35th European Photovoltaic Solar Energy Conference and Exhibition (EU PVSEC 2018)

Brussels, Belgium
24-28 September 2018

Volume 1 of 3

Editors:

P. Verlinden
P. Helm
R. Kenny

ISBN: 978-1-5108-7155-7

Printed from e-media with permission by:

Curran Associates, Inc.
57 Morehouse Lane
Red Hook, NY 12571

CURRAN ASSOCIATES INC. proceedings.com

Some format issues inherent in the e-media version may also appear in this print version.

Printed by Curran Associates, Inc. (2019)

WIP – Renewable Energies
Sylvensteinstr. 2
81369 Munchen
Germany

Phone: +49 89 72012735
Fax: +49 89 72012791

wip@wip-munich.de

Additional copies of this publication are available from:

Curran Associates, Inc.
57 Morehouse Lane
Red Hook, NY 12571 USA
Phone: 845-758-0400
Fax: 845-758-2633
Email: curran@proceedings.com
Web: www.proceedings.com

35th European Photovoltaic Solar Energy Conference and Exhibition

Proceedings of the International Conference held in Brussels, Belgium

24 September – 28 September 2018

Edited by:

P. VERLINDEN
Amrock Pty Ltd
Adelaide, Australia

R. KENNY
European Commission Joint Research Centre
Ispra, Italy

P. HELM
WIP Renewable Energies
Munich, Germany

Coordination of the Technical Programme:

European Commission Joint Research Centre
Via E. Fermi 1
21020 Ispra (VA)
Italy

Institutional Support:

European Commission

UNESCO – United Nations Educational, Scientific and Cultural Organization, Natural Sciences Sector

WCRE – World Council for Renewable Energy

Supporting Organisations:

ARE – Alliance for Rural Electrification

Energy Ville

EREF – European Renewable Energies Federation

ESA – European Space Agency

EUFORES –The European Forum for Renewable Energy Sources

EUREC – The Association of European Renewable Energy Research Centres

European Technology & Innovation Platform PV

IEA PVPS – International Energy Agency, Photovoltaic Power Systems Programme

imec

REN21 – Renewable Energy Policy Network for the 21st Century

Institutional PV Industry Cooperation:

SolarPower Europe

SOLARUNITED

Supporting Associations:

AIE – European Association of Electrical Contractors

AREA – African Renewable Energy Alliance

CABA – Continental Automated Buildings Association

EPIC – European Photonics Industry Consortium

MPIA - Malaysian Photovoltaic Industry Association

SASIA - Saudi Arabia Solar Industry Association

EU PVSEC realised by:

WIP
Sylvensteinstr. 2, 81369 München, Germany
Tel: +49 89 720 12 735, Fax: +49 89 720 12 791
Email: pv.conference@wip-munich.de
www.photovoltaic-conference.com
www.wip-munich.de

Proceedings produced and published by:

WIP
Sylvensteinstr. 2, 81369 München, Germany
Tel: +49 89 720 12 735, Fax: +49 89 720 12 791
Email: pv.conference@wip-munich.de
www.photovoltaic-conference.com
www.wip-munich.de

35th EUROPEAN PHOTOVOLTAIC SOLAR ENERGY CONFERENCE AND EXHIBITION
24 SEPTEMBER – 28 SEPTEMBER 2018
BRUSSELS, BELGIUM

EU PVSEC COMMITTEES

A. Ciesla, UNSW Australia, Sydney, Australia
A. Danel, CEA, Le Bourget du Lac, France
B. Dauksher, Arizona State University, Tempe, USA
P. Delli Veneri, ENEA, Portici, Italy
S. Dubois, CEA, Le Bourget du Lac, France
T. Dullweber, ISFH, Emmerthal, Germany
O. Eibl, Vienna University of Technology, Austria
M. Ernst, ANU, Canberra, Australia
P. Fath, RCT-Solutions, Konstanz, Germany
W. Favre, CEA, Le Bourget-du-lac, France
S. Gall, HZB, Berlin, Germany
I. Gordon, imec, Leuven, Belgium
C. Hagendorf, Fraunhofer CSP, Halle (Saale), Germany
J. Haschke, EPFL, Neuchâtel, Switzerland
K. Hesse, Wacker Chemie, Burghausen, Germany
J. Horzel, CSEM, Neuchâtel, Switzerland
J.-R. Huang, Motech Industries, Taoyuan County, Taiwan
G.J.M. Janssen, ECN part of TNO, Petten, The Netherlands
J. John, imec, Leuven, Belgium
S.D. Kim, SAIT, Yongin-si, Republic of Korea
J. Laurent, Vesuvius, Feignies, France
J. Levrat, CSEM, Neuchâtel, Switzerland
Y.-S. Lin, E-TON Solar Tech., Tainan, Taiwan
A. Meeder, Geo-En Energy Technologies, Berlin, Germany
J. Meier, Meier Technologies, Berlingen, Switzerland
A. Metz, Alzenau, Germany
J.W. Müller, Hanwha Q CELLS, Bitterfeld-Wolfen, Germany
R.C.G. Naber, Tempress, Vaassen, The Netherlands
Y. Ohshita, Toyota Technological Institute, Nagoya, Japan
S. Peters, Hanwha Q CELLS, Bitterfeld-Wolfen, Germany
K. Petter, Hanwha Q CELLS, Bitterfeld-Wolfen, Germany
R. Preu, Fraunhofer ISE, Freiburg, Germany
J. Rentsch, Fraunhofer ISE, Freiburg, Germany
H. Schlemm, Meyer Burger, Hohenstein-Ernstthal, Germany
J. Schmidt, ISFH, Emmerthal, Germany
M.C. Schubert, Fraunhofer ISE, Freiburg, Germany
B. Terheiden, University of Konstanz, Germany
M. Vetter, IPHT, Jena, Germany
A. Yamada, Tokyo Institute of Technology, Japan
D.L. Young, NREL, Golden, USA
J. Zhao, Meyer Burger, Hohenstein-Ernstthal, Germany

Topic 3 – Non Silicon-Based Thin Film Photovoltaics

C. Camus, ZAE Bayern, Erlangen, Germany
A. Redinger, University of Luxembourg, Luxembourg
A.N. Tiwari, EMPA, Dübendorf, Switzerland
S.C. Veenstra, ECN part of TNO, Eindhoven, The Netherlands
W. Wischmann, ZSW, Stuttgart, Germany

G. Bardizza, European Commission JRC, Ispra, Italy
V. Bermudez Benito, QEERI, Doha, Qatar
C.J. Brabec, FAU, Erlangen, Germany
S. Cros, CEA, Le Bourget du Lac, France
T. Dalibor, Avancis, Torgau, Germany
B. Dimmler, NICE Solar Energy, Schwäbisch Hall, Germany
D. Dimova-Malinovska, Bulgarian Academy of Sciences, Sofia, Bulgaria
K. Durose, Stephenson Institute for Renewable Energy, Liverpool, United Kingdom
C.J. Fell, CSIRO Energy Technology, Mayfield West, Australia
S. Huang, UNSW Australia, Sydney, Australia
S.J.C. Irvine, Swansea University, St. Asaph, United Kingdom
S. Ishizuka, AIST, Tsukuba, Japan

P. Kratzert, Solibro, Bitterfeld-Wolfen OT Thalheim, Germany
C.-F. Lin, NTU, Taipei, Taiwan
M. Lira-Cantú, CIN2, Barcelona, Spain
M.Ch. Lux-Steiner, HZB, Berlin, Germany
M. Meuris, imec, Genk, Belgium
R. Po, eni, Novara, Italy
M. Powalla, ZSW, Stuttgart, Germany
A. Romeo, University of Verona, Italy
J.R. Sites, Colorado State University, Fort Collins, USA
V. Sittinger, Fraunhofer IST, Braunschweig, Germany
W. Tress, EPFL, Lausanne, Switzerland
A. Virtuani, EPFL, Neuchâtel, Switzerland

Topic 4 – Concentrator and Space Photovoltaics

G. Flamand, imec, Leuven, Belgium
C. Signorini, European Space Agency, Noordwijk, The Netherlands

I. Antón Hernández, UPM, Madrid, Spain
R. Campesato, CESI, Milan, Italy
V. Khorenko, AZUR SPACE, Heilbronn, Germany
R. King, Arizona State University, Tempe, USA
G. Siefer, Fraunhofer ISE, Freiburg, Germany
G. Timò, RSE, Piacenza, Italy

Topic 5 – Photovoltaic Modules and BoS Components

M. de Wild-Scholten, SmartGreenScans, Groet, The Netherlands
R. Gottschalg, Fraunhofer CSP, Halle (Saale), Germany
W. Herrmann, TÜV Rheinland Energy, Cologne, Germany
N.M. Pearsall, Northumbria University, Newcastle upon Tyne, United Kingdom
M. Perrin, Oscarlab, BERNIN, France
K. Wambach, Wambach-Consulting, Aindling, Germany

K. Baert, KU Leuven, Heverlee, Belgium
G. Beaucarne, Dow Silicones, Seneffe, Belgium
C. Buerhop-Lutz, ZAE Bayern, Erlangen, Germany
R. Einhaus, Apollon Solar, Lyon, France
G. Friesen, SUPSI, Canobbio, Switzerland
V. Fthenakis, Columbia University, New York, USA
W.J. Gambogi, DuPont, Wilmington, USA
G. Graditi, ENEA, Portici, Italy
Y. Hishikawa, AIST, Tsukuba, Japan
U. Jahn, TÜV Rheinland Energy, Cologne, Germany
A.R. Lagunas, CENER, Sarriguren-Navarra, Spain
J. Lopez-Garcia, European Commission JRC, Ispra, Italy
M. Mittag, Fraunhofer ISE, Freiburg, Germany
C. Monokroussos, TÜV Rheinland, Shanghai, China
M.A. Muñoz-García, UPM, Madrid, Spain
H. Nagel, Fraunhofer ISE, Freiburg, Germany
H. Nussbaumer, ZHAW, Winterthur, Switzerland
S. Oktik, Sisecam, Kocaeli, Turkey
G. Oreski, PCCL, Leoben, Austria
W. Palitzsch, Loser Chemie, Zwickau, Germany
J.N. Roy, IIT Kharagpur, India
T. Sample, European Commission JRC, Ispra, Italy
R. Sinton, Sinton Instruments, Boulder, USA
M. Van Iseghem, EDF R&D, Moret-sur-Loing, France
E. Voroshazi, imec, Leuven, Belgium
A. Wade, First Solar, Mainz, Germany

Topic 6 – PV Systems - Performance, Applications and Integration

F.P. Baumgartner, ZHAW, Winterthur, Switzerland
F. Frontini, SUPSI, Canobbio, Switzerland
P. Lechner, ZSW, Stuttgart, Germany
O. Mayer, Munich, Germany
C. Nyman, Soleco, Porvoo, Finland
C. Protogeropoulos, EEPS, Athens, Greece
A. Sabene, ENEL Green Power, Rome, Italy
A. Scognamiglio, ENEA, Portici, Italy
W.G.J.H.M. van Sark, Utrecht University, The Netherlands
I. Weiss, WIP Renewable Energies, Munich, Germany

C. Alonso-Tristán, UBU, Burgos, Spain
P. Blanc, MINES ParisTech, France
P. Bonomo, SUPSI, Canobbio, Switzerland
F. Bourry, CEA, Le Bourget du Lac, France
E. Cunow, LSPV Consulting, Gröbenzell, Germany
I. de la Parra, UPNa, Pamplona, Spain
G.C. Eder, OFI, Vienna, Austria
B. Gaiddon, Hespul, Lyon, France
M. Grottke, WIP Renewable Energies, Munich, Germany
I.B. Hagemann, Architekturbüro Hagemann, Aachen, Germany
B. Herteleer, Ekistica, Alice Springs, Australia
T. Huld, European Commission JRC, Ispra, Italy
E. Kaijuka-Okwenje, Rural Electrification Agency, Kampala, Uganda
S. Koopman, Enterag South Africa, Cape Town, South Africa
K. Lappalainen, Tampere University of Technology, Finland
G. Makrides, University of Cyprus, Nicosia, Cyprus
P. Malbranche, CEA, Le Bourget du Lac, France
U. Muntwyler, BUAS, Burgdorf, Switzerland
N. Munzke, Karlsruhe Institute of Technology, Eggenstein-Leopoldshafen, Germany
G. Mütter, Alternative Energy Solutions, Vienna, Austria
K. Peter, ISC Konstanz, Germany
T. Reijenga, BEAR-iD, Gouda, The Netherlands
J. Remund, Meteotest, Bern, Switzerland
S. Rodrigues, M-ITI, Funchal, Portugal
K.T. Roro, CSIR, Pretoria, South Africa
M. Schroedter-Homscheidt, German Aerospace Center, Wessling, Germany
M. Sengupta, NREL, Golden, USA
K. Sinapis, SEAC, Eindhoven, The Netherlands
J.S. Stein, Sandia National Laboratories, Albuquerque, USA
D. Stellbogen, ZSW, Stuttgart, Germany
R.M.E. Valckenborg, SEAC, Eindhoven, The Netherlands
J.A.M. van Roosmalen, ECN part of TNO, Petten, The Netherlands
M. Zehner, Rosenheim University of Applied Sciences, Germany

Topic 7 – PV Economics, Markets and Policies

C. Breyer, Lappeenranta University of Technology, Lappeenranta, Finland
M. Getsiou, European Commission DG RTD, Brussels, Belgium
G. Masson, Becquerel Institute, Brussels, Belgium
T. Nordmann, TNC Consulting, Feldmeilen, Switzerland
S. Nowak, NET Nowak Energy & Technology, St. Ursen, Switzerland
H. Ossenbrink, Former European Commission, Italy

S. Caneva, WIP Renewable Energies, Munich, Germany
I. Kaizuka, RTS Corporation, Chuo-ku, Japan
P. Mazzucchelli, EUREC, Brussels, Belgium
D. Polverini, European Commission DG GROWTH, Brussels, Belgium
S. Szabó, European Commission JRC, Ispra, Italy
N. Taylor, European Commission JRC, Ispra, Italy

Awards Coordinators

Student Awards Coordinator
A.H.M. Smets, Delft University of Technology, Delft, The Netherlands

Poster Awards Coordinator
A.H.M. Smets, Delft University of Technology, Delft, The Netherlands

SUBJECT INDEX

New Materials and Concepts for Photovoltaic Devices

Sessions 1AP.1, 1AO.1, 1AO.2, 1AO.3, 1CO.1, 1CO.2, 1CV.4

Silicon Cells

Sessions 2BP.1, 2AO.4, 2AO.5, 2AO.6, 2BO.1, 2BO.2, 2BO.3, 2BO.4, 2CO.9, 2CO.10, 2CO.11, 2CO.12, 2DO.1, 2DO.2, 2AV.1, 2AV.2, 2AV.3, 2DV.3

Non Silicon-Based Thin Film Photovoltaics

Sessions 3CP.1, 3CO.6, 3CO.7, 3CO.8, 3DO.4, 3DO.5, 3DO.6, 3BV.2, 3BV.3

Concentrator and Space Photovoltaics

Sessions 4CP.2, 4CO.5, 4BV.4

Photovoltaic Modules and BoS Components

Sessions 5DP.1, 5BO.9, 5BO.10, 5BO.11, 5BO.12, 5DO.7, 5DO.8, 5DO.9, 5EO.1, 5CV.1, 5CV.3

PV Systems - Performance, Applications and Integration

Sessions 6DP.2, 6AO.7, 6AO.8, 6AO.9, 6BO.5, 6BO.6, 6BO.7, 6BO.8, 6CO.3, 6CO.4, 6DO.10, 6DO.11, 6DO.12, 6EO.2, 6BV.1, 6CV.2, 6DV.1

PV Economics, Markets and Policies

Sessions 7EP.1, 7DO.3, 7EO.3, 7DV.2

Topic Code

1 New Materials and Concepts for Photovoltaic Devices
2 Silicon Cells
3 Non Silicon-Based Thin Film Photovoltaics
4 Concentrator and Space Photovoltaics
5 Photovoltaic Modules and BoS Components
6 PV Systems - Performance, Applications and Integration
7 PV Economics, Markets and Policies

Session Type

P = Plenary Session
O = Oral Session
V = Visual Session

Day Codes

A = Monday, 24 September 2018
B = Tuesday, 25 September 2018
C = Wednesday, 26 September 2018
D = Thursday, 27 September 2018
E = Friday, 28 September 2018

e.g. 2BP.1 ⇒ 2= Silicon Cells, B=Tuesday, P=Plenary Session, 1=Session 1

TABLE OF CONTENTS
of 35th EU PVSEC 2018 Proceedings Papers

New Materials and Concepts for Photovoltaic Devices

Plenary SESSION 1AP.1 Routes to High Efficiency Photovoltaics

Oral PRESENTATIONS 1AO.1 Fundamental Studies

Oral PRESENTATIONS 1AO.2 Advanced Material Combinations for n-Terminal Multijunctions

Oral PRESENTATIONS 1AO.3 Advanced Materials for Solar Cells

Silicon Cells

Visual PRESENTATIONS 2AV.3 Heterojunction Solar Cells

Visual PRESENTATIONS 2DV.3 Thin Film and Foil-Based Si Solar Cells/Characterisation & Simulation Methods for Si Cells/Manufacturing & Production of Si Cell

Non Silicon-Based Thin Film Photovoltaics

Visual PRESENTATIONS 3BV.3 Perovskite, Organic and Dye-Sensitised Devices

Concentrator and Space Photovoltaics

Photovoltaic Modules and BoS Components

Visual PRESENTATIONS 5CV.1 PV Module Design, Manufacture, Performance and Reliability

Visual PRESENTATIONS 5CV.3 PV Module Design,Manufacture,Performance and Reliability/Inverters and Balance of System Components/Sustainability and Recycling

PV Systems - Performance, Applications and Integration

Oral PRESENTATIONS 6BO.8 Performance Analysis and Evalution of PV Systems

Oral PRESENTATIONS 6CO.3 Modelling for PV Systems

Oral PRESENTATIONS 6CO.4 Design and Calculations

Oral PRESENTATIONS 6DO.10 Solar Radiation

PV Economics, Markets and Policies

FOREWORD

The 35th EU PVSEC 2018 in Brussels concluded with the vision that TeraWatt by TeraWatt, PV will play a central role in a transformed sustainable energy economy with 100% decarbonized electricity generation to power, directly or indirectly, all Energy Sectors and Industrial processes. Research and innovation across the sector, together with fast deployment and numerous application options, are paving the way to putting PV at the centre-stage of sustainable energy systems. This year 2,037 participants from more than 80 countries took part in this most inspiring platform. It provided a vital forum for information exchange in the field of PV solar energy research, demonstration and applications. In the exhibition 48 companies from all parts of the world welcomed visitors and presented their products and services.

The 35th EU PVSEC 2018 Proceedings give a comprehensive overview of the photovoltaic solar sector, its current status and future prospects in science, research, innovation, development and deployment on almost 4,000 pages. Selection for inclusion in the conference was made by paper review experts and topic organisers (see the listing on pages III-V), to whom we express our sincere gratitude for their comprehensive review work and overall contribution to the success of the conference. In addition to the 468 submitted papers, the proceedings include 68 presentations (slides) shown during the plenary and oral presentations as well as 169 poster files of the visual presentations. In total this amounts to 705 publications.

The 2018 EU PVSEC Proceedings maintain our commitment to providing quick and open access to high quality scientific results. They constitute a powerful tool for targeted and quick information search and retrieval, enabling you to search by keyword, paper title and authors. We are sure that these features will help to simplify the use and exploit the full potential of this extensive source of information.

The Conference Proceedings are published as downloadable DVD file and are also fully accessible online. A DOI code (Digital Object Identifier) has been assigned to each paper. This ensures unequivocal and permanent identification and full citability. The papers can be viewed and downloaded in a full free open access from the EU PVSEC Proceedings website www.eupvsec-proceedings.com.

We are confident that these Proceedings will play an important role in providing a comprehensive overview of the current actors and activities in the global PV sector and that they will disseminate information on the state-of-the-art of technologies and applications. This can generate further research, add momentum to innovation and promote interest in PV worldwide.

We would like to cordially thank all authors and participants of the 35th EU PVSEC 2018 for their contributions and look forward to welcoming you in 2019 at the EU PVSEC 2019, 9 to 13 September, in Marseille, France.

The Editors

EU PVSEC 2018

35th European Photovoltaic Solar Energy Conference and Exhibition

Conference Highlights

Brussels
28 September 2018

General Figures

- 2,037 participants including visitors
- From more than 80 countries
- 884 presentations in the Scientific Programme
- 112 additional presentations in 9 parallel events

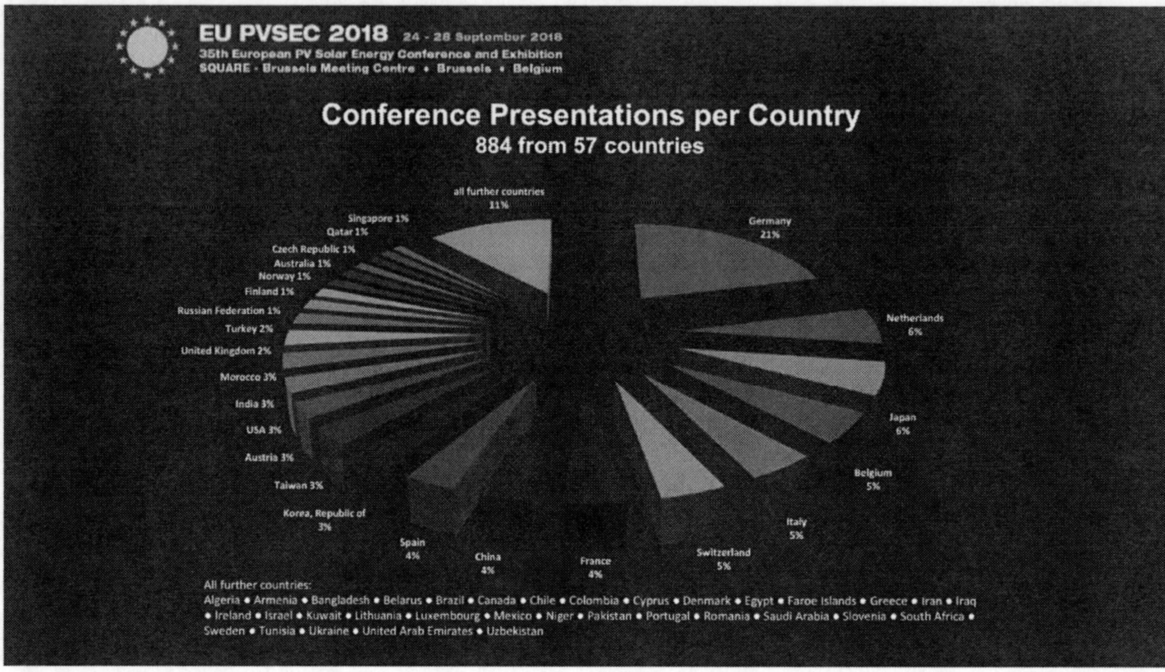
EU PVSEC 2018 24 - 28 September 2018
35th European PV Solar Energy Conference and Exhibition
SQUARE - Brussels Meeting Centre • Brussels • Belgium
Conference Presentations per Country
884 from 57 countries
all further countries 11%
Germany 21%
Netherlands 6%
Japan 6%
Belgium 5%
Italy 5%
Switzerland 5%
France 4%
China 4%
Spain 4%
Korea, Republic of 3%
Taiwan 3%
Austria 3%
USA 3%
India 3%
Morocco 3%
United Kingdom 2%
Turkey 2%
Russian Federation 1%
Finland 1%
Norway 1%
Australia 1%
Czech Republic 1%
Qatar 1%
Singapore 1%
All further countries:
Algeria • Armenia • Bangladesh • Belarus • Brazil • Canada • Chile • Colombia • Cyprus • Denmark • Egypt • Faroe Islands • Greece • Iran • Iraq • Ireland • Israel • Kuwait • Lithuania • Luxembourg • Mexico • Niger • Pakistan • Portugal • Romania • Saudi Arabia • Slovenia • South Africa • Sweden • Tunisia • Ukraine • United Arab Emirates • Uzbekistan

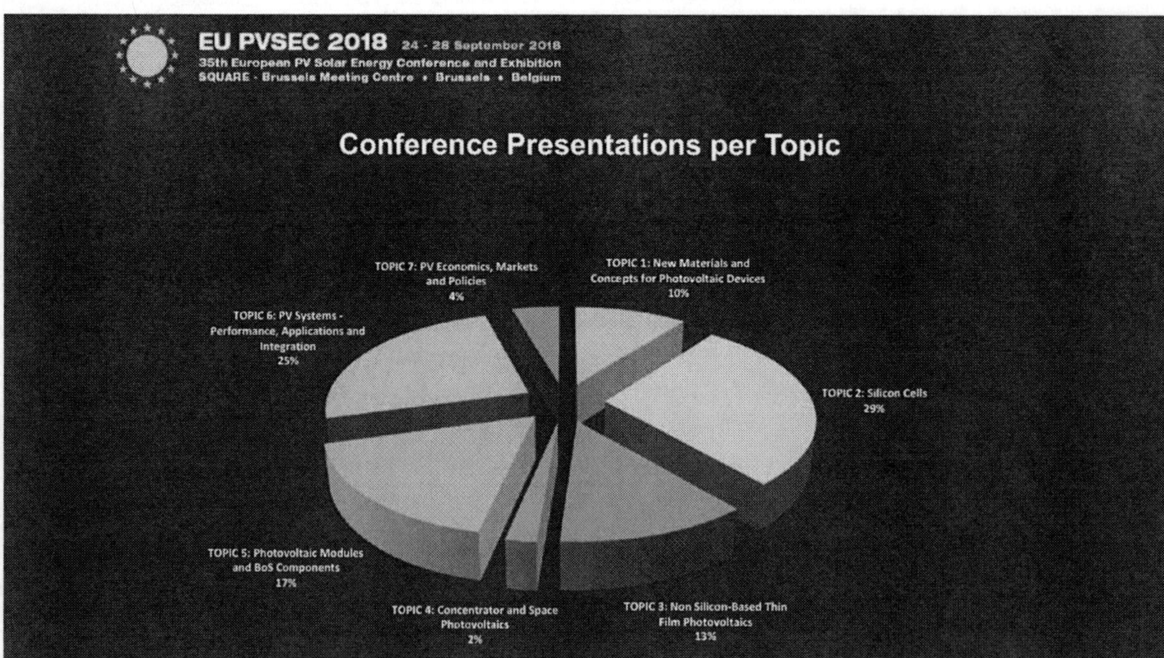
EU PVSEC 2018 24 - 28 September 2018
35th European PV Solar Energy Conference and Exhibition
SQUARE - Brussels Meeting Centre • Brussels • Belgium
Conference Presentations per Topic
TOPIC 1: New Materials and Concepts for Photovoltaic Devices 10%
TOPIC 2: Silicon Cells 29%
TOPIC 3: Non Silicon-Based Thin Film Photovoltaics 13%
TOPIC 4: Concentrator and Space Photovoltaics 2%
TOPIC 5: Photovoltaic Modules and BoS Components 17%
TOPIC 6: PV Systems - Performance, Applications and Integration 25%
TOPIC 7: PV Economics, Markets and Policies 4%

EU PVSEC is a huge team effort: Thanks to the members of the Scientific Committee, ExCo, ISAC, WIP & ETA, twitter feed team, and especially to all contributors and delegates for your dedication and enthusiasm

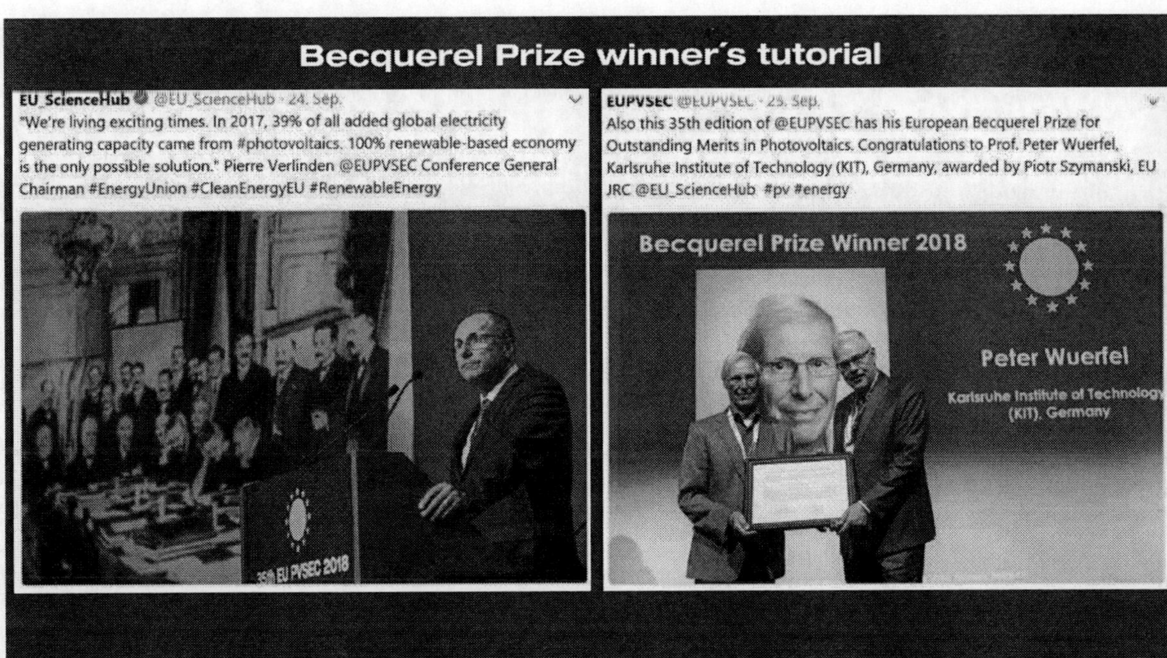
Becquerel Prize winner´s tutorial
EU_ScienceHub @EU_ScienceHub · 24. Sep.
"We're living exciting times. In 2017, 39% of all added global electricity generating capacity came from #photovoltaics. 100% renewable-based economy is the only possible solution." Pierre Verlinden @EUPVSEC Conference General Chairman #EnergyUnion #CleanEnergyEU #RenewableEnergy
35th EU PVSEC 2018
EUPVSEC @EUPVSEC · 25. Sep.
Also this 35th edition of @EUPVSEC has his European Becquerel Prize for Outstanding Merits in Photovoltaics. Congratulations to Prof. Peter Wuerfel, Karlsruhe Institute of Technology (KIT), Germany, awarded by Piotr Szymanski, EU JRC @EU_ScienceHub #pv #energy
Becquerel Prize Winner 2018
Peter Wuerfel
Karlsruhe Institute of Technology (KIT), Germany

1AP.1.2 Routes to high efficiency photovoltaics

1AP.1.2 High efficiency monolithic perovskite/silicon tandems

EUPVSEC @EUPVSEC · 24. Sep.
Christophe Ballif from @EPFL_en shows the many advances in and the exciting world of the perovskite-crystalline silicon tandem cells. @EUPVSEC 2018 #openingsession

EPFL/CSEM' 25.2% VERSUS OxfordPV's 27.3%

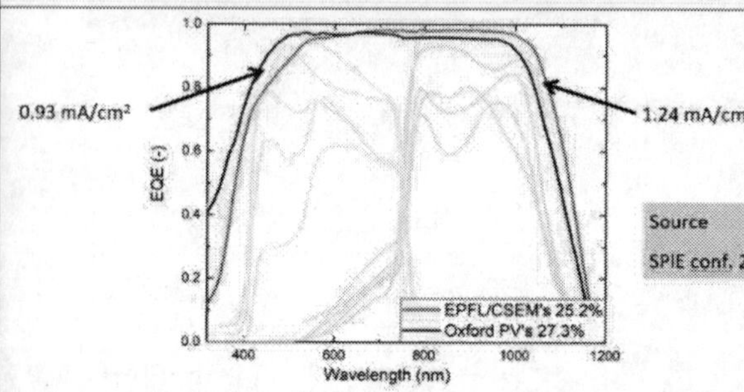

Source
SPIE conf. 2018

Total current from EQE:
EPFL/CSEM: 40.65 mA/cm²
Oxford PV: 40.35 mA/cm²

Conclusion:
EPFL/CSEM can gain 1 mA/cm2 at the front
Oxford PV can gain 1.2 mA/cm2 at the back

1AO.1.3 Use and abuse of Woc as a figure of merit

- Study of fundamental contributions to the Band gap offset voltage (Woc)
- Materials with low absorption at the bandgap can give a misleading low Woc - does not translate to cell performance

Summary

- Woc provides a useful, and memorable empirical figure of merit
- Relates directly to PV device measurements.
- Applies to a wide range of materials.
- Should not be used when the solar cell absorptivity near the band-gap is small !

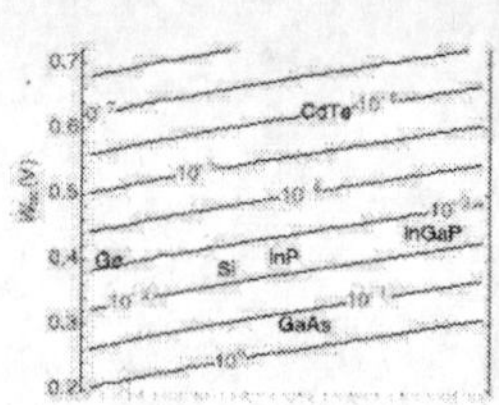

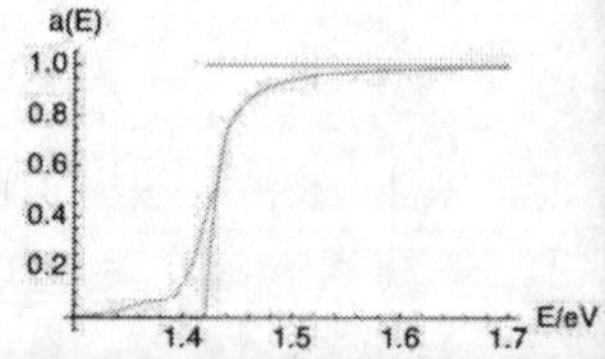

1AO.2.1 Nearly 30% High Efficiency Low Concentration InGaP/GaAs//Si 3-Junction Solar Cells

- Session presented options for II-V, CIGS, perovskite on Si
- Understanding role of selective contacts

Conclusion

supported by NEDO New Energy and Industrial Technology Development Organization

- We examined a low concentration InGaP/GaAs//Si 3-J solar cell with Smart Stack technology.
- We focused on the reduction of bonding resistivity for GaAs//Si interface.
- Using BHF treatment for Si surface, the bonding resistivity was reduced to be 1/10.
- The best efficiency for InGaP/GaAs//Si 3-J solar cell was 28.68 % at 10.3 suns.
- Obtained results suggest the possibility as new generation solar cells.

Ref 1) H. Mizuno, et. al., Appl. Phys. Lett., Vol. 101 (2012) 191111.
2) K. Makita, et. al., 29th European Photovoltaic Solar Energy Conference, (2014)3O.4.1.
3) H. Mizuno, et. al., Jap. J. Appl. Phys., 55, 025001_1-025001_7(2016)
4) H. Mizuno, et. al., Appl. Phys. Express, 10, 072301(2017)
5) K. Makita, et. al., 33rd European Photovoltaic Solar Energy Conference, (2017)1AO.3.1

Takeaways from Topic 1

Progress in materials understanding

Tandems on Si substrates

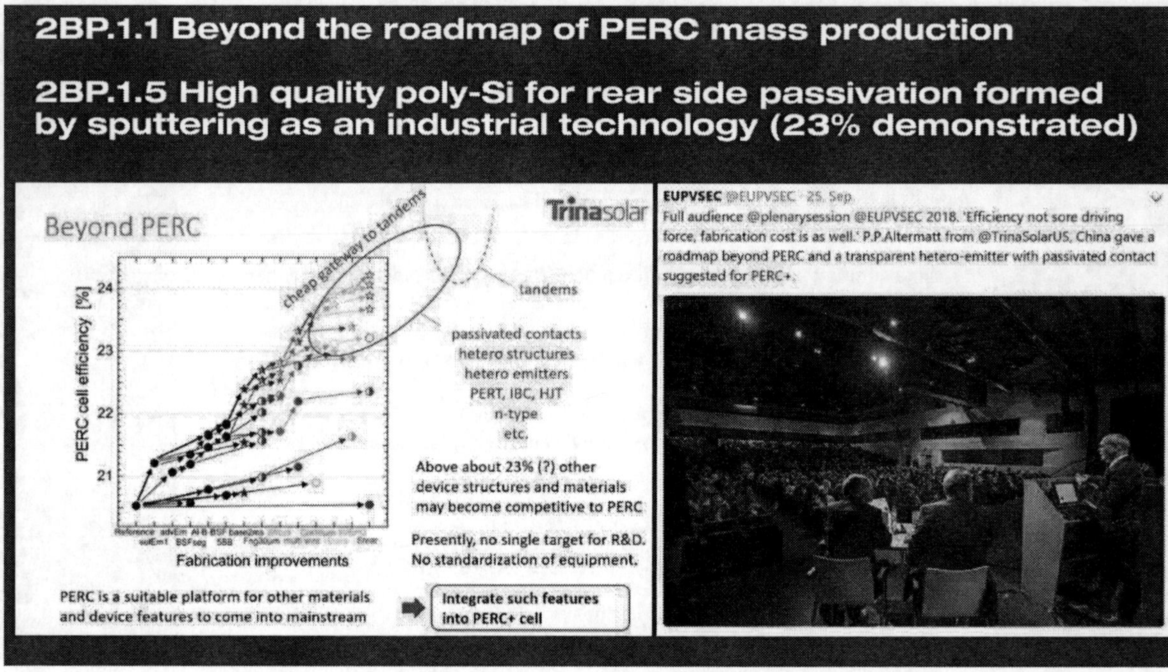

2BP.1.1 Beyond the roadmap of PERC mass production
2BP.1.5 High quality poly-Si for rear side passivation formed by sputtering as an industrial technology (23% demonstrated)
Beyond PERC
Trinasolar
cheap gateway to tandems
tandems
passivated contacts
hetero structures
hetero emitters
PERT, IBC, HJT
n-type
etc.
PERC cell efficiency [%]
Fabrication improvements
Above about 23% (?) other device structures and materials may become competitive to PERC
Presently, no single target for R&D. No standardization of equipment.
PERC is a suitable platform for other materials and device features to come into mainstream
Integrate such features into PERC+ cell
EUPVSEC @EUPVSEC · 25. Sep
Full audience @plenarysession @EUPVSEC 2018. 'Efficiency not sore driving force, fabrication cost is as well.' P.P.Altermatt from @TrinaSolarUS, China gave a roadmap beyond PERC and a transparent hetero-emitter with passivated contact suggested for PERC+.

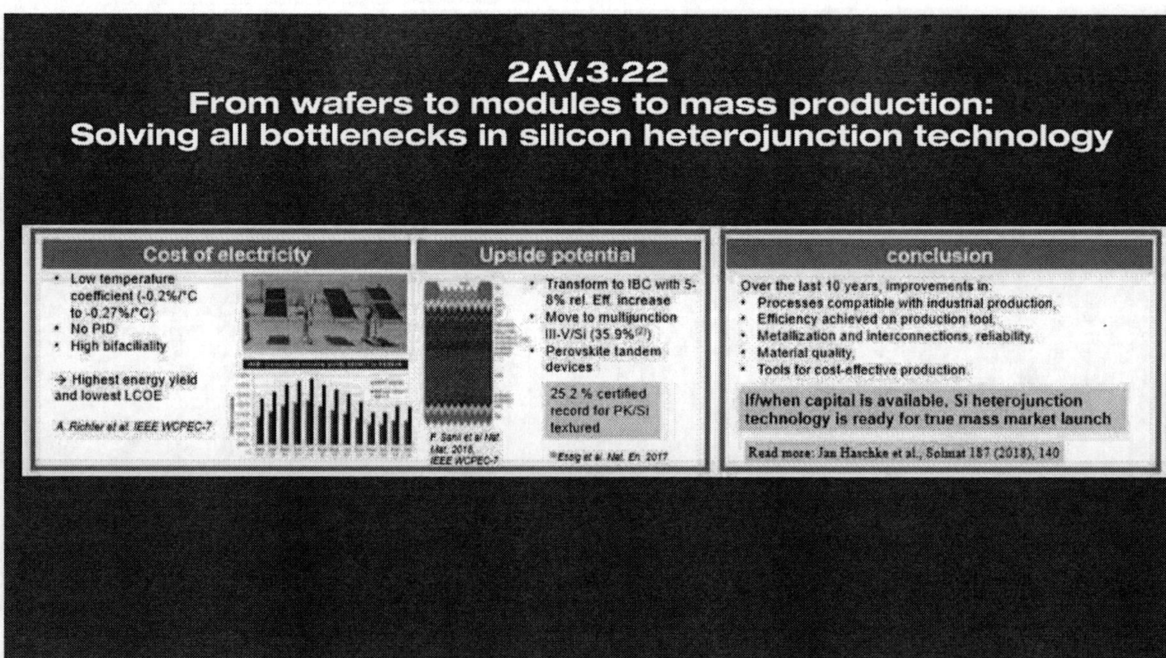

2AV.3.22
From wafers to modules to mass production:
Solving all bottlenecks in silicon heterojunction technology
Cost of electricity
Low temperature coefficient (-0.2%/°C to -0.27%/°C)
No PID
High bifaciliality
→ Highest energy yield and lowest LCOE
A. Richter et al. IEEE WCPEC-7
Upside potential
Transform to IBC with 5-8% rel. Eff. increase
Move to multijunction III-V/Si (35.9%)
Perovskite tandem devices
25.2 % certified record for PK/Si textured
F. Sahli et al Nat. Mat. 2018, IEEE WCPEC-7
Essig et al. Nat. En. 2017
conclusion
Over the last 10 years, improvements in:
Processes compatible with industrial production,
Efficiency achieved on production tool,
Metallization and interconnections, reliability,
Material quality,
Tools for cost-effective production.
If/when capital is available, Si heterojunction technology is ready for true mass market launch
Read more: Jan Haschke et al., Solmat 187 (2018), 140

2BO.1.4
Multicrystalline Informatics to realize ideal crystalline silicon for solar cells

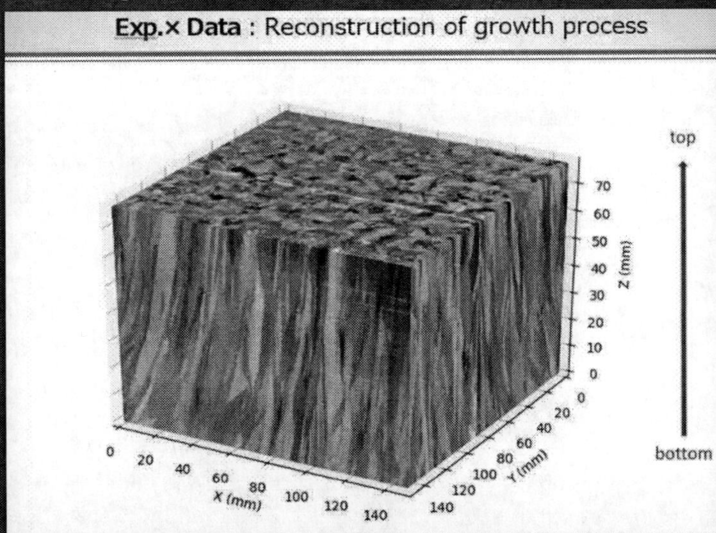

2CO.10.1
Intrinsic Poly-Crystalline Silicon Region in between the p+ and n+ POLO Contacts of an 26.1%-efficient IBC Solar Cell

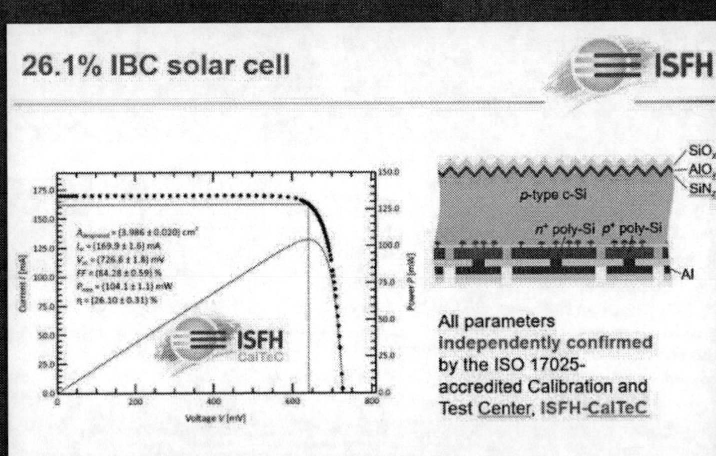

2BO.4.7
IBC-HJT @ 23.8% on Large Area with Low Cost Processes

- 23.8% cell based on tunnel junction technology on large area (100 cm²)
- First of its kind demonstration of an IBC solar cell with a single masking step on an industrial CZ wafer using an industrial compatible shadow masking technology for the PECVD deposition of microcrystalline silicon.

IBC HJT results

MEYER BURGER

IBC HJT efficiency starts where HJT ends!

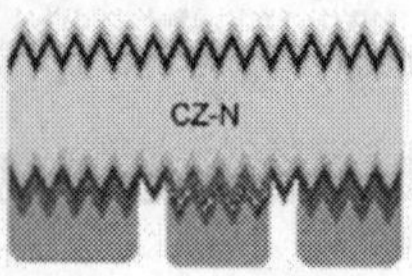

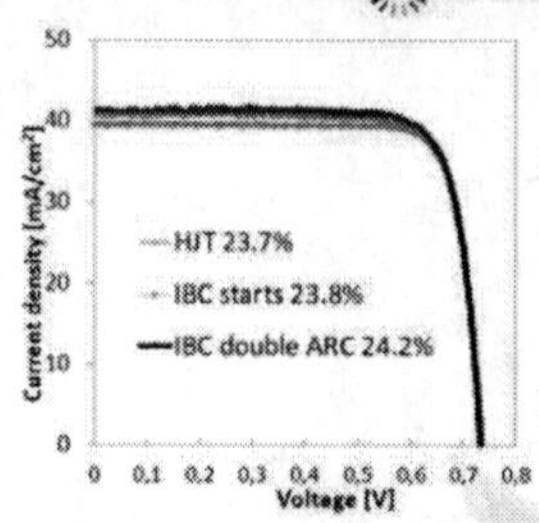

	eta (%)	FF (%)	J_{SC} (mA/cm^2)	Voc (V)
HJT (149cm²)	23.7%	81.9%	39.5	0.733
IBC (90.25cm²)	23.8%	79.6%	40.6	0.735
IBC (90.25cm²); x2 ARC	24.2%	79.8%	41.3	0.735

On CZ 155 μm thick wafer

Conclusions

23.8% Indium-free IBC HJT tunnel technology is deployed with only 10 process steps

Double ARC push IBC efficiency to **24.2%**

2BO.3.1

- Industrial rear-emitter nPERT cells with ~23% efficiency
- Over 700 mV Voc

Summary

- Several advantages to rear-emitter nPERT:
  1. easy upgrade from pPERC
  2. efficiency potential >24.0% + no LID
  3. easy to upgrade to bifacial (>22% with ~90% bifaciality already demonstrated)

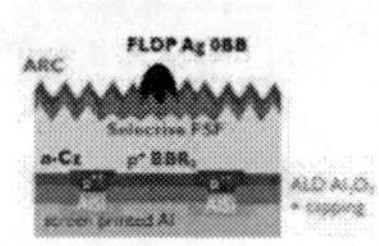

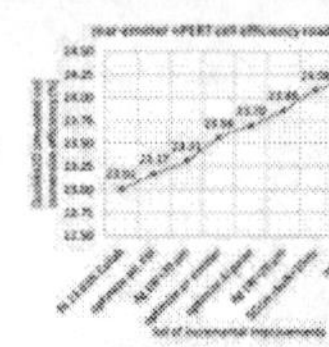

- Highlights of progress made at imec in recent years:
  - J_{0pass} <20 fA/cm² achieved for both $POCl_3$ and BBr_3 on textured surfaces
  - laser doped selective FSF from PSG with excellent J_{0metal}, J_{0pass}, and $R_{contact}$
  - 40 μm Ag linewidths applied in cell run using FLDP
  - <20 μm Ag linewidths demonstrated on test wafers using FLDP

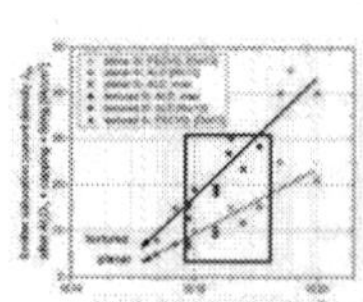

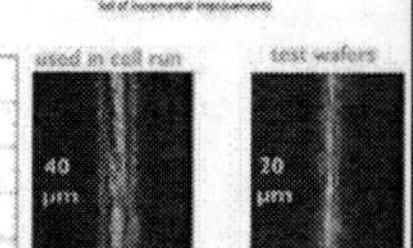

- Rear-emitter nPERT at imec:
  - average cell efficiency = 22.8±0.1%
  - champion cell = 23.03% (externally confirmed)

imec 17

2BO.2.4
Black Silicon to eliminate Light Induced Degradation
Black mc-Si PERC - suffers little or no LID

Black silicon provides several benefits

- **Negligible reflectance**
  - ➢ Improved I_{sc}
  - ➢ No need for AR-coating
- **Diamond wire sawing of mc-Si possible**[1]
  - ➢ Material savings
- **Enhanced impurity gettering**[2]
- **Applicable to industrial production (2AV.2.5)**
  - Long-term stability important

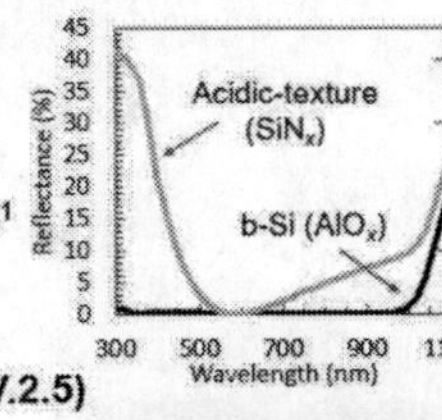

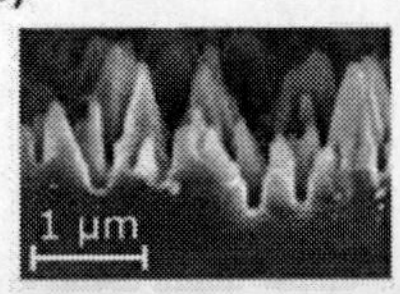

EUPVSEC @EUPVSEC · 25. Sep.
Toni Pasanen from @AaltoUniversity showed this afternoon his results on the
ck mc-Si PERC & the benefits of this technology emphasizing the
lls suffer little or no light induced degradation.
presentation #pv #solar

2BO.4.1 SiO2 passivation layers - From the past to the future

Conclusion

- SiO_2 is the easiest and one of the most efficient ways of surface passivation
- It allowed for breakthroughs in microelectronics (i.e. planar technology)
- First 20% silicon solar cells relied on SiO_2 surface passivation
- SiO_2 experiences a renaissance as the interfacial oxide for silicon-based passivating contacts
- → Perfect material for fixing the leaky bucket

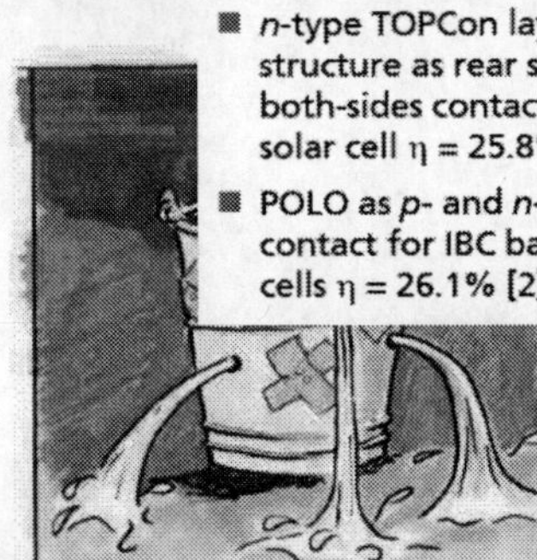

19

Fraunhofer ISE

Poly-Si/SiO$_x$ Contacts
Cell Results

- *n*-type TOPCon layer structure as rear structure for both-sides contacted n-type solar cell η = 25.8%[1]
- POLO as *p*- and *n*-type contact for IBC back contact cells η = 26.1% [2]

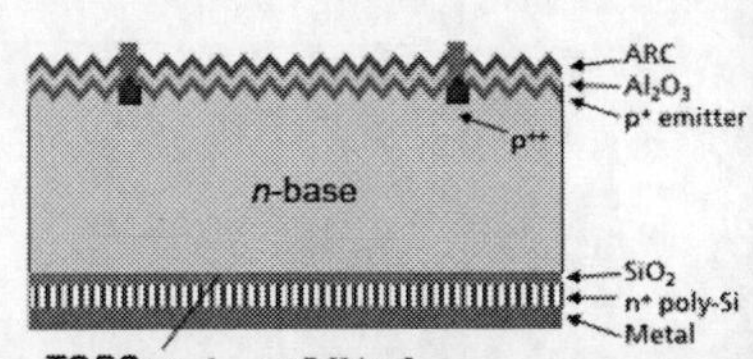

Takeaways from Topic 2

Steady Progress in Si cell technology

PERC well established (50%)

n-type

Passivated contacts increasingly important

Pathway to future tandem concepts

bifaciality

3DO.4.1
Standardization activity for emerging PV devices

EUPVSEC @EUPVSEC · 8 Min.
G. Bardizza @EU_Commission Commission JRC, Ispra, Italy: "We need standardized measurement procedures to correctly assess the performance of emerging #PV #technologies. @EUPVSEC 2018 #oralpresentation

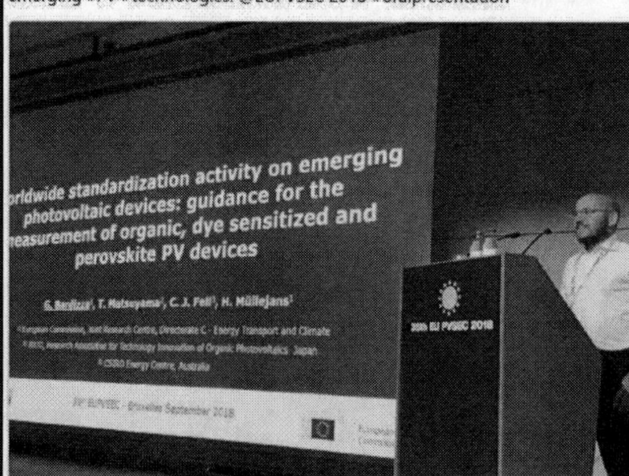

3CP.1.2 CIGS production technology above 18%
EUPVSEC @EUPVSEC · 22 Std.
#HappeningNow @EUPVSEC 2018 #plenarysession Philipp Kratzert from Solibro, Germany, presented exciting news of achieving world record efficiency of 18.7% for commercial full-size thin film module.
#solar #pv #photovoltaic #conference #research #conference #plenaries

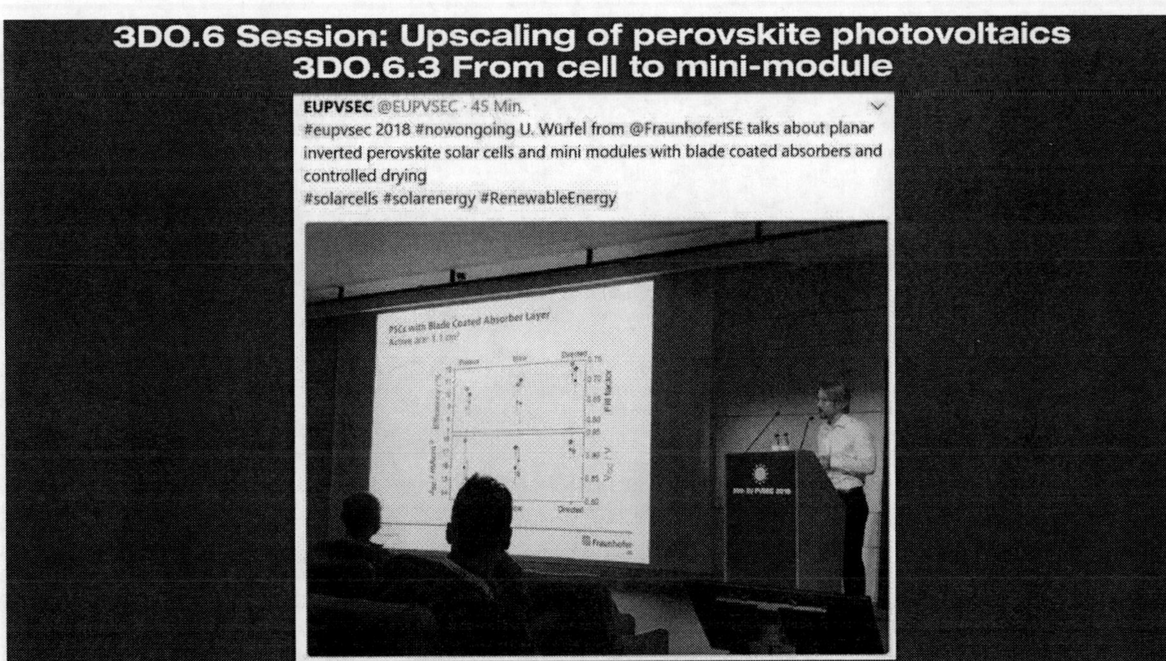
3DO.6 Session: Upscaling of perovskite photovoltaics
3DO.6.3 From cell to mini-module
EUPVSEC @EUPVSEC · 45 Min.
#eupvsec 2018 #nowongoing U. Würfel from @FraunhoferISE talks about planar inverted perovskite solar cells and mini modules with blade coated absorbers and controlled drying
#solarcells #solarenergy #RenewableEnergy

Takeaways from Topic 3

Record efficiencies continue to be obtained

Perovskites a very hot topic

Looking at tandem concepts

Upscaling and standards

4CP.2.1 Final results of CPVMatch - Concentrating Photovoltaic Modules using Advanced Technologies and Cells for Highest Efficiencies

Summary

- Four junction cells show potential for highest efficiencies towards 50 %, record efficiency 46 % on InP engineered substrate, 42% on Germanium
- Silicone on glass lenses suffer from chromatic aberration – mirrors and achromatic lenses to avoid these effects
- Full glass achromat proves potential for CPV module efficiencies exceeding 40 %
- Combining CPV and flat plate PV allows for highest efficiencies under global irradiance – 36.8 % measured outdoors!

37

5BO.12.2 TREND ANALYSIS OF PV MODULE FAILURE OCCURRENCE IN DIFFERENT CLIMATE ZONES

- Detailed failure tree map for PV system failure database
- With IEA Task 13 integrating more than 450 sites
- To assist climate and age dependent failure mode analysis

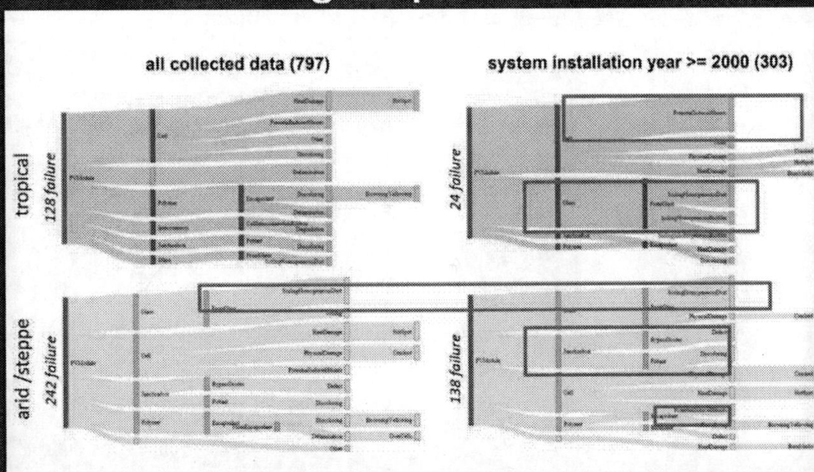

5DO.9.1 35 years of TISO 10kW solar plant

5DO9.2 The completed IEC 61853 standard series for Energy Rating

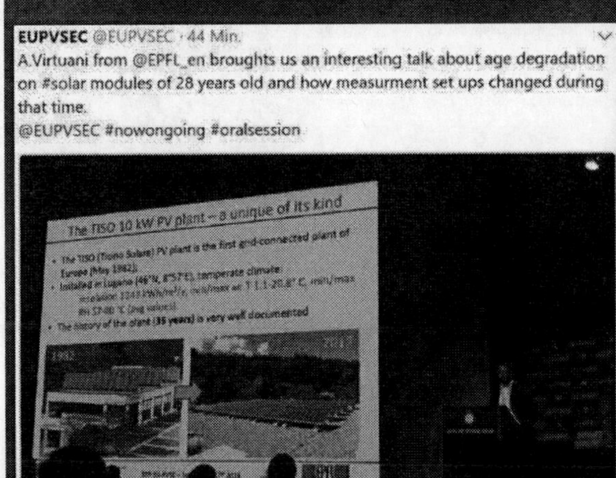

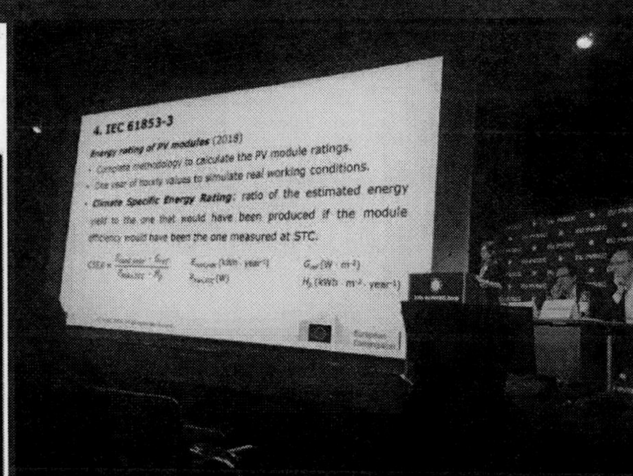

5BO.9

- Interesting session on bifacial modules and issues raised in their characterization and yield estimation
- 5BO.9.3 presented a shingle bifi design that generated much audience intervention

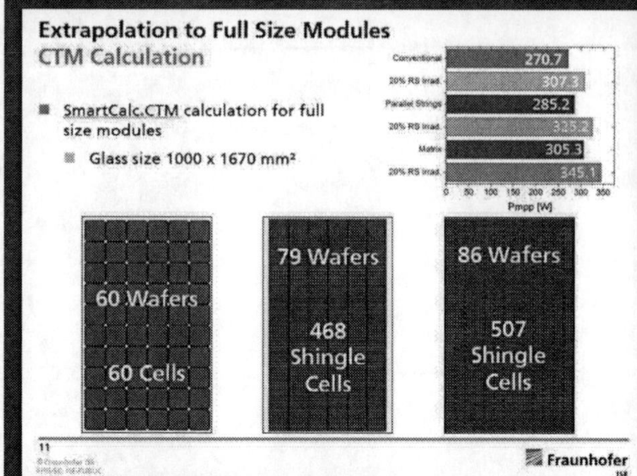

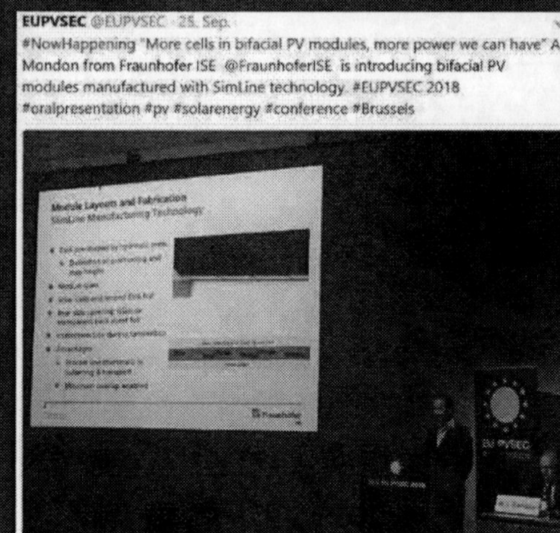

5BO.11

- Session: Innovative use of Electroluminescence
- 5BO.11.5 Magnetic field imaging shows defects that cannot be shown with EL. High resolution current image on PV modules

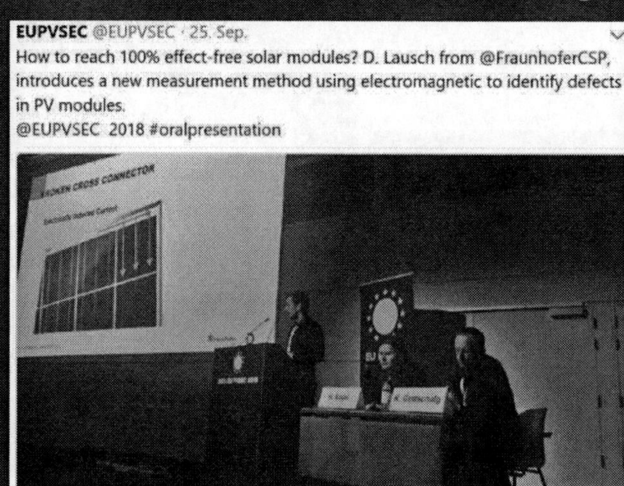

- **Interlaboratory comparison (8 labs) for measurement of angular dependency of cells**
- **Further progress is needed, especially for large incident angles**

DTU Fotonik
Department of Photonics Engineering

Conclusions

- **The 5 labs that met the <2% symmetry criteria at ±80° showed the lowest deviations to the weighted mean and better comparability.**
  - Suggests that it is a good practice to verify the symmetrical performance of a test system
- **Comparability is not dependent on the light source used.**
  - e.g. 4 labs used Xe flash lamps w/ very different results.
- **More frequent non-comparability observed beyond AOI ≈ 70°.**
  - Labs may need to consider a more conservative Uc at these angles.
- **Measurements at ±85° are problematic!**
  - High to low range of up to 75% observed among all 8 labs.
  - 4-5 of 8 labs are not comparable within stated or 'assumed' Uc.

5DP.1 Session

- **5DP.1.1 Overview of IEC Standards important to ensure confidence and reduce risk**
- **5DP.1.2 Overview of bifacial technologies**
- **5DP.1.3 Storage technologies for self-consumption – the DIY energy transition**
- **5DP.1.4 Environmental aspects and recycling**

History and future of PV, PERC, bifiPV

ISC International Solar Energy Research Center Konstanz

2000	2003	2006	2009	2012	2015		2021	2024	2027

PV against other energy sources (total installations)

0%	0%	0%	0%	0.1%	0.5%		4%	6%	8%

First they ignore you then they laugh at you then they fight you and then you win.

PERC against Al-BSF (market share)

0%	0%	0%	0%	0%	5%		70%	100%	100%

First they ignore you then they laugh at you then they fight you and then you win.

Bifacial against monofacial (accumulated installations)

0%	0%	0%	0%	0%	0.1%		5%	10%	40%

First they ignore you then they laugh at you then they fight you and then you win.

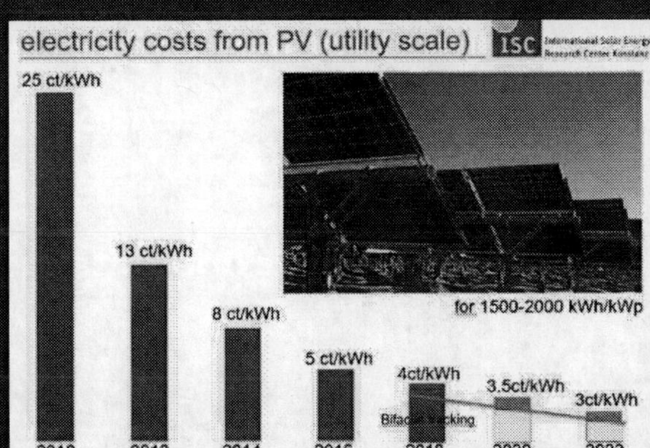

EUPVSEC @EUPVSEC · 18 Std.

@EUPVSEC #parallelevent Solar Industry Forum/Session 6 #SolarIndustryForum #SIF2018. Fabrizio Bizzarri, head of ENEL Green Power @EnelGroup presents a disruptive strategy for reducing the #energy generation cost based on EGP 3SUN PV Factory and high efficiency Si-HJT solar cells

Takeaways from Topic 5

Characterisation, lifetime, yield estimation, Standards

Bifacial modules

Storage

Environmental aspects/recycling

6DP.2.1 Interaction between curtailment, spinning reserve settings and generator limits on Australian projects with medium to high renewable energy fractions

How do other generators see & behave with PV?

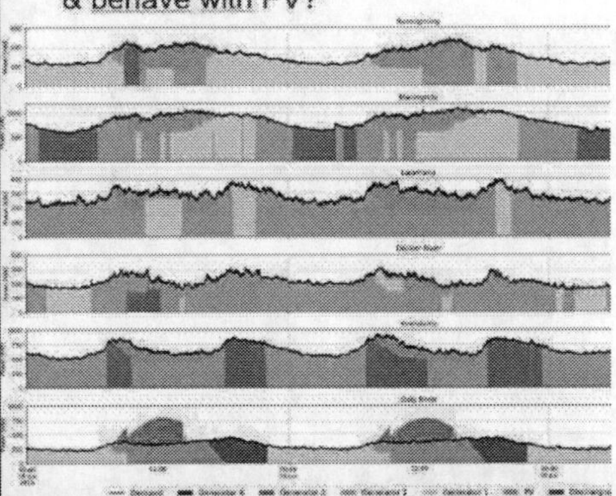

Cost changes

$$LCOE = \frac{\sum_{t=1}^{n} \frac{I_t + O_t + F_t}{(1+r)^t}}{\sum_{t=1}^{n} \frac{E_t}{(1+r)^t}}$$

Costs
Energy
Risk
Time

Licence to operate:

- Technical: PV needs to contribute more to grid & be valued for this
- Social: use of valuable land, benefit sharing (co-investment)
- Economic: creating & delivering value; stranded assets?

PV itself: $/W ↘ but…

Services currently not fully provided by PV:
Grid forming capabilities ⇒ PV alone, or with BESS? Anti-islanding…
Ancillary services ⇒ inverters need more capabilities

Co-funding of grid assets:

- Transmission
- Protection equipment

And yet likely still cheaper than current BAU … with the right business models

6CO.3.4 Predicting yields of bifacial PV Power Plants - What Accuracy is Possible?

RESULTS

	System 1	System 2	System 3	System 4
Description	Single n-type module with varying tilt, albedo of 25%, Germany	n-PERT modules, fixed-tilt 30° with 50% GCR, albedo of 30%, 7 months period, France	p-PERC modules, HSAT with 35% GCR, albedo of 32%, 6 months period, California	n-PERT modules, fixed-tilt 12° with 70% GCR, albedo of 30%, 1 year period, Israel
EDF results	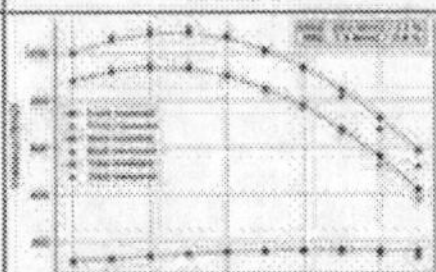	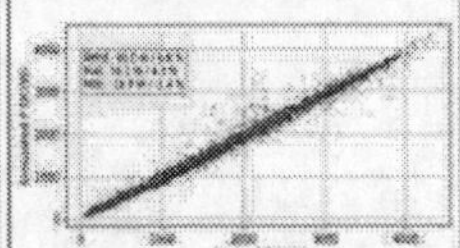	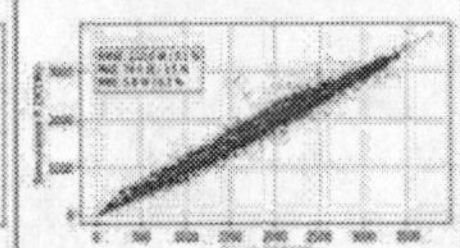	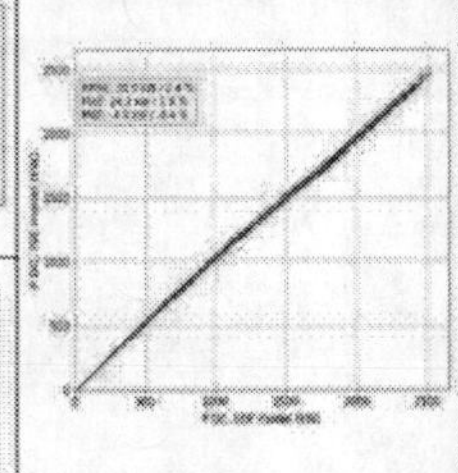
ISE results	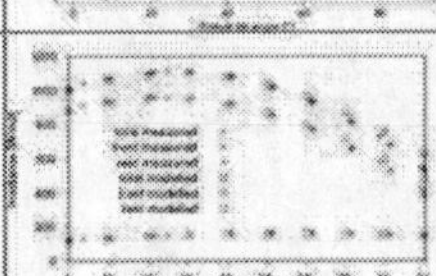	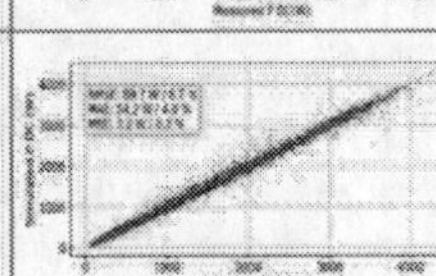	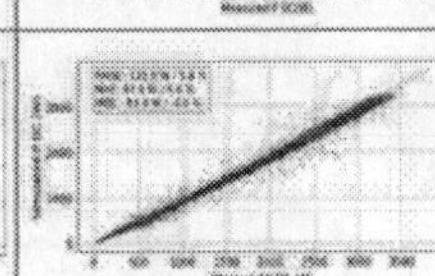	
Mean Bias (%)	-0.8% / 2.6%	-1.4% / 0.2%	0.3% / -4%	-0.4%

EDF Fraunhofer ISE

Predicting yields of bifacial PV Power Plants – What Accuracy is Possible? | 09/2018 | 16

6BO.8.2 Fully automated system modelling for low cost energy management applications based on power measurement data

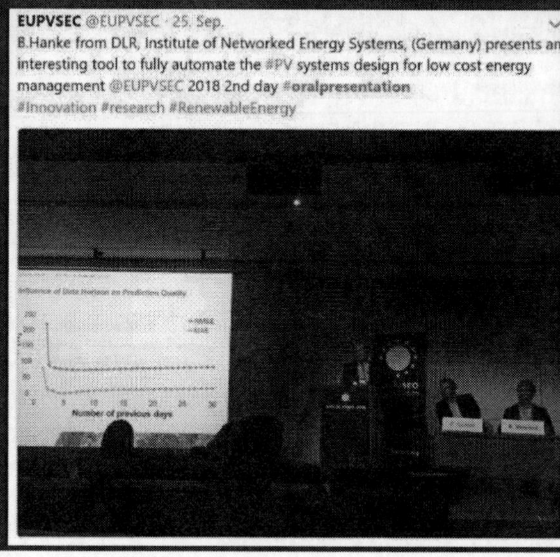

6BO.7

- Session describes different methodologies to handle local differences from laboratory assumptions in real PV plants
- Methods employed ranged from practical measurements to advanced simulation tools
- 6BO.7.4 Automated module failure detection

6AO.9

- PV goes beyond building and fields into:
- World´s largest solar noise barrier (bifacial)
- Combined tunnel lighting system with semi-transparent cells
- PV drone (student award)
- 6AO.9.6 Solutions for A Fully Integrated Solar Electric Car Roof

ECN › TNO innovation for life

TECHNICAL CHALLENGES FOR PV

- Maximum performance (Max. sun-facing area 5 m^2)
  - 20+% efficiency
  - Good shade performance (<30% shading losses)
  - Curved surface - 95% utility
- Lightweight
- Aerodynamics and Aesthetics
  - Needs to flexibly follow car roof
    - – reduced drag (30% less than current best in class vehicles)
  - Materials, coatings, and layout
- Reliability
  - 15 years minimum
  - High wind and speed damage

LIGHTYEAR

Takeaways from Topic 6

This area is growing rapidly which is indication of the continuing maturity of PV

Design, installation, monitoring of systems

Storage/grid integration concepts critical for continued growth

7DO.3.4 Impact of Batteries and Electric Vehicles on the Competitiveness of Solar PV

A 5 kWp PV system with 5 kWh storage becomes better investment than PV alone after 2025 in Germany

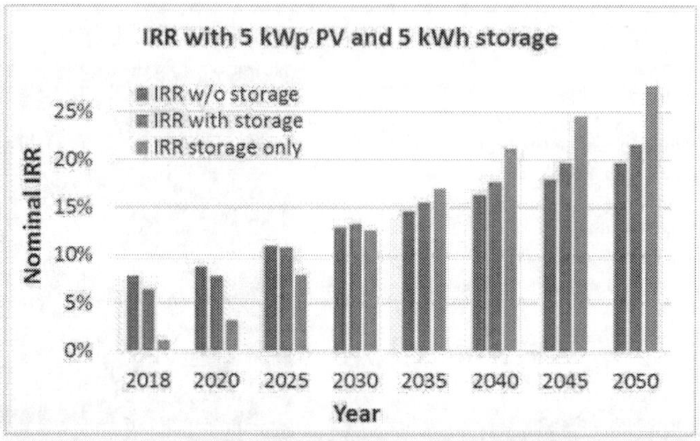

PV self-consumption:
39% with 0 kWh
58% with 5 kWh

Takeaways from Topic 7/Closing Session

- Plenary 7EP.1 indicated how sustainable transport and ambitious policymaking can ensure 2050 goals are met.
- A. Smets gave an entertaining presentation explaing why the PV community should think big.
- In closing, conference Chair P. Verlinden outlined his PV vision.
- The PV community urgently needs to broadcast these messages to a wider audience.

LIST OF EXHIBITORS
(in alphabetical order)

Abet Technologies, Inc.	USA
Agfa-Gevaert	Belgium
ArcelorMittal Europe – Flat Products	Luxembourg
Belgian Pavilion	Belgium
Bentham Instruments Ltd.	United Kingdom
CEA INES	France
CSEM Centre Suisse d'Electronique et de Microtechnique	Switzerland
Delft University of Technology	Netherlands
ECN part of TNO	Netherlands
Endeas Oy	Finland
EnergyVille	Belgium
ENGIE Laborelec	Belgium
Engineered Materials Systems, Inc.	USA
ETA - Florence Renewable Energies	Italy
EUREC	Belgium
European Commission JRC	Italy
exateq GmbH	Germany
Fraunhofer Institute for Solar Energy Systems ISE	Germany
h.a.l.m. elektronik	Germany
IEA PVPS (International Energy Agency Photovoltaic Power Systems Programme)	Switzerland
imec	Belgium
InnoLas Solutions GmbH	Germany
IOP Publishing	United Kingdom
ISC Konstanz e.V.	Germany
Jan Sehnoutek	Czech Republic
Jonas & Redmann Photovoltaics Production Solutions GmbH	Germany
Kopel/Kyoshin Electric Co., Ltd.	Japan
neonsee GmbH	Germany
Newport Spectra-Physics GmbH (MKS Instruments)	Germany
pv magazine group	Germany
PV-Vlaanderen	Belgium
RENA Technologies GmbH	Germany
Semilab Co. Ltd	Hungary
SINGULUS TECHNOLOGIES	Germany
Sinton Instruments	USA
SolarPower Europe	Belgium
SolarSwissConnect	Switzerland
SOLARUNITED	Germany
Solaxess SA	Switzerland
SOLTECH	Belgium
SVCS s.r.o.	Czech Republic
Tempress Systems	Netherlands
Thales Alenia Space in Belgium	Belgium
Ulbrich Solar Technologies Austria	Austria
University of Ljubljana, LPVO	Slovenia
VON ARDENNE GmbH	Germany
WIP - Renewable Energies	Germany
ZSW Zentrum für Sonnenenergie- und Wasserstoff-Forschung Baden-Württemberg	Germany

We thank the 35th EU PVSEC 2018 Sponsors

exateq
experience aided techniques

.h.a.l.m.

RENA

SINGULUS

VON ARDENNE

Electronic ratchets as necessary stepping stones for new PV concepts

Amaury Delamarre[1,2], Daniel Suchet[1,4], Nicolas Cavassilas[1,5], Zacharie Jehl[1,2], Yoshitaka Okada[1,2], Masakazu Sugiyama[1,2] and Jean-François Guillemoles[1,3,*]

[1]NextPV, LIA, CNRS-RCAST/U. Tokyo-U. Bordeaux, Tokyo, Japan; [2]RCAST, The University of Tokyo, Tokyo, Japan ; [3] CNRS, IPVF, UMR 9006 Palaiseau, France ; [4] Ecole Polytechnique, IPVF, UMR 9006 Palaiseau, France ; [5]Aix Marseille Université, IM2NP UMR 7334, Marseille, France

* Corresponding author: jf.guillemoles@cnrs.fr

ABSTRACT — Although promising, new solar cells concepts such as Intermediate Band Solar Cells have not shown efficiency improvements beyond single junction records so far. A potential path for increasing the performances would be to introduce electronic ratchets, which consists in a slight energy shift within one of the three bands of the system. Considering realistic properties such as sub optimal absorption and non-radiative recombination, we show that intermediate band can result in severe decrease of the performances, while a ratchet state can compensate those non-idealities. This allows us to consider new structures for the practical realization of intermediate band solar cells, such as quantum wells, which have been largely less considered than quantum dots so far.

I. INTRODUCTION

A ratchet is a concept popularized by Richard Feynman[1], in which a dented wheel is designed in such a way that it can rotate in one direction only. The allowed rotation involves a dissipation of ΔE so that the pawl can block the reverse rotation. The wheel can be coupled to blades, which, being sensitive to gas pressure fluctuations, could rotate the wheel so as to lift a weight, thereby producing work from a single heat source (Fig. 1).

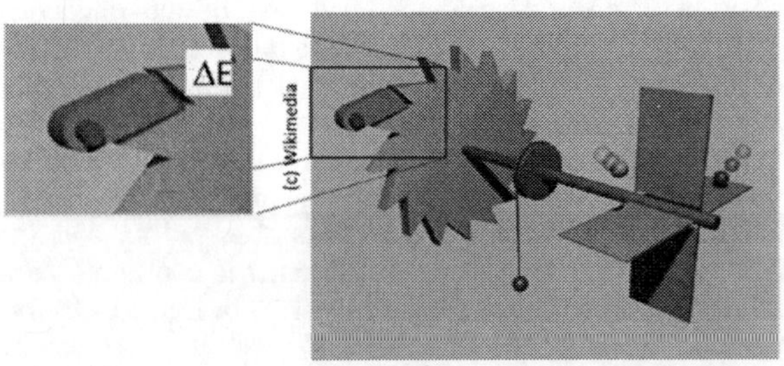

***Figure 1**: The principle of a Feynman Ratchet is illustrated. A dented wheel is designed in such a way that it can rotate in one direction only. The allowed rotation involves a dissipation of ΔE so that the pawl can block the reverse rotation. The wheel can be coupled to blades, which, being sensitive to gas pressure fluctuations, could rotate the wheel so as to produce work.*

This apparent paradox, which opposes the second law of thermodynamics, can be lifted by a closer look into the details of the device. The pawls has to be loose enough to allow the wheel to rotate at all; but then temperature induced fluctuations, can also lift the pawl, and allow the wheel to rotate backwards. At the end of the day, at constant temperature, the device works as an electrical rectifier where the mechanical torque plays the role of the electric field and the angular velocity that of the electrical current.
It is therefore not surprising to see that such a ratchet system can bring significant insights in the context of solar cells, where the same rectifying behavior allows charge separation and thereby energy conversion. And just as in the case of Feynman's ratchet, a closer look on the details of charge separation in a solar cell has led (and still leads) to lively debates concerning the behavior of the ratchet mechanism – in this case, the built-in potential [2], [3].

In this work, we highlight the role of a ratchet system for advanced concepts in PV. We will mainly focus on Intermediate level Solar Cell (IBSC), an advanced concept [4]–[7], that has been predicted to result in efficiencies of about 46.7% under one sun illumination [6]. We show that adding a ratchet system to such a device appears not as a mere improvement, but as a requirement to overcome the Shockley-Queisser limit [8] with materials currently available. Considering electrical figures of merit, we offer a qualitative and quantitative interpretation of the influence of the ratchet on such systems. This analysis can contribute to a better understanding of the physics at stake behind the ratchet mechanism.

II. THEORETICAL MODEL

The IBSC is a concept of device where, in parallel to the direct generation of carriers by absorption of photons, these can be generated in two absorption steps via an intermediate level (*Figure*). Nevertheless, photogenerated carriers can also recombine back using the intermediate level, that acts as a catalyst for this reaction. The theoretical performance of IBSC is close to that of triple junctions, series connected, but efficient devices have not been demonstrated yet because of current material limitations in absorption efficiency and bandwidth, and because of non-radiative recombination.
In nanostructures (quantum dots (QD) and quantum wells (QW)), which are one of the most considered path to the practical implementation of an intermediate band [9], this can be explained by two intrinsic properties: the reduced absorption energy width between the quantized states and the continuum, as well as increased non-radiative recombinations. As a consequence, a V_{oc} drop and a limited current increase are commonly observed. We have quantified [10] the impact of those two properties on the expected efficiencies, and highlighted that currently considered

structures are not compatible with efficiencies in excess of the SQ limit.
It has been suggested that a ratchet mechanism can increase the efficiency of an IBSC [10]–[15]. When implemented on the intermediate band (as in Figure (b), although implementation on the conduction or valance band can be also considered [14]), a ratchet consists in an additional state at a slightly lower energy, from which recombination to the valence band is forbidden. In case of reduced absorption strength, it has been further demonstrated that the efficiency reduction becomes less pronounced [14]. However, the effect on the non-radiative recombination has been given less consideration so far [12], despite a V_{oc} loss being regularly reported in practical realization of IBSC.

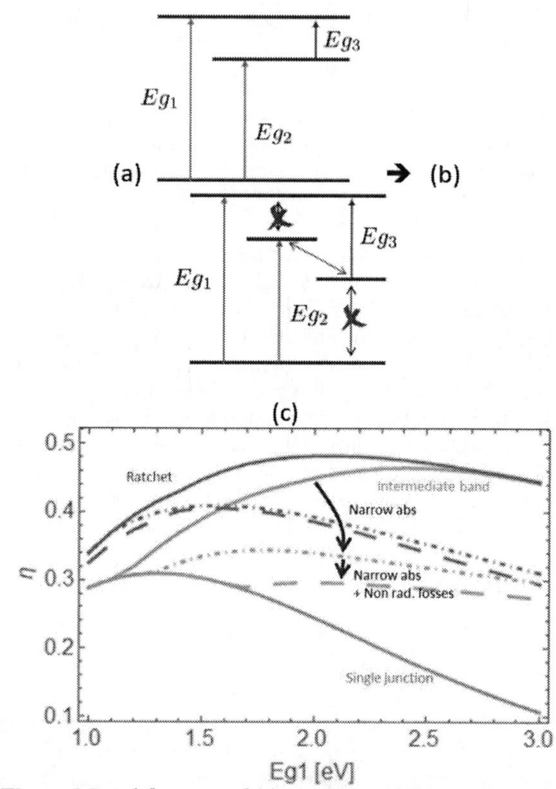

***Figure 2** Band diagram of (a) an IBSC and (b) an RBSC with a ratchet implemented in the IB. (c) Efficiencies of the single junction (full green line), IBSC and RBSC, as a function of the largest gap value (Eg_1), in the radiative limit and for an ideal absorption (respectively full orange and blue lines), with a narrow absorption (δE = 250 meV, orange and blue dotted lines), with a narrow absorption and non-radiative recombination ratio of 10^{-3} for the transition Eg_3 (dashed orange and blue lines).*

To account for these effects, the bandgap configuration and the notations of *Figure 2* (a) are considered. The limit efficiencies are calculated following a method similar to [13], to which the non-ideal behavior of the nanostructures will be added. For each transition i, the absorption G_i is given by the convolution of the absorption a_i and the sun spectrum, approximated by the black body radiation flux at a temperature T_{sun} = 6000 K. We consider the efficiencies in non-concentrated light. The radiative emission R_i for each transition i is given by the generalized Planck's law, for T_{cell} = 300 K and using the quasi-Fermi level splitting for each bandgap, $\Delta\mu_i$.

The transitions associated with the bandgap Eg_1 and Eg_2 are considered ideal, that is, a_1=1 strictly above Eg_1, and a_2=1 strictly between Eg_2 and Eg_1, whereas their recombination is purely radiative. Non-idealities will be considered for Eg_3. This will mimic the nanostructured devices in which the intermediate band is obtained thanks to quantized states in the conduction band, such as the widely studied InAs QDs in GaAs [9]. For this transition, both absorption and recombination are assumed only between Eg_3 and $Eg_3+\delta E$ where δE is the absorption energy width. Moreover the recombination is increased by considering a radiative recombination ratio r_{rad} such that $R_3^* = \frac{1}{r_{rad}} R_3$.
The full model can be found in [10], [13], [16]. It is firstly used to numerically compute I-V curves and efficiencies for given sets of bandgaps, which allows finding the best configurations. Another solution is to solve the system analytically, in the framework of Lagrange multipliers, which provides additional equations by explicitly considering the system constraints (matching of the currents through Eg_2 and Eg_3, $\Delta\mu_1 = \Delta\mu_2 + \Delta\mu_3$).

III. Results

On *Figure* (c) are represented the efficiency as a function of Eg_1, for the single junction, the IBSC and the RBSC. The position of the intermediate level and of the ratchet is optimized for highest efficiency. The full lines represent the efficiencies in the ideal configuration (r_{rad}=1, unlimited δE). Already in that configuration, it is observed that the implementation of a ratchet can result in an efficiency improvement. When the absorption width is reduced (δE = 250 meV), an efficiency reduction is observed for both the IBSC and the RBSC. When considering in addition a non-radiative recombination ratio of $r_{rad}=10^{-3}$ for Eg_3, we observe that the IBSC gives efficiencies lower than what can be achieved with the best single junction. Oppositely, we find that the introduction of the reduced radiative efficiency has a negligible impact on the RBSC performances.
A more detailed study [10] shows that a ratchet makes the cell tolerant to non-ideal characteristics, to that point that any value of the absorption width and radiative recombination ratio is compatible with efficiencies in excess of the Shockley-Queisser limit. This result gives new opportunities to realize IBSC with quantum structures. So far, QD have been preferred to QW for the practical implementation of IBSC, since the former are expected to exhibit larger radiative recombination ratios. We show here that when a ratchet system can be included, the reduced radiative efficiency of the QW can be acceptable to overcome the SQ limit, thanks to their larger expected absorption.

For a better understanding of the mechanisms at stake, the V_{oc}, J_{sc} and FF values are compared in *Figure* , instead of the aggregated efficiency value. As could be seen on Fig 2.c, for too small values of Eg_1, the optimal situation in the absence of ratchet aborts the intermediate band and behaves as a single junction. Above this threshold, we observe that the

V_{OC} of the cell including a ratchet keeps values close to that of the single junction, whereas the standard IBSC shows values reduced by about 10 %. Conversely, the RBSC gives a J_{sc} close to that of the IBSC, much above that of SJ, therefore taking advantage of the additional generation from the IB.

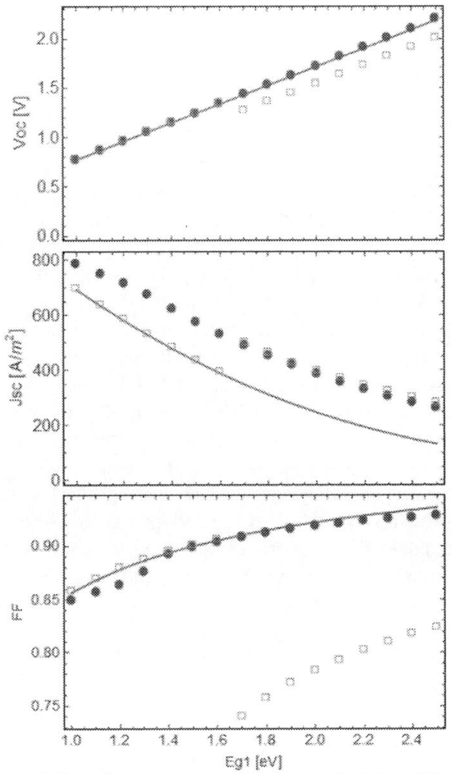

Figure 3 Top: Open-circuit voltage, as a function of the Eg1 bandgap, for an IBSC with a ratchet (filled blue symbols) and an IBSC w/o ratchet (open yellow symbols). A narrow absorption (δE = 250 meV) and non-radiative recombination ratio of 10-3 are considered for the transition Eg3 Middle: same for the short circuit current. Bottom: same for the FF.

These numerical observations have been supported analytically [16]. It has been shown that the introduction of a ratchet, by offering one additional degree of freedom to the system thanks to the ΔE adjustable parameter, allows for the simultaneous preservation of the open-circuit voltage and the current matching of the transition associated to Eg_2 and Eg_3. Oppositely, a standard IBSC cannot fulfill those two criteria, resulting in a trade-off providing lower performances.
This is illustrated in *Figure* , where the global open-circuit voltage, the generation through the sub-Eg_1 transitions (Eg_2 and Eg_3) and the energy conversion are represented as a function of the ratchet energy. For the energy conversion graph, we have differentiated the energy provided by the largest gap (Eg_1) and the energy provided by the two lowest gaps (together Eg_2 and Eg_3). In those graphs, we fixed $\delta E = 200$ meV and $r_{rad}=10^{-3}$. Eg_1 is also fixed at 1.86 eV, which is the value that optimizes the IBSC efficiency. We observe that the ratchet is at the same time able to achieve current matching (G_2 equaling G_3) and to optimize the recombination, the V_{oc} reaching a plateau, equivalent to the single junction at a similar Eg_1.

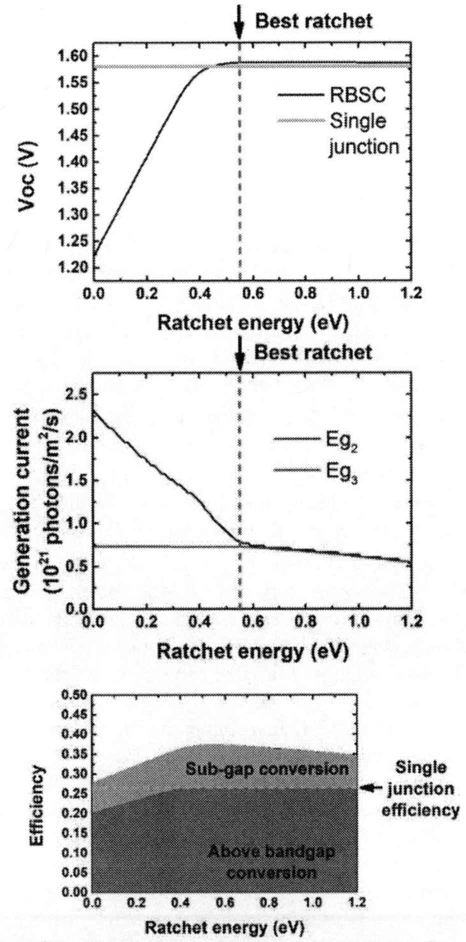

***Figure** 4 Evolution of recombination, generation and conversion efficiency as a function of the ratchet value, for Eg_1 fixed at 1.86 eV, which is the optimal value for an IBSC when the absorption width is 200 meV and the radiative efficiency 10^{-3}. Eg_2 is optimized for each ratchet value. Top: Voc as a function of ratchet energy ΔE. The green line represents the value of the single junction for the same Eg_1. Middle: photogenerated current via transition 2 (middle) and 3 (small gap) as a function of ΔE. Bottom: fraction converted by the Eg_1 transition and fraction converted with the help of the intermediate band (i.e. converted by Eg_2 and Eg_3). The green dashed line represents the efficiency of the single junction with the same Eg_1.*

For an implementation of a ratchet in real world, it is needed that some transitions must be forbidden (*Figure*), and this calls for specific selection rules. They may be spatial, in momentum space, or spin related [9]. Structure with quantum cascade, exhibiting spatial separation have been demonstrated experimentally recently [17] . Structures using spin selection rules could be realized in principle with dilute magnetic semiconductors [12], albeit in practice it is hard to obtain long minority carrier lifetimes and high Curie temperatures in these materials. Finally, another unexplored possibility is some 2D materials (MoS_2 nanoribbons) that could exhibit relevant selection properties in reciprocal space [18].

These conclusions were obtained with detailed balance calculations using the generalized Planck law. When more elaborate models are used, such as NEGF are used [19], an additional V_{oc} loss is found due to broadening of the final states that leads to increased recombination. Moreover, J_{sc} losses from lower optical absorption were also found.

IV. Discussion

Several interpretations are often considered to explain the influence of a ratchet. It is sometimes argued that the ratchet induces an energy dissipation ΔE as carriers relax from the top to the bottom of the shifted intermediate band. However, this picture does not hold when considering the full process driving a carrier from the valence to the conduction band. In an IBSC, when absorbing two photons with energies $E_{\gamma1}$ and $E_{\gamma2}$, the carrier will dissipate $(E_{\gamma1} - E_{g2})$ when relaxing after the first transition from VB to IB and $(E_{\gamma2} - E_{g3})$ when relaxing after the second transition from IB to CB. In an RBSC, the carrier would first dissipate $(E_{\gamma1} - E_{g2})$, then ΔE through the ratchet, then $(E_{\gamma2} - (E_{g3} + \Delta E))$ over the second transition. The total amount of dissipated energy is unchanged; what changes is the band in which this energy is dissipated, and the spectral range of absorbed photons.

It is also worth noting that, in some conditions, the optimal ratchet can be negative (see Figure 5). In this case, carriers brought to the intermediate band have to climb up a potential barrier, gathering energy from the ambient temperature, before they undergo a second transition to the conduction band. This situation occurs when photogeneration G3 is initially so low that its increase compensates the additional recombination induced by ΔE<0. While this situation is purely conceptual (an appropriate treatment would require considering the influence of populations on the absorptivity), it highlights the limits of an interpretation based on relaxation.

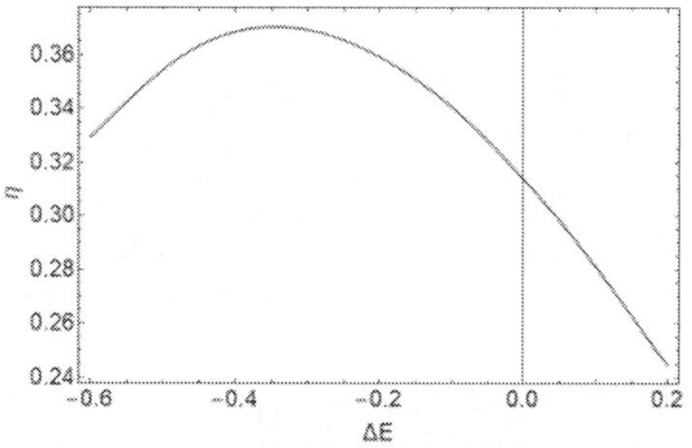

Figure 5 Power conversion efficiency as a function of the ratchet ΔE in the ideal case with Eg_1 = 2 eV and Eg_2 = 1.1 eV.

Ratchet have also been suggested to reduce the "Boltzmann" entropy creation [13], which is due to the geometrical mismatch between the absorption and emission optical étendues. As concentration is increased, it appears indeed that the role of the ratchet becomes more and more limited in the ideal case [13], even if a direct link with the absorption and emission étendues mismatch is not obvious. This numerical observation on the impact of concentration can be expanded with the other non-idealities considered above. We have found that the ratchet remains useful even under full concentration when non-radiative recombination (which produce additional entropy) are included, but not when considering a narrow absorption (which does not produce additional entropy). While there is a clear role of the ratchet energy to improve net current photogeneration through the IB while keeping current matching, the numerical observation above may also support the idea that the ratchet helps mitigating entropy production in the system.

V. Conclusion

We have discussed the ratchet analogy that has been proposed in the context of intermediate band solar cell. In this course, we have quantified the impact of non-radiative recombination and reduced absorption on the efficiency of nanostructured IBSC. The impact of those two intrinsic deficiencies result in strong performance reductions, unless an electronic ratchet is introduced, in which case efficiencies in excess of the SQ limit can be expected.

We have analyzed in depth the origin of the action of the ratchet on efficiency and concluded after careful examination of the proposed explanations that the main actions are (i) matching of the currents in the 2 step photogeneration process and (ii) reduction of the recombination losses.

These conclusions were obtained with detailed balance calculations using the generalized Planck law. When more elaborate models are used, such as NEGF, other factors affecting recombination and generation were found that also serendipitously pointed out to additional beneficial action of the ratchet.

References

[1] R. Feynman, R. Leighton, and S. Matthew, "Ratchet and pawl," in *The Feynman Lectures on Physics. Volume I: Mainly mechanics, radiation, and heat*, Addison–Wesley, 1964.

[2] U. Wurfel, A. Cuevas, and P. Wurfel, "Charge Carrier Separation in Solar Cells," *IEEE J. Photovolt.*, vol. 5, no. 1, pp. 461–469, Jan. 2015.

[3] P. Würfel and U. Würfel, *Physics of solar cells: from basic principles to advanced concepts*, 2nd, updated and expanded ed ed. Weinheim: Wiley-VCH, 2009.

[4] M. Wolf, "Limitations and Possibilities for Improvement of Photovoltaic Solar Energy Converters: Part I: Considerations for Earth's Surface Operation," *Proc. IRE*, vol. 48, no. 7, pp. 1246–1263, Jul. 1960.

[5] M. J. Keevers and M. A. Green, "Efficiency improvements of silicon solar cells by the impurity photovoltaic effect," *J. Appl. Phys.*, vol. 75, no. 8, pp. 4022–4031, Apr. 1994.

[6] A. Luque and A. Martí, "Increasing the efficiency of ideal solar cells by photon induced transitions at intermediate levels," *Phys. Rev. Lett.*, vol. 78, no. 26, p. 5014, 1997.

[7] J.-F. Guillemoles and S. Kettemann, "Limiting efficiencies of LDS intermediate band solar cells," in *Proceedings of the 13th European Photovoltaic Solar Energy Conference and Exhibition*, Nice, 1995, pp. 119–121.

[8] W. Shockley and H. J. Queisser, "Detailed Balance Limit of Efficiency of p-n Junction Solar Cells," *J. Appl. Phys.*, vol. 32, no. 3, pp. 510–519, Mar. 1961.

[9] Y. Okada *et al.*, "Intermediate band solar cells: Recent progress and future directions," *Appl. Phys. Rev.*, vol. 2, no. 2, p. 021302, Jun. 2015.
[10] A. Delamarre, D. Suchet, N. Cavassilas, Y. Okada, M. Sugiyama, and J.-F. Guillemoles, "An Electronic Ratchet Is Required in Nanostructured Intermediate-Band Solar Cells," *IEEE J. Photovolt.*, pp. 1–7, 2018.
[11] T. Trupke, M. A. Green, and P. Würfel, "Improving solar cell efficiencies by up-conversion of sub-band-gap light," *J. Appl. Phys.*, vol. 92, no. 7, pp. 4117–4122, Oct. 2002.
[12] P. Olsson, C. Domain, and J.-F. Guillemoles, "Ferromagnetic Compounds for High Efficiency Photovoltaic Conversion: The Case of AlP:Cr," *Phys. Rev. Lett.*, vol. 102, no. 22, Jun. 2009.
[13] M. Yoshida, N. J. Ekins-Daukes, D. J. Farrell, and C. C. Phillips, "Photon ratchet intermediate band solar cells," *Appl. Phys. Lett.*, vol. 100, no. 26, p. 263902, Jun. 2012.
[14] A. Pusch *et al.*, "Limiting efficiencies for intermediate band solar cells with partial absorptivity: the case for a quantum ratchet: Limiting efficiencies for intermediate band solar cells," *Prog. Photovolt. Res. Appl.*, vol. 24, no. 5, pp. 656–662, May 2016.
[15] M. A. Green and S. P. Bremner, "Energy conversion approaches and materials for high-efficiency photovoltaics," *Nat. Mater.*, vol. 16, no. 1, pp. 23–34, Jan. 2017.
[16] D. Suchet *et al.*, "Analytical optimization of intermediate band systems: Achieving the best of two worlds," *Prog. Photovolt. Res. Appl.*, vol. 26, no. 10, pp. 800–807, Oct. 2018.
[17] A. Vaquero-Stainer *et al.*, "Semiconductor nanostructure quantum ratchet for high efficiency solar cells," *Commun. Phys.*, vol. 1, no. 1, p. 7, Mar. 2018.
[18] S.-F. Chen and Y.-R. Wu, "A design of intermediate band solar cell for photon ratchet with multi-layer MoS2 nanoribbons," *Appl. Phys. Lett.*, vol. 110, no. 20, p. 201109, May 2017.
[19] N. Cavassilas *et al.*, "Beneficial impact of tunneling in nano-structured intermediate-band solar cell," *Submitt. EPJ Photovolt.*, 2018.

ANALYSIS FOR NON-RADIATIVE RECOMBINATION IN QUANTUM DOT SOLAR CELLS AND MATERIALS

Lin Zhu[1,2], Kan-Hua Lee[3], Masafumi Yamaguchi[3,*], Hidefumi Akiyama[2], Yoshihiko Kanemitsu[4], Kenji Araki[3] and Nobuaki Kojima[3]

[1] Sun Yat-sen University, Higher Education Mega Center, Guangzhou 510006, China

[2]University of Tokyo and OPERANDO-OIL, 5-1-5 Kashiwanoha, Kashiwa, Chiba 277-8581, Japan

[3]Toyota Technological Institute, 2-12-1 Hisakata, Tempaku, Nagoya 468-8511, Japan

[4]Kyoto University, Uji, Kyoto 611-0011, Japan

ABSTRACT: Quantum dot (QD) solar cells have drawn much attention of research because of their capability of tuning the band gap and potential of realizing many novel concepts such as intermediate transitions. However, high non-radiative recombination rates in the quantum dot layer stands in the way of realizing high-efficiency QD solar cells. In this paper, efficiency potential of QD solar cells is discussed based on external radiative efficiency (*ERE*), open-circuit voltage loss and fill factor loss and non-radiative recombination loss via current-voltage characteristics in detailed balance model. Intrinsic loss of QD solar cells is substantially increased with increase in binding energy and volume density of QDs. External radiative efficiency of QD solar cells is decreased with increase in binding energy and volume density of QDs due to non-radiative recombination.

Keywords: Quantum Dots, III-V Semiconductors, High-Efficiency, Recombination, Diffusion Length, Defects

1 INTRODUCTION

For creating new clean energy infra-structure based on photovoltaics, further development of science, technology and deployment of photovoltaics is necessary. Especially, development of high-efficiency, low-cost and highly reliable solar cells are very important. The authors have already discussed about efficiency potential of current high-efficiency solar cells [1,2]. In addition, future generation solar cells such as wide bandgap chalcopyrite, CZTS (Cu_2ZnSnS_4), and CZTSSe ($Cu_2ZnSn(S,Se)_4$) solar cells are very attractive because of high efficiency potential. The MQW (Multi Quantum Well) solar cells [3] and QD (Quantum Dot) solar cells [4], generally can achieve more photo-current by more infrared absorption, have also drawn much attention because of its capability of tuning the band gap to maximize the energy yields. However, most of experimental results [5-14] shows significant voltage drop in MQW and QD cells. In these results, while moderate photo-current add-on is achieved, the conversion efficiencies of the MQW and QD cells are often lower than their bulk reference cells. It is thus important to quantitatively identify the sources of their efficiency loss, which may be due to non-radiative recombination possibly induced by interfacial defects, dislocations, originated from lattice mismatch between QD and host materials. Brief analytical results for efficiency potential of future generation solar cells will be reported by the authors [15].

In this paper, efficiency potential of QD solar cells is analyzed in detail based on external radiative efficiency, open-circuit voltage loss, fill factor loss and non-radiative recombination losses. Effects of non-radiative recombination upon properties of QD solar cells are also discussed.

2 ANALYTICAL PROCEDURE FOR OPEN-CIRCUIT VOLTAGE AND EFFICIENCY

In order to realize higher efficiency of various solar cells, improvements in short-circuit density J_{sc}, open-circuit voltage V_{oc} and fill factor FF are substantially necessary.

From the ideal I-V curve of an illuminated solar cell in detailed balance model neglecting carrier generation or recombination in space-charge region and electric field in neutral regions [16], the open-circuit voltage (V_{oc}) is formulated by

$$V_{oc} = \frac{kT}{q}\ln\left(\frac{J_L}{J_0^*}+1\right) \approx \frac{kT}{q}\ln\left(\frac{J_L}{J_0^*}\right) \quad (1)$$

where k is Boltzmann constant, T is absolute temperature of cell, q is electronic charge, J_L is photon generated current density, and J_0^* is saturation current density, stemming from all of radiative and non-radiative recombination events in the cell. In order to increase V_{oc}, increase in J_L and decrease in J_0^* is essential.

Via separating out the part for non-radiative recombination, Eq. (1) can be re-expressed by [1,2, 15-17]

$$V_{oc} = V_{oc,rad} + V_{oc,non-rad} = V_{oc,rad} + \frac{kT}{q}\ln(ERE) \quad (2)$$

$V_{oc,non-rad}$ represents the voltage loss due to non-radiative recombination. It directly relates to cell's external radiative efficiency (*ERE*), that is, the ratio of external radiative recombination rate emitting from cell (R_{ext}) to the sum of that and non-radiative recombination rate (R_{ext} + R_{nr}). On the other hand, $V_{oc,rad}$ is open-circuit voltage in radiative limit, given by [1,2, 15-17]

$$V_{oc,rad} = \frac{kT}{q}\ln\left(\frac{J_L}{J_{0,rad}}+1\right) \approx \frac{kT}{q}\ln\left(\frac{J_L}{J_{0,rad}}\right) \quad (3)$$

$J_{0,rad}$ is reverse saturation current density in the case of only radiative recombination. According to the reciprocity relation [17], $J_{0,rad}$ can be evaluated by integral of the measured EQE data and 300-K blackbody thermal emission spectrum $B(\lambda)$ over the total wavelength region, given as

$$J_{0,rad}=\int EQE(\lambda)B(\lambda)d\lambda\cdot \quad (4)$$

Fill factor (FF) is dependent upon V_{oc} and ideal fill factor (FF_0) used in the calculation that is empirically expressed by [18-20],

$$FF_0=\frac{v_{oc}-\ln(v_{oc}+0.72)}{v_{oc}+1}, \quad (5)$$

where v_{oc} is normalized open-circuit voltage [18-20], given by

$$v_{oc}=V_{oc}/(kT/q)\cdot \quad (6)$$

FF is also decreased as increase in series resistance (R_s) and decrease in shunt resistance (R_{sh}) of solar cell, which is approximated by

$$FF\approx FF_0(1-r_S)(1-1/r_{sh})\approx FF_0(1-r_S-1/r_{sh}) \quad (7)$$

where r_s and r_{sh} are respectively normalized series and shunt resistance [18-20], given by

$$r_s=R_S/R_{CH}, \quad (8)$$

$$r_{sh}=R_{sh}/R_{CH}\cdot \quad (9)$$

R_{CH} is the characteristic resistance [18-20], expressed by

$$R_{CH}=\frac{V_{oc}}{I_{sc}}. \quad (10)$$

On the basis of the general expressions Eq.(1)-(3), the open-circuit voltage of QD solar cells (V_{oc}^{QD}) in comparison with the radiative V_{oc} of GaAs bulk cells ($V_{oc,rad}^{GaAs}$) can be expressed as,

$$V_{oc}^{QD}=V_{oc,rad}^{\mathrm{GaAs}}-\Delta V_{oc,rad}^{QD}(\text{intrinsic})-\Delta V_{oc,non\text{-}rad}^{QD}(\text{extrinsic}). \quad (11)$$

where intrinsic and extrinsic V_{oc} drops are given by,

$$\Delta V_{oc,rad}^{QD}(\text{intrinsic})=V_{oc,rad}^{\mathrm{GaAs}}-V_{oc,rad}^{QD}\approx-\frac{kT}{q}\ln\frac{J_{0,rad}^{\mathrm{GaAs}}}{J_{0,rad}^{\mathrm{QD}}} \quad (12)$$

$$\Delta V_{oc,non\text{-}rad}^{QD}(\text{extrinsic})=-\frac{kT}{q}\ln ERE. \quad (13)$$

$J_{0,rad}^{GaAs}$ and $J_{0,rad}^{QD}$ are the reverse saturation currents in radiative limit for GaAs and QD solar cells, respectively.

Table 1 shows the experimental records on voltage and related parameters for GaAs and QD solar cells [10-12,21]. To analyze their V_{oc} drops, the ERE and $V_{oc,rad}$ were evaluated by Eq. (3) from their measured V_{oc} and EQE results extracted from these works [10-12,21]. Table 1 exhibits these QD cells with significant V_{oc} drops and moderate J_{sc} add-ons, obtained much lower conversion efficiency than GaAs bulk solar cells.

Table 1: Voltage losses and other parameters for GaAs and QD solar cells [10-12,21]

Solar cell	V_{oc} (V)	J_{sc} (mA/cm2)	FF	efficiency (%)	$V_{oc,rad}$ (V)	$V_{oc,rad}$(GaAs) -V_{oc} (V)	ERE	Reference No.
GaAs	1.11	29.6	0.84	27.6	1.146	0.036	0.225	21
	1.03	29.8	0.86	26.4	1.146	0.116	1.26E-02	21
	1.04	28.2	0.89	27.4	1.15	0.110	1.00E-02	10
	1.04	22.5	0.89	23.1	1.15	0.110	1.47E-02	11
	0.88	17.5	0.81	12.4	1.146	0.266	4.03E-05	12
QD	0.84	27.66	0.89	26.1	1.063	0.306	1.79E-04	10
	0.991	23.3	0.89	23.5	1.128	0.155	5.08E-03	11
	0.994	23.3	0.89	23.5	1.128	0.152	5.70E-03	11
	0.88	23	0.89	23.2	1.129	0.266	6.81E-05	11
	0.868	17.7	0.81	12.4	1.095	0.278	1.55E-04	12
	0.783	18.7	0.77	11.3	1.089	0.363	7.50E-06	12
	0.671	19.7	0.74	9.8	1.077	0.475	1.54E-07	12
	0.643	21.5	0.72	10.0	1.071	0.503	6.62E-08	12
	0.616	22.7	0.67	9.4	1.063	0.530	3.24E-08	12
	0.616	23.6	0.63	9.2	1.059	0.530	3.77E-08	12

3. ANAYSIS FOR EFFICIENCY POTENTIAL OF QUANTUM DOT SOLAR CELLS

Figure 1 shows plots of calculated $V_{oc,rad}^{GaAs}$ - V_{oc} and V_{oc} drops due to non-radiative loss in QD and GaAs cells ($\Delta V_{oc,non\text{-}rad}^{QD/GaAs}$) as a function of ERE, based on the reported experimental record values for GaAs solar cells by Alta Devices [21], FhG-ISE [21,23,24] and QD solar cells reported Ioffe Inst [10], Rochester Inst. [11] and AIST [12]. They are respectively total V_{oc} drops with reference to radiative V_{oc} of GaAs cells ($V_{oc,rad}^{GaAs}$ = 1.15V) and the extrinsic V_{oc} drops estimated by using Eq. (2) and (13), where $V_{oc,rad}$ are achieved via experimental J_{sc} and evaluated J_0 by integral of 300-K blackbody emission spectrum and EQE records of those QD and GaAs cells [10-12,21]. In Fig.1, the total V_{oc} drop of GaAs solar cells originate from extrinsic drops, and are relative smaller than QD cells, because GaAs cells usually have much higher ERE, such as 22.5% and 1.26% for Alta Devices and FhG-ISE cells [21], as the results of high-quality thin-layer, reducing front and rear surface and interface recombination, and optical management [22]. In contrast, QD solar cells [10-12] generally have much larger extrinsic voltage drops or lower ERE values of less than 1%. This suggests that reduction of non-radiative recombination loss, such as interface recombination, is necessary. Besides extrinsic drops, quite obvious intrinsic drops co-exist in QD cells are clearly exhibited by the deviations between triangles and circles, especially at the lower *ERE* region less than 10^{-4}, which is also worth studying carefully, in order to suppress the total voltage drops further.

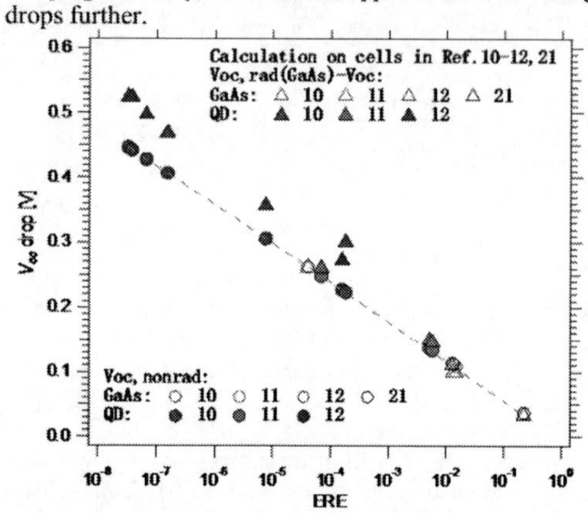

Figure 1: Open-circuit-voltage drops, $V_{oc,rad}$(GaAs) - V_{oc} and $V_{oc,nonrad}$, for GaAs and QD solar cells [10-12,21] as a function of *ERE*.

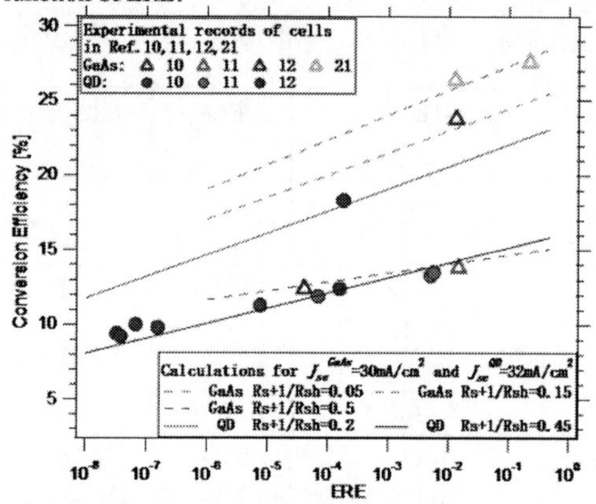

Figure 2: Calculated efficiency of QD solar cells as a function of *ERE* and r_s + $1/r_{sh}$ in comparison with efficiencies of GaAs solar cells reported by Alta Devices and FhG-ISE [10-12,21] and QD cells reported by Ioffe Institute [10], Rochester Institute of Technology [11], and AIST [12]

Figure 2 shows calculated efficiency (curves) and reported experimental efficiency values for GaAs solar cells [10-12,21] (triangles), and QD solar cells [10-12] (circles) as a function of *ERE*. It exhibits that the most QD solar cells with extremely low efficiency of less than 14% have inferior *ERE* of less than 10^{-2}. Only the QD solar cell of the Ioffe Inst. [10] (green circle) with a very inefficient *ERE* of 1.8E-4, however, has a relative high conversion efficiency of over 18%. This primarily stems from that its low voltage was compensated by its high J_{sc} of 27.7 mA/cm^2, which in the other QD cells [10-12] are less than 23.6mA/cm^2. Additionally, further improvements in efficiency are thought to be possible not only by improving photocurrent, *ERE*, minority-carrier lifetime and reducing non-radiative recombination losses in QD solar cells, but also by decreasing $r_s + 1/r_{sh}$. For example, if resistance loss ($r_s + 1/r_{sh}$) are close to 0.45, efficiencies of QD cells with J_{sc} of 32 mA/cm^2 increasing from 9% to 13% would be realized by improvements in *ERE* from 10^{-8} to 10^{-2}. On the other hand, one can realize efficiencies over 18% when *ERE* close to 10^{-4}, if reduced $r_s + 1/r_{sh}$ from 0.45 to 0.2, as shown in Fig.2. In the following section, intrinsic losses in QD solar cells are also discussed separately from non-radiative recombination losses.

4. DISCUSSION ABOUT INTRINSIC AND EXTRINSIC VOLTAGE LOSSDES IN QUANTUM DOT SOLAR CELLS

Intrinsic and extrinsic losses in QD solar cells were numerically analyzed on the basis of a detailed-balance limit theory by Zhu et al.[0,0]. Here, we analyze experimental data reported in previous works [10-12], by comparing with this theoretical calculations.

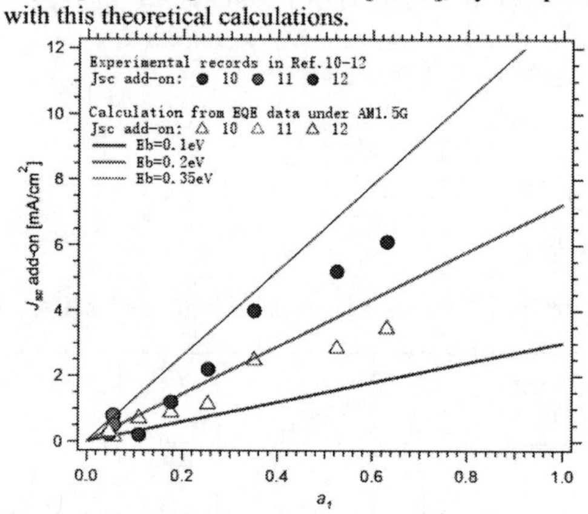

Figure 3: Boosts in short-circuit current density J_{sc} of QD solar cells compared with control cell as a function of binding energy E_b of QDs and QD absorptivity a_1 for QD solar cells [10-12] in comparison with analytical results by using the detailed-balance-limit theory [25,26]

Figure 3 shows the measured boosts in J_{sc} for the QD solar cells [10-12] compared with the corresponding control GaAs cells (circles) as a function of average absorption of QD (a_1). It is proportional to QD volume density and can be crudely estimated by

$$a_1 = \frac{\int_{E_g-E_b}^{E_g} EQE(E)S(E)dE}{\int_{E_g-E_b}^{E_g} S(E)dE}, \qquad (14)$$

where $S(E)$ is AM1.5G solar spectrum, E_b is binding energy of QDs and E_g is the bandgap energy for host materials, respectively. Clearly, Eq. (14) reveals the J_{sc} boosts of QD cells increase as increase in E_b and a_1. On the basis of two-step function absorptivity model [25,26], the calculated J_{sc} boosts against a_1 for E_b = 0.1, 0.2, 0.3 eV (curves), were also plotted in Fig. 3, in order to assess E_b of these QD cells. However, this calculation is probably with relatively low precision, because the absorptivity of practical QD cells differs with the two-step function absorptivity used in this calculation. For this reason, the J_{sc} boosts evaluated from practical EQE records [10-12,21] (triangles) rather than two-step function absorptivity, regarded as more accurate references, were also added into Fig.3. It demonstrates E_b of the QD solar cell fabricated by Rochester Inst. [11] (blue triangle) and that by Ioffe Institute [10] (green triangle) with smaller a_1 < 0.1 are regarded as 0.1eV and 0.2 eV, respectively. While E_b of AIST QD cells [12] with larger $a_I > 0.1$ are observed to locate between 0.1 eV and 0.2 eV. Additionally, it is also observed that these calculations (triangles) deviated from the experimental records (circles), especially at larger a_I region, which is probably caused by making a crude estimation on a_I in these QD solar cells with more QD densities. Even a rough calculation, however, still can exhibit the same tendency as experimental records. As expected[0], J_{sc} boost of those QD solar cells increase gradually as E_b and a_I increase.

Figure 4 shows the evaluated total V_{oc} drops of the QD solar cells [10-12] in reference to $V_{oc,rad}^{GaAs} = 1.15$ V (circles) plotted as a function of QD absorptivity a_I. Also shown in the figure are theoretical calculations for $E_b = 0.1$ eV via the detailed-balance-limit theory [25,26] (curves), where that of $ERE = 1$ (solid), regarded as the intrinsic drop, increases gradually as a_I increasing. While dashed curves for $ERE = 0.1, \ldots, 10^{-8}$ represent total V_{oc} drop containing increasing non-radiative recombination losses. In comparison with these curves, we can crudely assess that both of intrinsic and extrinsic drops in the three QD cells [10,11] (green and blue circles) with lower QD density ($a_I < 0.1$) may be much lower than those [12] (red circles) with more QD density ($a_I > 0.1$).

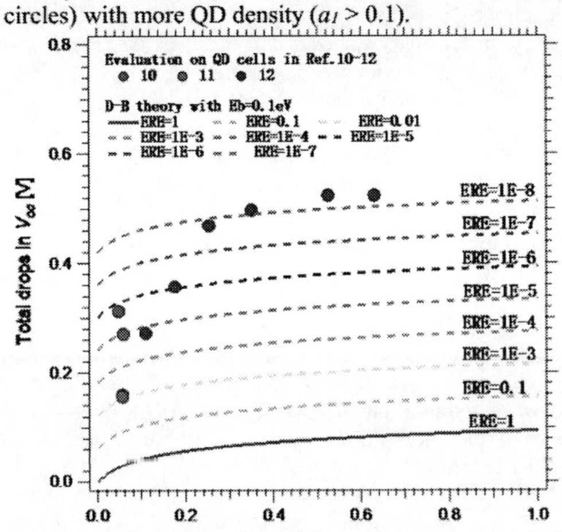

Figure 4: Total open-circuit-voltage loss, $V_{oc,rad}$(GaAs) - V_{oc}, of QD solar cells as a function of binding energy E_b of QDs and QD absorptivity for QD solar cells [10-12] in comparison with calculated results by using the detailed-balance-limit theory [25,26]

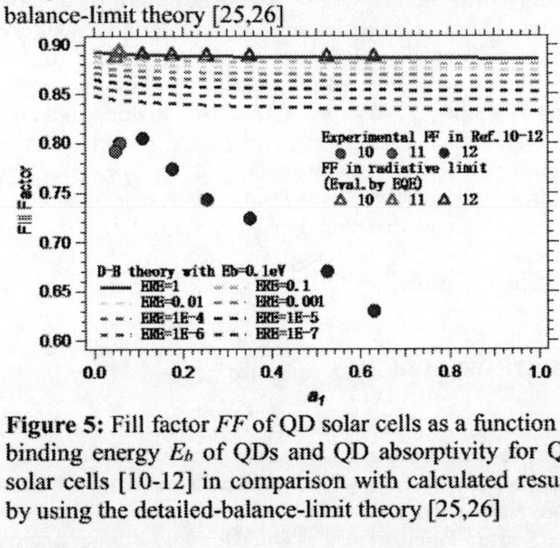

Figure 5: Fill factor FF of QD solar cells as a function of binding energy E_b of QDs and QD absorptivity for QD solar cells [10-12] in comparison with calculated results by using the detailed-balance-limit theory [25,26]

Figure 5 shows the experimental fill factor (FF) of QD solar cells[0-0] (circles) as a function of QD absorptivity (a_I) in comparison with analytical results with binding energy $E_b = 0.1$ eV and $ERE = 1 \sim 10^{-8}$ (curves) in the detailed-balance-limit theory [25,26]. Moreover, the radiative FF (in radiative limit) evaluated from those QD-cell EQE records [10-12] (triangles) are also plotted in Fig.5. Most of them [11,12] well agree with the curve of $ERE = 1$ and $E_b = 0.1$ eV, reflecting that FF drop caused by intrinsic reason is tiny, while non-radiative recombination loss is the primary reason leading to the larger FF drops. Also note that the experimental FF are obviously less than the calculations, even much less than that of extremely low ERE, which can be explained that apart from non-radiative recombination losses, QD cells may also have remarkable resistances losses and/or lower carrier mobility or carrier collection efficiency.

Figure 6 shows experimental conversion efficiency of QD solar cells [10-12] (circles) as a function of QD absorptivity (a_I) in comparison with calculation results of $E_b = 0.1$ eV and $ERE = 1 \sim 10^{-8}$ via the detailed-balance-limit theory [25,26] (curves). The conversion efficiency in radiative limit evaluated from EQE [10-12] (triangles) are also added in Fig. 6, regarded as more accurate assessment with only intrinsic drop than the solid red curve calculated by the two-step function absorptivity model [25-26]. Referenced to the triangles, even for these QD cells with perfect material qualities ($ERE = 1$), high efficiency of over 20% is obtained only for low binding energy of less than 0.1eV and a_I of less than 0.1 or greater than 0.3. In addition to the intrinsic loss, non-radiative recombination losses and/or other losses in QD solar cells are clearly shown as the deviation between the radiative conversion efficiency (triangles) and experimental records (circles). This exhibits these QD cells with the more QD volume density, for example, those with $a_I > 0.2$, almost have the larger extrinsic drops, which results in the lower conversion efficiency. If both intrinsic and extrinsic losses are also taken into account, QD solar cells with low E_b of less than 0.1eV and tiny a_I of less than 0.1 are more promising to achieve over 18% efficiency. For example, 18.3% efficiency has been achieved by QD solar cell [10] with E_b of 0.1 eV and a_I of 0.05, shown as the green circle in Fig.6.

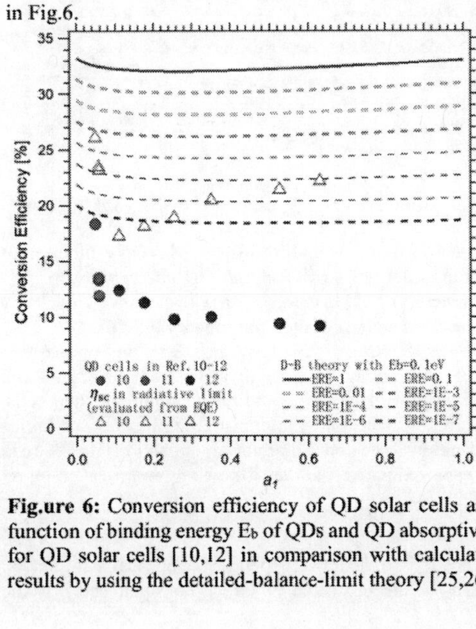

Fig.ure 6: Conversion efficiency of QD solar cells as a function of binding energy E_b of QDs and QD absorptivity for QD solar cells [10,12] in comparison with calculated results by using the detailed-balance-limit theory [25,26]

Figure 7 separately shows the evaluated intrinsic (circles) and extrinsic contributions (triangles) in the total V_{oc} loss for these QD solar cells [10-12] shown in Fig. 4. Meanwhile, the intrinsic part (E_b = 0.1eV, 0.2eV and ERE=1, marked as solid curves) and extrinsic part (E_b = 0.1eV and ERE<1, marked as dashed curves) in detailed-balance calculations [25,26] in Fig. 4 are also extracted and plotted in Fig.7. The evaluated intrinsic V_{oc} drop in the QD solar cells reported by Rochester Inst. and AIST [11,12] (green and blue circles) well matches with the calculations with E_b = 0.1eV (red curve), while that of QD cell from Ioffe Inst. [10] (green circle) is much close to that of E_b = 0.2eV (black curve), reflecting a larger intrinsic drop than other QD cells [11,12]. Compared to intrinsic loss, the extrinsic V_{oc} drops in QD solar cells are observed to be much higher, and are increased not only with increase in QD absorptivity (dot density), but also with decrease in ERE. For example, the AIST QD cells [12] (red triangles) with extremely inferior ERE (< 10^{-5}) and large a_I (> 0.1), have obvious extrinsic losses of greater than 0.3V and intrinsic drop of below 0.1V. The huge V_{oc} drop of 0.4V combining with the moderator J_{sc} of 17.7~23.6 mA/cm^2, lead to the extremely low conversion efficiency of below 13%. In contrast, the QD cell with the highest conversion efficiency of 18.3% [10], is realized by a largest J_{sc} of 27.7 mA/cm^2 accompanied with a moderate V_{oc} drop, including 0.09V of intrinsic loss (green circle) and 0.22V of extrinsic loss (green triangle). By decomposing V_{oc} loss of QD cells into intrinsic and extrinsic contributions, it is easily observed that the voltage loss in the QD solar cell with higher absorptivity a_I is mainly caused by their larger non-radiative recombination losses (extrinsic reason).

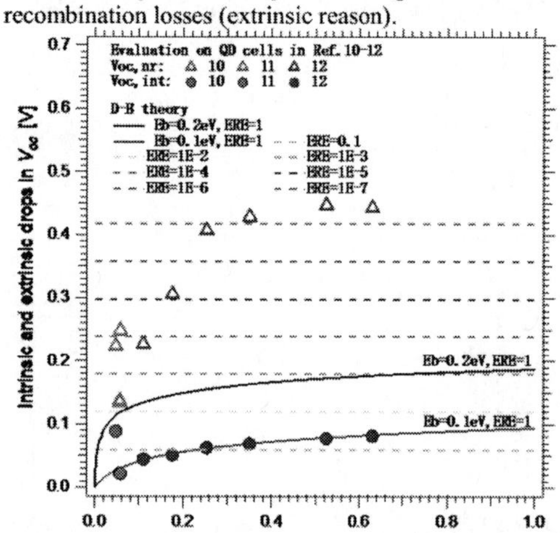

Figure 7: Extrinsic and intrinsic V_{oc} loss as a function of binding energy E_b of QDs and QD absorptivity for QD solar cells [10,12] in comparison with calculated results by using the detailed-balance-limit theory [25,26]

Extrinsic voltage loss mainly arises from the defects due to lattice mismatch growth in the quantum dot solar cells, including dislocations, interface recombination or lattice relaxations. The non-radiative recombination rate (R_{nr}) can be approximated as the following empirical relation [27,28]:

$$R_{nr} = \alpha \frac{\Delta a}{a_0} \frac{N}{h} \quad (15)$$

where α is a material dependent empirical parameters, $\Delta a/a_0$ is the mismatch of lattice constant, and h is the layer thickness, and N is the total number of QD stacks. For simplicity of later discussion, we can group the terms in the right-hand-side of Eq. (15) other than α as r_{nr}, namely,

$$r_{nr} = \frac{\Delta a}{a_0} \frac{N}{h} \quad (16)$$

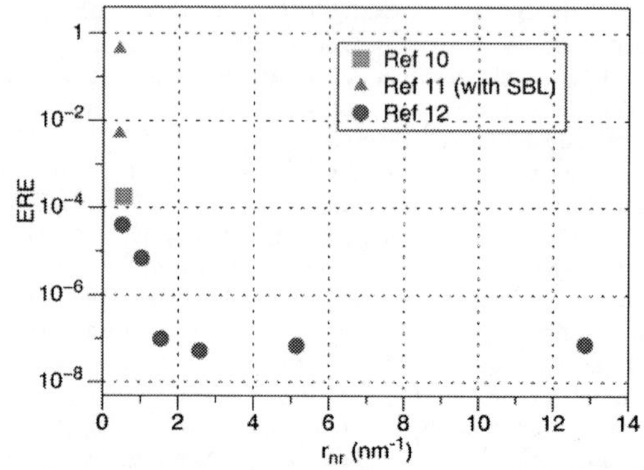

Figure 8: ERE of experimental quantum dot solar cells [6,10-12,28] against r_{nr} of Eq. (16)

Figure 8 plots ERE against r_{nr} of the extracted quantum dot data. It shows that the ERE decreases with r_{nr} in general, in other words, the ERE generally decreases with higher lattice mismatch and larger number of QD stacks. This comparison shows that the EREs of the QD cells with strain-balanced technique is higher than the EREs of QD cells with similar r_{nr}, suggesting that strain-balanced quantum dots is a promising pathway to further reduce the extrinsic loss of quantum dot solar cellss

5. SUMMARY

Efficiency potential of QD solar cells is discussed based on external radiative efficiency (ERE), open-circuit voltage loss and fill factor loss. QD cells with J_{sc} of 32mA/cm^2 have efficiency potential of 19.1%, 20.5% and 22.1% in the case of resistance loss of 0.2, and 21.5%, 23.3% and 24.8% in the case of resistance loss of 0.1, by improvements in ERE to 10^{-3}, 10^{-2} and 0.1, respectively. Effects of non-radiative recombination upon properties of QD solar cells are also discussed. Intrinsic loss of QD solar cells is substantially increased with increase in binding energy of QDs. The QD solar cells are clarified to have non-radiative recombination loss due to interface recombination in addition to intrinsic loss. ERE of QD solar cells are decreased with increase in binding energy and volume density of QDs due to non-radiative recombination. Further improvements in minority-carrier lifetime based on understanding defect behavior, in addition to improvements in front surface, rear surface and interface passivation, and decease in series resistance and increase in shunt resistance are necessary, in order to realize high efficiency QD solar cells.

ACKNOWLEDMENTS

The author thanks members in the NEDO, METI and JST, and to Prof. Y. Ohshita, Prof. A. Yamamoto and Dr. Y. Hayashi at Toyota Technological Institute for their cooperation and fruitful discussions. L. Zhu also thanks to the Startup Foundation for "One-Hundred Young Talents plan" supported by Sun Yat-sen University.

REFERENCES

[1] M. Yamaguchi, H. Yamada, Y. Katsumata, K.-H. Lee, K. Araki, and N. Kojima, J. Mater. Res. 32, 3446 (2017).

[2] M. Yamaguchi, H. Yamada, Y. Katsumata, K.-H. Lee, K. Araki, and N. Kojima, Progress in Photovoltaics, 26, 543 (2018).

[3] K.W.J. Barnham, and G. Duggan, J. Appl. Phys., 67, 3490 (1990)

[4] A. Luque, and A. Marti, Phys. Rev. Lett., 78, 5014 (1997).

[5] Y. Wang, Y. Wen, H. Sodabanlu, K. Watanabe, M. Sugiyam, and Y. Nakano, IEEE Journal of Photovoltaics, 1, 002674 (2011).

[6] Y. Okada, S. Yagi, and R. Ohshima, Oyo Butsuri, 79, 206 (2010) (in Japanese).

[7] M.A. Green, K. Emery, Y. Hishikawa, W. Warta, and E,D. Dunlop, Progress in Photovoltaics, 24, 905 (2016).

[8] B.M. Kayes, H. Nie, T. Twist, S.G. Sproytee, F. Reinhardt, I.C. Kizilyalli, and G.S. Higashi, Proceedings of the 37th IEEE Photovoltaic Specialists Conference, 2011, p. 4-8.

[9] Y. Okada, R. Ohshima, and A. Takata, J. Appl. Phys., 106, 024306 (2009).

[10] S.A. Blokhin, A.V. Sakharov, A.M. Nadtochy, A.S. Pauysov, M.V. Maximov, N.N. Ledentsov, A.R. Koush, S.S. Mikhrin, V.M. Lantratov, S.A. Mintairov, N.A. Kaluzhniy, and N.Z. Shvarts, Semiconductors, 43, 537 (2009).

[11] C.G. Bailey, D.V. Forbes, R.P. Raffaelle, S.M. Hubbard, Appl. Phys. Lett., 8, 163105 (2011)

[12] T. Sugaya, U. Numakami, R. Oshima, S. Furue, H. Komaki, T. Amano, K. Matsubara, Y. Okano, and S. Niki, Energy Environ. Sci. 5, 6233 (2012).

[13] K. Tanabe, D. Guimard, D. Bordel, and Y. Arakawa, Appl. Phys. Lett. 100, 193905 (2012).

[14] Y. Shoji, K. Akimoto, and Y. Okada, Proceedings of the 38th IEEE Photovoltaic Specialists Conference, (IEEE, New York, 2012).

[15] M. Yamaguchi, L. Zhu, H. Akiyama, Y. Kanemitsu, H. Tampo, H. Shibata, K.-H. Lee, K. Araki, and N. Kojima, Jpn. J. Appl. Phys. 57, 04FS03 (2018).

[16] A. Marti, J.L. Balenzategui, *and* R.F. Reyna, Journal of Applied Physics, 82, 4067-4075 (1997).

[17] U. Rau, Phys. Rev. B76, 085303, (2007).

[18] M. A. Green, Solar Cells 7, 337–340 (1982).

[19] M. A. Green. *Solar Cells*. Kensington: UNSW; (1998).

[20] M. Yamaguchi, K. H. Lee, K. Araki, N. Kojima, H. Yamada, and Y. Katsumata. Progress in Photovoltaics: Research and Applications, 26(8), 543-552, (2018).

[21] M. A. Green, Progress in Photovoltaics 20, 472 (2012).

[22] J. F. Geisz, M.A. Steiner, T. Garcia, S.R. Kurtz, D.J. Friedman, Appl. Phys. Lett. 103, 041118 (2013).

[23] M.A. Green, K. Emery, Y. Hishikawa, and W. Warta, Progress in Photovoltaics 18, 346 (2010).

[24] M. A. Green, K. Emery, Y. Hishikawa, and W. Warta, Progress in Photovoltaics 19, 84 (2011).

[25] L. Zhu, H. Akiyama and Y. Kanemitsu, "Intrinsic and extrinsic drops in open-circuit voltage and conversion efficiency in solar cells with quantum dots embedded in host materials". Scientific Report 8, 11704 (2018)

[26] L. Zhu, H. Akiyama, and Y. Kanemitsu, Proceedings of the 44th IEEE Photovoltaic Specialists Conference, (2017).

[27] M. Yamaguchi, Phys. Status Solidi, C12, 489 (2015)

[28] Y. Okada, N.J. Ekins-Daukes, T. Kita, R. Tamaki, M. Yoshida, A. Pusch, O. Hess, C.C. Phillips, D.J. Farrell, K. Yoshida, N. Ahsah, Y. Shoji, T. Sogabe, and J.-F. Guillemoles, Appl. Phys. Express, 2, 021302 (2015).

THE USE AND ABUSE OF W_{OC} AS A FIGURE OF MERIT

N. Ekins-Daukes and A. Pusch
School of Photovoltaic & Renewable Energy Engineering, UNSW Sydney, 2052 NSW, Australia

ABSTRACT: W_{oc} is defined as the difference between the semiconductor band-gap (E_g) and open circuit voltage (V_{oc}) of a solar cell. It has the virtue of being straightforward to determine and can be used to compare the quality of solar cells with different band-gap energies. Here we revisit the fundamental limits to W_{oc} and highlight some pitfalls where it can produce misleading data when applied to solar cells that have relatively low densities of states close to the band-edge; a situation commonly encountered with nanostructures.
Keywords: Electrical Properties, Characterisation, Quantum Well, Quantum Dot,

1 BACKGROUND

The open circuit voltage of a solar cell is bounded by fundamental limits that ensure that V_{oc} is always smaller than the band-gap energy [1]. The offset between the two quantities is known as W_{oc} and is defined as $W_{oc} = E_g/q - V_{oc}$ where E_g is the band-gap energy and V_{oc} is the open-circuit voltage. It is commonly used as a figure of merit for comparing the voltage performance of solar cells with different band-gaps and is convenient since it arises directly from a standard light-IV measurement from which Voc is readily measured.

A comprehensive survey of W_{oc} was performed by King et al. [2] and Yamaguchi [3], comparing W_{oc} for different component sub-cells in a III-V multijunction stack. As a figure of merit for bulk solar cells, W_{oc} is reliable and is approximately comparable between cells with different band-gap. However, W_{oc} is bounded by a lower, radiative limit which itself depends on the band-gap energy. It also depends strongly on the profile of the density of states and devices can be found where the W_{oc} appears to be anomalously low. Under these circumstances the effective band-gap threshold for the material is poorly defined and W_{oc} no longer provides a reliable measure of relative performance. In this paper we develop a simple model for W_{oc} and show examples of both where it is a reliable and where it is an unreliable figure of merit.

2 SIMPLE THEORETICAL MODEL FOR W_{oc}

To illustrate the behaviour of W_{oc} with different profiles for the electronic density of states we construct a simple photovoltaic model for W_{oc} [4]. Following the notation of Rau [5] the current density J delivered by a solar cell at voltage V and temperature T is given by:

$$J(V) = J_{sc} - (J_{em} + J_{ne}).\left(e^{qV/kT} - 1\right) \quad (1)$$

$$J_{em} = \frac{1}{4\pi^2\hbar^3c^2}\int_0^{\infty} EQE_{PV}(E)\frac{E^2}{e^{E/kT}}dE \quad (2)$$

$$J_{ne} = \frac{J_{em} - EQE_{LED}\,J_{em}}{EQE_{LED}} \quad (3)$$

Where J_{sc} is the short circuit current, J_{em} and J_{ne} represent the diode saturation current for radiative and non-radiative recombination respectively. Two distinct quantum efficiency terms are defined: $EQE_{PV}(E)$ corresponds to the external photovoltaic quantum efficiency at a photon energy E and provides a measure of the semiconductor joint density of states. EQE_{LED} corresponds to the external radiative quantum efficiency and accounts for non-radiative processes; taking the value of unity in the radiative limit.

V_{oc} is obtained when J(V)=0 so recognising that $(J_{em} + J_{ne}) \ll J_{sc}$ W_{oc} is given by

$$W_{oc} = \frac{E_g}{q} - \frac{kT}{q}ln\left(\frac{J_{sc}}{(J_{em}+J_{ne})}\right) \quad (4)$$

3 SURVEY OF TYPICAL W_{OC} VALUES

Values for Eg and Voc were taken for a variety of semiconductor materials [2,3,6] and used to produce the chart shown in Figure 1 illustrating the relationship between band-gap energy, W_{oc} and external radiative efficiency EQE_{LED}.

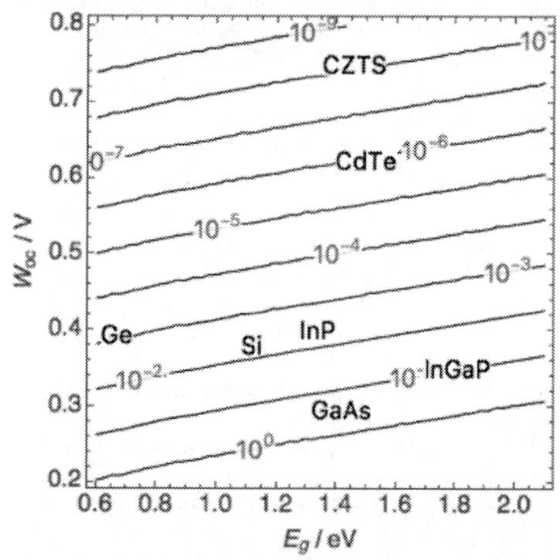

Figure 1: W_{oc} plotted as a function of band-gap energy (E_g). The contours correspond to the external radiative efficiency EQE_{LED}. The named semiconductors correspond to the position of present record solar cell devices.

GaAs holds the lowest W_{oc} of any device, as a consequence of the very high external radiative efficiency. By contrast CZTS devices have a W_{oc} that is more than double that of GaAs and an external radiative efficiency that is eight orders of magnitude smaller. W_{oc} therefore provides a forgiving, linear scale for comparing PV device performance, while external radiative efficiency provides an exponential and arguably less memorable measure. However, external radiative efficiency has the virtue of having a single limiting value of unity, whereas the radiative limit to W_{oc} (shown as the 10^0 line in Figure 1) spans 0.2V to 0.3V for the band-gap energy range shown.

4 ARBITRARY PHOTOVOLTAIC EQE PROFILE

An ideal Shockley-Queisser solar cell has $EQE_{PV}=1$ for all photon energies greater than the band-gap. This approximation is reasonable in the case of bulk semiconductor solar cells and has been assumed for the calculation of external radiative efficiency contours in Figure 1. However, the profile for EQE_{PV} can have a dramatic effect on the W_{oc} of a photovoltaic device. To illustrate this point we now consider three solar cells based around a GaAs single junction cell with different EQE_{PV} profiles shown in Figure 2:

1. Shockley-Queisser solar cell with a band-gap at 1.42eV shows an abrupt transition from $EQE_{PV}=1$ to $EQE_{PV}=0$ at 1.42eV
2. GaAs solar cell has $EQE_{PV}=0$ for E<1.42eV but rises more gently, saturating around 1.6eV.
3. InAs quantum dot device which has EQE_{PV} profile that matches GaAs at energies > 1.44eV and tapers slowly to $EQE_{PV}=0$ at 1.3eV. The profile is derived from data published by Bailey et al., [7]

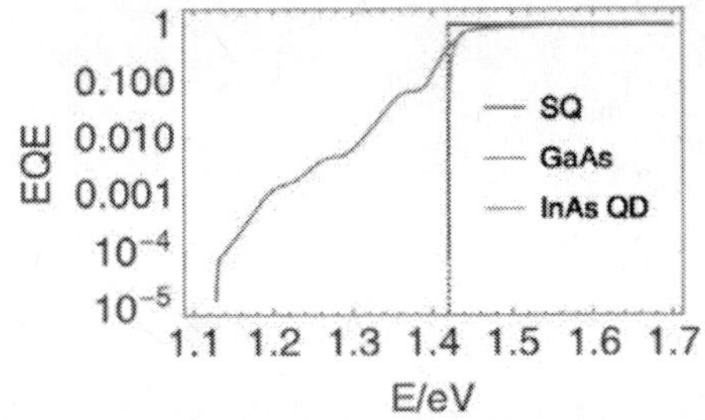

Figure 2: EQE profiles plotted on a logarithmic scale for three different solar cells: a Shockley-Queisser cell with band-gap 1.42eV, a bulk GaAs solar cell, also with band-gap of 1.42eV and an InAs quantum dot solar cell with a host band-gap of 1.42eV but with QD absorption extending to 1.3eV.

For clarity the W_{oc} for these EQE profiles is calculated assuming unity external radiative efficiency $EQE_{LED}=1$ and presented in Table 1.

Cell	Eg	Voc	Woc
SQ	1.42eV	1.17V	0.25V
GaAs	1.42eV	1.18V	0.24V
SQ	1.13eV	0.89V	0.23V
InAs QD	1.13eV	1.09V	0.04V

Table 1: Values for Eg, Voc and Woc for cells with different EQE profiles.

The abrupt onset in the EQE profile for the GaAs solar cell leads to a V_{oc} and hence W_{oc} that is similar to the equivalent Shockley-Queisser cell. Indeed, the V_{oc} for the GaAs cell is slightly higher than that of the SQ (1.42eV) cell on account of the finite gradient of the EQE profile; the integral in Eqn.2 will be maximum for a SQ step-function leading to the lowest value for J_{em}.
To make a meaningful comparison with the InAs QD cell another Shockley-Queisser cell with $E_g=1.13eV$ is considered (EQE profile is not plotted). Here we find the SQ limit to W_{oc} is 0.23V, comparable to the earlier values for W_{oc} and consistent with the radiative limit plotted in Figure 1. However, the W_{oc} for the InAs QD cell is unphysically small, 0.04V. This arises since although the "band-gap" has been taken at the lowest confined state of the QD (1.13eV), the EQE is extremely small at this wavelength. The open circuit voltage for the QD device is reported at 0.994V [7] which compares favorably with the radiatively limited voltage of 1.09V shown in Table 1. We emphasise that the voltage result of Bailey et al., [7] is sound (and they do not abuse W_{oc}) but their data provides a good example of how surprisingly high values of V_{oc} arise naturally from devices with a very low density of electronic states at the absorption edge.

To further illustrate this point further we approximate the EQE_{PV} profile for a confined semiconductor solar cell as two step-functions shown in figure 3.

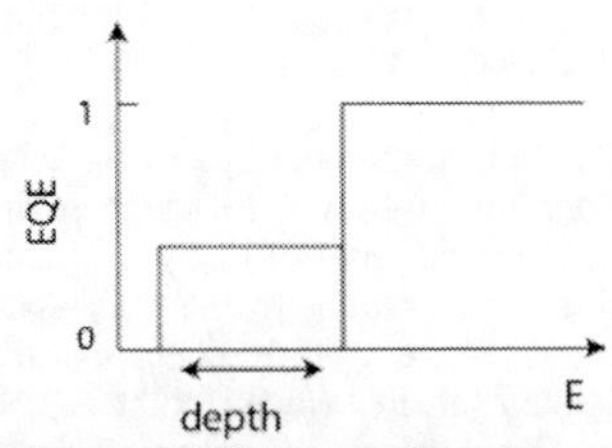

Figure 3: Stepped EQE_{PV} profile where a sub-unity EQE_{PV} region exists due to quantum confinement over an energy range indicated by 'depth'.

This stepped EQE_{PV} profile could correspond to a quantum well [8] or quantum dot [7] device where unity EQE_{PV} is obtained in the bulk barrier region of the device but the limited density of electronic states leads to sub-unity EQE_{PV} over an energy range given by the confinement depth. In a quantum well device, a good rule of thumb is that each confined layer and level contributes 1% to the EQE_{PV}, while in a QD device each confined layer and level contributes 0.2% to the EQE_{PV}.

Evaluating W_{oc} while varying the lower energy EQE_{PV} step shown in Figure 3, again with unity external radiative efficiency, results in the contour map shown in Figure 4.

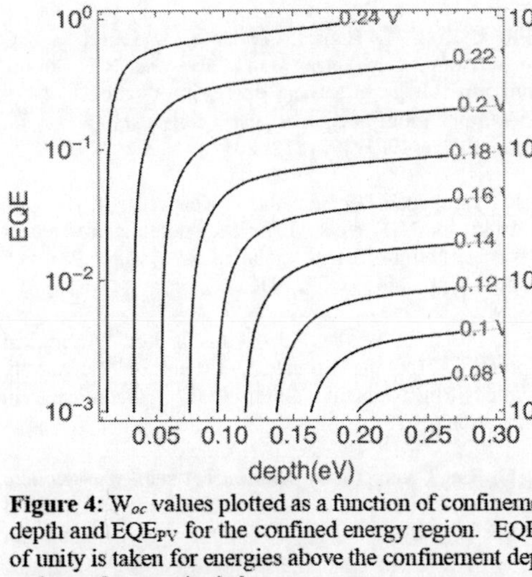

Figure 4: W_{oc} values plotted as a function of confinement depth and EQE_{PV} for the confined energy region. EQE_{PV} of unity is taken for energies above the confinement depth and zero for energies below.

Taking a cross-section section at 0.15eV we find that the limiting value for W_{oc} can vary from 0.24V down to less than 0.1V. It is clear that comparing W_{oc} for solar cells with different low energy EQE_{PV} profiles can give very different values. Under these circumstances W_{oc} is no longer a good figure of merit at all.

In devices with a strong absorption onset (e.g. GaAs), the rise in EQE_{PV} at the band-edge coincides with the band-gap energy and W_{oc}, as defined in eqn. 4, is valid. In QD and few QW devices this is no longer the case: the band-gap (as defined as the energy separation between occupied and unoccupied states) no longer coincides with the onset of significant absorption. As a result, W_{oc} should not be used as a figure of merit for QW or QD devices.

5 CONCLUSIONS

W_{oc} provides a reliable figure of merit for solar cells with similar absorption profiles. Efficient solar cells will generally have a rapid onset of EQE_{PV} at photon energies above the band-gap, similar to the bulk GaAs example described above. However, if the concept of W_{oc} is applied to cells with unusually low values for the EQE at the band-edge, such as QD and few-QW systems, values of W_{oc} can be obtained that approach zero. Under these circumstances, W_{oc} gives an overly optimistic and ultimately misleading measure of the relative recombination rate in the solar cell.

ACKNOWLEDGEMENTS

The authors gratefully acknowledge Professors Masafumi Yamaguchi and Martin Green for useful discussions.

REFERENCES

[1] L. C. Hirst and N. J. Ekins-Daukes, "Fundamental losses in solar cells," *Prog Photovoltaics*, vol. 19, no. 3, pp. 286–293, 2011.

[2] R R King, D Bhusari, A Boca, D Larrabee, X Q Liu, W Hong, C M Fetzer, D C Law, and N H Karam. Band gap-voltage offset and energy production in next-generation multijunction solar cells. *Progress In Photovoltaics*, 19(7):797–812, 2011.

[3] M. Yamaguchi, H. Yamada, Y. Katsumata, K.-H. Lee, K. Araki, and N. Kojima, "Efficiency potential and recent activities of high-efficiency solar cells," *J Mater Res*, vol. 32, no. 18, pp. 3445–3457, 2017.

[4] A. Pusch, P. Pearce, N.J. Ekins-Daukes, "Analytical Expressions for the Efficiency Limits of Radiatively Coupled Multi-Junction Solar Cells", manuscript in preparation (2018).

[5] U. Rau, "Reciprocity relation between photovoltaic quantum efficiency and electroluminescent emission of solar cells," Phys Rev B, vol. 76, no. 8, p. 085303, 2007.

[6] M. Green, Y. Hishikawa, E. D. Dunlop, D. H. Levi, J. H. Ebinger, and A. W. Y. Ho-Baillie, "Solar cell efficiency tables (version 51)," *Prog. Photovolt: Res. Appl.*, vol. 26, no. 1, pp. 3–12, 2018.

[7] Christopher G Bailey, David V Forbes, Ryne P Raffaelle, and Seth M Hubbard. Near 1 V open circuit voltage InAs/GaAs quantum dot solar cells. *Applied Physics Letters*, 98(16):163105, 2011.

[8] N. J. Ekins-Daukes, J. Barnes, K. Barnham, J. Connolly, M. Mazzer, J. Clark, R. Grey, G. Hill, M. Pate, and J. Roberts, "Strained and strain-balanced quantum well devices for high-efficiency tandem solar cells," *Sol Energ Mat Sol C*, vol. 68, pp. 71–87, 2001.

TRANSITION METAL OXIDES AS PASSIVATED HOLE-CONTACTS LAYER FOR SILICON SOLAR CELLS: INTRINSIC AND EXTRINSIC DEFECTS IN MoO_3 FROM FIRST-PRINCIPLES CALCULATIONS

Md. Anower Hossain[1,2*], Sergey Rashkeev[1], Vinod Madhavan[1], Tian Zhang[2], Chang-Yeh Lee[2], Bram Hoex[2], Nouar Tabet[1], and Amir Abdallah[1]
[1]Qatar Environment and Energy Research Institute, Hamad Bin Khalifa University, Doha 34110, Qatar
[2]School of Photovoltaic and Renewable Energy Engineering, UNSW Sydney, Sydney 2052, Australia
*Corresponding author: Md. Anower Hossain, Phone: +61 (0)468450858, e-mail: a.hossain@unsw.edu.au

ABSTRACT: In this work, we present first-principles calculations for different defects in MoO_3 oxide which is considered as a hole-selective contact for crystalline silicon solar cells. An attempt was made to understand the role of different defects in charge carrier transport. In particular, we consider intrinsic defects including oxygen vacancies (V_O) and molybdenum vacancies (V_{Mo}) as well as extrinsic defects (dopants) in MoO_3, namely, hydrogen (H), nitrogen (N), and silicon (Si). We show that some dopants have a shallow level in the MoO_3 bandgap close to the conduction band minimum (CBM) which could positively affect hole collection from silicon. The methodology used in this work should facilitate rational design of materials and processes for the charge carrier selective contacts for silicon solar cells.

Keywords: Transition metal oxides, Semiconductors, Defects, c-Si, Silicon solar cells, Doping, Passivation, Density functional theory, Modelling, Projected density of states, Charge carrier recombination

1 INTRODUCTION

The passivated emitter and rear contact (PERC) solar cell technology with its current record efficiency close to 25% on monocrystalline silicon, is predicted to dominate silicon solar cell technologies for mass production in the next decade.[1] It is generally expected that the efficiency of PERC silicon solar cells manufactured in high volume can go up to 25%.[2] To further improve the efficiency, silicon solar cells with so-called carrier-selective contacts are required. Currently, doped amorphous and poly-silicon contacts are the most mature carrier selective contacts that enabled silicon solar cells to reach efficiencies above 26%.[3, 4] An alternative approach is to use undoped transition metal oxides such as molybdenum trioxide (MoO_3), tungsten oxide (WO_3), and nickel oxide (NiO_x) as hole-selective contacts, which is particularly appealing as these layers show a significantly lower optical absorption compared to their doped silicon-based counterparts.[5-13]

MoO_3 is an emerging material which has shown its potential as hole-selective layer in silicon solar cells.[11-13] It has fascinating electrical, optical and band energetic properties that are known to be crucial for passivating hole-contact of silicon solar cells and for interfacial hole transporting layer of perovskite solar cells. Because of its high transparency in the visible part of the solar spectrum due to its relatively high bandgap of 3.2 eV,[14, 15] and its large work function of about 6 eV, MoO_3 was also used as a transparent contact in solar cells and organic electronics.[12]

As the electronic properties of semiconductors strongly depend on the presence of point defects, many of their electronic structure properties can be tuned by incorporating corresponding intrinsic defects and extrinsic dopants. While deep level defects act as minority charge carrier recombination centers and suppress the quantum efficiency of solar cells, incorporation of certain defects, especially in the carrier-collecting region, can assist the transport of majority carriers and reduce the concentration of minority carriers by suppressing the recombination in the carrier-collecting region.[16, 17] Therefore, defects with energy levels positioned in the region of carrier collection, *i.e.*, near the valence and conduction band edges of the interface between the contact material and Si could improve the performance of carrier selective contacts.

In this work, we tried to understand how different defects modify the electronic properties MoO_3. In particular, we considered intrinsic defects including oxygen vacancies (V_O) and molybdenum vacancies (V_{Mo}). The oxygen vacancy states are known as *n*-type defects for MoO_x ($x < 3$). It was shown that due to the high MoO_3 workfunction, the band alignment with silicon results in holes accumulating in silicon near the c-Si and molybdenum oxide interface and some defects in MoO_x film may facilitate hole transit through defect derived states. In particular, it was shown that dopant-free, substoichiometric molybdenum trioxide (MoO_x, $x < 3$) could be a hole-selective contact for silicon solar cells.[10]

In this work, we performed first-principles, density functional theory (DFT) based electronic structure calculations for a number of intrinsic defects (vacancies) as well as for extrinsic defects (dopants) including hydrogen (H), nitrogen (N), and silicon (Si) in MoO_3 and showed that both vacancies and dopants can positively affect the electronic properties of the material. Introduction of these defects typically occurs in the material fabrication process and depends on temperature, pressure, deposition environment, *etc*. These parameters may significantly affect the physical properties of the synthesized MoO_3 material. We calculated the defect formation energies and the defect level energies for the charged defects with respect to the MoO_3 bandgap. The defect level energies were found to significantly vary for different dopants which could provide a possibility to design MoO_3 contacts for solar cells with optimal transport properties.

2 COMPUTATIONAL DETAILS

The first-principles calculations were conducted using DFT based with projector augmented wave (PAW)[18] pseudopotentials as implemented in the Vienna Ab-initio Simulation Package (VASP).[19] The Perdew-Burke-Ernzerhof (PBE)[20] exchange-correlation density functionals as well as the Mo_pv ($4p^65s^14d^5$) and O

($2s^22p^4$) pseudopotentials were used in the calculations. Apart from the generalised gradient approximation (GGA) density functionals, the (GGA+U) methodology known as Dudarev's approach[21] (with U = 4.2 eV) was employed to take the on-site electronic correlations for localised d-orbitals of the transition metal atoms into account. A bulk 3×1×3 supercell of MoO_3 containing 36 formula units (144 atoms) was utilized in this study. A large supercell was considered to reduce errors related to finite size effects as described in Ref.[22] A plane wave basis set with a 520 eV energy cut-off was used to construct the electronic wavefunctions. The supercell was fully relaxed until the residual forces acting on all atoms became less than 0.01 Å eV^{-1}. The Brillouin zone was sampled using the Monkhorst-Pack scheme.[23] Van der Waals interactions between neighbouring MoO_3 bilayers were incorporated by the D3 method of Grimme which was reported to be successful in addressing structural and electronic properties.[24] The electronic density of states (DOS) was computed using a tetrahedron scheme; all performed calculations were spin-polarised.

Defect formation energy calculations provide useful information on defect stability as well as on thermodynamic transition levels between different charge states. The defect energy levels indicate whether a particular defect is a shallow donor/acceptor or deep defect. As the formation of a charged defect can be viewed as an exchange of electrons between the host atoms and the electronic reservoir, the formation energy of the defect (D) depends on the electronic Fermi energy (μ_e) and charge (q) as follows:[15, 25-27]

$$E^f(D,q) = E^D_{tot}(q) - E^{bulk}_{tot} + \sum n_i\mu_i + q(E_{VBM} + \mu_e) + \Delta^q \quad (1)$$

where $E^D_{tot}(q)$ is the total energy of the supercell containing the defect, and E^{bulk}_{tot} is the energy of the perfect supercell; n_i and μ_i are the number of atoms of a species i removed from/added to the supercell, and the corresponding chemical potential of the atom, respectively; the last term is the energy correction applied for charged defects to counter the self-interaction across the periodic boundary conditions of finite supercells. The zero of the electronic Fermi energy (μ_e) corresponds to the valence band maximum (E_{VBM}) in the defect free perfect supercell; the μ_e values are varied from 0 to the band gap energy (3.2 eV) in MoO_3. The corrections related to band alignments, and interactions among the charged defects originated from periodic boundary condition of supercell, were also conducted.[25, 28] It was previously reported by Lambert *et al.*[26] that for MoO_3, the chemical potential of Mo atom should be calculated using the BCC-Mo crystal, while the chemical potential of oxygen gas molecule should be calculated considering the standard formation energy of MoO_3:

$$\Delta G_f^{MoO_3}(p^0_{O_2}, T^0) = \mu_{MoO_3(s)} - \mu_{Mo(s)} - \frac{3}{2}\mu_{O_2(g)}(p^0_{O_2}, T^0) \quad (2)$$

where $\Delta G_f^{MoO_3}(p^0_{O_2}, T^0)$ and $\mu_{O_2(g)}(p^0_{O_2}, T^0)$ are the formation energy of MoO_3 and chemical potential of oxygen at standard condition. Because of its gaseous state, the chemical potential of oxygen strongly depends on pressure and temperature. However, the chemical potential of MoO_3 ($\mu_{MoO_3(s)}$) and Mo ($\mu_{Mo(s)}$) does not strongly depend on the partial pressure and temperature due to their solid nature. Therefore, the chemical potential of oxygen gas at room temperature was determined using the ideal gas law as follows:

$$\mu_{O_2(g)}(p_{O_2}, T) = \mu_{O_2(g)}(p^0_{O_2}, T^0) + \frac{1}{2} k_B T \log\left(\frac{p_{O_2}}{p^0_{O_2}}\right) \quad (3)$$

An assumption that the chemical potential of oxygen in MoO_3 is equal to that of oxygen gas corresponds to the case of O-rich conditions.

3 RESULTS AND DISCUSSIONS

3.1 Intrinsic defects: oxygen and molybdenum vacancies in MoO_3

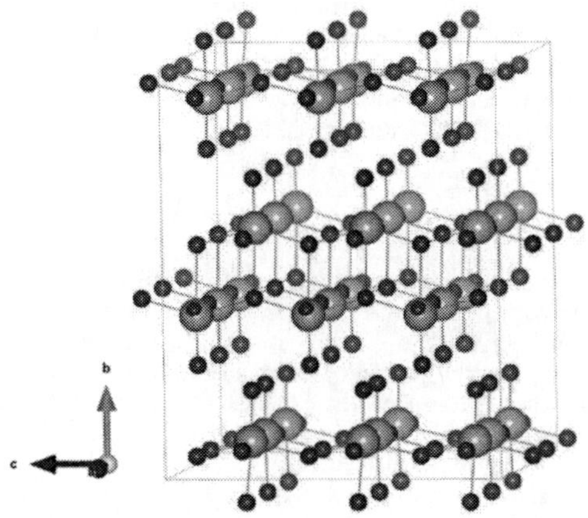

Figure 1. A schematic of fully relaxed 3×1×3 supercell of orthorhombic α-MoO_3 structure. Mo and O atoms are shown by purple and red spheres, respectively.

As shown in Figure 1, the layered α-MoO_3 is comprised of weak Van der Waals bonded layers of edge-sharing Mo–O_6 octahedra. The three types of symmetry-equivalent oxygen atomic sites are: symmetric (O_s) sites which connect the Mo atoms to form chains in the *c*-direction, asymmetric (O_a) sites which connect two Mo atoms and are oriented along the *a*-direction, and terminal (O_t) sites which face interlayer spacing along *b*-direction. The formation energy vs. the Fermi level are plotted for the oxygen vacancy (V_O) at all three inequivalent oxygen sites, for the charge states $q = 0, +1$, and $+2$ (Figure 2). The oxide growth condition was assumed to be oxygen-rich, meaning that the removed oxygen atoms from the MoO_3 structure will form oxygen molecules in the ambient atmosphere. Among the multiple possible charge states of the point defect, the lowest formation energy corresponds to the most stable charge state for a given Fermi level. A similar approach should be used for oxygen-poor conditions.

The oxygen vacancy formation energies for the obtained neutral states were 3.86 eV, 2.0 eV, and 1.56 eV for the symmetric, asymmetric, and terminal sites, respectively, suggesting that vacancy formation process is favourable for the terminal sites rather than for two other sites. A transition of +2 state to +1 state followed by the transition from +1 to neutral state was observed for the oxygen vacancies when moving through the band gap. The transition energy of +2/+1 transition is the lowest for V_{Oa}

(0.24 eV above VBM, Figure 2b). These result qualitatively agrees with results obtained earlier in Ref.[26]

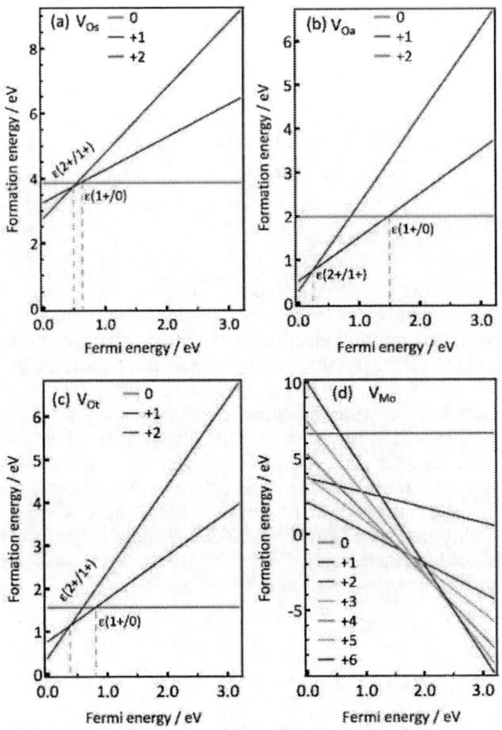

Figure 2. The vacancy formation energies at (a) O_s; (b) O_a, and (c) O_t sites of the MoO_3 supercells at charge states of $q = 0$, +1, and +2 at the O-rich condition. Figure (d) shows the formation energy of the Mo vacancies at different charge states and O-rich condition.

The projected density of states (PDOS) plots in Figure 3 (a) shows that the conduction band of MoO_3 is dominated by the Mo states. Also, the PDOS plots indicate the appearance of V_O defect states in the MoO_3 band gap [Figures 3(b,c,d)], suggesting that the electrons released from a corresponding oxygen vacancy after removing a neutral oxygen atom have a tendency to occupy empty Mo 4d orbitals positioned at the bottom of the conduction band. In particular, it is clear that oxygen vacancies created within the MoO_3 layer [at the symmetric (O_s) and asymmetric (O_a) sites] shows levels closer to the CBM (i.e., are more shallow) than vacancies formed at the transition (O_t) sites with deeper defect levels. This is understandable because removal of transition oxygen atom will inevitably form dangling bonds that typically are deep level defects. Also, this result concludes that charge carrier transport assisted by oxygen vacancy defects is anisotropic and will mainly occur within the layers.

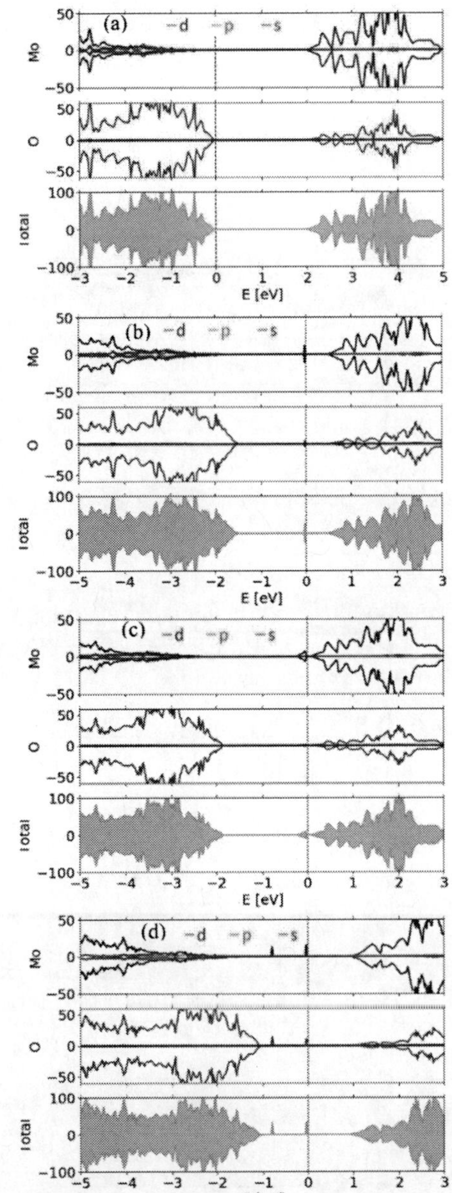

Figure 3. PDOS of: (a) pure MoO_3 supercell, (b) supercell with V_{Os} vacancy, (c) supercell with V_{Oa} vacancy, and (d) supercell with V_{Ot} vacancy. All vacancies are in their neutral charge state. The blue, green and red lines represent the d, p and s orbital contributions of corresponding atoms. The total DOS near the CBM is mainly composed of Mo 4d states, and the DOS near the VBM is mainly of O 2p states.

3.2 Extrinsic defects: substitutional Si(Mo), N(O), and interlayer H

Because of the observed high concentration of hydrogen in the as-deposited metal oxide by atomic layer deposition, we also considered hydrogen-induced defects as well as a few other extrinsic defects, such as Si(Mo) (silicon substituting molybdenum) and N(O) (nitrogen substituting oxygen) in the MoO_3 matrix. Some selected results are shown in Figure 4.

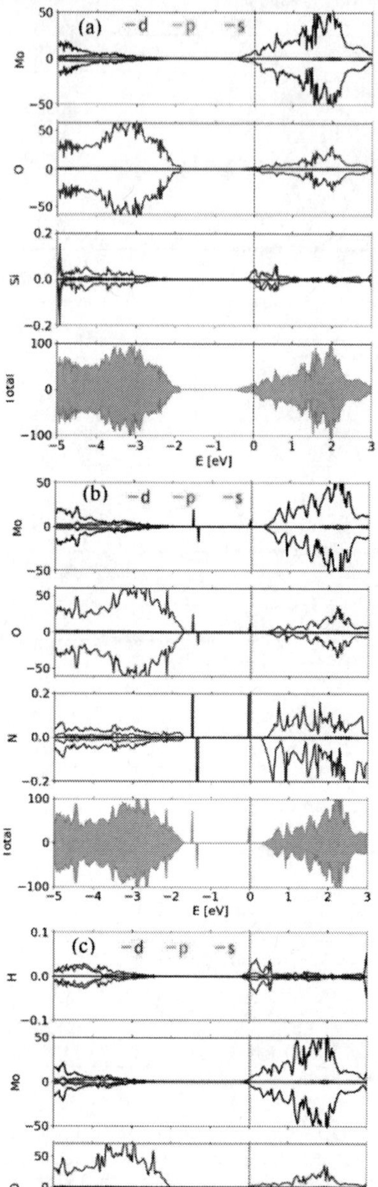

Figure 4. PDOS plots of the (a) Si (Mo), (b) N(O), and (c) interlayer H incorporated in the MoO_3 structure showing defect levels in the band gap. The blue, green and red colours represent the d, p and s orbital contributions of the respective elements, respectively.

The position of Fermi level for charge neutral defects indicates that Si(Mo) and interlayer H defects provide shallow levels positioned near the CBM in MoO_3 while N(O) levels are located deeper in the bandgap. It means that presence of nitrogen at the MoO_3 film fabrication stage may form deep dopant levels assisting carrier recombination in the device while the presence of hydrogen induced defects should not negatively affect the performance of solar cells containing MoO_3 passivating layer.

4 CONCLUSIONS

In summary, we considered the incorporation of different intrinsic and extrinsic defects in the MoO_3 which results in an appearance of defect levels within the band gap of the oxide. We suggest that some of these defects have shallow levels in the MoO_3 bandgap and, therefore, should suppress the recombination of charge carriers in the carrier-collecting region of the silicon solar cells thus increasing the solar energy conversion efficiency. Energy levels for other defects, however, are located deeper in the bandgap and may assist carrier recombination which is highly undesirable. The methodology used in this work should facilitate rational design of materials and processes for the charge carrier selective contacts in silicon solar cells.

5 ACKNOWLEDGEMENTS

This work was supported by a grant from the Qatar National Research Fund (a member of Qatar Foundation; Grant No. NPRP9-021-2-009).

6 REFERENCES

1. Green, M. A. *Sol. Energ. Mat. Sol. C.* **2015,** 143, 190-197.
2. Green, M. A.; Emery, K.; Hishikawa, Y.; Warta, W.; Dunlop, E. D.; Levi, D. H.; Ho-Baillie, A. W. Y. *PROG. PHOTOVOLTAICS* **2017,** 25, (1), 3-13.
3. Feldmann, F.; Bivour, M.; Reichel, C.; Hermle, M.; Glunz, S. W. *Sol. Energ. Mat. Sol. C* **2014,** 120, 270-274.
4. Yoshikawa, K.; Kawasaki, H.; Yoshida, W.; Irie, T.; Konishi, K.; Nakano, K.; Uto, T.; Adachi, D.; Kanematsu, M.; Uzu, H.; Yamamoto, K. *Nature Energy* **2017,** 2, 17032.
5. Melskens, J.; Loo, B. W. H. v. d.; Macco, B.; Black, L. E.; Smit, S.; Kessels, W. M. M. *IEEE Journal of Photovoltaics* **2018,** 8, (2), 373-388.
6. Zhang, T.; Lee, C.-Y.; Wan, Y.; Lim, S.; Hoex, B. *J. Appl. Phys.* **2018,** 124, (7), 073106.
7. Zhang, T.; Lee, C.-Y.; Gong, B.; Lim, S.; Wenham, S.; Hoex, B. *J. Vac. Sci. Technol. A* **2018,** 36, (3), 031601.
8. Chang-Yeh, L.; Mohammad Izzat Abdul, A.; Stuart, W.; Bram, H. *Jpn. J. Appl. Phys.* **2017,** 56, (8S2), 08MA08.
9. Hoex, B.; Dielen, M.; Lei, M.; Zhang, T.; Lee, C.-Y. *AIP Conf. Proc.* **2018,** 1999, (1), 040010.
10. Battaglia, C.; Nicolás, S. M. d.; Wolf, S. D.; Yin, X.; Zheng, M.; Ballif, C.; Javey, A. *Appl. Phys. Lett.* **2014,** 104, (11), 113902.

11. Battaglia, C.; Yin, X.; Zheng, M.; Sharp, I. D.; Chen, T.; McDonnell, S.; Azcatl, A.; Carraro, C.; Ma, B.; Maboudian, R.; Wallace, R. M.; Javey, A. *Nano Lett.* **2014,** 14, (2), 967-971.
12. Sacchetto, D.; Jeangros, Q.; Christmann, G.; Barraud, L.; Descoeudres, A.; Geissbühler, J.; Despeisse, M.; Hessler-Wyser, A.; Nicolay, S.; Ballif, C. *IEEE Journal of Photovoltaics* **2017,** 7, (6), 1584-1590.
13. Woojun, Y.; James, E. M.; Eunhwan, C.; David, S.; Nicole, A. K.; Erin, C.; Young-Woo, O.; Phillip, P. J.; Ajeet, R.; Robert, J. W. *Jpn. J. Appl. Phys.* **2017,** 56, (8S2), 08MB18.
14. Carcia, P. F.; McCarron, E. M. *Thin Solid Films* **1987,** 155, (1), 53-63.
15. Peelaers, H.; Chabinyc, M. L.; Van de Walle, C. G. *Chem. Mater.* **2017,** 29, (6), 2563-2567.
16. Richter, A.; Hermle, M.; Glunz, S. W. *IEEE Journal of Photovoltaics* **2013,** 3, (4), 1184-1191.
17. Liu, Y.; Stradins, P.; Deng, H.; Luo, J.; Wei, S.-H. *AIP Appl. Phys. Lett.* **2016,** 108, (2), 022101.
18. Blöchl, P. E. *Phys. Rev. B* **1994,** 50, (24), 17953-17979.
19. Kresse, G.; Joubert, D. *Phys. Rev. B* **1999,** 59, (3), 1758-1775.
20. Perdew, J. P.; Burke, K.; Ernzerhof, M. *Phys. Rev. Lett.* **1996,** 77, (18), 3865-3868.
21. Dudarev, S. L.; Botton, G. A.; Savrasov, S. Y.; Humphreys, C. J.; Sutton, A. P. *Phys. Rev. B* **1998,** 57, (3), 1505-1509.
22. Lany, S.; Zunger, A. *Phy. Rev. B* **2008,** 78, (23), 235104.
23. Monkhorst, H. J.; Pack, J. D. *Phys. Rev. B* **1976,** 13, (12), 5188-5192.
24. Grimme, S.; Antony, J.; Ehrlich, S.; Krieg, H. *J. Chem. Phys.* **2010,** 132, (15), 154104.
25. Walle, C. G. V. d.; Neugebauer, J.; E., P. W. *J. Appl. Phys.* **2004,** 95, (8), 3851-3879.
26. Lambert, D. S.; Murphy, S. T.; Lennon, A.; Burr, P. A. *RSC Advances* **2017,** 7, (85), 53810-53821.
27. Jiang, H.; Stewart, D. A. *ACS Appl. Mater. Interfaces* **2017,** 9, (19), 16296-16304.
28. Freysoldt, C.; Neugebauer, J.; Van de Walle, C. G. *Phys. Rev. Lett.* **2009,** 102, (1), 016402.

NEARLY 30% HIGH EFFICIENCY LOW CONCENTRATION INGAP/GAAS//SI 3-JUNCTION SOLAR CELLS USING SMART STACK TECHNOLOGY

Kikuo Makita[1], Hidenori Mizuno[1], Masaaki Baba[2], Ryuji Oshima[1], Takeshi Tayagaki[1], Taketo Aihara[1], Takashi Nakamoto[3], Noboru Yamada[2], Hidetaka Takato[1], and Takeyoshi Sugaya[1]
[1]National Institute of Advanced Industrial Science and Technology (AIST), Japan
[2]Nagaoka University of Technology, Japan
[3]Tokyo City University, Japan
AIST Tsukuba, Central 2, Umezono 1-1-1, Tsukuba, Ibaraki, 305-8568, JAPAN
Phone: +81-29-861-3263, Fax: +81-29-861-5615, E-mail: kikuo-makita@aist.go.jp

ABSTRACT: Low concentration multi-junction (MJ) solar cells have a practical solution to consist with high efficiency and low cost. We have demonstrated a low concentration InGaP/GaAs//Si 3-junction solar cell with 23.74% at 8.0 suns using Smart Stack technology, which is a unique mechanical technology with Pd nanoparticle alignment. In this report, it is focused on the reduction of the bonding resistance. Especially, the bonding resistance at Pd/Si interface had become an important issue for low concentration performance. We examined the surface treatment of Si cell using Buffered Hydrogen Fluoride (BHF) solution. As a result, the bonding resistance at the high current density is reduced, and cell performance for low concentration was improved drastically. Using this technology, we fabricated an InGaP/GaAs//Si 3-junction solar cell and observed the maximum efficiency of 28.68% at 10.3 suns (AM1.5g). These results strongly suggested the high potential of our Smart Stack technology to achieve high efficiency GaAs//Si MJ solar cells for practical use.
Keywords: Silicon, III-V Semiconductors, Nanoparticles, Multijunction Solar Cell

1 INTRODUCTION

Multi-junction (MJ) solar cells have enabled very high efficiencies over 30% due to the effective utilization of the sun's energy. The MJ solar cells using mechanical stacking technology have been particularly interested as next generation solar cell [1, 2]. The mechanical stacking technology enables high efficiency and low-cost because of flexible and most appropriate combination of different cells. Among them, GaAs/Si MJ solar cell is a useful combination. GaAs-based cell essentially has high efficiency. Si cell has high sensitivity for long wavelength region and holds the promise of low cost bottom cell. According to our theoretical prediction, the current-matched InGaP/GaAs/Si 3-junction solar cells have a potential to reach more than 30% efficiency. In addition, low concentration technique is a practical solution to obtain cost effective system. Low concentration system does not require a tracking system. According to our cost model, low concentration systems below 10 suns realize a competitive cost level.

Against this background, we have developed a unique mechanical staking method using a Pd nanoparticle alignment, which is named "Smart Stack" technology [3-6]. Using this method, we successfully reported an InGaP/GaAs//Si 3-junction solar cell with 25.1% efficiency at 1sun [7], and confirmed the long-term reliability under sever conditions as initial inspection [8]. On the other hand, we reported the low concentration characteristics with the InGaP/GaAs//Si 3-junction solar cells. However, the maximum efficiency of 23.74% at 8.0 suns [8] was not enough. In this report, we focused on the improvement of the bonding resistance, particularly for Pd/Si interface. As a result, we realized the nearly 30% high efficiency for low concentration performance.

2 DEVICE APPROACH

Figure 1 shows a schematic diagram of the fabricated low concentration InGaP/GaAs//Si 3-junction solar cell. As the top cell, two-junction cell consisting of p-InGaP (Eg-1.89 eV) and p-GaAs (Eg-1.42 eV) absorption layers connected with a tunneling layer were used. The GaAs-based cell was separated from the growth substrate by Epitaxial Lift-off (ELO) technique. As the bottom cell, the simplest and widely-used Si cell with Al-back surface field-type was used. Si (100) wafer was a single-side mirror polished with the thickness of 625 μm. These cells are connected using Smart Stack technology. In here, Smart Stack technology is the semiconductor bonding technology with the integration of conductive nanoparticle (Pd) alignments at the bonding interface. In our experiment, the bonding resistance for GaAs/Pd interface is enough low less than 1 Ωcm^2, which depends on being easy to form an ohmic-contact based on solid-phase-topotaxial reaction [9]. On the other hand, the bonding resistance at Si/Pd interface is relatively high, and thus we previously demonstrated the reduction of the bonding resistance using a partial ITO film [10]. In this report, we especially paid attention to the resistance depending on the handling of Si surface before the Pd processing.

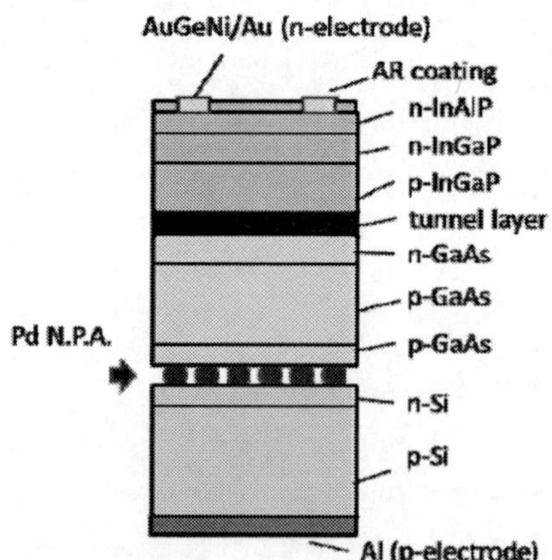

Figure 1: Schematic diagram of the fabricated InGaP/GaAs//Si 3-junction solar cells with Smart Stack technology [8].

3 EXPERIMENT AND DISCUSSION

3.1 Bonding Experiment

We prepared two kinds (group-A and group-B) of InGaP/GaAs//Si MJ samples with the different pre-treatment process for Si cell. For group-A, Pd alignment was formed on Si cell without special treatment. For group-B, the silicon surface was treated with Buffered Hydrogen Fluoride (BHF) solution before the Pd processing. All samples were measured for low concentration characteristics. Figure 2 shows the device resistance evaluated against the concentration factor (C). The reduction in the resistance of group-A samples is relatively slow. On the other hand, that of group-B samples is reduced steadily. Inclination against the concentration factor is approximately asymptotic to C^{-1}, which is ideal characteristic. In Fig. 3, the relative efficiencies evaluated against the concentration factor are shown. Although the group-B samples keep the relative high efficiency, while the group-A samples degrade at a significantly lower factor. These tendencies are mainly based on the bonding resistance, which means the superiority of the BHF-treated method.

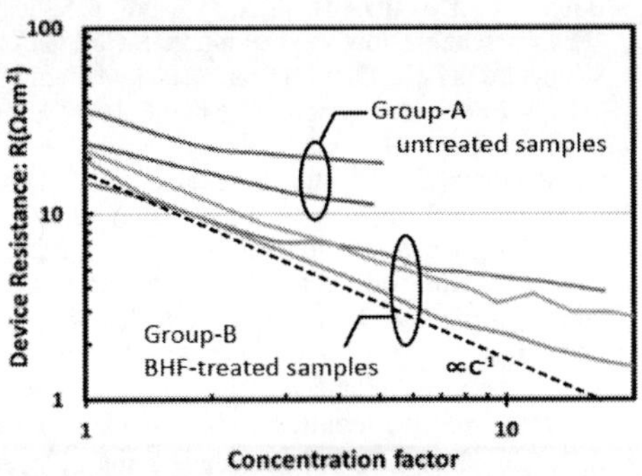

Figure 2: The device resistance of the InGaP/GaAs//Si 3-junction solar cells against the concentration factor.

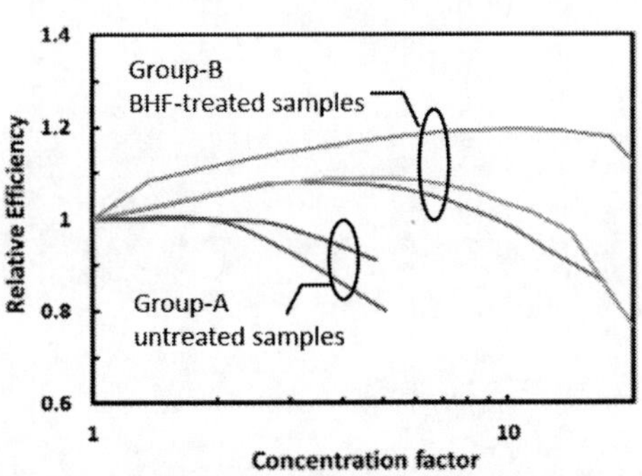

Figure 3: The relative efficiency of the InGaP/GaAs//Si 3-junction solar cells against the concentration factor.

Figure 4 shows a magnified view of the bonding interface, consisting of GaAs (the bottommost layer of the GaAs cell), Pd domain, and Si with the HAADF-STEM (High-angle Annular Dark Field Scanning TEM). The image (a) and (b) represents group-A sample and group-B sample, respectively. In HAADF-STEM, the existing area of Pd is observed as a relatively bright field of vision. The Pd domain contributing to the bonding looks like the aggregate of nanometer-order Pd. In group-B sample, it is observed that Pd is distributed relatively uniformly in the domain. Figure 5 shows the material distribution for the bonding interface with the TEM-EDX (Energy Dispersive X-ray Spectroscopy). This result suggests Pd distribution in group-B sample is uniform. On the other hand, that in group-A sample is segregated to the top GaAs cell side.

In Smart Stack technology, Pd nanoparticle alignment is formed on Si cell using the self-align process of the block copolymer (polystyrene-block-poly-2-vinylpyridine: PS-b-P2VP). In here, Pd^{2+} ions selectively fill in the patterned P2VP domains (hydrophilic region). By handling Si cell with BHF solution beforehand, the shape of Pd domains is different as shown in Fig. 4, and it is thought that Pd^{2+} ions became easy to fill in the domains. In addition, BHF solution treatment has the effect of removing an oxide film. By these effects, the bonding resistance between Pd and Si decreases and it brings a remarkable effect in concentration characteristics.

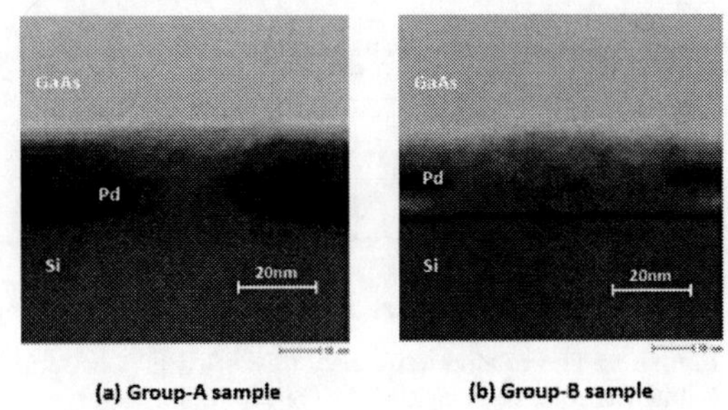

Figure 4: The magnified view of the bonding interface with the HAADF-STEM. (a) Group-A sample (untreated), (b) Group-B sample (BHF-treated)

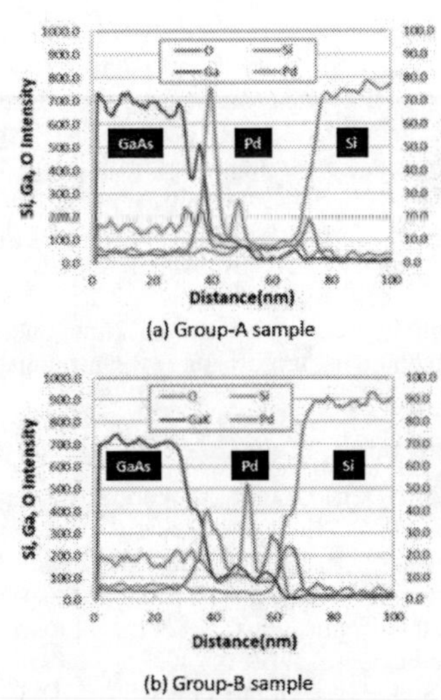

Figure 5: The material distribution of the bonding interface with the TEM-EDX. (a) Group-A sample (untreated), (b) Group-B sample (BHF-treated)

3.2 Device Characteristics

Using the BHF-treated method, we fabricated an InGaP/GaAs//Si 3-junction solar cell. Figure 6 shows the efficiency in dependence of the concentration factor. Under 1sun, the total efficiency (η) was 24.01% with open-circuit voltage (Voc), short-circuit current density (Jsc) and fill factor (FF) of 2.64 V, 11.84 mA/cm^2 and 0.77, respectively. The efficiency normally increases for increasing concentration factor. Maximum efficiency was 28.68% at 10.3 suns. The increase of efficiency is concerned in proper increase of open circuit voltage. The efficiency decreases along with the deterioration of fill factor in concentration factor 20 or more, which is originated in the tunneling characteristics of the top InGaP/GaAs cell. In here, the obtained efficiency largely updates our past result (23.74% at 8.0 suns) [8]. This depends on the reduction of the bonding resistance by applying the BHF-treated method. This performance is practical level for low concentration system below 10 suns.

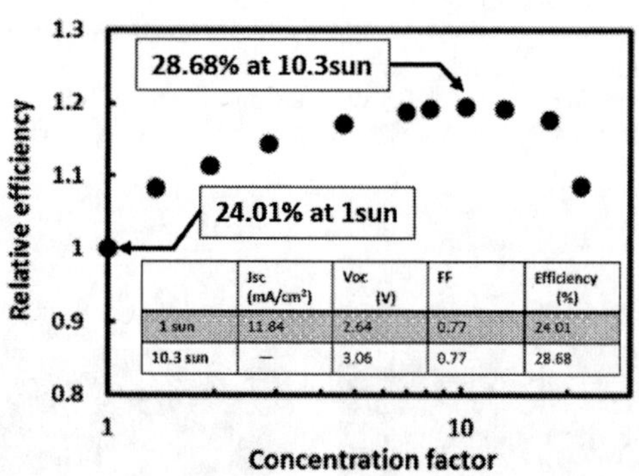

Figure 6: The relative efficiency of the InGaP/GaAs//Si 3-junction solar cells against the concentration factor.

4 CONCLUSIONS

We fabricated the low concentration InGaP/GaAs//Si 3-junction solar cell with Smart Stack technology. Appling the BHF-treated method, we confirmed the maximum efficiency of 28.68% at 10.3 suns. According to our theoretical prediction, the efficiency can be improved over 30%. These results suggest that Smart Stack technology is promising to obtain the low concentration InGaP/GaAs//Si MJ solar cells, as next generation solar cell.

This work was supported by the New Energy and Industrial Technology Development Organization.

References

[1] P. T. Chiu, D. C. Law, R. L. Woo, S. B. Stinger, D. Bhusari, W. D. Hong, A. Zakaria, J. Boisvert, S. Mesropian, R. R. King, and N. H. Karam, "Direct semiconductor bonded 5J cell for space and terrestrial applications", IEEE Journal of Photovoltaics, Vol. 4 (2014) 493.

[2] A. W. Bett, S. P. Philipps, S. Essig, S. Heckelmann, R. Kellenbenz, V. Klinger, M. Niemeyer, D. Lackner, and F. Dimroth, "Overview about technology perspectives for high efficiency solar cells for space and terrestrial applications", 28th European Photovoltaic Solar Energy Conference, (2013) 1AP.1.1.

[3] H. Mizuno, K. Makita, and K. Matsubara, "Electrical and optical interconnection for mechanically stacked multi-junction solar cells mediated by metal nanoparticle arrays", Appl. Phys. Lett., Vol. 101 (2012) 191111.

[4] K. Makita, H. Mizuno, H. Komaki, T. Sugaya, R. Oshima, H. Shibata, K. Matsubara, and S. Niki, "Over 20% efficiency mechanically stacked multi-junction solar cells fabricated by advanced bonding using conductive nanoparticle alignments", 2013 MRS Spring Meeting, (2013) FF2.08.

[5] K. Makita, H. Komaki, H. Mizuno, H. Sai, T. Sugaya, R. Oshima, H. Shibata, K. Matsubara, and S. Niki, "A 24.2%-efficiency III-V/CuInGaSe mechanical stacking multi-junction solar cells using semiconductor bonding method", 29th European Photovoltaic Solar Energy Conference, (2014) 3AO.4.1.

[6] H. Mizuno, K. Makita, T. Sugaya, R. Oshima, Y. Hozumi, H. Takato, and K. Matsubara, "Palladium nanoparticle array-mediated semiconductor bonding that enables high-efficiency multi-junction solar cells", Jap. J. Appl. Phys., Vol. 55 (2016) 025001.

[7] H. Mizuno, K. Makita, T. Tayagaki, T. Mochizuki, T. Sugaya, and H. Takato, "High-efficiency III–V//Si tandem solar cells enabled by the Pd nanoparticle array-mediated "smart stack" approach", Appl. Phys. Express, Vol. 10 (2017) 072301.

[8] K. Makita, H. Mizuno, R. Oshima, T. Tayagaki, M. Baba, N. Yamada, H. Takato, and T. Sugaya, "Highly reliable low concentration InGaP/GaAs/Si 3-junction solar cells with Smart Stack technology", 33rd European Photovoltaic Solar Energy Conference, (2017) 1AO.3.1.

[9] E. Yablonovitch, T. Sands, D.M. Huang, I. Schnitzer, T. J. GmitterS. K. Shastry, D. S. Hill, and J. C. C. Fan, "Van der Waals bonding of GaAs on Pd leads to a permanent, solid-phase-topotaxial, metallurgical bond", Appl. Phys. Lett., Vol. 59 (1991) 3159.

[10] M. Baba, K. Makita, H. Mizuno, H. Takato, T. Sugaya, and N. Yamada, "Reduction of bonding resistance of two-terminal III-V/Si tandem solar cells fabricated using smart-stack technology", Jap. J. Appl. Phys., Vol. 56 (2017) 122302.

Exploring new convergences between PV technologies for high efficiency tandem solar cells : Wide band gap epitaxial CIGS top cells on silicon bottom cells with III-V intermediate layers

Daniel Lincot[1,2], Nicolas Barreau[3], Ahmed Ben Slimane[1,4], Thomas Bidaud[4], Stéphane Collin[4], Markus Feifel[5], Frank Dimroth[5], Matteo Balestrieri[1,2], Damien Coutancier[1,2], Solène Béchu[1,6], Muriel Bouttemy[6], Arnaud Etcheberry[6], Olivier Durand[7], Marie Amandine Pinault-Thaury[8], François Jomard[8]

1.Institut Photovoltaïque Ile de France (IPVF), 30 Route Départementale 128, 91120 Palaiseau, France
2. CNRS, Institut Photovoltaïque Ile de France, UMR 9006, 30 Route Départementale 128, 91120 Palaiseau, France
3. Institut des Matériaux Jean Rouxel –UMR 6502, CNRS, Université de Nantes, 2 rue de la Houssinière BP 32229, 44322 Nantes cedex 3, France
4. Centre de Nanosciences et de Nanotechnologies (C2N), CNRS, Avenue de la Vauve, 91120 Palaiseau, France Palaiseau,
5. Fraunhofer Institute for Solar Energy Systems ISE, Heidenhofstrasse 2, 79110 Freiburg, Germany
6. Institut Lavoisier de Versailles(ILV),Université de Versailles Saint-Quentin enYvelines, Université Paris-Saclay, CNRS,45 avenue des Etats-Unis,78035,Versailles,France
7. Univ Rennes, INSA Rennes, CNRS, Institut FOTON - UMR 6082, Rennes 35000, France
8. Goupe d'Etude de la matière Condensée (GEMAC), Université de Versailles Saint-Quentin en Yvelines, Université Paris-Saclay, CNRS,45 avenue des Etats-Unis,78035,Versailles,France

Contact : daniel.lincot@ipvf.fr

ABSTRACT: A new concept is proposed to develop efficient tandem cells based on silicon bottom cells and wide band gap $Cu(In,Ga,Al)(S,Se)_2$ (noted CIGS) top cells. It consists in introducing an epitaxial III-V buffer layer at the interface. The benefit of this configuration is discussed in terms of lattice and band offsets matching between CIGS and III-V's. Selected combinations are proposed, with AlGaAs and GaAlP buffer layers. Preliminary results with $CuGaSe_2$ on various substrates confirm the influence of the substrate on the growth of CGS and PV properties. Epitaxial growth on GaP and possible selective back contact effect with GaAlAs are evidenced. Preliminary results on using advanced characterization methods are presented ((quantum efficiency, x-ray diffraction, secondary ion mass spectroscopy, nanoAuger, cathodoluminescence).

Keywords: CIGS, epitaxy, III-V, silicon, tandem cells.

1 OBJECTIVES

In the new road map of PV towards increased efficiencies beyond the Schockley Queisser limit for single junctions, tandem junctions are taking a leading role, which will impose an hybridation between crystalline silicon and thin film technologies (III-V viewed as thin film junctions, perovskites, chalcogenides). Very important efforts are presently devoted to develop tandem cells by using perovskites or III-V junctions on the top of silicon bottom cells. We propose to also consider chalcogenide junctions as top cells, since they refer to well established commercial technologies based on CdTe on one side and CIGS materials on the other side. Both technologies demonstrated record efficiencies beyond 22%, which would perfectly fit with more than 30% efficiency for tandems.

The aim of this work is to explore the potential of III-V/CIGS interfaces to fulfill the ideal back contact configuration shown in figure 1, which is lacking for present CIGS device structures (non selective Mo back contact)

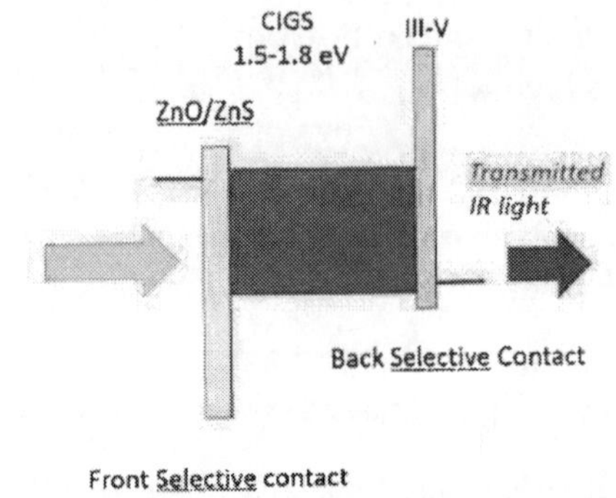

Figure 1. Scheme of an ideal CIGS solar cell, with front and IR transparent back selective contacts.

We will focus on the case of growing high efficiency wide band gap CIGS (1.6-1.8 eV) within the family of copper indium gallium selenosulfide (Cu(In,Ga, Al)(S,Se)$_2$) on top of existing III-V buffer layers on crystalline silicon. The very low cost of silicon wafers and cells makes possible to consider now them for CIGS tandem solar cells with real industrial development perspectives. The key hypotheses for improving the efficiency of the wide gap CIGS cell beyond the actual level (>15%) are : (1) beneficial aspect of epitaxial growth which reduces the detrimental grain boundaries effects in wide band gap CIGS, (2) possibility of higher deposition temperatures (>550°) beneficial for life time increase in sulfide based CIGS, (3) optimization of band offsets matching between III-V and CIGS absorbers as compared to classical Mo/MoSe$_2$ interfaces, (4) new tandem architectures with 2 or 4 terminals.

The device concept for the 2 terminal configuration is given in figure 2.

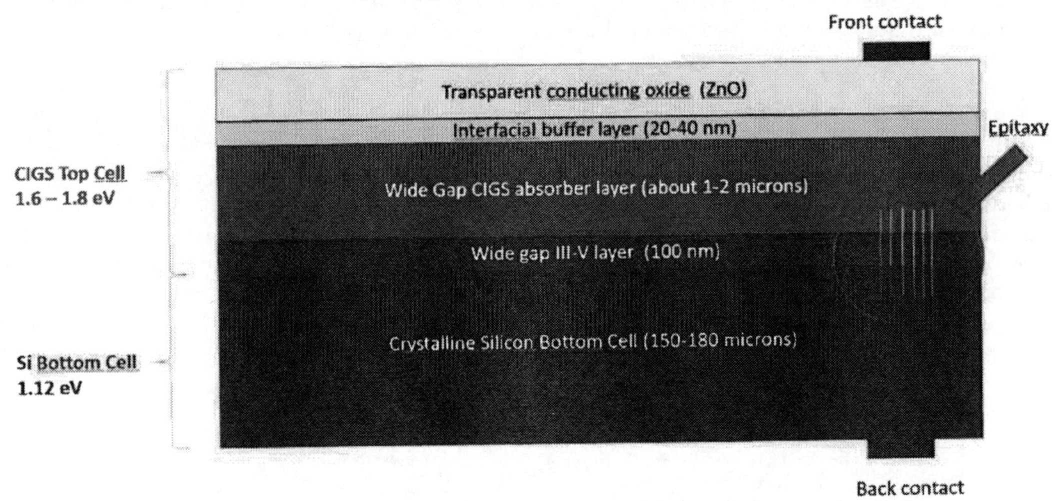

Figure 2. Scheme of the 2 terminal tandem Si-CIGS architecture concept with an interfacial III-V buffer layer.

2 FEASIBILITY ANALYSIS

CIGS solar cells are delivering record efficiencies (> 20%) in a limited zone of band gap values around 1.1 to 1.2 eV, corresponding to a Ga/(Ga+In) ratio (noted GGI) around 30% for $CuIn_{0.7}Ga_{0.3}Se_2$ pure selenides and $CuIn_{0.7}Ga_{0.3}(S,Se)_2$ selenosulfides. Moving to lower and larger GGI leads to reduced efficiencies, in particular towards larger GGI associated to wide band gaps in the range of 1.6-1.8 eV, which are needed for top cells. A first systematic study was carried out at NREL in 2012 showing that increasing the temperature allows to increasing significantly the efficiency up to 1.3-1.45 eV [1]. Recent breakthroughs occurred when using sulfides and selenosulfides, raising the record efficiencies up to 16% for band gaps about 1.55 eV and 11.9% for pure $CuGaSe_2$ at 1.7 eV [2] deposited on Mo/glass substrates. Very recently [3], a value of 15.7 % was obtained with epitaxial CIGSe cell on GaAs single crystal substrate, without Mo, with a GGI of 0.6 (Eg about 1.55). Figure 3 provides an overall view of possible combinations between CIGS absorbers and III-V's in terms of band gap and lattice matching. It appears that two domains exist which fulfill the required conditions. One concerns pure selenides on GaAlAs substrates (blue) and the other one sulfoselenides on GaAlP susbtrates (red zone). III-V nitrides are not included since the lattice constants are significantly smaller.

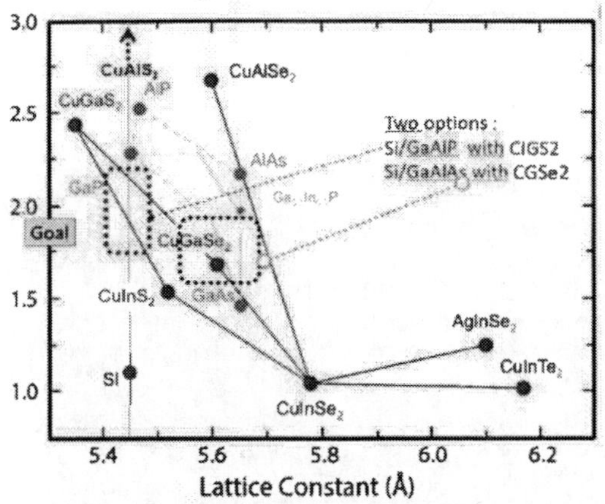

Figure 3. Energy gap (in eV) versus lattice constant for III-V's and chalcopyrites. The red and and blue rectangular zones are adapted for optimal lattice matching and band gap for CIGS top cells.

Considering now the selective contact criterion, the band offsets have to be considered. The conduction band edge energy of the III-V has to be higher than that of CIGS in order to act as an electron barrier/mirror, and the valence band has to be just higher (or slightly lower in case of a spike like configuration) than that of CIGS to provide an easy extraction of holes, giving the optimal open circuit voltage. The general figure given in figure 4 gives an overview of the band offsets for ternary CIGS sulfide and selenide compounds with Al, Ga and In, together with those of binary III-V compounds with N, P and Al. Silicon is included for comparison. The values are taken for the literature. It can be mentioned that some differences appear from one reference to the other as shown in the case of the selenide series (dotted line vs continuous line).

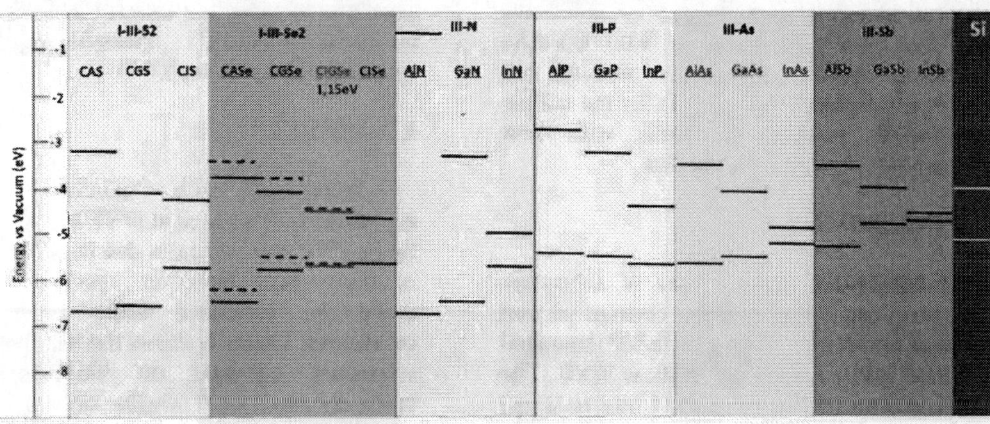

Figure 4. Band energy diagrams of ternary chalcogenides with selenium and sulfur, binary III-V's with nitrogen, phosphorus and arsenic, and silicon.

From both criteria, the combinations between a given CIGS absorber composition and the III-V can be evaluated. Figure 5 gives the example of $CuGaSe_2$.

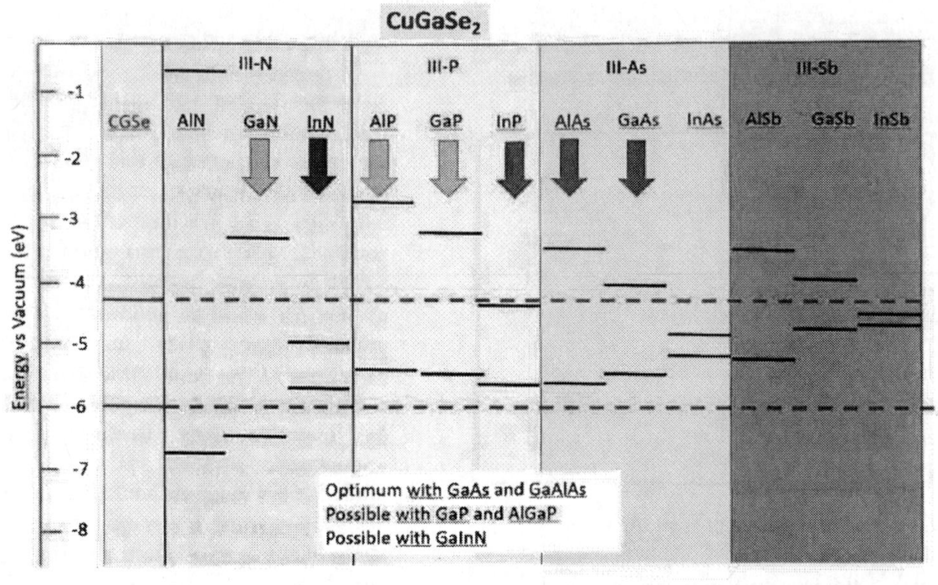

Figure 5. Prediction of the compatibility of III-V's buffer layers with $CuGaSe_2$.

The arrows give the level of compatibility with the III-V's. Best fit is with GaAlAs substrates, GaAlP are then also suitable, but less. Applying the same analysis for the sulfide $Cu(In,Ga)S_2$ reverse the order with best compatibility with the phosphides.

3 EXPERIMENTAL

III-V substrates. Three types of substrates have been used. The first one consists of thin epitaxial layers of GaAlAs or GaAlP deposited on GaAs (100) wafers by MBE at IPVF. The epitaxial films (about 300 nm to 1 micron thick) are p^+ doped (10^{18} cm^{-3}) and the GaAs is insulating. The second consists in thin GaP/GaPN epitaxial layers (230 nm) deposited on silicon (100) substrates by MOCVD at ISE Fraunhofer. The third one is a monocrystalline wafer of p type GaP (about 10^{17} cm^{-3}), with (111)B orientation. CIGSe layers have been deposited by single stage coevaporation at 570°C at IMN on these substrates. One reference sample with a standard three stage CIGSe absorber has been deposited on the GaP/Si(100) substrate. Cells have been completed for GaAs based substrates and the GaP single crystal with classical CdS/ZnO front contacts. In the case of GaAs substrate, back contact were taken on the III-V epilayers with silver dots. In the case of GaP it was simply done by contacting the back side of GaP by the silver dots. In other cases the CIGS surface was kept free to carry out specific characterization methods: XRD, NanoAuger, XPS, cathodoluminescence, SIMS.

4 RESULTS

Solar cells with $CuGaSe_2$ have been successfully fabricated at IPVF at IMN but with limited PV performances due to non-optimized configurations. However spectral responses could be measured under short circuit conditions. Figure 6 shows the relative spectral responses obtained on GaAlAs, GaInP, epilayers and GaP single crystal. We can observe that the best result is obtained with $Ga_{0.51}Al_{0.49}As$ with a good collection towards larger wavelengths. This could be an indication for reduced back contact recombination as expected from the analysis.SEM analysis shows that large grains (micron size) are observed contrary to usually small grain structure of gallium rich layers deposited on Mo/glass substrates. This demonstrates that the use of single crystalline III-V substrate influence markedly the structural and PV properties of the CIGS layer as expected.

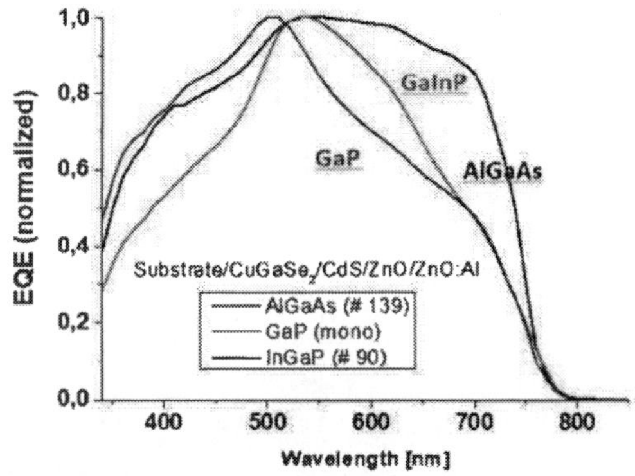

Figure 6. Relative spectral responses for $CuGaSe_2$ cells with different III-V's substrates.

XRD studies, both in θ–2θ and ϕ scan configurations, have been carried out to analyze the structure of the films and the possibility of epitaxy (IPVF, FOTON). We observed that the film were generally not epitaxial, excepted for the GaP (111)B substrate were the epitaxial component was clearly visible in the y scan configuration as shown in figure 7. The presence of 6 peaks can be attributed to a multipositionning effect (inset), as the (101) plane has only two possible orientations with respect to the (112) plane of CGS.

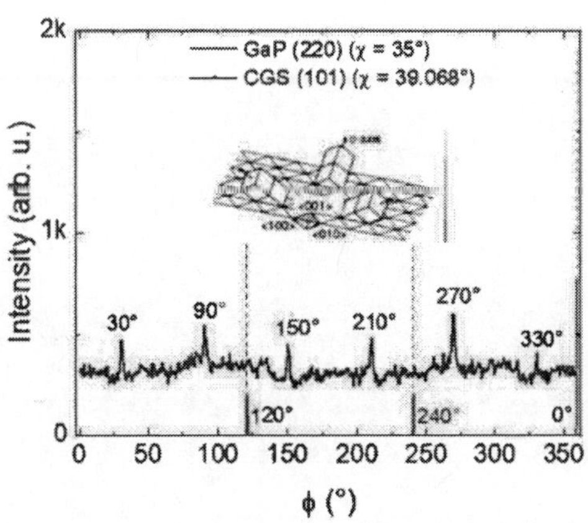

Figure 7. ϕ scans recorded on bare GaP (111) substrate and after deposition of a $CuGaSe_2$ layer.

Composition profiles have been addressed by SIMS (GEMAC) and NanoAuger (ILV) on Cross section. The interface appears sharp with no significant diffusion of copper across the interfaces. Figure 8 shows nanoAuger profiling on the cross section of the CIGS/GaP/silicon heterostructure.

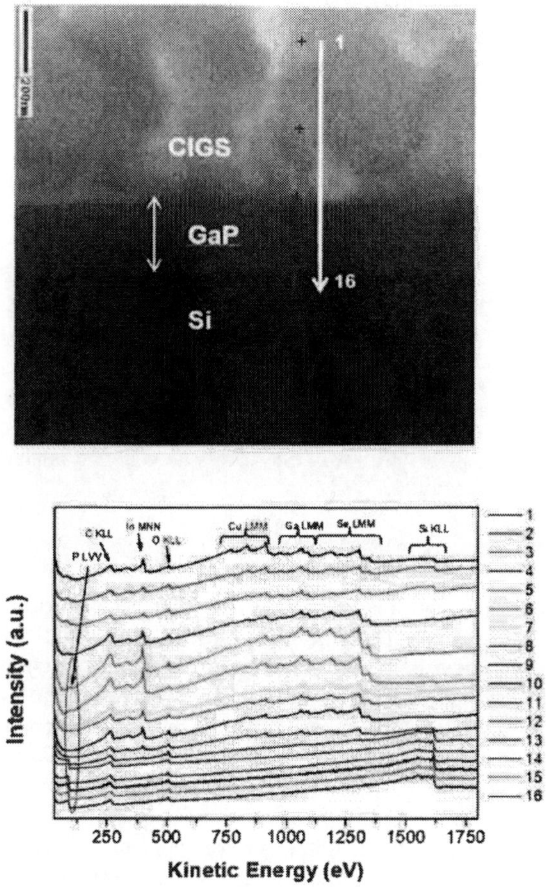

Figure 8. NanoAuger composition profiling across the cross section of a $Si/GaP/CIGSe_2$.

Cathodoluminescence experiments (C2N) show the presence of a sharp peak at about 1.68 eV on the CIGSe samples and high energies features related to possible high energy localized levels. As shown in figure 9, at room temperature the peak is sharper for the CGSe layer grown on the AlGaAs substrate, which is associated to the best spectral response. Some differences are thus observed for the different sample which may reflect different qualities of the layers.

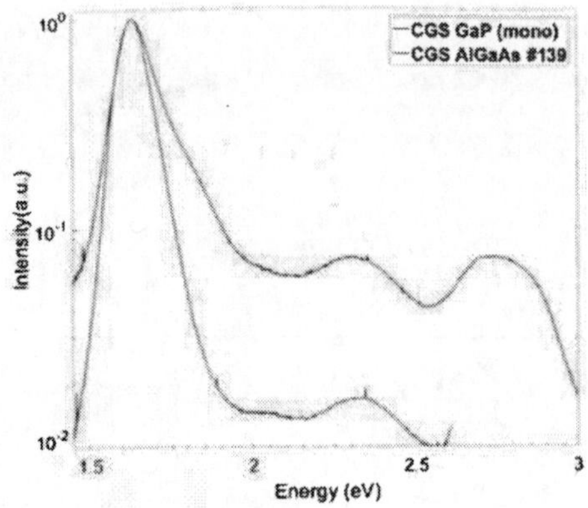

Figure 9. Cathodoluminescence spectra at room temperature for $CGSe_2$ layers on GaP and AlGaAs substrates.

5 CONCLUSION

A new concept for the realization of tandem cells between silicon and CIGS has been presented which consists in introducing a thin buffer layer of III-V layers between the two materials. First experiments seem to confirm the beneficial influence of this layer in the case of wide gap $CGSe_2$ absorber layers, in particular for GaAlAs buffers. Epitaxiãl growth has also been demonstrated. This is very promising for further development of this concept for efficient tandem cells combining silicon and CIGS. We can note that a recently published paper announced that an epitaxial cell of $CuInGaSe_2$ on GaAs (100) was able to deliver an efficiency of 20% [3]. This further supports the interest of developing III-V selective contact layers with CIGS and chalcogenides for high efficiency tandem solar cells.

6 ACKNOWLEDGEMENTS

This research is supported by IPVF in the frame of joint project C on wide gap CIGS solar cells for tandem applications, and by Agence Nationale de la Recherche (ANR).

7 REFERENCES

[1] M. Contreras et al., Progress in Photovoltaics, 20(2012)843
[2] F. Larsson et al., Prog. Photovoltaics, 25(2017)755
[3] J. Nishinaga et al., Applied Physics Express 11(2018)082302.

PERFORMANCE OPTIMIZATION OF A FOUR-TERMINAL CU_2O/C-SI TANDEM HETEROJUNCTION SOLAR CELL

Ørnulf Nordseth[1,*], Halvard Haug[1], Raj Kumar[2], Kristin Bergum[2], Florin Drăgan[3], Dan Crăciunescu[3], Laurențiu Fara[3,4], Irinela Chilibon[5], Edouard Monakhov[2], Sean Erik Foss[1], Bengt Gunnar Svensson[2]
[1]Institute for Energy Technology, Instituttveien 18, 2007 Kjeller, Norway
[2]Department of Physics/Center for Materials Science and Nanotechnology (SMN), University of Oslo, P.O. Box 1048, Blindern, NO-0316 Oslo, Norway
[3]University Politehnica of Bucharest, Spl. Independentei 313, RO-060042, Bucharest, Romania
[4]Academy of Romanian Scientists, Spl. Independentei 54, RO-030167, Bucharest, Romania
[5]National Institute of Research and Development for Optoelectronics (INOE-2000), Bucharest-Magurele, Str. Atomiștilor 409, Măgurele RO-077125, Romania.
*Corresponding author: ornulf.nordseth@ife.no

ABSTRACT: In this work, the electrical characteristics of a four-terminal Cu_2O/c-Si tandem heterojunction solar cell were investigated. Numerical modelling was used to analyze and optimize the main electrical parameters (open circuit voltage, short circuit current density, fill factor, and power conversion efficiency) of the individual subcells, in order to maximize the performance of the four-terminal tandem cell. The electrical and optical properties for aluminum doped ZnO (AZO) and undoped Cu_2O thin film layers synthesized by magnetron sputtering were determined from experimental characterization and used as input for the numerical modelling of the tandem cell. Hall effect measurements for the Cu_2O thin film show that the majority carrier (hole) mobility increases from 10 to 50 $cm^2/V\cdot s$ and the resistivity decreases from 560 to 200 $\Omega\cdot cm$ after post-deposition thermal annealing at 900°C. The complex refractive indices for the AZO and Cu_2O thin film layers were derived from spectroscopic ellipsometry measurements, and optical ray tracing method was used to calculate the reduced spectrum transmitted to the c-Si bottom cell, corresponding to a power density of ~44 mW/cm^2. For the ZnO/Cu_2O top subcell, a conversion efficiency of approximately 12.5 % was achieved when a buffer layer with an electron affinity χ = 3.65 eV was inserted between the n-type ZnO emitter layer and the p-type Cu_2O absorber layer, whereas for the c-Si bottom subcell a power density of 12.2 mW/cm^2 was achieved, yielding a combined power conversion efficiency of 24.7 % for the four-terminal Cu_2O/c-Si tandem heterojunction solar cell under 1 sun illumination.

Keywords: Cuprous oxide, heterojunction, silicon solar cell, tandem

1 INTRODUCTION

Metal oxides with semiconducting properties such as Cu_2O and ZnO are promising candidate materials for implementation in photovoltaic applications, since they are earth-abundant, non-toxic and have the potential for low production cost. The band gap of Cu_2O is 2.1 eV, dictating a theoretical maximum conversion efficiency of about 19 % for a ZnO/Cu_2O heterojunction solar cell under one sun illumination [1], however, the highest conversion efficiency currently achieved experimentally for a Cu_2O-based heterojunction solar cell based on thermally oxidized copper sheets is 8.1% [2], which suggests that further research and development is required in order to realize the full potential of Cu_2O as an absorber layer in photovoltaic applications. In general, the direct bandgap of approximately 2.1 eV makes Cu_2O attractive as a top cell absorber layer for tandem solar cells. By combining a ZnO/Cu_2O heterojunction solar cell with a crystalline silicon (c-Si) solar cell in a tandem cell configuration the high-energy photons can be utilized more efficiently, making it possible to increase the conversion efficiency of the device beyond the limit of conventional c-Si single-junction solar cells. The Cu_2O/c-Si tandem heterojunction solar cell has a theoretical potential to reach a conversion efficiency above 30 % [3].

The objective of this work is to evaluate the performance of a four-terminal Cu_2O/c-Si tandem heterojunction solar cell, as depicted in the schematic device design shown in Figure 1. The tandem cell is formed by combining a conventional c-Si cell with a ZnO/Cu_2O heterojunction cell. The rather large current mismatch between the two subcells in such tandem configuration, caused by the large band gap difference for the two absorbers, favors the arrangement of a mechanical stack of independently connected cells, i.e. a four-terminal configuration. Deposition of a ZnO/Cu_2O thin film stack on a transparent glass substrate (quartz) enabes low-energy photons to be transmitted through the ZnO/Cu_2O top subcell for subsequent absorption in the c-Si bottom subcell. A low-absorbing polymer material, such as ethylene vinyl acetate (EVA), may serve as an intermediate encapsulant layer in order to construct a Cu_2O/c-Si tandem solar cell. Numerical modeling was carried out in order to optimize the performance of the individual subcells for this four-terminal Cu_2O/c-Si tandem cell. The effect of parameter variation on the electrical characteristics of the subcells was investigated.

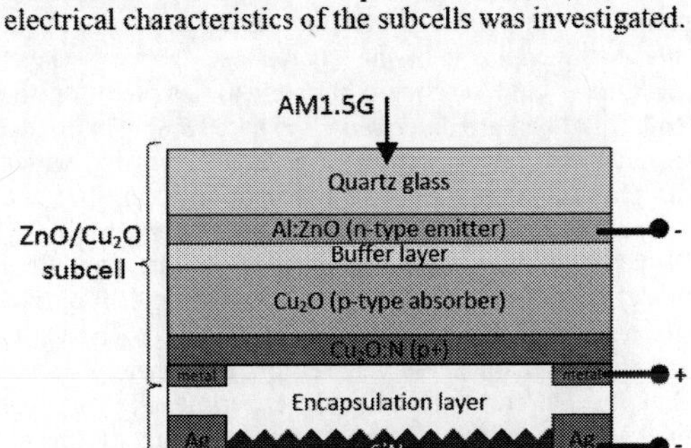

Figure 1: Schematic design of a four-terminal tandem heterojunction solar cell constituted of a ZnO/Cu_2O top subcell and a c-Si bottom subcell.

2 METHODOLOGY

2.1 Experimental details

Undoped Cu_2O and Al-doped ZnO (AZO) thin films were deposited on 20 x 20 x 0.5 mm^3 quartz substrates using a direct current/radio frequency (DC/RF) magnetron sputtering system (Semicore Triaxis) to form the ZnO/Cu_2O top subcell. Cu_2O was deposited by reactive sputtering of a 99.999% Cu target in O_2/Ar (6/49 sccm) with a substrate temperature of 400 °C. As-grown Cu_2O films were annealed at 900 °C for 3 minutes in vacuum (pressure ~10^{-1} Torr). AZO was deposited by co-sputtering of a 99.99% ZnO ceramic target and a 99.999% Al target in Ar with a substrate temperature of 400 °C, yielding an aluminum content of approximately 4 wt% in the deposited layers. The complex refractive indices of the AZO and Cu_2O thin films were derived from spectroscopic ellipsometry measurements using a Horiba Jobin Yvon UVISEL spectroscopic ellipsometer. Room temperature Hall effect measurements (LakeShore 7604) using the van-der Pauw configuration were carried out to determine the carrier mobility and concentration, and resistivity of the AZO and Cu_2O thin films.

A tube furnace (Tempress TS8603) was used to form n-type emitters in the surface of p-type doped (~3 ohm·cm) c-Si wafers using a $POCl_3$ source. The emitter doping profile was analyzed using secondary ion mass spectrometry (SIMS) and four-point probe measurements. For analysis of the carrier recombination in the emitter, the diffused c-Si samples were passivated by deposition of a 40 nm thick hydrogenated amorphous silicon (a-Si:H) layer onto the surface, using plasma enhanced chemical vapor deposition (PECVD). A Sinton WCT-120TS quasi-steady-state photoconductance (QSSPC) setup was used to evaluate the passivation quality and emitter saturation current pre-factor (J_{0e}) values were extracted using the approach of Kane and Swanson [4].

2.2 Device modelling of the ZnO/Cu_2O subcell

Silvaco Atlas was used to model the ZnO/Cu_2O heterojunction subcell, using materials parameters obtained partially from experimental characterization and partially from published literature. The full details of the Silvaco model and the parameter values (layer thicknesses and electronic properties) adopted for the ZnO/Cu_2O cell are described elsewhere [5,6]. The model assumes the heterojunction as several layers for which the physical properties were based on experimental and theoretical results previously reported in the literature [7]. Planar (flat) surfaces were adopted in the simulation model for the ZnO/Cu_2O subcell and the incident optical spectrum was air mass 1.5 global (AM1.5G). The Cu_2O absorber layer thickness was set to 2 μm, mainly due to experimental constraints, i.e. the experimentally feasible thickness range is typically a few micrometers as a result of the relatively low deposition rate for the sputtered layers. The optical characteristics (reflectance, absorptance, and transmittance) of the ZnO/Cu_2O heterojunction subcell were analyzed by optical ray tracing software (OPAL 2) [8] and used to calculate the reduced spectrum transmitted to the c-Si bottom cell.

2.3 Device modelling of the c-Si subcell

Numerical analysis of the c-Si subcell was carried out using PC1Dmod 6.2 [9] and Quokka 2 [10]. The c-Si bottom cell was simulated using Quokka 2, based on the template settings file for a commercial "passivated emitter and rear cell" (PERC) published in Ref. 11. The cell model was slightly modified with enhanced recombination properties on the rear side to increase the conversion efficiency to above 21%, to better reflect a higher efficiency solar cell which might be chosen for use in a tandem configuration. A range of different phosphorus dopant profiles were investigated for the emitter of the bottom cell, with sheet resistance R_{sheet} in the range between ~50 and ~280 Ωcm. The phosphorus dopant profiles used for the simulations were found by interpolating between a smaller set of experimental dopant profiles measured by SIMS. Because Quokka 2 does not solve the diffused regions explicitly, a selection of characteristic parameters for each emitter was found in separate calculations and used as input to the model. These include the emitter saturation current pre-factor J_{0e} for the passivated and metallized regions, the collection efficiency, and the contact resistance.

For each dopant profile, J_{0e} and R_{sheet} were simulated from a diode under moderate forward bias in the dark using PC1Dmod6.2, using the default set of physical models, as described in Ref. 12. The surface recombination velocity parameter S_{0p} was set as a function of the surface dopant concentration N_{surf}, using the recent parameterization by Kimmerle et al. [13], with parameters slightly adjusted to match experimental values measured on a selection of the same dopant profiles. The collection efficiency was simulated from the J_{sc} of a 1D solar cell model similar to the one used for the Quokka simulations. Finally, the specific contact resistance ρ_c was determined as a function of the surface dopant concentration using an analytical expression derived from literature data [14].

3 RESULTS

3.1 Electrical and optical properties of ZnO and Cu_2O

The majority carrier mobility, film resistivity and majority carrier concentration for Cu_2O (as-grown and annealed) and AZO thin films deposited on quartz, derived from room temperature Hall effect measurements, are presented in Table I.

Table I: Majority carrier mobility, film resistivity, and majority carrier concentration for AZO and Cu_2O thin films on quartz, derived from room temperature Hall effect measurements. (*) denotes the Cu_2O thin film after annealing at 900 °C.

Parameter	AZO	Cu_2O	Cu_2O*
Resistivity (Ohm·cm)	5e-4	560	200
Mobility (cm^2/V·s)	20	10	50
Carrier concentration (cm^{-3})	3e20	3e15	1e15

The data in Table I suggests that the electrical properties for the Cu_2O thin film are improved after annealing, i.e., the majority carrier (hole) mobility increases from 10 to 50 cm^2/V·s and the resistivity decreases from 560 to 200 Ω·cm after annealing. The increase in carrier mobility after annealing can, at least partly, be attributed to the increase in grain size and reduced grain-boundary scattering. The average grain size for a 500 nm thick Cu_2O thin film on quartz increases from about 70 nm for the as-grown film to about 600 nm for the annealed film, as revealed from

SEM images of the thin film surface [5].

Figure 2a shows the complex refractive indices of the AZO and Cu_2O thin films as obtained from spectroscopic ellipsometry measurements. The complex refractive indices for the AZO and Cu_2O layers were implemented in the optical analysis and used for calculating the optical transmittance for the ZnO/Cu_2O subcell [5]. The optical transmittance and absorption spectra for an AZO (200 nm)/Cu_2O (500 nm) thin film stack deposited on quartz are shown in Figure 2b. The data shows that high energy photons ($\lambda < 550$ nm) are absorbed in the Cu_2O absorber layer. Based on the calculated optical transmittance for the heterojunction top cell, the optical spectrum that is transmitted to the c-Si bottom subcell was estimated. Figure 3 shows the air mass 1.5 global (AM1.5G) spectrum along with the optical spectrum that is available for photocurrent generation in the c-Si bottom subcell. This transmitted spectrum corresponds to a power density of ~44 mW/cm^2, in comparison to the 100 mW/cm^2 power density for the full AM1.5G spectrum.

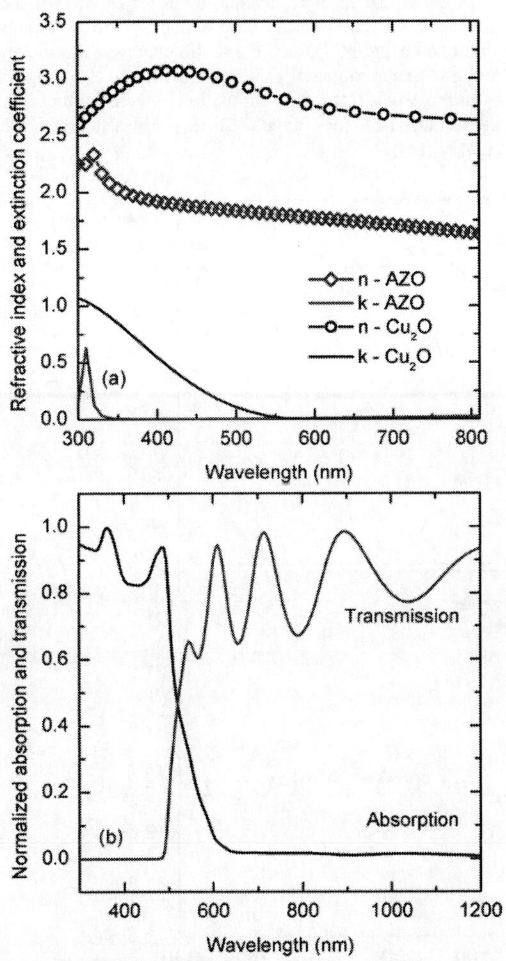

Figure 2: (a) Refractive index *n* and extinction coefficient *k* for the AZO and Cu_2O thin film deposited on quartz. (b) Calculated optical absorption and transmission spectra for a quartz/ZnO (200 nm)/Cu_2O (500 nm) sample.

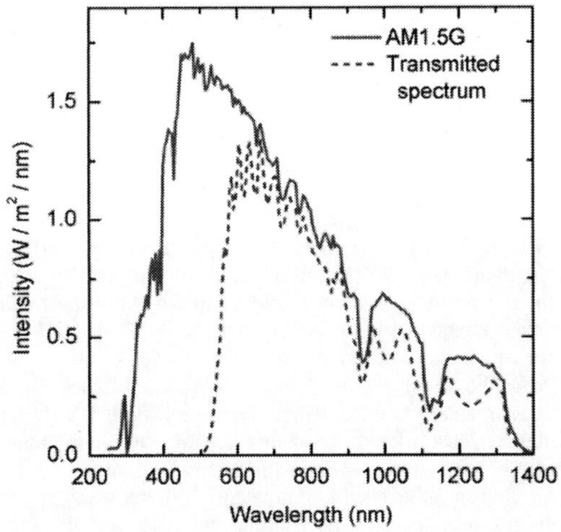

Figure 3: Air mass 1.5 global (AM1.5G) spectrum, along with the spectrum transmitted through the heterojunction top cell and is thus available for the Si bottom cell in the 4-terminal device. The transmitted spectrum corresponds to a power density of ~44 mW/cm^2.

3.2 Performance optimization of ZnO/Cu_2O top subcell

The electrical performance of the ZnO/Cu_2O subcell with incorporation of a buffer layer was evaluated based on device modeling in Silvaco Atlas. Table II summarizes the I-V parameters for the ZnO/Cu_2O heterojunction subcell with and without a buffer layer inserted between the AZO emitter layer and the Cu_2O absorber layer. The data shows that the electrical performance of the ZnO/Cu_2O subcell is substantially enhanced by the incorporation of a buffer layer [2,6].

Table II: I-V parameters for the ZnO/Cu_2O subcell without a buffer layer and with an ideal (optimized) buffer layer with electron affinity $\chi = 3.65$ eV. J_{sc}, V_{oc}, FF, and η denote the short circuit current density, open circuit voltage, fill factor and power conversion efficiency, respectively.

Buffer layer	J_{sc} mA/cm^2	V_{oc} V	FF %	η %
No buffer layer	9.9	1.05	69.1	7.9
Ideal buffer layer	10.3	1.48	82.4	12.5

For the ZnO/Cu_2O subcell the highest performance was achieved for an electron affinity of 3.65 eV for the buffer layer, yielding a fill factor (FF) of 82.4 % and a power conversion efficiency of 12.5 %, assuming an electron affinity of 3.2 eV and 4.4 eV for Cu_2O and AZO, respectively. Further enhancement of the performance can be achieved by improving the optical characteristics of the ZnO/Cu_2O subcell, for example by introducing a more advanced anti-reflection coating in combination with texturing of the front surface. The ZnO/Cu_2O subcell without a buffer layer displays S-shaped I–V characteristics, which limits the FF. Previously reported simulation results for ZnO/Cu_2O heterojunction solar cells have suggested that such distortion of the I-V curve may appear for conduction band offsets of more than 0.4 - 0.5 eV [15]. The origin of the S-shaped deformation might be space charge accumulation caused by restricted charge transport over the ZnO/Cu_2O interface, e.g., the conduction band offset can result in an energy barrier that limits the electron current flowing across the interface.

3.3 Performance optimization of c-Si bottom subcell

The electrical performance of the c-Si bottom subcell for different emitter dopant profiles was evaluated based on device modeling in Quokka 2. Figure 4 shows the emitter saturation current pre-factor J_{0e} as a function of the emitter sheet resistance, for a series of diffused phosphorus n-type emitters on p-type doped c-Si wafers, simulated using PC1Dmod 6.2. The dopant profiles used for the simulations were found by interpolating between experimental doping profiles obtained by SIMS and are shown in the inset in the top right corner of Figure 4. Also shown in Figure 4 are J_{0e} values obtained from experimental characterization for four different diffused emitters. The plot suggests that there is good agreement between the simulated and experimental J_{0e} values.

Figure 5 shows the simulated I-V parameters (open circuit voltage V_{OC}, short circuit current J_{SC}, fill factor FF, and power density at the maximum power point P_{MPP}) as a function of the emitter sheet resistance R_{sheet} for the c-Si bottom cell. The simulation results are shown for illumination with both the AM1.5G spectrum, i.e., with no ZnO/Cu_2O subcell on top, and the spectrum transmitted through the ZnO/Cu_2O top subcell, cf. Figure 3. We observe that V_{OC} and J_{SC} increases with increasing R_{sheet} due to reduced recombination and improved collect-

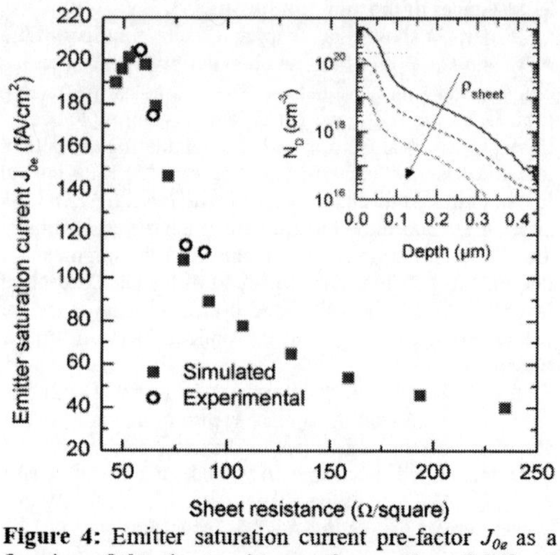

Figure 4: Emitter saturation current pre-factor J_{0e} as a function of the sheet resistance, for a series of diffused phosphorus n-type emitters on p-type c-Si wafers. Values simulated using PC1Dmod 6.2 (solid symbols) are shown along with experimental data (open symbols). The dopant profiles used for the simulations were found by interpolation of experimental SIMS profiles (inset in top right corner).

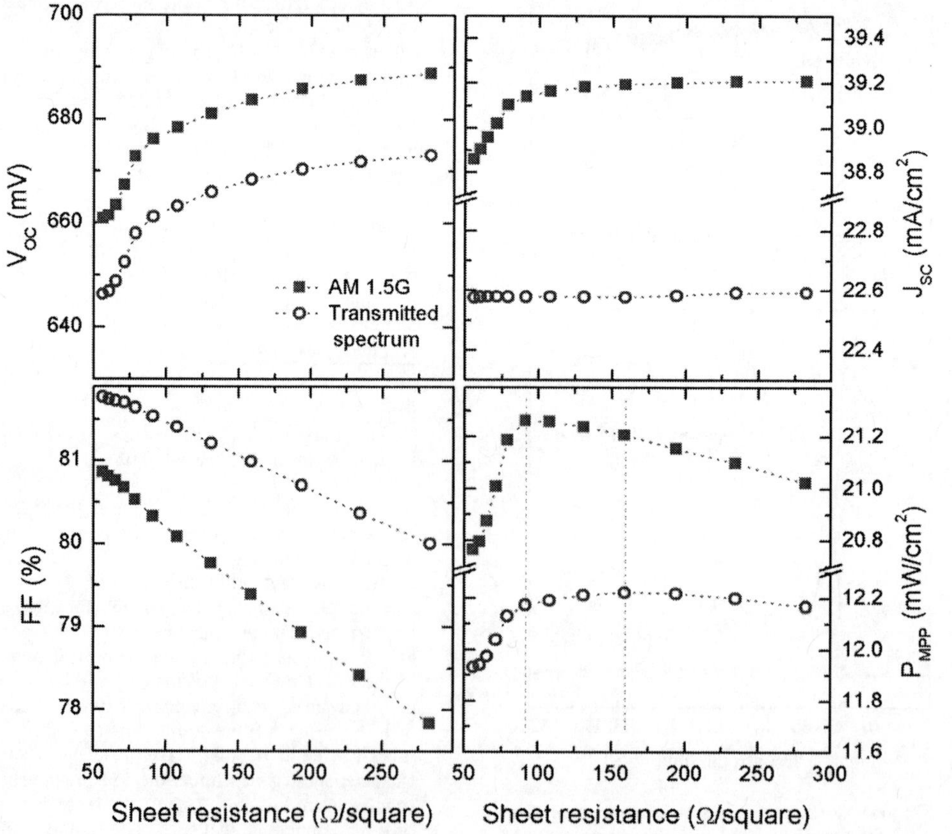

Figure 5: Simulated I-V parameters V_{oc}, J_{sc}, FF and P_{mpp} as a function of the emitter sheet resistance for c-Si bottom subcell, for both the full AM1.5G spectrum and the spectrum transmitted through the ZnO/Cu_2O top subcell (cf. Fig. 3). The specific contact resistance ρ_c, collection efficiency, and emitter saturation current pre-factor J_{0e} (for both the contacted and passivated part of the emitter) are changed accordingly for each simulation.

Table III: I-V parameters for the optimized subcells of the four-terminal Cu_2O/c-Si tandem heterojunction solar cell along with those for the stand-alone c-Si solar cell under 1 sun illumination.

Cell	Intensity mW/cm²	J_{sc} mA/cm²	V_{oc} V	FF %	η %	P_{MPP} mW/cm²
ZnO/Cu_2O top subcell	100	10.3	1.480	82.4	12.5	12.5
PERC c-Si bottom subcell	44.0	22.6	0.665	81.0	27.7	12.2
PERC c-Si cell (1 Sun)	100	39.2	0.677	80.3	21.3	21.3

ion efficiency, respectively, whereas the FF decreases with increasing R_{sheet} because of increased resistive losses. For this particular set of simulation parameters, the optimal sheet resistance for maximum cell performance, marked with vertical lines in the bottom right of Figure 5, is found at different R_{sheet} values for the AM1.5G spectrum and the transmitted spectrum. For the full AM1.5G spectrum the optimal R_{sheet} value is about 90 Ω/square, whereas for the transmitted spectrum the optimal R_{sheet} value is about 160 Ω/square. The initial hypothesis was that a lower R_{sheet} value would be beneficial for the bottom cell when illuminated with the transmitted spectrum since the cell in this case would be less sensitive for collection efficiency losses in the blue part of the wavelength spectrum. However, this effect is more than counteracted by the fact that the *FF* is less sensitive to the increased series resistance of a lowly doped emitter (high R_{sheet}) for the transmitted spectrum, due to the relatively low current density (~23 mA/cm²) in this case. The optimal values for the emitter sheet resistance will depend on the chosen simulation parameters, but this analysis illustrates the usefulness of device simulations for optimization of the c-Si bottom subcell in a four-terminal tandem solar cell. The simulation results presented in Figure 5 suggest that the optimized parameters for the PERC type c-Si cell under 1 sun illumination are not necessarily valid when the cell is used as bottom subcell in a tandem cell configuration.

The I-V parameters for the optimized subcells of the four-terminal Cu_2O/c-Si tandem heterojunction solar cell are summarized in Table III. Placing the ZnO/Cu_2O subcell on top of a c-Si solar cell in a tandem configuration (cf. Figure 1) will result in a current density of 22.6 mA/cm² and a power density of 12.2 mW/cm² for the c-Si bottom subcell, as a result of the short-wavelength photons being absorbed in the top subcell. The combined power conversion efficiency for the subcells in the four-terminal Cu_2O/c-Si tandem heterojunction solar cell is 24.7 % when this tandem cell is placed under 1 sun illumination. In comparison, the PERC type c-Si placed under 1 sun illumination has a power conversion efficiency of 21.3 %. The data suggests that for this particular tandem cell configuration with a commercially feasible PERC type c-Si bottom cell, the conversion efficiency of the ZnO/Cu_2O heterojunction cell needs to be higher than ~9 % for the additional power generated in the top subcell to outweigh the losses introduced by placing it on top of the c-Si solar cell.

4 CONCLUSION

The performance of a Cu_2O/c-Si tandem heterojunction solar cell was analyzed by numerical modelling. The rather large current mismatch between the two subcells in such tandem structure favors the arrangement of a four-terminal configuration, which allows for individual optimization of the top and bottom subcells. The electrical and optical properties for aluminum doped ZnO and Cu_2O thin film layers synthesized by magnetron sputtering were determined from experimental characterization and used as input for the numerical modelling of the photovoltaic device. Based on the complex refractive index for ZnO and Cu_2O, derived from spectroscopic ellipsometry measurements, the optical spectrum that is transmitted through the heterojunction top cell and onto the c-Si bottom subcell was calculated. This transmitted spectrum corresponds to a power density of ~44 mW/cm². For the ZnO/Cu_2O top subcell, a conversion efficiency of approximately 12.5 % was achieved when a buffer layer with an electron affinity χ = 3.65 eV was inserted between the n-type ZnO emitter layer and the p-type Cu_2O absorber layer. For the c-Si bottom subcell the optimal performance was achieved for an emitter sheet resistance of ~160 Ω/square, yielding a power density at the maximum power point of 12.2 mW/cm². The combined power conversion efficiency for the four-terminal Cu_2O/c-Si tandem heterojunction solar cell was estimated at 24.7 % under 1 sun illumination. The investigation suggests that for this particular tandem cell configuration, the conversion efficiency of the ZnO/Cu_2O cell should be higher than 9 % for the additional power generated in the top subcell to outweigh the losses introduced by placing it on top of the c-Si cell.

ACKNOWLEDGEMENTS

This work was conducted under the research project "High-performance tandem heterojunction solar cells for specific applications (SOLHET)", project 251789, financially supported by the Research Council of Norway (RCN) and the Romanian Executive Agency for Higher Education, Research, Development and Innovation Funding (UEFISCDI) through the M-Era.net program. RCN is also acknowledged for the support to the Norwegian Micro- and Nano-Fabrication Facility, NorFab, project number 245963/F50.

REFERENCES

[1]. W. Shockley and H.J. Queisser, "Detailed Balance Limit of Efficiency of PN Junction Solar Cells". J. Appl. Phys. 32 (1961) 510-519.

[2]. T. Minami, Y. Nishi and T. Miyata, "Efficiency enhancement using a Zn_{1-x} Ge_x -O thin film as an n-type window layer in Cu_2O-based heterojunction solar cells", Appl. Phys. Express 9 (2016) 052301.

[3]. F. Meillaud, A. Shah, C. Droz, E. Vallat-Sauvain, C. Miazza, "Efficiency limits for single-junction and tandem solar cells", Solar En. Mater. and Solar Cells 90 (2006) 2952.

[4]. D. Kane, R. Swanson, "Measurement of the emitter saturation current by a contactless photoconductivity

decay method", In Proc. of the 18th IEEE Photovoltaic Specialist Conference, New York (1985) 578-583.

[5]. Ø. Nordseth, R. Kumar, K. Bergum, L. Fara, S. E. Foss, H. Haug, F. Drăgan, D. Crăciunescu, P. Sterian, I. Chilibon, C. Vasiliu, L. Baschir, D. Savastru, E. Monakhov and B. G. Svensson, "Optical Analysis of a ZnO/Cu_2O Subcell in a Silicon-Based Tandem Heterojunction Solar Cell", Green and Sustainable Chemistry 7 (2017) 57-69.

[6]. Ø. Nordseth, L. Fara, R. Kumar, S. E. Foss, C. Dumitru, V.-F. Muscurel, F. Drăgan, D. Crăciunescu, K. Bergum, H. Haug, P. Sterian, I. Chilibon, C. Vasiliu, L. Baschir, D. Savastru, E. Monakhov, and B. G. Svensson, "Electro-optical modeling of a ZnO/Cu_2O subcell in a silicon-based tandem heterojunction solar cell", Proceedings of The 33rd European Photovoltaic Solar Energy Conference and Exhibition, Amsterdam, The Netherlands, 25-29 September 2017, 172-177.

[7]. Y. Takiguchi and S. Miyajima, "Device simulation of cuprous oxide heterojunction solar cells", Jpn. J. Appl. Phys. 54 (2015) 112303.

[8]. S.C. Baker-Finch and K.R. McIntosh, "OPAL 2: Rapid Optical Simulation of Silicon Solar Cells", Proceedings of the 38th IEEE Photovoltaic Specialists Conference, Austin, 3-8 June 2012, 265-271.

[9]. H. Haug and J. Greulich, "PC1Dmod 6.2 – Improved simulation of c-Si devices with updates on device physics and user interface," Energy Procedia 92 (2016) 60-68.

[10]. A. Fell, "A free and fast three-dimensional/two-dimensional solar cell simulator featuring conductive boundary and quasi-neutrality approximations," IEEE Transactions on Electron Devices 60 (2013) 733–738.

[11]. A. Fell, K.R. McIntosh, P.P. Altermatt, G.J.M. Janssen, R. Stangl, A. Ho-Baillie, H. Steinkemper, J. Greulich, M. Muller, B. Min, K.C. Fong, M. Hermle, I.G. Romijn, and M.D. Abbott, "Input parameters for the simulation of silicon solar cells in 2014", IEEE J. Photovoltaics 5 (2015) 1250-1263.

[12]. H. Haug, A. Kimmerle, J. Greulich, A. Wolf, and E. Stensrud Marstein, "Implementation of Fermi–Dirac statistics and advanced models in PC1D for precise simulations of silicon solar cells", Sol. Energy Mater. Sol. Cells 131 (2014) 30.

[13]. A. Kimmerle, M. Momtazur Rahman, S. Werner, S. Mack, A. Wolf, A. Richter, and H. Haug, "Precise parameterization of the recombination velocity at passivated phosphorus doped surfaces", J. Appl. Phys. 119 (2016) 025706.

[14]. P. N. Vinod, "Specific Contact Resistance Measurement of Screen-Printed Ag Metal Contacts Formed on Heavily Doped Emitter Region in Multicrystalline Silicon Solar Cells", J. Electron. Mater. 42 (2013) 2905-2909.

[15]. A. O. Pudov, A. Kanevce, H. A. Al-Thani, J. R. Sites, and F. S. Hasoon, "Secondary barriers in CdS-$CuIn_{1-x}Ga_xSe_2$ solar cells", J. Appl. Phys. 97 (2005) 064901.

THREE-TERMINAL TANDEM SOLAR CELLS COMBINING BOTTOM INTERDIGITATED BACK CONTACT AND TOP HETEROJUNCTION SUBCELLS: A NEW ARCHITECTURE FOR HIGH POWER CONVERSION EFFICIENCY

Jean-Paul Kleider,[1] Zakaria Djebbour,[1,2] Anne Migan-Dubois,[1] Walid El-Huni,[1]
Cyril Leon,[1] Marie Gueunier-Farret,[1] James P. Connolly[1]
[1]GeePs, Group of Electrical Engineering - Paris, UMR 8507 CNRS, CentraleSupélec, Univ. Paris-Sud, Université Paris-Saclay, Sorbonne Université, 91192 Gif-sur-Yvette Cedex, France
[2] Département des Sciences Physiques, UVSQ, 45 avenue des Etats-Unis, 78035 Versailles, FRANCE

ABSTRACT: We describe the working principle of a three-terminal tandem solar cell that associates a silicon interdigitated back contact bottom cell with a top heterojunction cell. Illustrations are given and calculations are performed in two examples: (i) the bottom silicon absorber is n-type and the top absorber is a III-V compound with a p-type top layer forming a p-i-n heterojunction; in this case the n+ back contact layer is common to both cells and allows the extraction of electrons generated in both absorbers (ii) the bottom absorber is p-type and the top absorber is a $MaPbI_3$ perovskite; in this case the p+ back contact common to both cells allows the extraction of electrons generated in both absorbers. Of particular importance in the proposed 3-terminal architecture using a heterojunction subcell is the band alignment at the heterointerface that provides selective carrier transport. It is shown that, in the case of an n-type silicon bottom absorber, a selective interface buffer having a large valence band offset and a zero or negligible conduction band offset can significantly improve the photovoltaic performance. In the case of a p-type bottom silicon absorber, misalignment of the valence band should be avoided while a large conduction band offset provides selective electron extraction. The concept is promising since it can avoid some drawbacks of the two-terminal and four-terminal tandem cells while providing equivalent high conversion efficiencies.
Keywords: tandem, silicon, 3-terminal, III-V semiconductors, perovskite

1 INTRODUCTION

The laboratory record efficiencies of single junction solar cells are getting closer and closer to the theoretical limit. This is especially true for silicon solar cells where the theoretical limit at one sun illumination taking into account intrinsic Auger recombination is 29%, while achievements of 26.7% have been reported [1,2]. Multijunction solar cells are being actively studied since they can overcome the limit of single junction cells, with one-sun theoretical values of 42% and 49% for dual (tandem) or triple junction cells, respectively [3]. Historically, progress in the field of multijunction solar cells has relied on the development of III-V compounds, with, for instance, the GaInP/GaAs/Ge triple cells that have been widely used in space applications. However the scarcity of some elements, or the cost of substrates (in this example, Ge), together with costly processing technologies are serious obstacles for their use for a large scale terrestrial production of electricity. This is why strong research efforts are currently devoted to the development of multijunction solar cells based on silicon, especially for tandem cells. Interesting results have been obtained recently combining silicon with III-V compounds like GaInP [4] or with perovskites [5, 6], with efficiencies above 30% and around 25%, respectively. Two-terminal (2-T) and various four-terminal (4-T) tandem cell architectures have been proposed [7], the most common ones being the monolithically integrated 2-T and the mechanically stacked 4-T. Our aim is to propose a route for tandem solar cells based on a three-terminal (3-T) concept that can offer several advantages: 1) to suppress the constraint of photo-current matching and of tunnel junctions required in the 2-T design; 2) to facilitate the access to the different contacts of the top and bottom cells without the need for etching and without having to align buried contact grids; 3) to be adapted to the well mastered silicon technology for the bottom cell.
In 2015, Marti and Luque [8] proposed theoretical calculations on a three-terminal double heterojunction bipolar transistor solar cell, showing that the limit efficiency was the same as for conventional tandem cells. However, only a few publications have been devoted to 3-T configurations so far (see [9] and references therein). We here propose a 3-T architecture that we recently patented [10]. It is built on an interdigitated back-contact (IBC) silicon bottom cell combined with a heterojunction front cell. Using the ATLAS module of SILVACO TCAD, we illustrate the operation principle and show how a suitable management of the hetero-interface can improve the performance of this 3-T cell. Using realistic calculations that take transport and recombination mechanisms into account, we demonstrate that efficiencies beyond 35% can be realistically attained.

2 THREE-TERMINAL CONCEPT

2.1 Basic architecture

The basic architecture is shown in Figure 1 for the two possible polarities of n-type (a) and p-type (b) c-Si bottom absorber. In case of an n-type c-Si bottom absorber, the top absorber must have a larger bandgap than c-Si and be either undoped or p-type and covered with a (p+) hole extracting layer.

The combination of the top layers with the bottom n-type c-Si absorber thus forms either a p+/p/n or a p+/i/n junction, with either a p/n or an i/n heterointerface. The back n+-Si region works as the unique electron extracting layer and thus as the common n-electrode for both subcells. It must collect all electrons whatever the region where they are photogenerated. Therefore, the top large bandgap semiconductor must be chosen so that there is no barrier for electrons at the heterojunction inhibiting electron transport to the silicon: the conduction band offset between the top absorber and the bottom absorber, $\Delta E_C = E_{C,top} - E_{C,bottom}$ should not be strongly negative. On the other hand, a positive value could lead to thermalization losses for the electrons photogenerated in the large bandgap layer since they will be collected in the lower bandgap c-Si. Thus, a small ΔE_C value is preferable. The electrons and holes photogenerated in c-Si will thereby be collected by the back electrodes on the (n+) and (p+) c-Si regions, respectively, like in a conventional 2-terminal interdigitated silicon back contact cell. However, holes generated in the top absorber should be collected by the top extracting layer in order to keep the advantage of the larger top bandgap.

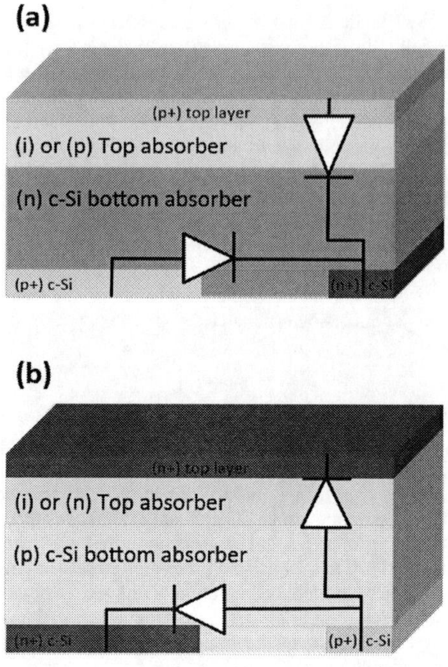

Figure 1: Schematic view of the 3-T architecture for (n) c-Si bottom absorber (a) and for and (p) c-Si bottom absorber (b). Electrodes are not shown.

In case of a p-type c-Si bottom absorber (Figure 1b), one just has to change all polarities. The back p+ c-Si region then works as the common hole extracting layer and common p-electrode. This implies that, in this case, the valence band offset $\Delta E_V = E_{V,bottom} - E_{V,top}$ should be adjusted close to zero.

2.2 Illustration using a III-V top absorber on (n) c-Si

An example of the proposed 3-T solar cell concept on an n-type c-Si absorber IBC bottom-cell is shown in Figure 2. The top p-i-n heterojunction cell is made of a 20 nm thick p-type GaP window cap layer, a 50 nm thick intrinsic $Ga_xIn_{1-x}P$ layer with graded bandgap (from 2.35 eV down to 1.78 eV), and a 1 µm thick intrinsic $Ga_{0.35}In_{0.65}P$ absorber, the bandgap of which is 1.78 eV. The graded bandgap layer has been chosen in order to smooth the band offsets between the window layer and the absorber.

GaInP is a suitable candidate for the top absorber in our structure since it has almost the same electron affinity (4.1 eV) as Si (4.05 eV). In addition one has to avoid collection of electrons at the front electrode, which is here ensured by the electric field produced through the absorber by the p-type GaP layer as well as by the front conduction band barrier, as can be seen in Figure 3.

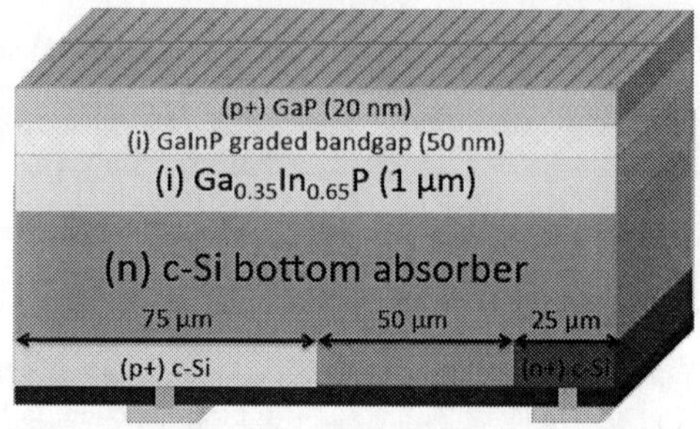

Figure 2: Sketch of the structure chosen to illustrate the working principle and performance of the 3-T architecture on an n-type silicon wafer. The top subcell is made of a p-i-n heterojunction with the p and i layers of p-type GaP and undoped i-$Ga_{0.35}In_{0.65}P$ absorber layer onto n-type crystalline silicon.

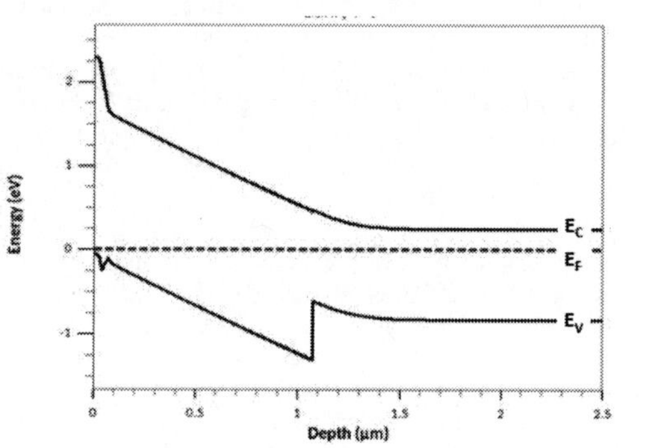

Figure 3: Energy band diagram at equilibrium across the p-i-n heterojunction of Fig. 2.

The maps of electron currents at short circuit conditions confirm that all electrons are collected at the common back n-electrode [9]. There is a pronounced S shape in the overall electron current flow. Our 3-T architecture can be viewed as a vertical/lateral cell, e.g. a combination of a subcell working mainly laterally (the bottom IBC cell) and a subcell working mainly vertically (the top heterojunction cell). This is clearly visible in the total current map (Figure 4), where some current lines are rather horizontal while others express the vertical and S shape flow from top to bottom.

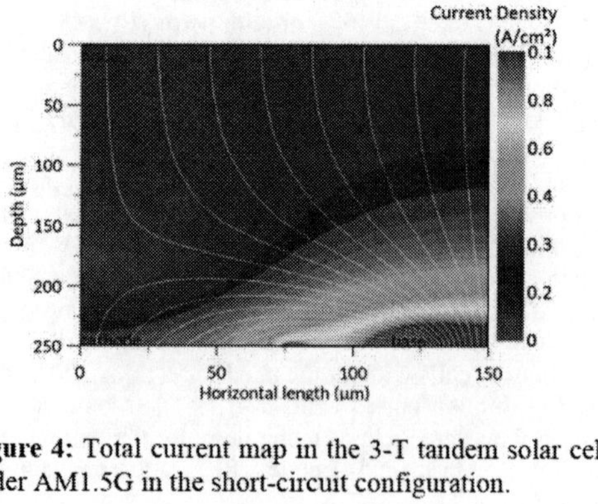

Figure 4: Total current map in the 3-T tandem solar cell under AM1.5G in the short-circuit configuration.

Due to the large valence band offset at the hetero-interface (Fig. 3), holes photogenerated in silicon cannot pass into GaInP, and they are collected at the back p-electrode, just as in the case of the traditional IBC cell. On the other hand, holes photogenerated in the GaInP layer are primarily collected by the front electrode due to the built-in field of the p-i-n heterojunction. However, when this junction is forward biased, the electric field is strongly decreased, as can be seen in Figure 5, so that diffusion begins playing an important role. Since holes have no barrier to pass from the large bandgap layer into the silicon, some holes will diffuse into silicon. Interestingly, these holes are not lost and can then be extracted by the back p-electrode, thus increasing the photocurrent in the IBC cell. However, due to the valence band offset when going from GaInP to silicon they thermalize, causing some power loss and limiting the open circuit voltage of the top cell to a value of 1.13 V, which is much less than the value obtained in single record GaInP cells (> 1.4 V) [9].

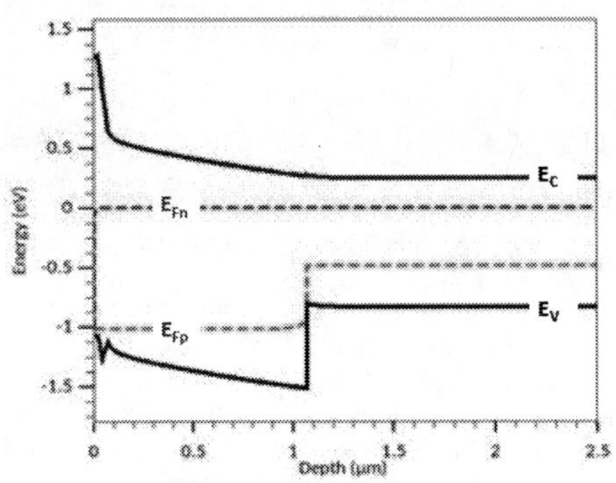

Figure 5: Energy band diagram in the dark across the p-i-n heterojunction of Fig. 2 when it is biased at 1.02 V (value at the maximum power point under AM1.5G). The silicon bottom diode is short circuited.

3 INTERFACE ENGINEERING FOR IMPROVED PERFORMANCE

In order to limit the thermalization loss due to holes flowing from the top absorber to the n-Si bottom absorber, one can adapt the heterojunction by inserting a thin layer of a larger bandgap semiconductor. A good candidate here can be GaN (Fig. 6).

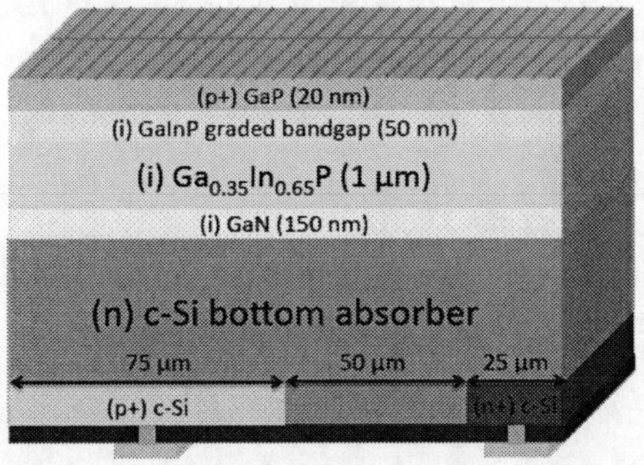

Figure 6: Sketch of the improved 3-T structure integrating a thin buffer GaN layer at the heterojunction with c-Si.

The GaN buffer plays the role of a selective band offset layer. Indeed the large valence band offset provides a barrier for holes in both directions, it prevents holes from entering into Si and favours their collection at the top electrode even for vanishing internal field, as can be seen in Figure 7, where the hole quasi Fermi level abruptly changes at the GaN/Si interface to join that of electrons in Si, the bottom cell being under short circuit conditions. In addition, the almost matched conduction band (electron affinity of GaN was taken equal here to 4.1 eV [11]) allows electrons to pass into Si in order to be collected at the common n-Si back electrode.

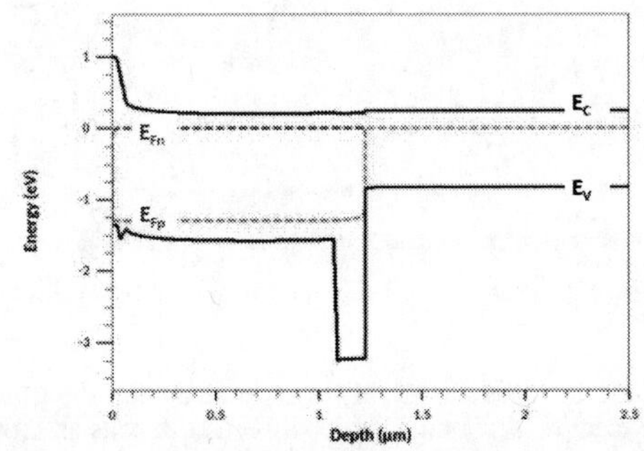

Figure 7: Energy band diagram in the dark across the heterojunction of Fig. 6, when it is biased at 1.32 V (value at the maximum power point under AM1.5G). The silicon bottom diode is short circuited.

Thanks to the introduction of the selective interface layer the open circuit voltage of the top cell is increased to 1.49 V and the two subcells are now operating fully independently of one another. The output characteristics of the two subcells are shown in Figure 8. The total power conversion efficiency can amount to 35% [9].

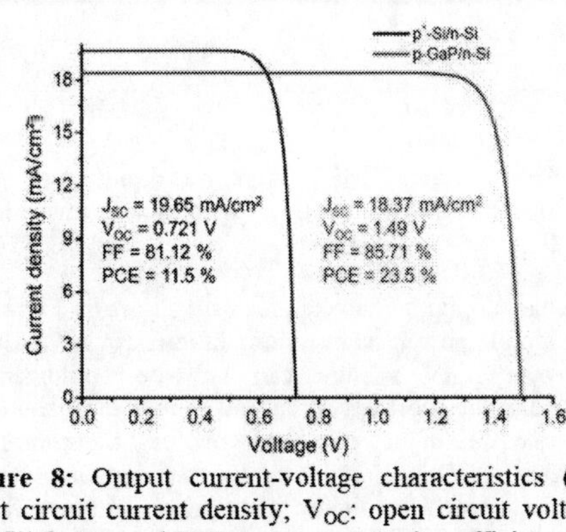

Figure 8: Output current-voltage characteristics (J_{SC}: short circuit current density; V_{OC}: open circuit voltage; FF: fill factor, and PCE: power conversion efficiency) of the two subcells in the 3-T architecture of Fig. 6.

4 PEROVSKITE/SILICON 3-T STRUCTURE

The proposed 3-T architecture is of particular interest for a perovskite/silicon combination since perovskites can be directly deposited onto Si. There are various possibilities for electron and hole transport layers in perovskite cells. Here we give an illustration on a p-type silicon bottom absorber (Figure 9). The modeled perovskite top cell is based on the work of T. Minemoto [12]. The absorber is the well-known methyl ammonium lead iodide ($MaPbI_3$) assimilated to an intrinsic semiconductor layer with a bandgap of 1.55eV. Hole transport layer (HTM) and electron transport layer (ETM) are modeled as p+ and n+ semiconductors with bandgaps of 3 eV and 3.5 eV, respectively.

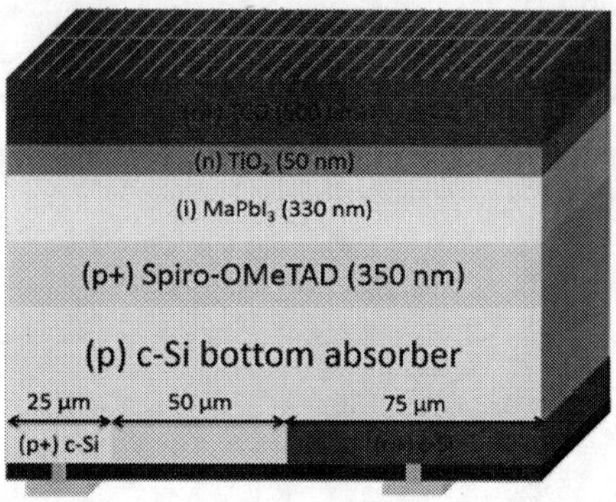

Figure 9: Example of 3-T structure with a perovskite top cell on p-type silicon.

In this example, the HTM layer also takes the role of the electron separation layer between both cells as shown in the energy band diagram (Fig. 10).

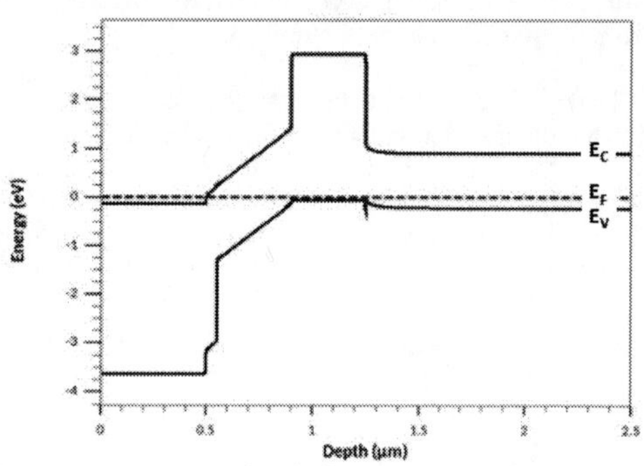

Figure 10: Energy band diagram at equilibrium across the top heterojunction cell of the structure presented in Fig. 9.

Calculated output characteristics are shown in Figure 11. The total power conversion efficiency is 27.9 %. However, the structure can still be optimized. In particular the short circuit current in both cells should be increased through enhanced photon absorption. In addition, $MaPbI_3$ does not have the suitable bandgap energy to reach the highest conversion efficiency for the tandem cell on silicon, and perovskites with bandgap around 1.7 eV should be rather considered in the future.

4 CONCLUSIONS

We have proposed a 3-T architecture using an IBC silicon bottom cell combined with a heterojunction top cell that opens new possibilities for high efficiencies. Realistic calculations for a combination of GaInP on silicon have shown that efficiencies of 35% can be attained. Preliminary modeling results have shown that the proposed 3-T concept can also be suitable for a perovskite/silicon combination. Future work should focus on the fabrication of prototype cells for the 3-T concept where the management of interfaces and optical losses should be the main issues.

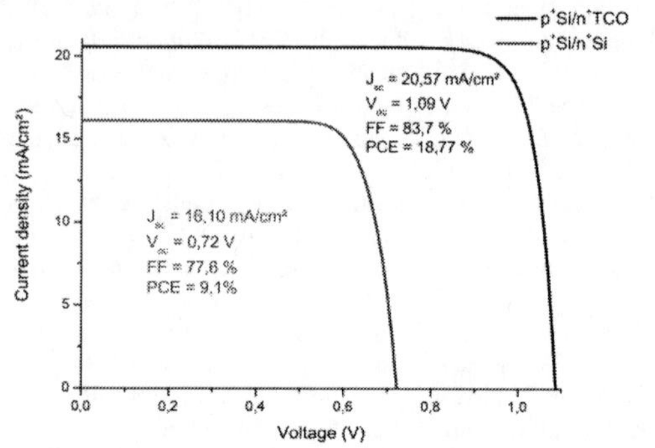

Figure 11: Output current-voltage characteristics (J_{SC}: short circuit current density, V_{OC}: open circuit voltage, FF: fill factor, and PCE: power conversion efficiency) of the two subcells in the 3-T architecture of Fig. 9.

REFERENCES

[1] K. Yoshikawa, H. Kawasaki, W. Yoshida, T. Irie, K. Konishi, K. Nakano, T. Uto, D. Adachi, M. Kanematsu, H. Uzu K. Yamamoto, Nature Energy 2 (2017) 17032.

[2] K. Yamamoto, D. Adachi, K. Yoshikawa, W. Yoshida, T. Irie, K. Konishi, T. Fujimoto, H. Kawasaki, M. Kanematsu, H. Ishibashi, T. Uto, Y. Takahashi, T. Terashita, G. Koizumi, N. Nakanishi, M. Yoshimi, Proceedings 33rd European Photovoltaic Solar Energy Conference (2017) 201.

[3] A. De Vos, J. Phys. D: Appl. Phys. 13 (1980) 839.

[4] S. Essig, C. Allebé, T. Remo, J. F. Geisz, M. A. Steiner, K. Horowitz, L. Barraud, J. S. Ward, M. Schnabel, A. Descoeudres, D. L. Young, M. Woodhouse, M. Despeisse, C. Ballif, and A. Tamboli, Nat. Energy 2 (2017) 17144.

[5] J. Werner, L. Barraud, A. Walter, M. Bräuninger, F. Sahli, D. Sacchetto, N. Tétreault, B. Paviet-Salomon, S.-J. Moon, C. Allebé, M. Despeisse, S. Nicolay, S. D. Wolf, B. Niesen, and C. Ballif, ACS Energy Lett. 1 (2016) 474

[6] K.A. Bush, A. F. Palmstrom, Z. J. Yu, M. Boccard, R. Cheacharoen, J. P. Mailoa, D. P. McMeekin, R. L. Z. Hoye, C. D. Bailie, T. Leijtens, I. M. Peters, M. C. Minichetti, N. Rolston, R. Prasanna, S. Sofia, D. Harwood, W. Ma, F. Moghadam, H. J. Snaith, T. Buonassisi, Z. C. Holman, S. F. Bent, M. D. McGehee, Nature Energy 2 (2017) 17009.

[7] J. Werner, B. Niesen, C. Ballif, Adv. Mater. Interfaces 5 (2018) 1700731.

[8] A. Martı and A. Luque, Nature communications 6 (2015) 6902.

[9] Z. Djebbour, W. El-Huni, A. Migan and J.-P. Kleider, Prog Photovolt Res Appl, to be published.

[10] Z. Djebbour, W. El-Huni, A. Migan and J.-P. Kleider, Patent WO 2017/093695.

[11] http://www.ioffe.ru/SVA/NSM/Semicond/index.html.

[12] T. Minemoto, M. Murata, Curr. Appl. Phys. 14 (2014) 1428.

SCREENING CARRIER SELECTIVE CONTACT COMBINATIONS FOR NOVEL CRYSTALLINE SI CELL STRUCTURES

R. Brendel[1,2], C. Kruse[1,2], A. Merkle[1], H. Schulte-Huxel[1], F. Haase[1], R. Peibst[1,3]

[1]Institute for Solar Energy Research Hamelin (ISFH), Am Ohrberg 1, 31860 Emmerthal, Germany
[2]Institute for Solid State Physics, Leibniz Universität Hannover, Appelstrasse 2, 30167 Hannover, Germany
[3]Institute for Electronic Materials and Devices, Leibniz Universität Hannover, Schneiderberg 32, 30167 Hannover, Germany

ABSTRACT: High efficiency crystalline Si solar cells require contacts with high carrier selectivity. This is ensured for contacts having low recombination currents as well as low contact resistances. A large variety of material systems for electron- and hole-selective contacts were measured in the literature. We screen a subset of electron- and hole-selective contacts to find promising combinations in terms of efficiency potential on the one hand and in terms of practical processes on the other hand. We use modelling of ideal Si cells with non-ideal experimental contact properties to determine the maximum efficiency and the optimized areal contact fractions for many contact combinations. Cells using a-Si and/or poly-Si contacts have the highest contact-limited efficiencies. Such cells are, however, quite different from today's PERC technology. We therefore also look for contact combinations that have one contact type equal to the current PERC technology and identify cell structures that combine a poly-Si(n) contact with a screen-printed Al-doped contacts (PAL cells) as an attractive upgrade for the PERC technology. We also report on experimental work on building blocks for various types of PAL cells.

Keywords: Carrier selective contacts, selectivity, loss analysis, silicon solar cell, laser processing, poly-Si

1 INTRODUCTION

Every piece of absorber material contributes to the recombination and to the series resistance of the solar cell. Dissipation of free energy by transport and by recombination reduces the cell efficiency. It thus depends on the recombinative and conductive material properties what level of efficiency can be achieved. For crystalline Si the conductivity is typically not limiting the cell efficiency. Instead it is carrier recombination that limits the cell efficiency to 29.6% under the illumination of one sun [1]. This limit only holds for ideal contacts that do neither contribute to the recombination nor to the resistance of the cell. Real contacts have, however, non-zero recombination currents J_c and contact resistances ρ_c.

Let us consider the rear side of a passivated emitter and rear cells (PERC) as an example. The contact recombination of a full-area Al-doped p^+-type back surface field (BSF) layer is large (high J_c). It therefore makes sense to reduce the contact area to a fraction f_c and to form partial contacts. This reduces the rear contact recombination to $f_c \times J_c$ and increases the contact resistance to ρ_c / f_c. The Al BSF cell becomes a PERC cell. The series resistance loss dominates if f_c becomes too small and therefore an optimum contact area fraction f_{max} exists that maximizes the cell efficiency.

We recently introduced the selectivity

$$S_{10} = \log_{10}\left(\frac{V_{th}}{J_c \times \rho_c}\right) \qquad (1)$$

for a carrier-selective contact [2], where V_{th} is the thermal voltage. [†] The maximum achievable efficiency

$$\eta_{max} = \left(\left(2.452\ S_{10} - 4.240\right)^{-19.52} + \left(29.21\right)^{-19.52}\right)^{-\frac{1}{19.52}} \% \qquad (2)$$

at optimum contact area fraction f_{max} was shown to increase linearly with the contact selectivity S_{10} until the efficiency saturates at above 29% due to intrinsic recombination (2 Ωcm n-type material, 110 μm wafer thickness) [2].

The literature reports many measurements of recombination currents J_c and contact resistance values ρ_c for various inorganic and organic material systems that do form carrier-selective contacts to Si. Figure 1 shows experimental (J_c, ρ_c) pairs of seven electron-selective contacts (red symbols) and seven hole-selective contacts (green symbols). Tables with the numeric values of (J_c, ρ_c) can be found in [3]. Knowing (J_c, ρ_c) we determine S_{10} with (1) and calculate the single-contact-limited maximum efficiency η_{max} using (2). The iso-efficiency lines η_{max} are shown in Figure 1 as solid lines. Excellent contacts are in the lower left corner. The dashed line shows the optimum contact area fraction f_{max} that maximizes the efficiency to η_{max}.

Now the important question arises which electron selective contact combined with which hole selective contact allows for a high efficiency in a mass producible way.

Please note that some parts of this contribution were already reported in workshops [4,5].

2 MODELLING CONTACT-LIMITED CELLS

We calculate the limiting efficiency for various contact combinations in an otherwise ideal cell using the approach described in [1]. We assume ideal Lambertian light trapping, one sun of AM1.5G illumination, 25°C cell temperature, and the intrinsic carrier lifetime of an n-type wafer with a resistivity 2 Ωcm and a wafer thickness

[†] Please note that we understand J_c to be equal to q R / [exp($V_{i,mpp}/V_{th}$) − 1] in case of a contact or surface recombination that does not follow the ideal Shockley diode equation (e.g. amorphous Si junctions). Here q is the elementary charge, R is the volume-integrated recombination rate, $V_{i,mpp}$ is the internal voltage at the maximum power point, and V_{th} is the thermal voltage.

of 110 μm. The intrinsic recombination is parametrized as in [6]. Grid shadowing and parasitic absorption by free carriers and metals are neglected. Flat quasi Fermi levels are assumed. We add the recombination and the resistances of the contacts and the recombination of the non-contacted surfaces in an obvious manner to the model given in [1]. All resistances contribute to a lumped series resistance. We considers the small scale limit [7] and thereby neglect lateral transport losses in the bulk and in the contact layers. The contacts have area fractions f_e and f_h for the electron- and the hole-selective contact, respectively. A contact area fraction larger than unity is possible if the Si surface is textured and/or the contact is not restricted to a single side of the cell.

If non-contacted surface regions exist, our model assumes a dielectric surface passivation layer with a recombination current that has the smallest among three possible values: The J_e and the J_h value of the two contact layers, respectively, and J_{Al2O3} = 10 fA/cm^2 for an Al_2O_3/SiN_x-passivated surface. We run a numerical optimization to find the pairs $f_{e,max}$ and $f_{h,max}$ that maximize the efficiency to η_{max}. In case of $f_e + f_h < 2$ a fraction $(2 - f_e - f_h)$ of the cell is not contacted. This is a three-parameter optimization problem. Varying f_e, f_h, and the internal voltage $V_{i,mpp}$ maximizes the power output.

The sole purpose of this crude modelling is to point to potentially interesting contact combinations without distraction by other physical limitations. Please note that the results depend on the experimental input parameters for J_c and ρ_c collected in Figure 1 and [3]. These parameters are, however, subjected to change due to technological improvements.

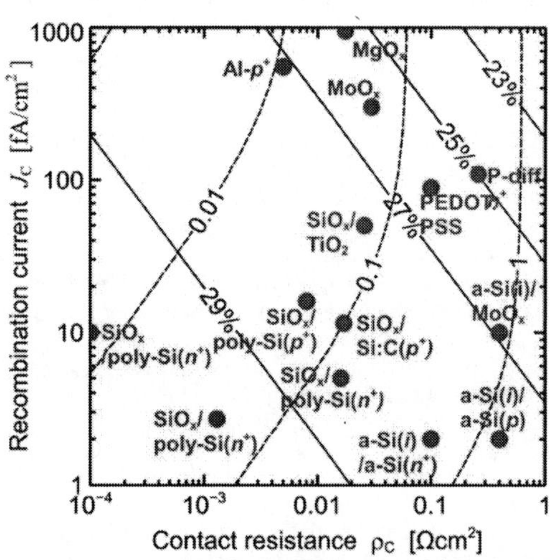

Figure 1: Experimental data pairs (symbols) of recombination current density parameters J_c and contact resistances ρ_c for carrier-selective contact systems reported in the literature. References for these data are given in Table I and [3]. Broken lines show the contact fraction f_{max} that maximizes the efficiency to η_{max}. Iso-η_{max} lines are shown as solid lines. The calculation assumes one real contact with (J_c, ρ_c) while the other contact is assumed to be ideal.

3 RESULTS FOR MODELLING CONTACT LIMITED CELLS

The seven electron- and seven hole-selective contacts that we selected from the literature allow for 49 contact combinations. We apply the above described model. Table I shows four calculated figures for all contact combinations: The (i) optimum contact area fraction $f_{e,max}$ of the electron-selective contact (top right in Table I) and (ii) the optimum contact area fraction $f_{h,max}$ of hole-selective contact (bottom left) that allow for (iii) the maximum efficiency η_{max} (lower right). We also calculate (iv) the combined selectivity $S_{10,max}$ for both contacts and the passivated and non-contacted surface if any (top left in Table I). The combined selectivity is determined from (1) with

$$J_c = f_{e,max} \times J_e + f_{h,max} \times J_h + \left(2 - f_{e,max} - f_{h,max}\right) \times \Theta\left(2 - f_{e,max} - f_{h,max}\right) \times J_{Al_2O_3} \quad (3)$$

using the Heaviside function $\Theta(x)$ and

$$\rho_c = \rho_e / f_{e,max} + \rho_h / f_{h,max} \, . \quad (4)$$

3.1 HIT benchmark case using a-Si:H(i)/a-SiH(p^+) and a-Si:H(i)/a-SiH(n^+)

This is the entry (2|2) in row 2 and column 2 of Table I and is encircled by a red box as a reference case. The amorphous Si contacts have low J_c values around 2 fA/cm^2. The p-type a-Si:H(i)/a-SiH(p^+) is the least recombination-active contact system among the hole-selective layer systems considered in this paper. The estimated contact resistances of 0.1 Ωcm^2 and above are, however, at the high end. The current world record for crystalline Si solar cell is held by an HIT-IBC cell using amorphous Si contacts. The experimental efficiency is 26.7% [8,9]. Our contact-limited cell model calculates a maximum efficiency η_{max} = 27.5%. Please note that the HIT benchmark combination is the only one with rather symmetric (76% and 152%) contact fractions. This is an advantageous and unique feature among all combinations considered here.

3.2 Combinations interesting because of efficiency and contact area

We are searching contact combinations that allow for highest efficiencies. All efficiency entries larger or equal to the HIT benchmark efficiency ($\eta_{max} \geq 27.5\%$) are marked with a red background. We find 8 such combinations that all apply amorphous Si and/or poly-Si on oxide (POLO)-type of contacts. This is a consequence of the low J_c values of these contacts as can be seen in Figure 1.

For simple processing it is advantageous if a third material for passivating the non-contacted cell surface is *not* required. This is the case for combinations that have $f_{e,max} + f_{h,max}$ approximately equal to 200%. As we did not precisely distinguish whether the reported J_c values apply to planar or textured surface we only require 150 % < $f_{e,max} + f_{h,max}$ < 250%. Combinations fulfilling this conditions have $f_{e,max}$ and $f_{h,max}$ fields marked with gray background in Table I. We find 32 of such combinations. Please note that many combinations have exactly $f_{e,max} + f_{h,max}$ = 200%. This happens if a non-contacted area has a recombination current density that is equal to that of one of the contacts. Enlarging the contacted area is then advantageous when optimizing the efficiency and thus the non-contacted area fraction ($2-f_{e,max}-f_{h,max}$) vanishes. We find eight combinations that have red and gray backgrounds marked. These combinations are encircled with a dashed black line in Table I; the HIT benchmark case (red box) is one of them. All eight combinations contain amorphous Si and/or poly-Si on oxide type of contacts. Among the combinations considered here it is SiO_x/poly-Si(n^+) as electron-selective contact layer combined with an a-Si:H(i)/a-Si:H(p) as hole-selective contact layer that reaches the highest efficiency limit of 27.9%. Two such combinations (2|4) and (2|5) exist in Table I. We mark (2|4), the one with the less asymmetric contact fractions, with a blue box.

3.3 Combinations interesting because of similarity to current industrial cells

For a future mass production it would be extremely helpful if one of the contacts of a new cell generation would be identical to one of the two sides of today's PERC cells. Today's PERC cells have a P-diffused emitter (with possibly reduced doping between the fingers) on the front and screen-printed local p^+-type Al-doped carrier-selective contacts on the rear. Contact combinations with a local Al-doped surface are listed in the first row; those with a P-diffused surface are listed in the first column of Table I. The two combinations with the highest modelled efficiency of η_{max} =26.7% are those marked by a green box (solid and dashed) in Table I. Screen-printed local p^+-type Al-doped carrier-selective contacts require a firing step that is not compatible with amorphous Si. That is why we will only consider the combination with the poly-Si in more depth later in this paper.

Table I: Calculation of two-contact-limited cell efficiencies. Combined selectivity S_{10} (top left entry), contact area fraction $f_{e,max}$ of electron-selective layer (top right), contact area fraction $f_{h,max}$ of hole-selective layer (bottom left), and efficiency η_{max} (bottom right) are tabulated for all contact combinations. Maximum efficiencies η_{max} > 27.5% are marked with a red background. Area fractions with 150 % < $f_{e,max}$+ $f_{h,max}$ < 250% are marked with a gray background. The marked combinations are discussed in the text.

$S_{10,max}$	$f_{e,max}$ [%]	Electron-selective contacts													
$f_{h,max}$ [%]	η_{max} [%]	P-diffused n [2]		a-Si:H(i) /a-Si:H(n) [10]		th-SiO_x/ poly-Si(n) PECVD [11]		th-SiO_x/ poly-Si(n) LPCVD POLO(n^+) [12]		chem-SiO_x/ poly-Si(n) LPCVD [13]		SiO_x/TiO_2 [14]		MgO_x [15]	
Hole-selective contacts	Al-doped (p) [16]	11.7	69	12.7	198.6	12.7	198.2	12.8	198.5	12.6	197.7	12.3	21.6	11.8	5.2
		4.1	24.3	1.4	26.7	1.8	26.4	1.5	26.7	2.3	26	2.6	25.6	3.7	24.6
	a-Si:H(i) /a-Si:H(p) [8]	11.8	50.4	13.2	75.9	13.3	20.5	13.5	8.6	13.5	1.1	12.8	9.6	12	3.6
		461.5	24.7	151.9	27.5	179.5	27.7	191.4	27.9	198.9	27.9	190.4	26.8	375.1	25.1
	SiO_x/ poly-Si POLO(p) [17]	11.8	59.4	13.5	191.7	13.5	187.4	13.7	190.6	13.5	177.5	12.7	18	12	4.5
		42.3	24.7	8.3	27.8	12.6	27.3	9.4	27.8	22.5	26.8	25.7	26.2	37.7	25
	SiO_x/ Si:C (p) [18]	11.8	59	13.4	185.2	13.4	176	13.6	183	13.7	135.1	12.7	17.8	12	4.4
		122.6	24.7	14.8	27.8	24	27.3	17	27.7	64.9	26.8	74.4	26.3	109.1	25
	a-Si:H(i)/ MoO_x [19]	11.7	58.9	12.6	105.6	12.7	60.6	12.7	80.4	12.8	3.1	12.4	16.2	11.8	4.3
		241.2	24.3	94.4	26.5	139.4	26.5	119.6	26.7	196.9	26.6	183.8	26	202	24.7
	MoO_x [20]	11.5	78.2	12.3	193.6	12.3	192.5	12.4	193.3	12.2	191.1	12	25.4	11.6	6
		15.5	24	6.4	25.9	7.5	25.7	6.7	25.9	8.9	25.5	10.1	25.1	14.1	24.2
	PEDOT :PSS [21]	11.6	76	12.4	178.4	12.3	174	12.4	177	12.3	168	12.1	24.5	11.7	5.8
		56.1	24	21.6	26	26	25.8	23	26	32	25.6	36.3	25.2	50.8	24.3

4 DISCUSSION OF NOVEL CELL STRUCTURES

Here we discuss possible realizations of the contact combinations marked in Table I.

4.1 Combination of poly-Si on oxide (POLO) and amorphous silicon a-SiH(*i*)/a-Si:H contact

The structure with the largest limiting efficiency in the above screening is a combination of an *n*-type poly-Si on oxide (POLO(*n*)) and a *p*-type amorphous silicon a-SiH(*i*)/a-Si:H(*p*) contact (blue box (2|4) in Table I). The calculated efficiency limit is 27.9% and thus 0.4% higher than the limit for the HIT cell. The POLO(*n*) contact covers 8% of the rear surface while all the rest of the rear and the front surface (192%) is left to the a-SiH(*i*)/a-Si:H(*p*) junction. Covering > 100% with one contact means surface texturing and/or a three terminal cell as sketched in Figure 2. This structure is of conceptual rather than practical interest for several technological reasons. Please note that a-Si contacts do not have a unity ideality factor that was implicitly assumed by our modelling.

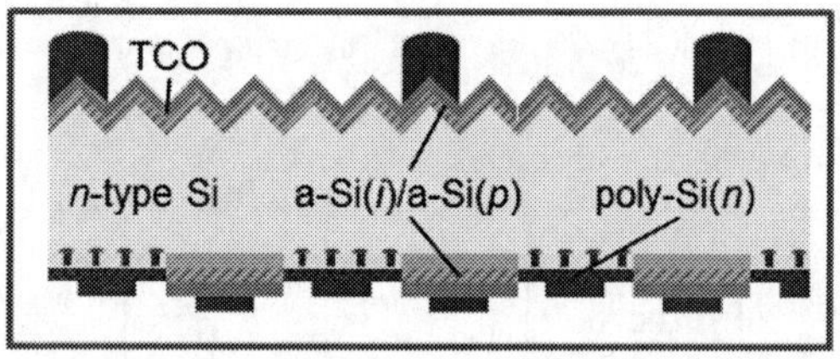

Figure 2: Sketch of a contact combination (2|4) marked with a blue box in Table I. Contact-limited efficiency is 27.9%. It combines a-Si and poly-Si contacts and realized in two-side contacted cell, in an IBC cell or in a three terminal cell.

4.2 Combination of poly-Si on oxide (POLO(*n*)) contacts with poly-Si on oxide (POLO(*p*)) contacts

The contact combination poly-Si(*n*) on oxide contacts with poly-Si(*p*) on oxide contacts in field (3|4) in Table I reaches a contact limited efficiency of 27.8% which is just 0.1% less than the structure discussed in 4.1. It also has highly asymmetric contact fraction with 191% for poly-Si(*n*) and 9% for poly-Si(*p*). This asymmetry is due to the smaller J_c value for the *n*-type POLO junction. Using symmetric contact areas of 100% poly-Si(*n*) on one side of the cell and 100% poly-Si(*p*)on the other side reduces contact limited the efficiency only by 0.3% to 27.5%. This is certainly a highly interesting structure because the high-temperature stable POLO junctions can be combined with screen-printing, the current industry standard for applying metal contacts. We term cells that have poly-Si at both contacts to belong to the PolyPoly cell family. Such cells can be realized in front-junction (PolyPoly-FJ), back junction (PolyPoly-BJ) of interdigitated back contact (PolyPoly-IBC) geometry.

Figure 3 shows an *n*-type PolyPoly-BJ cell as an example. Screen-printing on the front is easier for a thicker poly-Si layer. Reduced parasitic light absorption requires, however, a thin poly-Si layer. This thickness variation could be realized by masked etching and the poly-Si on the front could be fully removed. A new process to do this will be described below. Polarities of the poly-Si layers and the wafer that are inverted to those in Figure 3 are also attractive.

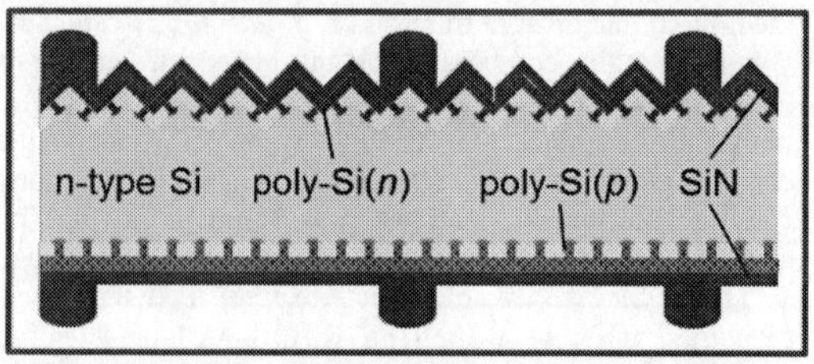

Figure 3: Sketch of PolyPoly-BJ cell with a contact combination (3|4) and marked with a violet box in Table I combining poly-Si(*n*) and poly-Si(*p*) contacts. The contact-limited efficiency is 27.8% for asymmetric contact areas and 27.5% for symmetric contact areas. Optics may require a thinned poly-Si(*n*) in between the front contacts. Inverse polarities f poly-Si and wafer and full removal of the poly-Si in between the front fingers is also attractive.

4.3 Combination of poly-Si on oxide (POLO(*n*)) contacts with local *p*-type Al-doped contacts

The contact combination marked by the solid green box in Table I has a large area poly-Si on oxide contact and a small area Al-doped contact formed by screen-printing (PAL cell). The PAL cell has a high contact-limited efficiency of η_{max} = 26.7 % in Table I. Figure 4 sketches three different geometries of a PAL cell: back junction (PAL-BJ), front junction (PAL-FJ), and PAL-IBC. These different members of the PAL family have specific advantages and disadvantages.

The PAL-IBC cell has the largest efficiency potential due to the advantages that are intrinsic to IBC cells such as no front side shadowing and using more metal on the rear side. For mass production the structuring of the rear side certainly is a challenge.

The PAL-FJ cell has a rear side that is identical to the rear side of the PERC+ cell. This is a big advantage when industrializing this concept. It replaces the diffused emitter of the PERC+ cell by a poly-Si(*n*) and thereby removes the bottleneck of future PERC cell development: The fabrication of an emitter with very low recombination. Fabricating this cell will, however also be challenging, since the poly-Si(*n*) on the front has to be very thin to reduce absorption of solar light in the poly-Si(*n*). Screen printing on the poly-Si(*n*) will, however, be the more challenging, the thinner the poly-Si(*n*) is. Also, the lateral conductivity of the POLO junction decreases with decreasing poly-Si thickness. A possibility to address these counteracting requirements is the restriction of the poly-Si just below the metal fingers, while utilizing a conventional *n*-type diffused c-Si emitter in the inter-finger regions. This concept requires a structuring of the poly-Si (as for PolyPoly-BJ and Poly-Poly-IBC). A new approach and first experimental results for structuring poly-Si are reported below.

The PAL-BJ cell has the poly-Si(*n*) emitter on the rear side. This allows for a thicker poly-Si layer and simplifies the screen printing on the rear. Local screen-printed Al-doped contacts are placed on the front-side of the cell. They are finger-shaped to allow light entering through an Al_2O_3/SiN_x passivation into the Si wafer. We choose a *p*-type wafer for sufficient lateral hole conductivity. This makes the front side of the PAL cell identical to a (rear side-textured) bifacial PERC+ cell [22] which is currently being rolled out by various companies. The PAL design has various potentially attractive production sequences, in particular when using a truly single-side poly-Si deposition technique. The PAL-BJ cell does not require lateral structuring except for the laser contact opening and screen printing.

The shadowing loss by the Al-printed fingers on the front is a challenge for this cell structure because the typical finger width on the rear of PERC+ cells is currently around 120 μm. However, the development of PERC+ cells towards high bifaciality is rapidly pushing the Al paste development towards thinner fingers. This makes the PAL-BJ design a candidate for upgrading PERC cells while avoiding a P-diffused selective emitter.

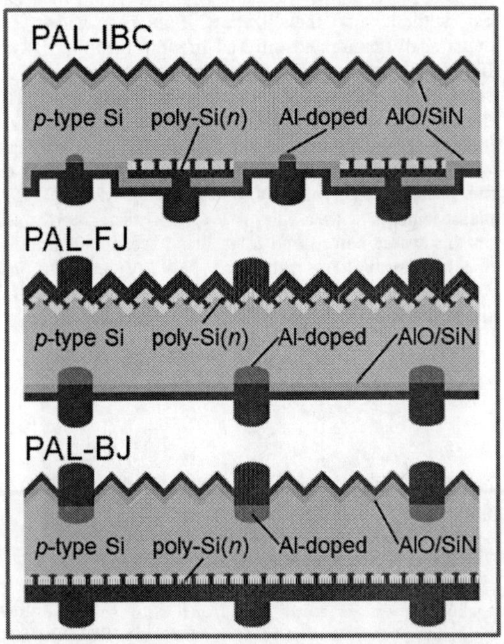

Figure 4: Three realizations of a PAL solar cell with a poly-Si(*n*) emitter and screen-printed local Al-doped contacts utilizing the contact combination (1|4) that is marked with a solid green box in Table I. The PAL cell can have a back junction (PAL-BJ), a front junction (PAL-FJ), or a PAL-IBC geometry.

5 MODELLING A PAL CELL WITH THE CONDUCTIVE BOUNDARY MODEL USING EXPERIMENTAL INPUT PARAMETERS

The above screening simulations only disclose the limits set by the contact properties while disregarding optical losses as well as resistive and extrinsic recombination losses in the wafer. For a more realistic evaluation of the expected performance of a PAL-BJ cell we conduct a Quokka 2 simulation [23]. Quokka has our Conductive Boundary (CoBo) model [24] implemented and thus accepts measured recombination currents J_c, sheet resistances $\rho_{\square}$, and contact resistances ρ_c as primary inputs. We take all these parameters from our experimental development work on PERC cells [25] and on POLO-IBC cells [26]. Using these parameters we optimize the geometry and calculate an efficiency of 23.6% when assuming *p*-type Si with a resistivity of 1 Ωcm, SRH lifetimes of τ_{n0} =1.25 ms and τ_{p0} = 12.5 ms, front fingers that are 50 μm wide, and an antireflection coating similar to that of our current PERC solar cell with a recombination current of 10 fA/cm^2.

Figure 5 shows a Synergistic Efficiency Gain Analysis (SEGA) [16] of this hypothetical PAL-BJ cell. The emitter recombination, which is the dominant loss mechanism in our PERC+ cell [25], becomes insignificant for the PAL cell with a rear side POLO(*n*) emitter. The dominating losses are finger shading losses (0.63%) and recombination at the Al_2O_3/SiN_x front surface passivation (0.51%). The shading loss will be smaller in a module. We consider 23.6% to be an attractive efficiency when compared to the efficiency of today's PERC cells.

Please note that we released the SEGA-GUI program for automatically generating plots as that in Figure 5 for many different cell structures at this conference [27].

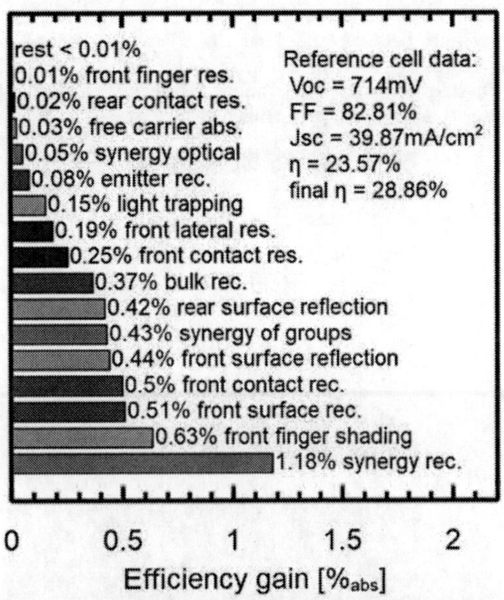

Figure 5: SEGA analysis of a PAL cell using experimental input parameters for the modelling. Finger shading and front surface recombination are the largest losses.

6 EXPERIMENTAL WORK TOWARDS PAL CELLS

We started experimental work on how to structure the front-side for a PAL-FJ cell, as well as of a process sequence for the PAL-BJ cell.

6.1 Laser structuring of poly-Si on oxide

The structuring of the poly-Si -for example for a PolyPoly-BJ cell or a PAL-FJ cell - could be performed by printing masks and subsequent etching or by laser ablation. The latter requires machining of ~97% of the surface area. It is more attractive to laser machine only the finger regions where the poly-Si should remain (~3% of the surface area). For this purpose, we introduce local oxidation of the poly-Si by a laser. Laser oxidation of Si is known from microelectronics [28,29] for growing of ultra-thin gate oxides and was considered in photovoltaics as a side effect occurring during the formation of laser contact openings [30]. Here we utilize the oxide that grows during laser machining in air as a local etch mask against KOH.

We use saw damage etched *n*-type Cz wafers with 90 nm SiN_x on both sides. On top of the SiN_x, we deposit 150 nm Si by low pressure chemical vapor deposition (LPCVD). Directly prior to the laser treatment, we perform an HF dip to remove native oxide. We use a picosecond laser with a wavelength of 355 nm. The laser fluence is below the ablation threshold of the Si. We laser machine finger regions with two overlaps of the single shot laser spots. Subsequently, the non-machined poly-Si regions are removed in diluted KOH. The SiN_x acts as an etch stop.

Figure 6 shows that the thin oxide grown during the laser treatment functions as an etch mask for the poly-Si underneath. While the non-machined Si in the inter-finger regions is completely removed (blue color), the poly-Si in the finger regions remains (pink color). Figure 6 (a) shows a finger pattern that we realized with a sufficient overlap between the laser spots as opposed to an insufficient overlap in Figure 6 (b). We are currently investigating the passivation quality of the POLO junctions in the laser-machined areas.

a) b)

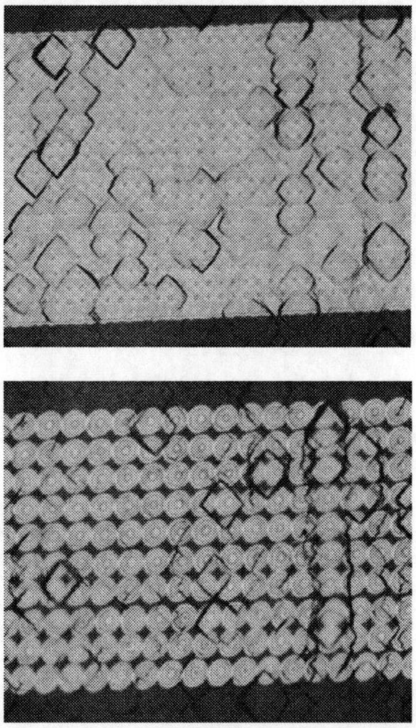

Figure 6: Optical microscope images of finger regions laser machined with (a) sufficient and (b) insufficient overlap of the laser spots. Pink regions corresponds to poly-Si which was protected during the KOH etch by the laser oxide. The blue layer underneath is SiN_x.

6.2 Compatibility of Al_2O_3/SiN_x passivation and POLO(*n*) junction

We have started the experimental realization of a 156.7 ×156.7 mm^2-sized pseudo-square PAL-BJ cell and briefly report the present status of this work. We use a B-doped *p*-type CZ Wafer of 1.7 Ωcm resistivity and 170 µm in thickness after the damage and the texture etching.

The front side is surface-textured and passivated by Al_2O_3/SiN_x. The cell has line-shaped laser contact openings with a width of 40 µm. The screen-printed Al fingers have a width of 150 µm and a pitch of 1600 µm. The five busbars are 500 µm wide each.

The rear side of our first PAL-BJ cell is planar (damage-etched) and passivated with a POLO(*n*) junction that has a nominal thickness of 100 nm plus an SiN_x layer. We use a firing-through Ag paste. The Ag fingers on the rear side are 45 µm-wide and have a pitch of 1200 µm. The five busbars on the rear side are 500 µm wide and are screen-printed using a non-firing-through paste.

We first tested the compatibility of the firing process with the passivation by Al_2O_3/SiN_x on the textured front and by poly-Si(*n*)/SiN_x on the rear. To this end we measured the carrier lifetime of co-processed cell precursors exhibiting the same coatings as the final cell but without any metallization. Figure 7 shows the dynamically calibrated infrared lifetime mappings (dyn-ILM [31]) before and after the firing process under 0.25 suns of illumination for such a cell precursor. The lifetime is spatially homogeneous and we observe a slight increase of the average lifetime from 905 µs to 1088 µs, which shows a positive impact of the firing process on the passivation. From photoconductance decay (PCD) measurements with a Sinton WCT120 tester performed on this wafer before and after firing we obtain implied open circuit voltage values of 717 mV and 722 mV respectively. As expected this value is larger than the V_{oc} simulated in Figure 5 including contact recombination.

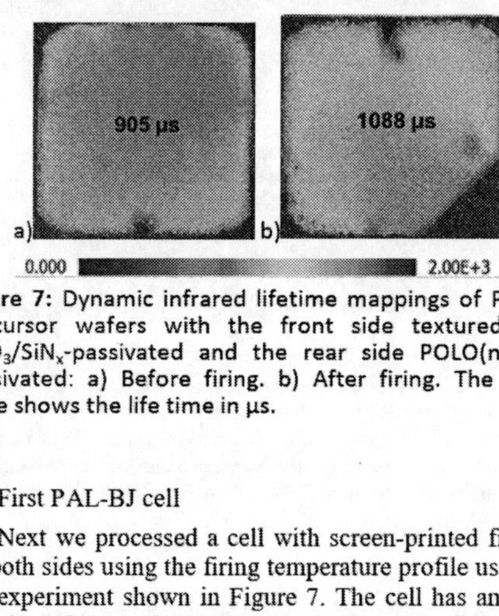

Figure 7: Dynamic infrared lifetime mappings of PAL-BJ precursor wafers with the front side textured and Al_2O_3/SiN_x-passivated and the rear side POLO(n)/SiN_x passivated: a) Before firing. b) After firing. The color scale shows the life time in µs.

6.3 First PAL-BJ cell

Next we processed a cell with screen-printed fingers on both sides using the firing temperature profile used for the experiment shown in Figure 7. The cell has an open circuit voltage of 675.7 mV, a short-circuit current density of 35.5 mA/cm^2, a fill factor of 77.3% and an efficiency of 18.5% as independently confirmed by the certified ISFH CalTeC [32].

The open circuit voltage is promising. However, due to various technological imperfections our first PAL-BJ cell is 40 mV behind the expectations from Figure 5. The small short circuit current density of 35.5 mA/cm^2 suffers from 11% shading by the 150 µm-wide Al fingers and the busbars. There is, however, a good perspective to significantly reduce this shading to the level assumed in the simulation. Figure 8 shows our recent Al printing result [33] using a paste developed by Toyo Aluminum K. K., Japan. The finger is only 50 µm wide.

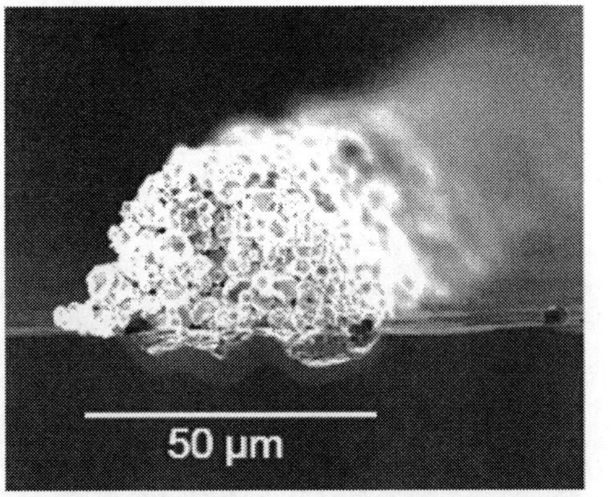

Figure 8: Scanning electron micrograph of a 50µm-wide Al print fabricated at ISFH. Source of image is Ref. [33].

7 SUMMARY

We calculated the contact-limited efficiencies of 49 combinations of electron- and hole-selective contact systems. The HIT combination using two amorphous Si-based contacts has the unique property of giving maximum efficiencies for approximately 100% contact area fractions for both polarities. Poly-Si on oxide junctions can improve the contact-limited efficiency compared to the reference contacts of amorphous Si by $0.4\%_{abs}$. Contact combinations of a-Si and poly-Si contacts and combinations of poly-Si(*n*) and poly-Si(*p*) offer the highest contact-limited cell efficiencies when using the experimental data collected in Figure 1. It is an important research task to find lean process sequences for realizing these options.

A further outcome of this screening is the PAL cell family combining poly-Si(*n*) junctions with the low-cost local Al-doped contacts fabricated by screen printing. Three different modifications are possible – the front junction cell (PAL-FJ), the back junction cell (PAL-BJ), and the interdigitated back-contacted cell (PAL-IBC).

A structuring of the poly-Si may be advantageous for the PolyPoly-BJ, the PAL-FJ and many other cell structures. For this purpose, we demonstrated the local oxidation of the Si by laser treatment in air. We showed that the oxide grown locally protects the Si in the finger regions in a KOH etch that removes the Si in the inter-finger regions.

The PAL-BJ cell uses a front side that is equal to the rear side of PERC+ cell and has a poly-Si on oxide emitter on the rear. The PAL-BJ structure is presumably easy to fabricate on a large scale and lifts major challenges of the current PERC development without sacrificing efficiency potential. We thus see the possibility of largely improving todays PERC cells while keeping major parts of the PERC cell and module process unchanged.

We also demonstrated first experimental steps towards the PAL-BJ cell and an open circuit voltage of 676 mV.

8 ACKNOWLEDGEMENTS

We thank U. Sonntag, S. Spätlich, B. Gehring and T. Friedrich for their help with sample processing, and T. Neubert for his contributions to the laser experiments. Parts of this work were supported by the State of Lower Saxony via the project vOx under contract No 74ZN1354.

REFERENCES

[1] S. Schäfer and R. Brendel, Accurate calculation of the absorptance enhances efficiency limit of crystalline silicon solar cells with Lambertian light trapping, IEEE Journal of Photovoltaics **8**, 1156 (2018).

[2] R. Brendel and R. Peibst, Contact selectivity and efficiency in crystalline silicon photovoltaics, *IEEE J. Photovolt.* **6**, 1413 (2016).

[3] J. Schmidt, R. Peibst and R. Brendel, Surface passivation of crystalline silicon solar cells: Present and future, Solar Energy Materials and Solar Cells 187, 39-54 (2018).

[4] R. Brendel and B. Lim, Silicon cell concepts - trends in research, Presentation held at *PV Module Technology & Applications Forum 2018* (TÜV Rheinland, Köln) on Jan. 29, 2018. Link to presentation: https://isfh.de/isfh-gibt-ausblick-auf-solarzellentechnologien-beim-pv-module-technology-applications-forum-2018/ , loaded Sept 19th, 2018.

[5] R. Brendel and C. Kruse, Screening carrier selective contact combinations for novel crystalline Si cell structures, in *Proc. 28th NREL Workshop on Crystalline Silicon Solar* Cells & Modules: Materials and Processes (Winter Park, CO, Aug 12-15, 2018), pp. 58-65., Link to paper: https://isfh.de/isfh-beitrag-zum-siliziumphotovoltaik-expertenworkshop-des-national-renewable-energy-laboratory-usa, loaded Aug 27th, 2019.

[6] A. Richter, S. W. Glunz, F. Werner, J. Schmidt and A. Cuevas, Improved quantitative description of Auger recombination in crystalline silicon, *Physical Review B* **86**, 165202 (2012).

[7] M. Schofthaler, U. Rau and J. H. Werner, Direct Observation of a Scaling Effect on Effective Minority-Carrier Lifetimes, *Journal of Applied Physics* **76**, 4168 (1994).

[8] K. Yoshikawa, W. Yoshida, T. Irie, H. Kawasaki, K. Konishi, H. Ishibashi, T. Asatani, D. Adachi, M. Kanematsu, H. Uzu, and K, Yamamoto, Exceeding conversion efficiency of 26% by heterojunction interdigitated back contact solar cell with thin film Si technology, *Sol. En. Mat. Sol. Cells* **173**, 37 (2017).

[9] M. A. Green, Y. Hishikawa, W. Warta, E. D. Dunlop, D. H. Levi, J. Hohl-Ebinger, and A. Ho-Baillie, Solar cell efficiency tables (version 50), *Prog. Photovolt. Res. Appl.* **25**, 668 (2017).

[10] S. Y. Herasimenka, W. J. Dauksher, and S. G. Bowden, >750 mV open circuit voltage measured on 50 µm thick silicon heterojunction solar cell, *Appl. Phys. Lett.* **103**, 053511 (2013).

[11] D. Yan, A. Cuevas, J. Bullock, Y. Wan, and C. Samundsett, Phosphorus-diffused polysilicon contacts for solar cells, *Sol. En. Mat. Sol. Cells* **142**, 75 (2015).

[12] M. Rienäcker, M. Bossmeyer, A. Merkle, U. Römer, F. Haase, J. Krügener, R. Brendel, and R. Peibst, Junction resistivity of carrier-selective polysilicon on oxide junctions and its impact on solar cell performance, *IEEE J. Photovolt.* **7**, 11 (2017).

[13] G. Kökbudak, R. Müller, F. Feldmann, A. Fell, R. Turan, S.W. Glunz, *Proc. 33rd European Photovolt. Solar Energy Conf.* (WIP, Munich, 2017), pp. 242.

[14] X. Yang, K. Weber, Z. Hameiri, and S. De Wolf, Industrially feasible, dopant-free, carrier-selective contacts for high-efficiency silicon solar cells, *Prog. Photovolt.: Res. Appl.* **25**, 896 (2017).

[15] Y. Wan, C. Samundsett, J. Bullock, M. Hettick, T. Allen, D. Yan, J. Peng, Y. Wu, A. Javey, and A. Cuevas, Conductive and stable magnesium oxide electron-selective contacts for efficient silicon solar cells, *Adv. En. Mat.* **7**, 1601863 (2016).

[16] R. Brendel, T. Dullweber, R. Peibst, C. Kranz, A. Merkle and D. Walter, Breakdown of the efficiency gap to 29% based on experimental input data and modeling, *Prog. Photovolt.: Res. Appl.* **24**, 1475 (2016).

[17] D. Yan, A. Cuevas, Y. Wan, and J. Bullock, Passivating contacts for silicon solar cells based on boron-diffused recrystallized amorphous silicon and thin dielectric interlayers, *Sol. En. Mat. Sol.* **152**, 73 (2016).

[18] G. Nogay, J. Stuckelberger, P. Wyss, E. Rucavado, C. Allebé, T. Koida, M. Morales-Masis, M. Despeisse, F.-J. Haug, P. Löper, and C. Ballif, Interplay of annealing temperature and doping in hole selective rear contacts based on silicon-rich silicon-carbide thin films, *Sol. En. Mat. Sol. Cells* **173**, 18 (2017).

[19] J. Geissbühler, J. Werner, S. M. Nicolas, L. Barraud, A. Hessler-Wyser, M. Despeisse, S. Nicolay, A. Tomasi, B. Niesen, S. De Wolf, and C. Ballif, 22.5% efficient silicon heterojunction solar cell with molybdenum oxide hole collector, *Appl. Phys. Lett.* **107**, 081601 (2015).

[20] J. Bullock, A. Cuevas, T. Allen, and C. Battaglia, Molybdenum oxide MoO_x: A versatile hole contact for silicon solar cells, *Appl. Phys. Lett.* **105**, 232109 (2014).

[21] D. Zielke, C. Niehaves, W. Lövenich, A. Elschner, M. Hörteis, and J. Schmidt, Organic-silicon solar cells exceeding 20% efficiency, *Energy Procedia* **77**, 331 (2015).

[22] T. Dullweber, C. Kranz, R. Peibst, U. Baumann, H. Hannebauer, A. Fülle, S. Steckemetz, T. Weber, M. Kutzer, M. Müller, G. Fischer, P. Palinginis and H. Neuhaus, PERC+: industrial PERC solar cells with rear Al grid enabling bifaciality and reduced Al paste consumption, *Progr. in Photovoltaics* **24**, 1487 (2016).

[23] A. Fell, A free and fast three-dimensional/two-dimensional solar cell simulator featuring conductive boundary and quasi-neutrality approximations, *IEEE Trans. Electron Devices* **60**, 73 (2013).

[24] R. Brendel, Modeling solar cells with the dopant-diffused layers treated as conductive boundaries, *Prog. Photovoltaics Res. Appl.* **20**, 31 (2012).

[25] C. Kranz, J. H. Petermann, T. Dullweber, and R. Brendel, Simulation-based efficiency gain analysis of 21.2%-efficient screen- printed PERC solar cells, *Energy Procedia* **92**, 109 (2015).

[26] F. Haase, F. Kiefer, S. Schäfer, C. N. Kruse, J. Krügener, R. Brendel, and R. Peibst, Interdigitated back contact solar cells with polycrystalline silicon on oxide passivating contacts for both polarities, *Jpn. J. Appl. Phys.* **56**, no. 8S2, 08MB15 (2017).

[27] C. Kruse , K. Bothe, B. Lim, T. Dullweber, and R. Brendel, Synergistic efficiency gain analysis: An easy to use SEGA simulation tool for the photovoltaik community, *Proc 35th European Photovltaic Specialsts Conf.* (WIP, Brussel, 2018), in press.

[28] I. W. Boyd, J. I. B. Wilson and J. L. West, Laser-induced oxidation of silicon, *Thin Solid Films* **83**, L173 (1981).

[29] B. J. Kailath, A. DasGupta, N. DasGupta, B. N. Singh and L. M. Kukreja, Growth of ultra-thin SiO_2 by laser-induced oxidation, *Semiconductor Science and Technology* **24**, 105011 (2009).

[30] Z. Du, C. Zhang, F. Li, R. Zhou and M. Hong, Impact of Laser-Induced Oxidation on Silicon Wafer Solar Cells' Performance, *IEEE Journal of Photovoltaics* **6**, 617 (2016).

[31] K. Ramspeck, S. Reissenweber, J. Schmidt, K. Bothe and R. Brendel, Dynamic carrier lifetime imaging of silicon wafers using an infrared-camera-based approach, *Applied Physics Letters* **93**, 102104 (2008).

[32] The ISFH CalTeC (Calibration and Test Center) is accredited according to ISO 17025 for the calibration of solar cells by the national accreditation body of the Federal Republic of Germany DAkkS (Deutsche Akkreditierungsstelle) under the registration number D-K-18657-01-00. https://isfh.de/en/dienstleistungen/isfh-caltec/solarzellen/.

[33] T. Dullweber, H. Hannebauer, B. Lim, P. Jäger, M. Rudolph, H. Schulte-Huxel, F. Haase, K. Bothe, M. Köntges, S. Kajari-Schröder, R. Peibst, R. Brendel, Bifacial PERC+ solar cells with conversion efficiencies above 22% and high energy yield, *Presentation held at SNEC 2018* (Shanghai, May 28th 2018). Link to presentation , loaded Sept 21, 2018.

SIMPLE YET EFFICIENT CHEMICALLY DEPOSITED AG REAR SIDE METALLIZATION ON ITO FOR HIGH-EFFICIENCY C-SI SOLAR CELLS

H. Nagel[1], D. Sontag[2], M. Glatthaar[1] and S. W. Glunz[1]

[1]Fraunhofer Institute for Solar Energy Systems ISE, Heidenhofstr. 2, 79110 Freiburg, Germany
[2]Meyer Burger Germany GmbH, An der Baumschule 6-8, 09337 Hohenstein-Ernstthal, Germany

ABSTRACT: A simple process for preparation of full-area Ag rear side contacts on indium tin oxide (ITO) coated heterojunction technology (HJT) Si solar cells was developed. It employs the well known Tollen's reagent to chemically deposit Ag. Investigations of the microstructure revealed that dense Ag layers without pores are formed. Compared with evaporated Ag the optical and electrical performance of HJT solar cells is practically the same if edge shunts can be avoided by thorough application of a polymer ink all around the edge before chemical Ag deposition. Hence, expensive high-vacuum deposition systems can be replaced. The process is well suited for lab R & D and could also find application in production when combined with smart wire connection technology.

Keywords: metallization, back contact, ITO

1 INTRODUCTION

There is growing interest in solar cells featuring heterojunction or passivated contacts because they are the leaders when it comes to high energy conversion efficiency [1]. For both sides contacted cells a transparent conducting oxide (TCO) is applied to the front and rear. TCOs provide low ohmic contact resistance on a-Si:H or polycrystalline Si and act as an efficient antireflection or reflection coating, respectively. Indium tin oxide (ITO) features high electrical conductivity and low light absorption and hence is a widespread TCO. On high efficiency level the bench mark for rear side metallization on ITO for cell optimization in the lab is full area physical vapor deposited (PVD) Ag. Ag forms a very efficient rear side reflector and simultaneously provides high lateral conductivity as well as low contact resistance to ITO. However, huge high-vacuum systems are needed in order to deposit Ag on a large quantity of industrial size solar cells. As a simple to use alternative we investigated the performance of chemically deposited Ag as a full area rear side metallization on silicon heterojunction solar cells. The same chemical process is utilized for industrial mirror production if high optical reflectance is aimed at. For this application the liquids are sprayed on large area glass panes in horizontal inline process machines, see Fig. 1. After silvering the surface is sealed with varnish in order to avoid yellowing. Note that if lower reflectance is sufficient, sputtered aluminum is used for industrial mirror making.

Fig 1: Photo of industrial Ag mirror production by spray deposition [2].

2 EXPERIMENTAL

6 inch *n*-type CZ Si solar cells featuring a rear side hetero-emitter were investigated. A schematic cross-sectional view is depicted in Fig. 2. The cells were alkaline textured and coated with ITO on both sides. On the front side Ag paste was screen printed in a 5 busbars layout and cured at 200 °C for 20 min. On the rear side we chemically deposited Ag using well known Tollen's reagent [3]. It is an aqueous solution of diaminesilver(I) complex ($[Ag(NH_3)_2]^+$). Deposition starts as soon as it is mixed with a reducing agent.

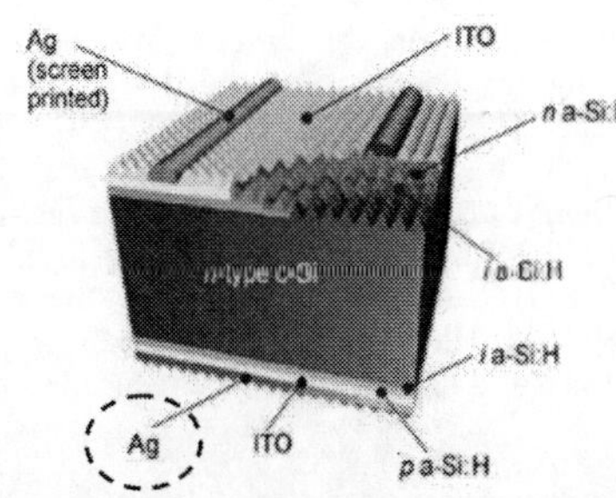

Fig. 2: Schematic representation of a Si heterojunction-technology (HJT) solar cell.

The chemical reactions are as follows [4]: Aqueous solutions of silver nitrate and sodium hydroxide are combined:

$$2\ AgNO_3 + 2\ NaOH \rightarrow 2\ AgOH\downarrow + 2\ NaNO_3 \quad (1)$$

Silver hydroxide precipitates and forms the more stable and also only slightly water soluble silver(I) oxide:

$$2\ AgOH \rightarrow Ag_2O + H_2O \quad (2)$$

Adding ammonia the solution becomes transparent due to the formation of easily water soluble diaminesilver(I) complex:

$$Ag_2O + 4\ NH_3 + H_2O \rightarrow 2\ [Ag(NH_3)_2]^+ + 2\ OH^- \quad (3)$$

This is Tollen's reagent. Note that the diaminesilver(I) complex breaks down to silver nitride (Ag_3N) within about 1 h at room temperature if the storage container is not perfectly sealed. Silver nitride is highly explosive leading to surprise detonations. At low mechanical pressure and elevated temperature it decomposes to metallic silver and nitrogen gas. Hence, for safety reasons promptly mix Tollen's reagent after preparation with reducing agent as described in the next paragraph. This way handling is perfectly safe. For industrial applications diaminesilver(I) complex is stabilized by additives and very long shelf life is obtained. Commercial solutions containing diaminesilver(I) complex are declared as stable.

As a reducing agent we chose an aqueous solution of glucose. It is readily available and not toxic. Glucose is not as effective as e. g. formaldehyde because in water it predominantly exists in closed ring forms [5], see Fig. 3, which don't act as a reducing agent. Nevertheless, in equilibrium 0.25 % of the glucose is in the desirable open-chain aldehyde form shown in Fig. 4.

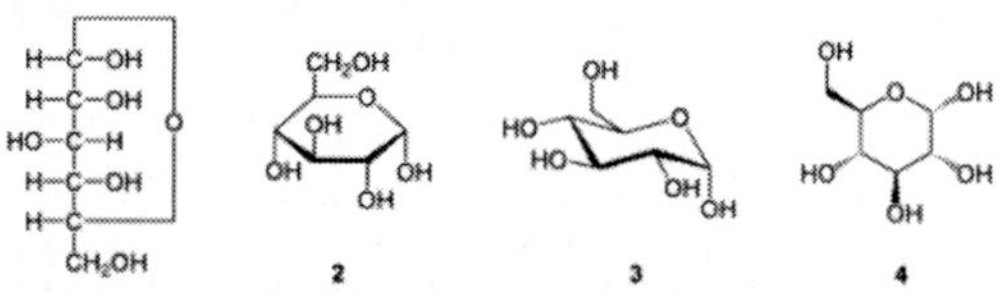

1 2 3 4

Fig. 3: Closed ring forms of glucose dissolved in water.

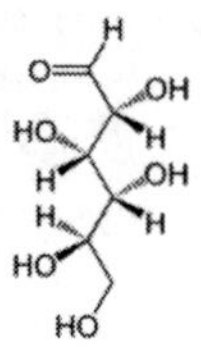

Fig. 4: Open-chain aldehyde form of glucose in water. It is a reducing agent for the diaminesilver(I) complex in Tollen's reagent.

Combining aqueous solution of glucose and Tollen's reagent the following reaction takes place in which the diaminesilver(I) complex is reduced by the aldehyde form of glucose. The aldehyde group oxidizes, gluconic acid is formed and the silver ions are reduced to metallic silver:

$$R\text{-}CHO + 2\,[Ag(NH_3)_2]^+ + 2\,OH^- \longrightarrow R\text{-}COOH + 2\,Ag\downarrow + 4\,NH_3 + H_2O \quad (4)$$

In order to speed up reaction rate the temperature can be increased to e. g. 55 °C so that a solid Ag layer is formed within about 2 min. See [3] and the hands-on instruction in reference [6] for more details.

As mentioned in the introduction, in industrial mirror manufacturing the liquids are sprayed on the glass panes. In contrast, in this work we simply placed the solar cells bottom up on a horizontally levelled plastic base and poured aqueous solution of glucose and Tollen's reagent simultaneously onto, see Fig. 5. It is important to note that we applied polymer varnish to the circumferential edge of the cell before silvering in order to minimize edge shunting. After silvering residual chemicals were rinsed with DI water and the cells were dried in N_2.

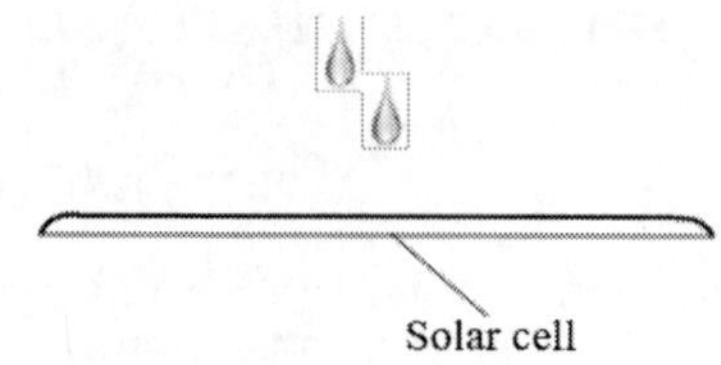

Fig. 5: Schematic representation of the process for silvering the rear side of HJT cells.

As a reference, full area rear side metallization was applied on the same type of cells by vacuum evaporation of Ag. All the cells were characterized by illuminated *IV*, dark *IV*, suns-V_{oc} and light reflection measurements.

Additionally, with the same process we chemically deposited Ag on *p*-type alkaline textured CZ Si wafers which were not coated with ITO before. The sheet resistance of the Ag layer was mapped by four point probe measurements and its microstructure was investigated by secondary electron microscopy (SEM).

3 RESULTS AND DISCUSSION

3.1 Thickness uniformity

Fig. 6 is a photo of a 6 inch Si wafer after full-area chemical silver deposition. No holes or cracks are visible in the Ag layer on a macroscopic scale. Because polymer varnish was not applied to the edge of this cell, Ag deposition extends up to the rim.

Fig. 6: Photo of a 6 inch alkaline textured CZ Si wafer (without ITO on the Si and without polymer ink at the edge) chemically coated with Ag.

Fig. 7 depicts the sheet resistance of the Ag layer measured on 9 spots on a 6 inch Si wafer. Note that the effect of the Si substrate on the measured values is negligible because its sheet resistance is high (about 50 Ω/sq). In a preceding experiment we found that within the measurement error the electrical conductivity of chemically deposited Ag and bulk Ag are equal. Using this value we calculated the thickness of the Ag layer from the measured sheet resistance considering the surface area of alkaline textured cells which is a factor of 1.7 larger compared to flat surfaces. The result is shown on the right hand side of Fig. 7. The average Ag thickness is 654 nm and the thickness variation is ± 13 %. Hence, reasonable thickness uniformity was obtained by the simple chemical deposition process investigated in this work.

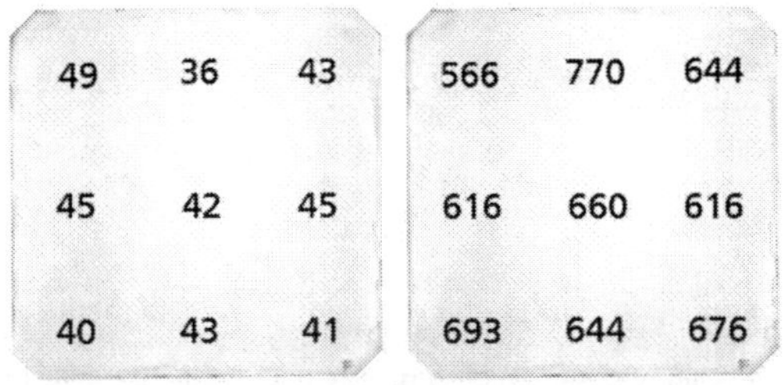

Fig. 7: Left: Photo of a 6 inch CZ Si wafer (without ITO in this case) chemically coated with Ag showing the measured sheet resistance of the deposited Ag in mΩ/sq. **Right:** The same photo showing the thickness map of the Ag layer in nm calculated from the measured sheet resistance.

3.2 Microstructure and adhesion of the Ag layer

Fig. 8 shows SEM pictures of an ion polished cross section of a chemically deposited Ag layer on an alkaline textured Si wafer. The Ag layer is formed by small grains that are densely packed. No gaps between Ag grains or between Ag and Si were found. The thin Ag layer very cosely follows the lines of the underlying alkaline textured Si surface. On a microscopic scale the Ag layer thickness variation is quite low.

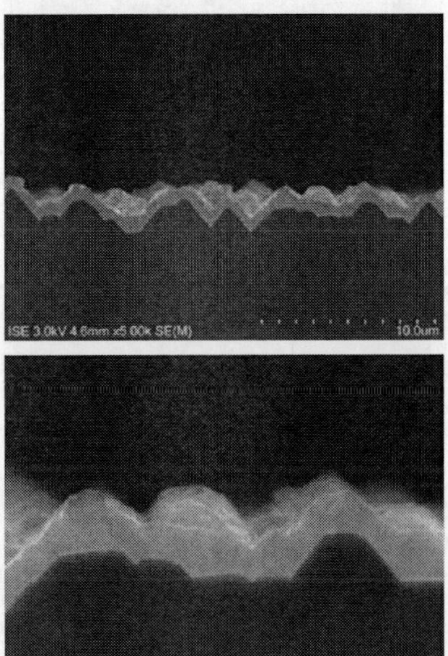

Fig. 8: Secondary electron microscope pictures taken of a cross section of electroless plated Ag on alkaline textured CZ Si (without ITO in this case).

To check adhesion of chemically deposited Ag on HJT cells we performed tape tests. They were passed. However, when it comes to soldering of tinned Cu connector ribbons onto the rear contact, Ag completely alloys with the Sn on the ribbons leading to low adherence. Hence, for this application another metal layer could be deposited on the Ag.

3.3 Optical performance

By means of an integrating sphere we measured the front side global reflection of Si HJT cells featuring full area chemically deposited Ag and thermally evaporated Ag on the rear, respectively. The Ag layer thickness was about 1 µm on both samples. Fig. 9 shows that the measured curves overlay exactly. Considering this result for wavelengths exceeding 1000 nm, which refers to reflection of infrared light from the rear of the cells, we conclude that the optical performance of chemically deposited Ag and PVD Ag is the same.

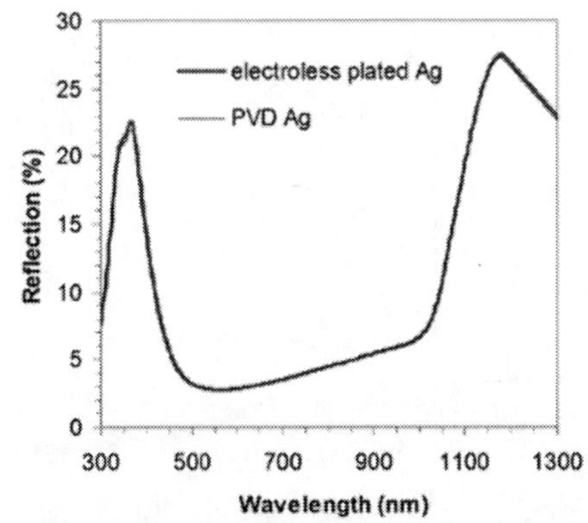

Fig. 9: Measured front side global reflection of Si HJT cells featuring about 1 µm thick full area chemically deposited Ag and thermally evaporated Ag on the rear, respectively.

For cost reasons it is desirable to reduce the Ag layer thickness as much as possible. However, very thin Ag films are optically transparent. To find out the minimum Ag thickness needed for high infrared light reflection we calculated the short circuit current density loss due to transmitted light ($J_{sc.escaped}$) as a function of Ag thickness for a 150 µm thick HJT cell. We used the ray tracing app 'SunSolve' [7] without change of the preset optical constants. It can be seen from Fig. 10, that $J_{sc.escaped}$ is negligible for Ag layers being thicker than 60 nm. Note that this thickness corresponds to 26 mg of Ag if the surface area enhancement factor of alkaline textured Si (1.7) is taken into account.

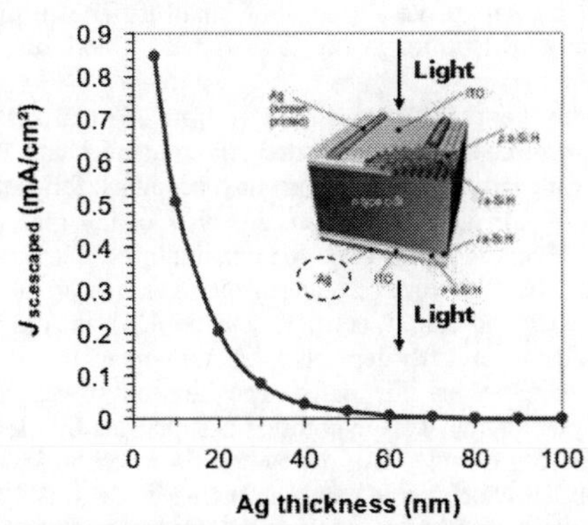

Fig. 10: Calculated short circuit current density loss owing to infrared light that escapes to the rear of a HJT cell as a function of the Ag rear contact's thickness.

3.4 Resistive power loss

With respect to minimum Ag thickness it is also important to investigate the lateral resistive power loss in the full-area Ag contact. As a meaningful example we assumed a 100 nm thick Ag layer on a 22 % efficient

HJT solar cell. Fig. 11 shows the calculated absolute efficiency loss caused by the lateral electrical resistance of the Ag as a function of the number of equally spaced contacting wires. The efficiency loss is less than 0.02 % absolute for 16 wires which is the number often used in smart-wire interconnection technology. Hence, in combination with the results of the optical assessment in paragraph 3.3 we conclude that thin full area Ag rear contacts are suitable as an efficient rear contact.

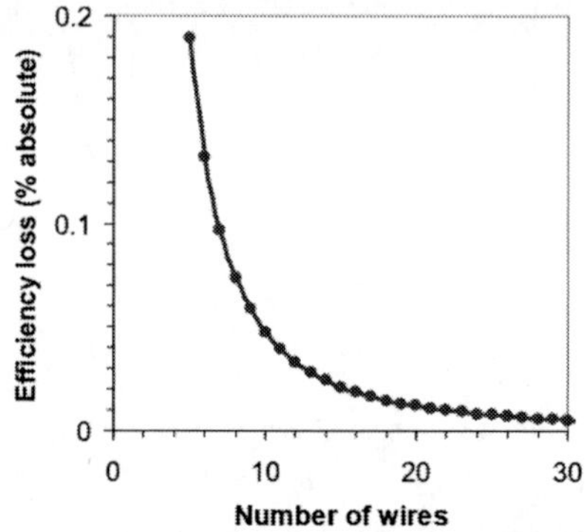

Fig. 11: Calculated efficiency loss of a 22 % efficient 6 inch Si cell caused by lateral electrical resistance of the Ag rear contact.

3.5 Solar cell results

A comparison of the measured electrical performance parameters of 6 inch HJT cells featuring chemical deposited and vacuum evaporated Ag rear contacts, respectively, is presented in Table 1. The thickness of the Ag layer was about 1 µm in both cases. The short circuit current densities (J_{sc}) as well as the open circuit voltages (V_{oc}) of the cells are practically equal. However, the fill factor (*FF*) of the cell featuring chemically deposited Ag is 0.7 % absolute lower compared to the cell featuring evaporated Ag. The pseudo fill factor (*pFF*) is 0.9 % absolute lower. Theoretically the reductions in *FF* and *pFF* should be the same because the series resistances are about the same. We attribute the small difference to the measurement errors of the illuminated *IV* and suns-V_{oc} curves.

The shunt resistance is reduced from 363 to 3 kΩ cm² in case of chemically deposited Ag compared with PVD Ag. Calculations reveal that this accounts for 0.4 % absolute fill factor reduction. Analysis of the measured dark *IV* curves showed that the remaining *FF* decrease of about 0.4 % absolute is caused by an increased second diode recombination current density (J_{02}) in the cell featuring chemically deposited Ag rear contact. However, we don't attribute this to Ag deposition but to variations in the cell batch. This hypothesis is supported by the fact that during chemical Ag deposition no mechanical force is applied which could cause an increase of J_{02}. A larger number of cells have to be manufactured in order to verify this and improve the statistics.

It is interesting to consider again the series resistances of the cells which are virtually identical. Because vacuum evaporated Ag is known to provide very low contact resistance on ITO, it can be concluded that the contact resistance of chemically deposited Ag on ITO is very low, too.

4 SUMMARY AND CONCLUSION

Dense Ag layers without pores are formed by Tollen's reagent combined with aqueous solution of glucose. No gaps between Ag layers and cells were found. This explains why the measured optical and electrical performance of the chemically deposited Ag is the same as that of vacuum evaporated Ag. We even found that the contact resistance of chemically deposited Ag on ITO is as low as that of vacuum evaporated Ag.

For HJT cells featuring full area chemically deposited Ag on the rear side, there is the danger of shorting front and back side across the cells' edges. Hence, extra measure has to be taken by thorough application of a polymer ink around the edge. All things considered, the chemically deposited Ag rear contact yielded 21.4 % efficiency compared with 21.6 % for the evaporated Ag.

When it comes to potential industrial applications, simulations showed that an Ag layer thickness of 100 nm is sufficient to reduce the efficiency loss caused by the full area rear contact's electrical resistance to 0.02 % absolute if 16 wires are used for cell contacting. In this case the Ag material costs are 1.7 euro cent per cell provided that the residual Ag in the aqueous solution after chemical deposition is recycled. In comparison, the standard metallization of Si heterojunction cells in industry is low-curing-temperature Ag paste. The price of 39 mg of this paste is approximately the same. However, due to its higher bulk resistivity it provides only about 1/3 to 1/2 of the electrical conductivity of a 100 nm thick pure Ag layer.

5 REFERENCES

[1] https://www.nrel.gov/pv/assets/pdfs/pv-efficiencies-07-17-2018.pdf.
[2] www.youtube.com/watch?v=OtUof0EH3-E.
[3] C. Heber, Publications of the Astronomical Society of the Pacific 23 (1911) p. 15–19.
[4] B. Tollens, Berichte der Deutschen Chemischen Gesellschaft 15 (1882) p. 1635–1639 (in German).
[5] P. C. Heinrich, M. Müller and L. Graeve, Biochemie und Pathobiochemie (2014) p. 27 (in German).
[6] www.wikihow.com/Make-Mirrors-by-Silvering-Glass.
[7] www.pvlighthouse.com.au.

Tab. 1: Comparison of the measured electrical performance parameters of 6 inch heterojunction CZ Si solar cells featuring chemically deposited and thermally evaporated full area Ag rear contacts, respectively.

	J_{sc} (mA/cm²)	V_{oc} (mV)	*FF* (%)	*pFF* (%)	R_{ser} (Ω cm²)	R_{shunt} (kΩ cm²)	Efficiency (%)
Evaporated Ag	37.8	728	78.5	83.0	0.94	363	21.6
Chemically deposited Ag	37.8	727	77.8	82.1	0.93	3	21.4

NANOABSORBERS FOR PV: NEW POTENTIAL APPLICATIONS

Volker Steenhoff, Norbert Osterthun, Kai Gehrke, Martin Vehse, Carsten Agert
DLR Institute of Networked Energy Systems
Carl-von-Ossietzky-Str. 15, 26129 Oldenburg, Germany

ABSTRACT: Nanoabsorber materials enabling strong absorption with thicknesses in the few-nanometer range have recently gained a lot of attention in the research community, and their utilization in photovoltaic devices has been proposed. However, with ever decreasing costs and increasing efficiencies of both c-Si based solar cells as well as existing thin film technologies, it is often argued that there is limited merit in developing further PV technologies. In this contribution, we present two novel applications in which nanoabsorber-based solar cells pose strong technical advantages over standard PV. In particular, we propose a spectrally selective solar cell design which transmits significant fractions of the blue and red light required for the photosynthesis of algae and higher plants, while the remaining parts are utilized for electricity generation. Another concept, which we call a "Smart PV Window", can be switched from transparent to shaded mode, while the shaded mode not simply blocks the light but instead harvests electric energy from it.
Keywords: Nanoabsorbers, Germanium, Building Integration, Nanocavity

1 INTRODUCTION

Semiconducting nanoabsorbers, i.e. absorber materials trapping light on the nanometer scale, have recently gained much attention in the research community because of their promising properties for application in optoelectronics. Particularly, some 2d semiconductors – mostly transition-metal dichalcogenides (TMDCs) – have been proposed for application in PV because they sustain high charge carrier lifetimes and mobilities even at nanometer thickness and when interfaced with other materials [1–3]. So far it seems hard to make high-quality layers on larger scales, so that only few working solar cells made on very small areas have been presented up to now. In the past years, we have shown that resonant-cavity-enhanced germanium (a-Ge:H) solar cells, which are easy to fabricate and relatively robust, can be used as a working model system for nanoabsorber-based PV. This, for example, helps to assess new light trapping concepts specifically designed for such thin devices [4–8].

However, due to impressive achievements in both wafer-based and existing thin-film technologies, a question often raised is, why should there still be research on new types of PV technologies? In this contribution, using our a-Ge:H based model system, we demonstrate two novel applications in which nanoabsorber-based solar cells will outperform existing technology.

2 SPECTRALLY SELECTIVE SOLAR CELLS FOR GREENHOUSES

Our first example is a solar cell specifically designed for combining growth of photosynthetically active species (i.e. plants / algae) with photovoltaics. For photosynthesis, mainly light in the blue and red spectral range is required. Remaining light can be utilized by a solar cell. While it might be possible to achieve this type of absorption characteristic using specialized organic materials, tuning the exact position of the transmission and absorption peaks in this case would require tailoring of the absorber material itself. This leads to strong interdependency between optical and electrical optimization, making it difficult to achieve at the same time high transparency in the relevant wavelength ranges and high efficiency. With classical solar cells based on thick semiconductor films, achieving high transparency in the blue wavelength range would also be difficult. As the absorption coefficient of semiconductors is normally higher at shorter wavelengths, strong absorption in the green and infrared spectral ranges would always imply strong absorption and low transparency in the blue and green range as well. Recently, an approach has been demonstrated which is based on a luminescent solar concentrator absorbing mainly light which is not used by plants [9]. In such devices, the light reradiated by the luminescent dye undergoes total internal reflection inside the glass, so that it is guided to the edges of the window where conventional solar cells are placed to absorb the light. However, while being an interesting approach, this requires very expensive and highly transmissive glass to avoid undesired attenuation of the light.

Here we present an alternative and cost-efficient approach exploiting the specific properties of nanoabsorber-based resonant-cavity-enhanced PV. Figure 1 depicts a schematic representation of such a device, which in principle works as follows: An ultrathin nanoabsorber is placed inside a nanocavity consisting of a transparent, semi-reflecting front electrode and a back mirror with spectrally selective transparency. The thickness of the absorber is chosen such that small but nonzero single-pass absorption is achieved. For wavelengths at which the back mirror is transmissive, light passes through the absorber only once so that only a small fraction of the incident light is absorbed. At wavelengths where the mirror is highly reflective, resonances form inside the cavity, strongly amplifying the absorption.

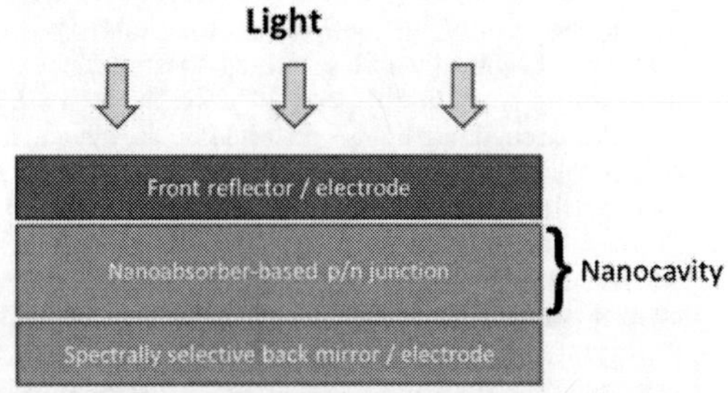

Figure 1: Schematic representation of a spectrally selective solar cell based on a resonant nanocavity with a spectrally selective back mirror.

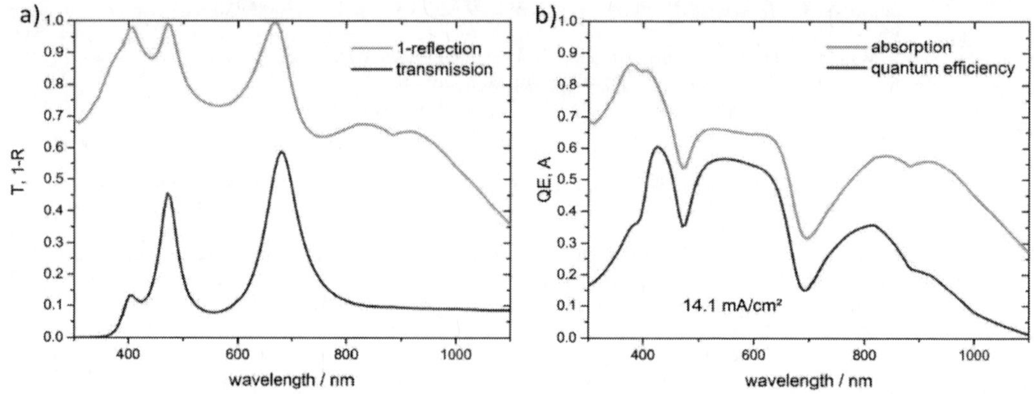

Figure 2: Simulation of a solar cell with spectrally selective transparency designed for combined photosynthesis and photovoltaics. (a) Transmission and reflection (b) total absorption and quantum efficiency. The quantum efficiency of the simulated cell corresponds to a calculated short circuit current density of 14.1 mA cm^{-2}.

This way, the solar cell selectively either transmits or absorbs the incoming light depending on its wavelength, and only small amounts are reflected. The advantage of this concept is that the absorption coefficient of the absorber itself does not need to possess the desired spectral selectivity but can be designed in a classical manner to have good optoelectronic properties. The spectral selectivity of the transmission is instead achieved independently by choosing a proper design of the back reflector.

Figure 2 shows first simulation results for a resonant-cavity-enhanced a-Ge:H nanoabsorber device making use of this concept. In the simulation the pin-junction is realized in the form of a SiC(n) / a-Ge:H(i) / SiC(p), each of these layers having a thickness of only 5 nm. The spectrally selective back reflector consists of an alternating sequence of thin silver and zinc oxide layers, which leads to interference patterns that can be tuned by changing the thickness of the oxide films. High-bandgap SiC layers were used to avoid parasitic absorption inside the doped layers, so that light is mainly absorbed inside the a-Ge:H layer. This was done to allow for a general feasibility test of the concept as well as to assess its potential optical limitations. The thickness of the a-Ge:H layer was chosen such that it absorbs a small fraction of the incoming light, but a significant amount is transmitted as long as only single pass absorption takes place. At wavelengths where the spectrally selective back mirror strongly reflects the light, however, standing waves are formed, which strongly amplifies the absorption inside of a-Ge:H film in these spectral ranges.

As shown in Figure 2a, when using a suitable design of the back mirror, the cell indeed transmits significant fractions of light in the blue and red spectral range, with peak values close to 50% and 60%, respectively. While part of the remaining light is reflected by the device, most of it is absorbed by the cell stack. Figure 2b demonstrates that the simulated absorption and quantum efficiency of the cell have distinct dips at about 480 and 680 nm, whereas at other wavelengths, light is trapped inside the a-Ge:H nanoabsorber efficiently. This way, a current density of about 14 mA cm^{-2} is still achievable with this type of device. More details about the design and experimental realization of the spectrally selective back reflector as well as first devices implementing this concept can be found elsewhere [10].

3 SOLAR CELLS WITH SWITCHABLE TRANSPARENCY

Our next example of an application made possible by nanoabsorber-based PV is a great opportunity for the glass industry: The color neutral semitransparent solar window. Even better, this concept can be easily extended to a reversibly switchable shading element. This smart window would be able to block light for indoor light management and would generate electric current at the same time, instead of just reflecting the light or generating heat.

A straightforward method to build this type of window is to use an absorber material featuring switchable absorption behavior. Solar cells implementing this concept have been demonstrated just recently [11]. These devices exploit the photothermal dissociation of methylammonium lead iodide to switch the solar cell to its absorbing state. However, to switch back to the transparent mode, exposure to CH_3NH_2 gas is required. While in this particular case the requirement of gas exposure makes the fabrication of large scale modules complex, there is a much more fundamental disadvantage of using switchable absorbers for PV: Again, optimization of the switching ratio and repeatability is strongly interdependent with optimizing the electronic properties of the cells.

Here again nanoabsorber-based PV is an alternative solution. The same idea described in the previous chapter can be employed, but the spectrally selective back reflector is replaced by a switchable mirror. Switchable mirrors are a technology that has been under intensive investigation for years already [12]. In the simplest case, they consist for example of gasochromic magnesium films or alloys that can be switched using a hydrogen-rich atmosphere [13,14], but more complicated structures allow electrical switching as well [15].

By using a switchable mirror instead of a switchable absorber, independent optimization of switching capabilities and electronic solar cell properties are achieved again. Figure 3a shows the simulated transmission and quantum efficiency of two a-Ge:H nanoabsorber-based solar cell stacks with and without using a back reflector, this way mimicking the behavior of a switchable mirror. While the first stack was simulated using idealized doped layers, the second is simulated using a layer sequence known to work

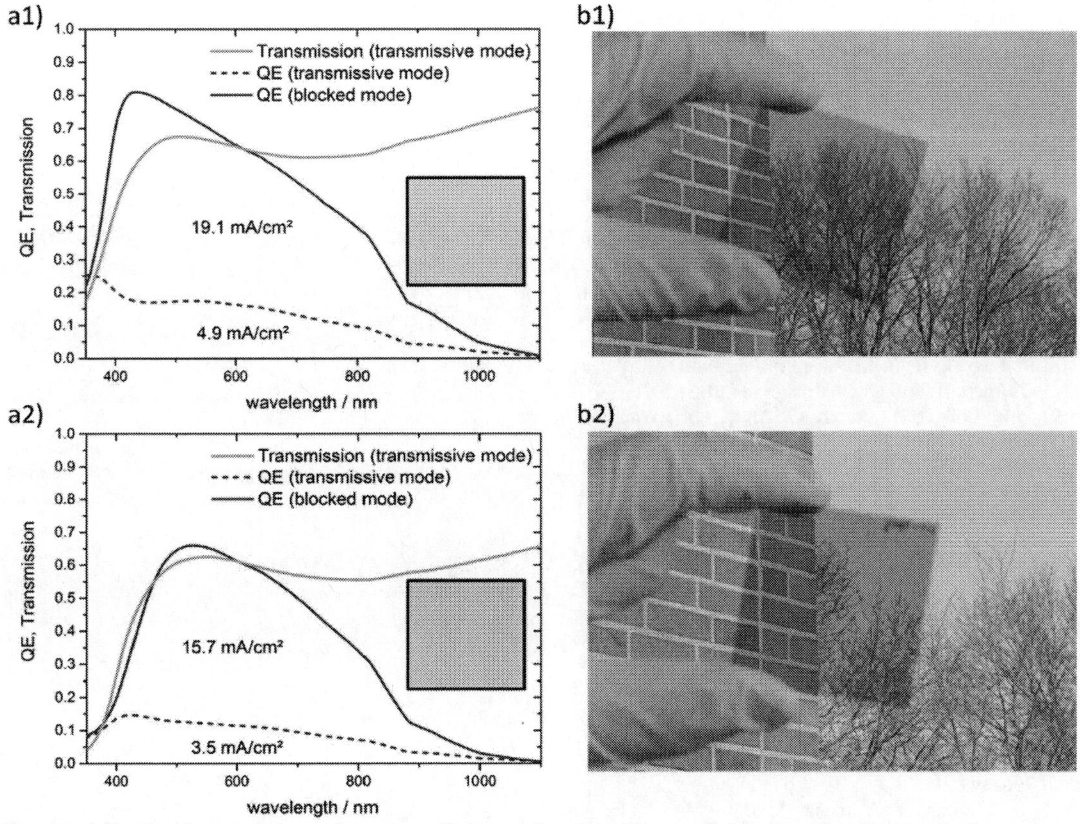

Figure 3: a) Simulated transmission and quantum efficiency of semitransparent a-Ge:H solar cells with neutral color appearance (transmissive mode) as well as of the same cells equipped with a silver back reflector (blocked mode) The insets show the simulated color appearance in transmissive mode. b) Photographs of samples implementing the simulated semitransparent layer stacks. (a1/b1 show results based on idealized doped layers, whereas a2/b2 use silicon-based doped layers in a solar cell stack known to work already.)

electrically.

Without a back reflector, both simulations yield a transparency of 50-60% in nearly the whole visible wavelength range. This leads to a remarkably neutral color appearance especially in the idealized case. Even when using the silicon-based stack, only a small fraction of the blue spectrum is underrepresented in the transmitted light, leading to a slightly greenish appearance only. Figure 2b shows photographs of first samples implementing the simulated layer stacks, showing that the actual color appearance is in very good agreement with the simulation results. At the same time short circuit current densities between 3.5 mA cm^{-2} and 4.9 mA cm^{-2} are predicted by our simulations. When integrating a back reflector into the same cell stacks to mimic the shading mode, up to 19.1 mA cm^{-2} can be achieved.

4 DISCUSSION AND CONCLUSION

In this contribution we have presented two novel solar cell applications in which nanoabsorber-based devices show strong advantages over classical PV approaches – the spectrally selective solar cell for greenhouses and the switchable PV smart window. We have used resonant-cavity-enhanced a-Ge:H solar cells as a model system for nanoabsorber-based PV to perform first simulations of these devices.

Our results indicate that spectrally selective solar cells with peak transmissivity close to 50% and 60% in the blue and red spectral ranges relevant for photosynthesis are possible, which at the same time allow for quantum efficiencies of more than 50% in the intermediate wavelength range and yield short circuit current densities of about 14 mA cm^{-2}. Furthermore, switchable solar cells with neutral color appearance and close to 5 mA cm^{-2} short circuit current density in the transmissive mode as well as 19 mA cm^{-2} in shading mode seem possible with a-Ge:H-based devices.

We estimate that color neutral windows with >60% transmission and yielding 1-2% photovoltaic efficiency under AM1.5g illumination can be realized by combining our concept with novel nanoabsorber materials. By adding a switchable back reflector, the same devices will turn into switchable shading elements that generate current from the blocked light with about 5-10% efficiency.

REFERENCES

[1] D. Jariwala, A.R. Davoyan, J. Wong, H. a Atwater, ACS Photonics. 4 (2017) 2962–2970.

[2] L. Wang, L. Huang, W.C. Tan, X. Feng, L. Chen, X.

Huang, K.-W. Ang, Small Methods. 1700294 (2018) 1700294.
[3] M.M. Furchi, A. Pospischil, F. Libisch, J. Burgdörfer, T. Mueller, Nano Lett. 14 (2014) 4785–4791.
[4] V. Steenhoff, M. Theuring, M. Vehse, K. von Maydell, C. Agert, Adv. Opt. Mater. 3 (2015) 182–186.
[5] V. Steenhoff, A. Neumüller, O. Sergeev, M. Vehse, C. Agert, Sol. Energy Mater. Sol. Cells. 145 (2016) 148–153.
[6] V. Steenhoff, M. Juilfs, R.-E. Ravekes, M. Vehse, C. Agert, Nano Energy. 27 (2016) 658–663.
[7] V. Steenhoff, M. Juilfs, R.E. Ravekes, M. Ahrlich, M. Kellermann, O. Siepmann, M. Vehse, C. Agert, IEEE J. Photovoltaics. 7 (2017) 3–10.
[8] C. Lattyak, R.-E. Ravekes, V. Steenhoff, M. Vehse, C. Agert, IEEE J. Photovoltaics. 8 (2018) 3–7.
[9] M.E. Loik, S.A. Carter, G. Alers, C.E. Wade, D. Shugar, C. Corrado, D. Jokerst, C. Kitayama, Earth's Futur. 5 (2017) 1044–1053.
[10] N. Osterthun, V. Steenhoff, N. Neugebohrn, K. Gehrke, M. Vehse, C. Agert, in: EUPVSEC Proc., 2018.
[11] L.M. Wheeler, D.T. Moore, R. Ihly, N.J. Stanton, E.M. Miller, R.C. Tenent, J.L. Blackburn, N.R. Neale, Nat. Commun. 8 (2017).
[12] C.G. Granqvist, Thin Solid Films. 614 (2016) 90–96.
[13] Y. Yamada, H. Sasaki, K. Tajima, M. Okada, K. Yoshimura, Sol. Energy Mater. Sol. Cells. 99 (2012) 73–75.
[14] Y. Yamada, M. Miura, K. Tajima, M. Okada, K. Yoshimura, Sol. Energy Mater. Sol. Cells. 117 (2013) 396–399.
[15] K. Tajima, Y. Yamada, S. Bao, M. Okada, K. Yoshimura, Appl. Phys. Lett. 91 (2007).

WHISPERONIC SOLAR CELLS

C. Sudakar, Das Tapan Kumar, and P. Ilaiyaraja
Multifunctional Materials Laboratory, Department of Physics,
Indian Institute of Technology Madras, Chennai-600036, India

ABSTRACT: Sensitized solar cells (SSC), including DSSC and QDSSC, are third generation solar cells, which aim to achieve low cost, highly efficient solar light to electricity conversion devices. The challenge that prevails till today is to find approaches to achieve high efficiency. Of several important factors that limit the efficiency in these SSC, including poor connectivity of the TiO_2 nanoparticles, charge separation and transport in the photoanode layer, inefficient light absorption by the dye/quantum dot, non-directional transport of charge carriers with significant recombination, and inefficient loading of the sensitizer on the photoanode, the non-optimal utilization of the light entering the devices is the most important one. A large fraction of the light that enters the solar cell gets lost or do not get converted into excitons. Here we propose a new type of solar cell called "whisperonic solar cells (WSC)". To make efficient use of the light entering the device, WSC, a modified version of Grätzel solar cell, uses mesoporous microspheres exhibiting whispering gallery modes as the photoanode. Enhanced power conversion efficiency of devices is achieved by the efficient scattering of light by way of intense optical whispering gallery modes that eventually gets absorbed by the sensitizer in an efficient way. Mesoporous micron-sized spheres in addition to exhibiting large Mie scattering (whispering gallery modes) can be used to load large fraction of quantum dot sensitizer. Further, using composite photoanode (mesoporous microspherical TiO_2 in combination with the nanocrystalline TiO_2) better connectivity among the particles leading to directional transport is achieved. The recombination of charge carriers is suppressed largely with these composite photoanodes. All these characteristics of the photoanode have been tested in the dye, quantum dot and thin film sensitized solar cells. Using CdSe and $CuInS_2$ QDs, and $CuInS_2$ thin film in the WSC devices we show significant enhancement in the power conversion efficiencies.
Keywords: Solar cells, whispering gallery mode, scattering, WSC.

1 INTRODUCTION

Reducing the specular reflection of light through anti-reflection coating and texturing the surface to increase diffuse reflection, which results in multiple scattering and hence leads to higher absorption by sensitizer, have been employed to enhance the harvesting of light [1, 2]. In this article we propose an efficient way to improve light harvesting by employing a composite photoanode, which mostly comprise of 80 wt.% mesoporous microspherical TiO_2 filled with 20 wt.% nanocrystalline TiO_2. Microspherical TiO_2 acts as a light-trapping medium through the optical whispering gallery modes (WGM).

The prevalence of WGM in microspheres and other microstructures is known in the literature [3-6]. Normally large-sized microstructures with size comparable to the wavelength of incident light are preferred as it enhances the light scattering in sensitized solar cells (SSC). When a photon is incident on a microsphere, it gets trapped inside the sphere by scattering and produces a resonance due to phase matching with the incident waves [7]. The total internal reflection of light at the oxide-air interface in microspheres leads to the confinement of light. When a particular wavelength couples into the microsphere-resonator, it leads to the propagation and circulation of scattered light in a ring-like path through the surface of the microsphere. When the wavelength of emitted light satisfies the constructive interference condition $\pi D = {}^{m\lambda}/_{n}$ (where D is the diameter of the sphere, m is the integer mode number, λ is the resonant wavelength and n is the refractive index of the microsphere), sharp resonating light emissions, WGMs, in the form of multiple peaks are observed on top of the regular PL spectra [4]. Such scattering is generally seen in the case of microspheres, nano-rods, nano-tubes, and photonic crystals. This concept of scattering mechanism, prevailing in the composite photoanode is used for making the effective light absorption by the sensitizers.

Approaches to enhance the absorption in the sensitizer are need to be addressed to enhance the power conversion efficiency (PCE). Various methodologies have been used including the concept of plasmonic light trapping mechanism to scatter the light as in "plasmonic solar cells"[8, 9]. We find using whispering-gallery mode (WGM) assisted scattering as a way to significantly enhance light absorption and hence the PCE in dye, quantum-dot and thin film sensitized solar cells. Significant increase in the power conversion efficiencies ~ 25 % in DSSC (using N719 dye), ~90 % in QDSSC (using CdSe and $CuInS_2$) and ~ 60% in TFSSC devices are demonstrated. This novel method of using WGM to scatter the light in the photoanode certainly provide a solution to enhancing the absorption in the sensitizer.

2 EXPERIMENTAL

Preparation of smooth mesoporous microspherical TiO_2 (SμS-TiO_2) particles by solvothermal method is described in detail elsewhere [10]. The composite photoanode is made by mixing the SμS-TiO_2 (80 wt.%) with the nanocrystalline P25-TiO_2 (20 wt.%). The mixing is carried out during the preparation of TiO_2 paste. TritonX-100 binder in ethanol solvent was used for mixing and coating the photoanode as detailed in Ilaiyaraja et al [10]. The photoanode coating is applied on a transparent FTO substrate. The sensitization of the photoanode with dye [11], quantum dot (CdSe or $CuInS_2$) [11-13] or thin film ($CuInS_2$)[14] is made by procedure described elsewhere. In all these devices, iodide/tri-iodide (I^-/I_3^-) electrolyte (0.5 M of LiI and 0.05 M of I_2 and 0.5 M of 4-TBP) was filled in the inter electrode space. The devices were tested under 1 Sun illumination (PET CT50AAA) with a Keithley 2400 current source meter.

3 RESULTS

The field emission scanning electron microscopy image of SμS-TiO_2 is shown in Fig. 1a. The size of the microsphere ranges between 1 to 6 μm with an average size of the sphere ~ 3 μm. The microspheres are all well-connected forming a network for the conductive path. The TiO_2 mesoporous microspheres exhibit whispering gallery modes as shown in Fig. 1b. Unlike the nanocrystallite TiO_2 particles, the smooth mesoporous microsphere TiO_2 exhibit strong WGM as evidenced from the photoluminescence (PL) studies carried out using a laser with an excitation wavelength λ_{exc}=325 nm. In contrast to the smooth broad PL spectra of P25-TiO_2, the SμS-TiO_2 show sharp resonance spikes. These resonant modes arise due to the repeated circulation of photons due to the total internal reflections [15, 16]. In an effort to harvest the benefit of WGM along with the high surface area available in the porous nanocrystalline P25-TiO_2, composite photoanode were made. Since our optimization studies suggested that the composite made of SμS-TiO_2 and P25-TiO_2 in the ratio of 80:20 wt. % show significant enhancement in PCE of DSSC and QDSSC devices, we have used this composite ratio for fabricating the photoanode for the device testing. The FESEM image of composite photoanode is shown in the Fig. 1c. The nanocrystalline P25-TiO_2 particles fill up the void regions between the microspheres. They are uniformly coated surrounding the particle as seen from the bright field TEM images shown in Fig. 1d.

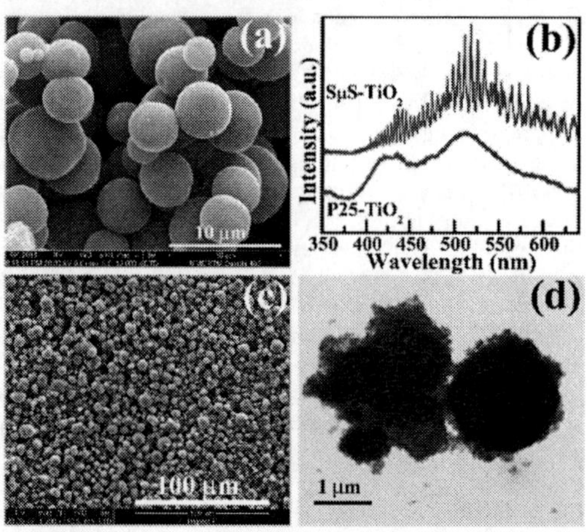

Figure 1: (a) Field emission scanning electron microscopy (FESEM) image of TiO_2 microsphere. (b) Photoluminescence (PL) spectra from nanocrystalline P25-TiO_2 and microspherical SμS-TiO_2 particles. (c) FESEM and (d) bright-field transmission electron microscopy (TEM) images of composite TiO_2 photoanode.

The benefit of having optical whispering gallery modes in the photoanode is learnt from the strong coupling between the resonant modes with the emission characteristics of the sensitizer. Fig. 2 shows this sort of coupling for various kinds of sensitizer loaded on the microspherical TiO_2 photoanode. A detailed account of the sensitization process for various sensitizers can be obtained from the references 11 to 13 [11-13]. These microspheres sensitized with N719 dye, CdSe/$CuInS_2$ QDs or $CuInS_2$ thin film are shown to exhibit emission characteristics with the presence of resonant spikes from the WGM. Such modes are not seen when these sensitizers are loaded on P25-TiO_2 photoanode (Fig.2). Further, in presence of microsphere, the PL emissions appear to be several folds stronger. This suggests a strong coupling of WGM with the sensitizer.

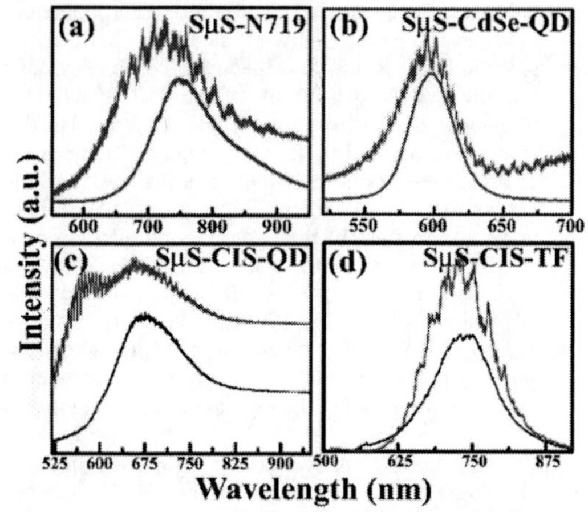

Figure 2: Photoluminescence spectra of (a) N719 dye, (b) CdSe QDs, (c) $CuInS_2$ QDs and (d) $CuInS_2$-TF coated on TiO_2 microsphere. The PL emission spectra are collected from an isolated microsphere using 488 nm excitation line from Ar-ion laser source. All of these PL spectra were collected from the sample dispersed on Si substrate.

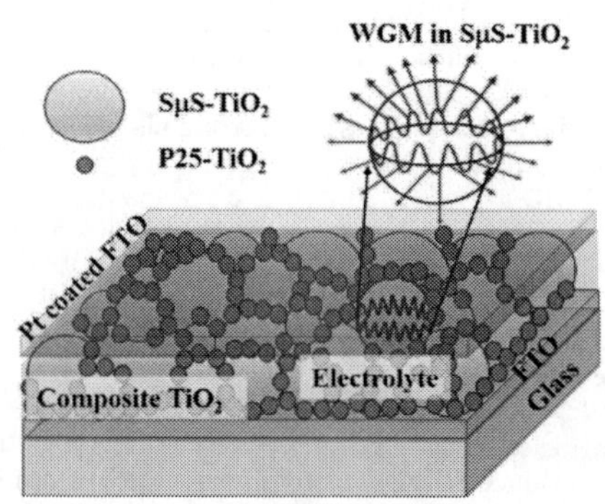

Figure 3: Schematic of whisperonic solar cells. The sensitizer is depicted by a thin red line around the surface of P25 and SμS-TiO_2. The picture also depicts the formation of whispering gallery mode in TiO_2 microsphere.

A typical device configuration for the whisperonic solar cell is shown in Fig. 3. The device is typical of Grätzel type solar cell [17], however with the use of composite photoanode. The solar cell device is thus configured by assembling the Pt coated FTO counter electrode above the sensitized TiO_2 composite photoanode. In all the devices discussed here the gap between the two electrodes were filled with iodide/tri-iodide (I^-/I_3^-) electrolyte. A detailed characterization on the material components used, detailed device performance and photovoltaic characteristics of these sensitized solar cells can be found in the references 11 to 14 [11-14]. Here we show the power conversion efficiency (PCE) of devices made with photoanodes fabricated with (I) P25-TiO_2, (II) SμS-TiO_2 and (III) the composite (P25+SμS)-TiO_2. Clearly the PCE measured on several test devices suggest that the composite photoanode exhibit highest efficiency for the devices made under similar conditions with same configuration. Increase in PCE of ~24 % in observed in DSSC devices. For QDSSC devices the increase is ~ 80 to 95 %. It should be noted that all these devices were tested with a cell area of 0.25 cm^2. Even in the case of thin film sensitized photoanode the increase is found to be ~ 60%. These studies clearly suggest that, TiO_2 composite exhibiting WGM can be used as a generic photoanode to achieve maximum PCE for Grätzel type sensitized solar cells. The increase is mainly attributed to the WGM enabled enhancement in the absorption.

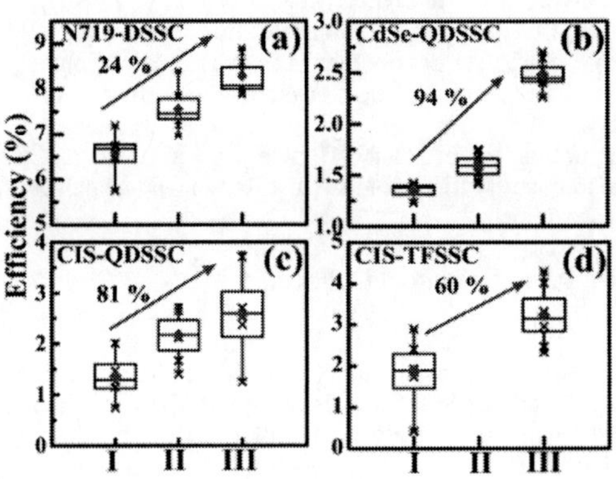

Figure 4: The efficiency of DSSC, QDSSC (with CdSe and $CuInS_2$ QD) and $CuInS_2$ thin film sensitized whisperonic solar cells (II and III) in comparison to Grätzel solar cells (I); The photoanode for solar cell device set (I) are made of P25-TiO_2, (II) are made of SμS-TiO_2 and (III) are made using the composite photoanode (P25+SμS)-TiO_2

4 CONCLUSION

In conclusion, we show that large Mie scattering from mesoporous micron-sized spheres in the form of optical whispering gallery modes could be used to enhance the sensitizer absorption and hence the PCE. The mesoporous TiO_2 microspherical particles in combination with the nanocrystalline TiO_2 are excellent photoanodes for sensitized solar cells. Significant loss in the photoconversion efficiency (PCE), which generally occurs due to the inefficient light absorption, is mostly avoided in these device configurations. While the devices tested employ a composite comprising of mesoporous TiO_2, micron sized microstructures of various shapes and sizes that exhibit whispering gallery modes can be employed. Other possible microstructures that will be suitable as WGM exhibiting Mie scattering structures include hexagonal platelets, circular discs, hollow spheres, hollow hexagons, etc. Various photoanode materials that are suitable for SSC devices including TiO_2, ZnO, SnO_2, In_2O_3, Fe_2O_3, or any metal oxides binary, ternary or quaternary oxides, in pure or doped forms of metal oxides can be employed as photoanode. Such composite photoanodes exhibiting WGM can also be used for various sensitized solar cells including perovskite based sensitizer. Such sensitized solar cell devices which employ optical whispering gallery resonant modes can generally be classified as "Whisperonic Solar Cells".

REFERENCES:

[1] T. Hong Nhung, N. Van Hieu, N. Bich Ha, V. Dinh Lam, Advances in Natural Sciences: Nanoscience and Nanotechnology, 7 (2016) 013001.
[2] Z. Tang, W. Tress, O. Inganäs, Materials Today, 17 (2014) 389.
[3] A. Paunoiu, R.S. Moirangthem, A. Erbe, physica status solidi (RRL) – Rapid Research Letters, 9 (2015) 241.
[4] A. Madani, L. Ma, S. Miao, M.R. Jorgensen, O.G. Schmidt, Nanoscale, 8 (2016) 9498.
[5] V.R. Dantham, P.B. Bisht, Chemical Physics, 388 (2011) 38.
[6] R.S. Moirangthem, P.-J. Cheng, P.C.-H. Chien, B.T.H. Ngo, S.-W. Chang, C.-H. Tien, Y.-C. Chang, Opt. Express, 21 (2013) 3010.
[7] M. Himmelhaus, S. Krishnamoorthy, A. Francois, Sensors, 10 (2010) 6257.
[8] K.R. Catchpole, A. Polman, Opt. Express, 16 (2008) 21793.
[9] W.R. Erwin, H.F. Zarick, E.M. Talbert, R. Bardhan, Energy & Environmental Science, 9 (2016) 1577.
[10] P. Ilaiyaraja, T. Kumar Das, P.S.V. Mocherla, C. Sudakar, Solar Energy Materials and Solar Cells, 169 (2017) 86.
[11] T.K. Das, P. Ilaiyaraja, C. Sudakar, ACS Applied Energy Materials, 1 (2018) 765.
[12] T.K. Das, P. Ilaiyaraja, C. Sudakar, Scientific Reports, 8 (2018) 9709.
[13] P. Ilaiyaraja, B. Rakesh, T.K. Das, P.S.V. Mocherla, C. Sudakar, Solar Energy Materials and Solar Cells, 178 (2018) 208.
[14] P. Ilaiyaraja, T. Kumar Das, P.S.V. Mocherla, C. Sudakar (manuscript under communication).
[15] D. Venkatakrishnarao, E.A. Mamonov, T.V. Murzina, R. Chandrasekar, Advanced Optical Materials, 6 (2018) 1800343.
[16] M.R. Foreman, J.D. Swaim, F. Vollmer, Adv. Opt. Photon., 7 (2015) 168.
[17] B. O'Regan, M. Grätzel, Nature, 353 (1991) 737.

ANALYSIS OF GRAIN-SIZE DISTRIBUTION AND YIELD STRENGTH OF INTERCONNECTOR RIBBONS AND WIRES AT DIFFERENT STRETCHING CONDITIONS USING COLOR ETCHING

Johann Walter, Jonas Stegmaier, Achim Kraft* and Ulrich Eitner
Fraunhofer Institute for Solar Energy Systems ISE, Heidenhofstraße 2, 79110 Freiburg, Germany
*Phone: +49 761 4588 5544, E-Mail: achim.kraft@ise.fraunhofer.de

ABSTRACT: In this paper we analyze the microstructure of solder coated copper ribbon and wire interconnectors for silicon solar cells from different manufacturers at different stretch levels in longitudinal and cross microsections by color etching and microscopy. The used etchant colors each grain according to its crystallographic orientation. This enables the extraction of the grain size and frequency information by image and data processing tools. Furthermore we measure the yield strength of the solar cell interconnectors for strain levels of 0.5 %, 5 % and 10 % and evaluate the impact on the microstructure of the interconnector.
We find a large variation in the copper microstructure, especially for wires and observe an inverse relation between yield strength and grain sizes corresponding to the Hall-Petch relation. The lowest measured yield strength for a wire interconnector is about 82 MPa (avg. grain size: 237 µm²), which is about 20 MPa higher compared to the lowest yield strength measured for ribbons (avg. grain size: 247 µm²). The wire with the highest yield strength of 148 MPa shows fine grains (avg. grain size: 29 µm²). In the ribbon analysis we find the same overall correlation between grain size and yield strength with some exceptions. This underlines that grain size distribution is not the only attribute which affects the yield strength. The analysis of copper ribbons at different stretch levels discloses a deformation or refinement of the copper grains associated with rising yield strength.
In general the results show that the approach of a color etching, optimized for solar cell interconnector cross and longitudinal sections, is a suitable, fast and cost-effective solution to quantify the grain size distribution and evaluate mechanical impacts like stretching or bending on the copper microstructure.

Keywords: Copper, Color Etching, Solar Cell Interconnector, Grain, Manufacturing and Processing

1 INTRODUCTION

According to the common practice, silicon solar cells are interconnected by 4-5 copper ribbons or 12-15 copper wires. The interconnectors consists of a copper core coated with solder (Sn60Pb40 is widely used). For five busbar interconnection the ribbon dimensions are typically 0.22 x 0.9 mm. For multi-wire interconnection wires with a diameter between 300 µm and 400 µm are used. To reduce mechanical stress in solder joints of interconnected solar cells, caused by the CTE mismatch between silicon ($\alpha_{Si} = 2.6 \cdot 10^{-6}$ K^{-1}) [1] and copper ($\alpha_{Cu} = 17.0 \cdot 10^{-6}$ K^{-1}) [2], the use of soft interconnectors with low yield strength is recommended [3-4]. The yield strength is strongly influenced by the manufacturing process of the copper interconnector resulting in different bulk material structures [5-6]. Typical yield strength values are around 60 MPa for ribbons and 80 MPa for wires. Before interconnecting the solar cells by soldering in a stringer the interconnectors are typically pre-stretched to enable a precise alignment to the solar cell.

In this work we investigate the copper grain structure of ribbons and wires with different yield strengths and at different stretching modes by color etching of metallographic cross and longitudinal sections prepared similar to literature [7].

Color etching, also called tint etching or stain etching, has been known for almost 100 years [8]. Many different reagents, called tint etchants or color etchants, were developed for different selective materials [9-11], such as copper [2,12] or alloy phases [13]. During the color etching process a stable fine film is deposited at the microsection surface [14]. The thickness varies for different grains depending on its crystallographic orientation. Due to interference effects these grains appear in different colors in the microscope image.

Due to improvements in image processing in the last decade, advanced tools and algorithms allow an efficient and quantitative analysis of color etched images and enable the examination of grain size information. Color etching enables a very fast grain boundary scanning of large areas making the technique a promising approach for the analysis of grain size distributions of solar cell interconnectors.

2 EXPERIMENTAL APPROACH

In this study we analyze ribbons and wires, in terms of yield strength and copper microstructure. The interconnectors are pre-stretched and the impact on grain size and yield strength is investigated.

A precise and reproducible pre-stretching of the solar cell interconnectors, as well as the determination of the yield strength is performed using a tensile tester according to ISO 6892-1 [15]. Table I shows the interconnector configurations and measured yield strength data. The interconnectors vary in cross section area, manufacturer (A – D) and initial yield strength. In terms of mechanical stress a low yield strength is beneficial for the interconnection of solar cells [3].

Table I: Measured yield strength of ribbon and wire interconnectors for different pre-stretching modes.

Pre-stretching mode [%]		Initial state	0.5	5.0	10.0
	Mfr.	Yield strength [MPa]			
Ribbon R1 (0.2 x 1.5 mm²)	A	61.1	68.8	137.5	182.3
Ribbon R2 (0.2 x 1.5 mm²)	B	65.3	73.4	136.3	178.2
Ribbon R3 (0.2 x 1.0 mm²)	B	75.7	83.9	146.7	190.0
Ribbon R4 (0.22 x 0.9 mm²)	C	78.0	85.6	145.7	186.5
Ribbon R5 (0.18 x 1.0 mm²)	C	110.0	117.6	162.2	193.6
Ribbon R6 (0.06 x 1.5 mm²)	C	163.8	171.2	218.2	248.5
Wire W1 (Ø = 350 µm)	A	81.9	89.4	165.5	203.5
Wire W2 (Ø = 350 µm)	B	106.7	112.6	164.1	198.4
Wire W3 (Ø = 300 µm)	D	148.1	158.2	201.6	231.9

We analyze the grain boundary by means of an optimized color etching process. For each sample group, we cut six randomized 20 mm pieces out of one 1.2 m interconnector strip for color etching. In this way we ensure to cover a large variety of microstructures within the interconnector in our analysis.

Figure 1 shows the process steps. After the very last very fine polishing, color etching is applied to the samples and a high resolution image is captured by digital microscopy. The image is analyzed by software to extract the grain boundary size of each grain. An algorithm detects the grain boundaries. Subsequently, the data are processed to enable a comparison of grain size distribution. Figure 1d) shows the cumulative relative grain area (a) and the cumulative relative frequency (f) for four different cross sections of one interconnector. For the analysis of the frequency a bin size of 0.2 µm^2 is used. Since the cumulative area curves often show a higher deviation, especially for larger grain sizes, we recommend to use the frequency for comparison of different interconnectors. In the result diagrams we show the averages of 3 to 6 curves for each interconnector (figure 1e). The number of evaluated samples depends on the quality of the obtained etched surfaces. Etching quality varies across samples and depends mainly on factors such as evenness, polishing, impurities, oxidation or etching process stability. All samples in one group are etched simultaneously.

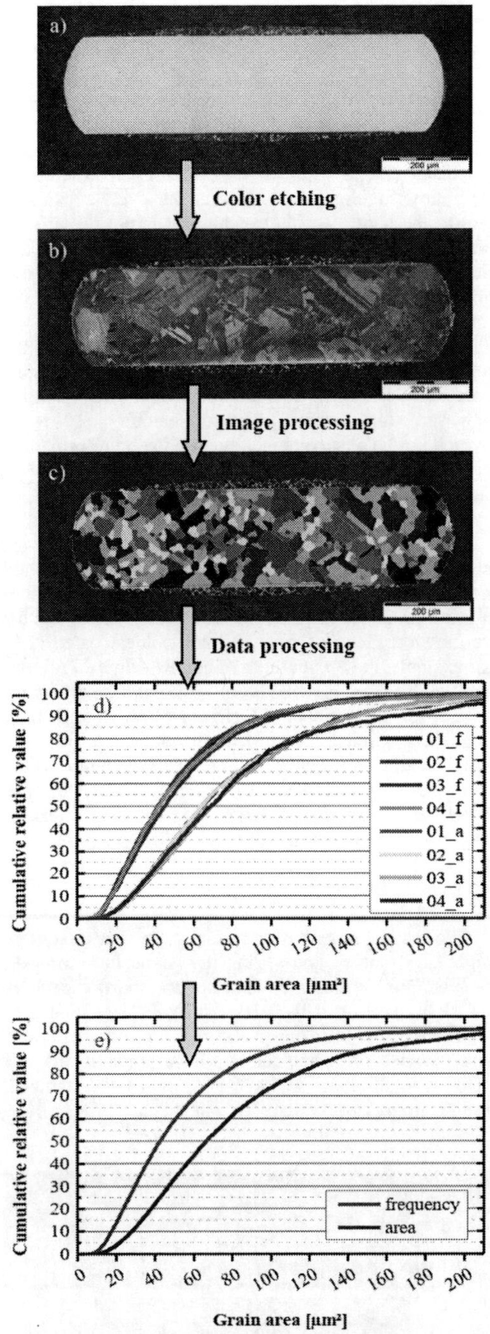

Figure 1: Microscope image of the metallographic cross section of a ribbon after grinding and polishing a), after color etching revealing the grain structure of the interconnector copper material b), after image processing c). The relative area as well as the relative frequency (bin = 0.2 µm²) of all grains are plotted cumulatively over the full grain boundary range d). By cumulating the values, the differences between interconnectors are more clearly visible. Curves are averaged respectively for area and frequency e).

3 RESULTS

3.1 Color etching

One representative color etched overview and detail image of each interconnector sample (cf. table I) at initial state (non-stretched) is shown in figure 2 to figure 10. The detail images show the magnified microstructure of the yellow highlighted sections in the overview image. Due to the nature of color etching, it is not possible to compare colors between images. The color result depends on parameters such as temperature, etching time, concentration of the color etchants, microstructure and polishing. Slight deviation can lead to different colors. However, the variety of colors gives an idea of the homogeneity in grain orientation since identical colors represents identical orientation within the sample. However for copper the orientation {100} is favored [16]. To extract the grain orientation information, methods such as electron backscatter diffraction (EBSD) can be used.

Depending on the manufacturing process the interconnectors are treated differently (temperature profile, system etc.). This is reflected by the differences in microstructure. In general, flat ribbon interconnectors are drawn, rolled flat, annealed and solder coated [17]. For example, the 60 µm thick ribbon R6 (figure 7) shows small grains (avg. grains: 53 µm²), whereas the grains for the 220 µm thick ribbon R4 (figure 5) are significantly larger (avg. grains: 247 µm²). Both ribbons are from the same manufacturer (table II). For wire W1 (figure 8) the color etching reveals very large grains whereas for wire W2 (figure 9) we observe larger grains in the center and smaller grains all around the edges. Wire W3 (figure 10) shows overall very small grains. Further analyses are shown in chapter 3.3.

We assume in case of very fine grain sizes that there might be some issues in the annealing process (temperature profile, interconnector transport) especially for wires because of their less mechanical resilience to deformation and less copper material. This correlates also with the fact that manufacturers have issues to achieve yield strength about 60 MPa for wires (cf. table I, manufacturer A).

Figure 2: Ribbon R1 (0.2 x 1.5 mm²), manufacturer A

Figure 3: Ribbon R2 (0.2x1.5 mm²), manufacturer B

Figure 4: R3 (0.2x1.0 mm²), manufacturer B

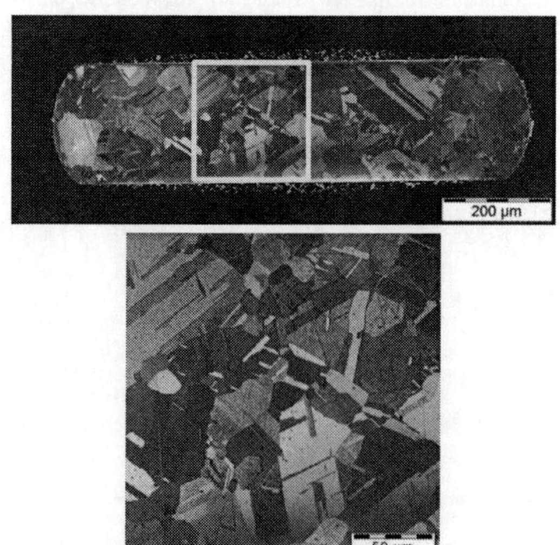

Figure 5: R4 (0.22 x 0.9 mm²), manufacturer C

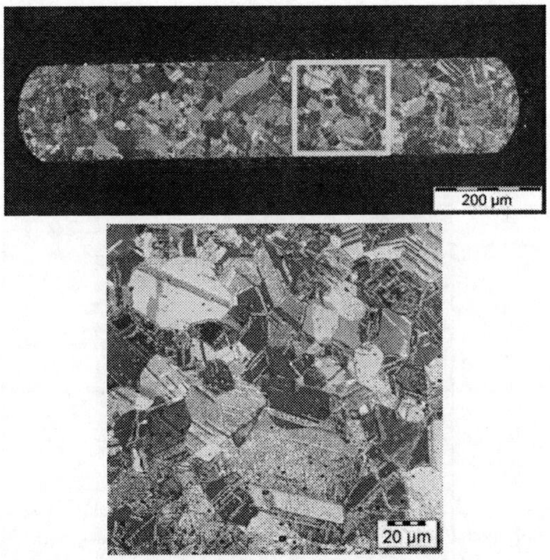

Figure 6: R5 (0.18 x 1.0 mm²), manufacturer C

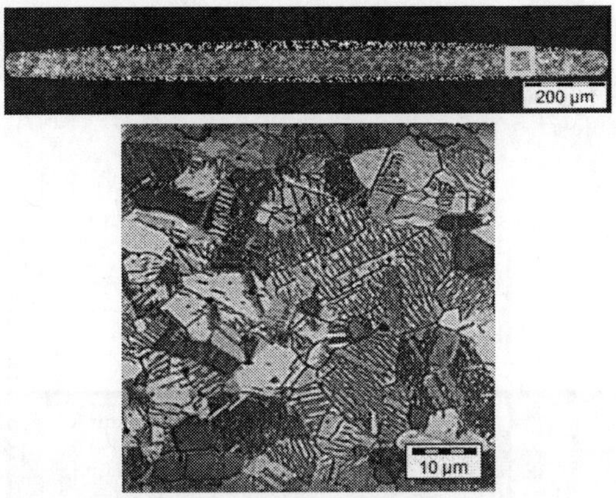

Figure 7: R6 (0.06 x 1.5 mm²), manufacturer C

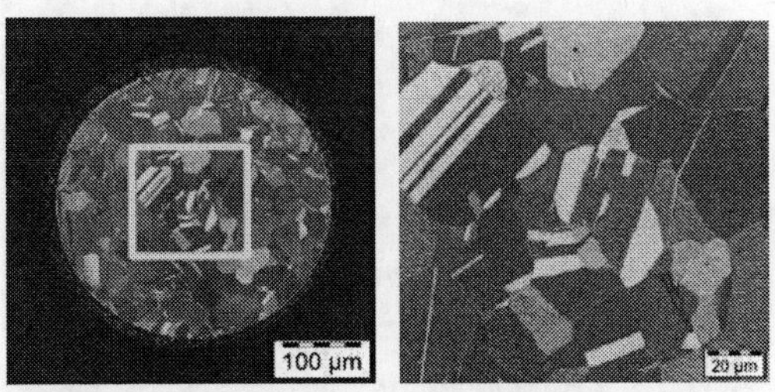

Figure 8: W1 (Ø =350 μm), manufacturer A

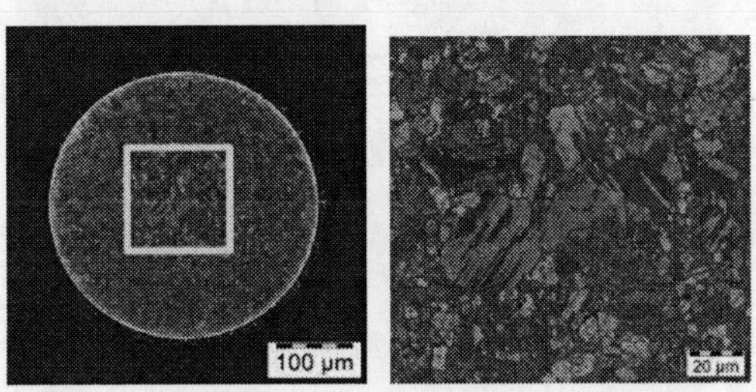

Figure 9: W2 (Ø =350 μm), manufacturer B

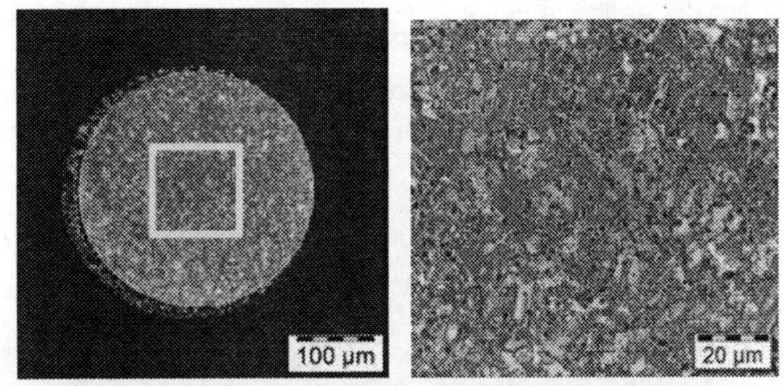

Figure 10: W3 (Ø =300 μm), manufacturer D. The black dots represent polishing artefacts.

3.2 Influence of stretching modes on microstructure

We apply the presented method (figure 1) to evaluate the grain size distribution of the interconnectors at initial state and after stretching. Figure 11 shows the deviation within a group of wire sample W3 non-stretched (initial state) and after 10 % stretching. In this case each group consists of four samples. The variation within a group is less than 5 $\%_{absolute}$. In general color etching shows a good agreement with the grain size distribution determined by EBSD [18].

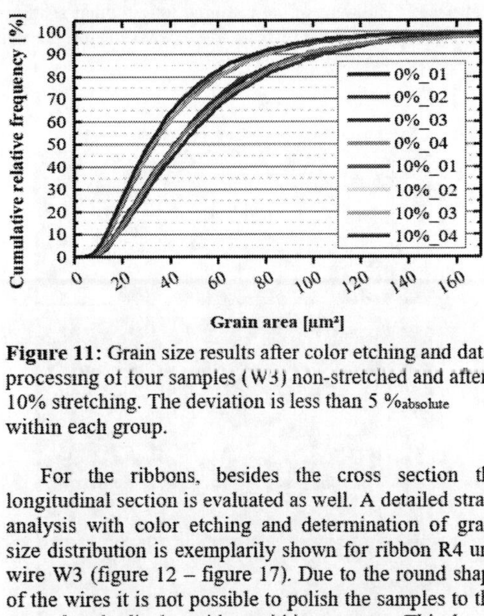

Figure 11: Grain size results after color etching and data processing of four samples (W3) non-stretched and after 10% stretching. The deviation is less than 5 $\%_{absolute}$ within each group.

For the ribbons, besides the cross section the longitudinal section is evaluated as well. A detailed strain analysis with color etching and determination of grain size distribution is exemplarily shown for ribbon R4 und wire W3 (figure 12 – figure 17). Due to the round shape of the wires it is not possible to polish the samples to the same longitudinal position within a group. This has a significant impact on the evaluated area and the results due to very inhomogeneous grain size distributions in the cross section e.g. center and edge. For this reason, we do not perform the analysis of longitudinal sections for wires.

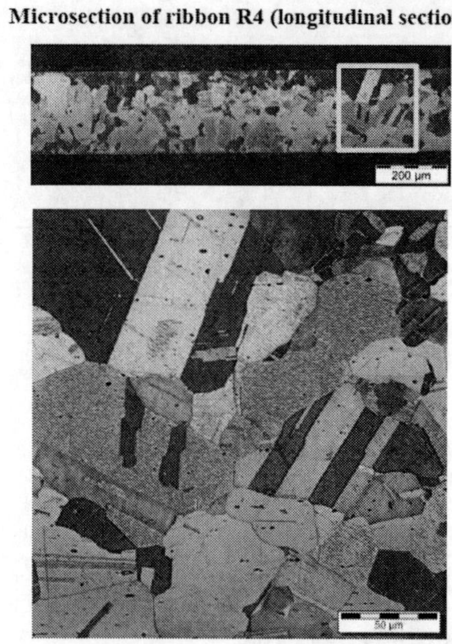

Figure 12: Half size of the captured longitudinal section and detailed image of ribbon R4, without pre-stretching ($R_{p0.2}$ = 163.8 MPa).

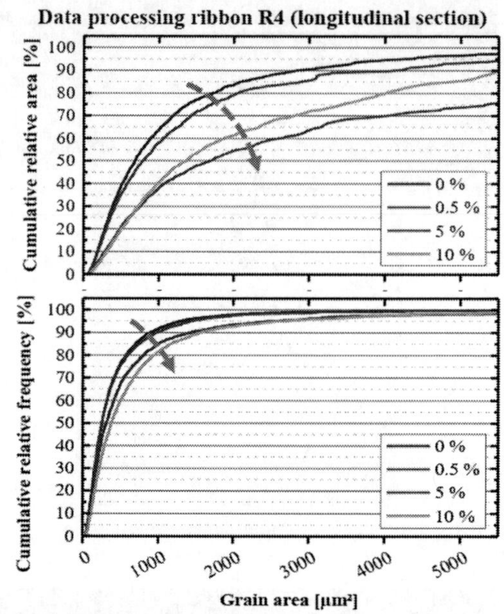

Figure 14: Distribution of grains of ribbon R4 (longitudinal section). The arrows indicated the change direction in grain distribution due to stretching.

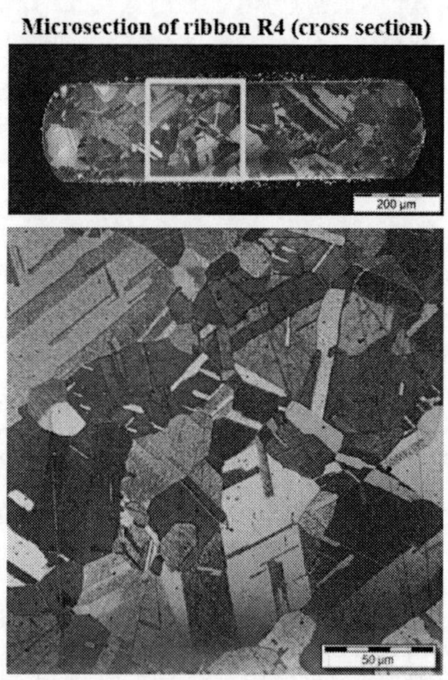

Figure 13: Cross section and detailed image of ribbon R4 after 0.5 % pre-stretching ($R_{p0.2}$ = 171.2 MPa).

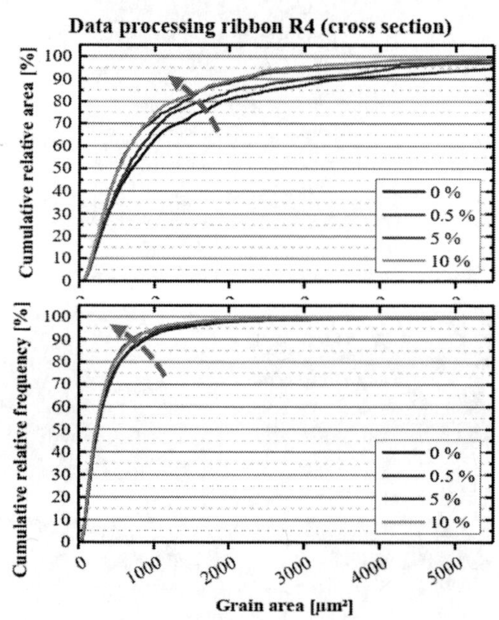

Figure 15: Distribution of grains of ribbon R4 cross section. The arrows indicated the change direction in grain distribution due to stretching.

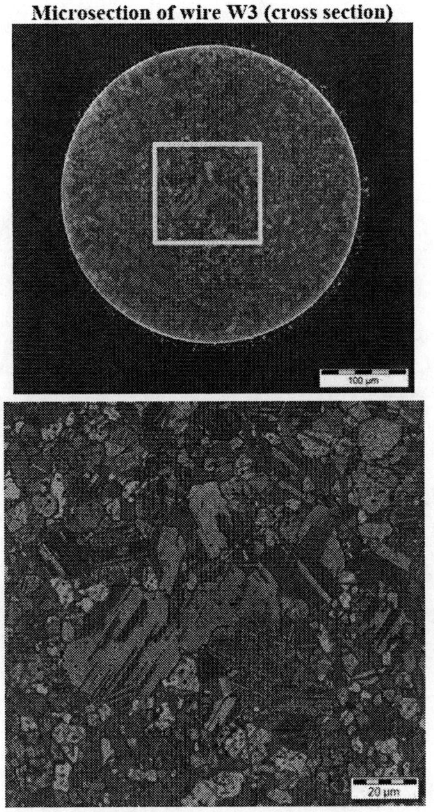

Figure 16: Cross section and detailed image of wire W3 after 0.5% pre-stretching ($R_{p0.2}$ = 158.2 MPa).

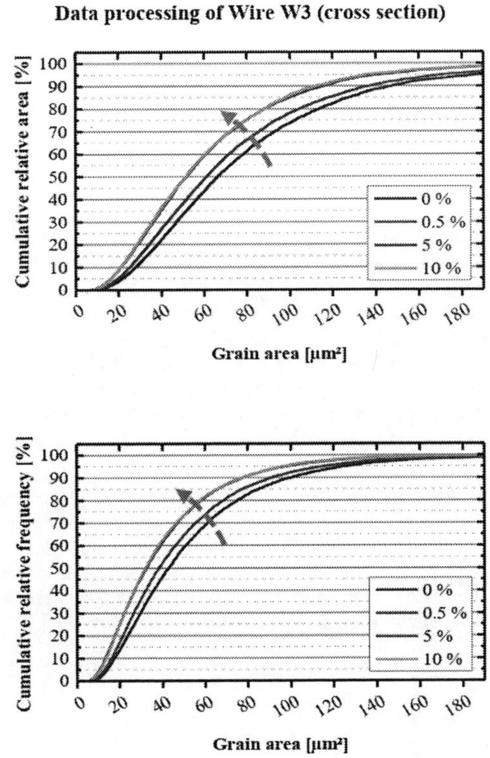

Figure 17: Distribution of grains of wire W3 cross section. The arrows indicated the change direction in grain distribution due to stretching.

Figure 14, 15 and 17 show the results after image and data processing of the color etched images (procedure cf. figure 1). Each curve represents the mean value of 3 - 6 evaluated microsections. The number of detected grains within a batch varies between 2.000 - 10.000 grains depending on the sample. The curves showing the relative area distribution present the area proportions of different grain sizes. The 50% threshold means 50 % of the area is covered by grains of the given grain sizes. Larger grains result in a flatter appearance of the curve.

Furthermore the curves show the cumulative relative frequency. This gives information about the frequency proportion of different grain sizes (bin size: 0.2 µm² grain area). A higher number of grains within the same grain bin leads to a steeper shape of the curve. A very similar number of grains of each grain size would result in a linear curve. The 50% threshold represents the median.

According to the cross section data both evaluated samples, ribbon R4 and wire W3, indicate a drift to smaller grains with increased pre-stretching.

In the longitudinal section analysis, higher pre-stretching levels result in larger grains. The curves become flatter (figure 14). The grain size appears smaller in cross sections but larger in the longitudinal sections. It seems that larger grains get deformed or refined into smaller ones during stretching. This explains the increased frequency for smaller grains. Due to literature stretching is most likely [19] but has to be investigated on further. For 0 % and 0.5 % stretching the grain size change seem to be too low to detect any difference.

3.3 Analysis of yield strength and different grain size distributions at initial state.

Based on the method as shown in figure 1 we analyze the microstructure of different ribbon and wires at initial state and calculate the cumulative relative frequency curves (figure 18 and figure 20) and the non-cumulative frequency curves (figure 19 and figure 21) for every group.

As shown in figure 14 and figure 15, stretching the grains of ribbon R4 does not e.g. doubling or halving any values in the grain size distribution curves, although the yield strength increases by a factor of 2.4 after 10 % stretching (table I). This means, grain size and yield strength does not change with same dynamic due to stretching. However, figure 18 to figure 21 disclose that the manufacturing process has a large impact on the grain size distribution curves.

Figure 19 shows that the relative frequency peak of W1 (81.9 MPa) is shifting from 52.2 µm² grain area to 15 µm² for W3 (148.1 MPa). The average grain size is changing from 237 µm² for W1 to 29 µm² for W3 (table II).

In figure 18 we observe a steeper curve shape with increasing yield strength. In figure 20 the same behavior can be observed for the ribbons R1 (61.1 MPa), R2 (65.3 MPa), R5 (110.0 MPa) and R6 (163.8 MPa).

R3 (75.7 MPa) and R4 (78.0 MPa) have larger grains compared to R1 and R2 but we measure a 10 - 20 MPa higher yield strength. This correlation is also confirmed by the average grain sizes in table II. These differences are caused by the fact that the yield strength depends not only on grain size, but also on material properties such as impurities, dislocations or grain orientation [20-21].

The grain size distribution is correlated to yield strength but its not the only influence parameter. We also find that different interconnector types, such as ribbon or wires, can not be compared directly since the resulting curve shape differs significantly for similar yield strengths as displayed for W2 and R5. However, for most of the interconnectors, we observe that larger grains result in lower yield strength, which is also proclaimed by the Hall-Petch relation [22-23]. It predicts a linear dependence of the yield strength or hardness on the average grain size ($\sigma_{\text{yield stress}} \sim D_{\text{grain size}}^{-1/2}$) [24]. With decreasing grain size less dislocation slipping can happen as the grain boundary represents an obstacle [25]. Since dislocation slipping is necessary for plastic deformation, material with smaller grains becomes harder, resulting in higher yield strength. Although grain refinement is not useful for interconnectors it is a typical technique for strengthening a material [26-27].

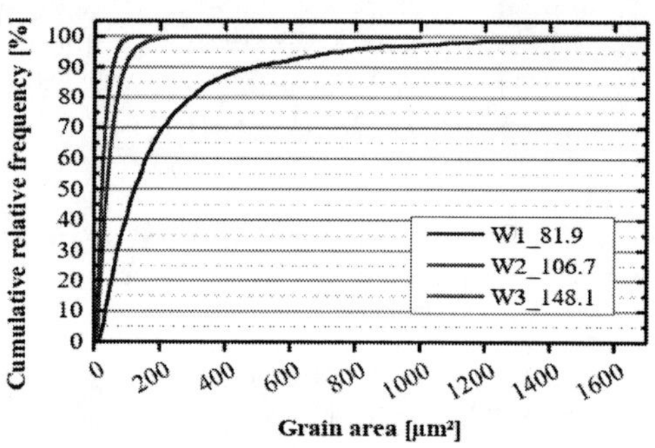

Figure 18: Distribution of the grain sizes at initial state of three different wires with different yield strength (cumulative).

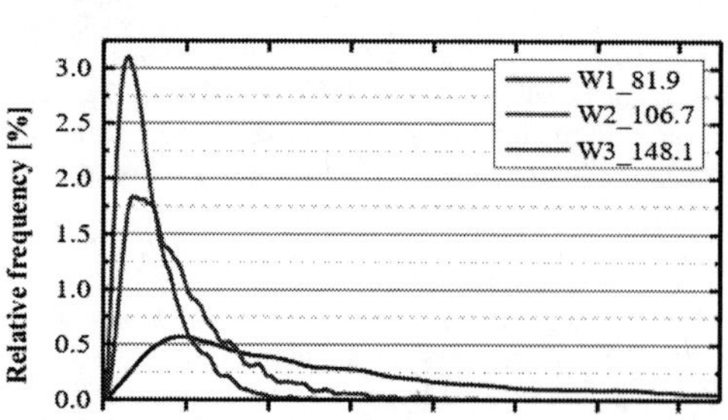

Figure 19: Distribution of the grain sizes at initial state of three different wires with different yield strength (relative).

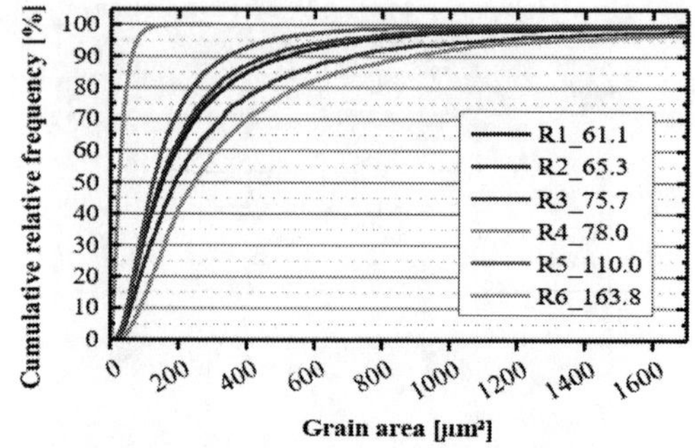

Figure 20: Distribution of the grain sizes at initial state of six different ribbons with different yield strength (cumulative).

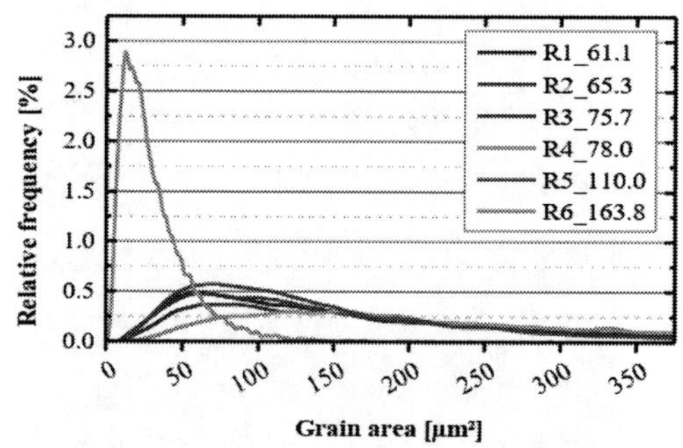

Figure 21: Distribution of the grain sizes at initial state of six different ribbons with different yield strength (relative).

Table II: Average grain size data after image and data processing from different non-stretched interconnectors and manufacturers.

	Mfr.	Initial yield strength [MPa]	Avg. grain size [µm²]
Ribbon R1 (0.2 x 1.5 mm²)	A	61.1	247
Ribbon R2 (0.2 x 1.5 mm²)	B	65.3	238
Ribbon R3 (0.2 x 1.0 mm²)	B	75.7	332
Ribbon R4 (0.22 x 0.9 mm²)	C	78.0	452
Ribbon R5 (0.18 x 1.0 mm²)	C	110.0	174
Ribbon R6 (0.06 x 1.5 mm²)	C	163.8	53
Wire W1 (Ø = 350 µm)	A	81.9	237
Wire W2 (Ø = 350 µm)	B	106.7	50
Wire W3 (Ø = 300 µm)	D	148.1	29

3.4 Analysis on wave-shaped interconnector

Wave shaped interconnectors, which are beneficial for low mechanical stress in solder joints [3], can also be analyzed by color etching. During the shaping process the grains get crushed in the inner radius and stretched with increasing radius as displayed in figure 22. On the one hand wave shaping lowers the effective yield strength [28] but on the other hand the process results in unfavorable mechanical properties in the bended area. Further analyses on this issue to enhance the bending process are in progress.

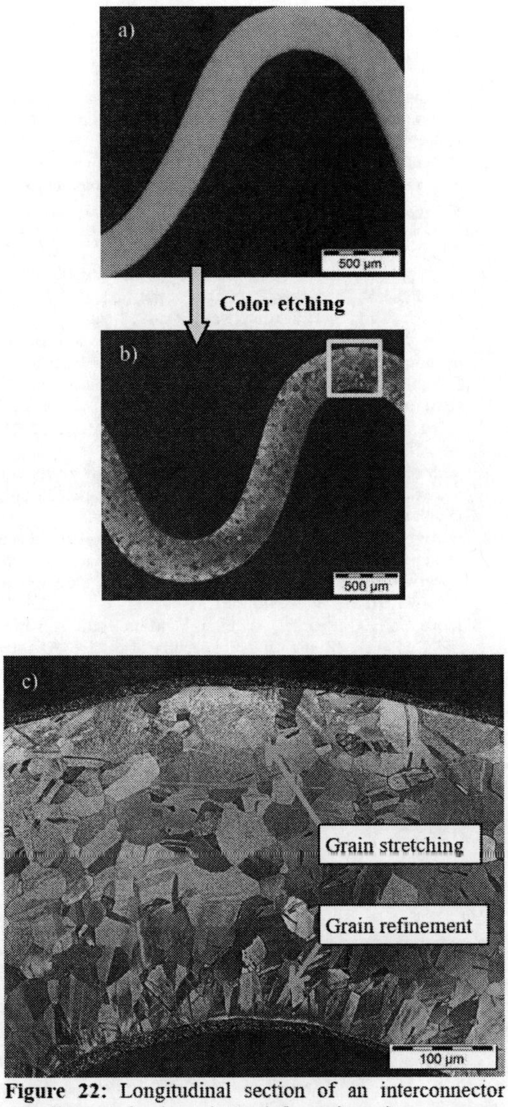

Figure 22: Longitudinal section of an interconnector revealing grain boundary deformation due to wave shaping. a): Original longitudinal section of a wave shaped interconnector. b): Wave shaped interconnector after color etching. c) Detailed view of the in b) highlighted bent area.

4 CONCLUSION

In this paper we analyze the grain boundary structure of 6 different ribbons and 3 different wires for solar cell interconnection by an optimized color etching process. The samples come from different manufacturers and vary in their initial yield strength and geometry. We analyzed them at non-strained and different strain levels and we complement the results with yield strength measurements. Color etching reveals the microstructure of the copper core of the ribbon by coloring the crystallographic orientation of each grain which is studied in longitudinal and cross microsections.

Compared to other methods - such as EBSD or LabDCT - color etching, combined with microscopy, delivers very fast grain boundary information of large areas, which can be used to extract the grain size distribution by image and data processing. The largest evaluated microstructure in our study is about 2270 x 220 µm². Typical etching times are within a few minutes. As the absolute color result is strongly dependent on the parameters of each individual sample (microstructure, temperature, color etchants concentration, etching time, etc.), quantitative color comparison across different samples is not possible. However, since identical colors represent identical grain orientation within a sample, knowledge about grain orientation homogeneity can be delivered but the exact orientation is unknown.

We showed that the grain size deviation of samples within the same group is < 5 % which is an acceptable variation for the purpose of this study. The strongest influence is the etching quality which depends on the copper surface quality (evenness, polishing, impurities, oxidation or etching process stability). For the analysis we recommend to use the frequency curves instead of the area curves since large single grains, which sometimes occur in samples, have a significant impact on the area curves, resulting in a stronger curve deviation at larger grains, but hardly affect the frequency curves.

The analyzed wires show a correlation between yield strength and grain sizes distribution, similar to the Hall-Petch relation, which expects microstructures with large grains to have a lower yield strength (table II bottom). W1 (82 MPa) has the largest grains (figure 8) and lowest yield strength (table I). W2 (107 MPa) has large grains in the center and small grains all around the edge (figure 9). W3 (148 MPa) has evenly distributed small grains (figure 10). This relation is also shown in the frequency curves (figure 18). However, for the ribbons we find the same overall tendency but also observe that yield strength is not only dominated by grain size. R3 and R4 have overall larger grains but a 10 -20 MPa higher yield strength compared to R1 and R2 (table II top). We assume this might be related to differences in grain orientation, dislocations or impurities, which also affect yield strength [20-21] and for which our color etching method is not sensitive. The highest yield strength is measured for R6 (164 MPa) which also exhibits the smallest thickness (60 µm) and smallest grains. It is notable that the lowest ribbon yield strength is about 60 MPa and for wires it is about 80 MPa. As the low yield strength for ribbons has been requested by the PV industry and is an established feature of solar ribbons today our analysis indicates that there might be optimization potential for processing round-shaped wires for an application in PV.

Stretching enlarges the grains in the longitudinal direction and reduces the grain size in the normal plane direction. The reduction of grain size is either caused by grain refinement or grain deformation. Due to the cold hardening the yield strength increases. For ribbon R4 at 10 % strain level (187 MPa) the yield strength raises by a factor of 2.4. The influence of stretching is also visible on grain size level in the longitudinal (figure 14) and cross sections (figure 15) for ribbon R4. Due to stretching not only the grain size but also the grain orientation can change [19] which cannot be detected by color etching. For this reason we suggest using this method on interconnectors in their initial state only, as differences in the manufacturing process can be observed easily (figure 18, figure 20, table II).

Our study shows that color etching is a suitable, fast and cost-effective alternative to reveal the copper microstructure of solar cell interconnectors and analysis of grain sizes and frequency over a large area.

ACKNOWLEDGEMENTS

This work has been supported by the Federal Ministry for Economic Affairs and Energy under the contract number 0324057B, acronym BACKBONE.

REFERENCES

[1] K. G. Lyon, G. L. Salinger, C. A. Swenson, and G. K. White, "Linear thermal expansion measurements on silicon from 6 to 340 K", Journal of Applied Physics, vol. 48, no. 3, 1977, pp. 865-868, doi: 10.1063/1.323747

[2] J. R. Davis, "Copper and copper alloys", ASM International, 2001, ISBN: 0-87170-726-8.

[3] R. Meier, M. Pander, R. Klengel, S. Dietrich, S. Klengel, M. Ebert, and J. Bagdahn, "Reduction of soldering induced stresses in solar cells by microstructural optimization of copper-ribbons", Proceedings of Society of Photographic Instrumentation Engineers, vol. 811206, 2011, pp. 1-13, doi: 10.1117/12.893519.

[4] B. Kang N. Park, S. J. Tark, W. W. Oh, S. Park, Y. D. Kim, H. Lee, and D. Kim, "Advanced yield strength of interconnector ribbon for photovoltaic module using crystallographic texture control", vol. 20, no. 2, 2014, pp. 229-232, doi: 10.1007/s12540-014-2005-x.

[5] A. Butz, D. Helm, and R. Meier, "Microstructural analysis and process chain simulation of copper-ribbons for solar cell Interconnections", 8th European Solid Mechanics Conference, 2012.

[6] A. Butz, R. Meier, and D. Helm, "Simulationsbasierte Untersuchung der Eigenschaften von Bauteilen aus Kupfer am Beispiel der Fertigungskette von Solarzellenverbindern", Metall, vol. 3, 2012, pp. 28-31.

[7] D. Eberlein, P. Schmitt, and P. Voos, "Metallographic solder preparation of soldered solar cells", Practical Metallography, vol. 48, 2011, pp. 402-408.

[8] H. S. Rawdon, and M. G. Lorentz, "Metallographic etching reagents", National Bureau of Standards, vol. 16, 1920.

[9] G. Pestrow, "Metallographic etching: techniques for metallotraphy, ceramography, plastography", ASM International, 1999, ISBN: 978-0871706331.

[10] G. F. Vander Voort, "Metallography and microstructures", vol. 9, ASM International, 2004, pp. 493-512, ISBN: 978-0-87170-706-2.

[11] E. Weck, and E. Leistner, "Metallographic instructions for colour etching by immersion", Deutscher Verband für Schweißen und verwandte Verfahren, vol. 2, 1995, ISBN: 978-3-87155-146-8.

[12] G. F. Vander Voort, "Copper color metallography", Advanced Materials & Process, vol. 158, no. 1, 2000, pp. 36-40.

[13] G. F. Vander Voort, "Color metallography", Microscopy Today, vol. 13, no. 6, 2005, pp. 22-27.

[14] G. F. Vander Voort, "Metallography: principles and practice", ASM International, 1984, reprinted by ASM International, 1999, ISBN: 978-0871706720.

[15] ISO 6892-1, "Metallic materials - Tensile Testing", 2009.

[16] D. P. Field, "Metallography and microstructures", vol. 9, ASM International, 2004, pp. 215-226, ISBN: 978-0-87170-706-2.

[17] R. Meier, M. Pander, S. Großer, and S. Dietrich, "Microstructural optimization approach of solar cell interconnectors fatigue behavior for enhanced module lifetime in extreme climates", Energy Procedia, vol. 92, 2016, pp. 560-568, doi: 10.1016/j.egypro.2016.07.020.

[18] T. Bernthaler, and G. Schneider, "Fortschritte in der Metallographie", Deutsche Gesellschaft für Materialkunde, 51. Metallographie-Tagung, 2017, pp. 79-84, ISBN: 978-3-88355-415-0.

[19] H. Oettel, and H. Schumann, „Metallografie: Mit einer Einführung in die Keramografie", Wiley-VCH Verlag, vol. 15, 2011, ISBN: 978-3527322572.

[20] J. Rösler, H. Harders, and M. Bäker: "Mechanisches Verhalten der Werkstoffe", Springer Vieweg, vol. 4, 2012, ISBN: 978-3835102408.

[21] B. V. Petukhov, "A Theory of the Effect of Impurities on the Yield Stress of Silicon Crystals", Semiconductors, vol. 38, no. 4, 2004, pp. 369-375.

[22] E. O. Hall, "The deformation and ageing of mild steel", Proceedings of the Physical Society, section B, vol. 64, 1951, pp. 747-753, doi: 10.1088/0370-1301/64/6/305.

[23] N. J. Petch, "The cleavage strength of polycrystals", Journal of the Iron and Steel Institute, vol. 174, 1953, pp. 25-28.

[24] M. A. Meyers, and K. K. Chawla, "Mechanical behavior of materials", Prentice Hall, 2009, ISBN: 978-0131395060.

[25] J. Schiotz, "Simulations of nanocrystalline metals at the atomic scale. What can we do? What can we trust?", Science of Metastable and Nanocrystalline Alloys, 2001, pp. 127-139, ISBN: 87-550-2916-7.

[26] C. M. L. James, "Mechanical properties of nanocrystalline materials", CRC Press, vol 51, no. 4, 2006, pp.133-162, ISBN: 978-9814241977.

[27] E. Arzt, "Size effects in materials due to microstructural and dimensional constraints: a comparative review", Acta Materialia, vol. 46, no. 16, 1998, pp. 5611-5626, doi: 10.1016/S1359-6454(98)00231-6.

[28] J. Walter, L. C. Rendler, A. Halm, V. Mihailetchi, A. Kraft, and U. Eitner, "Ribbon interconnection of 6'' BC-BJ solar cells", Energy Procedia, vol. 124, 2017, pp. 504-514, doi: 10.1016/j.egypro.2017.09.287.

ELECTRICALLY CONDUCTIVE ADHESIVES FOR CELL INTERCONNECTION IN SHINGLED MODULE TECHNOLOGY: THE IMPACT OF MATERIAL PROPERTIES ON MINI-MODULE PERFORMANCE

Marc Estruga, Liesbeth Theunissen, Antonio Ardizzone, Bob Willems, Anja Henckens
Henkel Belgium NV
Nijverheidsstraat 7. B-2260 Westerlo, Belgium

ABSTRACT: Modules manufactured with shingled crystalline silicon solar cells have become more prevalent as of late. For these structures, the electrical and mechanical properties of the electrically conductive adhesive (ECA) used to attach the shingles together plays a key role in the final performance of the assembled modules. In this work, ECAs based on different chemistry platforms are presented and the impact of their properties (such as adhesion, bulk electrical conductivity, contact resistivity to Ag ribbons and storage modulus) on module thermocycling reliability (-40/85°C) and static mechanical load performance is studied. In particular, full-sized modules were built with ECAs with a wide range of storage moduli -- from very flexible, to very rigid materials -- and it was found that materials with the highest storage modulus show a larger power loss upon standard mechanical load tests. In addition, one-cell mini-modules were built to study the effect of the contact resistivity on the reliability performance, with findings suggesting that ECA usage is the main contributor to the series resistance increase upon cycling. Lastly, the work also proves that other materials in the mini-module manufacture process, such as the backsheet, play a critical role in reliability.
Keywords: Shingling, Metallization, Contact, Electrical Properties, Stability

1 INTRODUCTION

Among the different available advanced technologies to improve the overall efficiency of crystalline silicon (c-Si) modules, shingling has become one of the most popular choices due to its ease of implementation for large volume manufacturing. Shingling technology is achieved by laser cutting each cell into several pieces (shingles), placing them one by one on top of each other, overlapping a certain area to form a string, so that no busbars are needed. In this configuration, an electrically conductive adhesive is applied to secure the shingle interconnection, providing both electrical conductivity and mechanical stability of the whole string assembly. Several strings can then be connected to the ribbon during the module assembly. This process can be carried out at the module manufacturing level by adding a cell lasering step and an ECA application step (dispensing, jetting or printing), thus it does not require additional changes upstream (at wafer or cell manufacturing level). For this reason, shingling technology is expanding rapidly and gaining market share versus conventional c-Si cells.

When compared to standard H-pattern ribbon attached cells, shingled modules have the advantage of lower Ohmic losses, due to lower current and higher voltage at which the shingled cells work. In addition, shingling provides a more compact module design since there is less inactive space between cells, which results in lower installation cost per watt. As opposed to soldering, shingling interconnections are compatible with thinner cells, providing lower breakage rates, and deliver better absorption of the mechanical stresses generated by coefficient of thermal expansion (CTE) mismatches in the module.

Commercial modules using ECAs as shingle interconnection are already on the market, as already demonstrated in the Sunpower® Performance Series [1] and Solaria PowerXT® solar panels [2], proving that there is great potential for the shingled module technology in terms of cost, performance, and long-term reliability.

In this work, the role of the ECA on the electrical and mechanical properties of shingled modules is studied. In particular, ECAs based on different chemistry platforms are presented and the impact of their properties (such as adhesion, bulk electrical conductivity, contact resistivity to Ag ribbon and storage modulus) on module thermocycling reliability (-40/85°C) and on the static mechanical load module performance is studied. Additionally, it has been proven that the backsheet plays a significant role in module reliability performance.

2 ECA PROPERTIES

In general, the chemical nature of the resins used to formulate the ECA determines the electrical, and especially the mechanical properties of the cured adhesive. Usually, acrylates are fast-curing, provide good adhesion and have a storage modulus which can be tuned from flexible to rigid, but are more sensitive to process conditions such as moisture and temperature fluctuations. On the other hand, silicones are more flexible, having a lower storage modulus, but might suffer from slow curing and low adhesion. Lastly, epoxies typically deliver medium to high adhesion, medium to high storage modulus, reasonable cure speed and are well-known for providing reliable electrical contacts. These are generic guidelines and there is a certain overlap between the three chemistries; these properties can be tuned by formulation work.

With these characteristics understood, ECAs from these three chemistry families are presented (Table I). They all show fast curing speed, achieving sufficient adhesion and electrical conductivity when cured for 90 s. at 180°C, although data in the table corresponds to a 15 min. 150°C curing profile, unless otherwise noted.

The ECAs from the *Chemistry 1* group show a wide range of mechanical properties, from low to medium to high storage modulus (1c, 1ab and 1d, respectively). It is also worth noting the difference in density between these four ECAs, since the low-density materials with similar electrical properties are advantageous in terms of lower material consumption, with the goal to reduce the total module cost. *Chemistry 2* ECA, exhibits very low storage modulus, thus a high flexibility. *Chemistry 3* ECAs show a wide formulation window, in this case, storage modulus ranges from medium to high, while keeping excellent electrical properties.

Table 1: ECA mechanical and electrical properties

Property	Units	1a	1b	1c	1d	2a	3a	3b
Bulk resistivity	Ohm.cm	$2{,}0\cdot10^{-4}$	$3{,}7\cdot10^{-3}$	$1{,}7\cdot10^{-4}$	$7{,}0\cdot10^{-4}$	$1{,}3\cdot10^{-3}$	$1{,}7\cdot10^{-4}$	$3{,}0\cdot10^{-4}$
Die shear strength (die/ceramic) 90 s 180˚C cure	MPa	15,4	17,2	6,1	32,6	3,3	4,4	20,6
Storage modulus at 25°C	MPa	1650	1390	145	4800	10	1680	4040
Contact resistivity (CR) ECA-Ag ribbon	mohm.cm^2	0,17	0,42	0,30	0,23	0,81	0,10	0,12
Reactivity (DSC peak)	˚C	72	67	76	75	149	160	130
Density	g/cm^3	4,3	1,9	2,1	2,1	2,4	2,3	3,0

3 MODULE ASSEMBLY

In order to understand the effect of the different ECA properties on the reliability and mechanical load module performance, five shingles per cell mini-modules, as well as one- or two-meter string full modules, were assembled (respectively). In both cases, monocrystalline PERC solar cells of engineering-grade quality were used.

The assembly of the one cell mini-modules for reliability tests (Section 4.1) was performed by stencil printing the ECA on each shingle and attaching them manually prior to cure (90 s. @ 180˚C). Then, for bussing the string, an Ag-coated Cu ribbon was used, and the ribbon was attached with ECA 1a. Subsequently, the strings were encapsulated with Ethylene-vinyl acetate (EVA) and laminated between glass and a layer of backsheet. Initial mini-module performance was measured with the Meyer Burger 'Pasan SpotLIGHT' cell tester and then the modules were subjected to thermo cycling from -40 to 85˚C with a dwell time of 30 min.

Assembly of the full modules for mechanical load tests (Section 4.2) was carried out with the Applied Materials tool [3]. The application of the ECA to the silver busbar of the cell was achieved via screen and stencil printing, depending on the ECA, with a design of 29 pads of 3 by 0,4 mm, spaced at 1,8 mm. The shingle overlap was set to 1,0 mm and the strings were cured at 150°C. For bussing these strings, SnPbAg ribbons were soldered. Polyolefin elastomer (POE) was used to encapsulate the strings between the glass and the backsheet.

For the static mechanical load test, the full modules were subjected to high pressure regulated by a pneumatic system with clamps. Initially, the IV curve and EL images of each module were recorded. Then, modules were subjected to five cycles of 1 h at 2400 Pa (pull/push) before a final push at 5400 Pa for 1 h. Additionally, IV and EL were recorded after the five cycles at 2400 and the final push at 5400.

4 RESULTS AND DISCUSSION

4.1 ECA and usage effect on thermocycling (TC) performance

In order to investigate the effect of the ECA properties on the reliability, mini-modules were assembled using different ECAs from Table 1. Additionally, coupons were built to measure the contact resistance (CR) between the ECA and Ag ribbon by the TLM method [4]. The initial CR values can be found in Table 1 for a contact area of 6 mm^2. Samples were subjected to the same thermocycling (TC) treatment as the mini-modules, and CR was measured after certain TC cycles (Figure 1).

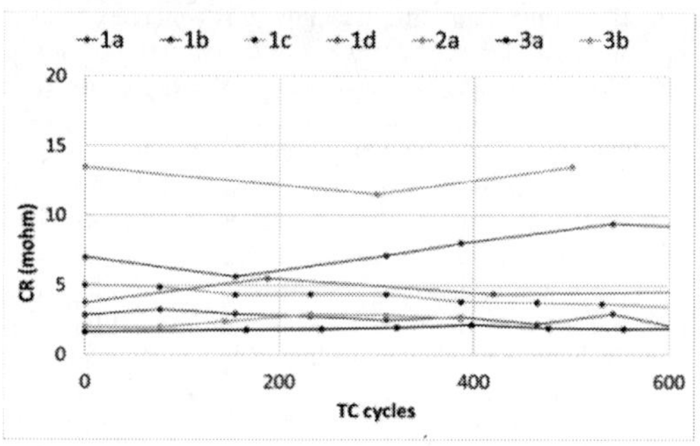

Figure 1: CR and TC reliability (-40/85˚C) of the different ECAs presented in this work.

Reliability results show that there are two groups of ECAs in terms of initial CR performance. While 1b and 2a show a relatively higher CR, all other materials show lower values, below 5 mohm. This is in good agreement with the bulk resistivity data from Table 1, which is also slightly higher for 1b and 2a. In addition, the two ECAs, especially 1b, are less stable in TC, as shown by the slight increase in resistance during cycling. All the other ECA show stable CR, which suggests that the electrical pathway within the ECA and at the interface with the ribbon is not degraded by temperature cycles from -40 to 85˚C.

In terms of IV performance, the evolution over thermocycling of the series resistance extracted from the IV curves is shown in Figure 2. Firstly, the ECA usage has an impact on the initial series resistance, as shown for 1c, which shows about 10% lower value for the minimodule with higher usage (3,8 mg per shingle vs 1,1). Secondly, the series resistance of 1c is stable over TC, both at lower and higher usage. This suggests that the shingle to shingle electrical contact reliability is not limited by the lower ECA usage for 1c.

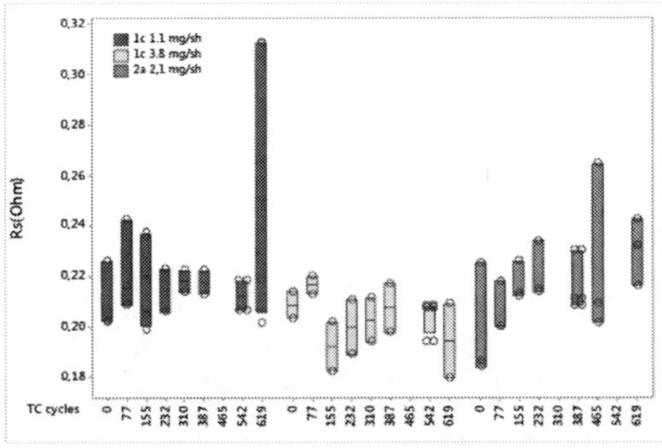

Figure 2: Series resistance evolution upon TC reliability (-40/85°C) of the mini-modules built with different ECAs.

In addition, the minimodule assembled with 2a shows a relatively slightly lower initial series resistance compared to 1c at high usage, although the amount of 2a per shingle is lower (2,1 vs 3,8 mg/shingle). This could be explained by the different ECA rheology causing a different contact area between the shingles, which could explain these small differences in series resistance.

In terms of mini-module reliability, 2a shows series resistance increasing over TC, as expected from the slight increase of CR between the ECA and Ag ribbon shown in Figure 1.

These results suggest that the TC performance is a combination of electrical and mechanical properties, low CR is needed to achieve initial low Rseries (and such, high maximum power), and certain flexibility (lower storage modulus) provides better stress absorption during TC cycling. The optimum balance point between these two will depend on the specific ECA pad design, bond line thickness and cell characteristics.

4.2 ECA effect on static mechanical load test performance

In addition to electrical performance, the effect of the ECA properties on module mechanical performance has been investigated. Full modules with different ECAs were built (1a, 1b, 1c, 2 and 3b). For ECAs 1a and 1b, modules were built using two different string orientations (Figure 3).

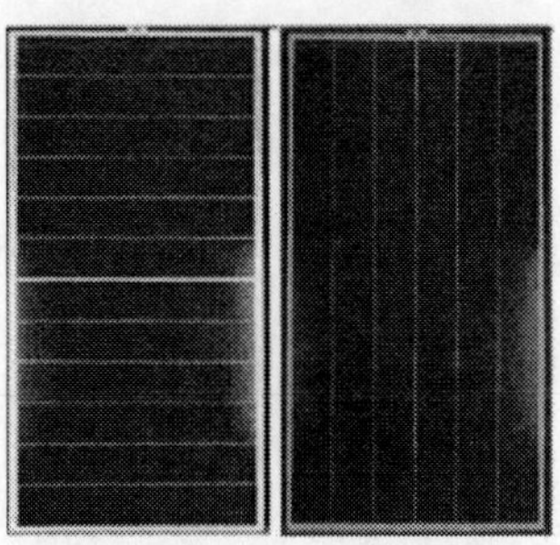

Figure 3: String orientation and layout for the 12 one-meter strings (left) and 6 two-meter strings (right) modules.

After measuring the initial IV and EL performance, each module was subjected to five 1 h cycles at 2400 Pa, IV and EL measured before a final 1 h cycle at 5400 Pa. The final IV and EL performance was also measured. The deflection (displacement) at the center of the modules was similar for all ECAs and string designs (averaging 31,8 mm at 2400 Pa, and 42,3 mm at 5400 Pa), so it can be assumed that the amount of stress to which the ECA is subjected is of the same magnitude in all cases.

The power drop after the static mechanical load tests (SMLT, Figure 4) shows that ECA 3b is performing worse than the rest, with a power drop of 4%. This can be attributed to the larger storage modulus at room temperature, which makes it too stiff to accommodate the stresses generated by the weight during the test. EL pictures reveal around 100 branched microcracks (Figure 5).

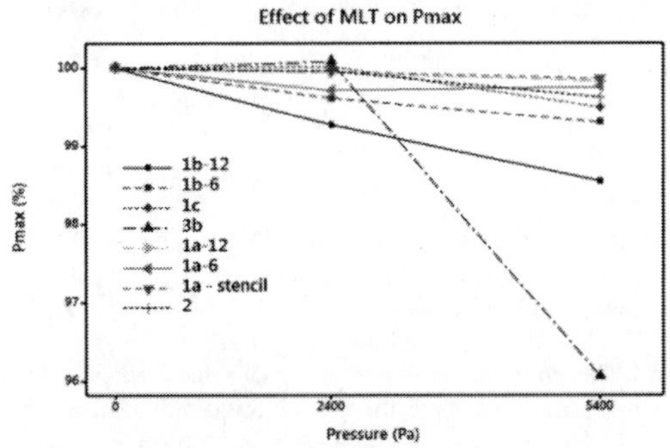

Figure 4: Power drop after SMLT tests of full modules.

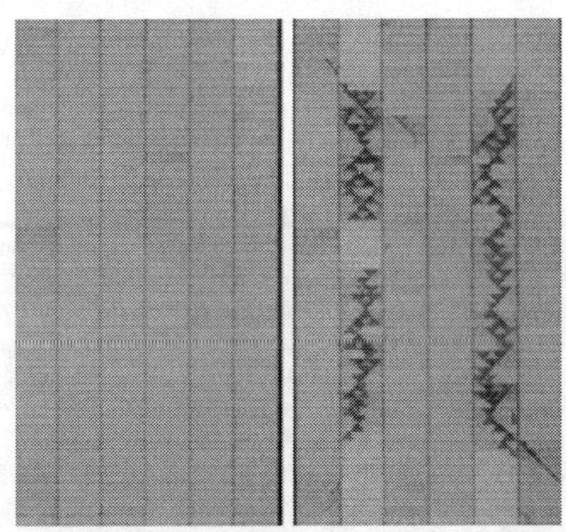

Figure 5: EL pictures of the modules built with ECA 3b, before (left) and after (right) SMLT.

The remaining ECAs perform similarly, with a power drop of 1% or better. There seems to be little or no effect of the string orientation for both 1a and 1b.

4.3 Effect of backsheet on TC performance

In the previous two sections, it has been described that different ECA parameters have an impact on the module electrical and mechanical performance. Indeed, due to the nature of shingling technology, the ECA plays a key role in reliability, but it is not the only modifiable parameter for improved module performance.

In order to prove this, two sets of mini-modules were assembled with an ECA very similar to 1b (optimized for dispensing, 3 mg per shingle in this case), using two different commercial backsheets available from the market, namely A and B.

The assembled mini-modules were subjected to TC aging under bias at -40/85°C and the resulting IV performance is presented in Figure 6. It can be seen that the minimodules assembled with backsheet B show a more significant degradation.

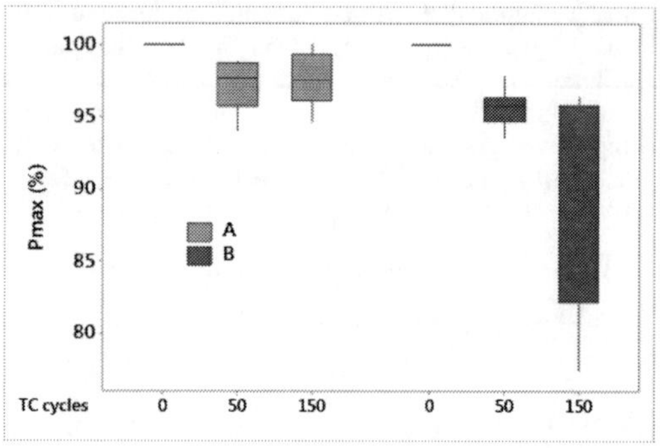

Figure 6: IV performance upon TC aging under bias at -40/85°C of the mini-modules assembled with two different commercial backsheets.

This could be attributed to the fact that backsheet B is causing more stress to the ECA-shingle bond during the cool down after lamination and during the cycling. When this stressed bond is subjected to thermocycling, it is more prone to fail electrically than a less stressed system.

5 CONCLUSIONS

ECAs for solar shingling can be based on different chemistry platforms, including acrylates, silicones and epoxies, and their performance depends on the material properties, the design of the cell assembly, and the bill of materials in the module. In terms of TC performance, this work suggests that the ECA usage appears to play a more critical role than the contact resistivity. Regarding the static mechanical load test performance, the storage modulus of the ECA at room temperature is the main driver of stability, as stiffer materials result in less capable performance. This is related to the ability of more flexible ECAs to accommodate the stresses that are built up during module manufacturing. String orientation has a more limited impact for the tested materials.

Lastly, ECAs can contribute to improved TC performance, but only to a certain extent. For example, backsheet selection can have a big impact on TC stability.

6 REFERENCES

[1] https://us.sunpower.com/solar-panels-technology/p-series-solar-panels/

[2] http://www.solaria.com/rooftops-utility/

[3] D. Tonini, G. Cellere, M. Bertazzo, A. Fecchio, L. Cerasti, and M. Galiazzo, "Shingling Technology for Cell Interconnection: Technological Aspects and Process Integration", Proceedings of the 33rd EUPVSEC, Amsterdam, The Netherlands, 2017, p.38-41.

[4] H. H. Berger, "Contact Resistance and Contact Resistivity" J. Electrochem. Soc. 1972, 119, 507-514.

PROGRESS IN ENCAPSULANT-INTEGRATED MULTI-WIRE INTERCONNECTION

Jonathan Govaerts[1], Tom Borgers[1], Philippe Nivelle[1,2], Rik Van Dyck[1,3], Ibrahim El-Chami[1,3,4], Ibrahim Issa[1,3], Tom Hoogewijs[1,3], Arvid van der Heide[1], Eszter Voroshazi[1], Jozef Szlufcik[1], Jef Poortmans[1,2,3]

[1]imec-Energyville, Thorpark 8320, 3600 Genk, Belgium
[2]UHasselt, Hasselt, Belgium
[3]KULeuven-ELECTA, Kasteelpark Arenberg 10, 3001 Leuven, Belgium
[4]Simon Fraser University Mechatronics Systems Engineering, Surrey, BC, Canada

ABSTRACT: In previous contributions, we presented already several multi-wire interconnection approaches, both for back-contact [1,2] and 2-side contacted cells [3]. In this paper we want to report our progress in further investigating and developing the latter approach.
We introduce our methodology to allow mapping of individual solder joint quality (that can be extended quantitatively) on commercially available solar cells. Applying the technology on 1-cell samples, we compare different encapsulants, lamination times and stitch patterns, and demonstrate a 2x2-cell module. We subject the fabricated samples to thermal cycling to get preliminary feedback on their performance reliability-wise, and to improve our understanding of the underlying mechanisms. Finally, we made a 9-cell module with busbarless cells and adapted metallization design, to show the full module-level interconnection potential of this technology.

1 DEADLINES AND DELIVERY

In the field of c-Si PV interconnection, a growing interest towards multi-wire approaches for 2-side contacted bifacial and busbarless cells is materializing [1,2,3,4]. This trend is in line with the predictions put forward in the ITRPV roadmap [5], and is attracting attention due to the potential it holds for a significantly improved trade-off between optical (finger shading) and electrical (resistive transport) losses, while the potential reduction in cell metallization is promising from a cost perspective. These benefits are valid on the cell-level, but also materialize into the module-level performance.
In this context, 2 such technologies are already in an advanced stage of development or even commercially available [1,2]. Though they are both implementing multi-wire interconnection, they are at the same time very distinct in technology:

- One approach effectively mimics the standard technology by soldering wires on finger solder pads, replacing the busbar. As in standard production, this step is then followed by a separate encapsulation process [1].
- Another approach applies a contact foil incorporating the interconnect wires directly onto the metallized cell followed by a lamination process [2] in which the wires of the contact foil are soldered directly to the metal fingers of the cell.

As a next-generation development, we propose an approach in which we immediately incorporate the wires into the encapsulant material for lamination. More details about the state-of-the-art and our approach are published in [3,4]. In this paper we want to elaborate on the progress we made in developing and characterizing this technology.

2 SAMPLE FABRICATION FOR DETAILED CHARACTERIZATION

At present, we fabricate the interconnect/encapsulant foils through manual stitching of encapsulant foils with wires. For the cells we use commercial bifacial cells with 3 busbars and a screenprinted front- and backside metallization. This adds to the industrial relevance of the results, but the presence of overdimensioned fingers and busbars at the same time makes it difficult to determine the quality of individual solder joints as the collected current will be transported to a neighbouring wire with a low-resistive contact, thus blurring the results. To resolve this, we interrupted the fingers between where the wires will end up after lamination through laser scribing, as illustrated in Figure 1.

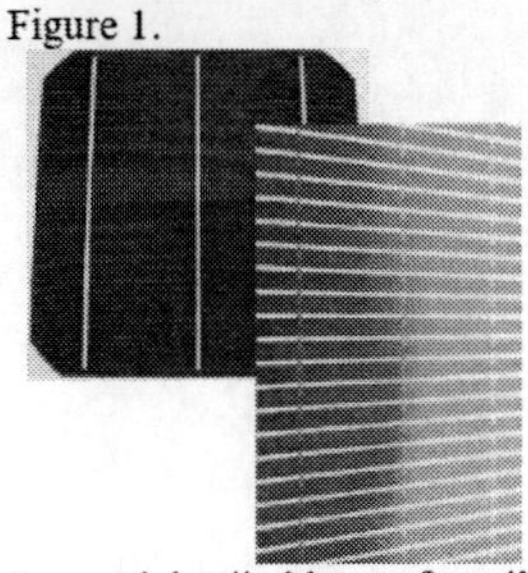

Figure 1: picture and detailed inset of a cell laser grooved at the rearside

This way, very little current will be collected from an area with a finger that is not contacted by the wire and will show up as a dark "pixel" in electroluminescence (EL) imaging, as illustrated in Figure 2. Thus it also allows to monitor degradation of the individual contacts, by checking the increase of dark pixels, as illustrated in Figure 3.

Figure 2: EL imaging of a multi-wire laminated 1-cell sample without (left) and with (right) laser grooving indicating a method of visualization of the presence or absence of individual wire-finger solder contacts

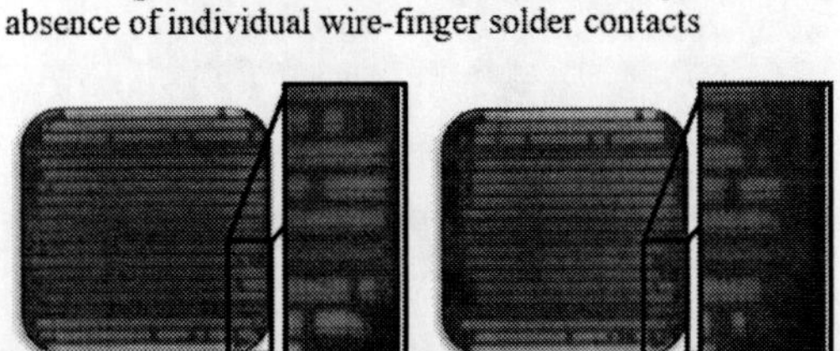

Figure 3: degradation of the individual contacts can be monitored throughout reliability testing: e.g. before (left)

and after (right) thermal cycling), including zoomed area

Obviously, we want to minimize any damage from the laser scribing process. Since we use bifacial cells, we have the freedom to choose between scribing either the front- or the backside fingers. Considering the used n-type cells with a diffused p-type emitter on the frontside, lasering on the front will more likely damage the emitter and induce more performance loss (increased shunting) than lasering the rear. This is confirmed experimentally after characterization with photoluminescence (PL) in Figures 4 and 5.

Figure 4: standalone cell before (left) and after (middle) front side laser grooving with detailed view of the grooved fingers (right)

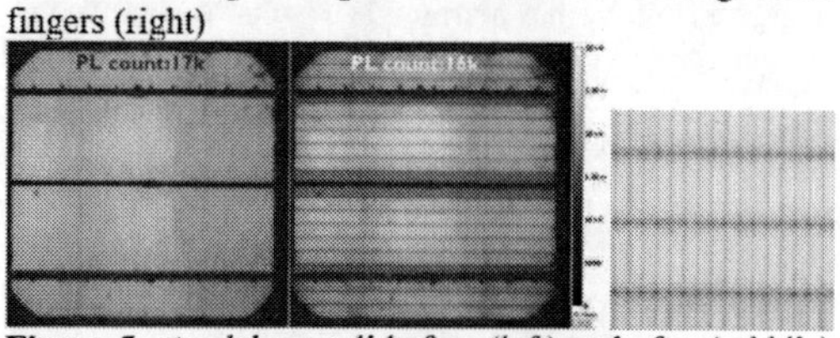

Figure 5: standalone cell before (left) and after (middle) rear side laser grooving with detailed view of the grooved fingers (right)

The average PL count drops from 19k to 11k for the front side grooving, while it only decreases from 17k to 16k for the rear side grooving. The busbar areas have a reduced intensity, which suggests they are not floating and so recombination is higher in those "stripes" due to a higher contact fraction.
Alternatively, the same principle can be used by changing the finger design towards interrupted fingers instead of laser grooving the cells, but this limits then the technique to dedicated samples.
As a side remark, such an approach with interrupted fingers can also be implemented to further improve current and voltage by reduced finger shading (no longer full lines), and contact fraction (increased passivation). On the downside, it obviously results in a reduced redundancy and alignment tolerance.

3 SAMPLE MATRIX AND PRELIMINARY RESULTS

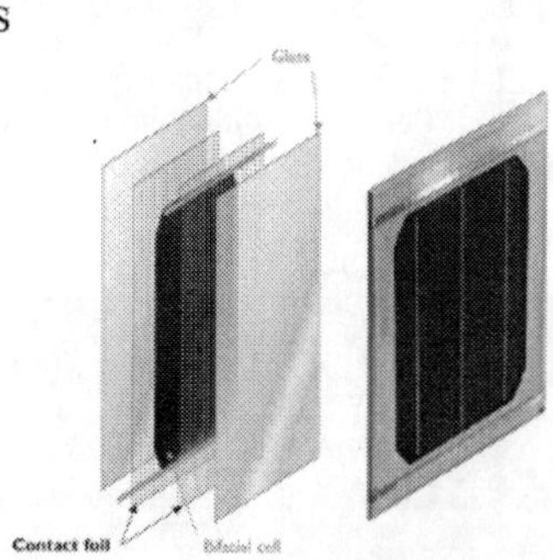

Figure 6: buildup and result after lamination of the fabricated samples

In the presented work, we wanted to compare the impact of different encapsulant materials, all non-crosslinking polyolefins, lamination times and stitching patterns. For all samples we used the aforementioned cells with grooved rear fingers, a lamination set temperature of 170°C and a lamination pressure of 1000 mbar. Figure 6 illustrates the buildup and the final result after lamination. Finally, the fabricated samples are characterized by EL, IV and optical inspection.
Electroluminesce (EL) imaging, together with the fill factor (FF) values determined by IV measurements, for the resulting samples is shown in Figure 7. Knowing that the bottom row is linked to one specific encapsulant material, it is very clear that the encapsulant can play, depending on the used conditions, a significant role.

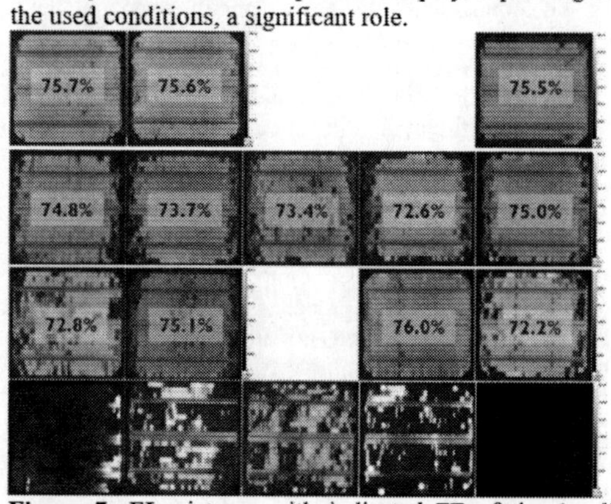

Figure 7: EL pictures with indicated FF of the 1-cell samples with different encapsulants, lamination times and stitching patterns

In [6] we deepen our understanding of the results for this encapsulant, in addition to a more detailed analysis of the solder joints. Although the variation in performance is quite large, the results indicate the promising potential of the technology. Currently we attribute the significant variation to significant non-reproducibility in the manufacturing of the foils based on manual stitching, as well as non-uniformities in the lamination process in terms of pressure and temperature (cf. later). This is still under investigation.

4 4-CELL PROOF-OF-CONCEPT MODULE

While we have published results of a first proof-of-concept 4-cell module [4], this was based on monofacial cells with a full Al BSF at the backside. To demonstrate the technology with more industry-relevant materials, we applied it using also bifacial cells. The available cell metallization (parallel fingers on front and back) still necessitate the use of bussing ribbons to interconnect the 2 strings. The resulting performance is shown in Figure 8, together with an optical impression of the module (front and backside) and the EL image, and indicates the potential of the technology, using 20 wires of 200 µm diameter and a SnBi-based solder coating, identical to the 1-cell samples discussed higher.

*	Pmpp [W]	Isc [A]	Voc [V]	FF [%]
Cells	4.629	9.312 ± 0.036	0.642 ± 0.003	77.4 ± 0.5
Module**	17.524 (4x 4.381)	9.330	2.522 (4x 0.631)	74.5
CtM	94.6	100.2	98.2	96.3

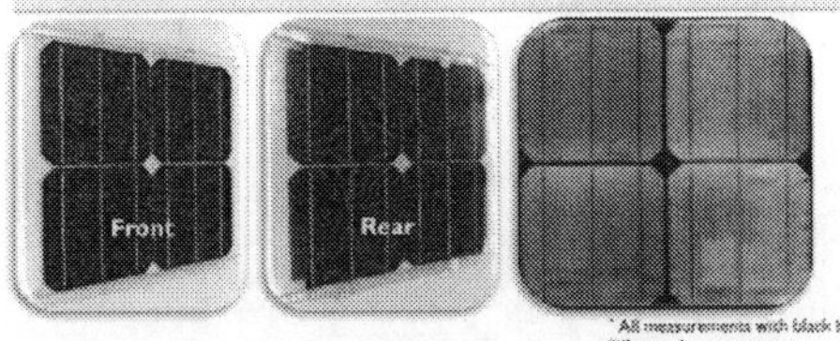

* All measurements with black background
** External measurements

Figure 8: IV-performance (top) and optical (bottom left) and EL (bottom right) images of 4-cell laminates

5 RELIABILITY TESTING

Of course critical for a new interconnection technology is reliability testing, in particular thermal cycling as that induces significant thermomechanical stresses due to mismatch in CTEs between the different materials involved (glass, encapsulant, metal interconnect wires, solder joints, cell metallization and Si wafers). The fabricated samples were subjected to 4h-cycles between 85 and -40°C, resulting in degradation, typically as shown in Figure 9. The EL mapping indicates the degradation is mainly coming from the glass edges, also for the 4-cell module.

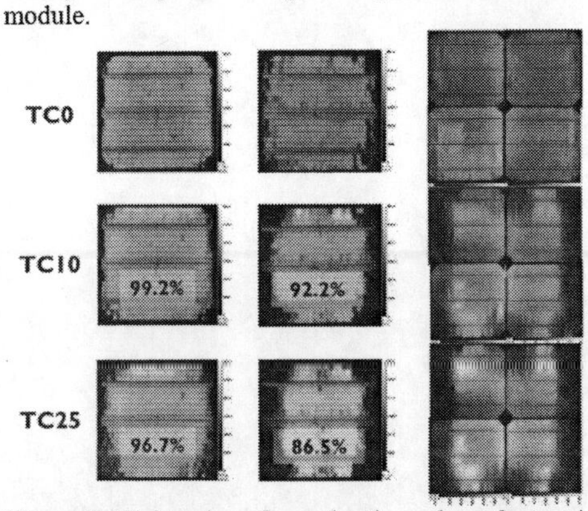

Figure 9: EL imaging of samples throughout the test (with relative FF degradation indicated for the 1-cell samples) at 0, 10 and 25 thermal cycles

At this point, the grooving of the fingers obviously imposes a worst case scenario for the testing, as no redistribution of current is allowed at all in this case. But even so, current samples are clearly very unlikely to pass the standard IEC test criterium of 200 cycles and further improvement is needed. Additional inspection indicates excessive solder reflow during lamination, resulting in local accumulation of solder. Based on this, we naturally expect non-uniform soldering behaviour.

Currently we are investigating the cause of these observations with 2 running hypotheses:

- The aforementioned manual stitching induces mechanical deformation to the wires, yielding non-uniformities (and non-reproducibilities) during lamination and thermal cycling
- During the lamination, a non-uniform pressure and temperature distribution likewise leads to non-uniform solder joint quality across the cell area. Figure 10 schematically illustrates how such non-uniformities may be introduced in the lamination process.

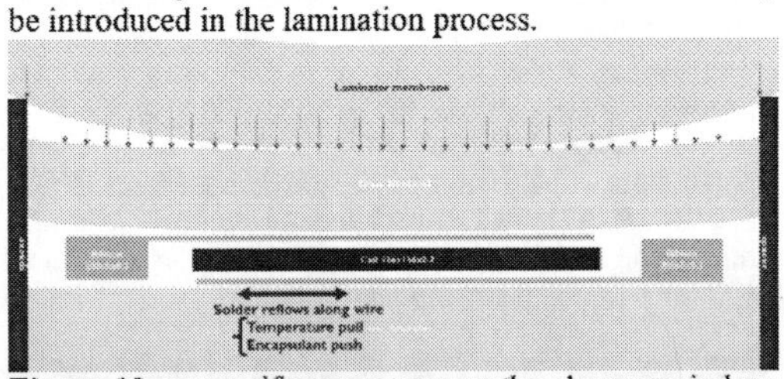

Figure 10: non-uniform pressure on the glass may induce non-uniformities in pressure and temperature distribution during solder reflow

The complexity of the combined soldering and lamination process entails solder reflow subjected to a pull from temperature gradients and a push from encapsulant pressure, and might be affected by process as well as material properties and interconnect foil preparation. As work in progress, we aim to investigate this with future samples.

6 9-CELL MODULES

While the above-mentioned results are based on a preliminary process optimization for the simultaneous soldering and lamination process, we implement the findings and knowledge from the more detailed analysis study of the solder joints [6] in the fabrication of a series of new 9-cell modules. Similar contact foils are used, but the foil layout is optimized for the new cell metallization layout.

A single cell laminate was made with the same process parameters as used in the 9-cell modules, however with thin 0.7 mm glass to enable and facilitate later cross-section preparation for solder joint evaluation [6]. Cross-sections in Figure 11 at different locations on the laminate demonstrate clearly the accumulation of solder between the wire and the cell finger and a good solder joint formation. However at the sides of the cell, less solder was accumulated, which is probably attributed to non-uniform pressure distribution during the lamination process, caused by the thin glass, resulting in solder flowing along the wires.

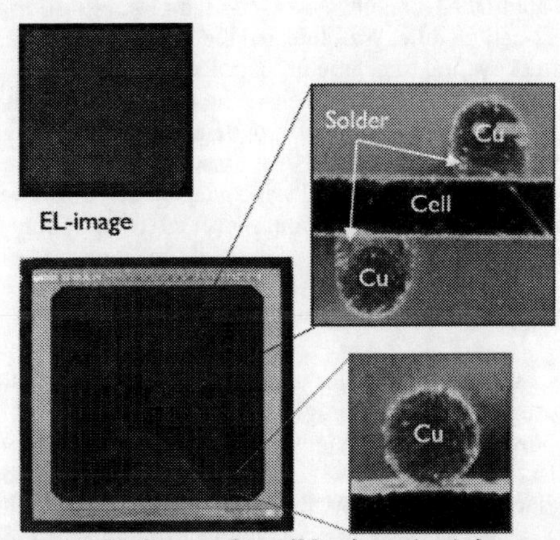

Figure 11: EL-image of 1-cell laminate (top), image of 1-cell laminate (bottom), cross-section images (left)

While the 4-cell proof of concept was based on bifacial cells with busbars, the 9-cell modules are based on industrial cells without busbars, having parallel fingers on front and diagonal fingers on backside. Hence no cell string bussing is needed as the contact foils can be used to interconnect cells strings by aligning perpendicularly to the cell string. Together with the 9-cell module with contact foils with 20 wires of 200 µm diameter and a SnBi-based solder coating, an additional module is made for comparison using the same cell type, but with a 5-busbar print for traditional tabbing and stringing.

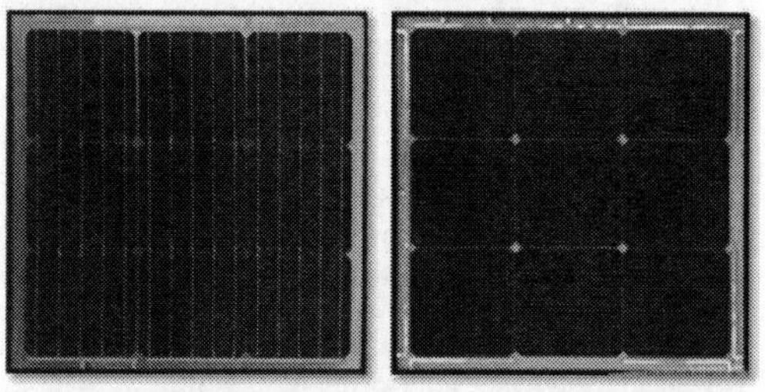

	I_{sc} [A]	V_{oc} [V]	FF (%)	P_{mpp} [W]
5BB	9.51±0.05	5.95±0.01	77.6±0.2	43.94±0.27
Twill	9.52	5.84	76.7	42.57

Figure 12: 9-cell module with 5-busbar cells (left), 9-cell module with multi-wire technology (right), table with module performance (bottom)

The total cell-to-cell copper interconnection cross-section is 0,63 mm² for the contact foil, and 0,99 mm² for the 5-busbar module. The performance of both modules is shown in Figure 14, together with an optical impression and the EL image. The module performance clearly indicates the potential of the multi-wire technology.

7 CONCLUSION AND OUTLOOK

In this paper we report on our progress made in developing our encapsulant-integrated multi-wire approach for 2-side contacted solar cells. We introduce our methodology to allow mapping of individual solder joint quality (that can be extended quantitatively) on commercially available solar cells.
We apply the technology on 1-cell samples, as well as on a 2x2-cell module. We subjected the fabricated samples to thermal cycling to get preliminary feedback on their performance reliability-wise. As the technology is still underperforming, we indicate our current understanding and our outlook to tackle these. Finally, we made a 9-cell module with busbarless cells and adapted metallization design, to show the full module-level interconnection potential of this technology.

ACKNOWLEDGMENT

The authors gratefully acknowledge imec's SiPV industrial affiliation programme and its partners, the support of the flemish government through the funded project SolsThore (IWT.155025), Eliosys for module measurements and DSM for supplying AR-coated glass. This project has also received funding from imec.ICON in the TWILL-BIPV project.

REFERENCE

[1] S. Braun et al., SiliconPV, 2013
[2] T. Söderström et al., EUPVSEC, 2013
[3] T. Borgers et al., IEEE PVSC, 2016
[4] T. Borgers et al., EUPVSEC, 2017
[5] ITRPV roadmap, 2016
[6] R. Van Dyck et al., EUPVSEC, 2018 (accepted)

GLUED SOLAR CELLS - A SOPHISTICATED TECHNOLOGY FOR PV MODULES

W. Mühleisen[1], L. Neumaier[1], J. Scheurer[2], B. Stoesser[2], W. Pranger[3], A. Schuetz[3],
F. Vollmaier[4], T. Fischer[5], R. Lorenz[5], M. Schwark[6], R. Ebner[6] ,C. Hirschl[1]

[1] CTR Carinthian Tech Research AG, Europastraße 12, 9524 Villach, Austria
Cor. Author: wolfgang.muehleisen@ctr.at, T: +43 4242 56300-238 (lukas.neumaier@ctr.at, christina.hirschl@ctr.at)
[2] Polytec PT GmbH, Ettlinger Str. 30, 76307 Karlsbad, Germany, (j.scheurer@polytec-pt.de, b.stoesser@polytec-pt.de)
[3] Ulbrich of Austria GmbH, Industriestr.1, 7052 Müllendorf, Austria, (wpranger@ulbrich.com)
[4] PVP Photovoltaik GmbH, Wernersdorf 111, 8551 Wies, Austria, (franz.vollmaier@kioto.com)
[5] Teamtechnik GmbH, Planckstraße 40, D-71691 Freiberg/Neckar, Germany (Robert.Lorenz@teamtechnik.com)
[6] AIT Austrian Institute of Technology GmbH, 1210 Vienna,Austria, (Michael.Schwark@ait.ac.at, Rita.Ebner@ait.ac.at)

ABSTRACT: Crystalline silicon PV modules are normally manufactured with lead-coated interconnectors in a soldering process. Due to the significant improvement of electronic conductive adhesives (ECA) in the last years, the interconnection of silicon solar cells by gluing has become possible as well. This young technology could help fulfil the European directive RoHS without exceptions and avoid lead in interconnectors and components in PV modules. Furthermore, the gluing concept is ideal for high efficiency solar cell technologies that cannot deal with high soldering temperatures. Scientific investigations of standard soldered solar cells and glued ones were done at laboratory and outdoor test platforms. Reliability and economic issues were taken into account. Finally, a 5 kWp PV plant consisting of 20 panels was manufactured and built for demonstration purposes.

Keywords: PV Module, Interconnection, Electronic Conductive Adhesive (ECA), Light Capturing Ribbons (LCR), Glued Strings, Lead-free PV Modules

1 INTRODUCTION

The standard procedure in 2018 to connect crystalline silicon solar cells in series is by using leaded ribbons with flux agent and a soldering process at temperatures around 220°C. Modern single track cell stringers can achieve 2100 cells per hour, which compares to 35 standard modules with 60 cells. In the last years, most electronic based industries have needed to comply with RoHS (Restriction of Hazardous Substances) regulations [1]. Due to many reasons this has not yet passed over towards the solar manufacturing industry, but it is a question of time until regulations will come. Lead-free ribbons are already on the market but they pose significant challenges especially for the solar industry since higher soldering temperatures result in more thermal stress and a higher risk of cell breakage for tin tin/silver alloys. Another disadvantage of lead-free soldering ribbons is the higher price.

An alternative to lead-free soldered ribbons are lead-free glued ribbons. Many research papers show that this concept could be an interesting alternative [2-8]. In addition, glued ribbons offer the advantage of possible usage of structured ribbons that compensate reflection losses. Besides this advantage, lower process temperatures can be used and high efficiency cell technologies would be workable. The electrically conductive adhesive (ECA) is based on a low silver particle formulation to have conductivity at moderate cost. Cell stringers compatible with the novel gluing technology are available on the market.

1.1 Current state of technology: Soldering

The interconnection of silicon solar cells in series is commonly done in several steps. First, a robot brings the cells in position (cell handling). Second, a dosed flux is applied in the area of the future contact to support the soldering and connection process. Third, the ribbons are placed in position and held in contact to the respective cells using a hold-down device. Fourth, a heating step over a time of a few seconds interconnects cell for cell to a string. After ten cells have been strung in series, a cutting unit separates the string and a robot picks up the string and places it on a prepared glass with embedding material, mostly ethylene vinyl acetate (EVA) foil. The operating temperature for the soldering process, mostly done by a non-contact infrared (IR) light unit, depends on the soldering material. A common standard soldering material is a lead-tin alloy (Sn60Pb40) with a melting temperature in the range of 183-188 °C. Lead-free tin-silver alloy materials like Sn96.5Ag3.5 are in the range of 221 °C. Special low temperature bismuth-tin based alloys like Bi58Sn42 have a melting point at 138 °C. With these three coating alloys for a copper ribbon, a wide range of cell technologies can be processed, for example, standard screen printed silicon solar cells, passivated emitter rear cells (PERC) or heterojunction technology cells (HJT) [Figs. 1, 2]. When producing PV panels, an empirical breakage rate value is found in the range of 0.1-0.3 % for standard cells and standard ribbons.

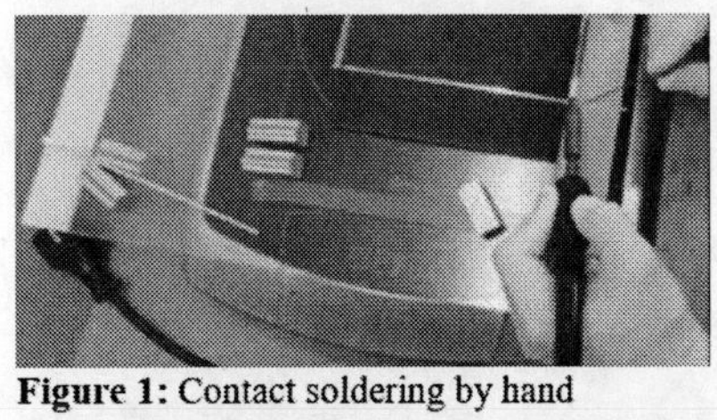

Figure 1: Contact soldering by hand

Figure 2: Contactless IR soldering by machine

1.2 Current state of law: Rules

In order to reduce the potential large amounts of toxic electronic waste, a restriction in the use of certain hazardous substances in electrical and electronic equipment was introduced more than 15 years ago. The Restriction of Hazardous Substances Directive 2011/65/EU (RoHS 2) restricts materials such as Lead (Pb), Mercury (Hg), Cadmium (Cd) and others – a total of ten toxic substances[1]. However, there are exceptions to this directive for some sectors that are still allowed to use, for example, leaded tin-solders in some applications for space, military, health and photovoltaic. The photovoltaic exception is stated as follows: "Photovoltaic panels intended to be used in a system that is designed, assembled and installed by professionals for permanent use at a defined location to produce energy from solar light for public, commercial, industrial and residential applications" [1]. But it could be a matter of time until this exception for leaded-tin solders or all lead containing components is also revoked for PV modules.

Figure 3: RoHS stickers to mark compliant products

1.3 Current state of technology: Gluing

Electronic conductive adhesives have been used for years especially for thin film PV modules. Gradual improvements in the formulation over the last years have opened new applications, especially the interconnection of silicon solar cells. The properties of the epoxy resin based adhesive are long pot life, no solvents and fast curing. Due to a progressive reduction of silver particles, a cost reduction was achieved at sufficient conductivity. Curing times of about five minutes at 150°C and a few seconds at higher temperatures are normal [Figs. 4, 5].

Figure 4: Manually applied ECA

Figure 5: Industrial screen printed ECA

2 MEASUREMENTS AND RESULTS

To analyze glued ribbons on silicon solar cells and modules, some indoor and outdoor tests were performed. The comparison of soldered standard flat ribbons, glued flat ribbons and glued structured ribbons (LCR) shows the potential of this novel environmentally friendly lead-free technology.

2.1 Indoor/ Outdoor Tests: Standard vs. Glued Ribbons for Mini Modules

The characterization of built single cells mini modules with the help of a Sinton Instruments FCT-350 cell flash tester was done with soldered and two types of glued ribbons. The more the Isc with help of structured glued ribbons, the higher the gain achieved compared to the standard soldered reference [Fig. 6]. It is worth mentioning that the Isc measurements are a relative comparison and the Isc values are higher in total. This is because of an elevated position to the light source in the cell flash tester due to the mini module dimensions and an adapted cell flash test system.

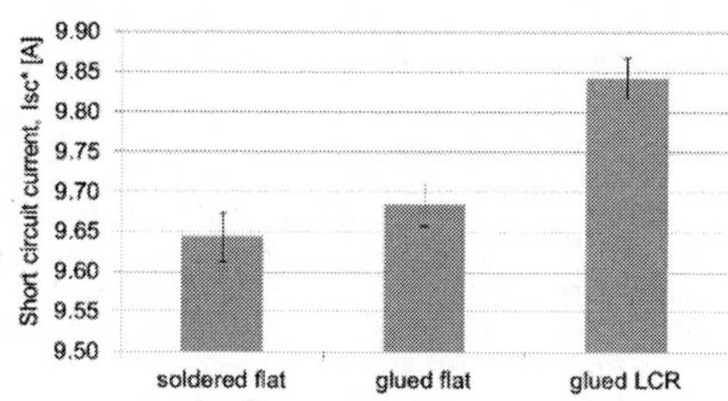

Figure 6: Relative gain in Isc between soldered flat and glued structured ribbons [2]

An averaged gain for Isc of 2 % (± 0.46 %) was measured for the glued structured ribbons compared to the soldered ones. Further measurements outdoor at natural sunlight and depending on the angle of light incidence were done and are shown in Figs.7, 8 [2].

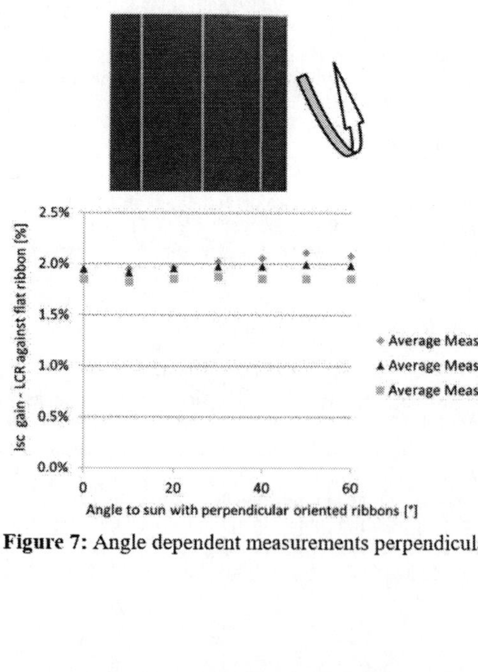

Figure 7: Angle dependent measurements perpendicular

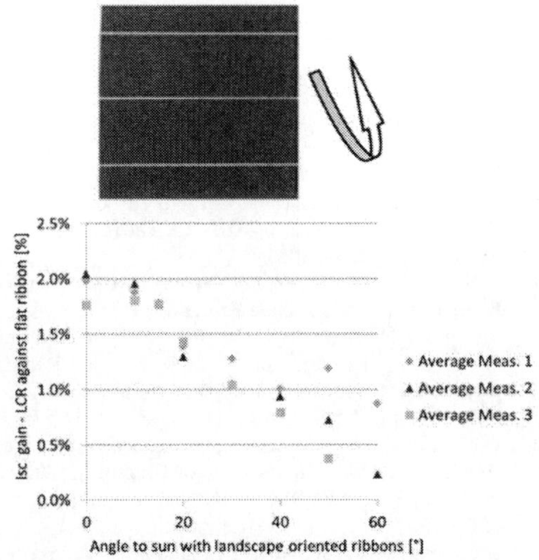

Figure 8: Angle dependent measurements landscape

Both indoor and outdoor data show similar results with regard to the gain of lead-free glued structured ribbons compared to soldered ones.

2.2 Indoor/ Outdoor Tests: Standard vs. Glued Ribbons for 60-cell PV Modules

Manufactured 60-cell modules with standard soldered ribbons as well as glued structured ribbons were power measured indoors and yield measured outdoors [Fig. 9]. It is worth mentioning that 60 cell PV modules are manufactured with production-related fluctuations. Therefore, 60 cell PV modules can only be compared for plausibility.

Figure 9: Outdoor yield measurements (left LCR, right standard soldered) [2]

The module flash test results showed a gain in power of about 1.75 % for glued to soldered variation. The yield measurements over the last four years showed seasonal yield gain fluctuations in the range of 0.3 – 2.3 %, but, on average, the higher gain is at 1.7 % for glued structured to soldered technology [Fig. 10]. After four years, these modules had to overcome the weather conditions in the "Cfb" climate zone with hot summer and snowy winter days.

So far, the repeating seasonal yield gain is an indication of good working PV modules with the novel technology.

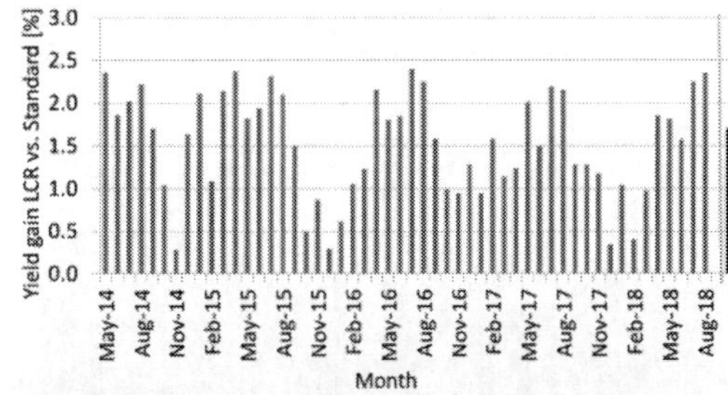

Figure 10: Relative yield gain for glued structured ribbon compared to standard soldered flat one

3 INDUSTRIAL MANUFACTURING AND TEST PLANT

Because of the promising results in creating a higher gain when using glued structured ribbons and the environmental idea of avoiding lead containing PV, 20 new PV modules were produced on an assembly line [Figs. 11, 12]. The city of Villach/Austria bought the lead-free and glued structured ribbon PV modules as flagship product and to promote this environmental friendly technology [Figs. 13, 14].

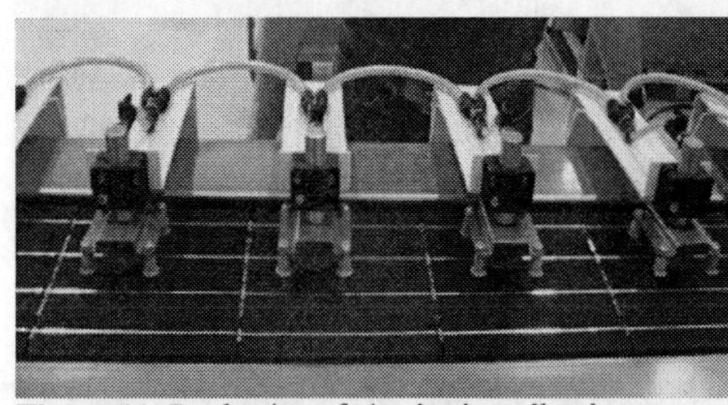

Figure 11: Production of glued solar cell strings

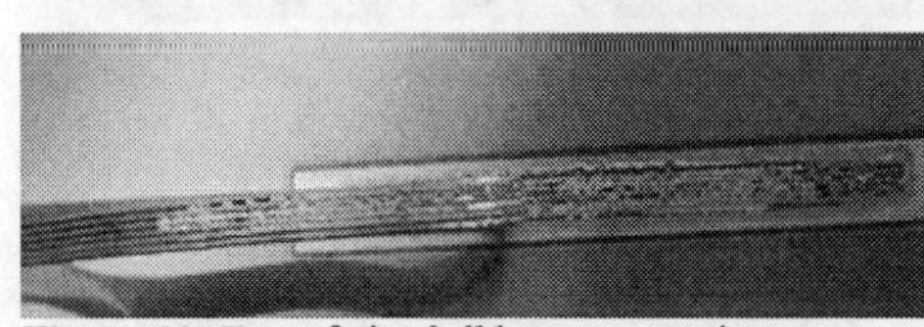

Figure 12: Test of glued ribbon as control step

Figure 13: Environmental friendly lead-free PV plant

Figure 14: Electroluminescence after one season

4 ECONOMICAL CONSIDERATIONS

When replacing environmentally toxic lead with silver, the costs could be higher than before. But there are some ideas to compensate for the additional costs due to higher amount of silver in a PV module. First, there will be a gain due to the structures in the ribbon that tolerates higher costs for the additional expenditure in manufacturing. Second, the conventional solar cells do not need a busbar because the electronic conductive adhesive can contact the fingers directly [Fig. 15]. The saved costs at the cell level can be transferred to the module level and reissued with the electronic conductive adhesive.

Figure 15: Changed cell concept for saving costs

5 SUMMARY AND CONCLUSION

A glued string of series connected solar cells is an alternative to a soldered one with the advantage of fulfilling RoHS requirements due to the lead-free concept. The usage of structured ribbons in combination with a gluing process results in a short circuit current gain of approximately 1.7 % compared to soldered flat ribbons. Two PV modules in comparison and over a time period of four years in real operating conditions showed no special occurrences. The municipality of Villach bought a commercial PV plant with the new technology to support the approach of the completely lead-free PV technology.

6 ACKNOWLEDGEMENTS

This work was conducted as part of the Austrian "Energy Research Program" project "INFINITY", using results from the predecessor project "InnoModu". Funded by the Austrian Climate and Energy Fund and the Austrian Research Promotion Agency (FFG), both are gratefully acknowledged.

7 REFERENCES

[1] Webpage of CE marking association https://www.cemarkingassociation.co.uk/restriction-of-the-use-of-certain-hazardous-substances-rohs/ , UK, 2018

[2] W. Mühleisen et al, "Comparison of output power for solar cells with standard and structured ribbons", EPJ Photovoltaics 7, 70701 (2016), DOI: 10.1051/epjpv/2016003

[3] J. Schneider, M. Turek, M. Dyrbal, I. Baumann, B. Koll, T. Booz, "Combined Effect of Light Harvesting Strings, Anti-Reflective Coating, Thin Glass and High UV Transmission Encapsulant to Reduce Optical Losses in Solar Modules", in Proc. 28th EU PVSEC, Paris, France, 2013, pp. 46–51

[4] G.C. Gläser, L. Hamann, L. Prönneke, "Structured Ribbons for Short Circuit Current Gain in Modules", in Proc. 25th EU PVSEC, Valencia, Spain, 2010, pp. 634–636

[5] E.M. Sachs, J. Serdy, A.M. Gabor, F.V. Mierlo, "Light-Capturing Interconnect Wire for 2% Module Power Gain", in Proc. 24th EU PVSEC, Hamburg, Germany, 2009, pp. 3222–3225

[6] Y. Zemen, M. Strasser, K. Turlapati, "Performance and Yield Evaluation of c-Si Modules Built with Reflective Interconnectors", in Proc. 29th EU PVSEC, Amsterdam, The Netherlands, 2014, pp. 3302–3304

[7] U. Eitner, T. Geipel, S.-N. Holtschke, M. Tranitz, "Characterization of Electrically Conductive Adhesives", Energy Procedia 27, 676 (2012)

[8] Md Zahidul Huq, "Interconnection of front side busbar-free c-Si solar cells using conductive adhesives", Master thesis, University of Freiburg, Germany, 2013, https://www.researchgate.net/publication/261510083
Interconnection of Front Side Busbar-Free c-Si Solar Cells Using Conductive Adhesives - Md Zahidul Huq

RESEARCH ON NEW MATERIALS FOR BUILDING INTEGRATED PHOTOVOLTAIC APPLICATIONS: AISOVOL PROJECT

O. González[1], A.B. Cueli[2], R. Castelo[1], P. Hernández[1], A. Álvarez[1], A. Linares[(1,3)], E. Llarena[1], C. Montes[1], D. Molina[1], A. Pío[1], L. Ocaña[1], C. Quinto[1], M. Friend[1] & M. Cendagorta[1]

[1]Instituto Tecnológico y de Energías Renovables, S. A. (ITER)
Pol. Industrial de Granadilla, s/n, E 38600 Granadilla de Abona, España
Tlf. +34 922 747 700, Fax +34 922 747 701
E-mail: iter@iter.es
[2]Centro Nacional de Energías Renovables (CENER)
Ciudad de la Innovación, nº 7, 31621 Sarriguren (Navarra), España
Tlf. + 34 948 25 28 00, Fax + 34 948 27 07 74
E-mail info@cener.com
[3]Agencia Insular de Energía de Tenerife (AIET)
Pol. Industrial de Granadilla, s/n, E 38600 Granadilla de Abona, España
Tlf. +34 922 747 700, Fax +34 922 747 701
E-mail: agenergia@agenergia.org

ABSTRACT: AiSoVol project consists on the development, manufacturing and testing of a new concept of photovoltaic modules, conceived for facilitating their use as architectural elements. Thus, diverse modules have been fabricated by encapsulating its constituent electrical elements with sandwich lamination techniques and using a transparent polycarbonate instead of tempered glass, as well as binding materials structurally strengthen by a grid made of high-tenacity yarns for avoiding the need for aluminum frames. As a backsheet, polycarbonate and polyester-based materials layers were tested. In this paper, the results of the visual inspection and the electrical characterization of the manufactured prototypes are shown. Various defects such as bubbles, lack of adhesion and buckling arose when using conventional lamination parameters. For this reason, a stepwise heating lamination process was implemented in order to cure the EVA, assure the integrity of the polycarbonate and so doing, avoid these defects. Two functionalities could be assigned to final module depending on its backsheet. Those manufactured with polycarbonate as frontsheet and backsheet, transparent and stiff, could work as active windows or skylights, whereas those using polycarbonate as frontsheet and polyester based materials as a backsheet, lighter but less resistant, would be screwed into any surface of a building.
Keywords: BIPV, encapsulation, polymer film, polycarbonate.

1 INTRODUCTION

One of the main demands of the photovoltaic sector is related to the availability of inexpensive solar panels that can be adapted for using as energy production elements in architectural projects. In that sense, replacement of conventional materials composing a standard solar panel for innovative elements to ensure that adaptation represents a significant advancement to this technology. To this end, AiSoVol project (photovoltaic generation solution as an alternative construction material) aims to develop a new concept of photovoltaic solar module, lighter and versatile, conceived for Building Integrated Photovoltaic (BIPV) applications.

The suitability of a wide range of materials for processes involved in photovoltaic module manufacturing was studied at earlier stages of the AiSoVol project [1], choosing as package constituent materials: a transparent thermoplastic layer, polycarbonate (PC), as superstrate instead as tempered glass; a commercial EVA (ethylene vinyl acetate) as encapsulant; polycrystalline silicon solar cells; a reinforcement grid made of polyester and HMPE (High Modulus Polyethylene) fibers embedded between two layers of polyester taffeta; and-transparent TPT (Tedlar-Polyester-Tedlar) as a backsheet.

At the present phase of the AiSoVol project, the focus has been put on manufacturing a set of prototypes based on mentioned study of materials that can be employed as benchmark in a later stage of commercialization. Thus, diverse modules have been fabricated maintaining the conventional lamination process with a view to adapt the equipment and systems of the Technological and Renewable Energy Institute, S.A. (ITER) factory and avoiding cost overruns, although adjusting the lamination parameters to the new materials. Besides, several characterization tests were performed at the facilities of the National Renewable Energy Center (CENER-CIEMAT Foundation), official accredited laboratory to perform tests according to standards UNE EN 12975, UNE EN 12976 and ISO 9806.

2 MATERIALS AND METHODS

Prototypes fabrication process was studied in two different stages. First one was focused on setting the prototypes configuration and lamination process parameters (time and temperature). To this end, a detailed visual inspection of the laminates obtained was carried out at the ITER factory to evaluate:

- Laminates appearance: lamination defects such as bubbles or wrinkles within the EVA and also adhesion between layers [1].
- Buckling degree: when laminating at high temperatures polycarbonate may bend around an axis [2].

Various configurations of prototypes were tested as shown in the sequence of the figure 1. Firstly, laminates were manufactured without cells and without reinforcement grid in order to adjust the lamination process to the polycarbonate and EVA features. Then, the reinforcement grid was introduced to verify its compatibility with the rest of the innovative materials. In the last step, solar cells were added.

It was employed polycarbonate sheets of two different thicknesses as frontsheet, namely 2 mm and 4 mm, and 2 mm thick polycarbonate or different layers of polyester-based materials -transparent TPT (Tedlar-Polyester-Tedlar) or standard white PYE- as backsheet. Regarding size, two different models were manufactured, both were 1 meter wide: one with 1.3 meters long and 48 solar cells and the other with 1.62 meters long and 60 solar cells.

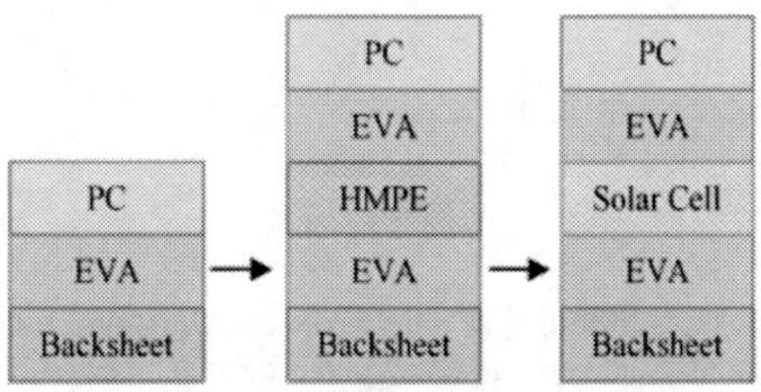

Figure 1: Sequence of prototype configurations

Altogether, over 100 prototypes were produced and the 10 most representative samples were selected for the study. Their layouts are reported in table I.

Table I: Layouts of the samples

Sample	Layout: Material	Layout: Thickness (mm)
1, 2, 3, 5	Polycarbonate	4
	EVA	0.5
	TPT	-
4	Polycarbonate	4
	EVA	0.5
	Reinforcement grid	-
	EVA	0.5
	TPT	-
6	Polycarbonate	4
	EVA	0.5
	60 solar cells	-
	EVA	0.5
	Polycarbonate	2
7	Polycarbonate	2
	EVA	0.5
	60 solar cells	-
	EVA	0.5
	Polycarbonate	2
8	Polycarbonate	4
	EVA	0.5
	60 solar cells	-
	EVA	0.5
	PYE	-
9, 10	Polycarbonate	4
	EVA	0.5
	48 solar cells	-
	EVA	0.5
	Polycarbonate	2

Regarding the lamination process, guidelines provided by EVA manufacturers are generally specific for conventional photovoltaic modules. There is hardly any information on laminating processes for innovative photovoltaic modules such as the ones involved in this project. Thus, laminating parameters, concretely time and temperature, must be reconsidered in order to find an accurate process because, in particular, difficulties arise when laminating polycarbonate. Table II compiles time, temperature and pressure in upper chamber used to laminate selected samples. Vacuum pressure is required in the lower chamber during the entire lamination process.

Table II: Parameters of the lamination process

Sample	Lamination process: Total time (min)	Time (min)	Temperature (°C)	Upper Pressure (mbar)
1	22.7	7.5	120	0
		15.2	120	1000
2	16.7	5.0	145	0
		11.7	145	1000
3	22.7	7.5	130	0
		15.2	130	1000
4	24.7	7.5	130	0
		17.2	130	1000
5	44.0	18.0	110	0
		13.5	120	1000
		6.5	130	1000
		6.0	145	1000
6, 7, 8	45.5	18.0	110	0
		13.5	120	1000
		6.5	130	1000
		7.5	145	1000
9, 10	43.0	10.0	110	0
		9.0	120	1000
		11.5	130	1000
		12.5	145	1000

Second stage of the research was focused on the characterization of prototypes containing solar cells, namely samples 6, 7, 8, 9 and 10. For this purpose, diverse laboratory tests were performed in CENER facilities:

- Electroluminescence imaging: technique with spatial resolution which allows the identification of various defects within the prototype such as broken cells, micro-racks or shunts.
- Electrical characterization: determination of the main electrical characteristics of samples in order to obtain its I-V characteristics curve and establish the output performance and the solar efficiency.

3 RESULTS AND DISCUSSION

3.1 Adjustment of the lamination process

Results of the visual inspection carried out in samples 1 to 8 to set lamination parameters are displayed in table III, where a numerical scale from 0 to 2 was assigned to quantify each of these defects: 0 means no defect, 1 means moderate defect and 2 means severe defect. Some examples of the worst defects appeared during the manufacturing process can be seen in figure 2.

Table III: Results of the visual inspection of the samples

Sample	Laminates appearance			Buckling
	Bubbles	Wrinkles	Adhesion	degree
1	0	2	2	0
2	2	1	0	2
3	1	1	1	1
4	1	1	2	1
5	0	1	1	1
6	0	0	0	2
7	0	0	0	0
8	0	0	0	1

Figure 2: Pictures exhibiting some of the most severe lamination defects: bubbles (top left), lack of adhesion (top right) and buckling (bottom).

3.1.1 Lamination process at constant temperature

Previous studies revealed that lamination of 16.5 x 16.5 cm samples of the same materials at constant temperature provided results close to those desired [1]. Under this premise, samples 1, 2 and 3 were laminated selecting a temperature of 120°, 145° and 130 °C, respectively.

It was found that lamination processes at 120 °C during 22.7 minutes (sample 1) resulted in laminates with a large amount of wrinkles of EVA and adhesion strength between the different layers did not fulfill minimum requirements, as can be noted in Table III. These factors could be caused by a low temperature, not enough for the EVA crosslinking reaction to take place. By contrast, a temperature of 145 °C and shorter time (sample 2) produced less wrinkles and a suitable adhesion, though the buckling degree of the sample was excessively high and a lot of bubbles appeared as a result of a too high temperature.

When choosing 130°C and same minutes as in 120 °C process (sample 3), number of bubbles and buckling degree decreased in comparison to 145°C, but a slightly lack of layer adhesion was noticed at the edges of the samples, so further adjustments on the lamination process were required.

3.1.2 Reinforcement grid

Efforts were made to introduce the reinforcement grid but results achieved to date demonstrated a absence of adhesion between the reinforcement grid and the TPT (sample 4). However, the reinforcement grid was not dismissed yet. Several attempts to improve this lack of adhesion were carried out, such as drilling the reinforcement grid in order to allow the EVA to flow through it, or finding another reinforcement grid with different features, but further researches need to be made within the lamination process in order to include the reinforcement grid in the prototypes' configuration, because of the stiffness it would provide to the final prototype.

3.1.3 Stepwise heating

To achieve better results, changes were introduced in the lamination process in order to reach 145 °C in a progressive way and remain so for a shorter time. This is meant to achieve a desirable degree of crosslinking within the EVA while avoiding both damage on the polycarbonate and buckling

This stepwise heating lamination process can be divided into four steps. The first and longest-lasting one takes place at 110 °C and allows the EVA to begin melting. In second and third step the temperature increases by 10 °C in each, allowing the EVA to melt completely and mitigating the impact on the polycarbonate. Finally, a temperature of 145 °C is reached to EVA crosslinking reaction to take place.

Sample 5 and subsequent were manufactured following this stepwise heating lamination process (see figure 3). Comparisons with samples laminated at constant temperature (samples 1, 2 and 3) led to believe that stepwise heating contribute to the general improvement in the lamination results.

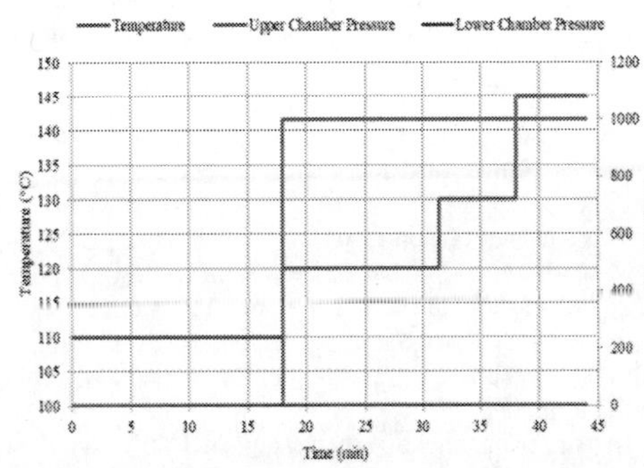

Figure 3: Stepwise heating lamination process of sample 5

In particular, time passed in 145°C was reduced by over 10 minutes in sample 5 when compared to sample 2. There were no visible bubbles and buckling degree was also lower (figure 4), though adhesion among layers slightly worsened. When laminates spent 7.5 min in 145°C (sample 6, 7 and 8), a suitable adhesion was accomplished but buckling differed depending on the sample configuration, as can be seen in table III. Aspects related to these variations will be discussed in the following paragraphs.

Figure 4: Absence of bubbles in sample 5 (above) and elimination of buckling in sample 7 (below).

3.1.4 Frontsheet comparison

Several tests were conducted employing polycarbonate of 2 mm and 4 mm of thickness as frontsheet, and polycarbonate of 2 mm of thickness as backsheet. Results found evidenced a significantly higher buckling degree in laminates with different thickness of polycarbonate as frontsheet and backsheet (see results of the visual inspection of sample 6 in Table IV) than in those using the same thickness of polycarbonate for both sheets (sample 7). This higher buckling could be partly attributed to the asymmetry of the laminate's layout, which produces a temperature gradient when samples cooling takes place.

Nevertheless, previous hail impact tests [3] indicated the inability of the 2 mm polycarbonate to work as frontsheet. Hence, further research is need on this subject.

3.1.5. Backsheet comparison

Regarding to the backsheet, samples using 2 mm thick polycarbonate and PYE or TPT were tested, appreciating a higher buckling degree in those laminated with PYE or TPT, presumably due the named asymmetry of the laminates.

In order to take advantage of both types of prototypes, different application could be assigned to each of them. On the one hand, there would be those samples manufactured with 2 mm thick polycarbonate as frontsheet and backsheet (see sample 7 in figure 5). Their usefulness lies in their transparency since natural light is able to go through the space between the cells of the module. These prototypes could be integrated directly into the structure of a building and work as an active window or skylight, providing both energy and light.

On the other hand, there would be those others laminated using 4 mm thick polycarbonate as frontsheet and TPT or PYE as backsheet (see sample 8 in figure 5). Their main advantage would be their lightness, but since they exhibit a slightly buckling degree, they could be screwed onto any kind of building panel, so that the buckling might be counteracted.

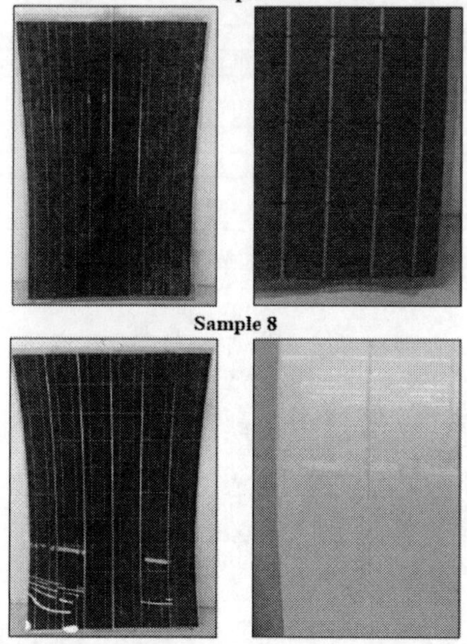

Figure 5: Pictures of sample 7 (above) and sample 8 (below). Frontside is shown on the right and enlarged backside on the left.

3.2 Samples characterization

A characterization of samples 6 to 10 was carried out selecting for this purpose those giving best lamination results.

3.2.1 Electroluminescence

Electroluminescence imaging test of all samples highlighted several defects such as broken cells and micro-racks, as can be seen in figure 6. Primarily, the widespread break of cells in sample 8 evinced the difficulty when manipulating and transporting samples with TPT or PYE as a backsheet, consequently a protocol for handing these types of prototypes should be implemented. Moreover, sample 7 presented fewer broken cells and micro-racks than sample 6. This could be caused by the higher buckling degree of samples employing different thickness of polycarbonate in the frontsheet and the backsheet.

Regarding size, laminates with 1.3 meters long and 48 solar cells (samples 9 and 10) displays fewer defects than those with 1.62 meters long and 60 solar cells (samples 6 and 7) ,especially in terms of broken cells. This may be because the smaller samples are easier to handle and also have less buckling degree, maintaining the integrity of the cells.

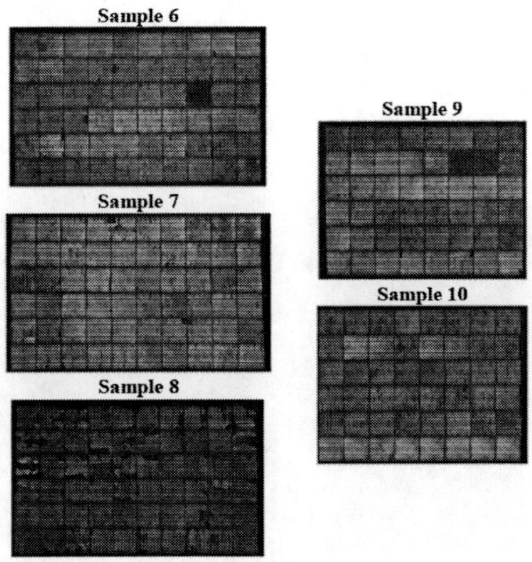

Figure 6: Electroluminescence imaging of samples 6 to 10.

3.2.2 I-V characteristics curve

Table IV displays the main electrical parameters of samples 6 to 10 extracted from their I-V characteristic curves (figure 7). As befits laminates with greater number of cells, samples 6, 7 and 8 provided higher power output than samples 9 and 10. The lower power output of sample 8 compared to samples 6 and 7 provides evidence of damage suffered by the solar cells of sample 8, as previously mentioned.

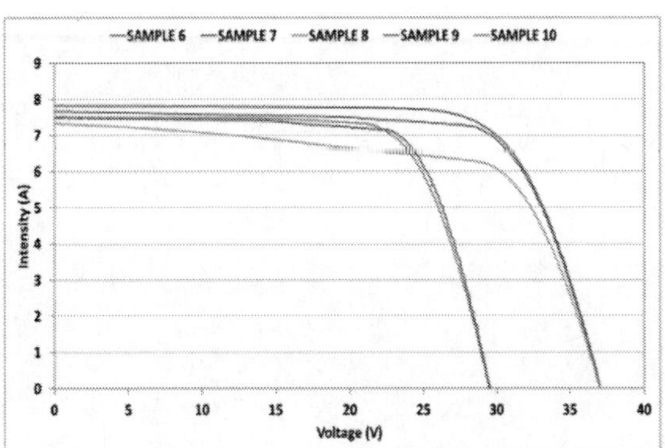

Figure 7: I-V Characteristics curve of samples 6 to 10.

Table IV: Electrical parameters of samples 6 to 10

Sample	I_{SC} (A)	V_{OC} (V)	V_{MP} (V)	I_{MP} (V)	P_{MAX} (W)	FF (%)
6	7.66	37.0	29.4	7.06	207.3	73.1
7	7.83	37.1	29.1	7.26	211.1	72.7
8	7.34	37.0	29.7	6.12	181.9	67.0
9	7.49	29.5	23.4	6.97	163.0	73.7
10	7.52	29.4	22.8	7.01	160.1	72.3

The theoretical power output for the 1.30 meter long and 48 solar cells samples is 190 W, while the theoretical power output for the 1.62 meter long and 60 solar cells samples is 240 W. Thus, the experimental power output of manufactured samples is close to their theoretical power output.

4 CONCLUSIONS

Conventional solar panel lamination process was adjust to fabricate the innovative prototypes involved in this research since several defects, such as bubbles, wrinkles, lack of adhesion and buckling, appeared when standard lamination guidelines were followed. To this effect, a stepwise heating method was developed and validated, which progressively melts the EVA in four steps at constant temperature (110-120-130-145 °C). This allows the EVA to melt and cure besides avoiding the buckling of the polycarbonate.

Two different types of prototypes were manufactured using a stepwise heating lamination method and satisfactory lamination results were obtained in all of them. The first type of prototypes was manufactured with polycarbonate layer as frontsheet and backsheet. Those consisting of 4 mm polycarbonate as frontsheet and 2 mm polycarbonate as backsheet were rejected since they showed a slight buckling degree. It was possible to eliminate that buckling when 2 mm polycarbonate was used as both frontsheet and backsheet, though these prototypes are not able to resist the hail impact test. The electroluminescence results were acceptable in both prototypes and the characteristic electrical parameters were close to the theoretical ones. This type of laminates is intended to be integrated into the building structure and work as an active window or skylight, providing both energy and light.

The second type consisted on a prototype with 4 mm polycarbonate as frontsheet and TPT or PYE as backsheet. The buckling degree was excessive although it could to be avoided by screwing the prototype onto a building panel. However, the electroluminescence imaging results revealed a generalized break of the cells, possibly due to the fragility of the laminate during its manipulation. Consequently, the characteristic electrical parameters were lower than those expected.

Subsequent researches will be focused on solving several issues, such as the resistance of the 2 mm polycarbonate to work as frontsheet, and finding a solution to include the reinforcement grid in the prototype.

5 ACKNOWLEDGEMENTS

AiSoVol project is funded by the Ministry of Economy and Competitiveness and co-financed by European FEDER Funds, through the State Programme for R&D&I dedicated to social challenges, specifically Challenge 3: Secure, Clean and Efficient Energy Challenge.

The project, which began on 1 October 2015 and will end on 30 September 2018, is a partnership between of the Technological Institute of Renewable Energies, SA (ITER) as coordinator and the National Renewable Energy Centre, CENER-CIEMAT Foundation.

6 REFERENCES

[1] A.B. Cueli, et al., Proceedings 32nd European Photovoltaic Solar Energy Conference (2016), 1BV.5.47.
[2] M.M. Hackmann et al., Solar Energy Materials & Solar Cells 84 (2004) 105–115
[3] R. García Navarro, Master's Thesis, Public University of Navarra (UPNA), 2017. https://hdl.handle.net/2454/25684

APPROACH FOR SIMULATING OUTPUTS OF PV MODULE/ARRAY OF DIFFERENT TECHNOLOGIES WITH HIGH ACCURACY

Sofiane Kichou and Petr Wolf
University Centre for Energy Efficient Buildings, Czech Technical University in Prague,
Třinecká 1024, 273 43 Buštěhrad, Czech Republic.
e-mails: sofiane.kichou@cvut.cz, petr.wolf@cvut.cz

ABSTRACT: Simulations of photovoltaic (PV) modules outputs are of primal importance for reproducing the real behaviour of the PV module/array under changing outdoor conditions. An accurate PV output estimation is crucial for many applications such as PV energy forecasting and fault detection in PV systems. The present work compares two different PV models based on numerical and metaheuristic algorithms used for estimating their unknown parameters. The unknown parameters of the PV models have been extracted from datasheet information, measured I-V curves and from daily profiles of the monitored outputs of the PV modules. The obtained results showed that the extraction of the models parameters from dynamics allows reproducing the real behaviour of the PV modules with root mean square error (RMSE) values less than 3%.
Keywords: Photovoltaic, one-diode model, double-diode model, parameter extraction.

1 INTRODUCTION

The accuracy of simulations depends on the models used for the main components of the photovoltaic (PV) system –especially on the PV module's model. Considering the one diode five parameters model, its precision is strongly dependent on the unknown parameters (series and shunt resistances, ideality factor, photocurrent and reverse saturation current) which are not provided in the PV module datasheet.

Several research works available in the literature discussed the topic of PV models parameter estimation by applying different methodologies (analytical, numerical, heuristic algorithms…etc). A method based only on datasheet information was presented in [1]. Another work presented a mathematical model developed to determine solar intensity and cell temperature dependency of PV module parameters and thermodynamic efficiency figures. A simple one-diode model is proposed there considering series resistance and shunt conductance [2]. Hybrid approaches combining numerical and analytical solutions based on minimal information from the module datasheet published in [3], [4] showed good accuracies.

The models unknown parameters can be estimated using data provided in the manufacturer's datasheet, measured I-V curve of module/array, or from dynamic behaviour considering daily profiles of PV module/array measured outputs together with measurement of solar irradiance and module temperature [5]. However, it is still fuzzy to conclude what is the best approach that can be used to simulate the output of PV module/array with high accuracy.

The aim of the paper is to suggest the best approach and input data to be used in order to reproduce the real comportment of PV module/array of different technologies under outdoor conditions. In addition, the present work provides a comparison in term of accuracy and complexity between the one diode and the double diode models considering two different PV technologies. The extraction of the models parameters is carried out using numerical and metaheuristic algorithms considering datasheet information, measured I-V curve and daily monitored outputs profiles of the PV module/array.

The remainder of the paper is organized as follow: the description of the experimental as well as the PV module technologies included in the study are given in Section 2. The followed methodology and the PV models used are presented in section 3. The obtained results are shown and discussed in Section 4. Finally, general conclusions are provided in Section 5.

2 EXPERIMENTAL SETUP AND DATA

The PV module technologies considered in this work correspond to polycrystalline (mc-Si) and cadmium telluride (CdTe). The PV modules are installed at the University Centre for Energy Efficient Buildings (UCEEB) situated in Buštěhrad, small city 20 km northwest of Prague (Czech Republic, Latitude: 50°09'24.2"N, Longitude 14°10'10.5"E). The mc-Si PV panel under study is mounted in the roof of UCEEB building facing the south with an azimuth angle of 35°. However, the considered CdTe PV array composed of three panels connected in parallel is installed in the south façade on an open rack with a tilt angle of 75° as it is shown in Fig. 1.

The nameplate parameters of the PV modules under standard test conditions (STC): G^{*}=1000W/m^2, T_c^{*}=25°C, are given in Table I.

Figure 1: PV module/string used in the study.

Table I: Main parameters at STC of the PV modules.

PV module	Holleis Lotus G2	Calyxo CX3-80
Technology	mc-Si	CdTe
P_m (W)	245	80
I_m (A)	8.19	1.72
V_m (V)	29.9	47
I_{sc} (A)	8.84	2.01
V_{oc} (V)	37.1	62.8
Temperature	P_m: -0.44 %/C	P_m: -0.25 %/C
Coefficients	V_{OC}: -0.30 %/C	V_{OC}: -0.24 %/C
	I_{SC}: +0.04 %/C	I_{SC}: +0.02 %/C

For simulations and the extraction of the models parameters in dynamics, real monitored profiles of irradiance, temperature and maximum power point (MPP) coordinates are need. Thus, the measured coordinates of the MPP of the PV module/string are provided by the optimizers every five minutes in the monitoring system based on SolarEdge web monitoring portal. The irradiance (G) data are measured at the same inclination of the PV modules through a Silicon Irradiance Sensors (Si-RS485TC-2T). The PV panel temperatures (T_c) are measured by a temperature sensor Pt1000 pasted to the backside of the PV modules.

However, for the extraction of the models parameters from a measured I-V curve, the I-V characteristics of the PV module/array are obtained through out an EuroLink Pro® I-V tracer.

3 METHODOLOGY

The aim of the present work is to clarify what is the best approach and input data to be used in order to reproduce the real comportment of PV module/array of the two different technologies with high accuracy.

First of all, after a good modelling and developing an efficient parameter extraction procedure. The chosen models as well as the developed parameter extraction technique are described in the following subsections.

In this study, three main approaches are considered:

- Case 1; the models parameters were extracted from the datasheet information of the PV panels considered in this study. Then, the extracted parameters are used in order to simulate the dynamic behaviours of the PV module/string under real weather conditions, taking into account daily profiles of G and T_c.
- Case 2; the extraction of the parameters of the models is carried out from a real measured I-V curves of the mc-Si module and the CdTe string. Then, the dynamic behaviour of the PV module/string is simulated the same way as in the case 1.
- Case 3; the extraction of the models parameters is carried out from dynamics. In this case, the daily monitored profiles of current, voltage, temperature and irradiance are used. Then, the extracted parameters from dynamic will be used by the PV models in order to simulate the outputs of the PV module/string based on measured profiles of G and T_c of another different day.

In order to conclude which is the best approach, the performance of the PV models in predicting the real behaviour of PV module/array under real weather conditions are evaluated considering the three different cases 1, 2 and 3.

3.1 PV modelling

The simulation of the PV module/array outputs has been carried out based on two well-known PV models one-diode model (ODM) and double-diode model (DDM) which offer good accuracy with different degree of complexity. The ODM of a solar cell, also known as five-parameter model includes a parallel combination of a photogenerated controlled current source I_{ph}, a diode, described by the single-exponential Shockley equation [6],[7], a shunt resistance R_{sh} and a series resistance R_s modelling the power losses as shown in Fig. 2(a). However, the DDM is similar to the ODM but consists of two diodes put in parallel instead of one as it could be seen in Fig. 2(b).

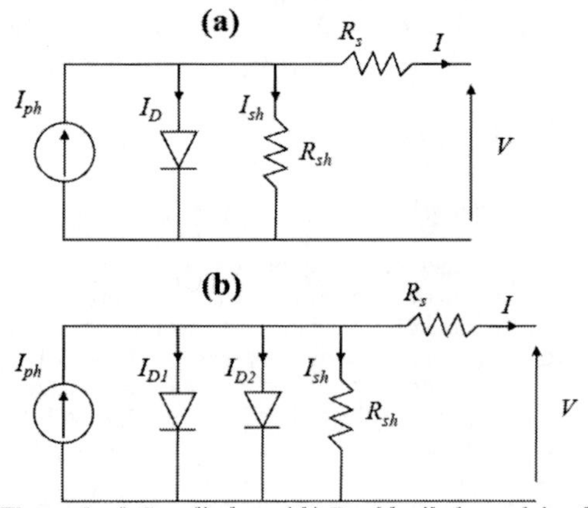

Figure 2: a) One diode and b) Double diode models of a PV cell.

The current-voltage (I-V) characteristic of a solar cell could be derived from Kirchhoff's current law and it is given by the implicit and nonlinear equations (1) and (2) for the ODM and DDM respectively as follows:

$$I = I_{ph} - I_o\left(e^{\left(\frac{V+R_sI}{nV_t}\right)} - 1\right) - \left(\frac{V+R_sI}{R_{sh}}\right) \quad (1)$$

$$I = I_{ph} - I_{o1}\left(e^{\left(\frac{V+R_sI}{n_1V_t}\right)} - 1\right) - I_{o2}\left(e^{\left(\frac{V+R_sI}{n_2V_t}\right)} - 1\right) - \left(\frac{V+R_sI}{R_{sh}}\right) \quad (2)$$

where I_o and n are the reverse saturation current and ideality factor of the diode respectively. V_t is the thermal voltage.

The scaling rules and equations to pass from a solar cell model to a PV module/array models are well detailed in [6] and [5].

3.2 Parameters extraction

To predict accurately the real outputs of a PV module/array operating under outdoor conditions, a good estimation of the model's unknown parameters is crucial. The parameter extraction procedure employed in the present work is the same presented before in [5]. However, in the present work only, one numerical algorithm based on Levenberg-Marquardt (LMA) and one metaheuristic algorithm based on artificial bee colony (ABC) have been used as optimization algorithms.

The parameters to be extracted are defined as $\theta = f(I_{ph}, I_o, n, R_s, \text{and } R_{sh})$ for the one diode model and $\theta = f(I_{ph}, I_{o1}, I_{o2}, n_1, n_2, R_s, \text{and } R_{sh})$ for the double diode model. Based on the chosen optimization algorithms, the parameter extraction procedure evaluates the unknown parameters (θ) of the two PV models described before in order to get the global minimum of the function $S(\theta)$ given bellow as:

$$S(\theta) = \sqrt{\frac{1}{N}\sum_{i=1}^{N}[Ii - I(Vi,\theta)]^2} \quad (3)$$

where, the N represent the number of measured data, Vi and Ii represent the input voltage and current of the data point i.

The extracted parameters based on the two algorithms (ABC and LMA) will be used for the simulation of the I-V characteristics of the PV module/string and for the reproduction of their dynamic behaviour based on real measured profiles of solar irradiance (G) and module temperature (T_c). Finally, the obtained results will be compared and evaluated through root mean square error (RMSE) indicator in order to see the effectiveness of the algorithms and the PV models compared to real measured data.

4 RESULTS AND DISCUSSIONS

I-V characteristics and dynamic behaviour of one mc-Si PV module and one string composed of three CdTe PV modules have been simulated in the present work. The results have been obtained by following the methodology presented in section before. The chosen PV cell models as well as the optimization algorithms were implemented on Matlab®/Simulink software based on the same equations previously published in [5].

For the evaluation of the models performance, I-V curve from datasheet and real measured one were used. The I-V curve from datasheet is reproduced based on the MPP coordinates as well as the current short-circuit and the voltage open-circuit values. However, for the evaluation of the PV models in dynamics, in order to avoid increasing the number of figures, only the output power of the PV module/string is shown.

4.1 Case 1

In this case, the ODM and DDM parameters were extracted from the I-V curve data at STC provided in the datasheet of the PV modules. The obtained results showing the extraction of the parameters from datasheet I-V curves using ABC algorithm are given in Figs. 3 and 4 for mc-Si and CdTe PV modules.

From Figs. 3 and 4, it could be seen that both PV models based on ABC algorithm provide high accuracy in reproducing the datasheet I-V curve of the PV panels considered in this study. The extracted parameters are summarized in Table II and III. The calculated RMSE vales are provided in Table IV.

From Table IV, it could be seen that the RMSE_datasheet values does not exceeds 0.5%. Which means that there is a good agreement between simulated I-V and real I-V curve provided in datasheets.

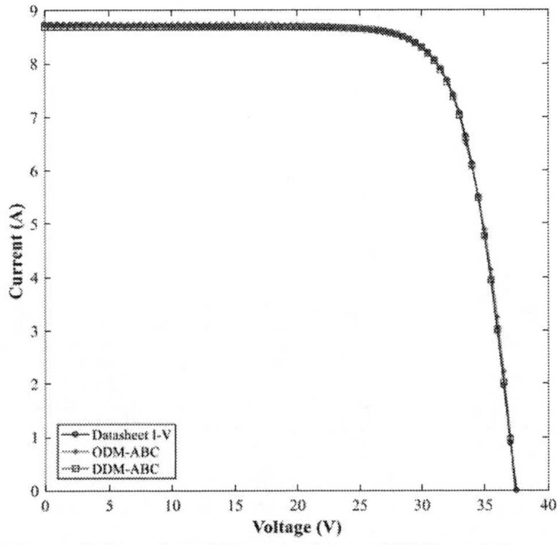

Figure 3: Datasheet I-V curve for mc-Si PV module.

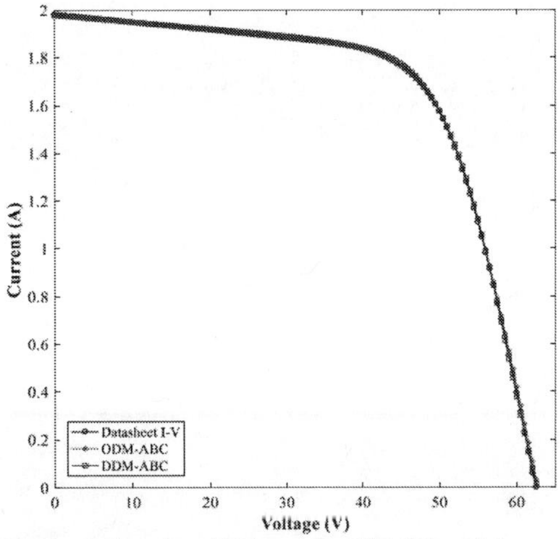

Figure 4: Datasheet I-V curve for CdTe PV module.

Based on the extracted parameters from the I-V curves given in Figs. 3 and 4. The output power of the PV module/string has been simulated using real daily profiles of G and T_c based on the two PV models. The obtained result is depicted in Fig. 5. As it can be seen, the simulated output power of the PV modules is over estimated by the models based on the parameters extracted from datasheet. Where, the calculated RMSE values showed that the error between measured and simulated curves is higher than 20% for the case of the mc-Si PV module, and even more than 30% for the case of CdTe PV string (see Table V). The huge difference between measured and simulated output power for the case of CdTe is that, the information present in the datasheet represent the healthy case of the PV modules before degradation. Moreover, it is already demonstrated that after a certain period of time the performance of CdTe PV modules degrade significantly by a degradation rate up to 3.5%/year [8], [5]. Thus, in order to simulate the outputs of these PV modules technology it is important to take in consideration the losses due to degradation.

Table II: Extracted parameters of the ODM for mc-Si PV module.

	Extracted from Datasheet		Extracted from measured I-V		Extracted from dynamic profiles	
Parameters	LMA	ABC	LMA	ABC	LMA	ABC
R_s (Ω)	0.236	1.043	1.133	2.289	0.789	1.13
R_p (Ω)	8.65 10^3	1.63 10^3	6 10^3	7.1 10^3	422	6 10^3
n	1.16	1.15	1.23	1.21	1.45	1.42
I_{ph} (A)	9.687	9.785	9.2	9.01	8.9	8.87
I_o (A)	4.2 10^{-8}	7.7 10^{-8}	6.07 10^{-6}	2.7 10^{-6}	2.3 10^{-7}	6.07 10^{-6}

Table III: Extracted parameters of the DDM for CdTe PV string.

	Extracted from Datasheet (one panel)		Extracted from measured I-V		Extracted from dynamic profiles	
Parameters	LMA	ABC	LMA	ABC	LMA	ABC
I_{ph} (A)	2.156	2.15	2.12	2.10	2.07	2.05
R_s (Ω)	6.66	9.43	12.13	2.289	20.97	15.11
R_p (Ω)	2.47 10^4	5.7 10^4	6.9 10^3	7.1 10^3	5.75 10^3	6 10^3
n_1	1.45	1.5	1.48	1.52	1.66	1.62
I_{o1} (A)	1.11 10^{-7}	7.9 10^{-7}	6.07 10^{-7}	2.7 10^{-7}	6.29 10^{-7}	6.07 10^{-7}
n_2	1.85	1.87	1.89	1.90	1.97	1.91
I_{o2} (A)	1.5 10^{-6}	5.17 10^{-6}	6.07 10^{-6}	2.7 10^{-6}	5.78 10^{-6}	6.07 10^{-6}

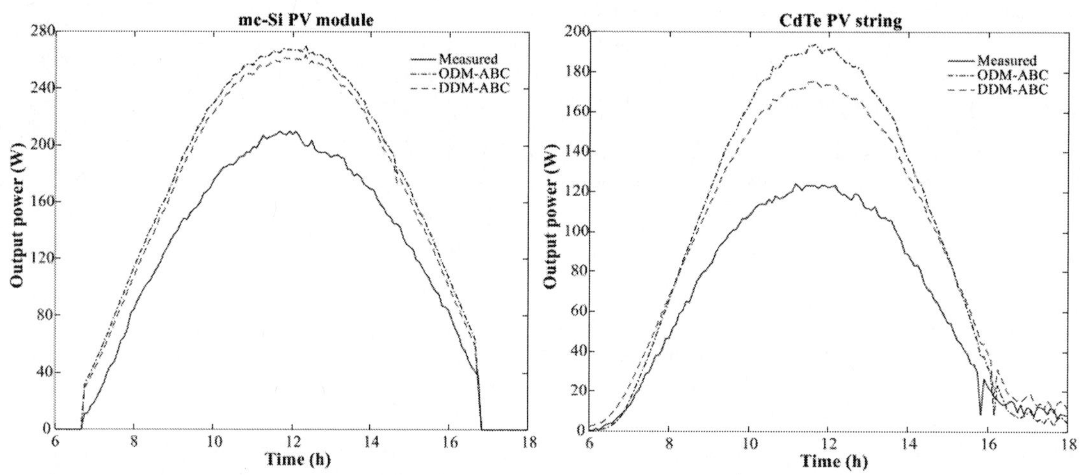

Figure 5: Dynamic behaviour of the PV modules based on the extracted parameters from datasheet.

4.2 Case 2

Similar to the case 1, the models parameters were extracted from an I-V curve, but in this case the I-V curve was measured using an EuroLink Pro® I-V tracer. The measured I-V characteristics as well as the simulated ones by the ODM and DDM based on the extracted parameters are shown in Figs. 6 and 7 for mc-Si PV module and CdTe PV string respectively.

From Fig 6 and 7, it could be seen that the two models based on ABC or LMA algorithms provide good fitting to the real measured curve. The calculated RMSE values considering each case are listed in Table IV. It can be observed that, the DDM model performs slightly better than the ODM. Moreover, the extracted parameters provided by the ABC optimization algorithm are more precise compared to the parameters provided by the LMA.

Here also from Table IV, it could be seen that, the RMSE are less than 2% in most of cases. The highest RMSE value does not exceed 2.5% and it is obtained from the ODM based on LMA.

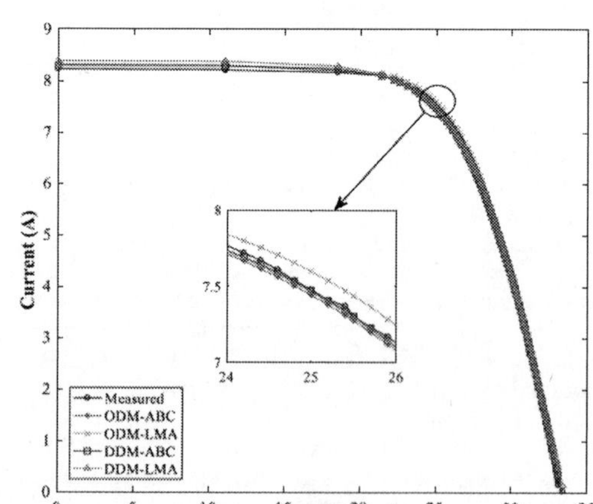

Figure 6: Measured I-V curve of mc-Si PV module under $G = 922$ W/m^2, $T_c = 56$ °C.

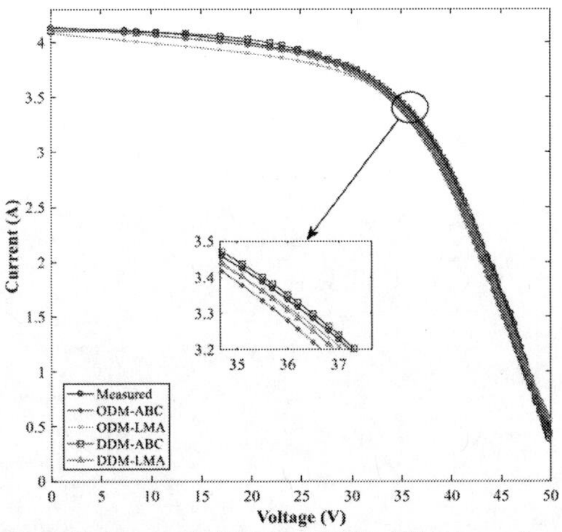

Figure 7: Measured I-V curve of CdTe PV string under G = 676 W/m^2, T_c = 50 °C.

The extracted parameters from measured I-V curves were used to reproduce the behaviour of the PV panels under the same day outdoor conditions simulated previously in case 1. The obtained results are shown in Fig. 8. From the obtained results, it could be seen that the ODM and DDM reproduce the output power of the mc-Si panel and CdTe PV string with good accuracy. The obtained RMSE reported in Table V are around 5% for the DDM based on ABC algorithm and 7% for the ODM based on the same algorithm. This amount of deviation between measured data and simulated ones is mainly due to the system losses (wiring, inverter efficiency, mismatch...) which are not included in the model.

However, in the present case, there is no significant over estimation of the output power of the CdTe PV string. This is due to the already included losses of degradation in the real measured I-V characteristic.

Table IV Calculated RMSE values between datasheet, measured I-V characteristics and simulated ones using the two PV models based on different optimization algorithms.

Module/String	CdTe				mc-Si			
PV Model	ODM		DDM		ODM		DDM	
Optimization algorithm	ABC	LMA	ABC	LMA	ABC	LMA	ABC	LMA
RMSE_datasheet (%)	0.31	0.42	0.26	0.36	0.30	0.49	0.21	0.33
RMSE_meas_IV (%)	1.35	2.4	1.16	1.59	0.59	0.99	0.55	0.91

Table V Calculated RMSE values from dynamic simulations of output power estimated by the two models based on the extracted parameters considering the three different scenarios.

Module/String	mc-Si		CdTe	
Optimization algorithm	ABC		ABC	
PV Model	ODM	DDM	ODM	DDM
RMSE_datasheet (%)	23.3	20.38	38.07	30.02
RMSE_meas_IV (%)	6.59	5.40	7.19	5.93
RMSE_dynamic (%)	2.85	2.10	3.69	3.27

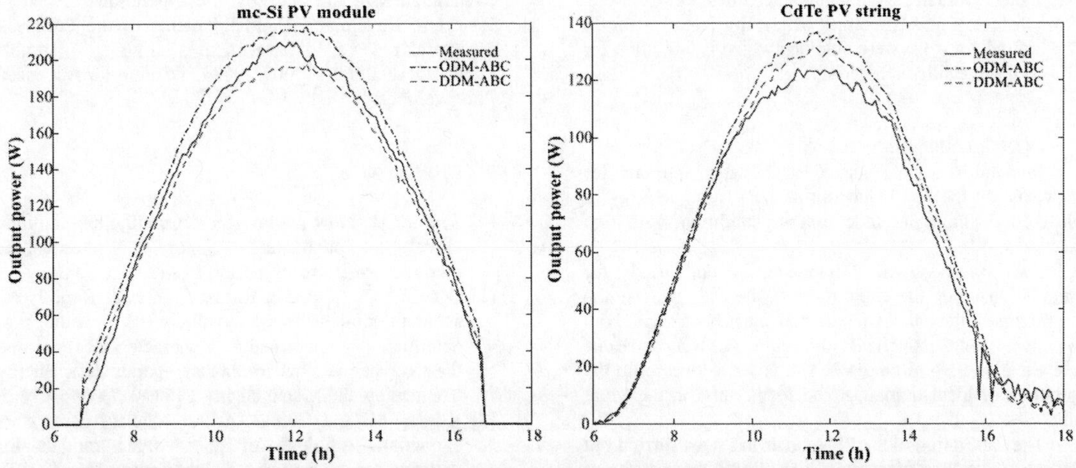

Figure 8: Dynamic behaviour of the PV modules based on the extracted parameters from measured I-V curves

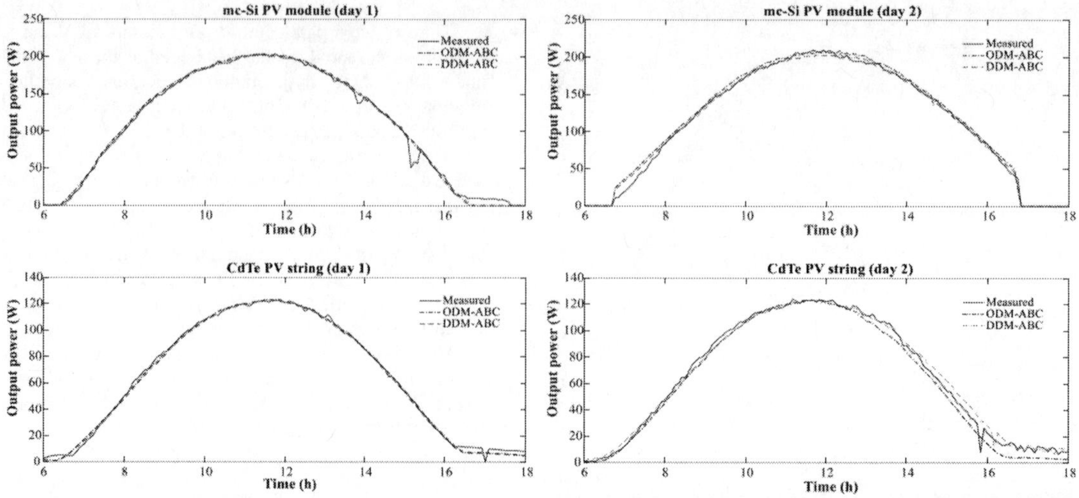

Figure 9: Dynamic behaviour of the PV modules based on the extracted parameters from real measured profiles I-V curves.

4.3 Case 3

In this third case, the extraction of the models parameters has been carried out directly from real measured profiles of voltage and current at MPP based on G and T_c profiles of the same day. Two different days were selected. The extraction of the parameters was from the measured profiles of the first day (day 1). The second day (day 2) has been used for the validation of the PV models based on the extracted parameters from the first day using ABC algorithm. The obtained result is shown in Fig. 9 and the calculated RMSE values corresponding to that case are reported in Table V.

From Fig. 9 it could be clearly seen that there is a good agreement between the measured and simulated output power profiles compared to the previous cases. The lowest RMSE values of 2.1% is obtained for the case of DDM based on the extracted parameters using ABC algorithm.

The extraction of the models parameters from real profiles results in high accuracy in simulating the real behaviour. All system losses are taken in consideration, and this explains the meaningless values of the unknown parameters extracted from dynamic profiles. For example, from Tables II and III, values of ideality factor of mc-Si PV around 1.4 – 1.6 were obtained which is too much for the crystalline silicon material.

5 CONCLUSIONS

In the present work it has been shown an approach for reproducing the real behaviour of a PV module/array of different technologies under outdoor conditions with high accuracy.

Two well-known PV models were used for simulations. For the calibration of the PV models, the extraction of the unknown parameters has been carried out by using one numerical algorithm (LMA) and one metaheuristic algorithm (ABC) in order to overcome the problem of local minimum and focus only on the input data.

The calibration of the PV models has been carried out by extracting their parameters considering three different set of input data; datasheet information, real measured I-V curves and daily monitored profiles.

The obtained results demonstrates that the simulation of the real behaviour of the PV module/array using datasheet information is inaccurate specially for the case of CdTe PV modules where their degradation is totally ignored. RMSE values up to 30% were found. Results obtained based on measured I-V curve provided an acceptable accuracy.

Finally, it could be concluded that the accuracy of simulating the outputs of a PV system is affected mainly by the input data used for the extraction of the models parameters. Moreover, the best approach to consider for simulating the real behaviour of a PV system with a very good accuracy is to use daily monitored profiles of its MPP outputs for the calibration of the PV models.

6 ACKNOWLEDGEMENTS

This work has been supported by the Ministry of Education, Youth and Sports within National Sustainability Program I (NPU I), project No. LO1605 – University Centre for Energy Efficient Buildings – Sustainability Phase and by the Operational Program Research, Development and Education of the European Structural and Investment Funds, project CZ.02.1.01/0.0/0.0/15_003/0000464 Centre for Advanced Photovoltaics.

7 REFERENCES

[1] Şentürk, A. New method for computing single diode model parameters of photovoltaic modules. Renewable Energy, 128 (2018), 30-36.

[2] Cuce, E., Cuce, P. M., Karakas, I. H., & Bali, T. An accurate model for photovoltaic (PV) modules to determine electrical characteristics and thermodynamic performance parameters. Energy Conversion and Management, 146 (2017), 205-216.

[3] Kumar, M, and Kumar, A. An efficient parameters extraction technique of photovoltaic models for performance assessment. Solar Energy 158 (2017): 192-206.

[4] Yahya-Khotbehsara, A., & Shahhoseini, A. A fast modeling of the double-diode model for PV modules

using combined analytical and numerical approach. Solar Energy, 162 (2018), 403-409.
[5] Kichou, S., Silvestre, S., Guglielminotti, L., Mora-López, L., & Muñoz-Cerón, E. Comparison of two PV array models for the simulation of PV systems using five different algorithms for the parameters identification. Renewable Energy, 99 (2016), 270-279.
[6] Castañer L, Silvestre S. Modelling Photovoltaic Systems using PCspice 2002.
[7] McEvoy AJ (Augustin J, Markvart T, Castañer L. Practical handbook of photovoltaics : fundamentals and applications. Academic Press; 2012.
[8] Silvestre, S., Kichou, S., Guglielminotti, L., Nofuentes, G., & Alonso-Abella, M. Degradation analysis of thin film photovoltaic modules under outdoor long term exposure in Spanish continental climate conditions. *Solar Energy*, *139* (2016). 599-607.

A PLANAR INDIUM-TIN-OXIDE THERMOPHOTOVOLTAIC EMITTER WITH HIGH-TEMPERATURE VERIFICATION

Dong-Han Wu[1], Parag Parashar[2], Hao-Ming Chou[2], Yi-Shiuan Lin[2], Yi-Chun Lai[3], PeiChen Yu[3], Yu-Bin Chen[1], Albert Lin[2*]
[1]Department of Power Mechanical Engineering, National Tsing Hua University, Hsinchu, Taiwan 30013.
[2]Department of Electronic Engineering, National Chiao-Tung University, Hsinchu, Taiwan 30010.
[3]Department of Photonics, National Chiao-Tung University, Hsinchu, Taiwan 30010.
*Corresponding author: hdtd5746@gmail.com, tel:+886 35712121~54267, fax: +886 35724361

ABSTRACT: Far-field thermophotovoltaics has been proposed in the early 1960s where the predicted efficiency can be as high as >80% with some idealized assumptions. The key part in thermophotovoltaic conversion is the design of the emitter spectral emissivity. The ideal spectral shape of the emission is a moderate bandwidth high emission at an energy that is just above the semiconductor diode bandgap. The long wavelength cutoff is required to prevent the diode becomes transparent to the incoming emitted photons. The short wavelength cut-off is to reduce the thermalization loss in the photovoltaic cell. While there have been tremendous amount of effort and publications in far-field TPV using various photonic nanostructures, here we show that in fact a simple planarized structure using refractory metal-indium tin oxide and an aperiodic dielectric stacking can achieve the desired emissivity for TPV. In addition, the emission wavelength can be tailored by the SiO_2/Si_3N_4 aperiodic stacking design and the ITO composition tuning.
Keywords: Selective Emitter, Thermophotovoltaics, TCO Transparent Conducting Oxides

1 INTRODUCTION

Thermophotovoltaics (TPV) has always been an extensive area of research due to its proposed efficiency that inherently exceeds the conventional photovoltaic efficiency limit [1-6]. This can be attributed to the effective utilization of photons energy lower than the semiconductor bandgap and the minimization of thermalization loss due the suppression of photons with energy much higher than semiconductor bandgap edge. A typical TPV system consists of an absorber/micro-combustor to heat up the TPV emitter and a photovoltaic cell beneath that absorbs the infrared photons from the emitter and generate electricity. Thus, designing an effective emitter design for tuning of the spectral response is a key part for the overall optimization of an efficient TPV system. To date, all of the experimentally realized TPV emitters and systems are far-field TPV systems. Extensive studies and proposals of efficient TPV emitters have conducted in the past includes Ta-based photonic crystals (PhC) [7, 8], tungsten (W) based metal-dielectric nanostructures [9], SiO_2-Si planar stacking [10], and a theoretical study of erbium-doped aluminum garnet (ErAG) as an emitter material based double side filtering structure [11].

In this present work, we propose the use of fully planarized indium tin oxide (ITO) emitter structure for the TPV applications. This structure is suitable for an efficient ideal emission in the 1.4-2.3μm spectral regime that is suitable for the most TPV applications. This planar structure eliminates the use of lithography and etching processes, thereby, adding cost and time efficiency to our design together with the possibility of large area scaling. Moreover, ITO based TPV emitter eliminates the fast oxidation of metal emitters, metal surface diffusion and deformation at high temperatures. This can promote the usage of TPV approach in ambient conditions instead of vacuum environment and thus facilitates the wide applicability of the TPV technologies.

In fact, TCOs have been an essential element in various near-field TPV applications as well due to their near-infrared plasmonic properties. Several remarkable theoretical efforts have been made in recent years in this field to exploit the plasmonic properties of the TCOs through the coupling of electromagnetic excitations to the plasmon at infrared frequencies [12-16]. This resonance can be thermally excited and thus can be coupled with the adjacent photovoltaic cell leading to the strong evanescent radiative heat transfer that matches its bandgap excitation energy. The calculation is based on fluctuational electrodynamics, somehow different from standard far-field thermal emission. Though the results have been very promising but experimental realization of these near-field TPV systems are quite challenging and need further considerations. Also, vacuum environment in an inevitable part of these systems. However as compared to these systems, our ITO emitter structure is experimentally attainable and could be operated in ambient environment with less constraints.

Lastly, it is worthwhile to point out that in our work, there are no surface plasmonic effects since it is a planar far-field structure. The ideal spectral emission at far-field is mostly due the near-infrared absorption characteristics in ITO dielectric response. This point is somehow different from the previous usage of ITO in near-field TPV where plasmonic resonance is the main physics behind the desired spectral shaping. Fig. 1(a) shows the schematic of our proposed structure with labeled dimensions. A 300nm thin ITO layer has been deposited on 300nm tantalum (Ta) metal ground plate. A three pairs of aperiodic dielectric staking of SiO_2/Si_3N_4 layers are proposed to be deposited on top of the Ta/ITO emitter for enhancing emission peak values and wavelength scalability of the device. Fig. 1(b) represents the emitter emissivity spectral curves for the ideal emitter for TPV applications. Long wavelength cut off is an imperative condition for ideal emitter spectral response as these photons are transparent to the PV cell beneath. And this cutoff should correspond to the bandgap excitation energy of the PV cell. Below the cutoff wavelength emitter spectrum ideally should resemble the blackbody radiation spectrum at a given temperature.

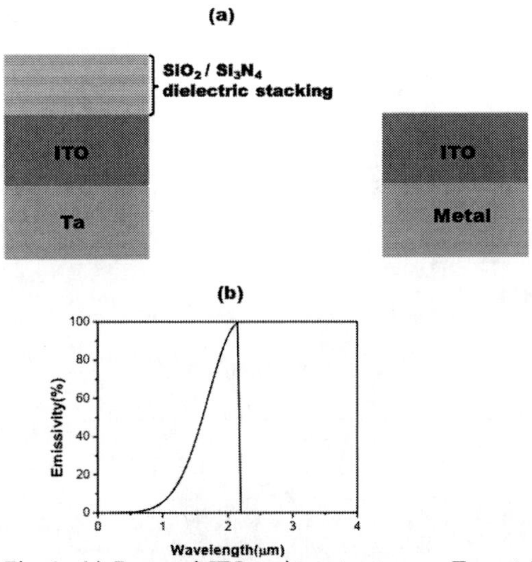

Fig. 1. (a) Proposed ITO emitter structure on Ta metal ground plate with aperiodic dielectric stacking (b) Ideal emitter characteristics. t_{ITO}= 300 nm, t_{Ta}= 300 nm, t_{Al}= 300 nm.

2 METHODS

The Fourier transform infrared (FTIR) spectroscopy has been used from λ=1.2μm to λ=10μm for the measurements of absorption/emission of the structure both at room temperature and at elevated temperatures. Bruker IFS 66v/S Fourier transform infrared spectroscopy (FTIR) is with 0.25 cm^{-1} resolution. Silicon carbide (SiC) glow bar has been used as the light source and a pyroelectric detector using deuterated lanthanum α alanine doped triglycine sulphate (DLATGS) with KBr window is used for detection purposes. Absorption/emission has been calculated via the formula Absorptivity (A)=1-reflectance(R) transmittance(T). And under thermal equilibrium, absorptivity is equivalent to emissivity due to Kirchhoff's law. For high temperatures, external source beam port of the machine is used for to collimate radiation from the sample.

In this work, rigorous coupled wave analysis (RCWA) method in Rsoft Diffractmod^{TM} with S-Matrix formulation has been used for simulations. RCWA is a semi-analytic numerical method in computational electromagnetics for the periodic dielectric structures to solve scattering problems. It implements eigenmode expansion method that incorporates the waveguide modes in the layered structure as the bases for the solution fields in the periodic nanostructures.

3 RESULTS

Fig. 2(a) represents the simulated spectral characteristics of our proposed structure with Al/ITO and Ta/ITO configurations in λ=1.2μm-10μm regime. Two metal ground plates (Al and Ta) have been adopted here for comparison purposes. Long wavelength cut off is clearly visible in both the configurations with the absorption peak at around 1.6μm. This long wavelength cut-off corresponds to non-absorption of long wavelengths into ITO thin layer and high reflection property of both Ta and Al metal ground plates. As a proof of concept, even higher absorption values and wavelength scalability can be achieved by incorporating only three pairs of SiO_2/Si_3N_4 aperiodic dielectric stacking on top of ITO/Ta as shown in Fig. 2(b). The thickness of the aperiodic stacking layers can affect the spectral response of the emitter. Thus, we can tailor the spectral response as well as the shift the emission peak by adjusting the geometry. With the inclusion of aperiodic dielectric stack on top of the emitter, enhancement of emissivity peak can be observed in Fig. 2(b) as compared to bare Ta/ITO and Al/ITO emitters in Fig. 2(a). Emission peak also shifts to nearly 2μm and 3μm from 1.6μm by varying thicknesses of the SiO_2/Si_3N_4 dielectric stack. Thus, wavelength tunability is not an issue with our proposed structure.

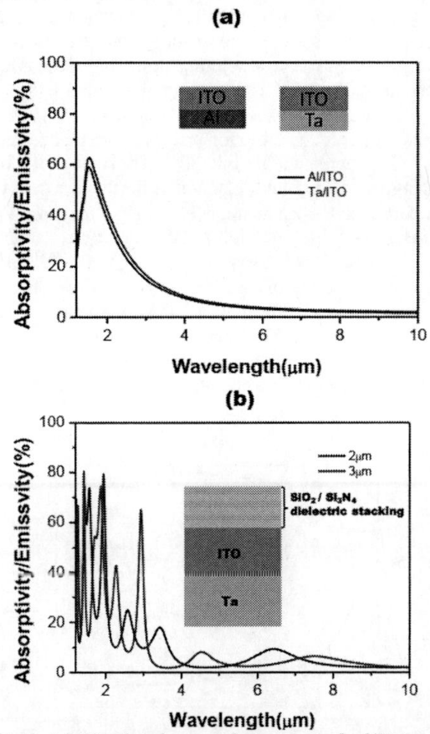

Fig. 2. (a) Simulated spectral response of Al/ITO emitter and Ta/ITO emitter (b) Simulated results of the Ta/ITO emitter with emission peak at 2μm and 3μm with 3 pairs of SiO_2/Si_3N_4 dielectric stack. Geometrical dimensions of 3 pairs of dielectric stack for 2μm emission: t_SiO_2 = 0.415μm, t_Si_3N_4= 0.626μm, t_SiO_2 = 0.392μm, t_Si_3N_4= 0.524μm, t_SiO_2 = 0.376μm, t_Si_3N_4= 0.439μm. Geometrical dimensions of 3 pairs of dielectric stack for 3μm emission: t_SiO_2 = 0.252μm, t_Si_3N_4= 0.657μm, t-_SiO_2 = 0.057μm, t_Si_3N_4= 0.15μm, t_SiO_2 = 0.314μm, t_Si_3N_4= 0.758μm.

Fig. 3(a)-(b) represents the experimental FTIR reflection, transmission and absorption measurement data at room temperature for the Al/ITO and Ta/ITO configurations, respectively, together with the corresponding simulated results. It could be seen experimental data is in well accordance with the simulated data. Emission peak has been observed at around 1.6 μm as expected. However, almost 20%

emissivity enhancement is achieved with ITO/Ta emitter as compared to ITO/Al emitter. This enhancement is due to the increased absorption in Ta metal ground plate. Sharp cut off emissivity at short wavelength can also be observed due to the intrinsic emission property of the ITO layer in both the configurations. Desired long wavelength suppression is also evident here, thereby, confirming the that applicability of our structure as an efficient TPV emitter. Selection of the metal ground plate is an important aspect in designing any optimized emitter for TPV applications. Ideal metal ground plate should possess high reflectance, compatibilty with present silicon IC technology, preferably low cost and high melting point depending on the high temperature oriented application such as micro combustor based TPV systems. Gold (Au), silver (Ag) and aluminum (Al) metals all have high reflectance values. Au and Ag also posses moderate melting points at 900°C and 1000°C, respectively, but these noble metals are not compatible with present Si IC technology. Besides, low melting point of Al at 660°C renders its usability at elevated temperatures in typical TPV applications. Nickel (Ni), titanium (Ti) also have high melting point suitable for practical micro combustor based TPV applications but the reflectance at long wavelengths are still not high enough for the practical use. Refractory metals such as tungsten (W) and tantalum (Ta) also posses very high melting point with high reflectance in the entire long wavelength regime. Although reflectance values of these refractory metal are still not comparable to the noble metals but high enough for the satisfactory results in TPV applications. Though, cost of Ta in still high but in our structure Ta has been used as the metal ground plate for the efficient emitter design.

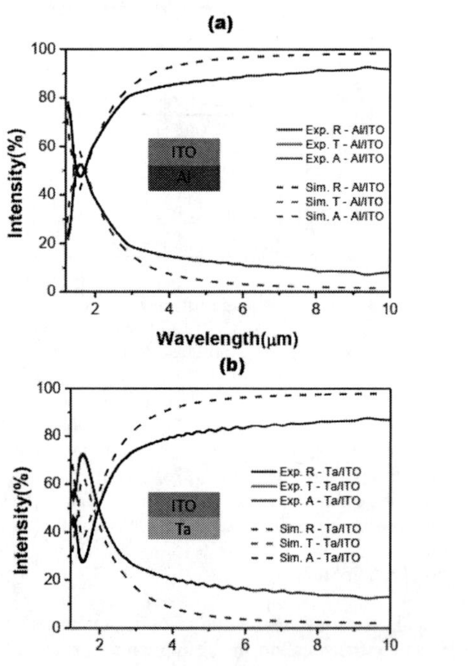

Fig. 3. (a) FTIR experimental spectral results of the ITO/Al emitter at room temperature (b) FTIR experimental spectral results of the ITO/Ta emitter at room temperature.

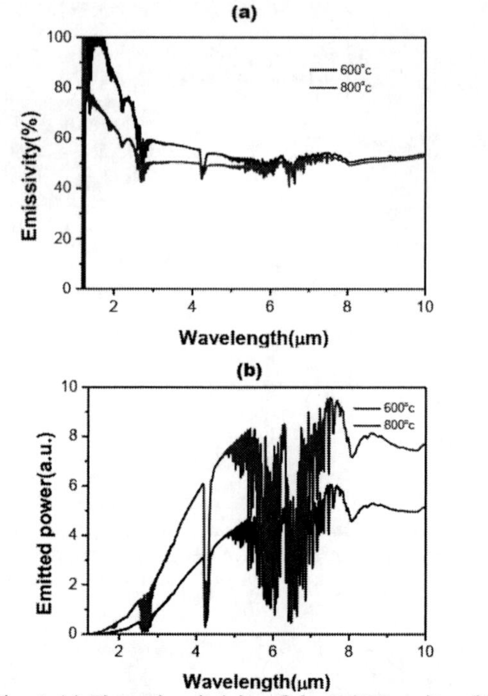

Fig. 4. (a) Thermal emissivity of the Ta/ITO emitter (b) Thermal emitted power of the Ta/ITO emitter.

Fig. 4(a)-(b) shows the thermal emissivity and emitted power data for the Ta/ITO emitter at 600°C and 800°C temperature, respectively. Thermal emissivity data shown here is consistent with the room temperature FTIR data presented in Fig. 3(b). An emission peak at around 1.45μm can be observed in Fig. 4(a) with suppression at long wavelengths and in turn again confirming the validation of the proposed structure as an effective TPV emitter at elevated temperatures. A typical practical temperature for TPV applications in the range of 800°C and a very few studies actually provide the experimental data at these elevated temperatures [17]. Although, at 800° C blackbody radiation peaks at around 3μm and our proposed emitter power spectrum does not resemble completely with the black body radiation at this temperature. But a typical TPV system normally use a solar cell with large band gap excitation energy due to associated lower dark current advantage. And this in turn leads to a higher efficiency. Thus, the high emissivity band at 1.4-2.3μm is a critical requirement for a high efficiency TPV system.

4 CONCLUSION

In this work, indium tin oxide (ITO) based planar TPV emitter has been implemented as a high effective TPV emitter. High temperature emissivity measurements have been included in this work and results have been very promising with emission peak near 1.45μm is obtained at 800°C with Ta/ITO emitter configuration. These elevated temperature results are well in accordance with FTIR results obtained at room temperature with emission peak around 1.6μm. The tailoring of the emission cutoff to longer wavelengths is further possible using the design of the aperiodic stacking on top of emitter structure as shown in Fig. 2(b). The λ=1.4µm-2.3µm range covers the entire potential TPV applications for elevated temperatures. Also, further shifting in the emission wavelength to 3µm can potentially be achieved by adjusting compound compositions of the TCO or selecting a different TCO altogether. Due the inherent emission property of the ITO layer at short wavelength, a sharp cut off emissivity can be observed here, thereby, the minimizing losses due to the thermalization. The planarized nature of the proposed emitter structure facilitates large area emitter design together with the elimination of lithography and etching processes. Besides, our functional oxide based TPV emitter eliminates the fast oxidation of metal emitters, metal surface diffusion and the deformation of the photonic nanostructure at elevated temperatures. Again, ITO based emitter design can be used in ambient conditions instead of vacuum environment, thereby, infusing less constraints and wider applicability of the TPV technologies.

REFERENCES

[1] Z. Zhou, Q. Chen, and P. Bermel, "Prospects for high-performance thermophotovoltaic conversion efficiencies exceeding the Shockley–Queisser limit," *Energ. convers. manage.*, vol. 97, pp. 63–69, 2015.

[2] H. Daneshvar, R. Prinja, and N. P. Kherani, "Thermophotovoltaics: Fundamentals, challenges and prospects," *Appl. Energy*, vol. 159, pp. 560–575, 2015.

[3] C. Ferrari, F. Melino, M. Pinelli, P. R. Spina, and M. Venturini, "Overview and Status of Thermophotovoltaic Systems," *Energy Procedia*, vol. 45, pp. 160–169, 2014.

[4] P. Bermel, M. Ghebrebrhan, W. Chan, Y. X. Yeng, M. Araghchini, R. Hamam, *et al.*, "Design and global optimization of high-efficiency thermophotovoltaic systems," *Opt. Exp.*, vol. 18, pp. A315-A334, 2010.

[5] R. E. Nelson, "A brief history of thermophotovoltaic development," *Semicond. Sci. Technol.*, vol. 18, p. S141, 2003.

[6] P. A. Davies and A. Luque, "Solar thermophotovoltaics: brief review and a new look," *Sol. Energ. Mat. Sol.*, vol. 33, pp. 11–22, 1994.

[7] Y. Nam, Y. X. Yeng, A. Lenert, P. Bermel, I. Celanovic, M. Soljačić, *et al.*, "Solar thermophotovoltaic energy conversion systems with two-dimensional tantalum photonic crystal absorbers and emitters," *Sol. Energ. Mat. Sol.*, vol. 122, pp. 287–296, 2014.

[8] V. Rinnerbauer, A. Lenert, D. M. Bierman, Y. X. Yeng, W. R. Chan, R. D. Geil, *et al.*, "Metallic photonic crystal absorber-emitter for efficient spectral control in high-temperature solar thermophotovoltaics," *Adv. Energy Mater.*, vol. 4, p. 1400334, 2014.

[9] C. Wu, B. Neuner, J. John, A. Milder, B. Zollars, S. Savoy, *et al.*, "Metamaterial-based integrated plasmonic absorber/emitter for solar thermo-photovoltaic systems," *J. Opt.*, vol. 14, p. 024005, 2012.

[10] A. Lenert, D. M. Bierman, Y. Nam, W. R. Chan, I. Celanovic, M. Soljacic, *et al.*, "A nanophotonic solar thermophotovoltaic device," *Nat. Nanotechnol.*, vol. 9, pp. 126–130, 2014.

[11] E. S. Sakr, Z. Zhou, and P. Bermel, "High efficiency rare-earth emitter for thermophotovoltaic applications," *Appl. Phys. Lett.*, vol. 105, p. 111107, 2014.

[12] R. St-Gelais, G. R. Bhatt, L. Zhu, S. Fan, and M. Lipson, "Hot Carrier-Based Near-Field Thermophotovoltaic Energy Conversion," *ACS Nano*, vol. 11, pp. 3001-3009, 2017/03/28 2017.

[13] A. Karalis and J. D. Joannopoulos, "Transparent and 'opaque' conducting electrodes for ultra-thin highly-efficient near-field thermophotovoltaic cells," *Scientific Reports*, vol. 7, p. 14046, 2017/10/25 2017.

[14] Y. Yang, J.-Y. Chang, P. Sabbaghi, and L. Wang, "Performance Analysis of a Near-Field Thermophotovoltaic Device With a Metallodielectric Selective Emitter and Electrical Contacts for the Photovoltaic Cell," *Journal of Heat Transfer*, vol. 139, pp. 052701-052701-9, 2017.

[15] A. Karalis and J. D. Joannopoulos, "Temporal coupled-mode theory model for resonant near-field thermophotovoltaics," *Applied Physics Letters*, vol. 107, p. 141108, 2015.

[16] A. Karalis and J. D. Joannopoulos, "'Squeezing' near-field thermal emission for ultra-efficient high-power thermophotovoltaic conversion," *Scientific Reports*, vol. 6, p. 28472, 07/01/online 2016.

[17] W. R. Chan, "High Effciency Thermophotovoltaic Microgenerators," PhD Thesis, Department of Electrical Engineering and Computer Science, Massachusetts institute of technology, Massachusetts, U.S.A., 2015.

RAPID CALCULATION OF THE BACKSHEET COUPLING GAIN USING RAY GROUPS

Andrea Pfreundt, Max Mittag, Martin Heinrich, Ulrich Eitner
Fraunhofer Institute for Solar Energy Systems ISE
Heidenhofstraße 2, 79110 Freiburg
andrea.pfreundt@ise.fraunhofer.de

ABSTRACT: Internal reflections within photovoltaic modules are known to contribute to power gains from cell to module. The module rear cover, usually a white backsheet, is one module component reflecting additional light onto the solar cell. A novel approach to model the effect of backsheet reflectance on the achievable coupling gain in solar modules is presented. Using a discrete ray optics approach, results can be calculated rapidly for arbitrary reflectance distributions using a partition of the emerging rays into groups. The model is fully wavelength resolved, using measured data to model optical material properties. It is therefore suitable for arbitrary material stacks in front of and behind the solar cell with a single diffusely scattering layer. We study the impact of layer thicknesses, incidence angle and distribution function on the coupling gain using the presented approach. Comparison to measurements of the coupling gain using single cell modules shows good agreement with the calculated results.
Keywords: PV Modules, Ray Tracing, Simulation, Module Efficiency, Coupling Gains

1 INTRODUCTION

Cell-to-module (CTM) efficiency losses in photovoltaic modules due to inactive module areas account for up to 2.5% abs. for a common module setup with 2 mm cell and string distance. Additionally, optical losses due to parasitic absorption and reflection contribute approximately 0.8% abs. [1, 2]. Light recycling from inactive module areas such as cell spacing and the outer module border contributes as coupling gain and reduces these losses [3, 4]. The magnitude of the coupling effect depends mainly on cell spacing, optical properties of the encapsulation materials and the reflectance of the module rear side [5][6]. Several methods to enhance the effect have been demonstrated, such as light redirecting ribbons and structured backsheets [7]. Experimental methods to investigate the coupling gain rely on short circuit current measurements of module samples with apertures or defined cell spacing. Either a single sample is built for each configuration [4, 5] or different backsheets are coupled to the same module stack by a liquid coupling agent to improve comparability [8].

In this paper we are presenting a computational approach based on data from transmission and reflection measurements of the individual encapsulation materials and geometric parameters of the module stack. The model enables rapid analysis of the coupling gain for parameter variations, including layer thickness, angular dependence and consideration of bifacial cells. It is compatible to the module simulation software SmartCalc.CTM and can be used as an integral part of the module simulation.

2 Simulation Methods

2.1 Cell-to-module loss analysis

The calculation method is compatible to the CTM methodology of the cell-to-module analysis tool SmartCalc.CTM. As published earlier, optical loss factors due to reflection and absorption for the front and rear encapsulation layers are calculated from measured material data [5]. The previously used model has been extended to incorporate the dependence of each of these optical factors on the angle of incidence (AOI). The relevant loss factors are listed in Table I.

Table I: Definition of optical loss factors

Symbol	Description
k_3	Air/cover reflection
k_3^{inv}	Cover/air reflection
k_4	Cover absorption
k_5	Cover/encapsulant reflection
k_5^{inv}	Encapsulant/cover reflection
k_6	Encapsulant absorption

The radiant flux (or power P) on the cell front through the encapsulation layers of the module P_{ref} is calculated from these loss factors and acts as a reference for the calculation of the backsheet gain. P_0 is the incident flux onto the module surface.

$$P_{ref}(d) = P_0 \cdot k_3 \cdot k_4 \cdot k_5 \cdot k_6 \quad (1)$$

The gain associated with the reflectance of the backsheet is denoted as k_{bs}. For a given cell spacing d it is defined as follows:

$$k_{bs}(\mathrm{d}) = 1 + \frac{P_{bs}(d)}{P_{ref}} \quad (2)$$

$$\Delta \mathrm{I}_{sc}(\mathrm{d}) = \frac{I_{sc}(d)}{I_{sc,0}} \quad (3)$$

The radiant flux $P_{bs}(d)$ describes the light that is redirected onto the cell via the surrounding backsheet. Similar definitions exist using the short circuit current as the defining parameter, which can be measured directly using module prototypes. We use the short circuit current gain, defined in (3), to characterize the backsheet coupling gain. Assuming a linear relationship between irradiance and short circuit current is valid for the considered range of irradiances [9], making the two values comparable but neglecting changes in voltage. We use (2) when analyzing module prototypes as described further down. Another option to evaluate backsheet reflection gains is to define the recovery probability of the backsheet, which considers the short circuit current gain per visible backsheet area [8]. We choose the first approach as it directly relates to current gain of the

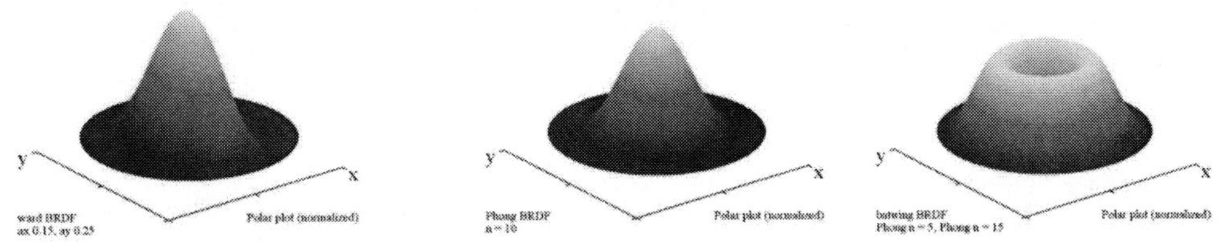

a) Ward model **b)** Phong model **c)** Phong model (batwing)

Figure 1: Different bidirectional reflectance distribution functions (BRDF) to model backsheet reflectance. All BRDFs are normalized to the same total reflectance. The ward parameters are $\alpha_x = 0.15$ and $\alpha_y = 0.25$. The Phong distribution is modeled with n = 10. The batwing model is based on two Phong distributions with n = 5 (base) and n = 15 (dip).

individual cell and makes it possible to consider cells with different surrounding geometries in one module and subsequently analyze the impact on electrical module losses.

2.2 Ray group model

2.2.1 Backsheet reflectance models

The backsheet reflectance distribution function (BRDF) is modeled by one of three approaches commonly used in ray tracing; Lambert, Phong or Ward distribution. The Lambertian reflectance model (not shown in the figure) considers a totally diffuse reflectance with a uniform distribution of the reflected light independent of the direction of the incoming/outgoing ray. The Phong model adds a specular peak (gloss lobe in rendering) to the Lambertian distribution, the width of which is determined by a parameter n that enters as exponent on the cosine of the angle between incident and reflected rays. The Ward distribution is capable of modelling an anisotropic distribution by adding a Gaussian specular glare with two main axes defined by their variances α_x and α_y. Examples of BRDFs based on these models are shown in Figure 1. It is possible to use other functions or measurement data to replace these models, for example to simulate structured backsheets or in order to optimize the reflectance for a specific module setup. We use hemispheric reflection data obtained from an integration sphere, which collects all the reflected light over the full hemisphere or with an additional opening of approximately 8° to exclude the direct reflection component. This results in a total hemispheric as well as a diffuse reflection spectrum. For mostly diffusely scattering backsheets we expect the direct component of the reflectance in air to disappear in the laminated module due to fusion between backsheet surface and encapsulation material.

2.2.2 Definition of ray groups

We partition the light reflected at the backsheet at each location x into ray groups. The x axis origin is located at the cell edge, pointing away from the cell. Each group is characterized by a predetermined trajectory through the encapsulation materials and a landing point on one of the areas of the solar cell (front, side, back). In this way, the pre-calculated optical loss factors of the encapsulation layers are reused to build an angle dependent loss function by matrix multiplication for each ray group. Loss function and BRDF are denoted with an M to indicate their discrete character as pre-calculated matrix in the implementation. For the zenith angle θ and azimuth angle φ the loss matrix for the cell front and back is calculated as follows:

$$M_{cf}^{k}(\theta) = [k_6(\theta)]^2 \cdot k_5^{inv}(\theta) \cdot k_4(\theta) \cdot (1 - k_3^{inv}(\theta)) \cdot k_4(\theta) \cdot k_5(\theta) \cdot k_6(\theta) \tag{4}$$

$$M_{cb}^{k}(\theta) = k_6(\theta) \tag{5}$$

The subscript indicates the location where the ray path terminates (e.g. cf = cell front). Since the encapsulation materials are assumed to be isotropic, M is independent of the azimuth angle φ.

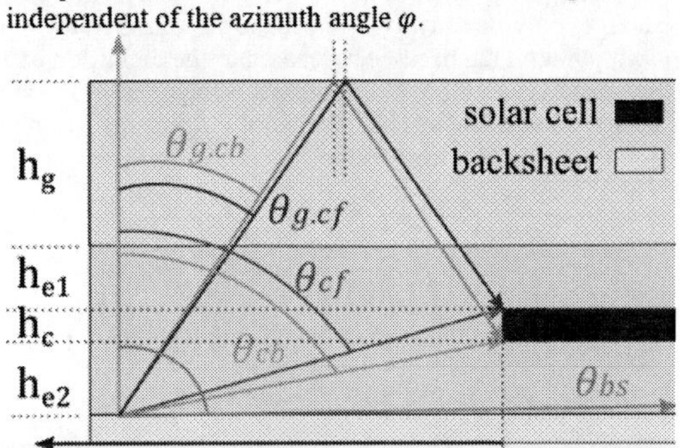

Figure 2: Illustration of the zenith threshold angles that divide the ray groups into sections on the reflected half sphere. For each azimuth angle φ and distance x from the cell edge these angles can be calculated directly from the geometry of the module.

The geometry of the module is modeled by defined layer thicknesses and a given cell spacing. For each distance from the cell edge, threshold angles can be defined that bound the ray groups and enable the calculation of their contribution without explicitly calculating each ray trajectory and interface intersection. The threshold angles are illustrated in Figure 2. The extent of the cell is not shown but considered in the calculation of the ray groups reaching the cell front and backside.

Calculation of the threshold angles is straightforward from geometrical considerations:

$$\varphi_0(x) = \tan^{-1}\frac{w_{cell}}{x} > 82°, \text{for relevant x} \tag{6}$$

$$\theta_{g.cf}(\varphi, x) = \tan^{-1}\frac{x/\cos\varphi}{2h_g + 2h_{e1} + h_{e2}} \tag{7}$$

$$\theta_{g.cb}(\varphi, x) = \tan^{-1}\frac{x/\cos\varphi}{2h_g + 2h_{e1} + h_{e2} + \frac{1}{2}h_c} \tag{8}$$

$$\theta_{cf}(\varphi, x) = \tan^{-1}\frac{x/\cos\varphi}{h_{e2} + \frac{1}{2}h_c} \quad (9)$$

$$\theta_{cb}(\varphi, x) = \tan^{-1}\frac{x/\cos\varphi}{h_{e2}} \quad (10)$$

$$\theta_{bs}(\varphi, x) = \tan^{-1}\frac{w_{cell} + x}{h_e \cos\varphi} \approx \frac{\pi}{2} \quad (11)$$

Table II: Definition of ray groups by reflectance angle

Index	Description	Condition for θ
g.bs	Backsheet via glass reflection	$\theta \le \theta_{g.cb}$
g.cs	Cell side via glass reflection	$\theta_{g.cb} \le \theta \le \theta_{g.cf}$
g.cf	Cell front via glass reflection	$\theta_{g.cf} \le \theta \le \theta_{cf}$
cs	Cell side directly	$\theta_{cf} \le \theta \le \theta_{cb}$
cb	Cell back directly	$\theta_{cb} \le \theta \le \theta_{bs}$

Since each side of the solar cell is a continuous surface, only one ray group needs to be considered from each point on the backsheet, leading to the definition of 5 ray groups not considering multiple reflections which are shown in Table II. The irradiance originating from a point at distance x from the cell edge is defined as

$$E_{group}(x) = E_{bs}(x) \cdot f_{bs.cf}(d) \quad (12)$$

$$f_{bs.cf}(d) = \frac{\int_{-\varphi_0(x)}^{\varphi_0(x)} d\varphi \int_{\theta_1(\varphi,x)}^{\theta_2(\varphi,x)} d\theta \, \sin\theta \; M_{BRDF}(\theta,\varphi) M_{cf}^{k}(\theta)}{\int_{-\pi}^{\pi} d\varphi \int_0^{\pi} d\theta \, \sin\theta \; M_{BRDF}(\theta,\varphi)} \quad (13)$$

The angle φ_0 is defined by the lateral extent of the cell along the cell edge. An average value of half the cell length is assumed. $\theta_1(\varphi, x)$ and $\theta_2(\varphi, x)$ are defined by the module geometry in x direction and layer thicknesses; they are unique for each of the ray groups (see Table II).

The total radiant flux (or power P) incident on the cell is determined by integration over all contributions in the given ray groups from the backsheet area corresponding to a specific cell distance.

$$\begin{aligned} P_{group}(d) &= dA \, E_{group}(d) \\ &= dy \int_0^d dx \, E_{group}(x) \quad (14) \\ &= dy \, P_{group}^{y}(d) \end{aligned}$$

2.2.3 Multiple reflections

For one setup of the model with a maximal cell distance all smaller cell distances are calculated simultaneously. Light that is scattered in backward direction (away from the cell) can contribute to the backsheet gain by way of multiple reflection. While the consideration of this additional reflection in detail would increase model complexity, the maximal contribution for each cell distance d is readily obtained in the simulation (group 1) and can be used to estimate the total contribution of multiple reflections. An average of the ray groups reaching the cell is used to scale the redirected light and calculate the additional contribution to the total power on each cell area. Exemplary for the cell front

$$P_{g.bs.cf}(d) = P_{g.bs}(d) \cdot \frac{f_{bs.cf}(d)}{d} \quad (15)$$

where integral in (13) is scaled with the total area to average the contributions over the full length. This is a simplification that neglects the landing point of multiply reflected rays on the backsheet.

2.2.4 Shading at non-normal incidence

Variation of the incidence angle leads to a reduction in incident light and shading of the cell within the cell spacing. This is considered in the model by shifting the integration boundaries of the integral in (14). The contributions from the four sides of the cell are accounted for individually (for varying azimuthal angles). As illustrated in Figure 3. For a change in zenith angle, x_{s1} and x_{s2} are added to/subtracted from the integration boundaries when integrating the contributions from the cell spacing area. Considering variations in azimuth angle requires the calculation of shading distances on all 4 sides of the cell.

$$x_{s,1}(\theta_{t2}, \varphi) = h_{EVA} \tan\theta \sin\varphi \quad (16)$$

$$x_{s,2}(\theta_{t2}, \varphi) = (h_{EVA} + h_{cell}) \tan\theta \sin\varphi \quad (17)$$

$$y_{s,1}(\theta_{t2}, \varphi) = h_{EVA} \tan\theta \cos\varphi \quad (18)$$

$$y_{s,2}(\theta_{t2}, \varphi) = (h_{EVA} + h_{cell}) \tan\theta \cos\varphi \quad (19)$$

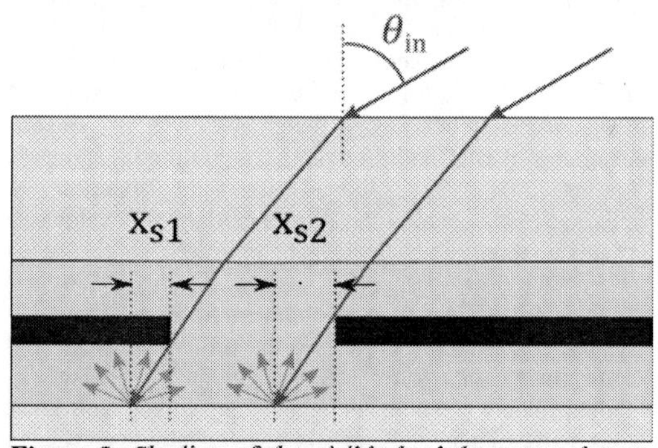

Figure 3: Shading of the visible backsheet area due to variations in the angle of incidence leads to additionally illuminated area at x_{s1} and shaded area x_{s2}.

2.2.5 Implementation of the ray group model

The model has been implemented in MATLAB and made compatible with the simulation software SmartCalc.CTM. The result of the optical simulation of the module layers, which is based on measurement data, is used as input to the ray group calculation. This makes it possible to incorporate the calculation method for backsheet coupling gains into the full cell to module loss calculation.

3 EXPERIMENTAL METHODS

3.1 Short circuit current measurements

Mini-modules consisting of a single solar cell embedded in the module stack are fabricated with different white and black backsheets. The hemispheric and diffuse reflectance of the backsheets is obtained using an integrating sphere, the direct contribution is calculated as the difference of the two measurements and corresponds to the reflected light into an opening angle of approximately 8° from the surface normal.

The open circuit current is measured at STC conditions using a Xenon Lamp calibrated to 1000 W/m². Aperture masks are used to limit the incident light to an area around the cell corresponding to different cell spacing in order to measure the contribution of this area to the individual cell current. A mask with a 135 mm opening is used in addition to the 156 mm mask as control to limit the influence of stray light. We estimate the error of cell distance due to placement of the mask and the influence of stray light to be around 0.25 mm. Examples of the used masks are shown in Figure 8 and the setup is illustrated in Figure 9.

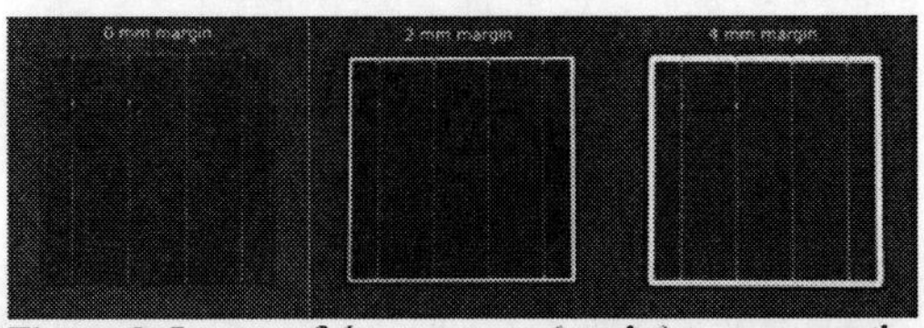

Figure 8: Layout of the apertures (masks) to measure the coupling gain for a specific margin where the backsheet is visible. The margin corresponds to the full cell distance in the simulations.

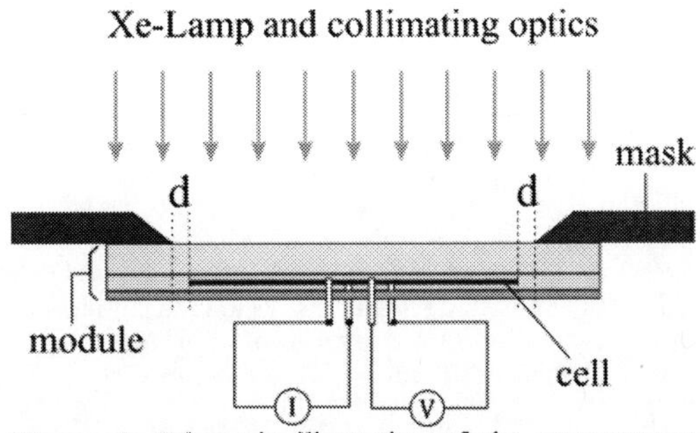

Figure 9: Schematic illustration of the measurement setup to measure the short circuit gain associated with backsheet coupling effects. A sun simulator with a total flash intensity of 1000 W/m² in the module plane is used for illumination.

4 RESULTS

4.1 Contributions of different areas on the cell to the backsheet coupling gain

By using ray groups it is straight forward to analyze the contribution of different cell areas to the backsheet

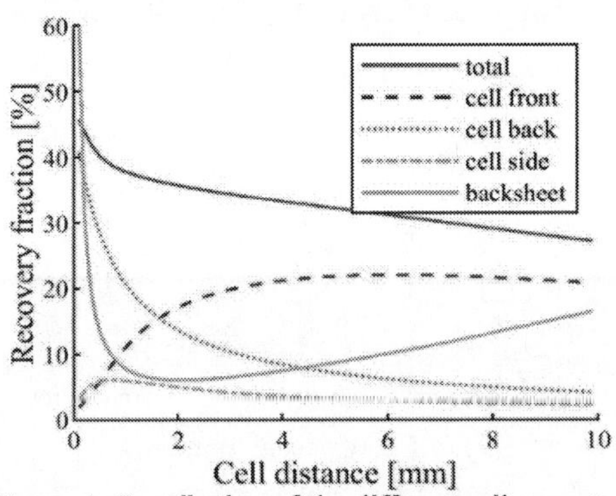

Figure 4: Contribution of the different cell areas to the backsheet coupling gain expressed as total recovery fraction of the light incident on the cell gap for each cell distance.

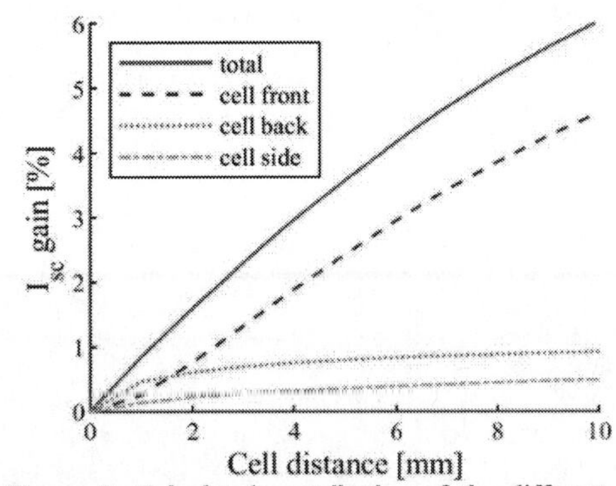

Figure 5: Calculated contribution of the different cell areas to the backsheet coupling gain expressed relative short circuit current gain of an individual cell surrounded by a cell gap of the specified size (cell distance).

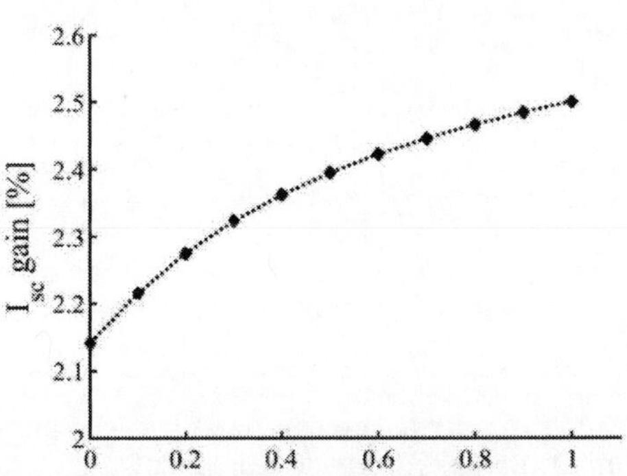

Figure 6: Backsheet coupling gain for a cell distance of 4 mm and different widths of the active area on the cell backside.

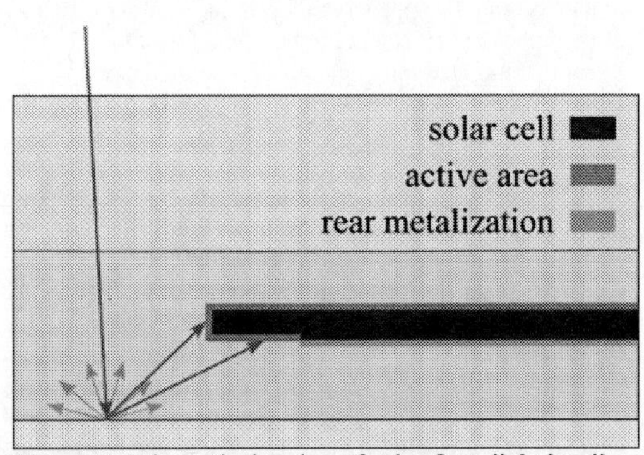

Figure 7: Schematic drawing of gains from light irradiant on the solar cell rear and side after backsheet reflection.

coupling gain. The collected contributions for each cell area are shown in Figure 5 for a typical solar module setup using bifacial solar cells and a totally diffuse reflecting backsheet with a reflectance of 82%, modeled using Lambertian reflection. The fraction that returns to the backsheet via the glass surface is also shown. This contribution as well as the contribution on the cell back side start at very high recovery values, since the impact of the lateral extent of the cell gap contributes significantly. Since the absolute irradiance entering the cell gap for small gap sizes is small, the absolute contributions will be small, as seen in the following results, which are shown in terms of as absolute current gain for an individual cell.

Today most industrial cells are "monofacial" which is a term used to describe that the rear side is usually blocking incident light and therefore light from the rear is not contributing to power generation. We find the assumption of complete photoelectrical inactivity of the solar cell rear side to be insufficient to describe effects of backsheet gain. Common cells feature a margin between the cell edge and blocking rear side cell metallization (i.e. aluminum) which leads to active cell areas on the cell rear side. Since these areas are located near the cell edge additional backsheet gains can be realized (Figure 7).

Measurements at Fraunhofer ISE CalLab Cells of the generated photocurrent at the edge and backside of monofacial cells have shown current density levels of 70% and 85% respectively compared to the front side. This significantly adds to the gain that can be obtained through backsheet coupling. Simulations show an increase from 2.15 % to 2.4 % for a cell distance of 4 mm and an active backside edge of 0.5 mm. Figure 6 shows the simulation result for a variation of the backside contribution for a fixed cell distance of 4 mm.

4.2 Impact of incidence angles on the relative backsheet coupling gain.

Variation of the zenith incidence angle between 0° and 80° has a very small impact on the backsheet gain. For cell distances up to 10 mm, the effect is less than 0.07 % across all angles. Variations in the azimuthal angle have no impact on the coupling gain as long as the reflection of the backsheet is isotropic. The different shading effects on the four cell edges cancel each other for a rectangular cell. The cell thickness has a small impact on the discrepancy between the shading distances of the opposite cell sides (Figure 3 and (16)-(19)), but the effect is found to be negligible in the studied cases.

4.3 Variation of geometrical parameters

4.3.1 Glass thickness

The thickness of the glass cover affects the coupling gain due to higher absorption of the reflected rays reaching the cell front, which shows especially at longer distances from the cell edge. The results in Figure 10 show that for thicker glasses the dependence of the coupling gain on cell distance changes from an exponential behavior towards a linear curve for distances up to 10 mm. For thicker glasses the exponential behavior is shifted to higher distances. Since the module standard for glass thicknesses moves towards thinner glasses, it is not sufficient to assume a linear model and it is necessary take the geometry into consideration.

4.3.2 Encapsulation thickness

The encapsulation thickness influences the coupling gain due to additional absorption in the polymer and by changing the light distribution in the ray groups due to the changes in geometry. For common EVA films we find a significant variation in coupling gain between 0.5 % and 1.2 % for common cell distances and for a range of EVA thickness between 150 µm and 900 µm corresponding to half respectively double the most common encapsulation sheet thickness. Figure 11 shows the variation with cell distance which is largest for common cell distances between 2 mm and 6 mm. Small variations in the encapsulation thickness which might arise due to differences in the lamination process will not have a significant influence on the backsheet coupling gain as we see in the curves in Figure 12, but will lead to uncertainties of approximately 0.1% per 100µm when values are compared with measurements.

4.3.3 Cell thickness

The influence of the cell thickness is investigated for common solar cells between 100 µm and 250 µm. Simulations show a variation of the coupling gain due to the increased surface area of the cell sides below 0.1 % for cell distances up to 10 mm. Compared to the impact of other parameters, this is a negligible contribution.

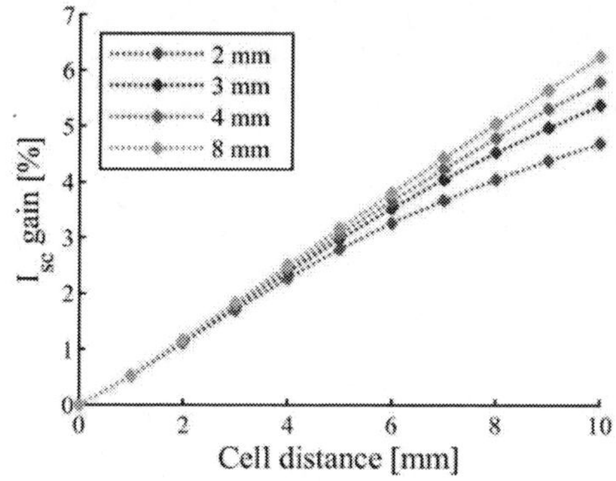

Figure 10: Backsheet coupling gain for identical module stacks with different glass thicknesses.

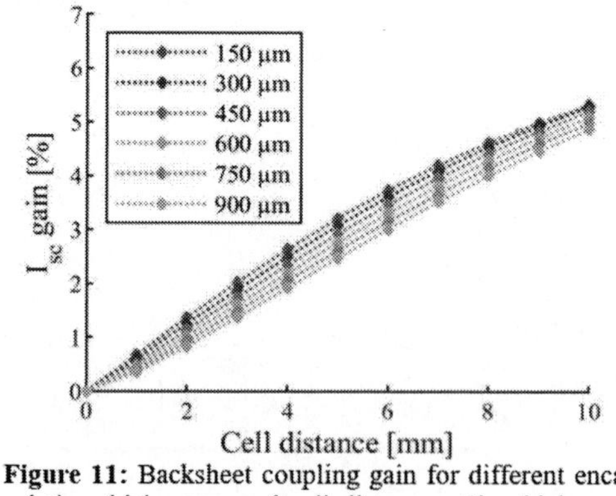

Figure 11: Backsheet coupling gain for different encapsulation thicknesses and cell distances. The thickness is the same on front and rear side of the cell.

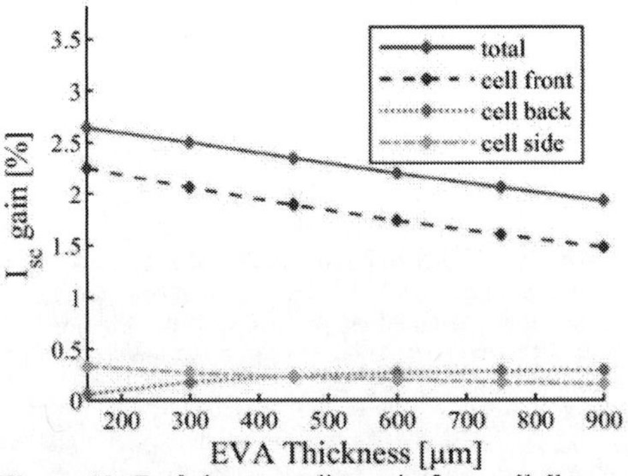

Figure 12: Backsheet coupling gain for a cell distance of 4 mm split into contributions from the different areas of the cell (ray groups) for a variation of the encapsulation thickness between 150 µm and 900 µm.

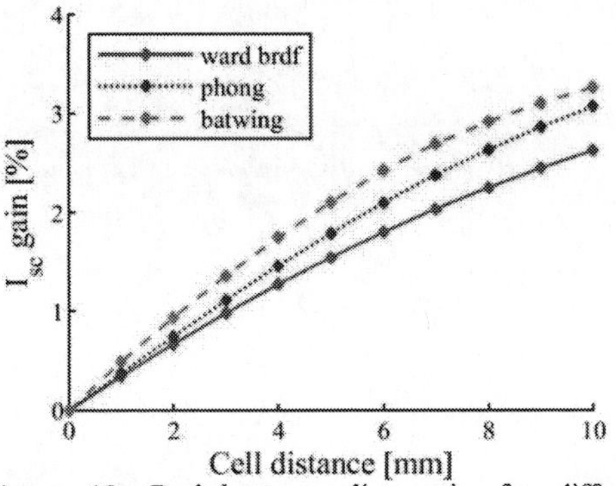

Figure 13: Backsheet coupling gain for different reflection distributions shown in Figure 9. For a total BRDF distributed reflectance of 75 % and 25 % diffusely distributed.

4.4 Comparison of BRDFs

Different bidirectional reflectance distribution functions (BRDFs) can be used to describe backsheet reflection. Ideally, precise measurements are performed to determine the distribution function. As these measurements are quite expensive and time consuming it is advisable to model the distribution function based on a few measurement parameters. Commonly the Lambertian distribution is used to model perfectly diffusive reflection behavior.

The shape of the reflection function influences the light distribution among the ray groups. Figure 1 shows three different light distributions, modeled with different approaches. a) shows an isotropic Phong distribution with n = 10, b) shows a non-isotropic Ward distribution with FWHMs of 0.15 and 0.25 in x (perpendicular) and y (parallel to cell edge) directions respectively, c) shows a batwing distribution. All BRDFs are normalized to the same total reflectance. The resulting coupling gain for different cell distances is shown in Figure 13.

4.5 Comparison of ray group calculation results with measurement data

The short circuit current gains associated with backsheet coupling are measured on single cell modules with different backsheets. The modules are built with monofacial solar cells, 450 µm EVA sheets and 3 mm solar glass without anti-reflective coating. For the simulation the reflection and transmission spectra of the individual materials are used to model their behavior. The cell surface and influence of metallization on the front side is not considered. The shape of the resulting curves shows an excellent agreement with the measured data while the absolute values show minor discrepancies (Figure 14). The individual contributions of the three cell areas are shown. We assume an active cell edge on the backside of 450 µm in the simulation, which is responsible for the contribution of the cell backside. A measurement tolerance in the effectively visible backsheet area due to stray light that enters the apertures at an angle is estimated to be of the order of 0.25 mm. Considering that this might lead to a shift in the measured data, we conclude that the presented model can be used to calculate the backsheet coupling gain with good accuracy.

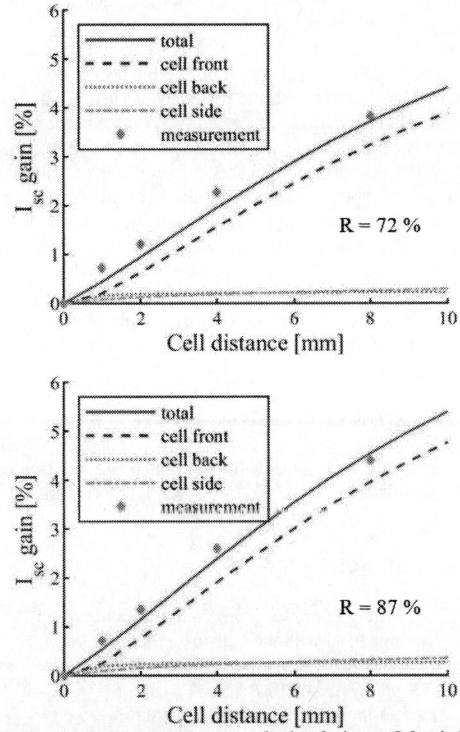

Figure 14: Measurement and simulation of backsheet coupling gain. The simulated gain is shown for different ray groups as well as the total gain. The measurement data is shown as diamonds. Data is shown for two backsheets with a reflectance of 72 % and 87 %.

5 SUMMARY AND OUTLOOK

We present an approach to rapidly calculate the backsheet coupling gain for individual cells in solar modules by using a model based on ray groups. The method is a simplified discrete ray optics approach, based on the explicit calculation of all the emerging rays. By dividing the reflected rays emerging from a diffusely scattering surface into ray groups by threshold angles, no ray tracing is necessary. This reduces the required number of calculations significantly. We show the simulation of backsheet coupling gains for a range of parameter variations such as geometrical parameters and reflection behavior of the backsheet. The presented method enables us to investigate new module setups with arbitrary backsheet reflectance functions in terms of achievable current gains. It simplifies the investigation of structured surfaces with complex reflection behavior, which can now be modeled to study their impact on module performance. Since the incidence angle in the module plane can be varied, the impact of the coupling gain on the yearly solar energy yield can be included in performance analyses.

We analyze the backsheet gain in terms of contributions from different reflection paths and find the reflection via the front cover to be the most important. We confirm a significant contribution of the cell side and rear, which is predominant for small cell distances. Our analysis of different module setups shows glass thickness to increase the backsheet coupling gain at long cell distances which is relevant for the effect on module margins. The thickness of encapsulants has a minor impact, while the effect of cell thickness is negligible. We investigate the dependence of incidence angle on the backsheet coupling gain and find that due to the symmetrical setup, shading effects balance resulting in virtually no dependence of the relative coupling gain on incidence angle. For non-isotroptic scattering backsheets such as structured surfaces we expect a stronger dependence on the angle of incidence that can have implication for annual yield analysis. Our present results confirm the importance of backsheet coupling for cell to module loss analysis. The method enables the quantification of the associated gains and their consideration for module optimization and evaluation of trade-offs between total module area and inactive area gains.

6 ACKNOWLEDGEMENT

We would like to thank the German Federal Ministry of Economic Affairs and Energy (FKZ 0324033) for their funding.

7 REFERENCES

[1] M. Mittag *et al.*, "Systematic PV module optimization with the cell-to-module (CTM) analysis software," *Photovoltaics International*, no. 36, pp. 97–104, 2017.

[2] *cell-to-module.com*. [Online] Available: http://www.cell-to-module.com. Accessed on: Feb. 28 2018.

[3] O. Dupré *et al.*, "Reassessment of cell to module gains and losses: Accounting for the current boost specific to cells located on the edges," in *Silicon PV 2018*, p. 90001.

[4] M. Mittag *et al.*, "Analysis of backsheet and rear cover reflection gains for bifacial solar cells," *33rd European PV Solar Energy Conference and Exhibition*, vol. 2017.

[5] I. Hädrich *et al.*, "Unified methodology for determining CTM ratios: Systematic prediction of module power," *Solar Energy Materials and Solar Cells*, vol. 131, pp. 14–23, 2014.

[6] P. Grunow *et al.*, "Modelling of the Encapsulation Factors for Photovoltaic Modules," 4th IEEE WCPEC, Waikoloa, Hawaii, USA, 2006, pp. 2152–2155.

[7] M. R. Vogt *et al.*, "PV module current gains due to structured backsheets," *Energy Procedia*, vol. 124, pp. 495–503, 2017.

[8] M. Köntges *et al.*, "Method to measure light recovery probability of PV module backsheets enabling 20.2% module efficiency with passivated rear emitter and rear solar cells," *32nd Eurpean Photovoltaic Solar Energy Conference (EUPVSEC)*, vol. 2016, 1532–1538.

[9] A. Schmid *et al.*, "IV Measurement of Bifacial Modules: Bifacial vs. Monofacial Illumination," 33rd EUPVSEC, Amsterdam, 2017, pp. 1624–1627.

GENERALIZED RECIPROCITY RELATION IN P-I-N JUNCTION SOLAR CELLS

Kasidit Toprasertpong[1*], Amaury Delamarre[1,2], Kentaroh Watanabe[1,2], Yoshiaki Nakano[1,2], Jean-François Guillemoles[2,3], and Masakazu Sugiyama[1,2]

[1] School of Engineering, the University of Tokyo, Japan
[2] NextPV, CNRS-RCAST, the University of Tokyo, Japan
[3] CNRS, Ile-de-France Photovoltaic Institute, France
*Address: #460, Eng. Bldg. 10, the Univ. of Tokyo, Yayoi, Bunkyo-ku, Tokyo, 113-0032, Japan
Phone: +81-3-5841-6733, E-mail: toprasertpong@mosfet.t.u-tokyo.ac.jp

ABSTRACT: The reciprocity relation, whose original version is not applicable to solar cells with wide depletion regions, is investigated and generalized to apply to p-i-n junction solar cells. By linearizing the transport equation in the i-region, we have derived the general form of reciprocity relation between electroluminescence and quantum efficiency at forward bias, not short circuit. This generalized reciprocity relation has been experimentally confirmed by comparing the electroluminescence and the voltage-dependent quantum efficiency of p-i-n multiple quantum well solar cells.
Keywords: Electroluminescence, p-i-n junction, Reciprocity relation, Quantum efficiency, Quantum Well, Voltage dependence

1. INTRODUCTION

The photovoltaic reciprocity relation [1] connecting electroluminescence (Φ_{EL}) spectrum and short-circuit external quantum efficiency (EQE) of solar cells (Fig. 1(a)) is a well-known theorem which helps enhance our understanding of the fundamental operation of solar cells and develops several new cell evaluation techniques based on electroluminescence measurements [2,3]. It is expressed by

$$\Phi_{EL}(\varepsilon,V) = \mathrm{EQE}^{(sc)}(\varepsilon)\Phi_{bb}(\varepsilon)\left(e^{qV/kT}-1\right), \quad (1)$$

where ε is photon energy, Φ_{bb} is the blackbody photon flux spectrum, V is the applied voltage at the device terminal, q is the elementary charge, k is Boltzmann's constant, T is the temperature and, the (sc) subscribe denotes the short-circuit operation. It has also been used to derive the well-known relation between the open-circuit voltage V_{oc} and the luminescence efficiency η_{ext} [1]:

$$V_{oc} = V_{oc}^{(rad)} + \frac{kT}{q}\ln\eta_{ext}, \quad (2)$$

where $V_{oc}^{(rad)}$ is the open-circuit voltage at the radiative limit (i.e. η_{ext} = 1). The reciprocity relation, however, is valid only if the deviation of device operation, e.g. carrier transport and recombination, from the thermal equilibrium is linear. The non-linearity of the system and consequently the failure of the original reciprocity relation in p-i-n junction solar cells (Fig. 1(b)) have been discussed [1,4] as the carrier dynamics in the i-region cannot be considered linear with the excess minority-carrier density. This leads to the limitation of cell evaluation methods that are based on the reciprocity relation in p-i-n junction solar cells. In this study, we examine how to modify the original reciprocity relation to account for the carrier dynamics in the i-region.

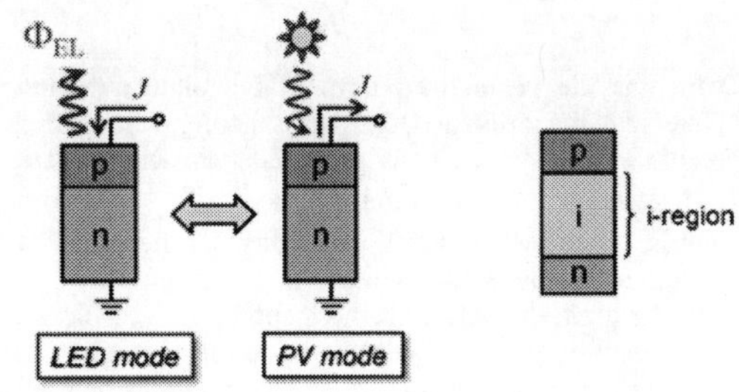

(a) (b)

Fig. 1. (a) Reciprocity relation describing the connection of the two device modes: the electroluminescecnce in the LED mode and the EQE in the photovoltaic mode. (b) p-i-n junction solar cells. Carrier dynamics in the i-region is not simply linear with the excess minority-carrier density, violating the assumptions in the previously proposed Donolato's theorem and reciprocity relation.

2. THEORETICAL APPROACH

By considering the carrier transport equation (current continuity equation), it can be confirmed that carrier transport and recombination in the i-region is not linear with the excess minority-carrier density as addressed in [1,4]. For instance, the equation for electron current continuity in the p-region can be written by

$$G = -\nabla\cdot(q\mu_n nE + qD_n\nabla n) + q\left(n-\frac{n_i^2}{p}\right)\left(\frac{1}{\tau_n}+Bp+C_p p^2\right), \quad (3)$$

where G is the photogeneration rate, n, p, and n_i are the electron, hole, and intrinsic carrier densities, respectively, μ_n is the electron mobility, D_n is the electron diffusion coefficient, τ_n is the electron non-radiative lifetime, B is the radiative recombination coefficient, and C_p is the 2-hole Auger recombination coefficient. In the p-region, p can be considered constant and Eq. (3) becomes linear relation. In the i-region, however, both p and n have spatial dependence as well as voltage dependence and thus Eq. (3) is not linear with the excess minority-carrier density $n - n_{V=0}$.

Analysis of the carrier transport equation (the detailed analysis and derivation will be published elsewhere [5]) provides a new finding that the carrier transport equation can be linearized in terms of u, which corresponds to quasi-Fermi level splitting $\Delta\varepsilon_F$

$$G = \mathcal{L}_V u, \tag{4}$$

$$u \equiv e^{\Delta\varepsilon_F/kT} - 1. \tag{5}$$

Here, $\mathcal{L}_V$ is a linear operator that has voltage V dependence. The carrier transport equation in Eq. (4) implies that carrier transport is therefore linear with variable u but not linear with voltage V. Mathematical treatment on Eq. (4) and (5) provides the new relation between the carrier injection efficiency $f_i(V)$ under voltage V and the *voltage-dependent* photocarrier collection efficiency $f_c(V)$ through [5]

$$f_i(V) = f_c(V), \tag{6}$$

which is the generalized form of Donolato's theorem. Note that the original Donolato's theorem claimed the relation between $f_i(V)$ and the photocarrier collection efficiency $f_c^{(sc)}$ at short circuit: $f_i(V) = f_c^{(sc)}$ [6]. In other words, it has been assumed in the original theorem that f_i and f_c are independent of voltage V.

The generalized Donolato's theorem in Eq. (6) can be confirmed by the agreement between the simulated $f_i(V)$ and $f_c(V)$ in a p-i-n junction solar cell at various operating voltages V shown in Fig. 2. Even though both $f_i(V)$ and $f_c(V)$ have clear voltage dependence in p-i-n junction solar cells, $f_i(V) = f_c(V)$ still holds, which has not been predicted by the original theorem.

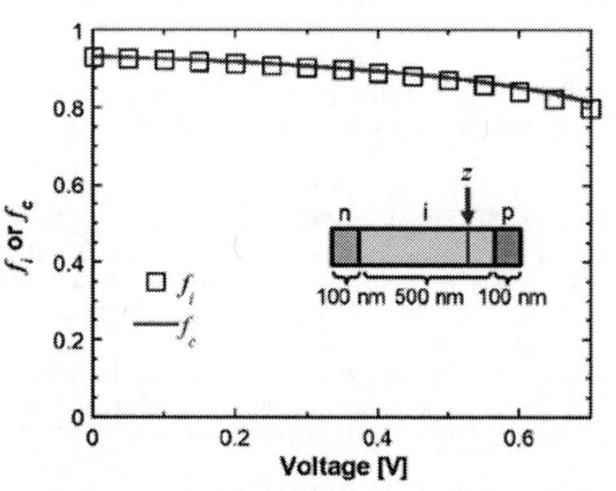

Fig. 2. Simulated dark-carrier injection efficiency f_i and photocarrier collection efficiency f_c of a p-i-n junction solar cell as a function of applied voltage V. The p-i-n junction solar cell simulated here had 100-nm p-region, 500-nm i-region, and 100-nm n-region. The simulated f_c here is the collection efficiency of carriers photogenerated at the position z (see the inset; z is at 100 nm from the p-region) and f_i is the efficiency of carrier injection to the position z. A mobility μ of 0.1 cm²/Vs and a lifetime τ of 100 ns have been assumed.

3. PHOTOVOLTAIC RECIPROCITY AND EXPERIMENTAL RESULTS

The original version of the photovoltaic reciprocity relation in Eq. (1) has employed the original version of Donolato's theorem [6] as one main assumption; therefore it becomes invalid if Donolato's theorem is modified. Fortunately, the modification of the photovoltaic reciprocity relation is straightforward: the voltage-dependent photocarrier collection $f_c(V)$ in Eq. (6) directly links to the voltage-dependent EQE. This leads to the general form of the photovoltaic reciprocity relation

$$\Phi_{\mathrm{EL}}(\varepsilon, V) = \mathrm{EQE}(\varepsilon, V)\Phi_{\mathrm{bb}}(\varepsilon)\left(e^{qV/kT} - 1\right) \tag{7}$$

connecting electroluminescence spectrum with *voltage-dependent* EQE (NOT short-circuit EQE in Eq. (1)).

The theoretical investigation described above has been experimentally confirmed by the electroluminescence measurement on p-i-n junction InGaAs/GaAsP multiple quantum well (MQW) solar cells with the same absorption spectrum but with different carrier non-radiative lifetime as shown in Fig. 3. The device preparation has been explained in [7]. The measured electroluminescence and voltage-dependent EQE are shown in Figs. 4(a)-(d), respectively. The generalized reciprocity relation in Eq. (7) were used to estimate electroluminescence from the measured EQE, which is plotted as open circles in Fig. 4(a). The agreement between the measured electroluminescence and the electroluminescence estimated from voltage-dependent EQE using Eq. (7) justifies the theoretical investigation on the reciprocity relation made in this study.

Furthermore, it has been shown in [8] that the open-circuit voltage V_{oc} for solar cells with voltage-dependent carrier injection and collection can be generalized and written by

$$V_{\mathrm{oc}} = V_{\mathrm{oc}}^{(\mathrm{rad})} + \frac{kT}{q}\ln\eta_{\mathrm{ext}} + \frac{kT}{q}\ln\frac{F_c(V_{\mathrm{oc}})}{F_i(V_{\mathrm{oc}})}, \tag{8}$$

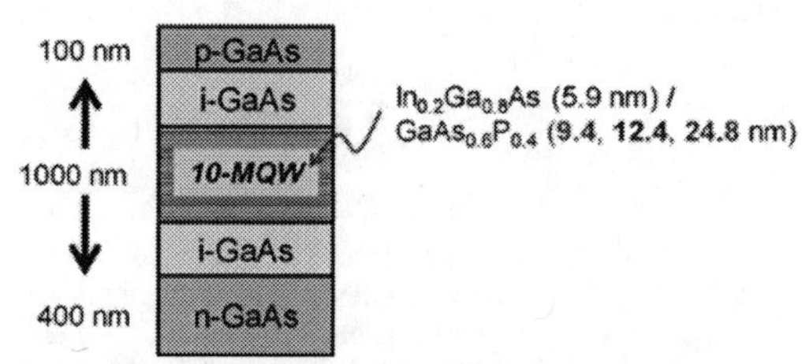

Fig. 3. InGaAs/GaAsP MQW solar cells with the same light absorption but different non-radiative lifetime [7]. Different thicknesses of GaAsP layers on GaAs substrates result in different net strains, which in turn varies the carrier lifetimes. On the other hand, the thickness variation of GaAsP, which has much larger bandgap than InGaAs, has negligible effect on light absorption near the abosprion edge, resulting in the similar light absorption among samples. This can be confirmed by the similar reverse-bias EQE in Figs. 4(b)-(d).

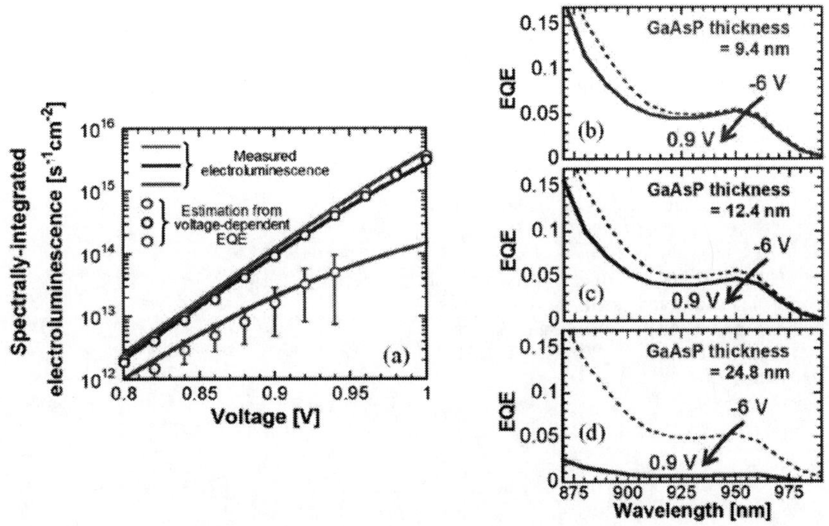

Fig. 4. (a) Measured (lines) and estimated (open circles) electroluminescence of InGaAs/GaAsP MQW solar cells, which have different carrier lifetimes but similar light absorption. The generalized reciprocity relation in Eq. (7) was used to estimate the electroluminescence from voltage-dependent EQE. Reverse-bias and forward-bias EQEs of MQWs with GaAsP thicknesses of (b) 9.4 nm, (c) 12.4 nm, and (d) 24.8 nm. The low signal-to-noise ratio of EQE in MQW cells with 24.5-nm-thick GaAsP results in large error bars for the estimated electroluminescence intensity from this EQE spectra.

where F_i and F_c are the spatial averages of the carrier injection efficiency f_i and the photocarrier collection efficiency f_c. Note that f_i and f_c are defined by the injection and collection of carriers at the specific position as one example is shown in Fig. 2. The third term comes from the fact that the total current J is not simply given by the so-called superposition of dark current J_{dark} and short-circuit current J_{sc} if the photocurrent has voltage dependence. Under the condition that the generalized Donolato's theorem in Eq. (6) holds, we obtain $F_i(V) = F_c(V)$ and thus

$$V_{oc} = V_{oc}^{(rad)} + \frac{kT}{q}\ln \eta_{ext}, \qquad (9)$$

implying that the V_{oc}-η_{ext} relation in Eq. (2) also holds in p-i-n junction solar cells regardless of the failure of the superposition principle.

4. CONCLUSION

In conclusion, we have investigated the validity of the reciprocity relation in p-i-n junction solar cells, where the device operation is not linear with the excess minority-carrier density. By properly linearizing the carrier transport equation, we propose the general form of the reciprocity relation which holds in p-i-n junction solar cells. The cell evaluation techniques based on electroluminescence measurement are found to be available in p-i-n junction solar cells as well, provided that voltage-dependent EQE must be used instead of short-circuit EQE to interpret the measured electroluminescence data. The generalized Φ_{EL}-EQE reciprocity relation is used to confirm that the well-known formula for V_{oc}-η_{ext} relation is still applicable in p-i-n junction solar cells regardless of their voltage-dependent EQE.

ACKNOWLEDGEMENTS

A part of this study was supported by the Research and Development of Ultra-high Efficiency and Low-cost III-V Compound Semiconductor Solar Cell Modules project under the New Energy and Industrial Technology Development Organization (NEDO).

REFERENCES

[1] U. Rau, "Reciprocity relation between photovoltaic quantum efficiency and electroluminescent emission of solar cells," *Physical Review B* **76**, p. 085303, 2007.

[2] T. Kirchartz, U. Rau, M. Hermle, A. W. Bett, A. Helbig, and J. H. Werner, "Internal voltages in GaInP/GaInAs/Ge multijunction solar cells determined by electroluminescence measurements," *Applied Physics Letters* **92**, p. 123502, 2008.

[3] A. Delamarre, M. Paire, J. -F. Guillemoles, and L. Lombez, "Quantitative luminescence mapping of Cu(In, Ga)Se2 thin-film solar cells," *Progress in Photovoltaics: Research and Applications* **23**, pp. 1305-1312, 2015.

[4] T. Kirchartz and U. Rau, "Detailed balance and reciprocity in solar cells," *Physica Status Solidi (a)* **205**, pp. 2737-2751, 2008.

[5] K. Toprasertpong, A. Delamarre, Y. Nakano, J. -F. Guillemoles, and M. Sugiyama, "Generalized reciprocity relations in solar cells with voltage-

dependent carrier collection: application to p-i-n junction devices," to be submitted.
[6] C. Donolato, "A reciprocity theorem for charge collection," *Applied Physics Letters* **46**, pp. 270-272, 1985.
[7] K. Toprasertpong, T. Inoue, A. Delamarre, K. Watanabe, J. -F. Guillemoles, M. Sugiyama, and Y. Nakano, "Electroluminescence-based quality characterization of quantum wells for solar cell applications," *Journal of Crystal Growth* **464**, pp. 94-99, 2017.
[8] T. Kirchartz, J. Nelson, and U. Rau, "Reciprocity between charge injection and extraction and its influence on the interpretation of electroluminescence spectra in organic solar cells," *Physical Review Applied* **5**, p. 054003, 2016.

SIMULATIONS OF SOME EFFECTS OF GRAIN BOUNDARIES IN SOLAR CELLS

Tor Oskar Saetre
University of Agder
Post Box 509, NO-4898 Grimstad, Norway

ABSTRACT: Grain boundaries in solar cells can give several important effects relevant for the performance of solar cells. The grain boundaries can give both negative and positive effects on the power generation. Hence, the study of these effects is an important topic. It is in the present paper suggested that Monte Carlo simulations may be a possible tool for simulating these effects. Some preliminary results are given to demonstrate the method. It is possible to develop this simulation concept into more advanced models which can be of importance in studying several effects of grain boundaries in solar cells.
Keywords: Simulation, Grain, Defects, Fundamentals, Recombination, Monte Carlo simulation

1 INTRODUCTION

The well-known detailed balance model [1] is a very powerful tool for calculating the thermodynamic limits of cell performance, but because of the idealizations involved in the analysis, more insight into the potential of practical photovoltaic devices can be given through alternative approaches.

Monte Carlo computer simulations are a standard tool in scientific fields such as condensed-matter physics, chemical physics and field theory [2].

Some examples of the application of Monte Carlo simulations in the field of solar power can be found in the study of solar concentrators, cf. [3]-[5]. The method has also been used to study light-emitting devices, cf. e.g. [6]-[7]. Cañizo et al. [8] have shown the great potential of the Monte Carlo method in their study of photon conversion for solar cells. They presented a physical model describing different photon conversion mechanisms and implemented a Monte Carlo ray-tracing model. Organic / polymer solar cells have also been modeled by the Monte Carlo method, cf. e.g. [9]-[10]. Both multijunction, hot carrier and intermediate band cells have recently been studied by Monte Carlo simulations [11]-[16].

The effects of grain boundaries in solar cells have been studied extensively for the last decades, cf. e.g. [17]-[29]. Since the majority of solar cells are made of multicrystalline silicon, it is of great importance to study the effects of these structural defects. The role of grain boundaries in the photovoltaic behaviour of the cells is, however, still not fully understood. By employing the Monte Carlo method, which is a new approach, it is expected that new insight and ways of optimizing the grain boundary structures will be found.

The present model is a simple and idealistic approach to studying some effects of grain boundaries in solar cells. The objective of the present work is to demonstrate the method. More advanced details can be implemented and thus give interesting simulation results.

2 MODEL DESCRIPTION

2.1 Photon generation

The data set provided by ASTM G173-03 for the direct normal AM1.5 spectrum have been employed in the Monte Carlo computer program to generate the distribution of photons. This is achieved by using an intrinsic function random number generator which generates a uniform distribution between 0 and 1. These pseudo-random numbers can be converted into any distribution with well-known Monte Carlo techniques. A typical generated distribution of photons giving an AM1.5 distribution of radiation is seen in Fig. 1. The photons are then individually accounted for in the device model. At every simulation run, 10^6 photons were generated and accounted for.

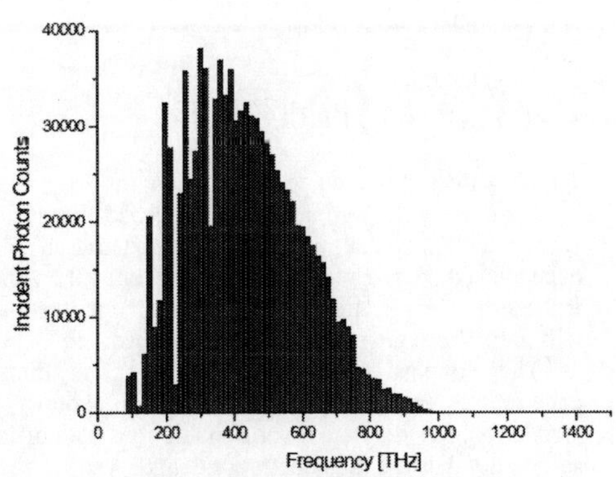

Figure 1: Typical photon distribution generated by employing the ASTM G173-03 data set for direct normal AM1.5 spectrum, downloaded from NREL. This sample contains 10^6 photons binned into 100 interval-bins for this plot.

2.2 Device model

The device model studied in the present paper is a simple and idealistic model, cf. Fig. 2.

Firstly, an ensemble of photons is generated. Any spectrum can be generated, but in the present work only the AM1.5 spectrum described above was employed. Secondly, the photons are distributed uniformly along the surface of the specimen. Thirdly, each successful photon is exciting an electron from the valence band to the conduction band, thus creating a hole in the valence band.

The effects of grain boundaries are introduced according to the following:

1. Grain boundaries are straight and normal to the surface.
2. The grain boundary width can be altered.
3. If a photon enters the grain boundary, it will be lost and not successfully create an electron-hole pair.
4. All photons have the same penetration length.
5. Surface effects like reflections are not considered.

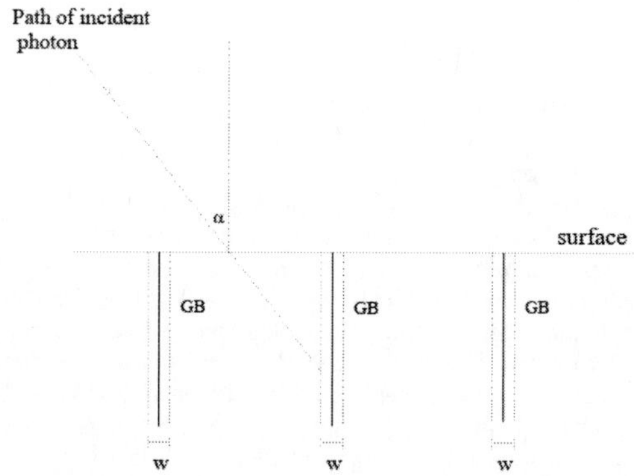

Figure 2: Model of a solar cell with a one-dimensional surface and grain boundaries 90 degrees to the surface. α is the angle of incidence of photons and *w* is the width of grain boundaries. The energy of photons which enter the grain boundary areas is lost.

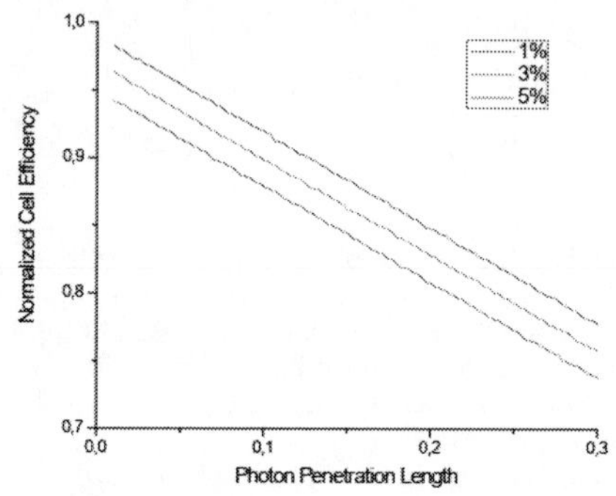

Figure 3: The effect of reduced cell efficiency due to the grain boundaries, where 1.0 is without grain boundaries. The effect of varying the penetration length of photons (arbitrary units), cf. Fig. 2. All photons were assigned identical penetration lengths at each simulation, giving the linear curves as the length is varied by multiple simulations. Results with different grain boundary areas at the solar cell surface are given by: black: 1%, red 3%, blue 5%.

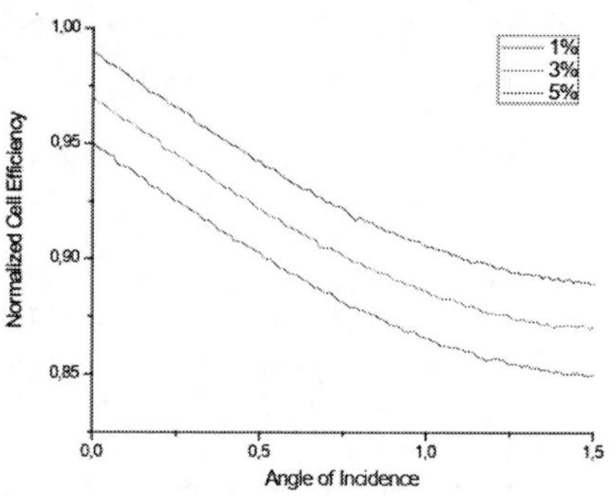

Figure 4: The effect of reduced cell efficiency due to the grain boundaries, where 1.00 is without grain boundaries. The angle of incidence of photons is the angle from the vertical, cf. Fig. 2. For every simulation run the angle of incidence is a constant for all the photons. The curves are given by many simulation runs, varying the angle of incidence. Results with different grain boundary areas at the solar cell surface are given by: black: 1%, red 3%, blue 5%.

3 RESULTS AND DISCUSSION

3.1 The effect of photon penetration length

Photons will normally be absorbed after some penetration into the material. The penetration length depends on the material properties and the photon frequency. The penetration length is the length the photon will move from the surface before it is absorbed.

The simplest modelling assumption is that the penetration length is a constant for all photons in a simulation. Results from several simulations with different assigned values of the penetration length, can be seen in Fig. 3. The resulting cell efficiencies are compared to a similar cell without grain boundaries which would give the value 1. It is seen that the three curves are linear. The black, red and blue curves are for 1%, 3% and 5% grain boundary volume of the cell material. The angle of incidence of the photons was $\pi/4$.

3.2 The effect of the angle of incidence of the photons

The angle of incidence of the photons is constant for each simulation. Results from several simulations with different angles of incidence can be seen in Fig. 4. The black, red and blue curves are for 1%, 3% and 5% grain boundary volume of the cell material. As the angle of incidence approaches 0, i.e. the vertical axis, the effect of grain boundaries is minimized.

3.3 Further work to obtain more realistic results

The preliminary results presented in the present paper demonstrate the potential of using the Monte Carlo method for the study of the effect of grain boundaries. More advanced models will be implemented in future research. The photon penetration length should be determined for each photon by the photon frequency and other factors. Moreover, the grain boundary area will also differ depending on the photon frequency. More complex mechanisms of the effect of the photons when entering the grain boundary area should also be included in order to give a more realistic model of the photon - grain boundary interaction. The strength of the Monte Carlo method is that each photon is individually accounted for and thus can be modelled with very complex behavior. Any photon spectrum can be generated and studied. The effects of other defects, e.g. dislocation clusters, can also be studied.

This work is an ongoing project and more details will be given in future publications.

4 CONCLUSIONS

It has been demonstrated that the Monte Carlo simulation method can be used for simulating some effects of grain boundaries in solar cells. Some preliminary proof of concept results have been given and the possibility of more advanced approaches has been discussed.

REFERENCES

[1] W. Shockley, H.J. Queisser, J. Appl. Phys. 32 (1961), 510.

[2] K. Binder (ed.), Applications of the Monte Carlo Method in Statistical Physics. Springer-Verlag, Berlin 1984.

[3] A. Kostro, B. Huriet, A. Schüler, Proceedings International Conference CISBAT, CD-ROM Edition, 107.

[4] S. Flores Daorta, M. Liscidini, L.C. Andreani, P. Scudo, R. Fusco, Proceedings 26th European Photovoltaic Solar Energy Conference, 2011. 10.4229/26thEUPVSEC2011-1CV.3.11.

[5] A. Schüler, A. Kostro, B. Huriet, C. Galande, J.-L. Scartezzini, Proceedings SPIE, Vol. 7046, 704609-1.

[6] A. Badano, J. Kanicki, J. Appl. Phys., 90 (2001), 4, 1827.

[7] S. Lee, A. Badano, J. Kanicki, IEEE Journal of Selected Topics in Quantum Electronics, 10 (2004), 1, 37; 1077-260X/04.

[8] C. Cañizo, I. Tobias, J. Péres-Bedmar, A.C. Pan, A. Luque, Thin Solid Films, 516 (2008), 20, 6757.

[9] P.K. Watkins, A.B. Walker, G.L.B. Verschoor, Nano Lett, 5 (2005), 9, 1814.

[10] B. Lei, Y. Yao, A. Kumar, Y. Yang, V. Ozolins, J. Appl. Phys., 104 (2008), 024504.

[11] T.O. Saetre, Proc. 29th European Photovoltaic Solar Energy Conf. and Exib., WIP Renewable Energies, 2014, 73.

[12] T.O. Saetre, Photovoltaic Specialist Conference (PVSC), 2015 IEEE 42nd, DOI: 10.1109/PVSC.2015.7356110.

[13] T.O. Saetre, Photovoltaic Specialist Conference (PVSC), 2015 IEEE 42nd, DOI: 10.1109/PVSC.2015.7355803.

[14] T.O. Saetre, Photovoltaic Specialist Conference (PVSC), 2016 IEEE 43rd, DOI: 10.1109/PVSC.2016.7749609.

[15] T.O. Saetre, Photovoltaic Specialist Conference (PVSC), 2016 IEEE 43rd, DOI: 10.1109/PVSC.2016.7749610.

[16] T.O. Saetre, Photovoltaic Specialist Conference (PVSC), 2017 IEEE 44rd, DOI: (*in press*).

[17] J.D. Zook, Appl. Phys. Lett. 37 (1980), 223. DOI: 10.1063/1.91832.

[18] J.G. Fossum, F.A. Lindholm, IEEE Trans. Electr, Dev., ED-27(4) (1980), 692.

[19] C. Donolato, J. Appl. Phys. 54 (1983), 1314.

[20] A. Agarwal, D. Bana, B. Prasad, C.M. Singal, Int. J. Electr., 58(6) (1985), DOI: 10.1080/00207218508939094.

[21] S. Banerjee, H. Saha, Solar Cells, 28(1) (1990), 77.

[22] W. Seifert, G. Morgenstern, M. Kittler, Semicond. Sci. Technol., 8(9) (1993), 1687.

[23] M.M. Deme, S. Mbodji, S. Ndoye, A. Thiam, A. Dieng, G. Sissoko, Revue des Energies Renouvelables, 13 (2010), 109.

[24] J.B. Li, V. Chawla, B.M. Clemens, Adv. Mater., 24(6) (2012), DOI: 10.1002/adma.201103470.

[25] D.P. Joshi, K. Sharma, Indian J. Pure Appl. Phys. 50 (2012), 661.

[26] H.C. Sio, S.P. Phang, H.T. Nguyen, D. Yan, T. Trupke, D. Macdonald, Proc. 31st European Photovoltaic Solar Energy Conf. and Exib., WIP Renewable Energies, 2015.

[27] B. Gaury, P.M. Haney, J. Appl. Phys., 120(23) (2016), DOI: 10.1063/1.4972028.

[28] Z. Chu, M. Yang, P. Schulz, D. wu, X. Ma, E. Seifert, L. Sun, X. Li, K. Zhu, K. Lai, Nature Comm., (2017), DOI: 10.1038/s41467-017-02331-4.

[29] K. Adamczyk, R. Søndenå, G. Stokkan, E. Looney, M. Jensen, B. Lai, M. Rinio, M.D. Sabatino, J. Appl. Phys., 123, 055705 (2018), DOI: 10.1063/1.5018797.

EVALUATING THE POTENTIAL OF OPTICAL MATERIALS AS SOLAR CELL ABSORBERS

B. Dzurnak*[1], L. Danos[2] and T. Markvart[1,3]
[1] – Centre for Advanced Photovoltaics, Czech Technical University, Prague, Czech Republic
[2] – Lancaster University, Lancaster, United Kingdom
[3] – University of Southampton, Southampton, United Kingdom
*address: Technicka 2, Prague 6, 166 27, Czech Republic,
phone: +420 773 484 211
email: branislav.dzurnak@fel.cvut.cz

ABSTRACT: We have developed a methodology to evaluate materials for potential use as absorbers in solar cells without the need for device fabrication. Using our expertise developed for fluorescence collectors we have fabricated relevant structures and characterised them for reabsorption and absolute fluorescence intensity. The latter then form a basis to deduce the chemical potential of the emitted light, closely related to open-circuit voltage of a solar cell where this material would act as an absorber. Detailed analyses are carried out of the relevant losses focusing on interplay between photon recycling, non-radiative quenching and absorptivity/emissivity. Our results present values of open-circuit voltage that can be achieved using easily available laser dyes.
Keywords: Concentrators, Dye-sensitised, Fundamentals, Optical Losses, Optical Properties, Organic Solar Cell, Photoluminescence, Spectroscopy

1 INTRODUCTION

1.1 Open-circuit voltage

Predictions of solar cell efficiencies usually relies on the Shockley-Queisser detailed balance model. The crucial assumptions within this model is a balance between absorbed and emitted light and perfect photon recycling (reabsorption).

Standard textbook equation for open-circuit voltage:

$$qV_{OC} = kT_0 \ln\left(\frac{I_{ph}}{I_0} + 1\right) \tag{1}$$

can be rewritten by considering the efficiencies for incident and emitted light (EQE_S and EQE_O respectively) and introducing quantum yield of photon emission Q and probability of reabsorption r [1]:

$$qV_{oc} = qV_{oc}^{SQ} + kT_o ln\left(\frac{EQE_s}{EQE_o}\right) + kT_o ln\frac{(1-r)Q}{1-rQ} \tag{2}$$

The first term is the limit based on SQ detailed balance, together with second term represents the radiative limit of SQ. The third term concerns effect of non-radiative recombinations and photon recycling. In materials with finite quantum yield, re-absorbed photons are delayed from being emitted but the probability of losing them due to non-radiative process rises. Therefore, higher photon recycling results in a lower open-circuit voltage.

1.2 Reabsorption

Dye fluorescent collectors are ideal materials for studying reabsorption effects, thanks to their high absorption and quantum yield. The total photon flux emitted by a collector consists of direct absorption of excitation illumination and reabsorbed photons of fluorescent emission. The presence of reabsorption alters the observed emitted spectra. Reabsorption depends on the overlap of absorption with emission spectra and can be characterized by probability of reabsorption r. After the first absorption and emission cycle, one would obtain first-generation spectra F_1 i.e. spectra not affected by reabsorption. After subsequent series of reabsorption and re-emission cycles the observed spectra F can be expressed as [2]:

$$F = \frac{F_1 . 10^{-A}}{1 - r.Q} \tag{3}$$

where A is the absorbance and Q is fluorescence quantum yield.

In a spectral region where absorption is weak (long-wavelength edge) the observed spectrum should be identical to the first-generation spectrum. Using appropriate scaling it is possible to determine the value of r by ([3], [2]):

$$r = \frac{\int \alpha F_1(\lambda)d\lambda - \int F(\lambda)d\lambda}{\int \alpha F_1(\lambda)d\lambda} \tag{4}$$

where α is scaling factor used to scale long-wavelength edges of first generation spectrum with observed spectrum.

1.3 Chemical potential of photon flux

If we assume a unit quantum yield, reabsorption and re-emission cycles bring photons within collector into thermal equilibrium with fluorescent species. The chemical potential of photon flux can then be introduced ([4], [5]) and fluorescence spectra can be approximated by black-body radiation in the form of ([5]):

$$F(T^*, E, \mu^*) = \frac{2\pi}{h^3c^2}E^3 exp\left(\frac{\mu^* - E}{kT^*}\right) \tag{5}$$

where μ^* is the chemical potential of the photon flux and T^* is the temperature given by the Kennard-Stepanov relation [6]. Equation (5) is valid for a spectral region where the absorption and emission bands of dye overlap and the reabsorption effect is the strongest (short-wavelength edge). This equation then allows fitting measured spectra of fluorescent collectors and calculation of the chemical potential.

It has been shown that experimental values of chemical potential of photon flux are close to values of dye chemical potential and to theoretical calculations [5]. Because the chemical potential of photons corresponds to free energy of recombined electron-hole pairs, it is directly related to open-circuit voltage by:

$$\mu = qV_{OC} \tag{6}$$

2 EXPERIMENTAL

We chose three BASF Lumogen F fluorescent dyes – Red R305, Orange O240 and Yellow Y083 due to their high quantum efficiencies (>90%, >90%, >85% respectively) [7]. The collectors were prepared by spin coating toluene solution of dyes and PMMA mixture (at 6:1 ratio) onto quartz glass slides of 20x20x1 mm size. Dye solutions were prepared at concentration 10^{-2} M and 10^{-4} M.

Spectroscopic measurements were performed using Bentham Series 200 Spectrometer equipped with TM300 monochromators and Hammamatsu R446 photomultiplier. Absorption spectra were recorded in the range of 400 – 800 nm. Fluorescence spectra were measured in the range of 450 – 900 nm, in two different setups (see Fig. 1).

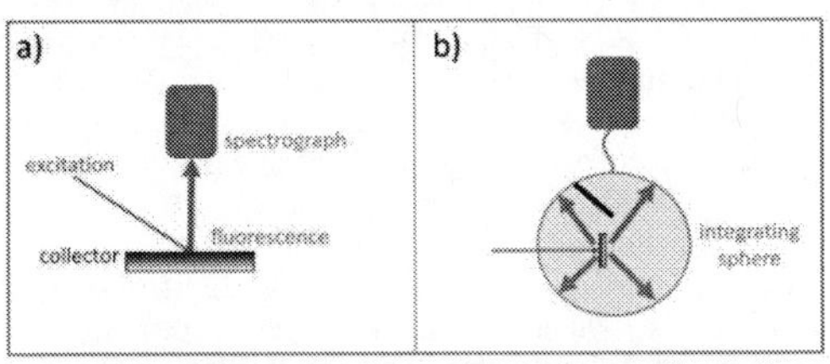

Figure 1: Schematic of experimental setup for a) front surface fluorescence measurements and b) measurement of absolute fluorescence irradiance

In "front surface fluorescence" setup (Fig. 1a), measured spectra of low-concentration collectors (10^{-4} M dye on quartz glass) represent the first-generation spectra, free from reabsorption. This can be confirmed by comparing front surface spectra to spectra of dilute solution (10^{-5} M) which are identical. The collector was excited with focused laser beam at 444 nm wavelength and 2 mW power, fluorescence signal was collected and focused on spectrograph slit by pair of lenses.

Previously, the fluorescence emerging from the edge of collectors has been used for studying reabsorption effects as the photons reaching the edge travel long optical path ([5], [3]). Here we use integrating sphere setup for such measurements. This allow us to observe reabsorbed fluorescence and at the same time to conveniently measure absolute irradiance of fluorescence photons. In "integrating sphere setup", collector was placed inside integrating sphere and excited by the same laser beam through the excitation port of the sphere (Fig. 1b). Signal was collected with optical fibre placed at the detector port of the sphere and guided towards spectrometer input. We used 445 nm longpass filter to supress the excitation beam in collected fluorescence spectra. The integrating sphere setup was calibrated with Bentham CL-2 halogen calibration lamp and fluorescence spectra were accordingly recalculated to absolute fluorescence power density units (mW.m^{-2}.ev^{-1}).

3 RESULTS AND DISCUSSION

The fluorescence spectra of three collectors are shown in Fig. 2. Each spectrum was obtained under the same excitation conditions. The absorbance spectra are also plotted in the graphs.

In the front fluorescence spectra two characteristic spectral regions can be distinguished – short-wavelength region where absorption edge overlaps with fluorescence emission edge, giving rise to re-absorption and long-wavelength region where reabsorption of fluorescent photons is negligible. Front fluorescence spectra are normalised and after scaling integrating sphere spectra it is apparent that shape of the long-wavelength spectra overlap. Using the equation (4) from above we calculated the probability of reabsorption for all collectors.

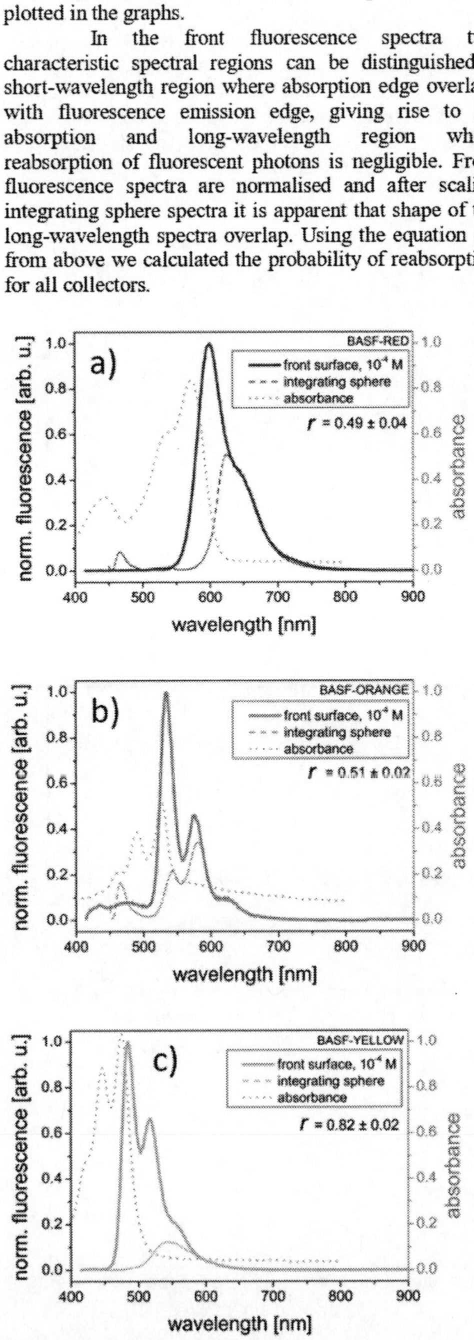

Figure 2: Scaled fluorescence of a) red b) orange and c) yellow collector. Solid line – front surface spectra of low-

concentration dye collector, dashed line – integrating sphere spectra, dotted line - absorbance

Reabsorption in short-wavelength part brings fluorescent photons into equilibrium with the fluorescent species. We therefore used equation (5) to fit the spectra in this region with black-body radiation. This is shown in Fig 3. Parameters of the fit are temperature T^* and flux chemical potential μ^*. It is worth noting that temperature T^* does not correspond to temperature of the collector (around 298 K). The discrepancy has been widely discussed in literature (see, for example, [8]). In contrast with many of these studies, however, the present paper relies on the measured temperature of the photon flux rather than extracting the temperature of the excited species directly using the Kennard-Stepanov relation.

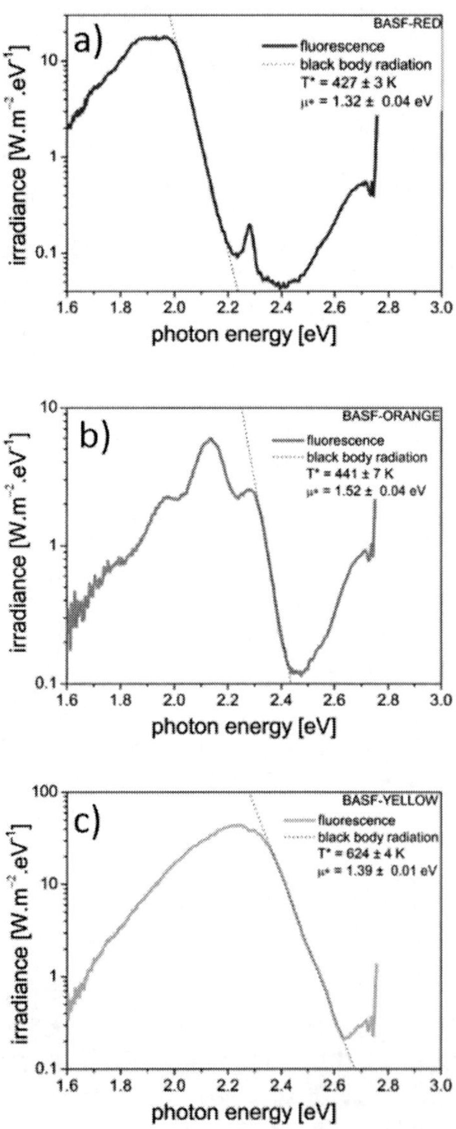

Figure 3: Absolute fluorescence of a) red b) orange and c) yellow collector. Solid line – fluorescence spectra, dotted line – black body radiation fit

Although equation (6) equates the chemical potential and the open-circuit voltage, it is worth noting that in this case, the expression is valid for the device temperature of T^*. For room temperature value of chemical potential, following transformation must be used:

$$\mu_0 = \mu_* \frac{T_0}{T^*} + E_0\left(1 - \frac{T_0}{T^*}\right) + kT_0 \ln\left(\frac{T^*}{T_0}\right) \qquad (6)$$

where μ_0, T_0 represent values at room temperature and E_0 is the energy of zero-phonon line. Values of μ^* obtained from fit and recalculated values μ_0 are summarised in following table:

Table I: Values of "hot" and room temperature chemical potential of collectors' emission

	μ^*	μ_0
Red	1.32	1.58
Orange	1.52	1.74
Yellow	1.39	1.82

The room temperature values exceed the values of hot photon flux since the hot photon flux is effectively cooled down and the excess energy is effectively converted into chemical potential and open-circuit voltage [9] whilst conserving the emitted photon flux. A more advanced concept using hot carrier solar cells where the cooling is performed in an adiabatic manner avoiding any heat dissipation would produce higher room-temperature chemical potentials than given in Table 1. We hope to return to this topic in a separate publication.

In conclusion, we showed the role of photon reabsorption in assessment of efficiency limits by means of Shockley-Queisser model. We proposed a methodology for evaluating reabsorption probability and chemical potential of solar cell absorbers by employing spectroscopy measurements of light emitted by fluorescent collectors. Our results also indicate acquiring of hot absorber which could be favourable for enhancing the values of open-circuit voltage of potential device.

ACKNOWLEDGEMENT

Centre for Advanced Photovoltaics is supported by the Czech Ministry of Education, Youth and Sport. CZ.02.1.01/0.0/0.0/15_003/0000464

REFERENCES

[1] T. Markvart, IEEE J. Photovoltaics, 8, 1, 2018
[2] T. Ahn, R. O. Al-Kaysi, A. M. Müller, K. M. Wentz and C. J. Bardeen, Rev. Sci. Instrum. 78, 086105, 2007
[3] P. Kittidachachan, L. Danos, T. Meyer, N. Alderman and T. Markvart, Chimia 61, 780-786, 2007W. Schockley and H. J. Queisser, J. Appl. Phys. 32, 510, 1961
[4] P. Würfel, J. Phys. C 15, 3967, 1982
[5] T. Meyer and T. Markvart, J. Appl. Phys. 105, 063110, 2009
[6] E. H. Kennard, Phys. Rev. 9, 58, 1917
[7] http://www2.basf.us/additives/pdfs/p3201e.pdf

[8] R. L. V. Metter and R. S. Knox, Chem. Phys 12, 333-340, 1976
[9] T. Markvart, Appl. Phys. Lett. 91, 064102, 2007

QUANTIFYING MINORITY CARRIER LIFETIME AND COLLECTION EFFICIENCY OF SOLAR CELLS BY COMBINED OPTICAL AND ELECTRICAL CHARACTERISATION TECHNIQUES

Vincent Tsai*, Christos Potamialis, Soňa Uličná, Martin Bliss, Thomas R. Betts, Ralph Gottschalg
Centre for Renewable Energy Systems Technology (CREST), Wolfson School of Mechanical, Electrical and Manufacturing Engineering, Loughborough University, Loughborough, Leicestershire, LE11 3TU, UK
*Tel.: +44 1509 635362, Email: v.tsai@lboro.ac.uk

ABSTRACT: This work presents the use of a combined measurement system for spectrally-resolved photoluminescence (PL), time-resolved photoluminescence (TRPL) and transient photocurrent decay (TPCD) to characterise the physical properties of solar cells and their materials. A physical model is proposed to quantify the localised carrier collection efficiency of solar cells from the measured localised minority carrier lifetime from TRPL measurements and the localised minority carrier diffusion time from TPCD measurements. A single excitation laser source is used to measure TRPL and TPCD at the same spot on the solar cell. Combined PL, TRPL and TPCD measurements are conducted on a CdS/CdTe and a CIGS sample. The resulting PL spectra for both samples show that the emission spectra can yield information on the material bandgap. TRPL and TPCD yield localised carrier lifetime and diffusion times of τ_{TRPL}=3.91ns and τ_{TPCD}=40.5ns respectively for the CdS/CdTe sample, and τ_{TRPL}=2.45ns and τ_{TPCD}=196.8ns respectively for the CIGS sample. The ratio between the τ_{TRPL} and τ_{TPCD} values is shown to be proportional to the localised carrier collection efficiency, yielding collection efficiencies of 21.97% and 7.93% for the CdS/CdTe and CIGS sample, respectively. The initial results show that the localised carrier collection efficiency may be affected by the sample's metal contact configuration. In short, this combined measurement approach can offer a novel and useful method of characterising the material quality of solar cells and the localised carrier collection efficiency of finished PV devices.
Keywords: carrier lifetime, diffusion, photoluminescence, CdTe, CIGS

1 INTRODUCTION

Characterisation of minority carrier lifetime in solar cells is important for the research and development of photovoltaic (PV) devices. Minority carrier lifetime usually indicates the quality of the semiconductor material in a solar cell and affects the overall efficiency of a PV device. However, there are other important factors which can limit the PV device efficiency such as carrier transport and collection at the contacts [1]. It is therefore important to experimentally characterise these parameters in order to gain a better understanding of the limiting factors.

Minority carrier lifetime can be directly measured by time-resolved photoluminescence (TRPL) through luminescence decay from radiative recombination [2]. Transient photocurrent decay (TPCD) measures the overall average effective carrier recombination, transport and collection processes [3][4]. Usually, different measurement systems are used for each individual characterisation technique with different excitation sources/methods. This can result in different areas of the sample being measured due to repositioning errors. Furthermore, different excitation source wavelengths can result in different layers of the sample being measured due to different penetration depths [5]. Therefore, it is difficult to directly compare and correlate the data from each measurement system to extract additional physical properties.

For example, the decay times measured from TRPL and TPCD using two separate measurement systems bear no direct correlation with each other if measured using different excitation sources or measurement spots. If TRPL and TPCD are measured at the same spot on the sample, a localised carrier collection efficiency can be extracted which yields additional information for the characterisation of solar cell performance.

As part of this work, a measurement system combining spectrally-resolved photoluminescence (PL), TRPL and TPCD has been developed [6]. Combined PL, TRPL and TPCD measurements were conducted on CdS/CdTe and CIGS solar cells. A physical model is proposed to quantify the localised minority carrier lifetime and carrier collection efficiency of solar cells from the combined measurements. The proposed model extracts the localised minority carrier lifetime from TRPL measurements and the localised minority carrier diffusion time from TPCD measurements at the same spot on the solar cell. This is conducted using a single excitation laser source. In addition, spectrally-resolved PL is measured for both samples to characterise the material bandgap.

2 METHODOLOGY

2.1 The combined measurement system

This section briefly details the measurement system used, an extended description is published in [6]. Figure 1 shows the overall measurement system schematic. The excitation light source is a pulsed picosecond laser with a wavelength of 640nm. A 10× objective lens focusses the laser onto the sample. The resulting luminescence from the laser excitation of the sample is collected using the same objective lens. The 650nm dichroic mirror and a 650nm long pass filter is used to remove any remaining laser light going into the monochromator used for wavelength scanning or selection.

The system uses three photodetectors: An amplified InGaAs photodiode with an extended wavelength range (500nm-1700nm) and two photomultiplier tubes (PMT); one extended visible range (230-920nm) PMT and one NIR (950-1700nm) PMT.

In previously presented work, the NIR PMT was not yet operational [6]. The NIR PMT allows for TRPL measurements of samples such as CIGS and CZTS. The PMTs are connected to a time-correlated single photon counting (TCSPC) board to measure TRPL, while the photodiode is connected to a data acquisition card to measure spectrally-resolved PL.

For transient electrical measurements, the sample is connected to a digital sampling oscilloscope with a 50Ω input impedance via a matching SMA coaxial cable and probes.

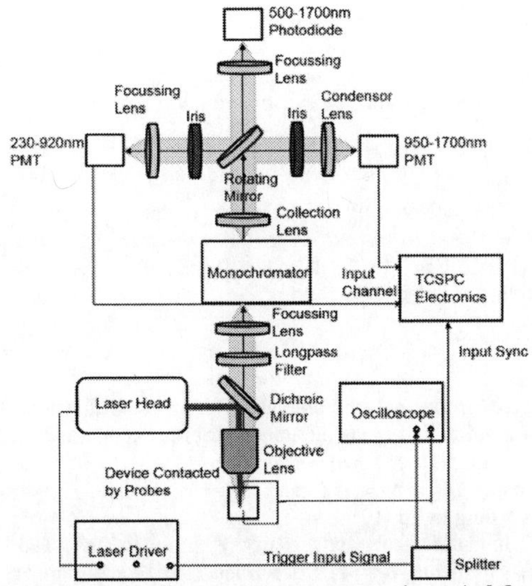

Figure 1: Overall measurement system schematic [6]

2.2 Sample fabrication

The test samples used during this work have been fabricated at CREST (see Figure 2). The CdS/CdTe sample is in the superstrate configuration. CdS was deposited on a 50×50mm substrate using a sono-chemical bath resulting in ~200nm thick films. CdTe was deposited by a close-space sublimation resulting in 4-6μm thick films. The CdS/CdTe devices were finished by depositing ~80nm of gold with an area of 5×5mm to act as the back contact. Further details on the CdS/CdTe device fabrication can be found in [7].

The solution-processed CIGS sample is in the substrate configuration using Mo/MoNx/Mo-coated soda lime glass as a back contact. The CIGS absorber layer thickness was ~3μm with a targeted composition of $Cu_{0.9}In_{0.7}Ga_{0.3}Se_2$. The CIGS device was completed by depositing CdS (~80 nm) using a chemical bath, followed by RF sputtering of intrinsic ZnO and Al doped ZnO (AZO) with a thickness of ~80 and ~500nm, respectively. Finally, a top contact silver grid was evaporated and cells with an area of 0.25cm^2 were separated by mechanical scribing. Further details on the CIGS device fabrication can be found in [8]–[10].

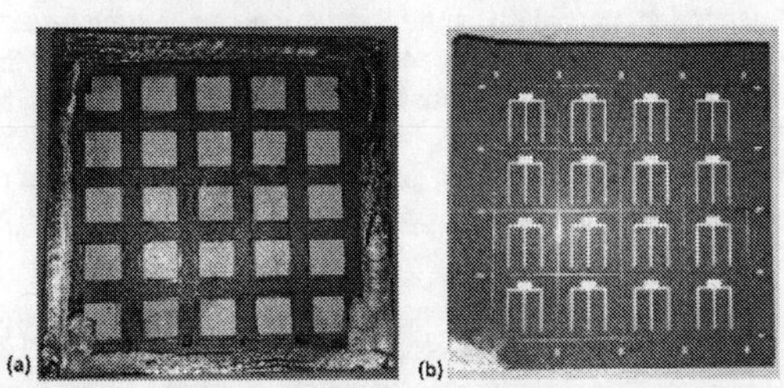

Figure 2: The (a) CdS/CdTe and (b) CIGS samples characterised in this work

3 THEOERETICAL MODELS

3.1 Carrier recombination

Recombination of minority carriers in semiconductors is due to three main processes, Shockley-Read-Hall (SRH), radiative and Auger recombination [11]. This relation can be represented by rate equations and so the overall recombination rate is as follows:

$$\frac{dn}{dt} = \left(\frac{dn}{dt}\right)_{\mathrm{SRH}} + \left(\frac{dn}{dt}\right)_{\mathrm{rad}} + \left(\frac{dn}{dt}\right)_{\mathrm{Auger}} \quad (1)$$

where n is the electron density and t is time. Eq.(1) can be calculated using the following equation [12][13]:

$$\frac{dn}{dt} = -\frac{np - n_0 p_0}{\frac{1}{A_n}(p + n_i) + \frac{1}{A_p}(n + n_i)} \quad (2)$$

$$-B(np - n_0 p_0) - C_n(np - n_0 p_0)n$$

$$-C_p(np - n_0 p_0)p$$

where n is the electron number density in the conduction band, p is the hole number density in the valence band, n_0 and p_0 are their values at equilibrium respectively, n_i is the intrinsic carrier density, A_n and A_p are the SRH recombination coefficients for electrons and holes respectively, B is the radiative recombination coefficient, and $C_n$and C_p are the Auger recombination coefficients for electrons and holes respectively. It is assumed in Eq.(2) that the energy level for the SRH recombination centre is located near the intrinsic Fermi level, which is usually located in the middle of the bandgap [14].

The p-type layer in the measured samples are thicker than the n-type layer by several orders of magnitude, as detailed in Section 2.2. Therefore, only the minority electrons in the p-type quasi-neutral region was considered in the physical model.

In the p-type layer, the majority carriers are holes which can be expressed as $p \cong p_0 - N_A$ [1], where N_A is the doping concentration of the p-type acceptors and p_0 is the hole number density at equilibrium. The hole number density p is greater than the electron number density n, (i.e. $p \gg n$) which can be represented by $n(t) = n_0 + \Delta n(t)$ [15] where n_0 is the electron number density at equilibrium and Δn is the excess electron number density. It is assumed that the excess electron number density is greater than the electron number density at equilibrium (i.e. $\Delta n(t) \gg n_0$). Therefore n_0 can simply be neglected (i.e. $n(t) \cong \Delta n(t)$) and Eq.(2) can then be simplified to [13]:

$$\frac{d\Delta n}{dt} = -\left(A_n + BN_A + C_p N_A^2 + C_n N_A \Delta n\right)\Delta n \quad (3)$$

Eq.(3) is a non-linear differential equation and can only be solved by numerical methods. Therefore, it is not suitable for fitting TRPL data because many physical parameters need to be known in advance. However, Eq.(3) can be simplified by assuming the low-injection condition, where the excess minority carrier concentration is less than the acceptor concentration (i.e. $\Delta n \ll N_A$). The $C_n N_A \Delta n$ term can be neglected and thus Eq.(3) can be simplified to:

$$\frac{d\Delta n(t)}{dt} = -\frac{\Delta n(t)}{\tau_n} \quad (4)$$

where τ_n is the minority carrier lifetime represented as [2]:

$$\tau_n = \frac{1}{A_n + BN_A + C_p N_A^2} \quad (5)$$

The general solution of Eq.(4) is:

$$\Delta n(t) = \Delta n(0)e^{-t/\tau_n} \quad (6)$$

It should be noted that the $C_n N_A \Delta n$ term cannot be neglected in the case of the high injection condition $\Delta n \approx p_0 = N_A$ (i.e. at the initial stage of the carrier decay process). In this case, if $C_n N_A \Delta n$ term is assumed to be a constant value, then Eq.(3) will become:

$$\frac{d\Delta n(t)}{dt} = -\frac{\Delta n(t)}{\tau_{\Delta n}} \quad (7)$$

where $\tau_{\Delta n}$ is the minority carrier lifetime at high injection conditions and is represented by:

$$\tau_{\Delta n} = \frac{1}{A_n + BN_A + C_p N_A^2 + C_n N_A \Delta n} \quad (8)$$

The general solution of this equation is:

$$\Delta n(t) = \Delta n(0)e^{-t/\tau_{\Delta n}} \quad (9)$$

If Eqs.(5) and (8) are compared, it can be concluded that the value of $\tau_{\Delta n}$ is less than τ_n:

$$\tau_{\Delta n} < \tau_n \quad (10)$$

In short, Eq.(8) describes the decay regime of the initial high injection conditions in the carrier decay process, while Eq.(5) describes the decay regime approaching equilibrium of the low injection condition. As a result, the overall decay process is the sum of these two regimes which can be described as follows:

$$\Delta n(t) \approx \Delta n_1 e^{-t/\tau_{\Delta n}} + \Delta n_2 e^{-t/\tau_n} \quad (11)$$

where Δn_1 and Δn_2 are the excess carrier densities for the high and low injection regimes, respectively. In short, there are two carrier lifetime constants which can be obtained from the TRPL measurement: $\tau_{\Delta n}$, the carrier lifetime under the initial high injection conditions, and τ_n, the carrier lifetime under low injection conditions. Since solar cells normally operate under low injection conditions, τ_n was interpreted as the carrier lifetime. The minority carrier lifetime τ_n is therefore represented as τ_{TRPL}:

$$\tau_{TRPL} = \tau_n = \frac{1}{A_n + BN_A + C_P N_A^2} \quad (12)$$

3.2 Carrier lifetime from TRPL

TRPL directly detects the photon flux which is generated from radiative carrier recombination [15] and can be expressed by the following equation:

$$\varphi_{PL} \propto B(np - n_0 p_0) \quad (13)$$

where B is the radiative recombination coefficient.

Since the n-type layer is much thinner compared to the p-type layer of the samples in this work, it can be assumed that the output photon flux seen in the PL emission signal is from the p-type quasi-neutral region. As discussed in Section 3.1, in the p-type quasi-neutral region, the majority carriers p can be expressed as $p \cong p_0 = N_A$. The minority carriers n can be expressed as $n = n_0 + \Delta n$, where Δn is the excess electron number density [15]. These two expressions can then be substituted into Eq.(13):

$$\varphi_{PL} \propto BN_A \Delta n(t) \quad (14)$$

The detected PL signal φ_{PL} is therefore linearly proportional to the excess minority electron density Δn. The resulting TRPL data can be fitted by a double exponential decay function:

$$\varphi_{PL}(t) = a_1 e^{-t/\tau_1} + a_2 e^{-t/\tau_2} \quad (15)$$

where a_1 and a_2 are parameters for the amplitudes, t is time and τ_1 and τ_2 are parameters for the exponential time constants. Eq.(15) is identical to Eq.(11), where $a_1 = \Delta n_1$, $a_2 = \Delta n_2$, $\tau_1 = \tau_{\Delta n}$ and $\tau_2 = \tau_n$, where $\tau_1 < \tau_2$ according to Eq.(10).

In short, the longer decay time in the TRPL measurement $\tau_2 = \tau_n$ describes the decay process approaching equilibrium under low injection conditions. This carrier lifetime is related purely to the material properties and is independent of the excess carrier density as indicated by Eq.(5). On the contrary, the shorter decay time in the TRPL measurement $\tau_1 = \tau_{\Delta n}$ describes the initial decay process under high injection conditions. Its value is dependent on the excess carrier density as indicated by Eq.(8). Therefore, $\tau_1 = \tau_{\Delta n}$ cannot represent the material properties and $\tau_2 = \tau_n$ represents the localised carrier lifetime of the TRPL measurement.

3.3 Carrier diffusion time from TPCD

It was given that the measured TPCD signal comes from the p-type quasi-neutral region by assuming that the contributions from the space charge region and n-type quasi neutral region are negligible due to the very thin n-type CdS layer used in the samples. The minority electron number density n obeys the diffusion-recombination continuity equation as follows [14]:

$$\frac{\partial n}{\partial t} = D_n \frac{\partial^2 n}{\partial x^2} + g - r = 0 \quad (16)$$

where D_n is the electron diffusion coefficient, g and r are the generation and recombination rates. The spatially dependent partial differential equation can be transformed into a spatially independent rate equation in which all the physical quantities discussed below are presumed to be spatially averaged.

Minority carriers are generated from the absorption of the input photon flux. Therefore, the generation rate is as follows [16]:

$$g = \alpha\varphi = \left(\frac{dn}{dt}\right)_{gen} \quad (17)$$

where α is the absorption coefficient and φ is the input photon flux. The recombination rate r is a combination of SRH, radiative and Auger recombination which can be described by a single value of the average minority carrier lifetime $\langle\tau_n\rangle$, as follows:

$$-r = \left(\frac{dn}{dt}\right)_{rec} = -\frac{n - n_0}{\langle\tau_n\rangle} \quad (18)$$

where n_0 is the minority carrier electron number density at equilibrium (i.e. no injection). It should be noted that $\langle\tau_n\rangle$ is the average carrier recombination lifetime along the diffusion path, which is different from the localised carrier lifetime measured from the TRPL. The diffusion process in Eq.(16) can be represented by the minority carrier diffusion time τ_{diff} as follows:

$$D_n \frac{\partial^2 n}{\partial x^2} \rightarrow \left(\frac{dn}{dt}\right)_{diff} = -\frac{n - n_0}{\tau_{diff}} \quad (19)$$

As a result, the partial differential equation of Eq.(16) can be represented by the following rate equations:

$$\frac{dn}{dt} = \left(\frac{dn}{dt}\right)_{gen} + \left(\frac{dn}{dt}\right)_{rec} + \left(\frac{dn}{dt}\right)_{diff} \quad (20)$$

It should be noted that in the actual TPCD measurement system, there are other factors which could possibly affect the measurement such as the PV device's junction capacitance and the measurement system capacitance in some cases. The advantage of using this rate equation approach is because it can be expanded further to accommodate other factors as follows:

$$\frac{dn}{dt} = \left(\frac{dn}{dt}\right)_{gen} + \left(\frac{dn}{dt}\right)_{rec} + \left(\frac{dn}{dt}\right)_{diff} \quad (21)$$
$$+ \left(\frac{dn}{dt}\right)_{SCR} + \left(\frac{dn}{dt}\right)_{RC} + \cdots$$

where 'SCR' represents the junction capacitance from the space charge region and 'RC' represents the measurement system capacitance which is the electronic response time of the measurement system. However, these two factors will be ignored in this work for simplicity. The '$diff$', 'SCR' and 'RC' terms in Eq.(21) are carrier transport processes and can summed to form an overall collection rate, which is different from the generation and recombination processes as follows:

$$\left(\frac{dn}{dt}\right)_{col} = \left(\frac{dn}{dt}\right)_{diff} + \left(\frac{dn}{dt}\right)_{SCR} + \left(\frac{dn}{dt}\right)_{RC} \quad (22)$$
$$= -\frac{n - n_0}{\tau_{col}}$$

where the collection time τ_{col} is introduced to account for the time for electrons travelling from the point of generation across the sample to the metal contacts and then the measurement system to become a detected current signal. Therefore, the detected current signal $J_{TPCD}(t)$ is linearly proportional to the carrier collection rate in Eq.(22) as follows:

$$J_{TPCD}(t) = qL\frac{n(t) - n_0}{\tau_{col}} \quad (23)$$

where q is the electron charge and L is the distance of the carrier diffusion path from the excited area to the metal contact.

The overall rate equation for the transient photocurrent decay is formed by combining Eqs. (17), (18) and (22):

$$\frac{dn(t)}{dt} = \alpha\varphi(t) - \frac{n(t) - n_0}{\langle\tau_n\rangle} - \frac{n(t) - n_0}{\tau_{col}} \quad (24)$$

Eq.(24) is the sum of the charge carrier generation, recombination, and collection processes. Since the excitation is a pulsed source, the photon flux $\varphi(t) = 0$ is assumed for $t \geq 0$. The decay from TPCD is measured and fitted after the excitation pulse so there will be no contribution to the generation at $t \geq 0$ in the decay process and the $\alpha\varphi(t)$ term can be omitted from Eq.(24):

$$\frac{dn(t)}{dt} = -\frac{n(t) - n_0}{\langle\tau_n\rangle} - \frac{n(t) - n_0}{\tau_{col}} \quad (25)$$
$$= -\frac{n(t) - n_0}{\tau_{decay}}, \quad \text{for } t \geq 0$$

where the decay time τ_{decay} is a combination of the carrier recombination and collection processes:

$$\frac{1}{\tau_{decay}} = \frac{1}{\langle\tau_n\rangle} + \frac{1}{\tau_{col}} \quad (26)$$

The solution of Eq.(25) is as follows:

$$n(t) = n(0)e^{-t/\tau_{decay}} + n_0 \quad (27)$$

where $n(0)$ is the carrier number density at $t = 0$ which depends on the injection level and τ_{decay} is the decay time in the TPCD. Eq.(27) can be substituted into Eq.(23) and the detected TPCD signal will become:

$$J_{TPCD}(t) = \frac{qLn(0)}{\tau_{col}} exp\left(-\frac{t}{\tau_{decay}}\right) \quad (28)$$
$$= J_{TPCD}(0) exp\left(-\frac{t}{\tau_{decay}}\right)$$

where L is distance of the carrier diffusion path from the excited area to the metal contact as shown by the diagram in Figure 3.

3.4 Localised collection efficiency

As indicated by Eq.(26) the decay time measured by TPCD τ_{decay} is determined by two factors: the average carrier lifetime $\langle\tau_n\rangle$ along the diffusion path is in the ns range so cannot currently be resolved from the TPCD measurement. The current TPCD measurement setup can only detect the slower process of carrier collection τ_{col} (10s to 100s ns range). Therefore:

$$J_{TPCD}(t) \approx J_{TPCD}(0) exp\left(-\frac{t}{\tau_{col}}\right) \quad (29)$$

If the minority carrier collection process is assumed to be dominated by the minority carrier diffusion process, the contribution of the 'SCR' and 'RC' terms in Eq.(21) can be ignored. Therefore the collection time τ_{col} is almost equal to the carrier diffusion time τ_{diff}:

$$\tau_{col} \approx \tau_{diff} \quad (30)$$

The measured decay time by TPCD τ_{TPCD} can therefore be represented by the minority carrier diffusion time τ_{diff}:

$$\tau_{TPCD} \approx \tau_{diff} \quad (31)$$

It can be shown that the minority carrier diffusion time is determined by the distance L of the carrier diffusion path from the excited area to the metal contact, and the minority carrier diffusion coefficient D_n as:

$$\tau_{TPCD} \approx \tau_{diff} = \frac{L^2}{2D_n} = \frac{L^2}{2L_n^2/\tau_n} = \frac{\tau_n}{2} \frac{1}{\eta_{col}^2} \quad (32)$$

It should be noted that the relation $L_n = \sqrt{\tau_n D_n}$ [1] is used in Eq.(32) where L_n is the minority carrier diffusion length, D_n is the diffusion coefficient and τ_n is the minority carrier lifetime. In Eq.(32), the localised collection efficiency η_{col} is defined as:

$$\eta_{col} \equiv \frac{L_n}{L} = \sqrt{\frac{\tau_{TRPL}}{2\tau_{TPCD}}} \quad (33)$$

In which $\tau_n = \tau_{TRPL}$ as described above in Section 3.3. It should be noted that the TRPL and TPCD measurements represent the localised physical quantities, which are measured at the same spot on the sample. The localised collection efficiency η_{col} is defined as the ratio between the diffusion length L_n and the distance of the diffusion path L, which characterises the percentage of the photogenerated carriers in the sample that have been transported and collected as a detected TPCD signal before the carriers recombine.

The merit of this combined measurement approach is thus demonstrated by Eq.(33) where the localised carrier lifetime is directly measured by TRPL and the localised carrier diffusion time is measured by TPCD at the same spot. The ratio between these two values will yield the localised carrier collection efficiency which can provide additional useful information on the PV device's performance. This localised carrier collection efficiency cannot be measured by separate measurement systems based on the physical model described above.

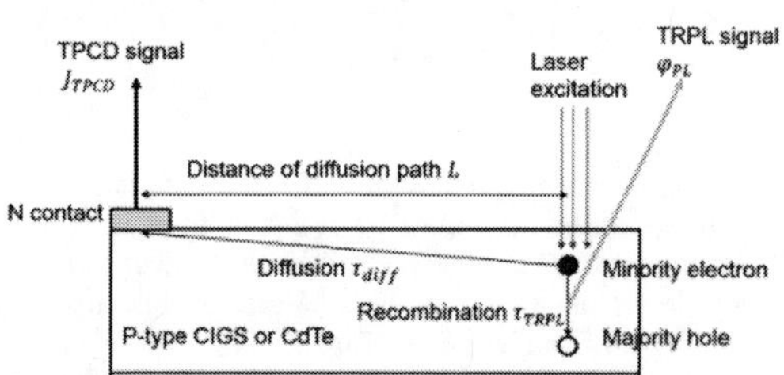

Figure 3: Illustration of the generation, recombination and collection processes of minority carriers in the measured samples.

4 INITIAL RESULTS AND DISCUSSION

Combined measurements of PL, TRPL and TPCD were conducted on the two test samples described in Section 2.2. The CIGS sample was measured to test the capability of the NIR PMT which was not previously operational. The CdS/CdTe sample has been measured in the previous work [6] and is used in this work to calculate the localised carrier collection efficiency.

4.1 Spectrally-resolved photoluminescence

Figure 4 shows the resulting PL spectra of the CdS/CdTe and CIGS sample. The laser excitation was pulsed at 2.5MHz for the CdS/CdTe sample and 40MHz for the CIGS sample to measure a detectable PL signal.

Qualitative analysis was conducted on the resulting PL spectra. The PL spectra have either one or two broad emission peaks which correspond to the material bandgaps, which are the main source of radiative band-to-band recombination.

The CdS/CdTe sample has two main emission peaks at ~1.47eV and ~1.50eV. The peak at 1.50eV is attributed to the CdTe layer while the peak at 1.47eV is attributed to the CdS_xTe_{1-x} layer. The formation of this CdS_xTe_{1-x} layer results from the diffusion of S into the CdTe layer, which is caused by the $CdCl_2$ annealing treatment as detailed in [17][18].

The CIGS sample has one main broad emission peak at around ~1.18eV which is attributed to the luminescence from the p-type CIGS absorber layer, and is in the expected range of CIGS solar cells (1-1.7eV) [19]. The CIGS p-type absorber layer has a tuneable bandgap by alloying $CuInSe_2$ with either S or Ga [10] which changes the bandgap depending on the material composition ratio. When the Ga/(Ga+In) ratio is between 25-30%, the bandgap is normally between 1.1 and 1.2eV [20] as is observed in Figure 4.

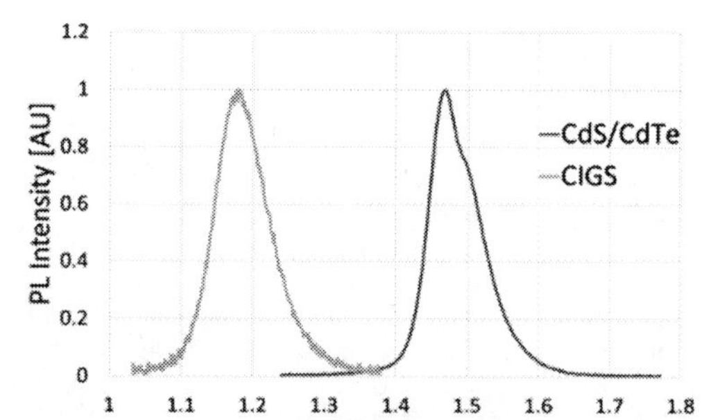

Figure 4: Normalised PL spectra of a CdS/CdTe and CIGS sample

4.2 Time-resolved photoluminescence

TRPL measurements were then conducted on the same spot as PL measurements without removing the samples. Figures 5 and 6 show the TRPL decay curves for the CdS/CdTe and CIGS sample respectively. The monochromator wavelength was set to the peak value of the PL emission spectrum: to 846nm (1.47eV) for the CdS/CdTe sample and to 1055nm (1.18eV) for the CIGS sample. The (230-920nm) PMT was used to measure the CdS/CdTe sample while the other (950nm-1700nm) PMT was used to measure the CIGS sample. The laser frequency was set at 2.5MHz for the CdS/CdTe sample and 20MHz for the CIGS sample.

A double exponential decay function was used to fit the two TRPL decay curves. The extracted values were ($\tau_1 = 1.39$ns, $\tau_2 = 3.91$ns) for the CdS/CdTe sample and ($\tau_1 = 0.63$ns, $\tau_2 = 2.45$ns) for the CIGS sample. As discussed earlier in Section 3.4, τ_1 is the lifetime at the initial high injection condition and τ_2 is the lifetime at low injection condition. Therefore, the value τ_2 will be used as the τ_{TRPL} value as it is the bulk recombination. It should also be noted that the TRPL decay curves did not have the instrument response function (IRF) deconvoluted which may result in slight overestimation of the measured carrier lifetime.

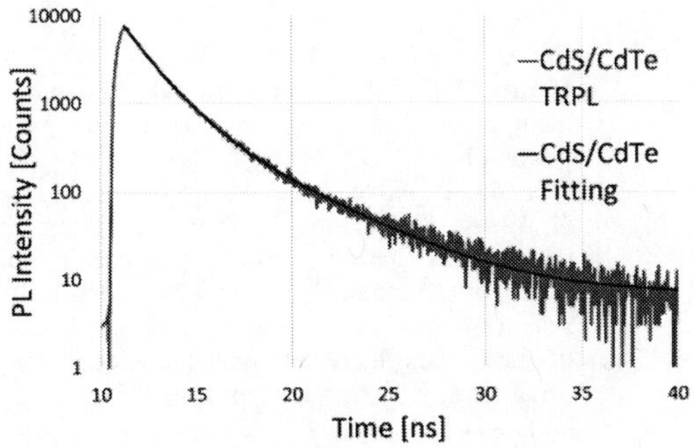

Figure 5 Corresponding room temperature TRPL decay curves of the CdS/CdTe sample (τ_1 = 1.39ns, τ_2 = 3.91ns) measured at 846nm with 30s integration time

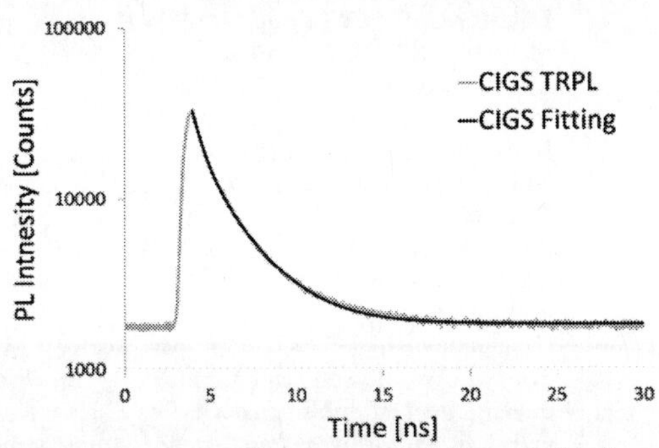

Figure 6 Corresponding room temperature TRPL decay curves of the CIGS sample (τ_1 = 0.63ns, τ_2 = 2.45ns) measured at 1055nm with 30s integration time

4.3 Transient photocurrent decay

TPCD measurements were conducted on the same area as the TRPL and PL on both samples (see Figures 7 and 8). During TPCD measurements, the PV device is contacted by probes which are connected to an oscilloscope. Each laser pulse creates excess carriers and thus generates a photovoltage. The subsequent decay is measured as a function of time and the minority carrier diffusion time τ_{TPCD} can be extracted through fitting. A single exponential decay function (as described by Eq.(29)) was used to fit the two decay curves. τ_{TPCD} values of 40.5ns and 196.8ns were extracted for the CdS/CdTe sample and the CIGS sample respectively.

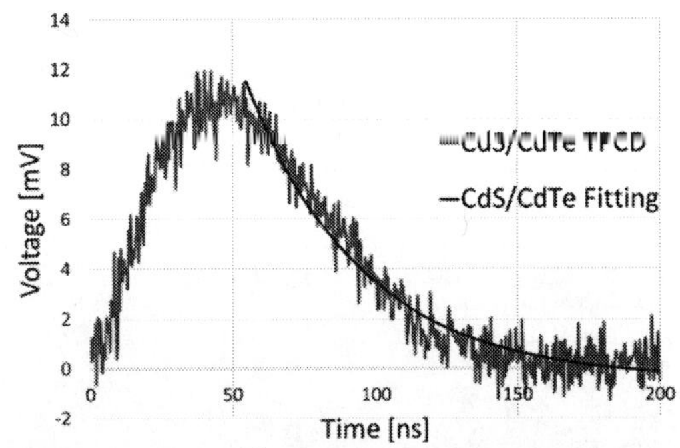

Figure 7 Corresponding room temperature TPCD curves of the CdS/CdTe sample ($\tau_{TPCD} = 40.5ns$) measured under 640nm wavelength and 2.5MHz frequency excitation without external biasing light

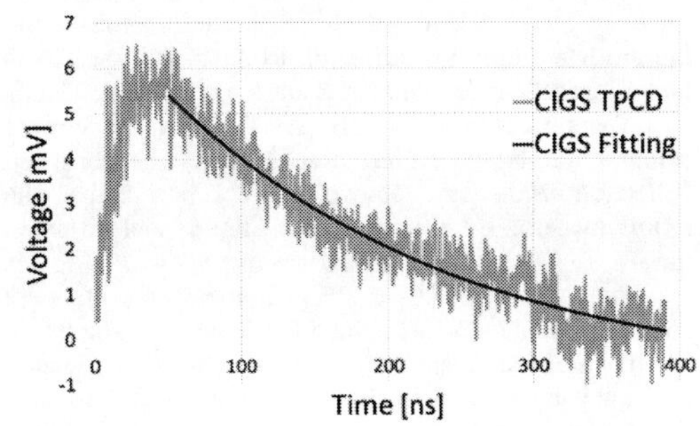

Figure 8 Corresponding room temperature TPCD curves of the CIGS sample ($\tau_{TPCD} = 196.8ns$) measured under 640nm wavelength and 2.5MHz frequency excitation without external biasing light

4.4 The localised carrier collection efficiency

The localised carrier collection efficiency η_{col} was calculated for both samples using Eq.(33) and the fitted values τ_{TRPL} and τ_{TPCD} from TRPL and TPCD, respectively (see Table I). η_{col} is the percentage of generated carriers collected by the metal contacts which do not recombine during the transport process.

It should be noted that the measured power conversion efficiency (PCE) η of a device is better related to the 'global' carrier collection efficiency where the whole sample is illuminated by the incident light. On the contrary, η_{col} measured from the 'localised' carrier collection efficiency is directly linked to the small spot of the laser excitation. The measurement system is currently only a localised point-based measurement and not a spatially-resolved measurement. This means that the measured localised carrier collection efficiency is not exactly proportional to the power conversion efficiency and is possibly more sensitive to the sample's contact configuration as will be discussed below. Therefore, the localised collection efficiency can only be compared with the overall PCE if the samples have the same contact configuration and also measurement setup.

The localised carrier collection efficiency η_{col} calculated by Eq.(33) show that the CdS/CdTe's η_{col} is 21.97% and 7.89% for CIGS. The CdS/CdTe's localised collection efficiency is more than double than that of CIGS. It is initially speculated that one of the reasons is because the front and back contacts of the CdS/CdTe sample have a larger area and are also more densely arranged than the CIGS sample. This means that there is a higher chance of the carriers being collected by the contact configuration of the CdS/CdTe sample measured here.

Table I: Fitted TRPL τ_{TRPL} and TPCD τ_{TPCD} values and calculated collection efficiencies η_{col}

Sample	τ_{TRPL} (ns)	τ_{TPCD} (ns)	η_{col} (%)
CdS/CdTe	3.91	40.5	21.97
CIGS	2.45	196.8	7.89

5 CONCLUSIONS AND FUTURE WORK

A combined measurement system of PL, TRPL and TPCD has been developed in this work. A physical model was proposed to quantify the localised carrier collection efficiency of solar cells from the measured localised minority carrier lifetime from TRPL measurements and the localised minority carrier diffusion time from TPCD measurements at the same spot on the solar cell. This is achieved by combined TRPL and TPCD measurements using a single excitation laser source. The localised collection efficiency is shown to be proportional to the ratio between the localised carrier lifetime and diffusion time.

Combined PL, TRPL and TPCD measurements were conducted on a CdS/CdTe and a CIGS sample. The initial results indicate that the localised carrier collection efficiency in this work is possibly affected by the sample's metal contact configurations and measurement setup. However, further investigation and measurements need to be conducted to verify this. This combined measurement approach can nevertheless offer a novel and useful method of characterising the material quality of solar cells and the localised carrier collection efficiency of the finished PV device.

The next steps in this work are to also examine the other factors which can affect the TPCD and TRPL measurements. For TPCD, external bias lighting will be applied to the sample during the TPCD measurement to observe the effect on the junction capacitance [21]. The signal-to-noise ratio and response time of the TPCD can also be improved. Furthermore, the IRF will be deconvoluted from the TRPL measurements to increase the accuracy of the extracted carrier lifetime values of TRPL.

Finally, the PL, TRPL and TPCD measurements are currently point-based measurements, an x-y scanning stage can be implemented for spatially-resolved measurements which in turn will provide further information on the point-by-point localised collection efficiency of the entire sample.

6 ACKNOWLEDGEMENTS

This work was supported by the UK Engineering and Research Council through the PVteam project EP/L017792/1: Photovoltaic Technology based on Earth Abundant Materials and by the EPSRC Centre for Doctoral Training in New and Sustainable Photovoltaics CDT-PV, EP/L01551X/1.

7 REFERENCES

[1] J. Nelson, *The Physics Of Solar Cells*. London: Imperial College Press, 2003.

[2] W. Metzger, R. Ahrenkiel, and P. Dippo, "Time-resolved photoluminescence and photovoltaics," in *Department of Energy Solar Energy Technologies Program Review Meeting*, 2004, vol. 1, no. January, pp. 1–2.

[3] S. Wood, D. O. Connor, C. W. Jones, J. D. Claverley, J. C. Blakesley, C. Giusca, and F. A. Castro, "Transient photocurrent and photovoltage mapping for characterisation of defects in organic photovoltaics," *Sol. Energy Mater. Sol. Cells*, vol. 161, no. November 2016, pp. 89–95, 2017.

[4] X. Wang, S. Karanjit, L. Zhang, H. Fong, Q. Qiao, and Z. Zhu, "Transient photocurrent and photovoltage studies on charge transport in dye sensitized solar cells made from the composites of TiO_2 nanofibers and nanoparticles," *Appl. Phys. Lett.*, vol. 98, no. 8, pp. 1–4, 2011.

[5] D. Kuciauskas, J. N. Duenow, A. Kanevce, J. V Li, M. R. Young, P. Dippo, and D. H. Levi, "Optical-Fiber-Based, Time-Resolved Photoluminescence Spectrometer for Thin-Film Absorber Characterization and Analysis of TRPL Data for CdS / CdTe Interface," in *Proceedings of the 38th IEEE Photovoltaic Specialist Conference*, 2011, pp. 1721–1726.

[6] V. Tsai, F. Bittau, C. Potamialis, M. Bliss, T. R. Betts, and R. Gottschalg, "Combined Electrical and Optical Characterisation of Recombination Mechanisms and Minority Carrier Lifetime in Solar Cells," in *14th Photovoltaic Science, Applications and Technology Conference*, 2018, pp. 57–60.

[7] C. Potamialis, F. Lisco, B. Maniscalco, M. Togay, J. W. Bowers, and J. M. Walls, "Process development of sublimated Cu-free CdTe solar cells," in *13th Photovoltaic Science, Application and Technology Conference*, 2017.

[8] S. Uličná, P. Arnou, C. S. Cooper, V. Wright, Lewis D.Malkov, J. M. Walls, and J. W. Bowers, "A comparison of different selenisation approaches for $Cu(In,Ga)(S,Se)_2$ solar cells," in *13th Photovoltaic Science, Application and Technology Conference*, 2017.

[9] S. Uličná, P. Arnou, A. Eeles, M. Togay, L. D. Wright, A. Abbas, V. Malkov, J. M. Walls, and J. W. Bowers, "Control of $MoSe_2$ formation in hydrazine-free solution-processed CIS / CIGS thin film solar cells," in *44th IEEE PVSC Conference*, 2017.

[10] P. Arnou, C. S. Cooper, S. Uličná, A. Abbas, A. Eeles, L. D. Wright, A. V. Malkov, J. M. Walls, and J. W. Bowers, "Solution processing of $CuIn(S,Se)_2$ and $Cu(In,Ga)(S,Se)_2$ thin film solar cells using metal chalcogenide precursors," *Thin Solid Films*, vol. 633, no. October, pp. 76–80, 2017.

[11] V. K. Khanna, "Physical understanding and technological control of carrier lifetimes in semiconductor materials and devices: A critique of conceptual development, state of the art and applications," *Prog. Quantum Electron.*, vol. 29, no. 2, pp. 59–163, 2005.

[12] D. Alonso-Álvarez, T. Wilson, P. Pearce, M.

Führer, D. Farrell, and N. Ekins-Daukes, "Solcore: a multi-scale, Python-based library for modelling solar cells and semiconductor materials," *J. Comput. Electron.*, vol. 17, no. 3, pp. 1–25, 2018.
[13] D. Schroder, *Semiconductor Material and Device Characterization*, 3rd ed. New Jersey: John Wiley & Sons Ltd, 2006.
[14] S. Sze and M.-K. Lee, *Semiconductor Devices: Physics and Technology*, 3rd ed. Wiley, 2013.
[15] W. K. Metzger, D. Albin, D. Levi, P. Sheldon, X. Li, B. M. Keyes, and R. K. Ahrenkiel, "Time-resolved photoluminescence studies of CdTe solar cells," *J. Appl. Phys.*, vol. 94, no. 5, pp. 3549–3555, 2003.
[16] W. K. Metzger, R. K. Ahrenkiel, J. Dashdorj, and D. J. Friedman, "Analysis of charge separation dynamics in a semiconductor junction," *Phys. Rev. B - Condens. Matter Mater. Phys.*, vol. 71, no. 3, pp. 1–9, 2005.
[17] D. Kuciauskas, A. Kanevce, J. N. Duenow, P. Dippo, M. Young, J. V. Li, D. H. Levi, and T. a. Gessert, "Spectrally and time resolved photoluminescence analysis of the CdS/CdTe interface in thin-film photovoltaic solar cells," *Appl. Phys. Lett.*, vol. 102, no. 17, 2013.
[18] A. Rohatgi, "A Study of Efficiency Limiting Defects in Polycrystalline CdTe/CdS Solar Cells," *Int. J. Sol. Energy*, vol. 12, no. 1–4, pp. 37–49, 1992.
[19] V. Buschmann, H. Hempel, A. Knigge, C. Kraft, M. Roczen, M. Weyers, T. Siebert, and F. Koberling, "Characterization of semiconductor devices and wafer materials via sub-nanosecond time-correlated single-photon counting," *J. Appl. Spectrosc.*, vol. 80, no. 3, pp. 449–457, 2013.
[20] M. Kemell, M. Ritala, and M. Leskelä, "Thin film deposition methods for $CuInSe_2$ solar cells," *Crit. Rev. Solid State Mater. Sci.*, vol. 30, no. 1, pp. 1–31, 2005.
[21] H. Cui, J. Wang, C. Wang, C. Liu, K. Pi, X. Li, Y. Xu, and Z. Tang, "Experimental Determination of Effective Minority Carrier Lifetime in HgCdTe Photovoltaic Detectors Using Optical and Electrical Methods," *Adv. Condens. Matter Phys.*, vol. 2015, pp. 1–6, 2015.

ENHANCED CVD-GRAPHENE SHEET TRANSFER ON A ZINC OXIDE ULTRATHIN FILM USING WET TRANSFER METHOD FOR SOLAR CELL APPLICATION

Ahmed F. Abdelaal[1], Mohammad K. Hossain[2], Ahmed Ibrahim[3], Billel Salhi[2], Tahar Laoui[1]
[1] Department of Mechanical Engineering, King Fahd University of Petroleum and Minerals, Dhahran 31261, Kingdom of Saudi Arabia
[2] Center of Research Excellence in Renewable Energy, King Fahd University of Petroleum and Minerals, Dhahran 131261, Kingdom of Saudi Arabia.
[3]Department of Mechanical Design and Production Engineering, Zagazig University, Zagazig 44519, Egypt

ABSTRACT: A hybrid structure of graphene sheets (GS) with ultrathin film of Zinc oxide (ZnO) is developed on glass substrates for thin film solar cell applications. ZnO is fabricated using RF sputtering technique. ZnO layer on glass was found to be very thin, ranging from 30 to 50 nm in thickness and transparent. Graphene film are synthesized on the Cu foils by using chemical vapor deposition method, and then transferred to the ZnO layer using wet transfer method. As-fabricated GS/ZnO hybrid structure was characterized using atomic force microscope and Raman spectroscopy to understand the effect of graphene layer on the ultrathin layer of ZnO.
Keywords: Graphene, Zinc oxide, solar cell applications

1 Introduction

Zinc Oxide (ZnO) has captivated immense consideration in many fields. More recently, well-known applications of ZnO are in polymers industry, ceramics, coating and specially nanodevices as an important semiconductor due to its unique optical, structural and electronic properties. It is especially significant as a semiconductor because ZnO's leading characteristics are a good photoconductivity, excellent transparency of (>90 %), wide band gap of (3.37 eV), high electron mobility (100 cm^2/Vs), large excitation binding energy of (60 meV)),, etc.[1]–[4]. ZnO has been a promising semiconductor material with many applications in transparent electrodes[5], gas sensing [6], light emitting diode [7] and photovoltaic cells, because of its binding energy that facilitate to collect carrier effectively and band gap. Its properties have been studied since the early days of semiconductor electronics, but the use of ZnO as transparent conducting oxide (TCO) in optoelectronic devices has been hindered because of low carrier transport characteristics. On the other hand, Graphene, single layer graphite with closely packed hexagonal lattices, has received a lot of attention in the field of research due to its unique carrier transport properties [8]. It is an excellent electrical conductor, and it has a high aspect ratio as well that make it a great material for use in optoelectronic applications. Considering a crucial challenge to improve the efficiency of solar cells by enhancing the way excitons are collected and transported, it would obviously infer to have a TCO of best carrier collection and transport capabilities. Therefore, a heterostructure of ZnO and graphene will be expected to serve the purpose.

The unique properties of graphene have been extensively investigated in the field of solar cell. There are some promising results that will enable the usage graphene for solar cell application, such as: conductive electrodes and active layer. Graphene has been used as a transparent conductive layer as well as an anode for flexible photovoltaic cell. The reported graphene film showed its sheet resistance down to 230 Ω/sq with transparency of 72%. The power conversion energy (PCE) was improved to 1.8 % Comparing with 1.27% without graphene [9]. Jiang and his team demonstrated the CdTe thin film solar cell using CVD-graphene as the front transparent electrode on top of ZnO layer. The resistance of the obtained graphene film was 220Ω /sq, with a transparency of 84% and PCE of 4.17% [10].

Here in this study, a very ultrathin layer of ZnO was fabricated on glass substrate followed by a further transfer of CVD-developed graphene atop. The thickness was kept as low as possible to optimize the transparency. Atomic force microscopy (AFM), Raman scattering spectroscopy, and optical microscopy were carried out to characterize ZnO and Graphene layer. It is indispensable to understand the inherent characteristics of ZnO-graphene heterostructures to be implemented as TCO in solar cell applications.

2 Methods and materials

An ultrathin layer of ZnO on a clean glass was fabricated by RF sputtering technique using automatic sputter coater (model NSC-4000). A high purity zinc (99.999 %) target received from Semiconductor Wafer, Inc. was pretreated following a standard procedure (ultrasonication for 15 min in acetone followed by 15 min ultrasonic cleaning in methanol). In addition, the zinc target was further cleansed before each experiment by a pre-sputtering process for 1 min. The base pressure in the chamber was kept ~2×10^{-6} torr and the working pressure was set to 7×10^{-3} torr by adjusting the Ar and O_2 gas flow at 80 sccm. The target substrate distance was maintained fixed at 10 cm.

Copper foil (25 µm thick from Alfa Aesar company) was used for graphene growth using the CVD method. As known, the principle behind CVD technique to grow graphene is that the substrate is exposed to gaseous precursors, which in contact with the substrate at high temperatures, deposit on its surface, creating a thin film [11]. The quality of the materials obtained by CVD is greatly dependent on the process parameters such as gas temperature, pressure, and time. To achieve single layer graphene by CVD with a quality similar to that of exfoliated graphene, these parameters have to be adjusted during the deposition process [12]. The grapheme was then transferred on the top of the ZnO file using wet transfer method. Graphene on the backside of copper substrate was etched by putting graphene/Cu on an ammonium persulfate (APS) bath for 7 min, then rinsing it in deionized (DI) water for 5 min. Next, the Copper substrate that holds the graphene on the top of it was etched away using Ammonium persulfate (APS) over the period of two hours [13]. The graphene was finally rinsed in 3 baths (for 10 min each) of DI water before being transferred it on ZnO/glass substrate as shown in Figure 1.

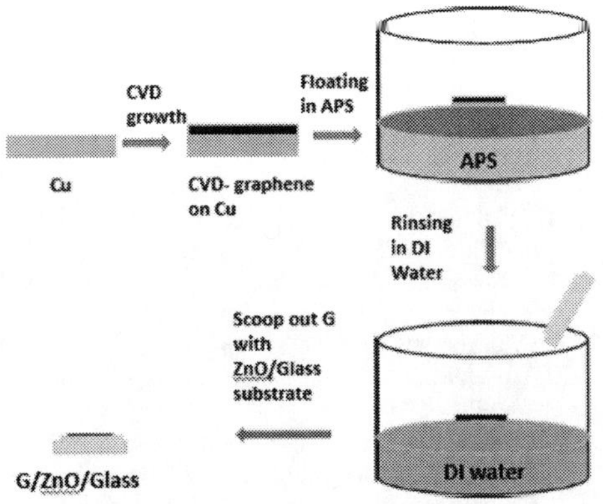

Figure 1: Stepwise schematic diagram for wet graphene transfer on ZnO/glass substrate.

Atomic force microscopy is a type of scanning probe microscopy with nanometric resolution. It is used to obtain topographic images of surfaces [17]. Dimension Icon- Bruker AFM machine was employed to identify the Zinc oxide and graphene superficial morphology. The tapping AFM mode was used to achieve the surface morphology. AFM was used to characterize graphene over copper and zinc oxide in order to investigate its continuity, uniformity, morphology and thickness of each layer.

Raman spectroscope (DXR™ 2 Raman Microscope) manufactured by ThermoFisher Scientific was utilized for characterization of graphene samples over zinc oxide.

3 Results and discussion

Graphene was successfully transferred to ZnO/glass substrate and glass substrate as explained in previous section. The total area of transferred graphene for both samples is 1X1 cm, as depicted in Figure 2.

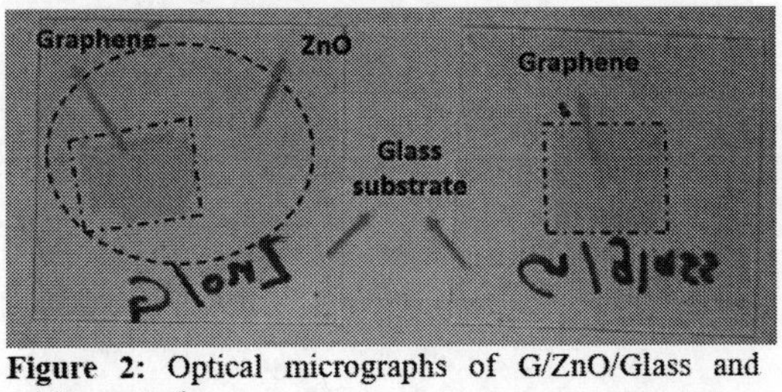

Figure 2: Optical micrographs of G/ZnO/Glass and G/glass samples.

Raman spectroscopy is one of the most popular and powerful non-destructive technique utilized to identify and characterize the materials and their structural properties. Raman phenomenon effect was discovered in 1928 by C.V. Raman [14]. One of the very prevalent characterization techniques for a carbon allotrope like graphene and carbon nano tube (CNT) which encompass basically of three characteristic peaks is Raman spectroscopy; the number of layer and quality of graphene and CNT can be identify from these peaks. Raman spectroscopy technique is nondestructive and very simple these are the main advantages of it, besides, Raman spectroscopy contributes to precise results.

A great deal of properties for graphene structure and defects can be extracted from the Raman spectra and typical graphene peaks. There are main Raman scuttaring peaks for grpahene expressed by two major Raman peaks and a few more negligible bands. The G-band is settled at a Raman shift of ~1600 cm^{-1} and at a Raman shift of ~2700 cm^{-1} there is a 2D-band . The D-band is the third band although often very feeble, is settled at ~1360 cm^{-1}. The quality of graphene, its defects, the doping effect in graphene layer and the number of graphene layer were deretrmined using these three Raman peaks and the realtion between them. [15][16]. Figure 3(a). shows the Raman peaks for ZnO/ glass substrate with the main peaks of ZnO which ZnO_{A1}(TO) and ZnO_{A1}(LO) beside the peak of glass substrate. Optical image of ZnO/glass is located inside, as shown in Figure 3(a) [18] .

Figure 3(b) shows Raman spectrum and optical images for G/Zno/glass. The Raman characterization of graphene shows the characteristic peaks around 1355, 1585, and 2700 cm $^{-1}$ corresponding to the D, G, and 2D peaks [15][16]. The ratio between intensity of 2D peak to intensity of G peak is less than 1.8 which shows that the graphene layer contains bilayer and multilayer regions because of free transfer. The graphene sheet folded on each other, resulting in a multilayer graphene [19][20].

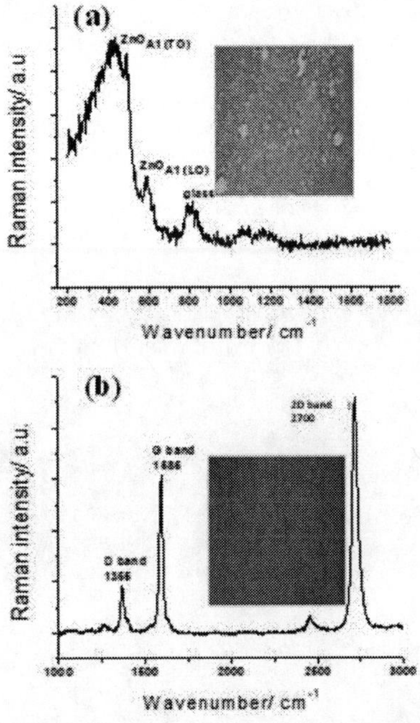

Figure 3: Raman scattering measurements of (a) ZnO on glass, inset: CCD image and (b) Graphene on ZnO on glass, inset: CCD image of each sample

Raman mapping measurements were conducted for G/ZnO/glass sample by investigating an area of 10x10 µm. A total of 100 μm^2 Raman spectra were collected and analyzed at Raman shift of ~ 2700 cm^{-1} which represent 2D peak of graphene. Figure 4 shows the mapping of an area where part of ZnO was covered by graphene to examine the difference between both areas. The blue area as shown in Figure 4(a) represents really low intensity of 2D peak, which represent a free graphene area. On the other hand, green, yellow and red areas dominated graphene layer. 3D Raman mapping represents the variety of 2D peak intensity on the G/ZnO/glass sample, as shown in Figure 4(b).

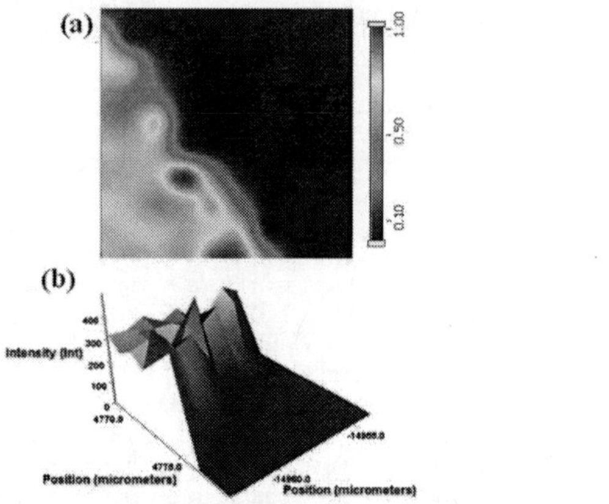

Figure 4: Raman mapping images of intensity of 2D graphene peak (a)2D image and (b) 3D image.

Figure 5 shows AFM imaging of sputtered ZnO on glass as well as Graphene on ZnO film. ZnO film was observed very smooth with maximum height of 5.6 nm as shown in Figure. 5(a). Further 3D imaging was shown in Figure 5(b). After the transfer of Graphene on as-obtained ZnO, it was noted that graphene layer contains bilayer and multilayer regions as shown in Figure 5(c). the dark spots are expected to be multilayers. This is simply because of free transfer and therefore the graphene sheet folded on each other, resulting in a multilayer graphene. Further 3D imaging confirmed maximum step height of 20.7 nm as shown in Figure. 5(d).

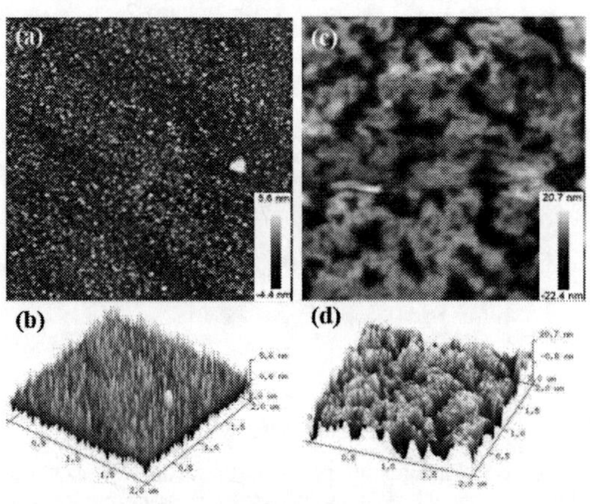

Figure 5: AFM spectroscopy of (a)-(b) ZnO on Glass and (c)-(d) Graphene on ZnO film.

4 Conclusion

An ultrathin layer of ZnO is fabricated on glass after the transfer of CVD-developed graphene using wet transfer method on the surface of ZnO. Further characterizations were carried out. Raman measurements confirmed the signature peaks of ZnO as well as those of Graphene atop.

5 Acknowledgement

Ahmed F. Abdelaal and MKH acknowledge Center of Research Excellence in Renewable Energy (CoRERE), Research Institute, King Fahd University of Petroleum & Minerals, (KFUPM), Dhahran 31261, Saudi Arabia.

6 References

[1] A. Kołodziejczak-Radzimska and T. Jesionowski, "Zinc Oxide-From Synthesis to Application: A Review.," *Mater. (Basel, Switzerland)*, vol. 7, no. 4, pp. 2833–2881, Apr. 2014.

[2] H. Morkoç and U. Özgür, *Zinc oxide : fundamentals, materials and device technology*. Wiley-VCH, 2009.

[3] A. Tereshchenko *et al.*, "Optical biosensors based on ZnO nanostructures: advantages and perspectives. A review," *Sensors Actuators B Chem.*, vol. 229, pp. 664–677, Jun. 2016.

[4] P. Zhu *et al.*, "Biomedical Applications of Functionalized ZnO Nanomaterials: from Biosensors to Bioimaging," *Adv. Mater. Interfaces*, vol. 3, no. 1, p. 1500494, Jan. 2016.

[5] S. Major and K. L. Chopra, "Indium-doped zinc oxide films as transparent electrodes for solar cells," *Sol. Energy Mater.*, vol. 17, no. 5, pp. 319–327, Aug. 1988.

[6] G. Biasotto, M. G. A. Ranieri, C. . Foschini, A. Z. Simões, E. Longo, and M. A. Zaghete, "Gas sensor applications of zinc oxide thin film grown by the polymeric precursor method," *Ceram. Int.*, vol. 40, no. 9, pp. 14991–14996, Nov. 2014.

[7] Y.-J. Lu *et al.*, "Improved performance of ZnO light-emitting devices by introducing a hole-injection layer," *Opt. Express*, vol. 22, no. 14, p. 17524, Jul. 2014.

[8] E. L. Lim, C. C. Yap, M. H. H. Jumali, M. A. M. Teridi, and C. H. Teh, "A Mini Review: Can Graphene Be a Novel Material for Perovskite Solar Cell Applications?," *Nano-Micro Lett.*, vol. 10, no. 2, p. 27, Apr. 2018.

[9] S. A. Kazmi, S. Hameed, A. S. Ahmed, M. Arshad, and A. Azam, "Electrical and optical properties of graphene-TiO2 nanocomposite and its applications in dye sensitized solar cells (DSSC)," *J. Alloys Compd.*, vol. 691, pp. 659–665, 2017.

[10] H. Bi, F. Huang, J. Liang, X. Xie, and M. Jiang, "Transparent Conductive Graphene Films Synthesized by Ambient Pressure Chemical Vapor Deposition Used as the Front Electrode of CdTe Solar Cells," *Adv. Mater.*, vol. 23, no. 28, pp. 3202–3206, Jul. 2011.

[11] J. Kong, A. M. Cassell, and H. Dai, "Chemical vapor deposition of methane for single-walled carbon nanotubes," *Chem. Phys. Lett.*, vol. 292, no. 4–6, pp. 567–574, Aug. 1998.

[12] A. Ibrahim, S. Akhtar, M. Atieh, R. Karnik, and T. Laoui, "Effects of annealing on copper substrate surface morphology and graphene growth by chemical vapor deposition," *Carbon N. Y.*, vol. 94, pp. 369–377, 2015.

[13] A. Ibrahim, G. Nadhreen, S. Akhtar, F. M. Kafiah, and T. Laoui, "Study of the impact of chemical etching on Cu surface morphology, graphene growth and transfer on SiO2/Si substrate,"

Carbon N. Y., vol. 123, pp. 402–414, Oct. 2017.
[14] L. F. Bates, "PHYSICS," *Science Progress in the Twentieth Century (1919-1933)*, vol. 23. Science Reviews 2000 Ltd., pp. 392–395, 1929.
[15] A. C. Ferrari *et al.*, "Raman Spectrum of Graphene and Graphene Layers," *Phys. Rev. Lett.*, vol. 97, no. 18, p. 187401, Oct. 2006.
[16] D. Stojanović, N. Woehrl, and V. Buck, "Synthesis and characterization of graphene films by hot filament chemical vapor deposition," *Phys. Scr.*, vol. 2012, no. T149, p. 014068, May 2012.
[17] B. M. Trafas, Y.-N. Yang, R. L. Siefert, and J. H. Weaver, "Scanning tunneling microscopy of Ag growth on GaAs(110) at 300 K: From clusters to crystallites," *Phys. Rev. B*, vol. 43, no. 17, pp. 14107–14114, Jun. 1991.
[18] S. Ben Yahia, L. Znaidi, A. Kanaev, and J. P. Petitet, "Raman study of oriented ZnO thin films deposited by sol–gel method," *Spectrochim. Acta Part A Mol. Biomol. Spectrosc.*, vol. 71, no. 4, pp. 1234–1238, Dec. 2008.
[19] D. Geng *et al.*, "Uniform hexagonal graphene flakes and films grown on liquid copper surface," *Proc. Natl. Acad. Sci.*, Apr. 2012.
[20] M. P. Levendorf, C. S. Ruiz-Vargas, S. Garg, and J. Park, "Transfer-Free Batch Fabrication of Single Layer Graphene Transistors," *Nano Lett.*, vol. 9, no. 12, pp. 4479–4483, Dec. 2009.

OPTIMAL BAND GAP ENERGIES FOR TWO-STEP PHOTON UP-CONVERSION SOLAR CELLS WITH PARTIAL ABSORPTIVITY

Yukihiro Harada, Tetsuhiro Matsuo, Shigeo Asahi, and Takashi Kita
Department of Electrical and Electronic Engineering, Graduate School of Engineering, Kobe University
1-1 Rokkodai, Nada, Kobe 657-8501, Japan

ABSTRACT: Photon up-conversion in a hetero-interface comprising different semiconductor band gaps can reduce the transmission loss without any complicated quantum structures, and is one of the promising concepts for improving the conversion efficiency of solar cell. In this work, we theoretically studied the conversion efficiency and the optimum band gap energies of the two-step photon up-conversion solar cell (TPU-SC) with partial absorptivity in the detailed-balance framework for offering a route to high-conversion-efficiency exceeding 50%. The optimal band gap energy of the narrow gap semiconductor is approximately 1.4 eV almost independent of the band discontinuity ratio in the hetero-interface at the absorptivity of 50% and the sunlight concentrations of 100–1,000. The conversion efficiency reaches 50% in the TPU-SC with the conduction- or valence-band discontinuity ratio of 0 at the absorptivity of 50% and the 1,000-suns illumination. The calculated results suggest that GaAs/AlGaAs and GaAs/InGaP heterostructures are promising candidates for demonstrating the high-conversion efficiency TPU-SC in the valence-band discontinuity ratios of 0.4 and 1, respectively.
Keywords: III-V Semiconductors; Intermediate Band; Solar Cell Efficiencies; Absorption

1 INTRODUCTION

Intermediate band solar cells (IBSCs) are expected to realize high conversion efficiency greater than the conversion efficiency limit of single-junction SCs which is so-called Shockley–Queisser limit [1]. Two-step photon absorption via an IB placed in the band gap of a host semiconductor plays a key role in the IBSC operation. Since quantum structures relax the optical selection rule for the intraband transition from the quantized states to the extended conduction band (CB), low-dimensional structures having quantum confinement in the growth plane yield stronger intraband photoabsorption compared with bulk crystals and quantum wells for the normal incident light. However, the absorptivity via the IB is still weak because of not only the optical selection rule but also a relatively short electron lifetime in the IB [2].

Recently, we have proposed a two-step photon up-conversion (TPU)-SC using a hetero-interface comprising different semiconductor band gaps [3]. The TPU-SC does not require any complicated quantum structures and is a single-junction SC containing a simple hetero-interface. In the TPU-SC fabricated on a *p*-type substrate, below-gap photons passing through the wide band gap semiconductor (WGS) layer excite the narrow band gap semiconductor (NGS), in which excited electrons drift towards the WGS/NGS hetero-interface and accumulate there, while holes reach the *p*-layer [3]. The long-lived electron is efficiently excited by another below-gap photon in the NGS and lifted to the CB of the WGS. This TPU boosts the output voltage at the hetero-interface [3]. We have demonstrated the up-conversion current two-order of magnitude larger than that in quantum dot IBSC, because the photon up-conversion process utilizes the bulk density-of-states in the TPU-SC [3].

On the other hand, the valence band (VB) discontinuity in the hetero-interface causes the energy loss compared with the ideal IBSC at a unity absorptivity [4]. Conversely, the unity absorptivity for the two-step photon absorption via the IB in the IBSCs and for the photon up-conversion in the WGS/NGS hetero-interface in the TPU-SCs are not a realistic assumption in the practical SCs. In this work, we theoretically studied the conversion efficiency and the optimum band gap energies of the TPU-SCs with partial absorptivity in the detailed-balance framework for offering a route to high-conversion-efficiency exceeding 50%.

2 CALCULATION MODEL

Our calculation is based on the detailed-balance framework which was originally proposed by Shockley and Queisser [5]. Here, we ignore nonradiative processes in the SC and spectral overlap between any transitions for predicting the ideal conversion efficiency limit. Figures 1(a) and 1(b) show the schematic calculation model of the IBSC and TPU-SC, respectively. The interband transitions, the transition between the VB and CB in the IBSC and the transitions in the WGS and NGS in the TPU-SC, can be regarded as a unity absorptivity. In this work, we assumed that the two-step photon absorption in the IBSC have same absorptivity, α, as shown in Fig. 1(a). The detailed calculation framework of the IBSC is reported in Ref. 6. Hereafter, we focus on the calculation framework of the TPU-SC used in this work.

The total photon emission flux, N, with the energy range between $E_{\min}$ and $E_{\max}$ can be calculated using the generalized Planck equation incorporating the effect of the chemical potential, μ:

$$N(E_{\min}, E_{\max}, T, \mu) = \frac{2\pi}{h^3c^2}\int_{E_{\min}}^{E_{\max}} \frac{E^2}{\exp\{(E-\mu)/k_BT\}-1}\mathrm{d}E\,,$$

where T is the temperature, h is Planck constant, c is the speed of light, and k_B is Boltzmann constant. By using this equation, the generation rates of G_{WGS} in the WGS, G_{NGS} in the NGS, and G_{up} for TPU can be expressed as follows:

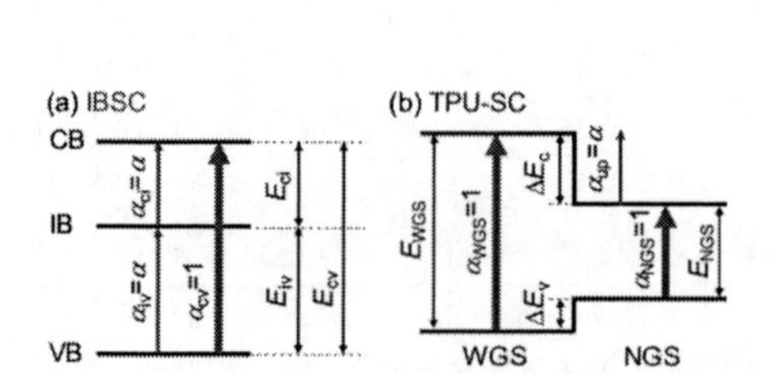

Figure 1: Schematic calculation model of the (a) IBSC and (b) TPU-SC.

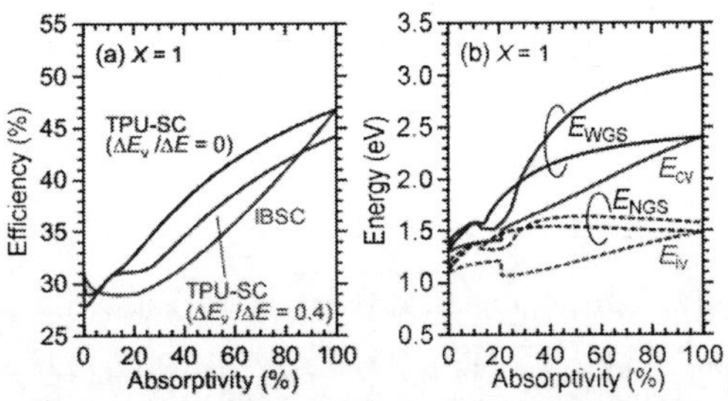

Figure 2: Absorptivity dependence of the (a) limiting conversion efficiency and (b) band gap energies in the IBSC and TPU-SCs at the one-sun illumination.

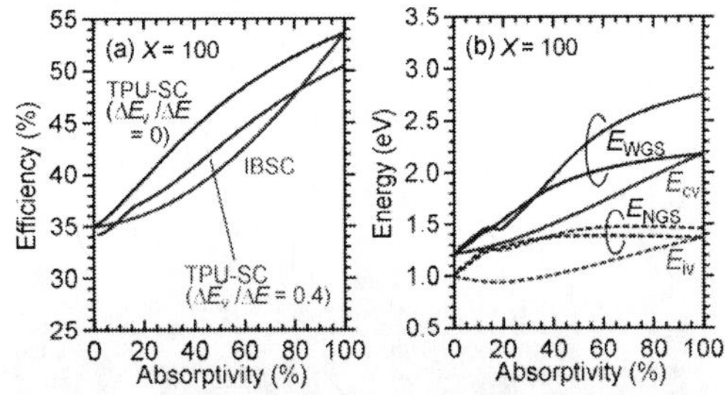

Figure 3: Absorptivity dependence of the (a) limiting conversion efficiency and (b) band gap energies in the IBSC and TPU-SCs at the 100-suns illumination.

$$\begin{aligned} G_{WGS} &= X f_{sun} N(E_{WGS}, \infty, T_{sun}, 0) \\ &\quad + (1 - X f_{sun}) N(E_{WGS}, \infty, T_{cell}, 0), \\ G_{NGS} &= X f_{sun} N(E_{NGS}, E_{WGS}, T_{sun}, 0) \\ &\quad + (1 - X f_{sun}) N(E_{NGS}, E_{WGS}, T_{cell}, 0), \\ G_{up} &= X f_{sun} \alpha N(\Delta E_c, E_{NGS}, T_{sun}, 0) \\ &\quad + (1 - X f_{sun}) \alpha N(\Delta E_c, E_{NGS}, T_{cell}, 0), \end{aligned}$$

where X is the solar concentration factor, f_{sun}=2.18×10^{-5} is the solid angle of the sun, α is the absorptivity for TPU as shown in Fig. 1(b), T_{sun}=6,000 K is the temperature of sun, E_{WGS} and E_{NGS} are the band gap energies of the WGS and NGS, respectively, ΔE_c is the CB discontinuity between the WGS and NGS, and T_{cell}=300 K is the temperature of the SC. The relation between E_{WGS}, E_{NGS}, ΔE_c, and the VB discontinuity, ΔE_v, can be described as follows:

$$E_{WGS} = E_{NGS} + \Delta E_c + \Delta E_v.$$

Here, we assumed that G_{up} takes place solely within the NGS and then the carriers are transferred to the WGS [4]. Each recombination rate can be given by:

$$\begin{aligned} R_{WGS} &= N(E_{WGS}, \infty, T_{cell}, \mu_{WGS}), \\ R_{NGS} &= N(E_{NGS}, E_{WGS}, T_{cell}, \mu_{NGS}), \\ R_{up} &= \alpha N(\Delta E_c, E_{NGS}, T_{cell}, \mu_{up}), \end{aligned}$$

where μ_{WGS} and μ_{NGS} are the quasi-Fermi level separations in the WGS and NGS, respectively, and μ_{up} is the quasi-Fermi level separation due to TPU. We do not consider the surface recombination at the WGS/NGS hetero-interface in this work. Here, we take into consideration electrons accumulated at the hetero-interface and the TPU occurring in the CB of the NGS. According to these relations, the total current, J, generated in the TPU-SC can be obtained by:

$$J/q = G_{WGS} + G_{NGS} - R_{WGS} - R_{NGS},$$

where q is the elementary charge. In the TPU-SC, the following current matching condition of TPU must be satisfied:

$$G_{NGS} - R_{NGS} = G_{up} - R_{up}.$$

The output voltage of the TPU-SC is given by:

$$qV = \mu_{WGS} = \mu_{NGS} + \mu_{up}.$$

Finally, the total electrical power generated in the TPU-SC is calculated as a product of VJ and, hence, the conversion efficiency can be estimated by the division of VJ by the total incident photon energy. In this work, the band gap energies are globally optimized. Here, the band gap energy difference between the WGS and NGS, $\Delta E = E_{WGS} - E_{NGS}$, is assumed to satisfy the following condition:

$$0.2\ \text{eV} \le \Delta E \le E_{WGS}/2.$$

Furthermore, $\Delta E_v/\Delta E$ was set to be 0 or 0.4 in this work.

3 RESULTS AND DISCUSSION

Figures 2(a) and 2(b) show the absorptivity dependence of the limiting conversion efficiency and the band gap energies, respectively, in the IBSC and TPU-SCs at the one-sun illumination. Red, blue, and green lines indicate the results for the IBSC, the TPU-SC ($\Delta E_v/\Delta E = 0$), and the TPU-SC ($\Delta E_v/\Delta E = 0.4$), respectively. Solid and dashed lines in Fig. 2(b) represent the E_{WGS} (E_{cv}) and E_{NGS} (E_{iv}) in the TPU-SCs (IBSC), respectively. In the IBSC, the calculated conversion efficiency decreases from 31% of the Shockley–Queisser limit with an increase in the absorptivity up to 20%. This is because the IB acts as an additional radiative recombination channel for the low absorptivity [6]. On the other hand, the conversion efficiency in the IBSC increases with increasing the absorptivity above 20% and reach 47% at the unity absorptivity [1]. In the TPU-SC ($\Delta E_v/\Delta E = 0$), the conversion efficiency is consistent with that in the IBSC at zero and unity absorptivity [4], however, the efficiency shows different absorptivity dependence as shown in Fig. 2(a) because of the unbalance in the generation (recombination) rates between the G_{NGS} (R_{NGS}) and G_{up} (R_{up}). On the other hand, the conversion efficiency in the TPU-SC ($\Delta E_v/\Delta E = 0.4$) is smaller than that in the TPU-SC ($\Delta E_v/\Delta E = 0$) for all the absorptivity because the VB discontinuity results in the energy loss occurring at the hetero-interface [3]. It should be stressed that the conversion efficiency in the TPU-SC ($\Delta E_v/\Delta E = 0.4$) is greater than that in the IBSC at the absorptivity in the range of 10 to 80% resulting from the unity absorptivity in the NGS. Conversely, the optimal band gap energies in the TPU-SC ($\Delta E_v/\Delta E = 0.4$) are larger than those in the IBSC and TPU-SC ($\Delta E_v/\Delta E = 0$) for satisfying the current matching condition at the absorptivity above 30%.

IBSCs require light concentration not only to reduce the device cost but also to ensure optimal IBSC operation for reducing the contribution of the recombination processes via the IB [7]. Figures 3(a) and 3(b) show the absorptivity dependence of the limiting conversion efficiency and the band gap energies, respectively, in the IBSC and TPU-SCs at the 100-suns illumination. The

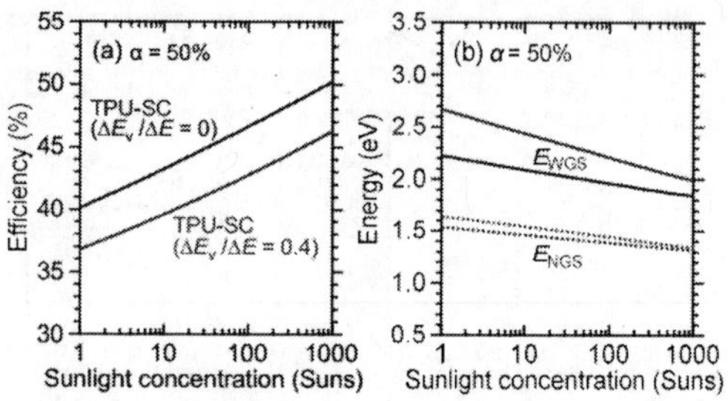

Figure 4: Sunlight concentration dependence of the (a) limiting conversion efficiency and (b) band gap energies in the TPU-SCs at the absorptivity of 50%.

conversion efficiency in the TPU-SC ($\Delta E_V/\Delta E = 0.4$) is still greater than that in the IBSC at the absorptivity in the range of 10 to 80%. On the other hand, the optimal band gap energies decrease with an increase in the sunlight concentration as shown in Figs. 2(b) and 3(b).

Figures 4(a) and 4(b) show the sunlight concentration dependence of the limiting conversion efficiency and the band gap energies, respectively, in the TPU-SCs at the absorptivity of 50%. The conversion efficiency monotonically increases with an increase in the sunlight concentration and reaches 50% in the TPU-SC ($\Delta E_V/\Delta E = 0$) at the 1,000-suns illumination. The optimal E_{NGS} is approximately 1.4 eV almost independent of $\Delta E_V/\Delta E$ at the sunlight concentrations of 100–1,000, which corresponds to the band gap energy of GaAs. On the other hand, the optimal E_{WGS} is 2.0–2.2 eV for $\Delta E_V/\Delta E = 0.4$ and 1.8–2.0 eV for $\Delta E_V/\Delta E = 0$ at the sunlight concentrations of 100–1,000. As $\Delta E_V/\Delta E$ of the GaAs/AlGaAs hetero-interface is approximately 0.4 [8], GaAs/AlGaAs is a promising candidate for demonstrating the high-conversion efficiency TPU-SC. On the other hand, $\Delta E_V/\Delta E$ of the partly ordered GaAs/InGaP hetero-interface is approximately 1 [8]. In the case of a TPU-SC fabricated on a *n*-type substrate, hole is accumulated in the WGS/NGS hetero-interface, and the long-lived hole is excited by another below-gap photon in the NGS and lifted to the VB of the WGS [3]. Therefore, GaAs/InGaP heterostructure is a promising candidate for achieving the high-conversion efficiency exceeding 50% in the TPU-SC with the partial absorptivity of 50%.

4 CONCLUSION

We theoretically studied the conversion efficiency and the optimum band gap energies of the TPU-SCs with partial absorptivity in the detailed-balance framework. The conversion efficiency in the TPU-SC ($\Delta E_V/\Delta E = 0.4$) is greater than that in the IBSC at the absorptivity in the range of 10 to 80% resulting from the unity absorptivity in the NGS. The optimal band gap energy of the NGS is approximately 1.4 eV almost independent of $\Delta E_V/\Delta E$ at the sunlight concentrations of 100–1,000 and the absorptivity of 50%. The calculated results suggest that GaAs/AlGaAs and GaAs/InGaP heterostructures are promising candidates for demonstrating the high-conversion efficiency TPU-SC in $\Delta E_V/\Delta E$ of 0.4 and 1, respectively.

REFERENCES

[1] A. Luque and A Martí, Phys. Rev. Lett. **78**, 5014 (1997).
[2] Y. Okada, N. J. Ekins-Daukes, T. Kita, R. Tamaki, M. Yoshida, A. Pusch, O. Hess, C. C. Phillips, D. J. Farrell, K. Yoshida, N. Ahsan, Y. Shoji, T. Sogabe, and J.-F. Guillemoles Appl. Phys. Rev. **2**, 021302 (2015).
[3] S. Asahi, H. Teranishi, K. Kusaki, T. Kaizu, and T. Kita, Nat Commun. **8**, 14962 (2017).
[4] S. Asahi, K. Kusaki, Y. Harada, and T. Kita, Sci. Rep. **8**, 872 (2018).
[5] W. Shockley and H. J. Queisser, J. Appl. Phys. **32**, 510 (1961).
[6] A. Pusch, M. Yoshida, N. P. Hylton, A. Mellor, C. C. Phillips, O. Hess, and N. J. Ekins-Daukes, Prog. Photovolt: Res. Appl. **24**, 656 (2016).
[7] T. Sogabe, Y. Shoji, M. Ohba, K. Yoshida, R. Tamaki, F.-F. Hong, C.-H. Wu, C.-T. Kuo, S. Tomić, and Y. Okada, Sci. Rep. **4**, 4792 (2014).
[8] S. Adachi, *Properties of Semiconductor Alloys: Group-IV, III–V and II–VI Semiconductors* (Wiley, 2009) Chap. 9.

INTERFACIAL BUFFER LAYER FOR THE INTEGRATION IN SWCNTS/A-SI:H HYBRID THIN FILM SOLAR CELLS

Hosni Meddeb, Oleg Sergeev, Hyo-Jei Cho, Martin Vehse, Pramod Rajanna*, Albert Nasibulin*, Sergei Bereznev**
DLR Institute of Networked Energy Systems, Carl von Ossietzky Str. 15, 26129 Oldenburg, Germany
* Skolkovo Institute of Science and Technology, Building 3, Moscow 143026, Russia
** Tallinn University of Technology, Ehitajate tee 5, 19086 Tallinn, Estonia

ABSTRACT: We report the efficient approach to protect the SWCNTs front window layer from hydrogen ion etching during deposition of the hydrogenated amorphous silicon absorber layer in the hybrid carbon nanotubes / thin film silicon solar cells. We studied the changes in SWCNTs films properties after deposition of different buffer layers based on low-hydrogen amorphous silicon and chemically inert molybdenum oxide. We defined a narrow process window for i-a-Si:H deposition that allow smooth coating of SWCNTs preventing the main anomalies such etching and protrusion. It is demonstrated that electron beam evaporation of MoO_x thin films can be advantageous over PECVD formation of "undiluted" a-Si:H buffer layers due to the preserved morphology and conductivity of grown films. A strong correlation between the morphological (network density) and structural (purity) properties of SWCNTs on one hand, and optical (transmission) and electrical (conductivity/mobility) properties on the other hand, has been established. Both optimized i-a-Si:H and MoO_x buffer layers can be recommended for efficient protection of carbon nanotubes. The optimization of our coating approach leads to the preservation of high conductivity of SWCNTs required for the integration as p-side layer, up to 750 S/cm and 800 S/cm for a-Si:H and MoO_x buffer layers respectively.
Keywords: Carbon nanotube, amorphous silicon, solar cells, heterojunction, buffer layers, plasma etching

1 INTRODUCTION

Hybrid thin film solar cells based on single walled carbon nanotubes (SWCNTs)/amorphous hydrogenated silicon (a-Si:H) are very promising PV devices. The potential of SWCNTs in amorphous silicon thin film solar cells consists on its capability to substitute both p-doped a-Si:H layer, and transparent conductive electrode [1, 2, 3, 4]. Thin SWCNTs films are advantageous with their high transparency due to their sub-band gaps, tunable photo-absorption from near infrared (NIR) to ultraviolet (UV) range, and high electrical conductivity.

The main challenge in the integration of SWCNTs in p-i-n superstrate configuration is the chemical etching of carbon nanotubes with hydrogen ions in the plasma during the deposition of the a-Si:H absorber. This effect leads not only to significant degradation of conductivity in the SWCNTs due to the deterioration of the network, but to the shunts in the solar cells due to the protrusion of released single carbon nanotubes through the absorber layer (Fig. 1a).

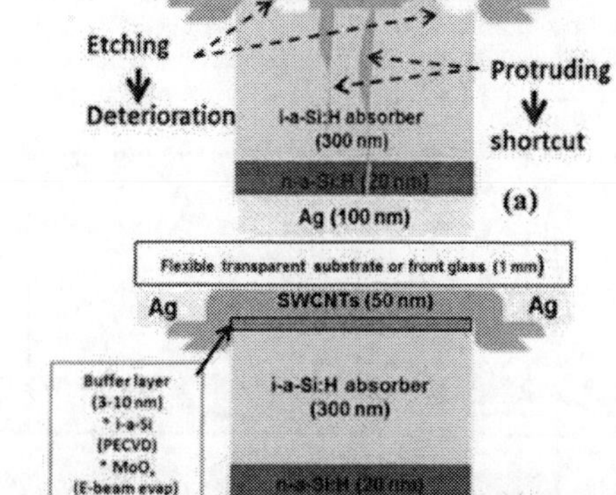

Figure 1: Schematic of main challenges in SWCNTs/ a-Si:H p-i-n solar cell structure (a) and proposed buffer layer solutions (b)

Commonly, for the fabrication of high quality defect-free absorber layers, the process gas composition is highly diluted with hydrogen. In this work, we developed "undiluted" intrinsic a-Si:H and molybdenum oxide (MoO_x) buffer layers for the SWCNTs protection. Thus, we tested two different approaches of SWCNTs protection, which are based on the gradient fabrication of the absorber layer, where at the first stage the reduction of hydrogen content during the deposition process will help the formation of a thin low-hydrogen buffer layer and on the fabrication of a very thin buffer layer of chemically inert material.

For the development of the most efficient protection we performed following studies:

- Determination of buffer layer process window inducing SWCNTs damage regime;
- Study of the structural, electrical and optical changes of SWCNTs films after buffer layer deposition.

2 EXPERIMENTAL

Within this study we fabricated different sets of samples. For the investigation of the changes induced by the hydrogen plasma species, the thin SWCNTs films of 50 nm thickness with optical transmittance of 65% at 600 nm and sheet resistance of 55 Ohm/sq [1] were dry-transferred [3] to a Corning Eagle XG glass substrate of 6.25 cm^2 size. The i-a-Si:H buffer layers with a fixed thickness of 10 nm were deposited by means of Plasma Enhanced Chemical Vapor Deposition (PECVD) at 13.56 MHz in a cluster system *Von Ardenne CS-400PS*. The reference i-a-Si:H layer was obtained at a pressure of 1 mbar, plasma power of 10 W and electrode distance of 20 mm. In order to optimize the i-a-Si:H buffer layer, we varied the silane pressure from 0.5 mbar to 2 mbar, plasma power from 7 W to 15 W, and the electrode distance from 15 mm to 25 mm. The heater temperature and silane flow were held constant at 200 °C and 100 sccm, correspondingly. To minimize the SWCNTs film degradation, no hydrogen was added during the deposition. Table 1 presents the starting process details of i-a-Si:H PECVD deposition. We vary one parameters at time.

Table 1: PECVD parameters of starting recipe of i-a-Si:H buffer layer

	T (∘C)	SiH4 (sccm)	Pr [mbar]	Pw [W]	d [mm]	T [nm]
Starting recipe	200	100	1	10	20	10

MoO_x Buffer layers with a thickness in the range between 3 and 5 nm were deposited using electron beam evaporation at room temperature using, 0.8% XTC power and 0,1 A°/s deposition rate.

Structural modifications of hydrogen bonding in i-a-Si:H thin buffer films and SWCNTs composition after coverage with buffer layers were analyzed with Fourier-transform infrared spectroscopy (*Perkin Ellmer Spectrum 400*). The total transmission (T) of thin film layers was measured with a photo-spectrometer (*Varian 5000*) with an integration-sphere from 300 to 2000 nm. Inspection of the changes in the characteristic peaks of SWCNTs films, pristine and protected with buffer layers, was performed with *Bruker Senterra Raman Scattering Microscope*. Dark conductivity and activation energy were measured with a self-developed system. The surface morphology of specimens were analysed with an atomic force microscope (AFM) *Nanosurf Mobile-S*, as well as scanning electron microscope (SEM) *Zeiss Leo VP*.

3 RESULTS AND DISCUSSIONS

The comparison of surface roughness values obtained with AFM (Fig. 2) clearly show that the deposition of i-a-Si:H buffer layer at higher energetic plasma conditions i.e. at lower pressure (0.5 mbar), higher power (15 W) and shorter electrode distance (15 mm), leads to an induced roughness. To shed more light on the change in SWCNTs morphology, we illustrate SEM images of some selected coating conditions (Fig.3). Pristine SWCNTs is characterized by dense network with 2 ~ 3 µm long nanotubes of 18 - 22 nm diameters (fig. 3a). Buffer i-a-SiH deposited at low pressure (0.5 mbar) results in more pronounced etching of SWCNTs with formation of porous networks, with larger voids (fig. 3b). High power (15 W) leads to aggravated protrusion (fig. 3c). Obviously, even low concentration of hydrogen ions in plasma discharge due to dissociation of undiluted silane is still sufficient to etch carbon nanotubes. In contrast, i-a-Si:H deposited at high pressure of about 2 mbar (fig. 3d) or MoO_x e-beam deposition (fig.3e) seems to be softer and smoother, preserving the morphology of pristine SWCNTs.

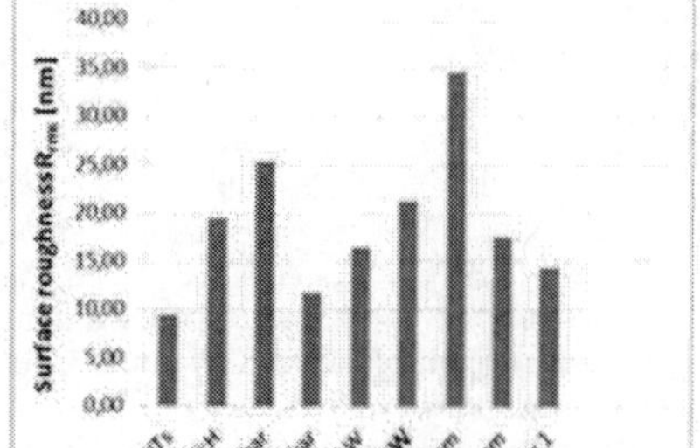

Figure 2: The RMS values of surface roughness of initial and coated SWCNTs films

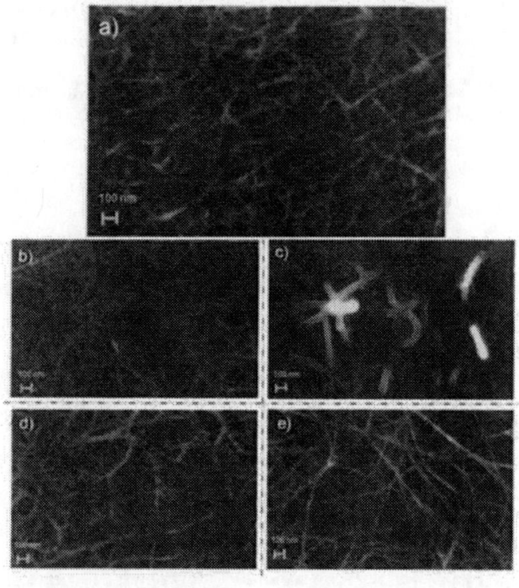

Figure 3: SEM image of pristine SWCNTs film (a) low pressure= 0.5 mbar (b) high power= 15 W (c) high pressure= 2 mbar (d) MoO_x 0 5 nm (e)

Figure 4 depicts the radial breathing mode (RBM) band (a) and the deconvolution of the G-band (b) of Raman spectra for pristine SWCNTs films and for coated ones with a-Si:H buffer layers deposited at different pressures. Based on the deconvolution of G-/G+ ratio, we confirm the coexistence in pristine SWCNTs bundles of metallic phase promoting the charge transport/collection and semiconducting nanotubes improving the charge separation at the interface with i-a-Si:H absorber [5]. An estimation of SWCNTs diameter is based on this relation within RBM bands:

$$\text{Diameter (nm)} = 248 / \text{Raman shift (cm}^{-1}\text{)}$$

for the following peaks at 135, 159, 187, 194 and 205 cm^{-1} that corresponds to the diameters of 1.85, 1.55, 1.32, 1.27 and 1.2 nm, respectively [6].

RBM bands indicate that high pressure preserves thin and thick SWCNTs, and G-band that semiconducting (G+) and metallic SWCNTs (G-) are not affected at high pressure. Evidently, buffer layer deposition at high pressure provides a high purity. The structure of SWCNTs network is severely affected at low pressure leading to a domination of a-Si:H coating at window layer. We speculate that smaller diameter tubes appear to be preferentially removed during buffer layer deposition

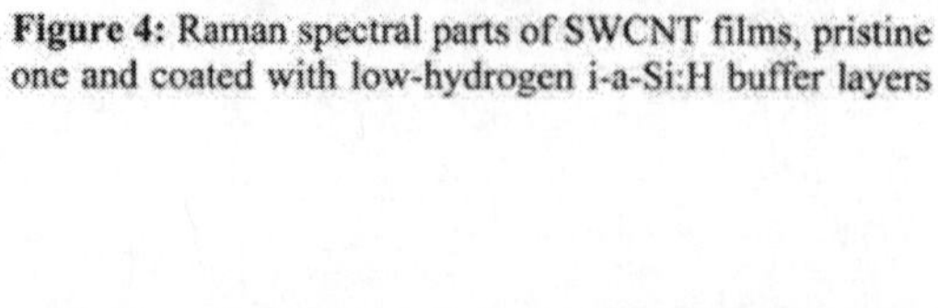

Figure 4: Raman spectral parts of SWCNT films, pristine one and coated with low-hydrogen i-a-Si:H buffer layers

fabricated at different process pressure

High plasma deposition power leads to SWCNTs network density deterioration, and both semiconducting (G+) and metallic (G-) SWCNTs reduction and thus to substitution of SWCNTs by i-a-Si:H areas at the window layer until the almost complete etching of the SWCNTs network at 15 W plasma power (Fig. 5).

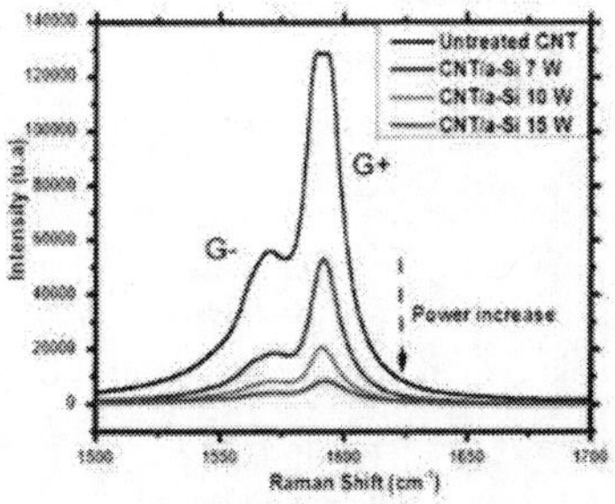

Figure 5: Raman related G-bands of SWCNT films, pristine one and coated with low-hydrogen i-a-Si:H buffer layers fabricated at different process power

On the other hand, decreasing electrode distance leads to SWCNTs thinning/etching (Fig. 6). Larger electrode distance is required to maintain the SWCNTs network.

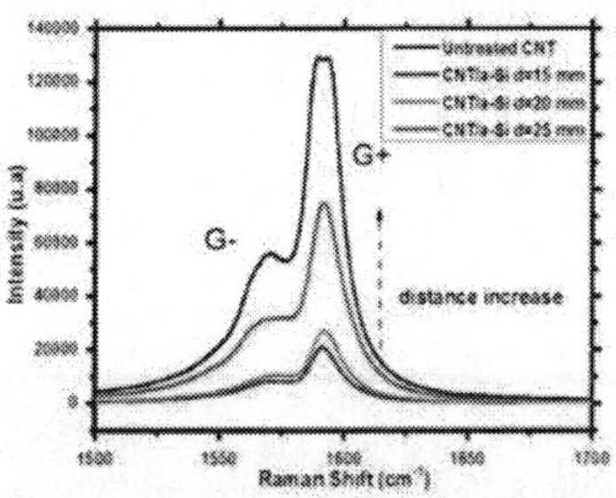

Figure 6: Raman spectral parts of SWCNT films, in pristine condition and coated with low-hydrogen i-a-Si:H buffer layers fabricated at different electrode distances.

For thin MoO_x layers (3 nm), a slight decrease in G-band as well as RBM peaks with low Raman shift related to thick SWCNTs bundles is observed (Fig. 7).

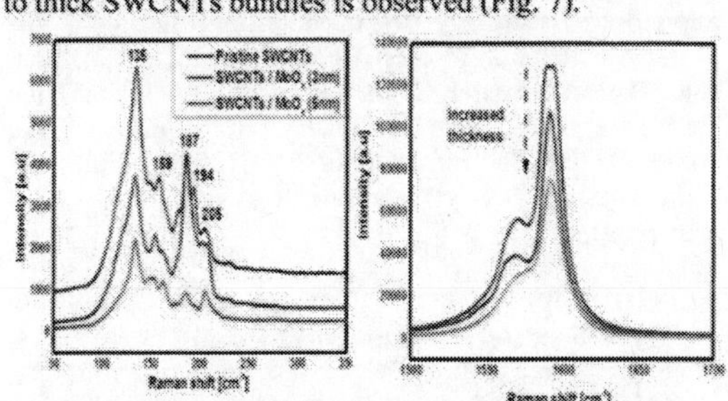

Figure 7: Raman spectral parts of RBM (left) and G-bands (right) of SWCNT films; untreated condition and coated with MoO_x buffer layers of different thicknesses

Thicker MoO_x layers (up to 5 nm) show drop in the mentioned peaks pointing out the degradation of SWCNTs bundles and the decrease of semiconducting and metallic SWCNTs network density. Therefore, a thin MoO_x buffer layers (~3 nm) are recommended for SWCNTs network protection coating.

For SWCNTs/undiluted i-a-Si:H stack fabrication, one can conclude that "soft" plasma employing high pressure (2 mbar), low power (7 W) and large electrode distance (25 mm) is recommended for reduction of damages in SWCNT-network.

Infrared spectra of SWCNTs films are displayed in figure 8. The existence of C-C stretching mode was observed.

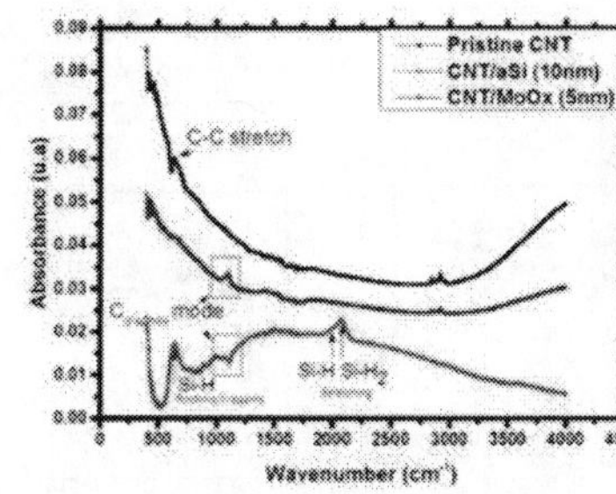

Figure 8: FTIR spectra of SWCNTs films, pristine and coated with i-a-Si:H and MoO_x buffer layers

The characteristic mode of graphite around 1100 cm^{-1} (diameter sensitive peak) corresponds to the calculated SWCNTS diameter:

$$D = 8.041 \times 10e - 4\lambda,$$

where the SWCNTs were protected with a-Si:H buffer. In the latter case the peaks related to Si-H bonding appeared. SWCNTs films protected with MoO_x show a similar range of characteristic graphite mode (~1100 cm^{-1}).

Infrared spectra of SWCNTs films protected with different a-Si:H buffer layers are compared to non-protected SWCNTs films in figure 9. Obviously, high power and lower electrode distance plasma conditions lead to the formation of Si-H bonding on the SWCNTs network.

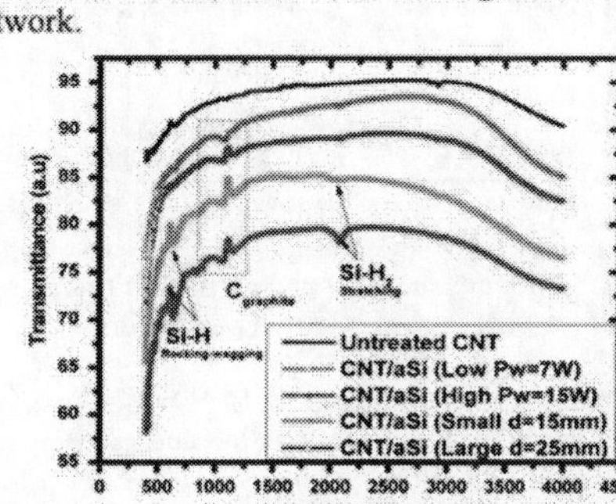

Figure 9: FTIR spectra of pristine SWCNTs films and coated with i-a-Si:H buffer layers obtained at different process parameters

"Soft" plasma regime with less damage due to the ion bombardment effect is recommended for the SWCNTs coating with i-a-Si:H buffer layer. Raman results consistently confirm that lower energetic plasma conditions lead to the preservation of SWCNTs network.

Based on UV-VIS spectra displayed in figure 10, lower energetic plasma conditions during formation of undiluted buffer layer also preserve the transmission degree of SWCNTs, whereas a high plasma process window leads to a decrease in transmission of SWCNTs/i-a-Si:H stacks as a window layers, due to coating with amorphous silicon.

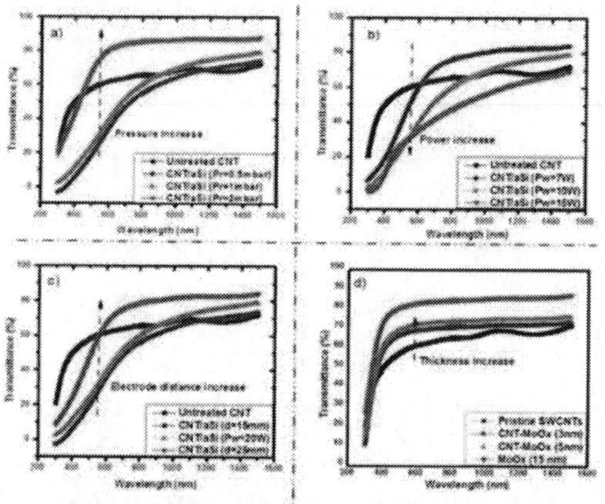

Figure 10: Impact of buffer layer deposition parameters pf i-a-Si:H : pressure (a), power (b), electrode distance (d) and MoO_x thickness (d) on transmission properties

For the SWCNTs films coated with MoO_x buffer layers the transmission improves with increased thickness of the MoO_x layer.

The results of dark conductivity and activation energy measurements corresponding to pristine and coated SWCNT films with MoO_x and undiluted i-a-Si:H buffer layers with different discharge powers are represented in Fig 11.

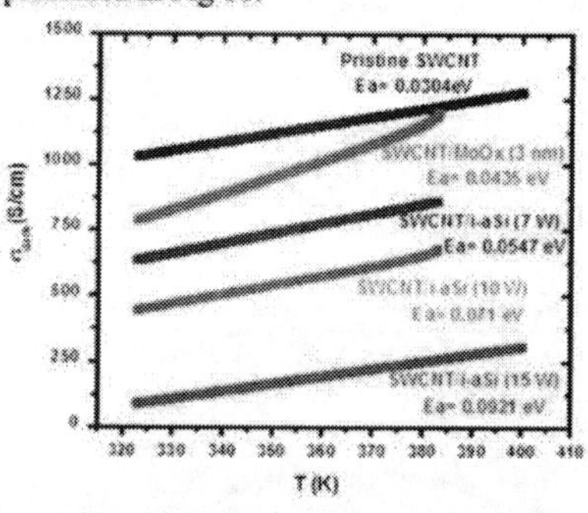

Figure 11: Dark conductivity and activation energy as a function of i-a-Si:H layer deposition parameters

It is observed that undiluted i-a-Si:H buffer layers on SWCNTs film induce higher activation energy compared to MoO_x layer. The conductivity is higher for MoO_x buffer layer compared to the i-a-Si:H counterpart. This can be explained with the p-doping tendency of MoO_x compared to the n-doping nature of intrinsic i-a-Si:H material. At high temperature up to 380 K, we speculate that Mo atoms can form alloy with C which further increase the conductivity.

Finally, the general trend in the properties of non-protected and protected SWCNTs films is represented in figure 12. The increase of G/D Raman bands ratio corresponds to the increased optical transmittance and sheet resistance. The increase of G+/G- Raman band ratio corresponds to the increased dark conductivity and reduced activation energy. Improvement of structural properties (high density of semiconducting SWCNTs networks at the SWCNTs/a-Si:H absorber interface with the high effective junctions area) leads to the increase of conductivity and of light transmission through the SWCNTs front window layer under the control of the buffer layers.

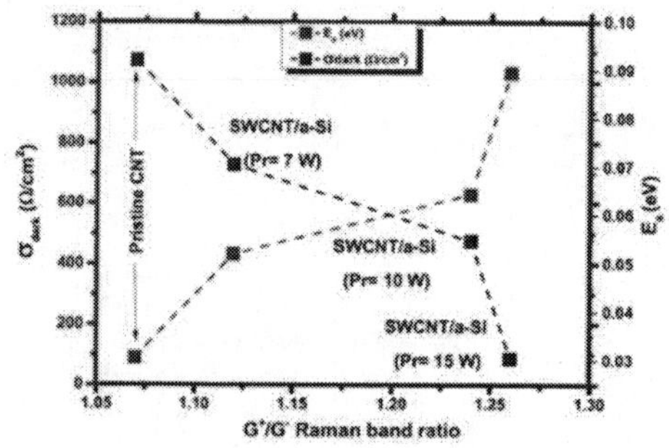

Figure 12: General trend and correlation of the properties of SWCNTs films, pristine and coated with different amorphous silicon and MoO_x buffer layers

4 CONCLUSION

To sum up, a general trend for the SWCNT films as a front window layer in p-i-n hybrid SWCNTs / hydrogenated amorphous silicon thin film solar cells is that deposition of protective layers of a few nanometers thickness is recommended to avoid damages related to the hydrogen etching. One can conclude that the improvement of structural properties (high density semiconducting SWCNTs networks at the SWCNTs/a-Si:H absorber interface (with highly effective junctions' area) protected with developed buffer layers on the basis of electron beam deposited molybdenum oxide leads to a high conductivity and to the increased light propagation through the window layer based on SWCNTs films. Further optimization of developed "undiluted" amorphous silicon buffer layers is proposed in order to combine it in the same process with the deposition of the absorber layer.

ACKNOWLDEGMENT

The authors gratefully thank Michael Kröner for the constructive discussions. We acknowledge the financial support from Ministry of Education and Science of the Russian Federation (project RFMEFI61815X0003), from the Estonian Research Council (project ETAG15028) and from German Federal Ministry of Education and Research (BMBF, ref. number 01DJ15030) within the joint project FLEXAPP in a framework of ERA.Net RUS Plus program.

REFERENCES

[1] A. Funde et al. Nanotechnology 401,185401, 6pp, 2016.

[2] Alekseeva et al., Phys. Status Solidi B, 255: n/a, 2017.

[3] Rajanna P. et al., Nanotechnology. 2018.

[4] Dosenovicova, D. et al Materials Sceince in Semiconductor Processing, 68, 1-5, 2017.

[5] A.C Dillon et al. J. Nanosci. Nanotech. Vol. 4, No. 7, 2004.

[6] A. Kumar et al. Journal Of Applied Physics 115, 084308, 2014.

RARE EARTH DOPED COMPOUNDS FOR ENHANCING THE EFFICIENCY OF SILICON BUILDING INTERGRATED CONCENTRATOR PV VIA SPECTRAL CONVERSION

J. Day*, T.K. Mallick, S. Senthilarasu
University of Exeter
Environment and Sustainability Institute, Penryn Campus, Penryn, Cornwall, TR10 9FE. United Kingdom.
* jd553@exeter.ac.uk

ABSTRACT: A detriment to the efficiency of silicon solar cells used in building integrated concentrator photovoltaics (BICPV) is their reduced electrical response to light of short or long wavelengths. In order to address this challenge, single cell static CPV modules were fabricated with some of the devices containing rare earth doped compounds dispersed into the system in differing concentrations and geometries. Under a solar simulator at 1 sun illumination, the power conversion efficiency (PCE) of devices improved up to 11.1% relative through the incorporation of these materials. At lower irradiances and compared to cells without concentrators, the relative efficiency gains were more pronounced and external quantum efficiency (EQE) measurements suggested spectral conversion was responsible for these enhancements. For a theoretical large scale BICPV system, a simple analysis showed cost per watt could fall by up to 8.1% and power output increased from 25.7 to 28.4 W/m^2 through this approach.
Keywords: Building integrated PV, Concentrator cells, Photoluminescence, Solar cell efficiencies, Spectral response.

1 INTRODUCTION

Optical concentrators have emerged as an area of interest in PV research, since the focusing of light allows for a smaller area of active cell to produce the same power output which could bring down module costs [1]. Many optical systems have been proposed to realize this (such as Fresnel lenses [2] and parabolic mirrors [3]) but 3-D static concentrators such as the square elliptical hyperboloid (SEH) design is an attractive option for low concentration silicon PV systems due to its ease of fabrication [4].

Meanwhile, building integrated PV systems have emerged which seek to incorporate PV into a building's architecture for aesthetic benefits. Luminescent solar concentrators (LSCs) with their semi-transparency and range of colors are strong candidates for one day replacing part of a building's window façade [5]. However, arrays of SEH concentrators coupled to silicon cells could also produce a semi-transparent effect and provide additional insulation in colder climates, so could be a viable alternative if their power per unit area is maximized.

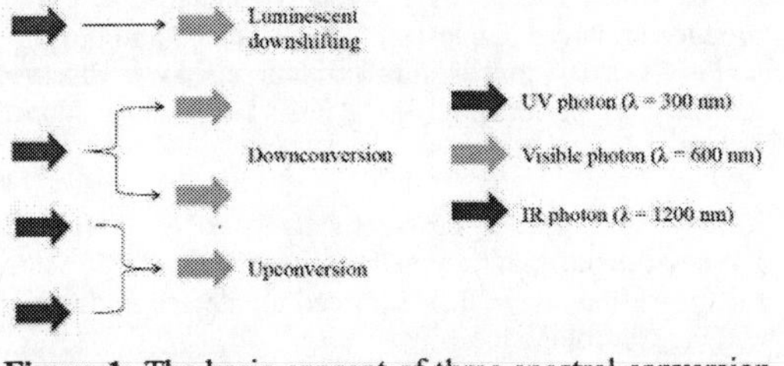

Figure 1: The basic concept of three spectral conversion processes: luminescent downshifting, downconversion and upconversion.

A fundamental source of loss for any single junction solar PV module is the spectral mismatch between the band gap of the cell and the incident sunlight. This results in high energy photons causing thermalization losses and low energy photons not generating any current. One way to overcome this is the concept of spectral conversion; whereby high energy photons can be converted into lower energy photons in processes known as downconversion and luminescent downshifting, or low energy photons into higher ones through upconversion [6].

In reality, spectral conversion can occur in rare earth doped compounds [7], organic dyes [8] or quantum dots [9] which are usually incorporated into the system in a layer above or below the cell. Hence, it was proposed to investigate the effect of the application of two rare earth doped compounds with spectral conversion properties to single cell SEH concentrator-silicon PV modules. The materials were embedded into the system in varying ratios (0.25%, 0.5%, 1% and 2% w/w in the optical coupling) and geometries (either 0.1% w/w homogenously throughout the SEH concentrator or confined to its top or bottom) and electrical performance compared with control devices to evaluate the impact of their addition (see Fig.2). PV cells without concentrators but rare earth compounds in a layer above them were also analyzed to determine the effect of the concentrator.

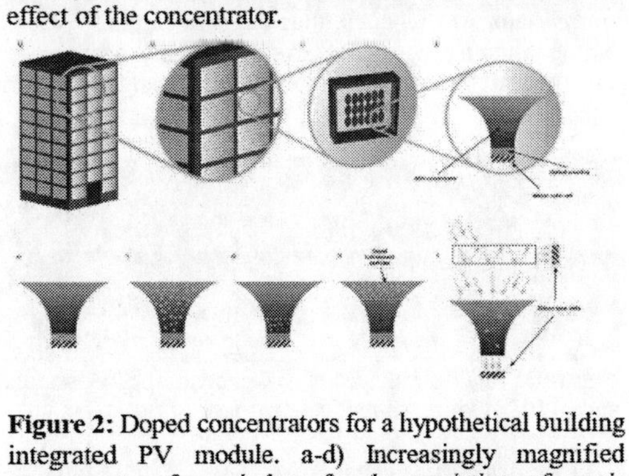

Figure 2: Doped concentrators for a hypothetical building integrated PV module. a-d) Increasingly magnified components of a window façade consisting of semi-transparent SEH-PV modules. e-h) Layout of how the rare earth doped compounds were incorporated into the concentrator PV modules in this experiment. i) Working principle of luminescent solar concentrator and SEH, both are based on total internal reflection.

2 MATERIALS AND METHODS

2.1 Materials

The two rare earth compounds used were $NaYF_4$: Er^{3+}, Yb^{3+} and $Sr_4Al_{14}O_{25}$: Eu^{2+}, Dy^{3+} of 99% purity (Sigma-

Aldrich). These were chosen because of their suitable excitation and emission profiles for spectral conversion; both can act as downconversion materials by absorbing UV-blue light and emitting in the blue-green [10] and $NaYF_4$: Yb, Er has additional upconversion properties by absorbing NIR photons and emitting them in the visible region [11]. The SEH concentrators were made from Crystal Clear® 200 casting resin which were attached to the solar cells by an optical coupling made from Sylgard®184 silicone elastomer. The silicon solar cells were provided by the National Renewable Energy Centre (NAREC) and had an active PV area of 1 cm^2.

2.2 Fabrication of concentrator PV devices

The SEH concentrators were made by mixing the two parts of the casting resin in the specified 10:9 ratio before degassing in a vacuum oven (to reduce air bubble formation) and leaving to set in a casting mold, before removal. If rare earth compounds were required within the concentrator, they were added to part B and dispersed with a magnetic stirrer before mixing with part A. If the particles were to be confined to the top or bottom of the concentrators, then the layers were prepared separately before pouring into the mold in the desired order. $NaYF_4$: Er^{3+}, Yb^{3+} were fine enough to disperse straight away but the $Sr_4Al_{14}O_{25}$: Eu^{2+}, Dy^{3+} was coarser, so was ground with by an agate mortar and pestle for 30 minutes to reduce particle size.

The optical coupling was made by mixing parts A and B of the elastomer in a 10:1 ratio. As with the casting resin, degassing was utilised to minimise bubble growth. However if rare earth compounds were to be added, they could be added after parts A and B were mixed and subject to ultra-sonication as well as magnetic stirring, due to the longer pot life of Sylgard and high viscosity of part A.

To put the device together, solar cells were placed on an acrylic base with a boundary which was laid by a glue gun to prevent elastomer leakage. Then, a small quantity of Sylgard was dropped on the center of the solar cell. If a concentrator was required, then one was carefully placed on top, aligning the square base with the PV active area. The system was then left to cure for 48 hours at room temperature.

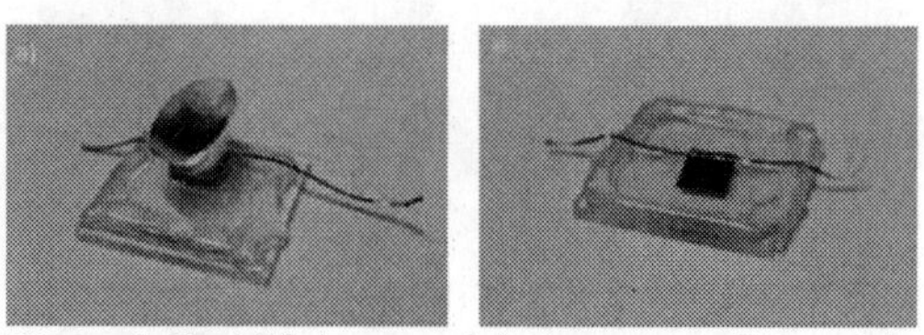

Figure 3: Real life images of a) Concentrator PV module and b) PV module without concentrator.

2.3 Electrical characterization

The devices were characterized, first, under a Wacom® solar simulator which can replicate the AM1.5G spectrum. The cells were placed on a horizontal platform beneath the lamp and connected by metal clips to a current-voltage tracer (EKO MP-160i) which could determine their electrical properties (after 30 seconds of illumination) including power conversion efficiency (PCE) and short-circuit current density (J_{SC})

Then, a Bentham PVE300, which uses a tunable laser as an illumination source, was used to determine the external quantum efficiency of the devices between 300 and 1100 nm (in 10 nm intervals). Both instruments were prior callibrated through reference silicon cells and the irradiance of the solar simulator could be modified using HeliCon software and reference short-circuit current measuremetns.

3 RESULTS

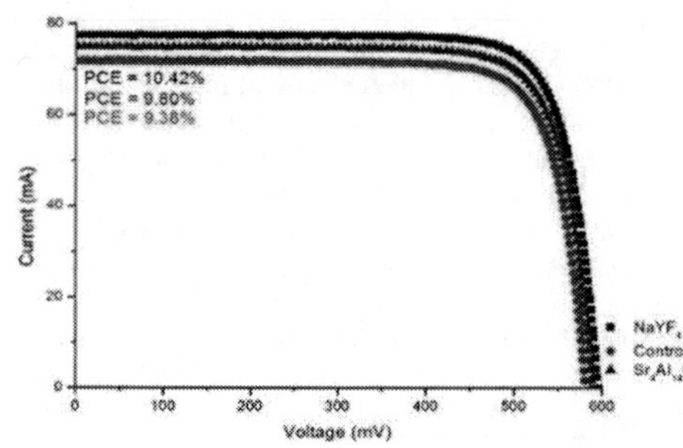

Figure 4: Current-voltage curves of the best performing $NaYF_4$: Er^{3+}, Yb^{3+} and $Sr_4Al_{14}O_{25}$: Eu^{2+}, Dy^{3+} doped devices and the control, obtained under the solar simulator.

At 1 sun illumination, the highest efficiency attained was 10.42% for $NaYF_4$: Er^{3+}, Yb^{3+} doped optical coupling (at 1% w/w) which is an 11.1% relative improvement on the control device. Whereas for $Sr_4Al_{14}O_{25}$: Eu^{2+}, Dy^{3+} at the same light intensity, the highest efficiency measured was 9.92% for the top doped concentrator (a 5.5% relative improvement on the control device, current-voltage curves in Fig. 4). From Table I, it can be seen $NaYF_4$: Er^{3+}, Yb^{3+} doped concentrators tended to perform weaker than $Sr_4Al_{14}O_{25}$: Eu^{2+}, Dy^{3+} concentrators. E.g. at 1 sun a 5.7% relative efficiency increase was observed for the top doped $Sr_4Al_{14}O_{25}$: Eu^{2+}, Dy^{3+} concentrator whereas the best improvement for a $NaYF_4$: Er^{3+}, Yb^{3+} doped concentrator was 1.7%. However, the $NaYF_4$: Er^{3+}, Yb^{3+} homogenous doped concentrators yielded better gains than their $Sr_4Al_{14}O_{25}$: Eu^{2+}, Dy^{3+} analogues, e.g. at 1 sun +1.7% for $NaYF_4$: Er^{3+}, Yb^{3+} vs -2.3% for $Sr_4Al_{14}O_{25}$: Eu^{2+}, Dy^{3+}. Furthermore, efficiency improvements were generally found to be more pronounced at lower irradiances as shown in Fig. 5. For example, the $NaYF_4$: Er^{3+}, Yb^{3+} doped device which had an 11.1% relative increase at 1 sun, produced relative increases of 15.5% at 0.8 and 0.6 suns and 17.7% at 0.4 suns. A similar relationship was observed for the best performing $Sr_4Al_{14}O_{25}$: Eu^{2+}, Dy^{3+} doped devices.

Table I: PCE and J_{SC} measurements for each concentrator PV device configuration at 1 sun irradiance. ΔPCE is the relative difference in PCE between the device in question and the control device.

$NaYF_4$: Er^{3+}, Yb^{3+}	PCE (%)	J_{SC} (mA/cm^2)	ΔPCE (%)
Control device	9.38	71.97	-
0.25% optical coupling	10.01	75.24	+6.7
0.5% optical coupling	10.03	76.54	+6.9
1% optical coupling	10.42	77.78	+11.1
2% optical coupling	9.88	74.63	+5.3
Homogenous concentrator	9.54	73.45	+1.7
Top of concentrator	9.47	72.30	+0.7
Bottom of concentrator	9.16	70.05	-2.3
$Sr_4Al_{14}O_{25}$: Eu^{2+}, Dy^{3+}	PCE	J_{SC}	ΔPCE

	(%)	(mA/cm²)	(%)
0.25% optical coupling	9.80	73.62	+4.4
0.5% optical coupling	9.78	74.95	+4.2
1% optical coupling	9.34	72.21	-0.7
2% optical coupling	9.38	73.30	+0.0
Homogenous concentrator	9.17	69.98	-2.3
Top of concentrator	9.92	75.17	+5.7
Bottom of concentrator	9.88	75.24	+5.3

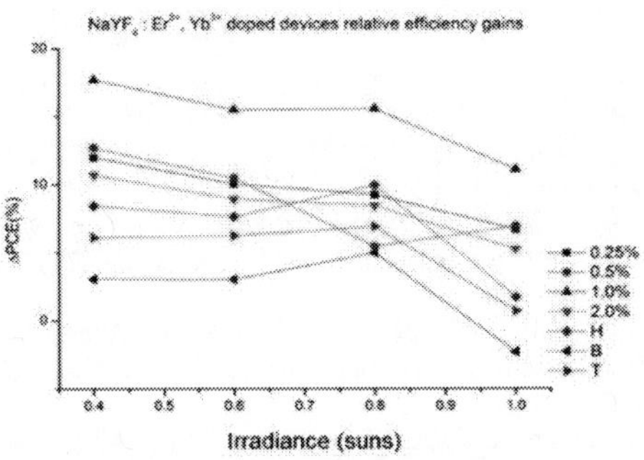

Figure 5: Efficiency gains relative to the control device for all configurations of NaYF4: Er^{3+}, Yb^{3+} containing devices at different irradiances.

Without concentrators, at 1 sun (see Table II), the greatest efficiency for $Sr_4Al_{14}O_{25}$: Eu^{2+}, Dy^{3+} doped sylgard was 16.31% at a doping concentration of 0.5% (a 2.1% relative improvement to undoped sylgard) whereas for for $NaYF_4$: Er^{3+}, Yb^{3+} doped sylgard was 16.75% at a doping concentration of 1.0% (a 4.8% relative improvement to un-doped sylgard). As shown by Fig. 6, the relative efficiency enhancements tended to be siginificantly bigger for both materials when concentrators were placed on top of the solar cells. These differences were more pronounced at lower irradiances. E.g. at 0.4 suns for 0.5% for $NaYF_4$: Er^{3+}, Yb^{3+} doped sylgard, with concentrator a relative increase in efficiency of 12.67% was observed and without concentrator a decrease of 3.46% was recorded, meaning the addition of an optic yielded a 16.13% greater relative enhancement in efficiency. For an appendix of all efficiencies at all irradiance levels, please contact the corresponding author.

Table II: PCE and J_{SC} measurements for each configuration of devices without concentrators at 1 sun irradiance. ΔPCE is the relative difference in PCE between the device in question and the control device.

$NaYF_4$: Er^{3+}, Yb^{3+}	PCE (%)	J_{SC} (mA/cm²)	ΔPCE (%)
Control device	15.98	36.33	-
0.25% optical coupling	16.10	36.79	+0.8
0.5% optical coupling	16.18	36.54	+1.3
1% optical coupling	16.75	36.97	+4.8
2% optical coupling	16.47	37.02	+3.1
$Sr_4Al_{14}O_{25}$: Eu^{2+}, Dy^{3+}	PCE (%)	J_{SC} (mA/cm²)	ΔPCE (%)
0.25% optical coupling	16.24	36.72	+1.7
0.5% optical coupling	16.31	37.55	+2.1
1% optical coupling	16.23	36.77	+1.6
2% optical coupling	16.07	36.98	+06

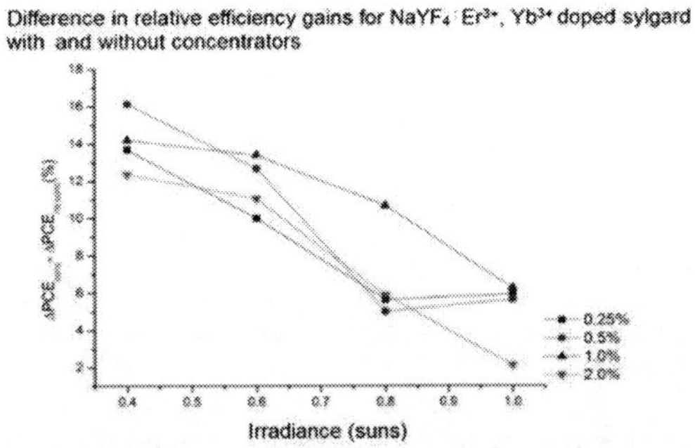

Figure 6: A key result; the difference in relative efficiency gains at different irradiances for $NaYF_4$: Er^{3+}, Yb^{3+} doped optical coupling concentrator-PV devices compared to the same doping level Sylgard topped PV cells but without concentrators

Short-circuit current density improvements tended to follow efficiency gains. At 1 sun the same device which had the highest efficiency also had the greatest J_{SC} at 77.78 mA/cm2 (an 8.08% relative increase). Although for $Sr_4Al_{14}O_{25}$: Eu^{2+}, Dy^{3+} doped devices the greatest J_{SC} was observed for the device with the 2nd highest efficiency. EQE measurements for these devices with the highest J_{SC} are presented (Fig. 7.) in which a clear increase of up to 4% for the $NaYF_4$: Er^{3+}, Yb^{3+} containing device at 450 nm is observed, suggesting downconversion or downshifting responsible for the enhanced performace.

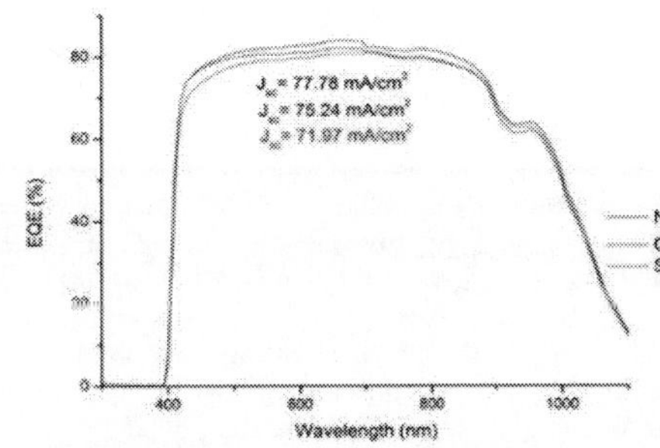

Figure 7: EQE spectra for the $NaYF_4$: Er^{3+}, Yb^{3+} and $Sr_4Al_{14}O_{25}$: Eu^{2+}, Dy^{3+} containing concentrator-PV devices with the highest short circuit current density, compared to the control.

4 DISCUSSION

It is important to gain insight into the physical reasons behind this data. The better performance of the $NaYF_4$: Er^{3+}, Yb^{3+} devices could be attributed to its wider absorption range, higher photoluminescent quantum efficiency or smaller particle size leading to less scattering losses. Although, this does not explain the relatively poorer performance of the $NaYF_4$: Er^{3+}, Yb^{3+} doped concentrators. Potentially this could have been due to self-absorption losses (from the overlap of the ion's excitation and emission spectra) which are more amplified over longer optical path lengths (i.e. when the particles are embedded in the concentrator as opposed to the optical coupling). The strong performance of $Sr_4Al_{14}O_{25}$: Eu^{2+}, Dy^{3+} when placed at the top or bottom of the concentrator is consistent with LSC studies which show improved results when the luminescent particles are confined to a thin film at the edge of the LSC [12].

The concentration ratio of the rare earth compounds within the optical components is another important factor for consideration. Clearly if the level of ions is too low then insufficient photoluminescent events will take place. Conversely, at too high concentrations energy transfer between ions will lead to efficiency losses [13]. In our investigation $NaYF_4$: Er^{3+}, Yb^{3+} shows this behaviour with an initial rise in efficiency gain with increasing particle concentration, reaching a maximum at 1% w/w before falling at 2% w/w. Light intensity also has been shown to affect spectral conversion rates [14] but in this study showed little impact, perhaps due to the relatively low variation between 0.4 and 1 sun.

Finally, a simple cost analaysis showed the cost per watt could fall up to 8.1% for the best performing $NaYF_4$: Er^{3+}, Yb^{3+} doped devices from \$51.81/W to \$47.59/W and to \$49.15/W for $Sr_4Al_{14}O_{25}$: Eu^{2+}, Dy^{3+}. These calculations were based on laboratory scale material prices and would significantly be reduced for studies conducted on a larger scale, to be brought in line with current building integrated PV costs of \$4.1-\$24.5/W [15]. A prototype BICPV module based on SEH concentrators would contain 24 PV cells in 0.031 m^2. Therefore incorporating $NaYF_4$: Er^{3+}, Yb^{3+} could theoretically increase the power per unit area from 25.7 W/m^2 to 28.4 W/m^2. Future work in this area could include optical models such as ray tracing to validate the experimental results and photoluminescent spectroscopy of the spectral conversion materials to determine their quantum efficiencies and underlying mechanisms.

5 CONCLUSION

In conclusion, rare earth doped compounds were successfully shown to improve the efficiency of single cell BICPV modules. At 1 sun, a relative PCE improvement of up to 11.1% was observed for $NaYF_4$: Er^{3+}, Yb^{3+} containing devices, the best performing compound. Both the relative PCE gains and difference between these values for concentrating and non-concentrating modules were more pronounced at lower irradiances; up to a 16.1% difference at 0.4 suns, highlighting the utility of spectral conversion for concentrating PV. EQE measurements indicated the downconversion and downshifting processes as potential reasons for these results but further theoretical and material investigations are required to fully understand the underlying mechanisms. It is hoped this work will lead to more spectral conversion studies for a range of BICPV geometries other than LSCs.

6 ACKNOWLEDGEMENTS

Funding for this work was provided by the Engineering and Physical Sciences Research Council (EP/J000345/2) and is gratefully acknowledged. The authors would also like to thank the following individuals: Mr James Yule, Dr Katie Shanks and Dr Shivangi Sharma (all from the University of Exeter) for their general help in ensuring high standard device fabrication.

7 REFERENCES

[1] K. Shanks, S. Senthilarasu, T.K. Mallick, Renewable and Sustainable Energy Reviews 60 (2016) 394.
[2] W.T. Xie, Y.J. Dai, R.Z. Wang, K. Sumathy Renewable and Sustainable Energy Reviews 15 (2011) 2588.
[3] A. Muthu Manokar, D. Prince Winston, M. Vimala, 2015, Procedia Technology 24 (2015) 4850491.
[4] N. Sellami, T.K. Mallick, D.A. McNeil, Energy Conversion and Management 64 (2012) 579.
[5] K. Wu, H. Li, V.I. Klimov, Nature Photonics 12 (2018) 105.
[6] J. Day, T.K. Mallick, S. Senthilarasu, Renewable Energy 132 (2019) 186.
[7] X. Huang, S. Han, W. Huang, X. Liu, Chemical Society Reviews 42 (2013) 173.
[8] R.T. Weigh, H. Donker, E.V.D. Van Loef, K.D. Oskam, A. Meijerink, Journal of Luminescence 87 (2000) 1017.
[9] S. Reineke, N. Seidler, S.R. Yost, F. Prins, W.A. Tisdale, M.A. Baldo, Applied Physics Letters 103 (2013) 093302.
[10] H. Sun et al, Dalton Transactions 43 (2014) 14936.
[11] B. Qu et al. Journal of Alloys and Compounds 658 (2016) 848.
[12] M.H. Hughes, D.A. Borca-Tasciuc, D. Kaminski, Solar Energy Materials and Solar Cells 171 (2017) 293.
[13] F. Vetrone, R. Naccache, V. Mahalingam, C.G. Morgan, J.A. Capobianco, Advanced Functional Materials 19 (2009) 2929.
[14] B. Ahrens et al, Physica Status Solidi (A) Applications and Materials Science, 205 (2008) 2827.
[15] R.J. Yang, P.X.W. Zou, International Journal of Construction Management 16 (2016) 39.

OPTICAL AND STRUCTURAL, PROPERTIES OF ZINC SUFIDE DOPED CHROMIUM THIN FILMS PREPARED BY RF-SPUTTERING

C. M. Samba Vall[1, 2], M. Chaik[1], H. El aakib[1], M.Elyaagoubi[1], A.Agdad[1], A.Tchenka[1], A.Oueriagli[1], M. Aggour[2], A. Outzourhit[1]

1Laboratory of Nanomaterials for Energies and Environments, Faculty of Sciences Semlalia, Cadi Ayyad University, Marrakech 40000, Morocco.

2Laboratory of Renewable Energy and Environment, faculty of Science Kenitra, Ibn Tofail University PO Box 133, Kenitra - Morocco.

Cheikhmohamed5@gamil.com

ABSTRACT: Zinc sulfide and Chromium doped Zinc sulfide thin films were grown on glass and Si substrates by cathodic radio-frequency sputtering. The deposition time was fixed at 2 hours for pure Zinc sulfide and was varied from 90 minute to 3 hours for Chromium doped Zinc sulfide while the radio-frequency power of both deposited was fixed at 200W. The optical, structural, of these thin films were investigated. The optical properties (mainly the thicknesses, refractive index, and optical band gap) were examined by optical transmission measurements in the ultraviolet-visible-near Infrared wavelength range. X-ray Diffraction analysis indicated that all the sputtered Zinc sulfide and Chromium doped Zinc sulfide films have the zinc blend structure with a preferential growth orientation along the (111) ,(2 2 0),(311) direction. The crystallite size was about 33.45nm for undoped Zinc sulfide films and varied from 16.5 nm to 29.5 nm for Chromium doped Zinc sulfide with a maximum value for a deposition time of 150 minutes. Ultra violet-Visible measurements showed that the Zinc sulfide film deposited for 2 hours have transmittance higher than 70% while the transmission of Chromium doped Zinc sulfide becomes less than 40% in the visible wavelength region. Moreover, it was observed that the band gap energy of the Zinc sulfide films is 3.48 eV, while that of Chromium doped Zinc sulfide films decreases from 2.77 to 1.68 eV as the deposition time increased. FT-IR data exhibited that absorption peaks at wave number 2730.65 cm-1 for ZnS due to the stretching vibration of O-H. Also, 1200/cm and 565.02/cm corresponding to asymmetric and symmetric Zn-S stretching bond.

Keywords: Structural Properties, Chromium doped Zinc sulfur, Thin Films prepared by Radio Frequency Sputtering.

1 INTRODUCTION

Thin films of doped and undoped chalcogenide materials like ZnS, ZnSe, and CdS [1] have potential applications in various areas of semiconducting industries such as photocatalysis, display panels, transistors[2], and solar cells[3],[4].

Cr-doped ZnS (ZnS:Cr) is very an interesting material that could be potentially used for various spin electronic and optoelectronic applications[5]. In addition, the insertion of magnetic transition metal ions in non-magnetic semiconductors will pave the way for the generation of diluted magnetic semiconductors (DMS) [6],[7]. In particular, ZnS:Cr is proposed as an intermediate band (IB) absorber thin films for use as in intermediate band solar cells (IBSCs)[8],[6]. zinc chalcogenides such as ZnS, ZnSe and ZnTe are semiconductors with zinc blende structure at ambient conditions. Chalcogenides such as ZnS, ZnSe have been widely used as important candidates, cathodoluminescence (CL), electroluminescence (EL), and photoluminescence (PL) devices[9] because of their better chemical stability.

The ionic radius of Cr is close to that of $Zn+^2$, which means Cr+3 can easily penetrate into ZnS crystal lattice and substitute for Zn2+ in the lattice[10].

There have been several reports on RF sputtering of wide-gap semiconductors doped with a few percent of 3d transition-metal ions. Amidst the most studied host materials, ZnS is a potential candidate for device applications due to its wide band gap (~3.68 eV), large exciton binding energy (40 meV), high index of refraction (2.27 at 1 µm) [11],[12].[13]. Another method to deposit films by PLD(pulsed-laser deposition) [14],[8],[15]. and molecular beam epitaxy[16], chemical co-precipitation route for synthesis of Cr doped ZnS nanocrystals[17]. Among the previously mentioned technique sputtering radio frequency RF cathodic sputtering, a relatively cost-effective deposition technique compared with those listed above, allows sufficient control over the stoichiometry and uniformity of the films[18]. Sputtering technique is widely used in industrial process , because high quality films can be obtained at low temperature substrates. However, to our knowledge there are no reports on the preparation of Cr-doped ZnS films by radio frequence technique sputtering [19].

In this work, we have prepared doped and undoped ZnS and ZnS:Cr thin films using RF cathodic sputtering to be used in intermediate band solar clls. The structural and optical properties as well as of the various films were investigated.

2 EXPERIMENTAL METHOD

ZnS and ZnS:Cr thin films were prepared by cathodic radio-frequency RF-sputtering in an Alcatel SCM 451 deposition system equipped with an Alcatel ARF 601 RF generator operating at 13.56 MHz, which was described in previous works . These films were deposited using a high purity ZnS target (99.99%) with a diameter of 10 cm[20],[18]. On which target of ZnS, chromnium Cr(99.99%) pieces had been placed with surface equal 4.5cm2. These pieces were uniformly spaced, and covered 5% of the ZnS disc surface.

The substrates used in these studies are Corning glass and silicium. They were chested with rinsed deionized water, isopropanol and acetone in an ultrasonic bath for 20 min, then put in roven for 15min at 70°C. After cleaning, the samples were immediately loaded into a chamber. Before deposition, the chamber was pumped down to a base pressure of 1.5×10-6mbar, The distance between the ta pert MPD difractometer using the CuKα radiation.. Optical transmittance measurements were performed using a Shimadzu UV-PC spectrophotometer and FI-TR VERTEX70 DTGS in the 200–3200 nm.rget and substrate was about 7.5 cm[21].

Table I: The deposition parameters of the undoped zinc sulfide and zinc sulfide doped chroiume thin films

.

Deposition parameter	Corresponding value	
Deposition time	ZnS:Cr	Varied from 90min to 180min
	ZnS	120min
rf-Power (input-reflected)	Fixed at 200W	
Target diameter	10cm	
Target-Substrate distance	8cm	
Bombardment gas	Argon (Ar)	
Gas pressure	0.01 mbar	
Substrates used	Glass, Si	
Substrate temperature	300K	

3 RESULTAT AND DISCUSSION

3.1. Optical properties of ZnS and ZnS:Cr thin films

3.1.1Optical Transmittance

The optical transmittance spectra in the wavelength range 200 to 3200 nm of ZnS, ZnS:Cr thin films at different times of deposit with RF power fixed are shown in Fig.1. All the spectra exhibited well contrasted interference fringes indicating a smooth surface. The observed fringes are attributed to interference between the waves reflected from the air-film and the film-substrate interfaces. The films formed at times between 90min and 150min have a relatively low average transmission in the visible and near IR regions (40%),and high transmission then 70% for pure ZnS deposited at 2h . The optical transmittance for our ZnS:Cr thin films is decreasing while the times of depositions increases and then transparency is comparable to that of other ZnS thin films deposited by RF sputtering[1],[23]. On the other hand, the average transmission of the films deposited at 180min dropped to an average of 30% and presents a small shoulder in the forte absorption region. Also a big shift in the absorption edge towards a higher wavelengths is also observed and can be assigned to the increase in the deposit times [24]. These results are in agreement with the fact that the transmittance of thin films depends on their thickness and their morphology [25].

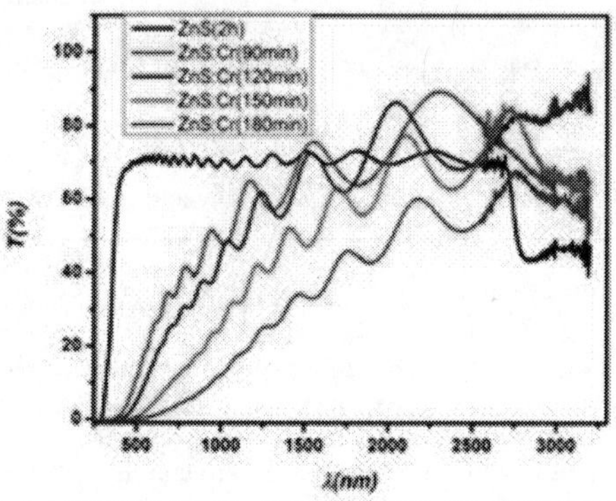

Figure 1: Wavelength dependence of the transmittance of ZnS and ZnS:Cr thin films deposited at different times.

3.1.2 Thickness and refractive index

The film thickness (d) and refractive index (n) were obtained from the measured optical transmission following the method described by N. Pimpabute[20],[26]. In the case where the film is thick on the one hand and smooth on the other hand, the reflections multiple of the light are between the lower surface in contact with the substrate and the free surface of the film, the result is the transmission spectrum with interference fringes with minima and maxima as a function of the wavelength. Whether λ1 and λ2 are the wavelengths of two consecutive maxima, TM1 and TM2, the respective transmissions, Tm is the transmission of the minima which found between its (Figure.2), fig.2 shows the maxima and minima of points using the upper and lower envelope. The thickness of the films is determined from of the relationship:

$$\mathbf{d} = \frac{\lambda_1\lambda_2}{2(\lambda_1 n_2 - \lambda_2 n_1)} \quad \text{... [1]}$$

Where n1 and n1 are the refractive indices of the films for the wavelengths λ1 and λ2.

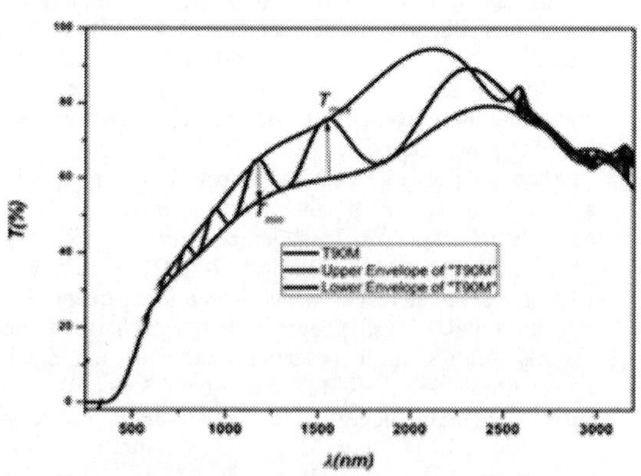

Figure 2: shows the transmittance of ZnS:Cr thin films at 90min with envelop.

Therefore the refractive index n is calculated from the interference fringes observed in the fully transmitting (transparency) region and the weak/medium absorption regions using Swanepoel's method. As mentioned above the interference fringes result from the interference of waves reflected from the upper and lower sides of the film (air-film and film-substrate interfaces). As demonstrated by E.R.Shaabaan[27]. The index of refraction of the film on a substrate is given by:

$$n = [M + (M^2 - n_s^2)^{\frac{1}{2}}]^{1/2} \quad \text{... [2]}$$

contrast and Tm is the maximum transmittance for a given fringe and equal is the $\left(\frac{T_{max}-T_{min}}{T_{max}*T_{min}}\right)$ ... [3]

And $$\mathbf{n_s} = \frac{1}{T_S} + \sqrt{\frac{1}{T_S} + 1} \quad \text{... [4]}$$

Is the substrate refractive index obtained from the measured transmittance Ts of the substrate.

The film thickness is then calculated from the following equation:

$$\mathbf{d} = \frac{\lambda_1\lambda_2}{2(\lambda_1 n_2 - \lambda_2 n_1)} \quad \text{... [5]}$$

The refractive index (n) and the thicknesses (d) of the pure and ZnS:Cr thin films are shown in figure (3), and figure(4) respectively. The refractive index decreasing from 2.7 to 1.19 while the times of deposition is increase, these decreasing may be is result of the increase of the thickness of films, that it's increased from 777,07 to 1555,94 nm for ZnS:Cr with 876nm for undoped ZnS as the times of deposited is increased from 90 to 180min. Higher film thickness could avoid damages caused by ZnO sputtering deposition process in making solar cells. Therefore, optimization of buffer layer thickness is essential to create an intermediate band between the valence band and the conduction band for improving the overall solar cell performance. The low refractive index of film deposited at 180min is accountable for the small fringe contrast observed in the transmittance spectrum.

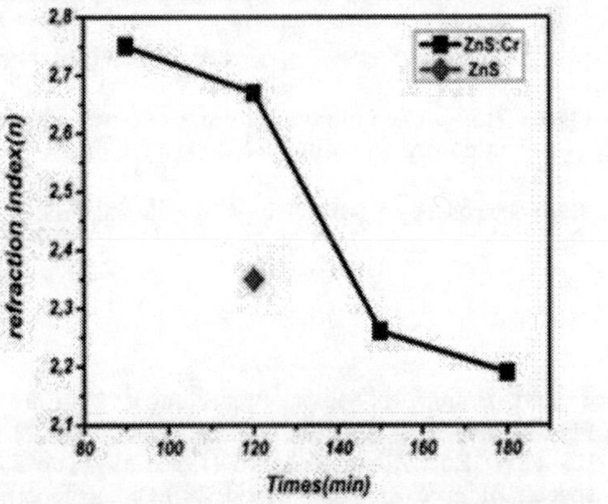

Figure 3: variant of the refractive index thin films with different times.

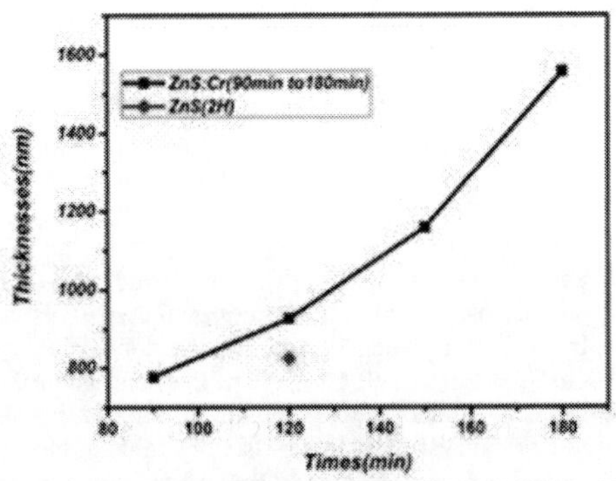

Figure 4: Variation of the thickness of pure and ZnS:Cr thin films with different times.

3.1.3 Band GapIllustrations

The values of the absorption coefficient α are calculated in the region of strong absorption [28],[29] by using the relation:

$$\alpha = \frac{1}{d} \times \ln\left(\frac{1}{T_n}\right) \quad \text{... [5]}$$

Where d is the thickness and T is the film transmittance in this region.

The optical band gap, Eg is determined using the Tauc relation, assuming a direct transition between valence and conduction bands[26],[30]. The optical absorption theory in semiconductors relates the absorption coefficient α to the photon energy hv as follows:

$$(\alpha h\upsilon)^{1/n} = A\,(h\upsilon - Eg) \quad \text{...[6]}$$

Where A is constant, h is Planck's constant, and Eg is the band gap width. n is a parameter that depends on the nature of the transition. For direct allowed transitions n = 1/2, while for indirect allowed transitions, n = 2 and for direct forbidden transitions n = 3/2.

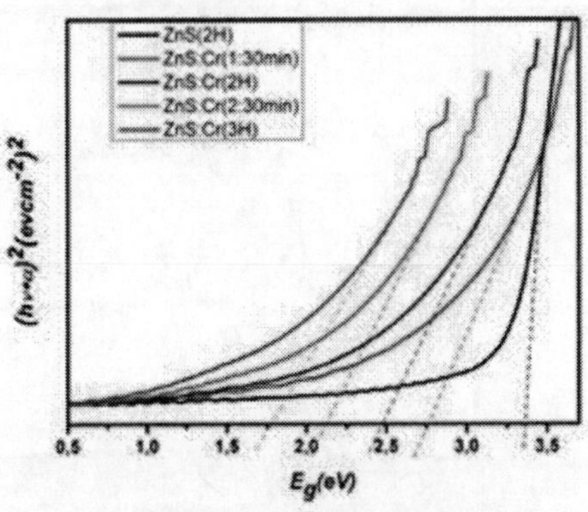

Figure 5: Direct band gap of the ZnS and ZnS:Cr thin films for different deposition times.

Figure.5 reveals the plot of (αhυ)2 versus the incident

radiation energy hυ(eV). The band gap values (Eg) obtained from the linear fitting and extrapolation are from 3.48 eV for undoped ZnS deposited at 120min and 2.77, 2.48, 2.35 and 1.6 8 eV for ZnS:Cr thin films grown at 90min, 120min, 150min and 180min, respectively. This supplementary band allows for better exploitation of both low and high energy photons in the solar spectrum, as sub band gap light can excite electrons from the valence band (VB) to the IB and from the IB to the conduction band (CB)[8]. The band gap decreased slightly as the deposition times increased. This is caused by the s-p and d interactions and the quantum confinement effect, the shift in the band gap affirms the uniform exchange of Cr ions in the ZnS lattice structure, and remains slightly lower than the value of bulk material (3.6 eV). It is well-known that the band gap in thin films tends asymptotically the band gap value of the bulk material as the thickness and the crystallinity of the films is improved[12],[31].

3.2.Structural properties of ZnS:Cr thin films

Figure.6. shows the XRD patterns of pure and doped ZnS:Cr thin films deposited by rf-sputtering at different times ranging from 90min to 180min. All these samples spectra exhibit one major peak at $2\theta \approx 28.08°$ with (111) plane, and intensity depending on the deposited time and RF power. The peak is attributed to the (111), (2 2 0), (311) planes of the zinc blende structure(JCPDS 05-0566, a = 0.5406 nm) for ZnS:Cr ,which suggests that the films have a preferential growth orientation in this (111) plane direction[3]. As the time of deposited is increased to 3h, the intensity of the peak increased significantly. Further increase in the Cr+ ion in the ZnS target as resulted in a slight decrease in the intensity of the peak. The higher peak intensity was observed for ZnS:Cr film grown at 2h, indicating that these films have the best crystallinity[32],[33].

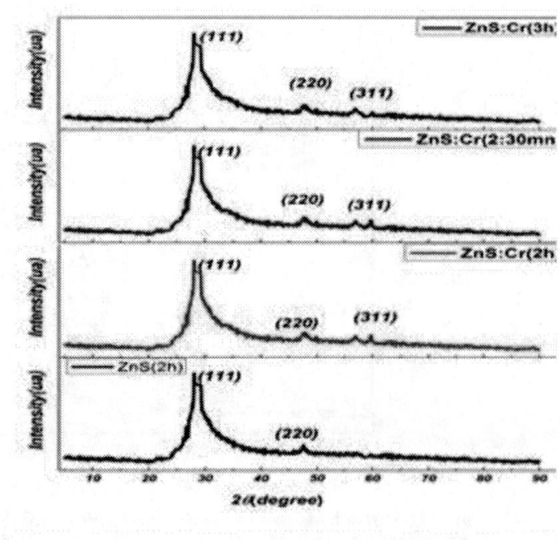

Figure 6: XRD diagrams of ZnS and ZnS:Cr thin films deposited at different times.

The mean crystallite sizes (L) of the films was calculated using Scherrer formula:

$$L = \frac{0.9 * \lambda}{\Delta(2\theta) * \cos(\theta')} \quad \ldots[6]$$

Where λ is the X-ray wavelength (0.15406 nm), $\Delta(2\theta)$ is the full width at half maximum [FWHM] of the film diffraction peak at 2θ, and θ is the Bragg diffraction angle. For the pure and doped ZnS:Cr films, the FWHM value decreased from 0.520° to 0.381° as the deposition times increased from 120min to 180min and subsequently decreased 0.154° at 90min, indicating a deterioration and improvement of the crystallinity of the films. The mean crystallite sizes, shown in Fig.7, were about 28.45nm for undoped ZnS at 120min , 16.7, 18.65, 25.8, and 33.6 nm for samples deposited at 90min, 120min, 150min, and 180min, respectively. These results suggest that the crystallinity of the films descended as the Cr+ ion incorporate in the target of ZnS at 90min in comparison with undoped ZnS, and then it began to increase according to time. This behavior is usually attributed to the increased mobility of the sputtered species on the surface of the substrate as the rf-power increase and consequently the kinetic energy of the sputtered species increase[34].The increased in the crystallinity, can be attributed to the growth rate of the particles pulverized , effect encountered at high thickness films because the kinetic energy of the sputtered species is high in this case[34].

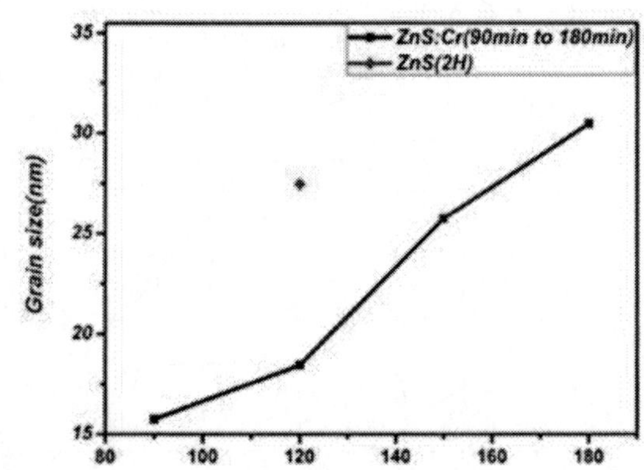

Figure 7: Evolution of crystallite size of pure and ZnS:Cr thin films with different deposition times.

3.3. Spectroscopy FT-IR analysis of ZnS and ZnS:Cr thin films

[1] M. Smith, A. Miller, Proceedings 17th European Photovoltaic Solar Energy Conference, Vol. I (2002) 903.
[2] K.W. Boer, Solar Cells 16 (1996) 591 The FT-IR spectra of pure ZnS and doped ZnS:Cr nanoparticles were obtained at wavenumber ranging from 4000 cm-1 to 400 cm-1 (Figure 10). It was performed to study the absorbance properties and the nature of bonds present in the synthesized samples. All samples illustrate

characteristic peaks, the weak peaks absorption band centered at 3800 cm-1 can be attributed to O-H stretching mode [35].

Fig 10. FTIR Spectrum of pure ZnS and ZnS:Cr thin films deposited for different times.

All the samples reveal a broad absorption band in the range of 3125 to 3400 cm-1 for ZnS:Cr and 2730.65 cm-1 for ZnS due to the stretching vibration of O-H. because of H2O adsorbed on the surface of the films while the water circle in the ancient when starting the deposit[36]. The wide peaks at 1362.66 to 1583.4 cm-1 may be attributed to S–S symmetric bond for pure ZnS. A broad band was observed at 1400 to 1725.4cm-1 attributed to the strong formation of ZnS nanoparticles, with some ions in the ZnS cationic surface replaced by Cr2+ ions. This observation suggested that in place of substituting Zn2+ ion[37]. IR absorption peak at 2465.66 cm-1 is corresponding the oxygen stretching and bending frequency, it may be because of the microstructure formation of the samples[38]. The peaks appearing from 436– 677 cm-1 are attributed to stretching vibrations of Zn- S bonds (corresponding to sulfides)[39].

4 CONCLUSION

In summary, ZnS:Cr thin films have been grown using RF cathodic sputtering and the effect of the deposits times on structural and optical properties of the films was investigated. XRD, studies showed that the particles crystallized in cubic Zinc blende structure with increasing of crystallite size. In the optical study it is observed that energy band gap decreases from 2.77eV to 1.68 eV with increasing of the depositions times; these results confirmed that, band gap of bulk ZnS is suitable material due to it large band gap and that band was created between the conduction band and the valence band with increasing of the thickness films. Thus, whatever the energy of the incident photon we have an excitation of pairs electron-hole. This supplementary band allows for better exploitation of both low and high energy photons in the solar spectrum.

REFERNECE

[1] M. Sreedhar, I. N. Reddy, P. Bera, and T. S. Shyju, "Studies of Cu-doped ZnS thin films prepared by sputtering technique," Surface and Interface Analysis, no. July, 2016, pp. 1–7.

[2] N. Hebalkar, A. Lobo, S. R. Sainkar, S. D. Pradhan, W. Vogel, J. Urban, and S. K. Kulkarni, "Properties of zinc sulphide nanoparticles stabilized in silica," JOURNAL OF MATERIALS SCIENCE, vol. 6, 2001, pp. 4377–4384.

[3] J. Hasanzadeh, A. Taherkhani, "Luminescence and Structural Properties of ZnS:Cu Nanocrystals Prepared Using a Wet Chemical Technique," CHINESE JOURNAL OF PHYSICS, vol. 51, no. 3, 2013, pp 540–550.

[4] J. Xie, "First-principles study on the magnetism in ZnS-based diluted magnetic semiconductors," Journal of Magnetism and Magnetic Materials, vol. 322, no. 19, 201, pp. L37–L41,.

[5] M. Nematollahi, X. Yang, U. J. Gibson, and T. W. Reenaas, "Pulsed laser ablation and deposition of ZnS : Cr," Thin Solid Films, vol. 590, 2015, pp. 28–32.

[6] X. Zeng, J. Zhang, F. Huang, X. Zeng, J. Zhang, and F. Huang, "Optical and magnetic properties of Cr-doped ZnS nanocrystallites," Journal of Applied Physics, vol. 111, 2012, pp. 123525–6.

[7] W. Q. Peng, S. C. Qu, G. W. Cong, X. Q. Zhang, and Z. G. Wang, "Optical and magnetic properties of ZnS nanoparticles doped with Mn 2 +," vol. 282, 2005 , pp. 179–185.

[8] T. Brakstad, B. R. Hope, M. Nematollahi, M. Kildemo, N. J. Podraza, K. Ghimire, and T. W. Reenaas, "Ellipsometric study of the optical response of ZnS:Cr for PV applications," Applied Surface Science, vol. 2016, 10, p. 157.

[9] A. Rahdar, H. A. Eivari, and R. Sarhaddi, "Study of structural and optical properties of ZnS : Cr nanoparticles synthesized by co-precipitation method," Indian Journal of Science and Technology, vol. 5, no. 1, 2012, pp. 1855–1858.

[10] M. Chattopadhyay, P. Kumbhakar, C. S. Tiwary, R. Sarkar, A. K. Mitra, U. Chatterjee, M. Chattopadhyay, P. Kumbhakar, C. S. Tiwary, R. Sarkar, and A. K. Mitra, "Multiphoton absorption and refraction in Mn 2 + doped ZnS quantum dots Multiphoton absorption and refraction in Mn 2 + doped ZnS quantum dots," Journal of Applied Physics, vol. 105, 2010, p. 024313.

[11] R. D. Mcnorton, T. M. Schuler, J. M. Maclaren, and R. A. Stern, "Systematic trends of first-principles electronic structure computations of Zn 1 − x A x B diluted," PHYSICAL REVIEW B, vol. 78, 2008,p. 075209.

[12] M. Bodke, H. Khawal, U. Gawai, and B. Dole, "Synthesis and Characterization of Chromium Doped Zinc Sulfide Nanoparticles," Open Access Library Journal, vol. 2, 2015, p. 1549.

[13] R. N. B. and D. Gallagher, "Optical Properties of Manganese-Doped of ZnS," vol. 72, no. 3, 1994,pp. 416–419.

[14] B. M. Kimpel, K. Lobet, H. Schulz, and E. Zeitler, "Generation of near-infrared light pulses from ZnS : Cr under," Meaa. Sci. Technol, vol. 6, 1995,pp. 1383–1388.

[15] N. A. Vlasenko, P. F. Oleksenko, M. A. Mukhlyo, Z. L. Denisova, and L. I. Veligura, "ZnS : Cr and ZnSe : Cr thin-film waveguide structures as electrically pumped laser media with an impact excitation mechanism," Annalender physik, vol. 525, no. 12, 2013, pp. 889–905.

[16] Kunio Ichino, Yasuharu Morimotol, "Molecular beam epitaxy and structural properties of ZnCrS.pdf,"vol.3. (2006).p. 776–779

[17] A. B. María Jadraque Evtushenko, A. David, M. Lo, V. Loriot, Y. G. Shukhov, L. S. Kibis, A. V Bulgakov, and M. Mart, "Co-Doped ZnS Clusters and Nanostructures Produced by Pulsed Laser Ablation," Journal of phy sical chemestery, vol. 117, , 2013 p. 5416–5423.

[18] E. Ech-chamikh a,*, E.L. Ameziane a, M. Azizan a, A. Bennouna a, J.L. Fave b, j. Cernogora b, y. Bounouh c and A, "Structural , electrical and optical properties of RF sputtered a-C films," Solar Energy Materials and Solar Cells, vol. 33, 1994,pp. 443–452,.

[19] A. Le Donne and D. Universit, "Study of the physical properties of ZnS thin films deposited by RF sputtering," Materials Processing, Semiconductor, no. October, 2017.

[20] L. Nkhaili, A. El, M. A. Ali, Y. Ijdiyaou, and

A. Elmansouri, "Effect of RF power on the structural and optical properties of RF-sputtered ZnO thin films," Eur. Phys. J. Appl. Phys, vol. 66, 2014, pp. 30302-p1-30302-p5.
[21] E. L. Amezianea, A. Bennounaa, M. Azizan, and T. A. N. Tanb, "Structural and optical properties of r.f.-sputtered Si, C, _ x :0 films," Thin Solid Films, vol. 259, 1995,pp. 18–24.
[22] A. Bennouna, E. L. Ameziane, A. Haouni, N. Ghermani, and M. Azizan, "X-RAY DIFFRACTION STUDY OF TEXTURE AND GROWTH OF RF-SPU'ITERED CdS FILMS *," Solar Energy Materials, vol. 20, 1990, pp. 405–415.
[23] M. S. Akhtar, S. Riaz, and S. Naseem, Optical properties of sol-gel deposited ZnS thin films: spectroscopic ellipsometry, vol. 2, no. 10. 2015, Elsevier Ltd.,.
[24] R. Zhang, B. Wang, and L. Wei, "Influence of RF power on the structure of ZnS thin films grown by sulfurizing RF sputter deposited ZnO," Materials Chemistry and Physics, vol. 112, 2008, pp. 557–561.
[25] S. Wook, G. L. Agawane, M. Gil, A. V Moholkar, J. Moon, J. Hyeok, and J. Yong, "Preparation and characteristics of chemical bath deposited ZnS thin films : Effects of different complexing agents," Journal of Alloys and Compounds, vol. 526, 2012, pp. 25–30.
[26] W. S. NUWAT PIMPABUTE, THANUSIT BURINPRAKHON, "Determination of optical constants and thickness of amorphous GaP thin film," Optica Applicata, vol. XLI, no. 1, 2011, pp. 257–268,.
[27] E. R. Shaaban and I. S. Yahia, "Validity of Swanepoel ' s Method for Calculating the Optical Constants of Thick Films Optical constants , ACTA PHYSICA POLONICA A, vol. 121, no. 3, 2012, pp. 628–635.
[28] Z. Y. Zhong, E. S. Cho, and S. J. Kwon, "Characterization of the ZnS thin fi lm buffer layer for Cu (In , Ga) Se 2 solar cells deposited by chemical bath deposition process with different solution concentrations," Materials Chemistry and Physics, vol. 135, no. 2–3, 2012, pp. 287–292,.
[29] C. D. Lokhande, E. Lee, K. Jung, and O. Joo, "Ammonia-free chemical bath method for deposition of microcrystalline cadmium selenide films," Mather chimy phy, vol. 91, 2005, pp. 200–204,.
[30] K. Deepa, K. C. Preetha, K. V. Murali, a. C. Dhanya, a. J. Ragina, and T. L. Remadevi, "The effect of various complexing agents on the morphology and optoelectronic properties of chemically deposited ZnS thin films: A comparative study," Optik - International Journal for Light and Electron Optics, vol. 125, no. 19, 2014,pp. 5727–5732.
[31] C. A. Rodríguez, M. G. Sandoval-paz, G. Cabello, M. Flores, H. Fernández, and C. Carrasco, "Characterization of ZnS thin fi lms synthesized through a non-toxic precursors chemical bath," Materials Research Bulletin, vol. 60, 2014, pp. 313–321, 2014.
[32] S. Ra, S. Wook, D. Sun, A. V Moholkar, J. Moon, and J. Hyeok, "Effect of pH on the characteristics of nanocrystalline ZnS thin films prepared by CBD method in acidic medium," Current Applied Physics, vol. 10, no. 3, , 2010, pp. S473–S477.
[33] P. K. Ghosh, S. Jana, S. Nandy, and K. K. Chattopadhyay, "Size-dependent optical and dielectric properties of nanocrystalline ZnS thin films synthesized via rf-magnetron sputtering technique," Materials Research Bulletin, vol. 42, no. 3, , 2007, pp. 505–514.
[34] M. Y. E. Azhari, M. Azizan, A. Bennouna, A. Outzourhit, E. L. A, and M. Brunel, "Structural properties of oxygenated amorphous cadmium telluride thin films," vol. 295, 1997, pp. 131–136,.
[35] and H. A. Rahizana Mohd Ibrahim, Masturah Markom, "Optical Properties of Ni 2 + - , Co 2 + - , and Mn 2 + -doped ZnS Nanoparticles Synthesized Using Reverse Micelle Method," ECS Journal of Solid State Science and Technology, vol. 4, no. 2, 2015, pp. R31–R37,.
[36] S. Cholan, N. Shanmugam, N. Kannadasan, K. Sathishkumar, and K. Deivam, "Effect of poly ethylene glycol (PEG) as surfactant on cerium doped ZnS nanoparticles," journal of Materials Research and technology, no. x x, 2014, pp. 2–7.
[37] G. Varughese and K. T. Usha, "Optical Studies in ZnS: Ce Nanocrystallite," Chemical Science Transactions, vol. 3, no. 4, 2014, pp. 1354–1359.
[38] V. D. Mote, Y. Purushotham, B. N. Dole, and A. Materials, "Structural , morphological and optical properties of Mn doped ZnS nanocrystals (Propriedades estrutural , morfológica e óptica," Cerâmica, vol. 59, 2013, pp. 614–619.
[39] M. Salavati-niasari, M. Ranjbar, and D. Ghanbari, "A rapid microwave route for the synthesis of ZnS nanoparticles," Journal of NanoStructures, vol. 1, 2012,pp. 231–235.

SYNTHESIS, STRUCTURAL AND PHOTO PHYSICAL PROPERTIES OF PEROVSKITE OXIDE $(KNbO_3)_{1-x}$+$(La_2NiMnO_6)_x$ FOR PHOTOVOLTAIC APPLICATION

Md Sariful Sheikh[1*], A. Dutta[2], T. K. Bhowmik[1], S. K. Ghosh[3], S. K. Rout[3], T. P. Sinha[1]
[1]Department of Physics, Bose Institute, 93/1, A. P. C. Road, Kolkata-700009, India
[2]Department of Condensed Matter Physics and Material Sciences, S. N. Bose National Centre for Basic Sciences, Block - JD, Sector - III, Salt Lake, Kolkata-700106, India
[3]Department of Physics, Birla Institute of Technology, Mesra, Ranchi-835215, India
*Corresponding Author: Phone No: +91 3323031194; Fax: (+91) 3323506790;
Email: sarifulsekh@gmail.com

ABSTRACT: Solid solutions of perovskite oxides $(KNbO_3)_{1-x}$+$(La_2NiMnO_6)_x$ (x=0, 0.1, 0.2 and 0.3) with a variation of band gap (1.33-3.6 eV) have been introduced as a promising photovoltaic absorber. The structural characterization of the prepared samples was carried out using X-ray diffraction (followed by Rietveld refinement) and Raman experiment. As the doping percentage of the monoclinic La_2NiMnO_6 in the solid-solution increases, the crystal structure of host $KNbO_3$ becomes more symmetric from orthorhombic to cubic. A large reduction in the particle size has also been observed in the solid solutions in comparison to the pure $KNbO_3$. The band gap (~ 1.33 eV) of the synthesized solid solution x=0.1 is found to be very close to the Shockley-Queisser band gap value of 1.34 eV, which suggests the promising photovoltaic possibility in this material. Photoluminescence (PL) emission spectra reveal a strong PL quenching in the solid-solutions in comparison to the $KNbO_3$. The overall structural and optical studies suggest the promising photovoltaic possibility in $KNbO_3$/ La_2NiMnO_6 solid solution.
Keywords: perovskite, characterization, magnetic field, band gap

1 INTRODUCTION

There is growing research interest to search for a narrow direct band gap solar cell absorber, that can be easily synthesized using the simple experimental techniques and low cost source materials [1-3]. In this regard, various multiferroic and ferroelectric perovskite oxides, like, $BiFeO_3$, $Pb(Zr.Ti)O_3$, $KNbO_3$ (KNO), etc., have shown their potential in the current photovoltaic research [4-7]. Ferroelectric domains in these materials produce a large internal electric field, which helps to separate the photo generated electron-hole pairs by dielectric screening and results in over band gap large photo voltage in multiferroic and ferroelectric solar cells. But, the band gap of these materials is too high to absorb the total visible region of the sun light, which restricts the photovoltaic performance of these materials [4-7]. Hence there is a significant research interest to search for a direct and narrow band gap multiferroic or ferroelectric solar cell absorber for optimal photovoltaic performance [7-11]. KNO, a widely studied ferroelectric material, have shown its potential in various light sensing semiconductor applications, like, photovoltaic, photo-catalytic dye-degradation, water splitting for H_2 production, etc [7, 9, 12, 13]. But like all other multiferroic and ferroelectric materials, band gap (E_g ~ 3.6 eV) of KNO is too large to obtain the optimal photovoltaic performance [7]. Recently, significant researches have been focused on the band gap tuning of KNO for photovoltaic application [7, 9, 14, 15]. However, still no significant performance has been reported on its photovoltaic behavior, and thus a large research efforts is highly desired in order to improve its photovoltaic performance. In this regard, we have tuned the band gap and enhanced the photo physical properties of KNO for high photovoltaic performance.

We have doped the narrow band gap ferromagnetic semiconductor La_2NiMnO_6 (LNMO) within the KNO lattice in order to synthesize narrow band gap multiferroic semiconductor material for the photovoltaic application. LNMO is a widely studied ferromagnetic semiconductor with its Curie temperature (T_c ~ 280 K) very close to the room temperature, narrow direct band gap (E_g ~ 1.08 eV), magneto-dielectric and photovoltaic effect [16-18]. The combination of wide band gap ferroelectric KNO and narrow band gap LNMO, are expected to posses the desired narrow band gap with the intrinsic ferroelectric property of KNO. A detailed structural, magnetic and optical properties of the synthesized KNO/LNMO solid solutions have been performed in order to elucidate its photo physical properties for photovoltaic application.

2 EXPERIMENTAL DETAILS

2.1 Material synthesis

First, KNO/LNMO solid solutions was prepared using standard solid state reaction technique. Powders of Nb_2O_5, K_2CO_3, La_2O_3, $NiCO_3.2Ni(OH)_2.4H_2O$ and MnO_2 were taken in stoichiometric ratios and mixed homogenously by grinding in acetone medium for 8 hours. Finally, the mixtures were calcined at 1073 K for 5 hours in an alumina crucible. The heating process was repeated to complete the all phase formation. The doping percentage of LNMO in the solid solution was 0, 10, 20 and 30 % in molar ratio.

2.2 Characterization

The room temperature X-ray diffraction (XRD) pattern of the prepared samples was carried out using a powder X-ray diffractometer (Rigaku Miniflex II diffractometer). The Rietveld refinement of the obtained XRD data was performed by using the Fullprof code [19]. The Raman spectrum of the prepared samples was obtained at an excitation wavelength of 488 nm (He-Ar laser) using a Lab–RAM HR 800 (Jobin-Yvon) Raman spectrometer. The particle size of the sample KNO was obtained from the field effect scanning electron microscope (FESEM) image. The morphology of the samples x=0.1, 0.2 and 0.3 were determined using transmission electron microscope (TEM) images (FEI

Tecnai G2, 200 KV). The UV-visible absorption spectrum of the prepared samples was taken by a Shimadzu UV-VIS spectrometer. Photoluminescence (PL) emission spectra measurement were performed using a fluorescence/luminescence spectrophotometer (LS 55, Perkin Elmer). The calcined powders were pelletized into discs using polyvinyl alcohol as a binder and sintered at 1123 K for the study of magnetism. The vibrating sample magnetometer (VSM; Lakeshore) was used to study the magnetization in the prepared samples.

3 RESULTS AND DISCUSSION

3.1 Structure and morphology

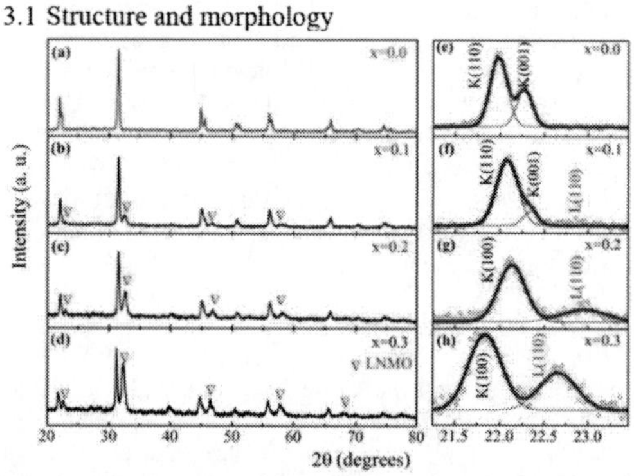

Figure 1: Room temperature XRD pattern of KNO/LNMO powder (a) x=0.0, (b) x=0.1, (c) x=0.2 and (d) x=0.3. Gaussian deconvolution of the XRD peaks of KNO/LNMO in the range from 21.3 to 23.5° of 2θ (e-h). The symbols, blue line, dark gray line and red line represent the experimental data, fit peak (KNO), fit peak (LNMO) and cumulative fit peak respectively. Suffix K and L represent the peaks originated from KNO and LNMO respectively.

The room temperature XRD patterns of the synthesized KNO/LNMO solid solutions are shown in Fig. 1 (a-d). From Fig. 1(a), it is observed that pure KNO sample crystallises in the orthorhombic phase with space group Bmm2 [JCPDS 04-007-9572]. The XRD peaks marked with the inverted delta symbol in Fig. 1(b-d) indicate the peaks originated from the monoclinic (*$P2_1/n$*) LNMO phase (JCPDS 04-013-0945). Fig. 1(e) displays the characteristic splitting of the (110) and (001) peaks for orthorhombic KNO. However it is clearly observed from Fig. 1(f-h) that in the sample x=0.1, this characteristic splitting of orthorhombic KNO reduces and disappears in the high doped KNO samples (x=0.2, 0.3). Disappearance of XRD peak splitting clearly indicates the recrystallization of KNO in the doped samples and points towards the possibility of more symmetric crystal structure of KNO. The XRD peaks of KNO in Fig. 1(c) matches very well with the tetragonal (P4mm) phase of KNO (JCPDS 71-0945). A significant shift of the (100) plane towards lower angle has been observed as shown in Fig. 1(d), which also indicates the possibility of recrystallization of KNO in the sample x=0.3. The XRD peaks of KNO in Fig. 1(d) may be due to the cubic (Pm-3m) phase of KNO [20]. Hence, to get a better understanding about the crystal structure of the solid solutions Rietveld refinement of the XRD data has been carried out and the corresponding refinement plots are shown in Fig. 2 (a-d). The values of the quality parameter (χ^2) of refinement for the samples x=0.0, 0.1,0.2 and 0.3 are 1.6, 1.27, 1.38 and 1.72, respectively, which indicate the goodness of the refinement.

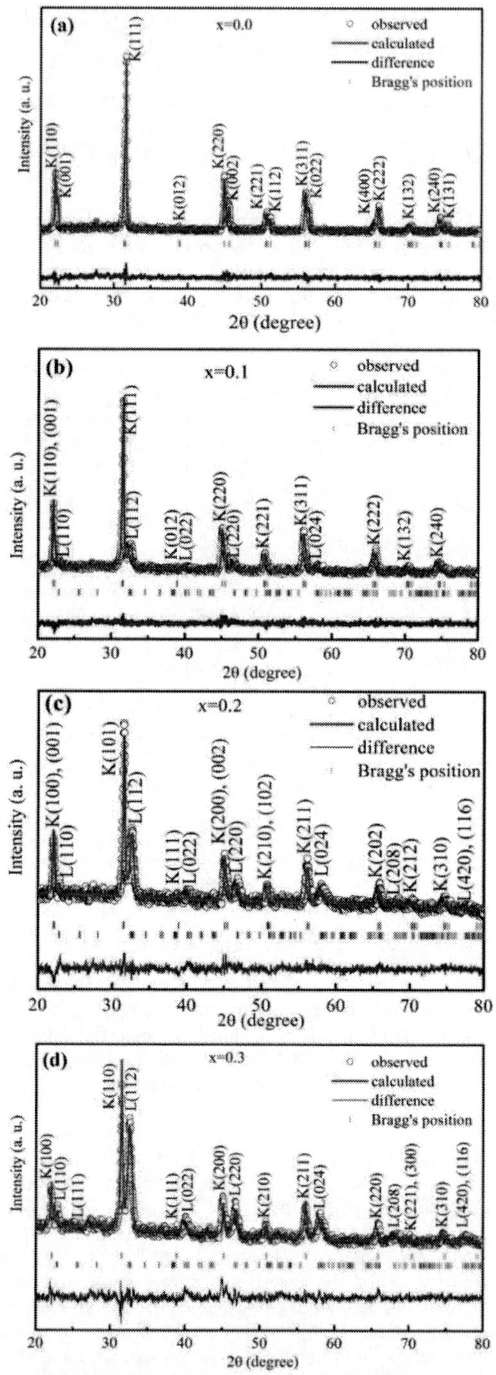

Figure 2. Rietveld refinement of the XRD data of solid solutions (a) x=0.0, (b) x=0.1, (c) x=0.2 and (d) x=0.3. Suffix K and L represent the crystal planes of KNO and LNMO lattice respectively.

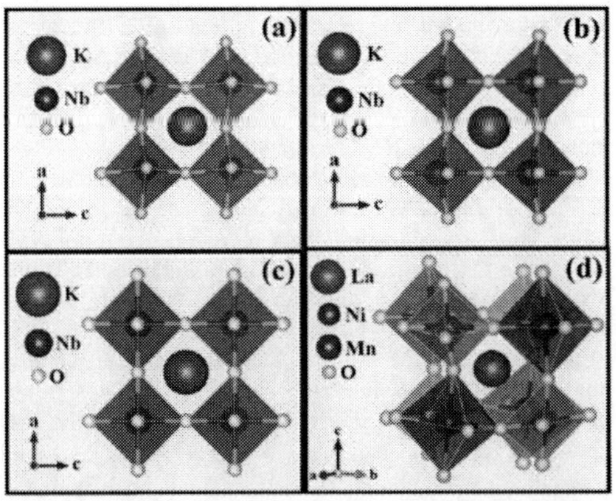

Figure 3. Unit cell representation of (a) orthorhombic, (b) tetragonal, (c) cubic KNO and (d) monoclinic LNMO.

The refined space groups and the lattice parameters of the samples are shown in Table 1, which are in good agreement with the previous reported lattice parameters of KNO and LNMO [20, 21]. It is clear that with the increase of LNMO percentage in the solid solution, the crystal structure of KNO becomes more symmetric from orthorhombic ($a \neq b \neq c$) to cubic ($a = b = c$). A schematic representation of the unit cells of KNO (orthorhombic, tetragonal and cubic) and LNMO (monoclinic) are shown in Fig. 3. Interestingly, due to the higher structural symmetry, cubic KNO shows better charge translation and photo reactivity than other less symmetric crystal structure of KNO [13]. Hence in respect of the crystal structure, KNO in the doped samples points towards the possibility of better charge transport as well as photovoltaic possibility in comparison to the pure KNO.

Table I: Refined lattice parameters

	KNO x=0.0	KNO x=0.1	KNO x=0.2	KNO x=0.3	LNMO x=0.3
Phase	Bmm2	Bmm2	P4mm	Pm-3m	P2₁/n
a (Å)	5.705	5.684	4.019	4.013	5.503
b (Å)	3.977	3.987	4.019	4.013	5.501
c (Å)	5.698	5.689	3.992	4.013	7.747
α	90°	90°	90°	90°	90°
β	90°	90°	90°	90°	89.28°
γ	90°	90°	90°	90°	90°

For a clear understanding about the structural properties of the solid solutions room temperature Raman spectra at an excitation of 488 *nm* has been carried out, which are shown in Fig. 4. In the orthorhombic (Bmm2) KNO phase (x=0.0) transverse and longitudinal phonon modes $B_1(TO_2)$, $B_1(TO_4)$, $A_1(TO_1)$, $B_1(TO_3)$, $A_1(TO_3)$ and $A_1(LO_3)$ are observed at 193, 278, 294, 532, 599 and 832 cm^{-1}, respectively [22].

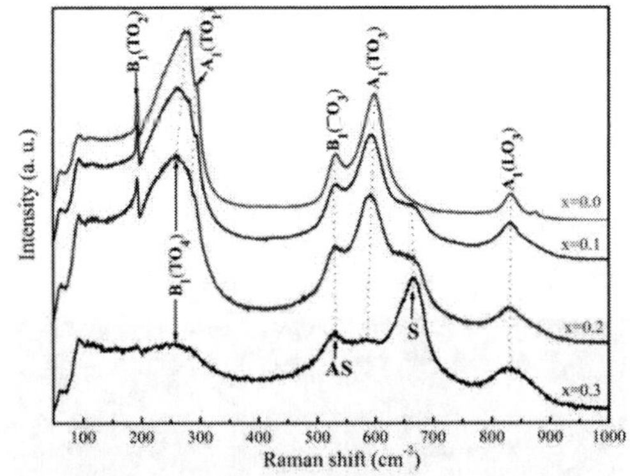

Figure 4. Raman Shift of the prepared samples.

But, in the doped samples the intensity of the lines $B_1(TO_2)$ and $A_1(TO_1)$ decreases as the doping percentage of LNMO increases. In the sample (x=0.3) the these two peaks become almost unobservable. This prominent change in $B_1(TO_2)$ and $A_1(TO_1)$ lines is related to the restructuring of the material and it represents the cubic phase of KNO in (x=0.3) [22]. In the cubic and tetragonal phase of KNO (x=0.2 and 0.3) the line $B_1(TO_3)$ at 532 cm^{-1} completely disappears. But in the doped samples (x= 0.2 and 0.3) a new line at 530 cm^{-1} appears. This may arise due to the appearance of the anti-symmetric stretching (AS) mode vibration of the $(Ni/Mn)O_6$ octahedra at 530 cm^{-1} [23]. The peak at 667 cm^{-1} arises in the doped samples, which is related to symmetric stretching mode vibrations (S) of the $(Ni/Mn)O_6$ octahedra [23]. The intensity of the peak at 667 cm^{-1} increases as the percentage of LNMO in the solid-solution increases. Hence, both the XRD and Raman experiment suggest the LNMO concentration dependent phase change in KNO host lattice.

Figure 5. (a) FESEM image of KNO particle; TEM image of the sample (b) x=0.1, (c) x=0.2 and (d) x=0.3.

FESEM image as shown in Fig. 5(a) shows that the synthesized KNO particles are cubic in shape and the calculated average particle size is 465 *nm*. However, the TEM images of the samples x=0.1, 0.2 and 0.3 as shown in Fig. 5(b-d) denote a large reduction in the average particle size in the doped samples. The average particle size of the samples x=0.1, 0.2 and 0.3 are 41, 34 and 25 *nm*, respectively. Thus bulk to nano transformation has been occurred upon chemical doping. Interestingly, nanomaterials exhibit improved performance in various light sensing semiconductor applications due to their high surface area, increased surface energy and increased light absorption in comparison to the corresponding bulk materials [24, 25].

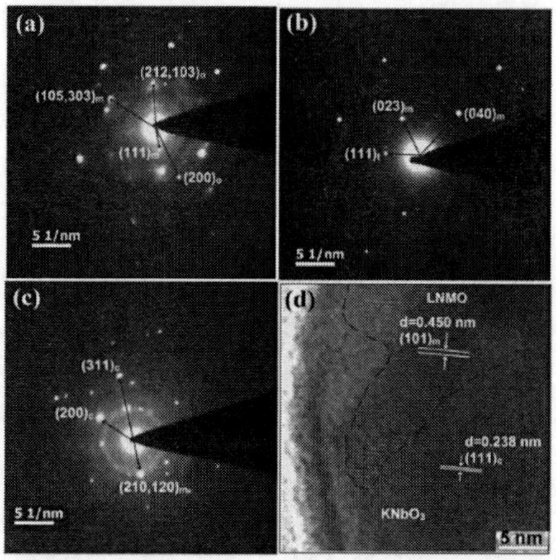

Figure 6. SAED pattern of sample (a) x=0.1, (b) x=0.2, (c) x=0.3; HRTEM image of sample (d) x=0.3. Prefix m, t, c and o represent the monoclinic, tetragonal, cubic and orthorhombic phases respectively.

Bright spots at regular positions in the selected area electron diffraction (SAED) patterns of the samples x=0.1, 0.2 and 0.3 as shown in Fig. 6(a-c) indicate the good crystalline nature of the doped samples. SAED patterns also confirm the coexistence of both KNO and LNMO lattice in the solid solutions. High resolution TEM (HRTEM) image of the sample x=0.3 as shown in Fig. 6(d) shows the formation of LNMO matrix within the KNO matrix. The large mismatch in the lattice parameters between KNO and LNMO, makes the LNMO matrix highly strained to accommodate within the KNO lattice and the crystal structure of KNO also changes from the orthorhombic to more symmetric cubic structure. The lattice misfit also prevents the host material KNO to form a larger lattice, which results a giant reduction in the average particle size of the doped samples in comparison to the pure KNO.

3.2 Magnetic property

Fig. 7 shows the temperature dependence of magnetic moment of the solid solutions in the temperature range from 80 *K* to 400 *K*. Though, KNO is a non-magnetic material, the solid solutions show ferromagnetic behavior and obviously, the magnetic contribution in the solid solutions originates from the ferromagnetic LNMO. Magnetic susceptibility (χ) of a ferromagnetic material in the paramagnetic region is governed by the Curie-Weiss law: $\chi=C/(T-T_C)$, where C, T and T_C represent the Curie constant, measuring temperature and Curie temperature, respectively. Here, T_C is estimated from the Curie-Weiss linear fit of the magnetic moment inverse (M^{-1}) versus temperature (T) plot (as shown in the inset of Fig. 7) in the paramagnetic region from temperature 250 to 350 K. The obtained value of T_C using Curie-Weiss law is found to be 230, 236 and 242 *K* for sample x=0.1, x=0.2 and x=0.3, respectively. A gradual increment in T_C has been observed as the LNMO concentration increases. Fig. 8 shows the magnetization versus field (*M-H*) curve at 80 *K*. The *M-H* curve as shown in Fig. 8, shows that the saturation magnetization also increases as the LNMO concentration in the doped samples increases and in all the cases magnetic moment tends to saturate above 5 *kG*, which indicates the ferromagnetic nature of the solid solutions. Inset of Fig. 8 shows that the remanant magnetization also increases with the LNMO doping percentage. However, coercivity almost remains the same in all the samples.

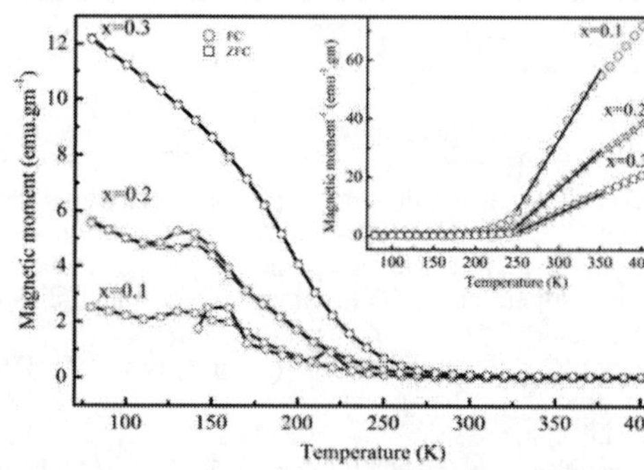

Figure 7. Temperature dependence of magnetic moment at a magnetic field of 2 *kG*. Inset shows the Curie-Weiss linear fit of M^{-1} vs. T in the temperature range of 250 to 350 *K*.

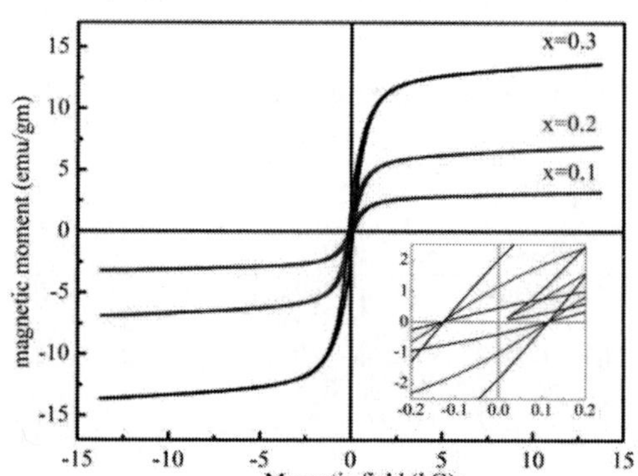

Figure 8. M-H loop of the KNO/LNMO samples at 80 K.

3.3 Optical property

The prime interest of this work was to reduce the band gap in the required range for the photovoltaic application of KNO as a promising light absorber. The optical band gap obtained from the Tauc plot as shown in Fig. 9 [26], is found to be 3.6, 1.33, 1.51 and 1.58 eV for the samples x=0.0, 0.1, 0.2 and 0.3, respectively. The band gap of the sample, x=0.1 matches very well with the Shockley-Queisser band gap value of 1.34 eV [27]. Shockley-Queisser band gap is the calculated value for which the theoretical power conversion efficiency of a single junction solar cell is normalized to the maximum value (> 33 %) [27]. The band gap of another sample, x=0.2 (~1.51 eV) is also similar to that of $CH_3NH_3PbI_3$ (~ 1.5 eV), which has shown high photovoltaic power conversion efficiency in recent years(>20%) [28, 29]. However, over doping of LNMO in the KNO crystal lattice increases the band gap of the solid solutions. The synthesized solid solutions (x=0.1 and 0.2) are able to able to absorb the total ultraviolet and visible region of solar spectrum, which is essential for optimal photovoltaic performance. Hence in respect of the band gap, we can conclude that the newly designed solid-solution, $(KNO)_{1-x}+(LNMO)_x$ (x=0.1 and 0.2) can be used as a promising light absorbing material for the high performing solar cell fabrication.

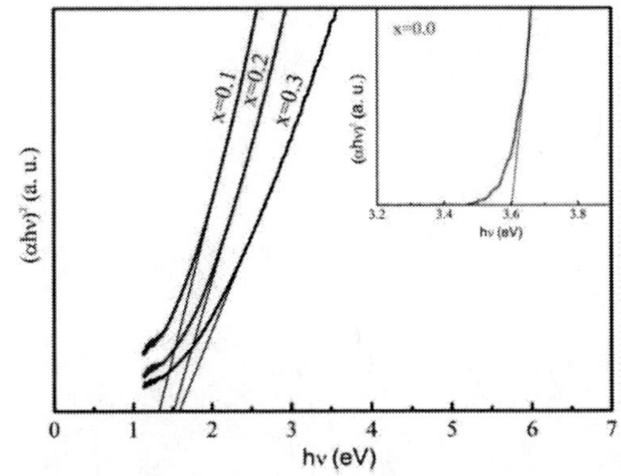

Figure 9. Tauc plot of the KNO/LNMO solid solution.

To further investigate the improved charge transport property in the materials, PL emission measurements were performed, which is shown in Fig. 10. It is observed that the doped samples present significant PL emission quenching in comparison to the pure KNO. Typically, a strong PL quenching of the light absorbing material provides an evidence of lower charge carrier recombination rate i.e. higher carrier life time in that material. Interestingly, high carrier life-time enhances the charge carrier extraction from the light absorber and hence improves the performance of a material in photovoltaic application. The enhanced charge carrier life time in KNO/LNMO may be due to the smaller particle size and better charge translation due to the more symmetrical crystal structure of KNO in the solid solutions.

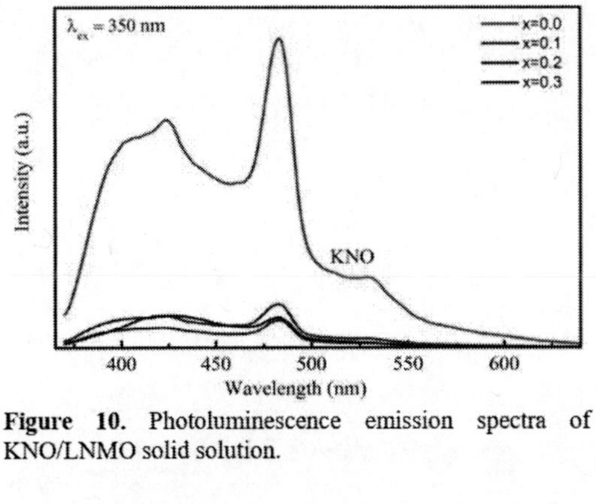

Figure 10. Photoluminescence emission spectra of KNO/LNMO solid solution.

4 CONCLUSION

A solid solution of KNO and LNMO has been introduced as a promising light absorber material for the photovoltaic application. The Rietveld refinement of the XRD data and Raman spectra of the synthesized materials show that with the increment in the LNMO doping percentage, the crystal structure of KNO becomes more symmetric from orthorhombic to cubic. Though KNO is a non-magnetic material, the doped samples show ferromagnetic behavior with high Curie temperature ($T_C \sim 242\ K$). The band gap of the solid solutions (x=0.1 and 0.3) is also very promising for high photovoltaic performance. A strong PL quenching confirms the lower carrier recombination rate and higher carrier life time in the doped samples in comparison to the KNO. Thus overall study shows that KNO/LNMO solid solutions posses many attributes suitable for high photovoltaic performance. KNO/LNMO solid solution may also find promising applications in photocatalytic dye degradation, water splitting, and other various light sensing semiconductor devices.

5 ACKNOWLEDGEMENTS

Md. S. Sheikh and T. K. Bhowmik would like to thank Department of Science and Technology (DST), Government of India for providing the financial support in the form of DST INSPIRE Fellowship (IF150220 and IF160418, respectively).

6 REFERENCES

[1] T. Omata, H. Nagatani, I. Suzuki, M. Kita, H. Yanagi, N. Ohashi, J. Am. Chem. Soc. 136 (2014) 3378.

[2] W. Taouali, M. E. Casida, A. A. M. H. M. Darghouth, K. Alimi, Comput. Mater. Sci. 150 (2018) 54.

[3] X. G. Zhao, J. H. Yang, Y. Fu, D. Yang, Q. Xu, L. Yu, S. H. Wei, L. Zhang, J. Am. Chem. Soc. 139 (2017) 2630.

[4] F. Zheng, Y. Xin, W. Huang, J. Zhang, X. Wang, M. Shen, W. Dong, L. Fang, Y. Bai, X. Shen, J. Hao, J. Mater. Chem. A 2 (2014) 1363.

[5] L. Wang, H. Ma, L. Chang, C. Ma, G. Yuan, J. Wang, T. Wu, Small 13 (2017) 1602355.

[6] M.-M. Yang, A. Bhatnagar, Z.-D. Luo, M. Alexe, Sci. Rep. 7 (2017) 43070.

[7] I. Grinberg, D. V. West, M. Torres, G. Gou, D. M. Stein, L. Wu, G. Chen, E. M. Gallo, A. R. Akbashev, P. K. Davies, J. E. Spanier, A. M. Rappe, Nature 503 (2013) 509.

[8] G. Zhang, H. Wu, G. Li, Q. Huang, C. Yang, F. Huang, F. Liao, J. Lin, Sci. Rep. 3 (2013) 1265.

[9] B. Song, X. Wang, C. Xin, L. Zhang, B. Song, Y. Zhang, Y. Wang, J. Wang, Z. Liu, Y. Sui, J. Tang, J. Alloys Compd. 703 (2017) 67.

[10] R. Nechache, C. Harnagea, S. Li, L. Cardenas, W. Huang, J. Chakrabartty, F. Rosei, Nat. Photonics 9 (2015) 61.

[11] D. Kim, H. Han, J. H. Lee, J. W. Choi, J. C. Grossman, H.n M. Jang, D. Kim, Proc. Natl. Acad. Sci. 201721503 (2018) 1. DOI: 10.1073/pnas.1721503115

[12] T. Zhang, W. Lei, P. Liu, J. A. Rodriguez, J. Yu, Y. Qi, G. Liu, M. Liu, J. Phys. Chem. C 120 (2016) 2777.

[13] T. Zhang, K. Zhao, J. Yu, J. Jin, Y. Qi, H. Li, X.

Hou, G. Liu, Nanoscale 5 (2013) 8375.
[14] F. Wang, A. M. Rappe, Phys. Rev. B 91 (2015) 165124.
[15] L. Yu, H. Deng, W. Zhou, P. Yang, J. Chu, Mater. Lett. 202 (2017) 39.
[16] D. A. B. Barbosa, M. W. Lufaso, H. Reichlova, X. Marti, M. V. S. Rezende, A. P. Maciel, C. W. A. Paschoal, J. Alloys Compd. 663 (2016) 899.
[17] N. S. Rogado, J. Li, A. W. Sleight, M. A. Subramanian, Adv. Mater. 17 (2005) 2225.
[18] Md. S. Sheikh, D. Ghosh, A. Dutta, S. Bhattacharyya, T. P. Sinha, Mater. Sci. Eng. B 226 (2017) 10.
[19] J. Rodriguez-Carvajal, Phys. B 192 (1993) 55.
[20] N. Kumada, T. Kyoda, Y. Yonesaki, T. Takei, N. Kinomura, Mater. Res. Bull. 42 (2007) 1856.
[21] C. L. Bull, D. Gleeson, K. S. Knight, J. Phys.: Condens. Matter. 15 (2003) 4927.
[22] I. S. Golovina, V. P. Bryksa, V. V. Strelchuk, I. N. Geifman, A. A. Andriiko, J. App. Phys. 113 (2013) 144103.
[23] S. Zhao, L. Shi, S. Zhou, J. Zhao, H. Yang, Y. Guo, J. App. Phys. 106 (2009) 123901.
[24] S. P. Kim, M. Y. Choi, H. C. Choi, Mater. Res. Bull. 74 (2016) 85.
[25] A. Punnoose, H. Magnone, M. S. Seehra, Phys. Rev. B 64 (2001) 174420.
[26] J. Tauc, R. Grigorovici, A. Vancu, Phys. Status Solidi. 15 (1966) 627.
[27] S. Rühle, Sol. Energy 130 (2016) 139.
[28] A. A. B. Baloch, M. I. Hossain, N. Tabet, F. H. Alharbi, J. Phys. Chem. Lett. 9 (2018) 426.
[29] W. S. Yang, B.-W. Park, E. H. Jung, N. J. Jeon, Y. C. Kim, D. U. Lee, S. S. Shin, J. Seo, E. K. Kim, J. H. Noh, S. I. Seok, Science 356 (2017) 1376.

SILICON NANOWIRE FOR SOLAR ENERGY APPLICATION

B. SALHI[1], M. K. HOSSAIN,[1], F. AL-SULAIMAN[1,2]*
[1]CENTER OF RESEARCH EXCELLENCE IN RENEWABLE ENERGY (CORERE), KING FAHD UNIVERSITY OF PETROLEUM AND MINERALS (KFUPM), DHAHRAN 31261, DHAHRAN, KINGDOM OF SAUDI ARABIA
[2]DEPARTMENT OF MECHANICAL ENGINEERING, KING FAHD UNIVERSITY OF PETROLEUM AND MINERALS
(KFUPM), DHAHRAN 31261, DHAHRAN, KINGDOM OF SAUDI ARABIA

ABSTRACT: Due to environmental and economic concerns associated with the use of fossil fuels, the need for cheap methods of generating energy from clean, renewable sources is becoming ever more urgent. Perhaps the most attractive renewable energy source is the sun, but finding materials that can efficiently convert the suns radiation to electricity is extremely challenging. The majority of solar cells currently in use are based on crystalline silicon, but the high fabrication costs for such cells hinder the competitiveness of solar cell technologies. Semiconductor nanowires are attracting intense interest as a promising material for solar energy conversion for the new-generation photovoltaic (PV) technology. In particular, silicon nanowires (SiNWs) are under active investigation for PV applications because they offer novel approaches for solar-to-electric energy conversion leading to high-efficiency devices via simple manufacturing. Thanks to their high surface area, SiNW arrays have strong optical absorption capabilities over the solar spectrum – less than 1% of the material containing in nanowire is required to obtain the same absorption efficiency compared to traditional planar wafer-based PV devices. Silicon nanowire solar cells with radial p-n junctions also offer a shorter collection length for charge carriers, so lower-quality silicon can be used. Even so, silicon nanowires can be produced with excellent electrical properties. All these advantages may reduce the production costs for SiNW-based solar cells without decreasing their efficiencies. Based on the previous results obtained by several authors, we propose a new combined mechanism and optimized process parameters for the formation of well-separated silicon nanowires. We believe that an improved solar cells performance can be achieved through further process optimization. Consequently, in the present study, we investigated MCEE process on silicon. The influence of the etching parameters (i.e., hydrofluoric acid concentration, and etching time) on the structure morphology, absorption, was studied.
Solar cells based on nanowires rather than crystalline films offer a promising route to reducing the cost of photovoltaic energy their increased surface area improves light absorption and carrier collection efficiency. Combining silicon nanowire with other material that absorbs light at different wavelengths may improve the sun radiation absorption.

Keywords: Fabrication, FDTD simulation, photovoltaic solar cell, Silicon nanowires.

1 introduction

Nanowires (NW) based solar cells are good candidates to decrease the cost of photovoltaic energy due to the increase of light absorption and carrier collection efficiency. There are two main approaches to elaborate the nanowires arrays: a bottom-up approach based on the growth of the nanowires usually by CVD (Chemical Vapor deposition) and a top-down approach based on etching methods. For photovoltaic energy production, it is preferable to use low cost and competitive technologies. Therefore bottom-up techniques are preferable to top-down methods that consume material and increase the cost of the device. In this study we have focus on bottom-up low cost technologies to grow nanowires arrays instead of more expensive etching techniques. Si NW arrays for photovoltaic applications are usually grown by CVD on top of silicon wafer, glass substrate or metal [1-5]. The most used technique is the vapor liquid solid (VLS) method which uses a metal catalyst to form a liquid eutectic with the desired nanowire material. Junction is preferably radial than axial for photovoltaic applications in order to increase collection probability (minority carriers only have to travel on a maximum distance corresponding to the radius of the nanowire). pn junction can be realized by doping but due to the small diameter of the nanowire, there might be a complete doping of the nanowire instead of just the surface. Another way is to deposit a conformal and doped polysilicon layer on the nanowire. However, the interface is usually highly recombinant. In order to passivate the surface of the nanowire and to create a junction, an interesting structure is the amorphous silicon / crystalline silicon (a-Si/c-Si) heterostructure, which has already demonstrated high efficiencies for first generation solar cells [6].

Nanowires solar cells are expected to outperform the thin film counterpart in terms of optical absorbance; however, the geometry of the array has to be optimized in order to maximize the carrier absorption for a given configuration of the media. Optimization of regular arrays involves the proper sizing of the heights, of the diameters as well as of the wire periodicity. As the dimensions of the investigated structures are on the same order of magnitude of the incident radiation wavelength, the optical properties of nanowires are influenced by diffraction effect. Interference and multiple reflection effects are also responsible of the absorption enhancement. Therefore the optical simulations of nanostructures must be performed by means of rigorous solvers of Maxwell equations like the Finite Element Method (FEM) [7], the Transfer Matrix Method (TMM) [8] and the Finite Difference Time Domain (FDTD) [9]. Methods like FEM and FDTD are remarkably cpu- and memory-demanding especially for applications involving the optical simulations of nano- and micro-structured solar cells. Recently the Rigorous Coupled-Wave Analysis (RCWA) method [10] based on the modal expansion of the electromagnetic field has been adopted for the numerical simulation of the light propagation in thin- film solar cells [11]. In RCWA the approximation is introduced by the number of Fourier expansion modes included in the calculation. However, RCWA allows to attain a good trade-off between accuracy and computational resources requirement, especially if compared with FDTD method [12].

2 Materials and Methodology

SINWs are prepared by metal assisted chemical etching of p- Si (100) wafer having resistivity of 1-10 ohm-cm. The wafer is treated by several chemicals, for cleaning and etching to produce SINWs. Specifically, the silicon sample is cleansed by Acetone, isopropanol, H2O 15 minutes each followed by piranha attack H2O2: H2SO4 (1: 1 v / v) for 10 minutes and then thoroughly rinsed and placed in the EDI (Deionized water) before drying with nitrogen. The sample is then directly introduced into the 5M HF 0.02N AgNO3 at different concentration mixture for a 1 minute after that the silicon wafer covered with Ag nanoparticles were immersed in a second etching solution of HF/H2O2 5M/ 30% followed by EDI rinsing and drying with nitrogen. 3D FDTD simulations were used to calculate the absorption, Poynting vector, electric field and exciton generation.

The nanowires obtained have a length range from 15- 20 μm. This length is homogeneous on the same substrate regardless of the sample size. Diameters between 10 nm and 100 nm and depending of the synthesis conditions the nanowires can be "linked" to each other and form rather 'nanosheets' or 'nanowalls'. As shown in fig1. In MCEE method, silicon is oxidized and etched preferentially by mixture solution of HF/H_2O_2 under thin layer of noble metal (e.g. Ag, Au, Pt, etc.), leaving bare non-etched silicon regions and the concentration of HF and H_2O_2 will have an effect on the structure and shape of the nanowires. The silver particles play a role of cathode, at which the reduction of the oxidant generates holes in these holes diffuse through the metal catalyst and then injected into the silicon, which is in contact with a noble metal, which causes silicon oxidized and dissolved by HF. This spontaneous bipolar electrochemical reaction generates an electric field drives the directional movement of the metal / Si and Si anisotropic dissolution. The nanowires obtained are crystalline and covered with an amorphous silicon layer often much finer than that of vapor-liquid-solid (VLS) nanowires synthesized. The nanowires obtained are crystalline and covered with an amorphous silicon layer often much finer than that of vapor-liquid-solid (VLS) nanowires synthesized. Furthermore, the method is not limited by the crystallographic orientation, the type or the concentration of the dopant, or the crystallinity of the substrate. Silicon nanowires are formed perpendicular to the surface for orientation <100>. The morphologies of these nanowires are reproducible and complementary (or different) from those of nanowires obtained by VLS or microporous silicon technology.

Figure1: SEM characterization of A) top view B) cross section of silicon nanowires produced by metal catalyzed electroless etching MCEE

Fig. 2a and Fig. 2b shows the absorption profiles simulation relative to the bulk Si and silicon nanowires grown on bulk silicon. We demonstrate that Fabray-Perot as absorption regime was held with regular pattern through only the bulk Si. It was observed that the absorption was confined within Si nanowires like Bloch-mode. We observed also that this earlier distribution depends on nanowire size and geometry. It is well known that the excitation of the exciton within the semiconductor is due the transfer energy resulted from the photons absorbed from solar radiation. Therefore, Poynting power flow, in fact, exhibits the connection between energy transfer and photon absorption. Poynting vectors for bulk crystalline Silicon and Si-NW grown on bulk silicon are shown in Fig. 2c and Fig. 2d respectively.

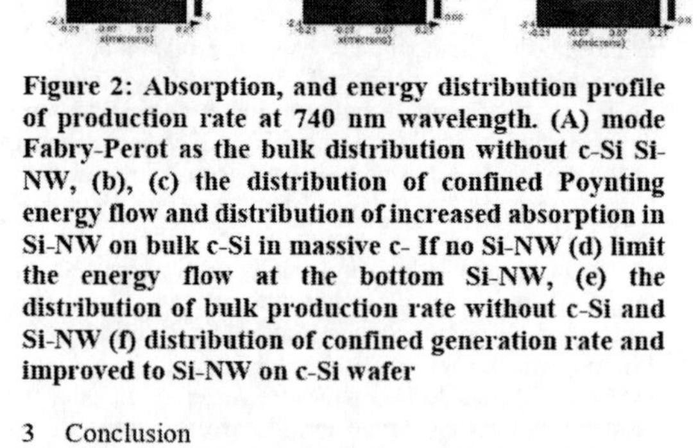

Figure 2: Absorption, and energy distribution profile of production rate at 740 nm wavelength. (A) mode Fabry-Perot as the bulk distribution without c-Si Si-NW, (b), (c) the distribution of confined Poynting energy flow and distribution of increased absorption in Si-NW on bulk c-Si in massive c- If no Si-NW (d) limit the energy flow at the bottom Si-NW, (e) the distribution of bulk production rate without c-Si and Si-NW (f) distribution of confined generation rate and improved to Si-NW on c-Si wafer

3 Conclusion

In conclusion we have demonstrated that the morphology of are affected by the fabrication parameters, we demonstrated that the HF concentration have a direct effect on the length and the density of the nanowires. We studied also the effect of the etching time on the optical absorption of the silicon nanowires, we showed that the light absorption increase with the increasing of the etching time. We are also conducted the simulation study of the optical absorption of the silicon nanowires. We observed

through FDTD simulation that silicon nanowires grown on bulk silicon strongly improved absorption, energy distribution, electric profile and generation.
The experimentation results demonstrate that the silicon nanowires produced by metal assisted chemical etching are very promising due to their potential application in solar cell or photovoltaic

References

[1] Erik C. Garnett, Mark L. Brongersma, Yi Cui, and Michael D. McGehee, Nanowire Solar Cells, Annual Review of Material Research, 41:11.1–11.27, 2011.

[2] Michael D. Kelzenberg, ShannonW. Boettcher, Jan A. Petykiewicz, Daniel B. Turner-Evans, Morgan C. Putnam, Emily L.Warren, Joshua M. Spurgeon, Ryan M. Briggs, Nathan S. Lewis and Harry A. Atwater, "Enhanced absorption and carrier collection in Si wire arrays for photovoltaic applications", Nature Materials, Vol 9, March 2010.

[3] Fan, Z., Razavi, H., Do, J., Moriwaki, A., Ergen, O., Chueh, Y.-L., Leu, P. W., Ho, J. C., Takahashi, T.,Reichertz, L. A., Neale, S., Yu, K., Wu, M., Ager, J. W., and Javey, A., Three-dimensional nanopillar- array photovoltaics on low-cost and flexible substrates, Nat Mater 8, (2009) 648-653

[4] Tsakalakos, L., Balch, J., Fronheiser, J., Shih, M.-Y., LeBoeuf, S. F., Pietrzykowski, M., Codella, P. J.,Korevaar, B. A., Sulima, O. V., Rand, J., Davuluru, A., and Rapol, U., Strong broadband optical absorption in silicon nanowire films, Journal of Nanophotonics 1, (2007) 013552

[5] O'Donnell, B., Yu, L., Foldyna, M., and Roca i Cabarrocas, P., Silicon nanowire solar cells grown by PECVD, Journal of Non-Crystalline Solids , (2011)

[6] Takahiro Mishima, Mikio Taguchi, Hitoshi Sakata, Eiji Maruyama, Development status of high- efficiency HIT solar cells, Solar Energy Materials and Solar Cells, Volume 95, Issue 1, (2011, 18-21)
[7] J-M. Jin, The Finite Element Method in Electromagnetics. John Wiley & Sons, ISBN-10: 0471438189, 2002

[8] Whittaker, D. M. and Culshaw, I. S., Scattering-matrix treatment of patterned multilayer photonic structures.Vol. 60, 4, pp. 2610-2618, 1999.

[9] K. S. Kunz and R. J. Luebbers. The Finite Difference Time Domain Method for Electromagnetics. CRC Press, 1993.

[10] M. G. Moharam and T. K. Gaylord. Rigorous coupled-wave analysis of planar grating diffraction. J. Opt. Soc. Am., 71 (7), pp. 811-818, 1981.

[11] M. Agrawal, M. Frei, Y. Bhatnagar, T. Repmann, K. Witting, J. Schroeder, and C. Eberspacher. Comprehensive experimental and numerical optimization of surface morphology of transparent th conductive oxide films for tandem thin film photovoltaic cells. Proc. Of 35 IEEE Photovoltaic Specialists Conference (PVSC), 301-304, 2010.

[12] I. Semenikhin, M. Zanuccoli, M. Benzi, V. Vyurkov, E. Sangiorgi, C. Fiegna, Computational efficient RCWA method for simulation of thin film solar cells. Optical and Quantum Electronics, Vol. 44, 3-5, 149-154, 2012

REDUCED OPERATING TEMPERATURE TO IMPROVE DURABILITY AND EFFICIENCY OF SOLAR MODULES

Yajie (Jessica) Jiang, Zibo Zhou, Supriya Pillai, Mark Keevers,
Jose Bilbao Bernales, Jinyi Guo, Nicholas Ekins-Daukes and Martin A. Green
Australian Centre for Advanced Photovoltaics, University of New South Wales (UNSW Sydney), Australia 2052
Emails: yajie.jiang@unsw.edu.au; m.green@unsw.edu.au

ABSTRACT: High operating temperature is detrimental to module performance and durability. Here we take a fresh look at four passive approaches to reduce module operating temperature: decrease the absorption of unproductive near-infrared sunlight, increase mid-infrared emissivity of the front and rear module surfaces, enhance convective losses by disrupting the normally laminar flow along the module rear surface, and increase lateral heat conduction to the module frame and support structure. We present initial simulation and experimental results, including outdoor measurements of a module with half of the rear covered by an array of small prototype vortex generators (VGs). A combination of these passive cooling approaches has the potential to reduce module operating temperature by 5-10°C, increasing module efficiency by 2-5% relative and module life by 50-100%, thus significantly improving lifetime energy production and the economics of PV.
Keywords: Silicon, Module, Thermal Performance, Optical Properties, Durability

1 INTRODUCTION

A silicon module typically operates 20-30°C above ambient in bright sunshine. This reduces efficiency and power output by 0.4-0.5% (relative) for every °C increase in temperature – up to 15% for a 30°C rise. It also accelerates module degradation and reduces module life below that otherwise achievable, as now evidenced in emerging hot markets such as India.

Generally, module generating capacity degrades 0.5% relative for each year in the field, with modules normally warranted to be above 80% of initial rating after 25 years of field exposure. As commonly observed in other engineering disciplines, lab testing of specific module degradation modes shows degradation rates approximately double for every 10°C increase in temperature [1]. On this basis, modules operating 10°C closer to ambient temperature would likely double their lifetime energy production, even not accounting for the additional durability benefits of reduced daily temperature cycles, with a correspondingly large positive impact upon the economics of energy production. A less ambitious 5°C reduction would increase output by 50%.

A new project in our group aims to markedly reduce solar module operating temperature, not only to increase module output, but also to greatly increase durability and lifetime energy production. We are investigating a range of passive approaches, some highly innovative [2] in order to increase the module radiative, convective and conductive losses, while also restricting the absorbed heat to be dissipated, all without major changes in module design. This paper describes these passive cooling approaches and presents our preliminary results.

2 FOUR APPROACHES: PRELIMINARY RESULTS

A schematic of the heat generation and loss processes in a solar module is shown in Fig. 1. A portion of the power Q_{IN} in incident sunlight is absorbed in the module, largely in the cells, normally encapsulated under an optically transparent glass coversheet with an opaque thin rear polymeric coversheet (reflection and electrical power extraction will reduce Q_{IN}). This heat is conducted to both module surfaces (small arrows) from where it is dissipated by free (Q_{FF} and Q_{FB}) and forced convection (Q_{WF} and Q_{WB}) as well as radiatively (Q_{RF} and Q_{RB}, note that glass is largely opaque at the radiative wavelengths). Some conduction also occurs perpendicularly to the module frame (Q_{Cf}, green arrows).

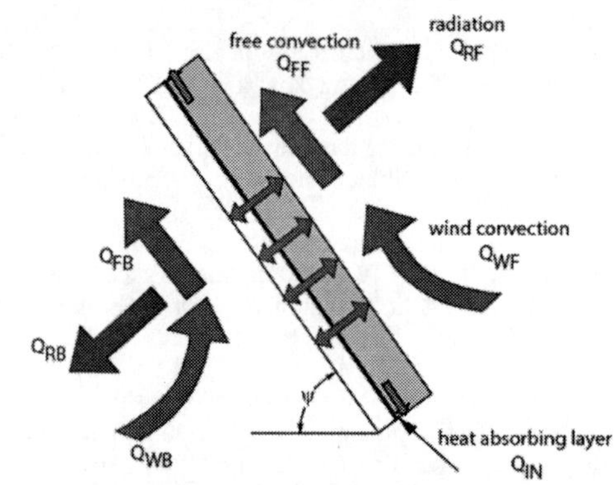

Figure 1: Module energy balance. Radiative, convective and conductive heat losses from the front (F) and back (B) surfaces can be significantly increased above present norms.

Prior to 2016, module thermal performance was characterised in terms of a single parameter, the NOCT (nominal operating cell temperature), defined as the temperature reached by the cells in a module under a defined set of operating conditions. With the introduction of an updated standard (IEC61215: 2016), NOCT has now been replaced by the NMOT (nominal module operating temperature), defined slightly differently [3]. Experimental differences are small, with NREL measurements [3] over 38 days for a single module giving NMOT = 48.7±1.7°C and NOCT = 47.9±1.3°C.

NMOT can be reduced by: (1) reducing Q_{IN}; (2) increasing radiative losses Q_{RF} and Q_{RB}; (3) increasing free convection losses Q_{FF} and Q_{FB} (and associated wind convection losses Q_{WF} and Q_{WB}); and (4) increasing conduction and convection losses to frame and support structures, Q_{Cf}.

2.1 Reject unwanted near IR (reduce Q_{IN})

Even the best silicon cells, such as PERC cells, now emerging as the commercial standard in PV manufacturing, respond only to wavelengths < 1200 nm. In the standard AM1.5G spectrum, there is 836 W/m^2 below this wavelength, with 164 W/m^2 above it, unusable by the cell. PERC cells reflect more of the latter than Al BSF cells and also extract more electrical power, giving 1.7°C lower NMOT [4]. However, if more near infrared light could be reflected, significantly reducing Q_{IN}, NMOT could be reduced by up to an additional 3.2°C [4].

Fan's group [5-7] recently designed a 45-layer dielectric coating to provide strong reflection in the 1200-2000 nm range and estimate 5.7°C lower MNOT. However, this approach is too expensive for practical use. We have investigated simpler approaches giving less ideal performance but still producing significant absorption reduction. One approach is to replace the single-layer antireflection coating (SLAR) now used routinely on modules by multilayer designs [8]. This has the advantage of reducing reflection at usable wavelengths but increasing it at wavelengths outside the design range. While a SLAR can have minimum and maximum reflectance of approx. 0.3% and 4.1%, assuming indices of 1.3 and 1.5 for coating and glass, a 2-layer 'V coating' [8] can have values of 0% and 28.3% (outside design range) assuming an index of 2.1 for the new layer. Saint Gobain "PV-Lite" glass may use such a design, although not for this purpose. Nanoporous SiO_2 is the most common AR coating in the solar glass industry. Our preliminary simulation results for a multilayer nanoporous SiO_2 AR design is shown in Fig. 2 as the baseline for further improvement. The unwanted reflection peak at 480 nm will be minimised in following work, and stronger reflection peak between 1.2-4 μm will be explored.

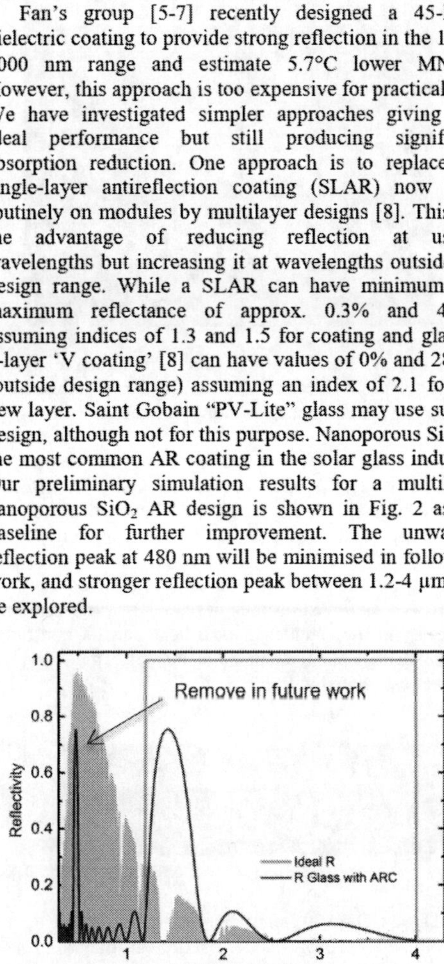

Figure 2: Reflectivity of glass for preliminary ARC designs compared with ideal reflectivity. Normalized AM1.5 solar spectrum is shown in grey.

A more innovative approach is to take advantage of 'resonant absorption' in the strong atmospheric absorption bands at 1.4 and 1.9 μm caused by a combination of atmospheric H_2O and CO_2. Fig. 3 shows our calculated refractive index if two impulses of absorption are introduced into the glass at these two wavelengths. This will not increase module solar absorption significantly if the region of strong absorption is constrained to the atmospheric absorption bands, but it disturbs the index significantly on either side of these two wavelengths, increasing reflection by undoing the effects of the AR coating. The calculated reflectivity of glass with this 'resonant absorption' feature is shown in Fig. 4 compared with ideal reflectivity.

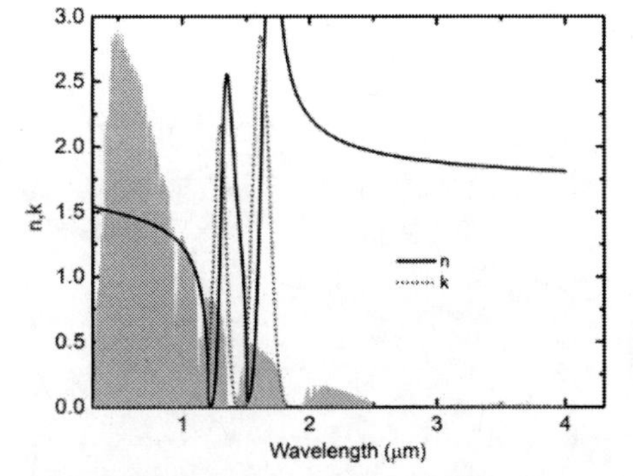

Figure 3: Effect of two resonant absorption impulses at the atmospheric absorption bands 1.4 μm and 1.9 μm on the refractive index n and extinction coefficient k. AM1.5 solar spectrum is shown in grey.

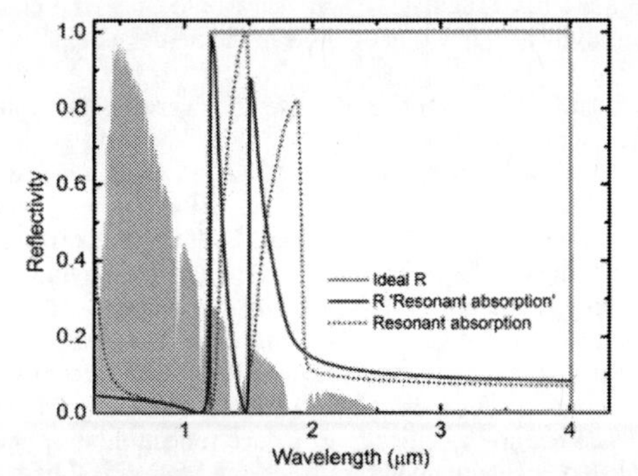

Figure 4: Reflectivity of glass with resonant absorption compared with ideal reflectivity. Normalized AM1.5 solar spectrum is shown in grey.

2.2 Increase radiative loss in the mid IR (Q_{RF} and Q_{RB})

For a module temperature around 40°C, the module will have peak radiative energy emission at about 9 μm, with significant emission over the 4-30 μm range. Our measured and calculated emissivity of bare glass over most of this range is shown in Fig. 5, evaluated using the optical constants of Rubin [9]. Glass changes from transparent to opaque beginning at 2.8 μm, with the transformation completed by 4.5 μm.

The features apparent in the near perpendicular emissivity (30°) curve at 9 and 21 μm are due to absorption by Si-O bond stretching and Si-O-Si bending, respectively, with weaker absorption by Si-Si bond stretching at 12 μm. This absorption causes similar disruption to the optical constants noted in Fig. 3, but the corresponding increased reflection at the present medium-to-long-wavelength IR region is a disadvantage (lower emissivity) in terms of increasing radiative losses.

Fan's group [5-7] has suggested increasing emissivity at these wavelengths by incorporating glass texture in the form of sub-wavelength upright pyramids, acting as a graded index layer to reduce reflection. Gentle and Smith [10] claim Fan's conclusions are overstated due to the large fraction of radiative emission that occurs at oblique angles (Fig. 5), where this approach would not be effective.

An alternative way of ensuring high emissivity is inclusion of doped transparent conducting oxides into the module AR coating to provide free carrier absorption of light at these long wavelengths before entering the glass, hence reducing glass reflection.

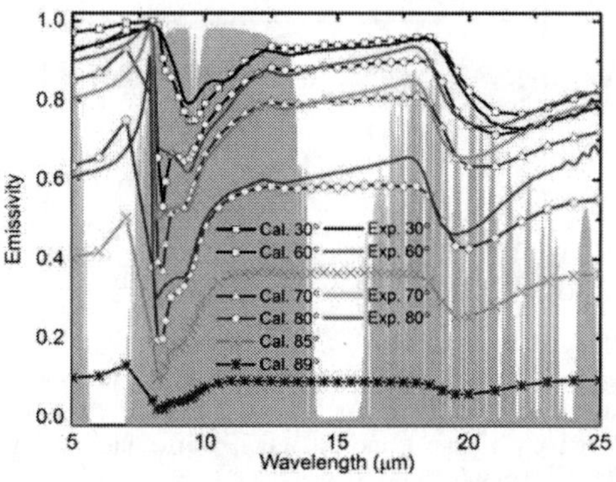

Figure 5: Calculated and measured planar glass emissivity as a function of angle to the perpendicular.

Another approach is the use of macro- rather than microscopic glass texture. Since most low-iron glass is rolled, such texturing can be incorporated at the mm scale at low cost, such as shown in Fig. 6 for standard solar glass manufactured at Pilkington, Sydney (the textured side generally faces away from the sun, for 'keying' of module laminations). By reducing the fraction of rays able to strike the glass at the most oblique angles of Fig. 5, the angular emissivity can be significantly increased. Some companies have marketed glass with deeper front surface texture specifically to reduce reflection at optical wavelengths, with good results reported for optical boosts [11] and for self-cleaning [12]. Our simulations of mm-sized upright pyramids using the COMSOL Multiphysics® platform have shown a 10% increase in radiative emission compared to a flat surface, similar to that expected.

Figure 6: Cross-sectional image of the shallow 'inverted pyramid' structure on solar glass.

2.3 Increase convective loss (Q_{FF} and Q_{FB})

The Heat Transfer Module of the COMSOL Multiphysics® platform is being used to evaluate and optimise alternative enhanced convection schemes prior to experimental evaluation. Fig. 7 shows our COMSOL simulation of free convection from the module front and back surfaces. Since the air is heated adjacent to the module, buoyancy causes it to rise, confining it to the underside of the module but allowing it to break free of the top surface. This increases the cooling rate of this front surface.

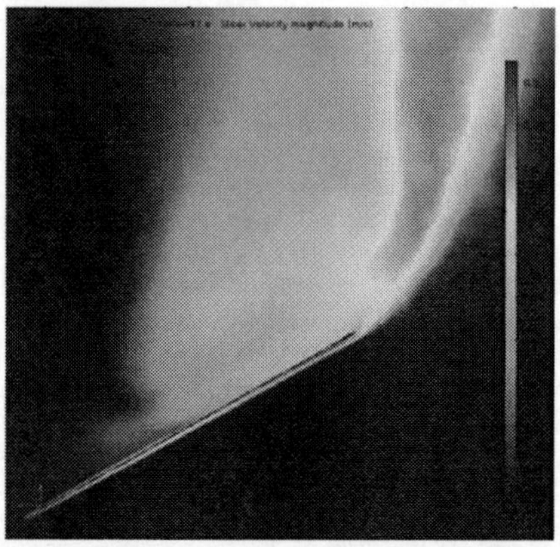

Figure 7: COMSOL simulation of convective module cooling. Rear convective loss can be greatly enhanced by disrupting the laminar flow along this surface.

Given the laminar flow along the back of the module, with associated boundary layer creating an effective 'thermal blanket', our efforts have focussed on attempting to disrupt this laminar flow by introducing longitudinal vortices using surface protrusions defined as vortex generators (VGs). Our investigation of triangular 'flap' VGs achieved promising results, giving significant increases in the calculated local heat transfer coefficient as shown in Fig. 8, despite the low effective Reynolds number associated with the flow.

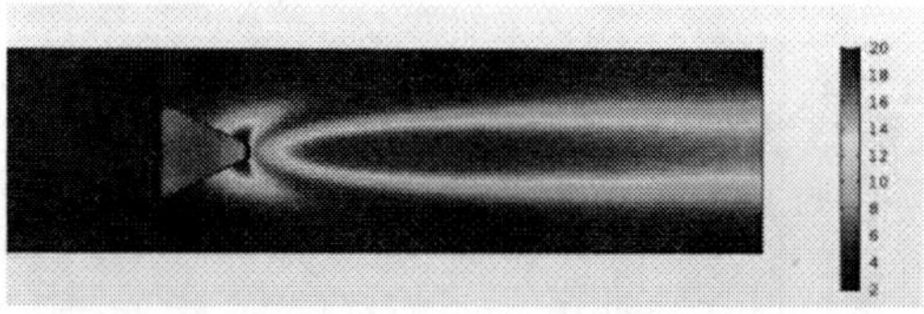

Figure 8: COMSOL simulation of the effect of a small flap on the local heat transfer coefficient (diagram rotated to horizontal). The flap produces a significant local enhancement of the heat transfer coefficient (units $W/m^2/K$) in its convective 'shadow'. Three dimensional geometries have shown even better performance under forced convection.

The layout of triangular flap VGs on the module rear was optimised [2], with improved designs doubling convective losses at the rear side under free convection conditions. The longitudinal temperature profile of the module simulated in COMSOL for free convection is shown in Fig. 9, indicating a significant temperature reduction up to 2°C.

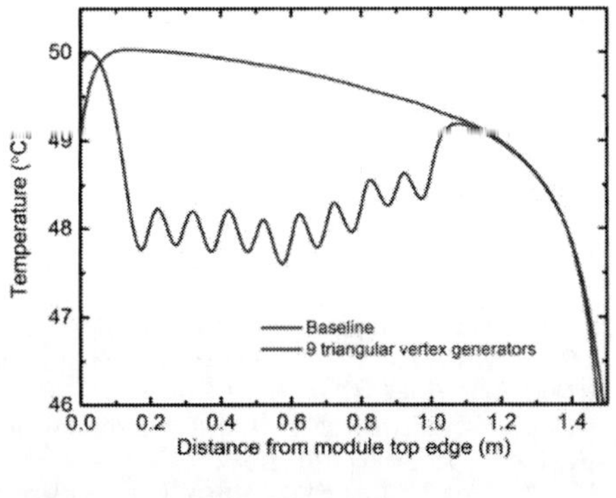

Figure 9: Longitudinal temperature profile of a standard solar module (purple) compared with our 'cooler' module with VGs attached at the rear (red).

We have established an outdoor module testing capability on the flat concrete rooftop of our building, using a standard 'ground mount' module rack, Fig. 10.

Figure 10: Module testing on the roof of our building.

For an initial proof of concept of our VG-enhanced convective cooling approach, we fabricated prototype VGs by 3D printing and attached them to one half of the rear of a standard Si module using double-sided tape, leaving the other half uncovered to serve as baseline. Temperature is measured by RTD probe, IR thermometer and FLIR thermal imaging camera (Fig. 11). Left-right temperature difference is also determined by proxy via Voc measurement of the left-most and right-most 20-cell strings (accessible in the junction box). To simulate free convection conditions we used corflute plastic sheets (not shown) to block wind from accessing the module rear whilst allowing hot air to escape from the highest points. All temperature measurements are consistent with 0.8°C temperature reduction due to the VGs, which is also consistent with our COMSOL simulations of this particular VG setup.

Figure 11: IR thermal and optical image of a module during outdoor testing of prototype VGs on one half of the module rear.

An intriguing aspect of the VG-enhanced convection approach is that VGs could be readily retrofitted to existing modules in the field, boosting energy production and increasing module life and project profitability.

Another issue that has been explored relates to the impact of module frames on convective losses and the redesign of frames to increase convective losses, such as by including slots in them at appropriate heights [2]. Also being investigated is the effect of wind convection on convective losses. In large fields, modules are shielded from wind by fencing and by adjacent rows of panels with wind possibly channelled along preferred directions. This effect and its exploitation are also being examined with COMSOL.

2.4 Increase conductive loss to frame (Q_{Cf})

Fig. 12 shows the approximate temperature distribution in two adjacent field-installed modules near the module frames (the near horizontal feature near the centre of the image). Temperatures within the module, ranging from 20.8 to 38.0°C, are reasonably accurately measured due to the high emissivity of glass over the detector's response range (7.5-14 μm), while the temperature of the frame, indicated as 26.1°C (circled area) is likely inaccurate (due to its different emissivity) since heat is clearly flowing to the frame from the module. The optical and thermal images are also not perfectly aligned due to the use of separate lenses.

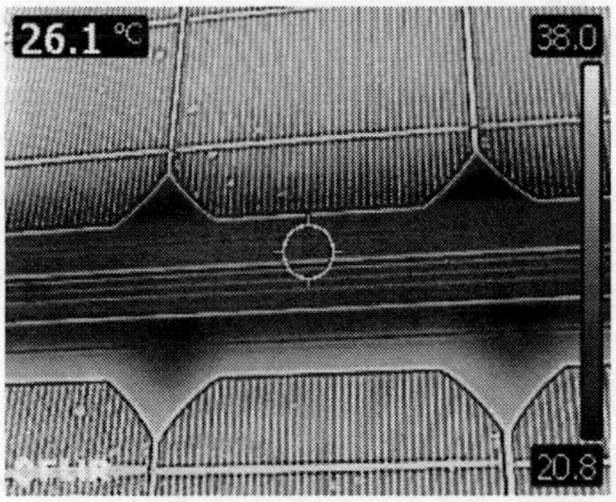

Figure 12: Infrared image showing the nominal temperature distribution near the frame of an operating solar module, superimposed on the outlines from an optical image.

Our analysis of this situation gives a diffusion length for the heat moving from the nearest cell to the frame given by the following expression:

$$L_{th} = \sqrt{\frac{\sum_i \kappa_i w_i}{H}} \approx \sqrt{\frac{1.1 \times 0.0032}{30}} = 0.011\ m = 1.1\ cm$$

where κ_i is the thermal conductivity of the layers (width w_i) giving lateral transport, mainly the glass coversheet, and H is the overall module heat transfer coefficient, typically 30 W/m^2/K. The total heat loss to the frame is then approximated by $P\ L_{th}^2\ H\ (t_{cell} - t_{frame})/min(S,\ L_{th})$ where P is the module perimeter and S is the cell to frame distance, or *34 W* for H=*30 W/m*^{2}*/K*, L_{th}=*1.1 cm*, *S~1 cm*, t_{cell} - t_{frame}=*18K* and *P=5.2 m*. Spacing the cells closer than L_{th} to the frame would increase this loss. Using a thermally conductive, electrically insulated metal back sheet as offered by some manufacturers or else laminating thermally conducting but electrically insulating layers such as tape-cast alumina of similar thickness to cells into the module near the edge would also substantially increase this loss.

We have also completed studies of the interaction of rear convective flows with the frame, showing frames reduce these by ~10%. This warrants much closer examination, due to the combination of flow disruption and cooling by the frame. Frame designs will be optimised to gain full benefit from these interactions. Both conduction and convection losses to the frame can be increased if the frame temperatures can be further reduced, as could be arranged by increasing the frame's infrared emissivity.

3 SUMMARY

Several innovative ideas and preliminary results are discussed in this work, with all contributing to increased module annual energy production and durability by lowering operating temperatures: (1) Use of a MLAR coating on glass, related to 'V' AR designs, to decrease reflection at optical wavelengths whilst increasing near IR reflection; (2) Use of 'resonant absorption' in the glass or adjacent layers at NIR wavelengths of atmospheric absorption bands to increase reflection at neighbouring wavelengths; (3) Use of free carrier absorption to improve glass mid IR emissivity and angular response; (4) Use of macroscopic texturing of glass to increase mid IR radiative losses whilst reducing reflection at optical wavelengths, with such texturing also useful on the module rear; (5) Use of flaps and other surface protrusions (vortex generators, VGs) to increase module convective losses, with different solutions possible for new modules and for retrospectively fitting to field installations to increase module output and life; (6) Redesign of module frames to maximise convective losses, including increased infrared emissivity to reduce frame temperature, while not compromising original design features; (7) Use of thermally conductive, electrically insulating layers integrated into the module encapsulation to increase heat loss through the module frame and support structure. Even though the potential temperature reductions identified are not necessarily cumulative, this work identifies a clear trajectory to reducing module operating temperature by 5-10°C.

REFERENCES

[1] O. Dupré, R. Vaillon, M. A. Green, *Thermal Behavior of Photovoltaic Devices: Physics and Engineering*. 2016: Springer International Publishing.

[2] M. A. Green, M. J. Keevers, Z. Zhou, Y. Jiang, N. Ekins-Daukes, *Australian Provisional Patent Applications 2018900639, 2018900640, 2018900641*. 2018.

[3] M. Muller, B. Marion, J. Rodriguez. *Evaluating the IEC 61215 Ed.3 NMOT procedure against the existing NOCT procedure with PV modules in a side-by-side configuration*. 2012 38th IEEE Photovoltaic Specialists Conference. 2012.

[4] M. R. Vogt, H. Schulte-Huxel, M. Offer, S. Blankemeyer, R. Witteck, M. Köntges, K. Bothe, R. Brendel, *Reduced Module Operating Temperature and Increased Yield of Modules With PERC Instead of Al-BSF Solar Cells*. IEEE Journal of Photovoltaics, 2017. **7**(1): p. 44-50.

[5] L. Zhu, A. Raman, K. X. Wang, M. A. Anoma, S. Fan, *Radiative cooling of solar cells*. Optica, 2014. **1**(1): p. 32-38.

[6] L. Zhu, A. P. Raman, S. Fan, *Radiative cooling of solar absorbers using a visibly transparent photonic crystal thermal blackbody*. Proceedings of the National Academy of Sciences, 2015. **112**(40): p. 12282-12287.

[7] W. Li, Y. Shi, K. Chen, L. Zhu, S. Fan, *A Comprehensive Photonic Approach for Solar Cell Cooling*. ACS Photonics, 2017. **4**(4): p. 774-782.

[8] A. Thelen, *Design of Optical Interference Coatings*. 1989: McGraw-Hill.

[9] M. Rubin, *Optical properties of soda lime silica glasses*. Solar Energy Materials, 1985. **12**(4): p. 275-288.

[10] A. R. Gentle, G. B. Smith, *Is enhanced radiative cooling of solar cell modules worth pursuing?* Solar Energy Materials and Solar Cells, 2016. **150**: p. 39-42.

[11] U. Blieske, G. Stollwerck, *Chapter Four - Glass and Other Encapsulation Materials*, in *Semiconductors and Semimetals*, G.P. Willeke, E.R. Weber, Editors. 2013, Elsevier. p. 199-258.

[12] W.Herrmann, *Auswirkung der Verschmutzung auf den Energieertrag von PV-Modulen mit verschiedenen Gläsern (Effect of pollution on the energy yield of PV modules with different glasses)*. Workshop "Photovoltaik-Modultechnik", 12-13 November 2015, Köln.

Effect of Annealing Process on Crystalline Silicon Solar Cells with Down-Conversion SiN_x : Tb^{3+}-Yb^{3+} Films

Yu-Chieh Lee[1], Shao-Chun Wu[1], Ing-Song Yu[1,*], Lucile Dumont[2], Julien Cardin[2], Christophe Labbé[2], Fabrice Gourbilleau[2]

[1] Department of Materials Science and Engineering, National Dong Hwa University, Hualien, 97401, Taiwan
[2] CIMAP, Normandie Univ, ENSICAEN, UNICAEN, CEA, CNRS, 6 Boulevard Maréchal Juin 14050 Caen Cedex 4, France.
*E-mail: isyu@gms.ndhu.edu.tw

Down-conversion (DC) is the process in which a photon with high energy can be converted into two photons of low energy by cooperative energy transfer to minimize the problem of thermalization of electron-hole pairs in solar cells. In this paper, SiN_x:Tb^{3+}-Yb^{3+} thin films were integrated on the top of c-silicon solar cells by reactive magnetron co-sputtering as spectrum converter and anti-reflection coating layers. After the process integration of monolithic c-Si solar cells with DC films, these films and the performances of solar cells were characterized and analyzed. Transmission electron microscope experiments show these layers uniformly on texturing Si surface. Photoluminescence spectroscopy depicts the energy conversion of DC films annealed at the temperatures of 650°C, 750°C, and 850°C, respectively. From the performance of solar cells, we can find the optimal annealing parameters to enhance the performance of monolithic c-Si solar cells with DC SiN_x:Tb^{3+}-Yb^{3+} films.

Keywords: Down conversion; Silicon nitride; Rare-earth ion; Anti-reflection coating

1. Introduction

Solar cells have been becoming the symbol of green energy in the past several decades. In particular, Si-based solar cells play a role in renewable energy generation. Recently, the efficiency of crystalline silicon (c-Si) solar cell (*SC*) has reached a high value of about 26% [1]. According to the Shockley–Queisser limit, the efficiency of a single junction solar cell with band gap energy of 1.1 eV is around 31% [2]. In other words, it is necessary to find a way out of breaking through the efficiency limit. The reason for limiting conversion efficiency comes from the energy losses. One of the main energy losses derives from the mismatch between solar spectrum and Si band gap. In the case of high-energy photons, the excess energy of photons will be released as heat in the *SC* [3]. To solve this problem, a promising approach called down-conversion (*DC*) has been investigated [4, 5].

DC process can convert a high-energy photon into two or more low-energy photons to minimize the problem of thermalization. As a result of conversion, the low-energy photons with E > 1.1 eV are absorbed by Si solar cells, and the energy loss can then be reduced. Rare earth (*RE*) couples such as Tm^{3+}-Yb^{3+}, Tb^{3+}-Yb^{3+}, Eu^{3+}-Yb^{3+}, and Pr^{3+}-Yb^{3+} have been investigated as sensitizers and activators. They have shown an efficient conversion of ultraviolet-visible (UV-Vis) light to near-infrared (NIR) light by energy resonance [6-9]. Besides, the band gap of host materials needs to be over 3.0 eV to avoid absorbing visible light, and it's necessary to choose a matrix that prevents multiphonon relaxations by reducing its phonon energy. Therefore, fluorides are commonly chosen as host matrixes due to their wide band gap and low-phonon energy [10]. These *RE*-doped materials are also used as light devices, fiber amplifiers and flat panel displays [11-13].

Among those materials mentioned above, Tb^{3+}-Yb^{3+} couples applied to DC materials in Si solar cells have attracted lots of attention in recent years. The DC mechanism has been reported in many previous studies [7, 10, 14, 15]. Upon a Tb^{3+} ion excited with high-energy incident light (wavelength < 490 nm), electrons transit to $^5D_3$ level and relax from $^5D_3$ to $^5D_4$ level by non-radiative de-excitation. After that, the DC process takes place through cooperative energy transfer from Tb^{3+} ion to two surrounding Yb^{3+} ions, resulting in NIR emission of Yb^{3+} ions due to the $^2F_{5/2}$-$^2F_{7/2}$ transition. Such emission around 980 nm (E ≈ 1.26 eV) is then absorbed by the Si-*SC* with a quantum efficiency of 188% [16].

Anti-reflection coatings (*ARC*) are widely used in solar cells to increase the absorption. Particularly, silicon nitride (SiN_x) has become the main material of ARC in industry due to its transmittance property and refractive index. With proper thickness of *ARC*, the reflectance of incident light can be reduced efficiently. Moreover, SiN_x is compatible with the fabrication of Si solar cells. With the above advantages, SiN_x is a promising candidate for the host material of *DC* layer. In this report, Tb^{3+}-Yb^{3+} doped SiN_x thin films are investigated as *DC*-Si technology compatible layer, combining *ARC* characteristics and frequency conversion process. Before depositing 70-nm thick SiN_x:Tb^{3+}-Yb^{3+} thin film by reactive magnetron co-sputtering, we grew 10-nm thick SiN_x thin film by plasma-enhanced chemical vapor deposition (PECVD) to improve the surface passivation. With the 10-nm thick SiN_x thin film between silicon and *DC* layer, the carrier recombination on the wafer surface can be reduced. After that, before the process of screen-printing and co-firing, the specimens are annealed at different temperatures ranging from 650°C to 850°C in order to activate the *RE* ions. We successfully fabricate the monolithic c-Si solar cell with *DC* layers and find out the optimal annealing process parameters for Si solar cells.

2. Experimental

200-µm thick (100) p-type c-Si wafers were cleaned in KOH solution to remove the surface contamination and saw damages. Then, the samples were textured by a dilute solution of KOH. Phosphorus doped Si were completed during thermal diffusion at 850°C for 30 mins. After that, the edge isolation process removed the phosphorus diffusion around the edge of the samples. We deposited a 10-nm thick SiN_x thin film by PECVD. Then, the SiN_x:Tb^{3+}-Yb^{3+} layers were deposited by reactive magnetron co-sputtering in nitrogen-rich plasma [7]. The RF power density applied on the Si target was fixed at 4.5 W/cm^2 while the nitrogen and argon fluxes were 2 sccm and 8 sccm, respectively. Besides, the plasma pressure was set as 3 mtorr, and the deposition temperature was fixed at 500°C. The doping concentrations of Tb^{3+} and Yb^{3+} ions were controlled by the RF power densities of Tb and Yb target and. were fixed at 8 at.% and 4 at.%, respectively. The deposition time was adjusted to obtain 70-nm film, and therefore, a total 80-nm thick *ARC* were completed. In order to investigate the properties of *DC* layers for different annealing process, the samples were annealed at 650°C, 750°C and 850°C by means of rapid thermal annealing (RTA) for 5 mins. Then, silver paste was printed on the surface of the samples through a patterned screen, and aluminum paste was covered on the rear by printing. Finally, the co-firing process was undertaken. Fig. 1 presents the schematic diagram of c-Si solar cell with *DC* layers.

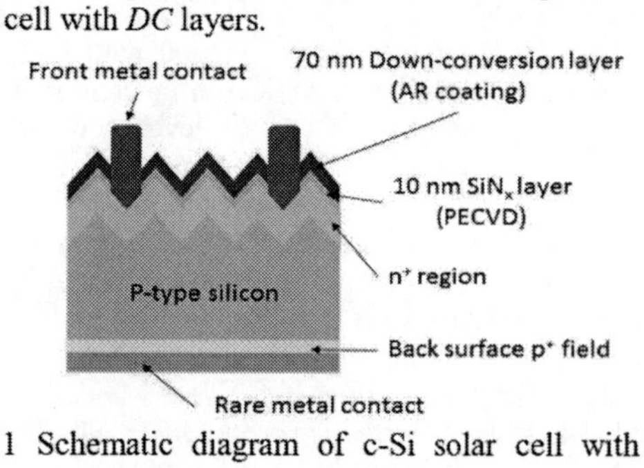

Fig. 1 Schematic diagram of c-Si solar cell with *DC* layers.

After the metallization process, the microstructure of *DC* layers was characterized using a JEOL-1400 transmission electron microscopy (TEM). The UV-Vis and NIR emission spectra were analyzed by photoluminescence spectroscopy (PL) equipped with 325 nm laser from the wavelength 300 nm to 1100 nm at room temperature. I-V curves of solar cells were measured by XES-40S1 solar simulator at room temperature.

3. Results and Discussion

3.1 Microstructure of DC layers on texturing surface

Fig. 2 presents the cross-section TEM image of c-Si solar cell with *DC* layers. From the TEM image, it is obvious to observe that the SiN_x:Tb^{3+}-Yb^{3+} layer and PECVD SiN_x thin film are successfully integrated on c-Si solar cells. The thickness of DC layer and SiN_x film is approximately 70 nm and 10 nm, respectively.

Fig. 2 Cross-section TEM image of SiN_x:Tb^{3+}-Yb^{3+} layer and SiN_x film on c-Si solar cells.

3.2 Emission spectra

Fig. 3 shows the visible emission spectra of SiN_x:Tb^{3+}-Yb^{3+} layers annealed at different temperatures of 650°C, 750°C, and 850°C under 325 nm laser excitation. The emission peaks at 488, 542, 586 and 624 nm can be found, and are assigned to the $^5D_4 \rightarrow {}^7F_J$ (J=6, 5, 4, 3) transitions of Tb^{3+} ions. While *DC* process takes place, Tb^{3+} ions transfer its energy to adjacent Yb^{3+} ions leading to the decrease of the PL emission in the visible range. Then, Yb^{3+} ions gain the energy and emit NIR photons around 980 nm. One can note here that the PL intensity of Tb^{3+} ions is reduced with the annealing temperatures.

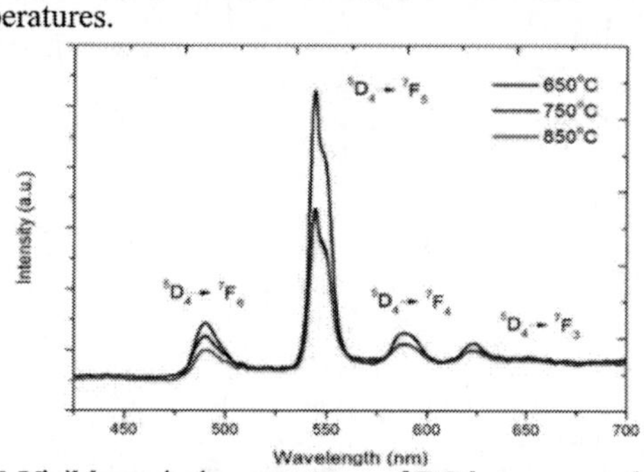

Fig. 3 Visible emission spectrum of DC layers annealed at different temperatures of 650°C, 750°C, and 850°C

On the other hand, the NIR emission intensity is increasing with the annealing temperatures. Fig. 4 presents the NIR emission spectrum of SiN_x:Tb^{3+}-Yb^{3+} layers annealed at different temperatures of 650°C, 750°C, and 850°C. The emission peak corresponding to the $^2F_{5/2} \rightarrow {}^2F_{7/2}$ transition is located at 980 nm. It is noticed that the sample of 850°C has a stronger PL intensity. The increase of the NIR concomitantly with the decrease of the visible counterparts is the signature of an efficient energy transfer from Tb^{3+} ions to Yb^{3+} ions.

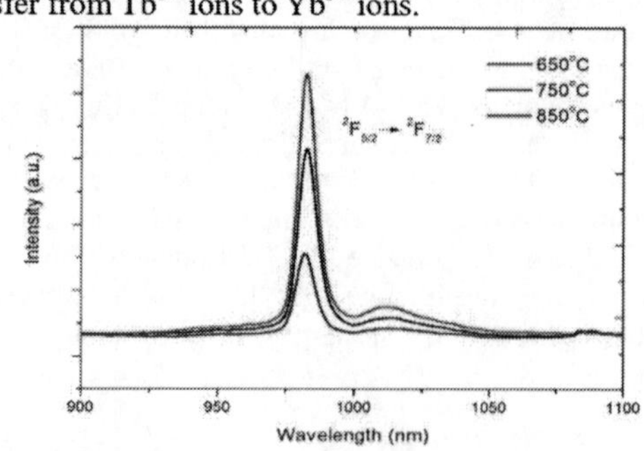

Fig. 4 NIR emission around 980 nm at different temperatures of 650°C, 750°C, and 850°C.

3.3 Performance of solar cells with DC layers

Fig. 5 displays the I-V curves of c-Si solar cell with DC layers annealed at different temperatures. The performance of samples is shown in Table 1. From the I-V curves, the higher annealing temperature favors the higher current density, which could be attributed to a higher conversion efficiency of the DC layers. However, the sample of 850°C has lower open-circuit voltage of 0.55V. Consequently, such a high annealing temperature is detrimental for the integrity of the SC p-n junction. Therefore, the annealing temperature of 750°C is more suitable for the c-Si solar cell topped with a DC layer.

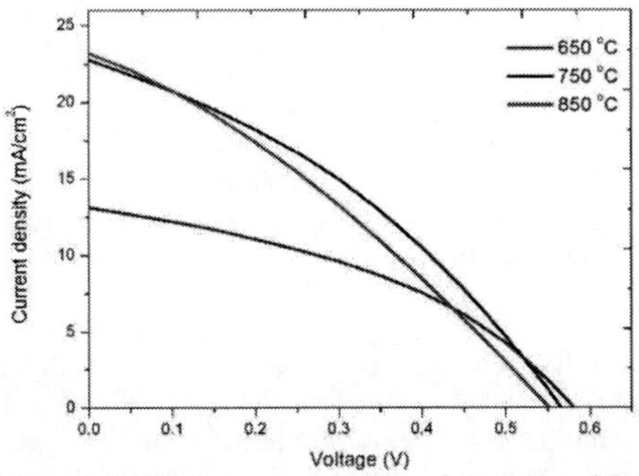

Fig. 5 I–V curves of c-Si solar cell with DC layers annealed at different temperatures.

Table 1. The performance of c-Si solar cell with DC layer annealed at different temperatures.

Sample	V_{oc} (V)	J_{sc} (mA/cm^2)	FF	η (%)
650°C	0.58	13.12	0.40	3.04
750°C	0.57	22.84	0.35	4.52
850°C	0.55	23.16	0.31	3.95

4. Summary

SiNx:Tb3+-Yb3+ thin films as down-conversion and anti-reflection coating layers have been integrated on c-Si solar cells by reactive magnetron co-sputtering. In order to improve the energy transfer of DC layers, high temperature annealing 850°C shows the better spectrum conversion from the results of PL spectra. However, the high temperature process could damage the performance of c-Si solar cells. In the report, we can find the optimized annealing temperature is 750°C from the performance of c-Si solar cells.

Acknowledgments

The authors acknowledge Taiwan Ministry of Science and Technology for financially supporting this study (105-2911-I-259-501 and 106-2221-E-259-010), and E-Ton solar Tech. Co. LTD for supporting c-Si solar cells. Part of this work has been supported by French Agence Nationale de la Recherche through the GENESE Project (ANR-13-BSS09-0020-01) and French Ministry of Research through the ORCHID Project n° 33572XF.

References

[1] K.Yashikawa et al. Silicon heterojunction solar cell with interdigitated back contacts for a photoconversion efficiency over 26%, Nature Energy 25 (2017) 3-13.

[2] W. Shockley, H. J. Queisser, Detailed balance limit of efficiency of p- n junction solar cells, J. Appl. Phys. 32 (1961) 510-519.

[3] B. S. Richards, Enhancing the performance of silicon solar cells via the application of passive luminescence conversion layers, Sol. Energy Mater. Sol. Cells, 90 (2006) 2329-2337.

[4] D. L. Dexter, Possibility of luminescent quantum yields greater than unity, Phys. Rev. 108 (1957) 630-633.

[5] T. Trupke, M. A. Green, P. Würfel, Improving solar cell efficiencies by down-conversion of high-energy photons, J. Appl. Phys., 92 (2002) 4117-4122.

[6] W. J. Park, S. J. Oh, J. K. Kim, J. Heo, T. Wagner, L. Strizik, Down-conversion in Tm^{3+}/Yb^{3+} doped glasses for multicrystalline silicon photo-voltaic module efficiency enhancement, J. Non-Cryst. Solids, 383 (2014), 181-183.

[7] L. Dumont, J. Cardin, P. Benzo, M. Carrada, C. Labbé, A. L. Richard, D. C. Ingram, W. M. Jadwisienczak, F. Gourbilleau, $SiN_x:Tb^{3+}-Yb^{3+}$, an efficient down-conversion layer compatible with a silicon solar cell process, Sol. Energy Mater. Sol. Cells, 145 (2016) 84-92.

[8] J. Y. Sun, Y. N. Sun, J. H. Zeng, H. Y. Du, Near-infrared quantum cutting in Eu^{2+}, Yb^{3+} co-doped $Sr_3Gd(PO_4)_3$ phosphor, Opt. Mater., 35 (2013), 1276-1278.

[9] X. J. Zhou, G. C. Wang, K. N. Zhou, Q. X. Li, Near-infrared quantum cutting in Pr^{3+}/Yb^{3+} co-doped transparent tellurate glass via two step energy transfer, Opt. Mater., 35 (2013), 600-603.

[10] Q. Y. Zhang, X. Y. Huang, Recent progress in quantum cutting phosphors, Prog. Mater. Sci., 55 (2010) 353-427.

[11] S. Tanabe, Rare-earth-doped glasses for fiber amplifiers in broadband telecommunication, Comp. Rend. Chem., 5 (2002), 815-824.

[12] C. X. Li, J. Lin, Rare earth fluoride nano-/microcrystals: synthesis, surface modification and application, J. Mater. Chem., 20 (2010) 6831-6847.

[13] J. McKittrick, L. E. Shea, C. F. Bacalski, E. J. Bosze, The influence of processing parameters on luminescent oxides produced by combustion synthesis, Displays 19 (1999) 169–172.

[14] W. X. Bai, Y. H. Liu, Y. P. Wang, X. H. Qiang, L. B. Feng, Cooperative down-conversion and near-infrared luminescence of $Tb^{3+}-Yb^{3+}$ co-doped $CaMoO_4$ broadband phosphor, Ceram. Int., 41 (2015) 12896-12900.

[15] X. J. Zhou, Y. J. Wang, G. C. Wang, and Q. X. Li, Cooperative down-conversion and near-infrared luminescence of Tb^{3+}/Yb^{3+} co-doped tellurite glass, J. Alloy. Compd., 579 (2013) 27-30.

[16] P. Vergeer, T. J. H. Vlugt, M. H. F. Kox, M. I. den Hertog, J. P. J. M. van der Eerden, A. Meijerink, Quantum cutting by cooperative energy transfer in $Yb_xY_{1-x}PO_4:Tb^{3+}$, Phys. Rev. B, 71 (2005) 014119.

CHARACTERIZATION OF A NOVEL PHOTOVOLTAIC BACKSHEET BASED ON POLYAMIDE-IONOMER TECHNOLOGY

Peter Perez[1], Darshana Hari[1], Jayant Kumar[1], Ramaswamy Nagarajan[1],
Andreas Rothacker[2], David Santoleri[2], Christopher Thellen[2]

University of Massachusetts, Lowell[1]
1 University Ave. Lowell, MA 01854 USA

Tomark-Worthen LLC[2]
11 McComb Avenue Glen Mills, PA 19342 USA

ABSTRACT: The use of low-performance, non-durable photovoltaic backsheet reduces the efficiency and safety of solar modules. These backsheets are highly susceptible to deterioration and degradation resulting from their construction. In an effort to improve the efficiency of solar energy, Tomark-Worthen has developed a novel photovoltaic backsheet known as PhotoMark® Reflections™. Produced via co-extrusion, PhotoMark® Reflections™ contains no adhesives, fluoropolymers, or poly(ethylene terephthalate) (PET). Instead, it is composed of a high performance polyamide/ionomer alloy developed to resist degradation. A comprehensive analysis of the performance of PhotoMark® Reflections™ indicates high stability of the backsheet after various accelerated aging tests, including Damp Heat, UV light exposure, and the Highly Accelerated Stress Test (HAST). Fourier-Transform-Infrared spectroscopy (FTIR) performed on PhotoMark® Reflections™ shows high resistance of the skin layers to degradation. Additionally, PhotoMark® Reflections™ and other laminate backsheets were subjected to the HAST. All laminate samples exposed to the HAST failed and delaminated prior to the 100-hour test completion time. Conversely, the PhotoMark® Reflections™ backsheet showed no inter-layer delamination after 100 hours of HAST exposure and retained its mechanical properties. The durability of PhotoMark® Reflections™ after extensive weathering indicates that the implementation of the backsheet will reduce the number of backsheet-induced module failures globally.
Key Words: PV Materials, Backsheet, Stability, Environmental Effect, Characterization

1 INTRODUCTION

Photovoltaic (PV) backsheets protect the interior components of the solar module from environmental ingress and serve as vital packaging materials [1]. They protect the interior components of the module, such as the solar cells, from damage due to environmental conditions [2,3]. The presence of reflected UV light, elevated temperatures, and high relative humidity are the most pertinent factors determining the degradation rate of current PV backsheets. [3 -15]. Premature degradation of the backsheet has proved highly problematic in modules warrantied for 25+ years. [2,4]. Despite making up only a fraction of the solar modules cost, backsheet failure has more of a negative impact on power output and safety than any other factor [5].

The degradation of traditional industry standard PV backsheets is a result of failure of the design as well as the materials utilized in their fabrication. These backsheets are typically composed of 3 polymeric layers: a polyolefin, polyethylene terephthalate (PET), and a fluoropolymer [2, 5-8]. These 3 unique polymer layers are laminated together using solvent based adhesives. The olefinic and polyester layers provide mechanical strength and high moisture barrier. The fluoropolymer layer is targeted at absorbing UV light and preventing degradation [2, 8, 9-12]. Figure 1 shows a schematic diagram of a typical industry standard PV backsheet [16]. The prominent degradation of backsheets with these compositions has spawned numerous research efforts aimed at determining modes of failure [1-15]. The most common modes of backsheet failure are cracking, delamination, and yellowing resulting from extensive exposure to both water vapor and UV light in the environment.

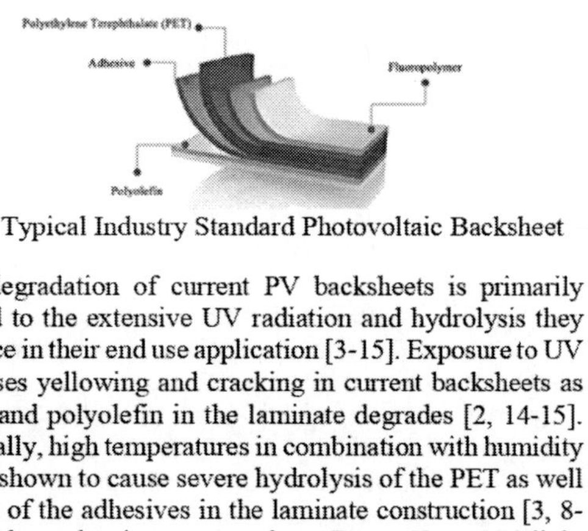

Fig.1 Typical Industry Standard Photovoltaic Backsheet

The degradation of current PV backsheets is primarily attributed to the extensive UV radiation and hydrolysis they experience in their end use application [3-15]. Exposure to UV light causes yellowing and cracking in current backsheets as the PET and polyolefin in the laminate degrades [2, 14-15]. Additionally, high temperatures in combination with humidity has been shown to cause severe hydrolysis of the PET as well as failure of the adhesives in the laminate construction [3, 8-11]. Accelerated aging tests such as Damp Heat, UV light Exposure, and the Highly Accelerated Stress Test (HAST), are often used to determine the durability and relative stability of PV backsheets [3-4, 13, 20-23]. These tests require exposing the backsheet samples to harsh environmental conditions in order to determine their modes and severity of failure in the field, without needing to wait decades for appropriate aging. The inadequate performance of PV backsheets based on traditional materials and adhesives has generated research into novel materials and processing techniques in their production. Tomark-Worthen has developed a PV backsheet novel in both its composition and means of production. Their PV backsheet, PhotoMark® Reflections™, is a 5 layer sheet produced via the coextrusion process. Figure 2 diagrams the coextrusion process [17]. The utilization of coextrusion removes the need for adhesives when bonding the different polymeric layers [4].

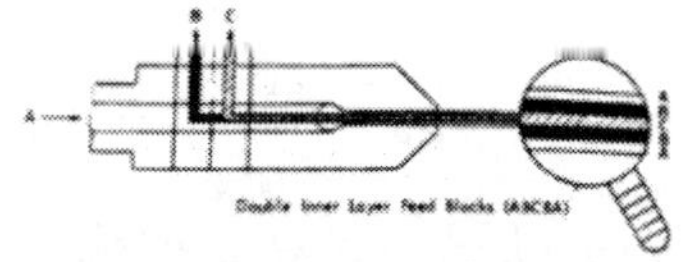

Fig. 2: Schematic Diagram of Coextrusion Process

Delamination of PV backsheets due to failure of the adhesive in environmental conditions is one of the most prominent modes of backsheet failure and leads to the deterioration of module performance [1, 5, 8, 13]. In the sheet coextrusion process, there could be molecular diffusion between the different polymers when they are in their molten state. The different polymers are essentially welded together as the melt exits the coextrusion die. The interlayer bond strength between two different polymers adhered in the coextrusion process is determined by their unique chemistries. A high interlayer bond strength indicates the polymer materials are compatible and able to diffuse into one another well at their interface. Conversely, dissimilar chemistries can cause different polymers to be immiscible and have a weak interlayer bond strength [17-19]. As a result, the interlayer bond strength of coextruded films is unique for each type of film/material structure. The interlayer bond strength of the current industry standard backsheet has been studied and is also reported by the manufacturer to be roughly 5 N/cm [9, 12]. However, the interlayer adhesion strength of Tomark-Worthen's PhotoMark® Reflections™ is much greater [4].

The novel polymer materials used in the construction of the PhotoMark® Reflections™ are targeted at improving the durability of the backsheet. Figure 3 shows a schematic diagram of the PhotoMark® Reflections™ backsheet. It is comprised of 3 different polymeric materials: a Polyamide/ionomer alloy outer layer, functionalized polyolefin core layer, and a stabilized polyamide tie layer [4].

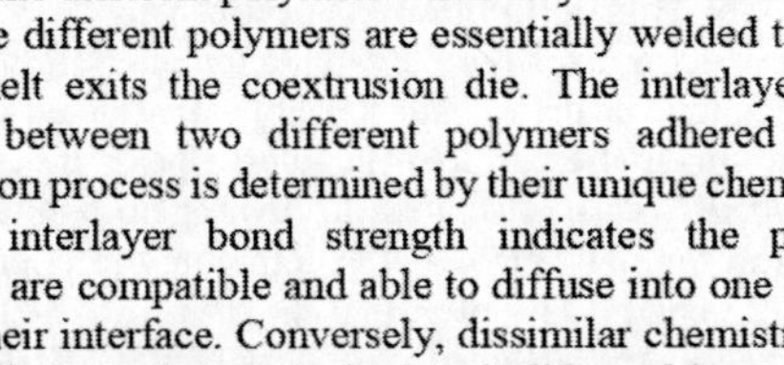

Fig. 3: Tomark-Worthen's PhotoMark® Reflections™ Co-extruded Photovoltaic Backsheet

Polyamides are susceptible to degradation due to hydrolysis and UV light, which lends concerns to their use in solar backsheets [23-28]. However, polyamide based backsheets have previously shown high stability under accelerated aging test conditions [4,13]. The outer polyamide/ionomer alloy is formulated to be durable despite extensive exposure to UV light and hydrolytic conditions. Furthermore, all five layers are stabilized with a comprehensive stabilization package including a Hindered Amine Light Stabilizer (HALS) and an antioxidant.

The goal of this research was to determine the durability and weatherability of Tomark-Worthen's PhotoMark® Reflections™ coextruded backsheet. While there is comprehensive literature on the degradation modes, mechanisms, and severity for current industry standard backsheets, Tomark-Worthen's films are largely uncharacterized. In this research, the chemical, physical, mechanical properties of industry standard and novel backsheet formulations were analyzed before and after accelerated ageing tests common in the photovoltaics industry.

2 EXPERIMENTAL

2.1 Materials and Backsheet Exposure

Samples of PhotoMark® Reflections™ polyamide/ionomer containing films were exposed to 1000 and 2000 hours of Damp Heat (DH). DH exposure requires constant conditions of 85°C and 85% relative humidity [3, 4, 8, 13]. Additionally, samples of PhotoMark® Reflections™ exposed to 2000 and 4000 hours of UV light were provided. The UV light exposure was performed by TUV SUD using: an Irradiance of (0.8 ± 0.08) $W/m^2/nm$ at 340 nm, equivalent to (81 ± 8) W/m^2 measured between 300 nm and 400 nm, using a xenon arc source with filters and complying with the latest version of ASTM D7869 "Standard Practice for Xenon Arc Exposure Test with Enhanced Light and Water Exposure for Transportation Coatings". The chamber air temperature was 65°C and the Black panel temperature was 90°C. The relative humidity in the chamber was 20% RH.

2.2 Fourier-Transform Infrared Spectroscopy

FTIR was used to determine the mechanism and severity of degradation for PhotoMark® Reflections™ samples after subjection to Damp Heat, UV light, and the HAST. FTIR spectroscopy has been routinely used to quantify changes in the material chemistry for backsheets exposed to different weathering conditions [2, 6, 10, 14]. The Thermo Scientific Nicolet 4700 FTIR-ATR was used to analyze the backsheets. Thirty two scans were taken for each sample. Each sample was characterized in the mid-infrared region between $4000cm^{-1}$ and $400m^{-1}$ with a resolution of $4cm^{-1}$.

2.3 Optical Microscopy

Optical microscopy was used to observe surface changes of industry standard backsheets and PhotoMark® Reflections™ after exposure to UV light. The Olympus System Microscope Model CX41 microscope was used with a magnification of 50x, 100x, and 200x. Both the cell and air side of each backsheet sample was analyzed. UV irradiation of the backsheet samples was performed on the Suntest CPS+, conforming to ASTM D6695 "Standard Practice for Xenon-Arc Exposures of Paint and Related Coatings" for 6000 hours.

2.4 Interlayer Adhesion Strength

The interlayer adhesion strength of PV backsheets and its deterioration has been extensively characterized [1, 4, 5, 8-9]. It is crucial to the performance and integrity of the solar module. However, neither the interlayer adhesion strength nor the durability of adhesion between layers have been assessed for PhotoMark® Reflections™. In this research, the interlayer adhesion strength of various multilayer polymer backsheets was evaluated via ASTM D903-98: "Standard Test Method for Peel or Stripping Strength of Adhesive Bonds". Figure 4 displays a specimen under test as per ASTM D9003-98. Ten samples of each multilayer film were prepared by cutting them in 1in x 12in strips (25.3mm x 304.8mm). These samples were then adhered to an Acrylonitrile-Butadiene-Styrene (ABS) "rigid member" with ethyl-cyanoacrylate adhesive. As per ASTM D903-98, the ABS rigid member was 8in x 1in x 1/8in (203.2mm x 25.4mm x 3.175mm). One film was adhered to each side of the rigid ABS member. The samples were then set aside for 24 hours in order to ensure that the adhesive had cured completely. The samples were placed in the tensile grips such that the rigid member was clamped in the stationary top grip and the multilayer polymer was clamped into the moving lower grip. For each sample, the force required to continue delamination plateaus, yielding a constant and consistent force reading for each sample tested.

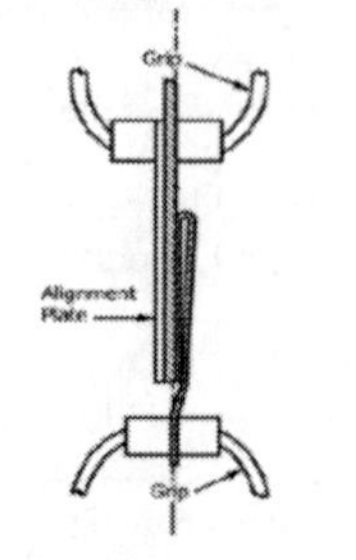

Fig 4: Specimen Under Test as per ASTM D903-98

2.5 Highly Accelerated Stress Test (HAST)

The Highly Accelerated Stress Test (HAST) is an accelerated ageing test which subjects samples to harsh temperature, pressure, and hydrolytic conditions. The HAST is commonly used in both the photovoltaic and electronics industry to determine the stability of individual components [20-23]. The HAST was carried out using the All American 25X Steam Sterilizer. The films were placed in the autoclave and treated for 100 hours with a pressure of 17 psi (1.17 bar) and temperature of 256°F (124.44°C). The autoclave was saturated with steam for the extent of the test.

2.5.1 Mechanical Properties after HAST

Backsheet films exposed to the HAST were tested as per ASTM D882: "Standard Test Method for Tensile Properties of Thin Plastic Sheeting". Tensile properties of backsheet and plastics materials provide insight into the severity of degradation in the polymers [4, 8, 10, 12, 14]. The tensile properties of each film were evaluated after 10-hour intervals of the HAST, up to 100 hours, using the Intron 6025 Universal Tester. All samples were conditioned for 48 hours at 21°C and 50% Relative Humidity prior to testing.

2.5.2 Thermal Properties after HAST

Differential Scanning Calorimetry (DSC) was used to determine the effect of the HAST on the polyamide/ionomer-based PV backsheets. DSC has been used extensively in determining the changes in thermal properties of polymer and backsheet materials [4-5, 8-9]. DSC was performed using the TA Instruments Q100 Differential Scanning Calorimeter.

2.6 Ultra Violet Light and Backsheet Yellowing

Backsheet samples were exposed to UV light and analyzed for color changes. Both the cell and air side of each backsheet sample was analyzed. UV irradiation of the backsheet samples was performed on the Suntest CPS+, conforming to ASTM D6695 "Standard Practice for Xenon-Arc Exposures of Paint and Related Coatings".

3 RESULTS

3.1 Fourier-Transform Infrared Spectroscopy

The spectrum for the polyamide/ionomer alloy exhibited several unique peaks whose intensity was affected subsequent to environmental exposure. The intense peak in the spectrum at 3300cm^{-1} was attributed to hydrogen bonded amines present in the polyamide and is commonly referred to as the amide A band [23, 29-30]. The peaks at 2920cm^{-1} and 2850cm^{-1} were assigned to $-CH_2-$ and $-CH_3$ groups, respectively, in both the polyamide and ionomer [23,32]. Additionally, both the polyamide and ionomer contain carbonyl groups. The Amide 1 peak is comprised mostly of hydrogen bonded C=O vibrations within the polyamide and ionomer; it is observed in the peak at 1635cm^{-1} [10, 13, 23-30]. Prior research on the polyamide/ionomer alloy indicate that these two polymers form co-continuous phases upon melt blending [31, 33-34]. The fingerprint region of the spectra was also of analyzed. The peak at 1450cm^{-1} was assigned to the aromatic vibrations of the polyphenol antioxidant [36-39].

There were no changes in the spectrum observed in samples exposed to 1000 hours of Damp Heat and samples exposed to 2000 hours of UV light. However, samples exposed to both 2000 hours of Damp Heat and the HAST experienced an increase in the intensity of the amide A peak. Figure 5 below shows the increased intensity of amide A peak in the films exposed to 2000 hours of Damp Heat. It has been postulated that an increase in the intensity of this peak for polyamides indicates chain scission at the amide bond under hydrolytic conditions. The presence of water causes hydrolysis of the amide bond, creating more amine end groups in the amorphous phase [26-31]. The Damp Heat testing

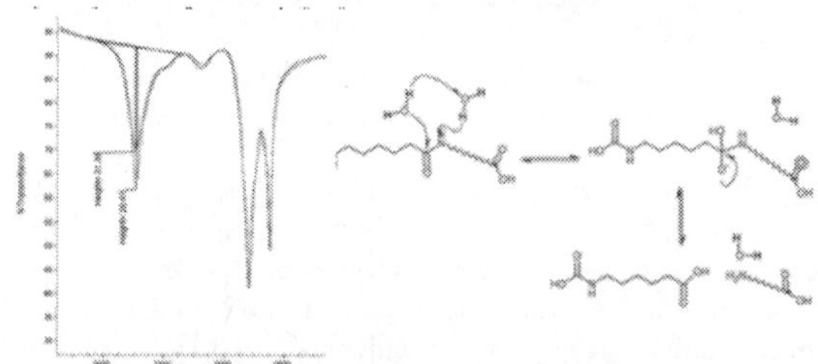

Fig.5: Change in Amide A intensity follow the HAST and Mechanism of Polyamide Hydrolysis

temperature of 85°C and HAST temperature of 125°C, gives these chain ends enough mobility to hydrogen bond. This explains the increased intensity of the amide A peak after Damp Heat Testing, as shown in figure 5 along with the mechanism of hydrolysis [35]. The hydrolysis and polycondensation of polyamides is also a competing reaction, even under harsh hydrolytic conditions [30, 32]. This can lead to an equilibrium in which the chain scission and polycondensation of the polymer negate each other. Mechanical testing of PhotoMark® Reflections™ films after both Damp Heat and the HAST indicate a retention of properties superior to that of current industry standard backsheets [4].

All backsheet samples exposed to 4000 hours of UV light exhibited a significant shift in peak intensity at 1450cm^{-1}. This peak was attributed to aromatic vibrations of polyphenols [36-39]. A shift in this peak after 4000 Hours of UV light exposure is consistent with the polyphenols being oxidized and reacting, potentially with the HALS present [36]. The UV package present in the backsheet is protecting the polymers from UV degradation. Figures 6a and 6b show the apparent shift in the spectrum and mechanism of oxidation, respectively. Photo-oxidation of the polymers in PhotoMark® Reflections™ is not apparent in the spectra. During the photo-oxidation of polyamides, the ($NH\text{-}CH_2$) bond is broken. This leads to the formation of chain ends in the amorphous phase. These chain ends have increased mobility and are able to chemi-crystalize, especially because of the elevated temperature present during UV light exposure. The chemi-crystallization would again manifest itself as an increased intensity of the amide A peak. . This is not observed in the spectra indicating high resistance of PhotoMark® Reflections™ to photo-oxidation.

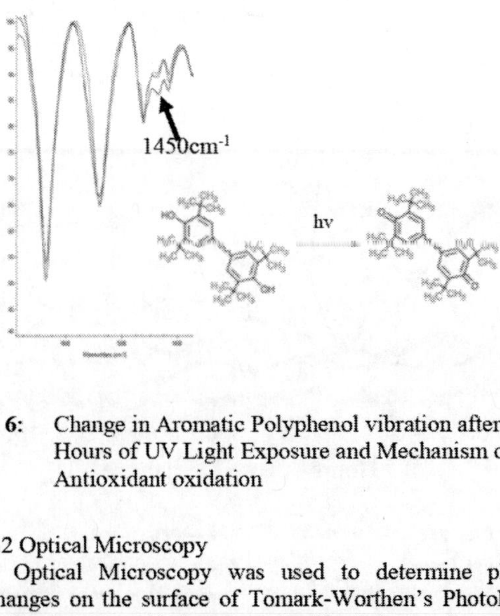

Fig. 6: Change in Aromatic Polyphenol vibration after 4000 Hours of UV Light Exposure and Mechanism of Antioxidant oxidation

3.2 Optical Microscopy

Optical Microscopy was used to determine physical changes on the surface of Tomark-Worthen's PhotoMark® Reflections™ and industry standard backsheets after exposure to 6000 hours of UV radiation. Figure 7 shows the images of the polyamide/ionomer skin layers and industry standard cell and air side layers after exposure. No changes on the surface of the film were present for PhotoMark® Reflections™, indicating high resistance to UV light degradation. Severe cracking was apparent on the cell side of the Industry standard films. The cell side of industry standard backsheets is composed of a polyolefin which is highly sensitive to UV light degradation. This layer experiences deep, widespread cracking. The air side of the TPE and KPE backsheets is composed of a UV light resistant fluoropolymer. The resistance of these polymers to UV degradation is apparent as neither exhibited cracking. Commonly, Delamination of the backsheet leads to exposure of the cell side polymeric layer to UV light, leading to backsheet failure and reduced module performance.

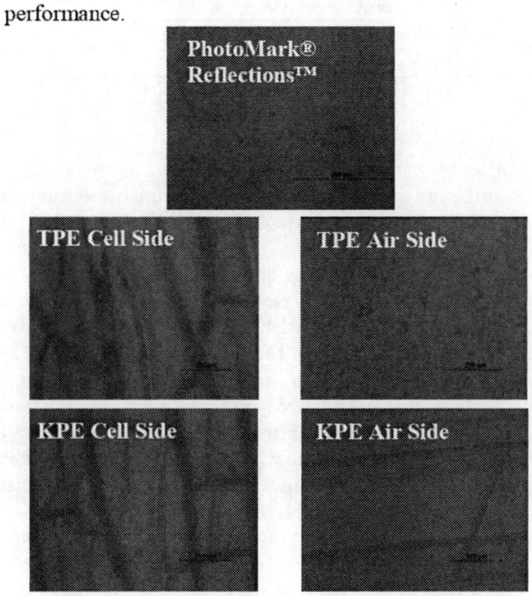

Fig. 7: PV Backsheet Surfaces after UV Light Exposure

Polyamides are sensitive to photo-oxidation and degradation. Chain scission, bond breaking, and photo-oxidation caused by UV light exposure occurs in the amorphous phase of the polymer as oxygen is able to more easily permeate this this region. The ($NH\text{-}CH_2$) bond can be broken in the presence of UV light and oxygen leading to chemi-crystallization, whereby highly mobile broken chain ends in the amorphous phase hydrogen bond and enter into the crystalline domain of the polymer matrix [40-42]. However, the lack of cracking on the surface of PhotoMark® Reflections™ and the FTIR analysis show that the antioxidant present in the skin layers is protecting the polymer and entire backsheet from photo-degradation and cracking. [43].

3.3 Interlayer Adhesion Strength

Previous work by Thellen et. al. showed the superior interlayer bond strength of PhotoMark® Reflections™ resulting from adhesion via co-extrusion. ASTM D903-98: "Standard Test Method for Peel or Stripping Strength of Adhesive Bonds", was determined to be the most appropriate method to testing the interlayer adhesion strength of all backsheet films. Table I shows the average peel strength for the industry standard as compared to PhotoMark® Reflections™. Backsheet delamination resulting from the weak strength of the laminate structure is one of the major modes of module degradation [1, 4, 5, 8, 9].

The utilization of the coextrusion increases layer adhesion strength while reducing the time, cost and number of steps associated with the manufacturing of the laminate films. Furthermore, PhotoMark® Reflections™ backsheet samples exposed to 100 hours of the Highly Accelerated Stress Test

PV Backsheet	Delamination Force (N/cm)	Std. Dev (N/cm)
PhotoMark® Black	*28.36*	*2.24*
PhotoMark® White	*28.25*	*1.62*
PhotoMark® HAST 100	*20.72*	*2.54*
TPE Control	*4.22*	*0.45*
FPE Control	*0.746*	*0.15*
PPE Control	*2.74*	*0.40*

Table I: Peel Strength of PV Backsheets

still maintained a superior interlayer adhesion strength while all laminate backsheets debonded. The improved layer adhesion shows a greater resistance to delamination during the weathering endured by solar modules.

3.4 Highly Accelerated Stress Test (HAST)

Hydrolytic conditions and elevated temperatures have been shown to accelerate the degradation of polymer materials used in PV backsheets [1-5, 8-10, 13, 19-23]. The HAST exposes the test specimen to: high temperature (124°C), high pressure (1.2 Bar), and harsh hydrolytic conditions (full steam saturation) [20-23]. Several films were exposed to the HAST to determine the extent and mechanism of degradation due to these conditions. Figure 8 shows the PV Backsheet films after exposure to 100 hours of the HAST.

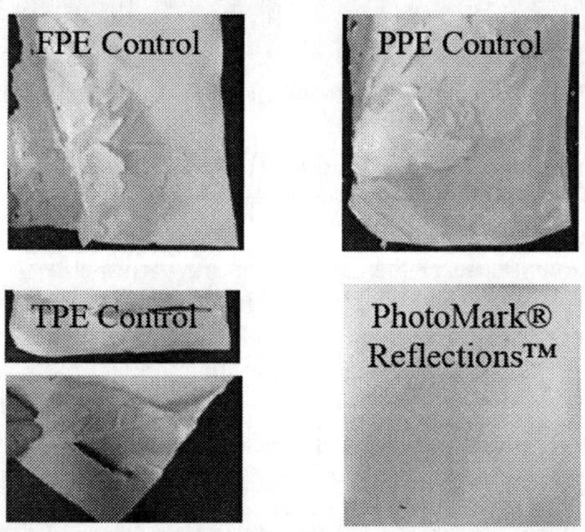

Fig. 8: PV Backsheets After 100 Hours of the HAST

There was significant delamination and degradation of the laminated PV backsheets. Both the adhesives and the polymer materials that made up these backsheets failed. The layers of these films were completely separated. Additionally, there was significant yellowing and hydrolysis of the polyester present in these films. Conversely, PhotoMark® Reflections™ experienced no delamination, cracking, or major yellowing. The TPE film completely decomposed after 20 hours of the HAST test, preventing samples after 20 hours from being collected for mechanical testing.

3.4.1 Tensile Properties after HAST

Backsheet samples were exposed to the HAST in 10-hour intervals in order to determine their performance trends. Figure 9 shows the trends in the elongation for the laminated and co-extruded backsheets. The laminate backsheets all experienced failure between 60 and 70 hours of HAST exposure. Severe delamination led the rapid deterioration in the yield strength and elongation of the backsheet samples. The significant hydrolysis of the polyester in the films caused this layer to break early in testing and cause total failure [3, 8, 10-13]. This early breakage greatly decreased the strength of the entire backsheet as this layer is the main source of strength in current PV backsheets.

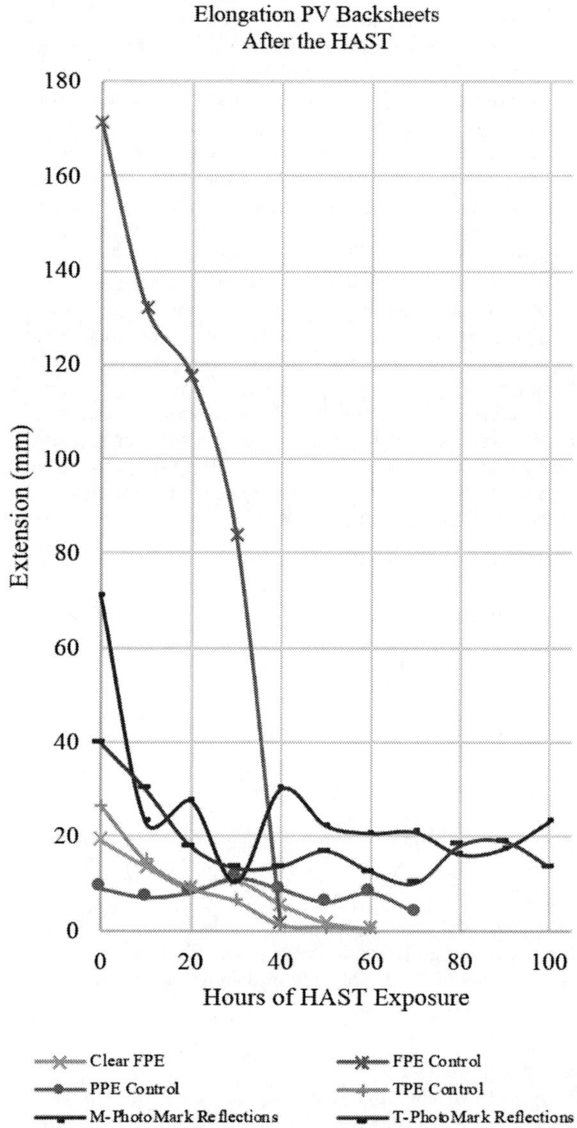

Fig 9: Elongation of PV Backsheets After the HAST

The co-extruded backsheets exhibited the opposite trends in performance as a result of HAST. The moduli and maximum stress of these films increased after exposure. Table II shows the moduli values for PhotoMark® Reflections™ after exposure to the HAST. Additionally, the ultimate elongation of PhotoMark® Reflections™ decreased as well after exposure. This resulted from crystallization of the polyamide and polyolefin present in PhotoMark® Reflections™.

Differential Scanning Calorimetry was used to quantify the degree of the crystallization subsequent to HAST exposure.

Hours of HAST Exposure	Young's Modulus (MPa) of PhotoMark® Reflections™ after the HAST	
	Machine	Transverse
0	*1356*	*1292*
10	*1459*	*1297*
20	*1679*	*1259*
30	*1684*	*1309*
40	*1661*	*1374*
50	*1748*	*1367*
60	*1583*	*1431*
70	*1420*	*1399*
80	*1540*	*1291*
90	*1478*	*1416*
100	*1499*	*1258*

Table II: Young's Modulus of PhotoMark® Reflections™ after the HAST

3.4.2 Thermal Properties after HAST

Differential Scanning Calorimetry (DSC) was used to determine the effect of the HAST on Tomark-Worthen's PhotoMark® Reflections™ Backsheet. Figure 10 shows the first heat for a control and backsheet sample exposed to 100 hours of the HAST.

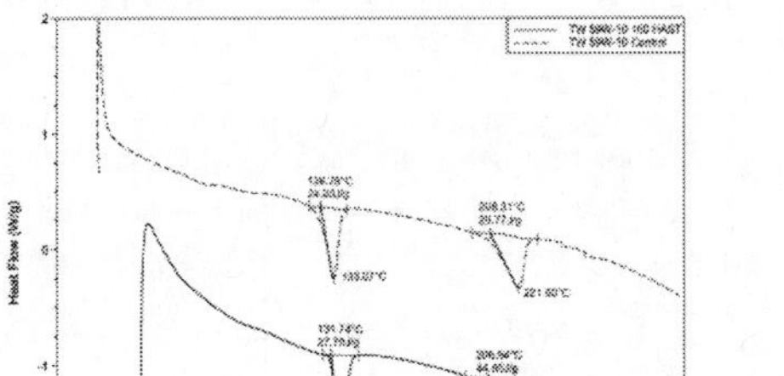

Fig 10: DSC First Heat of PhotoMark® Reflections™ control (red) and after 100 hours of HAST exposure (blue)

The endothermic peaks centered at 133°C and 137°C indicate the melting transition of the polyolefin present in the films. After the HAST, this endothermic peak increased in area and depth while its temperature shifted upward. The increased enthalpy of melting and melting temperature is indicative of the formation of larger crystals in the polyolefin due to annealing. The film was exposed to the HAST for 100 hours, allowing time for significant crystal growth in the polyolefin layer [44-45]. This increase in crystallinity of the polymer material also explains the increased modulus and max stress observed for PhotoMark® Reflections™ films exposed to the HAST.

The second endothermic peak indicates the melting transition of the polyamide present in the polymer film. The enthalpy of melting increases significantly for the melting transition experienced after the HAST. This endotherm broadens and experiences a lower onset melting temperature. This lower onset melting temperature is caused by crystallization of the polyamide during prolonged exposure to high temperature conditions. The elevated temperature of the HAST causes the formation of γ-crystals in the amorphous phase of the polymer [46-47]. These crystals form during extended exposure to temperatures below 130°C, as was present in HAST. The increase in crystallinity in the polyamide layers, resulting from the extended exposure to high temperature, also accounts for the increase in moduli observed for the co-extruded sheet, as shown in table 2.

3.5 Ultra Violet light and Backsheet Yellowing

Backsheet samples were exposed to UV light and analyzed for color changes. Both sides of each backsheet sample was subjected to UV light exposure. PhotoMark® Reflections™ contains the same outer layer on both sides of the film. Conversely, traditional laminate backsheets have a fluoropolymer air-side layer and olefinic cell-side layer. The fluoropolymer layer is highly resistant yellowing and degradation from UV light as is shown in figure 11 However, the olefinic side of these backsheets yellowed significantly.

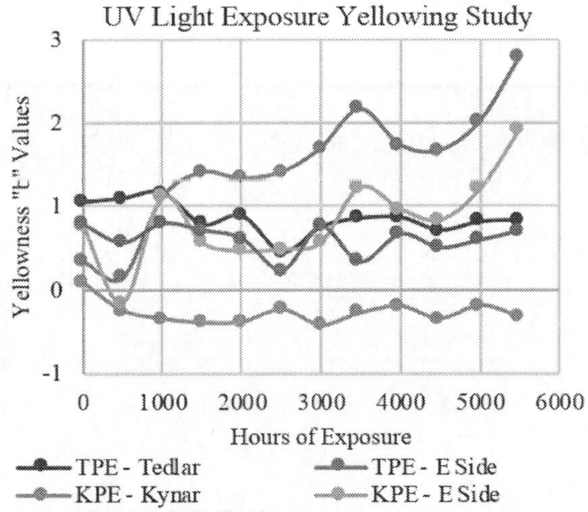

Fig 11: Degree of Yellowing for PV Backsheets exposed to UV light

4 Conclusions

With the implementation of solar energy increasing each year, it is vital to improve the longevity and effectiveness of solar modules. Currently, traditional industry standard Photovoltaic backsheets are prone to failure by means of cracking, delamination, and yellowing. This research has shown the superior performance of Tomark-Worthen's PhotoMark® Reflections™ backsheet as compared to the current industry standard. Tomark-Worthen's PhotoMark® Reflections™ backsheet experienced minimal degradation after extensive exposure to Damp Heat, UV Radiation, and the Highly Accelerated Stress Test. Additionally, the interlayer bond strength achieved via the co-extrusion process was 6.5 times greater than the interlayer bond strength of current industry standard laminate films. The implementation of PhotoMark® Reflections™ will reduce the number of backsheet related solar module failures and improve the efficacy of solar energy globally.

5 Acknowledgements

The authors would like to thank Dr. Jo Ann Ratto of the U.S. Army Natick Soldier Research, Development, and Engineering Center for her continued support and expertise throughout the entirety of the research. The authors would also like to thank the H.E.R.O.E.S initiative, "Soldier Lightweight Integrated Multifunctional Materials" - Extended Life Photovoltaic Backsheet and the U.S. Army Natick for contract number W911QY-17-2-0004 which made this research possible.

6 References:

1. [i]Dubey, Rajiv, et al. "Comprehensive Study of Performance Degradation of Field-Mounted Photovoltaic Modules in India." *Freshwater Biology*, Wiley/Blackwell, 1 Feb. 2017,
2. Fairbrother, Andrew, et al. "Differential Degradation Patterns of Photovoltaic Backsheets at the Array Level." *Solar Energy*, vol. 163, 2018, pp. 62–69., doi:10.1016/j.solener.2018.01.072.
3. Illya, G., et al. "Mechanical Studies of Solar Photovoltaics (PV) Backsheets Under Salt Damp Heat Environments." *Procedia Engineering*, vol. 215, 2017, pp. 238–245., doi:10.1016/j.proeng.2017.12.144.
4. Thellen, Christopher, et al. "European PV Solar Energy Conference and Exhibition." EU PVSEC Conference Proceedings, *CO-EXTRUSION OF A NOVEL MULTILAYER PHOTOVOLTAIC BACKSHEET BASED ON POLYAMIDE-IONOMER ALLOY SKIN LAYERS*, 2017, pp. 68–72.
5. Jordan, Dirk C., et al. "Photovoltaic Failure and Degradation Modes." *Progress in Photovoltaics: Research and Applications*, vol. 25, no. 4, 2017, pp. 318–326., doi:10.1002/pip.2866.
6. Liu, Fei, et al. "Ultra-Violet Degradation Behavior of Polymeric Backsheets for Photovoltaic Modules." *Solar Energy*, vol. 108, 2014, pp. 88–100., doi:10.1016/j.solener.2014.06.027.
7. Klinke, Addison G., et al. "A Non-Destructive Method for Crack Quantification in Photovoltaic Backsheets under Accelerated and Real-World Exposures." *Polymer Degradation and Stability*, vol. 153, 2018, pp. 244–254., doi:10.1016/j.polymdegradstab.2018.05.008.
8. Knausz, Marlene, et al. "Degradation of Photovoltaic Backsheets: Comparison of the Aging Induced Changes on Module and Component Level." *Journal of Applied Polymer Science*, vol. 132, no. 24, 2015,
9. Novoa, Fernando D., et al. "Environmental Mechanisms of Debonding in Photovoltaic Backsheets." *Solar Energy Materials and Solar Cells*, vol. 120, 2014, pp. 87–93., doi:10.1016/j.solmat.2013.08.020.
10. Voronko, Yuliya, et al. "Correlation of the Loss in Photovoltaic Module Performance with the Ageing Behaviour of the Backsheets Used." *Progress in Photovoltaics: Research and Applications*, vol. 23, no. 11, 2015, pp. 1501–1515., doi:10.1002/pip.2580.
11. M. A. Quintana, D. L. King, T. J. McMahon and C. R. Osterwald, "Commonly observed degradation in field-aged photovoltaic modules," *Conference Record of the Twenty-Ninth IEEE Photovoltaic Specialists Conference, 2002.*, 2002, pp. 1436-1439. doi: 10.1109/PVSC.2002.1190879
12. Li, Hua, et al. "Nondestructive Estimation of Strength Deterioration in Photovoltaic Backsheets Using a Portable near Infrared Spectrometer." *Solar Energy Materials and Solar Cells*, vol. 101, 2012, pp. 166–169., doi:10.1016/j.solmat.2012.01.017.
13. Geretschläger, Klaus J., et al. "Damp Heat Aging Behavior of a Polyamide-Based Backsheet for Photovoltaic Modules." *Journal of Solar Energy Engineering*, vol. 138, no. 4, 2016, p. 041003., doi:10.1115/1.4032977.
14. Lin, Chiao-Chi, et al. "A Novel Test Method for Quantifying Cracking Propensity of Photovoltaic Backsheets after Ultraviolet Exposure." *Progress in Photovoltaics: Research and Applications*, 2018, doi:10.1002/pip.3038.
15. Lu, T., et al. "UV Degradation Model forPolymers and Polymer Matrix Composites." *Polymer Degradation and Stability*, vol. 154, 2018, pp. 203–210., doi:10.1016/j.polymdegradstab.2018.06.004.
16. ENF Ltd. "ENF Ltd." *ENF Solar*, www.enfsolar.com/pv/backsheet.
17. Griff, Allan L. *Plastics Extrusion Technology*. Reinhold, 1962. p. 273.
18. "Extrusion." *Plastics Processing: Technology and Health Effects*, Noyes Data Corp., 1986, p. 137.
19. Schwartz, Seymour S., and Sidney H. Goodman. *Plastics Materials and Processes*. Van Nostrand Reinhold, 1982. p. 597-598.
20. Pei, Chun, et al. "Failure Analysis of Plastic Encapsulated Circuits after Autoclaving." *Engineering Failure Analysis*, vol. 66, 2016, pp. 421–431., doi:10.1016/j.engfailanal.2016.04.030.
21. Mehr, M. Yazdan, et al. "A Review on Discoloration and High Accelerated Testing of Optical Materials in LED Based Products." *Microelectronics Reliability*, vol. 81, 2018, pp. 136–142., doi:10.1016/j.microrel.2017.12.023.
22. Hagihara, Hideaki, et al. "Degradation Mechanism of Carbon Fiber-Reinforced Thermoplastics Exposed to Hot Steam Studied by Chemical and Structural Analyses of Nylon 6 Matrix." *Composites Part A: Applied Science and Manufacturing*, vol. 112, 2018, pp. 126–133., doi:10.1016/j.compositesa.2018.05.034.
23. Ogunsona, Emmanuel O., et al. "Accelerated Hydrothermal Aging of Biocarbon Reinforced Nylon Biocomposites." *Polymer Degradation and Stability*, vol. 139, 2017, pp. 76–88., doi:10.1016/j.polymdegradstab.2017.03.013.
24. Okamba-Diogo, Octavie, et al. "Molecular and Macromolecular Structure Changes in Polyamide 11 during Thermal Oxidation." *Polymer Degradation and Stability*, vol. 108, 2014, pp. 123–132., doi:10.1016/j.polymdegradstab.2014.05.028.
25. Forsström, Dan, and Björn Terselius. "Thermo Oxidative Stability of Polyamide 6 Films I. Mechanical and Chemical Characterisation." *Polymer Degradation and Stability*, vol. 67, no. 1, 2000, pp. 69–78., doi:10.1016/s0141-3910(99)00122-6.
26. Dong, Weifu, and Pieter Gijsman. "Influence of Temperature on the Thermo-Oxidative Degradation of Polyamide 6 Films." *Polymer Degradation and Stability*, vol. 95, no. 6, 2010, pp. 1054–1062., doi:10.1016/j.polymdegradstab.2010.02.030.
27. Singhal, Anu, and Leslie J. Fina. "Dynamic Two-Dimensional Infrared Spectroscopy. Part II: Morphological Assignments in Melt-Crystallized Nylon 11 Spectra." *Vibrational Spectroscopy*, vol. 13, no. 1, 1996, pp. 75–82., doi:10.1016/0924-2031(96)00031-8.

28. Jacques, B, et al. "Hydrolytic Ageing of Polyamide 11. 1. Hydrolysis Kinetics in Water." *Polymer*, vol. 43, no. 24, 2002, pp. 6439–6447., doi:10.1016/s0032-3861(02)00583-9.
29. "Spectroscopy and the Characterization of Chain Structure." *Fundamentals of Polymer Science: an Introductory Text*, by Paul C. Painter and Michael M Coleman, CRC Press, 1998, pp. 165–166.
30. Porubská, Mária, et al. "FTIR Spectroscopy Study of Polyamide-6 Irradiated by Electron and Proton Beams." *Polymer Degradation and Stability*, vol. 97, no. 4, 2012, pp. 523–531., doi:10.1016/j.polymdegradstab.2012.01.017
31. Krämer, R.h., et al. "Degradation of Poly(Ethylene-Co-Methacrylic Acid)–Calcium Carbonate Nanocomposites." *Polymer Degradation and Stability*, vol. 92, no. 10, 2007, pp. 1795–1802., doi:10.1016/j.polymdegradstab.2007.07.006.
32. Zhou, Qi, et al. "Substrate Effects on the Surface Properties of Nylon 6." *Applied Surface Science*, vol. 282, 2013, pp. 115–120., doi:10.1016/j.apsusc.2013.05.075.
33. Bose, Suryasarathi, et al. "Specific Interactions and Reactive Coupling Induced Dispersion of Multiwall Carbon Nanotubes in Co Continuous Polyamide6/ionomer Blends." *Macromolecular Symposia*, vol. 263, no. 1, 2008, pp. 11–20., doi:10.1002/masy.200850302.
34. Jaklewicz, Magdalena, et al. "Friction and Scratch Resistance of Polyamide 6 Modified with Ionomeric Ethylene/Methacrylic Acid Copolymer." *Journal of Applied Polymer Science*, vol. 91, no. 6, 2004, pp. 3866–3870., doi:10.1002/app.13587
35. Hocker, Samuel J.a., et al. "Graphene Oxide Reduces the Hydrolytic Degradation in Polyamide-11." *Polymer*, vol. 126, 2017, pp. 248–258., doi:10.1016/j.polymer.2017.08.034.
36. Müller, M., et al. "In-Situ ATR-FTIR for Characterization of Thin Biorelated Polymer Films." *Thin Solid Films*, vol. 556, 2014, pp. 1–8., doi:10.1016/j.tsf.2013.12.025.
37. Beißmann, Susanne, et al. "Analytical Evaluation of the Performance of Stabilization Systems for Polyolefinic Materials. Part I: Interactions between Hindered Amine Light Stabilizers and Phenolic Antioxidants." *Polymer Degradation and Stability*, vol. 110, 2014, pp. 498–508., doi:10.1016/j.polymdegradstab.2014.09.020.
38. Lee, Michael R. F. "Forage Polyphenol Oxidase and Ruminant Livestock Nutrition." *Frontiers in Plant Science*, vol. 5, 2014, doi:10.3389/fpls.2014.00694.
39. Pantoja-Castro, Mayra a., and HoraCio González-rodriGuez. "STUDY BY INFRARED SPECTROSCOPY AND THERMOGRAVIMETRIC ANALYSIS OF TANNINS AND TANNIC ACID." *Revista Latinoamericana De Química*, vol. 39, no. 3, Feb. 2012, pp. 108–112.
40. Thanki, P.n, et al. "Photo-Irradiation Induced Morphological Changes in Nylon 66." *Polymer*, vol. 42, no. 2, 2001, pp. 535–538., doi:10.1016/s0032-3861(00)00374-8.
41. Zhang, Jie, and Alina Adams. "Understanding Thermal Aging of Non-Stabilized and Stabilized Polyamide 12 Using 1H Solid-State NMR." *Polymer Degradation and Stability*, vol. 134, 2016, pp. 169–178., doi:10.1016/j.polymdegradstab.2016.10.006.
42. Cerruti, Pierfrancesco, et al. "Comparison of Photo-Oxidative Degradation of Polyamide 6,6 Films Stabilized with HALS and CuCl2+KI Mixtures." *Polymer*, vol. 46, no. 13, 2005, pp. 4571–4583., doi:10.1016/j.polymer.2005.03.065.
43. Awaja, Firas, et al. "Cracks, Microcracks and Fracture in Polymer Structures: Formation, Detection, Autonomic Repair." *Progress in Materials Science*, vol. 83, 2016, pp. 536–573., doi:10.1016/j.pmatsci.2016.07.007.
44. Moyses, Silvio Calafati, and Julio Zukermann-Schpector. "Annealing in Low Density Polyethylene at Several Temperatures." *Polymer Journal*, vol. 36, no. 9, 2004, pp. 679–683., doi:10.1295/polymj.36.679.
45. Wang, Long, et al. "Effects of Annealing on the Hierarchical Crystalline Structures and Mechanical Properties of Injection-Molded Bars of High-Density Polyethylene." *Polymer International*, vol. 63, no. 2, 2013, pp. 296–306., doi:10.1002/pi.4505.
46. *Fornes, T.d., and D.r. Paul. "Crystallization Behavior of Nylon 6 Nanocomposites." Polymer, vol. 44, no. 14, 2003, pp. 3945–3961., doi:10.1016/s0032-3861(03)00344-6.*
47. Zhao, Xiong-Yan, and Ming-Zhu Wang. "Structure and Thermal Behaviour of Nylon 6 Nanocomposites." *Journal of Applied Polymer Science*, vol. 100, no. 4, 2006, pp. 3116–3122., doi:10.1002/app.22530. doi:10.1002/app.42093.

GROWTH AND CHARACTERIZATION OF $(ZnSe)_{0.1}(SnSe)_{0.9}$ FILMS FOR USE IN THIN FILM SOLAR CELLS

Razykov T.M.[1], Bosio A.[2], Ergashev B.[1], Kouchkarov K.M.[1] Mavlonov A.A.[1],
Romeo A.[3], Artegiani E.[3], Romeo N.[2], Yuldoshov R.[1]

[1]Physical-Technical Institute, Uzbekistan Academy of Science, Bodomzor Yoli 2B, Tashkent 100084, Uzbekistan
Phone: +998-71-235-4103, Fax: +998-71-235-4291, E-mail: razykov@uzsci.net
[2]University of Parma, G.P Usberti 7/A - 43124 Parma, Italy
[3]Universita' di Verona, Ca' Vignal 2- Strada Le Grazie 15, 37134 Verona, Italy

ABSTRACT: $(ZnSe)_x(SnSe)_{1-x}$ films were fabricated from ZnSe and SnSe precursors using chemical-molecular beam deposition (CMBD) method at a substrate temperature of 500°C. The structural and morphological properties of $(ZnSe)_{0.1}(SnSe)_{0.9}$ films have been studied. The data from the scanning electron microscope showed that the grain sizes of the films were 5-6 μm and the films had a close-packed polycrystalline structure. The results of X-ray diffraction analysis of the samples revealed that the films have an orthorhombic structure. Structural parameters of the obtained films are given.

Keywords: $(ZnSe)_x(SnSe)_{1-x}$, CMBD based deposition, thin film, absorber layer

INTRODUCTION

At present, CdTe and $Cu(In,Ga)Se_2$ based absorber layers are one of the most suitable materials for thin-film solar cells due to their basic properties. These elements have an optimal bandgap at room temperature E_g = 1.5 eV for CdTe and 0.9-1.7 eV for $Cu(In,Ga)Se_2$, which cover the entire spectrum of the infrared and visible region of the solar radiation and have high absorption coefficient (~ 10^4 -10^5 cm^{-1}), allowing the effective absorption for incident solar radiation by a thickness of several microns. This significantly reduces the material consumption compared to silicon based solar cells.

The recent advances of the thin-film solar cells, i.e. obtained by various methods and scientific centers, are presented in Table 1. As shown, the efficiency of the thin film solar cells increases regularly. The leading companies in this field are First Solar and ZSW, where the highest efficiency values have been reported for CdTe and $Cu(In,Ga)Se_2$ based solar cells, being 22.1% [1] and 22.3% [2] in laboratory conditions and 16.1% [3] and 17% [4]in the modules, respectively.

Despite the high efficiency of solar cells based on these materials, their further application on a large scale is limited due to the limited In, Ga, Te in the earth's crust, as well as the high price of Ga.

Therefore, many research centers and laboratories conduct research to replace the abovementioned expensive materials with Earth abundant elements based compounds such as Cu_2ZnSnS_4 (CZTS). This absorber layer is not only cost effective and non-toxic, but also has similar optical properties to that of $Cu(In,Ga)Se_2$ absorber layer.

As depicted in Table 1, $Cu_2ZnSnS_xSe_{4-x}$ based solar cells has achieved the efficiency of 12.6% [5] reported by the IBM research center. However, this efficiency is still lower than that of $Cu(In,Ga)Se_2$. According to Ref. [6], low efficiency of CZTS is due to the complexity of sample preparation methods and controlling the film composition.

To date, novel absorber layers such as SnSe, SbSe, $Cu_2(Sn_{1-x}Ga_x)Se_3$ (CTGS), and Cu_2SnS_3 have been studied and developed, which are made up low cost and non-toxic elements. These materials are attractive due to their similar optical properties to $Cu(In,Ga)Se_2$, e.g. their bandgap lies in the range 0.87-1.7 eV [7.8]. The initial attempts to obtain thin film solar cells on their basis yielded an efficiency of 4-6% [8]. Currently, researchers are working to further increase the efficiency of such solar cells.

Despite the fact that in the world literature there is no information on zinc and tin selenide ($(ZnSe)_x(SnSe)_{1-x}$) thin films, we believe that this material is ideally suited for use in photovoltaics. It is consisted of Earth abundant and environmentally friendly elements and has the following excellent properties:

- The bandgap can vary over a wide range of (0.9 ÷ 2.7) eV, which is suitable for the absorber layer with an optimal bandgap of 1.45 eV.
- The absorption coefficient is in the range of (10^4 ÷ 10^5) cm^{-1}. This means that an absorber layer with a thickness of (0.5÷1.0) μm is sufficient for manufacturing highly efficient solar cells.

In previous study, we investigated the physical properties of SnSe thin films [11]. In this study, we report on structural and morphological properties of $(ZnSe)_x(SnSe)_{1-x}$ thin films with x=0.1.

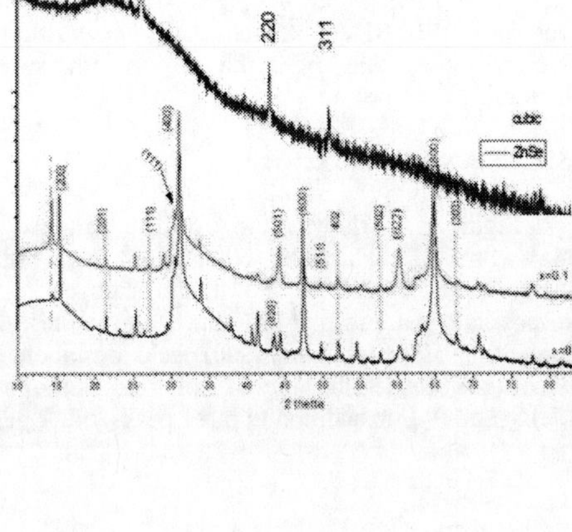

Figure 1. X-ray data of ZnSe and $(ZnSe)_x(SnSe)_{1-x}$ films for indicated x values, which were grown at a substrate temperature of 500 °C.

Table 1. Recent advances on CdS/CdTe, $Cu(In,Ga)Se_2$ and $Cu_2ZnSnS_xSe_{4-x}$ based thin film solar cells' efficiency obtained by various methods and companies [9,10]

Year	Structure	CdS and CdTe/ process	Activation process	V_{OC} (mV)	I_{SC} (mA/cm^2)	FF (%)	η (%)	Size (cm^2)	Research group
1993	glass/SnO_2:F/CdS/ CdTe	CBD/ CSS	$CdCl_2$ heat treatment	843	25.1	74.5	15.8	1.05	USF
2001	glass/CTO/ ZnO/CdS/CdTe	CBD/ CSS	$CdCl_2$ heat treatment	845	25.9	75.5	16.5	1.03	NREL
2013	glass/CTO/ ZnO/CdS/CdTe	CBD/ CSS	$CdCl_2$ heat treatment	875	28.9	78.0	19.7	1	First Solar
2014	glass/CTO/ ZnO/CdS/CdTe	CBD/ CSS	$CdCl_2$ heat treatment	872	29.5	79.5	20.4	1	First Solar
2014	glass/CTO/ ZnO/CdS/CdTe	CBD/ CSS	$CdCl_2$ heat treatment	903	30	79.8	22.1	1	First Solar
2013	glass/Mo/CIGSe /CdS/ ZnO/ZnO:Al			757	34.8	79.1	20.8	0,5	ZSW
2014	glass/Mo/CIGSe /CdS/ ZnO/ZnO:Al			746	36.6	79.3	22.3	0.5	ZSW
2008	glass/Mo/CZTS /CdS/ ZnO/ITO	Thermal evaporation		661	19.5	65.8	8.4	0.46	IBM
2012	glass/Mo/CZTSS /CdS/ZnO/ITO	Growth in solution		460	34.5	69.8	11.1	0.44	IBM
2013	glass/Mo/ CZTSS /CdS/ZnO/ITO	Growth in solution		513 .4	35.2	69.8	12.6	0.44	IBM

PM - printing method; CSS – closed-space sublimation; CBD - chemical bath deposition; USF - University of South Florida; NREL - National Renewable Energy Laboratory; ZSW – Zentrum für Sonnenenergie und Wasserstoff-Forschung (Center for Solar and Hydrogen Energy Research); IBM - International Business Machines. CTO – Cd_2SnO_4; CIGS – $Cu(In,Ga)Se_2$: CZTS – Cu_2ZnSnS_4; CZTSS – $Cu_2ZnSnS_{4-x}Se_x$

EXPERIMENTAL

The preparation of polycrystalline $(ZnSe)_x(SnSe)_{1-x}$ thin films was carried out according to the procedure described in [12]. The temperature of precursor is in the range (850-950) °C, whereas the substrate temperature is varied in the range of (500-600) °C. The carrier gas, i.e. hydrogen, flow was ~ 20 cm^3/min. The deposition time depends on the required thickness of the films and ranges from (30-60) minutes. Borosilicate glasses were used as substrates.

The crystal structure and the phase composition of the materials were studied by X-ray diffraction ("Panalytical Empyrean" diffractometer with CuKα radiation ,λ = 1. 5418 ˚A) with 2θ in the range of (20-80) ° and a step of 0. 010. Analysis of the phase composition was performed using the Joint Committee on Powder Diffraction Standard (JCPDS) database. Morphological studies were carried out with the scanning electron microscope (SEM-EVO MA 10). The resulting $(ZnSe)_{0.1}(SnSe)_{0.9}$ thin films had a smooth surface without cracks and pores.

RESULTS AND DISCUSSIONS

Figure 1 shows the XRD pattern for $(ZnSe)_{0.1}(SnSe)_{0.9}$ thin films obtained at a substrate temperature of 500 °C. As shown, the main peaks are corresponding to those of (400) and (800) planes. The total intensity of the (400) and (800) peaks was about 90-95% of the total intensity of all the film peaks $(ZnSe)_{0.1}(SnSe)_{0.9}$. In addition to these peaks, other peaks corresponding to the (200), (201), (111), (400), 600), (511), (402), (502), (303) and (323) planes are observed, although their intensity was extremely small compared to those of former peaks. These peaks also correspond to the SnSe peaks which is present in the phase (see Figure 1). According to XRD data, all films have an orthorhombic structure. The crystal lattice parameters for the samples were calculated by the following formula: $1/d^2 = h^2/a^2 + k^2/b^2 + l^2/c^2$, where d is the distance between the planes, h, k, l - Miller indices. The parameters of the lattice constant for films deposited from ZnSe and SnSe compounds at the substrate temperature of 500 °C have the following values: a = 11.48 Å, b = 3.78 Å, c = 4.47 Å. The structural parameters of all films are presented in Table 2. As shown, the values of a and b lattice parameters for $(ZnSe)_{0.1}(SnSe)_{0.9}$ films decrease, whereas the value of c increases. This phenomenon is explained by the fact that when the molar content of ZnSe in the vapor phase increases during the $(ZnSe)_x(SnSe)_{1-x}$ film deposition, tin atoms are replaced by zinc atoms. It is known that the ionic radius of zinc (Zn + 0.88 Å) is much smaller than the ionic radius of tin (Sn + 2 0.93 Å).

Figure 2 shows SEM images for $(ZnSe)_x(SnSe)_{1-x}$ samples deposited at a substrate temperature of 500 °C. As shown, the grain shapes of the samples had a flattened appearance with the grain sizes of about 5-6 μm. The shape and grain size of the samples changes depending on the film composition. Increased molar content of ZnSe in $(ZnSe)_x(SnSe)_{1-x}$ films led to a decrease in the grain size, and the shape of the grains retained their previous form corresponding to the composition film with x = 0.

This can be due to the fact that the formation energy of the ZnSe lattice is greater than that of SnSe.

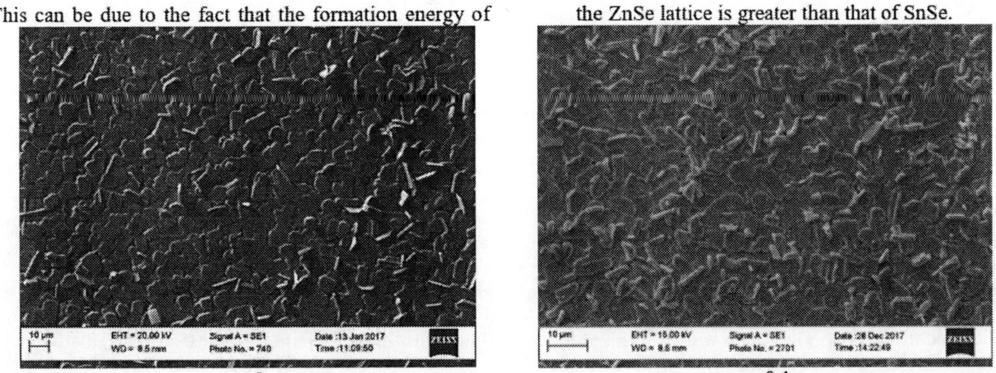

x=0 x=0.1

Figure 2. SEM images of $(ZnSe)_x(SnSe)_{1-x}$ films for indicated x values.

Table 2. Structural parameters of $(ZnSe)_x(SnSe)_{1-x}$ films with x=0.1 grown at substrate temperature of 500°C.

2Θ	(h k l)	d (Å)
15.4	(2 0 0)	5.75
25.3	(201)	3.51
27.14	(1 1 1)	3.28
31.06	(4 0 0)	2.87
47.4	(6 0 0)	1.9
49.6	(5 1 1)	1.8
51.9	(4 0 2)	1.75
57.7	(5 0 2)	1.59
60.26	(2 0 2)	1.5
64.86	(8 0 0)	1.4
67.9	(3 0 3)	1.3
84.02	(3 2 3)	1.15

SUMMARY

The results of the XRD and SEM data revealed that: 1) the films have an orthorhombic polycrystalline structure; 2) the grain sizes of the films are 5-6 μm and the films have a close-packed polycrystalline structure, and 3) increased molar content of ZnSe in $(ZnSe)_x(SnSe)_{1-x}$ films led to a decrease in the grain size, and the shape of the grains retained their previous form.

ACKNOWLEDGMENT

The work was carried out within the framework of the project Fundamental Research Project (Grant No. FA-F3-003).

REFERENCES

[1] http://www.pv-magazine.com/news/details/beitrag/first-solar-sets-new-cadmium-telluride-thin-film-cell-efficiency-record-at-22.1_100023341/#ixzz415kIS7YW

[2]http://www.renewableenergyworld.com/rea/blog/post/2014/03/two-thin-film-solar-efficiency-records-broken-this-week?cmpid=WNL-Wednesday-March5-2014

[3]http://www.businesswire.com/news/home/20130409006059/en/Solar-Sets-CdTe-Module-Efficiency-World-Record.

[4] M.A. Green, K. Emery, Y. Hishikawa, W. Warta and E.D. Dunlop. Solar cell efficiency tables. Prog. Photovolt: Res. Appl. 21, 1–11 (2013).

[5] W. Wang, M.T. Winkler, O. Gunawan, T. Gokmen, T.K. Todorov,Y. Zhu, and D.B. Mitzi. Device Characteristics of CZTSSe Thin-Film Solar Cells with 12.6% Efficiency. Adv. Energy Mater. 4(7), 1301465 (2014).

[6] T.K. Todorov, J. Tang, S. Bag, O. Gunawan, T. Gokmen, Y. Zhu, and D.B. Mitzi. Beyond 11% Efficiency: Characteristics of State-of-the-Art $Cu_2ZnSn(S,Se)_4$ Solar Cells. Adv. Energy Mater., 3, 34–38 (2013).

[7] J. Li, C. Xue, Y. Wang, G. Jiang, W. Liu, C. Zhu. Cu_2SnS_3 solar cells fabricated by chemical bath deposition–annealing of SnS/Cu stacked layers. Solar Energy Materials and Solar Cells, 144, 281-288 (2016).

[8] M. Umehara, Y. Takeda, S. Tajima, T. Motohiro, T. Sakai and R. Maekawa. Improvement of red light response of $Cu_2Sn_{1-x}Ge_xS_3$ solar cells by optimization of CdS buffer layers. J. Appl. Phys., 118, 154502 (2015).

[9] L. Kranz, S. Buecheler, A.N. Tiwari. Technological status of CdTe photovoltaics. Solar Energy Materials & Solar Cells 119, 278–280 (2013).

[10]http://onlinelibrary.wiley.com/doi/10.1002/pip.2573/pdf

[11] T.M. Razykov, E. Artegiani, G.S. Boltaev, A. Bosio, B. Ergashev, K.M. Kouchkarov, N.K. Mamarasulov, A.A. Mavlonov, A. Romeo, N. Romeo, R. Yuldoshev. Novel $Zn_xSn_{1-x}Se$ Absorber for Use in Thin-Film Solar Cells, Proc. Of the 33rd European Photovoltaic Solar Energy Conference and Exhibition, Amsterdam, Netherlands, pp. 141-145, (2017).

[12] T.M. Razykov . Chemical molecular beam deposition of II-VI binary and ternary compound films in gas flow. Applied Surface Science, 48/49(1), 89-92 (1991).

EFFECT OF CO AND CR DOPING ON THE OPTICAL BAND GAP OF ZNTE

M Chaik*, H. El Aakib, C. Sambeval, M. Elyaagoubi, A. Outzourhit
Laboratory of Nanomaterials, Energy and Environment (LN2E), Department of Physics, Faculty of Sciences Semlalia, Cadi Ayyad University, BP2390, Marrakech, Morocco
*Corresponding author: mounir.chaik@edu.uca.ma

ABSTRACT: Recently, there has been a growing interest in intermediate band solar cells as their expected efficiency as higher than that of traditional single junction cells. Efficiency limits as high as 63.2 % were predicted for this type of cells. This is mainly due to the possibility of sequential absorption of photons in a single material with intermediate bands (IB). Such IB can be formed through doping of large band gap semiconductors that otherwise do not absorbed the solar spectrum efficiently. In this work we will focus on ZnTe because its optical gap is 2.26eV and the fact that it can be easily doped.
In this work, we will focus studying the effect of doping ZnTe thin films by Cobalt and Chromium and the possibility to create an intermediate band in the band gap of this material.. For this purpose, several films Co and Cr-doped ZnTe thin films were deposited by RF co-sputtering from a pure ZnTe target, partially covered by pure Cr and Co chunks with various surface coverage ratio. The deposition was carried out in an Argon atmosphere under an RF-power of 150W for a fixed time of 1,5 hour. The structural, microstructural and optical properties of the obtained films were investigated. A decrease in the optical band gap of the films was observed as the doping level is increased.
Keywords: see enclosed list of keywords

1 INTRODUCTION

To increase the efficiency of photovoltaic solar cells via sequential absorption of photons in a single material has a relatively long history Starting with the impurity photovoltaic solar cell in 1960, where sequential absorption was suggested to proceed via defect levels, the concept gained renewed interest with the proposal of the intermediate band solar cell (IBSC) in the mid-1990s. The compelling difference between the impurity band and the IB concept is that forming an IB should, in principle, reduce the nonradiative transition rate that is likely to dominate when carriers are localized onto isolated impurities. However, the principle challenge to establish an IB within a semiconductor material that supports strong optical transitions with relatively low nonradiative recombination in comparison to the rate of photo-generation by solar photons has remained [1].

For that we chose telluride Zinc material as base on intermediate band solar cell. ZnTe is a Group II-VI compound semiconductor material with a direct band gap of 2.26 eV at room temperature [2].we finds this promoting material in many different applications such as in electric and thermoelectric devices [3], high efficiency multi-junction solar cells [4], photonics as light emission [5] or in terahertz (THz) devices [6]. The morphological, structural, or optical properties play big roles on the choice of the appropriate applications. For the structural properties in the literature Zinc telluride crystalized in cubic [7] or hexagonal wurtzite [8] structures.

We choose the Cobalt and Chromium as dopants to create deep defaults on zinc telluride in order to create intermediate band gaps. Also the deposition of ZnTe can be realized by many techniques such as chemical vapor deposition (CVD) [9], electro deposition method [10], pulsed laser ablation [11] and RF-sputtering [12-13]; as the RF-Sputtering technique gives stoichiometric, dense and thick thin films materials with an easy control of several parameters like the deposition power and Argon pressure.

Following all the reasons already mentioned above we deposited in this work Zinc telluride ZnTe, Cobalt doped Zinc telluride ZnTe:Co, and Chromium doped Zinc telluride ZnTe:Cr by RF-sputtering. The optical, structural and morphological properties results of our examples were investigated.

2 RESULTS AND DISSCUTIONS

2.1 Structural properties

The structural properties were performed by using the X-Ray diffraction (see Figure 1) from θ=5°to 70°. The films show the (111), (220) and (311) peaks characteristics of ZnTe, which indicates that our films crystalized in a cubic structure with an F-43m Space group. We remark also there is no presence of other peaks neither of Cobalt nor for Chromium.

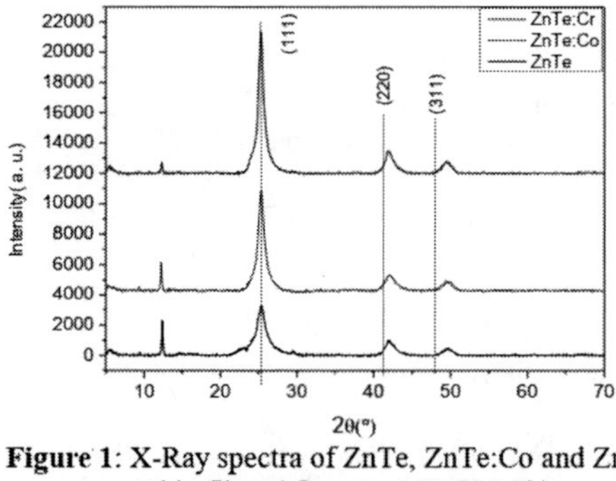

Figure 1: X-Ray spectra of ZnTe, ZnTe:Co and ZnTe:Cr thin films (rf-power 150 W;1,5h)

2.2 Morphological properties

The morphological properties of the films were investigated by scanning electron microscopy (SEM). Figure 2 shows the SEM images of ZnTe, the ZnTe:Co and ZnTe:Cr similar and show homogenous films with nano-sized grains.

Figure 2: SEM images of ZnTe film

2.3 Optical properties

The optical properties of the films namely the absorption coefficient and optical band gap and the refractive index, were obtained from the optical transmission measurements in the Ultraviolet-Visible-Near Infrared (UV-Vis-NIR) wavelength range. The ZnTe film showed a high transparency in the near-infrared, and an absorption edge around 700 nm(Figure 3). Interference fringes are also observed in the transparency region of this film. However, in the case of ZnTe:Co and ZnTe:Cr, the absorption edges shifts to higher wavelengths and the contrast of the fringes is reduced as a result of an increased absorption ZnTe:Cr start absorbing from 800nm to UV wavelengths and ZnTe:Co start form 1μm.

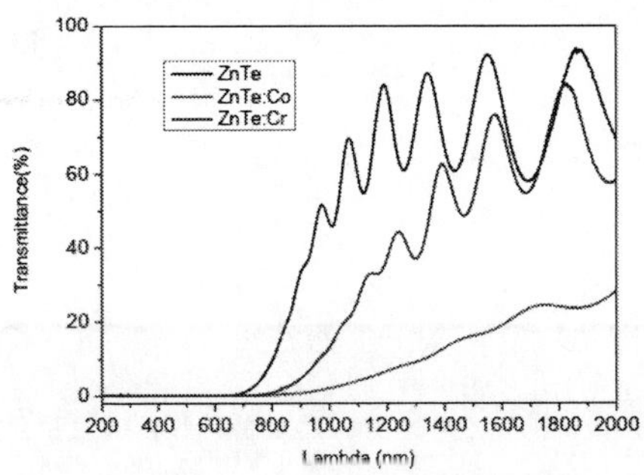

Figure 3: Transmittance spectra of ZnTe, ZnTe:Co and ZnTe:Cr thin films (rf-power 150 W;1,5h)

In order to estimate the optical ban gap Eg, we used the following Tauc formula [14]:

$$(\alpha h\nu)^{1/r} = k(h\nu - E_g)$$

With:

r = 1/2 for direct allowed transitions.
r = 3/2 for direct forbidden transitions.
r = 2 for indirect allowed transitions.
r = 3 for indirect forbidden transitions.

As known, ZnTe has a direct band gap, for that we used r=1/2 to plot the variations of the absorption coefficient α with photon energy hv (αhv)2 versus hv. The Figure 7 show calculated band gaps of the films. The value found for the ZnTe films was 1,82eV wish is very closed to the values reported by M. Shobana el al. they found a values between 2,4 and 1,85eV[15].

The optical band gap change in both cases it started by the value 1,82eV of ZnTe to 1,37eV of ZnTe :Co and 1,5 eV of ZnTe:Cr.

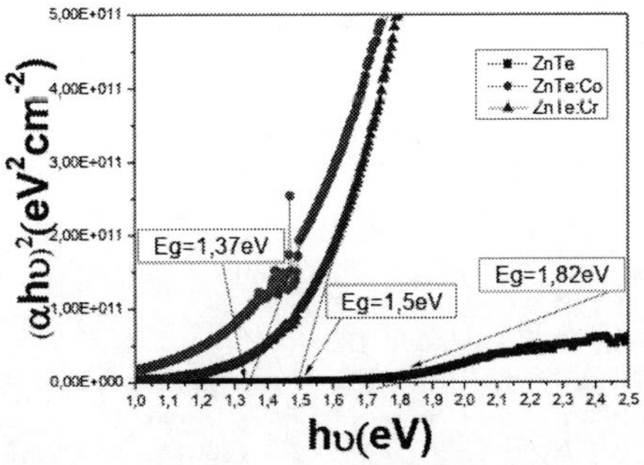

Figure 4: The optical band gap Eg of ZnTe, ZnTe:Co and ZnTe:Cr thin films (rf-power 150 W;1,5h)

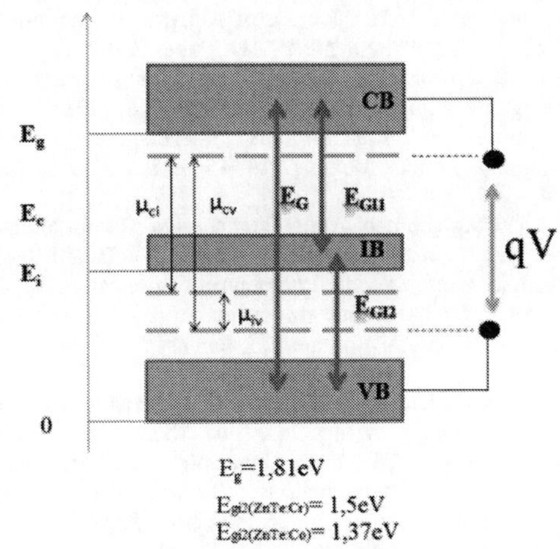

Figure 5: energy diagrammed of intermediate band solar cells.

Figure 5 illustrate the energy diagram of the intermediate band gap solar cells, the gap values founded of the chromium and cobalt doped Zinc telluride refers to the intermediate band gap Eg1.

3 CONCLUSION

Zinc telluride, chromium doped Zinc telluride and Cobalt doped Zinc telluride was deposited by RF-sputtering under argon plasma in borosilicate glass and silicon substrates. The films obtained is crystallized in cubic phase with F-43m space group with no appearance of characteristic peaks of cobalt or chromium phases, the films has a good surface morphologies as demonstrated the SEM images, and the optical band gap founds from the Tauc method is varied from 1,82eV for the ZnTe films, to 1,37eV of ZnTe :Co and 1,5 eV of ZnTe:Cr.

4 REFERENCES

[1] Okada et al., Intermediate band solar cells: Recent progress and future directions, Appl. Phys. Rev. 2, 021302 (2015).

[2] Muhammad Arshad Kamran, Novel low-temperature synthesis and optical properties of 1D-ZnTe nanowires, Journal of Science: Advanced Materials and Devices 3 (2018) 226-229

[3] P.Gupta, B.Maiti, S.Chaudhuri, A.K.Pal, ZnS□ZnTe alloy films: preparation and properties, Thin Solid Films, Volume 227, Issue 1, 5 May 1993, Pages 66-73.

[4] P. Saikia, P.K. Saikia, D. Saikia, Fabrication and characterization of ZnSe / ZnTe / CdTe / HgTe multijunction solar cell, 5 (2011) 204–207.

[5] N.Spalatu, I.Evtodiev, I.Caraman, S.Evtodiev, I.Rotaru, M.Caraman, D.Untila; Optical Properties of GaTe-ZnTe Nanolamellae Composite; Volume 84, December 2015, Pages 176-182.

[6] T. Löffler, T. Hahn, M. Thomson, F. Jacob, H.G. Roskos, Large-area electro-optic ZnTe terahertz emitters, Opt. Express. 13 (2005) 5353. doi:10.1364/OPEX.13.005353.

[7] M.Shobana S.R.Meher ; Experimental and ab-initio study of the structural, electronic and vibrational properties of ZnTe ; Journal of Alloys and Compounds; Volume 762, 25 September 2018, Pages 260-271

[8] Guoxiang Wang,Yawen Zhang, Peng Liu, Jiangxing Wang, Qiuhua Nie, Xiang Shen; Phase change behavior of pseudo-binary ZnTe-ZnSb material; Materials Letters; Volume 213, 15 February 2018, Pages 342-345.

[9] Mohammad Amin Baghchesara, Ramin Yousefi, Mohsen Cheraghizade, Farid Jamali-Sheini, Abdolhossein Sa'aedi; Photocurrent application of Cd-doped ZnTe nanowires grown in a large scale by a CVD method; Vacuum; Volume 123, January 2016, Pages 131-135.

[10] Kaifeng Qin, Lunjuan Li, Jian Huang, Ke Tang, Xiaotian Zhang, Meng Cao, Yue Shen, Linjun Wang; Influence of ZnTe intrinsic layer on the performance of p-ZnTe/n-Si heterojunctions; Surface and Coatings Technology; Volume 320, 25 June 2017, Pages 153-157.

[11] K.P.Acharya, A.Erlacher, B.Ullrich; Responsivity properties of ZnTe/GaAs heterostructures formed with pulsed-laser deposition; Solar Energy Materials and Solar Cells; Volume 93, Issue 1, January 2009, Pages 25-27.

[12] B.Späth, J.Fritsche, F.Säuberlich, A.Klein, W.Jaegermann; Studies of sputtered ZnTe films as interlayer for the CdTe thin film solar cell; Thin Solid Films; Volumes 480–481, 1 June 2005, Pages 204-207.

[13] H.U. Bellakhder, A. Outzourhit, E.L. Ameziane, Study of ZnTe thin films deposited by R. F. Sputtering, (2001) 30–33.

[14] A.En Naciri, F.Ahmed, M.Stchakovsky; Optical properties of ZnTe and ZnS nanocrystals by critical-points and Tauc–Lorentz models; Thin Solid Films; Volume 519, Issue 9, 28 February 2011, Pages 2843-2846.

[15] M. Shobana, S.R. Meher, SC, J. Alloys Compd. (2018). doi:10.1016/j.jallcom.2018.05.208.

NANOSPHERE COLLOIDAL COATING FOR IMPROVEMENT OF SOLAR CELL EFFICIENCY

Seyedehnasim Seyedpour Esmailzad,[1,2,3], İbrahim Murat Öztürk[3,4], Mona Zolfagharibora[1,4], Nardin Avishan[1,4] and Alpan Bek[1,3,4]
1. Micro and Nanotechnology Program, Middle East Technical University, Ankara 06800, Turkey
2. Department of Electrical and Electronics Engineering, TED University, Ankara, Turkey
3. Physics Department, Middle East Technical University, Ankara 06800, Turkey
4. Center for Solar Energy Research and Applications, Middle East Technical University, Ankara 06800, Turkey

ABSTRACT: CPV (Concentrated Photovoltaics) is an emerging technology for developing high-efficiency, low-cost solar cells. CPV uses optics such as mirrors or lenses to capture the sunlight in a more effective manner. In this study, closed-packed arrays of nanoscale dielectric spheres (nanospheres, NS) located on top of the cell, can be utilized as lenses which increase the absorption of c-Si solar cells. NS can reduce the reflection and suppress the transmission by changing the incident light into a set of collimated beams. NS can focus the light inside the photovoltaic layer which causes an increase in the absorption and consequently in the photovoltaic current. The fabrication of such light-trapping structure is low-cost and less complicated than standard alternatives such as vacuum evaporated multilayer anti-reflection coatings. We find that by using silica NS, the efficiency improvement is limited by the spherical shape of the NS causing premature focussing due to sphere shape. For that reason, we proceeded with polystyrene NS in which we were able to show further improvement in the solar cell efficiency by annealing based shape adjustment of NS.

Keywords: c-Si, solar cell, dielectric, nanosphere, CPV

1 INTRODUCTION

A considerable amount of incident light is reflected due to the high refractive index of silicon, which lead to a considerable decrease in the conversion efficiency of solar cells. To fabricate solar cells with high conversion efficiency, reduction of reflectivity at the surface of solar cells is very important. Anti-reflection (AR) layers with ordered micro/nanostructured arrays of uniform shapes and sizes have attracted growing interest recently because of their applications in light harvesting [1–3].

The inverted pyramid surface texturing has been experimentally proven to be very effective in trapping light by total internal reflection. However, most methods for texturing Si involve the use of dry or wet etching processes, either by a mask to obtain large ordered features or without a mask to create much smaller, random features. Most of the texturing methods are too costly and have low reproducibility. Moreover, a distinguished privilege of nanoparticle scattering is that, there is a smaller damaging effect on the electrical properties at the semiconductor surface.

Recently plasmonic nanostructures are employed in most of the devices due to their advantages [4], which are related to the excitation of the intense localized surface plasmon resonance (LSPR) in metallic nanostructures and strong suppression of light reflection in wavelength region close to the resonance. However, the application of plasmonic nanostructures is faced by many problems, such as, large ohmic losses of metals, which suppress nanoparticle resonances, or parasitic surface oxidizing, which changes the optical properties of nanostructures.

The enhancement of optical absorption is analysed in crystalline Si solar cell devices employing dielectric nanospheres that scatter incident light in a way proper for photocurrent generation. The size, shape, and concentration of dielectric nanosphere play a significant role in the transmission of light into the semiconductor. Experimental measurements show enhancements the silica or polystyrene nanosphere based scattering in photovoltaics.

In addition, dielectric nanospheres can be utilized as optical lenses in the structure of Concentrating Photovoltaic (CPV) technologies to concentrate a sunlight towards decreased areas of photovoltaic (PV) material [5]. In this manner, the noticeable reduction of the required solar cell area provides an opportunity to lower cost, as an expensive semiconductor material is replaced with inexpensive mirrors or lenses. However, novel designs, methods, and materials are still being investigated to reach an acceptable cost range in CPV systems. Solar concentrator technologies have many layers and factors to be considered when designing. This method is offered to simplify the CPV structures.

2 ELECTROMAGNETIC SIMULATIONS

Maxwell solutions are carried out numerically using finite elements method (FEM) for 880 nm silica (SiO_2) NS decorated (Figure 1) and, 500 nm (Figure 2) and 1μm diameter (PS) NS decorated c-Si with varying spheroid aspect ratio (Figure 3). The wavelength of the incident light is adjusted to 600nm.

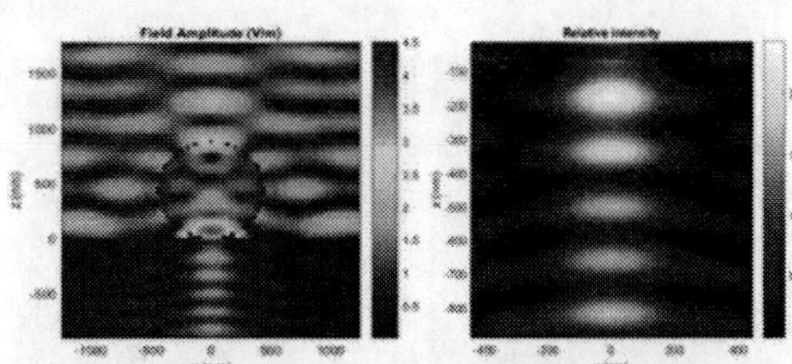

Figure 1: Maxwell simulation of Silica nanosphere with 880nm diameter.

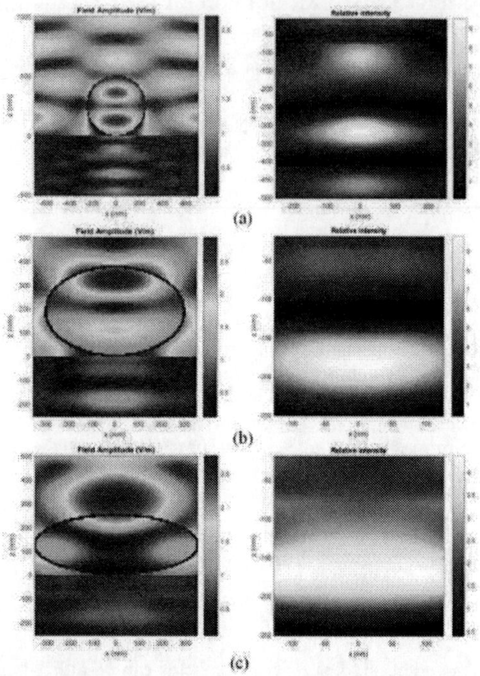

Figure 2: Maxwell simulation(a) Polystyrene nanosphere with 500nm diameter, (b),(c) Nanosphere with different annealing durations.

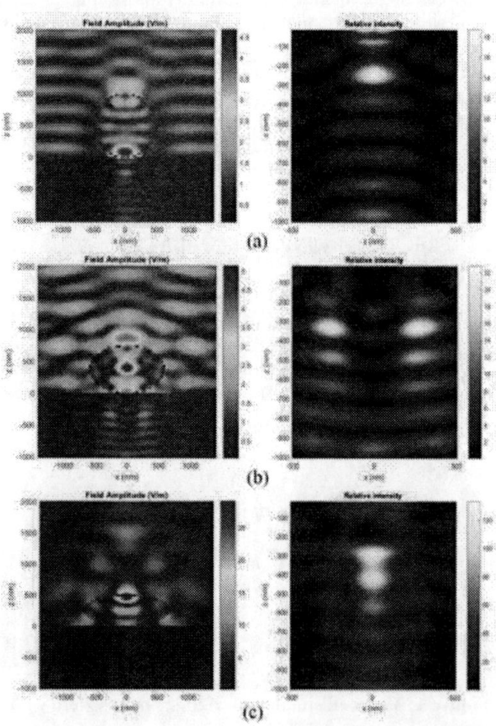

Figure 3: Maxwell simulation(a) Nanosphere with 1μm diameter, (b),(c) Polystyrene nanosphere with different annealing durations.

3 EXPERIMENTAL METHODS

The device structure is comprised of an n+/p crystalline Si solar cell. After a short KOH based saw-damage etching (SDE), the flat front side surface is deposited with SiN_x to prevent front side texturing after KOH etching. However, the back-side surface is textured by KOH etchant. After HF and RCA1 and RCA2 etching, we achieve a Si structure with SDE front side and textured backside. The surface of Si is doped with n-type dopant using $POCl_3$ diffusion in furnace.

To enhance the efficiency, surface passivation is performed with dry oxidation. After oxidation, the front surface is coated with very thin (6 nm) silicon nitride. To observe the photonic effects of dielectric nanospheres, the silicon nitride thickness is opted small in order not to have an ARC effect itself. In the metallization step, H shaped silver (Ag) grid is screen printed on the front surface and the rear side is aluminum deposited with thermal evaporation system.

Finally, laser edge isolation is applied to prevent shunt current between the front and rear surfaces caused by n-layer deposition. PS nanospheres are spin-coated on fabricated functional devices from solutions containing colloidal PS nanospheres. The schematic structure of the device is illustrated in Figure 4.

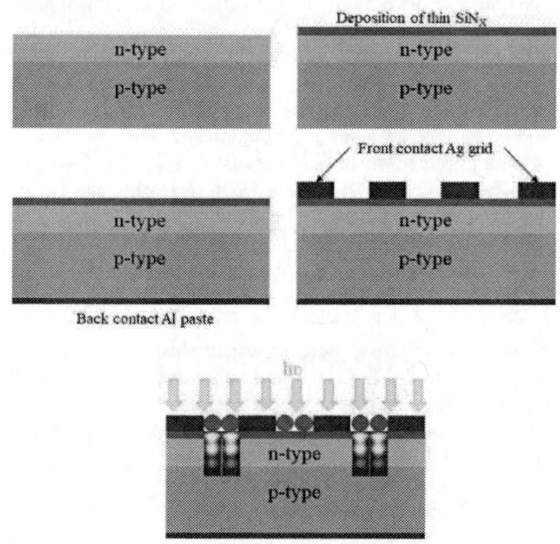

Figure 4: Cross-section schematics of a c-Si solar cell decorated with NS.

Scanning electron microscopy (SEM) images of the device surface before and after nanosphere annealing are shown in Figure 5. Zoomed-out view of solar cell device coated with 1μm diameter Polystyrene nanosphere is shown in Figure 5.a . Zoomed-in view of solar cell device coated with 1μm diameter Polystyrene nanosphere is shown in Figure 5.b. For changing the spheroid aspect ratio, Polystyrene nanospheres are annealed at 105℃ for 16 minutes. Figure 5.c demonstrate a closed view of solar cell coated with nanospheres after annealing.

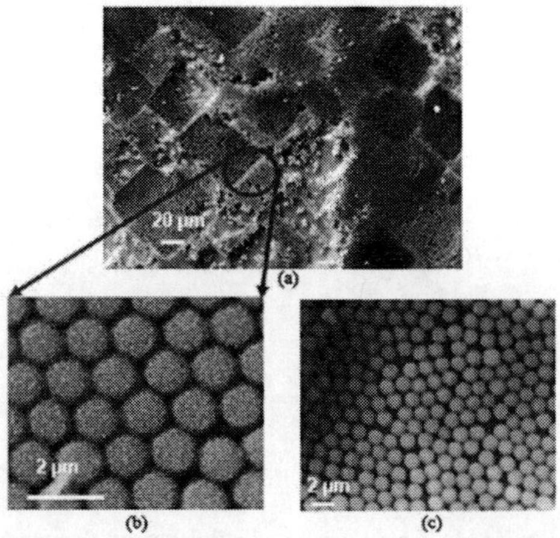

Figure 5: SEM images showing devicesurface (a) Zoomed-out view of solar cell coated with 1µm diameter nanosphere, (b) Zoomed-in view of solar cell coated with 1µm diameter nanosphere, (c) After annealing nanospheres.

4 RESULTS AND DISCUSSIONS

The photocurrent response of the samples is measured with the solar simulator at AM 1.5 condition. The device efficiency is given in the following Tables. of the same device before and after nanosphere coating.

Table I: I-V characteristics of solar cells coated with 880nm diameter silica nanospheres.

	V_{oc}	J_{sc}(mA/cm^2)	η (%)
Unprocessed PV	596	20.57	9.3
PV with 880nm Silica NS	599	20.49	10.02

Table II: I-V characteristics of solar cells coated with 500nm diameter polystyrene nanospheres

	V_{oc}	J_{sc}(mA/cm^2)	η (%)
Unprocessed PV	578	27.28	8.81
PV with 500nm NS	584	27.13	9.18
500nm NS after annealing	588	27.14	9.42

Table III: I-V characteristics of solar cells coated with 1µm diameter polystyrene nanospheres

	V_{oc}	J_{sc}(mA/cm^2)	η (%)
Unprocessed PV	580	23.38	8.34
PV with 1µm NS	580	23.64	8.69
1µm NS after annealing	580	23.45	9.01

The reflection measurement of the solar cell device before and after nanosphere coating is illustrated in Figure. 6, Figure.8 and Figure.10. Nanospheres act as an antireflection layer and decrease the reflection of the solar cell in the visible wavelength. Therefore, the uncoated sample has higher reflection than the samples deposited with nanospheres.

According to EQE measurements which is illustrated in Figure.7, Figure.9 and Figure.11, the photocurrent efficiency of silicon solar cell device is increased after nanosphere coating.

This increase is due to the focusing of light inside the silicon solar cell by nanospheres. By annealing the Polystyrene nanospheres, the spheroid aspect ratio is changed which cause an enhancement in the focusing of light. Therefore, external quantum efficiency is further improved after nanosphere annealing. However, PS nanospheres absorb the incoming light at the wavelength 400-600nm. Therefore, it is expected that at this wavelength region, the external quantum efficiency after nanosphere coating will be reduced (Figure.9 and Figure. 11).

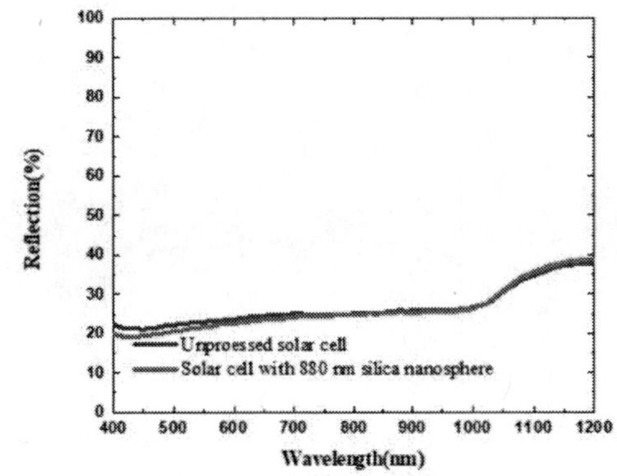

Figure 6: Reflection of PV before and after Silica nanosphere coating.

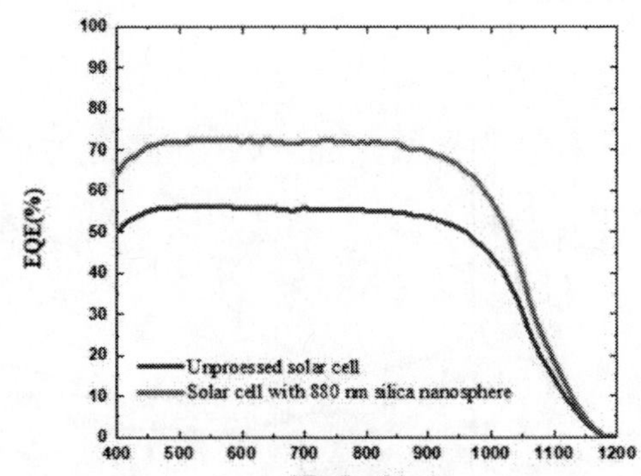

Figure 7: External quantum efficiency of PV before and after Silica nanosphere coating.

In order to survey the effect of annealing time, solar cells are annealed at different time duration of 16 minutes, 26 minutes and 31 minutes. The wavelength averaged EQE vs annealing time results is illustrated at Figure.12.

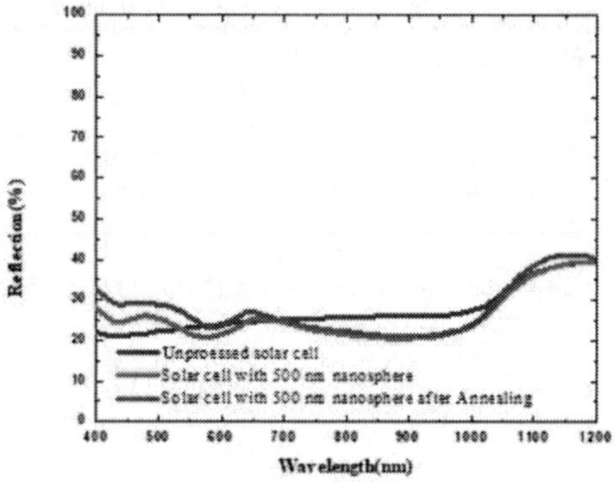

Figure 8: Reflection of PV before and after PS 500nm nanosphere coating and after PS annealing.

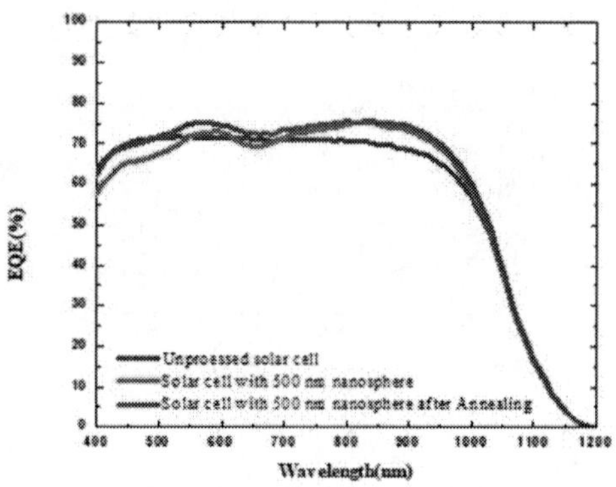

Figure 9: External quantum efficiency of PV before and after PS 500nm nanosphere coating and after PS annealing.

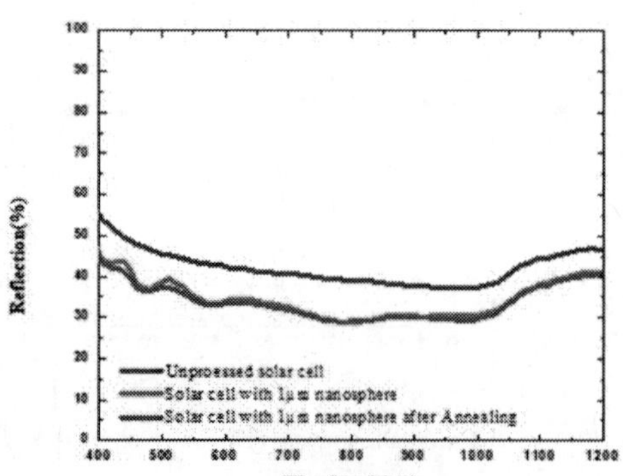

Figure 10: Reflection of PV before and after PS 1μm nanosphere coating and after PS annealing.

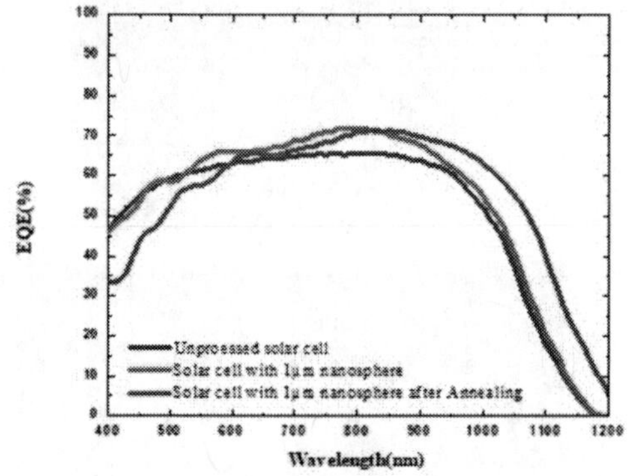

Figure 11: External quantum efficiency of PV before and after PS 500nm nanosphere coating and after PS annealing.

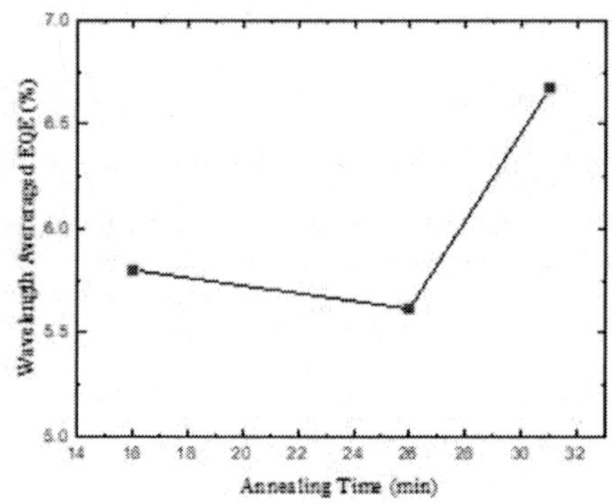

Figure 12: The wavelength averaged EQE vs annealing time.

5 ACKNOWLEDGEMENT

This research has been supported by Middle East Technical University under project number: TÜBİTAK 115F167

6 REFERENCES

[1] D. Wan, H. L. Chen, T. C. Tseng, C.Y. Fang, Y. S. Lai, and F. Yeh, "Antireflective Nanoparticle Arrays Enhance the Efficiency of Silicon Solar Cells", Journal of Adv. Funct. Mater, 20, (2010) 3064–3075.
[2] C.K. Huang, K.W. Sun , and W.-L. Chang, "Efficiency enhancement of silicon solar cells using a nanoscale honeycomb broadband antireflection structure", Optics Express A85, Vol. 20, No. S1 , (2012).
[3] K.-S. Han, J.-H. Shin, W.-Y. Yoon, and H. Lee, "Enhanced performance of solar cells with anti-reflection layer fabricated by nano-imprint lithography," Sol. Energy Mater. Sol. Cells 95(1), (2011) 288–291.
[4] M. Borra, S. Kayra Güllü, F. Es, O. Demircioğlu, M. Günöven, R.Turan, A.Bek,"A feasibility study for controlling self-organized production of plasmonic enhancement interfaces for solar cells", Applied Surface

Science, Vol. 318, (2014) 43-50.
[5] K. Shanks, S. Senthilarasu, K. Mallick, "Optics for concentrating photovoltaics: Trends, limits and opportunities for materials and design", Renewable and Sustainable Energy Reviews, V 60, (2016), 394-407.

SPECTRALLY SELECTIVE SOLAR CELLS FOR SIMULTANEOUS USE OF PHOTOSYNTHESIS AND PHOTOVOLTAICS

N. Osterthun, V. Steenhoff, N. Neugebohrn, K. Gehrke, M. Vehse, C. Agert
DLR Institute of Networked Energy Systems
Carl-von-Ossietzky-Str. 15, 26129 Oldenburg, Germany

ABSTRACT:
Photosynthesis of green algae and plants needs mainly red and blue light in chlorophyll. One possible way for combining photosynthesis and photovoltaics is to transmit this light, while the remaining light can be used for electricity generation. The straight forward solution to this problem is to use a high band gap absorber, so that only UV-light is utilized. However, this would imply that a great amount of photons of the solar spectrum remains unused. A solar cell concept that enables spectrally selective transmission could significantly increase efficiency. In this contribution, we present a solar cell design which transmits light around 480 nm and 680 nm for photosynthesis while generating electricity from the remaining spectrum. We demonstrate the concept of a spectrally selective solar cell utilizing a novel back reflector. This reflector is based on a tunable cavity determining the optical properties. Simulations as well as first experimental realizations of the reflector itself and of the solar cell are shown.
Keywords: spectrally selective, Germanium, cavity enhanced solar cell

1 INTRODUCTION

Combining photosynthesis and photovoltaics (PV) would allow a dual use of areas like agricultural fields or greenhouses. One known concept for this challenge is the agrivoltaic system [1, 2]. The suggestion is to build evenly spaced lines of PV modules on top of agricultural area. Although this leads to different shadow positions during the day, all plants on the field might get enough sun to grow.

Another idea is to use spectrally selective solar cells covering the complete area without full shadowing: For photosynthesis green algae and plants absorb red and blue light in chlorophyll (Figure 1) [3, 4]. The light within the green gap and the near infrared light is not absorbed in chlorophyll and, hence, can be used for photovoltaic power generation.

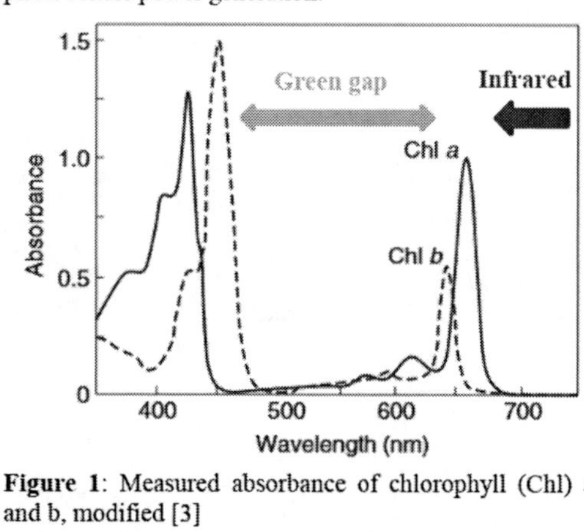

Figure 1: Measured absorbance of chlorophyll (Chl) a and b, modified [3]

Recently, an approach for wavelength-selective photovoltaic windows was published [5, 6]. It is based on luminescent solar cell technology and partly transmits light to growing plants. The remaining part is absorbed in thin solar panels along the windows to generate electricity.

We have realized another wavelength-selective solar cell without luminescent materials by using the existing technology of the "Ultrathin Resonant-Cavity-Enhanced Solar Cell", which was described in detail previously [7-9]. For our concept the most important points of this type of solar cell are as follows: The solar cell is based on a nip-doping-structure where all layers have a refractive index of 4-5. This stack has a small thickness of less than 60 nm with a nanoabsorber and is sandwiched between a semitransparent front electrode with a refractive index of approximately 2 and a highly reflective back electrode. The small thickness and the difference of the respective refractive indices of ~2 lead to broadband optical interference. This leads to an enhanced absorption within the nanoabsorber layer.

In order to design a spectrally selective solar cell, the back reflector has to be selective, i.e. reflect only at certain wavelengths. This can be realized with a multilayer reflector consisting of alternating metal and oxide thin films (Figure 2). A metal-oxide-metal-oxide stack (MOMO) shows optical interference fringes because of the difference in refractive index of the layers. These can be optimized to transmit blue and red light and reflect the remaining spectrum. By integrating this spectrally selective reflector in the ultrathin resonant-cavity-enhanced solar cell, single pass absorption for red and blue light is achieved, while cavity enhanced absorption for the remaining light of the solar spectrum remains.

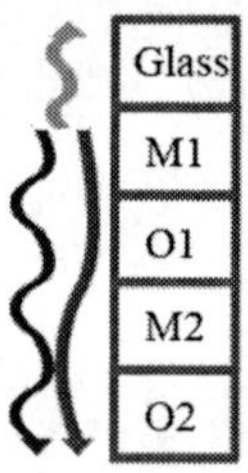

Figure 2: Concept of the spectrally selective reflector with alternating metal (M) and oxide (O) layers

This spectrally selective solar cell has the advantage, that it is produced with well-known technologies like PE-CVD and magnetron sputtering. Furthermore, the solar cell does not need ultra clean glass for wave-guiding like luminescent solar cell technology, and it can be utilized on the full area.

In this proceeding, the results of the experimental design of the novel spectrally selective back reflector are shown. Moreover, we present the integration in the solar cell and show the optical and electrical functionality.

2 METHODS

2.1 Optical simulation

1d optical simulations were performed with the software package Scout/Code using self-refined refractive index data. Due to the fact, that all layers of the stack are thin films, the light in the stack is assumed to be coherent. Refractive index data of the different materials were obtained by UV-VIS spectroscopy using a Cary 5000 spectrophotometer with integrating sphere on single layers and subsequent fitting.

2.2 Thin film preparation

For the deposition of the spectrally selective reflector via sputtering, a silver target was used as metal (M) source and a ZnO:Al target with 0.5wt% Al_2O_3 doping (AZO) as oxide (O) source. The films were deposited without breaking vacuum in two chambers being part of a von Ardenne cluster tool type CS400PS. The deposition method was DC magnetron sputtering at room temperature with 200 mm targets. The process parameters are summed up in table 1. As substrates Soda lime float glasses (Saint Gobain) were used (5x5 cm² and 10x10 cm²).

Table 1: Process parameters for MOMO fabrication

Process parameter	Ag	AZO
Gas flow Ar/O_2 [sccm]	60/-	100/10
Power [kW]	0.2	1
Pressure [mBar]	8e-3	6e-3
Distance target-substrate [mm]	70	75
Deposition rate [nm/s]	1	1.45

The solar cell stack consists of the following layer sequence:

a-Si:H(n)/a-Si:H(i)/a-Ge:H/μc-Si:H(i)/μc-Si:H(p)

All layers were fabricated in the same above-mentioned cluster tool by plasma-enhanced chemical vaper deposition on glass substrates with a transparent conductive oxide. Discussions concerning the cell stack and the production parameters can be found elsewhere [9]. The *jV*-curves were measured with a WACOM dual lamp solar simulator according to standard test conditions (AM1.5G-spectrum, 1000 W/m², 25°C).

3 SPECTRALLY SELECTIVE REFLECTORS

For designing the MOMO, optical simulations were performed (Figure 3). Up to 75 % of the red and up to 60 % of the blue light is transmitted, while the reflection between these two transmission maxima and in the near infrared region exceeds 80 %.

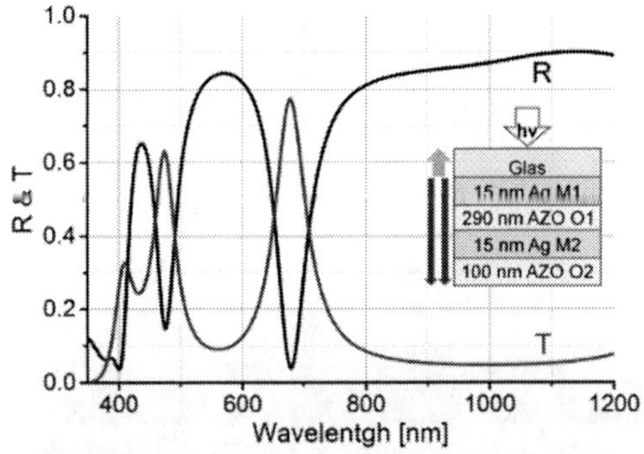

Figure 3: Simulated R&T spectra of a spectrally selective reflector. The stack in the figure depicts the simulated stack.

The silver layers are thin enough to transmit the light, still ensuring the conductivity of the reflector. The AZO layer O2 protects the silver against oxidation and supports the resonance of the AZO O1 layer, which mainly determines the peak positions.

To validate the simulated spectra, we fabricated three different MOMOs with an AZO O1 thickness d of 240 nm, 280 nm and 320 nm while keeping the dimensions of the other layers constant ($d_{M1}=d_{M2}=15$ nm, $d_{O2}=$ 100 nm). The results, shown in Figure 4, demonstrate that depending on the design, the position of the transmission peaks can be shifted over a broad wavelength range. This allows optimizing the transmission for highest overlap with the photosynthetically active absorption peaks.

In Figure 5 the measured transmission and its sum with reflection for the sample are presented, showing the best match with the photosynthesis spectrum shown in Figure 1, which had an AZO O1 thickness of 280 nm. With this device, high transmission of 70 % for red light and 55 % for blue light is reached. Simultaneously, up to 80 % of the remaining light is reflected and thus can be absorbed by a solar cell for electric current generation.

The next crucial step for realization of this concept is to combine the selective reflector with an ultrathin resonant cavity a-Ge:H solar cell.

4 SPECTRALLY SELECTIVE SOLAR CELLS

For the integration of the MOMO, optical simulations with a 5 nm a-Ge:H nanoabsorber were performed. To

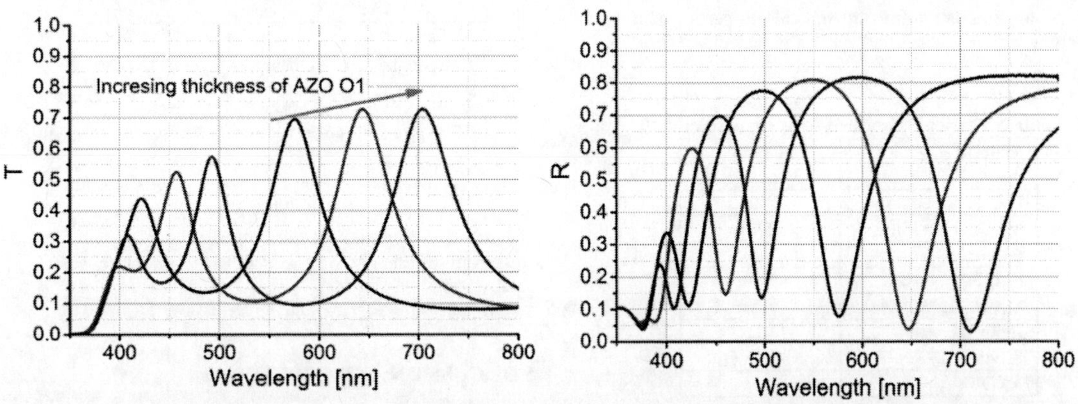

Figure 4: Measured transmission (T and reflection R for three different MOMO back reflectors with different AZO O1 thicknesses (240, 280 and 320 nm)

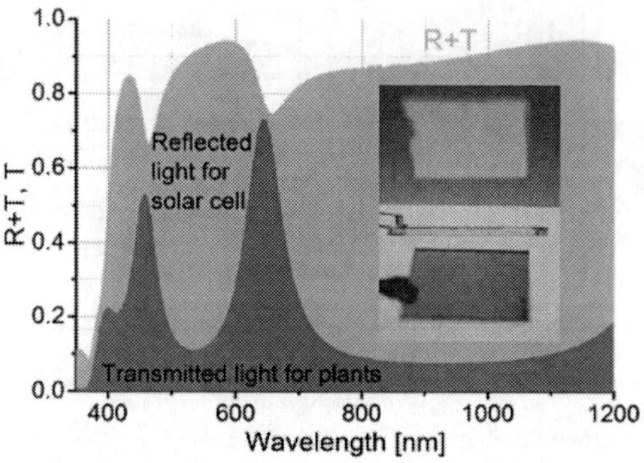

Figure 5: Final properties of the novel spectrally-selective reflector. The green area can be used in the solar cell, while the violet area is transmitted for the plants. The white area is parasitic absorption. Inset demonstrates the reflection and transmission of a MOMO.

minimize parasitic absorption, 5 nm silicon carbide was used for the doped layers. It has a higher bandgap compared to doped silicon and, therefore, absorbs less light in the blue, which is needed for photosynthesis.

The simulated spectra in Figure 6 show the theoretical feasibility of the concept. Transmission maxima close to 50 % for blue and 60 % for red light are calculated, respectively. A quantum efficiency of around 50 % for the remaining light can be expected, leading to a short circuit current density of 14.1 mA/cm².

To validate the simulation, a superstrate solar cell with MOMO reflector was realized. The cell stack is shown in table 2.

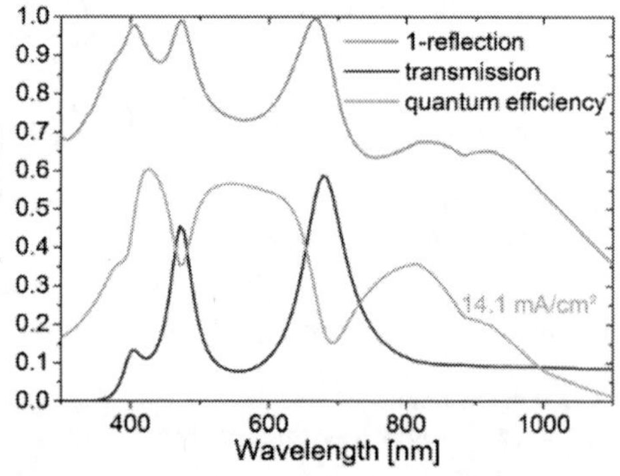

Figure 6: Simulation of the optical properties and quantum efficiency of the spectrally selective solar cell.

Table 2: Stack of fabricated superstrate solar cells

Material	Layer thickness[nm]
ITO Front contact	70
n a-Si:H/ i a-Si:H	15/15
i a-Ge:H	13
i μc-Si/ p μc-Si	7/7
Silver M1	15
AZO O1	322
Silver M1	15
AZO O1	83

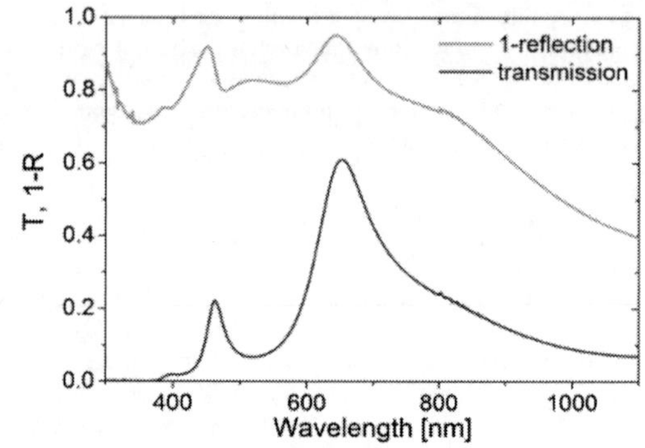

Figure 7: UV-VIS measurements of a spectrally-selective solar cell

Although amorphous and microcrystalline silicon was used instead of the simulated silicon carbide, the transmission of the red light reaches almost the predicted 60 %. However, the transmission of blue light is only at 20 %. This is most probably the result of parasitic absorptions in the doped silicon layers used in the experiment. Hence, the transmissivity can be improved in the future by employing doped layers with a wider bandgap. Due to the fact that the front contact is not optimized yet, this cell shows no suitable IV-curve as the high resistivity leads to a strong reduction of the fill factor

To show the electric functionality, glass substrates with 1 μm AZO as front electrode were used. The cell stack as described before in table 2 was built on top. Only the thickness of the AZO O1 was around 460 nm instead.

Although the thick AZO of the front contact alters the optical properties of the reflector with its own cavity effect, the solar cell still shows transmission in the red and blue light. The electric properties of the solar cell are shown in figure 8. Here the novel back contact is compared with a silver back contact on the same solar cell stack. The results demonstrate that apart from the expected reduction of the short circuit current density and a small increase in series resistance, there is no significant negative influence on the electronic properties of the device.

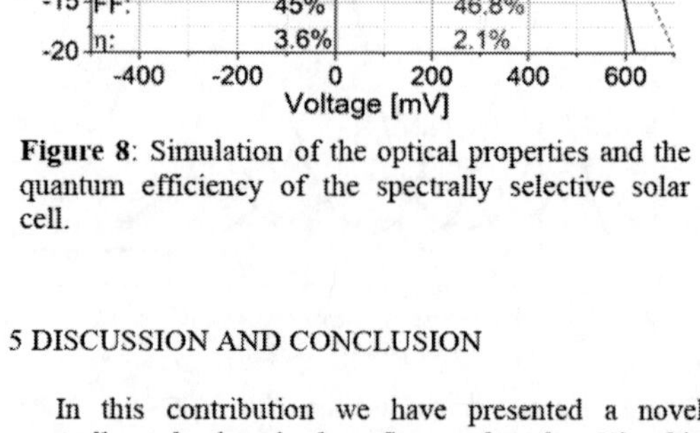

Figure 8: Simulation of the optical properties and the quantum efficiency of the spectrally selective solar cell.

5 DISCUSSION AND CONCLUSION

In this contribution we have presented a novel spectrally selective back reflector for the Ultrathin

Resonant-Cavity-Enhanced Solar Cell. The optical properties of the reflector on glass and the integration in the solar cell were demonstrated.

The first results indicate that it is possible to combine photosynthesis with photovoltaics in a way that red and blue light is transmitted. The remaining light of the solar spectrum was successfully used for photovoltaic energy conversation with no significant degradation of the open circuit voltage and fill factor of the device.

Several further steps are still needed to make combined PV and photosynthesis a reality: First, the simultaneous functionality of optical and electrical properties must be ensured by integrating an optimized front contact with sufficient conductivity. Furthermore, the parasitic absorption has to be reduced to improve the transmission in the blue spectral range. This can be achieved by employing high-bandgap materials inside the doped layers.

ACKNOWLEDGEMENTS
The authors thank Hosni Meddeb and Dennis Berends for experimental support. We acknowledge funding from the „Forschungsinitiative Zukunft Bau des Bundesinstitutes für Bau-, Stadt- und Raumforschung" (SWD-10.08.18.7-17.02)

REFERENCES

1 Dupraz, C., Marrou, H., Talbot, G., Dufour, L., Nogier, A., and Ferard, Y.: 'Combining solar photovoltaic panels and food crops for optimising land use: Towards new agrivoltaic schemes', Renewable Energy, 2011, 36, (10), pp. 2725-2732
2 Goetzberger, A., and Zastrow, A.: 'On the coexistence of solar-energy conversion and plant cultivation', International Journal of Solar Energy, 1982, 1, (1), pp. 55-69
3 Lichtenthaler, H.K., and Buschmann, C.: 'Chlorophylls and carotenoids: Measurement and characterization by UV-VIS spectroscopy', Current protocols in food analytical chemistry, 2001
4 Moss, R., and Loomis, W.: 'Absorption spectra of leaves. I. The visible spectrum', Plant Physiology, 1952, 27, (2), pp. 370
5 Corrado, C., Leow, S.W., Osborn, M., Chan, E., Balaban, B., and Carter, S.A.: 'Optimization of gain and energy conversion efficiency using front-facing photovoltaic cell luminescent solar concentratordesign', Solar Energy Materials and Solar Cells, 2013, 111, pp. 74-81
6 Loik, M.E., Carter, S.A., Alers, G., Wade, C.E., Shugar, D., Corrado, C., Jokerst, D., and Kitayama, C.: 'Wavelength-Selective Solar Photovoltaic Systems: Powering Greenhouses for Plant Growth at the Food-Energy-Water Nexus', Earth's Future, 2017, 5, (10), pp. 1044-1053
7 Steenhoff, V., Juilfs, M., Ravekes, R.-E., Ahrlich, M., Kellermann, M., Siepmann, O., Vehse, M., and Agert, C.: 'Optimized optical field profile in resonant-cavity-enhanced a-Ge: H nanoabsorber solar cells for tandem cell application', IEEE Journal of Photovoltaics, 2017, 7, (1), pp. 3-10
8 Steenhoff, V., Neumüller, A., Sergeev, O., Vehse, M., and Agert, C.: 'Integration of a-Ge: H nanocavity solar cells in tandem devices', Solar Energy Materials and Solar Cells, 2016, 145, pp. 148-153
9 Steenhoff, V., Theuring, M., Vehse, M., von Maydell, K., and Agert, C.: 'Ultrathin Resonant-Cavity-Enhanced Solar Cells with Amorphous Germanium Absorbers', Advanced Optical Materials, 2015, 3, (2), pp. 182-186

Electron selective La:$BaSnO_3$ thin films *via* Pulsed laser deposition (PLD) : Effect of deposition pressure

Akash Kumar, Ashish K. Singh, K. R. Balasubramaniam
Department of Energy Science and Engineering, Indian Institute of Technology Bombay,
Mumbai -400076, India.
144170001@iitb.ac.in

ABSTRACT: The effect of O_2 partial pressure during deposition on the properties of 1% Lanthanum doped Barium Stannate (LBSO) thin films is studied. X-ray diffraction revealed the growth of [001] oriented LBSO films at higher O_2 partial pressures. The transparency of the films also increases as we increase the deposition pressures. Around 300 nm thick films deposited at $P_{O_2} \geq 0.56$ mTorr, have ~80% transmittance in visible range (400-700 nm). It was observed that either extremities of deposition pressure were unfavourable for superior electrical properties. Values of electron concentration (n_e) and Hall mobility (μ) was found to be 6.09×10^{20} cm^{-3} and 13.30 cm^2/V-s, for the films deposited at 5.6 mTorr P_{O_2}. The high figure of merit (FoM: 2.08 mS) of LBSO is comparable to that of other commercially available TCOs and thus makes it a good candidate for the transparent electron selective layer in solar cells.

Keywords: Perovskite, Deposition, Electrical Properties.

1 INTRODUCTION

$BaSnO_3$ (BSO) is an oxide perovskite which crystalizes in cubic structure (a = 4.11 Å) and has a bandgap of 3.1 eV [1], [2]. Intrinsically, it is an insulator but external dopants such as La or Sb at Ba or Sn site make it an n-type degenerate semiconductor. Over the time, La has proven to be a good n-type dopant in several organic-inorganic and purely inorganic perovskites [3]–[6]. It was observed by Kim *et al.* that a 4% doped single crystal of LBSO gives a mobiltity (μ_e) of 300 $cm^2.V^{-1}s^{-1}$ with a carrier concentration (n_e) of 8.0×10^{20} cm^{-3} [7]. Similarly, a 30 nm thick film deposited by MBE on a $PrScO_3$ substrate shows a mobility of 150 $cm^2.V^{-1}s^{-1}$ at a carrier concentration of 7.2×10^{19} cm^{-3} [8]. For Sb doped films a moblity of 1.75 $cm^2.V^{-1}s^{-1}$ is found at a carrier concentration of 1.65×10^{21} cm^{-3} [9]. Owing to the fact that Sb doping doesn't compete well in improving the electrical properties, La doping is explored more frequntly.

In addition to external doping, oxygen vacancies (an inherent property of oxide perovskites) arised due to deposition pressure (P_{O_2}), plays a vital role in determing the carrier concentration and subsequntly, the electrical properties of thin films. In practical, one should observe an increase in carrier concentration as the deposition pressure reduces. But, several studies on pressure dependent ITO deposition have revealed that the films deposited in the range of 20-40 mTorr show the least resistivity [10]–[12]. The resistivity increases as the deposition pressure falls beyond this window. A severe deficiency of oxygen during deposition leads to highly disordered film and poor crystallinity which is unfavourable for electron mobility, resulting into an increment in resistivity below a certain P_{O_2}. Similarly, at higher partial pressure, oxygen vacancies are supressed due to the presence of ample amount of oxygen leading into an increased overall resistivity. Thus, an optimum window of deposition pressure is required for films to be enough conducting to be used in practical purposes like transparent conductors.

Recently, Liu *et al.* have done a pressure dependent study of LBSO growth on MgO substrates *via* Pulsed laser deposition (PLD), wherein, the deposition pressure was varied between 0.22-150 mTorr [13]. For the films deposited at 2.2 mTorr the resistivity was found to be minimum (0.8 mΩ-cm) and it increases towards both ends of the P_{O_2}. For example, the resistivity of the film deposited at 150 mTorr was found to be 1.03×10^8 mΩ-cm, which is significantly high to be conducting electrons through it. Similarly, the carrier cncentration decreases by 8 orders of magnitude as the deposition pressure is varied from 0.22 to 150 mTorr. Clearly, the films become inslulating when deposited at extremely high pressures in absence of oxygen vacancies and external dopants. It is noteworthy that the pressure in Liu's report was varied vastly from 0.22 to 150 mTorr, thus having a larger leap values of P_{O_2} (0.22, 2.2, 11, 37, 150 mTorr). In order to bring down the whole window of deposition pressure, this report summarizes the optical and electrical propeties of 1% doped LBSO thin films deposited *via* PLD in a relatively smaller pressure range (0.02-18 mTorr). Additionally, the films were deposited with the smaller intervals of P_{O_2} (0.02, 0.06, 0.18, 0.6, 1.8, 0.6, and 18 mTorr), to understand even the miniscule change in its properties within this range. In summary, the current work will aid us to identify the limits of oxygen partial pressure for LBSO thin film deposition.

2 EXPERIMENTAL DETAILS

1% doped LBSO thin films were grown on [001] oriented single crystal LSAT substrates by PLD. Single side polished LSAT substrates were surface treated by ultra-sonicating it in deionized water and 2-Propanol followed by annealing under flowing oxygen at 1000 °C for 2 hours [14]. KrF laser (λ=248 nm) was used for the deposition and the laser energy was kept constant as 2 $J.cm^{-2}$. Parameters such as target to substrate distance, substrate temperature and laser frequency were fixed at 5 cm, 800 °C and 5 Hz respectively. As mentioned earlier, the pressure was varied between from 0.02 to 18 mTorr while keeping the thickness of the film constant as 300 nm. The structural characterization of the films was done using X-ray diffraction (XRD) (Rigaku Smart Lab). The patterns were obtained between 2θ range of 20°- 50°. Room temperature electrical properties were measured using hall measurement system (Lakeshore 3404) in a van der Pauw geometry. Optical properties of the films were obtained by UV-Vis spectroscopy (Perkin Elmer Lambda 900).

3 RESULTS AND DISCUSSION

For structuaral analysis, XRD pattern of the films were taken in the 2θ range of 20-50°. LSAT being a perovskite substrate (a = 3.86 Å), is expected to fecilitate an epitaxial (00l) growth of LBSO thin films over it. All the films tend to show the desired (002) plane with its peak positioned at 2θ value of ~43.77°. However, quite interestingly, films also show an X-ray reflection corresponding to its (011) plane which is at a 2θ value of ~30.45°. From the XRD pattern of the films (shown in Fig. 1) it can be observed that (011) plane is not only present in all the films, it also becomes predominant as the partial pressure of oxygen is reduced during the deposition. Table-1 summarizes the intensity ratios of the (011) to (002) planes and a decresing trend can be observed. Moreover, the intensity of the peak corresponding to (001) plane of BSO (2θ = ~21.43) also gets inensified for $P_{O_2} \geq 0.60$ mTorr.

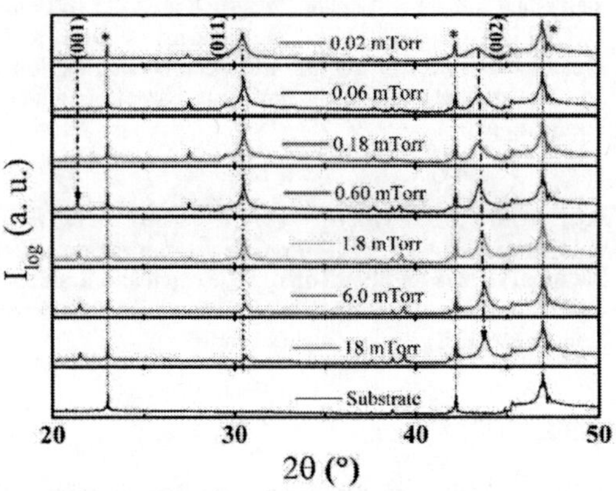

Fig. 1 X-ray diffraction of the films deposited at various oxygen partial pressures. Peaks indexed as * corresponds to the (001) and (002) planes of the LSAT substrates. Owing to the high crystallanity of the substrate, reflections from (002) plane of the substrate by Cu-K_β is also observed at 2θ value of 42.20°.

Table 1 Intensity ratios of the peaks corresponding to (011) and (002) planes in ascending P_{O_2} order.

Pressure (mTorr)	$I_{(011)}/I_{(002)}$
0.02	33.163
0.06	25.911
0.18	26.853
0.6	2.256
1.8	0.228
6.0	0.014
18	0.006

The peak position of (002) plane was used to calculate the lattice parameters of the deposited films. Fig. 2 shows the variation in the lattice constant as the deposition pressure of the films are varied. It is observed that the films deposited at 0.02 mTorr has a lattice constant of ~ 4.15 Å suggesting that the film is not properly relaxed. The value of lattice constant saturates for P_{O_2}between 0.06 to 1.8 mTorr to ~4.13 Å and then reduces eventually. Lattice constant for the film deposited at 18 mTorr is calculated to be ~4.12 Å which is close to the lattice constant of the bulk BSO (4.11 Å). This suggests that the films are more relaxed when deposited at higher pressures. The observed trend of the lattice constant could also be attributed to the poor crystallinity of the films at low pressures. Similar result was obtained by Liu *et al.* for the film deposited at 0.22 mTorr, wherein they suggest that the oxygen deficiency can lead to the deformity in the crystal structure by introducing coulombic repulsion between Sn and Ba cations [13]. Which, in turn, increases the volume of the unit cell and subsequenlty forces the lattice constant to expand. As a result, the films consist high amount of a dislocations and grain boundaries which eventaully inhibit the electrical properties. As more oxygen is introduced into the crystal, coulombic repulsion starts decaying and the unit cells begin to aquire their desired structure resulting into a smoother and relaxed epitaxial film.

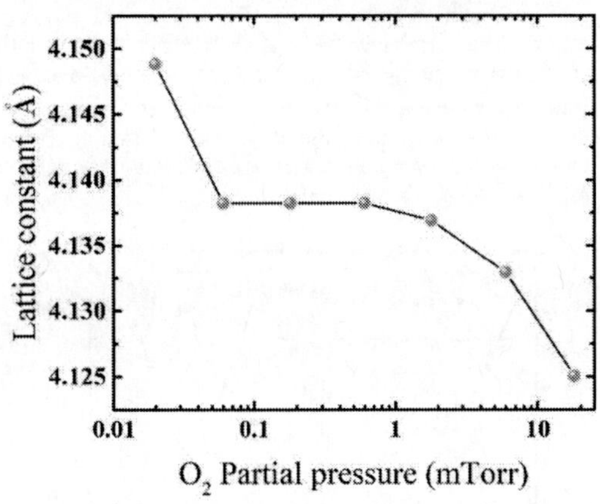

Fig. 2 Lattice constant of LBSO films as a function of partial pressure of oxygen.

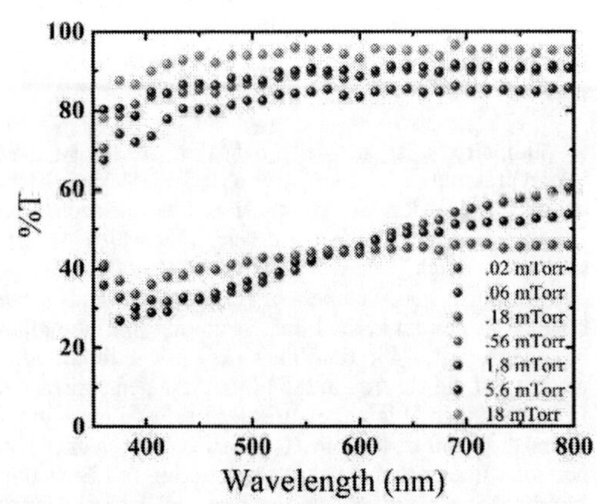

Fig. 3 Substrate corrected transmittance of the films deposited at various pressures.

UV-Vis characterization (shown in Fig. 3) of the films revealed that the films deposited at lower pressure had lower transparency in the visible range (400-700 nm). With ~50 % transparency, films deposited at pressures ≤ 0.18 mTorr, are not suitable for transparent conductor applications. This behavior is presumably because of the elevated defect densities in the films due to oxygen deficiencies, which further causes numerous sub band gap states accrediting to the high opacity of the films. On the contrary, the films deposited at pressures ≥ 0.56 mTorr had almost similar transparency (≥ 80%). A similar trend was observed by Kim *et al.* where the tranmittance of ITO films increased from 20 % to 90 % and saturated afterwards for the P_{O_2} values of 40 mTorr [11]. Likewise, it can be understood from the transmittance spectrum that a $P_{O_2} \geq 0.56$ mTorr is mandatory for LBSO films to have significant transmittance in visible range.

Room temperature electrical properties obtained by hall measurement of the films are shown in Fig. *4*. Film deposited at 0.02 mTorr was found to be extremely resistive to be measured. However, the n_e of the films deposited at 0.06 mTorr and was found to be 6.41×10^{19} cm^{-3}. Such high carrier concentration at room temperature can only occuer in a degenrate semiconductor. Thus, a sudden jump from insulating to metallic behaviour in the films was observed within a small leap of 0.02 to 0.06 mTorr. The carrier concentration further increases upto 6.02×10^{19} cm^{-3} and then staurates for the films deposited at ≥ 1.8 mTorr. This behaviour of enhanced carrier density with respect to ascending P_{O_2} is intruguing as one would expect it to follow the opposite trend due to oxygen vacancies. A possible reason could be the partial annihilation of charge carriers at the grain boundaries and dislocations which act as a trap center for free carriers.

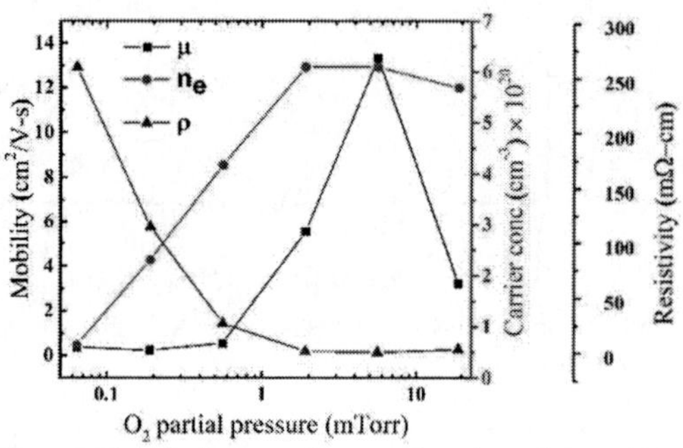

Fig. 4 Electrical properties as a function of deposition pressure.

Similalry, a strong dependence of resistivity over partial pressrue of deposition was observed. The resistivity changes two orders of magnitude and as the deposition pressure is increase upto 1.8 mTorr, after which it starts saturating. High values of resistivity could again be correlated to the number of dislocations and grain boundaries present in the films. Scattering form the grain boundaries and dislocations play a key role in limiting the mobility of the electron in the films. As stated earlier, the films deposited at low pressures tend to have more grain boundaries and disloctions. It has also been deduced in previous studies that the type of scattering in LBSO thin films can be classified on the basis of carrier density present in it. For $n_e < 10^{20}$ cm^{-3}, dislocation scattering is the dominant and for $n_e \geq 10^{20}$ it is mainly because of phonon electron scattering and ionized impurity scattering. Since our n_e values fall within the range dislocation scattering, we expect that it hinders the flow of electron, shooting up the resistivity values in the film. It is noteworthy that the μ was found to increase from 0.37 cm^2/V-s to 13.3 cm^2/V-s for the films deposited at 0.02 mTorr and 5.6 mTorr, respectively.

4 CONCLUSIONS

Role of oxygen partial pressure for LBSO thin films was studied. A [001] oriented growth can be observed only at high pressures. Films deposited at low pressures are highly disordered and contain plenty of deformity to limit the optoelectronic properties of the film. The films deposited at extremely low pressures (0.02 mTorr in our case), are almost as insulators and no electrical characterization could be performed. Additionally, the films deposited under ≤ 0.18 mTorr are also extremely resistant to be used in a practical use. These films are opaque in nature and in the visible range (400-700 nm) never show a transmittance above ~50%. However, as the deposition pressure is increased further (≥ 1.8 mTorr), the conductivity as well as transparency of the films improves in few orders of magnitudes. Film grown at 6 mTorr P_{O_2} has a transparency of ~90% in visible range and ne and μ for the same film was found to be 6.09×10^{20} cm^{-3} and 13.30 cm^2/V-s, respectively. The figure of merit of this film was calculated using the formula ($\varphi_{TS} = T^{10}/R_S$) and was found to be 2.08 mS which is comparable to any other well stablished transparent conductors such as AZO, IZO and FTO [15]. Owing to its high transparencey of ~90 % and a metallic behavior, LBSO thin films can be well exploited as a candidate material for transparent conductor applications.

ACKNOWLEDGEMENT

This work was made possible by the supported of National Center for Photovoltaics Research and Education (NCPRE), funded by Ministry of New and Renewable Energy (MNRE), Government of India.

REFERENCES

[1] A. Prakash *et al.*, "Wide bandgap $BaSnO_3$ films with room temperature conductivity exceeding 10^4 S cm^{-1}," *Nat. Commun.*, vol. 8, p. 15167, 2017.

[2] B. Hadjarab, A. Bouguelia, and M. Trari, "Optical and transport properties of lanthanum-doped stannate $BaSnO_3$," *J. Phys. D. Appl. Phys.*, vol. 40, no. 19, p. 5833, 2007.

[3] A. Spinelli, M. A. Torija, C. Liu, C. Jan, and C. Leighton, "Electronic transport in doped $SrTiO_3$:Conduction mechanisms and potential applications," *Phys. Rev. B*, vol. 81, no. 15, p. 155110, 2010.

[4] A. Kumar, K. R. Balasubramaniam, J. Kangsabanik, Vikram, and A. Alam, "Crystal structure, stability, and optoelectronic properties of the organic-inorganic wide-band-gap perovskite $CH_3NH_3BaI_3$:Candidate for transparent conductor applications," *Phys. Rev. B*, vol. 94, no. 180105, p. 1, 2016.

[5] R. P. Wang and C. J. Tao, "Nb-doped $CaTiO_3$ transparent semiconductor thin films," *J. Cryst. Growth*, vol. 245, no. 1–2, p. 63, 2002.

[6] T. J. Jacobsson, M. Pazoki, A. Hagfeldt, and T. Edvinsson, "Goldschmidt Rules and Strontium Replacement in Lead Halogen Perovskite Solar Cells: Theory and Preliminary Experiments on $CH_3NH_3SrI_3$," *J. Phys. Chem. C*, vol. 119, no. 46, p. 25673, 2015.

[7] H. J. Kim *et al.*, "High mobility in a stable transparent perovskite oxide," *Appl. Phys. Express*, vol. 5, p. 061102, 2012.

[8] S. Raghavan, T. Schumann, H. Kim, J. Y. Zhang, T. A. Cain, and S. Stemmer, "High-mobility $BaSnO_3$ grown by oxide molecular beam

epitaxy," *APL Mater.*, vol. 4, no. 1, p. 016106, 2016.
[9] Q. Liu, J. Dai, Z. Liu, X. Zhang, G. Zhu, and G. Ding, "Electrical and optical properties of Sb-doped $BaSnO_3$ epitaxial films grown," *J. Phys. D. Appl. Phys.*, vol. 43, no. 455401, p. 1, 2010.
[10] J. B. Choi, J. H. Kim, K. A. Jeon, and S. Y. Lee, "Properties of ITO films on glass fabricated by pulsed laser deposition," *Mater. Sci. Eng. B Solid-State Mater. Adv. Technol.*, vol. 102, no. 1–3, p. 376, 2003.
[11] H. Kim, J. S. Horwitz, G. P. Kushto, Z. H. Kafafi, and D. B. Chrisey, "Indium tin oxide thin films grown on flexible plastic substrates by pulsed-laser deposition for organic light-emitting diodes," *Appl. Phys. Lett.*, vol. 79, no. 3, p. 284, 2001.
[12] S. H. Kim, N. M. Park, T. Kim, and G. Sung, "Electrical and optical characteristics of ITO films by pulsed laser deposition using a 10 wt.% SnO_2-doped In_2O_3 ceramic target," *Thin Solid Films*, vol. 475, no. 1–2 SPEC. ISS., p. 262, 2005.
[13] Q. Liu *et al.*, "High electrical conductivity in oxygen deficient $BaSnO_3$ films," *Journal of Alloys and Compounds*, vol. 655. pp. 389, 2016.
[14] D. Kim *et al.*, "Roles of the first atomic layers in growth of $SrTiO_3$ films on $LaAlO_3$ substrates," *Appl. Phys. Lett.*, vol. 74, no. 15, p. 2176, 2012.
[15] G. Haacke, "New figure of merit for transparent conductors," *J. Appl. Phys.*, vol. 47, no. 9, p. 4086, 1976.

OPTICAL AND COMPOSITIONAL PROPERTIES OF ALD GROWN TiO_x

Akdemir Ozan[1,2,*], Nasser Hisham[1], Zolfaghari Borra Mona [1,2], Turan Raşit [1,2,3], Bek Alpan[1,2,3]

[1]The Center for Solar Energy Research and Applications (GÜNAM), Middle East Technical University, Dumlupinar Blvd. 1, 06800 Ankara – Turkey
[2]Micro and Nanotechnology Graduate Program, Middle East Technical University, Dumlupinar Blvd. 1, 06800 Ankara – Turkey
[3]Department of Physics, Middle East Technical University, Dumlupinar Blvd. 1, 06800 Ankara –Turkey

* akdemir.ozan@metu.edu.tr, Tel: +903122105071

ABSTRACT: TiO_x is an n-type semiconductor with a high dielectric constant, that has been frequently employed in photovoltaic solar cell applications. Thin films of TiO_x have been recently demonstrated to act as an efficient electron selective and passivation layers for n-type crystalline silicon (c-Si) solar cells. In this study, optical and compositional characterizations of TiO_x thin films deposited by thermal atomic layer (ALD) using tetrakis(dimethylamido)titanium (TDMAT) and H_2O as precursors on n-type c-Si substrates, were investigated by using spectroscopic ellipsometry (SE) and x-ray photoelectron spectroscopy (XPS) techniques. Effects of substrate and precursor temperature as well as post-deposition annealing on the optical and compositional properties of ALD-grown TiO_x thin films were investigated.
Keywords: Electron Selective Layer, TiO_2, Atomic Layer Deposition, Chemical Composition, Optical Properties

1 INTRODUCTION

Titanium dioxide (TiO_2) is an n-type semiconductor that possesses attractive physicochemical properties, and thus, commonly utilized in a wide range of applications such as high dielectric constant gate oxides [1,2], anti-reflection coatings [3,4], biocompatible coatings [5,6] and passivation layers for solar cells [7,8]. TiO_2 thin layers have also been used as electron selective layers in crystalline silicon (c-Si) solar cells [8,9].

The requirement of high quality ultra-thin layers for c-Si solar applications have led researchers to utilize ALD technique to deposit TiO_x with superior electrical and optical properties [7,10,11,12]. ALD is a thin film deposition method which is capable of depositions of uniform and pinhole-free ultra-thin films over large areas. It is based on sequential chemical reactions of two or more precursors which take place at the surface of a material, inside a reaction chamber. The precursor doses are separated by purging periods that remove the byproducts and surplus precursors using an inert gas, such as N_2. Repetition of these reactions and purge periods in a cyclic manner results in thin film growth on the surface of the material conformal with its contour. One of the most attractive properties of ALD is its self-limiting nature, which can be defined as the self-termination of the reaction of precursors with the surface; once all the active catalytic sites are reacted. The importance of the self-limiting property reveals itself with the control it gives on the thickness at sub-angstrom levels [13,14]. ALD usually requires low pressure inside the reaction chamber, however, depositions at even in atmospheric pressure are possible [15].

ALD grown TiO_x thin films with $x \sim 2$ have recently been demonstrated to act as an efficient electron selective and passivation layers for n-type c-Si solar cells due to their promising electronic properties and low fabrication costs [12,16]. The electron selectivity of TiO_2 contacts on c-Si originate from the large valance band offset and small conduction band offset that blocks holes while allowing electrons to pass through [17]. ALD grown, full-area electron selective TiO_2 layers that are deposited between c-Si and Al back contacts, are proven to reduce the carrier recombination and contact resistance, resulting in an increase in the open circuit voltage (V_{oc}), fill factor (FF) and thus, the power conversion efficiency (PCE) of n-type c-Si solar cells. Further increase up to 21.6% PCE was achieved by inserting a thin tunnel SiO_2 layer between c-Si and TiO_2 film and the potential of TiO_2-based contacts for high-efficiency, low cost c-Si solar cells was highlighted [8]. By decreasing the thickness of TiO_2 and Al metallization layers, the design was improved and PCE over 22% was achieved with the reduction of contact resistivity and absorption losses [16]. ALD grown TiO_x films have also been studied on dopant-free asymmetric heterocontact (DASH) solar cells, as electron selective passivating contacts [18]. As a thin passivating interlayer between c-Si and LiF_x, to reduce the carrier recombination, TiO_x and a-Si:H(i) thin films were compared. Although it was shown that both c-Si/TiO_x/LiF_x/Al and c-Si/TiO_x/ a-Si:H(i)/Al stacks could be effectively applied as full-area electron selective contacts, a-Si:H(i) gave better efficiency results. Nevertheless, the high potential of ALD grown TiO_x as electron selective and passivating layer was once again observed. All these studies indicate that ALD grown TiO_x is a very promising material for photovoltaic applications.

In this study, TiO_x thin films were deposited on n-type c-Si substrates via thermal ALD using tetrakis(dimethylamido)titanium (TDMAT) and H_2O as precursors. Samples deposited with different substrate and precursor temperatures were characterized via spectroscopic ellipsometry (SE) and x-ray photoelectron spectroscopy (XPS) techniques to observe the effects of substrate and precursor temperature on the optical and compositional properties of ALD grown TiO_x thin films. In addition, effect of post-deposition annealing on the prepared thin-films were investigated.

2 EXPERIMENTAL DETAILS

TiO_x thin films were deposited on c-Si substrates via thermal ALD. Nitrogen (N_2) was used as carrier and purging gas and its flow rate was set as 20 sccm, which resulted in ~0.5 mbar pressure inside the reaction chamber.

Before the deposition of TiO_x thin films, c-Si substrates were cleaned by RCA process and just before loading into ALD reaction chamber, they were treated with diluted hydrofluoric acid (10%) to remove the oxide on the surface. During one ALD cycle, pulse time of both TDMAT and H_2O was fixed to 15 ms and N_2 purging time after each dose was 15 s. TiO_x thin films were grown on each sample for 300 cycles. Two different set of growth parameters were considered. In the first set, the precursor temperature (Tpre) was set to 100°C while the substrate temperature (Tsub) was varied between 80-300°C. In the second set, Tsub was fixed to 230°C and Tpre was varied between 90-100°C. Moreover, three of the TiO_x thin films were annealed at 250°C for 30 minutes to investigate the effect of post-deposition annealing on the optical and compositional properties of ALD-grown TiO_x.

Elemental compositions of the thin films were measured with XPS (PHI 5000 VersaProbe) equipped with monochromatic AlKα X-ray source. To remove the impurities on the surface, the films were sputter cleaned.

Optical properties of the ALD grown films were characterized by SE (Sobra-GES5E) in the range of 1.23-4.00 eV at an angle of incidence of 70°. After the measurements, the data from TiO_x thin films were fitted using Tauc-Lorentz dispersion model.

3 RESULTS AND DISCUSSION

XPS measurements revealed the Ti2p and O1s core level spectra of the samples in the first set as shown in Figure 1. The peak positions were corrected using the adventitious carbon C1s binding energy 284.6 eV. For the first set, Ti2p and O1s peak positions shifted to higher binding energies with increasing Tsub from 80 to 230°C. Further increase of Tsub from 230°C to 300°C shifted the peak positions back to lower binding energies. However, it can be seen from Figure 2 that neither changing Tpre nor post deposition annealing resulted in a significant effect on the peak position of Ti2p and O1s.

For samples in both sets, the binding energy difference between Ti2p3 (high intensity Ti peak at lower binding energy) and Ti2p1 (low intensity Ti peak at higher binding energy) peaks were found to be ~5.7 eV. Every measured peak were separately analyzed in terms of area under the curve and intensity of the peak, to calculate the atomic concentration ratio of O to Ti ([O:Ti]). O1s peak includes oxygen from both Ti-O bonds and OH species that forms organic C-O bonds on the surface. Therefore, to extract the amount of oxygen bonded to titanium, O1s peaks were de-convolved. The calculated O/Ti ratio revealed that the ALD-grown TiO_x in this study is oxygen deficient with different x values for each deposition condition. Results are shown in Table I.

Optical characterization of ALD grown TiO_x thin films were made with SE measurements. Figure 3 shows the thickness and bandgap of deposited films in set I and II and the corresponding refractive index (n) as extracted from SE measurements are tabulated in Table I. Band gap energies of the ALD-grown TiO_x thin films were not affected by changing the Tsub and Tpre or by post-deposition annealing. Thickness of TiO_x thin films showed a gradual decrease with increasing Tsub, but Tpre did not have a significant effect on thickness.

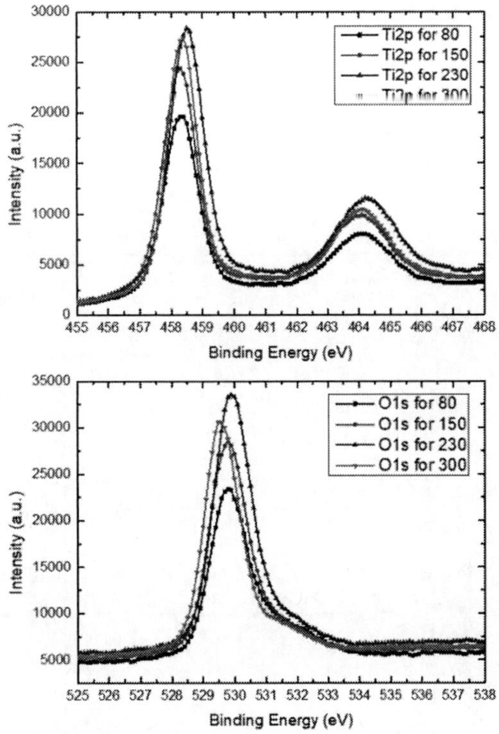

Figure 1: Ti2p (top) and O1s (bottom) core level spectra of the samples in set I (with fixed Tpre)

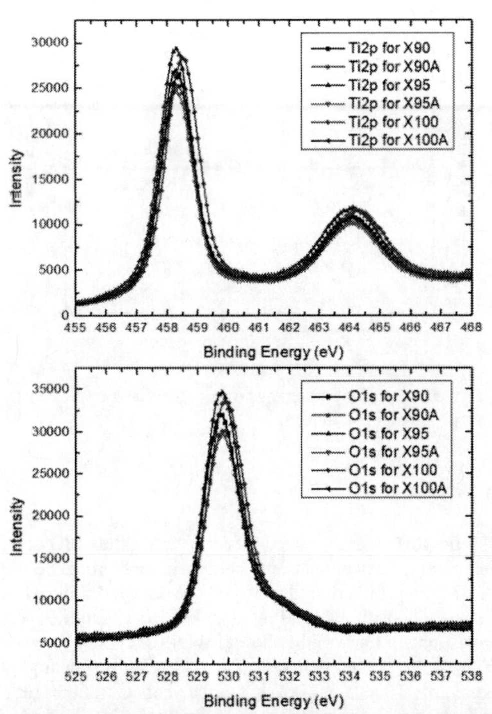

Figure 2: Ti2p (top) and O1s (bottom) core level spectra of the samples in set II (with fixed Tsub)

Table I: Process parameters and measurement results of ALD-grown TiOx thin films

Sample Set	Sample name	Tsub (°C)	Tpre (°C)	Post-Deposition Annealing	TiO_x	n @ λ=632.8 nm
I	80	80	100	-	$TiO_{1.70}$	2.210
	150	150		-	$TiO_{1.79}$	2.225
	230	230		-	$TiO_{1.91}$	2,220
	300	300		-	$TiO_{1.78}$	2,260
II	X90	230	90	-	$TiO_{1.90}$	2.320
	X90A		90	Yes	$TiO_{1.95}$	2.390
	X95		95	-	$TiO_{1.99}$	2.320
	X95A		95	Yes	$TiO_{1.89}$	2.300
	X100		100	-	$TiO_{1.91}$	2.220
	X100A		100	Yes	$TiO_{1.87}$	2.260

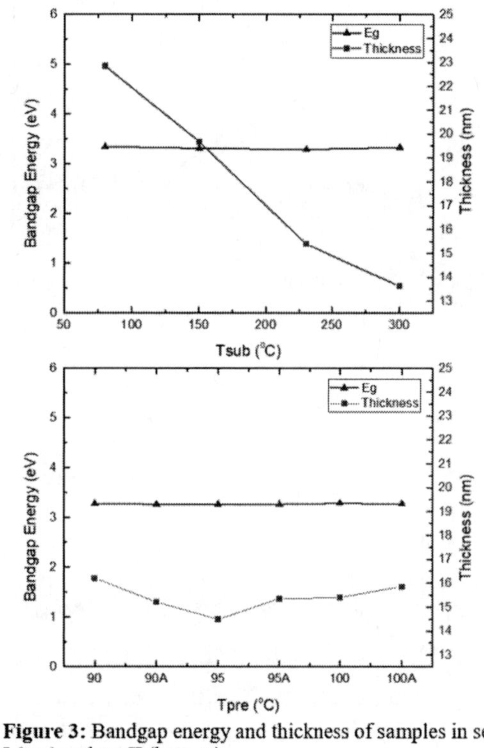

Figure 3: Bandgap energy and thickness of samples in set I (top) and set II (bottom)

4 CONCLUSION

In this work, effect of substrate and precursor temperature on optical and compositional properties of ALD grown TiO_x thin films was studied. In addition, effect of post-deposition annealing at 250°C for 30 minutes was investigated. Our results show that all the ALD deposited TiO_x were oxygen deficient. For the samples prepared at Tpre of 100 °C, [O:Ti] atomic concentration ratio increased with increasing Tsub from 80 to 230°C. Further increase in Tsub from to 300°C reduced [O:Ti] ratio. However, the films deposited at Tsub of 230 °C at different Tpre had high [O:Ti] ratio. Finally, no significant change in band gap was observed by changing Tsub and Tpre.

5 REFERENCES

[1] J. Robertson, Reports on Progress in Physics, **69 (2)** (2005) 327-396.
[2] K.F. Albertin, M.A. Valle, I. Pereyra, ECS Transactions, **4 (1)** (2006) 409-416.
[3] R. Prado, G. Beobide, A. Marcaide, J. Goikoetxea, A. Aranzabe, Solar Energy Materials & Solar Cells, **94** (2010) 1081–1088.
[4] S. Jeong, J. Kim, B. Kim, S. Shim, B. Lee, Vacuum, **76** (2004) 507–515.
[5] G.A. Crawford, N. Chawla, K. Das, S. Bose, A. Bandyopadhyay, Acta Biomaterialia, **3** (2007) 359–367.
[6] D.S. Koktysh, X. Liang, B. Yun, I. Pastoriza-Santos, R.L. Matts, M. Giersig, C. Serra-Rodriguez, L.M. Liz-Marzân, N.A. Kotov, Advanced Functional Materials, **12(4)** (2002) 255.
[7] B. Liao, B. Hoex, A.G. Aberle, D. Chi, C.S. Bhatia, Applied Physics Letters, **104(25)** 253903 (2014).
[8] X. Yang , Q. Bi , H. Ali , K. Davis , W.V. Schoenfeld, K. Weber, Advanced Materials, **28** (2016) 5891–5897.
[9] H. Kim, K. Ou, X. Wu, P.F. Ndione, J. Berry, Y. Lambert, T. Mélin, N.R. Armstrong, S. Graham, Journal of Materials Chemistry A, **3** (2015) 17332.
[10] G.T. Lim, D. Kim, Thin Solid Films, **498** (2006) 254 – 258.
[11] Q. Xie, Y. Jiang, C. Detavernier, D. Deduytsche, R.L. Meirhaeghe, G. Ru, B. Li, X. Qua, Journal of Applied Physics, **102** 083521 (2007).
[12] D. Ahiboz, H. Nasser, E. Aygün, A. Bek, R. Turan, Semiconductor Science and Technology, **33(4)** (2018) 045013.
[13] J. Iqbal, A. Jilani, P.M.Z. Hassan, S. Rafique, R. Jafer, A.A. Alghamdi, Journal of King Saud University – Science, **28** (2016) 347–354.
[14] E. Guziewicz, M. Godlewski, L. Wachnicki, T.A. Krajewski, G. Luka, S. Gieraltowska, R. Jakiela, A. Stonert, W. Lisowski, M. Krawczyk, J.W. Sobczak, A. Jablonski, Semiconductor Science and Technology, **27** (2012) 074011.
[15] J.S. Jur, G.N. Parsons, ACS Applied Materials & Interfaces, **3** (2011) 299–308.
[16] X. Yang, K. Weber, Z. Hameiri, S.D. Wolf, Progress in Photovoltaics: Research and Applications, **25(11)** (2017) 896-904.

[17] J. Jhaveri, S. Avasthi, K. Nagamatsu, J.C. Sturml, IEEE 40th Photovoltaic Specialist Conference (PVSC), (2014) doi:10.1109/pvsc.2014.6925206
[18] J. Bullock, M. Hettick, J. Geissbühler, A.J. Ong, T. Allen, C.M. Sutter-Fella, T. Chen, H. Ota, E.W. Schaler, S.D. Wolf, C. Ballif, A. Cuevas, A. Javey, Nature Energy, **1(3)** (2016) 15

INTEGRATION OF FLEXIBLE SOLAR CELLS IN PLASTIC AND COMPOSITE MATERIALS

Sanne Kristensen, Hugo de Moor, Martèn Driesser, Edwin Geldof and Karel Spee
Avans University of Applied Science
Onderwijsboulevard 215, 5223 DE 's-Hertogenbosch, The Netherlands
E-mail: hhc.demoor@avans.nl +31 6 5787 5291

ABSTRACT: Processes have been developed to integrate CIGS solar foils directly in plastic or composite based products. Using injection moulding and vacuum forming it turned out, that mechanical stress is applied to the solar foil, upon cooling. Using intermediate stress relief layers, such as polyolefin, this stress could be reduced considerably. The high temperatures needed in injection moulding have only a minimal effect on the performance of the solar cells. Making curved products is possible with the solar foils, leading to new markets for solar energy. Outdoor performance is attractive, but indoors, only applications that require little power have potential.
Keywords: Flexible Substrate, CIGS, Product Integration

1 INTRODUCTION

Application of solar energy is focused on large-scale applications to battle climate change and reduce greenhouse emissions. Forgotten is often the use of solar energy to power small autonomous devices. The value of the energy is much higher than bulk electricity. Therefore, it is worthwhile to look at the integration, physically and electrically, in appliances ranging from lighting to roof tiles and from sensors to road infrastructure. Producers of these products are interested in the energy harvesting and aesthetically nice integration while maintaining the process conditions of their manufacturing equipment.

This paper focuses at the physical integration of thin-film flexible solar foils, to be able to make curved products. New processes are developed to integrate flexible solar cells into plastic and composite materials, based on well-established techniques as injection moulding, vacuum forming and vacuum bagging. This creates a broad range of design and application possibilities. Using solar foils, it is possible to make PIPV (Product Integrated Photovoltaics) products, that need not to be flat. This gives new opportunities for PV and opens markets where crystalline silicon cannot be used.

As seamless integration of flexible solar cells in the production process is aimed for. Harsh production conditions such as high temperature and high pressure must be dealt with. Knowledge about the production processes of plastics and composites must be connected with the expertise on thin-film flexible CIGS. In this research, an important aspect is the adhesion of the cell to the plastic or composite material.

2 INTEGRATION TECHNIQUES

The industrial partners in the SEPAC project[1] are producers of plastic and composite based products, that are highly interested in the possibilities to use solar energy, to partially power their products. Commonly the solar modules are attached to the products in the final stage. From a process and aesthetical point of view, it is favoured to apply the solar cells already in the production process. This can be done with solar foils, as they can form to the product shape. The research question is, to what extent the temperature and mechanical stress can be dealt with, by the solar cells and whether good adhesion can be achieved.

2.1 Injection moulding

Most plastic products are manufactured with injection moulding. Plastic pellets are dried, and then fed through a hopper in a heated chamber, where they melt. With a hot screw-type plunger, the molten plastic is under high pressure, pressed in a heated mould. After solidifying, the mould is opened and the product removed. Moulds are very expensive, and the time to prepare the pellets considerable, so, it is mostly used for large series.

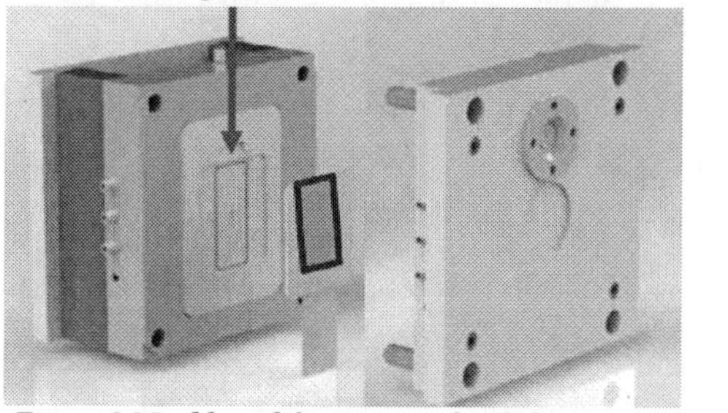

Figure 1 Mould used for testing solar foils

In Figure 1 can be seen, that thin plates are made with the solar foil underneath. Injection parallel to the solar foil is favoured, to minimize the force exerted on the solar cell, and gives a uniform plastic sample. This set-up is used to test the adhesion of the foil to the plastic.

2.2 Vacuum forming

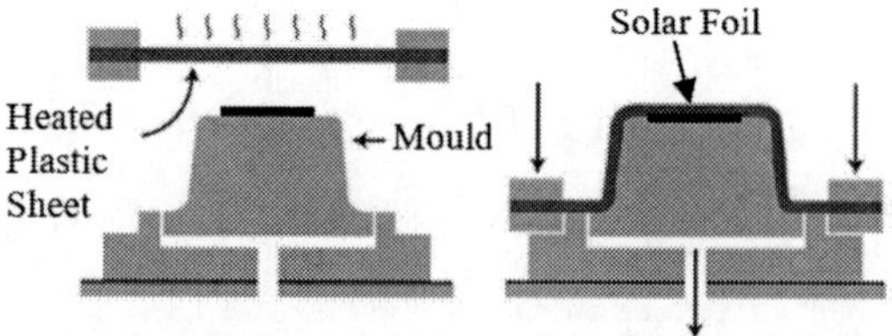

Figure 2 Vacuum Forming

In Figure 2 can be seen that a thermoplastic sheet, like ABS or PETG, gets hot using IR heaters. It softens and is stretched over a mould. In this work a solar foil is laid on a warm mould and pressed into the plastic sheet.

2.3 Composites

For structural materials, resins reinforced with fibers are used. The most commonly used resins are polyester, epoxy or vinyl ester. Fibres woven to give directional strength can be transparent, glass-based, enabling to be on the sunny-side of the solar cells or non-transparent, such as carbon or aramid.

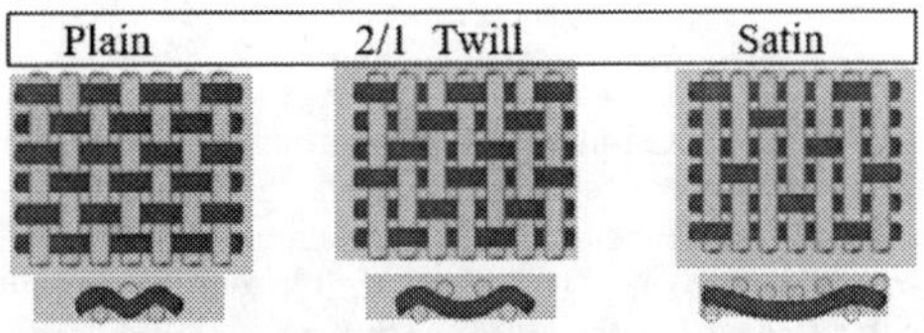

Figure 3 Fibres, different weave patterns, resin filled (red)

3 SOLAR FOILS

In this study CIGS flexible ICI solar submodules are used from Global Solar[2]. Six cells of 220 mm width, are connected in series via 5 dots to result in a 3.3 V module. The submodules can be cut in rectangular shapes to fit with the required product. This will lower the current and/or the voltage. Preliminary tests have shown that different forms are also possible.

Figure 4 CIGS ICI module
Horizontal cut lowers the current
Vertical cut gives a discrete number of cells

The modules, as received, are encapsulated in thin PET foil and have a limited lifetime. Depending on the application the encapsulation must be adapted. For outdoor use it is needed to laminate the modules with polyolefin and a 125 μm barrier PET foil to prevent moisture ingress.

3.1 Mechanical stress

During processing considerable forces can act on the solar foils. Therefore, tensile testing is performed to get an impression, to what extent a module can be stressed and where the weak points are.

Figure 5 Solar Foil in Tensile Testing Machine

In Figure 5 can be seen that the contact area, where the cells are series connected is deformed, left, and delamination takes place, right contact. In Figure 6 can be seen that eventually the whole cell is delaminated.

Figure 6 Delamination upon stretching

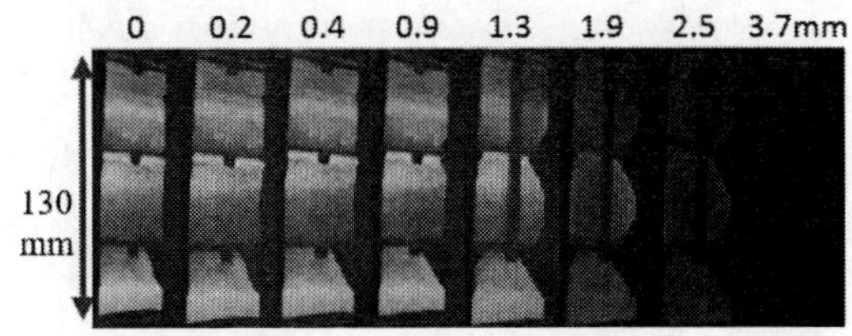

Figure 7 EL picture of a stressed solar foil

In Figure 7 can be seen that with an elongation of 1% (1.3 mm) or more the cell interconnects fail. The dark areas show that the electrical activity is lowered between the interconnects.

3.2 Thermal stress

ICI submodules should not be processed above 160°C[2]. Higher temperatures might be applied but only for a limited time. In chapter 4 will be shown that in injection moulding the solar foils can withstand process temperatures up to 250°C, though with a decreased efficiency.

3.3 Energy Yield

The first question to be answered is how much energy is harvested, in order to establish the storage capacity needed.

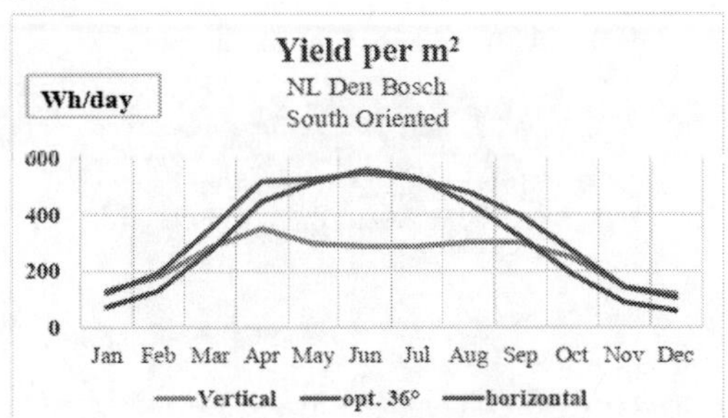

Figure 8 Energy yield of 1 m^2 ICI Modules in NL

The outdoor yield is calculated using PVGIS[3] for a 122 Wp system and a performance ratio (PR) of 85%. At maximum an average of 500 Wh/day can be harvested.

There is a great distinction between outdoor and indoor use. Indoors the light intensity is limited. In an office with only indoor lighting the light intensity is typically 500 Lux ≈ 5 W/m^2. The PV power is then 0.6 W/m^2, assuming an efficiency of 12%. For the present generation CIGS this is far too optimistic [4]. So, ultimately, in 8 hours the energy yield is at maximum 5 Wh for a 120 Wp system. The battery of a Samsung S8 has a capacity of 12 Wh. So, in the best case 1 m^2 of CIGS it takes 16 hours or two days to charge a smart cell phone. In case of a window, giving day light in a room, energy harvesting becomes far better.

4 INJECTION MOULDING

As a first step the adhesion of several foils to the injected plastic is studied. For 4 materials a plasma treatment, Surface Dielectric Barrier Discharge (SDSB) with N2, was used.

Table 1 Adhesion of foils on injected plastic

Foil / Mass	PC	PP	PO	PFA	PFA plasma	ETFE	ETFE plasma	PET	PET plasma	PEN	PEN plasma
PC											
PP											
ASA											
PBT											
PBT-PC											
PET											
Tritan											
PS											
ABS											

No adhesion
Limited (≈ 3 MPa), not reproducible adhesion
Strong adhesion (> 7 MPa)

In Table 1 can be seen that only the combination of the foils, polypropylene and polycarbonate, with their own injected plastic, give a proper adhesion, as tested according to ASTM D4541 with a Dolly. For PC and PP the glue to mount the dolly was the weak point, not the PC-PC or the PP-PP interface.

Adhesion on PET is at best poor. This means that adhesion on the solar foil, with PET as outer surface, needs a new approach. As can be seen in Table 1 PP-PP and PC-PC are promising combinations. So, when laminating the solar foils first PO, acting as adhesive, is added. then an additional layer of polycarbonate, 125µm (Lexan 8010 of Sabic), was added or polypropylene, 25µm (Innovia Propafilm RGP) or 300µm (Priplak Opaline).

Figure 9 Solar Foil with PC foil 125 µm on injected PC

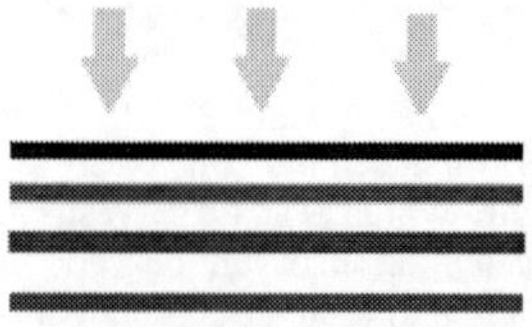

Figure 10 Lay-up for sample shown in fig.9 and fig.11

It can be seen in Figure 9 that a flat sample is obtained using polycarbonate. No detachment is observed after 800 hours Damp Heat (85°C, 85% humidity). The sample can be placed under or above the polycarbonate as the layers are transparent. In this simple structure it is assumed that the thick injected PC or PP layer protects the solar foil from water ingress.

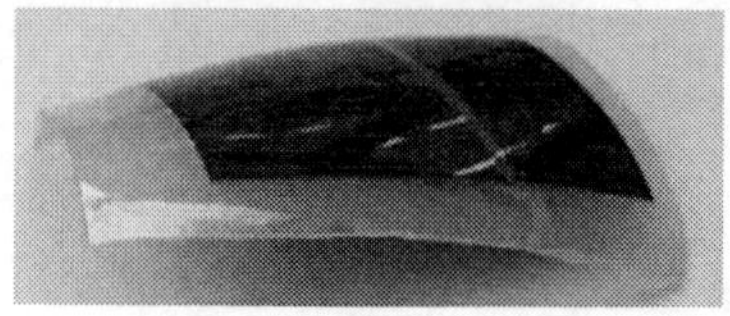

Figure 11 Solar Foil with PP foil 300 µm

The process applied for polycarbonate doesn't work for polypropylene as can be seen in Figure 11. The shrinkage of PP results in a bent sample, as the PET-encapsulated solar foil doesn't shrink with temperature. Such a curved sample cannot be put in the mould; Figure 1 So, a symmetric lay-up is made, as shown in Figure 12.

Figure 12 Symmetric lay-up

Figure 13 Solar Foil on PP with a symmetric lay-up

In a symmetric lay-up the bending of the PP-sample is prevented. This allows to sample to be put in the mould. However, the difference in thermal expansion between the inject PP and the PET foil still gives bending as can be seen in Figure 14.

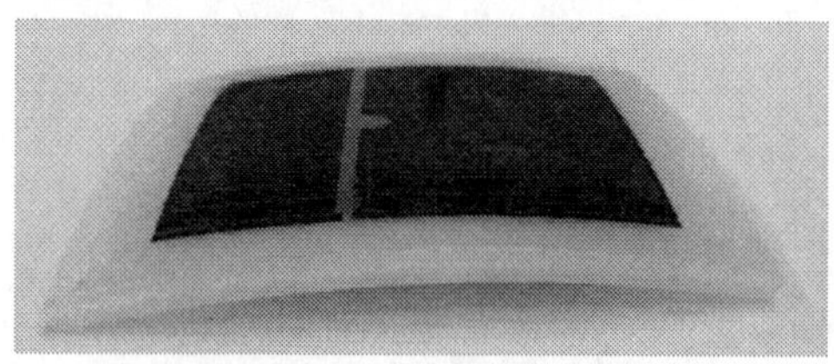

Figure 14 Solar Foil with PP foil 25 µm on injected PP

The pressures in the injection moulding process were 160 bar and temperatures up to 250°C. In Figure 15 can be seen that after injection moulding the cell efficiency drops significantly. For 200°C the cell performance restores to a large extent after 3 days. It was tested whether light soaking[5] might restore the cells. Light soaking (LS) for 20 hours, indeed restores the cell performance to a large extent. However, 64 hours of soaking, influenced the cell efficiency negatively. With a proper light soaking even temperatures up to 250°C can be used for injection moulding. It should be realised that the cells will not reach 250°C, because the mould is at a lower temperature. In addition, the total process time is only tenths of seconds.

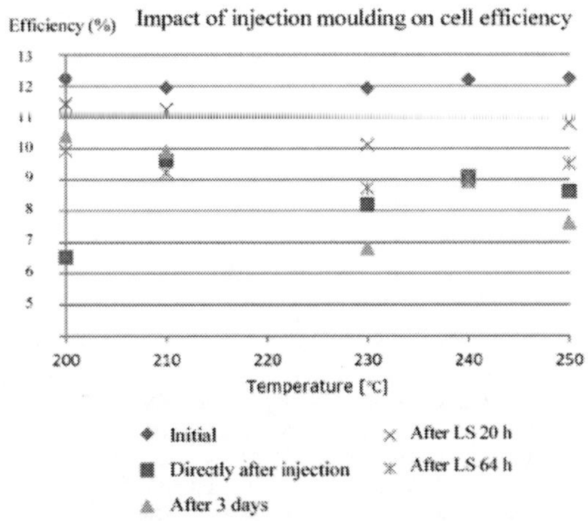

Figure 15 Impact of injection moulding temperature on cell efficiency (measured at STC)

5 VACUUM FORMING

The challenge when integrating solar foils in vacuum formed plastic products is handling the shrinkage of the plastic. In this work PETG and ABS are used. Both show similar behaviour, so only PETG is described in this paper.

PETG is heated to 130-140°C. The plate is very soft then and can be stretched easily. After removal of the IR-heaters, vacuum is applied and the PETG is stretched over the mould. After cooling down to 80°C, the plate with mould is taken out and the mould is removed.

Figure 16 Displacement of a solar strip

To understand the process, three encapsulated solar strips where marked on the PETG and the end of the strip. Filming the process, the marking didn't move the first seconds, so the strip moved with the PETG when the plate was stretched. In the final stage of stretching, after 7 seconds and a temperature of ≈100°C, the marking moved. During cooling down further the distance between the marking remained constant indicating that the solar strip was fixed to the PETG.

Initially small solar foils where used, but when a large product was made with a complete submodule, 29x22 cm, the module detached after a few hours, as the PETG continues to shrink even when at room temperature. Using an acrylic adhesive on PET gave better adhesion, but there still was considerable stress on the solar foil. Therefore, an intermediate stress relief layer was used.

Finite element calculations learn that a combination of polyolefin and polyurethane on the PET-barrier of the solar foil could be a solution to release the stress caused by the cooling of the PETG. PU diminishes the stress of the PETG, while PO has a small Young's modulus and converts the residual stress in deformation.

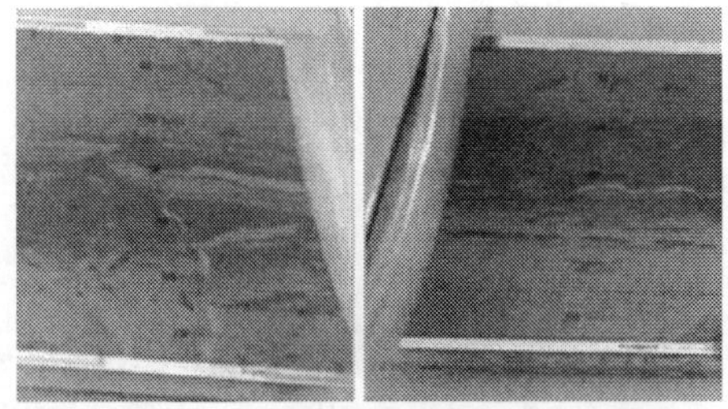

Figure 17 Solar Foils directly on PETG with intermediate PO/PU

In Figure 17 can be seen that most of the stress is relieved in the PO/PU-layer. Further progress is expected from an adhesive on top of the PU.

In Figure 18 can be seen that the solar module and its interconnects withstand the stretching in the process and that a curved surface (closer to the lamp) shows a good performance.

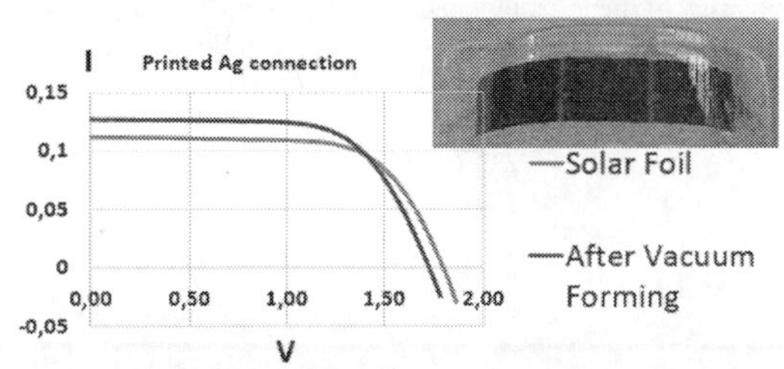

Figure 18 Performance before and after vacuum forming

6 COMPOSITES

With composite materials very large elements can be made such as bridges, pavement of solar roads and wind turbine blades. This is a limitation of injection moulding and vacuum forming. Another advantage is the absence of thermal stress as the processes are at room temperature or slightly above. Adhesion of PET to the resin is obviously still a requirement.

In this study an inventory was made of suitable techniques to integrate solar foils in composite products.

Table 2 Inventory of techniques to integrate solar foils with composites

Methods	Pro's	Con's
Hand Layup	- No high tech equipment - Simple method	- Lots of air bubbles enclosed in the resin
Vacuum Bagging	- Amount of air bubbles minimized - Easy control of resin distribution through the fibers	- Vacuum pump - Knowledge and experience
Vacuum Infusion	- Amount of air bubbles is minimized - The flow can be adjusted for different configurations	- Complicated process - Knowledge and experience - Vacuum pump and suitable resin

In Table 2 are the three most suitable techniques depicted. Experiments were performed to test the feasibility of these techniques. It turned out that with a hand layup too many air bubbles were enclosed in the product. This is aesthetically unacceptable and may have an influence on lifetime and strength of the composite.

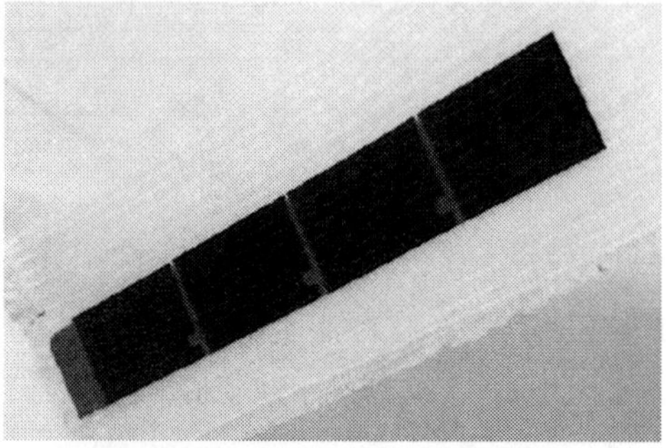

Figure 19 Solar Foil in composite using vacuum infusion
Back glass fiber and polyester
Front glass fleece and polyester

In vacuum infusion the resin is drawn through a dry stack of fibres using vacuum. The flow of resin through the fibres must reach the complete fibre package, so the position of the inlet, the quantity of resin and viscosity that increases in time must be well controlled. So, in this case a simple flat product is made on a flat surface.

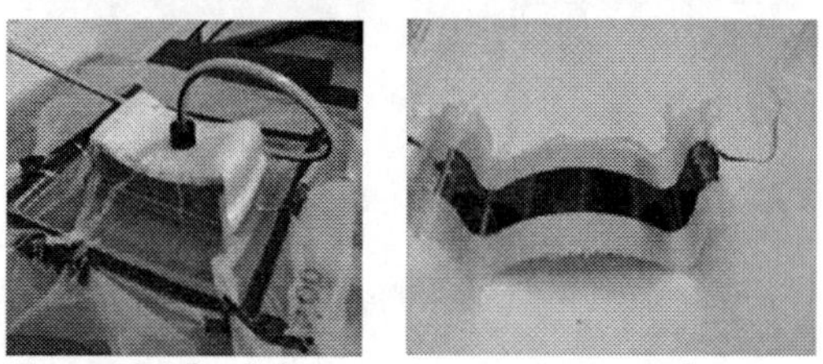

Figure 20 Solar foil in composite using vacuum bagging

In vacuum bagging the fibre layup is impregnated with resin. The vacuum is used to remove the excess of resin. Complex and large products can be made. In Figure 20 on the left the mould with the layup is shown. On the right the resulting product is shown. The mould is negative as this results in a smooth surface on the sunny-side.

In damp heat tests, 1000 hours, the solar module did not delaminate from the composite.

7 CONCLUSION

With flexible solar foils, curved products can be made of different sizes and plastic materials. These solar powered products can be made with injection moulding, polycarbonate and possibly polypropylene, with vacuum forming, using an intermediate flexible layer and composite, preferably with vacuum bagging. As the solar foils are usually laminated in PET-foil that has a very low coefficient of thermal expansion, shrinkage cannot be accommodated by solar foils. It is needed to find foils that allow shrinkage and simultaneously protect the solar foil. A limited lifetime of the solar foil is acceptable, as long it survives the product.

8 ACKNOWLEDGEMENTS

This work is carried out in the frame of the Dutch RAAK-mkb programme coordinated by SiA in the project SEPAC: Solar Embedded Polymers and Composites. The authors would like to thank M. van den Nieuwenhof and M. Koetse of Solliance and the industrial partners in SEPAC.

9 REFERENCES

[1] www.avans.nl/onderzoek/projecten/detail/sepac
[2] www. globalsolar.com/original-equipment-manufacturers
[3] http://re.jrc.ec.europa.eu/pvgis/
[4] A. Sacco et al. Characterization of photovoltaic modules for low-power indoor application, Applied Energy 102 (2013) 1295–1302
[5] J.A. del Cueto et. al. Progress Toward a Stabilization and Preconditioning Protocol for Polycrystalline Thin-Film Photovoltaic Modules, NREL/CP-520-47748

EVALUATION OF DIFFERENT MODULE DESIGNS AND DETERMINATION OF DIFFERENT PHYSICAL LOSS MECHANISMS BY MEANS OF A PRACTICAL MULTI-PHYSIC MODEL

Hamed Hanifi[1,2]*, Jens Schneider[1]
[1]Fraunhofer-Center for Silicon Photovoltaics CSP, Otto-Eissfeldt-Str. 12, 06120 Halle (Saale), Germany
[2]Anhalt University of Applied Sciences, Bernburger-Str. 55, 06366, Koethen, Germany
*Corresponding author: +49 345 5589 5515, hamed.hanifi@csp.fraunhofer.de

ABSTRACT: Incident light on solar modules is either lost in reflection or heat generation or generates desired electricity. In this work, we introduce and extended evaluation of PV modules by means of simulation and measurements and quantify optical reflection losses, heat sources and electricity values in PV modules with different bill of materials based on the material stack level. We introduce our optical and electrical loss analysis model which evaluates PV modules in spectral dependent optical and voltage dependent electrical analysis which can be used for any defined spectrums in any location in the word. We show that almost 75% of the incoming incident light turns into heat sources with different loss mechanism while 7% to 10% of incoming light can be reflected out of the stack. We also show that modules with anti-reflection coating can benefit about 2-4%rel. extra efficiency compared to the modules with similar glass and no AR properties. The reduced reflection in glass level increases the efficiency as well as the heat sources of the module. The Module with UV+ polymer (Polyvinylbutyral, PVB) shows a low UV (Ultraviolet) absorption in the polymer in UV range and decreases the degradation effects such as discoloration. Furthermore, modules with PVB encapsulation materials show about 2.5% extra electricity generation in module level compared to the similar module with encapsulant without UV+ properties (ethylene vinyl acetate, EVA).

Keywords: Partial cells, Half-cell module, Third-cell module, Desert Module, PV Module, Module Manufacturing, Ribbons, Design, Electrical Properties, Experimental Methods, Optical Losses, ARC, PVB

1 INTRODUCTION

Common characterization methods of PV modules are based on Standard Test Conditions. This characterization gives the electrical characteristics of the PV modules typically until 1200 nm and do not consider the heat sources of the module which influence the performance of the PV modules. More detailed PV module indoor characterizations also target the cell-to-module power ratios [1, 2], but do not consider the heat sources and wider range of wavelengths.

Extended characterization of PV modules under Standard Test Conditions enables us to estimate the detailed reflection losses and heat sources in module level [3, 4]. Spectral dependent optical analysis and voltage dependent electrical losses in a PV module determine the heat sources and optical reflection losses in each layer of the PV module as well as electricity output at the terminals of the PV module [4]. Figure 1 demonstrates the interaction of heat, optical reflections and electricity in a PV module which the model is based on.

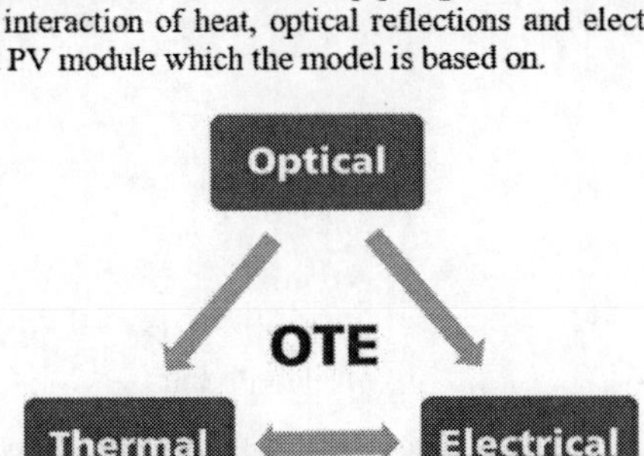

Figure 1: Trend of different physical processes which takes ina PV module and their interactions together.

In this work, we evaluate different stacks of module materials and their influence on module power and performance. For this purpose we chose glasses with different thicknesses and anti-reflection properties as well as two different encapsulations polymers. We determine the optical reflections and heat sources as well as electricity production for modules with different module stack properties.

2 EXPERIMENTAL SETUP

To evaluate the influence the module components on the performance of the PV modules, several module components were analyzed optically and electrically. Optical measurement is performed on each layers of the module stack by Perkin Elmer Lambda 1050 spectrometer equipped with an integrated sphere. Each module component is measured from 200 nm to 2200 nm. This translates to almost 977 Wm^{-2} input power density under AM 1.5g.

The electrical characteristics of the solar cells are measured by sun simulator and loss analysis tool LOANA to determine the quantum efficiency and power characteristics of the solar cells. The measurement device combines I-V measurement with spatially resolved analysis such as light-beam-induced-current and luminescence imaging.

Used materials		
Material	**Type**	**Specifications**
Glass	Float 1	No AR , 4 mm
		No AR , 2 mm
		AR , 2 mm
	Float 2	No AR , 3.2 mm
		AR , 3.2 mm
Polymer	EVA	No UV+
	PVB	UV+
Solar cells	Mono - Si	Selective emitter [4, 2]

Table 1: Overview of studied materials and their specifications for mini-module application

Different types of module stack materials are investigated by means of aforementioned methods. Table 1 demonstrates each module component type.

It should be mentioned that the glasses with AR properties are made from the similar related glass with no AR properties. This means that the glass bulk properties for AR and non-AR versions are almost similar. The electrical properties of the solar cells are similar to the cell measured in ref. [4, 2].

3 MODELLING AND SIMULATION

Modelling of the share of each module component to the heat sources, electricity and optical reflections is based on the model which is already developed in Fraunhofer CSP [4].

The loss analysis of a PV module is divided to spectral dependent loss analysis until generation of short-circuit current per each square centimeter and voltage dependent loss analysis after current generation. In this model [4], there are twelve loss factors to the incoming light to the module until power generation in the terminals. Equation (1) determines the output power of the module by considering the reduction factors,

$$P_{Electricity} = P_0 \times \prod_{i=1}^{m} f_i \qquad 1 \leq m \leq 12 \tag{1}$$

where $P_{Electricity}$ (W) is the output power of the module, P_0 (W) is the incoming incident power on top of PV module and f_i is the consecutive reduction factor which takes place by considering the travel path of light in the stack. For $m \leq 8, f_i$ determines the spectral dependent optical reduction factors and $m > 8, f_i$ refers to the constant reduction factors to determine the module power and efficiency. The consecutive reduction factors are

Reflection at air|glass interface $f_1(\lambda)$, Absorption in glass $f_2(\lambda)$, Reflection at glass|polymer interface $f_3(\lambda)$, Absorption in polymer $f_4(\lambda)$, Effective reflection from solar cell and cell|polymer interface $f_5(\lambda)$, Absorption of below band gap energy $f_6(\lambda)$, Thermalization losses $f_7(\lambda)$, Collection losses $f_8(\lambda)$, Reflection of interconnection f_9, Thermodynamic losses including non-radiative recombination, Carnot, angle mismatch f_{10}, Resistive losses in solar cell f_{11} and Resistive losses in interrconnection f_{12}.

The share of each power loss due to each reduction factor can be calculated by (2).

$$\Delta P_{rel.f_i} = \frac{\int (p_{i-1}(\lambda)(1 - f_i(\lambda))\, d(\lambda)}{\int p_0(\lambda)\, d(\lambda)} \times 100 \tag{2}$$

where $\Delta P_{rel.f_i}$ (%) is the share of each reduction factor from the incoming incident power and $(p_{i-1}(\lambda))$ ($Wm^{-2}nm^{-1}$) is the power density before a reduction factor.

Total optical reflection losses and heat sources are calculated based on (3),

$$P_{loss-Refl.} = \sum_{i\in\{1,3,5,9\}} P_{i,Refl.} \tag{3}$$

$$P_{loss-Heat} = \sum_{i\in\{2,4,6,7,8\}} P_{i,abs.} + \sum_{i\in\{10,11,12\}} P_{i,ele.} \tag{4}$$

where $P_{loss-Refl.}$(W) is the total power share of the incoming incident which produce neither heat nor electricity and leaves the module stack, $P_{loss-Heat}$ (W) is the total share of the incoming power which turns to heat either by spectral dependent absorption $P_{i,abs.}$ (W) in module stack and solar cells or voltage dependent electrical losses $P_{i,ele.}$ (W) in solar cells and interconnection.

4 RESULTS

The results are discussed based on the optical and electrical evaluation of the module components and loss analysis and evaluation of mini-modules with different module components.

4.1 Optical evaluation of module components

By using the optical measurement of module components, transmission and reflection coefficients of module components are determined. For this purpose, transmission and reflection results of glass, encapsulant polymers and solar cells in the stack are evaluated.

a. Air / Glass

The transmission and reflection coefficients of the glass are measured from 200-2200 nm for the given glasses introduced in Table 1. Figure 2 demonstrate the calculated transmission coefficient in the glass bulk and reflection coefficient of the incident light on air/glass interface.

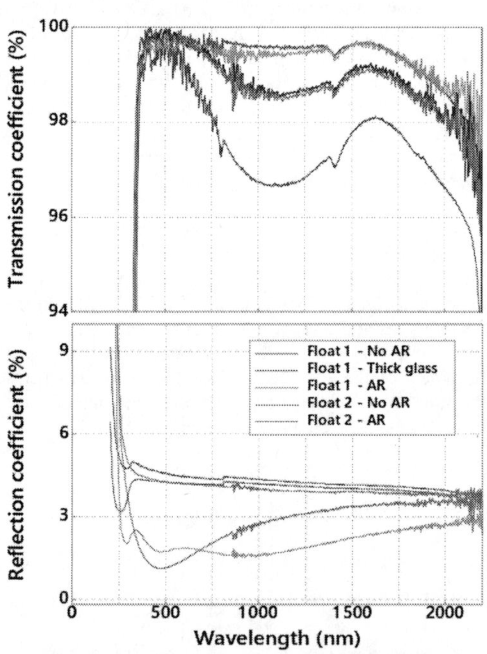

Figure 2: Calculated transmission and reflection coefficient of the glass samples in the air/glass interface based on the measurement result.

The results show that the losses due to the transmission in bulk are higher for thicker glasses. Float 1 glass with AR show considerably lower reflection on the glass compared to the similar glass without AR properties. Float 1-AR shows lower reflection in UV-IR range while Float 2-AR show better properties in visible range.

Power density loss due to the reflection in air/glass interface is almost similar for all glasses with no AR properties. However, the Float 2 glass with AR demonstrates better anti reflection properties compared to the Float 1 glass with relatively 6.7% less reflection compared to the Float 1 glass with AR in all UV-VIS-IR range.

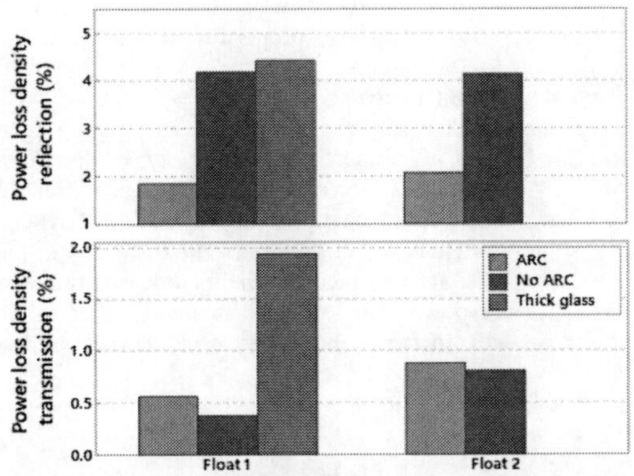

Figure 3: Total power loss density from incoming incident light due to transmission through the glass bulk and reflection in air/glass interface.

b. Glass / Polymer

Calculated transmission and reflection coefficients of polymer encapsulants are shown in Figure 4. The transmission coefficient curve of the UV+ polymer (PVB) demonstrates less power absorption in Ultraviolet (UV) range compared to EVA which does not have UV+ properties. This property is highly considerable when the modules are operating in extreme climates such as deserts or highlands with high amount of UV dose.

On the other hand, the reflection properties of both polymers in the interface between glass and polymer are quite similar and negligible.

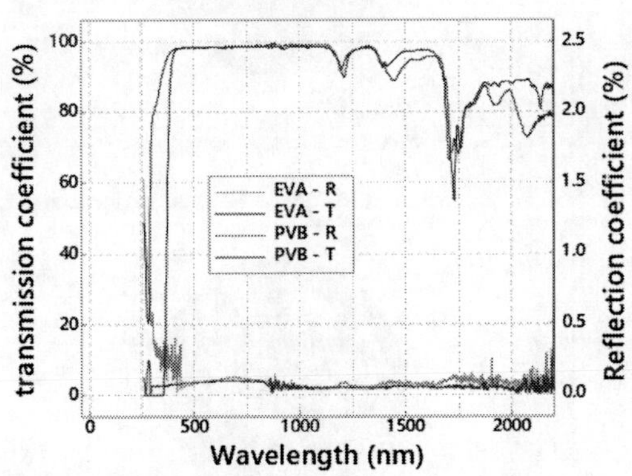

Figure 4: Calculated transmission and reflection coefficients of the polymer encapsulant samples based on the measurement results.

The power density loss due to the absorption of the incoming power in the UV+ polymer is more than half of the light which is absorbed in EVA without UV+ properties. Since most of the light absorbed in polymer is in UV range, UV+ encapsulants show are better choices in comparison to non-UV+ encapsulants for places with high UV doses which suffer from reliability issues such as delamination or discoloration of the polymers.

Since both polymers have quite similar refractive index to the glass, the reflection in the interface with glass is below 0.1% and can be neglected.

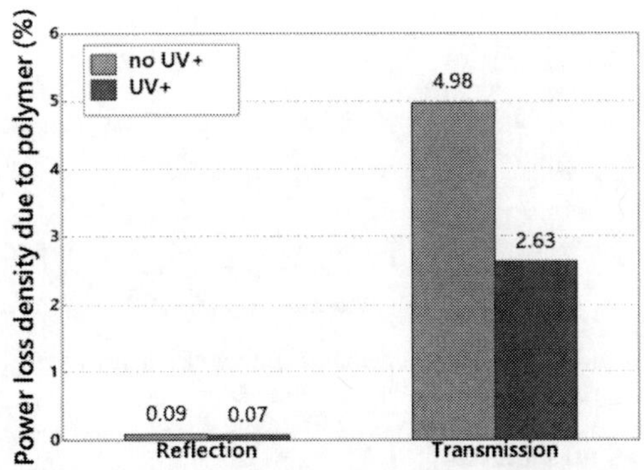

Figure 5: Total power loss density from incoming incident light due to transmission through the polymer bulk and reflection in glass/polymer interface. No UV+ represents EVA encapsulant without UV transparency properties and UV+ represents PVB with UV+ properties.

c. Polymer / Solar cell

The reflection in polymer/solar cell interface is quite similar for both polymers and the share of power density loss for EVA and PVB are 4.17% and 4.35% of the incoming incident respectively.

Figure 6 demonstrates the reflection coefficient of PVB and EVA in the interface between the polymers and solar cell.

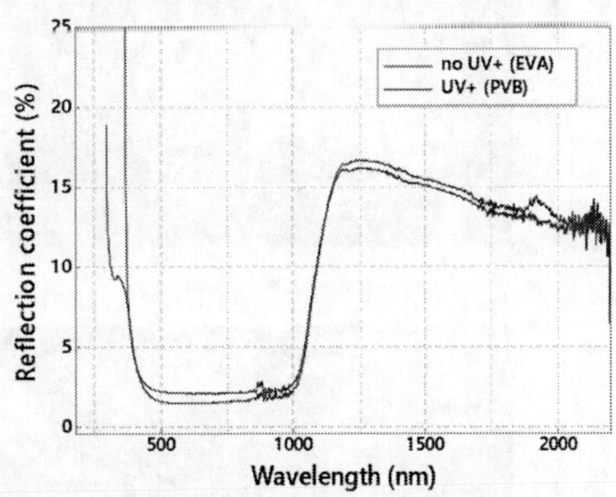

Figure 6: Calculated reflection coefficient of the solar cell in the solar cell/polymer interface based on the measurement result.

4.2 Electrical evaluation of module components

To determine the electrical properties of the solar cells, the cells are characterized electrically. Figure 7 shows the internal quantum efficiency of the solar cell which is used for evaluation.

Further calculation of the voltage dependent losses which includes thermodynamic losses and ohmic losses in solar cell and interconnection is based on Ref [4].

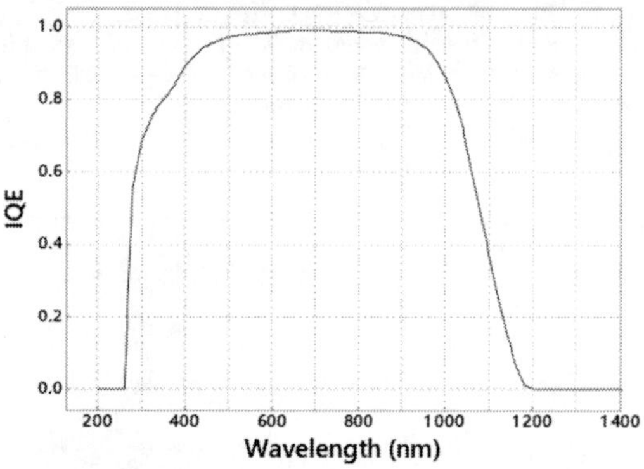

Figure 7: Measured internal quantum efficiency (IQE) of the solar cell by LOANA sun simulator.

4.3 Loss analysis

After measurement of all materials and determination of the reduction factors, the total spectral and voltage dependent loss map and the contribution of each loss mechanism from the incoming incident power on the module can be determined.

Figure 9 shows the spectral dependent and voltage dependent power densities of PV modules with Float 2 glass with and without AR properties and non-UV+ polymer as well as a module with Float 2-AR and UV+ polymer under AM1.5g spectrum from 200-2200 nm.

Module with Float 2-no AR and EVA shows 8.98% optical reflection, 73.96% heat sources and almost 17% efficiency. By adding AR properties to the glass, the reflection losses are decreased to 6.98% and heat and electricity are increased to 75.57% and 17.44% respectively. Changing the polymer from polymer with no low UV-cut off properties to UV+ polymer decreases the UV absorption. Decreasing the UV absorption in polymer does not change the optical refelctions considerably. However, it decreases the total heat in the module to 75.31% and increase the module efficiency to 17.86%. The power density which is not absorbed in the polymer will increase the other heat sources in the

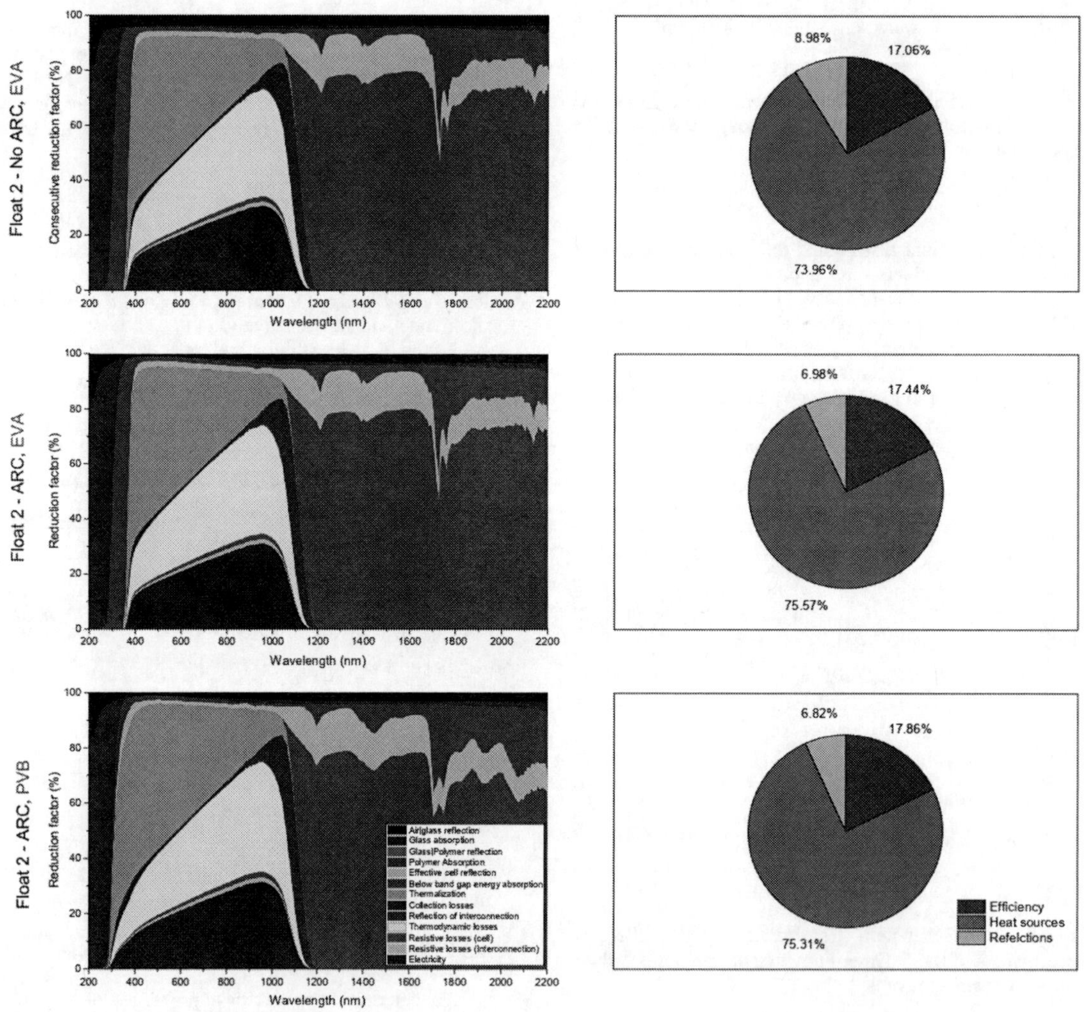

Figure 8: Calculated spectral dependent and voltage dependent reduction factors and total reflection and heat losses as well as electricity for a mini-module with Float glass 1 with AR and EVA (top), Float glass 2 with AR and EVA (middle) and Float glass 2 with AR and UV+ polymer (bottom).

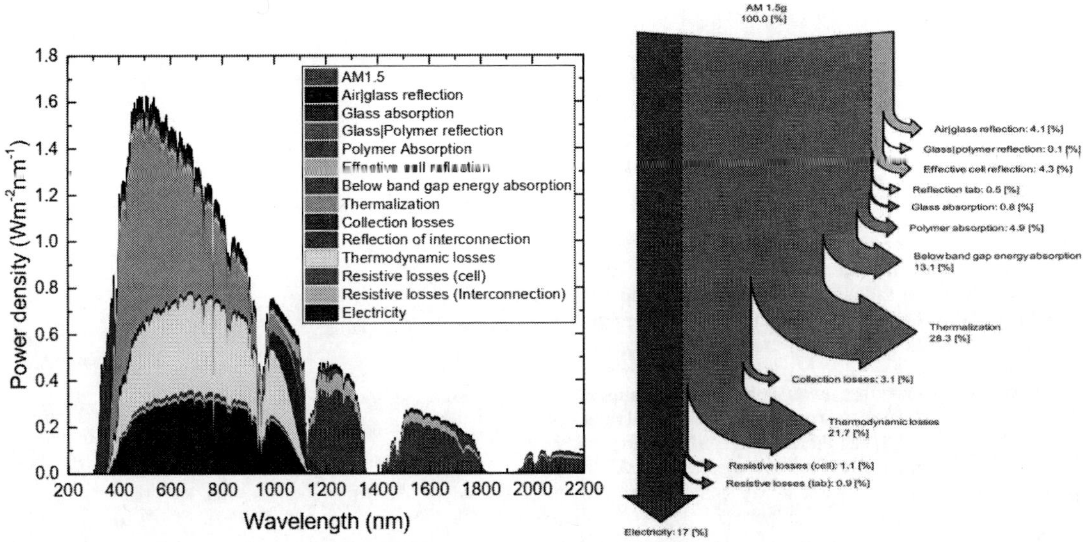

Figure 9: Calculated spectral and voltage dependent power density of a mini-module with Float 1 – No AR glass and EVA under AM1.5g spectrum from 200 to 2200 nm (left); and final share of each loss mechanism contribution from incoming light incident power and output electricity of the module.

module such a thermalization and therefore, the influence of considerable reduced absorption losses on module temperature is not significantly high.

Figure 10 shows the total reflection, heat and electricity in modules with different glass types in all UV-VIS-IR range.

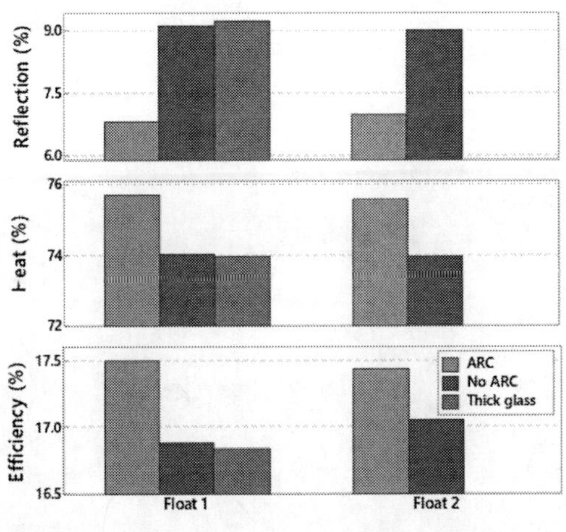

Figure 10: Total reflection, heat and electricity for modules with different concepts of glass in the UV-VIS-IR range.

By comparison of mini-modules with different glass types, the PV module with Float glasses 1-AR and 2-AR show the best results with almost 7%% and 6.8% reflection of total incoming light incident in different interfaces which is respectively 33%rel. and 29%rel. lower reflection compared to their related glasses without AR properties in the whole UV-VIS-IR range. Lower reflection means extra heat and electricity production. Float glass 1-AR shows 2.2%rel. extra heat and 3.7% extra electricity compared to the related glass with AR properties. Float glass 2-AR has the same increase of 2.2%rel. heat and electricity compared to the Float 2-No AR glass.

Finally, Float glass 1 generates 0.4% more electrical power, 0.1%rel. extra heat and 2.4% lower reflection in UV-VIS-IR range in comparison with the modules made from Float glass 2.

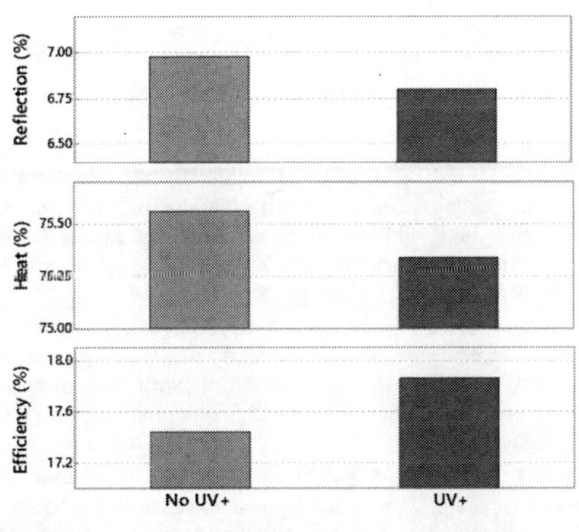

Figure 11: Total reflection, heat and electricity for modules with different concepts of encapsulant.

Figure 11 demonstrates total generated heat and electricity and optical reflections in PV modules with same glass and different polymer encapsulants.

Due to similar reflection properties, the total reflection in the module with different polymers is quite similar. The heat generated in the PV modules is also similar but with different mechanisms. PV module with no UV+ properties absorbs more energy in UV range while the UV+ polymer passes the UV due to the low UV cut-off properties. However, this extra incoming incident to the solar cell thermalizes in the solar cell level. The module with UV+ encapsulant shows 2.5% extra electricity generation compared to the similar module but with non-UV cut-off encapsulation material.

5 CONCLUSION

We introduced a spectral dependent optical and voltage dependent electrical loss analysis model. We evaluated the total heat sources, reflection and electricity of modules with different kind of stack materials in all UV-VIS-IR range.

It is shown that the reduced reflection in glass level increases the efficiency as well as the heat sources of the module under different mechanisms in module level. Float 1 and 2 with AR gain 3.7%rel. and 2.2%rel. extra efficiency compared to their related glasses with no AR properties.. We show that the absorption in glass does not play a considerable role in power generation or heat losses.

It is also explained that the UV+ polymer has almost twice less UV absorption compared to the polymer without UV+ properties and therefore benefit from higher lifetime in extreme climates. The PV module with UV+ polymer also gains 2.5% extra power compared to the module with no low UV cut-off properties.

6 ACKNOWLEDGEMENTS

The authors thank the Federal Ministry for Economic Affairs and Energy (BMWi) and Federal Ministry of Research and Education (BMBF) for financial support under contracts of "Wuestenmodul" with reference number of 03FH014IX4 and "CTM100+" with the reference number of 0324033.

7 REFERENCES

[1] I. Haedrich, U. Eitner, M. Wiese und H. Wirth, „Unified methodology for determining CTM ratios: Systematic prediction of module power," *Solar Energy Materials and Solar Cells,* pp. 14-23, 2014.

[2] J. Schneider, M. Turek, M. Dyrba, I. Baumann, B. Koll und T. Booz, „Combined effect of light harvesting strings, anti-reflective coating, thin glass, and high ultraviolet transmission encapsulant to reduce optical losses in solar modules," *Progress in Photovoltaics,* Bd. 22, pp. 830-837, 2014.

[3] M. Vogt, H. Holst, M. Winter, R. Brendel und P. P. Altermatt, „Numerical modeling of c-Si PV modules by coupling the semiconductor with the thermal conduction, convection adn radiation equations," *Energy Procedia,* Bd. 77, pp. 215-224, 2015.

[4] H. Hanifi, C. Pfau, D. Dassler, M. Turek und J. Schneider, „A practical optical and electrical model to estimate the power losses and quantification of different heat sources in Silicon based PV modules," *Renewable Energy,Vol 127, pp.602-612,* 2018.

INNOVATIVE SOLAR SPECTRAL BEAM SPLITTING CONCEPTS: COGENERATION AND PHOTOCHEMISTRY

Gur Mittelman[1*], Helena Vitoshkin[1], Beni Lew[1], Hadas Mamane[2], Avi Kribus[2]
[1]Agricultural Research Organization, POB 15159, Rishon LeZion 7528809, Israel.
[2]School of Mechanical Engineering, Tel Aviv University, Tel Aviv 69978, Israel
*e-mail: gur.mittelman@gmail.com tel: 972-52-9404011

ABSTRACT: The conversion efficiency from sunlight to electricity, for both solar thermal and solar photovoltaic converters, is usually in the range of 10—30%. More than two thirds of the solar energy, collected with considerable effort and capital investment, is rejected back to the environment. We present an analysis of novel spectral beam splitting systems with the applications of cogeneration and photochemical conversion. Significant advantages of these designs include (i) separate conversion mechanism for different radiation wavebands, such that typical efficiency limits are eliminated (ii) introduction of photochemical reactions driven by short waves, reducing photovoltaic thermalization losses and (iii) simple design with small number of elements and without a high level of radiation concentration, eliminating the requirements for accurate optics and active cooling. The results are very promising, with a considerable enhancement of the efficiency relative to photovoltaic and other reference plants.
Keywords: Bifacial, CdTe, High-Efficiency, Photovoltaic, Spectral Splitting, Medium Temperature, CHP, Cogeneration, Polygeneration, Photochemistry, Agrivoltaics

1 INTRODUCTION

The conversion efficiency from sunlight to electricity, for both solar thermal and solar photovoltaic (PV) converters, is usually in the range of 10—30%. More than two thirds of the solar energy, collected with considerable effort (are) and capital investment, is rejected back to the environment. Other conversion paths such as solar thermal power and solar fuels do not promise better efficiency as well. With solar spectral beam splitting, a separate conversion mechanism for each waveband is introduced, such that typical efficiency limits could be overcome. In this paper, several novel spectral beam splitting systems are analyzed. The applications of interest are cogeneration and photochemical conversion.

2 COGENERATION

The first application is solar Combined Heat and Power (CHP) generation (or cogeneration), with the opportunity to deliver medium grade heat. Here, Short Wave (SW) radiation is directed to a PV device, while Long Wave (LW) radiation is concentrated to a thermal receiver which is located elsewhere. Because the thermal receiver is not thermally coupled to the PV, it allows much higher operating temperatures compared to standard Photovoltaic/Thermal (PV/T) and CPVT [1] arrangements, where the temperature is typically well below 100°C. Spectral splitting systems that produce medium grade heat were proposed in the past, but the result was either a low overall solar-to-useful energy conversion efficiency, and/or too complex configurations [2-3].

The collector configurations proposed here include line focus concentrators, which greatly reduce complexity compared to point focus counterparts, while offering a reasonable conversion efficiency. In particular, linear Fresnel lens is considered due to the high acceptance angle (requires low tracking accuracy). To allow the implementation of commercial, single-junction PV modules, only the LW radiation is significantly concentrated, while keeping the PV receiver under low flux of SW radiation and low operation temperature. Two cogeneration configurations are shown in Fig. 1: (a) using 'cold mirror' beam splitter, and (b) using LW-transparent PV converters. Following the first design (Fig. 1a), the PV converter is receiving a radiation flux in the order of 1÷2 suns while the thermal receiver is subjected to a geometric concentration in the order of 20 suns. The wavelength cut-off can be, for example, 850 nm (e.g. CdTe), but can be optimized for different module materials. Due to the moderate concentration, the thermal energy can be provided at medium temperature (≈150-400°C), suitable for commercial and industrial process heat applications (e.g. steam). The difference in the second design (Fig. 1b), is that the spectral splitting is done within the PV converter itself, located between the concentrator and the thermal receiver. PV modules that are transparent to radiation above the bandgap are currently available for Bi-facial Silicon [4] but may be implemented in the future to higher bandgap materials such as CdTe and CIGS. The PV material directional transmittance is critical in the second configuration since radiation is focused towards the thermal receiver. Ongoing research is carried out towards optical characterization of such materials.

The input data for the energetic analysis is given in Table I for configuration (a). For CdTe modules with Standard Test Conditions (STC: 1000 W/m^2, AM1.5G, 25°C) efficiency of 17.6%, the conversion efficiencies are 16.0%, 22.2% and 27.5% for electricity, medium and low temperatures heat, respectively. The penalty in the electricity production is due to the optical losses in the lens and the beam splitter. If c-Si modules with the same STC efficiency are used instead, the conversion efficiency from solar to medium temperature heat reduces to 13.9% as less LW radiation is available for thermal conversion (cut-off at 1100 nm). In order to combine the electricity and medium grade heat products under the same thermodynamic value: work (or power/electricity), the thermal product is assumed to be converted into electricity by a heat engine with a second law efficiency of 50%. For a reveiver temperature of 265°C, the contribution of the thermal product to the electricity is 4.4%, making the overall solar-to-equivalent electricity (or exergy) efficiency 20.4% (the reference PV value is 17.6%).

The technology of "cold mirror" type spectral splitters (filters) is already available with several manufacturers worldwide e.g. the polymeric multi-layer film of 3M [5]. Using Physical Vapor Deposition (PVD), it is possible, in principle, to tune and optimize The spectral optical properties of the filters according to the system requirements.

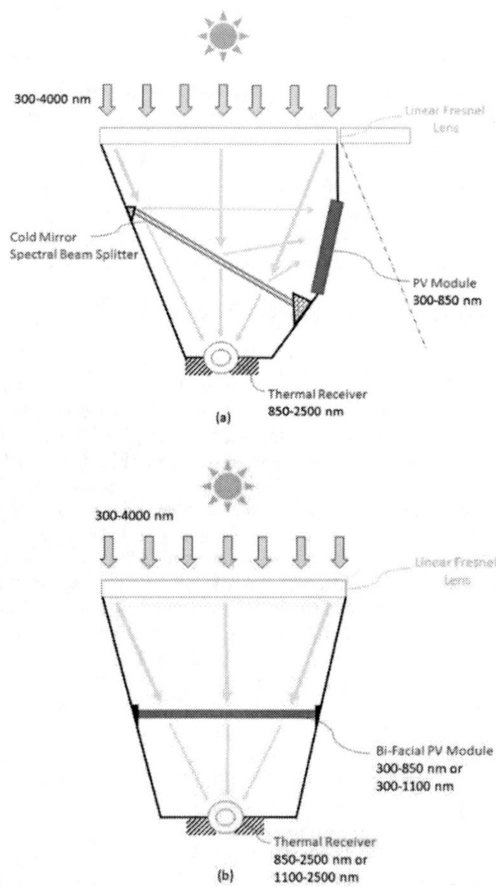

Figure 1: Cogeneration (a) with cold mirror beam splitter. The splitting is taking place at 850 nm, in accordance with the band-gap of CdTe. (b) with Bi-facial CdTe or Si modules.

Table I: Cogeneration data. C_{goem}= geometric concentration ratio.

Parameter	Value
Solar resource direct	AM1.5D
Fresnel lens transmittance	SOG [6]
Beam splitting cut-off	850 nm
Splitter transmittance (LW)	90%
Splitter reflectance (SW)	95%
PV module	CdTe
Thermal receiver C_{geom}	20
Thermal receiver $\tau\alpha$	90%
Thermal receiver emissivity	10%
Thermal receiver temperature	265°C
Heat engine low temperature	50°C
Heat engine 2nd law efficiency	0.5

3. PHOTOCHEMICAL CONVERSION

In the cogeneration systems shown above, due to the thermalization at the PV converter, a significant amount of energy is produced in the form of low grade heat (well below 100°C) in the PV cells rather than in the thermal receiver. This energy has a low thermodynamic value and a rather limited implementation potential (e.g. domestic hot water). A better use for the high-energy photons in the solar spectrum could be a photochemical reaction, where the product can be a chemical commodity, produced otherwise by a conventional, highly energy-consuming process. Thus, solar photochemistry is essentially saving energy, rather than producing it directly. Extensive solar photochemical conversion studies were carried out before with several applications suitable for SW radiation [7]. Hence, the two cogeneration configurations could be modified to include one more step before the PV: a photochemical conversion. This is shown on Fig. 2 where a photo-reactor is attached at the front of the PV device. The modified layouts are in fact polygeneation systems providing, simultanuosely, electriciy, heat and energy savings. Two applications will be discussed: the production of the chemical commodity ε-Caprolactam (section 3.1) and, water treatment (section 3.2).

3.1 CHEMICALS PRODUCTUION

The configuration in Fig. 2(a) was recently modeled and analyzed for the case of ε-Caprolactam production [8]. Within this process, a photo-reaction is initiated by activating NOCl which has a bandgap equivalent to approximately 600 nm. Manufacturing of ε-Caprolactam with photo-reactors is already in a commercial scale using coventional lamps [7, 8].

The energetic analysis considered the actual behavior of optical elements, thermal and PV collectors, as well as in-house optical measurements of the materials involved in the reaction. The output was evaluated by integration over all wavelengths and a specific mix of reactants and products was considered (i.e. the model is local). Not included in the analysis is the effect of the varying incidence angle at the reactor on the performance (normal beam was considered). The reference lamp performance data is based on state-of-the art blue LED (459 nm) and given in Table II. Apparantely, the net conversion efficiency from power to Quantum Yield (QY) in this lamp is only 44%, making any conventional photochemical plant potentially a big power consumer.

The resulting conversion efficiencies, normalized to the incident radiation, are summarized in Table III for all configurations. With photochemistry (Cog+PC), the power generated directly by the PV is decreased from 16% to about 7% due to the absorption in the reacor. However, the power saving in the photochemical conversion is remarkable, about 47% of the incoming sunlight. The total exergy conversion efficiency is close to 60%, significantly higher than the reference 17.6% and other current PV technologies.

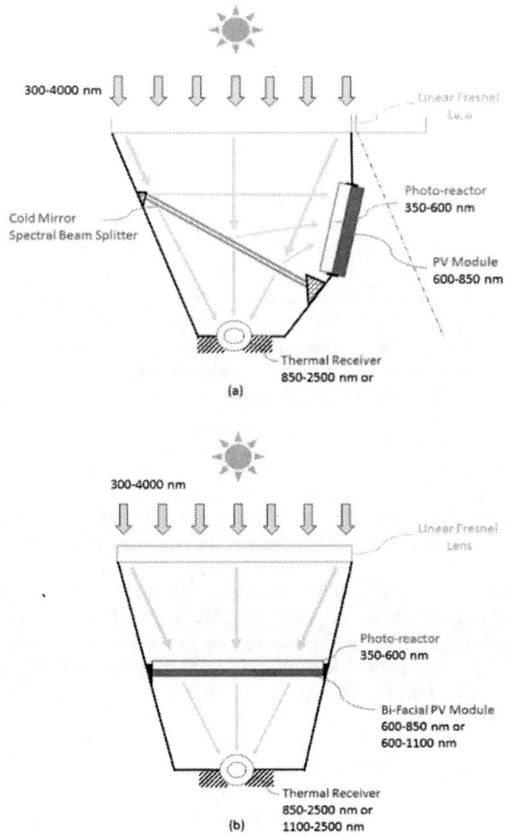

Figure 2: Cogeneration and photochemical conversion (a) with cold mirror beam splitter (b) with Bi-facial PV module.

Table II: Lamp performance data [9]. QY=Quantum Yield.

Parameter	Value
Lamp model	Blue LED
Lamp wall plug efficiency	66%
Lamp package efficiency	87%
Blue light QY, 600 nm	77%
Overall conversion from power to QY	44%

Table III: Efficiency – summary. Cog=Cogeneration (Fig. 1). COG+PC=Cogeneration + Photochemical (Fig. 2).

Configuration	PV only	Cog	Cog+PC
PV power	17.6%	16.0%	7.1%
Thermal energy	-	22.2%	22.2%
Power saving (lamp light)	-	-	46.8%
Solar to exergy efficiency	17.6%	20.4%	58.4%

3.2 WATER TREATMENT

The need for pure water is growing fast. However, water treatment (or disinfection) is a highly energy demanding process with significant environmental impacts such as air and ground pollution, as well as greenhouse emissions and other effects. Thus, water disinfection with solar photocemical processes was studied extensively [10]. In accordance with the current concept, water disinfection could be conducted by solar photo-chemical (or photo-catalytic) processes driven by the short wave part of the solar resource, up to the end of visible light, while near infra-red radiation is converted into electricity and medium grade heat. The electricity could be used to operate accessory devices such as membrane separators, pumps, etc., while the thermal energy could boost the disinfection rate as water photo-driven treatment processes are often accelerated at elevated temperatures. The idea could be implemented in greenhouse drainage water as well as wastewater, including oxidation of micro-organisms and oxidation of organic infects such as micro-infects. It is also compatible with distributed (or decentralized) systems, as the plant can be allocated close to the wastewater resource, and independent of the electrical grid (off-grid). Despite the wide scope of applications, our first study focused on Israeli greenhouse water. To validate the feasibility of the proposed concept, a set of preleminary testing was carried out and hereby described.

3.2.1 PRELEMINRAY EMPIRICAL RESULTS

In both of the envisioned configurations, the PV collector is located behind the photo-reactor and thus, the PV performance is sensitive to the optical properties of the water in the reactor (e.g. absorptivity). Hence, with the help of the Cary5000 spectrophotometer, we have tested several types of water including distilled (for the sake of calibration), tap, greenhouse drainage and buffer. The spectral transmittance for normal incidence light is given in Fig. 3. The test cell (cuvette) thickness is 10 mm and it is made of quartz. Apparently, the behavior of water containing salts typical for Israeli greenhouse is similar to distilled water expect for a narrow range from 300 to 500 nm. The transmittance between the end of the visible part and 1100 nm is good which allows the PV conversion. However, the transmittance of an actual photo-reactor could be different than these findings due to several reasons: (i) a reactor material different than quartz (e.g. Borosilicate) (ii) a reactor thickness different than 10 mm and (iii) actual water will include contaminants and catalysts.

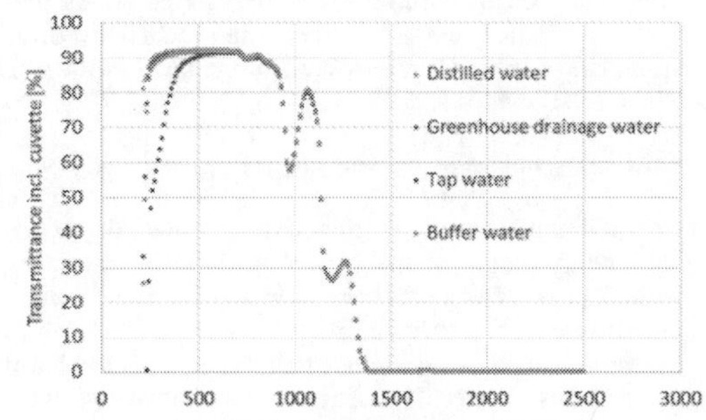

Figure 3: Spectral transmittance of different kinds of water, inserted in a 10 mm quartz cuvette.

Several preliminary experiments were carried out investigating the effect of photochemical treatment with common oxidants and catalysts on water contaminated by the Fusarium fungi, which is a frequent pathogen (disease) in tomatoes grown in Israeli greenhouses. The water were irradiated for 4 hours by a full spectrum solar simulator (a

150-W ozone-free xenon arc lamp, Sciencetech Inc., SS150 W, ON, Canada) while kept at 18°C. After the treatment, the water were added to an agar selective growth medium for Fusarium growth for a period of 3-5 days. In the first experiment, water including Fusarium (10^6 CFU/mL) and H_2O_2 oxidant (75 mg/Liter), was irradiated. The result was that the Fusarium concentration was reduced by 40% to 90%, in accordance with past work [11]. In the second experiment, water including Fusarium (10^6 CFU/mL) and TiO_2 catalyst (250 mg/Liter) oxidant, was irradiated. A reduction of 18% in the Fusarium concentration was observed (180,000 CFU/mL). Actually, the testing in [12] observed a much higher removal level, above 90%, after a same retention time of 4 hours. However, the initial Fusarium and TiO_2 concentrations were much lower, 10^3 CFU/mL and 50 mg/L, respectively, which could have a dramatic influence on the process since low titanium presence improve the water absorbance to light. Ongoing empirical work is carried out to investigate the effectiveness of the TiO_2 catalyst and visible light catalysts on Israeli greenhouse Fusarium.

3.2.3 PERFORMANCE ANALYSIS

The performance of the configuration in Fig. 2(a) for water treatment is expected to be at the same order of the ε-Caprolactam (section 3.1) due to the fact the envisioned catalysts are supposed to be sensitive to visible light (up to 700 nm). However, this has to be yet analyzed in detail as the water optical properties, as well as their spectral response, could be different.

3.3 AGRIVOLTAICS

Integrating PV collectors in agricultural crops ('agrivoltaics') has recently emerged as a fast-growing research area with the first agrivoltaic farm built in 2011 [13]. The agrivoltaic system provides two products from the same collection area: biomass (photochemical conversion) and electricity. Most of the open field agrivoltaics solutions considered so far are based on standard flat plate PV modules that block sunlight, including the visible light which is essential for photosynthesis. This leads to a significant reduction in crop yield and limits the module area that can installed in a given field. The productivity of an agrivoltaic farm is often measured by the Land Equivalent Ratio (LER), which is defined as

$$LER \equiv \left(\frac{Y}{Y_{ref}}\right)+\left(\frac{E}{E_{ref}}\right)$$

where Y and E are the crop yield (e.g. annual dry mass) and energy output of the agrivoltaic field, respectively, while Y_{ref} and E_{ref} are the crop yield and the energy output of equivalent crop only and solar PV only fields, each of the same area of the agrivoltaic field. Thus, a LER value of 2.0 indicates no mutual influence (or coupling) between the two products or in other words, no damage to the crops as a result of the PV installation. On annual basis, LER values between 1.19 and 1.53 were reported for several configurations of wheat and lettuce [14].

The fundamental problem is therefore, that crops and photovoltaics are actually competing on the same waveband: the visible light. Spectral beam splitting can resolve this conflict as the visible light (or Photosynthesis Active Radiation – PAR) can be separated from the Near Infra-Red (NIR) radiation while the latter, which accounts for more than 50% of the total insolation (AM1.5G), is converted into useful electricity. In this manner, a design based on a dish type, High Concentration Photovoltaics (HCPV) with a 'hot mirror' beam splitter integrated at the dish reflector, was recently suggested [15]. However, the engineering requirements involved in such a design are very demanding and include: very accurate 2-axis tracking, curved splitter, concentrator PV cells, cooling water system and other sophisticated elements.

A different configuration for open field agriculture is shown in Fig. 4. Sunlight is transmitted through a flat glass and hits a 'hot mirror' aligned at 45° with respect to the glass surface. The beam near infra-red irradiation is reflected towards a flat PV receiver while visible light (global) is transmitted towards the ground. The collector involves only planar, solid state elements and 1-axis tracking and because there is no radiation concentration, it is assumed that the requirement for high tracking accuracy could be relaxed.

The expected LER of the proposed design is between 1.48 and 1.74, depending on the choice of the PV materials. For multi-crystalline Silicon, the LER is 1.48 at Standard Testing Conditions (STC), which is below the maximum value of 2.0 due to electrical output loss in the visible light. If photovoltaic conversion of the 'green' light band say, from 460 to 630 nm, is introduced, the LER could be increased to 1.74. This is possible for plants which are less responsive to green, or if the green available from the part of the sky which is not blocked by the collectors is sufficient. A further improvement could be using Silicon/Germanium tandem cells [16], where because the Ge is responsive to light which is never consumed by crops (1100÷1900 nm), the LER could be increased from 1.48 to 1.57, for the case where the crops receive the entire visible spectrum. The LER results are summarized in Table IV.

The LER criterion suggests that electricity production and crop yield are of equal significance (equal contribution to the overall performance). However, farmers could be more sensitive to a damage in crops yield than to a reduction in electricity output, especially if we consider a high economic value of the crops, combined with a relatively low retail price for PV electricity (low tariff), which indeed reflects the current market status (tariffs has been dramatically declining over the last decade). In the proposed design, the contribution of the crop relative yield to the LER is assumed to be close to 1.0 (minor or no damage to crops). Thus, it might be more applicable than reference agrivoltaic systems.

Water evaporation and irrigation requirements are affected by the plant temperature. The shading caused by standard PV collectors (without spectral splitting) alleviate water evaporation due to reduced plant temperature and can result in water savings, which is beneficial especially during dry seasons and droughts. In the proposed design, considering that solar collectors cover about half of the field, then the collectors continuously block some of the solar NIR irradiation (which constitutes about 54% of the total solar AM1.5G). However, because the absorptivity of many plant leaves to NIR irradiation is small anyway (about 0.16, following the data of Ref. [17]), it only results in a slight negative (-) radiation gain of about 37 W/m^2 for a total insolation of 840 W/m^2. The plant energy balance ('energy budget') is also dictated by a Far Infra-Red (FIR) irradiation coming from the surroundings and sky [17]. For sky and surrounding temperatures of 253K and 273K, respectively, and a leaf absorptivity to FIR irradiation of

0.96, then because the collectors block some of the sky, a positive (+) radiation gain of some 90 W/m^2 is apparent at the plant. The overall effect is a net gain of (+) 53 W/m^2, which is assumed negligible considering the plant other heat transfer mechanisms (conduction and convection, and transpiration/evaporation).

The propsed idea is also apllicabale in green walls & closed plant factory. Here, integrating the collectors in the building vertical envelope can allow the penetration of natural visible light (or using this light for photosynthesis), while converting the NIR irradiation into electricity.

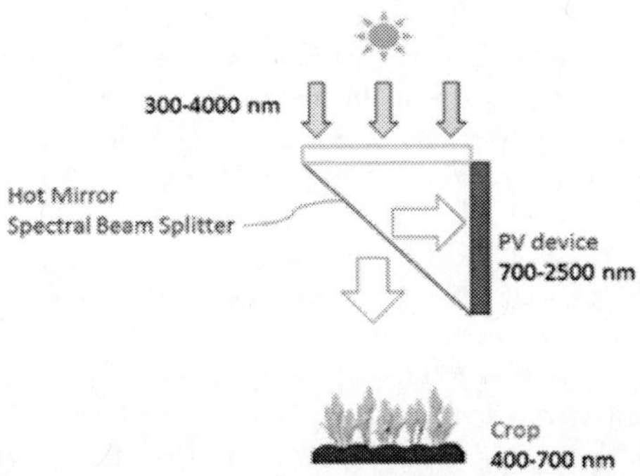

Figure 4: The proposed open field agrivoltaic system.

Table IV: Land Equivalent Ratio (LER)

Configuration	Ref. [14]	Proposed
LER	1.19-1.53	1.48-1.74

4 CONCLUSIONS

Novel configurations for the conversion of solar energy into electricity, heat and/or chemicals were presented, all based on spectral beam splitting. The approach is motivated by simple designs yet offering higher conversion efficiencies compared to available PV technologies. Due to the splitting, and in contrast to standard PV/T arrangements, medium grade heat could be produced at the thermal receiver, with a higher thermodynamic value. The significant thermalization energy loss in photovoltaic converters could be avoided by introducing a photochemical conversion. The proposed spectral beam splitting for agrivoltaics could be very valuable but may be implemented in buildings environmental control as well.

The implementation of the proposed designs in large industrial scale depends on several factors: (i) low cost, serial production of the spectral filters (especially for configuration (a)), (ii) validity of reactor materials optical properties under actual reaction and (iii) stable and controllable reaction behavior under real sun, dynamic conditions.

5 REFERENCES

[1] Ju Xing, Xu Chao, Liao Zhirong, Du Xiaoze, Wei Gaosheng, Wang Zhifeng, Yang Yongping, Science Bulletin 62 (2017) 1388-1426.

[2] Ju Xing, Xu Chao, Hu Yangqing, Han Xue, Wei Gaosheng, Du Xiaoze, Solar Energy Materials & Solar Cells 161 (2017) 305-327.

[3] Ju Xing, Xu Chao, Han Xue, Du Xiaoze, Wei Gaosheng, Yang Yongping, Applied Energy 187 (2017) 534-563.

[4] M. Rüdiger, S. Fischer, J. Frank, A. Ivaturi, B.S. Richards, K.W. Krämer, M. Hermle, J.C. Goldschmidt, Solar Energy Materials & Solar Cells 128 (2014) 57-68.

[5] V. Fthenakis, Third Generation Photovoltaics (2012) 183-200.

[6] K. Shanks, S. Senthilarasu, T.K. Mallick, Renewable and Sustainable Energy Reviews 60 (2016) 394-407.

[7] M. Oelgemöller, Chemical Reviews 116 (2016) 9664-9682.

[8] G. Mittelman, M. Epstein, A. Kribus, 2018 International Workshop on Thermal Chemistry and Solar Energy (2018IWTCSE). Harbin Institute of Technology (HIT), Harbin, China, August 2018.

[9] J. Brodrick, DoE SSL R&D Program (2017).

[10] D. Spasiano, R. Marota, S. Malato, P. Fernández-Ibáñez, I. Di Somma, Applied Catalysis B: Environmental 170-171 (2015) 90-123.

[11] C. Sichel, M. De Cara, J. Tello, J. Blanco, P. Fernández-Ibáñez, Applied Catalysis B: Environmental 74 (2007), 152-160.

[12] C. Sichel, P. Fernández-Ibáñez, M. De Cara, J. Tello, J. Blanco, Water Research 43 (2009), 1841-1850.

[13] C. Dupraz, H. Marrou, G. Talbot, L. Dufour, Y. Ferard, Renewable Energy 10 (2011) 2725-2732.

[14] B. Valle, T. Simonneaua, F. Sourdc, P. Pechiera, P. Hamarda, T. Frissonb, M. Ryckewaertb, A. Christophea, Applied Energy 206 (2017) 1495-1507.

[15] Wenjuan Huang, Hui Lv, Jun Liao, Yiyuan Shang, Wen Liu, Qinghua Lv, Chunfu Cheng, Yuehong Su, R. Saffa, Journal of Renewable and Sustainable Energy 9 (2017).

[16] V. Vijayakumar, P.B. Dunbar, Journal of Solar Energy Engineering 136 (2014) 1495-1507.

[17] P.S. Noble, Physicochemical and environmental plant physiology (2009).

Inorganic Cesium Carbonate Electron Transport Layer for High Efficiency Perovskite Solar Cells

M. I. Hossain[1], Iwan Zimmermann[2], Nouar Tabet[1], Mohammad Khaja Nazeeruddin[2], Abdelhak Belaidi[1]
[1]Qatar Environment & Energy Research Institute, Qatar Foundation,
Hamad Bin Khalifa University
[2] Institut des Sciences et Ingénierie Chimiques, École Polytechnique Fédérale de Lausanne, Valais Wallis, CH-1951 Sion, Switzerland
mhossain@hbku.edu.qa

ABSTRACT: In this work, we have explored the possibility to develop efficient double cation $(FAPbI_3)_{0.85}(MAPbBr_3)_{0.15}$ perovskite solar cells using inorganic n-type Cesium Carbonate (Cs_2CO_3) as electron transport material (ETM), which can be synthesized at low temperature and integrating this type of ETM with compact TiO_2 leads to 17% power conversion efficiency (PCE) under one sun illumination. External quantum efficiency (EQE) data yields around 85% of incident photon to electron conversion and integrated short circuit current extracted from aforementioned data confirms the short circuit current obtained from the IV test with a close agreement. Cs_2CO_3 (6wt%)/Perovskite/Spiro-OMETAD structures w/o any types of TiO_2 ETM has shown a short-circuit current density (J_{sc}) of 15.40 mA/cm^2, open-circuit voltage (V_{oc}) of 1.023 V, fill factor (FF) of 51.7%, and a PCE of 8.14%. Interestingly, c-TiO_2/Cs_2CO_3(0.2wt%)/Perovskite/Spiro-OMETAD structure w/o meso-TiO_2 boosts the performance with a short-circuit current density (J_{sc}) of 22.26 mA/cm^2, open-circuit voltage (V_{oc}) of 1.054 V, fill factor (FF) of 71.6%, and a PCE of 16.80% confirming Cs_2CO_3 as an alternative ETM to meso-TiO_2 modifies compact-TiO_2 surface. Also, high open circuit voltage and short circuit current values can be attributed to the passivation of defect states on TiO_2 surface resulting in effective charge transfer between perovskite and ETM layers. Hence, a thin layer of Cs_2CO_3 can replace thick and meso-TiO_2 with low process temperature (<200°C) with an improved efficiency.

1. INTRODUCTION

In this work, we have explored the possibility to develop efficient perovskite solar cells using inorganic n-type Cs_2CO_3 as electron transport material (ETM). Limited attempts have been reported to use with organic PV structure since such material can be synthesized at low temperature (<200 C). In the first stage, it is important to optimize the properties of Cs_2CO_3 films (good crystallinity, composition and structure and optoelectronic properties). The work function of Cs_2CO_3 can be modified from 3.45 to 3.06 eV by a low temperature below 200 °C annealing treatment. The affinity for such material is around 2.8 eV. Also, the valance band level (E_c=6.2 eV) could act as hole blocking layer (HBL) and, hence, suppresses any unwanted carrier diffusion. Aforementioned material properties certainly make Cs_2CO_3 suitable to replace the currently used ETM which requires a very high temperature processing. Also, both substrate and superstrate structures are possible due to its low processing temperature, which keeps the window for optimization. The thickness of Cs_2CO_3 will be varied between 1-5 nm to find the optimum. Also, it has been found that Cs_2CO_3-functionalized BP device presents almost 25 times enhancement of electron mobility from 1.1 cm^2 V^{-1} s^{-1} for pristine BP to 27.1 cm^2 V^{-1} s^{-1} at 10 nm Cs_2CO_3 coverage, revealing a remarkably improved electron transport in the BP channel. Cs_2CO_3 as a surface modification layer for perovskite solar cells has been fabricated, where promising results have been measured for the cells TiO_2/Cs_2CO_3 (0.2wt%)/Perovskite/Spiro with an efficiency of 17%, whereas, Cs_2CO_3 as electron transport material only shows an efficiency of 8.14% (V_{oc}=1.023V, J_{sc}=15.40mA/cm^2, and FF=51.7%) (it proves the usage of Cs_2CO_3 for perovskite solar cells). The take way message is-such material can be used as an alternative ETM, can alter the use of meso-TiO_2, which requires high temperature (450 C) to process, and thickness of 5 nm is good enough to fabricate high efficiency devices using Cs_2CO_3 as a surface modification layer. This is mainly due to the surface passivation character of Cs_2CO_3 on top of TiO_2 by reducing defect states and leakage current.

2. METHODOLOGY

Cs_2CO_3 layer was fabricated using solution process technique. 0.2 wt% (4 mg of Cs_2CO_3 powder and 2 ml of ethoxyethanol), 2 wt% (20 mg of Cs_2CO_3 powder and 2 ml of ethoxyethanol), 6 wt% (60 mg of Cs_2CO_3 powder and 2 ml of ethoxyethanol), and 20 wt% (200 mg of Cs_2CO_3 powder and 2 ml of ethoxyethanol) were used to prepare the solution and later coated by spin coating technique. The precursor solution of FA/MA perovskite solution was spin coated in a two-step program at 1000 and 5000 rpm for 10 and 30 s, respectively. During the second step, chlorobenzene (100 mL) was dropped on the spinning substrate 15 s prior to the end of the program. The substrates were then annealed at 100 C for 1 h in a nitrogen-filled glove box. Spiro-OMETAD solution as hole transport material was prepared inside the glove box.

3. RESULTS AND DISCUSSIONS

The structural analysis of thin films was performed by X-ray diffractometer with different diffraction angle 2θ from 10° to 80°. As shown in Figure 1 and Figure 2, both Cs_2CO_3 and Perovskite films show the preferential growth. Samples grown at a particular temperature showed better crystallinity. These changes attribute to the coalescence induced grain growth during the annealing process where smaller nuclei can easily rotate compare to the larger nuclei in order to minimize the interfacial energy. With higher temperature, the atoms will gain sufficient energy to be reoriented and will be much more stable with better crystallinity. UV-Vis absorption curves of perovskite layers with Cs_2CO_3 show no change in the absorption offset as shown in Figure 3.

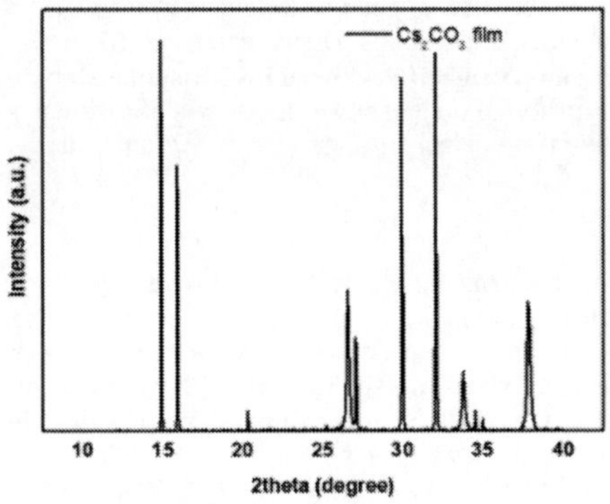

Figure 1: XRD results of Cs_2CO_3 films

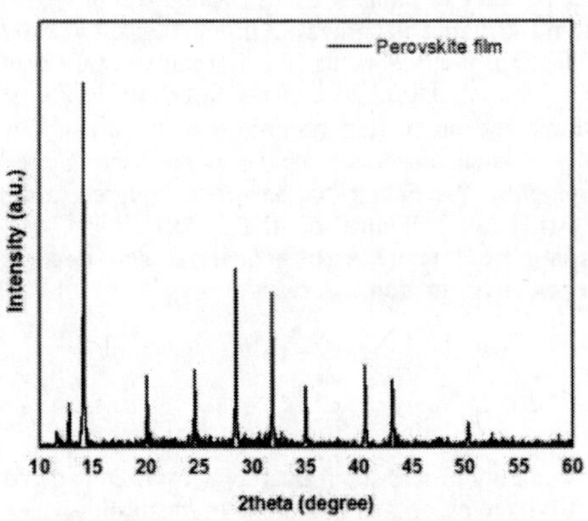

Figure 2: XRD results of Perovskite films

Huge improvement has been noticed, which can be attributed to the passivation phenomenon of Cs_2CO_3 to suppress the traps and current loss of TiO_2 surface

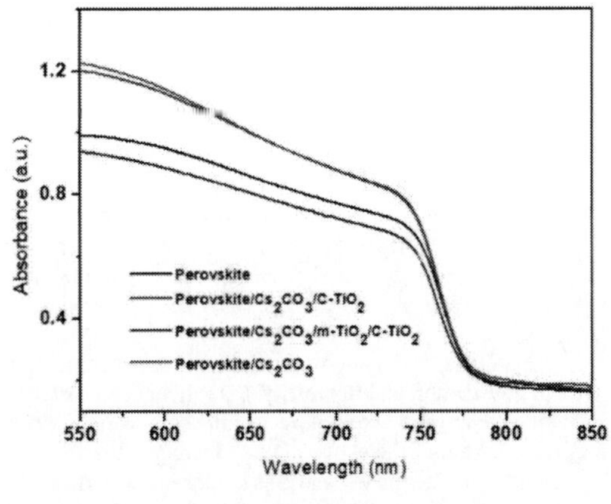

Figure 3: UV-Vis results of Perovskite films

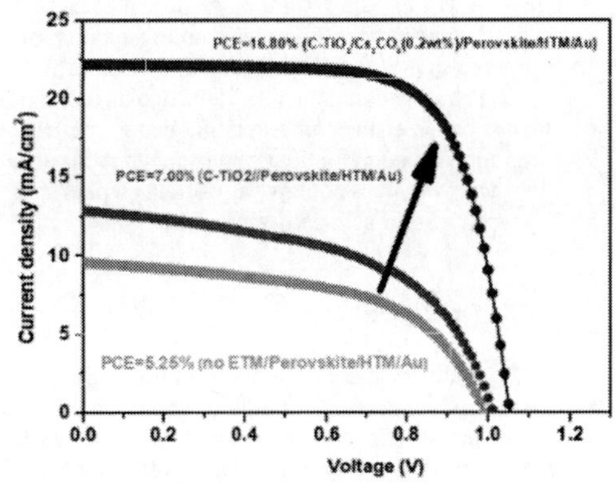

Figure 4: PCE results of Perovskite devices

As shown in Figure 4, promising results have been calculated for the cells TiO_2/Cs_2CO_3 (0.2wt%)/Perovskite/Spiro with an efficiency of 17%, whereas, Cs_2CO_3 as electron transport material only shows an efficiency of 8.14% (V_{oc}=1.023V, J_{sc}=15.40mA/cm^2, and FF=51.7%) (it proves the usage of Cs_2CO_3 for perovskite solar cells). Hence, such material can be used as an alternative ETM.

4. CONCLUSIONS

Electron transport material, hole transport material, and absorber layers have been optimized and shown preferential growth orientation. Cs_2CO_3 as a surface modification layer for the perovskite structure allows enhancing the efficiency to 16.80%, to the best of our knowledge, this is the best reported efficiency by one-step solution process.

SOLDER JOINT ANALYSIS OF TIN-LEAD AND BISMUTH BASED LEAD FREE PV RIBBONS IN HIGH THROUGHPUT MANUFACTURING

Narahari S Pujari[1*], Jayaprakash Sundaramurthy[1], Siuli Sarkar[1], Eric Poh[2], and Carl Bilgrien[3]
1: Alpha Assembly Solutions Alpha India Research Center Vaishnavi Bhavana 89/1 Industrial Suburb II stage Yeshvanthpur Bangalore-560094 (India)
2: Alpha Assembly Solutions, 14 Tuas Avenue 10, (Singapore) 639138
3: Alpha Assembly Solution, 109 Corporate Blvd South Planfield, New Jersey (USA) 07080
*: Phone:+91-80-4244312, Narahari.Pujari@alphaassembly.com

ABSTRACT: PV module manufacturers increasingly require higher throughput production lines with higher degrees of automation and flexibility. Establishing an electrical connection between solar cells in a module not only defines the final shape and output power of the solar panel, depending on the strategy applied it but also strongly impacts yield and throughput of the entire module factory. There are not too many reports on solder joint analysis in high manufacturing scenario. This paper analyses solder joints and factors affecting solder joints in high throughput manufacturing. Paper also report performance of new lead free solder, EcoSol™ along with standard tin lead ribbon in an automated tabbing machine. We found that under identical conditions, the solder joints in lead and lead free ribbons were comparable and peel strength of >2.5 N/mm on front and back side of the cell obtained. DOE results reveal that the cell metallization play an important role in determining quality of solder joints. Grade A PERC cells yielded higher peel strength compared to p-type c-Si cells due to the smooth surface topology. Similarly, flux having good wetting time performed better. Ribbon alignment was identified as one of the major causes of the weak solder joints. Due to decreased ribbon and bus bar size, this problem might occur frequently. Microscopic investigation yielded, absence of IMC (intermetallic compound) in either of the ribbons due to rapid production speed. Only diffusion of tin into silver matrix was observed. This, however, does not have any effect on peel strength and cohesive failure was evident from fracture surface analysis.
Keywords: Solar cell, soldering, solder joint

1 INTRODUCTION

The PV industry has evolved substantially during the last decade and witnessing continuous advancements with respect to throughput, efficiencies and materials used in the manufacturing of photovoltaic components. The growth is further fueled by the rising demand for energy, support extended by the government, sharp drop in module prices and the growing concern regarding environment [1].

In light of this growth, PV module manufacturers increasingly require higher throughput production lines with higher degrees of automation and flexibility. In the past the focus was on the cell level performance, now it has been shifted to the PV module performance itself. As the number of busbars in the photovoltaic cells increases, the interconnected welding process, becomes a bottleneck in many production lines. Moreover, due to new module designs and equipment, for which providers offer new machines to enable high-throughput production, the knowledge of solder joints becoming crucial and challenging. Establishing an electrical connection between solar cells in a module not only defines the final shape and output power of the solar panel, depending on the strategy applied it also strongly impacts yield and throughput of the entire module factory [2].

There are not too many reports on solder joint analysis in high manufacturing scenario. The effect of shrinking ribbon dimensions and narrower bus bars on reliability of joints is not known. Process parameters such as ribbon misalignment and its impact on peel strength is also not documented. The performance of standard and lead free ribbons in manufacturing conditions has not been studied simultaneously. This paper critically asses the solder joints and factor affecting the same in high throughput environment. Two types of ribbons, viz. tin lead (Sn-Pb) and lead free (Alpha's EcoSol™) were used for this study. EcoSol ribbon has liquidus temperature of 179°C. It is a non eutectic alloy with Bi content less than 40 wt.%, and performance enhancing additives to improve mechanical and thermal reliability [3]. Besides that, reduced silver (Ag) content in EcoSol makes it attractive in a cost sensitive industry such as PV. These ribbons were soldered on 4 bus bar (BB) c-Si (crystalline silicon) cells. The soldering was achieved in such a way that each busbar of the cell is soldered with lead (Pb) and lead free (EcoSol) ribbon alternatively. P-type and n-type PERC c-Si cells were used for comparison. Experiments were conducted on TeamTechnik machine TT2100, which is automated tabbing and stringing machine with production speed 2100 cells/h. Design of experiments (DOE) approach was used in which parameters such as cell metallization, flux type and volume and processing conditions were varied. The reliability of each process's output was measured by the peel strength of the ribbon that has been soldered to the cell metallization. Nature of the solder joints was investigated by fracture surface analysis and detailed microscopic investigation.

2 APPROACH

The soldering process (cell interconnect) is considered the most critical process in module manufacturing. A key point of a soldering process for photovoltaic (PV) modules is to increase an adhesive strength leading a low resistivity between ribbon and cell [4]. The interconnections provide electrical, mechanical and thermal contact between the solar semiconductor cell and electrodes. The main intention of this work is to assess the quality of solder joint in high throughput environment. A good solder joint is one where the solder wets the surface well and provides good adhesion between the two metals [4]. In the case of solar cells, the two metals of interest are Ag (from the cell busbar) and solder (from the interconnect). The soldering process has been reported as a preliminary source for degradation in PV modules.

For the DOE (Fig. 1), all the experiments were performed on TeamTechnik tabbing and stringing machine. Two spools of each, Sn-Pb and EcoSol were housed in a cabin and tabbing and stringing was done on cells in such a way that each ribbon is bussed alternatively (Table I, Fig. 2). Three strings of each DOE experiments were made (10 x 1 cells).

Table I: Processing parameter details

Process Parameter	Details
Machine	: Automated tabbing & stringing
Fluxing method	: Spray on cell
Soldering method	: IR soldering
Pre-heat temperature	: 170°C
Soldering time	: 800-1500 ms
Soldering temperature:	195°C-210°C
Peel strength	: Imada peel tester
Peel angle	: 180°
Observation	: Keyence microscope and SEM

Fig. 1 gives a set factors that affect quality of solder joint. We conducted DOE (design of experiment) around those factors to assess effect of each parameter and investigate optimum conditions for soldering and nature of joints formation in production. On material size, there are cell, ribbon and flux are the main players and process side, factors such as soldering time, temperature and ribbon alignment play an important role in determining the quality of a solder joint. In a DOE, each bus bar (BB) of 4 BB solar cell was tabbed with standard tin lead (Sn-Pb) and Ecosol alternatively.

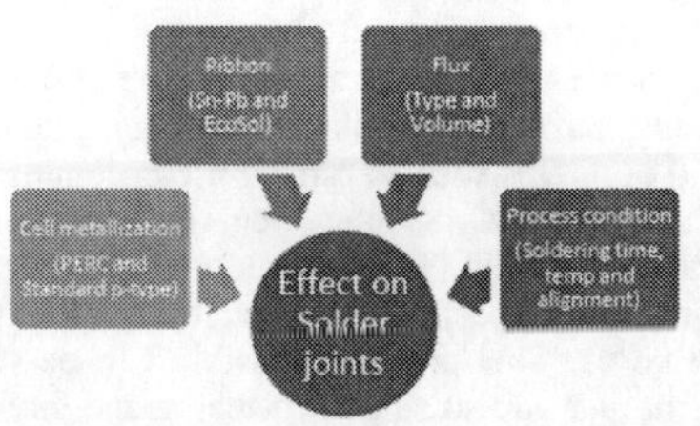

Figure 1: Factors affecting solder joints. Complete DOE was performed around these factors

The findings of this report may help PV professionals in taking some precautions in high speed manufacturing. They can have optimum conditions for tabbing and stringing and will be able to understand effect of ribbon camber and misalignment. In addition, companies may start to think of lead free option seriously.

3 RESULTS AND DISCUSSION

Currently, most PV ribbons used in the solar market are tinned with lead containing solders. However, lead and lead containing alloys are toxic. Disposal and recycling of lead containing modules which have been removed from service is quite difficult [3]. Expected changes in worldwide regulations will require that lead-based solders be replaced with less toxic alternatives. One of the objectives of this study was also to evaluate EcoSol in the production along with standard tin lead ribbon. This has not been done before. Any lead free (Pb-free) alternatives must exhibit acceptable soldering and reliability performance. EcoSol has liquidus of 179°C. This temperature is very close to the standard ribbon (183 and 192°C) and thus the ribbons can be processed at similar conditions to that of standard one. Fig. 2 shows the representative strings containing each bus bar soldered with tin-lead and EcoSol ribbon in an alternate fashion.

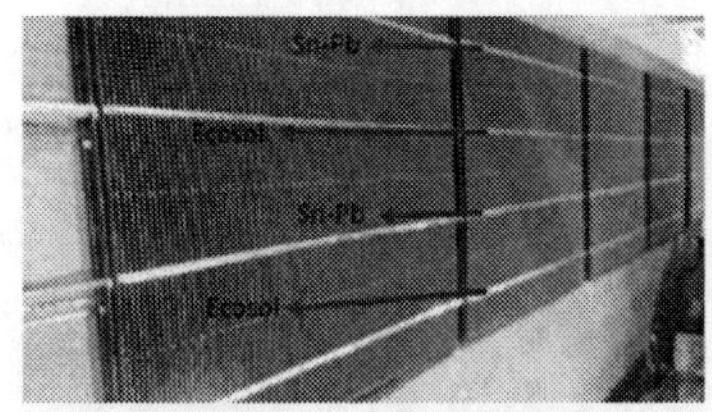

Figure 2: Strings of the soldered cells. Sn-Pb and EcoSol are bussed alternatively on each cell

3.1 Effect of cell metallization and ribbon

In the manufacture of solar cells, long-term reliability is of utmost importance. Peel strength is directly associated with reliability of the PV module. Any weak solder joint can severely affect the power output and module life. The joints are formed between silver and solder therefore, both factors are important. Silver metallization is a key for reliable joints [5]. In an industry's efforts toward reducing the silver consumption, the silver BB designs are getting narrower, BBs are patterned and the amount of silver that is used per cell is also reduced. This also means that there are very less contact points to form solder. In our experiment, n-type PERC cells and p-type c-Si cells were used. PERC type cells are made of telluride based silver paste with efficiency of 21 % and other p-type cells were 17.8 % efficiency. The design of the solder joint involves using appropriate parameters especially flux and soldering temperature. Fig. 3 shows the peel strength data of Sn-Pb and EcoSol ribbon processed at 210°C soldering temperature for 1.2 s. Fig. 3a shows results of n-type PERC cells while Fig. 3b shows p-type c-Si cells. As can be seen from the graphs, the peel strength values in both ribbons are excellent. Typically, on sunny side the values were >2.5 N/mm and on back > 3.0 N/mm. Both ribbons have comparable peel values, EcoSol being slightly less. EcoSol could be processed as same way as Sn-Pb without changing any parameters.

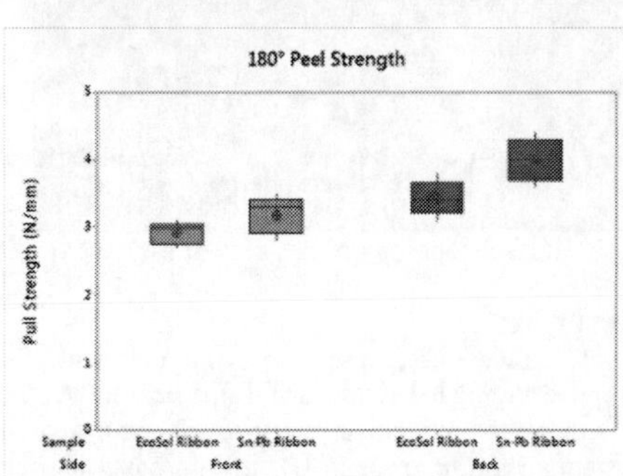

Figure 3a: Peel strength data of n-type PERC cell

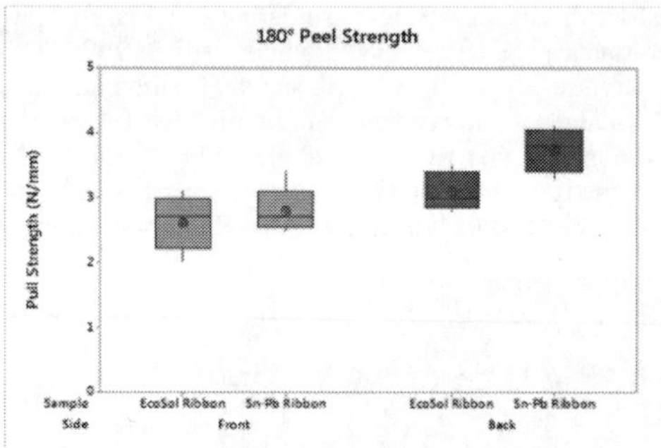

Figure 3b: Peel strength data of p-type c-Si cell

Close examination revealed, PERC cells give slightly better peel strength values and data is less scattered. Fig. 4a and 4b illustrate the result of the busbar profile from the cells used for the busbar profile experiment. As can be seen in the Fig. 4a PERC cell is smooth without much variation while in p-type cell (Fig. 4b), the edges are very high and there is a low lying saddle in the middle of the busbar with significant height variations. The difference between the highest (edge) and the lowest point in the middle of the busbar is in some places higher than 12 µm. The peaks at the edges will keep the ribbon off from the major part of the surface [6]. The consequence is an increase of thermal contact resistance between the busbar and the ribbon. In case of low quality cells where the variations are high and paste is not printed smoothly, impact is significant and this is the main reason for the failure mode. To evaluate effect of metallization further, some cells were oxidized and soldered under similar conditions. In all those cases non wetting or very less peel strength observed.

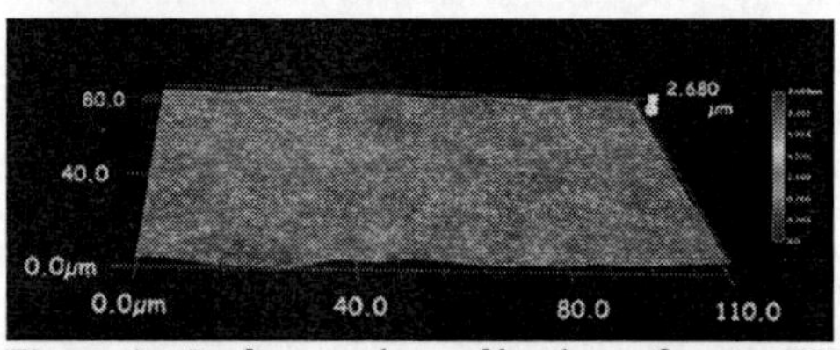

Figure 4a: Surface topology of bus bars of n type PERC cell

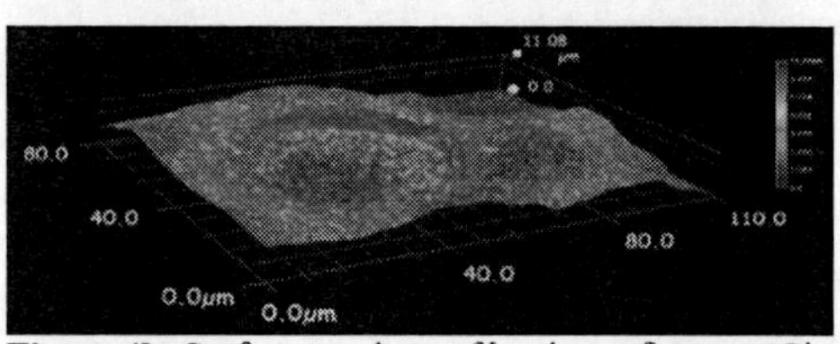

Figure 4b: Surface topology of bus bars of p-type cSi cell

3.2 Effect of flux

Due to higher throughput, the soldering time for tabbing and stringing has reduced and flux hass to wet the surface, remove the oxide and prepare surface for a good joint in such a short time span [7]. In full production run, only about 1-1.5 seconds are available for peak soldering. Many equipment have preheating zones, which sometime causes flux to evaporate making it less reactive.

Three fluxes, with solid content 1.1, 1.4 and 1.65 wt % were used for the study. The flux amount to be sprayed was adjusted by means of the cell conveyor speed. Slower the speed, more the flux volume per cell. The flux application at TeamTechnik takes place on both sides in one pass. Therefore, the specified amount applies to one cell (top and bottom side). Under controlled conditions and on dummy wafer, the flux amount was measured when sprayed at different speeds. Typically, 1000 mm/s is the desired speed so that TeamTechnik's Tabber and Stringer type 2100 is able to run 2100 cells/h. This translated to a calculated flux amount of 75 µl per cell. In addition, speeds of 900 mm/s = 84 µl and 700 mm/s = 109 µl were also tested.

The results were better with lower speed of cells but this in turn may create more contamination in the nozzle and other parts of the machine. Both, the flux selection and volume therefore, need to be carefully selected. Higher solid flux had a better wetting time and wetting force. For any lead free solder, wetting time and force matter a lot because Bi based alloys have low wetting capability than Pb based alloys. Higher solid % flux yielded better peel strength compared to other fluxes, however, the contamination was less with 1.1 % solid flux. Both, Sn-Pb and EcoSol could be soldered with standard flux. EcoSol does not require higher activator fluxes. Flux selection has a significant effect of increasing overall adhesion values and reducing effective soldering temperature. Use of an optimum flux and soldering condition will improve the quality of the solder joint so that failure mode is changed from non-wetting to optimum failure in fracture surfce analysis.

3.3 Effect of processing conditions

Processing conditions are important in determining the quality of solder joints. Effect of soldering time and temperature investigated. Typically, higher soldering time and temperature yielded better peel strength results. One of the important factors observed was the ribbon misalignment. As described above, as the ribbons and BBs are getting narrower, ribbons are becoming more like a wire and problems such as camber are becoming evident. Even in automated machine, the ribbon misalignment could not be avoided. We observed that even for slightest misalignment, the peel strength reduced by upto 40 % (Fig. 5 and Fig. 6). This problem possibly can be solved by making proper adjustments in machine and taking either of the one surfaces (ribbon or BB) wider than other. Slightly wider ribbon than the busbar has been preferred to increase the ribbon cross-section in order to enhance the module power output. Increasing ribbon thickness to increase the ribbon cross-section is in most cases not applicable. The thermo-mechanically induced cell stress in thicker ribbon is by far higher compared to the width of the ribbon [6].

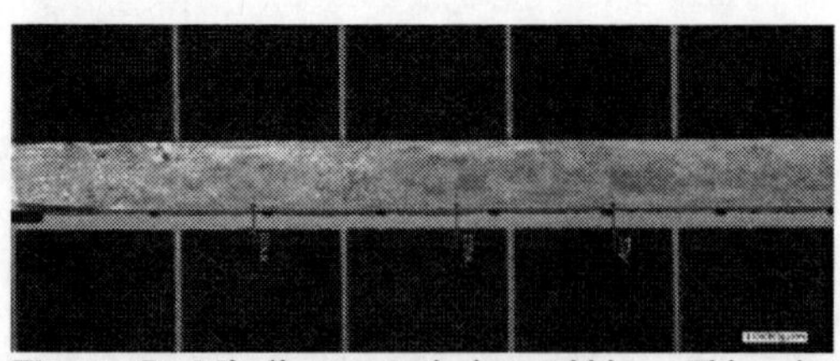

Figure 5: Misalignment during tabbing. This gives reduction in peel strength by upto 35 %

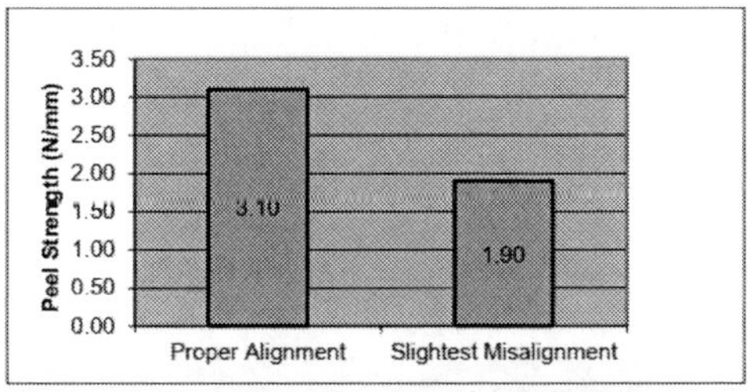

Figure 6: Peel strength comparision between properly aligned ribbon and misaligned ribbon over bus-bar

3.4 Fractured surface and microscopic investigation

Soldering defect usually won't be evident initially as they are more of long term concern. In high throughput manufacturing, under optimized conditions, peel strength was satisfactory in both Sn-Pb and EcoSol ribbons. However, peel strength is localized phenomenon and not distributed across the BB. In addition, BB is patterned and achieving peel strength per standard, EN50461 (>1 N/mm) is relatively easy. Analysis of fractured surface and failure mode is therefore, must be made to assess reliability of the solder joints. In theory, failure of the solder bond may occur at three different locations [3]: (1) at the Cu interconnect/solder interface; (2) at the Ag busbar/solder interface; or (3) within the solder itself. Failures at the Ag/solder interface, as well as within the solder itself appear to be the dominant modes. Our fractured surface analysis of cell and ribbon showed cohesive failures. In both case, silver from the BB transferred to the ribbon, however silicon cracking was not observed on front side in any case. In EcoSol, at some areas, bulk failure of the solder was observed. Cohesive failure can be seen in both cases (Fig. 7a and Fig. 7b).

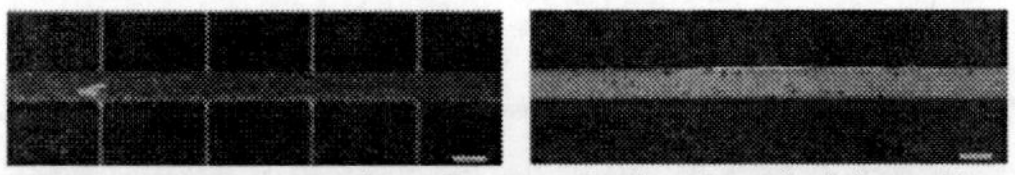

Figure 7a. Fractured surface analysis of peel ribbon (Sn-Pb soldered surface and ribbon)

Figure 7b. Fractured surface analysis of peel ribbon (EcoSol peel off soldered surface and ribbon surface)

Characterization by metallographic sample preparation and scanning electron microscopy did not reveal any intermetallic layer, although the samples showed good mechanical peel strength indicating good bonding (Fig. 8a and Fig. 8b). After mechanical polishing and ion milling the samples were imaged and analyzed using FESEM (Field emission scanning electron microscopy) and EDS (Energy dispersive X-ray spectroscopy). A systematic analysis of the cross-sections was carried out to verify the presence of bond formation and identify the intermetallic phase. Using these combined techniques of mechanical etching and surface smears removal, and high resolution imaging helped visualizing the various particles in the solder matrix, which were identified using EDS mapping.

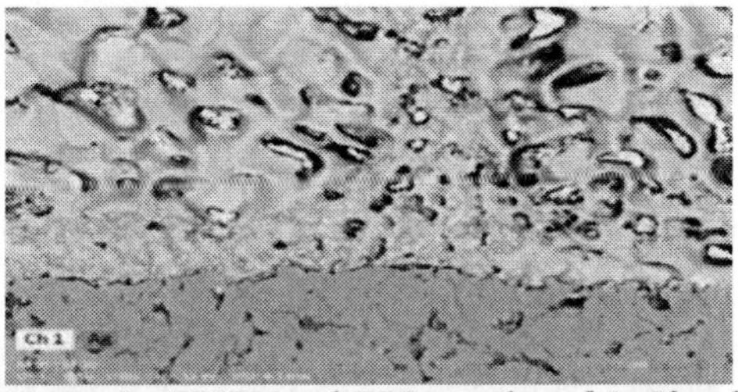

Figure 8a. FESEM and EDS mapping of Sn-Pb solder joint

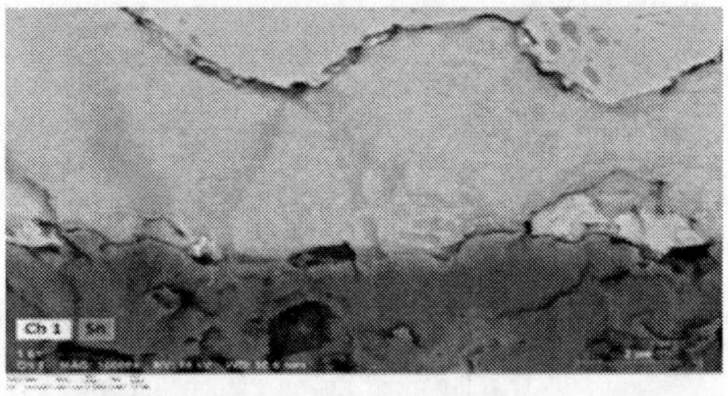

Figure 8b. FESEM and EDS mapping of EcoSol solder joint. Elemental distribution and Sn-Ag interdiffusion. No evidence of IMC compounds.

Despite the absence of a visible layer between the Ag busbar and the PV ribbon, EDS mapping was used to verify the elemental distribution of the soldered area. Fig. 8a shows that Ag particles diffused into the solder region. A further spot scan on the Ag rich region observed in the EDS mapping showed presence of Sn, Pb and Ag. Thus, we can assume that the Ag particles dispersed into the solder are starting to combine with Sn. However, due to the very short duration of heating during soldering, the time and temperature were not enough to ensure the formation of an actual intermetallic phase. Imaging of another area of the same sample confirmed these observations. Similarly, the cross-section analysis of EcoSol did not show presence of intermetallics formation, but EDS mapping of the Ag busbar/solder interface showed presence of a thin layer containing Sn and Ag. This was also confirmed by additional individual EDS mapping (Fig. 8b). An area scan determined that this thin layer has 85 wt.% Ag and 11.6 wt.% Sn. From the Ag-Sn phase diagram, we conclude that this layer was Ag and Sn in solid solution, and not the eutectic Ag_3Sn phase. Although we did not see formation of a clear intermetallic compound, we observed several areas with Ag-Sn inter-diffusion. This data is interesting and points out the situation in high throughput manufacturing. One could see nucleation sites but IMC layer is not formed. Although, in some instances, the IMC compounds can be noticeable under the microscope, observed IMC thickness is very thin and irregular in those cases. In absence of IMC, the strength of the bond can be realized by reduced surface tension, element interdiffusion and consequent disappearance of the metallization layer.

4 CONCLUSION

Solder joint analysis of standard tin lead and lead free, EcoSol in high throughput manufacturing is reported. Under identical conditions, the solder joints in lead and lead free ribbons were comparable and peel strength of >2.5 N/mm on front and >3.5 N/mm on the back side of the cell obtained. DOE results reveal that the cell metallization play an important role in determining quality of solder joints. Grade A PERC cells yielded higher peel strength compared to p-type c-Si cells due to the smooth surface topology. Similarly, flux having good wetting time performed better. Ribbon alignment was identified as one of the major causes of the weak solder joints. Due to decreased ribbon and bus bar size, this problem might occur frequently. Ribbon wider than BB can be suggested to alleviate this problem. Microscopic investigation yielded, absence of IMC in either of the ribbons due to rapid production speed. Only diffusion of tin into silver matrix was observed. This, however, do not have any effect on peel strength and cohesive failure was evident from fracture surface analysis. This study thus compare standard and EcoSol ribbons and further confirms lead free alternative can be used as an effective replacement to Pb containing ribbons. It looks possible to use lead free alternative as a "drop in" to the existing manufacturing set up.

4.1 Acknowledgement

We thank ORB solar for trials on their production line and Mrs. Divya K. for microscopic analysis.

5 REFERENCES

[1] World energy council. World energy resources 2016 solar. Link: https://www.worldenergy.org/wp-content/uploads/2017/03/WEResources_Solar_2016.pdf

[2] E. Schneller et al., Renew. Sust. Energ. Rev. 30 (2016) 992.

[3] N. Pujari, S. Telu, J. Sundaramurthy, M. Ribas, M. Murphy, 7th Workshop on Metallization and Interconnection for Crystalline Silicon Solar Cells, 2017. (Available at SSRN: https://ssrn.com/abstract=3152431 or http://dx.doi.org/10.2139/ssrn.3152431).

[4] M. Zarmai, N. Ekere, C. Oduoza, E. Amalu. Applied energy 154 (2015) 173.

[5] P. Schmitt, A. Russell, M. Tranitz, Proceedings 27th European Photovoltaic Solar Energy Conference, Vol. I (2012) 2058.

[6] Y. Zemen, H. Teusch, G. Kropke, S. Pingel, S. Held, Proceedings 27th European Photovoltaic Solar Energy Conference, Vol. I (2012) 2030.

[7] N. Pujari, E. Poh, S. Sarkar, M. Murphy and C. Bilgrien, Energy Procedia 2018 *Accepted.*

LEARNING FROM THE PAST TO LOOK BEYOND THE ROADMAP OF PERC SI SOLAR CELL MASS PRODUCTION

A Pietro P. Altermatt, Yang Yang, Yifeng Chen, Daming Chen, Xueling Zhang, Guanchao Xu, Zhiqiang Feng
Trinasolar - Key Laboratory for PV Science and Technology (SKL), Changzhou, Jiangsu Province, China,
pietro.altermatt@trinasolar.com

ABSTRACT: Past patterns of cell and module efficiency, fabrication cost, metallization, and growing independence from the IC industry are used to extrapolate progress into the future. It is concluded that the development of cheap Si bottom cells for tandems should be accelerated, scarce materials be avoided, and for passivated contacts to come into mainstream Si photovoltaics, it may be most advantageous to introduce them into the PERC fabrication process with the least required changes. Once both electron and hole contacts are passivated in the mainstream, it may be easier to switch from p- to n-type wafers to take advantage of higher bulk lifetimes, compared to switching to n-type cells like PERT and HJT now. In this sense, the million dollar question 'p-type or n-type?' may be posed too early, and concentrating the efforts on further developing PERC may be more successful. For this, a transparent passivated contact replacing the phosphorus emitter in PERC cells is suggested.
Keywords: PERC, Passivated Contacts, Emitter, ITO, Solar Cell Efficiencies

1 PATTERNS IN THE PAST

Figure 1 shows world record efficiencies [1] and in-house measurements [2] of mono-crystalline silicon solar cells over the years. Each evolution of the three basic cell designs progressed initially very quick, but eventually saturated in what can be called an 'efficiency hurdle', because each cell design has its limiting practical efficiency η_{limit}. The data can be fitted with the Goetzberger function [3]:

$$\eta(t) = \eta_{limit}\left[1 - \exp\left(\frac{t_0 - t}{c}\right)\right] \quad (1)$$

where the parameter c, quantifying progress, can be called efficiency loss decay constant. It is in units of years, required for reducing the efficiency loss by 1/e. The shorter c, the faster is progress. The starting time t_0, in units of years, has no practical significance because researchers didn't start at zero efficiency; hence, t_0 is usually a bit earlier than research activities historically started. From a least square fitting, c results as 2.5 years for the unpassivated emitter, 5.7 years for the passivated emitter (and rear), and 1.7 years for the passivated contacts. The passivated emitter progressed slowlier because passivating the front emitter alone would not have yielded the efficiency improvements; the rear had to be passivated as well, requiring work on both surfaces at the same time.

Because the initial progress is quick, t_0 is not far earlier than when researchers started with each cell structure. Hence, the values in Fig. 1 can be plotted counting from t_0 as shown in Fig. 2. This makes it apparent that progress has become quicker. One reason for this is that drawing from past experience is an advantage. Another reason may be that the research community has become larger over the years and attracted more research funds.

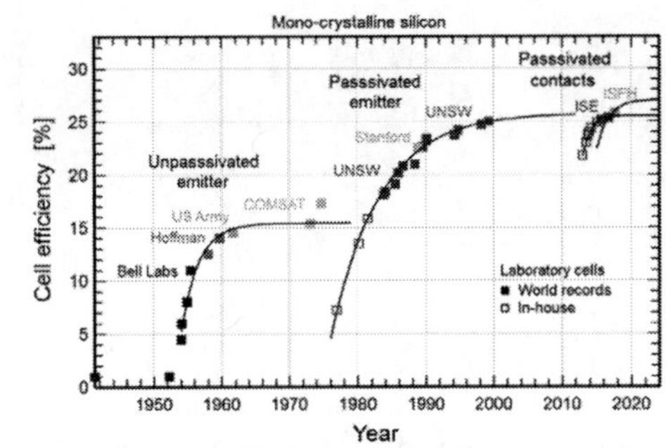

Figure 1: World record efficiencies and in-house measurements of mono-crystalline silicon solar cells over the years. The evolution of the three basic cell designs are fitted with Eq. (1).

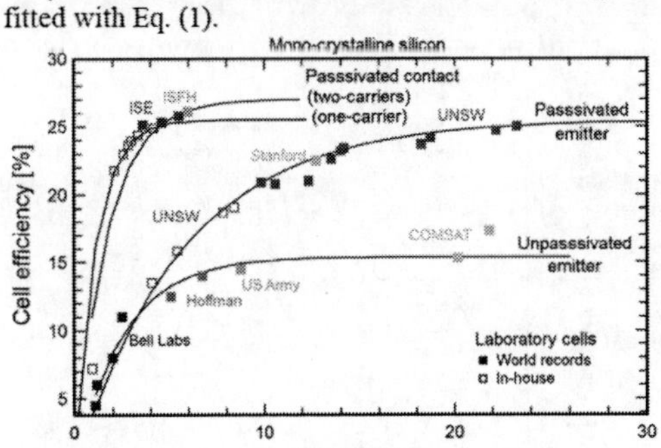

Figure 2: The efficiency values from Fig. 1 plotted counting from t_0 of Eq. (1) makes it apparent that progress in efficiencies has accelerated going to more advanced cell designs.

2 PATTERNS FOR THE NEAR FUTURE

This past pattern may be extrapolated to the future as a guide for what can be expected. For example, perovskite solar cells have been very actively developed in a large and diverse research community in recent years. Figure 3 shows the evolution of their world record efficiencies [4].

It progresses so fast that some researchers are tempted to write "such rapid progress has never been seen before in the history of solar cell development" [5]. However, fitting with the Goetzberger function yields c = 3.2 years for perovskite cells and 3.6 years for Si/perovskite tandems, while c of crystalline/amorphous silicon heterojunctions in the rear contact configuration is 1.8 years. Hence, progress in perovskite cells has been slower than in the

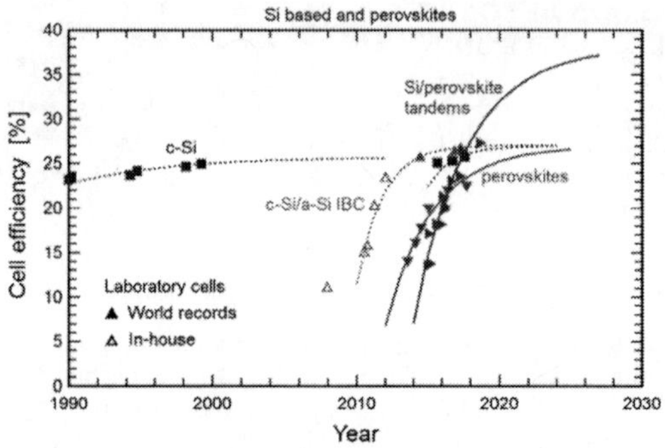

Figure 3: Evolution of world record efficiencies of crystalline/amorphous silicon heterojunctions in the rear contact configuration, of perovskite solar cells, and silicon/perovskite tandem configurations. Lines are least-square fits with the Goetzberger function, Eq. (1).

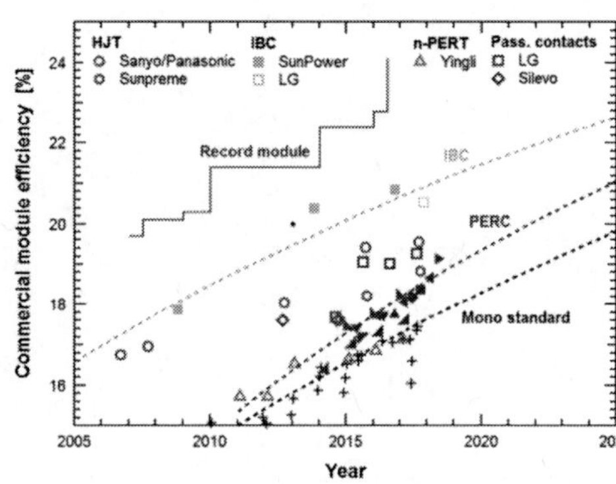

Figure 4: Efficiency of sold Si modules, calculated with data taken from companies websites (and therefore may lag behind up to one year). The dashed curves are fits with Eq. (1), the solid curve are record module efficiencies.

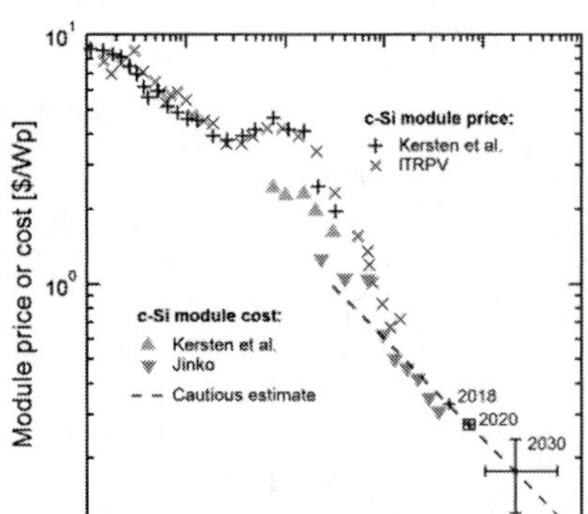

Figure 5: Silicon module fabrication cost (triangles) and sales price (crosses) show an exponential-like decline.

crystalline Si cell community. Understandably, because it is a new material with new complexities. Notwithstanding, the Goetzberger fit suggests that, typically, Si/Perovskite tandem may reach 35% efficiency in about 4 years (if not delayed by efficiency hurdles). Commercial type Si/perovskite tandems start passing stability tests [6]. They contain no spiro and no gold (which are expensive). Considering the expected fast progress, the crystalline Si cell community should accelerate its efforts to develop cheap Si bottom cells, otherwise, it will become the bottle neck of tandem development.

Other materials than perovskites are also candidates for tandem configurations and generally show better stability than perovskites. To assess how other materials can be used for mainstream production, it is helpful to be aware that photovoltaics will have a significant impact on climate change only once 10 TW_p are installed [7], and that 22 TW_p are necessary for 100% renewables in the power sector, globally [8]. In contrast, the present fabrication capacity is 0.1 TW_p/a, and needs to increase at least 10-fold to help meeting the climate goals of the Paris agreement. This required up-scaling causes undoubtedly mid- and long-term material supply risks. Most (but not all) III-V, II-VI, chalcopyrites and kesterites contain scarce elements with very high supply risks and hence are unavailable for mainstream photovoltaics, even if they are made feasible, for example by developing cheaper deposition techniques [9].Various recent papers [10-12] predict this, using various methods and data, particularly if the whole energy transition is considered [13 -18], like the case of nickel. The consequence of this is that, if your device should make a clear difference to our climate, do not use scarce elements like In, Cd, Te, Se, Pd, Mo, Ni, etc. as a bulk constitute (even in thin-films). Using these elements as dopants or for impurity engineering is feasible. For Ag [12], Ga [10,13], and Sn [14,18], price increase is expected. A consequence for mainstream silicon photovoltaics is that

1. Ag needs to be reduced to maximally 40 mg/cell; safer would be 20 mg/cell, for example by multibusbar and multiwire technologies. There is enough silver predicted that switching from silver to copper is not an urgent matter, but may come under pressure if other industries start using silver in large amounts.
2. Indium tin oxide (ITO) cannot be used for mainstream cells. The above-cited studies shows this undoubtedly. Already nowadays, the largest manufacturers in China are highly reluctant to introduce ITO in their cells considering the amounts of indium they (and their competitors) would need to buy each year, posing a price stability risks. In the long term, one may even need to replace SnO_2 by ZnO or other oxides.
3. Last but not least, an informed choice of materials for passivated contacts should be made. Even if traditional c-Si PV supplies all electricity in 2030, the required production growth rates for Si, Al, steel, etc. would still fall within the historical range. However, if scarce materials are chosen for passivated contacts, these cell designs won't be able to be mainstream.

3 MAINSTREAM BEYOND PERC

The efficiency evolution of sold Si modules is nearly linear, as shown in Fig. 4, calculated with data taken from companies websites (and therefore may lag behind up to one year). The dashed curves are fits with Eq. (1) with c between 20 and 30. The evolution is nearly linear because efficiency is not the sole driver; fabrication cost is at least as important. Silicon module fabrication cost (triangles in Fig. 5) show indeed an exponential-like decline. Mainstream decays fastest [19]. Crystalline silicon modules decay so fast for various reasons, including:

- there is a single target (the standard cell in the past, now PERC),
- enabling standardization in the whole value chain.
- Upscaling.
- High staff turnover in China fosters exchange of skills and knowledge.

Figure 6 shows how CAPEX (capital expenditure) per GW has declined in an exponential manner as well. PERC lines presently cost about 30 million Euro/GW in China. However, there is a small margin (as indicated in Fig. 5) and a fierce fight for gigawatts. CAPEX per GW declines partly due to higher throughput of most machines, as indicated by the blue symbols in Fig. 6.

The evolution of PERC module efficiency is shown in more detail in Fig. 7. The yearly efficiency increase is about 0.4 $\%_{abs}$. Typical cell (not module) efficiencies are presently up to 21.4% with homogeneous emitters, and from 21.5% to 22.2% for selective emitters. PERC roadmaps [20-23] predict that about 24% cell efficiency can be reached with present technology developed steadily further (i.e. without passivated contacts etc.). Combining this with the module efficiency evolution of Fig. 7 indicates that about 24% cell efficiency may be reached in typically 7 years in the industry median (if no major breakthroughs or disruptions occur). This is a rather cautious estimate, taking into account that old fabrication lines are not as clean; some factories most probably reach 24% quite a bit earlier. Combining this with Fig. 5 finally indicates that fabrication cost will approximately drop to half in the coming 7 years.

These PERC roadmaps are an important indicator how far present technology can go, but need not be followed to their end. For example, one of their last improvement steps is a front contact opening of only 10 µm. Instead, passivated contacts may be introduced into the PERC process flow. Process steps like this may become competitive at cell efficiencies above about 23%, depending rather sensitively on their additional cost and efficiency gain. However, presently it is unclear what kind of features or cell structures should be introduced into PERC cells, hence, there is no single target for R&D and no standardization of advanced equipment apparent. Large R&D activities are invested in n-type cells as an alternative to PERC by both public institutions and cell manufacturers, developing such as PERT, a-Si/c-Si heterojunction technology (HJT), interdigitated back contact designs (IBC), or combinations of those. Despite of this, PERC efficiencies seem to catch up with PERT efficiencies, while both HJT and IBC seem more expensive.

Our main point is that, for passivated contacts to come into mainstream Si photovoltaics, it may be most advantageous to introduce them into the PERC fabrication process with the least required changes. Once both electron and hole contacts are passivated in the mainstream, it may be easier to switch from p- to n-type wafers to take advantage of higher bulk lifetimes, compared to switching to n-type cells like PERT and HJT now. In this sense, the million dollar question 'p-type or n-type?' may be posed too early, and concentrating the efforts on further developing PERC may be more successful. Of course, it all depends on creativity, so future cannot be predicted, and we cannot be ultimately sure about our point we make. Presently, it looks to us that PERC is a suitable platform for other materials and device features to come into mainstream.

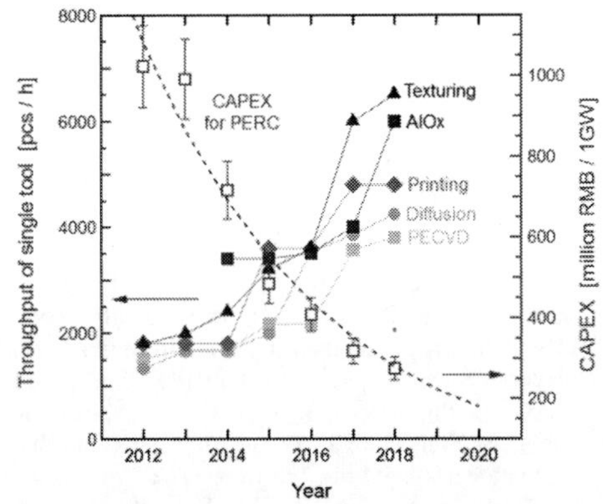

Figure 6: Capital expenditure (CAPEX) per GW in China for PERC lines (red curve) over time, partly caused by higher throughput (blue symbols).

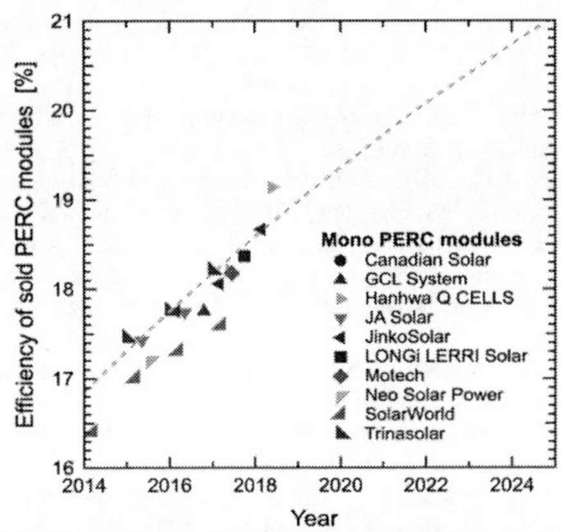

Figure 7: Efficiency of sold PERC modules from Fig. 4, calculated with data taken from companies websites (and therefore may lag behind up to one year). The dashed curve is a fit with Eq. (1).

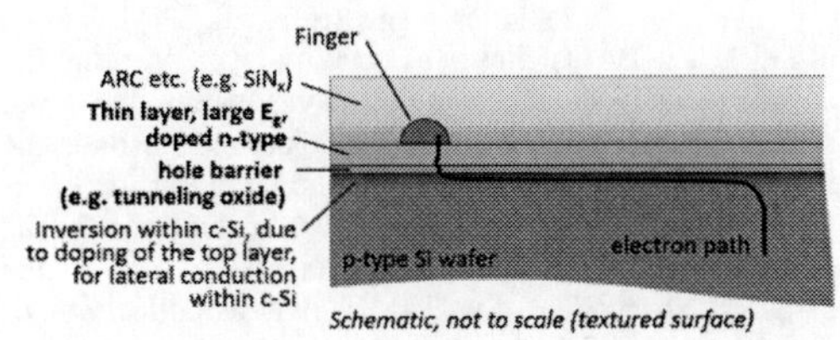

Figure 8: Proposed transparent hetero-emitter with a passivated contact.

4 PROPOSAL OF PERC EMITTER WITH PASSIVATED CONTACT

The recombination losses in the emitter are larger than at the rear in most industrial PERC designs. We therefore expect that more efficiency can be gained by introducing a passivated contact at the front. On the other hand, recombination at the front metal contacts is usually minor compared to the total recombination in the emitter. Hence, solely passivating the existing emitter contact would not increase PERC efficiency considerably. We therefore suggest to omit the phosphorus diffusion entirely and

1. to put a barrier that conducts electrons but blocks holes onto the front wafer surface, and that passivates the Si surface. This may be a thermally grown tunnel oxide with high quality, or sufficient may be any type of oxide (or other materials) fabricated by chemical means or by deposition or by other methods. See Fig. 8.
2. A layer comes on top of that tunnel barrier to define the electron quasi-Fermi level such that it bends the bands in Si downwards. We therefore call it 'bending layer' in this paper (its role differs from most window layers typically applied in thin-film PV). This layer can be as thin as metallization allows because there is no need that it conducts electrons laterally (its role therefore differs from the usual TCO's used in HJT and thin film cells). The band bending induces an inversion layer in the silicon wafer near the barrier, which depends on the amount doping within the bending layer and on the electron affinity of the bending layer.
3. The layer must not absorb a noticeable amount of light between 370 nm and 1200 nm. Due to the barrier, most absorption would be parasitic.
4. At this stage, the cell is a suitable bottom cell for a tandem configuration.
5. For a single-junction cell, the layer is contacted by fingers. Most electrons cross the hole barrier only near the finger because they cannot flow laterally in the bending layer for long distances. Hence, most electrons flow laterally in the inversion layer and in the rest of the silicon wafer. The fact that the bending layer does not need to provide lateral electron conduction poses less stringent requirements on it.
6. The bending layer may be covered by a standard nitride to provide an anti-reflection coating (ARC), release hydrogen in a firing step, and to protect against potential-induced degradation (PID). Because this ARC layer does not need to passivate a phosphorus-diffused emitter, its requirements may be somewhat relaxed as well and therefore may be processed differently or even with different materials.

An example simulation using Sentaurus [24] is shown in Fig. 9. ZnO:Al is chosen as bending layer. Varying the dopant density in the bending layer varies the band bending in the silicon. The resulting electron density in the inversion layer in p-type silicon is shown in Fig. 10.

In the following, we like to give optical criteria for the material choice of the bending layer. Crucial is the parasitic absorption. Fig. 11 shows a collection of absorption coefficients of various semiconductors and insulators, plotted over photon energy E in respect to their band gap, E-E_g. The upper and lower bounds can be fit with

$$\alpha = A\left(E - E_g\right)^2 \qquad (2)$$

using A = 7×10^3 and 3×10^5 cm^{-1} eV^{-2}, respectively, see grey lines in Fig. 11. We take these lower and upper bounds as input for the ray tracer SunSolve from PV lighthouse [25]. We use our optical model [26] for a standard PERC module. Hence, there is a standard EVA (absorbing at wavelengths below about 370 nm) and a nitride ARC that protects from PID. The ray tracing results for the parasitic absorption in the bending layer are shown in Fig. 12. From this, we conclude that typical amorphous semiconductors require $E_g > 2.75$ eV, while typical poly-

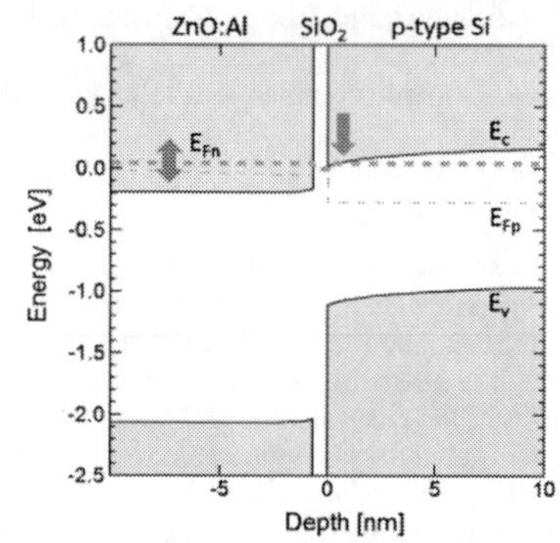

Figure 9: An example emitter simulated with Sentaurus. Band diagram with the quasi-Fermi energies for electrons and holes.

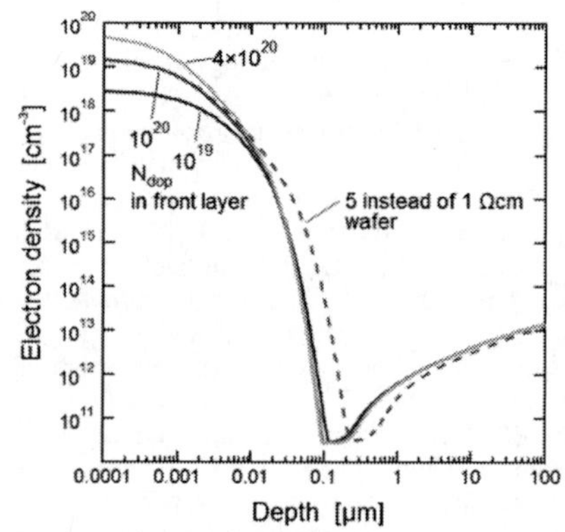

Figure 10: Simulated electron density in p-type silicon, affected by the band bending shown in Fig. 9 for the indicated dopant densities in the bending layer, and for a 5 Ωcm wafer instead of the usual 1 Ωcm wafer.

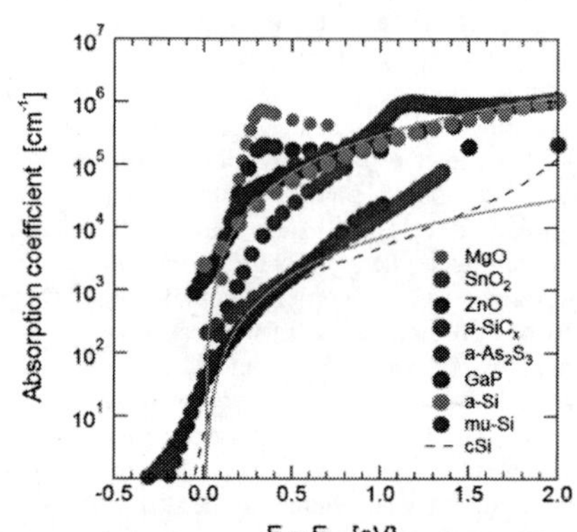

Figure 11: Absorption coefficient of various semi-conductors and insulators, plotted over photon energy E in respect to their band gap, E – E_g. Grey lines: chosen lower and upper bounds for parameter A in Eq. (2).

semiconductors and oxides require $E_g > 1.90$ eV. It is hard to find amorphous semiconductors with such a large band gap. For example, amorphous GaAs and amorphous GaP have both a band gap near 1 eV. Therefore, a semiconductor needs to be crystallized during the firing step at least into nano-crystallites or, instead, a metal oxide or other material having a wide band gap need to be

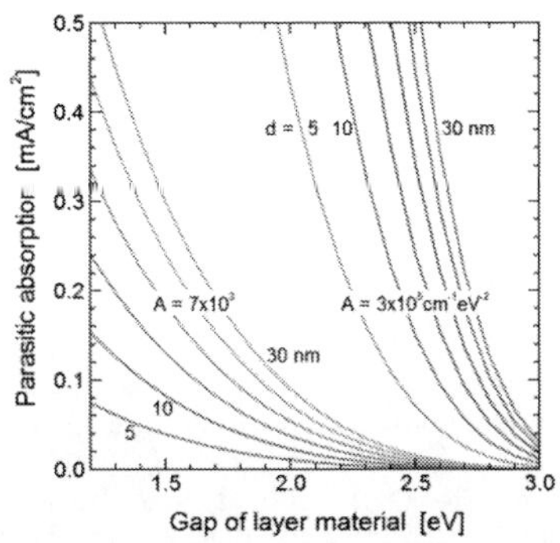

Figure 12: Parasitic absorption in the bending layer as a function of its band gap, for the indicated bending layer thicknesses and for the lower as well as the upper bound from Fig. 11. Calculated with ray tracing simulations in a module using SunSolve from PV Lighthouse.

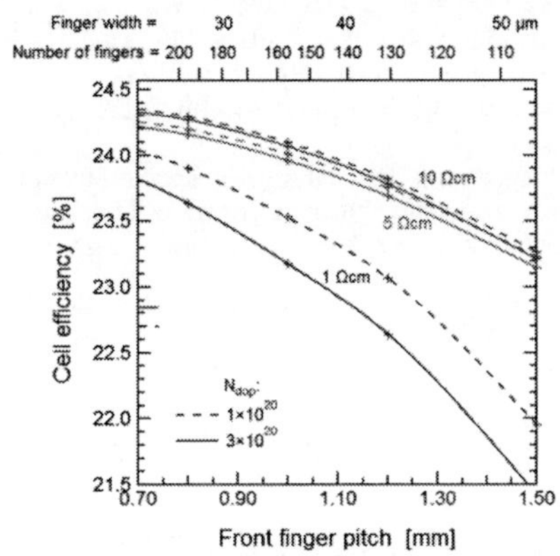

Figure 13: Cell efficiency in air (not in the module), simulated with Sentuaurs, in dependence of front finger pitch (bottom x-axis), which corresponds to the number of fingers on standard M2 wafers as indicated at the top x-axis. For keeping the shading of the fingers constant, the indicated finger widths are required. The wafer resistivity as well as the dopant density in the bending layer are chosen as indicated.

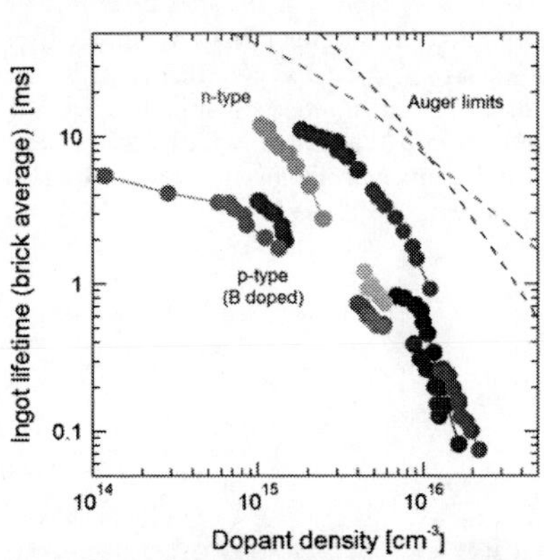

Figure 14: Ingot lifetimes (brick averages) in typical examples of Cz ingots sold in 2017 on a GW scale by Longi, China (the p-type ingots below a dopant density of 2×10^{15} cm^{-3} were custom-made).

chosen. Well-known candidates are ZnO and SnO_2, for prototypes there is ITO. These oxides can be easily doped n-type. But there exist also a range of other materials with suitable properties.

Using the optical and electronic band structure properties of ZnO:Al, cell efficiency in air (not in the module) is simulated as an example with Sentaurus and shown in Fig. 13. Three main features become apparent:

A. Optimum performance is obtained with high p-type wafer resistivities near 5 Ωcm,
B. where cell efficiency becomes rather insensitive to dopant density in the bending layer (as long as it is above about 1×10^{19} cm^{-3}, depending on the effective density of the conduction band, N_c).
C. An advanced metallization is required with finger width of about 35 µm, enabling to print about 150 fingers per cell on a M2 wafer. The simulations in Fig. 13 were done with Trina's present 12 busbar design.

The fact that high-resistivity p-type wafers perform optimum has a few crucial effects:

- The wafer lifetimes in presently commercially available boron-doped Cz ingots increase from about 300 µs in typical 1 Ωcm material to an expected 1.5 ms in 5 Ωcm material. This is deduced from the data shown in Fig. 14, obtained with a Sinton BCT400 at an injection density of 1×10^{15} cm^{-3} (p-type) and 5×10^{14} cm^{-3} (n-type) at 3 mm depth. The Auger limits for p-type [27] and n-type [28] were accordingly calculated at these injection levels. The simulations in Fig. 13 were therefore done with 1.5 ms lifetime.
- Soon, gallium-doped p-type wafers will be the standard material. This material can be more easily fabricated in the 5 Ωcm range.
- The rear surface passivation of 5 Ωcm wafers is better than in the presently usual 1 – 0.8 Ωcm range, possibly enabling simpler rear passivation procedures.

For evaluating the feasibility of advanced metallization, Fig. 15 shows the past behavior pattern of various printing techniques. There is linear progress towards 30 – 25 µm wide fingers having an area cross-section of about 300 µm^2. The above-mentioned sufficient target for the proposed emitter can be reached without the high aspect ratios some alternative printing techniques enable. The graph also shows that plating becomes a viable alternative to screen-printing due to short plating times. Hence, there seem to be two main scenarios:

1. If the bending layer is able to replace the nitride ARC (in terms of transparency, refractive index, hydrogen source, and anti-PID), the bending layer can be made sufficiently thick to function as the ARC. Its thickness may enable screen-printing without affecting the hole barrier layer. The bending layer's conductivity needs to be sufficiently high so the electrons can pass its thickness without significant voltage drop.
2. If the nitride ARC remains necessary for at least some of its above mentioned functions, the bending layer may be chosen as thin as plating does not affect the hole barrier layer. In that case, plating may need to be done after firing of the nitride for hydrogenation; the nitride layer would need to be carefully opened but can act as mask during plating. In case of plating, the

material of the bending layer needs to be resistant to the plating bath, which may favor SnO_2 or other materials over ZnO. Possibly, other metallization techniques than plating may be favorable.

5 R&D STRATEGIES

The idea of an inversion layer is not new. Nor is the thinning of a TCO and putting an ARC on top. Combining it with a transparent passivated contact to replace the phosphorus emitter may be new. Particularly the absence of an a-Si or nc-Si layer between the tunnel oxide and the TCO (by soft deposition methods such as PECVD), as well as avoiding lateral conduction in the TCO to lower the electron concentration and, with this, to lower parasitic absorption. However, this structure is not patented by us, as we like to foster its development in the wide PV community. The patterns of the past, outlined at the beginning of this paper, suggest that a whole community is required for successful integration into mainstream.

As R&D strategy, we suggest to first fabricate prototypes for testing whether the simulation model [29] is sufficiently realistic. We suggest that various research groups may use the materials they have most experience with, like ITO, their printing technique etc. Research groups who do not have advanced printing techniques available may use today's common finger pitch and expect low fill factors.

If the prototypes perform in a promising way, an efficient and necessary R&D target is to keep materials, fabrication tools and processes as close as possible to mainstream PERC production. Ideally, in the first scenario above, the phosphorus diffusion (including its single-side removal and edge isolation in the PERC process) can be replaced by fabrication of the hole blocking layer and the deposition of the band bending layer. For example, a thermal tunnel oxide may be grown on both sides of a single-side textured wafer, followed by rear passivation, front PECVD deposition of the bending layer (for soft deposition without disturbing the tunnel oxide, and for including hydrogen from precursors), rear laser opening,

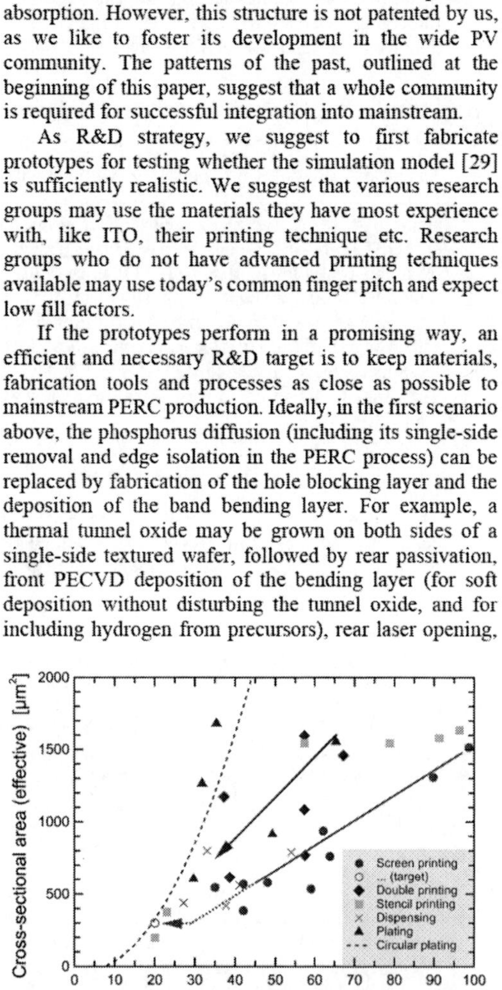

Figure 15: The cross-sectional area of front metal fingers is responsible for resistive losses and is plotted over finger width, responsible for shading losses. Data points are collected from literature over the years, the arrows indicate the trends over time, while the dashed curve describes increasing circular plating time. See Ref. [23].

and finally screen-printing of a multi-busbar or multi-wire design (with pastes possibly needing adaptation to the bending layer). This would be a significantly simplified process sequence compared to the typical PERC sequence nowadays. However, reliability tests of modules need to be passed.

If screen-printing does not work or the layer bending material cannot fulfill the important tasks of today's nitride, the second scenario would make it necessary to find alternative printing techniques such as plating. It is generally advantageous if progress in industrial fabrication is anticipated in R&D also in metallization (not only in wafer lifetime, which is usually done e.g. by choosing n-type or Fz). Because still some of today's processing steps would be unnecessary in the second scenario, alternative printing may be accomplished within today's PERC process complexity. However, the growing throughput shown in Fig. 6 must be considered. After all, a cell efficiency of about 24% with today's wafer material and today's rear structure is predicted in Fig. 13, which is about $2\%_{abs}$ higher than is nowadays fabricated. Such an improvement can alternatively be reached with other progress in material and technology outlined in Refs. [20-23].

Last but not least, for enabling a successful integration of this type of emitter into mainstream, a few important points need to be considered. The process sequence

- must not require long annealing times (should be done with firing),
- no separate hydrogenation process (should be done with firing),
- no prolonged, high annealing temperatures,
- and it should cause no significant parasitic absorption.

Most of today's R&D in both public institutions and in industry focus on the passivated contact made with poly-Si. This is somewhat surprising because most of these contacts do not fulfil the above points. It is questionable whether such poly-Si passivated contacts can enter mainstream, even when using n-type wafers. Recently, a passivated contact made by firing has been demonstrated [30], with its scope outlined in Ref. [31]. This opens possibilities to come a step closer to fulfilling the above points.

The poly-Si contacts were introduced into PV from the IC industry. Currently, the PV community may repeat a pattern of its past not mentioned so far in this paper: that it has stayed for (too?) long in line with IC fabrication and materials, and only developing its own requirements boosted its success.

REFERENCES

[1] M.A. Green, Prog PV 17, 183 (2009); M.A. Green et al, Prog PV, solar cell efficiency tables.
[2] Data from the first author for UNSW, F. Feldmann for ISE.
[3] A. Götzberger et al., Sol. En. Mater. Sol. Cells 74, (2002) 1.
[4] Data taken from NREL efficiency table.
[5] W.J. Yin, J. Phys. Chem. C 119 (2015) 5253.
[6] Henry Snaith (private communication), Oxford PV: 4 cm2 area with <5%rel eff. drop after: 85%RH/85°C for 3000 h, 200 thermal cycles from -40 to +85°C, or 60°C light soaking.
[7] D.B. Needleman, Energy Environ Sci 9 (2016) 2122.
[8] C. Breyer etal, Prog PV 26 (2018) 505.
[9] A.L. Greenaway et al, ACS Energy Lett 2 (2017) 2270.
[10] C. Helbig et al, Appl Energy 178 (2016) 422.
[11] G. Kavlak, Energy Environ. Sci. 8 (2015) 1651.
[12] A. Månberger et al, Energy Policy 119 (2018) 226.
[13] M. Frenzel et al, Resources Policy 52 (2017) 327.
[14] E. M. Harper et al, J Industrial Ecology 19 (2015) 628.
[15] S. Mohr et al, Resources 7 (2018) 17.
[16] K. Tokimatsu et al, Appl Energy 225 (2018) 1158.
[17] T. Watari et al, Minerals 8 (2018) 156.
[18] C. Yang et al, Sci Total Env 635 (2018) 1351.
[19] Y. Chen, et al. 7th WCPEC, Waikaloa, (to appear in IEEE J PV).
[20] B. Min et al, Proc. 31st EUPVSEC (2015) 473.
[21] H. Huang et al, Sol. Energy Mater. Sol. Cells 161, (2017) 14.
[22] Min et al, IEEE J. Photovolt 7 (2017) 1541.
[23] P.P. Altermatt et al, Photovoltaics International 41 (2018) 54.
[24] Sentaurus Device, Synopsys inc., Mountain View, CA.
[25] K.R. McIntosh, PV Lighthouse, Australia, www.plvlighthouse.com.au
[26] Y. Yang et al, Energy Procedia 124 (2017) 240.
[27] A. Richter et al., Phys Rev B 86 (2012) 165202.
[28] B.A. Veith-Wolf, J. Schmidt, PSS RRL (2017) 1700235.
[29] This cell model (Excel template for handling Sentaurus Device) can be downloaded from PV Lighthouse, www.pvlighthouse.com.au.
[30] A. Ingenito et al, Nature Energy 3 (2018) 800.
[31] P.P. Altermatt, Nature Energy 3 (2018) 718.

INLINE CHARACTERIZATION OF DIAMOND WIRE SAWN MULTICRYSTALLINE SILICON WAFERS

J. Haunschild[1], N. Bergmann[1], T. Hammer[1], K. Krieg[1], T. Kaden[1], O. Anspach[2], H. Schremmer[3], S. Rein[1]
[1] Fraunhofer Institute for Solar Energy Systems ISE, Heidenhofstraße 2, 79110 Freiburg, Germany
[2] PV Crystalox Solar Silicon GmbH, Gustav-Tauschek-Straße 2, 99099 Erfurt, Germany
[3] Hennecke Systems GmbH, Aachener Straße 100, 53909 Zülpich / Germany
Corresponding author: Jonas Haunschild | Phone: +49 (0)761 4588 5634 | e-mail: jonas.haunschild@ise.fhg.de

ABSTRACT: While for monocrystalline silicon ingot slicing diamond wire sawing processes are widely used, the general approach for slicing multicrystalline bricks has been based on slurry wire processes. Driven by a strong decrease in wafer prices recently, wafer manufacturers of multicrystalline materials switched rather quickly to the diamond-cut process, although several severe technological challenges were to be faced. Unfortunately, sawing is more difficult and an adaptation of solar cell production is essential for texturing processes considering diamond-cut wafer surface morphology. Also characterization methodologies must be adapted to these new material properties. Especially saw mark inspection and thickness topology measurements are key aspects of developments. General imaging methods also need to be evaluated for defect detection and wafer rating purpose. In this paper, an overview of inline characterization methods and their applicability on diamond-cut wafers will be given. Five out of eight methods need hardware or software adjustments to work on diamond-cut wafers. Special attention is paid to the photoluminescence analysis which allows a forecast of solar cell parameters if the relevant defect features are correctly extracted from the images. Due to the increased optical artifacts, this is not a trivial task.

Keywords: characterization, multicrystalline silicon, wire sawing, diamond wire

1 INTRODUCTION

The multicrystalline silicon (mc-Si) wafering market has rapidly turned from slurry cut to diamond-wire slicing processes. Respecting the rising cost pressure, the diamond cut process for multicrystalline wafers seems to prevail through its many benefits (especially the shorter sawing time) [1] and has become the next-generation slicing technique [2]. Especially in recent months, the prices for monocrystalline wafers have decreased a lot and were almost on the same level as for multicrystalline wafers. As the material quality of multicrystalline wafers is lower than of monocrystalline wafers, the manufacturing costs must be cheaper and thus the introduction of the diamond-cut technology on multicrystalline wafers is a pressing concern. However, still much research and development is necessary, as the slicing of multicrystalline wafers is more difficult than for monocrystalline wafers. Also, established texturing processes have reduced efficiency on the diamond-cut surface and the impact on solar cell efficiency is under investigation [3, 4].

But also inline metrology [5] needs to be adapted and therefore, in this paper, we will focus on the challenges which metrology faces when dealing with the diamond-cut surface on multicrystalline wafers. After summarizing the technological differences between slurry cutting and diamond cutting, we give an overview of state-of-the-art characterization techniques and of the common parameters which are addressed during inline inspection. Then, we discuss why those inspection methods have limits on the diamond-cut surface, which sawing-issues are problematic and where further development is necessary.

2 THE WIRE-SAWING PROCESS

The common standard for wafer manufacturing is to use a wire saw to cut the monocrystalline ingot or the multicrystalline brick into wafers [6]. The wire is spooled around wire guides forming a wire web, so that the whole brick is cut at once when pushed through the running wire web. An illustration can be seen in Figure 1.

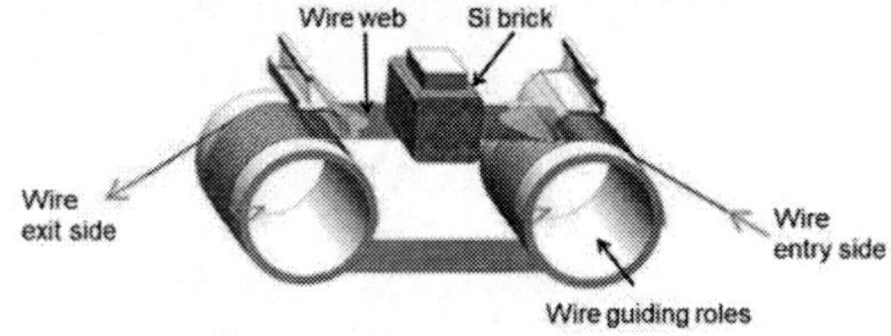

Figure 1: Scheme of the wire-sawing process.

The technological progress has brought various improvements to reduce the cutting time and increase overall quality and performance [7]. One of these improvements is the utilization of structured wire [8].

In recent years, many manufacturers have switched from plain-steel wires to wires coated with diamonds. When using the plain-steel wire, the cutting is done by small silicon carbide (SiC) particles, which are part of the cooling liquid called slurry. As the cutting performance is limited by the SiC particles, the time to cut a brick lies in-between 4-6 hours. With the introduction of the diamond wire process, the slurry can be replaced by a water based coolant. The cutting now is caused by fixed diamonds coated onto the wire. Obviously, the diamond wire is more expensive and therefore a bidirectional movement of the wire is applied with the wire being moved in forward direction and then moved back for a certain distance. This mode is called the pilgrim mode and reduces the usage of wire. Due to the better cutting efficiency of the coated diamonds compared to the SiC particles in the slurry, process time can be reduced down to 90 minutes for a monocrystalline ingot and a wire speed up to 33 m/sec [9]. For multicrystalline bricks, 2-3 hours are needed. Whether the cause for the reduced cutting efficiency lies in the presence of hard particles (SiN, SiC), different fracture strength of the grains or other multicrystalline properties, is still under investigation [10–12].

One important aspect from a metrologist's view is illustrated in Figure 2. While the brick is pushed through the wire web, the web is bent. In slurry cutting, this shape is also present, but hardly recognizable after the process. For diamond cutting with pilgrim mode being applied, we observe strong concave shaped variations of reflection and thickness on the wafers. These inhomogeneities are a challenge for inline wafer metrology, as will be pointed out in section 4 of this paper.

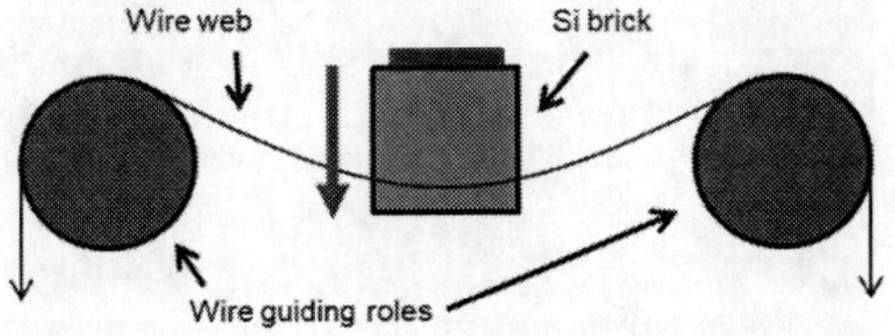

Figure 2: Side view illustration of the wire sawing process. The mc-Si brick is pushed through the wire web with constant speed. The pilgrim mode in the diamond-cut process thus forms concave-shaped marks on the wafers.

According to ITRPV, monocrystalline cutting is almost done exclusively with the diamond-wire process. For multi the situation is different. Not only is the diamond-cut process on multi more challenging than on mono, also the texture process in the beginning of the cell manufacturing process faces a technological challenge. For monocrystalline wafers, an alkaline etching solution is used forming a lowly reflecting pyramidal structure. The alkaline texture works fine with the rough surface from slurry cutting and the smooth surface from diamond cutting, especially as a saw-damage etching is applied beforehand. Unfortunately, the alkaline texture is not applicable on multi because of its varying grain orientations. The standard acidic texturing process needs a rough surface and performs poorly on the smooth surface of diamond-cut wafers. Both technological challenges have prevented the multicrystalline wafer manufacturing process from switching quickly to diamond cutting so far.

However, recent improvements and cost reductions in the monocrystalline wafer manufacturing have led to a significant decrease of wafer prices. According to our metrology partners, almost all multicrystalline production lines have switched from slurry cutting to diamond cutting in order to stay competitive. This sudden change seems to come along with non-optimal sawing conditions.

3 INLINE WAFER INSPECTION

3.1 Checking the specification

During incoming inspection for solar cell processing or outgoing inspection of wafer manufacturing, inline wafer inspection systems are used. The main purpose is to check the specifications of the wafers, to sort out wafers which do not meet the specifications, or to do a quality binning. The specifications cover different aspects and can be divided into surface, geometry, electric and material properties:

- Surface properties: stains, saw marks, chips, etc.
- Geometric properties: size, angels, thickness, total thickness variation (TTV), cracks, etc.
- Electric properties: resistivity, doping, lifetime, etc.
- Material properties: structural defects, crystallinity, oxygen, iron, inclusions, etc.

Not all of them are inline accessible. For example, there is no method up to now to measure oxygen, iron or other elemental concentrations inline within the wafers. Unfortunately, these impurities are decisive for a variety of meta-stable defects or degradation behavior. The charge carrier lifetime is a special case. While there are several inline methods to measure this parameter, it is limited in the as-cut state by surface recombination and thus holds little information on bulk recombination. Ways to overcome this limitation are to rate structural defects like dislocations or to calculate the bulk lifetime with a surface model. This topic will be covered in section 4.2.

3.1 Inline wafer inspection systems

In this work, the inline characterization tool HE-WI-06 from Hennecke Systems [13] has been used. Comparable machines are available from Semilab [14], Applied Materials [15] and others. Eight characterization modules are available in our HE-WI-06: (1) Thickness, (2) Resistivity, (3) Saw Marks, (4) Stain, (5) Micro Cracks (IR Transmission and Transflection), (6) Edge, (7) Geometry and (8) PL. These modules, their measurement types and the underlying SEMI standards [16] are depicted in Table 1.

Table I: Overview of the inspection modules of the Hennecke HE-WI-06 wafer inspection system and the related SEMI standards.

Module	Type	Standard
1) Thickness	capacitive *3 lines*	PV41
2) Resistivity	inductive *middle line*	PV28
3) Saw Marks	optical *laser line topology*	PV40
4) Stain	optical *diffuse illumination*	none
5) Micro Cracks	optical *IR transmission*	PV39
6) Edge	optical	none
7) Geometry	optical	PV46
8) PL	optical *IR luminescence*	PV51

It is important to note that all optical inspection modules needed an adjustment in terms of optical design, to work with the shiny diamond-cut surface. Existing inspections lines can be upgraded to be appropriate for the diamond-cut surface.

3.2 Sample set

A broad analysis of inline measurement methods applicable to diamond-cut as-cut wafers was carried out to investigate diversifying slicing effects. To learn from as many effects as possible, a large sample set of 2400 wafers was provided by PV Crystalox including 36 process variations on varying ingot materials [12]. Additional 1600 wafers with high quality from five different suppliers are part of the sample set. For comparison also wafers from a wafer set of 13000 slurry-cut wafers for production are taken into account.

All wafers have been thoroughly characterized and selected samples were textured to learn which surface issues remain visible after texturing.

4 RESULTS

4.1 Common issues

Before we deal with defects which are specific for mc-Si diamond-cut wafers, we will give some brief but important examples for incoming wafer inspection in general. The parameters to be addressed are listed in section 3.1 and the inspection modules are listed in table 1. In Figure 3, two examples are shown, a) microcrack detection and b) stain detection. A reliable detection and sorting out of (micro)cracks is very important, as cracked wafers can break in a later processing stage and cause machine downtime. Fortunately, the detection methods have improved a lot in the past years [17]. In infrared transmission images cracks are visible, but also grains and grain boundaries form similar strong contrasts. Thus, the detection rate is limited to avoid false positives. An improvement is the infrared transflection method, where the light source is on the same side as the camera. In this setup, the cracks act as scatter centers and reflected the light into the camera. Signatures of cracks with a higher contrast are the result. The downside is that in this setup, inclusions of particles cannot be detected anymore, so that both systems are part of the standard inspection modules of the HE-WI-06. In Figure 3a, an infrared transflection image is shown with a crack having a very distinct contrast and being easily detectable with inline image evaluation. In Figure 3b, an image of another module is shown. Here, a diffuse white light illumination is used to filter out reflection inhomogeneities and to find residuals from cleaning, handling or packaging. Depending on the follow-up processes, such stains need to be avoided and affected wafer to be sorted out.

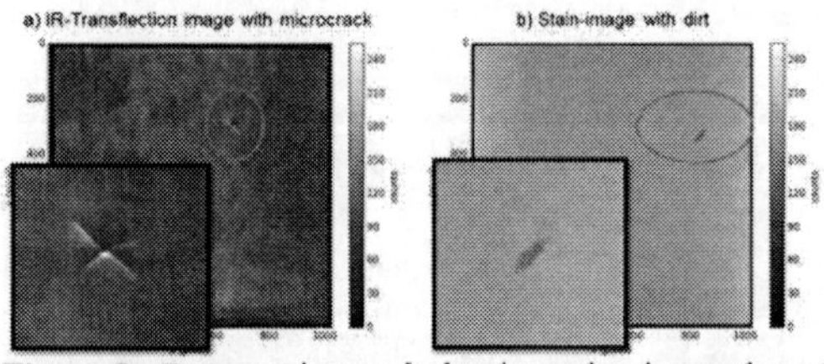

Figure 3: Common issues during incoming inspection of mono and multicrystalline wafers are the detection of (a) cracks and (b) stains. The images show diamond-cut wafers and for this surface the optics of the systems had to be adjusted to yield proper results.

4.2 Dislocations and impurities

One very important issue with multicrystalline wafers are structural defects. Grain boundaries and dislocations can be highly recombination-active and can significantly reduce efficiency of solar cells [18–20]. Also increased impurity concentrations in top- and bottom regions of bricks and edge regions of ingots can have a similar negative influence on efficiency. Both can be observed in PL images and the suppliers of wafer inspection system offer appropriate solutions.

In Figure 4, four Photoluminescence-Images (PL) are shown with different amounts of dislocations. Rating and highlighting here comes from the Hennecke PL-rating algorithm [21]. The given efficiency values are from the same wafers being processed to PERC solar cells. In this case, we observe $1.7\%_{abs}$ efficiency loss due to dislocations. Such a big impact of this material property can easily cover intentional process variations and thus needs to be monitored. Also, we recommend to perform a sorting of the wafers into different quality bins or sort out all wafers above a certain dislocation level. For the algorithm to work properly, we need a homogenous surface. Within the next section, we will discuss some examples where this condition is not fulfilled and will cause errors.

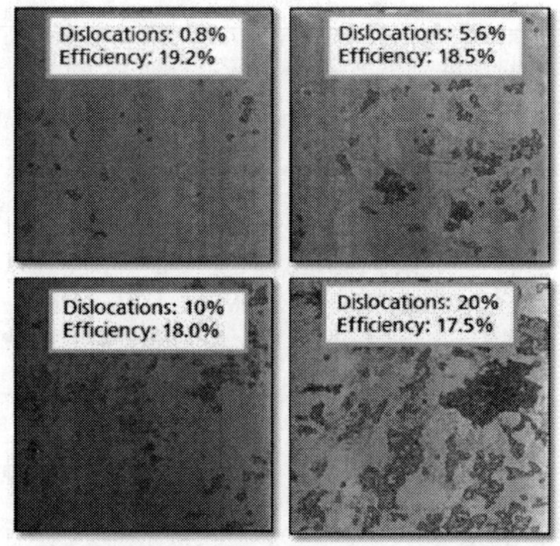

Figure 4: Four PL-Images of wafers with different dislocation (marked in red) values. Underneath the dislocation value, the efficiency of the solar cells is shown, which were made from these wafers.

One important note on this topic is that the manufacturers of PL-Systems follow different approaches and have different metrics for rating the images [20–23]. These metrics are based on a combination of image features, such as dislocations and impurities. For a more accurate rating, more image features like intensities, lines, patterns or even abstract filter responses are taken into account. While a rating for low-level Al-BSF solar cell processes can follow an analytic combination of the most common features, we observe that solar cell processes with high lifetime potential (e.g. PERC) require additional information from incoming inspection and a training of the rating model [20].

Other approaches to get information on the electrical material quality are: (i) The actual bulk lifetime can be calculated from the effective lifetime, which can be measured in as-cut state. A surface recombination model has to be taken into account for this approach [24, 25]. (ii) The effective lifetime can already be measured on brick level yielding significantly higher values due to lower surface recombination [26, 27]. (iii) A third approach is to apply a temporal wet chemical passivation (e.g. during texturing) to reduce the surface recombination velocity. In this case, the metrology tool needs to be placed within the wet chemical environment [28, 29].

4.3 Total thickness variation

After we covered some common issues and the topic of PL-Imaging, we now move to a basic wafer characteristic, which seems to be trivial, but needs to be handled with care on diamond-cut surface.

The total thickness variation (TTV) is calculated by the maximum value minus the minimum value. The characterization module used for this is the E&H MX152. It generates three line scans by means of capacitive sensors. Each scan gathers 500 data points over a whole 156x156 mm^2 wafer depending on the belt speed. In Figure 5, we see a visualization of the three traces for a slurry-cut wafer (a) and a diamond-cut wafer (b). The slurry-cut wafer shows its typical wedge-like structure with lower values at the upper side of the image. The cause for this structure is that the silicon carbide particles in the slurry reduce in size during the sawing process and thus the wafer thickness changes. The diamond-cut example features typical traces of saw marks, which not only cause a reflectivity variation from the pilgrim mode, but also a more inhomogeneous topology.

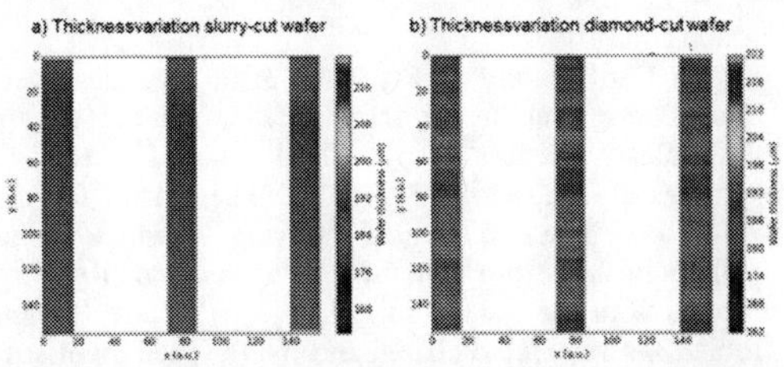

Figure 5: Visualization of the three-line measurement with capacitive thickness sensors. (a) The slurry-cut example features a smooth, but wedge-like surface. (b) The diamond-cut wafer shows a varying thickness due to saw marks.

In order to get a better impression of common TTV values, we analysed two datasets from the PVTEC laboratory. The first one contains 13,000 high-quality mc-Si wafers. This dataset (see Figure 6a) has TTV values roughly between 5-20 µm. In Figure 6b, a dataset of diamond-cut wafers is displayed and the colored histograms show different manufacturers. It has to be taken into account, that these also contain experimental wafers with sawing variations. So, we observe a bigger scattering but also lower TTV values. With technological improvements, we expect the TTV on diamond-cut mc-Si wafers to reach mean values well below 10µm.

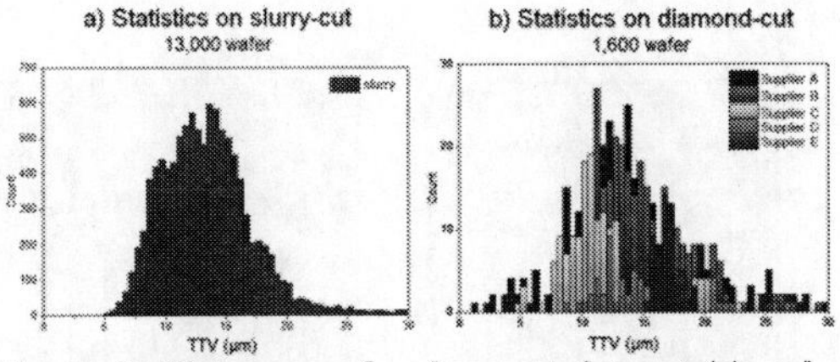

Figure 6: Histograms of a slurry-cut data set (a) and a diamond-cut wafer set (b). The histogram of total thickness variation of the slurry-cut wafers has its maximum between 10-20µm because of the wedge-like structure. The diamond-cut dataset is more experimental and this shows more variation. Five manufacturers are highlighted to show specific differences in TTV.

As pointed out in the beginning of this section, there is one very important point when measuring diamond-cut wafers. If the measurement traces are in line with saw marks, most of the wafers topology cannot be measured. Only if the measurement traces are perpendicular to the saw marks, the thickness variations can be covered. In Figure 7, this behavior is shown. Graph a) shows the histogram of a wafer box with 200 wafers which was measured in standard orientation (0°, with saw marks perpendicular to the direction of travel in the wafer inspection system) in blue and the measurement of the same wafers rotated by 90° with the saw marks then being parallel to the direction of travel in red. We observe the expected result for slurry-cut wafers: the histograms overlap. With diamond-cut wafers this is not the case. In Figure 7b, we see a strong deviation of the two histograms, the one measured in parallel to the direction of travel showing very low values. The origin of these low values is that only few saw marks are covered.

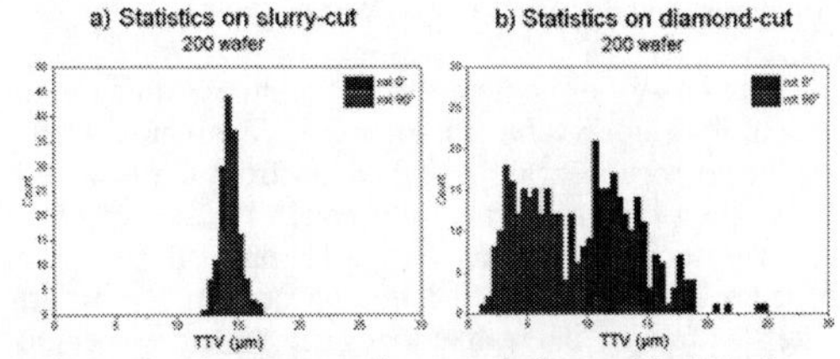

Figure 7: Histograms of a sample box with 200 slurry-cut wafers (a) and diamond-cut wafers (b) for a measurement parallel (red) and perpendicular (blue) to the saw marks. Due to the rather smooth surface of the slurry-cut process, the result is independent of the measurement direction. For the diamond-cut wafers, the measurement direction plays a decisive role.

As a consequence, the TTV of diamond-cut wafers (mono and multi) can only be measured accurately, if the wafers are measured with the correct orientation. Unfortunately, some wafer sawing- and sorting systems cannot maintain the rotation of the wafers and thus wafer stacks can be differently rotated in the wafer boxes.

4.4 Sawing issues

In the following, we discuss three examples of different surface conditions arising from different parameter setting of the cutting process or process problems.

a) Different sawing speed

As pointed out in the beginning, an increased speed (table or wire speed) in sawing will directly improve productivity and thus lower the costs. Typically, the pilgrim rate is kept at a constant level, so that by a faster pushing of the brick through the wire web, fewer pilgrim waves should be noticeable on the wafer and the reflection should be lower. This behavior can be seen in Figure 8. In this comparison of two typical diamond-cut mc-Si wafers, we observe that the pilgrim waves are clearly visible in the optical images, with 130 pilgrim waves on wafer A and 200 pilgrim waves on wafer B. The total reflection at 600 nm was measured offline with values of 31.7% and 33.5%, respectively. Such differences are no challenge for the metrology tools.

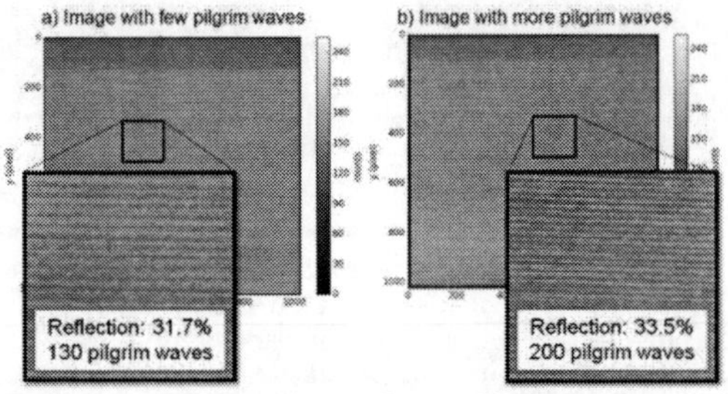

Figure 8: A comparison of two diamond-cut mc-Si wafers from optical inline inspection. The pilgrim waves are visible and influence the reflection value.

It is interesting to note that the multicrystalline grain structure is not visible in Figure 8. The cause is the higher proportion of direct reflection from these wafers. Visualizing the grain structure through the saw marks is rather difficult. This can be problematic if the grain structure needs to monitored, e.g. for quasi-mono wafers. After texturizing the wafers, the grain structure becomes visible, but the saw marks remain and the reflection is too high for solar cells with highest efficiencies. This can be observed in Figure 9. As pointed out in the beginning, better or improved versions of the acidic texture are needed.

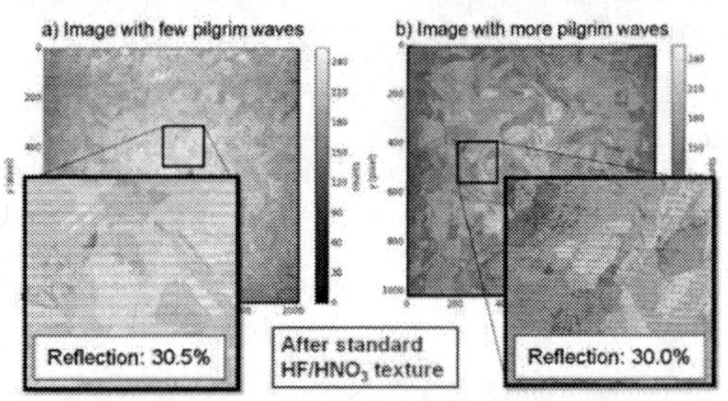

Figure 9: After texturizing the wafers from Figure 8, the grain structure becomes visible, while the pilgrim waves are still present. The lower reflection is not sufficient for making good solar cells.

b) Variation in sawing speed

In the second example, we encounter a more challenging artifact. Many wafers feature a more or a less reflecting part then the rest of the wafer's surface. The reason lies in the sawing process. Before the sawing process is finished, the pushing speed is reduced in order not to cut too deeply into the connection of the brick to the sawing machine. This leads to a more polished surface in this region. This feature is common for mono- and multicrystalline diamond-cut processes and can be more or less pronounced.

In Figure 10, we find the optical image of a wafer with strongly increased reflection at the bottom edge of the wafer. In the image, the higher reflecting part has a lower count rate. This change in contrast can cause problems for the automatic detection algorithms for surface defects (e.g. stains, chips, cracks, etc.).

After texture, we again can clearly observe the grains and saw marks. In addition we observe the region to be more reflecting which might cause a power loss on solar cell level or form an undesired pattern on module level.

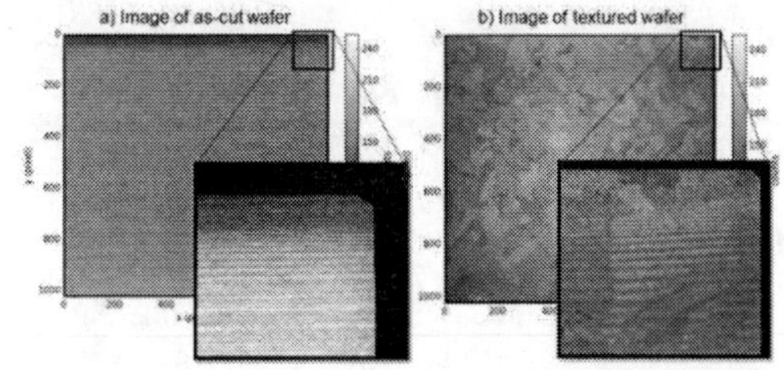

Figure 10: (a) A typical finding are wafers with a different reflectivity at one side. (b) After texturing the difference in reflectivity is reduced and grains and saw marks become visible.

c) Step contrast

The third example of surface artifacts causes very strong contrast at an arbitrary position on the wafers, being clearly visible also with the naked eye. Two examples are shown in Figure 11. The reason for this step-like contrast can have several origins, which cannot be distinguished from only the images. It can be a general problem with the sawing machine (wire rapture, coolant problem), a very rapid change in sawing speed (similar to the previous example), or it can be a silicon carbide particle forcing the sawing wire to shift. As previously stated, this strong contrast causes problems in the image evaluation (especially stain and PL). An example for false detection is given in Figure 12. After texture, the contrast remains visible (Figure 13).

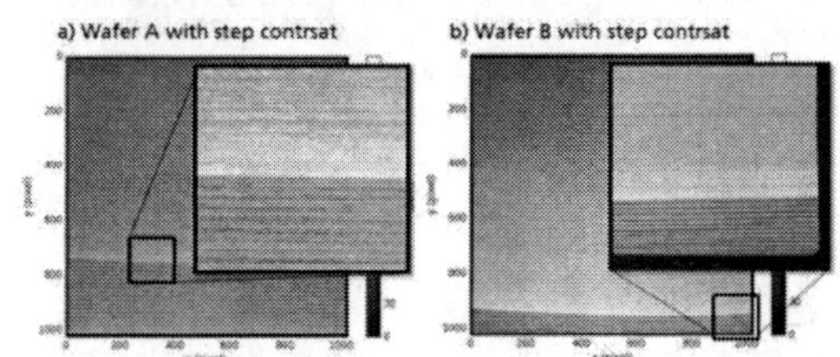

Figure 11: A very strong change in reflectivity on the wafer can lead to problems in the image evaluation. Such features remain visible after texture.

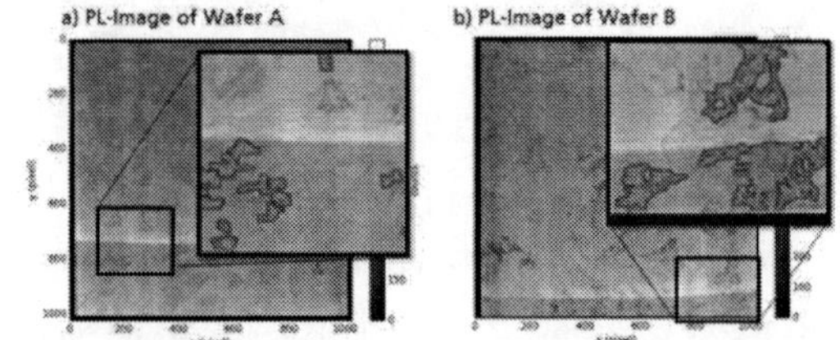

Figure 12: PL-Images of these wafers. The step contrast leads to a false detection of dislocations.

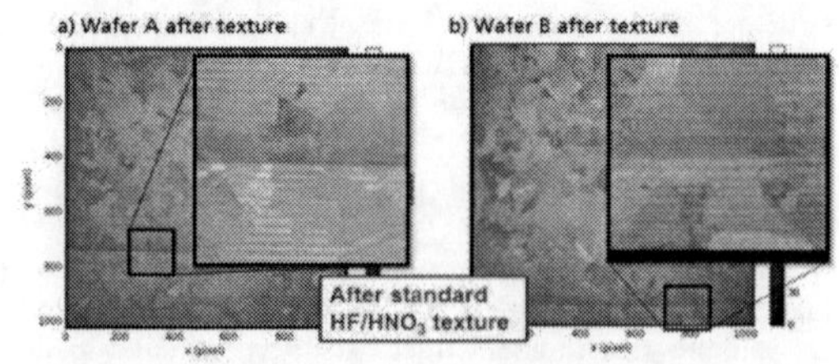

Figure 13: Visual images of these wafers after texture. The step contrast remains visible.

As a final remark, we can state that differences in reflectivity are a problem not only for texturing but also for inline metrology. For best error recognition results, the inspection modules and the image evaluation algorithms are optimized for a certain reflection. If the reflectivity changes (due to different sawing conditions), the accuracy in finding defects is reduced and by that false detections are more likely. Especially when the surface of the wafer is inhomogeneous (examples b and c), the detection of defects is challenging. For the algorithms to be fast enough with inline speed while dealing with high resolution, they need to be simple. Rather than time-consuming contrast-invariant methods, simple thresholding or grey-scale contrast algorithms are applied. If now, the reflectivity and thus the grey-values of the camera signal are changing, then errors in image evaluation can be the consequence.

The manufacturers of the inline inspection systems are working on that problem. One approach is to measure the reflectivity of the wafers and pass the result to the cameras in order to set the optimal visual conditions for image processing. However, if there is a sudden change of reflectivity on the wafer, this approach will not work.

5 CONCLUSION

In this work, we analysed the status of inline metrology when dealing with diamond-cut surfaces in general und multicrystalline wafers in special. Most visual inspection systems had to undergo a hardware modification in order to measure on the more reflecting diamond-cut surface. We observe that the common issues that need to be checked for wafer specification (e.g. size, thickness, cracks, doping, etc.) can be accessed with high accuracy. Differences or problems in the sawing process can cause strong inhomogeneities and variations in reflection on the wafer surface, which disturb the imaging methods and can cause problems in the detection algorithms.

For multicrystalline wafers, the detection of PL features such as dislocations and impurities is of major importance, because these parameters allow a rating of the electric quality in terms of IV parameters. The existing metrics from PL manufacturers are not comparable. Regarding the measurement of wafer thickness and the total thickness variation, we observe that the direction of measurement has to be taken into account, because saw marks parallel to the measurement direction cannot be detected.

Further work has to be invested in order to increase the quality of the diamond-cut surface, while inline metrology has to develop better algorithms and to be able to work with wafers which have a partially different reflecting structure.

7 AKNOWLEGMENTS

This work was funded by the German Federal Ministry for Economic Affairs and Energy within the project "DIADEM" (contract number 0324091B).

8 REFERENCES

[1] A. Bidiville, K. Wasmer, R. Kraft, and C. Ballif, "Diamond Wire-Sawn Silicon Wafers – from the Lab to the Cell Production," (eng), 2009.

[2] F. Cao *et al.*, "Next-generation multi-crystalline silicon solar cells: Diamond-wire sawing, nano-texture and high efficiency," *Solar Energy Materials and Solar Cells*, vol. 141, pp. 132–138, 2015.

[3] H. Seigneur, E. J. Schneller, N. S. Shiradkar, and W. V. Schoenfeld, "Effect of Diamond Wire Saw Marks on Solar Cell Performance," *Energy Procedia*, vol. 92, pp. 386–391, 2016.

[4] e. a. B. Kafle, "On the fabrication of high-efficiency mc-Si PERC-based solar cells on diamond wire-sawn surfaces using industrially viable etching technologies," *PV International*.

[5] J. M. Greulich *et al.*, "In-line quality control in high efficiency silicon solar cell production," *Photovoltaics International*, vol. 35, pp. 46–57, 2017.

[6] H.J. Möller, "Basic Mechanisms and Models of Multi-Wire Sawing," *advanced engineering materials*, vol. 2004.

[7] A. Kumar and S. N. Melkote, "Diamond Wire Sawing of Solar Silicon Wafers: A Sustainable Manufacturing Alternative to Loose Abrasive Slurry Sawing," *Procedia Manufacturing*, no. 21, 2018.

[8] O. Anspach, B. Hurka, and K. Sunder, "Structured wire: From single wire experiments to multi-crystalline silicon wafer mass production," *Solar Energy Materials and Solar Cells*, vol. 131, pp. 58–63, 2014.

[9] Meyer Burger, *Factsheet DW 291*.

[10] C. Yang, H. Wu, S. Melkote, and S. Danyluk, "Comparative Analysis of Fracture Strength of Slurry and Diamond Wire Sawn Multicrystalline Silicon Solar Wafers," *Adv. Eng. Mater.*, vol. 15, no. 5, pp. 358 365, 2013.

[11] T. Kaden, "Optimization potential of the wire sawing process for multicrystalline silicon," *Photovoltaic International*, vol. 37, 2017.

[12] K. Sunder, R. Buchwald and O. Anspach, "Origin of the periodic strucutre of diamond wire sawn wafers," in *31th EUPVSEC*.

[13] MeyerBurger. [Online] Available: http://www.meyerburger.com/de/en/meyer-burger/products-and-systems/detail/wis-06/.

[14] Semilab. [Online] Available: http://www.semilab.hu/category/products/u-pcd-carrier-lifetime-wafers.

[15] Applied Materials. [Online] Available: http://www.appliedmaterials.com/products/applied-vericell-solar-wafer-inspection-system.

[16] SEMI Standards for PV. [Online] Available: http://www.semi.org/en/industries/SolarPV.

[17] M. Demant *et al.*, "Microcracks in Silicon Wafers I: Inline Detection and Implications of Crack Morphology on Wafer Strength," *IEEE J. Photovoltaics*, vol. 6, no. 1, pp. 126–135, 2016.

[18] J. Haunschild *et al.*, "Quality control of as-cut multicrystalline silicon wafers using photoluminescence imaging for solar cell production," *Sol. Energy Mater. Sol. Cells*, vol. 94, no. 12, pp. 2007–2012, 2010.

[19] M. Demant, J. Greulich, M. Glatthaar, J. Haunschild, and S. Rein, "Modelling of physically relevant features in photoluminescence images," *Energy Procedia*, vol. 27, pp. 247–252, 2012.

[20] M. Demant *et al.*, "Inline quality rating of multi-crystalline wafers based on photoluminescence images," *Prog. Photovolt: Res. Appl.*, vol. 24, no. 12, pp. 1533–1546, 2016.

[21] H. Schremmer, A. Bergmann, J. Grohs, and N. Cüppers, "Precise cell performance estimation based on wafer PL measurement," in *31st EU PVSEC*, Hamburg, 2015.

[22] F. Korsós, Z. Kiss, Ch. Defranoux, S. Gaillard, "Inline PL Imaging Techniques for Crystalline Silicon Cell Production," Marseille, 2016.

[23] BT Imaging. [Online] Available: https://www.btimaging.com/products-c1g2d.

[24] K. Bothe, R. Krain, R. Falster, and R. Sinton, "Determination of the bulk lifetime of bare multicrystalline silicon wafers," *Prog. Photovolt: Res. Appl.*, vol. 18, no. 3, pp. 204–208, 2010.

[25] R. A. Sinton, J. Haunschild, M. Demant, and S. Rein, "Comparing lifetime and photoluminescence imaging pattern recognition methodologies for predicting solar cell results based on as-cut wafer properties," *Prog. Photovolt: Res. Appl.*, n/a, 2012.

[26] e. a. B. Mitchell, "PERC Solar Cell Performance Predictions From Multicrystalline Silicon Ingot Metrology Data," *IEEE JPV*, vol. 7, 2017.

[27] N. Schüler, B. Berger, K. Dornich, and J. R. Niklas, "High resolution inline topography of iron in P-doped multicrystalline bricks by MDP," *Energy Procedia*, no. 38, pp. 176–182, 2013.

[28] S. Al-Hajjawi, J. Haunschild, M. Zimmer, T. Dannenberg, R. Preu, "Predicting bulk lifetime values by applying wet Predicting bulk lifetime values by applying chemistry H-termination for inline quality control of silicon wafers," in *Silicon PV 2017*.

[29] H. Sugimoto and M. Tajima, "Photoluminescence Imaging of Multicrystalline Si Wafers during HF Etching,," *Jpn. J. Appl. Phys*, vol. 46, 2007.

STATUS OF THE EU H2020 DISC PROJECT: EUROPEAN COLLABORATION IN RESEARCH AND DEVELOPMENT OF HIGH EFFICIENT DOUBLE SIDE CONTACTED CELLS WITH INNOVATIVE CARRIER-SELECTIVE CONTACTS

B. Min[1]*, T. Wietler[1], S. Bordihn[1], R. Peibst[1], T. Desrues[2], P. Carroy[2], J. Jourdan[2], M. Hermle[3], F. Feldmann[3], J. Bartsch[3], C. Allebée[4], L. Ding[4], J. Horzel[4], A. Lachowicz[4], J. Diaz Leon[4], A. Ingenito[5], F. Haug[5], E. Schneiderlöchner[6], V. Linss[6], R. Köhler[6], K. Luedemann[6], A. Čampa[7], M. Bokalič[7], M. Topic[7], M. Zwegers[8], B. Hartlin[9], B. Field[9], B. Bénédicte[10], A. Richter[10], Z. Adam[11], J. Penaud[12], S. Filonovich[12], N. Harder[12], E. Marcon[13], J. Chupin[13], F. Tamini[13]

[1]ISFH Institute for Solar Energy Research Hamelin, [2]CEA-INES Commissariat à l'Energie Atomique et aux Energies Alternatives France, [3]FH-ISE Fraunhofer-Instituts für Solare Energiesysteme Germany, [4]CSEM Centre Suisse d'Electronique et de Microtechnique SA, [5]EPFL École Polytechnique Fédérale de Lausanne (EPFL), Institute of microengineering (IMT), Photovoltaics and Thin Film Electronics Laboratory, [6]VON ARDENNE GmbH, [7]University of Ljubljana, [8]Meco Equipment Engineers B.V., [9]ERM Environmental Resources Management Limited, [10]Meyer Burger AG, [11]Ecosolifer Modulgyarto Fotovillamos Technologiak Korlatolt Felelossegu Tarsasag, [12]TOTAL Gas Renewable and Power, [13]Ayming

*corresponding author: min@isfh.de, Phone: +49 5151 999 644, Fax: +49 5151 999 400

ABSTRACT: The DISC project addresses the development of key technologies for the next generation of high-performance photovoltaic solar cells and modules. Our approach is to fully exploit the potential of silicon to its maximum by the use of passivating contacts or carrier selective contacts and junctions. Such contacts allow for simple, non-patterned double-side contacted device architecture and an enhancement of the energy yield, which will be key elements for achieving very low levelized costs of electricity. In this paper, selected highlights are presented concerning the first 18 months in the development of cell and module components such as In-free transparent conductive oxide layers with low sputtering damage. The synergies and results created within DISC will be also interesting for a drop-in combination with other solar cell and module technologies.
Keywords: carrier selective contacts, passivated contacts, transparent conductive oxide, plating, multi wires

1 INTRODUCTION

The concept proposed by DISC is to develop double-side contacted silicon solar cells with carrier selective contacts (CSC) with target efficiencies above 25.5 % on large area cell (> 100 cm²) and module efficiencies above 22 %. We consider only the industry feasible technologies in order to be compatible with low manufacturing costs. Together with a reduction of non-abundant material consumption (Ag, In), and with an enhancement of the energy yield, we attempt a significant contribution to make the photovoltaic one of the cheapest electricity source. In addition, we apply a life cycle thinking approach to assess the environmental, social and economic impacts and benefits across the entire value chain of our prototype product.

Finally, we aim to strengthen the PV industry in Europe by helping our industrial partners VON ARDENNE, Meyer Burger and Meco to improve their portfolio with new technologies developed within this project, and Total and Ecosolifer to increase efficiency, reliability and durability of their products.

The results created within DISC will be also interesting for other cell concepts, since we address the major hurdles for a mass production of a-Si:H/c-Si heterojunctions (i.e., incompatibility with existing processing lines), by implementing high-temperature stable CSCs or by developing In-free transparent conductive oxides (TCO).

The DISC project started in October 2016. In this paper, we will present the main achievements concerning R&D activities on cell and module level in the first 18 project months.

2 APPROACH

We apply carrier-selective contacts on double-side contacted cells without sophisticated patterning as shown in Fig. 1. The both cell surfaces are passivated with CSCs with different polarities. The application on the entire front side requires sufficient transparency of the CSCs. For this purpose, we evaluate thin Si based layers, as well as transparent metal oxides. Both CSC candidates exhibit a low lateral conductivity. They are therefore covered with transparent conductive oxide layers in order to ensure the sufficient lateral conductivity between the metal fingers either realized with screen-printing or plating technologies.

For the interconnection in the module we apply the SmartWire technology to ensure compatibility of processes of different partners.

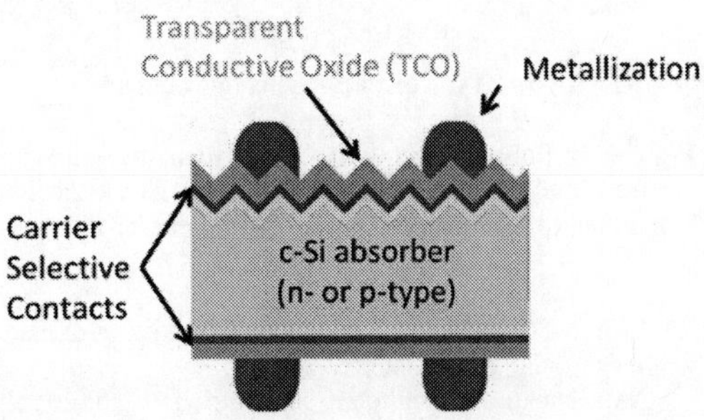

Figure 1: The common cell structure of the DISC project

3 MAIN ACHIEVEMENTS IN THE DEVELOPMENT OF INDIVIDUAL COMPONENTS

Within the first 18 project months, the project focuses on the development of individual components for high efficient solar cells and modules. Therefore, different candidates for CSCs and TCOs are evaluated and optimized regarding transparency, passivation and transport properties. In addition, different metallization technologies are evaluated for the compatibility with the CSCs and TCOs developed in this project. The following subsections present the main achievements in the technical work packages during the first half of the project.

3.1 Excellent passivation quality with poly-Si based CSCs

Figure 2 shows emitter saturation current density J_{0e} values of electron- and hole-collecting CSCs on planar and textured surfaces after hydrogenation.

Impressively low J_{0e} values beneath 9 fA/cm² are achieved reproducibly by ISFH, FH-ISE, EPFL and CSEM for electron-collecting CSCs on planar and textured surfaces. The corresponding values for hole-collecting CSC on planar surfaces are in comparably similar level but its performance on textured surfaces has still room for further improvement as demonstrated in Ref. [1].

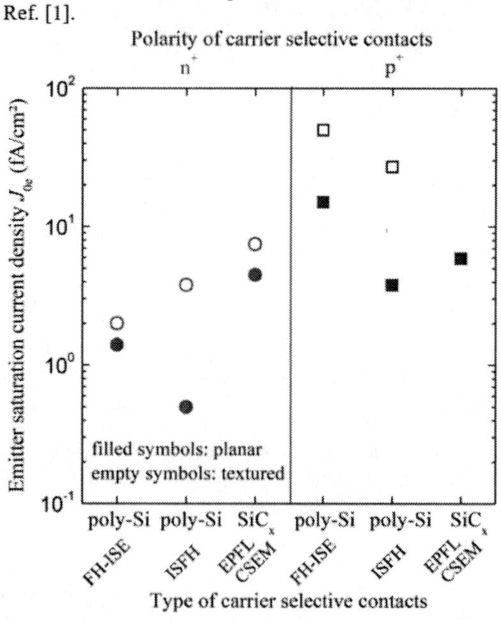

Figure 2: Emitter saturation current density values of electron- and hole-collecting carrier selective contacts on planar and textured surfaces after hydrogenations

3.1 In-free TCO with negligibly small sputtering damage while achieving high conductivity

As a cheap and abundant substitution of commonly used indium tin oxide, we develop aluminum-doped zinc oxide (AZO) with high conductivity (R_{sh} = 50 Ω/□ on glass substrate) prepared by DC magnetron sputtering combined with a post annealing step by 350 °C.

In order to investigate the influence of the AZO deposition step on the passivation quality, we monitor the iV_{oc} values on symmetrical lifetime samples with different type of electron-collecting carrier-selective contactss (CSC). In addition, a lifetime sample with a-Si:H/c-Si heterojunctions is used as reference. It is to note that the reference sample is annealed at a temperature below 250 °C (instead of 350 °C like other samples), since the passivation quality of the a-Si:H layers will be degraded for higher annealing temperature. The samples used for this study are n-type industrial M2-size Cz wafers ($\rho_b = 2 \sim 4$ Ωcm) with a thickness of 160 µm.

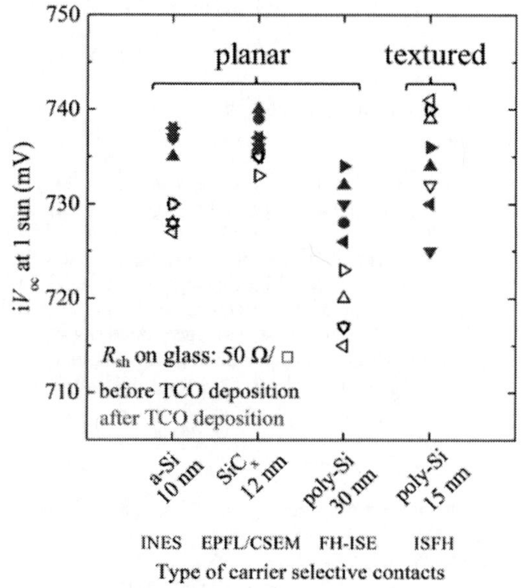

Figure 3: Influence of AZO sputter process on iV_{oc} of different CSCs. The values are measured before and after AZO deposition at different wafer position (▲: top, ▼: bottom, •: middle, ◀: left, ▶: right)

Figure 3 shows the iV_{oc} values of the lifetime samples with different CSCs fabricated by different institutes. The values are measured before and after the AZO deposition step at different wafer positions. The iV_{oc} values of the lifetime samples with a-Si:H/c-Si heterojunctions of INES increase up to 8 mV after the AZO deposition step. The samples with SiC_x and poly-Si based CSCs of EPFL-CSEM and FH-ISE show similar improvement of iV_{oc} values after the AZO deposition step. Contrary to this, the lifetime sample of ISFH shows a slight decrease in iV_{oc} values after the AZO deposition step. However, the iV_{oc} values are still in same level with the other samples, even though the sample is the only one in this experiment that has textured surfaces. This may be caused due to i) different surface morphology of those samples (textured instead of planar), ii) different crystallinity of the passivation layers (poly-Si instead of a-Si, SiC_x) or iii) different thickness of the poly-Si layer (15 nm instead of 30 nm).

These results highlight the higher thermal stability of the passivation quality of high-temperature processed CSCs compared to conventional silicon heterojunction.

3.3 Plating metallization with high aspect ratio enabling fill factors close to 83 % on 6 inch bifacial solar cells with CSCs

For high efficiency solar cells featuring novel CSCs, the metallization scheme should not compromise the passivation quality of CSC (below 1% iV_{oc} loss for grid metallization with an area fraction below 5%). Thus, the metallization scheme will essentially contribute to the efficiency potential > 25% by enabling a high *FF*, maintaining high V_{oc}, and inducing a very small amount of optical shading (high J_{sc}). Furthermore, the metallization has to be compatible with the new generation of multiple-wire interconnection technology with improved performance and decreased costs with respect to state-of-the-art SmartWire technology.

During the first 18 months of the DISC project, a Cu plating approach with high aspect ratio contacts (width < 30 µm, height > 15 µm, see Fig. 4) after bifacial co-plating is developed at CSEM in collaboration with FH-ISE and Meco. On different precursors with carrier selective contacts and TCO layers on both surfaces this plating approach demonstrated very low contact resistivity (ρ_c < 0.16 mΩcm²). Fill factors close to 83 % were achieved on 6 inch bifacial solar cells with CSCs [2, 3] while other precursors with CSCs and TCO layers showed lower fill factors despite the same low contact resistivity and the same metallization layout. These results show that the tested metallization approach does not limit the fill factor but other aspects of the CSC devices might do. Currently, a simplification of the process sequence is under work and a pilot line for bifacial Ni-Cu-Sn plating is established at Meco.

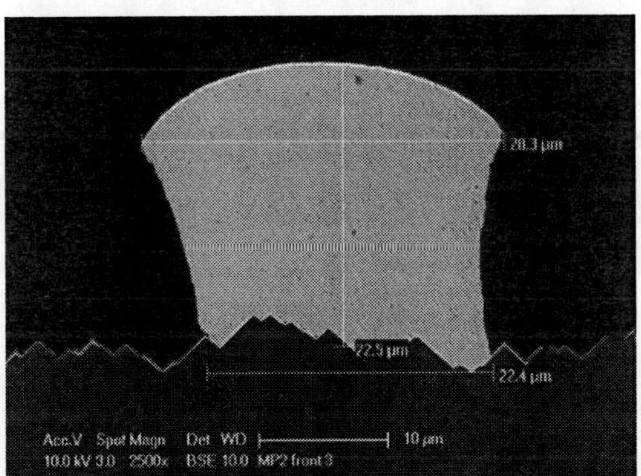

Figure 2: SEM cross section of plated finger fabricated by CSEM

3.4 In-free SmartWire interconnection with aging stability

In order to reduce module cost, In-free SmartWire interconnection is developed in collaboration with CSEM and Meyer Burger (see Fig. 5). It has been applied on SHJ and PERC cell interconnection for 60 cell modules and demonstrates high reliability (0.9 % and 0.5 % relative degradation after 400 thermo-cycles for SHJ and PERC cells, respectively) [4]. Hence, the modules clearly pass the IEC degradation test (IEC 61215, damp heat and thermal cycling).

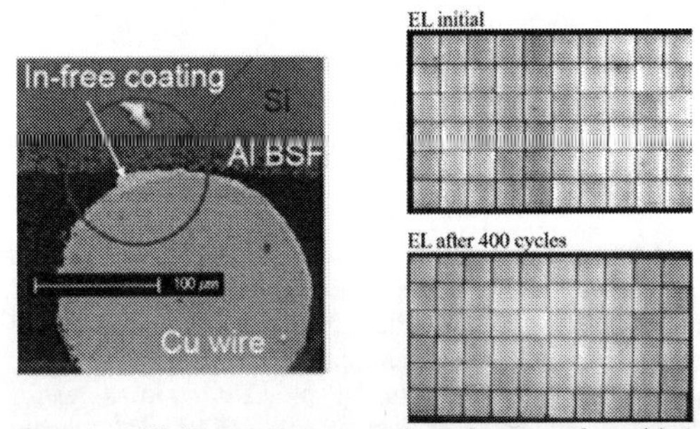

Figure 3: (left) SEM cross section of smart-wire with In-free solder and (right) electroluminescence of a 60 PERC cell module before and after 400 thermo-cycles

Already more than 10 different bill of materials for In-free SmartWire technology had been certified by TÜV Rheinland. Latest climate chamber tests (CSEM) show a high reliability for 72 cell glass/glass modules to pass several times IEC in damp heat and thermal cycling with very little power degradation.

4 FIRST CELL RESULTS MADE WITH M2-SIZE WAFERS

After the successful developments of single-components by different partners, the first integration run is lunched in order to evaluate the interplay of these components by applying the best known method for the each component.

Single-side textured n-type M2-size wafers are prepared by ISFH and the EPFL/CSEM deposited the CSCs with SiC_x layers for both polarities. Subsequently, the samples are splitted in two groups. One group is shipped to VON ARDENNE for ITO deposition on both sides and the other group is shipped to ISFH for AZO deposition on both sides. All samples are then metallized by CEA-INES with screen-printing technology. Finally, the samples are characterized by ISFH.

Table I presents the best cell results of different TCO types. Although the individual components are not optimized for the interaction, a cell efficiency of 21.2 % is achieved in the first integration run, if ITO is applied as TCO. The cell with AZO shows higher V_{oc} values than the one with ITO, but its efficiency is 0.6 $\%_{abs}$ lower, caused by the lower J_{sc} and *FF* values.

Table I: Best cell results in the first integration run

TCO	V_{oc} [mV]	J_{sc} [mA/cm²]	*FF* [%]	η [%]
ITO	708.1	37.5	79.85	21.2
AZO	719.4	36.9	77.69	20.6

Based on the first cell results, following main challenges need to be addressed in next project period:

1) Improvement of the J_{sc} values by reducing the parasitic absorption in the poly-Si based CSC layers (i.e. by reducing the thickness of poly)
2) Increase of the transparency of TCO layers while providing less sputter damage, low contact resistance between CSC and TCO and enough lateral conductivity

3) Application of the plating metallization on In-free TCO in order to reduce the shadowing due to metal fingers and improve the *FF* simultaneously.

5 SUMMARY

During the first 18 months, the DISC project is making good progress in the development and improvement of scientific understanding of cell and module components. Impressively high passivation quality is achieved reproducibly by different institutes for electron- and hole-collecting poly-Si based carrier selective contacts. The passivation quality remains nearly unchanged, even after the deposition of the In-free TCO, developed in this project. Regarding the metallization, the plating and In-free SmartWire technology show promising results to realize high efficient solar cells and modules with carrier selective contacts.

A cell efficiency up to 21.2 % is achieved although the individual components are not optimized yet for the interaction. While there is still some way to go to meet the final technical objectives, the results are promising for a first integration run involving four different partners in three different countries.

The DISC project is now in the integration phase, combining the best components from all partners targeting industry-feasible high efficient solar cells with carrier selective contacts.

6 ACKNOWLEDGEMENTS

This work was supported by the European Union's Horizon 2020 Programme for research, technological development and demonstration Grant Agreements no. 727529 (project DISC).

7 REFERENCES

[1] A. Mewe, M. Stodolny, J. Anker, M. Lenes, X. Pagès, Y. Wu, K. Tool, B. Geerlings, I. Romijn in AIP Conference Proceesings 1999, 040014, 2018.

[2] A. Lachowicz, J. Geissbühler, A. Faes, J. Champliaud, F. Debrot, E. Kobayashi, J. Horzel, C. Ballif, M. Despeisse, in Proc. of the European Photovoltaic Solar Energy Conference and Exhibition, Amsterdam, The Netherlands, 2017.

[3] A. Lachowicz, A. Faes, J. Champliaud, J. Hermans, J.-F. Lerat, P.-J. Ribeyron, C. Roux, G. Wahli, P. Papet, B. Strahm, J. Horzel, C. Ballif and M. Despeisse, in Proc. of the 7th Metallization Workshop, Constance, Germany, 2017.

[4] Moving from 5 busbar to multi-wire module interconnection: cost-reduction and performance benefits, oral presentation in conf. PVModule Tech, 8 nov. 2017, Kuala Lumpur, Malaysia.

SILICON SOLAR CELLS BY "DESIJN" (DEPOSITED SILICON JUNCTIONS)

Andres Cuevas, Di Yan, Sieu Pheng Phang, Yimao Wan and Daniel Macdonald

Research School of Engineering, The Australian National University (ANU)
Canberra ACT 2601, Australia

ABSTRACT: The deposition of silicon films to form the junctions required to make a silicon solar cell has permitted to achieve performance levels never seen before. Until now, such films have been deposited by chemical methods, both for the low temperature approach based on a-Si:H and the high temperature one based on polycrystalline silicon. We present here a novel approach for the latter, based on the physical deposition of silicon and an appropriate dopant element onto an ultrathin SiO_x interlayer. The electronic characteristics of these DEposited SIlicon JuNctions are, even at this early stage of development, essentially as good as those formed by chemical deposition in our laboratory. The initial focus of the novel PVD approach has been to develop p^+ poly-Si contacts doped in-situ with boron, for which a recombination current density $J_{0p+}\approx 20$ fA/cm^2, and a contact resistivity $\rho_c \approx 10$ mΩcm^2 have been achieved. These passivating contacts have been implemented as a full-area hole-selective contact on the rear side of p-type silicon wafers, in combination with a phosphorus diffused electron contact on the front side. The best devices fabricated so far have reached an open-circuit voltage of 701 mV and a conversion efficiency of 23%, with ample room for further improvement.

Keywords: silicon solar cells, passivated contacts, sputtering

1 INTRODUCTION

1.1 Preamble: how to design a solar cell

The design of a solar cell device hinges upon the following main premises: i) select a semiconductor material with two energy bands, one for free electrons and another for bound electrons (free holes), separated by approximately 1.1-1.4 eV, ii) maximise the absorption of photons and the transfer of their energy to bound electrons, iii) once generated, maintain as high as possible the concentration of free electrons and holes, iv) facilitate the exit of high energy electrons towards an external circuit and the re-entry of low energy electrons (i.e. the exit of holes).

Silicon is almost ideal to make the most of the energy spectrum of solar photons. The second condition includes a reminder that in some cases, when the concentration of free carriers is very high already in equilibrium (such as in highly doped regions), infrared photons can be absorbed by them, which is a non-productive process in terms of PV conversion, and hence should be avoided. The third condition means that recombination should be supressed everywhere in the device, and the fourth that two selective contacts need to be implemented, one for electrons and another one for holes. A material with a relatively low work function is necessary to let electrons out; its Fermi energy should be above or equal to the quasi-Fermi energy of electrons in the absorber. On the other side, a material with a relatively high work function is needed to facilitate the re-entry of electrons into the valence band of the absorber (i.e., the exit of holes), after they have released their energy to an external circuit. Note that in the block diagram of Figure 1 the details of the junctions between these two materials with extreme work functions and the absorber are not well described.

Doping permits to change the work function of silicon from a relatively low value of about 4 eV when doped n^+, so that E_F is slightly higher than E_C to a relatively high value of about 5 eV if doped p^+, which brings E_F close to E_V. An n^+ Si region contains a high concentration of electrons, which are readily available to take the energy from free electrons photo-generated in the absorber region and pass it on to electrons in the metal that forms the external circuit. Such high electron concentration also helps to form a low resistance contact between the metal and the n^+ semiconductor, even if, as it frequently happens, the work function of the metal is not lower than or equal to that of n^+ Si. The same can be said of p^+ Si.

Unfortunately, heavy doping comes with increased Auger recombination and does not fully stop minority carriers from flowing all the way to the outer surface, particularly to the fraction in contact with the metal terminal, where they recombine at high speed. It is therefore desirable to find carrier-selective contact structures that can also passivate the surfaces of the wafer, so that the third premise is still satisfied.

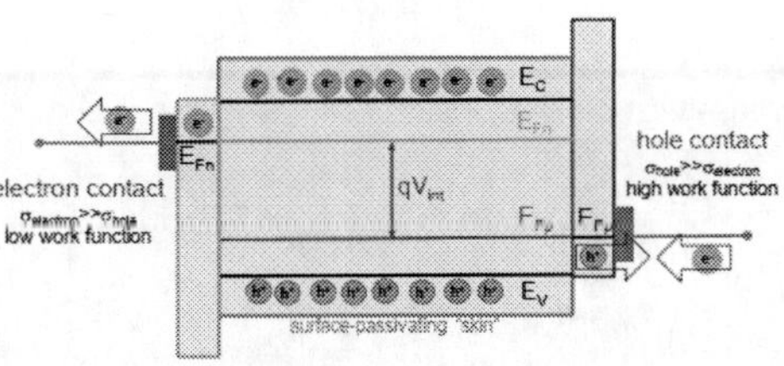

Figure 1. Conceptual diagram of a solar cell delivering electric power to an external circuit.

1.2 Passivating junctions based on deposited silicon

Among the many silicon solar cell device structures that are being developed, those based on deposited silicon junctions (DESIJN) have achieved the best performance so far. The silicon films can be amorphous or polycrystalline, or a mix of both; the selectivity to electrons or holes is conferred by adding dopants to them, complemented by the insertion of a surface passivating interlayer of a-Si:H or SiO_x. The combination of an ultrathin layer of SiO_x and doped polySi is gaining widespread acceptance, since it is based on techniques similar to those used in mainstream solar cell technology. In particular, the re-crystallisation of the deposited silicon layer and the ancillary activation of the dopant requires temperatures similar to those of conventional dopant diffusion processes. The corollary is that SiO_x/doped polySi structures can withstand subsequent thermal steps that may be needed to make the solar cells. Despite such commonality with conventional doping, SiO_x/doped polySi contacts can achieve much better levels of performance, particularly in terms of suppressing recombination. Figures 2 and 3 summarise pairs of values for the recombination current density and the contact resistance reported by different authors. It can be observed that in most cases people have achieved better levels of performance with phosphorus doped polySi contacts than with boron doped ones. Nevertheless, these two figures do not include reports of just one of the two parameters. It is worth mentioning that, in addition to ISFH (Peibst et al.), several laboratories have reported nearly perfect passivation, $J_0 \approx 1$ fA/cm², for both n^+ (Stodolny et al., Silicon PV 2017) and p^+ (Mack et al., pss-rrl 2017) polySi/SiOx contacts after hydrogenation by means of an anneal after capping the polySi with SiN_x. EPFL researchers (Nogay et al., IEEE PVSC 2018) have also used a hydrogenation treatment to achieve $J_0 \approx 4$-5 fA/cm² for both n^+ and p^+ polySi/SiO_x contacts.

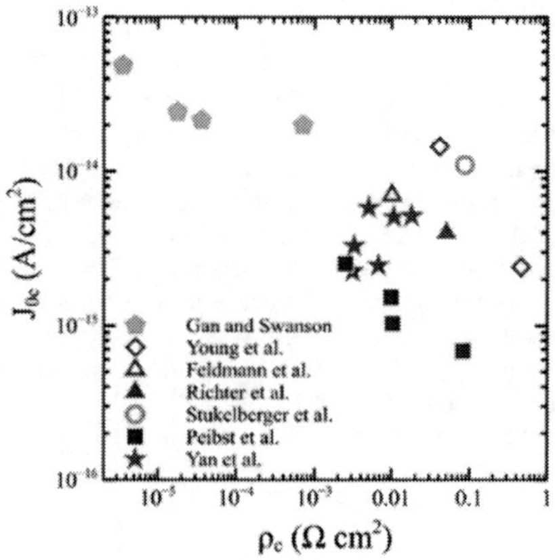

Figure 2. Recombination current and contact resistivity reported by various authors for phosphorus-doped n^+ polySi/SiO_x contacts.

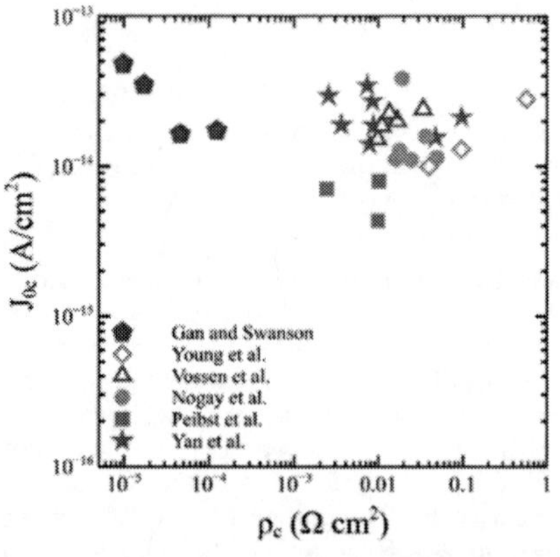

Figure 3. Recombination current and contact resistivity reported by various authors for boron-doped p^+ polySi/SiO_x contacts. The data points by Yan et al. include both PECVD and PVD polySi layers.

Commonly, the a-Si:H and poly-Si layers are formed by chemical vapour deposition (CVD), either plasma-enhanced (PECVD) or at low pressure (LPCVD), using gaseous precursors for Si. The Si layers may be doped in-situ by co-depositing boron or phosphorus from diborane or phosphine, which are category 1 (extreme) hazardous gasses. Alternatively, a subsequent doping step may be used, such as ion-implantation or conventional thermal diffusion. Although not shown in the table, excellent results have recently been achieved also using atmospheric pressure CVD, both for n+ and p+ polySi contact (ISFH and Schmid, 35th EUPVSEC, 2018). Like the other chemical deposition methods, APCVD needs to use hazardous gasses. Here we present an alternative technology based on the physical vapour deposition (PVD) of silicon films onto silicon wafers.

Sputtering, a common form of PVD, can be used to deposit both the silicon film and appropriate dopant elements using solid, non-toxic targets. The absence of hydrogen and the room temperature deposition mean that blistering of the thin silicon films is not an issue, contrary to PECVD techniques. Sputtering deposition is one-sided, which is advantageous for process integration. If so desired, metal deposition and TCO may also be performed by sputtering, as well as AR coatings. Sputtering is widely used in the fields of microelectronics, optics and large-scale coating of glass and plastics. Within photovoltaics, it is commonly used in thin-film solar cell technologies (CdTe, CIGS and a-Si) and to deposit the TCO layers needed for a-Si/c-Si heterojunction solar cells. Production-scale sputtering machines are capable of processing thousands of wafers per hour.

Whereas most institutions and companies have focussed on n^+ polySi applied to n-type Si wafers, it is important to address the largest segment of the PV market, which is based on p-type silicon. EPFL has reported a 22.4% p-type cell with rear p^+ SiC and front n^+ SiC layers deposited by PECVD [1]. The use of SiC is also mentioned in the "TOPCon" patent by Feldmann et al. as a means to reduce the occurrence of film blistering upon annealing, which originates in the relatively high concentration of hydrogen in the Si film resulting from the precursors used for PECVD [2]. More recently, ISFH have reported a 26.1% p-type IBC solar cell with both n^+ and p^+ poly-Si layers deposited by LPCVD followed by ion implantation of the dopants.

Table 1. Comparison between different polysilicon contact technologies.

method	LPCVD		PECVD		PVD
Si layer	2 sides		1 side		1 side
doping	diffusion	ion implant	diffusion	in situ	in situ
n-type cells	21.3% (ECN)	26.1% (ISFH)	24.7% (ANU)	25.8% (FhISE)	
p-type cells				22.4% (EPFL)	23.0% (ANU)

Although the PVD (sputtering) approach can produce n^+ doped poly-Si, our priority has been to develop boron doped p^+ poly-Si contacts and apply them to the rear side of p-type Si wafers. This would permit to not only boost performance very significantly compared to the incumbent p^+ regions formed by aluminium alloying, but also the use of any wafer resistivity, thanks to 1D current flow. High resistivity p-type silicon has been shown to be impervious to degradation related to the boron-oxygen defect. In the following sections we describe the development of p^+ poly-Si layers doped in-situ with boron and deposited by sputtering onto an ultrathin SiO_x interlayer, as well as their initial application to p-type DESIJN solar cells.

2 BORON-DOPED PVD P+ POLY-SI PASSIVATED CONTACTS

As an example of the ability to form boron-doped junctions onto silicon wafers using the PVD approach, Fig. 4 shows the concentration of electrically active (ionised) boron as a function of depth for passivated contact structures formed by simultaneous sputtering of Si and B onto a chemically grown SiO_x, followed by annealing at 900°C for 30 min, as measured by ECV.

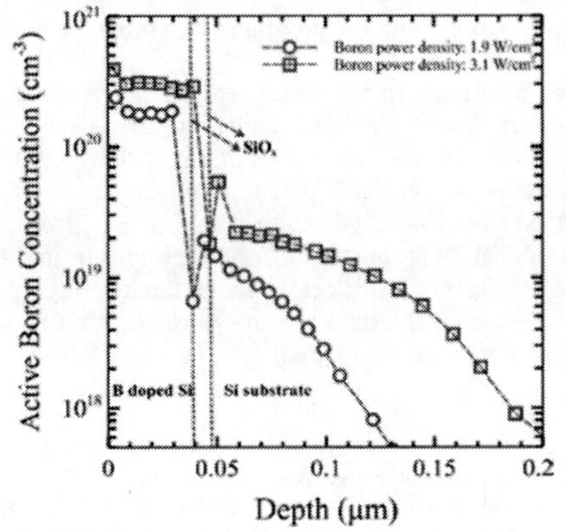

Figure 4. Ionised boron concentration as a function of depth for passivated contact structures formed by physical deposition of Si and B onto a chemically grown SiO_x and annealing at 900°C.

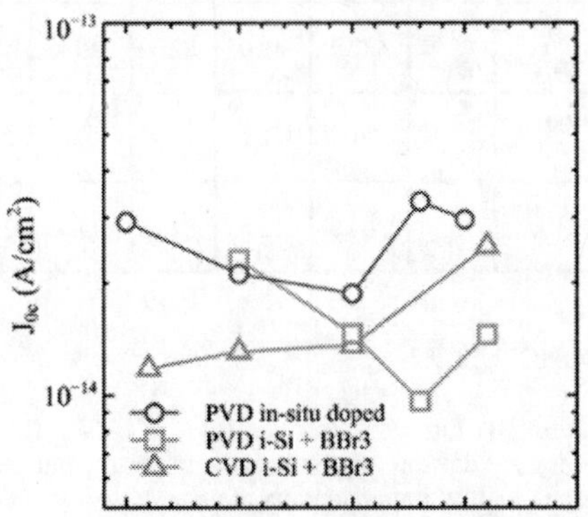

Figure 5. Recombination current parameter of in-situ doped PVD p^+ polySi/SiO_x contacts as a function of activation/crystallisation temperature compared to ex-situ doped PVD and PECVD polySi contacts as a function of boron diffusion temperature. After FGA.

Figures 5 and 6 show the recombination current J_{0c}, which establishes the level of "passivation", and the contact resistivity ρ_c, which determines the quality of the "contact", as a function of the temperature of the re-crystallisation/activation step (for in-situ doped PVD) or the diffusion of boron from BBr_3 (for ex-situ doped PVD and for PECVD) [3]. The measurements were taken after performing an anneal in forming gas at 400 °C; nevertheless, such anneal does not improve the passivation substantially for optimised contacts. The polySi layers were 40±10 nm thick and the chemical oxide interlayer about 1.4 nm thick. As can be seen in the figures, the three approaches can give sufficiently low J_{0c}<20 fA/cm² and $\rho_c \leq 30$ mΩcm² to enable high open-circuit voltages and high efficiency solar cells. The best passivation (lowest J_{0c}) in this set of experiments is for intrinsic silicon deposited by sputtering and recrystallised during a boron diffusion step at 930°C. The results for the PVD approach are, therefore, of the same order, or better that those achieved in our lab for PECVD Si films doped ex-situ by thermal boron diffusion.

The contact resistivity, shown in Fig.6, is higher for the ex-situ doped PVD than for the in-situ doped one, the latter being quite similar to that of the ex-situ doped PECVD films. Nevertheless, this should not be taken at this stage as a limitation, firstly because a $\rho_c \leq 30$ mΩcm² is low enough for a full-area contact, and secondly because further optimisation should produce even lower values.

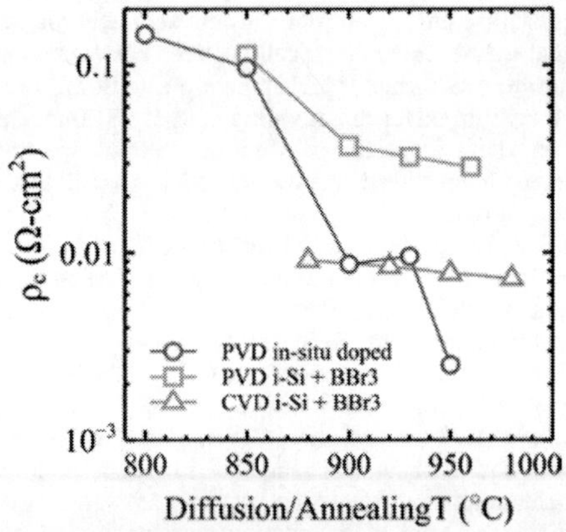

Figure 6. Contact resistivity of in-situ doped PVD p+ polySi contacts as a function of activation/crystallisation temperature compared to ex-situ doped PVD and PECVD polySi contacts as a function of boron diffusion temperature.

3 DESIJN SOLAR CELLS

The internal structure of the n^+pp^+ devices fabricated, shown in Fig. 7, is similar to the most common type of industrial solar cell, with a front textured surface coated with SiN_x, a front n^+ electron-selective conductor made by phosphorus diffusion from $POCl_3$, a p-type wafer and a p^+ hole-selective conductor at the back. The main peculiarity is that the latter is now a high performance PVD p^+ poly-Si/SiO_x passivated contact.

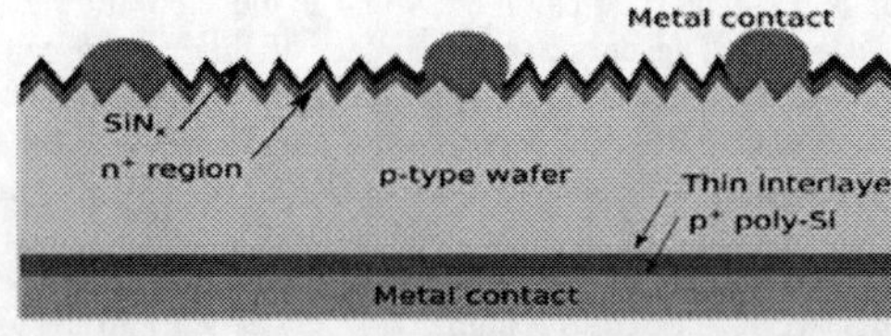

Figure 7. Cross-section of a p-type DESIJN silicon solar cell with a rear p^+ polySi contact and a conventional front phosphorus diffused region.

Monocrystalline p-type silicon wafers with a resistivity of approximately 1 Ωcm and a final thickness of 200 μm were used to form several 2x2 cm^2 cells on each 4-inch wafer. After random pyramidal texturing of the cell's active area, a dual n^+ front electron conductor was created within the active area of the cell by thermally diffusing phosphorus from $POCl_3$ twice: one to create a deep and highly doped n^{++} layer under the metal fingers and a second moderately doped n^+ layer over the rest of the front surface, onto which a silicon nitride antireflection/passivating coating was deposited by PECVD. Although complex, this selective diffusion scheme permits to reduce recombination losses at the front doped region to very low levels, in order to bring up the properties of the rear surface.

The p^+ poly-Si passivated contacts were formed by first growing an ultrathin (~ 1.4 nm) SiO_x on the wafers by chemical immersion in hot nitric acid and then depositing a 70 nm thick silicon/boron film on the rear surface by PVD (in-situ doping). Subsequently, the wafers were placed in a quartz furnace and annealed at 950 °C in order to crystallise the silicon film and activate the boron atoms so that they become electrically active. The rear metallic electrode was formed by evaporating a uniform 900 nm thick layer of silver directly onto the poly-Si film. Silver has excellent properties as a mirror, which helps to trap infrared photons inside the wafer and leads to a high short-circuit current.

Test samples were added to measure the most important parameters of the surface layers. The uniform n^+ region in Batch 1 had a sheet resistance of 70 Ohm/sq and a recombination current density J_{0n+} of 40 fA/cm^2, after SiN passivation (this measurement was taken on a flat surface and hence the value estimated for a textured surface is 1.73 times higher, that is, 70 fA/cm^2). The n^{++} region had a sheet resistance of 21 Ω/sq; adding its contribution (about 300 fA/cm^2 over a 3.25% area), we can estimate the total recombination current density of the front surface for this batch at $J_{0n+} \approx 80$ fA/cm^2). The p^+ poly-Si contact was measured to have a recombination current density $J_{0p+} \approx 20$ fA/cm^2, after FGA and metal deposition, of the same order as the optimised value shown in Fig. 5 (the poly-Si layer in Fig, 5 was slightly thinner, 50 nm, and the optimum annealing temperature was also slightly lower).

The finished solar cells were measured at standard testing conditions (AM1.5G, 25 °C) using a Sinton Instruments solar simulator and, as a reference, an n-type poly-Si contact cell measured by ISFH's calibration laboratory. A good repeatability was observed across the 24 cells fabricated in two batches, with an average efficiency close to 22%. The performance parameters of the best DESIJN solar cells, which reached a 23.0% conversion efficiency, are given in Table 2. For comparison, the parameters measured by ISFH CalTeC for the best n-type solar cell made so far in our laboratory are also given. The poly-Si rear contact of this n-type cell, which has an identical geometric design to the p-type cells, was made by PECVD deposition of a-Si:H followed by a phosphorus diffusion, and a double layer ARC was deposited on the front. The front n^+ layer of batch 1 of p-type cells had a sheet resistance of 70 Ω/sq, lower than targeted, which led to a higher J_{0n+}, lower V_{oc} and lower J_{sc} than expected. On the other hand, it helped to obtain a very good FF=82.3%. The series resistance of this device, measured by comparing the Suns-V_{oc} (pseudoFF=83.2%) and J-V characteristic curves (FF=82.3%), is very low, R_s=0.142 Ωcm^2, proving at the cell level the low contact resistivity of the PVD p^+ polySi passivated contacts, shown in Fig.6.

A subsequent batch was aimed at improving the open-circuit voltage by making the front phosphorus diffusion lighter, but we overshoot and achieved a sheet resistance of 400 Ω/sq, whereas the selective n^{++} region had a sheet resistance of 32 Ω/sq. As a result, V_{oc} increased by about 15 mV, reaching in one cell a value of 701 mV. Unfortunately, the excessively high sheet resistance of the front n^+ layer led to a lower fill factor of 79.9%, due to a series resistance of R_s=0.6 Ωcm^2. In addition, the SiN_x ARC was not optimal for these two batches, resulting in a lower than expected J_{sc}=41.1 mA/cm^2. Despite these shortcomings, the best cell in this batch 2 also achieved an efficiency of 23.0%.

Modelling with the program QsCell indicates that the best measured V_{oc}=701mV is consistent with recombination current parameters for the front and rear sides of J_{0n+}=30 fA/cm^2 and J_{0p+}=20 fA/cm^2, hence consistent with the characterisation of these parameters mentioned above. The 23.0% cell from batch 1 had a lower voltage of 684 mV, indicating that it suffered from higher recombination at the doped front n^+ regions or in the bulk of the wafer; if we neglect the latter, the measured voltage would be consistent with recombination current parameters of J_{0n+}=80 fA/cm^2 and J_{0p+}=20 fA/cm^2.

Table 2. Best p-type DESIJN solar cell incorporating a rear in-situ doped PVD polysilicon passivated contact. Also show is our best n-type solar cell, having a PECVD deposited, phosphorus diffused poly-Si rear contact, measured by ISFH CalTeC.

Si deposition method	wafer type	V_{oc} (mV)	J_{sc} (mA /cm^2)	FF (%)	R_s (Ωcm^2)	Effic. (%)
PVD batch 1	p-type	682.7	41.0	82.3	0.14	23.04
PVD batch 2	p-type	701	41.1	79.9	0.6	23.0
PECVD	n-type	704.8	42.4	82.6	0.34	24.7

5 CONCLUSION

Although further improvements of V_{oc}, J_{sc} and efficiency are definitely possible, the experimental results presented in this paper already demonstrate that in-situ doped, sputter-deposited polysilicon layers are suitable for making high efficiency silicon solar cells. The physical DESIJN approach is simple, safe and reproducible. It brings the fabrication of advanced silicon solar cells within the reach of any laboratory having a simple sputtering machine and an annealing furnace. High quality junctions may be formed on silicon wafers without detailed know-how on thermal dopant diffusion from liquid or gaseous sources, or without equipment for ion implantation or chemical vapour deposition. Perhaps more importantly, the PVDESIJN approach offers the PV industry an attractive technological option for manufacturing the next generation of very high performance silicon solar cells.

APPENDIX

As mentioned above, it is possible to produce n^+ doped poly-Si by sputtering onto a thin SiO_x layer, which in the experiments shown in Figure 8 was grown chemically in hot nitric acid. The recombination current J_{0c} was optimised as a function of the temperature of the re-crystallisation/activation step (for in-situ doped PVD) or the diffusion of phosphorus from $POCl_3$ (for ex-situ doped PVD and for PECVD). The measurements shown in this figure were taken after performing an anneal in forming gas at 400 °C, and the polySi layers were 40±10 nm thick. Only a few experiments could be performed with the single target that consisted of a mix of Si and P before the target became degraded, Nevertheless, it is straightforward to achieve very good passivation if an intrinsic silicon layer is deposited, either by PVD or PECVD, and subsequently doped by conventional phosphorus diffusion. In this group of experiments, the optimum diffusion temperature was found to be 860 °C, for which a very low $J_{0c}\approx 3$ fA/cm^2 was achieved. This result shows that there is practically no difference in terms of passivation between the physical and chemical deposition techniques. Nevertheless, in-situ doped n^+ poly-Si contacts by sputtering require more work.

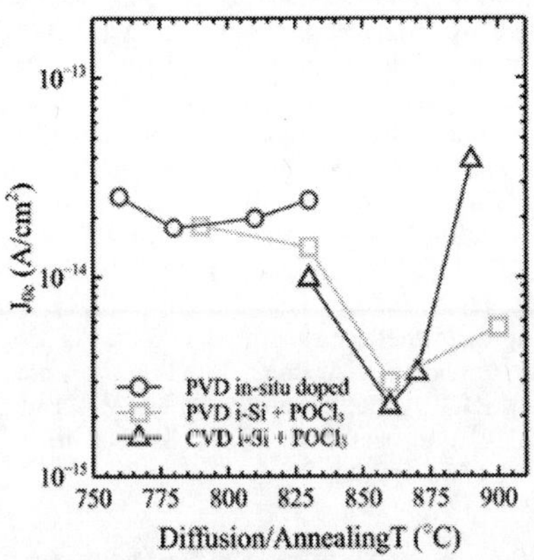

Figure 8. Recombination current parameter of in-situ doped PVD n^+ polySi/SiO_x contacts as a function of activation/crystallisation temperature compared to ex-situ doped PVD and PECVD n^+ polySi/SiO_x contacts as a function of phosphorus diffusion temperature. After FGA.

ACKNOWLEDGEMENT

This work has been partly supported by The Australian National University and the University of Canberra through the Discovery Translation Fund 2.0. Additional support has been received from the Australian Renewable Energy Agency (ARENA) via the Solar PV Research and Development Programme – Project AC/000001, and via the Australian Centre for Advanced Photovoltaics (ACAP)

REFERENCES

[1] G. Nogay, J. Stuckelberger, P. Wyss, E. Rucavado, C. Allebé, T. Koida, M. Morales-Masis, M. Despeisse, F.-J. Haug, P. Löper, C. Ballif, Interplay of annealing temperature and doping in hole selective rear contacts based on silicon-rich silicon-carbide thin films, Solar Energy Materials and Solar Cells, (2017).

[2] F. Feldmann, M. Hermle, Photoactive semiconductor component and method for producing a photoactive semiconductor component, in, Google Patents, 2015.

[3] D. Yan, A. Cuevas, Y. Wan, J. Bullock, Passivating contacts for silicon solar cells based on boron-diffused recrystallized amorphous silicon and thin dielectric interlayers, Sol. En. Mat. and Sol. Cells, 152 (2016) 73-79.

OPTO-ELECTRICAL MODELLING OF PERIODIC NANOSTRUCTURES, INTEGRATED INTO TWO-SIDE CONTACTED SILICON HETEROJUNCTION DEVICES

Arsalan Razzaq[1,2], Valérie Depauw[2], Miha Filipič[2], Ivan Gordon[2], Jozef Szlufcik[2] and Jef Poortmans[1,2,3,4]

[1] ESAT department, KU Leuven, Kasteelpark Arenberg 10, 3001 Leuven, Belgium
[2] imec, Kapeldreef 75, 3001 Leuven, Belgium
[3] U Hasselt, Wetenschapspark 2, 3590 Diepenbeek, Hasselt, Belgium
[4] EnergyVille, Thor Park 8320, 3600 Genk, Belgium

(corresponding e-mail: arsalan.razzaq@imec.be)

ABSTRACT: This work utilizes numerical simulations for evaluating 2D diffraction gratings as an alternative to randomly distributed pyramids for front-side texturing. The study is performed in the context of two-side contacted silicon heterojunction (SHJ) devices as a function of the absorber thickness. Optical simulation studies for a variety of nano-pattern shapes have been reported previously but these were seldom complemented by electrical modelling. Both aspects, eventually concluding to a detailed power loss analysis, are addressed in this work. Rigorous coupled wave analysis (RCWA) method is used for calculating the photo-generation rate profiles which are then coupled to the electrical model, defined in Sentaurus TCAD. The opto-electrical simulations in this work are performed for four different nano-pattern shapes that include nano-cylinders, nano-hemispheres, inverted nano-cones and inverted nano-pyramids. Nanostructured SHJ cells based on inverted nano-pyramids are found to reach the highest efficiencies as they combine the best optical and electrical characteristics.
Keywords: Silicon heterojunction cells, Rigorous Coupled-Wave Analysis, Nanotexturing, Modelling, Simulation

1 INTRODUCTION

Current densities of crystalline silicon (c-Si) solar cells are limited by poor absorption in the near-infrared part of the solar spectrum. This especially constraints the performance of thin c-Si solar cells that are meant to deliver high conversion efficiency at reduced production cost. With front-side texturing schemes, the incident illumination is made to in-couple at angles which allows light to make several passes via total internal reflection. This path length enhancement has conventionally been achieved by random pyramid texturing in wafer based cells where ~ 3-5 μm of c-Si is etched away, thereby making it non-viable for ultra-thin substrates below 10 μm. Alternatively by integrating periodic nanostructures, the infrared absorption in c-Si cells can be enhanced by diffraction with minimal material removal. Moreover, c-Si cells with superior infrared response are also potential candidates for bottom cell in tandem configuration. The resulting optical gain; however, usually comes at the expense of surface passivation which can potentially diminish cell efficiency [17, 22]. It therefore becomes imperative to relate the impact of nano-patterning to the corresponding electrical losses.

2 SIMULATION METHODOLOGY

2.1 Optical model

As the physical dimensions of nano-textures are comparable to the wavelength of light, numerical solutions of Maxwell's equations are required for determining photo-generated current densities. Frequency-domain based approach has been adopted in this work by using the rigorous coupled wave analysis (RCWA) method. The underlying principle is the Bloch theorem which states that the incident plane wave on a diffraction grating follows the same periodicity. This gives rise to periodic fields which can be expressed by Fourier series expansion, where each term represents a plane wave with a given wave vector $\vec{k}$ as depicted in Figure 1.

Influence of the geometry on light in-coupling is examined by considering four different nano-pattern shapes; nano-cylinders, nano-hemispheres, inverted nano-cones and inverted nano-pyramids, shown in Figure 2. Studies on analyzing the impact of the pattern shape on the optical absorption enhancement have been conducted in the past [1-12], without attempting to establish the impact on the electrical performance. The notable exceptions being [13, 21] where the photo-generation rate obtained from Finite-Difference Time-Domain (FDTD) method was coupled to the electrical model. With FDTD method however, only devices made on ultra thin substrates can be analyzed. RCWA method on the other hand, can also be applied for industrial c-Si solar cells that are typically more than 150 μm thick.

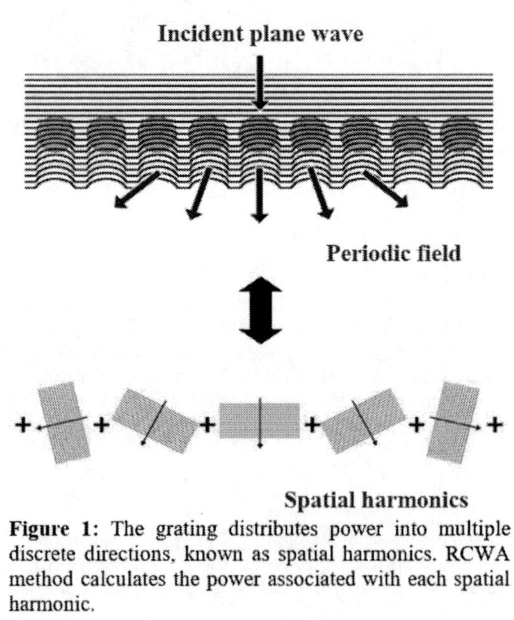

Figure 1: The grating distributes power into multiple discrete directions, known as spatial harmonics. RCWA method calculates the power associated with each spatial harmonic.

Besides the nano-pattern shape, the grating pitch, the area filling fraction and the aspect ratio of the openings also determine the efficiency of coupling light into the grating. Area filling fraction has been defined in this work as the ratio of the textured area to the total area [14] while the aspect ratio is defined by the depth and the diameter (base width in the case of inverted nano-pyramids) of the openings. We took the theoretical maximum filling fraction and a grating pitch of 900 nm for each pattern shape in our optical model. These critical parameters are listed in Table I. The absorption spectrum is calculated by assuming normally incident unpolarized light after averaging the results of transverse electric (TE) and magnetic (TM) modes. Sufficiently large number of plane waves is specified to ensure convergence of results. Refractive index data (n,k) for optical simulations was taken from either PV Lighthouse or from in-house ellipsometry measurements. Any dielectric layer stacks are assumed to be deposited conformally on the nano-patterned side. In reality, this may not be the case but depends on the deposition method.

Table I: Input parameters for optical simulation

Pattern shape	Pitch (nm)	Filling fraction	Aspect ratio
cylinders	900	0.78	$\sqrt{2}/2$
hemispheres	900	0.78	$1/2$
inverted cones	900	0.78	$\sqrt{2}/2$
inverted pyramids	900	1.00	$\sqrt{2}/2$

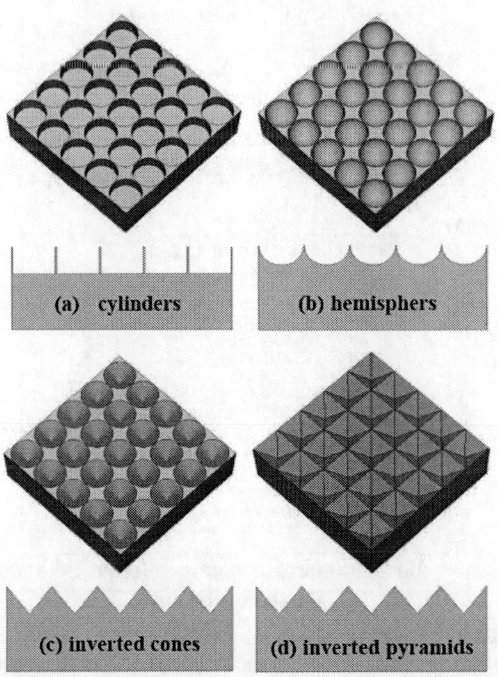

Figure 2: Top and cross-sectional view of the simulated nano-pattern geometries.

2.2 Electrical model

The photo-generation rate profiles from optical simulations are then applied on the electrical model which was implemented in Sentaurus TCAD. Use of the photo-generation rate profile greatly simplifies the computational complexity by allowing a flat unit cell in the electrical domain.

The electrical model was defined for a two-side contacted silicon heterojunction cell (SHJ), featuring emitter on the nano-patterned front side and a flat rear-side. This model, shown schematically in Figure 3, was implemented in 2D. The underlying assumptions for the electrical model are listed as follows:

a) Bulk lifetime is defined by intrinsic recombination model of Richter [15] and an injection independent SRH lifetime of 5 ms.

b) Material parameters for the intrinsic and doped a-Si:H layers are defined according to reference [16].

c) Interface defect density $D_{it,\ front}$ and contact resistance $R_{C,\ front}$ values on the nano-patterned front side are scaled by the surface area enhancement ratios for each shape. This assumption implies that the chemical passivation at the interfaces is identical for pattern shapes that offer the same area enhancement regardless of the processing steps. However, nano-patterning from the plasma etching (RIE) route generally causes a much higher surface damage [17, 22].

d) Lumped series resistance is defined for ohmic losses in the metal grid [18]. Besides line resistance, other components of the series resistance include resistance from the substrate $R_{subs.}$, lateral resistance of the front-side ITO and contact resistances of the front $R_{C,\ front}$ and rear side $R_{C,\ rear}$ contacts.

e) Grid shading loss is accounted by scaling the photo-generation rate with the metallization fraction.

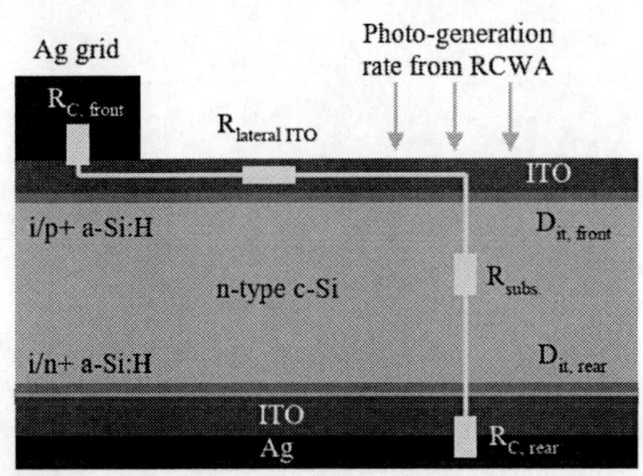

Figure 3: 2D electrical model for SHJ cell. Photo-generation rate from RCWA is coupled to the electrical model.

3 RESULTS AND DISCUSSION

The optical model was initially validated against measured reflectance results of a non-textured c-Si wafer (Figure 4(a)). Fabry-Perot interference fringes, obtained in simulation at infra-red wavelengths, are not experimentally observed. These interference fringes arise due to the assumed coherent wave propagation in RCWA. The assumption is invalid for substrate thicknesses much larger than the wavelength of the incident light. To circumvent the problem of light coherence, phase-elimination method has been proposed in [19] which is based on quarter-wavelength rule. The idea is to perform a first simulation at the original substrate thickness 'd' and a second simulation then follows at substrate thickness of 'd+λ/4'. Averaging the two simulation results filters out the fringes and leads to a direct comparison with the experimental data as shown in Figure 4(b).

The optical model was then further validated using the measured optical spectra of nano-patterned free-standing and bonded epitaxial foils (40 – 50 μm thicknesses) for two different pattern shapes and excellent agreement was found (Figure 5). Details on the fabrication of these nano-patterned epitaxial foils are available in [14].

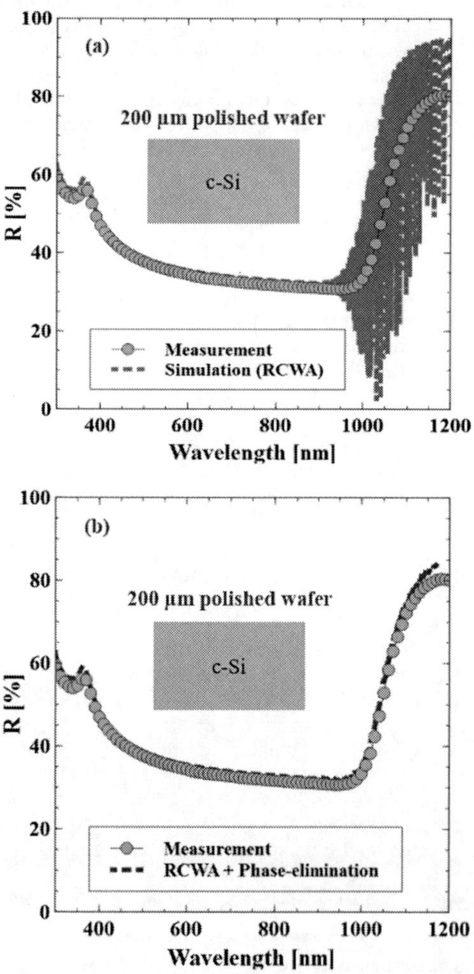

Figure 4: Measured and simulated reflectance results in the case of 200 μm thick non-textured wafer. (a) Fabry-Perot interference fringes at infra-red wavelengths are eliminated after applying the phase-elimination method (b).

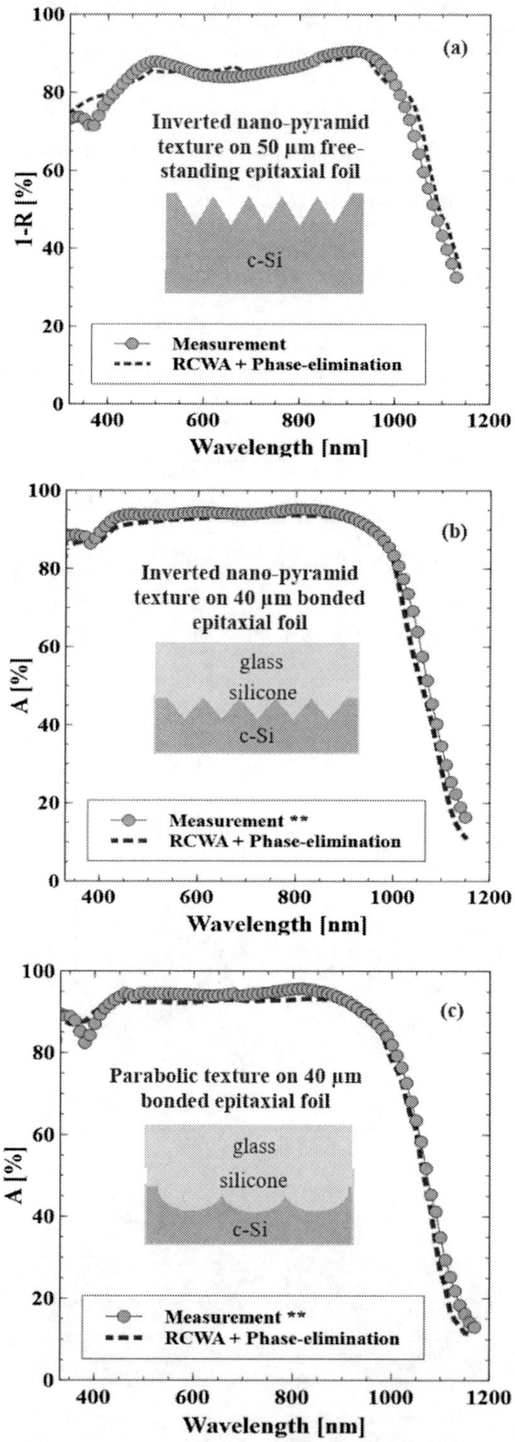

Figure 5: Measured and simulated reflectance results in the case of nano-patterned epitaxial foils. ** Measurement results from reference [14]. (a), (b) Inverted nano-pyramid pattern and (c) parabolic shaped pattern.

After incorporating the photo- generation rate determined by RCWA, SHJ cell efficiency is then obtained from the illuminated J-V curves as function of the absorber thickness (Figure 6).

The J_{SC} plots in Figure 6(a) reveal the loss in short-circuit current for all pattern shapes as the absorber is thinned down to below 10 μm. We observe that high current densities can be attained by inverted nano-pyramids and inverted nano-cones in contrast to nano-cylinders. This observation is explained by the graded index effect that is absent for cylindrical shaped openings. Light that is incident on a wavelength-scale cylindrical grating encounters an abrupt change in the refractive index as it couples from air to c-Si. The abrupt change in the refractive index is responsible for much higher reflection loss at the air/c-Si interface in the case of nano-cylinders. On the other hand, the transition in the refractive index from air to c-Si is more gradual for other pattern shapes. The transition is however relatively less gradual for nano-hemispheres and hence the reflection loss is higher compared to inverted nano-pyramids and cones. Another critical figure of merit for the optical absorption enhancement is the area filling fraction. Inverted nano-pyramids with a square base can theoretically reach much higher filling fraction than nano-patterns with circular openings. Consequently, they optically perform better than inverted nano-cones. Besides the reflection loss, J_{SC} is also strongly dependent on parasitic absorption in the dielectric stacks that provide surface passivation and allow for making electrical contacts. Since nano-patterning the substrate for minimizing reflection loss comes at the expense of increased surface area, pattern shapes that minimize the area enhancement are preferred for curtailing dielectric coverage. The area enhancement ratios for the investigated nano-pattern shapes are 3.20 for cylinders, 1.78 for hemispheres, 1.73 for inverted-pyramids and 1.57 for inverted-cones. This means that in addition to higher filling fraction and the graded index effect, reduced parasitic absorption also contributes to higher J_{SC} in SHJ cells that are based on inverted nano-pyramid texture.

The V_{OC} increases as the absorber gets thinner (Figure 6(b)) due to reduction in bulk recombination losses. SHJ cells are able to reach excellent open-circuit voltages, thanks to the chemical passivation provided by hydrogenated intrinsic a-Si:H layers. As the surface recombination losses scale up with surface area, SHJ cells featuring nano-cylindrical gratings reach lower V_{OC}.

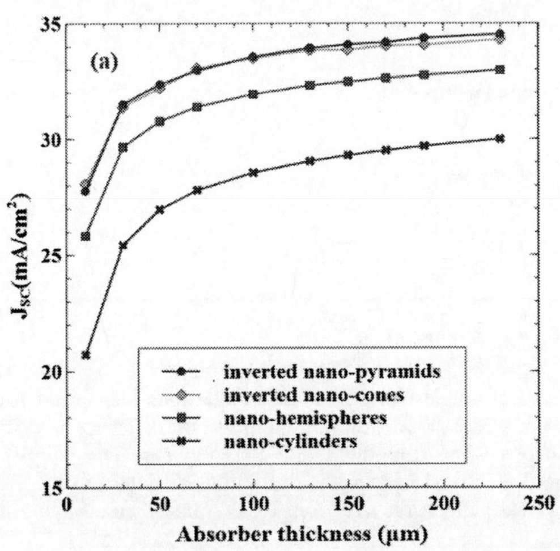

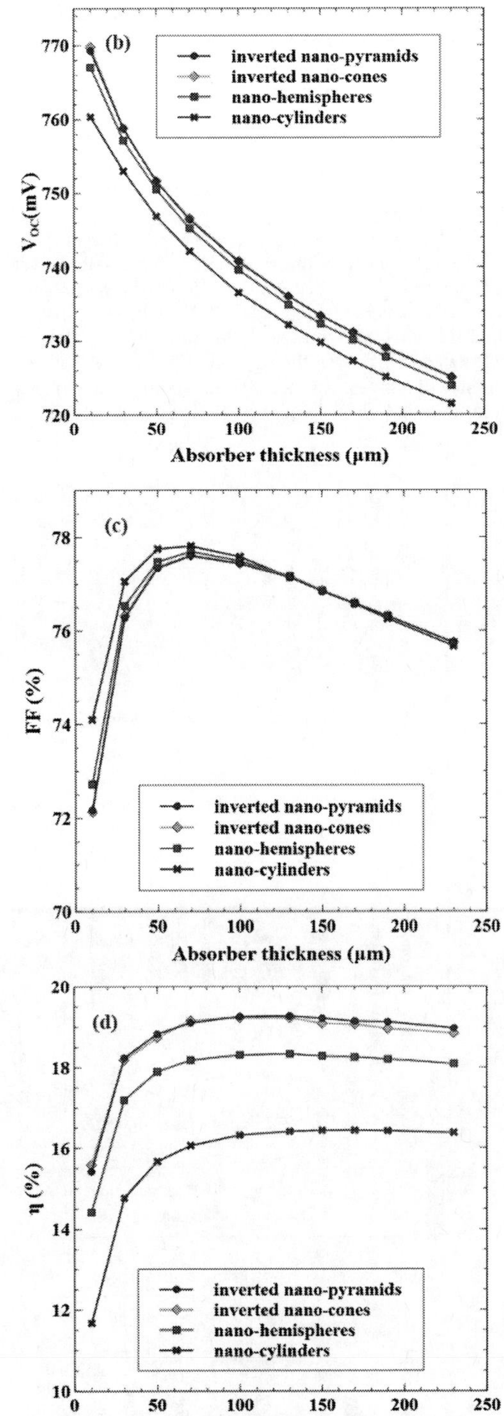

Figure 6: Illuminated J-V parameters as function of the absorber thickness for the investigated nano-pattern shapes.

The fill factors FF are limited by lower J_{SC} values for thin cells and by lower V_{OC} for thick cells. This results in maxima in the FF plots (Figure 6(c)). We observe marginal improvement in the FF with nano-cylinder texture for thin

SHJ cells. Again, this is explained by the much higher increase in the surface area that helps to lessen the contact resistance on the front-side $R_{C,\,front}$.

Overall, the cell efficiency η increases with the absorber thickness (Figure 6(d)) as the gain in J_{SC} outweighs the loss in V_{OC}. The higher J_{SC} and V_{OC} reached by inverted nano-pyramids and cones brings about a substantial improvement in η with respect to nano-cylindrical patterned cells.

Individual loss mechanisms can be quantified in terms of absolute efficiency loss via power loss analysis. Power loss analysis was carried out for 125 μm thick SHJ cells with nano-patterned front-side. Each loss mechanism is disabled, one at a time, and the cell efficiency is re-calculated. This gives the gain in cell efficiency in the absence of a given loss mechanism. Adding up efficiency gains from individual losses brings the final cell efficiency

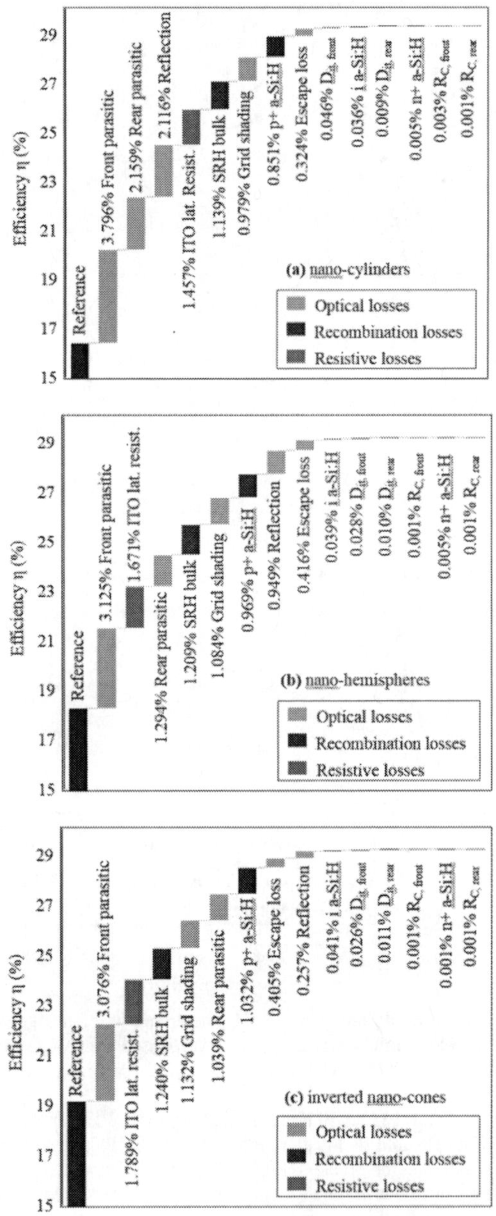

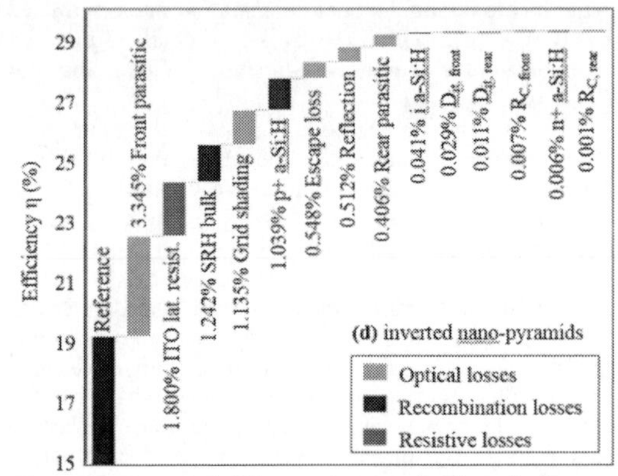

Figure 7: Pareto of all known loss mechanisms in the case of (a) nano-cylinders, (b) nano-hemispheres, (c) inverted nano-cones and (d) inverted nano-pyramids.

close to the theoretical limit of 29.8% [20] when the only loss mechanisms present are the intrinsic recombination and free-carrier absorption in c-Si. Individual loss contributions are then expressed in a Pareto chart (Figure 7). We find that for every nano-pattern shape, parasitic absorption in the front dielectric stack is the primary source of efficiency loss (> 3% abs.) in our simulated nano-patterned SHJ cells. Moreover, front-side parasitic absorption loss is the highest for nano-cylinder pattern that also has the highest surface area enhancement ratio. Besides parasitic absorption on the front, lateral conductance of front-side ITO is another major efficiency loss contributor. Therefore by optimizing the conductance of front-side ITO and the thicknesses of front-side dielectric stack, the efficiency of our simulated nano-patterned SHJ cells can be substantially enhanced. Nonetheless, SHJ cells based on inverted nano-pyramid texture have the potential of reaching higher efficiencies than the other investigated pattern shapes as they combine the best optical and electrical characteristics (Table II).

Table II: Summary of major performance indicators.
+ implies advantage and - implies disadvantage

Pattern shape	Area filling fraction	Graded index effect	Surface area enhancement
cylinders	-	-	-
hemispheres	-	-	+
inverted cones	-	+	+
inverted pyramids	+	+	+

4 CONCLUSIONS

Opto-electrical numerical simulations were used for comparing 2D diffraction gratings of different pattern shapes when integrated into SHJ devices. The coupled opto-electrical approach was built on determining the optical generation rate from RCWA and then applying it to the electrical model, defined in Sentaurus TCAD. The opto-electrical simulation results confirm previous findings that the pattern shape significantly influences light in-coupling. Inverted nano-pyramids give superior optical performance since the transition in the refractive index from air to c-Si is gradual and high area filling fractions are achievable. In view of electrical losses, pattern shapes with modest surface area enhancement ratios are preferred. This simultaneously allows for effective surface passivation along with reduced parasitic absorption. Power loss analysis show that the efficiency loss in our simulated nano-patterned SHJ cells is primarily coming from parasitic absorption in the front-side dielectric stack and the lateral conductance of front-side ITO. These two loss mechanisms need to be minimized for realizing considerable efficiency gains.

ACKNOWLEDGMENT

The authors gratefully acknowledge the financial support of the partners of imec's industrial affiliation program for Si PV (IIAP-PV) for part of this work. Samples used for optical measurements in Figures 5(b) and 5(c) were processed by Christos Trompoukis.

REFERENCES

[1] Wang, K. X., Yu, Z., Liu, V., Cui, Y., & Fan, S. (2012). "Absorption Enhancement in Ultrathin Crystalline Silicon Solar Cells with Antireflection and Light-Trapping Nanocone Gratings". *Nano Letters, 12(3), 1616-1619.*

[2] Han, S. E., & Chen, G. (2010). "Optical Absorption Enhancement in Silicon Nanohole Arrays for Solar Photovoltaics". *Nano Letters, 10(3), 1012-1015.*

[3] Zhu, J., Yu, Z., Burkhard, G. F., Hsu, C., Connor, S. T., Xu, Y., Wang, Q., McGehee, M., Fan, S., & Cui, Y. (2009). "Optical Absorption Enhancement in Amorphous Silicon Nanowire and Nanocone Arrays". *Nano Letters, 9(1), 279-282.*

[4] Mavrokefalos, A., Han, S. E., Yerci, S., Branham, M. S., & Chen, G. (2012). "Efficient Light Trapping in Inverted Nanopyramid Thin Crystalline Silicon Membranes for Solar Cell Applications". *Nano Letters, 12(6), 2792-2796.*

[5] Kowalczewski, P., Liscidini, M., & Andreani, L. C. (2013). "Light trapping in thin-film solar cells with randomly rough and hybrid textures". *Optics Express, 21(S5).*

[6] Bermel, P., Luo, C., Zeng, L., Kimerling, L. C., & Joannopoulos, J. D. (2007). "Improving thin-film crystalline silicon solar cell efficiencies with photonic crystals". *Optics Express, 15(25), 16986.*

[7] Yu, Z., Raman, A., & Fan, S. (2010). "Fundamental limit of light trapping in grating structures". *Optics Express, 18(S3).*

[8] Muller, J., Herman, A., Mayer, A., & Deparis, O. (2015). "A fair comparison between ultrathin crystalline-silicon solar cells with either periodic or correlated disorder inverted pyramid textures". *Optics Express, 23(11).*

[9] Brongersma, M. L., Cui, Y., & Fan, S. (2014). "Light management for photovoltaics using high-index nanostructures". *Nature Materials, 13(5), 451-460.*

[10] Meng, X., Depauw, V., Gomard, G., Daif, O. E., Trompoukis, C., Drouard, E., Jamois, C., Fave, A., Dross, F., Gordon, I., & Seassal, C. (2012). "Design, fabrication and optical characterization of photonic crystal assisted thin film monocrystalline-silicon solar cells". *Optics Express, 20(S4).*

[11] Ferry, V. E., Polman, A., & Atwater, H. A. (2011). "Modeling Light Trapping in Nanostructured Solar Cells". *ACS Nano, 5(12), 10055-10064.*

[12] Mallick, S. B., Agrawal, M., & Peumans, P. (2010). "Optimal light trapping in ultra-thin photonic crystal crystalline silicon solar cells". *Optics Express, 18(6), 5691.*

[13] Deceglie, M. G., Ferry, V. E., Alivisatos, A. P., & Atwater, H. A. (2012). "Design of Nanostructured Solar Cells Using Coupled Optical and Electrical Modeling". *Nano Letters, 12(6), 2894-2900.*

[14] Trompoukis, C. (2015). "Photonic Nanostructures for Advanced Light Trapping in Thin Silicon Solar Cells". *PhD Dissertation, KU Leuven.*

[15] Richter, A., Glunz, S. W., Werner, F., Schmidt, J., & Cuevas, A. (2012). "Improved quantitative description of Auger recombination in crystalline silicon". *Physical Review B, 86(16).*

[16] Fell, A., McIntosh, K. R., Altermatt, P. P., Janssen, G. J. M., Stangl, R., Baillie, A., Steinkemper, H., Greulich, J., Müller, M., Min, B., Fong, K. C., Hermle, M., Romijn, I. G., & Abbott, M. D. (2015). "Input Parameters for the Simulation of Silicon Solar Cells in 2014". *IEEE Journal of Photovoltaics 5(4).*

[17] Trompoukis, C., Daif, O. E., Sharma, P. P., Radhakrishnan, H. S., Debucquoy, M., Depauw, V., Nieuwenhuysen, K. V., Gordon, I., Mertens, R., & Poortmans, J. (2014). "Passivation of photonic nanostructures for crystalline silicon solar cells". *Progress in Photovoltaics: Research and Applications, 23(6), 734-742.*

[18] Brendel, R., Dullweber, T., Peibst, R., Kranz, C., Merkle, A., & Walter, D. (2015). "Breakdown of the efficiency gap to 29% based on experimental input data and modeling". *Progress in Photovoltaics: Research and Applications, 24(12), 1475-1486.*

[19] Campa, A., Krc, J., & Topic, M. (2013). "Two Approaches For Incoherent Propagation Of Light In Rigorous Numerical Simulations". *Progress In Electromagnetics Research, 137, 187-202.*

[20] Tiedje, T., Yablonovitch, E., Cody, G. D. & Brooks, B. G. (1984). "Limiting Efficiency of Silicon Solar Cells". *IEEE Transactions on Electron Devices 31(5).*

[21] Champory, F., Peretti, L., Lalouat, L., Gomard, G., Marion, L. C., Meng, X., Fave, A., Seassal, C., & Drouard, E. (2013). "Electro-Optical Optimization of c-Si Thin Solar Cells Patterened by Photonic Crystals". *28th Proceedings of EU PVSEC, 2634-2637.*

[22] Trompoukis, C., Stesmans, A., Simoen, E., Depauw, V., Daif, O. E., Lee, K., Gordon, I., Mertens, R., & Poortmans, J. (2015). "Photonic nanostructures for advanced light trapping in silicon solar cells: The impact of etching on the material electronic quality". *Physica Status Solidi (RRL) - Rapid Research Letters, 10(2), 158-163.*

OPTICAL INVESTIGATION OF HIGH-EFFICIENCY SILICON-BASED SOLAR CELLS WITH MULTI-SCALE INTERFACE TEXTURES ENABLED BY COUPLED MODELLING APPROACH

Benjamin Lipovšek[1], Žiga Lokar[1], Arsalan Razzaq[2,3], Valérie Depauw[2],
Ivan Gordon[2], Jef Poortmans[2,3,4], Janez Krč[1], and Marko Topič[1]

[1] University of Ljubljana, Faculty of Electrical Engineering, Tržaška 25, 1000 Ljubljana, Slovenia
[2] imec, Kapeldreef 75, 3001 Leuven, Belgium
[3] ESAT department, KU Leuven, Kasteelpark Arenberg 10, 3001 Leuven, Belgium
[4] Universiteit Hasselt, Martelarenlaan 42, 3500 Hasselt, Belgium
(corresponding e-mail: benjamin.lipovsek@fe.uni-lj.si)

ABSTRACT: In high-efficiency bifacial heterojunction silicon solar cells, microscale and nanoscale textures are utilized at different interfaces within the device structures to further improve their performance. We demonstrate the applicability of the coupled modelling approach (CMA) for accurate numerical analysis and optimization of such complex devices. We investigate the influence of surface texturization of the silicon wafer as well as light management properties of the encapsulation. The results show that texturization of the silicon surface alone boosts the short-circuit current density by 23 % with respect to a non-textured cell, however little differences between nanoscale or microscale textures are observed in this case. By introducing microtextured light management foil at the front and rear surface of the encapsulated solar cell, significant additional current gains of over 5 % can be achieved.
Keywords: Bifacial c-Si solar cell, Optical properties, Texturization, Modelling, Simulation

1 INTRODUCTION

The structures of state-of-the-art high-efficiency solar cells, such as heterojunction silicon (HJ Si), require an in-depth analysis to understand and optimize the optical and electrical effects in these devices [1,2]. With respect to the optical behavior, nanometer-thin and several micrometer-thick layers require consideration of coherent and incoherent light propagation through the structure, while various micro- and nanoscale textures at the interfaces require both refraction and diffraction effects to be accurately taken into account [3]. To analyze the operation of such complex devices and optimize their performance, numerical modelling is the key tool [4–9].

In this work, we demonstrate the applicability of the coupled modelling approach (CMA) for detailed and efficient optical analysis of complete encapsulated bifacial HJ Si solar cells. The developed CMA is based on the combination of 3D ray tracing (RT), 1D transfer matrix method (TMM), and 3D rigorous coupled wave analysis (RCWA). Therefore, the CMA enables us to include all the above mentioned optical effects related to the complex bifacial HJ Si solar cell structure within a single efficient simulation [10].

First, we describe the complete studied device and present the principle of its optical analysis using CMA. Validation of the method is demonstrated by comparing the simulation results with experimental measurements of fabricated samples. Then, we employ the CMA for detailed optical investigation end evaluation of various cases in which different textures are applied into the device structure. In particular, we study the influence of (i) surface texturization of the silicon wafer, for which both nanoscale and microscale textures are considered [11,12], and also of (ii) the light management properties of the front and rear surfaces of the encapsulation, namely the light management (LM) foil attached on top of the protective glass sheets [5,13,14]. For this case, flat, microtextured, and ideal anti-reflective LM foils are considered. We demonstrate how different combinations of textures affect the performance of the solar cell and indicate the optimal realization.

2 MODELLING

2.1 Structures

The photovoltaic device investigated in the scope of this work is the complete encapsulated bifacial HJ Si solar cell presented schematically in Fig. 1(a). Since we focus on optical effects related to various textures introduced at different places within the device structure, the metal contacts with their effects [4] are not included in this analysis.

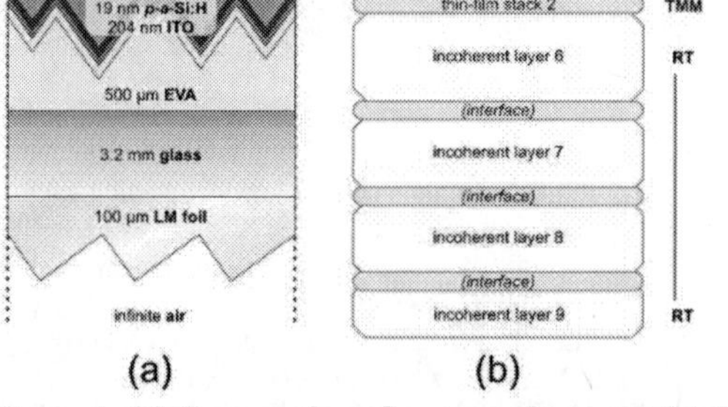

Figure 1: (a) Cross-section of an example encapsulated bifacial HJ Si solar cell, together with the layer thicknesses (dimensions not to scale). In this example, nano pyramids are realized on the top *c*-Si surface and micro pyramids on the bottom. Both LM foils feature the cornercube texture. (b) Segmentation of the structure into incoherent layers and thin-film stacks, together with the numerical method employed for their simulation within the CMA framework.

The basis of the cell presents the 200 µm thick monocrystalline silicon (*c*-Si) wafer that can be textured on any or both sides to enhance internal light management. Texturization can be either in the micrometer or nanometer range, as will be discussed later in a greater detail. The *c*-Si wafer is covered on both sides with thin amorphous silicon *i*-*a*-Si:H and *n*-/*p*-a-Si:H layers for surface fields, and indium-tin-oxide ITO layers for charge carrier collection. The respective thicknesses are given in Fig. 1(a).

The encapsulation on both sides consists of a layer of ethyl-vinyl-acetate (EVA) foil and thick glass. Additionally, a light management (LM) foil is applied on the surfaces of both glass sheets, for the purpose of external light management [5].

In the HJ Si structure presented in Fig. 1(a), textures can be introduced on the surfaces of the *c*-Si wafer and/or on the surfaces of the LM foil. For the *c*-Si wafer, two texture types were considered. The first are regular inverted nano pyramids that were experimentally realized by means of the UV nanoimprint lithography and alkaline etching [11]; their SEM image together with the primary texture parameters are shown in Fig. 2(a). The second are standard upright random micro pyramids realized by alkaline etching without masking [12]; their SEM image is shown in Fig. 2(b). For the LM foil, we considered different light management mechanisms (further details in section 3.3), one of which is the application of the cornercube microscale texture, which has been shown to offer superior optical properties [11,12]; the top-down image with the corresponding parameters is shown in Fig. 2(c).

In this paper, the following convention will be used to define different texture combinations: For the *c*-Si textures, letters "f" (flat), "m" (micro pyramids), and "n" (nano pyramids) will be used. For the textured LM foils, letters "F" (flat) and "C" (cornercubes) will be used. The entire combination of textures within the device will be given in the order from the top to the bottom (with respect to the primary source of illumination); the texture combination featured in Fig. 1(a), therefore, can be briefly given as "C-n-m-C".

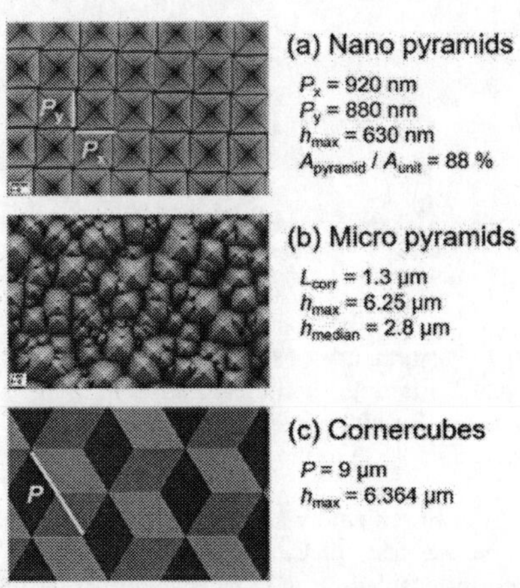

Figure 2: The three types of textures included in the analysis. The regular nano pyramids (a) and random micro pyramids (b) can be realized on the surfaces of the *c*-Si layer, while the cornercube texture (c) can be realized on top of the LM foil.

2.2 Coupled modelling approach (CMA)

In order to numerically simulate light propagation through the complex solar cell structure presented in Fig. 1(a), we employ a combination of different numerical methods, each targeted at different parts of the investigated solar cell [10]. The principles of this coupled modelling approach (CMA) are explained in the following.

The CMA algorithm first identifies all the thick layers in which incoherent propagation of light is assumed. In the example structure in Fig. 1(a), these are the *c*-Si wafer, the encapsulation layers (both glass sheets, EVA layers, and LM foils), as well as the incident medium and the medium in transmission (air in both cases). During the simulation, light propagation through these *incoherent layers*, as well as over the interfaces between them, is handled in three dimensions using geometric ray tracing (RT).

Next, any thin layers that may lie in between the incoherent layers are also identified. They can exist either as a single layer, or as a *thin-film stack* (e.g. EVA/*n*-*a*-Si:H/*i*-*a*-Si:H/*c*-Si). Coherent light propagation through such thin-film stacks is handled using wave optics. If the interfaces in the thin-film stack are locally flat, as is the case if the surfaces of the bordering incoherent layers are flat or microtextured, transfer matrix method (TMM) is employed [15]. If the thin-film stack is nanotextured, rigorous coupled wave analysis (RCWA) is employed [16]. In Fig. 1(b), segmentation of the structure between the incoherent layers and the thin-film stacks is indicated, together with the numerical methods employed to handle each segment.

In the beginning of the simulation, incident illumination is defined in form of a large number of incident rays within one of the incoherent layers, typically the incident medium (for front illumination) or the medium in transmission (for rear illumination). The wavelength, the power, the incident angle, the polarization etc. are all defined. Furthermore, the angular intensity distribution (AID) of the incident light can also be specified, which allows application of different illumination conditions [5]. For example, when assuming clear-sky conditions, direct (specular) illumination from the front side of the cell is assumed, while diffuse (Lambertian) illumination is assumed for the albedo illumination from the rear side.

The rays are traced through incoherent layers in three dimensions until they either reach the lateral borders, in which case the appropriate boundary conditions are applied (periodic in our case), or until they reach any of the thin-film stacks either above or below the incoherent layer. In this case, the *thin-film stack response* is determined – this means that based on the wavelength, the relative incident angle, and the local polarization with respect to the interface surface, the algorithm calculates (i) the polarization and the amount of the light reflected back into the incoherent layer, (ii) the polarization and the amount of the light transmitted over the thin-film stack into the next incoherent layer, and (iii) the amount of the light absorbed within each layer of the thin-film stack. The reflected and the transmitted ray(s) are then traced further through the respective incoherent layers. This iterative process takes place until the rays either escape the device at the front or rear side, or get absorbed in the device (their power drops below some predefined threshold).

Finally, it should be mentioned that while both TMM and RCWA can calculate the above mentioned thin-film stack response "on the fly" during each such event, RCWA used for nanotextured stacks is typically much slower compared to TMM used for locally flat stacks. This is why in the case of nanotextured thin-film stacks, a large amount of thin-film stack responses for different combinations of the incident angles and polarizations is preferably calculated in advance, prior to the simulation. During the CMA simulation, the pre-calculated response that best matches the current conditions is then applied, which helps to speed up the simulation considerably.

3 RESULTS AND DISCUSSION

3.1 Model validation

To validate the developed coupled modelling approach, different samples were fabricated, with different textures and/or additional thin layers realized on top of the *c*-Si wafer. The samples were optically characterized using a PerkinElmer Lambda950 spectrophotometer equipped with an integrating sphere.

The results for a perfectly flat *c*-Si wafer and for a *c*-Si wafer with one side nanotextured are presented in Fig. 3. The measured and the simulated total reflectance, R_{tot}, and (in the case of the textured sample) diffuse reflectance, R_{diff}, are shown in a broad wavelength range. The results confirm that the CMA simulations reliably predict the measured characteristics in all three cases. Only minor discrepancies can be observed which can mostly be related to the realistic imperfections of the fabricated regular nano pyramids that were not included in the simulations.

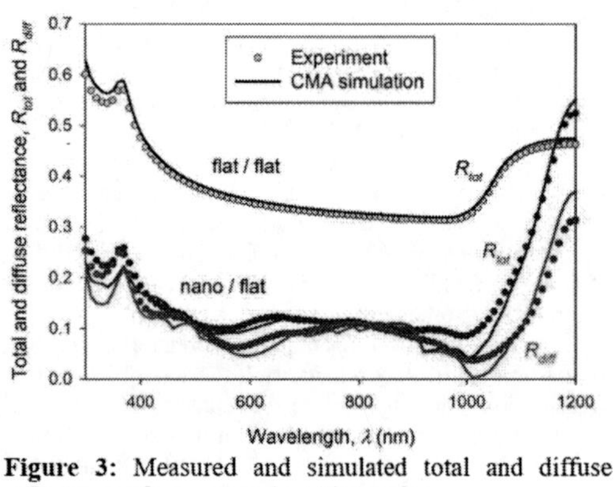

Figure 3: Measured and simulated total and diffuse reflectance of two example *c*-Si samples.

3.2 Light management by *c*-Si wafer

In the first phase of our investigation, we studied the influence of *c*-Si texturization only, while keeping the surface of the LM foil perfectly flat.

To mimic the operation of the bifacial solar cell, the front side was illuminated with direct (specular) illumination, and the rear side with diffuse (Lambertian) illumination. The spectral power distribution in both cases was assumed to be that of the reference AM1.5 solar spectrum. The total power of the rear illumination was assumed to be 30 % of that of the front illumination (albedo of 0.3) [4]. In the simulation range of 300 nm – 1200 nm, the short-circuit current density (J_{sc}) potential of the front direct illumination amounts to 46.53 mA/cm^2, and of the rear diffuse illumination to 13.96 mA/cm^2. The total J_{sc} potential of the total (front + rear) illumination thus amounts to 60.49 mA/cm^2.

The starting point of our study was the conventional F-m-m-F cell configuration in which both surfaces of the *c*-Si are microtextured. The total implied J_{sc} simulated using CMA in this case is 46.94 mA/cm^2, of which 36.65 mA/cm^2 comes from the front, and 10.29 mA/cm^2 from the rear illumination. It should be mentioned that this total J_{sc} of the F-m-m-F cell is almost 23 % larger compared to the perfectly flat F-f-f-F case (not shown here), which demonstrates the highly efficient light management properties of microtextured *c*-Si surfaces.

Next, we also simulated the various cases in which nanoscale textures (regular nano pyramids) are employed, either on their own or in a combination with the random micro pyramids. Namely, these were the cases F-m-n-F, F-n-m-F, and F-n-n-F. For each case, we calculated the J_{sc} increase/decrease with respect to the reference F-m-m-F cell. The results are presented in Fig. 4.

The first important observation is the fact that all four configurations yield similar results; the differences in the total J_{sc} in all cases are in the order of only 1 %. The optimal is the F-m-n-F cell (microtextured front surface, nanotextured rear surface), which enables the greatest total J_{sc}. Interestingly, both cases with nanotextured front surface show poorer performance, especially with respect to the front direct illumination. Therefore, for application to the front surface of the *c*-Si wafer, the random micro pyramids are evidently superior to the regular nano pyramids used in our study, although this conclusion may no longer hold true for other types and profiles of nanoscale textures and solar cell stacks.

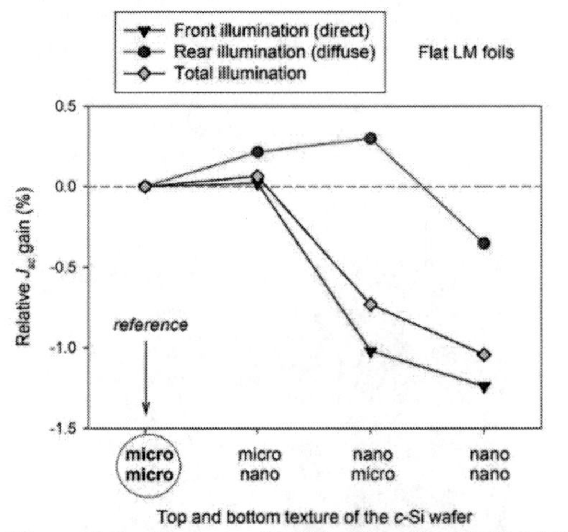

Figure 4: Performance of HJ Si solar cells with flat LM foils and different combinations of *c*-Si surface texturization. Results are expressed in terms of relative J_{sc} gains with respect to the reference F-m-m-F cell.

3.3 Light management by LM foil

In the second phase of our investigation, light management enabled by the LM foil located on the front and/or rear surface of the encapsulated HJ Si solar cell was also included in the analysis. For this purpose, two options were considered: The first was the (realistic) case of LM foil texturization, assuming the cornercube texture to be imprinted into the foil surface. In the second case, on the other hand, we assumed a perfectly flat LM foil, but such that it acts as an ideal omnidirectional (i.e. for all angles of incidence) anti-reflective coating (ARC). In our CMA simulations, this meant that there was zero reflection at the air/LM foil or LM foil/air interface, although refraction of light rays still occurred while crossing the interface. In the following discussion, this ideal ARC case will be denoted as "A" when specifying cell configurations.

The starting point was again the F-m-m-F cell for which the initial J_{sc} results were given in the previous section. Then, while keeping the micro-micro configuration of the *c*-Si wafer, we tested various combinations of the LM foil surfaces, e.g. C-m-m-F (cornercube texture at the front), A-m-m-F (ideal ARC at the front), etc. For each combination, we again calculated the relative J_{sc} gain with respect to the reference F-m-m-F cell. The results are presented in Fig. 5.

First, the results show that the performance of the device is significantly influenced by *both* sides of the encapsulated cell, even if considering illumination from one side only. For example, it is evident that the J_{sc} generated under direct illumination is improved greatly by applying the cornercube texture to the front side, which helps to reduce the initial reflectance and also trap some of the light reflected back from the internal interfaces in the front half of the structure [5]. However, even application of the cornercube texture to the rear side offers additional benefits, since the long-wavelength light that was not entirely absorbed in the *c*-Si wafer is now more efficiently reflected back at the rear internal textured LM foil/air interface. Therefore, the front and rear textured C-m-m-C case in total enables a 5.3 % J_{sc} boost compared to the reference F-m-m-F cell.

Second, the results demonstrate a surprisingly similar overall performance if ideal ARC is used instead of the cornercube texturization, even though the underlying light management principles are fundamentally different. While the cornercube texturization enables a reduced initial reflectance, prolonged optical paths and efficient internal light trapping [5], the ARC enables solely perfect transmission over the high-contrast air/LM foil interface, for all angles of incidence. While the former works better for direct illumination and the latter for diffuse illumination, the overall performances of C-m-m-C, A-m-m-A, A-m-m-C and C-m-m-A cases under total illumination are practically the same. This demonstrates that in this device, the (realistic) cornercube texture can indeed offer similar benefits as an (unrealistic) ideal flat omnidirectional anti-reflective coating.

Finally, it is evident that the cell performance under diffuse illumination is much more sensitive to various LM foil realizations. This is due to the fact that much of the diffuse illumination at larger incident angles cannot couple efficiently though the flat air/LM foil interface, which means that any texturization or application of ARC will greatly reduce the initial reflectance and, thus, enhance the J_{sc} of the device, even up to 10 %. However, since the power carried by the rear diffuse (albedo) illumination represents only 23 % of the total power shining onto the cell, the final results are mostly dominated by the behavior under direct illumination, which is also clear in Fig. 5.

The different realizations of the front and rear LM foil surface were tested also in combination with different surface texturizations of the silicon wafer. These results, shown for total illumination only, are presented in Fig. 6. For the most part, the general trends here follow the trend in Fig. 5, and also it is again evident that the different combinations of the silicon surface textures have only little significance, although some are more sensitive to certain LM foil realizations than others (which can be seen from the varying order of the symbols). However, it is important to emphasize that the order of the *absolute* values in all cases still follows that observed in Fig. 4. For the C-C case, for example, the C-m-n-C cell still generates the largest J_{sc} of 49.57 mA/cm^2, followed by C-m-m-C, C-n-m-C, and finally C-n-n-C. Therefore, the combination of micro pyramids at the front and nano pyramids at the rear of the silicon wafer remains the optimum for all cases studied within the scope of our investigation.

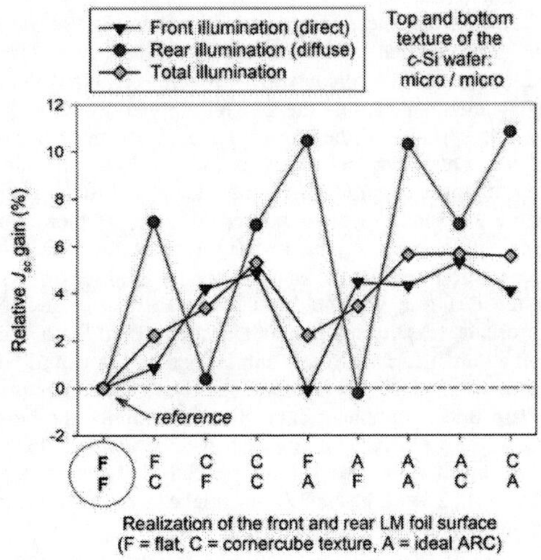

Figure 5: Performance of HJ Si solar cells with microtextured silicon wafer, as influenced by different realizations of the front and rear LM foil surface. Results are expressed in terms of relative J_{sc} gains with respect to the reference F-m-m-F cell.

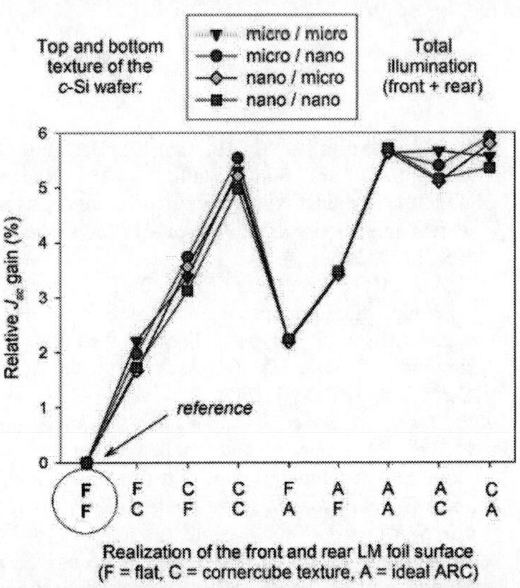

Figure 6: Performance of HJ Si solar cells with different texturizations of the silicon wafer, as influenced by different realizations of the front and rear LM foil surface. Results are expressed in terms of relative J_{sc} gains with respect to the reference F-F cell.

4 CONCLUSION

We demonstrated that the application of the coupled modelling approach, which combines different numerical methods, is crucial for accurate numerical analysis of complex bifacial heterojunction silicon solar cells. Using CMA, all the influential optical effects, such as incoherent and coherent light propagation, refraction at microscale textured interfaces and diffraction at nanoscale textured interfaces and thin-film stacks, can all be accurately taken into account. The approach was validated using experimental characterization results of fabricated samples.

To improve the performance of bifacial HJ Si solar cells, both internal light management enabled by surface texturization of the silicon wafer as well as external light management enabled by the LM foil applied to the front and rear surfaces of the encapsulated device need to be utilized. Our study showed that in the chosen device, without further optimization of the cell structure or texture features, microtextured front surface in combination with a nanotextured rear surface of the silicon wafer renders the optimal results in all cases. As for the LM foil, we found out that realistic cornercube microscale texture applied to the surfaces of both the front and rear LM foil offers similar benefits as would be in the case of an ideal omnidirectional anti-reflective coating. In the optimal combination of textures, the best total J_{sc} of 49.57 mA/cm^2 was achieved, which is a 30 % improvement with respect to the perfectly flat device, and represents 82 % of the total J_{sc} potential carried by the light illuminating the solar cell from both sides.

ACKNOWLEDGMENT

This work was supported by the Slovenian Research Agency (P2-0197). The authors gratefully acknowledge the financial support of imec's industrial affiliation program for Si-PV.

REFERENCES

[1] A. Descoeudres, Z. C. Holman, L. Barraud, S. Morel, S. D. Wolf, and C. Ballif, "> 21% efficient silicon heterojunction solar cells on n- and p-type wafers compared," *IEEE Journal of Photovoltaics*, Vol. 3 (2013) 83.

[2] L. Zhao, C. L. Zhou, H. L. Li, H. W. Diao, and W. J. Wang, "Design optimization of bifacial HIT solar cells on p-type silicon substrates by simulation," *Solar Energy Materials and Solar Cells*, Vol. 92 (2008) 673.

[3] M. Topič, M. Sever, B. Lipovšek, A. Čampa, and J. Krč, "Approaches and challenges in optical modelling and simulation of thin-film solar cells," *Solar Energy Materials and Solar Cells*, Vol. 135 (2015) 57.

[4] A. Čampa, A. Valla, K. Brecl, F. Smole, D. Muñoz, and M. Topič, "Multiscale modeling and back contact design of bifacial silicon heterojunction solar cells," *IEEE Journal of Photovoltaics*, Vol. 8 (2018) 89.

[5] B. Lipovšek, J. Krč, and M. Topič, "Microtextured light-management foils and their optimization for planar organic and perovskite solar cells," *IEEE Journal of Photovoltaics*, Vol. 8 (2018) 783.

[6] R. Santbergen, T. Meguro, T. Suezaki, G. Koizumi, K. Yamamoto, and M. Zeman, "GenPro4 optical model for solar cell simulation and its application to multijunction solar cells," *IEEE Journal of Photovoltaics*, Vol. 7 (2017) 919.

[7] N. Tucher *et al.*, "Optical simulation of photovoltaic modules with multiple textured interfaces using the matrix-based formalism OPTOS," *Optics Express*, Vol. 24 (2016) A1083.

[8] S. Altazin, L. Stepanova, J. Werner, B. Niesen, C. Ballif, and B. Ruhstaller, "Design of perovskite/crystalline-silicon monolithic tandem solar cells," *Optics Express*, Vol. 26 (2018) A579.

[9] M. Kovačič, P.-A. Will, B. Lipovšek, J. Krč, S. Lenk, S. Reineke, and M. Topič, "Combined optical model for micro-structured organic light-emitting diodes," *Informacije MIDEM - Journal of Microelectronics, Electronic Components and Materials*, Vol. 46 (2016) 267.

[10] Ž. Lokar *et al.*, "Coupled modelling approach for optimization of bifacial silicon heterojunction solar cells with multi-scale interface textures," submitted to *Optics Express*, 2018.

[11] C. Trompoukis *et al.*, "Passivation of photonic nanostructures for crystalline silicon solar cells," *Progress in Photovoltaics: Research and Applications*, Vol. 23 (2015) 734.

[12] D. Munoz *et al.*, "Optimization of KOH etching process to obtain textured substrates suitable for heterojunction solar cells fabricated by HWCVD," *Thin Solid Films*, Vol. 517 (2009) 3578.

[13] C. Ulbrich, A. Gerber, K. Hermans, A. Lambertz, and U. Rau, "Analysis of short circuit current gains by an anti-reflective textured cover on silicon thin film solar cells," *Progress in Photovoltaics: Research and Applications*, Vol. 21 (2013) 1672.

[14] S. Esiner, T. Bus, M. M. Wienk, K. Hermans, and R. A. J. Janssen, "Quantification and validation of the efficiency enhancement reached by application of a retroreflective light trapping texture on a polymer solar cell," *Advanced Energy Materials*, Vol. 3 (2013) 1013.

[15] B. Lipovšek, J. Krč, and M. Topič, "Optical model for thin-film photovoltaic devices with large surface textures at the front side," *Informacije MIDEM - Journal of Microelectronics, Electronic Components and Materials*, Vol. 41 (2011) 264.

[16] Ž. Lokar, B. Lipovšek, M. Topič, and J. Krč, "Performance analysis of rigorous coupled-wave analysis and its integration in a coupled modeling approach for optical simulation of complete heterojunction silicon solar cells," *Beilstein Journal od Nanotechnology*, Vol. 9 (2018) 2315.

SYNERGISTIC EFFICIENCY GAIN ANALYSES FOR THE PHOTOVOLTAIC COMMUNITY: AN EASY TO USE SEGA SIMULATION TOOL FOR SILICON SOLAR CELLS

C. N. Kruse[1,2], K. Bothe[1], B. Lim[1], T. Dullweber[1], R. Brendel[1,2]
[1] Institute for Solar Energy Research Hamelin (ISFH), Am Ohrberg 1, 31860 Emmerthal, Germany
[2] Dep. Solar Energy, Inst. Solid-State Physics, Leibniz University Hannover, Appelstr. 2, 30167 Hannover, Germany

ABSTRACT: A synergistic efficiency gain analysis (SEGA) determines the efficiency gain to be expected when idealizing a certain cell property. Moreover, it provides valuable information about synergistic effects due to the coupling of loss mechanisms, e.g. various recombination losses. Here we introduce an easy to use simulation tool to the PV community for performing a SEGA for PERC, PERC+, PERT, and IBC silicon solar cells. The simulation tool comes with a graphical user interface. This SEGA-GUI creates the required set of input files after entering all experimental parameters, e. g. the sheet resistance and recombination current of an emitter. The SEGA-GUI then performs the simulations by running Quokka 2 and analyzes the results. Programming knowledge is not required. The SEGA-GUI is designed to help the user in finding options for efficiency improvements. In this work we analyze our 22.1%-efficient PERC+ solar cell. The SEGA-GUI can be downloaded for free from the ISFH website.

1 INTRODUCTION

Identifying the major losses of an experimental solar cell is crucial for further improvements of the device. A detailed optical and electrical analysis is typically based on experimental parameters that allow modelling the cell and thereby quantify the losses. The free energy loss analysis [1] (FELA) allows for the extraction of all channels of power losses from a single simulation at the maximum power point. It is a popular approach in the PV community and was implemented in the widely used simulation tool Quokka 2 [2]. However, the FELA does in general not reveal the efficiency gain when closing a loss channel, because a single simulation cannot account for the change of the quasi Fermi level that goes along with reducing a loss [3]. The synergistic efficiency gain analysis [4], [5] (SEGA) overcomes this problem. It analyzes the efficiency gain of each loss channel and their synergies by modelling a series of hypothetical cells. This series of simulations breaks down the efficiency gap between an experimental cell (e.g. a passivated emitter and rear cell (PERC) that may have an efficiency in the range of 21% to 22%) and the theoretical maximum efficiency of Si solar cells of about 29%. A SEGA thus shows the experimentalists which loss channel to work on is most valuable in terms of efficiency improvements.

A SEGA requires about 20 separate simulations to be carried out separately. For each of these simulations input files have to be created, the simulations have to be run and checked for consistency and finally the results have to be collected. Performing 20 such simulations manually and then evaluating the potential efficiency gains and creating a graphical representation of the results is time consuming. This task could instead be automated in a program that offers a graphical user interface (GUI) allowing a SEGA without programming. Such a SEGA-GUI should cover common solar cell types such as PERC, PERC+, PERT and IBC cells.

In this work we present such a simulation tool (SEGA-GUI) to the photovoltaic community. The SEGA-GUI can be used on desktop computers and laptops in combination with the freely available program Quokka 2 by A. Fell [2]. The SEGA-GUI is programmed in MATLAB. The compiled program is available from the ISFH webpage at no cost:
http://www.isfh.de/tools/SEGA-GUI.

2 ANALYZING A PERC+ CELL WITH THE SEGA-GUI

2.1 Experimental input parameters

We demonstrate a synergistic efficiency gain analysis using the SEGA-GUI. For demonstration purposes we analyze our PERC+ solar cell that has an independently confirmed efficiency of 22.1% [6].

Figure 1 shows a schematic cross section of the cell. The Si bulk is Ga-doped and passivated on the rear with an AlO_x/SiN_y stack. The phosphorus-doped emitter is passivated with SiN_x that also serves as an anti-reflection coating. The emitter is contacted by screen-printed silver fingers while the rear side is contacted by screen-printed aluminum fingers. For a description of the cell process see [6].

The simulation of our PERC+ cell requires 18 input parameters: 5 of them are the geometrical parameters, 9 are electronic parameters and 4 are optical parameters. Table 1 lists all 13 electrical and optical parameters, together with the respective technique used to determine them.

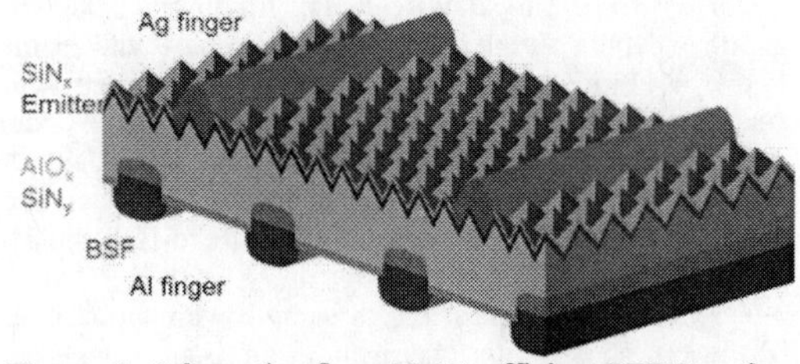

Figure 1: Schematic of our 22.1%-efficient PERC+ solar cell.

Table 1: Simulation parameters for the 22.1%-efficient PERC+ solar cell. The experimental data were measured by quasi steady state photo conductance (QSSPC), photoconductance-calibrated photoluminescence lifetime imaging (PC-PLI) and transmission line measurements (TLM).

Parameter	Value	Measurement technique
Bulk lifetime	1000 µs	QSSPC
Rear surface recombination	12 fA/cm²	QSSPC
Emitter recombination	100 fA/cm²	QSSPC
Front contact recombination	1000 fA/cm²	PC-PLI
Rear contact recombination	350 fA/cm²	PC-PLI
Bulk resistivity	0.8 Ωcm	4-point-probe
Emitter sheet resistance	95 Ω/□	TLM
Front contact resistance	1.5 mΩcm²	TLM
Rear contact resistance	3 mΩcm²	TLM
Front transmission	See Figure 2	Measured reflectance, extrapolated for near infrared
Rear reflection	0.90965	Fitted to reflectance
Lambertian factor front	0.335	Fixed
Lambertian factor rear	0.86435	Fitted to reflectance

The bulk lifetime as well as the recombination properties of the non-metallized surfaces are determined using a Sinton WCT-120 lifetime tester using the corresponding excel sheet [7], [8]. The injection dependence of the bulk lifetime is small. Consequently, we use a fixed Shockley-Read-Hall lifetime of 1000 µs. Further details on the bulk material can be found in the work of Lim et al presented at this conference. For materials with a strong injection dependence we could extract separate lifetimes for minority and majority carriers used by the SEGA-GUI that accepts electron and hole lifetimes of a mid-gap Shockley-Read-Hall (SRH) model.

Symmetric samples (e.g. a sample with the emitter diffusion and emitter passivation on both sides) are processed and measured to determine the surface recombination properties. We determine saturation current densities of 12 fA/cm² for the rear side passivation and 100 fA/cm² for the emitter.

Metallized samples cannot be analyzed with the QSSPC technique. We therefore determine the recombination properties of the contacted areas by photoconductance-calibrated photoluminescence lifetime imaging (PC-PLI) [9] of partially metallized samples [10]. It should be noted that these measurements need to be carried out using a short-pass filter to suppress light that is reflected the rear side metallization [11]. An alternative measurement setup could be the dynamic infrared lifetime mapping [12] which was also used to determine contact recombination current densities [10]. The resulting lifetimes measured for different metallization fractions are then used to determine the recombination current density of the contacted areas. We determine values of 350 fA/cm² for the aluminum contacts and 1000 fA/cm² for the silver contacts.

The resistive properties are determined by four-point probe measurements and transmission line measurements (TLM) [13]. For the extraction of the contact resistance of the aluminum contacts the TLM evaluation method described in reference [14] was used to account for the three dimensional current flow in the Si bulk.

The diamonds in Fig. 2 show the measured reflectance spectrum of the cell. The reflectance is measured using the LOANA cell tester by pv-tools and is then corrected for the reflectance of the metallization to extract the reflectance of the active cell area.

All the geometrical cell parameters are either measured using an optical microscope or they are known from the cell design.

3.2 Simulating the experimental cell (reference case)

The SEGA-GUI supports important cell structures such as passivated emitter and rear cells (PERC), passivated emitter rear totally diffused back junction cells (PERT-BJ), interdigitated back contact cells (IBC) and adaptions to these cell designs. Front and rear contacted cells may also be defined as bifacial cells.

All cell simulations are carried out using Quokka 2 [2], which is a widely used cell simulation program that applies the conductive boundary model [15]. In this model highly doped regions are treated as conductive boundaries to the Si bulk and thereby reducing the simulation time of a 3D unit cell compared to other approaches. Quokka 2 is required for running the SEGA-GUI. Quokka 2 is not included in the SEGA package and can be downloaded from the pv-lighthouse website: https://www2.pvlighthouse.com.au/resources/quokka2/Quokka 2.aspx.

For our PERC+ cell we use the measurement results from Table 1. The SEGA-GUI calculates the optical parameters and the photo-generation profile by fitting the measured reflectance spectrum in Figure 2 (blue diamonds) with the optical model described in [5]. Figure 2 also shows the extrapolated front surface reflectance in the near infrared spectral range (blue dotted line) and the fitted modelled reflectance (red line). The deviation to the measurement is negligible. Using the fitted parameters of the optical model [5] the SEGA-GUI calculates the photo-generation profile in the cell. The SEGA-GUI then starts Quokka 2 using the generation profile and all the preset geometrical, as well as all the electronic and resistive parameters listed in Table 1.

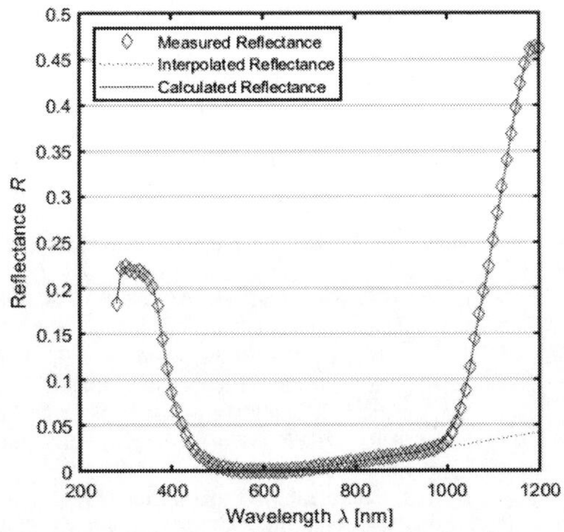

Figure 2: Measured, extrapolated and fitted reflectance of the example PERC+ solar cell.

3.3 Check consistency of *I-V*-parameters

It is crucial to check whether the input parameters reproduce the measured *I-V* characteristic of the analyzed solar cell correctly. For our PERC+ cell we simulate a conversion efficiency of 22.4% as compared to the measured experimental efficiency of 22.1%. J_{sc} and V_{oc} are reproduced correctly without any fitting of electronic, resistive and geometric parameters. The difference of the fill factor arises from a too high pseudo fill factor in the simulation. This is possibly due to the cell edges and lateral inhomogeneities that are neglected. Nevertheless, this reference simulation is a reasonable starting point for analyzing options for efficiency gains for this specific solar cell.

3.4 Results generated by SEGA-GUI

The SEGA-GUI runs the reference case plus one simulation per analyzed loss channel plus one simulation per synergistic effect. For our PERC+ cell this amounts in total of 21 simulations that are all performed at their respective maximum power points and take about 5 minutes on a typical laptop. Once these simulations are finished, the user may plot the results in various pre-defined representations (Figure 3 and 4) or use the comma-separated result files for creating customized plots.

Figure 3 shows the SEGA results of the analyzed PERC+ cell as it is automatically created by the SEGA-GUI: The red bars show potential gains due to recombination, green bars show potential optical gains and blue bars show potential gains due to an optimized carrier transport. The gray bar shows the synergistic gains for recombination, optics and transport losses as well as the synergy between these three groups. We find that the emitter recombination is the largest individual gain option (0.93%$_{abs}$) followed by optical gains due to shading (0.65%$_{abs}$), an imperfect anti-reflection coating and imperfect reflection at the rear side (0.42%$_{abs}$ and 0.4%$_{abs}$).

These efficiency gains would be achieved if e.g. the emitter had no recombination but still the same sheet resistance of 95 Ω/□. This is, of course, not possible but it allows cell developers to estimate the potential of cell optimizations.

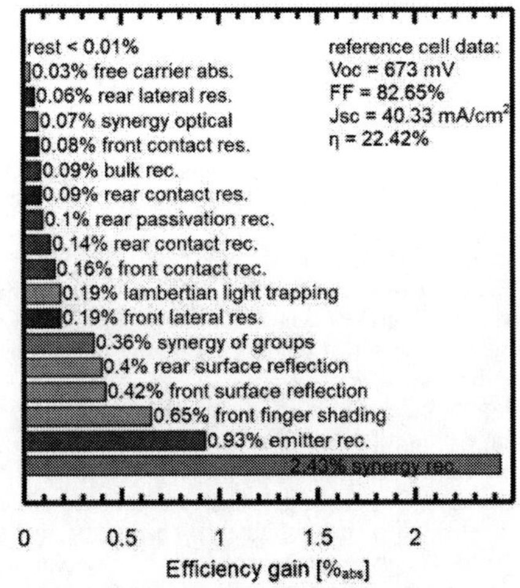

Figure 3: SEGA results of the example PERC+ cell

The advantage of these idealizing simulations is that the reference efficiency plus all efficiency gains shown in Figure 3 always add up to a value that is close to the ideal efficiency of 29%. In this case the sum is 28.8%. The theoretical limit is not exactly reached because the cell doping and the thickness remain unchanged and do not have the optimum values.

Figure 4 shows another automatically created representation of the SEGA results. In this pareto-style plot all losses are shown as bars each beginning at the top of the previous bar. The first bar (in this case the emitter recombination) starts at the efficiency of the reference cell. Together with the synergistic gains of the different groups (recombination, optical and resistive) and the synergistic gain between these groups this representation shows the full breakdown of the efficiency gap between the experimental reference case and the ideal Si solar cell with Lambertian light trapping.

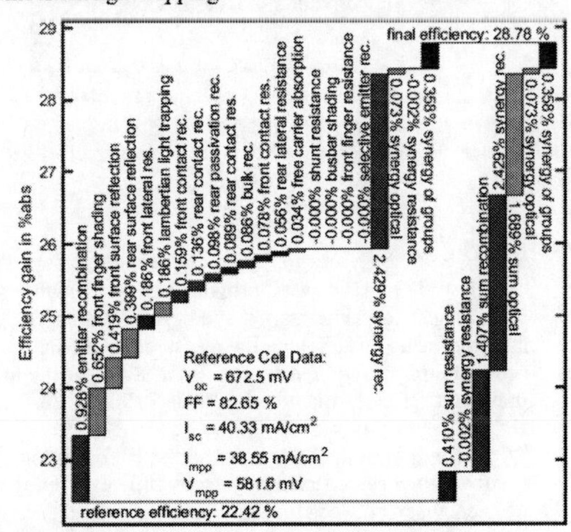

Figure 4: Pareto-style representation of the SEGA automatically created by the SEGA-GUI

The SEGA hints to the most severe loss channels. It does, however not suggest specific technological solutions how to reduce these losses.

3.5 Parameter variations

In the previous section we identified the emitter recombination and the shading by the silver fingers on the front surface as the two dominant bottlenecks for efficiency improvements.

Reducing the emitter recombination by optimizing the dopant profile and reducing the finger width by improving the screen printing process are, therefore important routes for improving the cell.

The SEGA-GUI also supports the user in simulating efficiency gains when gradually idealizing cell parameters in contrast to fully idealizing the parameters in section 3.4.

Figure 5 shows the efficiency change when varying the finger width (green), the emitter recombination (red) and the emitter sheet resistance (blue). We observe that a relative change of the finger width and the recombination rate has a larger impact than a change of the sheet resistance. This is in agreement with the observation that emitter recombination and front shading are more important bottlenecks of the efficiency than the emitter sheet resistance (compare Fig. 3). We also see that for relative improvements up to 30% reducing the finger width is as promising as reducing emitter recombination, despite the larger gain for a complete idealization of the emitter recombination. This rises from the non-linear effects when avoiding recombination.

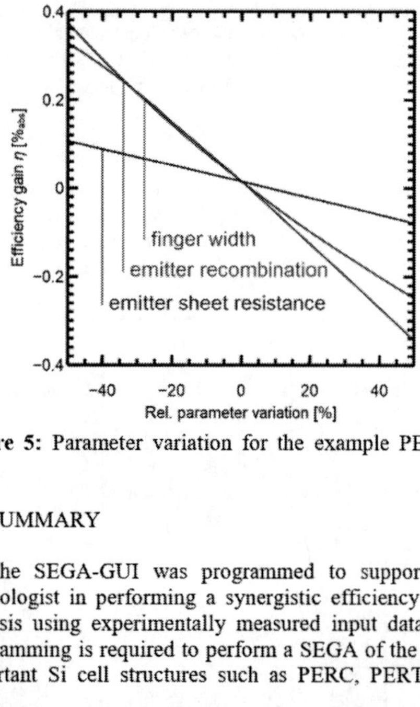

Figure 5: Parameter variation for the example PERC+ cell

4 SUMMARY

The SEGA-GUI was programmed to support the technologist in performing a synergistic efficiency gain analysis using experimentally measured input data. No programming is required to perform a SEGA of the most important Si cell structures such as PERC, PERT and IBC.

The output of the SEGA is a sorted list of options for improving the cell efficiency, thereby fully explaining the efficiency gap between the experiment and the theoretical limit.

The synergistic gains which result from the interaction of various losses are calculated with respect to freely definable groups of losses. The tool also allows evaluating efficiency gains when only partially closing specific loss channels.

The tool can be downloaded for free from the ISFH website: http://www.isfh.de/tools/SEGA-GUI.

5 REFERENCES

[1] R. Brendel, S. Dreissigacker, N. P. Harder, and P. P. Altermatt, "Theory of analyzing free energy losses in solar cells," *Appl. Phys. Lett.*, vol. 93, no. 17, pp. 3–5, 2008.

[2] A. Fell, "A Free and Fast Three-Dimensional/Two-Dimensional Solar Cell Simulator Featuring Conductive Boundary and Quasi-Neutrality Approximations," *IEEE Trans. Electron Devices*, vol. 60, no. 2, pp. 733–738, Feb. 2013.

[3] C. N. Kruse, K. Bothe, and R. Brendel, "Comparison of Free Energy Loss Analysis and Synergistic Efficiency Gain Analysis for PERC Solar Cells," *IEEE J. Photovoltaics*, vol. 8, no. 3, pp. 683–688, 2018.

[4] J. H. Petermann, "Prozessentwicklung und Verlustanalysen für dünne monokristalline Siliziumsolarzellen und deren Prozessierung auf Modullevel," Fakultät für Mathematik und Physik, Leibniz University Hannover, 2014.

[5] R. Brendel, T. Dullweber, R. Peibst, C. Kranz, A. Merkle, and D. Walter, "Breakdown of the efficiency gap to 29% based on experimental input data and modeling," *Prog. Photovoltaics Res. Appl.*, Oct. 2015.

[6] T. Dullweber, H. Schulte-Huxel, S. Blankemeyer, H. Hannebauer, S. Schimanke, U. Baumann, R. Witteck, R. Peibst, M. Köntges, R. Brendel, and Y. Yao, "Present status and future perspectives of bifacial PERC+ solar cells and modules," *Jpn. J. Appl. Phys.*, vol. 57, no. 8S3, p. 08RA01, Aug. 2018.

[7] R. A. Sinton and A. Cuevas, "Contactless determination of current-voltage characteristics and minority-carrier lifetimes in semiconductors from quasi-steady-state photoconductance data," *Appl. Phys. Lett.*, vol. 69, no. 17, pp. 2510–2512, 1996.

[8] A. Kimmerle, R. A. Sinton, J. S. Swirhun, and A. L. Blum, "An Updated Analysis to the WCT-120 QSSPC Measurement System Using Advanced Device Physics," *28th Eur. Photovolt. Sol. Energy Conf. Exhib.*, pp. 1521–1523, Nov. 2013.

[9] S. Herlufsen, J. Schmidt, D. Hinken, K. Bothe, and R. Brendel, "Photoconductance-calibrated photoluminescence lifetime imaging of crystalline silicon," *Phys. status solidi - Rapid Res. Lett.*, vol. 2, no. 6, pp. 245–247, Dec. 2008.

[10] J. Müller, K. Bothe, S. Gatz, F. Haase, C. Mader, and R. Brendel, "Recombination at laser-processed local base contacts by dynamic infrared lifetime mapping," *J. Appl. Phys.*, vol. 108, no. 12, p. 124513, 2010.

[11] J. Müller, K. Bothe, S. Herlufsen, T. Ohrdes, and R. Brendel, "Reverse Saturation Current Density Imaging of Highly Doped Regions in Silicon Employing Photoluminescence Measurements," *IEEE J. Photovoltaics*, vol. 2, no. 4, pp. 473–478, Oct. 2012.

[12] K. Ramspeck, S. Reissenweber, J. Schmidt, K. Bothe, and R. Brendel, "Dynamic carrier lifetime imaging of silicon wafers using an infrared-camera-based approach," *Appl. Phys. Lett.*, vol. 93, no. 10, p. 102104, 2008.

[13] H. Berger, "Contact resistance on diffused resistors," in *1969 IEEE International Solid-State Circuits Conference. Digest of Technical Papers*, 1969, vol. XII, pp. 160–161.

[14] S. Eidelloth and R. Brendel, "Analytical Theory for Extracting Specific Contact Resistances of Thick Samples From the Transmission Line Method," *IEEE Electron Device Lett.*, vol. 35, no. 1, pp. 9–11, Jan. 2014.

[15] R. Brendel, "Modeling solar cells with the dopant-diffused layers treated as conductive boundaries," *Prog. Photovoltaics Res. Appl.*, vol. 20, no. 1, pp. 31–43, Jan. 2012.

CAPACITIVE EFFECTS IN HIGH-EFFICIENCY SOLAR CELLS DURING I-V CURVE MEASUREMENT: CONSIDERATIONS ON ERROR OF CORRECTION AND EXTRACTION OF MINORITY CARRIER LIFETIME

Henri Vahlman*, Joonas Lipping, Jaakko Hyvärinen, Antti Tolvanen, and Sampo Hyvärinen
Endeas Oy
Ruukinkuja 1, FI-02330, Espoo, Finland.
*Corresponding author | Phone: +358 50 4523476, E-mail: henri.vahlman@endeas.fi

ABSTRACT: High-efficiency solar cells have a high internal capacitance that tends to distort I-V measurements during short voltage sweep times compatible with flash testing. Recently, it was shown that this distortion can be corrected with a weighted average of the measured output currents, using time derivatives of junction voltage as weighting factors. Although this procedure theoretically reproduces the steady-state I-V curve, errors from different sources exist. For example, time derivatives of measurement data are potentially sensitive to even low levels of noise in the original signal. In addition, the correction requires prior determination of the series resistance, which can be susceptible to an error. In this contribution, we investigate errors associated with the capacitance correction. We estimate the magnitude of additional noise that derives from using time derivatives of measurement data as weighting factors, as well as the significance of an error in the series resistance. Further, we provide indications that the physical lower limit for the voltage sweep length is dictated by minority carrier diffusivity. Additionally, we show that the effective minority carrier lifetime of the solar cell can be estimated based on output currents and time derivatives of the junction voltage during forward and reverse sweeps.
Keywords: Capacitance, Solar Cell Efficiencies, High-Efficiency

1 INTRODUCTION

Flash simulators enable a rapid and energy-efficient measurement of the I-V curve of solar cells and modules on device manufacturing lines. The length of light pulse of flash simulators can vary from few milliseconds to few hundred milliseconds, and a typical pulse length of many manufacturers is as low as 10 ms. To avoid unnecessary delays due to charging of the flash generators, the entire I-V measurement leading to the determination of the steady-state efficiency of the device is optimally completed within the duration of a single light pulse.

The photovoltaic industry is currently adapting high-efficiency solar cell architectures such as the passivated emitter and rear contact (PERC) cell, heterojunction technology (HJT) cell, and the interdigitated back-contact (IBC) cell into production. These architectures can store a considerable amount of charge in the quasi-neutral regions of the device, which is why their terminal current responds relatively slowly to changes in the applied terminal voltage [1]. This capacitive effect can considerably distort the power measurement of the solar cell during the relatively short voltage sweeps that are compatible with flash testing.

A method to correct the capacitive effects of solar cell I-V curve measurement was recently proposed by Sinton et al. [2]. This method is based on calculating a weighted average of the measured current during a forward (FW) and a reverse (REV) sweep of the terminal voltage (i.e. from short- to open-circuit and from open- to short-circuit, respectively), using time derivatives of the junction voltage as weighting factors. As capacitive currents are factored out during the calculation of the weighted average, theoretically only the steady-state components of the I-V curve remain after the correction.

Since time derivatives of the junction voltage are used in the correction of the capacitive effects with the weighted averaging method, the correction procedure involves numerical derivation of both voltage and current signals. However, the derivatives of measurement signals are prone to noise due to e.g. quantization upon digital signal processing and can notably exceed the noise level of the original signal. Hence, it has thus far remained unclear whether the application of the correction could also lead to increased uncertainty due to measurement noise as compared to a conventional I-V curve measurement.

In addition to the derivatives of the measurement signals, series resistance of the solar cell is an important quantity that is required for the capacitance correction. However, the significance of the error in the series resistance for the capacitance correction has not been assessed to date.

In this contribution, we investigate errors inherent to the correction of the capacitive effects of I-V curve measurement with the weighted averaging method. We write the expression of the capacitance correction in a generalized form such that it can be implemented with an arbitrary combination of two voltage sweeps in either FW or REV direction, or both. Consequently, we compare different combinations of directions and durations of the voltage sweeps and evaluate the magnitudes of the associated errors. Furthermore, we consider factors that dictate the ultimate physical lower limit for the duration of the voltage sweeps. With these investigations, we aim to provide insights for reliable I-V measurements of solar cells with a high internal capacitance.

In addition to the above considerations related to the accuracy of the capacitance correction, we also show how a fundamental physical quantity related to the performance of a solar cell, the effective minority carrier lifetime, can be estimated based on the capacitive effects of the I-V curve measurement.

2 EXPERIMENTAL

I-V measurements of solar cells were performed with the QuickSun 130E cell solar simulator using linear voltage sweeps. Two different solar cell types, listed in Table I, were characterized. The cell type no. 1 corresponds to a commercial state-of-the-art n-type solar cell with a high open-circuit voltage, V_{oc}, of over 0.72 V.

The cell type no. 2 is a 265 μm thick cell with an n-type base doping concentration, N_B, of ~2.3×10^{15} cm^{-3} (corresponding to a resistivity of ~2 Ωcm) and a V_{oc} of 0.66 V. The series resistances, R_s, of the two cell types were determined according to the IEC-60891 standard.

Table I: Solar cell types studied in this work.

Cell type	Thickness	N_B [cm^{-3}]	R_s (Ωcm^2)	V_{oc} (V)
1	~150 μm	Unknown	1.2	0.72
2	265 μm	~2.3×10^{15}	0.43	0.66

3 CAPACITANCE CORRECTION OF I-V CURVES

3.1 Correction of capacitive effects

The general mathematical form of the capacitance correction according to the weighted averaging method is given by

$$I_{s-s} = \frac{I_{out,1} \left.\frac{dV_j}{dt}\right|_2 - I_{out,2} \left.\frac{dV_j}{dt}\right|_1}{\left.\frac{dV_j}{dt}\right|_2 - \left.\frac{dV_j}{dt}\right|_1}, \quad (1)$$

where I_{s-s} is the capacitance-corrected steady-state current, $I_{out,1}$ and $I_{out,2}$ are output currents corresponding to two I-V curves measured using two different voltage sweep rates, t is the time, and V_j is the junction voltage whose time dependence is given by

$$V_j(t) = R_s I_{out}(t) + V_{out}(t), \quad (2)$$

where V_{out} is the output voltage.

Being based on solar cell equivalent circuit analysis, Eq. (1) factors out the current components that are proportional to the time derivative of V_j such as the capacitive current, which means that only the steady-state components remain in the I-V curve. This means that the correction of Eq. (1) is valid regardless of the type of capacitance (e.g. diffusion or junction capacitance) or the cell technology, that is, its application is not limited to crystalline silicon solar cells. Note that Eq. (1) takes all voltage sweep direction combinations into account (i.e. they can be either FW or REV sweeps, or both). Details on the derivation of Eq. (1) are elaborated in Appendix A.

4 ERROR OF CORRECTION

4.1 Error of time derivatives of measurement signals

It is noteworthy that the time derivatives of V_j used in the correction of Eq. (1) can be sensitive to noise in the measurement data. This is the case especially if the derivative is evaluated using discrete methods such as the central difference quotient, whose standard deviation is given by

$$\sigma_{derivative} = \frac{1}{\sqrt{2}} \frac{\sigma_{meas}}{\Delta t}, \quad (3)$$

where σ_{meas} is the standard deviation of the measured quantity, and Δt is the time interval between the measurement points. This means that, whereas the value of the time derivative depends on the total time of the measurement, its standard deviation also depends on the required density of measurement points around the region of interest. Using the central difference quotient, we have found that the noise level of the time derivative of V_j around the maximum power point (MPP) may rise to above 5 % relative despite a very low noise level well below 0.1 %$_{rel}$ in the originally measured data and is approximately constant at different voltage sweep times.

Curve fitting is often used to reduce noise related to the numerical derivation of experimental data. In this case, factors that limit the level of noise in the derivative are the suitability of the fitting function and the fit window width. On the other hand, the fit window should not be made excessively wide to avoid bias in the estimate of the derivative and to preserve good behaviour around discontinuities in the slope. In our case, e.g. local polynomial fitting restricts the measurement noise level of the time derivative of V_j around MPP to ~0.5 – 1 %$_{rel}$.

4.2 Theoretical evaluation of measurement noise

In this section, we assess the magnitude and the significance of measurement noise on the capacitance-corrected current of Eq. (1) based on its total derivative which is written out explicitly in Eq. (B1) of Appendix B. The evaluation at this stage is purely theoretical, and is performed with simulated I-V data obtained using the PC1D software [3]. This theoretical analysis will be followed by support from experimental data below in Sec. 4.3.

The PC1D simulations were performed using a cell thickness of 150 μm, an n-type base resistivity of 1.5 Ωcm, and a series resistance of 1.2 Ωcm^2. In addition, various bulk and surface loss mechanisms were minimized to obtain a V_{oc} close to 720 mV similar to the cell type 1 in Table I, and hence to reproduce capacitive effects of comparable magnitude. Two different linear output voltage sweep time combinations were used to model the I-V curves, which include: (1) a FW and a REV sweep with a sweep length ratio of 1:1 and (2) two FW sweeps with a sweep length ratio of 1:2. In addition, steady-state I-V curves were obtained by simulating a very long sweep.

In the error calculations based on Eq. (B1), we assume that the current measurement is associated with a constant relative noise level of 0.1 %$_{rel}$ around the MPP. In addition, based on the considerations of Sec. 4.1, we presume two different scenarios for the measurement noise level of the time derivatives of V_j. In scenario A, the noise level, $\Delta(dV_j/dt)$, equals a constant 5 %$_{rel}$ near the MPP. In scenario B, the voltage derivative data is obtained such that $\Delta(dV_j/dt)$ is reduced to 1 %$_{rel}$. We also neglect the covariance terms of Eq. (B1) since, as described further below in Appendix B, the covariance was found insignificant in our experimental analysis.

Fig. 1 illustrates the theoretical magnitude of the error of the capacitance-corrected current due to the measurement noise around the MPP in the case of the above-described two scenarios and voltage sweep combinations. The results are reported as a function of the summed total duration of both linear voltage sweeps, which corresponds to the minimum flash pulse duration that is required for the measurement. In general, using a FW and a REV sweep provides the smallest error, while the error in the case of two FW sweeps is notably higher. In addition, increasing the total sweep time decreases the calculated error considerably. These trends are directly attributable to the mathematical form of Eq. (B1) and its dependence on the time derivatives of V_j on the one hand, and the difference between the values of the output currents of the two sweeps on the other hand.

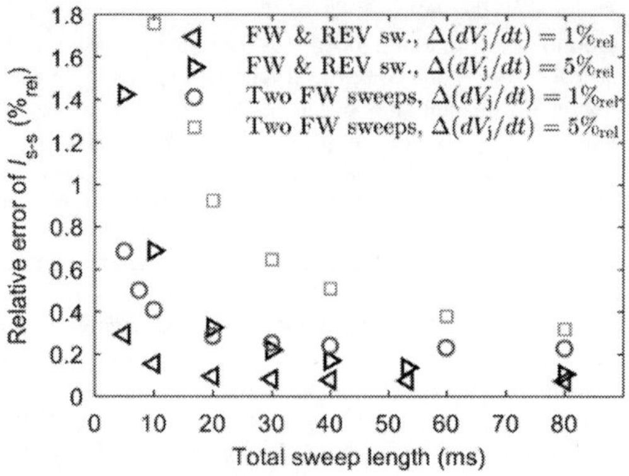

Figure 1: Measurement noise level of the capacitance-corrected current of Eq. (1) near the MPP as a function of the total sweep time in the case of different voltage sweep combinations. The calculation is based on simulated I-V curves and the total differential of Eq. (B1) in Appendix B assuming a constant relative noise level of 0.1 $\%_{rel}$ for the measured output current and either a 1 $\%_{rel}$ or a 5 $\%_{rel}$ noise level of the time derivative of V_j around the MPP.

It is noteworthy that the error at the shortest total sweep lengths is 3 – 20 times higher than the 0.1 $\%_{rel}$ error assumed for the original current data. In this specific example case, the total time of both sweeps that is required to decrease the noise-related error below 0.1 $\%_{rel}$ exceeds 80 ms in scenario A involving a 5 $\%_{rel}$ relative error of the time derivative of V_j, whereas for scenario B (1 $\%_{rel}$ error) the corresponding time decreases to 20 ms. On the other hand, the error related to using two FW sweeps does not descend below 0.2 $\%_{rel}$ even at the longest sweep times. Therefore, the measurement noise-related error of the capacitance-corrected current is expected to depend greatly on both the details of the utilized voltage sweeps and the method of calculating the time derivatives.

4.3 Experimental measurement noise

In this section, we evaluate the magnitude of measurement noise in capacitance-corrected experimental data that was measured from cell type no. 1 in Table I. This analysis is based on fitting a second order polynomial to the capacitance-corrected current near the MPP and determining the standard deviation between the data points and the fit. Two different methods were employed to obtain the time derivatives of V_j. In the first case, the time derivatives were calculated as a simple central difference quotient without any post-processing of the data, whereas in the second case they were calculated using local polynomial fitting to V_j data [see Eq. (2)].

Fig. 2 shows the relative measurement noise level of the experimental capacitance-corrected current as a function of the total sweep length in the case of the two different methods of obtaining the time derivative of V_j. Importantly, the main trends of the experimental data are consistent with the theoretical investigation of Sec. 4.2. These trends include: (i) In the case of the FW and REV sweep, the noise level is considerably lower than in the case of two FW sweeps. (ii) The error increases notably when the total sweep time decreases. Note that the slight increase of the error in the case of the polynomial fitting method with increasing sweep length derives from a gradual increase of the error of the derivative of V_j with increasing sweep length associated with this method.

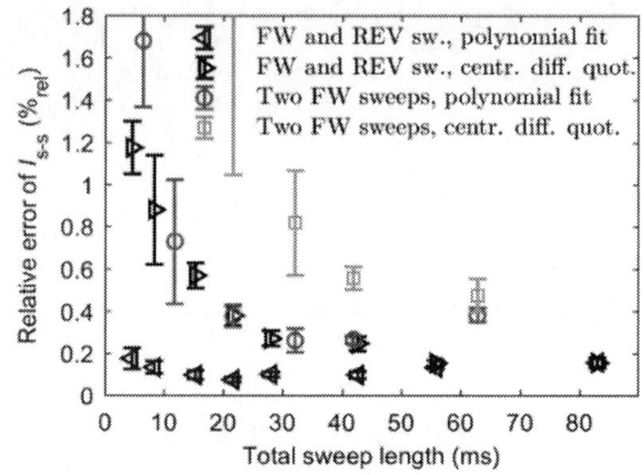

Figure 2: Error of the experimental capacitance-corrected current due to measurement noise as a function of the total sweep length of both sweeps in the case of different voltage sweep combinations and methods to obtain the time derivative of V_j. The error bars represent the standard deviation of at least six different measurements.

Generally, the overall magnitudes of the experimental errors in Fig. 2 agree with the theoretical errors in Fig. 1. For example, using the central difference quotient to obtain the time derivative of V_j corresponds well with the scenario A in Sec. 4.2. On the other hand, calculating the time derivative of V_j with the polynomial fitting method decreases the error and agrees with the scenario B. Therefore, the behavior of the experimental noise levels with respect to sweep length and direction are explained by the theoretical analysis of Sec. 4.2.

4.4 Error due to uncertainty in series resistance

Based on Eq. (2), both V_j and its time derivative depend on R_s, which must be known before applying the correction of Eq. (1). Hence, the accuracy of the power conversion efficiency, η, calculated from the corrected I-V curve depends on the precision of R_s determination and the associated error, ΔR_s. The relationship between the error of η after the capacitance correction and ΔR_s is evaluated here based on the PC1D simulations replicating the cell type 1 (see Sec. 4.2).

Fig. 3 illustrates the relative error of η when the value of R_s used in the capacitance correction is deviated from its actual value used in the simulations, as a function of the total sweep length of a FW and a REV sweep. Note that the relative error of η is defined here in relation to the η obtained with the correction at each sweep length by using the accurate value of R_s. According to the results, the positive ΔR_s values cause a one-sided deviation of η to the negative direction, whereas the negative ΔR_s values result in a positive deviation. In general, the error increases strongly toward shorter sweep lengths. On the other hand, above the total sweep length of 40 ms, the relative error of η has decreased below 0.1 $\%_{rel}$ at all the values of ΔR_s in Fig. 3. It was also found that using two FW sweeps for the correction resulted in overall lower errors (not shown in Fig. 3) compared to using a FW and a REV sweep due to the lower currents associated with the FW sweeps. Note that the relative error of η due to ΔR_s depends strongly on the V_{oc} of the cell and was found to reduce notably with decreasing V_{oc}.

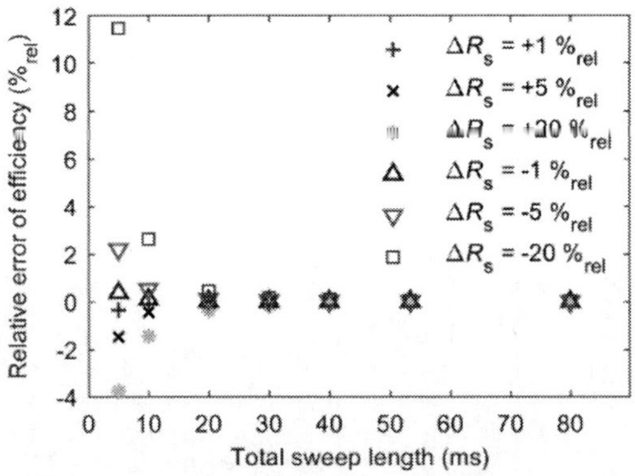

Figure 3: Relative error of the efficiency after the capacitance correction in the case of different relative errors of R_s, from the efficiency obtained with the accurate value of R_s, as a function of the total sweep time. The calculations are based on PC1D simulations replicating the cell type 1 in Table I (see Sec. 4.2).

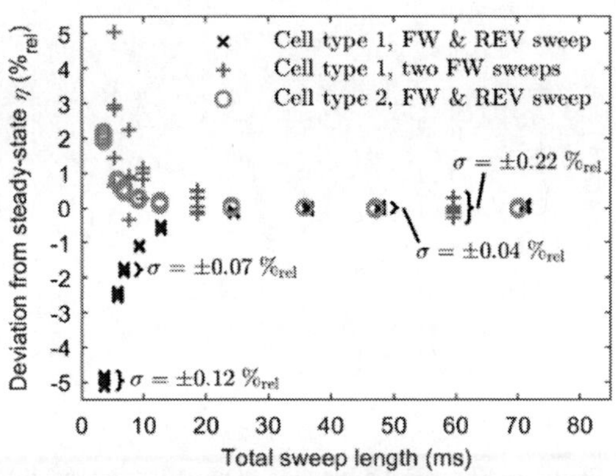

Figure 4: Deviation of the efficiency, η, (in $\%_{rel}$) determined from capacitance-corrected I-V curves from the steady-state η at different total durations of the FW and the REV sweep that were used for the correction. In the figure, each point represents an individual measurement event and σ is the relative sample standard deviation of at least four individual efficiency measurements.

4.5 Experimental efficiency after capacitance correction

Efficiencies of the two solar cell types of Table I were determined to investigate the error of η after the capacitance correction. The efficiencies of both cell types were determined using a FW and a REV sweep and the polynomial fitting method for the time derivative of V_j. For comparison, η of the cell type 1 was also determined using two FW sweeps.

Fig. 4 shows the relative deviation of the measured η values of the two cell types from their steady-state values. Note that we define the steady-state value of η as the value obtained when using a FW and a REV sweep at long sweep lengths. While calculating η, parabolic fitting is performed to the capacitance-corrected current around the MPP, which provides averaging and hence reduces the error due to the measurement noise. Hence, when using a FW and a REV sweep, the standard deviation (σ) of η at its maximum at short sweep lengths is only 0.12 $\%_{rel}$ in the case of the cell type 1 and decreases below 0.04 $\%_{rel}$ when the sweep time is increased. This means that in this case the noise-related error of η is acceptable at practically all sweep lengths. On the other hand, using two FW sweeps results in an increased noise that is visible as a high standard deviation of η of 0.22 $\%_{rel}$ at the longest total sweep length of 60 ms and increases further up to above 1.5 $\%_{rel}$ when the total sweep time is decreased. The high level of noise in this latter case is in qualitative accordance with the results of Secs. 4.2 and 4.3 and is generally considered too high for a reliable I-V measurement.

In addition to the noise component of the error, another, systematic deviation from the steady-state η arises in Fig. 4 at short sweep lengths. This deviation is manifested as a gradual decrease of η with decreasing sweep length in the cell type 1 when using a FW and a REV sweep, and a corresponding gradual increase in the case of cell types 1 and 2 when using two FW sweeps or a FW and a REV sweep, respectively. Notably, when using a FW and a REV sweep, the systematic deviation from steady-state exceeds 0.1 $\%_{rel}$ below a total sweep length of ~28 ms in the case of the cell type 1, and ~16 ms in the case of the cell type 2.

We noticed that the systematic deviation of η at the short sweep lengths can be decreased considerably by adjusting the value of R_s from its experimental value determined according to the IEC-60891 standard. This is a manifestation of the fact that, such as calculated in Sec. 4.4, the sensitivity of the efficiency after the capacitance correction to R_s increases considerably with decreasing sweep length. Consequently, in the case of the cell type 1, almost the entire systematic deviation of η at short sweep lengths disappeared when adjusting the R_s only ~20 $\%_{rel}$. From this perspective, it would seem possible that the deviation of η at short sweep lengths could be caused by inaccuracy of R_s determination. However, in the case of the cell type 2, over a 300 % increase of R_s would be required to remove most of the deviation. Since the value of R_s determined according to the IEC-60891 standard (see Table I) was in the case of the cell type 2 further confirmed using equivalent circuit fitting, it is unlikely that the deviation at short sweep lengths is caused by an erroneous value of this parameter. Hence, this deviation is likely related to another phenomenon, which is investigated further below in Sec. 5.

5 FUNDAMENTAL LOWER LIMIT OF VOLTAGE SWEEP LENGTH

5.1 Systematic deviation from steady-state efficiency at short voltage sweep lengths

To investigate the notable deviation of the capacitance-corrected η at short sweep lengths in Fig. 4, which is unexplainable by a possible error of R_s especially in the case of the cell type 2, we performed device simulations using the PC1D software. Importantly, the simulation approach allows removing the mentioned inaccuracies related to R_s. The simulation parameters were adjusted based on experimental details of the cell type 2 such that the simulated steady-state I-V curves would closely coincide with the corresponding experimental I-V curves.

Fig. 5 illustrates the relative deviation of η from the steady-state in the case of the simulated and capacitance-corrected I-V curves as a function of the total duration of the different voltage sweep combinations described above in Sec. 4.2. Note that, in the case of two FW sweeps, the deviation was negative but was nevertheless plotted on the positive axis to better compare with the FW and REV sweep combination. It is noteworthy that the magnitude of the deviation of the corrected η from the steady-state coincides well with the experimental results of the cell type 2 in Fig. 4. The deviation increases with decreasing total sweep time despite of the fact that R_s is known accurately and exceeds 0.1 $\%_{rel}$ below a total sweep time of ~15 – 20 ms.

5.2 Effect of minority carrier mobility on shape of I-V curves

The reason for the deviation of η from the steady-state at short sweep lengths described in Sec. 5.1 was investigated by modifying the minority carrier (hole) mobility parameter, μ_h, in the PC1D software. In addition, surface and bulk recombination losses in the simulations were minimized to reduce the effect of μ_h on I_{sc} and V_{oc}, and hence to better illustrate the impact on the shape of the I-V curves. Fig. 6 shows the effect of varying μ_h on the FW-swept, the REV-swept, and the capacitance-corrected I-V curve in the case of the cell type 2 when the total voltage sweep length of the two sweeps is 5 ms. While decreasing μ_h from 450 cm²/Vs corresponding to N_B = 2.3×10^{15} cm^{-3} at room temperature [4] deviates the corrected I-V curve further from the steady-state, increasing its value decreases the deviation. Also, modifying the mobility of the majority carriers was found to decrease the error but to a much lesser extent than when varying the mobility of the minority carriers.

Based on the I-V curves of Fig. 6, the current response of the solar cell upon the quickly varying voltage is delayed more and more when μ_h is decreased. This implies that, during short voltage sweeps, the cell is in a nonequilibrium state. The delay affects especially the slope of the current in the reverse direction, making it shallower, which consequently decreases the absolute value of the time derivative of V_j in the reverse direction [see Eq. (2)]. Subsequently, the relative weight of the current of the forward sweep is decreased in Eq (1), which leads to the observed increase of the capacitance-corrected current in Fig. 6. Therefore, the error in the simulated I-V curves at very short sweep lengths is dominated by the limited diffusivity of the minority carriers and is likely at least partially responsible for the systematic deviations of the experimental η values from their steady-state values at short sweep lengths in Fig. 4. This physical limitation dictates the ultimate lower bound for the voltage sweep lengths of the I-V curves used for the capacitance correction.

In Fig. 5, it is noteworthy that correcting the capacitive effects of the simulated data using two voltage sweeps in the FW direction generally gives a smaller error than a FW and a REV sweep. This trend is a manifestation of the fact that during the FW sweep the device is operating closer to steady-state than during the REV sweep [2], which relaxes the diffusion limitation of the minority carriers. As the error due to R_s is also reduced by utilizing two FW sweeps (see Sec. 4.4), purely physical considerations favor this approach in comparison to using a FW and a REV sweep. However, as illustrated in Fig. 4, the error due to measurement noise is higher in the case of two FW sweeps than the FW and REV sweep combination, which in this case practically negates the potential benefits of using two FW sweeps for the correction.

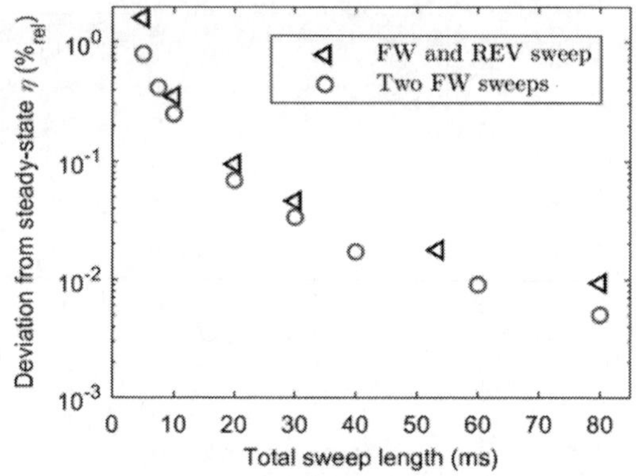

Figure 5: Relative deviation of the capacitance-corrected η of simulated I-V curves from the simulated steady-state as a function of the total duration of the different voltage sweep combinations. Note that the two FW sweeps reproduced a negative deviation which was plotted on the positive axis for comparison with the FW and REV sweep.

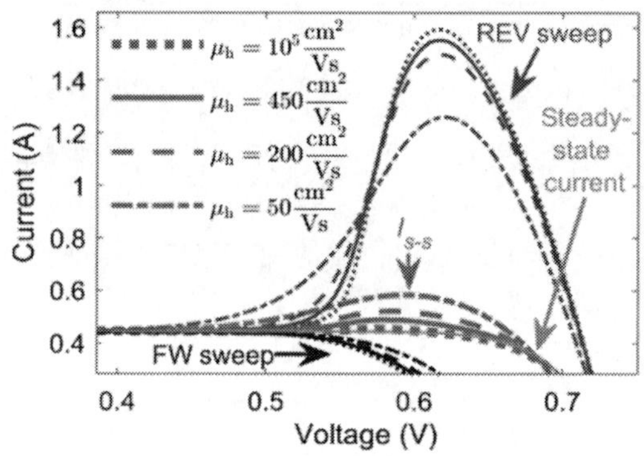

Figure 6: The effect of varying minority carrier mobility on the shape of the simulated FW- and the REV-swept I-V curves, as well as the capacitance-corrected I-V curve at the total sweep length of 5 ms. Note that the value μ_h = 450 cm²/Vs corresponds to the hole mobility in Si with n-type N_B = 2.3×10^{15} cm^{-3} at room temperature.

6 ESTIMATION OF MINORITY CARRIER LIFETIME FROM CAPACITIVE EFFECTS

6.1 Expression of minority carrier lifetime

In this section, we show how the capacitance of a solar cell, as determined from its I-V characteristics, can be linked to a fundamental quantity in view of the solar cell performance, the minority carrier lifetime.

In steady-state conditions, the charge stored in the solar cell base can be written as [5]

$$Q_B = \tau_{eff} I_{rec}, \tag{4}$$

where τ_{eff} is a base-centric effective lifetime of the solar cell as defined by Cuevas and Recart [5] and I_{rec} is the total recombination current of Eq. (A1) in Appendix A. Since the base charge dominates the total capacitance of the solar cell [2], the capacitance can be written as

$$C_{total}(V_j) = \frac{dQ_B}{dV_j} = \frac{d\tau_{eff}}{dV_j} I_{rec} + \tau_{eff}\frac{dI_{rec}}{dV_j}. \tag{5}$$

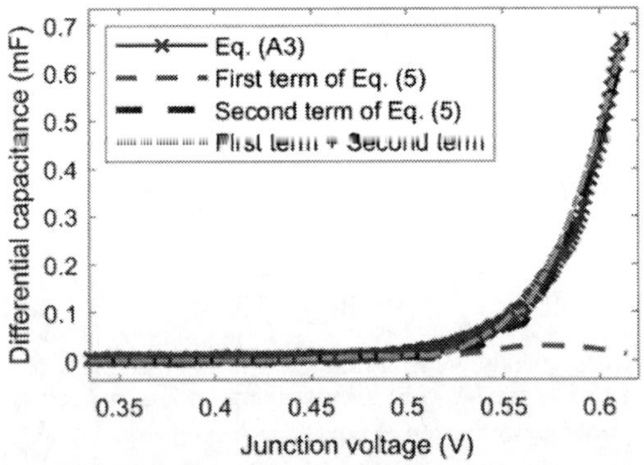

Figure 7: Comparison of the different terms of Eq. (5) for the solar cell capacitance with the capacitance as calculated with Eq. (A3).

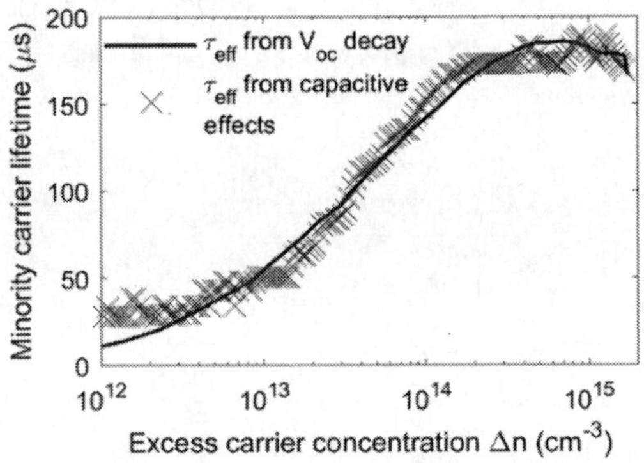

Figure 8: Comparison between τ_{eff} calculated with Eq. (8) based on capacitive effects of I-V measurement and τ_{eff} determined with the V_{oc} decay method.

As shown in Eq. (A3) in Appendix A, the capacitance, $C_{\mathrm{total}}(V_{\mathrm{j}})$, can also be calculated based on the output currents and the time derivatives of the junction voltage from a forward and a reverse sweep. Combining Eq. (A3) with Eq. (5) yields a differential equation for τ_{eff}:

$$\frac{d\tau_{\mathrm{eff}}}{dV_{\mathrm{j}}} I_{\mathrm{rec}} + \tau_{\mathrm{eff}} \frac{dI_{\mathrm{rec}}}{dV_{\mathrm{j}}} = \frac{I_{\mathrm{out,1}}(V_{\mathrm{j}}) - I_{\mathrm{out,2}}(V_{\mathrm{j}})}{\left.\frac{dV_{\mathrm{j}}}{dt}\right|_2 - \left.\frac{dV_{\mathrm{j}}}{dt}\right|_1}. \quad (6)$$

It is noteworthy that the numerical solution of Eq. (6) is complicated by a relatively high level of noise deriving from the derivative terms of both I_{rec} and V_{j} (see Sec. 4.1). As will be shown in the following section, Eq. (6) can be simplified further to provide an approximate estimate of τ_{eff} without a need to solve the full differential equation.

6.2 Experimental determination of lifetime

Fig. 7 shows the capacitance of the solar cell type 2 in Table I as a function of V_{j} as calculated with Eq. (A3) using FW- and REV-swept output currents and time derivatives of V_{j}. This capacitance is compared in Fig. 7 to another value of the capacitance that was calculated with Eq. (5) using τ_{eff} as determined independently with the V_{oc} decay method [6] and I_{rec} calculated based on Eq. (A1) as

$$I_{\mathrm{rec}} = I_{\mathrm{sc}} - I_{\mathrm{s-s}}, \quad (7)$$

where $I_{\mathrm{s-s}}$ is calculated with Eq. (1), and I_{sc} is the short-circuit current which provides a close approximation of I_{ph}. Based on Fig. 7, the second term of Eq. (5) dominates the total capacitance, and coincides relatively well with both the sum of the first and the second term of Eq. (5), and the capacitance as calculated with Eq. (A3). Consequently, Eq. (5) can be simplified by neglecting its first term. This approach provides an approximation of τ_{eff} that does not require solving the full differential equation of Eq. (6), yielding

$$\tau_{\mathrm{eff}} = \frac{I_{\mathrm{out,1}}(V_{\mathrm{j}}) - I_{\mathrm{out,2}}(V_{\mathrm{j}})}{\left.\frac{dV_{\mathrm{j}}}{dt}\right|_2 - \left.\frac{dV_{\mathrm{j}}}{dt}\right|_1} \left(-\frac{dI_{\mathrm{s-s}}}{dV_{\mathrm{j}}}\right)^{-1}. \quad (8)$$

Fig. 8 shows τ_{eff} as calculated with Eq. (8) as compared to τ_{eff} as determined with the V_{oc} decay method. In both cases, the value of V_{j} was converted to excess carrier concentration, Δn, as described in Ref. [6]. As mentioned above, the derivatives of Eq. (8) cause a relatively high level of noise, and hence data filtering was applied to smooth the data. Despite the noise, the overall excess carrier-dependent shape of τ_{eff} of Eq. (8) coincides approximately with that determined with the V_{oc} decay method. Hence, the capacitive effects of the I-V curve measurement can be utilized to obtain a rough estimate of the effective lifetime. Note that, unlike in the V_{oc} decay method, the base doping concentration does not need to be known for the calculation of τ_{eff} as a function of V_{j} from Eq. (8) (although both N_{B} and V_{j} are needed for the determination of Δn). Hence, the capacitive effects provide information on τ_{eff} without a necessity to know the value of N_{B}.

7 CONCLUSIONS

Correction of the capacitive effects of I-V measurement by weighted averaging of two I-V curves can be associated with three main sources of error. These include errors related to the series resistance, the limited minority carrier diffusivity and the measurement noise deriving from the time derivatives of the junction voltage.

Series resistance that is used in the calculation of the junction voltage and its time derivative represents an important source of error. Based on our theoretical evaluation, for example a 10 % relative error of the series resistance can cause a several % relative error in the efficiency after the capacitance correction at short voltage sweep lengths. On the other hand, even in the case of the highest efficiency solar cell type of this work and the highest investigated relative error of the series resistance of 20 $\%_{\mathrm{rel}}$, this error was observed to decrease below 0.1 % relative when the total sweep length of the linear sweeps was increased above 40 ms.

A systematic error of the efficiency was observed at short voltage sweep lengths, which could not be attributed to the series resistance. Simulations revealed that the origin of this phenomenon is the limited diffusivity of minority charge carriers in the solar cell bulk. In the studied cell types, this effect caused a greater than 0.1 $\%_{\mathrm{rel}}$ error when the total duration of both linear voltage sweeps was decreased below 15 – 30 ms and it dictates the physical lower limit for the usable sweep lengths.

Notably, both the error due to series resistance and the error due to the limited diffusivity of minority carriers increase in magnitude when the total duration of the two sweeps is decreased. Although using two forward sweeps for the correction could reduce both errors compared to using a forward and a reverse sweep, a higher measurement noise deriving from the mathematical form of the capacitance correction equation can negate these potential benefits. Hence, to obtain the efficiency at an accuracy greater than 0.1 $\%_{rel}$ compared to the steady-state, the use of a forward and a reverse sweep with a minimum total sweep length of 20 – 30 ms (linear sweeps) was required in this work when correcting the capacitive effects of high-efficiency solar cells with the weighted averaging method.

In addition to the error estimations, we derived an equation for calculating the effective minority carrier lifetime of a solar cell from the capacitive effects of I-V curve measurement. Experimental results show that despite measurement noise that derives from derivative terms that appear in the expression of the lifetime, the proposed method reproduces a reasonable estimate for the minority carrier lifetime.

ACKNOWLEDGEMENTS

The authors acknowledge Micro and Nanotechnologies research group (MNT) of the Departament d'Enginyeria Electrònica, Universitat Politècnica de Catalunya · BarcelonaTech (UPC) for providing solar cells for characterization.

REFERENCES

[1] S. Winter and J. Metzdorf, Proceedings 2nd World Conf. on PV Solar Energy Conversion (1998).
[2] R. A. Sinton, H. W. Wilterdink, and A. L. Blum, IEEE Journal of Photovoltaics 7 (2017) 1591.
[3] H. Haug and J. Greulich, Energy Procedia 92 (2016) 60.
[4] F. Schindler, M. Forster, J. Broisch, J. Schön, J. Giesecke, S. Rein, W. Warta and M.C. Schubert, Solar Energy Materials and Solar Cells 131 (2014) 92.
[5] A. Cuevas and F. Recart, Journal of Applied Physics 98 (2015) 074507.
[6] M. J. Kerr, A. Cuevas, and R. Sinton, Journal of Applied Physics 91 (2002) 399.

APPENDICES

Appendix A. Determination of solar cell capacitance from I-V data

The output current of a solar cell can be written as a sum of its steady-state and capacitive currents as

$$I_{out}(V_j) = I_{ph} - I_{rec}(V_j) - \underbrace{C_{total}(V_j)\frac{dV_j}{dt}}_{I_C}, \quad \text{(A1)}$$

where I_{ph} is the photocurrent, I_{rec} represents all steady-state current losses, I_C is the capacitive current, and C_{total} is the total differential capacitance of the solar cell.

Importantly, both I_{ph} and I_{rec} in Eq. (A1) are independent of $\frac{dV_j}{dt}$. This allows factoring the steady-state current components out by a junction voltage point-wise subtraction of two I-V curves. These two curves are measured with two different voltage sweep rates, yielding

$$I_{out,1}(V_j) - I_{out,2}(V_j) = C_{total}(V_j)\left(\frac{dV_j}{dt}\bigg|_2 - \frac{dV_j}{dt}\bigg|_1\right). \quad \text{(A2)}$$

Hence, C_{total} can be solved from Eq. (A2) as

$$C_{total}(V_j) = \frac{I_{out,1}(V_j) - I_{out,2}(V_j)}{\frac{dV_j}{dt}\big|_2 - \frac{dV_j}{dt}\big|_1}. \quad \text{(A3)}$$

Consequently, inserting Eq. (A3) in Eq. (A1) written for either of the I-V curves and solving for $I_{s-s} = I_{ph} - I_{rec}$ yields the correction of Eq. (1).

Appendix B. Total differential of the capacitance-corrected current with respect to noise components

An error estimate of I_{s-s} in Eq. (1) due to measurement noise can be obtained by calculating the total differential with respect to $I_{out,1}$, $I_{out,2}$, $\frac{dV_j}{dt}\big|_1$, and $\frac{dV_j}{dt}\big|_2$, yielding

$$\begin{aligned}\Delta I_{s-s} = \Bigg(& \frac{\Delta I_{out,1}^2 \left|\frac{dV_j}{dt}\big|_2\right|^2 + \Delta I_{out,2}^2 \left|\frac{dV_j}{dt}\big|_1\right|^2}{\left|\left(\frac{dV_j}{dt}\big|_2 - \frac{dV_j}{dt}\big|_1\right)\right|^2} \\ & + \left|I_{out,2} - I_{out,1}\right|^2 \times \frac{\Delta\left(\frac{dV_j}{dt}\big|_1\right)^2 \left|\frac{dV_j}{dt}\big|_2\right|^2 + \Delta\left(\frac{dV_j}{dt}\big|_2\right)^2 \left|\frac{dV_j}{dt}\big|_1\right|^2}{\left(\frac{dV_j}{dt}\big|_2 - \frac{dV_j}{dt}\big|_1\right)^4} \\ & + 2\frac{\left|\frac{dV_j}{dt}\big|_2\right|}{\left|\left(\frac{dV_j}{dt}\big|_2 - \frac{dV_j}{dt}\big|_1\right)\right|}\frac{\left|I_{out,1} - I_{out,2}\right|\left|\frac{dV_j}{dt}\big|_2\right|}{\left(\frac{dV_j}{dt}\big|_2 - \frac{dV_j}{dt}\big|_1\right)^2} \times \mathrm{Cov}\left(I_{out,1}, \frac{dV_j}{dt}\bigg|_1\right) \\ & + 2\frac{\left|\frac{dV_j}{dt}\big|_1\right|}{\left|\left(\frac{dV_j}{dt}\big|_2 - \frac{dV_j}{dt}\big|_1\right)\right|}\frac{\left|I_{out,2} - I_{out,1}\right|\left|\frac{dV_j}{dt}\big|_1\right|}{\left(\frac{dV_j}{dt}\big|_2 - \frac{dV_j}{dt}\big|_1\right)^2} \times \mathrm{Cov}\left(I_{out,2}, \frac{dV_j}{dt}\bigg|_2\right)\Bigg)^{\frac{1}{2}}, \end{aligned} \quad \text{(B1)}$$

where $\Delta\left(\frac{dV_j}{dt}\big|\right)$ is the error associated with the time derivative of V_j and $\mathrm{Cov}\left(I_{out,i}, \frac{dV_j}{dt}\big|_i\right)$, where i = 1 or 2, are the covariances between the output currents and the time derivatives of V_j.

Note that the covariance between the output current and the time derivative of the junction voltage was evaluated experimentally by fitting a second order polynomial to both the measured current and the time derivative of V_j near the MPP. This was followed by determination of the covariance using data points and fits of the output current and the time derivative of V_j at same values of t. Importantly, the magnitude of the covariance was found so small that, in our case, the covariance terms can be neglected while calculating the error of the capacitance-corrected current with Eq. (B1).

CHARACTERIZATION OF HETEROJUNCTION RIB-SI SOLAR CELLS BY EL AND DLIT IMAGING

Makoto Konagai [1], Rei Kondo[1], Yukimi Ichikawa[1], and Yasuaki Ishikawa[2]
[1] Advanced Research Laboratories, Tokyo City University, 8-15-1 Todoroki, Setagaya-ku, Tokyo 158-0082, Japan
[2] Graduate School of Materials Science, Nara Institute of Science and Technology, 8916-5 Takayama, Ikoma, Nara 630-0192, Japan

ABSTRACT: Heterojunction solar cells with a Rib structure are prepared aiming at ultra-thin Si solar cells. In the Rib structure, the Rib region of about 10% in area ratio mechanically supports the thinned region. In this paper, evaluation was carried out with a focus on the in-plane characteristic distribution of Rib solar cells. Both electroluminescence (EL) and dark lock-in thermography (DLIT) imaging were carried out to evaluate the local electronic properties of Rib solar cells. It was found that EL intensity from Rib-region is 2.2 times larger than EL from thin-region. The difference in EL intensity corresponds to the difference of *Voc*, about 23 mV. DLIT imaging also showed that the current density in the Rig-region is higher than the thin-region. In addition, evaluation by PL imaging is also carried out. And the relationship between the PL intensity distribution and the Rib wafer thickness distribution is discussed based on the thickness dependence of the effective carrier lifetime
Keywords: Electroluminescence, Heterojunction, Photoluminescence, Silicon Solar Cell

1 INTRODUCTION

Currently, Si is the material used in solar cells, and this trend is unlikely to change for some time. The energy conversion efficiency of Si solar cells is reported to have exceeded 26% as a result of the use of heterojunction and back-contact structures [1]. It is predicted that the thickness of the Si wafers used will decrease to approximately 100 μm in the future [2]. Furthermore, Richter *et al.* reported that a maximum theoretical efficiency of 29.43% are achieved for a 110-μm-thick solar cell made of undoped silicon [3].

In the HIT cell, junctions are formed at low temperatures using a method of plasma-CVD of amorphous Si, rather than dopant thermal diffusion. A characteristic of low-temperature formation is that, even if the wafer is extremely thin, warping does not occur after the process [4]. In future, Si solar cell wafers will become increasingly thin, and there are high hopes for the application of heterojunction technology in cells using thin wafers.

We are developing a heterojunction solar cell with a Rib structure aiming at ultra-thin Si solar cells [5, 6]. In the Rib structure, the Rib region of about 10% in area ratio mechanically supports the thinned region. There are thick regions (as thick as the original wafer) and thin regions (etched regions) in the Si wafer. Since the thickness is distributed in the plane, the characteristics of the whole cell are the averaged characteristics of the thick and thin regions. In this paper, evaluation was carried out with a focus on the in-plane characteristic distribution of Rib solar cells. Both electroluminescence (EL) and dark lock-in thermography (DLIT) [7] imaging were carried out to evaluate the local electronic properties of Rib solar cells.

Since the crystal quality of Si wafers is very uniform, and the carrier lifetime is as extremely large as 10 msec, the information obtained by EL mapping is purely thickness effect. It is well known that the *Voc* increases as the Si wafer is made thinner. If regions with different thicknesses are formed on one wafer, the influence of the thickness on *Voc* can be clearly shown. In this work, we obtain the in-plane distribution of EL intensity and measure the distribution of dark current density in the plane by DLIT. This verifies the validity of the EL and DLIT mapping technique in evaluation Rib-Si solar cells.

2 PREPARATION OF HETROJUNCTION RIB-SI SOLAR CELLS

2.1 Fabrication process of Rib wafers

Fig.1 shows the fabrication process of Rib wafers. Double-side polished n-type FZ (100) Si (100) substrate (~3 Ω-cm, ~280 μm , area of about 5 cm × 5 cm) coated with SiN_x was used for fabrication of the Rib structure. First, SiN_x of the region to be removed by etching is removed with a laser (Keithley, Laser Marker VK-X200). Next, anisotropic etching is carried out using KOH. As a result, the Rib region remains without being etched, and the remaining region is thinned. Utilizing the characteristics of anisotropic etching, it has been found that the area of SiN_x to be removed by laser can be about 50% of the required area. The width of Rib is typically 100 μm and the shape of the thin region is 1 mm x 2 mm.

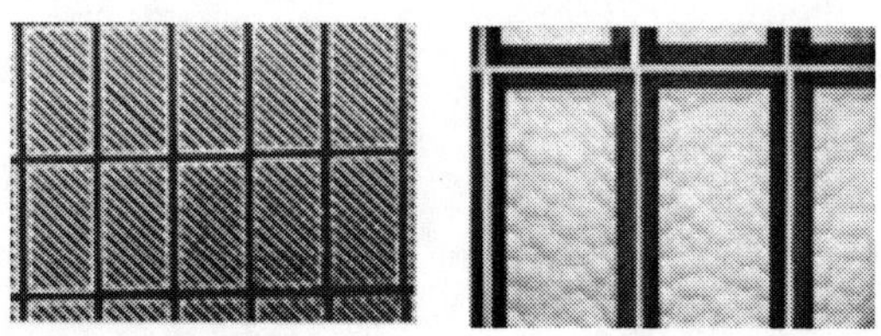

(a)Laser scribed (b) Anisotropic etching

Figure 1: Fabrication process of Rib wafers.
Rib pattern: 1mmx2mm, Rib width: 100 μm.

2.2 Preparation of Rib solar cells

After anisotropic etching to prepare the Rib structure, heterojunction solar cells are formed by SAMCO PD-2203L Plasma Enhanced CVD system. The intrinsic a-Si:H and a-$Si_{1-x}O_x$:H buffer layers were deposited with the mixture of the following gases: silane (SiH_4), hydorgen (H_2) and carbon dioxide (CO_2) as the oxygen source. Typical deposition conditions of each layer are described in references [8, 9].

In order to examine the passivation quality of an a-Si:H (i) buffer layer without impact of a doped layer, a wafer was passivated with identical buffer layers at both sides. Quasi-steady-state photoconductance (QSSPC, Sinton Consulting WCT-100) measurement was performed on passivation samples to evaluate the effective lifetime (τ_{eff}) as a function of excess minority-carrier density (Δn), and the implied V_{OC} was estimated under 1 sun illumination. As shown in Fig. 2, we have obtained a-Si:H(i) layer that provides excellent c-Si surface passivation. Excellent effective carrier lifetime of 10 msec was obtained for i-a-Si:H film thickness of 10 - 20 nm at the excess minority carrier density of $1\text{x}10^{15}/\text{cm}^3$.

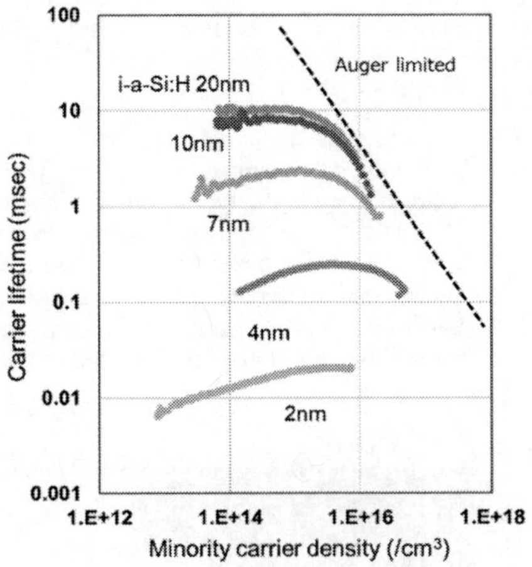

Figure 2: Effective carrier lifetime measured by QSSPC.

Fig.3 shows the Implied Voc of samples shown in Fig.2. Implied Voc of 0.72 V at 1 sun is obtained for a passivation film with a thickness of 10 - 20 nm.

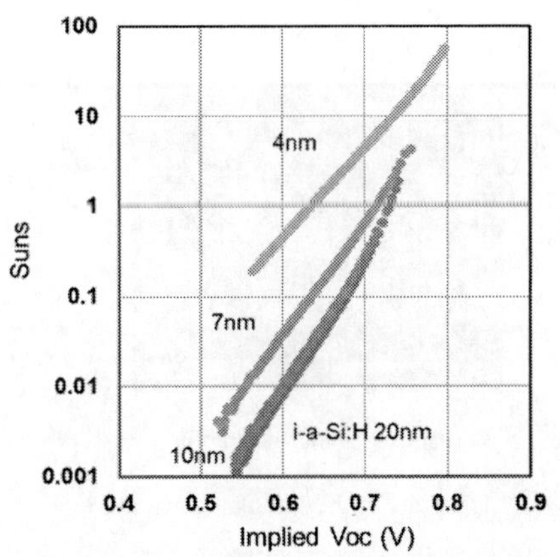

Figure 3: Implied Voc as a function of solar irradiance.

Fig. 4 shows the cross sectional structure of Rib solar cell. The front surface on the light irradiation side of heterojunction Rib-Si solar cell is specular, and the surface on the back side is composed of (100) and (111) planes. The (111) plane is inclined at an angle of 54.7 degrees to the (100) plane.

We fabricated a heterojunction cell having an μc-SiO:H(p)/a-SiO:H(i)/c-Si wafer(n)/a-Si:H(i)/a-Si:H(n) structure; the thickness of μc-SiO:H(p) and a-Si:H(n) layers was 25nm and that of a-SiO:H(i) and a-Si:H(i)layers was 2nm and 3nm, respectively.

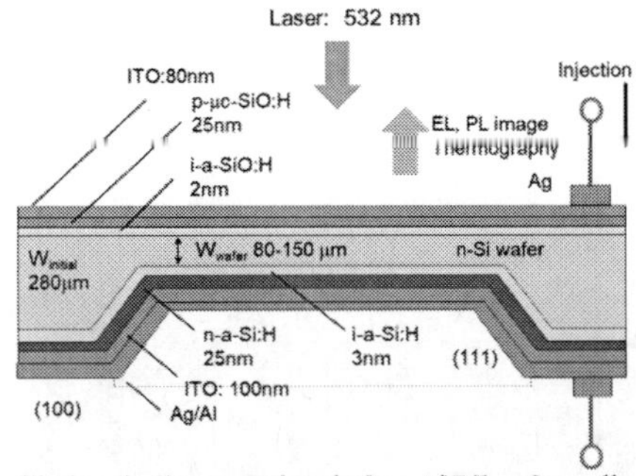

Figure 4: Cross sectional view of Rib solar cell.

3 CHARACTERISTICS OF RIB-SI SOLAR CELLS

3.1 Photovoltaic characteristics

We have fabricated heterojunction Rib-Si solar cells with a thickness from 80 to 150 μm using 280 μm thick n-Si wafers. Fig. 5 shows *I-V* characteristics of a 100μm thick Rib-Si solar cell with an efficiency of 18.8%.

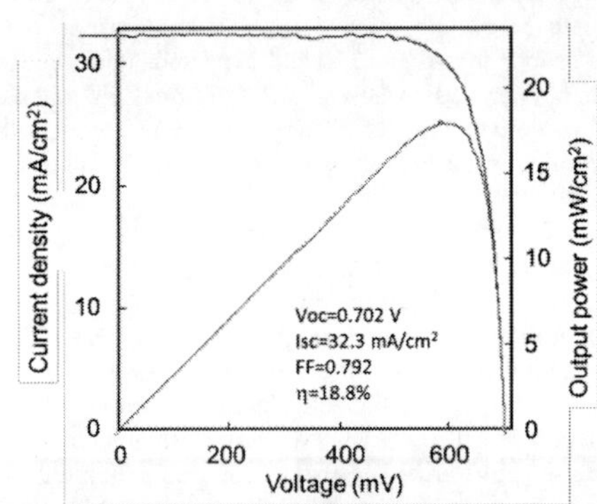

Figure 5: *I-V* characteristics of 100μm thick Rib-Si solar cell. AM1.5, 100mW/cm².

3.2 *Voc* and *EL-EQE*

Fig. 6 shows the *Voc* measured under 1 sun condition and EL external quantum efficiency of various samples as a function of thickness of thin region (W_{Wafer}). Since optical confinement structure is not used and the surface on the light irradiation side is specular, the *Isc* decreases by about 1 mA/cm² as W_{Wafer} decreases [9]. On the other hand, the *Voc* increased by 20 to 30 mV as the wafer was thinned from 280 to 80 μm.

For measurement of external quantum efficiency of EL, a Ge heterojunction solar cell was used as a photo-detector. Specifically, the Si solar cell as the light emitting device and the Ge solar cell as the photo-detector were separated from each other by about 10 mm. EL external quantum efficiency was just calculated from the ratio of the photocurrent flowing in the Ge solar cell and the current injected into the Si solar cell [10]

The EL external quantum efficiency of the Rib-Si solar cell is currently about 0.01-0.1%. A clear correlation was found between the *Voc*, the EL external quantum efficiency and the thickenss of thin region. Thus, it was found that the EL external quantum efficiency measurement can be one of measures to evaluate the performance of Si solar cells, in particular, *Voc*.

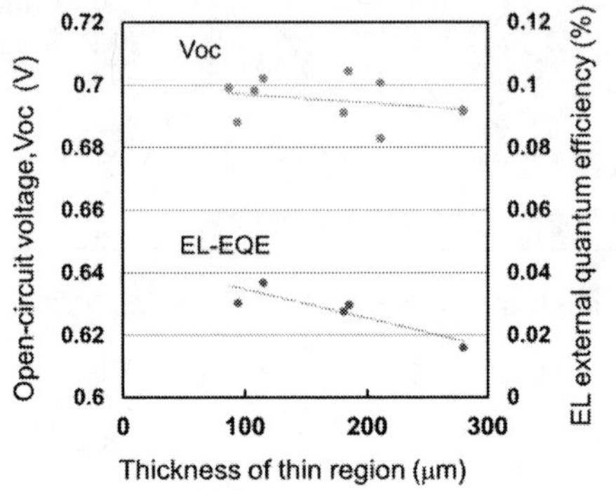

Figure 6: *Voc* and *EL-EQE* as a function of W_{Wafer}.

4 EL AND DLIT IMAGING

4.1 EL imaging

In order to improve the energy conversion efficiency of heterojunction Rib-Si solar cells, it is important to evaluate separately the electrical characteristics of Rib region and thin-region. It is expected that the local efficiency analysis deliver informations for obtaining higher conversion efficiency. In this work, local electronic properties were evaluated by EL and DLIT imaging.

Fig.7 shows EL image of heterojunction Ri-Si solar cell (W_{Wafer} =115 μm) at a forward current of 8.3 mA /cm^2. EL is measured from the p side as shown in Fig.2. In EL imaging, a very clear grid pattern appears. This is originated from the Rib structure formed on the back side. In the same figure, the EL intensity distribution from each region is also shown.Two large peaks are observed, but these peaks arise because the photon generated in the Rib region was reflected by the specular (111) plane etched with KOH. The important point in the figure is the difference in EL intensity between Rib region and thin-region (1 mm × 2 mm). EL intensity from Rib-region is 2.2 times larger than EL from thin-region.

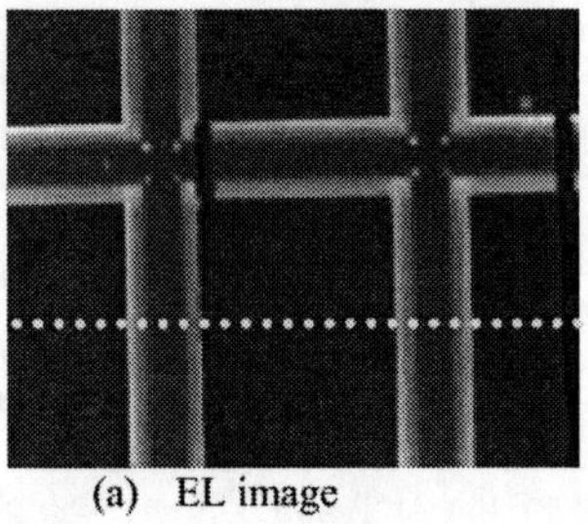

(a) EL image

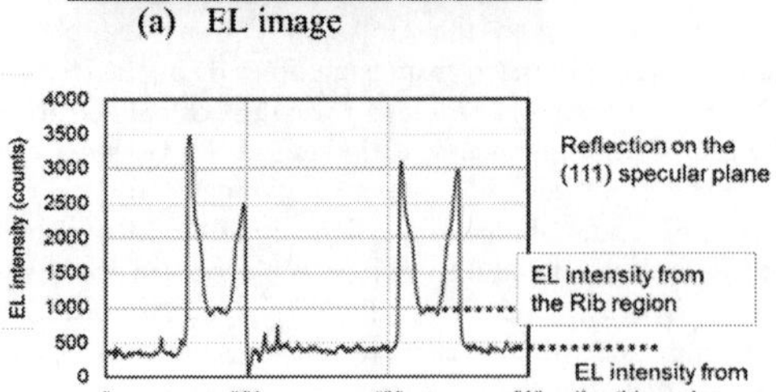

(b)EL intensity on the dotted line in (a).

Figure 7: EL image of Rib-Si solar cell.

In measuring the EL intensity, the same forward voltage is applied to both Rib-region and thin-region. In heterojunction solar cells, as shown in Fig. 6, *Voc* improves as wafer becomes thinner. That is, when the same voltage is applied to the junction, the current density *Id (thin)* flowing in the thin-region becomes smaller than the current density *Id (Rib)* flowing in the Rib region. Therefore, the difference in the concentration of the injected excessive minority carriers appears in the EL intensity distribution. Assuming that the diode quality factor (*n*-value) is the same in Rib-region and thin-region, the difference in EL intensity corresponds to the difference of *Voc*, about 23 mV.

4.2 DLIT

Fig.6 shows the measurement results of DLIT imaging. The color display in the figure shows a relative change, and it can't be converted to a change in the absolute value of the temperature at the present stage. However, the temperature of the light green region（Rib-region) is higher than that of the deep blue-green region (thin-region), which also shows that the current density in the Rig-region is higher.

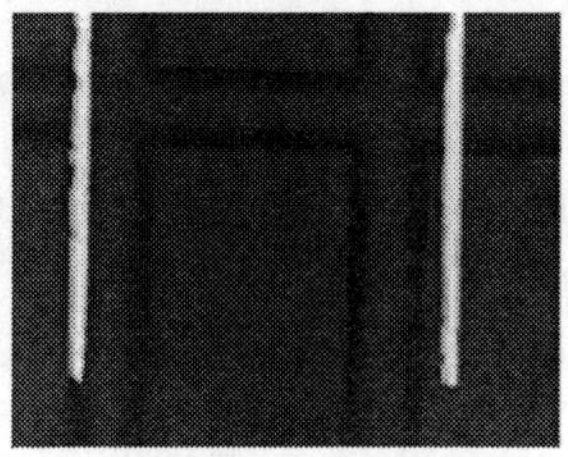

(a)Infrared microscope

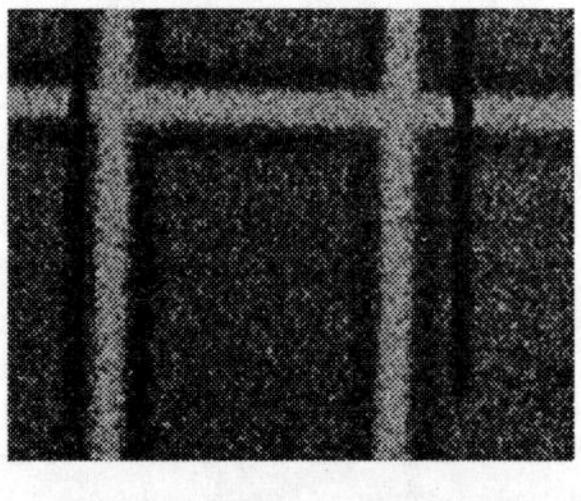

(b)Lock-in thermography

Figure 8: DLIT image of the sample shown in Fig.7.

Another point to consider when relating EL images to *Voc* is the change in effective carrier lifetime depending on the thickness of the Si layer. Fig.9 shows the relationship between the thickness of Si wafer and effective carrier lifetime measured by QSSPC method. Even though the passivation film is exactly the same, the effective carrier lifetime is different depending on the thickness of the wafer. The surface recombination velocity can be calculated based on the data in the figure. As a result of analysis, $S_{(100)}$ = 3.8 cm/s, $S_{(111)}$ = 6.1 cm/s in the structure of ni-a-Si /n-Si/in-a-Si and $S_{(100)}$ = 2.5 cm/s in the structure of pi-a-Si /n-Si/ip-a-Si on the (100) plane (Fig.9). By using these surface recombination velocities, we can clarify the relationship between EL intensity and *Voc* more accurately by conducting 2D device simulation.

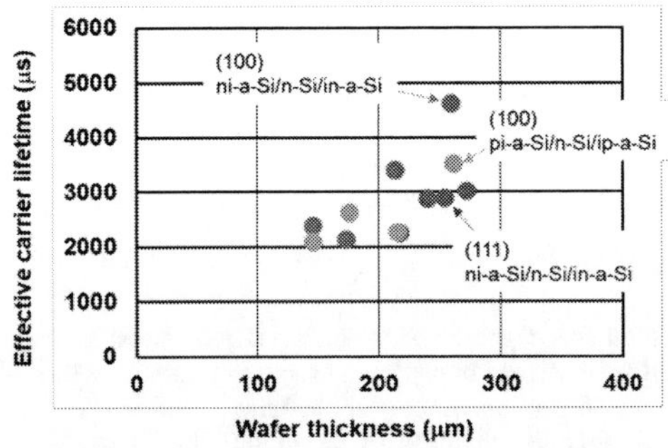

(a)Effective carrier lifetime measured by QSSPC.

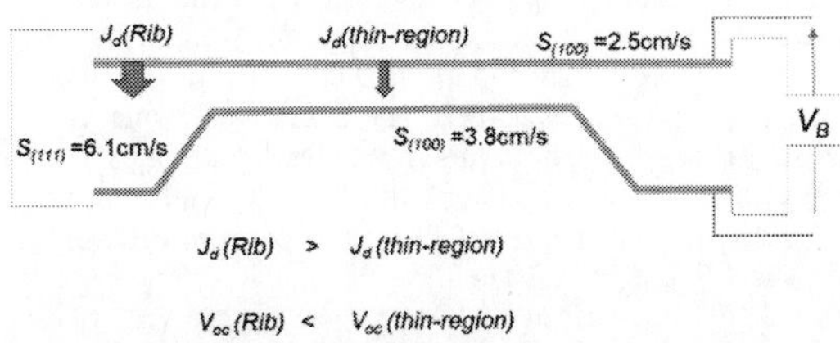

(b) Surface recombination velocity and dark current density in Rib region(*J_d(Rib)*) and in the thin-region(*J_d(thin-region)*).

Figure 9: Dark current density in the Rib and thin-region.

4.3 PL image

In the above, evaluation results mainly by EL and DLIT imaging are described. In addition, evaluation by PL imaging is also carried out as shown in Fig.10. The position of the line scan of PL intensity is the same as the position shown in Fig7.

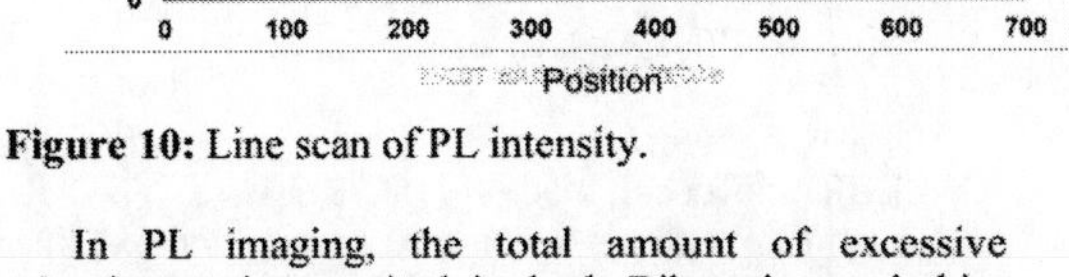

Figure 10: Line scan of PL intensity.

In PL imaging, the total amount of excessive minority carriers excited in both Rib-region and thin-region is just the same. On the other hand, due to surface recombination, the excess minority carrier concentration varies with the Si thickness. Furthermore, as the excess minority carrier concentration changes, the effective carrier lifetime changes, and so the anaysis becomes more complicated. By comprehensively analyzing data such as EL, PL imging and characteristic mapping such as Voc, Isc, it becomes possible to understand more clearly the device operation of the current Rib solar cell. Furthermore, detailed analysis will clarify factors limiting the conversion efficiency, and it is expected that important guidelines for achieving higher Voc will be obtained.

6 SUMMARY

In EL imaging, a very clear grid pattern originating from the Rib structure formed on the back side appears. EL intensity from Rib-region is 2.2 times larger than EL from thin-region. The difference in EL intensity corresponds to the difference of *Voc*, about 23 mV. DLIT imaging also showed that the dark current density in the Rig-region is higher than the thin-region.

In addition, evaluation by PL imaging is also carried out. In PL imaging, due to surface recombination, the excess minority carrier concentration varies with the Si thickness. As a result, it was found that PL image was very similar to EL image.

Acknowledgement
This study was supported by JST, MIRAI Project.

References

[1] K.Yoshikawa, W.Yoshida, T.Irie, H.Kawasaki, K.Konishi, H.Ishibashi, T.Asatani, D.Adachi, M.Kanematsu, H.Uzu, and K.Yamamoto, Solar Energy Materials and Solar Cells, 173(2017) 37
[2] International Technology Roadmap for Photovoltaic(ITRPV), Ninth Edition, March2018, 11
[3] A. Richter, M. Hermle, and S. W. Glunz, IEEE JOURNAL OF PHOTOVOLTAICS, VOL. 3, NO. 4, OCTOBER 2013, 1184
[4] M.Taguchi, A.Yano, S.Tohoda, K.Matsuyama, Y.Nakamura, T.Nishiwaki, K.Fujita, and E.Maruyama, IEEE Journal of Photovoltaics, 4 (2014) 96
[5] S. Yoshiba, M. Hirai, Y. Abe, M. Konagai, and Y. Ichikawa, AIP Advances 7(2), 025104 (2017).
[6] Y. Ichikawa, S. Yoshiba, M. Hirai, and M. Konagai, 2017 IEEE PVSC-44, Washington, D.C. (2017).
[7] O. Breitenstein, C. Shen, H. Kampwerth, M. A. Green, Energy Procedia, 38 (2013) 2-12
[8] He Zhang, Kazuyoshi Nakada, Makoto Konagai, Thin Solid Films, 628 (2017) 214
[9] M. Konagai, R. Kondo, T. Otani, S. Takahashi, K. Sawano, Y. Ichikawa, K. Nakada and Y. Kurokawa, WCPEC-7, Hawaii (2018)
[10] M.Konagai, R. Kondo, K.Sawano, and Y. Ichikawa, 27th Photovoltaic Science and Engineering Conf., Otsu, Japan (2017) 1ThPo.5.8

CHARACTERIZATION OF THE REVERSE BREAKDOWN INHOMOGENEITY OF ZEBRA IBC SOLAR CELLS

Stephan Großer[1,*], Martina Werner[1], Haifeng Chu[2], Christian Hagendorf[1]
1) Fraunhofer-Center for Silicon-Photovoltaics, Otto-Eißfeldt-Strasse 12, 06120 Halle (Saale), Germany
2) International Solar Energy Research Center Konstanz e.V., Rudolf-Diesel-Strasse 15, 78467 Konstanz, Germany
*) Corresponding Author: Tel: +49-345 5589-5112; Fax: +49-345 5589-5999; mail: stephan.grosser@csp.fraunhofer.de

ABSTRACT: The ZEBRA cell concept realizes low cost and large area interdigitated back contact (IBC) silicon solar cells. Furthermore, the concept allows a homogeneously distributed reverse bias breakdown at low voltages over the total cell area. Even without bypass diodes no severe hot spots will occur under partial shadings condition in a module. Hence, the homogeneity of the power dissipation is crucial for the reliability of the module under partial shading conditions. Root causes of non-uniform power generation under reverse condition must be understood and controlled. In this paper the macroscopic breakdown behavior and microscopic properties of the sample structure are compared by means of DLIT, reverse-bias EL and SEM/EBIC as well as ToF-SIMS. The microstructure and the local diffusion profiles of a IBC cell with non-uniform reverse breakdown have been analyzed separately in terms of correlation to the reverse breakdown. It is know that the local formation of high doping gradients at p^+n^+-junction leads to a low reverse breakdown voltage. Microscopic reverse-bias EL was used as a fast technique for identification of non-uniform p^+n^+-junction formation. On microscopic scales the structural roughness contributes to a spatially non-uniform generation of highly doped p^+n^+-junctions. However, on larger scales the breakdown behavior can be controlled through the dopant diffusion process. Our results show that this effect originates from different boron diffusion profiles which results in a variation of the boron sheet resistance on a cell.
Keywords: Solar Cell; Characterisation; Defects; Doping

1 INTRODUCTION

The ZEBRA interdigitated back contact (IBC) cell is a cell concept for high efficiency at low production costs [1]. Industrial processes proven in mass-production are used to produce on 6 inch n-type Cz-wafers IBC cells with efficiencies up to 23.1% [2]. Furthermore, the ZEBRA IBC cell concept includes the benefits of bifaciality with a bifaciality factor of 76% [1]. The IBC solar cell has highly doped p+ and n+ areas on the rear side. The contact between the highly doped areas form a p^+n^+-junction as shown in Fig. 1 for the ZEBRA cell concept. These junctions possess the property that at reverse-bias conditions a low breakdown voltage $|V_{bd}| < 6$ V can be achieved. This characteristic is attributed to band to band tunneling and/or avalanche breakdown [3].

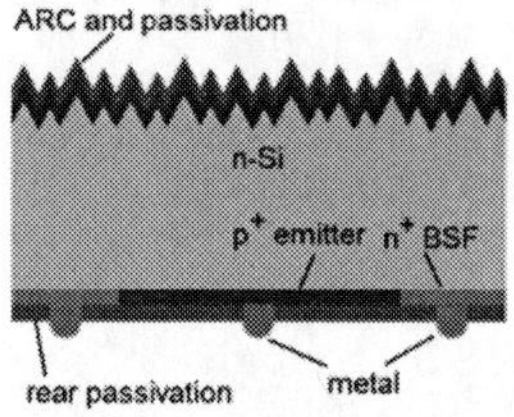

Figure 1: Schematic model of a ZEBRA IBC solar cell.

In other cell concepts like PERC a high $|V_{bd}|$ is desired to avoid hot spots due to e.g. defects [4, 5]. The IBC concept does not exhibit hot spots since the p^+n^+ junction is formed on the full rear side of the cell in a pattern which is determined by the structuring processes. Thus, power dissipation at reverse-bias conditions will be distributed over the full cell area resulting in low current densities and lower temperatures compared to a local hot spot. Accordingly for IBC cells the homogeneity of the reverse breakdown behavior plays an important role to guarantee hot spot free solar cells.

This paper focuses on the elucidation of the observed local inhomogeneity of the reverse breakdown property on a ZEBRA IBC cell. The first part deals with the investigation of macroscopic properties and the microstructure of the cell to reveal their impact on the p^+n^+-junction formation. Microscopy methods like scanning electron microscopy (SEM) show information about the formation of the interdigitated contact structure at the highly doped p^+n^+-junction area. Dark lock-in thermography (DLIT), reverse bias electro-luminescence (ReBEL) as well as electron-beam induced current measurements (EBIC) are performed to identify the position and the homogeneity of the formed p^+n^+-junction. In the second part of the paper the root cause of the inhomogeneous reverse breakdown will be analyzed by a comparison of areas with low and with high breakdown voltages on the same solar cell. Microstructure as well as local doping profiles will be utilized and allow clear conclusions on the break-down behavior for the investigated solar cell.

2 EXPERIMENTAL DETAILS

The experiments are conducted using a 6 inch n-type Cz ZEBRA solar cell made by ISC Konstanz. The solar cell was made with a non-standard process and exhibited a clear non-uniform reverse breakdown on the cell scale. Power dissipation imaging on the full solar cell scale is carried out by dark lock-in thermography (DLIT) measurements using a LOANA solar cell analysis system from pv-tools. Statistical information of the rear side microstructure on the full scale is measured by optical microscopy.

After finishing the analysis on the full scale solar cell regions of interest (ROI) with different reverse breakdown behavior, have been precisely separated from the solar cell by a laser cut process. This allows a detailed investigation on mini-solar cell formats for EBIC/SEM methods. Reverse bias electroluminescence images are recorded using a Keithley source meter unit and a home-build infrared-optimized microscope equipped with a Si-CCD camera from Andor. For visualization of the grid finger metallization a negligible illumination by external light is used which can be clearly distinguished from the ReBEL emission. Electron-beam induced current (EBIC) measurements are carried out on a scanning electron microscope (FE-SEM, Hitachi SU70) equipped with a DISS5 EBIC system (point electronic). The local doping profiles on the solar cell are measured by means of time-of-flight secondary ion mass spectrometry (ToF-SIMS) using a IONTOF TOF.SIMS 5.

3 RESULTS AND DISCUSSION

3.1 Homogeneity of the reverse breakdown

An ideal reverse breakdown for the ZEBRA cell is homogeneously distributed over the full cell area. For imaging of the reverse breakdown the ReBEL and the DLIT technique are commonly used [4, 6]. Bias-dependent measurements allow the evaluation of the homogeneity and localization of areas with low and high breakdown voltages (V_{bd}). Figure 2 represents DLIT measurements of the ZEBRA solar cell obtained at 2 different voltages where (a) is below and (b) around V_{bd}. Shunting defects at the solar cell edge dominate at voltages below V_{bd} (Fig. 2a) whereas at a bias around V_{bd} (Fig. 2b) the reverse breakdown at the highly-doped p^+n^+-junction occurs. For the given solar cell it can be clearly seen in Fig. 2b that the reverse breakdown is inhomogenously distributed. In comparison, the upper part of the cell is dominated by a low $|V_{bd}|$ and the lower part by a high $|V_{bd}|$.

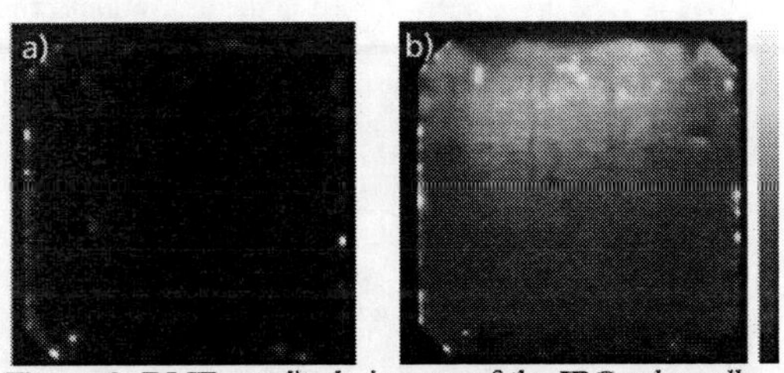

Figure 2: DLIT amplitude images of the IBC solar cell taken at a) -2V/-0.06 A and at b) -4.6 V/-8 A.

3.1 Impact of the microstructure on the reverse breakdown

Since the DLIT image taken from the full solar cell do not resolve the typical IBC pattern of the p^+n^+-junction (see Fig. 1), high-resolution voltage dependent ReBEL images from the rear side of the solar cell have been recorded. Figure 3 shows an ReBEL image from the lower part of the cell in figure 2 at the area with high $|V_{bd}|$. For better orientation, the grid finger structure was colored blue. The emission of light due to the reverse breakdown proceeds directly at the horizontally extended p^+n^+-junction. Furthermore, one can see that the light emission along the junction is interrupted and not uniform on the microscale. It is known from previous microstructure investigations (not shown here) that the roughness of the structure between base and emitter depends strongly on the production process parameters.

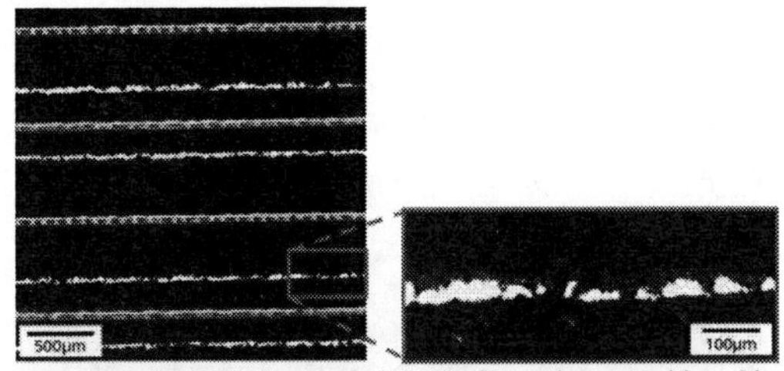

Figure 3: ReBEL image (-4.5 V) from the rear side with white colored light emission at the p^+n^+-junction (with magnification) and blue colored grid finger structure.

A detailed image of the microstructure and the corresponding p^+n^+-junction position can be obtained by SEM/EBIC technique. Figure 4 exhibits the overlay of a SEM image (grey colored) and the EBIC image (red). The brightness of the red color correlates with the EBIC current. On the lower left hand side the passivation layer was etched underneath (overlap) and suppressed the current generation locally. The expected trace of the p^+n^+-junction (see figure 1) is marked with a white dashed line. As can be seen, the junction position does not follow this simplified model. Thus, a potential impact on the p^+n^+-junction break-down behavior should be considered. Other measurements (not shown here) indicate that the under etching width (overlap) positively correlates with the deviation of the observed junction from the expected position. Accordingly, the microscopic width of the under etching overlap is measured as a potential parameter impacting the breakdown behavior as discussed below (Fig. 5b).

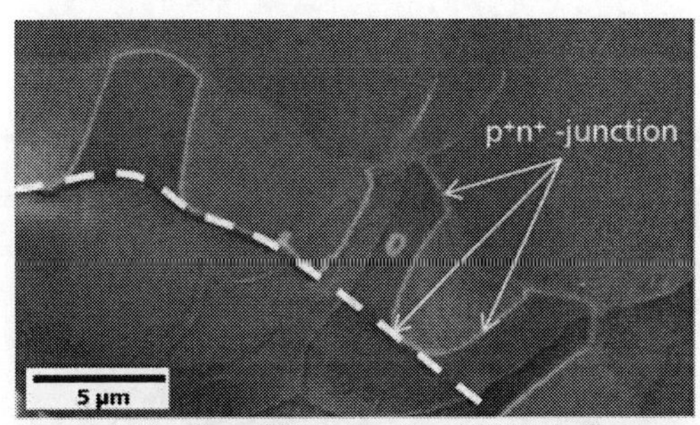

Figure 4: Overlay of SEM (grey) and EBIC (red) map from the rear side of the IBC. For better orientation the expected trace of the p^+n^+- junction is marked with a dashed line.

In order to correlate microstructural properties with the macroscopic homogeneity of the breakdown behavior, 9 regions of interest (ROI) were defined as shown in Fig. 5a with the blue frames. The ROIs represent areas of the solar cell with different breakdown voltages (low vs. high). The overlap width of the under etched passivation layer was measured at multiple positions for each ROI. Figure 5b shows the median of the overlap width for each ROI in a color value, assigned to the corresponding squared ROI on the total solar cell area. As can be seen, ROI 3/1 (column 3 vs. row 1) and ROI 1/3 show the same overlap width but a different reverse breakdown behavior in the DLIT image (Fig. 5a). Furthermore, along row 3 the overlap width increases from 22 µm to 31 µm but does not correlate with the observed reverse breakdown homogeneity for this positions.

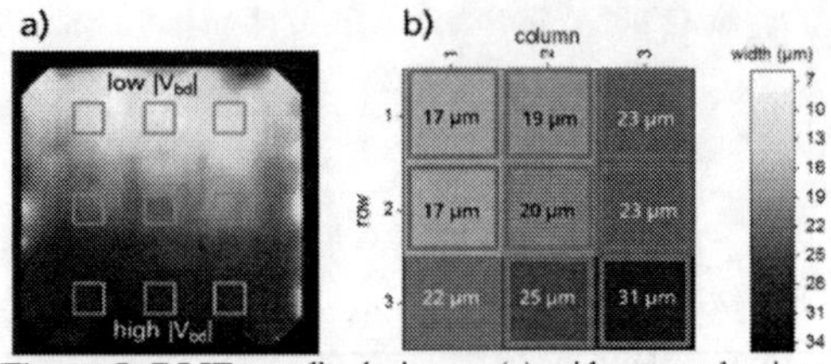

Figure 5: DLIT amplitude image (a) with area selections (blue boxes) for determination of the overlap width (b) from the passivation layer on 9 position on the cell.

Hence, it is concluded that the microstructural properties do not interact significantly with the reverse breakdown characteristic.

3.3 Impact of the boron diffusion on the reverse breakdown behavior

Rear contact patterns of p- and n-type doped areas on a IBC solar cell show lateral distances down to the sub-millimeter size. Therefore a very local measurement technique for doping profiles is needed to obtain the information precisely at the desired doped position. ToF-SIMS measurements on an area of ~100 µm x 100 µm have been conducted. Doping depth profiles in the p (boron doped) and the n (phosphorus doped) areas have been recorded separately for the ROI (1/1) showing low $|V_{bd}|$ and ROI (1/3) showing high $|V_{bd}|$ (see fig. 5). In Figure 6a the obtained doping profiles for phosphorus on both ROIs are presented in a logarithmic plot. The phosphorus concentrations versus the depth are identical for both ROI. In contrast to the measured boron depth profiles which are shown in Figure 6b.

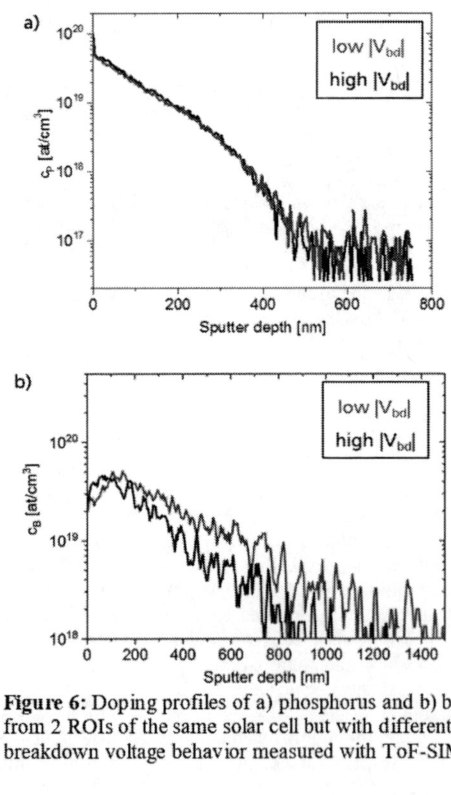

Figure 6: Doping profiles of a) phosphorus and b) boron from 2 ROIs of the same solar cell but with different breakdown voltage behavior measured with ToF-SIMS.

The profiles for both ROIs differ depending on the depth. A low breakdown voltage ROI exhibits a higher boron concentration for increasing depth. The surface boron concentrations are comparable. In terms of the related boron sheet resistance the area with the low $|V_{bd}|$ has a lower sheet resistance than the area with the high $|V_{bd}|$.

An independent experiment (not shown here) with ZEBRA IBC test structure manufactured with 2 different boron sheet resistances (by double boron-diffusion on the half of the cell) confirmed the observation from the ZEBRA solar cell that the sheet resistance of boron doped areas has a strong impact to the reverse breakdown voltage. Lower sheet resistances on the cell result in a lower breakdown voltage compared to areas with higher sheet resistance. The observed behavior is explained well by the theory that higher doping concentration gradients lead to lower breakdown voltages [3]. The breakdown mechanism by tunneling of charge carriers or avalanche multiplication depends on the doping concentration and the width of the space charge region. An increased doping concentration will enhance e.g. the tunneling probability at the p^+n^+-junction for low reverse voltages.

4 SUMMARY

The fundamental processes causing an inhomogeneous distribution of the reverse breakdown voltage on a ZEBRA IBC solar cells have been investigated by a detailed microstructure diagnostics approach using correlated high-resolution imaging (EBIC/SEM) and mass-spectrometry (TOFSIMS) techniques. Influences of the surface morphology on the shape of the p^+n^+-junction as well as from the boron doping process have been studied in detail. An impact of the ZEBRA IBC cell process induced undercut behavior to the inhomogeneity of the reverse breakdown voltage on cell scale can be excluded. In contrast, a correlation with the local boron depth-profile due to the diffusion process is shown. An increased boron concentration (deep emitter) was found in the area with low reverse breakdown voltage whereas a lower boron concentration (shallow emitter) was found in the area with higher reverse breakdown voltage. The findings confirm the theoretically expected behavior of strong doping gradients at highly-doped pn-junctions indicative for an avalanche or tunneling mechanism. The insight of the main impact parameter to the reverse breakdown voltage resulted in optimized boron-doping process for ZEBRA cell technology which can be verified through the uniformity of the sheet resistance values in production quality control.

5 ACKNOWLEDGEMENTS

We like to thank Susanne Richter, Volker Naumann, Sina Swatek and Stephan Eiternick for their contributions with measurements. Financial support by the German Federal Ministry for Economic Affairs and Energy within project "Processes for ZEBRA" (No. 0325840D) and Solar-Era-Net-Project "U-Light" (No: 0325886A) is gratefully acknowledged.

6 REFERENCES

[1] A. Halm, V. Mihailetchi, G. Galbiati, L. Koduvelikulathu, R. Roescu, C. Comparotto, R. Kopecek, K. Peter, J. Libal; In Proc. 27th Eur. Photovoltaic Sol. Energy Conf. (2012) 567-570
[2] G. Galbiati, H. Chu, V. D. Mihailetchi, J. Libal, and R. Kopecek, Presented in 7th World Conference on Photovoltaic Energy Conversion (WCPEC-7), Waikoloa, Hawaii, USA, (2018)
[3] S. M. Sze and K. K. Ng, Physics of Semiconductor Devices, John Wiley & Sons, Inc., New Jersey (2007)
[4] F. Fertig, S. Rein, M. C. Schubert, W. Warta, In Proc. 26th European Photovoltaic Solar Energy Conference and Exhibition (2011) 1168-1178
[5] S. Wendlandt, A. Drobisch, T. Buseth, S. Krauter and P. Grunow, In Proc. 25th European Photovoltaic Solar Energy Conference and Exhibition (2010) 4002-4006
[6] D. Lausch, K. Petter, R. Bakowskie, C. Czekalla, J. Lenzner, H. von Wenckstern and M. Grundmann, Applied Physics Letters 97 (2010) 073506

EVALUATION OF OXIDIZED SILICON SURFACES WITH LASER TERAHERTZ EMISSION MICROSCOPE (LTEM) AND CORONA CHARGING

T. Mochizuki[A], A. Ito[B], H. Nakanishi[B], K. Tanahashi[A], I. Kawayama[C], M. Tonouchi[C], K. Shirasawa[A], and H. Takato[A]

[A] Fukushima Renewable Energy Institute, AIST, Japan
[B] SCREEN Holdings Co. Ltd., Japan
[C] Institute of Laser Engineering, Osaka University, Japan

[A] 2-2-9 Machiikedai, Koriyama, Fukusima, 963-0298 Japan
[B] 322 Furukawacho, Hazukashi, Fushimi-ku, Kyoto 612-8486 Japan
[C] 2-6 Yamadaoka, Suita, Osaka 565-0871, Japan

ABSTRACT: Surface band bending evaluation by assessing Terahertz (THz) emission from silicon wafers using a laser terahertz emission microscope (LTEM) is systematically and quantitatively compared with the non-contact voltage-charge (V-Q) method which utilize a Kelvin probe. The amplitude of THz emission consistently changed with the surface barrier voltage Vsb of non-contact V-Q when the surface was depleted or weakly inverted. The deposited charge density required to zero THz amplitude was close to Q_{tot} of noncontact C-V, the charge needed to move the surface from initial condition to the flatband. When the surface had accumulation layer, the THz emission showed sensitivity to the surface charging, while V_{sb} = 0 and was insensitive to the surface charging. The amplitude of THz emission saturated when the surface was strongly inverted or accumulated. The results show that the LTEM can be a tool similar to the non-contact V-Q in combination with a corona charging. Furthermore, the sensitivity of THz emission to the both of depleted and accumulated surface suggests that the LTEM can be a tool to evaluate surface field of silicon surfaces to which corona charging is not applicable.
Keywords: Terahertz emission, Silicon photovoltaics, Characterization, Surface passivation, Silicon oxide

1 BACKGROUND

Surface Passivation is an essential part of design of high efficiency solar cells. Various materials including a-Si:H, SiO, SiN and AlO are deposited to fabricate passivation layer on the silicon substrate[1]. The quality of these layers as a part of solar cell is evaluated by effective surface recombination velocity. Combination of reduction of density of surface states which act as surface recombination centers and field-effect passivation determine the surface recombination velocity. When passivation layer of the solar cell is contaminated or faulty, the surface state density increases or the surface field do not repel the minority carriers. While the existence of the faults can be detected by many in-line capable evaluation tools including electroluminecence, photoluminescence, and LBIC, evaluation of surface field besides the surface state density is a good tool to investigate the nature of the contamination and contributes to the optimization of fabrication processes. Field-effect passivation is evaluated by measuring surface band bending with capacitance-voltage characteristics or electron-based spectroscopy. However, the former requires the passivation layers to be good insulators and the latter needs a complex setup. We have proposed using terahertz (THz) emission spectroscopy as a new, simple tool that can directly measure and map surface band bending[2–7]. Previously, we have compared the laser terahertz emission of a silicon metal-oxide semiconductor device with ITO semi-transparent electrode with the capacitance-voltage method and showed that the amplitude of THz emission sub-linearly depended on the surface potential and the waveform of THz emission inverted at the flatband condition[8]. With these results the THz emission from silicon surface is a measure of direction and relative strength of surface band bending. However, ITO electrode absorbs a significant part of THz emission and its emissivity cannot be accurately evaluated and the relation between the surface potential and the amplitude of THz emission cannot be used as a calibration curve for the measurements on the passivated silicon surfaces without ITO electrodes. In this work, we utilized corona charging to tune the surface band bending and measured the amplitude of THz emission from an oxidized n-type silicon substrate. The observed charge density dependence of the THz amplitude was consistent to all of the surface barrier voltage V_{sb}, which was measured with surface photovoltage (SPV) method, and effective minority carrier lifetime, showing that the change of surface potential of oxidized silicon surface without surface electrode can be measured by assessing THz emission.

2 EXPERIMENT

Figure 1 shows the schematic of the sample and the illustration of the measurements done in this work. The sample was a CZ n-type silicon monocrystalline substrate with a measured resistivity of 1.7 Ω cm, corresponds to 2.9 × 10^{15} cm^{-3} doping density. The substrate was thermally oxidized at 900 °C for 90 minutes and n = 1.48 and thickness of 79.8 nm were measured using an ellipsometer. Then the sample was annealed in 4% forming gas at 450 °C for 40 minutes. Negative or positive ion with various charge density was deposited by corona charging, and then the sample was transferred to the LTEM and the time-domain spectra of THz emission was measured. Initially, we can expect that positive fixed charge in the oxidized layer causes downward band bending and electron accumulation at the oxide-semiconductor interface of the silicon. When the sample is excited by laser pulse, the photocarriers are accelerated by the band bending to generate transient current toward the surface. This transient current is the source of the THz emission. When negative ion is deposited on the substrate by corona charging and the deposited charge density exceeded the positive charge density within the oxidized layer, the surface is depleted and the band bend upwards.

After the corona charging, photocarriers are accelerated to generate transient current against the surface, causing the waveform inversion of the THz emission. The amplitude of the THz emission is also dependent of the strength of the surface band bending i.e. the surface potential. The surface barrier voltage was measured using a kelvin probe as difference between contact voltage difference with and without strong illumination. The effective minority carrier lifetime $\tau_{eff.d}$ at 1-sun excitation was also measured with QSS-μPCD technique. LED-excited photoluminescence image was also measured to locate the center of the spot of the charge deposition and the motored sample stage was used to focus the excitation laser of THz time-domain spectroscopy measurements at the center of the charging spot.

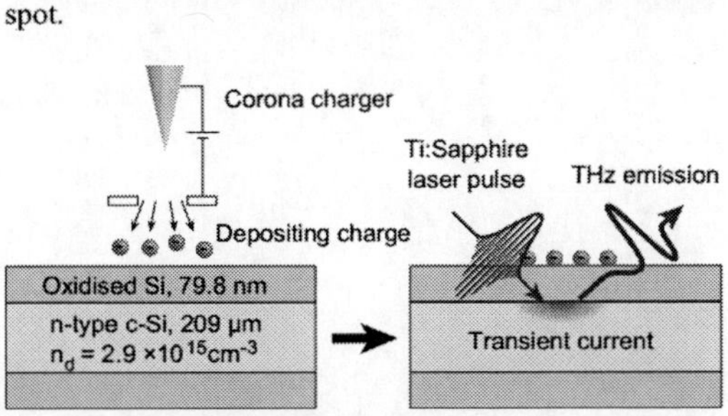

Figure 1: Illustration of the sample and the measurements. The THz time-domain spectroscopy measurements are done using the laser-terahertz emission microscope (LTEM).

3 RESULTS

Figure 2(a) shows the waveforms of THz emission from the sample with deposited charge density $\Delta Q_c$from -7.5 to 3.0 × 10^{11} cm^{-2}. The waveforms inverted showing that the initially accumulated surface changed to inverted surface by deposited negative charge. Figure 2(b) shows the THz amplitude averaged over the first peak, as indicated in the yellow rectangle in Fig. 2(a), as well as surface barrier voltage V_{sb}, measured with the corona-Kelvin non-contact V-Q. The THz amplitude crossed zero at -2.0 × 10^{11} cm^{-2}. This is consistent to Q_{tot} = -2.31 × 10^{11} cm^{-2} of the V-Q measurement. Fig. 2(c) shows the measured Vsb and the calculated surface potential s, neglecting the effects from interface trapped charge. V_{sb} is well fitted by 0.892 Ψ_s. This shows that the measured V_{sb} is a good measure of the surface voltage when the surface is depleted or inverted. Fig. 2(d) shows charge density dependence of the THz peak amplitude and minority carrier lifetime measured with the QSS-μPCD technique. We also observed a dip of minority carrier lifetime at -4.0 × 10^{11} cm^{-2} and we can expect that the lifetime dropped here because Fermi energy at the surface crossed the midgap under 1-sun illumination. This is reasonably consistent to s and V_{sb}, which indicate that Fermi level of the surface in the dark crossed the midgap at ΔQ_c = -3.0 × 10^{11} cm^{-2}. The observed results indicate that LTEM is a good measure of the surface voltage of silicon, especially if the surface is weakly inverted, depleted, or weakly accumulated. The amplitude of the THz emission was sensitive to both of positive and negative surface voltage, in contrast to the V-Q measurements.

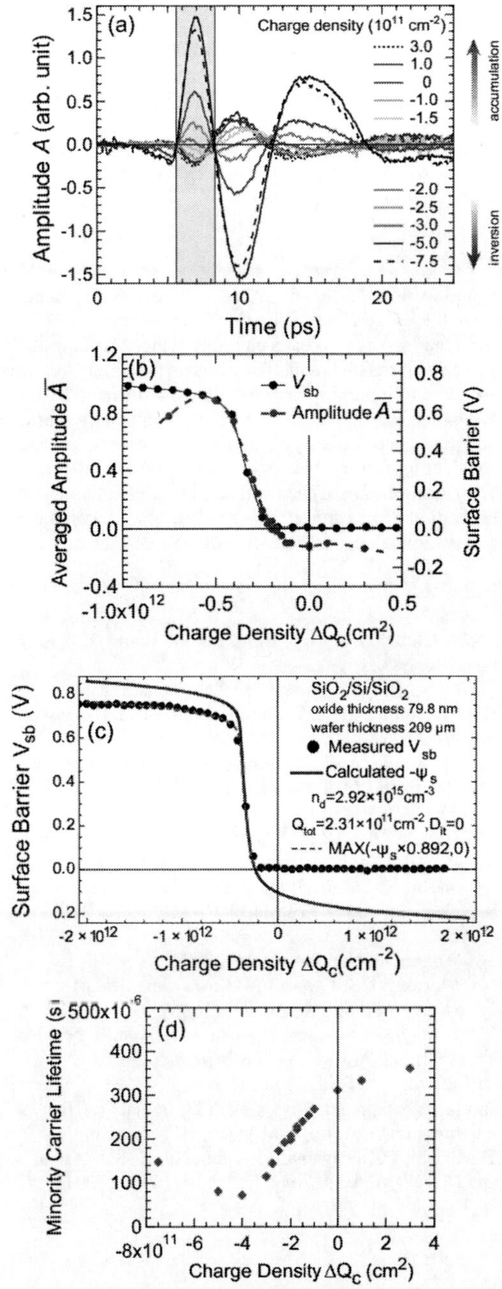

Figure 2: (a) Waveform of THz emission with various deposited charge density ΔQ_c. (b) THz amplitude averaged over the first peak and surface barrier voltage. (c) Measured surface barrier voltage and calculated surface potential at various ΔQ_c. (d) Charge density dependence of minority carrier lifetime.

4 DISUSSION

The observed amplitude of THz emission strongly correlated with the surface potential determined by the voltage charge method in the depleted or inverted region and was also strongly correlated with the calculated surface potential in the accumulation region. This suggests that LTEM can be a tool to evaluate surface potential besides the factors such as interface trap density and doping density. The obtained relationship can also be a very good reference when we observe silicon with other passivation layer such as a:Si with LTEM because the present measurements were done on the bare oxidized silicon without optical interference and can be quantitatively compared with the results from other silicon surfaces with passivation layers.

5 SUMMARY

To summarize, we measured waveforms of laser THz emission from Si-SiO interface without MOS structure, using Laser Terahertz Emission Microscope (LTEM) in combination with a corona charging setup. THz amplitude strongly correlated with the surface potential and was sensitive to to both of accumulated and depleted surface. We expect calibration of LTEM for evaluation of surface potential can be done by expanding the present experiment considering factors such as dope density and reflectance. With a calibration, LTEM can also be an effective tool for evaluation of surface potential with a conductive passivation layer, such as a-Si, silicon-rich SiN, or AlO.

REFERENCES

[1] S.W. Glunz, D. Biro, S. Rein, and W. Warta, J. Appl. Phys. **86**, 683 (1999).
[2] H. Nakanishi, S. Fujiwara, K. Takayama, I. Kawayama, H. Murakami, and M. Tonouchi, Appl. Phys. Express **5**, 112301 (2012).
[3] H. Nakanishi, A. Ito, K. Takayama, I. Kawayama, H. Murakami, and M. Tonouchi, AIP Adv. **5**, 117129 (2015).
[4] N. Ohta, K. Arimoto, M. Shiraga, K. Ishii, M. Inada, S. Yanai, Y. Nakai, H. Akiyama, T. Mochizuki, T. Takahashi, N. Takahashi, H. Miyagawa, N. Tsurumachi, S. Nakanishi, and S. Koshiba, J. Cryst. Growth **378**, (2013).
[5] M. Tonouchi, Nat. Photonics (2007).
[6] H. Murakami, S. Fujiwara, I. Kawayama, and M. Tonouchi, Photonics Res. **4**, A9 (2016).
[7] Y. Hotta, I. Kawayama, S. Miyake, I. Saiki, S. Nishi, K. Yamahara, K. Arafune, H. Yoshida, S. Satoh, N. Sawamoto, A. Ogura, A. Ito, H. Nakanishi, M. Tonouchi, and H. Tabata, Appl. Phys. Lett. **113**, 012103 (2018).
[8] T. Mochizuki, A. Ito, J. Mitchell, H. Nakanishi, K. Tanahashi, I. Kawayama, M. Tonouchi, K. Shirasawa, and H. Takato, Appl. Phys. Lett. **110**, 163502 (2017).

DETERMINING THE THERMAL BEHAVIOUR OF ENCAPSULANTS AND ITS IMPACT ON ENERGY YIELD

Ibrahim El-chami[1,2,3], Herman Oprins[4], Vladimir Cherman[4], Imre T. Horváth[2], Hans Goverde[2], Jonathan Govaerts[2], Tom Borgers[2], Eszter Voroshazi[2], Jef Poortmans[1,2,5]
[1]KULeuven-ELECTA, Kasteelpark Arenberg 10, 3001 Leuven, Belgium
[2]imec-Energyville, Thorpark 8320, 3600 Genk, Belgium
[3]Simon Fraser University Mechatronics Systems Engineering, Surrey, BC, Canada
[4]imec, Kapeldreef 75, 3001 Heverlee, Belgium
[5]also at UHasselt, Belgium

ABSTRACT: In this paper, the use of transient thermal measurements inspired by the micro-electronic industry is demonstrated. The transient thermal measurement method entails applying pulsed power through a thermal test chip, with integrated heaters and temperature sensors, laminated on top of a polyolefin elastomer encapsulant layer and a silicon substrate, mimicking the PV module buildup. Measurements of the transient temperature response show that the thermal response variations in the thermally insulating encapsulant material are proportional to the thickness of the material. In addition, introducing direct metal contact the cell surface causes a higher thermal response than only embedding metal in the encapsulant without contact to the cell. Following preliminary transient temperature response measurements of the polyolefin encapsulant, we ran simulations to represent the impact of thermal properties of encapsulant on the overall impact of energy yields in PV modules. Simulation results show that altering encapsulant material from polyolefin to EVA, as well as varying polyolefin thickness within the PV module, have minor impact on the overall energy yields in moderate and desert climates.

Keywords: thermal behaviour, encapsulant, multi-wire

1 BACKGROUND

Thermal effects within a PV module and its surrounding environment play a significant role in the performance of the solar cell as well as the overall energy output of the module. Thermal effects within the PV module include the thermal response of the solar cell itself as well as the thermal response of the material composing the module setup. The performance of solar cells decreases with the increasing temperature of the cell, as thermally excited electrons begin to dominate electrical properties of the semiconductor. The solar cell's open circuit voltage (Voc) and fill factor (FF) decrease substantially with temperature, while short circuit current (Isc) increases slightly [1].

In addition, a number of studies have shown thermal ambient effects on PV cells' performance [2,3] as well as thermal effects on degradation or discoloration of the encapsulant material and the subsequent effect on cells' performance [4]. Furthermore, there is an increasing interest in the use of multi-wire interconnection concepts and schemes in c-Si PV applications, mainly due to their demonstrated optical and electrical potential in decreasing shading due to metallization and resistive transport losses [5,6,7,8,9,10].

The transient thermal measurement method was initially developed to characterize the bulk thermal conductivity and the thermal contact resistance of underfill materials in die-die interfaces in 3D stacked structures for 3D-integrated circuits, as it was combined with thermal finite element simulations [11]. In this paper, we utilize the transient thermal measurements of the encapsulant, using a thermal test chip with integrated heaters and temperature sensors, to study the thermal response variations in the thermally insulating encapsulant material, polyolefin. The samples are prepared to represent the exact layup of solar cell modules, while varying parameters such as encapsulant thickness, in addition to varying the position of the wire interconnections within the encapsulant.

The method combined with thermal finite element simulations could potentially be used to characterize the bulk thermal conductivity of the ecanpsulant material and the thermal contact resistance at the different interfaces.

Following transient thermal measurements, we simulated the impact of thermal properties of encapsulant on the overall impact of energy yields in PV modules in moderate and desert climates. Oldenburg, Germany, is used to represent a city with moderate climate, and Kuwait City, Kuwait, is used to represent a desert climate. In this work, the impact of varying some encapsulant properties, such as altering encapsulant material from polyolefin to EVA as well as varying polyolefin thickness, on the overall impact of energy yields in PV modules is studied.

2 EXPERIMENTAL SETUP

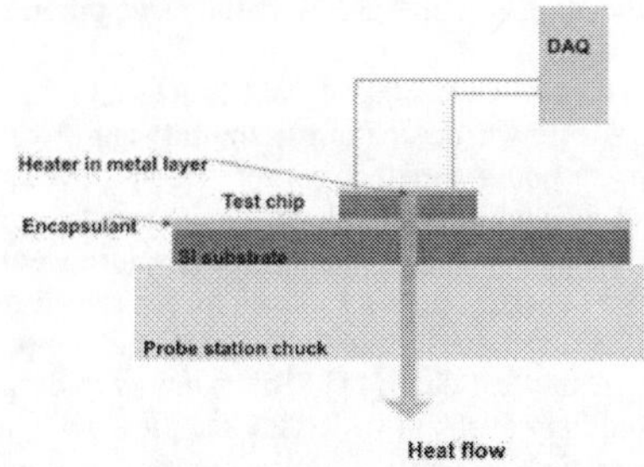

Figure 1: schematic of the experimental setup.

Figure 1 shows the experimental setup. Samples were prepared by laminating polyolefin elastomer encapsulant in between a thermal test chip and a silicon substrate(800μm thick). The samples are placed on the probe-station chuck. A probe station and a data acquisition microcontroller (National Instruments DAQ) (connected to a computer) are used to (1) apply a constant power to the heating elements of the chip, (2) supply a constant current through the temperature sensor resistors, and

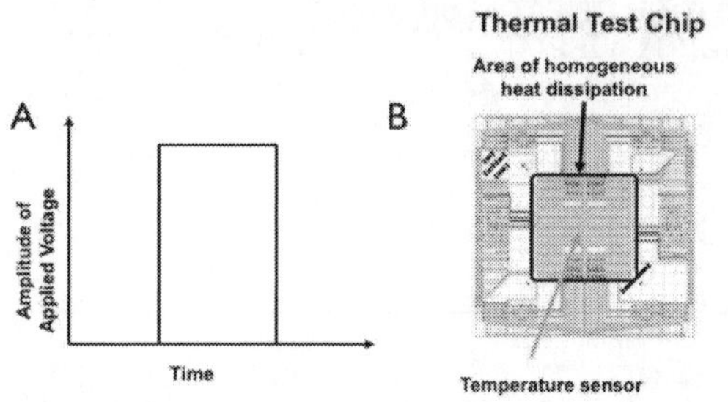

Figure 2: A. Square wave of constant-amplitude voltage is applied to samples at a period of 2τ, with τ being varied between one second and 30 seconds. B. The layout of the thermal chip showing heater resistors and meander resistors acting as temperature sensors.

(3) collect voltage measurements from the latter resistors, illustrated in Figure 2. Figure 3 shows the schematic cross-section of each sample: samples with bare encapsulant and varying thickness, figure 3.(A-C), encapsulant with metal wires completely decoupled from the samples, figure3.D, encapsulant with metal wire in contact with the cell figure 3.A, and metal wires in the glass fiber matrix, figure 3.F.

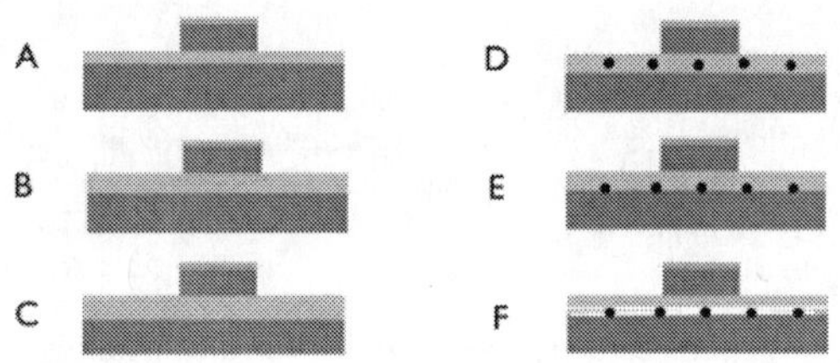

Figure 3: (A-C). Samples with increased encapsulant layer thicknesses. D. Metal interconnections are embedded in encapsulant. E. Metal interconnections are embedded in encapsulant while in contact with the substrate. F. Fiber-woven metal interconnections are embedded in encapsulant while in contact with the substrate. The interconnections samples E-F have the same pitch.

Voltage is applied as a square wave function to the heater resistors for power dissipation in the test chip, while the transient response of the sensor resistor voltage is measured (Figure 2.B). The voltage variations at the sensor resistor yield a local temperature, and the temperature coefficient of resistance for the sensor resistor material is used to calculate the average temperature increase in the test chip [11]. The samples undergo one heating-cooling cycle, in which the samples are heated up for a specified amount of time, figure 2.A, then, the samples cool down. The thermal impedance Zth is calculated as [11].

$$Z_{th} = \frac{V_{d,0} - V_d(T)}{\sigma . P} \quad (1)$$

where $V_{d,0}$ is the initial voltage on the sensor at the chuck temperature T0, Vd(T) is thesensor voltage at temperature T, σ is the resistor sensitivity, and P is the dissipated power. The dissipated power is obtained as follows:

$$P = V_{applied}\ I_{measured} \quad (2)$$

where $V_{applied}$ is the voltage applied to the heater and $I_{measured}$ is the measured current of the heater resistor.
In addition, the top side of the chip is covered with thermally insulating material, thus causing the heat generated at the chip to preferentially flow through the encapsulant towards the chuck, which acts as a heat sink. A reference measurement, where the test chip is directly placed on top of the substrate (i.e. with no encapsulant present) is used to determine ambient heat losses and convectional heat dissipations.

In order to simulate energy yield impacts of varying encapsulant material and properties, especially under a desert climate, accurate modelling of material properties and arrangements are essential in order to obtain accurate thermal management within the module. Thus, physics-based exploratory modelling of PV energy conversion was used for the simulation. The modeling takes into account previously measured environmental conditions, consisting of a time series of irradiation (global horizontal, direct normal, diffuse horizontal), ambient temperature, wind speed, and direction. In addition, cell electrical and thermal behavior, optical absorption and generation, as well as thermal conduction, convection and radiation are set by the material properties of the different layers. A full description and demonstration of this modelling approach can be found in [12].

3 LAYOUT SPECIFICATIONS

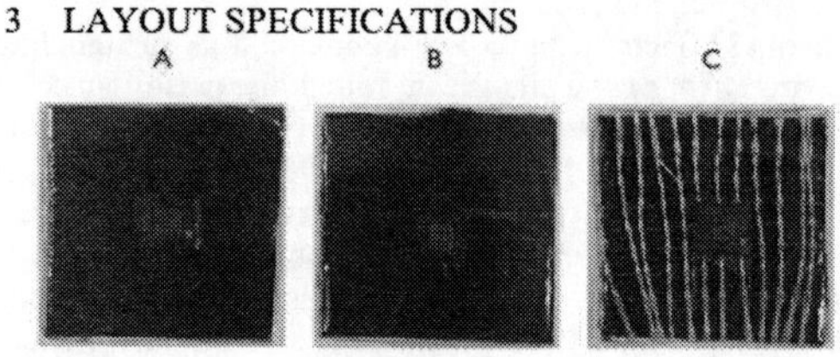

Figure 4: Some of the prepared samples. A. no metal interconnections. B. Metal interconnections embedded in encapsulant. C. Fiber-woven metal interconnections embedded in encapsulant.

Figure 4 shows some of the prepared samples, and Figure 5 shows the outcome of the transient measurements where the thickness of the encapsulant was increased (sample types A-C).

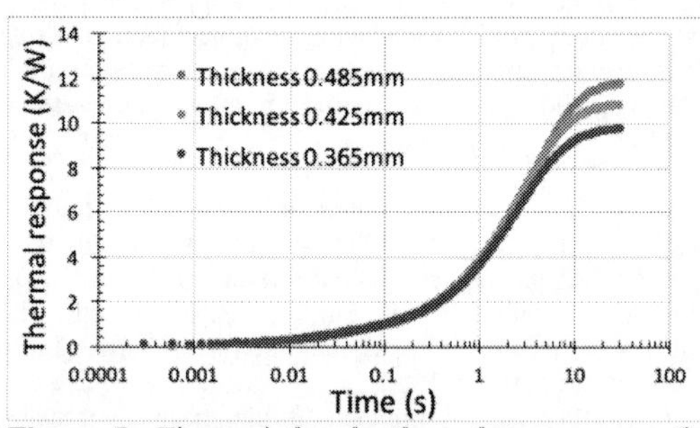

Figure 5: The variation in thermal response as the encapsulant thickness is varied.

The increase of steady-state temperature as the thickness of the thermally insulating encapsulant increases is expected as explained by equation 3, which states the relation between thermal resistance and material thickness.

$$R = \frac{\Delta T}{QL} \quad (3)$$

where ΔT is the change in temperature, Q is the heat flow rate, and L is the thickness of the material.

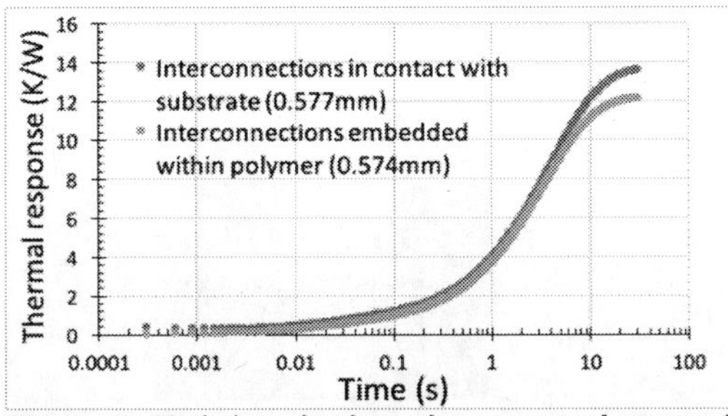

Figure 6: Variations in thermal response when metal interconnects are embedded in encapsulant (sample type D) or in contact with the substrate (sample type E).

Introducing metal interconnections results in different thermal responses depending on whether the metal interconnections are embedded within the encapsulant or are in direct contact with the Si substrate.

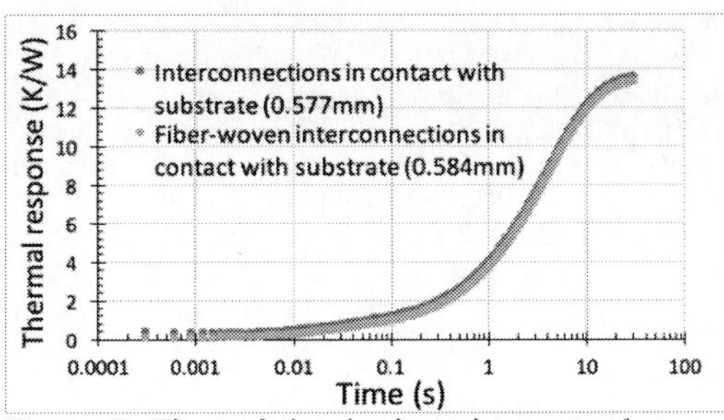

Figure 7: The variation in thermal response between samples.

The samples with metal interconnections (D-E) have a higher thermal response compared to the bare encapsulant (A-C), which is likely due to the increased thickness. Embedding interconnections within the encapsulant (D) causes a lower thermal response compared to the case when the metal interconnection wires are in contact with the Si substrate (sample type E), as shown by Figure 6. Intuitively, this could be explained by the heat sink effect induced by the wires, by drawing away heat laterally. The closer the wires to the heat source (with sample type D closer than E), the more they will sink heat from the source, and the lower the thermal response.

Alternatively, including glass fiber-woven metal interconnection wires results in a minimal alteration of the thermal response compared with non-fiber-woven samples as illustrated in Figure 7. However, when compared with an encapsulant of similar thickness with no interconnections, the fiber-woven sample indicates a slightly lower thermal response, thus suggesting that embedding fiber-woven interconnection helps to slightly lower the thermal resistivity.

Figure 8 shows the module layers are from back to front: back sheet (Polyvinyl fluoride), encapsulant (EVA), Si wafer, encapsulant (EVA), and glass.

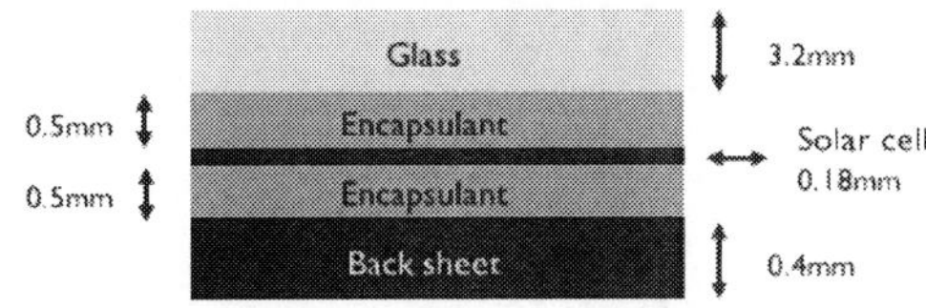

Figure 8. Simulated module layup with materials and layer thicknesses [12].

The simulated modules face South under both climates, where the modules in Kuwait City are tilted at 30° from horizontal, representing the optimal highest annual in-plane insolation. On the other hand, the module tilt angle in Oldenburg is 40°, less than optimum, corresponding to the higher latitude of the city [12], figure 9. Figure 10 shows module temperatures in Oldenburg and Kuwait city throughout the year due to measured environmental conditions.

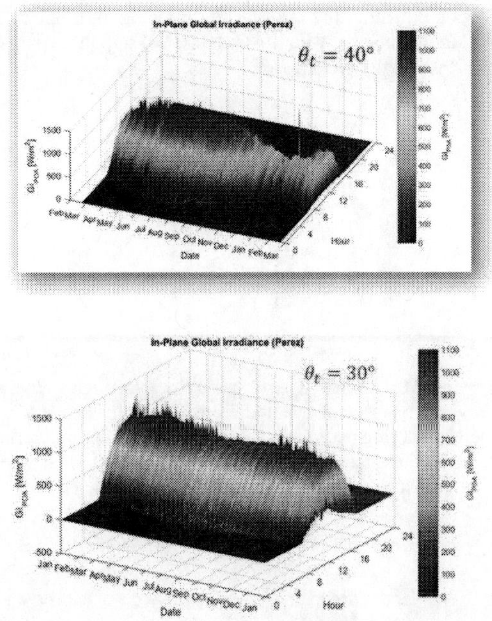

Figure 9. Measured irradiance quantities transformed into in-plane global irradiance values on the PV module surface, according to the Perez model [12], at tilt angles θ_t=40° in Oldenburg (top) and θ_t=30° in Oldenburg (bottom).

Figure 11 shows the annual energy yield in Oldenburg, as encapsulant material is varied from polyolefin to EVA. Encapsulant thermal behavior in the two encapsulant materials appear to have negligible impact on the energy yield in the module. On the other hand, increasing polyolefin thickness significantly from 1 layer to 5 layers, insignificantly decreases the annual energy yield of the module, as the energy yield gets reduced by 0.31%.

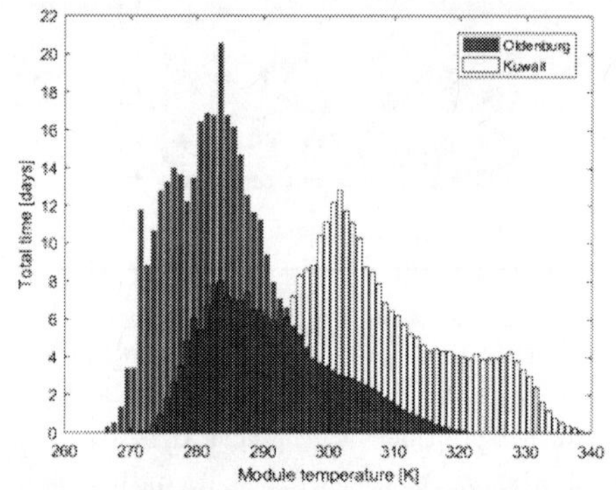

Figure 10. Module temperatures in Oldenburg and Kuwait city throughout the year due to measured environmental conditions.

The reduction in energy yield results are consistent with the transient temperature response measurements discussed earlier, as the encapsulant with the least thickness experienced the least thermal resistance, figure 6. Consequently, the material, surrounding the solar cell within the module, with the least thermal response causes the least energy yield losses.

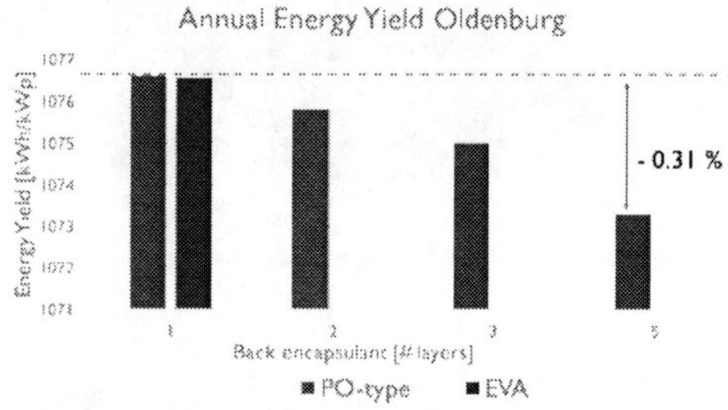

Figure 11. Annual energy yield in Oldenburg, while varying encapsulant material from polyolefin to EVA as well as varying polyolefin thickness. Increasing polylefin thickness reduces annual energy yield by 0.31%, and changing the encapsulant to EVA has negligible effects.

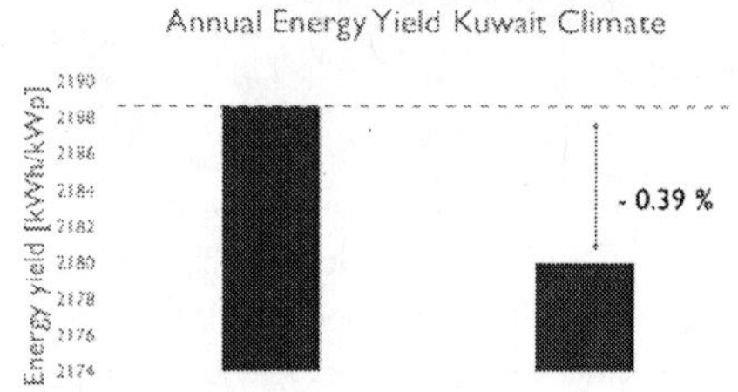

Figure 12. Annual energy yield in Kuwait. Increasing polyolefin thickness reduces annual energy yield by 0.39%.

On the other hand, Figure 12 shows the annual energy yield in Kuwait, as polyolefin thickness is significantly increased from 1 layer to 5 layers. Due to environmental conditions, the annual energy yield in Kuwait, more than doubled compared to Oldenburg, but still the encapsulant thermal effect on the overall energy yield of the module is insignificantly reduced by 0.39%.

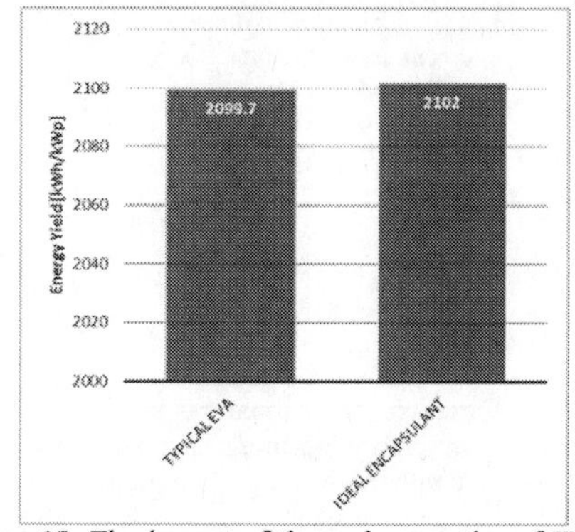

Figure 13. The impact of thermal properties of EVA on the overall energy yield in ideal conditions, at no thermal resistivity and no thermal capacity, vs in a typical case in Kuwait conditions and interdigitated back contact solar cells used. In the typical EVA case, the energy yield is reduced by 0.11%.

In addition, the impact of thermal properties of EVA on the overall energy yield in ideal conditions, at no thermal resistivity and no thermal capacity, is compared to the typical case in Kuwait conditions. Interdigitated back contact solar cells (IBC) type is used in the module, as in figure 13. Simulation results show that even with IBC, exhibiting the most absorption among solar cells, the reduction in energy yield is negligible at 0.11%.

4 CONCLUSIONS

A transient thermal measurement method used to extract thermal properties of materials, previously demonstrated for packaging and stacking in 3D-integrated circuit applications, is used to measure variations in thermal response in PV encapsulant materials due to variations in several parameters. We compared the impact of different thicknesses of encapsulant and of interconnection wires either directly or partly contacting the cells. We also included the impact of the glass fiber-woven metal interconnection wires. Furthermore, simulation results show that altering encapsulant material from polyolefin to EVA as well as varying polyolefin thickness has a negligible impact on the overall impact on energy yields in the PV module.

In future work, the transient thermal measurement technique, complemented with finite element simulations, as the modelled thermal responses are fitted to the transient temperature response measurements over the heating period, can be used to characterize the bulk thermal conductivity of the ecanpsulant material and the thermal contact resistance at the different interfaces of the module, including glass-encapsulant and encapsulant-back sheets interfaces.
Thus, the measurement technique, complemented with finite element simulations, has potential to be used in further studies and co-optimizations with respect to the thermal behavior of various encapsulant materials, novel materials and technologies, and interconnection schemes alongside with different embedding methods.

5 ACKNOWLEDGMENT

The authors gratefully acknowledge imec's SiPV industrial affiliation programme and its partners, the Kuwait Institute for Scientific Research (KISR) for providing climate data from Kuwait, the partial funding by the Kuwait Foundation for the advancement of Sciences under project number P115-15EE-01 and the support of the Flemish government through the funded project LENO PV (IWT.155025).

REFERENCES

[1] S. Dubey et al. Energy Procedia. 2013
[2] S. Krauter et al., Solar Energy Materials and Solar Cells. 1996
[3] B.J. Huang et al. Solar Energy. 2001
[4] A.W.Czanderna et al. Solar Energy Materials and Solar Cells. 1996
[5] T. Söderström et al., EUPVSEC, 2013
[6] S. Braun et al., SiliconPV, 2013
[7] T. Borgers et al., IEEE PVSC, 2016
[8] T. Borgers et al., EUPVSEC, 2017
[9] J. Govaerts et al., EUPVSEC, 2016
[10] T. Borgers et al., PiP, 2016
[11] H. Oprins et al. 13th IEEE ITHERM Conference
[12] I. Horváth et al. Solar Energy 173 (2018) 728–739.

IMPROVED CONTACTLESS METHOD OF IR REFLECTANCE UNDER GRAZING INCIDENCE FOR MEASUREMENT OF DOPING PROFILES

Jakub Holovský*[1,2], Zdeněk Remeš[1], Daniel Franta[3], Brianna Conrad[2], Lucie Abelová[2], David Bušek[2], and Aleš Poruba[1,4]

[1]Institute of Physics of the Academy of Sciences of the Czech Republic, Cukrovarnická 10, 162 00 Prague

[2] Centre for Advanced Photovoltaics, CTU in Prague, Faculty of Electrical Engineering, Technická 2, 166 27 Prague, Czech Republic

[3]Masaryk University, Faculty of Science, Kotlářská 2, Brno, Czech Republic

[4]Fill factory s.r.o., Televizní 2618, 756 61 Rožnov pod Radhoštěm, Czech Republic

*E-mail: holovsky@fzu.cz

ABSTRACT: We have elaborated contactless method of measurement and evaluation of doping profiles in silicon polished wafers based on infrared reflectance under high angle of incidence. We have found higher angle of incidence increases sensitivity, however approaching Brewster angle increases also experimental error, therefore 65 angle has been chosen. Moreover, to increase reproducibility we divide the measured spectra by reference spectra taken on an undoped sample, and further we rescale the spectra to fixed value in the region of 4000 cm^{-1}–7000 cm^{-1}. To reduce number of evaluated parameter, the carrier profile in boron-doped samples was parametrized by 3 parameters and that in phosphorous-doped samples was parametrized by 4 parameters, using additional empirically determined assumption that the first part of the profile is a constant plateau and that the following two exponential tails are joined at a value of $3x10^{19}$ cm^{-3}. As a physical model, the classical Drude equation is used and revised, leading to adjustment of effective mass and assuming concentration-dependent relaxation time and the contribution from band-to-band excitations. The relation of concentration-dependent relaxation time was represented in the range of interest by a power law that significantly deviates from typically used empirical data. The reason is that the empirical data had been obtained as a function of dopant density and not free carrier density. Because the doping efficiency drops at higher concentration this empirical data could not be used. This is one of the most important advancement of the free carrier absorption model. Finally, the model parameters were calibrated independently by infrared ellipsometry. Satisfactory agreement with the profiles measured by the electrochemical capacitance-voltage method was reached. The measurement using Fourier-transform infrared spectrophotometer takes few seconds and can be implemented as an in-line characterization technique in industry.

Keywords: infrared reflectance spectroscopy, diffusion profiles, free carrier absorption, in-line characterization

1 INTRODUCTION

Material conductivity can be detected destructively by methods such as electrochemical capacitance-voltage (ECV) or spreading resistance profiling (SRP), or in a contactless way by free carrier absorption of infrared light. Fundamentally, the presence of free carriers in a material modifies the dielectric function, ε, of the material as described by the Drude model. The dielectric function ε in turn determines the optical properties of a layer of the material. This optical approach to measuring the free carrier doping profile has been applied using numerous methods, in which the reflectance (intensity of reflected light) or ellipsometry (state of polarization of reflected light) were measured. Only reflectance at small angles of incidence (from normal) has so far been used [1-4]. However, this does not provide sufficient sensitivity for accurate determining of more than one parameter of the charge carrier profile – surface concentration. In this contribution we therefore use high-angle, polarization-resolved infrared reflectance to determine the free carrier profile in boron- and phosphorous-doped silicon. To calibrate all the parameters of our optical model, a reference optical method is necessary. We use infrared ellipsometry because this method has been proven to be highly sensitive, allowing for carrier profile evaluation. The number of parameters that define the carrier profile needs to be minimized by proper parametrization. We therefore use a parametrization that reduces the number of parameters describing the profile to either three or four, depending on the dopant.

2 OPTICAL MODEL

The optical response was simulated by a matrix-based optical model of layered media. The profile was divided into 80 layers of the same thickness and an infinite substrate of intrinsic silicon was added at the back, see Figure 1. For each layer, j, of the profile, the complex refractive index $n+ik$ was simulated by the Drude model as follows:

$$n=\sqrt{\tfrac{1}{2}\left(|\varepsilon|+\mathrm{Re}\,\varepsilon\right)},\quad k=\sqrt{\tfrac{1}{2}\left(|\varepsilon|-\mathrm{Re}\,\varepsilon\right)} \tag{1}$$

$$\varepsilon=\varepsilon_{Si}+\Delta\varepsilon_{\infty}-\frac{\omega_p^2}{\omega^2+i\omega/\tau} \tag{2}$$

where ε is dielectric function, ε_{Si} is dielectric function of pure silicon ω is optical frequncy, ω_p plasma frequency, τ is relaxation time.

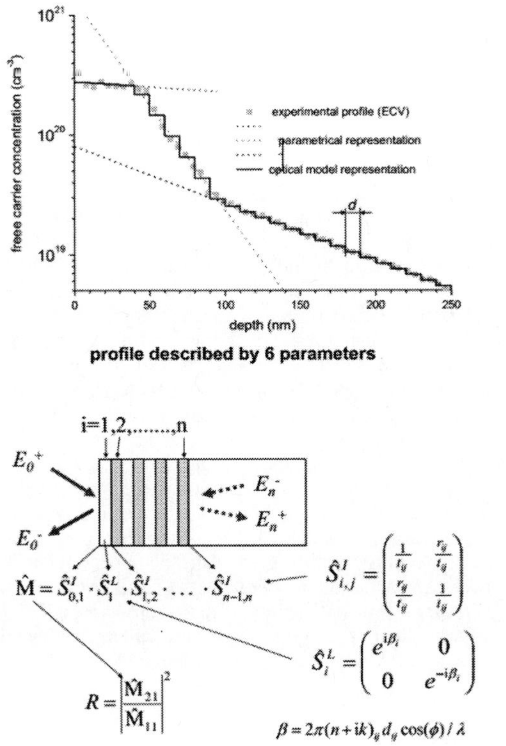

Figure 1: Example of parametrization of photosphorous carrier concenttration profile by three exponential. Mathematical reprezentation of the optical stack by a transfer matrix approach.

Optical frequency relates to wavenumber ν as ω[rad/s]=$2\pi \cdot 2.998\cdot 10^{10}\cdot \nu$[cm^{-1}]. Plasma frequency ω_p depends linearly on free carrier concentration N [cm^{-3}] as

$$\omega_p^2 = \frac{10^6 Ne^2}{\varepsilon_0 m_e}\frac{m_e}{m_{eff}} \tag{3}$$

where ε_0=8.854×10^{-12}A^{2}s^{4}kg^{-1}m^{-3} is vacuum permittivity, e = 1.602×10^{-19} is the electron charge, m_e=9.11×10^{-31} kg is electron mass and m_{eff} is effective mass of the charge carrier. The remaining parameters $\Delta\varepsilon_\infty$, m_{eff}, and τ_p are not strictly given and might even be a function of doping N. We assume the simplest case - linear dependence – because we assume that for zero concentration the dielectric function converges to the one of pure silicon:

$$\Delta\varepsilon_\infty = -\delta N \tag{4}$$

Dielectric function of pure silicon was assumed following:

Table 1: refraction index of pure silicon

n(cm^{-1})	0	2000	4000	6000	8000	10000
n_{Si}	3.552	3.563	3.583	3.615	3.655	3.716
n(cm^{-1})	12000	14000	16000	18000	20000	
n_{Si}	3.797	3.901	4.041	4.218	4.472	

The relaxation time τ is proportional to the mobility μ as $\tau=\mu m_{eff}/e$, while the mobility is known to depend on the carrier concentration N. In our approach we use our own simplified dependence described by equation with two free parameters, τ_0 and τ_1:

$$\tau = \tau_0 (N/10^{20})^{-\tau_1} \tag{5}$$

This dependence is shown on Figure 2 together with empirically calculated dependence [5] of mobility as a function of dopant concentration. The fitted dependence (5) significantly deviates from the empirical one because it accounts only free carrier concentration, i.e. only the active dopants. For high concentration the ratio of active dopants to the all the dopants can be quite low.

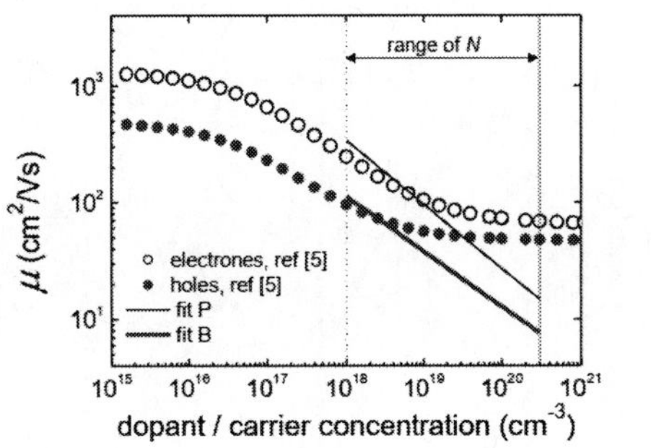

Figure 2: Dependence of mobility on dopant conentration (symbols) and on concentration of free carriers as simulated by equation (5).

3 EXPERIMENTAL DETAILS

Samples A and B were doped by boron implantation at 40 keV followed by activation annealing at 1050°C for 80 minutes. In the case of sample B, the last 20 minutes of annealing was performed under an oxygen atmosphere to create an oxidized layer on the top. Samples C and D were doped by phosphorous diffusion (from $POCl_3$). Optical measurements were taken in the range of 400cm^{-1} – 7000cm^{-1} with a Nicolet Nexus 870 Fourier-transform infrared (FTIR) spectrophotometer. For the polarized reflectance the Specac Grazing Angle Specular Reflectance Accessory was used. For each sample, the two reflectance spectra, R_P and R_S, were obtained for P and S polarizations, respectively. Additionally, reference spectra R_{P0} and R_{S0} were acquired with each measurement on an identical but undoped reference wafer.

3 OPTICAL MODEL CALIBRATION

For the purposes of calibration, the samples were analyzed by spectral ellipsometry with a FTIR VASE from J. A. Woollam Co. with an incident beam of 55, 65 and 75 degrees. ECV measurements were used for direct independent determination of the doping profile. The advantage in using ECV is that, since it measures electrical conductivity, it is insensitive to inactive dopants and directly gives the profiles of free carriers. We then simulated the ellipsometric spectra by our model based on known carrier profile and we vary the model parameters to best reproduce the experimental data. The obtained set of model parameters is given in Table 2. To double check this fitting procedure, we ran the procedure other way around with fixed model parameters and we vary the parameters of the carrier profile. The agreement between the profile found by fitting and the profile measured by ECV method is very good, see Figure 3 and Figure 4, and confirm sensitivity of ellipsometry.

Table 1: Optical model parameters

doping	τ_0 [s]	τ_1	m_{eff}/m_e	δ[cm^3]
B	6.30×10^{-15}	0.47	0.307	0.49×10^{-20}
B + oxid.	5.92×10^{-15}	0	0.254	3.52×10^{-20}
P	1.41×10^{-14}	0.55	0.325	0.38×10^{-20}

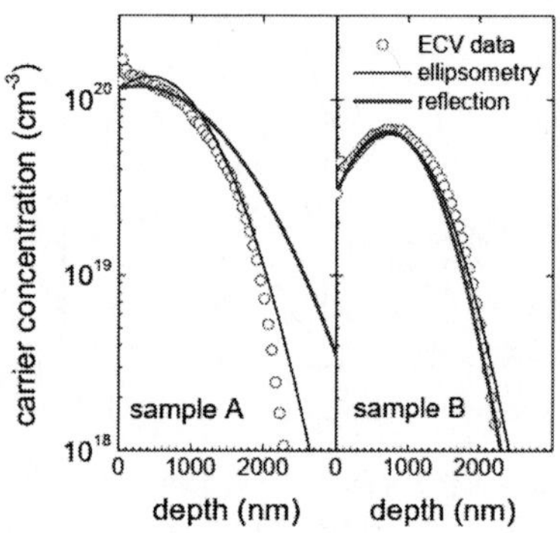

Figure 3: Carrier concentration profiles in boron-diffused samples determined by reflectance compared to those determined by ECV and ellipsometry.

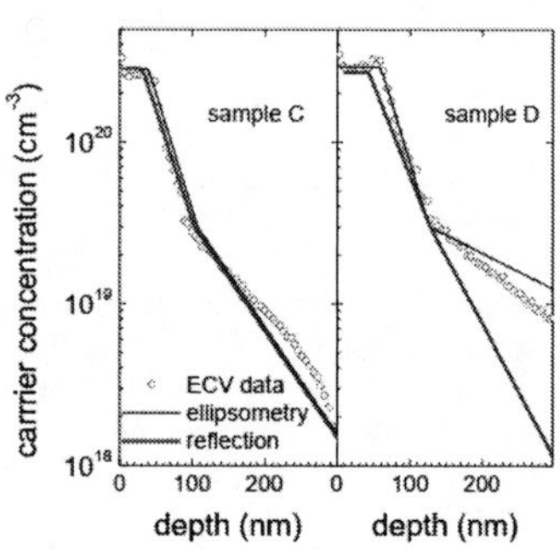

Figure 4: Carrier concentration profiles in phosphorous-diffused samples determined by reflectance, compared to those determined by ECV and ellipsometry.

4 RESULTS

To demonstrate functionality of our method of reflectance under grazing incidence, we ran the fitting procedure with fixed model parameters and again vary parameters of the carrier profile to best reproduce measurements of reflectanc, see Figure 3 and Figure 4. The agreement is fair, but slightly worse than for ellispometry, showing that reflectance is slighlty less sensitive compared to ellispometry.

5 CONCLUSIONS

This work investigated practical viability of doping profile measurement by the contactless method of infrared reflectance with a grazing angle of incidence. From the theoretical analysis and practical tests we chose to measure both s and p polarizations separately at a 65° incident angle, referenced to an undoped wafer. We normalized the curve to the reference curve in the 4000cm^{-1} - 7000cm^{-1} range. We used a Drude model that we first calibrated by comparing directly measured carrier profiles and additional ellipsometric measurements to describe the dependence of the dielectric function on doping. Four calibration constants were obtained for each type of doping. Doping profiles of boron doped samples were parametrized by three parameters and the profiles of phosphorous doped samples were parametrized by four parameters. With such simplified parametrization the method proved to be viable and well accurate at the surface where the concentration is high, though its accuracy decreases further from surface where concentration decreases. We prove here that a concept of using grazing angle has a potential to allow high accuracy of profile determination and we give plausible model including necessary calibration constants. Future work could investigate the applicability of these constants and might improve their applicability by testing on larger ensemble of samples. More details can be found in [6].

6 ACKNOWLEDGEMENT

This work was supported by Technology Agency of the Czech Republic through project no. TA01020972, by Ministry of Education Youth and Sports of Czech Republic through projects CZ.02.1.01/0.0/0.0/15_003 /0000464 Centre of Advanced Photovoltaics and no. LO1411 (NPU I), and by Czech Science Foundation through project no. 18-24268S

REFERENCES

[1] T. Abe, Y. Nishi, Non-Destructive Measurement of Surface Concentrations and Junction Depths of Diffused Semiconductor Layers, Japanese Journal of Applied Physics. 7 (1968) 397–403. doi:10.1143/JJAP.7.397.

[2] H.H. Wagner, R.R. Schaefer, Contactless probing of semiconductor dopant profile parameters by ir spectroscopy, Journal of Applied Physics. 50 (1979) 2697. doi:10.1063/1.326229.

[3] T. Abe, T. Kato, Infrared Reflectivity of N on N + Si Wafers, Japanese Journal of Applied Physics. 4 (1965) 742–751. doi:10.1143/JJAP.4.742.

[4] H. Engstrom, Infrared reflectivity and transmissivity of boron-implanted, laser-annealed silicon, Journal of Applied Physics. 51 (1980) 5245. doi:10.1063/1.327476.

[5] D.M. Caughey, R.E. Thomas, Carrier mobilities in silicon empirically related to doping and field, Proceedings of the IEEE. 55 (1967) 2192–2193. doi:10.1109/PROC.1967.6123.

[6] J. Holovský, et al. Measurement of doping profiles by a contactless method of IR reflectance under grazing incidence, Rev. Sci. Instrum. 89, 063114 (2018)

This is a pre-peer review version of the paper submitted to the journal Progress in Photovoltaics.
It has been selected by the Executive Committee of the 35th EU PVSEC 2018 for submission to Progress in Photovoltaics.

PREDICTION OF LOCAL TEMPERATURE DEPENDENT PERFORMANCE OF SILICON SOLAR CELLS

R. Eberle[1*], A. Fell[1], S. Mägdefessel[2], F. Schindler[1], M.C. Schubert[1]
[1]: Fraunhofer Institute for Solar Energy Systems ISE
Heidenhofstraße 2, 79110 Freiburg, Germany
[2]: University Freiburg, Hermann-Herder-Straße 3a, 79104 Freiburg, Germany

ABSTRACT: In this study, we present a method to predict the local temperature dependent performance of silicon solar cells from wafer lifetime images which enables local investigation of silicon solar cell parameters under realistic operation conditions. The multicrystalline silicon wafers investigated underwent high temperature steps equivalent to emitter diffusion and contact firing for the production of a PERC solar cell. Injection and temperature dependent lifetime images, gathered by calibrated photoluminescence imaging, are combined with numerical cell device simulations using Quokka3. Hereby, the spatially resolved open-circuit voltage V_{OC}, short-circuit current density j_{SC}, fill factor FF and the efficiency η of a virtual solar cell based on the characterized material are predicted for various temperatures. This data is used to compute temperature coefficients of the aforementioned cell parameters. Our predictions are validated by a comparison with measurement results of cells made from sister wafers, which are analyzed globally and spatially resolved via PL imaging, Lock-in Thermography and Light-Beam-Induced-Current measurements.
Keywords: c-Si, Modelling, Simulation, Thermal Performance

1 INTRODUCTION

The so-called "Standard Testing Conditions" (STC), a temperature of 25°C and an irradiation intensity equivalent to 1000 W/m^2 at an irradiation spectrum defined as AM 1.5G [1,2], are typically used to analyze solar cells. This enables an easy world-wide performance comparison of cells and modules in terms of their short-circuit current density j_{SC}, open-circuit voltage V_{OC}, fill factor FF and efficiency η at these conditions. However, these measurements do not reflect the real operation characteristics of solar cells as especially operating module temperature is usually much higher. In addition varying solar altitude and clouds change the irradiation spectrum and its intensity significantly.

In this study, the impact of varying temperature on cell performance is analyzed with spatial resolution. In operation, cell and module temperature can increase to around 70°C [3] which largely influences the module output: As the band gap energy E_g decreases with increasing temperature, a higher absorption of low energetic photons lead to increasing j_{SC} [1]. The fill factor FF reflects the ideal fill factor FF_0, reduced by injection dependent lifetime of excess charge carriers τ, and the cell's series resistance R_S [2, 3], but due to a complex temperature sensitivity of τ, temperature dependence of FF is non-trivial. Overall, the temperature sensitivity of V_{OC} has the strongest influence on a cell's temperature dependent behaviour [1]: V_{OC} decreases with temperature due to increasing intrinsic charge carrier density which leads to higher recombination.

The global temperature dependent characteristics of silicon solar cells has been studied in detail in the past and for global performance general equations could be derived [3,5,7-11]. More recently, some studies dealt with the temperature sensitivity of solar cells in a spatial resolution to analyze the local characteristics of temperature dependent cell performance [12,13].

In this study, we present an approach for temperature dependent efficiency analyses on wafer level which uses an improved "Efficiency Limiting Bulk recombination Analysis" *(ELBA)* [4, 5] approach for material characterization

2 MODELLING

In this study, we enhance the *ELBA* [4, 5] approach for material characterization for the analysis of temperature dependence. The *ELBA* approach is an established standard procedure to analyze the local and global efficiency potential of silicon wafers before metallization. Its principles and improvements are explained in the following:

Spatially resolved and injection dependent lifetime maps are gained via calibrated PL imaging for passivated silicon wafers [6]. An example of a lifetime map gathered at STC of 1 sun and 25°C is depicted in Fig. 1 a). A particular solar cell concept, in this work a Fraunhofer ISE PERC design similar to industry standard, is then used for simulation. We establish for this cell design a complete set of electro-optical device properties, except bulk lifetime, which is directly measured for the analyzed samples. The device simulation is first done for a broad range of τ values resulting in a look-up table for excess charge carrier density Δn, V_{OC}, j_{SC}, FF and η, and is then applied to the actual measured injection dependent lifetime maps. In an iterative procedure Δn and the bulk lifetime τ_{bulk} for a specific operating point are calculated, and via an interpolation V_{OC}, j_{SC}, FF and η are determined for each pixel.

The main achievement of this work is the enhancement of *ELBA* by temperature dependent sample characterization and cell performance prediction: measurements of effective lifetime are not only conducted at STC, but also at various temperatures from 15 to up to 70°C. An *ELBA* analysis is done at all measured temperatures, considering the actual sample temperature for evaluation. The gathered results are fitted with a linear approximation [4,12] to get insight into temperature coefficients *(TC)* of solar cells.

To predict reliable solar cell characteristics for various temperatures, the underlying cell device simulations have to take temperature dependence into account. Notably, some fundamental uncertainties in temperature dependent solar cell modeling still exist [7]. The device simulation in ELBA is changed from the previously applied software PC1D to Quokka3 [8]. By using the "lumped skin" approach based on the conductive boundary model [9], the near-surface regions ("skins", e.g. Back-Surface-Field (BSF) and emitter regions) are defined essentially by the input parameters dark saturation current J_0. While the required number of input parameters is simplified in this approach compared to the more detailed model of the skins as in PC1D, the overall sensitivity to varying bulk lifetime does not change due to its negligible effect on the lumped skin properties. For the temperature dependence of the lumped skin Quokka3 assumes that J_0/n_i^2 is temperature independent, mainly assuming a temperature independent effective minority carrier lifetime τ within those skins. In the bulk region the latest temperature dependent silicon material models are employed: band-gap and density of states, and the relation to intrinsic carrier density, by [10]; mobility by Klaasen [11]; band-to-band absorption by [12].

As it can predict the temperature dependence of the generation current, the lumped parameter optical model of Quokka3 is used in this work [13], where the light trapping parameters are fitted to match STC-j_{SC}. The recombination for measured samples in this study is dominated by the bulk and there is good agreement between simulation and measurements. Therefore, the used simplification of assuming that J_0/n_i^2 is temperature independent appears suitable for this work.

Most input parameters for electro-optical device simulation in Quokka3 are mainly gathered by independent measurement. The external series resistance R_S is determined via the difference between measured and simulated internal resistance at STC.

3 EXPERIMENTAL

Samples used in this study are industrial multicrystalline p-type wafers with a thickness of ~185 µm and a base resistivity of ~1.6 Ωcm. The samples (156x156 mm^2) for carrier lifetime measurements underwent high temperature steps for emitter diffusion and contact firing typical for the standard industrial "Passivated Emitter and Rear Cell" (PERC) concept [14,15]. Then the diffused layers are etched back and the wafers are passivated with Al_2O_3 and SiN_x. Sister wafers with similar crystal structure are used to produce solar cells with PERC at Fraunhofer ISE [16] to compare cell and wafer results: After acidic texturing and the phosphorus emitter diffusion, the passivation of the rear side is done by a stack of plasma-enhanced chemical vapor deposited (PECVD) aluminum oxide AlO_x and silicon nitride SiN_y, the front side by SiN_y. Contacts are applied via screen printing, the front contact is formed with a firing step in a fast firing oven and contact formation is done via a laser-fired contacts (LFC) process [16].

Several measurement tools are used in this study: Photoluminescence (PL) Imaging gains spatially resolved information about silicon lifetime samples which are irradiated with a 808 nm laser, a CCD camera captures photoluminescence of the samples and a stack of long pass filters with additional 1000 and 1050 nm short pass filter suppresses detection of reflected laser light and reduces optical blurring. Gathered PL images are calibrated to effective lifetime τ_{eff} with lifetime measurements via harmonically modulated PL [6].

Characterization of global cell parameters is done with the measurement system LOANA by PV Tools GmbH [17] which measures global η, V_{OC}, j_{SC} and *FF*, but also reflectance, external quantum efficiency *EQE* or grid parameters. The local performance of a solar cell is studied by a Total Cell Analysis *(TCA)* [18]. This method combines Coupled-Determination of Dark Saturation Current and Series Resistance *(C-DCR)* [19], a Local-IV analysis of Dark-Lock-In Thermography *(DLIT)* [20] and j_{SC} mapping via Light Beam Induced Current *(LBIC)* measurements [21] to map the spatially resolved η of a solar cell as well as V_{OC}, j_{SC} and *FF* (for details see cited literature).

To analyze the temperature dependent behaviour of silicon solar cells and lifetime samples, the aforementioned measurement approaches are enhanced. The measurement chucks are temperature controlled with Peltier elements by which measurements at various temperatures from 15 to 70°C are conducted.

4 RESULTS

4.1 PREDICTION OF TEMPERATURE SENSITIVITY

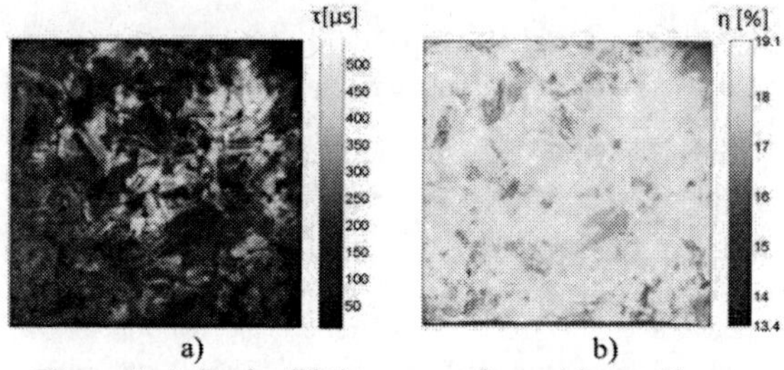

Figure 1 a) Carrier lifetime map of a multicrystalline p-type silicon wafer taken at STC-temperature of 25°C and monochromatic irradiation @808 nm with intensity equivalent to ~1 sun, b) With *ELBA* calculated efficiency map of the same sample at 1 sun and 25°C.

An example of a lifetime map gathered at STC-temperature of 25°C and 1 sun irradiation with 808 nm light is shown in Fig. 1 a). The resulting simulated η for this lifetime sample at 25°C and 1 sun is shown in Fig. 1 b). For that several lifetime-calibrated injection dependent PL images are combined with a cell simulation with Quokka3,. From measurements in the temperatures range between 15 and 70°C local temperature coefficients are calculated.

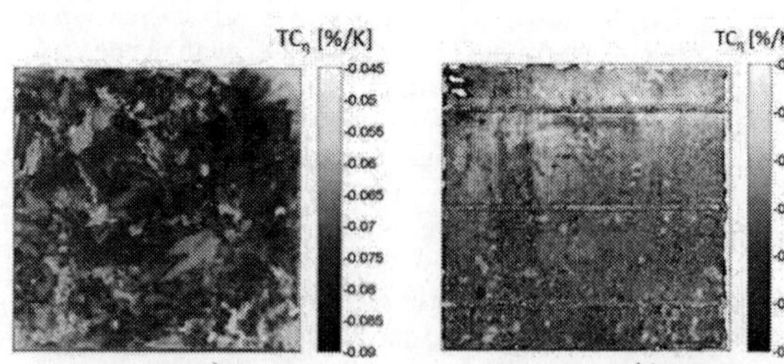

Figure 2 a) Temperature dependent *ELBA* results were used to fit local temperature coefficient *TC(η)* in spatial resolution, b) temperature sensitivity of η of a solar cell made from a sister wafer gained from temperature dependent TCA.

The spatially resolved change in η with temperature is analyzed in Fig. 2 a). The multicrystalline structure of the characterized sample is still visible in the temperature coefficient map, areas of dislocation clusters can clearly be distinguished from ones with good grains and high τ. A very similar spatial distribution can be seen for temperature dependent local V_{OC} in Fig. 3 a), which commonly dominates the temperature sensitivity of silicon solar cells. V_{OC} shows the same trends in temperature sensitivity for good areas and areas with dislocations as Fig. 2 a).

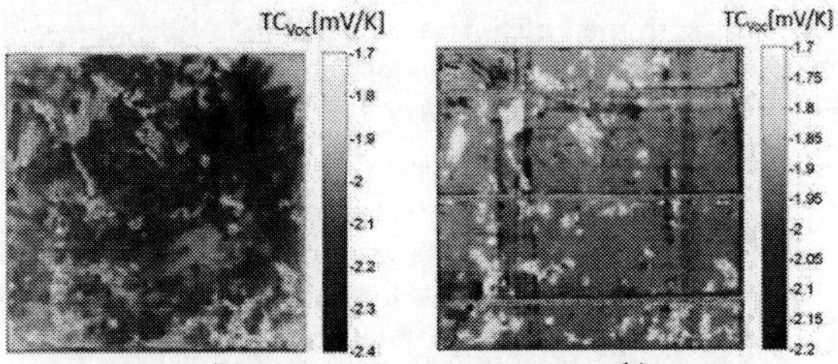

a) b)

Figure 3 a) Spatially resolved *TC(V_{OC})* calculated with *ELBA* results, b) measured spatially resolved temperature coefficient of V_{OC} of a sister PERC solar cell.

To assess the validity of our approach results are compared with other already established measurement techniques. For this, PERC cells are analyzed with two different characterization methods. The cells are made of sister wafers with very similar crystal structure and material properties, e.g. bulk resistivity, as the lifetime samples used for the temperature dependent *ELBA* analysis.

A global cell characterization via IV measurements with the measurement system *LOANA* [17], as well as a local *TCA* [18] is conducted with PERC samples. In Table 1 results are depicted for an irradiation intensity equivalent to 1 sun and a temperature of 25°C: *ELBA* prediction and *TCA* are averaged over the whole sample area (for *ELBA* according to [22], for *TCA* j_{SC} arithmetic and V_{OC} square-root harmonic) and compared to global measurement via *LOANA*. . The slight overestimation of *FF* by *ELBA* is explainable by the simulation neglecting edge effects, non-ideal surface recombination, as well as *FF*-losses caused by grain-boundary recombination [23].

Table 1 Comparison of performance parameters η, V_{OC}, j_{SC} and *FF* predicted with *ELBA*, gathered by global cell measurements and by spatially resolved *TCA* at STC.

Data gathered via	*η [%]*	*V_{OC} [mK]*	*j_{SC} [mA/cm^2]*	*FF [%]*
ELBA	17.7	640.9	36.17	76.4
TCA	17.3	639.2	36.22	74.7
Global	17.6	642.5	36.55	75.0

Measurements for all shown approaches are not only conducted at STC-temperature, but also at temperatures from 15 up to 70°C. As for ELBA predictions, for TCA and the global IV characterization a linear trend for the temperature sensitivity [1] is assumed to fit temperature coefficients of solar cell parameters. Arithmetically averaged results are shown in Table 2.

The satisfactory agreement between all approaches confirms that the assumptions made for cell device simulation in Quokka3 within the improved *ELBA* analysis are sufficient for a temperature sensitivity analysis and efficiency prediction of the investigated wafers.

Table 2 Comparison of temperature coefficients of η, V_{OC}, j_{SC} and *PFF* predicted with *ELBA*, gathered by global cell measurements and by spatially resolved TCA.

Data gathered via	*TC(η) [%/K]*	*TC(V_{OC}) [mV/K]*	*TC(j_{SC}) [mA/cm^2/K]*	*TC(FF) [%/K]*
ELBA	-0.072	-2.10	0.019	-0.11
TCA	-0.072	-1.95	0.022	-0.11
Global	-0.085	-1.93	0.017	-0.14

Besides global results, local cell parameters and their temperature sensitivity can be investigated for their spatial distribution. Figs. 2 and 3 compare temperature coefficients of η (Fig. 2 a)) and V_{OC} (Fig. 3 a)) gathered via ELBA with measured local cell results for both parameters (see Fig. 2 b) and 3 b) respectively).

Measured temperature coefficients (Figs. 2b) and 3b)) are distributed more evenly. We attribute this to lateral conduction of charge carriers over the whole cell area in the emitter present in the solar cells. Nevertheless, also for cells a different temperature dependent behaviour for areas of dislocation clusters is distinguishable. As already shown in [24] and [25], and also obvious from the local results presented here, areas of dislocation clusters with a low initial lifetime are characterized by decreased temperature sensitivity. This leads to a reduced change in IV characteristics with increasing temperature and to a more stable cell and module performance with increasing temperature for samples with high concentration of impurities than would be expected. The cause of the temperature sensitivity difference in dislocation and high lifetime areas is discussed in the following.

4.2 ROOT CAUSE FOR LOCAL TEMPERATURE COEFFICIENTS

According to theory of global temperature dependent behaviour [1], *TC(V_{OC})* increases with increasing V_{OC} and dominates the overall temperature sensitivity of silicon solar cells. However, some dislocation clusters of our multicrystalline silicon samples show the opposite trend. As can be seen in Table 3 by a comparison of actual *TC(V_{OC})* with predicted *TC(V_{OC})* by ELBA, the actual temperature sensitivity of V_{OC} in areas with low initial lifetime and respective low V_{OC} is contradictory to the behavior expected from text-book theory. To analyze the root cause of this anomaly, areas with high initial lifetime (green rectangle in Fig. 5 a)) and lower material quality (blue rectangle in Fig. 5 a)) are compared and further investigations are conducted.

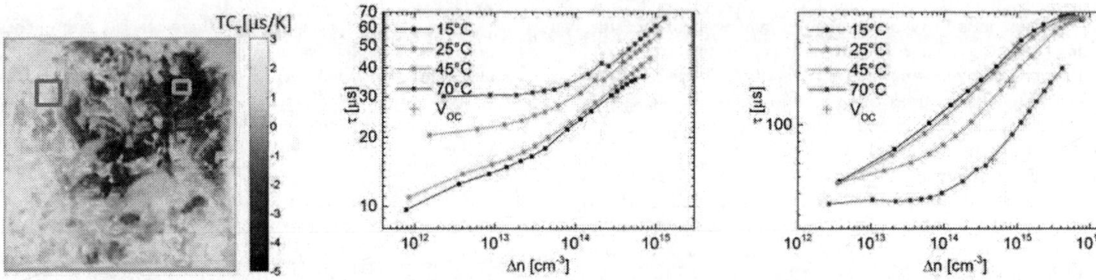

Figure 5 a) Spatially resolved temperature coefficient of the excess carrier lifetime assuming linear behaviour with temperature, images used for fitting were taken at irradiation intensity equivalent to ~1sun, b) injection level dependent mean lifetimes at measured temperatures for area with low lifetime (blue rectangle in Fig. 5 a)), c) with high lifetime (green rectangle in Fig. 5 a)).

Table 3 Comparison of temperature coefficients of τ and V_{OC} for whole sample and high (green rectangle in Fig. 5a)) as well as low lifetime areas (blue rectangle in Fig. 5 a)). For comparison of V_{OC} in the different areas, results from ELBA, TCA and calculations with the formula in [1] ($TC(V_{OC})_{Green}$) are used. Values are averaged harmonically for calculations with τ.

Averaged values for	*$TC(V_{OC})_{Green}$ [mV/K]*	*$TC(V_{OC})_{ELBA}$ [mV/K]*	*$TC(V_{OC})_{Cell}$ [mV/K]*	*$TC(\tau)$ [μs/K]*
Whole sample	-1.93	-2.10	-1.95	0.39
Good area	-1.90	-2.31	-1.92	-3.77
Bad area	-1.99	-1.98	-1.77	0.35

Fig. 5 a) depicts the local $TC(\tau)$ calculated with lifetime calibrated PL images taken at ~1 sun irradiation intensity showing two different areas, one with negative temperature coefficients of grains with high initial lifetimes (e.g. green rectangle) and a second type of area with positive temperature coefficients (e.g. blue rectangle). This map shows the same spatially resolved pattern as could be seen for ELBA predictions of $TC(V_{OC})$ and $TC(\eta)$. In addition, Fig. 5 b) and c) show temperature and injection level dependent lifetime for two different areas with low (Fig. 5 b)) and high effective lifetime at STC (Fig. 5c)), averaged values are listed in Table 3. Especially the difference in relative change of lifetime at low carrier injection for the good and bad area is significant. Again, we attribute deviation between ELBA and cell results to smearing of excess charge carriers due to the emitter. Nevertheless, ELBA correctly predicts the reverse temperature dependent behaviour of dislocation clusters.

One possible cause for the anomaly of temperature dependent behaviour in dislocations could be diffusion of excess charge carriers from good grains into bad areas. To assess this possibility, temperature and injection dependent lifetime maps, which were used for the ELBA analysis, were corrected for lateral charge carrier diffusion leading to diffusion corrected images [26]. This correction is not applied as a standard feature as diffusion of charge carriers also takes place in solar cells and is therefore important for local efficiency prediction [14].

Repeated ELBA analysis with diffusion corrected maps delivered the same results as for the original analysis which implies that charge carrier diffusion can be excluded as cause for improved $TC(V_{OC})$ and $TC(\eta)$ of dislocation clusters.

The analyzed multicrystalline sample contains a significant concentration of impurities. This is proven by measurement of local spatially resolved interstitial iron concentration in Fig. 6 according to [27]. However, it was shown in [12] that iron cannot be responsible for the relatively strong increase of lifetime in low quality areas as a negative $TC(\tau)$ for various iron concentrations was calculated.

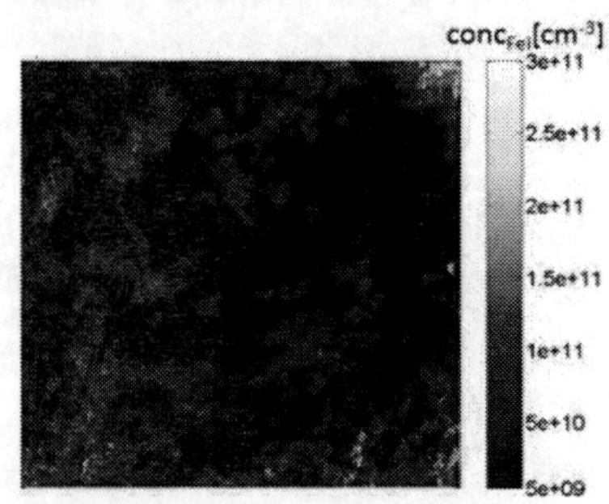

Figure 6 Spatially resolved measurement of interstitial iron concentration in investigated lifetime sample.

Dislocation clusters are usually characterized by a higher concentration of impurities, these can induce Shockley-Read-Hall (SRH) defects at an energy level in the middle of the band gap and change τ_{eff} significantly [28, 29]. To assess this possibility the injection dependent *SRH lifetime* τ_{SRH} is simulated for various *trap energy levels* E_t at different temperatures (see Fig. 7). Fig. 7 a) shows a calculation of τ_{SRH} for E_t-E_v=0.374 eV of interstitial iron [30]; there is no significant change in τ_{SRH} at high Δn values and only a weak one for low excess charge carrier densities. Both results confirm the statement of [24] that iron is most probably not responsible for the observed temperature sensitivity in dislocation clusters.

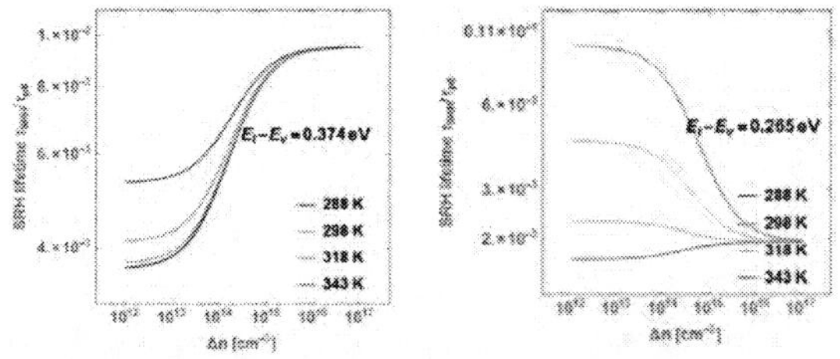

a) b)

Figure 7 Simulated SRH lifetimes for temperatures from 15 to 70°C and two different energy values of trap levels induced by impurities in the band gap of silicon leading to SRH defects, for the example of a) interstitial iron impurities, b) chromium boron pairs..

A trap level E_t of 0.27 eV (Fig. 7 b)), which represents chromium-boron pairs at these temperatures [31], results in a very different behaviour: With decreasing Δn a strong temperature dependence of τ_{SRH} is observed: τ_{SRH} decreases with declining Δn for low to medium temperatures, whereas τ_{SRH} strongly increases for decreasing Δn at 70°C. This is only an example for a possible impurity inducing *SRH* recombination with an energy level in the band gap of silicon. Simulations of defects with E_t-E_v of 0.2 to 0.3 eV demonstrate that lifetime strongly increases with temperature from the value at low Δn at low temperature to high temperatures and the related increased charge carrier density. This contributes positively to V_{OC}, thus the decrease of V_{OC} with increasing temperature is reduced which results in better $TC(V_{OC})$. However, the observed decrease in lifetime at high injection cannot be explained by SRH defects and might stem from a different mechanism as surface recombination which is still to be investigated.

5 CONCLUSION

In this work, we report an improved approach for the prediction of solar cell parameters at an early production stage. Temperature and injection dependent lifetime images are combined with a solar cell device simulation to assess the local and global characteristics of a solar cell and its temperature sensitivity before metallization. The proposed method can be applied to all kinds of silicon samples, and can be used to distinguish between effects induced by bulk material or cell production.

As there is reasonable agreement of the predictions with global and local measurements of solar cells made of sister wafers, the analysis approach, as well as the assumptions for electro-optical device modelling appear to account for all important effects with sufficient accuracy. The local results demonstrate that especially sample regions with low initial lifetime like dislocated areas are characterized by temperature dependent behavior which is opposed to expectations from established simpler analytical models.

The cause for this anomaly was investigated: Simulations of various *SRH* trap levels exhibit a strong increase in τ_{SRH} for increasing temperatures which results in increasing τ_{eff} for low excess charge carrier densities. These results imply that impurities leading to *SRH* defects with E_t of 0.2 to 0.3 eV may be part of the complex reason for decreased temperature sensitivity of dislocation clusters.

We conclude from our investigations that the disadvantage of multicrystalline silicon, which is the significantly decreased efficiency caused by a lower local crystal quality, is weakened as the gap in efficiency between bad and good areas decreases at real operation conditions.

ACKNOWLEDGEMENT

We thank Sven Wasmer for providing silicon lifetime and cell samples and detailed sample information for cell simulations. Rebekka Eberle would like to thank the "Studienstiftung des deutschen Volkes" for funding her dissertation project. This work was supported by the German Federal Ministry for Economic Affairs and Energy BMWi and by industry partners within the research project PROGNOSIS under contract No. 0324160. The authors are responsible for the content.

References

[1] M. A. Green, K. Emery, and A. W. Blakers, "Silicon solar cells with reduced temperature sensitivity," *Electron. Lett.*, vol. 18, no. 2, p. 97, 1982.

[2] O. Dupre, R. Vaillon, and M. A. Green, "Experimental assessment of temperature coefficient theories for silicon solar cells," in *2015 IEEE 42nd Photovoltaic Specialist Conference (PVSC)*, New Orleans, LA, 2015, pp. 1–3.

[3] Martin A. Green, *Silicon solar cells: Advanced principles & practice*. Sydney: Centre for Photovoltaics devises and systems, University of New South Wales, 1995.

[4] B. Michl, M. Rüdiger, J. A. Giesecke, M. Hermle, W. Warta, and M. C. Schubert, "Efficiency limiting bulk recombination in multicrystalline silicon solar cells," *Solar Energy Materials and Solar Cells*, vol. 98, pp. 441–447, 2012.

[5] B. Michl, M. Kasemann, W. Warta, and M. C. Schubert, "Wafer thickness optimization for silicon solar cells of heterogeneous material quality," *Phys. Status Solidi RRL*, vol. 7, no. 11, pp. 955–958, 2013.

[6] J. A. Giesecke, M. C. Schubert, B. Michl, F. Schindler, and W. Warta, "Minority carrier lifetime imaging of silicon wafers calibrated by quasi-steady-state photoluminescence," *Solar Energy Materials and Solar Cells*, vol. 95, no. 3, pp. 1011–1018, 2011.

[7] H. Steinkemper, I. Geisemeyer, M. C. Schubert, W. Warta, and S. W. Glunz, "Temperature-Dependent Modeling of Silicon Solar Cells—Eg, n i, Recombination, and VOC," *IEEE J. Photovoltaics*, vol. 7, no. 2, pp. 450–457, 2017.

[8] A. Fell, J. Schön, M. C. Schubert, and S. W. Glunz, "The concept of skins for silicon solar cell modeling," *Solar Energy Materials and Solar Cells*, vol. 173, pp. 128–133, 2017.

[9] R. Brendel, "Modeling solar cells with the dopant-diffused layers treated as conductive boundaries," *Prog. Photovolt: Res. Appl.*, vol. 20, no. 1, pp. 31–43, 2012.

[10] R. Couderc, M. Amara, and M. Lemiti, "Reassessment of the intrinsic carrier density

temperature dependence in crystalline silicon," *Journal of Applied Physics*, vol. 115, no. 9, p. 93705, 2014.

[11] D.B.M. Klaasen, "A unified mobility model for device simulation - Electron Devices Meeting, 1990. Technical Digest., International," *International Technical Digest on Electron Devices*, 1990.

[12] C. Schinke, K. Bothe, P. Christian Peest, J. Schmidt, and R. Brendel, "Uncertainty of the coefficient of band-to-band absorption of crystalline silicon at near-infrared wavelengths," *Appl. Phys. Lett.*, vol. 104, no. 8, p. 81915, 2014.

[13] A. Fell, K. R. McIntosh, and K. C. Fong, "Simplified Device Simulation of Silicon Solar Cells Using a Lumped Parameter Optical Model," *IEEE J. Photovoltaics*, vol. 6, no. 3, pp. 611–616, 2016.

[14] A. W. Blakers, A. Wang, A. M. Milne, J. Zhao, and M. A. Green, "22.8% efficient silicon solar cell," *Appl. Phys. Lett.*, vol. 55, no. 13, pp. 1363–1365, 1989.

[15] M. A. Green, "The Passivated Emitter and Rear Cell (PERC): From conception to mass production," *Solar Energy Materials and Solar Cells*, vol. 143, pp. 190–197, 2015.

[16] S. Wasmer, J. Greulich, H. Hoffler, N. Wohrle, M. Demant, F. Fertig, and S. Rein, "Impact of Material and Process Variations on the Distribution of Multicrystalline Silicon PERC Cell Efficiencies," *IEEE J. Photovoltaics*, vol. 7, no. 1, pp. 118–128, 2017.

[17] "pv-tools GmbH," http://www.pv-tools.de/.

[18] B. Michl, M. Padilla, I. Geisemeyer, S. T. Haag, F. Schindler, M. C. Schubert, and W. Warta, "Imaging Techniques for Quantitative Silicon Material and Solar Cell Analysis," *IEEE J. Photovoltaics*, vol. 4, no. 6, pp. 1502–1510, 2014.

[19] M. Glatthaar, J. Haunschild, M. Kasemann, J. Giesecke, W. Warta, and S. Rein, "Spatially resolved determination of dark saturation current and series resistance of silicon solar cells," *phys. stat. sol. (RRL)*, vol. 4, no. 1-2, pp. 13–15, 2010.

[20] O. Breitenstein, "Nondestructive local analysis of current–voltage characteristics of solar cells by lock-in thermography," *Solar Energy Materials and Solar Cells*, vol. 95, no. 10, pp. 2933–2936, 2011.

[21] M. Padilla, B. Michl, B. Thaidigsmann, W. Warta, and M. C. Schubert, "Short-circuit current density mapping for solar cells," *Solar Energy Materials and Solar Cells*, vol. 120, pp. 282–288, 2014.

[22] J. Isenberg, J. Dicker, and W. Warta, "Averaging of laterally inhomogeneous lifetimes for one-dimensional modeling of solar cells," *Journal of Applied Physics*, vol. 94, no. 6, pp. 4122–4130, 2003.

[23] F. Schindler, A. Fell, R. Müller, J. Benick, A. Richter, F. Feldmann, P. Krenckel, S. Riepe, M. C. Schubert, and S. W. Glunz, "Towards the efficiency limits of multicrystalline silicon solar cells," *Solar Energy Materials and Solar Cells*, vol. 185, pp. 198–204, 2018.

[24] R. Eberle, S. T. Haag, I. Geisemeyer, M. Padilla, and M. C. Schubert, "Temperature coefficient imaging for silicon solar cells," *IEEE J. Photovoltaics*, vol. 8, no. 4, 2018.

[25] R. Eberle, W. Kwapil, and M. C. Schubert, "Temperature dependent imaging of solar cell losses," *Energy Procedia*, 2018.

[26] S. P. Phang, H. C. Sio, and D. Macdonald, "Carrier de-smearing of photoluminescence images on silicon wafers using the continuity equation," *Appl. Phys. Lett.*, vol. 103, no. 19, p. 192112, 2013.

[27] D. Macdonald, J. Tan, and T. Trupke, "Imaging interstitial iron concentrations in boron-doped crystalline silicon using photoluminescence," *Journal of Applied Physics*, vol. 103, no. 7, p. 73710, 2008.

[28] W. Shockley and W. T. Read and Jr., "Statistics of the Recombinations of Holes and Electrons," *Physical Review*, vol. 87, no. 5, p. 835, 1952.

[29] R. N. Hall, "Electron-Hole Recombination in Germanium," *Physical Review*, vol. 87, no. 2, p. 387, 1952.

[30] A. A. Istratov, H. Hieslmair, and E. R. Weber, "Iron and its complexes in silicon," *Applied Physics A: Materials Science & Processing*, vol. 69, no. 1, pp. 13–44, 1999.

[31] C. Sun, F. E. Rougieux, and D. Macdonald, "Reassessment of the recombination parameters of chromium in n- and p-type crystalline silicon and chromium-boron pairs in p-type crystalline silicon," *Journal of Applied Physics*, vol. 115, no. 21, p. 214907, 2014.

CHALLENGES FOR THE QUANTIFICATION OF METAL INDUCED RECOMBINATION LOSSES

D. Herrmann*, S. Lohmüller (née Werner), H. Höffler, A. Fell, A. A. Brand, A. Wolf
Fraunhofer Institute for Solar Energy Systems ISE, Heidenhofstraße 2, 79110 Freiburg, Germany
*Corresponding author: David Herrmann | Phone: +49 (0)761 4588 5049 | E-mail: david.herrmann@ise.fraunhofer.de

ABSTRACT: One major loss mechanism for the currently highly relevant passivated emitter and rear cells (PERC) is locally enhanced recombination at the interface between semiconductor and front side metallization. For investigating these losses in detail, a reliable detection technique is crucial. A method we call calibrated photoluminescence imaging (PLI) method is a promising technique to extract the local dark saturation current density in the metallized area $j_{0,\mathrm{loc}}$. To investigate the sources of error of this method, metallized test samples are processed and the influence of sample specific parameters is considered in detail. Additionally, the difference in the resulting $j_{0,\mathrm{loc}}$ between the calibrated PLI method and a simulative approach using numerical PLI simulations (Quokka3) is evaluated. We find that the results from the calibrated PLI method strongly depend on the accurate knowledge of the base resistivity ρ_B (including the impact of thermal donors) and less strongly on the reflectivity R of the illuminated side of the sample. In addition, metastable defects in the Si bulk can falsify the results, due to changes in the PLI intensity as a function of the illumination time. The difference in the resulting $j_{0,\mathrm{loc}}$ between the calibrated PLI method and the simulative approach is within the error tolerances, which implies that the calibrated PLI method delivers accurate results despite the assumption of a uniform Δn throughout the sample. Here, it is important to mention that the non-uniformity of Δn is expected to be stronger for structures without a highly doped region at the investigated side (e.g. PERC rear contacts).
Keywords: Photoluminescence, Recombination, Metallization, Quokka3

1 INTRODUCTION

During the last years passivated emitter and rear cells (PERC) have gained more and more relevance for both, research institutes and industry [1]. Various authors identified recombination losses induced in the emitter region—especially at the front side metal contacts—to be currently one of the dominating loss mechanisms [2–5]. Although the fraction of metallized surface on the front side of a typical PERC solar cell is just a few percent, the share of recombination losses induced in this area is comparable with the losses in the passivated emitter area. Quantifying the recombination losses at the metal contacts in terms of the local dark saturation current density $j_{0,\mathrm{loc}}$ is essential for development of advanced emitters, metallization pastes as well as accurate device simulation. Hence, a reliable and accurate technique to extract $j_{0,\mathrm{loc}}$ is needed. A method we call from now on calibrated photoluminescence imaging (PLI) [6, 7] is a method which is frequently used for quantifying $j_{0,\mathrm{loc}}$ and which is of rising interest in the PV community. Using this approach the average detected PLI signal (Counts/Pixel*s) is calibrated to an implied open circuit voltage iV_{oc} measured by quasi steady state photoconductance (QSSPC). Thereby the PLI signal can be translated into an iV_{oc} and further with the one-diode equation into a total dark saturation current density j_0. Comparing the resulting j_0 values of metallized and non-metallized samples $j_{0,\mathrm{loc}}$ can be determined with the known metallization area fraction.

However, using this technique, several uncertainties strongly affect the final result. In the present work, the influence of sample parameters such as base resistivity ρ_B, metal impurities in the wafer bulk and surface reflectivity R are discussed in detail. Furthermore, the results using the calibrated PLI method are compared to the results using an approach which uses numerical PLI simulations to replicate PLI measurements as described in [8, 9].

2 SAMPLE PREPARATION

To investigate the influence of sample specific parameters on the determination of metal induced recombination losses, two samples types, which from now on are called both-sided emitter- and PERC-samples are fabricated, as shown in Fig. 1 a) and b).

Fig. 1 c) exemplary shows the process flow for the fabrication of both-sided emitter-samples. The characterization steps are highlighted in white letters. Sample preparation starts with pseudo-square n-type Czochralski-grown silicon (Cz-Si) wafers with an edge length of 156 mm and a base resistivity of $\rho_B \approx 4\ \Omega\mathrm{cm}$ determined after thermal donor anneal [10]. After alkaline texturing, a $POCl_3$-based diffusion process in an industrial tube furnace forms phosphorus-doped regions symmetrically on both sides with a sheet resistance of $R_{sh} \approx 130\ \Omega/\square$. Subsequently, the phosphosilicate glass (PSG) layer, which forms during the diffusion process, is removed in hydrofluoric acid (PSG etching). A wetchemical clean is performed, followed by an annealing step in N_2 [11] and subsequent deposition of a silicon nitride (SiN_x) anti-reflection coating (ARC) on both sides of the both-sided emitter-samples using plasma enhanced chemical vapor deposition (PECVD). Thereafter, a fast firing step is carried out in an industrial conveyor belt furnace to activate the passivation layers. Spatially resolved PLI measurements, using the Modulum tool at the Fraunhofer ISE, and QSSPC measurements, using a Sinton WCT-120 lifetime tester [12], follow. After screen printing a finger grid with 150 fingers of about 40 µm width on one side of the samples using a commercially available silver paste, and contact firing similar to the first fast firing step, an additional PLI measurement is performed using the procedure as described in section 3. PLI measurements on non-metallized reference samples, conducted after a first and a second firing step, confirm that the sample properties in the non-metallized areas are hardly affected by a second firing process. Additionally PERC-samples from p-type Cz-Si industrial precursors with an edge length of

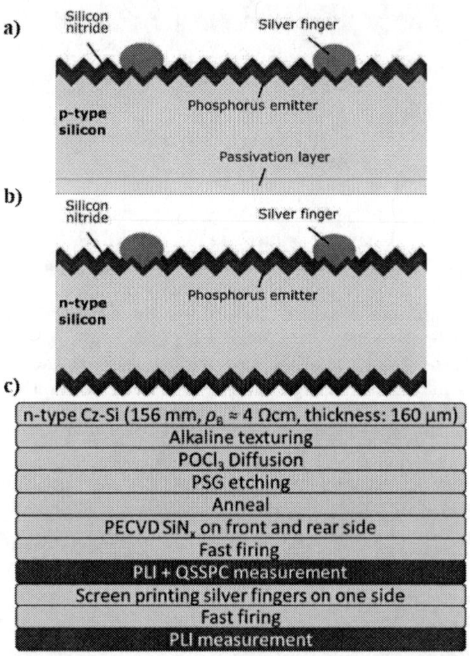

Fig. 1: a) Schematic draw of the PERC-samples and b) the both-sided emitter samples. c) Process flow for the fabrication of both-sided emitter-samples fabricated from n-type Czochralski-grown silicon (Cz-Si) wafers.

156 mm and a nominal resistivity of $1\ \Omega\text{cm} \leq \rho_B \leq 3\ \Omega\text{cm}$ are fabricated following a similar process route, with the ARC just on the front side.

For the comparison between the calibrated PLI method and the simulative approach, laser ablated samples are used. The samples are fabricated similar to the process flow shown in Fig. 1 and were already investigated in Ref. [13]. Instead of screen printing and the second fast firing step the passivation is locally ablated on one side of the sample via irradiation with a pulsed laser using a pulse duration of 15 ps and a wavelength of 355 nm. A similar ablation process is used for Ni-Cu plated metallizations [14] and is known to cause amorphization of the remaining silicon surface [15], thus inducing high surface recombination velocities, comparable to that of a screen printed metallization. The ablation geometry consists of parallel lines with an ablation width of $w \approx 16$ µm and eight different line distances d from $d = 100$ µm up to $d = 1500$ covering the entire sample area. Also reference samples are processed, for which no laser process is applied.

3 APPROACH

This section describes the approach for the determination of the local dark saturation current density used in this work.

A scheme of the measurement steps of the calibrated photoluminescence method is given in Fig. 2. The routine is shown exemplarily for *p*-type PERC-samples fabricated, as described in section 2. Starting from the upper left in the non-metallized state the samples are measured by QSSPC and PLI. The PLI measurement is performed with the passivated rear side pointing upwards to the illumination source and detector. Additionally, two short pass filters with cut-off wavelengths $\lambda_{\text{cutoff,1}} = 1050$ nm and $\lambda_{\text{cutoff,2}} = 1000$ nm are used to minimize the impact of optical variations of the non-illuminated surface. The averaged PLI-intensity $\bar{\varphi}_1$ in the area where the QSSPC measurement is performed, is linked to the implied open circuit voltage iV_{OC} determined by QSSPC, via the calibration constant $C = \bar{\varphi}_1 / \exp(\frac{iV_{OC}}{V_t})$ as derived in [16], where V_t is the thermal voltage at 25°C. In this step it is essential, that the iV_{OC} is determined at the identical generation rate G, at which the PLI measurement is performed, to ensure that the measurement values are determined under preferably similar conditions (similar excess carrier density Δn).

Using the sample specific C, the averaged PLI intensity $\bar{\varphi}$ can be related to an averaged implied open circuit voltage $i\bar{V}_{OC}$ in the non-metallized as well as in the metallized state. Assuming an ideal diode behavior $j_{rec} = j_0 \cdot \exp(iV_{OC}/V_t)$—the subtrahend -1 is neglected, since it is insignificant—, where j_{rec} is the recombination current density, the total dark saturation current density j_0 can be extracted with knowledge of j_{rec}. During a PLI measurement the samples are in steady state open circuit conditions, leading to equality of recombination current density and generation current density: $j_{gen} = j_{rec}$. The generation current density can be expressed as $j_{gen} = q \cdot (1 - R) \cdot j_\gamma$ where q is the elementary charge, R is the reflectivity of the illuminated sample side and j_γ is the photon flux of the illuminating laser on the sample per square centimeter.

Transmission losses can be neglected due to the short absorption length at the applied wavelength $\lambda = 790$ nm, compared to the wafer thickness. Thus, we can determine the total dark saturation current densities of the non-metallized samples $j_{0,\text{non-metall.}}$ as well as the ones of the metallized samples $j_{0,\text{metall.}}$ using C and the averaged PLI intensity in the same region of interest (ROI) for the non-metallized ($\bar{\varphi}_2$) and metallized PLI measurement ($\bar{\varphi}_3$). The difference in total dark saturation current density

$$\Delta j_0 = j_{0,\text{metall.}} - j_{0,\text{non-metall.}} = j_{rec} \cdot C \cdot (1/\bar{\varphi}_3 - 1/\bar{\varphi}_2) \qquad (1)$$

can be associated to the dark saturation current density in the metallized area $j_{0,\text{loc}}$ by

$$\begin{aligned} j_0 &= j_{0,b} + (1 - F) \cdot j_{0e} + F \cdot j_{0,\text{loc}} \\ &= j_{0,b} + j_{0,e} + F \cdot (j_{0,\text{loc}} - j_{0e}) \qquad (2) \\ \Rightarrow \Delta j_0 &= F \cdot (j_{0,\text{loc}} - j_{0,e}), \end{aligned}$$

where $j_{0,b}$ represents the dark saturation current density of the bulk and rear side and F is the metallization fraction of the metallized side. If j_0 is now plotted against F, $j_{0,\text{loc}}$ can be calculated from the slope of the graph $m = j_{0,\text{loc}} - j_{0e}$, if j_{0e} is known.

A further approach to quantify the locally enhanced recombination in a certain region is a comparison between numerical PLI simulations using Quokka3 [17] and PLI measurements, as similar proposed in Ref. [8, 9]. Further information about the detailed approach will be published [18]. The major benefit of this simulative approach in contrast to the calibrated PLI is the fact, that it accounts for a lateral and vertical non-homogeneous carrier injection density Δn in the sample. The assumption of homogeneous Δn is less fulfilled especially for high $j_{0,\text{loc}}$ values. Therefore, in this work

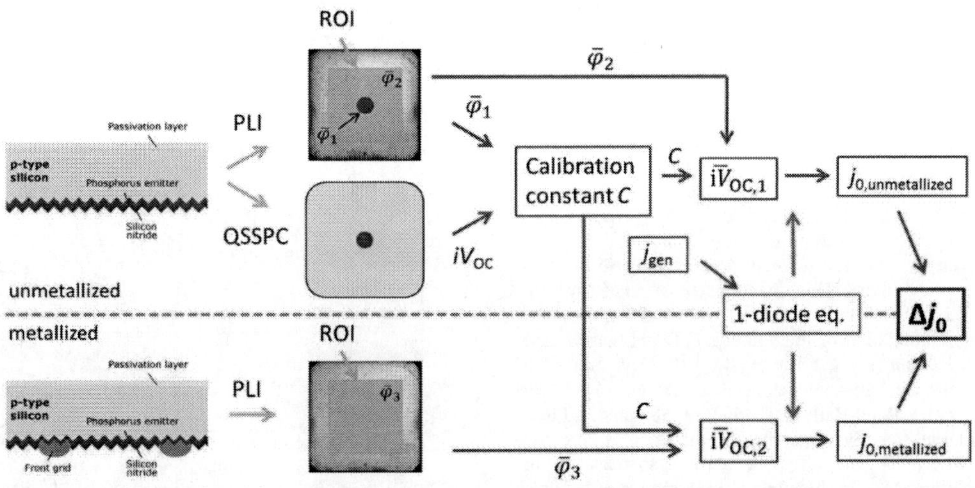

Fig. 2: Scheme of the measurement routine for the calibrated photoluminescence imaging (PLI) method exemplarily shown for *p*-type PERC-samples. Combining the resulting Δj_0 and the metallization fraction F, $j_{0,loc}$ can be determined. A detailed description of the single steps can be found in section 3.

we use this approach to estimate the influence due to a non-uniform Δn on the apparent $j_{0,loc}$ determined by calibrated PLI.

4 RESULTS

In this section the influence of sample-specific parameters as ρ_B, R and metastable defects in the wafer bulk on the calibrated PLI method are discussed in detail. At first the influence of ρ_B is discussed, which is shown in Fig. 3. On top, the influence of the input parameter ρ_B on the apparent iV_{OC} (left axis) measured by QSSPC in the non-metallized state (at $G_{QSSPC} = G_{PLI}$) and on the apparent Δj_0 between metallized an non-metallized samples (right axis) using calibrated PLI (see Fig. 2, assuming a rear side reflectivity of $R_{average} = 28\%$) is exemplarily shown for a *p*-type PERC-sample with $\rho_B \approx 1.85\ \Omega$cm (measured after thermal donor annihilation). The iV_{OC} measured by QSSPC is given by

$$\mathrm{i}V_{OC} = kT/q \cdot \ln(((\Delta n + N_D)\cdot\Delta n)/n_i^2), \quad (3)$$

where k is the Boltzmann constant, $T = 25$°C is the temperature at standard testing conditions, Δn is the minority excess carrier density, n_i is the intrinsic charge carrier density(here $n_i = 8.6\cdot10^9$ cm^{-3} as it is used in the Sinton Software Version 4.6.0) and N_D is the dopant concentration, which is defined by ρ_B.

The wafers are specified with $1\ \Omega\text{cm} \leq \rho_B \leq 3\ \Omega$cm. A variation from $\rho_B = 1\ \Omega$cm to $\rho_B = 3\ \Omega$cm affects the apparent iV_{OC} by more than 25 mV. As explained in section 3, iV_{OC} affects the calibration constant C exponentially and hence the extracted Δj_0 rises exponentially as a function of iV_{OC}. Therefore, the apparent Δj_0 changes from $\Delta j_0 \approx 70$ fA/cm^2 to $\Delta j_0 \approx 200$ fA/cm^2 if ρ_B is changed from $\rho_B = 1\ \Omega$cm to $\rho_B = 3\ \Omega$cm, respectively.

Usually, Cz-Si wafers contain thermal donors, making a representative measurement of ρ_B impossible without an additional thermal donor annihilation process [10] or the etch back subsequently to the process chain. At the bottom of Fig. 3 exemplary measurements of ρ_B

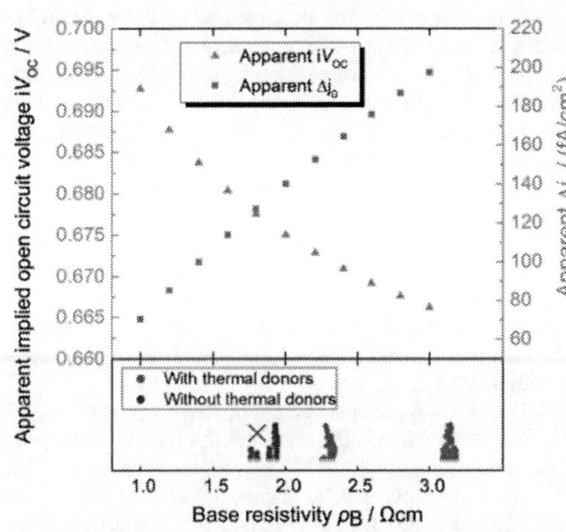

Fig. 3: Influence of the base resistivity ρ_B on the apparent iV_{OC} measured by QSSPC as well as on the apparent Δj_0 determined by the calibrated PLI method. ρ_B is varied from $\rho_B = 1\ \Omega$cm to $\rho_B = 3\ \Omega$cm similar to the manufacturer specifications of the samples. At the bottom the measured ρ_B values of the *p*-type PERC-samples are given before and after thermal donor annihilation. The base resistivity of the investigated sample is marked with a red cross.

with and without thermal donors are shown, using industrial *p*-type Cz-Si precursors from the same box as used for the experiment of this work. The base resistivity ρ_B decreases by up to more than $1\ \Omega$cm after thermal donor annihilation.

For the measurements shown in Fig. 3, the Δj_0 results taking into account the values for ρ_B without thermal donors differ by up to approximately 75% compared to the Δj_0 results using ρ_B measured before diffusion—with thermal donors. In consideration of the strong deviation of the Δj_0 results an accurate knowledge of ρ_B is crucial for a precise analysis of the metal induced recombination.

A further parameter, which influences the evaluation, is the density of metal impurities within the Si wafer. Fig.

4 shows the normalized averaged PLI intensity as a function of the illumination time at an illumination photon flux of $2.5 \cdot 10^{17}$ cm^{-2}s^{-1} (corresponding to 1 sun intensity, which is equivalent to 1000 W/m^2) preliminary to the measurement, exemplarily for a n-type and a p-type sample as described in section 2. The measurement was performed by repetitively illuminating the sample for $t_{ill} = 4$ s and then taking a PLI image with an exposure time of 1 s. The PLI intensity φ of the p-type sample increases to its maximum within the first 20 s of illumination. The initial φ of the exemplarily shown sample is $\varphi \approx 82$ %. This effect can be attributed to iron impurities which form iron-boron pairs (FeB) in the dark. During illumination of the sample, FeB dissociates into interstitial iron (Fe_i) and boron (B) [19]. For open circuit conditions with injection carrier densities $\Delta n > 10^{14}$ cm^{-3} the Shockley-Read-Hall (SRH) lifetime of interstitial iron τ_{Fei} is larger than the respective SRH lifetime of FeB τ_{FeB} [19]. Therefore, the effective lifetime increases after illumination (Verified by effective lifetime measurements using QSSPC). In contrast the PLI intensity for n-type samples stays constant and is independent on the time of previous illumination, since there are no FeB pairs present.

Using the calibrated PLI method it is important to perform the PLI measurements at a stable state of the samples, at which further illumination does not change the effective lifetime τ_{eff}. Otherwise the measurements before and after metallization are difficult to compare, especially if different exposure times are applied. Therefore, we suggest a measurement procedure—if a similar characteristic as for the p-type samples is present—where the samples are illuminated at least 20 s by the PLI laser preliminary to the measurement. Additionally, it has to be mentioned that the strong carrier injection dependency (due to FeB impurities) of the effective lifetime affects the validity of the one-diode equation.

The last sample-specific parameter discussed here, is the reflectivity R of the rear side illuminated during the PLI measurement. On the top of Fig. 5 the influence of R on the apparent implied open circuit voltage iV_{OC} (determined by QSSPC, assuming a fixed optical constant f_{Abs}, since the QSSPC measurement was performed with the textured front side pointing upwards, R was found to be almost constant for the front side from sample to sample) and on the apparent Δj_0 using calibrated PLI is exemplarily shown for one p-type sample. By increasing the reflectivity from $R = 0\%$ to $R = 40\%$ the apparent iV_{OC}

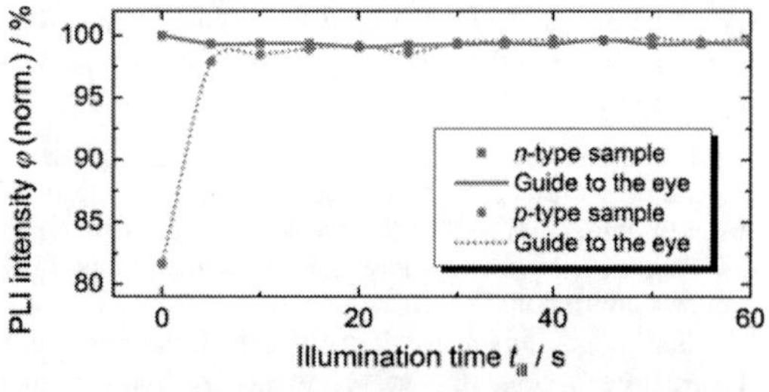

Fig. 4: The normalized PLI intensity φ in a certain region of interest as a function of the illumination time t_{ill}—at1 sun intensity—preliminary to the measurement is shown for an exemplary p-type and n-type sample. The samples are preliminary illuminated by the laser of the PLI setup.

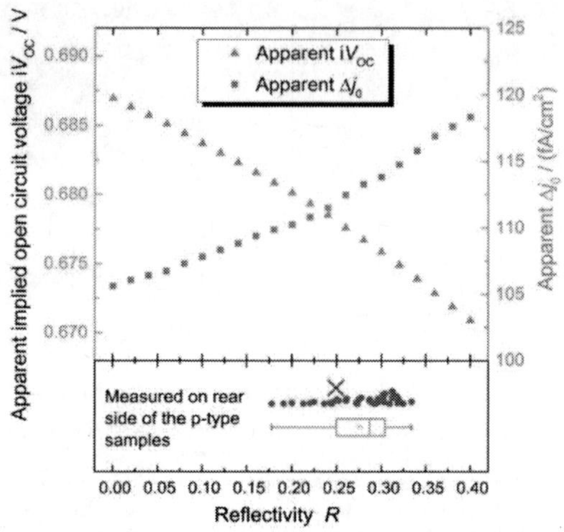

Fig. 5: On top the impact of the rear side reflectivity R on the apparent iV_{OC} as well as the apparent Δj_0, determined by the calibrated PLI method, is shown. The PLI measurement is performed with the rear side pointing upwards to the illumination source and detector. At the bottom the measured R values of the used set of p-type PERC-samples are demonstrated (one measurement per wafer in the wafer center). The R of the investigated sample is marked with a red cross.

drops from i$V_{OC} \approx 687$ mV to i$V_{OC} \approx 670$ mV and the apparent Δj_0 increases from $\Delta j_0 \approx 105$ fA/cm^2 to $\Delta j_0 \approx 120$ fA/cm^2. Two impacts have to be considered. On the one hand R affects the generation rate $G = j_Y \cdot (1-R)/W$, where W is the thickness of the sample. Increasing R leads to a decrease in G, since less photons enter the sample. As explained in section 3, the iV_{OC} measured by QSSPC is evaluated at G of the PLI measurement. Therefore, an overestimation of R leads to an underestimation of iV_{OC} and thus, to an overestimation of both, the calibration constant C and Δj_0 (see section 3). On the other hand, R affects the calculation of the recombination current density $j_{rec} = q \cdot (1-R) \cdot j_Y$ which is linearly connected to Δj_0 (see section 3). In this case an overestimation of R leads to an underestimation of j_{rec} and thus, to an underestimation of Δj_0. In our case, the latter aspect is weaker compared to the first, thus these two impacts do not fully compensate each other.

At the bottom of Fig. 5, the measured reflectivity R of the plain passivated rear side of the p-type samples (wafer center, one measurement per sample) at the wavelength of the PLI laser $\lambda_{Laser} = 790$ nm is shown. For the used set of identically processed wafers, the results vary within $18\% < R < 33\%$ resulting in a mean value $R_{mean} = (28 \pm 4)\%$. Although our samples show a quite large sample-to-sample variation in R the use of a wrong R-value for a specific sample would lead to an error in Δj_0 of about 5% in the worst case. Therefore the influence of R on the resulting Δj_0 is less significant than the influence of ρ_B.

In Reference [13], the locally enhanced recombination of the laser ablated samples is determined using the calibrated PLI method. The reported value in this publication is $j_{0,loc} = (3700 \pm 500)$ fA/cm^2. In our work the identical samples are investigated using the simulative method as described in section 3. The result using this approach is $j_{0,loc} = (4100 \pm 100)$ fA/cm^2. Further details of this evaluation will be published in Ref. [18]. It has to be mentioned that the error estimated for the calibrated PLI approach $\sigma_1 = 500$ fA/cm^2 includes also systematical errors, whereas the error for the simulative approach $\sigma_2 = 100$ fA/cm^2 only includes the standard deviations of the results from laser ablated samples with different area fractions. The results are in good agreement within the error tolerances.

Theoretically, it is expected that the calibrated PLI underestimates the resulting $j_{0,\mathrm{loc}}$. The underestimation of $j_{0,\mathrm{loc}}$ originates from the assumption of a uniform Δn. Especially, for high $j_{0,\mathrm{loc}}$ values or high metallization fractions, this condition is violated, as it is also discussed in [20]. Therefore Δn is overestimated at the highly recombinative surface and hereby $j_{0,\mathrm{loc}}$ is underestimated. In our comparison we identify a similar tendency.

The results further imply that the 'easy-to-perform' approach of the calibrated PLI delivers results in good accordance with the simulations, even the investigated $j_{0,\mathrm{loc}} > 4000$fA/cm^2 induces stronger Δn variations than most of the screen printed pastes, which normally induce $j_{0,\mathrm{loc}} \leq 3000$fA/cm^2 [21]. For lower $j_{0,\mathrm{loc}}$ values the deviation of Δn is expected to be smaller. Additionally, it is important to mention that the results in this work are generated on symmetrical samples with both-sided emitter. For structures without highly doped regions (e.g. PERC rear contacts) the Δn deviations are expected to be stronger and therefore the results for $j_{0,\mathrm{loc}}$ between the simulative method and calibrated PLI are expected to differ more significantly.

5 SUMMARY

In summary, we find a strong sensitivity of the calibrated PLI method on specific sample parameters. Especially the base resistivity ρ_B and metastable impurities can falsify the results significantly. For our samples a determination without considering thermal donors can falsify the result by approximately 75%. Excluding the effect of metastable impurities is important, since the measurements before and after metallization should be performed in a similar and stable state. Therefore, an accurate knowledge of the samples' parameters and properties is crucial.

Additionally, a comparison between the simulative approach using Quokka3 and the calibrated PLI method indicates just a minor difference within the error tolerances. We conclude that the results using calibrated PLI with accurate knowledge of the samples' parameters and proper realization delivers accurate results on a full area emitter.

ACKNOWLEDGEMENT

The authors would like to thank Sebastian Meier-Meybrunn and Jonas Huyeng for the support with the simulations. Further, David Herrmann would like to thank the "Deutsche Bundesstiftung Umwelt DBU" for funding his dissertation project. This work was supported by the German Federal Ministry for Economic Affairs and Energy BMWi and by the industry partners within the research POLDI under contract No. 0324079D. The authors are responsible for the content.

5 REFERENCES

[1] ITRPV, "International Technology Roadmap for Photovoltaic: 2016 Results," Jun. 2017. [Online] Available: http://www.itrpv.net/Reports/Downloads/.

[2] P. Saint-Cast *et al.*, "Analysis of the losses of industrial-type PERC solar cells," *Phys. Status Solidi A*, vol. 214, no. 3, p. 1600708, 2017.

[3] M. Müller *et al.*, "Loss analysis of 22% efficient industrial PERC solar cells," *Energy Procedia*, vol. 124, pp. 131–137, 2017.

[4] S. Wasmer, A. A. Brand, J. M. Greulich, "Metamodelling of numerical device simulations to rapidly create efficiency optimization roadmaps of monocrystalline silicon PERC cells," *Energy Procedia*, vol. 214C, 2017.

[5] H. Hannebauer, T. Dullweber, U. Baumann, T. Falcon, and R. Brendel, "21.2%-efficient fineline-printed PERC solar cell with 5 busbar front grid," *Phys. Status Solidi RRL*, vol. 8, no. 8, pp. 675–679, 2014.

[6] S. Herlufsen et al., "Photoconductance-calibrated photoluminescence lifetime imaging of crystalline silicon," *Phys. Status Solidi RRL*, vol. 2, no. 6, pp. 245–247, 2008.

[7] B. Hallam et al., "Photoluminescence imaging for determining the spatially resolved implied open circuit voltage of silicon solar cells," *Journal of Applied Physics*, vol. 115, no. 4, p. 44901, 2014.

[8] M. Ernst, A. Fell, E. Franklin, and K. J. Weber, "Characterization of Recombination Properties and Contact Resistivity of Laser-Processed Localized Contacts From Doped Silicon Nanoparticle Ink and Spin-On Dopants," *IEEE J. Photovoltaics*, vol. 7, no. 2, pp. 471–478, 2017.

[9] A. Fell, D. Walter, S. Kluska, E. Franklin, and K. Weber, "Determination of Injection Dependent Recombination Properties of Locally Processed Surface Regions," *Energy Procedia*, vol. 38, pp. 22–31, 2013.

[10] W. Götz et al., "Thermal donor formation and annihilation at temperatures above 500 °C in Czochralski-grown Si," *Journal of Applied Physics*, vol. 84, no. 7, pp. 3561–3568, 1998.

[11] S. Werner *et al.*, "Key aspects for fabrication of p-type Cz-Si PERC solar cells exceeding 22% conversion efficiency," in *33rd EU PVSEC*, Amsterdam, The Netherlands, 2017.

[12] R. A. Sinton and A. Cuevas, "Contactless determination of current–voltage characteristics and minority-carrier lifetimes in semiconductors from quasi-steady-state photoconductance data," *Appl. Phys. Lett.*, vol. 69, no. 17, pp. 2510–2512, 1996.

[13] H. Höffler et al., "Uncertainty in the Determination of Local Saturation Current Densities," Otsu, Japan, Nov. 14 2017.

[14] A. Grohe *et al.*, "Novel low temperature front side metallisation scheme using selective laser ablation of anti reflection coatings and electroless nickel plating," in *21st EU PVSEC*, Dresden, 2006, pp. 750–753.

[15] J. Bonse, K.-W. Brzezinka, and A.J. Meixner, "Modifying single-crystalline silicon by femtosecond laser pulses: An analysis by micro Raman spectroscopy, scanning laser microscopy and atomic force microscopy," *Applied Surface Science*, vol. 221, no. 1-4, pp. 215–230, 2004.

[16] P. Wurfel, "The chemical potential of radiation," *J. Phys. C: Solid State Phys.*, vol. 15, no. 18, pp. 3967–3985, 1982.

[17] A. Fell, J. Schön, M. C. Schubert, and S. W. Glunz, "The concept of skins for silicon solar cell modeling," *Sol. Energy Mater. Sol. Cells*, vol. 173, pp. 128–133, 2017.

[18] D. Herrmann et al., "to be published,"

[19] D. Macdonald, A. Cuevas, and J. Wong-Leung, "Capture cross sections of the acceptor level of iron–boron pairs in p-type silicon by injection-level dependent lifetime measurements," *J. Appl. Phys.*, vol. 89, no. 12, p. 7932, 2001.

[20] R. Dumbrell, M. K. Juhl, T. Trupke, and Z. Hameiri, "Extracting Metal Contact Recombination Parameters From Effective Lifetime Data," *IEEE J. Photovoltaics*, pp. 1–8, 2018.

[21] T. Fellmeth *et al.,* "Recombination at metal-emitter interfaces of front contact technologies for highly efficient silicon solar cells," *Energy Procedia*, vol. 8, pp. 115–121, 2011.

INFLUENCE OF EMITTER LAYERS ON LETID KINETICS IN MULTICRYSTALLINE SILICON

Alona Otaegi[1,2], Daniel Skorka[1], Andreas Schmid[1], Annika Zuschlag[1], Juan Carlos Jimeno[2], and Giso Hahn[1]
[1]University of Konstanz, Department of Physics, 78457 Konstanz, Germany
[2]Technological Institute of Microelectronics, UPV/EHU, 48013 Bilbao, Spain

ABSTRACT: The influence of emitter layers on LeTID kinetics in multicrystalline silicon is analysed in this work. Multicrystalline silicon samples with emitter layers with sheet resistivity of 60, 175, 200, 230 and 360 Ω/sq have been processed and exposed afterwards to 1 sun illumination at 75°C for more than 2000 h. The effective lifetime of the minority charge carriers of the samples has been measured repetitively by spatially resolved photoluminescence. The resulting lifetime maps are analysed and the effective defect concentration of the samples is calculated. This approach leads to different degradation rates, depending on the emitter sheet resistivity, and to the same defect regeneration rate, but differing in the onset of regeneration.
Keywords: Multicrystalline silicon; LeTID; degradation; regeneration; emitter layer

1 INTRODUCTION

The solar cell market is evolving towards a major implantation of PERC (passivated emitter and rear cell) cells over Al-BSF (back surface field) cells [1]. Efficiencies above 21% are obtained from PERC-type solar cells due to less recombination at the rear and the increase of rear surface reflection by an extra rear passivation [2-4], and production costs are close to standard Al-BSF technology [5]. However, this technology can be particularly sensitive to elevated temperatures under illumination in a phenomenon entitled LeTID (light and elevated temperature induced degradation), and the efficiency can drop by more than $10\%_{rel}$ [6-9].

The easiest way to describe the LeTID phenomenon in multicrystalline silicon (mc-Si) PERC solar cells is via a 3-state reaction scheme analogue to the effect observed for BO-degradation in Cz Si [10]; however, the inducing mechanisms are different for the different technologies, since BO complex formation and FeB pair dissociation have been found to be responsible for Cz Si degradation [11, 12], but have been discarded for LeTID [6-8].

While the main root causing LeTID remains unknown, changes in the processing steps are involved in the research for eliminating or reducing the effect. Gettering is shown to be efficient to diminish the degradation effect on good material quality [13]; a dark anneal after firing reduces LeTID [14]; exposing samples to elevated temperatures and high illumination accelerates the degradation and following regeneration process [15, 16]; and changing the temperature profile to avoid a fast cool cycle during firing avoids the effect [17, 18]. Up to now, most studies have shown the necessity to apply a thermal load after diffusion at temperatures well above 600°C (*e.g.* in form of a firing step) to activate the occurrence of LeTID (*e.g.* [19]). But latest results indicate that it is not the temperature load of the firing step that is needed, but the presence of hydrogen which is causing LeTID [20].

The relevance of the bulk lifetime and instabilities of the passivation layer have been subjects under study [21-24]; however, no mention of emitter layers and their influence on LeTID kinetics have been found in literature. This work aims to delve into a further characterization by analysing the influence of the presence of emitter layers on LeTID kinetics, which could help to understand the mechanisms involved.

2 EXPERIMENTAL DETAILS

Fig. 1 sketches the setup of the applied process sequence. Starting with 1.5 Ωcm boron doped mc-Si, wafers with comparable grain and defect structure (sister wafers) were used to study the influence of different emitter layers on LeTID.

The samples were chemically etched to remove saw damage. A standard gettering was applied afterwards by $POCl_3$ diffusion (60 Ω/sq). The emitter of some samples was then removed partially via wet chemical etch-back in order to analyse the influence of emitter layers' sheet resistivity R_{sheet} on the LeTID degradation and regeneration process. To avoid uncertainties of the etch-rate due to the different crystallographic orientations, R_{sheet} of the samples was measured on (100)-oriented Cz-samples. The measured R_{sheet} values correspond to 60 Ω/sq (no emitter removal), 175 Ω/sq, 200 Ω/sq, 230 Ω/sq and 360 Ω/sq (maximum emitter removal). A PECVD (plasma-enhanced chemical vapour deposition) SiN_x:H layer was then deposited on the samples and the manufacturing process was finished with a firing step in a belt furnace with peak wafer temperature of 786°C, monitoring the temperature on the wafer surface with a K-type thermocouple.

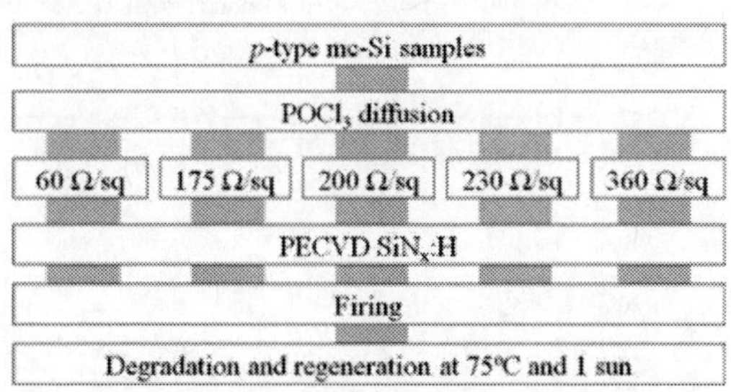

Figure 1: Process sequence of the investigated samples.

After firing, the samples are held at a temperature of approximately 75°C on a hot-plate under constant illumination with halogen lamps (0.9±0.1 suns). Effective minority charge carrier lifetime τ_{eff} is measured repetitively by the fast and self-calibrated transient photoluminescence imaging (TR-PLI) method [25 26] at room temperature. For each measurement, a spatially resolved lifetime map is obtained, as well as the harmonic average τ_{eff}. This method allows tracking the changes in τ_{eff} in different areas of the samples throughout the experiment.

3 RESULTS AND DISCUSSION

Fig. 2 shows τ_{eff} maps of three relevant stages throughout the experiment for the samples with R_{sheet} of 200 Ω/sq and 360 Ω/sq, respectively. The first stage corresponds to the initial situation, right after the firing with no degradation (left); the second stage corresponds to a full degradation of the samples with τ_{eff} below 20 µs after 400 h (middle). The third stage shows maps of the samples during regeneration (right).

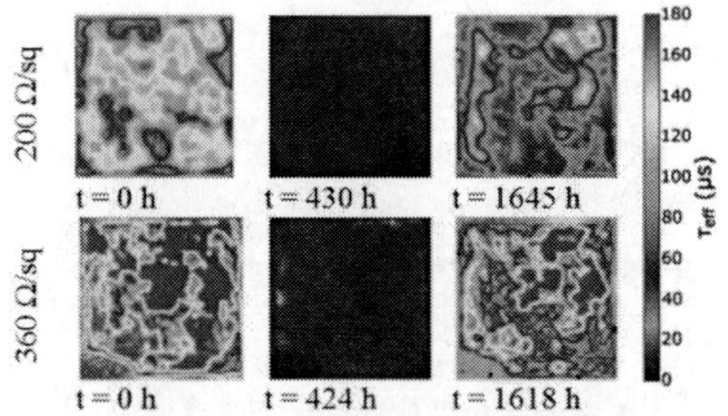

Figure 2: Effective lifetime mappings for samples with R_{sheet} of 200 Ω/sq (upper row) and 360 Ω/sq (lower row) directly after firing (left), around maximum degradation (middle) and after onset of regeneration (right).

Fig. 3 shows the evolution of the harmonic average τ_{eff} for all samples under study, from their initial state to above 2,000 h. In this figure, the initial τ_{eff} of the samples are different as expected, due to different recombination strengths in the emitter. Initially, the τ_{eff} distribution of all samples narrows down to a low lifetime level with different slopes. After around 100 h, the different samples show almost identical values indicating that the different emitter layers have no significant effect on maximum LeTID defect concentration. All the samples regenerate, although the onset of regeneration starts earlier for higher emitter sheet resistivities (inset in Fig. 3).

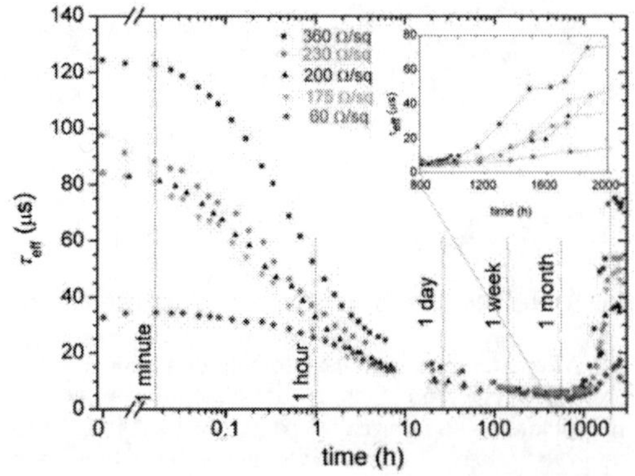

Figure 3: Evolution of the average τ_{eff} for the samples with different emitter layers (75°C, 0.9 suns).

The rainbow plot in Fig. 4 helps to recognize what is happening in each of the samples' regions; this rainbow plot represents the τ_{eff} behaviour during the experiment time for the sample. Each line in the graph represents an area of 150x150 µm^2 in the spatially resolved TR-PLI lifetime measurement and represents one of 2,500 small areas on the sample. The colour code applied is based on the initial $\tau_{eff}(0)$ of each sample at the beginning of the experiment. The dashed line represents the harmonic average τ_{eff} of the sample throughout the degradation-regeneration experiment. All investigated samples follow a similar tendency as shown exemplarily in Fig. 4 for the 200 Ω/sq sample. Regeneration sets in earlier in areas of high initial material quality (starting at 800 h in Fig. 4) and later in areas of lower initial material quality.

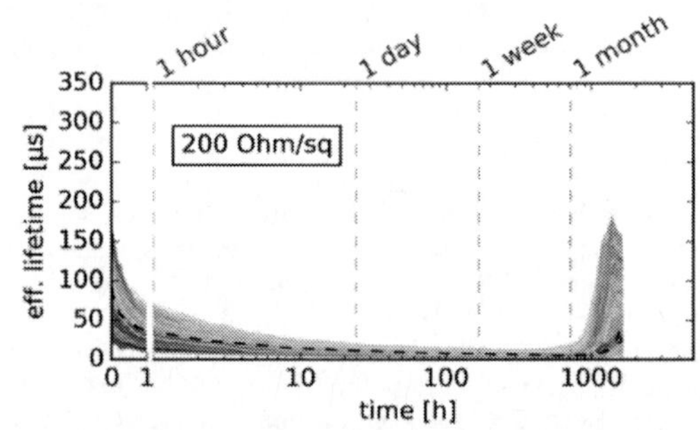

Figure 4: Effective lifetimes for a sample with emitter sheet resistivity of 200 Ω/sq (x-axis: 0-1 h linear scale, 1-2000 h log scale).

3.2 Analysis of the degradation

Degradation of the samples can be seen during the first few hundred hours of the experiment and all the samples show similar minimum average τ_{eff} values well below 10 µs after around 100 h of treatment time. However, two degradation behaviours can be detected from the data shown in Fig. 3; a *fast* degradation mechanism from the start of the experiment up to the first hour and a *slow* degradation mechanism from the first hour up to 100 h approximately. This evolution is consistent with [27].

To further quantify the degradation, the time-dependent effective defect concentration, $N^*(t)$, is determined as also used in, *e.g.*, [27-29].

—— —— (1)

For degradation, it is possible to fit $N^*(t)$ by a two exponential model as also used in [27]:

(2)

with a and b the pre-factors for the degradation rates R_{fast} and R_{slow}. Fig. 5 shows exemplarily the $N^*(t)$ values over treatment time for the two samples with 230 and 360 Ω/sq, respectively, using the fit parameters listed in Tab. I. The quality of the fits according to Eq. (2) (also included in Fig. 5 as solid lines) is generally very good and reliable as indicated by RMSE values close to 1. Nevertheless, fits to $N^*(t)$ using Eq. (2) for samples with 175 and 200 Ω/sq, respectively, resulted in slightly worse RMSE values, and fitting parameters obtained for these samples have a higher inaccuracy.

Table I shows the fitting parameters obtained with

MATLAB® for the pre-factors as well as for the degradation rates following the two exponential model.

Table I: Fitting of $N^{*}(t)$ according to the two exponential model for the samples under study.

Emitter	Fast degradation		Slow degradation	
	a	R_{fast} [1/h]	b	R_{slow} [1/h]
60 Ω/sq	0.0589	0.150	0.132	0.00574
175 Ω/sq	0.0965	0.207	0.131	0.00364
200 Ω/sq	0.0836	0.181	0.115	0.00316
230 Ω/sq	0.0417	0.454	0.139	0.00717
360 Ω/sq	0.0251	0.534	0.129	0.0110

The fast degradation rate R_{fast} depends on the emitter sheet resistivity; in fact, the higher the R_{sheet}, the faster the degradation seems to happen; however, the slow degradation rate R_{slow} seems to be slightly less dependent on the characteristics of the emitter.

The two exponential model for degradation has already been reported for mc-Si material [19, 27].

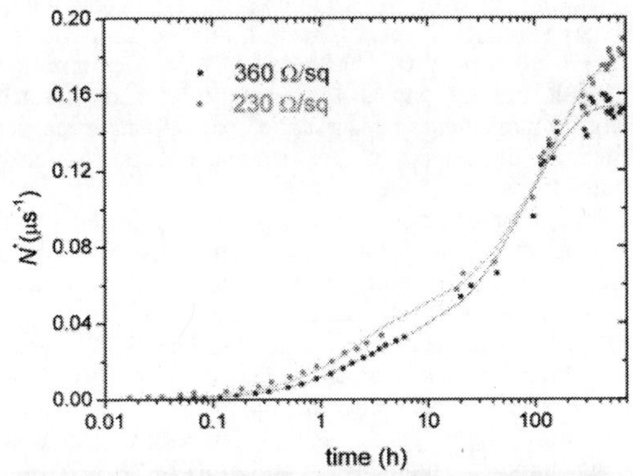

Figure 5: Evolution of N^{*} for the samples with R_{sheet} of 230 and 360 Ω/sq. The solid lines are numerically simulated by solving Eq. (2) using the coefficients from Tab. I.

3.3 Analysis of the regeneration

The inset of Fig. 3 shows the behaviour of the harmonic average τ_{eff} at the onset of regeneration for the samples under study, from 800 h up to the end. In this figure, it can be seen that regeneration of the samples starts around 1,000 h of treatment time.

The regeneration of the sample with maximum emitter removal (360 Ω/sq) begins earlier, and after >1,600 h the regeneration is more pronounced and areas with τ_{eff} of >180 µs can be found in regions far from the edge (from Fig. 2).

A more detailed sketch is shown in Fig. 6 where the evolution of N^{*} for all the samples is shown for the time interval of 800-2,500 h. In this figure, a fitting to a single exponential term is made for 60, 200 and 360 Ω/sq samples, shown with dash-dotted lines. According to the fitting, all the samples show the same regeneration rate, differing only in the moment at which regeneration starts. The higher the R_{sheet}, the earlier the regeneration starts. The three samples with R_{sheet} of 175, 200 and 230 Ω/sq show approximately the same N^{*} values over time.

The rainbow plot in Fig. 4 shows that even though the regions with the highest effective lifetime recover up to their initial value, regeneration is not complete as the average τ_{eff} is only around half of its initial value. This behaviour of incomplete regeneration has also been found in boron-oxygen defect experiments [30], where slow regeneration rates are related to incomplete regeneration processes. In addition, the SiN_x:H surface passivation degrades as well under these treatment conditions as shown, *e.g.*, in [23, 24].

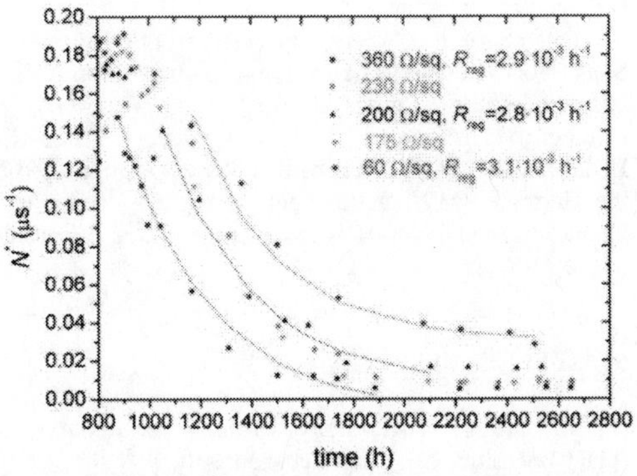

Figure 6: Evolution of N^{*}, from 800 h of experiment up to the end, for the samples under study.

It was speculated that regeneration rate is increased by higher excess carrier density Δn (*e.g.*, [8, 13]), but in our case Δn should be almost identical for all samples between 100-900 h of treatment. Nevertheless, the samples with lower emitter sheet resistivity show an earlier onset of regeneration leaving room for other explanations.

4 CONCLUSIONS

The influence of the emitter layers on LeTID kinetics in mc-Si was analysed in this work. The LeTID effect for samples with emitter sheet resistivities of 60, 175, 200, 230, and 360 Ω/sq, respectively, was studied at 75°C and 1 sun illumination for more than 2,000 h. The evolution of the samples was tracked via lifetime maps obtained by TR-PLI measurements.

In terms of degradation, two behaviours have been observed; a fast degradation rate, relevant up to the first hour approximately, and a slow degradation rate, from 1 hour up to 100 hours. The fast degradation rate seems to be dependent on the emitter nature, and is faster for higher emitter sheet resistivities. The slow degradation rate does not seem to be influenced by the nature of the emitter.

In terms of regeneration, similar regeneration rates have been found regardless of the emitter sheet resistivity, but the regeneration starts earlier for higher emitter sheet resistivities, although harmonic average τ_{eff} values are almost identical at maximum degradation in all samples. None of the samples is fully regenerated after ~1,800 h of treatment and the harmonic average τ_{eff} remains at about half its initial value. The regeneration seems to be better (more complete) in good quality material areas of the samples.

5 ACKNOWLEDGEMENTS

Alona Otaegi would like to gratefully acknowledge the entire team of the Department of Physics in the University of Konstanz for the assistance received during her postdoctoral stay, and especially K. Bayer, J. Lindroos, K. Weller, J. Fichtner, L. Mahlstädt, B. Rettenmaier, A. Dastgheib-Shirazi, and J. Engelhart for technical assistance.

This research has been supported by the Secretary of State for Research, Development and Innovation, attached to the Spanish Ministry of the Economy and Competitiveness, in its project TABACO ENE2014-56069-CA. Part of this work was supported by the German BMWi under contract numbers 0325763B, 0324204B, and 0324001. The content is the responsibility of the authors.

6 REFERENCES

[1] International Technology Roadmap for Photovoltaic (ITPV): Results 2017, ninth edition. July 2018. See http://www.itrpv.net/Reports/Downloads/

[2] A. W. Blakers, A. Wang, A. M. Milne, J. Zhao, and M. A. Green, "22.8% efficient silicon solar cell", Appl. Phys. Lett. vol. 55, pp. 1363-1365, 1989.

[3] M. A. Green, "The passivated emitter and rear cell (PERC): from conception to mass production", Sol. Energy Mater. Sol. Cells, vol. 143, pp. 190-197, 2015.

[4] O. Schultz, S. W. Glunz, and G. P. Willeke, "Multicrystalline silicon solar cells exceeding 20% efficiency", Prog. Photovolt: Res. Appl., vol. 12, pp. 553-558, 2004.

[5] P. A. Basore, "Understanding manufacturing cost influence on future trends in silicon photovoltaics", IEEE J. Photovolt., vol. 4, No. 6, pp. 1477-1492, 2014.

[6] K. Ramspeck, S. Zimmermann, H. Nagel, A. Metz, Y. Gassenbauer, B. Birkmann, and A. Seidl, "Light induced degradation of rear passivated mc-Si solar cells" Proc. of the 27th EU PVSEC, pp. 861-865, 2012.

[7] F. Fertig, K. Krauß, I. Geisemeyer, J. Broisch, H. Höffler, and J. Ove Odden, "Fully solderable large-area screen-printed Al-BSF p-type mc-Si solar cells from 100% solar grade feedstock yielding $\eta > 17\%$: challenges and potential on cell and module level", Proc. of the 27th EU PVSEC, pp. 1031-1038, 2012.

[8] F. Kersten, P. Engelhart, H. -C. Ploigt, A. Stelnikov, T. Lindner, F. Stenzel, M. Bartzsch, A. Szpeth, K. Petter, J. Heitmann, and J. W. Müller, "Degradation of multicrystalline silicon solar cells and modules after illumination at elevated", Sol. Energy Mater. Sol. Cells, vol. 142, pp. 83-86, 2015.

[9] F. Kersten, F. Fertig, K. Petter, B. Klöter, E. Herzog, M. B. Strobel, J. Heitmann, and J. W. Müller, "System performance loss due to LeTID", Energy Procedia, vol. 124, pp. 540–546, 2017.

[10] A. Herguth, G. Schubert, M. Kaes, and G. Hahn, "Investigations on the long time behavior of the metastable boron-oxygen complex in crystalline silicon", Progr. Photovolt. Res. Appl., vol. 16, pp. 135-140, 2008.

[11] J. Schmidt, A. G. Aberle, and R. Hezel, "Investigation of carrier lifetime instabilities in Cz-grown silicon", Proc. 26th IEEE PVSC, pp. 13-19, 1997.

[12] L. J. Geerligs, and D. Macdonald, "Dynamics of light-induced FeB pair dissociation in crystalline silicon", Appl. Phys. Lett., vol. 85, 5227, 2004.

[13] A. Zuschlag, D. Skorka, and G. Hahn, "Degradation and regeneration analysis in mc-Si after different gettering steps", Prog. Photovolt: Res. Appl., vol. 25, pp. 545-552, 2017.

[14] C. Chan, T. H. Fung, M. Abbott, D. Payne, A. Wenham, B. Hallam, R. Chen, and S. Wenham, "Modulation of carrier-induced defect kinetics in multi-crystalline silicon PERC cells through dark annealing", Sol. RRL, 1600028, 2017.

[15] K. Krauβ, A. A. Brand, F. Fertig, S. Rein, and J. Nekarda, "Fast regeneration processes to avoid light-induced degradation in multicrystalline silicon solar cells", IEEE J. Photovolt., vol. 6, No. 6, pp. 1427-1431, 2016.

[16] D. N. R. Payne, C. E. Chan, B. J. Hallam, B. Hoex, M. D. Abbott, S. R. Wenham, and D. M. Bagnall, "Rapid passivation of carrier-induced defects in p-type multi-crystalline silicon", Sol. Energ. Mat. Sol. Cells, vol. 158, pp. 102-106, 2016.

[17] R. Eberle, W. Kwapil, F. Schindler, S. W. Glunz, and M. C. Schubert, "Firing temperature profile on light induced degradation of multicrystalline silicon", Phys. Status Solidi RRL, vol. 10, pp. 861–865, 2016.

[18] R. Sharma, A. G. Aberle, and J. B. Li, "Optimization of belt furnace anneal to reduce light and elevated temperature induced degradation of effective carrier lifetime of p-type multicrystalline silicon wafers," Sol. RRL vol. 2, 1800070, 2018.

[19] D. Bredemeier, D. Walter, S. Herlufsen, and J. Schmidt, "Lifetime degradation and regeneration in multicrystalline silicon under illumination at elevated temperature", AIP Advances, vol. 6, 035119, 2016.

[20] M.A. Jensen, A. Zuschlag, D. Skorka, S. Wieghold, A.E. Morishige, G. Hahn, T. Buonassisi, "Evaluating root cause: the distinct roles of hydrogen and firing in activating light- and elevated-temperature induced degradation", J. Appl. Phys., vol. 124(8), 085701, 2018.

[21] A. Zuschlag, D. Skorka, and G. Hahn, "Degradation and regeneration analysis in mc-Si", Proc. 43rd IEEE PVSC, pp. 1051-1054, 2016.

[22] F. Kersten, J. Heitmann, and J. W. Müller, "Influence of Al_2O_3 and SiN_x passivation layers on LeTID", Energy Procedia, vol. 92, pp. 828-832, 2016.

[23] D. Sperber, A. Herguth, and G. Hahn, "Instability of dielectric surface passivation quality at elevated temperature and illumination", Energy Procedia, vol. 92, pp. 211-217, 2016.

[24] D. Sperber, A. Heilemann, A. Herguth, and G. Hahn, "Temperature and light induced changes in bulk and passivation quality of boron-doped float-zone silicon coated with SiN_x:H", IEEE J. Photovolt., vol. 7, No. 2, pp. 463-470, 2017.

[25] D. Kiliani, G. Micard, B. Steuer, B. Raabe, A. Herguth, and G. Hahn, "Minority charge carrier lifetime mapping of crystalline silicon wafers by time-resolved photoluminescence imaging," J. Appl. Phys., vol. 110, 054508, 2011.

[26] D. Kiliani, A. Herguth, G. Micard, J. Ebser, and G. Hahn, "Time-resolved photoluminescence imaging with electronic shuttering using an image intensifier unit", Sol. Energ. Mat. Sol. Cells, vol. 106, pp. 55-59, 2012.

[27] D. Bredemeier, D. Walter, and J. Schmidt, "Light-induced lifetime degradation in high-performance multicrystalline silicon: detailed kinetics of the defect activation", Sol. Energ. Mat. Sol. Cells, vol. 173, pp. 2-5, 2017.

[28] J. Lindroos, and H. Savin, "Review of light-induced degradation in crystalline silicon solar cells", Sol. Energ. Mat. Sol. Cells, vol. 147, pp. 115-126, 2016.

[29] M. A. Jensen, A. E. Morishige, J. Hofstetter, D. B. Needleman, S. Chakraborty, R. Sharma, H. S. Laine, B. Lai, V. Rose, A. Youssef, E. E. Looney, S. Wieghold, J. R. Poindexter, J. E. Correa-Baena, T. Felisca, H. Savin, J. B. Li, and T. Buonassisi, "Solubility and diffusivity: important metrics in the search of the root cause of light- and elevated temperature-induced degradation", IEEE J. Photovolt., vol. 8, No. 2, pp. 448-455, 2018.

[30] S. Wilking, M. Forster, A. Herguth, and G. Hahn, "From simulation to experiment: understanding BO-regeneration kinetics", Sol. Energ. Mat. Sol. Cells, vol. 142, pp. 87-91, 2015.

INTRODUCTION OF SMART TECHNOLOGIES IN PV PRODUCTION EQUIPMENT

M. Zimmer[1], L. Mohr[1], B. Broese[1], T. Krick[1], J. Rentsch[1], B. Mandlmeier[2], L. Papp[2], M. Menschick[2], R. Kogler[2], A. Strauch[3], T. Will[4], V. Meckelin[4], P. Mutz[5], D. Korner[5], M. Kremer[6]

1) Fraunhofer ISE, Heidenhofstrasse 2, 79100 Freiburg, Germany
2) Singulus Technologies AG, Fraunhoferstrasse 9, 82256 Fürstenfeldbruck, Germany
3) camLine Dresden GmbH, Königsbrücker Str. 28/30, 01099 Dresden, Germany
4) MIB GmbH, Bahnhofstraße 35, 79206 Breisach, Germany
5) SICK AG, Erwin-Sick-Str. 1, 79183 Waldkirch, Germany
6) JUMO GmbH & Co. KG, Moritz-Juchheim-Straße 1, 36039 Fulda, Germany

ABSTRACT: Increasing digitalization is a consequent development of the automation of PV manufacturing processes. A central concept in the digitalization of production processes is the digital twin of physical assets. This digital twin represents the asset in its properties and behavior. In this paper, it is shown, that a digital twin of a wet chemical process tool is suitable to improve processes and save production costs. Computational flow dynamics simulations were used to investigate and improve the flow properties in wet chemical process baths. Furthermore, a coupling with a chemical reaction simulation can predict the rounding process of random pyramids in ozone based cleaning baths. As a third use case, a concept for a predictive maintenance system is shown, which could reduce the unscheduled down time of the tool and therefore save maintenance costs.
Keywords: Cost reduction, Manufacturing and Processing, Simulation,

1 INTRODUCTION

The production of photovoltaic (PV) modules has to withstand a high cost pressure. The prices for photovoltaic modules decrease by approx. 23 % with every duplication of the cumulative PV module shipment [1]. This effect is shown in fig. 1, where the average module sales price is drawn over the cumulative PV module shipment.

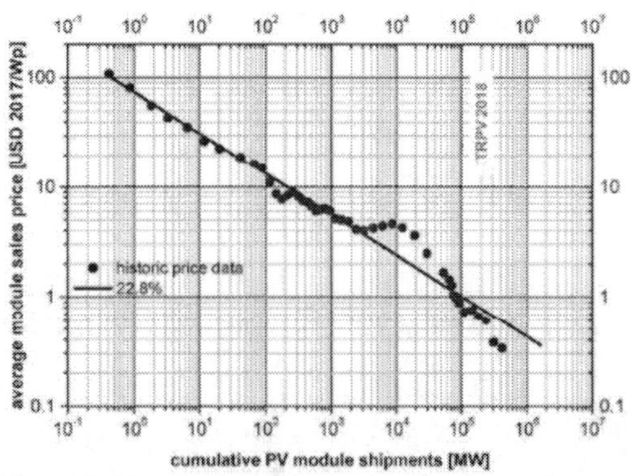

Figure I: Learning curve for module prices as a function of cumulative shipments [1]

One factor that leads to an increasing cost reduction is the degree of automation. In the economic history, several industrial revolutions far-reaching changed the production processes. The first industrial revolution was the usage of steam power and later of electricity. The second revolution was the introduction of belt production, what dramatically reduced the cost of mass production. The implementation of software, which assists the production process is called the third industrial revolution, software solutions like manufacturing execution systems (MES) are widely used to control the production process. Today, the fourth industrial revolution takes place. The rapid development of computer technology enables a decentralization of the production intelligence. Finally, this leads to more and more autonomous acting production assets.

In the beginning of the PV manufacturing, many processes were done manually, e.g. solar cell strings were soldered manually, the manual loading and unloading of production tools was common and the process quality control was conducted by offline measurements. This can be considered as "industry 2.0"

Today, high degree of automation during the PV module manufacturing is reached. Wafer loading and unloading as well as the logistic processes are mostly done completely automatically, while operators can handle several machines, where they have only a control task. The production is often controlled by a central MES. "industry 3.0" is reached in most PV fabs.

A further development of the automated fab is a completely digitalized fab according to "industry 4.0". To reach this goal, the tools will need a digital image, which represents their properties and behavior – a digital twin. This paper shows three use cases, how such a digital twin can help to improve the manufacturing tools and processes as well as decrease the manufacturing costs for solar cells.

2 THE DIGITAL TWIN

2.1 Concept of a Digital Twin

The digital twin is a concept of the so called "internet of things" and is a digital representation of a physical or virtual object. The digital twin contains all relevant data of an asset like geometries, electric and media flow, sensor data which describes the actual state of processes as well as the machine itself. Furthermore, it enables simulations, which are able to predict the behavior of the asset or its components in the future.

In this paper, the main focus is put on the content and usage of the data, which is stored in the digital twin. Details like file formats, data standards etc. are not subject of our investigations, but still an important open point in current industry 4.0 discussions.

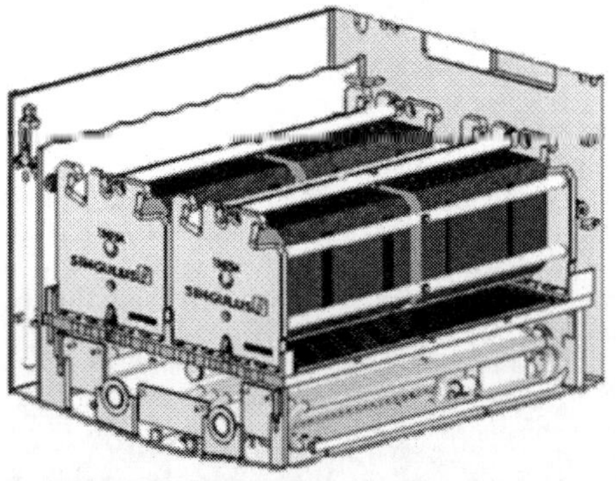

Figure II: Process bath with two carriers inside.

2.2 A Wet Chemical Process Tool in the Context of the Internet of Things

The tool consists of three alkaline process baths for alkaline preclean, saw damage etching and texturing. Furthermore, two ozone-based cleaning baths and one ozone-free cleaning bath for HF/HCl as well as quick dump and overflow rinses are available.

Figure II shows a process bath which is used in the wet chemical process tool. The process solution is supplied via supply pipes in the bottom of the bath, where heaters are also located in alkaline process basins. The solutions flow is then homogenized through a perforated plate, on which two carriers with each 100 wafers are placed.

The digital twin of such a system contains the properties like geometries, the electrical wiring scheme and the piping, but it should also represent the behavior of the system. Therefore, additional simulation models have to be developed, which describe the behavior of the asset during production and in case of maintenance.

3 USE CASES

As it was described above, the digital twin represents a physical asset during the whole life cycle. In the following use cases, it will be demonstrated, how the digital twin can help to improve the quality of process tools, of processes and to minimize the production costs of solar cells.

3.1 Use Case 1: Digital Engineering

Usually, the bath of a wet chemical process tool is circulated. The process media flow should be homogeneously distributed over the whole basin. Using the geometry and flow data from the digital twin enables to simulate the flow in the basin using computational flow dynamics (cfd) simulations with a finite element method

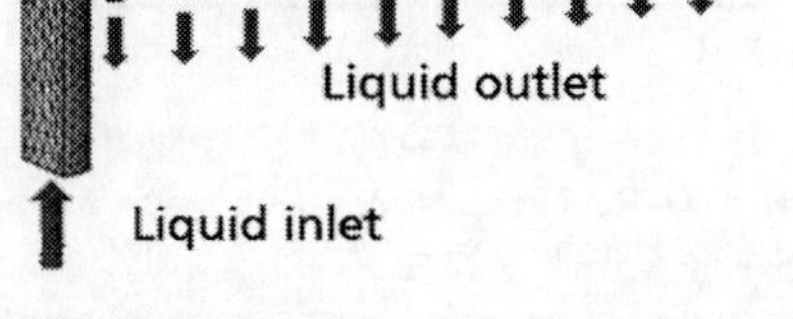

Figure III: Symmetry element of a T-type inlet pipe

The simulation of the whole bath is very time consuming, and therefore not well suited for the optimization of geometric elements in the process bath. The simulation can be divided in different parts, which can be optimized one after the other. In this case, it is assumed – and was proven [2] – that the process bath itself has no influence on the flow in the inlet pipes.

A symmetry element of a T-type inlet pipe is shown in Figure III, the inlet for the process solution is on the bottom, 10 holes that are directed to the bottom work as outlet into the basin. If all holes had the same area, the pressure and therefore the flow velocity out of the openings would vary over the pipe length. Hence, a continuous reduction of the opening area has to compensate that effect. Therefore, the ratio between the radius of the first and the last opening was varied from 100% (all radii equal) to 82%, shown in Figure IV.

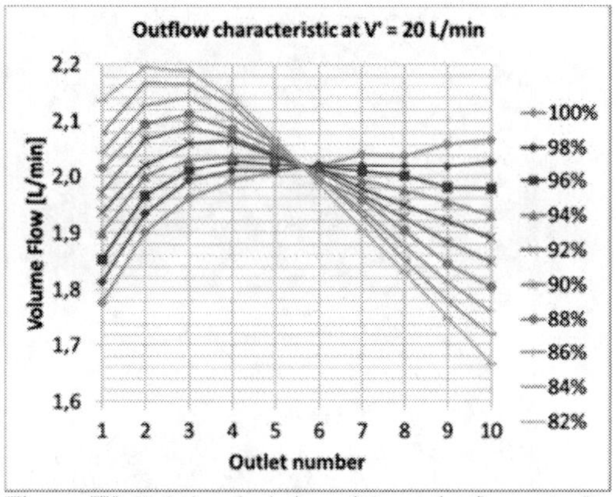

Figure IV: Relative deviation of the outlet flow velocity over the outlet opening position depending on the ratio of the opening radii r_1/r_{10}

3.2 Use Case 2: Process Simulation

The information of a digital twin is not limited to the engineering phase; Future process tools will more and more work autonomously, optimization of process recipes will be done by the tool itself. To reach this target, simulations, which predict the process results are necessary. An interesting process to simulate is the etching and rounding of pyramid tips in the ozone clean, where a diluted mixture of hydrofluoric acid (HF) and hydrochloric acid (HCl) are used for wafer cleaning. The presence of ozone has an oxidizing effect on the surface cleaning, while the HF removes SiO_2 immediately from the surface, what leads to a slow etching in this process that rounds the pyramid tip slightly (Figure V).

It is known from the literature [3], that these tip-rounding processes are useful for the production of heterojunction solar cells, since the round tip leads to a better passivation of the surface defects. A further understanding and optimization of the process in the given flow environment can help to optimize the lateral homogeneity of the process. Therefore, a process simulation for this etching process was conducted.

a)

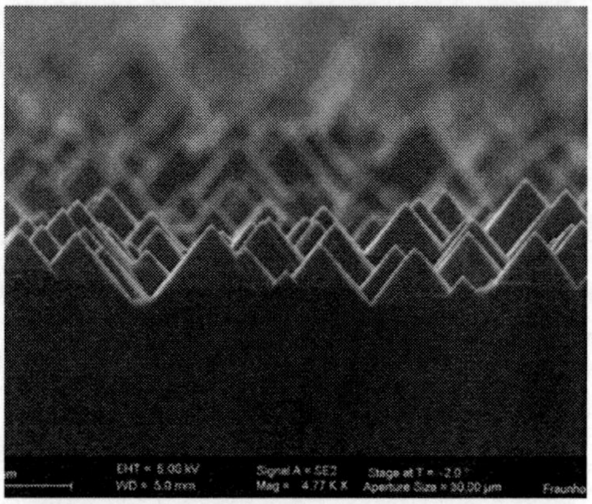

b)

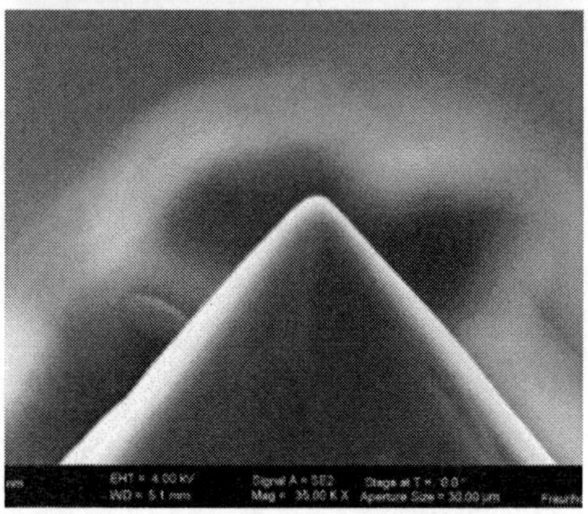

Figure V: a) random pyramids are created in an alkaline etching process on monocrystalline silicon. b) after an etching in HF/HCl/O_3, the pyramid tip is rounded.

The simulation model for the process simulation was built up from five pyramids which are located on the silicon surface [4]. The flow of the process media was directed parallel to the diagonal of the pyramid basis. (Figure VI, upper graph). The simulation consists of flow simulation in the considered volume, the transport of the diluted species in the circulated process solution as well as the surface reaction, where O_3 and HF are consumed, while hexafluorosilicic acid (H_2SiF_6) is generated.

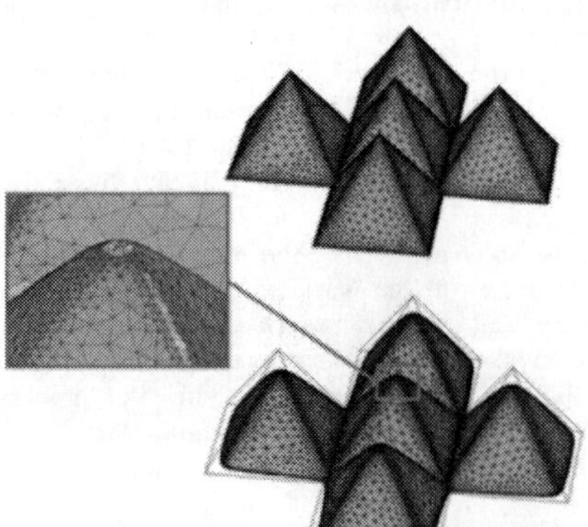

Figure VI: (upper part) Simulation model for the pyramid rounding in HF/HCl/O_3. (lower part) Result of the simulated etching on the pyramid tip.

The etching process is isotropic with respect to the crystal orientations. Therefore, the tip of the pyramid is attacked from several sides, while the pyramid surfaces are etched only from one direction. Therefore, the tip is rounded during this process (Figure VI, lower graph).

In Addition to the qualitative result, that the pyramid tips were rounded, quantitative results are also available. The dependence of the etching rate as well as that of the tip radius on the temperature and the volume flow can be predicted using such a process simulation. The etching rate increases with increasing temperature as well as with an increasing volume flow. Both effects also lead to a more pronounced rounding of the pyramid tips.

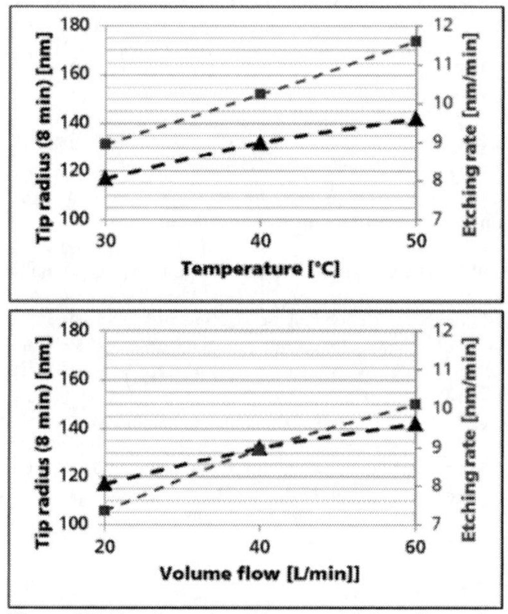

Figure VI: Development of the pyramid tip radius and etching rate with respect to the temperature (upper part) and volume flow (lower part).

3.3 Use Case 3: Predictive Maintenance

Predictive maintenance is a modern maintenance concept, which combines the advantages of the reactive maintenance and the preventive maintenance concept [5].

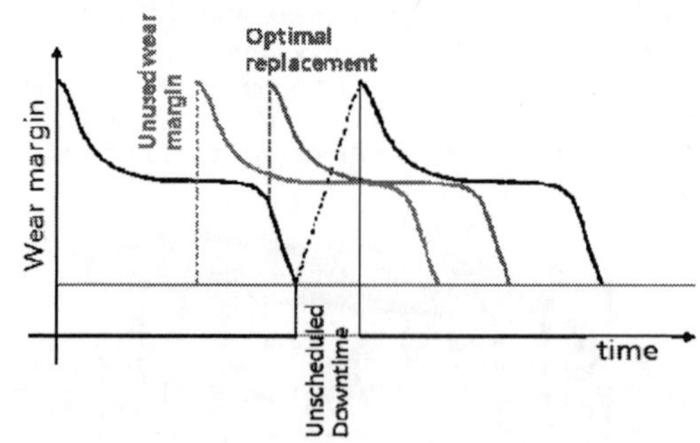

Figure VII: Replacement of spare parts in different maintenance concepts.

In the reactive maintenance concept, a tool is used until it breaks down due to a defect part. Immediately after the error detection, the maintenance team starts to repair or to replace the defect component (black line in Figure VII). This strategy leads to significant unscheduled down times, but a maximal use of the wear margin can compensate that in special cases, e.g. if the components are expensive, but quickly available. Another maintenance strategy is the preventive maintenance. Parts are replaced according to a schedule based on previous experiences (orange line in Figure VII). This means that the wear margin of the reference part is not completely used, what is uneconomic, but this strategy can avoid unscheduled down times, since all maintenance times can be planed und organized in the production process.

The most economic maintenance strategy is the predictive maintenance, where the breakdown of parts can be predicted via the evaluation of sensor data or via the modelling of the parts wear margin during its life cycle. The optimal time for the replacement is immediately before the occurrence of a defect. This can avoid down times and maximize the usage of the wear margin at once.

The implementation of such a system requires a huge knowledge of the breakdown mechanisms for each part.
On one hand, this knowledge is not yet available since most components do not have such sophisticated digital twins which describe the actual state of the physical part. Furthermore, breakdown models for simple components like pneumatic valves will probably be not much more precise than a mean statistical lifetime. On the other hand, tool manufacturers often have service personal with a large experience. This work experience can be drawn into business rules, which enable a knowledge-based system to help the service personal with trouble shooting.

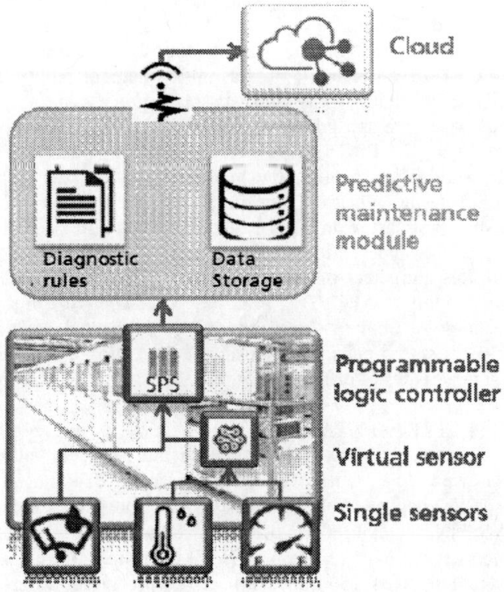

Figure VIII: Architecture of the Cheops Predictive Maintenance Concept.

A flexible smart maintenance system was designed which is able to make predictions from the development of the sensor data (Figure VIII). Therefore, all sensor data is read out of the programmable logic controller (plc) into a fifo (first-in-first-out) memory of the predictive maintenance module. This memory is continuously monitored by an inference engine, which applies a set of business rules to this memory. The business rules represent the knowledge of the tool manufacturer, hence, the existing knowledge can be used to avoid down times. Additionally, every system failure leads to a transfer of all data in the predictive maintenance module to a cloud memory, where the error as well as the relevant data history is available for a detailed analysis. From this analysis, new rules to predict system failures can be created after a successful trouble shooting. These new rules can be sent to the diagnostic rules memory of the predictive maintenance module.

4 DISCUSSION AND OUTLOOK

Digitalization is one of the most interesting topics in mechanical engineering industry. It offers new technologies, which can make the production more flexible and more cost efficient. A central topic in the digitalization of productions is the digital twin. In this paper, we describe use cases, where a digital twin can help to improve a wet chemical process tool and its processes. Furthermore, a digital twin can be included in a predictive maintenance concept and therefore reduce the unscheduled downtimes, which can be a significant cost factor, especially in highly linear production lines, where a system failure of a single tool can reduce the throughput of the whole line.

In further researches, the concept of the digital twin will be generalized and transferred to further assets like carriers, chemistry containers and finally to the solar cell and module.

ACKNOWLEDGEMENTS

This work was partly funded by the German Federal Ministry for Economic Affairs and Energy (Contract Number 0324056B).

REFERENCES

[1] International Technology Roadmap for PV, Ningh Edition 2018, http://www.itrpv.net/Reports/Downloads/
[2] L. Mohr, M. Zimmer, M.; AIP Conference Proceedings 1999(1), (2018) 110004
[3] A. Moldovan, T. Dannenberg, J. Temmler, L. Kroely, M. Zimmer, J. Rentsch, Energy Procedia 92 (2016) 374 – 380
[4] L. Mohr, T. Krick, *to be published*
[5] M. Austerjost, T. Hegmanns, A. Kuhn, 2016 (2016). Logistics Journal : Proceeedings, (urn:nbn:de:0009-14-43280)

FABRICATION OF APCVD PSG EMITTER-BASED INDUSTRIAL PERC SOLAR CELLS REACHING > 21.0% CONVERSION EFFICIENCIES

B. Kafle[1], P. Saint-Cast[1], U. Belledin[1], S. Lohmüller (née Werner)[1], H. Zunft[2], H. Knauss[2], P. Palinginis[3], C. Kusterer[3], R. Köhler[3], Thomas Zehl[3], A. Wolf[1], M. Hofmann[1]

[1]Fraunhofer Institute for Solar Energy Systems ISE, Heidenhofstr. 2, 79110 Freiburg, Germany
[2]Schmid, Gebr. Schmid GmbH, Robert-Bosch-Str. 32–36, 72250 Freudenstadt, Germany
[3]SolarWorld Industries GmbH, Am Junger-Löwe-Schacht 2, Freiberg, Germany

Correspondence: bishal.kafle@ise.fraunhofer.de, +49 (0) 761 4588-5499

ABSTRACT
Atmospheric pressure chemical vapour deposition (APCVD) of phosphosilicate glass (PSG) layer allows a separation of glass deposition and phosphorous diffusion step. This permits the application of the selective laser-doping process either before or after the thermal diffusion, which increases the freedom of tailoring the doping profiles of both lowly and highly doped regions of selective emitter structures. We fabricate industrial *p*-type PERC solar cells featuring APCVD PSG-based selective emitters and study the effect of laser-doping process on current-voltage characteristics. Mean conversion efficiency $\eta_{mean} = 21.1\%$ and an absolute gain $\Delta\eta_{mean} = 0.2\%$ to the homogeneous emitters is achieved for solar cells after applying the laser-doping process either before or after the thermal drive-in. The most promising selective emitter group is the one featuring the laser process before the thermal diffusion step, where a carrier concentration profile with depth $d = 0.9$ µm as well as a low specific contact resistivity ($\rho_{C,mean} \approx 3$ mΩ cm^2) are achieved. The major electrical loss of this group is related to the spectral response, where a higher front-reflection and an increased recombination at the illuminated fraction of the highly doped region need to be minimized in order to further push the efficiency level to $\eta > 21.5\%$.

Keywords: Doping, Atmospheric pressure, PERC solar cells

1. INTRODUCTION

With a continuous improvement in electrical properties and reduction of production cost, passivated emitter and rear cell (PERC)-type solar cell concept is set to become the standard cell architecture in industrial production [1]. However, still the losses in emitter region account for a significant proportion of the total recombination losses in a PERC solar cell [2]. In order to further minimize the emitter-related recombination losses, selective-emitter approaches are being increasingly adopted as it allows diffusion of an emitter with low surface concentration and thus, low saturation current density between the fingers ($j_{0,pass}$), whereas facilitating a lower contact resistivity (ρ_C) in contact areas. Additionally, another goal of the selective emitter approach is to minimize the recombination of minority charge carriers, i.e. lowering the saturation current density $j_{0,met}$ in the metallized area of the solar cell. Two of the major ways to achieve this are: a) formation of highly doped n^{++} areas under metal contacts for effective shielding of minority charge carriers towards reaching the contact area, b) minimizing the laser-induced damage in the Si lattice.

Meanwhile, the choice of the dopant source and the applied selective laser process can have a substantial effect on the resulting recombination as well as on the final contact resistance values of screen printed silver contacts. Typically, a phosphosilicate glass (PSG) layer formed by $POCl_3$-based tube diffusion process is used as the doping source to form laser-doped selective emitter regions [3]. In this study, we choose atmospheric pressure chemical vapour deposition (APCVD) as the technology to deposit phosphosilicate glass (PSG) layers on the Si surface. Apart from the economic potential of this technology compared to the conventional $POCl_3$-based tube furnace diffusion, its choice for this study is mainly due to: a) convenience of depositing single layer or stack of PSG glass with desired P concentration, and b) separation of glass (dopant source) deposition and thermal diffusion steps. The latter allows a freedom of tailoring doping profiles by independently varying the PSG deposition, laser and thermal diffusion steps, aiming to strike a balance between strong P diffusion and low laser-induced damage in the selective-emitter region.

One of the major barriers for the adoption of APCVD-based technology is a limited demonstration of high efficiencies using industrial-scale process steps. In this paper, we apply laser-based selective doping process on APCVD PSG layers, either directly after PSG deposition step or after the thermal diffusion step, and compare the electrical performances in the solar cell level. We start with electrical characterization of such selective n-type emitters on test structures using doping profiles and sheet resistance measurements. Afterwards, we apply APCVD PSG-based selective-emitter technology on large-area industrially fabricated *p*-type PERC solar cells and discuss the current-voltage characteristics of different groups.

2. EXPERIMENTAL

Figure 1 shows the experimental process plan followed to prepare PERC-type solar cells using APCVD-deposited PSG layer as the doping source. Alkaline textured *p*-type Czochralski (CZ) monocrystalline Si wafers with 1-3 Ω cm resistivity and edge length of 156 mm are used as precursor wafers. After HF-dip and DI-water rinsing, an inline APCVD deposition tool from SCHMID Thermal System Inc. is used to deposit a PSG layer stack on the textured side of the wafers. During the APCVD process, single doped layer or a stack of doped layer capped with a diffusion barrier layer (SiO_x) can be deposited in a single continuous run through the multiple deposition chambers. PSG layer deposition is performed by injecting N_2-diluted SiH_4, PH_3 and O_2 through the CVD injectors [4].

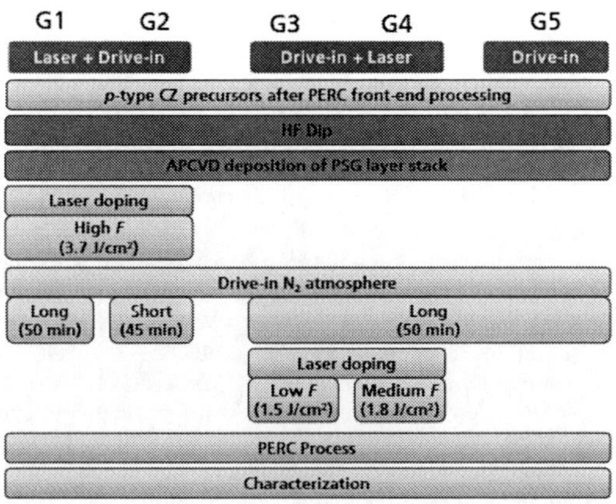

Figure 1: Experimental plan used to fabricate APCVD PSG-based CZ-PERC type solar cells. APCVD deposition of PSG layer stack is performed at Schmid GmbH, whereas laser doping, thermal diffusion and rest of the PERC processing steps are performed at SolarWorld Industries GmbH.

The concentration of P in the PSG layer mainly depends upon the gas flows and can be set as an input process parameter [5]. The gas flows are varied to form PSG layers of different P concentrations as well as to form un-doped barrier layers (SiO_x). Here, a stack of low concentration PSG/SiO_x/ high concentration PSG/SiO_x layers is deposited. The choice of phosphorous concentration in the PSG layers is made based upon our extensive study dedicated to the development of the homogeneous and selective emitters using APCVD PSG layer, and is discussed in another parallel conference paper [6].

Afterwards, highly doped regions are selectively formed either by applying laser-doping process before (*laser diffusion + drive-in*), or after (*drive-in + laser diffusion*) thermal diffusion at SolarWorld Industries GmbH. Groups G1 and G2 received *laser diffusion + drive-in* process, and differ only by the duration of the thermal drive-in process from each other. Groups G3 and G4 received *drive-in + laser diffusion* process, and differ only by the fluence (F) applied during the laser-doping process. Thermal drive-in process is performed under N_2 atmosphere in a tube furnace for 50 minutes in case of G1, G3, G4 and G5; whereas G2 received a shorter drive-in duration of 45 minutes.

In case of *laser diffusion + drive-in*, the laser process with a fluence of 3.7 J/cm^2 and pulse overlap of 100% is chosen based upon our past experience of this process leading to a low contact resistivity (ρ_C). For groups that received *drive-in + laser diffusion*, we choose two values of laser fluences (1.5 J/cm^2 for G3, 1.8 J/cm^2 for G4) for the solar cell precursors.

Please note that the selective laser-doping is applied only underneath the intended area for fingers and busbars for the solar cell precursors of G1-G4. For reference purpose, homogeneous emitter is formed for G5 without any laser doping. In parallel, characterization samples with laser fields are also prepared, where laser doping is performed in a larger area that can be used to measure doping profiles and emitter sheet resistances (R_{SH}) of both lowly and highly doped regions. Carrier concentration profiles are measured by using electro-capacitance-voltage (ECV) method on alkaline-textured samples and the profiles are corrected by adapting the surface factor that the sheet resistance determined from the charge carrier profile matches the sheet resistance measured by using four point probe (4pp) method [7,8].

After completing the diffusion processes, PSG etching is performed and the standard PERC fabrication sequence is followed for the solar cell precursors in the pilot line of SolarWorld Industries GmbH. It should be noted that the highly doped n^{++} region and the screen-printed metallization of the Ag grid are aligned for the selective emitter groups. In order to keep a tolerance for alignment, the selective-doping is performed in broader area than the width of the screen-printed fingers. The fraction of illuminated area with high doping to the total area of the solar cell is close to 6%, and the total metallization fraction is close to 4.5%. The screen-printing of rear and front sides is performed by using commercially available pastes, followed by a variation in the contact-firing process. *I-V* characteristics and spectral response of solar cells are measured after the regeneration process.

3. RESULTS AND DISCUSSION

3.1 Doping profiles of homogeneous and laser-doped emitters

Figure 2 shows the measured carrier concentration profiles of laser-doped emitters used in G1, G3 and G4, which received identical drive-in process (850°C, N_2 atmosphere, 50 min.). For comparison, doping profile of the homogeneous emitter that received no laser process (G5), is plotted as well. Please note that since G1 and G2 differ slightly in the drive-in duration, G2 is expected to have lower total phosphorous dose than G1 (profile not measured here). The corresponding emitter characteristics are listed in Table 1.

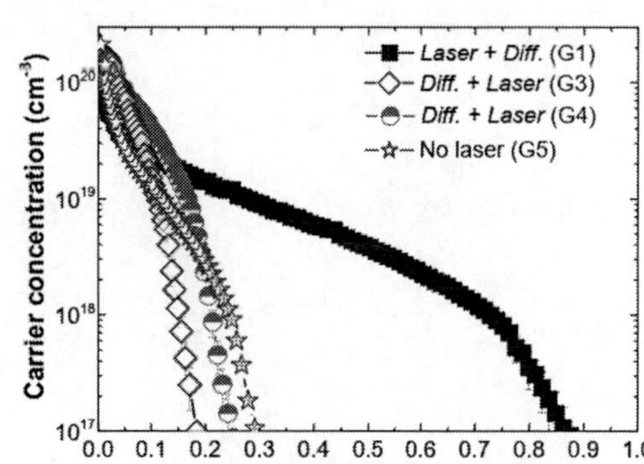

Figure 2: Carrier concentration measured by ECV for the homogeneous and selective doping processes applied in the solar cell batch measured on textured surface. The profiles are corrected by a surface factor to match the sheet resistance measured by 4pp.

For the interpretation of the carrier concentration profile, it should be noted that a significantly higher laser fluence is used for *laser diffusion + drive-in* group (G1,G2) in comparison to groups featuring *drive-in + laser diffusion* (G3,G4). In Figure 2, *laser diffusion + drive-in* group G1 leads to a lower N_S in comparison to the *drive-in + laser diffusion* selective doping processes (G3/G4). For G3, the depth measured by ECV is lower than that of the homogeneous emitter (G5), which could be related to the slight differences in PSG layer thicknesses within the test wafer on which laser fields were prepared to perform ECV measurements.

Table 1: Characteristics of selective (laser-doped) and homogeneous emitters used in the solar cell batch, which are subjected to identical drive-in duration of 50 minutes. Emitter depths (d) are measured at 10^{16} cm^{-3} after linear extrapolation.

Doping process	**Gr.**	R_{SH} **(Ω/□)**	N_S **(10^{20} cm^{-3})**	d **(nm)**
laser + drive-in	G1	67	0.72	1000
drive-in + laser	G3	95	1.7	225
drive-in + laser	G4	73	1.75	290
drive-in	G5	119	2.0	320

In fact, *drive-in + laser diffusion* mainly increases the the near-surface emitter doping in comparison to the emitter formed without any laser-doping. Nevertheless, the thermally diffused emitter also features a high N_S, although a very steep decline in P concentration in the depth of the emitter. Moreover, applying *laser diffusion + drive-in* process leads to a significantly deeper profile ($d \approx 1$ µm) with an extended depth of the highly doped region ([P] > 10^{19} cm^{-3}) in comparison to the highly doped areas formed by *drive-in + laser diffusion* groups and to the emitter formed only by thermal *drive-in*. The differences in profile of selectively doped regions based on whether the laser doping is applied before or after the laser doping process is expected to be related to an easier access to a rich P source (PSG layer with [P]=20%) during drive-in process in the former case. Further studies are needed in order to understand the exact mechanism of dopant diffusion from such a stack of PSG layers.

Nevertheless, in case of a thick stack like the one used in this experiment, significantly higher laser fluences are required to induce dopant diffusion into Si for the group *laser diffusion +drive-in* in comparison to the *drive-in + laser diffusion*. Therefore, collective optimization of the layer stack (thickness and P concentration) and the laser process is worthwhile to investigate for such a diffusion process. In case of *drive-in + laser diffusion*, lower laser fluences are required to reach an equivalent value of R_{SH}. A full investigation of applying *laser + drive-in* and *drive-in + laser* processes for various laser fluences and evaluating the resulting doping profiles and the corresponding electrical parameters is out of scope of this paper and will be discussed in a future publication.

3.2 Solar cell results

I-V characteristics of the PERC-type solar cells are plotted in Figure 3. Excellent open-circuit voltages (V_{OC}) above 670 mV and a gain in V_{OC} compared to the homogeneous emitter (G5) is achieved for best groups of both types of selective emitter cells.

For *laser diffusion + drive-in* group, an advantage of shorter drive-in duration (G2) is observed in both V_{OC} ($\Delta V_{OC} \approx 2$ mV) and short circuit current density (J_{SC}) ($\Delta J_{SC} \approx 0.1$ mA/cm^2) of the solar cells. It should be noted that for a shorter drive-in duration (45 minutes), a higher average sheet resistance $R_{SH,mean} \approx 134$ Ω/□ is measured for the illuminated emitter in comparison to the longer drive-in duration of 50 minutes ($R_{SH,mean} \approx 120$ Ω/□). Hence, we attribute the gain in V_{OC} for G2 to a lower Auger recombination in the emitter in comparison to G1.

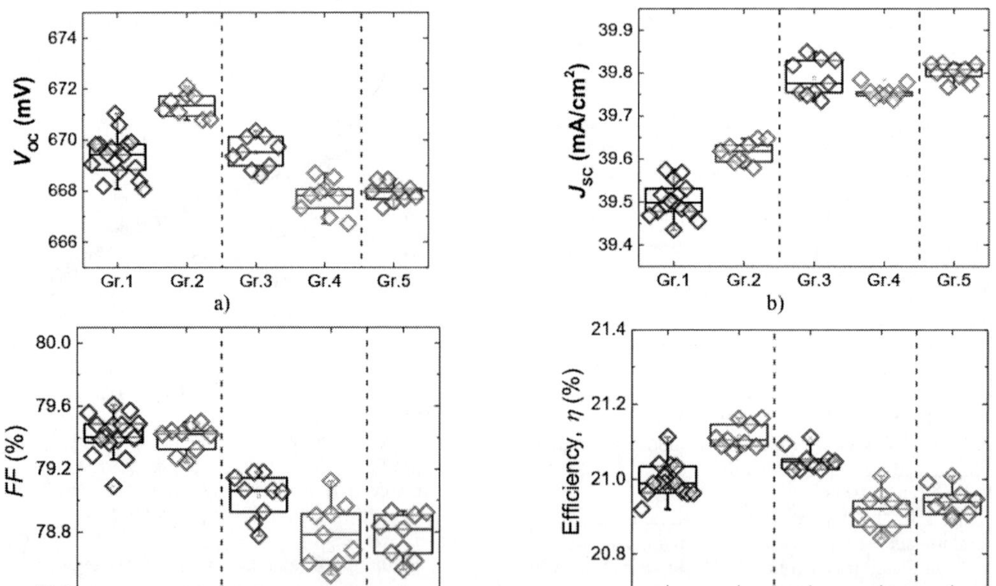

Figure 3: Illuminated *I-V* characteristics of the large area (156 mm edge length) *p*-type CZ PERC-solar cells measured in stable regenerated state, by using industrial cell tester with the reference cell calibrated to Fraunhofer ISE Callab.

In general, the solar cells that feature *laser diffusion + drive-in* show a significant loss in J_{SC} in comparison to other selective or homogeneous emitter solar cell groups. The reason for this will be discussed in the latter part of the paper. The solar cells featuring *laser diffusion + drive-in* (G1, G2) show a significantly higher fill factor (*FF*) in comparison to *drive-in + laser diffusion* (G3, G4) and the homogeneous emitter (G5). For *drive-in + laser diffusion* group, lower laser fluence of 1.5 J/cm^2 (G3) leads to a higher *FF* in comparison to a higher fluence of 1.8 J/cm^2 (G4). Overall, mean conversion efficiencies, $\eta_{mean} \geq 21.1\%$ is achieved for best selective emitter groups G1, G2 and G3. The champion solar cell from G2 reached 21.2% conversion efficiency. In comparison, homogeneous emitter solar cell group reached $\eta_{mean} = 20.9\%$.

3.3 Losses in selective-emitter solar cells

3.3.1 *FF* losses

The APCVD PSG-based selective-emitter groups reached higher *FF* than the homogeneous emitter group of solar cells. However, the *FF* values achieved in this solar cell batch are still lower than the expected values. Therefore, APCVD-based solar cells still have potential to reach higher *FF* and η if these losses can be minimized. To quantify the losses, a *FF* loss analysis is performed based upon the following measured parameters: a) *FF* from illuminated *I-V* curve, b) ideal value of fill factor FF_0 [9] that consists of only losses due to the first diode recombination parameter J_{01} and c) pseudo fill factor (*pFF*) from Suns-V_{OC} measurements that is free of series resistance (R_S) related losses [10]. By using these parameters, it is possible to identify the limiting loss-factors in *FF* of the solar cell. For example, *pFF-FF* provides a direct measure of R_S-related losses, whereas FF_0-*pFF* gives the extent of *FF* losses due to parallel resistance (R_P) and non-ideal recombination losses usually lumped in a second diode recombination parameter j_{02}. Furthermore, specific contact resistivity (ρ_C) is measured for two cells of each group (9 measurements per sample) using transmission line model (TLM) [11]. The parameters used to identify *FF*-losses in different solar cell groups are plotted in Figure 4.

Figure 4 a) shows the results for the *FF* loss analysis. It suggests that the selective emitter solar cells featuring *laser diffusion + drive in* are not limited by the R_S-related losses. It is also confirmed by a low mean value of specific contact resistivity $\rho_{C,mean}$=2.5 mΩ cm^2 for both G1 and G2 (see Figure 4 d)). In contrast, selective emitter solar cells featuring *drive-in + laser diffusion* are expected to be limited by R_S-related losses, mainly due to a higher value of ρ_C. We also confirmed that all groups have comparable values of Ag-grid related resistance losses.

Comparing Figure 4 d) with the carrier concentration profiles in Figure 2, one should notice that the behaviour is complex and cannot be reduced to a simple relation of ρ_C and N_S. In this study, the lowest ρ_C values are not obtained for emitter with high N_S value. Here, the shape of the profile and possibly the surface roughness left after the laser process is expected to have a large influence on the contacting property. In Figure 4 b), all groups show an equivalent value of FF_0-*pFF*, whereas as per Figure 4 a), losses due to R_P can be neglected for all groups (R_P > 10k Ω cm^2).

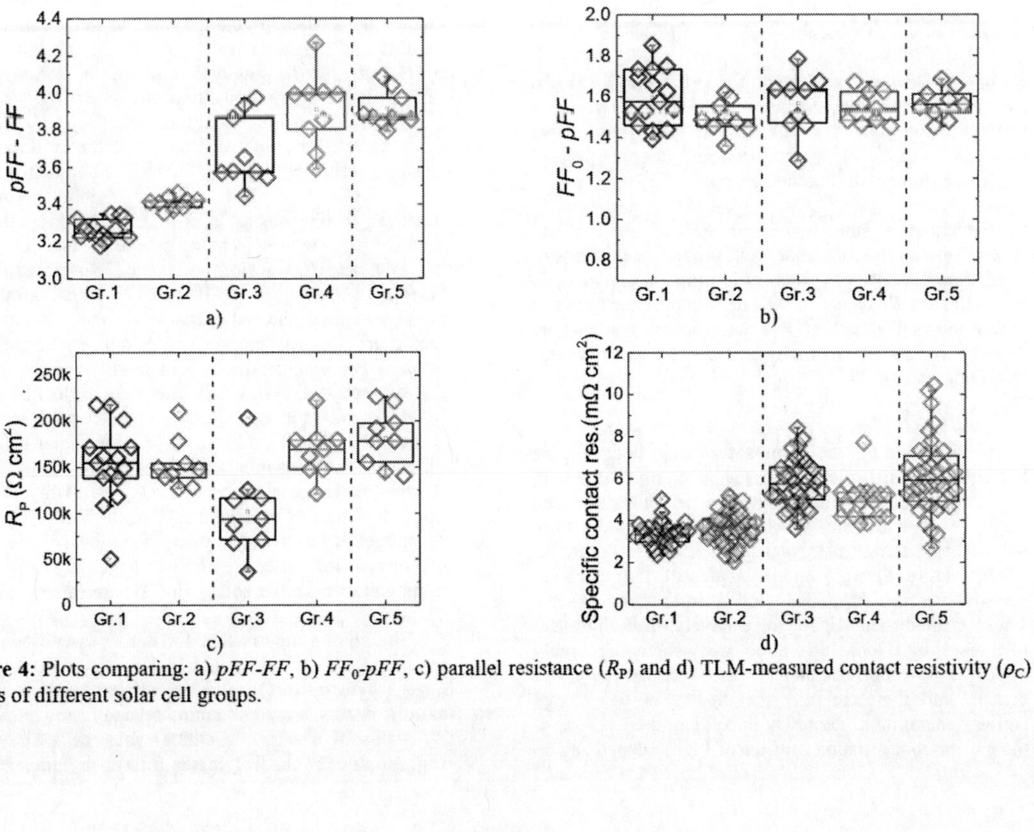

Figure 4: Plots comparing: a) *pFF-FF*, b) FF_0-*pFF*, c) parallel resistance (R_P) and d) TLM-measured contact resistivity (ρ_C) values of different solar cell groups.

In Figure 5, breakdown of j_{01}-limited *FF* (FF_0) of APCVD PSG-based solar cells, which followed either *laser + drive-in*, *drive-in + laser* or only *drive-in* processes are compared. APCVD-PSG based selective-emitter and homogeneous emitter solar cell groups clearly show high non-ideal recombination or j_{02}-related *FF* losses. Such j_{02}-related losses are generally related to the recombination losses in abrupt junctions, space charge region and other injection dependent recombination. In our case, origin of this loss in not yet completely understood, but expected to be related with impurity contamination during cell-processing steps and/or the wafer transport between different laboratories. As the rear-side of all solar cells are identically processed, the losses should be related to the front-side processing steps. In case of laser-doped emitters, this negative impact of j_{02} is expected to be compensated to some extent by the better shielding of minority charge carriers in the contact area, i.e. lower $j_{0e,met}$.

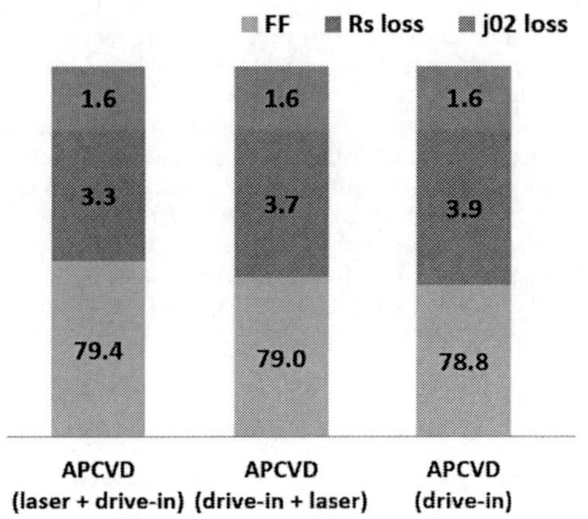

Figure 5: Breakdown of ideal *FF* (FF_0) into measured *FF* and losses due to series resistance (R_S) and recombination related to the second diode component (j_{02}). No significant losses occurred due to parallel resistance (R_P) on all solar cell groups.

Nevertheless, the j_{01}-limited *FF* or FF_0 of all APCVD-based PERC solar cell groups are identical. Furthermore, APCVD-group featuring *laser +drive-in* shows lowest R_S-related loss. The homogeneous emitter group shows the highest R_S-related losses that can be directly correlated to high contact resistivity values (see Figure 4 d)).

3.3.2 J_{SC} losses for *laser diffusion + drive in* group

In Figure 3 b), the groups featuring laser doping before thermal diffusion process show significant loss in J_{SC} in comparison to other groups In order to understand the losses, a schematic showing a unit area of the front side of a selective emitter solar cell is shown in Figure 6.

In Figure 6, w_{high} and w represent the width of selectively highly doped area and that of the screen-printed metal finger, respectively. The alignment tolerances lead to a laser-dope area where the highly doped area is not completely placed under the metal contact and therefore also participates in the charge carrier generation as the lowly doped emitter. Looking at the carrier concentration profiles of highly doped region for *laser + drive-in* group, one can expect a significantly higher Auger-related recombination in this region in comparison to the lowly-doped emitter. In combination with the much deeper junction, a notable loss in blue response is expected.

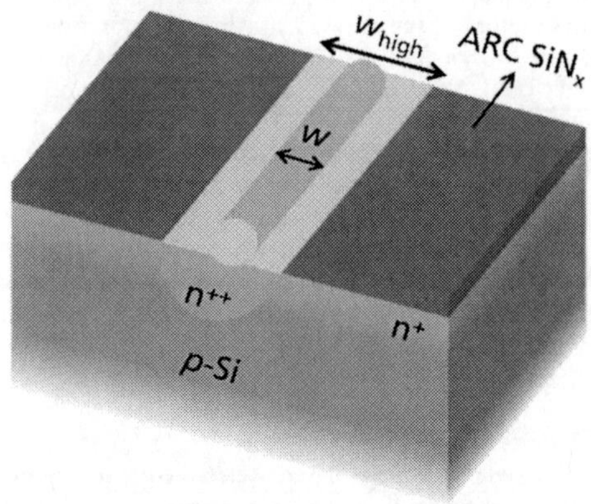

Figure 6: Schematic depicting the formation of selective emitter with a width w_{high} in the front-side of a solar cell, on which a Ag-grid with fingers of width w is printed.

In their studies, Jäger et al. argue that such Auger-related recombination losses forming dead layer in highly doped region is responsible for J_{SC} losses in selective-emitter solar cell that do not exhibit a perfect alignment between laser-doping and screen printing [12]. Apart from this, laser-induced defects can also activate Shockley Read Hall (SRH) recombination channel, although the consequent *drive-in* process is expected to anneal the defects to some extent. Meanwhile, higher and longer laser pulse energies could also significantly change the surface topography due to longer and deeper melting and crystallization phases of Si substrate. Thus, apart from the recombination losses that are typically visible in the internal quantum efficiency (*IQE*) of solar cells, losses in external quantum efficiency (*EQE*) due to a higher reflection of the illuminated highly doped area can also influence J_{SC}. This effect is often neglected for conventionally applied selective-doping processes; however, it demands a closer look if high laser fluences and overlaps are used as in the case of G1 and G2.

EQE and *IQE* measurements are performed for the solar cells from groups G1 and G5. These groups are chosen as they showed respectively the minimum and maximum J_{SC} of all groups, and therefore are expected to allow a better understanding of losses that are limiting J_{SC} of groups G1 and G2. The front reflection-related losses in J_{SC} for each group is calculated by using measured *IQE* and front-reflection of the solar cell using the AM 1.5 spectrum as described in [13,14]. Recombination related losses in J_{SC} are extracted based upon deviation of measured *IQE* to the ideal case, using the model of Fischer [15] as explained in [14]. J_{SC}-losses calculated for selective-emitter solar cell (G1) and homogeneous-emitter solar cell (G5) are shown in Figure 7.

The chart suggests that for higher laser fluences as used in the selective emitter solar cell group featuring *laser + drive-in* (G1), reflection-related losses can be as significant as the recombination related losses in the final J_{SC} value of a selective emitter solar cell. The model suggests higher shading losses for G1 in comparison to

G5. This could be related to a change in surface topography due to the laser-doping process, which is also suggested by a higher reflection-related J_{SC} loss for G1 in comparison to G5. Higher laser fluence is most likely to deform the surface texture due to a longer melting of Si [16], thereby changing the wetting and flow behaviour of the Ag-paste on Si surface, leading to a higher paste bleeding during the screen-printing process. Further characterization of the contact geometry on differently lasered surfaces is necessary in order to prove this hypothesis.

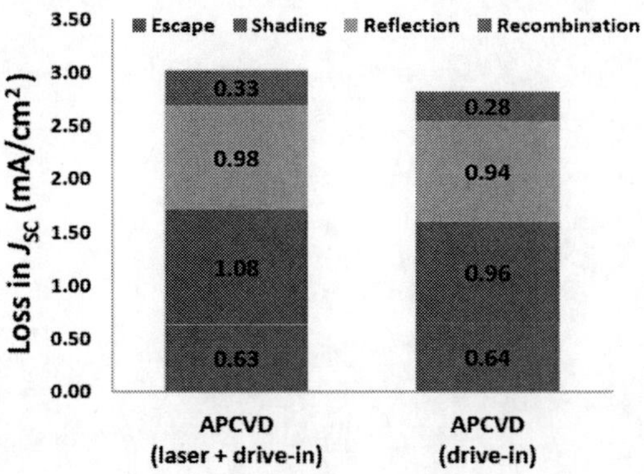

Figure 7: Loss in J_{SC} in the selective emitter solar cell group G1 (left) and the homogeneous emitter solar cell group G5 (right), due to recombination in the emitter region, front-reflection losses, shading due to metal contacts and escape reflection-related losses.

4. SUMMARY

We fabricated p-type CZ-PERC selective emitter solar cells using APCVD-PSG layer as the doping source in industrial pilot cell fabrication line of the high volume manufacturer SolarWorld Industries GmbH. An APCVD tool produced by SCHMID Thermal System Inc. is used to deposit PSG and barrier layers. The APCVD technology allows the separation of PSG deposition and drive-in step in comparison to conventional selective emitter cells that use $POCl_3$ as the doping source. This allowed us to perform selective doping either before or after the thermal drive-in step. In first demonstration of this technology in high-efficiency PERC solar cells, conversion efficiencies of up to $\eta_{max} \approx 21.2\%$ are achieved. To the knowledge of authors, this is one of the most systematic applications of this technology on PERC-type solar cells that have demonstrated with high conversion efficiencies.

Both selective emitter groups (*laser + drive-in* & *drive-in + laser*) show an average efficiency improvement of 0.2% to the homogeneous emitter group. Solar cells featuring *laser + drive-in* are measured to have low ρ_C values ($\rho_{C,mean} \approx 3$ mΩ cm^2) and a deep carrier concentration profile ($d \approx 0.9$ µm), hence show advantages in both V_{OC} and *FF* values to the homogeneous emitter group. In contrast, ρ_C-related R_S losses dominate the *FF* of *drive-in + laser* and further optimizations in layer properties and laser-doping are required. Significant non-linear recombination -related losses in *FF* and V_{OC} are observed for all APCVD-based solar cell groups. Finally, the J_{SC} loss of 0.2-0.3 mA/cm^2 for *laser + drive-in* groups is attributed to the illuminated fraction of the highly doped region of the selective-emitter solar cell. A loss analysis of J_{SC} using *QE* measurements suggests that for sufficiently higher laser fluences, contribution of front-reflection and metal shading is significant and should be also considered along with the usually reported Auger-related recombination in the illuminated fraction of the highly-doped emitter region.

5. ACKNOWLEDGEMENTS

This work is funded by the federal ministry for economic affairs and energy (BMWi) of Germany within the project APPI under the contract number 0325895A. Project APPI is supported under the umbrella of SOLAR-ERA.NET Cofund by Agence de l'Environnement et de la Maîtrise de l'Energie, Project management Jülich, Technology Strategy Board and Centro para el desarollo tecnológico. SOLAR-ERA.NET is supported by the European Commission within the EU Framework Programme for Research and Innovation HORIZON 2020 (Cofund ERA-NET Action, N° 691664). The authors would like to thank all co-workers at photovoltaics department of Fraunhofer ISE, especially Stefan Schmidt for ECV analysis, Sebastian Mack for helpful discussions, Eleni Miethig for help in solar cell processing; and Elisabeth Schaeffer, Felix Martin for their help in solar cell characterization.

6. REFERENCES

[1] International Technology Roadmap for Photovoltaic (ITRPV): 2017 Results including maturity report: Eighth edition, 2018.

[2] M. Müller, G. Fischer, B. Bitnar, S. Steckemetz, R. Schiepe, M. Mühlbauer, R. Köhler, P. Richter, C. Kusterer, A. Oehlke, E. Schneiderlöchner, H. Sträter, F. Wolny, M. Wagner, P. Palinginis, D.H. Neuhaus, Loss analysis of 22% efficient industrial PERC solar cells, Energy Procedia 124 (2017) 131–137.

[3] T.C. Röder, S.J. Eisele, P. Grabitz, C. Wagner, G. Kulushich, J.R. Köhler, J.H. Werner, Add-on laser tailored selective emitter solar cells, Prog. Photovolt: Res. Appl. 18 (7) (2010) 505–510.

[4] K.O. Davis, K. Jiang, C. Demberger, H. Zunft, H. Haverkamp, D. Habermann, W.V. Schoenfeld, Improved control of the phosphorous surface concentration during in-line diffusion of c-Si solar cells by APCVD, Phys. Status Solidi RRL 7 (5) (2013) 319–321.

[5] Felix Book, Holger Knauss, Carsten Demberger, Florian Mutter, Giso Hahn, Phosphorous Doping from APCVD Deposited PSG. in 32nd European Photovoltaic Solar Energy Conference and Exhibition (2016) 824–827.

[6] Pierre Saint-Cast, Udo Belledin, Elmar Lohmüller, Bishal Kafle, Julian Weber, Sven Seren, Andreas Wolf, Marc Hofmann, SELECTIVE EMITTER USING APCVD PSG LAYERS AS DOPING SOURCE, 34th EUPVSEC (this conference).

[7] Robert Bock, Pietro P. Altermatt and Jan Schmidt, Accurate extraction of doping profiles from Electrochemical capacitance voltage measurements, in 23rd EUPVSEC, Valencia, Spain (2008).

[8] Achim Kimmerle, Diffused surfaces for crystalline silicon solar cells - process development, characterization and modelling. Dissertation, Universtiy of Freiburg, Freiburg, Germany, 2015.

[9] M.A. Green, Solar cells: operating principles, technology, and system applications, Prentice-Hall, 1982.

[10] R. A. Sinton, A. Cuevas, A Quasi-Steady-State open-circuit voltage method for solar cell characterization, in 16th EUPVSEC, Glasgow, UK (2000).

[11] G. K. Reeves and H. B. Harrison, Obtaining the Specific Contact Resistance from Transmission Line Model Measurements, IEEE ELECTRON DEVICE LETTERS 3 (5) (1982) 111–113.

[12] U. Jäger, B. Thaidigsmann, M. Okanovic, R. Preu, Quantum Efficiency Analysis of Highly Doped Areas for Selective Emitter Solar Cells, Energy Procedia 8 (2011) 193–199.

[13] B. Thaidigsmann, A. Wolf, D. Biro, ACCURATE DETERMINATION OF THE IQE OF SCREEN PRINTED SILICON SOLAR CELLS BY ACCOUNTING FOR THE FINITE REFLECTANCE OF METAL CONTACTS, in 24th EUPVSEC, Hamburg, Germany (2009).

[14] B. Thaidigsmann, A. Wolf, F. Clement, D. Biro and R. Preu, COMBINING THE ADVANTAGES OF WRAP THROUGH METALLIZATION AND REAR SURFACE PASSIVATION INTO INDUSTRIAL MWT-PERC DEVICES. in 25th EUPVSEC, Valencia, Spain, 2010.

[15] B. Fischer. M. Keil, P. Fath, E. Bucher, Scanning IQE measurement for accurate current determination on very large area solar cells, Conference Record of the Twenty-Ninth IEEE Photovoltaic Specialists Conference, New Orleans, LA, USA (2002) 454–457.

[16] T. Röder, A. Esturo-Breton, S. Eisele, C. Wagner, J.R. Köhler, J.H. Werner, Fill Factor of Laser Doped Textured Silicon Solar Cells. 23rd European Photovoltaic Solar Energy Conference and Exhibition, Valencia, Spain 1-5 September (2008) 1740–1742.

OPTIMIZATION OF INLINE PROCESSES FOR THE PRODUCTION OF FREESTANDING EPITAXIALLY GROWN THIN FILMS FOR SOLAR CELLS

Alexey Ivanov, Ralf Sorgenfrei, Elke Gust, Philipp Barth, Kris Van Nieuwenhuysen*, Saskia Kühnhold-Pospischil, Stephan Riepe, Stefan Janz
Fraunhofer Institute for Solar Energy Systems (ISE), Heidenhofstrasse 2, D-79110 Freiburg, Germany
*imec v.z.w., Kapeldreef 75, B-3001 Leuven, Belgium

ABSTRACT: In this publication we discuss inline aspects of fabricating freestanding epitaxially grown thin silicon for solar cells. Especially the inline electrochemical porosification and its limitations are discussed and homogeneity improvements achieved with aid of electric simulations are presented. Different types of inhomogeneity for porous silicon layers lead to difficulties when forming a proper separation layer through reorganization and the detachment yield of the epitaxial thin film is affected. One of the solutions to ensure good detachment of the films is to form several local separation layers in a kind of puff pastry like structure where detachment of epitaxial film might proceed in any of the separation layers thus increasing yield. However, in case of too high porosity of the highly porous layer, large cavities without smaller supporting silicon pillars can collapse during reorganization and the surface layer can re-attach to the substrate making detachment of this area impossible. Additionally, local warping of the reorganized template before epitaxy can lead to detrimental defects like stacking faults.

KEYWORDS: Thin Film Solar Cell, Manufacturing and Processing, Porous Silicon, Epitaxy, electrochemical etching.

1 INTRODUCTION

The use of single crystalline flexible silicon thin films and foils is widespread throughout many applications like flexible IC chips [1], SOI [2] and solar cells [3] [4]. One of the technologies for those applications is the epitaxial growth on reorganized porous silicon substrates, which allows the subsequent detachment of the grown epilayer and reuse of the parent substrate. Especially for solar cells, this approach has been adopted to obtain large area epitaxial thin films in order to reduce material costs substantially while maintaining the efficiency potential of conventional mono wafers. A very important issue to lower the production costs is the consequent inline implementation of all single processes. At Fraunhofer ISE, the required core processes, inline porosification and inline epitaxy, have been developed and optimized in the last years.

In contrast to thick epitaxial wafers, the growth of detachable thin layers (EpiFoils) with thickness below 100 µm is more difficult due to handling issues and stronger influence from impurities of the substrate. We report the complete value chain from porosification, reorganization and epitaxy to characterization and detachment. The focus is to point out some limitations for production of high quality EpiFoils with high detachment yield. Finally, we present useful approaches to overcome the limitations.

2 EXPERIMENTAL AND CHARACTERIZATION METHODS

2.1 Fabrication steps

For the experiments, (100) silicon (Si) as-cut wafers of p-type with size 156×156 mm^2, thickness 550 µm, and resistivity of 10–15 mΩcm were used. Before porosification, standard damage etch (SDE) was performed. Alternatively in some cases chemical polishing was used to reduce surface roughness.

To form porous silicon (PorSi) layer, wafers were anodically etched in an experimental inline lab tool, capable to etch wafers up to 180 mm width and at least 156 mm length. The tool has been developed at Fraunhofer ISE in cooperation with RENA Technologies GmbH (Gütenbach, Germany). The schematic of the inline porosification tool is shown in Figure 1a. The setup consists of at least three electrolyte tanks, whereas the two side tanks are connected anodic and the middle one connected cathodic. As electrodes, glossy carbon cylindrical electrodes are used. The tanks of opposite polarity are isolated from each other by an air gap /

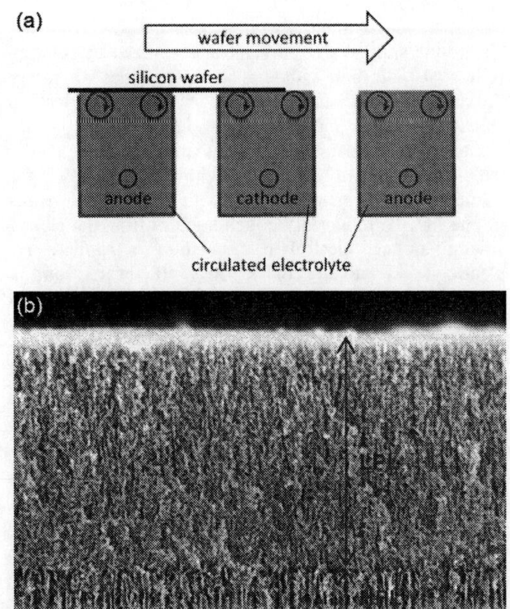

Figure 1: (a) Schematic of the inline porosification tool in a cross-section. (b) SEM image of a cross-section of fabricated porous silicon stack with lowly porous layer (LPL) and highly porous layer (HPL).

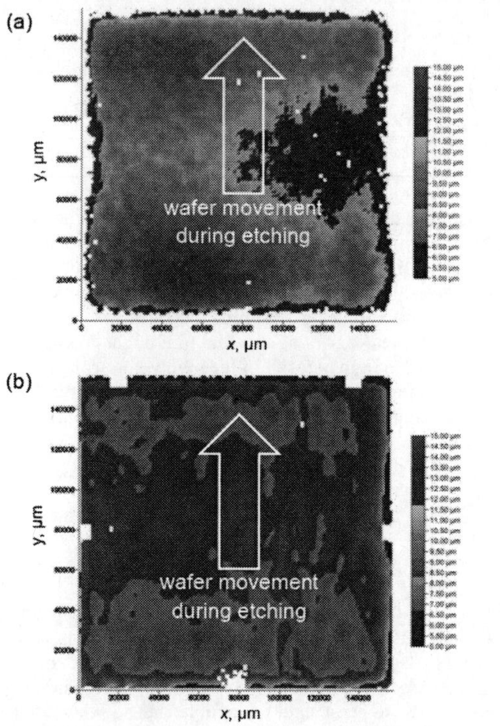

Figure 2: Thickness maps of PorSi layers etched in the inline porosification tool (a) before and (b) after process optimization. For the central region of 136×136 mm^2 the relative standard deviation of thickness is for (a) 9.26 % and for (b) 5.84 %.

electrolyte sink. Wafer movement is realized by rotating rollers. Anodic and cathodic contacts to the wafer are provided on the same wafer side as wet contact to electrolyte. Thus the wafer side facing down (wafer frontside) is porosified while moving over the anodic tank. The setup can be extended with additional tanks for inline etching of multiple layers with different properties in one run. To remove hydrogen bubbles from the etched silicon surface, electrolyte circulation is applied. As electrolyte, a mixture of 49 m% HF, water, and a surfactant was used resulting in approximately 20 m% HF solution. The process was conducted in current-control mode with computer-controlled current sources from Plating Electronic GmbH (Sexau, Germany). Due to limited range of possible wafer movement speed in the tool, lowly porous layer (LPL) of maximum 200–300 nm could be etched per run. To form LPL of total thickness of 1–2 µm, the wafers had to be processed multiple times for several (up to 20) runs to reach the desired thickness. For highly porous layer (HPL), several runs were also applied at higher wafer movement speed for more efficient removal of hydrogen bubbles and safety reasons. After porosification, the wafers were rinsed in deionized water and placed into IPA for 60 s to remove possible rests of surfactant from PorSi and reduce mechanical stress during drying of PorSi.

In order to optimize homogeneity of the inline porosification tool, numerical simulation (FEM) in Comsol 5.3a were conducted.

A cross section, characterized with Scanning Electron Microscopy (SEM) as one representative example of fabricated PorSi stack with LPL and HPL is shown in Figure 1b.

As a reference, wafers (8") from imec with LPL and HPL were used, which were stored after porosification up to several months and HF-dipped right before further processing.

The porosified Si wafers were reorganized at 1120 °C under 100 % H_2 atmosphere in a high throughput inline CVD tool [5]. The Si epitaxy was done at the same temperature in a chlorosilane / H_2 atmosphere leading to layers of 40–100 µm thickness. The sample doping was realized by inserting diluted PH_3 gas during the epitaxy process.

With a dicing saw the detachment area was defined followed by a lift-off process with a vacuum tool.

2.2 Characterization

Thickness of the wafers before porosification and after epitaxy was measured with CT250T optical thickness measurement device from cyberTECHNOLOGIES GmbH (Eching, Germany). The difference between the thickness measurements showed the thickness of the epitaxial layer. In order to evaluate the homogeneity of PorSi single layers and to optimize the current profile for the etching process, the same characterization procedure has been applied for few wafers with thickness measurement before porosification and after selective removal of PorSi in dilute KOH solution.

After separation, some sample cross-sections have been examined with SEM to examine the reorganization of the porosified template. For preparation of sample cross-sections, cleaving of samples with diamond pen provided sufficiently good results.

3 RESULTS AND DISCUSSION

3.1 Porosification

Due to the inline character of the porosification tool, PorSi layers showed some inhomogeneity. An example of PorSi thickness map with typical types of inhomogeneity is shown in Figure 2.

One type of lateral inhomogeneity results from connection of cathodes to the wafer frontside. Due to charge flowing through the wafer from side tanks to the middle tank, current density is inhomogeneous over the wafer frontside, with current crowding occurring close to the side walls of the anodic tank. Additionally, there is also the wafer movement over the anodic tank. Our electric simulations of the process showed that with simple current profile with ramp up / constant / ramp down parts the central part of the wafer (y in the range 60–100 mm in Figure 2a) gets more total charge density. This results in a thicker PorSi in the middle part of the wafer. With help of FEM simulations, a more complex current profile was developed to partially reduce this effect (s. Figure 2b).

Another source of lateral inhomogeneity of PorSi thickness was the voltage drop in the glossy carbon electrodes connected only from one side to the current source. This resulted in thickness and porosity reduction in x-direction from right to left (s. Figure 2).

The last but not least source of the lateral inhomogeneity in the used inline etching tool is the current crowding at the wafer edges resulting in thicker and less porous PorSi (s. Figure 2). This kind of inhomogeneity (edge effect) is typical for electrochemical setups where the size of workpiece and the tanks with electrodes are not matched. Supported with electric simulations, the edge effect on the wafer edges parallel to wafer movement could be partially reduced by reducing the length of the electrodes in the tanks (s. Figure 2b). Another possible solution could be introduction of additional objects in form of a tunnel or a blind into the anodic tank to modulate current flux. This however could interrupt circulation of electrolyte in the tank thus reducing the removal of hydrogen bubbles from the wafer frontside.

Besides homogeneity of PorSi layers over wafer area, the homogeneity of PorSi layers in depth is also important for correct development of the HPL to a separation gap during reorganization. Due to the movement of wafer during porosification, lateral inhomogeneity of current density results in transversal inhomogeneity of the PorSi layers. In Figure 1b this kind of inhomogeneity helps to distinguish two layers in HPL formed in two runs.

Porosity and thickness of PorSi layers define the degree of reorganization the PorSi layers can get at certain reorganization conditions. Inhomogeneity in PorSi layers make finding optimal process conditions for porosification and reorganization difficult and might reduce yield of EpiFoil detachment dramatically.

While etching PorSi multilayers, the properties of the subsequent PorSi layers (porosity, thickness) depend strongly on the properties of the upper PorSi layers which are etched first. Inhomogeneity of porosification process during etching of LPLs can result in not sufficiently developed or overetched HPLs. In the latter case the PorSi layer might detach during porosification or further processing making growth of homogeneous EpiFoils not possible. For the wafers etched in our inline tool the PorSi lift-off was especially strong on the wafer edges due to the edge effect which made full-area detachment more difficult. That is why wafers in this work were detached in a smaller area.

For formation of a good separation layer during reorganization, it is crucial to have well-defined porosity gradient (e.g. a clear double-layer structure of LPL+HPL) [6]. The transversal inhomogeneity of PorSi layers might make formation of a good separation layer during reorganization difficult. However, transversal inhomogeneities can be used on purpose to form several local separation layers in a kind of puff pastry like structure where detachment of EpiFoil might proceed in any of the separation layers thus increasing yield (s. Figure 3).

3.2 Reorganization and epitaxy

Reorganized PorSi layers with epitaxially grown Si layer of thickness in the range 40–100 μm were obtained in the inline CVD tool. For test and characterization purposes, also thicker epitaxial layers were fabricated.

(a)

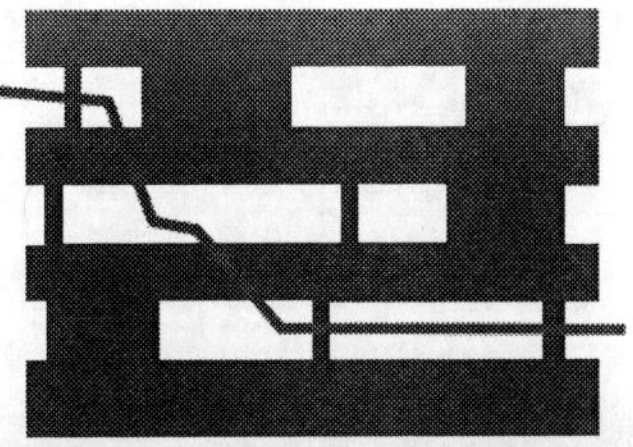

(b)

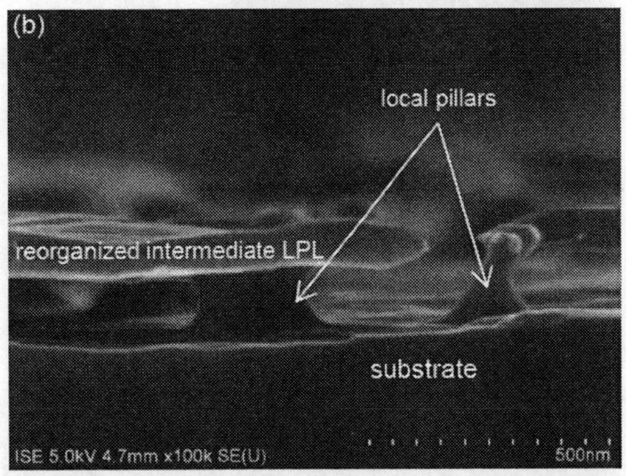

Figure 3: (a) Schematic cross-section of PorSi after reorganization with multiple separation layers. (b) SEM image of a cross-section of PorSi stack with multiple separation layers after reorganization and detachment.

For wafers with small number of HPLs, we identified thick interconnections like Si pillars (Figure 4) in the highly porous detachable layer after reorganization as main cause for incomplete or impossible EpiFoil lift-off. Several explanations are possible how these large pillars are built. Two examples of process related issues leading to large remaining interconnections are shown by SEM images in Figure 4.

The first example shows that selective etching [7] of the PorSi during the inline reorganization or too strong reorganization leads to the formation of such interconnections with large cavities in-between (Figure 4a). The formation of large cavities can also be the results of laterally inhomogeneous PorSi layer (s. section 3.1).

To compensate for this inhomogeneity, multiple HPLs were used (s. Figure 3), so that the detachment can proceed along multiple step heights. The drawback of having several separation layers is in possible contamination of the working area with pieces of PorSi layers flowing away from the substrate during the detachment in form of "Si-flakes". Additionally, PorSi designed for multiple separation layers tends more easy to lift off during porosification or reorganization, thus contaminating the CVD tool and other processed wafers. By applying PorSi layers with multiple LPLs and HPLs with reduced current, and by optimizing the number of HPLs iteratively, we achieved detachable EpiFoils with almost no "Si flakes".

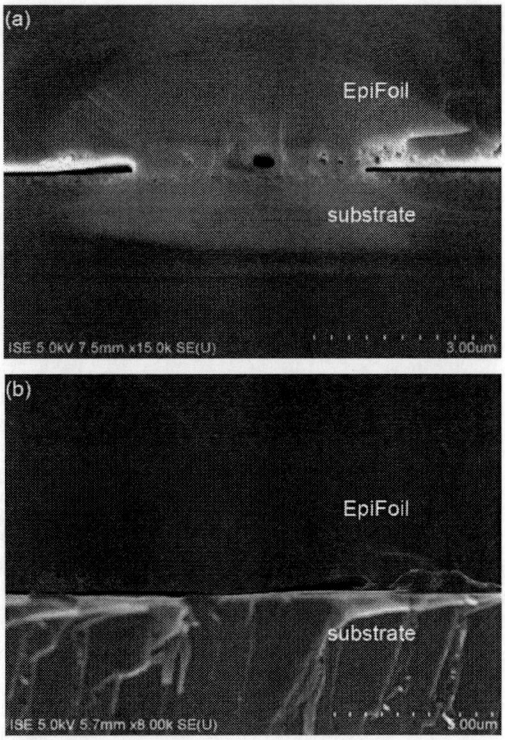

Figure 4: (a) Formation of a large diameter silicon pillar; (b) collapse of the cavity between substrate and epitaxially grown layer.

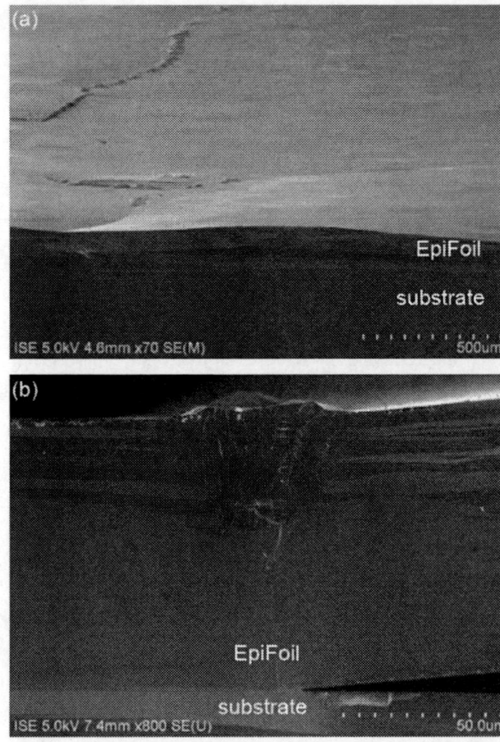

Figure 5: Local warping of the reorganized template leading to (a) sticking of the silicon EpiFoil at the edges of the warped area, and (b) formation of defects (stacking faults).

The second example shows that large cavities without smaller supporting Si pillars can collapse and the EpiFoil can re-attach to the substrate making detachment of this area impossible (s. Figure 4b). Additionally, local warping of the reorganized template before or during epitaxy can lead to detrimental defects like stacking faults (see Figure 5).

The adjustment of etch currents and the numbers of porosification runs for LPL and HPL also helped to offer sufficiently small supporting sites to prevent the collapse of the cavity during the H_2 anneal.

Through an improved process within our inline CVD equipment, selective etching of the upper LPLs through formation of SiO_x during reorganization [7] could be prevented. Although the main task was to improve the detachment yield of large area Si EpiFoils, there are still improvement steps necessary to further increase the quality of the EpiFoils, e.g. effective carrier lifetimes. Since it is known, that the pore size of the surface influences the quality of the reorganized template, a main trigger to increase the lifetime, is to manipulate the surface of the porous Si surface in the porosification process.

3.3 EpiFoil area definition and detachment

The saw dicing and detachment was done in tools with manual wafer handling. However, with state-of-the-art automatic wafer handling solutions, batch dicing and detachment of wafers is possible and already done in microelectronic fabs.

4 CONCLUSIONS

In this paper inline aspects of fabricating freestanding epitaxially grown thin silicon (EpiFoils) for solar cells are discussed. Movement of wafers during porosification introduces lateral and transversal inhomogeneity of PorSi layers which might make formation of good separation layer more difficult. Electric simulations can help to optimize inline porosification process.

One of the solutions to ensure good detachment of the films is to form several local separation layers in a kind of puff pastry like structure where detachment of epitaxial film might proceed in any of the separation layers thus increasing yield.

Improper process conditions during porosification or reorganization might lead to formation of large cavities without smaller supporting silicon pillars, which can collapse during reorganization making detachment of this area impossible. Additionally, this can lead to formation of stacking faults.

Further work has to be done to enhance yield and quality of fabricated EpiFoils.

ACKNOWLEDGEMENT

The authors would like to thank Michaela Winterhalder for sawing the EpiFoils before detachment with a dicing saw. The work was done in frame of the project CHEETAH financed by the European Commission (EC GA No. 609788).

REFERENCES

[1] J. Burghartz, "You can't be too thin or too flexible," *IEEE Spectrum*, vol. 50, no. 3, pp. 38-61, 2013.

[2] T. Yonehara and K. Skaguchi, "Cutting edge 2: ELTRAN (R); Novel SOI Wafer Technology," *Japan Society of Applied Physics International*, vol. 4, pp. 10-16, 2001.

[3] S. Janz, D. Amiri, E. Gust, S. Kühnhold-Pospischil, S. Riepe, F. Heinz and M. Drießen, "Origin and impact of crystallographic defects in epitaxially grown Si wafers," in *33rd European Photovoltaic Solar Energy Conference and Exhibition, EU PVSEC 2017: Proceedings of the international conference held in Amsterdam, The Netherlands, 25 September - 29 September 2017*, 2017.

[4] "Nexwafe GmbH," [Online]. Available: www.nexwafe.com.

[5] S. Reber, M. Arnold, D. Pócza and N. Schillinger, "ConCVD and ProConCVD: Development of high-throughput CVD tools on the way to low-cost silicon epitaxy," in *24th European Photovoltaic Solar Energy Conference and Exhibition, EU PVSEC 2009: Proceedings of the international conference held in Hamburg, Germany, 21 - 25 September 2009*, 2009.

[6] N. Ott, M. Nerding, G. Müller, R. Brendel and H. P. Strunk, "Evolution of the microstructure during annealing of porous silicon multilayers," *Journal of applied physics*, vol. 95, no. 2, pp. 497-503., 2004.

[7] V. Depauw, I. Gordon, G. Beaucarne, J. Poortmans, R. Mertens and J.-P. Celis, "Large-area monocrystalline silicon thin films by annealing of macroporous arrays: Understanding and tackling defects in the material," *Journal of Applied Physics*, vol. 106, p. 033516, 2009.

2BO. 1.1

A Proposal of Improved CZ Growth Technique of Monocrystalline Silicon for PV Cells

Tetsuo Fukuda[1], Yukichi Horioka[2], Kozo Fujiwara[3], Katsuto Tanahashi[1], Katsuhiko Shirasawa[1], and Hidetaka Takato[1]

[1]Renewable Energy Research Center, National Institute of Advanced Indutrial Science and Technology (AIST)
2-2-9 Machiike-dai, Koriyama, Fukushima 963-0215, Japan
[2]Frontier Technology Business Research Institute Co. LTD., Japan
[3]Institute for Materials Research, Tohoku University, Japan

[1]Phone/E-mail: +81-29-849-1547/tetsuo-fukuda@aist.go.jp

ABSTRACT: Currently, monocrystalline CZ silicon ingots manufactured as photovoltaic applications are all those having long straight bodies. After heat treatments in cell fabrication process, ring-type defects, well-known to lower cell efficiency, are often observed in wafers taken from seed-side position of the long crystals. On the other hand, lower density of defects in wafers taken from tail-side position is observed. The difference in the defect density results from that of thermal history during the crystal growth process in a CZ puller. We have developed special silica crucible (called liquinert crucible) whose inner surface has non-wettable property at high temperatures. Using the liquinert crucible, we have challenged the growth of "massive silicon" characterized by the diameter of more than 0.7 times the crucible diameter and the very short body-length with the expectaion of reduced thermal history. Currently, using a small CZ puller, we started to grow small "massive silicon" from the melt (3 kg) in a liquinert crucible with the diameter of 165 mm. In order to clarify the effect of thermal history, we measured lifetime of each of the seed-side and the tail-side in both conventional silicon having the diameter (~80 mm) and the "massive silicon" having the large diameter (116 – 120 mm).
Keywords: silicon, Czochralski, lifetime, photovoltaic

1 INTRODUCTION

In the photovoltaic (PV) industry, conversion efficiency is recognized as a key metric of solar cell technologies and high-quality substrates have been required.[1), 2)] However, most of monocrystalline silicon wafers are cut from long CZ silicon ingots, causeing a serious problem. While ring-shaped defects are easily generated in wafers cut from the position with small solidified fractions (seed-side position), the defects are hardly formed in wafers from the position with large fractions (tail-side position).[3)]

We[4), 5)] have developed barium (Ba)-doped silica glass crucibles for CZ silicon crystal growth. It has been known that the Ba-doped silica glass accelerates crystobalite (polycrystalline structure of SiO_2) formation, effectively suppressing so-called brownish rings at the inner surface and increasing the yield of monocrystalline silicon growth.[6)] Our method is characterized by both the use of highly pure silica crucible with bubble-free and the suppresion of crack occurrence after devitrification heat treatment. These resulted in a surprising effect that the inner surface of the crucible was almost non-wettable to molten silicon, causing less dissolution of the inner surface than conventional silica crucible.[4)] We gave our non-wettable crucible the name "liquinert (= liquid + inert) crucible"[4)] and named the liquinert-crucible applied CZ method the "LCZ technique".[4)] Also silicon crystals grown by the LCZ technique were called "LCZ silicon".[4)] Lifetime values of LCZ silicon were considerably longer than those of conventional CZ silicon due to the impurity reduction resulting from the decrease in the dissolution.[4)]

We found that liquinert crucible makes it possible to grow alternative LCZ silicon with large diameter ratio (DR) more than 0.7, where the DR is defined by the ratio of the grown crystal diameter to the crucible diameter. Since the large DR silicon is inevitably becomes a short body, it is called "massive silicon" from its shape. It can be expected that the massive silicon will be less susceptible to thermal history.

In this paper, we will show the status of current efforts on massive LCZ silicon growth and the crystal quality although the growth has been performed using a laboratory scale CZ puller and small liquinert crucibles.

Our goal is to grow a massive LCZ silicon with a 450 mm diameter and a short body length (~ 300 mm) from a 600 mm diameter liquinert crucible and to obtain four full-square (156.75 mm x 156.75 mm) bricks.

2 EXPERIMENTAL

2.1 Observation and temperature measurement of melt surface

After allowing the heater temperature to be adjusted so that seeding could be done at the center of the melt surface, photographs of the melt surface were taken in both the conventional crucible and the liquinart crucible. In either case, the rotation rate of the seed/the crucible is the same at +10/-8 rpm, respectively. The seeding in both melts means that the center temperatures of the melt surfaces are exactly the same.

After the photographing was finished, the seed crystal was pulled up away from the melt surface while keeping the heater power constant, and the surface temperature was measured from the melt center toward the crucible wall using a radiation thermometer. The observation port of our puller was so small that the slide length of the thermometer was limited to 0 (center) to 40 mm (the middle position between the melt-surface center and the wall). This is unavoidable since the puller is not designed to grow silicon with DRs over 0.5. The crucible rotation rate was kept at 8 rpm also during the measurement.

2.2 Crystal growth

We grew monocrystalline silicon with DRs of 0.70, 0.73 and 0.82, respectively, from a melt (3 kg) of each liquinert crucible with a diameter of 165 mm. Since our puller is not equipped with automatic diameter controller, the control of crown shaping, straight body growth and tail shaping was operated by manually adjusting growth speed and heater temperature. In all growth, the rotation rate of the seed/the crucible was the same at +10/-8 rpm, respectively. Using a conventional silica crucible with the same diameter, we also tried to grow a silicon crystal with a DR above 0.50.

The reason why large DR growth is possible will be described in the literature.[7)]

2.3 Lifetime and impurity measurement

We cut off the crown and tail of each ingot to make the straight body only, and cut 2-mm-thick wafers from both ends for measuring oxygen, carbon and resistivity. The 2-mm-thick wafers were polished.

We measured the as-grown lifetime at the center area of each straight body using Sinton BCT-400 system, which performs contactless measurement complying with the SEMI PV-13 standard. Using the polished wafers, we measured the oxygen content by FTIR spectroscopy, where 3.14 x 10^{17} cm^{-2} (IOC-88) was used as an oxygen-concentration calibration factor. The carbon content was measured by secondary ion mass spectroscopy (SIMS) and the resistivity by the four-point probe measurement method.

3 RESULTS AND DISCUSSION

3.1 Observation and temperature measurement of melt surface

Figure 1 shows the fluctuation of melts in a conventional silica crucible (a) and a liquinert crucible (b). As readily seen, the molten silicon in the liquinert crucible is more stable, indicating the temperature distribution in the melt is smaller than that in the conventional crucible.

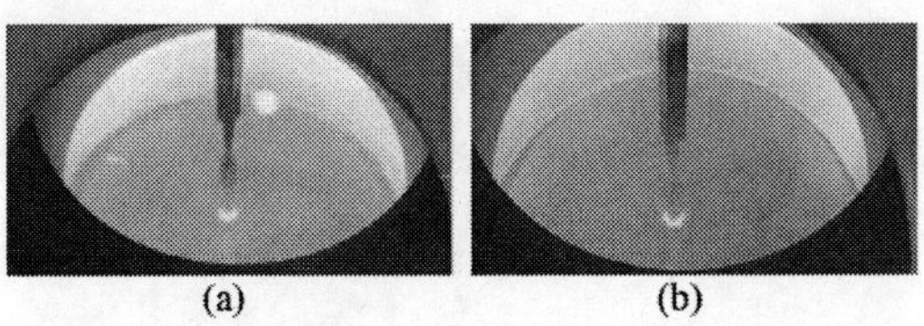

Figure 1. Photographs of melt surface. The melt in a conventional crucible (a) and that in a liquinert crucible (b).

Figure 2 represents temperature distributions of the melt surface from the center of the crucible to the crucible wall. The horizontal axis is relative position defined as the ratio of actual measurement position to the crucible radius. As predicted in the Figure 1, the surface temperature of the liquinert melt seems to be almost uniform compared to previous reports[8) - 10)].

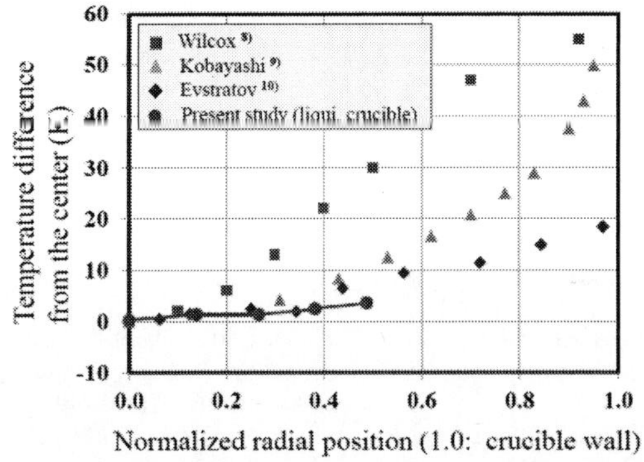

Figure 2. The temperature dependence of the melt surface on the radial position.

3.2 Crystal growth

Figure 3 shows massive silicon crystals with the DRs of 0.70 (=116/165) (a), 0.73 (=120/165) (b) and 0.82 (=135/165) (c).[11)] Figure 3 (b) demonstrates how much the massive silicon occupies the melt surface.

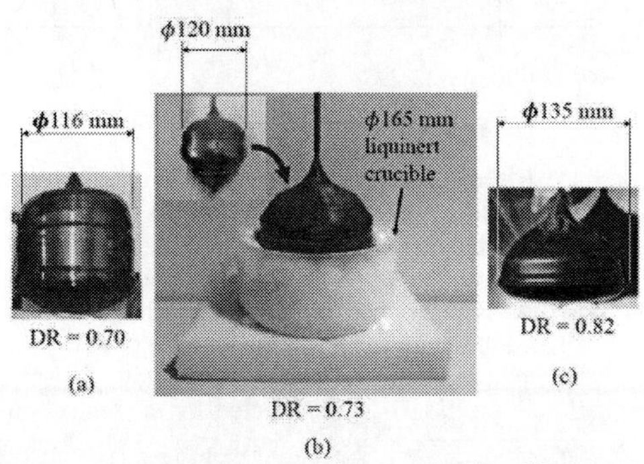

Figure 3. Various massive silicon crystals with the DRs of 0.70 (=116/165) (a), 0.73 (=120/165) (b) and 0.82 (=135/165) (c).[11)]

We could confirm that the growth conditions at large DRs more than 0.70 were quite stable using the liquinert crucible. On the other hand, in the growth from a conventional crucible, as DR approached 0.70, we had to give up further attempts to grow as the melt freezing started from the crucible wall.

The reason why massive silicon can be grown is shown in Figure 4.

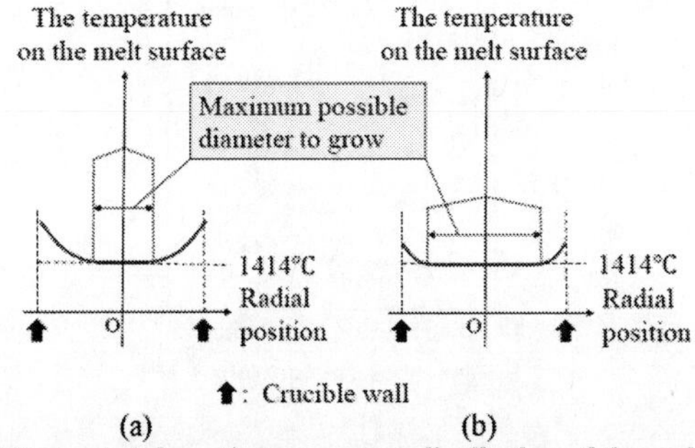

Figure 4. Schematic temperature distribution of the melt surface in a conventional crucible (a) and in a liquinert crucible (b).

Let's assume the temperature distribution of the melt surface of a conventional crucible as shown in Figure 4(a) and that of a liquinert crucible in 4(b). As can be seen immediately, the melt in the liquinert crucible has a larger uniform-temperature region, so that the maximum possible diameter to grow is larger.

3.3 The impact of thermal history on lifetime

Table I shows impurity concentration for conventional silicon-1, -2 and massive silicon-1. Since the impurity concentration differs depending on crystals, it is difficult to discuss crystal quality. Therefore, we developed a method for evaluating crystal quality using relative values of lifetime.

Table I: Impurity concentration.

Silicon ingot	Oxygen (10^{17} cm^{-3})	Carbon (10^{15} cm^{-3})	Phosphorus (10^{15} cm^{-3})
Conv. – 1			
seed-side	10.1	2.7	2.3
tail-side	9.1	2.7	3.5
Conv. – 2			
seed-side	10.1	1.6	2.5
tail-side	6.3	4.5	5.4
Massive – 1			
seed-side	11.2	4.7	2.5
tail-side	11.3	10.2	4.0

Since lifetime is influenced by both the cooling rate and temperatures after solidification at the growth interface, the relative difference in lifetime between the seed-side and the tail-side of an ingot is proportional to the magnitude of the thermal history to which the ingot was exposed in a puller. Taking the lifetime criteria to the tail-side value is justified by the fact that the lifetime at the tail-side position is little affected by thermal donors and oxygen precipitates[7)] but only by dopant concentration.

Our method for quatifying thermal history is shown in Figure 5. As shown by the dotted line, when there is no defect in the crystal, the lifetime at seed-side increases by I in proportion to the square of the phosphorus concentration ([P]).[7)] However, in reality, due to the various defects formed by thermal history, the lifetime decreases like a real curve. And let's define an impact factor of thermal history D as $D = d / I$, where d is the decrease in lifetime shown in Figure 5.

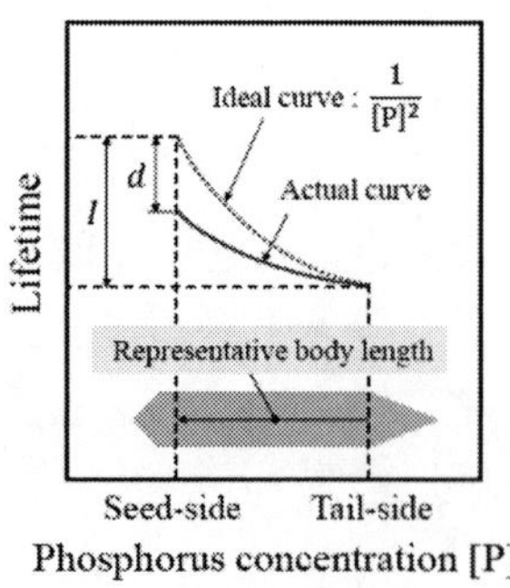

Figure 5. Our method for quatifying thermal history using lifetime.[7)]

Table II represents impact factors of thermal history The most strikeing feature is that the factor D of the massive silicon-1 is less than 1. This may seem natural, as the representative body length of the massive silicon-1 is shorter, but it is not so. Since the maassive silicon is thicker than conventional crystals, its part around the growth axis is hard to cool. Therefore, there was a risk that the thermal history would be lengthened accordingly, but the concern is likely to be denied as far as we grow massive silicon using the laboratory scale puller.

Table II: Each ingot size and the impact factor of thermal history D.

Silicon ingot	Representative body length (mm)	Diameter (mm)	D
Conv. - 1	150	~ 80	1.23
Conv. - 2	180	~ 80	1.16
Massive - 1	77	~ 116	0.91

4 CONCLUSION AND FUTURE PLAN

Using the laboratory scale CZ puller, we could demonstrate the growth of massive silicon crystals with the DRs exceeding 0.70. The reason why the growth is possible is because the uniform-temperature range of the melt surface is larger than that of conventional melt. Since the massive silicon has a short body part, it can be expected that it has a smaller thermal history than conventional crystals.

Using a 600 mm diameter liquinert crucible, we are planning to grow a 450 mm diameter silicon crystal (DR = 0.75) with a 300 mm length body which can provide four short bricks as shown in Figure 6(a1) and 6 (a2).[11)] For comparison, a brick obtained from a CZ silicon with 1470 mm body length is depicted in Figure 6(b). It is assumed that the weights of both crystals are equal (144 kg) in the as-grown states with crowns and tails. The amount of charged polycrystalline silicon is assumed 160 kg. Wafers with 156.75 mm-full-square are assumed to be obtained. Although the sum of the lengths of four short-bricks is 81.6% (= 300 x 4 / 1470) of the long brick, our massive LCZ growth will bring no cost-rising because the total estimated time of the growth will be shorter by 20% than that of conventional growth of long silicon.

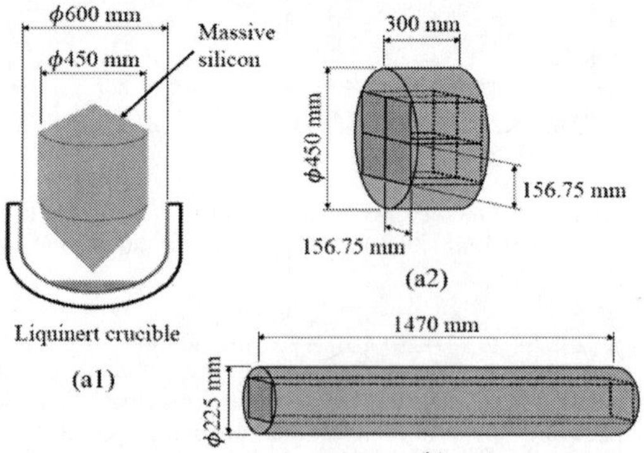

Figure 6. Our future plan of massive silicon growth, (a1) and (a2). The growth size target (a1), cutting four full-square-short bricks (a2) and a long full-square brick from a conventional CZ silicon.[11)]

References

[1] S. W. Glunz, J. Knobloch, D. Biro, and W. Wettling, Proc. 14th European Photovoltaic Solar Energy Conference, Barcelona (1997), pp. 392 - 395.
[2] J. Zhao, A. Wang, and M. A. Green, Prog. Photovol. Res. Appl. **7** (1999), pp. 471 - 474.
[3] T. Tachibana, K. Nakamura, A. Ogura, Y. Ohshita, T. Shimoda, I. Masada and E. Nishijima, AIP Advances 7, 045111 (2017).
[4] T. Fukuda, Y. Horioka, K. Fujiwara, K. Tanahashi, S. Simayi, K. Shirasawa, and H. Takato, J. Crystal Growth **438** (2016), pp. 76 - 80.
[5] USP 9, 593, 432 B2 (Date of Patent: Mar. 14, 2017).
[6] X. Huang, T. Hoshikawa and S. Uda, J. Crystal Growth **306** (2007), pp. 422 - 427.
[7] To be published in ECS J. Solid State Sci. and Tech. in 2018.
[8] W. R. Wilcox and L. D. Fullmer, J. Appl. Phys. **36** (1965), pp. 2201 – 2206.
[9] S. Kobayashi, S. Miyahara, T. Fujiwara, T. Kubo and H. Fujiwara, J. Crystal Growth **109** (1991), pp. 149 – 154.
[10] I. Y. Evstratov, V. V. Kalaev, A. I. Zhmakin, Y. N. Makarov, A. G. Abramov, N. G. Ivanov, A. B. Korsakov, E. M. Smirnov, E. Dornberger, J. Virbulis, E. Tomzig and W. von Ammon, J. Crystal Growth **237** - **239** (2002), pp. 1757 – 1761.
[11] T. Fukuda, Y. Horioka, K. Fujiwara, K. Tanahashi, K. Shirasawa, and H. Takato, Abstract Book, 10th Inter. Workshop Cryst. Silicon Solar Cells, CSSC-10 (2018), p. 21.

DEVELOPMENT OF HIGH PERFORMANCE MULTICRYSTALLINE SILICON WITH CONTROLLED SEEDING

A. Hess, P. Krenckel, T. Trötschler, T. Fehrenbach, S. Riepe
Fraunhofer Institute for Solar Energy Systems ISE
Heidenhofstrasse 2, D- 79110 Freiburg, Germany
Phone +49 761 4588-5644, adam.hess@ise.fraunhofer.de

ABSTRACT: High Performance multicrystalline silicon ingots are mostly produced with seeded growth on small size silicon material, which determines the beneficial initial grain structure and orientations of the crystallites. A drawback of this approach is the increased area of low minority charge carrier lifetime in the lower part of the ingot and thus a worse yield. Also, the solidification process time is increased as the seed has to remain solid in a well controlled manner. Both drawbacks can be addressed by using a thin seeding layer on the crucible bottom, which has a higher melting point than silicon. The area of low minority carrier lifetime at the ingot bottom is comparable to standard multicrystalline silicon, whereas the crystal structure is significantly improved. Several seeding layer configurations were investigated and SiO_2 with different grain sizes was chosen for G1-sized experiments. The resulting brick shows a grain and defect structure close to the High Performance multicrystalline silicon reference while reducing the width of the impurity indiffusion zone in the bottom. By optimizing the seeding layer in conjunction with the thermal process, a higher yield of high quality material is possible.
Keywords: Crystallization, Silicon, SiO_2, High Performance, Multicrystalline

1 INTRODUCTION

Wafer from multicrystalline silicon (mc-Si) are still the basis for approx. 50% of all sold solar cells [1]. In this fraction, the role of High Performance multicrystalline silicon (HP mc-Si) is dominating [2-3]. Due to the characteristic homogeneous and small grain structure with random orientations, dislocation creation is suppressed and induced ones tend to annihilate at random angle grain boundaries [4]. Thus, material quality limiting crystal defects can be reduced. But for this, the seed material has to remain solid during melting, leading to longer process times compared to standard mc-Si growth. Also, due to the seed preservation, the ingot yield is smaller as a bigger portion of the bottom part can't be used.

This work aims at the development of HP mc-Si ingots with low dislocation densities and optimized material quality based on the use of a thin seeding layer. In this approach, the standard mc-Si solidification process can be used leading to the improved material quality of HP mc-Si. Similar approaches have been tested with promising results [5-7], although most ones used an additional Si_3N_4-coating on top of the layers. Our experiments showed that for some seeding materials a Si_3N_4-coating leads to worse results than without. For that we used SiO_2-powder as a direct seeding layer with specific grain sizes to study their influence on the grain structure and the material quality of a resulting G1-ingot.

2 EXPERIMENTAL

In this work, we investigated the influence of different coatings as seeding layer for HP mc-Si growth. On small scale (G0, 800 g Si) various materials have been tested as nucleation layer with and without coatings on top. The samples have been prepared and analyzed for initial grain size, grain size homogeneity and evolution. SiO_2-powder without a coating was identified as best candidate for G1-experiments.
SiO_2-powders with a variation of initial particle size are used. The ingots were grown in a G1-sized crucible. Four ingots with 14.0 kg of high purity silicon feedstock were crystallized in a laboratory Vertical Gradient Freeze furnace (PVA TePla VGF632). The ingots were p-doped with boron for a target resistivity of 0.5 - 1.5 Ωcm.
We used an in-house developed, non-standard coating for the experiments (ISE coating) with and without nucleation layer and a standard industrial coating for a reference ingot without seed and one on granular Si.

The crucible bottom was covered with a 1 mm thick layer of high quality SiO_2-powders from two sizes, quartzpowder with a grain size distribution of 75-300 µm (Powder 1) on the left half and with 37-63 µm (Powder 2) on the right half (Figure 1). As a comparison a standard mc-ingot was grown with the same thermal process. Both materials have been compared with a reference HP mc-Si material grown on granular Si. For seed preservation, a different thermal process had to be applied.

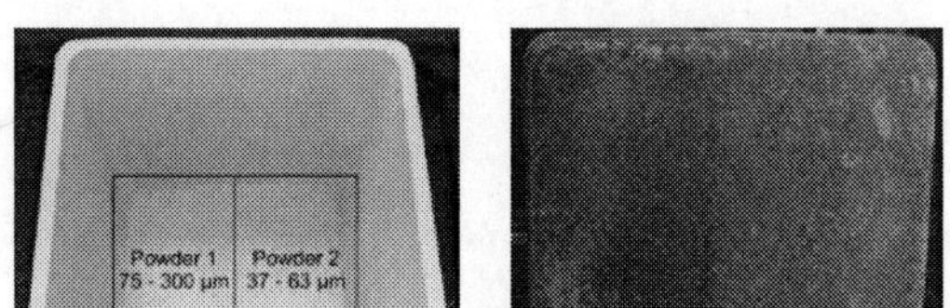

Figure 1: Experimental setup with two different grain size distributions on every half of the G1-sized crucible bottom (75 - 300 µm left and 37 - 63 µm right) and the bottom of the resulting Si-ingot.

A central brick with 156 mm edge length was prepared and analyzed for inclusions and brick minority carrier lifetime as well as electrical resistivity. Inclusions and other particles were identified using the Intego Orion HighRes IR block inspection system. Lifetime and electrical resistivity measurements were done with the Semilab WT-2000D. Fourier-transform Infrared Spectroscopy (FTIR) measurements were performed to identify oxygen concentrations using the Vertex 80v in combination with the Hyperion microscope (Bruker). The concentration of interstitial Oxygen (O_i) was measured every 5 mm along the brick height. For this measurement setup, the samples had to be divided into two parts, resulting in a data gap at 75-80 mm to avoid edge effects.

Wafers were produced from this ingot material by slurry based multi-wire sawing and characterized with respect to grain structure, dislocated area fraction, inclusions and material properties. To visualize the grain structure and the material quality Photoluminescence- (PL) and optical measurements were performed on the wafer using the Hennecke Wafer Inspection System.

3 RESULTS AND DISCUSSION

3.1 Inclusions

In a first step, the sawn bricks were polished and scanned for inclusions by IR inspection. The Intego Orion HighRes IR block inspection system allows a resolution down to 18µm to identify inclusion and other defects in the brick volume. This way, we analyzed if the SiO_2-powder may have been incorporated into the brick. As we used an experimental coating (ISE coating) on the crucibles, these two bricks were compared to a reference brick with standard industrial coating. In Figure 2, the Small Point Defect Density (SPDD) distribution over brick height is shown. There is no significant difference in SPDD in the bricks with and without SiO_2-powder, so the powder does not trigger increased particle inclusion. Whereas, in the brick grown with a standard industrial crucible coating, there are 10 times more small inclusions identified. In this case, the influence by the coating is dominant.

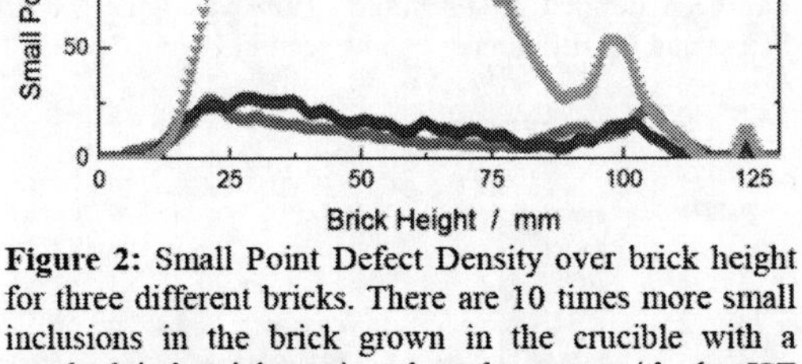

Figure 2: Small Point Defect Density over brick height for three different bricks. There are 10 times more small inclusions in the brick grown in the crucible with a standard industrial coating than the ones with the ISE coating.

3.2 Grain analysis

In Figure 3, we can see optical images of the wafer from three different brick heights (20, 60 and 100 mm). For the HP mc-Si brick grown on granules, the bottom-most wafer is at brick height of 40 mm due to the remained seed layer of 30 mm. It shows the typical small and homogeneous grain distribution. These characteristics and the random grain orientations are the key parameter to reach improved material quality by lowering mechanical stress and thus avoid the occurrence of dislocations. Even when dislocation clusters occur in some grains, they may be overgrown by other grains. This overgrowth is enhanced due to a small and homogeneous grain size distribution. The SiO_2-seeded brick shows a grain structure and size distribution close to the HP mc-Si reference. Although this brick has two different seeding layers, there is no obvious difference visible between the two halves. In the standard mc-Si brick, the undercooling at the start of the crystallization phase was higher. We can clearly see dendritic growth in the initial nucleation phase. As both experiments used the same thermal process, the combination of additional nucleation sites and slightly reduced thermal conductivity by the SiO_2-layer can avoid dendritic growth in this case.

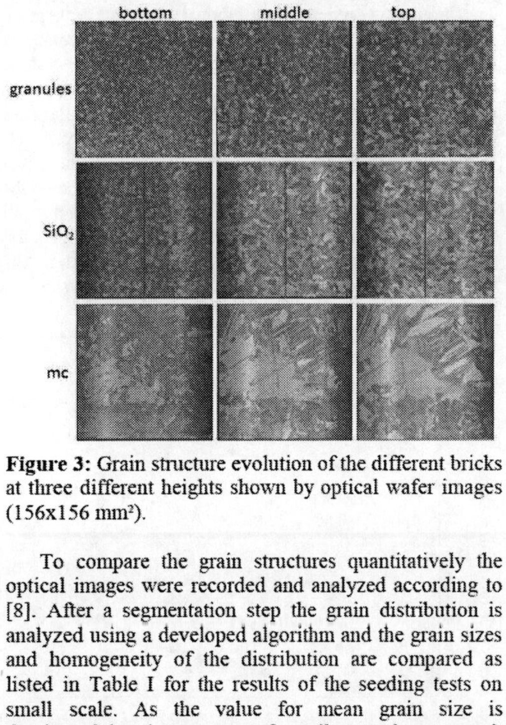

Figure 3: Grain structure evolution of the different bricks at three different heights shown by optical wafer images (156x156 mm²).

To compare the grain structures quantitatively the optical images were recorded and analyzed according to [8]. After a segmentation step the grain distribution is analyzed using a developed algorithm and the grain sizes and homogeneity of the distribution are compared as listed in Table I for the results of the seeding tests on small scale. As the value for mean grain size is dominated by the amount of smallest grains, a much better parameter to describe the grain size is the 50th weighted area percentile (A_{50}). The mean relative difference of the consecutive weighted percentiles showed to be a good measure for the inhomogeneity of the grain size distribution (GAI), where a smaller value means more homogeneity. Further explanations can be found in [8]. This way the best seeding layers were selected after various pretests in smaller crucibles, where SiO_2-powder without a Si_3N_4-capping proved to be the best option.

Table I: Grain analysis comparison of the bottom-most wafer of each brick from the small seeding tests using the procedure presented in [8]. 50th weighted area percentile (A_{50}), the mean relative difference of the consecutive weighted percentiles (GAI) are shown for experiments on SiO_2-powders with and without Si_3N_4-capping.

	Powder 1	Powder 1 + C	Powder 2	Powder 2 + C
A_{50} [mm²]	3.99	41.42	4.48	25.59
GAI	0.88	0.63	0.83	0.61

The evolution of the grain size over brick height can be found in Figure 4 and the initial values for the bottom-most wafer in Table II.

Table II: Grain analysis comparison of the bottom-most wafer of each brick using the procedure presented in [8]. 50th weighted area percentile (A_{50}), the mean relative difference of the consecutive weighted percentiles (GAI) are shown for experiments on granules, with SiO_2-powders, and without seeding layer.

	Granules	Powder 1	Powder 2	mc
A_{50} [mm^2]	1.44	3.50	3.78	66.91
GAI	0.51	0.46	0.44	1.41

The granular seeded brick shows the smallest grain distribution and a linear growth of the median over brick height. The SiO_2-seeded brick shows the same distribution and growth on both sides. Regarding initial grain size and grain size evolution, there is no difference between the two SiO_2-powders. The increase corresponds to the granules-seeded brick, but the overall median sizes are about 2 mm² smaller at the same height above the seeding position as the granules remain on the bottom of the crucible. The mc-brick with dendritic seeding has much bigger grains and a high inhomogeneity, too. Overall, the median stays at a value about 70 mm².

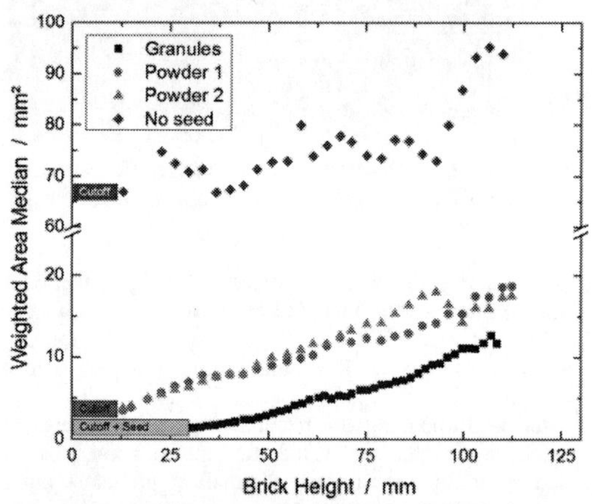

Figure 4: Weighted area median over brick height for the three bricks. The grain structure in the SiO_2-seeded brick shows a similar evolution like the one grown on granules, being about 2 mm² bigger in size.

3.3 Impurities and dislocations

In Figure 5 two PL-images are shown from the bottom and the top of the grown SiO_2-seeded brick. From bottom to top we can identify two different regions corresponding to the two seeding layers. Both show different effects on the PL-intensity. In the bottom wafer, the right side with Powder 1 as seeding layer shows a reduced PL-signal in the grains while the left one, grown on powder 2, doesn't. This effect is linked to the width of the indiffusion zone into the brick which is smaller here for the left side. On the other hand the dislocation density is higher on the left side in the middle wafer. However, both sides show the same grain structure and grain size in the bottom part of the brick as seen in Figure 3.

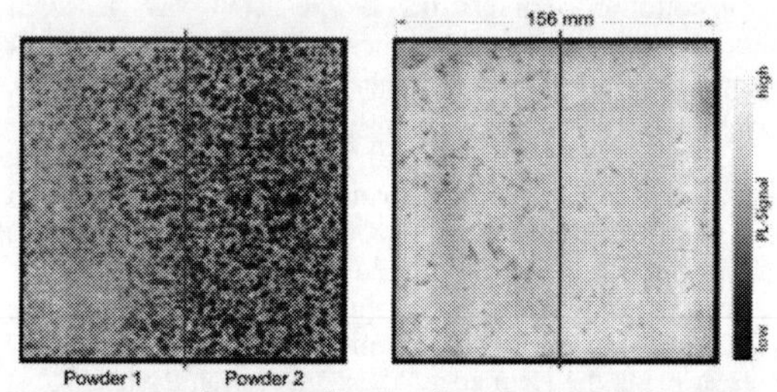

Figure 5: PL images of wafer from 10 mm (left) and 60 mm (right) ingot height. Different seed grain sizes were placed on each half. There is a clear difference in impurity indiffusion area and dislocation density for both sides.

In Figure 6, all wafer PL-images were stacked to create a virtual image of the brick vertical cross section to illustrate the different impurity indiffusion lengths. Bottom and top cutoffs are drawn in as these regions weren't wafered.

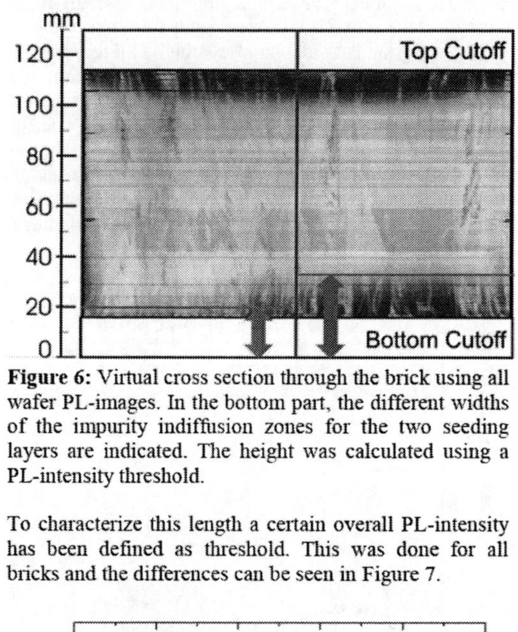

Figure 6: Virtual cross section through the brick using all wafer PL-images. In the bottom part, the different widths of the impurity indiffusion zones for the two seeding layers are indicated. The height was calculated using a PL-intensity threshold.

To characterize this length a certain overall PL-intensity has been defined as threshold. This was done for all bricks and the differences can be seen in Figure 7.

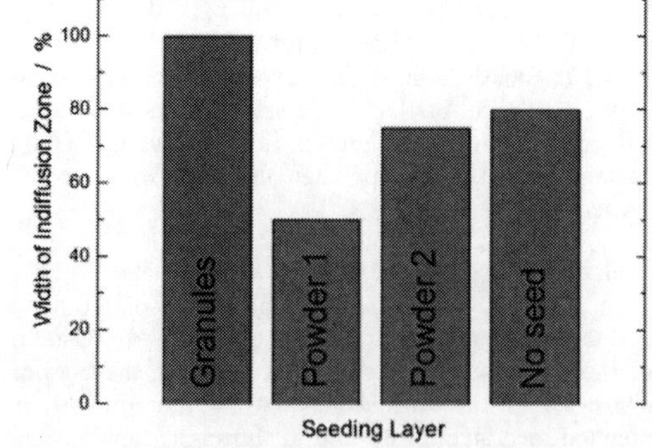

Figure 7: Width of impurity indiffusion zone extracted from PL-images for different seeding layers

For the ingot grown on Si granules, this value is higher as the seed remains solid and the process time is increased. For powder 2, the value is comparable to the standard mc-process whereas it's significantly reduced for the powder 1. The reason is either the SiO_2-material quality itself or an effect as diffusion barrier and has to be validated in further experiments.

The dislocation area fraction was calculated from the PL-images. The results are shown in Figure 8. The dislocation multiplication is high for the mc-material as grains are much bigger and stress relaxation by forming dislocations is common. The material grown on granules stays with only slight variations at 0.5 - 1 area%, which is a typical value for HP mc-Si. For the SiO_2-powder, there is a clear difference between the two grain sizes. On the coarser grains in powder 1, the value stays constant at about 2 area%. But especially on powder 2, dislocation density is reduced and in good agreement with HP mc-Si on granules.

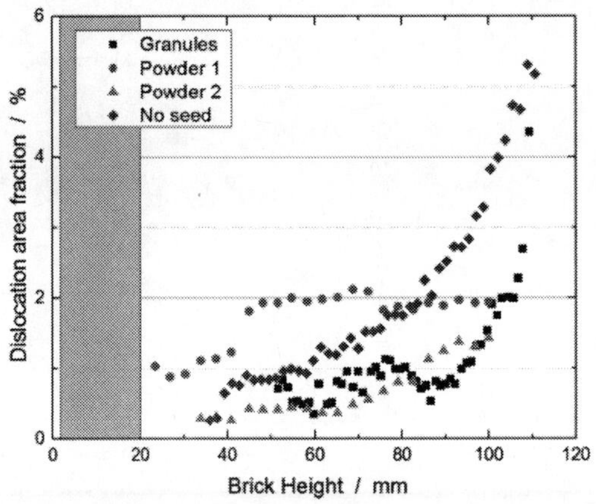

Figure 8: Dislocation area fraction over brick height for the seeded and not seeded bricks calculated from the PL-images. The indiffusion area was not analyzed due to the low PL-intensity.

3.4 Oxygen content

As the silicon melt is in direct contact with the SiO_2-powder, the Oxygen concentration could be a critical parameter. The interstitial Oxygen concentration over brick height is shown in Figure 9. Both curves show similar behavior. Only the first centimeters of the SiO_2-sample shows increased values which rapidly reach the same level at about 20 mm ingot height. Thus, the SiO_2 powder has no obvious negative effect on the O_i-concentration for most of the ingot volume.

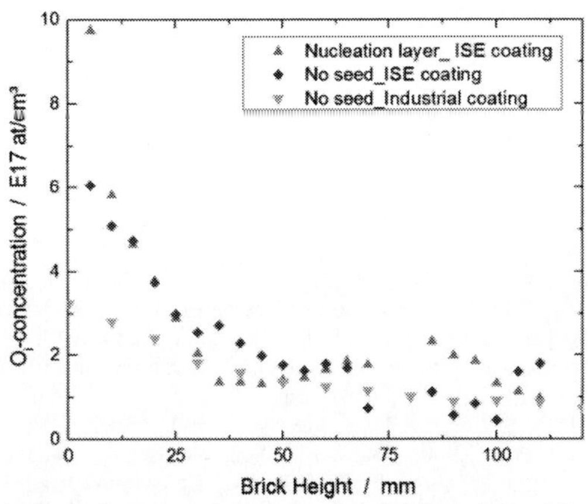

Figure 9: O_i-concentration over brick height; The distribution is comparable, only in the first centimeters the value is slightly higher in the sample with SiO_2-seed.

4 CONCLUSION

Different configurations of materials and particle sizes have been tested in a small crucible setup to identify the most promising candidates to act as a thin seeding layer for HP mc-Si growth in a standard mc-Si solidification process. These results were transferred to grow a G1-sized ingot. A HP mc-Si structure could be achieved with both seed grain sizes used at each ingot half. We also see a positive impact of the material used in PL wafer measurements in regard to lifetime for powder 1 and dislocation density for powder 2. Powder 2 shows a high potential as nucleation layer for HP mc-Si ingots with reduced width of impurity indiffusion zone compared to granular seeds. By optimizing the nucleation layer in conjunction with the thermal process, a higher yield of high quality material should is possible.

5 ACKNOLEDGMENTS

The authors would like to thank Wacker Polysilicon and AlzChem for crystallization materials and all colleagues of the crystallization and characterization teams at Fraunhofer ISE, especially F. Haas and P. Häuber for material processing.

This work was part of a SOLAR-ERA.NET project and was funded by the German Federal Ministry for Economic Affairs and Energy within the research project "HighCast" (grant number 0325894)

6 REFERENCES

[1] N. S. Pujari et al., "International Technology Roadmap for Photovoltaic (ITRPV)", Ninth Edition 2018, p. 40

[2] Lan, C. W.; Yang, C. F.; Lan, A.; Yang, M.; Yu, A.; Hsu, H. P.; Hsu, B.; Hsu, C., "Engineering silicon crystals for photovoltaics", CrystEngComm 18 (9) (2016), 1474-1485

[3] Lan, C. W.; Lan, A.; Yang, C. F.; Hsu, H. P.; Yang, M.; Yu, A.; Hsu, B.; Hsu, W. C.; Yang, A., "The emergence of high-performance multi-crystalline silicon in photovoltaics", Journal of Crystal Growth, vol. 468 (2017), 17-23

[4] G. Stokkan, A. Song, B. Ryningen, “Investigation of the Grain Boundary Character and Dislocation Density of Different Types of High Performance Multicrystalline Silicon”, Crystals 8(9) (2018), 341
[5] Huali Zhang, Da You, Chunlai Huang, Yihua Wu, Yan Xu, Peng Wu, “Growth of multicrystalline silicon ingot with both enhanced quality and yield through quartz seeded method”, Journal of Crystal Growth, vol. 435 (2016), 91–97
[6] Junjing Ding, Yunyang Yu, Wenliang Chen, Xucheng Zhou, Zhiyong Wu, “Effect of the fused quartz particle density on nucleation and grain control of high-performance multicrystalline silicon”, Journal of Crystal Growth, vol. 454 (2016), 186–191
[7] Yu, Yunyang; Ding, Junjing; Chen, Wenliang; Zhang, Zhaoyu; Zhou, Xucheng; Zhong, Genxiang; Huang, Xinming, “Growth of High-Quality Multicrystalline Silicon Ingot through the Cristobalite Seeded Method”, Journal of Materials Science and Engineering B (11-12) (2016), 304-310
[8] T. Strauch, M. Demant, P. Krenckel, S. Riepe, S. Rein, “Analysis of grain structure evolution based on optical measurements of mc Si wafers”, Journal of Crystal Growth, vol. 454 (2016), 147–155

MULTICRYSTALLINE INFORMATICS TO REALIZE IDEAL CRYSTALLINE SILICON FOR SOLAR CELLS

Noritaka Usami[1], Yusuke Hayama[1], Tetsuro Muramatsu[1], Kazuya Tajima[1], Soichiro Kamibeppu[1], Kentaro Kutsukake[2], Tetsuya Matsumoto[3], and Hiroaki Kudo[3]
[1]Graduate School of Engineering, [2]RIKEN Center for Artificial Intelligence Project,
[3] Graduate School of Informatics, Nagoya University, Japan

ABSTRACT: We introduce our multidisciplinary scientific approach, "multicrystalline informatics". The fundamental idea is to integrate data science with experimental science, theory, and computation in order to enable a paradigm shift in development of materials. Especially, we apply this methodology to realize ideal block-cast silicon by clarifying structural and electrical properties of crystal defects and controlling mechanisms in complicated multicrystalline silicon. The usefulness of multicrystalline informatics is shown by various aspects including analysis of nucleation sites using weighted Voronoi diagram and 3D visualization of generation, propagation, and annihilation of crystal defects.
Keywords: Multicrystalline Silicon, Crystallisation, Defects, Photoluminescence

1 INTRODUCTION

Since the announcement of Materials Genome Initiative in 2011, integration of experiments, computation, and theory with informatics attracts attention as a promising approach to enable a paradigm shift in materials development to accelerate the timeframe of the materials discovery to deployment process [1]. In various countries, national projects related with this methodology are initiated. As for Japan, cross-ministerial strategic innovation promotion program promotes such a multidisciplinary approach in research and development of structural materials for the future of Japanese aircraft industry [2]. In 2017, Japan Science and Technology Agency launched a new strategic basic research program on revolutional material development by fusion of experiments with theory/data science [3]. Since we believe that this approach must be useful also in research and development of materials for photovoltaics, we proposed a research project entitled "Multicrystalline informatics toward establishment of general grain boundary physics and realization of high-quality silicon ingot with ideal microstructures". The proposal was accepted as one of the 5 projects to start in 2017 [4].

In this paper, we introduce our attempt to integrate data science with experimental science, theory, and computation in order to realize ideal block-cast silicon by clarifying structural and electrical properties of crystal defects and controlling mechanisms in complicated multicrystalline silicon (mc-Si). The grain structures in mc-Si around the bottom of the ingot was reproduced using weighted Voronoi diagram to analyze nucleation sites. 3D visualization of crystal defects in mc-Si could reveal generation, propagation, and annhiration of defects, and collect a set of images for the source of dislocation clusters, which could be utilized to understand how to suppress generation of dislocations.

2 WEIGHTED VORONOI DAIAGRAM

The grain structures formed at the bottom of the crucible would affect the overall structures of mc-Si ingot. Therefore, it is of crucial importance to understand the nucleation sites as the first step for better control of microstructures. To analyze nucleation sites, we considered to utilize Voronoi diagram by reproducing grain structures around the bottom of the ingot [5]. Voronoi diagram is a partitioning of a plane into regions based on the distance to arbitrarily arranged generating points. If we assume simultaneous nucleation and isotropic growth rate, generating points of Voronoi diagram could be regarded as nucleation sites. Since these assumptions are not realistic, variation of the nucleation timing and anisotropic growth rate must be considered as weights to improve concordance rate between grain structures and the weighted Voronoi diagram.

The flow to maximize the concordance rate is as follows. First, experimentally obtained image of the grain structure is loaded followed by calculation of the center of the gravity of each grain. The center of the gravity is considered as initial generating points. The timing of the nucleation is estimated based on the area of each grain, and the crystal shape during growth was assumed as a cross section of an octahedral bounded by {111} planes [6,7], which permits to draw the weighted Voronoi diagram. Then, positions of generating points are moved to improve the concordance rate, and the weighted Voronoi diagram is redrawn with renewed weights. These process are repeated until none of the generating points move.

Figure 1 shows snapshots of the nucleation and lateral growth process reproduced by the weighted Voronoi diagram. It is seen that time-dependent nucleation and orientation-dependent crystal shape are well considered.

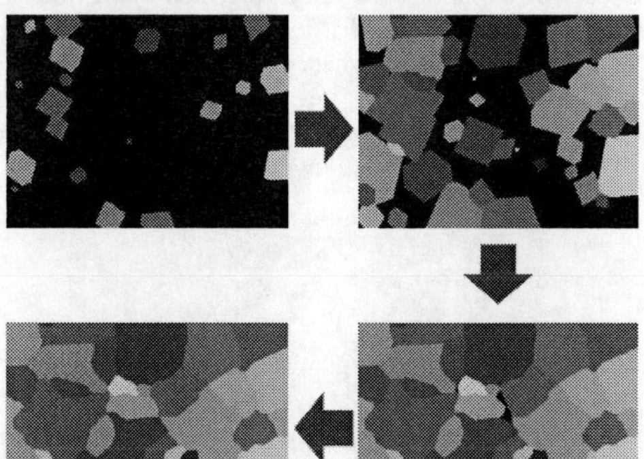

Figure 1: Nucleation and lateral growth process reproduced by the weighed Voronoi diagram.

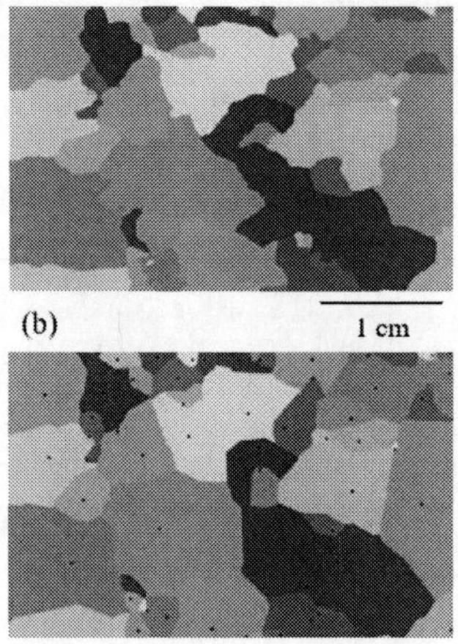

Figure 2: Comparison between (a) experimentally obtained grain structures and (b) weighted Voronoi diagram with generating points.

Figure 2 shows comparison between (a) the grain structure around the bottom obtained by electron back-scattering pattern analysis and image processing and (b) the weighted Voronoi diagram with generating points. They are in good agreement with each other except details, and the concordance rate amounts to 88.5%. It is noted that the initial concordance rate using Voronoi diagram without weights is only 58.3%. Due to the high concordance rate, the generating points could be regarded as most probable nucleation sites. Further improvement in the concordance rate might be possible by considering nonuniform temperature distribution, which could affect the anisotropic growth rate. This would be a target for study in the future.

Figure 3 shows superimposed image of generating points with grain boundaries and processed photo of the nucleation layer at the bottom of the crucible. In this particular sample, we intentionally introduced single layer silicon beads at the bottom of the crucible to give a millimeter scale roughness and enhance inhomogeneous nucleation for grain refinement [8]. However, the obtained grain size was found to be larger than the size of silicon beads. The superimposed image revealed that generating points are not randomly distributed, but are found around the concave part at higher probability, which act as preferential nucleation sites. This locally enhanced nucleation owing to inhomogeneous geometry would be the reason for disturbing grain refinement [5].

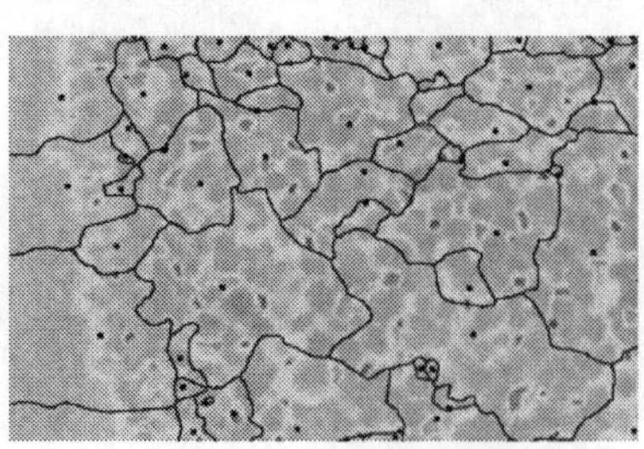

Figure 3: Superimposed image of the weighted Voronoi diagram and bottom of the crucible.

3 3D VISUALIZATION OF CRYSTAL DFECTS

After the nucleation at the bottom of the crucible, mc-Si ingot is produced by the directional solidification. During directional solidification, microstructures could change either kinetically or energetically. For example, orientation-dependent growth rate could modify the structure by grain competitions. Energy minimization would modify the grain boundary structures. In addition, generation of crystal defects takes place during directional solidification, which could affect macroscopic performance of mc-Si wafers produced from the ingot. Therefore, it is important to understand generation mechanisms of generation of dislocations to realize high-quality ingot with low dislocation density. For this purpose, we attempted to visualize crystal defects in mc-Si ingot [9].

A block of high-performance mc-Si ingot could produce ~1,000 wafers with fine crystal grains, and photoluminescence (PL) images of these wafers are supposed to contain data on structural and electrical properties of crystal defects. PL images of as-sliced wafers were measured under same conditions (laser wavelength: 940 nm) by EPL-100s, Hamamatsu Photonics K.K. Since the surface of as-sliced wafers by diamond wires is shiny with directional saw marks, raw data were affected by the saw marks and presence of crystal defects was not obvious. For better visualization, following image processing was applied.

Firstly, fast Fourier transform and inverse fast Fourier transform were used for the reduction of scratched patterns caused by the wafering. Then, gray level of PL images were normalized, because the crystal quality strongly depends on the ingot height due to the dopant density. Then, smoothing and unsharp-mask filter were used for the reduction of small noises and grain edge enhancement.

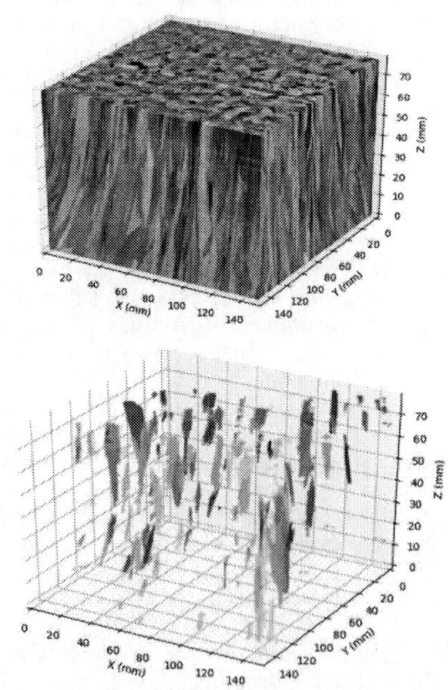

Figure 4: 3D visualization of evolution of microstructures (top) and crystal defect (bottom) in mc-Si ingot based on a set of image-processed PL images

Then, crystal defects, which are most likely assigned as dislocation clusters, were visualized by thresholding. By stacking these images, 3D visualization of directional solidification is possible.

Figure 4 shows compares 3D microstructures of mc-Si ingot and corresponding defect distribution inside the ingot. This manifests generation, propagation, and annihilation of dislocation clusters. Especially, annihilation process of dislocation clusters is considered to play an important role in keeping dislocation density low. Analysis of the direction of dislocations propagation were done by tracing the center of gravity of each dislocation cluster in each PL image. It was found that the slope of the path is strongly dependent on the ingot position, which would be closely connected with growth conditions. Therefore, this 3D imaging technique will be very useful to optimize crystal growth process.

It is of crucial importance to clarify the mechanism of generation of dislocations. We could collect a large number of PL images around the source of dislocations, which would contain important information. Figure 5 shows a typical example of image patches around the source of dislocations. Although it is difficult to recognize any features by human eyes, machine learning might clarify hidden features. It would be more probable if we could add information on orientation distribution. Optical images of mc-Si wafers with white light illumination could be utilized for fast evaluation of orientation distribution. Figure 6 illustrates how average luminance changes depending on the illumination angles, where the illumination directions are shown by arrows. It is seen that the luminance of each crystal grain and its dependence on illumination angles are different, which indicates that these could be utilized for prediction of crystal orientations. A set of training data is under preparation for machine learning so that we can predict orientation distribution simply by taking a photo unlike time-consuming conventional methods.

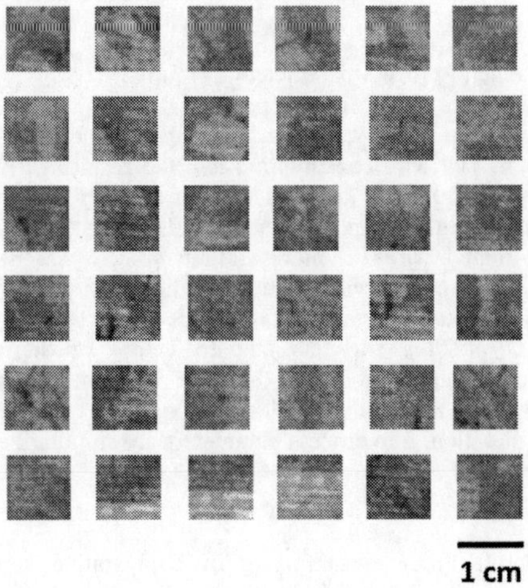

Figure 5: Example of image patches around the source of dislocations

Figure 6: Typical optical images of mc-Si wafers with different white light illumination angles. Illumination directions are shown by arrows.

4 SUMMARY

We reviewed our attempt to integrate data science with experimental science, theory, and computation in order to realize ideal mc-Si ingot for solar cells. The weighted Voronoi diagram was shown to be a powerful tool to analyze nucleation sites at the bottom of the crucible. 3D visualization of microstructures and crystal defects in mc-Si is possible by stacking a lot of PL images, which could be utilized to improve growth process of mc-Si ingot. Features of source of dislocations might be identified by machine learning. As illustrated by these examples, our multidisciplinary scientific approach, "multicrystalline informatics", is believed to be very promising approach to enable a paradigm shift in development of high-quality silicon ingot for solar cells.

This work was supported by Japan Science and Technology Agency (JST), CREST, JPMJCR17J1.

5 References

[1] "Materials Genome Initiative" https://www.mgi.gov/
[2] http://www.jst.go.jp/sip/k03/sm4i/dl/jst_pamphlet_Japan.pdf
[3] https://www.jst.go.jp/kisoken/crest/en/research_area/ongoing/areah29-3.html
[4] https://www.jst.go.jp/kisoken/crest/en/project/1111096/1111096_2017.html
[5] T. Muramatsu, Y. Hayama, K. Kutsukake, K. Maeda, T. Matsumoto, H. Kudo, K. Fujiwara, and N. Usami, J. Cryst. Growth 499 (2018) 62.
[6] X. Yang, K. Fujiwara, R. Gotoh, K. Maeda, J. Nozawa, H. Koizumi, and S. Uda, Acta Materialia 60 (2012) 3259.
[7] X. Yang, K. Fujiwara, K. Maeda, J. Nozawa, H. Koizumi, and S. Uda, Prog. Photovolt: Res. Appl. 22 (2014) 574.
[8] G. Anandha babu, I. Takahashi, S. Matsushima, and N. Usami, J. Cryst. Growth 441 (2016) 124.
[9] Y. Hayama, T. Matsumoto, T. Muramatsu, K. Kutsukake, H. Kudo, and N. Usami, Solar Energy Mat. and Solar Cells (2018) in press.

METAL CONTAMINATION IN THE DIAMOND WIRE SAWING PROCESS OF SILICON AND INFLUENCE ON THE SOLAR CELL EFFICIENCY

Thomas Kaden, Lydia Lottspeich
Fraunhofer Technology Center for Semiconductor Materials THM, Am St.-Niclas-Schacht 13, 09599 Freiberg
Corresponding author: Thomas Kaden, e-mail: thomas.kaden@ise.fraunhofer.de

ABSTRACT: In diamond wire sawing (DWS) of silicon wafers a steel core wire is used with diamonds embedded in a particular coating layer. This layer is either a nickel-based metallic layer or a resin-based layer. Due to the different properties of the layer material, the density of diamonds on the wire and the wire wear during the sawing process differ strongly between these two types of wire coatings. It was investigated, how the different coatings affect the metal contamination of monocrystalline silicon wafers produced in diamond wire sawing processes. The process parameters were chosen to obtain the same silicon contact length, which is a measure of the wire wear, for both types of wires. A rigorous analysis of the wires and the wafers with respect to the metal contamination was done. For Ni-coated wires, sum concentrations of Ni and Co in the range of 10^{16} at/cm³ were obtained in layers up to 12 µm below the wafer surface. For resin coated this concentration was significantly lower. PERC solar cells were fabricated from these wafers and the conversion efficiencies are compared.
Keywords: Silicon, Wafering, Diamond wire, Metal Contamination

1 INTRODUCTION

Multi-wire sawing of silicon ingots is still the standard process for the production of silicon wafers for solar cells. The diamond wire sawing (DWS) technique was introduced a decade ago in the photovoltaic industry [1] and has now become the dominant technology [2, 3]. There are two kind of coatings in which the cutting diamonds are embedded on the steel wire core. Either a nickel-based coating is used, which may contain also other metal components or the coating is based on a resin material. The market share of resin-bond diamond wire in the worldwide PV wafer production is currently assumed to be about 10% and is supposed to remain on that level within the next decade [3]. Resin-bond wires could have advantages compared to nickel-bond wires in terms of lower manufacturing costs, a lower environmental impact and an improved wafer surface quality [4, 5].

In this contribution a detailed analysis of the metal contamination from diamond wires with nickel as well as resin coating is presented. In comparison of both techniques for the embedment of cutting diamonds on core steel wires, the wires before and after the sawing process, the silicon kerf as well as the produced wafers are investigated. Extensive, stepwise layer etching of the wafers in combination with analysis of the sawing damage identifies the origin and impact of metallic impurities in the diamond wire sawing process.

While the influence of sawing parameters on the nickel concentration in silicon is demonstrated in a recent publication [6], the differences of metal contamination between nickel and resin coated wires is highlighted in this contribution. Since the share of resin-bond diamond wires is expected to increase in the future, the understanding of the effects of the wafering process on the solar cell parameters is of high relevance. The findings are also helpful to optimize the saw damage depth as first step of the solar cell production.

2 EXPERIMENTAL APPROACH

A schematic of the diamond wire sawing process is shown in Fig. 1, where also the notations of the respective ingot and wafer positions are given. The wire entry side of the ingot is the position, where wire from the fresh wire spool is fed into the wire web; correspondingly the wire exit side is the position towards the used wire spool. [6]

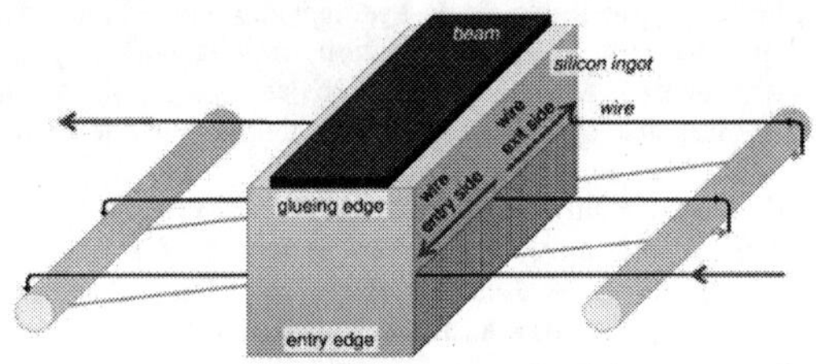

Figure 1: Schematic of diamond wire sawing process and used notations [6]

General process parameters in multi-wire sawing are the feed rate, which is the speed of the ingot movement through the wire web, the wire speed and the wire tension. The wire consumption and the resulting silicon contact length $l_c(Si)$ are particular for the oscillating wire movement in diamond wire sawing. To compare Ni- and resin-coated wires, monocrystalline silicon was sliced with 120 µm core wire diameter and process parameters that generate similar maximum silicon contact lengths $l_{c,max}(Si)$ of the wire at standard values of the wire consumption. Furthermore, the feed rate was chosen in order to get the similar resulting force of the wire web in feed direction. The process parameters are summarized in Table I.

Table I: Process parameters of comparative sawing processes for nickel and resin coated diamond wires

Wire coating	*Wire speed [m/s]*	*Feed rate [mm/min]*	*Wire consumption [m/wafer]*	$l_{c,max}(Si)$ *[m]*
nickel	18	1.4	3.9	4,500
resin	18	1.0	7.1	4,200

$l_c(Si)$ is a measure for the wire wear that defines, which length of silicon a particular piece of wire was cutting. $l_c(Si)$ increases from the wire entry side of the ingot to the wire exit side as well as from the ingot entry edge to the glueing edge. Since new wire is fed from the wire entry side in each reciprocation of the wire web, the lowest silicon contact length occurs along this side. The maximum silicon contact lengths ($l_{c,max}(Si)$) is obtained at the glueing edge on the wire exit side, see Fig. 2.

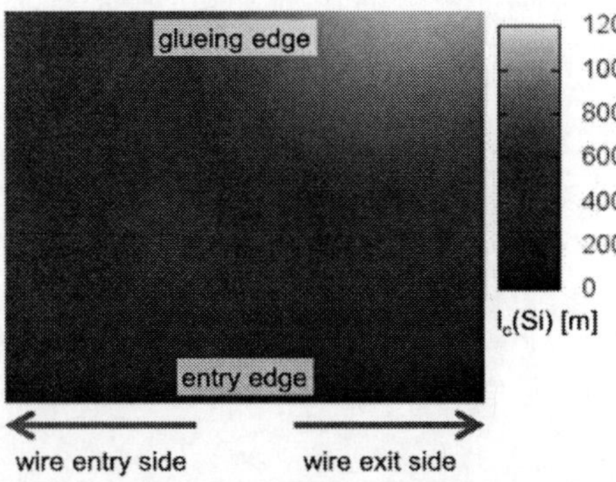

Figure 2: Simulation of silicon contact length $l_c(Si)$ [6]

3 RESULTS

3.1 Analysis of diamond wires

The detailed analysis of the wires reveals that the resin-coated wire features a higher diamond density compared to the Ni-coated wire, see Fig. 3. The resin layer itself is thicker (about 6-8 μm in average) compared to the standard nickel layer thickness of 2-3 μm in average.

Wire samples were analyzed before and after the sawing processes, using high quality cross section preparation and REM/EDX analysis. An unused wire corresponds to a silicon contact length of $l_c(Si)=0$, whereby the maximal used wire is described by $l_{c,max}(Si)$.

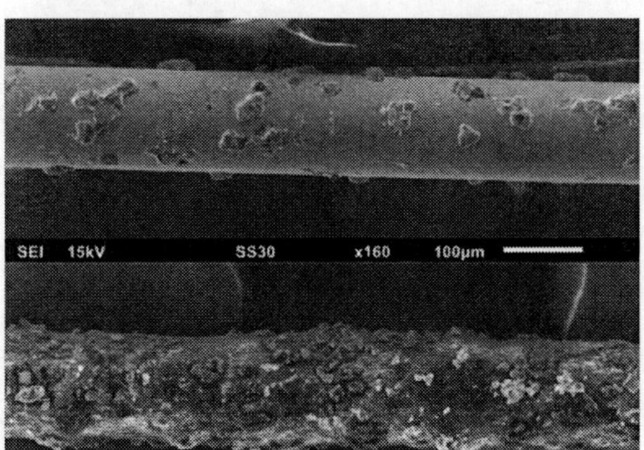

Figure 3: REM images of nickel- (top) and resin-coated (bottom) diamond wires with 120 μm core diameter

A cross section of an unused resin-coated wire is shown in Fig. 4 (*left*). As displayed in Fig. 4 (*right*), diamonds can be completely removed in the wire sawing process and the resin layer is stripped to a high extent. It should be emphasized that the diamonds used for the resin bonded wire are covered by a layer of nickel as well, which may be due to manufacturing issues but could generally be avoided.

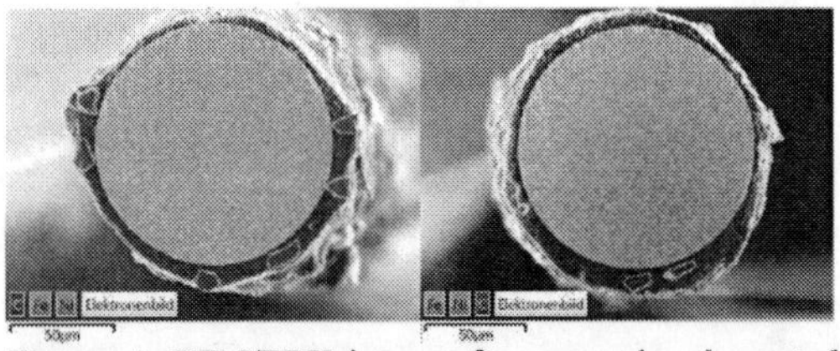

Figure 4: REM/EDX image of an unused resin-coated wire cross section *(left)*; resin-coated wire after the sawing process *(right)*

3.2 Metal concentration in silicon kerf

After the respective sawing processes, samples of the used coolant, containing silicon kerf were taken to determine the metal content in the kerf by means of ICP-OES (Inductively Coupled Plasma - Optical Emission Spectroscopy). The main metallic contaminations are nickel and cobalt, which are the main layer components. In the case of the Ni-coated diamond wires, a clear correlation was recently shown between the maximum silicon contact length $l_{c,max}(Si)$ and the metal concentration in the kerf [6].

In this work, the resulting metal concentrations from Ni-coated and resin-coated diamond wire sawing processes are compared for a similar maximum silicon contact length of $l_{c,max}(Si) \approx 4{,}200 - 4{,}500$ m. Surprisingly, a significantly higher Ni concentration was found in the kerf of the resin-coated diamond wire process, which is somewhat unexpected. As shown in Fig. 5, the concentration is twice as high compared to the Ni-coated diamond wire. These findings can be explained by the fact that the diamonds of the resin-coated wire are covered with a layer of nickel, too (see Fig. 4). Since the diamonds are removed to a high extend from the wire, the nickel in the kerf originates mainly from the diamonds.

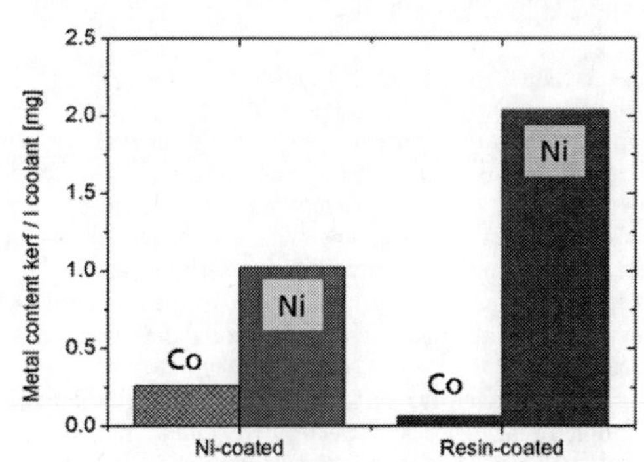

Figure 5: Concentration of cobalt and nickel in the silicon kerf from diamond wire sawing processes with nickel- and resin-coated diamond wires as measured by ICP-OES

3.3 Wafer geometry

The wafer quality was assessed by optical topography measurements. The total thickness variation (TTV) and average wafer thickness are displayed in Fig. 6 and 7.

The TTV is lower for wafers from the resin-coated wire process, which is reasonable with respect to the higher wire consumption in this process (7.1 m/wafer) and the higher diamond concentration of resin-bond wires. The thickness of the wafers is about 7 µm lower compared to wafers from the Ni-coated wire process. This is due to the higher coating layer thickness of the resin-coated wire, while using the same wire core diameter as for the Ni-coated wire, resulting in an increased sawing channel width.

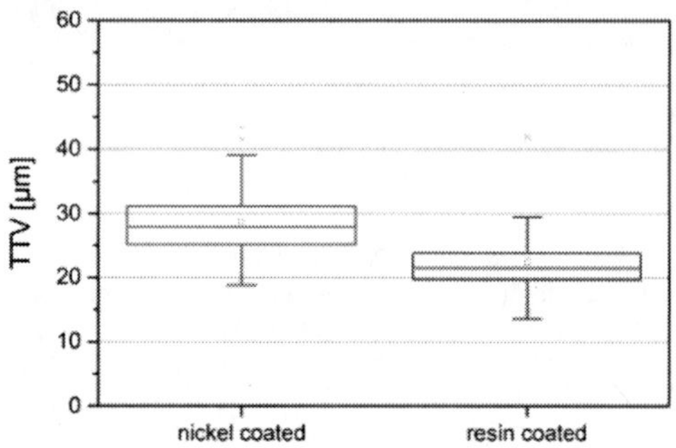

Figure 6: Total thickness variation (TTV) of wafers from nickel- and resin-coated wire processes

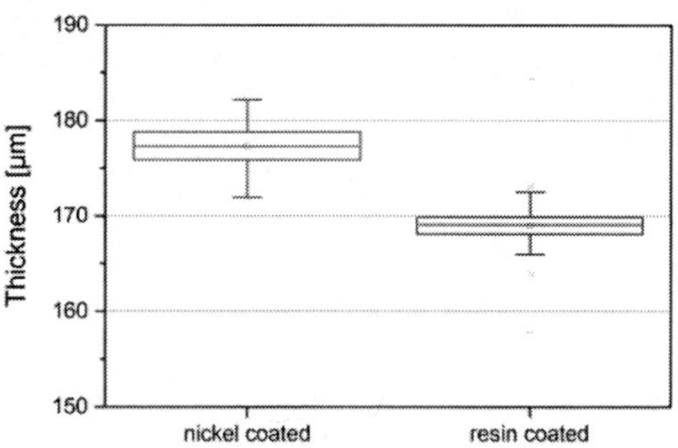

Figure 7: Average thickness of wafers from nickel- and resin-coated wire processes

3.4 Metal concentration in the wafers

Wafer samples were taken at the wire entry side (fresh wire - *fw*) and the wire exit side (used wire - *uw*) as defined in Fig.1. The metal concentration in different depths below the wafer surface was determined by step-wise etching of three layers with a thickness of 2 µm, respectively and 2 layers with 3 µm thickness each. Thus, a layer thickness of 12 µm below the as-cut wafer surface from both wafer sides was completely dissolved in an acidic solution.

The analysis was performed by ICP-MS (Inductively Coupled Plasma - Mass Spectrometry) measurements. As shown in Fig. 8, the sum concentration of Ni and Co generally decreases by three orders of magnitude from the surface-nearest layer (0-2 µm) to the deepest etched layer (9-12 µm depth). The metal concentration is lower for wafers from the resin-coated wire process, as it is expected.

The detailed analysis of the total metal concentration in the whole depth of 12 µm below the as-cut surface is shown in Fig. 9. There is a significant difference between the two types of wires. While for the Ni-coated wires the metal concentration correlates with the wire consumption [6], it is lower for the resin-coated wire even at a drastically higher wire consumption of 7.1 m/wafer. Furthermore, a significant difference between wafers from the wire entry side of the ingot and the wire exit side was detected. At the wire exit side, where the wire wear is higher, the incorporation of metals in the surface-near layers is higher. At the wire entry side, there is no difference between entry and glueing edge. However, a higher metal concentration was found near the glueing edge for the wire exit side of the ingot, which corresponds to the largest silicon contact length $l_{c,max}(Si)$.

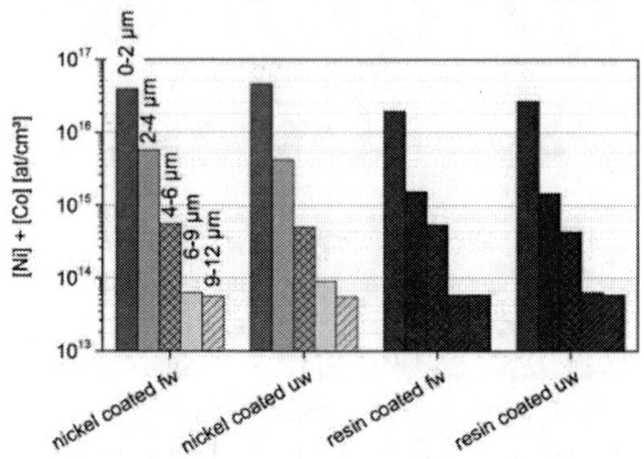

Figure 8: Sum concentration of nickel and cobalt in surface near layers of wafers from nickel-coated and resin-coated DWS processes as determined by ICP-MS

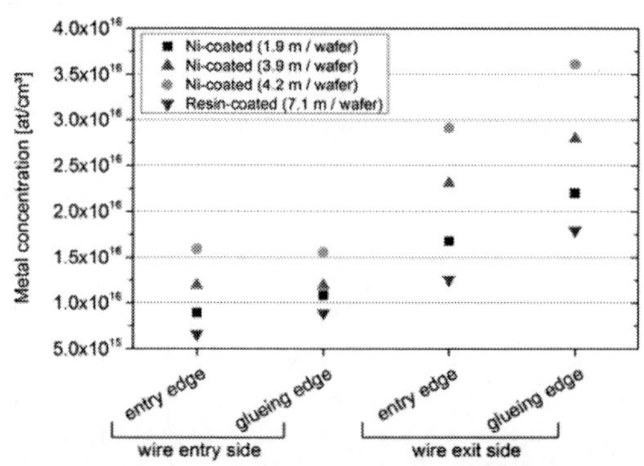

Figure 9: Sum concentration of Ni and Co in a layer depth of 0-12 µm below the as-cut surface for DWS processes with Ni-coated wires with different wire consumption [6] and resin-coated wire

3.5 Model for the incorporation of metals into the surface

From the results presented here and previous investigations on the surface damage in the DWS process [7-10], a simple model for the incorporation of metals into the as-cut wafer surface is derived. When the solid state diffusion of Ni and Co in silicon is calculated with the ingot temperatures of 30 - 40°C commonly achieved in the DWS process, a negligible increase of the metal concentration in surface near layers is obtained. It seems more convenient to assume a diffusion of metallic impurities into the cracks that are generated on the as-cut wafer surface with a depth up to 12 µm [7]. A schematic of this model is shown in Fig. 10.

Before the texturization step in solar cell production

the saw marks with a depth up to 5 µm [10] are flattened and most of the sawing damage is etched off. However, diffused metals in the remaining deepest cracks could affect the solar cell efficiency.

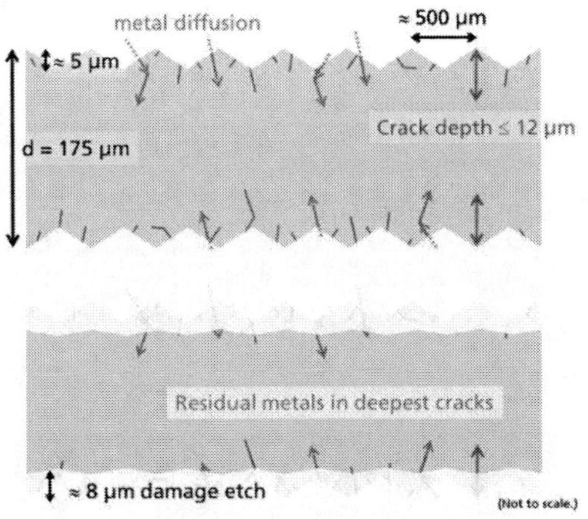

Figure 10: Schematic for the metal diffusion along cracks in the as-cut wafer surface of DWS wafers before *(top)* and after damage etching *(bottom)*

3.6 Influence on solar cell efficiency

PERC solar cells have been produced from wafers of the *fw*- and *uw*-side of the ingot. Wafers from both processes were labeled and mixed within the same batch for the solar cell processing. A layer of 8 µm thickness is etched off from both sides of the wafers at the beginning of the cell process (damage etch). Analyzing these PERC solar cells it was shown that the conversion efficiency directly correlates with the metal concentration introduced by the sawing process with nickel-coated diamond wires [6].

Since the nickel and cobalt concentration is lower for the resin-coated wire compared to the nickel-coated wires (Fig. 9), a beneficial influence on the solar cell efficiency is expected. However, the contrary effect was found. As shown in Fig. 11, the solar cell efficiency is lower for wafers from the resin-coated process although the concentration of Ni and Co is significantly lower compared to the Ni-coated wire processes. Furthermore, there is a reduction in cell efficiency from the wire entry side (*fw*) to the wire exit side of the ingot (*uw*).

With the considerations in Section 3.5 it could be possible, that deeper cracks are present in the surface of the wafers from the resin-bond process and the in diffusion of metals into these cracks is detrimental for the conversion efficiency of the solar cells.

In order to test this assumption, the crack depth on the as-cut wafer surfaces was determined for the DWS processes presented above. As shown in Fig. 12, the crack depth distribution is similar for all Ni-coated DWS process with a median of 4.0 - 4.2 µm, independent of $l_{c,max}(Si)$. For the resin-coated DWS process, the crack depth is even lower with a median value of about 3 µm. Thus it can be concluded that metal diffusion into deeper cracks is not the cause of efficiency reduction for the solar cells from resin-bond DWS wafers.

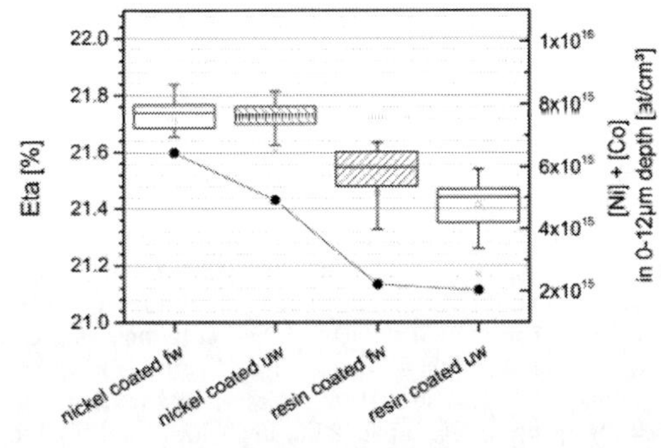

Figure 11: Efficiency of PERC solar cells fabricated from wafers of nickel-coated and resin-coated wire processes and sum of metal concentration between 0-12 µm below the wafer surface

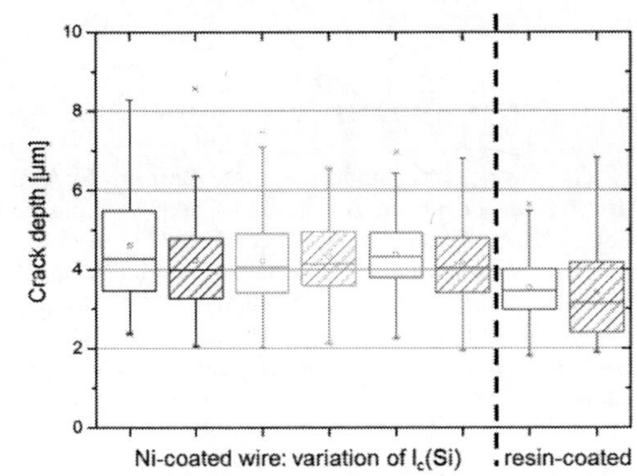

Figure 12: Crack depth distribution for wafers from nickel- and resin-coated wire processes

The analysis of the solar cell parameters shows that the reduction of the conversion efficiency is mainly caused by a reduction of J_{sc}, see Fig. 13. It is assumed that the higher concentration of residual organics originating from the resin coating is detrimental for the texturization step in the cell production. Furthermore, the reduced thickness of the resin-bond DWS wafers (see Fig. 7) partly contributes to the reduction of J_{sc}.

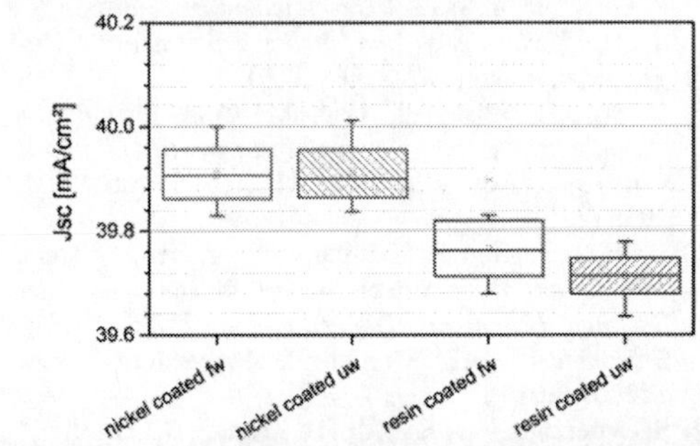

Figure 13: Short circuit current (J_{sc}) of the PERC solar cells fabricated from wafers of nickel-coated and resin-coated wire processes

4 CONCLUSION

Metal concentrations in surface layers with a maximum depth of 12 µm below the wafer surface and resulting conversion efficiency of PERC solar cells were compared for diamond wire sawn wafers, which were produced with nickel- and resin-coated diamond wires.

For a wire sawing experiment with similar silicon contact length $l_{c,max}(Si)$ an increase of the metal concentration in the silicon kerf by a factor of 2 was obtained. The total metal concentration, comprising Ni and Co, in a depth up to 12 µm is lower in wafers produced in resin-bond DWS processes. A simple model for the incorporation of metallic impurities by diffusion along cracks in the wafer surface was derived.

The lower conversion efficiency of wafers from the resin-bond DWS process were attributed to a non-optimal texturization of the wafers, caused by organic residuals on the wafer surface as well as the lower wafer thickness, which arises from a higher effective wire thickness of the resin-bond diamond wire.

5 ACKNOWLEGMENTS

This work was funded by the German Federal Ministry for Economic Affairs and Energy within the project "DIANA" (contract number 0324087B), which is gratefully acknowledged. The authors would like to thank for the support provided by the members of the Wafering Group at Fraunhofer THM and the teams of the analytical laboratory and the cell pilot line of SolarWorld Industries GmbH.

6 REFERENCES

[1] A. Bidiville, K. Wasmer, R. Kraft, C. Ballif, *„Diamond wire-sawn silicon wafers – from the lab to the cell production"*, Proc. of the 24th EU-PVSEC, pp. 1400-1405 (2009).

[2] A. Kumar and S. N. Melkote, *"Diamond Wire Sawing of Solar Silicon Wafers: A Sustainable Manufacturing Alternative to Loose Abrasive Slurry Sawing"*, Procedia Manufacturing 21, pp. 549 – 566 (2018).

[3] International Technology Roadmap for Photovoltaics (ITRPV), 9th Edition (2018)

[4] P. Ge, Z. Hou, S. Li, *"Development of resin bonded diamond wires, saw and slicing experiments"*, Key Eng. Mat. 416, pp. 321-326 (2009).

[5] D. Xu, H. Zhang, J. Gao, M. Cong, *„Study on manufacturing technology of resin bonded diamond wire saw"*, Adv. Mat. Research 602-604, pp. 2054-2058 (2013).

[6] L. Lottspeich, P. Müller, T. Kaden, *"Metal Contamination of Silicon Wafers in Diamond Wire Sawing Processes Depending on the Sawing Parameters"*, AIP Conference Proceedings 1999, 140003 (2018).

[7] S. Wuerzner, A. Falke, R. Buchwald, H.J. Möller, *" Determination of the impact of the wire velocity on the surface damage of diamond wire sawn silicon wafers"*, Energy Procedia, 77, pp. 881–890 (2015).

[8] S. Wuerzner, M. Herms, T. Kaden, H.J. Möller, M. Wagner, *"Characterization of the Diamond Wire Sawing Process for Monocrystalline Silicon by Raman Spectroscopy and SIREX Polarimetry"*, Energies, 10, 414 (2017).

[9] R. Buchwald, K. Fröhlich, S. Würzner, T. Lehmann, K. Sunder, H.J. Möller, *"Analysis of the sub-surface damage of mc- and cz-Si wafers sawn with diamond-plated wire"*, Energy Procedia 38, pp. 901-909 (2013).

[10] W. Chen, X. Liu, M. Li, C. Yin, L. Zhou, *"On the nature and removal of saw marks on diamond wire sawn multicrystalline silicon wafers"*, Materials Science in Semiconductor Processing 27, pp. 220-227 (2014).

DIAMOND WIRE WAFERING: A MODEL-BASED EVALUATION OF DIFFERENT CONTROL STRATEGIES

Daniel Treyer *[,1], Sebastian Gaulocher[1], Stefan Niederberger[1], Helge Rafacl[2], Gregor Frech[2], Alfons Ams[3]
[1]University of Applied Sciences and Arts Northwestern Switzerland, Institute of Automation, Windisch, Switzerland
[2]Meyer Burger AG, Gwatt (Thun), Switzerland
[3]TU Bergakademie Freiberg, Institute of Mechanics and Fluid Dynamics, Freiberg, Germany
*Corresponding author: daniel.treyer@fhnw.ch

ABSTRACT: In this work, the diamond wire wafering process for the production of crystalline silicon wafers is considered. Using different control strategies, the aim is to optimize the wire sawing process regarding productivity while maintaining an adequate wafer quality. The following approach has been chosen: assuming a given set of production resources, a dynamic model of the wire sawing process is established that simulates the relevant process variables. Then, reference values are defined for variables for which a reliable operation is expected. Using these reference values as constraints, different control strategies are subjected to a model-based evaluation. Simulation results indicate that a significant increase in productivity is possible using feed rate control in combination with alternative strategies to distribute wire performance.
Keywords: wafering, diamond wire sawing, crystalline silicon, control strategy, process optimization

1 INTRODUCTION

In the photovoltaic industry, the wafering process is used for the production of crystalline silicon wafers. Employing the fixed abrasive sawing process (FAS), a diamond-coated wire is wound over two wire guiding rollers to form a wire web. The silicon brick is fed perpendicularly through the moving wire web, while the relative motion between brick and abrasive wire causes material removal.

For mono-crystalline silicon, FAS has almost completely replaced the slurry-based loose abrasive sawing process (LAS), see e.g. [1]. Even for multi-crystalline silicon, FAS will most likely continue to gain market share. Compared to the slurry-based process, diamond wire wafering results in higher productivity, reduced kerf-loss, and reduced environmental impact. Even though diamond wire wafering is a highly sustainable process, wafer manufacturers are compelled to operate their wire saws under the most favorable conditions.

The diamond wire sawing process as a part of a photovoltaics (PV) production line may be optimized according to various criteria. First, the quality of the produced wafers is relevant. Then, an optimal utilization of production resources such as diamond wire and cutting fluid as well as a low energy consumption is important. Further, a robust, predictable process with a low wire failure rate and a high productivity is desirable. Ultimately, all these factors reduce to one single characteristic, namely the cost per wafer.

In the literature, different approaches for the optimization of the FAS process have been described. In [2], the effect of different pilgrim cycle lengths on wafer strength has been analyzed in an experimental study. In [3], the cutting force is controlled using the feed rate as control variable, with the aim of decreasing wire breakage probability and wafer collapse as well as obtaining a superior wafer surface roughness.

The work presented here employs a model-based approach. Models describing different aspects of diamond wire wafering have been presented by various authors. In [4], models based on experimental data have been established for the prediction of wire wear, wire bow and total thickness variation (TTV). A model for the estimation of the periodic structures on diamond wire sawn wafers as a function of the recipe parameters has been presented in [5]. In [6] and [7], the cutting forces are deemed as a main impact factor on wafer quality. Therefore, an analytical force model, considering feed rate, wire speed, and contact length between the wire and brick has been developed. Similarly in [8], a model for predicting the cutting forces considering the wire sawing parameters has been presented, taking into account a combination of ductile and brittle fracture material removal.

The aim of this work is to evaluate different strategies to run the wire sawing process with a high level of productivity and under favorable conditions. In contrast to other contributions, the model employed here pursues a comprehensive approach, considering the wire sawing recipe parameters and in particular the varying conditions along the wire web.

This paper is organized as follows: Section 2 gives an overview of the methodology and the process model employed. Also, the relations between process variables and wafer quality are discussed. Further, the control algorithm employed is presented. In Section 3, different control strategies are evaluated using the process model. The simulation results are discussed in Section 4, whereas the last section concludes the paper.

2 METHODOLOGY

Defining a global optimization criterion for the wire sawing process is a tedious task, even more is it to find a global optimum due to the vast amount of influencing factors. In this work, the following approach has been chosen: first, a given set of production resources consisting of diamond wire, cutting fluid and beam material is assumed. A dynamic model of the wire sawing process is established to represent the relevant process variables for the given production resources. Then, reference values are defined for a number of process variables, for which a reliable process with a high yield is expected. Subsequently, the recipe of the wire saw, defined by wire speed, feed rate and new wire feed is controlled to meet the previously defined reference values. Based on the insights provided by the model, different strategies for the distribution of the wire performance over the extension of the wire are designed, simulated and evaluated regarding the productivity. All examined strategies are compared to a reference scenario consisting of a standard diamond wire sawing recipe.

In the remainder of this section, an overview of the dynamic model of the wire sawing process is given; process variables relevant to process reliability and wafer quality are discussed and the control algorithm is presented.

2.1 Model of the Wire Sawing Process

The model employed for this study is presented in [9] and [10]. In the following, the most important characteristics are summarized.

The model describes process variables such as line load $q(x)$ between wire and silicon brick, wire bow $w(x)$ and wire performance $u(x_{wire})$, given in sawn area of silicon per unit length of wire over wire web and process time. Two coordinate systems are introduced, a machine-based one x as depicted in Fig. 1 and a wire-based one x_{wire} as in Fig. 2.

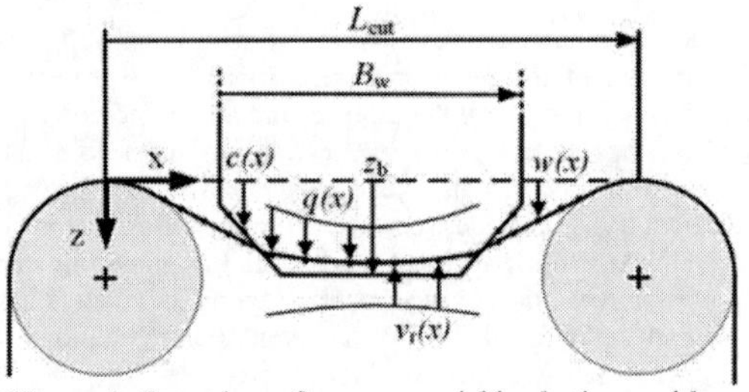

Figure 1: Overview of process variables in the machine-based coordinate system

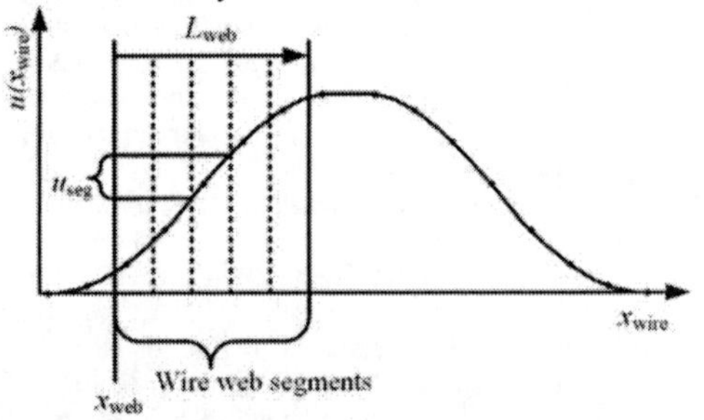

Figure 2: Overview of process variables in the wire-based coordinate system

The wire web, consisting of a multitude of wire loops, is partitioned into segments. Each wire web segment consolidates a number of real wire loops to one single simulated wire loop as shown in Fig. 1. By computing wire bow $w(x)$, line load $q(x)$ and wire performance u_{seg} for every segment, the model is able to represent the conditions along the entire wire web. Fig. 2 shows the wire performance in the wire-based coordinate system and the wire performance in the wire web segments u_{seg}.

The parametric material removal model given in (1) and (2) describes the interaction between the abrasive wire and the contour of the brick inside the kerf $c(x)$ for one specific wire web segment. The material removal rate $v_r(x)$ is formulated as a function of wire speed v_c, line load $q(x)$ and wire performance u_{seg} prevailing in a wire web segment. A number of parameters describing the influence of these process variables are identified using experimental data.

$$v_r(x) = \gamma v_{r,ref} \left(\frac{|v_c|}{v_{c,ref}}\right)^{\beta} \left(\frac{q(x)}{Dp_{ref}}\right)^{\alpha} f(u_{seg}) \quad (1)$$

$$f(u_{seg}) = \left(\frac{\delta k u_{seg} - k + 1}{\delta u_{seg} - k + 1}\right)^{\alpha} \quad (2)$$

The effects of the cutting fluid and thermal influences are not considered, and the wire management outside the wire web is not modelled, either.

For the simulations in this study, the parameters identified in [9] have been adopted. Further simulation parameters are the wire web length $L_{web} = 3525$ m, the number of wire loops $N_{loop} = 2501$, the wafer width $B_w = 156$ mm, the wire usage per cut $x_{wire,cut} = 1750$ m and a basis pilgrim length of $x_{wire,pilgrim} = 300$ m.

2.2 Process Variables Relevant to Wafer Quality

As mentioned above, reference values are defined for certain process variables. Using these reference values as constraints in the control algorithm, a reliable process with a high yield is expected.

In the literature, correlations between process variables and various wafer quality characteristics have been analyzed by a number of authors. In [11], the impact of feed rate, pilgrim motion and wire performance on periodic structures on the wafer surface is analyzed. The periodic structures inherent to the diamond wire sawing process are referred to as "pilgrim waves". In [12], periodic structures on the wafer profile are referred to as "pattern saw marks" (PSM). A correlation between PSM amplitude and PSM wavelength is established.

In [13], it was found that a low TTV is obtained by an appropriate matching of wire speed and feed rate. The depth of cut per abrasive grain is analyzed in [14]. The authors conclude that it is possible to increase process productivity by increasing feed rate in proportion to wire speed, in order to keep depth per cut constant.

A linear dependence of median TTV on the steady-state wire bow is observed in [15]. The correlation even holds for scenarios with increased wire bow due to higher table speed or decreased wire tension.

The effect of increasing diamond wire wear on surface roughness and other quality characteristics was investigated in [16]. As a consequence, the maximum wire performance has to be selected carefully.

Obviously, there are many correlations between wire sawing process variables and wafer quality characteristics. In this work, the following approach is chosen. The maximal wire bow in the wire web $\max(w(x))$ is limited to a given reference value w_{ref} in order to keep TTV low. In addition, the PSM wavelength psm is constrained to a maximal value psm_{ref} in order to keep the amplitude of periodic structures on the wafer below a given threshold. This method is chosen because the PSM amplitudes are not represented in the wire sawing model. Further, an upper bound for the wire performance u_{ref} is set.

$$\max(w(x)) \leqslant w_{ref} \quad (3)$$

$$\max(psm) \leqslant psm_{ref} \quad (4)$$

$$\max(u(x_{wire})) \leqslant u_{ref} \quad (5)$$

See Section 3.1 for the definition of the actual reference values.

2.3 Control Algorithm

For the control of the conditions given in (3), (4) and (5), wire speed, feed rate and new wire feed per pilgrim cycle may be used as control variables. As the wire speed v_c has a direct influence on the material removal rate $v_r(x)$, see (1), it is reasonable to operate at $v_c = v_{c,max}$. The amount of new wire feed per pilgrim cycle is kept constant to make different strategies comparable and to guarantee condition (5). As a result, a controller with feed rate v_f as the control variable has been implemented in the simulation environment to establish closed-loop control of the conditions (3) and (4).

The control law is given in (6). To guarantee stable saw in and saw out, the feed rate is set to constant values from the start of the cut until time t_{init} (saw-in stage) and after the first wire web segment has entered the wafer chamfer on the beam side of the silicon brick (saw-out stage). In (6), H_w and F_w denote the wafer height and the length of the wafer chamfer, respectively. During the sawing of the wafer bulk, two independent PI controllers with anti-windup are used for closed-loop control of maximum bow and maximum PSM wavelength. The equations for the control of bow and PSM wavelength are given in (8)-(9) and in (10)-(15), respectively. The minimum control value from these two controllers is used as control variable $v_f(t)$, see (7).

$$v_f(t) = \begin{cases} v_{f,init} & t < t_{init} \\ v_{f,final} & \max c(t) + z_b(t) > H_w - F_w \\ f_{w,psm} & else \end{cases} \quad (6)$$

$$f_{w,psm} = \min[v_{f,w}(t) \quad v_{f,psm}(t)] \quad (7)$$

In the simulation environment, the wire bow $w(x,t)$ is available at every time step for the evaluation of the control law. For the sake of clarity, the anti-windup is not shown in the equations below.

$$v_{f,w}(t) = k_w \left[e_w(t) + \frac{1}{T_w} \int_0^t e_w(\tau) d\tau \right] \quad (8)$$

$$e_w(t) = w_{ref} - \max(w(x,t)) \quad (9)$$

The computation of the PSM wavelength per wire web segment $psm(t_n)$ is based on the evaluation of the median wafer contour $\tilde{c}(t_n) = \text{median}(c(t_n, x)) + z_b(t_n)$ at the time of wire movement direction changes. These sampling points are denoted t_n. New values $psm(t_n)$ for the computation of the control variable $v_{f,psm}(t)$ are therefore available only after half a pilgrim cycle is completed. As a result, an event-based PI-controller for non-equidistant sampling times as presented in [17] is employed.

For the computation of $psm(t_n)$ in (14), the maximum value of the past two half pilgrim cycles is taken. This is because the PSM wavelengths vary considerably for wire movements in direction of new or old wire. The term $\max(psm(t_n))$ in (13) then refers to the maximum PSM wavelength of all wire web segments at time t_n.

In the equations below, $u_p(t_n)$ and $u_i(t_n)$ denote the proportional and integral terms of the PI controller, respectively.

$$v_{f,psm}(t) = u_p(t_n) + u_i(t_n) \quad (10)$$

$$u_p(t_n) = k_{psm} e_{psm}(t_n) \quad (11)$$

$$u_i(t_n) = u_i(t_{n-1}) + \frac{k_{psm}}{T_{psm}} \cdot T_s(t_n) \cdot e_{psm}(t_n) \quad (12)$$

$$e_{psm}(t_n) = psm_{ref} - \max(psm(t_n)) \quad (13)$$

$$psm(t_n) = \max \begin{bmatrix} \tilde{c}(t_n) - \tilde{c}(t_{n-1}) \\ \tilde{c}(t_{n-1}) - \tilde{c}(t_{n-2}) \end{bmatrix} \quad (14)$$

$$T_s(t_n) = t_n - t_{n-1} \quad (15)$$

3 EVALUATION OF CONTROL STRATEGIES

In this section, different strategies to distribute the wire performance over the extension of the wire are compared, whereas the process variables are controlled by the algorithm presented above.

In each scenario, a number of cuts are simulated until a quasi-stationary state is reached. Quasi-stationary in this context refers to the state where

i) the maximum wire performance $\max(u(x_{wire}))$ does not increase any further and
ii) one or several cuts are representative for the remaining cuts with the same wire spool, regarding wire performance at the start of the respective cuts and process time.

3.1 Reload Reference Scenario

First, a reference scenario using the common Reload mode with a given feed rate profile $v_f(t)$ is simulated and analyzed. In Reload mode, the position of the wire web is not changed between cuts.

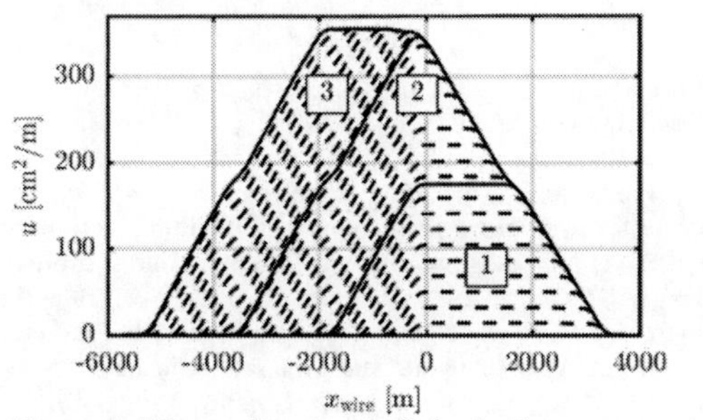

Figure 3: Wire performance in Reload reference scenario

Fig. 3 shows the wire performance in the wire-based coordinate system, solid lines refer to wire performance after every cut and dashed lines to wire performance during a cut in time equidistant sampling. After the end of the second cut, the wire performance does not increase anymore, as an equilibrium between new wire feed and growth of wire performance in the wire web is reached. Therefore, the third cut is assumed to be quasi-stationary and representative for the remaining cuts with the same wire spool.

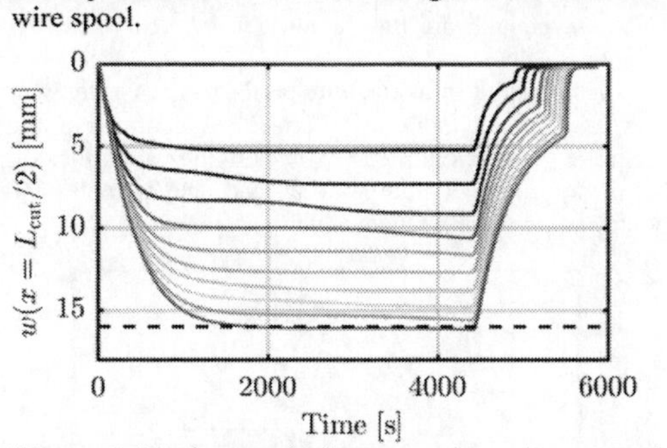

Figure 4: Maximum wire bow per wire web segment in Reload reference scenario (3rd cut)

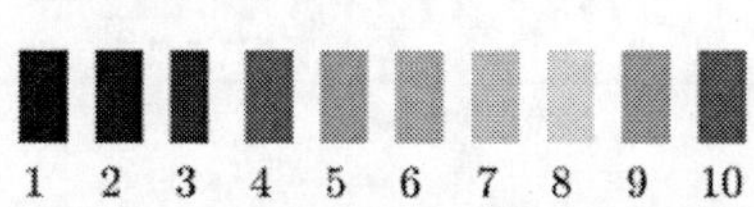

Figure 5: Color legend of wire web segments: '1' refers to the segment farthermost on the new wire side of the wire web, '10' to the segment farthermost on the old wire side (positive direction in wire-based coordinate system, compare to Fig. 2)

Fig. 4 shows the maximum wire bow in every wire web segment over process time in the third, stationary cut. The feed rate is reduced towards the end of the cut, which leads to a decreased wire bow. At the end of the cut, the wire bow decreases abruptly, which indicates that the wire in the respective wire web segment has entered the sacrificial beam.

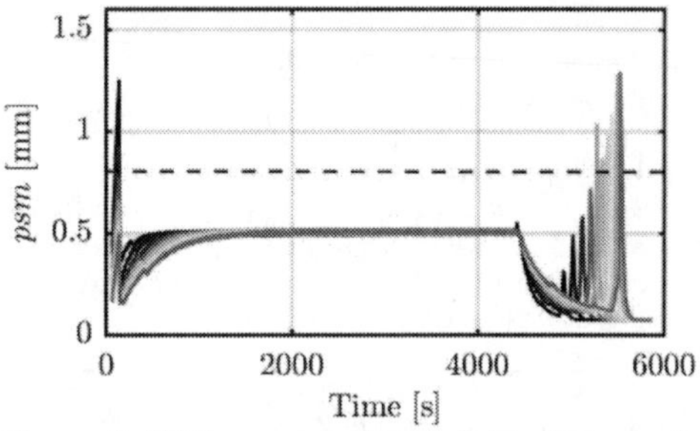

Figure 6: PSM wavelengths per half pilgrim cycle in Reload reference scenario (3rd cut)

Fig. 6 shows the PSM wavelength per half pilgrim cycle for every wire web segment in the third, stationary cut. To guarantee a stable saw in, the first pilgrim cycle is longer, leading to an increased PSM wavelength. Increased PSM wavelengths are also observed at the end of the cut, where the wire enters the sacrificial beam. Note that PSM wavelengths exceeding psm_{ref} are observed only during the saw-in and saw-out stages with constant feed rate according to (6).

Based on this scenario, the reference values for the process variables have been selected. The maximum wire bow should not exceed $w_{\mathrm{ref}} = 16$ mm, the wire performance should not exceed $u_{\mathrm{ref}} = 355\ \mathrm{cm}^2/\mathrm{m}$ and for the maximum PSM wavelength, a value of $psm_{\mathrm{ref}} = 0.8$ mm has been defined based on empirical knowledge.

3.2 Reload with Bow/PSM Control

Using the control law defined in (6), wire bow and PSM wavelength are now controlled to their respective reference values.

Figs. 7 and 8 show the wire performance and the wire bow for Reload mode with bow/PSM control. In this scenario, the maximum PSM wavelength is not a limiting factor for the feed rate, but instead, the wire bow is controlled to the reference value.

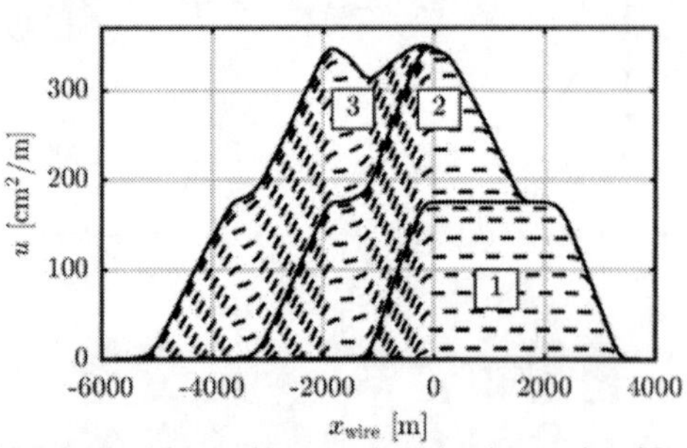

Figure 7: Wire performance in Reload scenario with bow/PSM control

3.3 Rewind with Bow/PSM Control

In the following, Rewind mode is used to distribute wire performance over the wire. In Rewind mode, the wire web is reset to initial position at the end of each cut, until

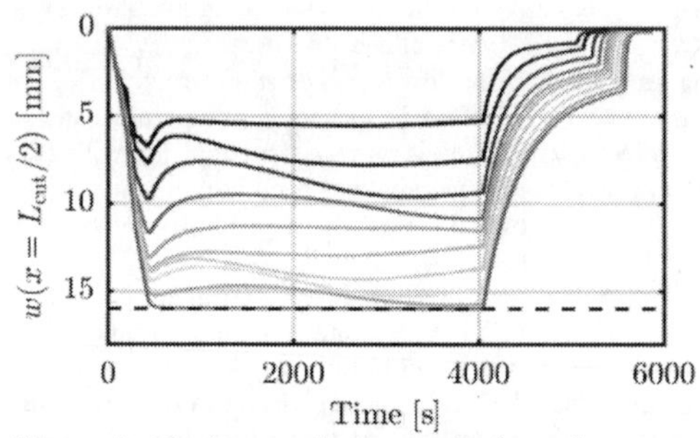

Figure 8: Maximum wire bow per wire web segment in Reload scenario with bow/PSM control (3rd cut)

maximum wire performance is reached and the entire wire web is shifted in direction of unused wire. Fig. 9 shows the resulting wire performance for the first four cuts. Cuts 3 and 4 are assumed to be quasi-stationary and are therefore evaluated concerning process duration.

Fig. 10 shows the wire bow for the fourth cut in Fig. 9. Wire web segments 1 and 2 are in the realm of new wire and exhibit a far lower wire bow than the rest of the wire web segments.

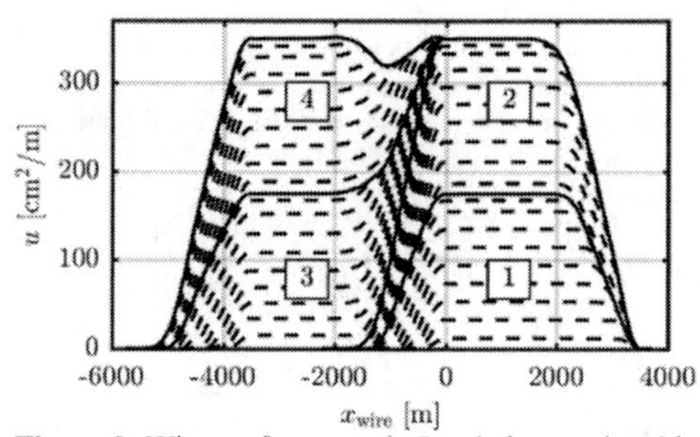

Figure 9: Wire performance in Rewind scenario with bow/PSM control

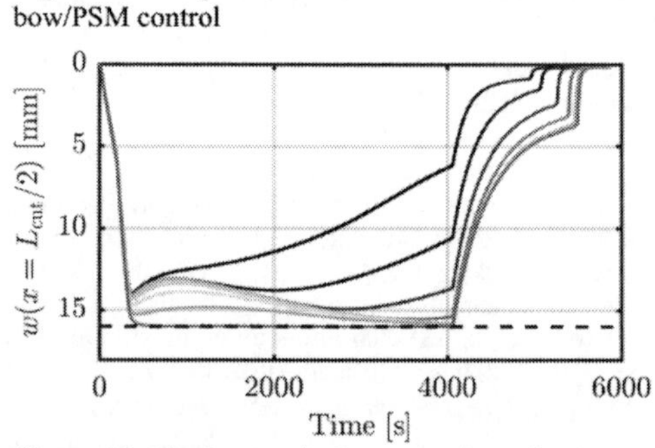

Figure 10: Maximum wire bow per wire web segment in Rewind scenario with bow/PSM control (4th cut)

3.4 Relocate with Bow/PSM Control

By using Relocate mode to distribute wire performance over the wire, two cuts starting with unused wire are performed first, leaving enough unused wire in between to place another cut which levels out the wire performance. The same procedure is applied for the "second floor" of wire performance. Starting a cut with equal distribution of wire performance over the wire web, or unused wire in particular, is favorable for short process durations using bow/PSM control.

Fig. 11 shows the resulting wire performance. In Relocate mode, there are four distinct types of cuts, namely cuts 2, 3, 6 and 7. As a result, the mean process duration is evaluated over these cuts. The fourth cut is placed only for the sake of demonstration; it should have been located far more in negative wire direction in a production scenario.

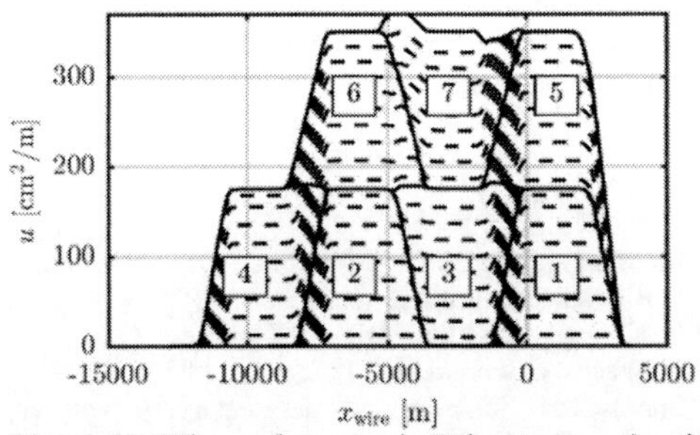

Figure 11: Wire performance in Relocate scenario with bow/PSM control

Fig. 12 shows the wire bow for the seventh cut in Fig. 11, second floor, level out wire performance. Particularly striking is the simultaneous saw-out of all wire web segments, caused by the equally distributed wire performance in the wire web.

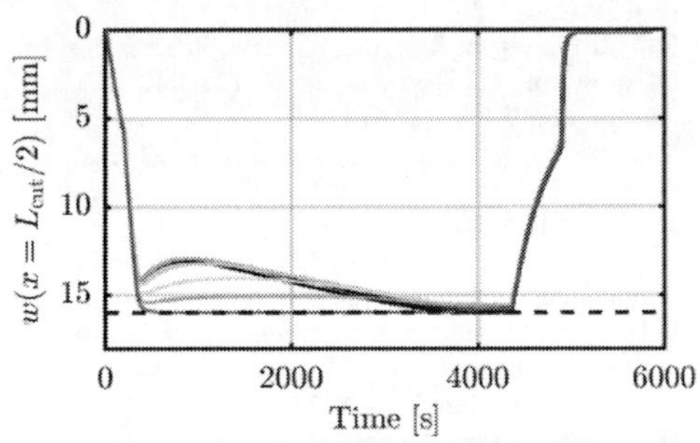

Figure 12: Maximum wire bow per wire web segment in Relocate scenario with bow/PSM control (7th cut)

Fig. 13 depicts the PSM wavelengths for the second cut in Fig. 11. In contrast to other cuts where the maximum wire bow is the limiting factor, here the maximum PSM wavelength is the factor that limits the feed rate during the bulk of the cut. This is due to the fact that unused wire with a high relative wire sharpness is used for this cut.

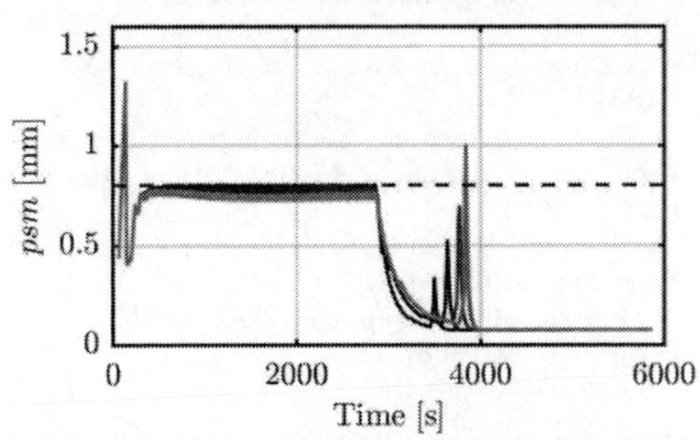

Figure 13: PSM wavelengths per half pilgrim cycle in Relocate scenario with bow/PSM control (2nd cut)

4 DISCUSSION OF SIMULATION RESULTS

In this section, feed rate profiles and simulated process durations for different control strategies are compared. Further, advantages and disadvantages of these strategies as well as implementation aspects are discussed.

4.1 Comparison of Feed Rate Profiles

Fig. 14 shows a comparison of the control variable feed rate in different scenarios. During the saw-in stage, the feed rate is equal in all scenarios. In contrast to the constant feed rate in the Reload reference scenario, the feed rate in the remaining scenarios is higher after the saw-in stage. This leads to faster increase of bow and therefore to a higher line load on the wire and ultimately to an increased material removal rate. In all scenarios, the feed rate is reduced to a constant value by the time the first wire web segment enters the wafer chamfer on the beam side of the silicon brick.

In cut 3 of the Rewind scenario, the feed rate is increased towards the end of the wafer bulk, because the wire web leaves the realm of high wire performance from the previous cuts due to the constant new wire feed.

In cut 2 of the Relocate scenario, a high feed rate is kept over the entire wafer bulk due to the operation with fairly unused wire.

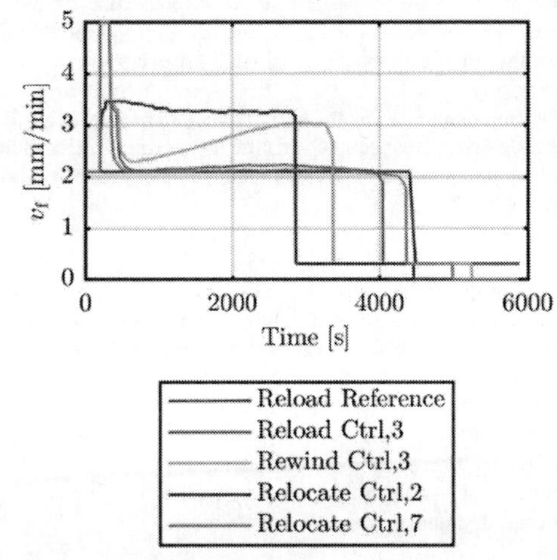

Figure 14: Comparison of feed rate profiles for different control strategies

4.2 Comparison of Process Durations

Table I gives an overview of simulated wafering process durations for different control strategies. Columns "First segment" and "Last segment" refer to the duration from process start to the saw out of the first and last segment, respectively. For recipe modes with varying processing times in the quasi-stationary case, values are given for a set of quasi-stationary cuts. Using Rewind or Relocate mode, the wire web is repositioned in between cuts, the time required is given in column "Mean spool time". For the repositioning, a displacement with maximum wire speed $v_{c,\max} = 30\,\mathrm{m/s}$ is assumed, acceleration and deceleration times are neglected.

4.3 Results

Comparing the total mean process durations of different recipe modes given in Table I, no gain in process duration is possible when using Reload with bow/PSM control instead of the standard Reload process with a predetermined feed rate profile.

A considerable gain in process duration in relation to Reload baseline is possible when using Rewind (-9.0%) or Relocate (-20.2%) with bow/PSM control.

Table I: Simulated process durations

Scenario	Stationary cuts [s]		Mean spool time [s]	Total mean time [s]
	First segment	Last segment		
Reload Baseline	4914	5518	0	5518
Reload Ctrl	5072	5580	0	5580
Rewind Ctrl	(3) 4064 (4) 4970	4422 5504	59 59	5022
Relocate Ctrl	(2) 3480 (3) 3374 (6) 4338 (7) 4908	3830 3390 4528 4926	236	4405

The decrease in process duration comes at the cost of increased process complexity for the operator and the wire saw control system, due to the varying process times and the necessary spool procedure between the cuts. In addition, increased loss of unused wire in case of wire failure is possible. These factors are difficult to model and must therefore be carefully evaluated for every production scenario.

Obviously, optimization of the wire sawing process is possible with respect to arbitrarily defined reference values for process variables due to the model-based approach.

4.4 Implementation Aspects

Regarding the implementation of these simulation results in the wire sawing process, two possibilities are feasible: First, the feed rate profiles generated by the PI controller in the simulation environment are quantized and used as an input to the traditional human-machine interface of the wire saw. Second, an observer-based implementation of the control algorithm in the wire sawing machine is possible, although associated with higher effort for implementation and sensor systems.

5 CONCLUSION

In this work, different control strategies for the diamond wire sawing process are compared using a dynamic model. Using other strategies to distribute wire performance over the wire may lead to a significant increase in productivity. Meanwhile, selected process variables such as wire bow and PSM wavelength are controlled to given reference values in order to guarantee a reliable process and to maintain adequate wafer quality.

The experimental validation of these control strategies is part of ongoing investigations, as well as their reassessment with models considering the decrease of wire sharpness due to the sawing in the sacrificial beam. In addition, using the dynamic model for online observer-based control of various process variables is also conceivable.

This study emphasizes the advantages of using a model of the diamond wire sawing process for optimization purposes. Rather than being an analysis tool, the model may be utilized to compare different strategies and recipe types and may therefore be highly relevant to PV manufacturers.

ACKNOWLEDGMENTS

This work was partly supported by the Swiss Commission for Technology and Innovation CTI. The authors would like to acknowledge Meyer Burger AG for many fruitful discussions and for sharing insights into the diamond wire sawing process.

REFERENCES

[1] International Technology Roadmap for Photovoltaics ITRPV 2017, Ninth Edition, March 2018.
[2] R. Koepge, C. Klute, F. Kaule, K. Sunder, O. Anspach, S. Schönfelder, Proceedings 31th European Photovoltaic Solar Energy Conference (2015) poster.
[3] S. Li, S. Du, A. Tang, R. G. Landers, Y. Zhang, Journal of Manufacturing Science and Engineering 137 (2015) 061003.
[4] K. Sunder, H. Uhle, S. Knöppel, O. Anspach, Proceedings of 28th European Photovoltaic Solar Energy Conference (2013) 1491.
[5] K. Sunder, R. Buchwald, O. Anspach, Proceedings of 31th European Photovoltaic Solar Energy Conference (2015) 595.
[6] S. Li, J. Wang, A. Tang, R. Landers, Proceedings of 2016 International Symposium on Flexible Automation ISFA (2016) 127.
[7] S. Li, S. Du, A. Tang, R. G. Landers, Y. Zhang, Journal of Manufacturing Science and Engineering 137 (2015) 061003.
[8] P. Wang, P. Ge, Y. Gao, W. Bi, Materials Science in Semiconductor Processing 63 (2017) 25.
[9] D. Treyer, S. Gaulocher, S. Niederberger, H. Rafael, A. Ams, Proceedings 33rd European Photovoltaic Solar Energy Conference and Exhibition (2017) 517.
[10] D. Treyer, S. Gaulocher, S. Niederberger, H. Rafael, A. Ams, Proceedings in Applied Mathematics and Mechanics PAMM (2017) 779.
[11] R. Brooker, H. Seigneur, W. Schoenfeld, Proceedings 43rd IEEE Photovoltaic Specialists Conference (2016) 618.
[12] M. Lanz et al., Proceedings 31st European Photovoltaic Solar Energy Conference and Exhibition (2015) 615.
[13] Y. Gao, P. Ge, T. Liu, Materials Science in Semiconductor Processing 56 (2016) 106.
[14] C. Chung, V. Le, Int J Adv Manuf Technol 80 (2015) 1337.
[15] K. Sunder, O. Anspach, Proceedings 29th European Photovoltaic Solar Energy Conference and Exhibition (2014) 588.
[16] A. Kumar, S. Kaminski, S. N. Melkote, C. Arcona, Wear 364-365 (2016) 163.
[17] K.-E. Årzén, Proceedings 14th IFAC World Congress (1999) 8687.

TEMPERATURE DEPENDENCE OF SLOW MINORITY CARRIER DE-TRAPPING IN CZOCHRALSKI SILICON WITH HIGH CONCENTRATIONS OF THERMAL DONORS

Manjula Siriwardhana[1], Y. Zhu[2], Z. Hameiri[2], F. Heinz[3], D. Macdonald[1] and F. E. Rougieux[2]
[1]Research School of Engineering, The Australian National University, Canberra ACT 2601, Australia
[2]School of Photovoltaic and Renewable Energy Engineering, University of New South Wales, Sydney, NSW, Australia
[3]Fraunhofer Institute for Solar Energy Systems ISE, Heidenhofstraße 2, D-79110, Freiburg, Germany

ABSTRACT: The slow de-trapping characteristics of defects related to thermal donors have been studied with transient photoconductance measurements in Cz monocrystalline silicon wafers. The temperature dependent photoconductance decay curves were analyzed to extract de-trapping time constants for a range of Cz-grown wafers. These included an as-cut wafer with a moderate concentration of pre-existing thermal donors, a wafer with a high concentration of thermal donors (High TD sample) which was annealed at 450ºC for 72 hours, and a sample with no thermal donors (No TD sample) annealed at 650ºC for 30 minutes to destroy the pre-existing thermal donors. The de-trapping time constants were measured at four temperatures (12ºC, 30ºC, 60ºC and 80ºC) using photoconductance decay measured up to 10 minutes after the illumination was turned off. We find that the de-trapping time constants of the as-cut sample are higher than those of the High TD sample for all the temperatures studied. Also, in both samples, the de-trapping time constants decrease with temperature, indicating that trapped minority carriers are released back into the bands more rapidly at higher temperatures. De-trapping was not observed in the samples without thermal donors. We also estimate the energy level of the significant traps using Arrhenius plots and found that in n-type Cz silicon, the as-cut wafers mostly contain traps close to the valence band, whereas the High TD sample has much deeper traps.
Keywords: Minority carrier trapping, Cz silicon, thermal donors, silicon defects.

1 INTRODUCTION

The use of injection-dependent lifetime data as a measure of the quality of silicon wafers is a common practice in the photovoltaic industry. However, with the presence of high concentrations of minority carrier trapping centres, lifetime measurements are known to be affected by an artifact that dramatically increases the apparent lifetime at low injection levels. The trapping centers capture a minority carrier and then release it back to the respective energy band where it was captured from. During the period of holding the minority carrier, the majority carrier density increases, creating the abnormal increase in the apparent lifetime measured by QSSPC [1]. These additional trapping centres do not contribute to recombination but only interact with minority carriers and the minority band [1]. Such trapping centres can be considered as 'fast' traps, as they have capture and release time constants that are faster than the recombination lifetime. 'Slow' traps can also affect photoconductance decay data as they are slowly released back into the bands. Such slow traps have been observed in Cz-Si samples [2]. Therefore, understanding slow trapping of minority carriers is useful in the characterization of silicon materials for photovoltaics. Different methods such as photoconductance [3], and Microwave Detected Photoconductivity [4, 5] have been used to study fast and slow trapping effects in silicon material.

At the same time, silicon heterojunction solar cells suffer from significant loss in efficiency due to the presence of thermal donors [6]. As such, it is important to understand the recombination and trapping properties of thermal donors. Although the recombination parameters of thermal donors [7] and the temperature dependence of the recombination lifetime in silicon wafers [8] have been studied recently, little is known regarding the trapping parameters of thermal donors and their dependence on temperature. Some studies have quantified the ratio of the de-trapping/trapping time constants [9] – [11]. Hornbeck and Haynes revealed the existence of two types of traps, present both in n-type [9] and p-type [10] silicon. However, recent work by Heinz et al. [2] and Zhu et al. [12] have shown that the trapping mechanisms of TDs are more complex than described by Hornbeck and Haynes, and little is known regarding the many thermal donor species (TD1...TDn) and how their structural configuration relates to their trapping properties. Previously, we have studied the effect of the concentration of thermal donors on slow de-trapping time constants [13], and found that the de-trapping time constants are inversely proportional to the concentration of thermal donors of silicon wafers [13]. Also, no slow decay of the photoconductance was observed when n-type Cz silicon wafers were annealed at 650ºC for 30 minutes to destroy the thermal donors.

In this study, we have investigated the temperature dependence of the slow de-trapping time constants in samples with different concentrations of thermal donors, allowing us to estimate the energy levels of the slow trap centers using Arrhenius plots. The research helps to improve our understanding of thermal donors especially in relation to their slow trapping properties.

2 EXPERIMENTAL METHODS

In this study, three sister wafers from an n-type monocrystalline Cz silicon ingot with an as-cut resistivity of 2 Ω.cm were used.

In order to create different concentrations of thermal donors, sample 1 was kept as-cut, sample 2 was annealed at 450 °C for 72 hours, and sample 3 was annealed at 650 °C for half an hour to remove any existing thermal donors in the sample. All three samples were then RCA cleaned and passivated with Al_2O_3 layers using thermal atomic layer deposition. The resistivity and thermal donor concentration was measured with a WCT 120 coil [14].

The lifetime was measured using a photoconductance based QSSPC lifetime measurement system from Sinton Instruments. Minority carrier trapping was identified in sample 1 and sample 2 from the artificially high lifetime at low injection levels. All the samples were then tested for changes in photoconductance during a slow de-trapping process at different temperatures (12ºC, 30ºC, 60ºC and 80ºC). The photoconductance measurements were done using a modified lifetime tester from Sinton Instruments [15]. The flash light source is a Xenon flash as used in QSSPC measurements with the standard Sinton Instruments WCT-120 lifetime tester. In order to analyze the de-trapping process, the samples were illuminated with a flash and the decay of the photoconductance was studied in each sample for up 1-10 minutes after the flash was terminated. The photoconductance decay curves, given by the voltage of the conductance coil, were then fitted with exponential curves to obtain the de-trapping time constants. The de-trapping rates were then fitted against the inverse temperature to generate Arrhenius plots, allowing and the energy levels of the slow traps to be estimated.

3 RESULTS AND DISCUSSION

3.1 Temperature dependent photoconductance decay of as-grown and high TD samples

During excitation, the traps are filled, and when the flash is terminated, the minority carriers are slowly released back into the band where they were captured from.

After terminating the flash, there is an immediate decline of the photoconductance due to recombination, and this part is not shown in Figure 1 and Figure 2. It is followed by a much slower decay which can be fitted with multi exponential curves, indicating the existence of more than one type of trap in a sample.

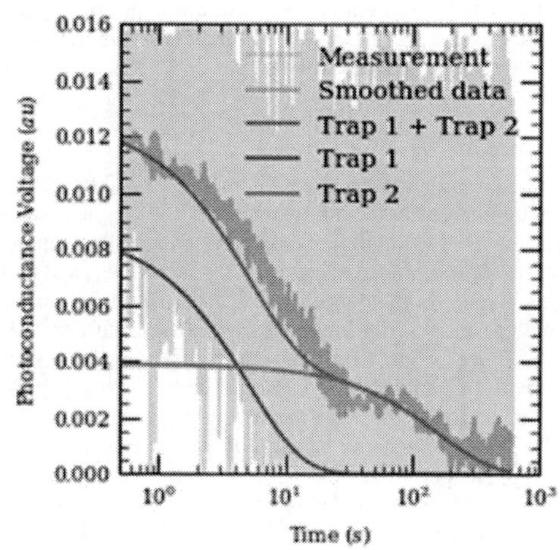

At 30ºC

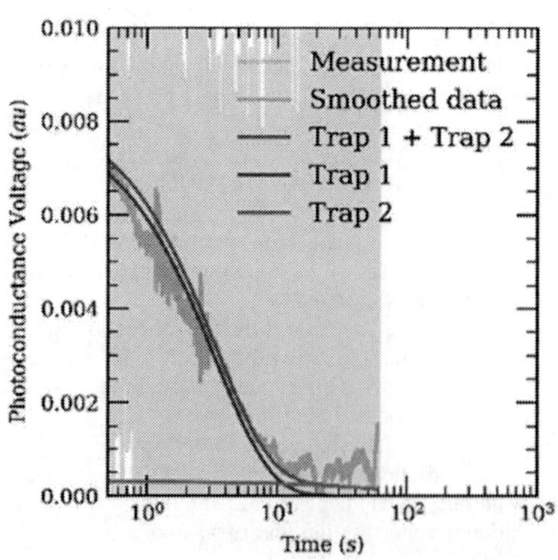

At 60ºC

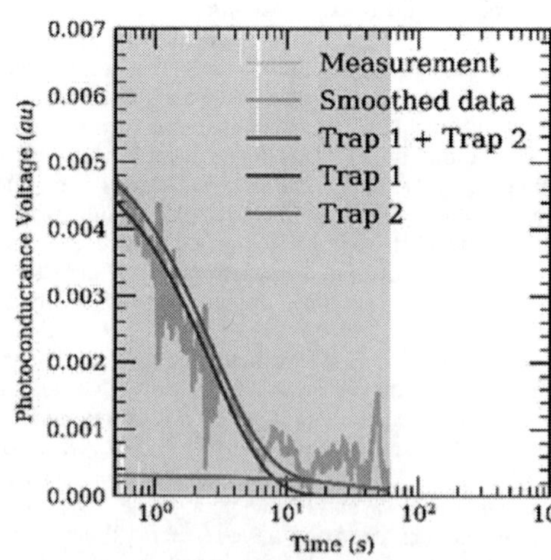

At 80ºC

Figure 1: Change in photoconductance coil voltage of an n-type Cz as-cut silicon wafer (sample 1) after termination of a strong illumination source, at different temperatures as specified under each Figure.

Figures 1 and 2 shows the measurements of the photoconductance coil voltage decay of sample 1 (as-cut sample) and sample 2 (High TD sample) after a flash light is terminated. The coil voltage is linearly proportional to the excess carrier density and serves as a proxy to measure the kinetics of excess carrier density decays. As shown in the figures, the photoconductance decay was measured for four different temperatures (12ºC, 30ºC, 60ºC and 80ºC). The de-trapping curves are fitted using a multi exponential function.

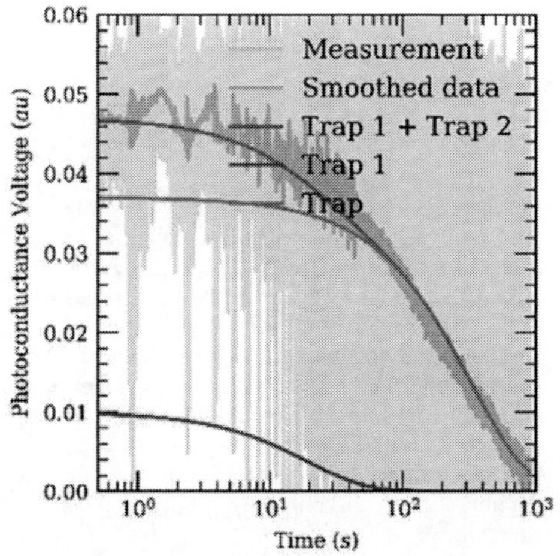

At 12ºC

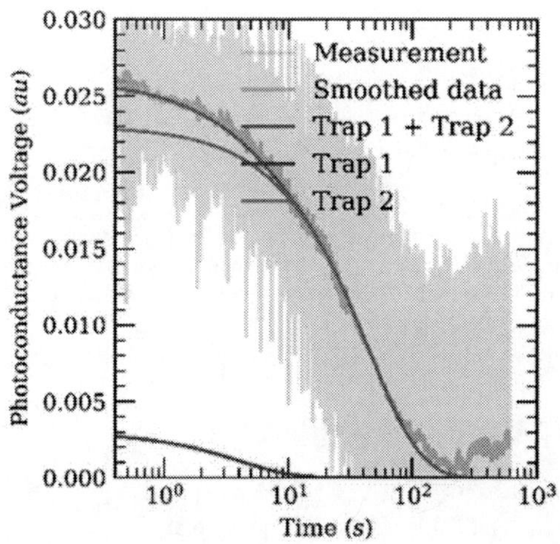

At 30ºC

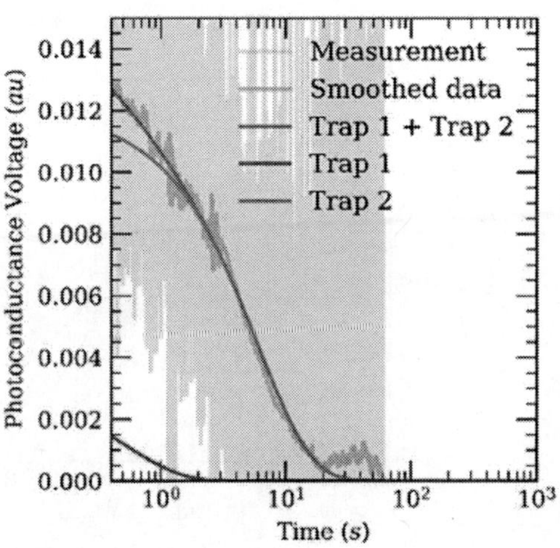

At 60ºC

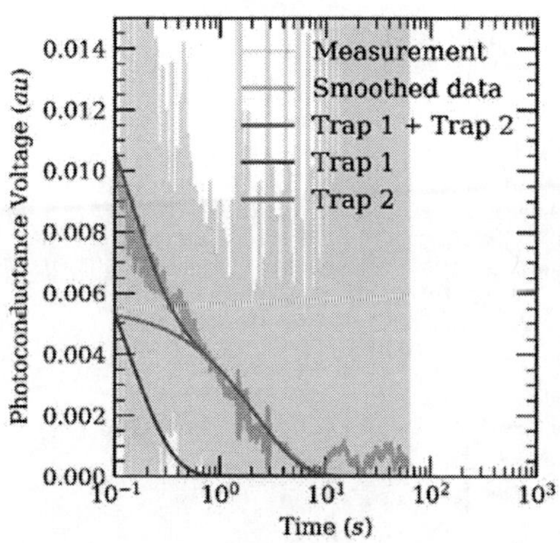

At 80ºC

Figure 2: Change in photoconductance coil voltage of an n-type Cz high TD silicon wafer which was annealed at 450ºC for 72 hours (sample 2) after termination of a strong illumination source, at different temperatures as specified under each Figure.

The de-trapping time constants at different temperatures are summarized in Table 1. The as-grown sample has longer de-trapping time constants compared to the sample with high [TD]. Further, it is obvious that the de-trapping is faster as the measurement temperature increases.

3.2 Estimation of trap energy levels

As demonstrated in Figures 1 and 2, the faster trap (trap 1) contributes more to the PC decay in the as-cut sample, and the slower trap (trap 2) contributes more to the PC decay in the High TD sample. Therefore, we further analyze trap 1 for the as-cut sample, and trap 2 for the High TD sample, in terms of Arrhenius plots.

Table I: De-trapping time constants for samples with different thermal donor concentrations at different temperatures.

Sample	As-grown (low [TD])		Annealed (high [TD])	
Resistivity (Ω.cm)	2.03		0.31	
Thermal donor concentration (cm^{-3})	$9.6x10^{14}$		$9.1x10^{15}$	
QSSPC lifetime (μs)	1388		22	
Temperature (°C)	τ_{1d} (s)	τ_{2d} (s)	τ_{1d} (s)	τ_{2d} (s)
12	-	-	20	330
30	5.1	165	4	44
60	3.3	80	0.55	6.0
80	2.7	50	0.13	2.2

Figure 3 shows the Arrhenius plot of trap 1 in the as-cut sample and Figure 4 gives the Arrhenius Plot of trap 2 in High TD sample. The energy level of each trap is determined using Equation 1, given by Dobaczewski et al. [16].

$$e_{n,p} = \sigma_{n,p}\langle \upsilon_{n,p}\rangle N_{c,\upsilon}\exp(-\Delta G_{c,\upsilon}/k_B T). \quad (1)$$

Where,

$e_{n,p}$ = Emission rate of electron or holes
$\sigma_{n,p}$ = capture cross sections,
$\upsilon_{n,p}$ = mean thermal velocities,
$N_{c,v}$ = effective density states
k_B = Boltzmann constant in eV/K
T = Temperature in K
$\Delta G_{c,\upsilon}$ = Energy Level

Therefore, the gradient of an Arrhenius plot of ln(1/tau) versus (1/T) gives $\Delta G_{c,\upsilon} / k_B$, from which the energy level can be estimated. The energy level of trap 1 in the as-cut sample was found to be 0.12 eV in agreement with previous studies (DOI 10.1002/pssa.2210560220) and that of trap 2 in the High TD sample was 0.64 eV.

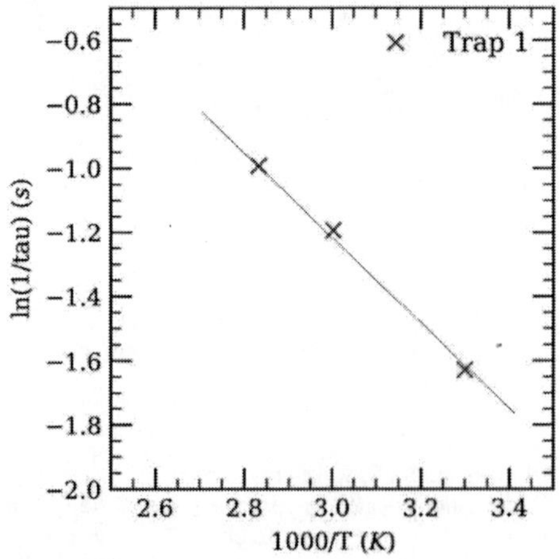

Figure 3: Arrhenius plot for Trap 1 of the as grown sample (sample 1)

In n-type silicon, holes, the minority carriers, are captured by traps and released back into the valence band during de-trapping. Based on the estimations of energy levels of the more significant trap in each sample, it is observed that the energy level of traps in the as-cut wafers are located at 0.12 eV above the valence band. In contrast, the dominant traps in the High TD sample are close to the middle of the gap, with an energy level of 0.64 eV above the valence band. The different de-trapping time constants may reflect the depth of the energy levels.

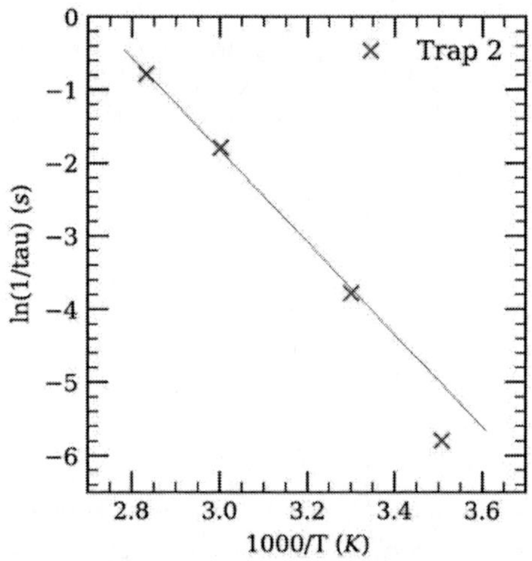

Figure 4: Arrhenius plot for Trap 2 of the high TD sample (sample 2)

4 CONCLUSIONS

The creation of additional thermal donors in Cz-Si wafers leads to lower measured carrier lifetimes. As-cut wafers with a low concentration of thermal donors have most trap levels closer to the valence band at 0.12eV above the valence band. Their de-trapping time is longer than the samples with a high concentration of thermal donors, for which the trap levels are located deep in the bandgap, closer to the mid-gap at 0.64eV above the valence band. In both the samples, the de-trapping time decreases as the temperature increases. The observed difference in de-trapping time constants may be associated with the depth of the traps, and the different trap parameters in each sample probably reflects the different species of thermal donors present.

Acknowledgements: This work has been supported by the Australian Research Council (ARC) Discovery Early Career Researcher Award (DECRA) DE160101368 and the Australian Renewable Energy Agency (ARENA) through project RND012 and the German-government BMBF funded CCPV project.

5 REFERENCES

[1] R. A. Bardos, T. Trupke, M. C. Schubert, T. Roth, Applied Physics Letters, 88(2006), 053504.
[2] F. D. Heinz, W. Warta, M. C. Schubert, Solar Energy Materials & Solar Cells, vol. 158(2016), pp. 107–114.
[3] J. Schmidt, K. Bothe, R. Hezel, 2002, Applied Physics Letters, 80(23), 2002.
[4] K. Lauer, M. Blech, A. Laades, A. Lawerenz, 24th EUPVSEC, Hamburg, Germany 2009.
[5] N. Schuler, T. Hahn, K. Dornich, J. R. Niklas, 25th EUPVSEC, Valencia, Spain, 2010.
[6] J. Li, X. Yu, S. Yuan, L. Yang, Z. Liu, D. Yang, Solar Energy Materials and Solar Cells, 179(2018), pp. 17-21
[7] M. Tomassini, J. Veirman, R. Varache, E. Letty, S. Dubois, Y. Hu, and Ø. Nielsen, Journal of Applied Physics, 119, 084508 (2016).
[8] M. Ichimaru, H. Tajiri, T. Ito, E. Arai, Journal of Electrochemical Society, 145(9), 1998.
[9] J. A. Hornbeck and J. R. Haynes, Physical Review 100(2), pp. 606-615, 1955.
[10] J. A. Hornbeck and J. R. Haynes, Physical Review 97(2), pp. 311-321, 1955.
[11] Y. Hu, H. Schøn, Ø. Nielsen, E. J. Øvrelid, and L. Arnberg, Journal of Applied Physics, 111, 053101 (2012).
[12] Y. Zhu, M. Juhl, G. Coletti, Z. Hameiri, IEEE Journal of Photovoltaics, vol. 8(2), pp. 421-427, 2018.
[13] M. Siriwardhana, D. Macdonald, F. D. Heinz, F. E. Rougieux, 7th World Conference on Photovoltaic Energy Conversion (WCPEC-7), 2018, In press.
[14] R.A. Sinton, A. Cuevas, Applied Physics Letters, vol. 69(17), pp. 2510-12, 1996.
[15] R.A. Sinton, A. Cuevas, M. Stuckings, 25th IEEE Photovoltaic Specialists Conference, 1996, pp. 457-460.
[16] L. Dobaczewski, A. R. Peaker, K. B. Nielsen, Journal of Applied Physics, 96(9), 2004.

INFLUENCE OF LOW-TEMPERATURE ANNEALING BEFORE FIRING ON LETID IN MULTICRYSTALLINE SILICON

Andreas Schmid, Jeanette Lindroos, Annika Zuschlag, Daniel Skorka, Jakob Fritz, Giso Hahn
University of Konstanz, Department of Physics, 78457 Konstanz, Germany

ABSTRACT: Light and elevated temperature induced degradation (LeTID) affects significantly the performance of multicrystalline silicon (mc-Si) solar cells. Even though the underlying mechanisms of LeTID and subsequent regeneration are still unknown, LeTID formation is known to be influenced by surface passivation, firing, and low-temperature annealing after firing. Since both the firing peak temperature and the cooling profile strongly influence LeTID, firing is presumed to dominate over any preceding low-temperature steps. However, lifetime measurements during illumination (1.0 ±0.1 suns) at 150°C show a clear difference in the LeTID defect density in samples annealed at 200-500°C *before* aluminum oxide deposition and *before* firing. Additionally, LeTID becomes worse when the pre-firing annealing is performed in an H plasma atmosphere. Therefore, the firing step does not erase the impact of low-temperature annealing or pre-firing hydrogenation on LeTID. Finally, thinner samples are shown to cause lower initial effective lifetimes (τ_{eff}) and less LeTID.

Keywords: Degradation, Multicrystalline Silicon, Annealing, Lifetime, Hydrogen, LeTID

1 INTRODUCTION

Multicrystalline (mc) Si PERC (passivated emitter and rear cell) solar cells, but also lifetime samples, show a strong degradation under illumination at elevated temperatures (*e.g.* [1-4]), known as light and elevated temperature induced degradation (LeTID). Previously known degradation mechanisms, such as the dissociation of FeB pairs and BO defects, cannot explain the observed degradation kinetics [1-3]. After degradation, solar cells [3] as well as lifetime samples [5] show a regeneration, but the underlying mechanisms of both effects are still unknown. However, the kinetics of both degradation and regeneration are known to be influenced by material properties, e.g. sample thickness [6], and treatment conditions, such as illumination and temperature (*e.g.* [3, 7]). Additionally, a low temperature dark annealing step *after* firing is known to impact LeTID formation in mc-Si [8]. However, as LeTID only forms at firing temperatures above 675°C [9], the firing step is presumed to dominate over any preceding low temperature steps. Interestingly, the surface passivation layer has been shown to also influence the LeTID defect density [10]. Passivation layers were deposited with different PECVD (plasma-enhanced chemical vapor deposition) tools, and it is still unknown whether the deposition temperature plays a role in LeTID formation. In this contribution, we subject sister wafers to different low temperature annealing steps (T_{var}< 500°C) in N_2 atmosphere *before* firing, after which we deposit the same Al_2O_3 layer on all samples.

Capture cross section analysis of the LeTID defect has hinted at a metal impurity [11, 12], whereas the impact of the PECVD passivation layer and the wafer thickness (*e.g.* [12-15]) indicate a direct involvement of hydrogen in the LeTID mechanism. Hence, we additionally apply an H plasma atmosphere during the low-temperature annealing step to investigate the influence of hydrogen on degradation and regeneration.

2 EXPERIMENTAL

To investigate the influence of a low temperature step before firing on LeTID and regeneration behavior, high quality B-doped p-type mc-Si sister wafers with comparable grain and defect structure from the middle of the ingot were processed to lifetime samples (5 x 5 cm²). To further investigate the influence of sample thickness, half of the wafers were etched to a thickness of 150 µm, the rest to 170 µm. A scheme of the applied process sequence is given in Fig. 1. Half of the wafers were gettered via $POCl_3$ diffusion (55 Ω/□), after which the emitter was removed. Next, all samples were deposited with thin PECVD SiO_x at room temperature to protect the samples from contamination during low temperature annealing. The samples were annealed for different times (0.5-2 h) in a tube furnace at different temperatures (T_{var} <500°C).

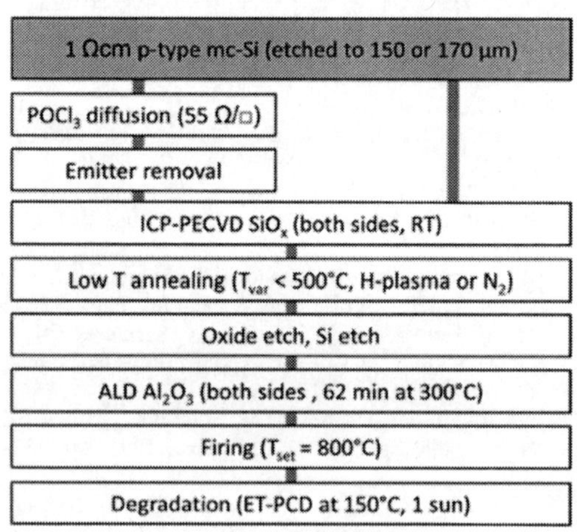

Figure 1: Process sequence of the investigated lifetime samples.

To investigate the influence of hydrogen on LeTID and regeneration behavior, half of the wafers were annealed in H plasma atmosphere, which was realized by microwave-induced remote hydrogen plasma (MIRHP). The rest of the wafers were annealed in N_2 atmosphere. After SiO_x and silicon etching, all samples were deposited with double-sided 30 nm Al_2O_3 at 300°C by atomic layer deposition (ALD) for a total of 62 min. Finally, all samples were fired in a belt furnace at a set temperature of 800°C.

For degradation, lifetime samples were held at a temperature of approx. 150°C under illumination with halogen lamps (1.0 ±0.1 suns). The effective minority charge carrier lifetime τ_{eff} was measured repetitively with photoconductance decay (PCD) at elevated temperature (ET) of 150°C. The lifetime was extracted at a fixed excess charge carrier concentration of $\Delta n = 1.5 \times 10^{15}$ cm^{-3}. Additionally, spatially resolved lifetime maps were measured at certain stages of degradation and regeneration (after firing, at the degraded state and after full regeneration) with the self-calibrated TR-PLI method [16] at room temperature (RT).

3 RESULTS

In the following either gettered or ungettered sample sets are shown and discussed. Nevertheless, all the observed results match qualitatively in each case.

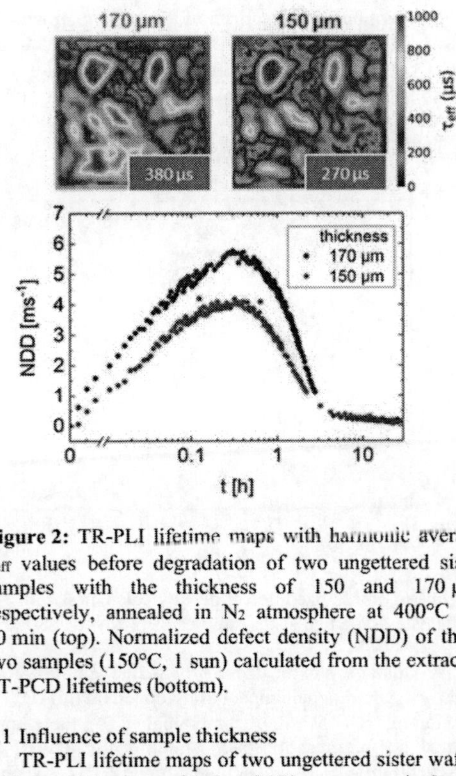

Figure 2: TR-PLI lifetime maps with harmonic average τ_{eff} values before degradation of two ungettered sister samples with the thickness of 150 and 170 µm, respectively, annealed in N_2 atmosphere at 400°C for 30 min (top). Normalized defect density (NDD) of these two samples (150°C, 1 sun) calculated from the extracted ET-PCD lifetimes (bottom).

3.1 Influence of sample thickness

TR-PLI lifetime maps of two ungettered sister wafers with the thicknesses of 150 and 170 µm, respectively, are shown in Fig. 2. Fig. 2 also shows the time-dependent normalized defect density (NDD), which is calculated using

$$\mathrm{NDD}(t) = \frac{1}{\tau_{eff}(t)} - \frac{1}{\tau_{eff}(0)} \quad (1)$$

where $\tau_{eff}(t)$ and $\tau_{eff}(0)$ are the extracted effective lifetimes measured with ET-PCD at any given time and the initial state, respectively. Both samples received a low temperature annealing step at 400°C in N_2 atmosphere before ALD Al_2O_3 passivation and firing. The thicker sample (170 µm) shows an initial harmonic average τ_{eff} of 380 µs, the thinner sample (150 µm) only 270 µs.

Regarding NDD, both samples show pronounced LeTID and a following regeneration phase. While the thicker sample shows stronger degradation, both samples regenerate completely to their starting values within a few hours. These observations are less pronounced but match qualitatively with [6], which also shows an influence of the sample thickness on LeTID and regeneration. In the model discussed in [6], a movement of impurities towards the wafer surface is assumed to be responsible for regeneration.

3.2 Low-temperature annealing before firing

To investigate the influence of a low temperature annealing step before firing, Fig. 3 shows the initial TR-PLI lifetime images of six gettered sister samples before degradation. The reference sample was only subjected to Al_2O_3 passivation before firing, whereas the other samples were annealed at 200, 300 and 400°C for 0.5-2 h before ALD and firing.

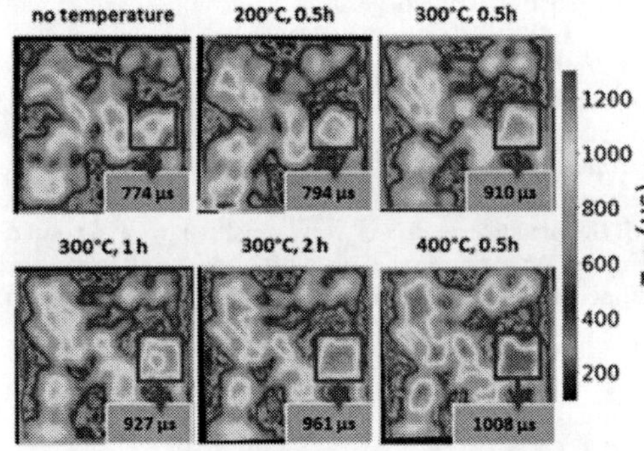

Figure 3: Initial TR-PLI lifetime maps (before illumination) of gettered sister samples subjected to different low temperature annealing steps in N_2 atmosphere before ALD and firing. The red squares highlight the change of one sample area with increasing temperature and annealing time. The harmonic average τ_{eff} values of this area are shown in the lower right corner.

It can be seen that a low temperature annealing step before firing improves initial τ_{eff}. As highlighted by the red squares in Fig. 3, adding a low temperature step of 200°C improves the lifetime from 774 to 794 µs. Elevating the temperature further to 400°C increases the lifetime up to 1008 µs. At 300°C, extending the annealing time from 30 min to 2 h increases τ_{eff} again from 910 to 961 µs. All of these improvements in the initial lifetime might be caused by low temperature gettering of impurities internally or towards the surface layer during the low-temperature annealing steps.

Fig. 4 displays the normalized defect density of the discussed samples. Within approx. 10 min of illumination, all the low temperature annealed samples have reached full degradation, but with significantly different defect densities. The reference sample without an additional low temperature annealing step shows weak LeTID, which remains at a similar level even when a 30 min anneal at 300°C is added before ALD and firing. Interestingly, extending the annealing time to 1 h at 300°C leads to the highest NDD. Regardless of the annealing temperature, LeTID becomes stronger by extending the step from 30 min to 1 h.

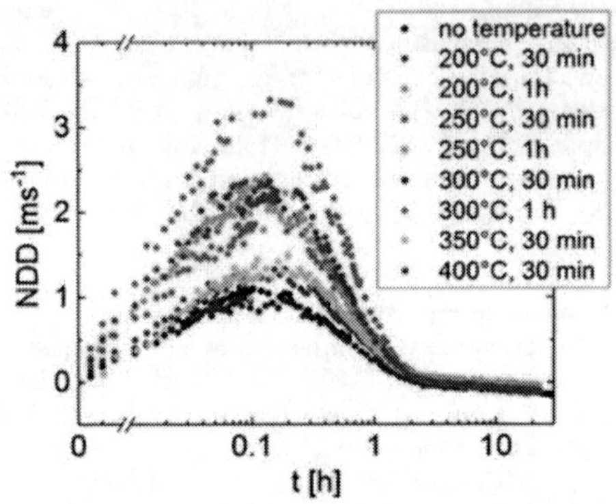

Figure 4: Normalized defect density of gettered sister samples annealed in N_2 atmosphere at 200-400°C for 0.5 and 1 h before Al_2O_3 deposition and firing.

While there is no obvious correlation between the temperature and the LeTID kinetics, the low temperature annealing before surface passivation and firing changes the maximum LeTID defect density reached in the degradation/regeneration cycle. This means that the LeTID precursor/s is/are very sensitive to even small temperature variations below 500°C, and the effects of this temperature load is not erased by the following firing step.

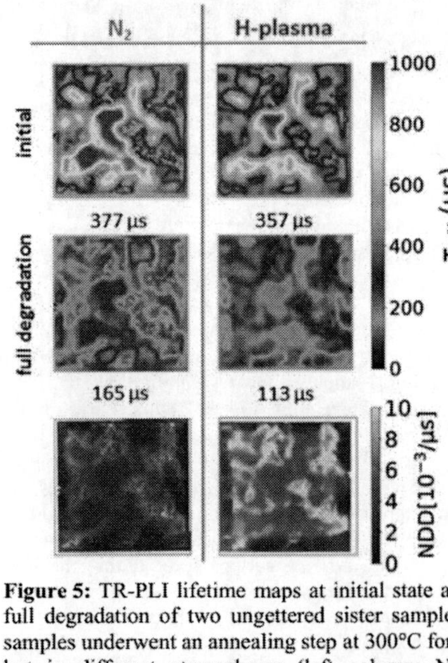

Figure 5: TR-PLI lifetime maps at initial state and after full degradation of two ungettered sister samples. Both samples underwent an annealing step at 300°C for 30 min but in different atmospheres (left column: N_2, right column: H plasma). The last row shows the NDD maps, calculated according to Eq. 1 with the respective lifetime maps above.

3.3 Influence of hydrogen

Fig. 5 shows the initial TR-PLI lifetime maps (directly after firing and after full degradation) of two ungettered sister wafers, which underwent an annealing step at 300°C for 30 min. The left sample was annealed in N_2 atmosphere, whereas remote hydrogen plasma was present during the annealing of the sample shown on the right. Comparing the lifetime maps at the initial state, the harmonic average τ_{eff} is clearly reduced from 377 to 357 µs by the presence of H plasma before ALD Al_2O_3 passivation. The spatially resolved maps reveal that areas with high τ_{eff} are decreasing, whereas areas with low τ_{eff} increase during H plasma annealing. This reduction in areas with high τ_{eff} is a first indication of the detrimental effect of pre-firing hydrogenation, which was already discussed in [15] and could lead to an issue regarding high performance mc-Si. The TR-PLI lifetime maps after full degradation shows that the sample treated in H plasma and therefore most probably containing more H in the Si bulk is affected more by LeTID, which was also recently observed in [17]. To further quantify these observations, the lifetime maps after firing and after full degradation were aligned to each other and NDD of each pixel was calculated according to Eq. 1. The resulting NDD maps are shown in the bottom row of Fig. 5. It can be seen that the sample with H-plasma present during annealing suffer much more from LeTID than the sample annealed in N_2 atmosphere.

Furthermore, a comparison of the NDD map with the initial TR-PLI lifetime map shows that areas with low initial τ_{eff} are more affected, but this is the case in both investigated samples. To further confirm these observations, time dependent NDD was calculated with ET-PCD data and is shown in Fig. 6.

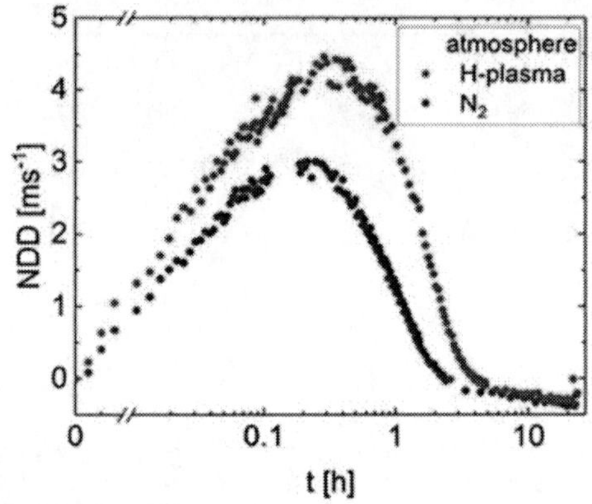

Figure 6: Normalized defect density calculated with ET-PCD data of the two sister samples displayed in Fig. 5.

A comparison of the differently treated samples shows clearly the detrimental impact of hydrogen on LeTID. The maximum NDD is significantly higher for the sample treated in H-plasma atmosphere. Additionally, a time shift of several minutes in reaching maximum NDD is visible. Interestingly, both samples regenerate completely. Whereas the sample annealed in N_2 atmosphere regenerates after approx. 2 h, the sample treated in H plasma atmosphere reaches the initial lifetime after 3 h. This might be due to the time shift in reaching maximum NDD.

This effect is not limited to the samples annealed at 300°C. All samples subjected to pre-firing annealing show higher LeTID in either the spatially resolved NDD maps measured at room temperature or the NDD plots calculated with ET-PCD data, when the annealing is performed in H plasma. The annealing step only influences the strength of this effect, which might be caused by a strong dependency on the local defect structure or a strong sensitivity to small experimental variations (*e.g.*, temperature ramps during unloading of samples after low temperature annealing). Note that PCD data only show weighted average values of part of the sample and the extracted data can be misinterpreted. So when trying to evaluate measurements of mc-Si samples, a combination of spatially resolved measurements (*cf.* Fig. 5) and PCD measurements (*cf.* Fig. 6) is necessary and seems to be a good solution.

4 SUMMARY AND DISCUSSION

The influence of sample thickness, low temperature annealing *before* firing, and the effect of an H plasma on LeTID and regeneration could be shown on mc-Si wafers using data from TR-PLI and ET-PCD. Spatially resolved τ_{eff} was combined with injection dependent τ_{eff} for detailed analysis of the effect of annealing and sample thinning. The effect of sample thickness on LeTID and regeneration which was already discussed in [6] could be confirmed, and its influence on initial τ_{eff} could additionally be demonstrated with spatially resolved lifetime maps.

A comparison of samples that underwent a low temperature anneal at 200-500°C for 0.5-2 h *before* surface passivation and firing showed an initial lifetime increase, which might be caused by low temperature gettering internally or towards the surface layer. The different annealing profiles clearly influenced the maximum defect density, but no obvious relation was found between the chosen temperature and the LeTID kinetics. Therefore, the LeTID precursor/s is/are observed to be very sensitive to even small temperature variations below 500°C *before* firing, and the effects of these steps are not erased by the following firing step.

In addition, we could show the combined detrimental and beneficial influence on the initial τ_{eff} of hydrogen introduced before ALD AlO_x surface passivation and firing. TR-PLI and ET-PCD measurements enabled a detailed comparison regarding LeTID and regeneration behavior of samples either annealed in N_2 or in H plasma atmosphere. While pre-firing hydrogen worsens LeTID, all samples regenerate to their initial τ_{eff} value after a few hours at 150°C. These results help explain why the surface passivation deposition step itself affects LeTID, and ultimately why LeTID varies so much in mc Si-PERC solar cells from different manufacturers.

5 ACKNOWLEDGEMENTS

Part of this work was supported by the German BMWi under contracts 0324204B, 0325763B and 0324001. The content of the publication is the responsibility of the authors. The authors would like to thank L. Mahlstaedt, B. Rettenmaier and J. Engelhardt for technical support.

6 REFERENCES

[1] K. Ramspeck, S. Zimmermann, H. Nagel, A. Metz, Y. Gassenbauer, B. Birkmann, A. Seidl, Proc. 27th EUPVSEC, Frankfurt/Main, Germany, 2012, 861.
[2] F. Fertig, K. Krauss, S. Rein, Physica Status Solidi RRL 9(1) (2014) 41.
[3] F. Kersten, P. Engelhart, H.-C. Ploigt, A. Stekolnikov, T. Lindner, F. Stenzel, M. Bartzsch, A. Szpeth, K. Petter, J. Heitmann, J. Müller, Solar Energy Materials and Solar Cells 142 (2015) 83.
[4] A. Zuschlag, D. Skorka, G. Hahn, Progress in Photovoltaics: Research and Applications 25(7) (2017) 545.
[5] A. Zuschlag, D. Skorka, G. Hahn, Proc. 43rd IEEE PVSC, Portland 2016, 1051.
[6] D. Bredemeier, D.C. Walter, J. Schmidt, Solar RRL 2 (2018) 1700159.
[7] J. Fritz, A. Zuschlag, D. Skorka, G. Hahn, Energy Procedia 124 (2017) 718.
[8] C.E. Chan, T.H. Fung, M.D. Abbott, D. Payne, A. Wellham, B. Hallam, R. Chen, S. Wenham, Solar RRL 1 (2017) 1600028.
[9] C.E. Chan, D.R. Payne, B.J. Hallam, M.D. Abbott, T.H. Fung, A.M. Wenham, B.S. Tjahjono, S.R. Wenham, IEEE Journal of Photovoltaics 6 (2016) 1473.
[10] F. Kersten, J. Heitmann, J.W. Müller, Energy Procedia 92 (2016) 828.
[11] A.E. Morishige, M.A. Jensen, D.B. Needleman, K. Nakayashiki, J. Hofstetter, T.-T.A. Li, T. Buonassisi, IEEE Journal of Photovoltaics 6(6) (2016) 1466.
[12] K. Nakayashiki, J. Hofstetter, A. Morishige, T. Li, D. Needleman, M. Jensen, T. Buonassisi, IEEE Journal of Photovoltaics 6(4) (2016) 860.
[13] T.H. Fung, M. Kim, D. Chen, C.E. Chan, B.J. Hallam, R. Chen, D.N.R. Payne, A. Ciesla, S.R. Wenham, M.D. Abbott, Sol. En. Mat. Sol. Cells 184 (2018) 48.
[14] D. Bredemeier, D.C. Walter, J. Schmidt, AIP Conference Proceedings 1999 (2018) 130001.
[15] D. Chen, P.G. Hamer, M. Kim, T.H. Fung, G. Bourret-Sicotte, S. Liu, C.E. Chan, A. Ciesla, R. Chen, M.D. Abbott, B.J. Hallam, S.R. Wenham, Solar Energy Materials and Solar Cells 185 (2018) 174.
[16] D. Kiliani, G. Micard, B. Steuer, B. Raabe, A. Herguth, G. Hahn, Journal of Applied Physics 110 (2011) 054508.
[17] M.A. Jensen, A. Zuschlag, D. Skorka, S. Wieghold, A.E. Morishige, G. Hahn, T. Buonassisi, Journal of Applied Physics 124(8) (2018) 085701.

KINETICS OF THE DEGRADATION AND REGENERATION OF P-TYPE MULTICRYSTALLINE SILICON UNDER DARK ANNEAL

Carlos Vargas[1], Gianluca Coletti[1,2], David Payne[1], Catherine Chan[1], and Ziv Hameiri[1]
[1] School of Photovoltaic and Renewable Energy Engineering, University of New South Wales, Sydney, Australia
[2] ECN Part of TNO – Solar Energy, Petten, the Netherlands

ABSTRACT:
In the last few years, it has been found that solar cells made using *p*-type multicrystalline silicon (mc-Si) wafers degrade under illumination at elevated temperatures. This degradation is commonly known as light- and elevated temperature-induced degradation (LeTID) or carrier-induced degradation (CID). CID is the main challenge for the introduction of mc-Si passivated emitter and rear cells (PERC) in mass production. Previous models of the CID kinetics are based on the presence of two defects (fast and slow), neglecting the simultaneous regeneration that may occur during the degradation. Recently, it has been reported that similar degradation can be observed under anneal at moderate temperatures, even *without* illumination. In this work, we developed and successfully validated a new model based on the presence of only one defect and on the simultaneous degradation and regeneration processes. The model is much simpler than the current available in the literature. We use this model to investigate the impact of thermal treatment in the dark on *p*-type mc-Si wafers. We find that high temperatures lead to faster degradation and regeneration rates, while the degradation extent is larger at low temperatures. The activation energies of the degradation and regeneration processes are obtained from the proposed model and found to be 1.08 ± 0.05 eV and 1.11 ± 0.04 eV, respectively. Interestingly, at the end of the dark anneal process, the samples treated at high temperatures show improvement of ~40% in their effective lifetimes.

Keywords: Multicrystalline silicon, Dark anneal, Degradation, Hydrogen

1 INTRODUCTION

In 2012 Ramspeck *et al.* [1] reported degradation of *p*-type multicrystalline silicon (mc-Si) cells under illumination and elevated temperature; since then, there has been a great interest to understand this degradation. Kersten *et al* [2] confirmed this degradation on cells and modules and named it light- and elevated temperature-induced degradation (LeTID). They also found differences in the degradation when open- or short-circuit conditions are used and that degradation could be induced using current-injection; therefore it is also known as carrier induced degradation (CID) [3], [4], since it is the injection of carriers that causes degradation. This is the term used in this study.

Recently, Chan *et al.* [5] reported modulation of the CID behavior when the wafers received an initial dark anneal (DA). In addition, they also reported a degradation of *p*-type mc-Si cells due to DA at moderate temperatures [5]. Fung *et al.* [6] presented an extended study of the effect of DA on the modulation of CID. They also concluded that DA induces degradation of *p*-type mc-Si. Chen *et al.* [5] then reported similar degradation and lifetime recovery due to anneal in the dark of boron-doped Cz and mc-Si wafers. These are the first reports to present a degradation of *p*-type mc-Si at moderate temperatures *in the dark.*

Lately, several models have been proposed to describe the time evolution of CID and to extract the rates of this process. Kwapil *et al.* [4] used a single exponential function to extract the degradation rates and found them to be proportional to the excess of minority carriers concentration (Δn). Bredemeier *et al.* [7] used two exponential functions to model the process, a model that was also supported by Kwapil *et al.* [4]. Based on degradation results obtained under illumination at elevated temperatures (75 °C to 120 °C), Bredemeier *et al.* assumed two defects (fast and slow) to explain the shape of the degradation curve [7]. Based on this model, they extracted the activation energies of both defects by an Arrhenius analysis and reported values of E_{fast} = 0.89 ± 0.04 eV (fast defect) and E_{slow} = 0.93 ± 0.06 eV (slow defect) [7].

In this contribution, we take into account that degradation and regeneration are happening simultaneously and propose a model to support this proposal. Furthermore, we investigate the temperature dependence, as well as the time evolution, of the degradation of *p*-type mc-Si wafers under DA.

2. EXPERIMENTAL METHOD

Nine six-inch neighboring high-performance boron-doped mc-Si wafers from a central brick with a thickness of 0.019 cm and resistivity 1.7 Ω.cm were used in this study. Symmetrical lifetime structures were prepared using the front PERC scheme (diffused layer passivated with SiN_x): RCA (Radio Corporation of America) cleaning, 60 Ω/□ phosphorous diffusion (both sides; carried out in a Tempress $POCl_3$ tube), hydrofluoric acid (HF) dip to remove the phosphosilicate glass, a second RCA clean before SiN_x deposition onto both sides using a Meyer Burger MAiA plasma enhanced chemical vapour deposition (PECVD) system [8]. The obtained 75 nm SiN_x layer (RI 2.08) was then fired in a belt furnace at a sample peak temperature of 800 °C.

The six-inch wafers were then laser cleaved into tokens of 3.9 cm × 3.9 cm. One of the tokens from each wafer underwent dark-anneal using a hotplate (IKA c-mag HS 10) at eight *wafer* temperatures (138 °C, 160 °C, 176 °C, 203 °C, 228 °C, 254 °C, 275 °C and 300 °C), as measured by a *k*-type thermocouple in direct contact with a similar token. One token was used as a baseline undergoing the laser-accelerated degradation of Payne *et al.* [3]. This process was done at a sample temperature of 140 °C using a 45 kW/m^2 laser intensity (wavelength of 938 nm).

After each process, effective lifetime measurements were done using a photoconductance (PC) based lifetime tester (WCT-120 from Sinton instruments at an actual sample temperature of 30 °C) using the generalized analysis method [9].

3. KINETICS MODEL

We define the normalized effective lifetime as $\tau_{norm} = \tau(t) / \tau_{fire}$, where $\tau(t)$ and τ_{fire} are the effective lifetimes at $\Delta n = 10^{15}$ cm^{-3} after a treatment time of t sec and immediately after firing, respectively. The normalized defect density (*NDD*) is defined as:

$$NDD(t) = \left(\frac{1}{\tau(t)} - \frac{1}{\tau_{fire}}\right). \quad (1)$$

The *NDD* time evolution was fitted using the following expression:

$$NDD(t) = NDD_{max}[1 - \exp(-r_{deg}t)] \\ -NDD_{max}(1 + A)[1 - \exp(-r_{reg}t)], \quad (2)$$

where NDD_{max} is the maximum possible *NDD* by the degradation, A is a factor to count for improvement in the effective lifetime after the treatment processes, r_{deg} and r_{deg} are the degradation and regenerations rates, respectively.

Compared with the model proposed by Bredemeier *et al.* [7], this one assumes only one defect, instead of two (slow and fast), the degradation and regeneration are simultaneously occurring, and the pre-factor NDD_{max} is the maximum amount of defects, similar to the constants a and b [7]. Note that the model is proposed to describe the kinetics of the degradation and regeneration, independently of the mechanism used to obtain them (via illuminated or dark annealing), as long as the rates can be considered constant.

4 RESULTS AND DISCUSSION

4.1. TEMPERATURE DEPENDENCE OF THE DEGRADATION AND REGENERATION.

The normalized effective lifetime for the different DA temperatures and the laser baseline process is presented in Figures 1 (a) and 1 (b). As the tokens are neighbors and were processed identically, they have a similar initial effective lifetime (after firing) of 75 ± 2 µs (at $\Delta n = 10^{15}$ cm^{-3}).

Figure 1 (a) indicates that higher temperatures lead to faster degradation and regeneration rates. Note that the wafer processed at 300 °C presents only improvement of the lifetime. The extent of the degradation increases when the temperature decreases. In addition, the effective lifetimes of the wafers treated by DA demonstrate a significant improvement, up to 40% compared to the initial lifetime, in agreement with Chen *et al.* [10]. The degradation extent of the laser degraded baseline [Fig. 1(b)] is similar to DA at 160 °C; however, the rates are much faster and only ~ 7% improvement of the effective lifetime was observed at the completion of the process (smaller than the > 20% improvement of the DA at 160 °C). The improvement in the minority carrier lifetime after DA is not stable under CID conditions [6]; therefore, the involved mechanism associated with the regeneration could be different for both situations. It is also possible that a portion of the observed lifetime improvement is associated with the recovery of pre-existing CID-related defects formed during the firing treatment, as suggested by Chen *et al.* [10].

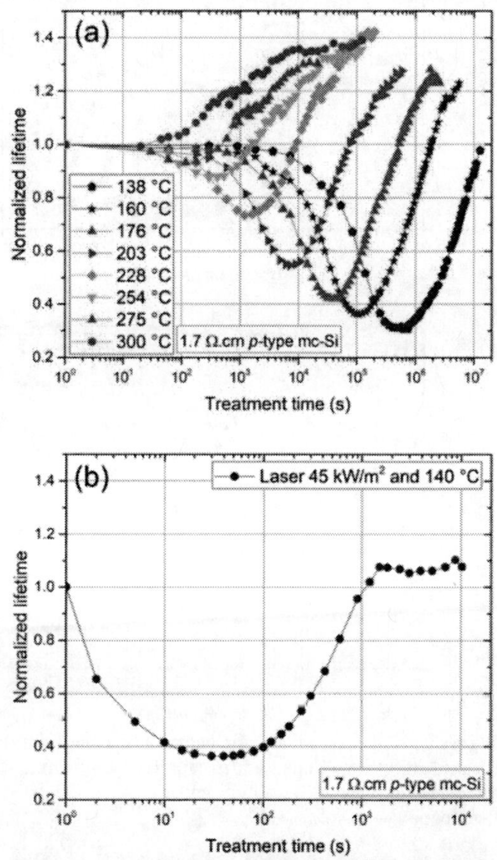

Figure 1: Evolution of the normalized effective lifetime of the wafers: (a) dark annealed at the different temperatures, and (b) the degradation under 45 kW/m^2 laser illumination and 140 °C. The lines are an aid for the reader.

4.2 TIME EVOLUTION AND FITTING MODEL.

Figure 2 shows the evolution of the *NDD* and the fits according to the suggested model of *NDD*(t) [Eq. (2)]. These fits have a minimum determination coefficient (r^2) of 0.96 (for the 275 °C case) indicating a good agreement between the experimental data and the model. Slight differences were found in the degradation part at low temperatures ($T < 200$ °C); Figure 2 (b) indicates that this fit is particularly good for the baseline case ($r^2 = 0.99$). The strong agreement between the model and the measurements supports our assumption that the degradation and regeneration occur simultaneously. This indicates that one defect, instead of two (fast and slow [7]), is sufficient to explain the degradation. Additionally, this model adequately describes degradation in both the dark and under illumination.

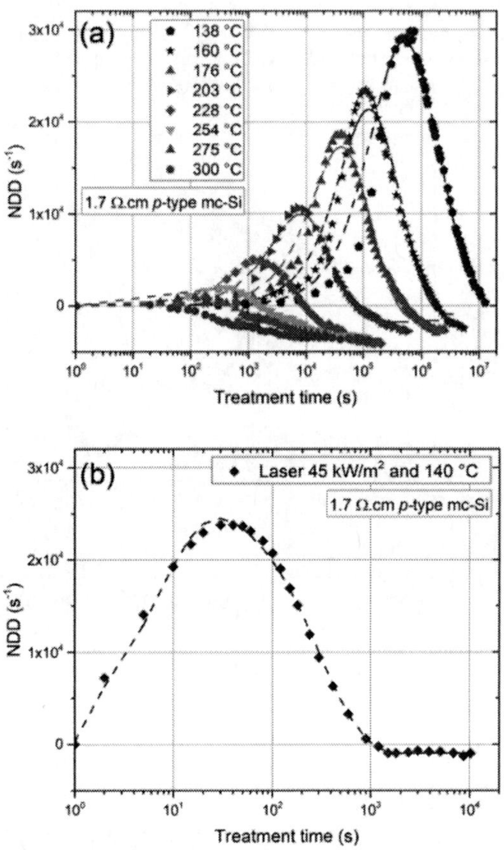

Figure 2: NDD evolution along with the fit using Eq. (2) (dashed lines): (a) Wafers under dark anneal at different temperatures, and (b) wafer under laser-accelerated degradation.

4.3 ARRHENIUS ANALYSIS AND ACTIVATION ENERGIES

Figure 3 (a) presents the degradation and regeneration rates as extracted from the proposed model [Eq. (2)] as a function of the inverse of the absolute temperature. The error bars of the *x*-axis are based on an uncertainty of 2 °C in the measured temperatures, while the error bars in the *y*-axis are associated with the uncertainty of the fit as calculated by the 95% confidence bounds method as implemented in Matlab [11]. Both the degradation and regeneration rates have strong temperature dependence as can be clearly observed in Figure 3 (a). The NDD_{max} is presented in Figure 3 (b). Higher temperatures lead to a smaller NDD_{max}.

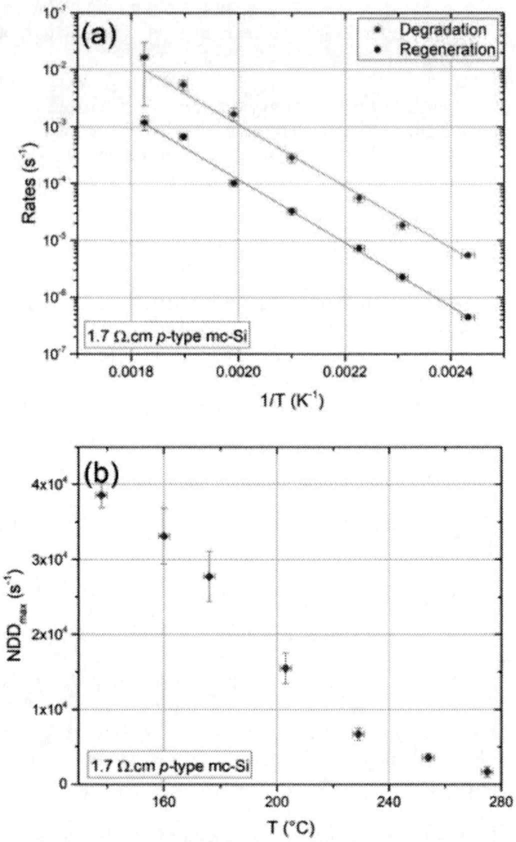

Figure 3: Fit parameters: (a) Degradation and regeneration rates for the different dark anneal temperatures, the lines corresponds to a fit to the Arrhenius analysis, and (b) NDD_{max} versus the process temperatures.

Based on the extracted degradation and regeneration rates, an Arrhenius analysis was performed [Figure 3 (a)]. The Arrhenius function is defined as:

$$r_{deg/reg} = C \times \exp\left[-\frac{E_{deg/reg}}{k_B T}\right], \quad (3)$$

where C is a constant, k_B is the Boltzmann constant, T is the temperature and $E_{deg/reg}$ are the degradation or regeneration activation energies, respectively. We determined activation energies of 1.08 ± 0.05 eV for the degradation process and 1.11 ± 0.04 eV for the regeneration process. These values are slightly higher compared to those reported by Bredemeier *et al.* for the fast and slow defects (E_{fast} = 0.89 ± 0.04 eV and E_{slow} = 0.93 ± 0.06 eV) [7]. A possible explanation of the observed difference is that Ref. [7] considers only the degradation, without taking into account the regeneration part of the process. Additionally, Bredemeier *et al.* [7] used illumination for the degradation. As discussed, illumination appears to significantly accelerate the degradation and regeneration processes (this can be seen by comparing the DA at 138 °C and the baseline laser process) and therefore, it can lead to an underestimation of the activation energies. Additionally, it is possible that the degradation mechanism in the dark is different to that in the light. It is highly likely that at least the recovery mechanisms in the light and dark are different, due to the difference in the resulting stability of samples that have undergone a degradation and recovery cycle in the light versus the dark [6]; in this case their activation energies could be different. The similarity between the activation energies for the processes indicates that both reactions may have a similar cause. This may support the suggestion of hydrogen involvement in the degradation (and regeneration) of *p*-type mc-Si, as suggested recently [5], [10], [12]–[14].

4 CONCLUSIONS

The time evolution of CID in mc-Si is described by a new model which has been developed based on the assumption of only one defect, and that degradation and regeneration occur simultaneously. Particularly, the temperature dependence of the degradation and regeneration of *p*-type mc-Si annealed in the dark has been studied. Differences in the rates and degradation extents were determined. It was found that high temperature leads to faster degradation and regeneration rates, while low temperature leads to higher degradation extent. Based on the extracted rates, the activation energy of the degradation processes was found to be 1.08 ± 0.6 eV and of the regeneration process 1.11 ± 0.4 eV. Since these values are similar, it is likely that both reactions have a similar cause. Furthermore, the lifetime increases up to 40% above the initial level.

ACKNOWLEDGMENT

The authors acknowledge support from the Australian Government through the Australian Renewable Energy Agency (ARENA, Project 2014/RND097 and ARENA Project 2017/RND010). The views expressed herein are not necessarily the views of the Australian Government, and the Australian Government does not accept responsibility for any information or advice contained herein. Z. Hameiri acknowledges the support of the Australian Research Council (ARC) through the Discovery Early Career Researcher Award (DECRA, Project DE150100268). The authors also acknowledge CN part of TNO – Solar Energy "Advanced hydrogenation and silicon materials" project and the help from the Solar Industrial Research Facility (SIRF) team of UNSW for their help in the sample preparation. Special thanks to Professor Stuart Wenham who sadly passed away recently for his many valuable comments, advice and suggestions.

REFERENCES

[1] K. Ramspeck *et al.*, "Light induced degradation of rear passivated mc-Si solar cells," *27th Eur. Photovolt. Sol. Energy Conf. Exhib.*, pp. 861–865, Oct. 2012.

[2] F. Kersten *et al.*, "Degradation of multicrystalline silicon solar cells and modules after illumination at elevated temperature," *Sol. Energy Mater. Sol. Cells*, vol. 142, pp. 83–86, Nov. 2015.

[3] D. N. R. Payne *et al.*, "Acceleration and mitigation of carrier-induced degradation in p-type multi-crystalline silicon," *Phys Status Solidi–Rapid Res Lett*, vol. 10, no. 3, pp. 237–241, Mar. 2016.

[4] W. Kwapil, T. Niewelt, and M. C. Schubert, "Kinetics of carrier-induced degradation at elevated temperature in multicrystalline silicon solar cells," *Sol. Energy Mater. Sol. Cells*, vol. 173, pp. 80–84, Dec. 2017.

[5] C. Chan *et al.*, "Modulation of carrier-induced defect kinetics in multi-crystalline silicon PERC cells through dark annealing," *Phys Status Solidi–Rapid Res Lett*, vol. 1, no. 2, p. 1600028, Feb. 2017.

[6] T. H. Fung, C. E. Chan, B. J. Hallam, D. N. R. Payne, M. D. Abbott, and S. R. Wenham, "Impact of annealing on the formation and mitigation of carrier-induced defects in multi-crystalline silicon," *Energy Procedia*, vol. 124, no. Supplement C, pp. 726–733, Sep. 2017.

[7] D. Bredemeier, D. Walter, and J. Schmidt, "Light-induced lifetime degradation in high-performance multicrystalline silicon: Detailed kinetics of the defect activation," *Sol. Energy Mater. Sol. Cells*, vol. 173, no. Supplement C, pp. 2–5, Dec. 2017.

[8] Z. Hameiri, N. Borojevic, L. Mai, N. Nandakumar, K. Kim, and S. Winderbaum, "Low-absorbing and thermally stable industrial silicon nitride films with very low surface recombination," *IEEE J. Photovolt.*, vol. 7, no. 4, pp. 996–1003, Jul. 2017.

[9] H. Nagel, C. Berge, and A. G. Aberle, "Generalized analysis of quasi-steady-state and quasi-transient measurements of carrier lifetimes in semiconductors," *J. Appl. Phys.*, vol. 86, no. 11, pp. 6218–6221, Dec. 1999.

[10] D. Chen *et al.*, "Evidence of an identical firing-activated carrier-induced defect in monocrystalline and multicrystalline silicon," *Sol. Energy Mater. Sol. Cells*, vol. 172, pp. 293–300, Dec. 2017.

[11] Mathworks, "PDF Documentation for Curve Fitting Toolbox." [Online]. Available: https://au.mathworks.com/help/pdf_doc/curvefit/index.html. [Accessed: 17-Aug-2018].

[12] C. Vargas *et al.*, "Influence of silicon nitride and its hydrogen content on carrier-induced degradation in multicrystalline silicon," *33rd Eur. Photovolt. Sol. Energy Conf. Exhib.*, vol. 33, pp. 561–564, Nov. 2017.

[13] D. Bredemeier, D. C. Walter, and J. Schmidt, "Lifetime degradation in multicrystalline silicon under illumination at elevated temperature: Indications for the involvement of hydrogen," *AIP Conf. Proc.*, vol. 1999, no. 1, p. 130001, Aug. 2018.

[14] T. Niewelt, F. Schindler, W. Kwapil, R. Eberle, J. Schön, and M. C. Schubert, "Understanding the light-induced degradation at elevated temperatures: Similarities between multicrystalline and floatzone p-type silicon," *Prog. Photovolt. Res. Appl.*, vol. 26, no. 8, pp. 533–542, Aug. 2018.

DEFECTS IN EPITAXIALLY GROWN SILICON WAFERS CAUSING LIFETIME PATTERNS

Marion Drießen, Patrick Beu, Friedemann D. Heinz, Tobias Fehrenbach, Elke Gust, Florian Schindler, Stefan Janz
Fraunhofer Institute for Solar Energy Systems (ISE)
Heidenhofstrasse 2, 79110 Freiburg, Germany

ABSTRACT: Epitaxially grown silicon wafers (EpiWafers) have the potential of gaining high electrical quality while production costs can be significantly reduced compared to conventional wafers. High efficiencies are already reported for solar cells made of EpiWafers yet defects originating from the specific production sequence are limiting their quality. Photoluminescence imaging shows line patterns of high recombination activity. A further reduced lifetime is visible in square shaped features often positioned at crossing points of the lines. Those features correspond mainly to stacking faults with polycrystalline inclusions, similar square shaped features beside lines to regular stacking faults. Micro photoluminescence measurements reveal that the lines consist of agglomerations of spots with high recombination activity. In the areas with those spots threading dislocations can be found on defect etched samples in a number exceeding the number of spots. We explain all observations by dislocation mediated impurity aggregation and precipitation.
Keywords: EpiWafer, Defects, kerf-less wafering

1 INTRODUCTION

Epitaxially grown silicon wafers (EpiWafers) are a promising alternative for standard wafers since material and energy consumption are significantly reduced leading to reduced overall costs per watt peak if the quality is comparable to standard wafers. High minority charge carrier lifetimes of up to 7 ms are already reported for EpiWafers [1]. Also efficiencies of up to 23 % for solar cells have been reported [2]. But still the specific production sequence leads to special defects that limit the quality of EpiWafers [3,4,5]. The fabrication contains epitaxial growth of the final wafer on top of a reorganized porous silicon double layer that acts as template for growth and predetermined breaking point for detachment. At nucleation sites on the template stacking faults grow along four {111} planes forming inverted pyramids within the epitaxial layer referred to SF in the following. Hence, apex of the SFs is at the side of the interface and the according base is exposed to the wafer's surface. The SFs can also contain polycrystalline inclusions bordered by the same {111} planes as SFs and therefore are called polySFs.

In [3] lifetime images exhibiting line patterns of reduced lifetime areas are reported. At crossing point of those lines of reduced lifetime often polySFs are positioned.

In the following the correlation of these lines and polySFs is investigated in more detail using photoluminescence measurements with a higher resolution. Afterwards the structure within these lines and the origin of the lifetime reduction are discussed. For this purpose micro photoluminescence measurements with an even higher resolution are compared with the defect structure in the region around polySFs. After looking at local strain around an exemplary polySF with Raman measurement a model is presented which can explain the origin of the lifetime pattern.

2 EXPERIMENTAL PROCEDURE

2.1 EpiWafer fabrication

Highly boron doped substrates with electrochemically etched porous silicon layers were provided by IMS (Insitut für Mikroelektronik Stuttgart). The 6 inch samples were sawn with a dicing saw with two horizontal cuts to a height of 10 cm. The native oxide was removed with a HF dip directly before mounting the samples in the CVD reactor RTCVD160 [6]. Subsequently, they were heated up to the process temperature of 1150 °C with a ramp of 100 K/min. Porous silicon layers were reorganized for 32 min in hydrogen atmosphere. The epitaxial layer with an average thickness of about 120 µm was grown in 75 min using trichlorosilane as precursor (5.5 g/min TCS in 11.9 slm H_2). Phosphine is used as dopant source and the flow is adjusted for a doping concentration of 3.5×10^{15} cm^{-3}. After growth samples were cooled with a ramp of 150 K/min. Edges of the final 5 x 5 cm^2 EpiWafers were defined with a dicing saw and detachment was prosecuted with a mechanical lift off tool developed in-house. Remaining porous silicon was removed with a chemical polish etch (CP33).

After RCA cleaning samples were passivated with Al_2O_3 layers on both sides. For defect investigation Al_2O_3 layers were removed with HF and defects were visualized using an adopted Sopori [7] solution[1].

2.2 Characterization

Average lifetimes were measured using quasi-steady-state photo conductance (QSSPC) measurements. Spatially resolved lifetimes were measured for passivated samples with photoluminescence lifetime measurements over the whole wafer area (integrated PL imaging setup, PLI [8]). On selected small areas µPL measurements [9] were applied for an investigation of the PL emission on the micron scale. Afterwards, Al_2O_3 layers were removed by an HF solution and dislocations as well as non-visible stacking faults were visualized with a Sopori etch solution [4]. Defects were investigated before and after defect etch with an optical microscope. A few defects were additionally investigated by a confocal laser microscope (Sensofar Plu S neox) to obtain structural and height information.

3 RESULTS AND DISCUSSION

To investigate the lines of reduced lifetime in EpiWafers samples with an effective lifetime of 30 to 60 µs (measured by QSSPC) and high defect densities (average density of polySF up to 11 cm^{-2}) were chosen.

[1] Here: 36 parts HF (50 %), 15 parts CH_3COOH (100 %), 2 parts HNO_3 (69 %).

3.1 Origin of lines of reduced lifetime

In Figure 1 a spatially resolved lifetime image of a 5 x 5 cm^2 sample is shown.

In the lifetime image black squares are visible corresponding to SFs mainly with polycrystalline inclusions (polySFs). Additionally, the lifetime image shows vertical and horizontal lines in <110> directions with reduced lifetimes that are not optically visible on the sample. As already discussed in [3], stacking faults are mainly situated on the crossing points of these lines. It seems likely that one is a result of the other. The question is which is cause or consequence.

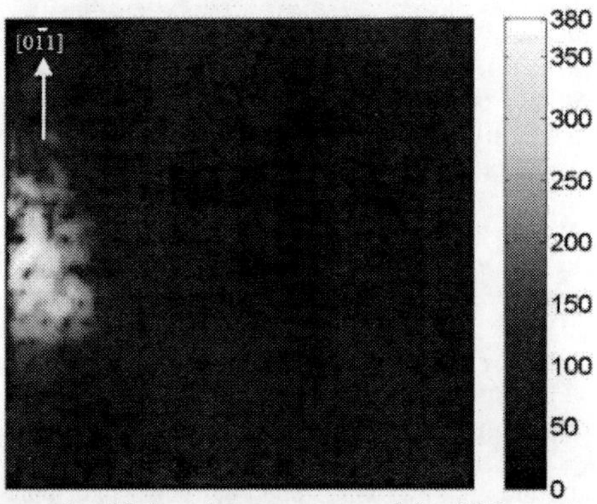

Figure 1: Lifetime image measured by PLI at ≈1 sun equivalent illumination intensity of a 5 x 5 cm^2 sample with a high defect density. Lifetimes are given in µs.

Lines of reduced lifetime could be caused by defects, particularly dislocations, that are induced by micro-scratches at the rim at elevated process temperature and move on {111} glide planes in <110> directions through the samples. This is a well-known glide system for silicon. Another explanation could be that stacking faults grow due to local disturbances of the template (e.g. crystal defects, holes or particles) and dislocations are generated at the boundary between regularly grown (100) silicon and stacking faults. Those defects could as well move through the sample on {111} glide planes leading to the line pattern of reduced lifetime.

In the first case, however, there should be more vertical then horizontal lines as only the horizontal edges of the samples are sawn causing possibly micro-cracks or other defects that in turn cause dislocations. Additionally, many lines do not extend to one of the edges of the samples as should be the case if the rim were the origin of the defects. Therefore, we conclude that defects within the epitaxially grown layer are the origin of the lines of reduced lifetime.

Taking a closer look, it can be seen that mainly polySFs are positioned at the crossing points of the afore-mentioned lines (Figure 2a). SFs without polycrystalline inclusions lead mainly to squares of reduced lifetime with the size of the SF but no surrounding line shaped features can be observed there (see Figure 2b). Thus, the following investigation focusing on the structure of the lines with reduced lifetime is restricted to the surrounding of polySFs.

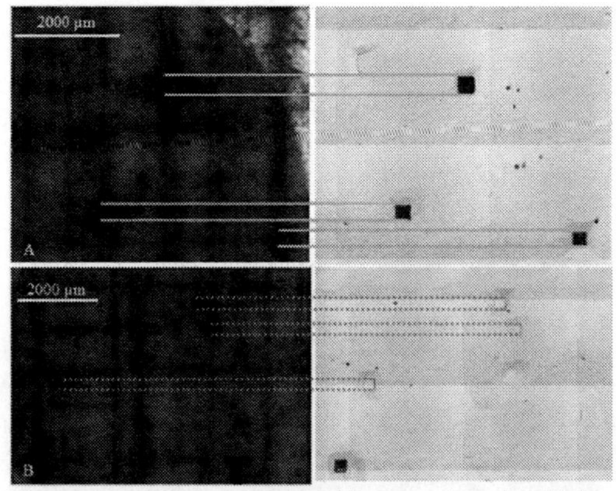

Figure 2: µPL images in arbitrary units with higher lifetime for brighter areas (left) and corresponding microscope images after defect etch (right) of selected areas of the sample shown in Figure 1. The brighter area in the upper right part of the µPL image A is a measurement artefact.

3.2 Origin for lifetime reduction at line patterns

To investigate the lines of reduced lifetime in more detail µPL images with a pixel size of 2 µm were recorded for selected polySFs and their surroundings (see Figure 3).

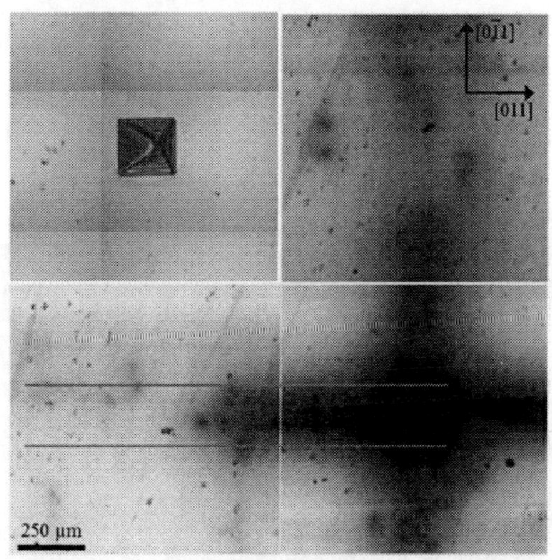

Figure 3: Picture and µPL image of a polySF and according surrounding including lines with reduced PL intensity the EpiWafer. Low PL intensity (dark) corresponds to high recombination activity. On the left side of the polySF the borders of the band in which locally reduced lifetime occurs is depicted exemplarily.

The lines visible with PLI (Figure 1) are not homogeneous in terms of lifetime. They are rather bands bordered by the edges of the polySF in which agglomerations of single dots with severely reduced lifetime are existent. Dots are often lying along two lines in <110> directions within one band. In Figure 4 the comparison of recombination activity and defect structure can be seen. Recombination within the polySF is drastically enhanced due to the polycrystalline structure (see Figure 4b) with a high defect density and most probably a high amount of impurities. The microscope images in the lower row are taken after etching in a Sopori solution to reveal dislocations. Dislocations cumulate in the same areas in which the lifetime is reduced: within bands limited by the edges of the polySF. Since the rear surface is planar, this is better visible there (Figure 4D).

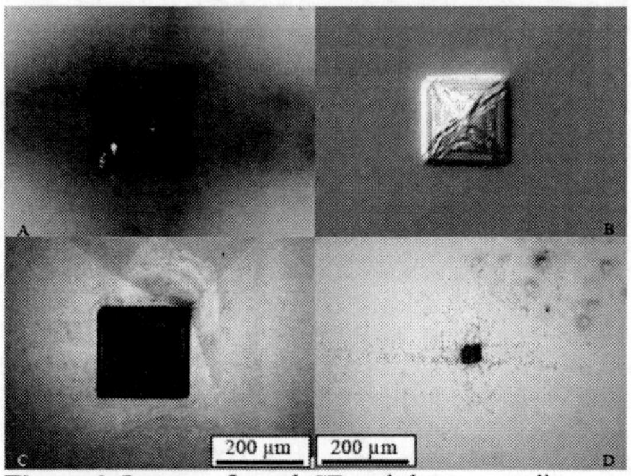

Figure 4: Images of a polySF and the surrounding area: µPL image (A), picture of the polySF taken with a confocal laser microscope (B), optical microscope pictures of front (C) and rear side of the EpiWafer after defect etch.

On front and rear side threading dislocations aligned along <110> directions are visible with increasing density near the polySF. Additionally, the shape of most etch pits is not round but with tails. Therefore, most dislocations don't break vertically through the surface. Those arguments lead to the assumption that many dislocations are lying on {111} glide planes. Such planes are bordered by an edge of the polySF on the wafer's front side and the tip of the polySF on the rear side (see scheme Figure 6).

In [4] impurities are found to be a major source for lifetime reduction in our current EpiWafer material. If an increased recombination due to a homogeneous decoration of the dislocations with impurities would be present, the dislocations should be visible as lines in the µPL image. Additionally, there seem to be many more dislocations in the microscope image than dots in the lifetime image. Therefore, we conclude that impurities agglomerate at some dislocations forming precipitates reasoning the dramatic local recombination.

3.3 Origin of dislocation structure

As already stated many dislocations are situated around polySF, most of them within bands in <110> directions bordered by the edges of the polySF. Many gather on {111} planes coinciding the edges of the polySF and the tip of the polySF. Dislocation densities around SFs without polycrystalline inclusions are much lower.

Impurities can agglomerate at grain boundaries and in areas with high defect densities which are present in the polySFs and at the interfaces between polySFs and regularly grown silicon with a (100) surface. These impurities lead to local stress during growth as well as during cool down. With µRaman measurements the Raman peak shift caused by strain in and around the polySF depicted in Figure 4 were measured (see Figure 5).

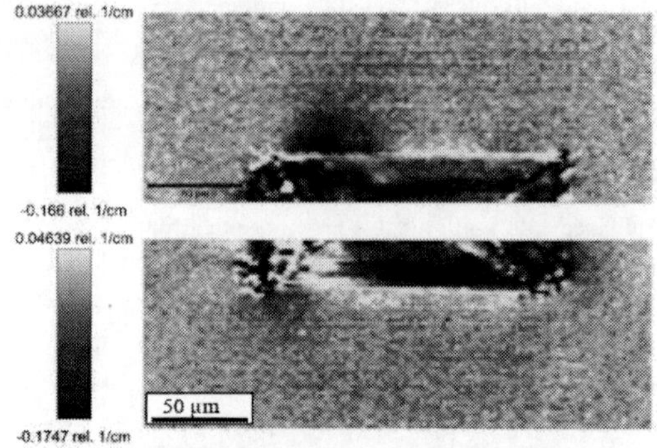

Figure 5: Local deviation from the Raman peak measured on upper and lower edge of the polySF depicted in Figure 4. A positive shift corresponds to tensile strain, a negative shift to compressive strain.

Within the polySF areas with compressive and tensile strain are visible. Within the regularly grown silicon around the polySF only small areas with compressive or tensile stress are visible. These areas are mainly located at the edges. At the corners were most dislocations seem to be generated no increased stress level can be observed. This leads to the assumption that stress is relieved by generation of dislocations. Most probable stress existing during the growth process and cool down until a certain temperature is relieved by generation of dislocations. Below that temperature thermal energy for dislocation generation is too low and strain generated during further cool down by different thermal behaviour of pure and decorated silicon cannot be relieved.

3.4 Model

Based on the observations detailed above the most probable formation mechanism of the lifetime patterns is described in the following.

At the beginning of the epitaxy process local disturbances cause growth of polySFs. Their structure and decoration with impurities lead to strain in surrounding areas during growth and cooldown. This strain is relieved by generation of dislocations that glide on {111} planes through the epitaxial layer. In Figure 6 a scheme of a polySF with {111} planes that pass through the sides of the polySF is shown. Dislocations are mainly positioned in the depicted planes and within the regions bordered by them. Therefore, bands of high dislocation density oriented in <110> directions as visible in Figure 4d can be observed. Impurities are gathered in the strain fields of these dislocations moving along the {111} planes by means of dislocation trails [10]. A part of these impurities form eventually precipitates. The high recombination activity of these precipitates arranged in bands in <110> directions primarily causes the lifetime patterns. A scheme of one poly SF with decorated areas in the (111) planes can be found in Figure 6.

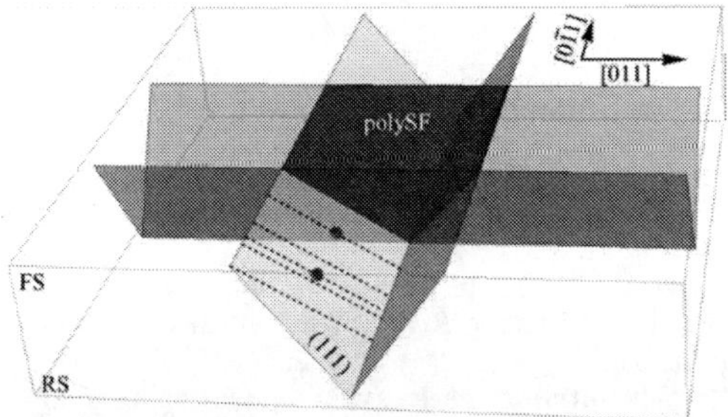

Figure 6: Scheme of a polySF with tip on the rear side (RS) and based bordered by four edges on the front side (FS). The {111} glide panes going through the sides of the polySF are depicted. Their projections on the surfaces (front and rear) are bands in <110> directions limited by the crossing of {111} planes and surfaces. On the (111) plane the assumed main position of dislocations and precipitates is shown exemplarily. Dislocations are indicated as dashed lines.

4 CONCLUSION

In lifetime images from epitaxial wafers deposited in the RTCVD160 reactor lines with reduced lifetime are visible. At the location of some crossing points of these lines the lifetime is further reduced. These locations correspond to stacking faults with polycrystalline inclusions, the most detrimental defect for local lifetimes. Close to the polycrystalline inclusions the lines are rather bands in which single spots with high recombination activity are agglomerated. In the same band high dislocation densities can be observed. We explain the observations by dislocations generated at edges and corners of the polycrystalline inclusions which move on {111} glide plains through the wafer. Impurities are gathered by the stress field of those dislocations. The high amount of dislocations and impurities around polycrystalline inclusions lead to the formation of precipitates. These precipitates with high recombination activity are aligned in <110> directions and cause the line patterns with reduced lifetime.

5 ACKNOLEDGEMENTS

The authors would like to express their gratitude to Philipp Barth, Michaela Winterhalder and Karin Zimmermann at ISE for their support and input in many valuable discussions. This work would not have been possible without the close cooperation with the NexWafe GmbH in particular with Nena Milenkovic, Kai Schillinger and Stefan Reber.
This work was funded by the German Federal Ministry for Economic Affairs and Energy (FKZ 020E-41V7452).

6 REFERENCES

[1] C. Gemmel, J. Hensen, L. David, S. Kajari-Schröder, R. Brendel, Jpn. J. Appl. Phys. 57, 041301 (2018).
[2] E. Kobayashi, Y. Watabe, R. Hao, T. S. Ravi, Prog. Photovolt: Res. Appl. 24, 1295-1303, (2016).
[3] S. Janz, D. Amiri, E Gust, S. Kühnhold-Pospischil, S. Riepe, F. Heinz, M. Drießen, Proceedings of the 33rd EUPVSEC, 343-347 (2017).
[4] P. Beu, F. Schindler, A. Fell, F. D. Heinz, D. Amiri, E. Gust, S. Janz, M. C. Schubert, IEEE Journal of Photovoltaics, under review.
[5] M.M. Kivambe et al., Journal of Crystal Growth 483, 57–64 (2018).
[6] S. Reber, C. Haase, N. Schillinger, S. Bau, A. Hurrle, Proceedings of the 3rd World Conference on Photovoltaic Energy Conversion, 1368–1371 (2003).
[7] B.L.Sopori, J. Electrochem. Soc. 131(3), p. 667-672 (1984).
[8] J.A. Giesecke, M.C. Schubert, B. Michl, F. Schindler, W. Warta, Solar Energy Materials & Solar Cells, vol. 95, 1011-1018 (2011).
[9] F.D. Heinz, L.E. Mundt, W. Warta, M.C. Schubert, Physica status solidi (RRL) 9 (12), 697–700 (2015).
[10]I.E. Bondarenko, V.G. Eremenko, B.Ya. Farber, V.I. Nikitenko, E.B. Yakimov, Phys. Stat. Sol. (b) 68, 53-60 (1981).

LARGE AREA MONOFACIAL SCREEN-PRINTED REAR-EMITTER NPERT CELLS APPROACHING 23% EFFICIENCY

L. Tous, J. Chen*, P. Choulat, S. Singh, A. van der Heide, M. Aleman, I. Kuzma, J. John, F. Duerinckx, J. Szlufcik
imec, Kapeldreef 75, 3001 Leuven, Belgium
*now with Jolywood, Kaiyang Rd #6, Jiangyan Economic Development Zone, Taizhou, China

ABSTRACT: In this paper, we present a simple processing sequence to fabricate large area (244.3 cm^2) n-type Czochralski-grown (Cz) monofacial screen-printed rear-emitter Passivated Emitter, Rear Totally diffused (nPERT) cells that can be implemented in production as an upgrade to conventional p-type Cz Passivated Emitter Rear Cell (p-Cz PERC) manufacturing. We use conventional $POCl_3$ and BBr_3 tube diffusions, selective n^{++} laser doping from phosphorous silicate glass (PSG), atomic layer deposited (ALD) Al_2O_3 emitter passivation, and front and rear screen-printing metallization using commercially available pastes and conventional calendered mesh screens. In a first experiment optimizing rear emitter contact pitch, we achieve average cell energy conversion efficiencies of 22.6±0.1% (internal measurement). In a second experiment implementing fine line double printing (FLDP) and double antireflective coating (DARC), we improve average conversion efficiencies to 22.8±0.1% (internal measurement) with the champion cell measured at 23.03% (externally confirmed). Finally, we identify main loss mechanisms using Free Energy Loss Analysis (FELA) and perform Quokka2D simulations to define a feasible roadmap towards >24% efficiencies.
Keywords: n-type, PERT, Screen-Printing.

1 INTRODUCTION

Photovoltaic (PV) modules made of Passivated Emitter and Rear Cells (PERC) processed on p-type Czochralski-grown (p-Cz) substrates are set to become the dominant technology in late 2018 driven by demand for high-efficiency PV modules both in China ("Top Runner Program") and globally [1-2]. Most advanced p-Cz PERC manufacturers are currently approaching average stable cell conversion efficiencies of 22% [3-5]. Importantly, p-Cz PERC treated to reduce/eliminate light-induced degradation (LID) can still degrade after module lamination [6]. Champion p-Cz PERC with efficiencies approaching 24% have been demonstrated [7-8] on Gallium (Ga) doped p-Cz but typical Ga-doped p-Cz production ingots still suffer from a large gradient in resistivities and bulk lifetime [9] which may delay their adoption in mass-production.

The purpose of this work is to develop a simple processing sequence for n-type Cz Passivated Emitter, Rear Totally diffused (nPERT) cells that can be implemented in production as an efficiency upgrade to conventional p-Cz PERC manufacturing. For this, we focus on a rear-emitter nPERT structure (see Fig. 1a) since: (i) it closely resembles the one of conventional p-Cz PERC with screen-printed silver (Ag) at the front and screen-printed aluminum (Al) at the rear (see Fig. 1b), (ii) it offers potential for efficiencies >24% as we will show in this work, and (iii) it doesn't suffer from LID [10]. A known disadvantage of nPERT cells compared to p-Cz PERC is the added complexity required to form diffused junctions at both the rear-emitter side and the front surface field (FSF) side. However, this issue is being addressed by R&D institutes and companies with various promising approaches currently under development to simplify the junction formation sequence in nPERT cells [10-13]. In this work, we focus on improving rear-emitter nPERT cell efficiencies and consequently make use of conventional $POCl_3$ and BBr_3 tube diffusions.

This paper is structured in three separate sections. First, we explain how know-how gained at imec in recent years helped us choose appropriate recipes for: (i) $POCl_3$ and BBr_3 tube diffusions and subsequent passivation, (ii) n^{++} laser doping from phosphorous silicate glass (PSG), and (iii) fine line screen printing of Ag. Then, we present and discuss results obtained, in this work, for large area (244.3 cm^2) rear-emitter nPERT cells resulting in a predictive roadmap towards 24% cells. Finally, we give a short summary and outlook for this work.

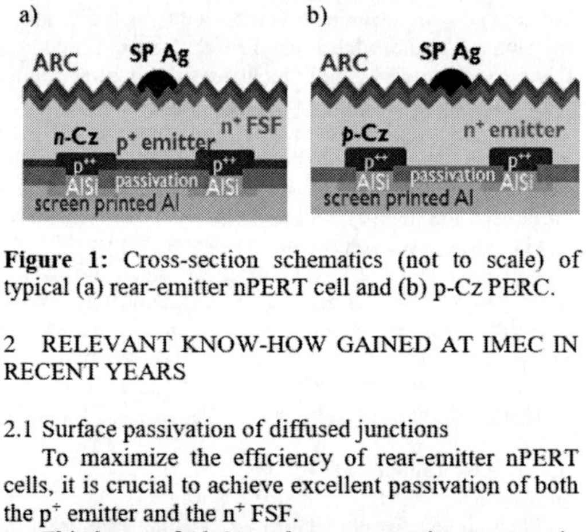

Figure 1: Cross-section schematics (not to scale) of typical (a) rear-emitter nPERT cell and (b) p-Cz PERC.

2 RELEVANT KNOW-HOW GAINED AT IMEC IN RECENT YEARS

2.1 Surface passivation of diffused junctions

To maximize the efficiency of rear-emitter nPERT cells, it is crucial to achieve excellent passivation of both the p^+ emitter and the n^+ FSF.

Thin layers of Al_2O_3 are known to passivate extremely well p^+ emitters, with excellent results demonstrated when using ALD or PECVD Al_2O_3 layers capped by PECVD SiN_x and fired [14-15]. Surface conditioning, ALD Al_2O_3 deposition parameters, PECVD capping, and firing are known to impact passivation results [15-17]. Therefore, all these parameters were optimized at imec using an industrial ALD Al_2O_3 system [16-17]. In addition, p^+ emitter doping profile and in particular boron p^+ peak doping concentration are known to impact emitter saturation current densities J_{0e} [14-15]. Thus, BBr_3 tube diffusion recipes were optimized at imec to minimize both J_{0e} and process time. As can be seen in Figure 2, J_{0e} values obtained experimentally at imec on planar and textured p^+np^+ test samples are in line with state-of-the-art values.

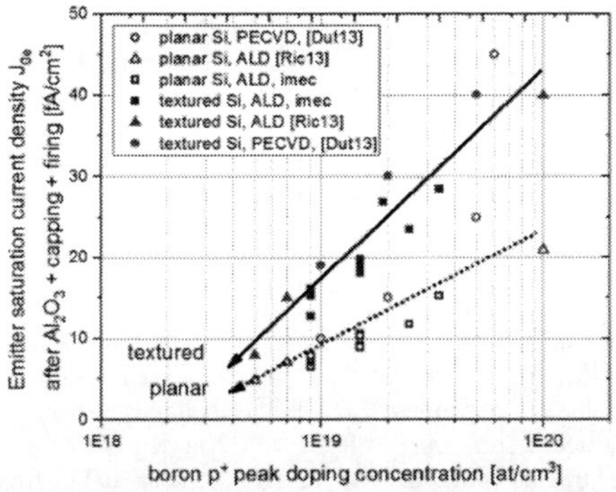

Figure 2: Experimentally obtained J_{0e} values as a function of boron p^+ peak doping concentration on textured (solid symbols) and planar surfaces (open symbols) using either ALD or PECVD Al_2O_3. Arrows are guide to the eye. Also shown, for comparison, are state-of-the-art J_{0e} values reported in the literature. [Dut13] values are taken from Ref. [14] and [Ric13] are taken from Ref. [15].

Similarly, we have optimized $POCl_3$ diffusion recipes and subsequent passivation to achieve saturation current densities on n^+ textured passivated surfaces ($J_{0n+pass}$) below 20 fA/cm² for n^+ sheet resistances >200 Ω/sq. As shown in Fig 3, results obtained at imec after firing with either PECVD SiN_x passivation or a stack of thermal oxide (SiO_2) and PECVD SiN_x are in line with state-of-the-art $J_{0n+pass}$ on textured surfaces.

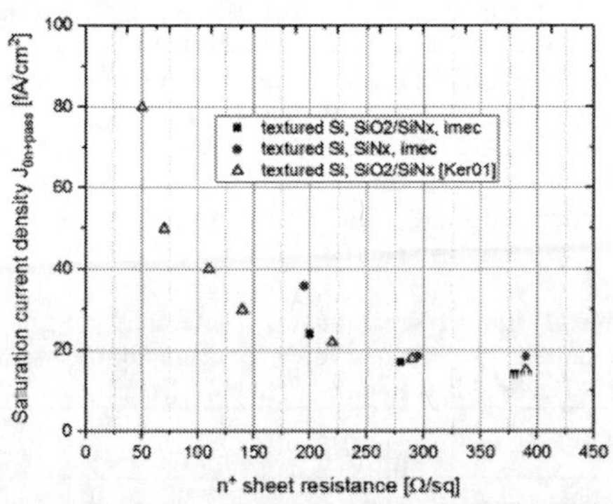

Figure 3: Saturation current density of $POCl_3$ n^+ doped passivated textured surfaces ($J_{0n+pass}$) as a function of n^+ sheet resistance. Passivation consists either of PECVD SiN_x or a stack of thermal oxide (SiO_2) and PECVD SiN_x. Also shown, for comparison, are state-of-the-art $J_{0n+pass}$ values reported in the literature. [Ker01] values are taken from Ref. [18].

2.2 n^{++} laser doping from phosphorous silicate glass

Using a selective FSF with n^{++} doping under the contacts and n^+ doping in between is an effective approach to maximize cell efficiencies of rear-emitter nPERT cells [11, 19]. We already demonstrated large area (239 cm²) rear-emitter nPERT cell efficiencies up to 22.9% (externally confirmed) using selective FSF formation, after ARC deposition, via laser doping from an external dopant source and subsequent self-aligned Ni/Cu/Ag plated metallization [19].

A more industrial approach is to use laser doping from phosphorous silicate glass (PSG) to form n^{++} regions prior to ARC deposition and to subsequently align to these regions screen-printed firing-through Ag metallization. This approach has been mostly employed to form selective emitters in p-type cells [20-21]. At imec, we have optimized n^{++} laser doping from PSG using the shallow (>200 Ω/sq) $POCl_3$ FSF profiles with $J_{0n+pass}$ < 20 fA/cm² presented earlier. In particular, we have optimized laser parameters, evaluated various Ag pastes, and optimized firing profiles. As detailed in Ref. [22], for n^{++} sheet resistances ($R_{sheet,n++}$) in the range of 80-160 Ω/sq (depending on laser doping conditions used) we have obtained excellent J_0 values in the n^{++} passivated regions ($J_{0n++pass}$ ~50-100 fA/cm²) and in the n^{++} metallized regions ($J_{0n++metal}$ ~500-1000 fA/cm²) while maintaining sufficiently low specific contact resistance values ($\rho_{c\text{-}Ag}$~1-3 mΩ.cm²). As a result, we are able to employ a wide range of laser doping conditions to form the selective FSF.

2.3 Fine line screen printing of Ag

Reducing linewidths of Ag fingers is a key enabler to improve cell efficiencies as it reduces both shading losses and contact recombination losses. In recent years, we have achieved drastic reductions in screen-printed Ag linewidths. This was made possible by optimizing both screen parameters (calendered mesh density/diameter, emulsion thickness, screen opening, etc.) and printing parameters (snap-off, speed, pressure, etc.) for fine line double printing (FLDP). As shown in Fig. 4, we have been able to reduce average linewidths down to 40 µm (as used in cell experiments described in section 3) and more recently down to 20 µm. Importantly, all results were obtained using conventional calendered mesh-emulsion screens which suffer much less from screen deformation than more advanced knotless screens [5]. This is a critical aspect in our rear-emitter nPERT cells since the Ag fingers need to be aligned precisely to the laser doped n^{++} regions

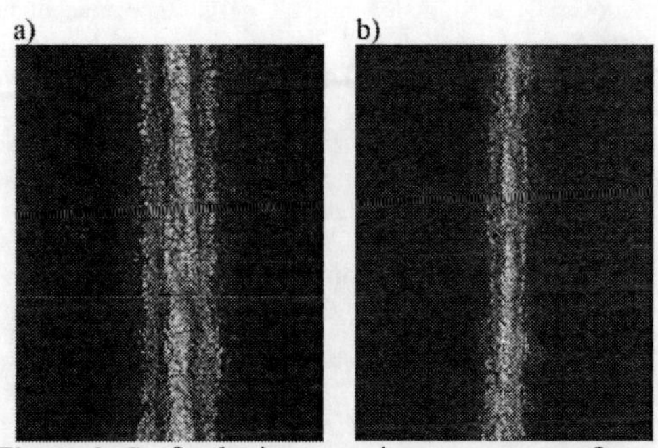

Figure 4: Confocal microscope images on test wafers of (a) 40µm wide and (b) 20 µm wide fine line double printed (FLDP) Ag fingers after firing.

3 RESULTS AND ANALYSIS

3.1 nPERT cells process sequence and I-V results

In this work, we used n-Cz wafers from Longi Solar (as-cut thickness ~190 µm, resistivity ~ 5 Ω.cm). Simple ozone-based cleaning recipes (no sulfuric acid, no hydrogen peroxide) were used throughout the process to reduce consumables costs. BBr_3 tube diffusion was performed after saw-damaged-removal (SDR) and cleaning. Following removal of the borosilicate glass (BSG), we deposited a PECVD layer which acted as a mask during both batch texturing (during which p^+ is removed at the front side) and subsequent $POCl_3$ FSF diffusion. Heavily doped n^{++} (selective FSF) regions were then formed by laser doping using PSG as dopant source.

Following cleaning (PSG and mask removal), front & rear side passivation was performed. In Fig. 5, we schematically grouped these passivation steps since it is possible to perform all of them in one equipment [23]. Since this equipment is not available at imec, we first did cleaning + front ARC deposition followed by another cleaning step + ALD Al_2O_3 (thickness $\leq$ 5 nm) + PECVD capping on the emitter side. Then, we used a laser to form dot contact openings on p^+ emitter side. Front & rear side screen-printing metallization (front: Ag fingers; rear: blanket Al) was performed using commercially available pastes and conventional calendered mesh-emulsion screens. Finally, cells were fired in a fast-firing infrared heated furnace and I-V parameters were measured using a Wacom, continuous-type, solar simulator (class AAA) in combination with a Gridtouch™ contacting setup.

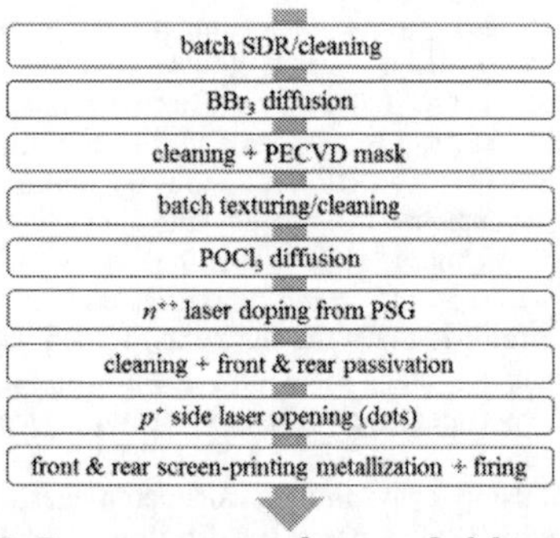

Figure 5: Process sequence for monofacial rear-emitter nPERT cells in this work.

Optimizing the rear side contact fraction is critical in monofacial rear-junction nPERT cells since inadequate local Al-p+ formation such as voids can lead to drastic efficiency losses [24-25]. Therefore, in a first experiment using single Ag printing at the front side (~40 µm linewidths), we fabricated four batches of cells (G1 to G4) with increasingly lower rear contact fraction. As shown in Fig. 6, the best group (G3) resulted in average (of 8 cells) efficiencies of 22.6±0.1%. For this group, we measured short-circuit current densities j_{sc} = 40.6±0.1 mA/cm^2, open-circuit voltages V_{oc} = 689.7±0.8 mV, and fill factors FF = 80.9±0.1 %.

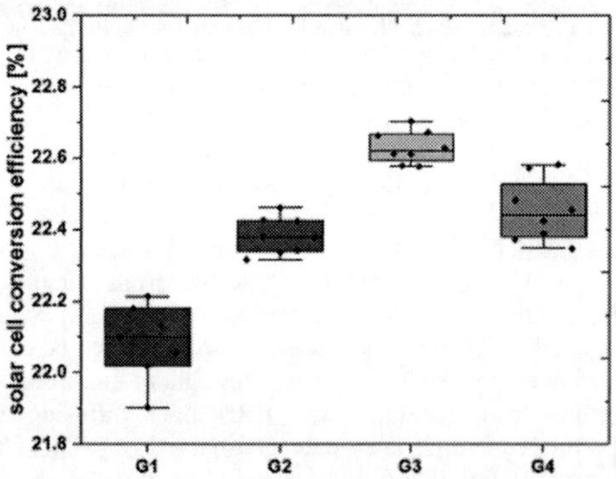

Figure 6: Experimentally obtained nPERT solar cell conversion efficiency for 4 batches (G1 to G4) with increasingly lower rear contact fraction. The boxes give the 25%, median, and 75% distribution thresholds.

In a second experiment, we implemented FLDP of Ag (~40 µm linewidths) and fabricated two batches of 8 cells with either single or double ARC at the front side. From the light I-V results given in Table I, it can be seen that cells with a double ARC feature gains in j_{sc} and V_{oc} of ~0.2 mA/cm^2 and ~2 mV respectively, and a loss in FF of ~ 0.2%. Based on spectral response and reflectance measurements (not shown), the j_{sc} gain can be explained by reduced reflection losses in the 300-500 nm range. Thick ARC layers have been shown to be more difficult to fire-through which reduces contact recombination losses and hence improves V_{oc} [26]. Based on measurements performed on test wafers with varying double ARC thicknesses (not shown), we can confirm that increasing double ARC thickness reduces $J_{0n^{++}metal}$ but it also increases $\rho_{c\text{-}Ag}$ which explains our I-V results.

Table I: Light I-V parameters (internal measurement) for two batches of rear-emitter nPERT cells (estimated cell area: 244.3 cm^2) with fine line double printing of Ag (FLDP Ag) and either a single or double anti-reflective coating (ARC).

ARC	average	j_{sc} [mA/cm^2]	V_{oc} [mV]	FF [%]	eta [%]
single	8 cells	40.5 ±0.0	685.1 ±1.1	81.6 ±0.1	22.6 ±0.0
double	8 cells	40.7 ±0.0	687.3 ±0.8	81.4 ±0.2	22.8 ±0.1

Two champion cells (A and B) with double ARC were sent to Fraunhofer ISE Callab for external calibration (spectral response + I-V measurement). As shown in Table II, the best cell (A) resulted in an externally confirmed efficiency of 23.03% with j_{sc}=41.1 mA/cm^2, V_{oc}=683.3 mV, and FF=82.0%. We attribute the significant difference in j_{sc} measured internally and externally to the reference cell used at imec being from an older batch of cells (spectral mismatch).

Table II: Light I-V parameters (*externally confirmed at Fraunhofer ISE Callab) of two champion rear-emitter nPERT cells. Measured cell area=244.1 cm^2.

ARC	cell	j_{sc} [mA/cm^2]	V_{oc} [mV]	FF [%]	eta [%]
double	A	41.1	683.3	82.0	23.03*
double	B	41.1	687.4	81.3	22.95*

3.2 Benchmark versus published nPERT cell results

To the best of our knowledge, the best champion cell obtained in this work has a higher efficiency than other published rear-emitter nPERT cells results on large area (>239 cm^2) Cz wafers (see Table III). We see further room to improve cell efficiencies as our cell V_{oc} remains significantly lower than the ones obtained on cells with plated Ni/Cu/Ag contacts at the front and PVD Al at the rear [13, 19]. The results given in Table III illustrate the importance of using a selective FSF design to reduce front side recombination losses in rear-emitter nPERT cells since best results published with an homogeneous FSF design are limited by either V_{oc} [10] or j_{sc} [12, 25]. This is particularly visible when employing homogenously n-doped polysilicon (n-poly) at the front side which very effectively suppresses contact recombination losses but results in large parasitic absorption losses [12].

Table III: Selected rear-emitter nPERT cell results (*externally confirmed) on large area (>239 cm²) Cz wafers. FSF type is selective (sel.) or homogeneous (hom.). Equipment used for emitter formation and metallization (metal.) front/rear are also indicated for reference. All cells reported in this table are monofacial except for the one from [12] which is bifacial.

Ref.	eta [%]	j_{sc} [mA/cm^2]	V_{oc} [mV]	FF [%]	FSF type	emitter formation	metal. front / rear
this work	23.03*	41.1	683	82.0	sel.	BBr_3	0BB SP-Ag / SP-Al
[19]	22.9*	40.9	705	79.4	sel.	BBr_3	0BB NiCu / PVD Al
[13]	22.5*	40.2	700	79.9	sel.	epitaxy wafer & p$^+$	3BB NiCu / PVD Al
[12]	22.3*	38.5	714	81.1	hom.	LPCVD n-poly	0BB SP-Ag /ITO+SP-Ag
[11]	22.0*	39.7	685	80.9	sel.	APCVD BSG	3BB NiAg / SP-Al
[10]	21.9	40.5	669	80.9	hom.	BBr_3	5BB SP-Ag / SP-Al
[25]	21.78*	39.5	686	80.4	hom.	BBr_3	5BB SP-Ag / SP-Al

For comparison, we also report in Table IV selected large area front-emitter nPERT cell results. Interestingly, applying homogeneous n-poly only at the rear side is an effective way to improve the V_{oc} without suffering from very large parasitic absorption losses [28-29]. However, all cells in Table IV are mainly limited by emitter contact recombination losses [27-30]. This is not the case in our rear-emitter nPERT structure as shown next.

Table IV: Selected front-emitter nPERT cell results (*externally confirmed) on large area (>239 cm²) Cz wafers. Back surface field (BSF) type is selective (sel.) or homogeneous (hom.). Equipment used for BSF formation and metallization (metal.) front/rear are also indicated. All cells reported in this table are bifacial.

Ref.	eta [%]	j_{sc} [mA/cm^2]	V_{oc} [mV]	FF [%]	BSF Type	BSF formation	metal. front / rear
[27]	22.8	40.5	694	81.1	sel.	$POCl_3$ + laser doping	0BB NiAg / NiAg
[28]	22.6*	40.8	698	79.3	hom.	PECVD n-poly	0BB SP-Ag / SP-Ag
[29]	22.4*	40.0	688	81.4	hom.	n.p. n-poly	SP-Ag / SP-Ag
[30]	21.8*	40.3	673	80.3	hom.	Implant	5BB SP-Ag / SP-Ag

3.3 rear-emitter nPERT cell efficiency roadmap

To define a realistic rear-emitter nPERT cell efficiency roadmap we first identify dominating losses in our champion rear-emitter nPERT cell and then conduct numerical simulations by considering incremental improvements based on data collected on test wafers.

Detailed characterization was performed on finished cells (measurements of spectral response, reflectance, series resistance, shunt resistance, J_{02}, contact resistance, finger resistance, etc.) and test wafers (sheet resistances, J_0 contributions, etc.) processed simultaneously to cells. We used these results as inputs in PVlighthouse's Quokka2D numerical simulations [31] to obtain a simulated champion cell with near-identical eta, j_{sc}, V_{oc}, and FF. Finally, we make use of the free energy loss analysis (FELA) provided in Quokka2D simulations results to identify the main electrical and recombination losses (note that optical losses are excluded from this FELA analysis). Based on the results given in Fig. 7, we identify the top 4 loss mechanisms as being: (i) recombination losses at the passivated FSF side, (ii) bulk resistive losses, (iii) bulk recombination, and (iv) recombination losses at the passivated emitter side.

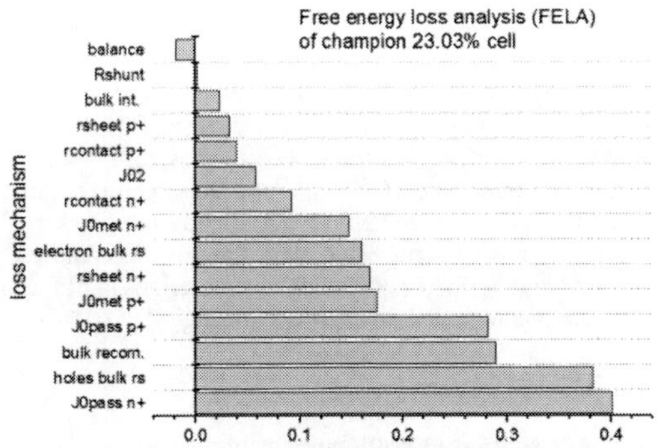

Figure 7: Free energy loss analysis (FELA) based on Quokka2D simulations of the champion 23.03% cell.

Starting from the Quokka2D 23.01% fit of the 23.03% champion cell (confirmed by Callab) as baseline, we performed Quokka 2D simulations (see Fig. 8) with the following list of incremental improvements:

- Optimized selective FSF with: 100 µm width (baseline=150 µm), $J_{0n+pass}$ = 15 fA/cm², $J_{0n++metal}$ = 500 fA/cm². Feasible as shown in Fig. 3 and 4a.
- Average Ag linewidth = 30 µm (baseline = 40 µm).
- Optimized p$^+$ emitter J_{0e} = 10 fA/cm² (Feasible as shown in Fig. 2).
- Optimized Al paste with $J_{0ep+metal}$= 500 fA/cm² and smaller contact radius (see Ref. [24]).
- Average Ag linewidth = 20 µm (see Fig. 4b).
- 5 Ω.cm effective bulk lifetime = 10 ms (See Ref. [9])
- no J_{02} (baseline = 1 nA/cm²)

Since we have already demonstrated most of these improvements on test wafers, we believe that efficiencies >23.5% are feasible in the short-term and that the efficiency potential for rear-emitter nPERT is >24%.

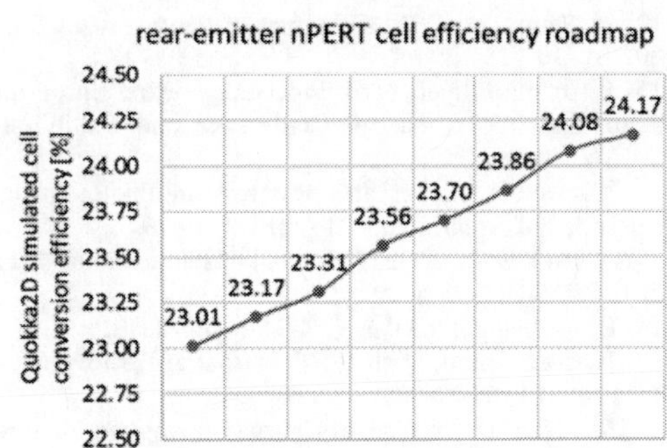

Figure 8: Rear-emitter nPERT cell efficiency roadmap.

4 SUMMARY AND OUTLOOK

In this work, we fabricated large area (244.3 cm^2) monofacial screen-printed rear-emitter n-PERT cells with energy conversion efficiencies approaching 23%. In a first experiment optimizing rear emitter contact pitch, we obtained average cell energy conversion efficiencies of 22.6±0.1% (internal measurement). In a second experiment implementing fine line double printing (FLDP) and double antireflective coating (DARC), we improved average conversion efficiencies to 22.8±0.1% (internal measurement) with the champion cell measured at 23.03% (externally confirmed). Finally, we identified main loss mechanisms using Free Energy Loss Analysis (FELA) and performed Quokka2D simulations to define a feasible roadmap towards >24% efficiencies.

5 REFERENCES

[1] M. Osborne, LONGI tripling monocrystalline wafer capacity to 45GW, published on Jan 24th 2018 on pv-tech.org.

[2] C. Lee, Panel predictions 2018, PV magazine, issue 11-2017

[3] F. Fertig, et al. Energy Procedia 124 (2017): 338-345.

[4] I. Höger, et al. AIP Conference Proceedings 1999, 110003 (2018); doi: 10.1063/1.5049312

[5] M. Galiazzo, Presentation at the 7th Workshop on Metallization & Interconnection for c-Si Solar Cells (2017).

[6] K. Lee, et al. Proceedings 31st European Photovoltaic Solar Energy Conference and Exhibition (2015): 1835–1837

[7] http://taiyangnews.info/technology/23-6-perc-mono-cell-world-record-from-longi/, accessed on 6th September 2018.

[8] https://jinkosolar.com/press_detail_1640.html, accessed on 6th September 2018.

[9] YC Wang, Presentation at the nPV workshop (2018).

[10] S. Bordihn et al. Energy Procedia Volume 124, (2017): 643-648.

[11] Y.S. Choi, et al. Proceedings 33rd European Photovoltaic Solar Energy Conference and Exhibition (2017): 586-588.

[12] R. Peibst, et al. IEEE Journal of Photovoltaics 8(3) (2018): 719-725.

[13] I. Kuzma-Filipek, et al. Proceedings 32nd European Photovoltaic Solar Energy Conference and Exhibition (2016): 560-564.

[14] S. Duttagupta, et al. Progress in Photovoltaics Research and Applications 21 (2013):760–764.

[15] A. Richter, et al. IEEE Journal of Photovoltaics 3(1) (2013): 236-245.

[16] B. Vermang. PhD thesis, KU Leuven, 2012.

[17] E. Cornagliotti, et al. IEEE Journal of Photovoltaics 5(5) (2015):1366-72.

[18] MJ. Kerr and A. Cuevas. Journal of Applied Physics 89, 3821 (2001).

[19] M. Aleman, et al. Proceedings of IEEE Photovoltaic Specialists Conference. (2017).

[20] G. Hahn, Proceedings of the 25th European Photovoltaic Solar Energy Conference and Exhibition. (2010): 1091-1096.

[21] M. Kim, et al Solar Energy Materials & Solar Cells 132 (2015): 215–220.

[22] S. Singh, et al. this conference.

[23] T. Grosse, et al. Proceedings of the 26th Photovoltaic Solar Energy Conference and Exhibition. (2016).

[24] V. Shanmugam, et al. Solar Energy Materials and Solar Cells. 186 (2018): 124-130.

[25] ZW. Peng, et al. AIP Conference Proceedings 1999, 100002 (2018); doi: 10.1063/1.5049309

[26] H. Chu, et al. Presentation at the 7th Workshop on Metallization & Interconnection for c-Si Solar Cells (2017).

[27] R. Russell, et al. Proceedings 33rd European Photovoltaic Solar Energy Conference and Exhibition (2017): 212 - 217.

[28] S. Duttagupta and N. Nandakumar. Presentation at SNEC Shanghai (2018).

[29] Y. Liu, Presentation at the nPV workshop (2018).

[30] F. Kiefer, et al. Solar Energy Materials & Solar Cells 157 (2016): 326–330.

[31] A. Fell. IEEE Transactions on Electron Devices 60.2 (2013): 733-738.

LID-FREE PERC+ SOLAR CELLS WITH STABLE EFFICIENCIES UP TO 22.1%

Bianca Lim[1], Agnes Merkle[1], Robby Peibst[1,2], Thorsten Dullweber[1], Yichun Wang[3], Rui Zhou[3]
[1]Institute for Solar Energy Research Hamelin (ISFH), Am Ohrberg 1, D-31860 Emmerthal, Germany
[2]Institute of Electronic Materials and Devices, Leibniz Universität Hannover, Schneiderberg 32, D-30167 Hannover, Germany
[3]LONGi Clean Energy Co., Ltd., No. 388, Middle Aerospace Rd., Chang'an District, Xi'an City, China
Corresponding author: phone: +49-5151-999-403, fax: +49-5151-999-400, email: b.lim@isfh.de

ABSTRACT: We evaluate the potential of advanced B-doped Cz-Si with extremely low oxygen concentration ([O_i]) of 2.6 ppma as well as Ga-doped Cz-Si to avoid light-induced degradation (LID) in state-of-the-art bifacial PERC+ solar cells. We compare these materials to current industry standard B-doped Cz-Si with [O_i] between 12 and 16 ppma.. We measure the solar cell efficiency and the lifetime of samples processed in parallel to the solar cells in three important states: as-processed, after illumination at room temperature (degraded), and after illumination at elevated temperature (regenerated). In the as-processed state, the low [O_i] as well as the Ga-doped solar cells yield 0.2%$_{abs}$ to 0.3%$_{abs}$ higher efficiencies than the industrial B-doped Cz-Si. In addition, their efficiency is stable under illumination at room temperature. Furthermore, the measured bulk lifetimes are used as input parameters in a device simulation. Subsequently, we compare the simulated solar cell efficiencies to the measured efficiencies. For the industry standard B-doped Cz-Si, the simulation predicts 0.7%$_{abs}$ loss due to LID, which fits to the experimental result. After regeneration, the device simulation predicts an increase by 0.4%$_{abs}$ compared to the as-processed state, whereas the measured PERC+ efficiency improves only to the as-processed level. We discuss possible reasons for this discrepancy.
Keywords: PERC, LID, Czochralski, Silicon, Simulation, Stability,

1 INTRODUCTION

Energy conversion efficiencies of industrial p-type PERC solar cells up to 22.8% have been reported by LONGi and JinkoSolar [1,2] and bifacial PERC+ solar cells with 22.1% efficiency by ISFH [3]. Such efficiencies are only possible with high-quality silicon wafers. However, current industry standard boron-doped Czochralski-grown silicon (Cz-Si) suffers from boron-oxygen-related light-induced degradation (LID) of the bulk carrier lifetime and, hence, the solar cell efficiency.

This problem has been known for a long time [4]. As early as 1997, it was shown that the degradation increases with increasing boron concentration [5] as well as with increasing oxygen concentration [6]. Consequently, early proposals to counteract BO-related LID already included reducing the interstitial oxygen concentration or replacing boron with gallium [5,7]. In 2006, it was also found that the BO-related defect causing LID can be permanently deactivated by illumination at elevated temperature [8], thus recovering the losses caused by LID. While first studies showed process times in the range of hours, it is now possible to permanently deactivate the BO-related defect – also referred to as regeneration of the lifetime or the solar cell efficiency – in a matter of seconds using industrial-type belt furnaces [9]. However, when using these industrial tools, the loss caused by LID cannot be fully recovered in most cases [9].

It is therefore still preferable to avoid LID in the first place. In the past, studies of low [O_i] or Ga-doped Cz-Si either used R&D-type Cz-Si or lab-type solar cell processes or both [7], thus leaving doubt about the feasibility of producing LID-free high efficiency solar cells on an industrial scale. In recent years, industrial-type PERC solar cells were fabricated on low oxygen B-doped magnetic Cz-Si [10,11] as well as Ga-doped Cz-Si [12] and LID in these cells was indeed very small. However, magnetic Cz-Si does not seem to be industrially viable and Ref. 12 did not state absolute efficiency values.

In this work, we combine mass-production capable Cz-Si (either B-doped with low interstitial oxygen concentration or doped with Ga) with an industrial PERC+ solar cell process. We demonstrate that these high-efficiency Si solar cells are completely LID-free. These results are compared to the performance of current industry standard B-doped Cz-Si. In addition to the PERC+ solar cells, we also fabricate dedicated samples to directly measure the bulk lifetime. By this, we assess whether the potential of the LID-free wafers is fully realized in the PERC+ solar cells.

2 EXPERIMENTAL DETAILS

2.1 Sample preparation

We study five different Czochralski-grown (Cz) silicon materials. One is R&D-type Ga-doped Cz-Si, which we use as an excellent, inherently LID-free reference. Two of the materials are industrial B-doped Cz-Si from LONGi with "standard" interstitial oxygen concentrations of 12 and 16 ppma. The other two materials from LONGi are assessed as industrial-scale LID-free options. The first is B-doped Cz-Si with an extremely low interstitial oxygen concentration of 2.6 ppma and the second is Ga-doped Cz-Si from an industrial-type puller. Table I summarizes the resistivity and interstitial oxygen concentration (measured by Fourier-transform infrared spectroscopy, FTIR, using the ASTM F121-83 standard) of the five materials.

Table I: Dopant species, resistivity ρ, interstitial oxygen concentration [O_i], and source of the five different Cz-Si materials investigated in this work.

#	Dopant species	Resistivity [Ω cm]	[O_i] [ppma]	Source
1	B	1.7	~16	LONGi
2	B	1.1	~12	LONGi
3	B	2.1	2.6	LONGi, R&D
4	Ga	1.0	~16	LONGi
5	Ga	0.8	n.a.	R&D

The extremely low oxygen concentration of 2.6 ppma (material 3) was achieved by using an advanced Cz pulling technology developed by LONGi which reduces the oxygen dissolution into the silicon melt during crystallization. This advanced pulling technology is developed as a potential candidate for future mass production for high-quality and low-cost Cz-Si wafers. The initial wafer thickness is about 180 µm for all wafers. The LONGi wafers are 156.75 × 156.75 mm^2 (M2), while the R&D Ga-doped Cz-Si (material 5) has a size of 156 × 156 mm^2.

The PERC+ solar cell process starts with saw-damage removal and wafer cleaning. We then apply a rear protection layer to act as etching and diffusion barrier during the following front surface texturing and phosphorus diffusion steps. Afterwards, the PSG and protection layer are removed and we apply a stack of aluminum oxide (AlO_x) and plasma-enhanced chemical vapor deposited (PECVD) silicon nitride (SiN_y) to passivate the rear side. PECVD-SiN_y is also deposited on the front to passivate the P-doped emitter and act as antireflection coating. To enable contact formation on the rear side, we locally remove the AlO_x/SiN_y stack using a picosecond laser with 532 nm wavelength before screen-printing Al fingers on the rear and Ag fingers on the front side. Most of the solar cells are processed with a 5 busbar grid design. A few PERC+ cells receive a busbarless front by printing only the Ag fingers, not the Ag busbars on the front side. At the end, the metal pastes are co-fired in a belt-firing furnace with a belt speed of 5.6 m/min. A schematic of the final solar cell structure is shown in Fig. 1.

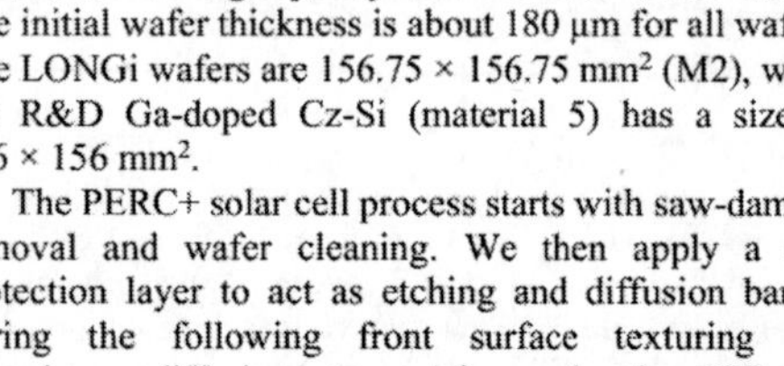

Figure 1: Schematic of a bifacial PERC+ solar cell.

In order to assess the combined quality of the bulk, the emitter, and of the passivated rear surface, we also process non-metallized solar cell precursors (so-called implied-V_{oc} samples). These are identical to the solar cells except for the laser contact opening and the screen-printing steps.

In addition, we process symmetrical test structures to measure bulk lifetime. These samples are processed in parallel to the solar cells: we remove the saw damage and clean the wafers followed by the same $POCl_3$ diffusion used for the solar cells. The resulting n^+ regions on both surfaces are subsequently etched off before we clean the wafers and deposit an AlO_x/SiN_y stack on the front and the rear for optimal surface passivation. The lifetime samples are then fired in a belt-firing furnace. Note that we have two groups, which are fired at two different belt speeds: 5.6 m/min, which is the speed at which the solar cells are fired, and 6.8 m/min, in order to obtain optimum lifetimes after regeneration [13].

After firing, both the solar cells and the lifetime samples are in the "as-processed" state. We then measure the *IV* curves of the 5-busbars solar cells with a LOANA from *pvtools* and the busbarless solar cells with a GridTouch from *Pasan*. The effective carrier lifetimes (for implied-V_{oc} samples) and the bulk carrier lifetimes are determined with a WCT-120 Lifetime Tester from *Sinton Instruments*, both in quasi-steady-state and photoconductance decay mode.

With regard to the bulk resistivity of up to 2.1 Ω cm, we then illuminate the samples for 24 h at about 0.1 suns and room temperature to achieve complete degradation of the carrier lifetime due to the formation of BO-related defects [14]. The regeneration treatment consists of annealing on a hotplate set to 185 °C and simultaneously illuminating with a halogen lamp at a light intensity of about 1 sun for 15 min. Please note that we verified that no further increase of the lifetime and the solar cell efficiency is observed when annealing under illumination for longer times.

2.2 Device simulations

For the device simulation we use the conductive boundary model [15] as implemented by Quokka2 [16]. The input parameters, which are based on extensive characterization of our baseline PERC+ solar cell process, are summarized in Table II. The injection-dependent lifetimes of the B-doped Cz-Si wafers, as measured on the lifetime test structures, can be described by a Shockley-Read-Hall (SRH) defect at midgap with a factor of 10 between τ_{n0} and τ_{p0}. The table lists the defect lifetimes for the as-processed state of the 1.7 Ω cm B-doped sample with 16 ppma [O_i] (material 1). Note that we used the lifetime measured on the designated structure that was fired at the same belt speed as the solar cells, i.e. 5.6 m/min.

The front contact shadowing is calculated from the optical finger width and the layout of the front grid. The saturation current density of the emitter J_{0e} is determined on symmetrical test wafers using the Kane and Swanson method. We do transmission line model (TLM) measurements to determine the specific contact resistances at the front and rear. The saturation current densities J_0 at the contacts are based on lifetime measurements done on test samples with varying metallization fraction, both for the emitter and the base contact. The depth-dependent generation profile is calculated by parameterizing the measured reflectance of a PERC+ solar cell according to Brendel *et al.* [17].

In addition to the PERC+ solar cell simulations, we also perform simulations of implied-V_{oc} samples. For this, we set the front metal recombination to be equal to that of the passivated emitter and the rear metal recombination equal to that of the passivated rear side. All other parameters are the same as for the solar cell simulation.

Table II: Input parameters for the 3D device simulations of PERC+ solar cells using the lifetime data of the B-doped Cz-Si with 16 ppma [O_i] in the as-processed state.

Parameter	Value
Wafer resistivity	1.7 Ω cm
Wafer thickness	170 µm
τ_{n0} for midgap SRH center	292 µs
τ_{p0} for midgap SRH center	2920 µs
Front contact shadowing	4.5%
Emitter sheet resistance R_{sh}	95 Ω/sq.
J_{0e} of SiN_y-passivated emitter	100 fA/cm^2
J_{0c} below Ag finger contacts	1000 fA/cm^2
Specific contact resistance, front	1.5 mΩ cm^2
J_0 at the passivated rear side	12 fA/cm^2
J_0 at the Al rear contact	350 fA/cm^2
Specific contact resistance, rear	3 mΩ cm^2

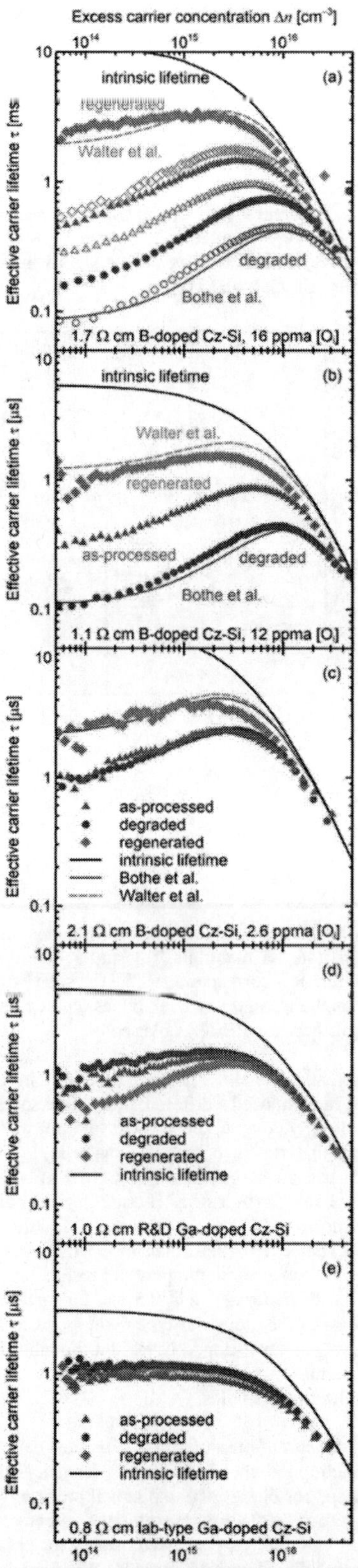

Figure 2: Effective lifetimes of the five different Cz-Si materials investigated in this work.

3 EXPERIMENTAL RESULTS

3.1 Bulk carrier lifetimes of different Cz-Si materials

The measured effective lifetimes of the five different materials are plotted in Fig. 2. Figures 2(a) and 2(b) show the significant changes that happen in B-doped Cz-Si with an industrially typical interstitial oxygen concentration. Figure 2(c) depicts the effective lifetimes of B-doped Cz-Si with an extremely low interstitial oxygen concentration of 2.6 ppma. The measured lifetimes of the industrial and R&D-type Ga-doped materials are shown in Figs. 2(d) and (e), respectively.

The blue triangles in the graphs correspond to the measured lifetime after the fast firing step (as-processed). The red circles give the lifetime after illumination at room temperature for 24 hours, i.e. in the case of the B-doped samples after light-induced degradation. The green diamonds refer to the lifetime after applying the regeneration treatment (annealing under illumination at 185 °C for 15 min). In addition, all graphs contain the respective intrinsic lifetime according to the model of Richter *et al.* [18] (solid black line).

For the B-doped 1.7 Ω cm sample [Fig. 2(a)], we plot the lifetimes measured after firing at different belt speeds. The open symbols correspond to the effective lifetimes measured after firing at a belt speed of 5.6 m/min and the filled symbols correspond to a belt speed of 6.8 m/min. For the slower belt speed the as-processed lifetime (open blue triangles) at an excess carrier concentration of $\Delta n = 10^{15}$ cm^{-3} is around 650 µs. After 24 h of illumination at room temperature, the lifetime drops to about 200 µs (degraded state, open red circles) and after application of the regeneration treatment, the lifetime increases to about 1.4 ms (regenerated state, open green diamonds). Using a faster belt speed during the fast firing process notably increases the effective lifetime in all three states. As-processed (filled blue triangles), we measure 1.1 ms at $\Delta n = 10^{15}$ cm^{-3}. In the degraded state (filled red circles) the lifetime is around 390 µs and regeneration results in an effective lifetime of 3.2 ms at $\Delta n = 10^{15}$ cm^{-3} (filled green diamonds).

Figures 2(a-c) also show the lifetime in the fully degraded state according to the parameterization of Bothe *et al.* (solid red line) [19]. As the samples underwent a P-diffusion as well as a fast-firing step, we assume an enhancement factor of 5.5 [20]. The dashed green lines in Figs. 2(a-c) correspond to the parameterization of the regenerated state proposed by Walter *et al.* [13].

For the 1.7 Ω cm sample the measured effective lifetime in the degraded state after firing at 5.6 m/min agrees well with the parameterization of Bothe *et al.* The lifetime of the sample that was fired at 6.8 m/min, however, lies well above the parameterization. The measured lifetime in the regenerated state fits well to the parameterization of Walter *et al* when firing at 6.8 m/min. For the 5.6 m/min firing step, however, the measured lifetime is notably lower. As mentioned above, such a dependence of the regenerated lifetime on the belt speed was already shown in Ref. 19.

It should be noted, however, that we do not observe such a strong impact of the belt speed during fast firing for all B-doped Cz-Si materials. The low [O_i] Cz-Si, for example, yields similar lifetimes after firing at 5.6 m/min and 6.8 m/min, respectively.

For the 1.1 Ω cm Cz-Si with an interstitial oxygen concentration of 12 ppma, we only have lifetime data after firing at 6.8 m/min. These lifetimes are comparable to those of the 1.7 Ω cm Cz-Si after firing at 5.6 m/min: about 600 µs in the as-processed state at $\Delta n = 10^{15}$ cm^{-3}, 210 µs after light-induced degradation, and around 1.4 ms after regeneration. Note that the lifetime in the degraded state again agrees with the parameterization of Bothe *et al.* Usually, one expects stronger degradation for Cz-Si with a higher boron content. However, in the present case, the reduced interstitial oxygen concentration offsets this increase. The measured lifetime after regeneration is only slightly lower than the parameterization of Walter *et al.*, proving the high quality of the material.

Figure 2(c) depicts the effective lifetimes of the B-doped material with a significantly lower interstitial oxygen concentration of 2.6 ppma. Here the changes are notably smaller. In particular, we do not observe any light-induced degradation of the bulk carrier lifetime between the as-processed and the degraded state. This shows that LID can be effectively suppressed when sufficiently reducing the oxygen concentration. Applying the regeneration treatment improves the lifetime from around 2 ms to about 3.8 ms at an excess carrier concentration $\Delta n = 10^{15}$ cm^{-3}.

The comparison with the parameterization of Bothe *et al.* (solid red line) reveals a significant deviation. This could be due to the fact that the present sample lies outside the range covered in Ref. 18. Bothe *et al.* used samples with an interstitial oxygen concentration within the range of 3 to 9×10^{17} cm^{-3}, whereas the present Cz-Si has an interstitial oxygen concentration of only 2.6 ppma, i.e. 1.3×10^{17} cm^{-3}. Another possible explanation could be the uncertainty of the measured [O_i] or rather slight deviations across the ingot. For example, a fit of the measured data to the parameterization of Bothe *et al.* by varying the interstitial oxygen concentration yielded good agreement at [O_i] = 2.3×10^{17} cm^{-3}.

Looking at the measured lifetimes of the Ga-doped Cz-Si [Fig. 2(d) and (e)], we see only small changes. The industrial material from LONGi [Fig. 2(d)] actually shows a slightly higher lifetime after 24 h of illumination at room temperature (red circles) than in the as-processed state (blue triangles). After the regeneration treatment, however, we see a slight decrease of the lifetime, especially at excess carrier densities below 5×10^{15} cm^{-3} (green diamonds). We are still investigating whether this decrease is a bulk effect or whether it is related to the surface passivation quality.

The R&D-type Ga-doped Cz-Si can be classified as completely stable at roughly 1 ms under the degradation and regeneration treatment (see Fig. 2(e)). In addition, we see virtually no injection-dependence for this material.

The fact that all lifetime curves are below the intrinsic lifetime published by Richter *et al.* (solid black line) can be attributed to the surface recombination. Assuming a surface recombination velocity of just 4 cm/s already closes the gap between measured lifetime and intrinsic model. This is a reasonable value for the used AlO_x/SiN_y stack.

3.2 Solar cell results

Figure 3 shows the PERC+ solar cell efficiencies for materials 1, 3 and 5 directly after processing. The whiskers indicate the minimum and maximum value, the horizontal line indicates the median, and the cross the mean. Note that there are two groups of solar cells: The first group (black boxes) has 5 busbars on the front side while the second group (red boxes) has a busbar-less front grid design. Such cells can be interconnected in a module using multiple thin wires, which reduces shadowing losses as well resistive losses in the front grid.

The B-doped cells with 16 ppma [O_i] and 5 busbars achieve an efficiency around 21.2%, whereas the Ga-doped and the B-doped Cz-Si with 2.6 ppma [O_i] yield efficiencies of 21.4%, i.e. 0.2%$_{abs}$ higher than the standard material. The solar cells without busbars on the front side achieve an efficiency of 21.7% and 22.0% for the standard B-doped Cz-Si and the Ga-doped Cz-Si, respectively. The highest efficiency obtained is 22.1% for a busbar-less Ga-doped PERC+ solar cell, as independently measured at Fraunhofer ISE CalLab [3].

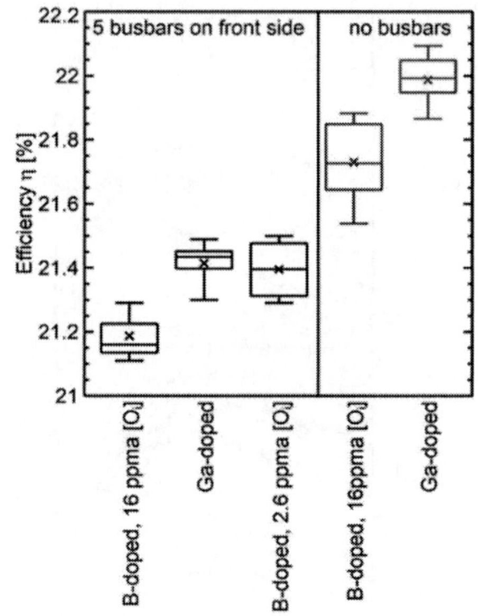

Figure 3: Measured solar cell efficiencies in the as-processed state for materials 1, 3, and 5. The three black boxes on the left correspond to PERC+ solar cells with 5 busbars on the front, the two red boxes on the right present data for busbar-less PERC+ solar cells.

Note that materials 2 and 4 are processed in a different solar cell run that used a different processing sequence and is thus not fully comparable to the results shown in Fig. 3. However, this run also includes material 1, which we therefore use as a reference point. In this second run, materials 1 and 2 (industrial B-doped) yield comparable efficiencies up to 21.0%, while material 4 (industrial-type Ga-doped) achieves efficiencies up to 21.2%.

Illumination at room temperature results in a decrease of 0.7%$_{abs}$ for material 1 and 0.5%$_{abs}$ for material 2. The efficiencies of the industrial-type Ga-doped Cz-Si as well as the low oxygen B-doped Cz-Si, on the other hand, are stable under illumination at room temperature, as expected from the lifetime samples.

3.3 Comparison of measured and simulated data

As mentioned above, the input parameters in Tab. II describe our baseline PERC+ solar cell process. We verify the simulation setup by inserting the effective lifetime measured in the as-processed state for material 1 (1.7 Ω cm B-doped with 16 ppma [O_i]) and comparing the result to measured *IV* parameters in the as-processed state. The results are summarized in Tab. III.

Efficiency, open-circuit voltage V_{oc}, and short-circuit current density J_{sc} match very well, while the measured fill factor FF is $0.45\%_{abs}$ higher than the simulated one. This is in part due to a higher pseudo fill factor pFF ($0.3\%_{abs}$) as well as a lower series resistance (~0.05 Ω cm²) but still it is a good match.

Table III: Simulated and measured *IV* parameters for the baseline PERC+ solar cell process and the 1.7 Ω cm B-doped Cz-Si with 16 ppma [O_i].

Parameter	Simulation	Measurement
Efficiency η [%]	21.49	21.53
V_{oc} [mV]	669.8	667.3
J_{sc} [mA/cm²]	39.54	39.53
FF [%]	81.16	81.62
R_s [Ω cm²]	0.468	0.420
pFF [%]	83.37	83.68

After verifying the simulation setup, we insert the lifetimes measured after process, after illumination at room temperature and after applying the regeneration treatment. Thus, we can compare the efficiency changes we would expect from the observed lifetime changes to the actual changes we observed on solar cells. We do this for materials 1, 3, and 5, i.e. the industry standard 1.7 Ω cm B-doped Cz-Si, the B-doped Cz-Si with extremely low interstitial oxygen concentration as well as the lab-type Ga-doped Cz-Si.

The results are shown in Fig. 4. We plot the solar cell efficiencies of the two B-doped samples in the as-processed state relative to the efficiency of the R&D-type Ga-doped Cz-Si (purple triangles). The change due to illumination at room temperature and the change after applying the regeneration treatment are plotted relative to the respective as-processed efficiencies. Values corresponding to measured solar cell efficiencies are plotted in filled symbols, the results from the simulation are depicted by open symbols.

As mentioned above, the actual Ga-doped Cz-Si solar cells (purple triangles) yield $0.2\%_{abs}$ higher efficiencies than the solar cells made on industrial B-doped Cz-Si (green diamonds). This difference is also predicted by the simulation when entering the respective effective lifetimes measured in the as-processed state. Based on the measured lifetime, the simulation predicts slightly higher efficiencies for the low [O_i] material (orange circles) than for the Ga-doped Cz-Si (purple triangles). Looking at the measured solar cell efficiencies, we find similar mean values and a spread of the data that is bigger than the expected difference (see Fig. 3).

Illumination at room temperature notably decreases the lifetime in the 16 ppma [O_i] B-doped Cz-Si. Inserting this degraded lifetime into the simulation yields a $0.7\%_{abs}$ loss compared to the as-processed state (open green diamonds). This is in good agreement with the degradation observed on actual PERC+ solar cells (filled green diamond). After regeneration, the measured PERC+ efficiency of the industrial B-doped Cz-Si improves to a similar level as "before LID", whereas the device simulation with the regenerated lifetime predicts an increase by $0.4\%_{abs}$ compared to before LID.

The measured PERC+ efficiency of the 2.6 ppma [O_i] B-doped material and the Ga-doped PERC+ solar cells, on the other hand, are stable both during the degradation and the regeneration treatment (within the measurement uncertainty). This is in accordance with the measured bulk lifetimes, which do not decrease after illumination at room temperature [see Figs. 2(c) and (e)]. While we do observe an increase of bulk lifetime for the 2.6 ppma B-doped Cz-Si after applying the regeneration treatment, this increase only translates to an efficiency gain of 0.02% in our device simulation, as the other recombination channels are much more dominant.

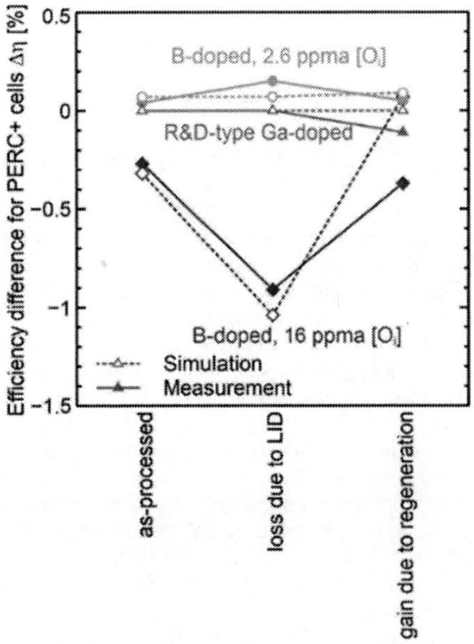

Figure 4: Measured (filled symbols, solid lines) and simulated (open symbols, dashed lines) difference in PERC+ cell efficiency Δη after illumination at room temperature and after applying the regeneration treatment.

4 DISCUSSION

The comparison between measured solar cell efficiencies and efficiencies simulated on the basis of measured bulk lifetimes revealed a notable discrepancy for the regenerated state of the industrial B-doped Cz-Si. The difference of the measured bulk lifetimes in the degraded and the regenerated state should translate to an efficiency gain of $1\%_{abs}$ after regeneration. However, on the actual solar cells we only observe an increase by $0.7\%_{abs}$.

This suggests that, when it comes to the regenerated state, there is a difference in the bulk lifetime in the designated test wafers and the PERC+ solar cells. As the designated test wafers were processed in parallel to the solar cells – as far as possible – this raises the question as to where this difference might stem from.

Looking at the detailed processing sequence used in this work, we note three major differences: (1) the solar cells have a rear protection layer during the P-diffusion step, (2) during the fast-firing process, the solar cells have an emitter on the front side, and (3) the solar cells are fired with metal pastes on both surfaces. In contrast, the n^+ emitter first forms on both sides of the lifetime samples. Subsequently, the emitter is removed and both surfaces are passivated with an AlOx/SiNy stack. And obviously we do not apply metal pastes to the lifetime samples.

Regarding the impact of a rear protection layer during P-diffusion, we have experimental results from a different study that makes it seem unlikely. In said experiment, the solar cells as well as the lifetime samples were diffused on both surfaces and the n^+ emitter on the rear was then removed by a rear polishing step. In that experiment, too, the lifetimes measured on designated test samples yielded a notably higher difference between degraded and regenerated state in the PERC+ simulation than what we observe on solar cells made on the same Cz-Si.

Regarding the presence of the n^+ emitter during the fast firing step, we measure the effective lifetimes of non-metallized implied-V_{oc} solar cell precursors. Unfortunately, this was not done as part of the original experiment and we did not measure the lifetime in the fully degraded state. Still, even though the as-processed state is not as clearly defined as the fully degraded or the regenerated state, we consistently measure higher lifetimes after regeneration than after processing on designated test wafers while solar cell efficiencies are always on the same level in both states.

Figure 5 shows the effective lifetime of such an iV_{oc} precursor. The blue triangles correspond to the as-processed state and the green diamonds to the regenerated state. The lines give the simulated effective lifetimes, i.e. the result of simulating an implied-V_{oc} structure, when assuming the bulk lifetime measured in the as-processed (solid blue line) and in the regenerated state (dashed green line), respectively.

In the as-processed state, the measured effective lifetime (blue triangles) matches the simulated effective lifetime (dashed blue line) very well. The measured effective lifetime on the non-metallized precursor after regeneration (green diamonds), however, is notably lower than the simulated effective lifetime (solid green line). It thus seems as if the presence of the n^+ emitter during the fast firing step negatively affects the bulk lifetime. We are currently investigating the possibility of so-called repoisoning, i.e. the re-emission of gettered impurities from the n^+ emitter into the bulk.

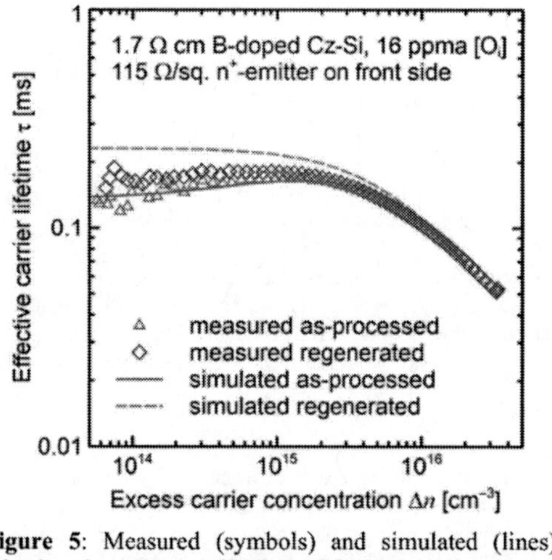

Figure 5: Measured (symbols) and simulated (lines) injection-dependent effective lifetimes of industry standard 1.7 Ω cm B-doped Cz-Si with a SiN_y-passivated 115 Ω/sq. n^+ emitter on the front side. Blue triangles and solid blue line correspond to the as-processed state, green diamonds and dashed green line to the regenerated state.

The absence of the metal pastes during the fast firing process (difference 3) might also add to the difference between simulation and measurement. As already mentioned above, different cooling rates during furnace firing may impact the carrier lifetime. The presence of the metal pastes and in particular the formation of the Al-Si eutectic on the rear side might change the cooling behavior of the silicon wafer. This, too, is currently under investigation.

5 CONCLUSIONS

Experimentally, we demonstrate for the first time LID-free PERC+ solar cells by using either boron-doped Cz-Si wafers with very low oxygen content or Ga-doped wafers. These two materials are currently developed by LONGi for industrial application and the combination with our industrial PERC+ solar cell process thus proves the feasibility of fabricating LID-free high-efficiency p-type Si solar cells.

As reference wafers, we also use 0.8 Ω cm R&D-type Ga-doped Cz-Si from a different source, which has a stable lifetime of 1 ms in all three states. Combining these Ga-doped wafers and a busbar-less front grid design, we demonstrate PERC+ solar cells with stable energy conversion efficiencies up to 22.1%.

Using experimentally determined input parameters for our basline PERC+ process and the software Quokka2, we also perform 3D device simulations. In particular, we simulate the solar efficiency in three distinctive states: as-processed, fully degraded, and after regeneration. For this, we use effective lifetimes measured on designated test samples processed in parallel to the solar cells. We find good agreement regarding the difference performance of the different materials in the as-processed state. In addition, the simulation yields the same extent of LID as observed on the PERC+ solar cells.

However, when comparing the predicted gain due to application of the regeneration treatment to the actually observed gain, we notice a deviation of $0.4\%_{abs}$. That is, the effective lifetime of the designated test samples seems to be higher than the bulk lifetime of the finished PERC+ solar cells. By further analyzing non-metallized solar cell precursors, we identify the presence of the n^+ emitter during the fast firing step as a possible source for this deviation.

6 REFERENCES

[1] Press release LONGi Solar, October 2017.
[2] Press release JinkoSolar, October 2017.
[3] T. Dullweber, H. Schulte-Huxel, S. Blankemeyer, H. Hannebauer, S. Schimanke, U. Baumann, R. Witteck, R. Peibst, M. Köntges, R. Brendel, and Y. Yao, Japanese Journal of Applied Physics **57** (8S3), 08RA01 (2018).
[4] H. Fischer and W. Pschunder, Proc. 10th IEEE Photovoltaic Specialists Conference, Palo Alto, CA, (IEEE New York 1973), 404.
[5] J. Schmidt, A. Aberle, and R. Hezel, Proc. 26th IEEE Photovoltaic Spec. Conf., Anaheim, CA, USA, 1997, 13–18.
[6] S. W. Glunz, S. Rein, W. Warta, J. Knobloch, and W. Wettling, Proc. of the 2nd World Conference on Photovoltaic Energy Conversion, Vienna, Austria (European Commission, Ispra, Italy, 1998), 1343.
[7] S. W. Glunz, S. Rein, J. Knobloch, W. Wettling, and T. Abe, Prog. Photovolt: Res. Appl., **7**, 463-469 (1999).
[8] A. Herguth, G. Schubert, M. Kaes, and G. Hahn, Proc. 21st European Photovoltaic Solar Energy Conference and Exhibition, Dresden, Germany, (Munich: WIP 2006), 530.
[9] T. Pernau, O. Romer, W. Scheiffele, A. Reichart, W.

Jooss, Proc. 31st European Photovoltaic Solar Energy Conference and Exhibition, Hamburg, Germany, 2015, 918-920.
[10] T. Dullweber, C. Kranz, U. Baumann, R. Hesse, D. Walter, J. Schmidt, P. Altermatt, and R. Brendel, Proc. 39th IEEE Photovoltaic Specialists Conference (PVSC), Tampa, FL, USA, 2013, 3074-3078.
[11] P. Saint-Cast, I. Reis, J. Greulich, S. Lohmüller, E. Lohmüller, H. Höffler, J. Haunschild, R. Preu, Proc. 32nd European Photovoltaic Solar Energy Conference and Exhibition, Munich, Germany, 2016, 1038-1043.
[12] Y. Zhang, Q. Li, R. Zhou, H. Xu, S. Turchetti, S. Keohane, Proc. 33rd European Photovoltaic Solar Energy Conference and Exhibition, Amsterdam, The Netherlands, 2017, 318-320.
[13] D. Walter, B. Lim, and J. Schmidt, Prog. Photovolt: Res. Appl. **24** (7), 920-928 (2016).
[14] D. W. Palmer, K. Bothe, and J. Schmidt, Phys. Rev. B **76**, 035210 (2007).
[15] R. Brendel, Prog. Photovolt: Res. Appl. **20** (1), 31-43 (2012).
[16] A. Fell, IEEE Transactions on Electron Devices **60** (2), 733–738 (2012).
[17] R. Brendel, T. Dullweber, R. Peibst, C. Kranz, A. Merkle, and D. Walter, Prog. Photovolt: Res. Appl. **24**, 1475–1486 (2016).
[18] A. Richter, S. W. Glunz, F. Werner, J. Schmidt, and A. Cuevas, Phys. Rev. B **86**, 165202 (2012).
[19] K. Bothe, R. Sinton, and J. Schmidt, Prog. Photovolt: Res. Appl. **13** (4), 287–296 (2005).
[20] D. C. Walter, B. Lim, K. Bothe, R. Falster, V. V. Voronkov, and J. Schmidt, Solar Energy Materials and Solar Cells **131**, 51-57 (2014).

Change of the bulk carrier lifetime of high quality silicon wafers during PERC solar cell processing

Matthias Müller*[1,2], Franziska Wolny[1,3], Gerd Fischer[1,4], Andreas Krause[1,5] Phedon Palinginis[1] and Holger Neuhaus[1,6]

[1] SolarWorld Innovations GmbH, 09599 Freiberg, Germany
[2] Technische Universität Bergakademie Freiberg, 09599 Freiberg, Germany
[3] Fraunhofer IPMS, 01109 Dresden, Germany
[4] Zittau/Goerlitz University of Applied Sciences, 02763 Zittau, Germany
[5] SGS Institut Fresenius GmbH, CRS Tech, 01109 Dresden, Germany
[6] Fraunhofer Institute for Solar Energy Systems, 79110 Freiburg, Germany

Abstract — **The change of bulk excess charge carrier lifetime of silicon wafers along the processing stages of PERC solar cells is investigated. As PERC solar cells efficiency improve currently very fast the material quality and its change during processing is becoming more and more important. We perform this work on Ga-doped Czochralski grown wafers which are not affected by boron-oxygen related light induced degradation. The bulk carrier lifetime reduces from an initial high level during the PERC process monitored after phosphorous diffusion and after the application of dielectric layers. However, after screen-printing and the following co-firing step the bulk carrier lifetime improves. A balanced temperature management during cell processing of temperature ramps, peak temperatures and times is necessary for high efficiency PERC cells which benefit more and more of high bulk carrier lifetime.**

I. INTRODUCTION

The highest PERC solar cell efficiency increased quickly in the last years and is expected to increase further [1]. High solar cell efficiency demand for high quality material and material stability during solar cell fabrication. In this work, we investigate the change of bulk excess charge carrier lifetime of Gallium-doped silicon Cz wafers along several processing stages of PERC solar cells.

II. EXPERIMENTAL DETAILS

For this work, we used Gallium-doped Czochralski (Cz) silicon wafers with a resistivity in a group of 1 Ωcm and 2 Ωcm. The interstitial oxygen concentration is about $(10\pm2)\times10^{17}$ at/cm³ which are typical values for wafers used for solar cell processing. Due to the Gallium doping, the resulting solar cells are not affected by boron-oxygen related light induced degradation (LID). This facilitates our investigation. The wafers are subjected to an industrial PERC solar cell process [2] as shown in Fig. 1.

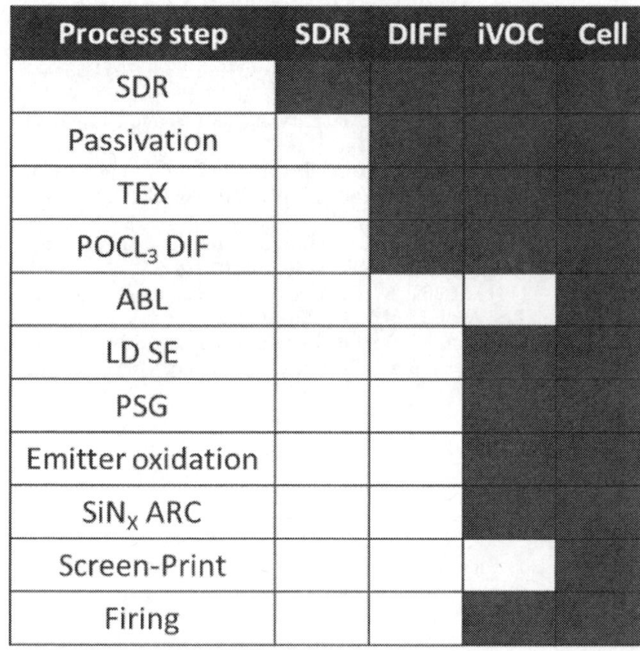

Fig. 1. Fabrication sequence of the PERC solar cell and its test structures.

Samples were drawn after the following processing steps:

- Saw-damage removal (SDR): as-cut wafer with approximately 12µm saw-damage removal from both sides,
- Diffusion (DIFF): after standard n^+ emitter diffusion,
- Antireflective coating (implied V_{OC}, short iVoc): after silicon nitride deposition on the front of the wafer, without laser ablation at the rear,
- Fast firing (Cell): after metallization and co-firing (the completed PERC process).

Schematics of the different sample types are shown in Fig.2.

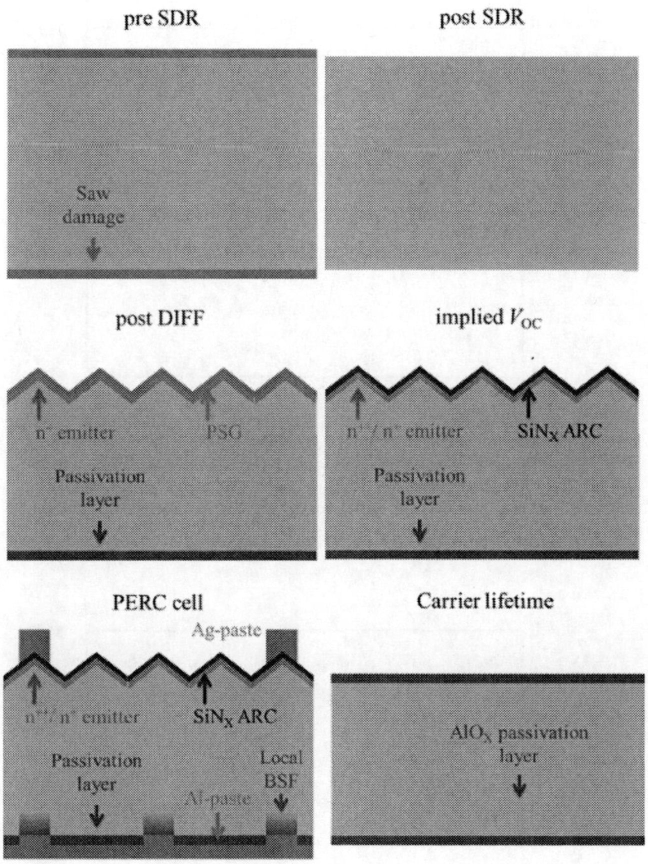

Fig. 2. Investigated PERC solar cell and its test structures.

The DIFF, iVoc and Cell samples are then subjected to different etching treatments in order to completely remove any metallization, passivation layers, anti-reflective coating or emitters/back surface-fields. The stripped wafers were then RCA cleaned and passivated with aluminum oxide Al_2O_3 using atomic layer deposition (see "Carrier lifetime" structure in Fig. 2.). The used passivation reaches a surface recombination velocity below 5 cm/s, i.e. the measured effective carrier lifetime is almost the bulk carrier lifetime. The excess carrier lifetime is measured using a Sinton WCT 120 lifetime tester. Here, we report the excess carrier lifetime at an excess carrier injection density of 5×10^{14} cm^{-3}. Using the calculations proposed in [3], we determine the material saturation current density $j_{0,mat}$ as an alternative, resistivity-independent parameter to describe the material quality.

III. Results and discussion

The fabricated PERC solar cell using the silicon, Ga-doped Cz wafers reached a high efficiency and open-circuit voltage level (compare Fig. 3 and 4) which allows relevant investigation on material effects.

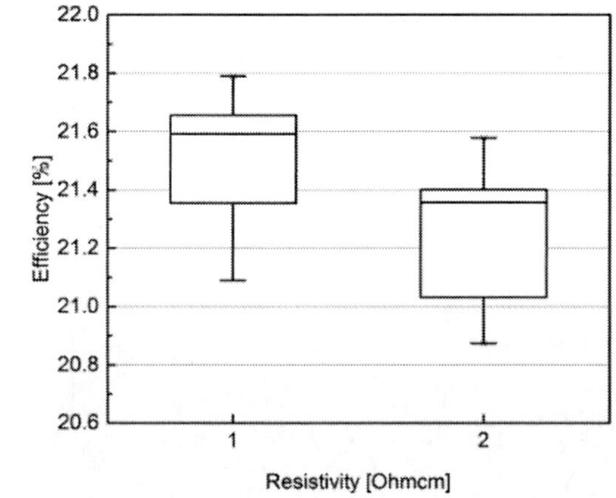

Fig. 3. PERC solar cell efficiency distributions for the 1 and 2 Ωcm resistivity groups.

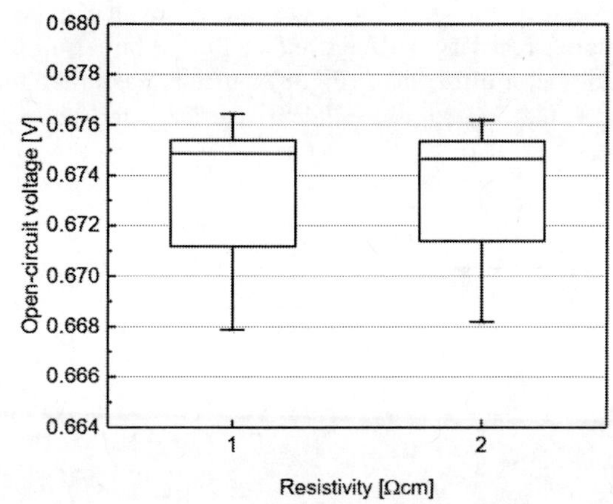

Fig. 4. PERC solar cell open-circuit voltage distributions for the 1 and 2 Ωcm resistivity groups.

Besides comparable open-circuit voltages, the short-circuit current density are also comparable for both resistivity groups (see Fig. 5)

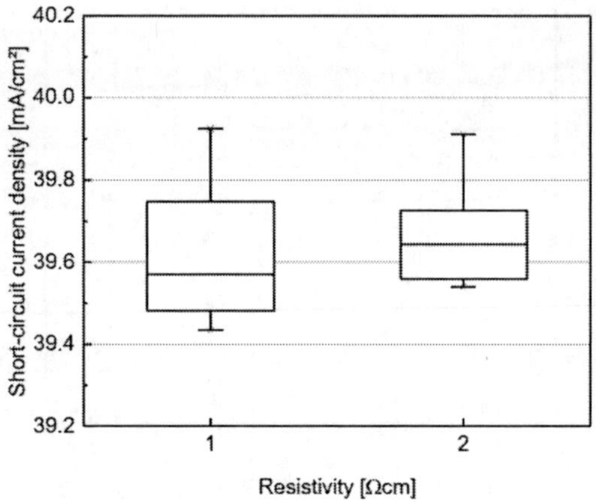

Fig. 5. PERC solar cell short-circuit current density distributions for the 1 and 2 Ωcm resistivity groups.

The Efficiency differences can be explained by fill factor differences (see Fig. 6). These are caused by the expected lumped series resistance difference (see Fig. 7) but also due to pseudo fill factor differences (Fig. 8) which are assumed to be related to the injection-dependent carrier lifetime curve differences.

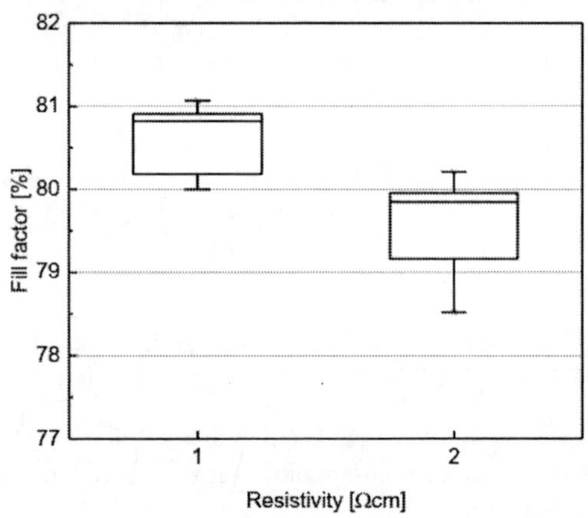

Fig. 6. PERC solar cell fill factor distributions for the 1 and 2 Ωcm resistivity groups.

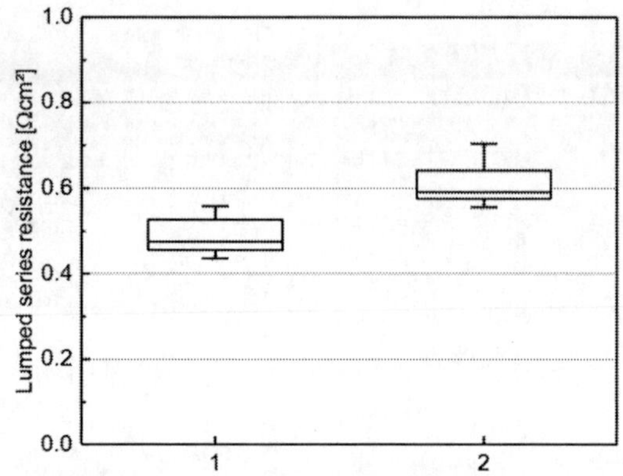

Fig. 7. PERC solar cell lumped series resistance distributions for the 1 and 2 Ωcm resistivity groups.

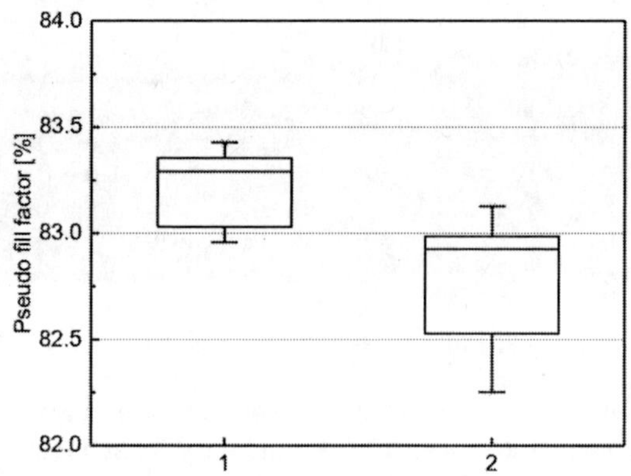

Fig. 8. PERC solar cell pseudo fill factor distributions for the 1 and 2 Ωcm resistivity groups.

The processed PERC solar cells and their test structures were etched back and cleaned and passivated. The effective carrier lifetimes which are almost the bulk carrier lifetimes after different process steps are shown in Fig. 9.

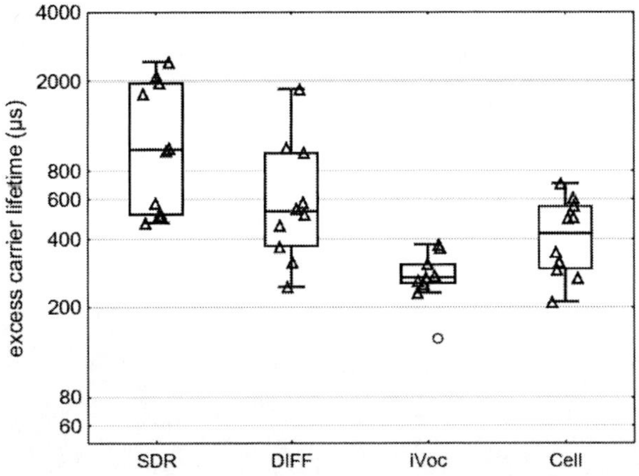

Fig. 9. Effective excess carrier lifetime after different process steps evaluated at $\Delta n = 5\times10^{14}$ cm^{-3}.

The bulk carrier lifetime reduces from an initial high level during the PERC process monitored after phosphorous diffusion (DIFF) and after the application of dielectric layers (iVoc). However, after screen-printing and the following co-firing step the bulk carrier lifetime improves. The possibility of the introduction of contamination into the silicon bulk during wet processing is tested by ICP-MS measurements of the metal contamination on the wafer surface. No significant metal concentration change is observed after saw-damage removal, texturing or phosphorous glass removal. Therefore a change in the wafer bulk is assumed as the reason for the carrier lifetime change. For better comparison between the two different resistivity wafer groups, the material saturation current density $j_{0,\mathrm{mat}}$ is derived and shown in Fig. 10. Thus, the widespread of carrier lifetime within groups can be mostly explained by the resistivity variation as the $j_{0,\mathrm{mat}}$ values are in a narrower distribution.

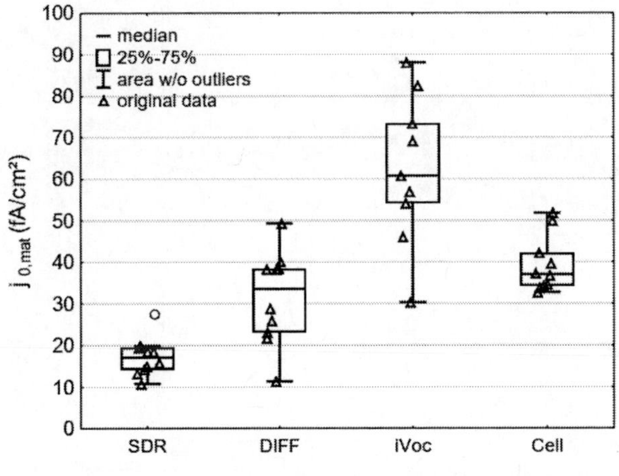

Fig. 10. Material saturation current density $j_{0,\mathrm{mat}}$ after different process steps.

The general tendency of bulk quality evolution within the PERC process is confirmed. Phosphorous diffusion, oxidation and deposition of dielectric layers, i.e. high temperature processing, reduces the bulk quality whereas the co-firing step improves it.

A reasonable explanation is the nucleation and/or growth of oxygen precipitates during high temperature steps which also impact the bulk carrier lifetime [4]. The improvement during the co-firing is possibly caused by the release of hydrogen from the dielectric layers into the wafer [5] which might passivate defects. A balanced temperature management during cell processing of temperature ramps, peak temperatures and times is necessary for high efficiency PERC solar cells.

III. SUMMARY

Ga-doped Cz wafers are used to fabricated PERC solar cells to reached a high efficiency level which allows for relevant investigation on material effects. The initial Cz wafers showed the highest carrier lifetime. Phosphorous diffused Cz wafers do not improve due to gettering. Implied Voc structures show lowest carrier lifetime. However, fired solar cells improve in carrier lifetime compared to Implied Voc structures. Prominent metals do not significantly vary on the wafer surface during solar cell processing. A balanced temperature management during cell processing of temperature ramps, peak temperatures and times is necessary for high efficiency PERC cells.

ACKNOWLEDGEMENTS

The authors would like to thank the colleagues at Solar World Innovation GmbH, especially the solar cell development group, pilot line team, characterization lab and analytical lab. We would also like to thank the colleagues at ISFH for carrier lifetime sample preparation.

REFERENCES

[1] B. Min, H. Wagner, M. Müller, Gerd Fischer, R. Brendel, P.P. Altermatt, H. Neuhaus, „A roadmap towards 24 % efficient PERC solar cells in industrial mass production," *IEEE J. Photovolt. vol. 7(6)*, p. 1541–1550, 2017.

[2] T. Weber, G. Fischer , A. Oehlke, C. Kusterer, K. Strauch, R. Schiepe, M. Mühlbauer, M. Müller, R. Köhler, F. Wolny, G. GruppMüller, E. Schneiderlöchner, K.-H. Stegemann, H. Neuhaus, „High Volume Pilot Production of High Efficiency PERC Solar Cells - Analysis Based on Device Simulation," *Energy Procedia vol. 38*, p. 474–481, 2013.

[3] M. Müller, „Reporting effective lifetimes at solar cell relevant injection densities," *Energy Procedia vol. 92*, p. 138–144, 2016.

[4] J. D. Murphy, R. E. McGuire, K. Bothe, V. V. Voronkov, R. J. Falster, "Minority carrier lifetime in silicon photovoltaics: The effect of oxygen precipitation." *Solar Energy Materials and Solar Cells vol. 120*, p. 402–411, 2014.

[5] J. Hong, W. Kessels, W. Soppe, A. Weeber, W. Arnoldbik, and M. Van de Sanden, "Influence of the high-temperature firing step on high-rate plasma deposited silicon nitride films used as

bulk passivating antireflection coatings on silicon solar cells". *Journal of Vacuum Science & Technology B vol. 21*(5), p. 2123–2132, 2003.

SELECTIVE EMITTER USING APCVD PSG LAYERS AS DOPING SOURCE

Pierre Saint-Cast[1], Udo Belledin[1], Elmar Lohmüller[1], Bishal Kafle[1], Julian Weber[1], Sven Seren[2], Sabrina Lohmüller[1], Andreas Wolf[1], Marc Hofmann[1]

[1]Fraunhofer Institute for Solar Energy Systems ISE, Heidenhofstraße 2, 79110 Freiburg, Germany
[2]Gebr. Schmid GmbH + Co., Robert-Bosch-Str. 32-36, 72250 Freudenstadt, Germany

ABSTRACT: Using atmospheric pressure chemical vapour deposition (APCVD) of phosphosilicate glass (PSG) allows the separation of the PSG deposition and the thermal drive-in, adding another degree of freedom and control to the emitter formation. A large survey is carried out varying the process parameters such as phosphorus concentration and layer thickness of the PSG layer and the temperature and duration of the thermal drive-in in a tube furnace. We show that the data of the survey can train a metamodel, which can predict the emitter sheet resistances R_{sh} based on the process parameters. An emitter with R_{sh} = 120 Ω/sq and an emitter dark saturation current density j_{0e} = 40 fA/cm² after firing (textured surface, SiN_X passivation) is demonstrated using this approach. For the formation of a selective emitter, laser diffusion is carried out. The laser process takes place between the PSG layer deposition and the drive-in process. A 1.4 µm deep profile with a peak concentration of 7×10^{19} cm^{-3} at the surface is fabricated using a long-pulse infrared laser. This process leads to a specific contact resistivity ρ_C = 4.2 mΩcm² using a commercially available screen-printed and fired silver paste. Specific contact resistivity down to ρ_C = 1 mΩcm² is reached using a short-pulse green laser process.

Keywords: APCVD; selective emitter; laser doping

1 INTRODUCTION

1.1 Context

The passivated emitter and rear cell (PERC) [1] has gained an important share of the solar cell market [2] within the last years. The improvement of the rear surface passivation compared to its predecessor, the aluminum back surface field (Al-BSF) solar cell, allows reaching efficiencies close to 22% in production. However, as the rear surface was greatly improved, the front of the cell became the main limiting factor [3]. Selective emitters allow decreasing front recombination and therefore increasing the efficiency limit of industrial PERC solar cells [4]. So far, selective emitter processes are rarely used [2] in the industry and unforeseen candidates might become major selective emitter technologies in the near future. This contribution presents the development of a laser-doped selective emitter based on the use of atmospheric pressure chemical vapour deposition (APCVD) of phosphosilicate glass (PSG) layers as doping source.

1.2 Approach

Figure 1 shows the process sequence needed for the formation of a homogeneous or a selective emitter using APCVD PSG as doping source. The "classical" approach using laser doping and $POCl_3$ diffusion is also shown for comparison [4]. A relatively simple solution is taken as point of reference [4]: first an HF dip is carried out, then the wafer undergoes a $POCl_3$ diffusion process, subsequently local laser doping is applied and finally the rear emitter and the PSG are wet-chemically etched. This selective emitter is therefore formed in four steps. The emitter between the fingers is obtained during the $POCl_3$ diffusion, the emitter under the front contacts is in addition modified by by laser diffusion. For the laser diffusion, the PSG layer is used as a doping source and in addition activation of inactive phosphorus in the diffused emitter might take place. Finally, because the $POCl_3$ diffusion process occurs on both sides of the wafer, a rear emitter removal and a PSG etching process is needed.

Forming a homogeneous emitter using APCVD PSG layers requires four steps as well, beginning with an HF dip followed by the deposition of the PSG layer by APCVD. The emitter is then formed by a thermal drive-in process, which allows the phosphorus atoms to diffuse from the PSG layer into the silicon wafer. Finally, because the deposition of the PSG layer is a one-side process, the rear emitter etch can be avoided in principle and only a simple PSG etching is required (this point is still under investigation and not fully demonstrated yet). Therefore, the additional process cost/complexity due to the separated PSG deposition step can be partially compensated by avoiding the rear emitter removal process. Furthermore, separating the PSG deposition from the thermal drive-in allows for stack drive in, where instead of filling one or two wafers per slot during the drive in, wafer packages of ten wafers or more are used. The throughput of the diffusion oven can be therefore increased significantly. In production, these both advantages combined might fully compensate or overcompensate for the increased cost due to the additional APCVD process.

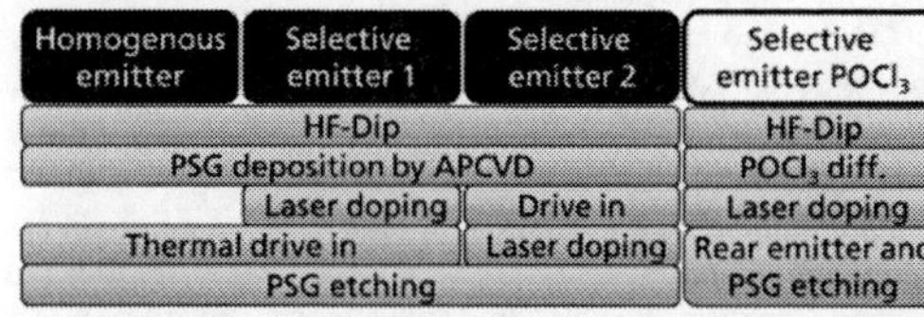

Figure 1: Schematics of the approach used to compare the analytical models and the numerical simulations.

For the formation of a selective emitter, two process sequences can be thought of. Within the approach "selective emitter 1", the local laser diffusion is carried out directly after the APCVD deposition and before the thermal drive-in. In this case, the drive-in process can be used to anneal possible laser damage and further decrease the sheet resistance. For the "selective emitter 2" sequence, the laser diffusion is carried out after the drive-in process; this sequence resembles the reference $POCl_3$ selective emitter process.

This study covers two major topics of selective emitter formation using APCVD PSG as doping source:

- The formation of the (illuminated) emitter: where the control of the sheet resistance R_{sh} and doping profiles $N(x)$ are presented as well as the emitter dark saturation current density j_{0e}.
- The formation of the locally diffused area (LDA) by laser doping, focusing on the "selective emitter 1" approach, where the laser diffusion process is performed before the thermal drive-in process. The doping profiles $N(x)$ as well as the specific contact resistivity (ρ_C) obtained are discussed.

In parallel to this contribution, first results of PERC solar cells presents another paper in this conference [5].

2 THE FORMATION OF THE EMITTER

2.1 Experimental

For the development of the illuminated emitter, *n*-type Czochralski-grown silicon (Cz-Si) wafers are used. The process begins with saw damage etching and a thermal process to deactivate the thermal donors. Then the wafers first undergo an alkaline texture process followed by an HF dip. Then, the wafers are coated on both sides with a PSG layer and a SiO_X capping layer using an APCVD deposition process performed with a Schmid Thermal Systems belt APCVD tool. The SiO_X layer protects the PSG layer from the humidity at ambient conditions. The thermal drive-in process is then carried out. In order to measure the emitter sheet resistance R_{sh}, an inductive sheet resistance measurement is performed before and after thermal drive in. Then, PSG etching is carried out followed by a wet-chemical cleaning (in figure 4: SC1 SC2 [6] in experiment 1-3, HNO_3 dip and HF dip in experiment 4. For experiment 4, an additional thermal process at 550°C for 5 min simulating the outgassing process needed for atomic layer deposited (ALD) AlO_X layers is applied. Then, a SiN_X layer deposited by plasma-enhanced chemical vapor deposition (PECVD) then passivates the wafers on both surfaces. Finally, the wafers are fired in a fast firing oven (FFO). A quasi steady state photo conductance (QSSPC) measurement yields the dark saturation current density j_{0e} using the procedure from Ref [7].

2.2 Parameter variation

Both, the characteristics of the APCVD PSG layer (phosphorus concentration [P], layer thickness and the parameters of the thermal drive-in process (temperature, duration)) determine the emitter sheet resistance R_{sh} and doping profile $N(x)$. Many combinations of PSG layers and drive-in processes might lead to the same R_{sh} but the doping profile might be completely different. An additional degree of freedom given by separating the PSG deposition and the drive-in allows a precise engineering of the doping profile for the aimed application. However, it increases the complexity as the number of parameters to explore increases. Our approach is to first investigate a very large range of process combinations regardless of the sheet resistance obtained in a large survey (see Figure 2). Then the results are implemented in a sophisticated fitting procedure called metamodel [8,9], which allows predicting sheet resistance from the process parameters.

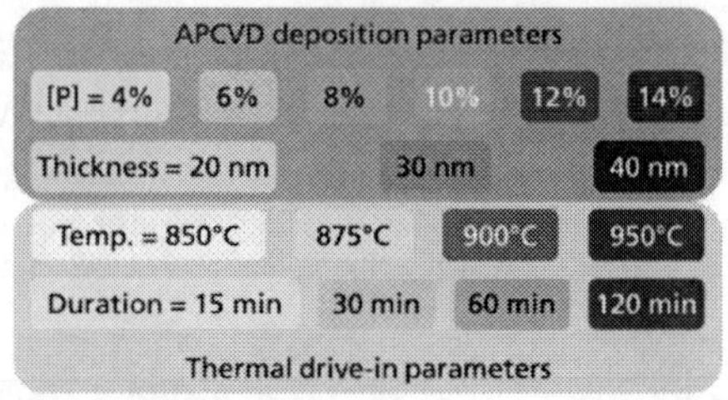

Figure 2: Schematic of the process parameters varied for the large sheet resistance survey.

2.3 Results

In Fig. 3, the sheet resistances obtained from the experiment are compared to the one obtained using the metamodel. The grey lines represent a ±25% deviation boundary. The accuracy obtained is not sufficient to pretend being able to make any whished emitter in a single try, however, it is a great tool to eliminate inappropriate process parameters. The metamodel has been also very helpful in order to transfer drive-in processes to three different diffusion furnaces, by calculating correction parameters to obtain the desired sheet resistance. The metamodel is further used to correct deposition parameters of the PSG layer after e.g. the change of a defect mass flow controller in the APCVD system, which altered the PSG layer.

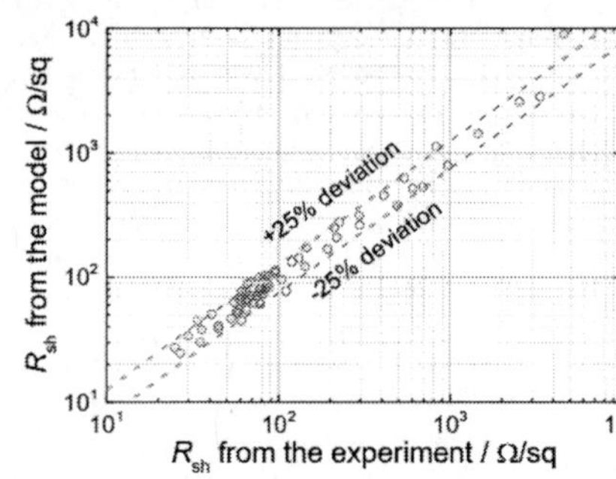

Figure 3: 10 fold cross comparison of the sheet resistance values predicted by the metamodel and measured in the experiment.

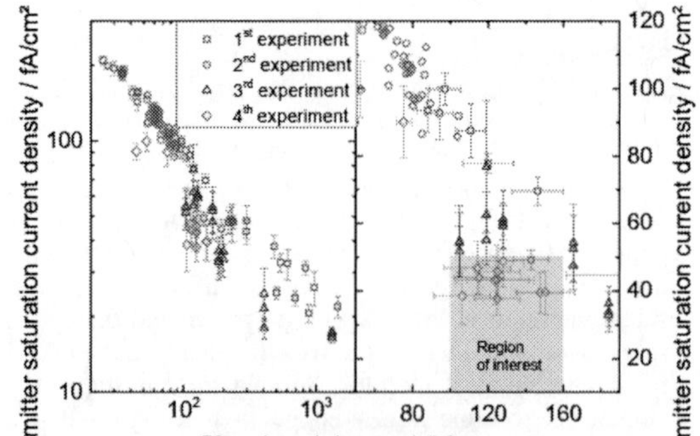

Figure 4: Emitter saturation current density j_{0e} as a function of the sheet resistance for four experiments. On the left hand side, the measured values are shown on a log/log scale. On the right, the scale is linear and focuses on the region of interest. All values belong to textured wafers. The passivation is PECVD SiN_X after firing (with an outgassing process for the 4th experiment).

The aim of this emitter development is to obtain a sheet resistance between $100\ \Omega/sq \leq R_{sh} \leq 160\ \Omega/sq$, while keeping j_{0e} lower than 50 fA/cm². In Figure 4, the j_{0e} for fired PECVD SiN_X passivation is plotted as a function of R_{sh} for four experiments. In the left part of the graph, the results are plotted in log/log scale covering a R_{sh} range from 10-2000 Ω/sq and a j_{0e} range from 10-300 fA/cm². After a first look, the results seem to line up, in a correlation where j_{0e} decreases, when increasing R_{sh}. This behavior can be explained as follows: as the sheet resistance decreases, the Auger recombination and the surface doping concentration decrease leading to an increase of the surface passivation quality. Unfortunately, this line does not cross the region of interest, the j_{0e} obtained for a sheet resistance from 100-160 Ω/sq are above 50 fA/cm². The goal of the process development is to reach points below this line, which means lowering j_{0e} without affecting the sheet resistance. In experiment 1 to 3, the focus has been given to variations of the PSG deposition and drive-in processes. In the 4th experiment, the passivation has been enhanced by adding an outgassing process (used for the outgassing of ALD layers) before the plasma-enhanced chemical vapor deposition (PECVD) of the SiN_X layer [10]. The region of interest is reached by many processes of the 4th experiment. This experiment demonstrates very low j_{0e} values below 40 fA/cm² for a sheet resistance of $R_{sh} \approx 120\ \Omega/sq$

2.4 Stacked drive in

The feasibility of stacked drive in is also tested. The wafer preparation and measurement is identical as described in section 2.1. The diffusion boat is filled with a stack of one wafer, three wafers, six wafers and nine wafers per slot. Each wafer stack is directly introduced in one boat slot. A stack of nine wafers is the maximum which can be introduced in one slot of the used quarts boat. Stacks with more than nine wafers would in principle be possible in a specially designed boat.

In Figure 5, the sheet resistance and its uniformity are shown for each wafer of each stack. The number of wafers in a stack seems to have very little influence on the sheet resistance and on the uniformity. In fact, between a stack of one wafer and nine wafers only an increase of 2 Ω/sq is observed and the uniformity is basically the same. Therefore, a stacking of up to nine wafers is demonstrated without major impact on the sheet resistance.

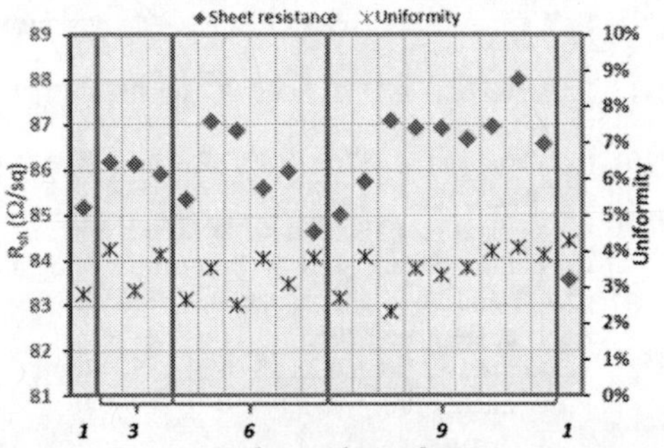

Figure 5: Emitter sheet resistance R_{sh} and emitter sheet resistance uniformity over the number of wafers in a slot for each wafer diffused. Each gray box represents a wafer stack.

3 THE FORMATION OF THE CONTACT EMITTER

3.1 Experimental

For the formation of the contact emitter area, we use *p*-type Cz-Si wafers. This study focused on the "selective emitter 1" sequence (see Figure 1), where the laser diffusion process is carried out directly after the PSG deposition and before the thermal drive-in process. Apart from the laser process, the wafer preparation is the same as described in section 2.1. Two lasers are used: a green nanosecond laser with 532 nm wavelength and 32 ns pulse duration and an infrared (IR) laser with 1064 nm wavelength and 1.4 µs pulse duration. For the contact resistance measurement, a silver grid resulting in 35 µm wide fingers is screen printed before the firing process. The sheet resistances and the contact resistances are measured using the transmission-line method (TLM). The charge carrier concentration profiles are measured using the electrochemical capacitance voltage (ECV) method, the profil were correct using the method described by Kimmerle and Werner [11,12].

3.2 Results

In Figure 6 (a) and (b), the specific contact resistivity and the sheet resistance are plotted as a function of the laser fluency for both laser types. The laser fluency increases from left to right.

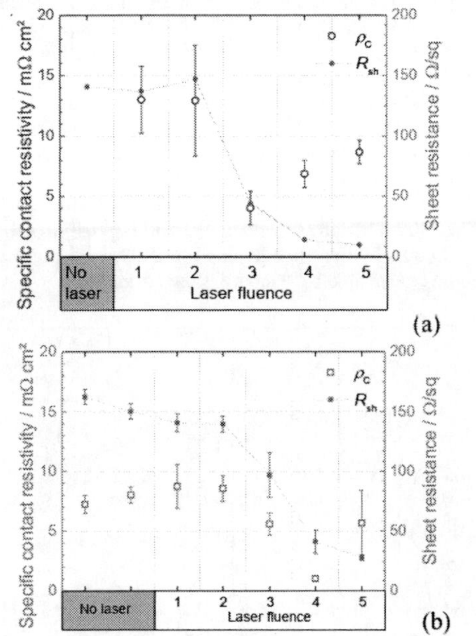

Figure 6: Specific contact resistivity and sheet resistance measured by transmission-line method (TLM) as a function of the laser fluency increase from left to right (the number does not refer to a particular unit) for the green laser (a) and the IR laser (b).

For low laser fluencies, the sheet resistance remains close to the sheet resistance of the as-diffused emitter (low fluency regime). Then the sheet resistance decreases (transition regime). After a quick decrease, the sheet resistance saturates (high fluency regime). For both lasers, the specific contact resistivity in the low fluency regime is close to the one measured for the non-lasered (as-diffused) emitter. In the transition regime, the specific contact resistivity decreases to reach a minimum value. In the high fluency regime, the specific contact resistivity increases again.

Both lasers yield ρ_C values below 5 mΩ cm² using standard screen-printing technology with commercially available silver paste. However, the lowest ρ_C value (~1 mΩ cm²) is reached using the short-pulse green laser. The sheet resistances obtained with this laser (R_{sh} = 42 Ω/sq) are comparable to the one obtained using the long-pulse IR laser (R_{sh} = 40 Ω/sq).

3.3 Doping profile and discussion

The doping profiles of the contact emitter and of the illuminated emitter are measured after drive in process and PSG etch using electrochemical capacitance voltage measurement (see Figure 7).

The light of the short-pulse green laser is absorbed surface-near. In addition, due to the short-pulse the thermal energy does not diffuse deeply into the wafer leading to high doping close to the silicon surface (see Figure 7). After drive-in process the surface doping concentration (N_S) increases by a factor 1.4. For the long-pulse IR laser, the light is absorbed much deeper in the silicon. Due to the long pulse, the thermal energy diffuses deeply into the wafer during the laser pulse, resulting in much deeper doping profiles with typically low surface doping concentrations (Gaussian fit figure 7). Such IR lasers are typically not used for laser diffusion due to the difficulty to reach high surface concentration and therefore low contact resistivity. However, the drive-in process leads to a peak in the doping concentration at the wafer surface. This peak – which appears after the drive-in process – is explained by a thermal diffusion of phosphorous atoms that are not actively incorporated into the emitter so far. The increase of the surface concentration corresponds to a factor 3.5±0.3. It is shown in section 2.3 that such doping profiles can be engineered to be contactable by standard screen printing.

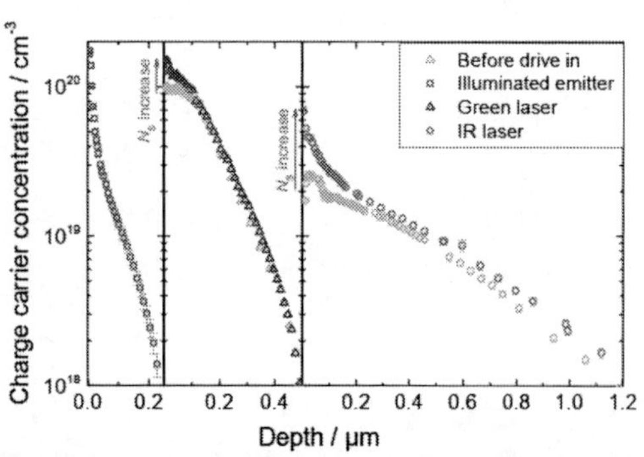

Figure 7: Doping profiles of the locally diffused area measured by electrochemical capacitance-voltage method for the short-pulse green laser and the long-pulse IR laser after laser doping before and after thermal drive in. The profiles have been corrected based on the method published by Kimmerle and Werner [11,12].

4 SUMMARY

The separation between the doping glass formation and the drive-in process, when using APCVD PSG deposition, adds another degree of freedom and control to the emitter formation. We show that a trained metamodel can be used to fit the emitter sheet resistances R_{sh} based on the process parameters (APCVD and thermal drive-in parameters). An emitter with sheet resistance R_{sh} = 120 Ω/sq and an emitter dark saturation current density j_{0e} = 40 fA/cm² (textured surface, using outgassing, PECVD SiN_X passivation, and firing) is demonstrated using this approach.

The emitter underneath the contacts is formed using laser diffusion that is conducted between the PSG layer deposition and the drive-in process. A 1.4 µm deep profile with a peak concentration of of 7×10^{19} cm^{-3} at the surface is fabricated using a long-pulse infrared laser. This process leads to a specific contact resistivity ρ_C 1-5 mΩ/cm² for a screen-printed and fired commercially available silver paste.

4 ACKNOWLEDGEMENTS

This work is funded by the federal ministry for economic affairs and energy (BMWi) of Germany within the project APPI under the contract number 0325895A. Project APPI is supported under the umbrella of SOLAR-ERA.NET Cofund by Agence de l'Environnement et de la Maîtrise de l'Energie, Project management Jülich, Technology Strategy Board and Centro para el desarollo tecnológico. SOLAR-ERA.NET is supported by the European Commission within the EU Framework Programme for Research and Innovation HORIZON 2020 (Cofund ERA-NET Action, N° 691664). The authors would like to thank all co-workers at photovoltaics department of Fraunhofer ISE.

6 REFERENCES

[1] A. W. Blakers, et al., Appl. Phys. Lett., vol. 55, no. 13, pp. 1363–1365, 1989.
[2] ITRPV, "International Technology Roadmap for Photovoltaic (ITRPV) - Results 2017".
[3] P. Saint-Cast, S. Werner, J. Greulich et al., Phys. Status Solidi A, 2016.
[4] U. Jäger, S. Mack, C. Wufka et al., IEEE J. Photovoltaics, vol. 3, no. 2, pp. 621–627, 2013.
[5] B. Kafle et al., this conference (2AO.6.3)
[6] W. Kern and D. Puotinen, RCA Rev. 31, 187–205 (1970).
[7] A. Kimmerle, J. Greulich, and A. Wolf, Sol. Energy Mater. Sol. Cells 142, 116 (2015).
[8] S. Wasmer, A. Brand, J. Greulich, Energy Procedia 124 (2017) 207-214
[9] F. Pedregosa, et al. J. Mach. Learn. Res. 12, 2825 (2011).
[10] S. Werner, et al. 33rd EUPVSEC proceedings pp. 406-412, 2017.
[11] A. Kimmerle, PhD thesis "Diffused Surface for Crystalline Solar Cells – Process Development, Characterisation, and Modeling", Shaker Verlag, Aachen, Germany, 2015
[12] S. Werner, E. Lohmüller, S. Maier, S. Mourad, A. Wolf, Energy Procedia 124 (2017) 936-946

NON MASS SEPARATION TYPE ION IMPLANTATION SYSTEM FOR BIFACIAL PERT CELL FABRICATION

Kyotaro Nakamura[1], Kazuo Muramatsu[2] Noboru Yamaguchi[3] and Yoshio Ohshita[1]
[1]Toyota Technological Institute 2-12-1 Hisakata, Tempaku-ku, Nagoya 468-8511, Japan
[2]NAMICS CORPORATION, 3993 Nigorikawa, Kita-ku, Niigata, 950-3131 Japan
[3]Institute of Semiconductor and Electronics Technologies, ULVAC, Inc., 1220-1 Suyama, Susono, Shizuoka, 410-1231 Japan
phone: +81-52-809-1877, fax: +81-52-809-1879, e-mail: kyotaro@toyota-ti.ac.jp

ABSTRACT: We developed crystalline Si solar cell fabrication process for both n- and p-bifacial PERT (Passivated Emitter and Rear Totally diffused) cell using non-mass separation type (NMS) ion implantation. At first, we fabricated n-bifacial selective emitter (SE) PERT solar cell and were able to control of Rsh of SE structure by controlling of conditions of ion implantation and activation annealing. Cell efficiency of n-bifacial PERT was obviously increased by introducing SE structure and the gain of η by SE structure was 0.4%$_{abs}$. Combination of SE and other optimization of cell processes improved all IV parameters, and we achieved 21%. Furthermore, we fabricated p-bifacial PERT cell using same process for n-bifacial PERT, and difference of η was just 0.7%$_{abs}$. Thus, we were able to confirm the applicability of our developed process to p-type cell. These results clearly show the effectiveness of our PERT cell process using NMS ion implanter and the feasibility of high efficiency low cost solar cell using our cell fabrication process.
Keywords: c-Si, Ion Implantation, PERT, Selective Emitter, Silicon Solar Cell

1 INTRODUCTION

Crystalline silicon solar cells are approaching a major technological turning point. New type high efficiency double-sided contact cells, i.e. PERC/PERT/PERL cells are rapidly developed and introduced to PV market in recent years and expected to gain significant market share over BSF cells in the near future [1]. One of key technologies to realize these high efficiency cells is to form high quality doping layer. As well known, ion implantation has huge potential to form high quality doping layer. However the ion implantation system has encountered some difficulties to becoming more mainstream. Actually, ion implantation is supposed to be applied in production but at with low market share of <5% [1]. Because the ion implantation system has some disadvantages, that is to say, relatively high equipment cost and low throughput. So, in recent years, a new ion implantation system, that is, NMS ion implanter are developed and proposed to apply for solar cell mass production, because it can lower the manufacturing cost. On the other hand, there are some concerns about NMS ion implantation. Other impurities may be implanted together with doping ion. Since bonded ion implantation, wafer damage might be significant. During activation anneal, ions may diffuse outward from the substrate.

In this work, we developed cell fabrication process of n-PERT solar cell using NMS ion implanter. For the improvement of cell efficiency, we applied SE structure for n-PERT cell. Additionally, one of features of our developed cell fabrication process is to be able to apply for both n- and p-type cells. Therefore, we investigated the applicability of our process by fabricating both n- and p-bifacial PERT cell using exact same process.

2 EXPERIMENTS

2.1 Non mass separation type ion implantation

The NMS ion implanter is expected to be introduced into mass production process for solar cells, because it can lower the manufacturing cost. Especially, ULVAC's NMS ion implanter has several advantages described as follows; (a) Control and reproduction of the amount of dopant, (b) Compact Footprint, (c) High accuracy mask implantation, (d) High throughput processing, (e) High accuracy alignment implantation by single wafer transfer, (f) Low cost compared to mass separation type ion implanter.

Figure 1(a) shows the schematic view of ULVAC's NMS ion implanter. BF_3 is used for Boron source and F, N, O, C etc. are implanted together with B. Thus, it was concerned that these ions affect cell performance. So, we investigated these issues in our previous work and confirmed these ions didn't provide negative impact on cell performance [2]. Therefore, in this work, we proceeded to the next step to improve cell performance using NMS ion implanter by formation of SE using mask as shown in Fig. 1(b).

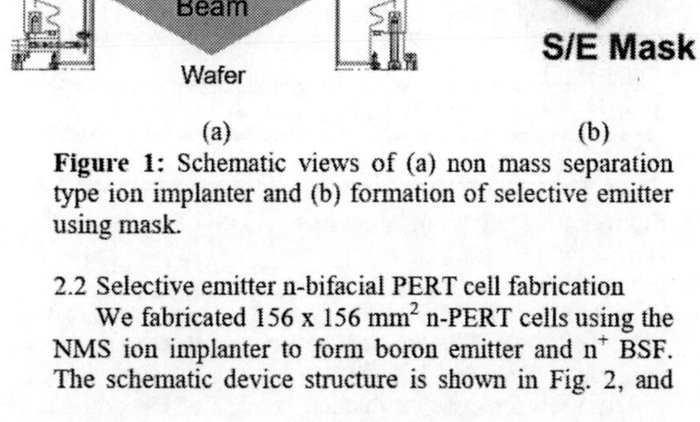

(a) (b)

Figure 1: Schematic views of (a) non mass separation type ion implanter and (b) formation of selective emitter using mask.

2.2 Selective emitter n-bifacial PERT cell fabrication

We fabricated 156 x 156 mm^2 n-PERT cells using the NMS ion implanter to form boron emitter and n$^+$ BSF. The schematic device structure is shown in Fig. 2, and the outline of our bifacial PERT solar cell fabrication process is shown in Fig. 3.

After damage removal and surface texture formation by an alkaline solution, boron ion implantation was performed using ULVAC's NMS ion implanter. Then, high temperature activation annealing was performed and then BSG was removed by HF etching. After that, phosphorus ion implantation and annealing were done, then phosphosilicate glass (PSG) was removed to form back surface field (BSF). Subsequent surface passivation and plasma-enhanced chemical vapor deposition (PECVD) formed the stacked passivation layer on the top of the boron emitter. At the same time, the rear surface was also covered by SiN_x layer. Finally, the contact formation steps of Ag/Cu dual contact cell were performed as follows; 1) Ag paste was printed as fingers on both side of wafer by screen-printing, 2) high temperature rapid firing by a belt furnace, 3) Cu paste was printed as busbars on both side (so called Dual printing), 4) low temperature curing by an oven.

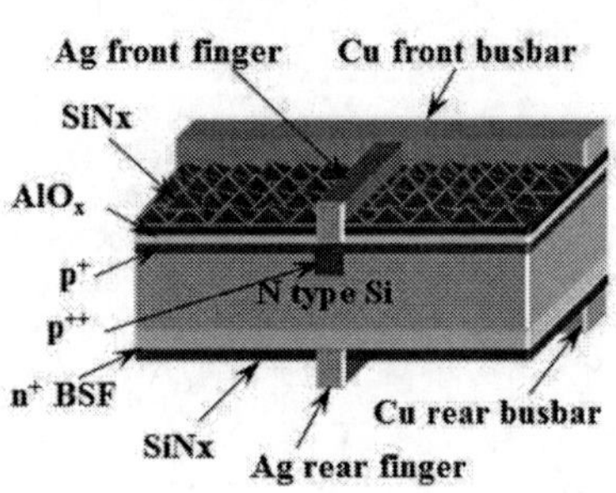

Figure 2: Schematic view of our n-bifacial SE PERT solar cell structure.

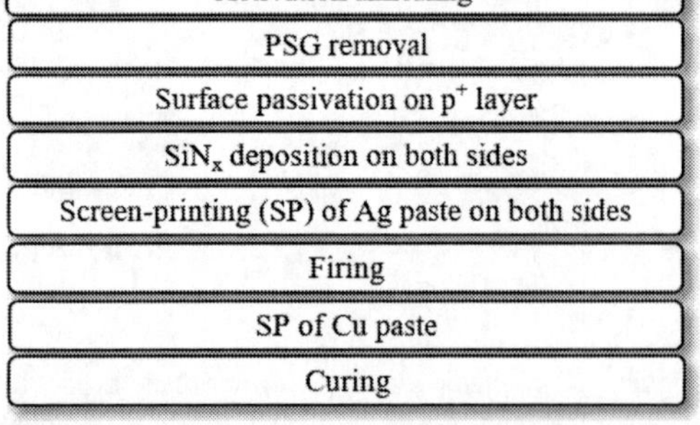

Figure 3: Outline of our bifacial PERT solar cell fabrication process.

Moreover, we introduced SE structure to improve cell efficiency. As well known, SE is very useful technique to improve cell conversion efficiency, however it isn't widely used for mass production because of the process complexity and increased process cost. However, we are able to make SE process simple by using the ion implanter because we can simultaneously dope and pattern the doped layer by using a graphite mask, as shown in Fig. 1(b). In the case of SE cells, we implanted highly doped pattern region shown as p^{++} in Fig. 2 at first, and then implanted lightly doped layer all over the wafer surface shown as p^+ in Fig. 2. The sheet resistances (Rsh) of each region were varied by controlling of conditions of ion implantation and activation annealing.

2.3 P-bifacial PERT cell using same process

One of features of our developed cell fabrication process is to be able to apply for both n- and p-type cell. Therefore, we investigated the applicability of our process by fabricating both n- and p-bifacial PERT cell using exact same process as described in Fig. 3. The schematic structure of our p-bifacial PERT solar cell is shown in Fig. 4.

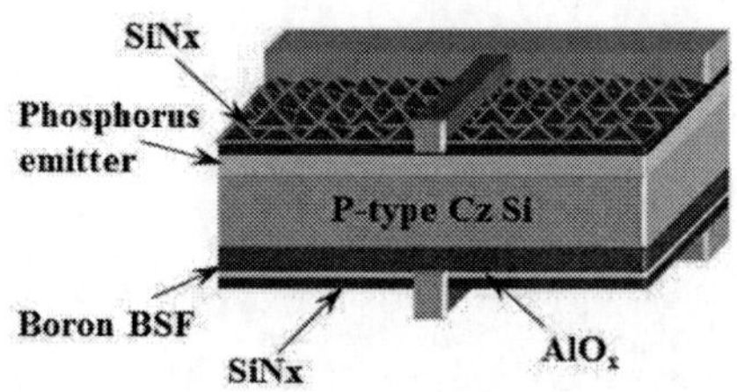

Figure 4: Schematic view of our p-bifacial PERT solar cell structure.

3 RESULTS AND DISCUSSIONS

3.1 Non mass separation type ion implantation

After several investigations, we fabricated homogeneous emitter n-type bifacial PERT cell by boron implantation using our improved conditions, compared with boron diffusion by APCVD. Table I shows the comparison of IV parameters of between APCVD boron diffusion and boron implantation solar cells. As a result, we succeeded efficiency 20.7% using low implantation energy and cap activation annealing conditions, and same as reference. Thus we were able to confirm that concerns about boron ion implantation don't have negative effect on cell performance.

Table I: Comparison of IV parameters of between APCVD boron diffusion and boron implantation solar cells

Emitter	Jsc (mA/cm^2)	Voc (V)	F.F.	η (%)
APCVD	40.6	0.651	0.772	20.4
Implant	40.5	0.652	0.782	20.7

After that, we investigated to apply phosphorus ion implantation instead of $POCl_3$ diffusion for BSF formation. As shown in Table II, when n layers of BSF are formed by using phosphorus ion implantation, the efficiency was 0.6% UP on average. Thus, we were able to establish n-type bifacial PERT solar cell fabrication process using both boron and phosphorus ion implantation.

Table II: Comparison of IV parameters of between $POCl_3$ diffusion and phosphorus implantation solar cells

Emitter	Jsc (mA/cm^2)	Voc (V)	F.F.	η (%)
$POCl_3$	38.4	0.628	0.773	18.6
Implant	39.2	0.641	0.765	19.2

3.2 Selective emitter n-bifacial PERT cell fabrication

Moreover, we introduced Selective Emitter structure to improve cell efficiency. Heavily doped emitter causes absorption loss of short wavelength light and recombination loss. In terms of this, selective emitter is a very useful technique to improve cell efficiency by reduction of these losses. However it isn't widely used for mass production because of the process complexity and increased process cost. In contrast with that, we are able to make the SE process simpler by using the ion implantation, as we can simultaneously dope and pattern the doped layer by using a graphite mask.

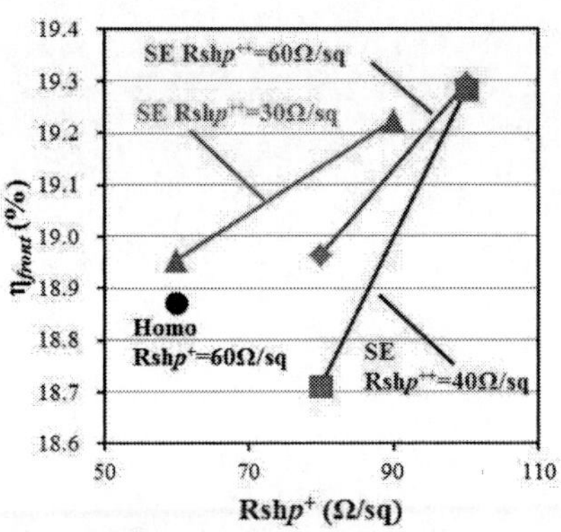

Figure 5: Efficiency dependence of n-bifacial PERT cells on Rsh_{p+} and Rsh_{p++}.

Conversion efficiencies of our n-PERT cells illuminated from front side (η_{front}) are shown in Fig. 4, as a function of Rsh of emitter. In Fig. 5, Rsh_{p+} means the Rsh of p^+ layer in Fig. 2, and Rsh_{p++} means Rsh of p^{++} layer. As you can see in Fig. 4, cell efficiency is obviously increased by introducing SE structure. Additionally, efficiency was increased with the increase in Rsh_{p+} in each Rsh_{p++}. In further detail, with decrease of Rsh_{p++} from 60 Ω/sq to 40Ω/sq, η_{front} tends to decrease. Then, with decrease of Rsh_{p++} from 40 Ω/sq to 30Ω/sq, η_{front} tends to increase.

Table III shows the comparison of IV parameters between homogeneous emitter and SE PERT solar cells. The increase of Jsc and Voc by SE suggests suppression of recombination loss by SE, and the gain of η by SE structure was $0.4\%_{abs}$. This result clearly shows the effectiveness of SE using the NMS type ion implanter.

Table III: Comparison of IV parameters between homogeneous emitter and SE PERT solar cells

Emitter structure	Rsh (Ω/sq)	Jsc (mA/cm^2)	Voc (V)	F.F.	η (%)
Homo-	60	38.4	0.625	0.786	18.9
SE	60/100	39.3	0.628	0.783	19.3

In addition, further optimization of our cell fabrication process such as surface passivation layer, printing and firing conditions of pastes, etc. is still under investigation. Table IV shows the PV performance of our current best SE n-PERT solar cell. Combination of SE and other optimization of cell processes improved all IV parameters, and we have achieved 21%. This result shows the feasibility of high efficiency low cost n-PERT solar cell using our cell fabrication process.

Table IV: PV performance of our current best SE n-PERT solar cell

Incident	Jsc (mA/cm^2)	Voc (V)	F.F.	η (%)	Bifaciality
p^+	40.2	0.652	0.803	21.0	0.924
n^+	37.9	0.642	0.797	19.4	

3.4 P-bifacial PERT cell using same process

Additionally, we fabricated p-bifacial PERT cell using exact same process for n-bifacial PERT to check the applicability of our developed process to p-type cell. Table V shows the comparison of IV parameters between n- and p-bifacial PERT cells. All parameters of p-PERT are slightly lower than n-PERT, however gaps are not large, and difference of η is just $0.7\%_{abs}$. Therefore, we were able to confirm the applicability of our developed process to p-type cell.

Table V: Comparison of IV parameters between homogeneous emitter and SE PERT solar cells

	Incident	Jsc (mA/cm^2)	Voc (V)	F.F.	η (%)	Bifaciality
n-PERT	p^+	39.5	0.654	0.787	20.3	0.949
p-PERT	n^+	38.7	0.644	0.784	19.6	0.852

4 CONCLUSIONS

We developed crystalline Si solar cell fabrication process for both n- and p-bifacial PERT cell using the NMS ion implantation.

At first, we fabricated n-bifacial SE PERT solar cell and were able to control of Rsh of SE structure by controlling of conditions of ion implantation and activation annealing. Cell efficiency was obviously increased by introducing SE structure. Additionally, efficiency was increased with the increase in Rsh_{p+}, and the gain of η by SE structure was $0.4\%_{abs}$. Combination of SE and other optimization of cell processes improved all IV parameters, and we achieved 21%.

Furthermore, we fabricated p-bifacial PERT cell using same process for n-bifacial PERT, and difference of η was just $0.7\%_{abs}$. Thus, we were able to confirm the applicability of our developed process to p-type cell.

These results clearly show the effectiveness of our PERT cell process using NMS ion implanter and the feasibility of high efficiency low cost solar cell using our cell fabrication process.

5 ACKNOWLEDGEMENT

This work was supported by the New Energy and Industrial Technology Development Organization (NEDO) under the Ministry of Economy, Trade and Industry (METI) in Japan. Authors are grateful to Mr. M. Morimura of TTI for his support for the experiment.

6 References

[1] International Technology Roadmap for Photovoltaic, Edition 2017, retrieved Aug. 23, 2017. From http://www.itrpv.net

[2] N. Yamaguchi, D. Hironiwa, H. Suzuki, K Muramatsu and K. Nakamura, "Development of N-PERT Solar Cell using Non Mass Separation Type Ion Implantation", in 27th International Photovoltaic Science and Engineering Conference, 2017, 1TuPo.16.

LASER-DOPED SELECTIVE EMITTER - PROCESS DEVELOPMENT AND SPEED-UP

J. Weber, S. Gutscher, S. Lohmüller, E. Lohmüller, A. A. Brand
Fraunhofer Institute for Solar Energy Systems ISE, Heidenhofstr. 2, 79110 Freiburg, Germany
Phone: +49 761 – 4588 5057; e-mail: julian.weber@ise.fraunhofer.de

ABSTRACT: This study pursues the development of a laser-doped selective emitter (LDSE) for p-type silicon passivated emitter and rear solar cells with screen-printed and fired silver contacts on the front. The LDSE is formed via local laser doping from the two-layer stack system of phosphosilicate glass and silicon dioxide that is located on the wafer surface after tube furnace diffusion using phosphorus oxychloride ($POCl_3$) as liquid dopant precursor. We aim for minimum emitter dark saturation current density at the LDSE-metal interface $j_{0e,met}$ and minimum specific contact resistance ρ_c. We use both an atmospheric pressure $POCl_3$ diffusion process and a high throughput low pressure $POCl_3$ diffusion process. Both $POCl_3$ processes are combined with a green nanosecond laser process at wavelength $\lambda = 532$ nm and pulse repetition rates 60 kHz $\leq f_{rep} \leq$ 100 kHz. Furthermore, we investigate high-speed infrared laser processes at $\lambda = 1064$ nm and $f_{rep} = 2$ MHz for which heat accumulation is expected to become relevant during LDSE formation. For some of the tested laser processes within this work, $\rho_c \approx 1$ mΩcm^2 is achieved. At the same time, simulations with the Quokka3 skin solver that are based on the LDSE doping profiles show that the LDSEs have the potential to lead to a $j_{0e,met}$ reduction by more than 50% compared to the not laser-doped emitter.
Keywords: Silicon Solar Cell, PERC, p-type, Selective Emitter, Laser Processing, Doping

1 INTRODUCTION

Within the field of passivated emitter and rear solar cell (PERC) technology, selective emitter concepts are again of great interest [1-4]. Gains of more than 0.3%$_{abs}$ in energy conversion efficiency are expected compared to the homogeneous emitter approach [2-5]. However, such a boost in energy conversion efficiency can only be realized, if both emitter dark saturation current density for the passivated emitter regions $j_{0e,pass}$ as well as for the metallized emitter regions $j_{0e,met}$ are reduced [5,6]. A decrease of $j_{0e,pass}$ can be achieved by reducing the surface doping concentration of the emitter within the photoactive area [6], whereas $j_{0e,met}$ can be reduced by heavily doping the emitter underneath the front metallization [4,7]. The heavily doped emitter should further allow for optimal electrical contact properties to allow for high fill factors and thus feature minimal specific contact resistance ρ_c. For p-type silicon solar cells, the most dominant technology for implementing a selective emitter [7] is local laser doping [8,9].

Within this study, laser-doped selective emitter (LDSE) processes are developed for p-type silicon PERC solar cells with screen-printed and fired silver contacts on the front. We aim for low $j_{0e,met} < 500$ fA/cm^2 and low $\rho_c < 2$ mΩcm^2. The LDSE is formed via local laser doping from a two-layer stack system consisting of phosphosilicate glass (PSG) and silicon dioxide (SiO_2). This layer stack system is located on the wafer surface after tube furnace diffusion using phosphorus oxychloride ($POCl_3$) as liquid dopant precursor. Thus, the interaction between the $POCl_3$ diffusion and the laser doping process is essential for the LDSE formation. In this work, three different process combinations are tested:

(i) We combine an atmospheric pressure (AP) $POCl_3$ diffusion process [3] and a green nanosecond laser process at wavelength $\lambda = 532$ nm and pulse repetition rate 60 kHz $\leq f_{rep} \leq$ 100 kHz (see section 4.1).

(ii) We switch from AP to low pressure (LP) $POCl_3$ diffusion [10-12] (see section 4.2). Within a LP system, the amount of processed wafers per cycle can be doubled, whereas the $POCl_3$ consumption is reduced [10]. Therefore, this technology is expected to play a major role in solar cell fabrication in the near future. In the long term, we thus aim for combining LDSE processes with a LP $POCl_3$ diffusion process.

(iii) We use the AP $POCl_3$ diffusion process from (i) for testing high-speed infrared (IR) laser doping processes at $\lambda = 1064$ nm and $f_{rep} = 2$ MHz (see section 5). When applying spatially overlapping laser pulses with such high repetition rates, consecutive laser pulses most likely interact with each other [13]. Heat is accumulated at the silicon surface which results in a drastic decrease of the absorption length, especially for IR laser irradiation [14]. The influence of this phenomenon on the LDSE formation has not yet been investigated. However, IR lasers are the most economical laser beam sources in terms of investment and operation. The approach (iii) allows for further speeding-up the LDSE formation, in order to achieve process times of clearly below one second per wafer. In this sense, the experiments performed at $f_{rep} = 2$ MHz not only address a new field of research, but are also relevant with regard to future LDSE developments.

2 APPROACH

For investigating LDSE processes, we use p-type Czochralski-grown silicon (Cz-Si) wafers with 156 mm edge length and an alkaline textured surface and apply the already mentioned different $POCl_3$ diffusion processes that are specially designed for the LDSE approach [3,12]. Laser test fields with an area of a few cm^2 are created consisting of overlapping laser lines. We vary laser process parameters, in particular laser pulse energy, pulse overlap, and pulse duration. Subsequent to laser doping, the wafers obtain a wet-chemical removal of the PSG layer and the emitter sheet resistance R_{sh} is determined via four-point-probe (4pp) measurements.

For a selection of test fields, charge carrier concentration profiles are recorded by electrochemical capacitance-voltage (ECV) measurements. Based on the 4pp measurements, the profiles are corrected as described in section 3. The corrected profiles are then used as input for the Quokka3 skin solver [15]. By setting the fundamental surface recombination velocity of electrons and holes to $S_n = S_p = 1 \cdot 10^7$ cm/s, we obtain an estimation for $j_{0e,met}$ [4]. However, $j_{0e,met}$ tends to be underestimated, since the formation of metal crystallites between the silver contacts and the silicon surface as well as potentially increased recombination due to residual laser damage are neglected.

On some of the wafers, a silicon nitride (SiN_x) passivation layer is deposited and metal contacts with a width of about 40 µm are applied by screen-printing a commercially available silver paste. Subsequently, contact firing is performed at different set peak temperatures. To determine ρ_c, transmission line measurements (TLM) [16] are performed.

3 CORRECTION OF ECV PROFILES

ECV profiling enables recording the charge carrier concentration $N(z)$ within a doped semiconductor as a function of depth z. An excellent review on this technique is given in Ref. [17].

Fig. 1 shows three exemplary ECV measurements of identically processed laser test fields. One of the samples received a wet-chemical PSG removal, whereas for two of the samples, a PSG removal and an additional wet-chemical emitter etch back (EEB) are performed. Shifting the ECV profiles against each other shows good agreement between all measurements. However, a decrease of N towards the silicon surface is observed. This pile-down effect leads to a mismatch of the profiles and is thus interpreted as an artefact introduced by the ECV technique. The same conclusion was drawn by Kimmerle [18] when analyzing ECV measurements of not laser-processed samples. Also, in theory, rather a pile-up is expected for phosphorous-doped silicon, whereas a pile-down cannot be explained physically [19]. In consequence, for all of the ECV measurements within this work, the pile-down artefact is removed and the profiles are straightened in the surface-near region. In addition, the part of the profiles in which $N(z)$ is flattened by the influence of the base doping is cut off. In Fig. 1 this would be $N(z > 650\text{ nm})$. However, correction is not yet completed.

A further correction step is necessary, since during the ECV measurement, it is assumed that the surface area of the electrolyte-semiconductor interface, denoted by A, is plane. This assumption is false, since the textured and laser processed silicon surface exhibits a rough topography. The enlargement of the surface can be

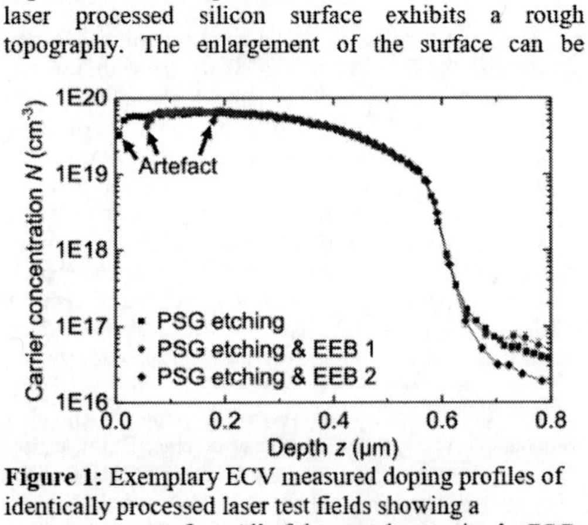

Figure 1: Exemplary ECV measured doping profiles of identically processed laser test fields showing a measurement artefact. All of the samples received a PSG removal. For two of the samples, an additional emitter etch back (EEB) is performed. The corresponding profiles are shifted to the right, so that they show the best possible match to the black squares.

accounted for by introducing the factor f in the sense of $A^{corr} = fA$ [18]. The factor f can be found by adjusting the emitter sheet resistance resulting from the ECV profile to the 4pp measured sheet resistance [18,20]. This adjustment eventually yields the corrected doping profile [18]. For the doping profiles corrected within this work, $1.0 \leq f \leq 1.5$ is found.

4 PROCESS DEVELOPMENT

4.1 AP $POCl_3$ diffusion & green laser at $f_{rep} \leq 100$ kHz

For the first group of wafers, we apply an AP $POCl_3$ diffusion process that is specially designed for the LDSE approach [3]. The resulting emitter yields $j_{0e,pass} \approx 40$ fA/cm^2 for textured and SiN_x-passivated surfaces [3]. For creating the laser test fields, a nanosecond laser is used at $\lambda = 532$ nm and 60 kHz $\leq f_{rep} \leq$ 100 kHz. The $1/e^2$-radius of the laser beam is $\omega_0 = (50.0 \pm 0.4)$ µm. The position of the test fields is varied for each wafer, in order to account for inhomogeneities of the PSG layer across the wafers. In total, eight different parameter sets are tested differing in laser pulse energy E_p, (full width half maximum) pulse duration τ_p, and pulse pitch d_p (distance of the laser pulses within the laser lines as well as from line to line). For fabricating the TLM samples, two different set peak firing temperatures are chosen: 840 °C and 880 °C.

Fig. 2 shows R_{sh} and ρ_c in dependence on E_p, τ_p, d_p, and the set peak firing temperature, whereas the errors for E_p and τ_p are about 3%. We compare the laser-doped test fields with the emitter that results from AP $POCl_3$ diffusion and therefore serves as reference ("Ref"). Concerning the emitter sheet resistance, a drop from $R_{sh} \approx 140\ \Omega$/sq to $21\ \Omega/\text{sq} < R_{sh} < 49\ \Omega$/sq after laser doping is observed. However, with increasing pulse duration τ_p, higher pulse energies E_p are necessary to achieve the same reduction in sheet resistance R_{sh}. In general, E_p shows a strong impact on R_{sh}. For the specific contact resistance ρ_c, an optimum occurs within the tested

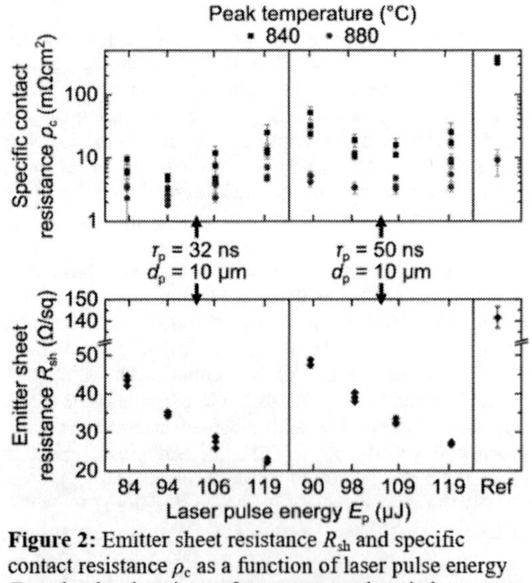

Figure 2: Emitter sheet resistance R_{sh} and specific contact resistance ρ_c as a function of laser pulse energy E_p and pulse duration τ_p for constant pulse pitch $d_p = 10$ µm. The emitter resulting from the AP $POCl_3$ diffusion process serves as reference ("Ref"). For ρ_c, a variation of the set peak firing temperature is shown.

range of E_p for both pulse durations. By increasing the set peak firing temperature from 840 °C to 880 °C, a decrease of ρ_c is obtained. For the shorter pulse duration $\tau_p = 32$ ns, ρ_c tends to be lower than for the longer pulse duration $\tau_p = 50$ ns. Minimal specific contact resistance $\rho_c \approx 2$ mΩcm² is achieved by $\tau_p = 32$ ns, $E_p = 94$ µJ and 880 °C set peak firing temperature. For this temperature, the reference yields $\rho_c \approx 9$ mΩcm².

For some of the laser parameter sets shown in Fig. 2, ECV profiles are recorded. Again, the emitter that results from AP $POCl_3$ diffusion serves as reference. Fig. 3 sums up the corrected doping profiles, the factor f that is used for correction, and the corresponding Quokka3 estimations for $j_{0e,met}$. The laser-doped emitters are about 0.7 µm deep, which is about twice as deep as the reference emitter. Doping depth grows with increasing laser pulse energy E_p and pulse duration τ_p. For the reference emitter, we find the surface-near doping concentration $N_{surf} \approx 4 \cdot 10^{19}$ cm^{-3}. Laser doping does not change N_{surf} drastically. However, for the longer pulse duration $\tau_p = 50$ ns, N_{surf} tends to be slightly reduced. Interestingly, both doping profiles obtained for the shorter pulse duration $\tau_p = 32$ ns show an increase with growing depth z for $z < 0.2$ µm, which is unexpected from a physical point of view. The reason for this unexpected finding is not clear yet. Possibly, the observed increase is an artefact of the ECV measurement similar to the pile-down artefact (see Fig. 1) described in section 3. It is also possible that laser doping results in a local inhomogeneous charge carrier concentration. Such inhomogenities cannot be resolved by the ECV method, since the carrier concentration is averaged over several mm². The factor f that quantifies the enlargement of the surface is found to be $f = 1.34$ for the reference. By laser-doping, f is reduced. This can be explained by the fact that the surface gets molten and therefore smoothened during the laser process [8]. For $\tau_p = 32$ ns, a decrease from $f = 1.14$ to $f = 1.09$ is found for increasing pulse

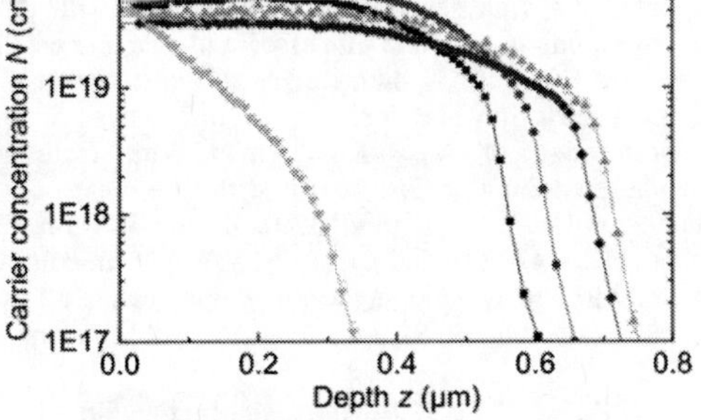

	E_p (µJ)	d_p (µm)	τ_p (ns)	f	$j_{0e,met}$ (fA/cm²)
■	94	10	32	1.14	271
●	106	10	32	1.09	230
◆	98	10	50	1.22	292
▲	109	10	50	1.17	257
▼	Ref			1.34	932

Figure 3: Emitter doping profiles measured by ECV and corrected as discussed for some of the laser parameter sets shown in Fig. 2. The emitter resulting from the AP $POCl_3$ diffusion process serves as reference ("Ref"). For the correction, the factor f is used. Via Quokka3, the dark saturation current density at the emitter-metal interface $j_{0e,met}$ is estimated for each profile.

energy from $E_p = 94$ µJ to $E_p = 106$ µJ. This is due to the increase in time in which the surface is molten [8]. The same effect is observed for $\tau_p = 50$ ns. Here a reduction from $f = 1.22$ ($E_p = 98$ µJ) to $f = 1.17$ ($E_p = 109$ µJ) is obtained. Concerning the estimated $j_{0e,met}$, the amount of electrically active phosphorous atoms per area, which we call the phosphorous dose D, plays the major role [4]. If doping depth and surface-near doping concentration are increased, the phosphorous dose D grows and $j_{0e,met}$ is reduced. However, no drastic differences between the tested laser processes occur, since the phosphorous dose is within the same range of $D \approx 2 \cdot 10^{15}$ cm^{-2}. Compared to the reference ($D = 4 \cdot 10^{14}$ cm^{-2}), a reduction from $j_{0e,met} = 932$ fA/cm² to 230 fA/cm² $\leq j_{0e,met} \leq$ 292 fA/cm² is achieved according to our estimation.

4.2 LP $POCl_3$ diffusion & green laser at $f_{rep} \leq 100$ kHz

For the second group of wafers, we use the same laser for creating the LDSE test fields as for the first wafer group. We stick to the short pulse duration $\tau_p = 32$ ns and vary the pulse pitch d_p instead. Furthermore, we replace the AP $POCl_3$ diffusion process by a LP $POCl_3$ diffusion process. This LP process is similar to the one introduced in Ref. [12]. It is specially designed for the LDSE approach and yields $j_{0e,pass} \approx 43$ fA/cm² for textured and SiN_x-passivated surfaces. For fabricating the TLM samples, two different set peak firing temperatures are chosen: 810 °C and 850 °C.

Fig. 4 shows the results obtained for R_{sh} and ρ_c. We find that laser doping results in a reduction from $R_{sh} \approx 150$ Ω/sq to 36 Ω/sq $< R_{sh} <$ 81 Ω/sq. The pulse energy E_p shows a strong impact on R_{sh}, whereas the pulse pitch d_p plays only a minor role. Interestingly, at $E_p = 85$ µJ, a decrease in R_{sh} is obtained for increasing the pitch from $d_p = 10$ µm to $d_p = 25$ µm, even though the applied energy per area is reduced. This counterintuitive finding could be due to variations within the properties of the PSG/SiO_2 layer system and the reference emitter that are caused by inhomogenities within the $POCl_3$ diffusion

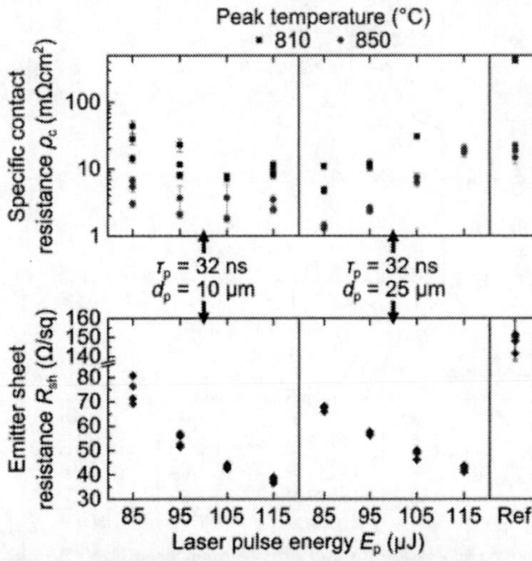

Figure 4: Emitter sheet resistance R_{sh} and specific contact resistance ρ_c as a function of laser pulse energy E_p and pulse pitch d_p for constant pulse duration $\tau_p = 32$ ns. The emitter resulting from the LP $POCl_3$ diffusion process serves as reference ("Ref"). For ρ_c, a variation of the set peak firing temperature is shown.

process. Concerning ρ_c, an optimum is found for $d_p = 10$ µm. This is in good accordance with the corresponding results obtained for the AP $POCl_3$ diffusion process ($\tau_p = 32$ ns and $d_p = 10$ µm in Fig. 2). For both AP and LP $POCl_3$ diffusion process, minimal specific contact resistance $\rho_c \approx 2$ mΩcm² is reached for $E_p \approx 100$ µJ. For the LP $POCl_3$ diffusion process and $d_p = 25$ µm (see Fig. 4), we expect the minimum of ρ_c to be reached at $E_p < 85$ µJ. However, within the tested pulse energy range, 1 mΩcm² $< \rho_c <$ 2 mΩcm² is achieved at 850 °C, whereas the reference yields $\rho_c \approx 19$ mΩcm². For 810 °C, ρ_c is found to increase.

For a selection of test fields, doping profiles are recorded and $j_{0e,met}$ is estimated. Fig. 5 sums up the results. If pulse energy is raised, the doping depth increases. Compared to the reference, profile depth is enlarged from about 0.3 µm to about 0.5 µm ($E_p = 85$ µJ) and 0.7 µm ($E_p = 115$ µJ), respectively. By changing the pulse pitch from $d_p = 10$ µm to $d_p = 25$ µm, the doping profiles are steepened and surface-near doping concentration is increased by more than $\Delta N_{surf} \approx 1 \cdot 10^{19}$ cm^{-3}. For $E_p = 85$ µJ, this increase yields $N_{surf} \approx 4 \cdot 10^{19}$ cm^{-3}, whereas for $E_p = 115$ µJ, $N_{surf} \approx 5 \cdot 10^{19}$ cm^{-3} is obtained. For the factor f that is used to correct the doping profiles, similar trends as in section 4.1 are observed. The reference yields $f = 1.50$, which is the largest value that is found. By laser-doping, f is reduced. This reduction strongly depends on the laser pulse energy E_p, whereas the pulse pitch plays only a minor role. For $E_p = 85$ µJ, $f \approx 1.32$ is found, whereas $E_p = 115$ µJ yields $f \approx 1.05$.

Concerning the $j_{0e,met}$ estimation, significant differences for the laser-doped test fields are obtained which reflects the differences in doping dose [4]. By all tested LDSE processes, however, a drastic reduction from $j_{0e,met} = 1070$ fA/cm² to $j_{0e,met} \leq 489$ fA/cm² is achieved according to the estimation. For $d_p = 25$ µm, a tradeoff between $j_{0e,met}$ and ρ_c exists within the tested range of E_p, since ρ_c but also the doping dose decreases when laser

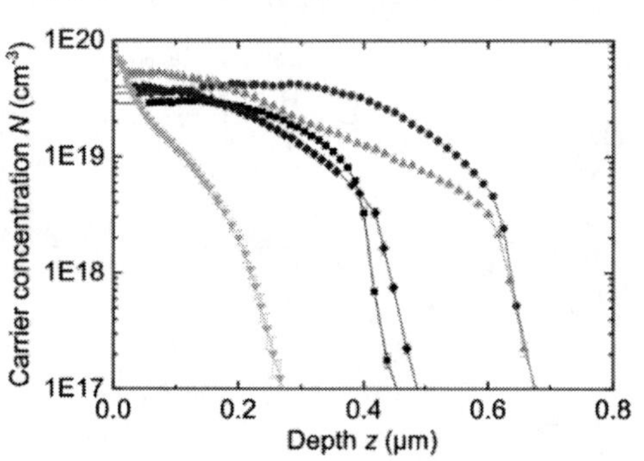

	E_p (µJ)	d_p (µm)	τ_p (ns)	f	$j_{0e,met}$ (fA/cm²)
	85	10	32	1.35	489
	115	10	32	1.08	281
	85	25	32	1.29	469
	115	25	32	1.02	313
		Ref		1.50	1070

Figure 5: Emitter doping profiles measured by ECV and corrected as discussed for some of the laser parameter sets shown in Fig. 4. The emitter resulting from the LP $POCl_3$ diffusion process serves as reference ("Ref"). For the correction, the factor f is used. Via Quokka3, the dark saturation current density at the emitter-metal interface $j_{0e,met}$ is estimated for each profile.

pulse energy is reduced.

Comparing the results of this section with the ones obtained for the AP $POCl_3$ diffusion process presented in section 4.1, we identify the most promising approach to be the combination of the LP $POCl_3$ diffusion process with the LDSE process at $\tau_p = 32$ ns, $E_p = 85$ µJ, and $d_p = 25$ µm. By this approach, the throughput of the $POCl_3$ diffusion process is increased and minimal specific contact resistance $\rho_c \approx 1$ mΩcm² is achieved. However, an experimental analysis of $j_{0e,pass}$ and $j_{0e,met}$ for the tested LDSE processes is required for a more sophisticated process development.

5 PROCESS SPEED-UP

For the third group of wafers, we apply the same AP $POCl_3$ process as for the first group. Laser doping, however, is performed using an IR laser ($\lambda = 1064$ nm) at constant repetition rate $f_{rep} = 2$ MHz, pulse duration $\tau_p \approx 30$ ns, and $1/e^2$-radius $\omega_0 \approx 28$ µm. A polygon scanner system [22] is used to create laser lines with scan speeds $v_s \geq 40$ m/s and a constant line pitch of 10 µm. For these laser parameters, we expect heat accumulation [13] to occur during the LDSE formation.

Within a first experiment, we test two laser processes A and B differing only in scan speed v_s. As Fig. 6 shows, after applying laser process A, no modification of the wafer surface is observed. In contrast, for process B for which the spatial overlap of the laser pulses within each laser line is increased, the threshold of modification is exceeded. This could be due to an enhanced heat accumulation resulting from the increase in overlap of the pulses. However, these observations can only be interpreted as an indication for the presence of heat accumulation.

We use process B as a starting point for testing potential high-speed laser processes for LDSE formation. In order to achieve significant doping, we raise pulse energy to $E_p = 44$ µJ and $E_p = 47$ µJ, respectively, while keeping the scan speed at $v_s = 40$ m/s. For both pulse energies, one test field is processed and characterized by 4pp and ECV. Fig. 7 summarizes the obtained results. Both doping profiles exhibit a surface-near doping concentration of $N_{surf} \approx 4 \cdot 10^{19}$ cm^{-3}. With increasing depth, the doping profiles start to differ from each other. For $E_p = 44$ µJ, the doping depth is below 1 µm. We obtain $R_{sh} = 47$ Ω/sq and $j_{0e,met} = 335$ fA/cm². In contrast, $E_p = 47$ µJ yields a doping depth larger than 1 µm and thus a reduction in R_{sh} of about $\Delta R_{sh} = 8$ Ω/sq and an

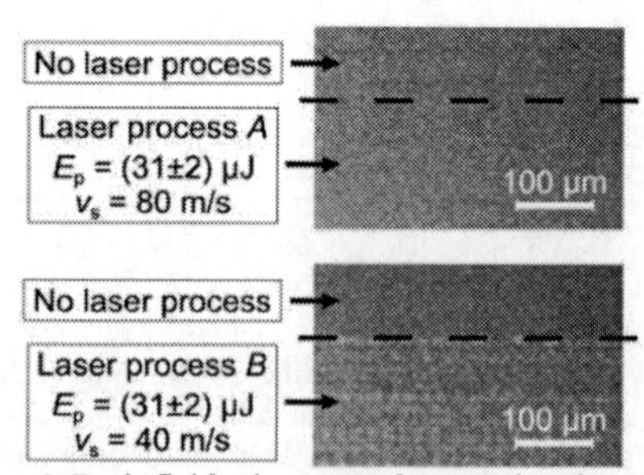

Figure 6: Dark-field microscope images showing parts of the wafer surface after performing the AP $POCl_3$ diffusion process and applying laser processes A and B with identical laser pulse energy E_p. A modification of the surface is observed by decreasing the scan speed v_s.

increase in $j_{0e,met}$ of $\Delta j_{0e,met} = 49$ fA/cm^2. Concerning the factor f, we find a slight decrease from $f = 1.10$ to $f = 1.05$ for increasing laser pulse energy. As already mentioned, this can be explained by the increase in time in which the wafer surface is molten during laser doping.

Compared to the doping profiles shown in Fig. 3 and Fig. 5, the high-speed laser processes yield steeper emitter profile and an increased doping depth. However, in terms of R_{sh}, f, estimated $j_{0e,met}$, and N_{surf}, similar results are achieved. Nevertheless, further research is necessary to investigate, if also for high-speed IR laser processes, low specific contact resistance $\rho_c < 2$ mΩcm^2 can be achieved. Also, deeper knowledge is required concerning the phenomenon of the heat accumulation.

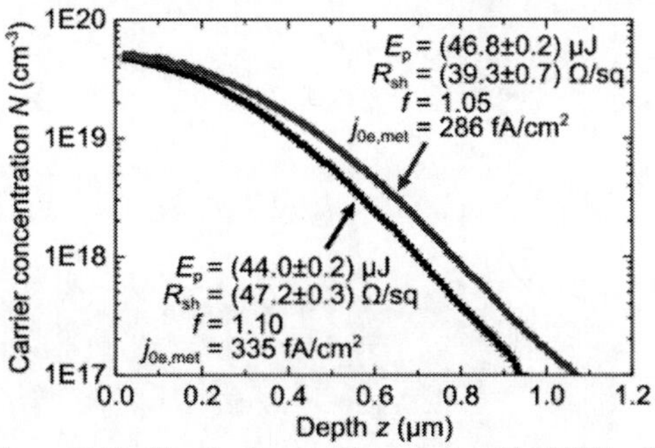

Figure 7: Emitter doping profiles measured by ECV and corrected by the factor f, emitter sheet resistance R_{sh}, and estimated dark saturation current density at the emitter-metal interface $j_{0e,met}$ for two laser processes with constant scan speed $v_s = 40$ m/s and differing laser pulse energy E_p. Laser doping is combined with the AP $POCl_3$ diffusion process.

6 SUMMARY AND CONCLUSION

Within this work, we develop laser-doped selective emitter (LDSE) processes for p-type silicon passivated emitter and rear solar cells with screen-printed and fired silver contacts on the front. For local laser doping, we use the two-layer stack system of phosphosilicate glass and silicon dioxide that is located on top of the silicon surface after phosphorus oxychloride ($POCl_3$) tube furnace diffusion. For the LDSE doping profiles that are recorded within this work, the emitter dark saturation current density at the LDSE-metal interface is reduced to 230 fA/cm$^2 \leq j_{0e,met} \leq 489$ fA/cm^2, which is a reduction of more than 50% compared to the not laser-doped emitter according to a Quokka3 estimation. For the combination of the atmospheric pressure $POCl_3$ diffusion process and the green nanosecond laser at wavelength $\lambda = 532$ nm, specific contact resistance $\rho_c \approx 2$ mΩcm^2 is achieved by using laser pulse energy $E_p = 94$ µJ, pulse duration $\tau_p = 32$ ns, and pulse pitch $d_p = 10$ µm. For the low pressure $POCl_3$ diffusion process with increased throughput, $\rho_c \approx 1$ mΩcm^2 is achieved by $E_p = 85$ µJ, $\tau_p = 32$ ns, and $d_p = 25$ µm. When applying high-speed infrared (IR) laser processes at pulse repetition rate $f_{rep} = 2$ MHz, we expect heat accumulation to occur. However, further research is necessary to understand under which conditions this phenomenon becomes relevant and how it affects the doping profile of the LDSE, the crystal structure, and the properties of the LDSE-metal interface, in particular the specific contact resistance.

ACKNOWLEDGEMENTS

This work was funded by the German Federal Ministry for the Environment, Nature Conservation and Nuclear Safety within the research project "POLDI" (Contract Number 0324079D).

REFERENCES

[1] M. Müller. G. Fischer, B. Bitnar et al., "22.61% efficient fully screen printed PERC solar cell," Energy Procedia, vol. 124, pp. 131-137, 2017.

[2] S. Wasmer, A. A. Brand, and J. M. Greulich, "Metamodeling of numerical device simulations to rapidly create efficiency optimization roadmaps of monocrystalline silicon PERC cells," Energy Procedia, vol. 124, pp. 207-214, 2017.

[3] S. Werner, E. Lohmüller, P. Saint-Cast et al., "Key aspects for fabrication of p-type Cz-Si PERC solar cells exceeding 22% conversion efficiency," in 33rd European Photovoltaic Solar Energy Conference and Exhibition, 2017, pp. 406-412.

[4] J. Weber, S. Lohmüller, E. Lohmüller et al., "Simulations on laser-phosphorous-doped selective emitters," in 7th World Conference on Photovoltaic Energy Conversion, to be published, 2018.

[5] A. Wolf, A. Kimmerle, S. Werner et al., "Status and perspective of emitter formation by POCl3-diffusion," in 31st European Photovoltaic Solar Energy Conference and Exhibition, 2015, pp. 414-419.

[6] S. Werner, E. Lohmüller, S. Maier et al., "Challenges for lowly-doped phosphorus emitters in silicon solar cells with screen-printed silver contacts," Energy Procedia, vol. 124, pp. 936-946, 2017.

[7] ITRPV, "International technology roadmap for photovoltaic (ITRPV) - 2017 Results - Ninth edition, March 2018," available online: http://www.itrpv.net/Reports/Downloads/, 2018.

[8] U. Jäger, "Selektive Laserdiffusion für hocheffiziente Solarzellen aus kristallinem Silicium," München: Verlag Dr. Hut, 2013.

[9] L. Ventura, A. Slaoui, and J. C. Muller, "Realization of selective emitters by rapid thermal and laser assisted techniques," in 13th European Photovoltaic Solar Energy Conference and Exhibition, 1995, pp. 1578-1581.

[10] M. Mühlbauer, A. Piechulla, C. Voyer et al., "Industrial low-pressure phosphorus diffusion for high performance and excellent uniformity," in 26th European Photovoltaic Solar Energy Conference and Exhibition, 2011, pp. 2028-2030.

[11] S. Lohmüller, S. Schmidt, E. Lohmüller et al., "Transfer of $POCl_3$ diffusion processes from atmospheric pressure to high throughput low pressure," AIP Conference Proceedings, vol. 1999, pp. 070002-1-070002-6, 2018.

[12] S. Lohmüller, M. Meßmer, S. Schmidt et al., "Analysis of phosphosilicate glass structures formed by atmospheric pressure and high throughput low pressure $POCl_3$ diffusion," in 7th World Conference

on Photovoltaic Energy Conversion, to be published, 2018.
[13] R. Weber, T. Graf, P. Berger et al., "Heat accumulation during pulsed laser materials processing," Optics Express, vol. 22, pp. 11312-11324, 2014.
[14] E. H. Sin, C. K. Ong, and H. S. Tan, "Temperature dependence of interband optical absorption of silicon at 1152, 1064, 750, and 694 nm," Physica Status Solidi (a), vol. 85, pp. 199-204, 1984.
[15] A. Fell, J. Schön, M. C. Schubert et al., "The concept of skins for silicon solar cell modeling," Solar Energy Materials and Solar Cells, vol. 173, pp. 128-133, 2017.
[16] H. H. Berger, "Contact resistance and contact resistivity", Journal of the Electrochemical Society, vol. 119, pp. 507–514, 1972.
[17] P. Blood, "Capacitance-voltage profiling and the characterisation of III-V semiconductors using electrolyte barriers," Semiconductor Science and Technology, vol. 1, pp. 7-27, 1986.
[18] A. Kimmerle, "Diffused surfaces for crystalline silicon solar cells - process development, characterization, and modeling," PhD thesis, University of Freiburg, Germany, 2015.
[19] A. S. Grove, O. Leistiko, and C. T. Sah, "Redistribution of acceptor and donor impurities during thermal oxidation of silicon," Journal of Applied Physics, vol. 35, pp. 2695-2701, 1964.
[20] R. Bock, P. Altermatt, and J. Schmidt, "Accurate extraction of doping profiles from electrochemical capacitance voltage measurements," in 23rd European Photovoltaic Solar Energy Conference and Exhibition, 2008, pp. 1510-1513.
[21] D. B. M. Klaassen, "A unified mobility model for device simulation - I. model equations and concentration dependence," Solid-State Electronics, vol. 35, pp. 953-959, 1992.
[22] R. De Loor, "Polygon scanner system for ultra short pulsed laser micro-machining applications," Physics Procedia, vol. 41, pp. 544-551, 2013.

DEVELOPMENT OF 4 NM-THIN PECVD ALUMINIUM OXIDE USING PLASMA ANALYSIS AND ITS APPLICATION TO PERC SOLAR CELLS AND MODULES

M. Hofmann[1], D. Wagenmann[1], C. Teßmann[1], M. Jäcklein[1], P. Saint-Cast[1], D. Eberlein[1], A. Kraft[1], T. Dippell[2], F. May[2], M. Dörr[2], B. Cord[2], T. Schütte[3], P. Neiß[3], L. Eichhorn[4], M. Klick[4], U. Richter[5], M. Siemers[6], P. Wiedemuth[7]
[1] Fraunhofer Institute for Solar Energy Systems ISE, Heidenhofstr. 2, 79110 Freiburg, Germany
[2] SINGULUS Technologies AG, Hanauer Landstr. 103, 63796 Kahl am Main, Germany
[3] PLASUS GmbH, Lechstr. 9, 86415 Mering, Germany
[4] Plasmetrex GmbH, Schwarzschildstr. 3, 12489 Berlin, Germany
[5] Sentech Instruments GmbH, Schwarzschildstr. 2, 12489 Berlin, Germany
[6] Fraunhofer Institute for Surface Engineering and Thin Films IST, Bienroder Weg 54E, 38108 Braunschweig, Germany
[7] TRUMPF Hüttinger GmbH + Co. KG, Bötzinger Str. 80, 79111 Freiburg, Germany

ABSTRACT: Very thin aluminum oxide passivation films (3.7 nm) are developed using ICP PECVD and applying plasma simulation (particle-in-cell Monte Carlo method), in-situ plasma characterization (spectroscopic plasma monitoring, self-excited electron plasma resonance spectroscopy, laser ellipsometry) and ex-situ layer analysis (lifetime and density of interface traps investigation). The thin (3.7 nm) AlO_x films are applied to PERC-type solar cells reaching 21% efficiency slightly outperforming solar cells applying standard 15 nm AlO_x layers. All solar modules manufactured applying the 3.7 nm thin AlO_x passivation layer successfully passed standard damp-heat and thermal-cycling tests proving the applicability of the novel technology.
Keywords: Passivation, PECVD, AlO_x, plasma monitoring, PERC, PV module, silicon solar cell

1 INTRODUCTION

In 2017 and 2018, more and more passivated emitter and rear cell (PERC) –type [1] solar cell production facilities were installed on industrial level. This trend is expected to go on and lead to PERC-type cells being the standard cell technology in the near future. One very attractive route for the implementation of the industrial rear passivation technology is the application of plasma-enhanced chemical vapor deposition (PECVD) of aluminum oxide (AlO_x) [2-4]. Its attraction is driven by the industrially proven PECVD technology already applied for decades for the front side anti-reflection coating (ARC) using silicon nitride (SiN_x) layers. PECVD will stay in use for the ARC application; therefore it is very logic to use the same technology for the AlO_x deposition as well. On the other hand, despite a successful proof in literature for a very high surface passivation quality possible by PECVD AlO_x the competition with atomic layer deposition (ALD) of AlO_x layers is still ongoing.

2 EXPERIMENTAL

ICP-PECVD allows very effective gas dissociation [5]. Thus, the gases can be used very efficiently. In this work, a SINGULUS Singular ICP-PECVD tool was applied for the fabrication of thin AlOx layers. Sketches and a more in-depth description of the tool can be found in [6]. The excitation of the plasma was performed using a frequency of 13.56 MHz. Additionally, further plasma process investigations were performed in a test chamber setup applying the same plasma excitation principle.
The deposition experiments and carrier lifetime investigations were carried out using float-zone (FZ) monocrystalline silicon wafers (p-type, boron doped) with a resistivity of 1…1.5 Ω cm and a thickness of about 250…300 µm. The solar cells were fabricated on p-type Czochralski-grown (CZ) monocrystalline silicon wafers of 156 mm x 156 mm format.
Plasma analysis applying spectroscopic plasma monitoring (SPM) provides deep time-resolved insights into the gas dynamics within the plasma. Therefore, one can recognize what kinds of species are present and how to control their ratios. Plasma characterization was in prior use in Si PV for example by Gabriel et al. [7] and us [6]. In our current work, a PLASUS EMICON 3MC system was applied. This is combined with self-excited electron plasma resonance spectroscopy (SEERS) to analyse electrical plasma properties. For this purpose, a Plasmetrex Hercules system was used. For a part of the investigations, an in-situ laser ellipsometer (LE) by Sentech was applied to observe the layer growth. Additionally, in-situ characterization must be combined with ex-situ analysis of the created layers to achieve a correlation of parameters and an optimization. Contactless capacitance-voltage (COCOS), spectroscopic ellipsometry and carrier lifetime measurements were performed besides others. Detailed characteristics of the plasma technology were elucidated by particle-in-cell Monte Carlo (PIC-MC) simulations of the ICP-PECVD process [8] providing local electron temperature, electron density and other distributions in the PECVD reactor that can facilitate an efficient deposition process. Finally, the layers are evaluated on PERC cells and in solar modules.

3 RESULTS

3.1 Plasma Process Simulation

Plasma deposition processes were modelled using particle-in-cell Monte Carlo (PIC-MC) simulations. Here, a code developed at Fraunhofer IST was applied [8] delivering additional insights into mechanisms within the deposition processes.
A first simulation example is shown in Figure 1: In (a) the reproduction of the chamber geometry is presented that is used for the simulations. The substrates are hanging in the upper part of the chamber, the gases and the radio frequency power are introduced to the chamber below the substrates.

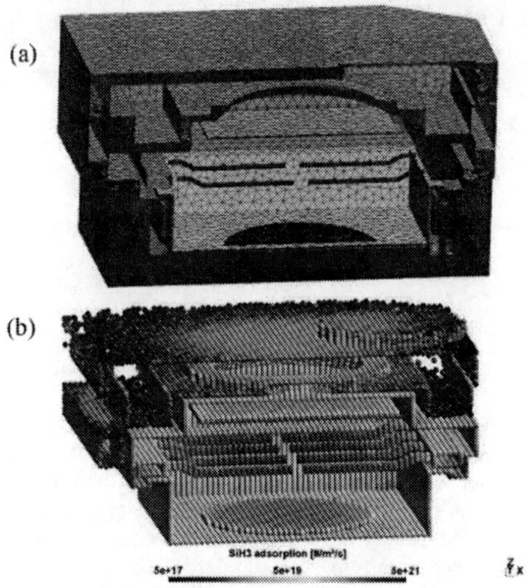

Figure 1: (a) Model of the deposition chamber. (b) simulated SiH_3 adsorption distribution in the chamber as one main factor for the layer homogeneity distribution.

Figure 1 (b) shows the distribution of SiH_3 molecules which were found to be one of the main parameters allowing the description of the SiN_x layer growth distribution. The square-shaped substrates in the upper part of the chamber are showing a homogeneous layer distribution.

3.2 in-situ Plasma Characterisation

The combination of plasma simulation, ex-situ layer (plasma process result) characterisation and the measurement of plasma process-describing in-situ parameters provides the best way to understand as much as possible about a plasma process. For the in-situ characterisation we investigated SiN_x and AlO_x deposition processes. Results of both processes are presented here.

SPM for chemical properties and SEERS for electrical properties of the plasma are used and combined with layer growth investigation applying LE. For one specific example of a cyclic SiN_x deposition process, Figure 2 provides first insights.

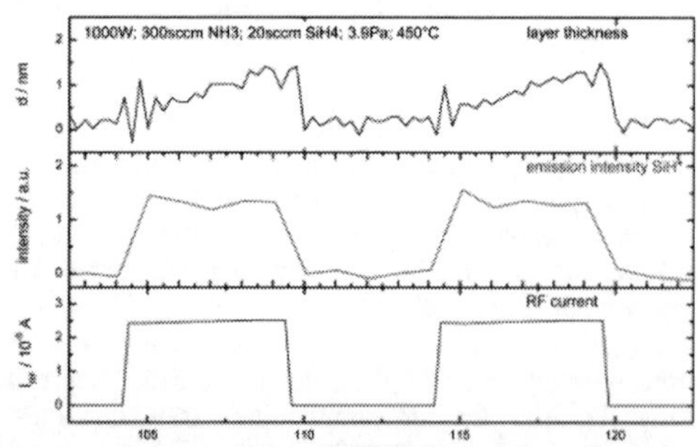

Figure 2: In-situ measurement results: layer thickness (laser ellipsometry LE, top), SiH optical emission intensity (spectroscopic plasma monitoring SPM, middle), RF current (self-excited electron plasma resonance spectroscopy SEERS, bottom).

Two deposition cycles are shown in Figure 2. The first deposition starts at ~104 s, the second at ~114 s. The RF current detected by SEERS is steeply increasing when the plasma is switched on. At the same time, the optical emission line of SiH increases (SPM-detected) and the layer thickness observed by LE is increasing.

As can be seen in the upper (LE) part of the figure, very small thickness changes are detectable. Within this small thickness change, also optical line intensities and variations of the electrical plasma properties are visible. Thus, a very powerful analysis of plasma processes is present.

Another example, also on a SiN_x deposition process, is exhibited in Figure 3. Here, results of SPM and SEERS are combined and shown over a time-dependent variation of the ICP power.

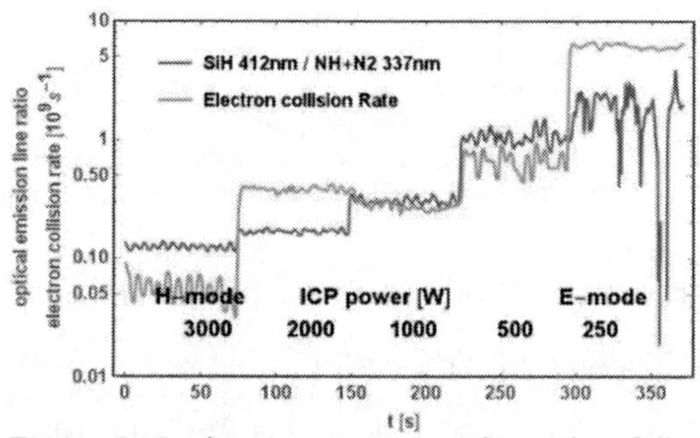

Figure 3: In-situ measurement results: ratio of line intensities SiH / (NH + N_2) of the SPM measurement, electron collision rate of the SEERS measurement. Variation of the plasma power over time.

Investigated here are, on the one hand, the optical emission line ratio of SiH at 412 nm wavelength and NH+N_2 at 337 nm and, on the other hand, the electron collision rate. Starting with at a high power of 3 kW, relatively low electron collision rates and SiH/(NH+N_2) ratios are detected. While stepwise decreasing the ICP power, both signals are increasing. Here, we are switching from the mainly inductive mode (H mode) at high powers to the mainly capacitive mode (E mode) for low power.

The surprising slight decrease in the electron collision rate when stepping from 2 kW to 1 kW is attributed to a combination of effects including a change in plasma skin depth. This depth is expected to decrease with increasing electron density und thus with increasing plasma power. The steady increase of the SiH/(NH+N_2) ratio with decreasing ICP power shows that the activation of SiH-type species happens more effectively at lower power compared to the activation of nitrogen-containing species.

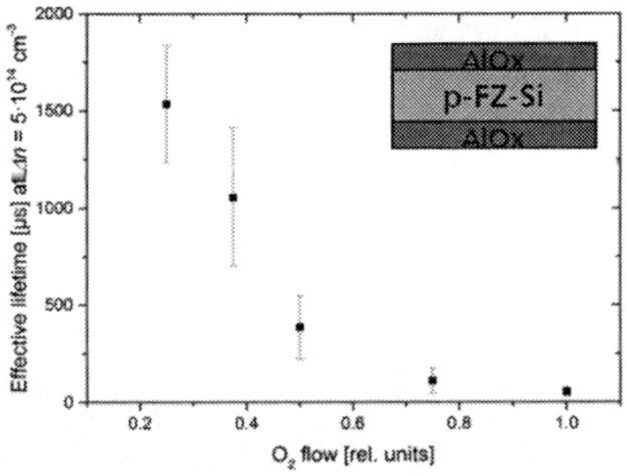

Figure 4: Dependency of the minority carrier lifetime on the O2 flow.

As already shown in [6], a significant effect of the oxygen flow on the passivation properties of a thin aluminium oxide layer on c-Si substrates can be observed for ICP PECVD deposition processes. Figure 4 shows this strong dependency. Here, symmetric samples were prepared using p-type float zone (FZ) Si wafers with shiny etched surfaces and a base resistivity of 1...1.5 Ω cm.

To analyse this phenomenon, SPM measurements were performed. Figure 5 shows AlO species in dependency of the O_2 flow. The AlO lines show a significant increase in case of a large AlO flow but quite stable conditions for lower flows.

In [6], we additionally pointed out that an increase in O_2 flow coincides with a decrease of hydrogen species.

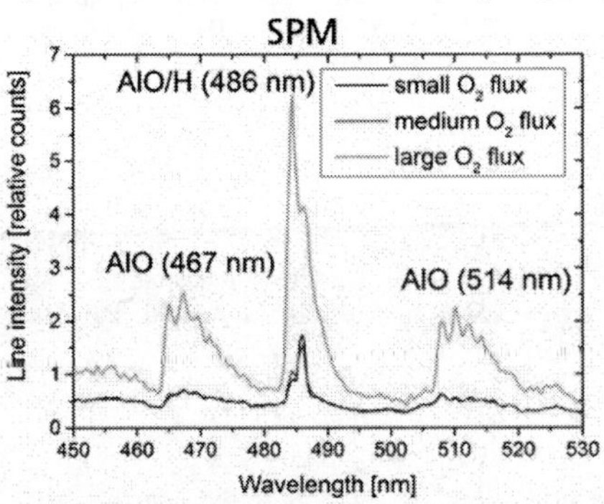

Figure 5: Plasma monitoring spectra of the AlO lines in dependency of the O_2 gas flux at constant flux of the aluminum precursor in the PECVD process, after actinometric data analysis.

3.3 Ex-situ Layer Analysis

COCOS measurements of the FZ Si wafers passivated with AlO_x further showed that the lifetime decrease is correlated with an increase in the density of interface traps (D_{it}). In Figure 6 this trend is visible with D_{it} values in the range between 1E11 to 1E12 cm^{-2} eV^{-1}.

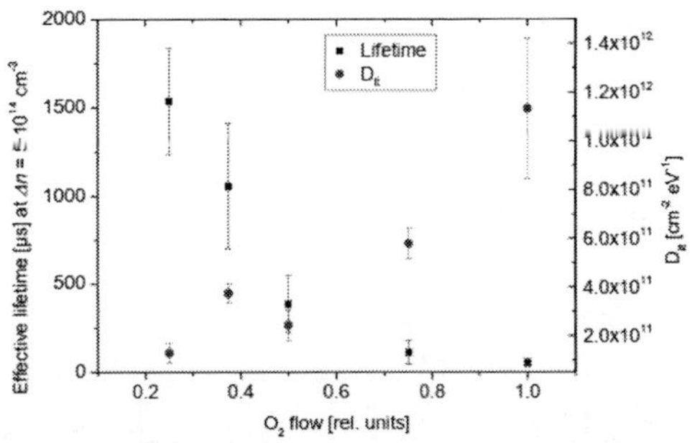

Figure 6: Carrier lifetime and interface trap density (D_{it}) over the O_2 flow.

Using the knowledge gained from the former investigations, we were able to decrease the layer thickness necessary for an excellent surface passivation using the ICP PECVD process as shown in Figure 7. This graph was already shown in [6], pointing out that excellent passivation down to 3.7 nm is possible with ICP-PECVD AlO_x layers which reach very thin layers formerly only achievable by ALD.

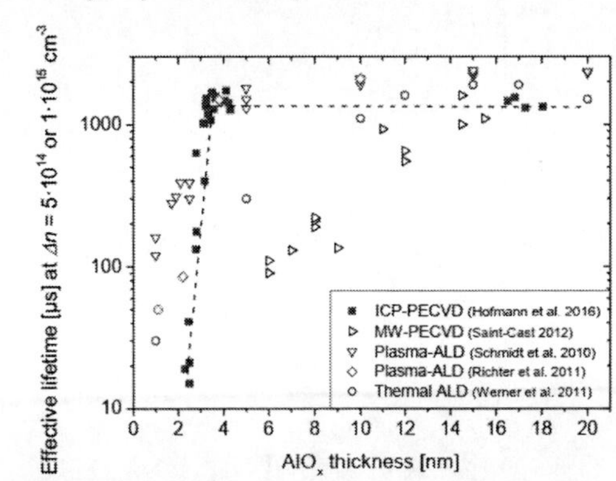

Figure 7: Carrier lifetimes of c-Si wafers with AlO_x layers of thicknesses in the range up to 20 nm on both sides after annealing (FZ-Si substrates 1.0…1.5 Ω cm, p-type, 250…300 µm). A slight variation of the O_2 flux provided the data behind the red ICP-PECVD symbols shown. It seems evident that the deposition process is quite robust in this parameter range. With changing O_2 flux, still excellent passivation properties were possible. Lifetimes of about 1.7 ms reflect a surface recombination velocity of about 5 cm/s. For comparison, also ALD-AlO_x and PECVD-AlO_x passivation layers on comparable substrate types from literature [9-12] are shown. Graph taken from [6].

3.4 Solar Cells and Modules

PERC-type solar cells were fabricated to show the applicability of the developed deposition processes. We would like to thank the company Hanwha Q-Cells for the pre-processing of the solar cells (texturisation and emitter diffusion). In the solar cell batch, a comparison between the current standard thickness for the AlO_x layer (15 nm) and the novel 3.7 nm AlO_x layer processes was performed. Additionally, a variation of the firing process peak temperature was used to find the optimum firing temperature conditions. Figure 8 gives an overview of the solar cell IV results.

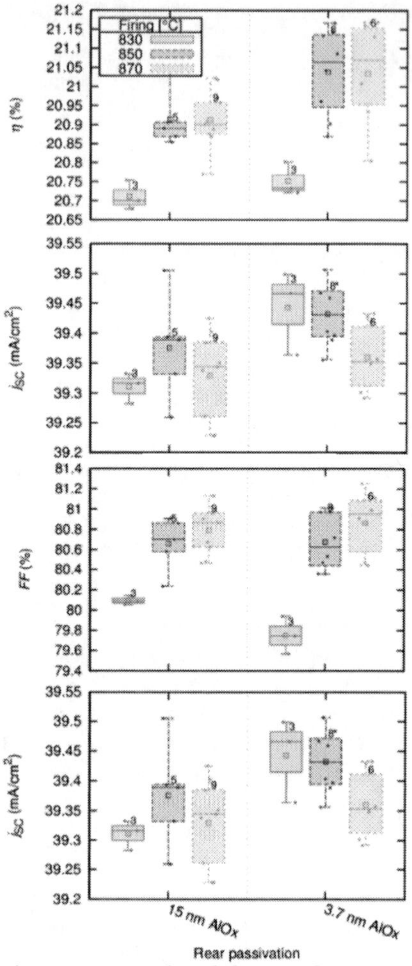

Figure 8: IV results of the manufactured PERC solar cells.

The fill factor (*FF*) is found to be increasing with increasing firing temperature whereas for the open-circuit voltage (V_{oc}), a peak voltage for the intermediate firing temperature was detected. No clear tendency was found for the short-circuit current density (j_{sc}) results. An increase for the energy conversion efficiency (η) of about 0.1 to 0.2% absolute was found for the solar cells applying the thinner (3.7 nm) AlO_x layers.

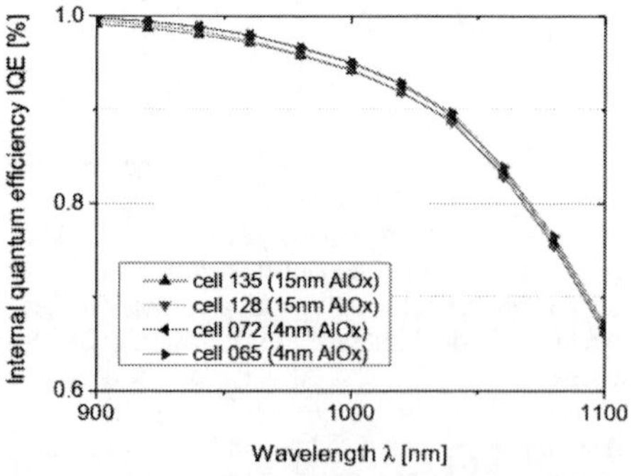

Figure 9: Exemplary internal quantum efficiency (IQE) curves of 2 PERC cells with thin (3.7 nm) and 2 cells with thick (15 nm) PECVD AlO_x layer.

This slight benefit of the 3.7 nm AlO_x layer in comparison to its thicker counterpart can also be found in internal quantum efficiency (IQE) measurements. Figure 9 shows the solar cell performances in the long-wavelength regime. The thin AlO_x layers slightly outperform the standard process.

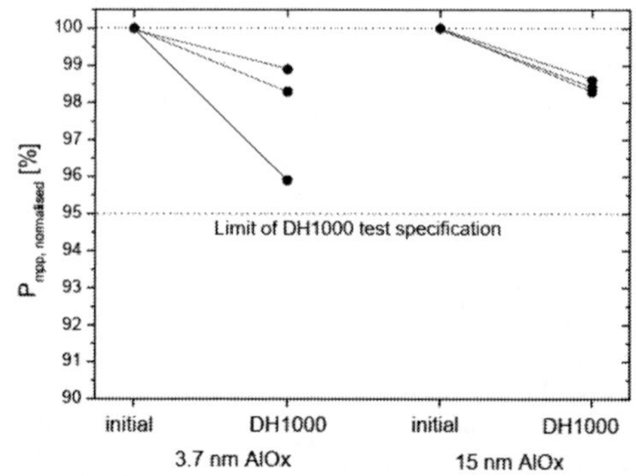

Figure 10: Solar module measurement results (power at the maximum power point P_{mpp}) before and after standard damp-heat test for 1000 hours (DH1000). Successful passing of DH1000 tests requires minimum of 95% of initial $P_{mpp,\ normalised}$ after test.

Small solar modules were manufactured out of the formerly investigated solar cells encapsulating either one or three solar cells in one small module. Three modules of each kind (3.7 nm AlO_x /15 nm AlO_x and one-cell / three-cell modules; in total 12 minimodules) were manufactured. The one-cell modules were used for damp-heat tests (1000 hours, STC tests initially and after the treatment), the three-cell modules were subjected to thermal-cycling tests (200 hours, STC tests initially and after the thermal treatment). All solar modules applying the novel thin (3.7 nm) AlO_x passivation layers successfully passed both tests. In Figure 10, the damp-heat test results are shown as an example.

CONCLUSION

Using the knowledge from the manifold analyses it was possible to deposit ICP-PECVD AlO_x layers with a thickness of only 3.7 nm that provide excellent surface passivation with a surface recombination velocity of ~ 5 cm/s which is even slightly outperforming thicker (15 nm) AlO_x layers allowing PERC solar cell efficiencies of 21%. Solar modules were manufactured proving the applicability of the novel technology by successfully passing standard thermal-cycling and damp-heat tests.

A relatively robust process was found that is capable of compensating variations in gas flow during the deposition process. In-situ plasma monitoring and ex-situ analyses revealed that increased lifetimes on rougher substrate surfaces can be achieved. Contactless CV measurements confirmed that low interface trap densities (D_{it}) at the c-Si / AlO_x interface lead to an excellent surface passivation.

Simulations of the plasma revealed that the layer thickness distribution in a SiN_x deposition process is depending on the distribution of SiH_3 molecules.

ACKNOWLEDGEMENTS

This work was partly funded by the German Federal Ministry for the Environment, Nature Conservation and Nuclear Safety (Contract Number 0325754D).
We would like to thank the company Hanwha Q-Cells for solar cell pre-processing.

REFERENCES

[1] Blakers, A. W.; Wang, A.; Milne, A. M.; Zhao, J.; Green, M. A.: 22.8% efficient silicon solar cell, Applied Physics Letters, 55(13), 1363–1365 (1989).

[2] Saint-Cast, P.; Kania, D.; Hofmann, M.; Benick, J.; Rentsch, J.; Preu, R. (2009): Very low surface recombination velocity on p-type c-Si by high-rate plasma-deposited aluminum oxide. Applied Physics Letters 95 (15), p. 151502. DOI: 10.1063/1.3250157.

[3] Veith, B.; Dullweber, T.; Siebert, M.; Kranz, C.; Werner, F.; Harder, N.-P.; Schmidt, J.; Roos, B.F.P.; Dippell, T.; Brendel, R. (2012): Comparison of ICPAlOx and ALD-Al2O3 layers for the rear surface passivation of c-Si Solar cells. In: Energy Procedia 27 (2012) 379-384.

[4] Pernau, T.; Fuchs, J.-U.; Nguyen, V. X.; Nickel, A.; Walk, U.; Jooss, W.; Hsu, K.; Chen, J.; Wiebecke, S.; Wu, H. H.; Hung, K.-H.; Lin, K. C.; Huang, W. K. W., 40 kHz PECVD of AlOx/SiNx Stacks Demonstrated in Industrial High Efficiency PERC Production, EUPVSEC32 (2016), 374-377.

[5] Lieberman, M. A.; Lichtenberg, A. J. (1994): Principles of plasma discharges and materials processing: John Wiley & Sons, New York.

[6] Hofmann, M.; Jäcklein, M.; Saint-Cast, P.; Wagenmann, D.; Cord, B.; Dippell, T.; Dörr, M.; Schütte, T.; Siemers, M., EUPVSEC32 (2016), 345-349.

[7] Gabriel, O.; Kirner, S.; Klick, M.; Stannovski, B.; Schlatmann, R., Plasma monitoring and PECVD process control in thin film silicon-based solar cell manufacturing, EPJ Photovoltaics 5, 55202 (2014)

[8] Pflug, A.; Siemers, M.; Melzig, T.; Schäfer, L.; Bräuer, G. (2014): Simulation of linear Magnetron Discharges in 2D and 3D, Surface and Coatings Technology 260 (2014) 411-416.

[9] Schmidt, J.; Veith, B.; Werner, F.; Zielke, D.; Brendel, R. (2010): Silicon surface passivation by ultrathin Al2O3 films and Al2O3/SiNx stacks, Proceedings of the 35th IEEE Photovoltaic Specialist Conference, pp. 885-890.

[10] Werner, F.; Veith, B.; Zielke, D.; Kühnemund, L.; Tegenkamp, C.; Seibt, M.; Brendel, R.; Schmidt, J.: Electronic and chemical properties of the c-Si/Al2O3 interface, Journal of Applied Physics 109, 113701 (2011)

[11] Saint-Cast, P: Passivation of Si Surfaces By PECVD Aluminum Oxide, PhD Thesis, University of Konstanz, 2012

[12] Richter, A.; Benick, J.; Hermle, M.; Glunz, Stefan W.: Excellent silicon surface passivation with 5 Å thin ALD Al2O3 layers: Influence of different thermal post-deposition treatments, Phys. Status Solidi RRL 5, No. 5–6, 202– 204 (2011)

WELL PASSIVATING AND HIGHLY TEMPERATURE STABLE ALUMINUM OXIDE DEPOSITED BY ATMOSPHERIC PRESSURE CHEMICAL VAPOR DEPOSITION FOR PERC AND PERT SOLAR CELL CONCEPTS

Josh Engelhardt, Benjamin Gapp, Florian Mutter, Giso Hahn, Barbara Terheiden
University of Konstanz, Department of Physics, 78457 Konstanz, Germany
Josh.Engelhardt@uni-konstanz.de, Benjamin.Gapp@uni-konstanz.de, Florian.Mutter@uni-konstanz.de, Giso.Hahn@uni-konstanz.de, Barbara.Terheiden@uni-konstanz.de

ABSTRACT: Atmospheric pressure chemical vapor deposited aluminum oxide films based on trimethyl-aluminum and oxygen as precursors reaching a surface passivation quality comparable to atomic layer deposited aluminum oxide layers are presented. Demonstrated are charge carrier lifetime values up to 8 ms on n-type Cz-Si for fired AlO_x/SiN_x:H passivation stacks, common to PERC and PERT solar cell designs. The APCVD AlO_x layers are highly temperature stable for set peak firing temperatures in the range of 700-920°C allowing for surface recombination velocities < 2 cm·s^{-1}. The influence of temperature during AlO_x deposition in the APCVD tool is investigated concerning Si material and fundamental passivation effect properties. The latter are investigated in comparison to ALD AlO_x determining the underlying causes for higher lifetime after firing for APCVD AlO_x films.
Keywords: APCVD; aluminum oxide; passivation; PERC; PERT

1 INTRODUCTION

PERC (Passivated Emitter and Rear Cell) solar cells based on p-type crystalline Si and PERT (Passivated Emitter, Rear Totally diffused) solar cells based on n-type crystalline Si gain influence in the commercial solar cell market. Since the market is highly competitive, a low cost production is necessary. Such processes are feasible with high throughput, low cost and low upkeep production tools. Atmospheric pressure chemical vapor deposition (APCVD) tools fulfill these requirements and can supply the necessary layer quality for a high-efficiency solar cell production. Based on trimethyl-aluminum and O_2 as precursors, aluminum oxide (AlO_x) layers [01] are deposited, reaching a surface passivation quality comparable to atomic layer deposition (ALD) based AlO_x layers [02, 03]. APCVD AlO_x layers are deposited at higher temperatures, leading to a higher temperature stability during contact firing, *i.e.* preventing blistering effects in a set peak firing temperature range of 700 to 900°C. Application of APCVD AlO_x layers [04, 05] in PERC and PERT crystalline solar cell concepts is shown to be advantageous, while investigating the influence of the APCVD technology on current Si materials such as n-type Cz-Si for high-efficiency silicon solar cells.

High-efficiency solar cells often use monocrystalline n-type Si substrates due to the high bulk lifetime and lower susceptibility to degradation by impurities. However, n-type Si is prone to other degradation effects as oxygen-related defect formation having a negative impact on material quality. These defects develop during the different high temperature steps such as the APCVD process. A pre-high-temperature (pre-HT) step, sometimes referred to as 'tabula rasa' process, applied prior to any other solar cell process step helps to mitigate the oxygen-related defect formation [06-09]. Therefore, an investigation of temperature effects during APCVD deposition and possible solutions are investigated.

Subsequently, the passivation mechanisms of APCVD AlO_x are investigated in detail concerning the major factors affecting the passivation quality and in comparison to common ALD AlO_x layers. Special focus is laid on the temperature impact and influence during APCVD processing.

2 EXPERIMENTAL

n-type 2-4 Ωcm Cz- and 2-200 Ωcm FZ-Si wafers (156x156 mm^2 and 150 mm diameter, respectively) were wet chemically treated before a pre-HT step was performed on some of them. Following a second HF-dip to prepare the Si surface, APCVD (roller system, SierraTherm) or ALD (FlexAL, Oxford Instruments) AlO_x layers were deposited on both sides of the substrates before a possible annealing step in N_2 atmosphere was performed. 40-90 nm PECVD (Plasma-Enhanced Chemical Vapor Deposition) SiN_x:H was subsequently deposited on some wafers to form the double layer passivation stack before firing in a belt furnace at a variety of set peak firing temperatures and charge carrier lifetime sampling by PCD (PhotoConductoance Decay) and PL (PhotoLuminescence) techniques. References included samples with a stack of thermal SiO_x/SiN_x:H fired in a belt furnace. Molecular bond density, sputtering rate and elemental concentrations, refractive index and fixed charge density were measured by FTIR (Fourier Transform InfraRed spectroscopy), GD-OES (Glow Discharge Optical Emission Spectroscopy), ellipsometry and CV (Capacitance Voltage) technique on pure AlO_x coated FZ-Si samples, respectively.

3 RESULTS AND DISCUSSION

Inherent to the APCVD technique, a high temperature deposition is necessary to form the respective dielectric films by chemical reaction of precursor gases on the heated silicon substrate surface, as it is moved through the gas curtain (see Fig. 1). It affects bulk Si properties as well as film properties, *i.e.*, AlO_x layer passivation characteristics.

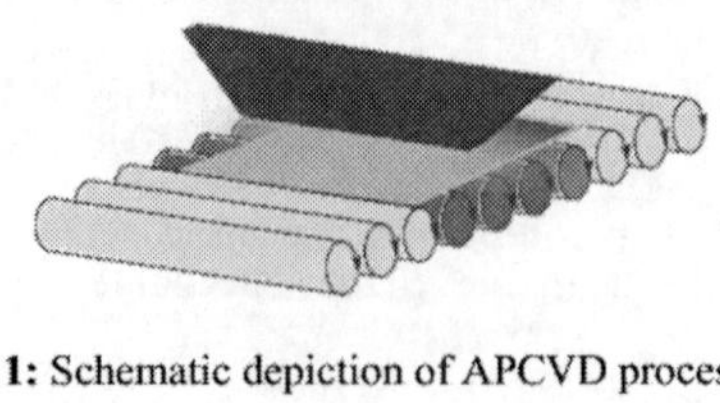

Figure 1: Schematic depiction of APCVD process.

3.1 Thermal impact from APCVD processing

The APCVD technique is based on the reactivity of precursor gases, *i.e.* trimethyl-aluminum (TMAl) and O_2 for AlO_x layer deposition. In contrast to PECVD, the reaction energy is only provided by the surface temperature of the Si substrate heated, in this case by infrared light (set process temperature). The temperature can be used to influence and optimize the properties of the dielectric layers. Typical process set temperatures range from 400 to 800°C.

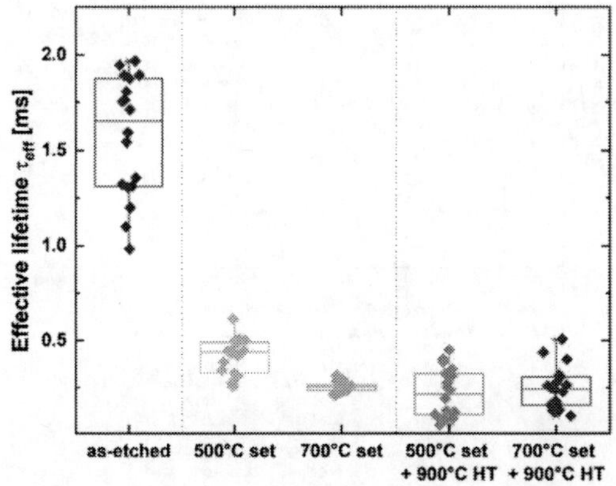

Figure 2: Box-plot diagram of effective lifetime τ_{eff} for as-etched (green), solely temperature treated (yellow/ light orange) at 500°C and 700°C in APCVD (N_2 gas flow) or additionally after APCVD temperature treatment HT treated samples (dark orange/red). All samples are passivated by fired SiN_x:H.

To investigate the effects of the heat treatment during AlO_x deposition in an APCVD tool at typical temperatures, Cz-Si wafers were processed at 500 and 700°C for about 30-60 s without any process gases but N_2 to simulate duration and temperature budget of an AlO_x deposition without the actual deposition of a film. A subsequent high temperature step (HT) at 900°C for about 10 min was carried out on some wafers of both groups. All wafers, including a non-temperature-treated group (as-etched), were passivated using fired SiN_x:H. Fig. 2 shows the effective lifetime of all five groups, none of which had an initial pre-HT step before APCVD treatment. The drop in lifetime from 1.6 ms before temperature treatment in the APCVD reactor to below 500 µs thereafter is dependent on the process temperature and cannot be reversed by a subsequent HT step at 900°C. Even SiN_x:H with hydrogen passivation has evidently no impact on the bulk lifetime degradation.

PL images show circular patterns common to (new) thermal donors activated by temperature treatment below 800°C for a sufficiently long time. Literature indicates that a HT step before low temperature treatment mitigates or even prevents the formation of thermal donor clusters (ring structures *i.e.*, in n-type Cz-Si) and subsequent degradation of bulk lifetime [06-09].

As can be seen in Fig. 3, during processing in the APCVD tool pre-existing or forming crystal defects (ring structure in left image) are visible after passivation. A possible explanation could be decoration of defect sites with impurities. Samples undergoing a subsequent HT step at 900°C show a further degradation of the inner part of the ring (middle). Using a pre-HT step before all following process steps prevents any occurrence of lifetime degrading effects (right).

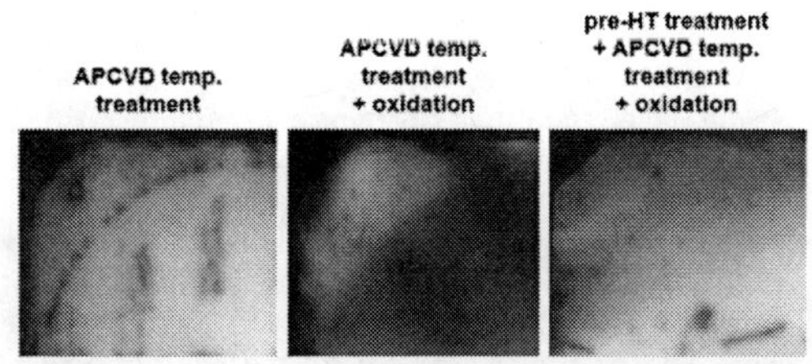

Figure 3: PL images of n-type Cz-Si wafers after APCVD heat treatment (left), APCVD heat treatment and HT step (middle), pre-HT step followed by APCVD heat treatment and HT step (right).

Since the pre-HT step has a significant influence for all further (H)T steps, the combination with a thermal oxidation was investigated. Some of the n-type Cz-Si wafers were pre-HT treated at 1000°C for 60 min before a low temperature treatment in the APCVD tool similar to the first experiment shown at around 500°C without deposition of AlO_x. Subsequent HT treatment to grow a thermal SiO_x at around 900°C in a tube furnace shows different effects on the effective lifetime for the four different groups in Fig. 4.

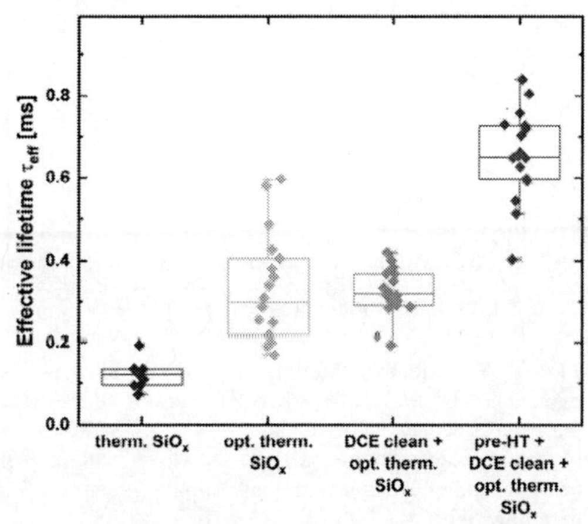

Figure 4: Effective lifetime of n-type Cz-Si wafers after low temperature treatment (APCVD) and subsequent passivation by fired thermal SiO_x/SiN_x stack. Shown are four different oxidation sequences with or without pre-HT step and/or DCE clean.

Passivation quality of thermal SiO_x layer depends on growth parameters during oxidation and thermal history of the Si substrate. In this case, a standard oxidation step in a tube furnace leads to a low lifetime of around 100 µs (green in Fig. 4) after subsequent SiN_x:H deposition and firing step. By applying a DCE (1,2-dichlorethane) clean or a different set of thermal oxide growth parameters (orange and yellow in Fig. 4), the lifetime can be tripled. Yet the highest impact has an initial pre-HT step before APCVD treatment, leading to up to 800 µs after thermal oxidation. Depending on n-type Cz-Si material quality and affinity to degradation, *i.e.*, thermal donor concentration, this effect of thermal donor activation prevention can vary.

The material shown here displayed a comparably strong thermal donor behavior.

3.2 Passivation for PERC and PERT solar cell concepts

APCVD AlO_x layer passivation quality highly depends on deposition parameters, *i.e.* deposition temperature, gas ratios, etc. as well as thickness (overall gas flow and/or belt speed dependent). Fig. 5 shows the effective lifetime τ_{eff} of FZ-Si samples passivated by APCVD AlO_x of various deposition parameters. The passivation quality increases with layer thickness up to about 15 nm and remains stable for thicker AlO_x layers. Overall, an annealing step (HT step, *e.g.* firing) increases the charge carrier lifetime of samples from 12 nm up to an AlO_x layer thickness of about 25 nm, whereas at this thickness the lifetime decreases slightly for some APCVD process parameter sets. FZ-Si wafers with a base resistivity of 200 Ω·cm show lifetimes up to 6.3 ms resulting in an effective surface recombination velocity of < 2 cm·s^{-1}. ALD AlO_x reference samples show a similar lifetime and surface recombination velocity level.

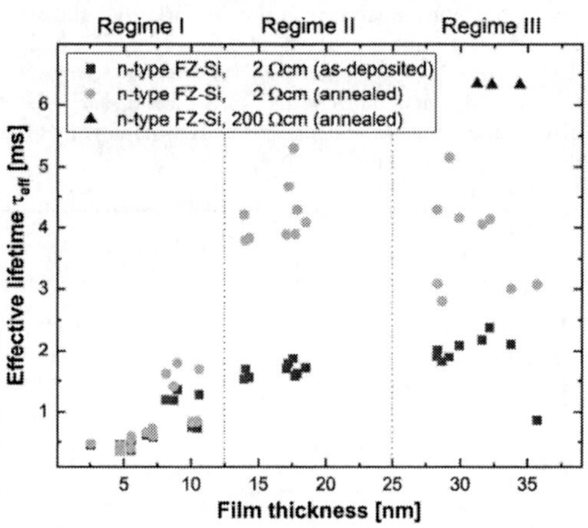

Figure 5: Effective lifetime τ_{eff} of FZ-Si samples passivated by APCVD AlO_x of various deposition parameters dependent on the layer thickness. Samples are shown after deposition (red), after an HT annealing step (yellow) and with a 200 Ω·cm base doping (green) for S_{eff} determination.

The overall level of up to 5 ms in effective lifetime and 2 cm·s^{-1} in surface recombination velocity is a sufficiently high passivation quality for PERC as well as PERT rear sides with all the advantages of APCVD technology.

Further, APCVD AlO_x displays different interesting properties, *e.g.*, high firing temperature stability broadening the available parameter range for contact formation. In this case, AlO_x/SiN_x passivated n-type Cz-Si wafers are fired at 810-930°C set peak firing temperature. Four different optimized deposition parameter sets are used for AlO_x deposition in the APCVD reactor and two for ALD AlO_x (Fig. 6). All APCVD AlO_x layers reach lifetime values above 2 ms (T < 900°C) and up to 8 ms for a common PERC and PERT cell set peak firing temperature of 810°C. In comparison, samples passivated with either a single PECVD SiN_x:H layer or an optimized ALD AlO_x layer show only effective charge carrier lifetimes of below 3 and 5 ms after firing, respectively.

Therefore, optimized AlO_x layers from APCVD are highly firing stable and allow for a firing parameter window between 700-900°C for PERC or PERT cell concepts. Furthermore, no blistering occurred during firing, allowing for homogeneous passivation without risking parasitic contact formation or loss in open circuit voltage, V_{OC}.

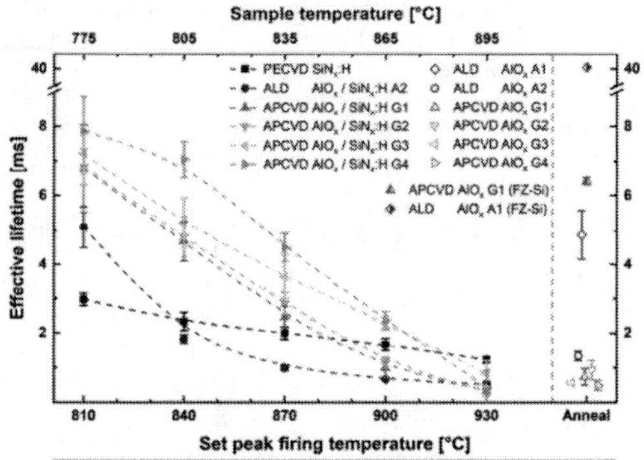

Group:	Dep.temp. [°C]:	Annealing conditions:
A1	170	30 min @ 450°C in N_2
A2	300	30 min @ 450°C in N_2
Group:	TMA-N_2 flow [slm] / O_2:TMA [%]:	Annealing conditions:
G1	0.75 / 100	20 min @ 470°C in N_2
G2	0.95 / 95	20 min @ 470°C in N_2
G3	0.87 / 90	20 min @ 470°C in N_2
G4	1.00 / 80	20 min @ 470°C in N_2

Figure 6: Effective lifetime τ_{eff} of Cz- and FZ-Si samples passivated by APCVD and ALD AlO_x with various deposition parameters. In case of fired samples (left side of dashed vertical line), SiN_x:H coating was applied before. The annealed samples (details see below graph) are shown on the right side.

In case of annealed AlO_x films, ALD surpasses APCVD layers on Cz-Si by 5 to 1 ms and on FZ-Si by 40 to 6.5 ms, respectively. There seems to be a difference in passivation behavior and effectiveness depending on contact firing or annealing to activate the passivation ability.

3.3 Investigation into passivation effects

Obvious reasons for a difference in passivation depending on temperature treatment and coating layers are on the one hand H diff-/effusion [10] and blistering effects [11] due to the same H diffusion effect and on the other hand differences in the underlying primary or rather limiting passivation type (chemical or field effect passivation).

To investigate the differences between ALD and APCVD techniques in that regard, Cz- and FZ-Si samples of the various parameter groups from Fig. 6 were fabricated as pure AlO_x films on the Si substrates in an as-deposited or accordingly annealed state. The effective lifetime depicted in Fig. 7 (graph A) is similar to the previously in Fig. 6 obtained values. Important to mention is the fact that annealing does not increase the effective lifetime of APCVD films in the same way as it does in case of ALD AlO_x layers. The reason is the in-situ annealing of APCVD films during deposition and shortly afterwards in the cool-down section of the tool. A subsequent anneal does only marginally change the layers further with respect to molecular structure and thus diffusion of H or formation of fixed charges as is the case for ALD layers. The refractive index ranges around 1.6 as expected from literature [04] and shows no difference between ALD and APCVD (graph D). Subsequent GD-OES measurements on the other hand show a significant difference in sputtering rate (graph B), indicating a lower film density for APCVD layers because of the higher sputtering rate. Thus, the AlO_x films are capable of better H diffusion allowing for less blistering in case of firing in a belt furnace with common SiN_x:H coating layers [05]. Furthermore, the GD-OES measurements show a slightly higher H concentration (atomic percentage) with a difference of up to $2.5\%_{abs.}$ in the APCVD AlO_x films compared to ALD (graph C). The Al/O ratio for APCVD is dependent on the total O_2 flow during deposition as is the H content. While the latter decreases with O_2 flow as expected considering the TMAl to be the H source during deposition, the Al/O ratio increases at the same time.

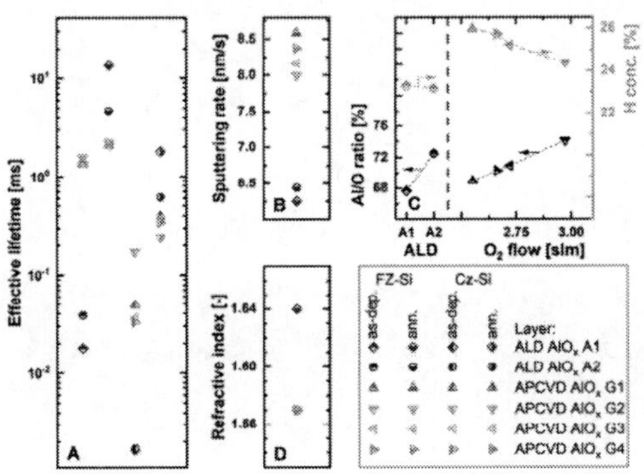

Figure 7: Effective lifetime τ_{eff} of Cz- and FZ-Si samples passivated by APCVD and ALD AlO_x with various deposition parameters in as-deposited and annealed state (graph A). Corresponding refractive index and sputtering rate in graphs D and B, respectively. Graph C shows the Al/O ratio and H concentration of the layers.

An explanation can be found in molecular bond density that changes with O_2 flow increasing the Al-O, Al-O_2 and Al-O_4 bonds in favor of Al-O_6 bonds to further an overall increase of incorporated Al in the APCVD AlO_x layer [15]. Fig. 8 shows the corresponding normalized molecular bond densities in the left graph and an example of an FTIR spectrum with identified bonds in the upper right graph.

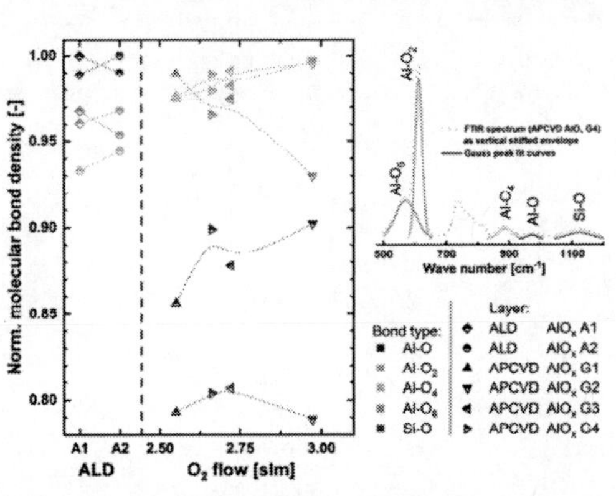

Figure 8: Normalized molecular bond density for all ALD and APCVD AlO_x layers (left graph) and example of an FTIR spectrum with identified bonds in the upper right graph.

A significant difference in bond density can be found for Si-O stemming from the interfacial silicon oxide layer between the AlO_x film and Si substrate. Previous publications indicate an importance of the growth and thickness of the interfacial SiO_x layer for chemical and field effect passivation quality [12, 13]. Since the relative bond density of Si-O is higher for ALD AlO_x, at least the chemical passivation properties of annealed ALD films are expected to be higher and explain the lack of the same in annealed APCVD layers. FTIR measurements further indicate the negative fixed charge density, known to depend on the density of Al on the surface of the interfacial SiO_x [12-14].

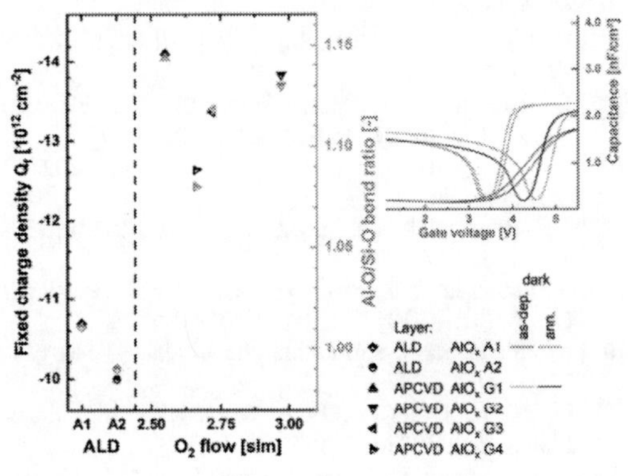

Figure 9: Negative fixed charge density from CV and Al-O/Si-O bond ratio from FTIR measurements (Fig. 8) for all ALD and APCVD AlO_x films. CV curves for dark measurements of one ALD and APCVD layer in as-deposited and annealed state in upper right graph.

CV measurements carried out with a Hg electrode setup show a 1.3 times higher negative fixed charge density in APCVD films compared to ALD AlO_x at a standard level of around $1{\cdot}10^{13}$ cm^{-2} (Fig. 9, left graph) [05]. Comparing the negative fixed charge density with the FTIR bond density, it turns out that the bond ratio of Al-O/Si-O is correlated with the negative fixed charge density, thus proving the aforementioned dependence on the Al density at the SiO_x interface. The CV curves in the upper right graph of Fig. 9 furthermore show the aforementioned change in the ALD layers during anneal (curve shape and slope changes from red to orange), while this does not occur in the in-situ annealed APCVD AlO_x films (CV curve does not change shape from light to dark green).

4 CONCLUSION

AlO_x layers by APCVD show high and stable passivation quality of up to 8 ms and above 2 ms even at high firing temperatures on n-type Cz-Si. With a surface recombination velocity of around 2 cm·s^{-1} they are very well suited for PERC and PERT cell application with all the benefits of APCVD tool processing. The negative impact of the processing temperature during APCV-deposition on n-type Cz-Si material quality can be reduced or even avoided by a pre-HT step or proper choice of material. A detailed characterisation investigating the passivation mechanisms of ALD and APCVD AlO_x films was presented to explain the differences between firing and lower temperature anneal on lifetime scale.

5 ACKNOWLEDGEMENTS

We would like to thank L. Mahlstaedt and B. Rettenmaier for their technical support. Part of this work was financially supported by the German Federal Ministry for Economic Affairs and Energy (FKZ 0324001 and 0324226A). The content is the responsibility of the authors.

6 REFERENCES

[01] R. Hezel et al., J. Electrochem. Soc. **136**, No. 2, 518-523 (1989).
[02] K.O. Davis et al., Phys. Status Solidi RRL **7**, No. 11, 942-945 (2013).
[03] B. Semmache et al., Proc. 29th EUPVSEC, Amsterdam, Netherlands, 859-862 (2014).
[04] L.E. Black et al., Appl. Phys. Lett. **100**, 202107 (2012).
[05] L.E. Black et al., Sol. Energ. Mat. Sol. Cells **120**, 339-345 (2014).
[06] V. LaSalvia et al., Proc. 43rd IEEE PVSC, Portland, Oregon, USA, 1047-1050 (2016).
[07] R. Falster et al., Solid state phenomena **57-58,** 123-128 (1997).
[08] D. C. Walter et al., Proc. 28th EUPVSEC, Paris, France, 699-702 (2013).
[09] E. E. Looney et al., Appl. Phys. Lett. **111**, 132102 (2017).
[10] A. Richter et al., IEEE J. Photovol. **3**, No. 1, 236-245 (2013).
[11] T. Lüder et al., En. Proc. **27**, 426-431 (2012).
[12] G. Dingemans et al., J. Appl. Phys. **110**, 093715 (2011).
[13] Y. Zhao et al., Nanoscale Res. Lett. **8**:114 (2013).
[14] V. Naumann et al., En. Proc. **27**, 312-318 (2012).
[15] J.M. Reyes et al., J. Electrochem. Soc. **160**, No. 10, B201-B206 (2013).

EVIDENCE OF REAR SURFACE RELATED DEGRADATION IN Cz-Si PERC-TYPE SOLAR CELLS

A. Herguth, C. Derricks, D. Sperber
University of Konstanz, Department of Physics, Universitätsstr. 10, 78457 Konstanz, Germany

ABSTRACT: A degradation phenomenon of unknown origin during an accelerated aging test comprising illuminated annealing is observed for boron-doped PERC-type solar cells with SiO_x/SiN_x:H rear surface passivation. By collating different measurement techniques and simulations, the rear surface passivation is identified as component responsible for the observed degradation. The observed degradation on solar cell level shows similarities to the degradation of similar passivation layers in lifetime studies.
Keywords: silicon solar cell, PERC, rear surface passivation, degradation, surface related degradation

1 INTRODUCTION

The efficiency potential of bifacially contacted silicon solar cells with front-side emitter strongly correlates with rear side passivation quality, quantified by the rear side surface recombination velocity S_{rear}, and material quality, quantified by excess carrier lifetime τ_{bulk}, as discussed with regard to open circuit voltage V_{oc} in the context of Fig. 3.

For quite a long time, a full area alloyed rear side contact was standard in p-type solar cell mass production. The alloying process with a typically screen-printed aluminum layer results in a highly Al doped region underneath the actual contact thus forming a high-low junction with the typical lowly p-type doped substrate. The concentration gradient of majority charge carriers and the resulting diffusion leads to the formation of an electric field which repels electrons, being the minority charge carrier species, from the back side (back surface field). Hence, the high-low junction acts as field-effect passivation for the otherwise strongly recombination active silicon-metal contact. Unfortunately, excess charge carrier lifetime in this highly-doped region is quite low and thus effective S_{rear} values for this technology typically exceed 200 cm/s with hardly any room for improvement.

Hence, further improvement requires a disruptive change in cell architecture. The passivated emitter and rear cell (PERC) architecture replaces the full area alloyed high-low junction as both passivation and contact component by a separated large area dielectric passivation (>90% of area) interrupted only by locally Al alloyed contacts (<10% of area). Even though high-low junction passivation quality of the locally Al alloyed contacts might be inferior to the full area alloyed scenario, the strongly reduced area fraction reduces the effective S_{rear} to values well below 100 cm/s thanks to the superior performing dielectric passivation with S being typically < 10 cm/s. Passivation by dielectric layers works by a combination of both chemical passivation of electronic states in the band gap at the silicon-dielectric interface and field effect passivation due to fixed charges in the layers preventing charge carriers in the substrate from recombination at the residual defect states at the silicon-dielectric interface.

However, passivation quality of different dielectric layers or stacks involving hydrogen-rich silicon-nitride layers were found in various lifetime studies to be instable during illuminated annealing even at low temperatures [1-3], hence raising the question if dielectric passivation in PERC-type solar cells might be affected as well, leading inevitably to a degradation of electrical entities like open circuit voltage V_{oc}, short circuit-current density j_{sc} and in consequence also of the conversion efficiency.

Within this contribution, long-term stability of PERC-type solar cells with a passivation layer stack consisting of silicon-oxide (SiO_x) capped by hydrogen-rich silicon-nitride (SiN_x:H) with regard to illuminated annealing is investigated.

2 SAMPLES AND SETUP

A commercially available PERC-type solar cell made of boron-doped Cz-silicon (1.3 Ωcm, 145 µm thickness) with SiO_x/SiN_x:H passivation layer stack at the rear side is used. Although the actual peak temperature of the solar cell during contact firing of the screen-printed contacts is unknown, the formation of Ag front and Al rear contacts implies an actual peak firing temperature above 700°C.

The sample is exposed to an accelerated aging test comprising illuminated annealing at 150°C and 1 sun halogen lamp illumination (j_{sc} equivalent [4]) on a hotplate. Treatment is interrupted every once in a while to measure a current-voltage characteristic yielding V_{oc} and j_{sc} under standard test conditions (STC, 25°C, AM 1.5g spectrum) and monitor the short- and long-term stability.

At the end of the measurement series the solar cell was etched back to the bulk substrate, cleaned in Piranha solution (H_2O_2, H_2SO_4, 80°C), HF-dipped and wet chemically re-passivated using a 0.8 molar iodine-ethanol solution. Effective lifetime was measured by a photoconductance decay (PCD) technique using a WCT120 lifetime tester from Sinton Instruments.

An additional lifetime sample was prepared with both-sided SiO_x/SiN_x:H passivation layer and fired at a peak temperature typical for screen-printed PERC-type solar cells. Again, this sample was exposed to the same illuminated annealing conditions and the effective lifetime was monitored by PCD.

3 RESULTS

Figure 1 shows the evolution of V_{oc} as well as j_{sc} for the solar cell during the accelerated aging test. V_{oc} as well as j_{sc} show a weak drop around 0.1 h which is attributed to a decrease in bulk lifetime due to boron-oxygen related degradation (BO-LID) as was expectable for the used oxygen-rich Cz-material [5,6]. Subsequently V_{oc} and j_{sc} recover from BO-LID as the conditions of the accelerated aging test (excess charge carrier injection at elevated temperature) promote a regeneration (BO-LIR) [6,7]. However, V_{oc} and j_{sc} drop for a second time in long-term leading to losses in V_{oc} of ~20 mV and j_{sc} of ~0.4 mA/cm².

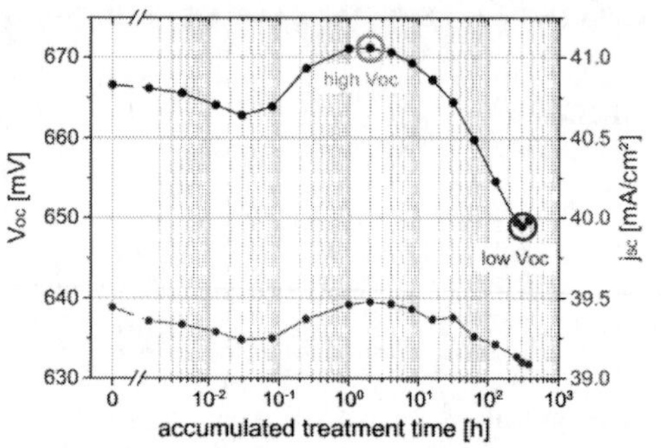

Figure 1: Open-circuit voltage V_{oc} and short-circuit current density j_{sc} measured under standard test conditions during the accelerated aging test.

Only based on these observations, it is hard to decide which cell component is responsible for this second degradation. Therefore, a detailed analysis was carried out which is described in detail elsewhere [8], whereas in this contribution only the major findings are presented. In the following, measurement data taken before the second degradation (~2 h) featuring still a high V_{oc} (and j_{sc}) and after second degradation (~300 h) featuring then a low V_{oc} (and j_{sc}) are compared.

In principle four cell components might be suspected to cause this degradation: defect formation in the emitter or deterioration of front side passivation, defect formation or in-diffusion of contaminants into the space charge region (SCR), defect formation in the bulk and deterioration of rear surface passivation. As the detailed analysis has shown [8], neither changes in the emitter or space charge region are likely and this part of the analysis is therefore not presented here.

Quantum efficiency measurements in the high V_{oc} and low V_{oc} state (shown in Fig. 2) prove that the observed degradation affects the long wavelength regime where light is absorbed deep in the bulk close to the rear surface as seen from the comparison of effective absorption length and cell thickness. Unfortunately, quantum efficiency analysis can virtually not distinguish between a degradation of bulk lifetime, meaning minority carriers more likely get lost in the bulk on their way to the emitter, and a deterioration of rear surface passivation quality, meaning these minority carriers get lost at the rear side [8].

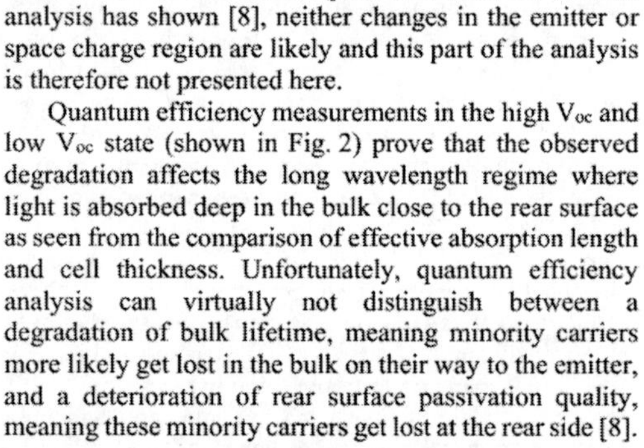

Figure 2: Wavelength-dependent internal quantum efficiency in the high V_{oc} and low V_{oc} state and effective absorption length in the long-wavelength regime.

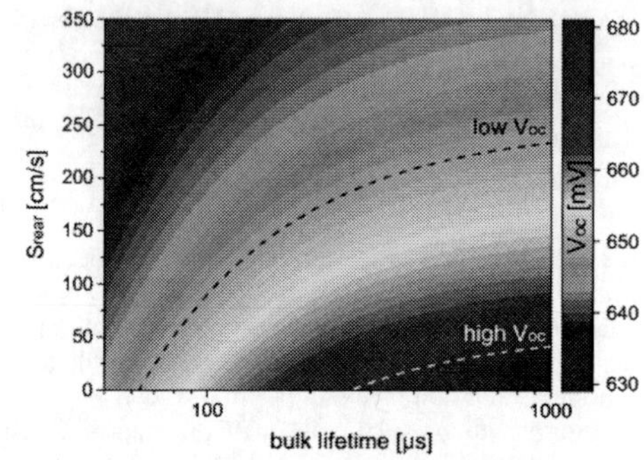

Figure 3: Calculated contour plot of open-circuit voltage V_{oc} in dependence of bulk lifetime (τ_{bulk}) and rear surface recombination velocity (S_{rear}) for the bifacially contacted, front emitter solar cell under investigation. Highlighted contour lines represent combinations of τ_{bulk} and S_{rear} that yield 'low V_{oc}' and 'high V_{oc}' of the investigated sample.

Figure 3 shows a contour plot of V_{oc} versus bulk lifetime τ_{bulk} and rear surface recombination velocity S_{rear} adapted to the specific cell properties. It should be highlighted that both $\tau_{bulk}(\Delta n)$ and $S_{rear}(\Delta n)$ probably depend on injection level Δn and that the specific τ_{bulk} and S_{rear} in Fig. 3 relate to whatever Δn is present at V_{oc}. The highlighted constant V_{oc} contour lines for high V_{oc} and low V_{oc} define possible combinations of τ_{bulk} and S_{rear}, but V_{oc} analysis is not capable of sorting out whether τ_{bulk} has decreased or S_{rear} has increased.

In order to resolve this issue, the solar cell was etched back to the substrate and wet-chemically re-passivated as described above. Figure 4 shows the obtained injection-dependent effective lifetime data. It should be noted that this effective lifetime data is influenced by both bulk lifetime and surface recombination limited lifetime and thus represents only a lower limit for actual bulk lifetime. From the measured low V_{oc} of ~650 mV in the degraded state it can be concluded that injection level is approx. $8.3 \cdot 10^{14}$ cm^{-3} thus the lifetime of ~600 µs at this injection is relevant for comparison with Fig. 3.

The high lifetime after re-passivation restricts possible changes between the high V_{oc} and low V_{oc} contour line in Fig. 3 almost completely to vertical changes suggesting a deterioration of surface passivation quality and excluding a degradation of bulk lifetime.

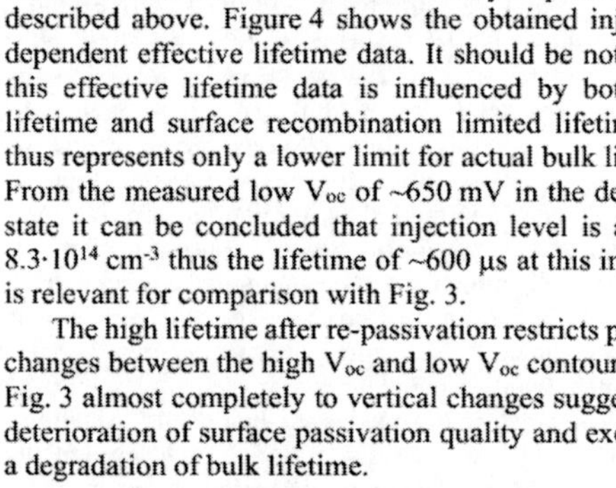

Figure 4: Injection-dependent lifetime data measured via PCD of the stripped and re-passivated solar cell. Low V_{oc} implies an injection level of $8.3 \cdot 10^{14}$ cm^{-3}.

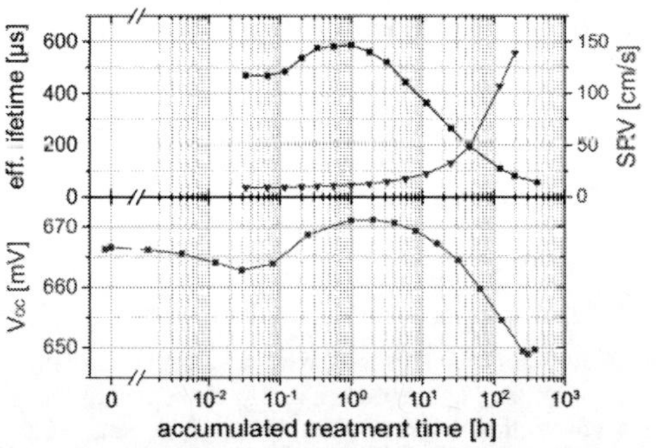

Figure 5: Comparison of the temporal evolution of open-circuit voltage (V_{oc}) of the PERC-type solar cell (bottom) as well as effective lifetime and surface recombination velocity (SRV) of the lifetime sample (top) during the accelerated aging test.

As mentioned in the sample description, boron-doped Cz-Si lifetime samples with both sided SiO_x/SiN_x:H passivation were prepared and fired. These samples were exposed to the similar accelerated aging test as the PERC-type solar cell. Figure 5 (top) shows the evolution of effective lifetime and effective surface recombination velocity (SRV). Even though the missing time resolution in the beginning conceals the ongoing BO-LID in the beginning, the increase in effective lifetime up to ~1 h is related to the regeneration from BO-LID. In the long-term, effective lifetime decreases significantly. In parallel, SRV is found to increase significantly thus the decrease in effective lifetime is caused by a deterioration of surface passivation quality. It is worth noting, that SRV did not show an increase due to BO-LID in the beginning, clearly demonstrating that the degradation in the long-term is caused by surface related degradation while degradation and subsequent regeneration in short-term are related to the bulk.

A comparison with the solar cell results (Fig. 5 bottom) shows the striking similarities of surface related degradation of the SiO_x/SiN_x:H passivation quality in the lifetime samples and the PERC-type solar cell suggesting that both suffer from basically the same phenomenon.

ACKNOWLEDGEMENTS

Part of this work was supported by the German Federal Ministry for Economic Affairs and Energy under Grants 0325877C, 0325763B and 0324001. The authors would like to thank A. Schwarz, L. Mahlstaedt, J. Lindroos, and M. Hagner for technical support as well as F. Huster for constructive discussions on loss mechanisms in solar cells. The content is the responsibility of the authors.

REFERENCES

[1] D. Sperber, A. Graf, D. Skorka, A. Herguth, G. Hahn *Degradation of surface passivation on crystalline silicon and its impact on light-induced degradation experiments* IEEE Journal of Photovoltaics **7** (2017) 1627 https://doi.org/10.1109/JPHOTOV.2017.2755072

[2] D. Sperber, A. Herguth, G. Hahn *On improved passivation stability on highly-doped crystalline silicon and the long-term stability of regenerated Cz-Si* Solar Energy Materials & Solar Cells **185** (2018), 277 https://doi.org/10.1016/j.solmat.2018.05.031

[3] D. Sperber, A. Schwarz, A. Herguth, G. Hahn *Enhanced stability of passivation quality on diffused silicon surfaces under light-induced degradation conditions* Solar Energy Materials & Solar Cells **188** (2018) 112 https://doi.org/10.1016/j.solmat.2018.08.019

[4] A. Herguth *On the meaning(fullness) of the intensity unit 'suns' in light induced degradation experiments* Energy Procedia. **124** (2017) 53 https://doi.org/10.1016/j.egypro.2017.09.339

[5] K. Bothe, R. Sinton, J. Schmidt *Fundamental boron-oxygen-related carrier lifetime limit in mono- and multicrystalline silicon* Progress in Photovoltaics: Res. Appl. **13** (2005) 287 https://doi.org/10.1002/pip.586

[6] T. Niewelt, J. Schön, W. Warta, S. Glunz, M. Schubert *Degradation of crystalline silicon due to boron-oxygen defects* IEEE Journal of Photovoltaics **7** (2017) 383 https://doi.org/10.1109/JPHOTOV.2016.2614119

[7] A. Herguth, G. Schubert, M. Kaes, G. Hahn *A new approach to prevent the negative impact of the metastable defect in boron doped Cz silicon solar cells* Proceedings of the 3rd World Conference on Photovoltaic Energy Conversion (2006) 940 https://doi.org/10.1109/WCPEC.2006.279611

[8] A. Herguth, C. Derricks, D. Sperber *A detailed study on light-induced degradation of Cz-Si PERC-type solar cells: Evidence of rear surface-related degradation* IEEE Journal of Photovoltaics **8** (2018) 1190 https://doi.org/10.1109/JPHOTOV.2018.2850521

DEVELOPMENT OF *P*-CZ PERC SOLAR CELLS APPROACHING 23% EFFICIENCY FOR GIGAWATT-LEVEL PRODUCTION

Benjamin G. Lee, Ingmar Höger, Tabitha Ballmann, Maximilian Kauert, Steffen Laube, Markus Neuber, Steffen Geißler, Tomasz Rudolph, Klaus Duncker, Ronny Lantzsch, Matthias Bartzsch, Fabian Fertig, Andreas Hubert, Matthias Köhler, Ronny Bakowskie, Sven Thormann, Susanne Schulz, Martin Schaper, Jörg W. Müller
Hanwha Q CELLS GmbH
Sonnenallee 17-21, 06766 Bitterfeld-Wolfen, Germany

ABSTRACT: Hanwha Q CELLS has developed high-efficiency PERC silicon solar cells with a selective emitter, achieving an efficiency of 22.7% for a 6-busbar cell, with open-circuit voltage above 684 mV. The selective-emitter cells are fabricated using industrially-proven processes, including conventional screen-printing, on B-doped, *p*-Cz silicon wafers. This advance builds on our Q.ANTUM technology, which combines leading efficiencies with a lean fabrication sequence and proven stability against light- (and elevated-temperature) induced degradation. Q CELLS currently produces an annualized 3 GW of Q.ANTUM cells on *p*-Cz wafers. In mass production, the efficiency of our 6-busbar cells now averages 22.15%. This enables high power ratings up to 395 W, for modules with 144 half-cells. With the results presented here, we demonstrate the capability of our Q.ANTUM technology to continue increasing the cell efficiency to 23% and beyond, within the next couple of years.
Keywords: PERC, Selective Emitter, Solar Cell, Module

1 INTRODUCTION

The Q.ANTUM technology of Hanwha Q CELLS has the key advantage of reaching high efficiencies using a low-cost, lean fabrication sequence [1,2]. Q.ANTUM adds only two cost-effective steps (dielectric rear passivation, local rear contact opening) to the process sequence of a standard Al-BSF silicon solar cell, combined with measures ensuring high yield, to obtain a PERC-like cell [3] with efficiencies of 22% and higher (Fig. 1).

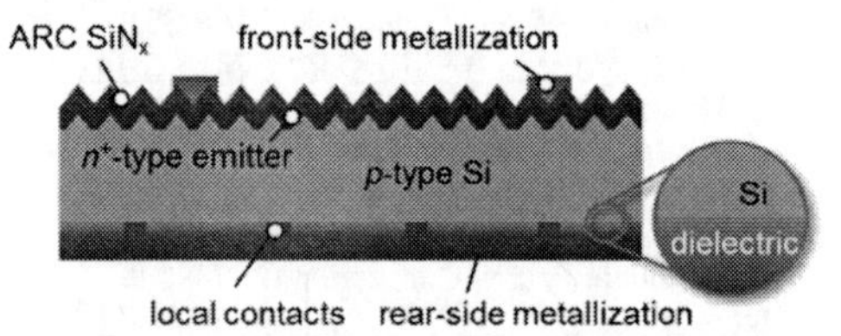

Figure 1: Q.ANTUM monocrystalline PERC-like solar cell.

Hanwha Q CELLS has maintained a learning curve that has cell efficiencies increasing at ~0.6% per annum, over the last 10 years (Fig. 2). In pilot mass production, we have reached cell efficiencies over 22.2%.

Moreover, Q.ANTUM technology shows robust stability against light-induced degradation (LID), including at elevated temperatures (LeTID) [4]. This is in contrast to conventional PERC cells, which can be susceptible to LID and particularly LeTID, even after B-O defect stabilization treatments [4]. The LID-stability and reliability of Q.ANTUM results in higher energy yield for installed modules.

With multi-GW in-house cell and module production, Hanwha Q CELLS is well-positioned for innovations optimizing both the cell and interconnection for top performance at the module-level. An example of this innovation is our Q.PEAK.DUO module product, which uses half-cut cells and 6-wire interconnection [5] instead of conventional ribbons. This design decreases series resistance losses through the half-cell and wire interconnection. The round wires also reduce effective shading at the module-level vs. flat ribbons, with light reflected from the wires mostly re-directed back onto the cell. Thus, the modules achieve an improved cell-to-module power ratio. Mass production data from cells and modules for Q.PEAK.DUO are presented in Section 2.

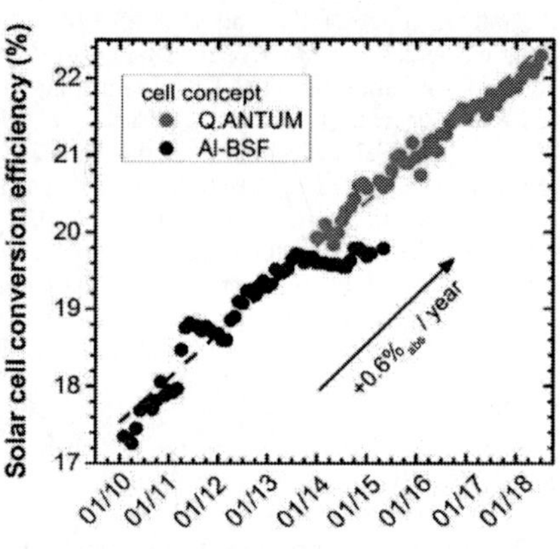

Figure 2: Progression of cell efficiency at Hanwha Q CELLS over time.

To enable even higher cell (and thus module) efficiencies, the remaining sources of recombination within the cell need to be addressed. In Section 3, we present an analysis of the main recombination pathways in the cell, and show how this can be improved by using a selective emitter design [3]. We have developed an industrially-viable process for PERC-like cells with a selective emitter, and discuss our cell results in this paper.

2 MASS PRODUCTION DATA

Hanwha Q CELLS has a production capacity of 8 GW of solar cells and modules, including 3 GW for monocrystalline Q.ANTUM cells (and also 3 GW multicrystalline Q.ANTUM PERC). From July 2018 mass production data, our Q.ANTUM cells used for Q.PEAK.DUO reached 22.15% average efficiency at GW-level production. Also, the distribution of cell efficiencies is fairly narrow, between 21.9 and 22.3%, as seen in Fig. 3. These cell efficiencies achieved at the GW-scale represent the highest-reported values for mass production of PERC solar cells.

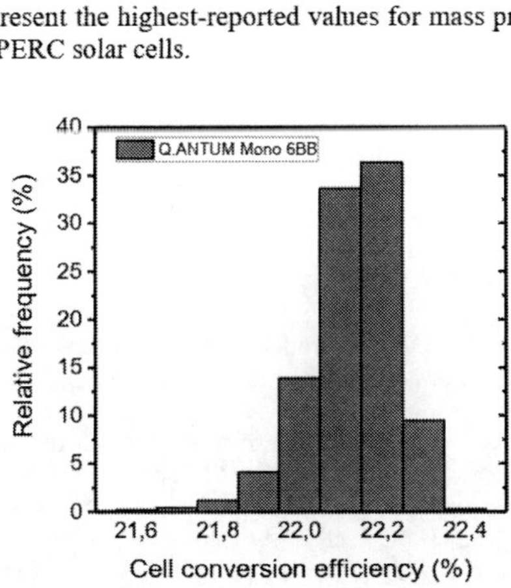

Figure 3: Histogram of cell efficiency for Q.ANTUM Mono 6BB cells. Data is from mass production during July 2018.

At the module-level, our 144 half-cell Q.PEAK.DUO L product has power ratings up to 395 Wp (Fig. 4). These modules have 6-wire interconnection; the round copper wires have lower shading losses as compared to flat ribbons with equivalent conductivity. Most of the Q.PEAK.DUO L modules have power classes of 385 and 390 Wp. This impressive performance is made possible by the high cell efficiencies for the Q.ANTUM cells, plus the narrow distribution of these efficiencies. These module power ratings are the best-in-class for PERC modules; indeed, only modules with significantly costlier heterojunction or rear-contacted cells have higher power ratings.

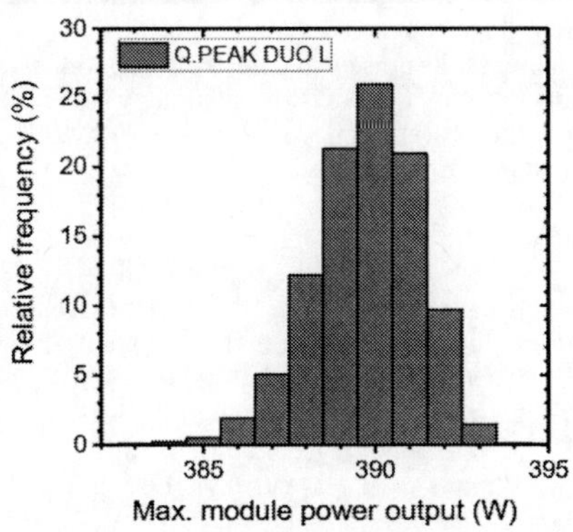

Figure 4: Histogram of module power for Q.PEAK DUO L modules with 144 half-cells and 6-wire interconnection. Data is from mass production during July 2018.

3 SELECTIVE EMITTER CELLS

In order to evaluate the potential for further efficiency gains, it is instructive to examine the loss mechanisms in the solar cell. Thus, we performed an analysis of the recombination losses of the Q.ANTUM cell in production, shown in Fig. 5a. This was done with simulations using the Quokka software [6]. The recombination associated with the homogeneous front emitter is the single largest contribution to losses in this cell. This includes recombination at the front metal contact, surface recombination of the passivated area, and Auger recombination. Since the front recombination is the dominant term, we seek to reduce it by adopting a selective emitter cell geometry (Fig. 6).

The selective emitter cell differs from the homogeneous emitter cell (Fig. 1) in that there is an extra n++ highly-doped emitter directly under the front metal grid. This provides added field-effect passivation that reduces the amount of recombination at the front metal contacts; our calculations and data show that the front metal recombination can be approximately halved. It can also decrease the contact resistivity, leading to an improved fill factor for the cell. Finally, it allows a separate optimization of the emitter in the non-contacted region for better surface passivation. This improved passivation can then reduce the surface recombination by about half, according to our calculations and data.

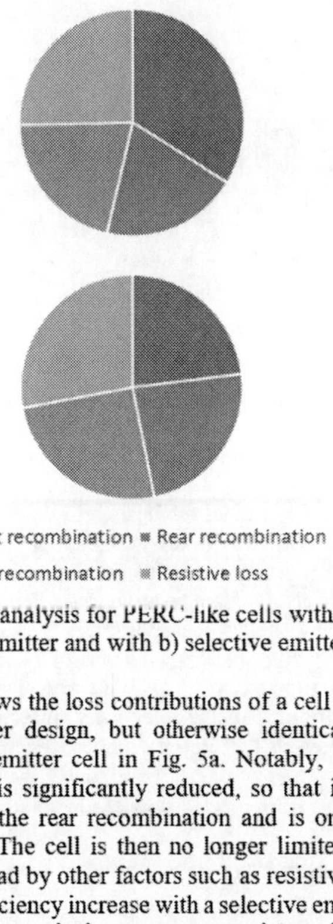

Figure 5: Loss analysis for PERC-like cells with a) homogeneous emitter and with b) selective emitter.

Fig. 5b shows the loss contributions of a cell having a selective emitter design, but otherwise identical to the homogeneous emitter cell in Fig. 5a. Notably, the front recombination is significantly reduced, so that it is now comparable to the rear recombination and is one of the smaller terms. The cell is then no longer limited by the front side, instead by other factors such as resistive losses. Overall, the efficiency increase with a selective emitter can be 1-2% relative vs. the homogeneous emitter cell.

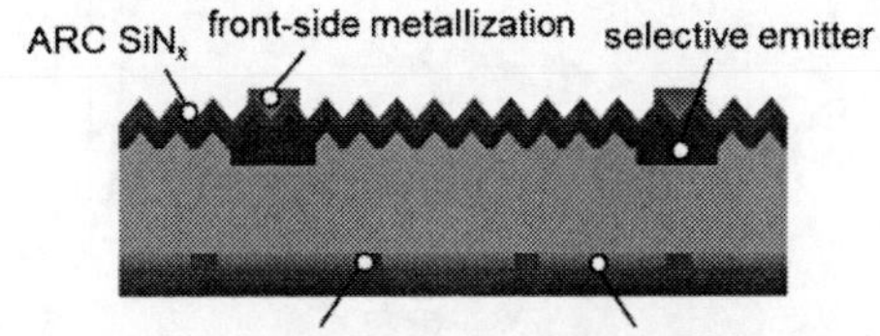

Figure 6: PERC-like cell with selective emitter.

In order for a selective emitter cell to be suitable for mass production, the additional processing costs need to be kept low. Therefore, we develop a process that only has one additional process step for realizing the selective emitter, keeping a lean process flow. The selective-emitters cells are fabricated using industrially-proven processes, including conventional screen-printing. Furthermore, an industrially-viable process needs to have good stability and reproducibility. As one example, the alignment between the n++ emitter and the printed front metal grid needs to be highly reproducible. Otherwise there can be efficiency loss and yield loss in production.

Here, we report results on PERC-like cells with a selective emitter; in Fig. 7, typical results from 50-cell test batches are shown. Standard 6-inch pseudosquare *p*-Cz wafers are used. The efficiencies shown are relative to a reference level of homogeneous emitter cells. The selective emitter cells have an efficiency level that is 1%rel higher than with the homogeneous emitter. Moreover, the efficiency level stays stable for the different process conditions A-D; this is meant to demonstrate a reasonably wide process window relevant to mass production.

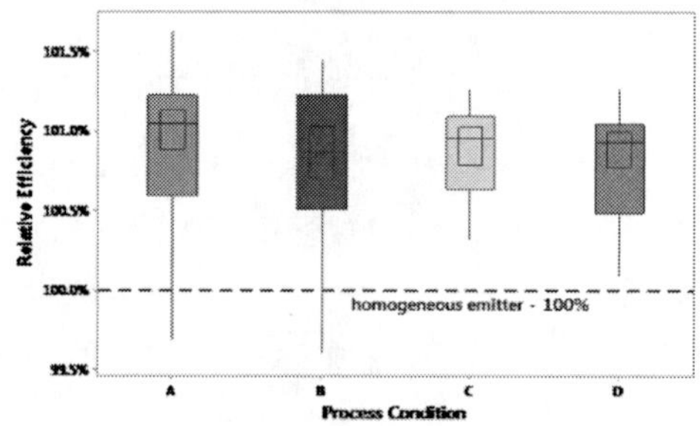

Figure 7: Batch performance of selective emitter cells with various process conditions. Compared relative to homogeneous emitter reference.

The efficiency of the best cell was measured in-house at 22.7% (Fig. 8) with 6BB contacting (including realistic shading); it also achieved an open-circuit voltage of 684.3 mV. This clearly demonstrates the efficiency potential of Q.ANTUM cells incorporating a selective emitter, benefiting from a strong increase in V_{oc} due to decreased recombination associated with the front emitter and metallization. Thus, a selective emitter could boost cell efficiencies by 1-2%rel, vs. a homogeneous emitter.

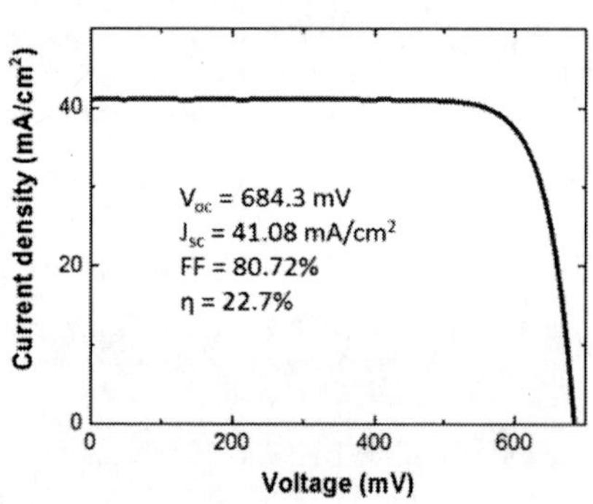

Figure 8: I-V characteristic of champion cell with selective emitter.

To reach 23% efficiency in mass production, we would need a few additional improvements. For example, by going to a busbarless cell design, which would decrease shading losses and allow for further advances in cell interconnection. Also, further optimization of the selective emitter design (i.e. emitter doping levels), front and rear dielectric passivation, and improvements to bulk lifetimes (e.g. material quality, gettering). These enhancements could boost cell efficiency by an additional 2%rel, and bring us to 23% efficiency.

4 SUMMARY

Hanwha Q CELLS produces an annualized 3 GW of PERC-like monocrystalline cells, using our Q.ANTUM technology, which combines leading efficiencies with a lean fabrication process and proven stability against light- (and elevated-temperature) induced degradation. In mass production, we have average efficiencies of 22.15% for 6-busbar Q.ANTUM cells. At the module-level, this translates to high power ratings up to 395 Wp for our Q.PEAK.DUO L product – a module with 144 half-cells and 6-wire interconnection. This represents the best-in-class power ratings for modules with PERC-like cells.

Currently only modules with *n*-type Si cells, and heterojunction or rear-contact technology, have higher power ratings. In the next 2-3 years, we anticipate that we can eliminate the remaining gap in performance. Notably, compared to the complex and thus costlier processing of those *n*-type Si cells, Q.ANTUM technology has a lean manufacturing process and extremely competitive manufacturing cost.

Q CELLS expects to continue on our roadmap of increasing cell efficiency by ~0.6% per annum. One route to higher efficiencies is to use a selective emitter; therefore, we have developed an industrially-viable process for PERC cells with a selective emitter. The selective emitter cells have efficiencies that are 1-2%rel higher than reference cells with homogeneous emitter. Moreover, the best cell achieved an efficiency of 22.7%, with cell voltage above 684 mV.

With the results presented here, we are confident in the capability of our Q.ANTUM technology to continue increasing the cell efficiency to 23% and beyond, within the next couple of years.

References

[1] M. Schaper et al., "The Q.ANTUM technology platform applied to p-Cz to maintain constant efficiency increase per year", 32nd EUPVSEC, 2016.
[2] B. Klöter et al., "Current Status of High-Efficiency Q.ANTUM Technology", 32nd EUPVSEC, 2016.
[3] A. W. Blakers et al., Appl. Phys. Lett. 55, p. 1363, 1989.
[4] F. Fertig et al., Energy Procedia 124, pp. 338–345, 2017.
[5] I. Höger et al., "Boosting module power by advanced interconnection and p-type Cz silicon solar cell efficiencies exceeding 22% in mass production", AIP Conference Proceedings 1999, p. 110003, 2018.
[6] A. Fell, "A free and fast 3D/2D solar cell simulator featuring conductive boundary and quasi-neutrality approximations", IEEE Transactions on Electron Devices 60, pp. 733–738, 2012.

Industrially feasible PERC cells on diamond wire sawing multi-crystalline silicon wafers textured by RIE towards 21.62% efficiency

Quanhua Ye, Wei Wang, Jianwen Dong, Wenwen Wei, Yadi Yang,Wenhao Cai, Jian Sheng, Chun Zhang

R&D Center, Solar BU, GCL System Integration Technology Co., Ltd., 288#, ChenFeng Road, YangShe Town ZhangJiaGang E&T Deve. Zone, SuZhou, JiangSu, China, 215600. yequanhua@gclsi.com

ABSTRACT: Cost reduction and cell efficiency improvement are the two keys to improving PV production processes. Diamond wire sawing (DWS) can significantly reduce the wafer processing cost and is gaining more market shares in multi-crystalline silicon (mc-Si) with advancements in texturing. Reaction ion etching (RIE) technology has been successfully applied to realize effective surface texturing in DWS mc-Si wafer regardless of grain orientations, with minimal chemical waste. Meanwhile, passivated emitter and rear cells (PERCs) and multi-busbar (MBB) are expected to be the mainstream cell technology with the largest gain in market share over the next few years, due to the high efficiency potential and compatibility with existing p-type materials and production lines. Combining the low cost material of DWS mc-Si with high efficiency structure of PERC with MBB, the final cell efficiency reached average of 21.29% with a batch size of 800 wafers and 21.62% for the best cell. Using 21% cells, the module average power reached 304W, and the cell to module (CTM) ratio reached 98.2%. It must be noted that the cell cost (per-watt) prepared by this route is actually decreased compared to the conventional technology of acidic texturing of DWS wafer with additives.

Keywords: DWS, RIE, PERC, MBB, multi-crystalline

1 INTRODUCTION

Several solar cell manufacturers have introduced PERC or RIE technology into *mc*-Si solar cells mass production in recent years, but few companies have combined these two technologies together. PERC technology improves the rear side of the cell by replacing the conventional Al-back surface field (Al-BSF) structure with passivated rear surface [1]. RIE solves the texturing problem of DWS *mc*-Si, and can achieve an efficiency gain of 0.4~0.6% in p-type DWS *mc*-Si solar cells with Al-BSF [2][3]. The combination of RIE and PERC technology can achieve an efficiency increase of >1.2%abs, while the advantage of low cost DWS *mc*-Si wafers can also be obtained. In addition, MBB technology can greatly reduce the silver paste consumption in PV cell production and improve the conversion efficiency, which can reduce costs and improve efficiency.

2 EXPERIMENTAL DETAILS

All solar cells were fabricated on 156.75×156.75 mm^2 p-type *mc*-Si wafers from GCL company, with a thickness of about 180~200 μm and a bulk resistivity of about 1.5 Ω·cm. The DWS *mc*-Si wafers were first damage-etched in an HNO_3/HF mixture solution, and then the nano-textures were fabricated by RIE. The subsequent damage etch was performed chemically and the reflectivity was optimized at the same time. After phosphorus diffusion, emitter with sheet resistance (R_{sheet}) of 100~130 Ω/□ was formed. On the rear side, the AlO_x and SiN_x layers were deposited by combination of atomic layer deposition (ALD) and plasma enhanced chemical vapor deposition (PECVD). On the front, the deposition of silicon nitride (SiN_x) antireflection coating was optimized to match the nano-textured surface. Laser ablation was applied on the rear side for contact opening. Finally, the cells were completed by conventional screen printing and optimized firing.

A QSSPC lifetime tester, Sinton Instruments WCT-

120, was used to monitor the passivation properties of the different front surface structures and the recombination parameter J_0 of the rear side. Contact resistivity was evaluated using the Transfer Length Method (TLM) [4]. A quantum efficiency measurement system was used to characterize the reflectance of different samples between the wavelength of 300nm and 1180nm.

Quokka, a free and fast multi-dimensional simulator developed by Andreas Fell [5], was used to simulate the series resistance and the saturation current density. The solar cell performance was measured on a production flash tester, using reference cells calibrated by Fraunhofer ISE CalLab.

3 RESULTS AND DISSCUSION

3.1 RIE process

RIE is a promising solution for DWS *mc*-Si wafer texturing. As shown in Fig. 1 (b), 300 nm to 500 nm nanostructures were successfully fabricated, which are smaller than that of wire slurry sawing (WSS) *mc*-Si wafer (3-4 μm) with acidic texturing, as shown in Fig.1 (a). The nanostructured surface shows lower reflectance even post ARC deposition, as shown in Fig. 2.

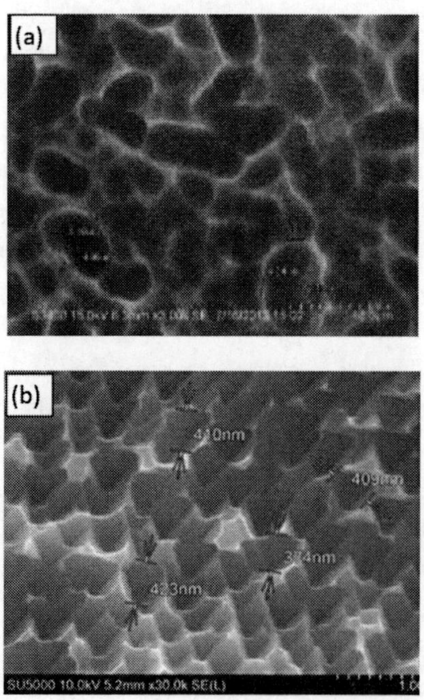

Figure 1: SEM images of (a) Regular acid texturing on WSS *mc*-Si wafer, (b) RIE on DWS *mc*-Si wafer

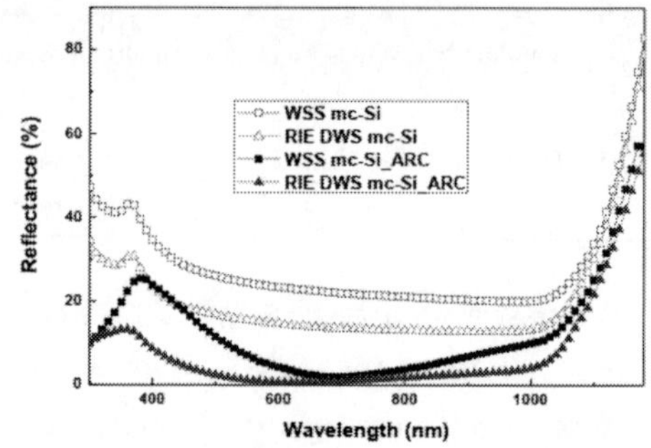

Figure 2: Reflectance curves of regular acid texturing WSS and RIE texturing DWS *mc*-Si wafers, with and without SiN_x anti-reflection coating (ARC).

The surface passivation of the front nanostructure is a challenge for industrial cell fabrication. In this paper, front surface J_0 of around 48.5 fA/cm^2 was achieved after optimization of the diffusion and passivation processes, while maintaining acceptable contact resistivity. Table 1 shows the J_0 of the baseline process (Group A) and the optimized process (Group B).

	Group A	Group B
	Jo_(fA/cm2)	Jo_(fA/cm2)
1	71.5	51.0
2	67.5	47.5
3	81.5	59.0
4	62.5	36.5
ave	70.8	48.5

Table 1 J_{0e} of baseline process (Group A) and optimized process (Group B).

3.2 Rear passivation

PERC cell aims to improve the rear surface recombination by replacing the conventional Al-BSF with a dielectrically passivated rear surface. As shown in Fig. 3 (b), the rear surface of WSS *mc*-Si wafer is rough after etch back of acidic texturing, with mounds of worm-like structures. In comparison, only the front surface is textured during RIE of DWS *mc*-Si wafer, while the rear surface was kept smooth even after etching back by standard industrial process. The smoother rear surface of RIE DWS *mc*-Si wafer is easier to be passivated by Al_xO_3/SiN_x layer.

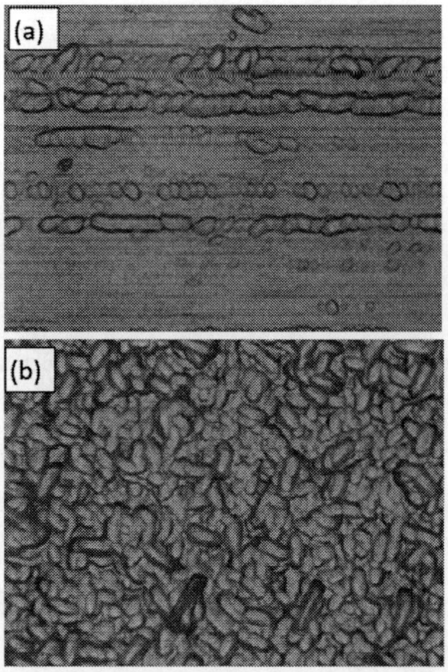

Figure 3: Surface morphology of (a) DWS and (b) WSS *mc*-Si wafers etched back by standard industrial process.

3.3 Multi-busbar (MBB)

The MBB technology reduces the surface shading of the cell, increases the light-receiving area, and shortens the finger length. It can effectively reduce the series resistance of the cell and the associated losses. According to our experimental data, the MBB technology can improve the conversion efficiency by about 0.2-0.4%, reduce silver consumption 25-35% , and increase the module power by about 5-10 W. In addition, the MBB technology has higher tolerance for cracks and disconnections, better long-term reliability, and ensures higher power generation overall. The MBB technology has wide adaptability, is compatible with all the mainstream cell technology, and the appearance is better.

3.4 Cell efficiency

Combining RIE, PERC, and MBB, the low cost DWS mc-Si wafers have been successfully transferred to mass production, and the IV performances are shown in Table 2. The final cell efficiency reached an average of 21.29% with a batch size of 800 wafers and 21.62% for the best cell (Fig. 4). The RIE+PERC cell efficiency is almost 2.59% absolute higher than DWS with conventional texturing technology. Further improvements in efficiency are achievable through further optimization of metallization. Efficiency distribution of pilot production is shown in Fig. 5.

Table 2: I-V performance of solar cells with different texturing

Group	Voc (mV)	Jsc (mA/cm²)	FF (%)	Eff (%)
DWS mc-Si with acid direct texture	636.9	36.69	80.02	18.70
DWS *mc*-Si with RIE+PERC (average)	667.1	40.01	79.77	21.29
DWS *mc*-Si with RIE+PERC (champion)	669.3	40.27	80.20	21.62

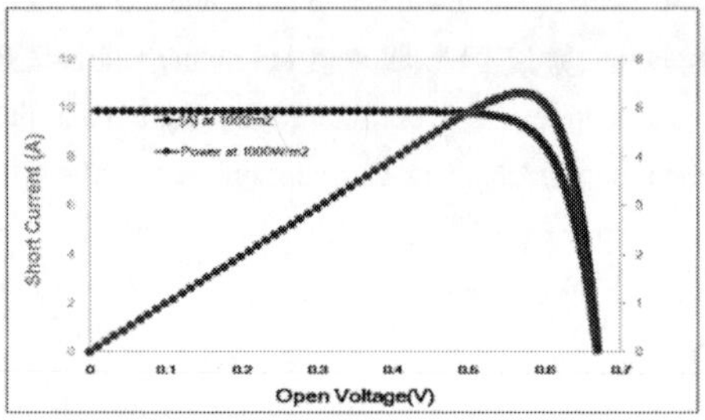

Figure 4: I-V curve of the champion 21.62% efficiency cell

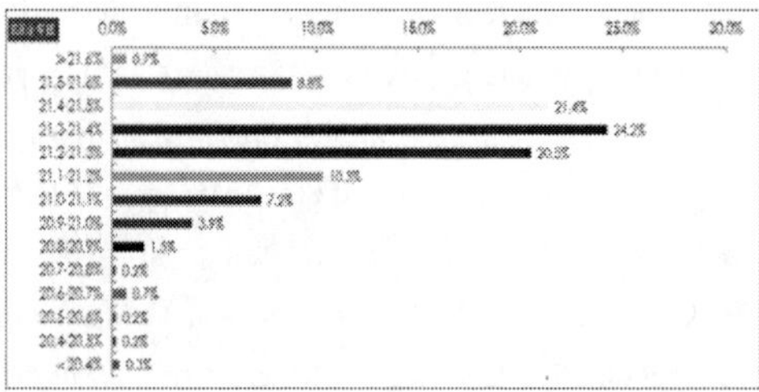

Figure 5: Efficiency distribution of pilot production

3.5 Module performance

In mass production, when the average efficiency of RIE+PERC DWS *mc*-Si solar cells is 20.2%, the module performance is as shown in Fig. 6. The main power grade is 290W, with 19.64% of modules showing less than 290W. After improving the average efficiency to 21.2%, the main power grade is expected to increase to 300W and the main power proportion to reach 96%.

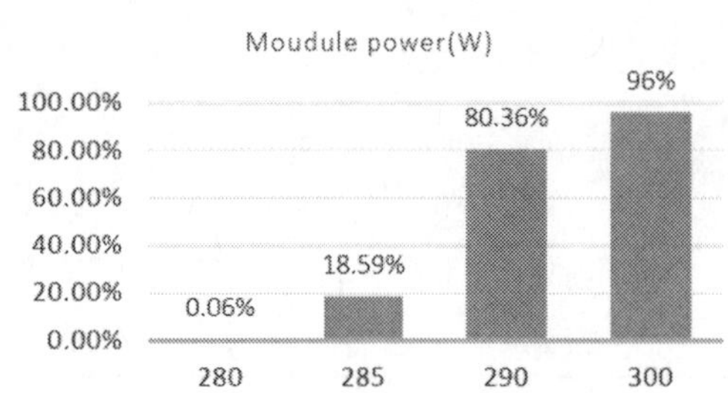

Figure 6: Module power distribution

4 SUMMARY

RIE has been successfully applied to realize the diamond wire cutting *mc*-Si wafer surface texturing regardless of grain orientations, resulting in efficiency improvement. The recombination in the Al-BSF in conventional cells limits the cell efficiency, while the PERC reduces rear surface recombination and improves the efficiency of the cell. Compared with the conventional PV cell technology, the MBB technology has lower series resistance, higher photocurrent, lower front silver paste consumption and higher reliability. Combining the high efficiency structure of RIE, PERC and MBB with low cost DWS *mc*-Si wafers, the cell efficiency has reached 21.29% in mass production and 21.62% for the best cell. With this process, the increase of cell efficiency can cover the increase of cost, resulting in a reduction in cost per watt.

5 ACKNOWLEDGEMENTS

The authors would like to thank Wei Wang, Jianwen Dong, Wenwen Wei, Yadi Yang,Wenhao Cai, Jian Sheng, Chun Zhang

6 REFERENCES

[1] A.W. Blakers, A.Wang, A. M. Milne, J. Zhao, and M. A. Green, "22.8% efficient solar cell," Appl. Phys. Lett., vol. 55, pp. 1363–1365, 1989.

[2] J.Yoo, G.Yu, J.Yi, Large-area multi crystalline silicon solar cell fabrication using reactive ion etching (RIE), Sol. Energy Mater. Sol.Cells 95(2011)2–6

[3] Y.D. Yang, W. Wang, Q.H. Ye, J. W. Dong, et al. Fabricating nano-texture on diamond-wire saw multi-crystalline silicon using RIE-etch, 32th European Photovoltaic Solar Energy Conference and Exhibition. Munich, Germany, 2016, pp.667-670

[4] D. L. Meier, D. K. Schroder, "Contact resistance: its measurement and relative importance to power loss in a solar cell," IEEE Trans. Electr. Dev. 31, 647(1984).

[5] A. Fell, "A free and fast three-dimensional/two-dimensional solar cell simulator featuring conductive boundary and quasi-neutrality approximations," IEEE Trans. Electr. Dev. 60, 733(2013)

[6] Y. W. Ok, A. D. Upadhyaya, Y. G. Tao, F. Zimbardi, K. Ryu, M. H. Kang, A. Rohatgi, "Ion-implanted and screen-printed large area 20% efficient

N-type front junction Si solar cells," Solar Energy Materials & Solar Cells. 123, 92(2014).

[7] J. Benick, B. Hoex, G. Dingemanns, A. Richter, M. Hermle, S. W. Glunz, "High-efficiency N-type silicon solar cells with front side boron emitter," 24th EUPVSEC, 2009, pp. 863-870.

[8] D. Suwito, U. Jäger, J. Benick, S. Janz, M. Hermle, S. W. Glunz, "Industrially feasible rear passivation and contacting scheme for high-efficiency N-type solar cells yielding a V_{oc} of 700mV" IEEE Trans. Electr. Dev. 57, 2032(2010)

PRODUCTION COMPATIBLE REMEDY AGAINST LeTID IN HIGH-PERFORMANCE MULTICRYSTALLINE SILICON SOLAR CELLS

Dennis Bredemeier[1,2], Dominic C. Walter[1], Thomas Pernau[3], Olaf Romer[3], and Jan Schmidt[1,2]
[1]Institute for Solar Energy Research Hamelin (ISFH), Am Ohrberg 1, D-31860 Emmerthal, Germany
[2]Leibniz University Hannover, Institute of Solid-State Physics, Appelstr. 2, D-30167 Hannover, Germany
[3]Centrotherm international AG, Württemberger Str. 31, D-89143 Blaubeuren, Germany

We examine the 'LeTID' (Light and elevated Temperature Induced Degradation) effect and the subsequent regeneration in high-performance multicrystalline silicon (mc-Si). We treat lifetime samples and PERC solar cells with a production-compatible inline regeneration furnace (c.REG) from centrotherm international AG and demonstrate that this industrial-type regeneration treatment is capable of effectively suppressing LeTID. On lifetime samples, we observe an increase in the lifetime by one order of magnitude compared to the untreated and fully degraded samples. On finished industrial-type PERC solar cells, the c.REG treatment results in a gain of 6 to 14 mV in the open-circuit voltage at the point of maximum degradation compared to untreated reference solar cells. This compares to a relative gain in conversion efficiency of up to 6.8%.
Keywords: multicrystalline silicon, light-induced degradation, regeneration

1 INTRODUCTION

In 2012, Ramspeck et al. reported that solar cells fabricated on multicrystalline silicon (mc-Si) wafers show a pronounced degradation upon illumination at elevated temperature (>50°C) [1]. They were not able to explain this effect by known degradation mechanisms such as the boron-oxygen defect activation [2] or the iron-boron pair dissociation [3] and hence attributed the efficiency loss to a new degradation mechanism. The increasing number of studies on this phenomenon in recent years underlines the practical and fundamental relevance of the effect, which is frequently denoted 'LeTID' (Light and elevated Temperature Induced Degradation) in the literature [4-10]. Since the degradation is particularly visible on mc-Si solar cells with an Al_2O_3/SiN_x-passivated rear surface, this degradation mechanism strongly interferes with the introduction of mc-Si passivated emitter and rear (PERC) solar cells into the market [1,4]. In this contribution, we present a production-compatible remedy against LeTID using the inline regeneration furnace c.REG from centrotherm international AG. We examine the stability of both regenerated lifetime samples as well as industrial-type PERC solar cells.

2 SAMPLE PREPARATION

In the first part of this study, we have processed lifetime samples from industrially available boron-doped high-performance (HP) block-cast multicrystalline silicon (mc-Si) wafers with a base resistivity of 2.1 Ωcm and a size of (156×156) mm². All experiments are conducted on samples from the same region of the ingot, so effects stemming from different defect structures do not affect the results. The as-cut HP mc-Si wafers are first cleaned with a surface-active agent and subsequently etched in a potassium hydroxide solution to remove the saw damage. A phosphorus diffusion is then performed in a quartz-tube furnace at a process temperature of 850°C resulting in n^+-layers on both wafer surfaces with a sheet resistance between 50 and 60 Ω/sq. The n^+-layers are chemically removed by a solution of hydrofluoric acid and nitric acid using a chemical polishing process. The surfaces are then passivated by an Al_2O_3/SiN_x stack, representing the PERC rear side, where the 10 nm thick Al_2O_3 layers are deposited by plasma-assisted ALD (FlexAL™, Oxford Instruments). The SiN_x layers with a thickness of 100 nm and a refractive index of 2.05 (measured at a wavelength of 633 nm) are deposited using an industrial-type plasma enhanced chemical vapor deposition (PECVD) process (SiNA, Meyer Burger). All samples receive a rapid thermal annealing (RTA) firing treatment at a set peak temperature of 900°C and a belt speed of 6.8 m/min using an industrial conveyor-belt firing furnace (centrotherm international AG, DO-FF-8.600-300). We have examined the actual wafer temperature during firing and measured that the actually reached peak temperature is (809±9)°C.

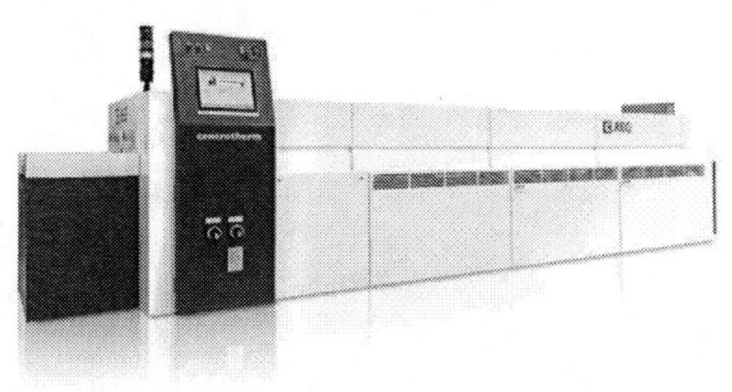

Figure 1: Photograph of the c.REG inline furnace from centrotherm international. The regeneration process takes place at intense illumination and at a temperature of 280°C.

In the second part of this study, we fabricate lifetime samples on boron-doped HP mc-Si wafers with a base resistivity of 1.2 Ωcm and a size of (156×156) mm² stemming from a different manufacturer. We use the same process steps as described above. We compare the behavior of our lifetime samples with the behavior of industrially produced full-square four-busbar PERC solar cells that have been fabricated on the same wafer material.
All samples with the exception of the non-treated reference samples receive a regeneration process in the c.REG furnace from centrotherm international AG. Figure 1 shows a photograph of the regeneration furnace.

Within this study, we vary the number of runs, i.e. the number of subsequent treatments in the furnace at maximum possible light intensity. One run takes a total time of 40 seconds. Please note that in the second part of this paper, we use another c.REG unit as in the first part, however, we adjust the temperature profile to match the profile used within the first part of our study.

3 EXPERIMENTAL DETAILS

Injection-dependent carrier lifetimes are measured using the photoconductance decay (PCD) technique (Sinton Lifetime Tester WCT-120) [11]. Spatially resolved lifetime measurements are carried out using the photoconductance-calibrated photoluminescence imaging (PC-PLI) method [12]. We extract the lifetime values at a fixed excess carrier concentration of $\Delta n = 10^{15}$ cm^{-3}. For the spatially resolved measurements, we calculate a lifetime image at a fixed Δn of 10^{15} cm^{-3} by interpolating a series of lifetime images, each recorded at a constant excitation intensity, resulting in a laterally varying Δn-distribution. In the first part of this study, we test the lifetime stability of the samples under illumination with a halogen lamp at a light intensity of 1 sun and a temperature of 75°C. The wafers from the second part of this study are tested under the same light source at 1 sun illumination and a temperature of 95°C to accelerate the experiments.

4 EXPERIMENTAL RESULTS

4.1 Regeneration of lifetime samples

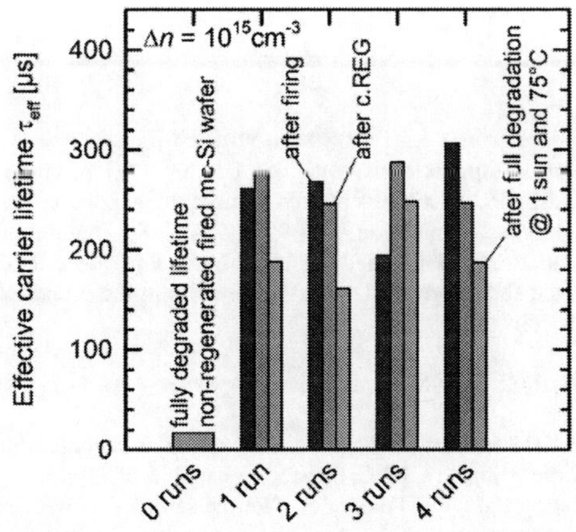

Figure 2: Effective carrier lifetimes of mc-Si samples treated in the c.REG regeneration furnace. The number of subsequent runs in the furnace varies between one single run and four runs. The lifetime during the subsequent stability test at 1 sun illumination intensity and 75°C are well above the lifetime of the untreated mc-Si reference sample.

Figure 2 shows the results using the maximum light intensity in the c.REG at a set peak temperature of 280°C and varying the number of runs between one single run and four subsequent runs in the c.REG. For one to three runs, the lifetime after regeneration treatment is comparable or even higher than the lifetime measured directly after the firing step. However, the sample that received four runs shows a decrease in lifetime from 306 μs to 247 μs. As can be seen in Fig. 3, the lifetime evolution shows a slight degradation and a subsequent regeneration during the stability test after the c.REG treatment. We observe that one run in the c.REG is sufficient to significantly reduce the LeTID extent compared to the fully degraded non-regenerated reference sample. However, we obtain the best results for three runs.

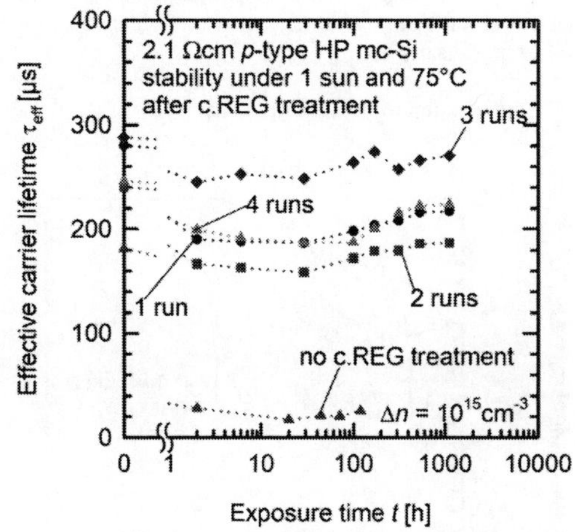

Figure 3: Effective carrier lifetimes as a function of the exposure time in hours to 1 sun light intensity (halogen lamp) at 75°C. Samples treated in the c.REG furnace show a slight degradation and regeneration. However, the lifetimes remain well above the lifetime of an untreated mc-Si reference sample (dotted lines are guides to the eyes).

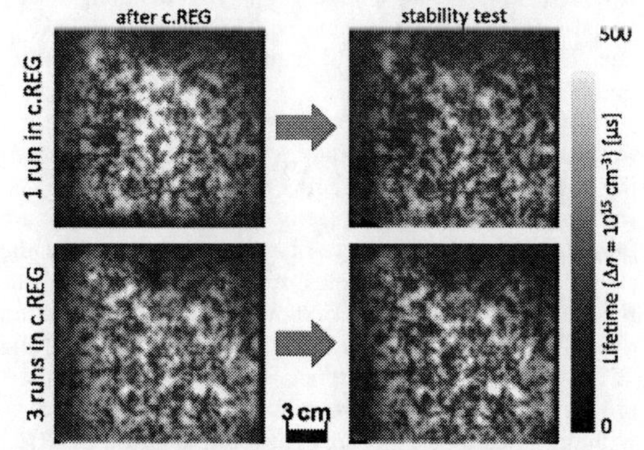

Figure 4: Lifetime images measured using the PC-PLI technique of two mc-Si wafers after c.REG treatment and after 314 hours of stability test at 1 sun, 75°C. The upper sample received one run and the lower sample three runs in the c.REG furnace. Both samples show an almost unchanged lifetime distribution during the stability test.

Figure 4 shows the spatially resolved lifetime images for two samples after 1 run and after 3 runs directly after regeneration in the c.REG and after 314 hours of illumination at 1 sun light intensity at 75°C. The LeTID is typically occurring homogeneously over a large area of the wafer. This effect can clearly not be seen in Fig. 4.

We conclude that LeTID is effectively suppressed by the c.REG treatment.

We monitor the surface passivation quality of float-zone silicon (FZ-Si) reference samples passivated with the same Al_2O_3/SiN_x stack as used on our mc-Si lifetime samples. After one run in the c.REG furnace, the lifetime of the Al_2O_3/SiN_x-passivated FZ-Si wafer decreases from 3.7 to 2.4 ms, the latter lifetime being equivalent to a surface recombination velocity of 3.3 cm/s. After 3 runs we measure that the lifetime decreases to 1.8 ms, still corresponding to a very low surface recombination velocity of 4.6 cm/s. Thus, the negligible increase of the surface recombination velocity does not limit the lifetime in our mc-Si samples.

4.2 Regeneration of PERC solar cells

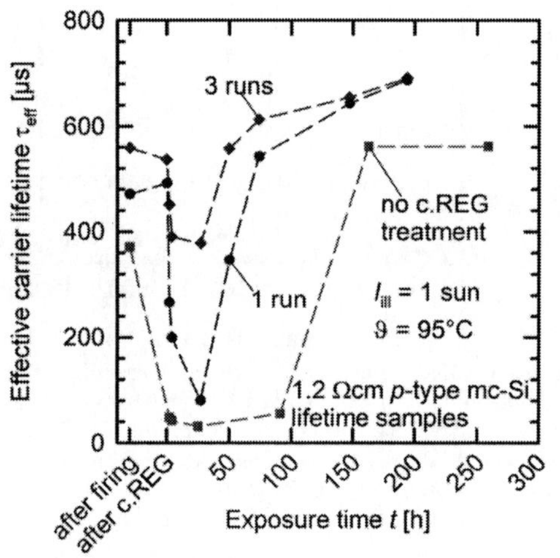

Figure 5: Effective carrier lifetimes as a function of the exposure time (1 sun halogen-lamp illumination, 95°C). The samples treated in the c.REG furnace show a decrease in the extent of the degradation with increasing number of c.REG runs.

In the second part of this study, we examine the impact of the c.REG regeneration process on industrially produced four-busbar PERC solar cells. Lifetime samples are fabricated on the same mc-Si material to compare the stability of lifetime samples with the stability of the finished solar cells. Figure 5 shows the lifetime evolution of the lifetime samples under illumination at 1 sun and 95°C. The untreated reference sample shows the most pronounced degradation from initially 371 to only 32 μs within 25 hours of exposure. The treatment in the c.REG furnace leads to a decrease in the extent of the degradation with increasing number of runs. The sample that received three subsequent runs shows a high lifetime of 378 μs in the point of maximum degradation. Also, note that the initial lifetimes and final lifetimes of the samples treated in the c.REG are higher than in the untreated reference sample. In contrast to the lifetime samples presented in the first part of this study, one run in the c.REG furnace seems to be not sufficient to effectively suppress the LeTID effect. This might be due to the different mc-Si material and/or the increased temperature of 95°C applied. Hence, we can conclude that the optimal c.REG regeneration conditions need to be adapted to the silicon material and the expected degradation conditions.

Figure 6 shows results on mc-Si industrially produced PERC solar cells fabricated on the same material as shown in Fig. 5. The PERC solar cell with no c.REG treatment shows the most pronounced degradation in terms of open-circuit voltage from initially 638 mV to only 595 mV after 4 hours of light exposure. At the point of maximum degradation, the solar cells treated in the c.REG furnace show a 14 mV higher open-circuit voltage than the untreated solar cells. Note that the treatment in the c.REG furnace alters the degradation and regeneration kinetics. In comparison to the untreated solar cell, the degradation of the regenerated solar cells is significantly slowed down. However, after about 100 hours of exposure, all solar cells reach open-circuit voltages close to their respective initial values. Note that at the point of maximum degradation, the c.REG-treated solar cells show an efficiency of up to 6.8% relative higher compared to the untreated reference cell.

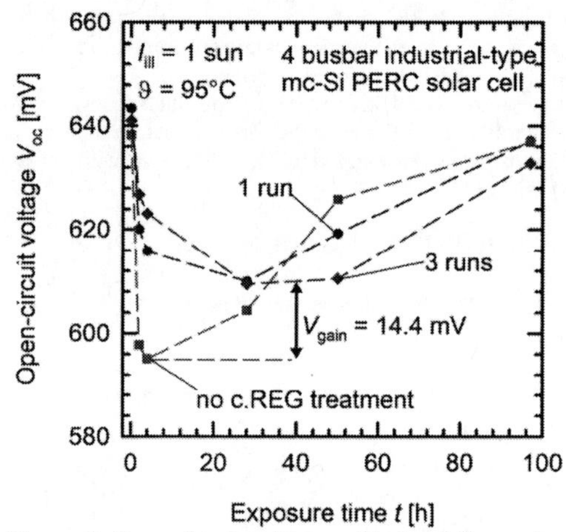

Figure 6: Open-circuit voltages of industrially produced mc-Si PERC solar cells plotted against the exposure time to 1 sun illumination at 95°C. At the maximum degradation point, the solar cells treated in the c.REG furnace show a gain of 14 mV compared to the untreated solar cells.

5 DISCUSSION

The data plotted in Fig. 3 show that the regeneration treatment in the c.REG furnace is capable of effectively reducing the LeTID extent on lifetime samples. However, the exact extent of the degradation depends strongly on the used mc-Si material. While one single run of c.REG treatment was sufficient to suppress degradation on the mc-Si material used in the first part of this study, the degradation is still very pronounced after one run on the material used in the second part of this study (see Fig. 5), but was strongly reduced after 3 runs. Therefore, it can be conjectured that the c.REG conditions need to be adapted to the mc-Si material used.

Figure 6 shows that the solar cells that underwent the c.REG regeneration treatment as well as the untreated reference cell feature a pronounced degradation in the open-circuit voltage under illumination at elevated temperature. The regeneration kinetics of the c.REG-treated solar cells was found to be much slower compared to the untreated cell. This is consistent with recent results demonstrating that the LeTID defect kinetics are modulated through dark annealingHow at elevated temperature or by applying a second rapid thermal annealing step [13,14]. At the point of maximum degradation, the c.REG-treated solar cells show an increase in open-circuit voltage that ranges from 6 to 14 mV compared to the untreated reference solar cell. Upon prolonged exposure, all solar cells show a regeneration. However, under typical field conditions the regeneration is too slow to take effect during the warranty period of a PV installation [4]. Concerning the efficiency, we measured a gain of up to 6.8% relative.

6 CONCLUSIONS

Within this contribution, we have shown that the pronounced lifetime degradation on mc-Si wafers under illumination at elevated temperature (the so-called 'LeTID' effect) can be significantly reduced using an industrially available inline regeneration tool (the c.REG furnace from centrotherm) which applies intense illumination at a set temperature of 280°C, in this study. We have examined the stability of the regenerated state of lifetime samples and PERC solar cells under typical LeTID conditions and found that the exact duration of the c.REG treatment needs to be adapted to the respective mc-Si material used. Nevertheless, a significant reduction of the extent of degradation was observed for both examined mc-Si materials stemming from different producers. The c.REG regeneration treatment also decreases the degradation extent on industrially produced mc-Si PERC solar cells. At the point of maximum degradation, we measured a gain of up to 14 mV in the open-circuit voltage in comparison to untreated reference solar cells. The efficiency at the point of maximum degradation was improved by 6.8% relative.

ACKNOWLEDGEMENTS

The authors thank C. Marquardt for sample processing. This work was funded by the German State of Lower Saxony and the German Federal Ministry of Economics and Energy (BMWi) and by industrial partners within the research project "LIMES" (contract no. 0324204D). The content is the responsibility of the authors.

REFERENCES

[1] K. Ramspeck, S. Zimmermann, H. Nagel, A. Metz, Y. Gassenbauer, B. Birkmann, and A. Seidl, *Proceedings of the 27th European Photovoltaic Solar Energy Conference* (WIP, Munich, 2012), p. 861.

[2] J. Schmidt, A. G. Aberle, and R. Hezel, *Proceedings of the 26th IEEE Photovoltaic Specialists Conference* (IEEE, New York, 1997), p. 13.

[3] S. W. Glunz, S. Rein, W. Warta, J. Knobloch, and W. Wettling, Proceedings of the 2nd World Conference on Photovoltaic Energy Conversion (WIP, Munich, 1998), p. 1343.

[4] F. Kersten, P. Engelhart, H.-C. Ploigt, A. Stekolnikov, T. Lindner, F. Stenzel, M. Bartzsch, A. Szpeth, K. Petter, J. Heitmann, J.W. Müller, Sol. Energy Mater. Sol. Cells 142 (2015) 83.

[5] F. Fertig, K. Krauß, and S. Rein, Phys. Status Solidi RRL 9, 41 (2014).

[6] A. Zuschlag, D. Skorka, and G. Hahn, Prog. Photovolt: Res. Appl. 25, 545 (2017).

[7] K. Nakayashiki, J. Hofstetter, A. E. Morishige, T.-T. A. Li, D. B. Needleman, M. A. Jensen, and T. Bounassisi, IEEE J. Photovolt. 6, 860 (2016).

[8] C. Chan, T. H. Fung, M. Abbott, D. Payne, A. Wenham, B. Hallam, R. Chen, and S. Wenham, Sol. RRL 1 (2017).

[9] R. Eberle, W. Kwapil, F. Schindler, M. C. Schubert, and S. W. Glunz, Phys. Status Solidi RRL 10, 861 (2016).

[10] D. Bredemeier, D. Walter, S. Herlufsen, and J. Schmidt, AIP Adv. 6, 035119 (2016).

[11] R. A. Sinton, and A. Cuevas, Appl. Phys. Lett. 69, 2510 (1996).

[12] S. Herlufsen, J. Schmidt, D. Hinken, K. Bothe, and R. Brendel, Phys. Stat. Sol. RRL 2, 245 (2008).

[13] C. E. Chan, D. N. R. Payne, B. J. Hallam, M. D. Abbott, T. H. Fung, A. M. Wenham, B. S. Tjahjono, and S. R. Wenham, IEEE J. Photovolt. 6, 1473 (2016).

[14] C. Chan, T. H. Fung, M. Abbott, D. Payne, A. Wenham, B. Hallam, R. Chen, and S. Wenham, Solar RRL 1, 160028 (2017).

BIFACIAL pSPEER SOLAR CELLS FOR SHINGLE MODULES

P. Baliozian, T. Fellmeth, N. Wöhrle, E. Lohmüller, R. Preu
Fraunhofer Institute for Solar Energy Systems ISE, Heidenhofstr. 2, 79110 Freiburg, Germany
Phone: +49 761 - 4588 5383; e-mail: puzant.baliozian@ise.fraunhofer.de

ABSTRACT: Shingling bifacial solar cells leads to higher output power densities p_{out} of silicon-based photovoltaic modules. This paper provides a short update on the current developments at Fraunhofer ISE in Czochralski-grown silicon bifacial "p-type shingled passivated edge, emitter, and rear (pSPEER)" solar cells. In the batch presented in this work, conventionally separated pSPEER cells are fabricated. Illuminated current-voltage measurements of the most efficient pSPEER cell results in a designated front side energy conversion efficiency of $\eta_f = 21.4\%$ for an irradiance $G_f = 1000$ W/m^2. The pSPEER cells show similar front side efficiencies independent of the pre-separation position within the 6-inch host wafer. Considering an additional rear side irradiance of $G_r = 100$ W/m^2, the most efficient pSPEER cell attains a total designated output power density $p_{out} = 22.8$ mW/cm^2.
Keywords: bifacial, p-type silicon, PERC, pSPEER, shingle solar cells

1 INTRODUCTION

Although the idea of shingling solar cells exists since the 1950s [1], it was only used for niche applications [2]. The aim towards achieving high output power density p_{out} modules has revived the concept of shingling, clearly seen in the increase in the number of recent publications [3], patents [4], and commercially available monofacial modules [5] based on shingling interconnection. In the case of shingling (overlapping) interconnection of solar cells, the rear side busbar of a cell overlaps the front busbar of the neighboring cell. Consequently, the active area within the module increases by omitting the spacing between the cells as well as eliminating the shadowing losses from the busbars covered by the neighboring cell's active area. Additionally, shingling interconnection decreases resistive losses on module level decreasing cell-to-module losses.

An additional gain in p_{out} can be attained by using bifacial solar cells that make use of the albedo light from the surroundings. Moreover, the p-type silicon passivated emitter and rear cell (PERC) [6] concept is of high interest for the industry, which is clearly shown by the recent and expected increase in market share [7]. Thus, joining the concepts of shingling, bifaciality and the state-of-the-art PERC technology provides industry relevant solutions for modules with high p_{out}.

This paper aims to portray recent developments in the bifacial "p-type shingled passivated edge, emitter, and rear (pSPEER)" solar cells that have been first introduced in Ref. [8]. The current fabrication process at Fraunhofer ISE and characterization results of pSPEER cells are discussed.

2 pSPEER SOLAR CELL FABRICATION AND CHARACTERIZATION

2.1 Fabrication Process

Currently, pSPEER solar cells are fabricated by using industrial 6-inch Czochralski-grown silicon (Cz-Si) precursors with rear and front side passivation layers that undergo the backend processes; see Fig. 1. In this work the 6-inch p-type Cz-Si precursors used feature a phosphorous oxychloride diffused n-type emitter. The rear and front sides are passivated by a industrial typical stack of aluminum oxide and silicon nitride. In a previously performed batch, similar precursors' mean thickness t is measured to be $t = 155$ µm and the base resistivity ρ_B is measured to be ranging between 0.3 - 0.9 Ωcm. The mean measured implied open-circuit voltage is i$V_{OC} = 702$ mV. For the formation of the local contact openings (LCO), an infrared laser process is used to locally ablate the rear side passivation layer. The rear and front side metallization layouts are printed by using commercial metal pastes in an industrial-type screen printing line. Initially, the rear side silver busbars are printed, followed by the rear side aluminum finger grid covering the LCO. Subsequently, the front side silver finger grid is printed and then contact fired in an industrial fast firing process at set peak temperatures around 840 °C. After firing, the rear side silver busbar features a mean measured width $w \approx 492$ µm. The mean rear side aluminum finger width is measured to be $w \approx 221$ µm, with a finger mid-to-mid pitch $f_p = 1.3$ mm. The mean front side silver busbar is $w \approx 800$ µm wide, while the measured mean finger width is $w \approx 37$ µm having the pitch $f_p = 1.3$ mm (details concerning the pSPEER metallization layouts can be found in Ref. [10].) Conventional laser scribing and manual cleaving is used to obtain 6 pSPEER cells from each host wafer. During this separation process approximately 1/3rd of the host wafer's thickness is scribed using a pulsed laser and the rest is manually cleaved. The pSPEER cell has the perimeter dimensions: 22 mm x 148 mm; see Fig. 2. These dimensions have been chosen to be compatible on the different industrially available wafer formats. In an industrial production of these cells, a format which utilizes the host wafer more efficiently would be used.

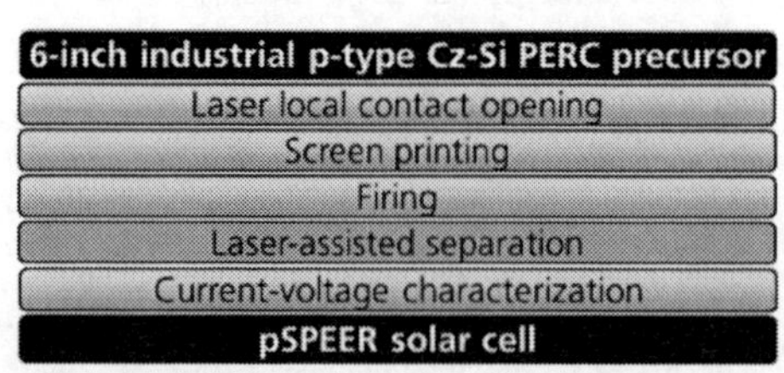

Figure 1: pSPEER solar cells backend fabrication processes starting from industrial 6-inch p-type Cz-Si PERC precurors (figure adapted from Ref. [9]).

2.2 Current-voltage measurement

The increased interest in cells ready for shingling interconnection also raises the questions of suitable current-voltage (*IV*) characterization for such shingle cells. Since the busbars of shingle cells are meant to be covered due to the overlap of the shingling interconnection, a designated area measurement (i.e. excluding busbar area) is also of interest. A total area measurement can be done and subsequently converted to a designated area. A calculation step by using the linear equation (1), converts total area short-circuit current $I_{SC,tot}$, into designated area short-circuit current density $j_{SC,des}$ by only subtracting the busbar area A_{BB} from the total area A_{tot} of the pSPEER solar cell.

$$j_{SC,des} = \frac{I_{SC,tot}}{A_{tot}-A_{BB}} \quad (1)$$

The measured total area *IV* parameters of shingle pSPEER cells can be converted into designated area *IV* parameters without inducing errors larger than usual measurement deviations in *IV* testing. An elaboration on the difference between a total area and a designated area measurement of pSPEER cells will be given in Ref. [11].

For the *IV* measurements of pSPEER cells shown in this work, an industrial cell tester is used. The pSPEER cells are measured at standard test conditions subsequently from either side with an irradiance of $G_f = 1000$ W/m^2 placed on a conductive copper chuck. The full area measurement is done by fully contacting the rear side by means of a conductive copper chuck. The front side is contacted by an *IV* pin array. The set-up is calibrated using pSPEER cells measured at Fraunhofer ISE CalLab PV Cells on a black non-reflective background. At an irradiance $G_f = 1000$ W/m^2, the output power density p_{out} in "mW/cm^2" and the energy conversion efficiency η in "%" have identical numerical values (for an explanation see Ref. [12]).

3 RESULTS AND DICUSSION

The peak designated area front side efficiency $\eta_f = 21.4\%$ is attained having an open-circuit voltage $V_{OC} = 666$ mV, short-circuit current density $j_{SC,des} = 40.2$ mA/cm^2, and fill factor $FF = 79.8\%$; see Table I. A front side pseudo-fill factor $pFF = 82.3\%$ is measured. $j_{SC,des}$ is calculated by using equation (1) where the total short-circuit current density $j_{SC,tot}$ is taken from the *IV* measurement results. The average total area of a pSPEER cell is measured to be $A_{tot} = 3263$ mm^2. The same pSPEER cell's rear side measurement shows a designated area rear side efficiency of $\eta_r = 13.7\%$. The rear side short-circuit current density is $j_{SC,des} = 26.2$ mA/cm^2, $V_{OC} = 654$ mV, and $FF = 79.7\%$. Finally, the bifaciality factor $\beta = \eta_r/\eta_f$ is calculated out of the respective designated efficiencies to be $\beta = 0.64$. Figure 3 shows the *IV* data measurement results of 54 measured pSPEER cells.

Considering an additional rear side illumination with an irradiance $G_r = 100$ W/m^2, the pSPEER cell with the highest p_{out} attains a designated total output power density $p_{out} = 22.8$ mW/cm^2.

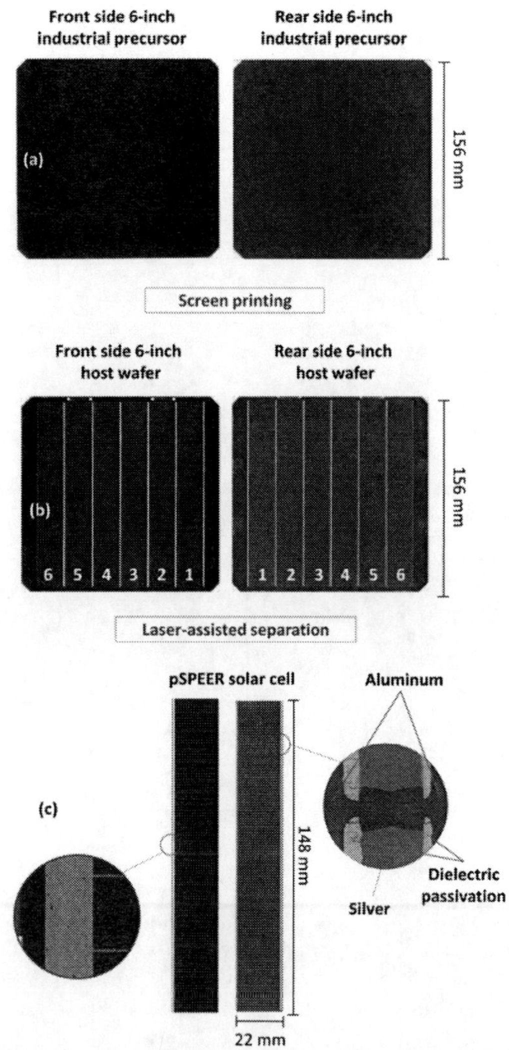

Figure 2: Images of the front and rear sides of (a) 6-inch industrial precursor (the rear side has a yellowish color since the precursor is optimized for monofacial use), (b) metallized host wafer after screen printing and (c) pSPEER solar cells of size 22 mm x 148 mm after separation.

Table I: *IV* data for the pSPEER solar cell with the highest output power density.

Meas. side	η (%)	V_{OC} (mV)	$j_{SC,des}$ (mA/cm^2)	FF (%)	β (1)
Front	21.4	666	40.2	79.8	0.64
Rear	13.7	654	26.2	79.7	

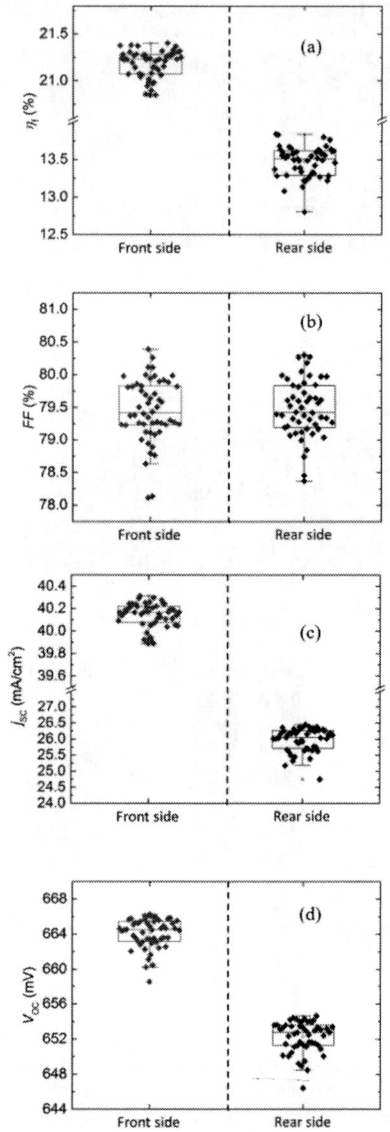

Figure 3: Front side and rear side *IV* measurement data of 54 pSPEER solar cells showing: (a) designated front side efficiency η_f, (b) fill factor *FF* (c) designated short-circuit current density $j_{SC,des}$, and (d) open-circuit voltage V_{OC}.

The dependency of η_f on their position pre-separated pSPEER cell within the host wafer is investigated. The designated front side efficiency η_f shows no significant difference in the range of results dependent on the position; see Fig. 4.

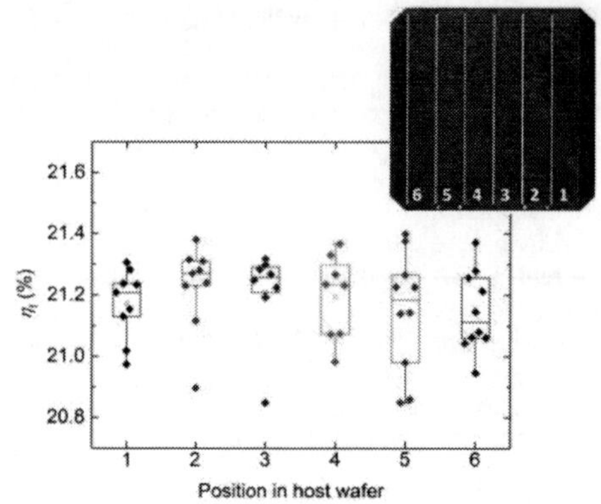

Figure 4: *IV* measurement results showing the designated front side energy conversion efficiency η_f of 54 pSPEER cells taken from different positions of the host wafers.

The currently used precursor is optimized for monofacial use. The lower V_{OC} for rear side measurment can be nearly attributed to the lower rear side $j_{SC,des}$. This can be verified using the one-diode model equation: The dark recombination current density j_{01} equals to $j_{01} \approx 250$ fA/cm^2, calculated out of the respective average front side values of j_{SC} and V_{OC} shown in Figure 3. Replacing j_{SC} front with j_{SC} rear yields a reduction of $\Delta V_{OC} \approx 12$ mV which equals to the measured difference.

Higher bifaciality values for pSPEER cells can be attained by adapting the rear side optical properties and optimizing the rear side metallization fraction towards decreasing shading. The fluctuations in the *FF* in Fig. 3(b), can be explained by the range of the measured base resistivity of the precursors used for the fabrication.

The increase in perimeter to area ratio would suggest the importance of supressing edge recombination [13]. Edge recombination can be shown by the rather lower pseudo-fill factor value $\Delta pFF \approx 1.2\%_{abs}$ compared to a 6-inch bifacial solar cell. In the near future, passivating the edges can improve pSPEER solar cell performance.

4 CONCLUSIONS

Bifacial shingle solar cells based on the p-type shingled passivated edge, emitter, and rear (pSPEER) cell concept are being developed and fabricated at Fraunhofer ISE. 6-inch industrial p-type Cz-Si passivated emitter rear cell (PERC) precursors are screen printed with special shingle metallization layouts to obtain six pSPEER cells per host wafer. The applied laser-assisted separation step leads to pSPEER cells of dimensions 22 mm x 148 mm. *IV* results of pSPEER cells show a peak designated front side energy conversion efficiency $\eta_f = 21.4\%$ and a bifaciality factor $\beta = 0.64$.

By including a rear side irradiance $G_r = 100$ W/m^2, the best pSPEER solar cell attains a total output power density $p_{out} = 22.8$ mW/cm^2.

ACKNOWLEDGEMENTS

The authors would like to thank all contributing colleagues at Fraunhofer ISE Photovoltaics Division who made the development of this work possible. Special thanks to Mohammad Al-Akash for measuring the pSPEER solar cells shown in this work. The work is conducted within the scope of "PV-BAT400" research project (contract no. 0324145) funded by the German Federal Ministry of Economic Affairs and Energy.

REFERENCES

[1] J.D.C. Dickson, "Photo-voltaic semiconductor apparatus or the like," US Patent 2,938,938, May 31, 1960.

[2] S. W. Glunz, J. Dicker, M. Esterle, M. Hermle, J. Isenberg, F. J. Kamerewerd, J. Knobloch, D. Kray, A. Leimenstoll, F. Lutz, D. Oßwald, R. Preu, S. Rein, E. Schäffer, C. Schetter, H. Schmidhuber, H. Schmidt, M. Steuder, C. Vorgrimler, and G. Willeke, "High-efficiency silicon solar cells for low-illumination applications," in *29th IEEE Photovoltaic Specialists Conference New Orleans*, New Orleans, 2002, pp. 450–453.

[3] D. Tonini, M. Bertazzo, A. Fecchio, M. Galiazzo, "Shingling Technology for Cell Interconnection: Technological Aspects and Process Integration," *33rd EUPVSEC, Amsterdam, The Netherlands*, 2017.

[4] R. Morad, G. Almogy, I. Suez, J. Hummel, N. Beckett, Y. Lin, D. Maydan, and J. Gannon, "Shingled solar cell module," US 9397252 B2 US 14/594,439, Jul 19, 2016.

[5] SunPower Corporation SunPower Introduces New Solar Panel: The Performance Series. https://us.sunpower.com/solar-panelstechnology/p-series-solar-panels/.

[6] A. W. Blakers, A. Wang, A. M. Milne, J. Zhao, and M. A. Green, "22.8% efficient silicon solar cell," *Appl. Phys. Lett.*, vol. 55, no. 13, pp. 1363–1365, 1989.

[7] VDMA, "International Technology Roadmap for Photovoltaic: Results 2017 - Ninth Edition," vol. 2018, http://www.itrpv.net/Reports/Downloads/.

[8] P. Baliozian, E Lohmüller, T. Fellmeth, N. Wöhrle, A. Krieg, R. Preu, "Bifacial p-type silicon shingle solar cells - the "pSPEER" concept," *Solar RRL*, 2018.

[9] P. Baliozian, E. Lohmüller, T. Fellmeth, N. Wöhrle, A. Krieg, R. Preu, "Bifacial Shingle Solar Cells on p-Type Cz-Si (pSPEER)," *SiliconPV 2018, AIP Conf. Proc. 1999*, 10002-1–110002-6.

[10] M. Al-Akash, P. Baliozian, T. Fellmeth, E. Lohmüller, N. Wöhrle, R. Preu, "Metallization Fraction of Bifacial pSPEER Shingle Solar Cells," *35th EUPVSEC, Brussels, Belgium (to be published)*.

[11] N. Wöhrle et. al (to be submitted) 2018.

[12] F. Fertig, S. Nold, N. Wöhrle, J. Greulich, I. Hädrich, K. Krauß, M. Mittag, D. Biro, S. Rein, and R. Preu, "Economic feasibility of bifacial silicon solar cells," *Prog. Photovolt: Res. Appl.*, no. vol. 24, 6, pp. 800–817, 2016.

[13] N. Wöhrle, T. Fellmeth, E. Lohmüller, P. Baliozian, A. Fell, R. Preu, "The SPEER solar cell Simulation study of shingled bifacial PERC technology based stripe cells.," *33rd EUPVSEC, Amsterdam, The Netherlands*, pp. 844–848, 2017.

[14] S. Werner, E. Lohmüller, P. Saint-Cast, J.M. Greulich, J. Weber, S. Maier, A. Moldovan, A.A. Brand, T. Dannenberg, S. Mack, S. Wasmer, M. Demant, M. Linse, R. Ackermann, A. Wolf, R. Preu, "Key aspects for fabrication of p-type Cz-Si PERC solar cells exceeding 22% conversion efficiency," in *33rd EU PVSEC*, Amsterdam, The Netherlands, 2017.

NOVEL SCHEMES OF P^+ POLYSI HYDROGENATION IMPLEMENTED IN INDUSTRIAL 6" BIFACIAL FRONT-AND-REAR PASSIVATING CONTACTS SOLAR CELLS

M.K. Stodolny[1], J. Anker[1], C.J.J. Tool[1], M. Koppes[1], A.A. Mewe[1], P. Manshanden[1], M. Lenes[2], I.G. Romijn[1]
[1] ECN part of TNO, Solar Energy, P.O. Box 15, NL-1755 ZG Petten, The Netherlands
[2] Tempress Systems, Radeweg 31, 8171 MD Vaassen, The Netherlands

ABSTRACT: In this paper we present our latest developments in bifacial 6" cells featuring polysilicon carrier-selective passivating contacts which can be seen as an upgrade to bifacial cell architectures such as PERC+ and PERT. In this study we present improvements of p^+ polySi passivation on industrially relevant textured surfaces by means of advanced hydrogenation schemes, involving novel dielectric stacked layers, and modification of the B-profile leading to a record passivation level of ~8 fA/cm² and 724 mV before firing and ~9 fA/cm² and 722 mV after firing. The results promote industrial implementation of p^+ polySi on 6" bifacial cells with excellent passivation properties and compatibility with screen-printed and fire-through metallization, especially as our latest results show $J_{o,contact}$ being reduced down to 200 fA/cm². Bifacial solar cells have been manufactured with industry compatible processing with front-and-rear passivating contacts with efficiencies above 20% in the first attempt and this development shows a potential of 23% with a bifaciality factor of 90%.

Keywords: Polysilicon, Passivating contact, Carrier selective contact, LPCVD, Bifacial

1 INTRODUCTION

Polysilicon carrier-selective passivating contacts, featuring an interfacial oxide between the wafer and polysilicon, currently receive the attention of many research institutes and industrial parties, due to their excellent passivating properties and industrial manufacturability. Application of n^+ polySi as rear contact for high-efficiency solar cells has resulted in impressive cell efficiencies both for non-industrial lab scale processes on *n*-FZ material (TOPCon) [1] as well as industrially relevant processes (PERPoly) [2-4] and very recently even in production [5].

Application of polysilicon to both the hole (p^+ polySi) and electron (n^+ polySi) contact shows the full (industrial) potential of carrier-selective contacts. The potential has been already demonstrated on small area interdigitated back contact cells with an efficiency of 25% [6]. However, the industrial implementation of p^+ polySi on 6" cells has until now not been reported for PERX (PERC, PERL and PERT) cell types. One of the reasons is the limited passivation level that is being reached for p^+ polySi on textured surfaces and limited compatibility with screen printed and fire-through metallization.

We present 6" bifacial cells with front-and-rear polysilicon passivating contacts, for which all processing has been realized using industrial equipment, and that are suitable for screen printed fire-through (FT) metallization. In this work we focus on improvements of p^+ polySi passivation on industrially relevant textured surfaces by means of advanced hydrogenation schemes and modification of the B-profile across the contact. To go one step further, the interaction of these passivating p^+ polySi layers with FT metallization is studied with respect to the contact recombination.

In this work we close the gap in passivation level between p^+ polySi and n^+ polySi by introduction of modified B-profiles and novel hydrogenation schemes using multi-layer dielectric stacks on top of the polySi that supply and maintain hydrogen passivation even during high-temperature steps. The obtained results promote industrial implementation of p^+ polySi on 6" cells with excellent passivation properties and compatibility with screen printed and fire-through metallization. This work yields bifacial solar cells with front-and-rear passivating contacts with efficiencies above 20% in the first attempt and shows a potential of 23% with bifaciality factor of nearly 90%.

2 RESULTS AND DISCUSSION

2.1 Hydrogenation schemes and modification of B-profile for improved p^+ polySi passivation

The passivation level of p^+ polySi is generally lower than that of n^+ polySi. Effective passivation is considered specifically difficult on the Si <111> planes [7], e.g. on standard wafers with random pyramid texture. The lower passivation quality of p^+ polySi is not completely understood yet, but is assumed to be related, amongst others, to the polysilicon doping level, boron segregation at the interfacial oxide [8], and a demanding interface hydrogenation.

In this contribution, we used symmetric planar and textured samples with boron doped polySi. We varied the B-profile across the p^+ polySi/SiO_x/c-Si by inserting an additional SiO_x interlayer in the polySi during the LPCVD run and we varied the composition of the dielectric ARC capping layer. Both novel approaches mitigate the interface passivation issues that are associated with p^+ polySi.

Fig. 1 shows the active B concentration profiles in the polySi layers *A* and *B* after the same BBr_3 diffusion measured with ECV analysis. In case of a typical structure of polySi/SiO_x/c-Si (*A*) the B-doping level at the interface reaches ~10^{20} cm^{-3} and an in-diffusion B-tail of 80 nm is observed. By inserting an additional oxide layer at 20 nm from the interface (structure *B* in Fig. 1) the B-doping at the interface is reduced to ~$6 \cdot 10^{19}$ cm^{-3} and the in-diffusion B-tail reaches only 40 nm. We propose that this leads to a reduced B segregation at the interface and thus to the improved passivation which is reported in Fig. 2.

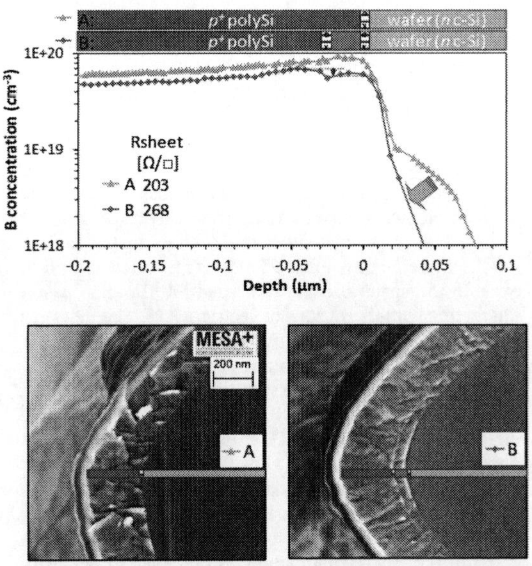

Figure 1: ECV of B doping profiles of investigated polySi/SiO$_x$/c-Si (A) and polySi/SiO$_x$/polySi/SiO$_x$/c-Si (B) structures with the corresponding R_{sheet} values. HR-SEM depicts the A and B structures.

By designing multi-layer dielectric stacks with different properties, we manage the hydrogen supply to the oxide interface. We improve the passivation levels and passivation stability of p^+ polySi on planar and especially on textured wafers. The results of this approach are presented in Fig. 2, where the implied V_{oc} and J_o of p^+polySi in combination with different capping layers, before and after firing (T_{set}=830°C), are shown. The SiN_x layer (*a*), in combination with p^+ polySi (*A*) on texture, only results in a proper interface passivation after a firing step (the heat treatment that is applied during firing of metal contacts), leading to J_o values of ~21 fA/cm^2. In contrast, a hydrogen-rich AlO_x layer inserted below SiN_x (stack *b*) dramatically improves the initial passivation to ~11 fA/cm^2, but a firing step reduces passivation to ~15 fA/cm^2. We ascribe this to the effusion of hydrogen during firing.

The solution to this problem is the addition of the top AlO_x capping layer. Such a multi-layer ($AlO_x/SiN_x/AlO_x$, called AnA) dielectric stack (*c*) on p^+ polySi (*A*) provides both the initial improved hydrogen supply and its preservation, resulting in an excellent passivation level of ~9 fA/cm^2 before firing (BF) and ~11 fA/cm^2 after firing (AF). Applying stack *c* on p^+ polySi with the modified B-profile (structure *B*, as shown in Fig. 1) leads to a record passivation level for p^+ polySi of ~8 fA/cm^2 (724 mV) BF and ~9 fA/cm^2 (722 mV) AF. On planar wafers, thus without the Si <111> planes, an excellent ~3 fA/cm^2 (735 mV) is achieved irrespective of the firing step. Interestingly, application of stack *c* to n^+ polySi also boosts passivation on the random pyramid textured samples to <1 fA/cm^2 (741 mV BF and 739 mV AF, not shown).

Until now the lower passivation quality of p^+ polySi has not been completely understood but in this study we provide clear insights that a combination of suppression of boron in-diffusion through the interface, supply of hydrogen as well as retaining the hydrogen is a key to obtain excellent passivation of p^+ polySi. Further improvements are feasible as demonstrated in Fig. 2 (orange symbols) by using further optimized AlO_x with respect to an additional increase in the supply of hydrogen.

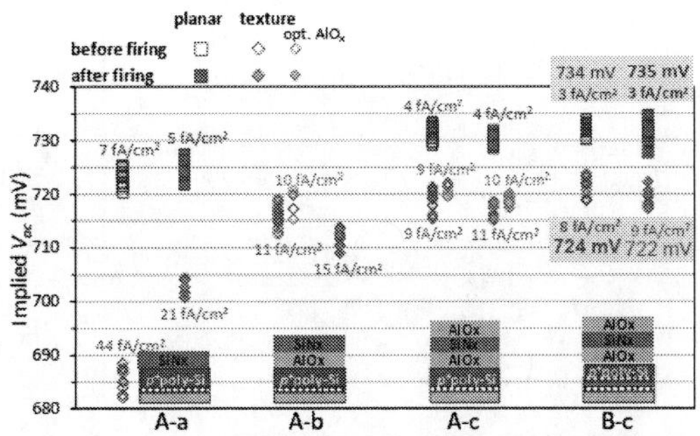

Figure 2: Implied V_{oc} results of hydrogenated p^+ polySi layers (*A*, *B*) on planar and textured n-Cz wafers with SiN_x(a), AlO_x/SiN_x (b) and $AlO_x/SiN_x/AlO_x$ (c), before and after firing (T_{set}=830°C). Best J_o values are indicated together with highest iV_{oc}.

2.2 FT metallization of p^+ polySi with various B-tails and dielectric stacks

We recently reported that for n^+ polySi metallized with FT paste the P-tail is essential to shield the Ag-intrusions that reach the c-Si below the polySi. $J_{o,contact}$ values as low as 200 and 80 fA/cm^2 were obtained with a P-tail of 0.3 and 1 µm depth, respectively [3]. Similarly, we anticipate that for the p^+ polySi with FT metallization a B-tail could be essential. To this end, we prepared a set of samples with varying depth of the pre-diffused B-tail, as depicted and described in Fig. 3.

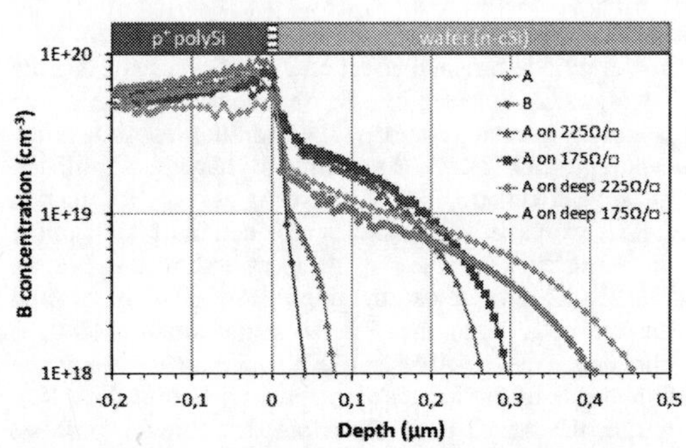

Figure 3: ECV of B-doping profiles of investigated p^+ polySi structures on various pre-diffused B-tails.

The addition of a B-tail can have a detrimental effect on the passivation due to increased Auger recombination which scales with the doping concentration of the B-tail. Fig. 4(a) shows that the J_o of p^+ polySi, e.g. with stack *c*, increases from ~10 to ~20 fA/cm^2 for samples with a pre-diffused B-tail. The tail depth is defined as the depth with 10^{18} cm^{-3} doping concentration. However, the presence and the depth of the B-tail has a strong beneficial effect on the $J_{o,contact}$ as also shown in Fig. 4(a).

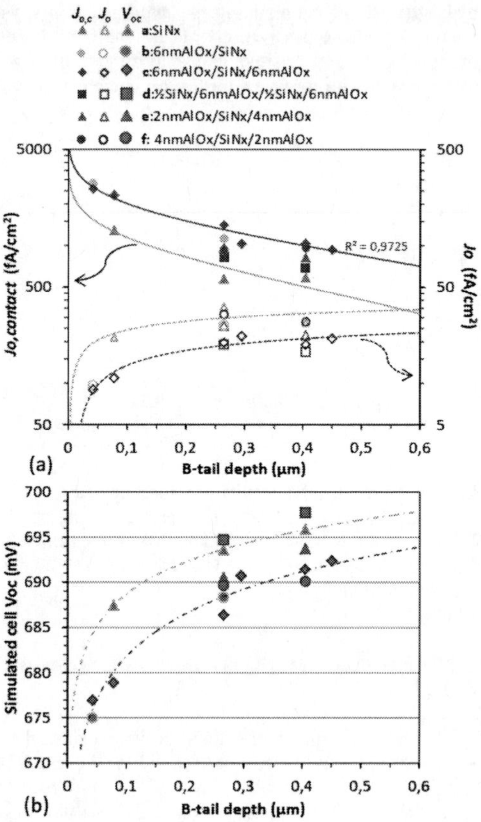

Figure 4: (a) J_o and $J_{o,contact}$ results of hydrogenated p^+ polySi layers with various dielectric stacks as a function of the B-tail depth. (b) Corresponding simulated V_{oc} of front-and-rear passivating contacts cells (p^+ polySi rear) with FT Baseline paste.

The composition of the dielectric stack can have however, a detrimental effect on $J_{o,contact}$. It seems that the lowest $J_{o,contact}$ values can be obtained with SiN_x at the p^+ polySi interface (stack *a*). The addition of AlO_x in the capping stack (stack *b*) seems to have a significant negative effect on $J_{o,contact}$. To investigate this phenomenon further we varied the amount and location of AlO_x in the dielectric stack (the *b,c,d,e,f* stacks as described in Fig. 4). The results indicate that the thicker the AlO_x layer is and the closer it is to the interface, the worse $J_{o,contact}$ is obtained. An HR-SEM study is ongoing to verify the hypothesis of the increased Al presence (from the AlO_x) during the Ag-Al paste (Baseline) fire-through process, leading to deeper and/or broader penetration of metal of the polySi/c-Si interface.

Because of these competing effects of the dielectric stack composition on the J_o and $J_{o,contact}$ it is interesting to simulate what would be the net effect on the solar cell V_{oc} with p^+ polySi passivating contact capped with various stacks (*a-f*) as a function of the B-tail depth. Fig. 4(b) shows the simulated V_{oc} of front-and-rear passivating contacts solar cells (type 1, shown in Table 1., i.e. with p^+ polySi rear (5% FT metal) and local front n^+ polySi fingers separated by an FSF). Indeed, the use of dielectric stack *c*, with the thickest AlO_x layer closest to the interface, results in the lowest V_{oc} for a given B-tail depth. Reducing the amount of AlO_x close to the interface improves the cell V_{oc}.

The specially designed stack *d* (40nmSiN_x/6nmAlO_x/40nmSiN_x/6nmAlO_x) with no AlO_x at the interface but still with excellent hydrogenation and ARC capabilities provides the best compromise of low J_o and fairly low $J_{o,contact}$. The cell simulation yields a V_{oc} of nearly 700 mV with the currently used FT Baseline paste.

Encouraged by this result we are currently investigating other dielectric stack compositions avoiding metallic elements but providing robust hydrogenation of p^+ polySi passivating contact used with FT metallization.

At the same time other FT pastes are tested as well to exclude the specific paste (i.e. used FT Baseline paste) interacting with the various dielectric stacks. The first very encouraging results using other FT pastes ('FT R&D 1' and 'FT R&D 2') are presented in Fig. 5. Most importantly we observed that $J_{o,contact}$ (same firing at T_{set}=830°C) can be reduced down to ~200-400 fA/cm² irrespective of the dielectric stack used. This provides an excellent opportunity to combine low J_o due to enhanced hydrogenation by improved dielectric stack with low $J_{o,contact}$ due to improved FT pastes. The resulting simulated cell V_{oc} (cell type 1 with a fixed B-tail of 0.265 µm) of such combinations is presented in Fig. 5(b). The results indicate that with excellent $J_{o,contact}$ reduced down to 200 fA/cm² combined with excellent passivation the cell V_{oc} can be boosted towards ~700-710 mV.

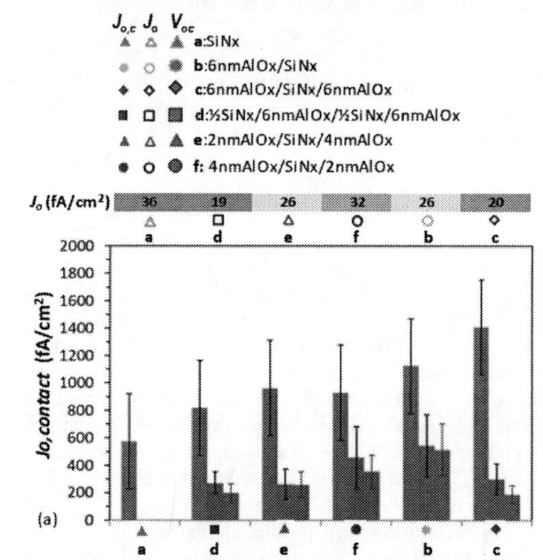

Figure 5: (a) J_o and $J_{o,contact}$ results of hydrogenated p^+ polySi layers with various dielectric stacks for a B-tail depth of 0.265 µm. (b) Corresponding simulated V_{oc} of front-and-rear passivating contacts cells (p^+ polySi rear) with different FT pastes.

Table 1: Parameters of textured front-and-rear passivating contacts solar cells

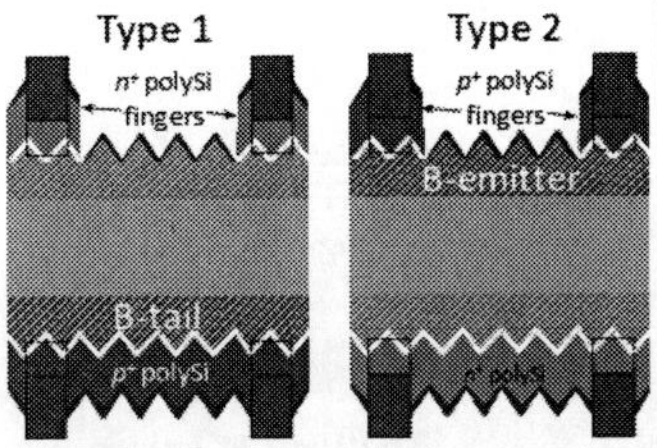

6" industrial solar cells with all FT metallization	0.265µm B-tail + p^+ polySi + dielectric stack	iV_{oc} BF (mV)	iV_{oc} AF (mV)	V_{oc} (mV)	$J_{o,contact}$ p^+ polySi (fA/cm²)	$J_{o,contact}$ n^+ polySi (fA/cm²)	J_{sc} (mA/cm²)	FF (%)	Eta (%)	Bifi. (%)
Type 1: p^+ polySi rear and n^+ polySi finger front	*A* with *a* (SiN_x)	678		677	626	214	38.2	77.7	20.1	79.1
Type 2: n^+ polySi rear and p^+ polySi finger front	*A (but 100 nm)* with *b* (AlOx/SiNx)	707	699	680		123	38.6	76.7	20.1	88.6
	A (but 100 nm) with *c* ($AlO_x/SiN_x/AlO_x$)	704	703	683	1495	234	39.0	76.5	20.4	88.2

2.3 Application of p^+ polySi in 6" bifacial solar cells with front-and-rear passivating contacts

In this section we present implementation of p^+ polySi on 6" bifacial *n*-type solar cells with front-and-rear passivating contacts. The process flow relies on industrially established steps like $POCl_3$ and BBr_3 diffusions to create the P- and B-tails and dope the LPCVD deposited polySi, and fire-through metallization. The front polySi is patterned into fingers under the metallization only, similar to ref [9]. One screen-printing patterning step is used to form these local front polySi fingers. Depending on the doping sequence, two types of front-and-rear passivating contacts solar cells are prepared: type 1 with a rear full-area p^+ polySi, or type 2 with front local p^+ polySi fingers as described in Table 1. In the first attempt (using FT Baseline paste and not yet FT R&D pastes for contacting p^+ polySi) in both cases efficiencies above 20% have been reached with excellent iV_{oc} of around 700 mV after firing, being reduced after FT metallization to ~680 mV. The influence of the dielectric stack on the J_o and $J_{o,contact}$ has been described above. For cell type 1, the SiN_x has been used yielding low $J_{o,contact}$ of ~600 fA/cm² but not satisfactorily good J_o (reflected in lower iV_{oc}). For cell type 2 the stack *b* and *c* showed improved hydrogenation thus better J_o (and higher iV_{oc}) but the $J_{o,contact}$ increased to around 1500 fA/cm² in line with Fig. 4.
In the next cell run we will implement the most promising dielectric stack *d* and other findings, as shown in Fig. 5 with excellent $J_{o,contact}$ reduced down to 200 fA/cm², to boost the cell V_{oc} towards ~700-710 mV. Including further cell processing optimization, we expect to reach 22-23% efficiency in the coming months.

3 CONCLUSIONS

The passivation properties of p^+ polySi on industrially relevant textured surfaces have been improved by means of novel hydrogenation schemes using dielectric multi-stacks (e.g. $AlO_x/SiN_x/AlO_x$) leading to J_o of ~9 fA/cm² before firing and ~11 fA/cm² after firing. Further improvements by modification of the B-profile across p^+ polySi lead to a record passivation level of ~8 fA/cm² (724 mV) before firing and ~9 fA/cm² (722 mV) after firing.

We studied the interaction of these passivating p^+ polySi structures with fire-through metallization with respect to the contact recombination ($J_{o,contact}$). We found that presence and the depth of the B-tail has a strong beneficial effect on the $J_{o,contact}$ and that the dielectric multi-stack needs to be carefully designed to not adversely affect $J_{o,contact}$. Recently excellent results were obtained on $J_{o,contact}$ being reduced down to 200 fA/cm².

Finally, we implemented p^+ polySi on 6" bifacial solar cells with front-and-rear passivating contacts, yielding efficiencies above 20% in the first attempt and showing a potential of 23% with bifaciality factor of 90%.

4 ACKNOWLEDGEMENTS

This work was partly funded by the Dutch Ministry of Economic Affairs as TKI Urban Energy Miracle project (TEUE116139).

5 REFERENCES

[1] A. Richter, J. Benick, F, Feldmann, A. Fell, M. Hermle, S.W. Glunz, "n-type Si solar cells with passivating electron contact: Identifying sources for efficiency limitations by wafer thickness and resistivity variation", Solar Materials and Solar Cells 173, p.96-105 (2017)
[2] Stodolny, M.K. et al. 2016, *Sol. Energy Mater. Sol. Cells*, Vol. 158, p. 24.
[3] M.K. Stodolny et al., "Material properties of LPCVD processed n-type polySi passivating contacts and application in PERPoly industrial bifacial solar cells", presented at 33th EUPVSEC, Amsterdam, 2017
[4] R. Peibst et al., "Building blocks for industrial, screen-printed two sides-contacted POLO cells with highly transparent ZnO:Al layers", presented at 33th EUPVSEC, Amsterdam, 2017.
[5] Chengfa Liu et al., Trina at Wuxi 4th PERC and bifacial forum 2018.
[6] F. Haase, F. Kiefer, S. Schaefer, C. Kruse, J. Kruegener, R. Brendel, R. Peibst, "Interdigitated back contact solar cells with polycrystalline silicon on oxide passivating contacts for both polarities", Japanese Journal of Applied Physics, 56, S. 08MB15 (2017)
[7] Y. Larionova, M. Turcu, S. Reiter, R. Brendel, D. Tetzlaff, J. Kruegener, T. Wietler, U. Hooehne, J.-D. Kaehler, R. Peibst, "On the recombination behavior of p+-type polysilicon on oxide junctions deposited by different methods on textured and planar surfaces", Phys. Status Solidi A 214, 8, 1700058 (2017)
[8] D.L. Young et al., "Gallium-Doped polySi:Ga/SiO_2 Passivated Emitters to n-Cz Wafers With iV_{oc}>730 mV" IEEE JOURNAL OF PHOTOVOLTAICS, VOL. 7, NO. 6, NOVEMBER 2017
[9] A. Ingenito, et al., "Silicon Solar Cell Architecture with Front Selective and Rear Full Area Ion-Implanted Passivating Contacts" Solar RRL 2017, 00, 1700040

ELECTRON BEAM EVAPORATION OF SILICON FOR POLY-SILICON/SIO$_2$ PASSIVATED CONTACTS

Jan Lossen[1*], Jan Hoß[1], Stephan Eisert[1],
Daniel Amkreutz[2], Martin Muske[2], Jonathan Plentz[3], Gudrun Andrä[3]
[1]International Solar Energy Research Center Konstanz (ISC), Rudolf-Diesel-Straße 15, 78467 Konstanz, Germany
[2] Institute for Silicon Photovoltaics / Helmholtz Zentrum Berlin (HZB), Kekuléstr. 5, 12489 Berlin, Germany
[3] Leibniz Institute of Photonic Technology (IPHT), Albert-Einstein-Str. 9, 07745 Jena, Germany
*Corresponding author: jan.lossen@isc-konstanz.de

ABSTRACT: We assess the use of electron-beam physical vapor deposition (EB-PVD) for the deposition of silicon layers to be used in doped poly-Si/SiO$_2$ passivated carrier selective contacts. It is shown that the crystallinity of the deposited layer can be tuned by the substrate temperature. Nano crystalline (nc-Si) intrinsic layers deposited at substantially different deposition rates of 25 nm/min and 500 nm/min show similar doping profiles after an ex-situ doping by POCl$_3$ diffusion and similar passivation quality with iV_{oc} > 715 mV for symmetrical lifetime samples on a n-type substrate. Best passivation characteristics with iV_{oc} of 732 mV and J_{0e} of 3 fA/cm^2 are achieved for in-situ doped layers with an active phosphorous concentration of $[P] = 2*10^{20}cm^{-3}$ after annealing. A low sheet resistance of 142 Ω/sq for a 100 nm thick layer make this stack eligible for ready integration into a screen printed solar cell. While layer properties are shown to be similar to those of parallel processed LPCVD layers, the single sided deposition characteristic EB-PVD constitutes a significant advantage for easy integration of the process step into a lean industrial process flow.

Keywords: Passivation, Polycrystalline, Silicon, Doping, High Deposition Rate

1 INTRODUCION

Recombination at surfaces and metal contacts are the dominant loss mechanisms in high efficient industrial crystalline silicon (c-Si) solar cells such as PERC and PERT cells. At lab scale, passivated carrier selective contacts based on a stack of a doped, at least partly crystalline silicon layer (poly-Si) and a thin silicon oxide (SiO$_2$) layer were demonstrated to be eligible for highest efficiencies [1, 2]. Such contact stacks are in principle compatible with a standard high temperature metallization scheme using firing through silver pastes, even though actual silver pastes still might deteriorate passivation properties [3].

A common deposition method for thin silicon layer is Low Pressure Chemical Vapor Deposition (LPCVD), which yields hydrogen-less silicon layers that can be annealed to poly-Si layers of high quality. Layers can be doped in-situ by addition of PH$_3$ during deposition or ex-situ by POCl$_3$ diffusion or ion implantation [3, 4, 5].

Main drawbacks of LPCVD, however, are a conformal growth characteristic, which does not allow for single-sided coating of a silicon wafer, and a comparatively low deposition rate. These make it difficult to include the LPCVD process into lean and cost effective production sequences. We investigate the applicability of electron-beam physical vapor deposition (EB-PVD) as single sided deposition method of silicon layers for passivated contacts. This method was used so far merely for the deposition of rather thick silicon layers (few micrometers up to several tens of micrometers) to act as absorber for solid- [6] or liquid phase crystallized solar cells [7, 8, 9]. Further, electron beam evaporation was used in combination with laser crystallization for deposition and epitaxial growth of doped and undoped poly-Si layers of thin film solar cells on glass [10].

2 EXPERIMENTAL

2.1 Electron beam evaporation of silicon layers

Intrinsic silicon layers are deposited at Helmholtz Zentrum Berlin (HZB) in a VON ARDENNE CS400PS cluster high rate electron beam evaporation system, as depicted schematically in Figure 1. The systems reaches a base pressure of $<1*10^{-7}$ mbar, which increases during deposition to $5*10^{-7}$ mbar and is operated with an acceleration of 10 kV. At the cluster, two evaporators are available: One chamber is configured for depositions on up to 4 wafers, while another chamber is configured for depositions on a single wafer. By changing the emission current from 190 mA to 540 mA, the deposition rate can be varied from 25nm/min up to 500nm/min. The morphology of the deposited film can be influenced by changing the substrate heater temperature. We use reference depositions on Corning Eagle XG glass substrate to determine the thickness of the layers as well as the microstructure using Raman spectroscopy.

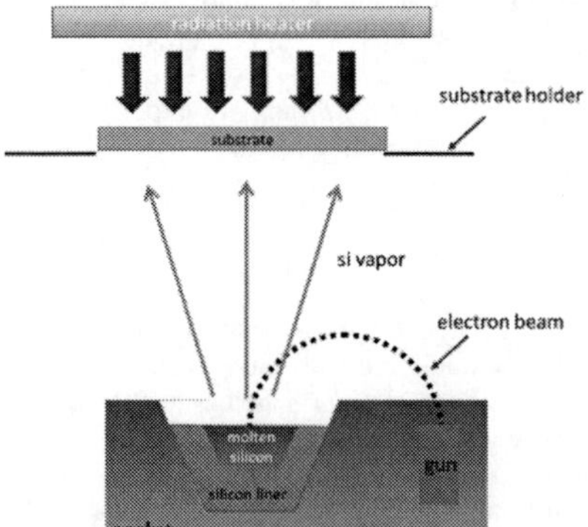

Figure 1: Schematic drawing of the high-rate silicon e-beam evaporators at HZB and IPHT

In-situ doped samples are deposited in a similar setup at Leibniz Institute of Photonic Technology (IPHT). This set-up offers the possibility of co-evaporating phosphorous from a gallium phosphide effusion source. At an acceleration voltage of 10.5 kV and an emission current of 100 mA a deposition rate of 12 nm/min is achieved. The low deposition rate was chosen to reach different doping levels easily with the existing equipment. The evaporation rate of the dopant is set by the temperature of the effusion source. The substrate temperature is kept below 200°C. We confirmed by XRD measurements that layers are fully amorphous under this conditions.

2.1 Sample preparation and characterizations

In order to assess the passivation quality we prepare symmetrical lifetime samples on commercial, PV-grade, diamond wire sawn n-type CZ wafers with base resistivity of 2 – 4 Ω*cm. Wafers are cut by laser to a substrate size of 10x10cm^2 in order to fit the substrate holders in the deposition systems.

After removing the saw damage in NaOH solution, wafers are cleaned and chemically oxidized in concentrated nitric acid at room temperature in the clean room of International Solar Energy Research Center Konstanz (ISC). In this state wafers are stacked with clean room paper and sent to IPHT and HZB, where the depositions are carried out in the different evaporation systems. Afterwards, wafers are sent back to ISC Konstanz, where annealing in a quartz tube furnace, deposition of SiNx by PECVD and fast firing in a belt furnace were carried out. Figure 2 displays the final sample structure. The passivation quality of the samples is assessed by quasi steady state photo conductance (QSSPC) measurements (Sinton WCT-120) after annealing/diffusion, after SiNx capping, and after the firing process.

The distributed processing might be a hindrance for achieving highest passivation quality, however homogeneity of passivation on the wafer is good, as confirmed by photoluminescence imaging. Furthermore, replications of processing conditions in different sample series yielded consistent results with iVoc values within the range of a few mV.

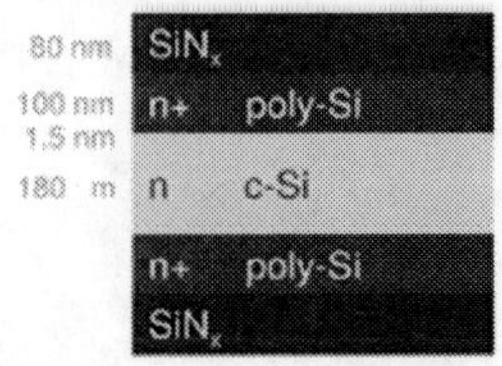

Figure 2: Schematic drawing of sample structure

Doping profiles are determined by electrochemical capacitance voltage (ECV) measurements after the annealing/diffusion step with a WEP CVP21 profiler.

Sheet resistances of the layers is measured by 4-point-probing on reference p-type samples, which are processed in the exact same manner.

3 RESULTS

3.1 Morphology of layers and influence of deposition rate

In a first experiment, silicon layers were deposited at different substrate temperatures on glass and on c-Si wafers. For heater temperatures of 450°C and 540°C layers are fully amorphous (a-Si), while for a heater temperature of 675°C the layer is partial crystalline after deposition (nc-Si), as can be seen from the Raman spectra in Figure 3. Layers deposited in another chamber with much high deposition rate and a heater temperature setting of 600°C are also partially crystalline. Table I lists the Raman crystallinity X_c, computed from the deconvoluted spectra as ratio of the peak areas according to the formula $X_c = (A_{510}+A_{520})/(A_{520}+A_{510}+A_{480})$.

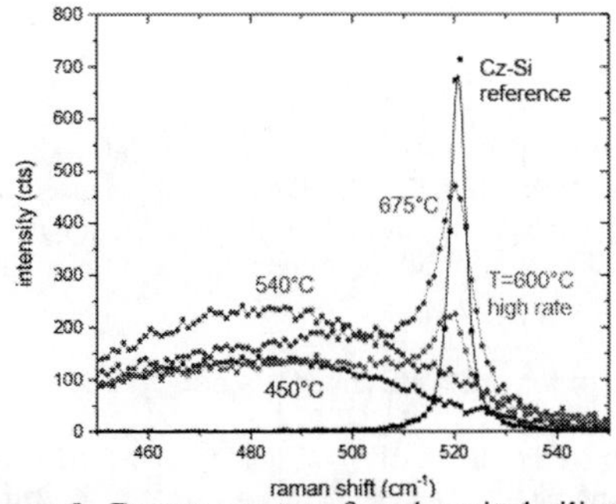

Figure 3: Raman spectra of as-deposited silicon layers produced at different substrate heater settings

Table I: Process conditions and resulting Raman crystallinity X_c

EBPVD System	Heater Temp [°C]	Emission current [mA]	Deposition Rate [nm/min]	Raman crys. X_c %
HZB A	675	190	25	30
HZB B	600	540	500	16
HZB A	540	190	25	0
HZB A	450	190	25	0
IPHT	200	100	12	(0)*

*amorphous phase confirmed by XRD

In order to assess the passivation quality of intrinsically deposited silicon layer, these were ex-situ doped in a combined diffusion and annealing process under POCl atmosphere (30min, 825°C).

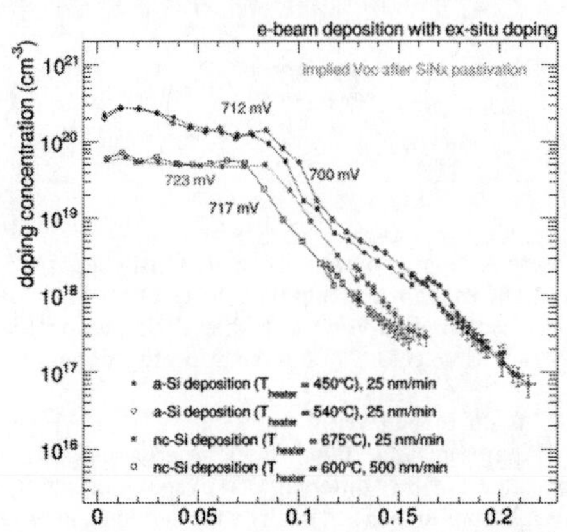

Figure 4: Doping profiles of ex-situ doped layer stacks of different crystallinity. Implied Voc values after SiNx passivation are given in same color close to the respective doping profile.

As shown in Figure 4, a crystalline fraction in the as-deposited state leads to a lower phosphorous concentration in the poly-Si layer and a steeper decrease at the depth of the oxide layer at approximately 100 nm. These observations are in accordance with findings of Stodolny et al. for LPCVD layers deposited at different temperatures [5], although the difference reported therein were smaller. As for the initially amorphous layers the oxide layer gets penetrated significantly by the stronger diffusion, implied Voc values decrease, as a result of an increased number of interface states and additional Auger recombination in the surface near region [4].

The nc-layer deposited at the 20-fold higher deposition rate of 500 nm/min. performs very similar to the nc layer deposited at 25 nm/min. Both iVoc value and the doping profile match quite well. We take this as evidence, that a very high deposition rate is not detrimental for a good layer quality, but further proof on a higher absolute iVoc value is desirable.

3.2. Passivation of ex-situ doped layers

To increase the passivation quality of poly-Si layers from amorphous precursor layers we developed an advanced ex-situ doping process, which is not based on the deposition of phosphorous silicate glass (PSG) in a $PCOl_3$ diffusion. In Figure 5, doping profile and corresponding iVoc value are given for the advanced doing process in comparison to the reference POCl3 diffusion process. A reduced penetration of the oxide explains the better passivation quality. However, this comes at the expense of low [P] in the poly-Si layer, reducing the field effect. Further optimization is ongoing.

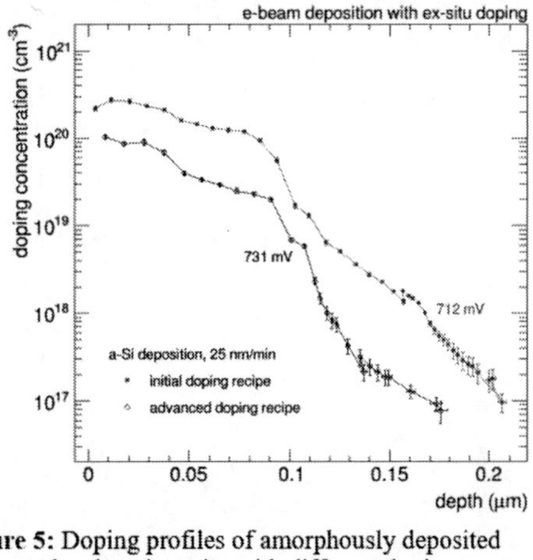

Figure 5: Doping profiles of amorphously deposited layer stacks, doped ex-situ with different doping processes. Implied Voc values after SiNx passivation are given in same color close to the respective doping profile.

3.2. In-situ doped layers

Approximately 100nm thick amorphous layers were deposited with different co-evaporation rates of phosphorous and subsequently annealed in N_2 ambient for 30 min at two different temperatures (825°C, 850°C).

The resulting concentration of active doping atoms in the poly-Si layer depends mostly on the co-evaporation rate, while the shape of the tail also depends on the annealing temperature, as depicted in Figure 6. Higher temperature leads to a more pronounced diffusion of dopants through theoxide. In Figure 6 also the sheet resistances of the layers are given, measured on identically processed p-type wafer by 4-point-probing. The values achieved with medium and high doping concentration range from 56 Ω/sq to 142Ω/sq, which is sufficiently low for including such layer into a solar cell with H-grid metallization.

Figure 7 shows implied Voc values after different processing steps. The best and most stable passivation is achieved for $[P] = 2*10^{20}$ cm^{-3} and annealing at 825°C. For these conditions, the doping profile exhibits a steep decrease of phosphorous concentration at the interfacial oxide. The layers with $4*10^{19}$ cm^{-3} also exhibit good passivation after SiNx deposition, but degrade in the firing process. Both layers with $3*10^{20}$ cm^{-3} and the layer with $2*^{20}cm^{-3}$ and annealing at 850°C suffer from substantial diffusion of phosphorous atoms through the oxide layer into the bulk, which limits their passivation.

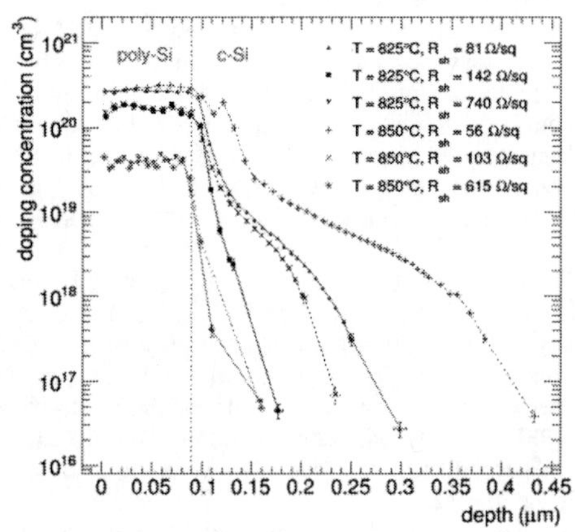

Figure 6: ECV doping profiles of in-situ doped EB-PVD deposited layers with different doping concentrations and two different annealing temperatures. R_{sheet} values in the legend are measured on equally prepared p-typ wafers.

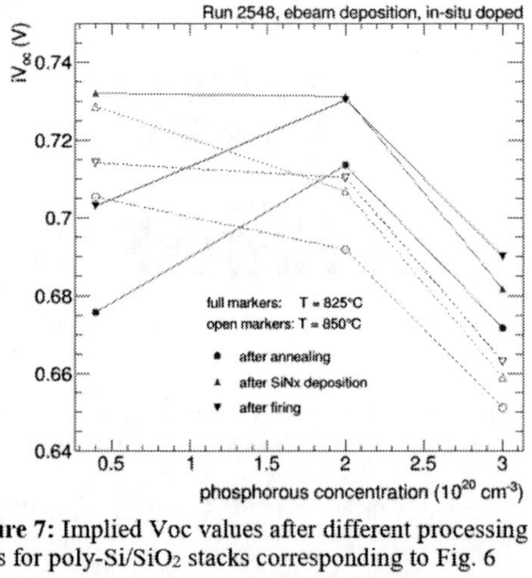

Figure 7: Implied Voc values after different processing steps for poly-Si/SiO_2 stacks corresponding to Fig. 6

In a last experiment we repeated the best settings and compared results with equally processed in-situ doped LPCVD layers with similar dopant concentration. Upon the same annealing process very similar doping profiles and implied Voc values are achieved as shown in Fig. 8.

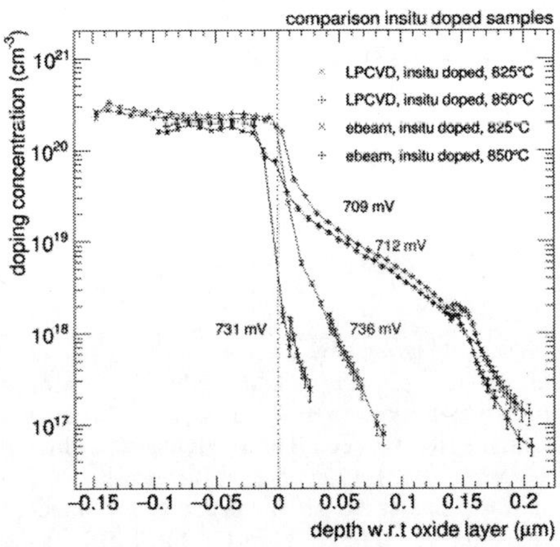

Figure 8: Doping profiles and implied Voc values of in-situ doped LPCVD and EB-PVD layers with similar doping densities.

4 CONCLUSIONS AND DISCUSSION

Electron beam evaporation was demonstrated to be an eligible and versatile technique for single sided deposition of silicon layers for the use in passivated carrier selective contact stacks.

The morphology of the as deposited layer can be tuned by the substrate temperature. Very high depositions rates are possible and seem not the have a negative impact on the layer properties. Ex-situ and in-situ doping are feasible and yield similar results, with in-situ doping yielding the best passivation so far. A passivation quality of 732 mV, corresponding to a Joe of 3 fA/cm2 per side has been achieved for in-situ doped layers, which is close to the values we achieved with equally processed LPCVD layers featuring a slightly higher doping concentration. Details are given in Table II.

Table II: Recombination parameters of the best layer stacks of this study

Methode	Doping	τ_{eff} [ms]	J_{0e} [fA/cm^2]	iVoc [mV]
EB-PVD	in-situ	1.7	3	732
EB-PVD	ex-situ	1.7	6	731
LPCVD	in-situ	3.0	2.6	736

We are confident that the passivation characteristics can be further improved by fine-tuning of the oxide properties as well as the hydrogenation and firing processes, which were not varied for this study.

The results show that poly-Si layers deposited by EB-PVD allow to achieve a similar surface passivation compared to LPCVD deposited layers. Because of the single sided deposition characteristic, EB-PVD offers the advantage of a much easier integration in a lean process flow for industrial solar cells.

Compared to PECVD deposition, EB-PVD has the inherent advantage of depositing silicon with a very low hydrogen content. This avoids problems of layer delamination or loss of passivation arising for thicker PECVD layers deposited at moderate temperatures.

A further advantage compared to both CVD methods is, that no silane infrastructure is needed, saving costs for gas supply, monitoring and abatement, as the material can be supplied in solid form.

ACKNOWLEDGEMENTS

The authors would like to thank our colleagues at HZB, IPHT and ISC who supported this work and the German Federal Ministry for Economic Affairs and Energy for funding the work of ISC Konstanz under contract number 0324198A (TuKaN).

REFERENCES

[1] A. Richter, et al., n-Type Si solar cells with passivating electron contact: identifying sources for efficiency limitations by wafer thickness and resistivity variation. Solar. Energy Mater. Sol. Cells 173, 96–105 (2017)

[2] R. Peibst et al., Implementation of n+ and p+ Poly Junctions on Front and Rear Side of Double-Side Contacted Industrial Silicon Solar Cells, Proc. 32st EU-PVSEC, Amsterdam, 2016

[3] H. E. Çiftpınara et al., Study of Screen Printed Metallization for Polysilicon Based Passivating Contacts, in 7th International Conference on Silicon Photovoltaics, Freiburg, Germany, 2017

[4] M. K. Stodolny et al., n-type Polysilicon Passivating Contacts for Industrial Bifacial n-PERT Cells, Proc. 32st EU-PVSEC, Amsterdam, 2016

[5] M. K. Stodolny et al., Material properties of LPCVD Processed n-type Polysilicon Passivating Contacts and its Application in PERPoly Industrial Bifacial Solar Cells, Energy Procedia, Volume 124, September 2017, Pages 635-642

[6] T. Sontheimer et al., Challenges and opportunities of electron beam evaporation in the preparation of poly-Si thin film solar cells, proc. 35th IEEE PVSC Honolulu, pp. 614-619, 2010

[7] Cham Thi Trinh et al., Potential of interdigitated back-contact silicon heterojunction solar cells for liquid phase crystallized silicon on glass with efficiency above 14%, Solar Energy Materials and Solar Cells 174 (2018) 187–195

[8] J. Haschke, D. Amkreutz, B. Rech, Liquid phase crystallized silicon on glass: Technology, material quality and back contacted heterojunction solar cells, Jpn. J. Appl. Phys 55 (4), no. 04EA04 (2016)

[9] J. Plentz, T. Schmidt, A. Gawlik, J. Bergmann, G. Andrä, D. Hauschild, V. Lissotschenko, Applicability of an economic diode laser emitting at 980 nm for preparation of polycrystalline silicon thin film solar cells on glass, Physica Status Solidi A 214 (2017) 1600882

[10] J. Plentz, G. Andrä, A. Gawlik, I. Höger, G. Jia, F. Falk, Polycrystalline silicon thin-film solar cells prepared by layered laser crystallization with 540 mV open circuit voltage, Thin Solid Films 562 (2014) 430-434

NANOCRYSTALLINE N-TYPE SILICON FRONT SURFACE FIELD LAYERS: FROM RESEARCH TO INDUSTRY APPLICATIONS IN SILICON HETEROJUNCTION SOLAR CELLS

A.B. Morales-Vilches[1*], D. Decker[2], L. Mazzarella[1], L.Korte[1], R. Schlatmann[1], D. Sontag[2], B. Stannowski[1]
[1]Helmholtz-Zentrum Berlin für Materialien und Energie GmbH, 12489 Berlin, Germany.
[2]Meyer Burger Germany, An der Baumschule 6-8, 09337 Hohenstein-Ernstthal, Germany

ABSTRACT: Nanocrystalline silicon and silicon oxide (nc-Si(O$_x$):H) layers grown by plasma-enhanced chemical vapor deposition (PECVD) have shown low parasitic absorption and excellent contact properties when implemented as n-type front surface field (FSF) contact in rear-junction silicon heterojunction (SHJ) solar cells [1-3]. In this contribution we present results from the successful process transfer from the lab at PVcomB at the Helmholtz-Zentrum Berlin (HZB), to the industrial pilot line at Meyer Burger Germany GmbH (MBG). Conversion efficiencies $\eta > 22.5$ % were demonstrated on SHJ cell (4 cm²) [2, 3]. The excellent cell performance in the lab and the potential to reduce parasitic absorption in the front stack by using nc-SiO$_x$:H motivated the process transfer from HZB to MBG. Initial cross processing experiments on 244 cm² wafers showed the benefit of using nc-Si:H as FSF layer. We here also emphasize the role of the Si texture on a fast nc-Si:H nucleation. After cross processing experiments a successful transfer of the nc-Si:H process and fine tuning resulted in a median cell efficiency of 23.4 %. This is in the same range as the MGB reference on 244 cm² cells, noteworthy, at the same throughput. Currently work is ongoing to further improve the optical performance of the cells by adding oxygen (CO_2).
Keywords: passivating contacts, transparent layers, silicon heterojunction solar cells.

1 INTRODUCTION

Silicon heterojunction (SHJ) solar cells have shown their efficiency potential and industrial relevance in the past years with world record efficiencies reported in all-back-contacted (26.7 % [4, 5]) and in 2-side contacted cells (24.5 % - 25.1 % [6, 7]). In these SHJ cells the p-type contact (junction) is usually placed at the front side. Placing it on the rear side instead allows to use n-doped nc-Si(O$_x$):H as the n-type contact, which exhibits excellent transparency and reaches a higher conductivity than its p-type counterpart. In recent papers we presented this, analyzing the optical benefit of (very thin) nc-SiO$_x$:H layers used on the cell's front side [1 - 3]. Now we incorporate our best processes in full-area cells in an industrial line in collaboration with MBG. In addition, the benefit of using these layers was shown in solar cells cross-processed at HZB and MBG, where most PECVD films were deposited at HZB and back-end processes done at MBG.

2 MATERIALS AND METHODS

Two different cell layouts were used to fabricate the rear junction silicon heterojunction solar cells (SHJ). Schematics of the device structure and pictures of the layouts are shown in fig. 1. A first layout of 14 small area (4 cm²) devices per wafer was used for the lab scale samples done at HZB. Detailed information about the different fabrication steps were shown elsewhere [2]. A second layout of full area cells (244.3 cm²) was used for the fab scale samples done both at HZB and at MBG. In this study we focus on the development of the front n-type silicon layers deposited by plasma enhanced chemical vapor deposition (PECVD).

2.1 Lab scale SHJ solar cells

Initially lab scale samples were used to analyze the different materials used as n-doped front contact (FSF) deposited by PECVD in an Applied Materials AKT1600 cluster tool for max. four wafers per run. Amorphous and nanocrystalline n-type layers have been already compared in a previous paper [2] demonstrating the benefit both optically and electrically when using the nanocrystalline material as FSF. The optical benefit of introducing oxygen in these layers has also been studied in the past in our group [1-3]. In this work we aimed to transfer the processes for these PECVD deposited layers to the industrial like HELiA PECVD tool from MBG with the capability of processing 6 inch wafers (56 wafers per run) at high throughput (>2400 wafers per hour).

2.2 Fab scale SHJ solar cells

The full-area SHJ solar cells were manufactured both at the lab at HZB and at the industrial like pilot line at MBG. In order to transfer the technology from "lab" to the "fab", cross-experiments between the two sites were done first to show the potential of these materials: front end processes at HZB (wet chemistry and PECVD) and back end processes at MBG (PVD for TCO and metal screen printing). The process transfer was then done first for the non-oxidic nc-Si layer. Once this layer was successfully implemented, oxygen (CO_2) was added to further optically improve the material.

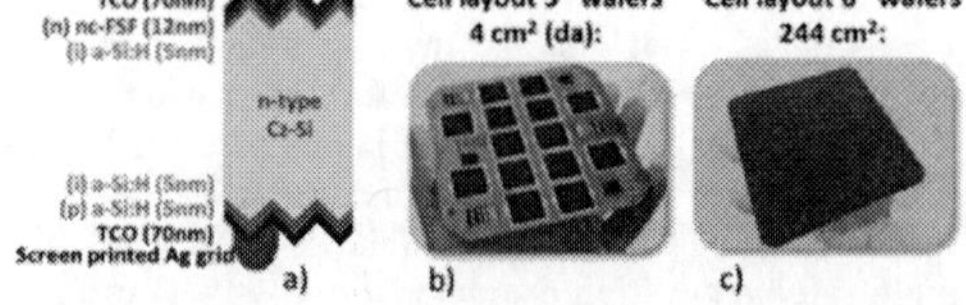

Figure 1: a) Cross section view of the bifacial rear-emitter silicon heterojunction (SHJ) solar cell studied in this work with ITO as the TCO and nanocrystalline Si front surface field layers; b) Cell layout of 4 cm² busbar-less cells on Cz-Si wafers and c) full-area busbar-less cells with a total area of 244.3 cm².

3 RESULTS AND DISCUSSION

Table 1 presents SHJ cell parameters obtained at HZB directly comparing nc-Si:H and nc-SiO_x:H FSF with efficiencies of 22.3% and 22.8 %, respectively. A net J_{sc} gain of +0.6 mA/cm² is found for cells with the oxidic layer. Full area cells (244 cm²) with nc-SiO_x:H FSF were then processed at HZB reaching 21.4 % [3]. Despite a comparable J_{sc} measured we could not yet reproduce the same efficiency level as obtained on small cells since the back-end processes for full size cells were not fully optimized. Therefore, in parallel we started cross processing experiments together with MBG, thereby, benefiting from the established full-size cell processes at MBG.

Table I: Nanocrystalline silicon (oxide) layer properties refractive index n at 632 nm wavelength and lateral conductivity σ, as derived from thick layers grown on glass and SHJ cell parameters using these layers as FSF. The J_{sc} results clearly show the optical benefit of the oxidic FSF due to lower parasitic absorption. Since the other electrical parameters keep the same excellent level, the optical benefit is transferred directly into the final conversion efficiency.

	Single layers on glass		Cell results (4 cm²)			
nc-	n	σ [S/cm]	J_{sc} [mA/cm²]	V_{oc} [mV]	FF [%]	η [%]
Si:H	3.7	62	38.3	727	80.0	22.3
SiO_x:H	2.7	0.8	38.9	730	80.1	22.8

To demonstrate the potential of the nc-Si(O_x):H layers developed at HZB the back-end processes for full area cells were done at MBG. Fig. 2 shows results of co-deposited cells on different wafer textures. Sample 1 with a larger pyramid size and a clear roughness on the pyramid facets reveals a detrimental effect in cell performance compared to Sample 2, which has a much smaller pyramid size and appears to have more smooth pyramid facets. These samples demonstrate that the surface topography has a critical influence on the cell performance when implementing nanocrystalline Si layers. We attribute this in particular to a certain degree of nano-roughness on the pyramid facets (compare Figs. 2b and 2c). This might disturb or delay the nc-Si:H nucleation resulting in lower crystalline fraction in very thin layers and, hence, increases parasitic absorption and lowers conductivity. As a result J_{sc} as well as FF and V_{oc} are reduced, the latter two due to an increased contact resistance and a reduced field-effect passivation.

After adapting the texturing process, a larger cross processing experiment with both nc-Si:H and nc-SiO_x:H FSF layers was carried out. Fig. 3 shows the results of four different experiments carried out between HZB and MBG, as described in the sketch in the lower part of the figure. The first three splits consist of 6 wafers each, while split 4 (dark blue) the reference devices with a-Si:H FSF, was fully processed at MBG and included about 100 wafers. Split 3 shows the same MBG reference but with the wet chemical processes performed at HZB. We ascribe the lower cell performance in this group to sample shipment between critical process steps, which probably led to Si surface deterioration in this split group. Therefore, in splits 1 and 2 both wet chemistry and PECVD were done at HZB within a very short time. In split 1 (orange) nc-Si:H was used as the FSF while the green group corresponds to the nc-SiO_x:H FSF samples. The high V_{oc} and FF of the nc-Si:H samples leads to an increase in the median η by 0.15% (split 1) compared to the reference in split 4. The optical benefit of the split 2 with the oxidic FSF is clearly visible in the J_{sc}, however, the FF is not yet as good as in the non-oxidic samples. Currently, the focus is on optimizing the deposition parameters to improve the FF of the cells using the nc-SiO_x:H FSF layers in 6-inch full area devices in the line at MBG.

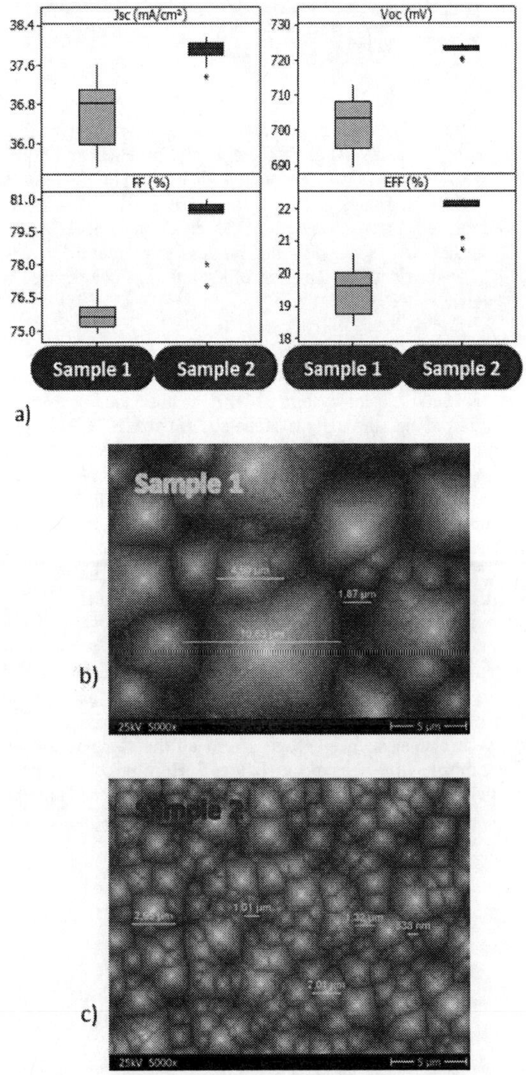

Figure 2: a) Solar cell characteristics of first cross-processed full-area size (244 cm²) SHJ solar cells using two texturing processes. The same nc-Si:H FSF was deposited at HZB in both samples to directly compare the texturing effect. Each boxplot includes the results of 6 wafers from the same sample group. b) and c) SEM images of the two differently textured wafers.

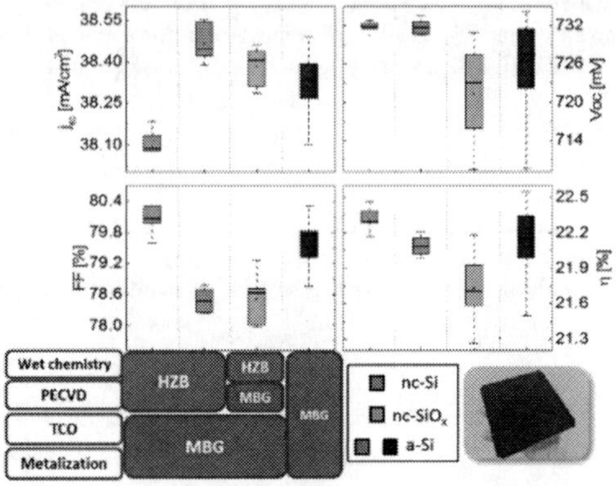

Figure 3: Solar cell characteristics of full area SHJ cells (244.3 cm²) from a cross processing experiment with four groups (split 1 – 4) carried out jointly by HZB and MBG on Cz-Si wafers with the optimized small-pyramid texture used for all groups (compare to sample 2 in Fig. 2c). In the lower left part a sketch of the different process splits clarifies where the different process steps were done. In the lower right part a picture of the full-area cell devices is shown.

After the cross experiments the challenge was to transfer the PECVD processes from the Applied Materials AKT1600 cluster system at HZB, to the industrial large-area and high-throughput Helia system at MBG (56 wafers per run). As a result the nc-Si:H process was successfully implemented obtaining efficiencies on the same level as the amorphous reference process at MBG, even maintaining the same throughput (fig. 4). The transfer for the oxidic layer (nc-SiO$_x$:H) was also started, obtaining efficiencies in the 23% range independent of the nc-layer used. There is still some room for improvement in the oxidic layer due to the lower FF, which prevents from obtaining the full potential of this material, in terms of optical gain. Additionally, both transferred nc-layers had an excellent homogeneity in all the tray, as can be observed from the solar cell efficiency and electroluminescence images from all the samples on the complete tray depicted in figure 5. Here only solar cells that use the nc-Si:H without oxygen are shown but the homogeneity was equally reproduced in the oxidic nc-layer.

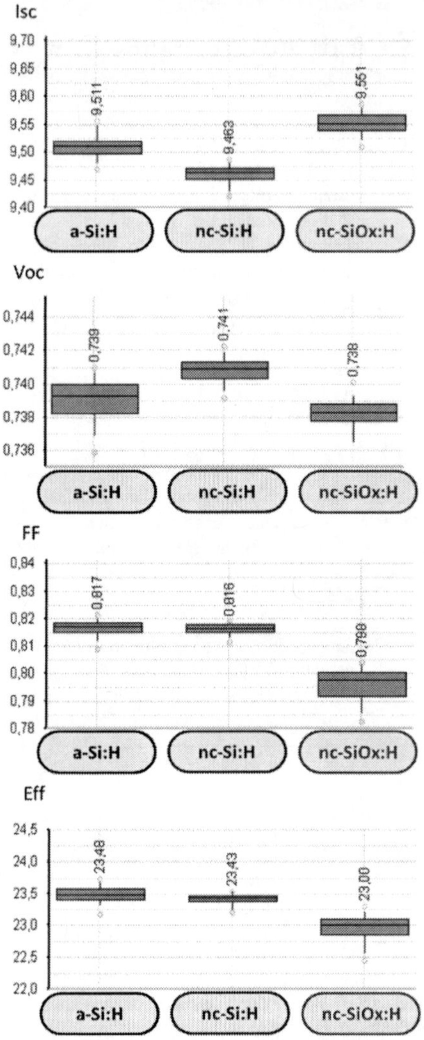

Figure 4: Solar cell characteristics of full area size SHJ cells (244.3 cm²) using three different n-type layers deposited at the MGB pilot line PECVD tool (Helia).

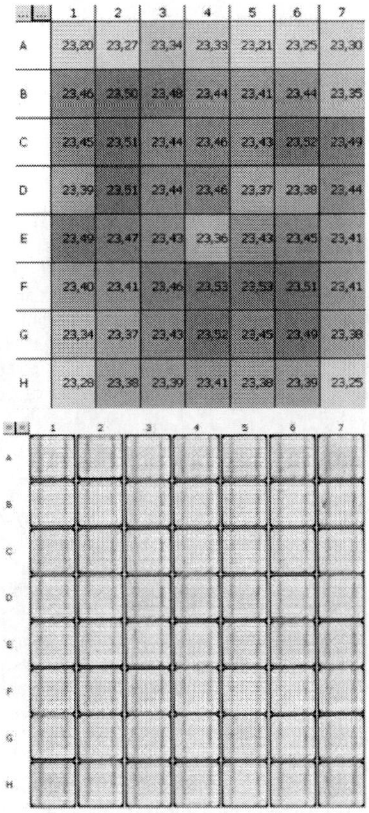

Figure 5: Solar cell efficiency and electroluminescence images of all the wafers deposited in the same PECVD tray. In that case using the nc-Si:H.

4. CONCLUSIONS

In this work we obtained excellent performance of full size (244.3 cm^2) cells with nc-Si:H as the FSF layer and a clear optical benefit of the nc-SiO_x:H FSF with a J_{sc} increase of +0.5 mA/cm^2. Different from the lab results, nc-SiO_x:H results in the cross experiment leads to a slight FF loss. The excellent results with the nc-Si:H FSF triggered the process transfer to the MBG pilot line which was done successfully, surpassing the cell efficiency of 23% for both nc-Si:H and nc-SiO_x:H n-type layers. We observed still a lower FF for the cells with the oxidic layer, which needs to be solved in future in order to benefit from the full potential of this material. Besides cell efficiency also the production requirements were fulfilled, such as high throughput (60s PECVD process time) and excellent cell homogeneity over the 56 wafers on the PECVD tray. Moreover, it was demonstrated in the MBG pilot line that the nc-Si:H PECVD process is very reproducible and stable over time.

ACKNOWLEDGMENTS

The authors would like to thank all colleagues at HZB and MBG that contributed to this work, especially Denise Debrassine, Manuel Hartig Tobias Henschel, Kerstin Jacob, Matthias Kubicki, Katja Mayer-Stillrich, Holger Rhein, Matteo Werth and Matthias Zelt for the valuable help in processing and characterizing. Special thanks to GP solar for providing the additive for the alkaline texturing solution. This work was partially supported by the German Ministry of Economic Affairs and Energy (BMWi) in the framework of the HERA project (reference #0325825I).

REFERENCES

[1] L. Mazzarella et al., "Nanocrystalline n-type Silicon Oxide Front Contacts for SHJ solar cells: Photocurrent Enhancement on Planar and Textured Substrates",IEEE J-PV, Vol. 8 (2017), 70-78.

[2] A. Morales et al., "Nanocrystalline vs. amorphous n-type silicon front surface field layers in SHJ solar cells: Role of thickness and oxygen content", Proc. of EUPVSEC 2017, Amsterdam.

[3] L. Mazzarella et al., "Ultra-thin nanocrystalline n-type silion oxide front contact layers for rear-emitter silicon heterojuntion solar cells", 179, 386-391 Sol. En. Mat, 2018.

[4] K. Yoshikawa et al., "Exceeding conversion efficiency of 26% by heterojunction interdigitated back contact solar cell with thin film Si technology", Solar Energy Materials and Solar Cells 173, December 2017, Pages 37-42.

[5] M. A. Green et al., "Solar cell efficiency tables (version 50)", Prog. Photovolt. Res. Appl. 2017; 25:668–676. https://doi.org/10.1002/pip.2909.

[6] M. Tauchi et al., "24.7% Record Efficiency HIT Solar Cell on Thin Silicon Wafer", IEEE J-PV, 4 (2014) 96-99.

[7] D. Adachi et al., "Impact of carrier recombination on fill factor for large area heterojunction crystalline silicon solar cell with 25.1% efficiency", Applied Physics Letters 107, 233506 (2015); doi: 10.1063/1.4937224.

PASSIVATING CONTACTS FOR SILICON SOLAR CELLS MADE OF AL_2O_3 AND TIO_X NANOLAYER SYSTEMS

David Tröger[1], Matthias Grube[1], Martin Knaut[2], Johana Reif[2], Johann W. Bartha[2] and Thomas Mikolajick[1,2]
[1]NaMLab gGmbH, Nöthnitzer Str. 64, 01187 Dresden, Germany
[2]Institute of Semiconductors and Microsystems, TU Dresden, Dresden, Germany

ABSTRACT: Local contacts for PERC solar cells reduce the cell efficiency by carrier recombination at the silicon-metal interface. This study investigates the possibility to avoid local contacts by using a full area contact with passivating behavior. The proposed structure is a double layer system consisting of a passivating Al_2O_3 and a conducting TiO_x. The work focuses on the enhancement of the conductivity of TiO_x and the entire contact stack. The resistance of the TiO_x could be decreased by doping the TiO_x with tantalum and interface engineering by employing a metallic titanium interlayer. The use of Ta:TiO_x led to a reduction of resistance of the contact stack by approximately 40 % compared to pure TiO_x. This could be enhanced further to 46 % by the addition of a 3 nm thick metallic Ti interlayer. The combined Al_2O_3/TiO_x and interlayer approach yielded a surface recombination velocity of 100 and 160 cm/s for TiO_x and doped Ta:TiO_x, respectively.
Keywords: surface recombination, passivating contacts, doped titanium dioxide, Al_2O_3, photovoltaic, silicon passivation, c-Si, homojunction, interfaces, PERC

1 INTRODUCTION

Passivated emitter rear cells (PERC) are increasing their market share continuously and this trend is expected to prevail for the next century [1]. In order to support this trend an uninterrupted improvement is needed. The development of passivating contacts is one of the main challenges for future high-efficiency solar cells. In state of the art PERC cells, Al_2O_3 and SiN_x:H are utilized for surface passivation of n-type and p-type Silicon. Those dielectrics need to be opened locally to contact the active areas and extract the charge carriers. In many process routes, this is done by fast firing at the front side and a local opening process at the back side. Two consequences arise from such an approach. First, the passivation is literally eliminated at the openings. Therefore, an increased recombination occurs. Second, the diffusion path of the charge carriers is mainly horizontal within the bulk Si due to the large distance of the local contacts compared to the cell thickness. A passivating full area contact will lead to an uninterrupted passivation layer and shorter diffusion paths. Thus, the carrier lifetime would be increased. One of the most successful full area contacts with industrial usage is known from the Silicon heterojunction cells. The modified band gaps of hydrogenated amorphous silicon lead to carrier selective contacts. Unfortunately, the implementation into PERC technology is not plug-and-play. There are obstacles like

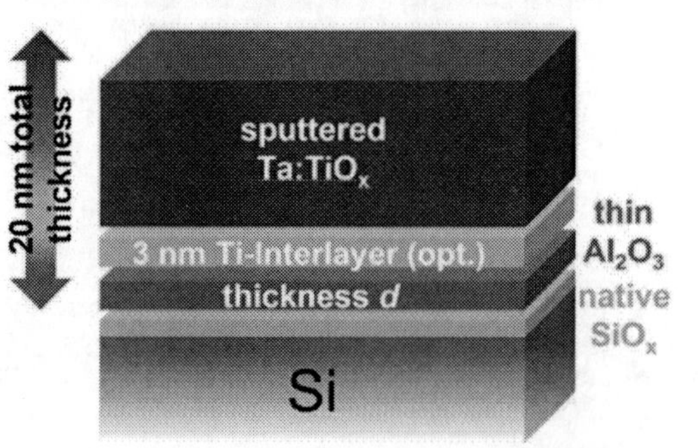

Fig. 1: Schematic of the used layer stack after deposition. A thin Al_2O_3 with thickness *d* acts as passivation, followed by an optional Ti interlayer and the conducting TiO_x, which is doped with Ta.

a temperature stability due to the metallization process, light management dealing with the backside reflection and equipment compatibility to existing factory equipment [2].

This study investigates a double layer system of Al_2O_3 as passivating tunnel oxide and TiO_2 as conductive oxide, which can fulfil the mentioned prerequisites [2]. The outstanding passivation performance of Al_2O_3 is accepted widely [3]. Additionally, anatase TiO_x is stable at temperatures up to 600 °C and known as a promising TCO layer [4]–[6]. An improvement of work published earlier [7], [8] will be shown. It reflects the replacement of the ALD deposition of the TiO_2 by sputtering. Additionally, the enhancements of the conductivity of the layer stack by Ta doping and interface engineering by Ti interlayers are demonstrated.

2 EXPERIMENTAL DETAILS

The final layer stack is shown in Fig. 1 schematically. A thin Al_2O_3 passivation is in direct contact with the Si substrate followed by an optional Ti interlayer and the conducting TiO_x. Two different types of Si substrates were used. On the one hand 250 µm thick boron doped *p*-type floatzone Si wafers with a resistance of 1-3 Ωcm were employed for the determination of the passivation behavior in order to exclude any bulk recombination. Theses samples were deposited symmetrically on both sides. On the other hand, the evaluation of the electrical characteristics was done with highly doped *p*-type Czochralski (Cz) Si wafers (< 0.005 Ωcm). The double layer consists of Al_2O_3, which acts as passivation layer and a TiO_x as conduction layer. The Al_2O_3 was deposited by thermal ALD (150 °C) using Trimethylaluminum as precursor and H_2O as oxidizing agent. The TiO_x was sputtered reactively using a metallic Ti target with an Ar:O_2 gas ratio of 10:1 and a pressure of $5 \cdot 10^{-3}$ mbar. The DC-power density at the target was 6.6 W/cm². A doping of the TiO_x was realized by a co-sputtering approach. A RF-plasma was ignited simultaneously with a power of 0…1.1 W/cm² at a metallic Ta target. This allowed a variation of the Ta content Ta/(Ta+Ti) from 0 to 0.2, which was verified by X-ray photoelectron spectroscopy (XPS). Specific samples got an additional Ti interlayer (Ti-IL) between TiO_x and Al_2O_3. The Ti-IL was deposited by electron beam evaporation or sputtering before the TiO_x deposition. In order to activate the passivation behavior, a 10 min post deposition anneal was performed in a forming gas atmosphere (N_2:H_2 = 9:1, 0.1 bar) at 350 °C (FGA) with all samples.

The physical characterization was done by grazing incidence X-ray diffraction (GIXRD) for evaluation of the crystallinity, by X-ray reflectometry (XRR)/ spectroscopic ellipsometry for thickness determination and microwave-detected photoconductivity (MDP) for verification of the effective carrier lifetime.

The resistance was investigated by current-voltage measurements. Therefore aluminum dots were deposited on top of the layer stack with a low resistive *p*-Cz Si-substrate (< 0.005 Ωcm). Due to the low resistance of the Si-substrate, the resistance is dominated by the oxide layers. In order to keep the efficiency of the PERC cell high, a voltage drop of 10 mV over the layer stack can be tolerated at a short-circuit current density of 40 mA/cm² [9]. Therefore, the normalized resistance of the stack has to be below 0.25 Ωcm².

3 RESULTS & DISCUSION

The maximum surface recombination velocity (SRV) needs to be about 100 cm/s to ensure a sufficient efficiency for PERC solar cells [10]. In previous works [7], [8] a SRV of 15 cm/s (lifetime of 800 µs) was demonstrated with a double layer stack of 5 nm Al_2O_3 and 15 nm TiO_x. The contact resistance of 20 Ωcm² reported, which was determined at 10 mV and 0.5 mA/cm². A typical short current of 40 mA/cm² could not be applied to the stack due to the large voltage drop and an accompanied dielectric breakdown. The dielectrics of that stack were both deposited by ALD. Especially the deposition of TiO_x is linked with low rates

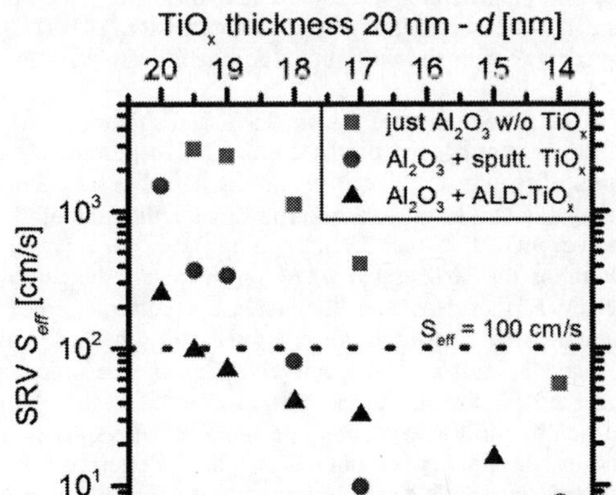

Fig. 2: Surface recombination velocity (SRV) vs. the Al_2O_3 thickness *d* deposited by ALD for a pure Al_2O_3 layer without TiO_x (green), a layer stack of Al_2O_3/sputtered TiO_x (red) and a layer stack of Al_2O_3/ALD-TiO_x. The total stack thickness was 20 nm except for pure Al_2O_3 layers. A 10 min FGA at 350 °C was done before the measurements. The horizontal dashed blue line shows the SRV limit for efficient PERC solar cells.

(≈ 0.3 nm/min) and not suitable to industrial solar applications.

In this study, the ALD-TiO_x is substituted by sputtered TiO_x. The SRV is calculated by

$$S_{eff} = \frac{W}{2\,\tau_{eff}}. \quad (1)$$

W represents the wafer thickness and τ_{eff} the measured effective lifetime. Equation (1) is valid for symmetrical passivation only. Therefore the layer stack in Fig. 1 was deposited on both sides of the *p*-FZ wafers.

Fig. 2 shows the passivation behavior of thin ALD-Al_2O_3 up to 6 nm without TiO_x on top. The SRV decrease with thicker Al_2O_3. 6 nm Al_2O_3 is necessary to achieve a SRV of $S_{eff} \approx 60$ cm/s. In general, the deposition of TiO_x with ALD or sputtering on top of Al_2O_3 improves the SRV. Sputtered TiO_x yields in lower SRV (≈ 10 cm/s) compared to ALD-TiO_x down to Al_2O_3 thicknesses of 3 nm. The origin might be a higher density of fixed charges in the Al_2O_3 during sputtering. This was previously observed for a double layer of thin ALD-Al_2O_3 (d < 5 nm) and sputtered Al_2O_3-layer on top [11]. Below 3 nm of Al_2O_3 the SRV of the stack with sputtered TiO_x is considerably higher than the one of ALD-TiO_x. Probably the sputtering process of TiO_x damages the Si/Al_2O_3 interface due to the bombardment by high energy atoms and ions [12], [13].

The improvement of the conductivity of TiO_x can be achieved by doping with Ta [14]–[17]. The substitution of a Ti^{+4} atom with a Ta^{+5} atom furnishes an additional electron to the conduction band. The activation energy of this electron is about 0.01 eV and is already excited at room temperature [14]. In Fig. 3, the resistance of a layer stack of Si/native SiO_x/Ta:TiO_x with total thickness of 20 nm is shown. The variation of RF-power at an additional Ta target provides various Ta doping levels in the TiO_x. The inset of Fig. 3 shows the Ta/(Ta+Ti)-atom ratio *A* for different RF-power at the Ta target. The RF-power between 20…50 W corresponds to a ratio *A* = 0.02…0.21. Once the layer is doped by small amounts of Ta, the stack resistance (Fig. 3) of the TiO_x

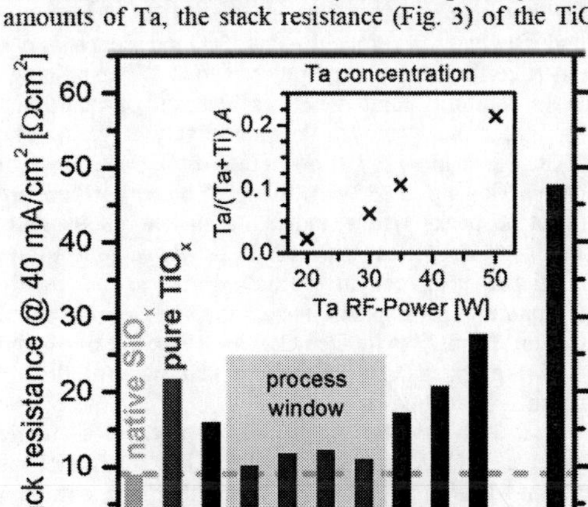

Fig. 3: stack resistance at 40 mA/cm² vs. RF-plasma power at the Ta-target (black bars) and pure TiO_x (without doping) in red, the total stack thickness was 20 nm, native SiO_x (≈ 1.5 nm) is shown in grey, **Inset:** Ta/(Ti+Ta)-atom-ratio *A* vs. RF-plasma power at the Ta-target measured with XPS, calculation was done with the XPS lines Ta 4d and Ti 3p

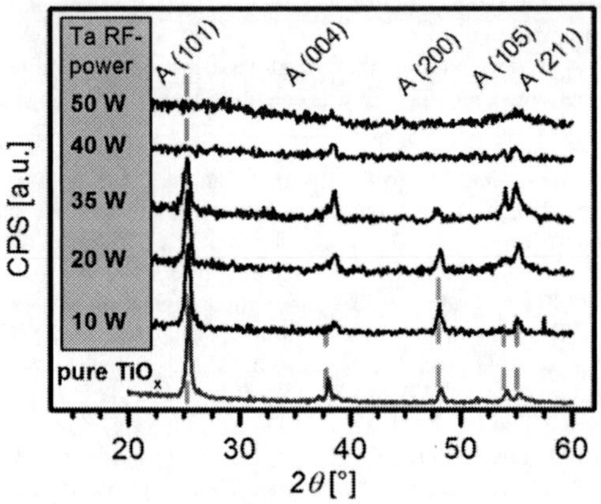

Fig. 4: GIXRD measurements of pure TiO_x (red) and $Ta:TiO_x$ (black) with different anatase TiO_x planes (blue dashed vertical), total stack thickness was 20 nm with 350 °C FGA

(red bar) decreases (black bars). A process window with almost constant resistance is observed for RF-power between 10 to 25 W. 20 W corresponds to an atom ratio $A = 0.02$. Since no XPS standards are used the measurement uncertainty for the XPS is about ± 0.01 of the atomic ratio. The minimum stack resistance is about 12 Ωcm^2 in this process window. Additionally, the GIXRD data in Fig. 4 exhibit an anatase phase for pure TiO_x and $Ta:TiO_x$ between 10 and 35 W RF-power at the Ta target. The stack resistance increases again at RF-power above 25 W. The solubility of Ta in TiO_x is about $A \approx 0.12$ for Ti-rich and O-poor conditions [14]. A higher oxygen content is expected to decreases the Ta solubility just slightly [14], [15]. However intermediate or O-rich films with higher Ta concentration tend to form Ti-vacancies V_{Ti}, which act as acceptor defects. This V_{Ti} is able to capture the free electrons and lower the conductivity [14]. Probably the V_{Ti} increase the stack resistance for a RF-plasma power above 25 W, although the Ta solubility limit is not reached with $A \approx 0.06$ and $A \approx 0.1$ for 30 W and 35 W, respectively (Fig. 3). The GIXRD data in Fig. 4 shows clear crystalline phase of anatase TiO_2 up to 35 W RF-power. For 40 W, there are almost no peaks visible, indicating an amorphous state. The layer deposited with 50 W RF-power shows no significant peaks anymore and seems to be totally amorphous. This explains the drastic increase by about 20 Ωcm^2 from 27 to 48 Ωcm^2 for an RF-power of 40 and 50 W (Fig. 3), since the amorphous TiO_x phase is already known as insulating [15].

Fig. 3 shows also the stack resistance of a simple Si/native SiO_x/Al-dot layer stack. The resistance is 9 Ωcm^2. This is just 3 Ωcm^2 lower than the optimum value of a $Ta:TiO_x$ layer stack in the process window (12 Ωcm^2). Therefore it is necessary to take the resistance of the native SiO_x into account. In [18], the removal of the SiO_x was attempted by an HF-dip before the deposition of the TiO_x layer, but no significant decrease in the stack resistance was observed. Probably the SiO_x is reformed during the FGA. Therefore, another approach to tackle the SiO_x was used here. An additional titanium interlayer (Ti-IL) should reduce the native SiO_x during a post deposition anneal. This technique is similar to oxygen scavenging known from high-k metal gates in

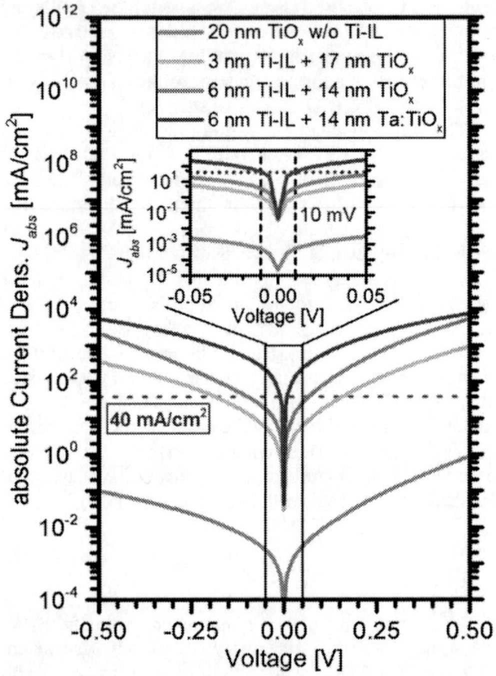

Fig. 5: Current vs. voltage characteristics of TiO_x layers with and without titanium interlayer (Ti-IL) in between and after a FGA at 350 °C, the dashed lines show the limit for an efficient PERC cell contact with 0.025 Ωcm^2, **Inset:** Current density between -0.05 V and 0.05 V to show the behavior at about 10 mV

CMOS transistors [19]. A Ti-IL was deposited by sputtering or evaporation on the substrate before the TiO_x deposition, whereby the total stack thickness is 20 nm. The I-V-measurements of the Si/native SiO_x/Ti-IL/TiO_x layer stack with and without Ta doping are shown in Fig. 5.

The introduction of a 3 nm Ti-IL leads to an increase of the current density by three orders of magnitude. This jump of the current density is not explainable by a 3 nm thinner TiO_x (17 nm). Furthermore, a reduction of the native SiO_x and the TiO_x capping layer is expected. Doubling the Ti-IL thickness (6 nm) increases the current density further. However, the increase in current seems to saturate. The measured current for 3 and 6 nm Ti-IL is 1.0 and 3.7 mA/cm^2, respectively. This is one order of magnitude below the target of 40 mA/cm^2 at 10 mV (blue dashed line in Fig. 5). Thus, the enhancements of doping and interlayers were combined. A 6 nm Ti interlayer was put together with a Ta doping of TiO_x by 20 W RF-power. It gained an additional order of magnitude (inset of Fig. 5). The achieved current density is 42.7 mA/cm^2 at 10 mV. This corresponds to a contact resistance of 0.23 Ωcm^2, which is below the target of 0.25 Ωcm^2.

Even so, the conductivity of TiO_x can be tuned up to a reasonable value, the SRV of those layers is above 1000 cm/s (not shown). Therefore, a double layer structure as presented in [7] is needed. A comparison of the approaches for conductivity tuning (sputtered TiO_x, Ta doping and the Ti-IL) in conjunction with the Al_2O_3 passivation are shown in Fig. 6. All presented layer stacks feature a 2 nm ALD-Al_2O_3. The stack with an ALD-TiO_x layer reaches the lowest measured SRV of

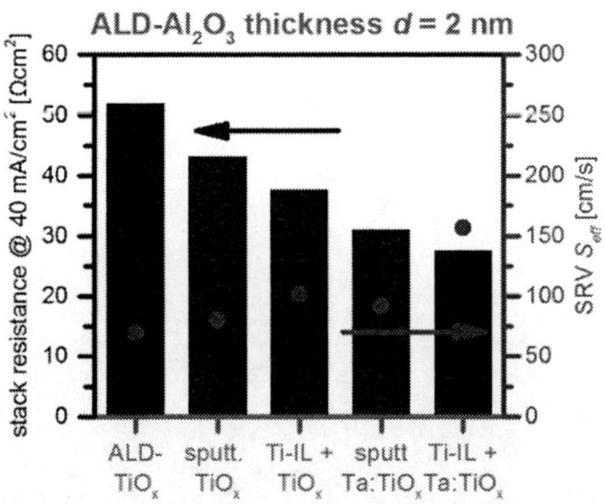

Fig. 6: Stack resistance at 40 mA/cm^2 (black bars) and minority carrier lifetime (red points) of samples with $d = 2$ nm Al_2O_3 deposited by ALD, TiO_x or $Ta:TiO_x$ was deposited on top by ALD and sputtering, 3 nm Ti-IL were evaporated before the sputtering of TiO_x and $Ta:TiO_x$ for two samples, the red line serves as guide for

70 cm/s. The SRV of sputtered TiO_x is slightly higher (80 cm/s). The stack resistance, which is calculated from the voltage drop over the $Si/SiO_x/Al_2O_3/TiO_x$-layer stack (Fig. 1) is slightly lower for the sputtered TiO_x (43 Ωcm^2) compared to the ALD-TiO_x (52 Ωcm^2). It is possible to decrease the stack resistance by inserting a thin 3 nm Ti-IL (38 Ωcm^2) or by doping the TiO_x with Ta (31 Ωcm^2). Both lead to a slight increase of SRV to 100 cm/s and 90 cm/s, respectively. A combination of the two approaches, Ta-doping and Ti-IL, leads to the lowest stack resistance of 28 Ωcm^2, however the SRV increased to approximately 160 cm/s. The Ti-IL exhibits a much higher influence on the current density without Al_2O_3. Probably, the inserted 2 nm Al_2O_3 hinders the scavenging effect. Additionally, the Al_2O_3 layer significantly contributes to the total stack resistance. Therefore, the engineering of the Al_2O_3 tunneling barrier is the crucial part in order to get an optimum of tunneling current and passivation behavior.

4 CONCLUSIONS

In this study, full area contacts were realized with Al_2O_3-TiO_x double layers. The surface recombination velocity (SRV) of the Al_2O_3/TiO_x deposited by ALD reaches 30 cm/s for 3 nm Al_2O_3. A sputtered TiO_x layer on top of 3 nm ALD-Al_2O_3 improves the SRV to about 10 cm/s. A thinner Al_2O_3 leads to more damage at the Si-surface during TiO_x-sputtering [12], [13] and a higher SRV (80 cm/s). In terms of full area contacts, there has to be a tradeoff between contact resistance and SRV.

Doping of the TiO_x by Ta demonstrated an improvement of the conductivity of the TiO_x. The Ta-doping can decrease the measured stack resistance at 40 mA/cm^2 by 40 % from 20 to 12 Ωcm^2. Probably the thin native SiO_x limits a further reduction in the Si/native $SiO_x/Ta:TiO_x$ layer stack (Fig. 3).

The deposition of an additional 3 nm or 6 nm titanium interlayer (Ti-IL) placed between the native SiO_x or the ALD-Al_2O_3 and the TiO_x reduces the conductivity by three orders of magnitude. It is expected, that the Ti-IL reduces the native SiO_x and the TiO_x. Doping in conjunction with a Ti-IL offers the possibility to reduce the resistance of the TiO_x layer to a value, which is below the target of 0.25 Ωcm^2.

Both approaches, the Ta doping and the Ti-IL, were used to improve the Al_2O_3/TiO_x passivation layer as full area contact. The stack resistance was halved from 52 to 28 Ωcm^2, however the SRV rises to 160 cm/s.

In summary, this study shows the limiting factor in terms of resistance is the Al_2O_3 tunneling barrier and the SiO_2 at the interface to the Silicon substrate. Additionally, a way is shown to passivate Si by a full area contact and substantiates the Al_2O_3/TiO_x layer as possible full area passivating contact.

4 ACKNOWLEDGEMENTS

We gratefully thank the Federal Ministry of Education and Research (BMBF) for the financial support within the framework of the TIBET project (FKZ: 03SF0538A) and the Free State of Saxony.

5 REFERENCES

[1] 'ITRPV Eighth Edition 2017- Maturity Report', 8th Edition, Sep. 2017.

[2] I. Dirnstorfer and T. Mikolajick, 'Dielectric Nanomaterials for Silicon Solar Cells', in *Nanomaterials for Sustainable Energy*, Q. Li, Ed. Cham: Springer International Publishing, 2016, pp. 41–94.

[3] D. K. Simon, P. M. Jordan, T. Mikolajick, and I. Dirnstorfer, 'On the Control of the Fixed Charge Densities in Al_2O_3 -Based Silicon Surface Passivation Schemes', *Appl. Mater. Interfaces*, vol. 7, pp. 28215–28222, Dec. 2015.

[4] M. M. Hasan, 'Effects of Annealing Treatment on Optical Properties of Anatase TiO_2 Thin Films', vol. 2, no. 4, p. 5, 2008.

[5] V. Legrand, O. Merdrignac-Conanec, W. Paulus, and T. Hansen, 'Study of the Thermal Nitridation of Nanocrystalline $Ti(OH)_4$ by X-ray and in Situ Neutron Powder Diffraction', *J. Phys. Chem. A*, vol. 116, no. 38, pp. 9561–9567, Sep. 2012.

[6] X. Yang, Q. Bi, H. Ali, K. Davis, W. V. Schoenfeld, and K. Weber, 'High-Performance TiO_2-Based Electron-Selective Contacts for Crystalline Silicon Solar Cells', *Adv. Mater.*, vol. 28, no. 28, pp. 5891–5897, Jul. 2016.

[7] I. Dirnstorfer *et al.*, '$Al_2O_3TiO_2$ Nanolaminates for Conductive Silicon Surface Passivation', *IEEE J. Photovolt.*, vol. 6, no. 1, pp. 86–91, Jan. 2016.

[8] D. K. Simon, P. M. Jordan, M. Knaut, T. Chohan, T. Mikolajick, and I. Dirnstorfer, 'ALD Al_2O_3 based nanolaminates for solar cell applications', 2015, pp. 1–6.

[9] S. Gatz, T. Dullweber, and R. Brendel, 'Evaluation of Series Resistance Losses in Screen-Printed Solar Cells With Local Rear Contacts', *IEEE J. Photovolt.*, vol. 1, no. 1, pp. 37–42, Jul. 2011.

[10] J. Schmidt, A. Merkle, R. Brendel, B. Hoex, M. C. M. van de Sanden, and W. M. M. Kessels, 'Surface passivation of high-efficiency silicon solar cells by atomic-layer-deposited Al_2O_3', *Prog. Photovolt. Res. Appl.*, vol. 16, no. 6, pp. 461–466, Sep. 2008.

[11] D. K. Simon, 'Realization and Characterization of Advanced Nanolayers for Photovoltaic Applications', Dissertation, Technischen Universität Dresden, Dresden, 2016.
[12] X. Zhang, A. Cuevas, and A. Thomson, 'Process Control of Reactive Sputter Deposition of AlO_x and Improved Surface Passivation of Crystalline Silicon', *IEEE J. Photovolt.*, vol. 3, no. 1, pp. 183–188, Jan. 2013.
[13] S. Chen, L. Tao, L. Zeng, and R. Hong, 'RF Magnetron Sputtering Aluminum Oxide Film for Surface Passivation on Crystalline Silicon Wafers', *Int. J. Photoenergy*, vol. 2013, pp. 1–5, 2013.
[14] J. Osorio-Guillén, S. Lany, and A. Zunger, 'Atomic Control of Conductivity Versus Ferromagnetism in Wide-Gap Oxides Via Selective Doping: V, Nb, Ta in Anatase TiO_2', *Phys. Rev. Lett.*, vol. 100, no. 3, Jan. 2008.
[15] M. Neubert *et al.*, 'Overcoming challenges to the formation of high-quality polycrystalline TiO_2:Ta transparent conducting films by magnetron sputtering', *J. Appl. Phys.*, vol. 114, no. 8, p. 083707, Aug. 2013.
[16] H. Anh Huy, B. Aradi, T. Frauenheim, and P. Deák, 'Comparison of Nb- and Ta-doping of anatase TiO_2 for transparent conductor applications', *J. Appl. Phys.*, vol. 112, no. 1, p. 016103, Jul. 2012.
[17] T. Hitosugi *et al.*, 'Ta-doped Anatase TiO_2 Epitaxial Film as Transparent Conducting Oxide', *Jpn. J. Appl. Phys.*, vol. 44, no. 34, pp. L1063–L1065, Aug. 2005.
[18] D. Tröger, M. Grube, M. Knaut, J. Reif, J. W. Bartha, and T. Mikolajick, 'Towards Full-area Passivating Contacts for Silicon Surfaces based on Al_2O_3/ TiO_x Double Layers', presented at the 7th World Conference of Photovoltaic Energy Conversion, Waikoloa Beach, Hawaii, USA, 2018, Manuscript nr. 665.
[19] M. M. Frank, 'High-k / Metal Gate Innovations Enabling Continued CMOS Scaling', *2011 Proc. ESSCIRC*, p. 9, 2011.

Electron-selective contact using i-a-Si:H/TiO_x/Yb/Ag stack for silicon heterojunction solar cells

*Jinyoun Cho[1,2], Maria Recaman Payo[2], Maarten Debuquoy[2], Hariharsudan Sivaramakrishnan Radhakrishnan[2], Ivan Gordon[2], Jozef Szlufcik[2], Moustafa Ghannam[3], Jef Poortmans[1,2,4,5]
[1] KU Leuven, Kasteelpark Arenberg 10, B-3001 Leuven, Belgium
[2] IMEC, Kapeldreef 75, B-3001 Leuven, Belgium
[3] Kuwait University, EE department, P.O. Box 5969, 13060, Safat, Kuwait
[4] Universiteit Hasselt, Martelarenlaan 42, B-3500 Hasselt, Belgium
[5] EnergyVille, Thor Park 8310, B-3600, Genk, Belgium
* Corresponding author

Abstract
Silicon heterojunction (SHJ) solar cells can be manufactured through relatively simple process steps. However, optoelectrical losses in doped a-Si:H and complex optimization of the doping process makes the cell fabrication difficult. Therefore, an electron contact without n-a-Si:H would be interesting and the i-**a**-Si:H/**T**i**O**$_x$/low work function **m**etal (ATOM) contact could be one of the alternatives. In this study, TiO_x grown by ALD and Yb (Φ: 2.5-2.6 eV) are used in the ATOM contact. Specific contact resistivities (ρ_c) and dark saturation current density at metalized area ($J_{0,metal}$) are investigated as a function of a TiO_x thickness. 1 nm thick TiO_x with Yb/Ag stack on 8 nm thick i-a-Si:H achieves the minimum ρ_c of 48 mΩ·cm^2 and a $J_{0,metal}$ below 15 fA/cm^2.

Introduction

Despite the high-efficiency potential and simple fabrication process sequence of SHJ cells, such cells incur optoelectrical losses related to parasitic absorption in the thick a-Si:H layers. Moreover, the complexity of the doping process optimization for doped a-Si:H makes it difficult to realize efficiency improvements. For these reasons, carrier-selective contacts without doped a-Si:H could enable high cell efficiencies.

However, while such doped layer-free contacts have often resulted in high ρ_c or poor contact passivation quality we have shown previously that our proposed ATOM (i-**a**-Si:H/**T**i**O**x/low work function **m**etal) contact structure could be a viable alternative to realize high-quality electron-selective contacts with both low ρ_c and high passivation quality [1,2]. The TiO_x layer in this ATOM contact reduces the metal Fermi level ($E_{F,metal}$) pinning at the a-Si:H surface and thereby results in a significant reduction in ρ_c by more than two orders of magnitude using i-a-Si:H/TiO_x/Ca/Al contact. Furthermore, TiO_x deposited by atomic layer deposition (ALD) proved to be suitable for this contact structure because no significant degradation of i-a-SI:H passivation quality was observed after TiO_x deposition, in contrast to TiO_x deposited by e-beam evaporation.

So far, a detailed thickness study of the ALD-TiO_x layer in the ATOM contact has not yet been conducted. Moreover, Yb is a very interesting for electron contacts because of the very low work function of 2.6 eV (vs 2.9 eV for Ca), and YbSi is also known to form a low Schottky energy barrier with Si [1].

For these reasons, the influence of TiO_x thickness in combination with Yb is studied in this work. The optimal thickness resulting from this study will be tested at device level.

Experimental details

For all test samples, double side textured Cz n-type Si wafers (3.3 Ω·cm, 182 μm) were used (see Fig.1 for the schematic cross sections). After HF:O_3:HCl and HF:HCl cleaning, intrinsic a-Si:H of 8 nm thickness was deposited on both sides. TiO_x layers were deposited on both sides by ALD using $Ti(OCH_3)_4$ and H_2O as a Ti precursor and an oxidant, respectively. TiO_x layer thicknesses of 1, 1.5, 2 and 3 nm were tested.

For ρ_c test samples, Yb (60 nm) and Ag (100 nm) were evaporated through a metal shadow mask on the front. On the rear, the full area was covered by the same metal stack. For $J_{0,metal}$ samples, Yb (5 nm) and Ag (5 nm) were evaporated only on the rear, in order to be able to measure carrier lifetime using quasi steady-state photoconductance (QSSPC). Lifetime measurements before and after one side metallization were used to determine the implied-V_{OC} and $J_{0,total}$ in order to extract the $J_{0,metal}$. $J_{0,bulk}$ of the Cz wafers was about 3 fA/cm^2, estimated by comparing $J_{0,total}$ of a Cz wafer after i-a-Si:H passivation to that of a high quality n-type FZ-Si wafers (3 Ω·cm, 200 μm). Detailed fitting method and sample preparation can be found in the authors' previous works [2,3]

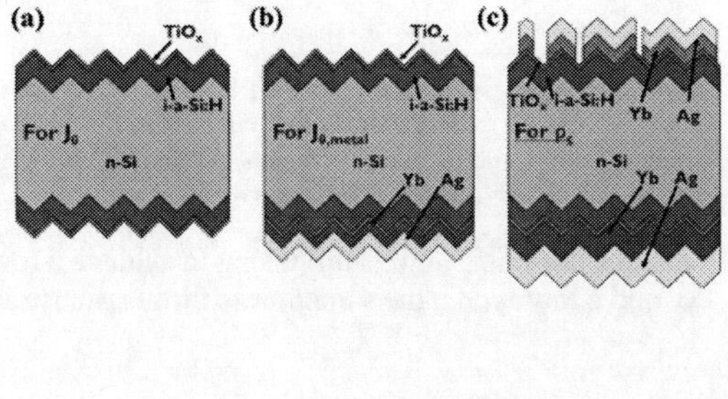

Figure 1. Test sample structures (a) for J_0 measurement, (b) for $J_{0,metal}$ measurement, and (c) for ρ_c measurement.

Results and discussion

The thickness of the insulator strongly influences ρ_c in a metal-insulator-semiconductor (MIS) contact. [4]. Indeed, from Fig. 2 (a), it can be observed that the thickness of the TiO_x layer in between i-a-Si:H and Ybstrongly influences the ρ_c. A TiO_x layer of 1 nm thickness reduces the $E_{F,metal}$ pinning and in that way the Schottky barrier at the interface and consequently the ρ_c. For a thickness above the 1 nm, ρ_c increases sharply due to the high resistance associated with tunneling through the TiO_x itself.

At the same time, $J_{0,metal}$ is maintained at a similar level below 15 fA/cm², regardless of the TiO_x thickness (Fig. 2 (b)). Presumably, the sensitivity of $J_{0,metal}$ to TiO_x thickness is low because in all cases the contact passivation is dominantly delivered by i-a-Si:H. Although wafer morphology is different, the trends for Yb-based ATOM contact were similar to those of Ca-based ATOM contact in terms of the ρ_c and $J_{0,metal}$ [3].

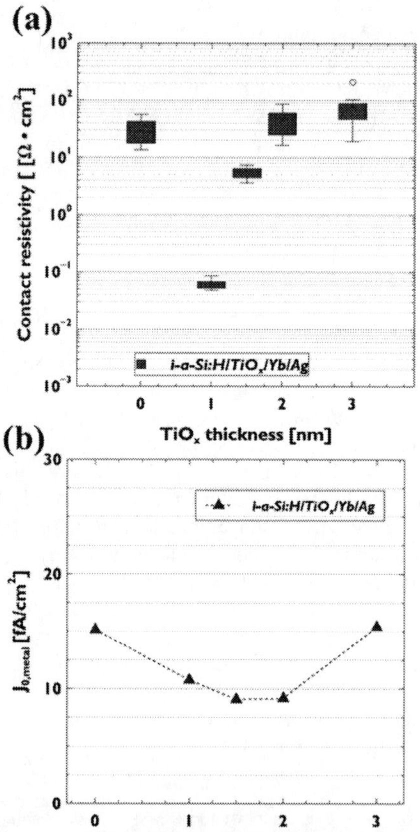

Figure 2. (a) ρ_c and (b) $J_{0,metal}$ in function of TiO_x thickness in i-a-Si:H/TiOx/Yb/Ag (ATOM) contacts.

Since 1 nm-thick TiO_x is promising to achieve a low ρ_c and a low $J_{0,metal}$, the sample was further analyzed using TEM and EDX to understand the phenomena occurring in these nanoscale layers.

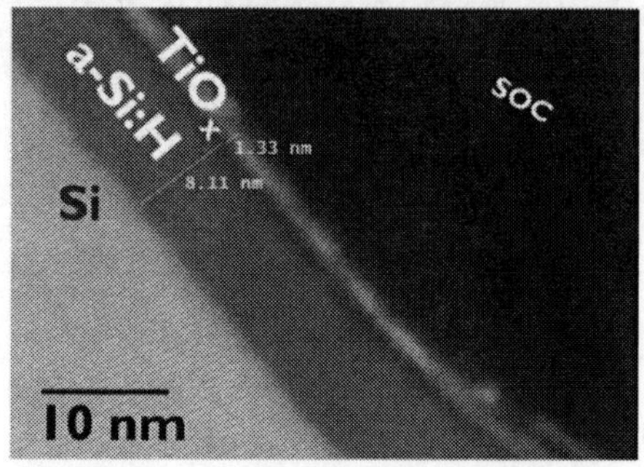

Figure 3. High-angle annular dark field (HAADF) image of n-Si/i-a-Si:H/TiO_x measured by Scanning Transmission Electron Microscopy (STEM). Spin on Carbon (SOC) is on the TiOx for the sample preparation for the TEM.

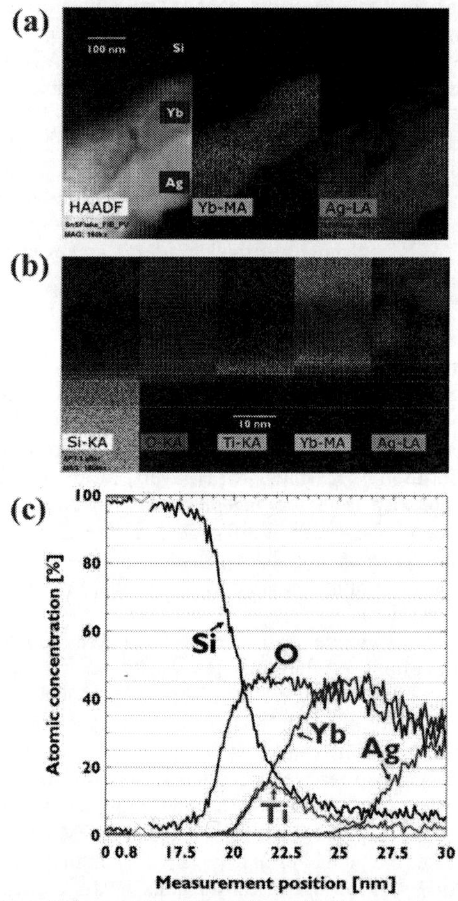

Figure 4. Cross-section Energy-dispersive X-ray spectroscopy images measured (a) at the Yb/Ag stack in large scale, and (b) at the c-Si surface in small scale. (c) Chemical line profile of the contact after Yb/Ag evaporation.

As can be seen in Fig. 3, TiO_x of 1.3 nm conformally covers the a-Si:H layer and the real thickness was very close to the target thickness of 1 nm. The small difference in nominal and real thickness is probably caused by the different growth rate for the first mono layer compared to that for the subsequent layers. Like other literature studies, amorphous SiO_x was formed underneath the TiO_x [2,3] and this may increase downward band bending by forming a dipole between the SiO_x/TiO_x interface. After Yb/Ag evaporation, the Yb/Ag stack was well formed and some Ag atoms appear to have diffused into the Yb layer (see, Fig. 4. (a)). However, high Yb concentration just above the TiO_x layer confirms that the resulting ρ_c is the result of Yb, and not due to Ag or Yb/Ag mixture. Moreover, the Yb also mixes into the TiO_x (see, Fig. 4 (b)) which may increase the conductivity of the TiO_x layer by Yb doping or defect-assisted conduction. Moreover, it has been reported that when TiO_x is doped by Yb, the density of the deep trap centers in the TiO_x is decreased and the Fermi level of the TiO_x is up-shifted [5]. From the chemical profiles shown in Fig.4 (c), the Yb stayed in TiO_x layer and the i-a-Si:H passivation layer was not contaminated by Yb, unlike the case for the ATOM contact with Ca [2]. TiO_x likely worked as a diffusion barrier for Yb which may help to maintain the a-Si:H passivation quality and might positively impact the stability of this ATOM contact, as also mentioned in literature for the stability of i-a-Si:H/TiO_x/LiF/Al contact [6].

Conclusion and future works

The combination of Yb with a thin TiO_x layer of 1nm on i-a-Si:H is promising to obtain a good contact performance. Contact resistivity below 0.1 $\Omega \cdot cm^2$, and $J_{0,metal}$ below 15 fA/cm^2 were achieved. The contacts will be evaluated in SHJ cells containing Yb-based ATOM contacts and compared to conventional SHJ cells. The ATOM contact structure may be a good alternative to achieve high-efficiency SHJ solar cells without the need for a doped a-Si:H.

Acknowledgment

The authors gratefully acknowledge the financial support of imec's industrial affiliation program for Si-PV. Moreover, the work in this paper was partially funded by Kuwait Foundation for the advancement of Sciences under project number P115-15EE-01.

References

[1] S. Zhu, J. Chen, M.F. Li, S.J. Lee, J. Singh, C.X. Zhu, A. Du, C.H. Tung, A. Chin, D.L. Kwong, N-type Schottky barrier source/drain MOSFET using Ytterbium silicide, IEEE Electron Device Lett. 25 (2004) 565–567. doi:10.1109/LED.2004.831582.

[2] J. Cho, M. Debucquoy, M. Recaman Payo, S. Malik, M. Filipič, H.S. Radhakrishnan, T. Bearda, I. Gordon, J. Szlufcik, J. Poortmans, Contact resistivity reduction on lowly-doped n-type Si using a low work function metal and a thin TiOx interfacial layer for doping-free Si solar cells, in: SiliconPV Conf. 2017, 2017: pp. 842–850. doi:10.1016/j.egypro.2017.09.356.

[3] J. Cho, J. Melskens, M. Debucquoy, M. Recamán Payo, S. Jambaldinni, T. Bearda, I. Gordon, J. Szlufcik, W.M.M. Kessels, J. Poortmans, Passivating electron-selective contacts for silicon solar cells based on an a-Si:H/TiO x stack and a low work function metal, Prog. Photovoltaics Res. Appl. 26 (2018) 835–845. doi:10.1002/pip.3023.

[4] A. Agrawal, J. Lin, B. Zheng, S. Sharma, S. Chopra, K. Wang, A. Gelatos, S. Mohney, S. Datta, Barrier height reduction to 0.15eV and contact resistivity reduction to 9.1×10−9 Ω-cm2 using ultrathin TiO2−x interlayer between metal and silicon, in: 2013 Symp. VLSI Technol. (VLSIT), 2013: pp. T200–T201. http://ieeexplore.ieee.org/document/6576652/.

[5] B. Roose, S. Pathak, U. Steiner, Doping of TiO2 for sensitized solar cells, Chem. Soc. Rev. 44 (2015) 8326–8349. doi:10.1039/C5CS00352K.

[6] J. Bullock, Y. Wan, Z. Xu, S. Essig, M. Hettick, H. Wang, W. Ji, M. Boccard, A. Cuevas, C. Ballif, A. Javey, Stable Dopant-Free Asymmetric Heterocontact Silicon Solar Cells with Efficiencies above 20%, ACS Energy Lett. 3 (2018) 508–513. doi:10.1021/acsenergylett.7b01279.

BENEFITS OF PATTERN TRANSFER PRINTING METHOD FOR FINGER METALLIZATION ON SILICON SOLAR CELLS

Adrian Adrian[1], Dominik Rudolph[1], Jan Lossen[1], Misha Matusovsky[2], Vinodh Chandrasekaran[3]
[1] International Solar Energy Research Center (ISC) Konstanz e.V., Rudolf-Diesel-Straße 15, 78467 Konstanz, Germany; Tel: +49-7531-36183-0; E-Mail: *adrian@isc-konstanz.de*
[2] Utilight Ltd., HaYarmuch St. 1, P.O.B 1341, 8122005 Yavne, Israel
[3] Heraeus, 24 Union Hill Road, PA 19428-2736 W. Conshohocken, USA

ABSTRACT: In this work, a new technology for the finger metallization on the solar cells, called Pattern Transfer Printing (PTP), is introduced and compared with the screen printing (SP) technology as reference. PTP is a contactless printing technology, which is based on laser induced deposition from a polymer substrate (called tape). Using trench width of 20 µm in the tape, the PTP technology is able to produce ultra-fine-line fingers (< 20 µm) with high aspect ratio (> 0.5). Therefore, the J_{sc} is increased significantly due to less shadowing and a cell efficiency of 21.8% for a finger cross section of 200 µm². This is a gain of +0.15%$_{abs}$ compared to the SP reference with 35 µm finger opening and simultaneously the silver paste consumption is reduced by 40 mg (-57%). As second reference group double printing with 20 µm finger opening screen is used to achieve a comparable aspect ratio. This results in a similar cell efficiency as for the single print with 35 µm finger opening but with higher paste consumption. The results show, that today the PTP technology is an attractive technology to replace SP for the finger metallization and enable to achieve ultra-fine-line finger printing below 20 µm.

Keywords: Pattern Transfer Printing, Metallization, fine-line printing, p-type, screen printing

1 INTRODUCTION

A look on the International Technology Roadmap for Photovoltaic (ITRPV) 2018 shows the importance the community attributes to the reduction of the finger width of the front metallization, which leads to lower production cost of the cells due to less silver consumption and also increase the efficiency because of less shadowing [1]. The commonly used technology for the finger metallization in the industrial production of silicon solar cells is screen printing (SP) due to its robustness and cost-effectiveness. Using screen printing technology with finger opening of 28-35 µm, it is possible to get finger lines with a width of 35-50 µm and reasonable aspect ratio. However, it becomes increasingly difficult to reduce finger width further due to limitations in the mesh wire thickness and emulsion material. Many studied alternative printing technology to overcome the current limitation of screen printing to print smaller fingers, such as dispensing [2] or flexographic printing [3].

We investigate a new printing method called Pattern Transfer Printing (PTP™) invented and commercialized by the company Utilight, which showed the ability to produce finer finger lines (< 30 µm) with a high aspect ratio. This technology is based on laser induced deposition from a polymer substrate (tape), which is not in contact with the wafer surface during the printing process. The tapes consist of many segments and each segment has many trenches. The silver paste is filled into these trenches with the help of two metal blades and is deposited by laser induced evaporation of the solvent in the interface area between paste and tape. A more detailed description about PTP printing method can be found in previous publication [4]. The PTP technology is already able to print in mass production capacity up to 3000 wafers/hour.

This paper will compare the status of the PTP method with state-of-the-art screen printing method as reference. The benefits and challenges, which need to be overcome in the future, are investigated.

2 EXPERIMENTAL SETUP

For this experiment precursors from the pilot line of Hanwha Q Cells in Germany were used [5]. These cells were made of 156.75 x 156.75 mm² Czochralski-grown Silicon (Cz-Si) with passivated emitter and rear cell structure (PERC). The full area aluminum metallization on the rear side and six bus bars on the front were screen printed already on the precursors. Only the finger metallization for the front side with screen printing technology (SP) and PTP technology were investigated in this work.

In this experiment 40 precursors were used and divided to four groups (see Table 1). A simulation work of Koduvelikulathu et al. [6] showed, that for cells with six bus bars the optimum finger width is between 20 to 30 µm depending on the finger aspect ratio that can be achieved. Hence, on the cells of the two PTP groups 112 fingers were printed from two different tapes, with different trench dimension. Opening width was 30 µm and 20 µm respectively, the further dimensions are given in Figure 1. In the printing position, the cells are placed at a gap distance of 200 µm from the tape surface. The paste in the trenches is deposited on the cell using a continuous wave infrared laser with a wavelength of 1064 nm.

The reference groups in this experiment were screen printed with two different screens containing both 100 fingers. The first screen has a 35 µm finger opening, 360 meshes per inch and wire diameter of 16 µm. The second screen is a development screen, which has a finger opening of 20 µm, 360 meshes per inch and wire diameter of 16 µm.

Table I: Experimental plan.

Group	Line opening	Remark
PTP	30 µm	
Screen Printing (SP)	35 µm	**Reference**
PTP	20 µm	
Screen Printing (SP)	20 µm	double print (DP)

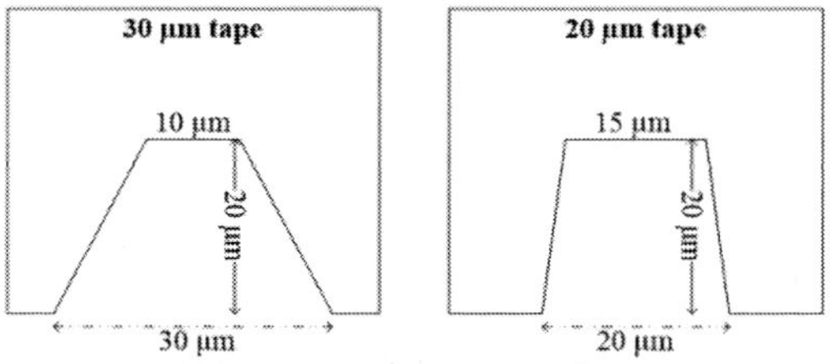

Figure 1: Trench dimension of the two tapes used in the study; (a) 30 µm tape and (b) 20 µm tape.

The reference silver paste for the screen printing group was SOL9641B from the company Heraeus, which currently is a frequent paste in the market and suitable for the ultra-fine-line printing [7]. A version of SOL9641B specifically modified and especially tailored for this application was used for the laser deposition process using PTP printer.

The preliminary test showed that the printability of the screen with 20 µm finger opening was insufficient, because there were many finger interruptions due to the mesh marks of the screen. We note that in this experiment, for the sake of simplicity, we used the single print paste SOL9641B for both single and double printing. There are other pastes available which are optimized for < 20 µm finger openings and double printing. To obtain a better printing result, the fingers of 20 µm SP group were double printed (DP), which means the fingers were printed, dried and then once more printed with the same alignment as the first print. After the finger metallization the cells from the whole groups underwent the drying and firing process. The cells were fired using a fast firing furnace (c.FIRE, Centrotherm AG) with a peak firing temperature of 905 °C and belt speed of 7 m/min.

3 RESULTS AND DISCUSSION

Using a 3D laser scanning microscope (LSM) from the company Olympus, the mean height and the mean width of the fingers after firing as well as the mean cross section for each group were determined. By dividing the mean height by the mean width of the finger, the mean finger aspect ratio for each finger was calculated (see Fig. 2). Fingers printed by PTP with the 30 µm tape have a mean height of 10 µm, a mean width of 25 µm at a cross section around 200 µm² and the finger aspect ratio lies between 0.3 and 0.7, which is slightly better compared to the screen printing reference with 35 µm finger opening.

Reducing the finger width further by using the screen with opening of 20 µm and double printed finger resulted in a higher aspect ratio compared to 35 µm finger opening, but resulted in a comparable cross section of around 500 µm². By using PTP with 20 µm tape, the fingers also have a higher aspect ratio. The average cross section is in similar range of around 200 µm². The average is 210 µm² for 30 µm tape and 200 µm² using 20 µm tape. The paste laydown for both PTP tapes was around 30 mg silver paste, while with screen printing the silver consumption was around 70 mg per cell. With double printing a higher finger aspect ratio was achieved, but the paste consumption was on the same level as for a 35 µm screen with single print.

The electrical parameters of all cells were measured using h.a.l.m flasher under the standard test condition (1000 W/m², 25 °C and AM1.5). Figure 3 shows the *I-V* results of each group. The reference single print screen printing group, using a finger opening of 35 µm, results in J_{sc} of 39.97 mA/cm², V_{oc} of 671.9 mV, *FF* of 80.47% and efficiency (*Eta*) of 21.61% averagely. By using PTP with 30 µm tape a gain in J_{sc} of +0.15 mA/cm² compared to the reference group was achieved, but V_{oc} and *FF* are slightly decreased by -0.4 mV and -0.29% respectively. The cells from both methods reached the same level of average cell efficiency of 21.6% with negligible difference (-0.01%).

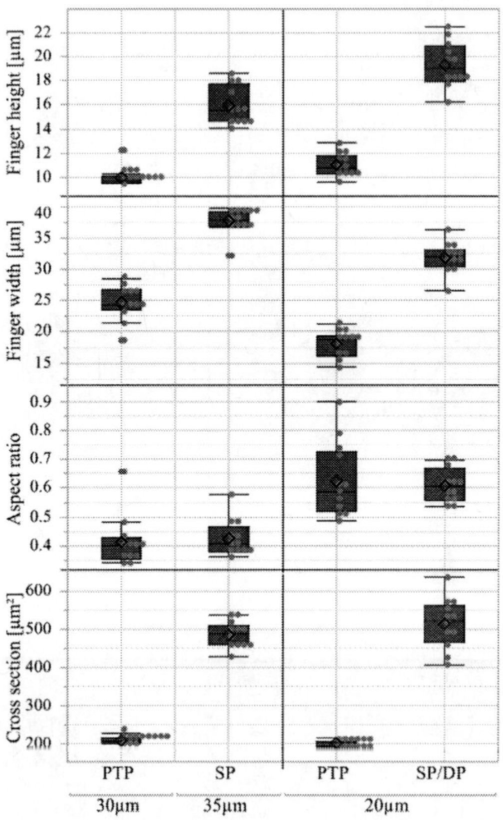

Figure 2: Average finger dimension measured by LSM for each group.

As compared to the reference SP group with 35 µm finger opening, reducing the finger width by using screen printing with opening of 20 µm and double printing resulted in a gain of J_{sc} by +0.13 mA/cm² with a slight decrease in V_{oc}. The *FF* is decreased by -1.11% due to losses on the series resistance. A slight gain in J_{sc} and at the same time a high loss on *FF* lead to a loss on the cell efficiency by average -0.25%. This lower cell efficiency showed the used 20 µm screen in this work did not work really well with double printing method. Using another screen design with different characteristics and silver paste designed for double print can achieve a higher cell performance than a single printed cell.

Reducing the trench width by using 20 µm tape results in higher J_{sc} of +0.38 mA/cm² (+1%) by average compared to the reference SP group with 35 µm opening. The mean V_{oc} value is slightly increased by +0.6 mV, but it is still in the measurement uncertainty. The mean efficiency is increased by +0.15%$_{abs}$ compared to the screen printing with finger opening of 35 µm.

The reduction on the finger width leads to an increase on J_{sc} due to less shadowing, but at the same time a decrease on *FF* due to higher series resistance. We found in a previous work [8], that a certain amount of glass frits in the silver paste is needed to enable a good contact formation between Si surfaces with the silver finger [8], which makes the paste needed for ultra-fine-line finger different as for standard finger. Same was shown earlier for the technique of aerosol printing, which results in very thin fingers. [9].

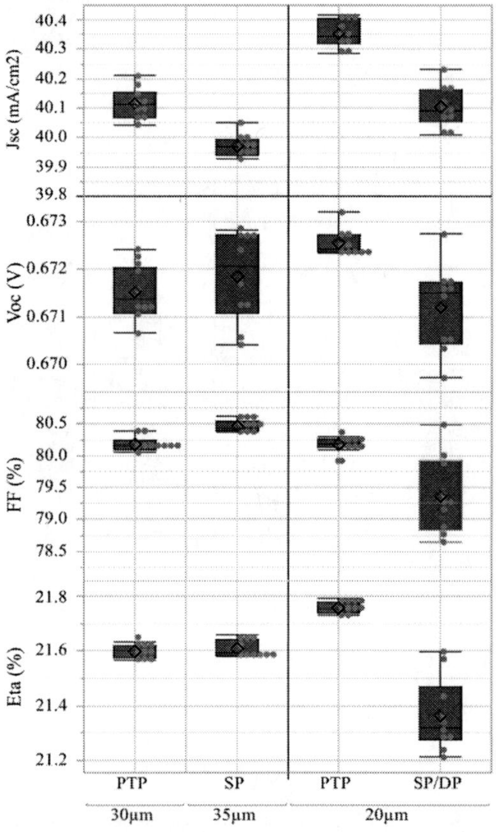

Figure 3: I-V results using h.a.l.m flasher under STC.

The contact resistivity for all groups were obtained with transfer length method (TLM) using a measurement device from GP Solar (GP 4-Test PRO). Two cells from each group were randomly selected and from each cell, six stripes of a width of 10 mm were cut out. As shown in Figure 4, the contact resistivity of the both PTP groups are slightly better (below 3 mOhmcm²) as the reference SP group with 35 µm finger opening. Hence, the contact resistance can be excluded as the reason for lower *FF*. The low contact resistivity also shows, that the commercial paste could be modified successfully to work with a finger cross section as low as 200 µm².

To measure the line resistance, two cells from each group were selected and cut into two stripes with a width of 22 mm. Then around 20 fingers distributed evenly were measured using the four-point-probe measurement method. Figure 5 shows the line resistance of all groups. The average line resistance of the PTP group with 30 µm tape is around 1.25 Ohm/cm and 2.5°times higher compared to SP group with 35 µm finger opening. The line resistance of 20 µm tape is slightly higher than the 30 µm tape probably because of a slightly small cross section or a higher number of constrictions. The higher line resistance of PTP groups due to the small finger cross section contributes to the higher electrical series resistance of the front metallization, which explains well the lower obtained *FF* (-0.3%) for PTP groups compared to the SP group with 35 µm finger opening.

The loss on *FF* due to a high line resistance could be reduced by a higher number of bus bars on the cells, like it realized in smart wire connection technology (SWCT). The combination of both technologies is expected to increase the individual gains.

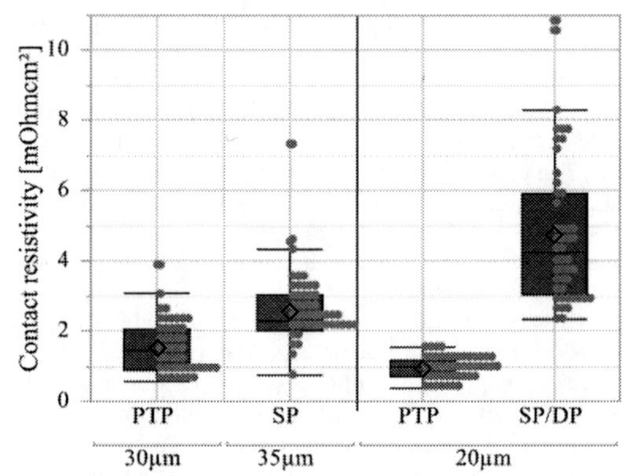

Figure 4: Contact resistivity measurement result achieved by TLM method.

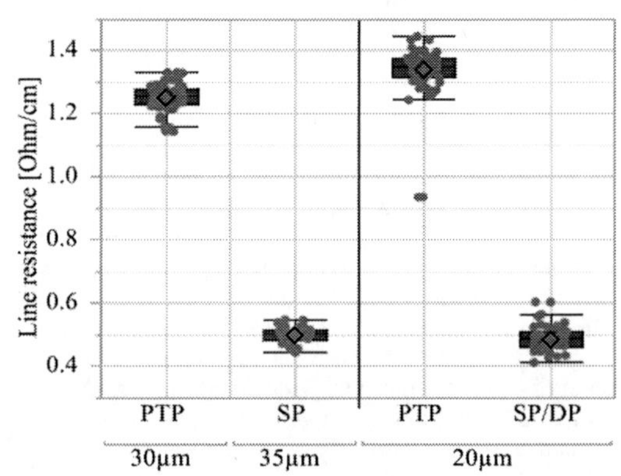

Figure 5: Line resistance of each group.

Figure 6 shows the microscopic finger image of PTP using 30 µm tape and SP with 35 µm finger opening with fiftyfold magnification (Fig. 6a) and its analysis (Fig. 6b) using surface imaging analysis software (MountainMap® Premium). As compared to the PTP finger, the height of the SP finger from top to bottom in the observation area varies due to the knots in the screen (connection between mesh) and the mesh itself. The highest finger height difference is around 10 µm. Both the knots and the mesh are the main problem for further reduction the finger opening, which increase the inhomogeneity of the finger and also interrupts the finger badly. Recently screen developments have shown very promising results with so called knotless screens [10]. However, use of this technique does result in higher screen costs.

On the other side, the finger of PTP is less straight compared to SP finger. The contactless printing process cause the paste to transfer not straight during the laser deposition process due to inhomogeneous release acceleration of the paste out of the trench. The finger of PTP can bent from one side to the other with a difference of 12 µm. This waviness of the PTP finger can be a challenge when printing cells with selective emitter.

The PTP finger is surrounded by debris, which is caused due to splashing effect of the organic components of silver paste during laser deposition. During the laser deposition, a laser beam illuminates the top of the trench and evaporates the solvent of the silver paste. This creates a high-pressure area between trench and silver paste and then releases the paste on to the cell. However, particles present in the surface near region where the solvent is evaporated, might be accelerated in diverging directions and leave fluid of paste. The influence of this debris to *I-V* parameter like J_{sc} due to shadowing and also V_{oc} due to higher recombination around the finger need to be investigated in the future experiments.

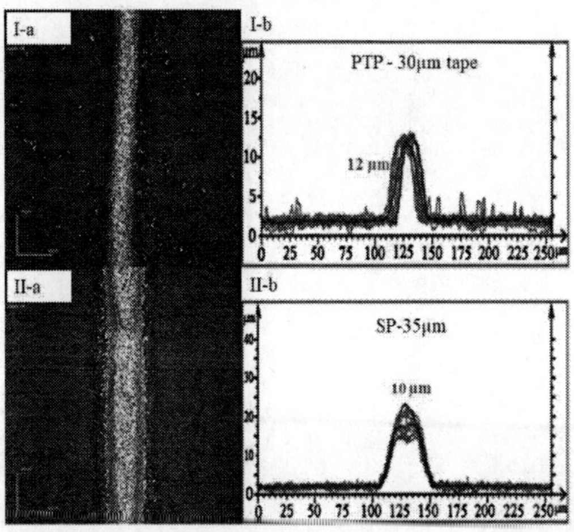

Figure 6: LSM picture (I-a, II-a) of the fingers and its analysis using mountain map (I-b, II-b) with different printing methods (I) PTP with 30 µm tape and (II) SP with 35 µm finger opening.

4 CONCLUSION

In this work, a new pattern transfer printing (PTP™) method for the finger metallization of the solar cell was introduced and compared with the screen printing method. The current screen printing technology has difficulties to print fine-line finger due to the problem of the knots in the mesh and the mesh marks on the finger, which can lead to interruption in the finger. In this experiment, double printing using 20 µm screen showed lower efficiency vs using a single print with wider opening. However, silver pastes designed for double printing can provide higher performance, depending on screen design.

The current PTP technology is able to print a fine-line finger (< 30 µm) with maintaining high aspect ratio and reduce the paste consumption to only 30 mg/cell (-57%). Therefore, the J_{sc} increases significantly but the *FF* is reduced due to higher losses on line resistance compared to the screen printing technology. The efficiency is increased by +0.15%$_{abs}$ by using PTP method with 20 µm tape. The contact resistivity of the PTP finger is slightly better compared to SP finger. Waviness and debris around PTP fingers are identified, but further experiments are needed to investigate their effects on the *I-V* parameters. However, the results of this work already show that PTP technology could be an interesting choice for replacing SP for the finger metallization in the quest of reducing line width further along the ITRPV roadmap. Good cells results for a finger width below 20 µm were demonstrated.

5 ACKNOWLEGMENT

This work was funded under the European Solar-era.net framework by the German Federal Ministry for Economic Affairs and Energy (BMWi) under the contract number FKZ 0324206 (RefinedPV). We are also thankful to Hanwha Q Cells and Heraeus for their support in this work.

6 REFERENCES

[1] ITRPV Working group, International Technology Roadmap for Photovoltaic (ITRPV) Results 2017 including maturity report 2018, 9th edition, September 2018. URL: www.itrpv.net/.cm4all/uproc.php/0/ITRPV%20Ninth%20Edition%202018%20including%20maturity%20report%2020180904.pdf?cdp=a&_=165a39bbf90 (access at 13.09.2018).

[2] M. Pospischil, et al., Process Development for a High-Throughput Fine Line Metallization Approach Based on Dispensing Technology, Energy Procedia vol. 43, pp 111-116, 2013.

[3] A. Lorenz, et al., Flexographic Printing – High Throughput Technology for Fine Line Seed Layer Printing on Silicon Solar Cells, in Proceedings 28th EUPVSEC, Paris, France, pp 1017-1023, 2013.

[4] J. Lossen, M. Matusovsky, A. Noy, Ch. Maier and M. Bähr, Pattern Transfer Printing (PTP™) for c-Si solar cell metallization, Energy Procedia vol. 67, pp 156-162, 2015.

[5] F. Fertig, et al., Mass production of *p*-type Cz silicon solar cells approaching average stable conversion efficiencies of 22 %, Energy Procedia vol. 124, pp. 338-345, 2017.

[6] L.J. Koduvelikulathu, J. Lossen, D. Rudolph, M. Matusovsky and G. Dishon, Why Multi-Busbars and Future Emtitters Require Further Shrinking of Finger Line Width, in Proceedings 32nd EUPVSEC, Munich, Germany, pp 550-554, 2016.

[7] Heraeus, Photovoltaic metallization paste: SOL 9641B Series, URL: https://www.heraeus.com/media/media/hpt/doc_hpt/metallization_pastes_downloads/SOL9641B_Series.pdf (access at 03.09.2018)

[8] D. Rudolph, A. Adrian, J. Lossen, P. Ferrada, C. Portillo, V. Campo, J. Correa, R. Sierpe and M.J. Kogan, Influence of the paste volume on the contact formation in fine line metallization, 7th World Conference on Photovoltaic Energy Conversion, Hawaii, USA, will be publish in the IEEE Journal of Photovoltaic, 2018.

[9] M. Hörteis, Fine-line printed contacts on crystalline silicon solar cells, Dissertation, Konstanz: University of Konstanz, 2009. ISBN: 978-3-86853-312-5

[10] Y. Zhang, L. Zhang, L. Jiang, L. Song, C. Guo, V. Dua, H. Yang, E. Kim and C. Chen, KNOTLESS SCREEN PRINTING FOR CRYSTALLINE SILICON SOLAR CELLS, presented at the 7th Workshop on Metallization and Interconnection for crystalline silicon solar cells, Konstanz, October 23, 2017.

PROGRESS ON BIFACIAL NI/AG PLATED NPERT CELLS FOR MODULE FABRICATION WITH SWCT

L. Tous[1], R. Russell[1], Y. Yao[2], B. Bonnet-Eymard[3], S. Jambaldinni[1], A. van der Heide[1], G. Doumen[1], F. Duerinckx[1], E. Voroshazi[1], J. Szlufcik[1]
[1]imec, Kapeldreef 75, 3001 Leuven, Belgium
[2]now with Sunyield Technology Co., LTD, 1288 Kanghe Road, Bulding 3, Level 3, Xiuzhou District,Jiaxing, China
[3]MEYER BURGER (SWITZERLAND) LTD, Schorenstrasse 39, 3645 Thun, Switzerland

ABSTRACT: In this work, we report on progress made when using busbarless (0BB) bifacial Ni/Ag plated n-type Czochralski-grown (Cz) Passivated Emitter, Rear Totally diffused (nPERT) cells and Smart Wire Contacting Technology (SWCT) for interconnection. We show that Ni/Ag glass-glass (GG) SWCT laminates with an adapted laser pattern display a limited average power loss of 1.7±0.0 $\%_{rel.}$ after 600 thermal cycles (TC) between -40 °C to +85 °C. Using this adapted laser pattern, we made a 60-cell GG SWCT module from a new batch of 0BB Ni/Ag nPERT cells using automated stringing equipment at Meyer Burger facilities. For this module, we measured a maximum power of 296.9 W_p under standard test conditions with a good cell-to-module ratio of 1.01. In addition, we also measured $BiFi_{100}$ and $BiFi_{200}$ values of 327.6 W_p and 356.7 W_p respectively. These results show the high potential of SWCT GG modules with 0BB Ni/Ag nPERT cells.
Keywords: n-type, PERT, bifacial, plating, SWCT

1 INTRODUCTION

This work is focused on the development of high-efficiency busbarless bifacial Ni/Ag plated n-type Passivated Emitter, Rear Totally diffused (nPERT) cells with reliable Smart Wire Contacting Technology (SWCT) as interconnection technology. High energy conversion efficiency is becoming increasingly important to minimize balance of system costs and achieve low levelized cost of electricity (LCOE). This is particularly relevant at utility-scale where higher-efficiency means fewer modules are required and consequently less land, racking, cabling, labor, and so forth [1]. Bifacial modules which can capture light from both sides offer a cost effective way to increase energy yield and hence further reduce LCOE [2]. Busbarless Ni/Ag plated nPERT cells (Fig. 1a), as being developed at imec, are interesting as: (i) cell metallization costs, which represent a large share of total cell processing costs, can be drastically reduced (we estimated a ~40% reduction versus a best case of 60 mg Ag/cell for bifacial screen printed nPERT), and (ii) extremely high cell bifaciality can be achieved (we have reported bifaciality values >97%) [3]. SWCT consists of several round Cu wires coated with a low-temperature solder alloy that are embedded in polymeric foil (Fig. 1b) that melts at low temperature enabling the formation of series-connected cell strings. The low-temperature solder alloy itself melts during module lamination, enabling the formation of solder joints between the round Cu wires and the cell metallization.

SWCT presents several benefits. First, part of the incident light hitting the round Cu wires is reflected back onto the cells increasing carrier generation [4]. Second, the high number of wires used (typically 18-24 for full size cells) relaxes cell metallization requirements (narrower and less conductive fingers can be used) [5]. Third, both stringing and lamination processes are low temperature and low stress making it particularly suitable for interconnection of thin (and hence cheaper) cells [6]. Fourth, no solder pads are required to achieve reliable results unlike with high-temperature soldering of round wires [7]. This last point, compatibility with busbarless cells, is particularly useful to: (i) reduce Ag/cell, (ii) reduce shading losses, (iii) reduce laser damage & process time in our nPERT cells, (iv) relax alignment requirements during stringing, and (v) achieve a uniform and aesthetically pleasing appearance.

Previously [3], we demonstrated that single cell Ni/Ag nPERT glass-glass (GG) SWCT laminates met the IEC61215 [8] pass criteria for damp heat (DH) (<5$\%_{rel.}$ power (P_{max}) loss after 1000 hrs. at 85°C, 85% relative humidity) and thermal cycling (TC) (<5$\%_{rel.}$ loss in P_{max} after 200 cycles between -40 °C to +85 °C).

In this paper, we repeat TC testing, this time up to 600 cycles (3x IEC61215 requirement) for three types of 1-cell GG SWCT laminates. Then, we fabricate two types of 60-cell GG SWCT modules and discuss I-V results. Finally, we give a brief summary and outlook for this work.

a) b)

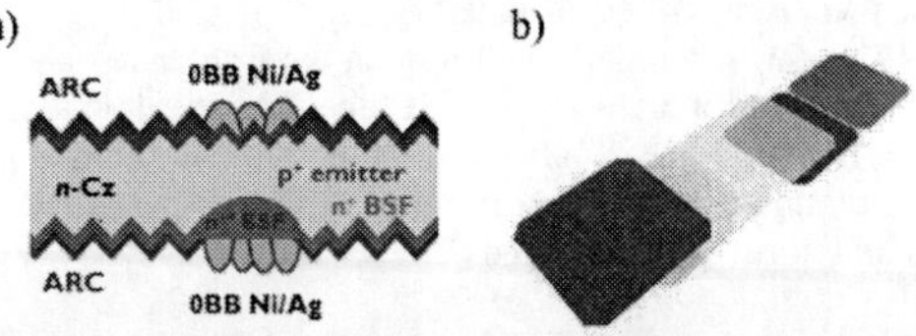

Figure 1: (a) Cross-section (not to scale) of busbarless (0BB) bifacial Ni/Ag nPERT cell [3]. (b) Illustration of stringing process using 0BB cells and SWCT [6].

2 EXPERIMENTS

2.1 Bifacial NiAg nPERT cell fabrication

The process sequence for busbarless (0BB) bifacial Ni/Ag plated nPERT cells in this work is shown in Fig. 2. First, a batch type tool was used for saw-damage-removal (SDR), random pyramid texturing, and wafer cleaning. This was followed by p^+ emitter formation in a BBr_3 tube furnace, removal of boron-silicate-glass (BSG) in diluted hydrofluoric acid (HF), and emitter isolation by means of single-side acidic etching. Then, a diffusion mask was deposited by plasma enhanced chemical vapor deposition (PECVD) onto the p^+ emitter side. Afterwards, the n^+ back-surface-field (BSF) was formed by means of $POCl_3$ diffusion. This was followed by removal, in diluted HF, of both phosphorous-silicate-glass (PSG) and diffusion mask and by a subsequent dry oxidation (SiO_2 formation). PECVD SiN_x was deposited on the BSF side resulting in a SiO_2/SiN_x n^+ passivation stack. Then, the p^+ passivation sequence was performed which consisted of: (i) SiO_2 removal on the p^+ emitter side in diluted HF, (ii) atomic-layer-deposition (ALD) of aluminum oxide (Al_2O_3), and (iii) PECVD SiO_x/SiN_x capping.

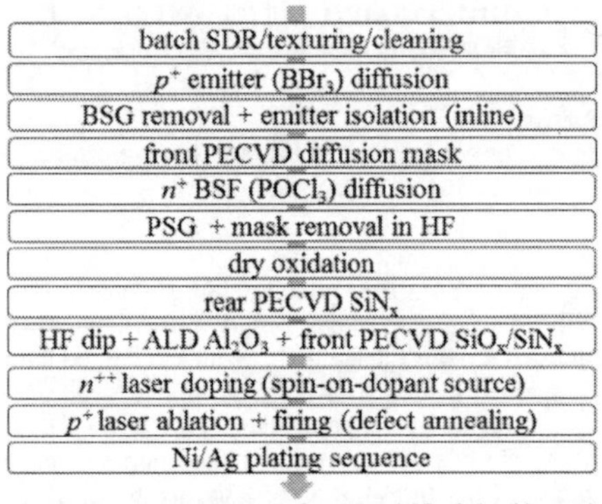

Figure 2: Process sequence for 0BB bifacial Ni/Ag plated nPERT cells in this work.

Laser patterning was done on both sides prior to the Ni/Ag plating sequence. At the rear side, a 45° finger pattern (to reduce amount of bussing ribbons needed in full-size modules [5]) was created by laser doping from a phosphorus containing spin-on-dopant source to simultaneously open dielectrics and form heavy n^{++} doping thereby reducing contact recombination losses [9]. At the front side, a vertical finger pattern was created by laser ablation. Importantly, surface roughness, which has been shown to be beneficial for finger adhesion [10], is much larger for the laser ablated p^+ surface (see Fig. 3a) than for the n^{++} laser doped surface (see Fig. 3b). Finally, a defect annealing was performed in a conventional fast-firing furnace prior to the self-aligned Ni/Ag plating sequence described in details in Ref. [3]. The resulting Ni/Ag finger widths are ~20 µm at the front side (see Fig. 3c) and ~25 um at the rear side (see Fig. 3d). In addition, the finger height is only ~4 µm on both sides (limited by Ni plating time for cost reasons) which is much thinner than typical screen-printed (SP) Ag fingers [11].

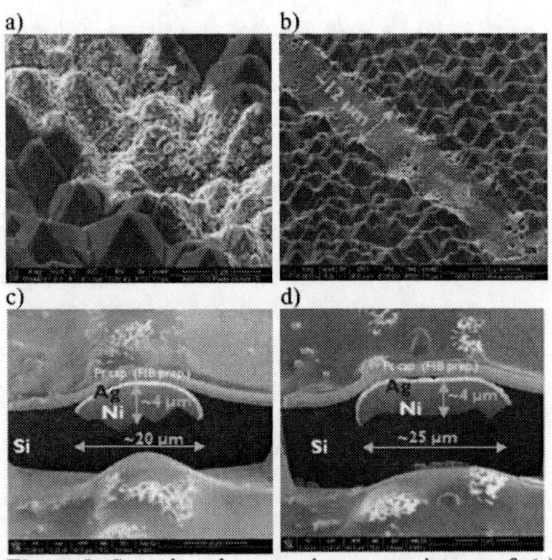

Figure 3: Scanning electron microscopy pictures of: (a) front side p^+ laser ablated surface, (b) rear side n^{++} laser doped surface, (c) front side Ni/Ag finger cross-section, (d) rear side Ni/Ag finger cross-section.

2.2 1-cell GG SWCT laminates fabrication

As shown in Table I, several 1-cell GG SWCT laminates were fabricated using nPERT cells with different metallization (plated Ni/Ag or SP-Ag) and origin (imec or industry). The sole reason for choosing 1-cell laminates was due to size limitations (illuminated area ~20x20 cm²) in the I-V setup installed at imec. In addition to the nPERT cells, materials used for laminate fabrication included: 2*2 mm glass (no ARC), thermoplastic polyolefin (TPO) encapsulant, butyl edge sealing, SWCT foils supplied by Meyer Burger (18*300 µm Cu wires, InSn solder coating), and bussing Cu ribbons (5*0.2 mm², Pb/Ag solder coating) for 4-wires I-V measurements. Lamination was done at imec using a single membrane laminator and optimized recipes.

Table I: Test matrix for 1-cell GG SWCT laminates with nPERT cells featuring different metallization and origin.

metallization	origin	number of 1-cell laminates
3BB SP-Ag	industry	3
0BB Ni/Ag laser v1	imec	5
0BB Ni/Ag laser v2	imec	11

Two issues were identified when making SWCT GG laminates with the original 0BB Ni/Ag nPERT laser v1 cells (Fig. 4a). Thus, we also fabricated SWCT GG laminates with 0BB Ni/Ag nPERT laser v2 cells featuring modifications in both the front (see Fig. 4b) and rear side (not shown but modifications were similar to the ones made on the front side) laser patterns.

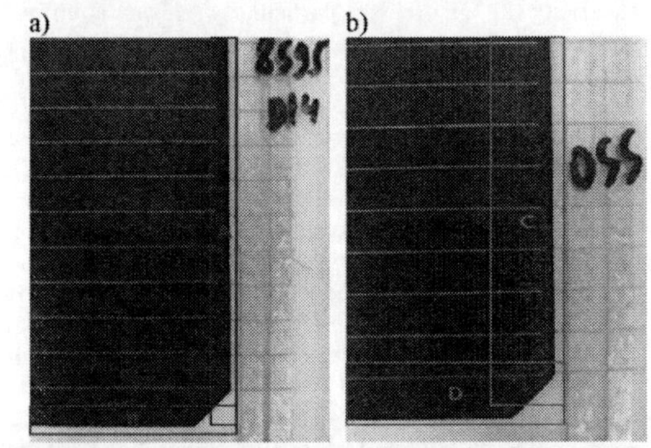

Figure 4: (a) Close-up image of front-side of GG SWCT laminate with 0BB Ni/Ag nPERT laser v1 cell. Due to absence of busbars, vertical fingers in region A are not electrically connected to horizontal Cu wires. In region B, distance to the first horizontal Cu wire is longer than in other regions of the cells, causing higher series resistance (r_s) losses. (b) Close-up image of front-side of GG SWCT laminate with 0BB Ni/Ag nPERT laser v2 cell featuring extra horizontal fingers in region C and extra vertical fingers in region D to reduce r_s losses.

2.3 60-cell GG SWCT modules fabrication

Two different 60-cell GG SWCT modules were fabricated as shown in Table II. Module fabrication was done at Meyer Burger facilities using automated stringing equipment and an industrial single membrane laminator.

Table II: Overview of 60-cell GG SWCT modules made.

Module name	type of 0BB nPERT cells used	Cu wires number/diameter in µm, solder coating
A	imec Ni/Ag	24x300, In-free
B	commercial SP-Ag	18x250, In-Free

Module lay-up was done by hand and materials supplied by Meyer Burger included: 2*2.5 mm tempered glass (front glass with ARC), TPO encapsulant, butyl edge sealing, SWCT foils (different wires were used for both modules as shown in Table II), bussing Cu ribbons, and decentralized (corner and middle mounting) edge junction boxes. An Aluminum frame was mounted on imec Ni/Ag module to avoid damage during transport between imec-Leuven (Belgium) and imec-Genk (Belgium) facilities while the commercial SP-Ag module was kept frameless and shipped directly to imec-Genk.

2.2 Thermal cycling (TC) testing and characterization.

1-cell GG SWCT laminates were subjected to thermal cycles (CTS chamber, -40 °C to +85 °C) without electrical load and characterized at regular intervals. Characterization included light I-V measurements (Wacom, continuous-type, class AAA solar simulator) using a Gridtouch™ (GT) contacting setup and series resistance (r_s) mapping using biased-pholuminescence (BTimaging).

2.3 Light I-V characterization of 60-cell SWCT modules.

Light I-V measurements of the 60-cell GG SWCT modules was performed using a Wavelabs Sinus 2100-p, pulse-type, class A+A+A+ solar simulator. Of note, the rear and surrounding surfaces in this tool are black to ensure that back side reflectivity is kept well below 3 W/m^2 as required accurate measurement of bifacial modules. Several measurements were performed consecutively to determine front STC, rear STC, $BiFi_{100}$, $BiFi_{200}$ values (section 3.2).

3 RESULTS AND ANALYSIS

3.1 TC results of 1-cell GG SWCT laminates

TC results for the various types of 1-cell GG SWCT laminates are shown in Fig. 5. After 600 cycles (3x IEC61215 requirement), 3BB SP-Ag laminates resulted in an average maximum power (P_{max}) loss of only 0.7±0.0 $\%_{rel}$ while 0BB Ni/Ag laser v1 and 0BB Ni/Ag laser v2 laminates resulted in average P_{max} losses of 3.0±0.0$\%_{rel}$ and 1.7±0.0$\%_{rel}$ respectively. The relative change in P_{max} remains sufficiently low for all laminates to easily pass the test (loss in P_{max} ≤5$\%_{rel}$). Unlike 3BB SP-Ag laminates, 0BB Ni/Ag laminates suffered from a limited progressive loss in FF as the number of TC increased (see Fig 5.b). The 0BB Ni/Ag laser v2 laminates degraded slower than the 0BB Ni/Ag laser v1 laminates which could indicate that degradation is related to effects at the perimeter of the cells. I-V data analysis (not shown) indicates that the FF degradation was solely the result of a progressive increase in series resistance. This discards any issues related to in-diffusion of Ni (shunt, junction damage) during TC testing. As shown in Fig. 5c and Fig. 5d, both short-circuit densities (I_{sc}) and open-circuit voltages V_{oc} remained relatively stable (<1$\%_{rel}$) during TC testing.

Series resistance (r_s) maps were taken at regular TC intervals for the three different types of laminates. Representative results for each type are shown in Fig. 6. At 0TC, it can be seen that 0BB Ni/Ag laser v1 laminates suffer from high r_s at the cell perimeter (Fig. 6) and that adopting a modified laser pattern (laser v2, see Fig. 4b) almost eliminates this issue. However, even with this laser v2 pattern, high r_s regions can be observed with increasing number of thermal cycles at the cell left/right edges. These regions correspond to the wires ends where stress during TC testing is most important. In addition, r_s degradation at the cell left/right edges appears to follow the 45° laser doping pattern of the rear side. Apart from a few finger breaks (also visible on SP-Ag laminates), the center of the cells do not show any signs of r_s degradation with thermal cycling. Further work is needed to understand why degradation is more pronounced at the cell left/right edges and along the 45° laser doping pattern of the rear side. The hypothesis for this is that finger adhesion is weaker at these locations due to non-uniformities issues caused by either PECVD deposition, laser doping process itself (spin-on-dopant + lasering), and/or Ni/Ag plating.

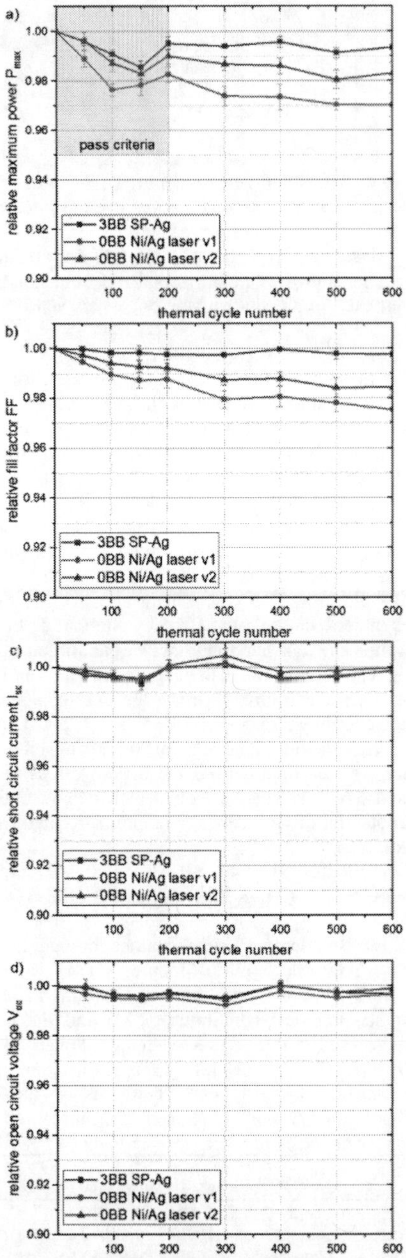

Figure 5: Thermal cycling testing results for three different types of 1-cell glass-glass SWCT laminates.

Figure 6: Series resistance mapping (light-to-dark scale: 0-2 Ω.cm^2) taken at various thermal cycling intervals for representative laminates of each type: (a) SP-Ag, (b) Ni/Ag laser v1, (c) Ni/Ag laser v2. In all images, interconnection wires are horizontal and front side fingers are vertical. For Ni/Ag, rear side fingers are at 45° angle.

3.2 I-V results of 60-cell GG SWCT modules

We made two 60-cell GG SWCT modules (for details see Table II). For module A with imec Ni/Ag cells, we measured a P_{max}=296.9 W_p under standard test conditions with a relatively good cell-to-module (CTM) ratio of 1.01 (see Table III). To calculate the CTM ratio we used the cell I-V parameters corrected for Gridtouch™ shading (see Ref. [12]) using the known wire number (24) and diameter (300 µm) in the module. Interestingly, the CTM for I_{sc} is 1.02 despite the fact that we suffer from: (i) absorption in the UV-blocking TPO encapsulant and (ii) transmission losses in between the cells. Thus, by eliminating this UV-absorption and transmission losses we can further improve P_{max} at STC. Finally, we also measured $BiFi_{100}$ and $BiFi_{200}$ values from the front side using the over-compensation method currently proposed in IEC 60904-1-2 [13]. Thanks to the high module bifaciality of 88.8% (despite the Al frame), we measured $BiFi_{100}$ and $BiFi_{200}$ values of 327.6 W_p and 356.7 W_p respectively which demonstrates the high potential of GG SWCT in combination with Ni/Ag nPERT cells.

For module B with SP-Ag cells, we measured P_{max}=287.2 W_p, $BiFi_{100}$=312.7 W_p, and $BiFi_{200}$=334.8 W_p.

Table III: Summary of results obtained for module A with imec Ni/Ag cells. Cell-to-Module (CTM) values are calculated from the ratio between module parameters under standard test conditions (STC) and the sum of individual cell powers corrected for Gridtouch (GTcor.) shading (for details, see ref. [12]). $BiFi_{100}$ and $BiFi_{200}$ measurements (meas.) were done using the over-compensation method [13] and the measured P_{max} bifaciality of 88.8% at STC.

meas.	I-V system	Intensity [W/m^2]	I_{sc} [A]	V_{oc} [V]	FF [%]	P_{max} [W_p]
∑60cells	Wacom+ GT	1000	9.8	40.4	77.9	308.7
∑60cells	Wacom+GTcor.	1000	9.4	40.4	77.9	294.4
STC	Wavelabs	1000	9.6	40.6	76.4	296.9
CTM	-	-	1.02	1.00	0.98	1.01
$BiFi_{100}$	Wavelabs	1088	10.6	40.8	76.1	327.6
$BiFi_{200}$	Wavelabs	1176	11.5	40.9	75.8	356.7

Table IV: Summary of results obtained for module B with commercial SP-Ag cells. $BiFi_{100}$ and $BiFi_{200}$ measurements (meas.) were done using the over-compensation method [13] and the measured P_{max} bifaciality of 80.0% at STC.

meas.	I-V system	Intensity [W/m^2]	I_{sc} [A]	V_{oc} [V]	FF [%]	P_{max} [W]
STC	Wavelabs	1000	9.4	39.4	77.6	287.2
$BiFi_{100}$	Wavelabs	1080	10.2	39.5	77.2	312.7
$BiFi_{200}$	Wavelabs	1160	11.0	39.6	76.8	334.8

4 SUMMARY AND OUTLOOK

In the first part of this work we demonstrated that 0BB Ni/Ag plated nPERT cells can suffer, after SWCT interconnection, from large r_s losses at cell perimeter if a non-optimized laser pattern is used. We showed that 0BB Ni/Ag GG SWCT laminates with an adapted laser pattern resulted in an average power loss P_{max} of only 1.7±0.0 $\%_{rel.}$ after 600TC (3x IEC61215 test duration). Based on r_s mapping, we found that degradation during TC testing mostly occurred at the cell left/right edges and that it appears to follow the 45° laser doping pattern of the rear side. This could possibly be due to non-uniformities issues during cell fabrication. Further work is needed to identify these issues before solving them.

In the second part of this work, we reported results obtained with two 60-cell GG SWCT modules: module A with imec Ni/Ag nPERT cells and module B with commercial SP-Ag nPERT cells. For module A, we measured P_{max} =296.9 W_p with a good cell-to-module ratio of 1.01, $BiFi_{100}$=327.6 W_p and $BiFi_{200}$=356.7 W_p. For module B, we measured P_{max}=287.2 W_p, $BiFi_{100}$=312.7 W_p, and $BiFi_{200}$=334.8 W_p. These results demonstrate the high potential of SWCT GG modules with imec Ni/Ag nPERT cells. In addition, further improvements should be possible at cell level, by reducing contact recombination losses [14]; and at module level by reducing both absorption and transmission losses while adopting half-cell technology [15].

5 AKNOWLEDGEMENTS

The authors gratefully acknowledge the financial support of the partners of imec's Industrial Affiliation Program for Photovoltaics.

6 REFERENCES

[1] M. Campbell, et al. Proceedings of the 34th IEEE Photovoltaic Specialists Conference, (2009): 421-426.
[2] R. Kopecek, et al. Photovoltaics International 26 (2014): 32-45.
[3] R. Russell, et al. Proceedings of the 34th European Photovoltaic Solar Energy Conference and Exhibition (2017): 212-217.
[4] K.R. McIntosh, et al. Proceedings of the 33rd European Photovoltaic Solar Energy Conference and Exhibition, (2016): 15-19.
[5] P. Papet, et al. Presentation at the 5th Workshop on Metallization of Crystalline Silicon Solar Cells. 2014.
[6] S. Hänni, et al. Presentation at SNEC. 2018.
[7] L.C. Rendler et al. Proceedings of the 32nd European Photovoltaic Solar Energy Conference and Exhibition, (2015): 84-88.
[8] IEC 61215: 2005 Edition 2.0.

[9] A. Urueña, et al. Progress in Photovoltaics: Research and Applications 24 (2016): 1149-1156.
[10] S. Kluska, et al. Energy Procedia 77 (2015): 733 – 743.
[11] P. Altermatt, et al. Presentation at the 6th Workshop on Metallization and Interconnection of Crystalline Silicon Solar Cells. 2016.
[12] P. Papet, et al. Presentation at the 7th Workshop on Metallization and Interconnection of Crystalline Silicon Solar Cells. 2017.
[13] V. Fakhfouri. Presentation at the 3rd Bifacial Photovoltaics Workshop. 2016.
[14] M. Recaman, et al. Proceedings of the 35th European Photovoltaic Solar Energy Conference and Exhibition, 2018.
[15] I. Haedrich, et al. Solar Energy Materials and Solar Cells. 131 (2014): 14-23.

SILICON HETEROJUNCTION SOLAR CELLS WITH OPEN-CIRCUIT-VOLTAGE ABOVE 750MV

A.Danel, S.Harrison, F. Gérenton, A. Moustafa, R.Varache, J.Veirman, C. Roux

CEA, LITEN, Department of Solar Technologies, F-73375 Le Bourget du Lac, France

ABSTRACT: In this paper we present silicon heterojunction (SHJ) cells with V_{oc} beyond 750mV fabricated on thin wafers on a semi-industrial pilot line. Once critical parameters for SHJ cell are insured, such as very high quality n-Cz material (2 Ohm.cm – 8ms typically), good wet cleaning along with right queue-time conditions, good PECVD a-Si:H passivation (S_{eff}<2 cm/s typically) and limited damage induced during ITO deposition and screen printing, the recipe development is not a first order work. The most important point is to insure cell processing below 90µm without extra defectivity. This was done in this work thanks to specific know how and practical adjustment of automation, handling, logistics of thin cells (40 to 90µm free standing 156x156 mm² wafers). The best V_{oc} of this work is 752mV and 751mV independently measured by ISFH-CalTeC on total area 244.3 cm² (M2 size), bifacial, rear emitter, 4 busbar cells. Accordingly, <1 g Si/Wp was acheived with a record at 0.52 g Si/Wp for a 44µm cell. Very promising implied V_{oc} up to 765mV on cell precursors post PECVD and losses analysis show possible cell V_{oc} >755mV and a <0.5 g Si/Wp figure of merite short term.

Keywords: silicon solar cells, heterojunction, industrial pilot line

1 INTRODUCTION

The amorphous/crystalline silicon heterojunction (a-Si:H/c-Si, SHJ) is one of the most attractive solar cell architectures combining high performance and industrial compatibility. Simple, low temperature, high throughput and cost effective processes, outstanding surface passivation, bifaciality, thin wafers (<100µm) and possible combination with rear-contact cell design are the SHJ cell main advantages, contributing to its competitiveness with regard to other architectures [1-3]. With SHJ in its interdigitated back contact version, Kaneka recently improved to 26.7% the world record of conversion efficiency for c-Si solar cells [4], and in standard double-side contacted configuration, Panasonic demonstrated a world record open-circuit-voltage (V_{oc}) of 750mV on a 97µm large area cell (100cm²) [1], and Arizona State University a 754.2mV on a 50µm 4cm² cell [5].

With the increased interest in thin wafer use for cost reduction, new module opportunities [6], higher efficiency potential [7] and new substrates [8], a focus of CEA SHJ activities performed on a semi-industrial pilot line [9] was put on thin wafers [10,11]. This study describes processes developed to optimize cell V_{oc}, presents our current record of 752mV achieved on 60µm thin 156x156mm² cells and discusses deviation between current cell V_{oc} and the high voltage potential of SHJ.

Increasing c-Si -based PV system performances by the use of ultra-thin substrates was calculated by A. Richter et al. with an optimum at 29% around 60µm for conventional 1 to 10 Ohm.cm n-type wafers [7]. Using similar approach, maximum implied V_{oc} (iV_{oc}) is plotted in figure 1 with band gap narrowing [12], intrinsic recombination with updated Auger contribution [13] and the ideal Lambertian case for light absorption [14]. This figure shows the V_{oc} potential >750mV of Si -based cells thinner than 100µm.

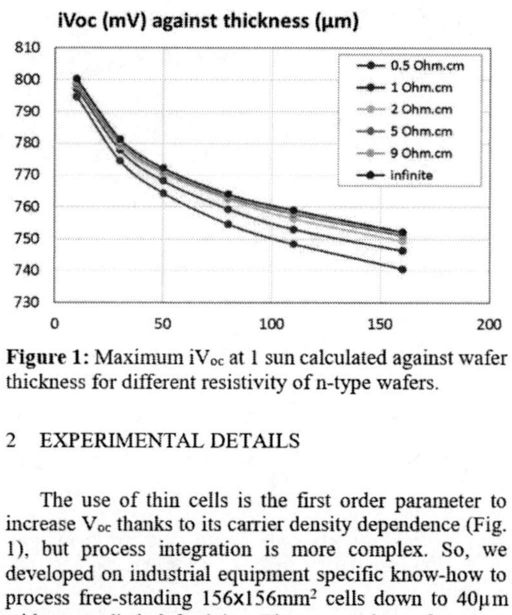

Figure 1: Maximum iV_{oc} at 1 sun calculated against wafer thickness for different resistivity of n-type wafers.

2 EXPERIMENTAL DETAILS

The use of thin cells is the first order parameter to increase V_{oc} thanks to its carrier density dependence (Fig. 1), but process integration is more complex. So, we developed on industrial equipment specific know-how to process free-standing 156x156mm² cells down to 40µm with controlled defectivity. Then, a review of process development specifically for thin cells was performed from substrate specification to screen printing step in order to optimize performances, especially Voc.

SHJ solar cells are fabricated with industrial processes on n-type Cz commercial wafers, M2 size, with a rear-emitter and 4 busbar bifacial design. All IV measurements are total cell area data under AM1.5G standard conditions using testers calibrated against sister cells independently measured by the FhG ISE CalLab. Implied parameters were measured by photoconductance decay on a WCT-120 Sinton instrument using the following setup: transient mode, 28°C, center point of cell precursors, iV_{oc} at 1 sun, precise thickness and doping taken into account from independent measurements cell to cell, band gap narrowing considered, optical constant of 0.97 to 0.93 estimated from cell short –circuit current density (J_{sc}).

The experiments against cell thickness were performed thinning down 180µm as-cut wafers by chemical etch prior process integration. Down to 80µm, cells were processed in full automatic mode on the pilot line with minor adaptations to insure low breakage rate. Below 80µm, cells were processed in semi-automatic mode, using some manual handling to by-pass the most critical points [10].

3 RESULTS AND DISCUSSION

3.1 Process development to enhance V_{oc}

Si wafers: In addition to the thickness variation, the electrical properties were investigated. As shown in Fig. 1, high resistivity n-type substrates show slightly higher potential for ideal cells. But in non-ideal practical cells with series resistance and non-null effective surface recombination velocity (S_{eff}), low resistivity material might behave well [15]. So, we selected 3 types of high quality material in this work: 1 Ohm.cm – 5ms minority carrier lifetime (Si bulk lifetime at $5.10^{14}/cm^3$ carrier density as measured by the wafer supplier) ; 2 Ohm.cm – 8ms ; 3 Ohm.cm – 12ms.

Wet processing: we studied the benefit of additional rounding applied after texturization and before final clean and HF last. As shown in figure 2, a few mV can be gained, but Isc decreases due to surface reflectance degradation. A few minutes long process was so selected, as highlighted in the chart.

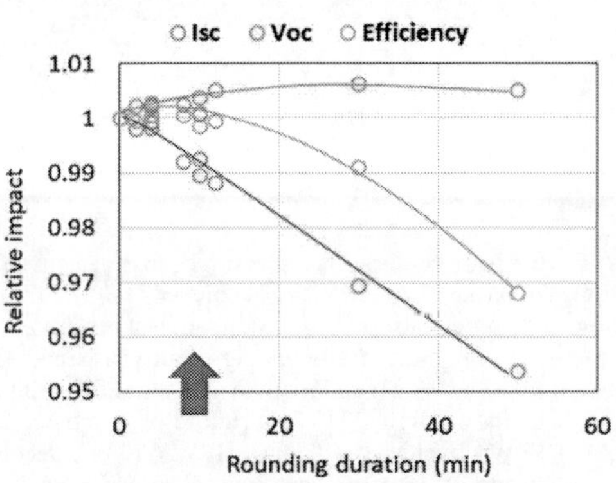

Figure 2: Relative impact of additional rounding on IV cell parameters.

PECVD: A major contribution for high V_{oc} was the optimization of plasma enhanced chemical vapor deposition (PECVD) of a-Si:H buffer layers, particularly for the emitter side holding boron-doped layers limiting the surface passivation. Figure 3 summarizes the optimization of recipes on 90µm cells. Recipe B is the improved reference process with 21.36% efficiency and 735mV V_{oc} (average). Recipe C with S_{eff}<1cm/s improved the Voc up to 741mV (average) – 746.4mV (best) and the overall efficiency to 21.69% (average) thanks to a specific tool seasoning and improved pre- and post- treatments before and after silane deposition steps. But this progress was made with a significant equipment throughput reduction.

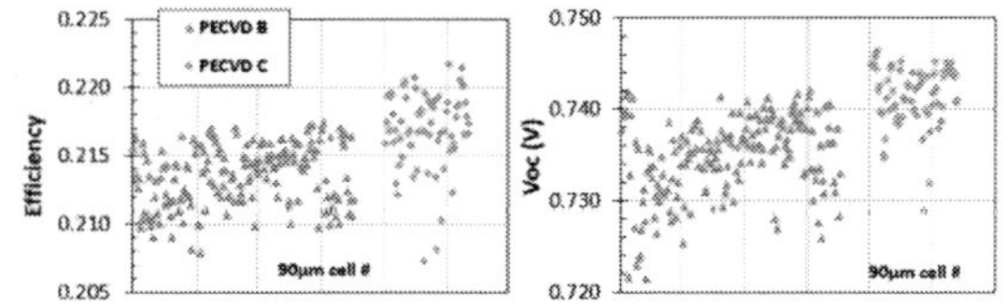

Figure 3: PECVD recipe improvement to enhance Voc.

PVD and screen printing: Indium tin oxide (ITO) deposition by PVD and Ag paste curing should degrade as less as possible the a-Si:H layers already deposited. This point cannot really be solved by recipe optimization since it mostly depends on hardware. Using a low energy ITO co-deposition equipment, the sputtering damage is limited to a few mV, as measured by iVoc difference of cell precursors post PVD compared to post PECVD. And a very low temperature high conductivity Ag paste was used in this work to avoid >200°C curing.

3.2 Thin cell fabrication and electrical results

It appeared that below 100µm the surface passivation can be significantly degraded. The loss of good surface passivation on very thin cells is firstly illustrated in figure 4 with transient effective lifetime measurement performed post PECVD on cell precursors. The fit of data using the well-known expression $\frac{1}{\tau_{eff.}} = \frac{1}{\tau_{Si-bulk}} + \frac{2S}{W}$, with W the cell thickness, gives a Si bulk lifetime of 8.7ms in good agreement with wafer supplier data and an outstanding S_{eff} of 0.52cm/s down to 110µm, then a significant S_{eff} increase is seen for thinner cells, here fabricated in production mode without automation adjustment.

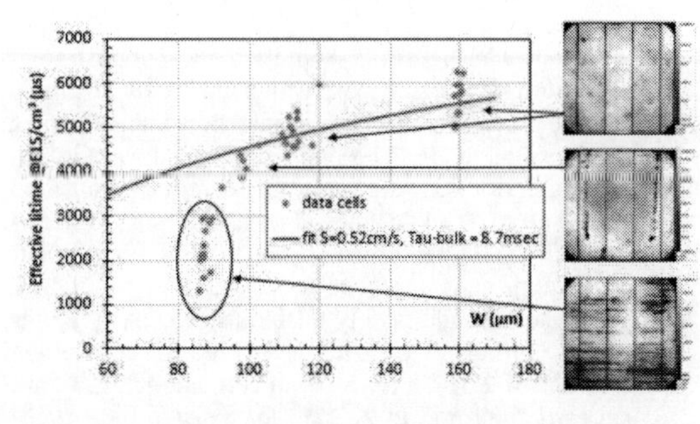

Figure 4: Effective lifetime at $10^{15}/cm^3$ apparent carrier density (Δp) against thickness of SHJ cells produced on CEA-INES pilot line. Highlight of localized defectivity occurring more frequently for thin cells as revealed by photoluminescence (PL) pictures.

Following this first statement, iterative improvement of automation and handling and right logistic of thin wafers was done on the CEA SHJ pilot line. Results are illustrated in figure 5 with normalized S_{eff} and PL Cells down to 80µm were manufactured in full automatic mode without significant loss of surface passivation related to defectivity. Down to 60µm, S_{eff} remains acceptable despite some degradation. Bellow this, the good surface passivation insured by a-Si:H layers is degraded and both V_{oc} and fill factor (FF) drop.

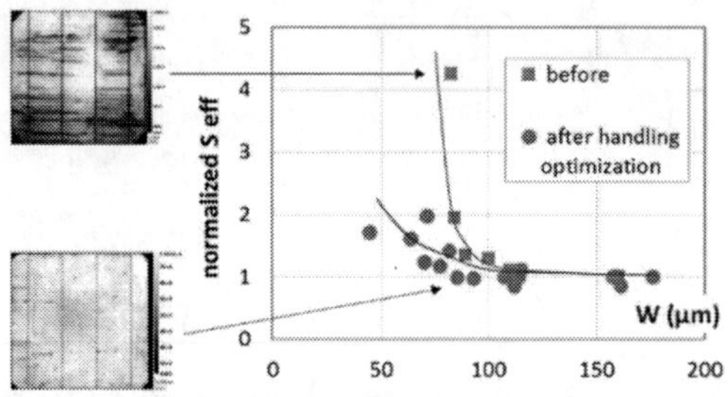

Figure 5: Normalized S_{eff} at $10^{15}/cm^3$ and PL pictures against cell thickness. Data 'before' issued from Fig. 1.

The best and more pro-V_{oc} recipes of each integration step were applied to several batches of cells of different thickness. Very good iV_{oc} was obtained post PECVD, without deviation from the iV_{oc} max trend, as shown in figure 6.

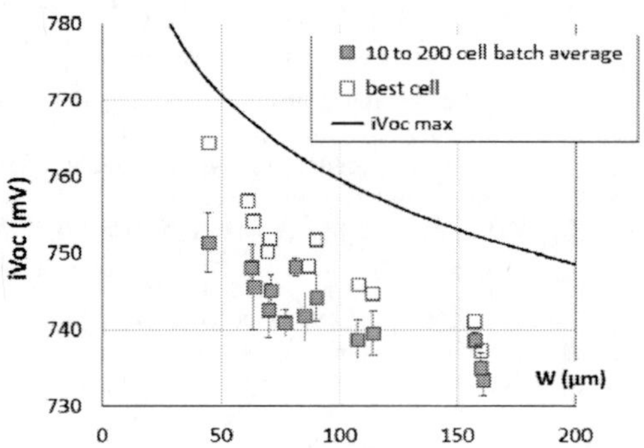

Figure 6: iV_{oc} at 1 sun post PECVD.

Results shown in figures 7, 8 and table 1 demonstrate a good control of surface passivation down to 80µm in automatic process mode and correct down to 60µm in semi-automatic mode. The V_{oc} difference to a Silvaco simulation performed with surface passivation parameters independent from cell thickness does not show a too strong increase from 160µm down to 60µm. But for ultimate standing alone cells of 40µm, the decrease of V_{oc} and FF is very important and mainly related to defectivity issues occurring during PVD. This gave record conversion efficiencies of 22.17% for a 89µm cell and 20.57% for a 61µm cell, compared to 22.32% for a sister 160µm cell (total area IV data). Corresponding record Voc are 750.3mV at 85µm and **752.1mV** at 61µm, improving the world record of total area cell, as far as we know from published data.

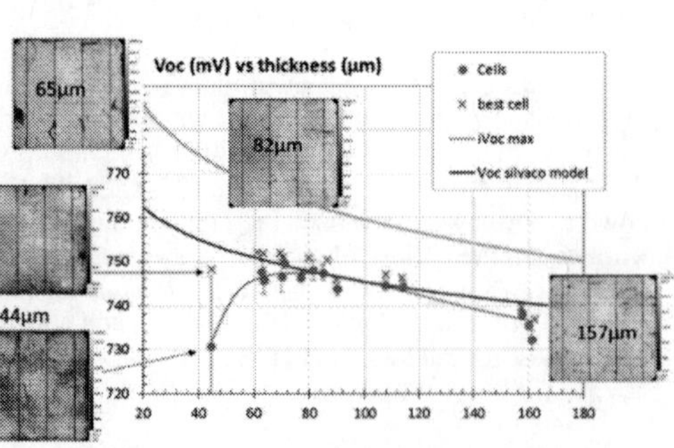

Figure 7: Cell V_{oc}, modeling and PL imaging.

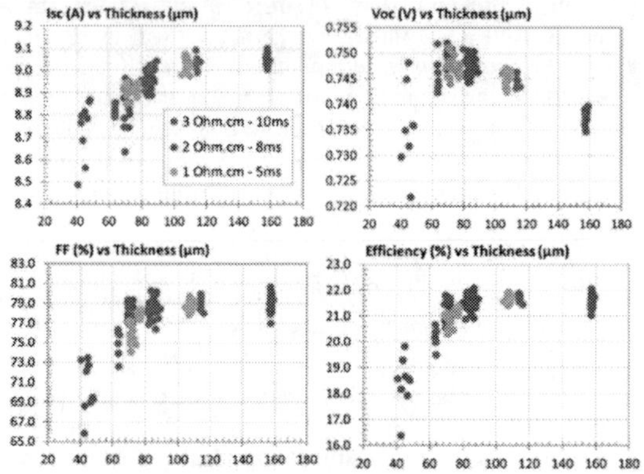

Figure 8: IV parameters of champion batches of cells (4 busbar, bifacial, 244.3 cm²) against thickness.

Table I: Summary of current CEA SHJ M2 cell best efficiency when applying pro-Voc integration. 63µm 751.1mV cell was also independently measured by ISFH-CalTeC.

W	Efficiency (%)	FF (%)	Jsc (mA/cm2)	Voc (mV)
40 µm	18.58	73.29	34.74	729.8
44 µm	19.84	73.44	36.09	748.2
61 µm	20.57	77.01	35.51	**752.1**
63 µm	20.33	76.02	35.61	751.1
CalTeC	20.36	75.18	36.06	751.1
86 µm	**22.09**	80.20	36.70	**750.3**
89 µm	**22.17**	81.21	36.58	746.4
160 µm	22.32	81.75	36.84	741.2

4 MODULES

With a rear-emitter, bifacial and 4 busbar design, cell efficiency in the 21 to 22% range achieved at 80 - 100µm allowed us to demonstrate under standard test conditions a breaking 1.02 gram of silicon per Watt-peak ratio (g Si/Wp) for a 313Wp 60-cell module with a standard glass-back sheet design [10, 11]. This should be compared to 1.66 g Si/Wp achieved with our 410Wp 72-cell record module made of 166µm champion cells. These module data are plotted in figure 9 with the g Si/Wp ratio of best cell. This demonstrates a possible outstanding valorization of the Si resource with a **0.52 g Si/Wp** record obtained with our 44µm champion cell.

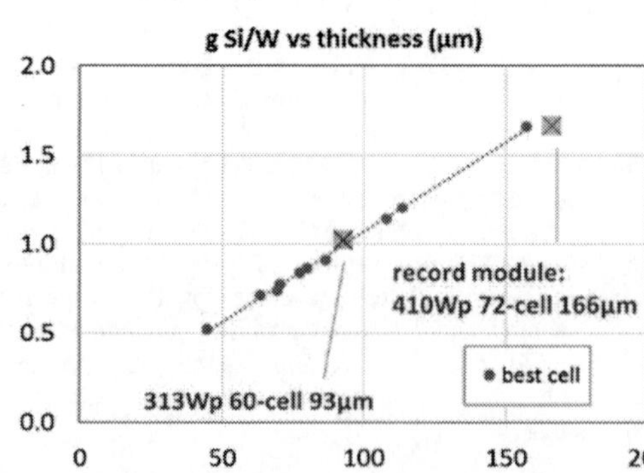

Figure 9: Best cell and module g Si/Wp ratio.

5 CONCLUSIONS AND PERSPECTIVES

We think that industrial process of thin SHJ cell could be performed short term down to 100µm and in the near future below 100µm. SHJ is also compatible with the potentially coming soon new thin substrates such as epitaxial foils.

On overall SHJ is natively a technology well adapted with the cost reduction roadmap based with thin wafers while keeping the highest cell efficiencies compared to other current mainstream technologies.

This work demonstrated outstanding Si resource / power ratio. A mid-term perspective of less than 0.7g Si/Wp in finished system is feasible considering bifacial modules manufactured on 21% 60µm cells. Because in this work process integration introduced surface passivation degradation on very thin cells (<60µm), to overpass the 752mV and 0.52 g Si/Wp record presented here, progresses on equipment parts with specific design for <60µm cells are needed.

ACKNOWLEDGEMENTS

Part of this work has received funding from the European Union's Seventh Program for research, technological development and demonstration under grant agreement No 609788 in the frame of the CHEETAH project, [16]. The authors kindly thank CEA and Meyer Burger teams for the work done throughout the year.

REFERENCES

[1] M. Taguchi et al, "24.7% Record Efficiency HIT Solar Cell on Thin Silicon Wafer", IEEE J. of Photovoltaics, Vol. 4-1 (2014), pp. 96-99.

[2] S. De Wolf et al, "High-efficiency Silicon Heterojunction Solar Cells: A Review", Green, Vol. 2 (2012), pp. 7-24.

[3] K. Yoshikawa et al., "Silicon heterojunction solar cell with interdigitated back contacts for a photoconversion efficiency over 26%", Nature Energy, Vol. 2-5 (2017), pp. 17032.

[4] K. Yamamoto et al., "Record-Breaking Efficiency Back-Contact Heterojunction Crystalline Si Solar Cell and Module, 2BP.1.1 in the 33rd European Photovoltaic Solar Energy Conference, Amsterdam, The Netherlands (Sept. 2017).

[5] T. Monga, "Development of Thin Heterojunction Solar Cells with High Open Circuit Voltage" Thesis at Arizona State University (Aug. 2015), https://repository.asu.edu/attachments/158030/content/Monga_asu_0010N_15243.pdf

[6] A. Augusto et al., "<1g/W solar cells on flexible silicon substrates" in the 27th PVSEC conference, Otsu, Japan (Nov. 2017).

[7] A. Richter et al, "Reassessment of the limiting efficiency for crystalline silicon solar cells", IEEE J. Photovolt., Vol. 3 (2013), pp. 1184–1191.

[8] H. Sivaramakrishnan Radhakrishnan et al., "Heterojunction IBC solar cells on thin (<50µm) epitaxial Si foils produced from kerfless layer transfer process" in: 33rd European Photovoltaic Solar Energy Conference, Amsterdam, The Netherlands (Sept. 2017), pp. 740-744.

[9] A. Danel et al., "Versatile pilot line to support the heterojunction solar cell industrial development: busbar and busbar-less configurations" in: 33rd European Photovoltaic Solar Energy Conference, Amsterdam, The Netherlands (Sept. 2017), pp. 447-450.

[10] S. Harrison et al., "Towards industrialization of heterojunction with thin and ultra-thin wafers" in the 27th PVSEC conference, Otsu, Japan (Nov. 2017).

[11] E. Gerritsen et al., "'Less is more': Ultrathin heterojunction cells offering industrial cost reduction and innovative module applications", PV International, March issue (2018).

[12] A. Schenk, "Finite-temperature full random-phase approximation model of band gap narrowing for silicon device simulation", Journal of Applied Physics, Vol. 84 (1998), pp. 3684.

[13] B.A. Veith-Wolf et al., "Reassessment of intrinsic lifetime limit in n-type crystalline silicon and implication on maximum solar cell efficiency", Solar Energy Materials and Solar Cells, Vol. 186 (2018), pp. 194–199.

[14] T. Tiedje et al., "Limiting efficiency of silicon solar cells", IEEE Transactions on Electron Devices Vol. 31 (1984), pp. 711-716.

[15] M. Bivour et al, "Silicon heterojunction rear emitter solar cells: Less restrictions on the optoelectrical properties of front side TCOs", Solar Energy Materials & Solar Cells, Vol. 122, (2014) pp. 120-129.

[16] J.M. Kroon, "Highlights from the FP7 project on photovoltaics CHEETAH: more power with less material", 7EO.3.5 paper in the Proc. of the 33rd European Photovoltaic Solar Energy Conference, Amsterdam, The Netherlands (2017), pp. 2844-2848.

Implementation of a novel silicon heterojunction IBC process flow using partial etching of doped a-Si:H with efficiencies close to 23%

Hariharsudan Sivaramakrishnan Radhakrishnan[1,*], MD Gius Uddin[1], Menglei Xu[1,2], Jinyoun Cho[1,2], Moustafa Ghannam[3], Ivan Gordon[1], Jozef Szlufcik[1] and Jef Poortmans[1,2,4]

[1]imec vzw, Leuven, Belgium; [2]KU Leuven, Belgium; [3]Kuwait University, Kuwait, [4]University of Hasselt, Belgium.

* Corresponding author: phone +32 16 28 1173, email sivarama@imec.be

Abstract

We present a novel process sequence to simplify the rear-side patterning of the heterojunction IBC cells. In this approach, interdigitated strips of a-Si:H (i/p$^+$) and a-Si:H (i/n$^+$) are achieved by partially etching a blanket a-Si:H (i/p$^+$) stack through a hard mask to remove only the p$^+$ a-Si:H layer and replace it *in situ* with an n$^+$ a-Si:H layer. This eliminates the *ex situ* wet clean after dry etch and also prevents re-exposure of the c-Si surface during rear-side processing. High-quality passivation and contact resistance comparable to reference translates into cell performance that is also similar to that of cells from the reference process. The best cell has an efficiency of 22.9% and V_{OC} close to 730 mV, demonstrating that the developed partial etch process sequence can be successfully implemented to achieve cell performance comparable to reference, but with a simpler process.

1. Introduction

The silicon heterojunction interdigitated back-contact (HJ IBC) solar cell technology have enabled the achievement of the world-record efficiency of 26.7% for Si solar cells [1]. Hydrogenated amorphous silicon (a-Si:H) used as passivating contacts in this technology is key to the high open-circuit voltages (V_{OC}) > 725 mV, while the IBC architecture ensures high short-circuit current densities (J_{SC}) > 41 mA/cm^2.

Since all processes are performed at low temperature (<200°C), the HJ IBC cell technology is also suitable for supported processing of thin (< 70 μm) Si wafers/foils bonded to glass superstrates using bonding agents such as silicones or EVA [2]. This is because, in such samples, all applied cell processes must be compatible with the entire material stack including the glass and the bonding agent. Since the bonding agent is typically a polymeric substance, high temperature processing is not possible, and hence the HJ IBC cell technology is used. A structural comparison between a freestanding and a glass-bonded HJ IBC cell is given in Fig. 1.

Our baseline HJ IBC process flow is described in Fig. 1. The main challenge for this cell technology is the simplification of the complex rear-side process sequence. One strategy to simplify this baseline process sequence into a simpler one that is litho-free, all-dry and inexpensive is shown in Fig. 1 (right). We have shown in our recent work the successful replacement of lithography by laser ablation [3] and wet etching by dry etching [2]. In this paper, the development of the partial dry etch route which eliminates the ex situ wet clean step will be discussed. Such a sequence is also more reliable for the processing of glass-bonded thin Si.

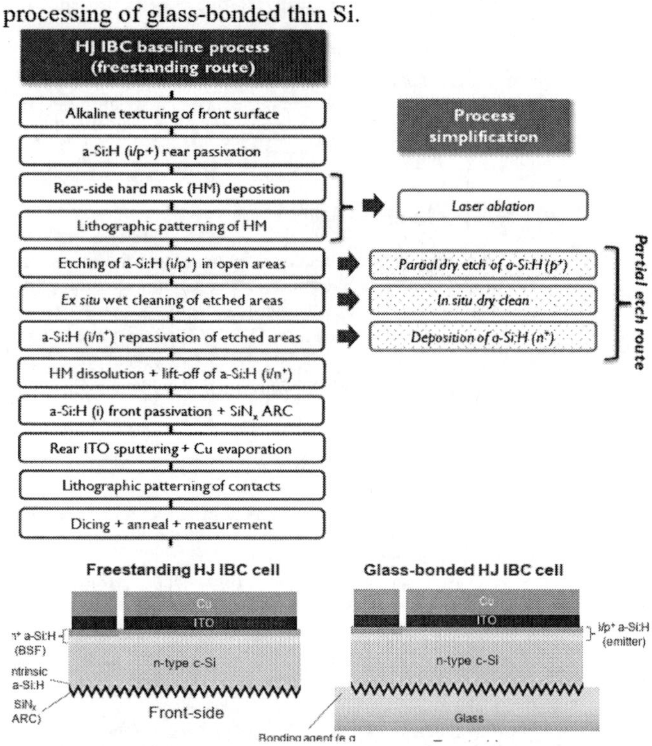

Fig. 1. The baseline HJ IBC process flow. Process simplifications leading to an all-dry, litho-free, cost-effective process flow are also indicated on the right. Cross-sectional schematics of a freestanding and a glass-bonded HJ IBC cell are given at the bottom.

2. Problem statements

A part of the HJ IBC rear-side process sequence is depicted schematically in Fig. 2. In the baseline process, after hard mask patterning, the areas with exposed a-Si:H (i/p$^+$) are completely etched to reveal the c-Si surface using NF_3/Ar plasma-based dry etching in the same PECVD tool used to deposit a-Si:H [2]. After dry etching, the surfaces are cleaned ex situ using a HF dip to remove residual contaminants on the surface before being inserted back into the PECVD tool for the deposition of a-Si:H (i/n$^+$), thereby producing interdigitated strips of a-Si:H (i/p$^+$) and a-Si:H (i/n$^+$).

When processing glass-bonded samples, achieving and maintaining excellent rear surface passivation in the presence of bonding agents is a challenging integration problem. This is because bonding agents contain volatiles which can outgas during the passivation process resulting in poor passivation. To address this, an extended outgassing step and in the case of silicones, an Ar plasma treatment step are applied to ensure reproducible high-quality passivation, as shown in Fig. 3 (b).

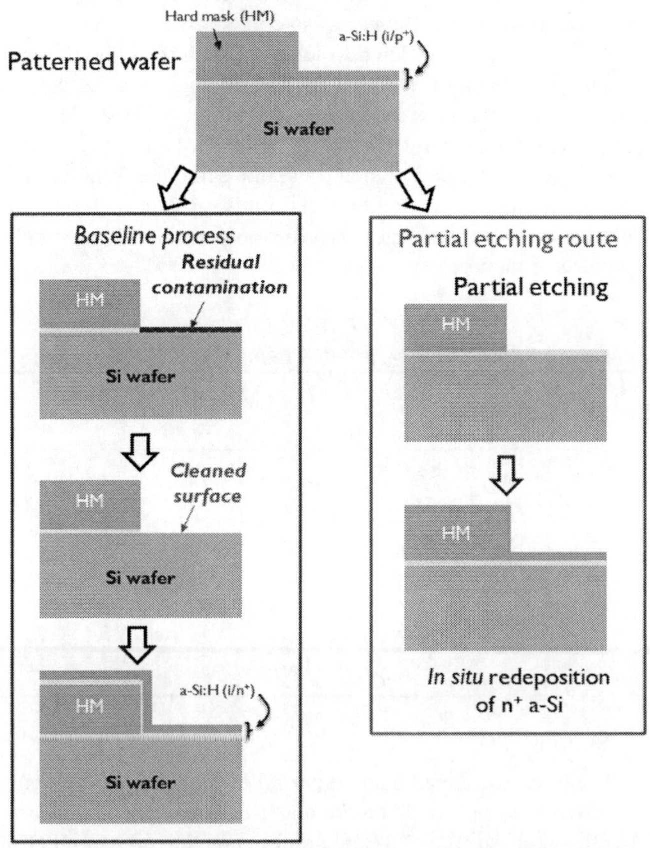

Fig. 2. Schematics depicting the steps after HM patterning till the realization of interdigitated strips of a-Si:H (i/p$^+$) and a-Si:H (i/n$^+$).

These processes, however, are not permanent fixes for the outgassing issue. Thus, any re-exposure of the c-Si surface during subsequent processes (e.g. after dry etch) can lead to problems with the repassivation of the re-exposed areas, due to the same organic contamination issue, as depicted in Fig. 3 (a). Figure 3 (c) shows the PL image after repassivation of dry-etched areas, having an outgassing pattern from the edges, which is indicative of organic contamination from the exposed silicone at the wafer periphery.

The partial etching route (Fig. 2 right) can solve these issues, whereby dry etching is carefully controlled to remove only the doped layer (p$^+$ a-Si:H), leaving behind the intrinsic a-Si:H layer from the first passivation and not re-exposing the c-Si surface. In the next step, n$^+$ a-Si:H is deposited in situ without vacuum break, in a single pump-down process. Skipping the ex situ wet clean can also reduce the process time and enable a fully in situ all-dry process sequence from dry etching till lift-off. This is particularly favorable for module-level cell processing, since wet processing of large-area glass is prohibitively cumbersome.

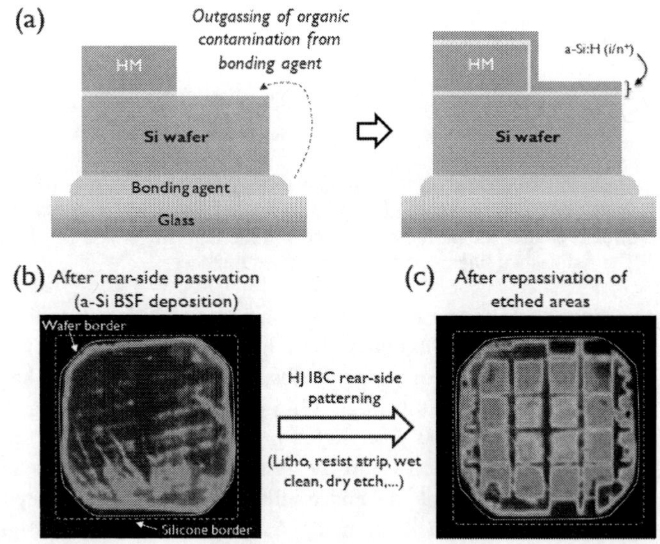

Fig. 3. (a) Schematics depicting outgassing of organic contamination from the bonding agent onto the re-exposed c-Si surfaces. (b) PL images showing the passivation quality during rear-side processing of HJ IBC cells on glass-bonded silicon.

3. Development of partial etching of a-Si:H

The standard dry etch process for the baseline flow is based on NF_3/Ar plasma with a flow ratio of NF_3:Ar = 200:100 sccm. This results in an etch rate (ER) of ~1.5 nm/s, which is too high to achieve a well-controllable partial etch process.

3.1 Selection of partial etch conditions

In order to reduce the ER to < 0.5 nm/s, the flow rates of NF_3 and Ar were individually varied, as shown in Fig. 4(a) and (b), respectively. It can be observed that the etch rate strongly depends on the NF_3 flow rate and only slightly on the Ar flow rate. Nevertheless, a high flow rate leads to better uniformity. Thus, we chose the flow ratio of NF_3:Ar = 25:500 sccm, which resulted in a well-controllable ER ≈ 0.32 nm/s.

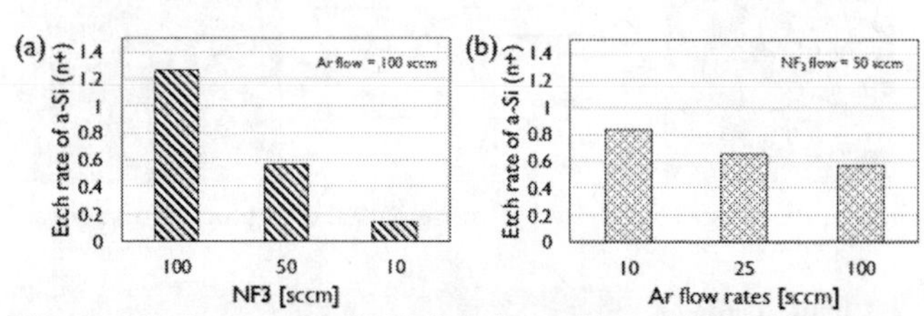

Fig. 4. Etch rates of a-Si:H when dry etching with NF3/Ar plasma with (a) different NF3 flow rates and (b) different Ar flow rates.

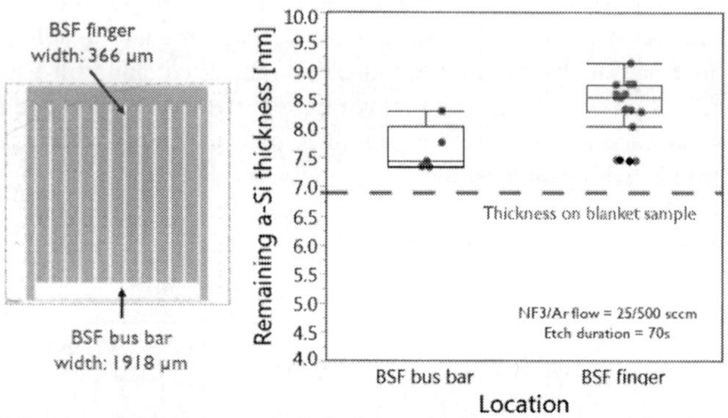

Fig. 5. Schematic on the left shows design of the rear-side of our small-area IBC pattern. Plot on right shows the remaining a-Si:H thickness after the partial etch process at the BSF finger and bus bar. The red dashed line indicates the expected thickness.

Since the above ERs were determined on blanket samples, the ER uniformity was checked using patterned samples having relevant features of different widths, namely, the BSF bus bar (1918 µm) and the BSF finger (366 µm), as depicted in Fig. 5 (left). The developed partial etch process was carried out on such patterned samples and the remnant a-Si thicknesses in different locations at the end of the process were estimated using spectroscopic ellipsometry. As shown in the plot in Fig. 5, the thicknesses measured in narrower features were thicker indicating a lower ER in such locations. This reduced ER was taken into account during device fabrication.

3.2 Surface passivation and contact resistance evaluation

Wafers (n-type, (100), Cz) passivated on both sides with a-Si:H (i/p$^+$) were subjected to the partial etching route i.e. partial dry etch of a-Si:H (p$^+$), followed by in situ deposition of a-Si:H (n$^+$). Steady-state photoluminescence (PL) and quasi-steady state photoconductance (QSSPC) measurements were performed on these wafers before and after the process to evaluate the minority carrier lifetime. As shown in Fig. 6 (a), the high initial lifetime (> 5 ms) is maintained after the partial etch process. Contact resistance measurements (not shown) were also performed and the contact resistance at the a-Si:H (i/n$^+$) contact from partial etch route was found to be comparable to that of the reference a-Si:H (i/n$^+$) contact.

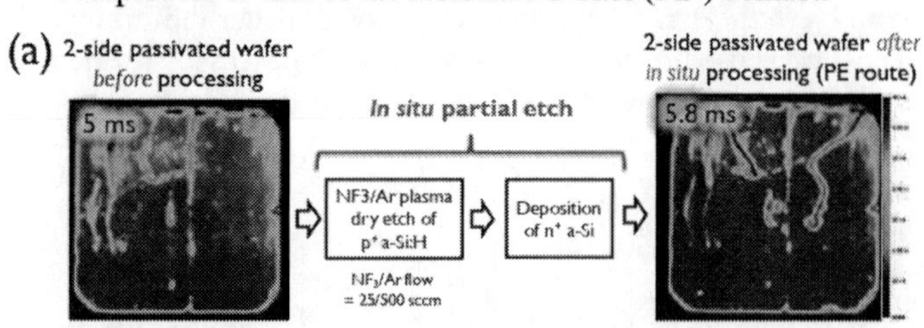

Fig. 6. PL images before and after the partial etch process sequence.

3.3 Cell results

The developed in situ partial etch route was implemented in our HJ IBC process flow on 200 µm-thick, n-type (100) FZ wafers, as described in Fig.1. For comparison, splits with ex situ clean were included as reference. Two different recipes of indium tin oxide (ITO) were also tested. Patterning was done using lithography. A total of 16 cells were produced from each wafer. The resulting cells (2.5×2.5 cm^2) were measured through an aperture area of 3.97 cm^2. The average and best results are tabulated in Table I.

The reference process resulted in excellent V_{OC}s close to 730 mV for both types of ITOs, indicating high quality surface passivation. Cells from the in situ partial etch split also showed high-quality passivation with V_{OC}s comparable to reference. As expected from contact resistance measurements, the fill factors of cells from all the splits are also comparable.

Overall, the best efficiency achieved with the reference split using the standard ITO is 23.3%, owing to a higher FF compared to the reference split with the new ITO, where the highest efficiency of 22.9% is reached. With the newly-developed in situ partial etch route, the same best efficiency of 22.9% is attained for the best cell, thus proving that a similar performance to reference can be achieved with a simpler process sequence.

TABLE I

CELL RESULTS OF 3 DIFFERENT SPLITS OF HJ IBC CELLS

Split		J_{sc} [mA/cm^2]	V_{oc} [mV]	FF [%]	Eta [%]
Reference (ex situ clean)	Best	41.9	729	76.2	23.3
	Avg.	41.9	728	75.3	22.9
Reference + new ITO	Best	41.6	728	75.7	22.9
	Avg.	41.4	726	75.1	22.6
Partial etch route + new ITO	Best	41.7	729	75.4	22.9
	Avg.	41.6	723	74.0	22.2

4. Summary

We have developed a novel partial etching process sequence whereby the doped a-Si on the rear-side is switched by etching off the exposed p$^+$ a-Si:H only and depositing an n$^+$ a-Si:H on top of the remaining intrinsic a-Si, without re-exposing the c-Si surface. This simplifies the rear-side patterning of HJ IBC cells by doing away with the need to perform the ex situ clean after dry etching in the baseline flow. In addition, since this route prevents re-exposure of the c-Si surface during rear-side patterning, this is advantageous for processing glass-bonded samples, where outgassing of organics from bonding agents can lead to poor passivation.

The partial etch process was based on NF_3/Ar plasma with high dilution of NF_3 in Ar of 25/500 sccm/sccm, which led to a controllable etch rate of ~0.32 nm/s on blanket samples. Lifetime and contact resistance measurements showed that high surface passivation quality and contact resistance comparable to reference can be achieved.

Finally, the partial etch route was implemented in the HJ IBC process flow, resulting in excellent V_{OC}s of ~730 mV. The FFs were also comparable to reference. Overall, a best efficiency of

22.9%, identical to the reference split was achieved, proving that the partial etching route can attain similar cell performance as the reference route, while simplifying the rear-side process sequence.

This process simplification implemented in this work eliminates the need to remove the wafers from the PECVD tool after dry etch and the need to use a cumbersome wet cleaning step in between dry etching and repassivation. Moreover, use of dangerous chemicals such as HF and the associated chemical waste disposal can be minimised. Such a process sequence is also beneficial for processing of glass-bonded samples (in module-level cell processing concepts), since c-Si surface re-exposure is avoided, thus maintaining high-quality surface passivation throughout the rear-side process sequence. This approach not only increases the processing simplicity but also improves the throughput.

A more detailed manuscript invited for publication in Progress in Photovoltaicss journal as part of the EUPVSEC is under review and will be published once accepted.

ACKNOWLEDGEMENTS

The authors gratefully acknowledge the funding received for this work from the EU for the H2020 project NextBase under grant agreement no. 727523, from Kuwait Foundation for the advancement of Sciences (KFAS) under project no. P115-15EE-01 as well as the imec's industrial affliliation program for Si-PV (IIAP-PV).

REFERENCES

[1] M. A. Green, et al., "Solar cell efficiency tables (version 50)" *Prog. Photovolt.: Res. Appl.* vol. 25, pp. 668-676, 2017.
[2] H. Sivaramakrishnan Radhakrishnan, et al., in *Proceedings of the 43th IEEE PVSC*, 2016.
[3] M. Xu et al., in *Proceedings of the 44th IEEE PVSC*, 2017.

OPTOELECTRONIC PERFORMANCE OF TCO ON SILICON HETEROJUNCTION REAR-EMITTER SOLAR CELLS

Alexandros Cruz[1], Sebastian Neubert[1], Anna B. Morales Vilches[1], Darja Erfurt[1], Stefan Körner[3], Florian Ruske[2], Bernd Stannowski[1], Bernd Szyszka[2] and Rutger Schlatmann[1]

[1]Helmholtz-Zentrum Berlin, PVcomB, Schwarzschildstr. 3, 12489 Berlin, Germany
[2]Technische Universität Berlin, Einsteinufer 25, 11087 Berlin, Germany

ABSTRACT: When designing silicon heterojunction (SHJ) solar cells with a rear-emitter (RE) configuration the requirements of the conductivity of the transparent conducting oxide (TCO) at the front side are relaxed due to the contribution of lateral current flow within the silicon wafer. In this study, we analyze two approaches that can be implemented to benefit from this. Firstly, the absorption of the TCO can be reduced by designing a thinner TCO layer than the single-layer anti-reflective optimum. In this case, a second anti-reflecting coating has to be deposited on top of the TCO to minimize reflection losses. Secondly, less conductive and more cost-effective materials can be used without strongly compromising the device's series resistance R_s and FF and, hence, the cell performance. To quantify the optoelectronic potential of these approaches in detail, we performed a comparative study of three different TCOs: indium tin oxide (ITO), aluminum doped zinc oxide (ZnO:Al), and hydrogenated indium oxide (IO:H) on rear-emitter SHJ solar cells. From simulations and experimental results, we conclude that, as expected, solar cells with IO:H as the front TCO reach the higheest efficiency. However, low-cost and higher resistive materials such as ZnO:Al can be implemented, without having a mayor efficiency penalty making them, moreover, competitive to the mainstream used ITO.

1 INTRODUCTION

Silicon heterojunction (SHJ) solar cells have gained significant relevance in the past years due to their high performance, with remarkable record efficiencies of 25.1% and 26.7% for two-side contacted and all-back-contact solar cells, respectively [1]–[3]. Main drivers for these achievements have been a very good passivation of the crystalline silicon wafer and the development of highly transparent contacting materials allowing current densities of over 40 mA/cm². By using a rear-emitter (RE) configuration the requirements of the conductivity of the front transparent conductive oxide (TCO) are relaxed due to the contribution of lateral current flow in the silicon wafer [4].

In order to profit from this RE configuration, we consider two approaches that can be adopted. Firstly, the optical absorption of the TCO can be reduced if the layer is designed thinner. Moreover, less conductive materials can be implemented without compromising the series resistance of the device.

To quantify the optoelectronic potential of these approaches in detail, we performed a comparative study of three different TCOs deposited by DC sputtering: indium tin oxide (ITO), a widely used TCO on SHJ solar cells; aluminum doped zinc oxide (ZnO:Al), an attractive TCO due to its high abundance resulting in lower costs and hydrogenated indium oxide (IO:H) as a TCO with very high carrier mobility.

With a thin IO:H as front TCO layer plus a second a-SiO_2 anti reflective coating (ARC), we achieve a generated current density calculated from the external quantum efficiency of the solar cell of 40.2 mA/cm² improving the reference solar cell by 0.8 mA/cm² with respect to 39.4 mA/cm².

From electrical simulations we see that IO:H has the potential to achieve the highest efficiency of 23%.Furthermore, the higher resistive ZnO:Al can also be very competitive when implemented into RE SHJ solar cells achieving 22.6% efficiency comparable to that of ITO with 22.7%.

2 EXPERIMENTAL DETAILS

2.1 TCO Deposition

TCO layers with thicknesses of 110±10 nm were deposited on 1.1-mm thick Corning Eagle glass in an in-line DC magnetron sputtering system from Leybold Optics (A600V7). Thicknesses were measured with a Dektak step profilometer. ITO layers were sputtered from a planar 97:3 In_2O_3:SnO_2 target, ZnO:Al from a rotatable 99:1 ZnO:Al_2O_3 target, and IO:H from a ceramic planar In_2O_3 target. ITO and ZnO:Al were deposited at a substrate temperature of about 150°C with oxygen flow ratios $(O_2)=q(O_2)/q(Ar+O_2)$ of 2.4% and 0.48%, respectively. IO:H was deposited at room temperature (without intentional heating) with an oxygen flow ratio of 2.0% and with an introduction of water vapor resulting in 1.7×10^{-6} mbar partial pressure. The depositions were carried out at a base pressure of 4×10^{-7} mbar. IO:H samples were solid-phase crystallized by annealing in vacuum at about 180°C for 1 h.

Charge carrier mobility μ_{Hall}, and carrier concentration, N_e, were determined for TCO layers on glass by Hall measurements with an Ecopia HMS 3000 system in the van de Pauw geometry. A Perkin Elmer Lambda 1100 spectrophotometer equipped with a 150 mm integrating sphere was used for optical characterization of the TCOs. Refractive indices n and extinction coefficients k of the TCO samples, were extracted from spectrophotometer and spectroscopic ellipsometry, measurements through fitting to the Drude-Tauc-Lorenz model.

2.2 Solar Cells

For solar cells preparation, we used n-type Czochralski (CZ) silicon wafers (c-Si) with 5 Ωcm resistivity. The as-cut c-Si substrates were etched to remove the saw damage. Its surfaces were then chemically textured to obtain random pyramids with 2-4 µm average height and <111> oriented facets The final wafer thickness was 125 µm. After RCA cleaning and just prior to silicon deposition a dip in 1% hydrofluoric (HF) acid solution was carried out to remove native formed oxide of the surfaces. Once the substrates were conditioned, intrinsic and doped thin-film silicon layers were deposited in an AKT1600 plasma enhanced chemical vapor deposition (PECVD) tool from Applied Materials operated at 13.56 MHz plasma excitation frequency. TCO (front) and ZnO:Al-silver (back) contacts were deposited through shadow masks to define 14 4-cm² size cells, three external quantum efficiency (EQE) spots, and three transfer-length-method (TLM) spots on 5-inch wafers. A busbarless 12-finger silver grid with a total shading of ~3% was screen printed and cured at 210°C for 7 minutes. Finally, selected devices were coated with a-SiO_2 as a second anti-reflective layer by PECVD.

All solar cells were characterized using current voltage (I-V) measurements under an AM1.5G solar-like spectrum when illuminated and standard test conditions in a Wacom WXS-155S-L2 dual source class AAA+ sun simulator. External quantum efficiency (EQE) and electrical measurements on the TLM structures to measure TCO-silver contact resistivity were made.

3 TCO PROPERTIES

3.1 Electrical properties

ITO is a polycrystalline material that presents carrier mobilities well over 30 cm²/Vs in a broad window of carrier concentrations going from 1 to 4 x10^{20} cm^{-3} when deposited at temperatures around 200°C [5].

The ZnO:Al presents poorer electrical properties showing mobilities between 10 and 20 cm²/Vs limited by the polycrystalline columnar growth nature of this material [18]. An important advantage for the ZnO:Al is the abundance of its components in the earth's crust in comparison to indium, positioning it as an attractive material substitute [6].

IO:H is a high-mobility TCO reaching $\mu_{Hall} > 100$ cm²/Vs.

Koida *et al.* demonstrated that this remarkable mobility is (partially) ascribed to the solid-phase crystallization leading to up to 400-nm sized grains with relaxed boundaries growing from the as-deposited amorphous material. The crystal growth is suppressed during deposition by the introduction of water (H_2O) vapor into the process chamber [7].

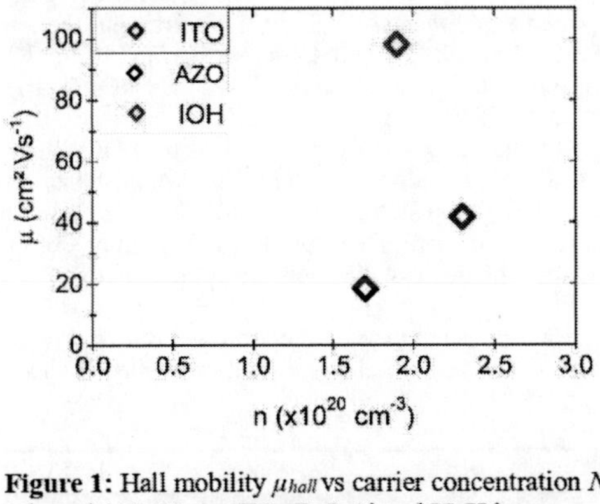

Figure 1: Hall mobility μ_{hall} vs carrier concentration N_e of selected 110±10 nm ITO, ZnO:Al and IO:H layers.

As we can see in Figure 1 the carrier concentration for the three different TCOs investigated, remains in a range of 2±0.5 x10^{20} cm^{-3}, hence the difference in conductivity between them is dominated by their mobility. For the layers implemented onto devices the mobility of the ITO is 35 cm²/Vs, the ZnO:Al presents a value oaround 20 cm²/s and IO:H reaches values of 100 cm²/Vs. The sheet resistances (R_{sh}) of these 110±10 nm TCO layers on glass are 30±10 Ω for the IO:H, 70±10 Ω for the ITO and 190±20 Ω for the ZnO:Al. From these reference layers, we calculate the R_{sh} of layers with different thicknesses by assuming that R_{sh} scales linearly with thickness, according to the equation $R_{sh}= \rho/t$ where ρ is the specific resistivity of the material and t is the thickness of the layer. The electrical properties of the devices, for relevant TCO thicknesses, was calculated with the Quokka2 program and will be discussed in the electrical simulations section.

3.2 Optical Properties

Figure 2 displays optical absorption spectra of the studied TCOs. We see that the differences is very subtle in the relevant range of the spectrum for the silicon absorber at the medium range and towards the near infrared region (600-1200nm), however the ZnO:Al as well as the IO:H present a slight advantage showing lower absorption than the ITO within this range. In the UV region, the IO:H shows noticeable lower absorption than the other materials. This is by combining a high optical bandgap, and a very low sub-bandgap absorption. The ZnO:Al on the other hand suffers from a higher wavelength fundamental absorption (lower band gap) positioned at 380 nm in comparison to the indium based TCOs at 350 nm.

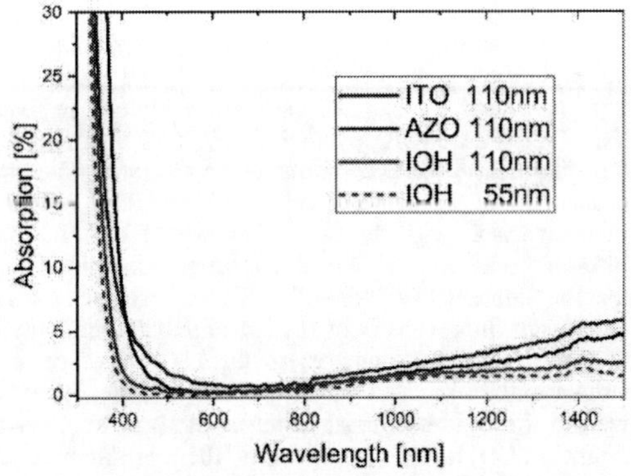

Figure 2: Optical absorption spectra of selected 110±10 nm ITO, ZnO:Al and IO:H layers on glass. An IO:H layer of 55±5 nm is presented for comparison.

By using a only 55±5-nm thick IO:H layer the absorption can be further reduced maintaining still a low enough sheet resistance of 80 Ω. This opens the possibility for a further improvement on the current density of devices. When using this thin IO:H layer a second anti-reflective layer is needed to reduce reflection losses [8]-[11]. In this case we used a-SiO_2. Reducing the layer thickness might, however, still result in higher resistive losses. In section 4.2, we quantify this effects in means of the solar cells opto-electrical performance.

4 RESULTS AND DISCUSION

4.1 Optical performance of solar cells

To analyze the optical performance of the materials in terms of generated current, simulations with the Matlab-based one-dimensional program GenPro4 [12] were carried out for our standard SHJ solar cell material stack.

The table in Figure 3 shows the simulated current density for the investigated TCOs. We can see improvements for ZnO:Al and IO:H of 0.3, 0.4 mA/cm² over the ITO. When introducing a thinner IO:H layer with a-SiO_2 second anti-reflective coating the current increase can be further enhanced up to 0.9 mA/cm² according to the simulated values. These improvements are consistent with the absorption spectra shown for the TCOs in section 3.2.

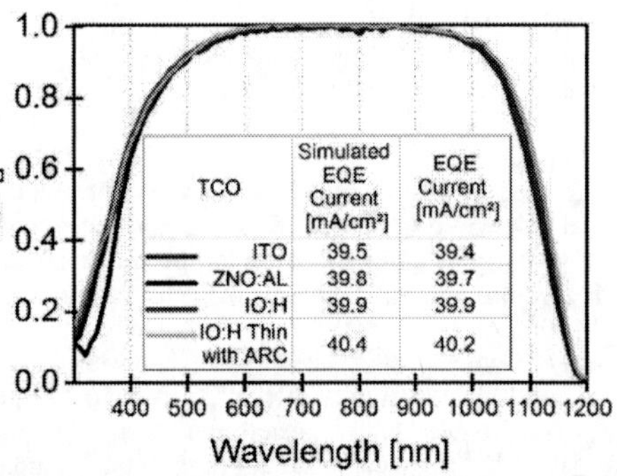

Figure 3: Internal quantum efficiency of experimental cells with different TCO (ITO, ZnO:Al, IO:H) materials with 75 nm nominal thickness. A solar cell with a thinner 35 nm IO:H layer is also presented. In the table simulated current from GenPro4 as well as calculated external quantum efficiency current densities values are presented.

To investigate the performance of the simulated solar cells on experimental devices, we processed SHJ solar cells with our standard material stack and variated the front TCO (ITO, ZnO:Al, IO:H). In Figure 3, the IQE curves for the different material combinations are shown. We see consistent improvements of the ZnO:Al and IO:H samples in comparison with the ITO reference of 0.3 and 0.5 mA/cm², respectively. We also confirm experimentally that the thinner IO:H layer with a-SiO_2 capping achieves a significant improvement of the generated current density of 0.8 mA/cm² as compared to the ITO reference and similar to the simulated values. Considering these results, we see that it is beneficial in terms of J_{sc} to implement thinner TCO layers. To quantify the performance of thinner layers for the three investigated materials further simulations were conducted to calculate the optimal TCO thicknesses for all TCOs. The resulting optimal TCO thicknesses are 40 nm for ITO and ZnO:Al and 55 nm for IO:H. For minimum reflection losses all layers must be capped with a 90-nm thick a-SiO_2 layer. In section 4.2, we consider this optimized layer thicknesses to assess their electrical performance on devices in comparison to standard 75 nm thickness layers with a respective optimal a-SiO_2 anti reflective coating.

4.2 Electrical performance of solar cells

To simulate electrical solar cell parameters, the Quokka2 program was used [13]. For this simulation, the geometry and material properties of our standard 4 cm² solar cell as it is described in the experimental methods section was introduced. For an n-type silicon wafer the Klaassen [14] model was considered. Auger recombination was modelled after Richter [15] and radiative recombination was set to $4.73x10^{-15}$ cm^{-3} according to Trupke et al. [16]. For the back ZnO:Al contacting layer a R_{sh} = 200 Ω was assumed. A saturation current density of J_0 = 15 $x10^{-15}$ mA/cm² and a recombination current density of J_{02} = 2.5 $x10^{-9}$ mA/cm² were determined as the input parameters from a two-diode model fit of IV curves of one of our typical ITO solar cells. The parameters were introduced and equally distributed to the rear and the front side of the device. For the n-type front contact, the R_{sh} was variated and set to the respective TCO value. The contact resistivity ρ_c for the metal-TCO interface was derived from TLM measurements and introduced to the simulation with a value of 3 mΩcm². For simplification the current generation was assumed to take place at the surface of the device, with a steady value at 40.7 mA/cm².

A lumped external R_s of 0.47 Ωcm² was added to account for all other R_s contributions, i. e. the grid-metal contact transport and the contact resistances at the remaining device interfaces.

Table I shows calculated R_{sh} of the TCOs at the relevant thicknesses and the simulated FF and efficiency of devices. For comparison, the best efficiency of experimental solar cells is shown too.

When analyzing the solar cells performance considering their simulated FF at a standard 75 nm TCO thickness, we can see that with a rear emitter configuration the ZnO:Al device will present a slightly lower FF than the ITO cell. However, through its improved optical properties the ZnO:Al will remain competitive showing same efficiency as the ITO of 22.5%. The IO:H device by the other side combines improved optical and electrical properties and can achieve a higher efficiency by 0.5% absolute than the other two materials. Considering the measured experimental devices we confirm that the less conductive ZnO:Al exhibits an efficiency of 22.8% comparable to the ITO with 22.5%. The IO:H cell's efficiency shows a lower value than expected being 22.4%, with reduced V_{oc} and FF . It was found that this is due to a lower pseudo Fill-Factor (pFF). We speculate that this deterioration originates from a silicon wafer passivation loss during the annealing of the sample for the IO:H solid-phase crystallization. By optimizing this

Table I: Simulated Fill-Factor FF and efficiency η for standard TCO thickness 75 nm as well as optimized layer thickness 40, 40 and 55 nm for ITO, ZnO:Al and IO:H respectively. For comparison, the best efficiencies from experimental solar cells η_{exp} is also shown.

	Standard TCO thickness				Optimized TCO thickness			
TCO	R_{sh} (Ω)	FF (%)	η (%)	η_{exp} (%)	Rsh (Ω)	FF (%)	η (%)	η_{exp} (%)
ITO	100	79.5	22.5	22.5	190	79.2	22.7	22.9
ZnO:Al	260	79.0	22.5	22.8	500	78.7	22.6	22.8
IO:H	60	79.7	23.0	22.4	80	79.6	23.0	22.4

process, it should be possible to exploit the full potential of this material. If we consider in Table I the optimized TCOs with reduced thickness we see a slight FF decrease due to the R_s increase as we expected. However, due to the improved optical properties of the solar cells through the thinner layers it is clear that the efficiency can be improved or in the worst case maintained equal as is the case of the IO:H. The experimental measured values confirm the slight increase or maintaining of the efficiency when designing devices with thinner TCO layers. We can also see that even for thinner TCO layers the ZnO:Al achieves comparable efficiencies as the ITO reference cell with 22.9% and 22.8%, respectively.

6. CONCLUSIONS

Through optical simulations and fabrication of experimental devices we quantitatively show the benefit that the implementation of a thinner TCO material layer in combination with a double layer anti-reflective coating with a-SiO_2 can achieve. By the introduction of such a material system in the example of a thinner IO:H layer an increase in current of 0.8 mA/cm² in comparison to an ITO standard solar cell is demonstrated.

Considering the opto-electrical performance of the solar cells we see, that the IO:H has a higher potential than the other materials increasing their efficiency by 0.5% absolute according to simulations. For the experimental devices, we do not see this performance improvement, probablydue to a damage of the Si passivation during the TCO annealing process that leads to reduced values for pFF and V_{oc}.

If we implement thinner TCO layers in SHJ solar cells, simulations show that the efficiency of the devices should slightly improve or in the worst case stay the same. This is confirmed by the experimental values. We also confirm that a cost effective less conductive material in the example of ZnO:Al can be implemented into rear-emitter solar cells achieving comparable efficiencies to ITO at 22.6% and 22.7% respectively.

7. ACKNOWLEDGMENTS

The authors would like to thank Manuel Hartig Tobias Henschel, Katja Mayer-Stillrich, Jonas Pogrzeba, Holger Rhein, and Matthias Zelt for technical support. We appreciate the many valuable discussions with Dr. Martin Dimer (Von Ardenne GmbH). This work was partially funded by the German Ministry of Economic Affairs and Energy (BMWi) in the framework of the HERA project (reference #0325825I).

8. REFERENCES

[1] K. Yoshikawa *et al.*, "Silicon heterojunction solar cell with interdigitated back contacts for a photoconversion efficiency over 26%," *Nat. Energy*, vol. 2, no. 5, p. 17032, May 2017.

[2] D. Adachi, J. L. Hernández, and K. Yamamoto, "Impact of carrier recombination on fill factor for large area heterojunction crystalline silicon solar cell with 25.1% efficiency," *Appl. Phys. Lett.*, vol. 107, no. 23, p. 233506, Dec. 2015.

[3] M. A. Green *et al.*, "Solar cell efficiency tables (version 50)," *Prog. Photovolt. Res. Appl.*, vol. 25, no. 7, pp. 668–676, Jul. 2017.

[4] M. Bivour, S. Schröer, M. Hermle, and S. W. Glunz, "Silicon heterojunction rear emitter solar cells: Less restrictions on the optoelectrical properties of front side TCOs," *Sol. Energy Mater. Sol. Cells*, vol. 122, pp. 120–129, Mar. 2014.

[5] A. Valla, P. Carroy, F. Ozanne, and D. Muñoz, "Understanding the role of mobility of ITO films for silicon heterojunction solar cell applications," *Sol. Energy Mater. Sol. Cells*, vol. 157, pp. 874–880, Dec. 2016.

[6] K. Ellmer, F. Kudella, R. Mientus, R. Schieck, and S. Fiechter, "Influence of discharge parameters on the layer properties of reactive magnetron sputtered ZnO:Al films," *Thin Solid Films*, vol. 247, no. 1, pp. 15–23, Jul. 1994.

[7] T. Koida, M. Kondo, K. Tsutsumi, A. Sakaguchi, M. Suzuki, and H. Fujiwara, "Hydrogen-doped In2O3 transparent conducting oxide films prepared by solid-phase crystallization method," *J. Appl. Phys.*, vol. 107, no. 3, p. 33514, Feb. 2010.

[8] D. Zhang *et al.*, "Design and fabrication of a SiOx/ITO double-layer anti-reflective coating for heterojunction silicon solar cells," *Sol. Energy Mater. Sol. Cells*, vol. 117, pp. 132–138, Oktober 2013.

[9] S. Y. Herasimenka, C. J. Tracy, V. Sharma, N. Vulic, W. J. Dauksher, and S. G. Bowden, "Surface passivation of n-type c-Si wafers by a-Si/SiO2/SiNx stack with <1 cm/s effective surface recombination velocity," *Appl. Phys. Lett.*, vol. 103, no. 18, p. 183903, Oct. 2013.

[10] J. Geissbühler *et al.*, "Silicon Heterojunction Solar Cells With Copper-Plated Grid Electrodes: Status and Comparison With Silver Thick-Film Techniques," *IEEE J. Photovolt.*, vol. 4, no. 4, pp. 1055–1062, Jul. 2014.

[11] A. B. Morales-Vilches *et al.*, "ITO-free silicon heterojunction solar cells with ZnO:Al/SiO2 front electrodes reaching a conversion efficiency of 23 %," *IEEE JPV accepted for publication.*

[12] R. Santbergen, A. H. M. Smets, and M. Zeman, "Optical model for multilayer structures with coherent, partly coherent and incoherent layers," *Opt. Express*, vol. 21, no. 102, pp. A262–A267, Mar. 2013.

[13] A. Fell, "A Free and Fast Three-Dimensional/Two-Dimensional Solar Cell Simulator Featuring Conductive Boundary and Quasi-Neutrality Approximations," *IEEE Trans. Electron Devices*, vol. 60, no. 2, pp. 733–738, Feb. 2013.

[14] D. B. M. Klaassen, "A unified mobility model for device simulation—I. Model equations and concentration dependence," *Solid-State Electron.*, vol. 35, no. 7, pp. 953–959, Jul. 1992.

[15] A. Richter, S. W. Glunz, F. Werner, J. Schmidt, and A. Cuevas, "Improved quantitative description of Auger recombination in crystalline silicon," *Phys. Rev. B*, vol. 86, no. 16, p. 165202, Oct. 2012.

[16] T. Trupke *et al.*, "Temperature dependence of the radiative recombination coefficient of intrinsic crystalline silicon," *J. Appl. Phys.*, vol. 94, no. 8, pp. 4930–4937, Sep. 2003.

HIGH MOBILITY TRANSPARENT CONDUCTIVE OXIDES FOR SILICON HETEROJUNCTION SOLAR CELLS DEPOSITED BY ROTATABLE MAGNETRONS

M. Dimer[1], A. Cruz[2], Er-Chien Wang[2], A.B. Morales-Vilches[2], J. Löhnert[1], U. Graupner[1], M. Thumsch[1], B. Stannowski[2], E. Schneiderlöchner[1]

[1]VON ARDENNE GmbH, Am Hahnweg 8, 01328 Dresden, Germany.
[2]Helmholtz-Zentrum Berlin, PVcomB, Schwarzschildstr. 3, 12489 Berlin, Germany.

ABSTRACT: Sputtered indium tin oxide (ITO) is a widespread used material for application as transparent conducting oxide as the front contact of rear-junction silicon heterojunction solar cells. Standard ITO films suffer from too high parasitic absorption in both the short- and the long-wavelength range. The aim of this contribution is to investigate lower absorbing TCOs deposited with industrial DC sputtering from rotatable targets. High V_{oc} values >735 mV demonstrate that in spite of the high-power industrial DC sputter process no significant sputter damage is observed, independent of the target material. Cell efficiencies up to 23.3% are obtained with titanium-doped (ITiO) and tin-doped indium oxide (ITO 99/1). The highest short circuit current values of 39.6 mA/cm^2 are reached with an ultimately low-carrier and high-mobility indium-oxide-based material. The cell efficiency in this case, however, is over-compensated by a too low FF, mostly due to a higher contact resistivity and TCO sheet resistances. We ascribe this effect to the very low carrier concentration.

1 INTRODUCTION

Silicon heterojunction solar cells have shown their ultimately high efficiency potential in the past few years with world record efficiencies for all-back-contacted cells of η = 26.7 % [1,2] and in both-side contacted cells with η up to 25.1 % [3,4]. The lean production process with only four main process steps makes this device concept, moreover, one of the most attractive candidates for mass production. Large scale and low cost producible high performance TCO is the key for further improving the conversion efficiency and reducing the costs of SHJ solar cells. A comprehensive overview on TCO materials was recently published in [5]. Indium oxide based transparent conducting oxides (TCO) are commonly used as contact layers in silicon heterojunction (SHJ) solar cells. High transparency in the spectral range of 400 – 1100 nm, low contact resistances to the n and p doped Si and the metal, processing at <200°C without deteriorating the c-Si passivation and cost-competitive producing are key and restrict the choice of possible TCO materials. The deposition method of choice is DC magnetron sputtering, which has been established in mass production on multi m^2-scale for instance in the glass industry. By sputtering from rotatable targets highest material utilization and scaling to large area both essential for lowest manufacturing costs can be achieved.

Sputtered indium tin oxide (ITO) is a widespread used TCO for this purpose. However, standard ITO films suffer from too high parasitic absorption in both the short and the long wavelength region. The aim of this contribution is to investigate lower absorbing TCOs deposited with industrial DC sputtering from rotatable targets. ITO with very low SnO_2 doping concentration (97/3 and 99/1) using hydrogen-containing sputter gases reaches a high carrier mobility up to 67 cm^2/Vs. With titanium-doped indium oxide (ITiO) 82 cm^2/Vs could be reached. Our approach here is to first optimize the opto-electronic properties of these TCO layers on glass, and then investigating the performance in state-of-the-art SHJ solar cells on Cz-Si wafers with screen-printed Ag contact grids as well as in transfer-length-method (TLM) structures for measuring the TCO sheet and metal-to-TCO contact resistances in the device. As a reference SHJ process we use standard (95/5) ITO for SHJ cells.

2 EXPERIMENTAL DETAILS

All TCO layers were deposited in an in-line DC magnetron sputtering system at temperatures between 100 and 200°C. Rotatable magnetrons with a high magnetic field were used to sputter from ceramic tube targets with 1 m length. Indium oxide based TCO materials with various doping, namely, ITO 97/3, ITiO, ITO 99/1 and "Y", indium-oxide material with unknown composition, were processed. Sputtering was done at industrial compatible high deposition rates at high DC power of > 4 kW/m target length. TCO thicknesses on glass were measured by a step-profilometer. The charge carrier mobility, μ_{Hall}, and the carrier concentration, N_e, were derived from Hall measurements done with a Phystech RH 2030 system in the van de Pauw geometry. A Perkin Elmer Lambda 950 spectrophotometer equipped with a 150 mm integrating sphere was used for optical characterization of the TCOs.

For solar cells preparation we used n-type Czochralski silicon wafers with 5 Ωcm dark resistivity. The as-cut c-Si substrates were etched to remove saw damage. The surface was textured by KOH to obtain random pyramids with 2-4 µm average height. The wafer thickness after wet-processing was ~130 µm. After RCA cleaning a dip in 1% hydrofluoric (HF) acid solution for three minutes was carried out to remove native formed oxide of the surfaces. Once the substrates were conditioned, intrinsic and doped thin-film silicon layers were deposited in an AKT1600 plasma enhanced chemical vapor deposition (PECVD) tool from Applied Materials at 13.56 MHz. A nanocrystalline silicon n-layer was used at the front side and an amorphous silicon p-layer for forming the p/n junction at the rear side. For bi-facial cells the TCO was sputtered through aligned shadow masks on the front side and the rear side to define 4-cm^2 cells as well as external quantum efficiency (EQE) spots and transfer length method (TLM) patterns. A 12-finger silver-grid busbarless metal contact design with total shading of ~3% was screen printed and cured at 210°C for 7 minutes.

Solar cells were characterized using current voltage (I-V) measurements under an AM1.5G spectrum when illuminated and standard test conditions in a Wacom WXS-155S-L2 dual source class AAA+ sun simulator. Electrical measurements on TLM structures were performed on the rear (p/n junction) side of the wafer to derive TCO/silver contact resistivity and the TCO sheet resistance (w/o influence of the Si substrate underneath).

3 RESULTS AND DISCUSSION

3.1 TCO layers on glass

The electrical properties of 110-nm thick TCO layers on borosilicate glass are summarized in table 1. Two TCOs ("Y1" and "Y2") have been deposited varying the process settings resulting in different opto-electric properties. Hall mobilities are between 57 and 89 cm²/Vs for all materials, with very similiar values before and after thermal annealing in ambient for all TCOs. The "Y" TCOs show a decrease in N_e upon annealing, resulting in a relatively high sheet resistance for "Y2" of 136 Ohm/sq. after annealing. The charge carrier density is decreasing from 2.7 to a very low value of 0.7 E20 cm^{-3} for "Y2". The sheet resistance of all materials is in a range of 40 – 70 Ω/sq before annealing, which is well low enough for application in SHJ solar cells. However, after annealing the sheet resistance of "Y2" reaches 136 Ω due to the reduced carrier density.

The optical TCO properties are shown in Figure 1 presenting the absorption spectra of layers on glass. In accordance with the decreasing carrier concentration, a decreasing (free-carrier) absorption in the near-infrared part of the spectrum (>800 nm) is observed, which is relevant for improved light in-coupling into the cell. In the short wavelength range of 400 - 600 nm, in contrary, we observe a less systematic behavior. Here TCOs ITO 99/1 and ITiO exhibit a higher absorption. This behavior can not easily be correlated with electrical properties.

Table 1. Properties of 110-nm thick TCO layers on glass. Values are given as deposited and, in parentheses, after thermal annealing (30 min at 200°C under ambient).

	ITO 97/3	ITiO	ITO 99/1	Y1	Y2
μ_{Hall} (cm²/Vs)	57 (57)	76 (79)	65 (69)	84 (89)	79 (79)
N_e (1E20 cm^{-3})	2.7 (2.7)	2.2 (2.0)	1.8 (1.8)	1.3 (0.9)	1.1 (0.7)
R_{sheet} (Ω)	41 (42)	34 (36)	47 (45)	51 (71)	68 (136)

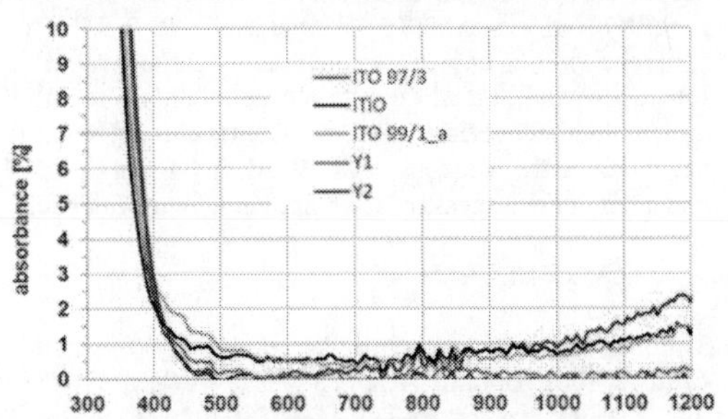

Figure 1. Absorption spectra of 110-nm thick TCO layers on glass.

3.2 Solar Cells

Figure 2 shows the IV parameters of solar cell with different TCOs at the front, and the same ITO 97/3 used at the rear side. In addition to the previously discussed TCOs, an ITO 95/5 reference, DC sputtered from a tube target at HZB, was added. For each TCO setting three wafers with in total 42 cell were prepared.

Except for few cells all V_{oc} values are in a narrow range of 735 to 740 mV. No TCO related variations can be found. For J_{sc}, in contrast, TCO related variations are observed. Generally, the J_{sc} values follow the trend observed in the absorption curves on glass (Fig. 1). ITO 97/3 results in the lowest value due to its high NIR absorption and both "Y" TCOs perform the best. Based on median values the total spread in J_{sc} is 0.5 mA/cm². Only the lab-reference exhibits lower J_{sc} values. When it comes to FF, the trend is vis-versa. Optically well-performing TCO lead to lower FF, finally dominating the cell efficiency. As the pseudo-FF (not shown), pFF, determined from dark and light IV curves, ranges from 83 to 85% for all cells without any correlation to the specific TCO used, differences in the cell's series resistance (R_s) are the dominating factor for the FF differences. From the good and constant V_{oc} and pFF trend we conclude that in all cases the DC sputtering does *not* result in a (significant) sputter damage. The samples with ITiO and ITO 99/1 show the best efficiencies above 22.5% reaching 23.3% in the best case thanks to the very good FFs combined with good J_{sc}'s. Inspite of the low J_{sc} the reference ITO 95/5, reaches efficiency values around 22.5% as well, thanks to good FF values.

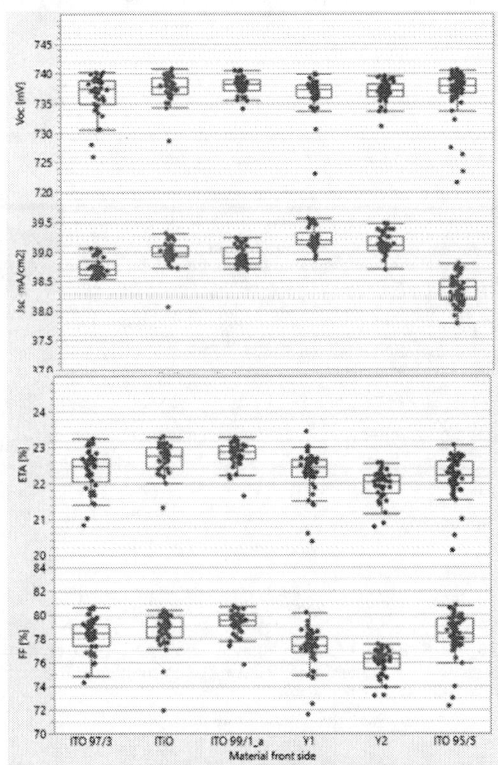

Figure 2. IV parameters of 4-cm² size bifacial silicon heterojunction solar cells with various front TCOs and ITO 97/3 on the rear side. ITO 95/5, DC sputtered from a tube target at HZB, was added for reference.

In order to elucidate the FF differences, in figure 3 we show the series resistance as derived from dark and light IV curves in the maximum power point. In addition, figure

4 shows the TCO sheet resistance and metal/TCO contact resistivity as derived from TLM measurements on the rear side of the wafer. The three samples on the left-side behave very similar with cell's R_s of 1.0 Ωcm^2, TCO sheet resistance around 70 Ω and a metal/TCO contact resistivity below 0.5 $m\Omega cm^2$, not significantly contributing to R_s. The sheet resistance is about 75% higher than was determined from the glass samples, which is mostly due to the rough Si surface resulting in a thinner layer and longer current path than on glass. For the "Y" TCOs, however, we find that both a high TCO sheet resistance and a high contact resistivity to the screen-printed silver are responsible for the high R_s and, hence, the low FF. We speculate that this is mostly due to the low N_e values.

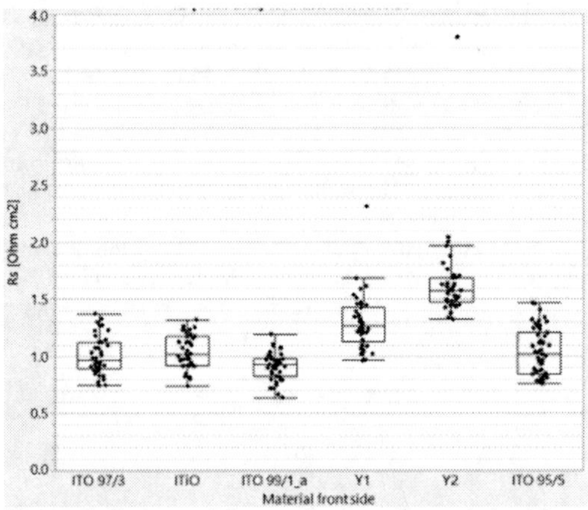

Figure 3. Series resistance of solar cells as derived from the dark and the light IV curves in the maximum power point.

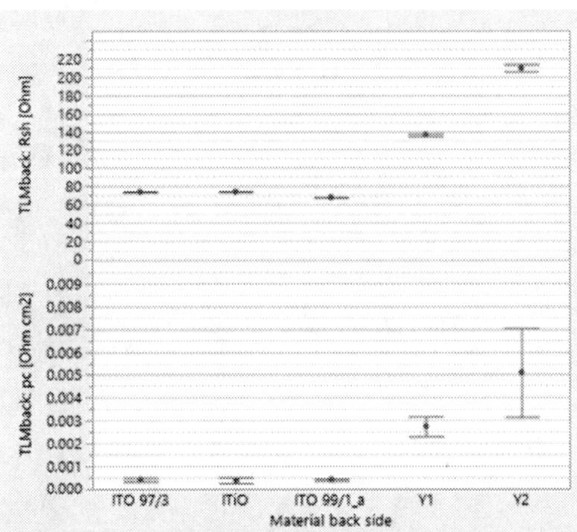

Figure 4. TCO sheet resistance and metal/TCO contact resistivity as derived from TLM measurements on the rear (junction) side of the wafer.

As we have shown the higher TCO sheet resistance as well as the metal/TCO contact resistivity both contribute to the lower FF in case of the "Y" TCOs. Additionally, we assume that a higher contact resistivity at the TCO/(n) nc-Si interface adds to the higher R_s. This, however, could not been quantified by measurements. In order to test the hypothesis we developed another process based on the ITO 99/1 target, but with lower carrier concentration. We here denominate the ITO 99/1 process condition used as "ITO 99/1-a" and the new process condition as "ITO 99/1-b". The latter one has a lower N_e = 1.2 E20 cm^{-3} and consequently higher R_{sh} = 80 Ω/sq vs. 45 Ω/sq with the same Hall mobility of 60-65 cm^2/Vs (compare Tab. 1). We implemented ITO 99/1-b in cells as single-layer front contact as well as a double-layer consisting of a 20 nm thick ITO 99/1-a contact layer and 55 nm ITO 99/1-b bulk layer on top. The resulting FF and R_s values are shown in figure 5. The trend in R_s and FF indicate that our assumption appears to be correct: With the ITO 99/1-a as a contact layer R_s is reduced slightly and the FF is a bit higher. These effects, however, are small.

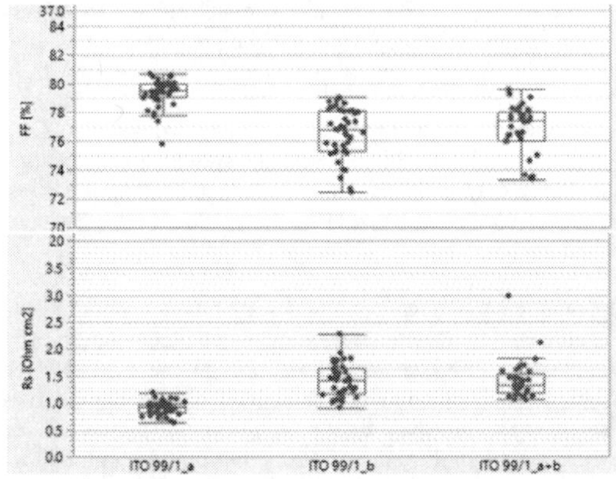

Figure 5. FF and series resistance of cell with different front TCO stacks. Left: ITO 99/1-a (= ITO 99/1 in Fig. 1 & 3). Center: ITO 99/1-b. Right: double layer with 55 nm ITO 99/1-b on top of 20 nm ITO 99/1-a.

4. SUMMARY AND CONCLUSIONS

In summary we demonstrated the effect of high-mobility indium-oxide-based front TCOs implemented in rear-junction silicon heterojunction solar cells. High V_{oc}'s over 735 mV for all TCOs and up to 741 mV in the best case demonstrate that in spite of the high-power industrial DC sputter process *no* significant sputter damage is observed, independent of the target material. The highest cell efficiency values up to 23.3% are obtained with titanium doped (ITiO) and tin-doped ITO 99/1 indium oxide as front TCO. Highest current values of J_{sc} = 39.6 mA/cm^2 are reached with a ultimately low-carrier and high-mobility indium-oxide-based material ("Y"). The cell efficiency in this case, however, is over-compensated by a lower FF value, mostly due to a higher contact resistivity and TCO sheet resistances. We ascribe this effect to the very low carrier concentration. Work is ongoing to study and quantify the specific R_s components in more detail.

5. ACKNOWLEGEMENTS

The R&D work was partially funded by the Federal Ministry for Economic Affairs and Energy (BMWi) within the R & D project PATOS (FKZ 0324074B). Special thanks to Tobias Henschel, Kerstin Jacob, Matthias Zelt and Stefan Janke for technical support and to Sebastian Neubert for support and fruitful discussions.

6. REFERENCES

[1] K. Yoshikawa *et al.*, "Exceeding conversion efficiency of 26% by heterojunction interdigitated back contact solar cell with thin film Si technology", Solar Energy Materials and Solar Cells 173, December 2017, Pages 37-42.

[2] M. A. Green *et al.*, "Solar cell efficiency tables (version 50)", Prog. Photovolt. Res. Appl. 2017; 25:668–676. https://doi.org/10.1002/pip.2909.

[3] M. Tauchi *et al.*, "24.7% Record Efficiency HIT Solar Cell on Thin Silicon Wafer", IEEE J-PV, 4 (2014) 96-99.

[4] D. Adachi et al., "Impact of carrier recombination on fill factor for large area heterojunction crystalline silicon solar cell with 25.1% efficiency", Applied Physics Letters 107, 233506 (2015); doi: 10.1063/1.4937224.

[5] M. Morales-Masis et al., "Transparent Electrodes for Efficient Optoelectronics", Adv. Electron. Mater. 2017, 1600529. DOI: 10.1002/aelm.201600529

ANALYSIS OF INFRARED LIGHT TRAPPING ON BIFACIAL SILICON HETEROJUNCTION SOLAR CELLS

Félix Gérenton*, Samuel Harrison, Perrine Carroy, Anthony Valla, Adrien Danel, Delfina Muñoz
Univ Grenoble Alpes, CEA, LITEN, INES, F-38000 Grenoble
*Corresponding author: felix.gerenton@cea.fr; Phone: +33.479.79.28.42

ABSTRACT: In silicon heterojunction (SHJ) solar cells, the transparent conductive oxide (TCO) layers used are designed to balance resistive losses and parasitic light absorption. This paper aims to determine the rear TCO optimization that can be made in the case of bifacial SHJ cells, taking into account both optical and electrical aspects. First, internal reflectance in bifacial cells is simulated and characterized for more or less transparent TCO layers, and compared to monofacial cells, in order to establish if the use of a more transparent TCO layer stays beneficial for bifacial cells as well. Then, the internal reflectance is further optimized by tuning the TCO thickness. In this work, we thus show that using a more transparent ITO layer remains an important path of improvement for the internal reflectance in bifacial cells, and that TCO thickness should be optimized differently than for monofacial cells. These results have been confirmed at cell level with a gain in short-circuit current without significant resistive losses observed, resulting in a final efficiency gain. Moreover, this optimization is proved to be also beneficial for the cell's bifaciality behavior.
Keywords: bifacial cells; heterojunction; transparent conductive oxide; light trapping; indium-tin oxide

1 INTRODUCTION

Silicon heterojunction (SHJ) solar cells have the advantages of both great conversion efficiencies [1] and high in-field performances, especially in the case of bifacial cells, where the sunlight reverberated behind a solar panel can be converted into electrical power as well. Moreover, the bifacial configuration does not require any additional step for SHJ cells fabrication [2], and uses much less metal paste.

In a bifacial SHJ cell, a metal grid is printed on both front and rear sides (Fig. 1a), and the pitch between two fingers should be optimized to limit both optical losses through shadowing, and resistive losses through lateral conduction. However, the impact of these two factors on conversion efficiency is quite different whether we consider front or rear cell side, as the amount of light received on the rear is much lower than the front (10 to 20% of the sunlight on typical field installations). In order to maximize the power output, it is thus preferable to use a smaller pitch on the rear side, resulting in a smaller photogenerated current at the rear but less resistive losses, which is more critical as it impacts also the cell power generated from the front side. Typical cell performances and associated bifaciality coefficient (ratio of efficiencies from the rear / from the front) with different rear grid pitches are shown in Fig. 1b.

The transparent conductive oxide layer (TCO), used for optical confinement and for carrier transport between the amorphous silicon (a-Si :H) contacts and the metal grid, needs to be optimized to balance its lateral conductivity and parasitic absorption in near-infrared (NIR) [3]. In the general case of a cell with a denser metal grid at the rear, front and rear TCO layers could be optimized differently. Indeed, the lateral conductivity of the rear TCO becomes less critical for the device, and a layer with a lower NIR absorption could reduce the optical losses, as described in [4] for monofacial cells, without critically increasing the resistive losses.

This study will first focus on the behavior of infrared light into a bifacial SHJ cell and will highlight the differences with monofacial cells, through optical simulations and characterizations, in order to understand the possible optimization paths for a rear TCO layer in a bifacial cell. Then, TCOs with different optical and electrical properties will be integrated into bifacial SHJ cells in order evaluate its impact on different cells parameters, especially short-circuit current (J_{sc}), fill factor (FF), conversion efficiency and bifaciality coefficient.

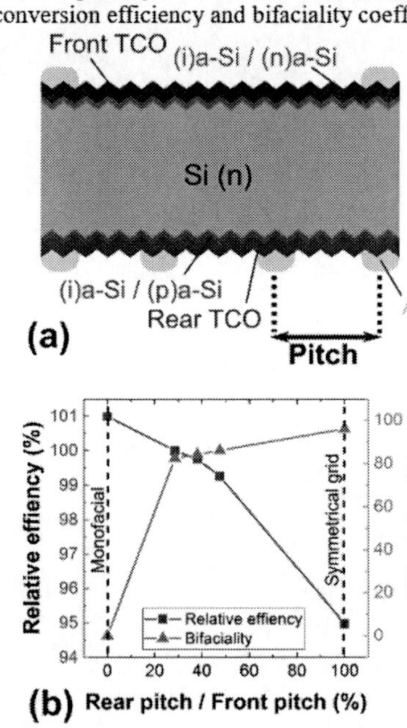

Figure 1: (a) Representation of a bifacial SHJ cell. (b) Typical SHJ cell relative effiency and bifaciality for different rear grid pitches.

2 MATERIAL AND SOLAR CELL PROCESSES

2.1 Fabrication of low-absorption TCO layers

This work focuses on the optimization of indium-tin oxide (ITO) layers, which is the most usual among the possible TCO materials used in SHJ cell process [5]. These layers are deposited by magnetron sputtering from an In_2O_3/SnO_2 target, under an Ar/O_2 gas mixture. The material properties are modified by changing the O_2 flow in the deposition chamber, from 8 to 24 sccm. The Ar flow stays fixed at 500 sccm. As the deposition rate is sensitive to the gas ratio in the chamber, the deposition time is adjusted to keep a fixed targeted thickness for all conditions (100 nm on polished substrate, unless stated otherwise). In order to reproduce the thermal step of metal paste curing in the cell process, all the layers are annealed at 205 °C for 9 min. Depositions are made simultaneously on polished silicon, glass and SHJ solar cells.

The different layers are characterized by spectroscopic ellipsometry using a combination of Tauc-Lorentz and Drude models [6], reflectometry (on glass samples), and Hall effect measurements. The absorbance vs wavelength, the carrier concentration (N) and the resistivity of the different ITO layers are summarized in Fig. 2.

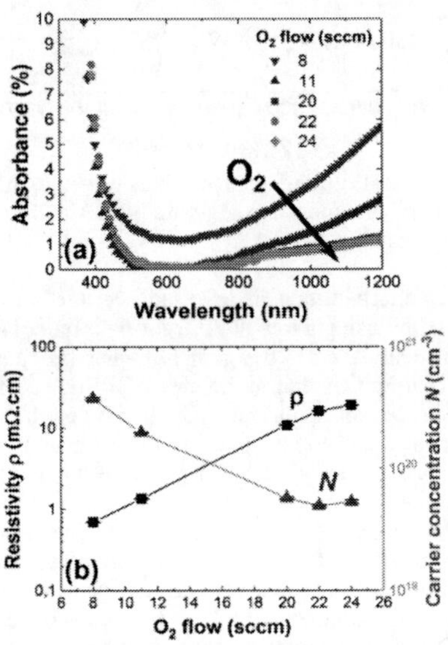

Figure 2: (a) Absorbance in ITO layer vs. wavelength for different O2 flows. (b) ITO resistivity ρ and carrier concentration N.

As usually observed [3-4], the absorbance in NIR range decreases as the O_2 flow increases. In this case, absorbance does not evolve from 20 to 24 sccm of O_2. The ITO resistivity still increases with O_2 until 24 sccm, even though carrier concentration stays quite constant from 20 to 24 sccm. The explanation stands in the mobility that keeps decreasing up to 24 sccm (not shown).

Therefore, the TCO layers deposited with an O_2 flow higher than 20 sccm are of no interest for integration in SHJ cells, as parasitic NIR absorption does not decrease anymore, and series resistance could increase because of the increased ITO resistivity. For the sake of simplicity, the following studies will focus on ITO layers deposited with an O_2 flow of 8 and 20 sccm, which will be called "high-N" and "low-N" respectively, representing their respective carrier concentrations N, of very different values.

2.2 Solar cell process

All the cells realized in this study are based on n-type wafers with a resistivity of 2.5 Ω.cm. Substrates are cleaned and textured, with a final thickness of 165 µm. Doped and intrinsic amorphous silicon (a-Si:H) layers are deposited on both sides, with n-type doping on the front side and p-type on the rear ("inverted emitter" architecture [3]). ITO is then deposited on the front and the rear side, with different sets of parameters for the rear as described in 2.1. Metal grids are screen-printed from a silver paste, with a 4 busbars pattern, a finger pitch of 2.1 mm on the front and 0.6 mm on the rear. After screen-printing, all cells are annealed at 205 °C for 9 min in order to cure the silver paste. Finally, all cells are characterized in current-voltage (IV) with probe bars on both sides, without a metal chuck in order to avoid parasitic reflections at the back of the cell. IV characterizations are made on both sides of the cell in order to extract bifaciality coefficients. More details about the cell process can be found in [2].

3 LIGHT PROPAGATION IN BIFACIAL SHJ CELLS

This part aims to give insights on the main differences in light interaction with the cell's rear side in the case of monofacial and bifacial cells. This analysis allows then to highlight the preferential optimization paths for the rear TCO layer in order to reduce the optical losses in a bifacial cell.

3.1 Internal reflectance: effect of material properties

In order to understand how the light propagates at the rear of mono- and bifacial cells for different TCO properties (high-N and low-N in this study), simulations of internal reflectance, absorbance and transmittance (R/A/T) are performed with a variable incident angle, on the online simulator OPAL 2 [7]. This calculator is based on the solving of the Fresnel equations [8] to compute R/A/T in a layer or stack of layers between two semi-infinite media. Here, we consider a layer of ITO with a complex refractive index measured with ellipsometry, and a thickness of 73.5 nm (thickness measured of 98.2 nm on flat substrate, converted to the thickness on textured substrate). a-Si:H layers are omitted, as they are transparent in the infrared range. The ambient media is silicon, and the external media is silver or air for mono- and bifacial cells respectively. Simulations are realized at a wavelength of 1200 nm and an angle varying from 0 to 90°.

Differentiation of light propagation regimes in the Si-ITO-air system occurs at two critical angles. First, the escape angle $\theta_e = sin^{-1}(1/n_{Si})$ where n_{Si} is the refractive index of silicon, above which light is totally reflected at the ITO/air interface. Then, the total reflection angle $\theta_r = sin^{-1}(n_{ITO}/n_{Si})$ with n_{ITO} being the refractive index of the ITO layer considered, above which light is totally reflected at the at the Si/ITO interface. One should note that even above θ_r, absorption in ITO is expected as light is still going through the ITO layer on the form of an evanescent wave [4]. For the Si-ITO-silver system corresponding to the monofacial case, the only angle separating different propagation regimes is θ_r. All cases are summarized in Fig. 3.

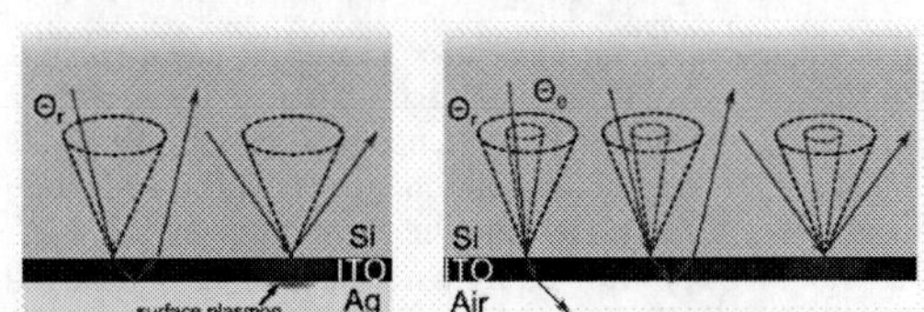

Figure 3: Representation of the different light propagation modes at the rear interface of mono- and bifacial SHJ cells.

R/A/T simulations results are presented in Fig. 4. The monofacial case presents, as already described in the literature [4], a peak of absorbance occurring at intermediate angles, above θ_r. This effect is known to come from the penetration of *p*-polarized waves interacting with the metal [9]. In our case, reflection is improved for all angles with low-*N* ITO. This is not a systematic rule, as the absorption peak shifts with the refractive index, the reflectance could be higher with a less transparent ITO on high angles. Moreover, in our case, absorption in silver is low compared to the ITO, which is not always the case.

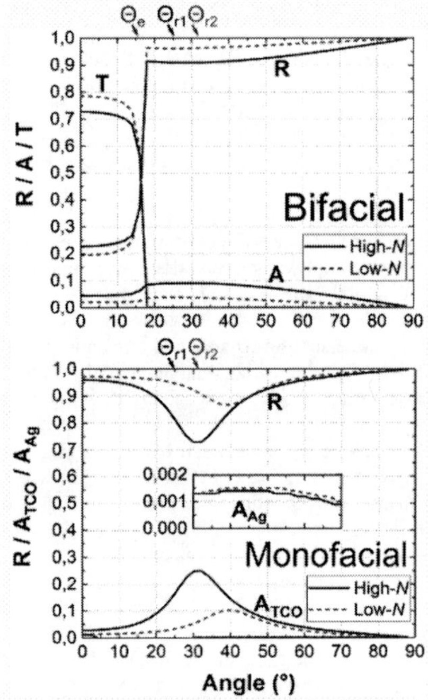

Figure 4: Reflectance, absorbance in ITO and transmittance (bifacial) or absorbance in silver (monofacial) versus incident angle for Si-ITO-air and Si/ITO/Ag systems. θ_{r1} and θ_{r2} are the total refraction angles for high-*N* and low-*N* ITO respectively.

The bifacial case behaves differently: from normal incidence to θ_e, where most of the light is transmitted into the air, reflectance is higher for the high-*N* ITO. This comes from a reduced transmittance to air, itself due to a lower refractive index value, increasing reflectance both at the Si/ITO and ITO/air interface. After θ_e, no light is transmitted outside of the cell, and parasitic absorption exists whether waves can propagate to the ITO (below θ_r) or are totally reflected (above θ_r), with an evanescent field decaying through the ITO layer. However, absorbance is lower for low-*N* ITO, which means that the value of θ_r, driven by n_{ITO}, has little influence on internal reflectance compared to the ITO absorption, driven by the extinction coefficient k_{ITO}. Moreover, the interaction with the air interface is very different from a metal, which is observed by the absence of the absorption peak present on the monofacial case.

Consequently, the ideal ITO layer for the rear side of a bifacial cell is ambiguous: depending on the incident angle with respect to θ_e, the reflectance can be improved with either a high-*N* or a low-*N* material. Therefore, the choice of material depends on the angular distribution of rays arriving at the rear interface. In order to determine the best material in this case, reflectometry is performed on cells, monofacial and bifacial, with high-*N* and low-*N* ITO on the rear (Fig. 5).

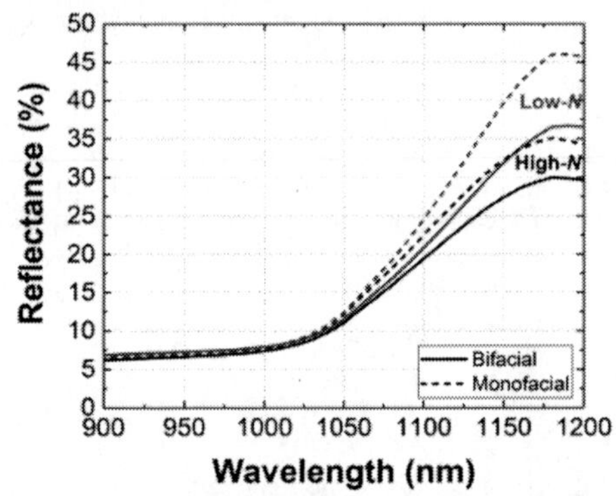

Figure 5: SHJ cell reflectance versus wavelength, for bifacial and monofacial cases, with high-*N* and low-*N* ITO on the rear side.

An important gain in reflectance at the cell level is observed by using a low-*N* ITO, for both monofacial and bifacial cells, even if the gain is higher for monofacial cells. This implies that in the case of a SHJ cell textured on both sides, the angles taken by the rays on the rear side are globally large enough to make the gain in internal reflectance for low-*N* ITO larger than the losses. This observation is verified at every wavelength in the 1000-1200 nm range, confirming a global reflectance improvement in the infrared. A consequence of this observation is that even if light can escape from the cell by the rear side, optimizing the TCO properties on the rear of bifacial cells still allows a large gain in reflectance.

Moreover, these reflectance gains translate well to external quantum efficiency (EQE) measurements in the infrared (Fig. 6), confirming a gain in J_{sc} by using a low-*N* ITO on bifacial cells.

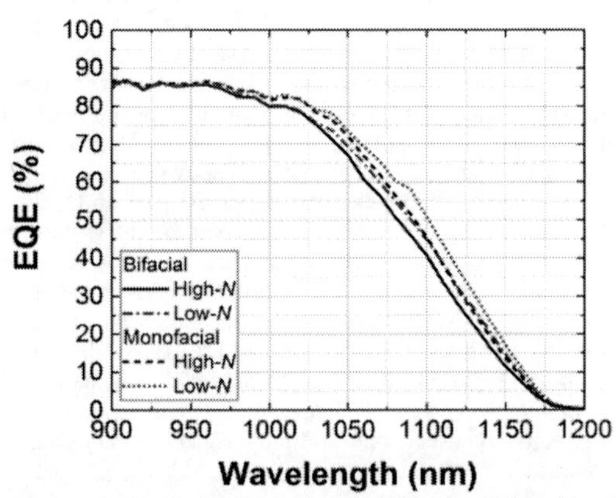

Figure 6: EQE in the infrared range for bifacial and monofacial cells with high-*N* and low-*N* ITO layers at the rear

3.2 Internal reflectance: effect of rear TCO thickness

On monofacial cells, optimizing the thickness of the rear TCO layer can be critical: evanescent waves can be absorbed through surface plasmons in the metal layer underneath if the ITO is too thin, or in the ITO itself with an excessive thickness [4]. In bifacial cells, no plasmon losses occur, and reflectance varies with angle in a very different way than monofacial cells, as already shown in Fig. 4. Therefore, the R/A/T values for different ITO thicknesses should be reassessed in the case of bifacial cells. For this purpose, we perform R/A/T simulations of a Si/ITO/air system on OPAL 2, focusing on a low-*N* ITO layer with three different thicknesses: 73.5, 147 and 220.5 nm (corresponding to the reference thickness of 73.5 nm used in 3.1, and this value multiplied by 2 and 3). Equivalent simulations for a monofacial case are performed as well. Results are shown in Fig. 7.

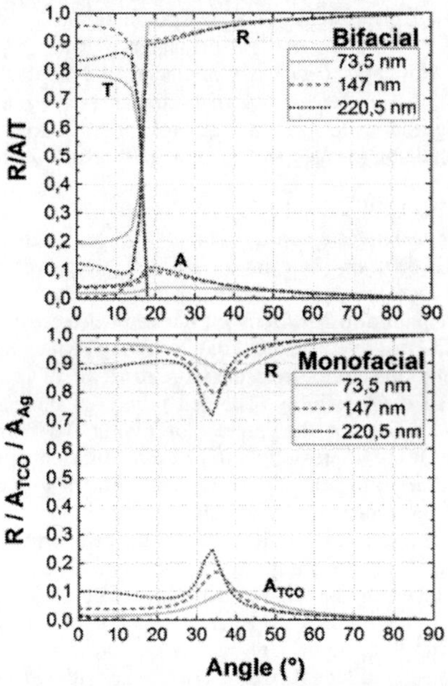

Figure 7: R/A/T versus incident angle for Si-ITO-air and Si/ITO/Ag systems.

As for the study on materials properties (see 3.1), monofacial and bifacial reflectors behave very differently with different ITO thicknesses. For the monofacial case, the absorption peak shifts to lower angles with increasing ITO thickness, making the best thickness for internal reflectance very dependent on the angular distribution. For the bifacial case, the reflectance is clearly improved at all angles for a thinner layer, decreasing both absorbance due to a smaller pathlength in the ITO, and transmittance at low incidence angles due to interference effects. Here again, these angle-dependent behaviors are now assessed on finished devices in order to analyze the TCO thickness impact on internal reflectance for a textured cell (Fig. 8).

For the monofacial case, the reflectance measurements confirm the high dependence of the response with the angle distribution: on a textured cell, the difference of global reflectance between different ITO thicknesses is quite small, and is not measurable on EQE.

For the bifacial case, the R/A/T graphs (Fig. 7) translate also well to global reflectance and EQE: with the thinner ITO layer, both reflectance and EQE are improved in the infrared, while the two higher thicknesses give very close responses.

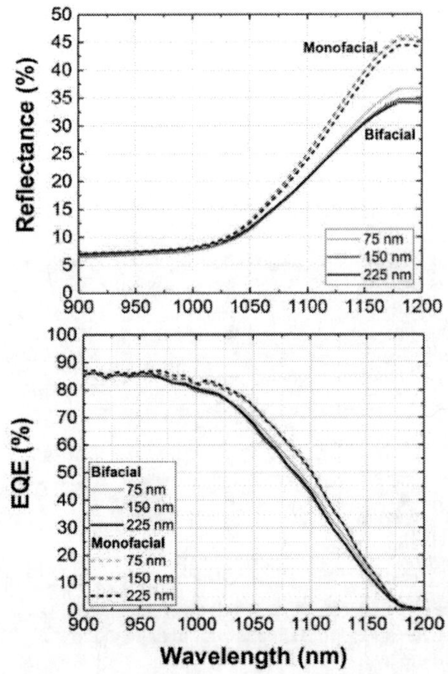

Figure 8: SHJ cell reflectance (top) and EQE (bottom) versus wavelength, for bifacial and monofacial cases, with ITO thicknesses of 75, 150 and 225 nm.

Finally, one can conclude that bifacial cells tend to have an improved internal reflectance with a thinner layer because of the very different light propagation mechanisms at the rear compared to monofacial cells. More precisely, reflectance above the escape angle θ_e is improved with a thinner layer because of the reduced pathlength into the TCO layer, while in monofacial cells, the absorbance peak shifts with thickness, making the response with angle more ambiguous. Therefore, even if some interference effects exist on bifacial cells, especially at angles below θ_e (Fig. 7), internal reflectance in bifacial cells is improved by a thinner layer on a wider angle range than monofacial cells, and consequently, on a textured device, internal reflectance rises with a thinner TCO layer. In our work, as the reference thickness used (~75 nm on textured substrate) demonstrated better results than the thicker layers, it will be chosen for the rest of the study.

4 PERFORMANCES OF BIFACIAL CELLS WITH OPTIMIZED REAR TCO

Two batches of 17 cells are fabricated with high-*N* and low-*N* layers. IV results and bifaciality coefficients are presented in Fig. 9.

V_{oc} does not evolve significantly between both sets of cells (1 mV of difference). J_{sc} increases by 0.2 mA/cm^2 for low-*N* ITO, confirming the gain in internal reflectance and quantum efficiency observed in 3.1. The fill factor (FF) drops in average by 0.2%$_{abs}$, but with a quite large variability within each batch, as the standard deviation is 0.8 and 1.2%$_{abs}$ for high-*N* and low-*N* ITO respectively. Finally, the efficiency rises by 0.1%$_{abs}$ by using a low-*N* ITO. It is worth noting that this result is potentially very dependent on the rear finger pitch, fixed at 0.6 mm here.

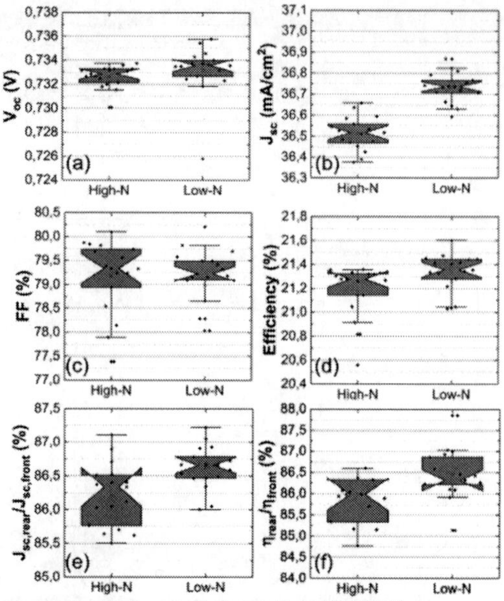

Figure 9: IV characteristics (a-d) and bifaciality coefficients (e-f) for SHJ cells with high-*N* and low-*N* ITO on the rear.

A second batch of cells realized in very similar conditions, with only slightly different ITO thicknesses (97 and 90 nm for high-*N* and low-*N* respectively) showed similar performances, in particular 0.3 mA/cm² in J_{sc}, +0.0%$_{abs}$ in FF and +0.1%$_{abs}$ in efficiency.

Finally, bifaciality slightly improves as well with a low-*N* ITO (Fig. 9f), going from 86.0% to 86.5% in average. This is due to the lower NIR parasitic absorption, which also improves current photogeneration when the cell is illuminated from the rear, as demonstrated in [4] for front-side TCO layers. This is confirmed by the improved bifaciality coefficient of J_{sc} (Fig. 9e), matching very well the improved global bifaciality.

5 CONCLUSION

The objective of this work was to assess if the possible optimizations of the rear TCO on monofacial SHJ cells stay valid for bifacial cells to some extent. In the practical case of bifacial cells with a smaller grid pitch on the rear side, the lateral conduction properties of the ITO are less critical, giving the possibility to improve J_{sc} by using a more resistive and more transparent layer.

The first objective was to verify how light propagates at the rear of a bifacial cell, and if a more transparent layer would improve internal reflectance. Through internal reflectance/absorbance/transmittance simulations, we showed that the propagation of light is very different on bifacial and monofacial cells. Especially, transmittance outside the cell is important at low incidence angles, and reflectance/absorbance are found quite different than monofacial cells for other angles. Moreover, in both cases, the internal reflectance of a device is dependent on the angle distribution at the rear. However, on a textured SHJ cell, we showed that internal reflectance is improved in the whole infrared range for a more transparent TCO. Consequently, even with the possibility of light escape at the rear of a bifacial cell, the internal reflectance can be improved in the same way than for a monofacial cell.

Another aspect to optimize was the TCO thickness. After similar investigations on internal reflectance, it appeared that monfoacial or bifacial configurations respond very differently with respect to the incident light angle. At cell level, the reflectance and quantum efficiency measurements showed almost no difference between ITO thicknesses in the monofacial case, due to this highly angular distribution dependent behavior. On the other hand, for bifacial cells, a clear trend appears in favor of our reference thickness, where reflectance was improved at all angles. This was confirmed at cell level with reflectance and EQE measurements as well. In conclusion, this very different behavior of light management with respect of TCO thickness between monofacial and bifacial cases leads to different optimal thicknesses for each configuration. In our case, our reference thickness (~75 nm on textured substrate) gave the best results for bifacial cells.

Finally, the simulation and characterization results of the optical investigations led on bifacial cells were applied on SHJ devices. The gains observed in internal reflectance for a more transparent TCO layer translated well to J_{sc}, with a gain of 0.2 mA/cm² on our cells. Fill factor loss is 0.2%$_{abs}$ in average, without knowing with certainty whether this comes from the more resistive TCO or from the process variability. The cell efficiency increased by 0.1%, validating the interest of tuning the rear TCO properties to improve light trapping on bifacial cells (with a smaller grid pitch on the rear). Bifaciality also showed a small but repeatable gain, confirming the interest of this TCO optimization for all aspects of the cell performances.

For a further study, this work could be extended to TCO layers thinner than our reference, who could improve further the internal reflectance because of the smaller pathlength of light into the TCO, but could also be detrimental to lateral conductivity, even with a dense rear metal grid. Another promising aspect will be the integration of the optimized rear TCO layers on thinner SHJ cells, in order to assess if the improvement observed could be enhanced on thinner cells where more photons reach the rear side.

REFERENCES

[1] K. Yoshikawa et al., Nature Energy 2 (2017) 17032

[2] A. Danel et al., Proceedings 31st European Photovoltaic Solar Energy Conference, (2013) 279-283.

[3] M. Bivour et al., Solar Energy Materials and Solar Cells 122 (2014) 120-129.

[4] Z. C. Holman et al., Journal of Applied Physics 113 (2013) 013107.

[5] Yu et al., Japanese Journal of Applied Physics 57 (2018) 08RB15.

[6] Y. S. Jung, Thin Solid Films 467 (2004) 36-42.

[7] K. R. McIntosh, S. C. Baker-Finch, Proceedings 38th IEEE Photovoltaics Specialists Conference (2012).

[8] H. A. Macleod, Thin-Film Optical Filters. Institute of Physics Publishing: London, 2001 (Chapter 2)

[9] Z. C. Holman et al., Light: Science & Applications 2 (2013)

EFFECTS OF MAGNESIUM-DOPING ON SILICON LEACHING FOR SOLAR GRADE FEEDSTOCK PRODUCTION

Mengyi Zhu[1], Adrian Murgau[2], Jafar Safarian[1]
[1] Department of Materials Science and Engineering, Norwegian University of Science and Technology (NTNU), Trondheim, Norway
[2] R&D Director, REC Solar Norway, Kristiansand, Norway

ABSTRACT: The rapid growth of photovoltaic industry has led to large demands of solar grade silicon (SoG-Si) feedstock production in recent years. Acid leaching as a low-cost and important part of the only available commercial process by the metallurgical route of SoG-Si feedstock production, has accordingly received significant attention. In this work, pure Mg_2Si is doped into a commercial metallurgical grade silicon (MG-Si) in molten state, solidified under controlled conditions, and leached by 10% HCl. The effect of Mg doping on the removal of impurities in leaching is studied at different conditions, at 60 °C and 80°C for different leaching durations from 5 minutes up to 1 hour. The results indicate high temperature and long leaching time are in favor of impurities removal. The eutectic Mg_2Si phase which contains the silicides of the impurities in association such as Al, Fe, Ca, Ti on the grain boundaries is leached. This causes the purification of silicon and it is shown that P and metallic impurities can be removed in large extents.

Keywords: Silicon, Magnesium, Phosphorus, Impurity, Acid leaching

1 INTRODUCTION

The development of cheaper processes for high purity solar grade silicon feedstock has become worldwide imperative demands and attrached significant attentions from both photovaltic industry and academia in recent years. Among the proposed alternative metallurgical purification methods, acid leaching is usually considered as one of the most promising candidates for the metallurgical grade silicon (MG-Si) refining due to its low-temperature operation feature and the ability to remove large amounts of impurities.

The main principle of acid leaching is based on the low segregation coeffiicent of the impurities in silicon, representative values of some selected impurites are listed below as Table I. It can be clearly seen that the segregation coefficient of most of metallic impurities such as Fe, Al, Ti etc. are between 10^{-3} and 10^{-6}. As a result, these impurities tend to segregate together as precipitate phase which can be leached away treated by acids while silicon is fairly noble. However it is also worth noting that the segregation coefficients of B and P are much higher compared to other impurites. This unqiue feature makes them, in principle, difficult to remove by MG-Si acid leaching process.

Table I: Representative segregation coefficient values of impurities at Si melting point[1]

Impurity	Segregation coefficient	Impurity	Segregation coefficient
B	8.0×10^{-1}	Ti	2.0×10^{-6}
P	3.5×10^{-1}	Zn	1.0×10^{-5}
Fe	6.4×10^{-6}	Mn	4.5×10^{-3}
Al	2.8×10^{-3}	Cu	4.0×10^{-4}

The research about acid leaching for silicon refining has been actively studied over decades. The first acid leaching of pulverized MG-Si was pointed out nearly hundred years ago in the pioneering work of Tucker in 1927[2]. After many years, Voos[3] obtained a patent in 1961 about a leaching process for pulverized MG-Si with different leaching agent. In addition Hunt et al.[4] later removed successfully more than 90% impurities from MG-Si by aqua regia leaching. In the 1980s, Dietl[5] systematically studied the optimal leaching purification conditions using a mixture of HF and HCl. Furthermore, Elkem A/S demonstrated higher leaching efficiency can be achieved after an alloying process of MG-Si with the association of precipitated $CaSi_2$ phase. Based on this method, leaching of alloyed MG-Si received increasing attention in the past few decades. Many additive refiner metals (or impurity getters) have been reported such as Ca[6]-[9], Fe[10],[11], Zr[12], Hf[13] and so on. As the most studied alloying elements, Ca is considered to have a strong affinity to impurities like P and some metallic impurities, it also have high reactivity with leaching acids. For example, according to the research of Shimpo et al [7], as much as 80% P can be removed by alloying 5.17% Ca. Based on the recent research conducted at NTNU in Norway[14]-[17], Mg also shows promising potential as the impurity getter for MG-Si doping due to the similar alkali earth element properties and high leaching efficiency.

In this work, the leaching of Mg doped MG-Si and original MG-Si were compared. Effects of leaching time and temperature on different impurities removal of Mg-doped Si was also studied in order to gain deeper understanding about the leaching mechanisms.

2 EXPERIMENTAL

2.1 Raw material and MG-Si doping

A commercial MG-Si was used in present study. The raw MG-Si was mixed in graphite crucible with certain amount of pure Mg_2Si was heated up to 1500 °C to form a Si-4.5wt% Mg melt in an induction furnace. After the Si-Mg alloy was fully melted and reach equilibrium, the melt was then slowly cooled down with an average cooling rate of 10 °C/min inside the induction furnace. The Si-Mg alloy obtained was then crushed and sieved to different particle sizes, and particles with 0.2-2.15 mm were further used. The same particle size for MG-Si was also used.

The chemical composition of raw MG-Si and obtained Mg-doped Si alloy were analyzed by high-resolution

Inductively Coupled Plasma-Mass Spectroscopy (ICP-MS) before acid leaching, results are shown as Table II.

Table II: Composition of MG-Si and Mg doped MG-Si by ICP-MS analysis.

Impurity	MG-Si	Alloyed MG-Si
B	46.2	42.4
P	15.8	15.6
Mg	9.6	44879.4
Ca	266.7	282.0
Ti	299.8	204.0
Fe	3167.5	2281.0
Al	2548.0	1744.0
Mn	70.4	53.9
Zr	10.7	7.0

2.2 Leaching and impurities analysis

The Mg doped MG-Si and raw MG-Si were leached by 10% HCl at the temperature of 60°C and 80°C within 1 hour. Totally 4 samples were made to understand the effects of different process parameters. All the leaching trials were performed in an ultrasonic bath. The samples after leaching were washed by distilled water and ethanol and then dried at 80°C.

Samples after leaching were also analyzed by ICP-MS for the impurities determination. Ultra-fine particles were separated and removed by a 50 microns sieve after drying in order to reduce the interference of metallic impurities.

3 RESULTS AND DISCUSSION

3.1 Effects of Mg doping on Si microstructure

The leaching effect of with and without Mg doping on Si microstructure was studied by SEM shown as Figure 1.

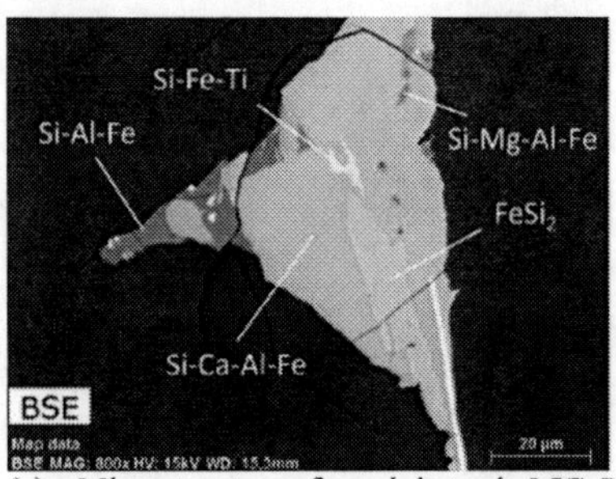

(a) Microstructure of precipitates in MG-Si

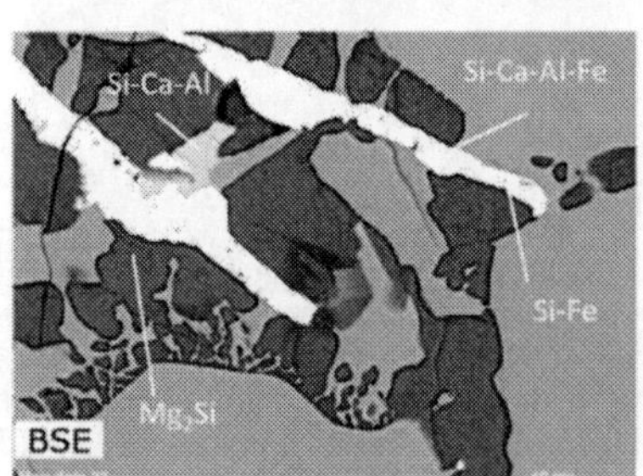

(b) Microstructure of precipitates in Mg-doped Si

Figure 1: SEM images of (a) MG-Si and (b) Mg-doped MG-Si.

It can be seen that the observed main phases in MG-Si are $FeSi_2$, Si-Fe-Ti, Si-Ca-Al-Fe, Si-Mg-Al-Fe, respectively. All these phases are Fe-bearing silicides and known as poor leachability. However, with the addition of Mg into MG-Si, it was found that the eutectic Mg_2Si becomes the dominant precipitated phase. The Fe-bearing silicides were embedded inside the Mg_2Si phase. It is known that Mg_2Si has a high reactivity with mineral acids. Thus, when the dominant Mg_2Si phase removed by acids during leaching, the poorly leachable Fe-bearing phases could be carried away at the same time. This feature offers an opportunity to enhance the leaching efficiency.

3.2 Effects of leaching time

The effects of leaching time on different impurities in Mg alloyed Si at 80 °C is shown as Figure 2. It can be seen that with the increasing of leaching time, the concentration of nearly all the impurities are decreased except B. More than a half of metallic impurities like Fe, Al, Mn, Ti were effectively removed after 1 hour leaching. It is also worth noting that the P removal is beyond 90%, which is problematical to be removed by conventional acid leaching of raw MG-Si. This result may suggest the formation of some complex magnesium phosphide compounds in the precipitate phase. The reason of no B removal during acid leaching might due to the high segregation coefficient of B in Si, which makes majority of B stay inside the Si grain and locked by the dislocations.

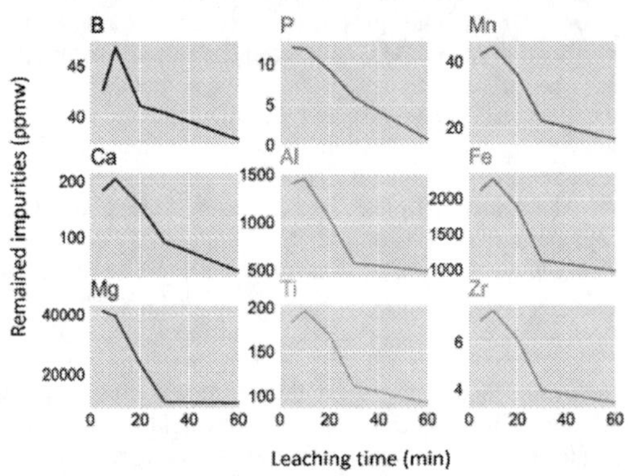

Figure 2: Time dependence of impurity concentration of Mg-doped Si during 10% HCl acid leaching at 80℃ in ultrasonic bath.

In addition, it can be observed that the removal of impurities can be divided into two stages; a fast removal stage in the first 30min, which indicates a chemical reaction control mechanism followed by a slow removal stage after 30 min, which is believed to be a diffusion controlled stage.

Detailed comparison of leaching performance of MG-Si and Mg-doped Si can be seen from Figure 3 in the following section.

3.2 Impurities after leaching

The remained impurity concentration of 4 different leaching conditions are shown as Figure 3, detailed data can be seen from Appendix Table I. It is clear that after Mg-doping, the remained concentrations of most impurities distinctly reduced. It is also worth noting that a great number of Mg impurity remained for Mg doped MG-Si leaching. This might be the reason that some amount of Mg_2Si precipitates close to Si grain boundaries were not completely removed within 1 hour leaching. In addition, it can be seen that for the conventional problematic P removal in MG-Si leaching, Mg-doping exhibits a promising potential to significantly remove P impurity amount as high as more than 90%.

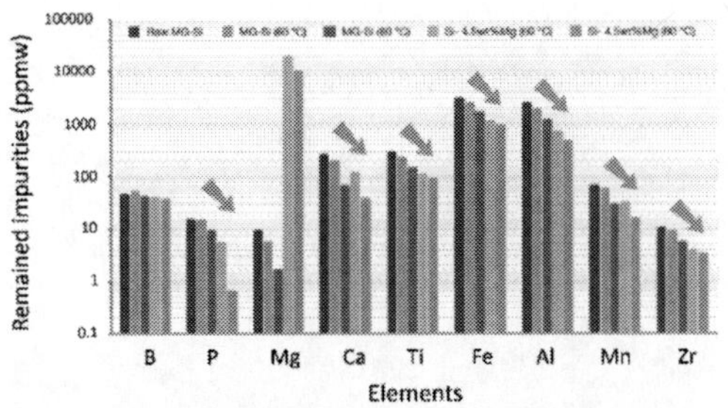

Figure 3: Remained impurities concentration after 1h leaching. The five bars from left to right respectively denotes the concentration of raw MG-Si, MG-Si after leaching at 60℃, MG-Si after leaching at 80℃, Mg-doped Si after leaching at 60℃, and Mg-doped Si after leaching at 80℃. The X-axis presents different impurity elements, while the Y-axis presents impurity level from 0.1 to 100000 ppmw.

The impurity removal degree is also calculated in order to compare the purification ability under different conditions. The results can be seen as Figure 4, detailed data seen from Appendix Table I. The calculation of specific impurity *i* based on ICP-MS analysis shown as:

$$Removal\ degree = \frac{c_i - c_f}{c_i} \times 100\%$$

where c_i and c_f denotes the initial impurity concentration and final concentration.

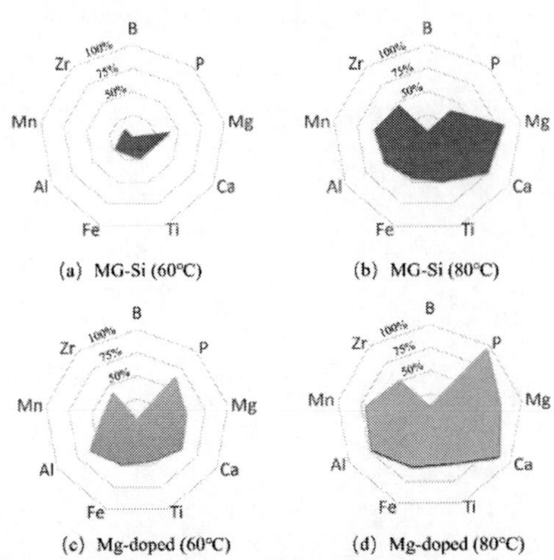

Figure 4: Remained impurities concentration of MG-Si and Mg-doped Si samples after 1h acid leaching by 10% HCl at different temperature in ultrasonic bath.

Owing to the Mg addition, more impurities are removed with the assistance of highly acid reactive Mg_2Si during leaching as well as the P removal. Additionally, higher temperature is in favor of impurity removal by enhancing the reaction kinetics. Therefore, a simple leaching efficiency order can be obtained as MG-Si (60℃)< MG-Si (80℃) <Mg-doped Si (60℃) <Mg-doped Si (80℃) from Figure 4, where the leaching of Mg-doped Si at 80℃ shows the highest impurities removal degree in present work.

4 CONCLUSIONS

A commercial MG-Si was doped with 4.5 wt% Mg and leached by 10% HCl with different leaching temperature. The leaching behaviour of different impurities were studied and the following conclusions have been obtained:

(1) Longer time and higher temperature benefits the impurities removal. A faster chemical control impurity removal stage and a slow diffusion control impurity removal stage were observed.

(2) Mg doping of silicon effectively improved the leaching efficiency for metallic impurities as well as the P removal due to the precipitated acid reactive Mg_2Si eutectic phase.

5 ACKNOWLEDGEMENT

This work was performed at NTNU within the Research Centre for Sustainable Solar Cell Technology (FME SuSolTech, project number 257639), co-sponsored by the Norwegian Research Council and industry partners.

6 REFERENCE

[1] K. Morita, and M. Takahiro. "Thermodynamics of solar-grade-silicon refining. " *Intermetallics*, vol. 11, pp. 1111-1117, 2003.

[2] N. P. Tucker, "Preparation of high purity silicon" *J. Iron Steel Ind.*, vol. 15, pp. 412–414, 1927.

[3] W. Voos, "Production of pure silicon" *U.S. Patent*, 2972521, 1961.

[4] L. P. Hunt, V. D. Dosaj, J. R. McCormick, and L. D. Crossman, "Production of solar-grade silicon from purified metallurgical silicon" in *12th IEEE Photovoltaic Specialists Conference*, 1976, pp. 125–129.

[5] J. Dietl, "Hydrometallurgical purification of metallurgical-grade silicon" *Sol. Cells*, vol. 10, no. 2, pp. 145–154, 1983.

[6] Y. V Meteleva-fischer, Y. Yang, R. Boom, B. Kraaijveld, and H. Kuntzel, "Intermetallics Microstructure of metallurgical grade silicon during alloying re fi ning with calcium" *Meta*, vol. 25, pp. 9–17, 2012.

[7] T. Shimpo, T. Yoshikawa, and K. Morita, "Thermodynamic study of the effect of calcium on removal of phosphorus from silicon by acid leaching treatment" *Metall. Mater. Trans. B*, vol. 35, no. April, pp. 277–284, 2004.

[8] M. D. Johnston and M. Barati, "Calcium and titanium as impurity getter metals in purification of silicon" *Sep. Purif. Technol.*, vol. 107, pp. 129–134, 2013.

[9] H. Lai *et al.*, "Leaching behavior of impurities in Ca-alloyed metallurgical grade silicon" *Hydrometallurgy*, vol. 156, pp. 173–181, 2015.

[10] F. Margarido, J. P. Martins, M. O. Figueiredo, and M. H. Bastos, "Kinetics of acid leaching refining of an

industrial Fe-Si alloy," *Hydrometallurgy*, vol. 34, no. 1, pp. 1–11, 1993.
[11] F. Margarido, M. H. Bastos, M. O. Figueiredo, and J. P. Martins, "The structural effect on the kinetics of acid leaching refining of Fe-Si alloys" *Mater. Chem. Phys.*, vol. 38, no. 4, pp. 342–347, 1994.
[12] Y. Lei, W. Ma, G. Lv, K. Wei, S. Li, and K. Morita, "Purification of metallurgical-grade silicon using zirconium as an impurity getter" *Sep. Purif. Technol.*, vol. 173, pp. 364–371, 2017.
[13] Y. Lei *et al.*, "Leaching behaviors of impurities in metallurgical-grade silicon with hafnium addition" *Hydrometallurgy*, vol. 169, pp. 433–439, 2017.
[14] J. Safarian and T. Gabriella, "Silicon purification through magnesium addition and acid leaching" in *32 European PV conference and Exhibition*, 2016, no. 7491, pp. 1011–1015.
[15] S. Espelien and J. Safarian, "Effect of acid leaching conditions on impurity removal from silicon doped by magnesium" *AIMS Energy*, vol. 5, no. 4, pp. 636–651, 2017.
[16] J. Safarian and S. Espelien, "Hydrometallurgical purification of Maganesium-doped silicon by difference acids" in *33 European Photovoltaic Solar Energy Conference and Exhibition*, 2017, pp. 480–482.
[17] S. Espelien, G. Tranell, and J. Safarian, "Effect of magnesium addition on removal of impurities from silicon by hydrometallurgical treatment" in *Energy Technology 2017*, Springer, 2017, pp. 355–366.

7 APPENDIX

Remained impurities content measured by ICP-MS and the calculated removal degree of each sample are listed in this appendix as Appendix Table I.

Appendix Table I: Leaching performance of 4 samples. The above line shows impurities content in unit ppmw while below indicates removal degree (%).

	MG-Si (60℃)	MG-Si (80℃)	Mg-doped (60℃)	Mg-doped (80℃)
B	52.5 (-13.6%)	42.5 (7.9%)	41 (3.3%)	37.7 (11.0%)
P	15.3 (3.2%)	9.6 (39.0%)	5.6 (64.1%)	0.7 (95.8%)
Mg	5.7 (40.6%)	1.7 (82.6%)	20609.0 (54.1%)	10344.8 (76.9%)
Ca	206.7 (22.5%)	69.4 (74.0%)	122.8 (56.5%)	39.5 (86.0%)
Ti	231.9 (22.6%)	152.8 (49.0%)	112.8 (44.7%)	92.1 (54.8%)
Fe	2602.6 (17.8%)	1730.5 (45.4%)	1173.2 (48.6%)	973.8 (57.2%)
Al	1975.0 (22.5%)	1215.5 (52.3%)	710.0 (58.3%)	487.8 (72.0%)
Mn	60.8 (13.6%)	30.1 (57.3%)	33.3 (38.2%)	16.3 (69.8%)
Zr	9.3 (13.1%)	5.7 (46.4%)	4.1 (41.4%)	3.4 (51.0%)

Phosphorus removal from Al-doped silicon by vacuum refining

Arman Hoseinpur [a*], Jafar Safarian [a,b]
Department of Materials Technology, Norwegian University of Science and Technology (NTNU), 7034 Trondheim, Norway.
b. Norwegian center for sustainable solar cell technology (www.susoltech.no)
*Contact E-mail: arman.h.kermani@ntnu.no, Phone: +47 92508688.

ABSTRACT: In production of solar grade silicon (SoG-Si), it is necessary to remove phosphorus (P) from Si to make the solar cells with proper efficiency. However, P is one of the most problematic elements to be removed from Si and cannot be removed by directional solidification because of its high segregation coefficient. Vacuum refining can be used as an effective method to remove P from Si due to the much higher vapor pressure of P than Si. In this research, the effect of Al addition on the P removal from a commercial Si is studied under controlled vacuum pressures. After refining process, the alloy is characterized by ICP-MS technique. Results show that P is removed from Si with higher rates when Al is added to the melt. In addition, it is seen that Al was removed from the Si considerably in the vacuum refining process.

Keywords: Vacuum, refining, Phosphorus, Silicon, Aluminum, Solar, Kinetics.

1 Introduction

Energy generation by photovoltaic (PV) panels is a clean and sustainable way for production of energy. The energy generation from solar panels increased 3 times over the past 4 years [1]. Currently 90% of the solar energy is produced by Si wafers . Silicon production for PV panels comprise 20% of the total cost for PV panel production. Thus low cost and sustainable production of Si feedstock is necessary for PV industries. Silicon is produced by the carbothermic reduction of SiO_2 in arc furnace [2], and the product of this process is called metallurgical grade silicon (MG-Si) and can have a purity about 99 % . However the Si that is used in PV panels must have at least a purity of 99.999 % (5N) and hence the refining of MG-Si becomes important.

Phosphorous (P) and boron (B) are two impurities that are had to be removed form Si due to their relatively large segregation coefficient (K_P = 0.35, K_B = 0.75). Solvent refining of Si is an effective technique to remove P from the Si, however, the mechanism of solvent refining is to decrease the segregation coefficient of the impurity. For P the segregation coefienct is 0.12, 0.085, and 0.06 when Si solidifies form Si-Al melt at 900, 1000, and 1100 °C, respectively. The best efficiency for P removal has been reported to be 98.5 % at 900 °C [3]. This means the final produced Si meets the required limit of SoG-Si only if the initial content of the P is lower than 10-15 ppm. However if the initial P content is more, the solvent refining process has to be repeated which increases the energy consumption of the process. In addition, it should be noted the Al consumption of the solvent refining increases at lower temperatures and according to Yoshikawa at 900 °C is 2.5 times of 1100°C [3].

Vacuum refining is an effective technique for removing volatile impurities from the melt [4]. As mentioned earlier it is not possible to remove P from Si if the initial concentration of P is high. But vacuum refining can remove P from silicon completely to the acceptable levels for SoG-Si. The driving force for vacuum refining is the difference between the vapor pressure of the volatile impurity and the melt [5–7]. In the case of P, the vapor pressure of P is millions of times higher than the vapor pressure of Si indicating the huge driving force for P removal from Si by vacuum refining.

Yuge et al [8] refined Si by vacuum refining using induction heating and in graphite crucible. The results achieved by Yuge was promising. Safarian et. Al. reported the vacuum refining P from Si in [9,10]. They discussed the various steps of P removal in vacuum refining as follows: of the P diffusion in bulk of the melt, diffusion in through the melt boundary layer, evaporation, and diffusion in the gas boundary layer. Their results showed the evaporation of P from the melt surface and in the gas phase are the slowest steps and control the whole process rate. Safarian [9] also showed Al and Ca evaporate efficiently and thus the vacuum refining applies on this elements too. The intension of this research is to apply vacuum refining for decreasing the P content in a Si-Al alloy to fix the solvent refining limitation on refining of high phosphorus Si.

2 Experimental procedure

In this research a commercial grade of Si (HQ-Silgrian®) and Al (99.99 wt %) were used as the initial materials. The experiments were performed in an induction furnace able to work in vacuum and Ar gas atmosphere conditions. The schematic representation of the furnace is shown in the Fig. 1. In order to make Si – Al alloy, 300 g of Si powder was charged in a graphite crucible. The crucible was wrapped by mica sheet and graphite wool and placed in the copper coil as shown in Fig. 1. The chamber was vacuumed to reach 1 pa and then was purged by Ar (5N) for 3 times before running the experiment. subsequently Ar gas was injected to the chamber and the heating process started when the pressure in the chamber reached to 1 atm. When Si was melted thoroughly Al was added to the melt. After doping, a sample was taken of the melt to determine the composition of the melt. Subsequently the melt

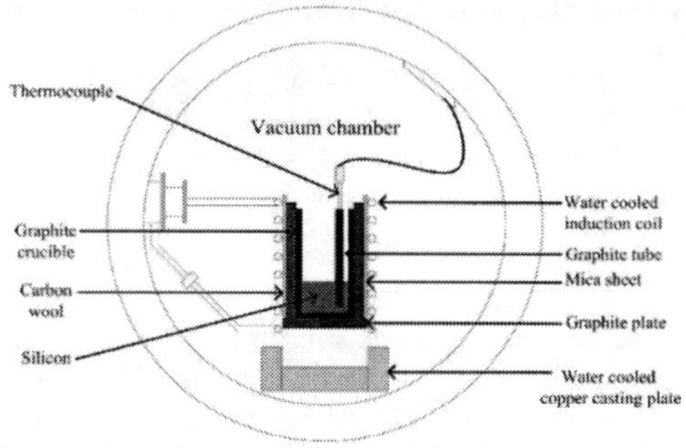

Fig 1. Schematic representation of the furnace chamber, courtesy from [11].

temperature was adjusted on the studying temperature and then the chamber was vacuumed. The pressure of the chamber was decreased immediately until reaching to 60 Pa and the pressure change to lower levels became gradual due to the evaporation of volatile species including of Al and P. Two samples were taken after 30 and 90 min of vacuum refining from the melt. The samples were characterized by ICP-MS to trace the impurities change in the Si – Al melt. Scanning electron microscopy (SEM) was used to investigate the interface of alloy-crucible.

3 Results and discussions

3.1. The challenge of adding Al to Si

Addition of Al to Si melt was challenging. The preliminary experiments showed that it is not possible to hold Al in a graphite crucible at high temperatures. It was observed that the crucible burst some minutes after adding Al, and the melt is drained out completely. Figure 2(a) shows the solidified melt that is drained off the cracks. The inner and outer side of the broken crucible is shown in Figure 2(b). This process was repeated three times and the crucible burst suddenly after adding Al. The formation of aluminum carbide (Al_4C_3) by means of the following reaction is suspected to be the cause of this phenomenon due to the significant volume expansion:

$$4Al + 3C = Al_4C_3 \qquad (1)$$

$$\Delta G°_{1500°C} = -95 \text{ kJ/mol}, \Delta V = +7 \text{ cm}^3\text{/mol}$$

The negative changes in Gibbs free energy at the process temperature shows the high feasibility of the reaction. In

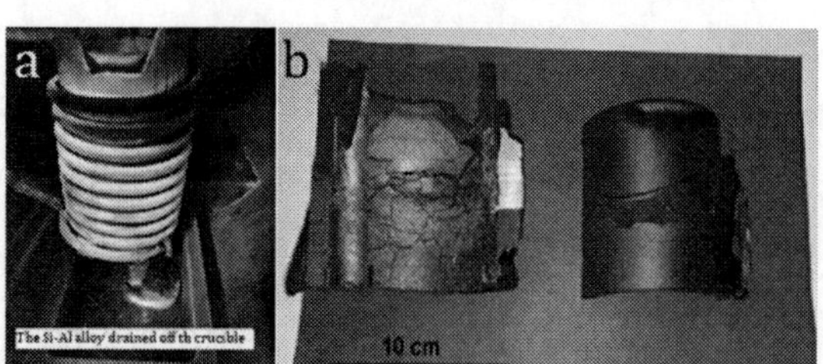

Figure 2: (a) a representation of the solidified alloy that is draned off the crucible due to the bursting of graphite. (b) a representation of the burst crucible

addition, the positive change in volume for the formation of Al_4C_3 leads to strain in the wall of the crucible. Since the graphite has low toughness, this strain leads to sudden crack propagation in the crucible's body. In order to tackle the bursting problem of the crucible it was attempted to form a layer of SiC on the surface of the crucible. SiC can be formed by holding Si melt in graphite crucible by means of the following reaction:

$$Si + C = SiC \qquad (2)$$

$$\Delta G°_{1500°C} = -55 \text{ kJ/mol}, \Delta V = -3.7 \text{ cm}^3\text{/mol}$$

This reaction has also a negative change in Gibbs free energy and is feasible to take place at 1500 °C. The volume change for formation of SiC is negative and thus there will not be expansion in the crucible body due to the formation of SiC. Thus in one experiment the Si melt was hold in the graphite crucible for 2 hours and then Al was added to the melt. This time the crucible withstand the Si-Al alloy and hold it until the end of vacuum refining.

Figure 3(a) shows the intersection of the crucible after the experiment. As can be seen the body of the graphite crucible is completely rigid and no crack can be detected in it. Figure 3(b) shows the interface of the alloy and the graphite crucible. It can be seen the primary SiC layer is formed on the graphite crucible evenly. The corresponding EDS characterizations reveal there is no Al_4C_3 in the graphite body. This layer functions as a barrier and prevents the diffusion of Al elements to the body of crucible. The corresponding EDS characterizations reveals that there is no Al_4C_3 in the graphite body.

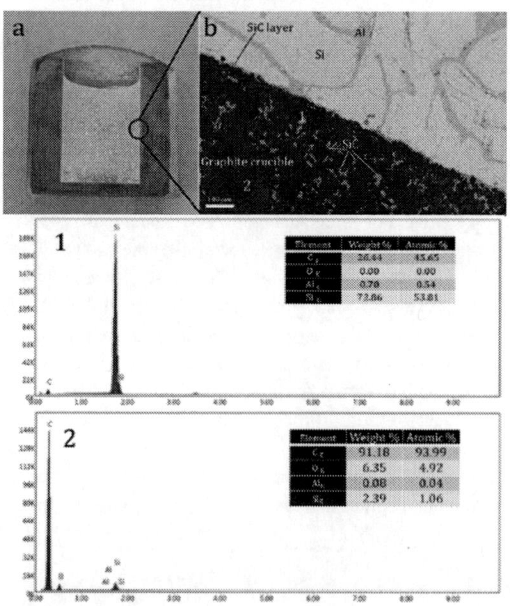

Figure 3: (a) the intersection of the cured crucible after 2 hours of vacuum refining. (b) SEM-BS image of the interface of Si-Al alloy and the treated crucible after vacuum refining. The corresponding EDS characterizations of the selected points are presented below.

In addition it is worthy to investigate the formation of Al_4C_3 by means of the reaction between SiC and Al as follows:

$$3SiC + 4Al = 3Si + Al_4C_3 \quad (3)$$
$$\Delta G°_{1500°C} = +69 \text{ kJ/mol of } Al_4C_3$$

Which shows the reaction (3) is not feasible due to the positive amount of ΔG°. Thus SiC would be a stable layer protecting the crucible during the refining process.

3.2. Phosphorus and Al removal

The effect of Al addition to melt on P removal was studied in the cured crucibles. Figure (4) shows P removal from Si-20 wt.% Al melt at two temperatures. This figure shows the feasibility of removing P by vacuum refining at lower temperatures of melting point of Si. In addition, it can be seen that the P removal was improved by increasing the temperature and more than 60% of the initial P is removed at 1500 °C.

Figure (5) compares the P removal in Si melt and Si-Al alloy at 1500 °C, showing the effect of Al addition. It can be seen that the addition of Al increases the rate of P removal significantly. This shows it is possible to remove P faster when Al is added to the melt.

The change in Al content in the melt at different temperatures is shown in Figure 6. It can be seen that Al evaporates during the vacuum refining significantly and it is also increased by increasing the temperature. The Al content of the melt is decreased to 13 and 11 at 1400 °C and 1500 °C, respectively. This result shows the vacuum refining could be considered as an appropriate technique for Al removal from Si. The positive effect of Al on the P removal could be related to the interaction coefficient of Al with P and no information can be found about it in the literature and it needs further investigations.

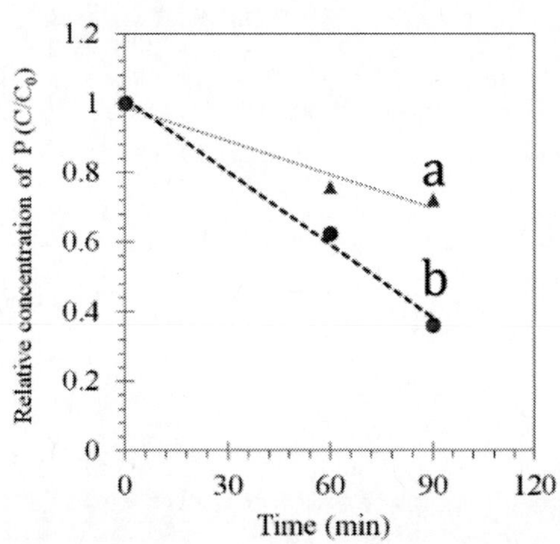

Figure 4: The change of P content in the Si-Al melt by vacuum refining at two different temperatures, (a): 1400 °C, (b) 1500°C.

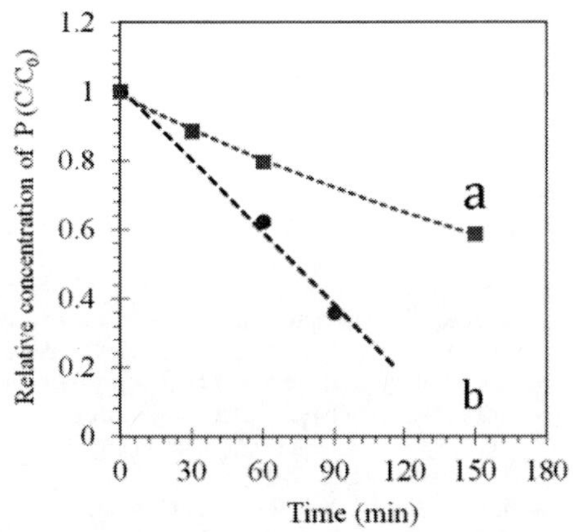

Figure 5: A comparison in change of P content of the melt in vacuum refining of, (a): Si melt, (b) Si-Al melt.

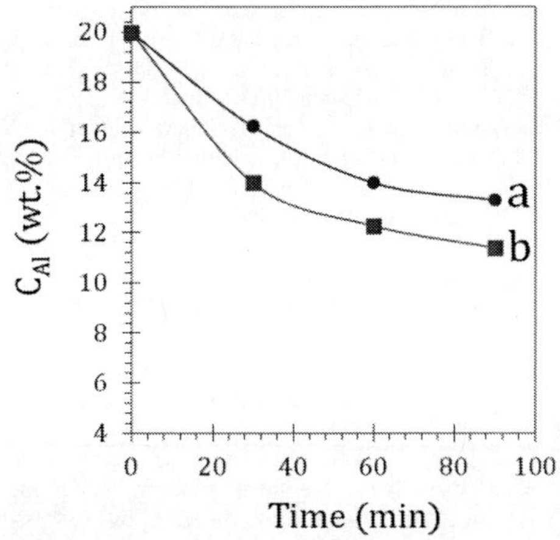

Figure 6: The change of Al content of Si-Al melt in vacuum refining at two different temperatures, (a): 1400 °C, (b) 1500°C.

4 Conclusions

The effect of addition of Al in the vacuum refining of Si was studied. The experiments were done in graphite crucible and it was necessary to treat the crucible before adding Al to Si melt. The treating consisted of holding Si melt for 2h at high temperatures prior adding Al to the melt. SEM investigations revealed the formation of SiC layer on the graphite protects the crucible body against Al diffusion. Results showed the addition of 20 wt. % Al to the Si melt improved the rate of P removal significantly. It was observed that 60 wt.% of the P was removed after 90 min of vacuum refining in 1500 °C when Al was added to melt. In addition, Al evaporated significantly from the melt. The Al content of the melt was decreased 11wt.% and 13wt.% at 1400°C and 1500 °C, respectively.

6 Acknowledgment

This work was financed by NTNU and has been done in cooperation with the Research Centre for Sustainable Solar Cell Technology (FME SuSolTech) in Norway. The support from Mr. Vegar Andersen for materials from Elkem Bremanger is acknowledged.

5 References

[1] Solar energy | Renewable energy | Statistical Review of World Energy | Energy economics | BP, (n.d.). https://www.bp.com/en/global/corporate/energy-economics/statistical-review-of-world-energy/renewable-energy/solar-energy.html (accessed May 4, 2018).

[2] M. Li, Y. Dai, W. Ma, B. Yang, Q. Chu, Review of New Technology for Preparing Crystalline Silicon Solar Cell Materials by Metallurgical Method, IOP Conf. Ser. Earth Environ. Sci. 94 (2017). doi:10.1088/1755-1315/94/1/012016.

[3] T. Yoshikawa, K. Morita, Removal of phosphorus by the solidification refining with Si – Al melts, Sci. Technol. Adv. Mater. 4 (2004) 531–537. doi:10.1016/j.stam.2003.12.007.

[4] E. Ozberk, R.I.L. Guthrie, A kinetic model for the vacuum refining of inductively stirred copper melts, Metall. Trans. B. 17 (1986) 87–103. doi:10.1007/BF02670822.

[5] K. Wei, D. Zheng, W. Ma, B. Yang, Y. Dai, Study on Al Removal from MG-Si by Vacuum Refining, Silicon. 7 (2015) 269–274. doi:10.1007/s12633-014-9228-9.

[6] J. Safarian, T.A. Engh, Vacuum Evaporation of Pure Metals, Metall. Mater. Trans. A. 44 (2013) 747–753. doi:10.1007/s11661-012-1464-2.

[7] C. Zhang, K. Wei, D. Zheng, W. Ma, Y. Dai, Phosphorus removal from upgraded metallurgical-grade silicon by vacuum directional solidification, Vacuum. 146 (2017) 159–163. doi:10.1016/j.vacuum.2017.08.037.

[8] N. Yuge, K. Hanazawa, K. Nishikawa, H. Terashima, Removal of Phosphorus, Aluminum and Calcium by Evaporation in Molten Silicon, J. Japan Inst. Met. 61 (1997) 1086–1093.

[9] J. Safarian, M. Tangstad, Kinetics and Mechanism of Phosphorus Removal from Silicon in Vacuum Induction Refining, High Temp. Mater. Process. 31 (2012) 73–81. doi:10.1515/htmp.2011.143.

[10] J. Safarian, M. Tangstad, Vacuum Refining of Molten Silicon, Metall. Mater. Trans. B. 43 (2012) 1427–1445. doi:10.1007/s11663-012-9728-1.

[11] B.S. Xakalashe, Removal of Phosphorus from Silicon Melts by Vacuum, Norwegian University of Science and technology (NTNU), 2011.

PERFORMANCE OF MODULES AND SOLAR CELLS MADE OF 100% SOLAR SILICON PURIFIED BY DIRECT ROUTE

E. Forniés [1,*], A. Souto[2], T. Vlasenko[3], A. Pérez Vázquez[2], M. Tojeiro[1], M. Anoshenko[4], G. Varela[5].

[1]Aurinka PV Group. Marie Curie Str. 19, 28521, Madrid, Spain
[2] FerroGlobe Group, Marie Curie Str. 19, 28521, Madrid, Spain
[3] FerroSolar OpCo, Ucrania Str. 6, 13500 Puertollano, Spain
[4]Pillar JSC, Severo-Syretskaya Str. 3, 04136, Kiev, Ukraine
[5]FerroGlobe Group, Polígono industrial de Sabón, Spain
* Phone: +34 914994197, E-mail: efornies@aurinkapv.com

After several years of research, FerroSolar has designed a novel but mature process that is able to purify the metallurgical grade silicon (MeG-Si) from 2 nines up to 6 nines of purity. An important step forward has been the capability of the process to reduce the phosphorous and boron concentration down to a level where compensation is enough to obtain a resistivity within the standard parameters of solar cell producers (ρ = 1-3 $\Omega\cdot$cm). This resistivity has been achieved with furnace charges of 100% solar silicon (or UMG). This paper describes the process followed by the solar silicon along the whole value chain, showing the outcomes in each step. MeG silicon has been purified in a multi-step process that involves, the usage of slags and vacuum purification, obtaining the solar grade silicon. Then multi-crystalline ingots were grown by high performance multi process (HPM). The wafers were sent to a solar cell and module producer where underwent through and standard process of manufacturing according to Al-BSF technology. Solar cells of up to 18.84% efficiency and modules of up to 333 W were obtained. This performance is particularly high, considering that no high efficiency technologies have been applied to the cell manufacturing.
Keywords: UMG, silicon purification, solar cells

1 INTRODUCTION

We are facing a moment when photovoltaic energy has become an energy which is able to compete with other feed in resources independently whether they are fossil or renewable. To get to this point an important reduction of prices has taken place in the last decade (see Fig.1).

In crystalline silicon PV in particular, the whole value chain has contributed to reach this goal by reducing the production costs of materials and a constant increase in the efficiency.

An important contribution to this costs reduction has been brought by the hand of the polysilicon producers, who have adapted their processes to the solar industry requirements, producing the so-called solar grade silicon (SoG-Si). The traditional way of producing polysilicon, indirect or chemical route, has reduced its ASP (see Fig. 2) by increasing the productivity and by the reintroduction of byproducts flows and energy wastes into the main stream. That reduction of polysilicon production costs reduced the costs percentage of polysilicon respect to module price from 40% approximately in 2007 to 20% (including MeG-Si) nowadays.

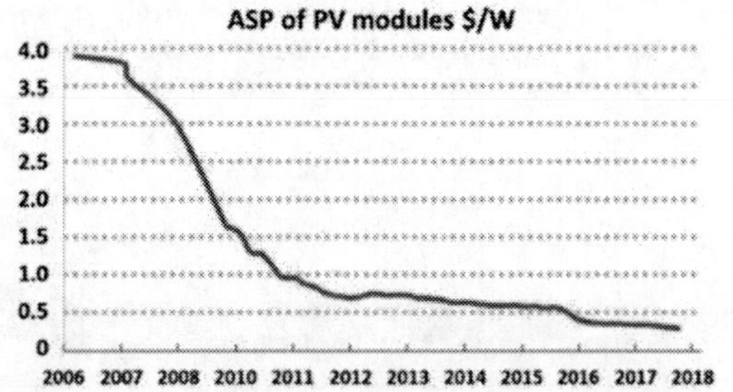

Figure 1: Average selling price (ASP) of PV modules since 2006.

Another contributor to the price dropping of SoG-Si has been the direct route industry [1], which main representatives are currently Elkem and FerroSolar. In mentioned direct route the silicon is purified by means of metallurgical processes. In particular, this paper assesses the silicon purified by FerroSolar route which has been used to manufacture wafers, solar cells and modules. All the processes used to transform the silicon into modules will be described as well as the results obtained in each step of the value chain.

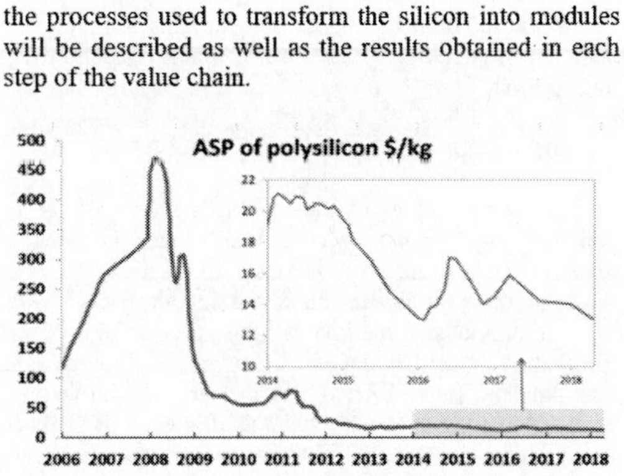

Figure 2: Average selling price (ASP) of polysilicon from 2006 to nowadays.

2 PROCESS DESCRIPTION AND RESULTS

2.1 Silicon purification

Metallurgical Grade Silicon (MeG-Si, chemical purity 98.5%-99.0%) is obtained by carbothermic reduction of silica (SiO_2) in submerged electric arc furnaces. Typically, Boron content in this silicon is in the range 10-40 ppm_w, Phosphorous 10-50 ppm_w and metals (Al, Ca, Fe, Ti, etc) 1500-2000 ppm_w.

The increasing demand of high purity silicon for photovoltaic industry (Solar Grade Silicon, SoG-Si) resulted in an intensive research in universities and companies for the search of cheaper and less polluting purification routes than traditional Siemens process. The result of these efforts is the development of different purification procedures at the industrial level, as directional solidification, slagging processes, acid leaching, treatments under vacuum [2] or reactive gases, among others. The efficiency of these procedures is different for each impurity, depending on their chemical properties [1].

Since 2000, Ferroglobe Group has been developing a purification process of MG-Si by the metallurgical route. The process starts with the selection and purification of raw materials used in the production of MG-Si. Following a series of successive steps allows to produce SoG-Si at industrial scale in a cheap and efficient way, without the environmental impact associated with the Siemens process. Basically, the developed process comprises of three stages. The first stage consists in the treatment of molten silicon with slags (slagging process) in induction furnaces. Once the slagging process is concluded, molten silicon is cast in suitable molds and subjected to controlled solidification (directional solidification). Last step of the purification process is a vacuum treatment to remove volatile impurities [2].

The process used for silicon purification is depicted in Fig. 3.

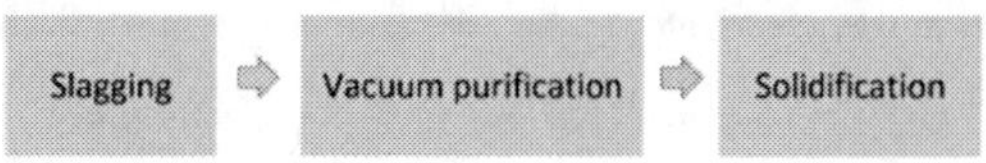

Figure 3: Sketch of the process of silicon purificacion by metallurgical route.

The result of this sequence of treatments is silicon with a purity higher than 99.9995%. Typical concentrations reached are:

$[B] < 0.2$ ppm_w, $[P] < 0.2\text{-}0.4$, $\sum [Me] < 0.5$ ppm_w.

Maximum values of total metals concentrations are below 5 ppm_w. The energy cost of this process is about 1/3 than Siemens, while the CO_2 emissions are about 1/4 -1/5. This silicon is suitable for photovoltaic industry.

In previous experiments the processing of 1st generation of this product with conventional methods in the industry (tests carried out in tier 1 manufacturer) resulted in solar cells with average efficiency of 18.15%, close to the 18.30% obtained with the reference polysilicon. With the same material and using special texturing techniques, average efficiencies of 19% have been reported. Maximum efficiency reached in this case was 19.2% [3].

2.2 Silicon growth

SoG-Si from the former process underwent a directional solidification to grow multi-crystalline ingots. As the silicon is already compensated only small concentrations of dopants were added to fine tune the resistivity profile. Fig. 4 shows the resistivity profiles of a representative G5 ingot measured at positions 5 (corner), 11 (side) and 13 (center). The technique used to determine the bulk resistivity is Eddy current (Semilab RT-1000). In the same picture the Klaassen model [4] has been used to calculate the resistivity with the input data of dopants concentration.

As it can be seen in the picture the model overestimates the resistivity from the half height of the ingot to the top.

The drop of resistivity at the bottom part of ingot is related to the method of seeding and nonuniform crystallization at first steps of growing.

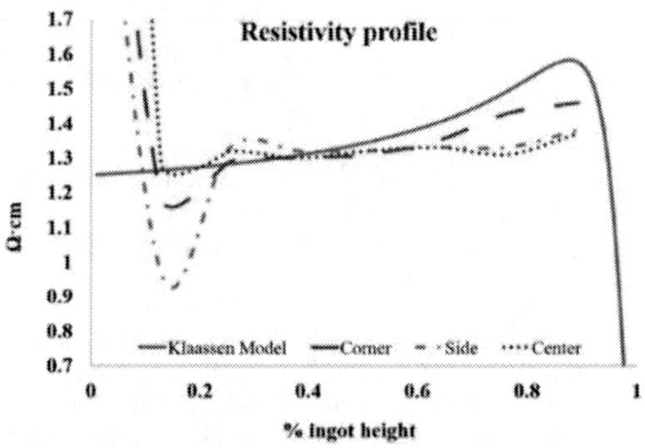

Figure 4: Resisitivity profile of one of the representative ingots grown meassured at three differen positions, Center, Side and Corner. Klaassen model has been included for comparison.

Several HPM silicon ingots were grown in identical industrial DS G5-sized furnaces using the same materials and consumables, with seeding on fine polycrystalline silicon chunks.

New solar grade silicon generation made by FerroSolar was used as 100% feedstock and 650 kg solar grade silicon was loaded into the silica crucibles with dimensions of 890 × 890 × 540 mm. Prior to loading, all crucibles were coated with Si_3N_4 to prevent sticking and to reduce silica erosion and impurity dissolution. Prior to growth, small quantities of dopants were added to ensure a resistivity ranging from 1.0 to 3.0 Ω·cm.

The growing technique used was HPM growth which gets specific structure with small grains and with random orientation along the ingot height. Between the different methods to obtain the HPM, seeding with polysilicon chunks has been used. The polysilicon chunks, with a predefined grain size in the range 0.5 – 5 mm, are loaded into the bottom part of the crucible, just below the charge of the UMG silicon feedstock [5].

2.3 Solar cells and modules

Standard Aluminum Back Surface Field (Al-BSF) solar cells were produced in a tier 1 solar cells supplier. All the cells were produced in the same line in one batch. In Fig. 5 the production sequence is depicted.

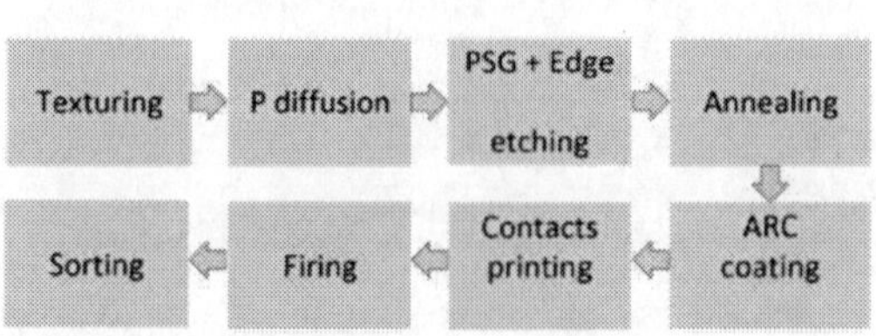

Figure 5: Sketch of the process of Al-BSF solar cells production.

Approximately 3000 solar cells were produced. Table I shows the average data for the cells made of FerroSolar silicon. The baseline represents the average data of solar cells obtained with standard polysilicon coming from indirect route. The baseline cells were produced in the same production line than the Ferrosolar's solar cells. The same materials and process parameters were used for both solar cells. The average efficiency of solar cells made of UMG silicon as base material were 18.52%, just 0.6 $\%_{rel}$ less than those made of polysilicon. Maximum cell efficiency achieved was 18.84%.

A total yield (breakage and rejection) of this run test was above 97.5%, which is inside the acceptance range of the production line.

Table I: Average electrical data of solar cells produced with UMG silicon feedstock and solar cells produced with polysilicon (baseline).

Type of wafer	Pmpp [W]	Uoc [V]	Isc [A]	FF %	Eff %	Rsh [Ω]	Rs [mΩ]	Total Yield %
FerroSolar	4.55	0.63	9.03	79.74	18.52	141	0.94	>97.6
Baseline (polysilicon)	4.58	0.63	9.06	79.69	18.64	131	0.90	--

As it is shown in Pic. 6, the efficiency distribution presents a peak at 18.50% classification.

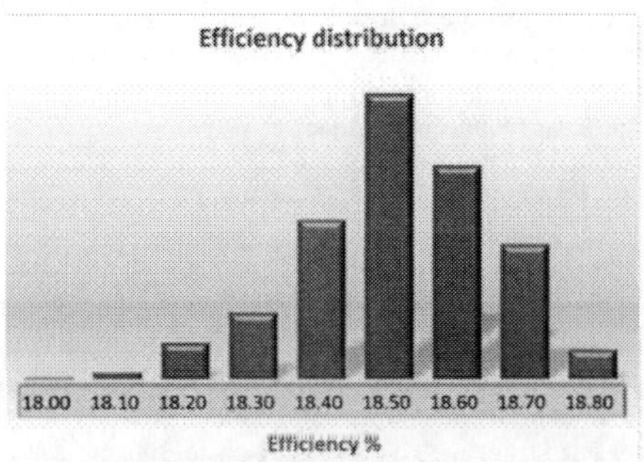

Figure 6: Efficiency distribution of the solar cells produced with UMG silicon material.

The solar cells underwent standard quality controls, as a light induced degradation test (LID test) and electroluminescence test (EL test). Pic. 7 shows a representative footage of one of the cells under EL test. Solar cells passed the LID test with less than 1% of degradation and EL test was also resulting in "pass marking". Additionally, adherence pulling test was carried out with acceptable results.

Figure 7: Picture of an EL test of a representative solar cell produced with UMG silicon material.

Finally, sample PV modules were produced in a standard process, which is described in Pic. 8. The electrical data of produced modules can be seen in Table II.

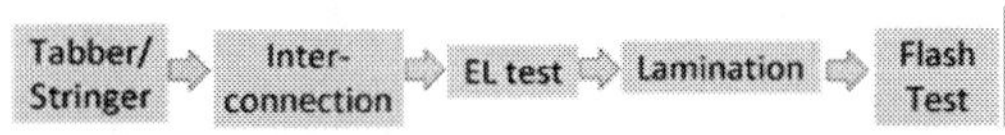

Figure 8: Sketch of the process of PV modules manufacturing.

Table II: Electrical data of the modules produced with solar cells made of UMG silicon feedsotck.

Cell Eff %	FF %	Isc [A]	Voc [V]	Ipm [A]	Vpm [V]	Module Pmpp [W]
18.80	76.76	9.37	46.30	8.81	37.80	333
18.70	76.76	9.36	46.17	8.81	37.61	332
18.60	76.62	9.35	46.03	8.79	37.48	330
18.50	76.54	9.29	45.81	8.76	37.18	326
18.40	76.32	9.25	45.90	8.70	37.24	324
18.30	76.43	9.28	45.60	8.70	37.16	323

3 CONCLUSIONS

And industrial test with UMG silicon feedstock has been performed involving all the steps of the value chain. Average solar cell efficiency of 18.52% and maximum efficiency of up to 18.84% were achieved. Results obtained in each step of the value chain reveals good performance results and acceptable LID confirming the UMG silicon from Ferrosolar's process as an alternative for multi-crystalline solar cells.

This test was performed with 100% solar-grade silicon produced by Ferrosolar demonstrating nearly the same efficiency than polysilicon material at least for Al-BSF technology.

Concerning solar cells, mass production test, PERC and black silicon tests are in process. Experiments of module and cell degradation and thermal coefficients determination are also being carried out.

4 REFERENCES

[1] E. Forniés, L. Méndez, M. Tojeiro, Photovoltaics International 31 (2016) 29-38.

[2] A. Souto, J. Bullón, R. Ordás, J. M. Míguez, Silicon for the Chemical and Solar Industry XII (2014) 67-76.

[3] V. Hoffmann, S. Zhou, J. M. Míguez, X. Su, 32nd European Photovoltaic Solar Energy Conference (2017), 613-616.

[4] D.B.M. Klaassen, Solid-State Electronics 35 (1992) 953-959

[5] I. Buchovska, O. Liaskovskiy, T. Vlasenko, S. Beringov, F. M. Kiessling, Solar Energy materials & Solar Cells 159 (2017) 128-135.

COST EFFECTIVE GROWTH OF SILICON MONO INGOTS BY THE APPLICATION OF THE MULTIPULLING TECHNIQUE COMBINED WITH ACTIVE CRYSTAL COOLING

F. Mosel[1], A.V. Denisov[1], B. Klipp[1], N. Sennova[1], M. Herms[2], R. Kunert[3], P. Dold[3]
[1] PVA Crystal Growing Systems GmbH, Im Westpark 10-12, 35435 Wettenberg, Germany
[2] PVA Metrology & Plasma Solutions GmbH, Am Naßtal 6/8, 07751 Jena-Maua, Germany
[3] Fraunhofer CSP, Otto-Eissfeldt-Str. 12, 06120 Halle/Saale, Germany
frank.mosel@pvatepla.com, phone: +49 64168690-125, fax: +49 64168690-822

ABSTRACT: The permanent demand for the improvement of solar cell efficiency and reduction of production costs has placed much focus on the multiple Cz-pulling techniques from a single crucible. PVA Crystal Growing Systems GmbH has developed a mobile recharge system for Czochralski pullers, the SiCharger MRS. This system can operate several Cz-pullers with in a consecutive mode and is characterized by low investment costs, reduced silicon consumption and higher output per crucible. This system and its potential were presented in detail at the 33rd EU PVSEC [1]. Another way to save costs due to an increased productivity is a higher pull rate. In order to enable a higher pull speed under stable crystal growth conditions, the hot zone was investigated and optimized by simulating the entire crystal growth arrangement with the commercial program package CGSim [2]. As a result, the pull rate could be increased significantly. In a next approach, an active crystal cooling system was implemented. This cooling device itself consists of a water cooled container, which is placed above the phase boundary of the growing crystal, in order to increase the axial temperature gradient in the growing crystal close to its solidification interface. This system is described in detail in [3]. With this arrangement the heat transport through the crystal is improved so that crystals with 8 inch diameter were grown with a pull rate of 1.8 mm/min resulting in an appreciable saving of process costs.
Keywords: c-Si, multipulling, active crystal cooling, melt replenishment, growth rate

1 INTRODUCTION

About 90% of the solar cells produced in the world are based on crystalline silicon (c-Si). The wafer market is already dominated by monocrystalline Si-material and the trend of increasing market share will continue as shown in fig.1. It is expected that in the next 10 years the market share of mono-Si will surpass the 60% mark. This market development is driven by the increasing demand for high-performance cells, but in presence of further increasing cost pressure. The cost reduction cannot be achieved by the wafer manufacturers only by reducing the consumption costs. A drastic increase in productivity while increasing product performance is essential.

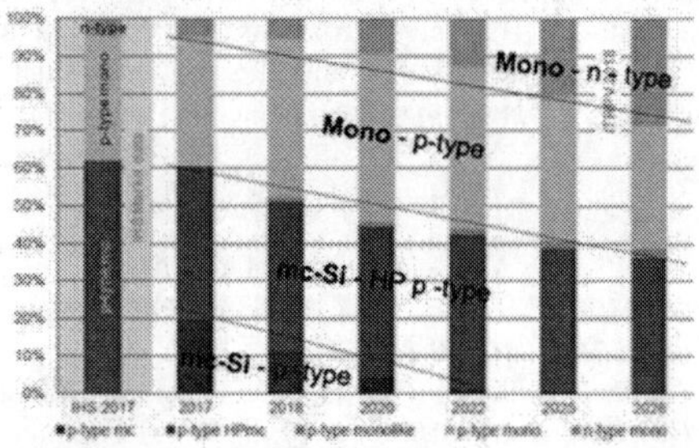

Figure 1: World market shares for wafer / ingot types. (International Technology Roadmap for Photovoltaic Results 2017 – Presentation)

Since the existing Czochralski pullers for the PV industry are limited regarding the melt volumes (crucible sizes) or the resulting crystal lengths, the pulling of several crystals from a single crucible (multipulling) is a necessary consequence. To increase productivity, maximizing the pull rate is the most powerful tool in the Cz-process. By a combination of both methods a maximum effect concerning cost savings can be achieved. In the following sections we describe the equipment for these low cost process conditions. The material properties of the crystals grown with this equipment are presented, an economic consideration of this process technique is given.

2 EXPERIMENTAL

2.1 Crystal growth equipment

The crystal growth experiments presented in this paper were performed in a SC24/26 Czochralski-puller from PVA Crystal Growing Systems GmbH. The crystal growth experiments were performed in a 24 inch hot zone. For larger melt volumes of up to 300 kg, a 26 inch hot zone will be available soon for this puller type.

2.2 Hotzone design with and without active crystal cooling

The physical limiting parameter for the growth rate of ingots in a Czochralski configuration is the dissipation of the latent heat of fusion of silicon at the interface crystal/melt by heat conduction through the growing crystal. The heat received by the ingot in contact with its melt can be determined as the heat of solidification plus the heat flux from the melt into the ingot. This situation can be written in a simple one dimensional model:

$$\lambda_s \cdot A \cdot \left(\frac{dT}{dx}\right)_s = \lambda_l \cdot A \cdot \left(\frac{dT}{dx}\right)_l + \rho \cdot A \cdot L \cdot v_p \quad (1)$$

in which λ is the thermal conductivity, A is the cross section of the ingot, (dT/dx) is the axial temperature gradient, v_p is the constant growth velocity, ρ is the crystal density, L is the latent heat of fusion and s,l are the subscripts for the solid and liquid phase. Assuming an isothermal melt i.e. $(dT/dx)_l = 0$ the maximum pull rate v_{pmax} can be estimated and related to the crystal radius r:

$$(2) \quad \lambda_s \cdot A \cdot \left(\frac{dT}{dx}\right)_s = \lambda_l \cdot A \cdot \left(\frac{dT}{dx}\right)_l + \rho \cdot A \cdot L \cdot v_p$$

Here σ is the Stefan-Boltzmannn constant, ε is the surface emissivity, m is the subscript for melting temperature. For a detailed derivation of the equations (1) and (2) see E. Billig [4].

To come closer to the theoretical limit of the maximum pull speed v_{pmax} the SC24-puller was equipped with an active crystal cooling (acc) system. Fig.2 shows a sketch of the active crystal cooling system which consists of a water cooled container in form of a coil which is placed between the inner heatshield and the growing crystal. The purpose of the cooling device is to alter the axial temperature gradient at the crystallization front in the crystal. The main influencing parameters of the cooling element are the cooling power given by the geometry and the cooling water flow, the vertical distance to the melt surface, the radial distance to the crystal and the emissivity of the cooling element. The crystal growth configuration including the active cooling system was developed and optimized by means of numerical simulation. calculations using the CGSim-

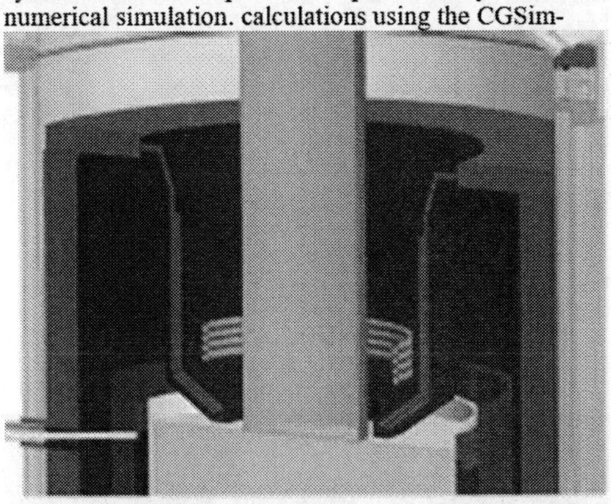

Figure 2: Arrangement of the active crystal cooling device in the hotzone

software-package from STR [2]. The most important evaluation criteria of the hotzone are the deflection of the phase boundary (concavity) and the radial temperature distribution on the melt surface. The phase boundary should be as plane as possible and the radial temperature gradient at the triple point for stable growth conditions should be steep. Fig.3 shows the interface curvature H of the phase boundary versus the average pull speed in the body phase at a body length of 500 mm for the different crystal growth configurations examined in this project (for details see [3]). Three different regions are shown in the diagram. In the stable growth region (green) the system is insensitive to changes in the average pull speed. This region is robust and suitable for industrial production. In the metastable growth region (yellow) all growth parameters have to be well tuned. Small changes can lead to unstable growth with loss of shape, i.e. spiral growth. In the unstable growth region (red) no regular crystal growth is possible. In the graph the points connected by the lines represent individual results of the simulation calculations for the examined crystal growth arrangements (V1-V5). The red circles in the graph represent experimental results. The configuration V1 is a standard configuration with a pull speed of 0.9 mm/min in the body phase. In the optimized configuration V2 an average pull speed of 1.3 mm / min could be established without an active crystal cooling. The configuration V5 with active crystal cooling shows the best values in terms of pull speed and energy saving by the optimized hot zone configuration (see tab.3). Configurations V3 and V4 were intermediate steps and will not be considered further.

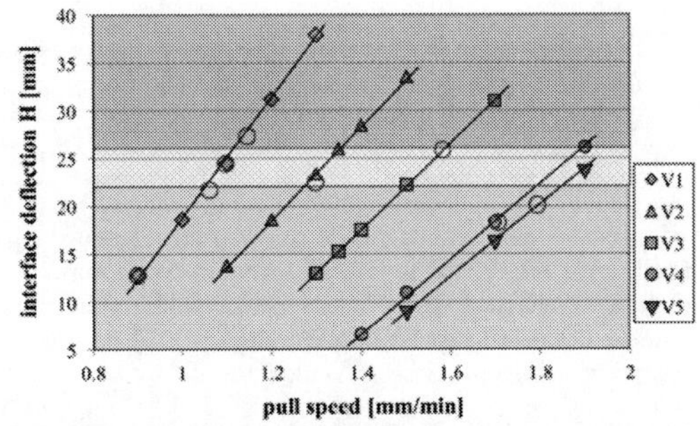

Figure 3: State diagram of the stable (green), metastable (yellow) and unstable (red) growth conditions for the different hotzone configurations. The main parameter: deflection H of the interface shape is plotted versus the pull speed and shown in fig.4.

Fig. 5 shows the dependence of the phase boundary deflection from the distance of the inner heat shield to the melt surface for different pulling rates in the growth configuration V5. In addition to the deflection of the phase boundary, the radial temperature distribution is an essential parameter for stable growth conditions. The radial temperature distribution on the melt surface depends on the cooling element in connection with the growth rate. As the pull rate increases, more heat of crystallization is released which reduces the radial temperature gradient and the formation of a supercooled area near the solidification front may occur, interfering with stable growth of the constant diameter crystal [5]. The dependence of the radial temperature distribution on the pull rate on the one hand and on the distance between the inner heat shield and the melt surface on the other hand is shown in figs.6 and 7. Together with the phase boundary deflection, these parameters must be reconciled to achieve optimal growth conditions.

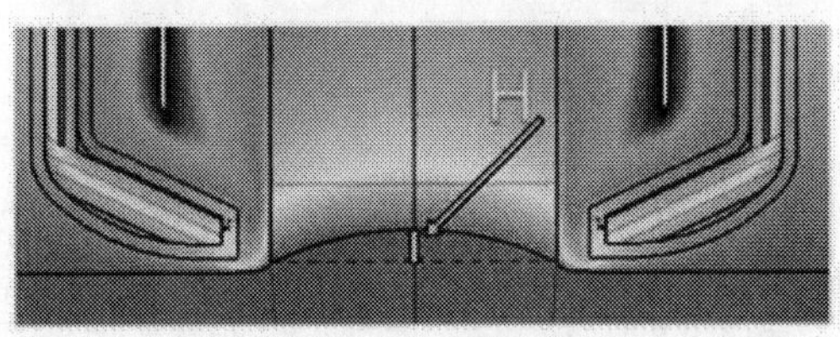

Figure 4: Sketch of the cooling device and the inner heat shield. H indicates the deflection of the phase boundary.

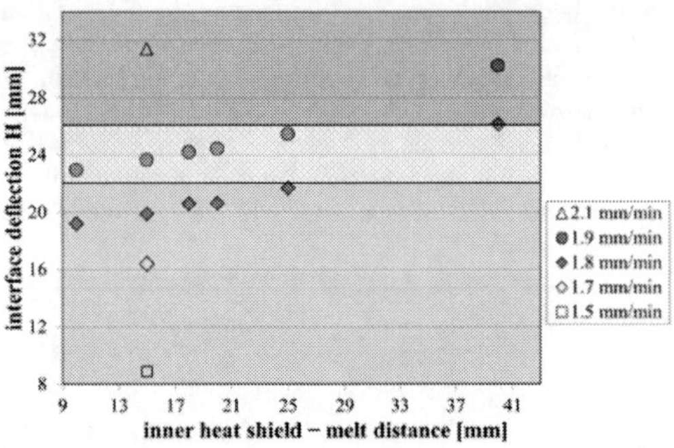

Figure 5: State diagram of the stable (green), metastable (yellow) and unstable (red) growth conditions for different distances of the inner heat shield to the melt surface in crystal growth configuration V5. The symbols represent results of numerical calculations.

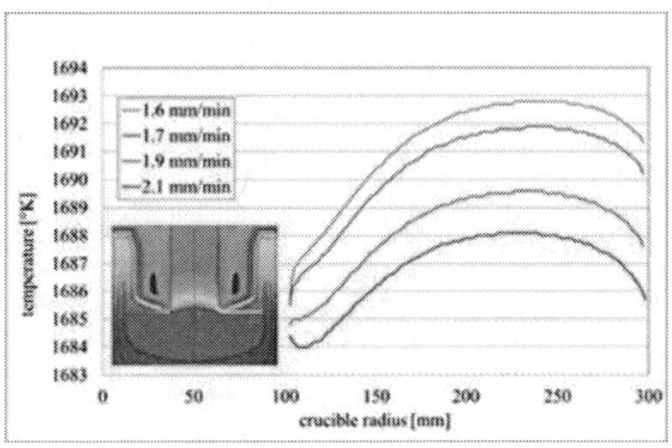

Figure 6: Radial temperature distributions (yellow marker in the inserted picture) on the melt surface for different growth rates in crystal growth configuration V5

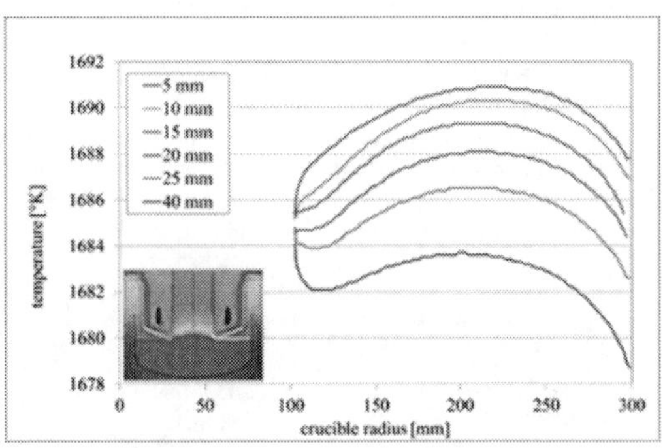

Figure 7: Radial temperature distributions (yellow marker in the inserted picture) on the melt surface for different distances between the melt surface and the inner heat shield in crystal growth configuration V5

The hotzone and crystal growth parameters used in this work were also evaluated for the calculated stresses in the crystals. Assuming isotropic material properties, the von Mises-stresses can be estimated. The von-Mises-stress is a theoretical value expressing a three-dimensional stress state as a scalar. The von-Mises-stress is therefore a viable criterion for judging crystal growth conditions with regard to the probability of dislocation formation [6]. A value of 80 MPa is considered as a critical value for the growth conditions of 8-inch silicon crystals [7]. The results, summarized in tab.2 later in this report, are all well below the critical value.

2.3 Mobile Recharge System

In the standard Cz batch process, one single crystal is pulled out of a single crucible. After this process is finished and the system is cooled down, the next process is prepared. This type of process management is very time-consuming and cost-intensive. In each process, one crucible is consumed and the hotzone is heavily stressed by the temperature cycles. In order to extend hotzone lifetime and to reduce crucible consumption, the multipulling process has been established, in which several crystals are sequentially pulled out of one crucible. Between two growth runs, the hot crucible is then refilled with silicon, which can be done in different ways [1]. The PVA Crystal Growing Systems GmbH has developed a mobile recharge system, which offers the unique possibility to serve several Cz-pullers time-delayed. To make the mobile feeding unit compact and easy to move, the hopper is separated from the passive conveyor, a quartz tube system. The connection of the hopper with the quartz tube is realized by a motor-driven active vibratory conveyor. As an advantage, the feeding rate can be predetermined and regulated, which in turn allows a high degree of automation of the feed process. Fig.8 shows the mobile recharge system in the process.

Figure 8: Mobile recharge system docked on Cz-puller

The interface from the feeder to the system is a special valve, which can be retrofitted to the special puller type. A more detailed description of the MRS system and the feed process can be found in [1].

One disadvantage of the multipulling technique may be the accumulation of impurities in the melt due to the segregation effect. An estimation of impurity accumulation in the melt due to the sequential melt replenishment has been published by Hopkins et al. [8].

The axial concentration profiles in Cz-ingots grown from replenished melts can be calculated as a function of the solidified fraction g, with C_0 is the initial concentration of the impurity in the liquid. The initial impurity concentration C_L^i (n) in the melt at the start of the n-th pull (3) can be calculated according to Hopkins with p = $(1\text{-}g)^{k_{eff}-1}$.

$$(3)\ C_L^i(n) = C_0\left[p^{n-1} + g\left(\frac{p^{n-1}-1}{p-1}\right)\right]$$

Fig. 9 shows the accumulation of impurities with a small segregation coefficient in the melt calculated for the sequential replenishment of the crucible. The refill is calculated after solidified fraction of 90% (g = 0.9).

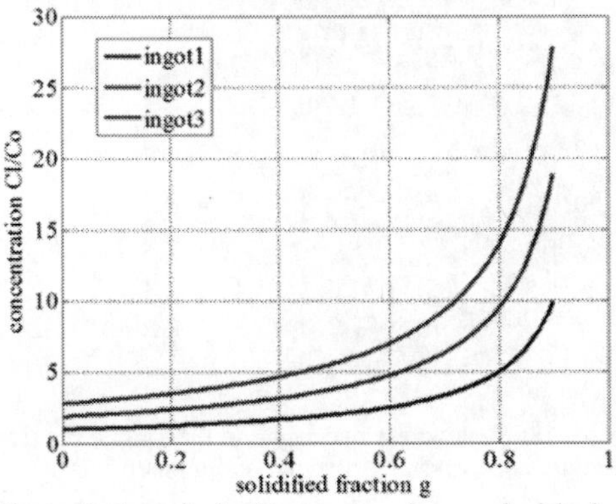

Figure 9: Impurity build-up curves for a replenished melt for the growth of 3 ingots in one crucible calculated according to Hopkins et al. [8].

For the calculation an effective distribution coefficient $k_{eff} = 10^{-3}$ was applied, which is in the range for metal impurities, having a strong impact on the solar cell performance [9]. It can be estimated that the maximum accumulation of impurities in the melt with small segregation coefficients is less than a factor of 3.

2.4 Crystal growth and characterization

In our study, crystals were grown in different crystal growth configurations under the aspect of a high pull rate. All crystals were n-type with an axial distribution of the resistivity in the range of 4-1 Ωcm from top to tail. Tab.2 gives an overview of the studied crystal growth configurations and the main growth parameters. From the crystals labeled V1, V2, V5 grown in the configurations V1, V2 and V5 five slices were prepared equidistantly over the length of the ingots according to fig10. Additionally a longitudinal section was cut out for the investigation of the interface shape by means of LPS technique [10].

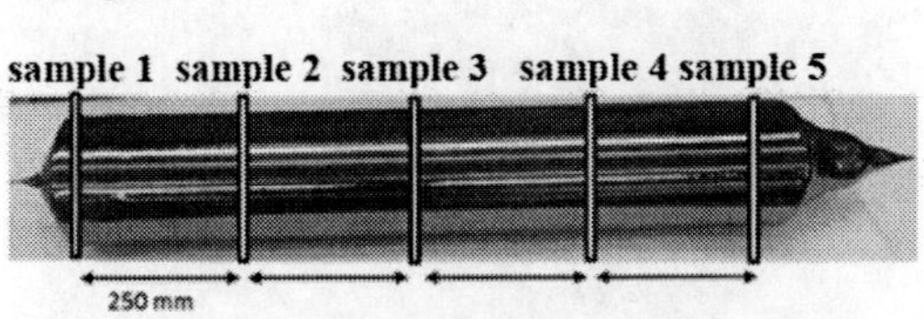

Figure 10: Position of the samples for characterization corresponding to a solidified fraction of g = 0.02 to 0.8.

The samples were subjected to FTIR measurements for the determination of interstitial oxygen and substituted carbon concentrations. The conversion factor is 3.14×10^{17} cm^{-2} for O_I and $1.0 \times 10^{17} cm^{-2}$ for C_S. Measurements were taken 7 mm from the edge, half radius and in the center. The samples are designated 1 to 5 from top to tail. The results for the O_I-concentrations are summarized in tab.1. The C_S-concentration in 95% of all samples were below detection limit ($<10^{16}$ cm^{-3}).

config. V1	sample 1	sample 2	sample 3	sample 4	sample 5	
center	10.8	9.1	8.2	7.5	6.5	x 10^{17} cm^{-3}
half radius	10.4	8.9	8.1	7.7	6.4	x 10^{17} cm^{-3}
edge	6.3	5.7	4.6	5.6	5.8	x 10^{17} cm^{-3}
config. V2	sample 1	sample 2	sample 3	sample 4	sample 5	
center	9.5	6.6	5.8	4.6	4.7	x 10^{17} cm^{-3}
half radius	9.2	6.6	5.6	4.0	4.3	x 10^{17} cm^{-3}
edge	3.6	3.2	2.3	1.6	3.6	x 10^{17} cm^{-3}
config. V5	sample 1	sample 2	sample 3	sample 4	sample 5	
center	9.8	6.9	6.7	7.8	9.3	x 10^{17} cm^{-3}
half radius	9.6	6.4	6.1	7.5	9.0	x 10^{17} cm^{-3}
edge	7.6	5.2	3.7	7.7	6.4	x 10^{17} cm^{-3}

Table 1: O_I-concentration in the crystals V1, V2, V5.

The measurement results do not have any statistical certainty. However, it can be concluded that the high pull speed in the configurations V2 and V5 does not have a negative effect on the distribution of oxygen and carbon content.

The minority carrier lifetime measurements on the crystal slices were measured with a Sinton BCT-400 / BLS-I (photoconductivity measurement) using the transient measuring method. The lifetime values were determined at a minority carrier density of 5×10^{14} cm^{-3}, which is usual for n-type material and are shown in figs. 11-13. The minority carrier lifetime is plotted versus the radial position on the different samples. The numbering of the samples (1 - 5) in tab.1 is replaced by the specification of the solidified fraction g (0.02 – 0.8).

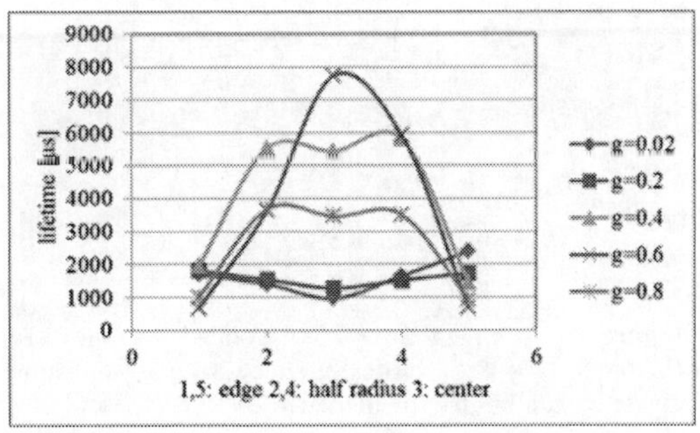

Figure 11: Minority carrier lifetime [µs] measured on the samples grown in crystal growth configuration V1

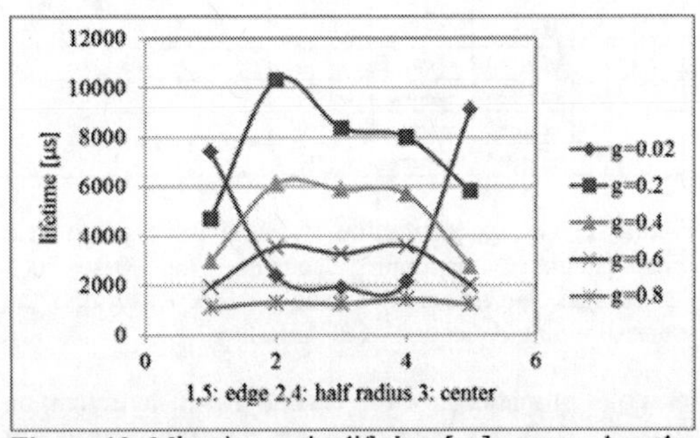

Figure 12: Minority carrier lifetime [µs] measured on the samples grown in crystal growth configuration V2

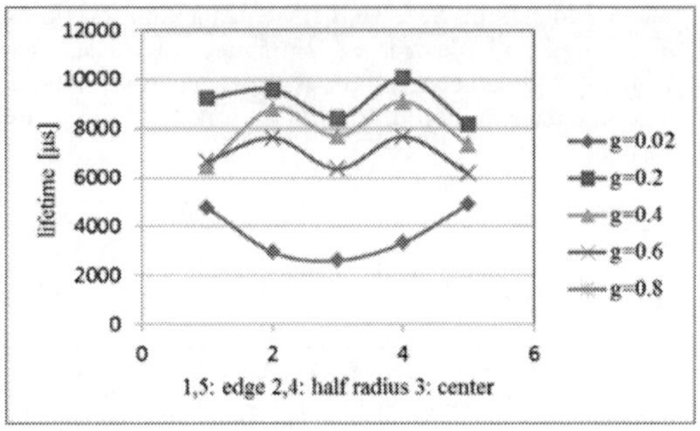

Figure 13: Minority carrier lifetime [µs] measured on the samples grown in crystal growth configuration V5. The sample 5 (g=0.8) was broken and could not be analyzed.

The distribution of the lifetime-reducing recombination centers should not be investigated or discussed in this context. However, it can also be stated here that the active crystal cooling in connection with the high pull speed does not appear to have a negative effect on the minority carrier lifetime distribution.

The phase boundary shape (concavity) was visualized on the longitudinal section using LPS technique at the Fraunhofer Center for Silicon Photovoltaics CSP. The results for the different crystal growth arrangements are shown in tab.2 and confirm the state diagram derived from the numerical simulation calculations in fig.3.

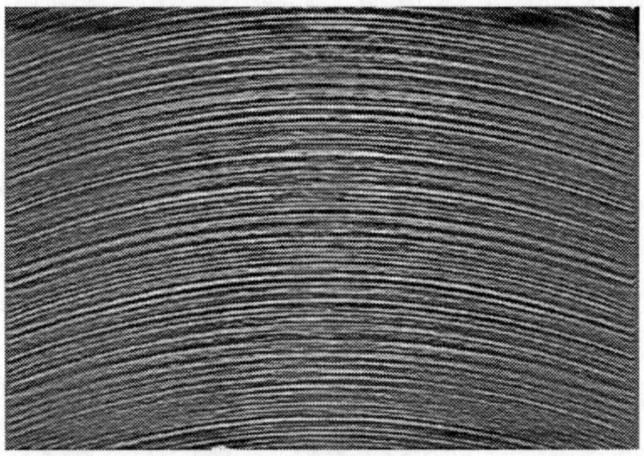

Figure 14: Growth striations revealed by the LPS technique [10] on a vertically sliced sample, indicating the deflection of the growth interface

growth configuration	hotzone	active crystal cooling	max. σ_{vM} [Mpa]	mean pull speed in body phase [mm/min]	measured deflection of interface in crystal center [mm]
V1	standard	no	$3.9*10^7$	0.9	13
V2	optimized	no	$4.6*10^7$	1.3	22
V3	optimized	standard	$4.8*10^7$	1.6	24
V4	optimized	optimized	$5.5*10^7$	1.7	21
V5	optimized	optimized	$6.5*10^7$	1.8	20

Table 2: Growth configurations with main parameters and calculated maximum occurring von Mises σ_{vM} stresses in the crystal at a length of 500 mm and the corresponding deflection of the interface.

Samples of ingots grown in crystal growth configuration V1, V2 and V5 were also evaluated by means of Scanning Infrared Stress Depolarization (SIRD). For this characterization technique we applied the SIRD system from PVA Metrology & Plasma Solutions GmbH. As silicon is optically isotropic in the unstressed state, residual stresses which may be present in the crystal due to the growth conditions lead to an optical anisotropy (stress birefringence), that can be evaluated with the SIRD system. A detailed description of this method can be found in [11]. Figs.15 -17 show three slices from the middle of the body of each examined crystal. For the characterization the samples were just etched. Therefore some saw marks can be seen.

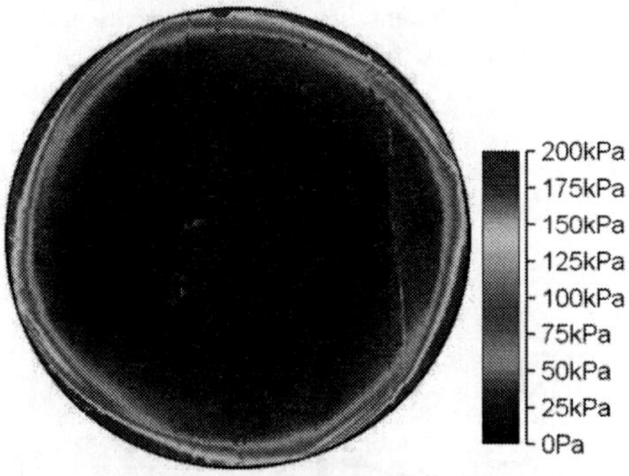

Figure 15: Shear stress maximum in the sample cut from the middle of a crystal grown in configuration V1

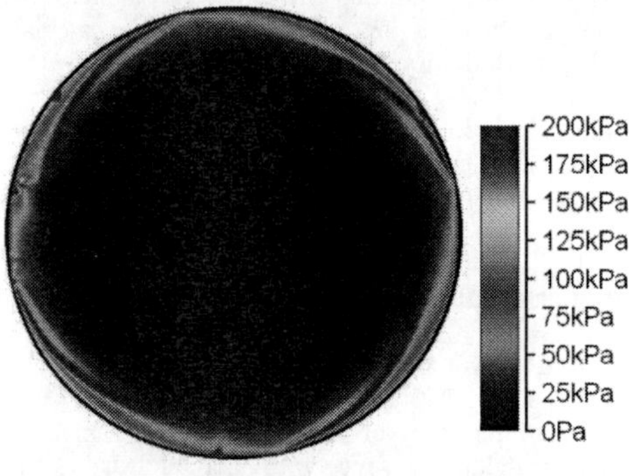

Figure 16: Shear stress maximum in the sample cut from the middle of a crystal grown in configuration V2

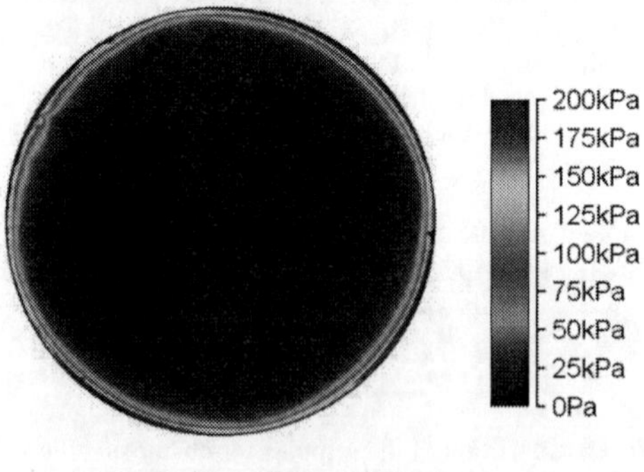

Figure 17: Shear stress maximum in the sample cut from the middle of a crystal grown in configuration V5

No differences between the examined samples were observed. All samples are stress-free except for the outer edge area. Along the circumference, the samples show slight tensile stresses, which may indicate a certain defect structure. A similar effect can also be observed in the lifetime distributions. Lifetimes measured on the as grown crystal surface show values lower than lifetimes measured 5 mm deep in the crystal volume. This effect was observed on several specially prepared crystals. An example is shown in fig.18.

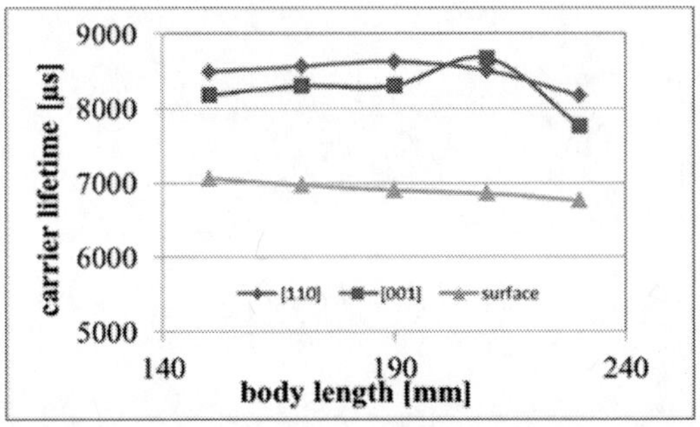

Figure 18: Minority carrier lifetime plotted versus a certain length in the body, measured on the as grown crystal and 5 mm beneath the surface

3 ECONOMIC ANALYSIS OF THE MULTIPULLING PROCESS COMBINED WITH ACTIVE CRYSTAL COOLING

An assessment of the economic potential cannot be presented quantitatively, because many factors depend on the technological condition of the process development as well as on the training level of the operational employees. Furthermore production costs may have different values depending on the location of the production plant and are subject to substantial fluctuations in the market. The cost savings of this growth technique is the reduced consumption of crucible material as well as the reduced amount of pot scrap (fig.21). The down-time of the multipulling process is less compared to a standard batch process. The biggest advantage is the drastic reduction of the process time and the resulting savings in process costs. This is shown in fig.19 where the power of the main heater and the bottom heater in a multipulling process of three ingots is compared with a standard batch process. The curves were derived from real crystal growth experiments performed at the crystal growth department of PVA CGS GmbH. The red curve represents a standard Cz-process with a pull speed of 0.9 mm/min in the body phase. The blue curve shows a multipulling process of three ingots grown with active crystal cooling and a mean pull speed of 1.8 mm/min in the body phase. Fig.20 shows a detailed section of Fig.19. The green line represents the individual crystal growth phases. Up to the first stage at about eight hours, the phases of pumping down, melting and stabilization of the melt take place. The steps represent the neck, shoulder, body and the end cone phase. After these steps the refilling phase takes place.

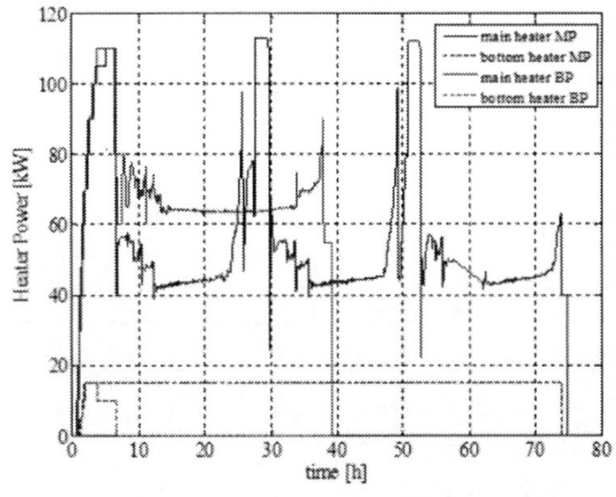

Figure 19: Power consumption of the main heater and the bottom heater versus process time for standard Cz (red curve) and multipulling process (blue curve).

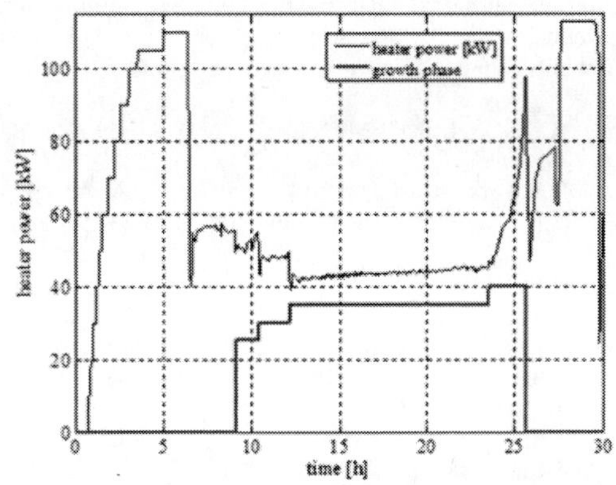

Figure 20: Detail of figure 19. The green line shows the different phases of a growth run (see text).

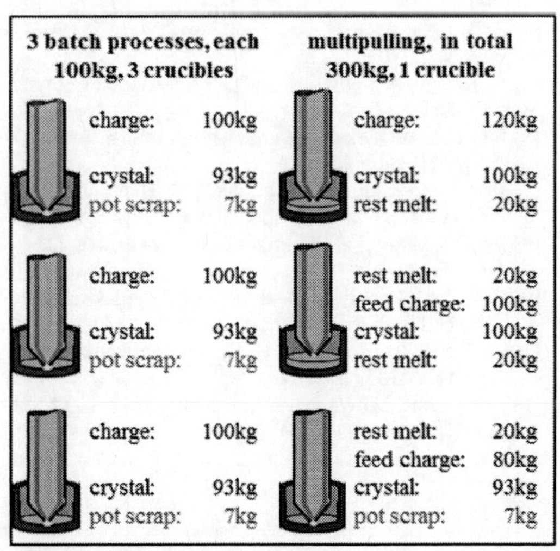

Figure 21: material consumption of a batch process compared to multipulling

Tab.3 shows the savings potential that results solely from the assumption of a higher pull rate. The main saving potentials are the process time and the reduced energy consumption. The potential savings in columns 2 and 3 relate only to the body phase and are related to the standard process in configuration V1.

growth config.	mean pull speed [mm/min]	Δ time body phase [%]	Δ energy [%]
V1	0.9	0	0
V2	1.3	29	32
V5	1.8	49	53

Table 3: Time and energy savings of the different crystal growth configurations

4 CONCLUSIONS

The results presented in this paper demonstrate that the combination of multipulling and active crystal cooling has a tremendous cost reduction potential. In addition, these techniques still have an optimization potential, e. g. by the use of larger hot zones. This probably requires a new generation of Cz-pullers.

5 ACKNOWLEDGEMENTS

This work was supported by the German Ministry of Economy and Energy under contract number 0325883A.

6 REFERENCES

[1] F. Mosel, A.V. Denisov, B. Klipp, R. Sharma, R. Kunert, P. Dold, Proceedings 32nd European Photovoltaic Solar Energy Conference, München, Germany, (2016), 1064-1068
[2] CGSim package, STR Group,Ltd.
[3] F. Mosel, A.V. Denisov, B. Klipp, R. Kunert, P. Dold, Proceedings 33rd European Photovoltaic Solar Energy Conference, Amsterdam, Netherlands, (2017), 495-500
[4] E. Billig, Proc. Roy. Soc. (London) 229 (1955) 346-363
[5] V. Kalaev, A. Sattler, L. Kadinski, J. Crystal Growth 413 (2015) 12-16
[6] I. Drikis, M. Plate, J. Sennikovs, J. Virbulis, Journal of Crystal Growth (2016)
[7] W. von Ammon, E. Dornberger, P.O. Hansson, Journal of Crystal Growth **198/199** (1999) 390-398
[8] R.H. Hopkins et al., Journal of Crystal Growth **42** (1977), 493-498
[9] G. Coletti, Impurities in silicon and their impact on solar cell performance, thesis
[10] H.J. Schulze, A. Lüdge, H. Riemann, Electrochem. Soc. **143**, 4105 (1998)
[11] M. Herms, M. Wagner, Phys. Status Solidi C **12**, No.8, 1085-1089 (2015)

MATHEMATICAL MODELING OF ELECTROMAGNETIC STIRRING OF SILICON MELT

Sergey M. Karabanov[1], Dmitry V. Suvorov[1], Dmitry Yu. Tarabrin[1], Andrey S. Karabanov[2], Evgeniy V. Slivkin[1], Andrey A. Trubitzin[1], Oleg A. Belyakov[2]

1 - Ryazan State Radio Engineering University, 59/1 Gagarina St., Ryazan 390005, Russia
2 - Helios Resource Ltd., 126 Proletarskaya St., Saransk 430001, Mordovia, Russia

ABSTRACT: The purpose of the work is the mathematical modeling of electromagnetic stirring of silicon melt using COMSOL Multiphysics software. In the calculations, the silicon stirring under the action of a traveling magnetic field are considered for 2D and 3D models. Using the model, distributions of the Lorentz force in silicon melt and the velocity fields of the melt movement under various conditions of electromagnetic stirring are obtained. The interrelation between electromagnetic stirring parameters and the rate of surface mass transfer is established.
Keywords: mathematical modeling, electromagnetic stirring, Lorentz force

1 INTRODUCTION

Currently mono- and multicrystalline silicon are the basic materials for solar cell production. High-purity silicon is produced by means of trichlorosilan technology (Siemens process, Union Carbide process). The main drawback of this technology is significant energy consumption, high production costs and environmental hazard of production. At present, there are direct purification methods that ensure the solar-grade silicon production of directly from metallurgical-grade silicon [1-4]. The most effective methods are vacuum refining [5-7], plasma-chemical purification [8-11], directional crystallization under the conditions of electromagnetic stirring [12,13]. These methods are based on impurity removal at the phase interface of "liquid silicon" – "gaseous medium" and "liquid silicon" – "multicrystalline silicon". Under the conditions of small purified silicon volumes, the effect of particle transport velocity to the melt surface on the purification efficiency can be disregarded. In case of industrial technology at purification of large silicon volumes, the problem of impurity transport to the melt surface becomes significant. To ensure the effective transport of impurities from the silicon volume to the phase interface, it is expedient to use electromagnetic stirring of melt [14, 15]. The paper presents the study of electromagnetic stirring of silicon melt by mathematical modeling for 2D and 3D calculation models and interrelation between electromagnetic stirring parameters and the rate of surface mass transfer.

2 MATHEMATICAL MODEL OF ELECTROMAGNETIC STIRRING OF SILICON MELT

Calculation of electromagnetic stirring of silicon melt was made by mathematical modeling using COMSOL Multiphysics software.

The mathematical model of electromagnetic stirring includes a set of basic Maxwell's equations in differential form:

$$rot\vec{E} = -\frac{\partial \vec{B}}{\partial t} \qquad (1)$$

$$div\vec{B} = 0 \qquad (2)$$

$$rot\frac{\vec{B}}{\mu\mu_0} = \vec{j} \qquad (3)$$

$$div\vec{j} = 0 \qquad (4)$$

and integral form:

$$\oint \vec{E}\cdot d\vec{l} = -\frac{\partial}{\partial t}\iint \vec{B}\cdot d\vec{S} \qquad (5)$$

$$\oiint \vec{B}\cdot d\vec{A} = 0 \qquad (6)$$

$$\oint \frac{\vec{B}}{\mu\mu_0}\cdot d\vec{l} = \iint \vec{j}\cdot d\vec{S} \qquad (7)$$

$$\oiint \vec{j}\cdot d\vec{S} = 0 \qquad (8)$$

Under the effect of an alternating magnetic field, the induced electromotive force is generated in the volume of silicon melt:

$$\varepsilon_i = -\frac{d\Phi}{dt} \qquad (9)$$

As a result of induced electromotive force, an induction current appears, and the Lorentz force acts on each elementary unit of silicon melt. The model includes the equation for the Lorentz force effecting silicon melt:

$$F_L = j \times B \qquad (10)$$

where B is the magnetic induction value, j – current density, the value of which can be found from the Ohm's law for isotropically moving media:

$$j_i = \sigma(E + u \times B) \qquad (11)$$

where σ is the liquid silicon conductivity, E – electric field strength, u – the flow velocity of a charged particle in silicon melt.

To describe the flows within liquid silicon, Navier-Stokes equation describing hydrodynamics of incompressible liquid is used:

$$\frac{\partial \vec{u}}{\partial t} = -\vec{u}\cdot(\nabla\vec{u}) + \frac{\eta}{\rho}\Delta\vec{u} - \frac{1}{\rho}\nabla p + \vec{F_m} \qquad (12)$$

where $\vec{u}$ is the vector velocity field, η – coefficient of dynamic viscosity, ρ – density, p – pressure, $\overrightarrow{F_m}$ – vector field of body forces.

For calculation of diffusion and particles mass transfer, the equation of continuity describing transportation of impurity particles in silicon melt is used:

$$\frac{\partial \overline{c_i}}{\partial t} + \nabla \cdot \left(-D_i \nabla c_i\right) + \overline{u} \cdot \nabla c_i = R_i \qquad (13)$$

where c_i is the concentration of i-impurity in silicon melt, D_i – the diffusion factor of i-impurity.

The system of the model boundary conditions included the condition of viscous friction at the melt-crucible boundary, given through the characteristic roughness of the crucible and the condition for ideal melt slip at the boundary above the melt surfaces.

To evaluate the motion type of the silicon melt (laminar, turbulent modes) under the effect of the Lorentz force, a criterion based on calculating the Reynolds number from the relation was used:

$$R_e = \frac{\rho u_0 L_0}{\eta} \qquad (14)$$

where u_0 is the characteristic flow rate, L_0 – characteristic length.

Figure 1 shows a computational grid of the axially symmetric numerical model. The geometrical dimensions of silicon melt correspond to the dimensions of a multi-crystalline silicon ingot of G5 size.

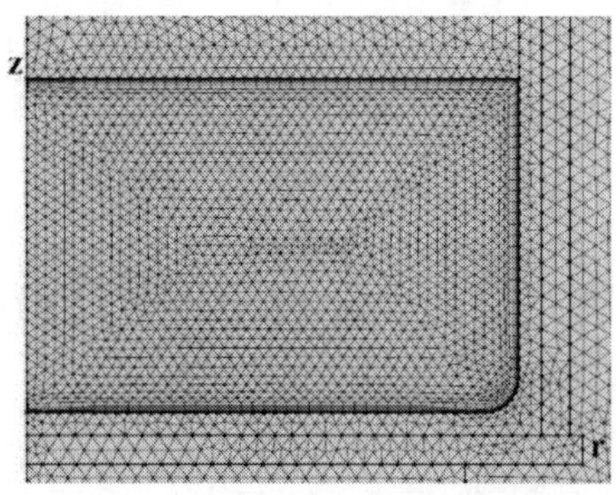

Figure 1: The computational grid in silicon melt.

A special feature of the model is the consideration of friction at the melt-crucible boundary. To improve the accuracy of calculating the stream flows on the silicon boundary the additional boundary layer condition is applied. This condition makes it possible to form a lot of boundary layers on the liquid phase boundary, that will ensure high accuracy of calculating the friction on the crucible walls, as well as swirling of liquid flows caused by melt turbulence.

The minimum thickness of the first boundary layer is 10 μm, that is half the value of the walls average roughness of the quartz crucible. At the upper boundary of silicon melt, ideal sliding condition is specified, since the friction at the boundary of the liquid phase-gas can be neglected.

3 CALCULATION RESULTS

3.1 Mathematical modeling conditions

The following parameters value of liquid silicon (for 1833 K) is used for electromagnetic stirring calculation:

- density, ρ – 2480 kg/m³;
- dynamic viscosity, η – 5.24·10⁻⁴ Pa·s
- conductivity, σ – 1.23·10⁻⁴ S/m;
- magnetic permeability, μ – 1.

The calculation of electromagnetic stirring modes conditions was performed for the geometry of the axisymmetric field formed by three ring inductors around the crucible. It is shown in [16] this geometry of the inductors arrangement is the most effective one. The amplitude of the inductor current was varied within the range of I = 100-1000 A, the magnetic field frequency, f = 20-200 Hz, the phase shift, φ = 0-120°).

3.2 Mathematical modeling results

As a result of modeling for various configurations and parameters of an electromagnetic stirring field, velocity distributions in silicon melt, Lorentz force distribution patterns in silicon melt, and flow rate distributions on the silicon surface were obtained.

Figure 2 shows the modeling results of axisymmetric electromagnetic stirring of silicon melt in the system of three ring copper inductors located around the crucible, with the basic initial parameters: I = 2 kA, f = 50 Hz, φ = 120°.

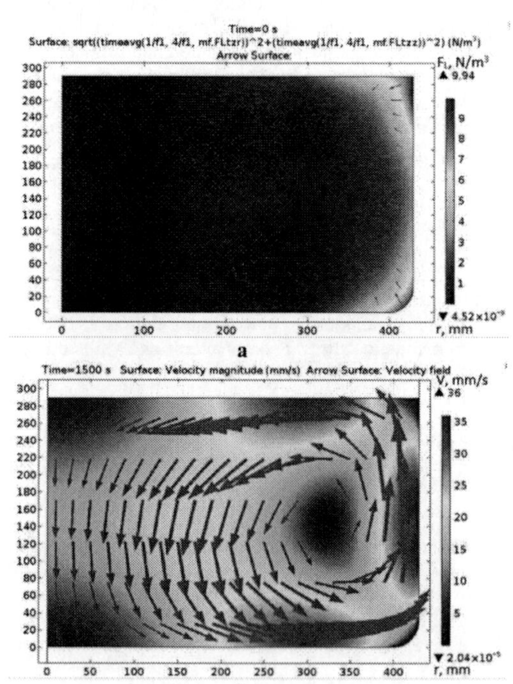

a

b

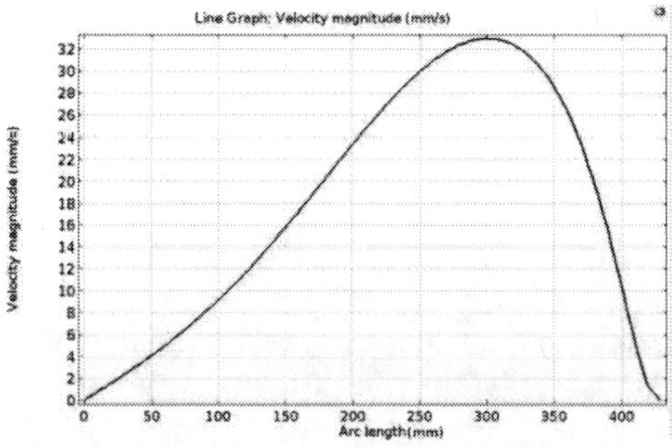

c

Figure 2: The Lorenz force distribution in silicon melt (a), the velocity field in silicon melt (b) and the velocity distribution on the melt surface (c) for axisymmetric electromagnetic stirring formed by ring inductors around the crucible.

To verify the calculation results obtained using a two-dimensional axially symmetric model, a 3D model was created. Figure 3 shows the distributions of the average Lorentz force in silicon melt for a 3D model.

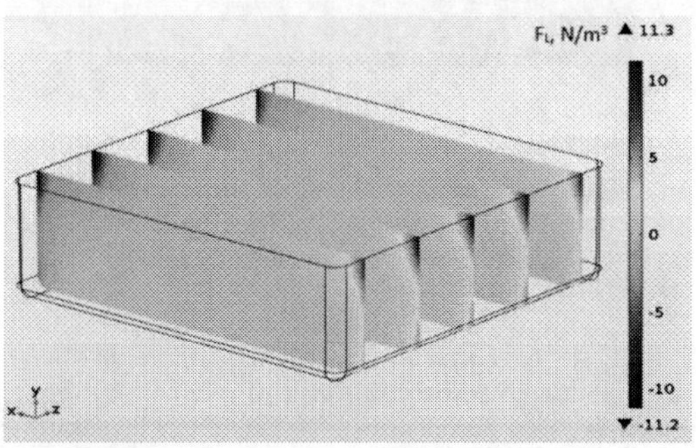

Figure 3: The Lorentz force distributions in silicon melt for a 3D model.

Comparison of the results obtained in 2-D and 3-D models shows that the local distributions of the Lorentz force are identical, and the maximum values correspond to each other within the permissible tolerance (9.94 N/m^3 for the axially symmetric approximation and 11.3 N/m^3 for a 3D model). The Lorentz force distribution (Figure 3) corresponds to the distribution of the Lorentz force obtained in the axially symmetric model (Figure 2a). Local distributions are identical, and maximum and minimum values of the Lorentz force have tolerance range due to errors at calculation (7.6 and 5.07 N/m^3 for 2D axially symmetric approximation and 7.63 and 5.62 N/m^3 for a 3D model). Thus, the results obtained in the course of modeling using the axially symmetric approximation can be considered valid.

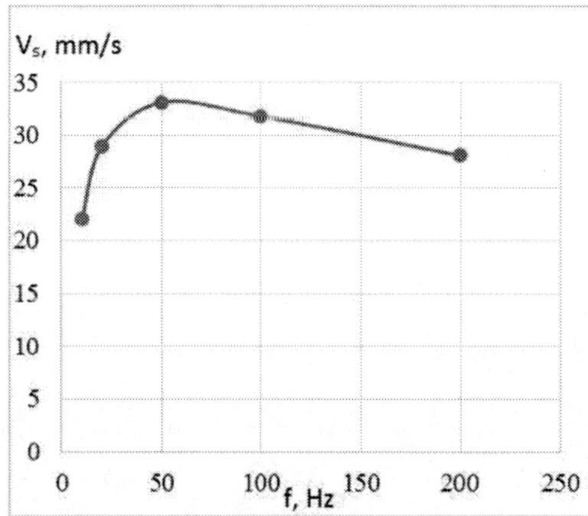

a

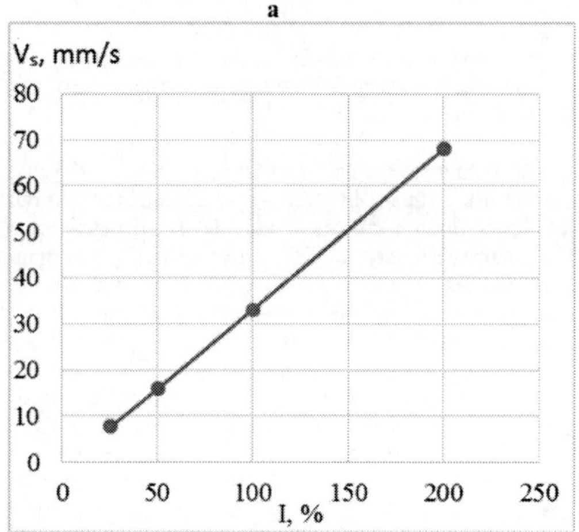

b

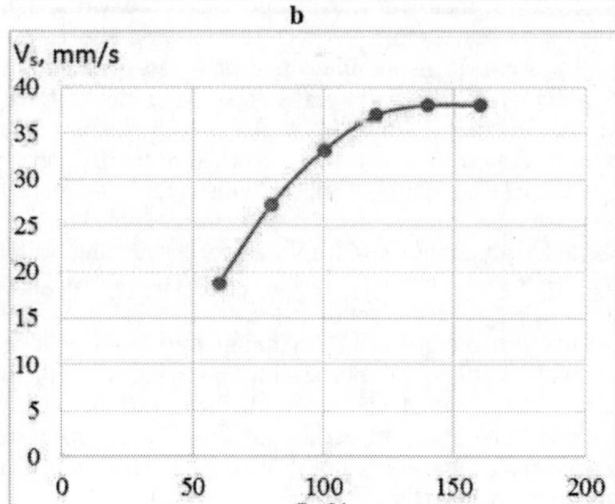

c

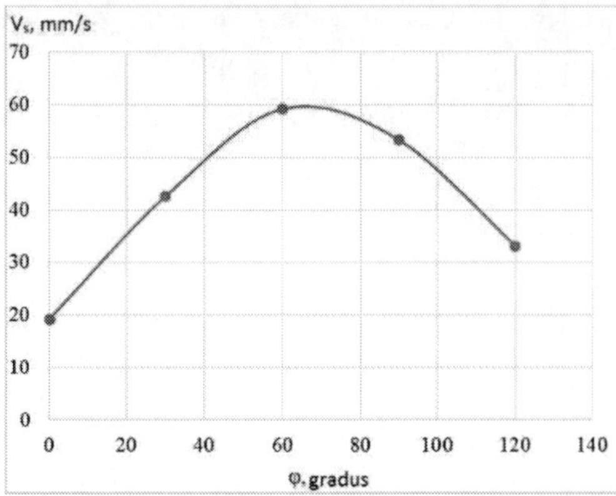

d

Figure 4: Dependences of the maximum rate on the silicon melt surface on frequency (a), the strength of current (b), the distance between inductors (c) and the phase shift (d), accordingly.

Modeling shows that the maximum rate on the melt surface is almost linearly depends on the current amplitude in the inductors (Figure 4). The dependence of the maximum rate on frequency has a maximum corresponding to the frequency of 50 Hz. The linear rate dependence on the current amplitude is explained by the Lorentz force increase proportional to the current effecting on the melt. The availability of the optimum frequency is explained as follows. In the low frequencies area (10 Hz or less), the field penetrates into silicon melt at great depth, and the Lorentz force is applied to a volume inside in melt near central axes. In this case the stirring efficiency is relatively small, since at mechanical trajectory of melt, circular by section, in areas near the ingot axis, the Lorenz force is directed opposite to the melt movement. As the frequency increases, the depth of the field penetration decreases and the Lorentz force vector becomes co-directed with the melt motion. The maximum stirring efficiency is reached at the frequency approximately equal to 50Hz. With further frequency increase, the area of the Lorentz force application is shifted to a thin near-surface layer near the crucible walls that leads to the influence increase of friction forces, and the intensity of the rotational motion decreases. The calculations show that the maximum stirring intensity is observed with a 60° phase shift between currents in inductors (Figure 4d). This is due to the availability of the optimal interaction efficiency of currents induced in silicon melt with the running wave of a magnetic field inducible by inductor currents.

Calculation of the impurity diffusion to the melt surface, where they are removed, shows that without electromagnetic stirring the transport process is caused only by the gradient of the impurity concentration and occurs extremely slowly. Figure 5 shows the distributions of the impurity concentration in silicon melt by the example of phosphorus, P, with and without electromagnetic stirring.

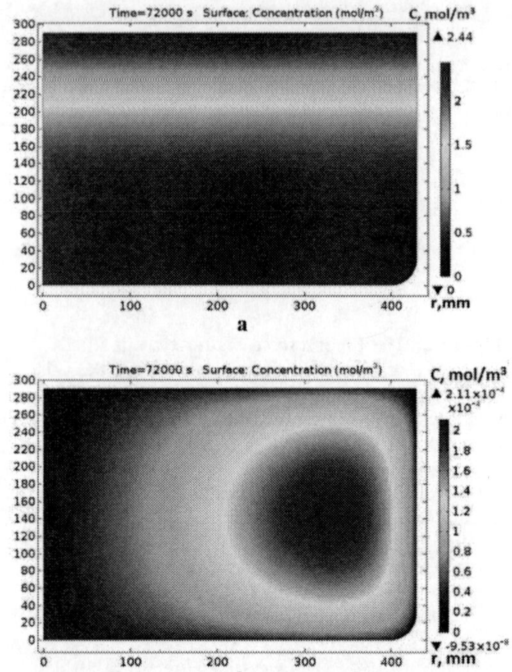

Figure 5: The distribution of the impurity (P) concentration, corresponding to the time interval of 72000 s without electromagnetic stirring (a) and with electromagnetic stirring (b).

Figure 6 shows the graphs of variance in the average impurity concentration (by volume) (by the example of phosphorus, P) in time without electromagnetic stirring and at different electromagnetic stirring conditions.

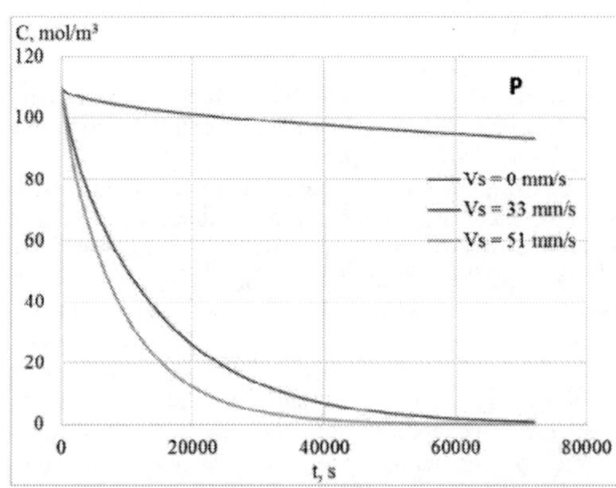

Figure 6: Time dependences of the average impurity concentration decrease (phosphorus) at different intensities of electromagnetic stirring: without electromagnetic stirring at a maximum surface velocity of Vs = 33 mm/s and 51 mm/s.

The obtained data show that the superposition of electromagnetic stirring leads to sharp rise in mass transfer intensity and to substantial increase of the removal rate of impurities from silicon melt. The purification intensity is proportional to the stirring rate.

4 CONCLUSION

1. The mathematical model of electromagnetic stirring of silicon melt is made.

2. Using the model,s the distributions of the Lorentz force in silicon melt, the velocity field of the melt motion under different conditions of electromagnetic stirring, the distributions of impurity concentration inside melt at different moments are obtained.

3. Based on the numerical calculations, the following regularities of electromagnetic stirring for 5G crucible are established:

- for the ingot dimensions corresponding to G5 size at axisymmetric electromagnetic stirring the area near the ingot axis is not sufficiently mixed;
- practically in all the considered electromagnetic stirring modes, the turbulent motion of melt in the crucible is provided, that contributes to a high rate of the surface mass transfer;
- optimal conditions for electromagnetic stirring (inductor frequency - 50 Hz, phase shift between inductor currents - 60°);

4. With electromagnetic stirring a sharp increase in the mass transfer rate with the surface occurs, that ensures a high efficiency of volatile impurities removal from silicon melt. The intensity of purification is proportional to the electromagnetic stirring rate.

5 ACKNOWLEDGMENT

The presented results have been obtained within the government contract No. 14.577.21.0263 of the Federal Target Program "Studies and developments of priority areas of development of Russian scientific and technological complex in 2014-2020". The project's unique identifier is RFMEFI57717X0263.

6 REFERENCES

[1] B. Ceccaroli, E. Ovrelid, S. Pizzini, Solar Silicon Processes: Technologies, Challenges, and Opportunities, CRC Press, 2016, ISBN 9781498742658.

[2] A.F.B. Braga, S.P. Moreira, P.R. Zampieri, J.M.G. Bacchin, P.R. Mei, New processes for the production of solar-grade polycrystalline silicon: A review, Solar Energy Materials and Solar Cells, 92. 2008. pp. 418-424.

[3] Y. Delannoy, Purification of silicon for photovoltaic applications, Journal of Crystal Growth, Vol. 360, 2012, pp. 61-67, ISSN 0022-0248.

[4] W. Lee, W. Yoon, C. Park, Purification of metallurgical-grade silicon in fractional melting process, Journal of Crystal Growth, Vol. 312, Issue 1, 2009, pp. 146-148.

[5] S.S. Zheng, W.H. Chen, J Cai, .et al., Mass transfer of phosphorus in silicon melts under vacuum induction refining. Metallurgical and Materials Transactions B. (2010) 41: 1268.

[6] J. Safarian and M. Tangstad, Vacuum refining of molten silicon. Metallurgical and Materials Transactions B. (2012) Vol. 43, Issue 6, pp. 1427-1445.

[7] W. Dong et al., Removal of phosphorus in metallurgical grade silicon using electron beam melting, Materials Science Forum, Vols. 675-677, pp. 45-48, 2011.

[8] N. Nakamura, H. Baba, Y. Sakaguchi, S. Hiwasa and Y. Kato, Boron removal in molten silicon with steam added plasma melting method, Journal of the Japan Institute of Metals, (2003) 67, pp. 583-589.

[9] C. Alemany, C Trassy, Bernard Pateyron, K.-I. Li, Y. Delannoy, Refining of metallurgical-grade silicon by inductive plasma (2002). Solar Energy Materials and Solar Cells, 72, pp. 41-48.

[10] S.M. Karabanov, D.V. Suvorov, D.Y. Tarabrin, E.V. Slivkin, V.A. Korotchenko, A.N. Vlasov, O.A. Belyakov, A.S. Karabanov, V.L. Dshkhunyan, Study of interaction of a plasma jet with the silicon melt surface under the conditions of its high turbulence, Conference Proceedings, 2017 IEEE International Conference on Environment and Electrical Engineering and 2017 IEEE Industrial and Commercial Power Systems Europe (EEEIC / I&CPS Europe), 6-9 June 2017, Milan, Italy, pp. 377-381, DVD-ROM, CFP1751I-DVD, ISBN 978-1-5386-3916-0.

[11] S. M. Karabanov, V. I. Yasevich, D. V. Suvorov, A. S. Karabanov, Mathematical modeling and experimental research of the method of plasma chemical purification of metallurgical-grade silicon, Conference Proceeding of 2016 IEEE, 16th International Conference on Environment and Electrical Engineering (EEEIC), 7-10 June 2016, Florence, Italy, pp. 939-943.

[12] P. Li, S. Ren, D. Jiang, J. Li, L. Zhang, Y. Tan, Effect of alternating magnetic field on the removal of metal impurities in silicon ingot by directional solidification, Journal of Crystal Growth, Vol. 437, 2016, pp. 14-19.

[13] C. Kudla, A.T. Blumenau, F. Bullesfeld, N. Dropka, Ch. Frank-Rotsch, F. Kiessling, O. Klein, P. Lange, W. Miller, U. Rehse, U. Sahr, M. Schellhorn, G. Weidemann, M. Ziem, G. Bethin, R. Fornari, M. Müller, J. Sprekels, V. Trautmann, P. Rudolph, Crystallization of 640kg mc-silicon ingots under traveling magnetic field by using a heater-magnet module, Journal of Crystal Growth, Vol. 365, 2013, pp. 54-58.

[14] N. Dropka, C. Frank-Rotsch, P. Rudolph, Comparison of stirring efficiency of various non-steady magnetic fields during unidirectional solidification of large silicon melts, Journal of Crystal Growth, Vol. 365, 2013, pp. 64-72.

[15] P. Rudolph, Travelling magnetic fields applied to bulk crystal growth from the melt: The step from basic research to industrial scale, Journal of Crystal Growth, Vol. 310, Issues 7–9, 2008, pp. 1298-1306.

[16] S.M. Karabanov, D.V. Suvorov, D. Yu. Tarabrin, E.V. Slivkin, A.E. Serebryakov, V.V. Klimakov, A.S. Karabanov, O. A. Belyakov, Mathematical modeling of vacuum refining of silicon melt under the conditions of electromagnetic stirring, AIP Conference Proceedings, 1999, 020012 (2018); https://doi.org/10.1063/1.5049251.

DEVELOPMENT OF METHODS FOR REDUCING THE RED-ZONE IN THE TOP REGION OF MC-SILICON INGOTS

T. Bähr*, M. Ghosh*, C. Kranert+, C. Reimann+,~, C. Morche°

* ACCESS e.V., Intzestraße 5, D-52072 Aachen, Germany
+ Fraunhofer THM, Am St.Niclas Schacht 13, D-09599 Freiberg, Germany
~ Fraunhofer IISB, Schottkystr. 10, D-91058 Erlangen, Germany
° ALD-VT GmbH, Otto-von-Guericke-Platz 1, D-63457 Hanau, Germany

ABSTRACT: In order to increase the high lifetime volume fraction in multicrystalline-silicon ingots for photovoltaic application by reducing the red-zone at their top, we tested three methods for melt removal from the top of the growing ingot: (1) Absorption by a porous medium, immersed into the melt, (2) vacuum suction of melt via pipes, and (3) pouring of melt by tilting the crucible. The experiments in our G1 lab furnace allowed to compare the different techniques and to identify their specific challenges. However, the final goal could not be reached yet. We hypothesize that an additional contamination via gas phase may be overlaying the positive effect of melt removal.
Keywords: multicrystalline silicon, crystallization, minority carrier lifetime, red-zone, impurities

1 INTRODUCTION

Multicrystalline silicon ingots produced by directional solidification exhibit a reduced minority charge carrier lifetime at their top, bottom and edge regions, known as red-zone [1, 2]. Specifically, the use of silicon from the top and bottom red-zone in solar cells would result in a strongly reduced conversion efficiency. Thus, these significant parts of the ingot are rejected from further processing into wafers and solar cells, leading to a substantial yield loss. The origin for the region of reduced minority carrier lifetime at the ingot's top is assumed to be caused by impurities (especially metals) with a distribution essentially determined by a combination of two mechanisms [3]:

- Enrichment of impurities in the melt at the top of the ingot by the segregation effect
- Back-diffusion of these impurities at high temperature after solidification from top downwards

The scope of the present work is to evaluate technological approaches preventing the impurity-enriched silicon melt at the top to become part of the ingot. By removing the impurity-enriched silicon melt (>1414°C) at the top shortly before the end of the solidification process (proceeding from bottom to top), the back-diffusion of segregated impurities is intended to be attenuated, therefore preventing the spread of a red-zone at the top. The advantage would be a potential minimization of yield loss.

Several ideas for reaching that aim have been developed and some of them are claimed in patents [4, 5], but no reports on technological realization of these approaches have yet been published. Our work aims to develop, implement, test, and evaluate such technologies in a G1 crystallization furnace.

2 METHODS FOR REMAINING MELT REMOVAL

2.1 Requirements

In order to reduce the ingot's top volume contaminated with segregated impurities, the following three requirements have to be satisfied by the melt removal technique:

- Fast:

The duration of the melt removal should be quick enough such that the concurring crystal growth during this time can be neglected. Therefore the removal process was targeted for no longer than a few seconds.

- Complete:

If some melt is left over after the removal procedure, the impurities present therein are still enough to (after solidification) eventually extend the red-zone by back-diffusion at high temperatures during the cool-down of the ingot.

- Sufficient Amount:

Regarding the goal to increase the yield of useable ingot material, the maximum amount of melt to be removed is given by the silicon amount forming the typical top red-zone, which would be ~15% of our G1-ingots (16.5 kg). The minimum amount of melt to be removed was estimated using the Scheil equation [6]. For an iron concentration in the initial melt of 10^{15} atoms per cm^3, which is a reasonable estimate for our ingots, and setting a red-zone carrier lifetime threshold limit of 10 µs, at least ~10% of the melt should be removed.

2.2 Melt Removal by Sponge Absorption

As described in [4], a porous body is immersed into

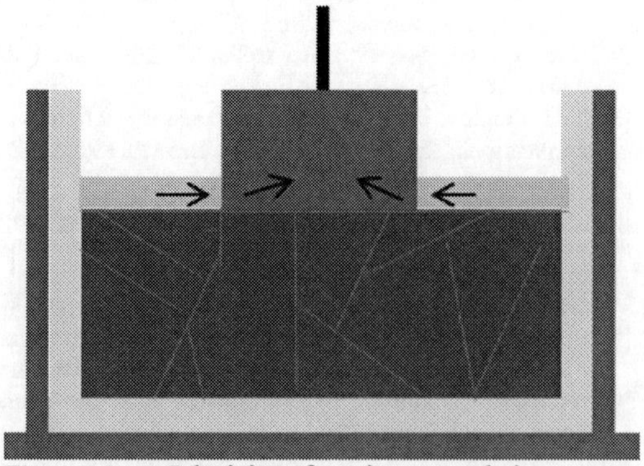

Figure 1: Principle of melt removal by sponge immersion.

the melt pool at the top of the growing ingot and will ideally absorb all the melt still present at that moment.

We implemented this approach by using a cubic body of graphite felt that was attached to a combined graphite/ceramic/metal rod through a hole usually used for a dip rod (see Figure 1 for a schematic illustration). The felt can be moved up and down through the rod.

2.3 Melt Removal by Pipe Suction

As described in [5], silicon melt is sucked through pipes into an extra container by vacuum application. In our realization, the container is placed on top of the hot box and can be moved up and down via the same mechanism as the graphite sponge. See Figure 2 for a schematic illustration.

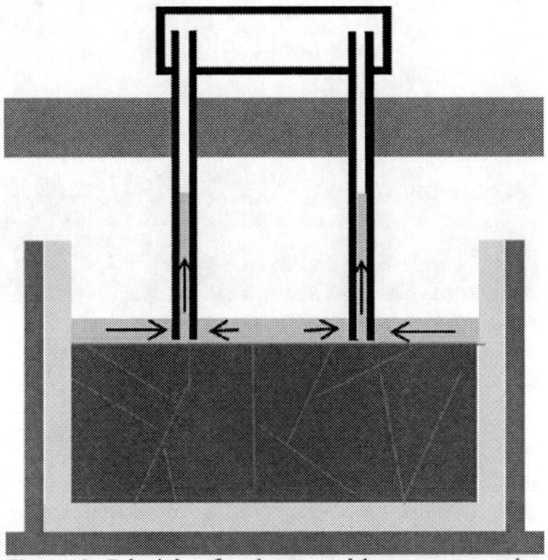

Figure 2: Principle of melt removal by vacuum suction via pipes.

2.4 Melt Removal by Tilting the Crucible

Another possibility to remove remaining melt from the mc-Si crystal surface is to pour the melt by tilting the crucible. As the implementation in our G1 furnace would be an extensive technical challenge, we decided to do experiments with G0-crucibles placed on a small graphite pedestal in G1-crucibles. A specially prepared rod (replacing the dip rod in our furnace) was used to tilt the G0-crucible and the overflowing melt was collected in a G1-crucible where it solidified (Fig. 3).

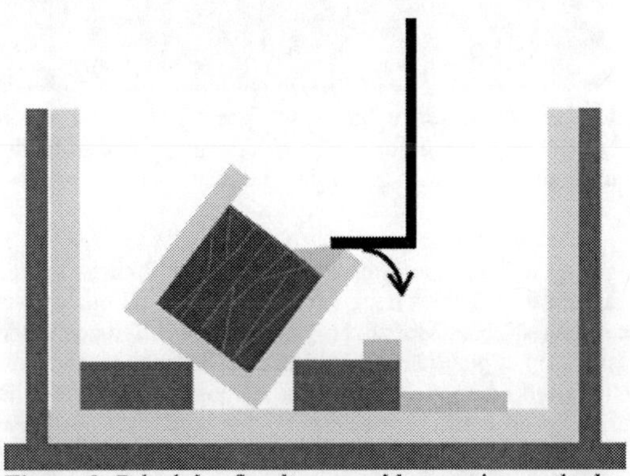

Figure 3: Principle of melt removal by pouring method.

3 CHARACTERISATION TOOLS

3.1 Minority carrier lifetime

For lifetime measurements of the minority carriers (in our case electrons), a 2 mm thick vertical cross section was cut from the center of each ingot. This sample was then measured in the as-cut state using an MDPmap tool by Freiberg Instruments, which is based on the principle of microwave-detected photocurrent (MDP). Because the regions of high minority carrier lifetime were not of interest for our investigations, we used a high-injection mode with a 4.8 W laser focused to a spot size of 1 mm. In this text all minority carrier lifetime mappings are scaled according to Figure 4.

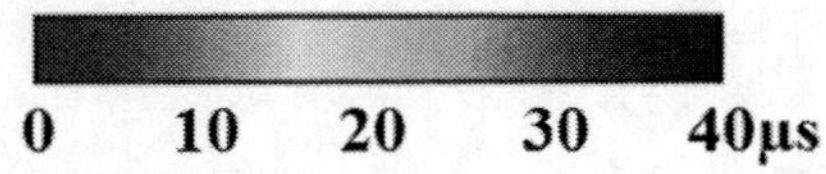

Figure 4: Scale bar for all minority carrier lifetime mappings in this text.

3.2 Infrared transmission microscopy

For IR transmission microscopy, cross-section samples adjacent to the samples of the lifetime measurement were polished on both sides. These were then mapped using the motorized stage of an Olympus transmitted light microscope and a 5x microscope objective. The focus area was extended by capturing five images across the sample thickness and combining them using the Extended Focus Image (EFI) feature of the Olympus analysis software.

4 REFERENCE PROCESS

Initially, a reference process for a 16.5 kg mc-Si ingot in a G1-crucible was defined. Relevant process parameters include a crystallization time of 14-15 h, a propagation speed of about 1 cm/h for the solid/liquid interface, and a cool-down time of 10 h. The solid/liquid interface is mostly flat or slightly convex. For feedstock, we used Wacker prime poly-silicon in the form of nuggets (size 3, 4) and granules (size 1). Boron dopant was added in the form of highly doped electronic grade silicon to reach a final resistance of ~2 Ωcm. The typical dimensions of the ingots are 22.5 cm in both horizontal directions and 14.5 cm in height.

Under these process conditions, the typical MDP values of the minority carrier lifetime are as depicted in Figure 5.

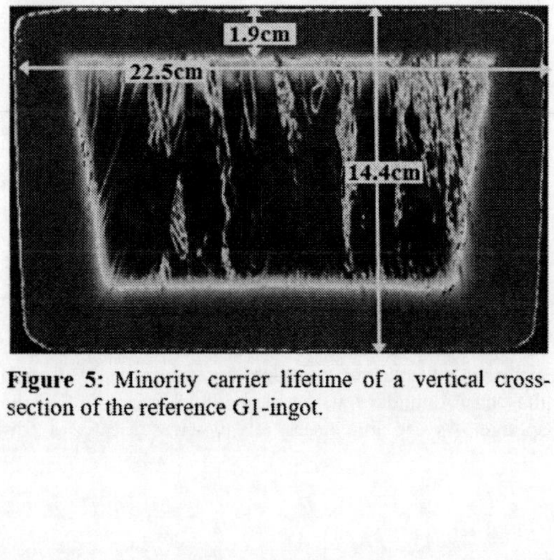

Figure 5: Minority carrier lifetime of a vertical cross-section of the reference G1-ingot.

The top red-zone has a slightly varying thickness of about 2 cm.

For the pouring experiments in G0-crucibles, the reference process conditions comprise a crystallization speed of 2 cm/h and a cool-down time of 8 h. The ingot mass was about 1.4 kg using Wacker chunks or about 1.8 kg if a massive polycrystalline Si cylinder of 10 cm in diameter and height was used as feedstock. A typical minority carrier lifetime distribution is given in Figure 6. The top red-zone has a height of about 1.8 cm to 2.0 cm.

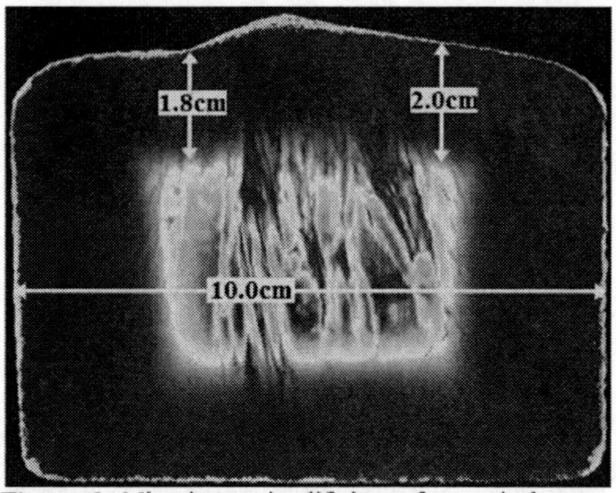

Figure 6: Minority carrier lifetime of a vertical cross-section of the reference G0-ingot.

5 RESULTS

5.1 Sponge Absorption Experiments

We carried out more than 15 experiments with sponge application. Typically, 2 – 12 % of the initial Si content was absorbed by the sponge (Figure7). Part of the absorbed silicon reacts with the graphite felt to SiC.

Figure 7: Sawed sponge with absorbed silicon.

Up to this point, we did not succeed to fulfill all the requirements defined in Chapter 2. Difficulties resulted especially from:

- Lack of knowledge on the actual vertical position of the phase boundary
- Relatively low speed of melt absorption by a sponge (some minutes for 0.5 kg)

Characterization of vertically cut samples by IR transmission microscopy nearly always showed a dark line near the top of the ingots, probably consisting of precipitates of SiC (see Figure 8). Based on our experience we suspect that this line marks the position of the phase boundary at the moment of immersion of the sponge. As the line nearly always was located a few millimetres below the ingot's surface, we could conclude that the phase boundary was not set free from the melt by the sponge method in these cases.

Figure 8: IR transmission of a vertical sample from the top of an ingot where a sponge was immersed in the melt.

Upon characterizing these ingots, we found no significant reduction of the top red-zone (see Figure 9), except in a few special cases where the sponge was frozen on the ingot's surface during its application.

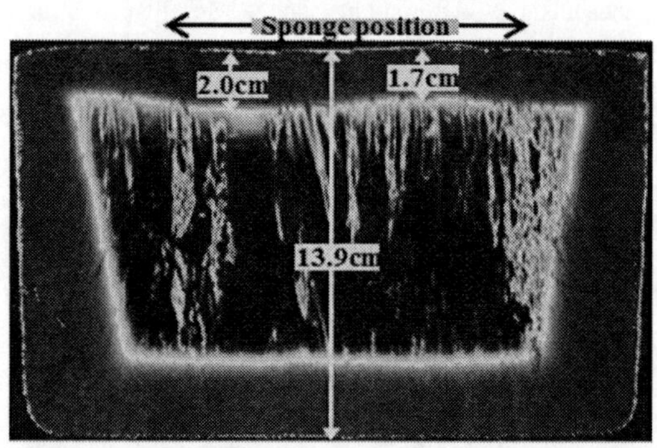

Figure 9: Minority carrier lifetime of a vertical cross-section of a G1-ingot where a sponge was immersed in the melt.

Figure 10 shows the minority carrier lifetime distribution of such an ingot. Already, the red-zones at the positions without sponge counterpart (positions C) are slightly thinner than the top red-zone of the reference G1-ingot. In the region next to the sponge, the melt was not removed completely by the sponge as also verified by a dark line below the ingot's surface in IR transmission microscopy.

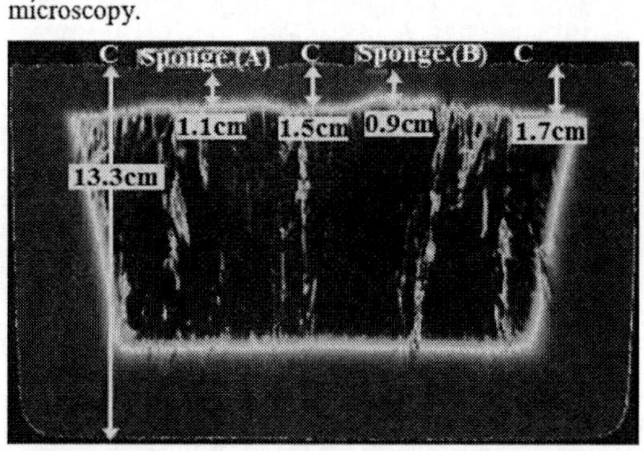

Figure 10: Minority carrier lifetime of a vertical cross-section of a G1-ingot where a sponge got frozen at the ingot during its immersion into the melt.

At the sponge positions A and B, the top red-zones are even thinner and the minority carrier lifetime values do not drop below 1 µs, which they do in all other parts of the red-zone. Most likely, the decisive difference to the situation at positions C is that below the sponge, the melt has been completely absorbed. At position B, where the sponge is entirely attached to the ingot, the red-zone thickness is minimal (0.9 cm). At position A, where the sponge is only locally attached at its edges to the ingot, the red-zone is slightly thicker (1.1 cm). This small difference might be due to a different shielding ability against the furnace atmosphere and a possible contamination therein.

With an ingot height of 13.3 cm in this experiment, instead of an average of 14.4 cm at our reference G1-ingot, the material loss at the ingot's top adds up to 2.0 cm at the sponge position B and to 2.2 cm at position A. This is similar to the top red-zone thickness of our reference G1-ingot (about 2 cm).

From this test and another test with the sponge frozen to the ingot's surface at the end of the crystallization, we conclude that in this special arrangement, the sponge absorbs the silicon melt underneath until the phase boundary is cleared from melt and that it can additionally act somehow as a sink for impurities in silicon. As a result, we observe a thinner top red-zone where the sponge gets permanently fixed to the ingot's surface as in the reference G1-ingot.

5.2 Pipe Suction Experiments

Up to this point, we made four G1-ingots with suction of silicon melt via pipes into a collect vessel. Through a viewing window in the furnace it was possible to study the melt flow and capture a video of it. During the suction process, which lasted only a few seconds, it can be observed how the melt drains towards the two pipes and the phase boundary eventually gets freed from the melt (Fig.11).

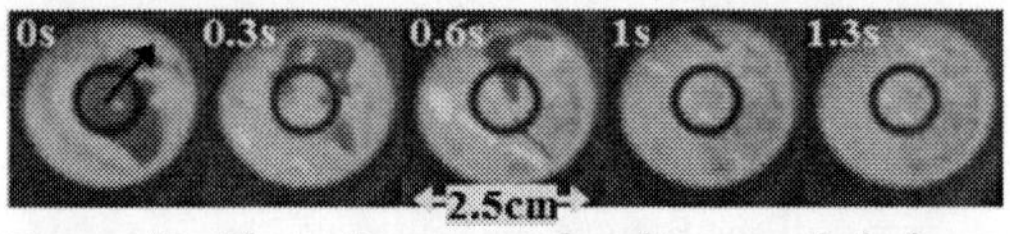

Figure 11: Photo Sequence of melt removal during a suction process. At the bright areas, silicon is already solidified, while darker areas still show melt. The black arrow in the left picture indicates the direction of melt flow. In the right picture, the phase boundary is fully cleared from melt.

Unfortunately, in the previous experiments, solidifying silicon caused a quick clogging of both pipes (Figure 12), which limited the amount of sucked silicon to a maximum of 2% of the initial silicon content.

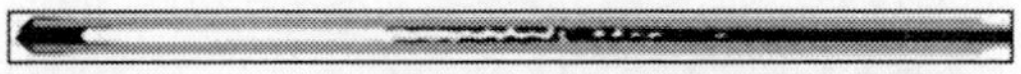

Figure 12: Computer tomography of a pipe (grey, inner diameter: 6 mm) clogged by sucked silicon (black).

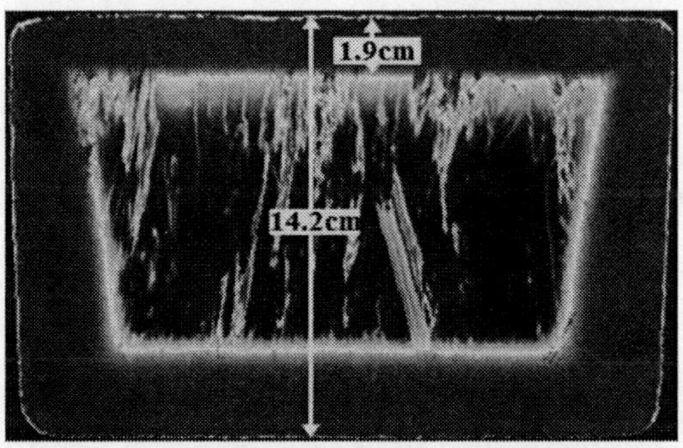

Figure 13: Minority carrier lifetime of a vertical cross-section of a G1-ingot where melt was sucked.

A reduction of the top red-zone was not detected and the minority carrier lifetime distributions of the ingots with suction applied (Figure 13) resemble the one of the reference G1-ingot. Improvements of the suction procedure to prevent, or at least delay, clogging in order to remove more melt are under development.

5.3 Pouring Experiments

Following our tests with melt removal by sponge absorption and the first experiments with pipe suction, we also started exploring melt pouring by tilting a G0-crucible placed on a small graphite pedestal in a G1-crucible. Towards the end of the crystallization process of the first two tests, the G0-crucibles were tilted to such an extent that the phase-boundary was in one case partly (Figure 14) in the other case completely cleared from melt (Figure 15). The thickness of the red-zone below these cleared surfaces, extending from 1.3 to 1.5 cm, is smaller than the top red-zone of the reference G0-ingot (1.8 cm to 2.0 cm).

However, despite the obviously eliminated contamination by back-diffusion, this still considerable extension of the red-zone is a further indication that an additional source of impurity contamination (e. g. via gas phase) is likely to hamper a further reduction.

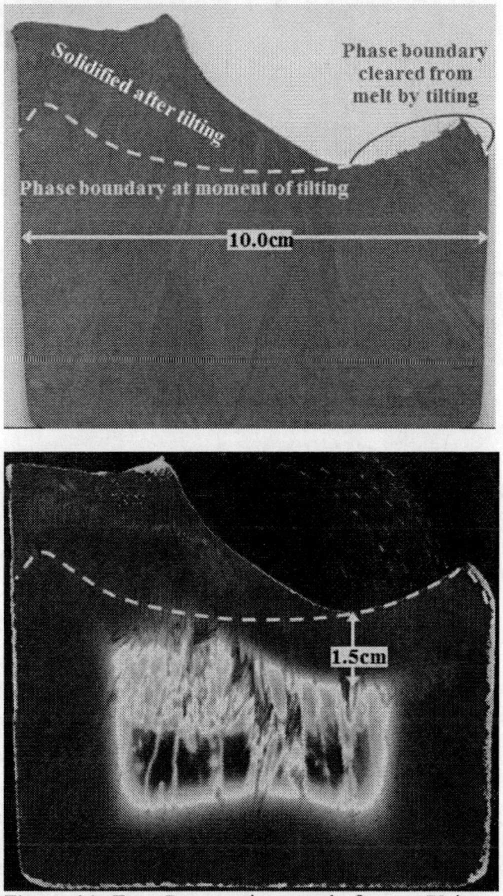

Figure 14: Top: Cross-section sample from a G0-ingot where melt was poured by tilting the G0-crucible. Bottom: Minority carrier lifetime mapping of this sample.

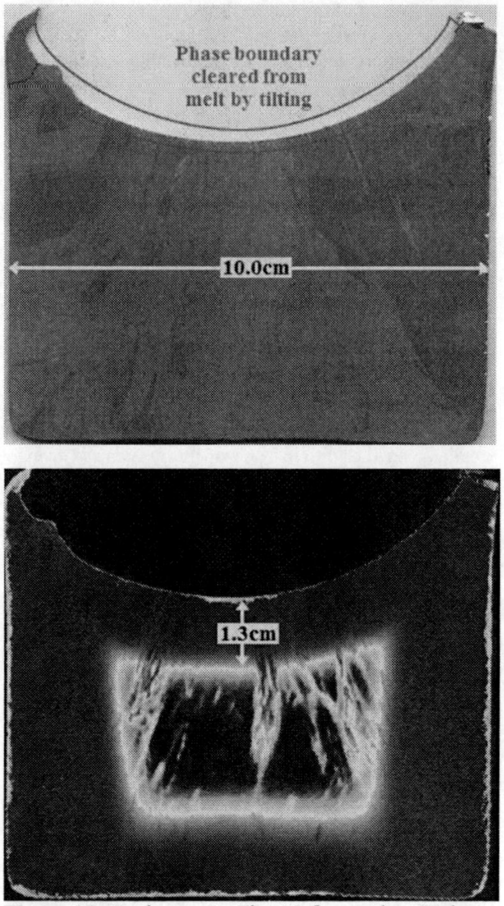

Figure 15: Analogous to Fig. 14 for another G0-ingot. The complete surface of this ingot was cleared from melt by tilting.

6 CONCLUSIONS

Three methods for the removal of the impurity enriched melt near the end of the crystallization process of mc-Si ingots were tested. Table I shows a summary of the main results.

Table I: Summary of main results

	Absorption by Sponge	**Suction**	**Pouring**
Number of ingots	16	4	2
Duration	few min	few sec	few sec
Silicon removed	≤ 12%	≤ 2%	≤ 21%
Phase boundary set free (area portion)	Only locally	≤ 25%	≤ 100%
Top Red-Zone reduced?	In special cases	No	Yes
Main Obstacle	Removal incomplete	Pipes clog	

The melt absorption by a sponge is usually not complete, so a reduction of the red-zone at the ingot's top is not observed.

In the first suction tests, both pipes were quickly clogged. This prevents a sufficient and complete melt removal. Improvement measures of the suction procedure are under development.

So far, only by pouring the melt, the phase boundary was set free and a reliable reduction of the red-zone at the ingot's top was achieved. In this case, it grew apparent that other sources of contamination are preventing further red-zone reduction.

REFERENCES

[1] F. Ferrazza: Large size multicrystalline silicon ingots. Solar Energy Materials and Solar Cells. Volume 72, Issues 1–4, p 77-81 (2002)

[2] Y. Boulfrad, G. Stokkan, M. M'hamdi, E. J. Øvrelid, L. Arnberg: Modeling of lifetime distribution in a multicrystalline silicon ingot. Solid State Phenomena, Volume 178-179. p. 507-512 (2011)

[3] R. Kvande, L. J. Geerlings, G.Coletti, L. Arnberg, M. Di Sabatino, E. J. Øvrelid, and C. C. Swanson, J. Appl. Phys. 104, 064905 (2008), doi: 10.1063/1.2956697

[4] Solar World America Inc., Patent Specification US 00000 893 6652 B2

[5] Solar World Innovations GmbH, Patent Specification DE 10 2009 039070 B4

[6] E. Scheil: Bemerkungen zur Schichtkristallbildung. Zeitschrift für Metallkunde 34, 70-72 (1942)

ACKNOWLEDGEMENTS

This project was funded by the German Ministry of Economics and Energy under contract 0324089 A, which is gratefully acknowledged.

Supported by:

Federal Ministry
for Economic Affairs
and Energy

on the basis of a decision
by the German Bundestag

EVALUATION OF A NEW HYBRID CRUCIBLE CONCEPT FOR CRYSTALLIZATION OF MC-SILICON INGOTS

C. Morche°, T. Bähr*, M. Ghosh#, C. Kranert+, A. Zimmermann°, H. Franz°

°: ALD-VT GmbH, Otto-von-Guericke-Platz 1, D-63457 Hanau, Germany
#: ACCESS e.V., Intzestraße 5, D-52072 Aachen, Germany
+: Fraunhofer THM, Am St.Niclas Schacht 13, D-09599 Freiberg, Germany

ABSTRACT: In order to overcome the well-known obstacles of silica crucibles in mc- and HPM silicon crystallization we tested a so called hybrid crucible consisting of a re-usable bottom plate with higher thermal conductivity and bottomless silica side walls. As a possible material for this bottom plate we considered silicon carbide as suitable. Crystallization experiments were performed in Access' SCU400-G1 furnace. In our tests the junction between bottom plate and side walls together with the choice of thermal process conditions revealed as critical parameters for success. The ingots produced so far resemble typical HPM ingots made under similar conditions in charge carrier lifetime and Oi and Cs content. So far the bottom plate was used for 3 following crystallization processes without significant wear.
Keywords: HPM-Si, crystallization, crucible

1 INTRODUCTION

Today, multicrystalline silicon ingots are produced by directional solidification in coated silica crucibles. The high thermal resistance of this material is an obstacle for higher crystallization rates, lower process times and a higher productivity in general. To overcome this the use of crucibles with narrower bottom and/or higher thermal conductivity has been the scope of former work [1,2].

One of the most promising ideas is to omit the silica crucible bottom completely and to crystallize a HPM ingot on the surface of an underlying re-usable plate directly [3] (figure 1).

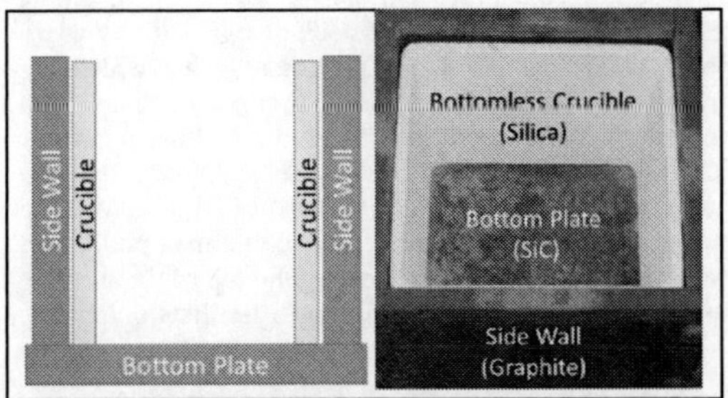

Figure 1: Sketch (left) and photography (right) of a bottomless silica crucible supported by a bottom plate and 4 side walls

Furthermore, mc-Si ingots exhibit a reduced minority charge carrier lifetime at their top, bottom and edge regions, the so-called red zone [4,5], which is to a large extend excluded from processing into wafers and solar cells. Here the absence of the silica crucible bottom may enable a new possibility for a reduction of the red zone at the bottom, and thus increase the yield substantially.

Additionally, the less needed crucible material offers potential in cost reduction for silicon crystallization and a mechanical less sensitive bottom plate can allow an automatized charging of the feedstock.

2 MATERIAL CHOICE FOR THE BOTTOM PLATE

According to the aim of the work, the relevant conditions for the new bottom plate and its material can be stated as follows:

- Temperature stable up to the melting point of Si (i.e. 1411°C, better: higher)
- Re-usable
- No reaction with Si, i.e. no sticking of Si crystal on the substrate
- No contamination of the Si crystal, especially with atoms which reduce charge carrier lifetime
- Thermal conductivity higher than that of the standard crucible material, SiO_2
- Mechanical strength

A first literature survey (using especially the sources [4] and [5]) revealed as condition fulfilling materials four material types: high melting metals, oxide ceramics, covalent ceramics and graphite. Table 1 on the following page gives an overview.

In most of the materials or material classes there is at least one drawback which would restrict them from a possible use. The named metals form eutectic phases with Si near 1400°C which means that even if they do not come into real contact with Si melt a potential risk of failure would exist. Additionally, a metal based bottom plate bears the risk of contaminating the silicon melt with fast diffusing metal atoms. Oxide ceramics have thermal conductivities not significantly higher than that of SiO_2 and give the risk of chemical reaction of oxygen to silicon. Furthermore, like a bottom plate made from metal, the potential of contamination with in-diffunding atoms from the plate material has to be considered. This is the reason why covalent ceramics like BN or AlN are not be suitable as they would affect the dopant level of the Si crystal.

At the end only two materials remained as a choice for the tests: carbon (graphite) and silicon carbide. So a bottom plate was made of graphite which was turned into SiC at its outer surface. The choice of such a bottom plate gives the additional advantage of an easy integration into standard crystallization furnaces and processes as such materials are used for other furnace components, too.

Table 1: List of possible materials for the bottom plate

Material	T-melt °C	Reaction/Contamination	Thermal conductivity	mechanics
Metals:				
Mo	2627	Eutectic phase 1400°C	90 W/mK	OK
W	3422	Eutectic phase 1390 °C	90 W/mK	OK
Nb	2477	Eutectic phase 1395°C	90 W/mK	OK
Ta	3017	Eutectic phase 1400°C	90 W/mK	OK
Oxide ceramics				
SiO_2 *	1200	Wetting, oxygen contamination	~2.5 W/mK	softening
Al_2O_3	2044	Dopant contamination	5 W/mK	OK
MgO	2825	?	10 W/mK	OK
Covalent ceramics				
BN	3000	No wetting, dopant contamination	30 W/mK	OK
AlN	2200	Dopant contamination	180 W/mK	OK
Si_3N_4	1600	No wetting, weak N contamination	30 W/mK	?
Carbon material				
graphite	3000	weak carbon contamination	20 W/mK	OK
SiC	2800	weak carbon contamination	120 W/mK	OK

*SiO_2 is the standard crucible material for mc-Si crystallization

3 EXPERIMENTAL SETUP

For the trials and tests of the hybrid crucible concept a graphite bottom plate with SiC coating in G1-size has been manufactured. The bottomless crucibles themselves have been generated from standard RTU G1 crucibles produced by Vesuvius (type XF 18310). For this purpose the bottom of the standard crucible has been just cut off. This means also slight loss in crucible height in comparison to a standard G1 crucible, because the rounding in the bottom corners needed to be removed.

For all trials the feedstock consists of a 35mm HPM seed composed of fine poly silicon material and a variation of different poly silicon chunk sizes with a total charge between 15 and 16kg. Additionally boron doped silicon wafers were added to the feedstock. The amount was calculated for a resistivity of 2 - 3 Ωcm.

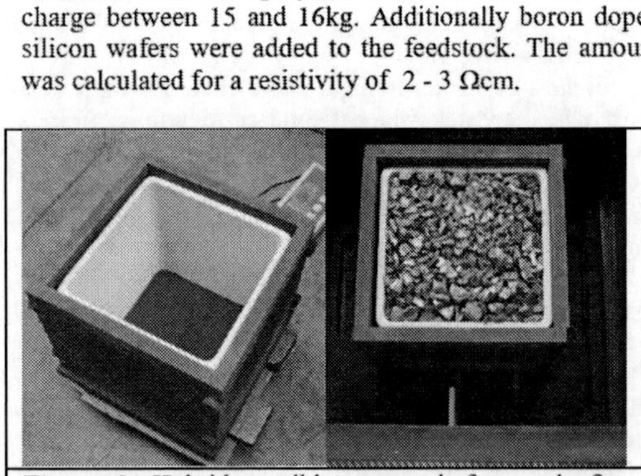

Figure 2: Hybrid crucible system before and after filling with feedstock.

The crystallization trials with the hybrid crucible have been all performed in Access' Tech-Center in Aachen with a special G1 version of a SCU400 furnace for solar silicon crystallization, produced by ALD. The furnace itself is equipped with a top- and a bottom heater and two side heaters in a hot zone of graphite hard felt. The power of each heater can be controlled individually and the crystallization can be controlled by power- or temperature based recipes. For HPM- and quasi-mono-crystallizations a boundary tracking device (BTD) is installed that allows the detection position of the solid/liquid interface during melting and crystallization [6].

3.1 Pre-Test

To see if the concept in general works a pre-test has been performed. In this trial the crucible setup should be tested with the described feedstock charge and a standard power-controlled HPM-crystallization recipe which has been used by Access for HPM crystallizations with normal standard crucibles. During the melting phase the melting behavior of the feedstock detected with the BTD was pretty off from previous experience with HPM. Also during melting and crystallization no real melt pool could be detected through a view port on the top of the furnace, so the trial has been aborted during crystallization phase.

Figure 3: Result of pretesting the hybrid crucible, remaining silicon after run out sticked to the bottom plate

This pre-test resulted in a nearly complete run out of silicon (fig.3). This shows that a simple overtake of existing HPM-process recipes to the new crucible system is not possible. Therefore a new temperature controlled recipe had to be developed with a more steep temperature gradient between bottom and top heater. Additionally, the remaining material inside the crucible had been fused together with the bottom plate, so a new bottom plate had to be manufactured.

3.2 Further tests

The new recipe was applied for 2 more crystallization experiments successfully. In both cases only a minor amount of Silicon melt penetrated the crucible bottom-side gap. The HPM like melting and crystallization process could be performed without damage to the furnace components. Fig. 4 shows the vertical position of the probed liquid solid interface during the process exemplarily. The resulting silicon ingots could be removed from the bottom plate easily. Fig. 5 shows the resulting bottom side of the ingot. Like in typical HPM ingots, the melt had infiltrated the primary silicon granules partially. The block height in all cases was around 140mm.

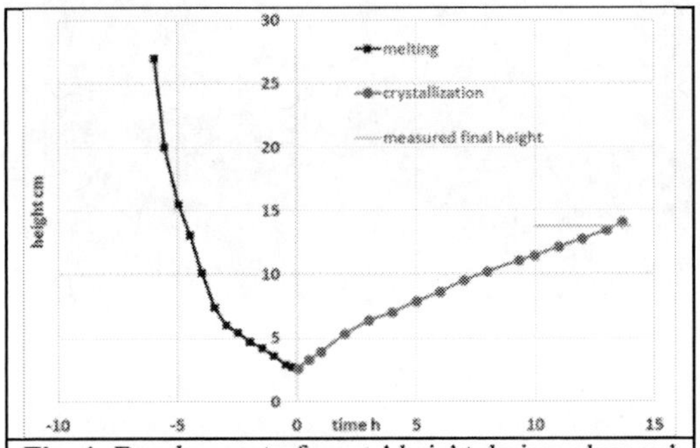

Fig. 4: Development of crystal height during advanced crystallization process, measured by BTD

Fig. 5: Bottom of a HPM Si crystal made with hybrid crucible

To establish a nearly complete inclusion of the melt during melting phase, in a third trial graphite felt at the outer side of the crucible bottom-side gap under the graphite side plates has been applied. The graphite felt prevents the thermal flow from the top heater to the bottom plate via side plates and provides a sharper separation between bottom and top temperature.

The new recipe in combination with modification of the hybrid crucible system with graphite felt enables to establish a process which is comparable in safety and behavior to a HPM crystallization with standard crucibles.

4 RESULTS

4.1 Sample preparation

All ingots of these trials were cut and analyzed by Fraunhofer THM regarding lifetime (MDP-measurement), resistivity, IR-scan and O- and C-content (FTIR-measurement). For the analysis along the center line of the ingot a thin board has been cut (figure 6) and the surface has been polished for measuring. The area for IR-scan, FTIR-measurement and resistivity is shown in figure 7.

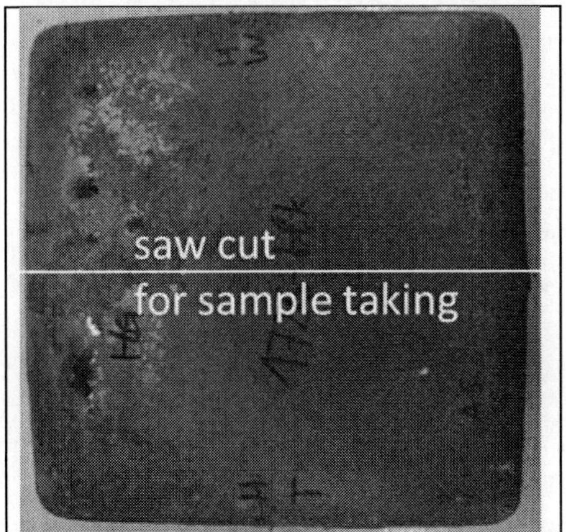

Fig. 6: Position of cutting on the ingot surface

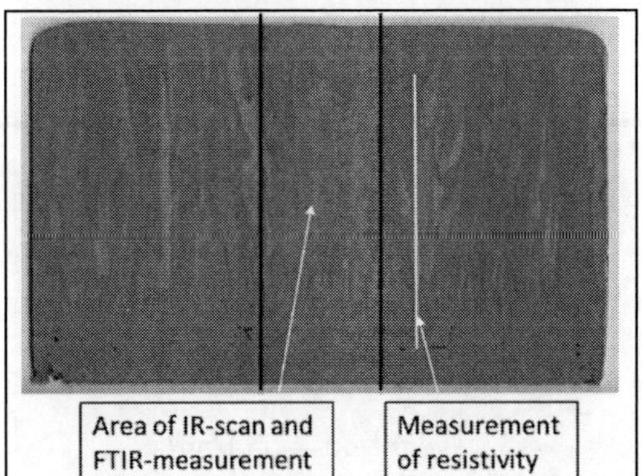

Fig. 7: Area for IR-scan, FTIR-measurement and resistivity

4.2 Carbon and oxygen content

The second ingot (the first with the new crystallization recipe) shows a high O-content in the seed layer (30mm block height) that decreases over the block height, while the C-content on the other hand increases fast (figure 8). At a height of around 60 mm the segregation level of carbon in silicon was reached which results in massive silicon-carbide clusters.

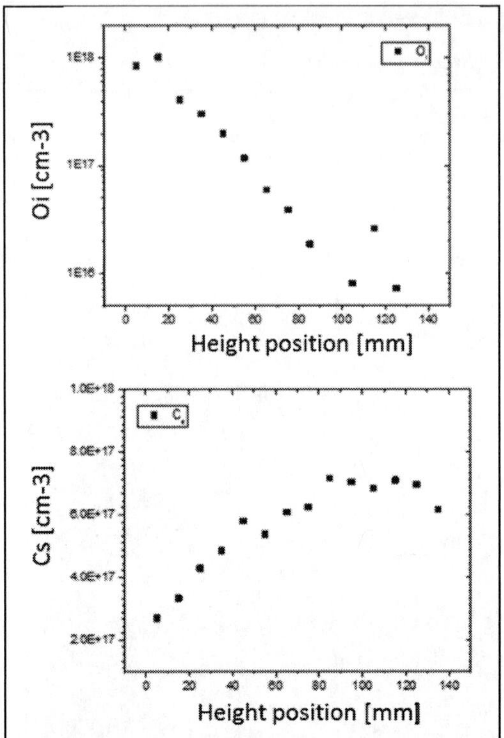

Fig. 8: C- and O-content of first ingot with power controlled recipe

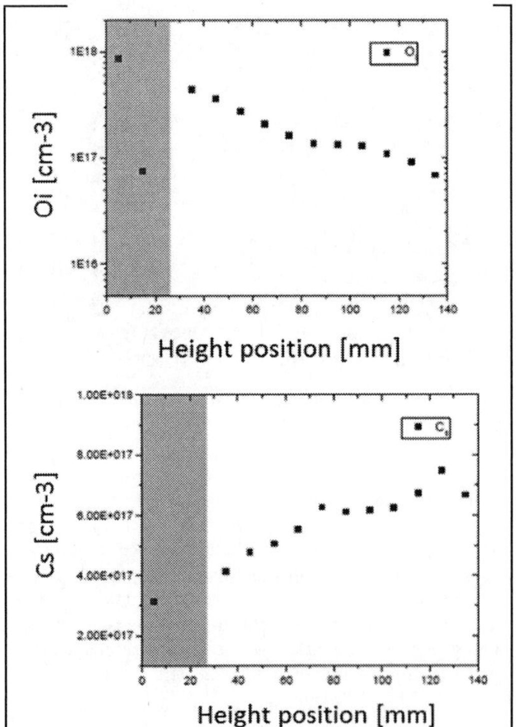

Fig. 9: C- and O-content of third ingot with temperature controlled recipe

Reasons for the strong difference in oxygen and carbon distribution could found in the crystallization speed of both ingots. The crystallization of the first ingot was finished after around 4h while crystallization of the third ingot took 14h. The extreme fast crystallization during the run for the first ingot was caused by the use of a crystallization recipe that was still too close for HPM-crystallization with a standard crucible.

Compared to a typical HPM process instandard crucibles in the used furnace we found similar oxygen and carbon concentrations.

4.3 Minority carrier lifetime (MDP)

An overlapping of IR- and lifetime-scan of the first ingot shows that the segregated carbon results in a massive reduction of the yield and the lifetime in the center of the ingot.

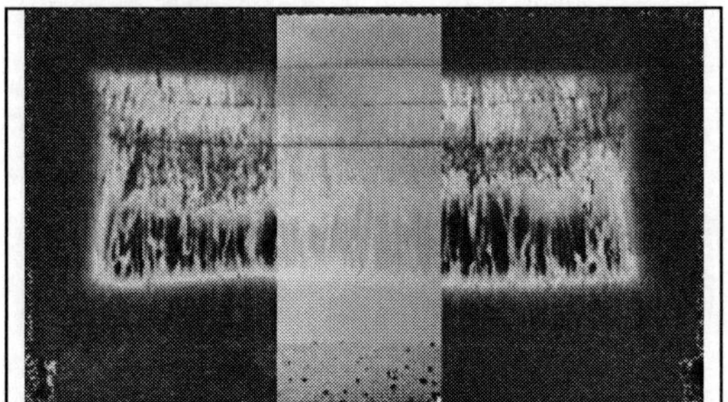

Figure 10: Overlapping lifetime scan and IR-scan for first ingot

Meanwhile the lifetime- and the IR-scan of the third ingot, shown in figure 11, present a red zone at top and bottom that is comparable with a HPM-ingot made in a standard crucible, exept for the carbon concentration at 120 mm ingot height.

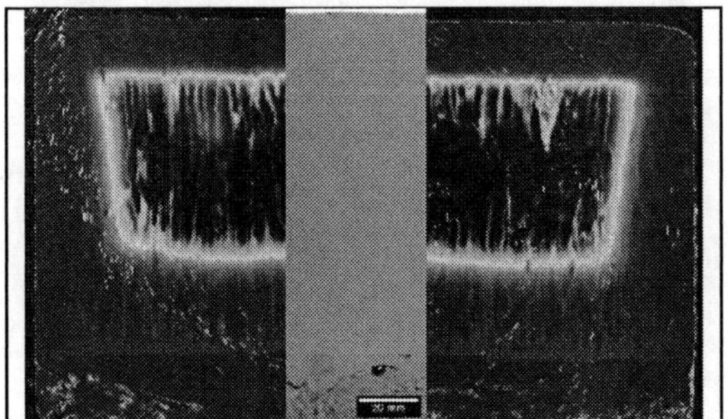

Figure 11: Overlapping lifetime scan and IR-scan for third ingot

Regarding the bottom red zone, in all cases there was no significant improvement in comparison to standard HPM. Obviously the main contamination source is not related to the crucible bottom itself. Detailed investigations are in progress.

4.4.Resistivity

We also recorded the resistivity of the ingots on their central cutting plane. Fig.12 and 13 give examples. The measured resistivities depend on the dopant content and are similar to those achieved with HPM ingots. The strong dropping resistivity in the lower half of the first ingot is mainly caused by the unusual extreme fast crystallization speed (see 3.1 pre-test).

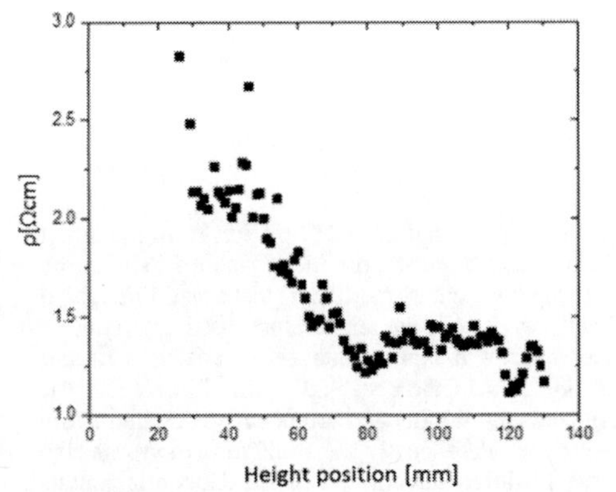

Figure 12: Resistivity first ingot with temperature controlled recipe

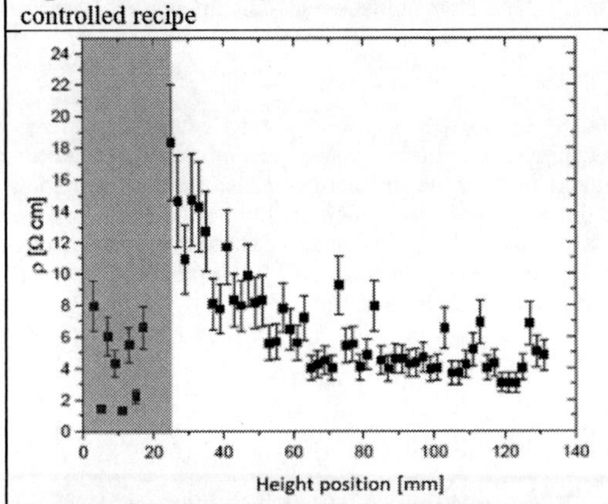

Figure 13: Resistivity third ingot with temperature controlled recipe

5 SUMMARY AND CONCLUSIONS

Regarding the usability of our new concept the tests have shown, that a HPM – crystallization process with the hybrid crucible system in G1-size can be handled safely and produce ingots that are close in quality to ingots which are processed with standard crucibles.

Regarding the process handling it has become clear that the melting and crystallization speed with the bottomless crucible is much more sensitive and reacts faster to the temperature gradient between top- and bottom heater and a flat uniform temperature profile over the height of the crucible system is much more of the essence than with a standard crucible, also regarding the safe inclusion of the melt. This means also that the heat transfer from top to bottom via the side plates of the crucible system has to be minimized as much as possible. So far the hybrid crucible has proved the theory that with the loss of the crucible bottom as a heat transfer resistance a faster and more accurate adjustment of the crystallization speed is possible.

Further trials will be performed to see if the new crucible concept can also provide the same ingot quality at a higher crystallization speed.

For the moment it is not exactly clear why even with a bottomless crucible it was not possible to reduce the red zone. Here additional investigation is necessary to clarify if the carbon or oxygen content in the seed layer is the source for the red zone or if the iron content has the main impact. For this, additional to the so far made analysis of the ingots examinations regarding the iron content shall be performed on existing samples.

As an alternative to a SiC-coated graphite plate, which is facing a continuous siliconizing during the process, Si_3N_4 will be also an option as material for the bottom plate that offers the possibility in reducing a source for carbon contamination into the melt.

6 REFERENCES

[1] M.P. Bellmann, G. Stokkan, A. Ciftja, R. Roligheten, J. Denafas, F. Buchholz, T. Kaden: Eco-Solar factory: Multicrystalline silicon ingot crystallisation from reusable silicon nitride crucibles, Proc. of 33. EU-PVSEC 2017 Amsterdam, p309-312 (2017)

[2] Lili Zhao, Tiezheng Lv, Qingsong Zhu: Si_3N_4/fused quartz composite crucible with enhanced thermal conductivity for multicrystalline silicon ingot growth, Journal of Crystal Growth 415 (2015), p 51-56

[3] ALD-VT GmbH, Patent specification DE 10 2014 102 980 B4

[4] B.Drevet, N.Eustathopoulos, J. Mat. Sci. 2012, DOI 10.1007/s10853-012-6663-0

[5] Y.Maeda et al., J. Electrochem. Soc. **133(2)**, 440-443 (1986)

[6] Systems and Methods for Monitoring a Solid-Liquid Interface, U.S. Patent specification 8,459,132

THERMAL STRESS MINIMIZATION IN SILICON RIBBON GROWTH PROCESSES BY THERMAL GRADIENTS MODULATION WITH 808 NM LASER SCANNING

David M. Pera*, M.C. Brito, A.M. Vallêra, João Manuel Serra, Jorge Maia Alves
Instituto Dom Luiz, Faculdade de Ciências, Universidade de Lisboa
Ed. C8 Campo Grande,1749-016 Lisbon, Portugal; *dmpera@fc.ul.pt

ABSTRACT: The temperature profile near the solid-liquid interface in the crystal growth direction is an important parameter since it will affect crystal defects structure. The bi-dimensional geometry of silicon ribbons reduces, by principle, the complexity of cooling gradients control when compared with traditional industrial directional solidification. In this work, we present a method to modulate and actively control the temperature distribution during growth of silicon ribbons. The method was explored in a silicon ribbon growth experimental set-up based on electric molten zone, EZ-Ribbon. Electric molten zone (EMZ) crystallization is a method that relies on silicon's electric conductivity increase with temperature. A positive feedback pattern leads to virtually all current collapsing into a thin molten capillary. The high power 808nm diode laser scanner used to stimulate locally the EMZ formation was also used to control the contiguous temperature distribution in the crystallization direction. The EZ-Ribbon technique configures a float zone method, featuring attractive kerfless, contamination-free and low energy budget crystallization mechanism. In particular, high crystalline quality may be, in the near future, its main issue regarding the present laser assisted thermal gradient modulation procedure.
Keywords: Ribbons, Silicon, Laser Processing, Crystallization

1 AIM AND APPROACH USED

Silicon based solar cells technologies has been so far dominated by ingot manufacturing processes. The associated costs are related to the high energy budget and the low silicon material yield inherent to the wafering process. In this context, Silicon Ribbon growing technologies use more efficiently silicon feedstock and energy, which makes them attractive for future of photovoltaic industry [1]. Nevertheless, ribbon shaped silicon growth still faces some challenges that must be overcome towards a successful deployment of ribbon solar cells, mainly due to both low grade crystallographic quality and throughput rates. Typically, these are limited by impurities contamination, contact with foreign materials like crucibles or shapers, high levels of thermally induced internal stresses and high dislocation densities.

In the last years, crystalline silicon ribbons growth techniques SDS Ribbons [2-5], gas to solid fast CVD process, and EZ-Ribbon [6,7], a liquid to solid process, were presented as virtually contamination free methods, promising particular advances in the reduction of thermo-mechanical stresses and dislocation densities in order to achieve high grade crystallographic quality.

The present work was developed in the context of the EZ-Ribbon technology (previously described [6]). This ribbon growth technique has a particularly low energy budget, since it avoids the necessity of large amounts of molten material and also because it uses a molten zone efficiently heated by Joule effect using an electric current directly through the liquid silicon. An electric molten zone may be stimulated by a very efficient, high power 808 nm diode laser scanning system that promotes the local increase of silicon electrical conductivity, and subsequent decrease of thermal conductivity, allowing the formation of a narrow and self-sustained molten capillary in a silicon sheet by Joule heating [8-10]. (Fig. 1).

The molten capillary is then fed by capillarity from a contamination free small molten reservoir (kept by 808 nm laser heating) while the crystalline shaper (original silicon sheet itself) is pulled. Earlier experimental results showed [6] a smooth cooling gradient characterized by a low magnitude and linear thermal second derivative profile along the plastic to elastic phase transition region promising potential high-quality crystalline silicon ribbons, since the thermal induced stress in a slim and long solid material is related to the established thermal distribution as fallows [17]:

$$\sigma \approx \alpha E \frac{\delta^2 T}{\delta y_i \delta y_j} L^2 \qquad (1)$$

α – linear expansion coefficient, E - modulus of elasticity, L- characteristic length

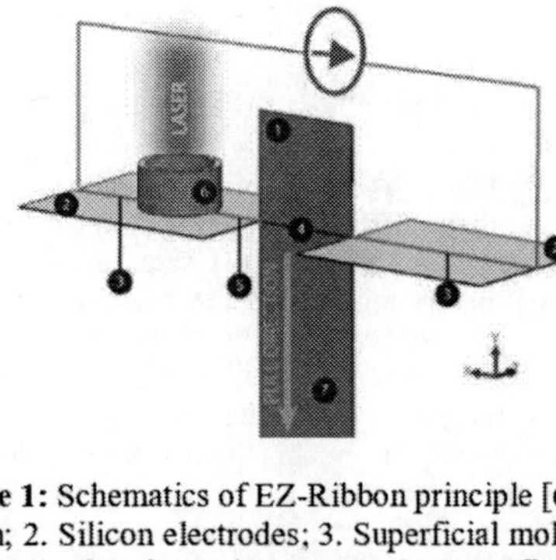

Figure 1: Schematics of EZ-Ribbon principle [6]: 1. Seed ribbon; 2. Silicon electrodes; 3. Superficial molten zones; 4. Linear electric molten zone; 5. Superficial silicon molten zone that conduces the molten feedstock material to the electric molten zone in the ribbon; 6. Granular silicon feedstock reservoir. The granules are molten by radiative heating; 7. Growing ribbon.

A stress-free cooling process will, in principle, avoid storage of mechanical stresses and the creation and/or multiplication of existing dislocations in the grown ribbons [12-16]. However, relaxation of existing dislocations will depend on the relation between pulling rate and the slope of the planar thermal gradient along the ductile to brittle transition region. The aim of the present work was to demonstrate the possibility of actively control the thermal distribution slope during the cooling in the crystallization of the EZ-Ribbon process using localized laser heating taking advantage of the scanning capabilities of the setup (Fig. 2 and Fig. 3).

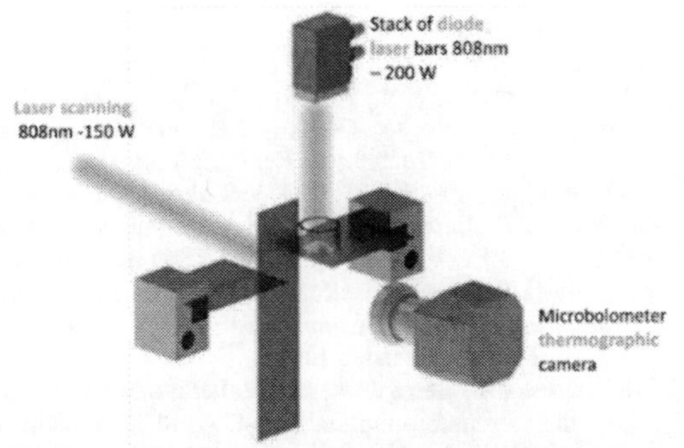

Figure 2: Schematics of EZ-Ribbon set-up. A fiber diode laser is used for EMZ stimulation; a stack bar diode laser is used to melt silicon granules feedstock; the temperature is monitored by a microbolometer thermal camera.

Figure 3: Picture of the interior of EZ-Ribbon furnace chamber.

For this purpose, an electric molten zone was stimulated transversely on a 25 mm wide (with 240 μm width) silicon ribbon using a high frequency scanning of a continuous wave 808 nm laser bidirectional line with 110 W optical power (Fig. 5. I) and 1mm spot diameter (Fig. 4). In this study we explored laser scanning schemes, using the same total optical power, such as rasters and vector patterns, in order to modulate the slope of the thermal cooling gradient preserving its planar characteristic and simultaneously guaranteeing the electric molten zone stimulation. The best results were achieved with parallel stripe patterns, such as the progressively spaced pattern (gap width = 0.5n mm; n = 1,2,3...) shown in Fig. 5. II and Fig. 6. The temperature distributions were monitored by an uncooled silicon micro-bolometer thermal camera, properly calibrated for the application.

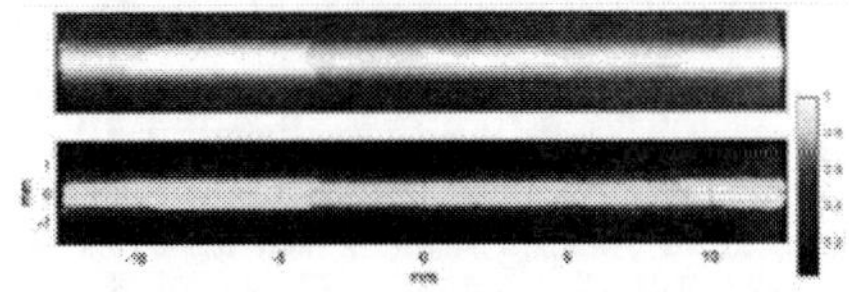

Figure 4: Fiber laser (808 nm) linear scanning on a 25 mm wide multicrystalline silicon ribbon. Top: CCD camera picture; Bottom: normalized intensity map. The laser spot diameter (FWHM) was estimated to be ~1mm.

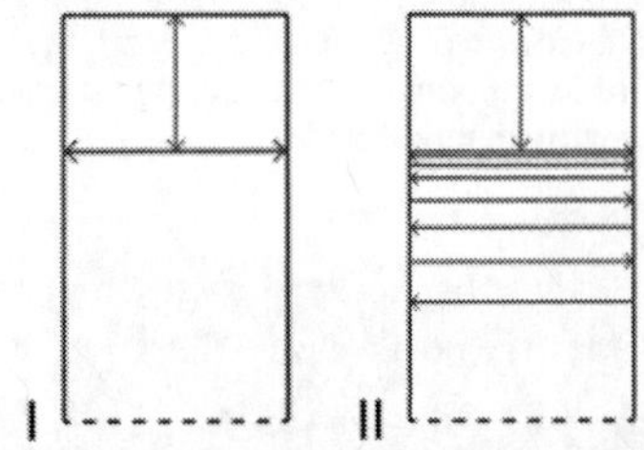

Figure 5: Schematics of the path vectors of laser scanning on a silicon sheet (red vectors). I – Standard configuration of EZ-Ribbon for radiative electric molten zone stimulation. II – Progressively spaced laser scanning pattern for thermal gradient modulation.

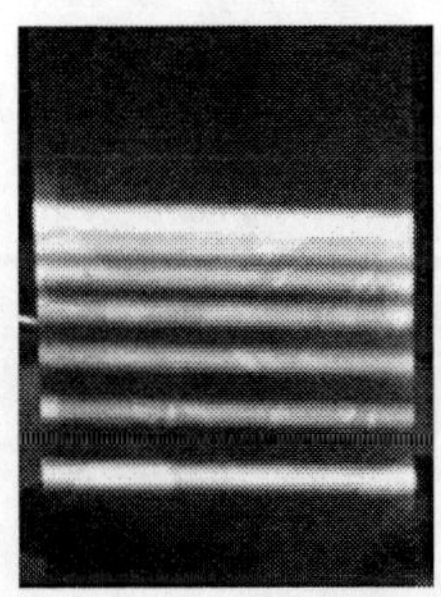

Figure 6: CCD camera picture of laser scanning pattern II of Fig. 5 on a 25 mm wide multicrystalline silicon ribbon.

2 SCIENTIFIC INNOVATION AND RELEVANCE

A new mechanism of thermal modulation of thermal gradients in a silicon ribbons growth process by diode laser heating was tested as a mean of reduce thermal induced defect densities in the grown crystal. The temperature profile along the growing direction is the most important parameter to control during the growth process since the temperature profile is related to the thermal stress magnitude and consequent mechanical stress and dislocations density in the grown ribbons [12-16]. Furthermore the proposed procedure doesn't increase the energy requirements of the growth process, once the input thermal power in the silicon ribbon is the same that was used in the previous electric molten zone stimulation, thus conserving the low energy budget feature of the EZ-Ribbon process.

3 RESULTS AND CONCLUSIONS

The thermograms (Fig. 7) show the temperature distributions on a silicon ribbon with a linear electric molten zone locally induced by laser radiation stimulated by laser, in the standard scanning scheme of EZ-Ribbon technique with just one bidirectional line, Fig. 5. I, and the specific case of the tested raster of Fig. 5. II. The second one exhibits a smoother and extended gradient.

As can be confirmed in the plot of Fig. 8, which refers to the thermal profile along the central line of the ribbon (white dashed line of Fig. 7) both cases are quite linear along the plastic to elastic transition regions of silicon fulfilling the requirement for a low second derivative as an indicator of low thermal stresses during the cooling of the crystalline growth process.

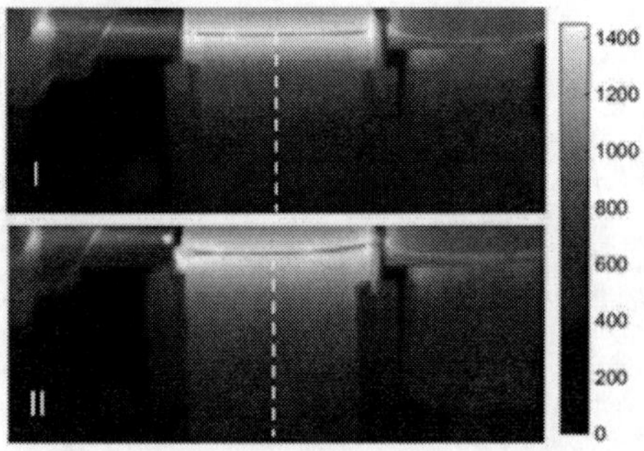

Figure 7: Temperature distributions of 25mm x100 mm silicon ribbons with a linear electric molten zone locally induced by laser radiation; Top and bottom images respect to laser scanning schemes I and II of Fig. 5 respectively.

Nevertheless, the smoother slope of the second profile offers an extended phase transition length which can be relevant in terms of dislocation density control.

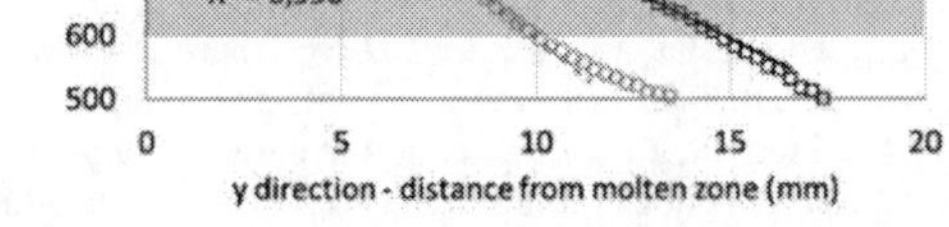

Figure 8: Cooling temperature profiles on a silicon ribbon along a 25 mm centreline (y = 0 corresponds to the canter of the linear molten zone.

A mechanism of thermal gradients modulation by laser scanning in a silicon ribbon growth process was thus successfully tested. This procedure may configure a future solution for dislocation engineering of high-quality crystalline silicon ribbons for solar photovoltaic application.

4 ACKNOWLEDGMENTS

That authors acknowledge the Portuguese Science & Technology Foundation (FCT) for funding through the R&D strategic project funding project UID/GEO/50019/2013

5 REFERENCES

[1] G. Hahn, S. Seren, M. Kaes, A. Schonecker, J.P. Kalejs, C. Dube, A. Grenko, C.Belouet, Review on Ribbon Silicon Techniques For Cost Reduction in PV, Proceedings of 4th WC PEC Waikoloa, 2006

[2] Pinto C.R., Serra J. M., Brito M. C., Gamboa R., Maia Alves J., Vallêra A. M.; "Zone melting recrystallization of self supported silicon ribbons obtained by fast cvd from silane"; Proceedings of the 21st EPVSEC, Dresden 2006

[3] Augusto A., Serra J.M., and Vallêra A.M., Progress on the Continuous Optical Fast CVD System to Grow Silicon Ribbons For Solar Cells Via SDS Process, Photovoltaic Specialists Conference (PVSC), 2010 35th IEEE

[4] Augusto A., Serra J.M., and Vallêra A.M., First multicrystalline silicon ribbons using the continuous SDS process, Photovoltaic Specialists Conference (PVSC), 2011 37th IEEE

[5] J. M. Serra, C. R. Pinto, J. A. Silva, M. C. Brito, J. Maia Alves and A.M. Vallêra, The silicon on dust substrate path to make solar cells directly from a gaseous feedstock, Semicond. Sci. Technol. 24 (2009) 045002 (5pp), doi:10.1088/0268-1242/24/4/045002

[6] D.M.Pera K. Lobato, M. C. Brito, J. Maia Alves, J. M. Serra, A. M. Vallêra, Linear Electric Molten Zone Stimulated by Laser for Silicon Ribbons Growth, Proceedings of the 28st EPVSEC, Paris 2013

[7] António M. Vallêra, et al. Patent: US 2007/0241981 A1 (227)

[8] António M. Vallêra, Jorge Maia Alves, João M. Serra, Miguel C. Brito, Roberto M. Gamboa, Appl. Phys. Lett. 90, 232111 (2007)

[9] M.C. Brito, A. Amaral, J. Maia Alves, J.M. Serra, I. Costa, A.M. Vallêra, Prog. Photovolt: Res. Appl., 17 (2009) 365–371

[10] R.M. Gamboa, M.C. Brito, J.M. Alves, J.M. Serra, A.M. Vallera, Journal of Crystal Growth 324 (2011) 26–30

[11] I. Costa, M.C. Brito, J.M. Serra, J. Maia Alves, A. Vallera, Journal of Crystal Growth, Volume 354, Issue 1 (2012) 198–201

[12] A.Augusto, D. Pera, H.J Choi, P. Bellanger, M.C. Brito, J. Maia Alves, A. M. Vallêra, T.Buonassisi, J.M.Serra, J. Appl. Phys. 113, 083510 (2013)

[13] M.C. Brito, J. Maia Alves, J.M. Serra, R.M. Gamboa, C. Pinto, and A.M. Vallêra, Sol. Energy Mater. Sol. Cells 87, 311 (2005)

[14] G. Sarau, S. Christiansen, M. Holla, and W. Seifert, Sol. Energy Mater. Sol. Cells 95, 2264 (2011).

[15] O.W. Dillon, Technical Report No. JPL 956571-1, University of Kentucky and Jet Propulsion Laboratory (NASA), 1985

[16] C. Donolato, J. Appl. Phys. 84, 2656 (1998)

[17] V. Indembom, Krist. Tech. 14, 493 (1979)

2AV.1.19

ECO-SOLAR FACTORY: UTILISATION OF KERF-LOSS FROM SILICON WAFER SAWING FOR THE MANUFACTURING OF SILICON NITRIDE CRUCIBLES

M.P. Bellmann[1], G. Noja[2], A. Ciftja[3]

[1]SINTEF Industry, Trondheim, Norway;
phone: +47 934 986 19, e-mail: martin.bellmann@sintef.no
[2]Garbo Srl, Cerano, Italy
[3]Steuler Solar Technology AS, Porsgrunn, Norway

ABSTRACT: H2020-FOF-13-2015 Eco-Solar submitted in 2015 envisions to increase resource and energy efficiency over the entire photovoltaic value chain, while simultaneously maximising recycling possibilities, by introducing design for recovery, repair and reuse, and collaborating for improvements in waste reduction.
Up to 50% of silicon is lost in form of fine powders (kerf-loss) during the mechanical processing of ingots to obtain wafers. The ability to recycle the kerf-loss has beneficial effect in terms of savings of precious poly-silicon consumption and of waste reduction. Due to the high energy requirement (150 kWh/kg) and CO2 emissions (40 kg CO2-eq/ kg Si) associated with poly-silicon production with the traditional Siemens route, significant impact in environmental sustainability is expected by the recycling of the kerf-loss as secondary feedstock material. In this work recent results on reusing recovered silicon kerf-loss for manufacturing of silicon nitride crucibles are presented.

Keywords: silicon crystallisation, silicon-kerf-loss, silicon nitride crucibles

1 AIM AND APPROACH USED

The European H2020-FOF-13-2015 Eco-Solar project that started in October 2015 envisions increasing resource and energy efficiency over the entire photovoltaic value chain, while simultaneously maximising recycling and remanufacturing efficiencies, by introducing design for recovery, repair and reuse, and collaborating for improvements in waste reduction. Reusing materials and reducing the consumption of raw materials will make solar cell panels both cheaper and greener.

The potential of reusing recovered and refined silicon kerf-loss for manufacturing of silicon nitride crucibles presents a fantastic opportunity to reduce material cost related to primary silicon fines. Silicon kerf-loss from diamond-wire sawing of monocrystalline silicon ingots has been used in manufacturing of silicon nitride crucibles. The material has been chemically treated to remove impurities that are introduced during the sawing process. These consists mostly of metal impurities from the steel wire used for cutting, with iron being the main pollutant, and from the glass plates used for fixing and holding the silicon block into the cutting device, where impurities such as Na, Ca, Al, Mg etc. are introduced.
Several green crucibles were manufactured with varying ratio of primary silicon to silicon kerf-loss and subsequently nitridation. Crucibles have been used for crystallisation of multi crystalline silicon ingots. The obtained material quality is compared to reference material obtained from conventional silicon nitride crucibles made entirely from primary silicon fines.

2 EXPERIMENTAL & RESULTS

2.1 Crucible manufacturing

Exhausted coolant from diamond wire sawing of monocrystalline ingots has been filtrated to obtain the solid fraction, the Si-kerf-loss filter cake. The Si-kerf-loss has been etched with HCl to remove surface contaminations, washed and finally dried in vacuum. After treatment the material has been analysed for carbon, oxygen and metallic impurities. The Si-kerf-loss contains around 1.8 wt% of carbon, 2.4 wt% oxygen and 15 ppmw surface metals.

Silicon nitride crucibles were manufactured, where the original silicon constituent in the recipe is only partly substituted with Si-kerf loss while the rest of the materials is not changed. Si-Kerf represent 10 % of the total mass. This mixture is used to cast a G1 (11") crucible (Fig. 2). As it can be seen the crucible maintained the shape and resulted with seemingly good mechanical properties. However, a few silicon flakes or spots appear at the bottom suggesting for a less than perfect nitridation process.

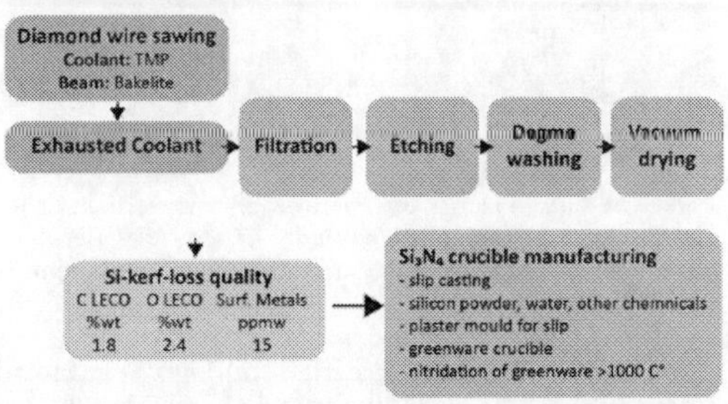

Figure 1: Treatment of silicon-kerf-loss from diamond wire sawing and subsequent silicon nitride crucible manufacturing process.

Figure 2: G1 crucible made from 10 wt.% Si-kerf-loss before (left) and after (right) nitridation

2.2 Crystallisation experiments

G1 HPmc-Si ingot (p-type) were directionally solidified from the silicon nitride crucible with 10 wt% Si-kerf-loss. Colloidal silica is used as crucible coating agent. Cross sectional cut of the obtained ingot is shown in Fig. 3.

Figure 3: Cross sectional cut of the grown HPmc-Si ingot. In the bottom area the remaining seeding layer can be seen.

Charge carrier lifetime mapping using µPCD has been conducted on the cross sectional cut and the results are shown in Figure 4.

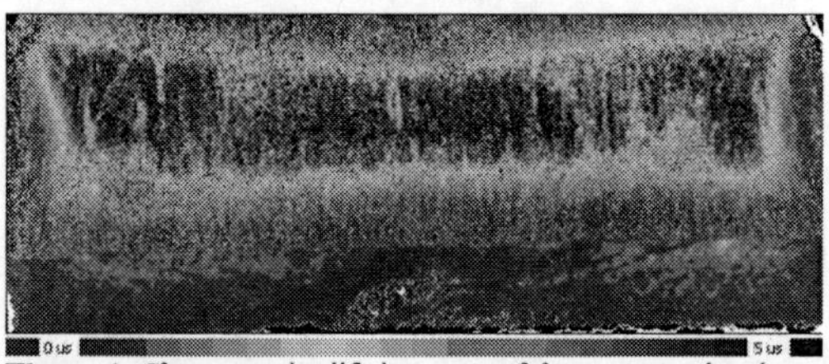

Figure 4: Charge carrier lifetime map of the cross sectional cut. Scale of the lifetime is from 0 to 5 µs.

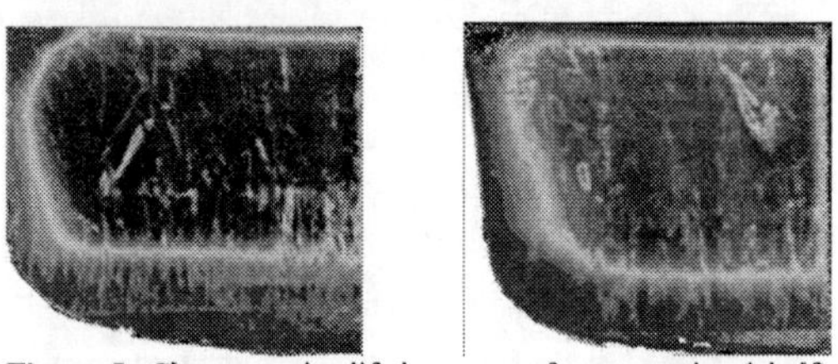

Figure 5: Charge carrier lifetime map of cross-sectional half-cuts for material grown in high purity EG-crucibles (left) and standard silicon nitride crucibles (right). Scale of the lifetime is from 0 to 5 µs.

Compared to reference materials in Fig. 5 (standard silicon nitride crucible, high purity EG-crucible), the red zone is more pronounced and average carrier lifetime is lower. Main contributor to lower carrier lifetime are metallic impurities from the Si-kerf-loss (15 ppmw).

Furthermore, the material has been analysed for carbon and oxygen by means of FTIR, results are shown in Fig. 6 and 7. Pronounced carbon levels can be observed for material grown in silicon nitride crucibles, no matter if silicon kerf-loss or standard material is used for crucible manufacturing. SiC impurities in the crucible are an additional source of carbon, contributing to increased carbon levels in the ingot.

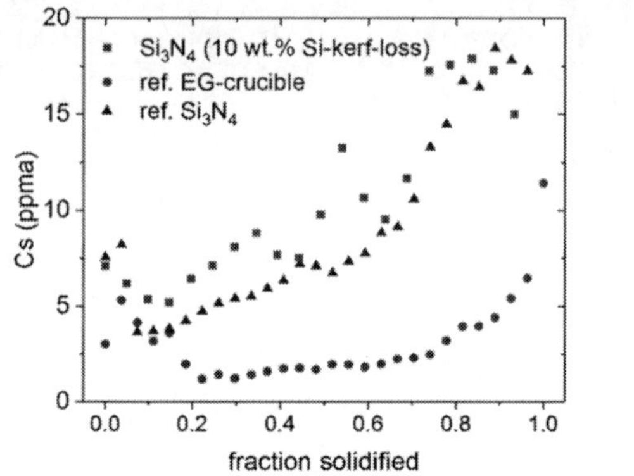

Figure 6: Carbon segregation profiles by means of FTIR measurements.

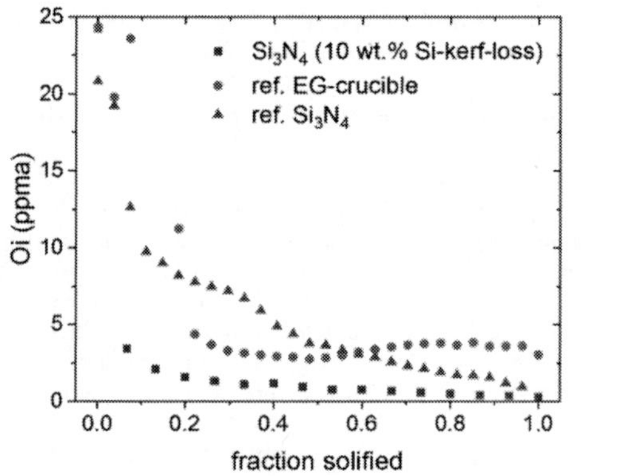

Figure 7: Oxygen segregation profiles by means of FTIR measurements.

Lowest oxygen levels (Fig. 7) are observed for the material grown in the silicon nitride crucible with 10 wt% Si-kerf-loss. Lower oxygen levels can be explained by the usage of a different coating agent, colloidal silica instead of silicon oxynitride. In addition, the crucible has been pre-treated together with the coating in inert, instead in oxidizing atmosphere.

3 SUMMARY

Silicon nitride crucible based on 10 wt.% Si-kerf loss was produced. Crucible was coated with colloidal silica and a HPmc-Si ingot was directionally solidified.

Material quality is suitable, but underperformance in comparison to reference ingots grown in high purity EG-crucibles and standard silicon nitride crucibles.

Due to SiC impurities in the silicon nitride crucible, carbon levels are more pronounced. Due to crucible pre-treatment in inert atmosphere, oxygen levels are significantly lower, below 4 ppma. Further purification of silicon kerf-loss is suggested to qualify as a secondary raw-material for silicon nitride crucible manufacturing.

ACKNOWLEDGEMENT

This project has received funding from the European Union's Horizon 2020 research and innovation programme. under grant agreement No. 679692.

THE IMPACT OF WAFER THICKNESS (210 AND 140 μm) FOR PHOTOVOLTAIC USE ON THE FRACTURE STRENGTH

Halubai Sekhar*, Tetsuo Fukuda*, Katsuto Tanahashi, and Hidetaka Takato

Renewable Energy Research Center (RENRC)
National Institute of Advanced Industrial Science and Technology (AIST), Japan.

*E-mail: halubai-sekhar@aist.go.jp, halubaisekhar@gmail.com, tetsuo-fukuda@aist.go.jp
Phone/Fax: +81-29-849-1547/+81-24-963-0824

ABSTRACT: We developed a diamond wire characterized by the core steel-wire diameter of 80 μm and the size of diamond abrasives ranging from 6 to 12 μm (labelled as 80d-M6/12). The diamond particles on the wire are designed to have lower density and less agglomeration than those on conventional wires. As-developed diamond wire is introduced into multi wire saw to cut monocrystalline silicon bricks into wafers with a thickness of 210 and 140 μm. The above wire allows us to reduce the silicon dust or kerf to 100 μm per wafer. After as-sawn wafers are separated from fresh and worn-out wire sides of bricks and labelled as fresh and worn-out wire side wafers. In three-line bending test, a variable load is applied in perpendicular direction to wire saw marks on the wafer surface. The fracture strength distributions are evaluated on fresh and worn-out wire side wafers. In thick wafers, the fracture stress distributions are approximately the same on both sides. On the other hand, in thin wafers, the fracture strength is extremely lower than that of thick wafers. And particularly, the strength of fresh wire side wafers is lower compared to that of worn-out wire side wafers. In thin wafers, a small variation in the surface and subsurface damage is a crucial factor to decide their fracture strength of both-side wafers. In case of thick wafers, a small variation in the surface and subsurface damage may not have strong influence on their fracture strength of both-side wafers.
Keywords: Silicon, Photovoltaic, Wafer, Three-line bending, Fracture strength, Diamond wire, Kerf

1 INTRODUCTION

In photovoltaic (PV) research and development, there has been enormous pressure to improve efficiency and to reduce the production cost per watt at every stage of the manufacturing chain by implementing innovative concepts and techniques [1-4]. Si is a brittle material, sawing silicon (Si) bricks into wafers is the first processing step to manufacture Si solar modules. There are two current ongoing research problems in the wafer sawing technology. First problem is to reduce sawing loss (Si kerf) per wafer, and the second problem is to saw thinner wafers [5-7]. The Si kerf or saw dust per wafer is calculated as the summation of wire diameter and abrasive size. This trend leads to use thinner wires with smaller abrasives than before. By sawing thinner wafers allows to increase number of wafers from the same amount of materials than before. In device fabrication process, the damage layers removing adds an additional material loss. In traditional cutting (slurry), more than half of the silicon brisk is consumed as a material loss (kerf + damage removal). To address this issues in wafer sawing industry looks for new technologies. Diamond wire wafer sawing technology gives an opportunity to address these issues and allow to reduce the material loss compare to their traditional slurry wire wafer sawing technology [8-10]. To cut mono-crystalline Si brisk into wafers, diamond wire technology rapidly gaining the market shares by replacing slurry technology [5].

Sawing is a mechanical process and there are some limitations in loose-abrasive slurry sawing technology on wafer thickness and Si kerf. This trend leads to shift the sawing technique from adopting loose-abrasive slurry to fixed-diamond abrasives on wire and as well, leads to saw wafers by thinner wires with smaller abrasives than before [8,9]. Since Si is a brittle material, the material removing or chipping in ductile mode can allow to reduce the wafer breakage and makes it possible to saw thinner wafers with a higher production yields in industrial scale. In diamond wire sawing (DWS) process the cuspidate diamond-particles fixed on the core wire moves to-and-fro motion and scratches the wafer surface with high compressive stress simultaneously results an indentation. This allows the sawing process to undergo in ductile mode. In the DWS, the material removing or chipping occurs in ductile mode leads to produce less density of micro cracks on the surface and subsurface regions [9-10]. This makes it possible to saw thinner wafers without sacrificing the fracture strength and to increase in process yield than before. In literature there are several other advantages were reported in DW sawing, allows to saw thinner wafers, easy to recycle silicon kerf, higher productions yields, less consumables costs and environmentally friendly [8-14].

In the present study, we have developed a diamond wire (80d-M6/12) and silicon blocks sawn into wafers with a thickness of 210 μm and 140 μm. As sawn wafers fracture stress evaluate using simple three line bending setup.

2 EXPERIMENTAL

To reduce the Si kerf per wafer (100 μm) in multi wire wafer sawing, we have developed a diamond wire and the wire labelled as 80d-M6/12. Where 80d represents wire diameter (80 μm) and M6/12 represents diameter of diamond particles (6 to 12 μm) on the wire. Using Keyence 3d laser microscope diamond wire was observed using a 20x objective lens shown in Fig. 1(a). From Fig.1, one can clearly observe that density of diamond particles on the wire is low and particles are uniformly dispersed with less agglomeration. In multi wire diamond saw, the optimized process parameters were as follows. The tension on the diamond wire was 13 N, the wire speed was 1000 m/min, and the brick feeding speed was 0.53 mm/min. Using above processing parameters in multi wire diamond saw, the pseudo-square monocrystalline Si block (of size 156 × 156 mm^2) sawn in to wafers with a thickness of 210 and 140 μm as shown in Fig. (b). The wafers were separated and labelled as fresh wire (wafers near entrance of diamond wire) and worn-out (wafers near exit of diamond wire) wire side wafers. As separated Si wafers were cleaned in soap water then rained in clean water and wafers are dried using air blow.

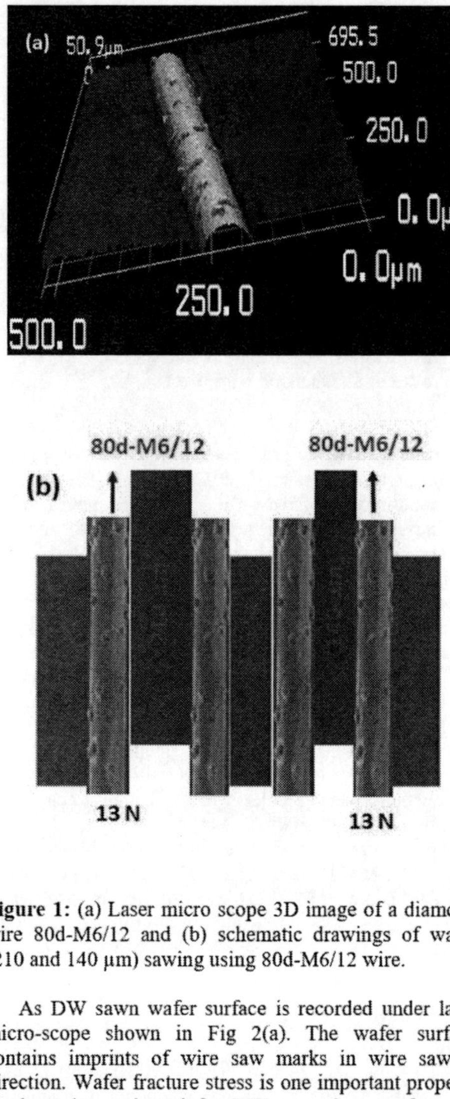

Figure 1: (a) Laser micro scope 3D image of a diamond wire 80d-M6/12 and (b) schematic drawings of wafer (210 and 140 μm) sawing using 80d-M6/12 wire.

As DW sawn wafer surface is recorded under laser micro-scope shown in Fig 2(a). The wafer surface contains imprints of wire saw marks in wire sawing direction. Wafer fracture stress is one important property needs to be evaluated for DW sawn bare wafers. To measure mechanical strength of as sawn wafers are introduced in bending experiment. In the present we have employed a three-line bending tester (Shimadzu Autograph AG-X plus) to measure the fracture stress. The wafer was inserted on two supporting bars and variable load attached to third line bar and loaded on the middle of the wafer. To address the impact of surface subsurface damage on fracture strength, loads are applied perpendicular to wire saw marks in three line bending test. The schematic of the three line bending tester is shown in Fig. 2 (b). In three-line bending wafer experience a gradient stress and maximum on the middle of the wafer as shown in Fig. 2(b). By measuring the loads at the moment of wafer breakage, we have evaluated its fracture strength using below equation

$$\text{Fracture strength: } \sigma_{frac} = 3P_{frac}L\,/\,(2bh^2)$$

where P_{frac} is the fracture load, L the distance between the two supporting bars (80 mm), b the width of the sample wafer (156 mm), and h the wafer thickness.

We have taken extra care in the bending experiment to collect shattered pieces from the middle of the wafer not to introduce any external artifacts on the samples. To understand fracture mechanism, shattered pieces from the middle of the wafer are recorded using a scanning electron microscope (SEM), JCM-6000PLUS NeoScope Benchtop.

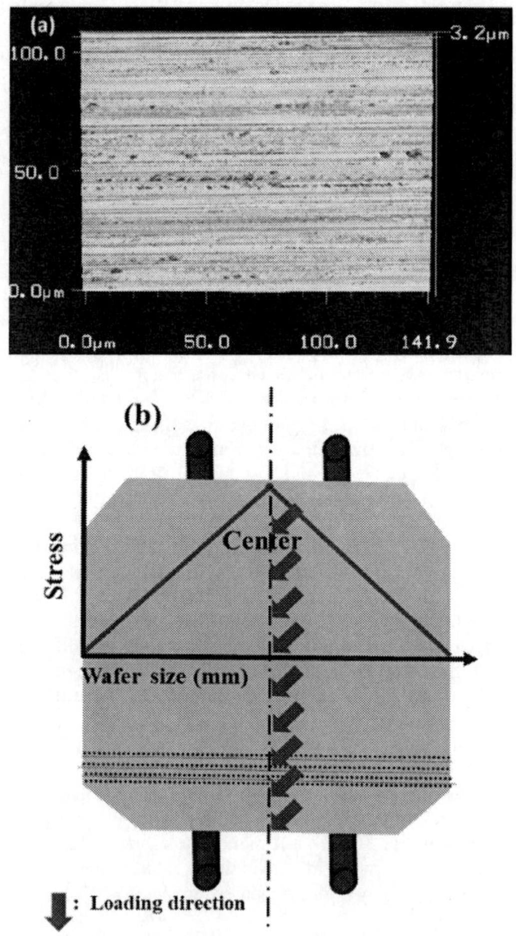

Figure 2: (a) Laser microscope image of diamond wire sawn wafer surface contains imprints of wire saw marks (b) schematic of three-line bending.

3 RESULTS AND DISCUSSION

Single crystal Si is a diamond lattice structure, brittle and anisotropic material [15-20]. The Si wafers contains no defects on surface, subsurface and bulk regions are loaded to fracture, resists to higher loads and cleave along crystallographic planes guided by their inter-atomic bonds. The fracture strength of Si is strongly depends on size of surface flaws and its location. In diamond wire sawing surface flaws were generated on silicon wafer surface and subsurface regions in the forms of terrace and pits as shown in Fig. 2(a). By applying tension 19N on a diamond wire (100d-M6/12) in our previous study, we have sawn a Si block into thin wafers (120 μm) having a higher fracture strength, [8, 9]. To reduce kerf loss in DW sawing, the diameter of the core wire lowered from 100 to 80 μm. Thinner wires become rupture to higher tensions, so it makes us to reduce (19 to 13 N) the tension on the wire. The fracture stresses of thick and thin wafers are measured and shown in Table 1. From this table one can clearly observe that fracture strengths of thick wafers on fresh (336±10 MPa) and worn-out (327±6 MPa) wire sides are approximately same.

Table 1: Fracture strength of as-sawn wafers.

Wafer thickness (μm)	Fracture strength (MPa)	
	Fresh wire side	Worn-out wire side
210	336±10	327±6
140	209±57	260±27

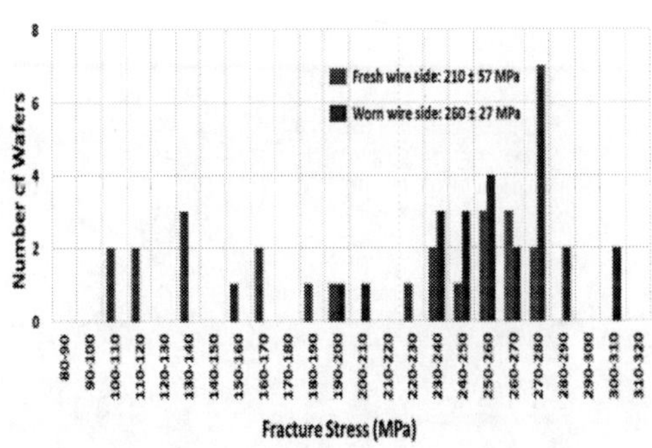

Figure 3: Fracture strength of as-sawn wafers with a thickness of 140 μm from the fresh and worn-out wire sides.

On the contrary in thin (140 μm) wafers a large number of fresh side wafers fractures at a lower strength compare to their worn wire side wafers and fracture strength are 209±57 MPa and 260±27 MPa. The fracture strength of thin (140 μm) wafers on both sides dramatically reduced compared to those of thick (210 μm) wafers. The fracture strength distributions of thin wafers taken from fresh and worn wire sides are shown Fig. 2(b). From this figure one can clearly observe that a large number of fresh wire side wafers are broken at lower fracture strength compare to their worn-wire side. The explanation for lower fracture stress is as follows. Sawing thin wafers using thin wires, both wire and wafer

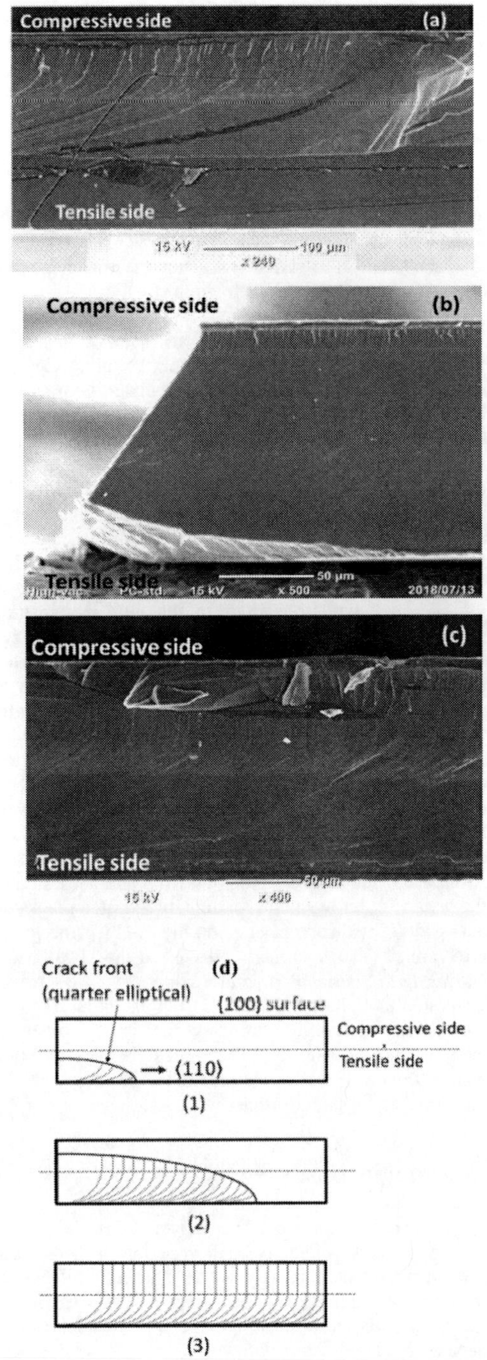

Figure 4: Fracture surfaces of (a) thick wafers (210 μm), (b) and (c) thin wafers (140 μm). (d) (1) shows the propagation as a quarter elliptical crack, (2) hackle of the advancing crack and (3) the deflection of the crack advance and the completion of the fracture.

vibrate and cause more damage by hitting the wire on wafer surface.

The explanation for lower fracture stress on fresh wire side is as follows: when sawing is in progress the sharp tips of the diamond abrasive on the fresh-wire side, scratch the silicon surface and generate microscopic cracks, steps and phase transformation on the wafer surface [8, 9]. These defective Si areas are responsible for the reduction in fracture strength of the wafer. When the wire passes from the fresh to the worn-out side, the sharp tips of the diamond particles are smoothened. The damage of wafers cut from the worn-out side caused by smoother diamond particles might be less than that to the wafers cut from the fresh-wire side [8, 9]. In thin wafers, a small variation in surface and subsurface damage is a crucial factor to decide their fracture strength.

To address the difference in fracture strength, fractured surfaces are observed under scanning electron microscope. Under dynamic loading in brittle silicon, a crack can propagate in several ways such as mirror, mist, hackle, and branching [15-20]. Mirror fracture is a low velocity, low energy mechanism, with little or no surface perturbations. The remaining fractures (mist, hackle, and branching) are categorized as high velocity, high energy mechanisms, with surface perturbations. The fractured faces of thick and thin wafers are shown in Fig. 4.

In perpendicular bending to the wire saw marks, scratching pits are considered as individual small defects. In thick wafers cracks propagated in a tortuous way from top to bottom surface choosing longer paths as like as shown in Fig. 4 (a). In both thick and thin fractured wafer cross sections shows quarter elliptical shape crack paths. The crack deflection mechanism is responsible to form curved nature of perturbations or quarter elliptical shape crack front in Si. The quarter elliptical crack fronts shape can be explained using Sherman's model [16, 18]. DW sawn bare wafers surface contains pre-existed cracks on surface and subsurface regions on front and back side of the wafer. Under dynamics loading, pre-existed cracks on back side of wafer propagates on the {111} plane to the direction of ⟨110⟩ and then deflected to the {110} plane according to Sherman's fracture model (Fig. 4 d). It was suggested that crack deflection or branching is the basic mechanism controlling the way the crack consumes energy while propagating. The results on thick and thin wafers fracture mechanism are under progress will be published as a separate article.

5 CONCLUSIONS

Two conclusions can be drawn from the above results. In thick wafers, a small variation in the surface and subsurface damage may not have strong influence on their fracture strength. In thin wafers, a small variation in the surface and subsurface damage is a crucial factor to decide their fracture strength.

REFERENCES

[1] C. Battaglia, Energy Environ. Sci. 9 (2016) 1552.
[2] N. Yamada, Prog. Photovoltaics 24 (2016) 846.
[3] M. M. Rahman, Jpn. J. Appl. Phys. 55 (2016) 032303.
[4] M. M. Rahman, Prog. Photovoltaics 24 (2016) 774.
[5] ITRPV, "International Technology Roadmap for Photovoltaics" 2018, 9th edition.
[6] B. Weber, C. Bierwisch, R. Kübler, G. Kleer, 23rd European Photovoltaic Solar Energy Conf. Exhib. 2008, p. 1285.
[7] B. Weber, and S. Riepe, 27th European Photovoltaic Solar Energy Conf. Exhib. 2012, p. 1060.
[8] H. Sekhar, Jap. J. Appl. Phys. 57 (2018) 08RB08.
[9] H. Sekhar, Jap. J. Appl. Phys. 57 (2018) 095501.
[10] A. Kumar, Procedia Manuf. 21 (2018) 549.
[11] N. Watanabe, Prog. Photovoltaics 18 (2010) 485.
[12] A. Kumar, Wear 364-365, (2016)163.
[13] X. Yu, Sol. Energy Mater. Sol. Cells 98 (2012) 337.
[14] H. Wu, Precis. Eng. 43 (2016) 1.
[15] Lv Zhao, Nature Communications 9 (2018) 1298.
[16] D. Sherman, J. Mater. Sci. 38 (2003) 783.
[17] I. Beery, J. Appl. Phys. 93 (2003) 2429.
[18] D. Sherman, Int. J. Fract. 140 (2006) 125.
[19] S. K. Bhaumik, Eng. Failure Analysis 9 (2002) 255.
[20] R. Pérez, Phys. Rev. Lett. 84 (2000) 5347.

EPITAXIAL GROWTH OF SILICON BY ELECTRON BEAM EVAPORATION DEPOSITION

M. Stange[1]*, T.O Sunde[1], R. Dahl-Hansen[1], A. S. Azar[1], J. S. Graff[1], J. Nørgaard[2], A. Ulyashin[1]

[1]SINTEF Industry, Box 124, Blindern, 0314 Oslo, Norway
[2]Polyteknik AS, Moellegade 21, DK-9750 Oestervraa

ABSTRACT: In this paper we report epitaxial growth of thin p-doped Si films at Si (100) substrates at deposition temperatures $T_s = 600–1000$ °C under non-ultra-high vacuum conditions using high rate electron beam evaporation. It is established that fully crystalline Si based layers (~ 50 μm thick) can be fabricated at a deposition rate of ~1.5 μm/min. The crystallinity and stress were analyzed by Raman spectroscopy, which revealed a peak at 520 cm^{-1}, confirming crystalline and stress-free silicon. Electron Backscatter Diffraction (EBSD) shows that the film growth proceeds epitaxially on the crystalline Si supporting substrate with excellent crystallinity at 800 and 1000°C. Hall effect measurements show that the film deposited at 1000 °C has a resistivity of 5.01 Ωcm, carrier concentration of 4.93^{15} cm^{-3} and mobility of 253 $cm^2V^{-1}s^{-1}$. These values comparable to those obtained for c-Si material grown using the same Si feedstock (solar grade Si). Secco etching tests were also performed to analyze the defect density in the films. The films deposited at higher temperatures have lower etch pits density. These results demonstrates that high temperature electron beam deposition can be considered as a promising and cost-effective alternative to conventional CVD based growth of epi-Si layers.

Keywords: silicon, epitaxy, cost reduction, high deposition rate, electron beam deposition,

1 INTRODUCTION

Today, high-quality monocrystalline silicon (Si) ingots are typically grown by the Czochralski (CZ)-method and cut to thin wafers creating the basis for the different devices. Fabrication of thin wafers (around~100 μm) is of increasing importance to reduce material consumption and to enable processing of novel devices with increased functionality, such as flexible solar cells. However, due to physical limitations of the sawing technology used today, especially the ratio between the saw and the wafer-thickness, an increasing amount of Si-material is lost in the cutting process as the wafer thickness continues to decrease. In fact, the thickness of wafers produced today is comparable to the thickness of the saw, meaning more than 50% Si material loss (kerf) per wafer produced. Although some material can be regained through kerf re-melting, the achieved yield is strongly limited by issues such as material contamination, increased energy usage, handling and breakage of the cut wafers and increasing number of processing steps. For these reasons new approaches and methods must be developed in order to reduce the associated energy and material costs (and the environmental footprint), and enable the continued technological development.

There are several different methods for producing "kerfless/kerffree" Si wafers without casting and sawing: solidifying melt either against a substrate (includes direct wafer) or as a vertical ribbon, ion implantation and cleaving, exfoliation, and epitaxial growth by CVD or physical vapor deposition (PVD) such as electron beam (e-beam) deposition followed by laser or e-beam crystallization processes [1-3]. Growth of vertical ribbon was commercialized but was not able to stay competitive due to very low throughput, non-standard geometry, and high dislocation density resulting in lower cell efficiencies and increased brittleness.

The last few years, at least three companies are investing heavily in potentially "kerffree" technologies. 1366 Technologies is planning to produce multi-Si substrates directly from silicon melt. In August 2017, Hanwha Q CELLS reached a 20.3% efficiency for PERC cells using 1366's wafers. NexWafe, a spin-off of ISE Freiburg, deposits a thick crystalline epi-Si layer (n-type, 50 – 180 μm thick) on a reusable template using trichlorosilane as a gaseous precursor (CVD method). The fully-grown wafer is then detached from the seed wafer aiming at replacement for traditional CZ wafers, and ~20% efficiency has been reached [4]. According to NexWafe, their technology could save up to 60% of Si-loss during sawing, reduce energy consumption during manufacturing by up to 80%, and require 70% less investment cost for its scrap-free wafer production. Crystal Solar in collaboration with Scatec and IMEC (Belgium) are working on a similar n-PERT cell with built-in junction demonstrating an efficiency of 22.5% [5] (also CVD method). Crystal Solar claims that their CVD-process uses 50 % less Si compared to the conventional process with less than 50 % of the capital cost. In addition, as will be shown, the epitaxial wafers have excellent material quality as evidenced by the lifetime measurements and cell efficiencies. Another benefit is lower O_2 content which leads to lower light induced degradation. These efficiencies clearly demonstrate that kerfless approaches have the potential to compete with conventional wafers.

Although being the most investigated method to grow Si films both for PV and electronics (CMOS, CCD), CVD methods have several drawbacks such as the use of toxic and explosive gases imposing the need for complex and expensive infrastructure and HSE-related considerations. Recent technological advances have enabled the possibility of epitaxy through e-beam evaporation of Si which offers several advantages over CVD based methods including the elimination of toxic gases (see Figure 1), feedstock flexibility and possibility for direct growth of p/n junctions using p/n Si feedstock. While both epi-Si and poly-Si with high quality have been grown using CVD at high temperature, no epi-Si with solar cell or electronic grade quality has been reported using PVD so far.

The results have been generated in the Research Centre for Sustainable Solar Cell Technology, FME SUSOLTECH, financed by the Norwegian Research Council and industry partners in Norway

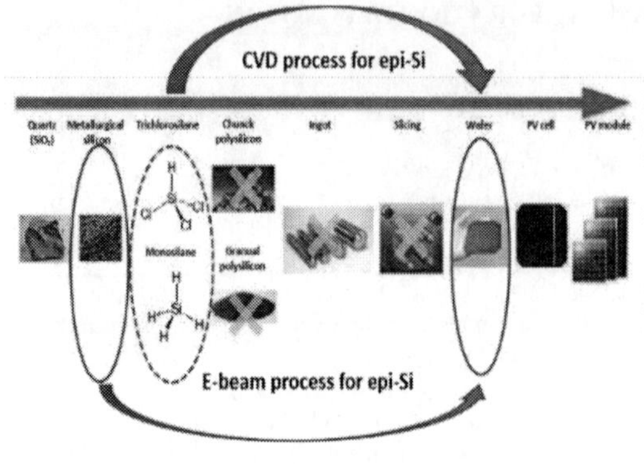

Figure 1: Comparison of eliminated capex and opex steps for Epi-Si production by e-beam and CVD. (Adapted from https://www.slideshare.net/MicheleBina/a-solution-to-the-crisis-in-photovoltaics-manufacturing-profitability)

2 ELECTRON BEAM DEPOSITION OF SILICON

2.1 Description of the high temperature e-beam

The high-temperature e-beam is a part of the new Flextura 200 cluster deposition tool series developed by Polyteknik, Denmark. The thin film deposition system is equipped with three independent process vacuum-chambers; two magnetron sputtering chambers and one dedicated e-beam chamber for high-rate fabrication of Si for solar cells from Si-feedstock on up to 6" substrates heated to 1000°C. The substrates are on ground potential. The e-beam power-supply is 10 kW, and the e-beam crucible size is 500 cm^3, with a scanning area of 5 x 5 cm. A RHEED system is installed in order to be able to monitor epitaxial growth. Plasma treatment of the substrates can be done in the magnetron sputtering chamber to remove substrate oxide layers, while introduction of hydrogen in the e-beam chamber can be done to reduce O_2 and H_2O partial pressures during deposition. The system is shown in Figure 2, while the e-beam chamber in operation is shown in Figure 3.

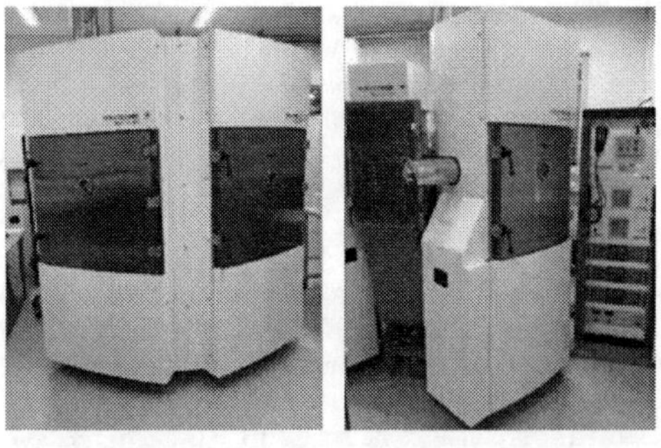

Figure 2: The Polyteknik Flextura 200 system. (a) two sputtering chambers, (b) e-beam chamber, the loadlock chamber and one of the control racks.

Figure 3: E-beam chamber in operation at 1000°C.

2.2 deposition of p-type Si on monocrystalline Si wafers

Thin p-Si layers have been deposited using e-beam deposition on n-doped monocrystalline 6" Si wafers with (100) orientation (Siegert wafer, resistivity >3500 Ωcm). The substrate temperature was varied between 600 °C and 1000 °C. The feedstock used in the depositions was p-doped Si from an ingot made by Elkem in the frame of the national project SolarUnited (resistivity of 5.4–5.7 Ω cm (top-cut)). The substrates were treated in 2% HF-solution for 30 s immediately prior to transferring into the load-lock of the deposition system. The depositions are done with 1 sccm of H_2 flow. The presence of H_2 has a positive impact on the crystallinity of the deposited layers. The layers deposited at 600 and 800 °C appear only partly crystalline while the layer deposited at 1000 °C appears fully crystalline. The deposition rate is ~1.5 µm/min, and the thickness of the deposited epi-Si layers is ~50 µm. The base pressure of the chamber was in the range of 1×10^{-7} mbar and during the depositions in the 10^{-5} mbar range.

3 CHARACTERISATION OF DEPOSITED FILMS

The properties of the deposited films were characterized by Raman spectroscopy to determine crystallinity, Scanning Electron Microscopy (SEM) for micro structural studies and Electron Backscatter Diffraction (EBSD) for grain orientation studies. Secco etch was used to determine defect density, and Hall measurements was used for resistivity measurements.

3.2 RAMAN

The deposited layers at different temperatures were investigated with Raman spectroscopy for their state of crystallinity. The laser light with wavelength of 532 nm was used at 0% attenuation, corresponding to 120 mW. The acquisition time for each spectrum was set to 30 seconds to maximize the peak to background noise ratio. A Raman detector at 2400 grating resolution was selected for acquisition to increase the resolution of the spectra. Figure 4 show that in all three selected deposition temperatures, the peak is located at 520 cm^{-1}, which is the representative wave number for crystalline and stress-free silicon material.

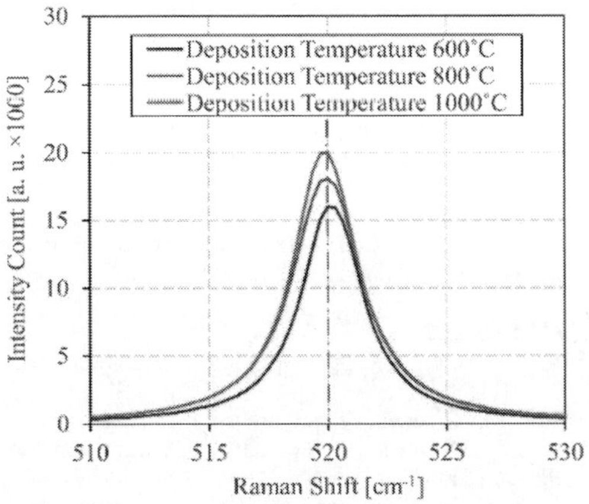

Figure 4: Raman shift at 520 cm^{-1} showing a crystalline and stress-free silicon.

3.1 SEM and EBSD

SEM images of the surface of the crystalline areas of the deposited layers are shown in Figure 5. Some defects are observed, many of which appear to originate around silicon droplets/particles in the deposited layer, and the number of these defects appear to increase with increasing deposition temperature. The defects in the 1000°C film is accompanied by formation of twin structures. Despite these defects, most of the surface area is flat, homogeneous and smooth. On-going work investigates measures to reduce the number of defects by a more careful control of the process parameters.

(a) (b) (c)

Figure 5: SEM images of the surface of the deposited films (a) 600°C, (b) 800°C, and (c) 1000°C.

EBSD images taken from the surface and cross-section of the films, shown in Figure 6 and 7, clearly demonstrate the epitaxial nature of the films. Most of the surface area clearly has a single (100) orientation, which is the same as the substrate it was deposited on. Many of the defects on the surface did not influence on the surface crystal orientation, but it is obvious that the crystal orientation is disturbed around some of the larger defects. The number and size of such defects increase with increasing temperature. It is not clear why these defects seem to depend more upon substrate temperature than electron beam energy, which is identical for all depositions, and the origin of the defects must be examined in more detail.

Investigations of the cross section shows that the deposited layer is ~50 µm thick, see Figure 7. EBSD show that there is no change in orientation from the substrate to the deposited layer, verifying that the layer is epitaxial. The orientation of the layer is (100) normal to the surface.

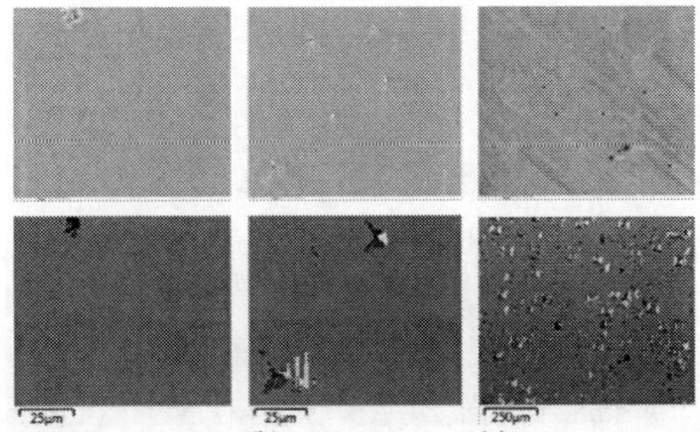

(a) (b) (c)

Figure 6: Top images: Electron image, Bottom: Inverse pole figure of same area. (a) 600 C, (b) 800 C, (c) 1000 C. Images for 600 and 800 °C depositions taken for 100µm area, while image for 1000°C taken for 1000 µm wide area.

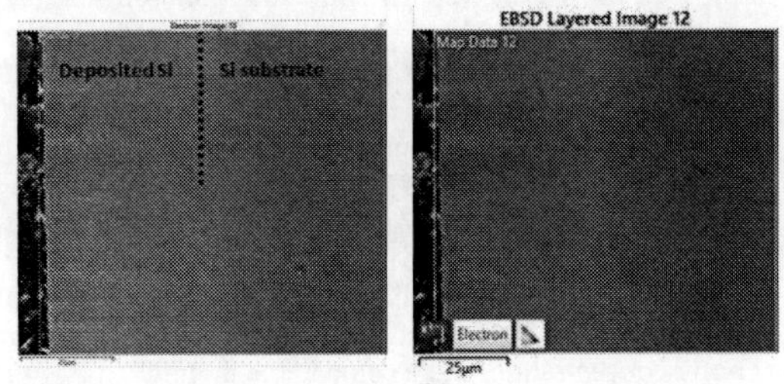

(a) (b)

Figure 7: Cross section of layer deposited at 1000°C in the left part of the image on top of the substrate to the left. (a) SEM and (b) EBSD. The interphase between deposited film and substrate is marked with a dotted line.

3.2 Secco etch

Secco etching was performed for 5 min to reveal the extended defects of the films. Figure 8 shows etch pits on films deposited at all temperatures. Only the defects which affected the crystal orientation can be observed in EBSD (Figure 6), and it appears as if the film deposited at 1000 °C had the highest defect density. The Secco etching, however, also reveals other defects not visible in the EBSD. It is obvious that the structural quality of the films depends significantly on T_s, and that the film deposited at the highest deposition temperature (1000 °C) actually has the lowest density of defects. The dislocation densities were determined to be ~2×10^7 cm^{-2} and ~5×10^6 cm^{-2} respectively at 800 and 1000°C. The film deposited at 600 °C had a higher number of dislocations, as seen from the large number of small pits in Figure 8 (d).

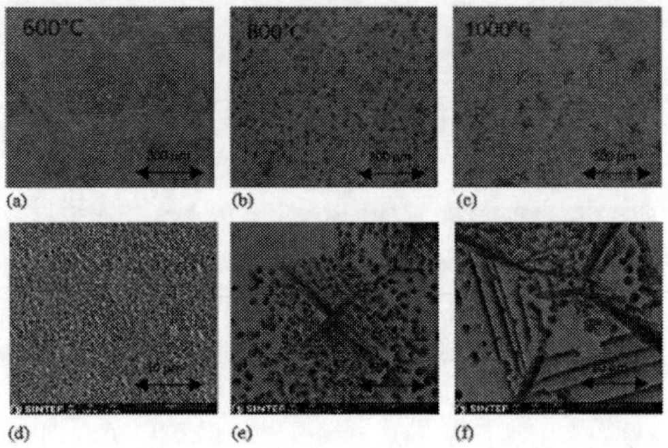

Figure 8: SEM images of Secco-etched surfaces of the Si films at $T_s = 600 - 1000$ °C.

3.3 HALL EFFECT

Hall effect measurements were performed on the deposited films in the van der Pauw configuration, and the results are given in Table 1. The feedstock Si used for the deposition has a resistance of 5.5 – 5.7 Ω*cm, and a boron concentration of $7{,}02^{15} \pm 0{,}07$ B-atoms/cm³. The specific resistance and carrier concentration of the film deposited at 1000 °C, 5.01 Ω*cm and 4.93^{15} 1/cm³, respectively, are close to the values of the feedstock, indicating an excellent transfer of the dopant to the film. The p-type hole mobility was 253 cm²/(V*s). The film deposited at 800 °C had about five times higher resistance, mainly due to a lower carrier concentration, and a comparable mobility.

Table I: Resistivity, carrier density and mobility from Hall effect measurements.

Sample	Resistivity [Ω*cm]	Carrier density [1/cm³]	Mobility [cm²/(V*s)]
800 °C	28.4	9.17^{14}	240
1000 °C	5.01	4.93^{15}	253
Reference values for feedstock	5.5-5.7	7.02^{15*}	-

*For the feedstock the value is the measure boron concentration, not the charge carrier concentration

4 CONCLUSIONS

Epitaxial growth of p-doped Si films on Si (100) substrates at T_s = 600–1000 °C under non-ultra-high vacuum conditions using high rate e-beam evaporation is demonstrated. Fully crystalline Si based layers (~ 50 μm thick) can be fabricated at 1000 °C at a deposition rate of ~1.5 μm/min. It should be mentioned that although epitaxial growth of Si upon e-beam deposition has been observed at 650 °C earlier [6], to our best knowledge, Si deposition at 1000°C with the deposition rate above 1 μm/min has not been reported so far. It is important to emphasize that the properties of the e-beam deposited layers at 1000°C are reasonably good at this initial stage of the developments, and the technology can be considered as a promising and cost-effective alternative to conventional CVD based growth of epi-Si layers and potentially - epi-Si wafers. It should be noted that deposition rate for e-beam deposited layers is comparable to that for the atmospheric pressure CVD based processes and can be as high as above 30 μm/min [7-9]. However, high density of extended defects was revealed by Secco etch experiments at T_s of 800 and 1000 °C, which indicates that further optimizations of the e-beam based growth of Si layers and epi-wafers and a comparative analysis with CVD based processes are required.

ACKNOWLEDGEMENTS

The authors acknowledge B. Karlsen for performing the Secco etch and H.N. Riise for Hall measurements. The Norwegian Research Council is greatly acknowleded for financial support of the Norwegian Research Centre for Sustainable Solar Cell Technology and for providing fundings for Polyteknik Flextura 200 system.

REFERENCES

[1] Rosenits, P., F. Kopp, and S. Reber, Thin Solid Films, 519 (2011).
[2] Becker, C., et al., Energy Procedia, 10 (2011).
[3] Kim, D.S., et al., Solar Energy Materials and Solar Cells, 90 (2006).
[4] https://www.ise.fraunhofer.de/en/press-media/press-releases/2015/nexwafe.html.
[5] http://www2.imec.be/be_en/press/imec-news/crystal-solar-si-pv.html.
[6] Gorka, B., et al., Thin Solid Films, 515 (2007).
[7] Heinß, J.-P., et al., in Proceedings of 31st European Photovoltaic Solar Energy Conference, (2015).
[8] Saager, S., et al., in Proceedings of 31st European Photovoltaic Solar Energy Conference, (2015).
[9] Saager, S. and B. Scheffel, to be published in Proceedings of 35th European Photovoltaic Solar Energy Conference, (2018).

SLURRY SAWING OF MULTICRYSTALLINE SILICON WITH LOW-VISCOSITY CARRIER LIQUID

Thomas Kaden[a], Christian Look[a], Vladislav Ischenko[b], Matthias Gröschel[b], Oliver Anspach[c]
[a]Fraunhofer Technology Center for Semiconductor Materials THM, Am St.-Niclas-Schacht 13, 09599 Freiberg
[b]SiC Processing (Deutschland) GmbH, Neuteichnitzer Str. 46, 02625 Bautzen
[c]PV Crystalox Solar Silicon GmbH, Gustav-Tauschek-Straße 2, 99099 Erfurt, Germany
Corresponding author: Thomas Kaden, e-mail: thomas.kaden@ise.fraunhofer.de

ABSTRACT: The wafering of crystalline silicon is industrially realized by wire sawing with loose abrasive particles dispersed in a carrier medium (slurry process) or by fixed diamond particles embedded on the wire (diamond wire sawing - DWS). DWS demonstrates cost advantages by the usage of a cheap cooling fluid, which consists of water with about 2% of additives. In the slurry process, SiC particles are used as abrasive typically dispersed in polyethylene glycol (PEG200) as carrier liquid. A cost reduction of this process by saving electrical energy can be achieved by the usage of a lower-viscosity carrier liquid. In this contribution, PEG200 is substituted by the lower-viscosity organic substance propylene carbonate. In four consecutive sawing processes on multicrystalline silicon bricks, the cutting performance of this liquid was tested. The results are compared with the standard sawing process using PEG200. It was found that the average power of the wire saw main drive can be reduced by about 60%.
Keywords: Silicon, Wafering, Mechanical properties, Recycling

1 INTRODUCTION

Silicon wafers for photovoltaic applications are nowadays mainly produced by the so-called diamond wire sawing (DWS) process. Compared to the slurry process, which uses a free SiC particles dispersed in an organic carrier liquid, typically polyethylene glycol (PEG200), DWS offers cost advantages because water can be used as a cheap cooling liquid. Further advantages are the usage of diamond particles fixed on the wire, allowing for shorter process times and the availability of very thin wires (60-70 µm core diameter), which increases the productivity significantly. [1]

In this contribution, a measure to reduce costs in the slurry sawing process was tested. The standard carrier liquid PEG200 was substituted by a low-viscosity organic liquid. The results are relevant for remaining capacities in the photovoltaic water production that are based on slurry sawing as well as other industries like the microelectronic wafer production with a high share of slurry-based processes.

2 EXPERIMENTAL APPROACH

PEG200 has a dynamic viscosity of about 60 mPa·s at room temperature. On the one hand, this could be beneficial to produce wafers with a high quality, for example in terms of low thickness variations or a low number of saw marks on the wafer. On the other hand, with such high viscosity the shear stress in the sawing channel is increased leading to significant higher machine power consumption [2] and furthermore the cyclic pumping of PEG200 in the wire saw requires a significant amount of energy.

For the substitution of PEG200 with a low-viscosity cooling liquid, possible organic liquids were pre-evaluated in terms of chemical and physical properties, non-toxicity and reactivity with silicon.

Propylene carbonate (PC) was identified as an appropriate substance. PC has the sum formula $C_4H_6O_3$ and a molecular weight of 102 g/mol. Relevant properties of PEG200 and PC are compared in Table I. While the density of PC is slightly higher, the dynamic viscosity of 2.8 mPa·s is significantly lower compared to PEG200. Also the specific heat capacity of PC is lower, possibly affecting the temperature regulation of the slurry in the process.

Table I: Properties of polyethylene glycol (PEG200) and propylene carbonate

Liquid	Density [$kg \cdot m^{-3}$]	Dynamic Viscosity [mPa·s]	Specific heat capacity [$kJ \cdot kg^{-1} \cdot K^{-1}$]
Polyethylene glycol (PEG200)	1.124	60-67	2.43
Propylene carbonate (PC)	1.205	2.8	1.71

Four consecutive sawing processes on multicrystalline silicon (mc-Si) were done using the same propylene carbonate based slurry. Bricks from the edge and center ingot positions were used. Selected process parameters are shown in Table II. A sawing process with the same parameters using PEG200 was used as the reference process.

The forces on the mc-Si ingot, the average power of the slurry saw main drive and the slurry temperature were measured during the processes. The resulting wafer quality was assessed in terms of total thickness variation (TTV) and roughness of the wafer surface.

Table II: Process parameters for the evaluation of propylene carbonate (PC) slurry

Parameter	Value
Wire	115 µm structured wire
SiC	F800
Slurry density at start	1.57 $g \cdot cm^{-3}$
Wire speed	15 $m \cdot s^{-1}$
Feed rate	0.6 $mm \cdot min^{-1}$

3 RESULTS

3.1 Process results

For the evaluation of the sawing processes, the forces on the wire web have been measured in three dimensions throughout the whole process time. The force in the crystal feed direction (F_z) is a measure of the cutting efficiency; the force in wire direction (F_x) is proportional to the friction of the wire in the developing sawing channel. With a low-viscosity cooling liquid, a lower F_x is expected. As shown in Fig. 1, the force in wire direction is significantly lower when the PC slurry is used. Compared to PEG200 slurry, F_x is reduced by about 60%. Furthermore it can be seen that F_x increases in each subsequent process, which is due to the loading of the slurry with silicon kerf. In the fourth process, the maximum silicon load capacity is nearly reached, corresponding to 14 wt% silicon in the slurry at a density of 1.7 g/cm³.

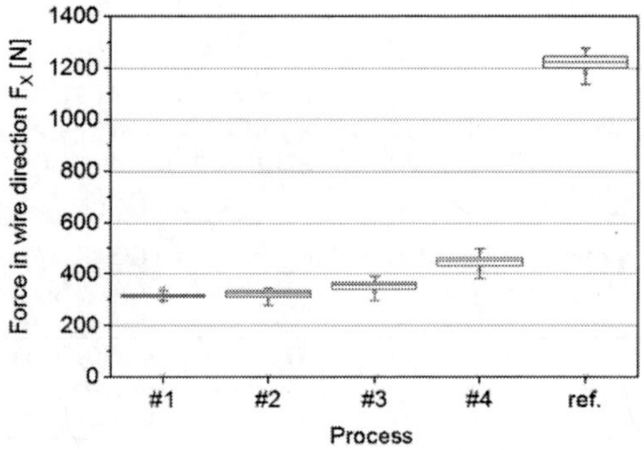

Figure 1: Force in wire direction (F_x) is significantly reduced with the PC slurry (processes #1 to #4) compared to the PEG200 slurry (reference)

A minor effect on F_z was observed, the force in feed direction with the PC slurry is about 10% higher than for PEG200 slurry, see Fig. 2. The higher F_z indicates a reduction of the cutting ability when PC slurry is used. An explanation can be found regarding the amount of free abrasive particles (SiC) that are transported into the sawing channel. With the lower viscosity it seems reasonable that a higher fraction of small SiC particles is transported by the wire into the sawing channel, excluding larger abrasive particles. Thus, the cutting ability is lowered as recognized by the increase of F_Z. [3]

Within the four consecutive processes with PC, F_z does not clearly increase, which would be an indication for a reduction of the cutting ability. The observed variation in F_z is mainly due to the usage of bricks from different ingot positions that contain inclusions of silicon carbide and silicon nitride in different sizes and concentrations [4].

The lower friction force directly results in a significantly reduced average power of the wire saw main drive, see Fig. 3. In the first process with PC slurry, a reduction of 60% compared to the PEG200 slurry was found. The energy consumption increases with increasing silicon load of the slurry in the consecutive processes. The slurry density values measured at the end of each process support this result. While the density was 1.61 g/cm³ after process #1, it was increasing to 1.71 g/cm³ after process #4, which requires a higher pump performance.

It should also be mentioned that the slurry temperature was kept constant at 23°C in the PC processes; the lower specific heat capacity did not affect the process stability.

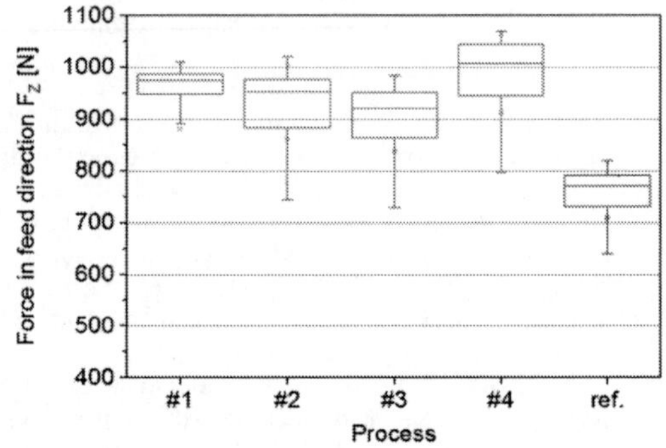

Figure 2: Force in feed direction (F_z) with PC slurry (processes #1 to #4) is slightly higher compared to the PEG200 slurry (reference), showing a minor reduction of the cutting ability

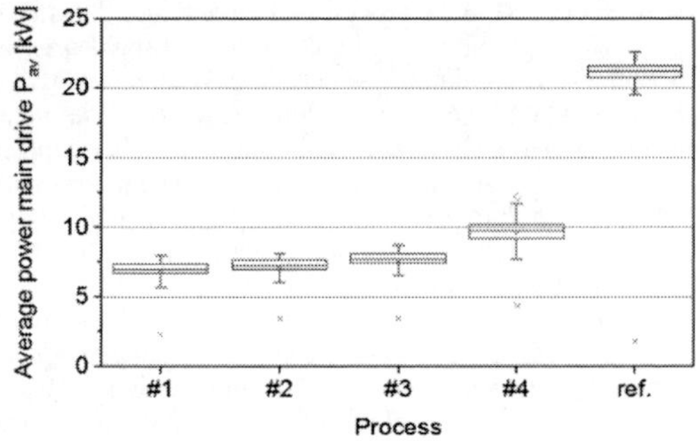

Figure 3: Average energy consumption of the wire saw main drive; a drastic reduction with the PC slurry (processes #1 to #4) is obtained in comparison to the PEG200 slurry (reference)

3.2 Wafer results

The wafers produced in the PC slurry processes were characterized in terms of geometry and roughness. The thickness of the PC slurry wafers exhibits median values of about 205 µm, which is equal to the wafers obtained from the PEG200 process, see Fig. 4.

However, the thickness of the PC slurry wafers shows a broader distribution, which is also reflected in the total thickness variation (TTV), see Fig. 5. Parameter optimizations were done within the four processes; with the parameter set used in process #3, the best wafer quality was obtained so far.

Further process optimization is necessary in order to produce wafers with a TTV of about 10 µm as demonstrated in the reference process with PEG200. Especially the slurry nozzle system needs to be adapted for the low-viscosity coolant.

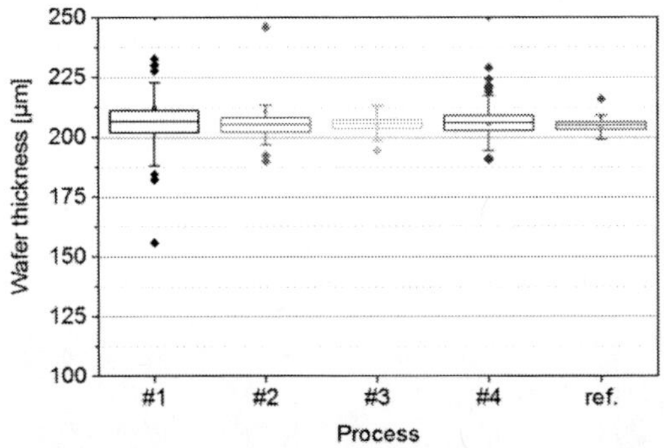

Figure 4: Wafer thickness obtained from four PC slurry processes and the PEG200 process as reference

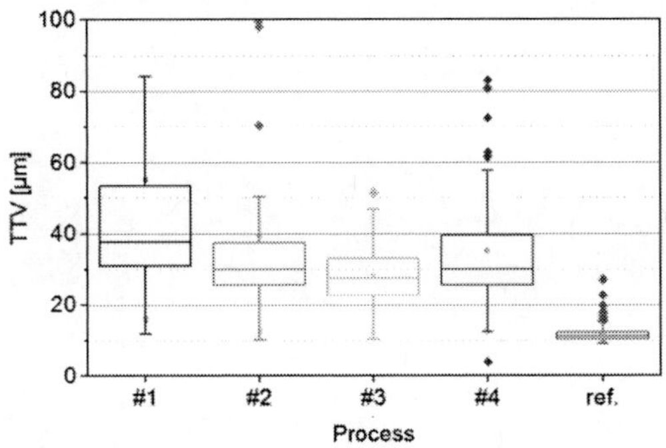

Figure 5: Total thickness variation for wafers obtained from four PC slurry processes and the PEG200 process as reference

As shown in Fig. 6, the roughness R_a of PC slurry-sawn wafers is about 15 % lower compared to wafers sawn with PEG200-based slurry, which could turn out to be beneficial for the mechanical strength of the wafers [5]

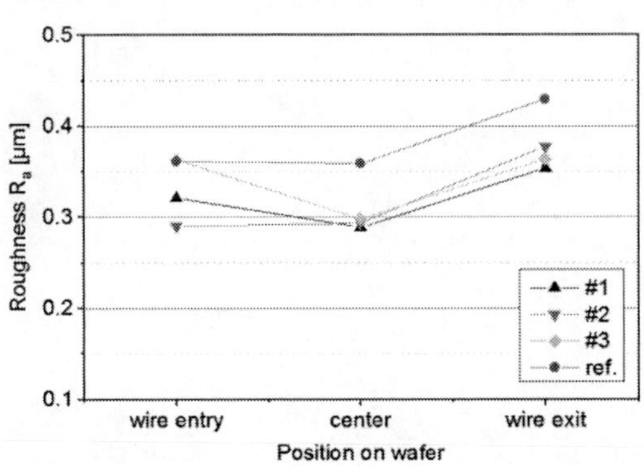

Figure 6: Roughness R_a at wire entry side, center and wire exit side of the wafer

The mechanical strength was determined by a ball on three ball (B3B) test on wafers of the same average thickness. The B3B test concentrates the load on the center of the wafer and excludes fracture effects on the wafer edges. Assuming a Weibull distribution of the fracture forces, the PC slurry sawn wafers exhibit a slightly higher critical fracture force of 3.68 N compared to 3.53 N for the PEG200 wafers, see Fig. 7. This result is in good agreement with the lower roughness of the PC-slurry sawn wafers.

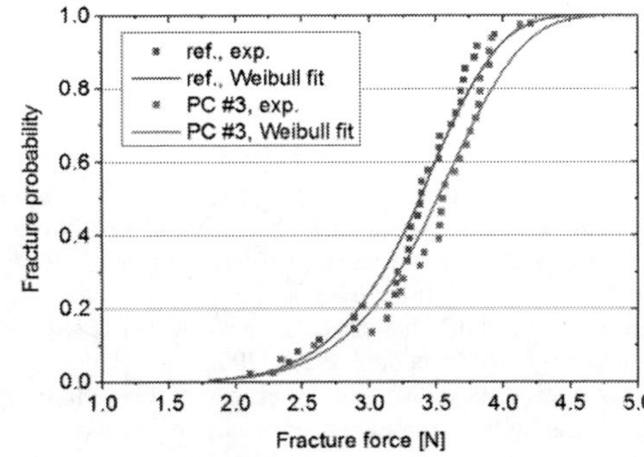

Figure 7: Weibull plots of the fracture forces in B3B test

4 RECYCLING OF USED SiC/PROPYLENE CARBONATE SLURRY

Propylene carbonate (PC) has been separated from solids by means of filtration. The moisture content of the used PC amounts to 1.8 wt%. While other carrying fluids, as for example PEG200, are prone to partial oxidation and formation of small amounts of carbon acids under conditions of the wafering process and, consequently, to higher corrosion rate of wire metals, PC remains completely stable.

This stability is reflected in the relatively high pH value (8.13) and low conductivity (13 µS/cm) of the used PC, as well as low amount of solved wire metals (Fe < 20 ppm, Cu < 5 ppm; Zn < 20 ppm). As the result, the recycling of PC is essentially simplified, limited to the filtration and drying to the required moisture content.

The recycling process of the SiC from the used propylene carbonate slurry is quite similar to the recycling of SiC from the conventional SiC/PEG200 slurry and comprises classification of particulate solids by means of hydro cyclones, as well as chemical removal of remained metals.

The yield of recycled SiC corresponded to 53.1%. There are indications of higher mechanical degradation of Si and SiC grains under conditions of wafering process in propylene carbonate, compared to PEG200, as shown in Fig. 8 by the shift of the particle size distribution after wafering to smaller particle sizes.

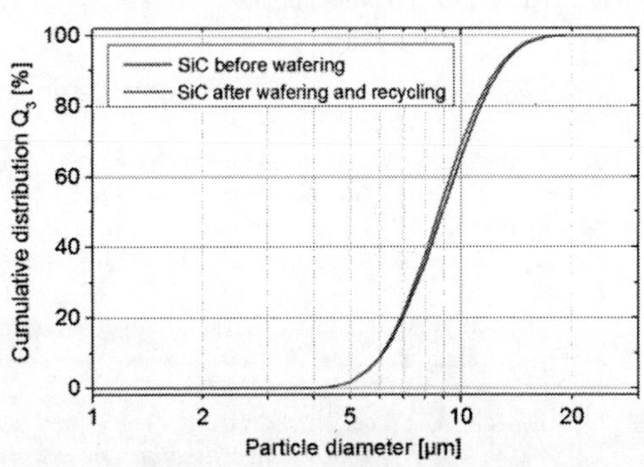

Figure 8: Particle size distribution of silicon carbide abrasives before wafering and after wafering/recycling

Beside the lower yield of the recycled SiC, this results in much lower retention rates of solids in the standard filtration step and in the higher content of fine Si particles in the underflow of the hydro cyclones, resulting in 1.7 wt% elemental silicon in the recycled SiC.

5 CONCLUSION

The substitution of the standard carrier liquid PEG200 in slurry sawing of crystalline silicon with a low-viscosity organic substance was investigated in this work. We found propylene carbonate (PC) to be a promising substance having a viscosity, which is reduced by a factor of 20 compared to PEG200.

In four consecutive sawing processes the suitability of propylene carbonate as carrier liquid was demonstrated. The lower viscosity resulted in a reduced force developing in wire direction and, consequently, in a reduction of energy necessary for the wire sawing process. The average power of the wire saw main drive could be reduced by about 60% compared to PEG200, which is a measure to reduce electrical energy costs.

Compared to wafers produced in slurry sawing processes with PEG200, wafers propylene carbonate based processes show the same thickness with a higher variation and thus with higher values of the total thickness variation. This may be improved by adapting process parameters as for example the slurry volume flow or the geometry of the slurry nozzles. Furthermore, a lower roughness and higher mechanical strength of the wafers produced with PC slurry were obtained. Although this is generally advantageous it is also a hint that the larger abrasive SiC particles are partially not transported into the sawing channel and thereby the cutting efficiency is slightly reduced. This was proved by an increase of the force measured in feed direction by about 10%.

Propylene carbonate remained absolutely stable during the wire sawing process which reduces the recycling procedure to filtration and drying steps. Advantageous is the content of dissolved metals in the used slurry. However, the recycling yield of SiC particles is lower compared to PEG200 which may be due to a higher mechanical degradation of SiC.

6 ACKNOWLEGMENTS

This work was funded by the German Federal Ministry for Economic Affairs and Energy within the project "REWASI" (contract number 0325938C).

7 REFERENCES

[1] A. Kumar and S. N. Melkote, *"Diamond Wire Sawing of Solar Silicon Wafers: A Sustainable Manufacturing Alternative to Loose Abrasive Slurry Sawing,"* Procedia Manufacturing 21, pp. 549 – 566 (2018).

[2] O. Anspach, F.W. Schulze, *Glycol to water-based slurry, investigation on the wire sawing process*, Proc. of the 22nd EU-PVSEC (2007).

[3] O. Anspach, A. Stabel, A. Lawerenz, S. Riesner, R. Porytskyy, F.-W. Schulze, *Investigations on single wire cuts in silicon*, Proc. of the 23rd EU-PVSEC, pp. 1098-1103 (2008).

[4] T. Kaden, E. Ershova, L. Lottspeich, M. Fuchs, *The Influence of Material Properties on the Wire Sawing Process of Multicrystalline Silicon*, Proc. of the 33rd EU-PVSEC, pp. 537-540 (2017).

[5] M. Oswald, T. Loewenstein, O. Anspach, J. Hirsch, D. Lausch, S. Schoenfelder, *On the Correlation of Surface Roughness to Mechanical Strength and Reflectivity of Silicon Wafers*, Proceedings of the 29th EU-PVSEC, pp. 764-768 (2014).

WAFERING PERFORMANCE OF STRUCTURED WIRE IN CORRELATION TO THE WIRE GEOMETRY

Ringo Koepge[1, a)], Kjell Buehler[1], Felix Kaule[1], Stephan Schoenfelder[1,3], Oliver Anspach[2]

[1]Fraunhofer Center for Silicon Photovoltaics CSP, Otto-Eissfeldt-Strasse 12, 06120 Halle (Saale), Germany
[2]PV Crystalox Solar Silicon GmbH, Gustav-Tauschek-Strasse 2, 99099 Erfurt, Germany
[3]Leipzig University of Applied Sciences, Karl-Liebknecht-Strasse 134, 04277 Leipzig, Germany

[a)] Corresponding author: ringo.koepge@csp.fraunhofer.de

ABSTRACT: Nowadays two main multi wire wafering technologies are established, the loose and the fixed abrasive process. The latter process, which is known as diamond wire technology, seems to dominate the market compared to the conventional slurry based wafering process over the next years. However, over the last years the usage of structured wire within the slurry based process was established and a significant decrease of process time and consumable's consumption was achieved. In this study an extensive DoE was performed for a closer investigation of these process parameter correlations. The structured wire diameter, the silicon carbide (SiC) particle size and the table speed were varied which results in a complex process behavior. Additionally, wire samples are analyzed by a high resolution topography measurement with a specific analysis software. The wire data are correlated to the wafering performance for each parameter setup. Twenty-seven process configurations are investigated. In conclusion, the best setup for an optimum of wafering performance is detected as well as a correlation to the wire structure.

1 INTRODUCTION

The dominating slurry based wafering technology for cutting multi crystalline silicon into photovoltaic (PV) wafers has started to be replaced by the diamond wire technology within the last years. In 2018 approximately 95 percent of mono crystalline wafers are manufactured by diamond wire technology - for multi crystalline wafers the share of diamond wire technology is about 50 percent or even higher. For PV applications the market share for slurry based processes is predicted to vanish over the next four years [1]. Though, the slurry technology will be still in use for manufacturing semiconductor and photovoltaic wafers, if wafer manufacturers are forced by different reasons to use their current slurry based wire saws by different reasons. Strong process improvements can be reached by changing the wire geometry from straight to structured wire. In comparison to straight wire the process efficiency for the slurry process has doubled over the last years [2] and there is still potential for improvements like higher throughput or less consumption of consumables.

A quantification of the process performance needs to consider the cutting performance, the process time, consumable consumption and the resulting wafer geometry. A decrease of power consumption and process time are the major parameters. Both improvements lead to a decrease of wire consumption [2, 3]. Furthermore, less usage of SiC slurry indicates a more efficient wafering process [2]. Unfortunately, the wafer geometry is not independent from these process modifications. Even stronger process improvements would be possible if the wafer geometry was negligible. The major wafer parameters are defined by the solar cell manufacturers. Typically, the total thickness variation, the mean thickness and the saw marks are focused for quality inspection. Mostly, one can find an analysis of the improvement of the wafer geometry [4, 5, 6] and the process performance [7, 8] independent from each other. But in some cases a correlation between both, wafer geometry and process performance, is desired. Thus, the influence of the wire tension to the total thickness variation was analyzed. A loss of wire tension leads to higher TTV values in a diamond wire process [9]. Moreover, the wire usage and the correlation to the wafer geometry was analyzed. Detailed investigations show a significant increase of the TTV by an increase of the wire usage [10]. Besides a comprehensive analysis method was presented for a quantification of the wire cutting performance by considering the wafer geometry [11]. In this study a performance limit of the three existing wafering technologies were determined and can be extended to each unique sawing wire.

The next step in process innovation for the structured wire wafering technology is to predict an ideal shape of structured wire, which leads to even higher process performance while keeping the wafer's quality constant. The target of this study is to analyze the impact of initial wire structure and the loss of wire structure during wafering in relation to the particle size distribution of the slurry to improve the current process performance on a level that is comparable with diamond wire processes

2 MATERIALS AND METHODS

2.1 Wafering and Process Data

An investigation of the influence of wafering consumables and process setting to the wafer quality was started by planning and performing a Design of Experiment (DoE). A silicon carbide slurry based process was chosen and nine wafering runs were performed at a Meyer Burger multi wire saw DS264. The wire speed was kept constant at 12 m/s and the wire tension was set to 1800 MPa for unwinding and 720 MPa for recoiling the wire spool. The wire pitch was set to 340 µm. The number of silicon bricks was also set constant for each wafering run (four bricks) at a total length of 760 mm.

Table 1: Overview of the performed wafering runs.

Class of Wire Diameter	Slurry Particle Size (FEPA)		
	F1000	F800	F600
101 µm	Run #1	Run #4	Run #7
105 µm	Run #2	Run #5	Run #8
115 µm	Run #3	Run #6	Run #9

For all runs multi crystalline silicon bricks were used with a 156 mm by 156 mm dimension. The different structured wire (eq. crimped wire) core diameters and the particle sizes of silicon carbide can be seen in Table 1. The silicon carbide slurries were processed with an equal viscosity the ensure a comparable particle transport by the polyethylene glycol. For each wafering run a virgin slurry was mixed. In addition, the table speed (eq. feed speed) was changed during the sawing run. It started at 440 µm/min and changed to 550 µm/min, then to 640 µm/min and back to 440 µm/min. These four process sequences were set to see any influence of the table speed. Figure 1 shows the schematic view of one of two glass beams that are inside the multi wire saw and the two silicon bricks that were mounted at each beam. The figure illustrates the four process sequences and the corresponding block position that changes from zero to 156 mm. A force sensor was mounted to measure the force in table movement direction during the entire wafering process.

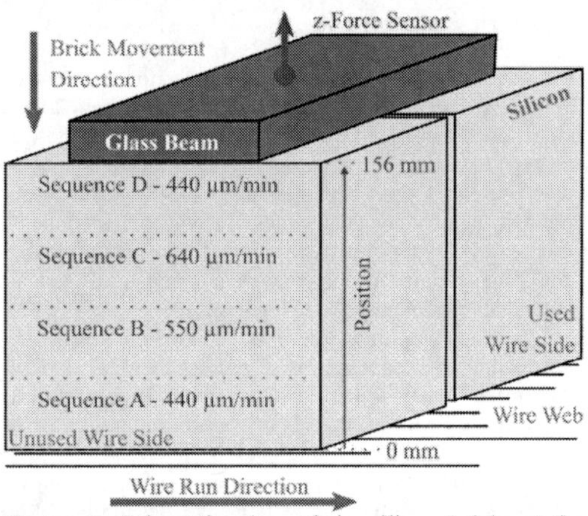

Figure 1: Schematic view of the silicon bricks at the beginning of the wafering process. Each wafering run was subdivided into four table speed sequences that were equal for all nine wafering runs.

For a quantification of the wafering performance two major parameters were defined. On the one hand a parameter was needed that describes the cutting capability of the wire in combination with the SiC particle size and the wafering process at all. Thus, the Preston coefficient [12, 13] was used to quantify the cutting efficiency. The Preston coefficient c_p [m²/N] is a proportionality factor between the material removal rate $v_{removal}$ [m/s], the contact pressure p [N/m²] and the relative slicing velocity v_{rel} [m/s] between the two interacting materials, in this case silicon and the steel wire or the abrasive. Then the removal rate can be written as

$$v_{\text{removal}} = c_p \, p \, v_{rel} \quad (1)$$

On the other hand one was looking for a parameter that will correlate to the quality of the wafer. The wafer geometry is the most important parameter, especially the wafer thickness and the thickness variation. For these investigations the wafer thickness deviation was selected. In Table 2 the brick positions are quantified and named Region of Interest (RoI).

Table 2: Process parameters that are collected during and after each wafering run for calculation of the major process parameters, the Preston coefficient and wafer thickness deviation.

Position [mm]	Calculated Parameters in the Region of Interest (RoI)			
	14 to 34	50 to 70	91 to 111	127 to 147
Median Force per Wire Median Kerf Width Wire Speed Table Speed	Avg. Preston Coefficient			
	Median Preston Coefficient	Median Preston Coefficient	Median Preston Coefficient	Median Preston Coefficient
Median Wafer Thickness	Avg. Thickness Deviation			
	Median Thickness Deviation	Median Thickness Deviation	Median Thickness Deviation	Median Thickness Deviation

Each RoI is located in only one process sequence. The input parameters that are necessary for calculating the process parameter can be seen at the left column of the table. For each RoI the median values were calculated. Further analysis was focused on the data of the process sequence A to C. Thus, an average of value the first three sequence parameters was calculated. Then, the influence of the changing table speed will be seen by the scattering of the data points.

2.2 Wire Analysis

All wafering runs were performed by using a structure steel wire. A structured wire is usually formed by two periodical bending shapes, which have to be characterized separately. Each shape is described by a period T_i and an amplitude a_i (see Figure 2).

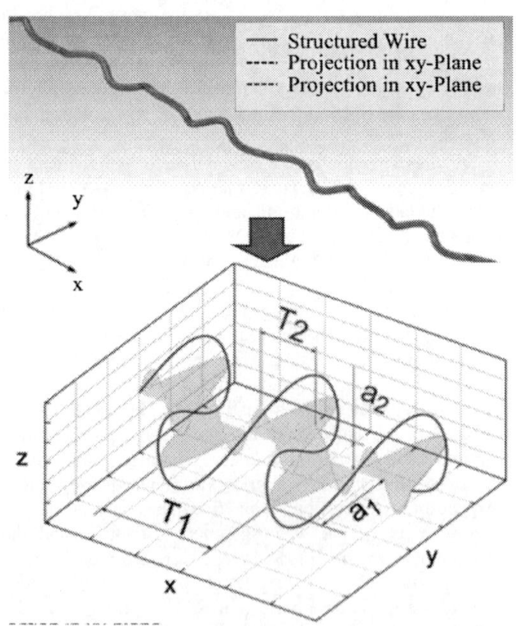

Figure 2: Schematic view of a structured wire and the four major wire parameters.

The wire surface topography was detected optically with the sentronics SemDex 301 system in a length of 100 mm and a measuring grid of 200 µm along the wire direction and in 1 µm perpendicular to the wire direction. During the optical measurements the wire was loaded with a tensile stress of 900 MPa to assure proper geometry data during the measurement. By use of a specifically developed evaluation software [14] the two bending periods (T_1 and T_2), the two amplitudes (a_1 and a_2) and the wire core diameter were determined from the 3D geometry.

The wire samples were taken at the beginning of the wire spool, after each wafering sequence for each wafering run and at the end of the wire spool. For Run #1 to #3 three measurements were performed for each sequence and the virgin wire. For Run #4 to #9 one wire was measured for each sequence and the virgin wire.

Because the wire parameters T_i and a_i are dependent on the tensile stress, a simulation model [15] was used to transfer the measured wire parameters into an unloaded stress state. Afterwards the wire parameters are correlated to the process settings and the DoE parameters.

2.3 Loss in Wire Tension Force

One part of the process analysis is the investigation of the wire's geometry. In this section, its significance for the wafering process is explained. Due to plastic elongation and abrasion of the sawing wire during the process its tension force reduces gradually [16]. This reduction leads to changing process conditions, which directly affect the quality of sawn wafers.

The difference between the period of the used wire and the period of the virgin wire represents the plastic elongation (eq. plastic strain) during the sawing process in this study. The plastic strain $\varepsilon_{pl,used}$ is calculated as

$$\varepsilon_{pl,used} = \frac{\sum_{i=1}^{2} \frac{T_{i,used} - T_{i,virgin}}{T_{i,virgin}}}{2} \qquad (2)$$

The axial stress σ of a wire can be described by

$$\sigma = E\,\varepsilon_{el} = \frac{F}{A} = \frac{4\,F}{\pi\,d^2} \qquad (3)$$

where E is the Youngs modulus, ε_{el} is the elastic strain, F is the force, A and d being the cross-sectional area and the diameter of the wire, respectively.

The total strain of a wire consists of an elastic and a plastic part

$$\varepsilon = \varepsilon_{el} + \varepsilon_{pl} \qquad (4)$$

It is assumed that the total wire strain (ε_{total}) (Equation (4)) remains constant inside the wire web regarding the unstressed virgin wire length and that the guiding rollers are guiding the wire perfectly, that means without any slip. Furthermore, the virgin wire is assumed to have no plastic elongation when it enters the wire web, so Equation (4) can be written as

$$\varepsilon = \varepsilon_{total} = \varepsilon_{el} + \varepsilon_{pl}$$

$$\varepsilon_{total} = \varepsilon_{el,virgin} = \varepsilon_{el,used} + \varepsilon_{pl,used} \qquad (5)$$

Considering Equations (3) and (5), the wire force $F_{used,pl}$ in consequence of a plastic wire elongation of the virgin wire (tension force F_{virgin}, cross-sectional area A_{virgin}) can be written as

$$F_{used,pl} = E\,A_{virgin}\left(\varepsilon_{el,virgin} - \varepsilon_{pl,used}\right) \qquad (6)$$

The wear of a wire changes its diameter but the elastic strain remains constant ($\varepsilon_{el,used} = \varepsilon_{el,virgin}$), plastic strain is neglected in this case ($\varepsilon_{pl,used} = 0$). As a result, the wire force $F_{used,wear}$ can be calculated with the cross-section area after abrasion A_{used} and the strain $\varepsilon_{el,virgin}$ and Equation (3)

$$F_{used,wear} = E\,A_{used}\,\varepsilon_{el,virgin} \qquad (7)$$

To consider both force reducing mechanisms, namely the plastic elongation of the wire and its wear, Equations (6) and (7) lead to the expression

$$F_{used,total} = E\,A_{used}\,\varepsilon_{el,used}$$

$$F_{used,total} = E\,A_{used}\left(\varepsilon_{el,virgin} - \varepsilon_{pl,used}\right)$$

$$F_{used,total} = E\,\frac{\pi\,d_{used}^{\;2}}{4}\left(\frac{4\,F_{virgin}}{E\,\pi\,d_{virgin}^{\;2}} - \varepsilon_{pl,used}\right) \qquad (8)$$

Where F_{used} and d_{used} are the force and the diameter of the used wire (plastically elongated and worn), respectively. The diameters of fresh and used wire are measured as described in section 2.1. The Youngs modulus of steel is E = 210 GPa which is used for all calculations.

3 RESULTS

3.1 Cutting Performance

The Preston coefficient was chosen for an independent quantification of the cutting performance. Process data of sequence A to C are taken for the calculation of an average value of the three median Preston coefficients. The influence of the changing table speed is represented by the scattering of the data points. Figure 3 shows the Preston coefficient versus the wire core diameter classes for different silicon carbide particle size distributions.

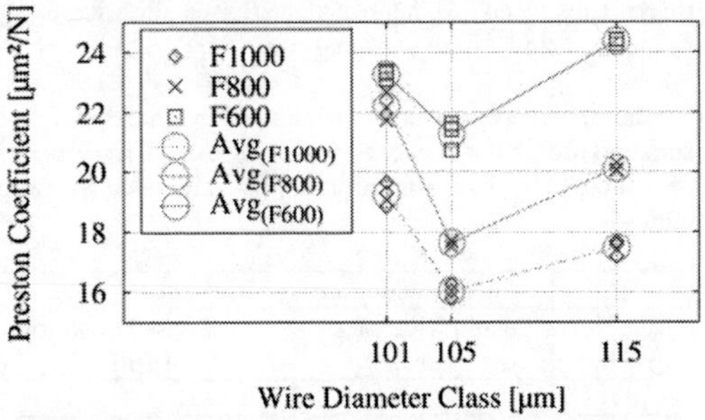

Figure 3: The diagram shows the correlation between the wire core diameter class and the Preston coefficient for different silicon carbide particle size distributions.

The influence of the silicon carbide particle size distribution can be clearly seen by an increase of the Preston coefficient when the particle size increases (from F1000 to F800 to F600). This effect is independent from the wire core diameter and takes place for all wire diameters. Looking at the wire diameter, a minimum of the wafering efficiency was detected at a diameter of 105 µm for all particle sizes. This special behavior triggers a closer look at the process data, especially at the applied wire tension. The wire tension was set to 1800 MPa to ensure comparable wafering efficiency. The wire tension was calculated for a wire diameter that was given by the wire manufacturer. Afterwards, the real wire diameter were determined by the geoemetry measurement system and the real applied wire tension was calculated (see Figure 4). The wire tension of the unwinding spool side is shown in correlation to the classified wire diameters.

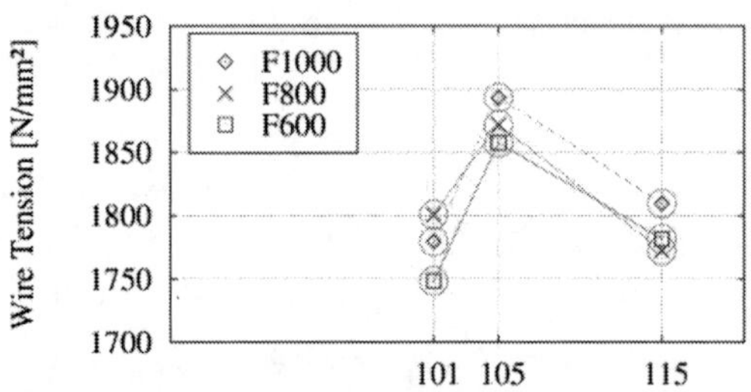

Figure 4: Wire tension for unwinding in correlation to the wire core diameter.

A maximum wire tension can be seen in all cases, using the 105 µm diameter of steel wire. The maximum correlates to the minimum of the Preston coefficient, so an influence of the wire tension can be assumed. An increasing wire tension seems to cause a decrease of sawing efficiency. This correlation has to be proved because in this study there was no intentional variation of the wire tension for the same wire diameter.

3.2 Wafer Quality

The second major process parameter is represented by the wafer geometry. Different wire diameters were used, so the wafer thickness differs becuase the wire pitch was set constant. Therefore, the wafer thickness is not useful for analysis, but the deviation of the wafer thickness can be taken for the quantification of the wafer quality. Table 3 summarizes the average values of wafer thicknesses and standard deviations for wafering sequence A to C of all wafering runs.

Table 3: Average values of the wafer thickness and standard deviation for each wafering run. The process sequences A to C are represented by calculated average values.

	Mean Parameters of Sequence A, B, C	
Run No.	Wafer Thickness	Thickness Deviation
[]	[µm]	[µm]
#1	212.0	8.9
#2	212.6	6.9
#3	199.1	8.1
#4	206.4	13.6
#5	200.1	5.7
#6	189.7	6.2
#7	196.0	11.1
#8	194.8	12.4
#9	181.8	11.2

The mean wafer thickness correlates to a changing wire thickness and silicon carbide particle size distribution. The wafer thickness deviation is related to process issues and can be seen versus the wire diameter and silicon carbide particle size distribution in Figure 5. The F1000 and F800 particle sizes cause a low thickness deviation apart from the F800 particle size in combination with a wire diameter of 101 µm. The F600 wafering runs show a high level of thickness deviation for all wire diameters. An optimized process should show a high Preston coefficient and a low thickness deviation, but both parameters are interacting with each other.

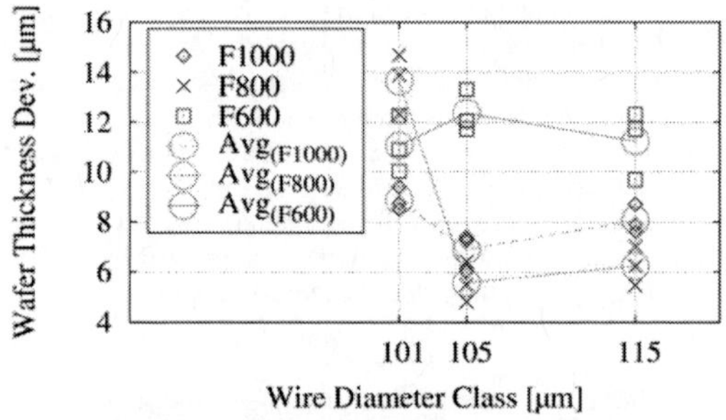

Figure 5: Wafer thickness deviation versus the wire core diameter and silicon carbide particle size distributions.

3.3 Process Optimum Quantification

Due to the fact that an optimum of two major process parameters needs to be found, a performance ratio r_p between the Preston coefficient c_p and the thickness deviation s_{thk} is defined by

$$r_p = \frac{c_p}{s_{thk}} \tag{9}$$

The process efficiency (quantified by the Preston coefficient) is converted to a thickness deviation independent parameter that also can be defined as a relative Preston coefficient. Figure 6 shows the ratio in correlation to the wire diameter and the silicon carbide particle size distribution. The advantage of a high Preston coefficient is degraded by a high thickness variation. The highest ratio will now show the best wafering efficiency for an equal thickness deviation or the best quality for an equal efficiency. Finally, the F800 run using a 115 µm or 105 µm wire diameter (oval marking) gives the best performance ratio. The performance ratio of the other wafering runs are up to 50 percent smaller.

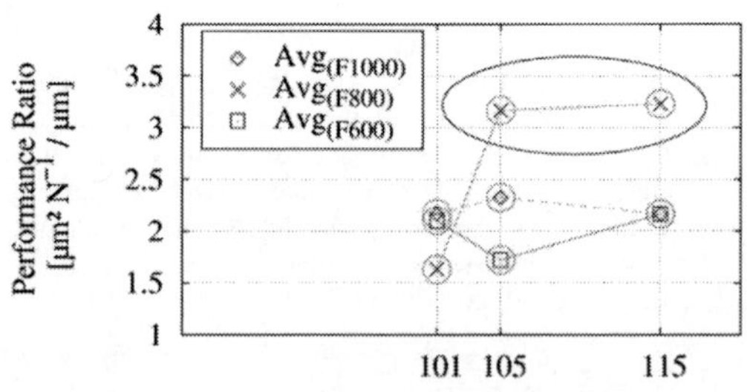

Figure 6: Relative performance ratio was defined as quotient of the Preston coefficient and the wafer thickness deviation. The quotient is shown in correlation to the wire thickness and the silicon carbide particle size distribution.

3.4 Change of Wire Structure

The wire parameters are determined to see any process issues that are not visible by analyzing the process data. The major wire parameters are measured before and after wafering and can be seen in Figure 7 and Figure 8. All measured wire data were transformed by calculation into a non-loaded wire parameter set for better comparison.

Having a look at the average values of the wire amplitudes (Figure 7) a huge difference between the virgin wire and used wire parameters was measured. A decrease up to 50 percent was observed for both amplitudes (related to large and short period). A unique influence of the silicon carbide particle size distribution in combination with a specific wire diameter was not observed. Generally, the wafering process decreases all amplitudes for each wafering run. The period, of the structured wire varies due to the change of the process parameters (see Figure 8). In general, the period increases by running a wafering process compared to the virgin wire. Having a closer look, some periods are almost near the initial virgin period length, i.e. 115 µm wire at F800. Thus, a minimum of wire extension occurs at this point. An average wire extension is calculated (see Equation (2) and used to subsequently calculate the wire force of the used wire leaving the wire web, following Equation (8). A percentage loss in wire tension force compared with the force of the unwound virgin wire was calculated and is shown in Table 4. The correlation to the DoE parameters is shown in Figure 9.

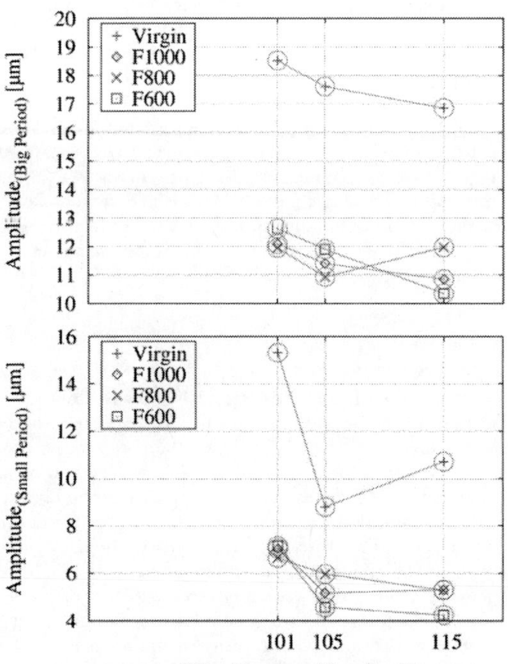

Figure 7: Wire amplitudes before and after the wafering process in correlation to the wire diameter class and silicon carbide particle size distributions. The amplitudes are calculated for an unloaded wire stress state.

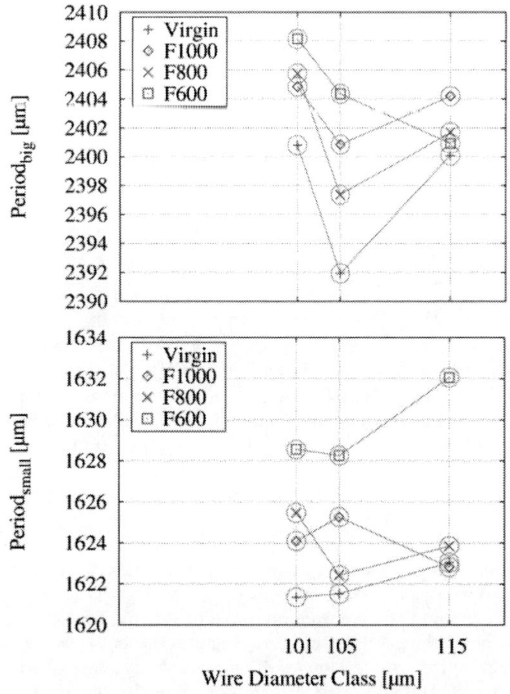

Figure 8: Wire wavelengths (eq. periods) before and after the wafering process in correlation to the wire diameter class and silicon carbide particle size distributions. The periods are calculated for an unloaded wire stress state.

The F800 at 115 µm and 105 µm wire diameter as well as the F1000 at 115 µm cause a wire force loss of less than 30 percent. All F600 runs and the F1000 at 105 µm show the highest force loss, from 40 to more than 50 percent. Correlating these results to the wafer geometry (see Figure 5) and the relative performance ratio (see Figure 6) an almost clear relationship was observed. Low wire force losses correlate to a low thickness deviation and a high cutting performance.

Table 4: Wire extension and the corresponding force loss caused by wire wear and plastic elongation.

Run No.	Diameter Class	Plastic Strain ε_{pl} [%]	Force Loss by Diameter [%]	Force Loss by Plastic Strain [%]	Total Force Loss [%]
1	101	0,17	-19,95	-16,21	-32,93
2	105	0,30	-33,92	-16,56	-44,86
3	115	0,09	-10,83	-16,10	-25,19
4	101	0,23	-27,21	-14,22	-37,56
5	105	0,14	-15,92	-15,85	-29,25
6	115	0,06	-6,81	-12,09	-18,08
7	101	0,38	-44,34	-13,30	-51,74
8	105	0,47	-52,45	-16,56	-60,32
9	115	0,29	-34,63	-12,25	-42,64

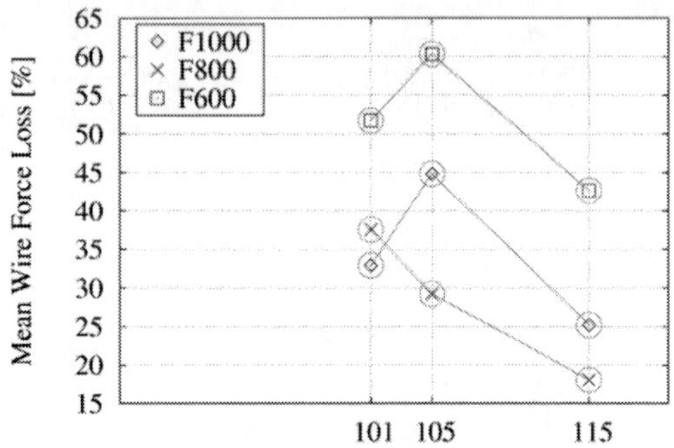

Figure 9: Wire force loss caused by the wire wear and plastic elongation in correlation to the silicon carbide particle size distributions and the wire diameter class.

4 DISCUSSION

This study presents correlations between the process settings and the wire structure. Looking especially at the wire tension that was applied at the virgin wire in the beginning of each wafering process, it could be seen that the cutting efficiency (see Figure 3) correlates to a changing wire tension (see Figure 4). An increase of wire tension by approx. 5 percent correlates to a decrease of the Preston coefficient of approx. 10 percent. The wire tension has an important influence to the wire structure, a detailed analysis was performed by A. Haendler [15]. The correlation to the Preston coefficient that was detected in this study was not independent from the wire structure. Too many parameters are affecting this correlation, the wire tension as well as the wire diameter and the size of silicon carbide particle. A verification of a direct correlation between the wire tension and the Preston coefficient needs further research.

The wire diameter was subdivided into classes due to the real wire diameter that was different to the manufacturer specification. Table 5 shows the diameter classes and corresponding diameters that were measured. All wires lose the initial diameter in case of usage during the wafering process. In addition, a decrease of diameter with an increase of the table speed (Sequence A to C) was measured. All parameter settings of sequence D are equal to sequence A. Thus, one can see the impact of a higher level of silicon kerf. Obviously, there is an effect (see Figure 10). The analysis of the process was focused at the changing cutting efficiency, changing wafer geometry, the wire amplitude and period.

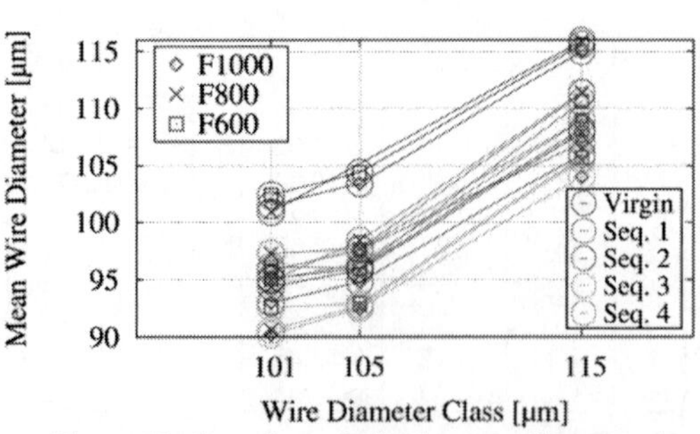

Figure 10: Overview of all wire core diameters that were determined before and after each wafering sequence.

The loss of wire diameter was considered for all calculation, but for the data plots, the diameter classes are used to see major impacts of the DoE parameters. In further investigations the decrease of wire diameter will be analyzed in detail.

Table 5: The table show the wire diameter classes and the corresponding wire diameters of each wafering run that were measured before and after each wafering sequence.

Wire Diameter Class	Run No.	Mean Wire Diameter [µm]				
		Virgin	Seq. A	Seq. B	Seq. C	Seq. D
101	1	101.5	95.8	93.0	90.2	95.0
105	2	103.4	97.6	94.8	92.5	96.4
115	3	115.0	107.0	106.0	104.0	109.0
101	4	100.9	97.3	94.4	90.6	94.9
105	5	N/A	97.7	95.8	92.7	98.4
115	6	115.9	111.3	107.9	105.5	111.4
101	7	102.4	96.1	95.1	92.6	95.8
105	8	104.4	96.0	96.0	92.9	97.6
115	9	115.6	110.6	108.2	105.7	109.0

The main driver for an improvement of the cutting performance and the wafer geometry is still defined by wafering costs. Actually, 18 percent of the entire PV module costs is the share of multi crystalline silicon wafers[1]. An increase of the wafer yield will be helpful for decreasing the wafering costs at all. Here, the most efficient wafering process was figured out for a F800 slurry and a 115 µm and 105 µm wire diameter (see Figure 6). The parameter wafer costs for each wafering run was not considered, but is not less important and may change the optimum process set. However, this correlation has to be done by the wafer manufacturer regarding their own cost structure.

5 SUMMARY AND CONCLUSION

The quantification of the performance for wafer manufacturing is still needed for an ongoing process improvement. This study included the performance of a DoE for a structured wire slurry process. The core wire diameter, the silicon carbide particle size distribution and the table speed were varied in order to investigate the impact to the wafering result. Two objects were investigated afterwards, the wafer geometry and the wire geometry. The wafer geometry represented by the wafer thickness deviation was correlated to the cutting efficiency (Preston coefficient). An optimum for wafer cutting was detected for a F800 slurry and a wire diameter of 105 µm and 115 µm. The analysis of the wire structure indicated an increase of wire length that results in a decrease of wire tension during the wafering process and an increasing wafer thickness deviation. Furthermore, the wire structure was analyzed regarding its resistance against structure loss and wire tension loss.

The wafer geometry and the cutting efficiency will remain in focus for all wafer manufacturers. The detailed wire structure measurement and a smart analysis software in combination with the FEM calculation model provides a high process understanding and enables the prediction of wire structure transformation. The introduced performance ratio parameter can be applied at any wafering technology process, e.g., diamond wire processes. Finally, the presented analysis shows a huge potential for the quantification of the process performance in multi wire sawing.

6 ACKNOWLEDGEMENT

This work was supported by the German Federal Ministry of Education and Research within the research project MechSi (contract no. 03IPT607X), which is gratefully acknowledged.

REFERENCES

[1] International Technology Roadmap for Photovoltaik (ITRPV), March 2018

[2] O. Anspach, B. Hurka, K. Sunder, "Structured wire: From single wire experiments to multi-crystalline silicon wafer mass production", Solar Energy Materials and Solar Cells, 2014

[3] D. Heppner, R. Sueldia, A. Daridon, P.M. Nasch, M. Yin, New Cost Saving c-Si PV Wafer Slicing Process by Advanced Wire Technology, 26th European Photovoltaic Solar Energy Conference and Exhibition, 2011

[4] M. Lanz, J. Heiber , Wafer Shape Measurement Investigations, 28th European Photovoltaic Solar Energy Conference and Exhibition, 2013

[5] B. Weber, S. Riepe, H.J. Möller, Increasing Cutting Efficiency of Slurry Based Sawing Process by Use of Structured Wire, 29th European Photovoltaic Solar Energy Conference and Exhibition, 2014

[6] B. Weber, S. Riepe, Challenges of the Multi Wire Sawing Process for Thin Wafers below 120 µm Thickness, 27th European Photovoltaic Solar Energy Conference and Exhibition, 2012

[7] D. Treyer, S. Gaulocher, S. Niederberger, H. Rafael, A. Ams, A Comprehensive Dynamic Model of the Diamond Wire Sawing Process, 33rd European Photovoltaic Solar Energy Conference and Exhibition, 2017

[8] X. Wang, L.-T. Lin, Modeling the Wiresawing Yield from SiC Shape Measurements, 29th European Photovoltaic Solar Energy Conference and Exhibition, 2014

[9] K. Sunder, O. Anspach, "Evaluation of Diamond Wires for Silicon Wafer Production", 29th European Photovoltaic Solar Energy Conference and Exhibition, 2014

[10] K. Sunder, H. Uhle, S. Knöppel, O. Anspach, Prediction of Wire Wear, Wire Bow and Wafer Total Thickness Variation in the Diamond Wire Wafering Process, 28th European Photovoltaic Solar Energy Conference and Exhibition, 2013

[11] R. Koepge, S. Brinnig, F. Kaule, H. Schwabe, S. Schoenfelder, „Advanced Analysis of Multi Wire Wafering Processes", 44th IEEE Photovoltaic Specialists Conference (PVSC-44), Washington D.C. (U.S.A.), 2017

[12] F. W. Preston, The Theory and Design of Plate Glass Polishing Machine, Journal of The Society of Glass Technology, Vol. 11, pp. 214-256, 1927

[13] J. F. Archard, Contact and Rubbing of Flat Surfaces, Journal of applied physics, 1953

[14] H. Uhle, Master Thesis Geometrische und mikrostrukturelle Bewertung von strukturierten Sägedrähten zur Herstellung von Siliziumwafern, Mittweida University of Applied Sciences, 2014

[15] A. Händler, Master Thesis Bewertung des mechanischen Verhaltens von struktruierten Sägedrähten für das Drahtsägen von Siliziumkristallen, Leipzig University of Applied Sciences Leipzig, 2016

[16] D. Meißner, S. Schoenfelder, B. Hurka, J. Zeh, K. Sunder, R. Koepge, T. Wagner, A. Grün, H.-J. Hagel, H.J. Moeller, H. Schwabe, O. Anspach, "Loss of wire tension in the wire web during the slurry based multi wire sawing process", Solar Energy Materials & Solar Cells 120, 2014

LIFETIME EVOLUTION DURING REGENERATION IN BORON-DOPED CZOCHRALSKI-SILICON

D. C. Walter[1], L. Helmich[1], D. Bredemeier[1], R. Falster, V. V. Voronkov[2] and J. Schmidt[1,3]
[1]Institute for Solar Energy Research Hamelin (ISFH), Am Ohrberg 1, D-31860 Emmerthal, Germany
[2]Sun Edison Semiconductor, Via Nazionale 59, 39012 Merano, Italy
[3]Leibniz University Hannover, Institute of Solid-State Physics, Appelstr. 2, 30167 Hanover, Germany
Corresponding author: phone: +49-5151-999-424, fax: +49-5151-999-400, email: d.walter@isfh.de

ABSTRACT: We measure the evolution of the carrier lifetime in boron-doped Czochralski-grown silicon wafers for the first time in-situ during permanent deactivation of the boron-oxygen defect under the applied conditions, i.e. illumination with a halogen lamp at elevated temperatures. Applying illumination intensities ≥1 sun on standard 1.5 Ωcm *p*-type Cz-Si, the lifetime (measured at the regeneration temperature) changes only negligibly during this regeneration process. As expected, in this case, the time dependence of the defect concentration (measured at room temperature) follows a single-exponential decay function during regeneration. For light intensities << 1 sun on the same material, the lifetime shows a significant change during the regeneration conditions and the evolution of the defect concentration does no longer follow a single-exponential decay curve. In addition, on 0.5 Ωcm *p*-type Cz-Si we observe a non-exponential decay of the defect concentration for a regeneration treatment performed at 1 sun illumination intensity and a single-exponential decay for a higher illumination intensity of 2.9 suns. These observations are well compatible with a deactivation rate increasing proportionally with the excess carrier concentration.
Keywords: boron-oxygen defects, regeneration, Czochralski-silicon

1 INTRODUCTION

The minority-carrier lifetime in boron-doped, oxygen-rich silicon, such as the typical *p*-type Czochralski-grown silicon (Cz-Si) material frequently used in solar cell production, is limited after illumination by a recombination-active defect center consisting of boron and oxygen [1–4]. This BO defect center exists in two structural configurations: a latent, recombination-inactive, and a highly recombination-active form which is activated during illumination or excess carrier injection [5]. The latent, recombination-inactive form of the BO defect is present after annealing in darkness at temperatures around 200°C. The degradation of the carrier lifetime, which is a consequence of the transition of the BO defect center from its recombination-inactive to its recombination-active form in the presence of excess carriers, is often referred to as light-induced degradation (LID). This LID effect poses a fundamental problem when using boron-doped Cz-Si wafers for electronic devices which require large diffusion lengths for best performances, such as silicon solar cells [6].

Fortunately, in 2006 Herguth *et al.* [7] observed that the degraded lifetime can be permanently regenerated when the material is exposed to illumination (or more generally, to the injection of excess carriers) at *elevated* temperatures. Over the past decade, several research groups have investigated this regeneration process [8–11]. However, the underlying physical mechanisms have still not been fully understood.

Within this study, we analyze the dynamics related to the lifetime regeneration by measuring the lifetime τ not –as typically done– at room temperature, but at the physically relevant conditions, i.e. at the temperature and the illumination intensity applied during the regeneration treatment.

In order to examine the time-resolved evolution of the BO defect during the regeneration treatment, typically, the effective defect concentration $N_t^*(t) = \frac{1}{\tau(t)} - \frac{1}{\tau_0}$, is characterized, with $\tau(t)$ being the lifetime at a fixed excess carrier density Δn during the regeneration process and τ_0 being the initial lifetime after annealing in darkness. From previous studies, it is known that the evolution of the effective defect concentration N_t^* follows a single-exponential decay during regeneration [7]. Based on experimental results [12,13] and on the theory presented in [14], it is expected that a higher excess carrier density Δn accelerates the regeneration process and, hence, leads to an increased deactivation rate constant R_{de}. As during the regeneration, the lifetime measured at room temperature usually changes by more than one order of magnitude, the excess carrier density Δn during the regeneration is expected to change by about one order of magnitude as well. Assuming a linear dependence between R_{de} and Δn as predicted in [14], one would expect a strongly non-exponential decay of N_t^* during regeneration. In contradiction to these expectations, the time-resolved evolution of the defect concentration follows an almost perfect mono-exponential decay with a fixed value of R_{de}. In order to solve this apparent contradiction, in this contribution, we examine the defect evolution for the first time in-situ under the actually applied regeneration conditions.

2 EXPERIMENTAL DETAILS

2.1 Sample preparation

Lifetime samples are prepared on boron-doped Czochralski-grown silicon (Cz-Si) wafers with a resistivity of 0.5 Ωcm featuring a final thickness of 200 µm and on wafers with 1.5 Ωcm resistivity and a final thickness of 160 µm. The sample processing consists of the following process steps: First, the silicon wafers are etched in an aqueous potassium hydroxide solution (KOH) to remove the saw damage from the surfaces. Afterwards, the samples undergo a standard RCA cleaning sequence before a phosphorus diffusion at 850°C for 51 min is performed. The diffusion process results in n^+-regions on both surfaces with a sheet resistance of 100 Ω/□. The n^+-regions are removed again using KOH and, after another RCA cleaning, an aluminum oxide layer (film thickness *d*=10 nm) is deposited by plasma-assisted atomic layer deposition in a FlexAL™ system (Oxford Instruments) on both wafer surfaces. Subsequently, a silicon nitride capping layer (*d*=70 nm, refractive index *n*=2.04 measured at a wavelength of 633 nm) is deposited by plasma-enhanced chemical vapor deposition (SiNA, Roth&Rau) on both surfaces. Finally, the samples undergo a fast-firing step (centrotherm DO FF–8.600–300) at a set-peak temperature of 850°C and belt speeds of 6.0 or 6.8 m/s. Before the lifetime measurements are performed, the wafers are laser-diced into a sample size of (5×5) cm².

In-situ lifetime measurements are conducted using the temperature-controlled Sinton lifetime tester WCT-120TS, either by measuring the lifetime at room temperature or at the applied regeneration temperature (up to 200°C). For our experiments, we combine the setup with an additional halogen lamp, which allows the illumination of the sample directly on the measurement stage. For this reason, we are able to perform the regeneration process of the lifetime sample directly on the setup. A picture of the WCT-120TS lifetime tester in combination with the displaceable halogen lamp is shown in Fig. 1. To perform a lifetime measurement, the halogen lamp is quickly removed (the handling takes approximately 10 seconds) to allow the flash lamp to illuminate the sample.

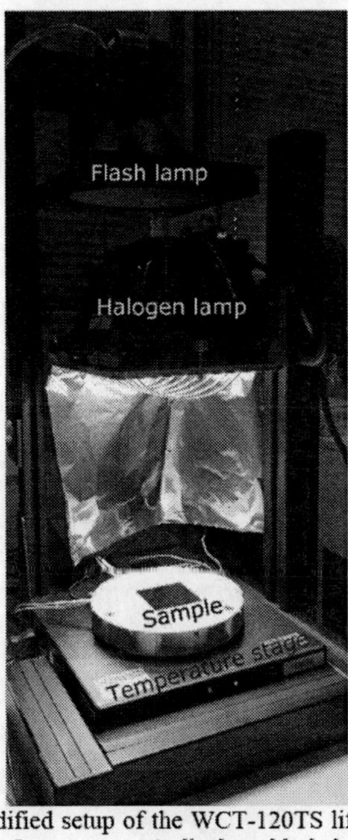

Fig. 1: Modified setup of the WCT-120TS lifetime tester from Sinton Instruments. A displaceable halogen lamp is mounted on top of the temperature stage to perform the regeneration treatment directly on the setup.

3 EXPERIMENTAL RESULTS AND DISCUSSION

3.1 Exponential and non-exponential regeneration

Figure 2 shows the evolution of two lifetime samples with a resistivity of 1.5 Ωcm, regenerated on a standalone hotplate under illumination with an halogen lamp. The lifetime, in this experiment measured at 30°C, is shown at a fixed excess carrier density of $\Delta n = 10^{15}$ cm^{-3}. We start with the regeneration process directly after annealing in darkness at 200°C for 10 min, i.e. the in-active state of the BO defect. For both regeneration conditions, at $\vartheta_{reg} = 170$°C and I=1.3 suns light intensity (orange squares) as well as for regeneration at $\vartheta_{reg} = 160$°C and I=0.12 suns (blue circles), we observe the typical degradation followed by the regeneration of the lifetime. Comparing the minimum and the maximum lifetimes, we can see that in both cases the lifetime changes strongly, from 350 to 2900 µs and from 150 to 2600 µs, for high (I=1.3 suns) and low intensity (I=0.12 suns) regeneration, respectively. Accordingly, as stated above, one would expect a change in Δn during regeneration of about one order of magnitude.

From these measured lifetimes, we calculate the effective defect concentrations as mentioned above. Figure 3 shows the evolution of the defect concentration N_t^* of both samples. The dashed lines in Fig. 3 are exponential fits to the data using an equation of the form:

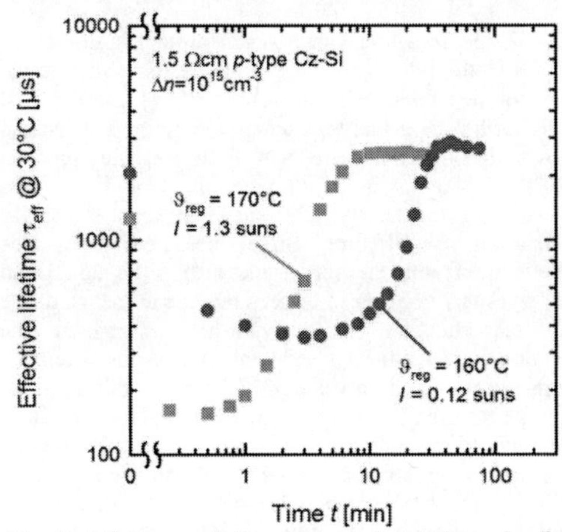

Fig. 2: Lifetime evolution, measured at 30°C and a fixed excess carrier density $\Delta n = 10^{15}$ cm^{-3}, of two lifetime samples with a resistivity of 1.5 Ωcm. The regeneration conditions are either $\vartheta_{reg} = 170$°C and I=1.3 suns (orange squares) or $\vartheta_{reg} = 160$°C and I=0.12 suns (blue circles).

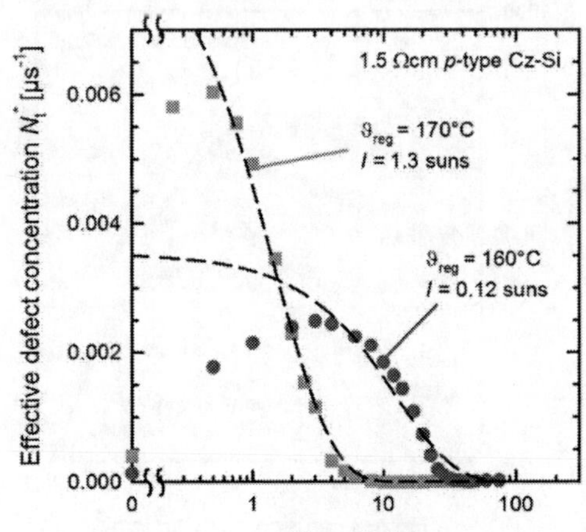

Fig. 3: Evolution of the effective defect concentration N_t^* calculated from the lifetime values shown in Fig. 2. Orange squares correspond to the regeneration at $\vartheta_{reg} = 170$°C and I=1.3 suns light intensity and blue circles correspond to the regeneration at $\vartheta_{reg} = 160$°C and I=0.12 suns. The dashed black lines are single-exponential fits.

$y = a \times e^{-b \times t}$, with b being identical with the rate constant of the regeneration R_{de}. Having a closer look at the data, one can see that for the regeneration under high illumination intensity the evolution of N_t^* can be well described using this single-exponential decay function. However, from the comparison of the dashed line with the data of the low intensity regeneration (blue circles) it becomes obvious that this single-exponential function describes the evolution of N_t^* insufficiently. The reason for the fundamentally different regeneration behaviors (single-exponential at high intensities, non-exponential at low intensities) is the hugely different changes in the excess carrier concentrations at the two different illumination levels during the actual regeneration treatment, as will be shown in the next section.

3.2 Lifetime measurements at regeneration conditions

Using the WCT-120TS setup we are able to determine injection-dependent lifetime curves $\tau(\Delta n)$ for the first time during the actual regeneration conditions. Figure 4 shows the injection-dependent lifetimes, measured at 170°C, during a regeneration treatment at 170°C and 1.3 suns light intensity after different time steps. By comparing the lifetime curve measured after 72 s regeneration with the curves measured after 234 s and 630 s, respectively, it can be seen that the influence of the BO recombination center vanishes throughout the regeneration treatment. Red stars show the lifetimes extracted at a fixed Δn value of 10^{15} cm^{-3}. In this case, the lifetime measured at 170°C changes strongly from 320 to 3300 µs during the regeneration treatment. However, extracting the lifetime at a constant illumination intensity of I=1.3 suns (red diamonds), the lifetime changes only slightly within the first 4 min before the intrinsic lifetime limit is reached (black solid line in Fig. 4). Analyzing the lifetime at constant illumination intensity further, we determine from the slope of $\tau(\Delta n)$ at 1.3 suns illumination the generation rate G to be 1.57×10^{19} cm^{-3}s^{-1}. Please note, this value is only an estimate for G under the regeneration

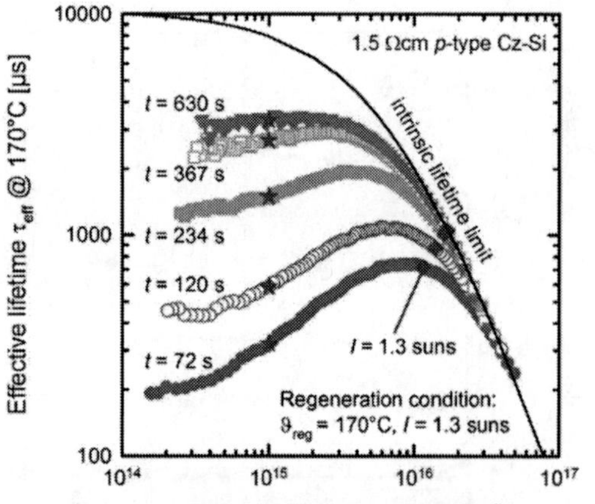

Fig. 4: Injection-dependent lifetime measured at 170°C during the regeneration process. Differently colored curves correspond to different time steps during regeneration, as indicated. Red stars show the lifetime extracted from these curves at a fixed injection of $\Delta n = 10^{15}$ cm^{-3}. Red diamonds show the lifetime at a constant illumination intensity of I=1.3 suns. The black solid line is the intrinsic lifetime limit according to Richter *et al.* calculated for 170°C [15].

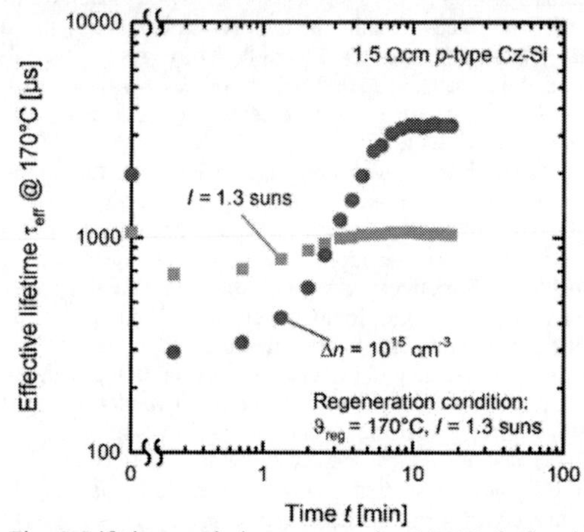

Fig. 5: Lifetime evolution, measured at 170°C, during the regeneration process at 170°C and 1.3 suns illumination intensity. The lifetimes are extracted from the injection-dependent lifetime curves at constant $\Delta n = 10^{15}$ cm^{-3} (blue circles) and at constant illumination intensity of I=1.3 suns.

condition, as the different light spectra from the Sinton flash lamp and the halogen lamp are applied for measurement and regeneration, respectively.

Figure 5 shows the evolution of the lifetime, measured at the regeneration temperature, during the regeneration process at 170°C and 1.3 suns light intensity. The lifetime is shown either at a constant excess carrier concentration of Δn=10^{15} cm^{-3} (blue circles) or at a fixed illumination intensity of I=1.3 suns (orange squares). Comparing both curves, we observe that while the lifetime at the constant Δn value changes by about one order of magnitude, the lifetime changes at I=1.3 suns are comparatively little, from about 700 to 1000 µs. Accordingly, also the excess carrier concentration $\Delta n = \tau \times G$, with the constant

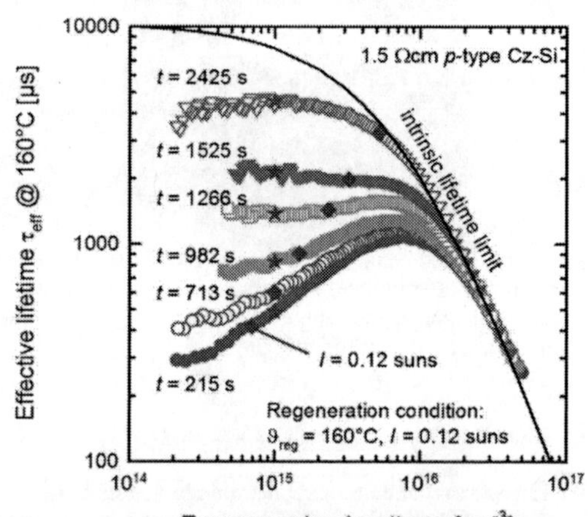

Fig. 6: Injection-dependent lifetime measured at 160°C during the regeneration process. As indicated, differently colored curves correspond to different time steps during regeneration. Again, red stars show the lifetime extracted from these curves at a fixed excess carrier density of $\Delta n = 10^{15}$ cm^{-3} and red diamonds the lifetime at a constant illumination intensity of, in this case, I=0.12 suns. The black solid line is the intrinsic lifetime limit according to Richter *et al.* calculated for 160°C [15].

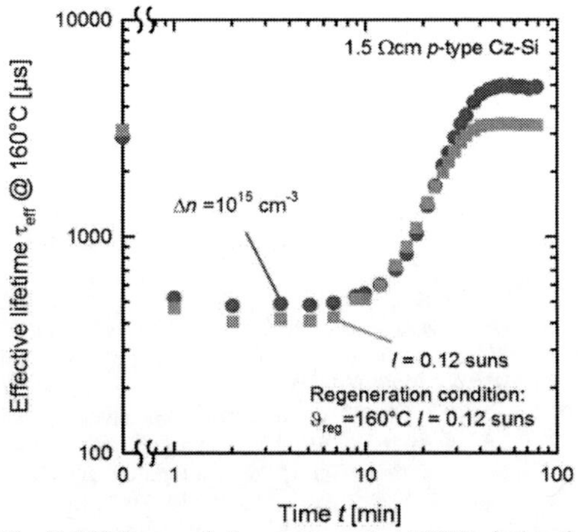

Fig. 7: Lifetime evolution, measured at 160°C, during the regeneration process at 160°C and 0.12 suns light intensity. Blue circles correspond to the lifetime extracted at $\Delta n = 10^{15}$ cm^{-3}, while orange squares correspond to the lifetime extracted under constant illumination at *I*=0.12 suns.

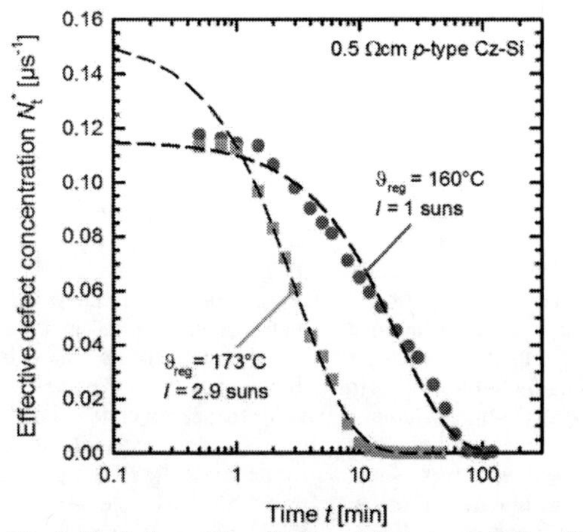

Fig. 8: Evolution of the effective defect concentration N_t^* of two samples with a resistivity of 0.5 Ωcm, determined from room temperature lifetime measurements. The regeneration was performed either at a regeneration temperature of ϑ_{reg} =173°C and an illumination intensity of *I*=2.9 suns (orange squares) or at ϑ_{reg} =160°C and *I*=1 sun (blue circles). Black dashed lines are fits using a single-exponential decay function.

generation rate *G* under constant illumination, shows only a minor change during the regeneration process. This small variation of about 30%, leads in case of a proportional increase of R_{de} with Δn to only 30% variation in R_{de}. From the single-exponential decay of N_t^* observed under these conditions in Fig. 3 (orange squares), we conclude that a 30% variation in R_{de} would be too small to be detectable in our experiments.

Additionally, the observation of a constant or only slightly changing lifetime under the applied regeneration conditions agrees well with the assumptions made by Voronkov *et al.* [14] that the lifetime at the regeneration conditions is no longer affected by the defect-limited lifetime, but largely by the intrinsic lifetime determined by the Auger recombination process.

As shown in Fig. 3, for the regeneration under low illumination intensities, the evolution of the effective defect concentration N_t^* can no longer be described by a single-exponential decay function. To further analyze this behavior, we measure again the injection-dependent lifetime at the regeneration temperature of 160°C. Figure 6 shows the corresponding lifetime curves.

At a fixed $\Delta n = 10^{15}$ cm^{-3} (red stars) the lifetime changes again strongly from 490 to 4600 μs during the regeneration treatment. However, this time, at the applied relatively low illumination intensity of *I*=0.12 suns, the lifetime changes also significantly from 410 to 3300 μs. In this case, we determine a generation rate $G = 1.61\times10^{18}$ cm^{-3}s^{-1}. Compared to the curves shown in Fig. 4, the intrinsic lifetime limit is only reached after about 40 min (open green triangles) compared to 4 min in Fig. 4 at the higher illumination intensity of 1.3 suns.

In Fig. 7, the evolution of the lifetime measured at 160°C during the regeneration treatment at a regeneration temperature of ϑ_{reg} =160°C and an illumination intensity of *I*=0.12 suns is displayed. From the graph it is evident that at these low illumination intensities the lifetime at the regeneration conditions changes strongly, in this case comparable to the lifetime extracted at $\Delta n = 10^{15}$ cm^{-3}. Consequently, the excess carrier density during the process changes also by about one order of magnitude.

Taking the linear dependence between the rate constant R_{de} and the excess carrier concentration Δn into account, the non-exponential decay of the defect concentration shown in Fig. 3 can be consistently explained.

3.3 Highly boron-doped Cz-Si material

In the previous section, we have discussed the evolution of the effective defect concentration on a relatively lowly doped Cz-Si material, typical for PV applications. However, it is worth noting that the same observations also apply to highly doped material with a resistivity of only 0.5 Ωcm. As in such materials, the lifetime is usually much lower, higher illumination intensities are needed to get closer to the limitations of either the surface passivation or the intrinsic lifetime, which would cause constant lifetime during the regeneration process. Accordingly, from the results presented in the previous section, we would expect that high illumination intensities are needed to observe a single-exponential decay of the effective defect concentration N_t^* in the 0.5 Ωcm Cz-Si material.

Figure 8 shows the evolution of the effective defect concentration N_t^* of two samples with a resistivity of 0.5 Ωcm. Orange squares correspond to the regeneration conditions of ϑ_{reg} =173°C and *I*=2.9 suns, while blue circles correspond to ϑ_{reg} =160°C and *I*=1 sun. The dashed lines are fits to the data using a single exponential decay function, as already shown in Fig. 3. Comparing the dashed lines to the data, we observe the same trend as before. High illumination intensities (*I*=2.9 suns) lead to a lower deviation of the fitted curve to the measurement data. In contrast, the exponential decay curve corresponding to the low illumination intensity of *I*=1 sun describes the measurement data not sufficiently. From these observations we can conclude that the change in the excess carrier density Δn during the regeneration process at *I*=2.9 suns is too low to cause a significant deviation from the single exponential decay. The contrary holds for *I*=1 sun illumination.

4 CONCLUSIONS

Within this study, we measured for the first time the change of the minority-carrier lifetime during the permanent deactivation of the boron-oxygen-related defect center under the applied regeneration conditions, i.e. at elevated temperatures and at the applied, constant illumination intensity. For a typical illumination intensity of 1 sun, we observe on typical 1.5 Ωcm *p*-type Cz-Si, that in contrast to the pronounced change in lifetime at a fixed excess carrier density Δn measured at room temperature, the lifetime at the regeneration temperature and at 1 sun illumination intensity remains almost unchanged as it is largely limited by intrinsic recombination. As expected, in this case the evolution of the effective defect concentration N_t^* can be described with a single-exponential decay function. We further demonstrate that if the excess carrier density Δn changes significantly at the applied regeneration conditions, the evolution of N_t^* is no longer sufficiently described by a single-exponential decay function.

On highly boron-doped Cz-Si with a resistivity of 0.5 Ωcm we observe a comparable behavior. While at 1 sun the change in N_t^* does not follow a single-exponential decay, at a higher illumination intensity of 2.9 suns a single-exponential decay is observed.

Finally, the observations here confirm the assumption of Voronkov and Falster [14] that under typical regeneration conditions, the lifetime is not affected by defect-related recombination, but by intrinsic recombination. Additionally, our observations are consistent with a deactivation rate constant increasing proportionally with the excess carrier concentration.

Acknowledgements

The authors would like to thank Cornelia Marquardt of ISFH for her help with sample processing. The work was funded by the German State of Lower Saxony and the German Federal Ministry of Economics and Energy (BMWi) within the research project "LIMES" (contract no. 0324204D). The content is the responsibility of the authors.

5 REFERENCES

[1] H. Fischer, W. Pschunder, Investigation of photon and thermal induced changes in silicon solar cells, Proceedings 10th IEEE PVSC, Palo Alto, CA (IEEE, New York, 1973) (1974) 404.

[2] J. Schmidt, A.G. Aberle, R. Hezel, Investigation of carrier lifetime instabilities in Cz-grown silicon, Proceedings 26th IEEE PVSC, Anaheim, CA (IEEE, New York, 1997) (1997) 13.

[3] S.W. Glunz, S. Rein, W. Warta, J. Knobloch, W. Wettling, On the degradation of Cz-Silicon solar cells, Proc. 2nd WC-PVSEC, Vienna, Austria (WIP, Munich, 1998) (1998) 1343.

[4] J. Schmidt, K. Bothe, Structure and transformation of the metastable boron- and oxygen-related defect center in crystalline silicon, Phys. Rev. B 69 (2004) 24107.

[5] V.V. Voronkov, R. Falster, Latent complexes of interstitial boron and oxygen dimers as a reason for degradation of silicon-based solar cells, J. Appl. Phys. 107 (2010) 53509.

[6] J. Knobloch, S.W. Glunz, D. Biro, W. Warta, E. Schaffer, W. Wettling, Solar cells with efficiencies above 21% processed from Czochralski grown silicon, Proceedings 25th IEEE PVSC, Washington DC (IEEE, New York, 1997) (1996) 405–408.

[7] A. Herguth, G. Schubert, M. Kaes, G. Hahn, A new approach to prevent the negative impact of the metastable defect in boron doped Cz silicon solar cells, Proc. of the 32nd IEEE PVSC (4th WCPEC), Waikoloa, USA (2006) 940–943.

[8] B. Lim, K. Bothe, J. Schmidt, Deactivation of the boron–oxygen recombination center in silicon by illumination at elevated temperature, phys. stat. sol. (RRL) 2 (2008) 93–95.

[9] S. Wilking, S. Ebert, A. Herguth, G. Hahn, Influence of Short High Temperature Steps on the Regeneration of Boron-Oxygen Related Defects, 28th EUPVSEC, Paris, France (2013) (2013) 34–38.

[10] D.C. Walter, B. Lim, K. Bothe, V.V. Voronkov, R. Falster, J. Schmidt, Effect of rapid thermal annealing on recombination centres in boron-doped Czochralski-grown silicon, Appl. Phys. Lett. 104 (2014) 42111.

[11] B.J. Hallam, S.R. Wenham, P.G. Hamer, M.D. Abbott, A. Sugianto, C.E. Chan, A.M. Wenham, M.G. Eadie, G. Xu, Hydrogen Passivation of B-O Defects in Czochralski Silicon, Energy Procedia 38 (2013) 561–570.

[12] V. Steckenreiter, D.C. Walter, J. Schmidt, Kinetics of the permanent deactivation of the boron-oxygen complex in crystalline silicon as a function of illumination intensity, AIP Advances 7 (2017) 35305.

[13] S. Wilking, S. Ebert, C. Beckh, A. Herguth, G. Hahn, Of Apples and Oranges: Why Comparing BO Regeneration Rates Requires Injection Level Correction, Proceedings 32nd EU PVSEC, Munich, Germany (2016) 487–494.

[14] V.V. Voronkov, R. Falster, Permanent deactivation of boron–oxygen recombination centres in silicon, physica status solidi (b) 253 (2016) 1721–1728.

[15] A. Richter, S.W. Glunz, F. Werner, J. Schmidt, A. Cuevas, Improved quantitative description of Auger recombination in crystalline silicon, Phys. Rev. B 86 (2012) 165202.

DEFECT ANALYSIS OF APCVD GETTERED MULTICRYSTALLINE SILICON

Martin Fleck, Jeanette Lindroos, Annika Zuschlag, Giso Hahn
University of Konstanz, Department of Physics, 78457 Konstanz, Germany
martin.fleck@uni-konstanz.de, Tel: +49 7531 882082, Fax: +49 7531 883895

ABSTRACT: The effects of atmospheric pressure chemical vapor deposition (APCVD) gettering on material quality of industry standard mc-Si wafers has been studied on a microscopic level and compared to $POCl_3$ gettered samples. The interaction of the gettering efficacy with the microscopic defect structure has been studied by a combination of measurements of the effective minority charge carrier lifetime (τ_{eff}), interstitial iron (Fe_i), optical microscope images of defect structures revealed with Secco etching, a subsequent high resolution etch pit density (EPD) analysis as well as Electron Beam Induced Current (EBIC) and Electron Backscatter Diffraction (EBSD). Findings include a similar gettering quality of APCVD- and $POCl_3$-based glasses and a reduction of EPD after the gettering step.
Keywords: Multicrystalline Silicon, Gettering, APCVD, Lifetime, Defects

1 INTRODUCTION

Gettering is known to strongly increase material quality of multicrystalline silicon. With APCVD (atmospheric pressure chemical vapor deposition)-based glasses, doping sources and getter sinks can easily and cost effectively deposited on one side only, making this technology valuable for passivated emitter, rear totally diffused (PERT) and other advanced solar cell concepts. Here we make a comparison between APCVD and $POCl_3$ based gettering and the interaction of the gettering efficacy with the microscopic defect structure, for which measurements of effective minority charge carrier lifetime (τ_{eff}), interstitial iron concentration ([Fe_i]), optical microscope images of Secco etched wafers, etch pit density (EPD), electron beam induced current (EBIC) and electron backscatter diffraction (EBSD) are combined.

2 EXPERIMENTAL

2.1 Sample Processing

Industrial multicrystalline silicon sister wafers with boron doped bulk and a resistivity of 1 Ωcm from middle ingot height were, after typical saw damage removal and cleaning steps, subjected to either of the following (Fig. 1): Either an APCVD BSG (boron-silicate glass) and PSG (phosphosilicate glass) deposition followed by a high temperature diffusion step at 840°C (referred to as APCVD gettering in the following) or else solely the previously mentioned high temperature step at 840°C or an industry standard $POCl_3$ diffusion. Hereafter, emitter and BSF (back surface field) were removed followed by PECVD (plasma-enhance chemical vapor deposition) deposition of SiN_x:H and firing.

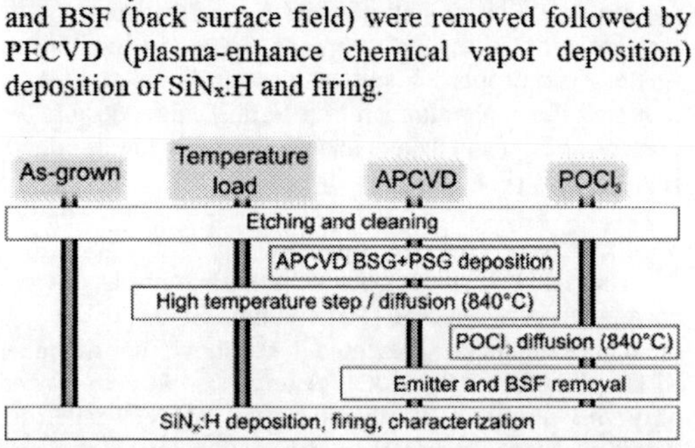

Figure 1: Process flow that the various samples have been subjected to.

2.2 Etching and etch pit density counting algorithm

Etch pits were formed by application of the Secco etch [1] on mechanically polished wafers for an etching time of 180 s. During the etching process, the beaker was agitated in an ultrasonic bath.

The etched wafer surface is recorded with an optical microscope. The etch pit density counting algorithm is based on [2]. The outline of the algorithm that was used is roughly as follows:

1. Thresholding of the image (black and white conversion)
2. Removal of small specks
3. Identification and characterization of remaining structures
4. Separation of the largest structure(s)
5. Counting of small structures – these are free standing, individual etch pits
6. Distinguishing between grain boundaries and high density clusters of etch pits
7. Estimation of the density of etch pits in regions where the etch pits are so dense that they cluster together to a compound structure (etch pit cluster). Based on the size distribution of the free standing etch pits of the current sample, an estimator for the density of etch pits in such clustered regions is calculated.
8. Filling of thus acquired spatial information of free standing and clustered etch pits into a single histogram results in the EPD maps as shown in this publication

The algorithm outlined above will be discussed and in detail and released in a future publication.

3 RESULTS

On the macroscopic scale, effective minority carrier lifetime (τ_{eff}) and interstitial iron (Fe_i) maps are accessible via QSSPC (quasi-steady-state photoconductance) calibrated photoluminescence (PL) imaging. The PL and Fe_i images shown in Fig. 2 and Fig. 3 are a superposition of the optical image of the Secco etched wafer with the respective map of the measurement variable, either τ_{eff} or [Fe_i]. Hence, defects revealed by the Secco etch are visible as black structures on top of the map.

3.1 Fe_i and τ_{eff} correlation

Macroscopic measurement variables for gettering strength used in this study are τ_{eff} and [Fe_i] (Fig. 2, Fig. 3). Temperature treatment slightly, if at all, decreases the harmonic mean of τ_{eff} from 60 µs in the as-grown state to 59 µs in accordance with the slight increase in [Fe_i]. APCVD ($POCl_3$) based gettering results in mean values for τ_{eff} of 158 µs (170 µs), respectively. Comparably strong gettering effects are observed for APCVD and $POCl_3$ gettered samples as already described in [3]. The observed increase in τ_{eff} (Fig. 2) is similar for both gettering methods whereas the reduction in interstitial iron in the depicted sample is stronger for the APCVD gettered sample. However, it should be noted that the comparison of τ_{eff} and [Fe_i] (Fig. 3) for various column heights and block positions is sometimes in favor of APCVD and sometimes in favor of $POCl_3$ gettering. Therefore, no clear tendency can be detected.

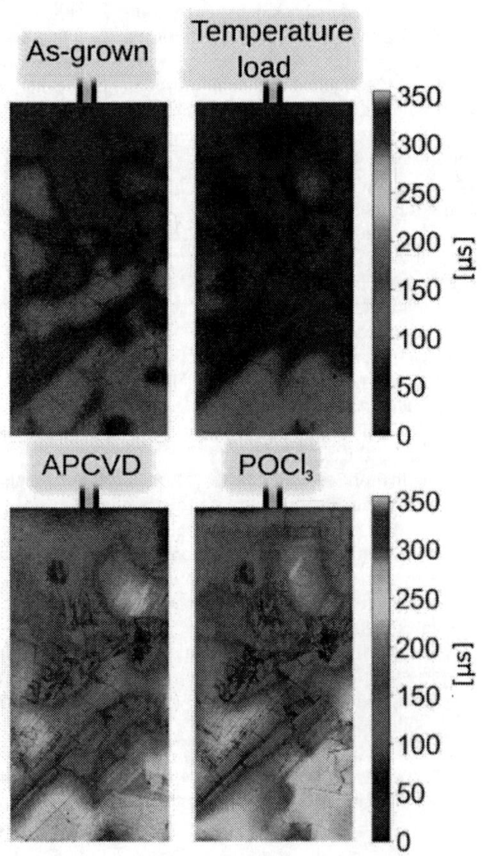

Figure 2: τ_{eff} maps of as-grown, temperature treated (840°C), and APCVD as well as $POCl_3$ gettered samples. Sample size: 12x25 mm².

When isolating the temperature load of the diffusion step (840°C) in absence of an external getter sink, a redistribution of Fe_i is observed, *i.e.* a slight overall increase in the as-grown [Fe_i] in regions of low and intermediate contamination strength, combined with a decrease in size of highly contaminated regions. Consequently, a slight reduction of τ_{eff} is observed. In absence of an external getter sink, no effective gettering effect is observed.

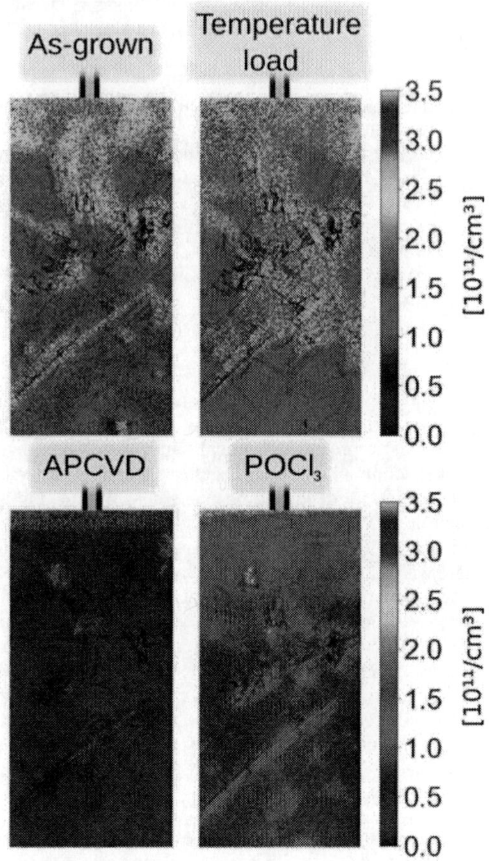

Figure 3: [Fe_i] maps of as-grown, temperature treated (840°C), and APCVD as well as $POCl_3$ gettered samples. Sample size: 12x25 mm².

A clear correlation between τ_{eff} and [Fe_i] is observed on the scale of millimeter sized structures within each of the differently treated samples. Regions of high Fe_i typically perform poor in τ_{eff} and *vice versa*.

Gettering is most effective, *i.e.* relative improvements are strongest, in regions of high [Fe_i]. This suggests the interpretation that the prepared gettering conditions (getter sink properties and temperature) are capable of reducing the contamination to a certain value, despite the presence of defect clusters that are typically found in high [Fe_i] regions (Fig. 3).

3.2 Extended crystal defect density

The EPD of the as-grown sample reveals a clear correlation between high [Fe_i] and high EPD (Fig. 4).

The sample set presented here shows considerable EPD reduction for the $POCl_3$ gettered sample (Fig. 4) and only very slight EPD reduction after APCVD gettering. Additional sample sets from different ingot heights and positions have shown an APCVD EPD reduction that is relatively stronger than for its sister $POCl_3$ gettered sample while at the same time exhibiting an absolute EPD reduction comparable to the $POCl_3$ sample depicted here. There is currently no indication for a correlation between the strength of EPD reduction and [Fe_i] results.

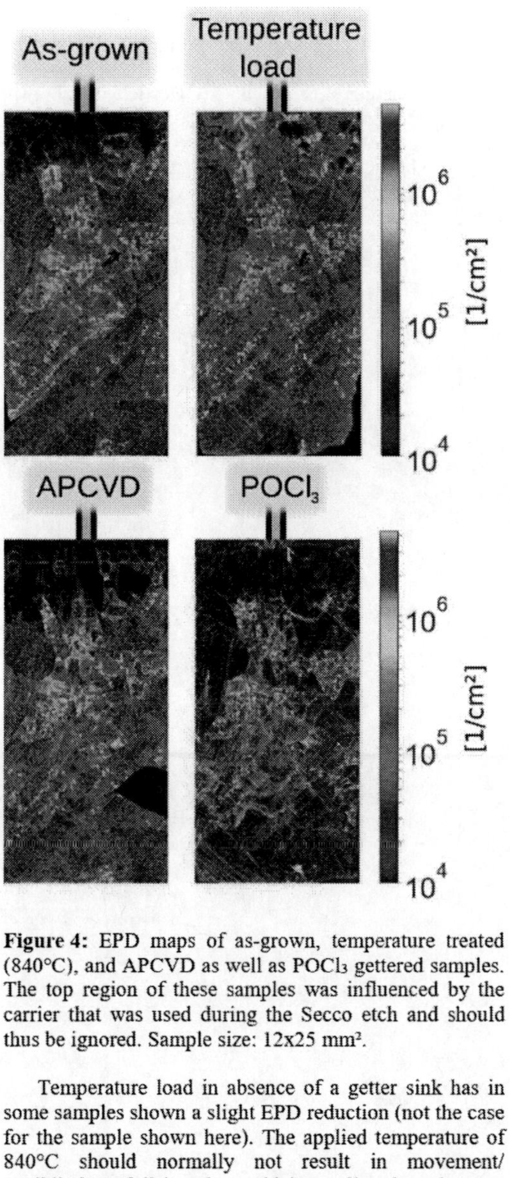

Figure 4: EPD maps of as-grown, temperature treated (840°C), and APCVD as well as $POCl_3$ gettered samples. The top region of these samples was influenced by the carrier that was used during the Secco etch and should thus be ignored. Sample size: 12x25 mm².

Temperature load in absence of a getter sink has in some samples shown a slight EPD reduction (not the case for the sample shown here). The applied temperature of 840°C should normally not result in movement/ annihilation of dislocations which usually takes places at significantly higher temperatures only. Nevertheless, similar effects have been observed before [4] and the underlying mechanism is still unclear. Further investigations are ongoing.

3.3 Recombination activity

EBIC measurements offer a unique possibility to analyze spatially resolved recombination activity with a similar high resolution as the microscopic etch pit density analysis.

The EBIC measurement of the temperature load sample (Fig. 5) features a more homogeneous EBIC contrast in comparison to the as-grown sample: Clouds of higher recombination activity in the vicinity of the most dense regions of recombination active defect lines and grain boundaries are not visible or less pronounced in the temperature load sample.

The comparison of EBIC signals of the as-grown and APCVD gettered samples with the corresponding EBSD map of the APCVD gettered sample (Fig. 6) reveals that most of the recombination active structures remaining after gettering are defect lines. Most grain boundaries show negligible recombination activity after gettering for both gettering methods. Only small parts of a few grain boundaries retain their recombination activity. This was observed for large angle as well as small angle grain boundaries.

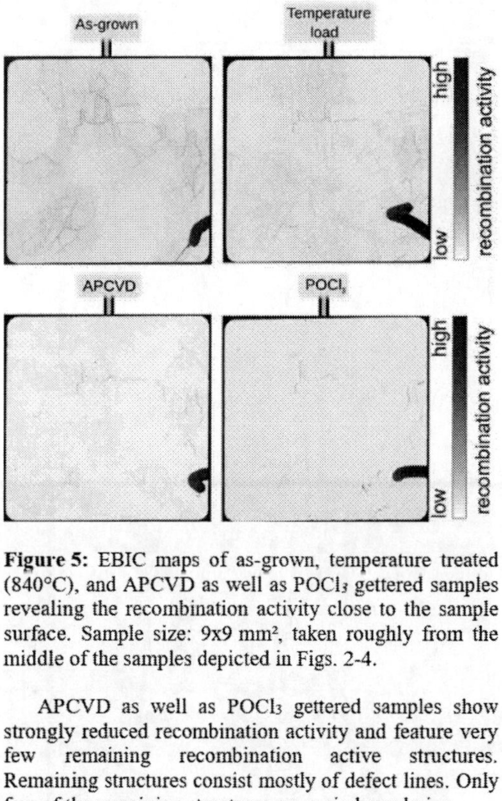

Figure 5: EBIC maps of as-grown, temperature treated (840°C), and APCVD as well as $POCl_3$ gettered samples revealing the recombination activity close to the sample surface. Sample size: 9x9 mm², taken roughly from the middle of the samples depicted in Figs. 2-4.

APCVD as well as $POCl_3$ gettered samples show strongly reduced recombination activity and feature very few remaining recombination active structures. Remaining structures consist mostly of defect lines. Only few of the remaining structures are grain boundaries.

5 CONCLUSION

Similarly strong effects for APCVD and $POCl_3$ based gettering were observed. Gettering is most effective in high [Fe_i] regions, despite high EPD in these areas. All samples show a strong correlation in local structures of τ_{eff} and [Fe_i] maps.

The gettering step introduces a reduction of EPD for APCVD as well as $POCl_3$ based gettering. Furthermore, APCVD and $POCl_3$ based gettering result in strong improvement of the EBIC signal. Recombination active structures that remain after the gettering steps are some defect lines as well as (segments of) large and small angle grain boundaries.

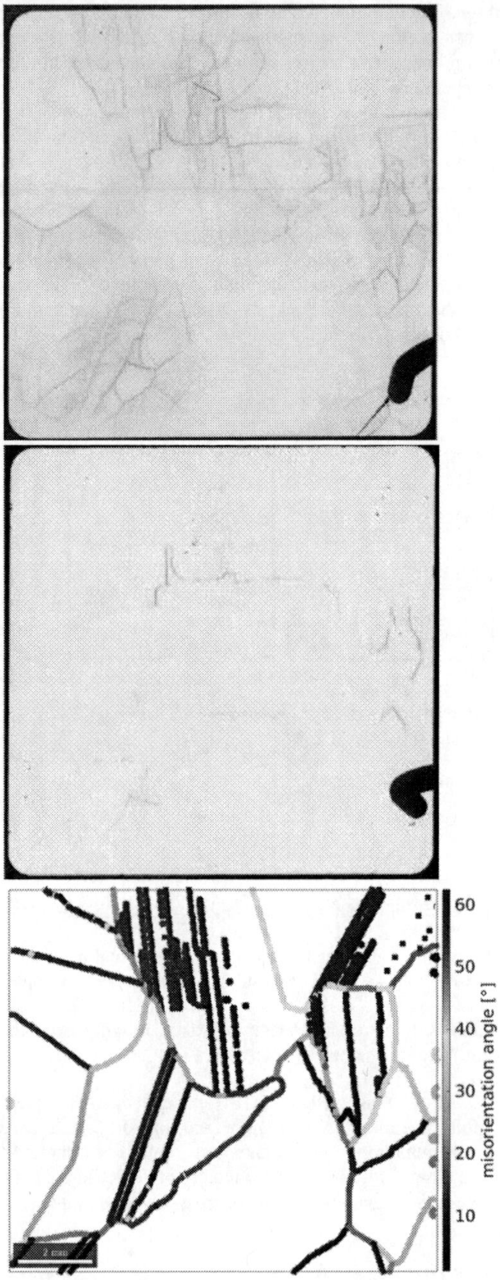

Figure 6: Comparison of EBIC results for the as-grwon (top) and APCVD gettered sample (middle) with the misorientation angle between grains at grain boundaries measured via EBSD. The APCVD gettered sample was used for the EBSD measurement.

6 ACKNOWLEDGEMENTS

Part of this work was supported by the German Federal Ministry of Economic Affairs and Energy under contract no. 0324041B and 0324001. The content is the responsibility of the authors.

7 REFERENCES

[1] F. Secco d' Aragona, Journal of the Electrochemical Society 119(7) 1972 948.

[2] D.B. Needleman, H. Choi, D.M. Powell, T. Buonassisi, Physica Status Solidi Rapid Research Letters 7(12) (2013) 1041.

[3] D.P. Fenning, A.S. Zuschlag, J. Hofstetter, A. Frey, M.I. Bertoni, G. Hahn, T. Buonassisi, IEEE Journal of Photovoltaics 4(3) 866.

[4] J. Fichtner, H. Zunft, A. Zuschlag, H. Knauss, G. Hahn, IEEE Journal of Photovoltaics (2018) in press; DOI: 10.1109/JPHOTOV.2018.2865509

GETTERING AND PASSIVATION OF ADVANCED HIGH PERFORMANCE MULTICRYSTALLINE SILICON MATERIAL

Christian Fischer, Jeanette Lindroos, Annika Zuschlag, and Giso Hahn
University of Konstanz, Department of Physics, 78457 Konstanz, Germany
christian.fischer@uni-konstanz.de, Tel: +49 7531 884995, Fax: +49 7531 883895

ABSTRACT: High performance multicrystalline (HP mc) silicon material with local effective minority charge carrier lifetime above 1 ms is a promising pathway to increase efficiency in mc silicon solar cells. Critical for the success of the material is the behavior of the material quality during solar cell processing. Phosphorus gettering during emitter formation for example is known to reduce contaminations and thus to enhance carrier lifetime in multicrystalline material. In contrast, also an increase of charge carrier recombination was recently reported for P-gettering in HP mc-Si. Therefore, a detailed analysis of different P-gettering profiles and te mperature effects combined with several surface passivations (SiN_x:H, AlO_x and iodine-ethanol) is performed to gain further understanding of this phenomenon. Especially changes in τ_{eff} due to different thermal load and in-diffusing hydrogen from SiN_x:H are investigated.
Keywords: Gettering, High Performance Multicrystalline Silicon, Lifetime, Passivation

1 INTRODUCTION

In 2017 the market share of high-performance multicrystalline silicon (HP mc-Si) material already exceeded standard p-type mc silicon [1]. This material is a promising pathway to increase efficiency in solar cells like recently projected for (PERC) passivated emitter rear contact cells [2].

On one hand effective minority charge carrier lifetimes τ_{eff} >1 ms with AlO_x passivation are reported [3], on the other hand changes in recombination activity are observed in HP mc-Si materials during solar cell processing. Phosphorus (P) diffusion is normally known to enhance material quality through gettering effects in mc-Si. A contrary influence of P-gettering was observed on HP mc-Si [4-7]. An increase in recombination was reported after P-gettering especially at grain boundaries [5, 7].

In this study, the influence of different thermal loads with and without presence of P-doping layers on material quality is investigated and effects of SiN_x:H and AlO_x deposition and surface passivation are compared by using wet-chemical iodine-ethanol (IE) passivation.

2 EXPERIMENTAL

156x156 mm² B-doped HP mc-Si wafers (~1-1.5 Ωcm) from an advanced crystallization run and mid ingot height were used with defined orientation of each wafer in the block and processed as 50x50 mm² samples. A total of 17 samples were processed and measured according to the process sequences shown in Fig. 1.

After wafer cleaning and saw damage removal, P-diffusion gettering was performed with several temperature profiles (samples 4-12). A $POCl_3$-diffusion used for solar cell processes resulting in a 50 Ω/□ emitter with a peak temperature of about 840 °C was compared to two diffusions with adjusted temperature profiles (Fig. 2). One was performed with an about 30 K elevated peak temperature resulting in a 27 Ω/□ emitter. The other diffusion with a peak temperature of 840 °C had a slow cool-down ramp before unloading (55 Ω/□). This should improve precipitation of defects and contaminations.

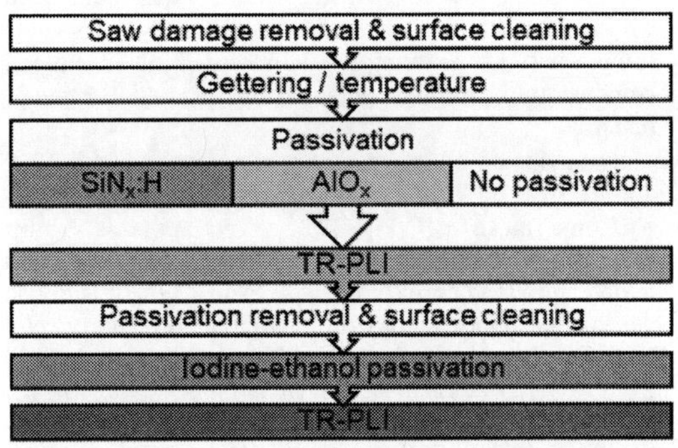

Figure 1: Process sequence of sample preparation and characterization of investigated lifetime samples.

Gettering \ Passivation		SiN_x:H	AlO_x	IE
untreated		1 & 1*	2	3 & 3⁺
$POCl_3$ gettered	840°C	4	5	6
	870°C	7	8	9
	840°C with CD	10	11	12
Temperature profile $POCl_3$		13 broken	14	15

Figure 2: List of all lifetime samples with P-gettering or temperature step, respectively, on the left side and passivation method in top line. P-gettering at about 840 °C with slow cool-down is labeled with CD. 1* unfired sister wafer. 3^+ fired sister wafer.

To distinguish between gettering efficacy and temperature-induced material changes, a set of samples received the temperature profile of P-gettering at T_{peak} 840 °C without PSG(phosphorus-silicate glass)-layer deposition. Effects of surface passivation were investigated with samples 1-3 (Fig. 2). Those samples were passivated without previous gettering or temperature step.

PSG and diffused layers were removed by chemical etching and, as all other samples received thickness relevant etching steps, a similar final wafer thickness was achieved for all samples. One group of samples was passivated with direct plasma-enhanced chemical vapor deposition (PECVD) SiN_x:H on both sides (deposition temperature 450 °C) and fired in a belt furnace (peak set-temperature 800 °C). One group was passivated with AlO_x from atomic layer deposition (ALD) and annealed at around 400 °C (30 min). The last group was directly wet chemically passivated with IE. τ_{eff} was measured for all samples by time resolved photoluminescence imaging (TR-PLI) [8]. Finally, SiN_x:H and AlO_x passivation were removed and the samples were wet chemical passivated with IE for lifetime measurement. Afterwards, all samples can be compared with similar surface passivation.

Untreated sister wafers not included in the list were CP4 etched and Piranha cleaned. Afterwards the samples were polished and Secco etched [9] for etch pit density analysis.

3 RESULTS AND DISCUSSION

Local lifetimes exceeding 1 ms and a harmonic mean τ_{eff} of 654 µs were observed on the AlO_x passivated sample without gettering as shown in Fig. 3 (sample 2) demonstrating the high potential of the used HP mc-Si material.

τ_{eff} maps of all 5x5 cm² samples passivated with AlO_x are shown in Fig. 3. P-gettering strongly reduces harmonic mean τ_{eff} (Fig. 3: 2 vs 5, 8, 11). As this reduction is also present in ungettered sample 14 with temperature load corresponding to P-diffusion, it is due to detrimental temperature load. The areas remaining on a high lifetime level after temperature treatment benefit from P-gettering and can even surpass $\tau_{eff,max}$ of the ungettered sample. The differences between several P-gettering profiles indicate that a slow cool-down after gettering at high temperatures results in slightly reduced negative effects in highly defected areas on the one hand and on the other hand reduces maximum τ_{eff}.

PECVD SiN_x:H and IE passivation result in lower lifetimes than AlO_x. This is due to better surface passivation of AlO_x. But also a temperature related effect induced by deposition and annealing of AlO_x cannot be excluded. P-gettering reduces drastically lifetimes in AlO_x passivated samples. This effect is also observed in samples with SiN_x:H and IE passivation, but is less pronounced due to a generally lower lifetime level (Fig. 4).

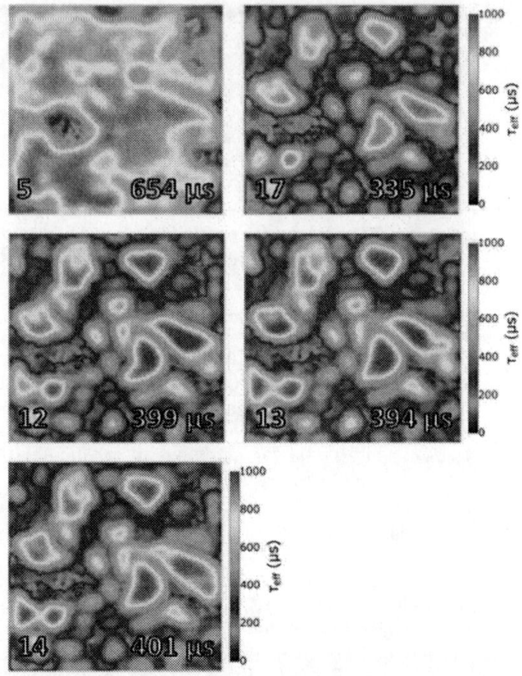

Figure 3: τ_{eff} maps of the 5x5 cm² samples passivated with AlO_x with sample codes according to Fig. 2 and harm. mean of τ_{eff}. 2: untreated sample, 14: sample with temperature profile of P-gettering without diffusion, 5: sample with P-gettering at 840 °C, 8: sample with P-gettering at 870 °C, 11: sample with P-gettering at 840 °C with slow cool-down ramp.

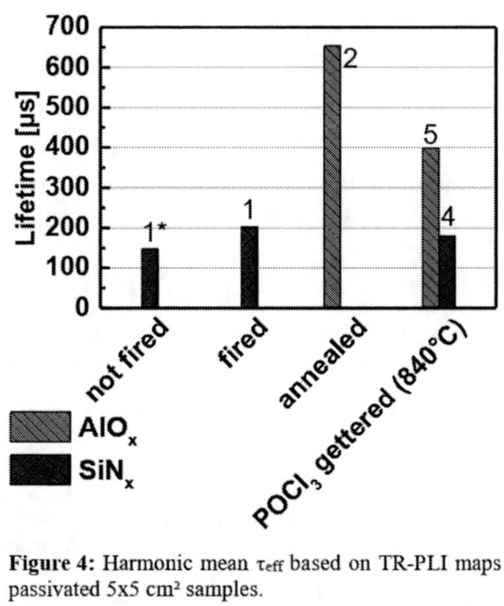

Figure 4: Harmonic mean τ_{eff} based on TR-PLI maps of passivated 5x5 cm² samples.

The final wet-chemical IE passivation after removal of all functional layers allows comparing all samples with similar surface passivation quality. Although general lifetime level is reduced due to limited IE passivation quality in areas with highest τ_{eff}, differences between τ_{eff} values for different sample (Fig. 5) can be attributed to the bulk. This reveals the negative effect of firing, as lifetimes of fired samples are in all cases below unfired samples (*e.g.*, Fig. 6: 3 vs 3^{+}). SiN_x:H passivated samples show expected behavior with low passivation quality of the unfired sample 1* (Fig. 6 1 vs. 1*). However, after removal of this layer and IE passivation the opposite behavior is observed (Fig. 6 1 IE vs. 1* IE). Sample 1^{*} with SiN_x:H deposition and no firing step shows with IE passivation higher lifetimes than sample 3 without any high temperature or passivation steps. A possible explanation can be in-diffusion of hydrogen during deposition or an internal gettering effect induced by deposition temperature of SiN_x:H. This may also explain more homogenous τ_{eff} maps of SiN_x:H samples and an improvement in low lifetime areas.

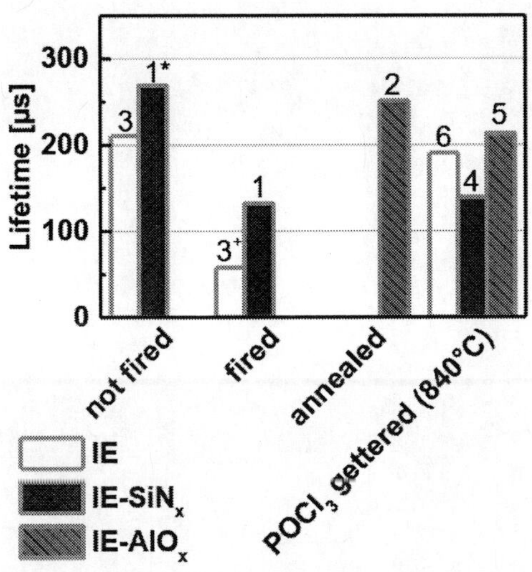

Figure 5: Harmonic mean τ_{eff}, based on TR-PLI maps of IE passivated 5x5 cm² samples.

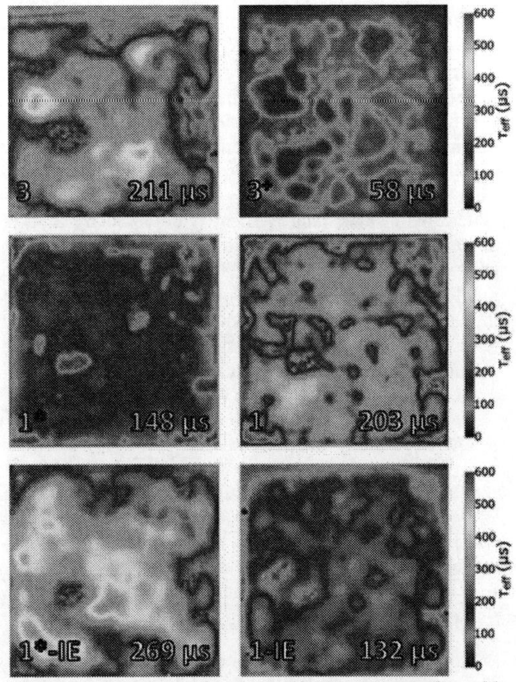

Figure 6: τ_{eff} maps of ungettered 5x5 cm² samples with sample codes according to Fig. 2 and harm. mean of τ_{eff}. 3: ungettered IE passivated, 3*: fired IE passivated, 1*: SiN_x:H-layer deposited not fired, 1: SiN_x:H-layer deposited and fired, 1*-IE: sample 1* with IE repassivation, and 1-E: sample 1 with IE repassivation.

In addition to TR-PLI analysis, sister samples were Secco-etched and microscopic pictures show etch-pit density in comparison to material quality as determined by TR-PLI. In Fig. 7 a detail of the TR-PL image of sample 3 with grain boundary overlay is compared with a microscopic picture of the etched region. A high etch-pit density corresponds to a low lifetime region. This could be the effect of dislocations like discussed in [2].

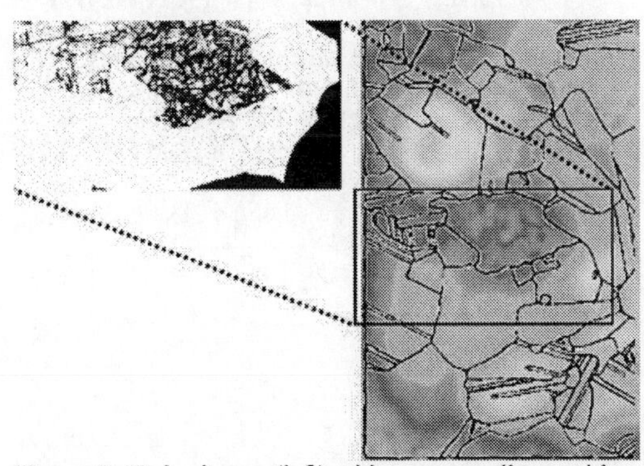

Figure 7: Etch-pit map (left) with corresponding position on a lifetime sample with a grain boundary overlay (right). The wafer broke during processing on right side and black parts in the picture result from missing edges.

4 CONCLUSION

In conclusion, this study gives improved insight in the behavior of high quality HP mc-Si material regarding P-gettering, temperature treatment and different surface passivation processes. Temperature steps above 800 °C are drastically decreasing material quality in most areas. This negative effect cannot be compensated by P-gettering. A slight improvement is observed by SiN_x:H deposition without firing. But altogether prediction of end-of-process material quality remains difficult to predict for HP mc-Si, as not only very high initial τ_{eff} in the as-grown state are important. Instead, final lifetime after process-relevant high temperature steps has to be considered when optimizing the crystallization process.

5 ACKNOWLEDGEMENTS

Part of this work was supported by the German BMWi under contracts 0324204B and 0324001. The content of the publication is the responsibility of the authors. The authors would like to thank M. Fleck, B. Gapp, and J. Rinder for technical support.

6 REFERENCES

[1] International Technology Roadmap for Photovoltaic (ITRPV), 2017 Results (Ninth Edition, March 2018)
[2] P. P. Altermatt *et al.*, Solar Energy (2018) DOI: 10.1016/j.solener.2018.01.073
[3] C. Winter, A. Zuschlag, D. Skorka, G. Hahn, AIP Conference Proceedings 1999 (2018) 130020.
[4] K. Adamczyk, R. Søndenå, G. Stokkan, E. Looney, M. Jensen, B. Lai, M. Rinio, M. D. Sabatino, Journal of Applied Physics 123 (2018) 055705.
[5] M. S. Wiig, K. Adamczyk, H. Haug, K. E. Ekstrøm, R. Søndenå, Energy Procedia 92 (2016) 886.
[6] G. Stokkan, D. S. Marisa, R. Søndenå, M. Juel, A. Autruffe, K. Adamczyk, H. V. Skarstad, K. E. Ekstrøm, M. S. Wiig, C. C. You, H. Haug, M. M'Hamdi, Physica Status Solidi A 214 (2017) 1700319.
[7] H. Sio, D. Macdonald, Solar Energy Materials & Solar Cells 144 (2016) 339.
[8] D. Kiliani, G. Micard, B. Steuer, B. Raabe, A. Herguth, G. Hahn, Journal of Applied Physics 110 (2011) 054508.
[9] F. Secco d' Aragona, Journal of the Electrochemical Society 119(7) (1972) 948.

QUANTIFYING THE IMPACT OF GRAIN BOUNDARIES ON STANDARD AND HIGH PERFORMANCE MC-SILICON SOLAR CELLS

Aleo Paolo Pacho, Beppe Petrelius and Markus Rinio
Department of Engineering and Physics, Karlstad University,
Universitetsgatan 2, 651 88 Karlstad, Sweden

ABSTRACT: Crystal defects such as grain boundaries affect the overall performance of a solar cell. The light beam induced current method allows for the localized quantification of the impact on the internal quantum efficiency of such defects. This work presents a method to estimate the separate impact of grain boundaries on the internal quantum efficiency (IQE) of multicrystalline silicon solar cells by correlating LBIC topographs with optical images of etched samples. Segmenting the impact of the grain boundaries on the IQE against those of other defects in our samples showed that the grain boundaries remain the most detrimental. The average IQE at 826 nm was reduced by up to 1.29 % (vs 0.25 % for other defects) absolute for standard multicrystalline and up to 1.15 % (vs 0.28 % for other defects) absolute for high performance multicrystalline silicon.
Keywords: LBIC, Multicrystalline Silicon, Defects

1 INTRODUCTION

Multicrystalline silicon, having an effective compromise between conversion efficiency and production cost, continues to be the most common material in commercial solar cell production. Inherent to this type of material are crystal defects such as grain boundaries and dislocations which may serve as recombination centers for generated carriers. Efforts towards developing crystallization methods for better ingot quality conventionally aimed to grow mc-Si ingots with larger grains and therefore lower density of grain boundaries in the material. However, recent progress in this area showed that the opposite approach may be better. Material commonly referred to as 'high performance multicrystalline silicon' (HPmc-Si) contains smaller grains and lower density of dislocations [1]. With the emergence of HPmc-Si in the industry, an interesting question arises as to why the presence of more grain boundaries in this material does not deter it from having better solar cell performance.

Light Beam Induced Current (LBIC) mapping is a technique based on measuring the locally induced short circuit current for all points within a region of interest. When properly calibrated and corrected for reflection, these measurements give a map of the internal quantum efficiency (IQE) of the solar cell for a particular wavelength. Such a map allows one to identify regions with high recombination rates. In combination with defect etching and standard image processing techniques, LBIC can give us an insight on the impact of grain boundaries on multicrystalline silicon solar cells in terms of the IQE.

2 EXPERIMENTAL METHODOLOGY

2.1 LBIC measurements

Solar cell samples fabricated from standard and HPmc silicon wafers, obtained from the top and bottom positions of each ingot, were used in this work. The IQE maps at wavelength of 826 nm were obtained using a custom-built LBIC system [2]. IQE maps of full size wafers were obtained at spatial resolution of 250 microns with incident laser power of about 1.8 μW.

2.2 Defect etching and imaging

After the LBIC measurements, the cells were cut into 25 mm x 25 mm samples. 4 samples from each of cell corresponding to the top and bottom positions of a standard multicrystalline and high performance multicrystalline silicon ingot were selected and further processed.

The samples were then prepared using a combination of mechanical and chemomechanical polishing to ensure that they are free from scratches and subsurface deformation. They were then subjected to a 1 min Secco etch [3] to reveal dislocations and grain boundaries. Imaging of the samples was then carried out using an optical microscope with a moveable stage with resolution of about 0.34 μm/pixel.

The acquired array of images for each sample is analyzed using ImageJ [4] to detect the grain boundaries. The relative pixel coordinates of each detected grain boundary is translated into real microscope coordinates and finally transformed into the LBIC map coordinates via a topographical correlation method [5]. Since the actual effect of each grain boundary on the IQE may extend to its immediate surroundings, the grain boundary positions on the IQE maps were taken to be the pixels within a radius of 100 μm from the translated position coordinates. The positions of the contact fingers and busbars were identified using the reflection maps which were also obtained using LBIC. In particular, positions in the IQE maps where the total reflection exceeded 8 % were masked (marked in black in figure 3) and excluded from the analysis. Having identified the LBIC map coordinates of the grain boundaries in the samples, the IQE at those positions are extracted and compared with the IQE at the other positions where defects other than grain boundaries, i.e. dislocations, may be present and at regions where the IQE of the sample was greatest.

3 RESULTS

Figure 1 shows the IQE map at 826 nm of a full size solar cell processed from an HPmc Si wafer. Areas with low IQE and therefore high bulk and back surface recombination can be identified in such a map. Also shown in this figure is the region where one of the 25 mm x 25 mm samples was taken for etching and grain boundary detection.

IQE:

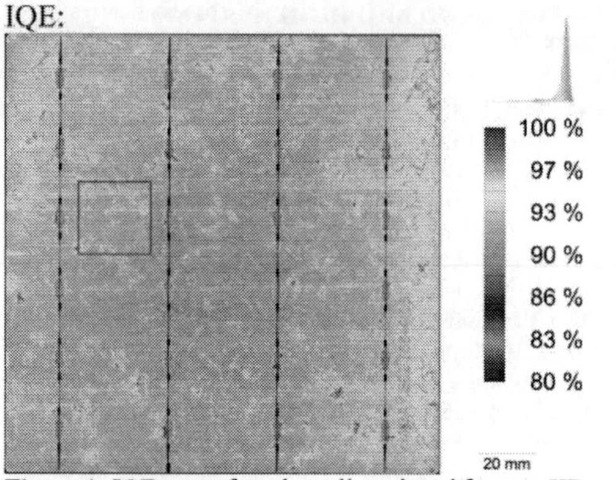

Figure 1: IQE map of a solar cell produced from an HPmc Si wafer. The region marked by a square indicates the position of one of the samples taken from this cell for further processing and analysis.

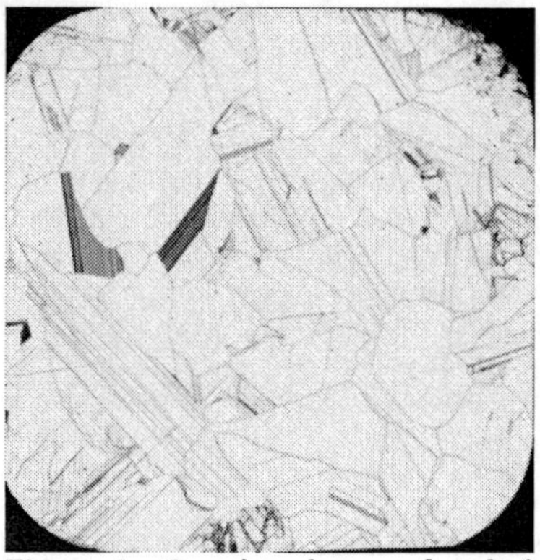

Figure 2: Overview image of etched sample. The brightness and contrast of the image shown here was adjusted to improve visibility.

Figure 2 shows an image of the etched sample taken from the region shown in figure 1. This overview image was stitched from about 4900 microscope images, scaled down and adjusted to be suitable for viewing here. In the actual image anlaysis the individual high resolution microscope images were used and not this scaled down version.

Figures 3A – 3C show a typical result of the IQE and grain boundary position matching. Figure 3A shows the IQE map of the entire region marked in figure 1. A defect free region in the sample with the greatest average IQE is marked by a small blue square. Figure 3B shows the same IQE map but with the positions of the grain boundaries marked in white. We observe in this image that almost all areas with relatively lower values of IQE have been masked. Figure 3C shows the IQE at the positions of the grain boundaries. Again we may note that the regions with the worst values of the IQE are those that correspond to grain boundary positions. However, we also see that not all these positions have a low IQE. While it is well known that not all grain boundaries are recombination active [6,7], we make no distinction in different types or orientations of grain boundaries in this work. Instead, the impact of all detected grain boundaries on the IQE at 824 nm is estimated and compared with the combined impact of the all other defects that may be present in the sample.

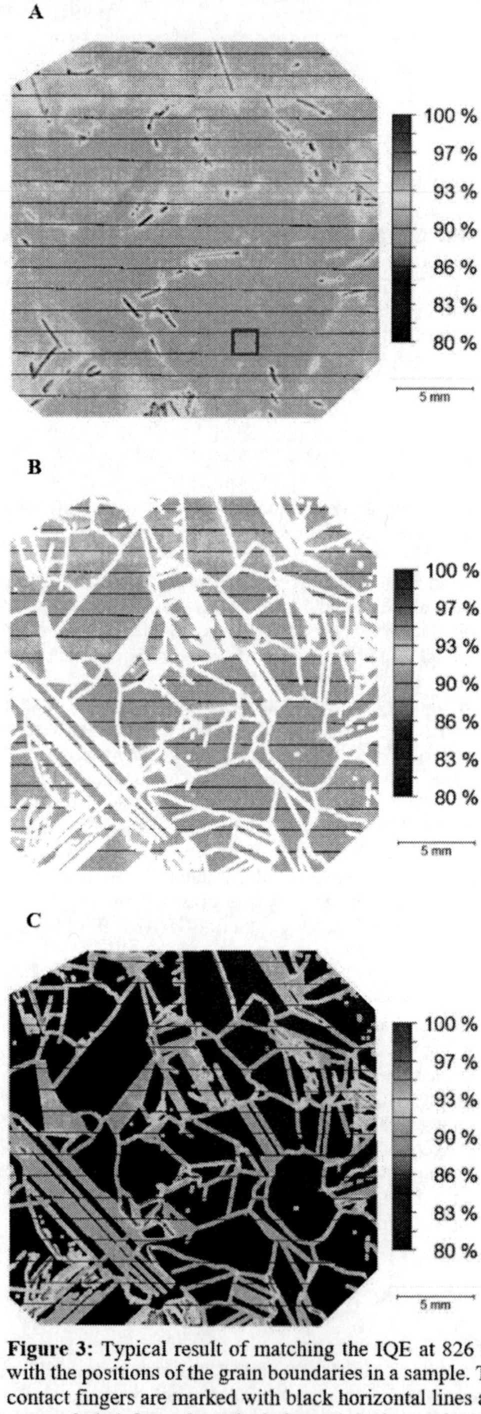

Figure 3: Typical result of matching the IQE at 826 nm with the positions of the grain boundaries in a sample. The contact fingers are marked with black horizontal lines and are excluded from the calculations. (A) shows the IQE map of the entire sample. The small region enclosed by the blue square is an area with the highest average IQE. (B) shows the same IQE map but with the positions of the grain boundaries marked in white. (C) shows the IQE at the positions of the grain boundaries.

The absolute reduction in IQE due to grain boundaries is calculated as the difference between:

(1) the expected average IQE if the entire cell was free from defects, which is taken to be the average IQE in the defect free region marked in figure 3A, and
(2) the average IQE of the cell if all sites other than those of grain boundaries (marked in black in figure 3C) are filled with points with IQE equal to the average IQE of the defect free region.

The absolute reduction in IQE due to defects that may be present in all other positions are calculated in a similar way, replacing the values of the grain boundary sites with the average IQE of the defect free region. In this manner, the comparison between the impact of grain boundaries and the combined impact of all other defects inherently incorporate both the area covered by the defects and the recombination activity in the corresponding areas.

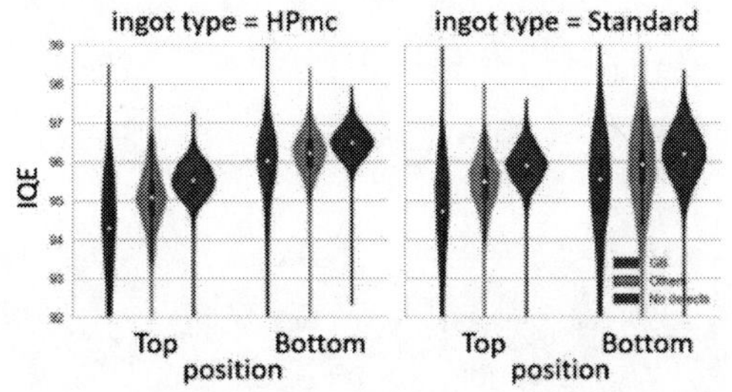

Figure 4: Internal quantum efficiency distributions for the positions of grain boundaries, other defects and the regions with the greatest IQE.

			Ave IQE
HPMC	Top	GB	93.29
		Others	94.94
		No defects	95.52
	Bottom	GB	95.38
		Others	96.09
		No defects	96.47
Standard	Top	GB	93.56
		Others	95.33
		No defects	95.90
	Bottom	GB	94.45
		Others	95.31
		No defects	95.92

Table I: Average values of the segmented internal quantum efficiency distributions shown in figure 4.

Figure 4 shows the distribution of the segmented IQE measurements. In this figure, the width of each shape at each IQE value corresponds to the frequency of the measurements such that each shape can be viewed as a histogram of the IQE values for the corresponding distribution. The thick black bars represent the interquartile range, the thin black lines represent the 95 % confidence interval and the white dots are the respective medians of the distributions. The average values of the distributions are tabulated in table I. We observe that the median of the internal quantum efficiency distributions at the positions of the grain boundaries are always lower than the median of the IQE distributions at the other positions where we may have other defects such as dislocations. This is despite the fact that we did not make a distinction between recombination active and inactive grain boundaries. Furthermore, this observation holds true for samples from both standard multicrystalline and high performance multicrystalline silicon ingots, regardless of whether the sample was from the bottom (about 3 mm from the bottom of the ingot) or the top of the ingot (about 3.6 mm from the top of the ingot).

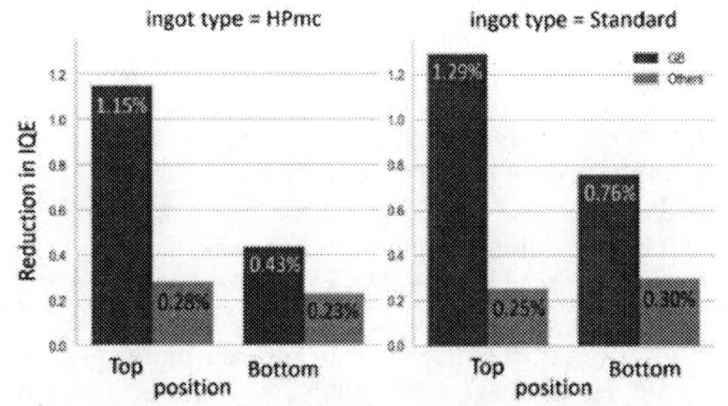

Figure 5: Absolute reduction in average internal quantum efficiency due to grain boundaries and other defects.

As mentioned earlier, all grain boundaries detected in the etched samples were marked and included in the segmentation of the IQE distribution regardless if they were recombination active or not. As a result we see a wider spread in the IQE distributions for the grain boundary positions as compared to the areas without grain boundaries. We also note that the absolute reduction in the average IQE due to the grain boundaries are larger compared to the reduction due to all other defects as depicted in figure 5. The reduction of the IQE due to grain boundaries is greatest for samples near the top of the ingot. The grain boundaries reduce the average IQE of standard multicrystalline samples by up to 1.29 % (vs 0.25 % for other defects) and at the bottom by up to 0.76 % (vs 0.30 % for other defects) absolute. For the high performance multicrystalline samples, the reduction of IQE due to grain boundaries at the top is 1.15 % (vs 0.28 % for other defects) and at the bottom is 0.43 % (vs 0.23 % for other defects) absolute.

4 CONCLUSION

A method to quantify the separate impact of grain boundaries on the average internal quantum efficiency of mc silicon solar cells has been presented. The segmentation of the grain boundaries from the other crystal defects in our samples showed that grain boundaries are more detrimental to the IQE at 826 nm compared to the combined impact of all other defects. The results suggest that grain boundaries in multicrystalline silicon can degrade the cell performance due to their recombination activity and must therefore be accounted for when designing crystallization techniques. The method presented in this work can be extended to account for the other defects such as dislocations and can be used to observe how defects on different solar cells from different crystallization processes affect the overall IQE.

5 ACKNOWLEDGEMENTS

This work was supported by the Solar-Era.Net project "HighCast" via the Swedish Energy Agency under project number 40184-1.

6 REFERENCES

[1] Y. M. Yang, A. Yu, B. Hsu, W. C. Hsu, A. Yang, and C. W. Lan. Development of high-performance multicrystalline silicon for photovoltaic industry. Progress in Photovoltaics: Research and Applications, 23(3):340–351, 2015. PIP-13-141.R1.

[2] M. Rinio, H. J. Möller, and M. Werner, Solid State Phenom. 63–64, 115 (1998).

[3] F.Secco d'Aragona. Dislocation etch for (100) planes in silicon. Journal of Electrochemical Society, 119(3), 1972.

[4] Johannes Schindelin, Ignacio Arganda-Carreras, Erwin Frise, Verena Kaynig, Mark Longair, Tobias Pietzsch, Stephan Preibisch, Curtis Rueden, Stephan Saalfeld, Benjamin Schmid, et al. Fiji: an open-source platform for biological-image analysis. Nature methods, 9(7):676, 2012.

[5] M. Rinio, A. Hauser, and H. J. Moller. Topography correlation - a powerful tool applied to the visualisation of remote plasma hydrogen passivation effects. In Photovoltaic Energy Conversion, 2003. Proceedings of 3rd World Conference on, volume 1, pages 118–121 Vol.1, May 2003.

[6] G. Stokkan, S. Riepe, O. Lohne and W. Warta. Spatially resolved modeling of the combined effect of dislocations and grain boundaries on minority carrier lifetime in multicrystalline silicon. Journal of Applied Physics 101(5), 053515, 2007

[7] Krzysztof Adamczyk, Rune Søndena, Gaute Stokkan, Erin Looney, Mallory Jensen, Barry Lai, Markus Rinio and Marisa Di Sabatino, Recombination activity of grain boundaries in high-performance multicrystalline Si during solar cell processing. Journal of Applied Physics 123, 055705, 2018.

An efficient optimized RTP process to minimize light induced degradation phenomenon and their effect on surface roughness in p-type Cz-Si wafers

Y.Kouhlane[a], D. Bouhafs[a*], N. Khelifati[a], S. Mezghiche[a], A. Guenda[a], W. Hetatache[b], F. Derkaoui[c]

[a)]Research Center in Semiconductor Technology for the Energetic (CRTSE), 02,BP.140, les sept mMerveilles 16038, Algiers, Algeria.
[b)]Faculty of Sciences, University of Sétif, El Bez 19000, Sétif, Algeria.
[c)]Faculty of Technology, University of Blida, Ouled Yaïch 09000, Blida, Algeria.
*Corresponding author: **bouhafsdjoudi@crtse.dz**

ABSTRACT: In this study, the carrier lifetime variation of p-type boron-doped Czochralski silicon (Cz-Si) wafers was investigated after a direct rapid thermal processing (RTP). Two wafers were passivated by silicon nitride (SiNx:H) layers, deposited by plasma-enhanced chemical vapor deposition (PECVD) system on both surfaces. Then the wafers were subjected to an RTP cycle at a peak temperature of 620°C. The first wafer was protected (PW) from the direct radiative heating of the RTP furnace by placing the wafer between two as-cut Cz-Si shield wafers during the heat treatment. The second wafer was not protected (NPW) and followed the same RTP cycle. The carrier lifetime τeff was measured by the quasi-steady-state photoconductance (QSSPC) technique, before and after an illumination step at 0.5 suns for 5 hours duration. The immediate results of measured lifetime τ_{RTP} after the RTP process have shown regeneration in the lifetime for the two wafers, but the PW wafer exhibited an important enhancement in τ_{RTP} compared to the NPW wafer. The QSSPC measurements after 5h of illumination have indicated a good stabilization of lifetime (τd) and a weak degradation effect was observed for the PW wafer compared to their initial lifetime. Additionally, to improve the correlation between the RTP heat radiation stress and carrier lifetime behavior, a simulation of the thermal stress and temperature profile by the finite element COMSOL software on the wafers surface at RTP peak temperature of 620°C was performed. The results confirm the reduction of the thermal stress and less heat loss for the protected wafer. Finally, the proposed method can lead to improving the lifetime of wafers by an RTP process with minimum energy costs.

Keywords: c-Si, Light-induced degradation, Rapid Thermal Processing, Carrier lifetime.

1. INTRODUCTION

RTP provides a suitable controlled environment for thermally activated processes. In addition, RTP is fully integrated into the solar cells production. However, an unwanted side effect of rapid thermal processing technique is the potential to generate large thermal stresses in the wafer. If the thermal stresses are large enough, crystallographic dislocations can occur, caused by temperature non-uniformity on the wafer [1]. These dislocations can be very harmful to the electrical proprieties of the treated wafers. In this work, two boron-doped Cz-Si wafers treated in an RTP furnace at a peak temperature of 620°C, were studied. Also, the effect of radiation shield during the RTP process on the stability of the carrier lifetime before and after light-induced degradation (LID) steps, were systematically investigated using the quasi-steady-state photoconductance (QSSPC) method. Furthermore, a simulation using COMSOL software was done to predict the thermal stress and temperature profile on the studied samples. Finally, a possible way to minimize the thermal stress during the RTP process and enhance the carrier effective lifetime are discussed.

2. EXEPERIMATEL SETUP

The material used for this investigation was boron-doped Cz-Si wafers of 380 µm thick, surface of 100 mm^2 and resistivity ρ between 1.5 to 3 Ω.cm. and an interstitial oxygen concentration [Oi]= $6x10^{17}cm^{-3}$. The wafers were etched in NaOH solution to remove saw damage. After etching, all wafers were cleaned in a standard RCA

solution. In order to evaluate the effect of hydrogen in the regeneration of the wafers carrier lifetime; an 80nm thick silicon nitride (SiN_x:H) layer was deposited via a PECVD (SEMCO) system onto both front and rear surfaces of all the samples. After deposition of the SiN_x:H passivation layer, two samples underwent a short temperature cycle process in an RTP furnace (AccuThermo AW610). The RTP cycle started with a temperature ramp from 25°C to 500°C, where the firing took place to peak temperature of 620°C. Then, the temperature was reduced to room temperature, where the furnace was shut down. In order to investigate the effect of the RTP process on wafers, one of the samples was isolated from the direct radiative heating of the RTP furnace by two as-cut Cz-Si shield wafers (Fig. 1) and denoted as PW. The other sample (NPW) was not protected and follows the same RTP cycle. The effective minority carrier lifetimes of wafers are measured using a WCT-120 tester before and after a light exposure under a 500Watt halogen lamp during 5 hours.

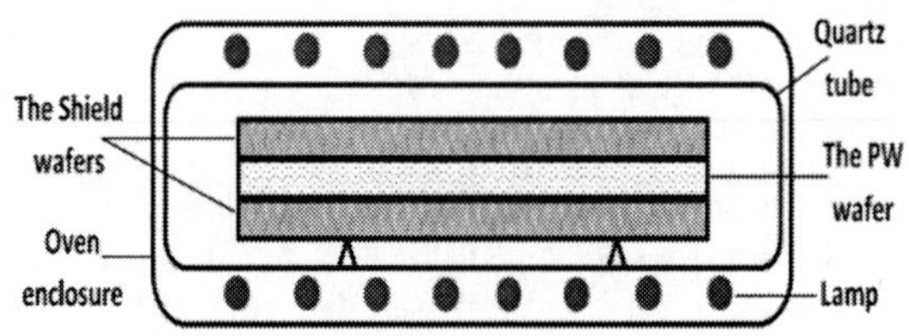

Fig.1: Diagram showing the method used to protect the wafer during the RTP process.

3. PHYSICAL MODEL

To simplify the numerical analysis, a single-sided modeled heating apparatus was built using COMSOL software based on the work of A.T. Fiory [2].

The model simulates the lamp as a solid object with a volume heat source of 25 kW. It is insulated on all surfaces except for the bottom, which faces the silicon wafer; an additional wafer can be added to create the shield between the heat source and the PW wafer.

At the surface, heat leaves the lamp as radiation only. The equation of the heat conduction is given as follows:

$$\rho C_p \frac{\partial T}{\partial T} + \nabla.(-k \nabla T) = Q \qquad (1)$$

Where ρ is the density, C_p is the heat capacity, k is the thermal conductivity, and Q is the heat source per unit volume as a function of position and time. Also, the convection and radiation into the ambient were given through the equation:

$$-n.(-k \nabla T) = h(T_{inf} - T) + \frac{\varepsilon}{1-\varepsilon}(J_0 - \sigma T^4) \qquad (2)$$

Where n is the surface normal vector; T_{inf} equals the temperature of the convection cooling gas; ε denotes the surface emissivity; J_0 is the expression for surface radiosity and σ is the Stefan-Boltzmann constant.

The components are contained in a chamber with temperature-controlled walls with a set point of 25°C. These results in a closed cavity and the geometry of the chamber walls can be neglected. Furthermore, the model assumes that this physical system is dominated by radiation and convection cooling. The model simulates the transient temperature field for 10s of heating to 620°C. The initial temperature is 25°C for all objects.

4. RESULTS AND DISCUSSION

Effective lifetime measurements were performed using QSSPC technique and denoted as τ_0 before thermal treatment, τ_{RTP} directly after RTP and (τ_d) after illumination steps of 5h under a halogen lamp of 0.5 suns energy. The lifetime values given in Figure 2 were extracted at a fixed injection level equal to $\Delta n = 10^{16}$ cm^{-3}. The τ_{RTP} lifetime increased for both wafers (Fig. 2). The immediate augmentation after the RTP process is due to the additional hydrogen diffusion in the wafer bulk from the SiN_x:H layers [3]. It was observed as an important enhancement in the value of τ_{RTP} for the PW wafer. Good temperature uniformities across the PW wafer can lead to enhancing the hydrogen passivation in the RTP process. Furthermore, the degradation of τ_d is pronounced for the NPW wafer. The degradation can be linked to the presence of a high concentration defect such as dislocations generated from the heat stress during the RTP process [1].

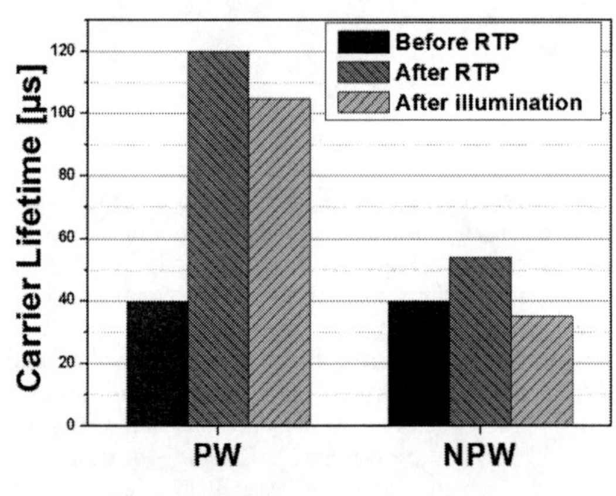

Fig.2: QSSPC lifetime results for PW and NPW wafers.

Simulation results of the principal stress (von mises stress) σv profiles on the surfaces of silicon wafers in the modeled RTP furnace is given in Figure 3. The maximum stress is located at the edge of the wafers; this is in agreement with previous experimental results in RTP process [1]. Furthermore, the PW wafers show clearly a reduction in the stress compared to the NPW wafers. The probability of sleep and dislocation formation that create a new sites for carrier traps is less important for the shielded wafer, which can partially explain the significant carrier lifetime regeneration detected by QSSPC system for the PW wafers.

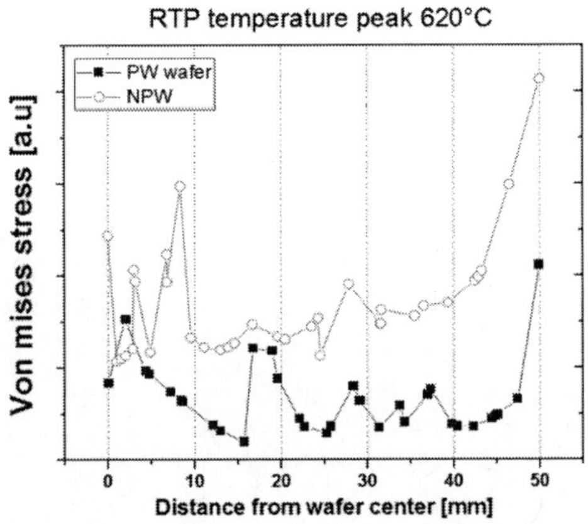

Fig.3: Simulation of the principal stress (von mises stress) σv, at temperature peak of 620°C on the surface of NPW and PW wafers.

Figure 4 illustrates the simulated temperature distribution on the silicon wafers surfaces at RTP peak of 620°C. Experimental results of temperature profile deviation in RTP process between the center and the edge of the wafer was reported previous by H.A. Lord [1]. Moreover, the visible enhancement of edge's temperature in the PW wafer (Fig. 4) can consolidate the assumption that the hydrogen optimized diffusion and bulk passivation in the PW wafer can reduce the light induced degradation generated by boron-oxygen related defect.

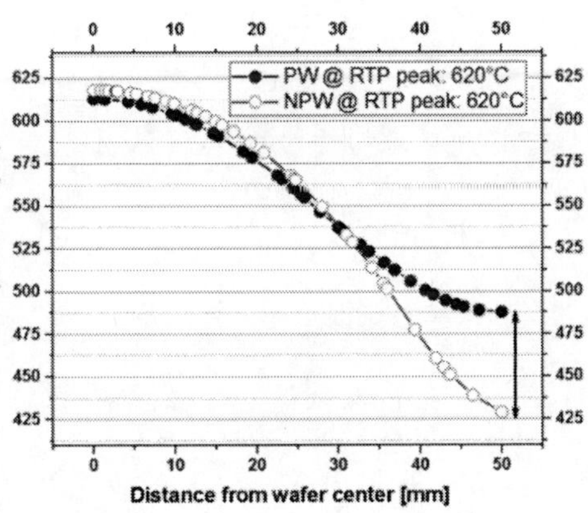

FIG.4: Simulation of the temperature distribution, at RTP peak of 620°C on the surface of wafers.

The experimental silicon surface profiling results obtained from the interferometer microscopy are illustrated in figure 5 in which we can observe the variation in the surface morphology of the treated samples by the RTP process as compared to the sample reference one. The roughness (Ra) factor, extracted from the surface profiling, is resumed in Table 1. The value of Ra is slightly higher for the NPW wafer in comparison with that of the PW wafer. This can be due to the shield protection as well as to a uniform temperature distribution on the surface of the protected wafer. The variation in Ra can affect the wafer surface quality and increase the density of surface dangling-bond states and contribute to increasing the electrical degradation of the wafer.

Table 1: Measured wafer surface roughness (Ra)

	Ref. Wafer	NPW	PW
Ra [nm]	264.5	333.9	329.4

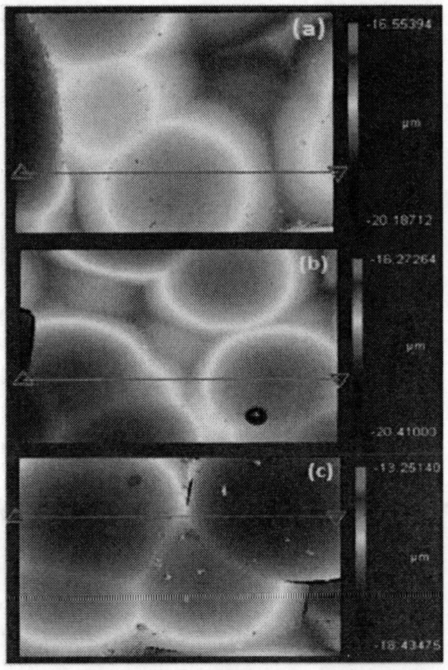

Fig.5: Interferometry surface mapping of the samples for reference wafer (a), protected RTP processed wafer (b) and RTP non-protected wafer (c).

5. CONCLUSION

The study of the RTP thermal stress impact on the carrier lifetime degradation has been carried out. Carrier lifetime shows an important enhancement after RTP process for the wafer with a shield protection. In addition, lifetime results after prolonged illumination present a good stability for PW wafer. These results can be related to the reduction of thermal stress for the PW wafer, leading to an optimization of the hydrogen passivation and reduction of defect concentration in the RTP process. Furthermore, numerical modeling confirms the reduction of heat stress and more homogeneity in temperature on wafers with shield integration in the RTP process. Finally, combining the numerical results with experimental observations reveal that it is possible, to improve the carrier lifetime by protecting the Cz-Si wafer from the direct heat radiation stress in the RTP process. However, further investigations are necessary, especially varying the shield physical proprieties. Also, experimental analysis of the stress impact and temperature profile in the wafers must be carried out to verify these findings.

Acknowledgments

This work is realized at the DDCS-Division/CRTSE and financed by the National Fund for Research DGRSDT/MESRS_Algeria.

REFERENCES:

[1] H.A. Lord, "Thermal and stress analysis of semiconductor wafers in a rapid thermal processing oven", IEEE Transactions on Semiconductor Manufacturing 1 (3), 105-114 (1988).

[2] A.T. Fiory, "Methods in Rapid Thermal Annealing", Proceedings of RTP 2000, 8th International Conference on Advanced Thermal Processing of Semiconductors, Gaithersburg, USA, 2000, pp. 15-25.

[3] G. Krugel, W. Wolke, J. Geilker, St. Rein, R. Preu, "Impact of hydrogen concentration on the regeneration of light induced degradation", Energy Procedia 8, 47–51(2011).

THE PERFORMANCE OF CAST MONO WAFER, CELL AND MODULE

Xusheng Wang*,1 Shuai Zhou, Yuanyuan Gu, Guoqiang Xing
Thierry Galvez[2]
1. Canadian Solar Inc., 199 Lushan Road, SND, Suzhou, China, 215129
2. Photowatt, 33 rue Saint-Honoré, Bourgoin-Jallieu, France
*Corresponding Author: Xusheng WANG, Xusheng.wang@canadiansolar.com

ABSTRACT: Cast mono (quasi mono or mono-like) material is gaining attention in photovoltaic industry due to its potential to reach high efficiency close to p-type mono while keeping the cost level close to multi. Nevertheless, there still remain challenges to industrialize the technology. One is how to control the brick quality, to keep the mono percentage as high as possible, another is how to treat the outlier wafer during cell texturing which will result in very shiny parts and be rejected by the end customers. The cost should also be considered nowadays as photovoltaic is a cost driven industry. The good news are we developed a suitable casting process which the mono percentage reaches 90% and we are using metal catalyzed chemical etching (MCCE) texturing process to create black silicon for the outlier wafer. The solutions we carried out paves the way for the cast mono technology. G5 cast mono brick is grown using ECM Crystalmax furnace by a special seed arrangement and corresponding recipe. The brick quality is greatly improved due to lack of grain boundaries (GB). The defect multiplication is also carefully controlled. The brick is squared, cropped, polished and cut into wafers using diamond wire sawing (DWS). The wafers are sorted by the center, the edge and the corner location to study different behavior. For the center wafers, as it is very close to mono, the alkaline texture and standard PERC process is applied. The cell efficiency reaches 20.9% average due to better lifetime. For the wafers from the edge and the corner, the MCCE texture and standard PERC process is using. The efficiency reaches 20.2% for edge and 19.7% for the corner due to lower light trapping. The cells have very uniform outlook without any shiny parts thanks to the black silicon treatment on the surface. Modules were made using the produced center cells. Max module power output reaches 370 watts using 144 half-cell designs. The power is quite promising which makes the product competitive on the market. Characterization from brick to wafer, cell and module were also carried out in this paper to verify the performance and give possible explanations.
Keywords: cast mono, dislocation, grain boundary, MCCE

1 Introduction

High power output and low cost are the key competency for photovoltaic industry. Mono-Si product is normally rated high power and dominates the high-end application market such as rooftop while multi-Si product is the mainstream of the field application due to low cost. Recently, several technical solutions are carried out to greatly improve the multi-Si cell performance, such as PERC (Passivated Emitter and Rear Cell) technology contributes 1% absolute efficiency increase in the mass production of multi-Si solar cell. 20% average efficiency for multi-Si cell is common for leading PV manufacturers. Of course, the multi-Si wafer still keeps the low manufacturing cost level thanks to the casting technology (directional solidification), but the method itself also restrains the multi-Si cell efficiency limit due to grain boundaries and large array of dislocations existed. Nowadays, the cast mono Si grown by the casting technique is deeply attracting the PV companies. The Si ingot quality will be obviously improved due to less grain boundaries and lower dislocation during the crystal growth by this approach [1.2]. Thus the cell efficiency can be estimated to increase to an even high level, which may be close to mono-Si. The cast mono wafer owns a very unique visual outlook due to large area of <100> crystal area, which is close to mono-Si wafer. Regarding the texture process in cell fabrication, the MCCE (Metal Catalyzed Chemical Etch) is more suitable rather than the conventional alkaline treatment for mono-Si wafer [3]. It successfully covers the shiny parts on non <100> area while still keeps a low reflection on light trapping. Combining with the low dislocation, the cell efficiency can reach as high as 20.9% average.

The objective of this article is to describe a detailed method from ingot growth to bricks selected, then to wafer, cell and module fabrication, which provides you with attracted results using cast mono Si wafer.

2 Process flow and experiment

2.1 Ingot:

Cast mono G5 ingot with dimensions of 855mm*855mm*360mm, and the weight of 500kg, was produced by directional solidification using ECM furnace. The bottom seed is specially laid comparing with the standard mc-Si. <100> thin plate Si seeds are placed flatly on the crucible bottom, with the top filled with the poly Si feedstock. A detailed thermal field temperature control recipe is used to precisely control the growing procedure.

The ingot was cut into the bricks according to the lifetime mapping.

2.2 Wafer:

The bricks are polished and cut into wafers with a square area of 156.75mm*156.75mm, and a thickness of 180 μm by DWS (diamond wire sawing). State of the art 70um diamond wire is using in the process to guarantee the surface quality.

2.3 Cell:

All the wafers are processed into cells using optimized PERC recipe. The main steps of cell are shown in Table I.

Table I: Main steps in cell process.

MCCE to form inverted-pyramid texture
Phosphorus diffusion to form a p-n junction emitter
Wet etch to remove edge, back pn junctions and PSG layer
ALD Al_2O_3 is applied as rear-surface passivation dielectric layer
PECVD SiN_X is applied as rear-surface capping layer
PECVD SiN_X is applied as front-surface passivation dielectric and AR layer
Locally open the back Al_2O_3/SiN_X stack by laser ablation
Screen printing to fabricate the rear/front electrodes, back surface field and sintering

2.4 Module:

Cells selected from different bricks are fabricated into modules with 144 half-cell design. Table II gives the process flow of the module fabrication.

Table II: Main steps in module process.

Laser cutting: preparation of half-cut cells
High precision half-cell automatic soldering
Layout & soldering: completion of low series resistance ribbon connection
Lamination: completion of cell encapsulation
Frame up: strengthening the module to resistant wind & snow load
Set up junction box: power output & circuit protection
EL & power Testing： ensure module intact & nominal component power

2.5 Characterization:

Various characterizations from brick to wafer, cell and module are carried out during the product processing.

3 Results and discussions

3.1 Brick results:

The top surface of the G5 ingot has a mono percentage as high as 90%, see Figure 1. There are almost no grain boundaries on the surface except the border area. We pick up part of the bricks from the G5 ingot for benchmark. Their locations shown by Fig1 are carefully chosen to reflect the whole ingot quality. B3, C3, D4 come from the center, while A4, C1 come from the edge and A5 comes from the corner.

	1	2	3	4	5
A				A4	A5
B			B3		
C	C1		C3		
D				D4	
E					

Figure 1: Location of bricks chosen for benchmark.

The resistivity for all the bricks is well controlled from 1.09ohm · cm on the bottom to 1.62 ohm · cm on the top illustrated by Figure 2. There is no much difference on the resistivity among the brick locations. The resistivity range is perfect for PERC cell structure.

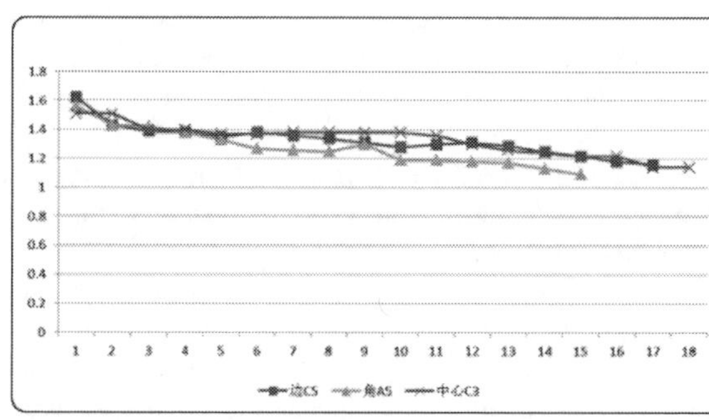

Figure 2: The resistivity from the top to the bottom of bricks from different locations.

The oxygen and carbon contents are important factors for PV products. The oxygen dominates the LID (Light Induced Degradation) performance on cell and module level, while carbon is more relatively connected with the reverse current behavior on cell. Table III gives the number we measured from the top and bottom of bricks from different locations. Oxygen content is little higher comparing with HPM (High Performance Multi), while the carbon content keeps on the same level. The cast mono process needs to improve on the oxygen content control.

Table III: O and C contents from different brick locations.

Brick	Location	O/10^17	C/10^17
C5	Top	0.8665	5.5633
	Bottom	1.9575	1.5819
A5	Top	1.1442	4.7317
	Bottom	2.2864	1.4418
C3	Top	1.0479	6.1315
	Bottom	1.8612	1.4498

MCLT (Minority Carrier Life Time) is another key characteristic of the brick. It determines the potential device performance. Figure 3 reports the mapping results for the bricks from the edge, the corner and the center (Left to Right). The average MCLT are 5.4 μs for the edge, 5.35 μs for the corner and 5.95 μs for the center. The mapping pictures show a good relevance with the MCLT values. There are lots of defects as dislocations and/or impurities on the edge and the corner comparing with the center part which leads to lower MCLT.

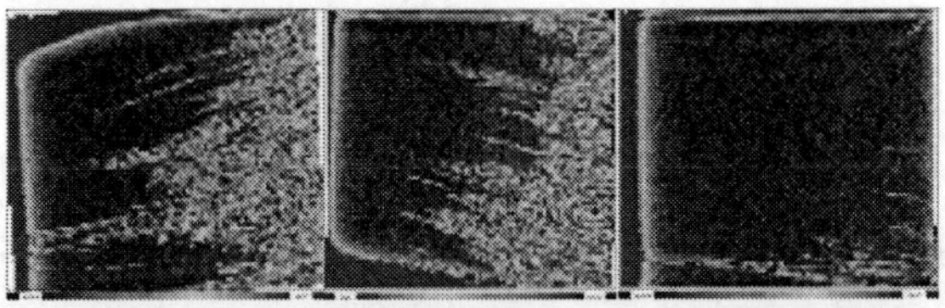

Figure 3: MLCT mapping for the bricks from the corner, the edge and the center.

IR (Infra-Red) detections were also carried out to check the internal brick quality. Dislocations are clearly observed in Figure 4 from the brick on the left (the corner) and the middle (the edge), while the right (the center) is perfectly good. The results reflect the same quality trend with the MCLT mapping. The center part has the best quality, while the edge and corner are much worse due to dislocation growth in the casting process.

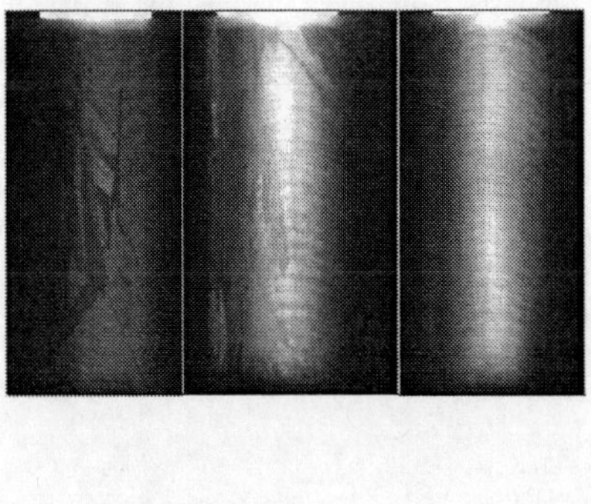

Figure 4: IR detection of the brick from the corner, the edge and the center.

We measured the brick hardness before DWS (Diamond Wire Sawing). It seems that the cast mono brick is much harder than the traditional HPM brick. It needs more wire and more time during DWS which results in higher manufacturing cost.

3.2 Cell Results:

All the bricks chosen were sawed into wafers using the diamond wire. Cells are processed using the optimized PERC recipe. Figure 5, 6, 7 report the efficiency distribution from the corner, the edge and the center bricks.

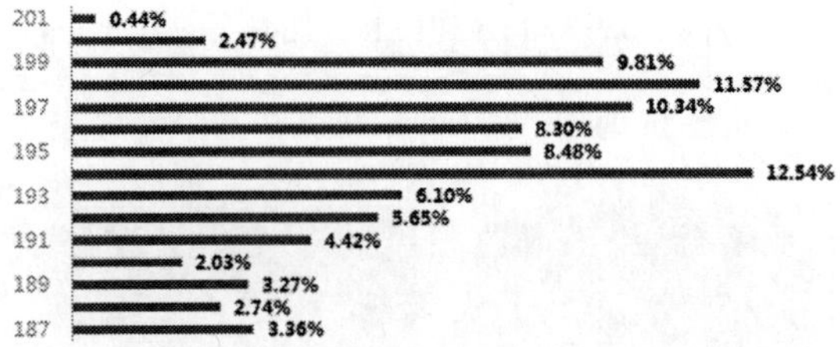

Figure 5: Cell efficiency distribution on the corner brick.

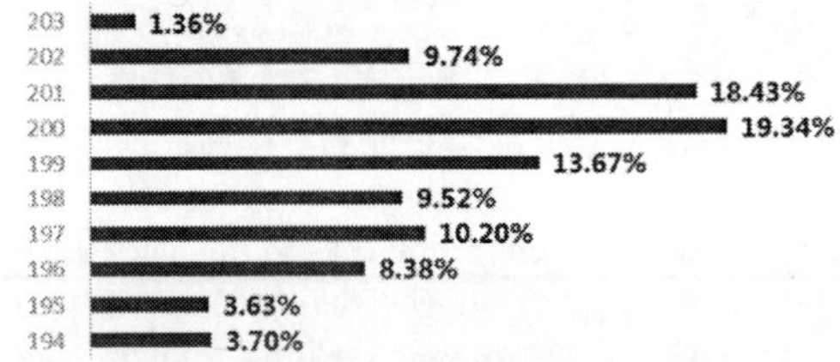

Figure 6: Cell efficiency distribution on the edge brick.

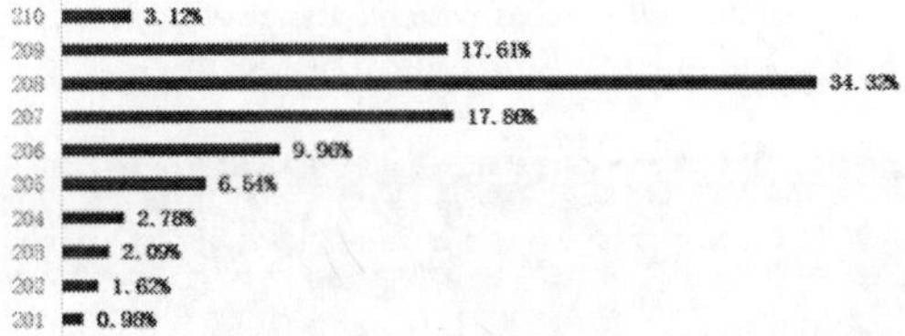

Figure 7: Cell efficiency distribution on the center brick.

The distribution is much tighter for the center brick, while the edge brick is scattering. The worst is the corner brick which has multiple peaks and a long tail because of the much more dislocations. Efficiency loss in the corner bricks is mainly caused by structural defects: grain boundaries, and dislocation which introduce a large incorporation of impurities. It is know that impurities such as Ti, Cu, Cr and Fe are originated from the crucible and the silicon nitride coating. It was seen that seed junction is a source of defect, and the density of these defect is increasing from bottom to the top of bricks during the growth. There is also huge difference from the average efficiencies. The center brick reaches an average of 20.94%, the edge brick reaches 20.19% and the corner only has a value of 19.66%. Considering the actual numbers of these bricks in a whole ingot, the estimated average cell efficiency for the whole ingot will be 20.38%. The reported cell efficiency is higher than HPM PERC, but still far away from mono PERC. Most of the low efficiency cell has lots of non-(100) orientation grains and grain boundaries which can be clearly seen on the cell surface due to texture treatment. It coincides well to the cell EL picture, as shown in Figure 8. Dislocations on the boundaries and inside grains limit the cell efficiency. LID on the cell level is also measured. The average value is around 2-3%, which is higher than mono and HPM PERC. Higher oxygen content on the brick level may be the reason and the casting process needs to be further improved.

Figure 8: Cell outlook and EL picture.

The cell outlook is uniform thanks to the black silicon texture process. Figure 9 gives the typical cell visual picture using MCCE treatment for the brick from the center, the edge and the corner. Shiny part will be seen if using alkaline treatment due to reflection difference from different grain orientations. MCCE helps the cast mono to find a way in the cell process which balances the efficiency and the visual outlook.

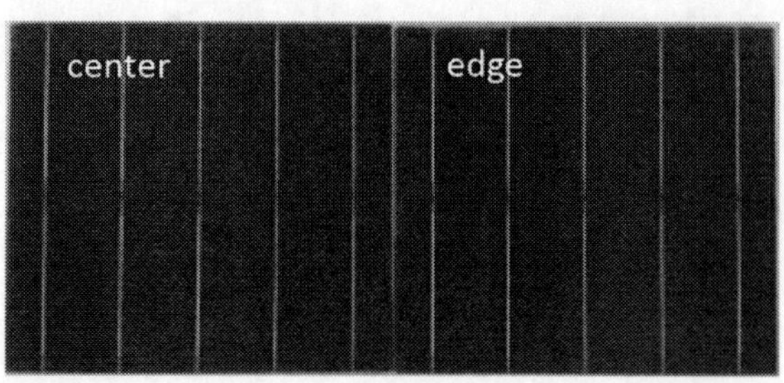

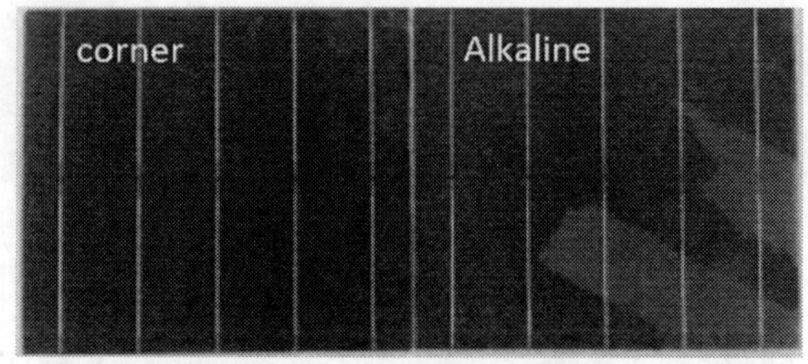

Figure 9: Cell outlook using black silicon from the center, the edge, the corner and alkaline treatment.

3.3 Module results:

We choose the cells from the main bins of the all the bricks to make standard 144 half-cell modules. Average module power output is 367.7 watts for 20.5% cell from the center, 364.7 watts for 20.0% cell from the edge and 361.8 watts for 19.6% cell from the corner as reported by Table IV.

Table IV: Module power output for main bins from the center, the edge and the corner.

	Main bin	Power output
center	20.5%	367.7W
edge	20.0%	364.7W
corner	19.6%	361.8W

The module power output lies between the standard mono PERC and the HPM PERC. The values are very promising. The product will be competitive if the manufacturing cost is controlled well in mass production.

4 Conclusions

Cast mono ingot is grown by a special developed furnace and a carefully tuned recipe. Then the bricks are sliced using DWS and all the wafers are processed into cells and finally modules are made. Cell efficiency and module power output are measured. Related characterizations are carried out and reasonable explanations are given. Cast mono seems to be a promising technology in the near future due to high performance between mono and HPM while keeping the manufacturing cost close to multi. The performance/cost ratio will be higher than the current existing product. MCCE black silicon process helps cast mono to solve the visual problems on cell and module level which get rid of the hurdles on product marketing. Nevertheless, cast mono still needs a long way to go into industry. The cell efficiency distribution shall be strictly controlled. This leads to further R&D works on the ingot growing process and/or furnace design. Unless we controlled the process on the origin of the manufacturing chain, cast mono product can be a good choice for the future potential PV users.

References

[1] Sergey Beringov, Timur Vlasenko, Sergiy Yatsuk, Oleksandr Liaskovskiy, Iryna Buchovska," *MONO-LIKE INGOT/WAFERS MADE OF SOLAR-GRADE SILICON FOR SOLAR CELLS APPLICATION*", *27th European Photovoltaic Solar Energy Conference and Exhibition.*

[2]C.Y. Lan, Y.C. Wu, A. Lan, C.F. Yang, C. Hsu, C.M. Lu, A. Yang, C.W. Lan, "*Control of ingot quality and solar cell appearance of cast mono-like silicon by using seed partitions*",*Journal of Crystal Growth 475(2017)136-143.*

[3] YuChen Wu, MingYen Chen, HsiHao Huang, PoSheng Huang, and LiWei Chen," *Analysis of Light-induce Degradation on Mono Cast Cells", 28th European Photovoltaic Solar Energy Conference and Exhibition.*

OPTIMIZATION OF SURFACE PASSIVATION FOR BLACK SILICON BASED ON THERMAL OXIDATION

Shude Zhang[1,2], Hongqiang Qian[2], Jiaqi Peng[2,3], Jiansheng Jie[1]*, Xiaohong Zhang[1], Qingzhu Wei[2,3]*, Zhichun Ni[2,3]
[1] Institute of Functional Nano & Soft Materials (FUNSOM), Jiangsu Key Laboratory for Carbon-Based Functional Materials & Devices, Soochow University, Suzhou, Jiangsu, 215123, P. R. China
[2] Suzhou Talesun Solar Technologies Co., Ltd., Changshu, Jiangsu, 215542, P. R. China
[3] Nanjing University of Aeronautics & Astronautics, Nanjing, Jiangsu, 210016, P. R. China
* Email: jsjie@suda.edu.cn; qingzhu.wei@talesun.com

ABSTRACT: Because of the large emitter specific surface area of black silicon solar cell, conventional silicon nitride coating deposited by PECVD can't provide conformal passivation, which results in severe surface recombination and low open circuit voltage. Through introducing thermally grown silicon dioxide between the emitter and antireflection coating, the serious surface recombination can be suppressed effectively without too much cost. Except for the improved open circuit voltage, short wavelength response can also be enhanced due to the reduced surface recombination velocity on the front side. Hence, the efficiency potential of black silicon solar cell can be further exploited on account of this low-cost silicon dioxide introduced, which is definitely beneficial for the reduction of LCOE (Levelized Cost of Electricity) and achievement of grid parity.
Keywords: Black Silicon, Surface Passivation, Thermal Oxidation

1 INTRODUCTION

Compared to traditional BSF (Back Surface Field) multicrystalline silicon solar cell, black silicon solar cell possesses a higher short circuit current density, resulting from the lower reflectance of submicrometer or nanometer scale surface texture. However, along with the enhanced light absorption, the defect state density rises significantly on account of the much larger specific surface area [1]. The conventional silicon nitride antireflection coating can't provide sufficient passivation probably owing to the PECVD (Plasma Enhanced Chemical Vapor Deposition) technology adopted, which is incapable of conformal deposition achieved by ALD (Atomic Layer Deposition) [2]. As a result, the surface recombination can't be effectively suppressed, which leads to a lower open circuit voltage. In this report, effort is made to address this passivation issue in order to further exploit the efficiency potential of black silicon solar cell.

In consideration of the cost, ALD technology isn't an optimal choice for mass production, therefore another method should be chose to deposit the conformal passivation layer. In this report, thermal oxidation [3] was adopted to achieve the conformal passivation, and a thermally grown silicon dioxide thin layer was inserted between the emitter and antireflection coating to improve the surface passivation of black silicon.

2 EXPERIMENTS

In this work, solar cells and lifetime samples were fabricated according to the corresponding process flows shown in Fig 1. All samples were fabricated with p-type DWS (Diamond Wire Sawing) multicrystalline silicon wafers with side length of 156.75 mm, thickness of 200 μm and resistivity of 1–3 Ω·cm.

For solar cells, acid etching or RIE (Reactive Ion Etching) [4] was employed to texture the wafer surface. Subsequently, phosphorus was diffused to form emitter with $POCl_3$ as the phosphorus source. After rear emitter removal, edge isolation and PSG (Phosphosilicate Glass) removal, the passivation and antireflection coating (i.e. SiN_x:H or SiO_2/SiN_x:H stack) was deposited on the emitter surface. Finally, the metal pastes were screen-printed onto both sides and the metallization was completed after sintering.

For lifetime samples, both wafer surfaces were textured with acid etching or RIE, and diffused with phosphorus to form symmetrical emitter structure. Then, the PSG was removed by dipping in the dilute hydrofluoric acid solution. After the passivation and antireflection coating (i.e. SiN_x:H or SiO_2/SiN_x:H stack) was deposited onto both sides, sintering was also carried out to undergo the same thermal history as solar cell.

Because DWS multicrystalline silicon wafers were adopted, a specific additive was added into the acid etching solution to reduce the texture reflectance. On the other hand, when RIE was applied as the texturization method, the surface damage caused by cutting was removed by alkaline solution before the reactive ion etching.

During the deposition of SiO_2/SiN_x:H stack, the silicon dioxide thin layer was grown on the emitter surface by thermal oxidation first, then the silicon nitride was stacked by PECVD.

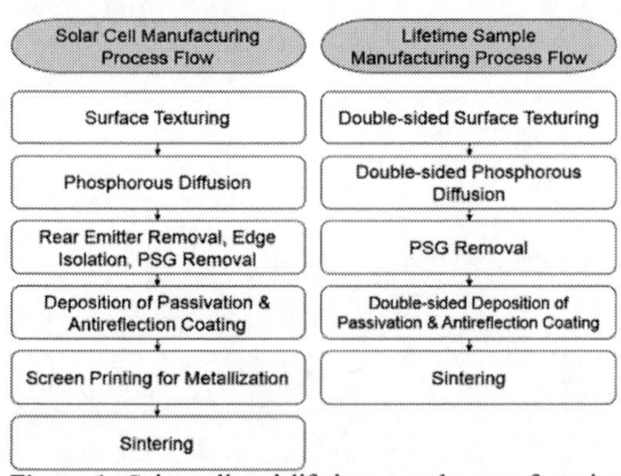

Figure 1: Solar cell and lifetime sample manufacturing process flows.

After the texturization, the surface morphology was observed by optical microscope and SEM (Scanning

Electron Microscope). And after the preparation of solar cells and lifetime samples, a few characterization techniques were applied, such as IV test, reflectance measurement, QE (Quantum Efficiency) measurement and QSSPC (Quasi Steady State Photoconductance) technique.

3 RESULTS AND DISCUSSION

DWS multicrystalline silicon wafers were adopted because of the lower cost than slurry sawing. The texturization was done with two different methods, namely additive-assisted acid etching and RIE combined with alkaline damage removal etch. The corresponding optical and scanning microscope images are shown in Fig 2. The typical worm-like micrometer scale structures can be seen on the surface of acid-etched wafer (Fig 2a & 2c), which verifies the validity of additive. As for the RIE-treated wafer (Fig 2b & 2d), there are only submicrometer scale holes on the surface without the presence of worm-like structures owing to the adoption of alkaline damage removal etch.

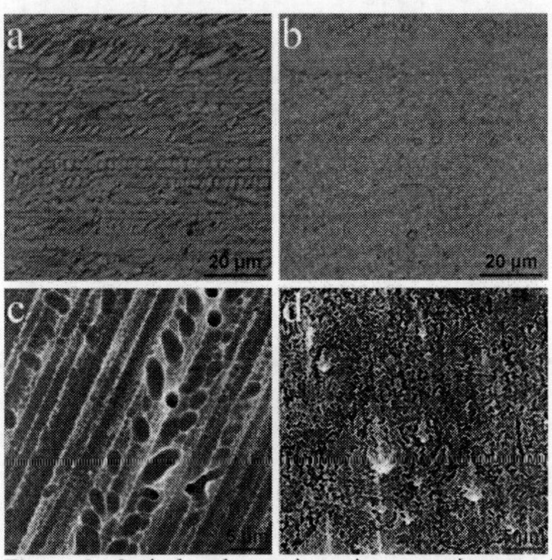

Figure 2: Optical and scanning microscope images of DWS multicrystalline silicon wafers textured with additive-assisted acid etching (a & c) and RIE combined with alkaline damage removal etch (b & d) respectively.

Corresponding solar cells, namely AE-BSF (BSF solar cell with acid etching) and black silicon (BSF solar cell with RIE), were prepared with the textured wafers, and the photovoltaic performance parameters are summarized in Table I. It can be found that the black silicon solar cell possesses a much higher efficiency than the traditional AE-BSF solar cell primarily due to the much higher short circuit current density, which benefits from the submicrometer scale light trapping structure. However, the open circuit voltage of black silicon solar cell is lower than the AE-BSF solar cell on the contrary, which may be attributed to the insufficient emitter surface passivation provided by silicon nitride.

Table I: Photovoltaic performance parameters of AE-BSF and black silicon solar cells. Mean values of about 100 samples for each kind of solar cell are listed.

	V_{oc} (mV)	J_{sc} (mA/cm^2)	FF (%)	Eff (%)
AE-BSF	640.0	36.31	80.14	18.62
Black Silicon	637.3	37.14	80.37	19.02

Both kinds of solar cells were submitted to the quantum efficiency measurement, and the resulting EQE (External Quantum Efficiency), reflectance and IQE (Internal Quantum Efficiency) curves are shown Fig 3. It is obvious that the spectral response below about 600 nm of black silicon solar cell is much more superior to the AE-BSF solar cell, mainly because of the much lower reflectance. Nevertheless, according to the IQE data, the much lower values below about 600 nm of black silicon mean that there is severe surface recombination on the front side, which leads to a decline in open circuit voltage.

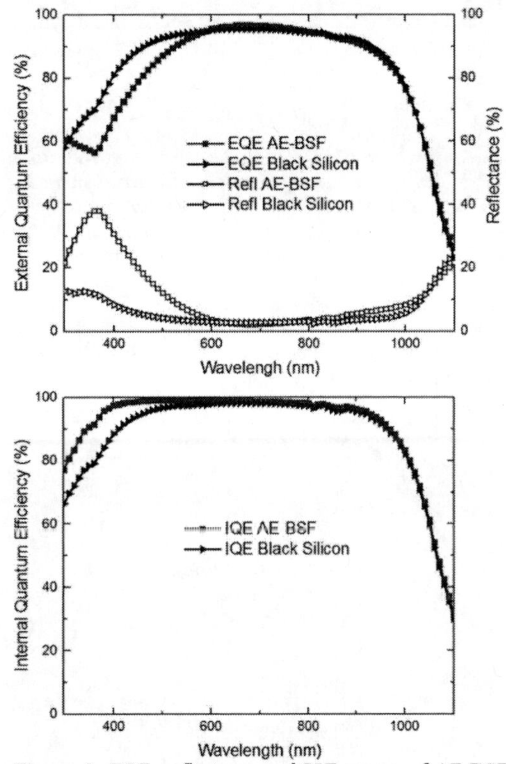

Figure 3: EQE, reflectance and IQE curves of AE-BSF and black silicon solar cells.

For this reason, the black silicon solar cell structure was optimized with a thermally grown silicon dioxide thin layer as shown in Fig 4 for the sake of surface passivation improvement.

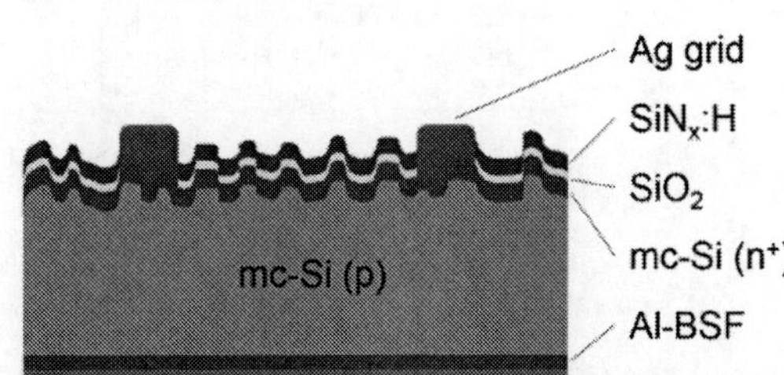

Figure 4: Schematic diagram of black silicon solar cell optimized with silicon dioxide.

According to the photovoltaic performance parameters of black silicon solar cells with SiO_2 listed in Table II, the open circuit voltage gets increased, which can be ascribed to the improved surface passivation provided by the addition of silicon dioxide. What's more, the short circuit current density gets increased as well. However, the fill factor suffers a decline, possibly because the emitter contact resistance is influenced by the introduced SiO_2. Eventually, an efficiency gain of 0.15% is achieved after the introduction of thermally grown SiO_2.

Table II: Photovoltaic performance parameters of black silicon solar cells without and with silicon dioxide. Mean values of about 100 samples for each kind of cell are listed.

	V_{oc} (mV)	J_{sc} (mA/cm^2)	FF (%)	Eff (%)
Black Silicon	637.3	37.14	80.37	19.02
Black Silicon with SiO_2	641.3	37.39	79.96	19.17

On the basis of EQE and reflectance curves shown in Fig 5, despite the higher reflectance of black silicon solar cell with SiO_2 below about 500 nm, the spectral response below about 400 nm gets higher on the contrary, which explains the higher measured short circuit current density. According to the IQE curves shown in Fig 5, it can be deduced from the much higher values below about 500 nm that the emitter surface passivation quality of black silicon solar cell is markedly enhanced with the introduction of thermally grown silicon dioxide.

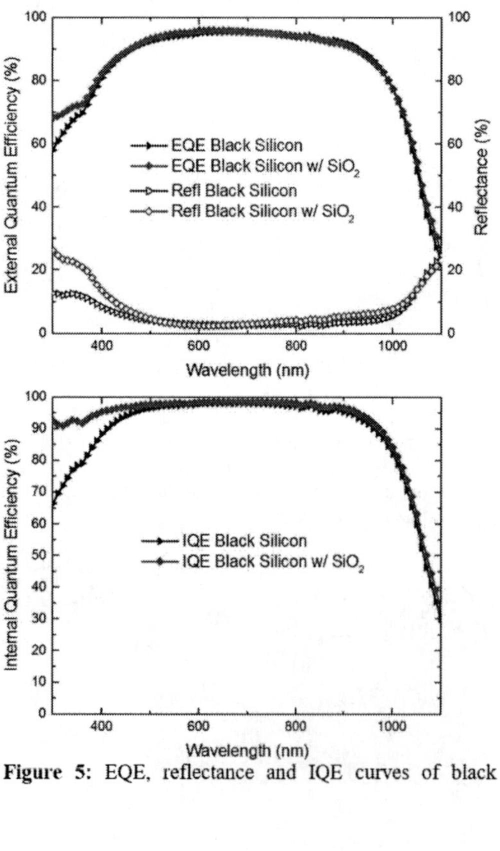

Figure 5: EQE, reflectance and IQE curves of black silicon solar cells without and with silicon dioxide.

The lifetime samples with symmetrical emitter structures were prepared for the evaluation of emitter surface passivation quality. The corresponding lifetimes, J_{0e} (Emitter Saturation Current Density) at the injection level of 5×10^{15} cm^{-3} and iV_{oc} (implied Open Circuit Voltage) were extracted from the QSSPC (Quasi Steady State Photoconductance) data, and summarized in Fig 6. It can be found that without thermally grown SiO_2, the lifetime of black silicon sample is lower and the J_{0e} is higher than the AE-BSF sample, which implies that black silicon solar cell possesses more serious emitter surface recombination than AE-BSF solar cell. And the lower iV_{oc} of black silicon lifetime sample without SiO_2 explains the lower measured V_{oc} of corresponding solar cell than AE-BSF. When thermally grown SiO_2 is inserted between emitter and silicon nitride, the lifetime, J_{0e} and iV_{oc} of black silicon lifetime sample are significantly improved, which means that the emitter surface recombination is effectively restrained and explains the higher measured V_{oc} of black silicon solar cell with SiO_2. Interestingly, the AE-BSF lifetime sample also benefits from the introduction of SiO_2, which further verifies the effectiveness of thermally grown SiO_2 for the emitter surface passivation improvement.

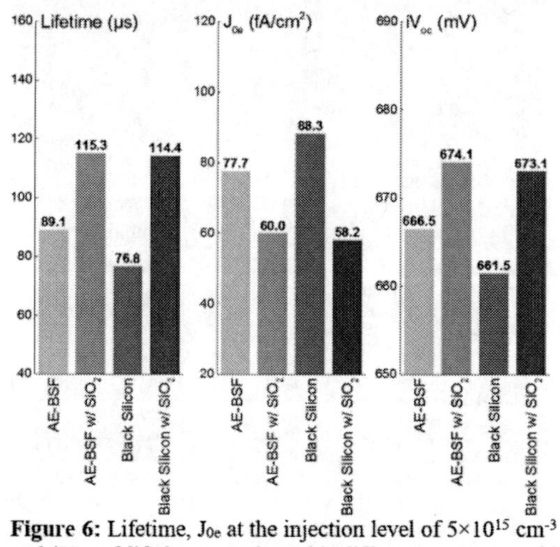

Figure 6: Lifetime, J_{0e} at the injection level of 5×10^{15} cm^{-3} and iV_{oc} of lifetime samples with different textures (acid etching and RIE, i.e. AE-BSF and Black Silicon) and passivation coatings (without and with SiO_2).

4 CONCLUSION

Because of the large emitter specific surface area of black silicon solar cell, conventional silicon nitride coating deposited by PECVD can't provide conformal passivation, which results in severe surface recombination and low open circuit voltage. Through introducing thermally grown silicon dioxide between the emitter and antireflection coating, the serious surface recombination can be suppressed effectively without too much cost. Except for the improved open circuit voltage, short wavelength response can also be enhanced due to the reduced surface recombination velocity on the front side. Hence, the efficiency potential of black silicon solar cell can be further exploited on account of this low-cost silicon dioxide introduced, which is definitely beneficial for the reduction of LCOE (Levelized Cost of Electricity) and achievement of grid parity.

5 ACKNOWLEDGEMENTS

This work was supported by the Research and Development Department of Talesun.

6 REFERENCES

[1] M. Otto, M. Kroll, T. Käsebier, R. Salzer, A. Tünnermann, R.B. Wehrspohn, Applied Physics Letters 100 (2012) 191603.
[2] R.W. Johnson, A. Hultqvist, S.F. Bent, Materials Today 17 (2014) 236.
[3] J. Oh, H.-C. Yuan, H.M. Branz, Nature Nanotechnology 7 (2012) 743.
[4] S. Winderbaum, O. Reinhold, F. Yun, Solar Energy Materials and Solar Cells 46 (1997) 239.

INDUSTRIAL APPLICABILITY OF ANTIREFLECTION-COATING-FREE BLACK SILICON ON PERC SOLAR CELLS AND MODULES

Toni P. Pasanen[1*], Ville Vähänissi[1], Franziska Wolny[2], Alexander Oehlke[2], Matthias Wagner[2], Mikko A. Juntunen[3], Ismo T. S. Heikkinen[1,4], Emma Salmi[4], Sami Sneck[4], Henri Vahlman[1,5], Antti Tolvanen[5], Jaakko Hyvärinen[5], and Hele Savin[1]
[1]Aalto University, Department of Electronics and Nanoengineering, Tietotie 3, 02150 Espoo, Finland
[2]SolarWorld Industries GmbH, Martin-Luther-King-Str. 24, 53175 Bonn, Germany
[3]Naps Solar Systems Inc., Ruosilankuja 4, 00390 Helsinki, Finland
[4]Beneq Oy, Olarinluoma 9, 02200 Espoo, Finland
[5]Endeas Oy, Ruukinkuja 1, 02330 Espoo, Finland

ABSTRACT: Black silicon (b-Si) is of particular interest within the photovoltaic community due to its ability to texture diamond wire-sawn multicrystalline silicon wafers. In this work, we apply a deep dry-etched black silicon nanotexture, which possesses less than 1 % solar-weighted reflectance with no antireflection coating (ARC), to industrial Passivated Emitter and Rear Cells (PERC). Additionally, the cells are processed further into small solar modules, which are characterized by their electrical and optical performance. The fragile nanostructures remain intact during industrial cell and module fabrication, maintaining the excellent optical properties until the final product. We show that b-Si modules with a typical cover glass retain their performance until incident angles larger than 60°, whereas the heavily increased reflectance of acidic-textured modules decreases their efficiency already after a 30° tilt. Furthermore, at an incidence angle of 70°, the efficiency of b-Si modules has reduced only 7 %, while that of the acidic-textured equivalents has decreased more than 25 %. Hence, the excellent optical properties of ARC-free b-Si are maintained also at module level. The results demonstrate that deep dry-etched b-Si nanostructures are fully applicable to current industrial PERC production facilities.
Keywords: black silicon, Passivated Emitter and Rear Cell, module, multicrystalline silicon, incidence angle

1 INTRODUCTION

Black silicon (b-Si) is currently highly topical for the PV industry due to its ability to eliminate reflective losses of diamond wire-sawn (DWS) multicrystalline silicon (mc-Si) wafers [1]. Indeed, the potential of b-Si as the next mainstream technology in the PV industry has been demonstrated as impressive b-Si cell results by several research groups [2-5]. A few b-Si fabrication methods, such as metal-assisted chemical etching (MACE) [6], have already demonstrated their industrial viability [7]. Although MACE clearly outperforms standard acidic texturing on DWS mc-Si wafers, the textured surfaces still typically require an antireflection coating (ARC) to decrease the surface reflectance to a sufficiently low level for high-efficiency solar cells [8, 9]. This is the case also with another promising b-Si fabrication method, atmospheric dry etching (ADE), where an excessively porous surface layer is removed after the ADE process by an additional polishing step [10], which changes surface morphology and increases reflectance.

The need for any ARC can be eliminated, and the optical properties improved further, by making b-Si nanostructures deeper, i.e., increasing their aspect ratio [11, 12]. Such surfaces can be fabricated e.g. via deep reactive ion etching (DRIE) [11], which is able to produce lattice defect-free deep b-Si nanostructures [13], eliminating the need for any post-etching polishing steps. In addition to increased performance, the application of deep b-Si would reduce cell manufacturing costs due to the exclusion of ARC deposition [14]. However, the larger height of b-Si increases the fragileness of the nanostructure, and only little effort has been made to study the applicability of deep dry-etched b-Si nanostructures to current industrial solar cell and module production lines. It remains ambiguous, whether the fragile nanostructure is damaged by wafer grippers or inline equipment, if screen-printing is applicable to deep b-Si, or how the b-Si affects the module encapsulation process. Moreover, the benefit from the excellent optical properties of deep b-Si may be partly diminished by ethylene vinyl acetate (EVA) and module cover glass [15].

In this contribution, we fabricate dry-etched b-Si Passivated Emitter and Rear Cells (PERC) at an industrial production line of SolarWorld Industries following the standard PERC process flow. The cells are processed further into small modules at the solar module factory of Naps Solar Estonia. The survival of b-Si is evaluated after various cell and module process steps, and the cells and modules are finally characterized by angle-dependent I-V measurements at the testing facility of Endeas.

2 EXPERIMENTAL METHODS

Industrial-sized (156 × 156 mm^2) mc-Si wafers were used as substrate material for the fabrication of b-Si PERC solar cells. The process flow followed the standard PERC process of SolarWorld Industries as applicable. First, a deep b-Si nanostructure (see inset in Fig. 1) was etched by DRIE on more than one hundred wafers using a process based on that reported in [11]. Due to clamping against the bottom electrode, approximately 5 mm from each wafer edge remained unetched. For reference purposes, some of the wafers were textured with the standard acidic texturing process in hydrofluoric-nitric acid (HF/HNO_3) solution [16].

Phosphorus emitter was formed by $POCl_3$ diffusion applying the same diffusion parameters for all wafers. After phosphosilicate glass (PSG) removal in HF, backside was passivated with a chemical-vapor-deposited (CVD) aluminum oxide/silicon nitride (AlO_x/SiN_x) stack. The acidic-textured cells were passivated with CVD SiN_x on the front according to the industry standard. Instead, the b-Si cells received an atomic-layer-deposited (ALD) AlO_x front surface passivation, since the conformality of CVD layers is insufficient on deep b-Si [17]. Furthermore, AlO_x is the most widely studied ALD material [18] and has been demonstrated to passivate b-Si surfaces efficiently due to the high density of negative fixed charges and the high-quality interface with silicon [11, 13, 19]. The thin film conformality and the efficiency of surface passivation on deep dry-etched b-Si over the large 156 × 156 mm^2 wafer area was investigated by passivating a few non-diffused b-Si wafers with ALD AlO_x on both sides using the process reported in [20]. The lifetime samples were characterized by minority charge carrier lifetime using the microwave-detected photoconductance decay (µ-PCD) method. Finally, the cells were finished by an industry-standard screen-printing process of silver (Ag) metal contacts, which was followed by firing at 880 °C. Screen dimensions were adjusted to match with the b-Si area to have a uniform metal contact throughout the active cell area.

The prototype cell fabrication was kept as close to the industrial standard as possible at the industrial production line of SolarWorld Industries. However, b-Si was etched with a single-wafer equipment (Oxford Instruments Plasmalab System 100) at Aalto University, and AlO_x thin film for front surface passivation was deposited using a large-scale batch ALD tool (Beneq P400) at Beneq.

The efficiency of the finished cells was determined by current-voltage (I-V) measurements. The slightly smaller active area in the b-Si cells was taken into account in the data analysis. Additionally, the survival of the b-Si nanostructure was evaluated visually and by SEM after various process steps. Specifically, SEM characterization was focused on the evaluation of the metal-Si interface of screen-printed contacts. In addition, the optical properties of b-Si cells were examined by reflectance spectroscopy.

The cells, both b-Si and acidic-textured, were processed further into single-cell modules at the solar module factory of Naps Solar Estonia. All modules were fabricated using the same procedure. The untextured edges of the b-Si cells were removed by laser cutting prior to module encapsulation to have fully black solar modules. The acidic-textured cells were cut to equal size, which was approximately 146 × 146 mm^2. Electric contacts were established by regular tab wires, which were soldered manually to ensure better positioning of the wires on cut wafers. The encapsulation was performed simultaneously to both b-Si and acidic-textured modules using an industry-standard process. The modules had a standard PV glass, which was single-side textured and had no antireflection coating. Additionally, the strong cover glass was replaced in one module by a plastic cover, which is typically used in e.g. boat installations. This enabled the cutting of the module and cross-sectional inspection of the EVA – b-Si interface with SEM for the possibility of damaged nanostructure or air gaps inside the module. Finally, module efficiencies were measured as a function of incidence angle between 0° and 70°. The efficiency was calculated taking only the cell area into consideration to enable direct comparison between cell and module efficiencies. In addition, the effect of reducing solar irradiance with increasing incidence angle was cancelled out by calculations in the angle-dependent data.

3 RESULTS AND DISCUSSION

3.1 Optical and electrical properties of ARC-free b-Si

Figure 1 compares the measured reflectance spectra of ARC-free b-Si and standard acidic-textured cells with SiN_x ARC. While the reflectance of acidic-textured surfaces can be minimized efficiently at a certain wavelength with a proper ARC, deep dry-etched b-Si has negligible reflectance across the whole relevant solar spectrum range. This is contrary to the shallower "black silicon" surface textures currently utilized in the PV industry for DWS mc-Si, which still require separate polishing and ARC steps [8, 9]. The solar spectrum-weighted average reflectance of ARC-free b-Si cells is only 0.8 %, which is significantly lower than 8.3 % of AR-coated acidic-textured cells. Furthermore, the excellence of deep b-Si is even more emphasized in the ultraviolet range, where its reflectance is tens of percents lower than in standard cells (Fig. 1). Indeed, deep dry-etched b-Si studied here appears truly black to naked eye (Fig. 2a) due to the negligible reflectance, which eliminates the need for any ARC, and hence, reduces cell manufacturing costs [14].

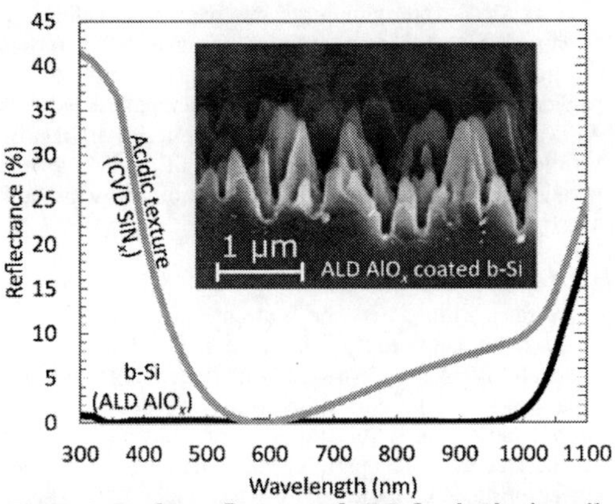

Figure 1: Surface reflectance of ARC-free b-Si solar cells with ALD AlO_x surface passivation and acidic-textured cells with standard SiN_x ARC. The solar spectrum-weighted average reflectance between 300 nm and 1000 nm of b-Si and acidic-textured cells is 0.8 % and 8.3 %, respectively. The inset presents a SEM image of the deep b-Si nanostructure coated conformally with ALD AlO_x.

Fortunately, the excellent optics do not have to come at the expense of electrical properties. Figure 2b shows a minority charge carrier lifetime map of a non-diffused b-Si wafer passivated with ALD AlO_x showing a high average lifetime of 250 µs. Separate grains are clearly visible in the map, indicating that the lifetime is limited by the bulk and the b-Si surface is efficiently passivated. Furthermore, the boundary of the b-Si area is not visible in the lifetime map, which further supports successful passivation throughout the 156 × 156 mm^2 area realized by a commercial batch ALD tool. The inset in Fig. 1 confirms that the deep b-Si nanostructures are indeed coated conformally with AlO_x, which is a prerequisite for efficient passivation. However, the efficiency of AlO_x surface passivation in the b-Si cells cannot directly be concluded from the lifetime results, since the cells have a phosphorus emitter at the front contrary to the lifetime samples, which will be discussed later in more detail.

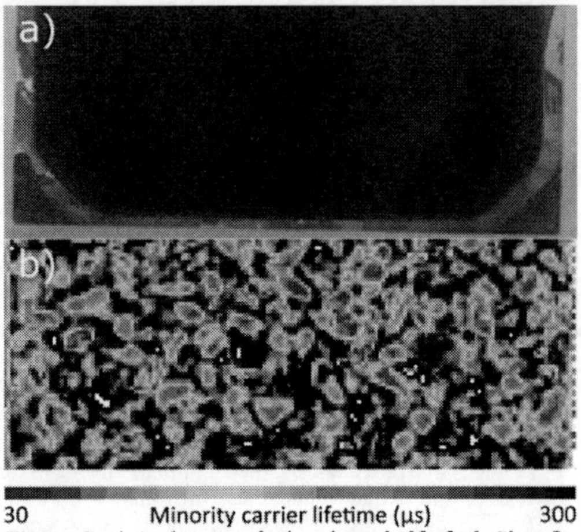

Figure 2: a) A photograph showing a half of a b-Si wafer with no ARC. Approximately 5 mm from each wafer edge remained unetched due to clamping against the bottom electrode during b-Si fabrication. b) Minority charge carrier lifetime map of the same b-Si wafer passivated with ALD AlO_x (average ~250 µs). Separate grains are clearly visible in the map, while the boundary of the b-Si area is invisible, indicating that the lifetime is limited by the bulk and b-Si surface is efficiently passivated.

3.2 Screen-printed contacts

Screen-printing is the current industry standard process for metal contact fabrication [1]. However, its applicability on deep b-Si is not obvious due to the high aspect ratio of the nanostructures. Nevertheless, earlier studies have promisingly reported that textured surfaces are in fact more compatible with contact formation by screen-printed Ag pastes than flat surfaces [21]. This is due to an increased available surface area for contacting and more optimal wetting of the dielectric, Si and Ag by the glass frit in the paste [21, 22]. The glass layer tends to concentrate in the valleys of the texture, enabling direct contact between the metal finger and Ag crystallites that form during the metallization process, which promotes low contact resistance [21, 22].

Figure 3a presents a SEM image of a screen-printed front contact finger of a b-Si cell fabricated in this work. Closer inspection reveals that the valleys between the nanostructures are indeed filled with the glass frit, whereas some b-Si tips are free of the thick insulating layer. This agrees with a previous study, which concluded that direct contacts between Si and Ag are the dominant current paths on b-Si surfaces [23]. Furthermore, the screen-printed metal finger and the Si – Ag interface on deep b-Si look rather similar to those on an acidic-textured surface (Fig. 3b), which suggests that the industrial metallization process is applicable also to b-Si. However, since screen-printing is performed through the front surface passivation film, the AlO_x layer needs to be thin enough (<10 nm) to avoid increased contact resistance [24]. Although the AlO_x film is too thin to be visible in the SEM image in Fig. 3a, it is expected that there is an insulating layer between Ag and Si, which increases contact resistance. Additionally, the performance of contacts is affected by the amount of inactive dopants and the depth of the highly-doped emitter region [23], which needs to be taken into consideration in the optimization of b-Si emitters [2, 12, 20].

Figure 3: Screen-printed front contact finger on (a) b-Si coated with ALD AlO_x and (b) acidic-textured surface coated with SiN_x. The inset in figure (a) shows a zoomed view of the b-Si – metal interface.

3.3 Mechanical durability and module encapsulation

Visual and SEM inspection after various process steps reveals that the deep b-Si nanostructure remains intact during cell and module processing despite its fragile nature. No visible mechanical damage is produced to the b-Si by inline equipment or wafer grippers, which is an essential finding for the industrialization of the new technology.

Figure 4 presents a cross-sectional SEM image of an ARC-free b-Si module without cover glass. Mainly, no air gaps were observed at the EVA – b-Si interface, as preferred. However, a few micrometer-sized gaps were located, which allowed closer inspection of the nanostructures. The inset in Fig. 4 presents a SEM image taken from such gap, showing intact b-Si covered with EVA. Thus, the ARC-free nanostructure survives the encapsulation process, which is observed also as a deep black appearance of the finished modules (inset in Fig. 6).

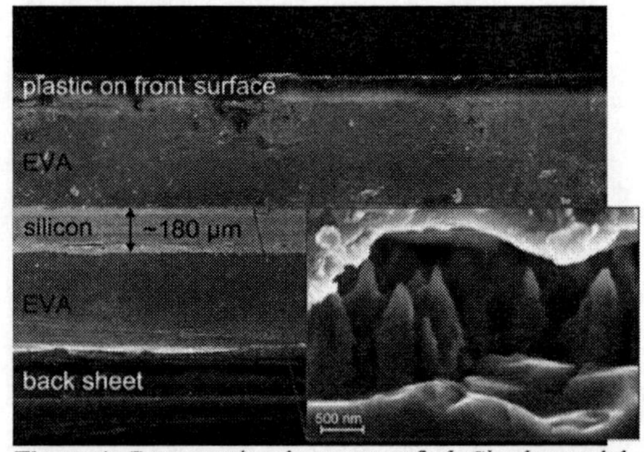

Figure 4: Cross-sectional structure of a b-Si solar module without glass. The inset presents a SEM image taken from a µm-scale gap at the ethylene vinyl acetate (EVA) – b-Si interface, which shows intact b-Si nanostructure covered with EVA.

3.4 Cell and module performance

The efficiencies of deep dry-etched b-Si and standard acidic-textured cells and modules are presented in Fig 5. The b-Si cells likely suffer from increased emitter losses due to an excessive amount of phosphorus in the heavily-doped layer [2, 12, 20]. This is a result of the application of the standard emitter diffusion parameters to b-Si, which has significantly larger surface area. Moreover, the efficiency of the b-Si cells is partly limited due to the non-optimal charge polarity in ALD AlO_x for phosphorus emitter passivation [20, 25, 26]. Nevertheless, despite the above-mentioned hindrances, the excellent optics of deep b-Si raise the efficiency of the nanostructured cells to the same level with acidic-textured equivalents. Hence, higher efficiencies should be expected from deep b-Si cells after further optimization of emitter diffusion process for high aspect ratio nanostructures [2, 12, 20] and replacement of AlO_x e.g. with positively charged SiO_x/AlO_x stacks [20, 26, 27].

Although cell efficiencies heavily determine the module performance, the efficiency of modules is a more direct quantity for comparison between photovoltaic devices under real operating conditions. Figure 5 reveals that the efficiencies of both b-Si and acidic-textured devices are reduced by 2-3 $\%_{abs}$ at module level compared to the corresponding cell performance. This reduction can be mainly attributed to two factors. First, the modules have a single-side textured cover glass with no antireflection coating, which slightly increases the reflectance. Moreover, manual tab wire soldering caused additional losses in the metal contacts and the process generated cracks to both acidic-textured and b-Si cells. Consequently, the b-Si modules show slightly lower efficiency than the acidic-textured equivalents.

Nevertheless, efficiency measured immediately after fabrication may not directly reflect the device performance under real operating conditions. Indeed, a recent study showed that after prolonged usage, the b-Si cells should maintain their performance, whereas standard cells would likely suffer from Light- and Elevated Temperature-Induced Degradation [28]. Furthermore, the performance gain obtained by the application of b-Si could be emphasized when using dirtier substrate material, since b-Si getters detrimental impurities more effectively than flat surfaces [29]. Thirdly, the efficiency measured for perpendicular incident light neglects the ability of b-Si to absorb light that arrives at large incidence angles efficiently [4], which will be shown in the next section in more detail.

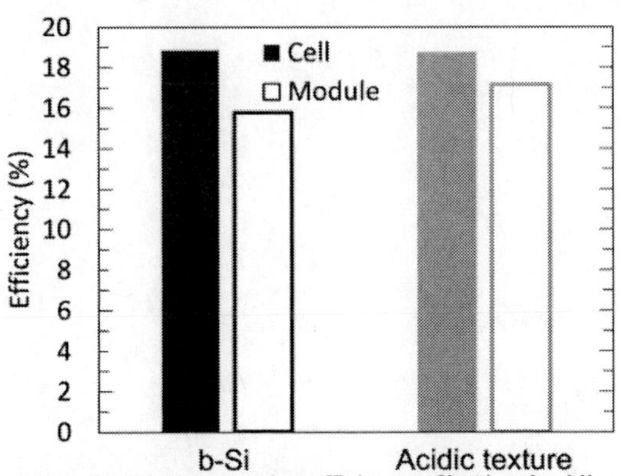

Figure 5: Power conversion efficiency of b-Si and acidic-textured PERC cells and modules. The efficiencies of b-Si cells are calculated considering only the active b-Si area.

3.5 Impact of incidence angle

Figure 6 presents the efficiency of b-Si and acidic-textured PERC modules as a function of light incident angle normalized to that of normal incidence. The reducing amount of solar irradiance arriving to the cell surface with increasing incidence angle (sketched in the inset) is cancelled out in the data by applying the equation presented in Fig. 6. The b-Si modules retain their performance until incident angles larger than 60°, whereas the efficiency of acidic-textured modules starts to decrease already after a 30° tilt due to heavily increased reflectance. This agrees well with earlier results reported at cell level [4]. Furthermore, at an incidence angle of 70°, the efficiency of b-Si modules has reduced only 7 %, while that of the acidic-textured equivalents has decreased more than 25 %. Hence, the results demonstrate that the excellent optical properties of deep ARC-free b-Si are maintained also at module level. This is an essential finding, since solar cells are always encapsulated, and hence, the cell level optics do not necessarily reflect the performance under real operating conditions.

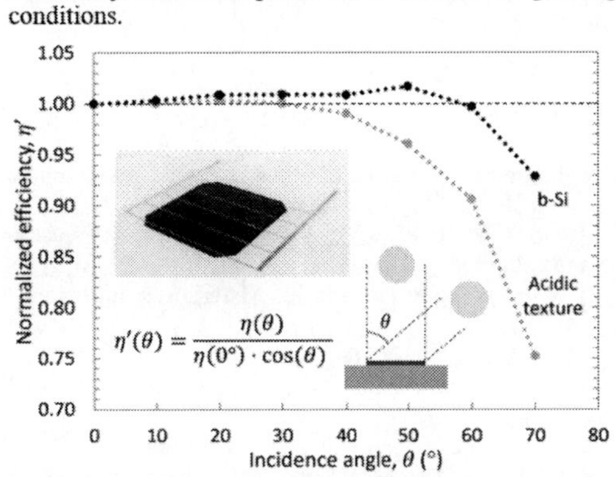

Figure 6: Incidence angle-dependent power conversion efficiency of b-Si and acidic-textured modules normalized to that of normal incidence. The effect of reducing solar irradiance with increasing incidence angle (sketched in the inset) is cancelled out in the data, as per the presented equation. The dotted lines act as a guide for eye. Values larger than unity can be accounted for measurement inaccuracy. The inset shows a photograph of a single-cell ARC-free b-Si solar panel.

4 CONCLUSIONS

This contribution demonstrated that deep dry-etched b-Si nanostructures, which possess less than 1 % solar-weighted reflectance with no ARC, survive well in industrial solar cell and module production, which has not been self-evident due to their fragile nature. Despite the lack of any polishing steps after b-Si etching, the nanostructure was efficiently passivated by a conformal ALD AlO_x thin film on the whole industrial-sized wafer area. Furthermore, the high aspect ratio of deep b-Si did not notably affect screen-printing. Instead, direct contact was formed between the emitter and silver finger on some b-Si tips, promoting good electrical contact. Finally, angle-dependent I-V measurements revealed that the b-Si modules retain their performance until an incidence angle of 60°, whereas the heavily increased reflectance of acidic-textured equivalents decreased their efficiency already after a 30° tilt. The results demonstrate that ARC-free b-Si is fully applicable to current PERC production facilities, and that the excellent deep b-Si optics are maintained also at module level. The main development area of the ARC-free b-Si technology is the upscaling of the dry etching process to meet the demanding throughput requirement set by the PV industry.

ACKNOWLEDGEMENTS

The authors would like to thank Ander Pukk and Pavel Tšukrejev from Naps Solar Estonia for guidance in module assembly. The authors acknowledge the provision of facilities by Aalto University at OtaNano – Micronova Nanofabrication Centre. This work was funded by project "BLACK" (project No. 2956/31/2014), which was supported under the umbrella of SOLAR-ERA.NET by Business Finland (former Tekes – the Finnish Funding Agency for Innovation). T. P. Pasanen acknowledges the Aalto ELEC Doctoral School, Jenny and Antti Wihuri Foundation, and Walter Ahlström Foundation for the financial support.

REFERENCES

[1] International technology roadmap for photovoltaic: Results 2016. 8th Edition (2017).
[2] J. Oh, H.-C. Yuan, H. M. Branz, Nature Nanotechnology 7 (2012) 743.
[3] W.-C. Wang et al., Applied Materials & Interfaces 5 (2013) 9752.
[4] H. Savin et al., Nature Nanotechnology 10 (2015) 624.
[5] J. Benick et al., IEEE Journal of Photovoltaics 7 (2017) 1171.
[6] F. Toor et al., Nanoscale 8 (2016) 15448.
[7] B. Beetz, PV Magazine (2018) Available online: https://www.pv-magazine.com/2018/01/04/suntechs-black-silicon-solar-cells-enter-mass-production/ (accessed on May 30, 2018).
[8] Z. G. Huang et al., Solar Energy Materials & Solar Cells 143 (2015) 302.
[9] G. Su et al., Solar Energy 170 (2018) 95.
[10] P. Saint-Cast et al., Energy Procedia 124 (2017) 260.
[11] P. Repo et al., IEEE Journal of Photovoltaics 3 (2013) 90.
[12] B. Kafle et al., Solar Energy Materials & Solar Cells 152 (2016) 94.
[13] G. von Gastrow et al., Solar Energy Materials & Solar Cells 142 (2015) 29.
[14] C. Modanese, H. S. Laine, T. P. Pasanen, H. Savin, J. M. Pearce, Energies 11 (2018) 2337.
[15] J. Gjessing, E. S. Marstein, Energy Procedia 38 (2013) 348.
[16] A. Hauser, I. Melnyk, P. Fath, S. Narayanan, S. Roberts, T. M. Bruton, Proceedings of 3rd World Conference on Photovoltaic Energy Conversion, Osaka, Japan, 2 (2003) 1447.
[17] Y. Liu et al., Small 8 (2012) 1392.
[18] G. Dingemans, W. M. M. Kessels, Journal of Vacuum Science & Technology A 30 (2012) 040802.
[19] [Otto2012] M. Otto, M. Kroll, T. Käsebier, R. Salzer, A. Tünnermann, R. B. Wehrspohn, Applied Physics Letters 100 (2012) 191603.
[20] T. Pasanen, V. Vähänissi, N. Theut, H. Savin, Energy Procedia 124 (2017) 307.
[21] E. Cabrera et al,. Journal of Applied Physcis 110 (2011) 114511.
[22] E. Cabrera et al., IEEE Journal of Photovoltaics 3 (2013) 102.
[23] B. Kafle et al., IEEE Journal of Photovoltaics 7 (2017) 136.
[24] A. To, W. M. Li, X. Li, B. Hoex, Energy Procedia 124 (2017) 914.
[25] B. Hoex, J. J. H. Gielis, M. C. M. van de Sanden, W. M. M. Kessels, Journal of Applied Physics 104 (2008) 113703.
[26] B. W. H. van de Loo et al., Solar Energy Materials & Solar Cells 143 (2015) 450.
[27] B. W. H van de Loo, A. Ingenito, M. A. Verheijen, O. Isabella, M. Zeman, W. M. M. Kessels, Applied Physics Letters 110 (2017) 263106.
[28] T. P. Pasanen et al., Progress in Photovoltaics: Research and Applications, in peer review.
[29] T. P. Pasanen, H. S. Laine, V. Vähänissi, J. Schön, H. Savin, Scientific Reports 8 (2018) 1991.

THE BLACK-SIN METHOD: A NOVEL APPROACH TO REDUCE THE REFLECTION OF SOLAR CELLS

Jens Hirsch[1,2], Alena Okhorzina[1], Stephan Großer[2], Norbert Bernhard[1], Dominik Lausch[2]
[1]Anhalt University of Applied Sciences, Faculty EMW, Bernburger Str. 55, DE-06366 Köthen (Anhalt)
[2]Fraunhofer Center for Silicon Photovoltaics CSP, Otto-Eissfeldt-Str. 12, DE-06120 Halle (Saale)
Jens.Hirsch@hs-anhalt.de

ABSTRACT: The Black-Si method is a promising alternative technique to reduce the reflection of c-Si solar cells. However, drawbacks of this technique are (I) the ion bombardment of the silicon surface, (II) the plasma induced impurities and (III) the increase of the silicon roughness. The Black-SiN method bypasses these drawbacks by a maskless plasma texturing of a SiN-layer. This approach separates the optical properties of the SiN-layer and the electrical properties of the silicon bulk and p-n junction. The Black-SiN texture is etched by a capacitively coupled power supply with a SF_6/O_2 gas mixture in a direct glow discharge. First results show a parabolic like Black-SiN texture in a scale of approx. 100 nm. A simulation of the Black-SiN optical properties revealed a global reflection minimum of 7.40 % between 300-1200 nm wavelength with respect to the air mass 1.5 spectrum on a plane silicon surface. Hence, the Black-SiN method can contribute to an improved solar cell efficiency.

Keywords: Black-Silicon, Plasma Etching, Effective Medium Theory, Silicon-Nitride, Black-Silicon-Nitride

1 INTRODUCTION

In 1995 Jansen et al. published for the first time the so-called 'Black-Silicon' method [1], which is a promising alternative technique to reduce the reflection of silicon solar cells. Typically, a Black-Si texture has a parabolic like morphology in the sub-µm scale, which fulfills the requirement of a so-called zero-order effective medium. In such a case, the optical properties of a texture can be approximated by the effective medium theory in combination with a gradual transition of an effective refraction index [2]. The gradual increase of the effective refraction index between air and silicon is the reason for the excellent optical properties of Black-Si.

However, the Black-Si method has some challenges, which prevent an successful application in an industrial environment. These challenges are (I) an ion bombardment of the silicon surface from the glow discharge [3], (II) an accumulation of plasma induced impurities on the silicon surface [4,5] and (III) a challenging surface passivation of the Black-Si texture by an industrial standard SiN-layer [6]. In particular, a SiN passivation is a serious problem for an industrial application of the Black-Si method, because a SiN-layer thickness of 80 nm can be in the same range as the plasma texture width. Further, all subsequent solar cell process steps, i.e. p-n junction formation or contact formation have to optimized to the rough and defected silicon surface.

To overcome these problems, our group developed the so-called 'Black-SiN' method. In this approach, the maskless plasma texturing of a silicon surface is replaced by a maskless plasma texturing of a SiN-layer on top of a silicon substrate. This can be performed even after the solar cell has been completely processed. The major benefit of this solution is no contact between the silicon surface and the glow discharge during the plasma etching. Hence, the sensitive silicon surface is protected against incoming ions and plasma induced impurities by the SiN-layer. The Black-SiN method does not apply an enlargement of the effective silicon surface area. Therefore, the solar cell process has not to be optimized. Furthermore, the textured SiN-layer acts as an effective medium which opens the opportunity to optimize the SiN-layer towards other parameters, i.e. higher refractive index for the Si/SiN interface. In this work, we present the concept of the Black-SiN method for the first time and demonstrate first results by SEM and TEM image acquisition. Further, we show an optimization of the Black-SiN method for an industrial application due to simulations by the effective medium theory [7] in combination with the Transfer-Matrix method [8].

2 THE BLACK-SiN METHOD

For the Black-SiN method, we used an adapted version of the Black-Si method. This version consists of a capacitively coupled plasma source with a SF_6/O_2 gas mixture in combination with a high ion acceleration. The ion acceleration was provided by a self-bias due to a capacitively power supply in combination with asymmetrical electrodes. Further, we used a low process pressure ($p < 100$ µbar) to ensure an enhanced ion path length. Figure 1 compares the Black-Si (a) and the Black-SiN (b) method by two schematic representations.

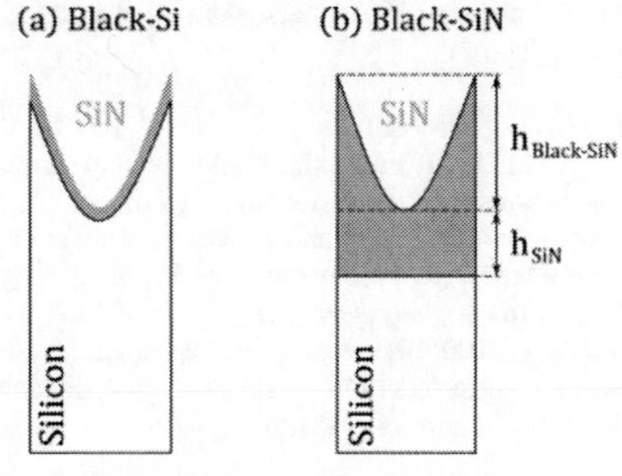

Figure 1: Schematic representation of the Black-Si method (a) and the Black-SiN method (b).

Figure 1 (a) shows a schematic representation of a standard Black-Si texture. The structure consists of a parabolic like texture with a width and height of approx. 300-500 nm, which causes a silicon surface enlargement but also the superior optical properties. There is an approx. 80 nm thick SiN-layer on this texture, which is usually deposited by a PECVD process for surface passivation. This SiN-passivation is a challenging task for an industrial application, because the SiN layer thickness and the Black-Si structure width are in the same range.

In comparison, figure 1 (b) shows a schematic representation of the Black-SiN method. The structure consists of a plane silicon surface with a thick SiN-layer on top. However, the Si surface can also be structured during the solar cell process using standard wet chemical etching. In the Black-SiN approach, the SiN-layer is partially plasma textured by a maskless capacitively coupled plasma process with SF_6 and O_2. Therefore, the SiN-layer consists of a volume fraction (h_{SiN}) and a textured fraction ($h_{Black\text{-}SiN}$). The thickness of both fractions must be optimized for a solar cell production, because the reflection can increase through constructive interference at the interfaces. This will be discussed in the optimization section of this contribution. By using the Black-SiN method, the silicon surface is completely protected during the plasma etching by the SiN-layer. This prevents contaminations of the silicon surface by metals.

3 EXPERIMENTAL

Starting point for the Black-SiN approach were commercial available as-cut, p-type, Czochralski wafers. In a first step, a 500 nm thick SiN-layer was deposited on the samples by a PECVD system from Oxford Instruments (PlasmaPro 100 System). Process gases were SiH_4, NH_3 and N_2 as carrier gas. A layer thickness of 500 nm was chosen to avoid a through-etching of the SiN-layer by the Black-SiN plasma etching. In a second step, the Black-SiN texture was plasma etched by a SF_6/O_2 gas mixture with a high ion acceleration in a RIE system from Oxford Instruments (PlasmaLab 100 ICP 65 System). Afterwards, SEM and TEM image acquisitions were performed on the plasma textured Black-SiN samples to elucidate the impact on the surface morphology.

3.1 Material

Commercial available Czochralski, p-type, as-cut, 15.6*15.6 cm² solar wafers were used with a 1-5 Ωcm boron doping.

3.2 SiN deposition

A PECVD PlasmaPro 100 system from Oxford Instruments was used for the SiN deposition. Process gases were SiH_4, NH_3 and N_2 as carrier gas. The process parameters were a temperature of 400°C, a SiH_4/NH_3 gas flow ratio of 1, a process pressure of 650 mTorr, a power supply of 100 Watt and a process time of 50 minutes. These parameters are equal to a deposited SiN-layer thickness of approx. 500 nm.

3.3 Black-SiN

A RIE PlasmaLab 100 ICP 65 system from Oxford Instruments was used for the Black-SiN plasma texturing. Process gases were SF_6 and O_2. The process parameters were a capacitively coupled power supply with 110 Watt, 20 minutes process time, 50 µbar process pressure, a SF_6/O_2 gas flow ratio of 0.72, 5 °C process temperature and a He backside cooling with 6.7 mbar pressure.

3.4 SEM/TEM image acquisition

SEM images were acquired with a Hitachi SU-70. TEM images were acquired with a FEI Tecnai G2 transmission electron microscope.

4 RESULTS

Figure 2 shows SEM and TEM images of a non-textured SiN-Reference (a) and a Black-SiN sample (b) in top-view and cross-section view. The reference images (a) show a typical flat SiN-layer on top of a silicon bulk material. The reference cross-section shows a plane SiN-layer with 500 nm thickness and indicates no additional texture in the sub-µm scale. In contrast, the Black-SiN images (b) show a rough surface in the top-view and cross-section. In particular, the cross-section shows a parabolic like texture at the SiN surface with a width of approx. 100 nm. Thus, these images indicate that the Black-SiN method is technically possible.

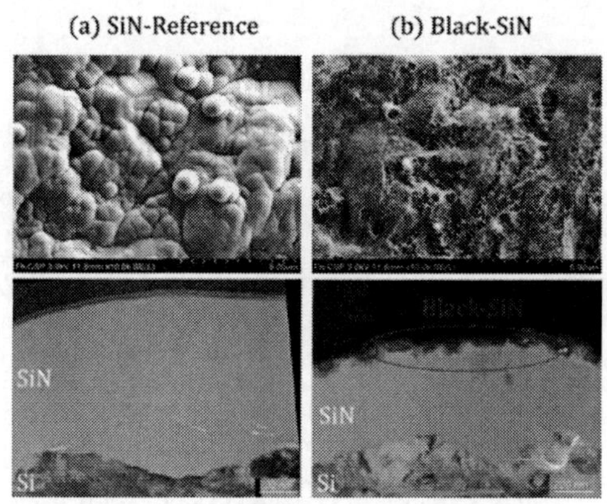

Figure 2: SEM and TEM images of a non-textured SiN-Reference (a) and a Black-SiN sample (b) in top-view (upper) and cross-section view (lower). The Black-SiN sample shows a parabolic like plasma texture in the sub-µm scale.

After the successful results shown in figure 2, we applied the Black-SiN method for the first time to an optimized SiN-layer with an initial height of 200 nm. This Black-SiN layer thickness optimization will be explained in more detail in the next section. Afterwards we used the Black-SiN method for a maskless plasma texturing of this SiN-layer. According to figure 1, we etched a parabolic like SiN texture with a height of 150 nm and a remaining SiN bulk height of 50 nm. The optical properties of this experiment are shown in figure 3.

Figure 3 shows the global reflection of an industrial standard 70 nm SiN-layer (a) and the optimized Black-SiN texture (b) as a function of the wavelength under a perpendicular angle of incidence. A comparison of both results show a reduced global reflection of the Black-SiN sample in the short wavelength region ($\lambda < 600$ nm). In the long wavelength region ($\lambda > 600$ nm), the Black-SiN sample shows a slight gain of the reflection. The two insets in figure 3 show a photo of the SiN reference and the Black-SiN sample.

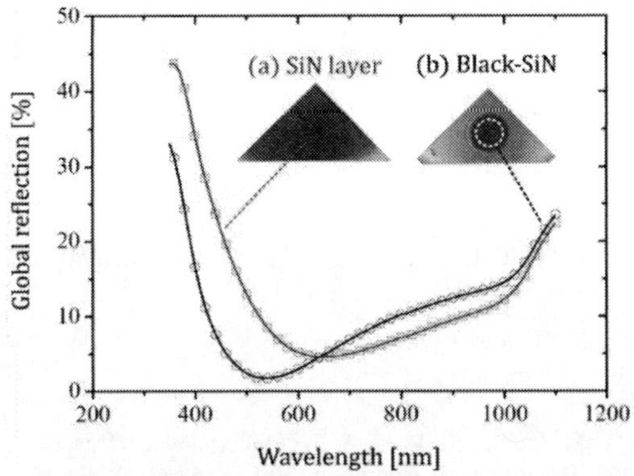

Figure 3: Global reflection of crystalline silicon with an industrial standard SiN-layer (a) and a Black-SiN antireflection layer (b) as a function of the wavelength under perpendicular illumination.

5 OPTIMIZATION

If we assume a Black-SiN texture with a structure width in the sub µm-scale, the optical properties can simulated by a combination of the effective medium theory and the Transfer-Matrix method. The following equation shows the critical ratio between a Black-SiN texture width w and the incident wavelength λ to fulfill the so-called zero-order requirement of an air-to-SiN interface [9].

$$\frac{w}{\lambda} < \frac{1}{n_s + n_i * sin(\Theta_i)}$$

With the refraction indexes of the substrate n_s and the incident medium n_i and the angle of incidence Θ_i. If the Black-SiN structure width fulfills this ratio, the optical properties can simulated by the effective medium theory in combination with the Transfer-Matrix method. If the structure width is above this ratio, the optical properties must simulated by wave optics. According to the equation above, the critical structure width of a Black-SiN texture as a zero-order effective medium in air is plotted in figure 4.

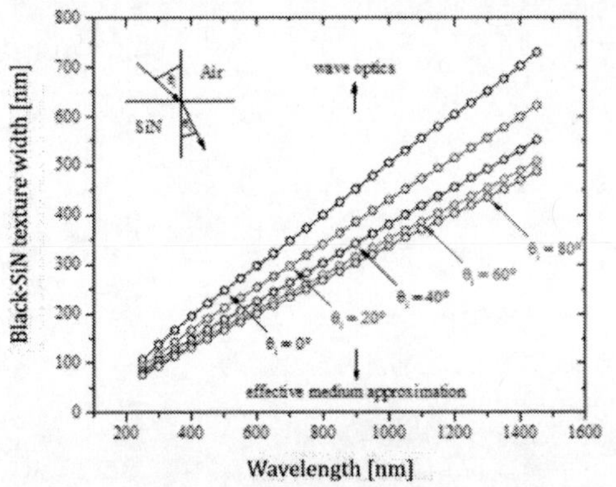

Figure 4: Maximum structure width of a zero-order effective medium from air to silicon-nitride as a function of the wavelength. If the Black-SiN texture width is below this critical ratio, the optical properties can simulated by the effective medium theory.

Figure 4 shows the maximum structure width of Black-SiN as a zero-order effective medium as a function of the incident wavelengths between 250-1450 nm. The figure shows an increase of the critical width from the short wavelength region ($w \approx 100$ nm at $\lambda = 250$ nm) to the long wavelength region ($w \approx 725$ nm at $\lambda = 1450$ nm). This gain is caused by the diffraction limit of the incident light and indicates if the light can resolve the Black-SiN texture as mentioned above. Hence, a Black-SiN texture width should be less than 100 nm, if we want to use the texture as a zero-order effective medium in air over the entire spectrum. Therefore, the optimal Black-SiN texture width as a zero-order effective medium is 100 nm.

Afterwards, this critical texture width can used for an optimization of the Black-SiN optical properties by a simulation with the Transfer-Matrix method. In this approach, the Black-SiN texture and bulk volume are sliced in thin layers with a thickness of 1 nm. For this, the so-called volume weighted effective refraction index is used. Then, the optical properties of this pseudo multi-layer stack are simulated by the Transfer-Matrix method. The refractive indices of silicon and silicon-nitride needed for this simulation were taken from the PV-Lighthouse database [10,11].

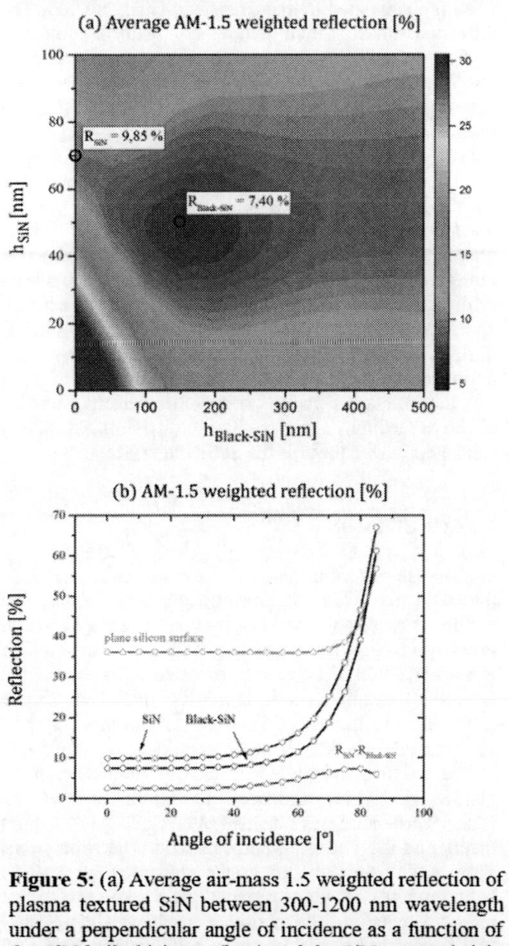

Figure 5: (a) Average air-mass 1.5 weighted reflection of plasma textured SiN between 300-1200 nm wavelength under a perpendicular angle of incidence as a function of the SiN-bulk thickness (h_{SiN}) and the SiN texture height ($h_{Black\text{-}SiN}$) simulated by the Transfer-Matrix method. (b)

Average air-mass 1.5 weighted reflection between 300-1200 nm wavelength of an optimized SiN and Black-SiN layer as a function of the angle of incidence simulated by the Transfer-Matrix method. The red curve shows the difference between both layers and indicates a benefit of the Black-SiN layer, in particular at angles of incidence above 60°.

According to Figure 1, the plot of figure 5 (a) shows the results of these Transfer-Matrix simulations as a function of the SiN-bulk thickness (h_{SiN}) and the Black-SiN texture height ($h_{Black\text{-}SiN}$). The simulations were performed at a perpendicular angle of incidence between wavelengths of 300-1200 nm with regard to the air mass 1.5 spectrum. The plot shows two points of interest. These are (I) a standard 70 nm SiN-antireflection layer with a reflection of 9.85 % and (II) the Black-SiN solution with a global minimum in the reflection of 7.40 %. This global minimum is achieved with a SiN-bulk thickness of 50 nm and a Black-SiN height of 150 nm. This result shows the possibility to reduce the reflection of a plane SiN-antireflection layer by the Black-SiN method.

Figure 5 (b) shows the reflection of the optimized SiN and Black-SiN textures as a function of the angle of incidence of the light with wavelengths between 300-1200 nm simulated by the Transfer-Matrix method. The reflection values of both textures are identical compared to figure 5 (a) in a perpendicular angle of incidence. Further, figure 5 (b) shows the reflection difference between both textures (R_{SiN}-$R_{Black\text{-}SiN}$). This difference rises from a vertical angle of incidence (R_{SiN}-$R_{Black\text{-}SiN}$ = 2.45 %) to tilted angle of incidences and has a maximum of 7.5 % at 80° incidence. This result shows the advantage of the Black-SiN texture at an oblique incidence of light. The cause of this benefit is the gradually increase of the Black-SiN effective refractive index from air to silicon-nitride. This increase provides a refraction of the incoming light to the surface normal, which results in a reduced reflection. Note, the silicon-nitride parameters, i.e. the refractive index was not optimized for these simulation. One should keep in mind that the interface between air and SiN performs like an effective medium. Therefore, the refractive index of SiN can be optimized towards the Si/SiN interface.

6 CONCLUSION

The Black-SiN method is a new approach to reduce the reflection of crystalline silicon solar cells. The method is based on a maskless plasma texturing of a SiN-layer by a SF_6/O_2 gas mixture in combination with a high ion acceleration. First results revealed a parabolic like Black-SiN texture in a scale of 100 nm. This texture fulfills the requirement of a so-called zero-order effective medium. Hence, the optical properties are improved compared to standard SiN-layers, in particular under oblique angles of incidence. An optimization of the Black-SiN texture was obtained by the effective medium theory and the Transfer-Matrix method. The simulations show a global minimum in the reflection of 7.40 % between 300-1200 nm wavelength with respect to the air mass 1.5 spectrum. The minimum is achieved by a Black-SiN texture height of 150 nm in combination with a 50 nm height of the SiN volume fraction.

7 ACKNOWLEDGEMENTS

This work was carried out within the joint research project "Black-SiN" of Anhalt University of Applied Sciences and the Fraunhofer Center for Silicon Photovoltaics CSP, which is funded by the European Union and EFRE program "Sachsen-Anhalt WISSENSCHAFT Schwerpunkte" with the funding code ZS/2017/7/86931.

8 REFERENCES

[1] H. Jansen, M. Boer, R. Legtenberg, M. Elwenspoek, J. Micromech. Microeng., Vol. 5, (1995), 155-120
[2] S. J. Wilson, M. C. Hutley, Optica Acta,Vol. 29, (1982), 993-1009
[3] J. Hirsch, M. Gaudig, N. Bernhard, D. Lausch, Appl. Surface Science, Vol. 374, (2016), 252-256
[4] S. J. Fonash, J. Electrochem. Soc., Vol. 137, (1990), 3885-3892
[5] D. Lausch, J. Hirsch, S. Wahl, S. Meyer, M. Gaudig, N. Bernhard, 31st EU-PVSEC, (2015)
[6] M. Gaudig, J. Hirsch, T. Schneider, A. N. Sprafke, J. Ziegler, N. Bernhard, R. B. Wehrspohn, J. Vac. Sci. Technol. A, Vol. 33, (2015)
[7] D. G. Stavenga, S. Foletti, G. Palasantzas, K. Arikawa, Proc. R. Soc. B, Vol. 273, (2006), 661-667
[8] H. A. Macleod, Taylor & Francis Group, Boca Raton, USA, (2010)
[9] D. H. Raguin, G. M. Morris, Appl. Optics, Vol. 32, (1993), 2582-2598
[10] M. Green, Solar Energy Materials & Solar Cells, Vol. 92, (2008), 1305-1310
[11] K. R. McIntosh, T. C. Kho, K. Fong, S. Baker-Finch, Y. Wan, S. Zin, E. T. Franklin, D. Wang, M. D. Abbott, N. E. Grant, 40th IEEE Photovoltaic Specialist Conference, Denver, (2014), 115-123

SILICON WAFER REFLECTION REDUCTION WITH MASKLESS PLASMA ETCHING BY CHF_3 AND H_2

Alena Okhorzina[1,2], Jens Hirsch[1,2], Norbert Bernhard[1], Dominik Lausch[2]
1 Anhalt University of Applied Sciences, Bernburger Str. 55, 06366 Köthen
2 Fraunhofer Center for Silicon Photovoltaics CSP, Otto-Eißfeldt-Straße 12, 06120 Halle (Saale)

ABSTRACT: Plasma etching can be applied to silicon and its compounds used in solar technology. Plasma texturing of Si surface allows getting a moth-eye surface texture which has a light trapping effect. This will increase efficiency of solar modules, due to a decrease of the reflectance. In this work, silicon samples were etched using fluorine-based plasma (CHF_3/H_2). First, screening experiments were conducted, aimed at identifying the main factors that affect the resulting surface structure. Next, the experiment is planned to obtain a black surface, which is a nanostructure of columns and cones.

Keywords: Black-Silicon, Plasma Etching, fluorine-based plasma

1 INTRODUCTION

The standard industrial texturing method for increasing the light absorption in crystalline silicon is a wet chemical etch process. Generally, pyramidal-shaped or similar structures are created.

The texturing surface has antireflective properties also in the short-wave limit, at wavelengths much smaller than the characteristic size of the texture. Texturing surface manufactures various kinds of lithographic techniques can be used, holographic patterning of the photoresist followed by etching, and plasma etching.

Plasma etching can be applied to silicon and its compounds used in solar technology silicon, silica, silicon nitride and glass. Plasma texturing of glass surface allows getting a moth-eye surface which has a light trapping effect. This will increase efficiency of solar modules.

To reach much higher aspect ratios and thus to gain more light absorption, the so-called black silicon method is a promising process. Further advantages of this dry, maskless plasma etch process are less silicon waste, no usage of fluid chemicals, better controllable etch treatment and the possibility to texture multicrystalline silicon with the same process.

Depending on the parameters of the plasma process, two modes are possible: deposition or etching. Usually, sulfur hexafluoride (SF_6) with argon (Ar) or Oxygen (O_2) are used for glass or Si etching. Interesting alternative way is using gas mix based on fluoroform (CHF_3) with hydrogen (H_2) [1].

2 MIXED PLASMA CHEMISTRIES

As a useful indicator of the predominance of etching (F) over deposition (CF_x) the F/C ratio of the discharge is often used [2]. The F/C ratio is four for CF_4, three for C_2F_6, etc. The ratio is lowered when extra Si which consumes F atoms is added or when CF_4 is mixed with H_2 or CH_4. The turn-over from deposition to etching is stimulated by ion bombardment. For a Si substrate, deposition is observed for small F/C ratios (< 2) [2].

The strongest halogen–carbon bond is the F–C bond. The most widely used mixed gas molecule is CHF_3. It dissociates primarily into CF_2 and HF species. The CF_2 intermediates are responsible for polymer formation, C_xF_y, on surfaces [3].

3 EXPERIMENTS

To create textured silicon surfaces, an etch chamber of Oxford Instruments (Plasma Lab System 100) is used. It includes a capacitive coupled power (CCP) generator as well as an inductive coupled power (ICP) generator.

Samples are monocrystalline, p-type CZ wafers with as-cut surface; wet chemical etching sample is made NaOH/IPA-process.

3.1 Pre-scanning experiments

The pre-scanning experiments series was conducted to determine the magnitude and direction of the parameters. Pre-scanning plasma etching process parameters are gas-mix is based on CHF_3/H_2: gas flow proportion 1:1; 2:1 and 1:2; pressure is between 20 and 60 mbar, CCP power is 0-200 W, ICP is 0-600 W, process temperature is 5 – 15 °C.

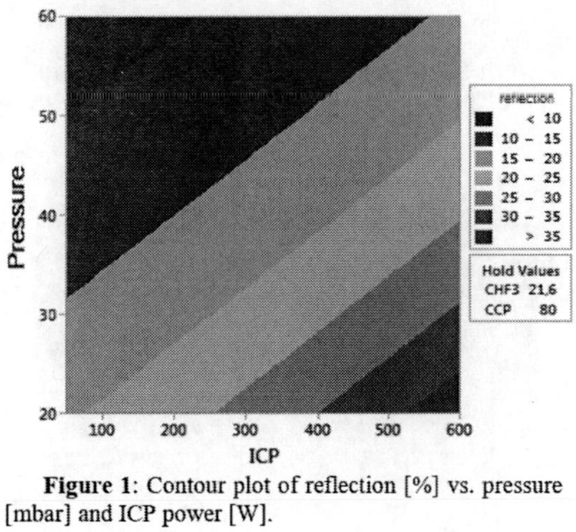

Figure 1: Contour plot of reflection [%] vs. pressure [mbar] and ICP power [W].

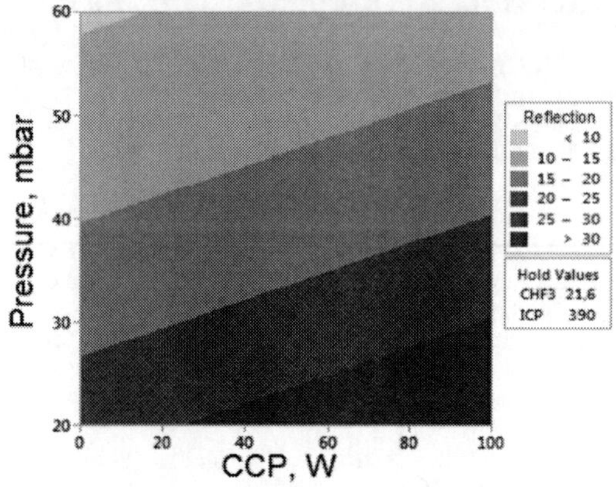

Figure 2: Contour plot of reflection [%] vs. pressure [mbar] and CCP power [W].

Pre-scan results (fig.1 and 2) show that the etching process requires a sufficiently high pressure of 40-60 mbar, high CCP power, but low ICP power. Magnitude and the ratio of gases did not give a clear understanding.

3.2 Design of experiments (DoE)

Based on the data obtained in the first experiments, a design of experiments was created. This plan is based on a DoE-algorithm for a full-factor experiment with 5 factors, 2 levels and 1 central point. The investigated response of the DoE is the wafer reflection.

Table I: Design of experiments

	low level	high level
Flow CHF_3, sccm	12	20
Flow H_2, sccm	4	10
Pressure, mbar	20	60
CCP power, W	10	100
ICP power, W	50	300

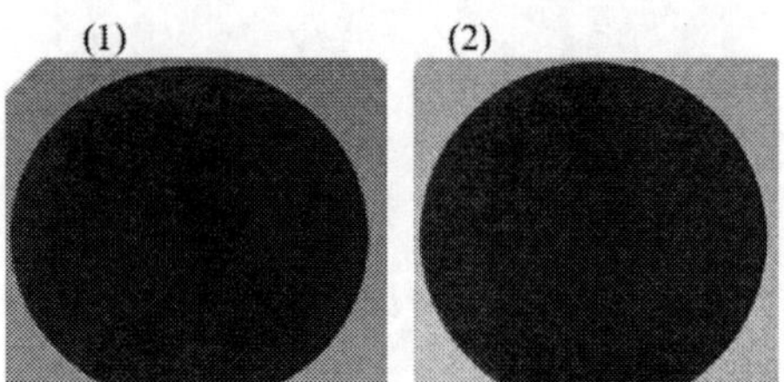

(1)	(2)
Flow CHF_3 ↓	Flow CHF_3 ↑
Flow H_2 ↑	Flow H_2 ↓
Pressure ↓	Pressure ↑
CCP power ↓	CCP power ↑
ICP power ↑	ICP power ↓

Figure 3: Photo of two samples (1) and (2) plasma etched by CHF_3/H_2.

The difference reflection of sample (1) and (2) is seen well in figure 3. The corresponding recipes of the plasma process are also different. The influence and interaction of the input factors were estimated using the DoE.

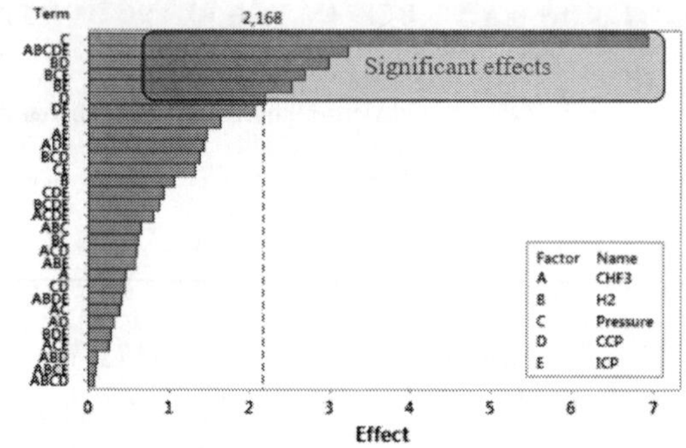

Figure 4: Pareto chart of the input effects to the wafer reflection after plasma etching (α=0,05; only 30 effects shown). High values are indicating a strong impact.

The Pareto diagram (fig. 4) is an effective means of determining which effects have the greatest contribution to the dependent variable of interest to us. Consider the Pareto effect diagram. This diagram shows the standardized coefficients, sorted by absolute value. Main parameters are pressure, CCP power and the interaction with H_2, pressure and CCP power. The influence of the amount of H_2 gas in the plasma process is very large and agrees with the theory, since the gas can enhance the deposition of the film [2].

4 RESULTS OF EXPERIMENTS

All samples can be divided into the main groups: (I) no impact to reflection, (II) brown surface after etching (sample 1) and (III) blue surface after etching (sample 2).

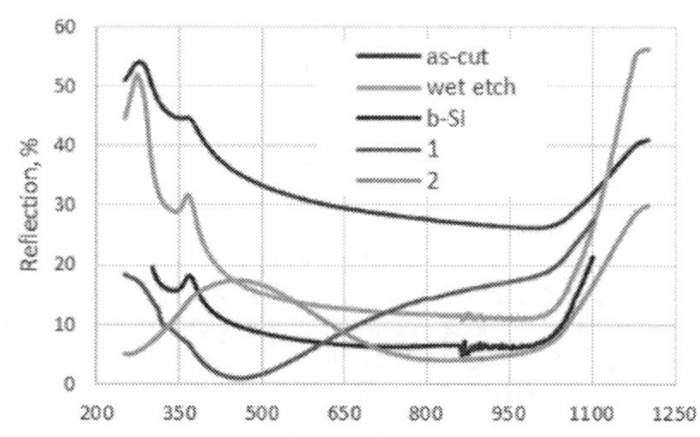

Figure 5: Global reflection of Black-Silicon etched by CHF_3/H_2 (sample 1 and 2) compared to Black-Silicon etched by SF_6/O_2 (red) and NaOH/IPA wet chemical etched silicon.

Figure 5 shows the global reflection of the different samples. The grey line is a reference as-cut wafer; orange line is etched by alkaline wet chemicals; red line is Black-Silicon etched by SF_6/O_2; blue line is sample 1 and green line is sample 2. The reflectance was measured with Perkin Elmer Lambda 1500 spectrometer.

The reflection of the two CHF_3/H_2 samples is less than the reflection of alkaline wet chemically textured sample, but is greater than the b-Si sample (red).

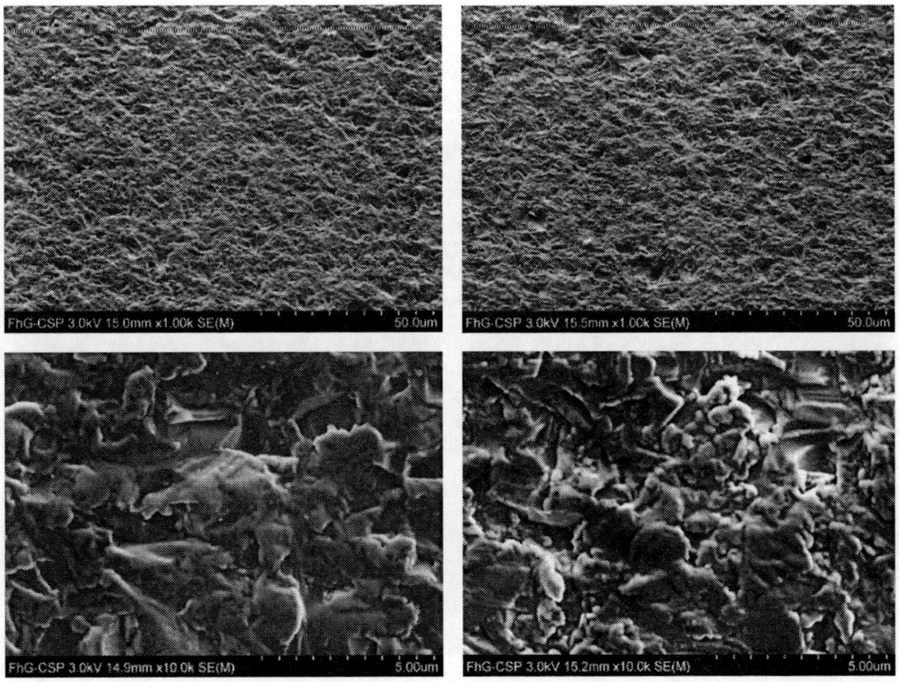

Figure 6: SEM top-view images of sample 1 and sample 2 after CHF_3/H_2 plasma texturing.

Figure 6 shows the results of the microscopy for samples 1 and 2. The image data indicate that the surface has many channels and ruts. The second sample has a rougher surface, but this surface is not yet black silicon. We can say that the greater the roughness, the lower the reflection. This statement corresponds to the method of black silicone and the eye moth effect.

5 CONCLUSION

In this work pre-scanning experiments were obtained for an maskless plasma texturing of silicon by CHF_3/H_2 within a design of experiments. Results of this investigation were the main impact parameters of the plasma etching according to the wafer reflection. The main impact parameters are (I) the CHF_3/H_2 fraction, (II) the value of capacitive and inductive coupled power, and (III) the pressure in the plasma chamber.

The reflection of two silicon samples after CHF_3/H_2 plasma texturing was investigated in this work. Sample one shows a reflection of 3-5 % in the short wavelength region (< 500 nm) and a reflection of 10-20 % in the long wavelength region (> 500 nm). The second investigated sample shows similar results, but has a reflection of approx. 20 % in the short wavelength region.

Thus, this works shows the opportunity of reduced silicon reflectance made by a maskless plasma texturing with CHF_3/H_2. This could be a promising approach to substitute SF_6/O_2 as process gases for the Black-Silicon method.

In upcoming studies, we will optimize the CHF_3/H_2 plasma texturing of silicon, based on the obtained results in this work.

6 REFERENCES

[1] Anderson H M et al 1994 Plasma Sources Sci. Technol. 3 302
[2] Coburn J W and Winters H F 1979 J. Vac. Sci. Technol. 16 391;
[3] Jancen J. Micromech. Microeng. 6 (1996) 14–28.

7 ACKNOWLEDGEMENTS

This work was carried out within the joint research graduate school "StrukturSolar II" of Anhalt University of Applied Sciences and Martin-Luther-University Halle-Wittenberg which is funded by the German Ministry of Education and Research under identification code 03EK3570A.

METALLIZATION FRACTION OF BIFACIAL pSPEER SHINGLE SOLAR CELLS

M. Al-Akash, P. Baliozian, E. Lohmüller, T. Fellmeth, N. Wöhrle, R. Preu
Fraunhofer Institute for Solar Energy Systems (ISE), Heidenhofstr. 2, 79110 Freiburg, Germany
Phone: +49 761 - 4588 5011; e-mail: mohammad.al-akash@ise.fraunhofer.de

ABSTRACT: We present the investigation of metallization fraction of bifacial p-type silicon shingled passivated edge, emitter and rear (pSPEER) solar cells intended for shingled module integration. For the first group G_1 (cell dimension: 23 mm x 148 mm), ten fabricated pSPEER cells, five each with a different busbar layout (single and double busbar), are utilized to measure the contact widths and calculate the metal coverage. Taking into account the shingling interconnected cell is overlapped, the reduced and therefore designated area metallization fraction is represented by the metal coverage of fingers and redundant line. The rear side metallization fraction for G_1 for single and double busbar layouts based on designated area remain for both busbar types almost the same with around 25.9%. The same applies to the front side resulting in around 3.0%. For the second group G_2 (cell dimension: 22 mm x 148 mm), five fabricated pSPEER cells are used. Metallization fraction for this group is lower than G_1. The designated area metal coverage fraction on the rear side is 19.8%. The front side features metallization fraction of 2.8%. A complete overlap should be ensured precisely to avoid any variations in metallization fraction affecting the expected short-circuit current density values.
Keywords: Bifacial shingle cells, pSPEER, metallization fraction, designated area.

1 INTRODUCTION

Shingling solar cells are realized by overlapping the front side busbar of a cell by the rear side busbar of neighboring cell leading to a visually busbarless front and rear side of the final shingle solar module. A complete overlap, attained by entirely covering the busbar, causes a reduction in the cell area and therefore a designated area which is the difference between total cell area and overlapped cell area [1]; see Fig. 1. Shingling interconnection of bifacial shingle solar cells is an approach to increase the output power density due to the reduction of the inactive area which is covered significantly by busbars on front and rear sides. Hence, the shading metal is then replaced by its adjacent neighbor's active area being valid also for the rear side for bifacial solar cells. The overlap provides a mechanical and electrical junction to adjacent cells. In addition, shingling decreases the series resistance on the interconnection level.

The interest in shingle cell and module technology is rising. Market share of bifacial solar cells as well as shingled interconnection technology is expected to rise in the upcoming years [2]. Monofacial shingle modules are available in the market in addition to an increase in the number of publications in shingling concepts [3–5].

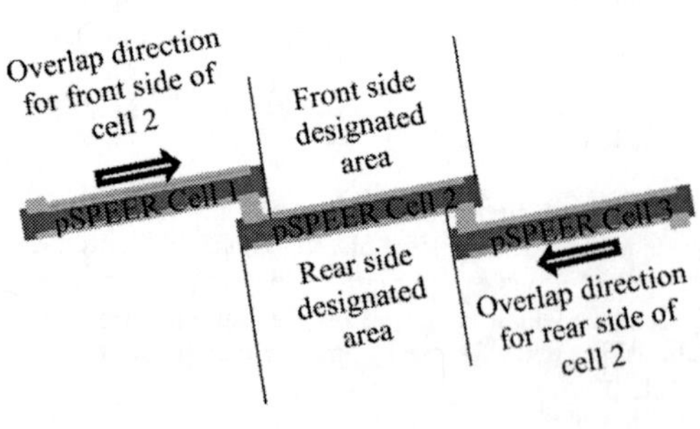

Figure 1: Labeled side-view illustration of shingled solar cells. The rear side busbar of pSPEER cell 1 overlaps the front side busbar of cell 2, whereas the rear side of cell 2 overlaps the front side of pSPEER cell 3 obtaining front and rear side designated areas of pSPEER cell 2.

The bifacial p-type silicon shingled passivated edge, emitter and rear (pSPEER) solar cells [6] are based on the p-type Czochralski-grown silicon (Cz-Si) passivated emitter and rear cell (PERC) approach. The 6-inch PERC host wafers exhibit the shingle metallization layouts on front and rear side and the single shingle cells are separated from the host wafer after contact firing. From each host wafer, six pSPEER cells are obtained.

This paper introduces a general overview of the metallization of two groups G_1 and G_2 of pSPEER solar cells with different geometries. It also investigates the change of visible metallization fraction of different overlap scenarios of pSPEER cells by taking into account scenarios meant for shingled module integration. It is crucial to examine the busbar area on both sides in order to precisely determine the overlap distance needed to maximize interconnection area and short-circuit current simultaneously.

2 METALLIZATION OF pSPEER SOLAR CELLS AND INVESTIGATION METHOD

Industrial 6-inch p-type Cz-Si PERC precursors with front and rear side dielectric passivation are used for the fabrication of pSPEER solar cells. Laser rear local contact opening is ensured by means of a pulsed infrared laser. Firstly, the metallization of the rear busbar is screen printed using a non-firing-through commercially available tabbing silver paste. Subsequently, the rear aluminum contact grid is screen printed on the local contact openings. The aluminum is connected to the priory printed silver busbar; see Fig. 2(a). Following that, a silver finger grid including busbars is screen printed on the front side. After fast firing, laser-assisted separation is used to obtain six pSPEER solar cells from each host wafer; see Fig. 2(b). For more details concerning the fabrication process, please refer to Ref. [7].

The pSPEER solar cells of the first group G_1 (cell dimension: 23 mm x 148 mm) feature single and double busbar formats, whilst the pSPEER cells of the second group G_2 (cell dimension: 22 mm x148 mm) only feature a single busbar format. Both groups feature different passivation layers. For G_1, ten fabricated pSPEER cells after separation from their host wafers [6] are utilized to investigate the metallization coverage f_{met} by light microscopic measurements; five of them have a single busbar format, while the other five feature the double busbar format. For G_2, five fabricated pSPEER cells [7] are used to determine f_{met}.

The work flow, see Fig. 3, starts by measuring busbars, fingers, pads, and redundant lines through light microscopic images taken at different locations of the cells for each sample and each side. 40 measurements are performed at different positions of the front side for each cell, whereas rear side measurements are done at 70 different positions on each cell. The median of the measured metallization dimensions is considered to avoid any outliers. Subsequently, f_{met} is determined for each group and busbar layout for front and rear sides and, additionally, for different overlap cases. The considered overlap ranges from 0 mm to 2.9 mm from the cell edge in the direction of overlap, see Fig. 1, with increment steps of 0.1 mm. f_{met} is calculated at each point using equation 1;

$$f_{met}(\%) = \frac{\sum A_m}{A_{non\text{-}ov.}} \quad (1)$$

A_m refers to the metallization area covering active area and $A_{non\text{-}ov.}$ is the difference between total area and overlapped area.

In order to compare the results based on designed and measured dimensions, f_{met} is also determined at the same overlap points using the CAD metallization layout to observe the deviation of f_{met} based on design and measurement.

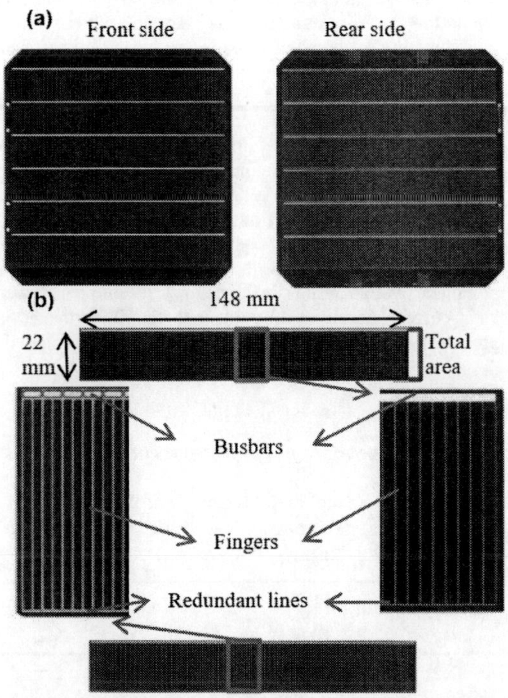

Figure 2: Photographs of (a) a host wafer of group G_2 with single busbar layout, (b) front and rear sides separated single busbar pSPEER solar cells from the G_2 host wafer. The metallization covering the active area is labeled.

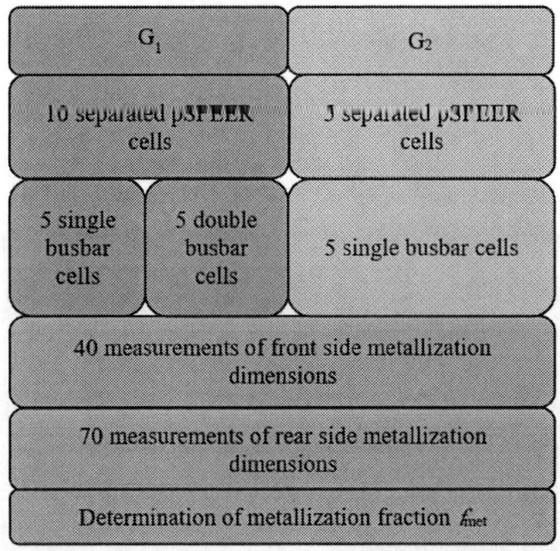

Figure 3: Flow chart of the investigation done in this work.

The rear side aluminum finger widths of the CAD layout for G_1 and G_2 are $w_{Al.rs,G1} = 250$ µm and $w_{Al.rs,G2} = 150$ µm, respectively. The rear side silver single busbar for G_1 and G_2 are $w_{Ag.rs,G1} = 600$ µm and $w_{Ag.rs,G2} = 500$ µm, respectively. For the front side, the finger width for G_1 and G_2 is designed to be $w_{Ag.fs} = 35$ µm. G_1 features a wider busbar than G_2 which reaches $w_{Ag.fs,G1} = 1300$ µm and $w_{Ag.fs,G2} = 800$ µm for G_2. The rear side redundant line of G_1 and G_2 widths are $w_{rl.rs,G1} = 100$ µm and $w_{rl.rs,G2} = 150$ µm. Whereas for the front side both groups have the same redundant line width $w_{rl.fs} = 40$ µm.

The rear side of each group and busbar layout has higher f_{met} than the front side due to the wider finger contacts on the rear side. In addition, the rear side has more fingers since the finger pitch is lower compares to the front side. Both groups have 148 fingers on the rear side. The front side of G_1 has 100 fingers compared to 114 fingers for G_2. The rear side of G_2 features lower f_{met} than the rear side of G_1 based on the finger width mentioned before.

3 RESULTS AND DISCUSSION

3.1 Measured metallization dimensions of pSPEER cells

The measured metallization dimensions on the rear and front side of each busbar layout for G_1 and G_2 is shown in Table I. In addition, the standard deviation is calculated. The standard deviation of metallization dimensions on the rear side of G_2 is higher than G_1. This can be explained due to using precursors with different rear surfaces causing a different paste spreading for both groups affecting contact geometry.

3.2 f_{met} based on design and measurements

The metal coverage differs based on designed or measured metallization dimensions, see Table II. Where f_{met} is calculated using $f_{met,tot}$ referring to the total cell size, whereas it is defined as designated area $f_{met,des}$ referring to the non-overlapped area after complete overlap of the busbars; see Fig. 5. $f_{met,des}$ of G_1 for the double busbar decreases less than for the single busbar layout, due to the existence of active area between each

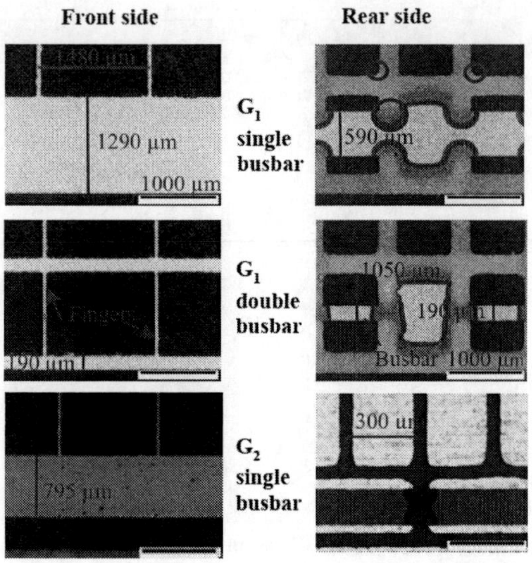

Figure 4: Microscopic images illustrating measured metallization dimensions of front and rear sides of pSPEER cells for both groups with each busbar layout. The rear side features a higher f_{met}. The aluminum fingers on the rear side smudges through the silver busbar at joining points. The highlighted joining points of fingers on the rear side feature a curved shape; however, its effect on f_{met} is negligible.

busbar on the front side and narrower busbar on the rear side. For G_1, the difference between f_{met} based on design and measured does not exceed 1%, while for G_2, it exceeds 4%. After taking the median of measured values, the single busbar finger width of the rear side based on measured data for G_1 and G_2 is $w_{Al,rs,G1} \approx 260$ µm and $w_{Al,rs,G2} \approx 195$ µm, respectively. The front side for G_1 and G_2 features a finger width of $w_{Ag,fs,G1} \approx 43$ µm and $w_{Ag,fs,G2} \approx 35$ µm, respectively.

3.3 f_{met} dependency on overlap

Taking into account the median of the metallization dimensions, f_{met} is calculated for each group. The rear side of G_1 pSPEER cells has higher $f_{met,rs}$ than the rear side of cells of G_2 mainly due to the wider fingers, see Fig. 6. While $f_{met,des}$ is determined after a complete overlap by entirely covering the busbar meant for shingled module integration as shown in the graph when f_{met} is at minimum and remains constant. After a complete overlap, the double busbar cell has lower f_{met} than the single busbar cell due to the existence of active area between each busbar on the front side, in addition to narrower busbar on the rear side compared to the single busbar. After a complete overlap, the metal coverage of the designated area is represented by the fingers area. Single and double busbar layouts of each side feature almost the same $f_{met,des}$, due to the similar finger width. A complete overlap is defined for each case in Table III. The curve shape of f_{met} vs overlap of each case is dependent on the metallization layout. Complete overlap should be precise when interconnecting shingle modules since a small variation of overlap before a complete overlap results in higher f_{met} resulting in lower short-current density. To point out the importance of precise complete overlap, two cases are considered.

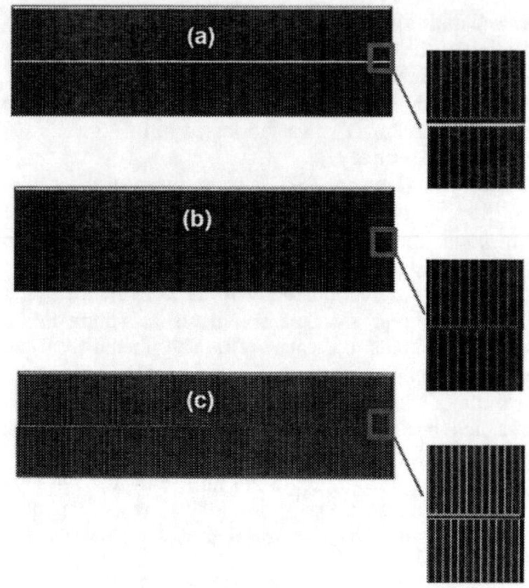

Figure 5: Photographs of (a) front side of single busbar pSPEER cells from G_2 after partial overlap, (b) front side of single busbar pSPEER cells after complete overlap and (c) rear side of single busbar pSPEER cells after complete overlap. Complete overlap from an adjacent cell is depicted to observe metallization coverage.

Table I: Measured metallization dimensions for both groups with each busbar layout of front and rear sides of pSPEER cells. Dimensions are measured through light microscopic images.

Gr.	Busbar layout	Meas. side	Finger width (µm)	Standard deviation (µm)	Finger pitch (µm)	Standard deviation (µm)	Busbar width (µm)	Standard deviation (µm)	Redundant line width (µm)	Standard deviation (µm)
G_1	Single	Front	43	4.0	1480	3.2	1290	11.3	43	6.0
		Rear	260	9.6	1010	9.4	590	10.8	180	5.2
	Double	Front	41	2.5	1480	2.7	(x2) 190	14.4	41	2.5
		Rear	260	8.1	1005	6.5	190	11.1	180	8.5
G_2	Single	Front	35	2.3	1300	2.9	795	9.5	40	2.8
		Rear	195	13.8	1005	4.0	480	11.1	210	14.2

Table II: Metallization fraction of total $f_{met,tot}$ defined as f_{met} before overlap and designated area $f_{met,des}$ defined as f_{met} after complete overlap by entirely covering the busbar for both groups of front and rear sides with different busbar layouts. f_{met} is calculated using equation 1.

Gr.	Busbar layout	Meas. side	$f_{met,tot}$ based on design (%)	$f_{met,tot}$ based on meas. (%)	$f_{met,des}$ based on design (%)	$f_{met,des}$ based on meas. (%)
G_1	Single	Front	7.9	8.3	2.5	3.0
		Rear	28.4	29.2	25.0	25.9
	Double	Front	4.1	4.4	2.5	2.9
		Rear	27.1	28.1	25.1	25.9
G_2	Single	Front	6.2	6.2	2.8	2.8
		Rear	18.1	22.5	15.4	19.8

The first case is to show the effect of overlap deviation by 0.5 mm from the complete overlap for G_1 of the rear side single busbar. The complete overlap for this case is 1.9 mm from the cell edge, measured $f_{met,des} \approx 25.9\%$. Assuming an overlap deviation of 0.5 mm causing an overlap of 1.4 mm with $f_{met} \approx 26.8\%$, resulting in $\Delta f_{met} \approx 0.9\%_{abs}$. The second case is realized by applying the same but on the rear side single busbar of G_2. The complete overlap is 1.3 mm with $f_{met,des} \approx 19.8\%$. After an assumed overlap deviation of 0.5 mm, the new overlap is 0.8 mm with $f_{met} \approx 21.4\%$, resulting in $\Delta f_{met} \approx 1.6\%_{abs}$.

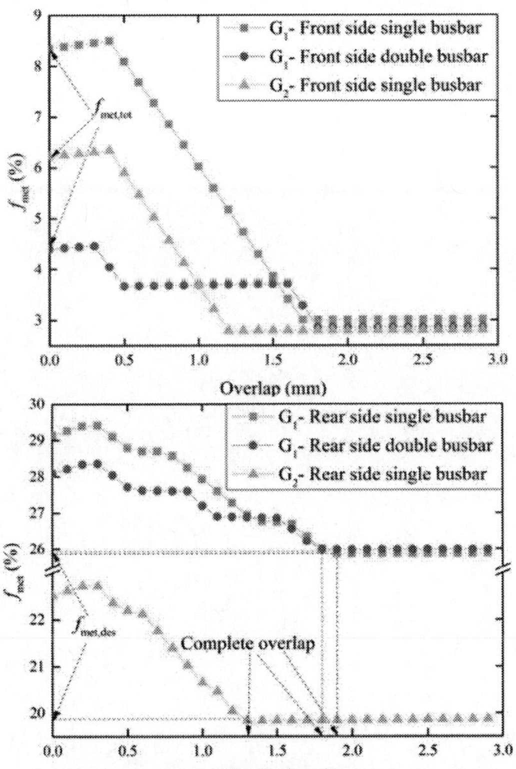

Figure 6: Calculated f_{met} versus overlap for the front and rear sides from the cell edge of each group and busbar layout. $f_{met,tot}$ is when the case of no overlap while $f_{met,des}$ is defined when f_{met} is at minimum which is shown after a complete overlap.

Table III: Complete overlap from the cell edge for both groups with each busbar layout of front and rear sides of pSPEER cells based on measured dimensions. The corresponding f_{met} is $f_{met,des}$.

Gr.	Busbar layout	Meas. side	Complete overlap (mm)
G_1	Single	Front	1.7
		Rear	1.9
	Double	Front	1.8
		Rear	1.8
G_2	Single	Front	1.2
		Rear	1.3

4 CONCLUSION AND OUTLOOK

We provide an overview of applied metallization layouts for bifacial p type silicon shingled passivated edge, emitter and rear (pSPEER) solar cells meant for shingled module integration at Fraunhofer ISE. The study shows the determined metallization fraction before overlap $f_{met,tot}$ and after complete overlap $f_{met,des}$ for two different groups of pSPEER solar cells: group G_1 with an area of 23 mm x 148 mm and G_2 with an area of 22 mm x 148 mm. Before a complete overlap, different considered overlap cases show that a small variation of overlap causes a significant change in metallization fraction. Hence, the intended overlap needed for shingle module interconnection should be precise to avoid any changes in visible f_{met}.

For G_1, after a complete overlap and therefore a designated area, $f_{met,des}$ for single and double busbar layout of each side is almost the same, as the active area is covered only fingers and redundant line. It reaches a front $f_{met,G1,fs,des}$ around 3.0% compared to rear side $f_{met,G1,rs,des}$ of around 25.9%.

For G_2, the metallization layout is represented only by a single busbar format. After a complete overlap, f_{met} for the rear and front sides is determined to be $f_{met,G2,rs,des} = 19.8\%$ and $f_{met,G2,fs,des} = 2.8\%$, respectively.

ACKNOWLEDGEMENTS

The authors would like to thank their colleagues at Fraunhofer ISE who contributed to this work. This work is supported by the German Federal Ministry of Economic Affairs and Energy within the research project "PV-BAT400" (contract no. 0324145).

REFERENCES

[1] M. A. Green, K. Emery, D. L. King, and S. Igari, "Solar cell efficiency tables (Version 15)," Prog. Photovolt: Res. Appl., vol. 8, no. 1, pp. 187–195, 2000.
[2] ITRPV, "International Technology Roadmap for Photovoltaic", 2017 Results, 2018.
[3] D. Tonini, M. Bertazzo, A. Fecchio, M.Galiazzo, Ed., Shingling Technology for Cell Interconnection: Technological Aspects and Process Integration, 33rd ed., 2017.
[4] "'Shingle cell' P-Series solar panel from SunPower maximizing polysilicon efficiencies,"
[5] H. Gochermann and J. Soll, "Shingle-type solar cell generator prodn. - allowing formation of curved or domed product," DE3942205C2, Germany 3942205, Feb 1, 1996.
[6] P. Baliozian et al., "Bifacial p-Type Silicon Shingle Solar Cells – the "pSPEER" Concept," Sol. RRL, vol. 55, p. 1700171, 2018.
[7] P. Baliozian, N. Wöhrle, E. Lohmüller, T. Fellmeth, R. Preu, Ed., Bifacial shingle pSPEER solar cells for shingle modules, 35th ed., 2018.

MODIFYING THE RATIO BETWEEN HIGHLY AND LIGHTLY DOPED EMITTERS FOR PERC WITH A SELECTIVE EMITTER STRUCTURE USING WET CHEMICAL ETCH-BACK PROCESS

Supawan Joonwichien*, Yasuhiro Kida, Masaaki Moriya, Satoshi Utsunomiya, Katsuhiko Shirasawa, and Hidetaka Takato
National Institute of Advanced Industrial Science and Technology (AIST),
Fukushima Renewable Energy Institute (FREA), 2-2-9, Machiikedai, Koriyama, Fukushima 963-0298, Japan

ABSTRACT: This paper demonstrates an improvement of a passivated emitter and rear cell (PERC) with a selective emitter (SE) structure by modifying the ratio between lightly (n^+) and highly doped (n^{++}) emitters. To produce the SE structure, screen-printed resist masking combined with wet chemical etch-back was introduced into the conventional PERC process. An adjustment of the ratio of the n^+/n^{++} regions can be achieved by modifying the areas between the unmasked and masked regions through screen-printing on an emitter. The ratio of the n^+/n^{++} regions was increased from a reference of 5 to ratios of 10 and 13, respectively, aiming to decrease the recombination losses at the front surface of the cell. As a result, the SE structure had a considerable impact on the performance of the PERC cell. Upon increasing the ratio of the n^+/n^{++} regions from 5 to 13, the open-circuit voltage (V_{oc}) increases considerably, except for the fill factor (*FF*). The decline in *FF* for SE PERC was due to the higher sheet resistance in the area for the n^+ regions during the etch-back. The improved V_{oc} is due to a lower Auger and Shockley–Read–Hall recombination at the front surface of a cell, and the enhanced performance of the PERCs. This confirms that the range of n^{++} regions should be narrower to increase V_{oc} and reduce the Auger recombination, which is a dominant recombination mechanism. The experiment results of the PERCs were modeled using the PC1D simulation program to calculate and compare the influence of the local contact pattern and SE structure on V_{oc}. The calculation using a two-diode model indicates that the recombination current for SE PERCs occurs in the depletion region of the pn-junction. The results suggest that an improvement of the emitter through an SE structure with a narrower range of n^{++} regions is essential for improving the conversion efficiency of industrial-sized PERC solar cells.
Keywords: Etch-back process, PERC, PC1D, Selective emitter, Two-diode model

1 INTRODUCTION

Saint-Cast et al. [1] conducted a loss analysis of p-type passivated emitter and rear cells (PERCs), and reported that more than half of the recombination events occur in the emitter and under the front contacts, and that the second important loss channel is associated with passivation at the rear surface and the local aluminum-back surface field. The recombination on the emitter and front metal grid can be reduced through the implementation of a selective emitter (SE) structure [2], [3]. The concept of an SE structure is based on removal of the highly doped layers in areas not intended for front metallization. The lightly doped emitter areas (n^+) showing a high sheet resistance (R_{sheet}) lead to a reduced Auger recombination. A lightly doped region ensures a better blue response, resulting in high internal quantum efficiency (IQE). However, a high contact resistance occurs, resulting in a high series resistance (R_s) [4]. In contrast, the highly doped emitter areas (n^{++}) underlying the contact regions reduce the contact resistance; however, the carrier lifetime easily degrades owing to the enhanced Auger recombination and Shockley–Read–Hall (SRH) recombination [2] [3]. Several different processes for SEs have been introduced for use in crystalline silicon solar cells, including the PERC family [5]. These processes include etch-back emitters [6]-[11], laser-doped SEs [12] [13], and the screen-printing of a phosphorus-dopant paste [14].

To produce an SE structure, we are interested in the etch-back process, which was first introduced by Zerga et al. [6], and this processcan be easily integrated into existing large-scale production lines [15]. The etch-back process allows a well-controlled reduction of the emitter surface concentration and a morphology preservation of the pyramid surface structure for maintaining a high short-circuit current density (J_{sc}). In this study, we demonstrate the development of the SE technology for use in industrial-sized PERC processing by modifying the ratio between the n^+ and n^{++} emitters. This aims to reduce the electrical losses of the phosphorus emitter, which have been reported to be major factors limiting the performance of standard PERCs. The ratio adjustment of the n^+/n^{++} regions can be achieved by modifying the areas between the unmasked and masked regions using screen-printing on the emitter. A narrower range of the n^{++} regions was designed to reduce the recombination losses at the front surface of the cell. This idea was expected to improve the open-circuit voltage (V_{oc}) and the short-circuit current density (J_{sc}) of the cell owing to a reduction in the emitter saturation current density (J_{0e}) in the n^+ regions.

As a result, an increase in the n^+/n^{++} ratio led to an improvement, which can be attributed to the reduced Auger and SRH recombination at the front surface of a cell, thereby enhancing the performance of the PERCs. In addition to conducting related experiments, we also simulated the resistance, including two saturation current densities of J_{01} and J_{02} through a curve fitting of the *I-V* data with a two-diode model, to investigate the origin of the losses owing to the SE structure.

2 EXPERIMENT DETAILS

Figures 1 and 2 show the corresponding process flow diagram and the cell structure for the PERC solar cells fabricated in this work, respectively. Solar cells were fabricated on industrial-sized (156 mm × 156 mm × 200 µm thick) 2.0 Ω.cm p-type Czochralski silicon wafers. After standard texturing and cleaning, n^{++} emitters with a R_{sheet} value of ~60 Ω/sq were initially formed in an industrial tube furnace using $POCl_3$ as a precursor gas, followed by the removal of phosphosilicate glass layers.

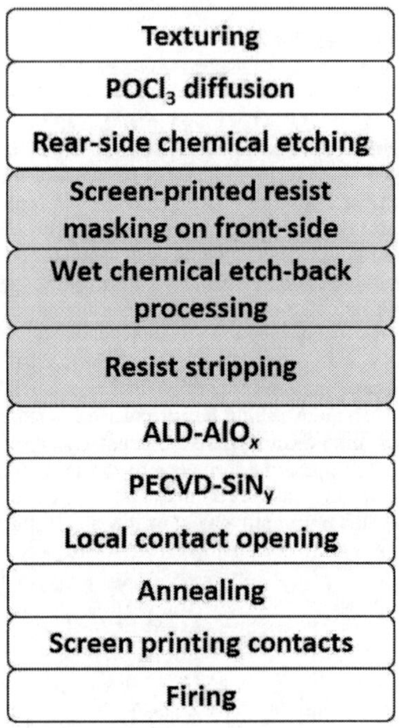

Figure 1: Process flow diagram for the PERC solar cells fabricated in this study.

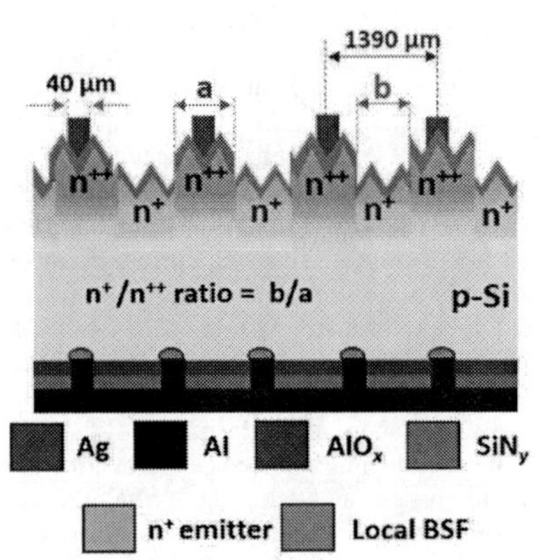

Figure 2: SE PERC structure fabricated in this study.

To produce an SE instead of a homogeneous emitter, screen-printed resist masking combined with a wet chemical etch-back was introduced into the standard p-type PERC process. A screen-printed acid resist mask was then printed on the areas intended for front Ag metallization, aiming to protect the n^{++} regions during the wet chemical etch-back process. The widths of the n^+ and n^{++} regions were controlled through the use of different screen-printed masking patterns. The ratio of the n^+/n^{++} regions was increased from the reference ratio of 5, which is normally used in the industry for inkjet printing, to ratios of 10 and 13, respectively. This idea is expected to improve the V_{oc} and J_{sc} values of the cell owing to a reduction in J_{0e} in the n^+ regions. The etch-back process can be achieved by immersing the samples into an etching solution, and the final R_{sheet} of the n^{++} regions depends on the etching times. All unmasked areas after an etch-back are at the same R_{sheet} level of 140–150 Ω/sq with an immersion time of 30 s, and 47-nm removal of the highly doped emitters. Note that the etching solution for the etched-back process was developed at Nippon Kasei Chemical Co. Ltd, and the ingredients mainly consist of HF and HNO_3 solutions. The acid-resistant barrier layer was subsequently stripped through soaking in isopropyl alcohol, and further extensive cleaning was applied in a standard SC1-SC2 sequence [16]. From here, the rear surface was passivated using a 10-nm-thick atomic-layer-deposited aluminum oxide (ALD-AlO_x) layer. Following this, a 190-nm-thick SiN_y diffusion barrier was deposited using plasma-enhanced chemical vapor deposition (PECVD). An HF-dip was used to remove the oxide layer prior to the deposition of an 80-nm PECVD-SiN_y film on the front side above the n^+ emitters. The rear passivation is locally removed through laser ablation to form line-shaped rear contacts with a local contact opening (LCO) line width of 76 µm and a pitch of 1 mm. The cells were subsequently annealed at 450 °C for 30 min to obtain a high level of surface passivation [17] [18]. Prior to metallization, the samples were briefly dipped into diluted KOH and HF solutions. The rear sides of the PERCs were metalized using a commercial Al PERC paste, and Ag printing was applied on the top with a finger length of 1.4 mm. Finally, the prepared PERCs were fired at the set firing temperature.

3 RESULTS

As shown in Table I, the resulting SE structure significantly impacted the performance of the PERC, showing an increase in V_{oc} of ~10 mV, but a decrease in the fill factor (*FF*) of ~1.8% with the other parameters remaining constant, compared with homogenous emitter (HE) PERCs. Note that the main J_{sc} losses for PERCs are due to rear recombination and absorption [1], which can be reduced through an improvement in the passivation scheme.

Table I: Cell parameters for industrial HE PERCs and SE PERCs under standard testing conditions with a cell area of 239 mm^2.

	LCO pattern	R_{sheet} (Ω/sq)	n^+/n^{++} ratio	J_{sc} (mA/cm^2)	V_{oc} (mV)	*FF* (%)	*Eff* (%)
HE 1	Line	95	–	38.6 ± 0.1	646 ± 0.8	80.8 ± 0.1	20.2 ± 0.1
HE 2	Dash	95	–	39.5 ± 0.1	651 ± 1.2	79.7 ± 0.3	20.5 ± 0.1
SE 1	Dash	95/140	5	38.6 ± 0.1	658 ± 1.6	77.8 ± 0.9	19.8 ± 0.3
SE 2	Line	60/140	5	38.7 ± 0.1	655 ± 1.2	79.0 ± 0.3	20.1 ± 0.1
SE 3	Dash	60/140	10	39.4 ± 0.1	661 ± 1.3	77.9 ± 0.2	20.3 ± 0.1
SE 4	Dash	60/140	13	39.4 ± 0.1	663 ± 0.6	77.6 ± 0.3	20.3 ± 0.1

When dash-shaped local contacts were introduced to the rear of the PERCs instead of line-shaped contacts, the results of which are shown in Table I, the values of V_{oc}, namely, ~5 mV, and that of J_{sc}, ~0.7 mA/cm^2, increased considerably. This can be understood by considering the increments in the area for the rear passivation layer and the internal reflection inside the cell. It should be noted that the small change in J_{sc} that occurred when the SE structure was applied is due to a trade-off between the competitive mechanisms for the Auger recombination in the n^{++} regions (R_{sheet} of 60 Ω/sq) and SRH recombination in the n^+ regions (R_{sheet} of 140 Ω/sq). Upon increasing the ratio of the n^+/n^{++} regions from 5 to 10, all *I-V* parameters increase, except for the *FF*.

4 DISCUSSION

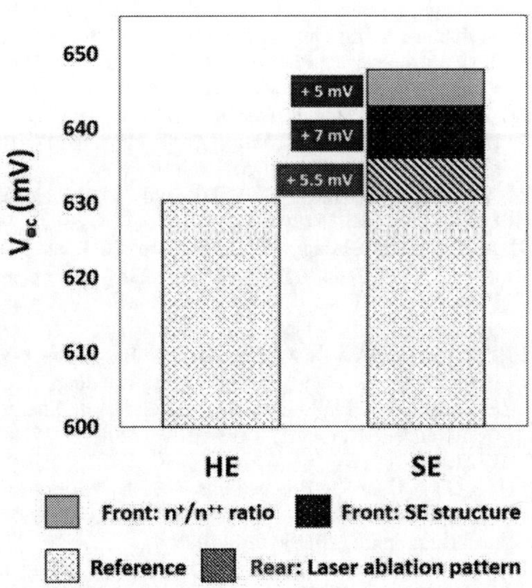

Figure 3: Simulated V_{oc} using PC1D.

As summarized in Table I, the results show significant improvements in V_{oc} for the SE PERC. However, the procedure producing the final R_{sheet} is different; that is, 95 Ω/sq emitter samples were obtained simply after phosphorus diffusion, whereas the other samples with a higher R_{sheet} were achieved after etching in a chemical solution. To compare and determine the influence of the LCO pattern, SE structure, and n^+/n^{++} ratio on V_{oc}, a simulation was carried out using the PC1D program [19]. To ensure a realistic simulation, the experimental data were used to calculate V_{oc} by fitting the calculation data with the measurements of the IQE and the reflectance of the reference cell. The results are shown in Fig. 3. Good agreement is found between the simulated V_{oc} results and the experimental data shown in Table I.

As the results of PC1D indicate, the increased V_{oc} of 7 mV is due to an SE structure with an n^+/n^{++} ratio of 5. Upon increasing the n^+/n^{++} ratio from 5 to 13, the improved V_{oc} can be attributed to the reduced J_{0e} in the n^+ regions because the SRH recombination at the passivation layer and the Auger recombination in the emitter are both low, as was previously investigated [20]. The gain in V_{oc} of 5.5 mV is due to the dash-shaped LCO. This can be understood by considering the increments in area for the rear passivation layer and internal reflection inside the cell. The value of V_{oc} increases with an increase in the thickness of the aluminum-back surface field, [21].

Table II: TLM measurements.

n^+/n^{++} ratio	5 (SE 2)	10 (SE 3)	13 (SE 4)
Actual R_{sheet} (Ω/sq)	169	166	165
Contact resistivity (mΩ.cm^2)	0.24	0.38	0.55
Finger pitch (mm)	1.39	1.39	1.39

Table III: Results of estimated parameters using two-diode model.

	R_s (mΩ)	J_{01} (A/cm^2)	J_{02} (A/cm^2)
HE 1	1.0	4.4×10^{-13}	2.7×10^{-8}
HE 2	1.0	3.7×10^{-13}	4.0×10^{-8}
SE 2	2.0	2.5×10^{-13}	3.5×10^{-8}
SE 3	2.3	1.8×10^{-13}	4.9×10^{-8}
SE 4	2.5	1.6×10^{-13}	4.9×10^{-8}

In this study, the SE PERC did not reach its potential because of a diminished *FF*. In particular, when the ratio of the n^+/n^{++} regions increased from 5 to 13, the

FF remarkably decreased by 1.4%. Table II shows the contact resistivity of the finished cells, measured using the transmission line method (TLM). It was found that the actual R_{sheet} values of the finished cells increased, presumably owing to the chemical treatment processes, i.e., the HF treatment, as well as phosphosilicate glass removal during cell fabrication. From a TLM analysis, the contact resistivity increases with an increase in the n^+/n^{++} ratio. In addition, we further calculated the resistance, including two saturation current densities of J_{01} and J_{02}, through the curve fitting of the *I-V* data with a two-diode solar cell model; the detailed principles of these calculations are described in [22]. The results are shown in Table III. The value of R_s increased from 1 mΩ to greater than 2 mΩ for the SE PERCs, showing a higher J_{02} for SE compared with the HE cells. This indicates a high contact resistance, and thus a high value of R_s, where the loss from the recombination current described by the increased J_{02} occurs in the depletion region of the pn-junction.

Therefore, we can conclude that the diminished *FF* is due to the increase in R_s upon the increase in the total area of the n^+ regions through the etch-back process. In addition, the observed improvement in the actual R_{sheet} values of the finished cells by TLM (Table II) could be one of the reasons for the reduced *FF* for the SE PERCs as compared to the HE PERCs.

Based on our calculations, to maintain an *FF* of more than 80%, the finger pitch should be reduced to ~0.78 mm for the SE PERC with a R_{sheet} value of the n^+ regions of 140–160 Ω/sq. This means that metallization technology of a fine-line finger front is required. It should be noted that the small change in J_{sc} that occurred when the SE structure was applied is due to a trade-off between competitive mechanisms for the Auger recombination in the n^{++} regions (R_{sheet} of 60 Ω/sq) and the SRH recombination in the n^+ regions (R_{sheet} of 140 Ω/sq). The modification of the n^+/n^{++} ratio does not alter the value of J_{sc} because the main J_{sc} losses for the PERCs are the rear recombination and absorption, which can be reduced by improving the LCO pattern and the passivation scheme.

5 SUMMARY

We reported our attempt to improve the PERC performance through the introduction of the SE technology using screen-printed resist masking combined with a wet chemical etch-back process. Upon applying the SE structure, the recombination losses at the front surface of the cell can be decreased by modifying the ratio between n^+ and n^{++} emitters, namely, the n^+/n^{++} ratio. Indeed, an increase in the n^+/n^{++} ratio from 5 to 13 led to an improved V_{oc} owing to the lower Auger and SRH recombination at the front, and enhanced the performance of the SE PERCs. Experiment results of the SE PERCs were modeled using the PC1D simulation program to calculate and compare the influence of the local contact pattern and SE structure on V_{oc}. In this work, a diminished *FF* was observed for the SE PERCs, which is attributed to the increase in the area of the n^+ regions showing a high R_{sheet} value, and thus a high R_s value, as calculated by fitting the *I-V* data with a two-diode model. The recombination current occurs in the depletion region of the pn-junction, showing a higher J_{02} compared with J_{01}. These results suggest that the improvement of the emitter by an SE structure with a narrower range of n^{++} regions is a very practical technique for improving the conversion efficiency of industrial-sized PERC solar cells.

ACKNOWLEDGEMENTS

This work was supported by the New Energy and Industrial Technology Development Organization (NEDO) of Japan.

REFERENCES

[1] P. Saint-Cast, S. Werner, J. Greulich, U. Jäger, E. Lohmüller, H. Höffler, and R. Preu, Phys. Status Solidi A 214 (2017) 1600708.

[2] A. Rohatgi, M. Hilai, D.L. Meier, A. Ebong, C. Honsberg, A.F. Carrol, and P. Hacke, Proceedings 17th European Photovoltaic Solar Energy Conference, Munich, Germany (2001) 22–26.

[3] B. Min, H. Wagner, A. Dastgheib-Shirazi, and P.P. Altermatt, Energy Procedia 55 (2014) 115–120.

[4] M.J. Kerr and A. Cuevas, J. Appl. Phys. 91 (2002) 2473–2480.

[5] M. Green, Sol. Energ. Mat. Sol. Cells 143 (2015) 190–197.

[6] A. Zerga, A. Shaoui, J.C. Muller, B. Bazer-Bachi, D. Ballutaud, N. Le Quang, and G. Goaer, Proceedings 21st European Photovoltaic Solar Energy Conference, Dresden, Germany (2006) 865–868.

[7] Y. Tao, K. Madani, E. Cho, B. Rounsaville, V. Upadhyaya, and A. Rohatgi, Appl. Phys. Lett. 110 (2017) 021101-1–021101-5.

[8] Y. Schiele, S. Joos, G. Hahn, and B. Terheiden, Energy Procedia 55 (2014) 295–301.

[9] A. Dastgheib-Shirazi, H. Haverkamp, B. Raabe, F. Book, G. Hahn, Proceedings 23rd European Photovoltaic Solar Energy Conference, Valencia, Spain (2008) 1197–1199.

[10] F. Book, A. Dastgheib-Shirazi, B. Raabe, H. Haverkamp, G. Hahn, P. Grabitz, Proceedings 24th European Photovoltaic Solar Energy Conference, Hamburg, Germany (2009) 1719–1722.

[11] G. Hahn, Proceedings 25th European Photovoltaic Solar Energy Conference, Valencia, Spain (2010) 1091–1096.

[12] R.S. Davidsen, H. Li, A. To, X. Wang, A. Han, J. An, J. Colwell, C. Chan, A. Wenham, M.S. Schmidt, A. Boisen, O. Hansen, S. Wenham, and A. Barnett, Sol. Energ. Mat. Sol. Cells 144 (2016) 740–747.

[13] K. Hirata, T. Saitoh, A. Ogane, E. Sugimura, and T. Fuyuki, Appl. Phys. Express 5 (2012) 016501-1–016501-3.

[14] S. Zhong, W. Shen, F. Liu, and X. Li, Sol. Energ. Mat. Sol. Cells 117 (2013) 483–488.

[15] H. Haverkamp, A. Dastgheib-Shirazi, B. Raabe, F. Book, and G. Hahn, Proceedings 33rd IEEE Photovoltaic Specialists Conference, San Diego, CA, USA (2008) 430–433.

[16] W. Kern and D.A. Puotinen, RCA Rev. 31 (1970) 187–206.

[17] B. Hoex, S.B.S. Heil, E. Langereis, M.C.M. van de Sanden, and W.M.M. Kessels, Appl. Phys. Lett. 89 (2006) 042112-1–042112-3.
[18] B. Hoex, J. Schmidt, R. Bock, P.P. Altermatt, M.C.M. van de Sanden, and W.M.M. Kessels, Appl. Phys. Lett. 91 (2007) 112107-1–112107-3.
[19] D.A. Clugston and P.A. Basore, Proceedings 26th IEEE Photovoltaic Specialists Conference, Anaheim, Calif, USA (1997) 207–210.
[20] S. Joonwichien, S. Utsunomiya, Y. Kida, M. Moriya, K. Shirasawa, and H. Takato, IEEE J. Photovolt. 8 (2018) 703–709.
[21] S. Joonwichien, S. Utsunomiya, Y. Kida, M. Moriya, K. Shirasawa, and H. Takato, IEEE J. Photovolt. 8 (2018) 54–58.
[22] M.R. AlRashidi, K.M. El-Naggar, and M.F. AlHajri, International Journal of Electrical, Robotics, Electronics and Communications Engineer 7 (2013) 93–96.

EFFECT OF LIGHT ABSORPTION FROM REAR SIDE IN BIFACIAL INTERDIGITATED-BACK-CONTACT (IBC) CRYSTALLINE SILICON SOLAR CELL

Tomihisa Tachibana, Katsuto Tanahashi, Toshimitsu Mochizuki, Katsuhiko Shirasawa, and Hidetaka Takato
National Institute of Advanced Industrial Science and Technology (AIST)
Fukushima Renewable Energy Institute, AIST (FREA), Koriyama, Fukushima 963-0298 Japan
E-mail: t-tachibana@aist.go.jp

ABSTRACT: In this study, effects of light absorption from rear side were investigated in bifacial interdigitated-back-contact (IBC) silicon solar cell. Our bifacial IBC solar cell has a front surface field (FSF) and screen-printed electrodes. A conversion efficiency of 19.7% was obtained from front side illumination with black sheet under the cell. When white sheet was placed under the cell, a conversion efficiency of 20.6% was obtained. These results indicate that bifacial IBC solar cells have the potential to fabricate two types of modules according to the condition of installing the modules. In the case of low absorption expectancy from the rear side such as residential space, a high reflectance back sheet under the cell would lead to a high-output as a one-sided absorption module. When set in a place with high absorption expectancy from the rear side such as PV plant, our bifacial module will lead to high output.

The internal quantum efficiency (IQE) spectrum and mapping analysis showed that further improvement could be obtained to suppress the carrier collection loss at the BSF region, especially busbar area. The busbar regions in both emitter and BSF act as low carrier collection centers. Thus, the Jsc of our cell is still low. The ratio of busbar electrodes on emitter and BSF region is approximately 7 % with respect to the cell area. By eliminating the busbar region, we expect to increase the Jsc by approximately 1 mA/cm^2, which should lead to over 21% conversion efficiency.

Keywords: c-Si, Back Contact, Bifacial

1 INTRODUCTION

Interdigitated-back-contact (IBC) crystalline silicon (c-Si) solar cell can achieve high conversion efficiencies because of the lack of optical shading losses due to the absence of front side metal electrodes. There are several reports of IBC solar cells with high cell efficiency exceeding 26%, with high quality passivation [1-3]. However, their production cost is still high due to the metallization process. Almost all IBC silicon solar cells with high cell efficiency are fabricated with contacts using physical vapor deposition (PVD) technology in combination with electroplating. However, the production cost by PVD technique is still high. For the future photovoltaics industry, it is necessary to reduce the production cost. To decrease the production cost, we applied a screen-printing and firing method to form electrodes for IBC c-Si solar cells [4]. In addition, our IBC c-Si solar cell has a bifacial structure due to the narrow width of the finger electrodes [5, 6]. A bifacial concept solar cell can absorb light on both sides, leading to a high output in the module.

The schematic cross-sectional image of our bifacial IBC c-Si solar cell is shown in figure 1. The bifacial IBC has a front surface field (FSF). The front side was passivated using plasma-enhaced chemical vapor deposition (PECVD) silicon nitride (SiN_x). The interdigitated emitter, back surface field (BSF), and metallization structure are located on the rear side. The electrodes were formed by screen printing and firing technology. In this study, we investigated the bifacial characteristics of our bifacial IBC c-Si solar cell, such as IV characteristics and internal quantum efficiency (IQE). In both analyses, a black sheet and white sheet were placed under the solar cell. The solar cells were fabricated with equipment used in standard c-Si solar cell production lines.

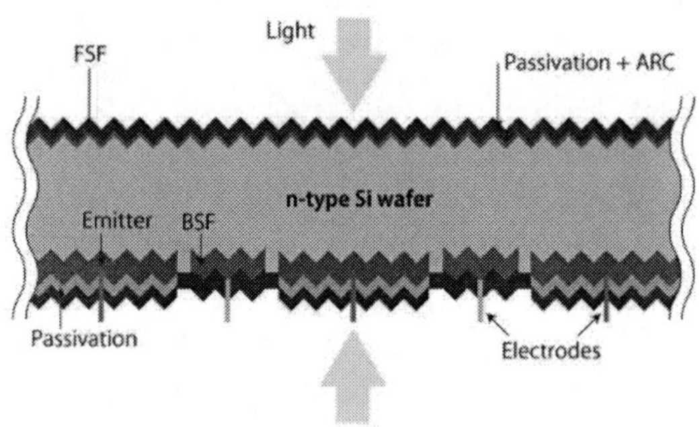

Figure 1. Schematic cross-sectional image of our bifacial IBC c-Si solar cell. The rear side can absorb light.

2 EXPERIMENTS

As starting material, n-type Cz silicon wafers with a thickness of 200 μm were used. The solar cells consist of an interdigitated boron-doped emitter and phosphorus-doped BSF and FSF. The emitter was passivated using an aluminum oxide (AlO_x)/SiN_x stacking layer and BSF and FSF were passivated using a SiN_x layer. An undiffused gap region separates the emitter and BSF regions. Al-Ag paste and Ag paste were used for emitter and BSF electrode, respectively.

On rear side, the widths of the emitter and BSF region were 1.65 mm and 0.53 mm, respectively. The undiffused gap region was 0.01 mm width. We call BSF and gap regions together and call it Base region. The ratio of the emitter and Base region was 3:1. The widths of finger electrodes were 0.06 mm. The fabricated cell size was 125 x 125 mm^2.

For the IQE analysis, we used a spectra mapping system [7]. The individual IQE spectra of the area above the emitter or the BSF region could be evaluated. The diameter of the excitation spot used in the experiment was 0.2 mm. The pitch of the measurement points was 0.1 to 0.2 mm for the mapping and line scan profile measurement.

3 RESULTS AND DISCUSSION

The IV characteristics of our fabricated bifacial IBC c-Si solar cells are shown in table 1. The cell area was 156.25 cm^2. A conversion efficiency of 19.7% was obtained from front illumination with a black sheet under the cell. A bifacial ratio ($\eta_{(rear)}/\eta_{(front)}$) of 0.90 was achieved. In addition, a conversion efficiency of 20.6% was obtained with a white sheet under the cell. These results indicate that our bifacial IBC solar cells have the potential to fabricate two types of modules: one-sided absorption or bifacial module. In the case of low absorption expectancy from the rear side, a high reflectance back sheet under the cell would lead to a high-output as a one-sided absorption module. When set in a place with high absorption expectancy from the rear side, our bifacial module will lead to high output.

The IQE spectrum and reflectance data of the bifacial IBC c-Si solar cell with a black sheet or a white sheet are shown in figure 2. The green curves are taken at the area above the emitter region and the red curves are from above the BSF region. The solid lines and dashed lines correspond to the spectra taken with a black sheet or a white sheet under the cells, respectively. Due to electrical shading effects [8, 9], there is the difference between the IQE values above emitter and BSF region. The IQE values are reduced above 950 nm with the black sheet under the cells. On the other hand, with the white sheet under the cell, the IQE spectrum increased at the long wavelengths in both emitter and BSF region. This improvement of IQE leads to an increase in J_{sc} and V_{oc}. Nevertheless it is clear from the IQE analysis that further improvement could be obtained to suppress the carrier collection loss at the BSF region.

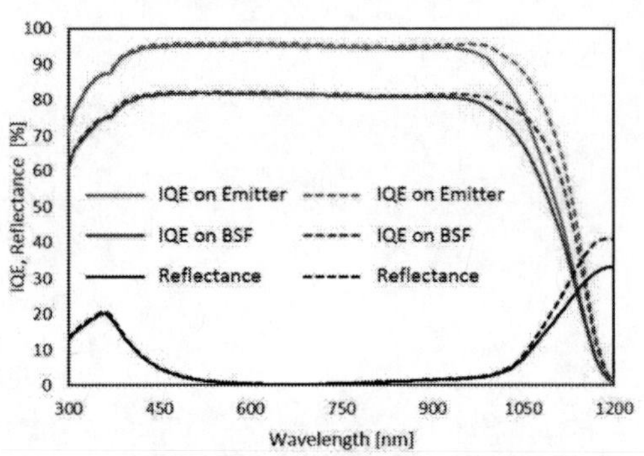

Figure 2. IQE spectrum and reflectance of bifacial IBC c-Si solar cell. Solid lines correspond to that taken with black sheet under cells, dashed ones to with white sheet.

The IQE mapping image is shown in figure 3. Low IQE values can be observed at the emitter busbar, BSF finger, and BSF busbar. The IQE values at the emitter and BSF busbar are approximately 80% and 25%, respectively. The busbar regions in both emitter and BSF act as low carrier collection centers. Thus, the J_{sc} of our cell is still low.These results agree well with the analysis data from 2D-simulation [10, 11], laser beam induced current measurement [12, 13], and photoluminescence [14].

The ratio of busbar electrodes on emitter and BSF region is approximately 7 % with respect to the cell area. By eliminating the busbar region, we expect to increase the J_{sc} by approximately 1 mA/cm^2, which should lead to over 21% conversion efficiency.

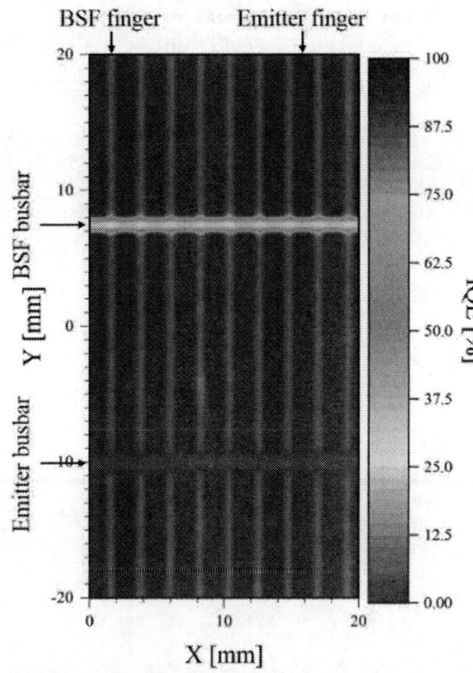

Figure 3. IQE mapping at 950 nm.

4 CONCLUSION

In this study, effects of light absorption from rear side were investigated in bifacial IBC silicon solar cell. A conversion efficiency of 19.7% was obtained from front illumination with black sheet under the cell. When white sheet was placed under the cell, a conversion efficiency of 20.6% was obtained.

The IQE spectrum and mapping analysis showed that further improvement could be obtained to suppress the electrical shading loss at the BSF region, especially

Table 1. Bifacial IBC performance measured with front or rear side illumination with black sheet under the cell. Front (white) was measured with white sheet under the cell.

Illumination	J_{sc} [mA/cm^2]	V_{oc} [mV]	FF [%]	*p*FF [%]	η [%]
Front (Black)	39.0	652.8	77.6	82.3	19.7
Rear (Black)	34.5	654.6	78.9	-	17.8
Front (White)	39.6	660.3	78.9	82.4	20.6

busbar area. The busbar regions in both emitter and BSF act as low carrier collection centers. Thus, the J_{sc} of our cell is still low. The ratio of busbar electrodes on emitter and BSF region is approximately 7 % with respect to the cell area. By eliminating the busbar region, we expect to increase the J_{sc} by approximately 1 mA/cm^2, which should lead to over 21% conversion efficiency.

REFERENCE

[1] D.D. Smith, P. Cousins, S. Westeberg, R. De Jesus-Tabajonda, G. Aniero, and Y.-C. Shen, IEEE J. Photovoltaics **4** (2014) pp. 1465-1469.

[2] K. Yoshikawa, W. Yoshida, T. Irie, Y. Kawasaki, K. Konishi, H. Ishibashi, T. Asatani, D. Adachi, M. Kanematsu, H. Uzu, and K. Yamamoto, Solar Energy Material and Solar Cells, **173** (2017) pp. 37-42.

[3] F. Haase, C. Hollemann, S. Schäfer, A. Merkle, M. Rienäcker, J. Krügener, R. Brendel, and R. Peibst, Solar Energy Materials and Solar Cells **186** (2018) pp. 184-193.

[4] T. Tachibana, K. Tanahashi, T. Mochizuki, K. Shirasawa, and H. Takato, Jpn. J. Appl. Phys., **57** (2018) pp. 040315-1 - 040315-4.

[5] G. Galbiati, H. Chu, V.D. Mihailetchi, J. Libal, and R. Kopecek, Proc. WCPEC-7.

[6] N. Guillevin, I. Cesar, A.R. Burgers, P. Venema, Z. Wang, J.Y. Zhai, and D. Liu, Proc. EU-PVSEC 2017, pp. 205 – 211.

[7] T. Mochizuki, S. Joonwichien, K. Tanahashi, K. Shirasawa, I. Sakata, H. Takato. Appl. Phys. Express, **11** (2018) pp. 086601-1 – 086601-5.

[8] C. Reichel, F. Granek, M. Hermle, and S.W. Glunz, J. Appl. Phys. **109** (2011) pp. 024507-1 - 024507-12.

[9] S. Kluska, F. Granek, M. Rüdiger, M. Hermle, and S. W. Glunz, Sol. Energy Mater. Sol. Cells **94** (2010) pp. 568-577.

[10] C. Reichel, F. Granek, M. Hermle, and S.W. Glunz, J. Appl. Phys., **109** (2011) pp. 024507-1 - 024507-12.

[11] S. Kluska, F. Granek, M. Rüdiger, M. Hermle, and S. W. Glunz, Sol. Energy Mater. Sol. Cells, **94** (2010) pp. 568-577.

[12] M. Hermle, F. Granek, O. Schultz-Wittmann, and S.W. Glunz, Proc. 33rd IEEE Photovoltaic Specialist Conf., pp. 1-4 (2008).

[13] M. Padilla, B. Michl, B. Thaidigsmann, W. Warta, and M.C. Schubert, Solar Energy Materials & Solar Cells **120** (2014) pp. 282-288.

[14] M. Padilla, H. Höffler, C. Reichel, H. Chu, J. Greulich, S. Rein, W. Warta, M. Hermle, and M.C. Schubert, Solar Energy Materials & Solar Cells **120** (2014) pp. 363-375.

X-RAY PHOTOELECTRON SPECTROSCOPY (XPS) STUDY OF THE PRINTED-SIO_X DL CAPIN PERC-TYPE SOLAR CELL APPLICATION

Yung-Sheng Lin[1], Jui-Yi Hung[2], Chen-Hao Ku[1], Jung-Ching Wang[3], Jusfong Yu[1*], Shih-Lung Lee[1], Tsung-Cheng Chen[1], Ting-Wei Guo[1], and Ching-Chang Wen[1], Yu-Chieh Lee[4], Ing-Song Yu[4]
1. E-ton Solar Tech Co., LTD., No.498, Sec. 2, Bentian Rd., Annan Dist., Tainan City 709, Taiwan, R.O.C.
2. New E Materials Co., LTD., 2F, No. 3, Luke 1st Rd., Luzhu Dist., Kaohsiung City 82151, Taiwan, R.O.C.
3. Eternal Chemical Co., LTD., No. 22, Changhsing Road, Luchu District, Kaohsiung 821, Taiwan, R.O.C.
4. Department of Materials Science and Engineering, National Dong Hwa University, Hualien 97401, Taiwan, R.O.C.

ABSTRACT: In our previous work, we demonstrated that an efficiency gain of 0.13 % can be achieved by implementing a dual capping (DL CAP, SiO_x/SiN_x) layer on our unique printed-AlO_x passivated emitter and rear cells (PERC) technology. The SiO_x layer was deposited by using screen printing and was confirmed to be a composite of SiO_2 and $Si(OH)_x$ by using Fourier-transform infrared spectroscopy (FT-IR) analysis. The $Si(OH)_x$ was speculated to be an oxidizing-agent provider which decomposes during firing, and then the released oxidizing agents diffuse to and oxidize the AlO_x/Si interface as a result of the reduction in the interface trap density (D_{it}) and better passivation quality In this work, we further analyzed the elemental composition and the electronic state of the elements of DL CAP and baseline (BSL) by using X-ray photoelectron spectroscopy (XPS) depth profile analysis. XPS result confirmed that DL CAP has a higher degree of oxidation than BSL do at the AlO_x/Si interface (O atomic %: 24.3 in DL CAP vs. 18.1 in BSL). Moreover, the binding energy of O 1s is about 534.1 eV in the printed SiO_x layer and shifts towards lower bonding energy of 532.7 eV as approaching the AlO_x/Si interface, meaning transformation from hydroxide to oxide as approaching the AlO_x/Si interface. These XPS results are consistent with the FT-IR results and are as the supporting evidence for the passivation mechanism we speculated..
Keywords: PERC, AlO_x, SiO_x, paste

1 INTRODUCTION

PERC originated in the 1980s and now PERC solar cell has become the mainstream technology in solar cell industry [1]. According to ITRPV's 2017 annual report, PERC solar cell is estimated to exceed 50 % of the global market share in the next 10 years [2].

The major differences between PERC solar cell and conventional aluminum-back-surface-field (Al-BSF) one are the passivation of dielectric film on the back surface and the locally rear metal contacts, which significantly reduces the surface recombination velocity and enhances the light reflectivity of the back surface. In our previous work, we developed a low-cost AlO_x passivation paste and applied the printed AlO_x layer to our PERC product, named ExcelTon™-Cell-III [3-7]. Instead of commonly used expensive vacuum coating equipment to deposit the rear AlO_x layer, such as atomic layer deposition (ALD) or plasma enhanced chemical vapor deposition (PECVD), E-ton's printed-AlO_x technology has advantages of low cost and high safety to grow the AlO_x layer for PERC solar cell application.

Recently, we have been developing a novel technology by applying a printed SiO_x layer to improve the passivation quality of PERC cells since 2016 [4]. DL CAP was used the same printed SiO_x technology and was designed for rear-side improvement [6]. The conventional single capping layer (usually SiN_x, for protection and hydrogenation) is replaced by a dual capping layer (SiO_x/SiN_x) where the silica layer is deposited by screen printing and the silicon nitride layer is deposited by PECVD. In our previous report, DL CAP contributed to an efficiency gain of 0.13 %, which was mostly ascribed to its 2 mV improvement in open circuit voltage (V_{oc}). However, the reason for this V_{oc} improvement is still not clarified yet [6].

Some research has indicated that the alumino-silicate layer formed at the Si/Al_2O_3 interface could induce negative charge at the interface, and the interface negative charge contributed to the formation of field effect passivation at back side of P-type silicon solar cells [8-9]. Thus the back side surface recombination velocity was reduced. In 2016, Wei Zhang et al. found the change of interfacial oxide layer after heat treatment, and proposed that the presence of interfacial alumino-silicate layer (Si-O-Al) was the reason for passivation quality improvement by wet annealing Al_2O_3 films [10].

Our previous FT-IR analysis confirmed that the absorption of O-H bonds significantly reduced, and Si-O related bonds and Al-O related bonds significantly increased after high-temperature firing of DL CAP [7]. Moreover, the absorption of Si-O related bonds in DL CAP also slightly shifted toward higher frequency after firing, implying interfacial Si-O-Al bridges increase after firing.

In this paper, XPS is utilized to investigate the elemental composition at each layer including the chemical state and electronic state of the elements of DL CAP and to provide supporting evidence to clarify the passivation mechanism of DL CAP

2 EXPERIMENTAL

This work is to investigate the chemical structure and clarify the passivation mechanism with DL CAP structure by scanning electron microscopy (SEM), life time, and XPS measurement. **Figure 1** shows the corresponding fabrication sequence and the rear-side cross-sectional SEM images of cell precursors. For approximating to the realistic solar cell performance, these cell precursors were conducted regular processes from texturing to firing in our production line except for skipping rear opening and metal deposition. As for more details about Eton's PERC process, please refer to our previous work [3-7].

In this study, an extra screen-printing SiO_x deposition, which is similar with the AlO_x deposition, was carried out before the SiN_x capping layer deposition, i.e. the rear-

AlO_x passivation layer was capped by the SiO_x/SiN_x layer to form an $AlO_x/SiO_x/SiN_x$ stack layer.

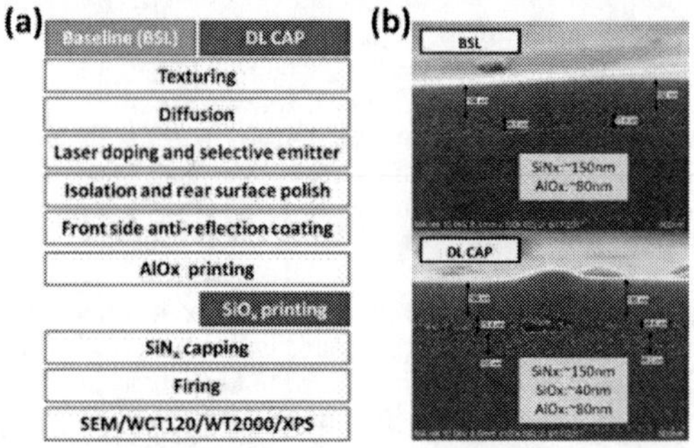

Figure 1. (a) The fabrication sequence of cell precursors with and without DL CAP structure. (b) The cross-sectional SEM images comparison between BSL and DL CAP structure.

The SiO_x layer for DL CAP structure was implemented by screen printing method and the printable silica paste was manufactured by New E Materials Corporation and Eternal Chemical Corporation. The silica paste was composed of SiO_2 precursor, additive, and organic solvent. After the paste was printed, a belt furnace was used to remove organic solvent and to provide an essential thermal budget for SiO_x formation.

WCT-120 and microwave photoconductivity decay (μ-PCD) lifetime mapping (WT-2000, Semilab) was used for minority carrier lifetime characterization. XPS (JPS-9010 MX, JEOL) analysis was performed to investigate the composition of the $AlO_x/(SiO_x/)SiN_x$ stack layer.

3 ESULTS AND DISCUSSION

The lifetime measurement by WCT-120 and μ-PCD mapping are shown in **Figure 2**. The lifetime of cell precursor was enhanced from 161 μs to 208 μs by applying DL CAP technology. The increase in lifetime implies that rear passivation quality was further improved. This result suggests that the printed-SiO_x layer in DL CAP layer plays a crucial role to enhance the passivation at the Si/AlO_x interface.

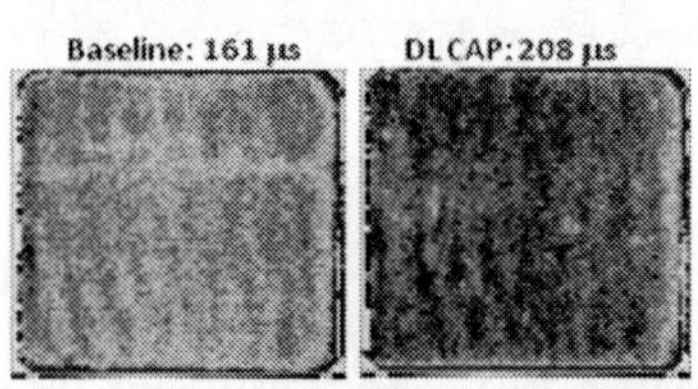

Figure 2. The lifetime comparison between BSL and DL CAP.

For clarifying the mechanism of this passivation enhancement, XPS analysis was used to obtain the chemical composition and electronic state of our printed-AlO_x/SiO_x dielectric stack. **Figure 3 (a)** and **3 (b)** show the depth profile of Si, O, Al and N in BSL and DL CAP, respectively. The corresponding thickness of each layer indicated in the figures was calculated from the etching rate of each layer. The chemical composition at some representative depths (87, 159, 209, 231, and 270 nm) was also shown in **Table 1** for the following discussion

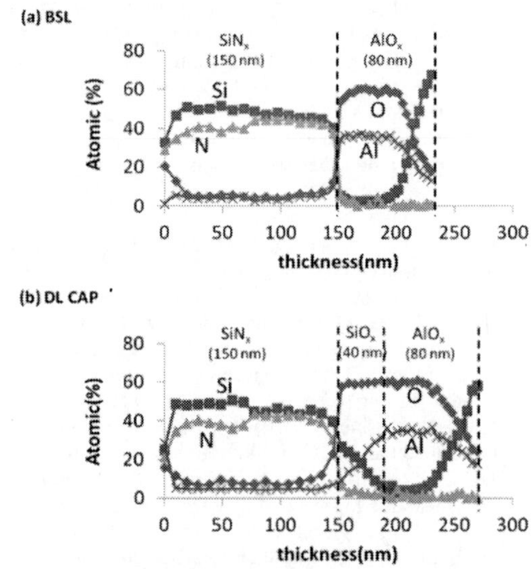

Figure 3. XPS depth profile of (a) BSL and (b)DL CAP.

In **Figure 3(a)**, the thickness of the SiN_x (applied as the capping layer) and the AlO_x (applied as the passivation layer) in BSL structure were 150 nm and 80 nm, respectively. Therefore the total thickness of the rear stack layer of BSL was about 230 nm. In DL CAP structure (**Figure 3(b)**), the SiN_x and SiO_x are applied as the capping layers stack, and the thicknesses were 150 nm and 40 nm respectively; the AlO_x is applied as the passivation layer and remained 80 nm. Therefore the total thickness of the rear stack layer of DL CAP was about 270 nm. The results of the thickness of each layer from XPS analysis are consistent with the results of the cross-sectional SEM observation in **Figure 1(b)**.

Table 1. The chemical composition (atomic %) at some representative depths of the capping layer obtained from the XPS analysis.

		SiNx	SiOx	AlOx		
	depth (nm)	87	-	159	209	230
compositions (atom%) of BSL	Si	47.2	-	3.2	22.2	67.6
	O	4.9	-	58.5	47.7	18.1
	Al	3.6	-	36.7	27.9	13.4
	N	44.3	-	1.6	0.0	0.9
	depth (nm)	87	159	209	231	270
compositions (atom%) of DL CAP	Si	44.3	23.9	3.7	8.0	57.8
	O	8.6	59.3	59.6	54.3	24.3
	Al	4.3	13.5	34.5	36.6	17.8
	N	42.8	3.2	2.2	1.1	0.0

As shown in **Figure 3** and **Table 1**, BSL decreased its oxygen content ratio from about 58.5 % to 18.1 % as it approached the AlO_x/Si interface (at a depth of 230 nm) from AlO_x layer (at a depth of 159 nm). Whereas the oxygen content of DL CAP decreased from 59.6 % to 24.3 % as it approaches the AlO_x/Si interface (from a depth of 209 to 270 nm). It is confirmed that the oxygen content of DL CAP is higher at AlO_x/Si interface than that of BSL. In addition, the oxygen content of DL CAP is higher than that of BSL for the entire depth range of XPS analysis (BSL: 0-230 nm, DL CAP: 0-270 nm). This result is consistent with our previous research [7]: FT-IR showed that the more intensive oxygen-associated signals of the DL CAP than those of the BSL. Both XPS and FT-IR results indicate that the DL CAP film has a higher degree of oxidation than the BSL do, especially at the AlO_x/Si interface where significantly affects the passivation quality. This higher oxidation ratio at the AlO_x/Si interface implies that the more oxygen-rich interfacial alumino-silicate layer formed in DL CAP after firing.

To clarify the mechanism and the effect of back side capping and passivation on various interfaces by applying printable-silica and AlO_x paste, we further investigated the electronic state of oxygen at different depths of DL CAP. The O 1s signal of a typical Al-O or Si-O bonding is 532 eV and the O 1s signal of the more electron-rich oxygen species such as Al-OH or Si-OH is greater than 532 eV [11-12]. **Figure 4** shows the evolution of the bonding energy of O 1s within the depth range of 150 to 230 nm. The binding energy of O 1s at the SiO_x/AlO_x interface remained, which was about 534.1 eV. However, the binding energy of oxygen shifted towards lower bonding energy of 532.7 eV as approaching the AlO_x/Si interface. This is consistent with our speculation from FT-IR analysis [7]: The printed-silica layer is a composite of SiO_2 and $Si(OH)_x$ and is mainly composed of $Si(OH)_x$. This $Si(OH)_x$ in DL CAP can serve as an oxidizing-agent provider which decomposes during firing, and then the released oxidizing agents diffuse to and oxidize the AlO_x/Si interface as a result of the reduction in D_{it}.

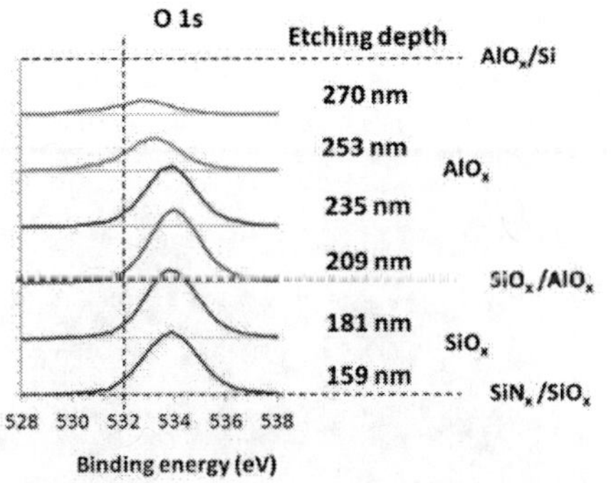

Figure 4. The O 1s XPS signal of DL CAP at different etching depth.

4 CONCLUSION

By applying E-ton printable-silica paste with the same organic system as the AlO_x paste, the DL(SiO_x/SiN_x) capping is realized and has better lifetime and passivation quality. In consistent with previous FT-IR analysis, the XPS analysis find that DL CAP film has a higher degree of oxidation than the BSL do at the AlO_x/Si interface, which helps the passivation quality. This improvement is suggested to be ascribed to the decomposition of Si(OH)x in DL CAP and the AlO_x/Si interface oxidation by the decomposition-released oxidizing agents during firing. Finally, this interface oxidation results in the reduction in D_{it} and better passivation quality

5 REFERENCE

[1] Andrew W. Blakers, Aihua Wang, Adele M. Milne, Jianhua Zhao, and Martin A. Green, "22.8 % efficient silicon solar cell," *Appl. Phys. Lett.* 55, 1363 (1989).

[2] Markus Fischer, "C-Si PV: Manufacturing trends & challenges - ITRPV 8th edition update (incl. maturity report)," World Solar Congress, Shanghai, September 5-6, 2017.

[3] Jui-Yi Hung et al., "Back-Side AlO_x Passivation Material and Technology for the Application of High Efficiency (20 %) and Low Cost PERC Solar Cells", n *41st IEEE Photovoltaic Specialist Conference*, 2014.

[4] Yung-Sheng Lin et al., "Improving the Passivation and the reflectance uniformity of SiN_x anti-reflection coating by using the SiOx passivation paste," in *43rd IEEE Photovoltaic Specialist Conference*, 2016.

[5] Tsung-Cheng Chen, Jui-Yi Hung, Yung-Sheng Lin, Chun-Fu Lin, Chen-Hao Ku, Jung-Ching Wang, S.-W. Chen, Cheng-Shun Hu, Ching-Chang Wen, "E-Ton's printed-AlO_x PERC Cells: efficiencies beyond 21 % with a next-generation AlO_x paste," in *32nd European Photovoltaic Solar Energy Conference and Exhibition*, 2016.

[6] Yung-Sheng Lin et al., "Toward 21.4 % efficiency by implementing industrially feasible technolgies in printed-AlO_x PERC technology," in *33rd European Photovoltaic Solar Energy Conference and Exhibition*, 2017.

[7] Tsung-Cheng Chen, Chen-Hao Ku, Ing-Song Yu, and Zu-Po Yang, "Passivation Mechanism of the Printed-SiO_2 DL CAP in PERC-type Solar Cell Application", submitted to *45th IEEE Photovoltaic Specialist Conference*, 2018.

[8] B. Hoex, S.B.S. Heil, E. Langereis, M.C.M. van de Sanden, and W.M.M. Kessels, *Appl. Phys. Lett.*, 89, 042112 (2006).

[9] B. Hoex, J. Schmit, R. Bock, P.P. Altermatt, M.C.M. van de Sanden, and W.M.M. Kessels, *Appl. Phys. Lett.*, 91, 112107 (2007).

[10] S. Joonwichien, K. Shirasawa, S. Simayi, K. Tanahashi, and H. Takato, "Wet oxidation effects on the electrical and interface properties of ALD Al_2O_3 and ALD-AlOx/SiNx passivation stacks for PERC solar cells," in *32nd European Photovoltaic Solar Energy Conference and Exhibition*, 2016.

[11] H. Li, A. Belkind, F. Jansen, Z. Orban, "An in situ XPS study of oxygen plasma cleaning of aluminum surfaces", *Surface and Coatings Technology*, 92, 171-177 (1997).

[12] M. E. Simonsen, C. Sønderby, Z. Li., E. G. Søgaard, "XPS and FT-IR investigation of silicate polymers", J. Mater. Sci. 2009, 44, 2079–2088.

LASER DOPING FROM PSG FOR SELECTIVE FSF OF SCREEN PRINTED REAR-JUNCTION N-PERT CELLS

Sukhvinder Singh[1], Loic Tous[1], Patrick Choulat[1], Jia Chen[2], Ronglin Liu[2], Liming Ma[2], Xinghua Wu[2], Jianming Wang[2], Zhifeng Liu[2], Filip Duerinckx[1], Ivan Gordon[1], Jozef Szlufcik[1], Jef Poortmans[1,3]
[1] imec, Kapeldreef 75, 3001 Leuven, Belgium
[2] Jolywood, NO.6 Kaiyang Rd., Jiangyan economic development zone, Taizhou 225500, Jiangsu, China
[3] UHasselt, B-3590 Diepenbeek, Belgium

ABSTRACT: We report detailed studies on laser doping from phosphorus silicate glass (PSG) to form the selective front surface field (FSF) for screen printed n-type passivated emitter, totally diffused (nPERT) solar cells. All relevant parameters for the laser doping process have been characterized at different laser conditions. These include measurements of sheet resistance, specific contact resistivity, recombination losses at the passivated laser-treated and metallized surface (measured by $J_{0n++pass}$ and $J_{0n++metal}$ respectively), optical and scanning electron microscopy SEM investigations. State of the art and very low dark saturation current densities have been achieved on passivated n^+, passivated laser doped (n^{++}) and metalized surfaces while maintaining low specific contact resistivity. The developed process for selective front surface field (FSF) with chosen doping parameters has been implemented in the monofacial nPERT devices with screen printed contacts resulting in independently confirmed efficiencies up to 23%.

Keywords:, n-type, Laser Processing, Doping, Screen Printing, Metallization

1 INTRODUCTION

According to the ITRPV roadmap of 2018, global production capacity of PV modules was estimated to have more than 90 % share of crystalline silicon solar cells by the end of 2017 [1]. It is mentioned that the module prices continued to decrease in 2017 but at slower rate than in 2016. The rapid decrease in module prices is required to bring down the levelized cost of PV electricity (LCOE). The lower LCOE due to balance of system costs translates in the need for cheaper solar cells with an increase in conversion efficiency [2]. Presently, the majority of PV modules are made using front junction solar cells based on p-type Czochralski (Cz) substrate with full aluminum BSF and more recently with p-type cells with local rear contacts (PERC cells) [3]. This transition is propelled by the need to reduce metallization related recombination and optical losses on the rear side of the cell thereby improving the efficiency. It has been stated that the most efficiency limiting parameter of the screen–printed metallization is the open–circuit voltage of the cells [4]. In order to reduce recombination losses from Ag front metallization, various selective emitter methods have been proposed [5,6]. These selective emitter technologies with heavy doping under screen printed contacts and shallow doping elsewhere offers the advantages of having both, a low specific contact resistivity and low recombinations for passivated surface and metal contacts ($J_{0n++metal}$) without compromise.

The next technology transition will be towards cells based on n-type substrates with several advantages such as absence of light induced degradation, higher tolerance to many transition metal impurities (including Fe, Ti, V, Cr) and much lower sensitivity to the presence of the laser induced dislocations [7,8]. The share of n-type silicon based solar cells is expected to increase in the coming decade [1]. Rear junction n-type Passivated Emitter, Rear Totally diffused (nPERT) cells could use many of the learnings from p-type PERC solar cells and allow an easier transition from p to n-type. For such a rear junction device, it is even more important to have selective FSF for low minority carrier recombination on front surface [9]. Different routes for selective doping have been reported in literature such as etch back by using mask, laser doping from spin-on-dopant or PSG, doping from a printed paste combined with $POCl_3$ diffusion and double diffusion [5,6,10]. Among these, laser doping from PSG to form selective FSF is one of the simplest routes since it doesn't require any masking, extra doping source or etching chemicals. Schematic of front side of n-type rear junction solar cell with selective FSF using laser doping is shown in Fig. 1. As depicted in the figure the laser doped regions (n^{++}) are wider than screen printed fingers. This is needed to have sufficient alignment margins between laser doped regions and screen printed fingers, due to the deformations and alignment repeatability of screen printed and laser doped patterns.

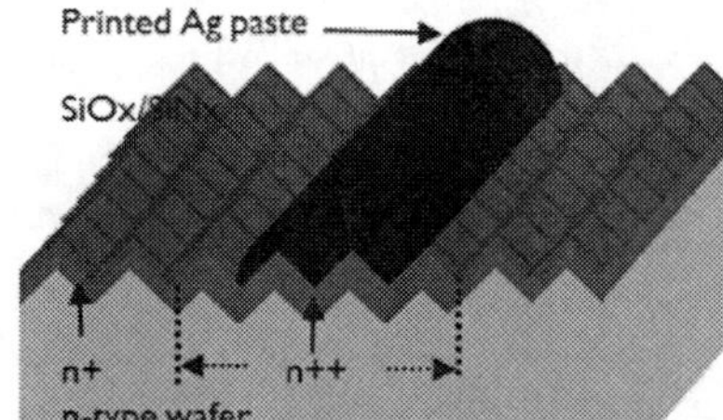

Figure 1: Schematic of front side of n-type rear junction solar cell with laser doped (n^{++}) selective FSF

Therefore, in order to benefit from the selective doping using laser, it is essential to have low surface damage on the textured wafer to limit reflectance and recombination losses at the laser-treated passivated surface ($J_{0n++pass}$) and have high enough doping to shield the surface from metallization induced recombination losses ($J_{0n++metal}$). Development and detailed characterization of such a selective FSF using laser doping from PSG obtained from homogeneous and shallow $POCl_3$ FSF (>200 Ω/sq) is presented in this work.

2 PROCESS FLOW

The process sequence for investigation of laser doped FSF for screen printed contacts is listed in Fig. 2. In this work, we use commercially available n and p-type Cz wafers (as-cut thickness ~200 µm, resistivity ~ 5 Ω.cm for n-type wafers). After saw-damaged-removal (SDR) and texturing, an ozone (O_3) based cleaning recipe was used prior to shallow $POCl_3$ FSF diffusion. Subsequently heavily doped n^{++} (selective FSF) regions were formed by laser doping using PSG as dopant source. Different sets of laser parameters were used to achieve varying sheet resistance at laser doped regions. Following PSG removal and cleaning, front and rear side passivation was performed using thermal oxide and PECVD SiN_x. Afterwards, screen printing of commercially available Ag fingers was carried out on part of the wafers. Some of the wafers underwent multiple printing to obtain varying metal contact fractions, allowing $J_{0n++metal}$ measurements. Then wafers were fired in a fast-firing infrared belt furnace using different firing profiles. In order to extract $J_{0n++metal}$ using QSSPC measurements, Ag paste was etched off in a diluted mixture of HCl and HNO_3. Detailed characterization was carried out at different stages of the process flow. These include, measurement of sheet resistance, J_0 measurements on homogeneously n^+ doped passivated regions ($J_{0n+pass}$), passivated laser doped ($J_{0n++pass}$), and on metallized regions ($J_{0n++metal}$) using QSSPC measurements, specific contact resistance ($\rho_{contact}$) and SEM measurements. Results of these characterizations are listed in the sections below.

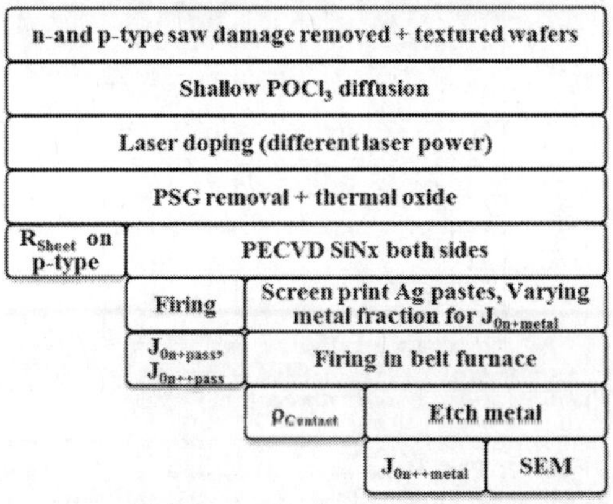

Figure 2: Process flow for investigation of laser doped FSF

3 RESULTS AND DISCUSSIONS

3.1 Sheet Resistance measurements

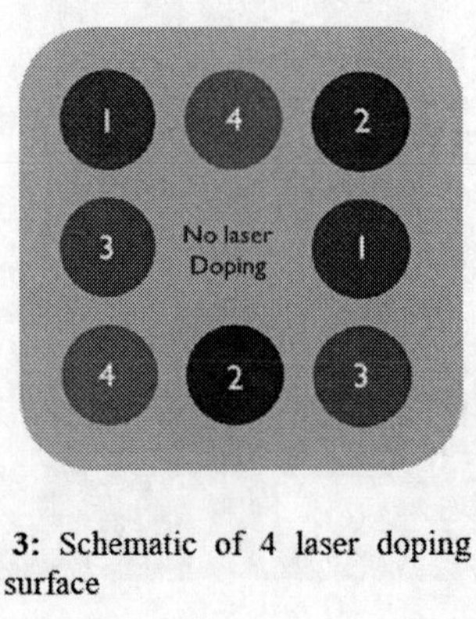

Figure 3: Schematic of 4 laser doping conditions on sample surface

Table I Average sheet resistance measured on laser doped p-type samples.

Laser doping parameters set	Average R_{sheet} (Ω/sq)	Standard deviation (±)
1	71	4
2	102	9
3	116	9
4	155	21
No laser	308	18

Fig 3 shows the schematic representation of laser doped regions on the test samples. Four different laser doping conditions were used on the PSG layer to achieve different sheet resistance. Homogeneous laser doped regions allow the measurement of sheet resistance, contact resistance and QSSPC measurements. Average R_{sheet} values along with standard deviation measured on p-type wafers before ARC coating are listed in table I. Starting from homogeneous shallow diffusion of around 300 Ω/sq, values ranging from 70 to 155 Ω/sq were achieved for different laser parameters. It was noted for high sheet resistance 155 Ω/sq, standard deviation increases drastically. Values ranging from 140 Ω/sq to 175 Ω/sq were observed at two different locations of the same sample. This is most likely related to the use of very low laser power being at the threshold of doping from PSG. Thus homogeneous doping up to 120 Ω/sq could be achieved starting from 300 Ω/sq $POCl_3$ diffused surface.

3.2 Specific contact resistivity measurements

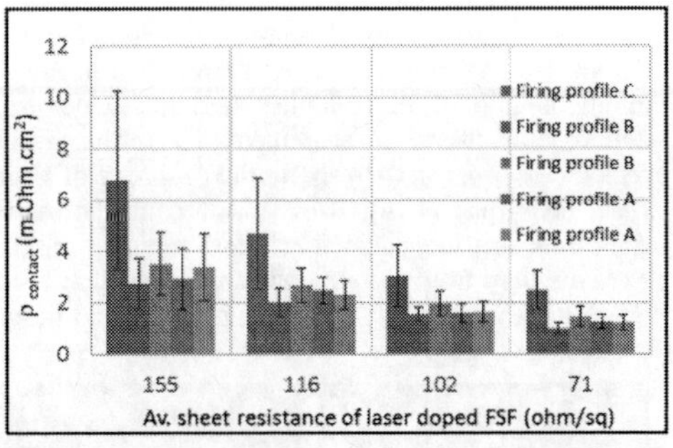

Figure 4: Contact resistivity vs different firing profiles

The objective of the present work is to find an optimum process window for laser doping process to achieve a low contact resistivity, low J_0 and low surface damage. Firstly, specific contact resistivity ($\rho_{contact}$) measurements were carried out using TLM method on screen printed and fired samples. For each sheet resistance, two locations on each sample (Fig. 3) were measured. Three different firing profiles namely A, B and C and two different commercial fire though Ag pastes were used. Fig 4 shows specific contact resistivity measured on screen printed samples, fired with different firing profiles. Except for the case of fast profile C, both firing profiles result in similar $\rho_{contact}$ values for each of the sheet resistance. It is observed that $\rho_{contact}$ values between 1.5 and 2.5 mΩ.cm² could be achieved for sheet resistance of 100-120 Ω/sq. For 155 Ω/sq, $\rho_{contact}$ values increase slightly above 3 mΩ.cm² with larger standard deviation. Similar values could be achieved for the other paste (not shown here). The larger standard deviation is due to inhomogeneity in laser doping for 155 Ω/sq at different area of the surface as mentioned above. Nevertheless, the $\rho_{contact}$ values are comparable to value of 4 mΩ.cm^2 reported in literature for homogenous FSF of 155 Ω/sq [11]. As expected, contact resistivity was too high to be measurable at wafer regions without laser doping (300 Ω/sq).

3.3 Photoluminescence and dark saturation current density (J_0) measurements

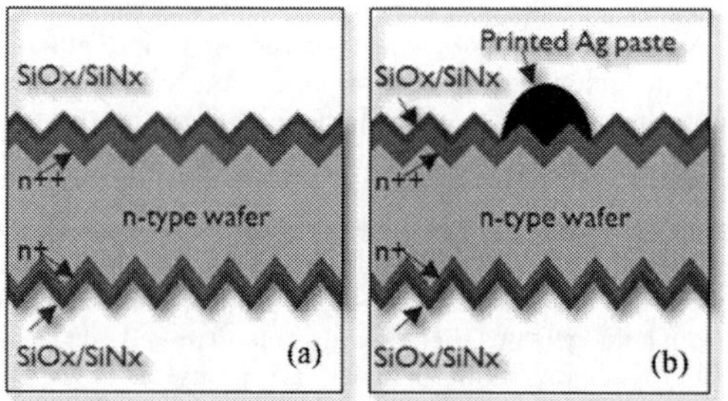

Figure 5: (a) Schematic of test samples for $J_{0n++pass}$ and (b) $J_{0n++metal}$ measurements

Test samples for measuring $J_{0n++pass}$ and $J_{0n++metal}$ measurements are schematically represented as in Fig 5 (a) and (b). The samples feature a laser doped surface with heavy n^{++} doping on the front and shallow n^{+} (of 300 Ω/sq) on the rear. For $J_{0n++metal}$ measurements, the metal contact fraction (C_{fmetal}) was varied between 3 % to about 10 % by screen printing. Typical uncalibrated PL images of test samples (measured from non-lasered side) with laser doping are shown in Fig 6. As represented in Fig 2, the middle region of the samples does not undergo laser doping, allowing the measurement of J_0 of the passivated region ($J_{0n+pass}$). Fig 6 (b) shows the PL image of a sample with screen printed Ag. The $J_{0n++pass}$ could be measured from tests structures shown in Fig 5 (a). $J_{0n++metal}$ values were mesuerd from test structures represented in Fig 5 (b) by plotting total J_0 vs C_{fmetal}, using the method described by Deckers et.al [12].

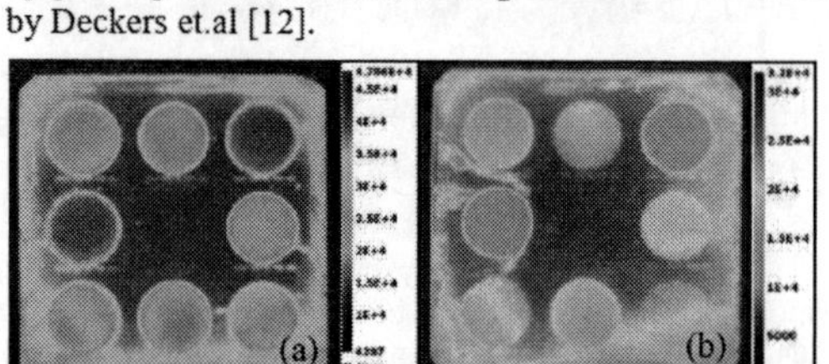

Figure 6: Un-calibrated PL images of (a) sample with laser doping and (b) with laser doping and screen printing.

Measured values of $J_{0n++pass}$ and $J_{0n++metal}$ are plotted against the n^{++} sheet resistance in Fig. 7. Very low and state of the art $J_{0n++pass}$ values for passivated n^{++} doped regions were measured for all conditions. $J_{0n++pass}$ values decrease with increase in sheet resistance as expected, partly due to lower Auger recombination. For sheet resistance values ranging from 100-120 Ω/sq, $J_{0n++pass}$ values down to 50 fA/cm^2 could be obtained. As expected, $J_{0n++metal}$ values (also shown in the Fig 7), increase with increase in sheet resistance. This is due to diminishing of shielding effect of metal contacts at higher sheet resistances. Very high $J_{0n++metal}$ value of 1000 fA/cm^2 is obtained for samples without laser doping and thus having only shallow (300 Ω/sq) diffusion underneath Ag contacts. This also emphasizes the importance of having higher doping under screen printed Ag contacts, and hence the need for a selective FSF. Since low contact resistance is a must for choosing a condition for laser doping, sheet resistance values up to 120 Ω/sq could be selected for device integration. This corresponds to $J_{0n++pass}$ values of < 70 fA/cm^2 and $J_{0n++metal}$ values of around 500 fA/cm^2. The values of $J_{0n++metal}$ are comparable to state of the art values reported in literature [4,13,14]. It has been stated that $J_{0n++metal}$ values <200 fA/cm^2 could be achieved if the screen printing pastes do not etch the emitter [13], However, no such values are reported in literature for fire-though pastes yet. Furthermore, very low $J_{0n+pass}$ values of around 14 fA/cm^2 were obtained for passivated region without laser doping.

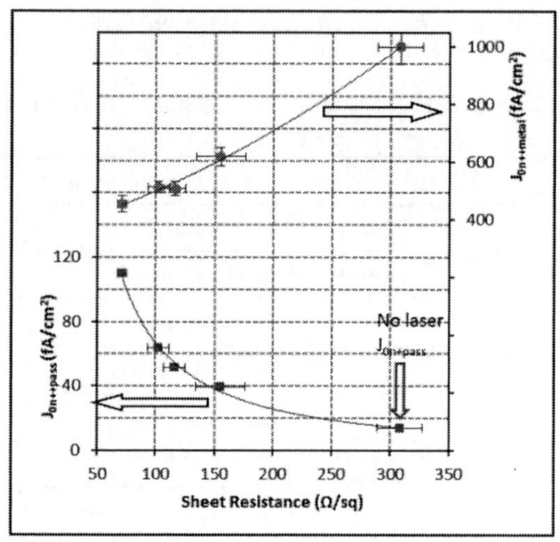

Figure 7: Plot of $J_{0n++pass}$ and $J_{0n++metal}$ for all samples using chosen firing profile and screen printing paste.

3.4 Surface damage analysis SEM analysis

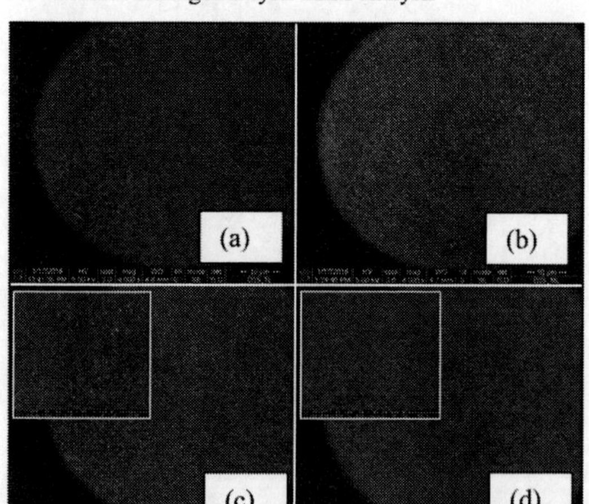

Figure 8: SEM images of laser doped surfaces.

It is intended to have minimal damage to the textured surface to avoid increase in front surface reflectance. SEM investigations of laser doped surfaces, carried out to check surface damage due to laser doping are shown in Fig 8. From SEM images of 70 Ω/sq laser doped sample (Fig 8 (a)), significant damage to the textured surface is observed. However, for other cases (100-155 Ω/sq), shown in Fig 8 (b) and Fig 8 (c) only melting of pyramid tips is observed, similar to such phenomena reported in literature [15]. A SEM image of the reference surface without laser is shown in Fig 8(d). Thus laser doping resulting in a sheet resistance of around 100-120 Ω/sq is chosen for device integration.

3.5 I-V result with process integration

The developed laser doped FSF process sequence was integrated in large area (244.3 cm^2) monofacial screen-printed rear-emitter nPERT busbar less cells. Laser doped regions of 150 μm and screen printed fingers ~ 40 μm wide were used. The details of the device structure and processing will be presented at this conference [16]. Light I-V results for the best cells are listed below.

Table II: Light I-V parameters (*externally confirmed at Fraunhofer ISE Callab) of two champion rear-emitter nPERT cells. Measured cell area=244.1 cm^2.

cell	j_{sc} [mA/cm²]	V_{oc} [mV]	FF [%]	eta [%]
A	41.1	683.3	82.0	23.03*
B	41.1	687.4	81.3	22.95*

A very high open circuit voltage of 687 mV with good FF > 81 % has been achieved. This corroborates well with the state-of-the-art saturation current densities (for laser, passivated and metalized regions) and low contact resistances measured in laser doped selective FSF process presented here. Further process developments are ongoing to reduce specific contact resistance, and linewidth of screen printed and laser doped regions.

4 CONCLUSIONS AND OUTLOOK

The selective front surface field (FSF) for screen printed n-type passivated emitter with rear totally diffused (nPERT) solar cells has been investigated in detail. Very low saturation current densities have been achieved in all regions while maintaining low specific contact resistivity and minimal damage to laser doped surface. We measured state-of-the-art saturation current densities of <15 fA/cm², <70 fA/cm² and ~ 500 fA/cm² for the passivated n^+ ($J_{0n+pass}$), n^{++} ($J_{0n++pass}$) and metallized n^{++} ($J_{0n++metal}$) FSF regions respectively, while keeping specific contact resistivity in the range of 1.5-2.5 mΩ.cm². The developed laser doped FSF process was integrated in the process sequence for large area (244.3 cm^2) monofacial screen-printed rear-emitter nPERT cells, resulting in externally confirmed efficiencies up to 23 % with open circuit voltage of 687 mV and fill factor > 81 %. Further process developments are ongoing to reduce screen printed and laser doped area linewidth.

5 ACKNOWLEDGEMENTS

Imec would like to acknowledge the financial support from Jolywood for part of this work.

6 REFERENCES

[1] ITRPV roadmap ninth edition 2018.
[2] P. Verlinden et al., presentation for the 28th EU PVSEC (2013).
[3] F. Colville, Photovoltaics International Volume 33 (2016) https://www.pv-tech.org/technical-papers/perc-solar-cell-production-to-exceed-15gw-in-2017
[4] A. Edler, V. D. Mihailetchi, L. J. Koduvelikulathu, C. Comparotto, R. Kopecek, R. Harney, Prog. Photovolt: Res. Appl. 23 (2015) 620.
[5] G. Hahn, 25th EU PVSEC / WCPEC-5 Valencia (2010).
[6] F. Colville, Photovoltaics International Volume 5 (2009) https://www.pv-tech.org/technical-papers/laser-assisted-selective-emitters-and-the-role-of-laser-doping
[7] J. E. Cotter, J. H. Guo, P. J. Cousins, M. D. Abbott, F. W. Chen, and K. C. Fisher, IEEE Trans. Electron Devices, 53 (2006) 1893.
[8] B Singha and C. S. Solanki, Mater. Res. Express, 4 (2017) 072001.
[9] M. Rüdiger, C. Schmiga, M. Rauer, M. Hermle, and S. W. Glunz, IEEE Trans. Electron Devices, 59 (2012) 1295.
[10] J.M. Bovatsek, A. Sugianto, B. Hallam, B. Tjahjono, S. Wenham, R.S. Patel, 26th EUPVSEC (2011) 895.
[11] B. Lim, T. Brendemuhl, T Dullweber and R. Brendel, IEEE JPV 6 (2016) 447.
[12] J. Deckers, X. Loozen, N. Posthuma, B. O'Sullivan, M. Debucquoy, S. Singh, M. Aleman, M. R. Payo, I. Gordon, P. J. Verlinden, R. Mertens, J Poortmans, 28th EUPVSEC (2013) 806.
[13] D. Inns, Energy Procedia 98 (2016) 23
[14] U. Jager, S. Mack, C. Wufka, A. Wolf, D Biro and R. Preu, IEEE J. Photovoltaics 3 (2013) 621.
[15] M. Kim, D. Kim, D. Kim, Y. Kang, Solar Energy Materials SolarCells132 (2015) 215.
[16] L. Tous, J. Chen, P. Choulat, S. Singh, M. Aleman, I. Kuzma, J. John, F. Duerinckx, J. Szlufcik, 35th EUPVSEC Brussels (2018).

THE IMPACT OF ADVANCED TEXTURING ON SATURATION CURRENT DENSITY IN N-TYPE PERT SILICON SOLAR CELL PROCESSING

J. John*[1,2], S. Jambaldinni[1,2], M. Haslinger[1], M. Gocyla[1], A. Hajjiah[3], I. Kuzma Filipek[1,2], L. Tous[1,2], R. Russell[1,2], F. Duerinckx[1,2], J. Szlufcik[1,2], J. Poortmans[1,2,4]

[1]Imec, Kapeldreef 75, 3001 Leuven, Belgium
[2]Energyville, Genk, Belgium
[3]Kuwait University, College of Engineering and Petroleum, Electrical Engineering Department, Safat, Kuwait
[4] Katholieke Universiteit Leuven, KUL, Leuven, Belgium
*Corresponding author: Phone: +32-16-28 1903; Fax: +32-16-28 1501; e-mail: joachim.john@imec.be

ABSTRACT. We report on the progress of wet chemical process development for silicon surface texturing of n-type passivated emitter with rear totally diffused (n-PERT) solar cells. A KOH-based texturing process sequence has been developed which provides potentially both improved conversion efficiency and reduction in cost of ownership (CoO). The improvement in conversion efficiency is attributed to decreased saturation current density after diffusion and laser processing, whereas, reduction of CoO is reached by reduced material consumption and process time. We have determined the saturation current density after subsequent process steps like Boron emitter diffusion, Phosphorous back surface field diffusion, passivation, laser ablation and selective laser doping for small pyramid (pyramid base length <2μm) and standard texturing (pyramid base length 3-5μm), respectively.

Keywords: KOH texturing, small Pyramids, n-type silicon, PERT, saturation current density

1 INTRODUCTION

Silicon solar cells based on crystalline silicon (c-Si) are the dominating technology in photovoltaics (PV) industry and have a market share of more than 85% of the modules produced for roof top installations [1]. The PV Industry continues to demand even lower cost per watt solar cells so that PV electricity can ultimately compete with conventional sources of energy. One viable approach is to increase the conversion efficiency and simultaneously reduce the manufacturing cost of the device [2]. The need for cheaper levelized cost of PV-originated electricity (LCOE), i.e. cost per KWh, translates in a need for cheaper solar cell processes with an increase in conversion efficiency [3]. In order to reach this goal, novel cell concepts and novel fabrication methods [4] have to be developed for implementation in industrial process lines [5]. In this context, wet chemical texturing processes should also evolve towards easy to implement, well controlled, and cost-effective steps [6-7].

Process development using wet chemical texturing is triggered by two main cost driven objectives: first, reducing the amount of chemicals to the minimum needed to reach the specified performance target and second, reducing the processing time to increase the throughput required in mass fabrication. On the performance side the requirements are, first, a high surface passivation potential and, second, a homogeneous textured silicon surface. In order to reach the first cost objective, low cost chemicals with increased chemical bath lifetimes and reduced chemical consumption by using highly diluted solutions were applied. Moreover, in order to reach the second cost objective, it is essential to reduce the process time for the sequential wet chemical steps by means of optimizing the wet chemical processes themselves. The performance of a wet chemical texturing can be benchmarked by determining the emitter saturation current density.

2 EXPERIMENTAL

This work is based on a bifacial Passivated Emitter Rear Totally diffused cell process [8]. The cell structure is schematically shown in Fig. 1 and the bifacial n-PERT process flow is depicted in Fig. 2 [9]. Emitter saturation current density extraction is performed after several steps highlighted in figure 2. In Fig. 3 the detailed advanced texturing process is shown. In contrast to standard texturing this process uses a KOH pre-Texturing step at low temperature and a KOH texturing step at 70°C. Fig. 4 shows the pyramid size distribution in a SEM picture with magnification 8000 times. Fig.4a shows small pyramids with an average base length less than 2μm, Fig. 4b shows the standard texturing process with pyramid base length of 3-5μm in average.

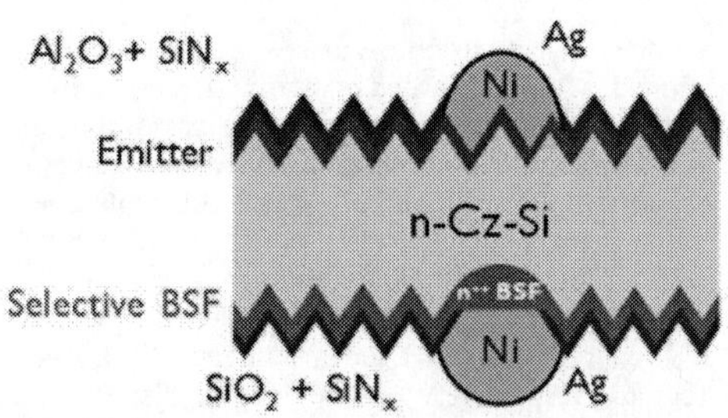

Figure.1 Schematic drawing of nPERT cell structure

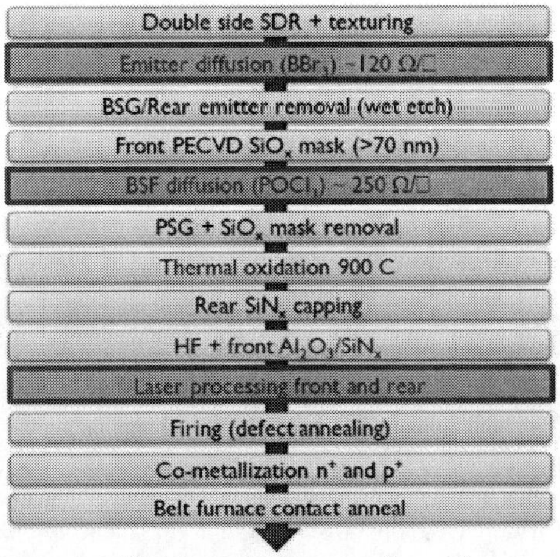

Figure.2: Bifacial n-PERT process flow. The processes where emitter saturation current density is extracted are highlighted.

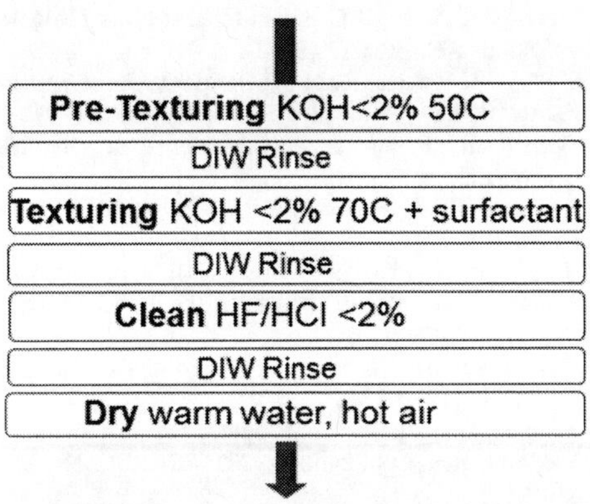

Figure.3: Detailed process flow for advanced texturing. Process time dry-in to dry-out ~30min.

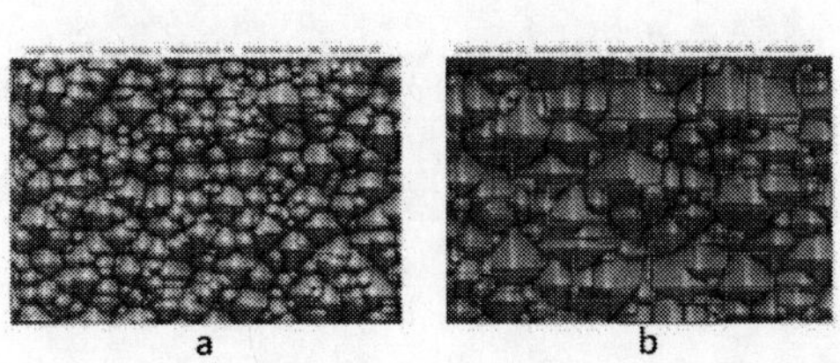

Figure.4: SEM (x8000 magnification) picture of pyramid distribution,
4a: Average pyramid base length less than 2µm
4b: Average pyramid base length between 3-5µm

3 RESULTS AND DISCUSSION

3.1 Impact of Small pyramids texturing on diffusion processes

Fig.5 shows the extracted saturation current density after boron and phosphorous diffusion, respectively. The boron diffused samples are passivated with ALD Al_2O_3 and PECVD SiO_x/SiN_x. The phosphorous diffused samples are passivated with a PECVD SiO_x/SiN_x layer. The J_{0pass} for small pyramids is measured to be 6 fA/cm^2 for the P-doped BSF and 25 fA/cm^2 for the B-doped emitter. We could confirm an enhanced passivation effect due to small pyramids like reported in [10]

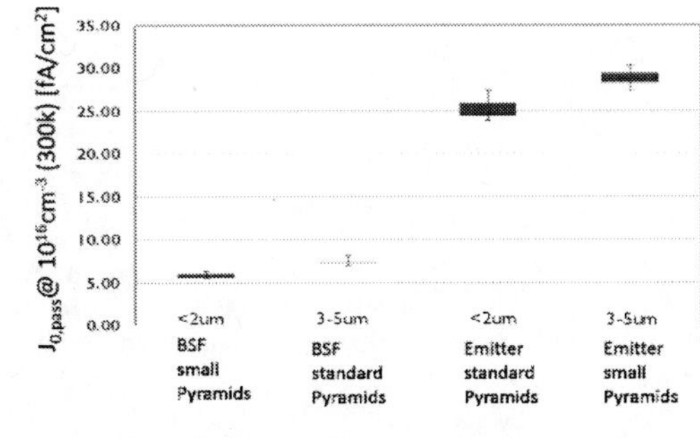

Figure.5: Saturation current density ($J_{0,pass}$) after the boron emitter formation and the phosphorous back surface field formation for silicon surface textured with small and standard size pyramids.

3.2 Impact of small pyramids texturing on laser processes

The saturation current density after laser processing of the silicon surface is extracted for the two process steps laser ablation and laser doping. Laser processes are a source of damage to the silicon surface and lead to an increase in J_0. It is expected that for small pyramids the power of the applied laser processes could be less damaging due to reduced topographical difference. The result of this approach is depicted in figure 6 and 7, where the reduced laser power could be translated into less damage and thus in lower J_0 values for the case of small pyramids.

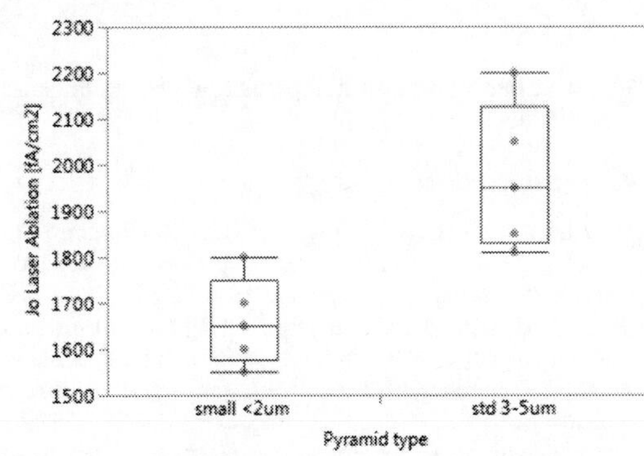

Figure.6: Impact of the small pyramid texturing on laser ablation on p$^+$ emitter. Please note that the laser energy for full opening passivation ablation on small pyramid textured surfaces could be reduced by 25% compared to standard texturing.

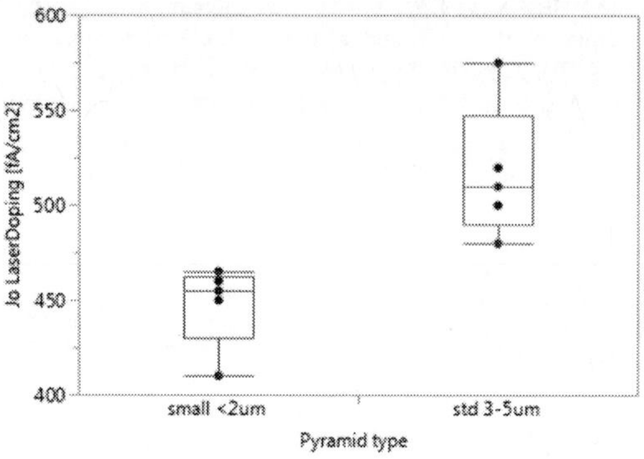

Figure.7: Impact of the small pyramid texturing on n^{++} laser doping. For small pyramid textured surfaces, the laser power for achieving the same doping effect could be reduced.

3.3 Impact of small pyramids texturing on solar cell efficiency

We have processed nPERT silicon solar cells using a co-plating metallization scheme. On the cells with small pyramid texturing solar cell efficiencies up to 21.7% are achieved.

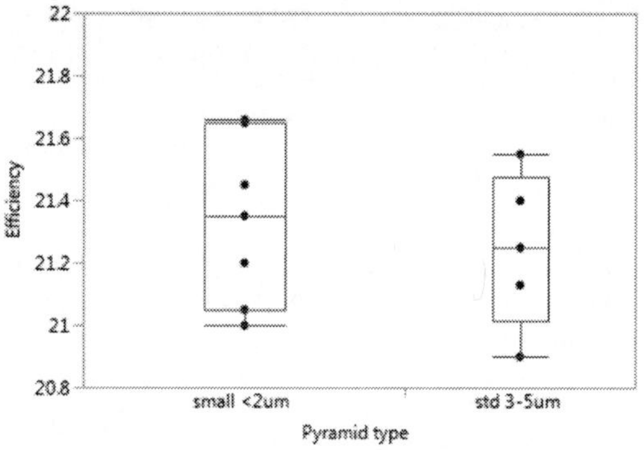

Figure.8: Impact of the small pyramid texturing on solar cell efficiency.

4 CONCLUSION

Industrial n-type Cz-silicon wafer (M2, 3-5 Ohm.cm) are processed with the imec nPERT process flow. Two different texturing sizes (pyramid base length <2µm and 3-5µm) are applied and compared during the cell process on their impact on the emitter saturation current density. After Boron emitter diffusion an emitter saturation current density J_{oe} of 25 fA/cm^2 is measured on the surface textured with small pyramids and passivated with ALD Al_2O_3 and PECVD SiO_x/SiN_x stack. Front surface field (FSF) Phosphorous diffusion and passivation with thermal SiO_x and SiN_x led to $J_{0,FSF}$ values down to 6 fA/cm^2 on small pyramids. After laser ablation we extracted a $J_{0,abl}$ of the small pyramid textured surface down to 1600 fA/cm^2. For laser doping on small pyramid surfaces $J_{0,dop}$ down to 420 fA/cm^2 could be measured. Cell results with small pyramid texturing in combination with different metallization schemes (Ni/Cu plating and Ag screen-printing) are in progress. First results on bifacial nPERT silicon solar cells are manufactured. Implied V_{oc} before metallization up to 695 mV and solar cell conversion efficiency up to 21.7% have been measured on small pyramids using a co-plating process which is still under development.

We have shown that a wet chemical advanced texturing process leading to pyramid sizes below 2µm base length could be beneficial for the passivation of silicon solar cell surfaces after applying different process steps like diffusion and laser treatment. This has been demonstrated in an industrial n-PERT process flow. In a next step the small pyramid approach is tested for different metallization schemes like fine line screen printing and Ni/Ag based plating for smart wire interconnection.

4 ACKNOWLEDGEMENTS

The authors would like to acknowledge the funding agencies and the industrial partners of the imec Industrial Affiliation Program in SiPV, and KFAS under project code: P115-15EE-01.

5 REFERENCES

[1] Jäger-Waldau, A. JRC Scientific and Policy Reports-PV Status Report 2012 (2012).
[2] C. T. M. International Technology Roadmap for Photovoltaics- Results 2012. ITRPEV. net 2 (2012).
[3] Verlinden, P. et al., presentation for the 29th EUPVSEC (2013)
[4] Recamán Payo, M. et al., ,Proceedings of the 30th EUPVSEC (2014)
[5] John, J. et al., EPJ Photovolt., 3 (2012), p. 35005
[6] Rentsch, J. et al., Proceedings of the 25th EUPVSEC (2009), p1113
[7] Schweckendiek, J. et al., Proceedings of the 11th UCPSS (2012), p283
[8] Urueña, Angel, et al., Progress in Photovoltaics: Research and Applications, 2016, 24. Jg., Nr. 8, S. 1149-1156.
[9] R. Russell et al., Proceedings of 33rd EUPVSEC (2017)
[10] LI, Guorong; ZHOU, Yuqin; LIU, Fengzhen., Journal of Non-Crystalline Solids, 2012, 358. Jg., Nr. 17, S. 2223-2226.

IMPACT OF SURFACE MORPHOLOGY AND INTERFACIAL OXIDE THICKNESS ON PASSIVATION QUALITY OF P+ POLYSILICON PASSIVATING CONTACTS

S. Mack[1], F. Feldmann[1], A. Moldovan[1], M. Lenes[2], J. M. Luchies[3], A. Wolf[1]
[1]Fraunhofer Institute for Solar Energy Systems ISE, Heidenhofstraße 2, 79110 Freiburg, Germany
[2]Tempress Systems, Radeweg 31, 8171 MD Vaassen, The Netherlands
[3]Amtech Systems Inc., Radeweg 31, 8171 MD Vaassen, The Netherlands
Phone: +49 761 - 4588 5048; e-mail: sebastian.mack@ise.fraunhofer.de

ABSTRACT: The work aims at increasing the knowledge on p-type passivating contacts, especially the impact of the surface morphology and interfacial oxide thickness. Here, p-type passivating contacts are formed by low pressure chemical vapour deposition (LPCVD) of a 240 nm thick boron doped polysilicon layer, which is deposited on top of an in-situ grown interfacial oxide. In the experiment, three different oxide thicknesses are compared on top of five different surface types, i.e. an alkaline textured surface, three subsequently wet-chemically treated alkaline textured surfaces, as well as a planar alkaline saw damage etched surface. After a thermal anneal and SiN_x deposition on both sides, which yields symmetrical samples, a contact firing step is used for hydrogenation purposes. The results show a strong correlation between surface morphology and passivation quality, and the highest passivation quality is achieved for saw-damage etched surfaces with implied voltages of up to 737 mV. In addition, the thickness of the interfacial oxide layer has a weaker impact on passivation quality than the morphology.
Keywords: LPCVD, passivated contact, oxide, polysilicon, passivation, recombination

1 INTRODUCTION

Passivating contacts are considered a follow up technology to the passivated emitter and rear cell (PERC) design, which is already in production by many cell manufacturers.

So far, the majority of solar cells with passivating contacts under investigation are fabricated from n-type silicon substrates and these solar cells feature also a n-type passivating contact, which is placed on the cell's rear side [1–5]. To facilitate a possible transfer to industry, the use of p-type passivating contacts is preferred, as large parts of the known PERC process sequence could be left unchanged. For this purpose, it is also important to investigate, how the surface morphology affects passivation properties. For transfer to industrial production, the interdependency of passivating contact layers with typical Si surface morphologies from industrial processing (e.g. alkaline texture or after rear emitter removal) needs to be investigated. Larionova et el. performed a similar study, however, only for selected samples, an additional hydrogenation process was performed to further reduce minority carrier recombination [6], which underlines the need for this investigation. Other groups also investigate p-type passivating contacts and related cell structures, either applying screen printed [7] or physical vapour deposited [8, 9] metallization.

Here, passivating contacts are formed by low-pressure chemical vapour deposition (LPCVD) of boron doped polysilicon layers on top of in-situ thermally grown interfacial oxides. A SiN_x layer acts as a hydrogen source during fast firing, to resemble a possible process sequence, which incorporates fire through contacts.

2 EXPERIMENTAL

2.1 Sample preparation

Fig. 1 shows the experiment layout. Phosphorous doped silicon wafers of 1.65 Ωcm resistivity (measured after thermal donor deactivation by thermal treatment) with edge length 156 mm are used in the experiment. After texturing the wafers of groups 1 to 4 in alkaline solution, the surfaces of groups 2, 3, and 4, are subject to a wet chemical post treatment. For groups 2 and 3 (denoted "O_3 short" and "O_3 long"), the pyramids tips are smoothed by wet chemical etching in $O_3/H_2O/HF/HCl$ mixtures, where group 3 is etched longer than group 2. For group 4 (denoted chemical edge isolation, CEI) an industrial chemical edge isolation process in an inline tool using HF/HNO_3 is applied on both sides. A standard saw damage etch (SDE) step serves as a reference in group 5. The wet chemical cleaning step is followed by tunnel oxide formation in situ in the poly-Si furnace preceding the deposition of the 240 nm thick in-situ boron doped polysilicon layer, as determined by SEM on planar samples. Here, a variation in the process parameters yields tunnel oxides of three different thicknesses, which are expected to be in the range of 1 to 2 nm. The next steps consist of thermal annealing at 900 °C for 30 min to activate the dopant, double sided hydrogen-rich SiN_x deposition (75 nm layer thickness) by direct-plasma enhanced chemical vapour deposition (PECVD) and thermal activation of the passivation layers in a contact firing step. The resulting sample structure is depicted in Fig. 2.

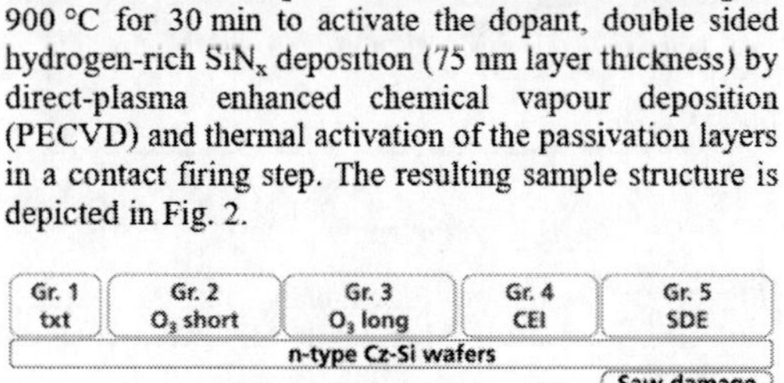

Figure 1: Process flow of the experiment.

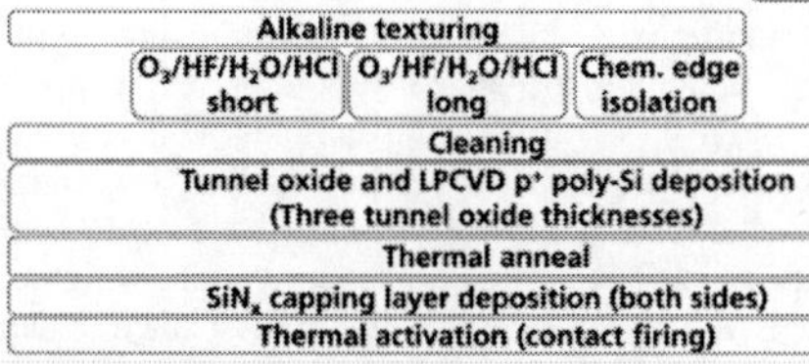

Figure 2: Schematic drawing of the cross section of the resulting non metallized test structure. The actual surface morphology is not shown.

2.2 Characterization

Following wet chemical processing, we measure the reflection of the different surfaces. This technique is supported by scanning electron microscopy analysis for a more detailed view of the surface. Additionally performed confocal microscopy reveals information about the surface enlargement factor f_{surf} of the different morphologies. The factor f_{surf} describes the ratio of actual surface area and the 2D projection (macroscopic) area.

After passivation, we apply the photoconductance decay (PCD) technique for determination of the minority carrier lifetime and the implied open circuit voltage iV_{oc} at one sun as well as the implied fill factor iFF.

3 RESULTS

3.1 Reflection

Fig. 3 shows the reflection of the different surface after wet chemical processing. The two different processes featuring O_3 result in a similar reflection, only marginally higher than for the textured sample. Compared to this, the chemical edge isolation process strongly increases the reflectance, as can be seen from Fig. 3.

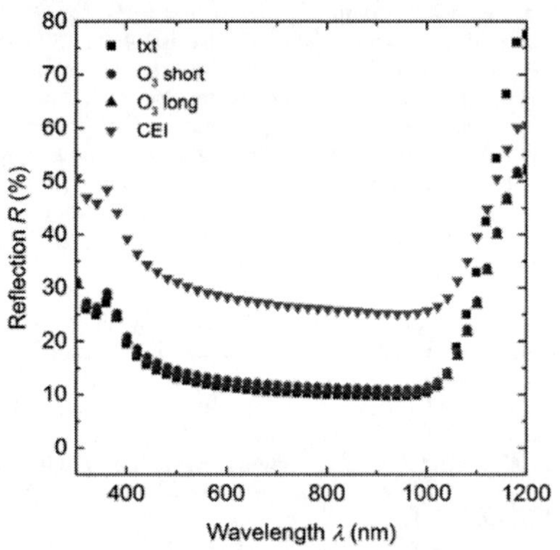

Figure 3: Reflection for initially alkaline textured surfaces, measured after different wet chemical treatments.

An additional reflection measurement after processing, including the poly layers and SiN_x capping layers, see Fig. 4, shows that the samples with textured and O_3 treated surfaces feature the more or less same reflection, and considerably less than the sample with CEI treated surface. The highest overall weighted reflection is found for sample with SDE surface. Thus, for a bifacial application with a high bifaciality and a poly-Si/SiN_x stack placed at the cell's rear side, a textured or O_3 treated surface is preferred due to the lower overall reflection.

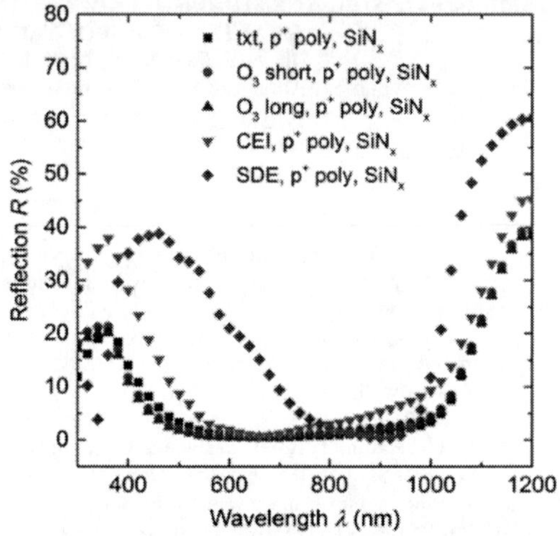

Figure 4: Reflection for several morphologies for the finished samples after poly-Si deposition, thermal anneal and SiN_x deposition.

3.2 Scanning electron microscopy

Fig. 5 shows exemplary scanning electron microscopy images of different surfaces after wet-chemical post treatment. The in-view measurements reveal that O_3 based processing indeed leads to a smoothing of pyramid tips, whereas the surface enlargement factor f_{surf}, is hardly affected by such a treatment, and longer O_3/HF treatment results in a decrease of $f_{surf} = 1.34$ (O_3 short) to $f_{surf} = 1.31$ (O_3 long), as is extracted from confocal microscopy measurements. In contrast to this, the pyramid tips of the wafer that was subject to a CEI process are even more concave than the textured reference without such a treatment. Thus, the applied edge isolation process leads to a surface with many sharp structures, which might contradict a high quality surface passivation by dielectric layers. The surface enlargement factor is decreased more heavily by CEI treatment, to values of around $f_{surf} = 1.17$.

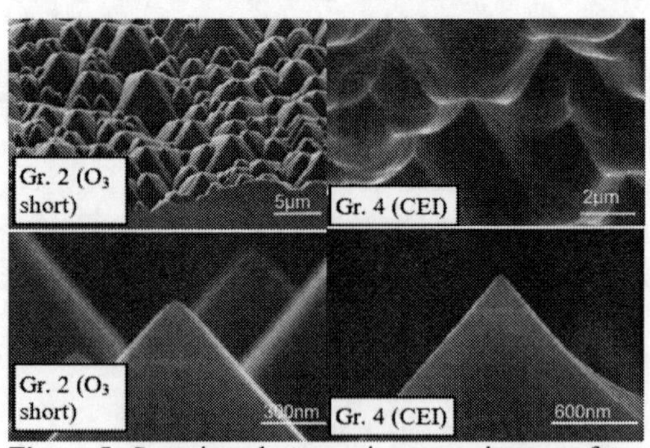

Figure 5: Scanning electron microscope images of two initially alkaline textured surfaces after different wet chemical treatments. Left: short O_3 etch (group 2), right chemical edge isolation (group 4). Top tilted view from above, bottom side view of an exemplary pyramid tip. The rounding edges leads to a smoothed pyramid tip, whereas the chemical edge isolation yields sharper structures.

3.3 Surface passivation quality

Fig. 6 shows the results from QSSPC measurements, performed after contact firing of the samples. The first graph 6a) focuses on the implied open circuit voltage iV_{oc} at one sun of the symmetric samples, as a representative of the passivation quality of the SiO_x/polysilicon/SiN_x stack for the five different surfaces and three investigated oxide thicknesses.

It is apparent that the passivation quality strongly depends on the surface morphology. The highest passivation quality is found for the most planar SDE surface, with highest values of 737 mV, including the contribution of the wafer itself to carrier recombination. This translates to a dark saturation current density of 4 fA/cm^2, which is higher than reported last year, due to a lower applied peak firing temperature [10].

The samples that were subject to CEI allow for iV_{oc} up to 715 mV, thus much lower compared to the SDE surface but still higher than the textured or O_3 treated surfaces which yield iV_{oc} of up to 705 and 707 mV, respectively. These findings are in accordance with literature [11]. With respect to scattering of the measurement data there are no differences in passivation quality for the textured samples with or without O_3 post treatment. Larionova *et al.* contributed this reduced iV_{oc} for textured surfaces to an increased surface recombination velocity compared to their planar counterparts [6]. Compared to planar SDE surfaces, two effects lead to enhanced recombination: 1) the increased surface factor and 2) the change in crystal orientation from (100) to (111), which presumably leads to increased interface trap density at the Si/SiO_x interface. A more detailed investigation will be published elsewhere.

Having a look now at the impact of interface oxide layers of different thickness, it is obvious that in case of the SDE surface, the oxide thickness has only a marginal impact on the passivation quality, at least for the investigated range of layer thicknesses, and all oxide layers give the more or less same passivation quality of above 730 mV. However, for the other four surfaces, there is a clear correlation and the thinnest oxide layer always shows the lowest passivation quality within each group, to values even below 690 mV. Here, increased minority carrier recombination due to a presumed highest surface recombination velocity at the Si/SiO_x interface might be the origin of the low iV_{oc}. Electro-chemical profiling showed no difference in the boron diffusion profile in the crystalline wafer beyond the interface oxide. For the other two oxide layer thicknesses, there seems to be no significant difference in passivation quality at one sun illumination. Comparing our results to that of others [6, 7, 11], we achieve slightly higher iV_{oc} on planar and textured surface, even for in-situ boron doped poly-Si layers.

Fig. 6b) shows the implied fill factor *iFF*, extracted from the QSSPC measurements again after contact firing. As for iV_{oc}, the highest values in the graph are determined for the SDE surface. However, here, there are quite significant differences visible for the three different oxide layer thicknesses. The highest *iFF*=85.8% results from the oxide layer with medium thickness and the lowest *iFF* from the thinnest oxide layer. For the wafers with textured or O_3 treated surface, *iFF* is on a similar value between 84 and 85%, and again there is a trend, that applying the thinnest oxide layer results towards the lowest *iFF* values.

Interestingly, the samples with CEI surface have the lowest *iFF* globally, and again, the samples with medium oxide thickness seem to have the highest values. The low *iFF* reflects a strong injection dependence of the surface recombination with reduced carrier lifetime in low injection. We assume that this behavior originates from local defects at the sharp-edged surface structures of the CEI samples, as this is the most obvious difference compared to the textured samples.

Although the iV_{oc} values for the CEI surface are promising, the low *iFF* already indicates necessary process improvements for this sequence to achieve a smoother surface morphology. The authors would like to point out, that this only holds true for the investigated CEI process and processes done at other tools or applying other chemicals might yield different results.

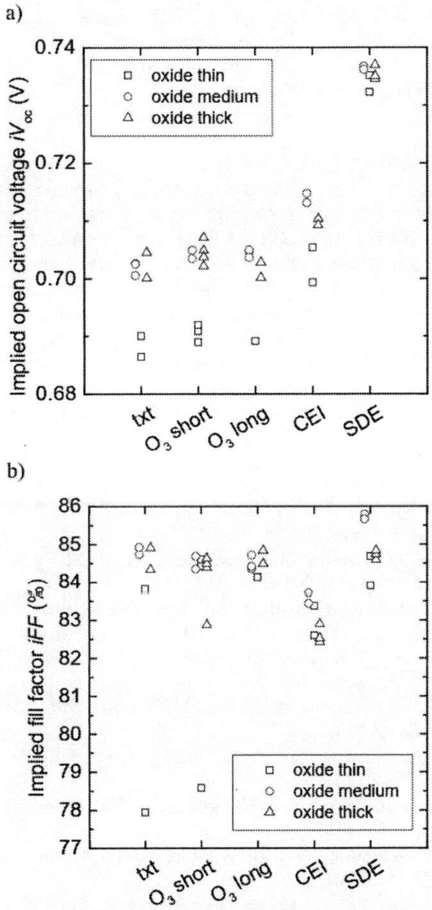

Figure 6: a) Implied open circuit voltage iV_{oc} at 1 sun and b) implied fill factor *iFF* of the investigated symmetric samples, measured by PCD.

4 SUMMARY

In this paper, the surface passivation quality of SiO_x/p^+ polysilicon/SiN_x stacks has been investigated for five different and typically used morphologies and three different oxide thicknesses. The highest passivation quality has been found on saw-damage etched surfaces, and medium or thick oxide layers with iV_{oc} = 737 mV and *iFF* = 85.8 %. The impact of the morphology strongly exceeds that of the thickness of the interface oxide layer, at least for the investigated range. Thus, the thickness can be selected such that it allows for an efficient carrier transport of the interface oxide.

The surface after wet chemical edge isolation here falls short. Although iV_{oc} is on an acceptable level, a strong decrease of *iFF* is observed, which at this point is not yet fully understood and currently attributed to the observed sharp-edged surface structures of the samples. The textured and ozone treated surfaces exhibit more or less the same passivation quality. Here however, the impact of a too thin interface oxide layer becomes notably.

The results show that in situ boron doped polysilicon layer formed by LPCVD combined with PECVD SiN_x layers are a promising and firing stable technology for an industrial realization of p-type passivating contacts. For a high passivation quality, the results of this survey indicate, that an alkaline saw damage etched surface might be necessary at this point.

ACKNOWLEDGEMENTS

The authors would like to thank their colleagues at Fraunhofer ISE for processing and characterisation. Project HIPPO is supported under the umbrella of SOLAR-ERA.NET Cofund by German Federal Ministry of Economic Affairs and Energy under contracts 0324086A and 0324086B, as well as Business Finland under contract 58/31/2016. SOLAR-ERA.NET is supported by the European Commission within the EU Framework Programme for Research and Innovation HORIZON 2020 (Cofund ERA-NET Action, N° 691664).

REFERENCES

[1] A. Richter, et al., Solar Energy Materials and Solar Cells, vol. 173, no. 173, pp. 96–105, 2017.

[2] R.C.G. Naber, et al., Proceedings 32nd EU PVSEC, München, Germany, 2016, pp. 430–433.

[3] S. Duttagupta et al., Solar Energy Materials and Solar Cells, vol. 187, pp. 76–81, 2018.

[4] Y. Tao, et al., Prog. Photovolt: Res. Appl., vol. 24, no. 6, pp. 830–835, 2016.

[5] M. K. Stodolny, et al., Energy Procedia, vol. 124, pp. 635–642, 2017.

[6] Y. Larionova, et al., Phys. Status Solidi A, vol. 214, no. 8, p. 1700058, 2017.

[7] K. Tool, et al., in 2018 WCPEC 7, Hawaii, USA, 2018, pp. 1188–1192.

[8] M. Nicolai, et al., IEEE J. Photovoltaics, vol. 8, no. 1, pp. 103–109, 2018.

[9] A. Ingenito, et al., Sol. RRL, vol. 1, no. 7, p. 1700040, 2017.

[10] S. Mack, et al., Physica Status Solidi Rapid Research Letters, 1700334, 2017.

[11] Y. Larionova, et al., Proceedings 32nd EU PVSEC, München, Germany, 2016.

WETALOX: A NOVEL, COST EFFECTIVE, NEGATIVE CHARGE PASSIVATION METHOD FOR CRYSTALLINE SILICON SOLAR CELLS

Elina Schmid[1] (1), Stefan Schmitt[1] (2), Tim Boescke[1] (3), Eckard Wefringhaus[1] (4), Florian Buchholz[1] (5), Christoph Peter[1] (6), Rafael Marczak[1] (7), Arun Ramakrishnan[2] (8), Markus Mateescu[2] (9), Peter Kunze[2] (10)
[1]ISC Konstanz e.V., Rudolf-Diesel-Str. 15, 78467 Konstanz, Germany
[2]GP Solar GmbH, Konstanz, Germany
(1) Corresponding author: elina.schmid@isc-konstanz.de, Tel.: +49 7531-36 183 552,
(2) stefan.schmitt@isc-konstanz.de, (3) t.boescke@gmail.com, (4) eckard.wefringhaus@isc-konstanz.de,
(5) florian.buchholz@isc-konstanz.de, (6) christoph.peter@isc-konstanz.de, (7) rafael.marczak@isc-konstanz.de, (8) arun.ramakrishnan@gpsolar.com, (9) markus.mateescu@gpsolar.com,
(10) peter.kunze@gpsolar.de

ABSTRACT: We present the development and evaluation of a novel negative charge passivation with cost effective aluminum oxide (Pascal). Instead of relying on vacuum processes, a thin, charge inducing aluminum oxide layer is generated by a novel wet-chemical process ("WetAlOx" process) on the silicon surface. This novel passivation layer was evaluated for state-of-the-art p-type PERC and n-type PERT solar cell concepts. The "WetAlOx" concept was investigated on cell level and precursor level with symmetrical lifetime samples. It was possible to generate negative surface charge densities (Q_{tot}) from $-4.0\cdot10^{11}$ to $-1.4\cdot10^{12}$ cm^{-2} depending on process conditions. The surface charge density was mainly controlled by pH of the aluminium salt solution. By increasing the pH the surface charge density is becoming more negative. An undisered side effect of the increasing pH is the increased adsorption rates of undisired ions, which contaminate the surface. When applying a "WetAlOx" passivation stack to PERC cells the "WetAlOx" shows worse performance compared to bare silicon oxide / silicon nitride and ALD AlOx (atomic layer deposition aluminum oxide) passivation stacks. In the n-PERT cell concept the "WetAlOx" passivation stack shows comparable performance to conventional passivation methods such as silicon oxide / silicon nitride stack.
Keywords: PERC, PERT, Passivation, Al_2O_3, Impurities, Surface Charge Density

1 INTRODUCTION

In the last years solar cell efficiencies were highly improved. Since crystalline silicon solar cells become thinner and thinner the solar cell efficiency is more and more determined by surface recombination losses [1]. Blakers et al. [2] introduced the first rear surface passivated silicon solar cell in 1989 called PERC (Passivated Emitter and Rear Cell). The first PERC cell used a thin silicon oxide layer for rear side passivation. Figure 1 (top) shows a cross section of an enhanced PERC solar cell with an aluminum oxide (AlOx) passivation layer. Similar to PERC on p-type a concept for passivated n-type silicon solar cells called PERT (Passivated Emitter Rear Totally Diffused, Figure 1, bottom) has been developed. N-type material has a lot of benefits compared to p-type material, such as a bigger tolerance to common impurities and it does not suffer from the boron-oxygen related light induced degradation [3].

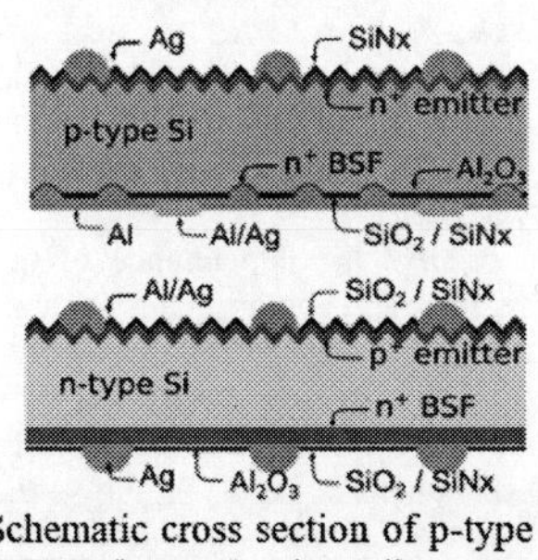

Figure 1: Schematic cross section of p-type PERC (top) and n-type PERT (bottom) solar cell concepts [4]. In the PERC solar cell concept the AlOx layer is deposited at the rear side before SiO_2/SiN_x deposition. For n-PERT the AlOx passivation layer is located between the n^+ region and the SiO_2/SiN_x anti reflection coating.

In general dielectric materials are necessary for the passivation of silicon surfaces. Since thermally grown silicon oxide (SiO_2), the initially used passivation method in PV industry, has several disadvantages, for example high production costs and instability regarding UV radiation, silicon nitride (SiN_x) layers grown by PECVD (Plasma Enhanced Chemical Vapor Deposition) were investigated. Beside the field effect passivation provided by positive charges, the plasma SiN_x layer features a lot of useful hydrogen. An alternative is the double-layer stack SiO_2-SiN_x [5]. More recently it was shown, that AlOx provides an excellent passivation layer. AlOx generates negative charges at the interface to silicon and induces a very effective field effect passivation and has a low defect density at the interface [6]. The common deposition techniques are ALD (Atomic Layer Deposition) and PECVD. ALD is a very precise method to deposit monolayers of AlOx by a gas phase chemical process. The film thickness is controlled by the cycle times. Thus the growth rate of conventional ALD is limited to approximately 2 nm/min. The PECVD method is faster and thus more useful as an industrial process. However, the PECVD AlOx is not stable in high temperature processes [7]. Given that wet chemical processes are very cost effective, a new method for the generation of an AlOx passivation layer was evaluated: "WetAlOx".

2 EXPERIMENTAL SECTION / SOLAR CELL PROCESS

2.1 Surface charge and lifetime samples

As surface charge and lifetime samples p-type wafers were used. Saw damage was removed (15 µm/s) in 22 % NaOH at 80°C. After a cleaning sequence with SPM (Sulfuric Peroxide Mixture) and 2 % HF, AlOx was deposited as described in section 2.4 (Figure 4). Different

Al salts ($Na[Al_{13}O_4(OH)_{24}(H_2O)_{12}](NO_3)_8$, $AlCl_3$ and $Al(NO_3)_3$) were tested in different concentrations (0,025 - 5 mmol/L). Furthermore, the pH in the aluminum salt solution was varied (pH = 4.0 - 7.5) by adding TMAH (tetramethylammoniumhydroxide). For reference the "WetAlOx" process was skipped. For another reference, samples were dipped in iron (25 and 250 mmol/L) instead of aluminum solutions. In order to determine concentrations of Al and contaminants on the silicon surface, wafers were taken from the process flow and analyzed by sandwich-etch ICP-OES similar to sandwich-etch ICP-MS described in [8]. On top of the thin AlOx layer a SiO_2/SiN_x passivation layer was deposited by PECVD. Afterwards the blue wafers were annealed. Implied V_{OC} and AlOx surface charge density were measured by QSSPC and PV 2000 (corona charge), respectively.

2.2 p-type PERC

In order to evaluate the "WetAlOx" passivation layer for the PERC solar cell concept, p-type wafers were treated as described in Figure 2. As reference to the "WetAlOx" passivated samples, cells with bare silicon oxide/silicon nitride and ALD AlOx passivation stack were produced.

Figure 2: p-type PERC process flow.

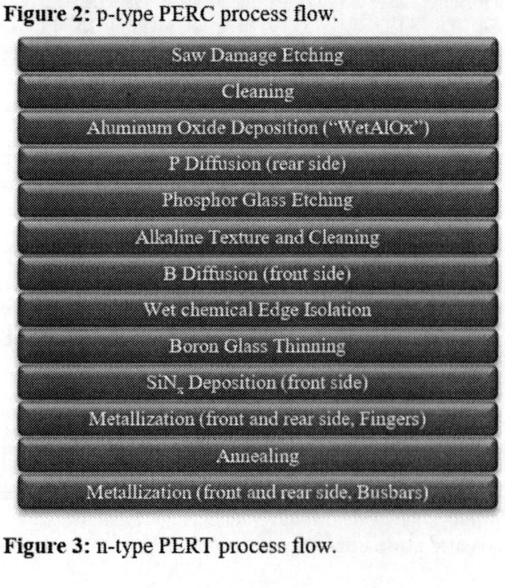

Figure 3: n-type PERT process flow.

2.3 n-type PERT

For the n-PERT cell concept the process flow described in Figure 3 was used. Similar to the experiments for the PERC cells two different bare SiO_2/SiN_x passivation stacks were used as reference. First a borosilicate glass (BSG)/SiN_x passivation stack and second a SiO_2/SiN_x passivation stack, where the SiO_2 was generated wet chemically. As a matter of course n-type wafers were used.

2.4 "WetAlOx"

The "WetAlOx" process flow is shown in Figure 4. The method is based on adsorption. Therefore the silicon surface needs to be activated by generating a thin SiO_2 layer, which is done by a wetchemical oxidation in hot sulfuric peroxid mixture. When dipping the activated surface into an aqueous aluminum salt solution, the surface will self-saturate by aluminum ions. Ideally one "WetAlOx" process cycle creates a monolayer of AlOx. To be sure self-saturation has finished, the wafers were dipped for 10 minutes. After aluminum adsorption, the wafers were cleaned by dipping into DI water. The AlOx is formed completely while drying. In combination with a subsequential SiO_2 and SiN_x deposition (PECVD), a passivation layer with properties comparable to an ALD AlOx passivation is expected.

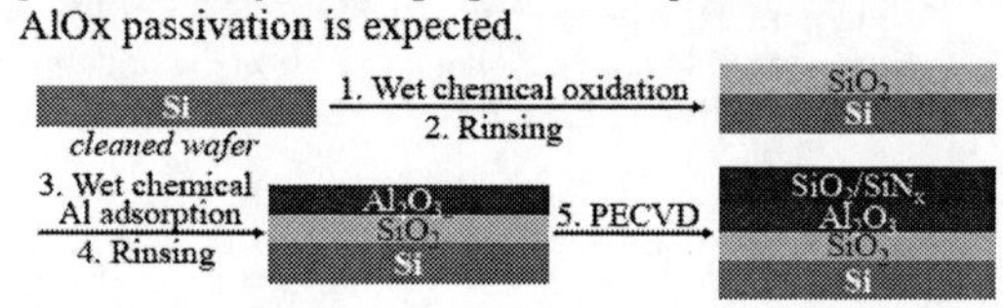

Figure 4: „WetAlOx" process flow and wafer cross section for the „WetAlOx" passivation method. In the first step a reactive surface (with OH-groups) is created by surface oxidation, followed by a rinsing step. In the third step Al ion adsorption takes place in an Al ion solution, followed by a second rinsing step. The passivation is completed by a PECVD SiO_2/SiN_x stack.

3 RESULTS

3.1 Silicon surface analysis: Surface Charge Density (Q_{tot}) and concentration of aluminum/contaminants

Negative surface charge from $-4.0 \cdot 10^{11}$ to $-1.4 \cdot 10^{12}$ cm^{-2} depending on process conditions were generated by the thin AlOx layer (Figure 5). The reference SiO_2/SiN_x passivation stack delivered a positive surface charge of $3.3 \cdot 10^{11}$ cm^{-2}. Figure 5 shows that the negative charge increases with pH due to shift of equilibrium towards Al ion adsorption (cf. also [9]). By increasing the aluminum salt concentration the surface charge does not increase significantly. Furthermore the kind of the anion does not play a role.

Figure 6 depicts the dependence of the amount of adsorbed aluminum and contaminants such as iron, copper and titanium on $AlCl_3$ concentration and pH. The amount of adsorbed Al increases with increasing pH and aluminum salt concentration. Comparison of Figure 5 and 6, which are combined in Figure 7, shows that only a low fraction of adsorbed aluminum is activated as negative charge. Namely only the first monolayer of AlOx contributes to the field effect passivation. If a corundum structure is assumed, a monolayer (of Al_2O_3) corresponds to $6 \cdot 10^{14}$ Al atoms/cm^2 [10].

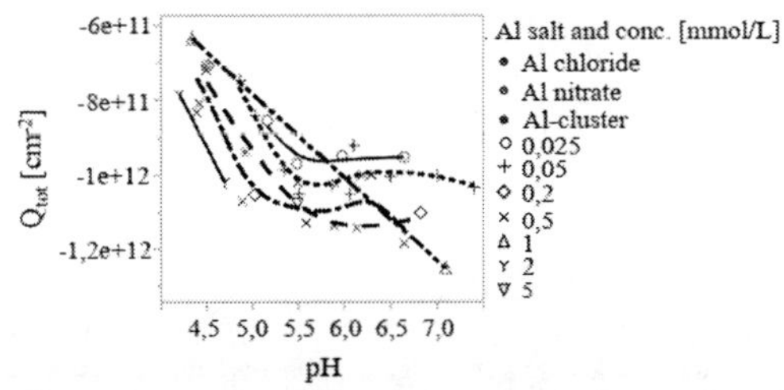

Figure 5: Resulting surface charge (Q_{tot}) of the "WetAlOx" passivation method using different and differently diluted Al salts vs. pH of the aluminum solution.

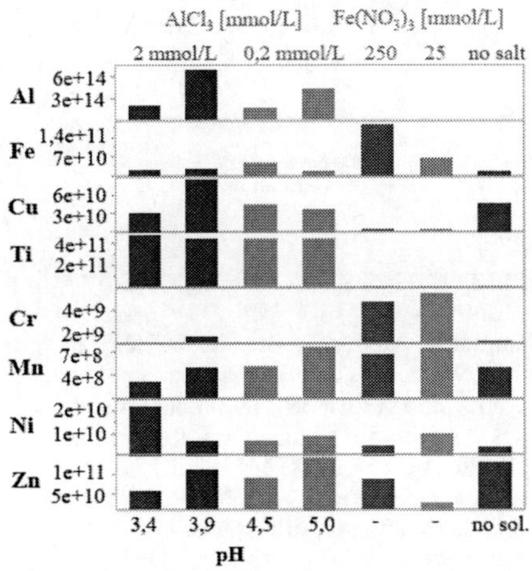

Figure 6: Adsorbed aluminum and adsorbed contaminants on activated silicon wafers (in atoms/cm^2) plotted versus pH and concentration of aluminum solutions (blue bars), and iron solutions (green bars). For comparison, a cleaned wafer was investigated (red bars).

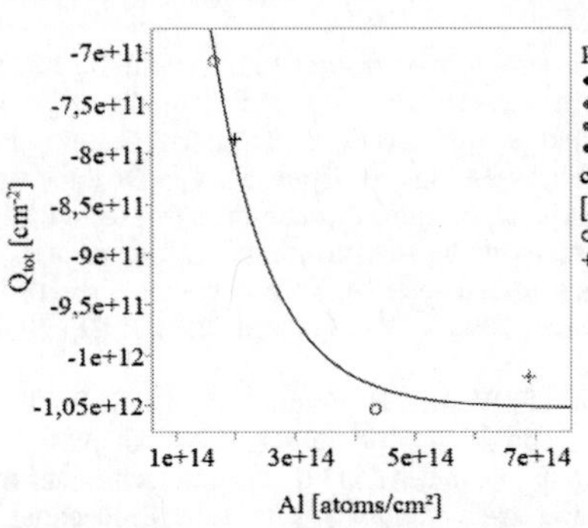

Figure 7: Fraction of activated aluminum as surface charge density (Q_{tot} vs. Al atoms). The caption lists the pH and concentration of the $AlCl_3$ solution, which was used in the "WetAlOx" process step.

As an undesired side effect, increasing pH also increases adsorption rates of other metals (Figure 6), which pollute the surface und act against the required passivation effect. Here the main impurities are iron, copper and titanium. Transition metal contaminants badly influence diffusion length and recombination both, in bulk and emitter [11]. Figure 8 shows the dependence of implied Voc (open circuit voltage) on pH in solution. By increasing pH the implied Voc decreases due to increasing metal contamination.

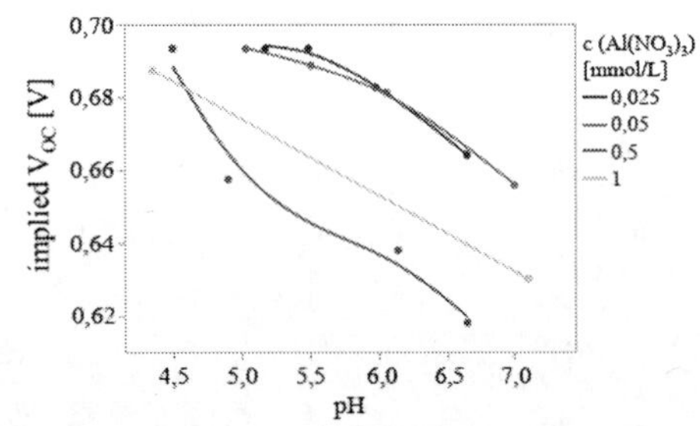

Figure 8: Implied Voc plotted against pH of aluminum nitrate solution. The colors refer to different concentrations of $Al(NO_3)_3$ (see caption).

3.2 p-type PERC

By applying the "WetAlOx" passiviation stack to p-PERC cells, the results presented in Figure 9 / Table I were achieved.

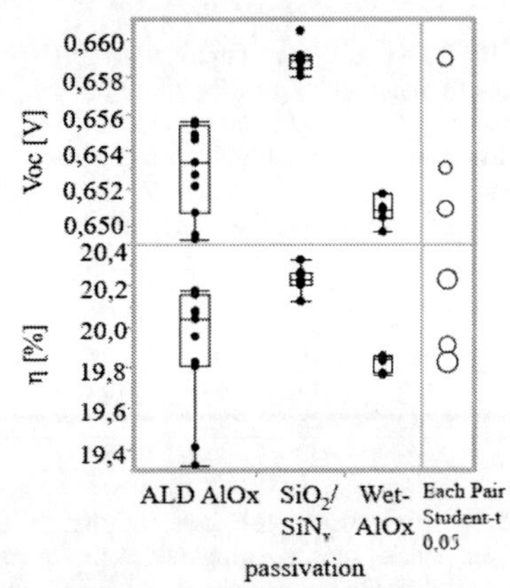

Figure 9 / Table I: Open circuit voltage (Voc, figure top) and efficiency (η, figure bottom) as well as short circuit current (Isc) and fill factor (FF) of p-type PERC cells with different passivation stacks, processed according to the procedure shown in Figure 2.

passivation	# of cells	Voc [V]	Isc [A]	FF [%]	η [%]
Silicon oxide/ nitride	7	0,659	9,57	78,31	20,21
WetAlOx	7	0,651	9,57	77,79	19,82
ALD AlOx	9	0,653	9,59	78,10	20,03

"WetAlOx" performs substantially worse compared to bare SiO_2/SiN_x and ALD AlOx passivation stacks. The mean cell efficiency was 19,82 % compared to 20,21 % for the SiO_2/SiN_x passivation stack. The reason is a lower open circuit voltage Voc (651 mV compared to 659 mV). We speculate that the Voc loss of 8 mV is related to contaminants, in particular iron, which compete with aluminum during ion adsoption in solution as mentioned in section 3.1.

3.3 n-type PERT

By applying the "WetAlOx" passiviation stack to n-PERT cells, the results presented in Figure 10 / Table II were achieved.

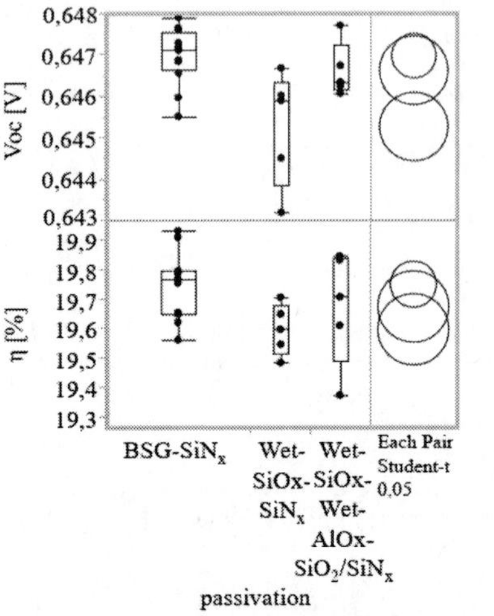

Figure 10/ Table II: Open circuit voltage (Voc, figure top) and efficiency (η, figure bottom) as well as short circuit current (Isc) and fill factor (FF) of n-type PERT cells with different passivation stacks (P: BSG-SiNx (a), WetSiOx-SiNx (b) and WetSiOx-WetAlOx-SiOxSiNx (c)), processed according to the procedure shown in Figure 3.

P	# of cells	Voc [V]	Isc [A]	FF [%]	mean η [%]	η best cell [%]
a	12	0,647	9,30	78,43	19,75	19,93
b	5	0,645	9,28	78,23	19,60	19,70
c	5	0,647	9,32	78,01	19,67	19,84

In the n-type PERT cell concept the "WetAlOx" passivation stack shows comparable performance to conventional passivation methods such as SiO_2/SiN_x stack. While the contaminants lead to a degradation of p-type material (see section 3.2), in case of n-type material contaminantion does not play a decisive role as Guillevin et al. 2009 [12] confirm.

4 CONCLUSION

In this study a new method for AlOx passivation called "WetAlOx" for n-PERT und p-PERC solar cells was presented. The method is based on adsorption of aluminum ions onto an activated silicon surface from a wet chemical solution. Surface charge densities up to $-1.4\cdot10^{12}$ cm^{-2} were reached. The surface charge increased by increasing pH of the aluminum salt process bath, which is due to better aluminum adsorption conditions. Only the first layer of AlOx on top of the thin SiO_2 layer contributes to negative charge densities. Unfortunately metal contaminants, which compete with Al adsorption, lead to implied Voc loss and hence to downgraded solar cell performances. This effect is clearly visible by applying the "WetAlOx" passivation to p-PERC cells. For the n-PERT cell concept this effect did not occur since n-type material has an excellent tolerance to many metal impurities. But still the n-PERT cells with "WetAlOx" did not perform better than cells passivated with a SiO_2/SiN_x stack.

In summary, by applying "WetAlOx" in combination with a SiO_2/SiN_x capping stack, a passivation layer with properties comparable to an ALD AlOx passivation is obtained.

5 ACKNOWLEGMENT

This work was supported by the Federal Ministry for Economic Affairs and Energy (BMWi) as a part of the project "Pascal" FKZ 0325837A.

Supported by:

Federal Ministry for Economic Affairs and Energy

on the basis of a decision by the German Bundestag

6 REFERENCES

[1] J. Schmidt, A. Merkle, R. Brendel, B. Hoex, M.v. de Sanden, W. Kessels, Surface passivation of high-efficiency silicon solar cells by atomic-layer-deposited Al2O3, Progress in Photovoltaics: Research and Applications 16(6) (2008) 461-466.

[2] A.W. Blakers, A. Wang, A.M. Milne, J. Zhao, M.A. Green, 22.8% efficient silicon solar cell, Applied Physics Letters 55(13) (1989) 1363-1365.

[3] J. Benick, B. Hoex, M. Van de Sanden, W. Kessels, O. Schultz, S.W. Glunz, High efficiency n-type Si solar cells on Al 2 O 3-passivated boron emitters, Applied Physics Letters 92(25) (2008) 253504.

[4] F. Buchholz, Metal surface contamination in c-Si solar cell processing, PhD thesis, University of Konstanz (2016).

[5] A.G. Aberle, Surface passivation of crystalline silicon solar cells: a review, Progress in Photovoltaics: Research and Applications 8(5) (2000) 473-487.

[6] J. Schmidt, A. Merkle, B. Hoex, M. Van De Sanden, W. Kessels, R. Brendel, Atomic-layer-deposited aluminum oxide for the surface passivation of high-efficiency silicon solar cells, Photovoltaic Specialists Conference, 2008. PVSC'08. 33rd IEEE, IEEE, 2008, pp. 1-5.

[7] J. Schmidt, F. Werner, B. Veith, D. Zielke, R. Bock, V. Tiba, P. Poodt, F. Roozeboom, A. Li, A. Cuevas, Industrially relevant Al2O3 deposition techniques for the surface passivation of Si solar cells, Proceedings of the 25th European Photovoltaic Solar Energy Conference, Valencia, Spain, 2010, pp. 1130-1133.

[8] F. Buchholz, E. Wefringhaus, S. Eisert, H. Egenolf, A. Groß, Measurement and impact of surface transition metal contamination of textured multicrystalline silicon wafers, Proceedings of the 26th European Photovoltaic Solar Energy Conference and Exhibition, 2011, pp. 1187-1190.

[9] M. Miyashita, H. Kubota, Y. Matsushita, R. Yoshimura, T. Tada, Quantum chemical analysis of metal adsorption mechanism onto silicon surface in cleaning solution, Japanese Journal of Applied Physics 34(3A) (1995) L288.

[10] I. Levin, D. Brandon, Metastable alumina polymorphs: crystal structures and transition sequences, Journal of the American Ceramic Society 81(8) (1998) 1995-2012.
[11] G. Coletti, P.C. Bronsveld, G. Hahn, W. Warta, D. Macdonald, B. Ceccaroli, K. Wambach, N. Le Quang, J.M. Fernandez, Impact of metal contamination in silicon solar cells, Advanced Functional Materials 21(5) (2011) 879-890.
[12] N. Guillevin, R. Naber, L. Geerligs, A. Weeber, High efficiency n-type multicrystalline solar cells, 19th Workshop on Crystalline Silicon Solar Cells & Modules, Materials and Processes, 2009, pp. 9-12.

INDUSTRIAL BIPERC SOLAR CELLS WITH VARIED REAR SIDE CHARACTERISTICS UNDER BIFACIAL ILLUMINATION

N. Wöhrle[1], T. Fellmeth[1], A. Krieg[1], P. Palinginis[2], T. Weber[2], S. Steckemetz[2], K. Ramspeck[3], J. Greulich[1], S. Rein[1]
[1]Fraunhofer Institute for Solar Energy Systems ISE, Heidenhofstraße 2, 79110 Freiburg, Germany
[2]SolarWorld Industries GmbH, Am-Junger-Löwe-Schacht 2, 09599 Freiberg, Germany
[3]h.a.l.m. elektronik gmbh, Friesstraße 20, 60388 Frankfurt/Main, Germany
Telephone: +49 761 4588 5964, Fax: +49 761 4588 7621, e-mail: nico.woehrle@ise.fraunhofer.de

ABSTRACT: Bifacial solar cells pose the challenge of optimizing a solar cell's rear side towards two opposing aspects: The front side efficiency and the rear side efficiency. After a previous work of the authors has dealt with the aspects of optimizing the grid this work focusses on the passivation thickness and the rear side roughness in optical and electrical aspects. A solar cell batch with 20.6% front and 13.5% rear peak efficiency has been produced. It contains a variation of rear side capping thickness, i.e. colors, as well as planarized and textured rear surfaces which are compared under different bifacial illumination scenarios with a pure front as well as a pure rear irradiance of 1000 W/m², as well as 1000 W/m² front and added rear irradiance of 100, 300 and 500 W/m². It has been found that a textured rear surface is capable of capturing more light shining onto the rear surface, but also exhibits enhanced rear surface recombination. Under 1000 W/m² front irradiance, the threshold for rear irradiance necessary to reach parity in terms of output power between a planarized and a textured rear surface PERC is found to be between 300 and 500 W/m² on cell level using an optimal SiN_x capping thickness of 60 nm for the planar and 80 nm for the textured surface. In compound with module encapsulant the rear texture's optical trapping advantage for rear irradiated light diminishes further, increasing the required rear irradiance for output power parity above 500 W/m². The development of PERC cells with bifacialities exceeding 70% could also increase the influence of rear side optics favoring textured rear sides.
Keywords: bifacial solar cell, PERC, rear passivation, rear texture, bifaciality

1 INTRODUCTION

Bifacial PERC solar cells are still in an early stage of industrial adaption and continue to gain attraction due to their potential of a bifacial light collection [1–6]. The aim of this work is to improve the rear side collection by optimizing the surface features in terms of grid layout, surface morphology and passivation properties. After having published a simulation-optimized and verified rear side grid layout a year ago [7], which included first hints on the impact of a textured versus a planarized PERC rear side, we now want to intensify the focus on the rear side morphology and passivation.

When optimizing a bifacial solar cell's rear surface, several opposing trade-off factors have to be considered. On the optical side these are transmission of front-illuminated light through the cell's rear versus light trapping of the rear illumination into the cell. This affects the choice of metal coverage [7], surface morphology and passivation thickness. On the electrical side we have again the metal coverage [7], surface morphology (i.e. surface size) and passivation thickness (stability). By changing these factors with systematic process variations we find an optimum. Due to the bifacial character of the cells, reliable information about the results of the rear side adjustments can only be gathered with appropriate bifacial measurement. We use a current-voltage (*IV*) measurement system for solar cells with independent top and bottom light sources from h.a.l.m. elektronik GmbH offering these realistic conditions. It is described in detail in another submission [8]. By building single-cell-modules from the cell batch we follow up with an assessment of cell characteristics with the inclusion of effects of cell encapsulation for a variation of rear encapsulants.

2 EXPERIMENT

A batch of bifacial PERC solar cells on Cz p-type silicon was produced at the facilities of SolarWorld Industries yielding efficiency levels of roughly $\eta_{front} = 20.6\%$ at 1000 W/m² front and $\eta_{rear} = 13.5\%$ at 1000 W/m² rear illumination. Cells and cell precursors for characterization were provided by SolarWorld's research line and represent cell technology and performance as of 2016. The general cell parameters, except for the now varied rear side, remain the same as in Ref. [7]. This time the experiment contained two different rear side morphologies: a) A planarization by means of a chemical etch isolation and polishing process and b) a pyramid texturing process (equivalent to the

Figure 1: Visualization of the different rear cappings in a combined image (overlay) on planarized surface of the processed PERC solar cells. The SiN_x-capping thickness is (from left to right): 40, 50, 60, 80, 100 nm.

front side texture). On the rear side varying silicon nitride capping layers between 40 and 100 nm thickness have been deposited (see Figure 1). The cell batch is accompanied by characterization samples such as unmetallized precursors of each variation to determine the intrinsic open-circuit voltage (iV_{oc}) (cell area average) using the established combination [9] of photoluminescence (PL) imaging and a carrier lifetime measurement by quasi-steady-state photoconductance (QSSPC). As different trade-off mechanisms were to be expected depending on the illumination conditions, the current-voltage parameters of the produced cells were determined in a bifacial cell flasher under a combination of constant 1000 W/m² (1 sun) front irradiance and varied rear irradiance of 100, 300, and 500 W/m². The 100 W/m² rear irradiance plot is not shown in this paper to keep the length reasonable. To separate front from rear side effects, pure 1000 W/m² irradiance scenarios from front and rear side only are shown as well. Furthermore some of the cells of the optimal rear side capping thickness were incorporated into single-cell-modules to review the effects of rear side morphology on module level, which have been observed on cell level.

3 RESULTS

First of all a short overview of the results is given: Figure 2 shows the iV_{OC} results. The cell results for different illumination conditions are shown over the following pages with the *IV* parameters given in Figure 3 to Figure 6. Figure 7 finally introduces the mini-module results. All boxplots show the upper and lower quartile with the box, which is separated by the horizontal line of the median. The antennas reach to the maximum and minimum value, the arithmetic mean value is given by the open square. The little dots or diamonds give the individual sample's result. Above each boxplot the number of samples is given.

3.1 Characterization sample results

The increased recombination due to enlarged rear surface area of the pyramids on lifetime samples was accounted to $\Delta iV_{OC} = 4$ mV at the peaks, but 10-15 mV in median (see Fig. 2). Layers thinner than 100 nm show an increasing scattering indicating unstable surface passivation. While the planarized rear side shows a median loss of $\Delta iV_{OC} = 5$ mV for 60 nm thickness and $\Delta iV_{OC} = 10$ mV for 40 nm, the thin capping on textured surfaces loses large parts of its passivating property.

3.2 Cell results

Compared to planarized rear surfaces, rear side pyramids have a higher transmission loss of the front side irradiated light due to the altered angle of incidence for the light passing through a solar cell's bulk [7]. Therefore the planarized rear cells reached a peak efficiency of $\eta = 20.6\%$ ($j_{SC} = 39.3$ mA/cm²) while textured rear cells reached a peak efficiency of $\eta = 20\%$ ($j_{SC} = 39.0$ mA/cm²) for pure front irradiance of 1000 W/m² (see Figure 3). The surface enlargement loss is indicated by the pyramid rear sides being $\Delta V_{OC} = 10$-15 mV down to the planarized surfaces for all capping thicknesses between 40 and 100 nm. An additional effect is the decreasing V_{OC} for thinner capping layers, which is caused by increased surface recombination, probably due to less effective

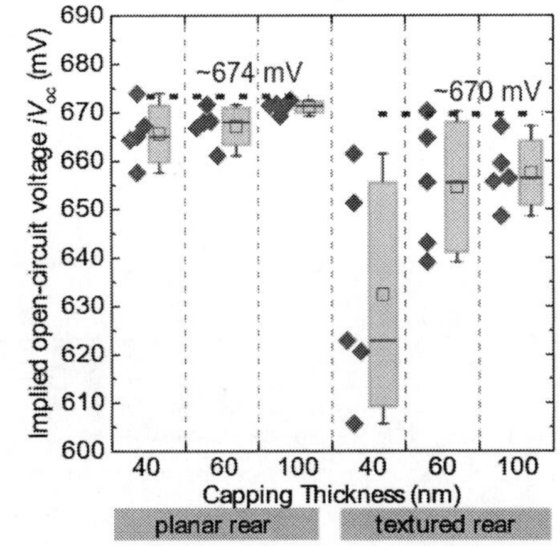

Figure 2: iV_{OC} measurements of five samples each for chosen rear layer thicknesses (40, 60, 100 nm) for planar and textured rear side. The projected upper limit for the reference thickness of 100 nm is indicated with a red dashed line.

hydrogenization and/or protection of the passivation layer by the capping layer against the rear metal. A simultaneous reduction of the fill factor is as expected observable as the maximum achievable fill factor FF_0 depends on V_{OC} [10]. Additionally it is visible that the textured rear surface has enormous spreads in passivation quality especially for reduced capping thickness. This is reflected in all three *IV* parameters indicating that a reduced passivation thickness causes problems in passivation (as already seen in the characterization samples before), optics, and capping against metal paste penetration of the passivation.

For the planarized rear surface this spread is significantly lower and capping thicknesses of 80 and 100 nm lead to a stable passivation, thicknesses down to 50 nm induce a 10% fraction of low-η cells.

To sum up, for a 1000 W/m² front illumination the planarized rear cells have a peak advantage of $\Delta\eta = 0.6\%$, moreover the textured rear side is susceptible towards unstable passivation if its SiN_x-capping thickness is reduced below 100 nm.

For pure rear side illumination of 1000 W/m² (Figure 4) a significant additional factor joins the equation: Rear light trapping. While it is obvious for pure rear illumination it plays a crucial role in bifacial illumination. The rear texture improves the light trapping of rear irradiated light, hence antagonizes the mentioned disadvantages of a textured rear against a planarized surface.

For the pure rear illumination (Figure 4) this leads to an optical advantage of $\Delta j_{SC} \approx 2$ mA/cm² for an optimal rear capping thickness of 60-80 nm. Despite the still reduced V_{OC} (5-10 mV less) this leads to an advantage for the textured rear over the planarized rear with $\eta_{text} = 13.5\%$ and $\eta_{planar} = 12.6\%$. Nevertheless the strong fluctuation inside the groups prevails. This means we have to assume that with process adjustments it is possible to produce a more solid capping layer with more consistent passivation results.

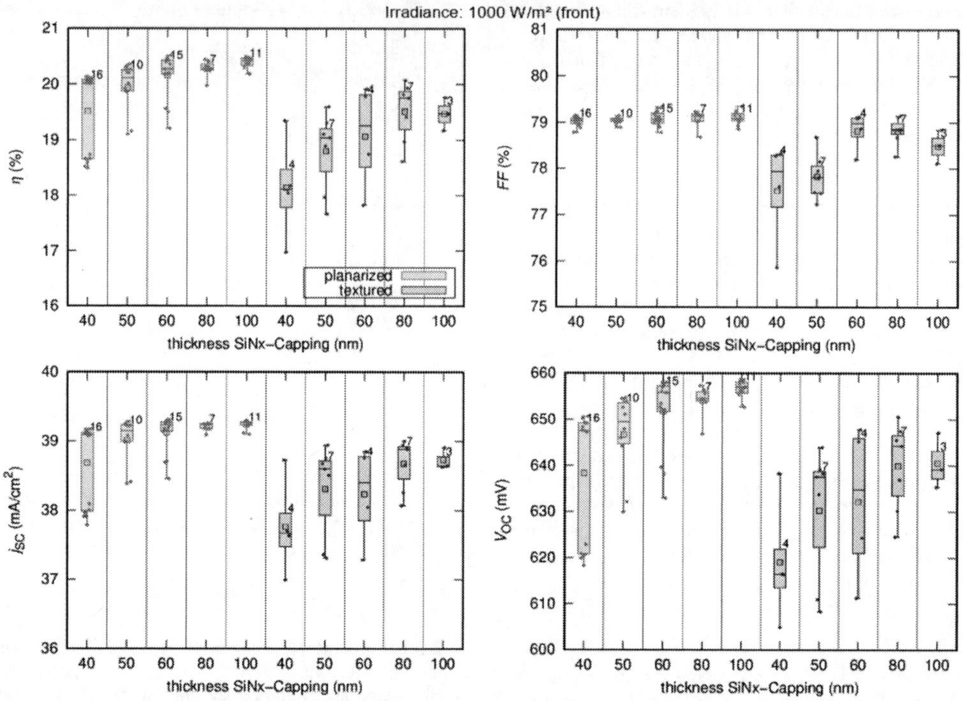

Figure 3: Measured front *IV* parameters for the biPERC cells with varied rear capping (SiN_x) thickness from 40 to 100 nm. The orange box plots show the results on chemically planarized rear, the blue boxes on random pyramid textured rear. The pure front illumination shows the optical advantage for the planarized rear side in short-circuit current by $\Delta j_{SC} \approx 0.5$ mA/cm² mainly due to less transmitted light. The scattered values, especially with the textured rear have their origin in rear surface recombination. The loss in open-circuit voltage ΔV_{OC} is approximately 10 mV for the textured rear surface against the respective planarized counterpart, if outliers are ignored. This is mainly caused by enhanced rear surface recombination.

The relevant follow-up question is if there's a relevant illumination scenario where a cell with textured rear side can overcompensate the deficits it has at front side illumination with rear side collection gains.

The flipping point for optical advantage in the simulation in the previous work [7] was determined to be at 300 W/m² rear irradiance additionally to the 1000 W/m² front irradiance. The bifacial measurements of our cells (Figure 5, Figure 6) show the break-even rather between 300 and 500 W/m² for an optimized capping thickness of 60-80 nm. As presumed this equality for the textured rear against the planarized rear cell is achieved by a balance of a superior optics (higher j_{SC}) and inferior passivation quality and stability (lower V_{OC} & *FF*). This also implies a consistent deposition process avoiding the strong fluctuations we still see. In a scenario where the rear irradiation surpasses the 300 W/m², Figure 6 displays measured 1000 + 500 W/m² bifacial irradiance, the balance finally flips towards an advantage (however a small one) for the peak of the textured rear cell, again driven by gains in j_{SC}.

3.4 Mini module results

For the mini modules the results for 1000 W/m² front and for 1000 W/m² rear illumination are shown in Figure 7. The plots start with the selected cell results, which then have been encapsuled in modules with different types of backsheets (glass, transparent foil, white foil) to provide different optics. The orange boxes represent planarized rear, the blue boxes textured rear surface. The cells all feature 80 nm rear capping thickness.

Unfortunately the glass-glass module assembly suffered from a contacting issue so its *FF* results are hard to compare with the other module results. Nevertheless it is visible that for front irradiation the rear side sheet type (transparent vs. white) makes no significant difference in cell efficiency and the loss on cell level from the textured rear against the planarized rear cell is directly reflected in module compound, both yielding $\Delta\eta \approx 0.5\%$.

For the case of 1000 W/m² rear illumination however the picture is more differentiated. The planarized rear cells have $j_{SC} = 24$ mA/cm² both on cell and module level (with glass as well as transparent backsheet). The textured rear cell's superior current of $j_{SC} = 26.8$ mA/cm² diminishes to 26.2 mA/cm² with a glass encapsulant and to 25.5 mA/cm² with a sole transparent backsheet. This kind of effect is known [11] and can be explained by external light reflection at the air-glass or air-backsheet interface and light trapping within the module, i.e. internal reflection.

As a result the advantage of the improved rear collection of the textured rear side is generally weakened in module encapsulation. The found flipping point at 1000 + 500 W/m² on cell level will thus not be sufficient on module level. Since the advantage on cell level is still small at 1000 + 500 W/m², this can be considered as a minimum rear irradiation for a bifacial cell with textured rear showing higher output power on module level.

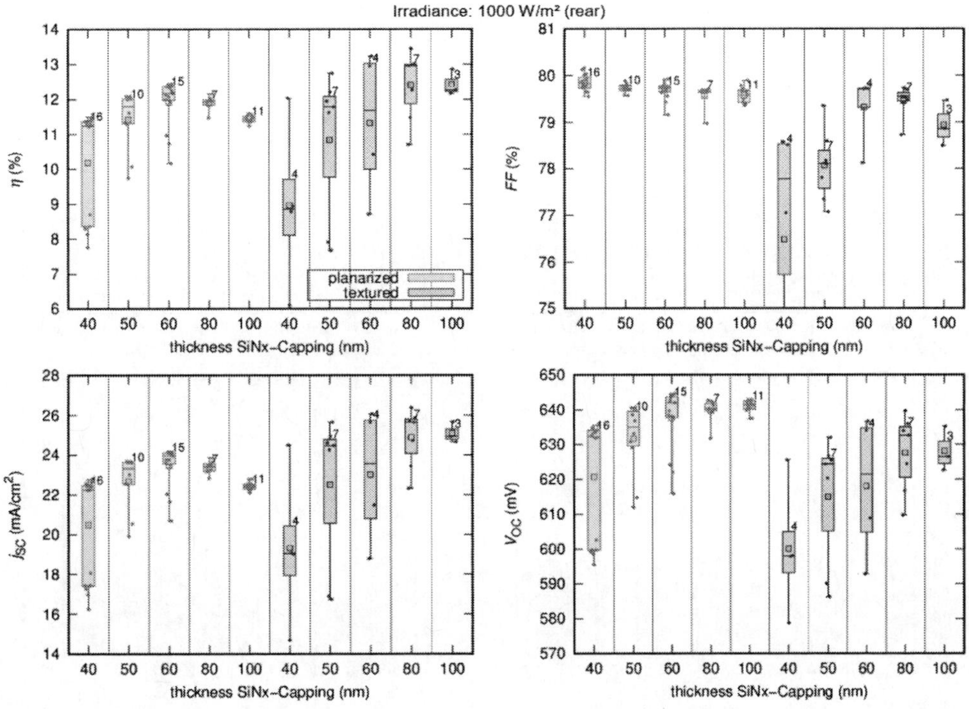

Figure 4: Measured rear *IV* parameters for the biPERC cells at 1000 W/m² rear illumination with varied rear capping (SiN_x) thickness from 40 to 100 nm. The orange boxplots show the results on chemically planarized rear, the blue boxes on random pyramid textured rear. The pure rear illumination shows the optical advantage of the textured surface of $\Delta j_{SC} \approx 2.8$ mA/cm² due to improved light trapping. V_{OC} is still in disadvantage by 10 mV, yielding a remaining efficiency advantage of $\Delta\eta \approx 1\%$. The optimum capping thickness is at 60 nm for planar and 80 nm for textured rear sides.

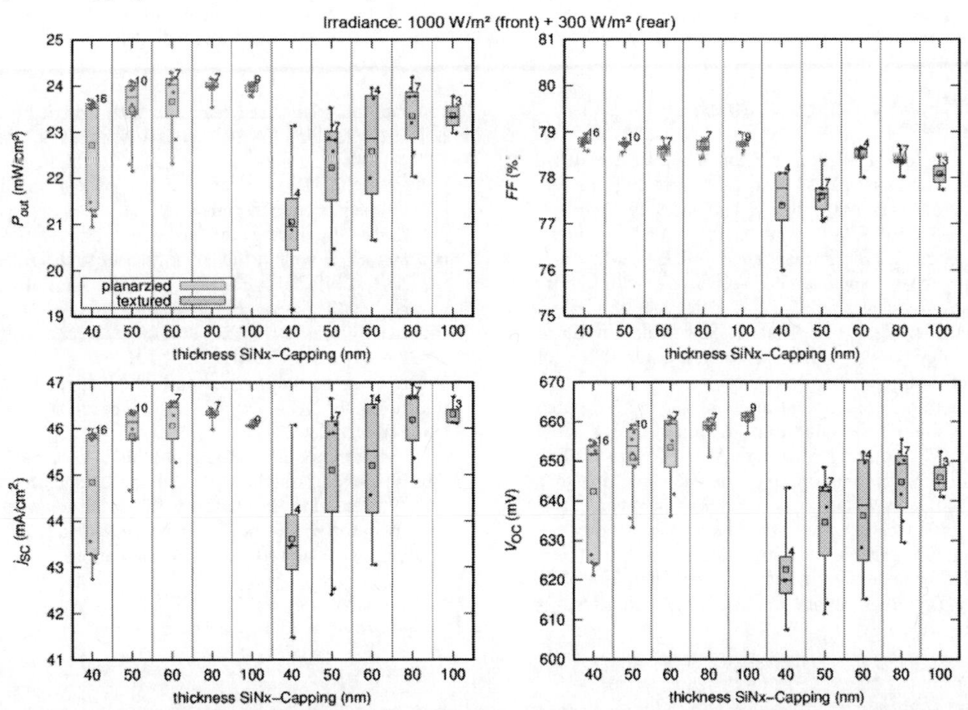

Figure 5: Measured *IV* parameters for the biPERC cells with varied rear capping (SiN_x) thickness from 40 to 100 nm for bifacial irradiance of 1000 W/m² (front) + 300 W/m² (rear). The orange boxes show the results on chemically planarized rear surface, the blue boxes on random pyramid textured rear. With this irradiance the improved light trapping of the textured rear compensates the losses by transmission of front irradiated light (compare maximum j_{SC} of the blue group against the orange group). However the higher recombination loss of the textured rear (*FF* and V_{OC}) still slightly discriminates the texture against the planarized rear in maximum output power (at 60 and 80 nm thickness) with 24.4 against 24.1 mW/cm².

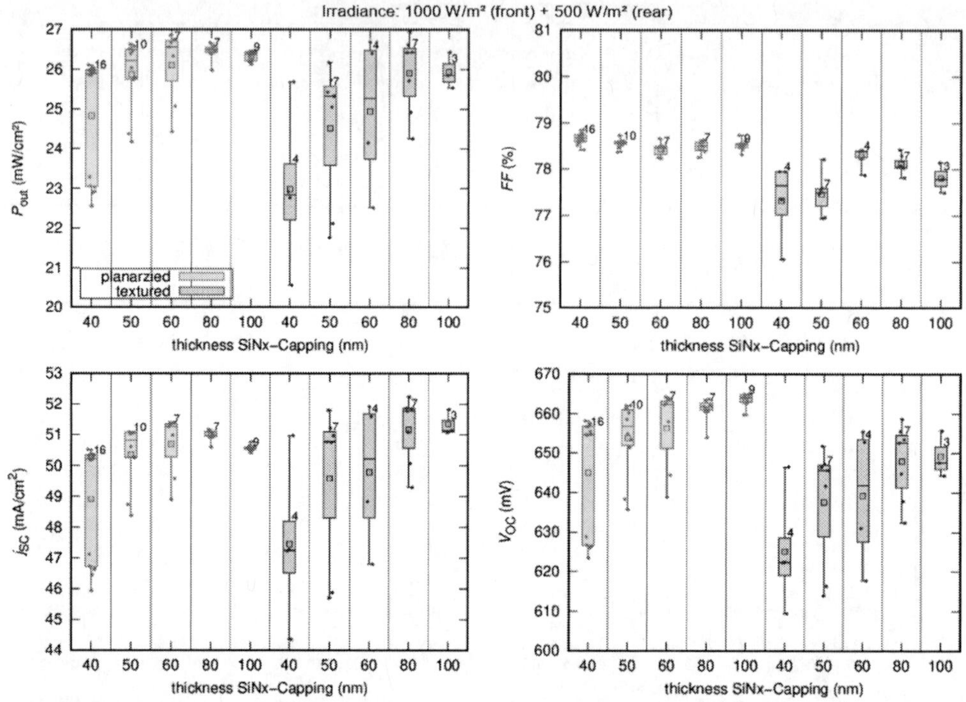

Figure 6: For 1000 W/m² (front) + 500 W/m² (rear) irradiance the improved light trapping of the texturized rear side yields 1 mA/cm² increased maximum j_{SC} over the planarized surface. This compensates the recombination losses and the planarized and texturized rear yield the same maximum p_{out} = 27 mW/cm² for the ideal rear capping thickness of 60 and 80 nm. However the efficiency scattering of the groups, especially the texturized rear sides imposes the need for cautious observation of the capping stability and a process tuning for the deposition step. Further an improvement in process stability for thinner layers would also reduce the general disadvantage of the textured rear in *FF* and V_{OC}, as the recombination beneath the rear fingers could be confined. In effect the textured-rear cell might reach rear parity to the planarized-rear cell at lower rear irradiance than 500 W/m².

4 DISCUSSION AND CONCLUSION

For our solar cell setup we conclude that a textured rear side on bifacial PERC solar cells can unfold its rear light trapping advantage only at comparably high rear irradiances of 500 W/m² or higher (in combination with a simultaneous front irradiance of 1000 W/m²). This exceeds most application scenarios with front illumination and rear albedo. It also exceeds the values discussed for a draft of a bifacial measurement norm of the IEC 60904-1-2, which currently considers 100 or 200 W/m² additional rear irradiance.

Furthermore, the optimization of the rear capping thickness follows the same rules as on the front side (if the expected spectrum is similar to AM1.5G) thus leading to 60 nm SiN_x-capping-optimum in our case (due to a second thin passivation layer beneath). However this is heavily influenced by the stability of the passivation, which tends to degrade for thicknesses lower than 80 nm, especially for textured or rough surfaces. Therefore, if no exceptionally high rear irradiance is expected, a conventional planarized rear side with a dark blue anti-reflection capping is the optimal solution for bifacial PERC cells and modules yielding the highest overall power output. This has been shown by our simulations [7] and the present experiment.

These findings are however limited to cell configurations with sufficient similarity to the solar cells we used. A bifaciality (ratio of rear and front efficiency) around 60% is rather low by today's standards. Industrially feasible PERC studies with bifacialities of 70-90% have been shown [5, 7, 12] giving the rear side optics a higher significance and which again lowers the threshold for rear the irradiance necessary favoring textured rear sides.

Furthermore alternative illumination scenarios are thinkable which might also shift the balance towards a textured rear side. In particular these can be weak light scenarios for the front side with less than 1000 W/m² irradiance such as vertical panel installations.

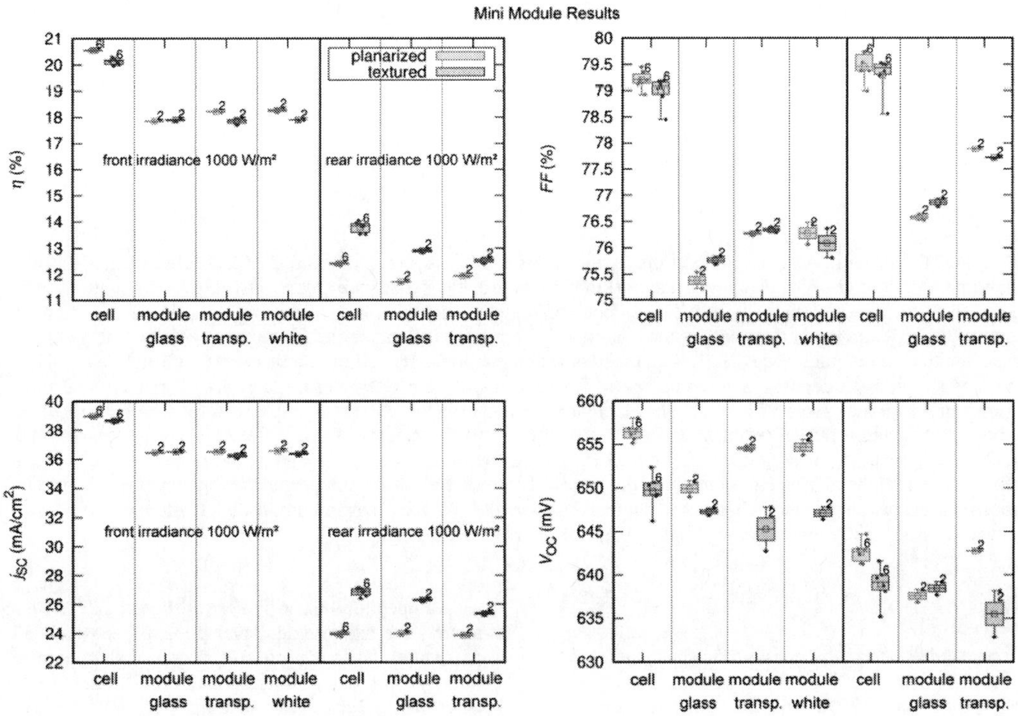

Figure 7: Mini module results for cells with 80 nm rear capping thickness each. The *IV* plots each show the cell results first, followed by the same cells built into 1-cell-mini modules with a glass rear encapsulation, a transparent backsheet or a white backsheet. The columns for rear side illumination lack the white backsheet for obvious reasons. The orange boxes show the cells with planarized rear side, the blue boxes show the textured rear side.

ACKNOWLEDGEMENTS

This work was conducted within the project "BiZePS" (FKZ: 0325909), supported by the German Ministry for Economic Affairs and Energy.

We thank .h.a.l.m elektrinik and SolarWorld Industries for the fruitful cooperation and wish all SolarWorld employees the very best for their future.

REFERENCES

[1] J. Parnell, *Intersolar Europe: SolarWorld to launch glass-glass bifacial modules.* [Online] Available: https://www.pv-tech.org/news/intersolar_europe_solarworld_to_launch_glass_glass_bifacial_modules. Accessed on: Mar. 09 2018.

[2] T. Dullweber *et al.*, "PERC+: Industrial PERC solar cells with rear Al grid enabling bifaciality and reduced Al paste consumption," *Prog. Photovolt: Res. Appl.*, vol. 24, no. 12, pp. 1487–1498, 2016.

[3] R. Kopecek and J. Libal, "Bifaciality: still an advantage for n-type?," Freiburg, 2017.

[4] K. Krauß, F. Fertig, J. Greulich, S. Rein, and R. Preu, "biPERC silicon solar cells enabling bifacial applications for industrial solar cells with passivated rear sides," *Phys. Stat. Sol. (a)*, vol. 213, no. 1, pp. 68–71, 2016.

[5] E. Lohmüller *et al.*, "Bifacial p-type PERL solar cells with screen-printed pure Ag metallization and 89% bifaciality," in *33rd EU PVSEC*, Amsterdam, The Netherlands, 2017.

[6] T. Fellmeth *et al.*, "Co-diffused bi-facial PERT solar cells," *Energy Procedia*, vol. 124, pp. 875–880, 2017.

[7] N. Wöhrle *et al.*, "Understanding the Rear-Side Layout of p-Doped Bifacial PERC Solar Cells with Simulation Driven Experiments," *Energy Procedia*, vol. 124C, pp. 225–234, 2017.

[8] A. Krieg *et al.*, "IV-measurements of bifacial solar cells in an inline solar simulator with double-side illu-mination," in *35th EU PVSEC*, Brussels, submitted, 2018.

[9] B. Hallam, B. Tjahjono, T. Trupke, and S. Wenham, "Photoluminescence imaging for determining the spatially resolved implied open circuit voltage of silicon solar cells," *J Appl Phys*, vol. 115, no. 4, p. 44901, 2014.

[10] M. A. Green, "Accuracy of analytical expressions for solar cell fill factors," *Solar Cells*, vol. 7, no. 3, pp. 337–340, 1982.

[11] J. Greulich *et al.*, "Optical simulation and analysis of iso-textured silicon solar cells and modules including light trapping," *Energy Procedia*, vol. 77, pp. 69–74, 2015.

[12] T. Dullweber *et al.*, "Present status and future perspectives of bifacial PERC+ solar cells and modules," *Jpn. J. Appl. Phys.*, vol. 57, no. 8S3, 08RA01, 2018.

IN-SITU PHOTOLUMINESCENCE STUDY ON EVOLUTION OF PASSIVATION PROPERTIES DURING PLASMA EXPOSURE

Mengkoing Sreng [1,2]*, François Silva [1,2], Pere Roca i Cabarrocas [1,2]
[1] Institut Photovoltaïque d'Ile-de-France, 30 Route Départementale 128, 91120 Palaiseau, France
[2] LPICM, CNRS, Ecole Polytechnique, Université Paris-Saclay, 91128 Palaiseau, France
*corresponding author: mengkoing.sreng@ipvf.fr

ABSTRACT: Vacuum UV light emitted from various plasma processes (e.g. deposition of amorphous silicon nitride by plasma-enhanced chemical vapor deposition (PECVD) or growth of transparent conductive oxides by sputtering) is known to cause a degradation of crystalline silicon surface passivation. A better understanding of the evolution of passivation properties under different plasma conditions is required to attenuate such detrimental effects. Using our in-house in-situ photoluminescence (PL) set-up, the passivation quality of crystalline silicon coated with aluminum oxide grown by atomic layer deposition or hydrogenated amorphous silicon grown by PECVD is characterized in real-time during argon-hydrogen plasma exposure. The PL signals emitted from both types of samples exhibit different behaviors at the very beginning of the treatment and also after turning off the plasma. These differences are linked with the passivation mechanism of both materials as well as the interaction between plasma emission and different types of chemical bonds at the interface. Furthermore, the relation between the passivation degradation induced by plasma and various plasma parameters (e.g. chamber pressure, RF power and substrate temperature) is also studied.
Keywords: in-situ photoluminescence, surface passivation, plasma exposure

1 INTRODUCTION

In crystalline silicon solar cells, surface passivation is widely used to neutralize the dangling bonds, leading to the increase of minority carrier lifetime as well as the improvement of open-circuit voltage. Hydrogenated amorphous silicon (a-Si:H) grown by plasma-enhanced chemical vapor deposition (PECVD), and aluminum oxide (Al_2O_3) grown by atomic layer deposition (ALD) are two of the most commonly used passivation materials. a-Si:H is known for its excellent chemical passivation [1], while Al_2O_3 provides both chemical and field effect passivation even for very thin layers [2]. In order to enhance the efficiency of the solar cells, additional capping layers, e.g. hydrogenated amorphous silicon nitride (a-SiN_x:H) or transparent conductive oxide (TCO), are required for better anti-reflection coating [3] or to provide lateral transport of photo-generated carriers, as can be found in high efficiency solar cells [4, 5]. Plasma processes such as PECVD and sputtering are commonly used to deposit these capping layers and generally a degradation of the passivation is observed afterward [6]. Many ex-situ studies of plasma effects on passivation quality provided by Al_2O_3 or a-Si:H have been reported [7, 8]; however, in-situ studies can provide more insights in order to understand the interactions between plasma species and the passivation layer. In previous works, we have shown that plasma emission, mainly vacuum UV light (VUV), is the main factor causing a degradation of the passivation properties [9]. Here, we report on the in-situ photoluminescence (PL) study of the evolution of surface passivation of c-Si provided by Al_2O_3 or a-Si:H during Ar/H_2 plasma exposure. Furthermore, the relation between passivation degradation and plasma parameters (e.g. chamber pressure, applied RF power and substrate temperature) is also discussed.

2 EXPERIMENTS

High quality double-side polished 4 inch n-type (100) FZ c-Si wafers with a resistivity of ~3 Ω.cm and a thickness of ~280 μm were used as substrates. All wafers were dipped for ~30s in a 5% hydrofluoric acid in order to remove the native oxide layer prior to deposition of the passivation layer. Thin films (~9 nm) of aluminum oxide were grown using the thermal ALD technique in a Picosun™ R200 reactor. During the growth process at 150°C, Trimethyaluminium (metal precursor) and water vapor (oxidant) were alternatively introduced to the chamber. Hydrogenated amorphous silicon thin films were grown by PECVD at 200°C in an OCTOPUS reactor from INDEOTEC. A mixture of silane and hydrogen was used as the gas precursor. After deposition of the passivation layers, all samples were transferred to another PECVD reactor fitted with an in-situ photoluminescence system, where the Ar/H_2 plasma treatment (see Table I) were performed.

Table I: Baseline plasma conditions

Parameters	Value
Ar/H_2 flow rates	10/10 sccm
Pressure	300 mTorr
RF power	15 W
Inter-elec. distance	25 mm
Substrate temperature	Room temperature

Thanks to in-situ photoluminescence set-up integrated into the PECVD reactor (a detailed description of the in-situ PL tool is reported elsewhere [10, 11]), the passivation properties of the samples can be characterized in real time during the experiments.

3 RESULTS AND DISCUSSION

3.1 Evolution of PL signal during plasma exposure

Figure 1 shows the evolution of PL signal emitted from both types of samples during the Ar/H_2 plasma exposure. For sample coated with Al_2O_3, a quick increase of PL signal just after plasma ignition followed by a gradual degradation is observed. However, only degradation of PL signal can be found on sample passivated by a-Si:H.

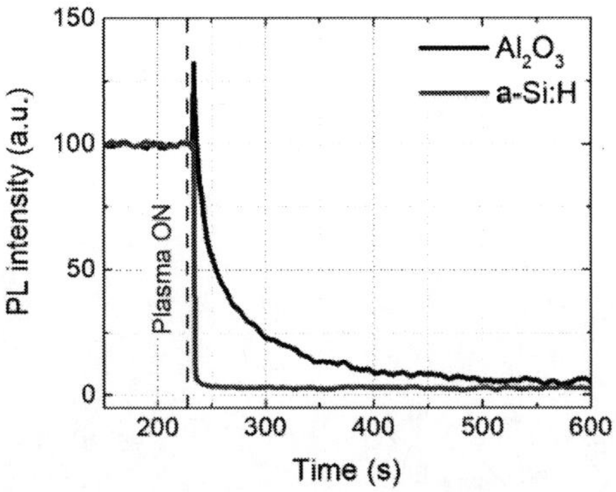

Figure 1: Evolution of PL signal from c-Si wafers passivated by Al_2O_3 and a-Si:H during the Ar/H_2 plasma exposure.

The complex evolution of the PL signal observed here is due to the interaction between the VUV light emitted by plasma and the chemical bonds near/at the interface between c-Si and passivation layers. The quick increase of PL signal observed on Al_2O_3-passivated sample is probably caused by the dissociation of hydroxyl radicals (Al–OH) near the c-Si/Al_2O_3 interface, which leads to an increase of fixed charge density in the Al_2O_3 layer (enhancement of field effect passivation). A more detailed discussion can be found elsewhere [9]. On the other hand, the slow decrease of PL signal may result from breaking of different chemical bonds (e.g. Si-H, Si-O, etc.) at c-Si/Al_2O_3 and c-Si/a-Si:H interfaces, inducing the degradation of chemical passivation.

One may notice that for the same layer thickness and under the same plasma conditions, the passivation quality provided by a-Si:H degrades much faster than that provided by Al_2O_3. The explanation of the difference between both results may be related to the strength of chemical bonds at the c-Si/Al_2O_3 and c-Si/a-Si:H interface. To find out, additional plasma experiments with an optical window (glass, $\lambda_{cut\text{-}off}$ = 275 nm or 4.5 eV) coverage were conducted. The idea is to prevent high energy photons (VUV) from reaching the passivated sample, but allow those with energy less than 4.5 eV to pass through.

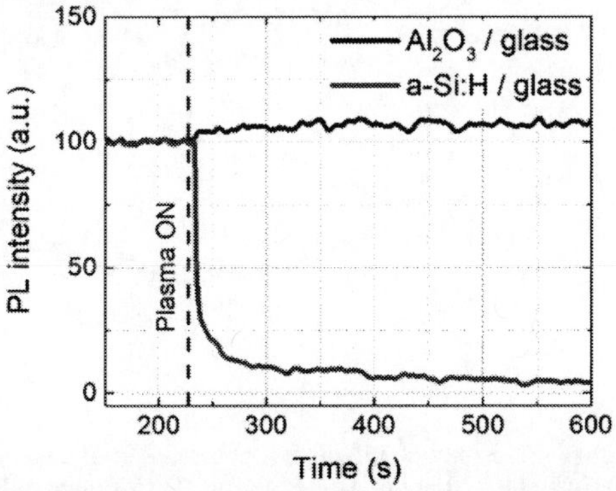

Figure 2: PL signal recorded during the Ar/H_2 plasma exposure of samples covered by an optical window (glass).

Figure 2 shows the PL signal emitted by samples covered by a glass window during the Ar/H_2 plasma exposure. We clearly see that by blocking the photons with energy higher than 4.5 eV, the degradation of surface passivation only occurs on sample coated with a-Si:H. These results suggest that the chemical bonds at the c-Si/Al_2O_3 interface are stronger than those at c-Si/a-Si:H interface.

3.2 PL signal recovery after plasma exposure

The evolution of passivation properties of each sample was also characterized after turning off the plasma. As shown in Figure 3, the PL signal from samples coated by Al_2O_3 remains stable after the plasma was turned off, while a slight increase is observed on the samples coated with a-Si:H. These results suggest that defects induced by plasma on the Al_2O_3-passivated samples remain stable at room temperature. However, the passivation provided by a-Si:H can partially recover by itself even after a significant deterioration caused by plasma. Further study is required to explain the mechanism behind this result.

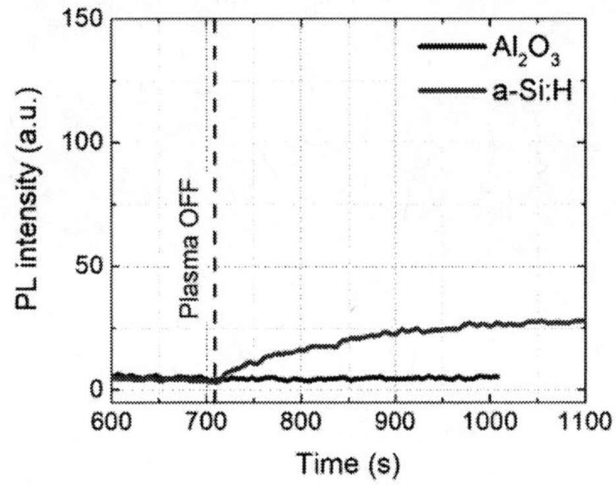

Figure 3: Evolution of the PL signal from both types of samples after turning off the plasma (experiments were conducted at room temperature).

3.3 Effect of layer thickness and process conditions

a. Effect of passivation layer thickness

In an attempt to reduce the detrimental effects of plasma processes on the passivation quality, the thickness of the passivation layers is increased from 9 nm to 40 nm.

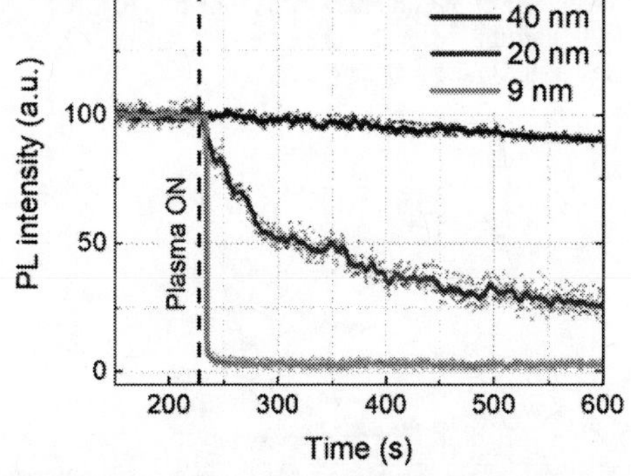

Figure 4: Evolution of PL signal emitted from sample passivated by a-Si:H of different thickness

As shown in Figure 4, the degradation of PL signal is significantly attenuated by increasing the a-Si:H thickness from 9 nm to 20 nm. Furthermore, this plasma induced degradation is hardly observed when the a-Si:H layer reaches 40 nm. The reasons behinds these thickness dependent results lies on the fact that a-Si:H is a direct band semiconductor with a moderate bandgap (~1.7 eV). As its thickness increases, the layer tends to absorb more high energy photons, preventing them from deteriorating the interface. However, it is worth mentioning that this behavior is not found on samples passivated by Al_2O_3 with a thickness from 9 nm to 50 nm. The thing is Al_2O_3 has a large bandgap (~7 eV); therefore, its photo-absorption property is less pronounced compared to the absorption of a-Si:H layer.

Although the increase of a-Si:H thickness can strongly reduce the detrimental effects of plasma, the a-Si:H layer also causes parasitic absorption that will lead to a decrease of solar cells efficiency. So other options should be considered.

b. Effect of pressure and RF power

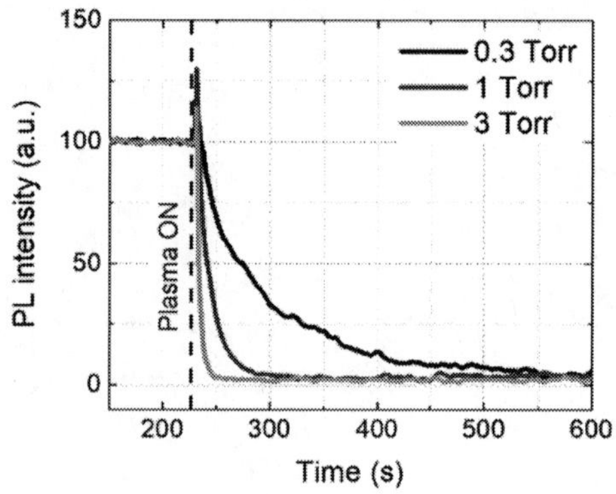

Figure 5: PL signals emitted by Al_2O_3-passivated samples under the Ar/H_2 plasma exposure at different pressure.

As plasma properties are strongly depended on the process parameters, variation of the chamber pressure and applied RF power may allow a diminution of plasma detrimental effects. Figure 5 shows the PL signal emitted by samples passivated by Al_2O_3 under the plasma exposure at different pressure. The results suggest that the degradation of the passivation quality can be reduced by lowering the chamber pressure, which results in a decrease of the plasma emission intensity as shown in Figure 6. One can see that the lower the pressure (3 Torr – 0.3 Torr) is, the less intense the plasma emission.

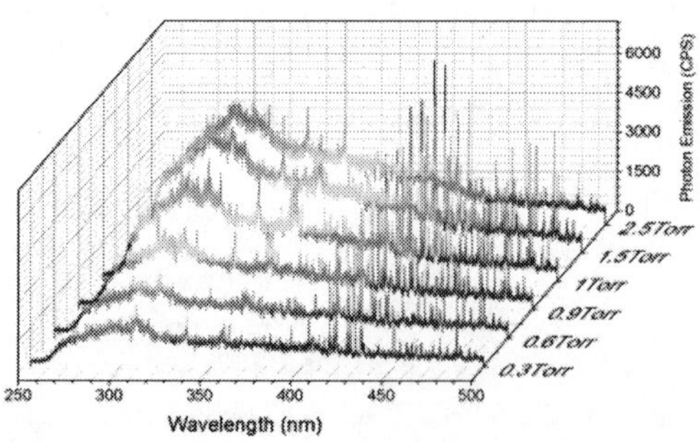

Figure 6: Plasma emission intensity as a function of wavelength for different value of the chamber pressure.

These results also hold true for the variation of applied RF power. The lower the RF power is, the slower the degradation of the PL signal. One can understand that by lowering the applied RF power, free electrons in the plasma will have less energy and therefore plasma emission is less intense.

It is also important to notice that the influence of chamber pressure and applied RF power observed on sample coated by Al_2O_3 above is also found on sample passivated by 20 nm of a-Si:H.

c. Effect of substrate temperature

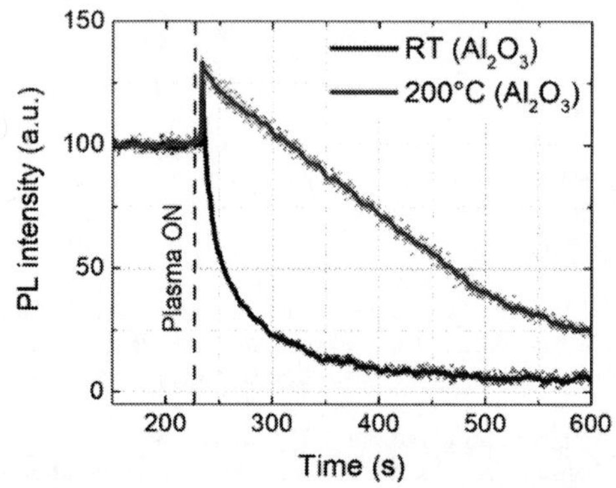

Figure 7: Emitted PL signals from Al_2O_3-passivated samples during the Ar/H_2 plasma exposure at RT and at 200°C.

Substrate temperature may have less influence on the plasma species; however, it has a great influence on the properties of the passivation materials, mainly the diffusion of atomic hydrogen. Figure 7 compares the PL signal obtained from the samples passivated by Al_2O_3 during the plasma exposure at RT and at 200°C. We can see that by increasing the temperature from RT to 200°C, the degradation of passivation quality is strongly decreased. The obtained result can be explain by two competing mechanisms occur simultaneously: the degradation caused by plasma emission and the annealing of defects induced by diffusion of atomic hydrogen (originating from Al_2O_3 layer), leading to a slower overall degradation of the passivation quality.

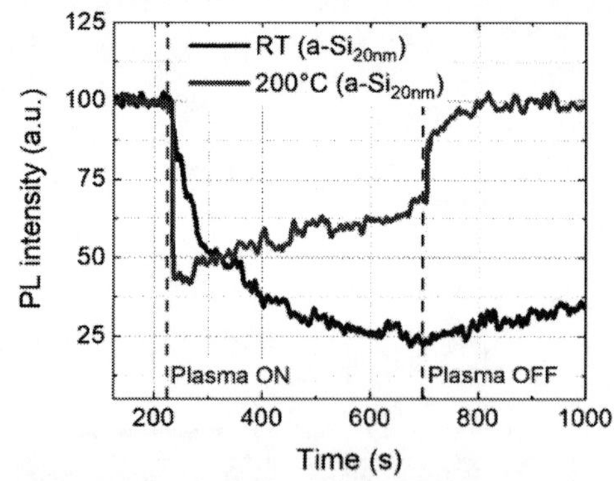

Figure 8: Emitted PL signal obtained from samples passivated by 20 nm of a-Si:H during the treatment at RT and at 200°C.

The plasma exposure at 200°C is also conducted on samples passivated by 20 nm of a-Si:H. Figure 8 compares the PL signals obtained during the treatments at RT and at

200°C. A complex behavior of the evolution of surface passivation is observed during treatment at 200°C.

(i) Just after plasma ignition, the PL signal quickly drops by more than 50%, much faster than the one treated at room temperature. A possible explanation of these results is that at high temperature the vibration of the chemical bonds increases, reducing the energy required to break them. As the bonds become weaker, the number of photons (UV or near UV) that can break these interfacial bonds will increase, and therefore many dangling bonds can be rapidly created.

This feature cannot be observed in the treatment of sample passivated by Al_2O_3 because the chemical bonds at the c-Si/Al_2O_3 interface are mostly strong bonds. So even though these interfacial bonds are weaken by the vibration, they still require high energy photons (VUV) to break them. In addition, the increase of fixed charge in the Al_2O_3 layer that leads to increase of PL signal may prevents the detection of such feature.

(ii) After the quick drop, emitted PL signal slowly increases, indicating a recovery of the passivation properties even under plasma exposure. To understand this feature, it is important to notice that only photons with energy equal or less than 3 eV can pass through the 20 nm a-Si:H. From the first stage of the experiment, we learn that most of weak bonds at the c-Si/a-Si:H interface are broken, leading to a large amount of dangling bonds at the interface. In this second stage, two competing mechanisms should be taken into account: interfacial bonds breaking caused by photons from plasma and healing of dangling bonds due to creation of both strong and weak bonds at the interface. While reforming the interfacial bonds, the weak ones may get broken again by the photons that pass through the 20 nm a-Si:H layer. However, these photons may not able to break some newly-generated strong bonds (e.g. Si–Si) which required energy more than 3 eV to break them. So as the number of strong interfacial bonds is gradually increased, the passivation is slowly recovered, leading to a slow increase of PL signal.

These mechanisms could also occur on samples passivated by Al_2O_3. However, as there is nothing to block the high energy photons (VUV) from reaching the interface, both strong and weak regenerated bonds will got broken.

(iii) When the plasma is turned off, the PL signal quickly returns to its initial value. In this last stage, the degradation caused by plasma emission is terminated. Only the annealing of dangling bonds occurs until reaching a steady state.

4 CONCLUSION

Using our in-house in-situ photoluminescence set-up, the evolution of passivation properties provided by Al_2O_3 and a-Si:H has been studied in real-time during and after an Ar/H_2 plasma exposure. On samples passivated by Al_2O_3, the PL signal quickly increases just after plasma ignition due to a quick dissociation of hydroxyl radicals (Al–OH), resulting in fixed charge generation in Al_2O_3 layer. The degradation of PL signal, observed on both samples, is linked to the breaking of other chemical bonds at the interface between c-Si and the passivation layers. It is important to note that photons with energy less than 4.5 eV do not cause degradation on samples coated with Al_2O_3, but with a-Si:H. A self-recovery of passivation properties at room temperature is observed on samples passivated by a-Si:H after turning off the plasma, and further study is required to explain the mechanisms behind this result. The thickness of a-Si:H layer could be increased to reduce plasma deterioration effects, but one should remember that increase of a-Si:H thickness also raises the parasitic absorption. Modifying plasma parameters can be an option to reduce the plasma induced passivation degradation. By decreasing the chamber pressure and applied RF power, the plasma emission become less intense and therefore the degradation of the passivation quality is reduced. Increasing the substrate temperature provides a competing annealing process which may cancel out the degradation process and restore the initial passivation properties when the plasma is turned off.

ACKNOWLEDGEMENTS

This project has been supported by the French Government in the frame of the program of investment for the future (Programme d'Investissement d'Avenir - ANR-IEED-002-01). The authors acknowledge the support from P. Bulkin for providing samples, J.C Vanel for software development and the BEER team from LPICM for technical support.

REFERENCE

[1] J. A. Schüttauf, K. H. M. van der Werf, I. M. Kielen, W. G. J. H. M. van Sark, J. K. Rath, R. E. I. Schropp, Applied Physics Letters, 98 (2011) 153514.

[2] D. Schuldis, A. Richter, J. Benick, P. Saint-Cast, M. Hermle, S. W. Glunz, Applied Physics Letters, 105(2014) 231601.

[3] D. Wu, R. Jia, W. Ding, C. Chen, D. Wu, W. Chen, H. Li, H. Yue, X. Liu, Journal of Semiconductors, 32(2011) 9.

[4] F. Feldmann, M. Bivour, C. Reichel, M. Hermle, S. W. Glunz, Solar Energy Materials & Solar Cells, 120(2014).

[5] M. Taguchi, A. Yano, S. Tohoda, K. Matsuyama, Y. Nakamura, T. Nishiwaki, K. Fujita, E. Maruyama, IEEE Journal of photovoltaics, 4(2014).

[6] Y. P. B. Mouafi, T. Lüder, G. Hahn, proceedings 28th European Photovoltaic Solar Energy Conference and Exhibition, (2013).

[7] F. Wang, X. Zhang, L. Wang, Y. Jiang, C. Wei, J. Sun, Y. Zhao, ACS Applied Materials & Interfaces, 17(2014).

[8] M. Mews, E. Conrad, S. Kirner, N. Mingirulli, L. Korte, Energy Procedia, 55(2014).

[9] M. Sreng, F. Silva, P. Roca i Cabarrocas, Physica Status Solidi A, submitted 2018.

[10] F. Lebreton, S. N. Abolmasov, F. Silva, P. Roca i Cabarrocas, Applied Physics Letters, 108(2016) 051603.

[11] S. N. Abolmasov, P. Roca i Cabarrocas, Journal of Vacuum Science & Technology A, 33(2015) 021201.

EMITTER FORMATION AND PASSIVATION DEPENDENCE ON CRYSTAL GRAIN ORIENTATIONS AFTER ATMOSPHERIC PRESSURE DRY NANOTEXTURING

A. I. Ridoy[1], B. Kafle[1], P. Saint-Cast[1], S. Lohmüller (née Werner)[1], M. H. Norouzi[1], L. Clochard[2], E. Duffy[2], M. Hofmann[1], J. Rentsch[1], R. Preu[1]
[1]Fraunhofer Institute for Solar Energy Systems ISE, Heidenhofstr. 2, 79110 Freiburg, Germany
[2]Nines Photovoltaics, IT Tallaght, Dublin 24, Ireland
Correspondence: ahmed.ismail.ridoy@ise.fraunhofer.de, +49 (0)761 4588 2183

ABSTRACT: In this work, we investigate the emitter performance of the multicrystalline silicon (mc-Si) solar cell as a function of the crystallographic orientation of the grains and the associated texturing level. Here, we applied a plasma-less nanotexturing process by atmospheric pressure dry etching that enables low reflectivity, followed by a short anisotropic alkaline etch. It is seen in our investigation that grains with lower reflection exhibit lower emitter sheet resistance ($R_{sh} \approx 72$ Ω/sq.) than that of grains with higher reflection ($R_{sh} \approx 79$ Ω/sq.). We show that with our current etching process flow, there is a linear correlation between charge carrier lifetime and weighted surface reflection of different grains in a mc-Si wafer.
Keywords: Atmospheric pressure dry etch, nanotexture, multicrystalline silicon, crystal grain orientations, emitter performance

1 INTRODUCTION

Alternatives to the incumbent wet-chemical texturing method have been sought after by the photovoltaic community for quite some time [1, 2]. More recently, nanotexturing processes (also often referred to as "black silicon") have generated renewed interest due to their ability to reach low reflectivities and higher solar cell efficiencies using slurry-sawn multicrystalline silicon (mc-Si) wafers [3, 4]. The plasma-less atmospheric pressure dry etching (ADE) process recently demonstrated efficiencies above 20% using mc-Si wafers [5]. ADE is also suitable for monocrystalline silicon (c-Si) wafers [6]. For mc-Si wafers, the texture varies depending on the crystallographic orientation of the grains and the degree of anisotropy of the etching/texturing steps that are used. Anisotropic etching causes reflection variations among the mc-Si grains due to the formation of structures different in morphology. It is observed that this difference in morphology of the crystal grains cause variations in emitter diffusion, affecting the carrier concentration and the emitter sheet resistance.

As a continuous improvement in the ADE technique, in this work, we investigate various crystal orientations within mc-Si wafer and their effects on specific cell processes, such as surface reflection, surface passivation, emitter diffusion, carrier concentration and lifetime. The purpose of this work is to understand these variations in a quantitative way and to analyze the distribution of results of the ADE-nanotextured mc-Si solar cells, which will allow further optimization of the surface texture and the subsequent cell manufacturing process steps to achieve higher conversion efficiencies.

2 APPROACH

2.1 Experiment design

The process workflow is presented in Figure 1. 156x156 mm² boron-doped *p*-type mc-Si wafers (base resistivity (ρ_B) ≈ 1.6 Ωcm) were pre-treated by saw-damage etching using wet-chemical process (alkaline), subsequently the ADE texturing was performed in order to form black silicon structures on wafer front sides [5–9]. The wafers then went through a post-treatment process in order to lower the surface roughness to allow better integration of this texture into common solar cell processing steps, most notably emitter diffusion and surface passivation [7]. During this post-treatment process the wafers were subjected to a short anisotropic wet-chemical treatment (alkaline) adjusted to reach an average value of surface reflection at 600 nm, $R_{600} \approx 15\%$. Some wafers were characterized for surface morphology, weighted surface reflectivity (R_w) [9] and scanning electron beam microscopy (SEM). Emitter was formed on the textured surface by industry-type $POCl_3$-based tube diffusion furnace, followed by chemical edge isolation including phosphosilicate glass (PSG) etching. Some wafers were characterized for emitter diffusion properties by using electrochemical-capacitance-voltage (ECV) and four-point-probe (4pp) measurement techniques. The wafers were first passivated by applying conformal layer of AlO_x both on front side (2 nm) and rear side (6 nm) by using fast atomic layer deposition (ALD) technique [8], followed by a low-temperature annealing, so called "outgassing" process [10] and were finally passivated by applying SiN_x depositions using plasma-enhanced chemical vapor deposition (PECVD) both on front side (75 nm) as anti-reflection coating (ARC) and on rear side (150 nm) as capping layer. Afterwards, the samples were fired at a set temperature of 850 °C.

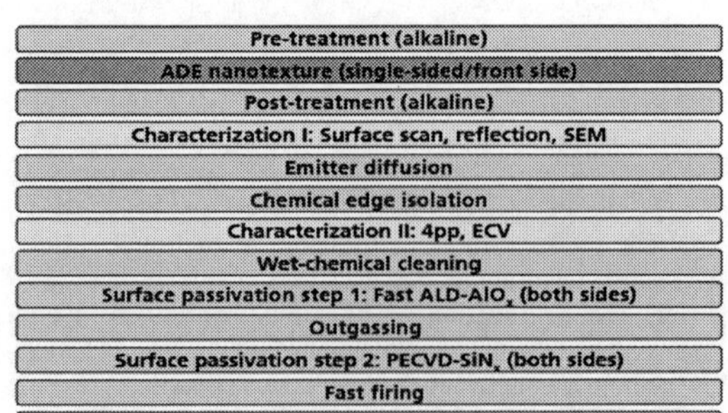

Figure 1: Schematic process workflow.

Hence, the precursors went all steps that are typically applied to fabricate passivated emitter and rear cell

(PERC) architecture apart from metallization. Finally, the processed samples were characterized by measuring surface reflection and photoluminescence (PL).

2.2 Characterization

In order to investigate the influence of the etching steps on mc-Si grains, at first, scanned image of an ADE-textured surface was characterized for its intensity distribution of the surface darkness. One dark grain (a) and one light grain (b) were selected for further characterizations, such as R_w and SEM.

Following emitter diffusion process, emitter sheet resistance (R_{sh}) of the dark grain (a) and the light grain (b) for identical emitter diffusion process was then characterized before surface passivation by using four-point-probe (4pp) technique. Furthermore, electrochemical-capacitance-voltage (ECV) method as per [11] was performed in order to obtain various emitter profile characterization values by characterizing the dark grain (a), the light grain (b) along with the flat surface for identical emitter. As the surface roughness influences the ECV measurement (the surface enlargement factor (f) correlates with the charge carrier concentration (N) and the depth (d)), we corrected the measured carrier concentration profile, that the R_{sh} determined from the profile matches the local R_{sh} determined by 4pp technique [12–14]. After surface passivation and firing, the samples were characterized with PL method for lifetime-calibrated PL-imaging.

3 RESULTS AND DISCUSSION

3.1 Reflection distribution over ADE-textured mc-Si

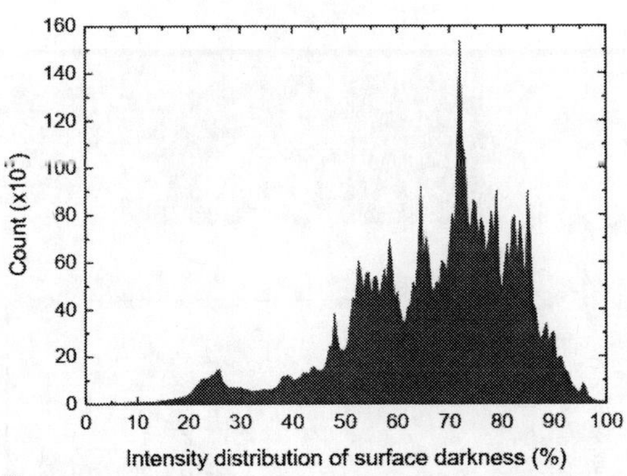

(i)

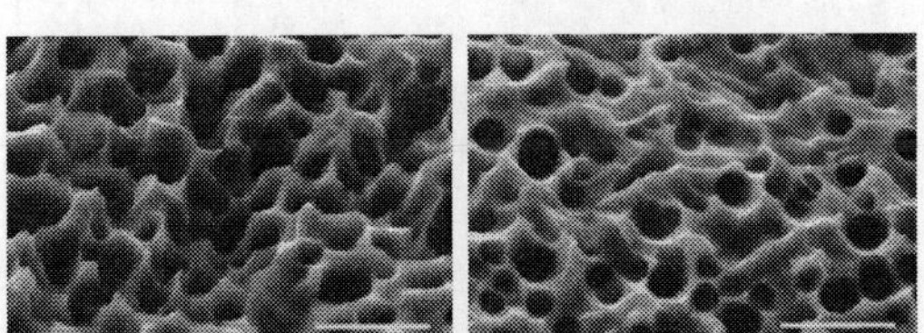

(ii) (iii)

Figure 2: (i) Histogram of an ADE-textured mc-Si wafer after surface scanning representing counts over intensity distribution of the surface darkness. (ii) and (iii) Inclined top-view of scanning electron microscopy (SEM) images of dark grain (a) and light grain (b) of the selected ADE-textured wafer respectively.

Figure 2(i) presents correlation between counts over surface darkness distribution for a scanned surface of an ADE-textured mc-Si wafer. Such distribution of surface darkness can be translated as a distribution function of surface reflection over the textured surface.

The SEM images of the lowest reflective grain of the wafer, namely the dark grain (a) ($R_w \approx 8.8\%$) and the highest reflective grain of the wafer, namely the light grain (b) ($R_w \approx 16.5\%$) are presented respectively in Figure 2(ii) for the dark grain (a) with densely situated porous structures, and in Figure 2(iii) for the light grain (b) with comparatively less porous structures than the dark one.

3.2 Emitter diffusion

Figure 3(i) presents the corresponding R_{sh} distributions for dark grain (a) and light grain (b) measured by 4pp technique. The emitter diffusion characterized by ECV method and fitted for R_{sh} measurement by 4pp technique is presented in Figure 3(ii) exhibiting a possible trend with highest carrier concentration (N) for lower reflection grains to lower N for higher reflection grains.

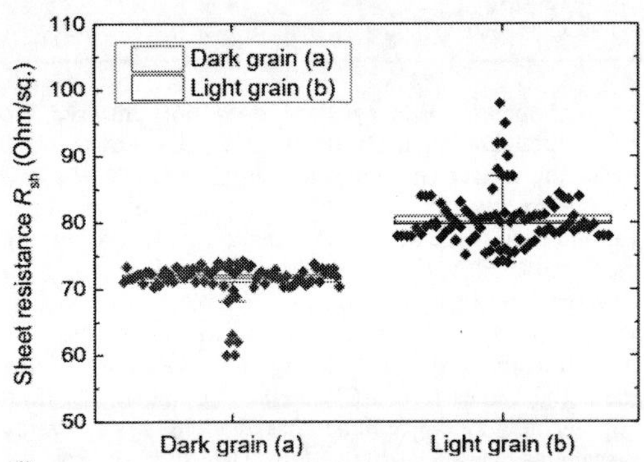

(i)

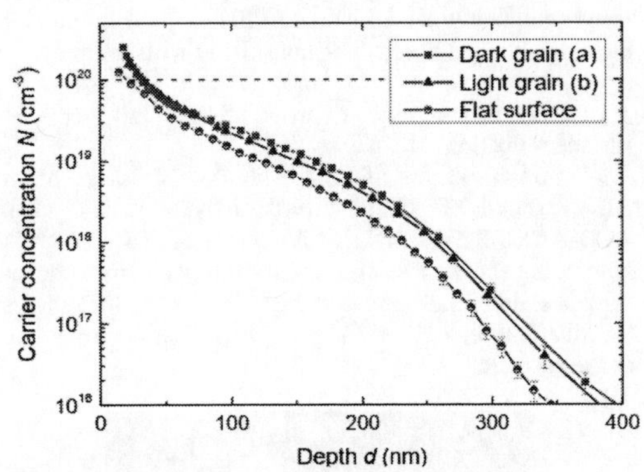

(ii)

Figure 3: (i) Emitter sheet resistances (R_{sh}) for dark grain (a) and light grain (b) measured by four-point-probe (4pp) technique. (ii) Carrier concentration (N) in comparison between ADE-textured dark grain (a) and light grain (b) to flat surface for carrier concentration (N) measured by electrochemical-capacitance-voltage (ECV) method and fitted for R_{sh} from Figure 3(i).

Surface enlargement is found to be higher for the dark grain (a) (factor of 2.2) due to higher texturing than for the light grain (b) (factor of 1.7), which is the reason behind higher N for dark grain (a) and comparatively lower N for light grain (b). In comparison, pertaining to a low surface enlargement, N is lower for the flat surface for the diffusion of the identical emitter.

Different emitter profile characterizations are, namely maximum carrier concentration at the surface (N_s), depth of the highly doped region (d_{high} at $N = 10^{20}$ cm^{-3}), depth of the lowly doped region (d_{low} at $N = 10^{16}$ cm^{-3}), emitter sheet resistance (R_{sh}) (see table I).

Table I: Characteristics of the emitter diffusion profile measured with ECV method and adjusted with 4pp technique for the dark grain (a), the light grain (b) and the flat surface after emitter diffusion and PSG etching.

Grains/ Properties	Dark grain (a)	Light grain (b)	Flat surface
N_s (x10^{20} cm^{-3})	(2.5±0.2)	(1.6±0.0)	(1.3±0.0)
d_{high} (nm)	35	30	20
d_{low} (nm)	400	390	350
R_{sh} (Ω/sq.)	72	79	88
R_w (%)	8.8	16.5	30.0

Nonetheless, ECV method does not preserve the surface roughness inherently obtained from texture itself, hence the surface enlargement is argumentative over two factors: mostly the texture and to some extent the ECV method itself. Moreover, ECV method only provides a mean value of N for the whole measurement area as it does not consider diffusion profile/emitter performance separately at the tip of the nanostructures or inside the nanostructures. Although during the adjustment of the measured values from ECV method (by using R_{sh} from 4pp technique) we assume that the emitter is evenly distributed over the whole texture. Therefore, it can be concluded that although the ECV method provides insight to the emitter diffusion profile to good extent, yet the data obtained needs to be considered with caution.

3.3 Comparison of charge carrier lifetime and weighted surface reflection

Figure 4(i) shows lifetime-calibrated PL image of an ADE-textured wafer passivated with fast ALD-AlO_x/PECVD-SiN_x stacks on front and rear sides after firing showing a variation of minority charge carrier lifetimes along the wafer surface area. Figure 4(ii) shows the distribution of the charge carrier lifetime over the entire surface.

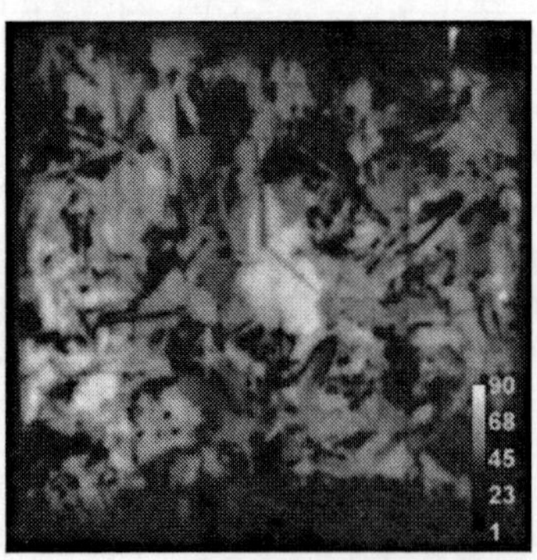

(i)

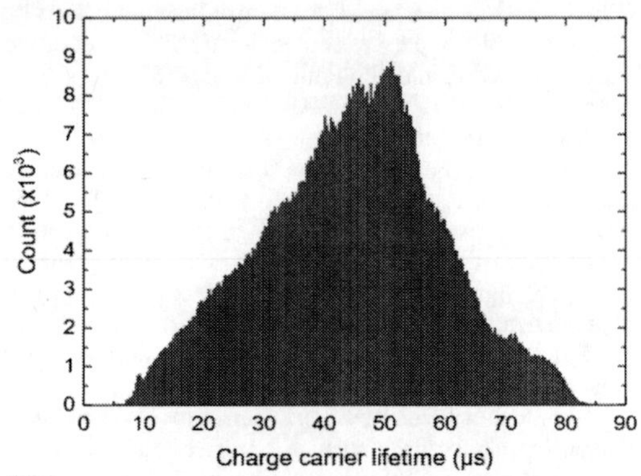

(ii)

Figure 4: (i) Lifetime-calibrated PL image of passivated ADE-textured surface. (ii) Histogram of the charge carrier lifetime representing counts over lifetime for an entire ADE-textured surface.

We can see from 4(i) and 4(ii) that the carrier lifetime varies throughout the surface. Although the carrier lifetimes in various mc-Si grains are known to be influenced by their inherent bulk lifetimes (due to grain boundaries, defect/dislocations, etc.), we expect that a variation in surface morphology in different grains also significantly affects the effective lifetime due to a difference in emitter formation and surface passivation.

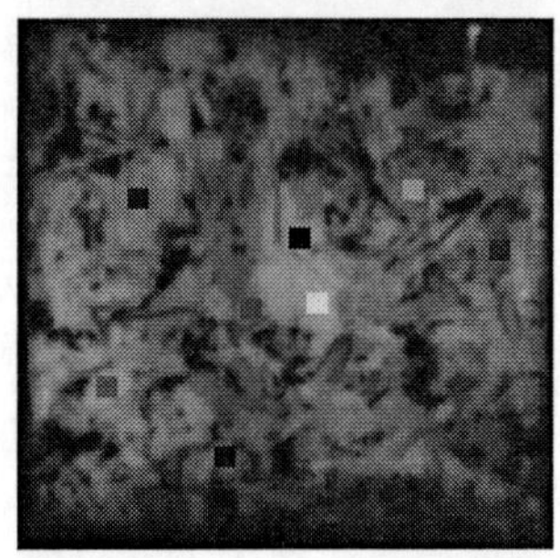

(i)

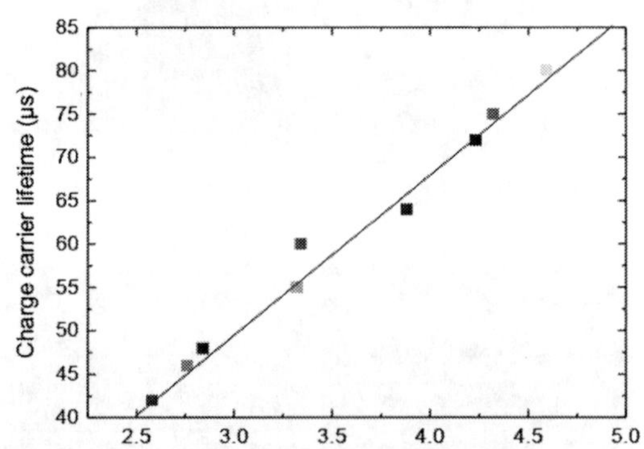

(ii)

Figure 5: (i) PL image of ADE surface presented in Figure 4(i) with arbitrary points for measuring their weighted surface reflections in order to correlate to their corresponding lifetime values obtained from Figure 4(i). (ii) Correlation between weighted surface reflection (R_w) and the corresponding charge carrier lifetime of the arbitrary points.

From the lifetime-calibrated PL image nine arbitrary points were identified for measuring their corresponding R_w by avoiding areas with bulk-inherited surface defects, dislocations and impurities. The correlation between carrier lifetime and weighted reflection is linear (Figure 5(ii)) within the investigated parameter range; with increasing weighted reflection the lifetime also increases.

4 CONCLUSION

We investigated how the electrical properties relevant for the solar cell performance may vary over the surface due to difference in surface morphologies in different grains, after using ADE-textured mc-Si wafers followed by a short anisotropic alkaline wet etch.

We find that the grains with lower surface reflection, pertaining to a higher surface enlargement, show higher carrier concentration and consequently a lower emitter sheet resistance in comparison to the grains with higher surface reflection. We also highlighted a linear relationship between weighted surface reflection and charge carrier lifetime.

This study provides an insight on methods through which the influence of a large distribution of surface reflection on electrical properties of a nanotextured solar cell can be studied. We expect that with further investigations, this approach could also facilitate the optimization of the emitter formation process on diamond-wire sawn mc-Si solar cells textured by various nanotexturing methods.

5 ACKNOWLEDGEMENTS

The authors would like to thank KIC InnoEnergy for the financial support of this work within the research project ADE-GLOBAL. The authors would also like to thank all contributing colleagues at the Division Photovoltaics of the Fraunhofer ISE who made the development of this work possible. A. I. Ridoy would like to thank the Deutscher Akademischer Austauschdienst (DAAD) for funding the work within the scope of his dissertation.

6 REFERENCES

[1] J. Rentsch, N. Kohn, F. Bamberg, K. Roth, S. Peters, R. Lüdemann, R. Preu, "Isotropic plasma texturing of mc-Si for industrial solar cell fabrication," Proceedings of the 31st IEEE Photovoltaic Specialists Conference, pp. 1316–1319, 2005.

[2] K.-s. Lee, M.-H. Ha, J. H. Kim, J.-W. Jeong, "Damage-free reactive ion etch for high-efficiency large-area multi-crystalline silicon solar cells," Solar Energy Materials and Solar Cells, vol. 95, no. 1, pp. 66–68, 2011.

[3] J. Oh, H.-C. Yuan, H. M. Branz, "An 18.2%-efficient black-silicon solar cell achieved through control of carrier recombination in nanostructures," Nature nanotechnology, vol. 7, no. 11, pp. 743–748, 2012.

[4] X. X. Lin, Y. Zeng, S. H. Zhong, Z. G. Huang, H. Q. Qian, J. Ling, J. B. Zhu, W. Z. Shen, "Realization of improved efficiency on nanostructured multicrystalline silicon solar cells for mass production," Nanotechnology, vol. 26, no. 12, p. 125401, 2015.

[5] B. Kafle, A. I. Ridoy, P. Saint-Cast, L. Clochard, E. Duffy, M. Hofmann, J. Rentsch, "Atmospheric pressure dry texturing enabling > 20% conversion efficiency on multicrystalline silicon PERC solar cells," AIP Conference Proceedings 1999, pp. 50003 (1–7), 2018.

[6] B. Kafle, J. Seiffe, M. Hofmann, L. Clochard, E. Duffy, J. Rentsch, "Nanostructuring of c-Si surface by F_2-based atmospheric pressure dry texturing process," Phys. Status Solidi A, vol. 212, no. 2, pp. 307–311, 2015.

[7] B. Kafle, T. Freund, A. Mannan, L. Clochard, E. Duffy, S. Werner, P. Saint-Cast, M. Hofmann, J. Rentsch, R. Preu, "Plasma-free dry-chemical texturing process for high-efficiency multicrystalline silicon solar cells," Energy Procedia, vol. 92, pp. 359–368, 2016.

[8] P. Saint-Cast, B. Kafle, R. Pandey, A. I. Ridoy, M. Hofmann, L. Clochard, T. Schwarze, M. Pittroff, J. Rentsch, R. Preu, "Rear passivated mc-Si solar cells textured by atmospheric pressure dry etching," Energy Procedia, vol. 124, pp. 260–266, 2017.

[9] B. Kafle, "Mask-less dry texturing of crystalline silicon solar cells in atmospheric pressure conditions," Dissertation, Albert-Ludwigs University of Freiburg, Freiburg, Germany, pp. 272, 2016.

[10] S. Werner, E. Lohmüller, P. Saint-Cast, J.M. Greulich, J. Weber, S. Schmidt, A. Moldovan, A.A. Brand, T. Dannenberg, S. Mack, S. Wasmer, M. Demant, M. Linse, R. Ackermann, A. Wolf, R. Preu, "Key aspects for fabrication of p-type Cz-Si PERC cells exceeding 22% conversion efficiency," 33rd European Photovoltaic Solar Energy Conference, Amsterdam, The Netherlands, 2017.

[11] B. Kafle, T. Freund, S. Werner, J. Schön, A. Lorenz, A. Wolf, L. Clochard, E. Duffy, P. Saint-Cast, M. Hofmann, J. Rentsch, "On the nature of emitter diffusion and screen-printing contact formation on nanostructured silicon surfaces," IEEE J. Photovoltaics, vol. 7, no. 1, pp. 136–143, 2017.

[12] R. Bock, P. P. Altermatt, J. Schmidt, "Accurate extraction of doping profiles from electrochemical capacitance voltage measurements," 23rd European Photovoltaic Solar Energy Conference, Valencia, Spain, 2008.

[13] A. Kimmerle, "Diffused surfaces for crystalline silicon solar cells – process development, characterization and modelling," Dissertation, Albert-Ludwigs University of Freiburg, Freiburg, Germany, 2015.

[14] S. Werner, E. Lohmüller, S. Maier, S. Mourad, A. Wolf, "Challenges for lowly-doped phosphorus emitters in silicon solar cells with screen-printed silver contacts," Energy Procedia, vol. 124, pp. 936–946, 2017.

OPTIMIZING TCO LAYERS FOR NOVEL BIFACIAL CRYSTALLINE SILICON HOMOJUNCTION SOLAR CELLS INTEGRATING PASSIVATED CONTACTS

Elise Bruhat[1,a], Thibaut Desrues[1], Danièle Blanc-Pélissier[2], Bernadette Grange[1] and Sébastien Dubois[1]
[1]Univ. Grenoble Alpes, INES, F-73375 Le Bourget du Lac, France CEA, LITEN, Department of Solar Technologies, F-73375 Le Bourget du Lac, France
[2]Université de Lyon, Institut des Nanotechnologies de Lyon INL - UMR5270, CNRS, Ecole Centrale de Lyon, INSA Lyon, Villeurbanne, France

ABSTRACT: This study explores the needed properties of both Transparent Conductive Oxides (TCOs) and the emitter region for bifacial homojunction solar cells with potentially passivated contacts. TCO layers with different electrical and optical properties have been tested on both sides of n-type homojunction cells for different resistivity of the boron doped region. The high lateral conductivity provided by the diffused emitter and back surface field (BSF) greatly reduces the constraints on TCO electrical properties in such structures. An understanding of the required properties of TCO for advanced homojunction applications is given. All in all moderately conductive TCOs on resistive junctions are shown to be suitable for such applications, allowing better optical properties and limitation of the recombinations losses without inducing resistive losses in the devices.
Keywords: Homojunction, solar cells, TCO, implantation, passivated contacts

1 INTRODUCTION

Most crystalline silicon (c-Si) homojunction solar cells present highly doped regions at the c-Si surfaces with a direct contact between the c-Si and the metal grid thus promoting electron-hole recombinations (see Fig. 1a). In order to obtain higher open circuit voltage (V_{OC}) and lower saturation current densities (J_0) "lightly doped homojunctions" [1], [2] approaches are being considered. However to overcome recombination issues due to the direct contact between the doped c-Si and the metal grid, and to counterbalance the higher sheet resistance (R_{Sheet}) of lowly doped emitters, TCO layers may be used (see Fig. 1b)[3], [4]. In addition to being conductive and antireflective, TCO layers can act as passivating layers. Such passivating properties can be obtained thanks to either hydrogenation of the TCO or in combination with a passivating ultra-thin underlayer[5], [6] (see Fig. 1c). The aim of this preliminary study is to get new insights into the resistive losses due to the interaction of TCO with homojunction dopings, therefore no passivating layers are introduced in this study.

High resistivity collectors (a-Si:H, thin poly-Si) require the use of TCO layers featuring high conductivity values to improve both the lateral charge transport and the charge collection by metal electrodes. However TCO layers with high conductivity values usually feature poor optical properties (i.e, light absorption).

For homojunction applications, high lateral conductivity is already achieved thanks to the c-Si over-doping at the surfaces. Thus it could be asked if TCOs with high conductivity are still needed or if another optimum between optical and electrical properties of the TCO layer could be considered. Previous studies show that a highly conductive junction beneath the TCO leads to a lowering of the TCOs electrical properties without affecting the cell performances [7]. However the influence of the doping profiles of the emitter region on the performance of homojunction solar cells is still to be studied. Thus, two different Indium Tin Oxides (ITO) layers were tested on 4 different boron profiles in order to investigate any resistive losses due to such electrical interfaces. It should be noted that no surface passivation layer (e.g, tunnel layer) between the TCO and the c-Si wafer is used in this study, thus reducing the short circuit current density (J_{SC}) and V_{OC} performances.

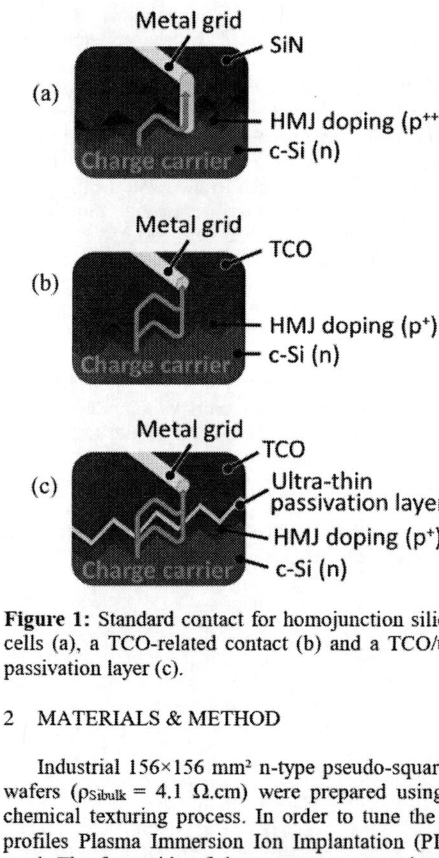

Figure 1: Standard contact for homojunction silicon sola cells (a), a TCO-related contact (b) and a TCO/ultra-thi passivation layer (c).

2 MATERIALS & METHOD

Industrial 156×156 mm² n-type pseudo-square Cz-Si wafers (ρ_{Sibulk} = 4.1 Ω.cm) were prepared using a wet chemical texturing process. In order to tune the emitter profiles Plasma Immersion Ion Implantation (PIII) was used. The front side of the precursors were doped with Boron and the rear side with Phosphorus. The precursors were then annealed under N_2 to diffuse and activate the dopants. Magnetron Sputtering (MS) ITO layers (80 nm) were deposited at 200°C on both sides of the textured Si wafers. Two different ITO layers were deposited: a standard one (Oxygen pressure pO_2 = 4.5 sccm) and a highly resistive one. ITO properties were tuned thanks to different pO_2/pAr ratios (fixed pAr at 350 sccm). The coated c-Si wafers were then metallized using Ag screen printing (4 bus bars, 1.8 mm pitch, and line resistivity of 7 μΩ.cm) and annealed 20 minutes at 300°C under air. To avoid residual shunt resistance (R_{Shunt}) the junctions were open thanks to laser scribing of the cells on the front side. In order to investigate their optical and electrical properties, ITO layers were also deposited on glass samples and annealed under the same conditions as the c-Si cells.

The R_{Sheet} values of the ITO layers were measured using a 4-point probe on glass substrates. To compare cells performances, illuminated I-V measurements were performed to extract the electrical photovoltaic parameters of the cells (V_{OC}, J_{SC}, Fill factor (FF). The conversion efficiency (η) was measured under 1-sun illumination (front emitter configuration). Suns-V_{OC} measurements were also performed, allowing the extraction of R_{Shunt} and series resistance (R_{Series}) losses within the cells. Finally, to check the quality of the contact between the TCO layer and the metal grid, Tranfer Length method (TLM) based measurements were performed[8] on Ag/ITO/thick thermal-SiO_2/c-Si samples. In order to provide accurate results, all characterizations were made on at least two samples, and standard deviation was calculated to show value dispersion on graphs.

3 RESULTS

3.1 Evolution of ITO with pO_2

As expected, the electrical analyses of the different ITO layers on glass show that increasing the amount of O_2 from 1.4 to 3.3% in the MS process increases the ITO layer R_{Sheet} from 150 to above 400 Ω/sq. The effective absorptions in the visible light and the infrared region also show an increase of the transparency of ITO with the percentage of O_2 during the MS process as shown in previous studies [9], [10].

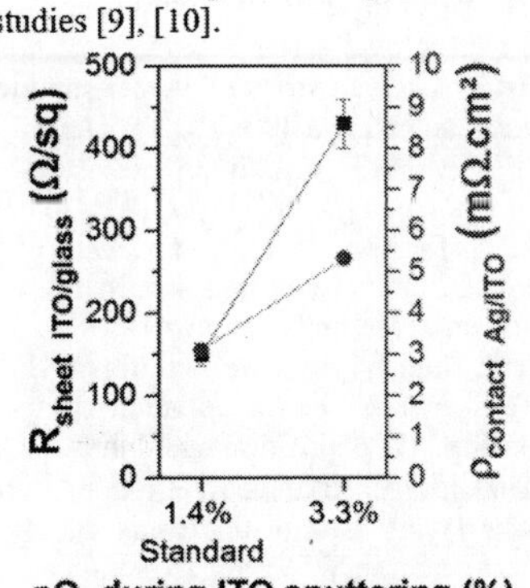

Figure 2: Variation of ITO sheet resistance on glass and contact resistance with screen-printed Ag.

Contact resistivity ($\rho_{contact}$) between the metal grid and the ITO for all pO_2 concentrations are extracted form TLM measurements (Fig. 2). After the metal grid deposition, high $\rho_{contact}$ are measured. However, a 20 minutes annealing step at 300°C allows contact resistivity to drop thus helping current collection for all samples. A longer annealing or at higher temperature might further decrease resistive losses due to the metal/ITO contact for the cells if $\rho_{contact}$<3mΩ.cm² is achieved.

3.2 Characterization of the doped region

For this study only the influence of the emitter profile is considered. Thus the same dose of Phosphorus was implanted for the BSF fabrication. Variations of the phosphorus profile on the rear are limited and $R_{Sheet\ BSF}$ = 44 ± 4 Ω/sq for all high-low junctions.

Changes in the emitter region are however more significant. The dopant profiles were tuned thanks to both the dose of implanted ions and the annealing temperature. Two doses were used : d_1=7.4.10^{15} at.cm^{-2} and d_2=3.8.10^{15} at.cm^{-2}, giving two different surface dopant concentration (SDC) in the profiles. Three annealing temperatures were as well tested in order to modify the junction depth. All profiles obtained are to be seen on Fig. 3.

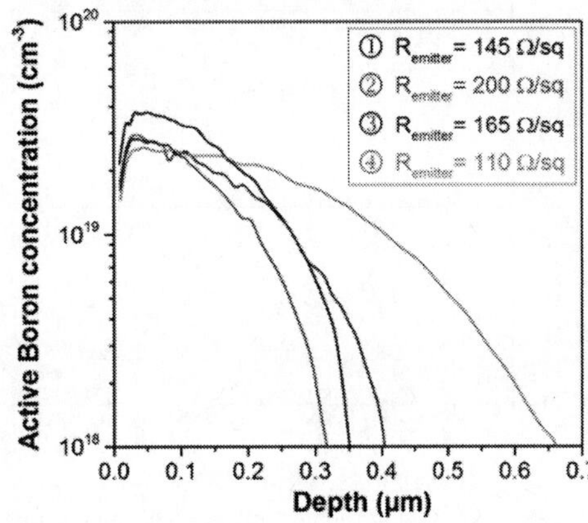

Figure 3: Variation of the active boron concentrations for the different cells studied (ECV measurements).

It is to be noted that comparing profile ① and ② gives insight to te influence of SDC with a decreasing amount of dopants from ① to ②. While comparing profile ①, ③ and ④ gives insight to te influence of the profile's depth, with an increasing amount of dopants from ② to ④.

3.3 Solar cells results

All solar cells results are shown on Fig. 4. Both the influence of the doping profile (x-axis) and the influence of the ITO resistivity (shift between both curves) are depicted. It should be noted that no surface passivation layer (e.g, tunnel layer) between the TCO and the c-Si wafer is used in this study, thus reducing the short circuit current density (J_{SC}) and V_{OC} performances.

3.3.1 Impact of ITO's resistivity

Enhancement of the J_{SC} and V_{OC} with the increase of pO_2 concentration during MS ITO deposition on the front side of the cell is noticeable. The increase in V_{OC} might be related to a change in work function (W_f) with the increased amount of [O] in the ITO layer. The increase in J_{SC} is probably due to the lower absorption of ITO at high O_2 concentrations.

The fill factor is mostly influenced by the resistive losses. Therefore the increase of the ITO resistivity (higher O_2 concentrations) should lead to a decrease of the FF. Surprisingly, an increase is noticeable with ITO's resistivity leading to think that the conductivity from the dopant-diffused regions to the overall conduction seems sufficient to reduce the constraints on the front TCO conductivity.

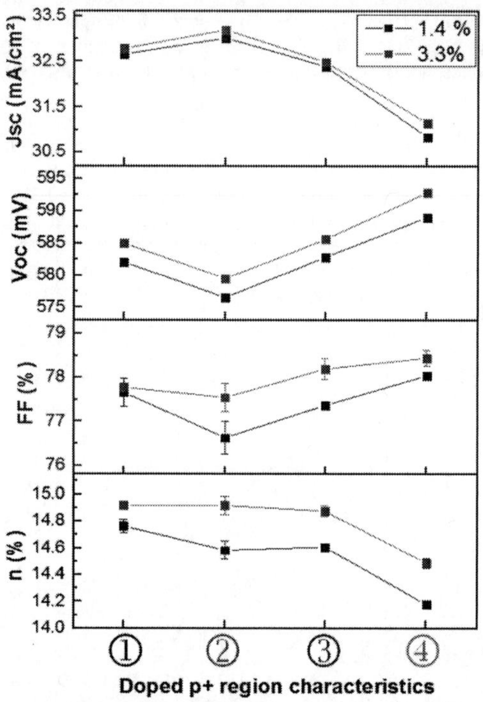

Figure 4: Variation of cell performances for two different ITOs for bifacial cells with different boron profiles on the front side.

To understand more the increase of FF with pO_2 Suns-V_{OC} measurements were performed in order to determine both the R_{Series} and the R_{Shunt} losses of the cells. First, as Suns-V_{OC} measurements are unsensible to R_{Series} changes, shifts in R_{Series} follow the variation of pFF-FF. From Fig. 5, it is clear that R_{Series} increases as expected with pO_2. Such changes should lead to an overall decrease of the FF values. However, analysis of the R_{Shunt} (Fig.5) show a huge increase with pO_2. Even though laser junction opening was performed on all cells, the high conductivity of the ITO lead to high shunt losses that can be completely avoided thanks to highly resistive TCOs.

Thanks to better short circuit current performances, best cell are obtained with $pO_2 = 3.3\%$. It is possible that a longer or higher temperature annealing might increase the properties of cells with higher O_2 concentrations as the contact between the metal grid and those ITOs is not optimal with a 20 minutes annealing steps.

3.3.2 Impact of the emitter profile

Enhancement of the J_{SC} and V_{OC} with lowering the concentration of dopant on the front side of the cell is noticeable. A higher concentration of dopants increases the amount of defects within the Boron junction. Thus an increase in Auger recombinations should lead to a decreased J_{SC}.

However degradation of the V_{OC} with lowering the concentration of dopant on the front side of the cell is observed. Most likely, increasing the amount of dopants leads to an increase of the field effect and limit Shockley Read Hall (SRH) recombinations.

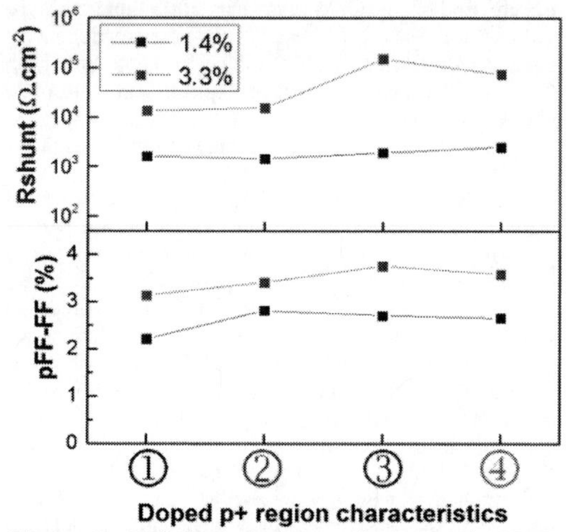

Figure 5: Variation of shunt resistance and pFF-FF characteristic for two different ITOs for bifacial cells with different boron profiles on the front side.

The fill factor is mostly influenced by the resistive losses. Therefore a more conductive emitter (higher activated boron concentrations) leads to an increase of the FF values.

Thanks to better short circuit current performances, best cell are obtained with thinner Boron junction. High FF values are obtained even on lowly doped junction (200 Ω/sq) contacted with a highly resistive TCO layer

4 CONCLUSION & PERSPECTIVES

- The TCO layer conductivity has a low influence on the global series resistance of homojunction devices, as shown by FF values superior to 77.5% obtained using a 400 Ω/sq ITO layer
- If for high resistivity collectors the sheet resistance of the ITO layer needs to be below 150 Ω/sq, for homojunction devices ITO layers with lower conductivity and absorbance can be used and increase cells performances even combined with highly resistive emitters. The best cell performances being obtained for $pO_2 = 3.3\%$ during ITO deposition and thin p^+ junctions.
- The poor surface passivation properties of the ITO layers used in this study clearly limits cell performances (V_{OC} and J_{SC}), indicating a need to develop different TCO materials and/or stacks with "tunnel" oxides inserted between the c-Si and the TCO layer and/or hydrogenation processes.

5 AKNOWLEDGMENTS

The team gratefully acknowledges Auvergne-Rhône-Alpes Region for financially supporting the project through the ARC4 program. This project has received funding from the French National Research Agency (Programs "Investment for the Future" ANR-10-ITE-0003 and "Oxygen" ANR-17-CE05-0035), and has also received funding from the European Union's Horizon 2020 research and innovation program under grant agreement No 727529.

REFERENCES

[1] R. Müller *et al., Sol. Energy Mater. Sol. Cells*, vol. 120, Part A, (2014) 431–435.
[2] G. G. Untila, T. N. Kost, A. B. Chebotareva, and E. D. Kireeva, *Sol. Energy Mater. Sol. Cells*, vol. 137, (2015) 26–33,.
[3] M. Kim *et al., Mater. Lett.*, vol. 116, (2014) 258–261.
[4] B. Min, J. Krügener, M. Müller, K. Bothe, and R. Brendel, *Energy Procedia*, vol. 124, (2017) 126–130.
[5] J. Panigrahi, Vandana, R. Singh, and P. K. Singh, in *2016 IEEE 43rd Photovoltaic Specialists Conference (PVSC)*, (2016) 2964–2966.
[6] A. Rehman *et al., Sol. Energy*, vol. 166, (2018) 90–97,.
[7] E. Bruhat, T. Desrues, B. Grange, D. Blanc-Pélissier, and S. Dubois, AIP Conference Proceedings 1999, 040004 (2018)
[8] H. H. Berger, *J. Electrochem. Soc.*, vol. 119, no. 4, (1972) 507–514.
[9] M. Bender, W. Seelig, C. Daube, H. Frankenberger, B. Ocker, and J. Stollenwerk, *Thin Solid Films*, vol. 326, no. 1, (1998) 72–77.
[10] H. Askari, H. Fallah, M. Askari, and M. C. Mohmmadieyh, "*ArXiv Prepr. ArXiv14095293* (2014).

INVESTIGATION ON POST CLEANINGS ON MODIFIED SURFACE USING LASER TEXTURING

Behrad Radfar [1,2,*], Firat Es [1], Rasit Turan [1,2,3]
[1]The Center for Solar Energy Research and Applications (GÜNAM), Middle East Technical University
[2]Micro and Nanotechnology Graduate Program, Middle East Technical University
[3]Department of Physics, Middle East Technical University
Dumlupinar Blvd. 1, 06800 Ankara – Turkey

Email: esfirat@metu.edu.tr, Tel: +90 312 210 5090
Email: turanr@metu.edu.tr, Tel: +90 312 210 5069
*Corresponding Author, behrad.radfar@metu.edu.tr, Tel: +90 312 210 4303

ABSTRACT: The surface of crystalline silicon (c-Si) solar cells can affect their performance as the surface can alter both electrical and optical properties. The c-Si texturing is carried out to increase the absorption; therefore, the optical properties will be enhanced. The anisotropic alkaline etchants are commonly used to form pyramids on the surface of monocrystalline Si wafers. However, this method is not usable for multi-crystalline silicon (mc-Si) because of its grains with random crystallographic orientations. The laser texturing, as an alternative method for texturing mc-Si wafers, is getting attention among texturing techniques. In this study, post-texture cleaning is investigated since it is a vital step after the laser texturing. This approach includes preparing the cleaning solution and performing the cleaning on the laser processed mc-Si wafers' surface. The different solution parameters, such as cleaning duration, temperature, and concentration, are justified by performing a series of experiments. To determine the effectiveness of the process, the weighted reflection and Scanning Electron Microscopy (SEM) images of the samples are considered. The SEM images show the evolution of the surface morphology versus cleaning duration, temperature, and concentration. Moreover, the measurements indicate that the samples' weighted reflection can change dramatically by changing the cleaning solution parameters.
Keywords: Multi-crystalline Silicon, Texturization, Post-texture cleaning

1 INTRODUCTION

With growing energy usage, expanding the energy sources is obligatory. The conventional energy sources like oil and coal are depleting and they cannot supply the required energy. Thus, finding new energy sources is crucial. One of the most important energy sources is solar energy which can be directly converted into electricity using solar cells.

Currently, most of the solar cells are fabricated from the crystalline silicon (c-Si) wafers. There are two types of c-Si wafers: multi-crystalline silicon (mc-Si) wafers and mono-crystalline silicon wafers. Multi-crystalline silicon wafers are more favorable than mono-crystalline silicon wafers because of their lower production expenses. However, mc-Si solar cells have lower efficiency than mono-crystalline silicon solar cells because of their lower carrier lifetime and their surface and interface qualities [1-2].

In general, enhancing the surface structure leads to higher light absorption which can be done by texturing. Texturing of mono-crystalline silicon wafers can be done using anisotropic alkaline chemicals like KOH and IPA that make pyramid shapes at the surface. These pyramids are typically 5-10 μm high with the angle of 54.7° and randomly located across the surface. In contrast, using anisotropic method is not possible for mc-Si because of the grains with random crystallographic orientations at the surface. Laser texturing is an isotropic method for texturing mc-Si surfaces [3-4]. The laser beam interacts with the surface and ablates vertical and horizontal lines. The laser processing parameters are investigated to have the favorable surface morphology [5]. The shape, depth, and spacing of these laser-ablated grooves are the key factors in determining the light absorption. In addition, the surface after laser texturing contains lots of ablated silicon and other residues which decrease electrical performance. Thus, post-texture cleaning is a crucial step to remove residues and ablated Si and decrease surface roughness to regain the electrical performance.

In the following sections, post-texture cleaning parameters like concentration, temperature, and cleaning duration are justified and their effects on the surface morphology are investigated by analyzing the weighted reflection and the SEM images of the samples.

2 METHODS AND EXPERIMENTS

In the current study, p-type multi-crystal silicon wafers with 200μm thickness and the size of 15.6cm x 15.6cm are used. Texturing of samples is done using a nanosecond infrared laser with the wavelength of 1064 nm. In our previous work, different patterns with different laser scanner parameters, the spacing between laser lines and different laser parameters such as laser pulse frequency and laser power were investigated [5]. The effects of the laser texturing on the surface morphology and the reflection were also presented [5]. With further experiments, the optimum texturing parameters are selected for the current investigation. The selected laser parameters are shown in Table I.

Table I: Laser parameters used for the laser texturing

Frequency	Power	Draw Step	Step Period	Scan Number	Space between laser lines
30kHz	100%	25 LSB	120 μs	1 iteration	75 μm

As it is shown in Fig. 1, the surface of the laser-textured samples consists of lots of silicon ablated residues. Thus, the post-texture cleaning step is crucial.

To perform post-texture cleaning, KOH solutions with different specifications are prepared which are shown in Table II.

Table II: KOH solution parameters

Concentrations	Temperatures	Post-texture Cleaning Durations
2%		30 sec
4%	60°C	1 min
8%		2 min
16%		4 min
24%	80°C	6 min
32%		8 min

3 RESULTS AND DISCUSSIONS

Figure 1 shows the uniformity of the laser texturing of the full mc-Si wafer. In this figure, the optical images of the untextured wafer (a) and the textured wafer (b) and the SEM image of the middle point of the wafer (c) are presented. The SEM images of other areas are the same as the presented one. It is good to mention that the laser texturing takes 35 minutes to texture the full-size wafer.

Figure 1: Optical and SEM images of full wafer sample; a) optical image of the untextured wafer, b) optical image of the textured wafer, c) SEM image of the textured wafer.

The effectiveness of the cleaning step is determined by measuring a weighted reflection of samples which are shown in Fig. 2. In Fig. 2, graph (a) shows the results of the 60°C solution and graph (b) shows the results of the 80°C solution.

In addition, the reflection curves of the reference sample and the best samples, which are 2% and 4% KOH solution with 0.5min, 1min, and 2min cleaning durations, are shown in Fig. 3 for temperatures of 60°C and 80°C in (a) and (b) respectively.

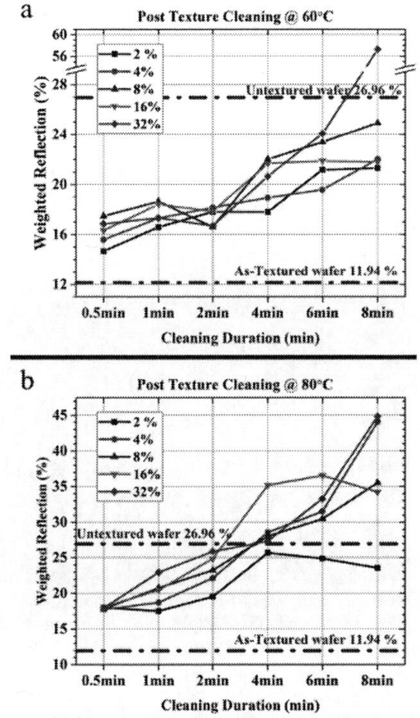

Figure 2: Weighted reflection of post-texture cleaned samples a) at 60°C, b) at 80°C.

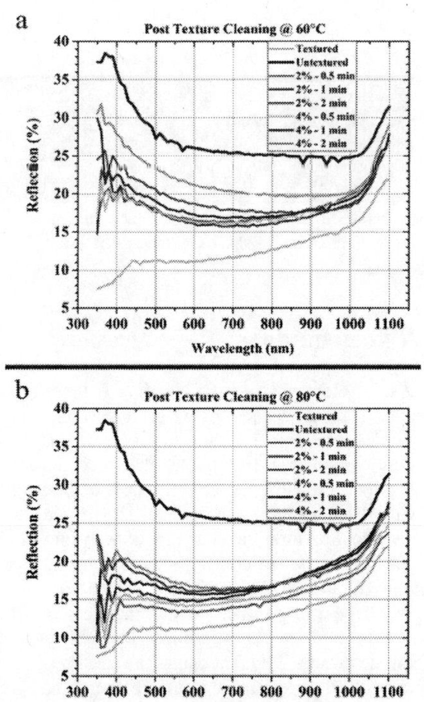

Figure 3: Reflection curves of post-texture cleaned samples at two different concentrations (2% and 4%) and three different durations (0.5 min, 1 min, and 2 min) for a) 60°C and b) 80°C.

Lastly, the development of the samples surface structure versus post texturing KOH cleaning is illustrated in Fig. 4 and Fig. 5. In these figures, the SEM images for the selected samples are shown. Fig. 4 shows the results of the solution with the temperature of 60°C and Fig. 5 shows the results for the solution with the temperature of 80°C. Each row indicates the changes over increasing the cleaning duration. Each column indicates the changes versus the KOH solution concentration.

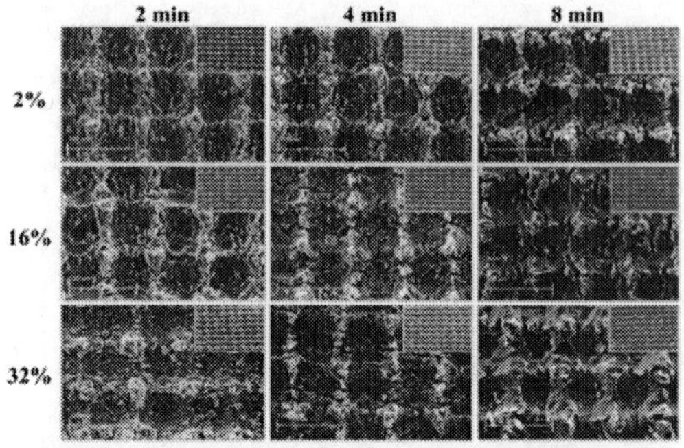

Figure 4: SEM images of KOH cleaned samples at 60°C with 2%, 16%, and 32% KOH over different cleaning durations.

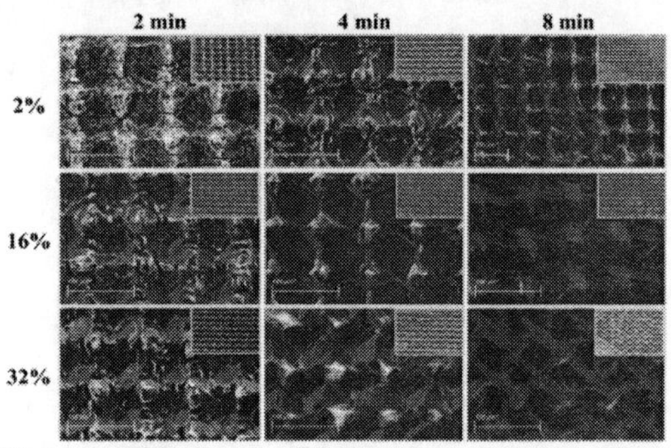

Figure 5: SEM images of KOH cleaned samples at 80°C with 2%, 16%, and 32% KOH over different cleaning durations.

4 CONCLUSION

In this study, post-texture cleaning is investigated as a critical step proceeded by laser texturing. Since the surface morphology of the mc-Si wafer after laser texturing, which is an isotopic texturing technique, is not suitable for fabricating solar cells, post-texture cleaning is required to improve the properties of the textured surface. Post-texture cleaning is done using a KOH solution with different concentrations, temperatures, and cleaning durations. The weighted reflection and SEM images show strong effects of the chemical solution on surface morphology. The optimum cleaning parameters can be determined by choosing the desired weighted reflection and corresponding surface morphology based on the SEM images.

5 REFERENCES

[1] D. Mulati, T. Fuyuki, H. Matsunami, "Determination of Minority-Carrier Lifetime in Multicrystalline Silicon Solar Cells using Current Transient Behaviours," JAP, vol. 38, no. 3R, pp. 1408, 1999.

[2] J. C. Zolper, S. Narayanan, S. R. Wenham, and M. A. Green, "16.7% efficient, laser textured, buried contact polycrystalline silicon solar cell," Appl. Phys. Lett., vol. 55, no. 22, p. 2363, 1989

[3] A. Goetzberger and V. U. Hoffmann, Photovoltaic Solar Energy Generation, vol. 112. Berlin/Heidelberg: Springer-Verlag, 2005.

[4] L. A. Dobrzański, A. Drygała, P. Panek, M. Lipińki, and P. Zieba, "Development of the laser method of multicrystalline silicon surface texturization," Arch. Mater. Sci. Eng., vol. 38, no. 1, pp. 5–11, 2009.

[5] B. Radfar, F. Es, and R. Turan, "Investigation on Different Surface Modifications Using Laser Texturing," in 33rd European Photovoltaic Solar Energy Conference and Exhibition, 2017, pp. 706–709.

Evolution of Contact Formation on P-type Crystalline Silicon Solar Cells

Ryan W. Mayberry, Vinodh Chandrasekaran, L. Karpowich

Heraeus Precious Metals North America Conshohocken LLC, W. Conshohocken, PA, 19428, USA

***Abstract* — This paper will examine the evolution of contact formation with temperature of a silver paste on a crystalline silicon solar cell. Detailed evaluation of the damage due to the front side grid metallization and its relation to contact formation and temperature will be carried out. Generally, a multi component silver paste is screen printed on a solar cell, then fired at high temperatures (750-800°C) to form an electrical contact to the base silicon material. The inorganic component, apart from silver (usually glass and/or crystalline additives), of the paste is responsible for reactions with the silicon nitride anti-reflective coating (ARC), silicon and silver that may lead to a low contact resistance between the metallization and bulk silicon. Knowledge of the detailed chemistry that controls the thermodynamic and kinetic temperature dependence of these reactions is generally vague and constantly changing. Both the formation and nature of silver crystallites, colloids, and precipitates formed at the silicon emitter region and in the glass layer will be investigated in detail over a wide range of temperatures. This will shed light on the evolution of contact formation with time and temperature for different metallization pastes.**

I. Introduction

Screen-printed thick film metallization on crystalline silicon solar cells is the predominant processing technology in the PV industry today. Rapid advances in cell structure call for simultaneous advances in metallization paste to ensure compatibility with limiting cell parameters. This could include processing temperature, adhesion requirements, new reaction limitations, etc. Examples of this are seen regularly with new passivation techniques, different stacking layers, new emitter structure, altered surface treatments, and so on [1]. In order to quickly develop new metallization solutions for these advancements, the entire metallization process must be understood. In this way, knowledge of the reactions taking place on the surface of the wafer will lead to the efficient development of metallization technology which directly follows, or even better leads, advances in wafer technology.

II. Experimental Approach

Typical thick film metallization pastes used on c-Si solar cells consist of 85-90% Ag powder, 5-10% organics, 1-5% glass frit and/or other inorganic additives. The silver pastes in this study were produced by thoroughly mixing each component, then homogenizing the paste through the use of a three roll mill.

2.1 Once milled, the pastes were screen printed using a 4-busbar pattern onto full Al-BSF p-type mono-crystalline (Cz) silicon wafers with a sheet resistance of ~120 Ohm/sq. The wafers used in this study have been pre-fired using an IR belt furnace before application of the front-side Ag paste in order to reach sufficient temperatures to form the back-side passivation, or back surface field (BSF). After the front-side metallization was printed, the wafers were dried at 150°C for 10 minutes in a box oven. After drying, the wafers were re-fired using the same IR belt furnace at varying peak temperatures. At this point, functional solar cells are made and will be characterized using various techniques which will be discussed below.

For this study, one commercially available front side Ag paste - SOL9641B, hereto referred as 9641B, and one internally modified experimental Ag paste – HG01, are used as test subjects. Once all wafers are fully processed, the formation of a functional solar cell will be analyzed at each separate firing temperature for each paste and wafer type. This will yield a step by step look into contact formation throughout the firing process. The overall conversion efficiency will be measured for each cell, although this will only be used as secondary information due to the number of temperature dependent factors effecting this value.

To truly examine the contact formation, implied Voc and direct contact resistance measurements will be the most valuable source of information. In this case, the many effects temperature has on the overall cell will be narrowed down to the effect at the metallization-wafer interface. Here we can infer the Ag paste's reaction with silicon nitride, silver, and silicon to further group these reactions by temperature. It will be seen when each process occurs and how it can be altered using a new Ag paste chemistry. Through the transmission length method (TLM) using a GP 4-Test v1.0, the contact resistance versus temperature plot can be studied. With this data, we will be able to determine when contact is made, how quickly the resistance drops with temperature, and how this effects the implied Voc values. As opposed to other cell measurements which can indirectly suggest the how well or poor the contact resistance is, such as fill factor (FF) or series resistance (Rs), the TLM method will separate the front-side metallization effects from the bulk wafer and back-side effects which are included in FF and Rs.

2.2 The second portion of this study will look at implied Voc and Nitrogen removal data. The substrates used for implied Voc are specialized PERC wafers. After removing the wafer temperature dependence through baseline measurements, the results are strictly due to the front side metallization paste; in effect showing the level of recombination at the paste-wafer interface as a result of firing temperature. Initial baseline studies were done that show the temperature dependence of the wafer itself (without metallization) assumed to be largely due to hydrogenation. Using the information seen in Figure 1, we are able to pre-fire the wafers at sufficient temperatures (760°C) for complete hydrogenation followed by an additional firing step to process the front-side metallization which proves to be important in order to separate metallization from all other effects. This will give us a detailed understanding of the impact temperature has on the deterioration of the front surface passivation under the metallization. Also measured is Nitrogen % removal by the firing process. This is an indirect measurement of ARC removal by assuming it is proportional to the Nitrogen % decrease as measured by SEM-EDX.

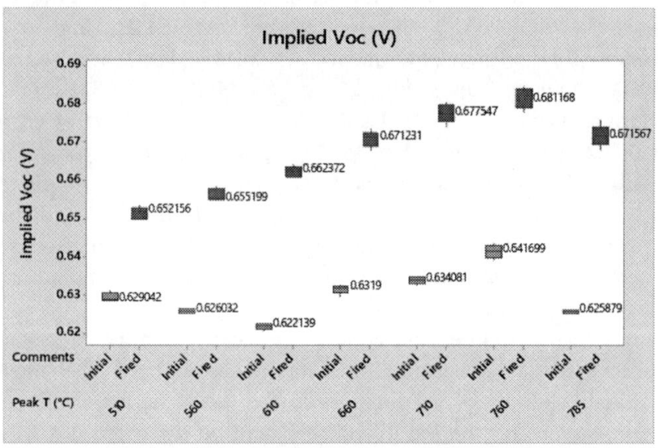

Figure 1: Implied Voc data for unfired vs fired wafers (without metallization) for baseline data

III. RESULTS AND DISCUSSION

Initial electrical results show the difference in firing behavior of 9641B and HG01. The overall conversion efficiency, which includes all effects such as front side metallization, back side metallization, and wafer influence is obtained first. Using pre-fired wafers, it was possible to get a general idea of the firing behavior. Without pre-firing of the un-metallized wafers, firing temperatures below a peak temperature of around 750°C would be dominated by insufficient processing of back side aluminum metallization and bulk wafer influence, as opposed to the ability of the front side metallization to make a good contact.

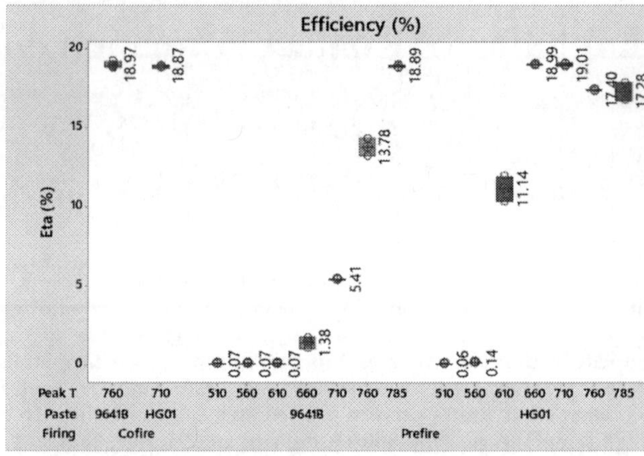

Figure 2: Efficiency values for each paste on cells that have been first pre-fired as well as two control groups of the same pastes at their optimal firing temperature printed on cells that have not been pre-fired (labeled Cofire)

Included in the electrical data is a control for each paste which is only fired once, in other words, co-fired with the backside paste. The temperatures chosen were found from previous studies to yield the optimum efficiency for each individual paste when the cell is only fired once. Here it can be seen why pre-firing is critical for HG01 which forms a good contact at lower temperatures. HG01, at its optimum co-fired temperature, yields a Voc that is over 8mV lower than its maximum Voc obtained using pre-fired wafers at a lower firing temperature. At even lower firing temperatures Voc data is not shown due to the inability to measure a functional value.

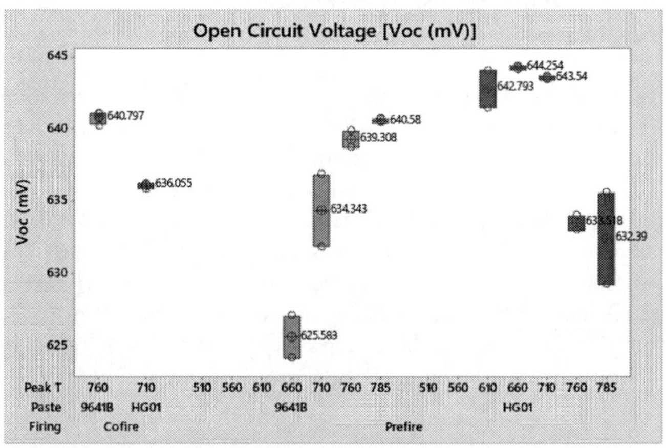

Figure 3: Voc data of each paste under the same conditions as previously reported on efficiency

Measured separately is the contact resistivity (rhoC) for both pastes. As seen in Figure 4, the rhoC data shows the same trend as does the efficiency. This again demonstrates the effectiveness of pre-firing the wafers, which satisfies the conditions needed for the bulk wafer passivation and the back-side aluminum paste while allowing for the front-

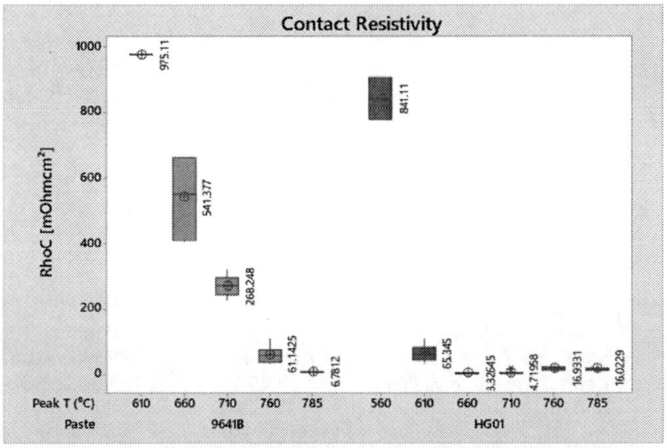

Figure 4: TLM data of measured contact resistivity over a range of firing temperatures for each silver paste

side silver metallization to be the dominate factor in changes to cell performance. RhoC values show that for 9641B, the 785°C or preferably even higher, is needed for the best contact. Alternatively, only 660°C is needed for HG01 to make sufficient contact; while already at 760°C effects of a decreasing shunt resistance at high temperatures leads to an increased measured contact resistance. Results of implied Voc versus firing temperature have also been obtained for each paste. The data here shows minimal impact of metallization on implied Voc values until 650°C. An important implication of this is that there is a temperature range for HG01 that has a good balance between sufficiently low contact resistance and minimal front surface damage. High implied Voc at 600°C and below indicates the silver paste metallization has not yet caused any significant amount of front surface recombination. Comparing the implied Voc values of 9641B and HG01 provides insight on the cause of decreases in implied Voc, which seem to be not so dependent on making a good contact.

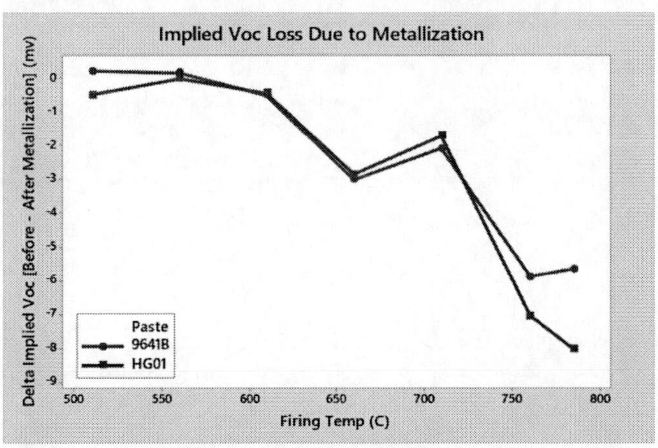

Figure 5: The red line shows the implied Voc for the HG01 paste while the blue line shows that of 9641B. Overall the two paste systems show that the loss in implied Voc is not dependent on when each paste makes a good contact.

As inferred through implied Voc measurements, there is essentially no damage to the wafer at temperatures of 600°C and below. Once 650°C is reached, there is the first large decrease in implied Voc (~3mV). At the highest firing temperatures there is a total voltage drop relative to the un-metallized cell of approximately six and nine millivolts for 9641B and HG01, respectively. In addition to these characterization methods, an attempt is made to quantify the amount of silicon nitride passivation that has been removed at each temperature. This can be seen below in Figure 6.

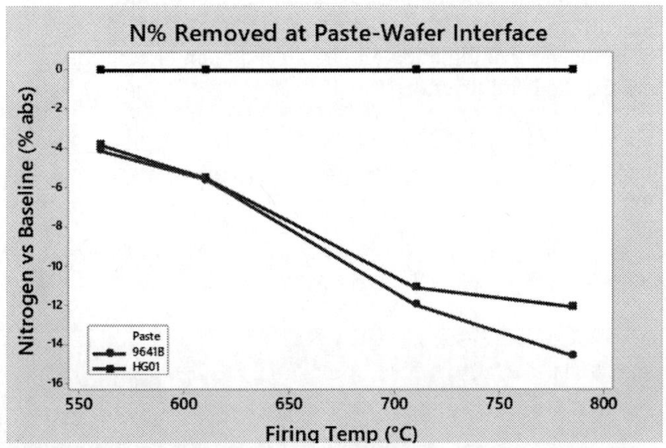

Figure 6: Absolute % of nitrogen removed versus baseline (non-metallized area) with at each firing temperature. This is done using SEM-EDX measurements of the wafer surface after removing the metallization.

This data shows that again, even though the two paste systems are drastically different in terms of when they are able to form a good contact, they differ only very little in this characterization method. Only at higher temperatures (> 700°C) is a difference seen. Overall, in this study it has not been concluded exactly where metallization induced recombination (resulting in decreased open circuit voltage) comes from. However, it may be just as valuable to conclude which phenomenon are not connected to a loss in Voc in the finished cell electrical measurements, which has been started here.

IV. Conclusion

The primary objective of this work is to evaluate contact resistance between the finger line and silicon wafer as a function of temperature and see how it is correlated to open circuit voltage losses due to metallization induced recombination. This is important because after contact formation, which is heavily studied [2-3], the next most important step is to reduce metallization induced recombination [4]. Therefore, determining which steps in the process of forming a good contact contribute most to recombination losses is crucial. In this paper, it has been shown that there is a high loss in Voc at increasing temperatures, regardless of when a good contact has been made. The conclusion here is that creating a metallization and wafer system that can be fired at lower temperatures should be a primary aim of those looking to immediately reduce metallization induced (and presumably most other forms of) recombination. Of course, this requires both front and backside metallization as well as the bulk wafer to meet all requirements (high grid conductivity, low contact resistance, good passivation, etc.) in a decreased temperature range. Here, only feasibility of the front-side metallization has been studied. Further research needs to be completed on backside metallization and wafer passivation to unlock the full potential of reduced recombination and increased open circuit voltage at the finished cell level.

REFERENCES

[1] S.W. Glunz, "High Efficiency Crystalline Silcon Solar Cells," *Advances in OptoElectronics*, vol. 2007, Article ID 97470, 2007.

[2] Kim et al, "Electrochemical nature of contact firing reactions for screen-printed silicon solar cells: origin of 'gray finger' phenomenon," *Prog. Photovolt: Res. Appl.* 2016; 24:1237–1250.

[3] Ching-Hsi Lin, Shih-Peng Hsu and Wei-Chih Hsu (2011). "Silicon Solar Cells: Structural Properties of Ag-Contacts/ Si-Substrate," *Solar Cell Silicon Wafer-Based Technologies*, Prof. Leonid A. Kosyachenko (Ed.), ISBN: 978-953-307-747-5, InTech.

[4] Edler et al, "Metallization-induced recombination losses of bifacial silicon solar cells," *Prog. Photovolt: Res. Appl.* 2015; 23:620-627.

PASSIVATION OF CRYSTALLINE SILICON SURFACES WITH ULTRA-THIN SILICON NITRIDE FILMS FORMED BY CATALYTIC CHEMICAL VAPOR DEPOSITION

Hao Song and Keisuke Ohdaira
Japan Advanced Institute of Science and Technology
1-1 Asahidai, Nomi, Ishikawa, JAPAN
Phone: +81-76-51-1563 E-mail: ohdaira@jaist.ac.jp

ABSTRACT: We attempt to replace tunneling silicon dioxide (SiO_2) films in tunnel-oxide passivated contact (TOPCon) solar cells by ultra-thin SiN_x films formed by catalytic chemical vapor deposition (Cat-CVD). Ultra-thin SiN_x films (<5 nm) can be formed with good controllability by tuning deposition duration and H_2 flow rate. Ultra-thin SiN_x films with lower refractive indices tend to have better passivation ability. Atomic hydrogen treatment onto the ultra-thin SiN_x films enhances the effective minority carrier lifetime (τ_{eff}) of crystalline Si (c-Si) wafers up to >1 ms, indicating excellent passivation quality.
Keywords: Silicon-Nitride, c-Si, Thin Film

1 INTRODUCTION

Crystalline Si (c-Si) solar cells are one of the most important photovoltaic (PV) technologies because of their domination of present PV market share. The most common structure of c-Si PV cells, Al-BSF, has a large semiconductor/metal contact area on their rear side, which induces significant minority carrier recombination and thus limits open-circuit voltage (V_{oc}). The structure of the c-Si PV cells has been shifting to those with a passivation structure to minimize or eliminate direct semiconductor/metal contact, which leads to the suppression of the rear-side surface recombination of minority carriers.

Recently, c-Si solar cells with a rear-side surface passivated by an ultra-thin (<2 nm) tunneling Si dioxide (SiO_2) film, tunnel oxide passivated contact (TOPCon) cells, have been developed [1–7]. The SiO_2 film must be sufficiently thin so that majority carriers can effectively pass through the thin SiO_2 by tunneling. The ultra-thin SiO_2 also contributes to reduce interface state density by the termination of dangling bonds. A field-effect passivation induced by an n^+ Si layer formed on the ultra-thin SiO_2 also leads to improvement in high V_{oc} and conversion efficiency.

Since SiO_2 has a quite large band offset against Si in the conduction band [8], the precise control of the thickness of the SiO_2 is necessary for an effective electron tunneling. This issue is particularly critical for industrial-sized cells in which the formation of SiO_2 with uniform thickness is more difficult. If a passivation film with less band offset can be prepared, a larger thickness or less uniformity for the passivation films for the effective carrier tunneling will be allowed.

Silicon nitride (SiN_x) formed by chemical vapor deposition (CVD) is one of the best materials for such a passivation film since it has lower conduction band offset against c-Si [9] and has good passivation quality particularly for an n-type c-Si surface. Furthermore, CVD SiN_x can be formed only on one side of a c-Si wafer, unlike in the case of SiO_2 formation through solution-based processes or thermal oxidation. Ultra-thin SiN_x films can thus be utilized as an alternative material for the TOPCon-like cells.

Catalytic CVD (Cat-CVD) is a proper method to prepare SiN_x films with good passivation property due to its plasma-damage-less nature [10–18]. The application of Cat-CVD SiN_x films to TOPCon-like solar cells has not been attempted yet. In this study, we have investigated the formation of ultra-thin (<5 nm) SiN_x films by Cat-CVD and improvement in the passivation quality of the ultra-thin Cat-CVD SiN_x films.

2 EXPERIMENTAL PROCEDURE

We first cleaved 290-μm-thick, mirror-polished n-type c-Si(100) wafers with a resistivity of 1–5 Ω·cm into 20×20 mm^2-sized pieces. The wafers were cleaned in acetone, ethanol and water for 5 min, respectively, in this order. The native oxide layers of the wafers were removed by dipping in 5% hydrofluoric acid. We then deposited SiN_x films by Cat-CVD on both sides of the c-Si substrates. A tungsten catalyzer temperature (T_{cat}), substrate temperature (T_{sub}), pressure, SiH_4 and NH_3 flow rates were fixed to 1800 °C, 200 °C, 1 Pa, 3 sccm, and 50 sccm, respectively. To control the thickness of the SiN_x films, a deposition duration were systematically changed in a range of 30–420 s, and we found that the thickness of SiN_x films can be ~5 nm when a deposition duration is ~30 s. The refractive index (n) of SiN_x films were tuned by varying a H_2 flow rate, based on a vapor-phase cracking of NH_3: NH_3+H → NH_2+H_2 and so on [19]. Atomic hydrogen treatment was performed on part of the SiN_x-coated c-Si substrates to introduce hydrogen into SiN_x films and SiN_x/c-Si interfaces and to enhance the passivation ability of the SiN_x films. We performed the atomic hydrogen treatment at a pressure of 0.2 Pa, at a H_2 flow rate of 20 sccm, and at T_{cat} and T_{sub} of 1800 and 200 °C, respectively. On both sides of the SiN_x films, n^+ amorphous Si (a-Si) films were deposited at SiH_4 and He-diluted (2.25%) PH_3 flow rates of 20 and 10 sccm, respectively, at a T_{sub} of 250 °C. The schematic structure of a c-Si sample for τ_{eff} measurement is shown in Fig. 1.

The thickness and n of SiN_x films were characterized by spectroscopic ellipsometry. We evaluated the passivation quality of SiN_x films by measuring effective minority carrier lifetime (τ_{eff}) by microwave photoconductivity decay (μ-PCD).

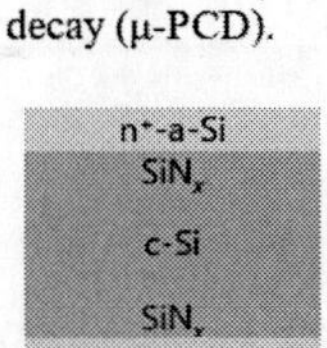

Figure 1: Schematic of a c-Si sample used for τ_{eff} measurement.

3 RESULTS AND DISCUSSION

Figures 2(a) and (b) show the thickness and *n* of Cat-CVD SiN_x films as a function of H_2 flow rate during the deposition. *n* of SiN_x films can be systematically controlled in a wide range from ~2.0 to >2.8 by changing a H_2 flow rate, with roughly keeping the thickness of SiN_x to be ~5 nm. We can thus choose ultra-thin SiN_x with proper *n* for the passivation of c-Si surfaces.

Figure 3 shows the τ_{eff} of the c-Si substrates passivated with ~8-nm-thick n^+-a-Si/SiN_x (*n*~2.3) stacks as a function of SiN_x thickness. Sufficiently high τ_{eff} of ≳1 ms can be obtained for the samples passivated with SiN_x films with thicknesses of >15 nm. On the contrary, the c-Si substrates with thin (~5 nm) SiN_x films show much worse τ_{eff}. Since sufficient tunneling conduction cannot be expected in thick (>10 nm) SiN_x films, we have to improve the passivation quality of thin SiN_x films.

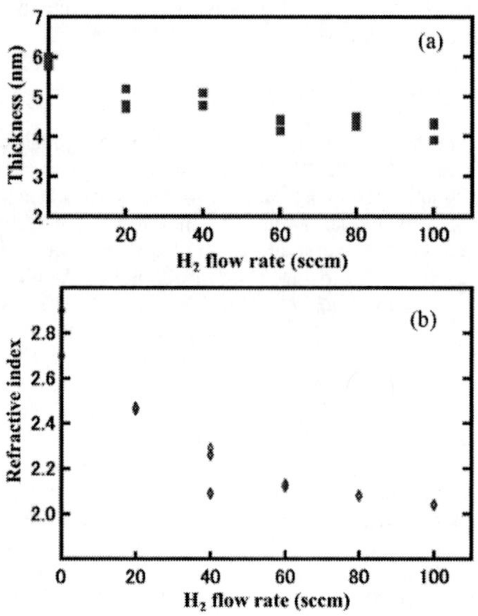

Figure 2: (a)Thickness and (b)*n* of Cat-CVD SiN_x films as a function of H_2 flow rate.

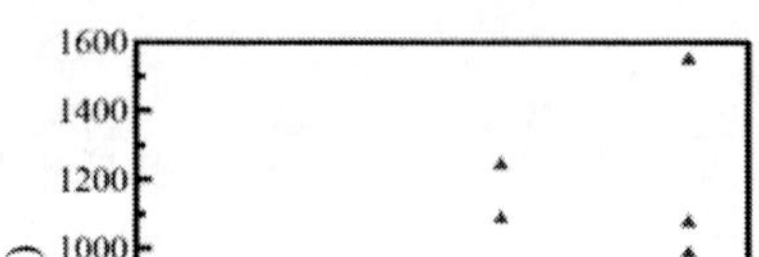

Figure 3: τ_{eff} of the c-Si samples passivated with SiN_x layers as a function of SiN_x thickness.

Figure 4 shows the τ_{eff} of the c-Si samples passivated with 5-nm-thick SiN_x layers with various *n* as a function of atomic hydrogen treatment duration. Low τ_{eff} of ≤100 µs is obtained for all the samples without atomic hydrogen treatment. On the other hand, τ_{eff} dramatically increases by performing hydrogen treatment. Note that the samples passivated with SiN_x with lower *n* shows larger improvement in τ_{eff}. This means that the passivation ability of near stoichiometric SiN_x films can be improved more effectively than Si-rich ones. We separately confirmed that the hydrogen treatment does not etch off the ultra-thin SiN_x with low *n*.

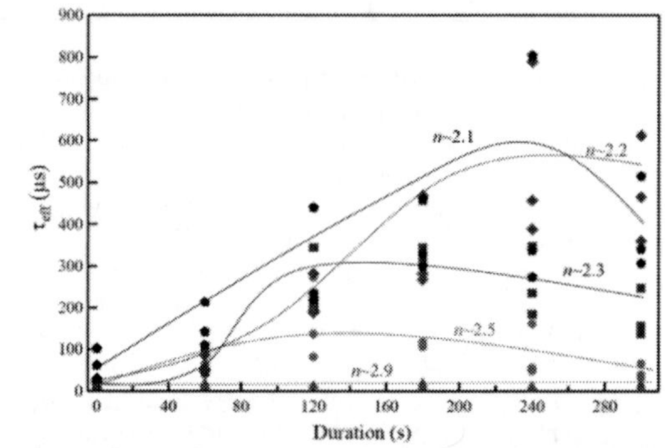

Figure 4: τ_{eff} of the c-Si samples passivated with 5-nm-thick SiN_x layers with various *n* as a function of atomic hydrogen treatment duration. The solid lines are guides to the eye.

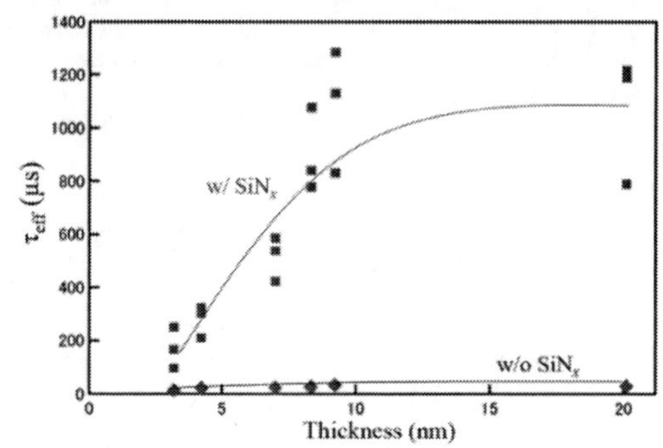

Figure 5: τ_{eff} of the c-Si samples with and without ultra-thin SiN_x layers as a function of the thickness of n^+-a-Si films. The solid lines are guides to the eye.

Figure 5 shows the τ_{eff} of the c-Si samples with and without the SiN_x layers as a function of the thickness of n^+-a-Si films deposited on the SiN_x. τ_{eff} increases with the thickness of n^+-a-Si films and then saturates. The improvement in τ_{eff} may be due to the enhancement of field effect and resulting stronger downward band bending. It should be emphasized that the samples without SiN_x show any remarkable improvement in τ_{eff} even with an increase in the thickness of n^+-a-Si. This clearly indicates that such a high-quality passivation of c-Si surfaces is realized mainly by the ultra-thin SiN_x films. By using n^+-a-Si with sufficient thickness, τ_{eff} of as high as ~1 ms can be obtained by SiN_x with a thickness of only ~5 nm. The utilization of the ultra-thin Cat-CVD SiN_x passivation films for TOPCon-like solar cells is expected.

4 SUMMARY

We have developed a way of forming ultra-thin (<5 nm) Cat-CVD SiN_x films with good passivation ability on n-c-Si surfaces. As-deposited SiN_x films do not show sufficient passivation ability, while τ_{eff} can dramatically improve up to ≥1 ms by performing atomic hydrogen treatment and using sufficiently thick n^+-a-Si layers. We have found that SiN_x with smaller n tends to have better passivation ability by performing atomic hydrogen treatment.

REFERENCES

[1] H. Steinkemper, F. Feldmann, M. Bivour, M. Hermle, IEEE J. Photovoltaics 5 (2015) 1348.

[2] S. W. Glunz, F. Feldmann, A. Richter, M. Bivour, C. Reichel, H. Steinkemper, J. Benick, M. Hermle, Proc. 31st European Photovoltaic Solar Energy Conference (2015) 259.

[3] S. Richter, K. Kaufmann, V. Naumann, M. Werner, A. Graff, S. Großer, A. Moldovan, M. Zimmer, J. Rentsch, J. Bagdahn, C. Hagendorf, Sol. Energy Mater. Sol. Cells 142 (2015) 128.

[4] F. Feldmann, M. Simon, M. Bivour, C. Reichel, M. Hermle, S. W. Glunz, Sol. Energy Mater. Sol. Cells 120 (2014) 270.

[5] Y. Tao, V. Upadhyaya, K. Jones, A. Rohatgi, AIMS Mater. Sci. 3 (2016) 180.

[6] A. Moldovan, F. Feldmann, M. Zimmer, J. Rentsch, J. Benick, M. Hermle, Sol. Energy Mater. Sol. Cells 142 (2015) 123.

[7] A. Richter, J. Benick, R. Müller, F. Feldmann, C. Reichel, M. Hermle, S. W. Glunz, Prog. Photovolt: Res. Appl. 26 (2017) 579.

[8] W.-C. Lee, C. Hu, IEEE Trans. Electron Devices 48 (2001) 1366.

[9] I. Kobayashi, T. Ogawa, S. Hotta, Jpn. J. Appl. Phys. 31 (1992) 336.

[10] H. Matsumura, Jpn. J. Appl. Phys. 25 (1986) L949.

[11] H. Matsumura, H. Tachibana, Appl. Phys. Lett. 47 (1985) 833.

[12] K. Koyama, K. Ohdaira, H. Matsumura, Thin Solid Films 519 (2011) 4473.

[13] K. Koyama, K. Ohdaira, H. Matsumura, Appl. Phys. Lett. 97 (2010) 082108.

[14] T. C. Thi, K. Koyama, K. Ohdaira, H. Matsumura, J. Appl. Phys. 116 (2014) 044510.

[15] T. C. Thi, K. Koyama, K. Ohdaira, H. Matsumura, Jpn. J. Appl. Phys. 55 (2016) 02BF09.

[16] T. C. Thi, K. Koyama, K. Ohdaira, H. Matsumura, Jpn. J. Appl. Phys. 53 (2014) 022301.

[17] T. Oikawa, K. Ohdaira, K. Higashimine, H. Matsumura, Curr. Appl. Phys. 15 (2015) 1168.

[18] K. Ohdaira, T. C. Thi, H. Matsumura, Jpn. J. Appl. Phys. 56 (2017) 102301.

[19] H. Umemoto, K. Ohara, D. Morita, T. Morimoto, M. Yamawaki, A. Masuda, H. Matsumura, Jpn. J. Appl. Phys. 42 (2003) 5315.

FULLY ION-IMPLANTED IBC SILICON SOLAR CELL WITH GAP STRUCTURE BETWEEN EMITTER AND BSF BY SELF-ALIGNED PROCESS

Katsuto Tanahashi, Tomihisa Tachibana, Masaaki Moriya, Yasuhiro Kida, Katsuhiko Shirasawa, and Hidetaka Takato
Renewable Energy Research Center, National Institute of Advanced Industrial Science and Technology (AIST)
2-2-9, Machiikedai, Koriyama, Fukushima 963-0298, JAPAN

ABSTRACT: We propose a fully implanted interdigitated back contact (IBC) silicon solar cell with a gap structure between the emitter and back surface field (BSF) fabricated using a self-aligned process with a single photolithography step. Boron was implanted on the entire rear side. After post-implantation annealing, the oxide layer was grown on the front and rear sides of the silicon wafer. The BSF region was patterned on the rear oxide layer via photolithography. A hollow structure with an overhang of the oxide layer was formed at the boundary between the emitter and BSF during the removal of the oxide and boron-implanted layer on the BSF region via reactive-ion etching and wet chemical etching. The overhanging oxide layer became the mask for subsequent phosphorus implantation on the BSF region.

1 INTRODUCTION

Interdigitated back contact (IBC) silicon solar cells have an optimal architecture to achieve high conversion efficiency owing to the absence of optical shading loss from metallization of the front side [1], [2]. In spite of their high conversion efficiencies, IBC silicon solar cells face the significant problem of complex fabrication processes. A gapless IBC silicon solar cell is suitable for simplifying the process sequence because of the self-aligned processing of the emitter and back surface field (BSF) [3], [4]. Moreover, ion implantation has been developed as an alternative to the diffusion process in order to reduce the number of processing steps required in the fabrication of IBC silicon solar cells [5]. Müller *et al.* reported a gapless IBC silicon solar cell incorporated with a phosphorus-implanted BSF and subsequently a boron-diffused emitter using the counter-doping technique [6]. They reported a large leakage current in the reverse electrical behavior in a gapless IBC silicon solar cell owing to direct contact between the emitter and BSF [7]. In contrast, an IBC silicon solar cell with a gap structure effectively inhibits the reverse current [7]. However, the fabrication of IBC silicon solar cells with a gap structure requires patterning of the emitter and BSF sequentially, thus increasing the required number of processing steps [8].

In this paper, we propose a fully implanted IBC silicon solar cell with a gap structure fabricated using a self-aligned process. We introduce the process sequence used to fabricate the separated emitter and BSF regions with a single photolithography step. Our process is characterized by the fabrication of a hollow structure with an overhanging oxide layer over the gap, which serves as an implantation mask. We show the boron distribution of the emitter and the phosphorus distributions of the BSF and gap observed using scanning capacitance microscopy (SCM). We also evaluate the reverse leakage current of the IBC silicon solar cell fabricated in this study.

2 EXPERIMENTAL

Fig. 1 shows the processing steps undertaken during the fabrication of a fully implanted IBC silicon solar cell with a gap structure. The IBC silicon solar cell was fabricated on an *n*-type (phosphorus doping) magnetic-field-applied Czochralski silicon wafer with a resistivity of 2.4 Ω·cm and a thickness of 180 μm. After the wafer was textured on the front and rear sides, the rear side was etched off with an acid solution to obtain a quasi-mirror surface. A screen oxide layer of thickness 20 nm was grown on the surface at 900 °C for 66 min in a dry oxygen atmosphere. In this study, a beamline implanter equipped with a mass analyzer was used. Elemental boron was implanted on the entire rear side with an acceleration voltage of 10 keV, dose of 3×10^{15} atoms/cm^2, tilt angle of 7°, and rotation angle of 90°. Isothermal annealing was performed at 1050 °C for 30 min in a nitrogen atmosphere as post-implantation annealing (PIA). After removing the screen oxide by dipping the sample in hydrofluoric acid, the oxide layer was also grown on the front and rear sides at 900 °C for 120 min in a wet oxygen atmosphere. The BSF on the rear side was patterned via photolithography. Elemental phosphorus was implanted on the BSF and subsequently on the front surface field (FSF). The post-implantation oxidation (PIO) of phosphorus implantation on the front and rear sides was simultaneously performed at 900 °C for 10 min in a wet oxygen atmosphere [9]. The emitter and BSF were passivated with a stack of Al_2O_3/SiN on the rear side, and the FSF was passivated with SiN. Al/Ag and Ag were metallized on the emitter and BSF, respectively, via screen-printing, and co-firing was performed.

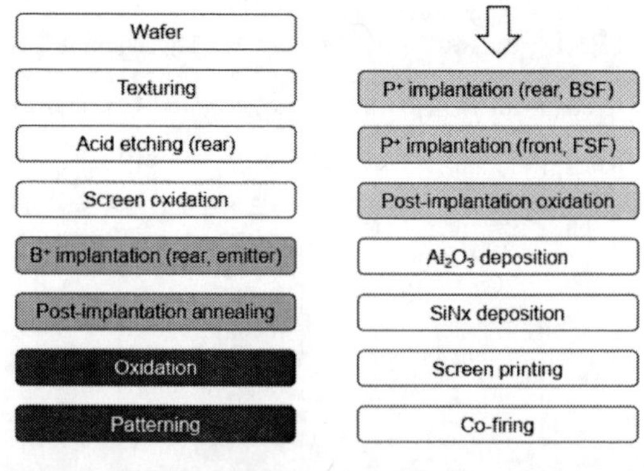

Fig. 1. Process flow for the fabrication of a fully implanted IBC silicon solar cell with a gap structure between the emitter and BSF with a single photolithography step.

The sample preparation and the experimental details of SCM observation are described in ref. [9]. The

SCM samples were cleaved at the center of the cells with a thickness of 800 μm. The cleavage plane was mirror-polished with colloidal silica and subsequently, the sample was observed using SCM. SCM images were observed with a DC bias voltage of −0.5 to +0.3 V, AC bias voltage of 1.0–2.0 V, bias frequency of 100 kHz, and scanning velocity of 1.0 kHz. In the SCM observation, the identification of a depletion state or an accumulation state under the probe depends on the AC bias voltage. This reflects a phase shift between the AC bias voltage and the AC output voltage. A phase shift in the resonant circuit is detected as the SCM signal. We determined the polarization of the observed position by noting whether the SCM signal was positive or negative.

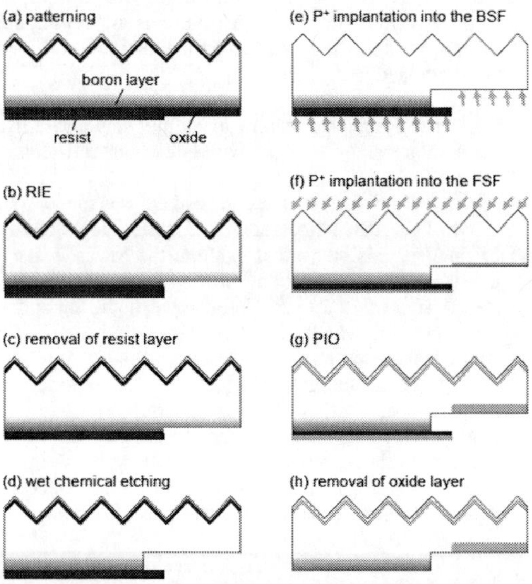

Fig. 2. Process sequence and the structure of the fully implanted IBC silicon solar cell with a gap structure between the emitter and BSF with a single photolithography step.

3 PROPOSED PROCESS SEQUENCE OF IBC SILICON SOLAR CELL WITH GAP STRUCTURE

Fig. 2 shows our proposed process sequence and the structure of a fully implanted IBC silicon solar cell with a gap structure fabricated using the self-aligned process. Fig. 2(a) shows the patterning of the BSF on the rear side. The resist layer on the BSF was removed using photolithography. Wet chemical etching with hydrofluoric acid is not suitable to remove the oxide layer on the BSF owing to peeling off of the resist layer and subsequently the oxide layer on the emitter. Therefore, the oxide layer on the BSF was removed via reactive-ion etching (RIE) using CH_4/CHF_3 plasma (Fig. 2(b)). The concentration of boron beneath the rear surface was in the order of 10^{19} atoms/cm^3. It is difficult to etch off a high-concentration boron-doped layer with an alkali solution such as KOH [10]. Therefore, the high-concentration boron-doped layer on the BSF was also removed via RIE using CF_4/O_2 plasma (Fig. 2(b)). The resist layer prevented the emitter from being etched off during RIE. After removing the resist layer on the emitter (Fig. 2(c)), the residual boron-doped layer was etched off with 24% KOH solution at 70 °C (Fig. 2(d)). Wet chemical etching shown in Fig. 2(d) was also used to remove the layer damaged by RIE. The boron-doped layer was etched off laterally and in the depth direction, resulting in a hollow structure with an overhang of oxide layer as shown in Fig. 2(d). Subsequently, the oxide layer on the front side was removed using a spin etcher. Elemental phosphorus was implanted on the BSF with a tilt angle of 7° and rotation angle of 90° (Fig. 2(e)). In this step, phosphorus ions were not implanted on the lower side of the overhanging oxide layer of the hollow structure. Consequently, the boundary between the emitter and BSF was un-implanted. Elemental phosphorus was implanted on the FSF with a tilt angle of 35° and rotation angle of 15° (Fig. 2(f)) [11]. The PIO of phosphorus implantation on the front and rear sides was simultaneously performed at 900 °C for 10 min in a wet oxygen atmosphere as shown in Fig. 2(g) [9]. The oxide layers on the front and rear sides of the silicon wafer were removed by dipping the wafer in hydrofluoric acid (Fig. 2(h)).

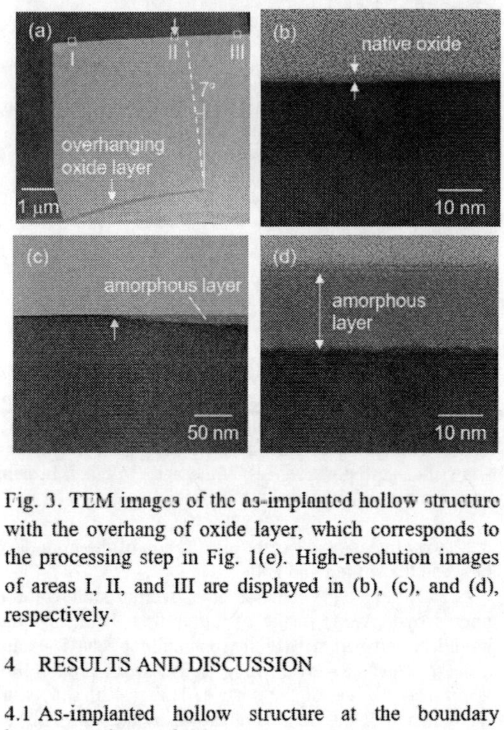

Fig. 3. TEM images of the as-implanted hollow structure with the overhang of oxide layer, which corresponds to the processing step in Fig. 1(e). High-resolution images of areas I, II, and III are displayed in (b), (c), and (d), respectively.

4 RESULTS AND DISCUSSION

4.1 As-implanted hollow structure at the boundary between emitter and BSF

Fig. 3 shows the bright-field cross-section transmission electron microscopy (TEM) images of the as-implanted hollow structure, which also corresponds to Fig. 2(e). The overhanging oxide layer and vertical wall are observed in Fig. 3(a). There is a possibility that the oxide layer was bent downward during preparation of the TEM sample. Higher-resolution images of areas I, II, and III in Fig. 3(a) are shown in Figs. 3(b), (c) and (d), respectively. The position of the yellow arrow in Fig. 3(a) corresponds to that in Fig. 3(c). An amorphous layer of thickness 17 nm is observed to the right side of the yellow arrow in Fig. 3(c). This amorphous layer was formed owing to the implantation of phosphorus, i.e., phosphorous-implanted region [9]. In contrast, the surface of the silicon wafer to the left side of the yellow arrow corresponds to the crystalline layer in Fig. 3(c). As shown in Fig. 2(e), phosphorus was implanted on the

BSF with a tilt angle of 7°. As the overhanging oxide layer serves as a protection layer for the implantation of phosphorus, the ion beam was not irradiated on the area to the left side of the yellow arrow, resulting in the un-implanted region. During PIO, an amorphous layer was recrystallized via solid-phase epitaxial growth and phosphorus atoms diffused into the bulk [9]. As described later, the diffusion length of phosphorus is approximately 0.7 μm under wet oxidation at 900 °C for 10 min. Therefore, the un-implanted region becomes a gap between the emitter and BSF.

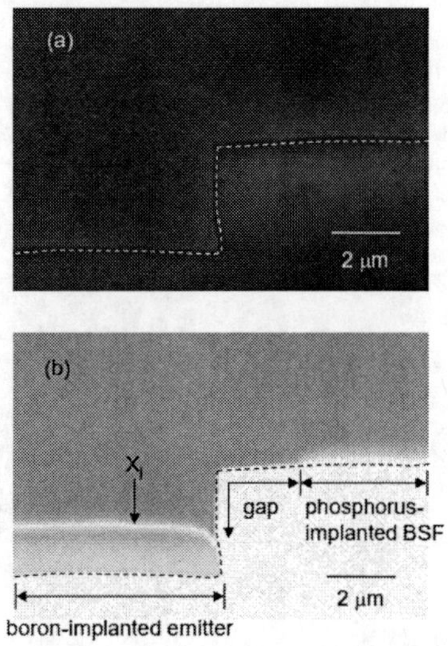

Fig. 4. Cross-section (a) AFM and (b) SCM images of the IBC silicon solar cell fabricated via the process sequence in Figs. 1 and 2. The emitter, BSF, and the boundary between them are displayed in the images. Both images show the same position. (a) White and (b) black broken lines are the rear surface of the solar cell.

4.2 Two-dimensional carrier distribution of the boundary between the emitter and BSF

Fig. 4(a) shows a cross-section atomic force microscope (AFM) image of the emitter, BSF, and the boundary between them in the solar cell fabricated via the process sequence shown in Fig. 2. The white broken line is the rear surface of the solar cell. Fig. 4(b) shows an SCM image of the same area shown by the AFM image in Fig. 4(a). The detailed explanation of the SCM image is given in ref. [9]. The broken line in Fig. 4(b) is the rear surface, which corresponds to the white broken line in Fig. 4(a). In Fig. 4(b), the orange contrast is due to the p-type polarity i.e., boron-implanted emitter. The orange contrast becomes dark and subsequently weak with the increase in depth from the rear surface. It is known that the intensity of the SCM signals becomes weak at high and low carrier concentrations [12]. In the emitter, the concentration of boron decreases with the increase in depth from the rear surface owing to PIA, oxidation at 900 °C for 120 min, and PIO. The white line denoted by Xj is the p–n junction depth where the SCM signal equals zero, whereas the blue contrast is due to the n-type polarity, i.e., phosphorus-implanted BSF and n-type silicon wafer. In the BSF, white contrast is observed immediately beneath the rear surface. Moreover, blue contrast becomes strong with the increase in depth from the rear surface. The tail of dark blue contrast is approximately 0.7 μm in depth. This contrast variation, i.e., phosphorus distribution, was formed during PIO. The area between the emitter and BSF, i.e., the gap region, has almost the same blue contrast as the n-type silicon wafer. The width of the gap between the emitter and BSF, as shown by the double-headed arrow in Fig. 4 (b), is determined by the depth of the vertical step except for the boron-doped layer, width of the overhanging oxide layer, tilt angle of phosphorus implantation on BSF, lateral diffusion length of phosphorus during the PIO of implantation on BSF, and diffusion length into the bulk of boron. The resulting width of the gap in this study is approximately 3.1 μm. In Fig. 4(b), the p–n junction denoted by Xj is properly formed along the rear surface on the emitter. Thus, the distributions of boron and phosphorus show that the separated emitter and BSF were fabricated in the solar cell.

4.3 Electrical characterization of a fully implanted IBC silicon solar cell with a gap structure using self-aligned Process

Fig. 5 shows the measured reverse current density of the IBC silicon solar cell with a gap structure fabricated in this study that exhibits the lowest leakage current of 6.32 × 10^{-6} A/cm^2 at the bias voltage of 2 V with the area of 12.5 cm^2. The inserted figure magnifies the vertical axis, which is of the order of 10^{-5} A/cm^2. The average leakage current of ten solar cells is (2.75 ± 2.25) × 10^{-5} A/cm^2 at the bias voltage of 2 V. As reported for IBC silicon solar cells with a gap structure [7], these cells exhibit a very low leakage current. This is due to the large space charge region at the boundary between the emitter and BSF [7]. Thus, a gap structure fabricated via the self-aligned process shows good reverse *I–V* characteristics of the solar cells.

Fig. 6 shows the *I–V* curve of the IBC silicon solar cell that exhibits the best performance among 10 solar cells (in-house measurement under standard testing conditions; AM 1.5G, 100 mW/cm^2, 25 °C). The short circuit current J_{sc}, open circuit voltage V_{oc}, fill factor *FF*, conversion efficiency η, and active area of the IBC silicon solar cell in Fig. 6 are listed in Table I. We further attempt to improve the electrical characteristics via improvement of bulk lifetime and optimization of the boron (emitter) and phosphorus (BSF and FSF) profiles, passivation, and metallization.

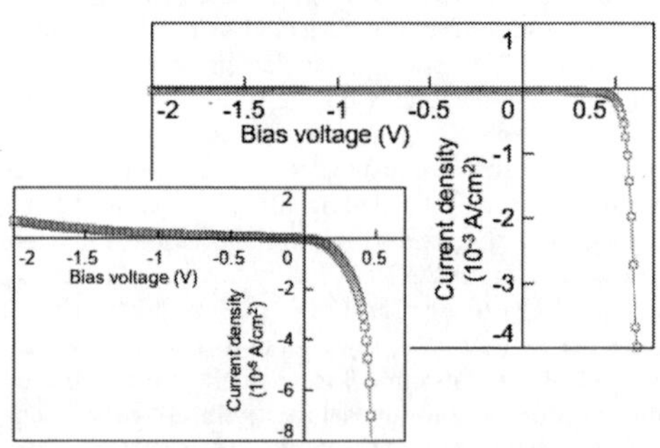

Fig. 5. Measured reverse current density of the IBC silicon solar cell with the gap structure that exhibits the lowest leakage current density among 10 cells fabricated using our proposed process sequence.

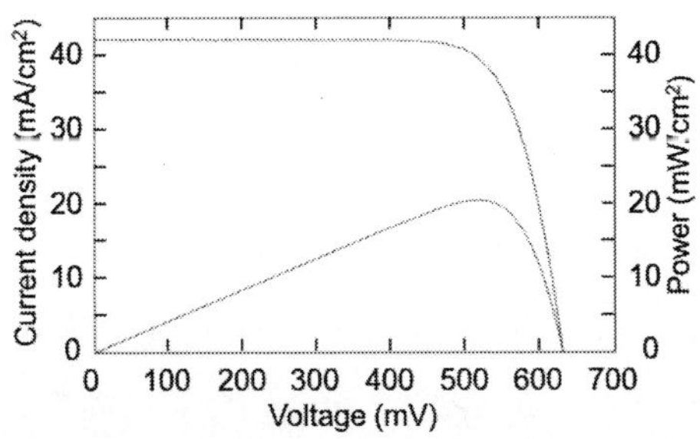

Fig. 6. *I–V* curve of the IBC silicon solar cell that exhibits the best performance among 10 solar cells.

TABLE I. *I–V* PARAMETERS OF IBC SILICON SOLAR CELL IN FIG. 6.

J_{sc} (mA/cm^2)	V_{oc} (mV)	*FF* (%)	η (%)	Active area (cm^2)
42.1	630.4	77.3	20.5	6.35

5 CONCLUSIONS

In conclusion, we proposed a fully implanted IBC silicon solar cell with a gap structure between the emitter and BSF fabricated via a self-aligned process. A hollow structure with an overhanging oxide layer was formed during the etching of the boron-implanted region on the BSF. The overhanging oxide layer at the boundary between the emitter and BSF acted as a mask for subsequent phosphorus implantation. TEM observation revealed an un-implanted region on the lower side of the overhanging oxide layer. The SCM image showed the carrier distribution of the separated boron- and phosphorus-implanted regions, i.e., the gap structure between the emitter and BSF. The IBC silicon solar cells fabricated using our proposed process showed a low leakage current density with reverse *I–V* characteristics.

ACKNOWLEDGEMENT

This work was supported by New Energy and Industrial Technology Development Organization.

REFERENCES

[1] K. Yoshikawa, H. Kawasaki, W. Yoshida, T. Irie, K. Konishi, K. Nakano, T. Uto, D. Adachi, M. Kanematsu, H. Uzu, and K. Yamamoto, "Silicon heterojunction solar cell with interdigitated back contacts for a photoconversion efficiency over 26%," *Nature Energy*, vol. 2, no. 17032, 2017.

[2] D. D. Smith, P. Cousins, and S. Westerberg, "Toward the Practical Limits of Silicon Solar Cells," *IEEE J. Photovolt.*, vol. 4, no. 6, pp. 1465–1469, 2014.

[3] N. E. Posthuma, J. Robbelein, S. Singh, M. Debucquoy, M. Aleman, K. Wostyn, B. J. Pawlak, X. Loozen, J. Fernandez, K. Baert, P. J. Verlinden, and J. Poortmans, "Development and analysis of small area high efficiency interdigitated back contact silicon solar cells," in *Proc. 27th European Photovoltaic Specialist Conf. and Exhibition (EuPVSC)* 2011, pp. 571–575.

[4] N. Zin, A. Blakers, K. McIntosh, E. Franklin, T. Kho, J. Wong, T. Mueller, A. G. Aberle, Z. Feng, and Q. Huang, "19% Efficient N-Type All-Back-Contact Silicon Wafer Solar Cells With Planar Front Surface," in *Proc. 49th Annual Conference of the Australian Solar Energy Society, Sydney* 2011.

[5] A. Merkle, R. Peibst, and R. Brendel, "High efficiency fully ion-implanted, co-annealed and laser-structured back junction back contacted solar cells," in *Proc. 29th European Photovoltaic Specialist Conf. and Exhibition (EuPVSC)* 2013, pp. 954–958.

[6] R. Müller, J. Schrof, C. Reichel, J. Benick, and M. Hermle, "Back-junction back-contact n-type silicon solar cell with diffused boron emitter locally blocked by implanted phosphorus," *Appl. Phys. Lett.*, vol. 105, no. 10, pp. 103503-1–103503-4, 2014.

[7] R. Müller, C. Reichel, J. Schrof, M. Padilla, M. Selinger, I. Geisemeyer, J. Benick, and M. Hermle, "Analysis of n-type IBC solar cells with diffused boron emitter locally blocked by implanted phosphorus" *Sol. Energy Mater. Sol. Cells*, vol. 142, pp. 54–59, 2015.

[8] E. Franklin, K. Fong, K. McIntosh, A. Fell, A. Blakers, T. Kho, D. Walter, D. Wang, N. Zin, M. Stocks, E.-C. Wang, N. Grant, Y. Wan, Y. Yang, X. Zhang, Z. Feng, and P. J. Verlinden, "Design, fabrication and characterization of a 24.4% efficient interdigitated back contact solar cell," *Prog. Photovolt.*, vol. 24, pp. 411–427, 2016.

[9] K. Tanahashi, M. Moriya, Y. Kida, S. Utsunomiya, T. Fukuda, K. Shirasawa, and H. Takato, "Impact of Post-implantation Annealing Conditions on Electrical Characteristics of a Phosphorus-implanted Emitter Crystalline Silicon Solar Cell," *IEEE J. Photovolt.*, vol. 7, no. 3, pp. 741–746, 2017.

[10] S. D. Collins, "Etch Stop Techniques for Micromachining," *J. Electrochem. Soc.*, vol. 144, no. 6, pp. 2242–2262, 1997.

[11] K. Tanahashi, M. Moriya, Y. Kida, S. Utsunomiya, T. Fukuda, K. Shirasawa, and H. Takato, "Phosphorus-implanted Emitter Crystalline Silicon Solar Cell with AL-BSF," in *Proc. 21th International Conference on Ion Implantation Technology (IIT)* 2016, pp. 1–4.

[12] J. Smoliner, B. Basnar, S. Golka, E. Gornik, B. Löffler, M. Schatzmayr, and H. Enichlmair, "Mechanism of bias-dependent contrast in scanning-capacitance-microscopy images," *Appl. Phys. Lett.*, vol. 79, no. 19, pp. 3182–3184, 2001.

POTENTIAL OF CHEMICAL ROUNDING FOR THE PERFORMANCE ENHANCEMENT OF A PYRAMID-TEXTURED BIFACIAL SI BOTTOM CELL

Hyunju Lee[1]*, Inseol Song[2], Sang-Won Lee[3], Soo Hyun Bae[3], Ji Yeon Hyun[3], Yoonmook Kang[2], Hae-Seok Lee[2], Atsushi Ogura[4], Donghwan Kim[2,3], Yoshio Ohshita[1]
[1]Toyota Technological Institute, 2-12-1 Hisakata, Tempaku-ku, Nagoya 468-8511, Japan
+81 52 809 1877, leehyunju@toyota-ti.ac.jp
[2]KU-KIST Green School, Graduate School of Energy and Environment, Korea University, Anam-dong, Seongbuk-gu, Seoul 136-713, Korea
[3]Department of Materials Science and Engineering, SERC, Korea University, Anam-dong, Seongbuk-gu, Seoul 136-713, Korea
[4]School of Science and Technology, Meiji University, Kawasaki 214-8571, Japan

ABSTRACT: We have investigated the effects of chemical rounding on the performance of bifacial *n*-PERT silicon solar cells with a p^+-emitter on a pyramid textured front surface. From the study, we found that solar weighted reflectance increases from ~2.5 to ~3.7% for AlO_x/SiN_x:H stack-passivated pyramid textured emitter samples according to chemical rounding time. Consequently, compared to the non-rounded cells, the *J*sc of the 200-sec-rounded cells with ~60 and ~100 Ω/sq p^+-emitters is reduced by 0.8 and 0.6 mA/cm^2, respectively. However, the loss in *J*sc is fully offset by the increased *FF* by 0.8 and 1.5% for the 200-sec-rounded cells with ~60 and ~100 Ω/sq p^+-emitters, respectively. The efficiency of the 200-sec-rounded cells with a ~100 Ω/sq p^+-emitter can be enhanced as a result, compared to the non-rounded cells. Finally, we investigated effects of chemical rounding on the optical property of an indium tin oxide layer widely used as a recombination layer for the fabrication of a monolithic perovskite/silicon tandem cell. From the results, it can be expected that efficiency of a bifacial *n*-PERT silicon bottom cell and its monolithic tandem cell with a perovskite top cell could be improved without additional complicated processes if chemical rounding and boron-doping processes can be properly optimized.
Keywords: p^+-Emitter, Bifacial *n*-PERT Cell, Chemical Rounding, ITO, Perovskite/Silicon Tandem Cell

1 INTRODUCTION

Recently, silicon heterojunction (SHJ) cells demonstrated the very high power conversion efficiency of 26.7% [1], therefore, only marginal improvements in silicon (Si) solar cell efficiency are still possible, considering the 29.4% theoretical limit for crystalline silicon (c-Si) solar cells [2]. One of the most investigated cell concepts is a perovskite/Si tandem cell to overcome the theoretical limit for the efficiency of Si solar cells.

In the perovskite/Si tandem cell, theoretically, the cell efficiency above 30% is feasible [3]. Currently, the highest cell efficiency reported have been 25.2% for a monolithic perovskite/Si tandem cell [4]. Recent study shows that Si bottom cells have been identified as the limiting cell mainly. This is because the Si bottom cells can only absorb the solar spectrum with energy ranging from 1.1 to 1.6 eV in the monolithic tandem cell though it is essential to match the currents of the top and bottom cells by balancing the photon distribution in each subcell in the monolithic tandem cell [4,5]. Therefore, for further improvement of the efficiency of the monolithic perovskite/Si tandem cell, it is necessary to increase the current of a Si bottom cell. Two pathways could be suggested to increase the current of a Si bottom cell. One is the reduction in the optical and electrical losses of a Si bottom cell and its interconnection with a perovskite top cell, and the other is the employment of a higher bandgap material for a perovskite top cell. In this study, we suggest employment of a *n*-PERT bifacial Si solar cell as a Si bottom cell for the fabrication of low-cost high-efficiency monolithic perovskite/Si tandem cells. Because bifacial solar cells simultaneously collect photons from incident and albedo radiation reaching both the front side and backside of a cell [6], which may alleviate strict requirements to develop new higher bandgap materials for a perovskite top cell.

Meanwhile, although n-PERT bifacial Si solar cells generate more current than monofacial Si solar cells, the formation of a pyramid textured surface on both sides of the cells is necessary to obtain the highest current [7]. In addition, n-type Si cells inevitably require the formation of a boron doped p^+-emitter on their pyramid textured surface [8]. Unfortunately, a boron-doped p^+-emitter on a pyramid textured surface have proved to demonstrate inferior performance to a phosphorus doped n^+-emitter on a random pyramidal textured surface even though n-type Si solar cells have demonstrated higher cell performance, compared to p-type Si solar cells [8]. Such lower performance of pyramidal textured p^+-emitters is partly attributed to 1) the enlarged surface area, formation of stressed and non-uniformly doped convex and concave regions and edges by pyramid-texturing, 2) the generation of higher defect density on {111}-oriented facets by boron doping as shown in the previous studies [9]. In addition, random pyramidal textured surfaces are very problematic for monolithic tandem cells because the conformal, pinhole-free deposition of the top cell layers on random pyramidal textured surfaces with many sharp features is highly challenging [10].

One of the possible ways of reducing such problems on pyramidal textured emitters and/or surfaces is chemical rounding (CR) of pyramid corners, which, however, demonstrates contradicting dual effects on the light-trapping performance and level of passivation at textured emitter surfaces, making quantitative understanding of the rounding effects important to optimize the degree of the rounding for processing of *n*-PERT bifacial Si bottom cells with a pyramidal textured p^+-emitter. In addition, there are few systematic studies to evaluate the effects of CR on the performance of amorphous aluminum oxide (AlO_x) passivation and indium tin oxide (ITO) recombination layers on pyramidal textured surfaces and p^+-emitters of *n*-PERT bifacial Si solar cells. Therefore, we have investigated effects of CR on the performance of AlO_x passivation on differently doped p^+ emitters and the performance of *n*-PERT bifacial Si solar cells with rounded p^+-emitters. In addition, we have also investigated effects of CR on the optical property of ITO thin layers deposited on AlO_x/SiN_x:H-passivated pyramidal textured p^+-emitters.

2 PREPARING THE MANUSCRIPT

2.1 Investigation of the effects of chemical rounding on the passivation and reflectance of p^+-emitters on pyramid textured surfaces

Chemical rounding (CR) experiments were carried out on *n*-type 125 × 125 mm^2 pseudo-square Cz mono-crystalline Si(100) wafers with a final thickness of ~170 µm and a resistivity of ~3.3 Ω·cm. The wafers were subjected to the conventional standard cleaning process of the Radio Corporation of America (RCA), saw damage etching (SDE) in acidic solution, and random pyramidal texturing in an alkaline (KOH) isopropanol solution containing a surfactant additive. CR of pyramidal surface textures was performed with a 1:50 (% (v/v)) solution of 49 % HF and 70 % HNO_3 at room temperature. The CR time (t_{CR}) varied from 0 to 200 sec.

Following the CR process, for investigating effects of CR on passivation performance of AlO_x on pyramidal textured emitter samples with different boron doping concentrations, symmetrical $p^+/n/p^+$ lifetime samples were fabricated by boron diffusion of ~ 60 or ~100 Ω/sq with boron silicate glass (BSG) as a diffusion source. Non-diffused and diffused carrier lifetime samples were passivated by ~10-nm-thick atomic layer deposited (ALD) AlO_x layers. All passivated samples experienced post-deposition annealing (PDA) in an industrial-type quartz tube furnace at 500 °C in N_2 for 15 min.

For investigating effects of CR on the reflectance of AlO_x and AlO_x/SiN_x:H stack-passivated pyramidal textured emitter samples with different boron doping concentrations, single-side pyramidal textured wafers with a textured p^+-emitter (p^+/n) were fabricated by boron diffusion of ~60 or ~100 Ω/sq.. Diffused reflectance samples were given an RCA clean, HF dip, and ~ 10-nm-thick AlO_x-passivation in the same way as for the previous lifetime samples. Selected AlO_x-passivated samples were capped by ~ 70-nm-thick plasma-enhanced chemical vapor deposition (PECVD) SiN_x:H layers having a refractive index of ~ 2.1 for enhancing antireflection performance. The AlO_x-passivated samples experienced PDA in an industrial-type quartz tube furnace in N_2 at 500 °C for 15 min and the AlO_x/SiN_x:H stack-passivated samples experienced PDA in an infrared (IR) lamp-heated belt furnace in air at 800 °C for a few seconds. Some annealed AlO_x/SiN_x:H stack-passivated samples were coated by an ~100-nm-thick remote plasma sputtering deposited ITO layer to investigate effects of CR on the reflectance of AlO_x/SiN_x:H /ITO passivated reflectance samples.

The morphological properties of the random pyramid textured and chemically rounded samples were characterized by a field-emission scanning electron microscope (FE-SEM). Reflectance and solar-weighted reflectance (SWR) with the AM 1.5G spectrum in the wavelength range from 400 to 1100 nm of the samples were characterized by a UV-Vis-NIR spectrophotometer with an integrating sphere. The effective excess carrier lifetime τ_{eff}, emitter saturation current density J_{0e}, and implied V_{OC} of the carrier lifetime samples were measured by using a WCT-120 Sinton Instruments tool in quasi-steady-state (QSS) and transient modes from the relation proposed by Kane and Swanson [11]. The schematic diagram of the fabricated sample structures are shown in Fig. 1.

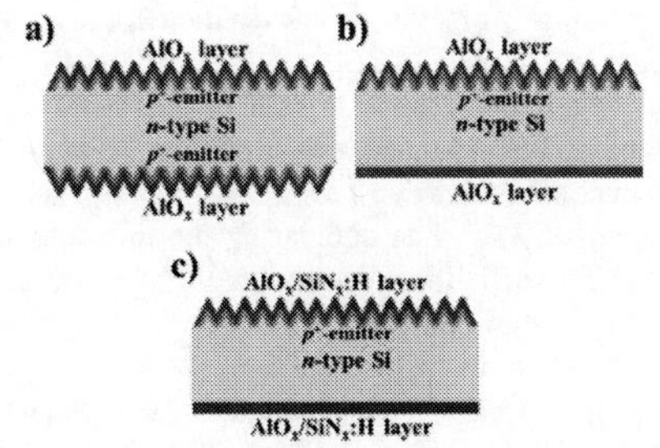

Figure 1: Schematic structures of (a) symmetrical lifetime and (b) reflectance samples passivated with AlO_x layers, and (c) reflectance samples passivated with AlO_x/SiN_x:H stacks.

2.2 Investigation of the effects of chemical rounding on the performance of bifacial *n*-PERT Si cells

To investigate effects of CR on the performance of *n*-PERT bifacial Si cells, we fabricated the cells on *n*-type pseudo-square 156 × 156 mm^2 Cz monocrystalline Si(100) wafers with a final wafer thickness of ~200 µm and a resistivity of ~3.5 Ω·cm. The cell structure is schematically shown in Fig. 2. After the RCA cleaning, SDE, and random pyramidal texturing of the wafers in the same way as for the previous CR samples, CR was performed on a front side of the wafers only. After CR of the wafers and removal of all oxide layers in HF, p^+-emitters and an n^+-back surface field (BSF) were formed by diffusion from BSG and $POCl_3$, respectively. After removal of all oxide layers in HF, surface passivation was performed by a stack of AlO_x/SiN_x:H deposited on the p^+-emitter and by a stack of thermal SiO_2/SiN_x:H on the n^+-BSF. Finally, the wafers were printed with an Ag-Al paste on a front surface and an Ag paste on a rear surface, and co-fired to form metal contacts. The electrical parameters of the solar cells were analyzed by current density–voltage (*J–V*) measurement under the AM1.5G, 1-sun illumination using a WACOM solar simulator.

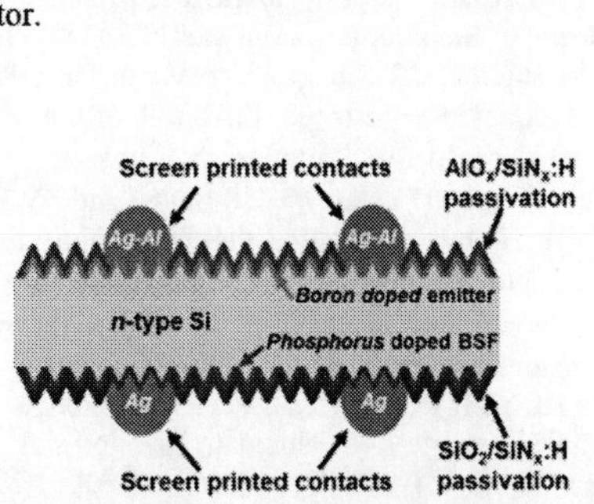

Figure 2: Schematic structure of fabricated *n*-PERT bifacial Si cells with a random pyramid textured or chemically rounded front boron-diffused p^+-emitter, and random pyramid textured rear phosphorus-diffused n^+-BSF.

3 RESULTS AND DISCUSSION

3.1 The effects of chemical rounding on the passivation and reflectance of p^+-emitters on pyramid textured surfaces

We measured τ_{eff} of AlO_x-passivated symmetric $p^+/n/p^+$ lifetime samples to evaluate the effects of CR on a level of surface recombination of pyramid textured p^+-emitters with different boron-doping concentrations. As shown in Fig. 3(a), 200-sec CR resulted in the reduced J_{0e} from ~570 to ~236 fA/cm^2 (i.e., ~59% reduction in J_{0e}) when comparing the non-rounded and 200-sec-rounded lifetime samples with ~60 Ω/sq p^+-emitters, and the reduced J_{0e} from ~54 to ~39.5 fA/cm^2 (i.e., ~27% reduction in J_{0e}) when comparing the non-rounded and 200-sec-rounded lifetime samples with ~100 Ω/sq p^+-emitters. In addition, Fig. 3(b) shows that 200-sec CR also resulted in the enhanced $V_{OC,impl}$ from 622 to 642 mV (i.e., ~20 mV increase in $V_{OC,impl}$) when comparing the

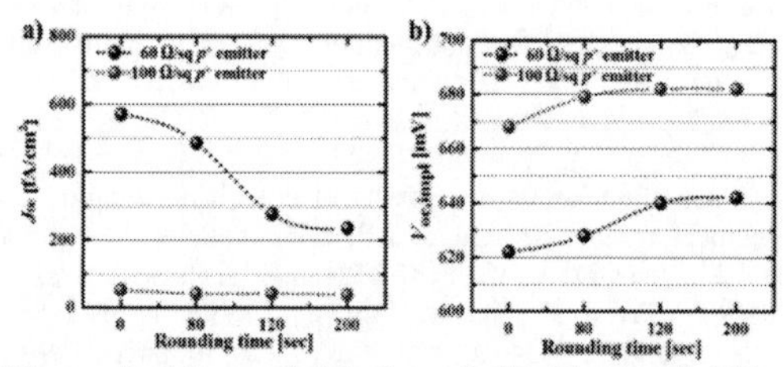

Figure 3: Measured (a) J_{0e} and (b) $V_{OC,impl}$ of AlO_x-passivated lifetime samples with pyramid textured ~60 and ~100 Ω/sq p^+-emitters as a function of t_{CR}.

non-rounded and 200-sec-rounded lifetime samples with ~60 Ω/sq p^+-emitters, and the enhanced $V_{OC,impl}$ from 668 to 682 mV (i.e., ~14 mV increase in $V_{OC,impl}$) when comparing the non-rounded and 200-sec-rounded lifetime samples with ~100 Ω/sq p^+-emitters. From the results, we could conclude that CR is the very efficient method to reduce surface recombination at pyramid textured p^+-emitters.

Fig. 4 shows the effects of CR on the reflectance of non-rounded and rounded reflectance samples passivated by a ~10-nm-thick AlO_x with or without a ~70-nm-thick SiN_x:H antireflection layer as a function of t_{CR}. It is noticeable that the difference in doping concentration does not result in any marked difference in the reflectance of the samples in our study. As expected from previous studies, CR leads to increase in the reflectance of the samples. For instance, the SWR of the ~10-nm-thick AlO_x-passivated samples significantly increased from ~9.8 to ~27% (i.e., ~2.8 times) and SWR of the AlO_x/SiN_x:H stack-passivated samples slightly increased from ~2.5 to ~3.7% (i.e., ~1.5 times) after 200-sec CR in an etch mixture of ~1:50 (v/v) HF/HNO_3, respectively, as shown in Table 1.

We investigated the surface morphology of the reflectance samples to elucidate the origins of the changes in passivation performance of AlO_x and reflectance of the samples according to t_{CR}. As shown in Fig. 5, FE-SEM images show that the increase in the passivation performance of AlO_x and the SWR of the rounded samples seems to be owing to CR of concave as well as convex areas of pyramidal textures. For instance,

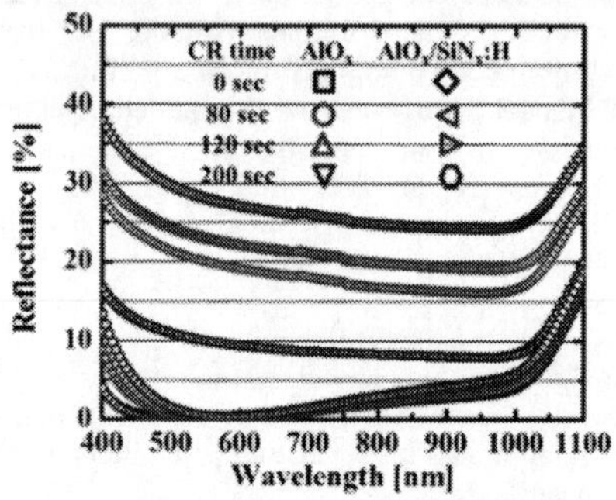

Figure 4: Measured reflectance of reflectance samples passivated by AlOx layers with or without a SiN_x:H antireflection layer as a function of t_{CR} by UV–Vis–NIR spectroscopy in the wavelength range between 400 and 1100 nm.

Table I: Determined solar weighted reflectance (SWR) of passivated emitter samples as a function of CR time by UV-Vis-NIR spectroscopy between 400 and 1100 nm.

t_{CR} (sec)	0	80	120	200
SWR of AlO_x sample	9.8%	19%	22%	27%
SWR of AlO_x/SiN_x:H sample	2.5%	2.9%	3.2%	3.7%

cross-sectional FE-SEM images shown in Fig. 5(a)-(d), and Table 2 show that pyramid vertex angle α of the reflectance samples are different from the ideal value of 70.52° expected from c-Si crystallography and the pyramid geometry strongly depends on t_{CR}. In particular, the observed α increased from ~76.4° for the non-rounded reflectance samples to 80.8° for the 200-sec-rounded reflectance samples. We calculated geometrical factor P=1/sin (α/2) given as a ratio between the surface area of the pyramid and its corresponding horizontal projection from the FE-SEM images. The calculated P only slightly decreased from 1.617 for the non-rounded reflectance samples to 1.543 for the 200-sec-rounded reflectance samples without considering the effects of CR on concave regions of the pyramidal textures. However, the surface recombination of the AlO_x-passivated lifetime samples with ~60 Ω/sq p^+-emitters is largely reduced after 120- and 200-sec CR as shown in Fig. 3(a) and (b). Therefore, the effects of CR on concave areas of pyramid textures should need to be considered and we found that the curvature of concave features largely decreases and some small pyramids in concave regions of pyramidal textures are significantly rounded after 120-sec CR, and even almost removed after 200-sec CR as shown in Fig. 5(c) and (d), respectively. As mentioned above, pyramidal texturing significantly increases surface recombination because an ideal pyramid textured surface has ~1.73 times larger effective area compared to a planar (100) surface. Furthermore, the resulting (111) facets of pyramids are associated with a higher defect density than (100) planes, and convex and concave surface features are supposed to cause process-induced defects during boron diffusion and plasma-based surface passivation processes, resulting in increased surface recombination. Therefore, from the experimental results, it could be concluded that CR in concave regions of random pyramidal textures efficiently reduces the effective area of pyramid textured surfaces and the surface defects generated during texturing, boron diffusion, and passivation processes. However, the degradation of convex and concave surface features of pyramidal textures and the removal of small pyramids significantly increase the reflectance of the samples as shown in Fig. 4 and Table 1.

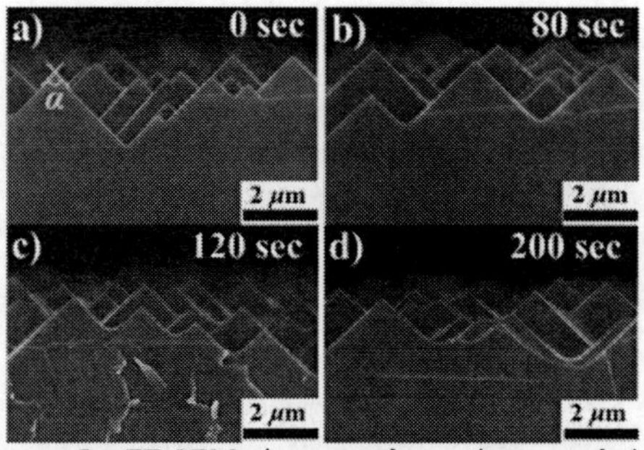

Figure 5: FE-SEM images show the morphology changes of non-rounded and rounded reflectance samples with a ~60 Ω/sq p^+-emitter as a function of t_{CR}. The FE-SEM images show cross-section of (a) the non-rounded sample, (b) 80-sec-rounded sample, 120-sec-rounded sample, and 200-sec-rounded sample, respectively.

Table 2: Measured pyramid vertex angles, calculated geometrical factors, and determined curvatures in concave regions of pyramidal textures as a function of t_{CR}. The determined curvature is the inverse of curvature radius of the measured samples.

t_{CR} (sec)	0	80	120	200
Pyramid vertex angle α (°)	76.4	78.8	80.0	80.8
Geometrical factor P	1.617	1.576	1.556	1.543
Curvature	–	0.27	0.17	0.09

3.2 The effects of chemical rounding on the performance of bifacial n-PERT Si cells

To investigate the effects of CR on solar cell performance, full size (156×156mm^2) non-rounded and rounded bifacial n-PERT Si cells with a ~60 or ~100 Ω/sq p^+-emitter (i.e., $n^+/n/p^+$ structured cells) were fabricated and their cell parameters were investigated. As shown in Table 3, for front p^+-emitter side illumination, the increased reflectance by 200-sec CR slightly reduces the J_{SC} of the fabricated bifacial n-PERT Si cells with ~60 and ~100 Ω/sq p^+-emitters by 0.8 and 0.6 mA/cm^2, respectively. However, the loss in J_{SC} is offset by the increased FF for the 200-sec-rounded cells. In particular, for the 200-sec-rounded cells with a ~100 Ω/sq p^+-emitter, the loss in Jsc was fully offset by the largely enhanced FF of 76.8% and slightly enhanced V_{OC} of 635 mV. The cell efficiency could be enhanced as a result, compared to that of the non-rounded cells. The enhanced V_{OC} seems to be due to the reduced surface recombination by 200-sec CR as shown in Fig. 3. To reveal the origin of the enhancement of FF by 200-sec CR, we investigated boron-doping profiles of the p^+-emitter samples with FE-SEM using secondary electrons. As shown in Fig. 6, the boron-doped emitter layer of the non-rounded ~100 Ω/sq sample is much thinner than that of the 200-sec-rounded ~100 Ω/sq sample in concave regions of the pyramidal textures. It has been well known that thinner boron-doped emitters are vulnerable to shunting problem caused by Al spiking in a space charge region during a metallization process with an Ag/Al paste. This shunting problem causes power losses in solar cells by introducing an alternative current path for the light-generated current. As a result, reduced FF is observed in the solar cells [12]. Therefore, we could conclude that CR seems to enhance FF by enabling boron dopants to diffuse deeper into concave regions of the pyramidal textures and consequently, reducing shunting problem during a front-side metallization process.

Table 3: Measured cell parameters of non-rounded and 200-sec-rounded bifacial n-PERT Si cells (238.8 cm^2) under front p^+-emitter side illumination.

Cell condition	V_{OC} (mV)	Jsc (mA/cm^2)	FF (%)	η (%)
60 Ω/sq 0-sec CR	633	38.0	75.1	18.1
60 Ω/sq 200-sec CR	633	37.2	75.9	17.9
100 Ω/sq 0-sec CR	633	38.1	75.3	18.2
100 Ω/sq 200-sec CR	635	37.5	76.8	18.3

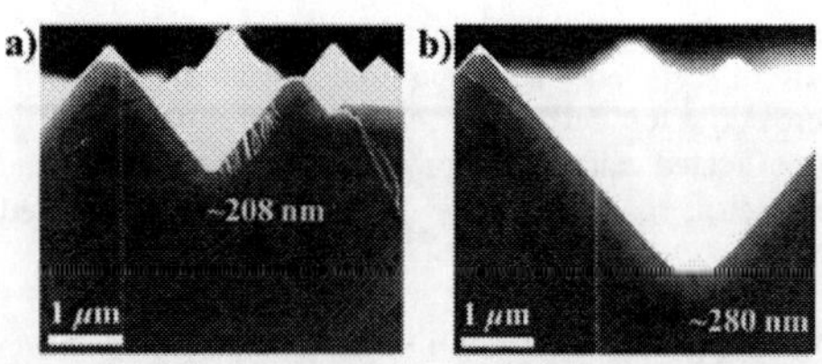

Figure 6: FE-SEM images of ~100 Ω/sq boron-diffused layers on (a) non-rounded and (b) 200-sec-rounded pyramid textured surfaces. The images were obtained by cross-sectional FE-SEM using secondary electrons.

3.3 Effects of chemical rounding on the reflectance of AlO_x/SiN_x:H/ITO-stack passivated emitter samples

Finally, the reflectance of non-rounded and 200-sec-rounded ~100 Ω/sq emitter samples passivated by an AlO_x/SiN_x:H/ITO stack was measured to investigate effects of CR on the reflectance of pyramid textured emitter samples. As shown in Fig. 7, the deposition of an ~100-nm-thick ITO layer results in the pronounced reduction in the SWR of the AlO_x/SiN_x:H-stack passivated non-rounded emitter samples from 4.7 to 1.5% and rounded emitter samples from 6.1 to 2.4 % in the wavelength ranging between 1.1 and 1.6 eV. As mentioned above, a Si bottom cell can only absorb the solar spectrum with energy ranging from 1.1 to 1.6 eV in a monolithic tandem cell and therefore deposition of an ITO thin layer is expected to be beneficial to enhance the efficiency of monolithic perovskite/textured Si tandem cells.

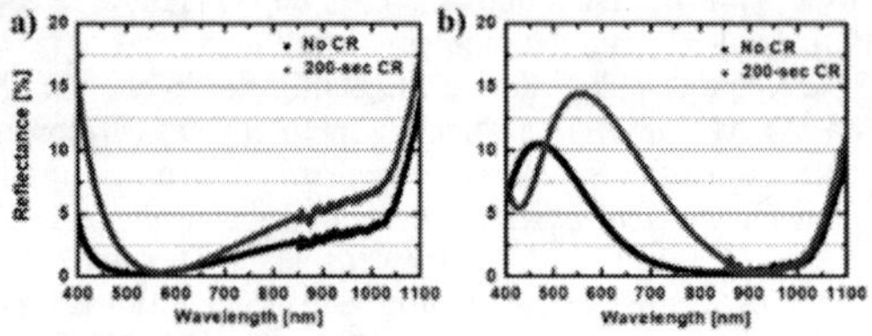

Figure 7: Measured reflectance of non-rounded and 200-sec-rounded reflectance samples passivated by (a) an AlO_x/SiN_x:H stack or (b) an AlO_x/SiN_x:H/ITO stack.

4 CONCLUSION

We have investigated the effects of CR on the performance of bifacial *n*-PERT Si solar cells with a p^+-emitter on a random pyramid textured front surface. From experimental results, we found that although CR increases the reflectance of the random pyramidal textured emitter samples and leads to the loss in *J*sc of fabricated *n*-PERT bifacial Si cells, the loss in *J*sc is fully offset by the increased *FF* for the 200-sec-rounded cells with a ~100 Ω/sq p^+-emitter and the cell efficiency of the 200-sec-rounded cells with a ~100 Ω/sq p^+-emitter can be enhanced as a result, compared to the non-rounded cells. Finally, we have also investigated the effects of CR on the optical property of an ITO thin layer deposited on an AlO_x/SiN_x:H-stack passivated textured p^+-emitter. We found that the deposition of an ~100-nm-thick ITO layer results in the pronounced reduction in the SWR of the AlO_x/SiN_x:H-stack passivated non-rounded and 200-sec-rounded emitter samples in the wavelength ranging between 1.1 and 1.6 eV. Based on our results, it can be expected that cell efficiency of a bifacial *n*-PERT Si bottom cell and its monolithic tandem cell with a perovskite top cell could be improved without additional complicated and costly processes if chemical rounding and boron doping processes can be properly optimized.

5 ACKNOWLEDGEMENTS

The authors thank the staffs of Collaborative Clean Room at Toyota Technological Institute for experimental support on sample fabrication. This work has been partly supported by the Research Center for Smart Energy Technology at Toyota Technological Institute and the New & Renewable Energy Core Technology Program of the Korea Institute of Energy Technology Evaluation and Planning (KETEP) granted financial from the Ministry of Trade, Industry & Energy, Republic of Korea (No. 20163030014020). This study is also partly supported by New Energy and Industrial Technology Development Organization (NEDO).

6 REFERENCES

[1] M. A. Green, Y. Hishikawa, E. D. Dunlop, D. H. Levi, J. Hohl-Ebinger, A. W.Y. Ho-Baillie, Progress in Photovoltaics: Research and Applications 26 (2018) 3.
[2] A. Richter, M. Hermle, S. W. Glunz, IEEE Journal of Photovoltaics 3 (2013) 1184.
[3] I. Almansouri, A. Ho-Baillie, M. A. Green, Japanese Journal of Applied Physics 54 (2015) 08KD04.
[4] F. Sahli, J. Werner, B. A. Kamino, M. Bräuninger, R. Monnard, B. Paviet-Salomon, L. Barraud, L. Ding, J. J. Diaz Leon, D. Sacchetto, G. Cattaneo, M. Despeisse, M. Boccard, S. Nicolay, Q. Jeangros, B. Niesen, C. Ballif, Nature Materials 17 (2018) 820.
[5] J. Werner, L. Barraud, A. Walter, M. Bräuninger, F. Sahli, D. Sacchetto, N. Tétreault, B. Paviet-Salomon, S.-J. Moon, C. Allebé, M. Despeisse, S. Nicolay, S. De Wolf, B. Niesen, C. Ballif, ACS Energy Letters 1 (2016) 474.
[6] R. Guerrero-Lemus, R. Vega, T. Kim, A. Kimm, L.E. Shephard, Renewable and Sustainable Energy Reviews 60 (2016) 1533.
[7] J. Frank, M. Rüdiger, S. Fischer, J.C. Goldschmidt, M. Hermle, Energy Procedia 27 (2012) 300.
[8] W. Jellett, C. Zhang, H. Jin, P. J. Smith, K. J. Weber, Proceedings of the 33rd IEEE Photovoltaic Specialists Conference, (2008) 1.
[9] K. R. McIntosh, L. P. Johnson, Journal of Applied Physics 105 (2009) 124520.
[10] J. Werner, B. Niesen, C. Ballif, Advanced Materials Interfaces 2017, 1700731.
[11] D. E. Kane, R. M. Swanson, Proceedings of the 18th IEEE Photovoltaic Specialists Conference, (1985) 578.
[12] F.D. Heinz, M. Breitwieser, P. Gundel, M. König, M. Hörteis, W. Warta, M.C. Schubert, Solar Energy Materials and Solar Cells 131 (2014) 105.

NANO-ROD ANTIREFLECTION HF-DOPED IN_2O_3 FILMS AND ITS APPLICATION TO SILICON HETEROJUNCTION SOLAR CELLS

G.H.Wang[a,b], L.Zhao[a,b], H.W.Diao[a], W.J.Wang[a,b]
[a] Key Laboratory of Solar Thermal Energy and Photovoltaic System of Chinese Academy of Sciences, Institute of Electrical Engineering, the Chinese Academy of Sciences, Beijing 100190, P R China
[b] University of Chinese Academy of Sciences, Beijing 100049, P R China

ABSTRACT: Nano-rod antireflection Hf-doped In_2O_3 films (IHFO) are prepared at a low wafer temperature of 200 °C by glancing angle radio frequency magnetron sputtering for the application to silicon heterojunction solar cell. The structural, electrical and optical properties of nano-rod IHFO films are investigated. Nano-rod IHFO film is applied to a-Si/c-Si heterojunction solar cells. The significant reduction of reflection loss is observed. The circuit current of solar cell is improved. We aim to improve the efficiency of solar cell in the future.
Keywords: Antireflection Coating, Nano-rod Hf-doped In_2O_3, Glancing-angle Magnetron Sputtering, Heterojunction Silicon Solar Cell

1. INTRODUCTION

Compared to other crystalline silicon cells, the short circuit current (J_{sc}) is low for amorphous silicon/crystalline silicon heterojuncion solar cells (HIT)[1]-[2]. The anti-reflective (AR) coating plays an important role in improving the J_{sc}. Periodic or random nano-structures have emerged as promising candidates for AR coating. The fabrication techniques involve electron beam[3], thermal annealing treatment[4], or chemical vapor deposition[5]. However, the resulting surface states could also degrade the device performance, making applications of nano-structure in commercial solar cells unrealistic. In addition, the techniques are not suitable for low temperature deposition of HIT solar cell.

In this work, nano-rod Hf_2O_3-doped In_2O_3 (IHFO) films were fabricated for the applications to HIT solar cells, where it serves as antireflection on front surface of HIT solar cells. Thus optical, structural and electrical properties were simultaneously investigated.

2. EXPERIMENTS AND MEASUREMENTS

The Hf_2O_3-doped In_2O_3 (IHFO) films were deposited on quartz or polished silicon wafer by glancing-angle radio frequency magnetron sputtering from a sintered ceramic Hf_2O_3-doped In_2O_3 (1 wt.%) in an argon atmosphere. Wafers are heated to 200 °C during the sputtering process. The thickness of the films was measured by Veeco Dektak 200 surface profiler. The scanning electron microscopy (SEM) (ZEISS SIGMA) was employed to characterize SEM images of the section. The crystalline structure was monitored by X-ray diffract meter (XRD) using Cu-K_α irradiation source. The optical reflection and transmission of thin films was measured by Varian Excalibur HE 3100 UV-VIS-NIR Spectrophotometer. The photo current-density versus voltage (J-V) characteristics of the fabricated solar cells was measured at 25°C under 1-sun (AM1.5, 100mW/cm^2) solar simulator radiation.

3. RESULTS AND DISCUSSION

The nano-rod columnar structure perpendicular to the polished silicon wafer is shown in the cross sectional SEM image of Fig.1. The film is deposited uniformly on the silicon wafer.

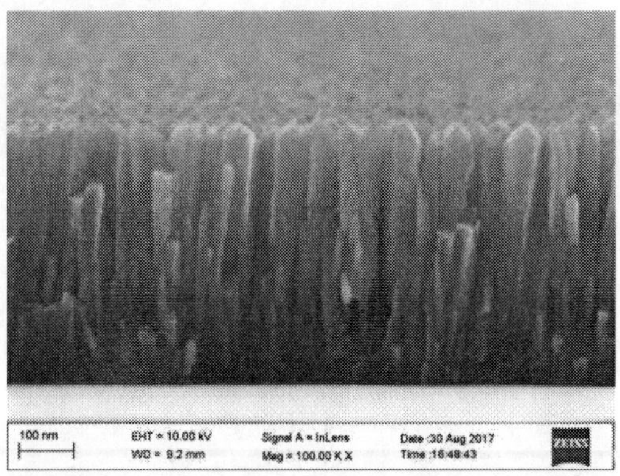

Figure 1: Cross-sectional image of the IHFO film on polished silicon wafer.

Fig.2 shows the XRD pattern of 450 nm thickness nano-rod IHFO film deposited on quartz wafer. The film shows typical polycrystalline structure with the weak diffraction peaks of (122), (222), (431), (440) and (622) together. The film is dominated by a strong peak at 2θ near 51°. The oxygen vacancies might lead to preferred orientations with (222) or (400) for conventional IHFO films deposited at different Ar flow rate[6].

Fig.3 shows the optical transmittance spectra of the 450 nm thickness IHFO film. The average transmittance in the range of 300-2000 nm is over 80%. It is higher than 90% in the range of 1100-2000 nm.

We applied the conventional and nano-rod IHFO films as the transparency electrodes and AR coating of HIT solar cell. The schematic diagram is shown in Fig.4. The thickness of conventional and nano-rod one is 70 and 10nm respectively.

Fig.5 shows reflectance spectra of conventional, conventional and nano-rod IHFO films on doped layer surface of light incidence plane for HIT solar cell. The minimum weighted mean reflectance in the range of 400-1100 nm is shown on surface with conventional and nano-rod IHFO films, which shows the significant reduction of reflection loss.

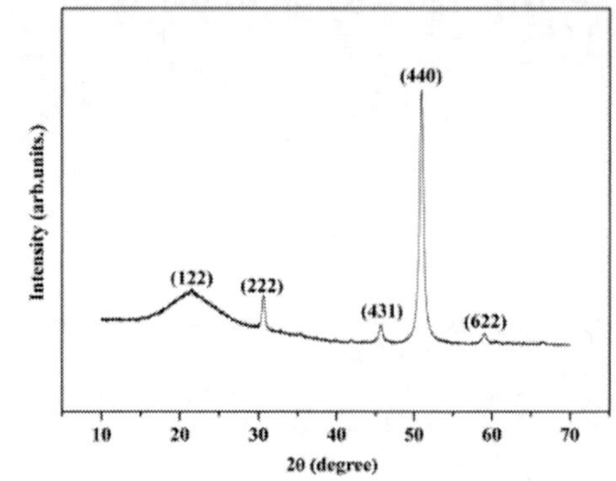

Figure 2:. XRD spectra of 450 nm thickness nano-rod IHFO film deposited on quartz wafer.

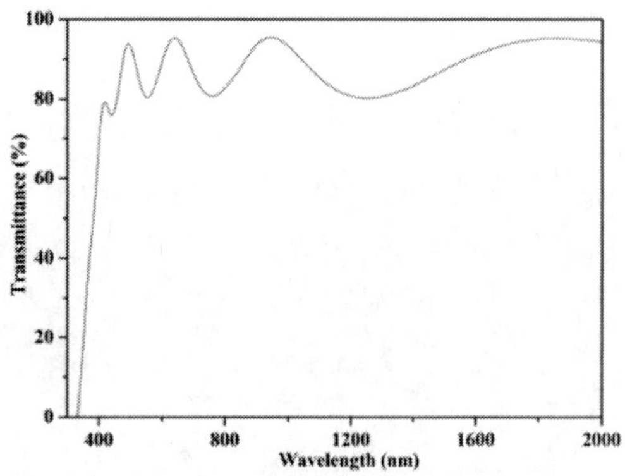

Figure 3: Transmission spectra of the 450 nm thickness nano-rod IHFO film.

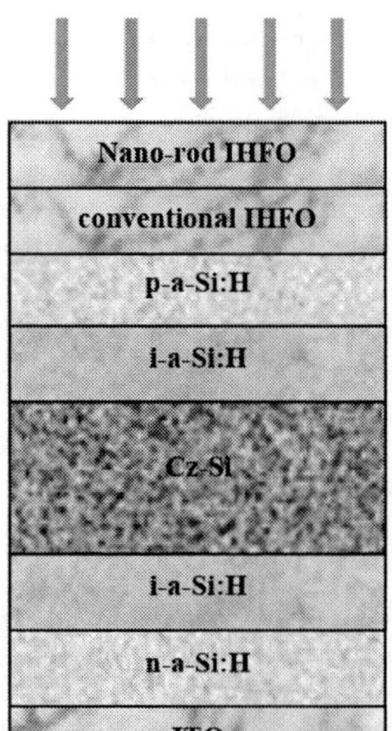

Figure 4: The schematic diagram of HIT solar cells.

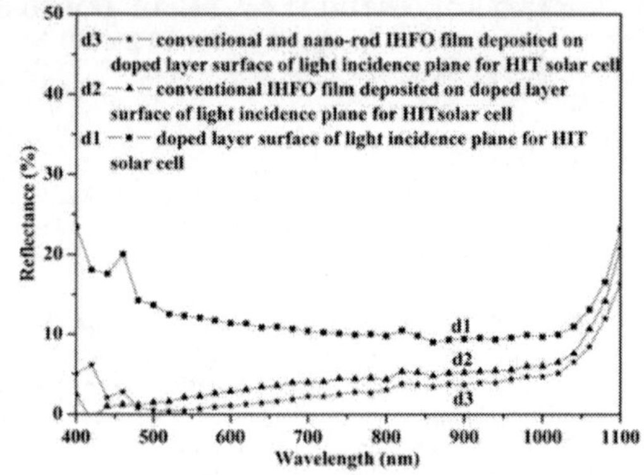

Figure 5: Reflectance spectra of conventional, conventional and nano-rod IHFO films on doped layer surface of light incidence plane for HIT solar cell.

The conventional and nano-rod IHFO films are applied as conductive and antireflection layers on front surface of HIT solar cells. Fig.6 shows current density-voltage characteristics of HIT solar cells. The J_{sc} is increased by 1.46 mA/cm^2 utilizing nano-rod IHFO film. The efficiency is improved by 0.21%.

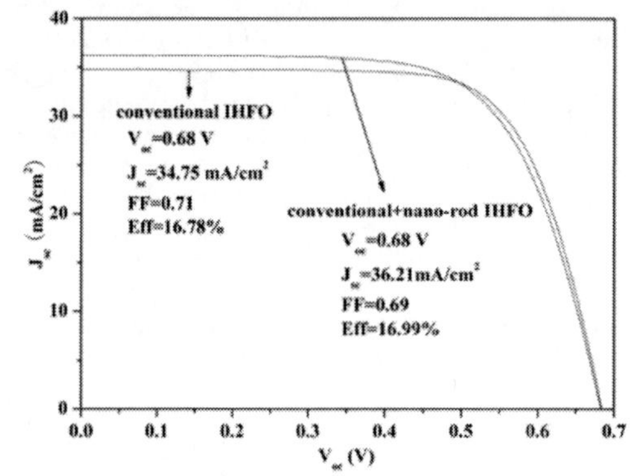

Figure 6: Current density-voltage characteristics of HIT solar cell utilizing a nano-rod IHFO layer as antireflection layer.

4. SUMMARY

In this study, the nano-rod IHFO films were fabricated by glancing-angle radio frequency magnetron sputtering. The cross sectional scanning electron microscope image shows columnar structure perpendicular to the wafer. The film shows typical polycrystalline structure by XRD. The maximum Hall mobility of 57 cm^{-2}/Vs and minimum resistivity of 4.3×10^{-4} Ω cm for double layers formed with conventional and nano-rod IHFO films is obtained. The average transmittance of 450 nm thickness IHFO film in the range of 300-2000 nm is over 80%. It is higher than 90% in the range of 1100-2000 nm. The minimum weighted mean reflectance in the range of 400-1100 nm is shown on surface with conventional and nano-rod IHFO films. The J_{sc} of HIT solar cell is increased by 1.46 mA/cm^2 utilizing nano-rod IHFO film as antireflection layer. The efficiency is improved

ACKNOWLEDGEMENTS

We are grateful for financial support from the National Key Research and Development Program of China (2017YFF0211902), National Natural Science Foundation of China (Grant No. 61504150 and 61674151), Strategic Priority Research Program of the Chinese Academy of Sciences (Grant No. XDA21060500), and Beijing Municipal Science and Technology Project (Z181100004718003).

REFERENCES

[1] M.A. Green, Prog. Photovolt: Res. Appl. 17 (2009) 183.
[2] K. Masuko, M. Shigematsu, T. Hashiguchi, D. Fujishima, M. Kai, N. Yoshimura, T. Yamaguchi, Y. Ichihashi, T. Mishima, N. Matsubara, T. Yamanishi, T. Takahama, M. Taguchi, E. Maruyama, S. Okamoto, IEEE JOURNAL OF PHOTOVOLTAICS 4(6) (2014) 1433.
[3] P. Yu, C.H. Chang, C.H. Chiu, C.S. Yang, J.C. Yu, H.C. Kuo, S.H. Hsu, Y.C. Chang, Adv. Mater. 21 (2009) 1618.
[4] D. Maestre, A. Cremades, J. Piqueras, L. Gregoratti, J. Appl. Phys. 103 (2008) 093531.
[5] N. Kumar, O. Parajuli, Feng M., J. Xu, J.-I. Hahm, Appl. Phys. Lett. 96 (2010) 053705.
[6] G.H.Wang, C.Y. Shi, L.Zhao, H.W.Diao, W.J.Wang, Appl. Surf. Sci. 399 (2017) 716.

STACKS OF A-SIO$_X$:H/A-SI:H PASSIVATION LAYER FOR LOW PARASITIC ABSORPTION AND HIGH PASSIVATION IN SILICON HETEROJUNCTION SOLAR CELLS

Kazuhiro Gotoh[1], Min Cui[1], Ryushiro Akaishi[1], Yasuyoshi Kurokawa[1] and Noritaka Usami[1]
1 Department of Materials Process Engineering, Graduate School of Engineering, Nagoya University, Japan
Furo-cho, Chikusa, Nagoya, 464-8603 Japan

ABSTRACT: Silicon heterojunction (SHJ) solar cells employing hydrogenated amorphous Si (a-Si:H) have been intensively studied owing to the extremely high conversion efficiency, since remarkable suppression of carrier recombination is achieved. For further improvement of SHJ solar cells, parasitic absorption should be minimized. A promising candidate is hydrogenated amorphous silicon oxide (a-SiO$_x$:H), which shows higher bandgap in comparison with that of a-Si:H. Although low parasitic absorption is realized by employing a-SiO$_x$:H, passivation performance is sacrificed in some case. Recently, stacking of a-SiO$_x$:H and a-Si:H have been proposed to achieve low parasitic absorption and surface recombination velocity. In this paper, we studied on optical and structural properties of a-SiO$_x$:H single layer and implied V_{OC} of a-SiO$_x$:H/a-Si:H stacking layer on c-Si. With increasing gas flow ratio f_{CO2} = [CO_2]/([SiH_4] + [CO_2]), optical bandgap monotonically increased. From Fourier transform infrared (FT-IR) spectroscopy, Si-O-Si stretching mode at around 980 cm^{-1} was observed for a-SiO$_x$:H, indicating oxygen atoms are successfully incorporated. in a-Si:H network. The implied V_{OC} decreased with increase in f_{CO2} regardless of a-SiO$_x$:H/a-Si:H stacking and a-SiO$_x$:H single passivation layer. The higher implied V_{OC} of a-SiO$_x$:H/a-Si:H stacking layer are obtained in low f_{CO2} and thicker a-Si:H layer of the stacks possibly due to low defect density at the a-Si:H/c-Si interface in a-SiO$_x$:H/a-Si:H stacking layer.
Keywords: a-Si:H, c-Si, Heterojunction, Passivation

1 INTRODUCTION

Silicon heterojunction (SHJ) solar cells are the leading solar cells showing high conversion efficiency [1-3]. The high conversion efficiency is attributed to extremely high open circuit voltage (V_{OC}) thanks to excellent passivation by hydrogenated amorphous silicon (a-Si:H). In standard SHJ soalr cells, the high V_{OC} is obtained by the sacrifaice of short-circuit current density (J_{SC}). Generally, photogenerated carriers in a-Si:H–paricularly doped a-Si:H–cannot contribute to current, since the lifetime of (doped) a-Si:H is very short [4,5]. Therefore, absorption in a-Si:H layer leads to current loss of SHJ solar cells, which is so called parasitic absorption. Although higher bandgap materials such as a-SiO$_x$:H have been studied to reduce parasitic absorption of standard heterojunction solar cells [6-9], Deligiannis *et al.* reported that the passivation performance of a-SiO$_x$:H is poor due to increase in the defect density at the a-SiO$_x$:H/c-Si interface and/or the bulk a-SiO$_x$:H layer compared with standard a-Si:H [10,11].

Recently, comparable passivation performance to standard a-Si:H is achieved by stacking a-SiO$_x$:H and a-Si:H layer [12,13]. In addition to the improved passivation performance, the stacks of a-SiO$_x$:H/a-Si:H shows higher fill factor (FF) as a by-product in comparison with a-SiO$_x$:H single layer. Larger valence band offset is generally created at the a-SiO$_x$:H/c-Si interface and it impedes hole transport [14]. With decreasing layer thickness of a-SiO$_x$:H, defect-assisted tunneling in a-SiO$_x$:H would be enhanced and thus transportation of holes is facilitated. Thus, the a-SiO$_x$:H/a-Si:H stacking passivation layer can enhance absorptance in bulk c-Si, passivation and efficient carrier extraction. In ref. [13], the layer thickness of underling a-Si:H was fixed to 2 nm. The ratio of layer thickness in the stacking structure would be crucial for understanding passivation by a-SiO$_x$:H/a-Si:H stacking layer and further improvement of device performance.

The passivation scheme of a-SiO$_x$:H/a-Si:H stacking layer is not fully understand. The ratio of layer thickness in the stacking structure would be important to clarify the impact of the stacking structure on the passivation performance. In this contribution, we investigated the effect of the ratio of layer thickness and gas flow rate of carbon dioxide on the passivation performance.

2 EXPERIMENTS

Double-side mirror-polished, floating-zone (Fz) grown n-type c-Si substrates were used. The resistivity and wafer thickness are about 2-3 Ω·cm and 280 μm, respectively. After cleaning the substrates using Semicoclean-23 (Furuuchi Chemical Co.) in ultra-sonic bath for 6 minutes, oxides on c-Si were removed by 5% HF. Then, the Si substrates were immersed into ozonized, deionized water (DI-O_3) at room temperature for 10 minutes to form protective ultra-thin oxide. Prior to depositing a-Si:H and a-SiO$_x$:H, the substrates were dipped into 2.5% HF. Then, substrates were immersed into 1.5% H_2O_2 for preventing a-Si:H from epitaxial growth. Intrinsic-(i-)type a-Si:H and a-SiO$_x$:H were deposited by a plasma enhanced chemical vapor deposition (PECVD) system. The growth conditions are summarized in Table I. The stacks of a-SiO$_x$:H/a-Si:H is prepared by systematic change of layer thickness of a-SiO$_x$:H and a-Si:H. The total thickness was fixed to 7 nm. No post-annealing was carried out.

The implied-V_{OC} (i-V_{OC}) was characterized by a Sinton WCT-120TS lifetime tester using quasi steady-state photoconductance (QSSPC) method. All lifetime measurements was carried out at room temperature. Furthermore, the optical bandgap was measured for a-SiO$_x$:H single layer on quartz substrate by a spectroscopic ellipsometry (SE). The Tauc-Lorentz model was used as a oprical model. Additionally, Fourier transform infrared (FT-IR) spectroscopy in transmittance mode was employed to obtain structural information of the a-SiO$_x$:H single layer. For SE and FT-IR measurements, about 200-nm-thick a-SiO$_x$:H film was deposited on quartz and high resistivity c-Si of over 100 Ω·cm, respectively.

Table I: Film deposition conditions

Parameter	a-Si:H (i)	a-SiO_x:H (i)
Growth temperature [°C]	180	180
[SiH_4] [sccm]	40	5, 10, 20
[CO_2]/([SiH_4] + [CO_2])	0.00	0.33-0.91
Pressure [Pa]	25	25
Power density [mW/cm^2]	32.5	32.5
Layer thickness [nm]	0.0/2.5/3.5/ 4.5/7.0	7.0/4.5/3.5/ 2.5/0.0

3 RESULTS AND DISCUSSION

Figure 1 shows E_{04} optical bandgap of a-SiO_x:H single layer on quartz substrates as a function of gas flow ratio f_{CO2} = [CO_2]/([SiH_4] + [CO_2]). The E_{04} optical bandgap is determined from a SE. The optical bandgap monotonically increased with increasing f_{CO2}. Figure 2 shows FT-IR spectra of the a-SiO_x:H deposited at different f_{CO2}. The transmittance at 640, 980, 2000, 2100 cm^{-1} represents Si-H wagging mode, Si-O-Si stretching mode, and Si-H low and high stretching modes, respectively. The Si-O-Si stretching mode were observed in region of f_{CO2} from 0.33 to 0.67, which indicates oxygen is successfully incorporated into a-Si:H network. The band at 2100 cm^{-1} became pronounced with increasing f_{CO2}. The high stretching mode is attributed to dihydrides and/or monohydrides at the surface of nanosized voids [10], which indicates more defective a-SiO_x:H layer is formed using higher f_{CO2}. On the other hand, it is known that Si-H stretching mode is shifted to higher wavenumbers due to backbonded O atoms [15]. From SE measurements, the optical bandgap increased with increase in f_{CO2}, suggesting the oxygen content in a-SiO_x:H monotonically increased with f_{CO2}. Therefore, the higher stretching mode could be caused by the shift of Si-H stretching mode and nanosized voids in a-SiO_x:H layer.

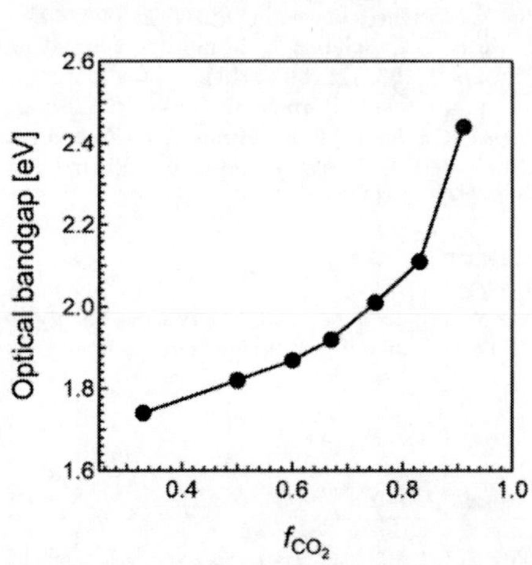

Figure 1: Optical bandgap of a-SiO_x:H single layer as a function of f_{CO2}.

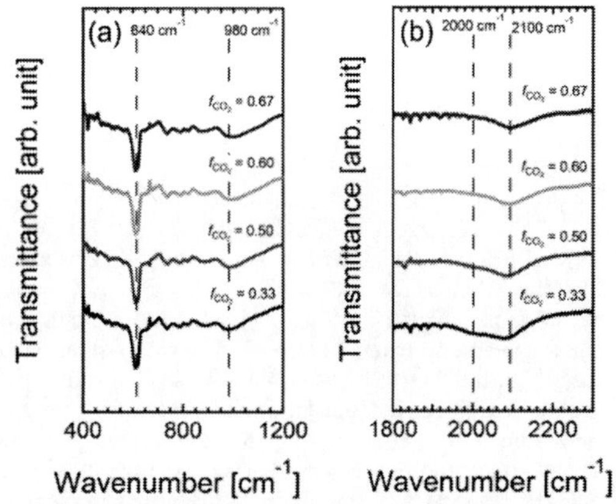

Figure 2: FT-IR spectra of a-SiO_x:H single layer grown at different f_{CO2}.

Figure 3 shows i-V_{OC} of a-SiO_x:H single and a-SiO_x:H/a-Si:H stacking layer as a function of f_{CO2}. The solid lines in Fig. 3 are visual guides. The i-V_{OC} depends on the photogenerated carrier density Lower i-V_{OC} was observed for a-SiO_x:H single layer, suggesting high defect density at the a-SiO_x:H/c-Si interface and in a-SiO_x:H layer. The i-V_{OC} was improved by a-SiO_x:H/a-Si:H stacking layer possibly due to reduced interface defects. With increasing f_{CO2}, the i-V_{OC} decreased. Inaba et al., reported that valence band offset affects to effective carrier lifetime of i-a-SiO_x:H passivated n-type c-Si and larger band offset results in longer effective lifetime [16]. The larger band offset is created by higher f_{CO2}, however the i-V_{OC} in Fig. 3 decreased with increasing f_{CO2}. This indicates that contribution of valence band offset to the i-V_{OC} is negligible in the stacks of a-SiO_x:H/a-Si:H. Furthermore, the i-V_{OC} increased with increasing layer thickness of a-Si:H ($t_{a\text{-}Si:H}$), i.e. decreasing layer thickness of a-SiO_x:H ($t_{a\text{-}SiOx:H}$). Beyer *et al.* addressed that increased oxygen content in a-SiO_x:H

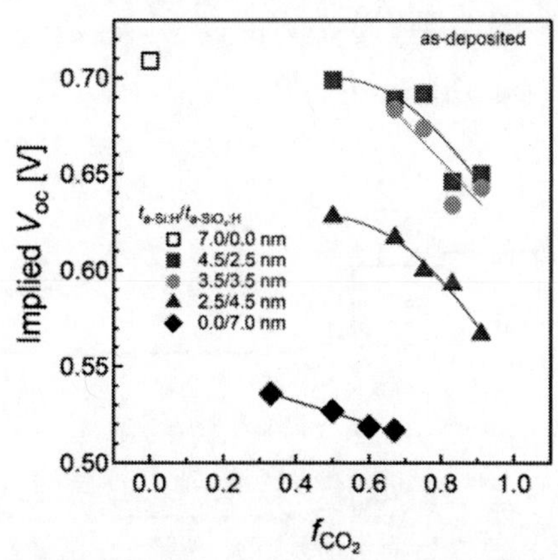

Figure 3: Implied V_{OC} of a-SiO_x:H single and a-SiO_x:H/a-Si:H stacking passivation layer as a function of f_{CO2}. The solid lines are visual guides.

leads to increase in interconnected voids [17]. From FT-IR measurements, the peak at 2100 cm^{-1} become stronger with increasing f_{CO2}, which implies the increase in nanovoids, i.e. defective a-SiO_x:H layer. Photogenerated carriers in c-Si would recombine at the voids or the a-SiO_x:H surface by enhanced defect-assisted tunneling through the increased defects in a-SiO_x:H. Therefore, reduction of i-V_{OC} with increase in f_{CO2} and $t_{a\text{-}SiOx:H}$ is attributed to defects in a-SiO_x:H layer.

From these results, schematic band lineup of the a-SiO_x:H/a-Si:H stacking layer are depicted in Fig. 4. In case of lower f_{CO2} and thicker a-Si:H layer, defects at the stacks/c-Si interface and in the stacks are few, which results in higher i-V_{OC} since concentration of photogenerated carriers become high. On the other hand, defects at the interface and in the stacks are too many to maintain high carrier concentration in case of higher f_{CO2} and thinner a-Si:H layer, which leads to lower i-V_{OC}. Hence, realization of nanovoids-lean, i.e. high quality, a-SiO_x:H layer is key issue for further improvement in passivation performance of a-SiO_x:H/a-Si:H stacking layer. H_2 dilution during deposition [18] and hydrogen plasma treatment after deposition are promising to improve the quality of a-SiO_x:H film [19,20].

4 SUMMARY

we investigated the effect of the ratio of layer thickness and gas flow rate of CO_2 on the passivation performance of a-SiO_x:H/a-Si:H stacking layer. Optical bandgap of a-SiO_x:H layer increased with increaseing f_{CO2}. Furtermore, Si-O-Si stretching mode were observed in region of f_{CO2} from 0.33 to 0.67. These results indicate oxygen is successfully incorporated into a-Si:H network. From lifetime measurements, low i-V_{OC} of about 0.5 V is obtained by single a-SiO_x:H passivating layer. On the other hand, a-SiO_x:H/a-Si:H stacking layer shows higher i-V_{OC} in comparison with single a-SiO_x:H passivating layer. The FT-IR spectra shows the peak at 2100 cm^{-1} become stronger with increasing f_{CO2}, which implies the increase in nanovoids, i.e. defects in a-SiO_x:H layer. In addition, higher i-V_{OC} was obtained by thicker a-Si:H layer. Therefore, the passvation performance is dependent on defects in a-SiO_x:H layer the f_{CO2}. Development of nanovoids-lean a-SiO_x:H layer is key for further improvement in passivation performance of a-SiO_x:H/a-Si:H stacking layer.

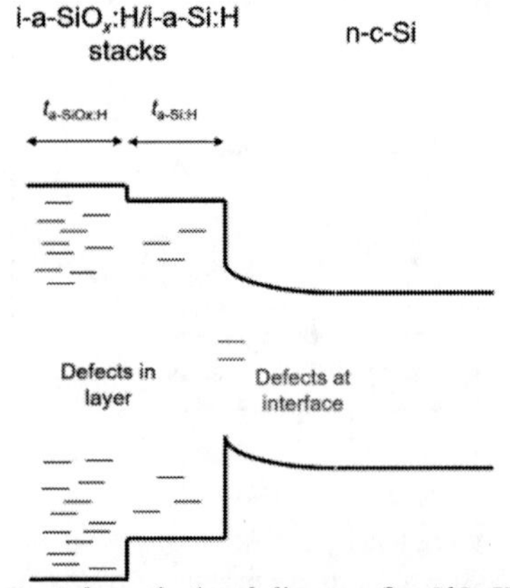

Figure 4: Schematic band lineup of a-SiO_x:H/a-Si:H passivated c-Si.

REFERENCES

[1] K. Yoshikawa, W. Yoshida, T. Irie, H. Kawasaki, K. Konishi, H. Ishibashi, T. Asatani, D. Adachi, M. Kanematsu, H. Uzu, and K. Yamamoto, Sol. Energy Mater. Sol. Cells **173** (2017) 37-42.
[2] D. Adachi, J. L. Hernández, and K. Yamamoto, Appl. Phys. Lett. **107**, 233506 (2017).
[3] M. Taguchi, A. Yano, S. Tohoda, K. Matsuyama, Y. Nakamura, T. Nishiwaki, K. Fujita, and E. Maruyama, IEEE J. Photovolt. **4** (2014) 96-99.
[4] S. De Wolf, A. Descoeudres, Z. C. Holman and C. Ballif, Green **2** (2012) 7-24.
[5] Z. C. Holman, A. Descoeudres, L. Barraud, F. Z. Fernandez, J. P. Seif, and S. De Wolf, and C. Ballif, IEEE J. Photovolt. **2** (2012) 7-15.
[6] B. Zhang, Y. Zhang, R. Cong, Y. Li, W. Yu, G. Fu, Sol. Energy **155** (2017) 670-678.
[7] K. Nakada S. Miyajima, M. Konagai, Jpn. Appl. Phys. **54**, 082301 (2015).
[8] K. Ding, U. Aeberhard, F. Finger, and U. Rau, J. Appl. Phys. **113**, 134501(2013).
[9] H. Fujiwara, T. Kaneko, and M. Kondo, Appl. Phys. Lett. **91**, 133508 (2007).
[10] D. Deligiannis J. van Vliet, R. Vasudevan, R. A. C. M. M. van Swaaij, and M. Zeman, J. Appl. Phys. **121**, 085306 (2017).
[11] M. Liebhaber, M. Mews, T. F. Schulze, L. Korte, B. Rech, and K. Lips, Appl. Phys. Lett. **106**, 031601 (2015).
[12] T. Krajangsang, S. Inthisang, J. Sritharathikhun, A. Hongsingthong, A. Limmanee, S. Kittisontirak, P. Chinnavornrungsee, R. Phatthanakun, K. Sriprapha, Thin Solid Films, **628** (2017) 107-111.
[13] J. P. Seif, A. Descoeudres, M. Filipič, F. Smole, M. Topič, Z. C. Holman, S. De Wolf, C. Ballif, J. Appl. Phys. **115**, 024502 (2014).
[14] M. Mews, M. Liebhaber, B. Rech, and L. Korte, Appl. Phys. Lett. **107**, 012902 (2015).
[15] K. Nakada S. Miyajima, M. Konagai, Jpn. Appl. Phys. **53**, 04ER13 (2014).
[16] M. Inaba, S. Todoroki, K. Nakada, and S. Miyajima, Jpn. J. Appl. Phys. **55**, 04ES04 (2016).
[17] W. Beyer, J. Non-Cryst. Solids **266-269**, 845 (2000).
[18] Y. Yamaguchi, T. Watahiki, and M. Konagai, 26th EUPVSEC conference proceeding (2011) 2138-2140.
[19] M. Mews, T. F. Schulze, N. Mingirulli, and L. Korte, Appl. Phys. Lett. **102**, 122206 (2013).
[20] A. Descoeudres, L. Barraud, S. De Wolf, B. Strahm, D. Lachenal, C. Guérin, Z. C. Holman, F. Zicarelli, B. Demaurex, J. Seif, J. Holovsky, and C. Ballif, Appl. Phys. Lett. **99**, 123506 (2011).

Acknoledgment

This work was financially supported by New Energy and industrial Technology Development Organization (NEDO).

TRANSPARENT MoO_X AND SiO_X WINDOW LAYERS FOR HETEROJUNCTION SILICON SOLAR CELLS

F. Menchini[1*], L. Serenelli[1,2], L. Martini[1,2], M. Izzi[1], G. Stracci[1], P. Mangiapane[1], E. Salza[1], S. Lombardo[3], R.C. Milazzo[3] and M. Tucci[1]
[1] ENEA Casaccia Research Center, Via Anguillarese 301, 00123 Rome, Italy
[2] DIET University of Rome "Sapienza", Via Eudossiana 18, 00184 Rome, Italy
[3]IMM-CNR, Strada VIII n. 5, Zona Industriale, 95121 Catania (CT), Italy

ABSTRACT: Emitter and buffer layers more transparent than amorphous Silicon is one of the improvement needed in amorphous/crystalline silicon heterojunction technology to reach the limit efficiency of silicon based solar cell. Recent studies have demonstrated the suitability of sub-stoichiometric Molybdenum Oxide as hole extracting layer in heterojunction solar cells based on n-type crystalline Silicon. However, up to now it has mainly been used in combination with amorphous Silicon, so limiting its potential as transparent emitter. This work investigates on a hole-selective contact based on the combination of a passivating hydrogenated amorphous Silicon Oxide and Molybdenum Oxide which, with their high band gaps (respectively 1.9 and 3.5 eV), should lead to best cell performances if compared to heterojunction devices based only on amorphous Silicon. Solar cells have been produced and optimized in terms of layers thicknesses. The widely reported problem of s-shape in the light current-voltage characteristics, due to undesired barrier formation against photo-carrier collection, is addressed and solved. The temperature response of the devices is also studied, and the Molybdenum Oxide film stability under thermal treatment up to 130°C is demonstrated when addressing the appropriate thermal treatments sequence.
Keywords: heterojunction; passivation; transparent oxides.

1 INTRODUCTION

One of the main issues in the fabrication of efficient solar cells is charge separation at the electrodes, which can be accomplished in two ways. The first one is the formation of charge-selective contacts, which establish a potential gradient and lead to holes/electrons collection while simultaneously blocking the complementary carriers. The second one is the surface passivation of crystalline silicon (c-Si), in order to minimize carrier recombination during charge transmission towards the outer electrodes. In amorphous/crystalline silicon heterojunction (HJ) solar cells both these effects are attained by doped and intrinsic hydrogenated amorphous silicon (a-Si:H) layers. However, the use of a-Si:H films is one of the major limits to high photocurrent generation due to the a-Si:H bandgap restricted to 1.7 eV [1], which introduces parasitic absorptions. Therefore the idea of replacing a-Si:H films by wider bandgap passivating layers is becoming of relevant interest [2]. To this aim this work investigates hole-selective contacts based on the combination of passivating hydrogenated amorphous Silicon Oxide (a-SiO_x:H) and sub-stoichiometric Molybdenum Oxide (MoO_x, $2<x<3$) thin films.

MoO_x and other transition metal oxides have been studied for many years due to their potential applications as emitters (or, more appropriately, hole-selective contacts [3]) for silicon based [2,4-6] solar cells. Thermal evaporation of MoO_3 powders leads to films of a slightly sub-stoichiometric material (MoO_x), in which the reduced Mo oxidation state produces an oxygen vacancy defect band slightly below the conduction band [7]. The material is characterized by high transparency and wide bandgap (3-3.8 eV), typical resistivity around 10^6 Ω·cm and work function between 4.5 and 6.7 eV [5,8,9]. MoO_x has already been successfully integrated in silicon HJ cells achieving high efficiency [2,10,11]. Although non-passivated MoO_x has shown encouraging performances in solar cells [10], passivation of the c-Si surface still remains the key factor for the best cell performances [4].

On the other hand, a-SiO_x:H produced by RF-PECVD (Radio Frequency-Plasma Enhanced Chemical Vapor Deposition) can be considered appealing in mass production of HJ solar cells because it has already demonstrated its good passivation capabilities [12-14] and has been successfully integrated in Si-based solar cells [1,13,15,16] as buffer or emitter layer. With respect to a-Si:H films, which are typically characterized by a band gap energy (Eg) of 1.7 eV [17], a-SiO_x:H has the advantage of improved thermal stability [14,18] and higher transparency due to its Eg larger than 1.8-1.9 eV [1,12,19].

The aim of this study is to test the feasibility of the a-SiO_x:H/MoO_x stacked window for hole extraction in HJ cells.

2 EXPERIMENTAL

Flat, n-type doped c-Si wafers with <100> orientation, 1-5 Ω resistivity and 250 μm thickness were used as substrates for the deposition of the samples. The substrates were cleaned according to standard SC1 and SC2 procedures [20], then dipped in 1% Hydrofluoric Acid (HF) for 60 seconds before the subsequent deposition steps. Thin layers of a-SiO_x:H were deposited for surface passivation on both sides of several n-type c-Si wafer by RF-PECVD in a parallel plate reactor. The deposition temperature was 250°C, while the RF power density was 18 mW/cm^2. The reactive gas mixture flow rates were 120 sccm of silane (SiH_4) diluted in Ar at 5%, 1.5 sccm of carbon dioxide (CO_2) and 200 sccm of hydrogen (H_2), at a total pressure of 2000 mTorr. The ratio between the CO_2 and (CO_2+SiH_4) flow rates was $f_O = \frac{[CO_2]}{[CO_2]+[SiH_4]} = 20\%$. MoO_x films with various thicknesses were grown by thermal evaporation from MoO_3 pellets at a 10^{-7} mbar base pressure. The film thickness was controlled by a calibrated quartz balance. A Transparent Conductive Oxide (TCO) layer of Indium Tin Oxide (ITO) for efficient lateral contact formation was sputtered over the MoO_x film at different substrate temperatures. The top contact was completed by a Silver grid with 1 mm-spaced 200 μm wide and 1.5 μm thick fingers.

Thermal treatments were carried out on single films and complete cell structures in a programmable ATV static furnace, in a nitrogen flux.

Reflectance (R) and Transmittance (T) measurements

were performed in a Lambda 950 Perkin Elmer UV-VIS-NIR Spectrophotometer equipped with a 15 cm integrating sphere. Lifetime measurements were carried out with a flash lamp in a WCT 120 Sinton Quasi-Steady State Photoconductance Decay measuring apparatus at a minority carrier density injection level in the order of 10^{15} cm^{-3}. The valence band region of MoO_x was investigated by Ultra-Violet Photoemission Spectroscopy (UPS) to determine the work function via excitation by the 21.2 eV radiation of the HeI line. Current-Voltage (J-V) characteristics of the cells were measured under dark and light standard conditions (1-sun, 25°C, AM 1.5G) in a calibrated Class A solar simulator. Spectral Internal Quantum Efficiency (IQE) was evaluated by measures of External Quantum Efficiency (EQE) and Reflectance.

Cross sectional TEM samples were prepared in the (110) orientation and analyzed with High Resolution TEM using a JEOL 2010f HR TEM operating at 200 keV.

3 RESULTS AND DISCUSSION

3.1 a-SiO_x:H layer

The hydrogenated amorphous silicon suboxide layer used in this work was extensively studied and characterized [19]. Average optical constants were evaluated for 12 nm thick a-SiO_x:H layers, finding for n and k at 630 nm values of 3.6 and 0.004 respectively, and Eg=1.9 eV, revealing the transparency of the material. The average effective lifetime of these samples was 2.6 ms, with an Implied V_{OC} of 736 mV. After thermal annealing at 300°C for 10 minutes, the optical parameters and V_{OC} remained substantially unchanged, while the lifetime increased to 3.6 ms. The oxygen inclusion in a-SiO_x:H films was evident from FTIR-ATR measurements in the 1850-2300 cm^{-1} range, not shown here, where the peaks related to the stretching of the Si-H bond are located.

A particular characteristic common to other amorphous thin films is the c-Si surface passivation achievable as a function of a-SiO_x:H thickness [21]: the passivation capabilities are best fulfilled when thicker films are deposited, reaching effective lifetimes well above the millisecond and Implied V_{OC} values exceeding 700 mV. Below 8 nm both values of lifetime and Implied V_{OC} rapidly decrease; a thickness of 5 nm constitutes the lower limit to obtain an adequate passivation of the c-Si surface with an effective lifetime of 200 μs, corresponding to 490 μm of diffusion length in n-type silicon. It is worth to remark, however, that when a 5 nm thick a-SiO_x:H layer was introduced in a complete HJ cell, the effective lifetime of the structure reached 1.7 ms after a thermal annealing at 300°C for 10 minutes.

3.2 MoO_x layer

Before implementing MoO_x films as hole-collecting layers in HJ solar cells, the electro-optical characteristics of samples grown on glass were measured.

The linear extrapolation of the UPS measurements, reported in Fig. 1, provided a work function value of 5.5 eV for the as-deposited sample, while after thermal annealing at 180°C the value dropped to 5.2 eV.

Optical R and T of films with different thicknesses are shown in Fig. 2. The optical bandgap Eg was deduced by the approximated Tauc Plot for indirect transitions (inset in Fig. 2), and it was around 3.8 eV, which ensured a very low absorption of the high-energy solar radiation.

MoO_x film resistivity measured on thick layers decreased from 5×10^5 Ω·cm to 10^2 Ω·cm after thermal annealing. The measured values were in accordance with literature and confirmed the sub-stoichiometric structure and the semiconductive behavior of the MoO_x layers even after thermal annealing [8].

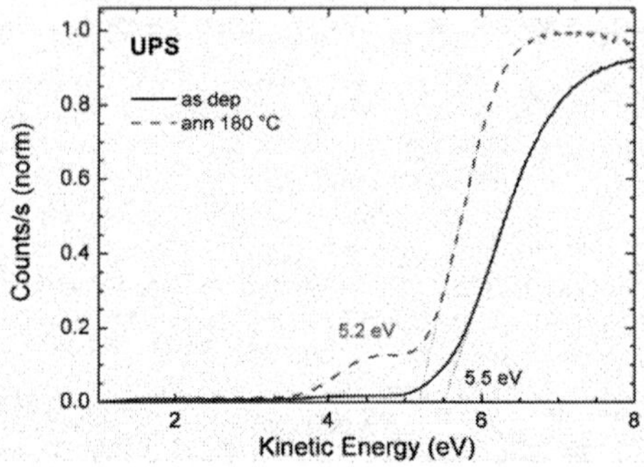

Figure 1: UPS measurements on a MoO_x film before and after annealing at 180°C

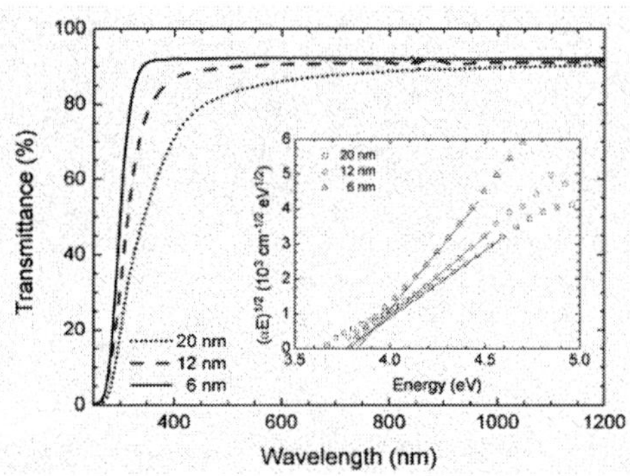

Figure 2: Transmittance of MoO_x films with different thickness. The inset shows the Tauc plot and the optical bandgap of the samples

3.3 Optimization of the a-SiO_x:H/MoO_x bilayer

In the initial proof-of-concept cell, the back side contact was not optimized. To obtain the base contact, the back a-SiO_x:H passivation layer was partially removed by a mechanical scratch to allow an ohmic contact of the n-type c-Si substrate with an Indium Gallium (InGa) alloy [22].

As a starting point, a 6 nm thick front buffer a-SiO_x:H layer was adopted and different thicknesses of MoO_x were tested in the range between 6 nm and 20 nm, as suggested by literature [4,5,10,23]. An ITO film was deposited over the MoO_x layer to enhance the lateral collection of the emitter layer. The initial passivation level, associated to a 600 μs effective lifetime, was partially reduced after the MoO_x layer deposition and subsequent ITO sputtering process at 100°C [24]. The effective lifetime value of 360 μs, measured on the cell structure without metal contacts, ensured an Implied V_{OC} of 675 mV. Nevertheless it is well known that further thermal processes could recover the initial passivation level [24] or even improve it [25].

The light J-V characteristics of all devices (shown in Fig. 3) proved efficient current generation capabilities, although the cells fill factors (FF) were affected by an unwanted s-shape clearly influenced by the thickness of the MoO_x layer, as already found in literature [4]. The presence of the s-shape is the evidence of a hole-collection issue, which limits the FF of the devices below 70% [4]. It has been suggested that possible causes for this effect can be found in an interlayer between MoO_x and TCO layers [11] or in the high reactivity of the MoO_x films to the environment that changes its oxidation state and lowers its work function. Thus the key to obtain better performances is to avoid MoO_x exposure to ambient atmosphere. To exclude any contribution coming from the MoO_x/ITO interface we produced a p-c-Si/MoO_x/ITO structure (12 nm thick MoO_x) avoiding prolonged MoO_x ambient exposure. Then we tested the current-voltage behavior of this structure obtaining a linear J-V characteristics, that confirmed both the behavior of MoO_x layer as hole extractor and the ohmic nature of the MoO_x/ITO contact. Therefore the s-shape of Fig. 3 did not depend on the MoO_x/ITO interface.

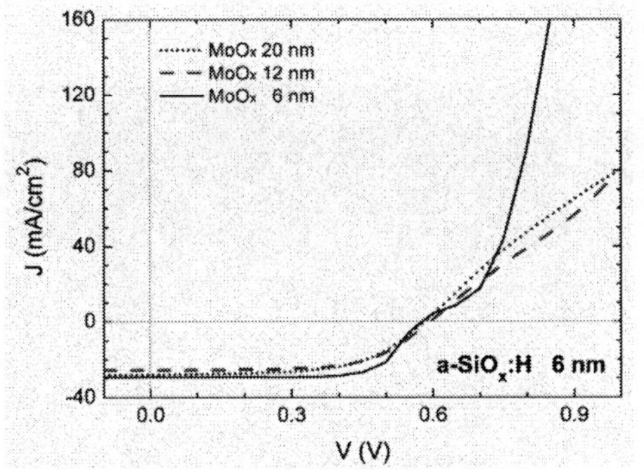

Figure 3: Light J-V characteristics of test cells with 6 nm a-SiO_x:H buffer layer and variable MoO_x layer thickness

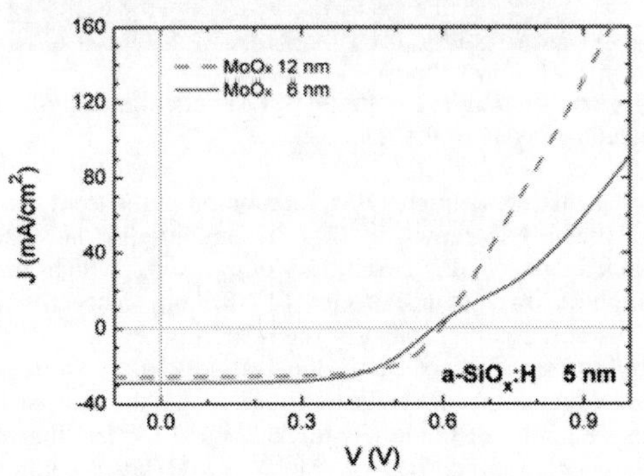

Figure 4: Best light J-V characteristics of test cells with a 5 nm a-SiO_x:H buffer layer and variable MoO_x layer thickness

One more reason to explain the s-shape could be the excessive thickness of the buffer a-SiO_x:H layer, as suggested by literature [26,27]. Thus we reduced the buffer thickness to 5 nm. New heterostructures were produced with 5 nm thick a-SiO_x:H buffer layers and both 6 nm and 12 nm thick MoO_x layers, respectively. From Fig. 4 it can be seen that the combination between 5 nm of a-SiO_x:H and 12 nm of MoO_x gave rise to J-V curves without any s-shape, while the cell with the 6 nm thick MoO_x layer is again characterized by an s-shape.

Despite the s-shape disappearance, the FF still resulted oddly low (66%), suggesting the presence of some other detrimental effects on the cells, besides the series resistance. To evaluate this subject, we investigated the a-SiO_x:H/MoO_x interface, as discussed in the following sections, also with the aid of band structure simulation.

3.4 Band structure of the cell

To have a detailed view of the c-Si/a-SiO_x:H/MoO_x/ITO junction and how it works, the complete heterostructure device was described with the help of a numerical finite difference model [28,29].

In Fig. 5a the band diagram of the entire device obtained from simulation is reported at the equilibrium condition in the dark. Looking at the front side of the cell (left side in the picture), in which the most relevant part of the heterostructure under investigation lays, the effect of the MoO_x layer on the whole interface has to be carefully considered. Even though the MoO_x layer can in principle be considered as an n-type semiconductor, its electron affinity, higher than the surrounding layers, strongly bends the c-Si/a-SiO_x:H interface. Since at the equilibrium the Fermi level must be kept constant everywhere along the band diagram, the n-type c-Si depletes and it is even inverted at the interface, thus forming the junction. For the n-type c-Si this behavior is quite similar to what happens in a typical p-a-Si:H/a-Si:H/n-c-Si heterojunction device, but in this particular case the band bending within the c-Si is induced by the n-type MoO_x layer instead of the p-type a-Si:H layer.

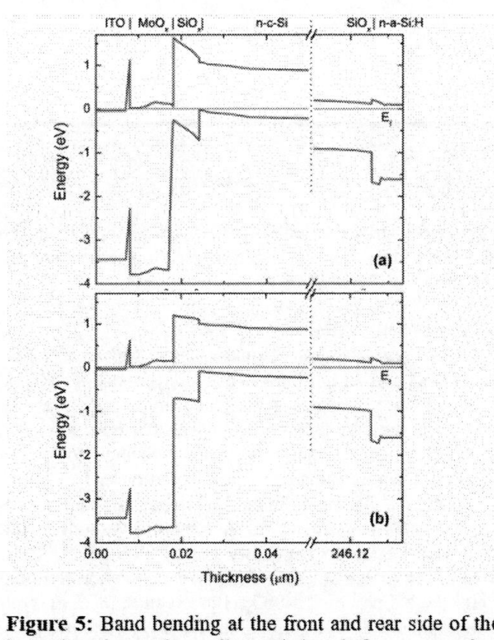

Figure 5: Band bending at the front and rear side of the heterojunction solar cell as deduced from numerical simulations. The value of MoO_x work function adopted in (a) is reduced by 0.5 eV in (b)

Thanks to the high electron affinity and work function of the MoO_x layer, there should be no need to be concerned about the depletion of this layer at the equilibrium that, instead, in certain conditions affects the p-type a-Si:H emitter layer of the heterojunction. The relevant concern is mainly on the MoO_x work function.

Since it is lowered after thermal annealing over 180°C, as suggested by the experimental results, the band bending induced within the c-Si would be less pronounced, thus producing a lower built in voltage and some difficulties to collect carriers over the valence band mismatch at the c-Si/a-SiOx:H interface. This effect is illustrated in Fig. 5b, where the work function and the electron affinity of the MoO_x layer were reduced by 0.5 eV.

Experimentally the effect of work function reduction produces a fill factor drop, thus remarking the difficulties of hole collection. Also the c-Si/a-SiO_x:H interface plays a role in this equilibrium. Indeed, the defect density (D_{it}) at the interface should be much reduced, as assumed in the simulation using a $D_{it}= 10^{11}$ cm^2, to avoid the pinning of the Fermi level toward mid-gap position, thus affecting again the carrier collection. Indeed the a-SiO_x:H layer has been already demonstrated suitable for effective c-Si surface passivation like a-Si:H film; moreover the chemical compatibility of this sub-oxide layer with any oxide growth over it favors the stability of the MoO_x layer during its growth. Since the simulations demonstrate that the investigated heterostructure effectively works as efficient solar cell, in the following paragraphs we show how to experimentally achieve similar results.

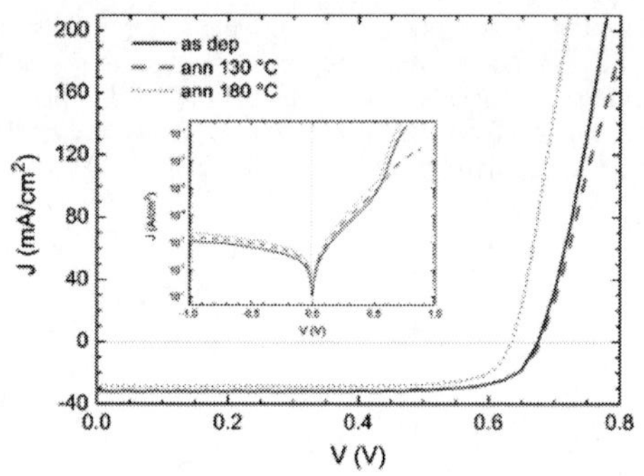

Figure 6: Dark (inset) and light J-V characteristics of the final heterojunction solar cell before and after thermal annealing at 130 and 180°C

3.5 Solar Cell performances

The optimized front structure consisting in a 5 nm-thick a-SiO_x:H/12 nm-thick MoO_x bilayer was introduced in a complete solar cell, paying additional care to minor aspects in the fabrication process. This included attention to environmental conditions to preserve the MoO_x free surfaces from unwanted parasitic oxidation and moisture adsorption, and a tuning in the ITO deposition parameters to avoid thermal stresses to the structure without losing transparency and conductivity of the ITO layer itself. According to the findings described in Section 3.3, the cells had a 5 nm thick a-SiO_x:H passivation layer on both sides of the c-Si wafer. In agreement with literature [30], a-SiO_x:H layers could achieve a moderate surface oxidation when exposed to the air. For this reason, as a precaution, we treated the passivated wafer with a fast dipping in HF 1% solution before the subsequent depositions. The back side of the cell was completed by a 10 nm thick n-type amorphous silicon film grown, as reported in [25], to selectively extract electrons. The optimized 12 nm thick MoO_x layer was evaporated on the front side of the cell, which was kept in nitrogen atmosphere before the deposition of an 80 nm thick ITO layer. This thickness for ITO was chosen for better front side reflectance reduction. Also the back contact was covered by an 80 nm thick ITO layer. At this stage the effective lifetime value of the cell was of 560 μs and the Implied V_{OC} had increased to 690 mV.

The cell was then completed by a homogeneous Silver contact on the rear side and a silver grid on the front side, with 1 mm-spaced, 200 μm-wide and 1.5 μm-thick fingers which were thermally evaporated through a physical mask, thus determining an effective area of 4.8 cm^2. Figure 6 shows the light and dark J-V curves measured on the cell. The undesired s-shape was completely removed and the superposition principle of the dark and lighted curves was respected, even if not shown. In this way the FF was boosted to a value of 77.4%. These measurements confirmed that the issue limiting the FF and efficiency (η) of the previously studied samples was located at the a-SiO_x:H/MoO_x interface. The photovoltaic parameters of the best cell were: V_{OC}=675 mV, J_{sc}=31.7 mA/cm^2, FF=77.4% and η=16.6%.

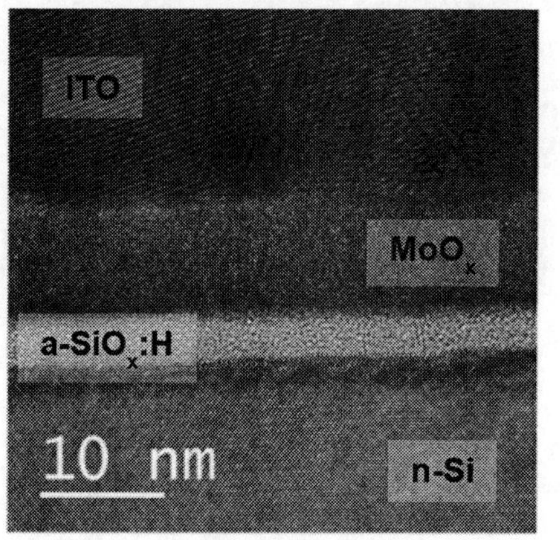

Figure 7: High resolution TEM micrograph of the window layers of the final cell.

A high resolution TEM micrograph of the front layers of the cell is shown in Fig. 7, and reveales the actual thcknesses of the a-SiO_x:H/MoO_x layers, which were slightly lower than expected (4.3/8.9 nm, respectively). This could partially justify the relatively low V_{OC} of the cell which did not reach the state-of-the-art value for heterojunction solar cells, considering that the passivation is a function of the buffer thickness and the fact that the passivation properties of the a-SiO_x:H layer could in principle be further improved after simultaneous ITO deposition and thermal treatment [25], with best effects at 300°C, as already discussed in Section 3.1. On the other hand, detrimental effects of temperature on the work function of MoO_x are well known [8,31]. For these reasons, the thermal stability of the cell was investigated with the aim of both increasing the passivation quality offered by the a-SiO_x:H layer and testing the MoO_x resistance to thermal treatments.

Figures 6 and 8 report the J-V characteristics and quantum efficiencies of the final cell after sequential thermal annealing at increasing temperatures of 130°C and 180°C in a nitrogen flux for 10'. The performances of the cell slightly improved when thermally annealed up to 130°C. In particular, the V_{OC} increased up to 680 mV because of c-Si surface hydrogenation promoted by the thermal treatment of the a-SiO_x:H layer. Nevertheless 180°C thermal annealing temperatures negatively affected both J_{sc} and V_{OC}, which lowered to 28.5 mA/cm^2 and 635 mV, respectively, resulting in an efficiency of 14.1 %. We suppose that this effect was related to energy band alignment modifications due to a reduction in the work function of the MoO_x layer, although the quantum collection efficiency on the front side was greatly increased thanks to the higher ITO transparency in the short-wavelength region after thermal annealing. These values were actually lower with respect to c-Si/a-Si:H heterojunction cells [32-34] despite the transparency of the a-SiO_x:H/MoO_x emitter. This was mainly due to the absence of c-Si surface texturing as demonstrated by the mirror-like surfaces of these cell leading to an effective reflectance of 19.8%. Moreover, the ITO layer absorption still limited the photon transmission towards the c-Si substrate. Nevertheless the obtained results are remarkable for a structure involving for the first time to our knowledge the a-SiO_x:H/MoO_x combination as HJ window layer.

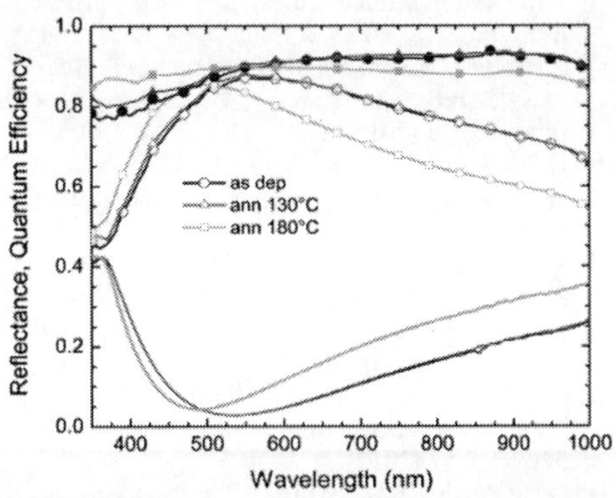

Figure 8: Reflectance (line), Internal (symbol) and External (open symbol) Quantum Efficiencies of the final heterojunction solar cell before and after annealing at 130 and 180°C

The IQE measurements on the non-annealed cell (Fig. 8) proved, as expected, the reduced absorption of all front layers and the fair passivation level provided by a-SiO_x:H. Therefore the aim of manufacturing a window layer more transparent than the present state of the art of HJ solar cell was demonstrated. The effect of annealing was to improve both front passivation and transparency, as deduced by the increase of the response in the short-wavelength range, while the signal in the near-infrared range was lowered because of the modified optical properties of ITO.

In conclusion, even though a high temperature treatment would be desirable to enhance the passivation of the c-Si surface and the transparency of the cell front side, the overall behavior of the device results strongly deteriorated after a thermal treatment above 130°C.

It can be noticed (Fig. 6) that no s-shape arose, even at the highest thermal annealing temperature. These results differ from other reported in literature [11], in which the s-shape appeared and worsened upon annealing over 130°C, and was hypothetically attributed to the thickening of an interlayer between MoO_x and TCO layers. Battaglia et al. [4] reported an initial s-shape which "straightened out" but did not completely disappear after low temperature annealing up to 60°C, and interpreted the results as a thermionic emission barrier which the holes had to overcome. In the experiments reported in this work, the s-shape issue was definitely solved thanks to the HF treatment of the a-SiO_x:H surface before MoO_x deposition, which removed unwanted surface oxidation [30].

Even though the a-SiO_x:H/MoO_x bilayer is more transparent than the amorphous counterpart and the s-shape issue was completely removed, the cells efficiencies were not as good as those reported in literature [2] for silicon-based devices containing MoO_x. Optimization of the front texturing, of the thermal processing sequence, of the metal grid thickness and of ITO transparency and conductivity will help in filling the gap.

3.6 Carrier transport at interface

Focusing on carrier transport, it is worth to notice that the carrier collection by the MoO_x hole-collecting layer has two contributions. The first one arises from the thermionic emission over the c-Si/a-SiO_x:H barrier in the valence band. When the holes reach the a-SiO_x:H/MoO_x interface, they immediately recombine with electrons coming from the MoO_x layer and supplied by the ITO antireflection coating. The second contribution to hole collection arises from a multi-step tunneling mechanism [35] of electrons which takes place at the a-SiO_x:H/MoO_x interface with the help of band tails and defect states within the a-SiO_x:H layer, and extends up to the c-Si/a-SiO_x:H interface. This assumption is demonstrated by the exponential trend of the low forward bias (0.3 V) dark current of the heterojunction as a function of temperature reported in Fig. 9.

The interface between ITO and MoO_x is less critical. The energy barrier evident in Fig. 5a is kept very narrow if the MoO_x layer does not deplete, and therefore it can be easily overcome by a tunneling mechanism.

Finally the rear side contact of the cell, where a-SiO_x:H works as a passivating buffer layer, as also simulated in Fig. 5, did not show any issue on the carrier collection.

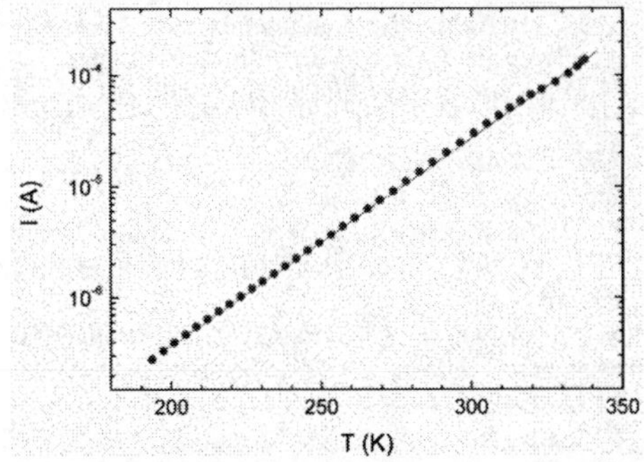

Figure 9: Exponential trend of the low (0.3 V) forward-bias dark current of the heterojunction as a function of temperature, revealing a multi-step tunneling mechanism

4 CONCLUSIONS

In this work MoO_x as hole selective layer was successfully used in combination with an a-SiO_x:H passivation buffer layer in an n-type silicon based solar cell. It was successfully verified that barriers at the a-SiO_x:H/MoO_x/ITO interfaces strongly depend on care taken during the fabrication processes. Once the formation of these barriers was avoided, the optimization of layer thicknesses, process flow sequence and thermal processing led to good performances of the solar cells. A V_{OC} of 675 mV and a J_{sc} of 31.7 mA/cm^2 were obtained together with the complete removal of the undesired s-shape on the J-V characteristics, which boosted the fill factor to 77.4% and the efficiency to 16.6%. Although the fabricated devices have quite modest performances in terms of V_{OC} and J_{sc}, FF values above 77% are nevertheless interesting, as this is generally the weak point of devices featuring MoO_x layers.

The thermal stability of the structures was investigated, recording an improvement of performances after annealing up to 130°C. Further heating brought to a general deterioration of the cell, in particular in terms of V_{OC} and FF, because of electronic modifications in MoO_x which were not counterbalanced by improvements in ITO conductivity and a-SiO_x:H passivation.

5 ACKNOWLEDGEMENTS

This work was supported by the Italian Ministry of Economic Development in the framework of the Operating Agreement with ENEA for Research on the Electric System.

6 REFERENCES

[1] M. Izzi, M. Tucci, L. Serenelli, P. Mangiapane, M. Della Noce, I. Usati, E. Esposito, L.V. Mercaldo, P. Delli Veneri, Appl. Phys. A 115 (2014) 705.

[2] J. Bullock, M. Hettick, J. Geissbühler, A.J. Ong, T. Allen, C.M. Sutter-Fella, T. Chen, H. Ota, E.W. Schaler, S. De Wolf, C. Ballif, A. Cuevas, A. Javey, Nature Energy 1 (2016) 15031.

[3] A. Cuevas and D. Yan, IEEE J. Photovoltaics 3 (2013) 916.

[4] C. Battaglia, S.M. de Nicolás, S. De Wolf, X. Yin, M. Zheng, C. Ballif, A. Javey, Appl. Phys. Lett. 104 (2014) 113902.

[5] S.I. Park, S.J. Baik, J.-S. Im, L. Fang, J.-W. Jeon, K.S. Lim, Appl. Phys. Lett. 99 (2011) 063504.

[6] M. Bivour, J. Temmler, H. Steinkemper, M. Hermle, Sol. Energy Mater. Sol. Cells 142 (2015) 34.

[7] J.W. Rabalais, R.J. Colton, A.M. Guzman, Chem. Phys. Lett. 29 (1974) 131.

[8] Z. Zhang, Y. Xiao, H.-X. Wei, G.-F. Ma, S. Duhm, Y.-Q. Li, J.-X. Tang, Appl. Phys. Express 6 (2013) 095701.

[9] V. Nirupama, M. Chandrasekhar, P. Radhika, B. Sreedhar, S. Uthanna, J. Optoelectron. Adv. Mater. 11 (2009) 320.

[10] C. Battaglia, X. Yin, M. Zheng, I.D. Sharp, T. Chen, S. McDonnell, A. Azcatl, C. Carraro, B. Ma, R. Maboudian, R.M. Wallace, and A. Javey, Nano Lett. 14 (2014) 967.

[11] J. Geissbühler, J. Werner, S.M. de Nicolas, L. Barraud, A. Hessler-Wyser, M. Despeisse, S. Nicolay, A. Tomasi, B. Niesen, S. De Wolf, C. Ballif, Appl. Phys. Letters 107 (2015) 081601.

[12] T. Mueller, S. Schwertheim, and W. R. Fahrner, J. Appl. Phys. 107 (2010) 014504.

[13] J.P. Seif, A. Descoeudres, M. Filipič, F. Smole, M. Topič, Z. C. Holman, S. De Wolf, and C. Ballif, J. Appl. Phys. 115 (2014) 024502.

[14] F. Einsele, W. Beyer, U. Rau, J. Appl. Phys. 112 (2012) 054905.

[15] M. Tanaka, M. Taguchi, T. Matsuyama, T. Sawada, S. Tsuda, S. Nakano, H. Hanafusa, Y. Kuwano, Jpn. J. Appl. Phys. 31 (1992) 3518.

[16] H. Fujiwara, T. Kaneko, M. Kondo, Appl. Phys. Lett. 91 (2007) 133508.

[17] R.A. Street, Hydrogenated amorphous silicon (Cambridge University Press, Cambridge, 1991)

[18] L. Martini, L. Serenelli, L. Imbimbo, F. Menchini, M. Izzi, R. Asquini, M. Tucci, Proc. 32nd European Photovoltaic Solar Energy Conference and Exhibition (2016) 651.

[19] L. Serenelli, L. Martini, L. Imbimbo, R. Asquini, F. Menchini, M. Izzi, M. Tucci, Appl. Surf. Sci. 392 (2017) 430.

[20] W. Kern, J. Electrochem. Soc. 137 (1990) 1887.

[21] D. Deligiannis, V. Marioleas, R. Vasudevan, C.C.G. Visser, R.A. van Swaaij, M. Zeman, J. Appl. Phys. 119 (2016) 235307.

[22] S.M. Sze, Semiconductor devices, physics and technology (J. Wiley & Sons, New York, 1985)

[23] J. Ziegler, M. Mews, K. Kaufmann, T. Schneider, A.N. Sprafke, L. Korte, R.B. Wehrspohn, Appl. Phys. A 120 (2015) 811.

[24] B. Demaurex, S. De Wolf, A. Descoeudres, Z.C. Holman, C. Ballif, Appl. Phys. Lett. 101 (2012) 171604.

[25] L. Martini, L. Serenelli, F. Menchini, M. Izzi, R. Asquini, G. de Cesare, D. Caputo, M. Tucci, Proc. 33rd European Photovoltaic Solar Energy Conference and Exhibition (2017) 2AV.3.30.

[26] B. Nemeth, Q. Wang, W. Shan, Proc. 38th IEEE Photovoltaic Specialists Conference, NREL/CP-5200-56957 (2012).

[27] K. Ghosh, C.J. Tracy, S. Herasimenka, C. Honsberg, S. Bowden, Proc. 35th IEEE Photovoltaic Specialists Conference, PVSC.2010.5614387 (2010).

[28] M. Tucci, G. De Cesare, J. Non-Cryst. Solids 338 (2004) 663.

[29] M. Tucci, M. Della Noce, E. Bobeico, F. Roca, G. De Cesare, F. Palma, Thin Solid Films 451 (2004) 355.

[30] A.C. Bronneberg, A.H.M Smets, M. Creatore, M.C.M. van de Sanden, J. Non-Cryst. Solids 357 (2011) 884.

[31] M.T. Greiner, L. Chai, M.G. Helander W.-M. Tang, Z.-H. Lu, Adv. Funct. Mater. 22 (2012) 4557.

[32] D. Adachi, J.L. Hernández, K. Yamamoto, Appl. Phys. Lett. 107 (2015) 233506.

[33] L.V. Mercaldo, E. Bobeico, I. Usatii, M. Della Noce, L. Lancellotti, L. Serenelli, M. Izzi, M. Tucci, P. Delli Veneri, Sol. Energy Mater. Sol. Cells 169 (2017) 113.

[34] A. Richter, V. Smirnov, A. Lambertz, K. Nomoto, K. Welter, K. Ding, Sol. Energy Mater. Sol. Cells 174 (2018) 196.

[35] S.J. Fonash, Solar cell device physics (Academic press, New York, 1981).

TEMPERATURE EFFECTS ON SPUTTERED ITO

F. Menchini[1], L. Serenelli[1,2], G. Stracci[1], M. Izzi[1], E. Salza[1], D. Caputo[2], G. de Cesare[2] and M. Tucci[1]
[1]ENEA, Casaccia Research Centre, Via Anguillarese 301, 00123 Rome (Italy)
[2]DIET, University of Rome "Sapienza", Via Eudossiana 18, 00184 Rome (Italy)

ABSTRACT: Indium Tin Oxide (ITO) is widely used in solar cell devices for its excellent electrical and optical characteristics, such as high transparency in the Ultraviolet-Visible range and good conductivity (around 10^4 $\Omega^{-1}cm^{-1}$). In this work we have compared thin (70-150 nm) ITO layers deposited by Direct Current or Radio Frequency sputtering. We have used different substrate temperatures during film growth and have afterwards thermally annealed the samples at different temperatures up to 300°C to investigate the effects on the electrical and optical properties of the material. We have found out that the different growth/annealing conditions induce changes in the optical properties of the samples as well as in the conductivity and carrier concentration.
Keywords: ITO; annealing

1 INTRODUCTION

In heterojunction solar cells, a key point for achieving high efficiency is the deposition of a Transparent Conductive Oxide (TCO) layer as full-area contact on both base and emitter side of the solar cell. Various TCOs have been studied through the years as alternatives to Indium Tin Oxide (ITO), to circumvent its relative high cost and Indium shortage in nature. Aluminum Oxide, Tungsten Oxide and Cerium Oxide [1-3] have attracted interest in the recent past, but ITO still remains the most widely used as TCO in heterojunction (HJ) solar cells [4,5] manufacturing because of its excellent electrical and optical characteristics, such as high transparency in the ultraviolet-visible range and good conductivity, typically 80% and 10^4 S·cm-1 respectively for a thin film also useful as antireflection coating [6].

Among the available deposition techniques for ITO growth, the most efficient and widely used is Physical Vapor Deposition (PVD). Both Radio Frequency (RF) and Direct Current (DC) magnetron sputtering are suitable to obtain high-quality films for photovoltaic applications. The DC process is very simple and easy to be scaled up on large area, with relatively high growth rates. In turn, the RF process requires the use of a matching network between the power generator and the target, and results in lower deposition rates. If this latter could be a drawback for scalability, it could represent an advantage because of the lower ion bombardment, which, in principle, should reduce stress on the underlying films. In this work we have studied the intrinsic properties of ITO for Photovoltaic applications in silicon heterojunction solar cells [7], where thin films are used to work also as antireflection coating. In particular, we have grown thin (<150 nm) ITO films by in-line, large area DC and RF magnetron sputtering, at different power densities and substrate temperatures. The samples have been subsequently thermally annealed up to 300°C to explore the effects of temperature during and after deposition. The results obtained on DC- and RF-grown samples have been compared and discussed.

2 EXPERIMENTAL

ITO samples have been deposited on Corning glass in an in-line, large area deposition system by DC Magnetron Sputtering at a constant pressure of 1.1×10^{-3} mbar in Argon (Ar) without Oxygen addition. The target was an Indium–Tin alloy with a 90%:10% In:Sn weight ratio and 170 cm^2 area. The power has been fixed at 200 W and the substrate temperature during deposition has been varied between Room Temperature (RT) and 180°C. Similar samples have been grown by 13.56 MHz RF sputtering from a 6 inches diameter round target of same ITO alloy at 100W power and 3.3×10^{-3} mbar pressure, at various substrate temperatures between RT and 180°C.

The samples have been characterized in terms of Reflectance (R) and Transmittance (T) by a Perkin Elmer Lambda 950 UltraViolet-Visible-NearInfrared (UV-Vis-NIR) spectro-photometer. Bulk film resistivity has been evaluated by measuring the sheet Resistance (R_{sh}) with a Four Point Probe tool and thickness by a Tencor Alphastep contact profilometer. After deposition, the samples have been thermally annealed in a static IR furnace under Nitrogen flux for 10 minutes at different temperatures from 100°C to 300°C, and again characterized. The samples have been also characterized in terms of mobility (μ), free carrier density (N) and resistivity (ρ) in a Biorad Hall profiler according to the Van der Paw configuration.

3 RESULTS AND DISCUSSION

We have started by characterizing the DC-grown samples. The absorption coefficient α has been calculated by the approximated formula

$$\alpha = \frac{1}{t} ln \frac{(1-R)}{T}$$

where t is the film thickness in cm, R is the Reflectance and T the Transmittance measured as a function of wavelength.

Figure 1 shows the absorption coefficient before and after thermal treatment. The as-dep samples show noticeable differences, and the higher the deposition temperature, the lower the absorption. After a thermal treatment, the absorption decreases for all samples in the UV-Vis range.

In Fig. 2 we report the effects on bulk resistivity (ρ) of thermal annealing performed in several temperature steps on the same samples reported in Fig. 1. All samples show a step-like decreasing trend of ρ with increasing annealing temperature, and the lower the deposition temperature, the higher the resistivity. The transition temperatures for the three samples, defined as the temperature at which the slope of the curves in Fig. 2 is maximum and the ITO films change drastically their characteristics, are very different for the three samples, and decrease from about 195°C for the samples grown at RT to 160°C and 155°C for the samples grown at higher temperatures. The ρ values after thermal annealing at 250°C become comparable for all the ITO films grown at different temperatures.

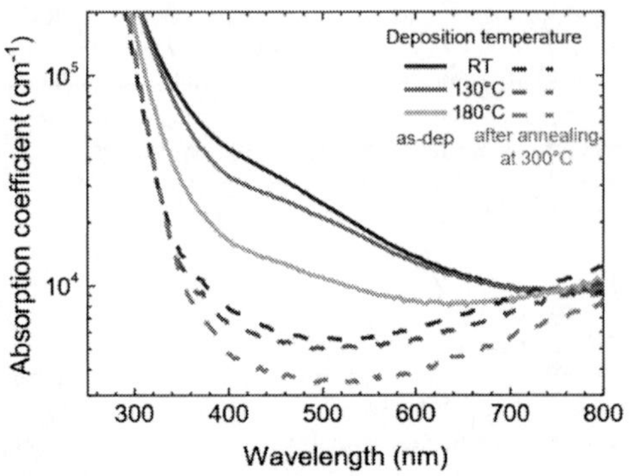

Figure 1: Absorption coefficient before (solid line) and after (dashed line) thermal annealing at 300°C in samples grown at different substrate temperatures and 200 W. Different color intensities refer to different deposition temperatures

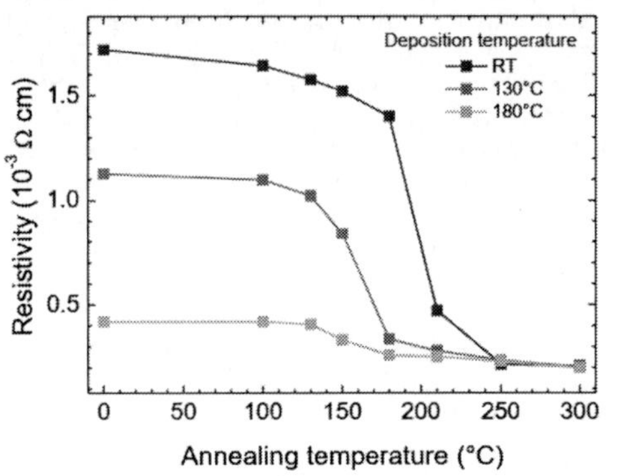

Figure 2: Resistivity as a function of thermal annealing temperature in samples grown at 200W and different substrate temperatures. Lines are only a guide for the eye.

On the basis of the previous observations, we have selected two deposition conditions, i.e. 200 W power and RT and 180°C substrate temperature, to be compared with two analogous of samples deposited by RF sputtering. The samples have been thermally annealed at different temperatures and optically characterized in terms of Reflectance and Transmittance in a range extended to 1200 nm, so as to cover the one of interest for crystalline silicon based heterojunction solar cells. Electrical characterizations have been carried out by Hall measurements

Figure 3 shows the absorption coefficient of the samples grown at RT and 180°C by both DC and RF process. It is evident that the DC samples are less transparent than the RF ones in the whole spectral range. For DC samples, a thermal annealing at 180°C induces only small changes on α, because the transition temperature has not been reached yet. On the other hand, growing the sample directly at 180°C leads to a higher improvement of transparency in the 400-650 nm range.

Concerning the RF samples, their absorption coefficients are much lower than those of the DC samples, and the higher the deposition temperature, the better the transparency in the 400-1200 nm range. Samples grown at RT show a slight increase in transparency after a thermal annealing at 180°C.

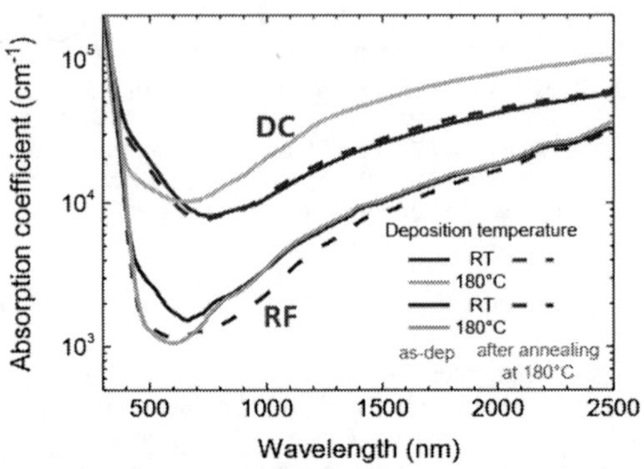

Figure 3: Absorption coefficient in the UV-Vis-NIR spectral range for samples grown by DC and RF sputtering at different temperatures.

Table I summarizes the extinction coefficients of the samples in the Vis-NIR range, calculated from the α values, clearly showing the higher transparency of the RF-grown samples.

Electrical characterizations have been performed on the samples in terms of mobility (μ), free carrier density (N) and resistivity (ρ), and the results are reported in Fig. 4. From Fig. 4 (top) it is seen that the resistivity of the as-dep RF samples (around 4×10^{-4} Ω·cm) is less influenced by the deposition temperature with respect to that of DC samples, which varies from 1.2×10^{-4} to 0.4×10^{-4} Ω·cm for RT and 180°C deposition, respectively. In both RF samples ρ is lower than that of the RT grown DC sample, but comparable to that of the DC sample deposited at 180°C. The effect of thermal annealing on resistivity is quite different for the 4 sets of samples. While for both DC samples a step-function decrease is observed, confirming what shown in Fig. 4, the RF samples show different behaviors depending on the deposition temperature. Indeed, for ITO deposited at 180°C (RF2), a small increment with thermal annealing temperature is recorded. On the other hand, for the films deposited by RF power at RT (RF1) a considerable increase is observed.

TABLE I: Vis-NIR extinction coefficients of DC and RF samples grown at different temperatures

Sample	633 nm	1000 nm
DC RT	0.06	0.09
DC 180°C	0.05	0.16
RF RT	0.01	0.03
RF 180°C	0.005	0.03

The free carrier density variation after thermal treatment are also shown in Fig. 4. The N trends can be directly correlated to those of ρ values. Indeed the relative differences in N values of the DC samples before thermal annealing reflect those of ρ, with values of 1.6×10^{21} cm^{-3} for the DC sample grown at RT, and lower but similar values (around 6.5×10^{20} cm^{-3}) for the DC sample grown at 180°C and for both the RF samples. After annealing, the N values for the DC samples have raised up to over $2.2\times10^{2}1$ cm^{-3}, while the RF samples show a much lower variability.

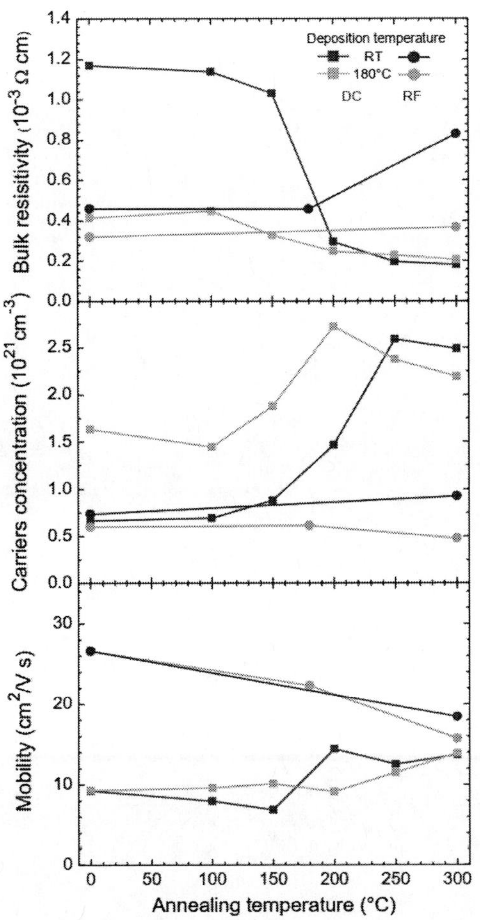

Figure 4: From top to bottom: resistivity, carriers concentration and mobility as a function of annealing temperature for samples grown at different substrate temperatures, by both DC and RF sputtering. Lines are only a guide for the eye.

For what concerns mobility, in the bottom of Fog. 4, it is immediate to note that just after deposition the RF samples are characterized by a higher mobility (26.6 cm^2/V s) with respect to the one of the DC samples (10 cm^2/V s). The mobilities of RF and DC samples show a decreasing and increasing trend, respectively, with increasing annealing temperature, with a convergence of the values for all samples around 16 cm^2/V s.

The coupled analysis of N and μ can explain the resistivity variations reported in Fig. 4. For the DC samples, the global increase of both N and μ with increasing annealing temperature results in a corresponding decrease of ρ. For the RF sample grown at RT, the annealing causes a decrease in both μ and N, with an enhancement of resistivity, especially at 300°C. In the RF sample grown at 180°C a compensation between the increase in N and the decrease in μ with increasing annealing temperature does not produce large variations in the film resistivity.

Summing up all these considerations, it can be established that the RF technique is able to produce ITO films having lower resistivity, higher mobility and lower carrier concentration with respect to the DC samples grown at RT. DC samples grown at 180°C are comparable to RF samples in terms of electrical parameters, even if their transparency remains lower. However the resistivity of films deposited by RF can be negatively affected by a thermal treatment, especially when carried out at temperatures above 180°C.

Layers deposited via DC sputtering can suffer from very high free carrier absorption, affecting the transparency in the NIR part of the spectrum, especially when deposited at 180°C, due to N concentrations in the order of 10^{21} cm^{-3}. Thermal treatments further increase this concentration as well as the mobility, which corresponds to an improvement in transport properties.
The illustrated different electro-optical properties of ITO films grown by DC or RF techniques can be exploited when designing the cells manufacturing process sequence, by choosing between the different deposition conditions depending on the properties to be privileged, i.e. transparency or conductivity, and on the thermal process the device has to undergo.

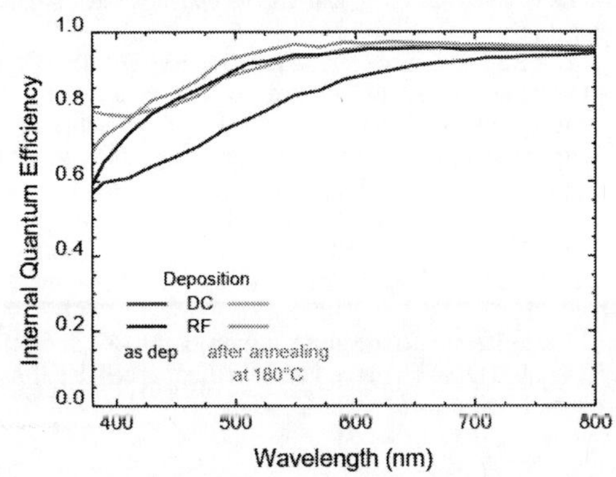

Figure 5: Internal Quantum Efficiencies of complete cells th different top ITO layers, before and after annealing.

To better asses the performances of ITO films, they have been used as top window layers in complete solar cells which were produced in parallel and differentiated only for the ITO layer. The cells structure starts from n-type c-Si wafer passivated on both sides by a buffer layer of intrinsic hydrogenated amorphous silicon (i-a-Si:H). The back contact is constituted by a layer of n-type amorphous hydrogenated Silicon Oxide (n-SiO_x:H) covered by Al-doped Zinc Oxide (AZO). The front side is completed by a p-type a-Si:H layer finally covered by the different ITO layers. Ag evaporated grids complete the cell on both sides as electrodes.

For this comparison, we have focused on ITO films deposited at RT for both DC and RF sputtering, and we have subsequently annealed the cells at 180°C. Since the different ITOs should only affect optical and electrical properties of the cells in terms of J_{sc} collection, we have measured the cells Quantum Efficiencies and Reflectances, and we have computed the Internal Quantum Efficiencies (IQE). The results are reported in

Fig. 5. As expected, the cell containing ITO grown by RF sputtering is the best performing one in the measured range, with a further slight improvement after thermal annealing. On the contrary, the cell containing the DC grown ITO layer shows much lower response, increasing after thermal treatment. This latter behavior is the consequence of enhanced transparency and conductivity of DC ITO after thermal annealing as already shown in Fig. 4.

4 CONCLUSIONS

In this work we have shown the effect of temperature on DC and RF sputtered ITO thin films. We have considered both substrate heating during the deposition and subsequent thermal treatments, finding out that in all cases higher temperatures lead to better results in terms of transparency.

For DC sputtered ITO, mobility has been found to be slightly dependent on substrate deposition temperature and subsequent thermal treatments. The free carrier concentration has resulted higher for samples deposited at 180°C, with an increase after thermal annealing for both deposition temperatures. These effects are reflected on the absorption coefficient, which is lower in the IR part of the spectrum for samples deposited at higher temperatures.

The RF sputtering technique produces ITO film with better characteristics just after deposition with respect to samples grown by DC sputtering at RT. In particular, the absorption coefficient is lowered by almost one order of magnitude in the 450-1200 nm range, while the resistivity is less than one half. Nevertheless the conductivity of these films is worsened by post-deposition thermal treatments, especially when performed at temperatures above 180°C.

The different temperature responses of DC and RF sputtered ITO layers have been verified at cell level by means of IQE measurements.

5 REFERENCES

[1] M. Wimmer, F. Ruske, S. Scherf, B. Rech, Thin Solid Films 520 (2012) 4203.

[2] F. Meng, J. Shi, Z. Liu, Y. Cui, Z. Lu, Z. Feng, Solar Energy Materials and Solar Cells 122 (2014) 70.

[3] E. Kobayashi, Y. Watabe, T. Yamamoto, Y. Yamada, Solar Energy Materials and Solar Cells 149 (2016) 75.

[4] S. De Wolf, A. Descoeudres, Z.C. Holman and C. Ballif, Green 2 (2012) 7.

[5] N. Kim, H.D. Um, I. Choi, K.H. Kim, and K. Seo, Applied Materials & Interfaces 8 (2016) 11412.

[6] Z.O. Adurodija, H. Izumi, T. Ishihara, H. Yoshioka, M. Motoyama, Solar Energy Materials and Solar Cells 71 (2002) 1.

[7] W.G.J.H.M. van Sark, L. Korte, F. Roca, Physics and technology of amorphous-crystalline heterostructure silicon solar cells. Berlin, Springer-Verlag (2012).

REAR DEVICE ARCHITECTURES FOR EVALUATING PASSIVATED ORGANIC/SILICON HYBRID CELLS

James Hack[1], Abhishek Iyer[2], Meixi Chen[3], Robert Opila[123]
1. University of Delaware, Department of Materials Science & Engineering,
2. University of Delaware, Department of Electrical Engineering & Computer Science,
3. University of Delaware, Department of Chemistry & Biochemistry
Newark, DE 19716

ABSTRACT: Hybrid organic-silicon devices were fabricated making use of benzoquinone-methanol passivation of silicon, which provides minority carrier lifetimes of 1500 microseconds at 10^{15} cm^{-3} minority carrier density (MCD), and an implied 1 sun Voc of .708V on textured silicon. Initial devices built on this structure show a large deficit between this implied Voc and the voltage of completed devices, indicative of recombination losses. To address these losses, 3 different rearside contact schemes were evaluated. Implementation of a BSF layer yielded a 58mV Voc gain, while making use of an a-Si stack yielded a 128 mV gain. This provides experimental quantification of the extent to which backside structures with differing passivation quality can effect the measurement of novel frontside structures. These tests clearly demonstrate the important role that optimization of rear-side passivation and contact schemes make in the performance of Si-organic hybrid solar cells. Devices constructed show excellent low wavelength quantum efficiency.
Keywords: Hybrid, Back-Contact, Passivation

1 INTRODUCTION

Hybrid organic/Silicon photovoltaic devices have attracted increased interest within the past 5 years. [1]-[4] Such devices hold promise to combine the excellent bulk absorber quality of silicon with the ease of processing provided by organic materials. If fully stabilized and developed, organic/silicon hybrid cells could unlock processing cost savings versus conventional high temperature and high vacuum processed devices. Our group has demonstrated in the past devices exceeding 9% efficiency, making use of benzoquinone (BQ) as an interfacial passivation layer between Si and a hole conducting poly(3,4-ethylenedioxythiophene) polystyrene sulfonate (PEDOT:PSS) layer. [5] In our current work, the benzoquinone passivation treatment of textured n-type wafers after H-Si termination is shown to give a minority carrier lifetime of 1500 microseconds at 10^{15} cm^{-3} minority carrier density (MCD), and an implied 1 sun Voc of .708V. If such high passivation levels can be maintained in a completed solar device structure, this provides potential for a highly-performing alternative silicon-based device architecture. Initial devices built on this structure show a large deficit between this implied Voc and the voltage of completed devices, indicative of recombination losses. To isolate the effects of voltage loss from the front and rear of the device, different backside devices architectures were compared. For KOH textured wafers with the same frontside BQ treatment and PEDOT:PSS recipe, three backside treatments were compared: (1) the textured wafer with no further treatment before Al metallization (TEX devices), (2) a phosphorous doped layer to create an n+ layer providing a back surface field (BSF devices), and (3) a stack of intrinsic a-Si layer and n-doped a-Si layer to create a heterojunction style backside architecture (a-Si devices). The a-Si structure proved a better test platform for novel frontside treatments, eliminating a large portion of the initially observed voltage deficit.

2 METHODS

2.1 Device Fabrication

N-type silicon (100) KOH-textured samples (doped with phosphorous, CZ) with and without a BSF layer were fabricated at Arizona State University. These wafers are 145 micron thick with 1-5 Ohm cm resistivity. Random texturization was performed on both sides by potassium hydroxide (2% KOH yielding pyramids of about 3-5 um base size) alkaline etching. These wafers were either used as-is (TEX devices), treated with an n+ backside layer (BSF devices), or with an intrinsic and n-doped a-Si stack (a-Si devices). For the a-Si devices, a 9nm i-Si layer and a 30nm phosphorous doped n-Si layer were subsequently deposited via PECVD. For the BSF devices, a 55 Ohm/square sheet resistance back surface field was created by a 15 minute POCl3 diffusion at 820° C, followed by a buffered oxide etch to remove the phosphosilicate glass layer. Before cleaning, a passivation solution of 0.01M of BQ with methanol is prepared in a sealed bag. All wafers (scribed down to 1.5cm by 1.5cm) were cleaned following the procedures described by Opila and Teplyakov et al.[6] Substrates were cleaned using a acid Piranha etch solution (H2SO4:H2O2 ratio of 4:1) for five minutes, followed by a five minute de-ionized (DI) water rise and a two minute immersion in hydrofluoric acid (HF, 2 wt%) to obtain hydrogen terminated surfaces[6].

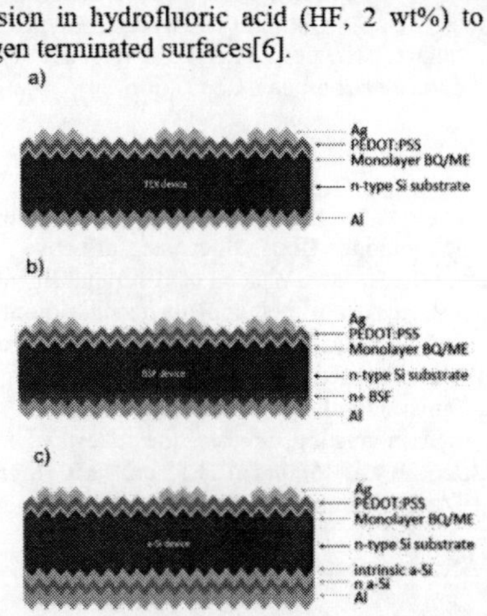

Figure 1: Device architecture for the three different

backside structures under investigation. a) The simple TEX devices. b) BSF devices with a n+ diffused back surface field. c) The a-Si device structure, featuring an i-Si and n-Si stack. All 3 structures make use of the same wafers and frontside processing.

After cleaning, the substrates were quickly DI water dipped and dried with a nitrogen gun and immediately placed in the passivation solution. The wafers are allowed to stay in the sealed bag containing passivation solution for 2 hours in ambient light conditions. PEDOT:PSS PH1000 (Heraeus Clevios, USA) was prepared with 7% ethylene glycol (EG) and 0.25% Triton X-100 (non-ionic surfactant). After 2 hours immersion, the wafers were removed from the sealed bag, rinsed with methanol and dried with a nitrogen gun. Wafers underwent spin coating at 4000rpm for 300 seconds to ensure good surface coverage and drying of the substrate. This was followed with 15 minutes in a hot-plate at 130 °C to drive out any remaining moisture. After spin coating and drying the sample, wafers were loaded into the electron beam physical vapor deposition tool for back and front contact deposition. 2 microns of aluminum was deposited on the backside followed by 400 nanometers of silver (through a shadow mask) on the frontside.

2.2 Device Characterization

Following immersion in the BQ methanol solution, the minority carrier lifetime of the wafers in solution was measured via photoconductance decay using a Sinton WCT-120 in the Generalized Mode. For electrical characterization of solar cell devices, the current density–voltage response was measured by means of a dc source meter (Keithley 2400 sourcemeter, USA) under illumination using an ELH bulb calibrated to 1 sun illumination intensity at air mass 1.5G standard. External quantum efficiency (EQE) of devices was measured using an Oriel monochromator from 300 to 1400 nm using an AC method with a lock-in amplifier.

2.3 Contact Resistivity

Contact resistivity tests were performed using a simple 1mm radius contact pad pattern, separated by a distance of 6mm. While this method does not enable a full calculation of sheet and contact resistivity, it can give a measure of the relative resistance of different contact techniques. On 25 Ohm-cm CZ wafers, n-type Si, three metal contact schemes, Al, TiPdAg, and Ni were deposited via electron-beam deposition.

3 RESULTS

Passivation studies in the Opila group have suggested that benzoquinone (BQ) does an effective job of passivating the surface. We can confirm that BQ abstracts hydrogen atoms from methanol to become quinhydrone (QH•), and both QH• resulting in an improvement of minority carrier lifetime (up to lifetime=4500μs, with SRV=1.6cm/s on polished surfaces). [7] For the textured wafers used in making devices, the measured minority carrier lifetime was 1500μs at 1E15 cm^{-3} MCD, giving an implied 1 sun open circuit voltage of 0.708V.

3.1 JV Measurements

Full JV characteristics are shown in Table 1 and Figure 2 for high-performing devices of each type. The BSF devices and a-Si devices yield a higher Jsc, shifting the JV curves upward. The BSF devices showed a modest improvement in Voc relative to the basic TEX devices. Devices with a-Si backsides showed a significantly increased Voc performance relative to the TEX and BSF devices, indicating that the rear treatment is a limiting factor in the performance of this class of devices. For the a-Si device shown, some shunting of PEDOT:PSS to the back contact is observed, however this effect can be reduced by edge isolation techniques. Series resistance is a significant limiting factor for the a-Si and other devices.

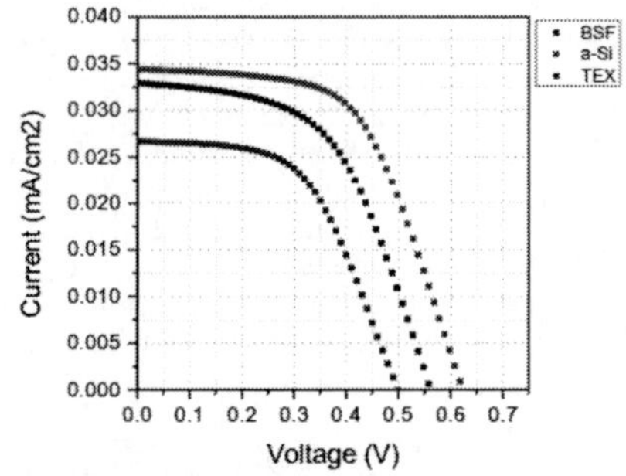

Figure 2: Light JV curve results

Device type	Jsc (mA/ cm²)	Voc (V)	η (%)	FF (%)	R_s (Ohm cm²)	R_{sh} (Ohm cm²)
TEX	26.7	.500	7.3	54	6.8	780
BSF	32.9	.558	9.9	54	6.1	286
a-Si	34.4	.628	12.4	57	5.9	645

Table 1: Device Results for three backside treatments

3.2 Quantum Efficiency Measurements

Improvements in these devices were further examined via quantum efficiency measurements, shown in Figure 3. All devices show a high quantum efficiency at low wavelengths which are absorbed near the front surface of the device. This is indicative of low frontside carrier recombination and low optical losses from reflection and parasitic absorption in the PEDOT:PSS at these wavelengths. A slight boost in overall QE is seen with the progression from TEX devices to BSF devices to a-Si devices. The most significant change is the shift in shape of the QE curve at around 950nm. The a-Si devices

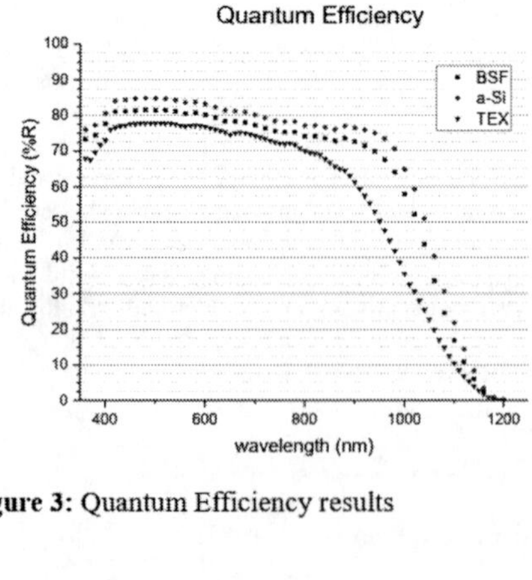

Figure 3: Quantum Efficiency results

reliably show a more square shape in this region with a sharper dropoff near the Si bandgap than the TEX devices and even the BSF devices which show a slightly more gradual rollover.

These results can be explained by differences in the surface recombination velocity (SRV) of the different backside structures. Unpassivated metal -silicon interfaces can provide extremely high surface recombination velocities (>1000cm/s), while high-low junctions similar to those used for backside passivation of Al-BSF devices can reduce the SRV to around 200cm/s. Stacks of a-Si layers used in silicon heterojunction devices can reach low SRV values of 2cm/s. [8][9] It is expected a secondary effect is observed in the EQE due to differences in backside reflectivity, with the a-Si layers boosting the rearside reflection.

3.3 Contact Resistivity Measurements

For the basic textured wafer (TEX), a variety of metals were evaluated for rear-side contact schemes. Due to the processing sequence, it was desirable to make use of a contact which could be deposited via electron-beam physical vapor deposition with no subsequent high temperature annealing steps. Al, Ni, and a TiPdAg stack deposited in this way gave measured relative contact resistances of 74 Ohms, 15000 Ohms, and 860 Ohms respectively. Aluminum was selected for use in backside contacts, displaying the most ohmic behavior, with a linear IV relationship and the lowest overall resistance (Table 2). Reported values are measured at 2V. The a-Si device still exhibits a high series resistance of 5.9 Ohm cm2 as measured at Voc, so it appears that optimization of the metallization could give a sizable effect on FF and the overall device efficiency.

Material	Contact Resistance
Al	74 ohms
TiPdAg	860 ohms
Ni	15000 ohms

Table 2: Measured resistivities of metal contacts

4 CONCLUSIONS

Carrier selective solar cells based of PEDOT:PSS and benzoquinone solution passivation on n-type silicon were fabricated and 12.4% power conversion efficiency was demonstrated. There is room for more improvement in series resistance of the a-Si devices, so alternative contact schemes are currently under investigation. The a-Si architecture gave the highest performance due to improved rearside reflection and lower surface recombination velocity, with a clear difference evident in QE measurements. Total efficiency results for the conventional (TEX) device gave 7.3% efficiency, 9.9% for the device with BSF, and 12.4% for the device with a-Si backside. For devices which cannot withstand high annealing temperatures, a structure consisting of a rear a-Si stack was shown to give a much improved base structure for evaluating novel front contact schemes. It was demonstrated that the backside architecture of the device was a significant contributing factor in losses of both short circuit current and open circuit voltage. This work gives a clear experimental indication of the extent to which backside device architecture can affect achievable open circuit voltages for novel front side treatments, with a 58mV and 128mV Voc gain for the BSF and a-Si structures, respectively. Of the 208mV deficit in iVoc and initial device results, a large portion was able to be attributed to the rearside passivation and contact scheme.

5 ACKNOWLEDGEMENT

Work at the ASU Solar Power Lab by Som Dahal, Rameshwari Ghimire and Stuart Bowden was supported by the Nanotechnology collaborative Infrastructure. Southwest (NCISW) with funding from NSF-ECCS1542160. Quantum efficiency measurements were measured at the Institute of Energy Conversion (IEC) under the guidance of Steven Hegedus, Patrick Conroy and Chris Thompson. PECVD was performed under the guidance of Ugochukwu Nsofor and Ujjwal Das at the IEC. Electron beam deposition was done at UD Nanofabrication facility under guidance of Scott McCracken. This material is based upon work primarily supported by the Engineering Research Center Program of the National Science Foundation and the Office of Energy Efficiency and Renewable Energy of the Department of Energy under NSF Cooperative Agreement No. EEC-1041895. Any opinions, findings and conclusions or recommendations expressed in this material are those of the author(s) and do not necessarily reflect those of the National Science Foundation or Department of Energy.

6 REFERENCES

[1] Thomas, J. P., & Leung, K. T. (2014). Defect-minimized PEDOT:PSS/planar-si solar cell with very high efficiency. Advanced Functional Materials, 24, 4978–4985. http://doi.org/10.1002/adfm.201400380

[2] Nagamatsu, K. A., Member, S., Avasthi, S., Jhaveri, J., & Sturm, J. C. (2014). A 12 % Efficient Silicon / PEDOT : PSS Heterojunction Solar Cell Fabricated at <100 ◦ C. IEEE Journal of Photovoltaics, 4(1), 260–264. http://doi.org/10.1109/JPHOTOV.2013.2287758

[3] Mahato, S., Gerling, L. G., Voz, C., Alcubilla, R., & Puigdollers, J. (2016). PEDOT:PSS as an alternative hole selective contact for ITO-free hybrid crystalline silicon solar cell. IEEE Journal of Photovoltaics, Accepted, 1–6. http://doi.org/10.1109/JPHOTOV.2016.2557072

[4] Li, S., Pei, Z., Zhou, F., Liu, Y., Hu, H., Ji, S., & Ye, C. (2015). Flexible Si/PEDOT:PSS hybrid solar cells. Nano Research, 8(10), 3141–3149. http://doi.org/10.1007/s12274-015-0814-y

[5] Hack, J., Iyer, A., Chen, M., Kotulak, N., Sriharan, A., Opila, R., Hybrid Organic-Inorganic Solar Cells with a Benzoquinone Passivating Layer. PVSC 44 Proceedings. 2017

[6] F. Tian, D. Yang, R. L. Opila, A. V. Teplyakov, "Chemical and electrical passivation of Si(111) surfaces, Applied Surface Science, 258, 3019, 2012.

[7] M. Chen, et al, J. Phys. Chem. C, 2017, 121 (39), pp 2136

[8] A. G. Aberle, "Surface passivation of crystalline silicon solar cells: a review," Prog. Photovoltaics Res. Appl., vol. 8, pp. 473–487, May 2000.

[9] F. Kløw, H. Haug, and S. Erik, "Surface recombination velocity measurements of metallized surfaces by photoluminescence imaging," Energy Procedia, vol. 43, no. 1876, pp. 18–26, 2013.

[10] M. Taguchi, A. Yano, S. Tohoda, K. Matsuyama, Y. Nakamura, T. Nishiwaki, K. Fujita, and E. Maruyama, "24.7% Record efficiency HIT solar cell on thin silicon wafer," IEEE J. Photovoltaics, vol. 4, no. 1, pp. 96–99, 2014.

Electron selective contacts based on $Al_2O_3/TiO_2/ZnO$ stacks for crystalline silicon solar cells

L.A. Zafoschnig, P. Ortega, I. Martín, G. Masmitjà, G. López and R. Alcubilla
Departament d'Enginyeria Electrònica, Universitat Politècnica de Catalunya.
C/ Jordi Girona 1-3, Mòdul C4, 08034 Barcelona, Spain.
Phone.: +34 93 405 41 93, Fax: +34 93 401 67 56, e-mail: isidro.martin@upc.edu

ABSTRACT: In the last years, our research group has focused on the development of Interdigitated Back Contacted (IBC) solar cells based on transition metal oxides. In particular, a very thin Al_2O_3/TiO_2 stack deposited by ALD has demonstrated excellent results. The point of this work is to try to develop a transparent conductive oxide that could be inserted in between the passivating layers and the metal contact in order to improve rear reflector properties. As a good candidate, we focus on Zinc Oxide (ZnO) deposited from DEZ (Ditehylzinc) and water as gas precursors. A deposition rate of 1.76 A/cycle is found for layers deposited at 125 °C indicating a self-limited reaction, i.e. atomic layer deposition process. This film has an optical band gap of 3.26 eV demonstrating good optical properties to work as a back reflector. Unfortunately, contact quality with these layers is very poor. In order to investigate the origin of such a poor performance, we measure ZnO workfunction by Scanning Kelvin Probe resulting in 4.85 eV. This value suggests that the developed ZnO films could work better in a hole selective contact. To demonstrate this point, contacts on MoO_x layers were fabricated leading to reasonable contact resistance values of 43 mΩ cm. Passivation properties of these contacts will be addressed in the future in order to consider their introduction into the final IBC solar cell.
Keywords: ZnO; selective contacts; surface passivation; c-Si.

1 INTRODUCTION

In recent years, Interdigitated Back-Contacted (IBC) c-Si solar cells have demonstrated their high efficiency potential holding the world record conversion efficiency at 1 sun-illumination [1]. Simultaneously, strong efforts have been paid to the development of selective contacts based on transition metal oxides, like MoO_x, V_2O_5 or TiO_2 [2-4]. In particular, our research group has demonstrated the excellent properties of an atomic-layer-deposited Al_2O_3/TiO_2 stack as electron transport layer (ETL). The Al_2O_3 film is kept very thin (~0.6 nm, 6 cycles) reducing the interface state density and allowing carrier transport by tunnel, while TiO_2 is promoting the electron transport (~1.2 nm, 20 cycles). Finally, this film stack is contacted by Magnesium which is capped by Aluminum. This film stack has been applied to IBC c-Si solar cells leading to efficiencies beyond 19 % [5]. Keeping the focus on IBC structures, our objective is the development of an ETL that could improve rear reflectance increasing the available photogenerated current in the cell. To reach this goal, we introduce a ZnO layer before metal deposition. This material is a transparent conductive oxide that can be deposited by ALD and can easily fit into the ETL fabrication process [6-7]. In this work, we report on the passivation and contact properties of the proposed film stack.

2 EXPERIMENTAL

As substrates, we used (100) 280 µm-thick n-type crystalline silicon wafers with a resistivity of 2.5±1 Ω cm. All wafers were cleaned with a 1% HF dip for 1 minute before any film deposition. Films of ZnO, TiO_2 and Al_2O_3 were deposited using Atomic Layer Deposition technique with deposition tool Savannah S200 from Cambridge Nanotech was used. Precursors were TMA (Trimethyl-aluminium), TDMAT (Tetrakis (dimethylamino) titanium) and DEZ (Ditehylzinc), for Al_2O_3, TiO_2 and ZnO layers respectively. In all cases, we used H_2O as the oxidant agent. Times for precursor injection and purge during the ALD deposition were kept constants to the values shown in Table I. Additionally, Al_2O_3/TiO_2 was deposited at 100 °C and films thicknesses were fixed to 6 and 20 cycles respectively [5].

Table I: Deposition conditions for Al_2O_3/TiO_2 stack

	Cycles	$t_{precursor}$ (s)	t_{purge} (s)	t_{H_2O} (s)	t_{purge} (s)
Al_2O_3	6	0.05	10	0.015	10
TiO_2	20	0.8	5	0.015	5
ZnO	-	0.06	2.5	0.015	2

Effective lifetime (τ_{eff}) vs. excess carrier density (Δn) measurements of symmetrical samples were performed using Sinton Consulting WCT-120. For the calculation of effective surface recombination velocity (S_{eff}), we apply the relation $1/\tau_{eff} = 1/\tau_i + 2S_{eff}/w$, where w is the sample thickness and τ_i is the intrinsic lifetime as modeled in ref. [8]. We use τ_{eff} value at Δn = 10^{15} cm^{-3} for direct comparison between samples.

Ellipsometry at λ= 632 nm was used to analyze thickness and refractive index of ZnO films with a Plasmos SD2100 ellipsometer. Further optical characterization of ZnO films was carried out by depositing films on borosilicate glass. Then, absorption of ZnO layers was determined by measuring the reflectance and absorptance spectrum of these samples in the range of 300-700 nm using Shimadzu UV-3600 UV-VIS-NIR Spectrophotometer. Finally, ZnO layer conductivity was investigated using a four-point measurement.

3 RESULTS AND DISCUSSION

3.1 ZnO material characterization

Our first objective is to determine ALD deposition conditions that lead to conductive ZnO films with good optical properties. As a first approach, we measure the dependence of ZnO conductivity and deposition rate on deposition temperature ranging from 100 to 200 °C. The results are shown in Table II.

Table II. Deposition rate and film conductivity of ZnO films deposited at different temperatures.

Deposition temperature (°C)	Depositon rate (A/cycle)	Resistivity, ρ (Ω·cm)
100	1.54	NA
125	1.76	$2.83 \cdot 10^{-2}$
150	1.76	$9.57 \cdot 10^{-3}$
200	1.49	$9.33 \cdot 10^{-3}$

A maximum deposition rate of 1.76 A/cycle is measured for 125-150 °C indicating an autolimited ALD process. Additionally, except for the film deposited at 100 °C, the rest of the films show reasonable low resistivities in the range of 10^{-2}-10^{-3} Ω·cm. This parameter does not need further optimization, since current in the IBC cell would vertically flow through the ZnO layer and consequently the ohmic losses introduced by any of these conductive films would be negligible.

Taking into account that Al_2O_3/TiO_2 films are deposited at 100 °C, we focus on ZnO deposited at 125 °C so as not to jeopardize passivation conditions. Optical characterization of this layer shows a refractive index of 1.98 at $\lambda = 632$ nm indicating that could work as a good back reflector on silicon substrates. Additionally, absorption coefficients (α) for $\lambda = 300$ to 700 nm of this material were also calculated from reflection and transmission measurements of layers deposited onto borosilicate glass. Once α is known, the optical band gap could be determined using Tauc's plot [9] which is shown in Fig. 1. As it can be seen, an optical band gap of 3.26 eV is obtained. This value agrees well with other measurements using the same technique reported in the literature [10-11]. This relatively high optical badgap indicates very low absorption in the infrared spectrum and, thus, that this film is a good choice for a back reflector of c-Si solar cells.

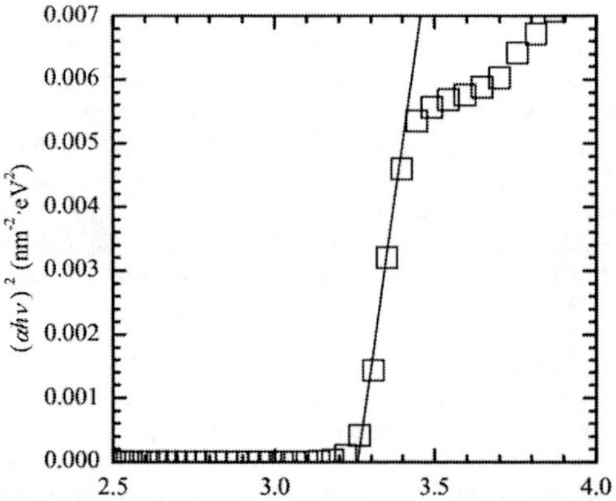

Figure 1: Tauc's plot of ZnO layer deposited at 125 °C

3.2 Passivation properties

Once we have fixed the deposition conditions of the ZnO films, we focus on their passivation properties. To do so, symmetrical samples were fabricated and lifetime measurements were carried out as explained in the experimental section. In this case, we would like to explore the effect of the ZnO thickness on passivation properties of Al_2O_3/TiO_2 stack. Three thicknesses were explored: 20 cycles (~3.5 nm), 100 cycles (~18 nm) and 420 cycles (~74 nm) the latter being optimum thickness for back reflector properties. The obtained results are shown in Fig. 2.

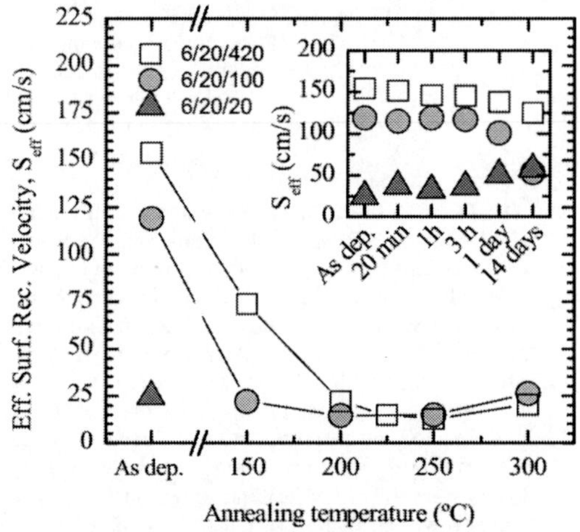

Figure 2: Effective surface recombination velocity as a function of annealing temperature; (inset) stability of the obtained passivation with time.

As it can be seen, surface passivation deteriorates from 22 to 154 cm/s with ZnO thickness increasing from 20 to 420 cycles. On the other hand, we also measured the evolution of S_{eff} with time resulting in the plot shown in the inset of Fig. 2. Interestingly, the passivation obtained with the thinnest ZnO film degrades with time while in the case of thick films it improves. Further research is needed to clarify the origin of this opposite behavior. In any case, thick ZnO films are the interesting ones from an optical point of view. In order to improve their passivation properties, we explore the effect of 5 minutes annealing in forming gas (N_2/H_2) at different temperatures whose results are shown in Fig. 2. A significant improvement is obtained even for the lowest explored temperature of 150 °C. A plateau is observed for 225-250 °C with S_{eff} values as low as 14 cm/s for both cases corresponding to an implied open-circuit (iV_{oc}) value of 688 mV. In addition, passivating properties of these annealed layers were monitored remaining constant for 5 weeks. This result is comparable to the Al_2O_3/TiO_2 stacks directly contacted by Mg/Al [5] and it demonstrates that excellent surface passivation could also be obtained with the ALD stack proposed hereby.

3.3 Contact properties

In order to get a complete picture of the contact properties, contact resistance (ρ_c) characterization is needed. Metal contacts were made of Al or Mg/Al deposited by thermal evaporation with a thickness of 1.5 µm for Al and 250 nm for Mg. Substrates symmetrically covered by the layer stack were fully contacted at the rear side while round dots with a diameter of 4 mm were defined on the front surface by means of shadow mask. Dark current-voltage (*I-V*) curves were measured at room temperature and the obtained curves are plotted in Fig. 3. Although the Magnesium contact performs better, a clear rectifying behaviour for both cases is found indicating very poor contact properties (notice the very low current in the range of 10^{-5} A).

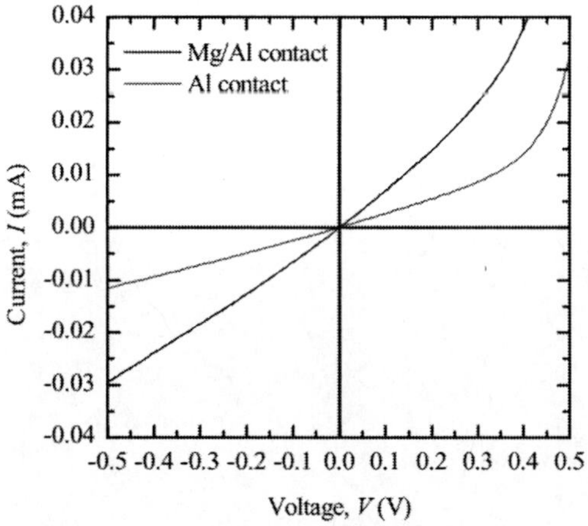

Figure 3: I-V curves of $Al_2O_3/TiO_2/ZnO$ stacks contacted by Al or Mg/Al.

In order to get a deeper insight of the origin of such a poor performance, we measured the ZnO work function using Scanning Kelvin Probe technique (SKP SPV E from KP Technology Ltd.). The obtained value is 4.85 eV which agrees with values reported in the literature [12]. This relatively high value indicates that this material is more suitable for a hole transport layer in c-Si contacts.

In order to confirm this point, we fabricate contacts on p-type substrates. In this case, the Al_2O_3/TiO_2 stack is replaced by a thermally evaporated MoO_x layer that has demonstrated good hole transport layer properties [3][13-14]. To quantitatively determine ρ_c, we use the Transfer Length Method [15] where contact bars are defined at different distances. A cross-section of the fabricated sample can be seen in Fig. 4.

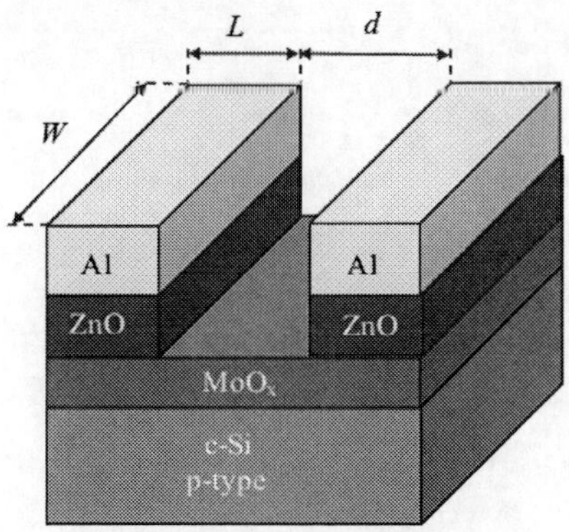

Figure 4: Cross-section of the fabricated samples to measure ρ_c on p-type substrates.

With this technique, the measured resistance shows a linear trend with the distance between bars (d) that can be modeled with the following expression:

$$R_T = \frac{R_S}{W} \times (d + 2L_T) \quad (1)$$

where R_s is the sheet resistance of the substrate, W is the width of the contacts and L_T is the transfer length that could be obtained from the x-axis cut. This magnitude indicates the average distance that a carrier travels in the silicon beneath the contact before it flows up into the contact. In case that L_T is shorter than the physical contact, then the contact geometry that plays a role in ρ_c is $W \times L_T$, whereas in the opposite case the full bar length should be considered, i.e. $W \times L$. The contact resistance (R_c) is determined as two times the y-axis cut and ρ_c is then calculated applying the corresponding effective area.

After checking out that the I-V curves of the measured samples were linear in the range of -0.5 to 0.5 V demonstrating that ohmic contacts were fabricated, we measured the resistance between bars defined at different distances leading to the plot shown in Fig. 5.

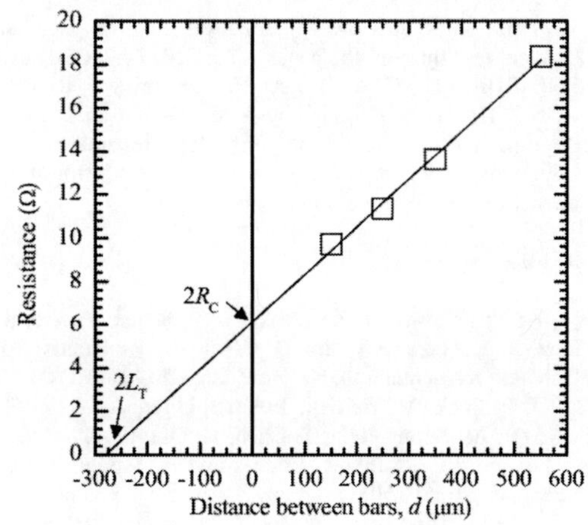

Figure 5: Measured resistance as a function of contact spacing.

As it can be seen, a value of 140 μm is obtained for L_T (notice the factor 2 in equation (1)) which is shorter than the physical contact length of 200 μm. As a consequence, knowing that W = 1 cm an area of 0.014 cm^2 should be considered for ρ_c calculation. From the y-axis we get a contact resistance of 3.068 Ω and thus a ρ_c of 43 mΩcm^2. This value is reasonable and comparable with the ones obtained for other transition metal oxide contacts like MgO_x/Al [16]. Furthermore, this result indicates that ZnO layers developed in this work have more potential if they are applied at the hole contact in our IBC structure. Then, once we have demonstrated that good quality contacts can be created on MoO_x, in future works we will address their surface passivation properties, i.e. contact selectivity, in order to introduce them into finished solar cells.

4 CONCLUSIONS

In this work, we report on the development of ZnO layers deposited by ALD. The original idea was to introduce these layers in between the Al_2O_3/TiO_2 stack, that works as an electron transport layer, and the metal contact in order to improve rear reflector properties. Using DEZ (Ditehylzinc) and water as gas precursors, we found self-limited reactions for deposition temperatures in the range of 125-150 ºC with deposition rate of 1.76 A/cycle. Since the Al_2O_3/TiO_2 stack is deposited at 100 ºC, we chose ZnO deposited at 125 ºC as a good candidate for contact formation. Additionally, an optical band gap of 3.26 eV was determined demonstrating good optical properties to work as back reflector. Surface passivation of the contacts is measured with surface recombination velocities as low as 14 cm/s after an annealing in the range of 225-250 ºC. Unfortunately, contact quality on electron transport layer is far away from the expected values. ZnO workfunction was measured resulting in 4.85 eV which suggests a higher potential as a hole contact. Then, the same ZnO layer is used to contact a hole transport layer based on MoO_x leading to a reasonable ρ_c value of 43 mΩcm. Taking into account this result, exploration of passivation properties of such contacts will be addressed in future works to determine the viability of the introduction of ZnO layers in the final IBC solar cell.

ACKNOWLEDGMENTS

This work has been supported by the Spanish Government through the projects TEC2017-82305-R and ENE2016-78933-C4-1-R, and by the project REFER COMRDI15-1-0036 funded by ACCIÓ and the European Regional Development Fund (FEDER). The authors also acknowledge the help of L.G. Gerling and J. Martorell for the help in the SKP measurements at ICFO.

REFERENCES

[1] K. Yoshikawa, H. Kawasaki, W. Yoshida, T. Irie, K. Konishi, K. Nakano, T. Uto, D. Adachi, M. Kanematsu, H. Uzu and K. Yamamoto, Nature Energy 2 (2017) 17032.
[2] J. Bullock, M. Hettick, J. Geissbühler, A.J. Ong, T. Allen, C.M. Sutter-Fella, T. Chen, H. Ota, E.W. Schaler, S. De Wolf, C. Ballif, A. Cuevas and A. Jarvey, Nature Energy 1 (2016) 15031.
[3] L.G. Gerling, S. Mahato, A. Morales-Vilches, G. Masmitja, P. Ortega, C. Voz, R. Alcubilla and J. Puigdollers, Solar Energy Materials and Solar Cells 145 (2016) 109.
[4] S. Avasthi, W.E. McClain, G. Man, A. Kahn, J. Schwartz and J.C. Sturm, Applied Physic Letters 102 (2013) 203901.
[5] G. Masmitjà, P. Ortega, J. Puigdollers, L. G. Gerling, I. Martin, C. Voz and R. Alcubilla, Journal of Materials Chemistry A 6 (2018) 3977.
[6] E. Guziewicz, M. Godlewski, L. Wachnicki, T.A. Krajewski, G. Luka, S. Gieraltowska, R. Jakiela, A. Stonert, W. Lisowski, M. Krawczyk, J.W. Sobczak and A. Jablonski, Semiconductor Science and Technology 27 (2012) 074011.
[7] T. Tynell and M. Karppinen, Semiconductor Science and Technology 29 (2014) 043001.
[8] A. Richter, S.W. Glunz, F. Werner, J. Schmidt, A. Cuevas, Physical Review B 86 (2012) 165202.
[9] J. Tauc, R. Grigorovichi, A. Vancu, Phys. Status Solidi 15 (1966) 627.
[10] Chen-Hui Zhai, Rong-Jun Zhang, Xin Chen, Yu-Xiang Zheng, Song-You Wang, Juan Liu, Ning Dai and Liang-Yao Chen, Nanoscle Research Letters 11 (2016) 407.
[11] J. Lopez, E. Solorio, H.A. Borbon-Nuñez, F.F. Castillon, R. Machorro , N. Nedev, M.H. Farías and H. Tiznado, Journal of Alloys and Compounds 691 (2017) 308.
[12] Min Wei, Chun-Fu Li, Xue-Ran Deng, Hong Deng, Energy Procedia 16 (2012) 76.
[13] J. Geissbühler, J. Werner, S. M. Nicolas, L. Barraud, A. Hessler-Wyser, M. Despeisse, S. Nicolay, A. Tomasi, B. Niesen, S. De Wolf and C. Ballif, Appl. Phys. Lett. 107 (2015) 081601.
[14] J. Bullock, C. Samundsett, A. Cuevas, D. Yan, Y. Wan and T. Allen, IEEE J. Photovoltaics 5 (2015) 1591.
[15] D.K. Schrode, "Semiconductor Material and Device Characterization", 3rd Edition, New York: Wiley-IEEE Press (2015).
[16] Y. Wan, C. Samundsett, J. Bullock, M. Hettick, T. Allen, D. Yan, J. Peng, Y. Wu, J. Cui, A. Javey and A. Cuevas, Advanced Energy Materials 7 (2017) 1601863.

AMORPHOUS SILICON-FREE METAL-OXIDES BASED CARRIER-SELECTIVE CONTACTS TO CRYSTALLINE SILICON SOLAR CELLS

Suren Patwardhan[1, 2], Sandeep Maurya[1], Akash Kumar[1] and Balasubramaniam Kavaipatti[1]
[1]Department of Energy Science and Engineering, Indian Institute of Technology Bombay, Mumbai 400076, India
[2]K J Somaiya College of Engineering, Mumbai 400077, India

ABSTRACT: Heterojunction carrier selective contact silicon solar cells are fabricated on both, Czochralski and Float-zone silicon wafers. Molybdenum trioxide and titanium dioxide are used as hole and electron selective layers respectively. The main intention of this work is to achieve dopant-free heterojunction solar cells without using intrinsic amorphous silicon passivation and any traditional anti-reflecting coating. Both the metal oxides are tested for their carrier selectivity and surface passivation. Maximum temperature used during entire cell fabrication process was 150°C. The total number of steps involved are limited to seven. The techniques used are well suited for large area fabrication over an industrial scale. Since this route avoids use of toxic and flammable gases, it results in low overall budget, which otherwise involves critical maintenance and safety measures for these gases. Preliminary results are promising showing ~ 8.7% active area efficiency at 1 Sun simulated radiation. Short circuit current density for the best cell exceeded 31 mA/cm^2 with open circuit voltage ~ 514 mV. This route has a potential to become a truly cost-effective, low thermal budget, simple and toxic gases-free process for Si cell production without losing on the current efficiency levels achieved.
Keywords: Heterojunction, c-Si, MoO_X, TiO_X

1 INTRODUCTION

Realization of carrier selectivity using heterojunction contacts to silicon has triggered research on a variety of materials, mainly metal oxides and fluorides. High work function transition metal oxides of molybdenum (MoO_X), tungsten (WO_X), vanadium (V_2O_X) [1] etc. have been demonstrated as hole selective contacts. Electron selective contacts include oxides of titanium (TiO_X) [2], magnesium (MgO_X) [3], tantalum (TaO_X) [4] and fluorides of lithium (LiF), potassium (KF), cesium (CsF) [5] etc. MoO_X and TiO_X are the most extensively studied materials with MoO_X based device reaching 22.5% [6] and TiO_X based device reaching 22.1% [7] efficiency. These are however, one-sided metal oxide-Si carrier-selective contact silicon (CSC-Si) solar cells as they include conventional n-doped/p-doped amorphous silicon (n^+ *a*-Si/p^+ *a*-Si) or boron/phosphorous diffused (p^+Si/n^+Si) layer on the other side. A recent trend is a structure involving both un-doped/asymmetric work function metal oxides/fluorides on silicon wafer. Many combinations have been demonstrated in this category with a recent record of 20.7% using MoO_X and TiO_X/LiF combination [8]. Still, they utilized passivation by intrinsic *a*-Si. An *a*-Si free CSC-Si cell with 19.1% efficiency is recently reported using aluminium oxide (Al_2O_3)/TiO_X stack for electron selectivity and V_2O_X for hole selectivity [9].

Our work is aimed at heterojunction CSC-Si cell fabrication process, which can be expanded on industrial scale. On industrial scale, it is always preferred to have lesser number of cell fabrication steps and to use cost-effective techniques. Therefore, we have set four guidelines for our work. First is to keep low overall thermal budget, second is to have a simple cell structure, third is to avoid use of intrinsic and doped *a*-Si and forth is to prepare cell for possible integration with another cell in a tandem solar cell configuration. The first guideline intends to avoid use of high temperature processes such as diffusion and thermal oxidation. Next, in order to keep a simple cell structure, the number of layers deposited on Si wafer should be minimal. This may result in some loss of efficiency but the rise in cell production cost per improvement in efficiency will be more relevant parameter, which ultimately decides the overall budget. The third guideline refers to the methods used for intrinsic as well as doped *a*-Si layers. They involve complex deposition techniques to get good interfaces with Si surface and need highly toxic and hazardous gases, which significantly increase cost due to their maintenance and safety measures. Moreover, there are problems such as parasitic absorption, degradation over time and dissociation due to UV-exposure. Following the forth guideline, we need to make hole and electron selective contacts on opposite faces of Si wafer with a suitable recombination layer instead of going for an Integrated Back Contact (IBC) type cell structure.

2 EXPERIMENTAL

2.1 Device fabrication

The device configuration as shown in Figure 1 consisted of n-type 1-5 Ω-cm single side polished (1 0 0) Czochralski (Cz)/Float-zone (Fz) silicon wafers with atomic layer deposited (ALD) TiO_2 film on one side and thermally evaporated or ALD grown MoO_3 film the other side. An ITO overlayer covered the MoO_3 film and they formed the front surface for the illumination. Evaporated Al at the back and Ag on the front served as metal contacts. The back contact was full area while the front contact consisted of finger-busbar pattern. Cell areas varied from 0.5 x 0.5 cm^2 to 1.1 x 1.1 cm^2.

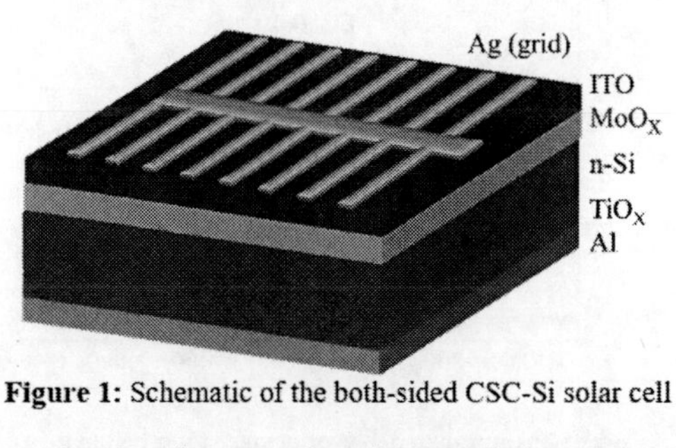

Figure 1: Schematic of the both-sided CSC-Si solar cell

The Si wafers were cleaned by standard RCA I and RCA II solutions followed by 2% Hydrofluoric acid (HF) dip. Then they were immediately transferred to the ALD chamber for TiO_x deposition. TiO_x films of thickness 3.5 nm were deposited in the ALD chamber using Tetrakis (Dimethylamido) Titanium (TDMAT) and water vapour (H_2O) as precursors for Titanium and Oxygen respectively. The process was carried at 150°C. A 300 nm thick Al layer was evaporated on the top of it. Prior to MoO_x deposition, the wafers were etched by 5% HF on front side to remove native oxide and TiO_x deposition on the front side from ALD process. MoO_x films of thickness 15 nm were deposited by thermal evaporation. Molybdenum trioxide powder source (99.97% pure, Sigma Aldrich) was used as source. The powder was taken in a closed Mo-boat having small openings for controlled evaporation. The evaporation was done by keeping the substrate at Room Temperature. For the ALD route, molybdenum hexacarbonyl [$Mo(CO)_6$] (Gelest, USA) and ozone (O_3) were used as precursors for Mo and oxygen respectively. Mo precursor sprayed using a dispenser and O_3 was obtained using an ozoniser. The process was carried at 170 °C in a flow type custom build reactor using N_2 as carrier and purging gas. Base pressure of the reactor was kept at around 1 Torr for all the experiments. MoO_x films of thickness 7-10 nm were grown. After MoO_x deposition, 80 nm thick ITO was RF sputtered at substrate temperature of 150°C. Front metallization was done by 300 nm Ag evaporated through a shadow mask bearing finger and busbar pattern.

2.2 Characterization

Dark and light I-V measurements of solar cells were taken on class AAA solar simulator from ABET Technology (Model: Sun 3000 Solar Simul). Quantum Efficiency measurements were done using QE tool from Bentham Instruments (Model: PVE 300). Thicknesses of MoO_x, TiO_x and ITO films were determined using Ellipsometry.. Suns-V_{OC} and carrier lifetime data was taken using carrier lifetime and Suns-Voc tool (Make/Model: Sinton Instruments WCT-120)

3 RESULTS AND DISCUSSIONS

Illuminated J-V plot of the best cell is shown in figure 2. This cell achieved full area J_{SC} value of 31.94 mA/cm^2 with the Fz-Si wafer and V_{OC} of 514 mV under 1 Sun simulated light.

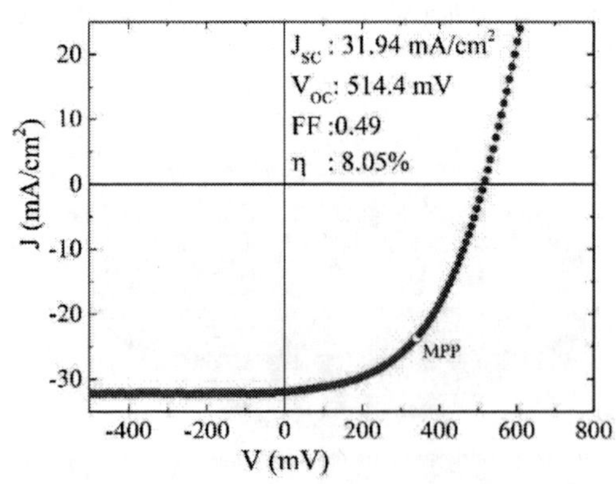

Figure 2: Illuminated J-V of best CSC-Si cell (Fz-Si)

Figure 3 shows EQE and reflectance plot of the 8.05% efficient cell shown above. Following shadowing losses, the effective area efficiency was 8.7%. The integrated J_{SC} as obtained from EQE data was 33 mA/cm^2, close to effective J_{SC} value of the cell measured from J-V curve. This value is significant since there was no texturiation or antireflecting film applied in the cell design.

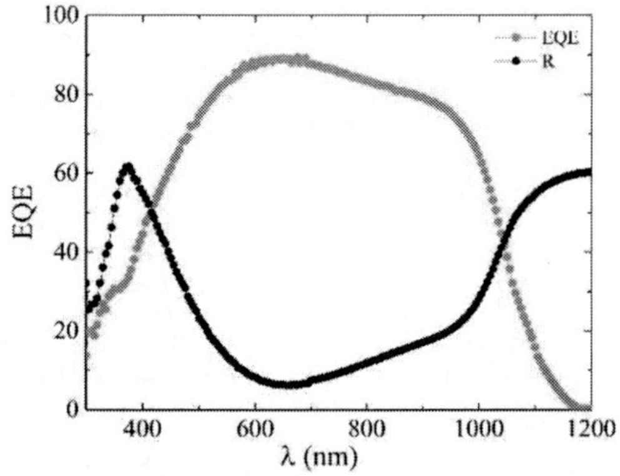

Figure 3: EQE and Reflectance of the best CSC-Si cell

The reflectance dropped down below 10% in the mid-wavelength region, which is a promising result to engineer ITO and MoO_x bilayer for effective antireflecting property in addition to carrier selectivity. The 20% EQE near short wavelengths suggests that there is much scope for improvement possible by achieving effective passivation at the front. The passivation quality achieved using MoO_x (or TiO_x) primarily depends upon the oxide state/s retained after deposition of these films. For MoO_x in particular, this becomes very critical as its oxide state is easily affected by exposure to air and humidity. Preserving the as-grown MoO_x state where $2 < x < 3$ leads to higher V_{OC}s. We have confirmed this by achieving $V_{OC} \sim 608$ mV using some experimental procedures with our reference MoO_x/Si cells. EQE in the bulk region can improve further if texturization and dedicated antireflection film such as Si_3N_4 or Si_3N_4/SiO_2 stacks are applied.

The cell fabricated on a Cz-Si wafer yielded best J_{SC} value of 27.6 mA/cm^2 and V_{OC} of 485 mV. Although, the estimated bulk lifetimes of minority carriers differed by an order of magnitude for the Fz and Cz Si wafers (~ 2500 μs for Fz-Si and 100 μs for Cz-Si), the corresponding V_{OC}s only marginally differed from each other. This indicates that the device parameters are largely governed by the properties of deposited MoO_x and TiO_x layers and the starting wafer type is less important. The measured carrier lifetime (mcd @ 1E15/cc) for MoO_x-passivated Fz-Si wafers was 55 μs. This was achieved for evaporated MoO_x. Our ALD-grown MoO_x devices yielded lower carrier lifetime and V_{OC} values. The J_{SC} values were almost same for MoO_x deposited by both techniques. Large variation (over 100 mV) in V_{OC} values was observed between wafers treated with HF just prior to MoO_x deposition and for wafers deposited with MoO_x with a delay of few hours after HF treatment (where, the samples were stored in a desiccator). The device characteristics were completely deteriorated when MoO_3 powder source stored in ambient conditions was used for evaporation. The best V_{OC} values are obtained using MoO_3 powder source stored in N_2 filled glove box permanently. In general, for CSC-Si cells without using *a*-Si interlayers, the pre-processing surface treatments of Si wafers and the deposition techniques will play the utmost important roles in the cell fabrication process.

Fill Factors (FF) for the cells were in the range 0.46-0.5 and have room for further improvement. Main reason for this low FF is the higher series resistance (R_S) observed in these devices, which was typically in the range of 700-1000 mΩ-cm^2. Since the ITO layer had reasonable conductivity (~ 1.25 x 10^3 S/cm), the interfaces namely, MoOx/ITO, ITO/Ag and TiOx/Al could be significant contributors to the total R_S. Reducing the interface contact resistance would be an important aspect of future work. Authors of this article have done some optimization mainly regaring the ITO-MoOx interface. Suns-V_{OC} measurement of cell which showed lowest R_S yielded a pseudo FF of 0.74, almost 50% more than the measured FF. This is an indication of improvisation in metal contacts itself. Thickening the busbars-fingers, back metal deposition and short low temperature annealing treatments would help to increase FF.

One peculiar observation for these cells was that they offered different resistances in dark and illuminated conditions. Due to this, the dark and illuminated curves crossed over each other as it can be clearly seen from the dark and illuminated J-V curve of figure 4. This observation was consistent for all the 44 cells fabricated in all.

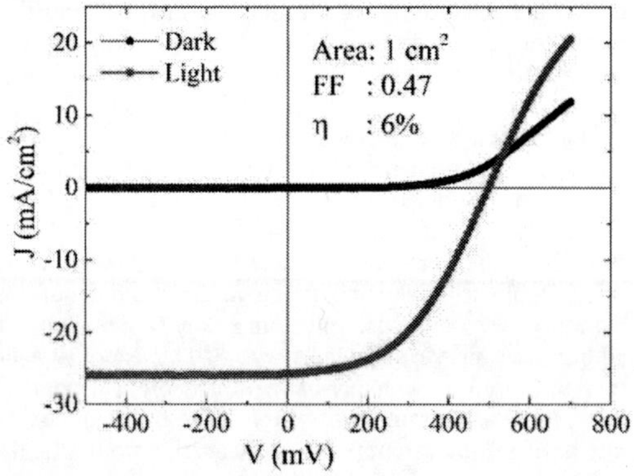

Figure 4: Violation of current superposition observed in both side CSC-Si cells

This type of behavior is however, common in a-Si/c-Si heterojunction and organic photovoltaic (OPV) cells. Such behaviour is generally attributed to non-idealities such as voltage dependent photocurrent. Moreover, this behaviour disappeared and the J-V curves followed normal current superposition in a test cell structure formed by replacing the back TiOx film with heavily doped phosphorous diffused (n^+) region. Dark and illuminated J-V characteristics of this test cell is shown in figure 2 (c) for comparison. As the test structure still had MoO_x-Si heterojunction, the voltage-dependent photocurrent might be due to TiOx-Si heterojunction. One more point to support this claim is the observed higher J_{SC} for cells with both heterojunction (nearly 40% higher) as compared to cells with only one heterojunction (that of MoOx-Si). One impact of the voltage dependent photocurrent is loss of fill factor. Apparently, our cells with both heterojunction had much lower fill factors (0.46-0.5) as compared to cells with only one heterojunction, which had fill factors in the range 0.58-0.64.

4 CONCLUSION

Carrier selective contacts to Si using MoOx and TiOx without intrinsic *a*-Si or any other additional means of passivation are demonstrated achieving preliminary efficiency of 8.7%. It involves a total of only seven processing steps keeping the fabrication simple. The V_{OC} values still exceeded 500 mV and can be increased further by optimizing MoOx and TiOx thicknesses and growth parameters. An optimized metallization would lead to higher fill factor as seen from Suns-V_{OC} measurements. Finally, almost all of the techniques used in the fabrication process are scalable on larger area cells and are cost-effective.

5 REFERENCES

[1] L. G. Gerling et. al., Transition metal oxides as hole-selective contacts in silicon heterojunctions solar cells, Sol. Energy Mater. Sol. Cells, vol. 145, pp. 109–115, Feb. 2016.
[2] X. Yang et. al., Silicon heterojunction solar cells with electron selective TiOx contact, Sol. Energy Mater. Sol. Cells, vol. 150, pp. 32–38, Jun. 2016.
[3] Y. Wan et. al., Conductive and Stable Magnesium Oxide Electron-Selective Contacts for Efficient Silicon Solar Cells, Adv. Energy Mater., vol. 7, no. 5, p. 1601863, Mar. 2017.
[4] Y. Wan et. al., Tantalum Oxide Electron-Selective Heterocontacts for Silicon Photovoltaics and Photoelectrochemical Water Reduction, ACS Energy Lett., vol. 3, no. 1, pp. 125–131, Jan. 2018.
[5] J. Bullock et. al., Efficient silicon solar cells with dopant-free asymmetric heterocontacts, Nat. Energy, vol. 1, no. January, 2016.
[6] J. Geissbühler et. al., 22.5% efficient silicon heterojunction solar cell with molybdenum oxide hole collector, Appl. Phys. Lett., vol. 107, no. 8, Aug. 2015.
[7] X. Yang, K. Weber, Z. Hameiri, and S. De Wolf, Industrially feasible, dopant-free, carrier-selective contacts for high-efficiency silicon solar cells, Prog. Photovoltaics Res. Appl., vol. 25, no. 11, pp. 896–904, Nov. 2017.
[8] J. Bullock et. al., Stable Dopant-Free Asymmetric Heterocontact Silicon Solar Cells with Efficiencies above 20%, ACS Energy Lett., vol. 1, pp. 508–513, Jan. 2018.
[9] G. Masmitja et. al., Interdigitated back-contacted crystalline silicon solar cells with low-temperature dopant-free selective contacts, J. Mater. Chem. A, vol. 6, no. 9, pp. 3977–3985, 2018.

6 ACKNOWLEDGEMENTS

This work was carried under National Centre for Photovoltaic Research and Education (NCPRE), a project funded by Ministry of New and Renewable Energy (MNRE), Government of India. We are thankful to them. We are also thankful to technical staff of NCPRE, IIT Bombay central facilities and Centre for Excellence in Nanotechnology (CEN). We express gratitude towards Dr. Ashok Sharma and Mr. Guru Burkule from NCPRE, Mr. Sunil Kale, Ms. Neha Ravrane, Mr Pijush from CEN and our colleagues Mr. Ashish Singh, Mr. Chandan Das and Ms. Garima Aggrawal for their help in experimentation, analyses and valuable suggestions from time to time.

CONTACTLESS INVESTIGATION OF THE P-TYPE DOPING CONCENTRATION LEVEL OF SINGLE MICROMETRIC SIZE GaAs CRYSTALS GROWN ON SILICON FOR MULTIJUNCTION SOLAR CELLS

Alexandre Jaffré[1], Jose Alvarez[1], Hung-Ling Chen[2], Hajer Makhloufi[2], Charles Renard[2],
Florent Loëte[1], Stéphane Collin[2], James Patrick Connolly[1], Jean-Paul Kleider[1], Denis Mencaraglia[1]
[1]GeePs, Group of Electrical Engineering Paris, CNRS, CentraleSupelec, Univ. Paris-Sud, Université Paris-Saclay, Sorbonne Université, 3&11 rue Joliot-Curie, Plateau de Moulon, 91192 Gif-sur-Yvette CEDEX, France
[2]C2N, Centre de Nanosciences et de Nanotechnologies, CNRS, Univ. Paris-Sud, Université Paris-Saclay, Avenue de la Vauve, 91120 Palaiseau, France

ABSTRACT: In previous work, we have demonstrated the perfect integration on silicon of micrometric GaAs crystals without any structural defects nor stress, using Epitaxal Lateral Overgrowth on Tunnel Oxide from nanoseeds (ELTOn). Then, it would be very interesting to integrate the crystals in a regular way to have a quasi-complete covering of the Si substrate, without coalescence of the GaAs microcrystals to maintain their very good electronic properties, avoiding then detrimental grain boundaries. The main focus of this work is to address the issue of the doping determination in a single micrometric size GaAs crystal which is a pre-requisite to develop the future design and technology of multijunction solar cells based on an array of non-coalescent GaAs crystals free of structural defects. In a previous work, local electrical characterizations by CP-AFM revealed a rather high non intentional p-type doping. To obtain a quantitative estimation of the doping level, we propose a contactless method based on the photoluminescence measurement of the evolution of the bandgap versus temperature and its modeling.
Keywords: Epitaxy, III-V Semiconductors, Doping, Photoluminescence, Multijunction Solar Cell, Silicon Solar Cell.

1 INTRODUCTION

For thirty years, researchers have been attempting to integrate III–V semiconductors, in particular, GaAs on silicon. However, three major intrinsic problems have to be overcome: the lattice parameter mismatch generating a high density of threading dislocations, the polar or non polar nature of GaAs or Si respectively, and the difference of thermal expansion coefficient, which is at the origin of cracks for thick GaAs layers [1]. Several approaches to these problems have been investigated to date, but until now, no one succeeded in achieving the defect-free and cost-effective integration of 2D continuous GaAs layers. One of these approaches, selective area epitaxy (SAE) of GaAs on patterned Si, has shown significant improvements, yielding mostly defect-free epitaxial layers. However, the relaxation process occurring at the GaAs/Si interface leads to a high density of misfit dislocations at levels, which are prohibitive for applications involving carrier transport through the interface area. We have recently demonstrated by a new method called ELTOn (for Epitaxial Lateral overgrowth on Tunnel Oxide from nano-seed) the defect- and strain-free heteropitaxy of micrometric GaAs crystal on Si, yielding an effective electrical contact of the epitaxial layer to the substrate [2].
An interesting consequence of this is that unlike many optoelectronic devices where continuous layers are desirable, the micrometric GaAs crystals form a textured surface which can be exploited to improve light management by reducing front surface reflection and contributing to light trapping. The GaAs array of non coalescent micro-crystals, free of structural defects, providing a quasi complete covering of the Si substrate, is therefore a promising route to III-V on Si mulltijunction solar cells.
Cost efficient multijunctions solar cells could be based on this approach if the doping of the GaAs microcrystals is fully controlled. However the doping level investigation on a single crystal is difficult by I(V) or C(V) measurements because in the first case the determination is too indirect and for the second, the determination is direct but the capacitance values will be too small to be evaluated.
It is then mandatory to develop a method sensitive enough to detect the doping level in one unique GaAs micrometric size crystal. The purpose of this work is to present such a contactless method based on the photoluminescence (PL) of a unique crystal which allows a precise measurement of the material bandgap energy. We have performed this measurement at different temperatures to investigate the bandgap evolution. Then modelling the bandgap dependence with temperature using published data and parametrized models of the literature, we could determine the doping density of a unique micrometric size GaAs crystal. This is not such an easy task both on an experimental or theoretical point of view. Indeed, experimentally even if the bandgap energy can be determined with a great accuracy from PL, the precise knowledge of the crystal lattice temperature is not always straightforward and can be somewhat different from the requested temperature. Some gradient can exist between the sample and the temperature probe, and also any significant sample heating by the laser excitation has to be avoided. Moreover, theoretically the semiconductors bandgap dependence with temperature is a complex physical problem for which up to now only semi-empirical relations have been proposed. Nevertheless a lot of work has already been performed on this topic and accurate parameter data sets are available, in particular for GaAs. To assess the validity of our method, we will first present its implementation on a nominally undoped GaAs bulk wafer (denominated here GaAs Ref sample), crosschecking the results with complementary experiments such as cathodo-luminescence for the bandgap determination, or a contactless Multi Frequency Eddy Current System (MFECS) that has been developed in GeePs to determine the GaAs Ref sample conductivity. Raman LO and TO peak positions were also measured with the same experimental conditions as for the PL measurements and compared to data of the literature for a strain free GaAs in order to have a better estimation of the lattice temperature of the samples under test.

2 EXPERIMENTAL SETUPS AND PROCEDURES

2.1 Micrometric size GaAs crystals growth

The micrometric size GaAs crystals were grown on silicon through a silica mask structured with nanoscaled openings (Fig. 1). The selective hetero-epitaxy ELTOn (for Epitaxial Lateral overgrowth on Tunnel Oxide from nano-seed) was realized by chemical beam epitaxy (CBE) at the growth temperature of 575 C using the gaseous precursors Trimethylgallium (TMGa) and Tertiarybutylarsine (TBAs). For more details of the growth procedure, and structural properties of our micrometric size GaAs crystals, the reader is referred to the previous studies [2 and our references therein].

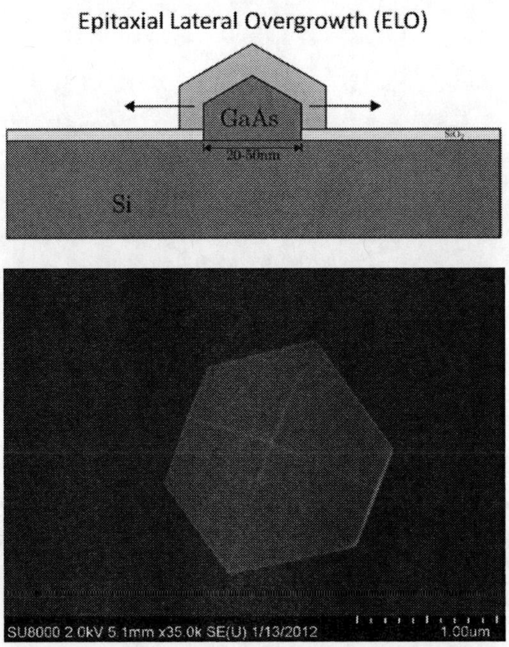

Figure 1: Top: Schematic view of Epitaxial Lateral overgrowth on Tunnel Oxide from nano-seed (ELTOn) through a silica mask structured with nanoscaled openings ; Bottom : Top SEM view of a micrometric size GaAs crystal grown on silicon using the ELTOn process.

2.2 µ-Photoluminescence/µ-Raman confocal microscope

µ-PL and µ-Raman measurements were performed on a WITec alpha-300R confocal microscope in a backscattering configuration. Samples were under vacuum at 10^{-3} mbar in a MMR cryostat and probed at a requested temperature imposed by the temperature controller from 100K to 450K, each 50K. Samples were excited at 532nm through an Olympus 20x long working distance objective, with 6.5mW laser power (82µW/µm²). This injection power is high but we verified that it does not affect the PL maximum peak position nor make the Raman peak shift. We choose this power to have enough Raman signal and to be able to compare PL and Raman in the same injection conditions. Raman scattered light is guided through a 50µm (pinhole) optical fiber to an UHTS300 spectrometer using a 1800 grooves/mm diffraction grating and a CCD silicon camera. For µ-PL analysis, we use a 200µm core diameter fiber, a Princeton SP-2300 spectrometer with a 150grooves/mm grating and an InGaAs CCD camera. Both cameras are cooled down to -70°C via 3-stages Peltier cooler.

Luminescence data are corrected by removing the spectral transfer function of the microscope. To achieve this we use an Ocean Optics HL-2000-CAL halogen calibration lamp. We place it as a point source under the microscope objective. Knowing the spectral irradiance of the lamp and measuring its spectra through the whole optical path (from the objective to the CCD camera), we are able to correct the luminescence data. The bandgap value of the sample is extracted for each temperature by evaluating the energy position of the maximum of the luminescence spectra.

Raman peak position and Full Width at Half Maximum (FWHM) are calculated by a Voigt function fit for each temperature step. Raman LO and TO peak positions for a strain free GaAs (ω(LO)=297.6cm^{-1} at 0K) are represented as a function of the requested temperature (not shown here) with an empirical model described in [3]. We compare our experimental data to this model and extract the local lattice temperature (denominated here Raman extracted temperature, T Raman extracted) by matching our Raman peak position values to the ones of the empirical model.

For PL and Raman measurements at each temperature step, we proceed as follows : we first reach 100K, then heat up the sample to 450K by taking spectra each 50K. For each step we wait five minutes before recording a spectrum to let the sample thermalize. We make two heat-up cycles with the same timing to be accurate, the first to collect PL, the second to record Raman scattering.

2.3 Nano-Cathodoluminescence

The nano-CL setup from Attolight brand is located at C2N. Cathodo-luminescence principle is based on the emission of light of a solid material which is excited with an electron beam. This tool has a spatial resolution down to 10nm. Samples were kept under vaccum in a He cryostat. CL measurements were performed at different temperature stages from 25K to 300K. The electron beam was accelerated at 6keV in order to have the same order of penetration depth in GaAs (~80nm) as for the 532nm PL excitation laser. Backscattered luminescence photons are injected in a spectrometer via a parabolic mirror. Spectral data were diffracted with a 150 grooves/mm grating and analyzed on a 3-stages Peltier cooled Silicon EMCCD camera. The Cathodo-luminescence data are also corrected with the spectral transfer function of the setup.

2.4 Conductance measurements

Conductance measurements have been achieved on our nominally undoped GaAs Ref sample in order to extract its conductivity and have an estimation of the doping concentration due to active residual impurities using a Multi Frequency Eddy Current System (MFECS) that has been developed in GeePs [4]. This patented contactless technique is based on eddy currents. The eddy current probe, constituted of a coil connected to a transmission line, interacts with the semiconductor wafer under inspection. The coil induces eddy currents in the semiconductor wafer that in return involve the creation of an induced magnetic field and a modification of the total magnetic field inside and in the vicinity of the wafer. The coil impedance is measured by reflectometry using a broadband multicarrier test signal, i.e. containing multiple frequencies. An analytical electromagnetic model of the coil-wafer interaction is then used to estimate the conductivity of the wafer [5, 6].

3 MODELLING METHOD

For many years, the Varshni's semi-empirical relation has been used to describe semiconductors bandgap dependence with temperature [7]. However the inadequate analytical structure of Varshni's formula was explained by Pässler in 1999 who proposed an analytical four-parameters expression capable of providing better numerical fittings and estimations of physical parameters [8]. Pässler refined further this work in 2003 [9] but considering our investigated temperature range, for the present purpose it is sufficient to use the previously proposed analytical four-parameters expression written as

$$E(T) = E(0) - \frac{\alpha\theta_p}{2}\left[\sqrt[p]{1+\left(\frac{2T}{\theta_p}\right)^p} - 1\right], \qquad (1)$$

with, in the GaAs case, the following values of the four parameters :
$E(0) = 1.519$ eV, $p = 2.44$, $\alpha = 0.472$ meV/K, $\theta_p = 230$ K and T being the sample lattice temperature.
This four-parameter Pässler's data set has been determined by fitting the high precision PL data of Grilli et al. [10] measured on a high-quality nominally undoped GaAs sample grown by molecular beam epitaxy.
For heavily p-type doped samples, we take into account the band gap narrowing (BGN) substracting to the right hand-side of Eq. (1) the expression of ΔE_{BGN} given by

$$\Delta E_{BGN} = AP^{\frac{1}{3}} + BP^{\frac{1}{4}} + CP^{\frac{1}{2}}, \qquad (2)$$

where P is the active dopant concentration, A, B, and C having the following constant values:
$A = 9.83\times10^{-9}$ eV/cm, $B = 3.90\times10^{-7}$eV/cm$^{3/4}$, $C = 3.90\times10^{-12}$ eV/cm$^{3/2}$ for p-type GaAs if ΔE_{BGN} is expressed in eV and P in cm^{-3} [11].
We will show that a least-square fit of our experimental data, without any change in the parameters of Eqs. (1) and (2) allowed us to have a good estimation of the active dopant density. The quality of the fits will be compared using either the requested temperature or the Raman extracted one defined in section 2.2.

4 RESULTS AND DISCUSSION

4.1 Validation of the accurate bandgap extraction

To check the accuracy of our experiments we first worked on a GaAs reference bulk and have done several cross-checks on it. The first step was to compare PL and CL spectra at 300K in equivalent experimental conditions (6keV electron acceleration for the CL so that the penetration depth is around 80nm in GaAs, equivalent to the 532nm PL Laser penetration depth). Figure 2 shows normalized PL and CL spectra in semi-log scale for our GaAs reference wafer. The curves are well superimposed around the bandgap value. On the PL spectrum we notice some noise at higher energies. This is due to the fact that PL detector is an InGaAs CCD camera, so the noise is important in this low CCD sensitivity region. We also notice that the slope of the high energy part is not constant on the PL spectra compared to the CL ones. This could be the consequence of hot carriers effect during the PL experiment.

In the low energy region, the noise is higher on the CL results. The reason is the same than previously but this time, it is correlated to low sensitivity of the Si CCD camera in this energy range. This comparison shows that the spectral acquisition is well calibrated. We also performed this study on the crystal and the curve behaviors are the same. From this point we extract the bandgap value of our sample by evaluating the energy position of the maximum of the luminescence spectra. At 300K, we extract a bandgap value of 1.422eV for the GaAs reference bulk which is in agreement with the literature [12].

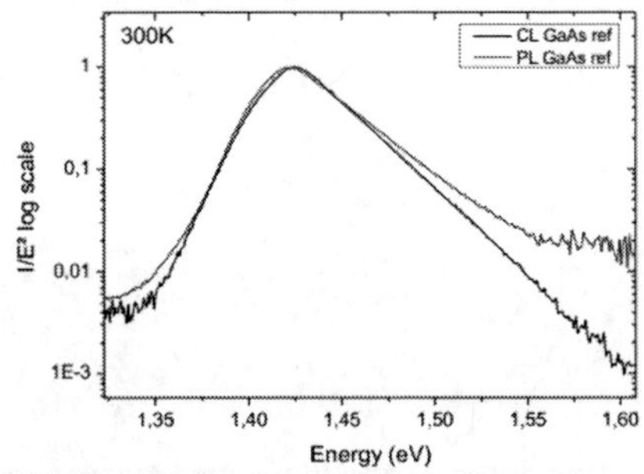

Figure 2: PL and CL spectral comparison at 300K on the reference bulk

4.2 Assessment of the undoped GaAs bulk reference quality

We performed Raman spectroscopy measurements at 300K of our undoped GaAs (100) reference bulk (Fig. 3). TO and LO modes position and FWHM are consistent with values given in the literature for an undoped GaAs bulk [13].

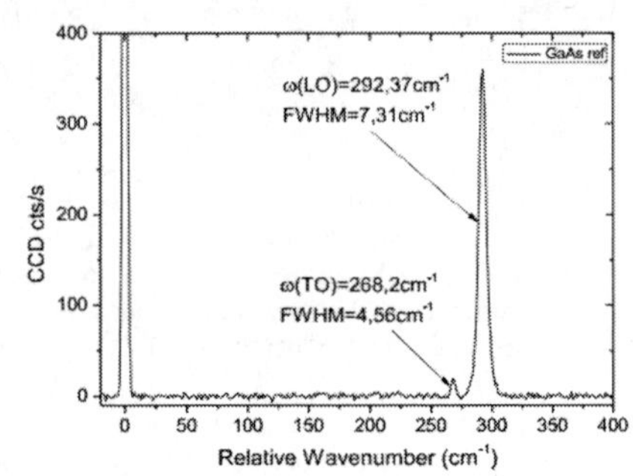

Figure 3: Raman spectrum of the undoped GaAs (100) reference bulk at 300K.

Our Multi Frequency Eddy Current System (MFECS) setup was used to derive the conductivity at room temperature of our 444 μm thick undoped GaAs reference sample. The MFECS data and its fit are shown in Fig. 4. The normalized coil impedance variation $\delta Z / X0$ due to the presence of the sample can be modeled as a function of the layer properties (conductivity, thickness) [6]. Knowing the thickness of the wafer, its conductivity was derived to be equal to $5.\Omega^{-1}$cm^{-1}, probably due to residual electrically active impurities. Taking into account the mobility dependence versus doping content, such a conductivity can be attributed either to a n-type doping around 5.10^{15} cm^{-3}, or to a p-type one around 2.10^{17} cm^{-3}. However such a rather high p-type doping content would have led to a bandgap narrowing effect easily detectable with our PL measurement. This is not the case as we will see later in this section (refer to Fig. 5). Then the only hypothesis left is a light n-type doping density around 5.10^{15} cm^{-3}. This shows that our unintentionally doped GaAs Ref behaves as a lightly n-type doped sample due to a low residual doping level that does not affect significantly the temperature dependence of the bangap as we will see in the next paragraph.

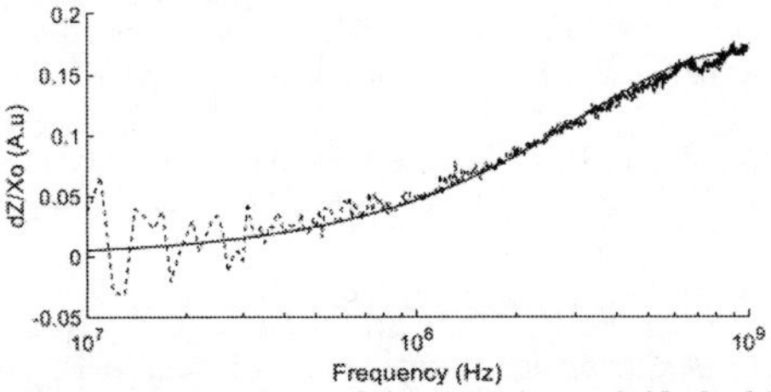

Figure 4: Comparison of the experimental (dashed line) and fitted (full line) normalized impedance variation of the coil in presence of the GaAs sample, as a function of frequency

Figure 5 shows the temperature dependence of the bandgap of our GaAs reference wafer derived from PL measurements (symbols) compared to the Pässler's analytical four-parameters set of data (full line) calculated with Eq. 1.

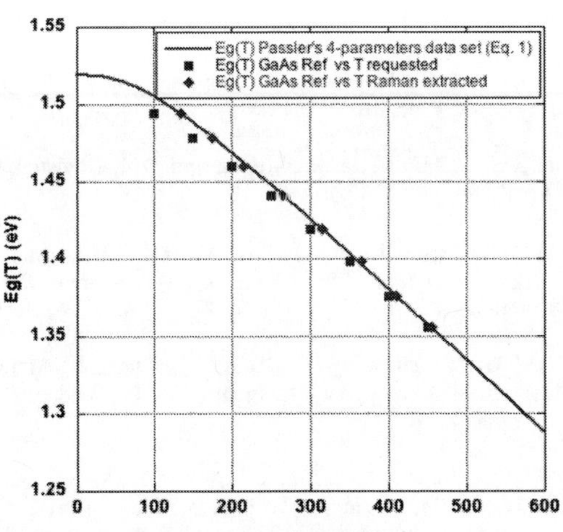

Figure 5: Temperature dependence of the bandgap of our GaAs reference wafer derived from PL measurements (symbols) compared to the Pässler's analytical four-parameters set of data (full line) calculated with Eq. 1.

We can note that using the Raman extracted temperature to deduce the lattice temperature of the sample under test, a perfect agreement is obtained between the bandgap temperature evolution of our GaAs Ref sample and the one modeled with Pässler's Eq.1 for a high quality undoped GaAs sample. This comparison also attests for the very good quality of our GaAs Ref sample and permits to validate the better estimation of the lattice temperature when using the Raman extracted temperature instead of the requested temperature.

4.3 Extraction of the doping density on a single GaAs crystal grown on silicon

Figures 6 and Figure 7 show the temperature dependence of the bandgap of a GaAs crystal derived from PL measurements, when the lattice temperature has been taken equal to the requested one or to the Raman extracted one, respectively.

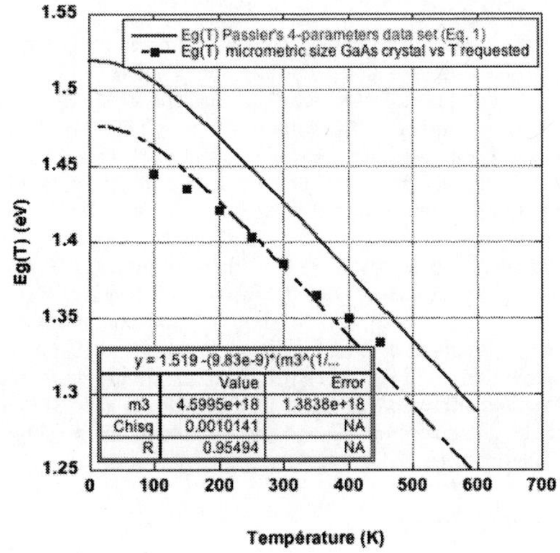

Figure 6: Temperature dependence of the bandgap of a micrometric size GaAs crystal grown on Si derived from PL measurements (symbols); curve fit (dashed line) using Pässler's Eq.1 (full line) corrected for the bandgap narrowing (Eq.2). The temperature in this figure corresponds to the requested one.

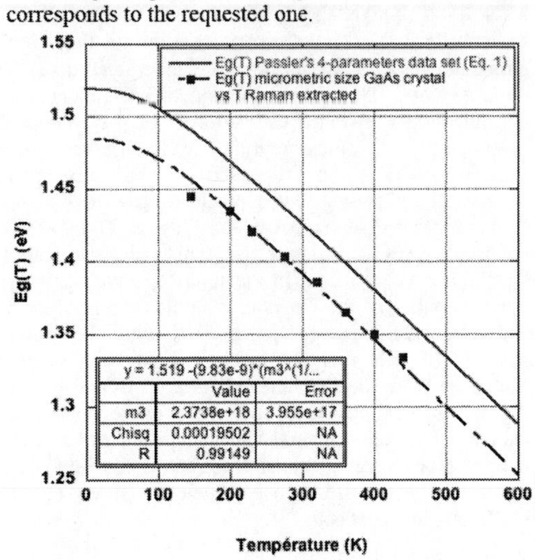

Figure 7: Temperature dependence of the bandgap of a micrometric size GaAs crystal grown on Si derived from PL measurements (symbols); curve fit (dashed line) using Pässler's Eq.1 (full line) corrected for the bandgap narrowing (Eq.2). The temperature in this figure corresponds to the Raman extracted one.

In both figures, the full line represents the Pässler's analytical four-parameters set of data calculated with

Eq.1 for a nominally undoped GaAs sample and the dashed line corresponds to the fit using Pässler's Eq.1 corrected for the bandgap narrowing (Eq.2). The quantitative result of the fit with the doping concentration as the only unknown parameter is given in the inserted tables of both figures, where m3 (representing P of Eq. 2) is the fitted variable. We can note the better quality of the fit in Fig. 7 when the lattice temperature has been taken equal to the Raman extracted one, leading to a p-type doping density comprised between $2.0x10^{18}$ cm^{-3} and $2.8x10^{18}$ cm^{-3}.
Such a rather high doping content has been attributed to carbon incoporation during the Epitaxal Lateral Overgrowth on Tunnel Oxide from nanoseeds (ELTOn) process using TriMethyl-Gallium (TMGa) and TertiaryButyl-Arsine (TBAs) as gas precursors [14]. This high carbon doping has been observed in the literature, and one of the proposed solutions to control the carbon incorporation is to adjust the concentration by the simultaneous use of TEGa (TriEthyl-Gallium) and TMGa together with TBAs. Using TEGa and TMGa in variable proportions is expected to allow a good control of the level of doping over a range from 10^{14} cm-3 to 10^{20} cm-3 [15]. This will be investigated in the future with the method presented in this paper which we have shown able to determine the doping content in a single micrometric size GaAs crystal.

5 SUMMARY

In this paper, contactless methods (PL, CL, Raman thermography) were investigated to determine the doping level on a single micrometric size GaAs crystal from the bandgap evolution with temperature. We have seen that the precise knowledge of the lattice temperature is a critical point to derive a good estimation of the doping level. Raman Stokes shift analysis versus temperature has been used to extract a more accurate determination of the lattice temperature of the sample under test than the requested one. The refined method presented here has permitted to have a better estimation of the p-type doping level due to carbon incorporation already observed in our previous works and attributed to an incomplete decomposition of the TMGa precursor gas used in our Epitaxal Lateral Overgrowth on Tunnel Oxide from nanoseeds (ELTOn) process. This characterization technique will be useful in the future to investigate the doping control of the GaAs crystals using a mixture of TMGa and TEGa as gas precursors during the growth process. If such a doping control during our ELTOn growth process can be achieved for both n- and p-type, the GaAs array of non coalescent micrometric size crystals, free of structural defects, providing a quasi complete covering of the Si substrate, could be a very promising route for cost efficient III-V on Si mulltijunction solar cells.

Acknowledgments
This work was supported by the ANR (project MULTISOLSI No. 2011 PRGE 00901). The authors also thank the Centrale de Technologie Universitaire MINERVE and RENATECH for technological backup.

References

[1] S. F. Fang, K. Adomi, S. Iyer, H. Morko, H. Zabel, C. Choi, and N. Otsuka, J. Appl. Phys. 68, R31 (1990)

[2] C. Renard, T. Molière, N. Cherkashin, J. Alvarez, L. Vincent, A. Jaffré, G. Hallais, J.P. Connolly, D. Mencaraglia, and D. Bouchier, Sci. Rep. 6, 25328 (2016)

[3] A. M. Ardila, O. Martínez, M. Avella, J. Jiménez, B. Gérard, J. Napierala, and E. Gil-Lafon, J. Appl. Phys. 91, 5045 (2002)

[4] D. Mencaraglia, Y. Le Bihan, F. Loëte, Contactless measurement of the conductivity of semiconductors, Patent FR 15 53392, 16/04/2015; WO2016/166449A1

[5] F. Loete, Y. Le Bihan and D. Mencaraglia, IEEE sensors journal, vol. 16, no. 11, pp.4151-4152, 2016

[6] C. V. Dodd and W. E. Deeds, J. Appl. Phys. 39, 2829 (1968)

[7] Y. P. Varshni, Physica 34, 149 (1967)

[8] R. Pässler, Phys. Stat. Sol. (b) 216, 975 (1999)

[9] R. Pässler, Phys. Stat. Sol. (b) 236, No. 3, 710 (2003)

[10] E. Grilli, M. Guzzi, R. Zamboni, and L. Pavesi, Phys. Rev. B 45, 1638 (1992)

[11] S. C. Jain, J. M. McGregor, and D. J. Roulston, J. Appl. Phys. 68, 3747 (1990)

[12] O. Madelung et al., Group III Condensed Matter 41A1b - Group IV Elements, IV-IV and III-V Compounds; p. 1174

[13] O. Madelung et al., Group III Condensed Matter 41A1b - Group IV Elements, IV-IV and III-V Compounds; p. 1470

[14] T. Moliere, A. Jaffre, J. Alvarez, D. Mencaraglia, J.P. Connolly, L. Vincent, G. Hallais, D. Mangelinck, M. Descoins, D. Bouchier and C. Renard, J. Appl. Phys. 121, 035704 (2017)

[15] H. Lüth, *Solid Surfaces, Interfaces and Thin Films*, Springer Science & Business Media, 2 sept 2010, p. 166

COMPARISON BETWEEN AMORPHOUS SILICON LAYERS DEPOSITED FOR HETEROJUNCTION SOLAR CELLS AT 13.56 MHZ AND 140 MHZ EXCITATION FREQUNECY

B. Leszczynska*[1], C. Strobel[1], S. Leszczynski[1], M. Albert[1], F. Stahr[2], J. Kuske[2], J. W. Bartha[1]

[1] Dresden University of Technology, Semiconductor and Microsystems Technology Laboratory, 01062 Dresden, Germany, *Phone: +49 351 463 33151, e-mail: barbara.leszczynska@tu-dresden.de
[2] Forschungs- und Applikationslabor Plasmatechnik GmbH, Gostritzer Str. 67, 01217 Dresden, Germany

ABSTRACT: The plasma enhanced chemical vapor deposition (PECVD) process is used for the deposition of thin-film silicon. By varying the deposition conditions, it is possible to produce silicon layers with different material structures. The increase of the excitation frequency leads to a decrease of the ion bombardment energy on the growing layer and a reduction of the surface damage during the deposition is possible [1]. The recombination losses at the interface between absorber and emitter layer strongly affects the heterojunction solar cells (HIT) and excellent interfacial passivation is required [2]. For this purpose, the application of very high plasma excitation frequencies (VHF) for the deposition of passivation layers in comparison with standard radio frequency (RF) is investigated. The VHF-process enables a highly productive deposition of a-Si:H with a significant enhancement of the deposition rate. Thus, the influence of the deposition rate on layer properties for 13.56 and 140 MHz frequency will be presented.

Keywords: PECVD, amorphous silicon, solar cell, very high frequency,

1 INTRODUCTION

The silicon heterojunction solar cells (HIT) based on crystalline silicon (c-Si) wafers and amorphous silicon layers (a-Si:H) has a great potential for cost reduction [2-4]. However, the recombination losses at the interface between absorber and emitter layer strongly affect the solar cell parameters [2]. In order to reduce the damages of the silicon wafer surface during the a-Si:H deposition the very high frequency plasma enhanced chemical vapor deposition technique (VHF-PECVD) has been introduced. An application of the very high excitation frequencies promises a reduction of the plasma induced surface damages through a reduction of the ion bombardment on the growing layers.

It has already been shown that an enhancement of the excitation frequency up to 140 MHz leads to a deposition rate increase without deterioration of the a-Si:H material [5]. In this study, the electrical and structural properties of a-Si:H layers deposited with varied deposition parameters like hydrogen concentration and plasma power for the 13.56 MHz and 140 MHz process will be compared. A correlation between material density and infrared spectral analysis will be carried out. Thereby, the infrared spectroscopy will be applied to investigate the influence of the Si-H bonding configuration on the material properties for different deposition conditions. Furthermore, the correlation between the excitation frequency and the interface properties between amorphous and crystalline silicon will be investigated by minority carrier lifetime.

2 EXPERIMENTAL

2.1 VHF-PECVD deposition system

A capacitively coupled inline VHF-PECVD system with linear plasma sources developed by FAP GmbH Dresden was used for the fabrication of thin-film silicon layers (Figure 1). The VHF input signal is generated using a signal generator and an amplifier with the frequency range of 100 kHz to 250 MHz. For working in the VHF range a new electrode with low electrode capacity has been specially designed to provide a suitable power coupling for plasma ignition [5].

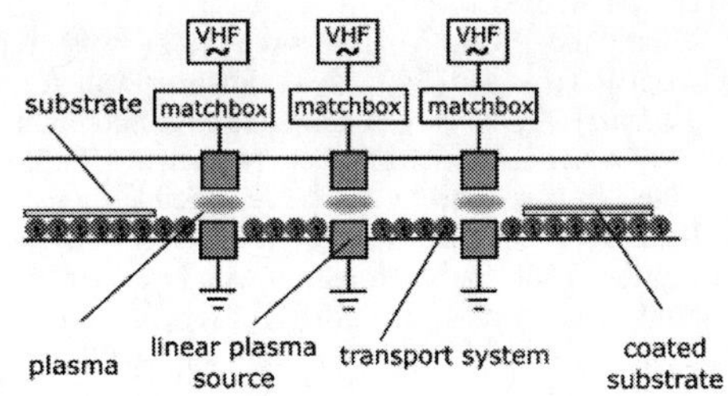

Figure 1: Inline deposition system- principle.

2.2 Experimental details

The a-Si:H intrinsic single layers were deposited on Corning Eagle XG glass substrates to analyze the deposition rate. Additionally, doubled sided polished c-Si wafers were used for structural measurements. For passivation studies the a-Si:H layers were deposited on 250 µm thick n-type double sided polished float zone silicon wafers with 1 - 5 Ωcm resistivity.

The substrate temperature during the deposition process was 170 °C. The process gases used for the deposition were silane (SiH_4) and hydrogen. The amorphous silicon was deposited at 30 Pa process pressure at varied silane concentration (SC) and power densities.

First, the thickness of the a-Si:H layers was measured by a surface profilometer from Veeco. Then, to exam the structural properties of the amorphous material the infrared spectroscopy was applied. The samples were measured from 380 to 3000 cm^{-1} using a Bruker Vertex 80v spectrometer in transmission mode and a DTGS detector. The influence of the Si-H bonding configuration on the material properties for different deposition parameters was studied. The behavior of the 840 cm^{-1} bending mode and the 2100 cm^{-1} stretching mode was investigated.

As the next step, the mass density (ρ) of the deposited films was calculated using the Clausius–Mossoti equation [6]:

$$\rho = \left(\frac{n^2-1}{n^2+2}\right) \cdot \frac{3m_{Si}}{4\pi \cdot \left[2\alpha_{Si-Si} + \frac{C_H}{1-C_H} \cdot \left(\alpha_{Si-H} - \frac{1}{2}\alpha_{Si-Si}\right)\right]}$$

where:
m_{Si} - mass of the Si atom
C_H - total hydrogen content
α_{Si-Si} and α_{Si-H} - bond polarizability in the amorphous phase of the Si–Si and Si–H bond

Finally, to characterize the passivation properties of amorphous silicon layers the effective carrier lifetime t_{eff} was measured with microwave detected photoconductivity (MDP) at the injection level of 10^{15}cm^{-3} [7].

3 MATERIAL PROPERTIES OF THE SILICON THIN-FILM DEPOSITED AT 13.56 AND 140 MHz

To get further insight into the influence of the excitation frequency on the material properties a correlation between the process parameters (silane concentration, power density) and properties of the a-Si:H layers was studied.

First, the silane concentration (SC) was varied between 10 % and 100 % at low deposition rates (< 0.4 nm/s). An increase in the silane concentration leads to lower material densities for both frequencies (Figure 2). At 100 % silane concentration a remarkable increase of Si-H_2 and (Si-H_2)$_n$ chains has been observed. The strong absorption at 840 cm^{-1} is correlated with near-neighboring pairs of Si-H_2 groups and (Si-H_2)$_n$ chains in the material structure, whereas the 2100 cm^{-1} mode is due to isolated Si-H_2 groups [8].

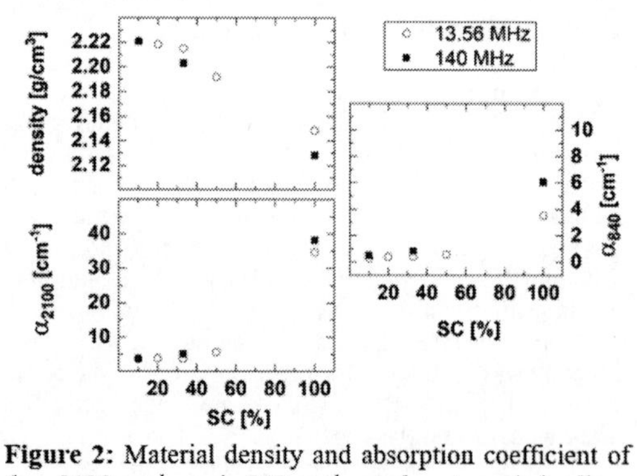

Figure 2: Material density and absorption coefficient of the 2100 cm^{-1} and 840 cm^{-1} mode at varied silane concentration.

Next, the influence of plasma power and deposition rate on the material properties was investigated. It was observed, that the deposition rate strongly depends on the power density and excitation frequency (Figure 3). Furthermore, at the same process parameters the deposition rate at the 140 MHz frequency is five times greater than at 13.56 MHz

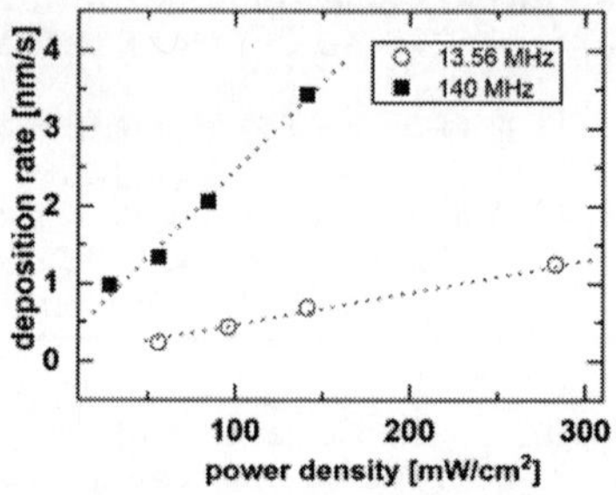

Figure 3: Deposition rate of a-Si:H as a function of the power density for 13.56 and 140 MHz frequency.

As a next step, the structural properties of a-Si:H were studied (Figure 4), whereby the deposition rate was varied by the power density. An enhancement of the deposition rate due to an increased power leads to lower material densities for both 140 and 13.56 MHz frequency. Additionally, a remarkable increase of both Si-H_2 and (Si-H_2)$_n$ polymers was observed. At higher deposition rates the density of the RF-deposited material is significant poorer than for the VHF-deposited one. Additionally, a higher amount of (Si-H_2)$_n$ chains has been observed for the high rate RF-process.

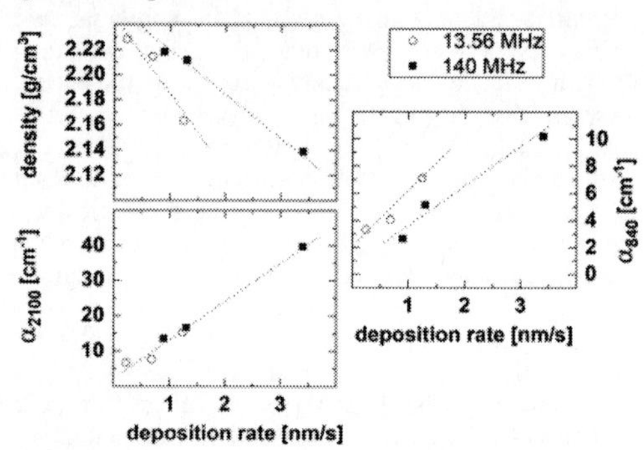

Figure 4: Material density and absorption coefficient of the 2100 cm^{-1} and 840 cm^{-1} mode as a function of deposition rate.

The hydrogen incorporation in the growing layer can be divided in two different processes [8]. The Si-H and Si-H_2 chains appears at the lower silane concentrations (below 10%) and lower deposition rates. For the higher concentration of silane in process gas and high-rate materials the enhanced formation of near-neighboring pairs of Si-H_2 groups and (Si-H_2)$_n$ polymers is detected. Hydrogen bound as Si-H_2 groups increases the structural disorder due to bond-angle distribution and consequently leads to the higher density of states in the gap. Enhanced amount of longer chain segments (Si-H_2)$_n$ cause the formation of empty microcavities where hydrogen is preferentially incorporated, which increase hydrogen content in deposited material.

4 PASIVATION PROPERTIES OF THE SILICON THIN-FILM DEPOSITED AT 13.56 AND 140 MHz

Subsequently, the effective carrier lifetime of amorphous silicon layers deposited on n-type FZ wafers was measured (Figure 5). The best lifetime values (7 ms) for the low-rate process (0,3 nm/s) was obtained at 100 % silane concentration due to H_2-plasma post-treatment. The reduced lifetime for lower silane concentration values can be explained through epitaxial like growth on the crystalline silicon substrate. However, for the VHF-PEVCD process the lifetime remains high even if epitaxial growth has probably started (50 % silane concentration). The increase of the excitation frequency leads to a decrease of the ion bombardment energy on the growing layer. Thus, at the 140 MHz frequency the interface crystallites are better passivated compared to the 13.56 MHz process, where the crystallites are probably damaged and defect rich.

At higher deposition rates, significantly better lifetime values were obtained for the VHF-process in comparison with the RF-process. This corresponds to material properties of high rate RF-process-layers. Without the H2-plasma post-treatment, a 2 ms lifetime value at 140 MHz frequency and 1.5 nm/s deposition rate was achieved.

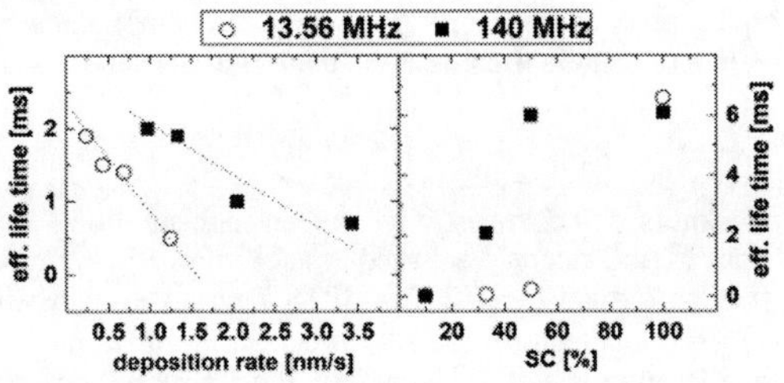

Figure 5: Effective carrier lifetime of a-Si:H layers deposited on n-type FZ wafers at varied deposition rates and silane concentration at 13.56 and 140 MHz.

5 CONCLUSIONS

The main goal of this study was an investigation of VHF process for a-Si:H passivation layer deposition.

First, the influence of the plasma excitation frequency on intrinsic a-Si:H layer properties was studied. The deposition process of the passivation layers was optimized to achieve high lifetime values. Thereby, the silane concentration was varied between 10 % and 100 % at low plasma powers and low deposition rates (< 0.4 nm/min). High lifetime values (7 ms) for both RF- and VHF-deposited a-Si:H have been measured.

An enhancement of the deposition rate due to an increased power leads to lower material densities for both 140 MHz and 13.56 MHz. This results in a remarkable increase of Si-H_2 groups and (Si-H_2)$_n$ polymers. At high deposition rates the passivation properties of VHF-grown a-Si:H were significantly better than those grown at 13.56 MHz frequency, whereby a higher amount of (Si-H_2)$_n$ chains for the high rate RF-process has been observed. Thus, the high process power leads to the enhanced formation of voids in deposition layer and production of very porous material.

Finally, the results show that working in the VHF-range allows an increase of the deposition rate without deteriorating the effective carrier life times.

6 ACKNOWLEDGEMENTS

This work is supported by the Bundesministerium für Bildung und Forschung (BMBF) under contract number 03SF0538.

7 REFERENCES

[1] F. Finger, U. Kroll, V. Viret, A. Shah, W. Beyer, X.M. tang, J. Weber, A. Howling, CH. Hollenstein, J. Appl. Phys. 71 (1992) 5665

[2] S. De Wolf, A. Descoeudres, Z. C. Holman, Ch. Ballif, Green, Vol. 2 (2012), 7–24

[3] Y. Tsunomura, Y. Yoshimine, M. Taguchi,T. Baba,T. Kinoshita, H. Kanno, H. Sakata, E. Maruyama, M. Tanaka, Solar Energy Materials & Solar Cells 93 (2009) 670–673

[4] M.Green, K. Emery, Y. Hishikawa, W. Warta, E. D. Dunlop, Solar cell efficiency tables (version 39), Prog. Photovolt: Res. Appl. 20 (2012) 12–20

[5] B. Leszczynska, C. Strobel, M. Albert, J.W. Bartha, U. Stephan, J. Kuske, 27th European Photovoltaic Solar Energy Conference and Exhibition (2012) 2507 – 2510

[6] Z. Remes, M. Vanecek, P. Torres, U. Kroll, A.H. Mahan, R.S. Crandall, J. Non-Cryst. Solids 227–230 (1998) 876–879

[7] F. Benner, P. Jordan, M. Knaut, I. Dirnstorfer, J. W. Bartha, T. Mikolajick, Investigation of the c-Si/Al2O3 interface for silicon surface passivation, 27th European Photovoltaic Solar Energy Conference and Exhibition (2012) 1793 – 1796

[8] C. Manfredotti, F. Fizotti, M. Boero, P. Pastorino, P. Polesello, E. Vittone, Phys. Rev. B, 50 (24) (1994) 18046-18053

BLACK SILICON WITH TUNNEL OXIDE PASSIVATED CONTACTS

Maren E. Kloster, Maria-Louise Witthøft, Denver C. Shearer, Alireza Hajijafarassar, Rasmus S. Davidsen, Dirch H. Petersen, Ole Hansen and Beniamino Iandolo
Department of Micro- and Nanotechnology, Technical University of Denmark, Ørsteds Plads, 2800 Kongens Lyngby, Denmark

ABSTRACT: We investigate surfaces of black silicon (bSi) fabricated by reactive ion etch (RIE) on n-type Si substrates and passivated by *in-situ* doped polycrystalline Si (poly-Si) deposited by low pressure chemical vapor deposition (LPCVD). We achieved full surface coverage of bSi surfaces for poly-Si thickness of 20 nm. We determined sheet resistance on p-type and n-type poly-Si by means of micro four-point probe measurements. Effective lifetime mapping on symmetrically passivated samples shows that n-type poly-Si offers excellent surface passivation after hydrogenation, reaching effective lifetime values of almost 4 ms on non-textured substrates. P-type poly-Si shows values of effective lifetime lower than 800 µs and requires further improvement. We calculated i-V_{oc} of up to 711 mV and 609 mV for asymmetrically passivated lifetime samples without texturing and with bSi, respectively. The effective lifetime is limited by the less-than-optimal passivation of the p-type poly-Si. Preliminary solar cell measurements indicate that both cell voltage and fill factor require major improvement on textured surfaces. Current work is directed towards the following goals: improving the quality of the p-type poly-Si; testing possible replacement of wet chemistry to grow the tunnel oxide using dry furnace oxidation; fabricating and comparing cells with p-n junction at the front or at the bottom of the cell.

1 INTRODUCTION

The tunnel oxide passivated contact (TOPCon) is a class of selective contacts that enables excellent charge carrier selectivity on crystalline Si. [1,2] It consists of an ultra-thin tunnel silicon oxide (SiO_2) layer in combination with a poly-Si layer, and it offers almost as high quality of surface passivation as its sister architecture HIT (heterojunction with intrinsic layer) based on hydrogenated amorphous Si (a-Si:H) [3], with the advantage of being compatible with high temperature processing (up to around 900 °C) [4]. Less-than-maximum short circuit current J_{SC} currently limits the power conversion efficiency of TOPCon solar cells. Black silicon (bSi) [5,6] has demonstrated great potential as texturing method for Si thanks to its excellent intrinsic antireflective properties [7]. Power conversion efficiencies between 18 and 22% have been achieved in the lab using laser-doped selective emitters [8] and interdigitated back contact (IBC) cells [9]. These results have been obtained using maskless reactive ion etch (RIE) for the Si texturing. Importantly, RIE works on any Si surface irrespective of the crystallinity and the method used for wafering, including diamond-wire. The increased surface recombination due to increased surface area and process-induced damage currently limits the open circuit voltage (V_{OC}) and thus the efficiency of RIE textured solar cells. TOPCon seems therefore an appropriate candidate to passivate a notoriously challenging surface such as RIE-textured Si. Here, we present our first attempt at combining RIE bSi with TOPCon passivation. We fabricated *in-situ* doped poly-Si layers by LPCVD, a method that offers higher compatibility with industrial production than commonly used ion implantation for doping. We demonstrate excellent passivation by n-type poly-Si, while further work is required to improve the passivation by p-type poly-Si and thus to obtain significant device efficiency.

2 EXPERIMENTAL DETAILS

2.1 *Fabrication*

100 mm n-type Czochralski (100) Si (350 µm thickness, 4 Ω cm resistivity, oxygen concentration < 11.5 ppa) from Okmetic were used as substrates. BSi texturing was performed by RIE in a SPTS Pegasus system with process temperature of -20 °C, process time of 14 min, SF_6 and O_2 plasma with 7:10 gas flow ratio, chamber pressure of 38 mTorr, 3 kW coil power and 10 W platen power. Wafers were then RCA cleaned and a tunnel silicon oxide with thickness of 1.5 nm was grown in 68 wt% HNO_3 at 95 °C for 10 min. *In-situ* doped poly-Si passivating contacts were deposited by LPCVD at 620 °C with SiH_4, B_2H_6 and PH_3 as precursors for Si, B and P respectively. Deposition time was varied in order to obtain nominal thickness ranging between 10 and 40 nm. All wafers where annealed in N_2 for 20 min at 850 °C. Symmetrical lifetime samples were prepared by depositing one poly-Si layer on both sides of the wafers during the same LPCVD run. Asymmetrically passivated wafers were prepared with the following steps: protection of the back side by depositing 300 nm of SiO_2 by plasma-enhanced CVD (PECVD, SPTS system); first tunnel-oxide/p-type poly-Si LPCVD run; RIE and wet etch in buffered HF solution to remove poly-Si and SiO_2 on the back side, respectively; protection of the front side by 300 nm PECVD SiO_2; second tunnel-oxide/n-type poly-Si LPCVD run; RIE (SPTS) and wet etch in buffered HF solution to remove poly-Si and SiO_2 on the front side, respectively; annealing of the poly-Si. Hydrogenation of lifetime samples was performed using the following steps: deposition of 75 nm SiN_x:H at 300 °C by PECVD; drive-in of hydrogen by annealing for 1 h at 400 °C in a tube furnace with N_2 atmosphere; removal of SiN_x:H by buffered HF. Solar cells with area of 6 cm^2 were fabricated from these asymmetrically passivated samples with following cross-section: front grid with 500 nm thick Al fingers/75 nm Al-doped zinc oxide (AZO)/15 nm p-type poly-Si/SiO_2/c-Si/SiO_2/40 nm n-type poly-Si/75 nm AZO/250 nm Al full back contact. Al was deposited by physical vapor deposition in a Temescal system. AZO was deposited by atomic layer deposition (ALD) at 200 °C in a Picosun system using Tetramethylammonium (TMA), Diethylzinc (DEZ) and water vapor as precursors for Al, Zn and O respectively.

2.2 *Characterization*

The thickness of the poly-Si was measured with a VASE ellipsometer on planar Si substrates. Cross-section scanning electron microscopy was performed in a Supra v40 SEM (Zeiss) at an accelerating voltage of 5 kV. Sheet resistance was measured on poly-Si layers using a micro-four point probe (μ4PP) CAPRES M300-tool with equidistant, straight 10 μm pitch micro four-point probes (M4PP). The n-type poly-Si was measured using 50 μA AC current at 13 Hz, using a punch-through current of 500 μA for 0.01 s. The p-type poly-Si was measured using 7 μA AC current at 13 Hz, using a punch-through current of 500 μA for 0.05 s. Minority carrier effective lifetime was measured on the full 100 mm area of the wafers using the microwave detected photoconductance method with a MDP mapper from Freiberg in steady-state configuration, with laser wavelength of 977 nm and photon flux of 4×10^{17} cm^{-2} s^{-1}. Current-voltage characteristics of solar cells were measured under 1 Sun using a solar simulator from Newport. Quantum efficiency and reflectance measurements were performed in a QEXL set-up equipped with an integrating sphere.

3 RESULTS

Figure 1 shows cross-section SEM images of bSi surfaces after deposition of poly-Si layers with nominal thickness of 10, 20 and 30 nm. Owing to the interest in minimizing the thickness of the front passivating contact to limit parasitic absorption, it is of interest to investigate what thickness of poly-Si is necessary to achieve complete surface coverage of the nanostructures. Less-than-complete surface coverage may result in degraded surface passivation. Some voids are visible in the poly-Si layer with nominal thickness of 10 nm, which implies that such thickness is not sufficient to provide full coverage of the hillock-like structures of bSi. Complete surface coverage is achieved for nominal thickness of 20 and 30 nm.

Next, we present measurements of sheet resistance of two 40 nm thick poly-Si layers, performed by μ4PP. The μ4PP is needed to avoid penetration of ultra-shallow junctions, which usually occurs when using macroscopic four-point probes. Penetration of the junction causes measurement errors, since the sheet resistance of the substrate is measured in parallel to that of the thin top-layer. Figure 2 shows line-scan measurements across the 100 mm wafer for p-type poly-Si and n-type poly-Si (top and bottom, respectively). A certain non-uniformity can be seen in the values of sheet resistance, likely due to non-uniformity of the LPCVD processing. The sheet resistance of the p-type poly-Si is rather high for a passivating contact, with lowest values of 4 kΩ close to the edge of the wafers. The n-type poly-Si, instead, is characterized by a by a more suitable sheet resistance between 600 and 800 Ω (apart from 1 mm or less from the edge of the wafer, where is drops to about 200 Ω).

Lifetime measurements on symmetrical samples (i.e. with poly-Si with the same polarity on both surfaces) are summarized in Figure 3. Both p-type and n-type poly-Si benefit from the hydrogenation induced by deposition of a sacrificial SiN_x:H followed by annealing. Importantly, the higher effective lifetime is retained in all cases after removing the sacrificial SiN_x:H layer. The lowest average effective lifetime of the p-type poly-Si passivated samples

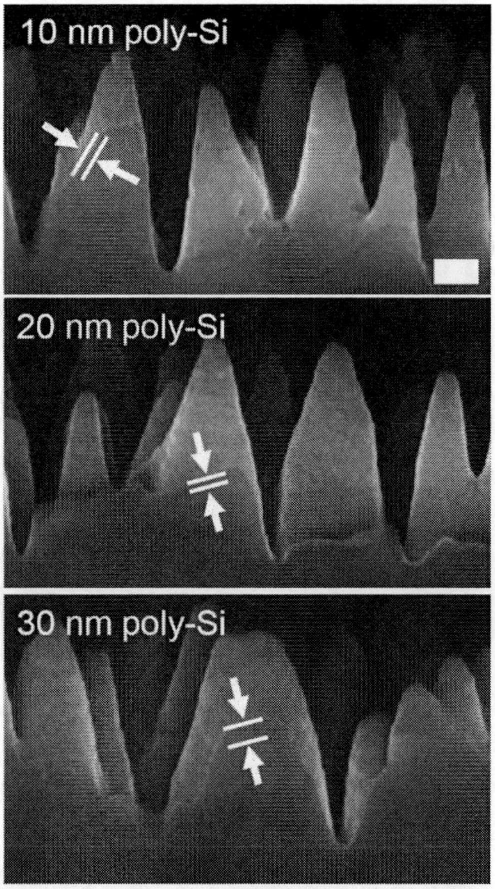

Figure 1: Cross-section SEM images of Si surfaces textured by RIE and coated with poly-Si layers of increasing nominal thickness. The poly-Si layers are indicated by the white arrow. The scale bar represents 100 nm.

is 162 ± 24 μs before hydrogenation. Both p-type poly-Si wafers show effective lifetime of around 750 μs after hydrogenation. Nonetheless, the quality of the passivation offered by our poly-Si is still far less-than-optimal and further work is needed to improve it. In addition to improve the quality of the p-type poly-Si, we are currently investigating the possibility to replace the SiO_2 tunnel layer grown by oxidation in HNO_3 with dry furnace oxidation. The quality of passivation afforded by as-deposited n-type poly-Si is an order of magnitude better than that of p-type poly-Si, with average lifetime between 2.8 and 3 ms. The average lifetime increases further upon hydrogenation, reaching 3.8 ms for the best sample.

Figure 4 summarizes measurements of effective lifetime mapping performed on asymmetrically passivated samples, i.e. with p-type poly-Si on the front and n-type poly-Si on the bottom (layer thickness of 40 nm in both cases). Two of the wafers were textured with bSi and this surface was passivated with p-type poly-Si. For these wafers, calculating the implied V_{oc} (i-V_{oc}) gives a more realistic indication of the voltage that the final cells will sustain. i-V_{oc} was calculated using the equation

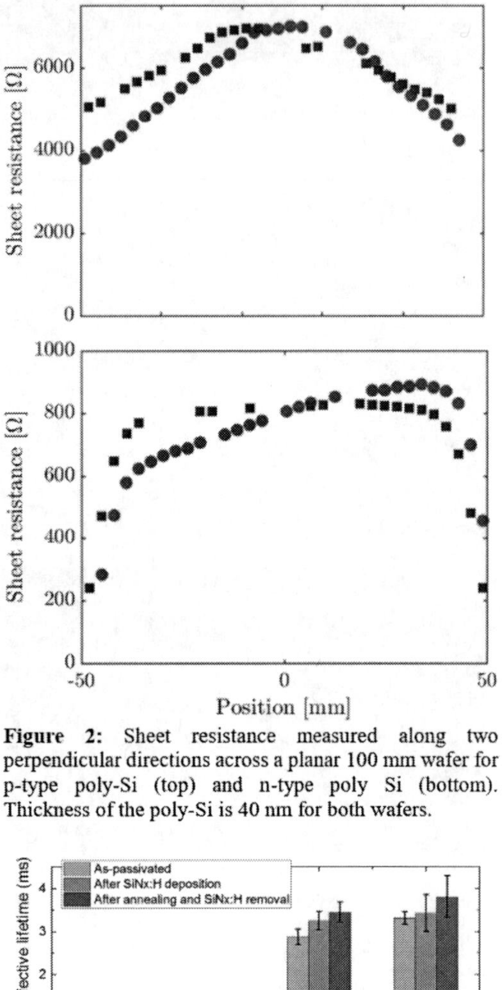

Figure 2: Sheet resistance measured along two perpendicular directions across a planar 100 mm wafer for p-type poly-Si (top) and n-type poly Si (bottom). Thickness of the poly-Si is 40 nm for both wafers.

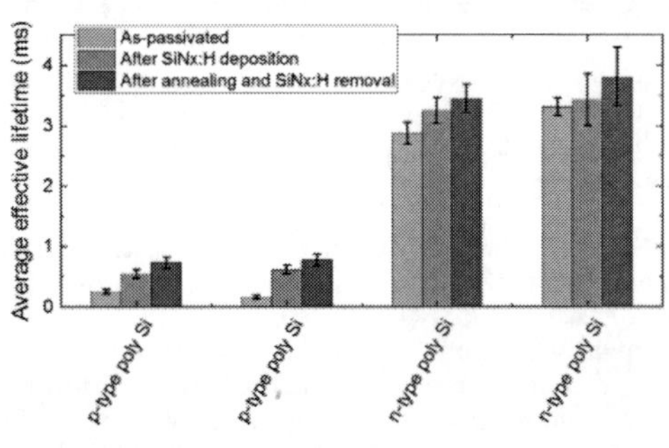

Figure 3: Effective lifetime averaged on full 100 mm wafers passivated symmetrically with 40 nm of either p-type or n-type poly-Si.

$$\text{i-}V_{OC} = \frac{kT}{q}\ln\left[\frac{\Delta n(\Delta n + N_D)}{n_i^2}\right]$$

where kT/q is the thermal voltage (25.85 mV at room temperature), N_D is the concentration of donors in the substrate, n_i is the intrinsic concentration of carriers in Si, and Δn is the concentration of light-generated carriers. Δn was calculated as the product of τ_{eff} and the photon generation rate G, which is known from measurements settings. The average lifetime before hydrogenation for planar wafers is between 333 ± 72 µs and 583 ± 62 µs. As expected, the lower quality of passivation afforded by the p-type poly-Si dominates the effective lifetime.

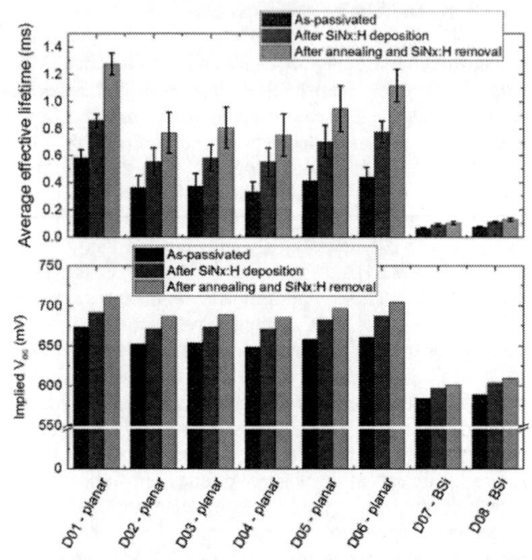

Figure 4. Top: effective lifetime averaged on full 100 mm wafers passivated asymmetrically with 40 nm p-type and n-type poly-Si on either side. Bottom: effective lifetime data converted in implied V_{oc} data for the same samples.

These values translate in i-V_{oc} of 648 and 673 mV, respectively. All samples show an increase of effective lifetime upon hydrogenation, with lifetime values after SiN_x:H removal of 1.28 ± 0.08 ms and 0.754 ± 0.16 ms. This translates into i-V_{oc} of 711 mV and 685 mV, respectively. The two bSi textured wafers show worse results, with average effective lifetime of 105 ± 12 µs and 130 ± 10 µs even after hydrogenation. This results into values of i-V_{oc} of 609 and 601 mV, respectively.

Finally, we present results from our first batch of solar cells. Here, the hydrogenation procedure was not carried out and thus lower performance is to be expected. Results of IV and spectral characterization are summarized in Figure 5. Our process flow, where the poly-Si is doped in-situ during LPCVD, allows for decoupling thickness of the front and back contact poly-Si, at the cost of a high number of process steps. Limiting the thickness of poly-Si front contact would be beneficial in terms of minimizing parasitic absorption. Here, we used a nominal thickness of 15 nm for the p-type poly-Si front contact, which should be sufficient to achieve full surface coverage even on textured substrates. The planar, non-textured cells show the highest performance, with maximum power conversion efficiency of 15.2%. The efficiency of cells with textured surfaces is considerably lower, 9.6 % for KOH texturing and 8.6 % for bSi texturing, respectively. The KOH texturing results in slightly higher J_{sc} than that of the planar reference (38.0 and 36.7 mA cm^{-2}, respectively). However, texturing results in considerably lower V_{oc} and lower fill factor. The latter may also be due to issues with metallization, which is currently under improvement. Finally, it is interesting that bSi texturing results in lower J_{sc} as well, as compared to the other cells. This is caused by the poorer passivation currently offered by the p-type poly-Si. The results in terms of J_{sc} are corroborated by measurements of quantum efficiency (QE) in the bottom panel of Figure 5. The data confirm that bSi textured cells, although with the highest potential for current generation thanks to the lowest front surface reflection, display the poorest QE and in particular very low QE for wavelength below 600 nm, which points once again at poor passivation by the front contact. Besides the required improvement in the quality of our p-type poly-Si, it would also be of interest to passivate textured surfaces with n-type poly-Si and place the p-n junction at the bottom of the cell. Fabrication and testing of cells with this structure are underway.

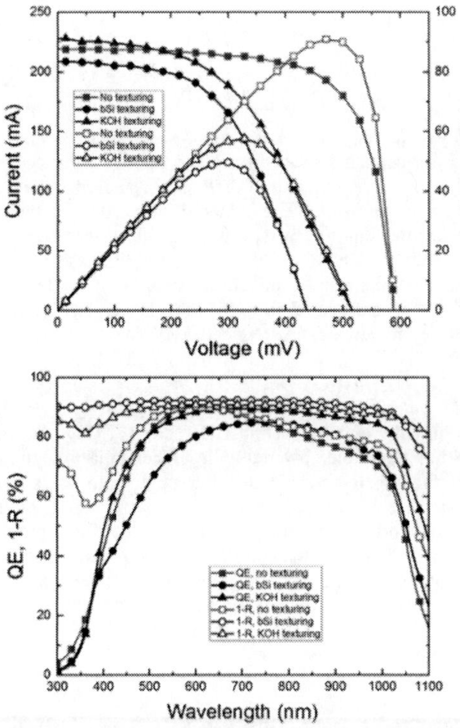

Figure 5: Characterization of solar cells with different surface texturing. Top: current-voltage characteristics under 1 Sun illumination. Bottom: quantum efficiency and front surface reflection subtracted from unity as function of photon wavelength.

4 CONCLUSIONS

We presented passivating contacts of *in-situ* doped poly-Si, fabricated by low pressure chemical vapor deposition, on planar and black Si textured surfaces. SEM characterization confirms that full surface coverage of black Si surfaces is achieved for thickness of poly-Si of 20 nm. We achieved implied V_{oc} values of up to 711 mV for asymmetrically passivated lifetime samples without texturing and after hydrogenation using a sacrificial SiNx:H. The effective lifetime is limited by the less-than-optimal passivation of the p-type poly-Si. The values of i-V_{oc} measured on bSi textured surfaces are lower (609 mV). Preliminary solar cells measurements indicate that both cell voltage and fill factor require major improvement on textured surfaces. Current work is focusing on: improving the quality of the p-type poly-Si; testing possible replacement of wet chemistry to grow the tunnel oxide with dry furnace oxidation; fabricating and comparing cells with p-n junction at the front or at the bottom of the cell.

5 ACKNOWLEDGMENTS

This work was supported by funding from EUDP (project number 64016-0030). We would like to thank CAPRES for assistance in performing electrical measurements with M4PP's on their microRSP-M300 tool.

6 REFERENCES

[1] F. Feldmann, M. Bivour, C. Reichel, M. Hermle and S. W. Glunz (2014), Solar Energy Materials and Solar Cells, 120, 270-274.
[2] F. Feldmann, C. Reichel, R. Muller, M. Hermle (2017), Solar Energy Materials and Solar Cells, 159, 265-271.
[3] K. Yoshikawa, H. Kawasaki, W. Yoshida, T. Irie, K. Konishi, K. Nakano, T. Uto, D. Adachi, M. Kanematsu, H. Uzu and K. Yamamoto (2017), Nature Energy, 2, 17032.
[4] J. Liu, Y. Yao, S. Xiao, X. Gu (2018). Review of status developments of high-efficiency crystalline silicon solar cells, J. Phys. D: Appl. Phys., 51, 123001.
[5] H. Jansen, M. de Boer, R. Legtenberg, M. Elwenspoek (1995), Journal of Micromechanics and Microengineering, vol. 5(2), 115, 1995.
[6] M. Otto *et al.* (2015), Adv. Optical Mater. 3, 147–164.
[7] R.S. Davidsen, J. Ormstrup, M. L. Ommen, P. E. Larsen, M. S. Schmidt, A. Boisen, and O. Hansen. (2015), Solar Energy Materials and Solar Cells, 140, 134-140.
[8] R. S. Davidsen, *et al.* (2017), Solar Energy Materials and Solar Cells, 144, 740-747.
[9] H. Savin, P. Repo, G Von Gastrow, P. Ortega, E. Calle, M. Garín, and R. Alcubilla (2015), Nature nanotechnology, 10(7), 624-628.
[10] D. H. Petersen, O. Hansen, T.M. Hansen, P. Bøggild, R. Lin, D. Kjær, P.F. Nielsen, T. Clarysse, W. Vandervorst, E. Rosseel, N.S. Bennett, N.E.B. Cowern (2010), Journal of Vacuum Science & Technology B, 28, C1C27--C1C33.

APPROACH FOR DEVELOPING AMORPHOUS SILICON PASSIVATION LAYERS AND P-TYPE MICROCRYSTALLINE LAYERS FOR HIGHLY EFFICIENT HIT SOLAR CELLS USING A DYNAMIC VHF-PECVD PROCESS WITH HIGH DEPOSITION RATES

S. Leszczynski*[1], C. Strobel[1], B. Leszczynska[1], M. Albert[1], F. Stahr[2], J. Kuske[2], J. W. Bartha[1]
[1] Dresden University of Technology, Semiconductor and Microsystems Technology Laboratory, 01062 Dresden, Germany, *phone: +49 351 463 32448, e-mail: sebastian_sylwester.leszczynski@tu-dresden.de
[2] Forschungs- und Applikationslabor Plasmatechnik GmbH, Gostritzer Str. 67, 01217 Dresden, Germany

ABSTRACT: In this study, the properties of the intrinsic amorphous silicon (a-Si:H) and p-doped microcrystalline silicon (μc-Si:H) materials were investigated in a wide range of the deposition rates using the plasma enhanced chemical vapour deposition (PECVD) technique at the very high excitation frequency (VHF) and standard radio-frequency (RF). Thereby, the high-rate and low-rate regimes were compared. The material properties of the passivation layers as well as carrier lifetime were studied. Furthermore, the new p-type μc-Si:H electrodes were introduced. The initial growth process of the p-layer on top of the a-Si:H material was developed. Additionally, the influence of the ion bombardment energy on the deposition process is studied using simulation methods. At the final stage of the development, the dynamic VHF-PECVD process for the passivation and electrode layers has been introduced and applied for a highly productive fabrication of the heterojunction silicon solar cells (HIT).
Keywords: PECVD, Solar Cell, High Deposition Rate

1 INTRODUCTION

Deposition techniques and process regimes have a significant impact on the thin-film silicon properties. It has been reported, that using the HIT solar cells very high efficiencies can be achieved (26.7% by Panasonic) [1]. Thereby, the hydrogenated amorphous silicon acts as a passivation layer and strongly influences the device properties. However, up to now, these highly efficient HIT solar cells are not cost competitive compared to other crystalline silicon-based technologies (AL-BSF, PERC). Therefore, a compromise between a high interface quality and low production costs has to be found. Thus, the high deposition rate processes for the silicon passivation layers and electrodes are required.

For the commonly used RF-PECVD technique at the 13.56 MHz frequency an enhancement of the deposition rate can be achieved through an increased plasma power. However, the high-power regime leads to an increase of the ion bombardment energy and therefore to a deterioration of the a-Si:H layer properties and a-Si:H/c-Si interface. Thus, we propose an innovative solution for the high-rate process without negative impact on the c-Si surface during the a-Si:H deposition. By applying the very high plasma excitation frequencies (VHF-regime, >100 MHz) a low ion bombardment energy and a high silicon dissociation can be combined. Therefore, the high deposition rates and the high interface quality were simultaneously achieved and will be introduced in this study. The developed deposition process has also been transferred for deposition of enhanced p-type electrode layers.

2 EXPERIMENTAL

2.1 In-line dynamic deposition system

To ensure the very thin layer thickness in the HIT-device (5-20 nm) and the high-rate deposition a dynamic fabrication technique in the VHF-regime with the high substrate velocities is introduced. The inline deposition system used in this study is based on the linear plasma source concept developed by Dresden University of Technology and FAP GmbH Dresden [2] (Fig. 1). This ensures a suitable power coupling for the plasma ignition and the homogenous deposition up to 140 MHz. The linear plasma source concept is optimized for a highly productive fabrication in a continuous deposition process with a moving substrate carrier. The excitation frequency can be further extended up to 200 MHz. The length of the plasma source can be also increased for the dynamic deposition of the thin silicon layers on the large area substrates.

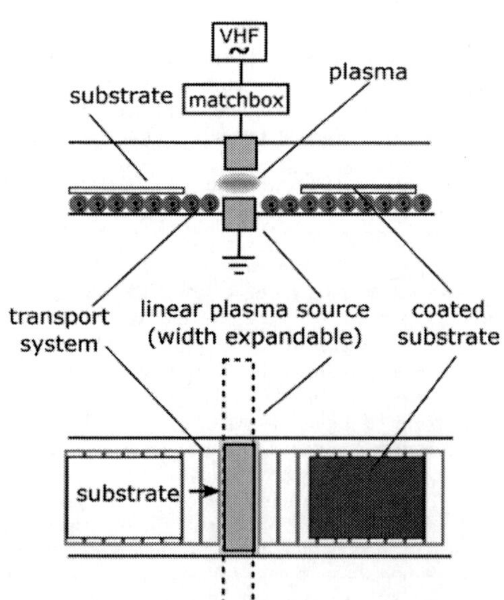

Figure 1: In-line deposition system with a moving substrate carrier and a linear plasma source.

The linear electrode used in this study is equipped with a showerhead gas delivery system to provide the homogenous gas flow in the deposition area. The input signal is generated using a signal generator and an amplifier with the 13.56 and 140 MHz frequency. During the dynamic deposition the substrates are placed on the carrier and moved horizontally inside the tunnel under the plasma source. The carrier velocity depends on the desired thickness of the fabricated layer.

2.2 Experimental details

To characterize the material properties at the 13.56 and 140 MHz frequencies the deposition of the a-Si:H layers was performed. The intrinsic and p-doped thin silicon films were fabricated on 1.1 mm thick Corning Eagle XG glass substrates. Additionally, doubled sided polished c-Si wafers were used for the structural measurements. The passivation studies were performed using the a-Si:H layers deposited on 250 µm thick n-type double sided polished float zone silicon wafers.

To investigate the deposition rates the thickness of the layers was measured by a surface profilometer from Veeco. Whereby, the deposition rate was varied by the plasma power values.

For the electrical measurements the a-Si:H layers were covered with the aluminum contacts and the dark conductivity was measured under vacuum conditions.

Additionally, a Fourier transform infrared spectroscopy (FTIR) was introduced to exam the structural properties of the amorphous material. The measurements were performed from 380 to 3000 cm^{-1} using a Bruker Vertex 80v spectrometer. The microstructure factor was defined by the ratio of integrated absorption bands of the higher stretching mode (HSM) at 2090 cm^{-1} and the lower stretching modes (LSM) at 2000 cm^{-1}. The HSMs correspond to hydrogen bonded to internal void surfaces and the LSMs correspond to the mono-hydrides in the amorphous network.

The interface properties between amorphous and crystalline silicon was investigated by the effective minority carrier lifetime measurements using microwave detected photoconductivity (MDP) [3].

Finally, the HIT solar cells were fabricated on the textured 300µm n-type float zone silicon wafers (n-FZ) and the solar cell parameters were obtained.

3 RESULTS AND DISCUSSION

3.1 Intrinsic amorphous silicon passivation layers

For a better understanding of the influence of the excitation frequency on the material properties the correlation between the deposition rate, silane concentration (SC) and the properties of the a-Si:H layers was investigated (Fig. 2). The a-Si:H material deposited at the low deposition rates shows very good structural parameters and no deterioration of the material properties. However, the low silane concentration ratio and very low deposition rate regime leads to an epitaxial growth of the a-Si:H films on the crystalline silicon surface which effects the surface passivation.

Next, the structural properties of the high rate a-Si:H were studied. The deposition rate was varied by the power density (Fig. 2 below). An increase in the a-Si:H deposition rate leads to higher microstructure factor values (increased porosity) for both frequencies. An increased porosity correlates with a larger volume fraction of silicon-di-hydride and tri-hydride in the material.

At the next step, carrier lifetime measurements were carried out (Fig. 3). Both, the RF-PECVD and VHF-PECVD process leads to good life time values (7 ms) for 100 % silane concentration (low deposition rate regime with plasma post-treatment). This can be explained with annealing-induced silicon passivation and transfer of hydrogen from higher silicon-hydride states in the layer to the monohydride states on the c-Si surface.

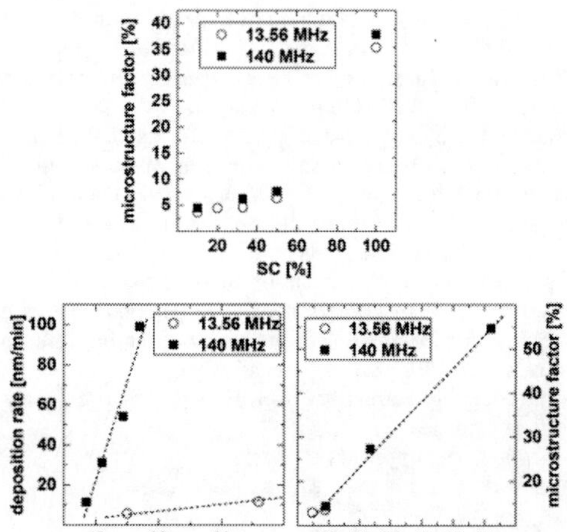

Figure 2: Microstructure factor as a function of the silane concentration (above), deposition rate as a function of the working power and microstructure factor as a function of deposition rate (below) for the VHF-PECVD and RF-PECVD process.

Furthermore, differences between material growth by the VHF- and RF-process were observed. For the RF-deposited a-Si:H layers the lifetime values decrease abruptly in the epitaxial growth regime (low SC) [4]. The VHF-PEVCD leads to the higher lifetime values even for the material grown with low SC ratios. It can be speculated, that for the 140 MHz process with the low ion bombardment energies the epitaxial grown crystallites are well passivated, compared to the damaged interfaces for the high energetic ion bombardment (RF-PECVD).

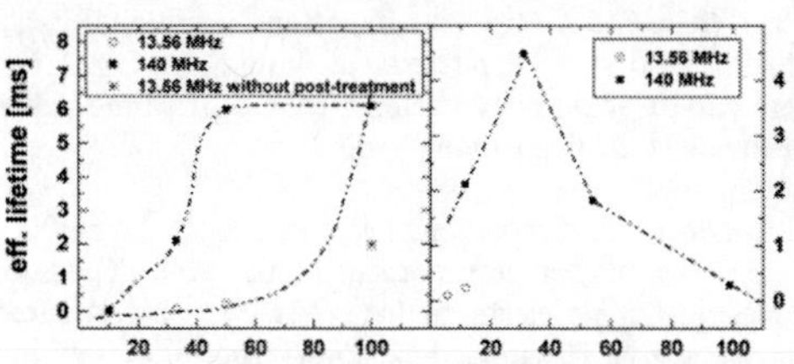

Figure 3: Carrier lifetime of the VHF-PECVD and RF-PECVD material as a function of silane concentration (left) and deposition rate (right).

The VHF-PECVD technique enables an enhancement of the deposition rate at a lower power level and ensured good lifetime values up to 60 nm/min (Figure 3 right). In comparison, for the RF-process due to an application of the very high plasma power values (up to 300 mW/cm^2) only 11 nm/min deposition rate with the 0.5 ms carrier lifetime has been achieved. The differences in the effective lifetime value between RF- and VHF-deposition could be explained by the ion bombardment damages of the c-Si surface.

Therefore, at the next step, the influence of the ion bombardment energy on the deposition process was studied (Fig. 4). A calculation of the projected ranges of the hydrogen ions (H^+) on the surface of c-Si was performed using the Monte-Carlo simulation program TRIM (Transport of Ions in Matter) developed by Ziegler et al. [5]. The results show a larger penetration depth for the higher ion energies. Furthermore, the ion bombardment processes at 45 eV energy for the a-Si:H layers with the different material densities were simulated, whereby a larger penetration depth for porous materials was observed. The surface passivation depends on the thickness and quality of the a-Si:H material and the surface termination of the c-Si wafer. The growth of the a-Si:H passivation layer on the c-Si surface can be explained as a replacement of Si-H, weak Si-Si or Si dangling bonds by strong Si-SiH_x (x <4) bonds. The high energetic ion bombardment processes at the high plasma powers (RF-PECVD) can negatively influence the c-Si surface quality during the a-Si:H deposition and cause surface damages.

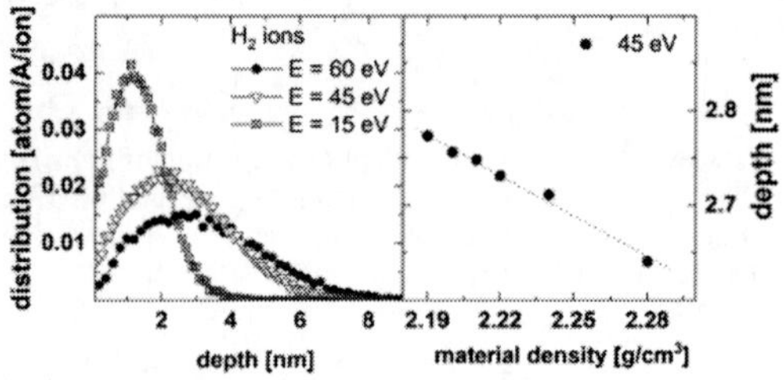

Figure 4: Monte-Carlo modelling of the H_2 ion bombardment on the surface of c-Si (left) and a-Si:H (right): ion distribution as a function of depth in the material for different ion energies (left), ion average range as a function of material density (right).

In the next development step, the passivation layers were further optimized. Using a new layer configuration, high effective carrier lifetimes of 13 ms and high open-circuit voltage (V_{oc}) of 727mV (300 μm n-FZ wafer) were achieved. For thinner wafers (approx.100μm) V_{oc}'s larger than 740mV can be expected. Additionally, the dynamic and static deposition of the a-Si:H thin layers were compared without a deterioration of life time values for the dynamic deposition process

3.2 p-doped microcrystalline silicon electrode layers

For the further development of the HIT deposition process different electrode materials were investigated. An innovative dynamic deposition process of the thin microcrystalline silicon p-type films was studied (Fig. 5). For this purpose, the conductivity of the very thin layers (< 20nm) fabricated by the static and dynamic RF- and VHF-process was investigated. Thereby, a good seeding of the microcrystalline material on the intrinsic a-Si:H films was very challenging and the different seeding possibilities were studied.

The VHF-PECVD technique shows the high conductivity values (> 1E-03 S/cm) for both dynamic and static deposition process at the high deposition rates. In comparison, the low conductivities (< 1E-09 S/cm) were achieved for the dynamically deposited RF-layers. This demonstrates the advantages of the deposition of the microcrystalline silicon at the high excitation frequencies.

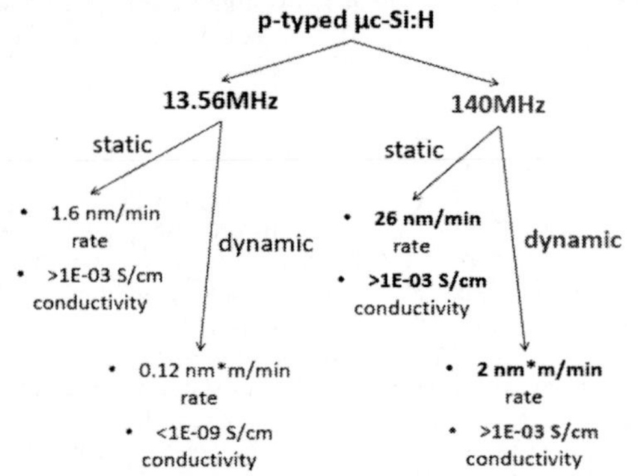

Figure 5: Diagram of the different deposition regimes and techniques related to the deposition rates and layers conductivities.

3 SUMMARIES

The properties of the VHF-PECVD process in comparison with RF-PECVD were investigated. The individual steps for development of the a-Si:H passivation process with the high carrier lifetimes and high V_{oc} values for the HIT solar cells were presented. Furthermore, an innovative process for the deposition of the highly transparent microcrystalline silicon electrodes using the dynamic VHF-PECVD technique at the high deposition rates was discussed.

Finally, the developed i- and p- layers were incorporated in HIT solar cells giving significant enhancement of the solar cell properties (see Fig. 6). The new dynamic deposition of the microcrystalline silicon electrodes leads to 2 mA/cm^2 enhancement of short-circuit current (J_{sc}).

Due to the very high current values (39.6 mA/cm2) the 2% increase in the solar cell efficiency was achieved (up to 21.8 %).

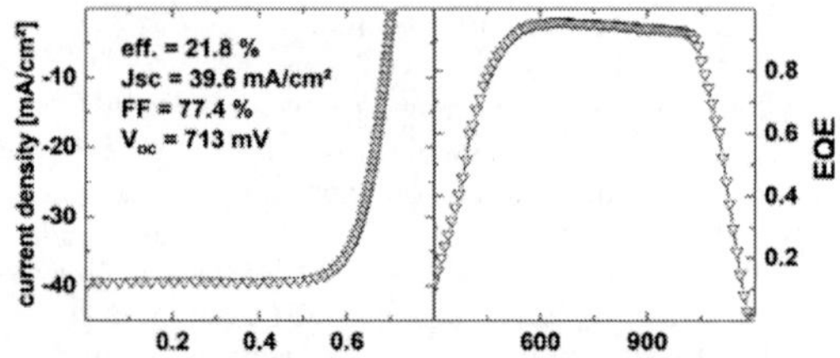

Figure 6: U/I characteristics and parameters of the HIT solar cell (left), and the corresponding external quantum efficiency (EQE).

4 ACKNOWLEDGMENT

This work is supported by the Bundesministerium für Bildung und Forschung (BMBF) under contract number 03SF0538.

5 REFERENCES

[1] K. Yoshikawa, H. Kawasaki, W. Yoshida, T. Irie, K. Konishi, K. Nakano, T. Uto, D. Adachi, M. Kanematsu, H. Uzu, K. Yamamoto, Nat. Energy 2 (2017) 17032.

[2] J. Rüdiger, H. Brechtel, A. Kottwitz, J. Kuske, U. Stephan, Thin Solid Films 427 (2003) 16–20.

[3] T. Mikolajick, J.W. Bartha, I. Dirnstorfer, M. Knaut, P.M. Jordan, F. Benner, 27th Eur. Photovolt. Sol. Energy Conf. Exhib. (2012) 1793–1796.

[4] H. Fujiwara, M. Kondo, Appl. Phys. Lett. 90 (2007) 013503.

[5] J.F. Ziegler, J.P. Biersack, U. Littmark, The stopping and range of ions in solids, Pergamon, 1985.

4-TERMINAL AND 2-TERMINAL TANDEM MODULES IN BIFACIAL OPERATION: MODEL ANALYSIS AND COMPARISON

L.J. Geerligs[1], D. Zhang[2], G.J.M. Janssen[1], S.L. Luxembourg[1]
[1]ECN part of TNO, Solar Energy, Westerduinweg 3, 1755 LE Petten, The Netherlands
[2]ECN part of TNO and Solliance, High Tech Campus 21, 5656 AE Eindhoven, The Netherlands

ABSTRACT: This paper models and compares the energy yield to be expected from bifacial 2-terminal (2T) and 4-terminal (4T) hybrid perovskite/crystalline silicon tandem modules. Following the work by Asadpour *et al.* the comparison includes 2T tandem cells employing a thick perovskite absorber layer with standard band gap, which in monofacial operation have significant mismatch of short circuit currents of top and bottom cell, and therefore low efficiency. The perovskite stack properties are taken from the current Solliance high-performance semi-transparent perovskite cells, and the c-Si model represents current high-efficiency industrial nPERT cells. As shown by Asadpour *et al.*, the combination of thick absorber and standard band gap with bifacial operation is a promising approach for 2T bfiacial tandem modules, resulting in higher performance than the c-Si bottom cells. We find that such a 2T bifacial tandem module should be able to outperform even an optimized 2T module with wide band gap perovskite. Additionally, employing a planar c-Si front side in combination with textured module glass can potentially perform nearly as well as employing a textured c-Si front side. These results support the feasibility of important simplifications for production of 2T tandem modules, as long as these will be operated bifacially. The dependence of energy yield of 2T bifacial tandems on albedo is qualitatively different from that of 4T and single junction c-Si bifacial modules. This reduces the competitiveness of 2T bifacial tandems, compared to 4T and SJ c-Si, with increase of the albedo of the environment.
Keywords: bifacial, tandem, perovskite, crystalline silicon

1 INTRODUCTION

A polysilicon (polySi) passivated contact as front junction on a crystalline silicon (c-Si) solar cell enables the manufacturing of a monolithic series-connected tandem cell (i.e., 2-terminal or 2T tandem cell) on low-cost industrial c-Si cell architectures such as PERC and n-type PERT [1]. Since these c-Si cells, as well as the silicon heterojunction cells (SHJ) which are so far more commonly used for 2T tandems, are suitable for bifacial module operation, i.e. for modules collecting light from rear as well as front, it raises the question of possibility for bifacial operation of 2T tandem cells. Additionally, a 2-sided glass module layout is very suitable for moisture-proof encapsulation of perovskite cells, as was recently demonstrated in damp heat and thermal cycling stability tests [2].

Top and bottom cells of 2T tandem cells operate by definition at the same current, but the carrier generation in the bottom cell can vary a lot in bifacial operation (with time and day, weather, and location). Therefore, one might suppose that 2T tandem cells will not perform well in bifacial operation.

However, Asadpour, Chavali, Khan and Alam [3] have brought forward the very interesting concept that a 2T tandem cell can in fact benefit from bifacial operation. They considered in particular perovskite/c-Si tandems and proposed that the 2T tandem can employ a top cell with an absorber layer of low band gap and/or relatively large thickness, which would absorb too much light for optimal monofacial operation. In bifacial operation, the short-circuit current density J_{sc} of the bottom cell would be increased by the rear irradiation. As soon as this is large enough to match the top cell current, the performance of the tandem cell starts to flatten out at an operating current close to the top cell's J_{sc}. By modeling, they showed that, with such a top cell, the 2T tandem cell can exceed the performance of the single junction (SJ) c-Si bottom cell significantly, for a wide and relevant range of albedos.

In this paper we will investigate by modeling how the performance of such bifacial 2T tandem modules with high absorption in the top cell, resulting in mismatch in monofacial operation, will compare to alternatives such as a 2T tandem cell with wider band gap top cell optimally matched for monofacial operation, or a 4-terminal (4T) tandem cell. We will pay particular attention to a further simplified top cell, not only using a standard low band gap perovskite absorber, but also using a polished front side of the c-Si bottom cell [4], where the effect of the absence of texture on the wafer is partially compensated by using textured module glass. It turns out that in bifacial operation, this simplified 2T module can even perform better than a 2T tandem module with textured c-Si cell and optimized wide band gap perovskite. To estimate practical benefits, we model the performance of the architectures not only under standard irradiation conditions but also for several field conditions.

2 MODELING DETAILS

2.1 Perovskite top cell and c-Si bottom cell

The perovskite top cell stack is modeled based on a high performance "pin" semi-transparent perovskite cell stack developed and characterized at Solliance [5]. The standard (experimental) band gap E_g is 1.6 eV. In the modeling in this paper, the perovskite absorber layer thickness is varied. The experimental approximate shunt resistance of 10 kΩcm^2 and series resistance of 3 Ωcm^2 were used both for 2T and for 4T. With a diode factor of 1.8, and prefactor of the dark recombination current J_o of 10^{-12} A/cm^2, the experimental V_{oc} and FF are quite well approximated. For the (hypothetical) wide band gap perovskite, with E_g=1.7 eV, the absorption curve versus photon energy was shifted by 0.1 eV, and J_o was reduced to 1.2×10^{-13} A/cm^2 to achieve a V_{oc}-gain of 100mV.

The c-Si cell model was based on a typical industrial performance nPERT cell, including homojunctions with representative doping density and free carrier absorption. Note that in practice tandem cells will benefit strongly from the use of a polySi passivated contact on at least the front, but for simplicity such a polySi contact was not specifically included in the present model. With shunt resistance of 7 kΩcm², series resistance of 0.7 Ωcm², diode factor equal to 1, and J_o of 1.5×10^{-13} A/cm² a typical cell is well approximated.

Both in 2T and 4T models, the p-type side of the perovskite stack is on the light-incident side. In reality, the stack order might be reversed in 2T compared to 4T. The present model approach has the advantage, however, of a simpler interpretation of differences between 2T and 4T results, since parasitic optical effects such as UV absorption in charge transport layers, will be similar in both architectures.

2.2 Module architecture

Fig. 1 shows schematically the two module architectures modeled in this work. The variations for which results will be presented in more detail are shown in Table 1.

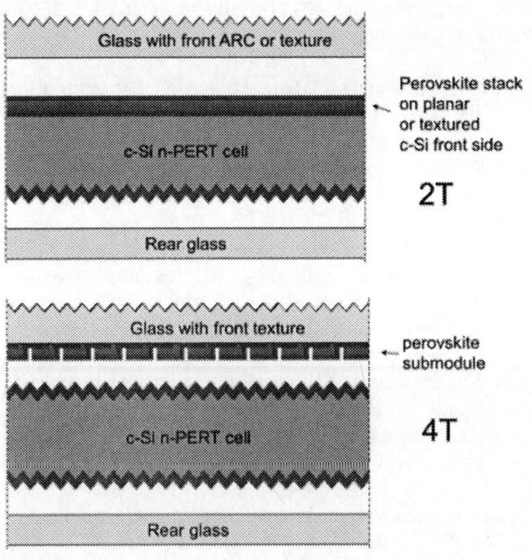

Figure 1: 2T and 4T module architectures modeled in this work. Encapsulant is present between cells and glass. For 2T, several variations for front of glass and front of c-Si cell were modeled, as listed in Table I (this figure shows the design of variations 2T-3 and 2T-4).

Table I: Module architecture variations. The variations concern presence of texture (text.) on cell or module glass front side, presence of ARC on module glass front side, perovskite band gap (E_g), thickness of absorber layer in the perovskite solar cell (t_{PSC}). The resulting J_{sc} of the perovskite solar cell (PSC) and the c-Si bottom cell are shown.

	module	cell	E_g (eV)	t_{PSC} (nm)	$J_{sc,PSC}$ (mA/cm²)	$J_{sc,c\text{-}Si}$ (mA/cm²)
2T-1	ARC	text.	1.6	360	19.6	19.6
2T-2	ARC	text.	1.7	850	19.6	19.6
2T-3	text.	planar	1.6	520	20.6	17.5
2T-4	text.	planar	1.6	1000	22.1	16.1
4T	text.	text.	1.6	520	20.6	17.8
c-Si	ARC	text.	n/a	n/a	n/a	40.7

Short circuit current generation in the absorber layers (perovskite and crystalline silicon) was calculated with ray-tracing simulations in the optical simulation software GenPro4 [6] for AM1.5 incident radiation, assuming a fixed internal quantum efficiency of 1 for perovskite as well as c-Si absorbers.

Module model 2T-2, based on a wide band gap perovskite on a textured c-Si cell, is representative for the technology that is mostly pursued in R&D nowadays.

In the subsequent efficiency and energy yield calculations, the J_{sc}'s listed in Table I were reduced by 3% to account for metal grid shading. The feasible thickness for the perovskite absorber layer (t_{PSC}) is limited in practice, since too thick layers result in reduced V_{oc} and FF. A thickness between 500 and 1000 nm seems a realistic target. For such thickness, the absorption of photons with near-band-gap energy in the perovskite absorber layer is significantly improved by increase of optical path length, which can be achieved by texture of wafer (and the perovskite following conformally the wafer texture), or by texture of module glass. In case of textured module glass we have modeled a random pyramid texture similar to wafer texture.

2.3 Performance modeling

The performance of the tandem modules was calculated in two ways: simply as power output under 1000 W/m² front irradiation and varying rear irradiation, or as annual energy yield for ground-mounted field conditions in Amsterdam (the Netherlands) and Doha (Quatar) with varying surface albedo. Spectral variations were not included, only intensity variations were included in the energy yield modeling, with the light generated J_{sc}'s scaling linearly with the irradiation power. Also temperature effects were omitted, noting that at this stage nothing is known of the perovskite cell temperature coefficients. The I-V characteristics of both sub-cells were calculated using a routine from PVLIB [7]. The front and rear irradiance over the year, discretized in 1 hour periods, was calculated for a module in the middle of a long string, using the model software BIGEYE [8].

3 RESULTS

3.1 Variation of texture and ARC

A polished front side of the c-Si cell will significantly reduce the challenge of applying the perovskite stack for a 2T tandem cell, and may result in more correspondence of the deposition technology with single junction (and 4T) perovskite process technology. These are both important aspects for industrialization. Therefore, module design variations 2T-3 and 2T-4, which are set up with industrial feasibility in mind, have a planar front side of the c-Si cell. The rear side of the c Si cell remains textured in these designs. Table II compares this design with other design variations that can be used to increase optical path length or reduce reflection. The values in Table II depend somewhat on perovskite thickness: the thicker the polySi layer, the less pronounced the benefit of texture.

We observe only small differences between texture on glass and texture on c-Si. However, note that we have assumed relatively ideal and perhaps difficult to produce texture features on the glass. Reports from field tests of textured glass are limited, but in dry climates need for more frequent cleaning would be expected [9]. The performance of industrially available textures would have to be assessed in future model work and field tests, to prove the feasibility and performance of glass texture for tandem modules.

Table II: Variations in design and their effect on J_{sc} relative to 2T-3.

module		c-Si text. front		c-Si planar front	
		$\Delta J_{sc,PSC}$	$\Delta J_{sc,c\text{-}Si}$	$\Delta J_{sc,PSC}$	$\Delta J_{sc,c\text{-}Si}$
ARC	text.	(mA/cm²)			
no	no	-0.1	-0.2	-1.3	-1.3
yes	no	+0.4	+0.1	-0.9	-1.0
no	yes	+0.5	+0.2	2T-3	
yes	yes	+0.6	+0.2	+0.1	0

Note that table II gives no guidance for effects on 4T. In case of 4T tandem device architecture, the J_{sc} of the planar perovskite layer on the glass superstrate benefits greatly from texture of the glass. Adding ARC to the glass texture gives insignificant additional improvement.

3.2 Power output under standard conditions

Fig. 2 shows the output power relative to the front-incident irradiation power of 1000 W (AM1.5) of the modules in bifacial operation:

$PCE^* = P_{out}/P_{in,front}$

This parameter PCE*, at a value for the rear/front incident power ratio of 20%, is often taken as a measure for the performance of a bifacial module.

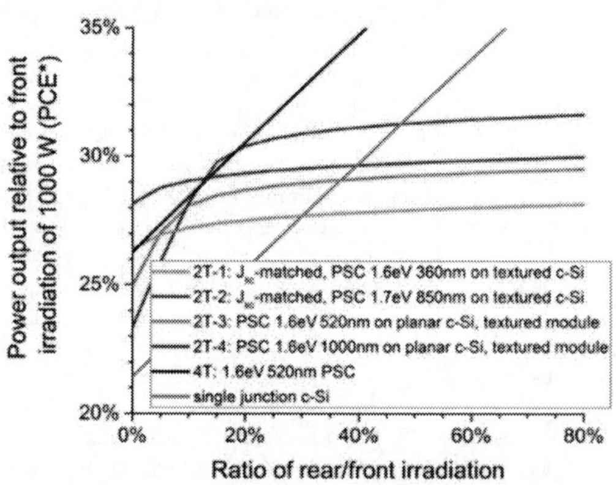

Figure 2: Power output relative to a fixed 1000 W front incident irradiation, as a function of rear incident irradiation (a ratio of x% meaning an additional rear incident irradiation power of x * 1000 W).

The behavior of the 2T devices is significantly different from the 4T and SJ c-Si devices: instead of the linear increase of PCE* with the fraction of rear irradiation observed for 4T and SJ c-Si, the PCE* of the 2T tandems rapidly increases at low fractions of rear irradiation, and then saturates. Fig. 3 shows the reason behind this (already explained in [3, 4]): for the 2T devices 2T-3 and 2T-4, the initial J_{sc}-mismatch without rear irradiation is rapidly reduced with increasing rear irradiation and PCE* rapidly increases. As soon as the J_{sc} of the bottom cell (with rear irradiation) exceeds the J_{sc} of the top cell, the J_{mpp} and output power of the tandem saturates. For the 2T tandem device stacks optimized for monofacial use, 2T-1 and 2T-2, there is much less room for increase of J_{mpp} and power output with increasing rear irradiation [10].

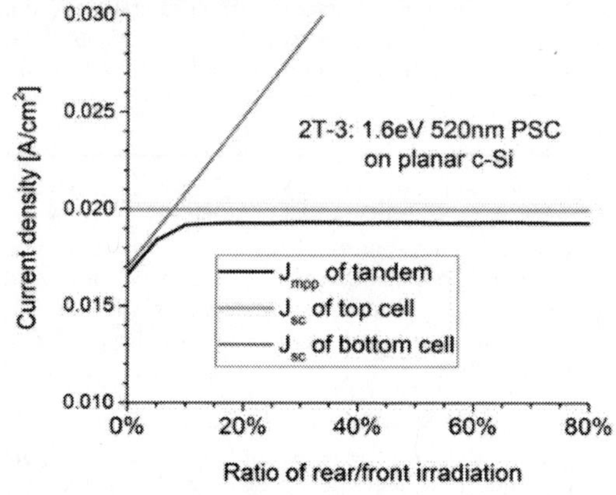

Figure 3: $J_{mpp,tandem}$, J_{sc,top_cell} and $J_{sc,bottom\_cell}$ as a function of rear irradiation, at fixed 1000 W front irradiation, for the 2T-3 tandem (1.6 eV 520 nm perovskite cell stack on planar c-Si).

Fig. 2 shows that 2T tandem designs that favor a high J_{sc} of the top cell (for example, 2T-4) perform very well in bifacial operation, potentially even significantly better than the optimized wide band gap tandem 2T-2.

The 4T tandem clearly outperforms the 2T designs (apart from the wide band gap 2T-2, because the 4T model is based on a standard band gap perovskite). Similarly, the PCE* of the single junction nPERT bifacial module approaches that of the 2T tandems when ratio of rear/front irradiation increases to around 0.4.

In the comparisons in Fig. 2 and the next section it should be kept in mind that all perovskite cell stacks for the 2T and 4T models in the present work include the same TCO layers, while in reality the 2T designs could do with thinner, more transparent, rear TCO.

3.3 Modeled annual energy yield of bifacial tandems

In this section, the efficiencies under standard conditions are translated to annual energy yield of bifacial modules versus ground albedo. As mentioned earlier, the modeling of annual energy yield in this paper does not take into account temperature coefficients of cell performance. In fact, for perovskite solar cells, temperature coefficients are not well established at this moment.

Fig. 4 shows the normalized annual energy production, calculated as the annual energy production (e.g., in Wh/m²) divided by the cumulative annual front irradiation on the module (similarly in Wh/m²), for ground-based module setups at two geographical locations. The main feature observed in Fig. 4 is that the horizontal axis is "stretched" compared to Fig. 2. Due to a limited "view factor" of the module's rear side on the ground surface, and due to shade cast by the module itself, the rear irradiation fraction is often significantly less than the ground surface albedo, in particular for high front irradiation values. Thus the advantage of 2T over single junction c-Si is maintained to higher albedos than Fig. 2 would seem to suggest if one would assume that albedo and rear/front irradiation ratio are somewhat similar, and the advantage of 4T over 2T builds up more slowly. Also, the advantage of the 2T tandem with high top cell J_{sc}, 2T-2, builds up more slowly than in Fig. 2.

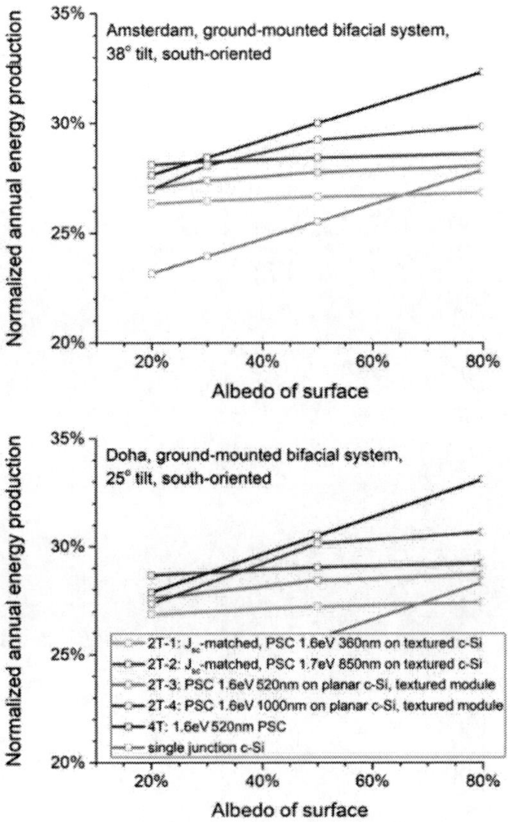

Figure 4: Normalized annual energy production of the various tandem designs for a simulated ground-based, free-standing module, landscape orientation,in the middle of a long row, south facing. Ground cover ratio 0.35; ground clearance 0.5 m. Normalized annual energy production is calculated as the annual energy production (Wh/m^2) divided by the cumulative annual front irradiation on the module (Wh/m^2). Top: in Amsterdam, Netherlands (modeled for albedos 0.2, 0.3, 0.5, 0.8). Bottom: in Dohar, Quatar (same legend, modeled for albedos 0.2, 0.5, 0.8). Lines are guides to the eye.

The significant stretching of the horizontal axis of Fig. 4 compared to Fig. 2 is important, because reported practical albedos (except for snow and purposely modified surfaces) are rather low; apparently, typical values are 20-35% [11].

The 2T tandem module with the top cell with the largest J_{sc}, 2T-2, performs best amongst the 2T designs for albedos larger than about 30%.

4 CONCLUSIONS

In line with the work by Asadpour *et al.*, standard band gap perovskite cell technology offers good opportunity for 2T tandems, if the cell and module are operated bifacially. Even optimal monofacial 2T designs with wide band gap will be matched or outperformed for high albedos of the ground surface.

The advantage of 2T bifacial tandems over the bifacial single junction c-Si technology used in the tandem (nPERT in this study) decreases as a function of albedo, and cost effectiveness of the 2T tandem architecture will become more critical for very high albedos.

4T tandems can of course also be based on standard band gap perovskite, and have an energy yield advantage over bifacial single junction c-Si modules which is constant as a function of albedo. They outperform the 2T tandems by a fraction that typically increases with albedo, in particular for very high albedos. Taking into account differences in series resistance and parasitic absorption in TCO between 2T and 4T tandems (not done in this study) may give some adjustment in this comparison.

Finally, we considered performance with textured module glass and front-polished c-Si wafers for 2T tandems, versus front-textured c-Si wafers with the perovskite stack conformally following the texture features, and these perform quite similarly.

Thus, there seems good opportunity for a 2T tandem design much simplified compared to present R&D efforts: standard perovskite band gap, thick absorber layer thickness, and deposited on a planar c-Si cell front side, as long as this is applied in bifacial operation. To verify this, further work should include modeling, and tests in reality, of the performance of industrially available glass textures.

REFERENCES

[1] S.L. Luxembourg *et al.*, "perovskite/crystalline silicon tandems: impact of perovskite band gap and crystalline silicon cell architecture", 33rd European Photovoltaic Solar Energy Conference (2017) 1176-1180; S.L. Luxembourg *et al.*, "Crystalline silicon solar cell with front and rear polysilicon passivated contacts as bottom cell for hybrid tandems", Energy Procedia 124 (2017) 621–627; D. Zhang *et al* "Four-terminal perovskite/c-Si tandem PV technology", Photovoltaics International Edition 40 (2018) 66-76.

[2] R. Cheacharoen *et al.*, "Encapsulating perovskite solar cells to withstand damp heat and thermal cycling", Sustainable Energy Fuels (2018), DOI: 10.1039/C8SE00250A, in press.

[3] R. Asadpour, R.V.K. Chavali, M.R. Khan and M.A. Alam, "Bifacial Si Heterojunction-Perovskite Organic-Inorganic Tandem to Produce Highly Efficient Solar Cell", Appl. Phys. Lett. 106 (2015) 243902.

[4] H. Chung *et al.*, "Modeling and designing multilayer 2D perovskite / silicon bifacial tandem photovoltaics for high efficiencies and long-term stability", Optics Express 25 (2017) A311-A322.

[5] D. Zhang *et al.*, "High efficiency 4-terminal perovskite/c-Si tandem cells", Solar Energy Materials and Solar Cells 188 (2018) 1–5.

[6] Santbergen *et al.*, "Minimizing optical losses in monolithic perovskite/c-Si tandem solar cells with a flat top cell", Optics Express 24 (2016) 1288-1299.

[7] W.F. Holmgren, C.W. Hansen, and M.A. Mikofski. "pvlib python: a python package for modeling solar energy systems." Journal of Open Source Software, 3(29), 884, (2018). https://doi.org/10.21105/joss.00884

[8] G.J.M. Janssen *et al.*, "Impact of inhomogeneous irradiance at the rear of bifacial panels on modelled

energy yield", 33rd European Photovoltaic Solar Energy Conference (2017) 1618-1623; and this conference.

[9] W. Hermann *et al.*, "Soiling and self-cleaning of PV modules under the weather conditions of two locations in Arizona and South-East India", 42nd IEEE PV Specialists Conference (2015), DOI: 10.1109/PVSC.2015.7356131.

[10] aThe front-incident J_{sc} of the monofacial tandems 2T-1 and 2T-2 were modeled with a back reflector, which results in slight increase compared to when no back reflector is present. For the comparisons in this paper, this effect is practically insignificant.

[11] P.P. Altermatt, on-line lectures, www.pvlighthouse.com.au, .section 4.5, with reference to P. Hupfer and W. Kuttler, Witterung und Klima, 11th edition (in German), B. G. Teubner, Stuttgart, Germany, 2005

OPTIMIZATION OF SILICON HETEROJUNCTION CELLS: A RECIPE FOR MORE THAN 26% EFFICIENCY

Moustafa Ghannam
EE department, Kuwait University, P.O. Box 5969, 13060, Safat, Kuwait
moustafa.ghannam@ku.edu.kw

ABSTRACT: Optimum conditions for high efficiency silicon heterojunction solar cells are investigated. It is shown that a large difference between the work functions of the p-type doped amorphous silicon layer (a-Si:H) and that of the n-type crystalline silicon substrate is required but subject to constraints imposed by the valence band offset and by doping concentrations. The optimum cell design when such constraints are taken into consideration should use a-Si:H with a band-gap in the 1.6-1.63 eV range with an electron affinity in the range 3.85-3.9 eV. The p-type doping concentration in a-Si:H should be around 10^{20} cm^{-3} with a maximum allowed peak dangling bond density half this concentration, and a substrate doping concentration in the $8x10^{16}$ to 10^{17} cm^{-3} range with a carrier lifetime in the ms range. With such a design, the predicted efficiency ranges from 25.6-26% for the full wafer thickness SHJ cell, and with effective light trapping reaches 26.1-26.4% for the thin 100 µm SHJ cell and 26.3-26.9% for the silicon thin film 50 µm SHJ cells.
Keywords: Over 26% SHJ cell efficiency, work function difference, valence band offset, thin silicon SHJ cell

1 INTRODUCTION

Extensive and continuous research and development devoted to the technology of a-Si:H/c-Si hetero-junction solar cells (SHJ) [1-5] led recently to the achievement of a record efficiency exceeding 26% [5]. In parallel, a substantial progress has been achieved in clearly understanding the principle of operation of this type of cells and in identifying the fundamentals constraints imposed on the cell parameters [6-10]. It is now ascertained that the high boost in the cell open circuit voltage, which is the main advantage provided by the SHJ cell, is associated with the presence of a strongly inverted p^+ inversion layer at the a-Si:H/c-Si hetero-interface in the n-type c-Si side. Although the passivation of the hetero-interface by an intrinsic a-Si:H(i) layer is and must be effective, it cannot by itself justify the very high open circuit voltage achieved in SHJ cells. It has been suggested that band-gap narrowing is expected to be smaller in the inversion layer emitter than in the p^+ doped emitters of c-Si cells, which is probably responsible to a great extent in limiting the enhancement of the emitter dark saturation current due to heavy doping, which allows V_{OC} to increase significantly [6]. Furthermore, it has been reported [e.g. 11,12] that, due to hole quantum confinement, band-gap widening rather than band-gap narrowing occurs in the inversion layer, which reduces the intrinsic carrier concentration and leads to further boost in V_{OC}.

Strong inversion, which must be maintained throughout the whole voltage range of cell operation, is mainly controlled by the amount of band bending at the hetero-interface. For better cell performance, the latter should be maximized which, however, may possibly generate trade-offs. An attempt is made in the present paper to understand the possible ways to enhance the band bending in the SHJ cells to ensure strong inversion, and resolve any possible trade-offs that may arise. An optimum structure could then be designed providing the highest possible efficiency using available technology and material.

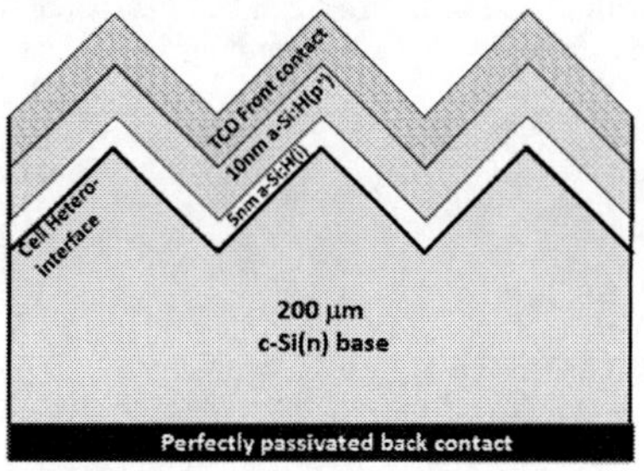

Figure 1: Standard SHJ cell structure

2. LINKING WORK FUNCTION DIFFERENCE, VALENCE BAND DISCONTINUITY AND CARRIER CONCENTRATIONS

The standard SHJ cell structure sketched in Fig. 1 consists of a c-Si n-type substrate with ultrathin doped/undoped a-Si:H(p^+/i) double layer front contact. The latter is topped with a thin transparent conductive oxide (TCO) forming a perfect ohmic contact with a-Si:H(p^+) layer, and not absorbing any light. The separate band diagrams of the a-Si:H(p^+) and c-Si(n) regions before contact is shown in Fig. 2a, with $\chi_{a\text{-}Si}$, $\chi_{c\text{-}Si}$, $E_{G,a\text{-}Si}$, $E_{G,c\text{-}Si}$, and $WF_{a\text{-}Si}$, $WF_{c\text{-}Si}$ representing the electron affinity

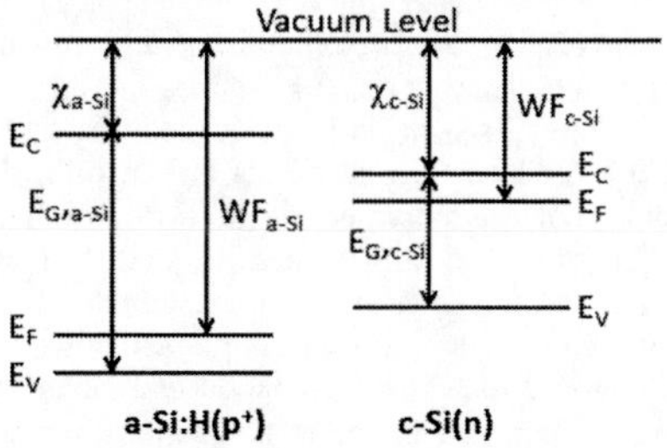

Figure 2a: Separate band structure of a-Si:H(p^+) and c-Si(n)

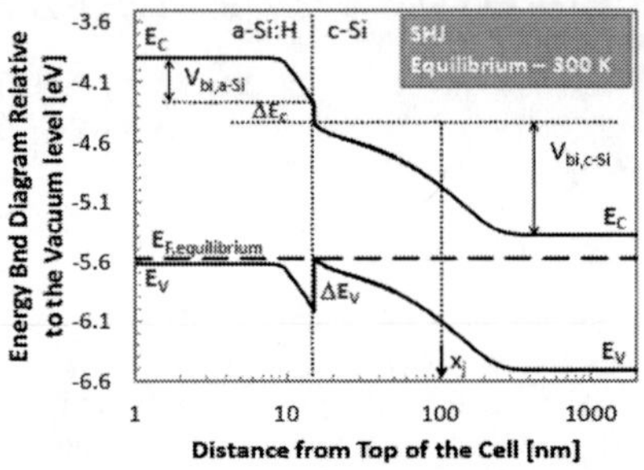

Figure 2b: Simulated band diagram of the SHJ at equilibrium

energy band-gap and the work function of a-Si:H and of c-Si, respectively, and E_C, E_V, and E_F represent the conduction, valence band edges and the Fermi energy, respectively. In principle, the equilibrium energy band diagram after setting the device and contacting the a-Si:H layer with the c-Si substrate can be built by applying the general affinity and band line-up rules [13,14] such that

$$\Delta\chi = \chi_{c-Si} - \chi_{a-Si} \quad (1a)$$

$$\Delta E_G = E_{G,a-Si} - E_{G,c-Si} \quad (1b)$$

$$\Delta E_c = \Delta\chi \quad (1c)$$

$$\Delta E_V = \Delta E_G - \Delta E_c = \Delta E_G - \Delta\chi \quad (1d)$$

$$\Delta E_V = (E_{G,a-Si} + \chi_{a-Si}) - (E_{G,c-Si} + \chi_{c-Si}) \quad (1e)$$

where ΔE_G is the difference between the energy band-gaps of a-Si:H and of c-Si, ΔE_C and ΔE_V are the conduction and valence band discontunities (offsets) appearing at the heterojunction, respectively. A typical band diagram of a SHJ cell simulated using AFORS-HET device simulator [15] is displayed in Fig.2b, which strictly obeys the electron affinity rule stated in equations (1a) to (1e), and in which flat band is ensured at the front surface to avoid any influence of the front contact on the band bending at the heterojunction [8].

As mentioned earlier, to achieve a high V_{OC} strong inversion at the hetero-interface must be ensured which requires that the valence band edge is the closest possible to the Fermi energy at the hetero-interface. As deduced from the band diagram of Fig.2b, such a situation requires maximum band bending in c-Si (maximum electric field at the hetero-interface), which can be achieved by increasing the difference between the work functions of a-Si:H and of c-Si, ΔWF. Since the latter is given by the total built-in potential in the cell, that is by the sum of the band bending in a-Si:H and in c-Si, the value of ΔWF could be increased by increasing the free carrier concentrations in a-Si:H and in c-Si through increasing the doping concentration in both layers. Higher limits for the doping concentrations, however, will be set to avoid material degradation and enhanced carrier recombination. On the other hand, ΔWF and the band bending could also be increased by increasing the values of $E_{G,a-Si}$ and χ_{a-Si}, which will also affect the valence band discontinuity ΔE_V at the hetero-interface as stated in (1e). However, with increasing ΔE_V, the thermionic emission barrier at the hetero-inetrface increases which degrades the cell Fill Factor [9]. To determine the optimum ΔWF, the analytical expression that governs the relationship between ΔWF, ΔE_V and the free carrier concentrations is established. The work function WF_{a-Si} in a-Si:H and W_{Fc-Si} in c-Si can be expressed as

$$WF_{a-Si} = \chi_{a-Si} + E_{G,a-Si} - kT\ln\frac{N_{V,a-Si}}{p_{a-Si}} \quad (2a)$$

$$WF_{c-Si} = \chi_{c-Si} + E_{G,c-Si} - kT\ln\frac{N_D N_{V,c-Si}}{n_{i,c-Si}^2} \quad (2b)$$

where $N_{V,c-Si}$ and $N_{V,a-Si}$ represent the effective density of states in the valence band of c-Si and of a-Si:H, respectively, N_D is the doping concentration in the n-type c-Si substrate, p_{a-Si} is the free hole concentration in a-Si:H(p^+) which is proportional to the chemical doping concentration N_{a-Si} , n_i is the intrinsic concentration of c-Si, k is Boltzmann constant, and T is the absolute temperature. Hence the work function difference is given by

$$\Delta WF = WF_{a-Si} - WF_{c-Si}$$

$$\Delta WF = -\Delta\chi + \Delta E_G + kTlnR_{NV}\frac{N_D p_{a-Si}}{n_{i,c-Si}^2} \quad (2c)$$

Substituting (1d) in (2c) we finally get

$$\Delta WF = \Delta E_V + kTlnR_{NV}\frac{N_D p_{a-Si}}{n_{i,c-Si}^2} \quad (2d)$$

with R_{NV} represents the ratio $N_{V,c-Si}/N_{V,a-Si}$ estimated to 0.2685 in AFORS-HET. Equation (2d) expresses the reationship between ΔWF and ΔE_V, and as expected confirms that the former could be increased by increasing ΔE_V, by increasing N_{a-Si} in a-Si:H(p^+), and by increasing N_D in the substrate. As mentioned earlier, the optimum cell should take into account the trade-offs associated with increasing ΔE_V at the hetero-interface, and by increasing the doping levels. These trade-offs impose the range for ΔWF variation within which the described losses could be avoided.

Optimization is carried out here starting with a SHJ cell built on an n-type 200 µm thick c-Si substrate with a doping concentration $N_D = 1.5\times10^{16}$ cm^{-3}, and having a 10 nm thick doped a-Si:H(p^+) layer on top of a 5 nm thick intrinsic a-Si:H(i) layer. The peak dangling bond density at the a-Si:H/c-Si hetero-interface is set at 10^{10} cm^{-2}eV^{-1} which ensures very good passivation. Since controlling accurately the chemical doping concentration N_{a-Si} and p_{a-Si} is not trivial, only two hypothetical values for N_{a-Si}, namely 10^{20} cm^{-3} and 5×10^{18} cm^{-3} for the high and low limits, respectively, are considered. The value of p_{a-Si} is determined by the default model implemented in AFORS-HET assuming proper dangling bond density in either case.

2.1 Leverage of ΔWF on the cell performance through varying ΔE_V

At first, the value of ΔWF is varied through changing ΔE_V, that is varying χ_{a-Si} or $E_{G,a-Si}$ or their sum, as suggested in (2d). The second term in (2d), however, is unaffected by changing ΔE_V, but depends on the doping level in a-Si:H, and is typically estimated to 0.89 eV for the heavily doped cell and 0.723 eV for the lowly doped cell. As depicted in Fig.3a, the efficiency η in both cells increases with increasing ΔWF and band bending, reaches a maximum, which has approximately the same value (24%) for both cells, then drops sharply. The dependence of η on ΔWF in the heavily doped cell is shifted to a higher ΔWF range due to the dependence of the second term of (2d) on the doping level in a-Si:H. Such a shift is removed and the interpretation becomes easier when η is plotted versus ΔE_V rather than versus ΔWF, as depicted in Fig.3b. Since the inversion becomes stronger as the band bending increases through increasing ΔE_V, the open circuit voltage V_{OC} in both cells also increases, as depicted in Fig.4a. Beyond $\Delta E_V = 0.2$ eV the rate of increase of V_{OC} in the heavily doped cell decreases which reflects on the cell efficiency that reaches a maximum and saturates at 24%, as depicted in Fig.3b. Since the band bending is smaller in the lowly doped cell, however, V_{OC} is smaller, as depicted in Fig.4a. The difference between V_{OC} in both cells decreases, however, with increasing ΔE_V, hence band bending, such that it becomes only 5 to 10 mV when ΔE_V reaches 0.3 eV, as depicted in Fig.4a. At this point the efficiency in the lowly doped cell also reaches a maximum and saturates at approximately 24%.

When ΔE_V exceeds the range within which η saturates, (beyond 0.36 eV and 0.4 eV for the lowly and heavily doped cells, respectively), the efficiency drops very sharply, as depicted in Fig.3b. This drop is associated with the thermionic emission barrier created by ΔE_V which becomes significant and hinders the transport of the photo-generated holes across the hetero-interface leading to a severe degradation of the cell Fill Factor (FF) [9,10], as depicted in Fig.4b.

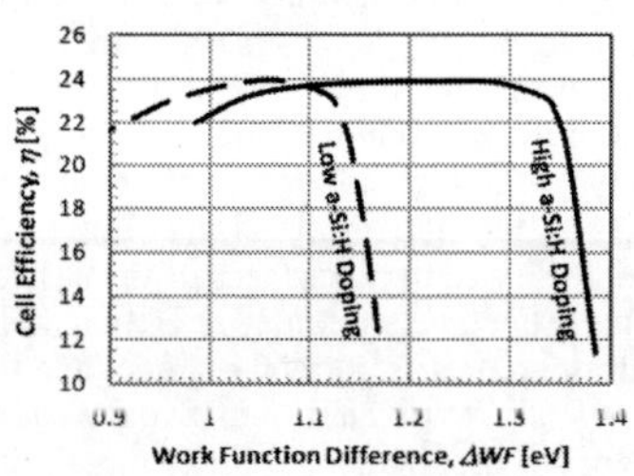

Figure 3a: SHJ cell efficiency vs ΔWF

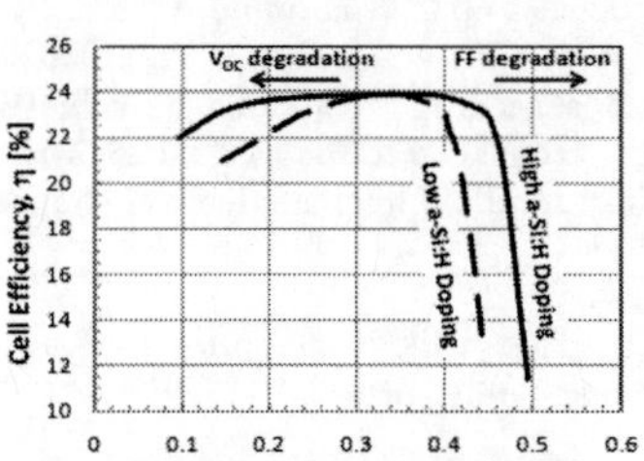

Figure 3b: SHJ cell efficiency vs ΔE_V

Consequently, it can be concluded that the optimum energy band-gap of a-Si:H that results in the highest efficiency corresponds to the ΔE_V range 0.2-0.4 eV *(ΔWF in the range* 1.09-1.3 eV) for the heavily dopd cell, and to the ΔE_V range 0.3-0.36 eV *(ΔWF in the range* 1.02-1.09 eV). Assuming that the conduction band offset ΔE_C is equal to 0.15 eV obtained assuming the electron affinity of a-Si:H $\chi_{a\text{-}Si}$ to be equal to 3.9 eV as often reported

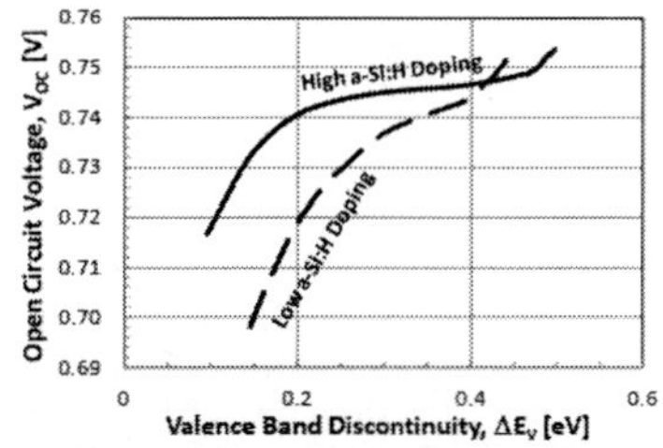

Figure 4a: Open circuit voltage vs ΔE_V

[e.g. 16,17], and with ΔE_V set to 0.33 eV, which is in the middle of the optimum range for the lowly doped cell but also perfectly falls in the optimum range of the heavily doped cell, ΔE_G could be evaluated using (1d) to 0.48 eV. Consequently, the average optimum band-gap of a-Si:H would be equal to 1.6 eV. With such an optimum value for $E_{G,a\text{-}Si}$, the resulting maximum efficiency amounts to 24%, only which is still far below the projected 26% efficiency, suggesting the necessity of further cell optimization.

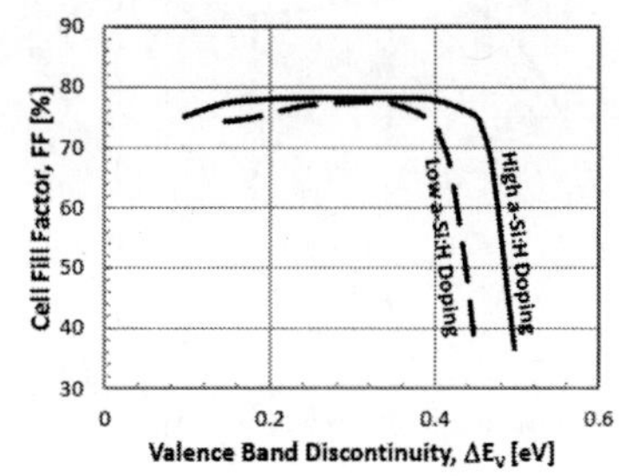

Figure 4b: Cell Fill Factor vs ΔE_V

.

2.2 Leverage of ΔWF on the cell performance through varying the doping concentrations

2.2.1 Impact of the doping concentration $N_{a\text{-}Si}$ in a-Si:H

As predicted in (2d) ΔWF increases with increasing the free hole concentration $p_{a\text{-}Si}$ in a-Si:H, that is with increasing the total (chemical) doping concentration $N_{a\text{-}Si}$ in a-Si:H(p$^+$). With increasing $N_{a\text{-}Si}$, the peak hole concentration p_{peak} of the inversion layer also increases, which reduces the emitter dark current and hence boosts V_{OC}. In addition, effective heavy doping is also necessary in a-Si:H(p$^+$) to avoid the layer depletion causing enhanced series resistance under illumination [10], and to avoid merging of depletion regions of the TCO/a-Si:H and of the a-Si:H/c-Si interfaces when flat band is not assumed at the front. It is well known, however, that the active or effective doping concentration $p_{a\text{-}Si}$ represents only a small percentage of $N_{a\text{-}Si}$ due to hole trapping by dangling bond defects [18]. Moreover, the density of the latter increases with increasing the chemical doping concentration [19], and becomes detrimental to the cell efficiency when the value of the peak defect density N_{DB0} (cm^{-3}eV^{-1}) exceeds half-to-full the doping concentration [cm^{-3}] [6]. Therefore, N_{DB0} will be assumed to be equal to 5x10^{19} cm^{-3}eV^{-1} in the cell with $N_{a\text{-}Si} = 10^{20}$ cm^{-3}. In such a case, the resulting free hole concentration $p_{a\text{-}Si}$ in a-Si:H(p$^+$) is estimated by to 2.1x10^{19} cm^{-3}, and the peak hole concentration p_{peak} at the hetero-interface to 4.8x10^{19} cm^{-3}. Although the maximum efficiency of the cell with such concentrations and optimized a-Si:H energy bandgap has been estimated here to only 24%, as depicted in Fig.3b, it will not be probably possible to beneficially achieve better cells with higher doping levels. Therefore, the optimum value of $N_{a\text{-}Si}$ is assumed to be around 10^{20} cm^{-3}.

2.2.2 Impact of the doping level N_D in the c-Si substrate

As stated by the second term of (2d), ΔWF increases also when N_D increases. In such case, the dark current injected into the substrate decreases which boosts V_{OC}. In addition, the bulk series resistance decreases which helps boosting the cell FF. As depicted in Fig.5, the optimum value for N_D is found to be in the range $7x10^{16}$ cm^{-3} to 10^{17} cm^{-3}, with a maximum efficiency of 25.7% when N_D = $9x10^{16}$ cm^{-3}, mainly due to the improvement of the FF from 78.2% to 81.9% as well as a slight increase in V_{OC} from 744 mV to 750 mV. Above 10^{17} cm^{-3}, however, bulk recombination starts to be important which reduces V_{OC} and the short circuit current and leads to the decline of the cell efficiency, as depicted in Fig.5.

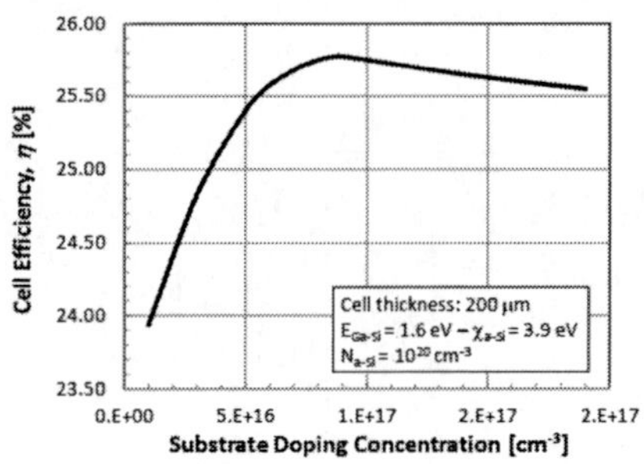

Figure 5: SHJ cell efficiency vs substrate doping concentration

3. OPTIMUM CELL THICKNESS

So far the optimization concerned a 200 μm thick cell, which does not necessarily provide the maximum possible effciciency. The trend is to go for thinner SHJ cells and therefore the efficiency will be investigated for: 1) 180 μm full thickness cell, 2) 100 μm thin Si cell and 3) 50 μm Si film cell. As the cell gets thinner, however, light trapping becomes a must, and will be considered in the simulations by adjusting the number of bounces given by the factor f at the cell surfaces. It is found that 25 bounces are needed for the optimum 180 μm full thickness cell to achieve the same current generated in the 200 μm thick cell treated above with 10 bounces. Therefore f = 25 is assumed to be standard for all cells. However, this factor should be higher in thinner cells to achieve the same current generated by the full thickness cell. The results displayed in Fig.6 indicate an efficiency of 25.6%, with V_{oc} = 752 mV and FF = 80.9% for the optimum full thickness cell, an efficiency of 26.12% with Voc = 764 mV and FF = 81.4%, for the 100 μm thick cell, that increases to 26.4 % with f boosted to 80 instead of 25. On the other hand, the efficiency of the optimum 50 μm silicon film cell is equal to 26.3%, with Voc = 777 mV and FF = 81.6% and may reach 26.9 % when f is boosted to 120.

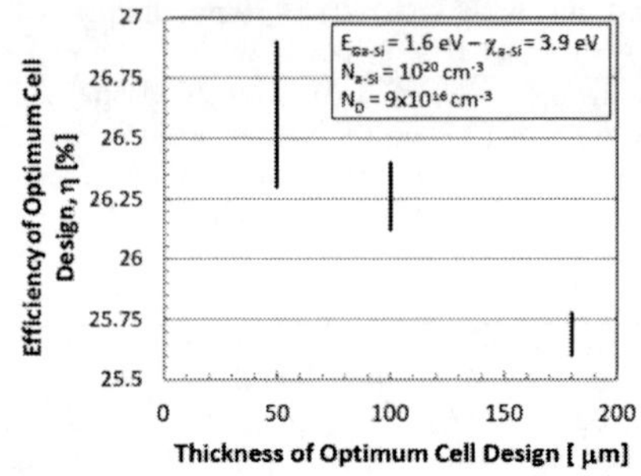

Figure 6: Efficiency of optimum design for: Wafer thickness (180μm), thin (100μm), Si thin film (50 μm), SHJ cells

4. CONCLUSIONS

A relationship is derived for the difference between the work functions ΔWF of a-Si:H and of c-Si, the valence band offset in p^+/n SHJ cells, and the doping concentrations in both materials. Two extreme doping concentrations in a-Si:H: 1) low doping with $N_{a\text{-}Si}$ = $5x10^{18}$ cm^{-3} and 2) high doping with $N_{a\text{-}Si}$ = 10^{20} cm^{-3}, have been investigated. The same maximum is achieved by the efficiency in both cases but in a smaller and restricted range of a-Si:H energy band-gaps 1.6-1.63 eV for the lowly doped cell, which extends to 1.675 eV for the heavily doped cell. With $E_{G,a\text{-}Si}$ = 1.6 eV and $\chi_{a\text{-}Si}$ = 3.9 eV, the optimum SHJ cell design, has an n-type high quality substrate, with a carrier lifetime longer than 1 ms, and a doping concentration N_D= 8-$9x10^{16}$ cm^{-3}. The doping concentration in the p-type a-Si:H should be around 10^{20} cm^{-3} and the maximum of the peak dangling bond defect density of $5x10^{19}$ $cm^{-3}eV^{-1}$ is afforded. The peak defect density at the hetero-interface is limited to 10^{10} $cm^{-2}eV^{-1}$. The front and back surfaces of the cell are textured to ensure effective multiple reflection. The TCO-a-Si:H(p^+) front contact is perfectly ohmic, and the back side has a perfectly passivated ohmic contact with an effective surface recombination velocity not larger than 10 cm/s. The thickness of the a-Si:H(p^+) and a-Si:H(i) layers are set to 10 nm and 5nm. The efficiency may gain about 0.025% to 0.05% absolute if the latter is reduced to 2 nm. With such a description, an efficiency close to or slightly exceeding 26% could be achieved, which can be further enhanced to the range 26.5-27% in thin Si cells and Si film cells if light is effectively trapped in the cell.

Acknowledgements: This work is supported by Kuwait University/KFAS/*imec* funded project P115-15EE-01

5. REFERENCES

[1] K. Wakisaka, M. Taguchi, T. Sawada, M. Tanaka, T. Matsuyama, T. Matsuoka, S. Tsuda, S. Nakano, Y. Kishi, Y. Kuwano, Proceedings 22nd IEEE Photovoltaic Specialists Conference (1991) 887.

[2] M. Tanaka, M. Taguchi, T. Matsuyama, T. Sawada, S. Tsuda, S. Nakano, H. Hanafusa, Y. Kuwano, *Jpn. J. Appl. Phys.* 31 Part 1 (1992) 3518.

[3] M. Taguchi, A. Yano, S. Tohoda, K. Matsuyama, Y. Nakamura, T. Nishiwaki, K. Fujita, E. Maruyama, *IEEE J. Photovolt.* 4 (2014) 96.

[4] K. Masuko, M.M. Shigematsu, T. Hashiguchi, D. Fujishima, M. Kai, N. Yoshimura, T. Yamaguchi, Y. Ichihashi, T. Mishima, N. Matsubara, T. Yamanishi, T. Takahama, M. Taguchi, E. Maruyama, S. Okamoto, *IEEE J. Photovolt.* 4 (2014) 1433.

[5] K. Yoshikawa, W. Yoshida, T. Irie, H. Kawasaki, K. Konishi, H. Ishibashi, T. Asatani, D. Adachi, M. Kanematsu, H. Uzu, K. Yamamoto, *Sol. Energy Mater. Sol. Cells* 173 (2017) 37.

[6] M. Ghannam, G. Shehada, Y. Abdulraheem, J. Poortmans, *Sol. Energy Mater. Sol. Cells* 132 (2015) 320.

[7] K. Ghosh, C. Tracy, S. Herasimenka, C. Honsberg, S. Bowden, Proceedings 35th IEEE Photovoltaic Specialists Conference (2010) 1383.

[8] M. Ghannam, Y. Abdulraheem, G. Shehada, *Sol. Energy Mater. Sol. Cells* 145 (2016) 423.

[9] M. Ghannam, Y. Abdulraheem, *Sol. Energy Mater. Sol. Cells* 171 (2017) 228.

[10] M. Ghannam, Y. Abdulraheem, to appear in special issue on Next Generation Photovoltaic Solar Cells, Applied Siences (2018).

[11] S.A. Hareland, S. Jallepalli, G. Chindalore, W.K. Shih, A.F. Tasch, C.M. Maziar Jr., *IEEE Trans. Electron Dev.* ED-44 (1997) 1172.

[12] M.J. Van Dort, P.H. Woerlee, A.J. Walker, *Solid-St. Electron.* 37 (1994) 411.

[13]. A.G. Milnes, D.L. Feucht, Heterojunction and Metal-Semiconductor Junctions, ScienceDirect, New York, 1972; ISBN 9780124980501.

[14] J.P. Kleider, In *Physics and Technology of Amorphous-Crystalline Heterostructure Silicon Solar Cells*; Sark W.J.H.M., Korte, L. Roca F. Eds.; Springer-Verlag, ISBN 978-3-642-22274-0, Ch. 12 (2012) pp. 405-414.

[15] R. Stangl, C. Leendertz, J. Haschke, *Solar Energy*; Rugescu R.D., Ed.; Intech: Vukovar (2012) 319.

[16] L. Korte, M. Schmidt, Doping type and thickness dependence of band offsets at the amorphous/crystalline silicon heterojunction, *J. Appl. Phys.* 109 (2011) 063714.

[17] M. Tucci, L. Serenelli, S. De Iuliis, M. Izzi, G. de Cesare, D. Caputo, In *Physics and Technology of Amorphous-Crystalline Heterostructure Silicon Solar Cells*; Sark W.J.H.M., Korte, L. Roca F. Eds.; Springer-Verlag, ISBN 978-3-642-22274-0 (2012) 337.

[18] D. Caputo, G. de Cesare, F. Irrera, A. Nascetti, F. Palma, *J.Non-Cryst. Solids* 266–269 (Part1) (2000) 565.

[19] M. Stutzmann, D. Biegelsen, R. Street, Phys. Rev. B35 (1987) 5666.

COMPOSITION-DEPENDENT HYDROGEN EFFUSION AND DIFFUSION MECHANISM WITHIN a-SiNx:H

S. Jafari[1,2,*], Chaitanya[1,2], J. Dwan[2], J. Hirsch[1,2], N. Bernhard[1], D. Lausch[2]
[1]Anhalt University of Applied Sciences, Köthen, Germany
[2]Fraunhofer Institute CSP, Center for Silicon Photovoltaics, Halle (Saale), Germany
*sahar.jafari@hs-anhalt.de, +49 345 5589 5124

The effect of the film structure of SiN_x on the hydrogen evolution at elevated temperatures was investigated by an effusion measurement system at constant heat rate. Control of hydrogen diffusion-limited effusion has been achieved by providing three different structures of SiN_x: (i) Si-rich (ii) N-rich and (iii) layer with identical content of N and Si. The hydrogen evolution peak shifts to higher temperature up to 900 °C by increasing the N content suggesting that N-rich SiN_x is a perfect compact layer for limiting hydrogen effusion due to firing. On the other hand, the Si-rich layer shows good surface passivation with large amount of hydrogen, however the hydrogen was almost effused out at temperatures lower than 800 °C indicating a void-rich structure. The desorption energy of hydrogen has been calculated out of the measurements at high temperature hydrogen evolution peaks and reveals a proportional relation to the N content.

Keywords: SiN_x , PECVD, hydrogen passivation, firing treatment, Hydrogen effusion

1 INTRODUCTION

Due to the current trend in PV industry towards higher quality solar cells having increased efficiency, it is of utmost importance to achieve a high quality of surface passivation since surface recombination is the dominating recombination loss within silicon solar cell technologies. Two strategies have been introduced for eliminating recombination losses on the surface: (i) reduction of interface defect density at the c-Si surface, which induces high surface recombination, by chemical passivation of dangling bonds [1] and (ii) application of an electrical field caused by fixed charges within the passivation layer, which reduces either the hole or electron concentration near the surface (field effect passivation) [2]. Passivation of crystalline silicon solar cells by plasma deposition of a-SiN_x:H at low temperature is the state of the art technology in industrial mass production and passivates the surface using both strategies. For the surface passivation, hydrogen plays a key role, which is due to high sticking coefficient S~0.6 of atomic hydrogen on a clean silicon surface [3]. It has also been reported that the presence of hydrogen in a-SiN_x:H accelerates the lifetime regeneration process under illumination and elevated temperature (LeTID), which is currently under intensive investigation [4,5]. Even though a-SiN_x:H layers on Si surfaces is a highly investigated layer system, little attention was paid to the firing process in solar cell production used for contact formation.

Effusion measurements in general provide a structural characterization of thin films, since the effusion of gas atoms and molecules from the film interior depends on the atomic density and bond structure and is strongly modified by the presence of empty spaces (voids) [6]. The focus of this article is a comparison study of different film composition regarding hydrogen diffusion and effusion using an effusion measurement system the FTIR measurements.

2 EXPERIMENTAL APPROACH

Planar 250 μm 2-5 Ω.cm p-type FZ wafers were used for the deposition of SiN_x. Wafers were cut by laser in 5 cm x 5 cm and followed by standard saw damaged removal. They were then etched by 10% HF and HNA each for 60 s prior to loading in the PECVD system. Amorphous hydrogenated SiN_x has been deposited on both sides of the samples at substrate temperature of 400 °C for about 2 min. The gas flow ratios of $[NH_3]/[NH_3+SiH_4]$ were 0.1, 0.5 and 0.9 which results in three types of SiN_x: (i) Si-rich (ii) N-rich and (iii) the identical concentration of Si and N. The N/Si ratio of the deposited films were estimated by the X-ray photo-electron spectroscopy XPS. Other details of the impact of deposition parameters on the film deposition of SiN_x are reported elsewhere [7]. N_2 gas was used as diluent in the gas mixture and all the processes were performed at a constant RF power of 50 W and pressure of 750 mTorr. The effective carrier lifetime after deposition of SiN_x was monitored at excess minority carrier density (Δn) of $1x10^{15}$ cm^{-3} using Sinton quasi-steady-state photo-conductance (QSSPC). The surface recombination velocity S_{eff} was then calculated using Schockley-Read-Hall formalism knowing the thickness of substrate and assuming very high bulk carrier lifetime.

$$\frac{1}{\tau_{eff}} = \frac{1}{\tau_{bulk}} + \frac{S}{2W}$$

Post SiN_x deposition in order to determine the thermal stability and the effect of co-firing process on the passivation quality of silicon nitride, samples have been fired at 800 °C in air for about 10 s. The firing process has been carried out in the standard rapid thermal processing belt furnace of an industrial partner. The concentration of different bonds in SiN_x and at the interface of Si/SiN_x was calculated by the fraction of peak areas in IR absorption spectra measured by Fourior-transform IR spectroscopy (FTIR). In addition, the thicknesses and the refractive index of the films were investigated by spectral ellipsometry,

3 HYDROGEN EFFUSION MEASUREMENT

The hydrogen effusion measurement takes place in a quartz tube at high vacuum 10^{-8} mbar in constant heat rate with 20 °C /min up to a temperature of 1200 °C. The release of molecules and atoms from the material can be detected by quadrupole mass spectrometry. The absolute quantity of hydrogen in the samples can be provided in this method [6]. The effusion rate is integrated over time

$$N_H = \int_0^t \int \left(\frac{dN}{dt}\right) dt$$

The typical low and high temperature effusion peaks are expected for the molecular (H_2) and atomic hydrogen (H) evolution in Si substrates passivated by amorphous SiN_x:H, respectively. The diffusion-limited effusion temperature highly depends on the film thickness and the diffusion coefficient. The diffusion process can be described by the Fick's diffusion law in an Arrhenius equation form,

$$D = D_0 exp(-E_D/kT)$$

Where D is the effusion rate, E_D is the energy of diffusion, D_0 is the diffusion coefficient and K is Boltzmann constant. The hydrogen effusion experiments are a series of experiments to determine the stability of hydrogen and how the hydrogen is distributed non-bounded or in bonded states (molecular state or saturating the dangling bonds) in the amorphous network of the SiN_x film.

PEAK ANALYSIS: The effusion peaks may be explained due to the hydrogen coming out of a void-rich film or compact film. At a higher temperature, the film tends to densify and it needs a much higher energy to get out the hydrogen of the film. Figure 1 shows a picture of the effusion measurement system contains of a quartz tube in which the samples are loaded, an oven mounted on a railway for tunable annealing time with a temperature controller, a quadrupole mass spectrometer to detect the different gas species and the vacuum units.

Fig. 1: Effusion system at Fraunhofer CSP.

4 RESULTS AND DISCUSSION

The concentration of various bonds in SiN_x films such as [Si-N], [Si-H] and [N-H] at varying gas flow ratio of $[NH_3]/[NH_3+SiH_4]$ was estimated by fitting the peak areas from IR absorption spectra by Lanford method [8]. Figure 2 shows the variation of the bond densities after the deposition and after firing process in three types of SiN_x as mentioned before.

Regarding the film composition, the highest concentration of [Si-H] was observed in the Si-rich layer and highest [N-H] in the N-rich layers about $2x10^{22}$ cm^{-3} and $7x10^{21}$ cm^{-3} respectively. The [Si-N] bond density shows the maximum value of $1x10^{24}$ cm^{-3} in the layer with the approximately identical amount of Si and N which drops to $9x10^{23}$ cm^{-3} in N-rich and to $6x10^{23}$ cm^{-3} in Si-rich SiN_x suggesting that the Si-rich SiN_x layer is void-rich while the two other layers have quite dense structure. The highest amount of hydrogen was achieved in Si-rich SiN_x up to $2x10^{22}$ cm^{-3}. Results are in good agreement with those reported previously [9].

The FTIR measurements after the firing shown in Fig. 2 with filled triangles present the film densification according to the increase of [Si-N] for all three layers. The loss of hydrogen concentration occurs due to activation energy provided during the firing process. The maximum release of hydrogen was observed in N-rich SiN_x due to the reduction of N-H bond concentration. The surface recombination velocity of the SiN_x passivated samples decreased from initial value of ~700 cm/s in bare reference Si wafer to the minimum of 10 cm/s in the Si-rich SiN_x shown in Fig.3. N-rich SiN_x shows higher surface recombination velocity of 70 cm/s indicating the increase of N content is destructive for the surface passivation likely due to the mismatching of the bond structure at the Si surface. The noisy curve at the low injection levels can be due to the defective interface of Si/SiN_x.

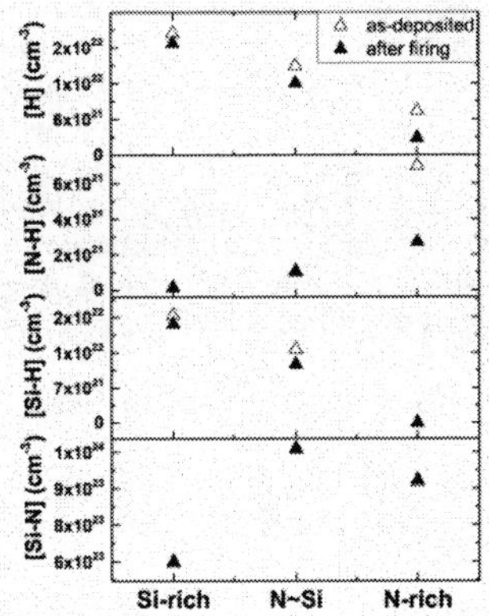

Fig. 2: The bond density of [Si-N], [Si-H], [N-H] and [H] in SiN_x deposited samples for three different film structure as deposited and after firing deduced from FTIR absorption peak areas.

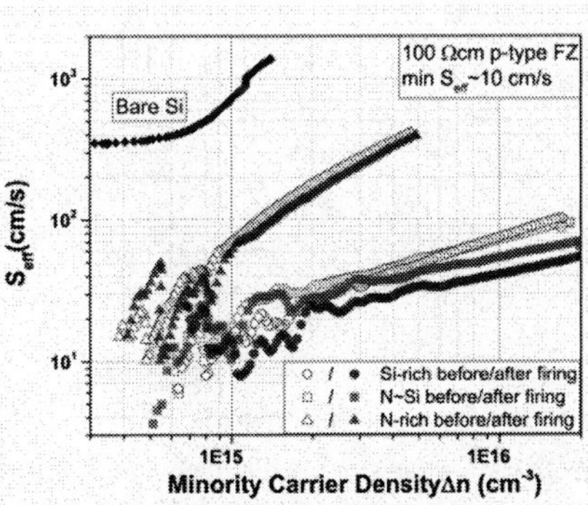

Fig. 3: Carrier injection dependent effective surface recombination velocity of three different film structures of SiN_x as deposited and after firing.

Although the different composition of SiN_x shows various quality of surface passivation, the surface recombination velocity reduced in all three types of silicon nitride after firing process. It can be attributed to the release of hydrogen from SiN_x to Si resulting in the passivation of more dangling bonds at the surface of Si. Furthermore, the increase of [Si-N] which shows the densification of the film due to the high temperature may cause the formation of K^+ centers in the SiN_x structure resulting in a higher amount of fixed charges and improvement of field-effect passivation [2].

To understand more about the mechanisms of the hydrogen evolution, in particular the effusion and diffusion due to the firing process, we measured the effusion of hydrogen by using the effusion measurement system. Samples were loaded in the quartz tube and transferred by a nickel boat to the heating area of the oven. Each sample was then heated separately with the constant heat rate of 20 °C/min up to temperature of 1200 °C.

Fig. 4 shows the hydrogen evolution curves as a function of temperature for three different types of SiN_x. Si-rich silicon nitride with high concentration of [Si-H] shown by the black curve in Fig. 4 shows the maximum effusion peak at lower temperature of ~ 650 °C comparing to N-rich at temperature of ~ 900 °C. It means that the N-rich SiN_x limits the release of hydrogen at firing temperature which is typically at 800 °C. The red curve in Fig. 4 shows SiN_x layer with the identical amount of Si and N with two maximum hydrogen peaks appear at temperature of 570 °C and 850 °C which determine that the hydrogen effusion temperature strongly depends on the bond structure of SiN_x. Fig. 2 and Fig. 4 both show that the higher hydrogen concentration can be achieved by deposition of the Si-rich SiN_x.

In addition, the structure of deposited film can be determined by the effusion measurements. It means the low temperature effusion peak shows a void-rich structure and the N content is increasing the density of SiN_x films. Furthermore, Beyer et al. [11] have reported that the effusion peaks shifts to lower temperature during the firing due to the densification of the film in which the transition of volume-bonded hydrogen to surface-bonded hydrogen takes place in a void-rich material.

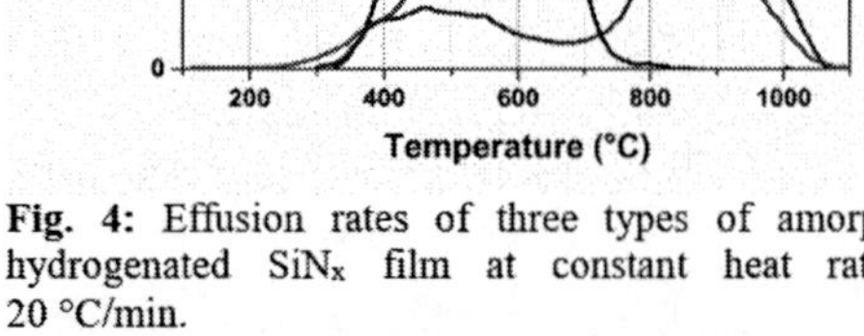

Fig. 4: Effusion rates of three types of amorphous hydrogenated SiN_x film at constant heat rate of 20 °C/min.

Two effusion peaks of hydrogen at low (LT) and high temperatures (HT) have been observed for each of the passivation structures. It is still not clear if the two LT and HT peaks are indicating the release of hydrogen from the two hydrogen bonds with different binding energies, [Si-H] (3.1 eV) and [N-H] (3.6 eV), respectively [12]. In a solid, usually these processes are not simply a breakage like Si-H → Si + H but rather (1) Si- (dangling bonds) + Si-H → Si-Si + H for hydrogen diffusion and (2) 2Si-H → 2Si + H_2 for H_2 surface desorption [6].

The desorption energy of hydrogen from two main bonds at the constant heat rate can be calculated from the Gibbs energy by the formulism proposed by Redhead [10] for the first order of reaction,

$$\frac{\Delta G}{kT} = Ln\left(\frac{v_1 T}{\beta}\right) - 3.64$$

where the β is the heat rate, v_1 is the reaction rate constant ($v_1 = kT/h$) and h is the Planck constant.

Table 1 presents the results for three different types of silicon nitride regarding in low and high temperature effusion peaks. The result shown in Table 1 reveals a large dependency of high temperature (HT) effusion peak on the composition of silicon nitride. The increase of N content shifts the peak to the higher temperatures. It can be concluded that there is a large correlation between the film density and the hydrogen effusion temperature. The desorption energy at HT effusion peaks are comparable with those reported for the N-H binding energy for N-rich and N~Si layers, the Si-rich shows almost no N-H bond concentration, thus desorption energy at HT in Si-rich is attributed to rupture of Si-H bond. The values of desorption energy at LT peaks are almost in same range, however for N-rich is relatively low. There is no clear correlation between the desorption energy at LT peaks and the film composition.

Table 1: Desorption energy of the low and high hydrogen effusion peaks in three different types of SiN_x

	Si-rich	N~Si	N-rich
ΔG at LT peak (eV)	2.24	2.54	2.15
ΔG at HT peak (eV)	2.62	3.34	3.4

Furthermore, released hydrogen from the bond breakage of either bonds (Si-H or N-H) will diffuse in a form of atomic or molecular hydrogen within the SiN_x depending on the film composition. For instance, the formation of H_2 before effusing out of the layer occurs more likely in a void-rich material. However the detected hydrogen in the effusion measurement system is always in molecular form.

5 CONCLUSION AND OUTLOOK

Direct experimental evidence of the correlation between the silicon nitride density regarding the [Si-N] concentration as a major bond and the hydrogen effusion temperature has been presented by comparing two characterization methods, FTIR analysis and hydrogen effusion measurement. We introduced three types of SiN_x (i) Si-rich, (ii) N-rich and (iii) the SiN_x with identical content of Si and N with quite different film structures which all of them show high thermal stability during the firing process resulting in reduction of S_{eff}. The temperature-dependent effusion curves describe the microstructure of the deposited film which is the key factor for the control of hydrogen effusion and diffusion at high temperature.

The N-rich SiN_x with high density of [Si-N] shows the hydrogen effusion at higher temperature comparing with the Si-rich layer promising that the N-rich SiN_x is the optimal barrier layer for the effusion of hydrogen and can act as a capping layer for the stacking structures. The Si-rich SiN_x is however an excellent candidate for the surface passivation with large concentration of hydrogen shown by both FTIR and effusion measurements. The low surface recombination of 10 cm/s has been achieved by deposition of Si-rich layers. The film structure of the deposited films have been also determined by the effusion measurement regarding the hydrogen effusion temperature. Results show that the Si-rich films are void-rich providing fast release of hydrogen at the temperature lower than typical firing temperature.

The two H effusion peaks appear in the effusion rate of silicon nitride assumed to be originating from Si-H bonds at low temperature and from N-H bonds at high temperature. The desorption energies calculated from the effusion measurements were in good agreements with reported binding energies of Si-H and N-H bonds.

6 AKNOWLEDGEMENT

This work was carried out within the joint research graduate school "StrukturSolar II" of Anhalt University of Applied Sciences and Martin-Luther-University Halle-Wittenberg which is funded by the German Ministry of Education and Research under identification code 03EK3570A. Especial thanks to Dr. Wolfhard Beyer from Jülich research center for his kind support and knowledgeable advices.

7 REFERENCE

[1] U. Hansen and P. Vogl, Physical review B, Vol 57(20), 13295, (1989).
[2] L. Maeckel et al., Journal of Applied Physics, p. 2602, (2002).
[3] A. Aberle, Solar Energy Materials and Solar Cells, 65, 239 (2001).
[4] K.A. Münzer, Proceedings 24th European Photovoltaic Solar Energy Conference, 1558 (2009).
[5] J. Hong, W. M. M. Kessels, W. J. Soppe, Journal of Vacuum Science & Technology, B 21, 2123 (2003).
[6] W. Beyer, Physica status solidi a, 1-14 (2016).
[7] S. Jafari et al. Proc. 33rd EUPVSEC, p. 553 (2017).
[8] W. A. Lanford, M. J. Rand, Journal of Applied Physics, 49, 2473 (1978).
[9] T. Lauinger, J. Moschner, A. Aberle, R. Hezel, Journal of Vacuum Science & Technology A: Vacuum, Surfaces, and Films **16**, 530 (1998).
[10] P. A. Redhead, Vacuum 12, 203 (1962).
[11] W. Beyer, H. Mell, Disordered Semiconductors, p. 641 (1987).
[12] J. Robertson, Philosophical Magazine B, Volume 69, p. 307 (1994).

A TRANSITION TO THINNER SI WAFERS AT HJT MASS PRODUCTION: AHEAD OF ITRPV SCHEDULE

D. Andronikov[1], A. Abramov[1], S. Abolmasov[1], K. Emtsev[1], G. Ivanov[1], I. Nyapshaev[1], D. Orekhov[1], A. Semenov[1], G. Shelopin[1], E. Terukova[1], E. Terukov[1]
N. Belkova[2], A. Dubrovskiy[2], P. Ishmuratov[2], A. Ivanov[2], D. Saykin[2], A. Smirnov[2], N. Saymurzanov[2], E. Sokolov[2], V. Tarasov[2],

[1]R&D Center of Thin Film Technologies in Energetics, Polytechnicheskaya str. 28,
194064 St. Petersburg, Russia
[2]Hevel LLC, Shorshelskiy 12, 429950, Novocheboksarsk, Russia

ABSTRACT: Si wafer price constitutes ca. 50% of the solar cell production costs resulting in a global trend in Si solar cell and module industry towards application of thinner as-cut wafers. In this paper, we report a full transition of Hevel HJT module production to 150 um as-cut wafers 4 years ahead of predicted ITRPV schedule. We also update the current technological status of Hevel production.

Keywords: silicon heterojunction, HJT modules mass production, thin silicon wafers

1 INTRODUCTION

Since Si wafer price constitutes around 50% of the solar cell production costs, there is a clear global trend in Si solar cell and module industry towards application of thinner as-cut wafers, which results in significant savings estimated as 0,015 USD/wafer for each 10 um of mono n-type Si wafer thickness [1]. Si heterojunction cell technology has an inherent advantage in terms of thin Si wafer application due to simple low temperature process resulting in low cell breakage rate even for smallest available wafer thicknesses. ITRPV reports minimum as-cut mono-Si wafer thickness of 170 um for the year 2017 and predicts transition to 150 um wafers for the year 2022 [2].

We have previously reported start of production for 160 MW/year HJT solar cells and modules at Hevel [3]. In this paper, we demonstrate a transition of Hevel HJT mass production line to application of 150 um as-cut CZ-Si 6" n-type wafers successfully accomplished in Q1 2018 – 4 years ahead of predicted ITRPV schedule.

2 TRANSITION OF HEVEL MASS PRODUCTION TO THINNER AS-CUT SILICON WAFERS.

First test runs with 150 um as-cut wafers have been performed in Q3 2017 followed by limited mass-production campaigns in Q4 2017 and full transition in Q1 2018.

Table 1. Average production cell characteristics corresponding to 150 um and 180 um as-cut wafer thicknesses at the time of transition averaged for 60 000 cells measured by GridTouch

Cell characteristics	150 um	180 um
Eff, %	21,94	22,03
Voc, V	0,723	0,721
Isc, A	9,287	9,327
FF, %	79,75	79,97

Decrease of wafer thickness in general resulted in slight decrease of cell efficiencies, with average production values going down from 22% to around 21,9% (Table 1) mainly due to reduction of short circuit current caused by thickness related change of spectral response in the IR-range (see Figure 1) partially compensated by increase of open circuit voltage.

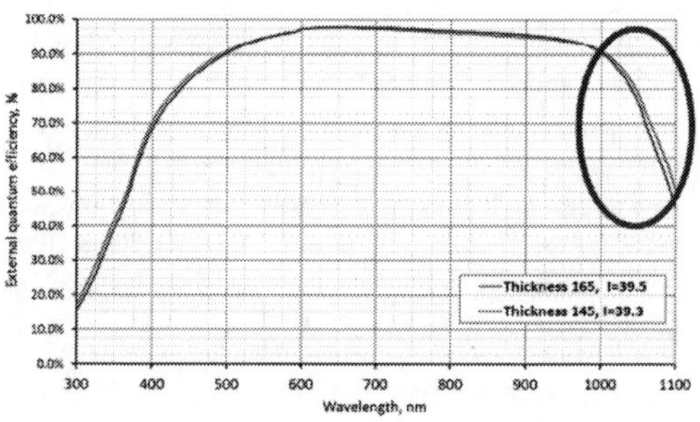

Figure 1: Comparative EQE for cells produced from 150 um (red color) and 180 um (blue color) as-cut Si wafers

Application of thinner Si wafers resulted in a drop of cell to module transition (CTM) losses by approximately 10% relative, compensating a slight drop of average cell efficiencies and yielding a 1-1,5W increase of average module power in production.

At the same time, we are facing a challenge of an increased cell breakage rate both at cell and module levels. These yield losses somewhat influences the benefit of Si wafer price reduction when switching to thinner wafers.

3 OUTLOOK

As the Si prices are constantly going down it is very important to consider the balance between material cost savings and increasing cell breakage rate in real production conditions causing decrease of an overall production yield. There are two strategies to attack this issue first of them being a direct increase of breakage rate by tuning the cell processing (e.g. modifying the wet chemistry recipe to decrease the etch removal thickness obtaining thicker cells) and carefully considering the optimal wafer thickness by taking into account the balance between the current Si wafer price and the observed breakage rates in production.

4 HEVEL PRODUCTION STATUS UPDATE

A significant progress has been achieved at Hevel production for the past year. The mean cell efficiency reached 22,3-22,4% with a champion production cell reaching 23% (see Figure 2)

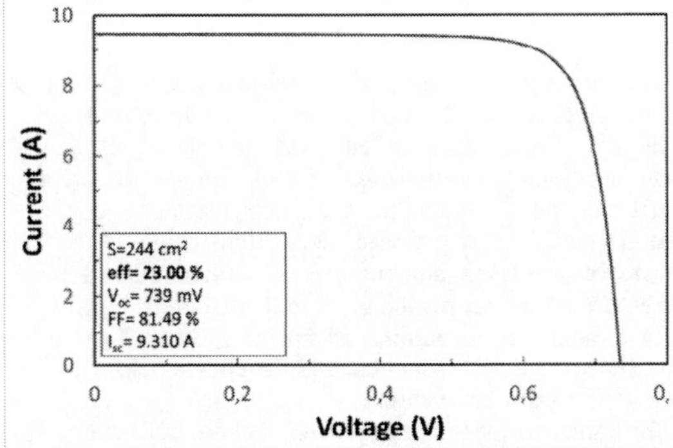

Figure 2: Champion production cell IV curve and characteristics. GridTouch measurements.

A lot of effort has been devoted to decreasing the CTM losses by implementing optimized materials for Smart Wire Connection Technology (SWCT) cell interconnection. As a result, we have achieved at Q3 2018 an average 60 cell glass/backsheet module power of 313W (Figure 3) with the best module power of 320W (table 2) resembling a 28W average module power increase compared to start of production in Q2 2017.

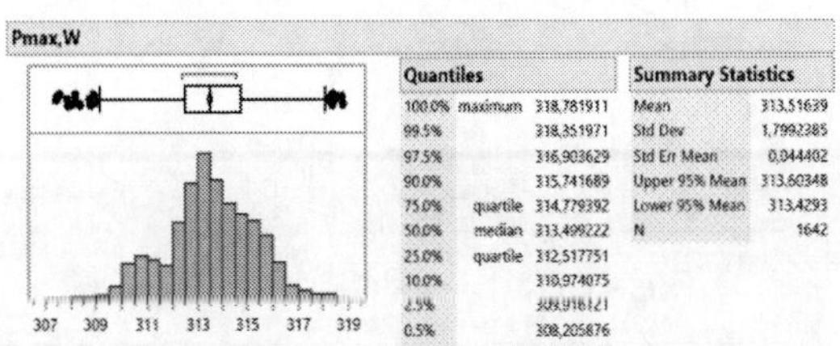

Figure 3: Distribution of Hevel 60 cell glass/backsheet SWCT IV-characteristics for a typical production day based on measurement of 1642 modules.

Table 2. Best production module IV-characteristics

Power, W	320
Voc, V	44,13
Isc, A	9,296
FF, %	78,00

4 CONCLUSIONS

Four years ahead of predicted ITPRV schedule Hevel has accomplished full transition from 180 um to 150 um as cut wafers. As a result, we have achieved a production costs reduction at the same time benefiting from lower CTM losses and average module power increase. At the same time, as the Si wafer prices are significantly decreasing on the one hand the optimal wafer thickness may have to be reconsidered taken into account the production yield loss when using thinner wafers and on the other hand additional process modifications have to be implemented to decrease the breakage rate coming from reduced Si wafer thickness.

The authors wish to thank Longi for constant support of Hevel production and adapting their products to our needs

REFERENCES

[1] Y.C. Wang, nPV workshop, 2018;
[2] ITRPV 9th edition 2018;
[3] D. Andronikov et. al, Proc. of the 33rd EUPVSEC 2AV.3.12, 2017

STUDY OF CHANGES IN INTRINSIC A-SI:H PASSIVATION LAYER INDUCED BY THE GROWTH OF N-DOPED NANOCRYSTALLINE LAYER

Hosni Meddeb, Oleg Sergeev, Hyo-Jei Cho, Otman Werfaly, Martin Vehse
DLR Institute of Networked Energy Systems,
Carl-von-Ossietzky-Str. 15, 26129 Oldenburg, Germany

ABSTRACT: In silicon heterojunction solar cells (SHJ), the passivation quality of an intrinsic hydrogenated amorphous silicon (i-a-Si:H) layers can be either harmfully or beneficially affected by the impact of subsequent employed doped layers as back (BSF) or front (FSF) surface field. In this work, we investigated the induced changes of the i-a-Si:H layer properties exposed to nanocrystalline regime plasma conditions with high Hydrogen flow and plasma power. Moreover, we studied the change of the i-a-Si:H passivating layers due to the deposition of n-doped nanocrystalline silicon layers (nc-Si:H) at high rate. It was found that the low pressures deposition regime ($Pr < 3$ mbar) of nc-Si:H leads to the etching of underlying i-a-Si:H passivation layer inducing lifetime degradation with a loss of about 90%. In contrast, the deposition at high pressures ($Pr > 6$ mbar) results in the degradation of lifetime due to induced disorder. Thus, moderate working pressures (4-6 mbar) are recommended for the growth of the n-doped nc-Si films on an i-a-Si:H passivation layer, to allow further defects reduction and beneficial field effect passivation, which provide up to 50% gain in terms of effective lifetime absolute value.
Keywords: Hydrogenated amorphous silicon; Passivation: Lifetime; Doped nanocrystalline silicon; Silicon heterojunction solar cells

1 INTRODUCTION

In silicon heterojunction solar cells (SHJ), the collection of charge carriers is commonly achieved by doped amorphous silicon (a-Si:H) layers of a few nanometers, deposited on the opposite sides of the crystalline silicon wafer [1]. A tendency towards the employment of thin doped nanocrystalline silicon layers (nc-Si:H) has been adopted to minimize the parasitic optical absorption and to promote better electrical contact with the adjacent electrodes [1,2]. It was demonstrated that Hydrogen atoms play a crucial role during the growth of the nc-Si films via a dominant etching process and selective removal of the amorphous phase [3, 4]. During the growth of nc-Si:H at high power and high Hydrogen dilution regime, a competition between etching and deposition can be manifested [3].

Hence, the growth of n-nc-Si:H doped layers at high deposition rates can lead to the degradation of the underlying i-a-Si:H passivation layer which is detrimental for open-circuit voltage (V_{oc}) in solar cell characteristics [1].

In this work, we report on the combined monitoring of in-situ plasma diagnostics, structural, morphological and chemical bonding to study the changes in the i-a-Si:H passivation layer which are induced by the deposition of the n-doped nc-Si:H films. Moreover, we figured out links between the passivation quality, growth mechanism, and structural material properties. In the light of this study, an optimum working pressure regime, and thus a lifetime improvement, was demonstrated.

2 EXPERIMENTAL DETAILS

The samples for lifetime measurement consist of a n-nc-Si:H (15 nm)/i-a Si:H (10 nm) symmetric structure on textured (111) n-type Czochralski (CZ) silicon wafers (CZ-c Si, 180 μm, resistivity of 5 Ω·cm). For single layer characterization, thin (25 nm) intrinsic a-Si:H films were deposited on polished crystalline float zone silicon wafers (FZ c-Si) as well as on Corning Eagle-XG glass substrates by means of Plasma Enhanced Chemical Vapour Deposition (PECVD) at 13.56 MHz in a cluster system Von Ardenne CS-400PS, as depicted in figure 1.

The deposition i-a-Si:H layers was performed at a temperature of 200°C with a SiH_4:H_2 ratio of 40:100 sccm, at a pressure of 2 mbar and a power of 15 W. The i-a-Si:H layers are exposed to plasma treatment at the same regime (high plasma-high power) as for n-nc-Si:H layer growth without addition of silane precursor gas, as presented in table 1.

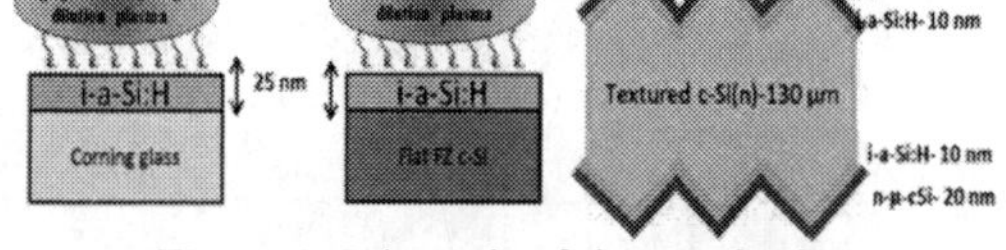

Figure 1: Schematic of characterization test structures

	T [°C]	SiH_4 [sccm]	H_2 [sccm]	PH_3 [sccm]	Pr [sccm]	Pw [W]	d [mm]
Plasma exposure	220	0	1000	2	1-8	125	15
n-nc-Si:H deposition	220	5	1000	2	1-8	125	15

Table 1: Detailed parameters for both etching and deposition processes

The effective carrier lifetime τ_{eff} was measured by quasi-steady-state photoconductance (QSSPC) using equipment of *Sinton instruments WCT-120* [5]. The reported values were obtained at room temperature and at an excess carrier density of $n = 10^{15}$ cm^{-3}. Using an amphoteric defect model for the a-Si:H/c-Si interface recombination [6], a correlation between effective lifetime values, the surface-state density (N_S) and surface charge density (Q_s) was obtained. Spectroscopic Ellipsometry (SE) measurements (using a *SENTECH SE850-ST* ellipsometer), were carried out to determine the thickness variation. To assess the Hydrogen bonding in the layers, Fourier Transform Infrared spectroscopy (FTIR) measurements using *Perkin Ellmer Spectrum 400* system were performed. The structure of the grown films was assessed by a *Bruker SENTERRA* Raman spectroscopy. Atomic Force Microscopy (AFM) was applied to analyze the change in surface morphology after plasma exposure. Furthermore, time-resolved

Optical Emission Spectroscopy (OES) was employed to monitor the plasma emission. The spectra were measured through a view port at the chamber wall. The intensity of Si* (288 nm), SiH* (414 nm), H_β (486 nm) and H_α (656 nm) as typical peaks in a Hydrogen dilution plasma were probed [7].

3 RESULTS AND DISCUSSIONS

3.1 Lifetime results

Figure 2 (a) shows the injection-level dependent effective charge carrier lifetime for different passivation test structures: i-a-Si:H layer as well as a film stack with different nc-Si:H n-doped layers deposited at different pressures from 1 to 8 mbar. The deposition time is tuned to get the same layer thickness of about 20 nm. The corresponding extracted lifetime at an injection level of $1x10^{15}$ cm^{-3} are shown in figure 2 (b), whereas, the extracted surface defect density N_s and surface charge density Q_s were plotted in figure 2 (c). We can easily distinguish three different pressure regimes:

i) Low pressures range 0-2.5 mbar: A severe drop of the lifetime values compared to as-passivated i-a-Si:H layer (down to ~90% drop) is obvious. This region is characterized by the highest defect density N_s at the a-Si:H/c-Si interface induced due to detrimental etching and deterioration of the underlying passivation layer.

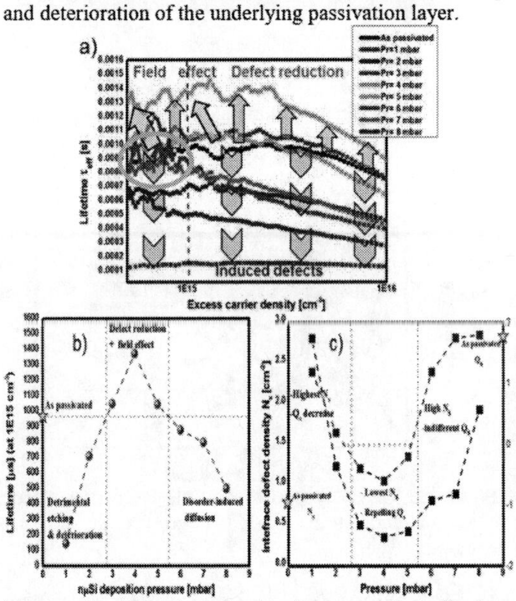

Figure 2: The injection-level dependent effective minority carriers lifetime for different n-nc-Si:H deposition pressures (a). Extracted lifetime at 1E15 cm^{-3} (b) and N_s and Q_s values (c) determined based on amphoteric model [6]

ii) Medium pressures range 2.5-5.5 mbar: It is the optimum lifetime region with an enhanced chemical passivation mechanism with the lowest defects density by further reducing the recombination center density (dangling bonds at a-Si:H/c-Si interface). The increase of lifetime due to defect reduction occurs for almost all injection levels (figure 2 (a) green arrows) The field effect passivation promoted through the reduction of one carrier type is reflected by the drop of Q_s manifested by creating an electric field due electrostatic charges with the overlaying nc-Si:H doped layer [8]. This impact of the latter mechanism is mainly weighted for low injection levels (below 1E15 cm^{-3}). Both enhanced passivation schemes lead to a gain of about 50% in terms of effective lifetime (absolute value).

iii) High pressures range 5.5-8 mbar: The lifetime value return to a decreasing trend. This is accompanied with a rise in recombination rate as expressed by high defect density N_s. This lifetime drop is mainly due to the fact that for relatively thin i-a-Si:H passivation layer (below 10 nm), high power and high pressures regime can lead to a diffusion mechanism towards the a-Si:H/c-Si heterointerface [9,10]. Such 'cascade' of Hydrogen atoms with larger diffusion length tend to alter the i-a-Si:H network creating further disorder and inducing not passivated dangling bonds which act as carrier recombination center [11]. It is worth to note that the drop of lifetime is more pronounced at high injection level (above 1E15 cm^{-3}), whereas for a low injection level, the lifetime level is relatively comparable to medium range pressures, as can be seen for the case of 6 and 7 mbar (highlighted circle).

3.2 OES plasma diagnostics and thickness measurements

Figure 3 (a) presents OES spectra corresponding to species dissociation in the plasma exposure at different pressures. In figure 3 (b) a zoom in of the H^*_α related peak is depicted. We notice a progressive decrease of the emission intensities of H_α^* with decreasing working pressure. While, low pressures regime results in the rise of Hydrogen ions concentration and radical formation in the plasma, susceptible to induce excessive etching process [12]. Furthermore, a slight appearance of SiH* related peaks at low pressure could be an indication of the possible i-a-Si:H layer etching and ion bombardment.

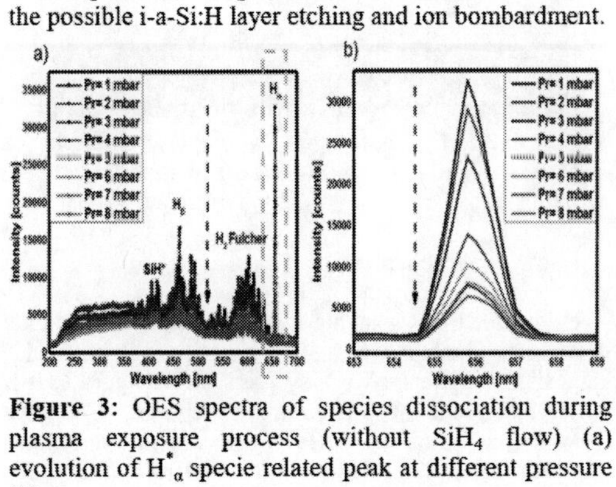

Figure 3: OES spectra of species dissociation during plasma exposure process (without SiH_4 flow) (a) evolution of H^*_α specie related peak at different pressure (b)

Figure 4 displays the evolution of i-a-Si:H layer etching rate upon plasma exposure and n-nc-Si:H films deposition rate as a function of the pressure. It is clearly seen that the etching rate of i-a-Si decrease monotonously with increasing pressure. Furthermore, the deposition rate of nc-Si:H presents almost a linear rise with increasing pressure. It is worth to note that only using RF-frequency at 13,56 MHz, we achieved considerably high deposition rate up to 0.3 A°/s, which can be competitive with others very high frequency (VHF) techniques. On the other hand, the deposition rate is very close to the etching rate at very low pressure (~1 mbar). These observations are consistent with the OES results showing a drop in the emission of H-related species, which are responsible for etching mechanism, with decreasing pressure.

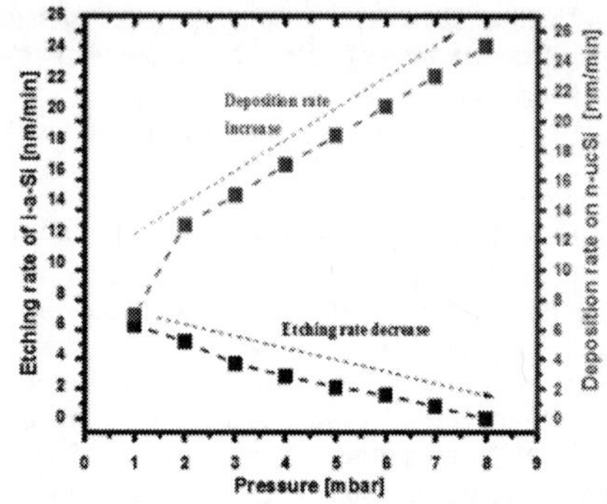

Figure 4: i-a-Si:H layer etching rate during plasma exposure and n-nc-Si:H films deposition rate at various pressure

3.3 Morphological properties

In figure 5, we show the change in i-a-Si:H surface morphology after plasma exposure at different pressures. The treatment at high pressure of 3 mbar (a) results in an i-a-Si:H layer with small grains and quite dense network. While plasma exposure at low pressure below 2 mbar leads to a partial etching of the film network and the creation of large pores, hence, a deterioration of the film network.

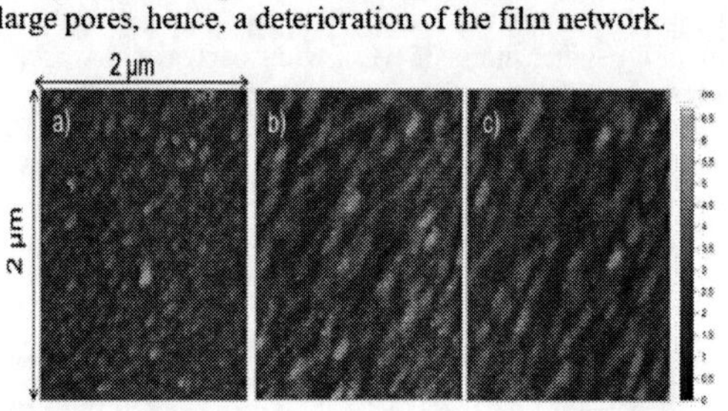

Figure 5: AFM images of thin i-a-Si:H layer after plasma exposure simulating n-µc-Si:H growth at different working pressure: 1 mbar (a), 2 mbar (b) and 3 mbar (c)

3.4 Structural properties

We illustrate FTIR wagging-rocking mode related to total Hydrogen content (C_H) for n-nc-Si:H layers [13] deposited at different working pressures in figure 6. It is observed that the low pressur growth regime (at Pr < 3 mbar) leads to low Hydrogen content in the films. This can be explained by the dominant role of H in the selective etching of amorphous network during the growth [4]. At average pressures, a slight increase of Hydrogen content is observed. Due to the fact that the nc-Si film consists of nano-sized crystalline grains mixed in amorphous matrix, the H atoms are capable to diffuse into intra-grains and the amorphous network. Such beneficial incorporation of H would provide better defects reduction. For high pressures plasma exposure, significant drop in Hydrogen content is noticed. By increasing the deposition conditions is closer to the silane depletion regime and Hydrogen is consumed in the plasma glow reactions rather than being incorporated in the films structure [14]. These finding shed more light on the role of Hydrogen in the completion between selective etching and the deposition growth mechanism.

In figure 7, Raman spectra of the i-a-Si:H layer after plasma exposure display n-nc-Si:H growth at different working pressure. We marked the typical bands of amorphous silicon. The most intense peak is located at 480 cm^{-1} (Transversal Optic TO mode) and another at 150 cm^{-1} (Transversal Acoustic TA mode), which are sensitive to variations in bond angle in the a-Si network. We can notice a rise in I_{TA}/I_{TO} ratio at extreme low and high pressures which reflects a lowering in the short range order by i-a-Si:H treated bulk material. A plasma exposure at medium range around 4 mbar provides the closest result to an "as deposited" i-a-Si:H layer, hence, Applying such deposition regime results in less induced disorder and minimized bulk film degradation.

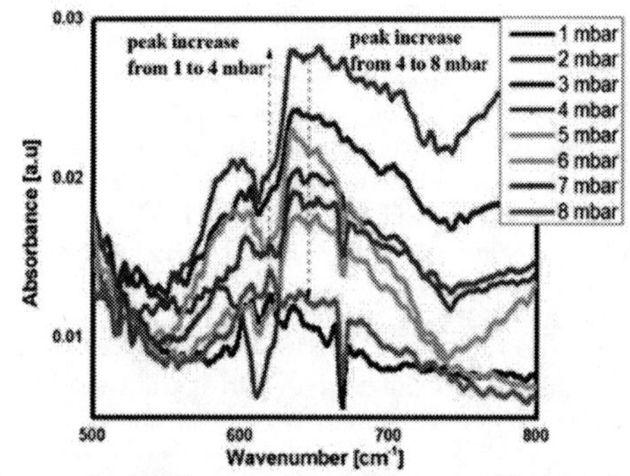

Figure 6: FTIR wagging-rocking mode related to Hydrogen content in the n-nc-Si:H layers deposited at different working pressures

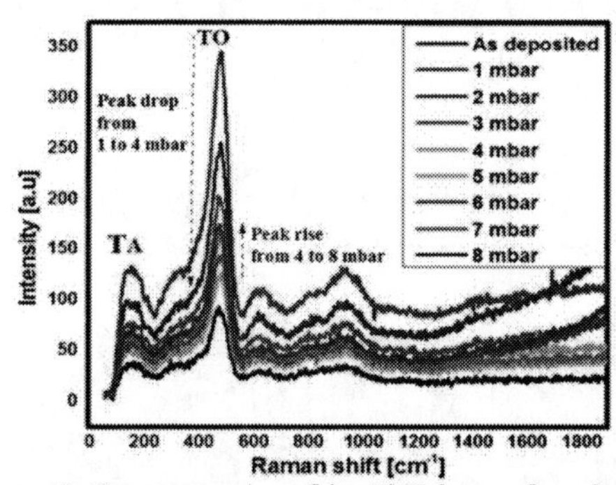

Figure 7: Raman spectra of i-a-Si:H layer after plasma exposure simulating n-nc-Si:H growth at different working pressure

4 CONCLUSIONS

We have studied the changes induced by the growth of n-doped nc-Si:H films in i-a-Si:H passivation layer for heterojunction solar cell integration. We found that for the low pressure deposition regime, lifetime degradation is caused by the etching of the amorphous network and by microvoid creation at the i-a-Si:H / n-nc-Si:H interface. At high pressur processes, the drop of lifetime is related to induced disorder and creation of further recombination site in the i-a-Si:H film. It is concluded that a moderate working pressure (4-5 mbar) is recommended for the beneficial growth of n-doped nanocrystalline films on i-a-Si passivation layer, with a gain of about 50% in terms of the absolute lifetime value.

5 ACKNOWLEDGMENTS

This work was supported by the German Federal Ministry of Economics and Technology (BMWi, reference number 0325825H, HERA).
We thank the company Meyer Burger and Fraunhofer-ISE for the wafer delivery. The authors acknowledge the insightful scientific discussions and experimental support of Michael Kröner, Martin Kellermann, Volker Steenhoff, Norbert Osterthun and Maximilian Götz.

6 REFERENCES

[1] M. O. Sergeev et al., Energy Procedia 124, 371-378, (2017)
[2] J.P. Seif et al., IEEE Vol. 6 pp 1132-1140, (2016)
[3] O. Madani et. al, Proceeding 27th European Photovoltaic Solar Energy Conference and Exhibition,Vol I, (2010)
[4] P. Cabarrocas et al., Solid State Phenomena Vol. 93 pp 257-268, (2003)
[5] R. Sinton et al. Appl.Phys.Lett.69, 2510–2512, (1996)
[6] S. Olibet et al, Phys. Rev. B 76, (2007)
[7] H. Meddeb et al., J. Phys. D Appl. Phys. 48, 415301 (2015).
[8] S. De Wolf et al., Phys. Rev. B 85, 113302, (2012)
[9] H. Meddeb et al., Phys. Status Solidi RRL 9 (No. 1) 53-56, (2015).
[10]A. Descoeudres et al., Appl. Phys. Lett. 99,123506 (2011).
[11] J. Geissbuhler et al., Appl. Phys. Lett. 102,231604 (2013).
[12] J. K. Rath, et al., Solar Energy Materials and Solar Cells 90, 3385 (2006)
[13] S. Lebib et al., Journal of Applied Physics 97, 104334, (2005)
[14] A.Matsuda et al., Sol. Energy Mater. Sol. Cells 78 3–26, (2003)
[15] H. Meddeb et al., Energy Proc. 55 818–26, (2014)

IMPROVEMENT OF NANOCRYSTALLINE DOPED LAYER PROPERTIES WITH ARGON AND HYDROGEN PLASMA TREATMENTS

Hosni Meddeb, Oleg Sergeev, Hyo-Jei Cho, Martin Vehse
DLR Institute of Networked Energy Systems, Carl-von-Ossietzky-Str. 15, 26129
Oldenburg, Germany

ABSTRACT: In silicon heterojunction solar cells (SHJ), the collection of charge carriers is commonly achieved by doped silicon layers of a few nanometers, deposited at opposite sides of the crystalline silicon wafer. In this work, we point out the potential application of Hydrogen (HPT) and Argon (APT) plasma treatments for the improvement of n-doped nanocrystalline layer (n-nc-Si:H) to be applied as window layer in SHJ solar cells. Moreover, we conducted a comparative study about the impact of HPT and APT on the structural, optical and electrical properties of n-nc-Si:H. It was found that both APT and HPT are beneficial for the enhancement of doped nc-Si:H layers properties. It was demonstrated that the integration of a combined Argon and Hydrogen treatment (AHPT) in bifacial SHJ, leads to more pronounced improvement, yielding a high value of short-circuit current density J_{sc} of about 39.23 mA/cm^2 with a gain of more than 2 mA/cm^2 and pseudo-efficiency up to 24%.

Keywords: Doped nanocrystalline silicon; Plasma post-treatments, Hydrogen, Argon

1 INTRODUCTION

Heterojunction silicon solar cells SHJ have attracted a considerable attention from PV community thanks to their high conversion efficiencies and their advantageous low thermal budget [1, 2]. Commonly, amorphous doped layers (a-Si:H) are employed as a back surface field (BSF) or front surface field (FSF) in the solar cells on n-type c-Si wafers [3]. Such allotropic forms present several drawbacks such as high defect density, low doping efficiencies, parasitic optical absorption and poor conductivity [4]. Hence, a tendency towards the substitution by thin doped nanocrystalline silicon layers nc-Si:H has been adopted to minimize the parasitic optical absorption and to promote electrical conductivity [4,5]. In our previous studies we explored and explained the positive influence of APT onto the passivation quality of silicon wafers with i-a-Si:H layers and have shown its benefits compared to commonly used Hydrogen plasma treatment (HPT) [6].

In this work, we report on the impact of plasma post-treatments on n-doped nc-Si:H. Moreover, a comprehensive in-situ plasma diagnostics, structural, morphological and electrical study of separate and combined APT and HPT was addressed. Along this report, a considerable enhancement in nc-Si:H doped layers, and thus an outstanding current density, were demonstrated.

2 EXPERIMANTAL DETAILS

A schematic illustration of the reference bifacial SHJ solar cells with nc-Si:H doped n and p layers used in this work is shown in Fig. 1. For the SHJ solar cells fabrication, we used 6-inch n-type randomly textured monocrystalline Czochralski silicon wafers with a similar thickness of about 130 μm and resistivity of 1-5 Ohm.cm. The wafers were already passivated with i-a-Si:H layers on both sides, stored in a N_2 box and delivered from the company *Meyer Burger*. Both n- and p-doped nc-Si:H layers were grown in a multi chamber plasma enhanced chemical vapor deposition (PECVD) *von Ardenne CS400PS Clustertool* at 13.56 MHz. A Silane (SiH_4) and Hydrogen (H_2) gas mixture was used with diborane (B_2H_6) and phosphine (PH_3), both diluted to 3% in H_2, added as dopant precursors. The deposition p-doped nc-Si:H layers was performed at a temperature of 200°C with a SiH_4:H_2 ratio of 5:1000 sccm and B_2H_6 dopant gas of 0.5 sccm, at a pressure of 5 mbar, a power of 125 W and electrode distance of 15 mm. The same set of parameters was adopted for n-doped nc-Si:H, except PH_3, which was applied with a flow of 2 sccm. The ITO layers were deposited symmetrically onto both sides of the cell with DC sputtering in the same Clustertool. A 80 nm-thick ITO layers was sputtered through a shadow mask to define a cell area of 2 x 2 cm^2 with an applied power of 0.3 kW, pressure of 3x10^{-3} mBar, and flows of Ar and O_2 gases at 50 and 1.1 sccm, respectively.
Single layers of 20 nm-thick n-doped nc-Si:H were deposited on *Corning Eagle-XG* glass substrates for further material characterization. Different plasma post-treatments were applied to n-doped nc-Si:H layers, as shown in table 1.

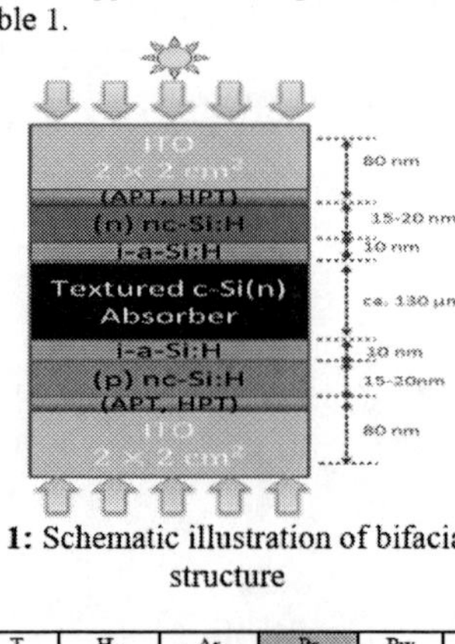

Figure 1: Schematic illustration of bifacial solar cell structure

	T [°C]	H_2 [sccm]	Ar [sccm]	Pr [sccm]	Pw [W]	d [mm]	t [s]
APT	230	200	0	1-3	20	20	90
HPT	230	0	100	1-3	20	20	90
AHPT	230	200	100	1-3	200	20	90

Table 1: Detailed parameters for different plasma post-treatment

Time-resolved OES was employed to monitor the plasma emission. The intensity of H_α (656 nm) and Ar (750 nm) as typical peaks in hydrogen and argon plasma, were probed [7]. The structure of the grown films was assessed by *Bruker SENTERRA* Raman spectroscopy. To calculate the crystallinity factor (Φ_c), the Raman spectra were deconvoluted to 3 Gaussian peaks - centered at 480 cm^{-1} for amorphous peaks and 510 and 520 cm^{-1} for crystalline peaks. Then, the crystallinity fraction was defined as the ratio of the sum of crystalline peaks height over the total peaks height of crystalline peaks and the amorphous peak at 480 cm^{-1}. Atomic Force Microscopy (AFM) was applied in *Nanosurf* setup to analyze the change in surface morphology of nc-Si:H after plasma treatment. The dark conductivity (σ_{dark}) and activation energy (E_{act}) were measured under direct current (DC) in vaccum, without light illumination and with increasing temperature. The current was activated by the temperature following an Arrhenius trend. E_{act} is the activation energy for dark conduction and can be exptracted from the slope of $\ln(\sigma_{dark}(T))$ versus 1/T. Hall measurements were conducted using Van der Pauw square geometry. The values of Jsc were calculated from external quantum efficiency measurements (EQE) calibrated with a c-Si photodiode. Since serial resistance (R_s) can considerably affect the solar cells without metal contacts, we obtained the overall solar cell performance based on Suns-Voc measurements (*Sinton Instruments*), which ignores the influence of R_s effect, giving the values of pseudo-fill-factor (pFF) and pseudo-efficiency (pEff).

3 RESULTS AND DISCUSSIONS

3.1 OES plasma diagnostics

Figure 2 presents the OES spectra corresponding to species dissociation during plasma treatments at different pressures. It is clearly seen that the emission intensities decrease gradually with increasing pressure for all plasma processes. This can be due to recombination reactions or mutual neutralization in the plasma [8]. At low pressures the energy of electrons is sufficient for species excitation, leading to higher emission intensity and high excitation temperatures [8]. Interestingly, Ar dissociation is ten orders of magnitude higher than its H counterpart. We speculate that Ar related species will have a dominant impact in the plasma glow and surface reaction. Excited Ar* atoms are susceptible to transfer their deexcitation energy into the film matrix, causing weak bonds breaking and network reconstruction [4].

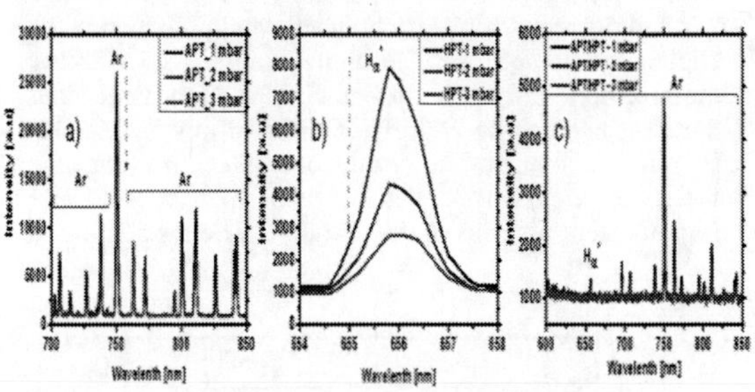

Figure 2: OES spectra for HPT, APT and combined AHPT plasma processes at different pressures

3.2 Morphological properties

Figure 3 presents the surface morphology of thin nc-Si:H before and after HPT and APT. It is clear that the initial nc-Si:H layer is characterized by small grains (~30 nm) and a dense network. After HPT, we notice a preservation of the nc-Si:H layer network, without noticeable change in morphology. This is mainly related to the fact that hydrogen reacts chemically with the film's surface. consisting on diffusion and insertion of hydrogen in the dangling bonds and intra-grain defects [8]. In the case of APT, a tendency of coalescence and formation of larger size clusters is manifested leading to film densification. It is well known that APT can lead to network restructuring and mechanical compaction accompanied with a strain release [6].

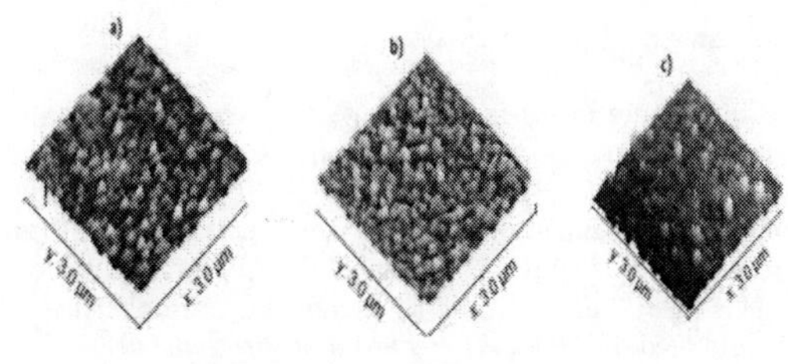

Figure 3: AFM images of thin nc-Si:H: before (a), after HPT (b), and after APT (c)

3.3 Structural properties

In figure 4, we depict Raman spectra for untreated and treated nc-Si:H with HPT and APT. Both treatments lead to an increase of crystallinity fraction from 63% to 66% for the case of HPT and to 67% for APT. This can be explained by the selective hydrogen etching of the amorphous network for HPT [9] in contrast to the network reconstruction and grains coalescence in the case of APT [6] .

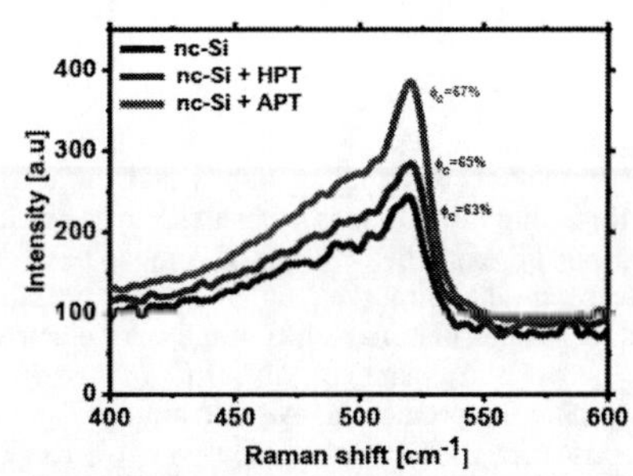

Figure 4: Raman spectra for thin nc-Si:H films before and after HPT and APT treatments

3.4 Optical properties

In figure 5, we present UV-VIS transmission spectra of nc-Si:H layers after post-treatments at different pressures..

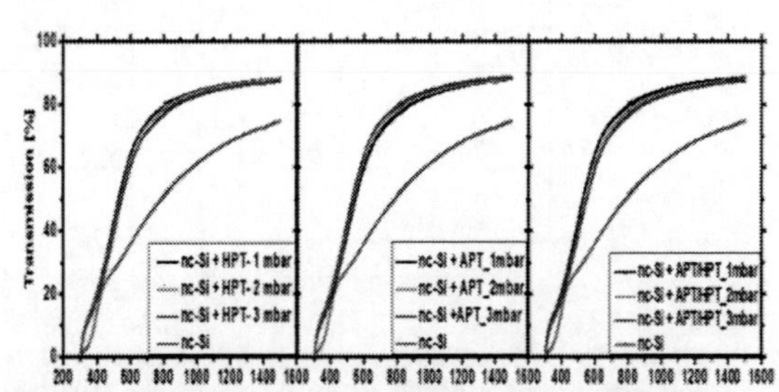

Figure 5: UV-VIS transmission spectra of nc-Si:H layers after post-treatments at different pressures

It is clearly seen that light transmission is enhanced over almost the full range of wavelengths for all treated layers. This is likely due to the combination of different phenomena i.e. the band gap widening, the improvement of crystallinity and the thinning effect. It is worthy to note that even thinner n-nc-Si:H layers don't lead to such a light transmission level.

3.5 Electrical properties

Figure 6 illustrates the changes in terms of resistivity, carrier density and mobility for thin n-nc-Si:H before and after HPT and APT treatments. Both HPT and APT lead to an enhancement of mobility accompanied with a drop in resistivity, whereas, free carrier density remains almost unchanged after plasma treatment, which keeps minimized the level of free carrier absorption.

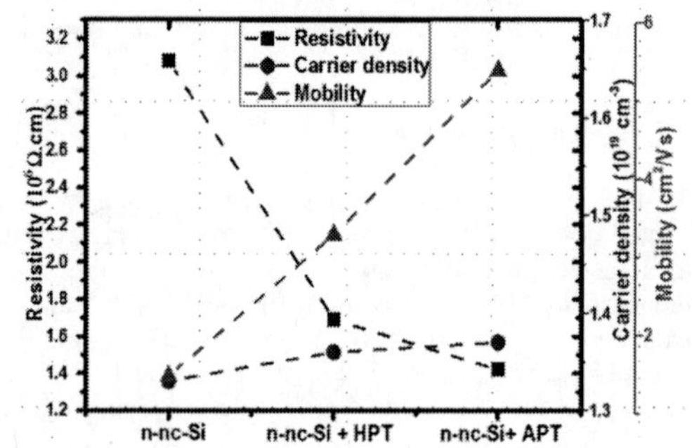

Figure 6: Hall measurements on n-doped nc-Si:H layers after HPT and APT

In the case of HPT, the improvement of electrical properties is mainly due to hydrogen diffusion and chemical passivation of dangling bonds and intra-grain defects. Regarding APT, these changes in electrical properties can be explained by better compactness and strain release in the film yielding a dense crystalline phase with better percolation path continuity for electrons transport between Si nanocrystals [9]. In terms of electrical properties promotion, APT treatment is more advantageous over HPT samples. However, we can get profit of both beneficial effects by applying combined AHPT.

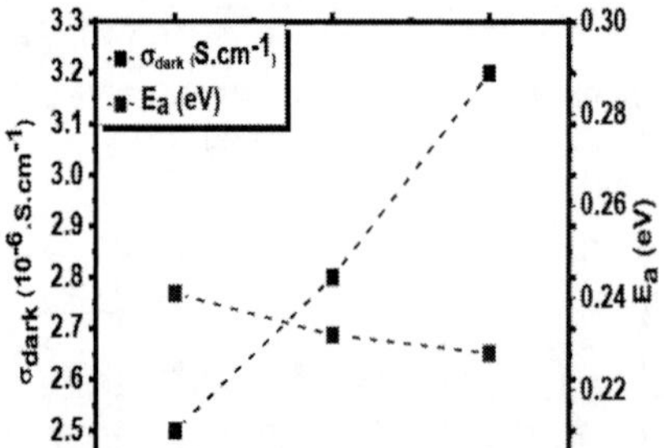

Figure 7: Dark conductivity and activation energy E_a measurement of n-doped nc-Si:H layers after HPT and APT

We illustrate the change in σ_{dark} and E_a of n-nc-Si:H layers after different post-treatments in fig. 7. Also improvement of conductivity shown is consistent with the increase of mobility. Moreover, the slight change in activation energy (E_a) confirms that the enhancement of the electrical properties of n-nc-Si after plasma treatments is related to the improvement of structural properties rather than a change in the doping level or free carrier density.

3.6 Solar cell results

In figure 8, we illustrate EQE curves of SHJ based on untreated and treated nc-Si:H using combined AHPT. Combined AHPT results in superior current density due to the outright improvement of structural and optoelectronic properties of nc-Si:H doped layers as well as the reduction of recombination rate at the TCO/nc-Si:H interface [5]. The enhancement of EQE is manifested over almost all visible and infrared wavelengths . Applying an optimized AHPT treatment on n-nc-Si:H, an outstanding short-circuit current density up to 39.29 mA/cm^2 with a gain of more than 2 mA/cm^2 is achieved.

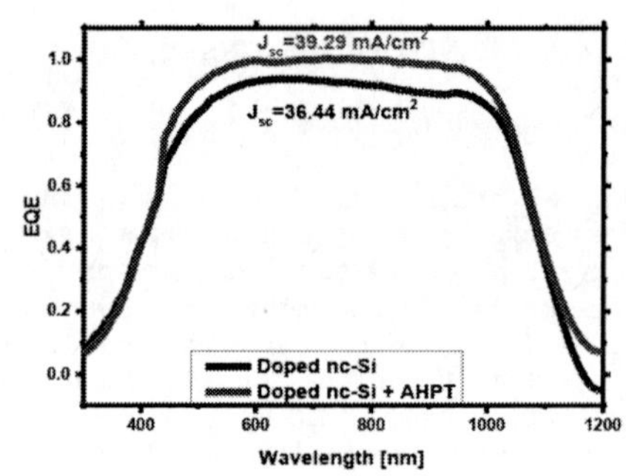

Figure 7: EQE of SHJ solar cells with untreated and plasma treated doped layers

Figure 8 depicts the pseudo I-V curve of SHJ based on untreated and treated nc-Si:H using combined AHPT. In table 2, we summarize SHJ characteristics with and without plasma post-treatments on nc-Si:H doped layers. A small decrease in V_{oc} is obtained, which can be due the slight alteration of i-a-Si:H/c-Si heterointerface. A further optimization in terms of precise and pressure and duration need to be addressed to avoid any V_{oc} losses. However, a compensation of the overall cell performance thanks to high current density leads to a considerable improvement in terms of the pseudo-efficiency pEff up to 24%.

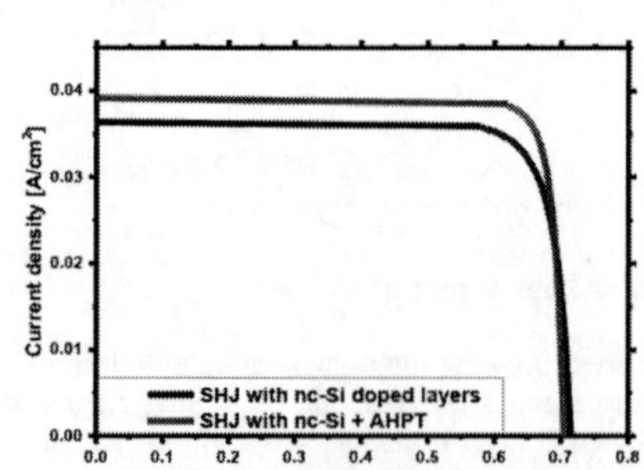

Figure 8: Pseudo-IV of SHJ solar cells with untreated

and plasma treated doped layers

	J_{sc} [mA/cm^2]	V_{oc} [mV]	pFF	pEff [%]
SHJ with untreated doped nc-Si:H	36.44	717	0.82	21.4
SHJ with doped nc-Si:H + AHPT	39.32	707	0.86	24

4 CONCLUSIONS

We have studied the effect of plasma post-treatments of n-doped nc-Si:H layers integrated in SHJ bifacial solar cells. We find that impact of plasma treatment on thin nc-Si:H layer is manifested as follow: HPT act via structural rearrangements of the network (mainly in the amorphous phase) through the combination of bond breaking and insertion of Hydrogen in the weak sites, dangling bonds and intra-grain defects. Whereas APT acts on basis of mechanical compaction and the grains coalescence accompanied with strain release. It was demonstrated that both APT and HPT enable further enhancement of the optical, electrical and structural properties of doped nc-Si:H layers. The integration of such combined AHPT treatment in SHJ process flow results in an outstanding value of J_{sc} (39.23 mA/cm^2), yielding an improved pEff of 24%.

5 ACKNOWLEDGMENTS

This work was supported by the German Federal Ministry of Economics and Technology (BMWi, reference number 0325825H, HERA).
We thank the company *Meyer Burger* and *Fraunhofer-ISE* for the wafer delivery. The authors acknowledge the insightful scientific discussions and experimental support of Michael Kröner and all our colleagues in at the DLR Institute of Networked Energy Systems. .

6 REFERENCES

[1] M. Green et al. Progress in Photovoltaics, https://doi.org/10.1002/pip.3040 , 2018
[2] Y. Tsunomura et al., Sol. Energy Mater. Sol.Cells 93, p.670, (2009)
[3] O. Sergeev et al., Energy Procedia 124, 371-378, (2017)
[4] J.P. Seif et al., IEEE Vol. 6 pp 1132-1140, (2016)
[5] O. Madani Ghahfarokhi et al., Appl. Phys. Lett. 104, 113901 (2014).
[6] A.Neumüller et al., Nano Energy, 43, 228-35, (2017)
[7] O. Gabriel et al. EPJ Photovoltaics 5, 55202 (2014)
[8] H. Meddeb et al. Phys. Status Solidi RRL 9 (No. 1) 53-56, (2015)
[9] A. Filonovich et al., Sci. Technol. Adv. Mater. 13, 045004 (8pp), (2012)

FRONT/BACK POLY-SI/SIO$_2$ PASSIVATED CONTACT DEVICE WITH V_{OC} >720 MV

David L. Young, Vincenzo LaSalvia, Bill Nemeth, San Theingi, Abhijit Kale, Dawn Findley,
Jake Atkins, Matthew Page, Pauls Stradins
National Renewable Energy Laboratory, Golden Colorado 80401 USA

ABSTRACT: We report on front/back poly-Si/SiO$_2$ passivated contact solar cells with iV_{oc} > 735 mV and V_{oc} > 720 mV. The devices were grown on n-Cz wafers using PECVD to deposit doped a-Si layers which were then annealed at 850 °C. Our best device has an efficiency of 21% which is limited by: (1) low current density from a large grid shadow loss and parasitic absorption in the front poly-Si:P layer; and (2) high series resistance from a high Al-poly-Si:P contact resistivity. We discuss the role of high temperature anneals throughout the process flow to preserve and improve the bulk lifetime of the n-Cz wafers.
Keywords: solar cells, passivate contacts, Tabula Rasa.

1 INTRODUCTION

Solar cell efficiency continues to be a main driver to lower the levelized cost of electricity in photovoltaic (PV) installations. As the industry aggressively moves from Aluminum back surface field (BSF) cells to higher efficiency PERC cells, the value of higher voltages, made possible by improved surface passivation, will begin to be realized. While the PERC cell architecture[1] utilizes selective area contact schemes to minimize poorly passivating metallized regions, fully passivated contacts, like a-Si/a-Si:B(P)[2] and poly-Si:B(P)/SiO$_2$ layers[3] offer lower recombination current densities, J_o, in the unmetallized *and* metallized regions. Passivated contact IBC-type cells have proven their record setting ability[2], but the simpler processing steps and potential bifacial advantage of the front/back design is more likely to be adopted by the greater PV industry.[4,1]

Passivating contacts formed from Polycrytalline silicon and thin tunneling oxide layers (poly-Si/SiO$_2$) offer a much higher temperature processing window than do a-Si-based contacts, typically up to 950 °C as compared with 250 °C. The high thermal stability of these poly-Si/SiO$_2$ contacts makes them compatible with other beneficial anneals to reform oxygen clusters, diffuse dopants, getter impurities and fire/anneal metal contacts.

In this contribution we report our progress on forming rear junction, front/back passivated contact cells utilizing poly-Si/SiO$_2$ layers on n-Cz wafers. We have found that a V_{oc} improvement can be realized by including a fast, high temperature annealing step to the wafer before adding passivation layers to the device. This anneal is similar to a tabula rasa (TR)[5,6] anneal known to dissolve oxygen precipitates which improves beneficial gettering of impurities by limiting the number and lowering the capture cross-section of recombination sites associated with metal/oxygen clusters. The anneal, along with other process improvements, allows iV_{oc} values over 735 mV and V_{oc} values of 722 mV on front/back passivated contact cells.

2 EXPERIMENT

2.1 Device preparation

n-Cz double-side textured Woongjin wafers with bulk lifetimes over 1 ms were laser scribed to 40 mm x 60 mm coupons. The samples were cleaned with a piranha, RCA regimen and then etched in 1% HF until hydrophobic before loading into a clean tube furnace. Next, a KOH etch-stop layer was grown. This was accomplished by either annealing the samples to 1100 °C for 15 mins to mimic a TR anneal and to grow an SiO$_2$ layer, or a SiNx layer was deposited on one side of the sample by plasma enhanced chemical vapor deposition (PECVD) at 320 °C. For samples with the SiO$_2$ thermal oxide layer grown on both sides, a remote plasma etch (RIE) in CF_4 and SF_6 was used to remove the protective SiO$_2$ layer on one side.

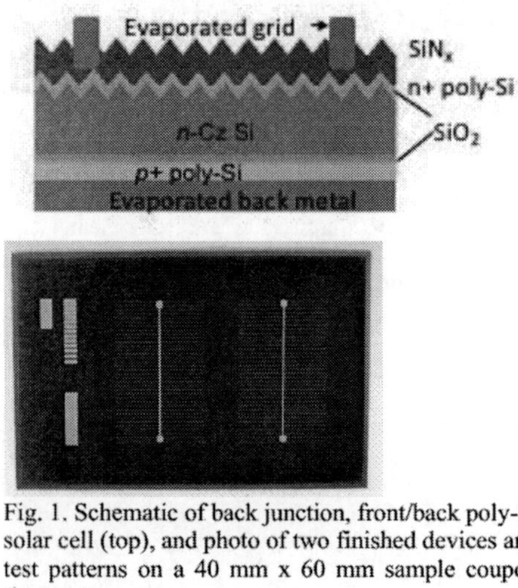

Fig. 1. Schematic of back junction, front/back poly-Si solar cell (top), and photo of two finished devices and test patterns on a 40 mm x 60 mm sample coupon (bottom).

Samples were then etched in KOH to remove the pyramid texturing on the unprotected side, resulting in single side textured samples. Following another cleaning, a low temperature oxide was grown on both sides of the samples at 700 °C for 5 mins in a clean tube furnace. Next, p/n a-Si:H layers were grown over the oxide using PECVD and SiH_4. The textured side of the wafer received 40 nm of P-doped a-Si:H over the full area, while the non-textured side of the wafer received 40 nm of B-doped a-Si:H deposited through a 4 cm^2 opening in a mask held between the sample and the plasma. Two devices and separate test patterns were deposited onto each sample. Next, samples were annealed in a tube furnace at 850 °C for 30 mins to crystallize the a-Si into poly-Si and to diffuse dopants. A 15 nm Al_2O_3 layer was deposited on both sides of the wafer using atomic layer deposition of trimethylaluminum

(TMA) and water followed by a 400 °C anneal in forming gas to diffuse H from the Al_2O_3 layer into the poly-Si/SiO_2 layers. Thermally evaporated Ti/Ag or Al formed the back contact and the front grid. Finally, a SiN_x antireflection coating was added to the textured side of the device. The final back junction, front/back poly-Si/SiO_2 solar cell configuration is shown in Fig. 1.

2.2 Device Characterization

Current vs voltage measurements were tested under a calibrated JV tester using the AM1.5G spectrum. To optically and electrically separate the two devices on each sample we electrically isolated each emitter mesa from the other using a thin dielectric sheet between the non-measured device and JV temperature-controlled chuck. The chuck made two independent contacts to the back of the cell under test. A shadow mask was aligned over the cells to block light from illuminating areas outside of the emitter mesa. Figure 1 shows an image of the two devices on each sample coupon. Quantum efficiency measurements were made using an Oriel IQE 200 instrument. Reflection measurements were captured using a Cary 5000i spectrometer with an integrating sphere. Contact resistivity measurements were made using a Keithley source meter and a transfer length measurement (TLM) pattern metal deposited onto the coupon (see Fig. 1) V_{oc} as a function of light intensity data and photoconductive decay measurements were taken with Sinton instruments SunsV_{oc} and WCT-120, respectively. Optical modeling of the device was accomplished with SunSolve using experimentally measured n and k data from all deposited layers. The generation profile output from SunSolve was then input to Quokka2 to model the electrical nature of the devices. Again, experimental data were used as input values to the model where ever possible.

3 SYMMETRIC STRUCTURE RESULTS

40 mm x 60 mm sample coupons were prepared from 156 mm randomized n-Cz wafers and processed through either the TR oxide growth step or the SiNx KOH etch-stop layer deposition. Samples given the TR annealing step grew a thin SiO_2 layer on both sides of the wafer. This layer was then removed on one side of the wafer using RIE. Samples that were not given the TR anneal were given a single side SiNx deposition. All wafers were then ramid were

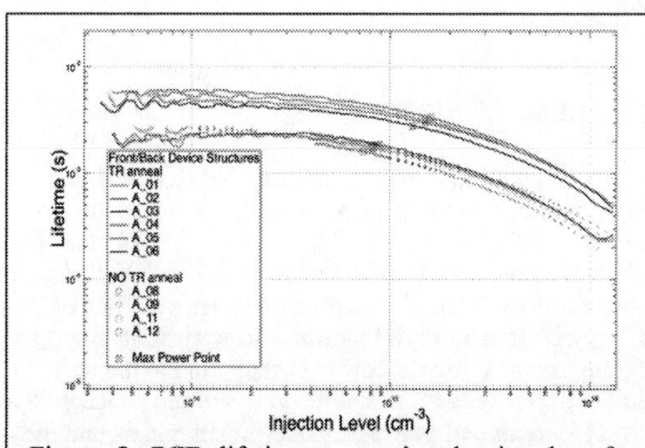

Figure 2. PCD lifetime vs carrier density data for unmetallized devices with and without TR anneal.

processed identically following the procedures outlined in section 2.1 to form back junction, front/back devices with textured front surfaces. Figure 2 shows lifetime vs carrier density data from these samples just after Al_2O_3 passivation, but before metal was added. Other than two outliner curves (A01 and A06), the data are bimodal with an increase in bulk lifetime and improved J_o, associated with the surface passivation, for samples subjected to the TR anneal. Samples that were not given the TR treatment showed bulk lifetime values of 2-3 ms near an injection level of 5E15 cm^{-3} and $2J_o$ values over 40 fA/cm^2, while those samples that did receive TR show bulk lifetimes of 4-6 ms and $2J_o$ ~20 fA/cm^2. Baring unknown factors concerning the RIE process and/or the PECVD of SiNx,

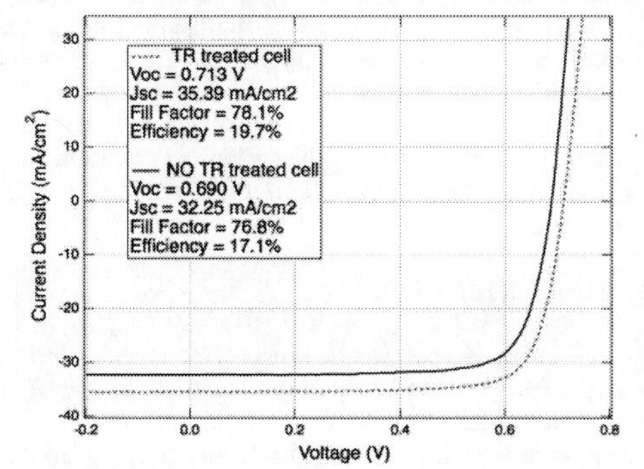

Figure 3. JV data for cells processed with and without TR treatments.

such as surface plasma damage, it appears that the TR anneal is beneficial for the bulk lifetime of the wafers and the ability to passivate the surface with respect to our device process flow and thermal protocol. The lifetime of the non-TR samples is similar to the bulk lifetime of the wafers before being subjected to high-temperature processing. The encouraging results of Figure 2 for bulk lifetime and surface passivation improvement with TR translated to device improvement as well, with TR-treated cells giving V_{oc} values of 713 mV and efficiencies of 19.7%, while non-TR-treated cells gave V_{oc} values of 689 mV and efficiencies of 17.1% (See Fig. 3)

Next, we investigated if either the n-type side or p-type poly-Si side was affected more by the TR treatment by growing symmetric structures and measuring *tau_bulk* and J_o using optimized grow parameters. For symmetric n-type samples processed identically to device cells we found that both TR-treated (oxide blocking layer grown on both sides) and non-TR treated samples (SiNx blocking layer grown) gave *identical* iV_{oc} values between 732 – 737 mV with corresponding J_o values between 3.75 - 2.3 fA/cm^2. Note that even though we grew KOH-blocking layers, the samples were symmetrically textured to mimic the surface of the devices. These highly passivated n-type textured surfaces gave near state-of-the-art iV_{oc} values for n-Cz poly-Si/SiO_2 contacts.

Next, symmetric P-type contacts were investigated using identical process steps as above except that the TR oxide and the NO-TR SiNx KOH-blocking layers were removed to produce symmetric planar surfaces. From the data sets obtained so far, we have not found a significant difference in iV_{oc} between symmetric p-type samples that were given a TR treatment and those that were not given the treatment. For example, the average iV_{oc} for NO-TR samples is 710.9 ± 6.3mV (J_o ~11 fA/cm^2) whereas for TR samples the average iV_{oc} is 708.5 ± 7.3mV (J_o ~13 fA/cm^2).

Given that neither symmetric n-type or p-type passivated contact samples revealed that the TR treatment improved the passivation of the wafer, we are lead to conclude that the combination of the n-type surface and the p-type surface act together during the TR treatment of devices to account for the data of Fig. 2. As an example, we recall that the TR treatment for n-type surfaces alone gives J_o values of about 3 fA/cm^2, and p-type surfaces give J_o values of 13 fA/cm^2. Using our best cell data (see below) which more closely mimics the exact processing of the symmetric samples used in this study, we find that the combination of n-type and p-type surfaces in a front/back device configuration give total $2J_o$ values of 12 fA/cm^2 ($J_{o\_n\text{-}type} + J_{o\_p\text{-}type} = J_{o\_total}$). This implies that the p-type surface J_o is lowered an additional 4 fA/cm^2 when TR annealed with an n-type surface opposite to the p-type surface. The presence of the n-type surface with the p-type surface drastically improves the surface passivation, J_o, of the p-type surface and the bulk lifetime (see Fig. 2).

4 DEVICE RESULTS

Following the process flow using TR anneals outlined above we grew several devices to show the potential of the treatments. Figure 4 shows an unofficial full area JV curve for an optimized device using a shadow mask aligned with the emitter area of 4 cm^2. The V_{oc} of 722 mV is quite high and shows the benefit of using the TR treatment to improve both J_o and *tau_bulk* of the n-Cz wafer. The J_{sc} value is consistent with a 40 nm n-poly-Si front layer and a nonoptimized grid/busbar coverage of 5.5 %. The FF is close to 80% but could be improved with lower contact resistivity values. Currently our evaporated Al contacts make low contact resistivity values (~ 8 mOhm-cm^2) to the p-type side (metal-poly-Si/SiO_2-wafer – measured separately on a p-type wafer), but high (~ 60 mOhm-cm^2) contact resistivity values to the n-type side (metal-poly-Si/SiO_2-wafer). Overall, the 21% efficient device is encouraging and represents one of the highest values for a device of this configuration, thermal history and on an n-Cz wafer.[7-9]

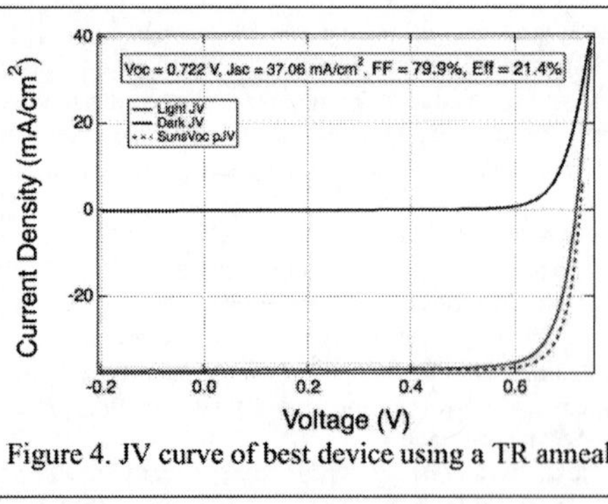

Figure 4. JV curve of best device using a TR anneal.

5 DISCUSSION

TR anneals are known to improve gettering in n-Cz wafers by dissolving oxygen precipitates that would otherwise grow in size during thermal treatment between 700 – 1000°C and provide attachment points for metal impurities on the precipitates, leading to bulk recombination sites and lower wafer lifetimes. Additionally, n-type poly-Si/SiO_2 contacts are known to getter impurities at elevated temperatures (>800 °C) and is one of the benefits of poly-Si/SiO_2 heterointerface contacts over a-Si:H contacts.[10] The high temperature (850 °C) crystallization anneal also acts as a gettering step during the device process flow. Thus, the additional step of a TR anneal simply improves this benefit and acts to improve both the lifetime of the wafer and the ability to passivate its surfaces.

Our results for symmetric p-type surfaces not showing effective gettering and improvement of passivation with TR treatments are consistent with those found for ion implanted poly-Si/SiO_2 samples where a surface boron-rich layer is not present to getter impurities.[10] Our *in situ* B-doped a-Si:H layers do not show an obvious boron rich surface (similar to ion implanted contacts) unlike contacts doped by BBr3 diffusion which use the BRL as an effective gettering surface.[10] Thus, the TR and gettering anneal only improve the p-type contact when it is coupled with an effective gettering n-type surface on the opposite side of the wafer. This could imply that the TR anneal improves the p-side surface for passivation by removing impurities from the surface during oxide growth at 1,100 °C, which are then gettered away during the 850 °C crystallization anneal to the n-type side.

For experiments with symmetric n-type surfaces showing no improvement in iV_{oc} with TR treatments, we suspect that the strong gettering process for n-type poly-Si is effective for high lifetime (> 1 ms) n-Cz wafer impurities regardless of the TR treatments for wafers with low impurity concentrations. TR treatments may be effective for higher impurity wafers too.

6 CONCLUSIONS

We show that a TR anneal of an n-Cz wafer improves the surface passivation, J_o, effective lifetime and device performance of a rear junction, front/back poly-Si/SiO_2 cell. We conclude that during the high temperature poly-Si crystallization anneal the TR treatment enables better gettering of impurities from the bulk or from surfaces which are benignly collected at the n-type contact, but not in the p-type contact. P-type poly-Si/SiO_2 contact passivation improves when coupled with an effective gettering n-type poly-Si/SiO_2 contact in a device configuration. This process has produced a device with V_{oc} of 722 mV with an efficiency of 21%.

7 ACKNOWLEDGEMENTS

This work was authored by the National Renewable Energy Laboratory, operated by Alliance for Sustainable Energy, LLC, for the U.S. Department of Energy (DOE) under Contract No. DE-AC36-08GO28308. Funding provided by the U.S. Department of Energy Office of Energy Efficiency and Renewable Energy Solar Energy Technologies Office. The views expressed in the article do not necessarily represent the views of the DOE or the U.S. Government. The U.S. Government retains and the publisher, by accepting the article for publication, acknowledges that the U.S. Government retains a nonexclusive, paid-up, irrevocable, worldwide license to publish or reproduce the published form of this work, or allow others to do so, for U.S. Government purposes.

8 REFERENCES

1 Green, M. A. The Passivated Emitter and Rear Cell (PERC): from conception to mass production. Solar Energy Materials & Solar Cells 143, 190-197 (2015).
2 Green, M. A., Hishikawa, Y., Dunlop, E. D., Levi, D. H., Hohl-Ebinger, J. & Ho-Baillie, A. W. Y. Solar Cell Efficiency Tables (Version 52). Progress in Photovoltaics: Research and Applications 26 (2018).
3 Haase, F., Hollemann, C., Schäfer, S., Merkle, A., Rienäcker, M., Krügener, J., Brendel, R. & Peibst, R. Laser contact openings for local poly-Si-metal contacts enabling 26.1%-efficient POLO-IBC solar cells. Solar Energy Materials and Solar Cells 186, 184-193 (2018).
4 Glunz, S. W., Feldmann, F., Richter, A., Bivour, M., Reichel, C., Benick, J. & Hermie, M. in EU- PVSEC 2015 (Hamburg, Germany, 2015).
5 Falster, R., Cornara, M., Gambaro, D., Almo, M. & Pagani, M. Effect of high temperature pre-anneal of oxygen preciptitates nucleation kinetics in Si. Solid State Phenomena 57-58, 123-128 (1997).
6 LaSalvia, V., Youssef, A., Jensen, M. A., Looney, E. E., Nemeth, W., Page, M., Nam, W., Buonassisi, T. & Stradins, P. Tabula Rasa for n-Cz silicon-based photovoltaics. Journal of Photovoltaics, 1-8, doi:DOI: 10.1002/pip.3068 (2018).
7 Ingenito, A., Limodio, G., Procel, P., Yang, G., Dijkslag, H., Isabella, O. & Zeman, M. Silicon solar cell architecture with front selective and rear full Area ion-implanted passivating contacts. RRL Solar 1, 1700040 (2017).
8 Nicolai, M., Zanuccoli, M., Feldmann, F., Hermle, M. & Fiegna, C. Analysis of silicon solar cells with poly-Si/SiOx carrier-selective base and Emitter contacts. IEEE Journal of Photovoltaics 8, 103-109 (2018).
9 Feldmann, F., Nogay, G., Löper, P., Young, D. L., Lee, B. G., Stradins, P., Hermle, M. & Glunz, S. W. Charge carrier transport mechanisms of passivating contacts studied by temperature-dependent J-V measurements. Solar Energy Materials & Solar Cells 178, 15-19 (2018).
10 Liu, A., Yan, D., Phang, S. P., Cuevas, A. & Macdonald, D. Effective impurity gettering by phosphorus- and boron- diffused poly-Silicon passivateing contacts for silicon solar cells. Solar Energy Materials & Solar Cells 179, 136-141 (2018).

NEWLY DEVELOPED AG COATED CU PASTE FOR SI HETERO-JUNCTION SOLAR CELL

Kyotaro Nakamura[1], Kazuo Muramatsu[2] Aki Tanaka[2] and Yoshio Ohshita[1]
[1]Toyota Technological Institute 2-12-1 Hisakata, Tempaku-ku, Nagoya 468-8511, Japan
[2]NAMICS CORPORATION, 3993 Nigorikawa, Kita-ku, Niigata, 950-3131 Japan
phone: +81-52-809-1877, fax: +81-52-809-1879, e-mail: kyotaro@toyota-ti.ac.jp

ABSTRACT: In this work, we have developed a new low cost low temperature curing (LTC) paste which is suitable for Si hetero-junction (SHJ) solar cell. It is made of Cu particle coated with Ag, that is to say, Ag coated Cu paste. To ensure the enough performance of LTC paste, Ag particles were blended with Ag coated Cu particles. We estimated cost reduction rate due to Ag coated Cu paste, and investigated conductor and contact resistance of Ag coated Cu paste electrode. Finally, we fabricated SHJ solar cells using the Ag coated Cu paste, and compared with Ag paste, the difference of cell efficiency was relatively just only 0.4%. In conclusion, we were able to confirm that the newly developed LTC Ag coated Cu paste replaces LTC Ag paste, and it can reduce about 30% of electrode cost of SHJ solar cell.
Keywords: c-Si, Heterojunction, Metallization, Silicon Solar Cell

1 INTRODUCTION

Low temperature curing (LTC) Ag paste is commonly used for Si hetero-junction (SHJ) solar cell in present. However, in general, LTC Ag paste is more expensive than high temperature firing (HTF) Ag paste which is commonly used for conventional solar cell. Additionally, the conductor resistance of LTC Ag paste is much higher than that of HTF Ag paste, thus, high aspect ratio of finger is required for LTC Ag paste electrode to realize high F.F. In addition to that, because Ag paste must be printed on both sides of SHJ solar cell, the consumption of Ag paste of SHJ cell is much larger than the conventional cell. Therefore, the ratio of electrode on the fabrication cost of SHJ cell is very high, and the cost reduction of electrode is very serious issue for SHJ cell. Therefore, many investigations for the cost reduction of electrode of high efficiency SHJ solar cell have been done [1]-[4].

In this work, we have developed a new low cost LTC paste which is suitable for SHJ cell. It is made of Cu particle coated with Ag, that is to say, Ag coated Cu paste. Needless to say, the market quotation of Cu is much lower than Ag, so, Ag coated Cu paste could drastically reduce the cost of LTC paste. We estimated cost reduction rate due to Ag coated Cu paste, and investigated conductor and contact resistance of Ag coated Cu paste electrode. Finally, we fabricated SHJ solar cells using Ag coated Cu paste, and compared with Ag paste one.

2 EXPERIMENTS

2.1 Ag coated Cu paste

Figure 1 shows schematic views of SHJ solar cells using LTC Ag paste and Ag coated Cu paste electrode. Ag coated Cu paste is able to be used as an alternative for LTC Ag paste in the same way of that. However, because the resistivity of Cu is higher than Ag, the conductor resistance of electrode made of 100% Ag coated Cu paste should increase.

To ensure the enough performance of LTC paste, it will be the realistic approach to blend Ag coated Cu particles and Ag particles with appropriate blend ratio.

On the other hand, the blend ratio of Ag particles will affect the cost of electrode.

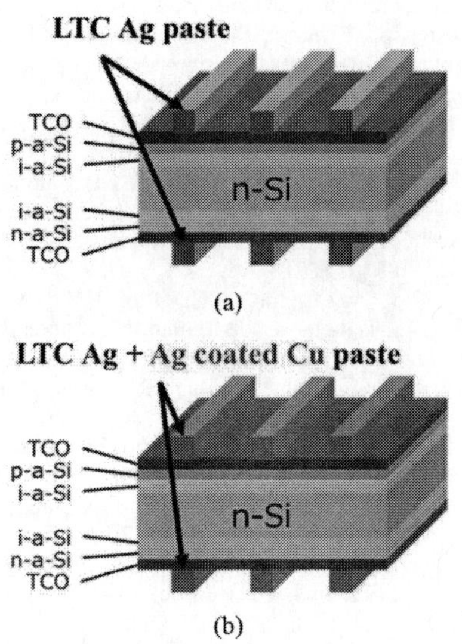

Figure 1: Schematic views of SHJ solar cells using (a) low temperature curing Ag paste and (b) Ag coated Cu paste electrode.

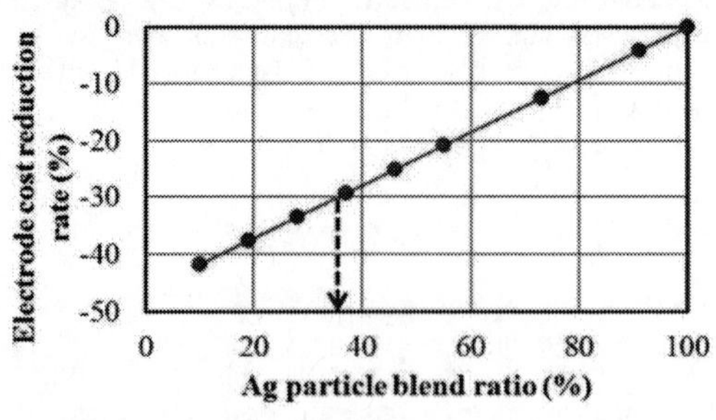

Figure 2: Electrode cost reduction rate as a function of Ag particle blend ratio into Ag coated Cu paste.

Figure 2 shows the electrode cost reduction rate of LTC paste as a function of Ag particle blend ratio into Ag coated Cu paste. As shown in Fig. 2, the increase of Ag particle blend ratio decreases cost reduction rate, that is, cost advantage. As one of the guidelines, to reduce 30%

cost, the blend ratio should be limited up to 35%.

2.2 Contact and conductor resistance

Ag particle blend ratio also impacts the performance of LTC paste, in particular, the contact and conductor resistance. To investigate the contact resistance between electrode and TCO (Transparent Conductive Oxide) at first, we made TEG (Test Element Group) samples. We printed Ag particle blended Ag coated Cu pastes on the top of TCO coated on Cz Si wafers. After curing at 200°C for 45 min, we measured the contact resistance using TLM (Transmission Line Model) method.

Then we evaluated the conductor resistance of Ag particle blended Ag coated Cu paste. After printing and curing of Ag coated Cu paste on Cz wafers as same way as described above, we measured the conductor resistance of electrode using LCR meter and confocal microscope.

2.3 SHJ solar cell fabrication

After the determination of appropriate Ag particle blend ratio as described above, we fabricated SHJ solar cells using Ag LTC paste and Ag particle blended Ag coated Cu LTC paste, as shown in Fig. 1.

After saw damage removal and texturization, amorphous i/p and i/n thin layer were deposited on both sides of the wafer using PECVD (plasma-enhanced chemical vapor deposition). Then, ITO was deposited on the both sides of wafer using RPD (Reactive Plasma Deposition). Finally, H pattern electrodes were formed as described above. IV characteristics were measured using a solar simulator under Standard Test Conditions. Cross sectional SEM images of electrodes also were observed.

3 RESULTS AND DISCUSSIONS

3.1 Contact and conductor resistance

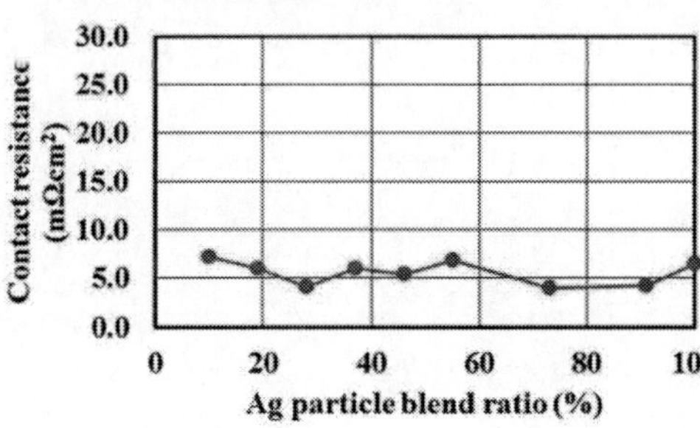

Figure 3: Contact resistance between TCO and Ag coated Cu paste electrode as a function of Ag particle blend ratio.

Figure 3 shows the results of contact resistance measurements between TCO and Ag coated Cu paste electrode as a function of Ag particle blend ratio. As shown in Fig. 3, the contact resistance doesn't largely depend on blend ratio, and the value is lower than 8 mΩcm^2, which is low enough for electrode of SHJ solar cell.

On the other hand, figure 4 shows the results the results of conductor resistance measurements of Ag coated Cu paste electrode as a function of Ag particle blend ratio. The conductor resistance heavily depends on blend ratio in the case under 40%, and sharply increases with decrease of blend ratio. However, up to 37%, the conductor resistance keeps lower than 10 $\mu\Omega$·cm, which is low enough for electrode of SHJ solar cell.

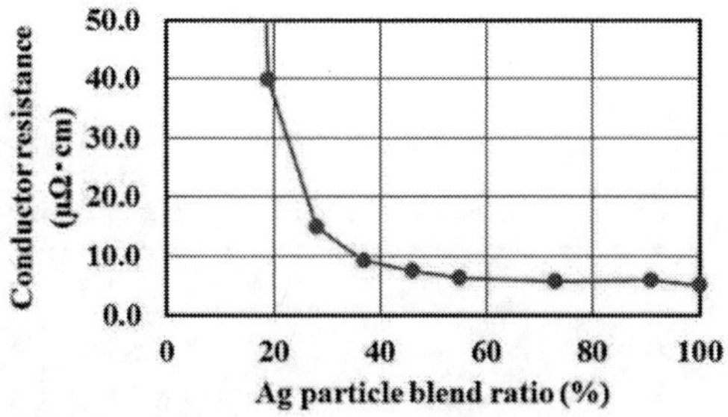

Figure 4: Conductor resistance of Ag coated Cu paste electrode as a function of Ag particle blend ratio.

Therefore, we determined the appropriate Ag particle blend ratio 37%.

Finally, we fabricated SHJ solar cells using 100% Ag LTC paste and 37% Ag particle blended Ag coated Cu LTC paste.

3.2 Cross sectional SEM image

Figure 5 show the cross sectional SEM images of 100% Ag paste electrode and 37% Ag particle blended Ag coated Cu paste electrode of SHJ solar cells. As shown Fig. 5(a), in the case of 100% Ag paste, small Ag particles are connected each other and contacted to the substrates, too. On the other hand, in the case of 37% Ag particle blended Ag coated Cu paste, as shown in Fig. 5(b), small Ag particles are filled into the space among the large Ag coated Cu particles. And it helps the Ag coated Cu particles to contact to the substrate.

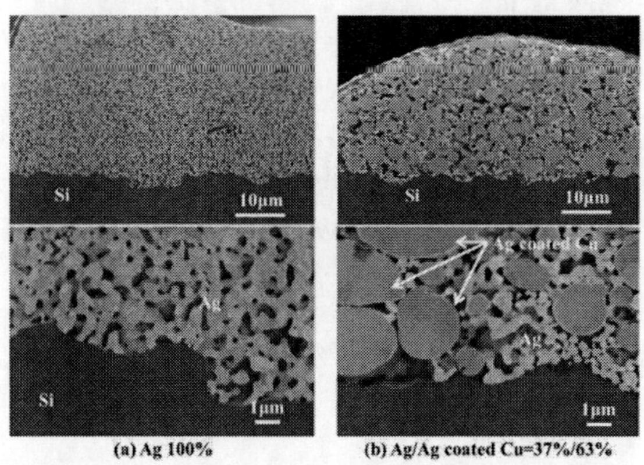

Figure 5: Cross sectional SEM images of (a) 100% Ag paste electrode and (b) 37% Ag particle blended Ag coated Cu paste electrode.

3.3 Cell fabrication results

Table I and Figure 6 show the comparison of PV performances of SHJ solar cells between 100% Ag paste electrode and 37% Ag particle blended Ag coated Cu paste electrode. The differences of Isc, Voc and F.F. between Ag paste and Ag blended Ag coated Cu paste are small, thus, the difference of cell efficiency is relatively just only 0.4%.

As a result, we were able to confirm that the newly

developed LTC Ag coated Cu paste replaces LTC Ag paste, and it can reduce about 30% electrode cost of SHJ solar cell.

Table I: Comparison of PV performances between Ag paste and Ag coated Cu paste

Blend ratio	Isc (A)	Voc (V)	F.F.	η (%)
Ag 100%	8.869	0.726	0.785	20.96
Ag/Ag coated Cu=37%/63%	8.999	0.723	0.773	20.89

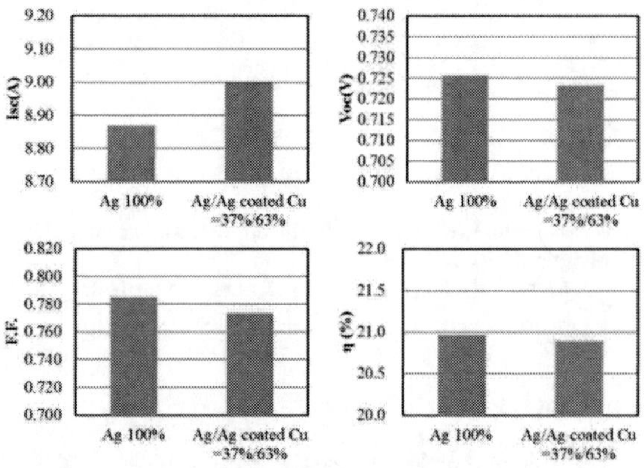

Figure 6: PV performances of SHJ solar cells using Ag paste and Ag coated Cu paste.

4 CONCLUSIONS

We have developed a new low cost LTC paste which is suitable for SHJ cell. It is made of Cu particle coated with Ag, that is to say, Ag coated Cu paste. Because the resistivity of Cu is higher than Ag, it is the realistic approach to blend Ag coated Cu particles and Ag particles with appropriate blend ratio to ensure the enough performance of LTC paste. We estimated cost reduction rate due to Ag coated Cu paste, and investigated conductor and contact resistance of Ag coated Cu paste electrode, then, determined Ag particle blend ratio 37%. Finally, we fabricated SHJ solar cells using the Ag coated Cu paste, and compared with Ag paste, the difference of cell efficiency was relatively just only 0.4%.

In conclusion, we were able to confirm that the newly developed LTC Ag coated Cu paste replaces LTC Ag paste, and it can reduce about 30% of electrode cost of SHJ solar cell.

5 ACKNOWLEDGEMENT

This work was supported by the New Energy and Industrial Technology Development Organization (NEDO) under the Ministry of Economy, Trade and Industry (METI) in Japan. Authors are grateful to Mr. M. Morimura of TTI for his support for the experiment.

6 References

[1] A. Rodofili, W. Wolke, L. Kroely, M. Bivour, G. Cimiotti, J. Bartsch, M. Glatthaar, J. Nekarda, Proceedings 33th European Photovoltaic Solar Energy Conference, Amsterdam, Netherlands, 2017 (WIP, München, 2017) pp. 402-405.

[2] Jochen Rentsch, Martin Bivour, Denis Erath, Frank Feldmann, Andreas Fischer, Sebastian Mack, Anamaria Moldovan, Jan Temmler, Martin Hermle, Stefan W. Glunz, Ralf Preu, Proceedings 33th European Photovoltaic Solar Energy Conference, Amsterdam, Netherlands, 2017 (WIP, München, 2017) pp. 441-446.

[3] A. Lachowicz, J. Geissbühler, A. Faes, J. Champliaud, F. Debrot, E. Kobayashi, J. Horzel, C. Ballif, M. Despeisse, Proceedings 33th European Photovoltaic Solar Energy Conference, Amsterdam, Netherlands, 2017 (WIP, München, 2017) pp. 753-756.

[4] Ankit Khanna, Vinodh Shanmugam, Gabby Alonzo de Luna, Jaffar Moideen Yacob Ali, Ge Jia, Thomas Mueller, Proceedings 7th World Conference on Photovoltaic Energy Conversion (WCPEC-7), Waikoloa, Hawaii, (2018) 320.

HOMOGENEOUS DEPOSITION OF HIGH PURITY SILICON THIN FILMS WITH HIGHEST RATES ABOVE 30 µm/min

Stefan Saager & Bert Scheffel
Fraunhofer Institute for Organic Electronics, Electron Beam and Plasma Technology (FEP)
Winterbergstr. 28, 01277 Dresden, Germany, phone: +49 351 2586 316, email: stefan.saager@fep.fraunhofer.de

ABSTRACT: We demonstrate currently reached results for depositing silicon thin films by crucible-free electron beam physical vapor deposition (EB-PVD). The feasibility of a crucible-free EB-PVD could successfully be demonstrated in experiments with different configurations of evaporation material. The process could be improved regarding the critical aspects, e.g. occurrence of cracks during heating up and spill out of the melt on the melting pool edges. Deposition rates above 500 nm/s, corresponding to 30 µm/min, have been reached. By using a double-ingot arrangement and superposing vapor from multiple sources the relative deviation of layer thickness could be reduced to $< \pm 5$ % on a substrate width of 200 mm (8″). Furthermore, we investigated impacts of process parameters on layer stress.
Keywords: Cost Reduction, High Deposition Rate, Silicon Thin Films, Electron Beam Evaporation, Crucible Free

1 INTRODUCTION AND PURPOSE

Due to ongoing breakthroughs in enhancing solar cell performance within the last decade, crystalline silicon thin film technology is becoming more and more attracting for commercialization. The fabrication technology requires an economic method for depositing silicon thin films of several micrometer thickness including following conditions:

- high throughputs, which corresponds to high depositions rates
- high purity layer without any metallic contamination
- low process temperature to minimize external entry of contamination impact
- layer thickness homogeneity
- dense and amorphous layer with low layer stress for subsequent crystallization

To fulfill these challenges the electron beam physical vapor deposition (EB-PVD) is a predestinated technique, because it offers the highest deposition rates and is well established in industries. Additionally EB-PVD can be combined in-situ with subsequent manufacturing processes, such as crystallization [1-2], metallization [3] and contact formation [4].
Because Si purity is the key for PV thin film application, a crucible-free evaporation from an electronic-grade purity Si single crystal was experimentally applied.

2 EXPERIMENTAL

Coating experiments have been carried out in the high vacuum system VERSA with vacuum lock evacuated by a cryogenic pump system, reaching a base pressure of $2 \cdot 10^{-5}$ Pa (Fig 1). For pretreatment the substrates were cleaned in situ by a plasma process using a hollow cathode arc discharge [5]. The Si layers were deposited on boron doped Si wafers ($\rho = 1...20$ Ωcm) which could be heated up to defined substrate temperature (50...200 °C) by a rear side radiation heater. Substrate temperature measurement was realized by rear side contact thermocouples. To avoid contamination, the coated substrates were handled in a glovebox and transferred in inert gas atmosphere.

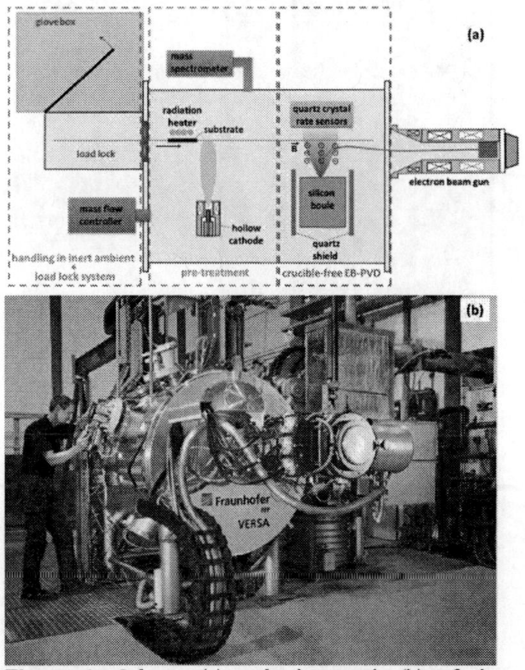

Figure 1: Scheme (a) and photograph (b) of the Fraunhofer FEP pilot plant VERSA used for crucible-free EB-PVD experiments at a process pressure of $4 \cdot 10^{-4}$ Pa.

For crucible-free EB-PVD Si evaporation, material (bricks or boules) is arranged in a special manner for avoiding direct contact to any contaminating material such as from a crucible (Fig 2). Applying the fast deflected EB on the rod's top face the silicon material melts and evaporates. Critical aspects for this method are heating up the brittle Si rod without the occurrence of cracks, the limitation of the melting pool to the rod's perimeter without spill out of the melt, the avoidance of splash formation from the melting pool and the cooling down the rod without stress cracks for reuse.

Figure 2: Crucible-free electron beam evaporation setup before (a, c) and after (b, d) evaporation experiment.
Configuration A: 4 multicrystalline silicon bricks, which are placed on a water-cooled plate.
Configuration B: one single crystalline silicon boule, shielded by a quartz tube.

3 RESULTS

The feasibility of a crucible-free EB-PVD could successfully be demonstrated in experiments with different configurations of evaporation material. The process could be improved regarding the critical aspects mentioned above. Deposition rates above 500 nm/s, corresponding to 30 µm/min, have been reached (Fig 3). Because the diameter of the Si melting pool increases by applying higher electron beam power, the maximum deposition rate was reached with configuration A, consisting of multicrystalline Si bricks.

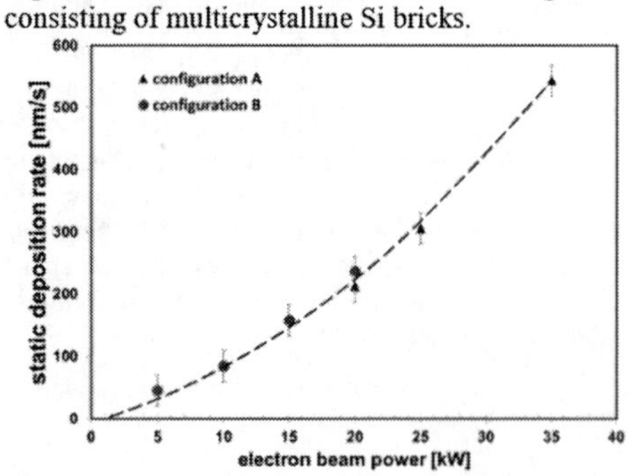

Figure 3: Static deposition rate vs. applied electron beam power for investigated crucible-free EB-PVD configuration.

However, the highest layer purity was reached by using an electronic-grade purity Si single crystal as evaporation source in configuration B. Layer purity has not been analyzed during the actual experiment period. However, SIMS analysis of prior work [7] shows a metallic impurity content below detection limit for iron ($< 5 \cdot 10^{14}\,\mathrm{cm}^{-3}$) and copper ($< 3 \cdot 10^{14}\,\mathrm{cm}^{-3}$). A comparable or even better purity is expected for current deposition experiments.

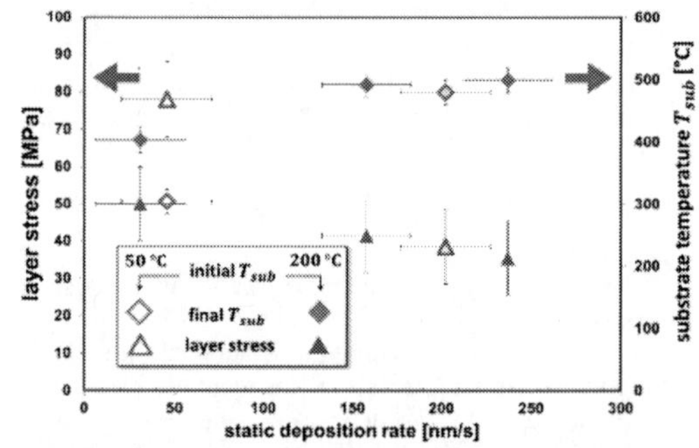

Figure 4: Layer stress vs. static deposition rate and final substrate temperature after depositing a 10 µm thick silicon layer in dependency on static deposition rate.

For further crystallization processing, layer stress and layer thickness are critical parameters for reaching reproducible results [8]. We analyzed layer stress of 10 µm thick Si thin films in dependency of applied deposition rate and initial substrate temperature (Fig. 4) by profilometer determination of differences in substrate curvature. Tensile layer stress reduces from +70 MPa down to +50 MPa by increasing initial substrate temperature T_{sub} up to 200 °C, whereas no influence of deposition rate on layer stress was observed.

Substrate area located directly over vapor source was coated with the highest deposition rate $\dot{d}_0$. However, the vapor flux density of source depends on the angle ϑ between the vapor propagation direction and the normal of the vapor-emitting surface and is proportional to a $\cos^m \vartheta$ term. As shown in Fig. 5, the deposition rate $\dot{d}(x_S)$ from one vapor source depends on lateral position x_S on substrate, leading to an inhomogeneous distribution of layer thickness. The symbols show experimental results, while the lines were obtained by simulation of vapor distribution on a plane substrate.

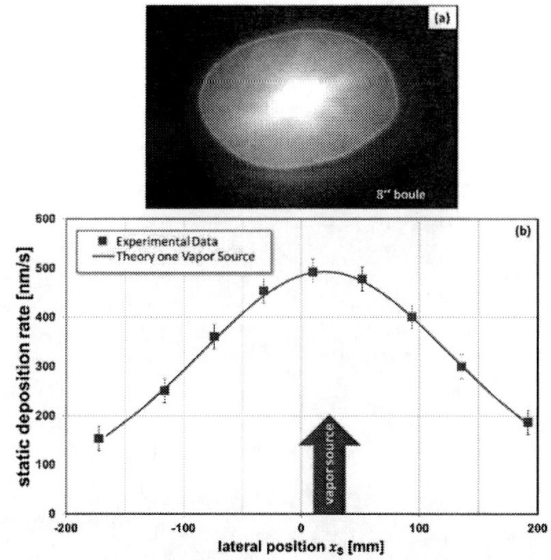

Figure 5: (a) Photograph of a 8″-boule during crucible-free EB-PVD experiment. (b) Lateral distribution of static deposition rate for a single high-rate evaporation source.

By using a double-ingot arrangement and superposing vapor from multiple sources the relative deviation of layer thickness could be reduced to < ±5 % on a substrate width of 200 mm (8″) (Fig. 6). Due to the smaller diameter of the used 4″ Si boules a minor electron beam power was allowed only, leading to a reduction in total deposition rate.

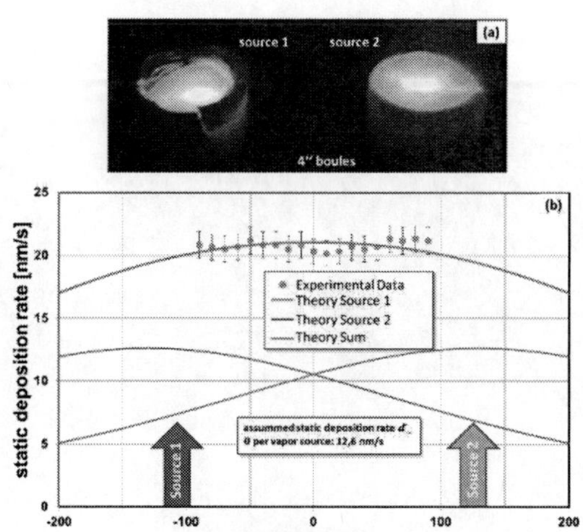

Figure 6: (a) Photograph of the two 4″-boules during crucible-free EB-PVD experiment. (b) Lateral distribution of static deposition rate for two vapor sources.

4 DISCUSSION

With enhancing deposition rate, an increase in layer purity is expected, because the ratio of vapor particle and residual gas impact rates is increased. Further magnifying of deposition rate is limited by the edges of melting pool. At coating experiments with medium deposition rates (~50 nm/s), layer stress could be manipulated greatly by varying substrate temperature. This dependency was already investigated by Kim *et al.* [9]. Due to higher heat flux at high deposition rate (>200 nm/s) substrate temperature rises to the same final value of ~500 °C nearby the thermal equilibrium. This forms a comparable low tensile stress of +40 MPa. To restrict substrate temperature at high deposition rates an effective method for substrate cooling has to be implemented [10]. For layer growth processes over 525 °C [11] the beginning formation of microcrystalline grains has to consider, which has a great influence of layer stress. Analyzing layer morphology of deposited Si thin films are in progress.

Thickness deviation over whole substrate width can be homogenized by applying a dynamic deposition process with linear substrate motion through superposing vapor stream from multiple sources.

5 CONCLUSIONS AND OUTLOOK

Encouraging high rate deposition experiments by crucible-free electron beam physical vapor deposition, ensuring high purity conditions, were demonstrated successfully. High purity Si-thin films have been deposited in the thickness range of 10-20 µm. For substrate temperature range up to 500 °C a low tensile layer stress in the rage of 50-40 MPa was observed. The possibility of excellent homogeneity in layer thickness was validated experimentally by using multiple Si-vapor sources. Crystallization experiments of deposited Si layers by an electron beam process are under investigation

6 REFERENCES

[1] G. Jia *et al.*„Bifacial multicrystalline silicon thin film solar cells," *Solar Energy Materials and Solar Cells,* vol. 167, pp. 102-108, 2017.

[2] C. T. Trinh *et al.*, „Progress in and potential of liquid phase crystallized silicon solar cells," *Solar Energy (In Press)*, 2018.

[3] J.-P. Heinß *et al.*, "In-Line High-Rate Deposition of Aluminum Onto RISE Solar Cells by Electron Beam Technology" , In *Proceedings of 26th European Photovoltaic Solar Energy Conference*, Hamburg, Germany, 2011, 2121-2124

[4] B. Graffel *et al.*, "Electron Beam-Fired Contacts: A Novel Approach for Rear Contact Formation in Crystalline Solar Cell Production", 2nd Silicon PV 2012, Leuven, Belgium

[5] J.-P. Heinß *et al.*, "Noch schneller zu sauberen Oberflächen im Vakuum", *Magazin für Oberflächentechnik,* vol. 66, no. 10, pp. 24-27, 2012.

[6] S. Saager *et al.*, "Density Determination and Gas Absorption Measurements in Ambient Nitrogen of Silicon Thin Films Deposited by Crucible-Free Electron Beam Evaporation" in *Proceedings of 31st European Photovoltaic Solar Energy Conference and Exhibition*, Hamburg, 3BO.6.3, 1036-1041 ,2015.

[7] J.-P. Heinß *et al.*, "Depositon of Silicon Layers with Rates up to 300 nm/s" in *Proceedings of 31st European Photovoltaic Solar Energy Conference and Exhibition*, Hamburg, 3BO.5.4, 1022-1025, 2015.

[8] S. Saager, "Electron Beam Crystallization of Amorphous Silicon Thin Films in the Solid Phase Regime and Assisted Simulations by Finite Element Method," in *Proceedings of 32nd European Photovoltaic Solar Energy Conference and Exhibition*, Munic, 3DV.1.6, 1256 –1259, 2016.

[9] J. Kim *et al.*, "Cracking behavior of evaporated amorphous silicon films," *Thin Solid Films,* vol. 518, no. 17, pp. 4908-4910, 2010.

[10] J.-P. Heinß *et al.*, "Temperature control of metal strip during high-rate vacuum coating", *Surface and Coatings Technology,* 290 (2016): 39-42.

[11] N. Annamalai *et al.*, "Properties of polycrystalline silicon grown on insulating substrates by electron beam gun evaporation", *Thin solid films,* vol. 155, no. 1, pp. 97-113, 1987.

7 ACKNOWLEDGEMENTS

The project was funded by the European Union and the Free State of Saxony (funding reference 100276968).

Europa fördert Sachsen.
EFRE
Europäischer Fonds für regionale Entwicklung

VALIDITY ANALYSIS OF THE TEXT BOOK LUMPED SERIES RESISTANCE APPROACH FOR SOLAR CELLS

Arvid van der Heide[1], Jef Poortmans[1,2,3]
[1]: imec (Energyville campus), Thor Park 8320, 3600 Genk, Belgium
[2]: KU Leuven, department ESAT – ELECTA, Kasteelpark Arenberg 10, 3001 Leuven, Belgium
[3]: UHasselt, Martelarenlaan 42, 3500 Hasselt, Belgium
arvid.vanderheide@imec.be

ABSTRACT: In this paper, the influence of the voltage variations due to sources of series resistance on the solar cell cell J-V curve is investigated. At first, the text book lumping approach for series resistance with its assumptions is explained. Then, the local diode equation is expanded as a Taylor series to analyse what happens for small voltage variations. Afterwards, the concept of a series resistance calculation based on voltage averaging is introduced. The voltage averaging approach is more intuitive and easier to grasp, and is proven to be exactly equivalent to the text book approach. Finally, the average current density in between two fingers will be calculated with a correct method for current density averaging and compared to the value predicted by the text book approach to check its validity. It is found that the text book approach is quite accurate for voltage variations that stay below the thermal voltage, but starts to deviate for higher values.
Keywords: series resistance, lumped series resistance, distributed series resistance

1 INTRODUCTION

The aim of this contribution is to check the validity of the textbook lumped series resistance approach to calculate the effect of series resistance on a solar cell J-V curve. In this approach, the specific series resistance ρ_{series} of a solar cell is a lumped value, in which the contributions of all sources of series resistance are summed. The term 'specific' is used to indicate that this parameter has been made area-independent, by multiplication of the series resistance with the area that generates the current through the series resistance. The different sources of series resistance in a typical silicon solar cell are the sheet resistance ρ_s of doped layers, the silicon bulk resistance, the contact resistance between front side metallisation and the silicon, the contact resistance between back side metallisation and the silicon, the resistance of the narrow metal lines on the front side, the resistance of the busbars and the resistance of the back contact, either a full metal layer or narrow metal lines and busbars.

Neglecting shunt resistance losses (which is reasonable for good quality cells), in the one diode model the J-V curve of a solar cell is given by

$$J(V) = J_L - J_0(\exp\left(\frac{V + J\rho_{series}}{mV_T}\right) - 1) \qquad (1)$$

where J is the total current density, V the external cell voltage, J_L the light generated current density, J_0 the diode saturation current density, m the diode ideality factor and V_T the thermal voltage, defined by $V_T \equiv kT/q$. In this equation, k is the Boltzmann constant, T the absolute temperature and q the elementary charge.

Since the local diode currents depend exponentially on the local voltage, it is not obvious at all how the voltage variations due to series resistance have to be taken into account in the calculation of the total diode current of a solar cell. The text book approach is to put a term $\rho_{series}J$ in the exponent as shown in (1), and to obtain ρ_{series} as a sum of ρ_{series} contributions that are calculated based on Joule heat integration, assuming a constant J everywhere.

To simplify the explanation of the calculation of contributions to ρ_{series}, it will be replaced by its area-dependent equivalent R_{series} for now, it is given by

$$R_{series,Joule} = \frac{P_{Joule}}{I_{tot}^2} = \frac{1}{I_{tot}^2}\int I^2(x)dR \qquad (2)$$

which is the integration of the locally dissipated Joule power $dP_{Joule} = I^2(x)dR$ for all locations relevant for a certain resistance source (e.g. all points along a finger), divided by I_{tot}^2, the square of the total current at the exit of the resistance source.

In this approach, it is assumed that J is constant everywhere, so that $I(x)$ in each resistive element is linearly increasing. It should be noted that this is not completely true in practice, due to local voltage variations.

It would be surprising that the method just described, and suggested to be correct in the text books, would be really exact, and in fact it is not, as has been pointed out before [1]. It should be noted that the integration of all local Joule powers across a solar cell, using the real local currents, does in fact give the exact total power loss due to series resistance in a solar cell [2], but that is something different than the approach described before.

The ρ_{series} values obtained by equation (2) for the different contributions are given in text books like [3]. An example of such a contribution that will be investigated in more detail in this work is that from the emitter to ρ_{series}:

$$\rho_{series,Joule,e} = \tfrac{1}{12}\rho_s s^2 \qquad (3)$$

where s is the separation between the fingers, from finger edge to finger edge. The subscript 'Joule' has been used to discriminate between this $\rho_{series,Joule}$ and an alternatively calculated ρ_{series} that will be introduced in the next section.

After that, for a region between two fingers the difference in average current density for the usual text book approach and a precise averaging will be calculated, showing the error made when using the text book approach.

2 OBTAINING ρ_{series} BY VOLTAGE AVERAGING

Before introducing the alternative way to calculate ρ_{series} contributions it is necessary to take a close look at the change in local current density due to an extra voltage $\Delta V(x)$ induced by local series resistance. This $\Delta V(x)$ is calculated with respect to the voltage V_0, which is defined as the voltage at the place where the current leaves the resistive element. Neglecting the current through shunts, the change in current density is only due to the change in the diode current J_d, which depends on $\Delta V(x)$ in the following way:

$$J_d(x) = J_0[\exp((V_0 + \Delta V(x))/mV_T) - 1] \approx J_{d0}\exp(\Delta V(x)/mV_T) \quad (4)$$

where $J_{d0} \equiv J_0\exp(V_0/mV_T)$ has been used to simplify the equation. The -1 has been omitted at the right hand side, since it can be safely neglected for $V_0 >> mV_T$. This condition is certainly met at the maximum power point voltage V_{mpp}, the most interesting point to evaluate solar cell losses, since $V_{mpp} \sim 20mV_T$.

The function exp(x) can be written as a Taylor series that converges for all x:

$$\exp(x) = 1 + x + \frac{x^2}{2} + ... + \frac{x^n}{n!} + ... \quad (5)$$

Using (5), the equation for $J_d(x)$ can be rewritten as:

$$J_d(x) = J_{d0}\left(1 + \frac{\Delta V(x)}{mV_T} + \frac{1}{2}\left(\frac{\Delta V(x)}{mV_T}\right)^2 + ... + \frac{1}{n!}\left(\frac{\Delta V(x)}{mV_T}\right)^n + ...\right) \quad (6)$$

Since $J_d(x)$ is position dependent, for the performance of a region it is the average $J_d(x)$ across the region that counts:

$$\overline{J_d(x)} = J_{d0}\overline{\exp\left(\frac{\Delta V(x)}{mV_T}\right)} = J_{d0}\left(1 + \frac{1}{mV_T}\overline{\Delta V(x)} + \frac{1}{2m^2V_T^2}\overline{\Delta V^2(x)} + ...\right) \quad (7)$$

Under illumination, the average total current density is obtained by adding the average J_L.

As long as $\Delta V(x) < mV_T$, the non-linear terms in (7) can be neglected and we can approximate the average J_d by a linear approximation:

$$\overline{J_d(x)} \approx J_{d0}\left(1 + \frac{1}{mV_T}\overline{\Delta V(x)}\right) \quad (8)$$

When this approximation is accurate enough (so for small $\Delta V(x)$), the averaging of the diode current in (7) can also be done by averaging only the $\Delta V(x)$ that is in the exponent instead of averaging the whole exponential function across the interval. To distinguish it from the correct but more complex averaging in (7), this average is called $J_{d,va}$ (*va* stands for voltage averaging):

$$J_{d,va} = J_{d0}\exp\left(\frac{\overline{\Delta V(x)}}{mV_T}\right) = J_{d0}\left(1 + \frac{1}{mV_T}\overline{\Delta V(x)} + \frac{1}{2m^2V_T^2}\left(\overline{\Delta V(x)}\right)^2 + ...\right) \quad (9)$$

This equation differs from (7) in the non-linear terms since the square of an average is not equal to the average of a square (and similar for the higher order terms). But for small $\Delta V(x)$ equation (7) can be replaced by (9).

Using (9), it is now possible to introduce a $\rho_{series,va}$ based on voltage averaging in the exponent, by choosing the following definition:

$$\rho_{series,va} \equiv \frac{\overline{\Delta V(x)}}{J} \quad (10)$$

where $\Delta V(x)$ refers to the voltage difference within one of the resistive contributions (finger, doped layer, ...), and J is taken over the region where the voltage averaging is done. In this way, we also get $\rho_{series,av}J$ in the exponent (similar as for the text book approach), but with a different definition for the ρ_{series} than usually.

Note that $\Delta V(x)$ is a function of the distribution of the total current density, in the usual situation of 1 sun illumination this includes a J_L that is typically (much) larger than J_d, except when the voltage is near open circuit condition.

Now we will determine $\rho_{series,va}$ for the emitter under illumination, assuming that the cell is working at V_{mpp}. A schematical drawing of this, for a cell based on p-type silicon, is shown below:

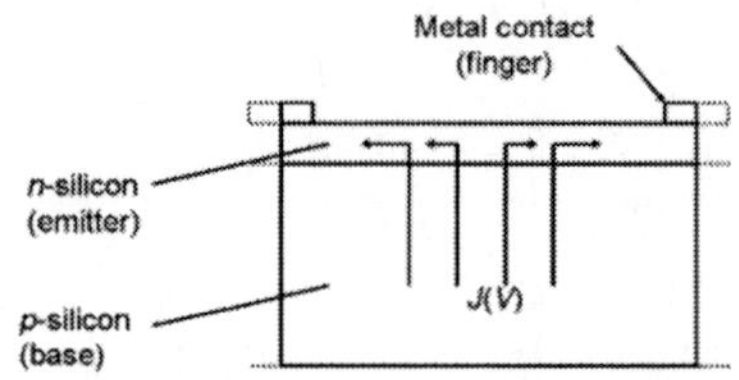

Figure 1: Current flow in base and emitter between fingers of a (p-type) solar cell. The horizontal current in the emitter increases towards the fingers, this is indicated by increasing length of the horizontal arrows.

When the potential on both fingers is equal and the contact resistances on both fingers are equal, the current flow to each finger will linearly increase from 0 at the centre between the fingers towards each finger, assuming that the total J between the fingers, J_f, is constant. This is reasonable at V_{mpp}, since the total J is then still dominated by J_L. To simplify the calculations, the position x=0 is now defined to be the centre between the fingers. The current flow towards a finger, per unit length of finger, is indicated by i, which increases linearly with x:

$$i = J_f x \quad (11)$$

The current flow in the emitter induces a voltage gradient dV/dx, the relation between the values is given by

$$i = -\frac{1}{\rho_s}\frac{dV}{dx} \quad (12)$$

Substituting in the previous equation gives

$$-\frac{dV}{dx} = \rho_s J_f x \quad (13)$$

When the fingers are at the same potential and the contact resistances are equal, the solution is a symmetric parabola with the maximum at the centre between the fingers, so at a distance $s/2$ from the fingers. The height of this parabola can be calculated to be $V_{max} = 1/8\ \rho_s J_f s^2$.

The average height of the parabola can be obtained by integration of $V(x)$ between the fingers and dividing by s, the result is 2/3 V_{max}.

In Figure 2, a graph of the parabolic potential between the fingers is shown, as well as a horizontal line that indicates the value of the average voltage.

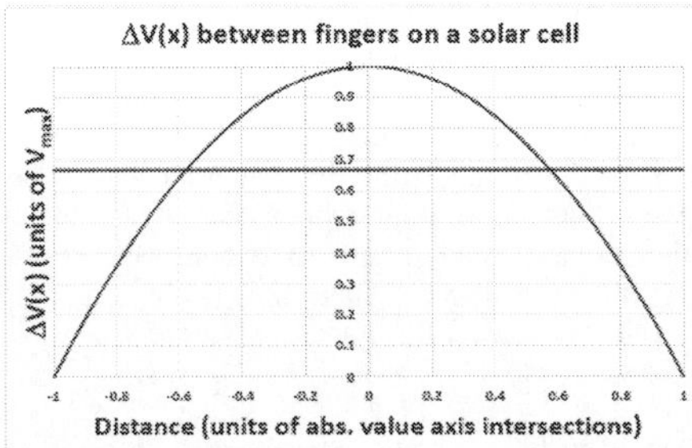

Figure 2: Parabolic potential in between fingers, including a horizontal red line indicating the average voltage of $2/3\,V_{max}$

So the emitter series resistance as obtained by voltage averaging $\rho_{series,va,e}$ is:

$$\rho_{series,va,e} = \frac{\overline{\Delta V(x)}}{J_f} = \frac{\frac{2}{3}V_{max}}{J_f} = \frac{\frac{2}{3}\frac{1}{8}\rho_s J_f s^2}{J_f} = \frac{1}{12}\rho_s s^2 \tag{14}$$

where $\Delta V(x)$ is in this case the emitter potential with respect to the fingers, that are at a certain potential V_0. This value for $\rho_{series,va,e}$ is exactly the same as $\rho_{series,Joule,e}$ calculated using the text book approach. In the next section it will be proven that both approaches are also equivalent in general.

3 EQUIVALENCE OF ρ_{series} CALCULATIONS

To simplify the proof for the equivalence of the two approaches, in the following area dependent variables R_{series}, P_{Joule} and I_{tot} will be used, as done before in the introduction. To start with, equation (2) used in the text book approach can be rewritten as:

$$R_{series,Joule} = \frac{P_{Joule}}{I_{tot}^{2}} = \frac{P_{Joule}/I_{tot}}{I_{tot}} \tag{15}$$

If it can be shown that P_{Joule}/I_{tot} is equal to the average of $\Delta V(x)$, the equation above would be the area dependent equivalent of the equation (10) for $\rho_{series,va}$.

To check this, we consider a linearly increasing current like we typically find in solar cell resistance sources. It starts from 0 at the beginning of the resistive element to I_{tot} at the end of it. For each length of the element dx the extra contribution to the current equals $i\,dx$, where i is the current flow per unit length into the element (assumed to be constant). The Joule power can then be calculated in the following way:

$$P_{Joule} = \int dP_{Joule} = \int \Delta V(x) i dx \tag{16}$$

This can be explained in the following way: for each amount of current idx entering the resistance, the Joule power generated when flowing to the exit of the resistance is equal to idx times the voltage difference $\Delta V(x)$ between the entrance point of the current idx and the exit point of that current. It should be noted that this integration is not written as the usual summation of the locally produced Joule powers in resistance pieces with length dx, but as the summation of Joule powers that current contributions dI produce over the total length of the resistance that they flow through. However, the outcome should be the same for both cases, namely the total Joule power generated along the resistance.

Now P_{Joule}/I_{tot} can be written as:

$$\frac{P_{Joule}}{I_{tot}} = \frac{\int \Delta V(x) i dx}{\int i dx} = \frac{\int \Delta V(x) dx}{\int dx} = \frac{1}{L}\int \Delta V(x) dx = \overline{\Delta V(x)} \tag{17}$$

where L is the total length of the resistance. Substituting in (15) gives

$$R_{series,Joule} = \frac{\overline{\Delta V(x)}}{I_{tot}} \tag{18}$$

This is exactly the area dependent equivalent of equation (10), so that the Joule heat method and the potential averaging method are in fact the same.

So the text book approach is equivalent to voltage averaging in the exponent, so using equation (9) as an approximation of equation (7). Although the approximation might work quite well in most cases (reasonably good cells), it is still unfortunate and incorrect that the approximation is usually suggested to be exact in the text books.

4 CALCULATION OF ERROR MADE IN LUMPING

In the following, it is assumed that the cell is illuminated at 1 sun intensity while it is operated at its V_{mpp}. Under these conditions, the emitter potential can be safely assumed to be parabolic, since the total current density J_L-$J_d(x)$ can be assumed constant (the variations in $J_d(x)$ are small compared to J_L-J_{d0}). For this case of a parabolic voltage, the error that is made in the lumping approach will now be calculated as a function of the maximum voltage of the parabola.

The parabolic voltage can be described in general by its intersections with the x-axis at $x = -a$ and $x = +a$, and a maximum height V_{max}, reached for $x = 0$. Then the equation for this parabola is:

$$V(x) = \frac{V_{max}}{a^2}(a-x)(x+a) = V_{max}\left(1-\frac{x^2}{a^2}\right) \tag{19}$$

Substituting the right hand side of equation (19) for $\Delta V(x)$ in equation (4) gives the following for $J_d(x)$:

$$J_d(x) = J_{d0}\exp\left(\frac{V_{max}}{mV_T}\left(1-\frac{x^2}{a^2}\right)\right) = J_{d0}\exp\left(v_{max}\left(1-\frac{x^2}{a^2}\right)\right) \tag{20}$$

using the normalised maximum voltage $v_{max} \equiv V_{max}/(mV_T)$ to simplify the equation.

In Figure 3, $J_d(x)$ is plotted (the 2 lower curves) for v_{max} = 1 and v_{max} = 2. Also the total $J(x) = J_L - J_d(x)$ is plotted for completeness, these are the upper curves. Typical values were used as input: J_L = 40 mA/cm^2, J_{d0} = 2.2 mA/cm^2 at $V_0 = V_{mpp}$.

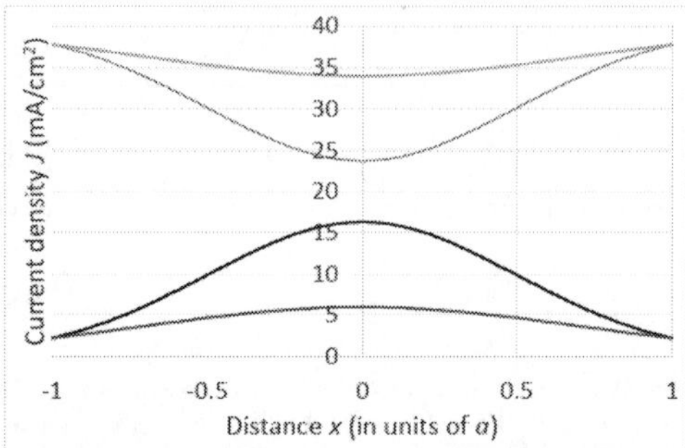

Figure 3: Graphs of $J_d(x)$ between two fingers for v_{max} = 1 (dark green) and v_{max} = 2 (dark blue). Also the corresponding curves for the total $J(x) = J_L - J_d(x)$ are shown, in light green and light blue. Typical input values were used: J_L = 40 mA/cm^2, J_{d0} = 2.2 mA/cm^2 at $V_0 = V_{mpp}$.

Figure 3 clearly shows the influence of the parabolic potential between the fingers on $J_d(x)$ and the total J. It will also be clear that the assumption that the varation in the total J in between the fingers are small (needed to obtain a parabolic potential) gets violated with increasing v_{max}. The assumption is still reasonable for v_{max} = 1, is not so good anymore for v_{max} = 2, and will be unacceptable for higher values of v_{max}. So this is another approximation in the text book approach that breaks down for higher $\Delta V(x)$.

To continue with the error calculation, the accurate average J_d (assuming that it is safe to use parabolic potential) is obtained by averaging of (20):

$$\overline{J_d(x)} = \frac{J_{d0}}{a}\int_0^a \exp(v_{max}(1-\frac{x^2}{a^2}))dx \qquad (21)$$

Taking the constant term outside the integration, the equation simplifies to

$$\overline{J_d(x)} = \frac{J_{d0}\exp(v_{max})}{a}\int_0^a \exp(-\frac{v_{max}x^2}{a^2})dx \qquad (22)$$

Substitution of $v_{max}x^2/a^2 = t^2$ finally gives

$$\overline{J_d(x)} = J_{d0}\frac{1}{\sqrt{v_{max}}}\exp(v_{max})\int_0^{\sqrt{v_{max}}} \exp(-t^2)dt \qquad (23)$$

wich can be written in terms of the normalised error function erf(z) in the following way:

$$\overline{J_d(x)} = J_{d0}\sqrt{\frac{\pi}{4v_{max}}}\exp(v_{max})\mathrm{erf}(\sqrt{v_{max}}) \qquad (24)$$

The text book approach, that was shown to be equivalent to the voltage averaging approach, has $J_{d,va}$ as the average current density between the fingers. This is given by equation (9), using that the average voltage for this parabolic $\Delta V(x)$ is 2/3 of V_{max}:

$$J_{d,va} = J_{d0}\exp\left(\frac{\overline{\Delta V(x)}}{mV_T}\right) = J_{d0}\exp(\tfrac{2}{3}v_{max}) \qquad (25)$$

Now the correct average current density between two fingers can be compared with the text book/voltage averaging approximation, the curves as a function of v_{max} are shown in Figure 4. The curves are only shown up to v_{max} = 2, since the parabolic potential assumption is violated too much for higher values (when the potential at the fingers V_0 is at maximum power point voltage at least, for significantly lower voltage the parabola assumption will even stay correct for higher v_{max}, since $J_d(x) \ll J_L$ in that case).

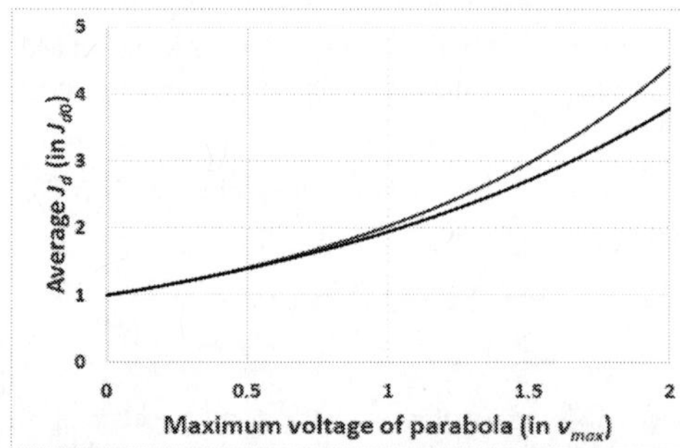

Figure 4: Curves of the average diode current density for the text book/voltage averaging approach (black line) and for the correct averaging (red line) as a function of the normalised maximum of the parabolic potential.

Both curves in Figure 4 have a linear start with a slope of 2/3 at v_{max} = 0, as they should have according to (8), taking into account the factor 2/3 for the average voltage of a parabola. At increasing v_{max}, the non-linear terms in (9) and (7) get more important, and since these terms are differing in these equations, an error gap is growing between correct diode current averaging and the text book approach. It can be observed that certainly up to v_{max} = 1 the text book is still quite accurate, and even at v_{max} = 2 the relative error is still only 16 %.

A typical value for v_{max} for good cells is around 0.5, so that in practice the error is usually quite small for the emitter. To draw conclusions for entire cells, it has to be taken into account that different sources of series resistance on a cell add up, causing total voltage differences across the cell that are larger than the voltage differences in the individual sources. Taking that into acoount, for good cells the voltage differences are still less than 1.5 mV_T so that the errors across the cell are also still acceptable. If there are local series resitance problems on the cell, like interrupted fingers, the voltage differences can get as large as several times mV_T so that deviations can get larger on a local scale.

5 CONCLUSIONS

It has been shown that the text book lumping approach to calculate the influence of series resistance on the average diode current density is equivalent to the distance averaging of voltage within the exponent of the diode equation. In principle, this is not correct since the distance averaging should be done by taking into account the whole exponential function instead of the exponent only.

However, since the methods only differ in the non-linear terms, the text book aproach is an acceptable approximation if the voltage variations are small.

In the last part, the error in the text book approach was quantitatively determined for the region between two fingers, as a function of the magnitude of the voltage variations. This error is the difference between correct current density averaging and voltage averaging within the exponent. It was found that the text book approach is performing quite well for v_{max} up to 1 (so maximal variations up to mV_T) while errors can get considerable for $v_{max} > 2$.

In practice, this means that the text book approach performs quite well for uniform good cells. However, care should be taken since the text book approach can also get incorrect if cells locally have large differences in voltage due to local series resistance problems (like finger interruptions or locally increased contact resistance).

6 ACKNOWLEDGEMENTS

The authors would like to thank Jan-Martin Wagner from the Christian-Albrechts-Universität in Kiel for valuable comments on this work.

7 REFERENCES

[1] Araújo GL, Cuevas A, Ruiz JM. "The effect of distributed series resistance on the dark and illuminated current-voltage characteristics of solar cells", IEEE Transactions on Electron Devices 1986: **33**(3): 391-401

[2] Jan-Martin Wagner, Andreas Schütt, Jürgen Carstensen, Rainer Adelung, "Linear-response description of the series resistance of large-area silicon solar cells: Resolving the difference between dark and illuminated behavior", 6th SiliconPV 2016, Energy Procedia **92** (2016): 255-264

[3] Green MA. *Solar cells: operating principles, technology and system applications.* Prentice-Hall: Englewood Cliffs (NJ,USA), 1982

AUTHOR INDEX

AUTHOR INDEX

AUTHOR INDEX

AUTHOR INDEX

AUTHOR INDEX

AUTHOR INDEX

AUTHOR INDEX

AUTHOR INDEX

AUTHOR INDEX

AUTHOR INDEX

AUTHOR INDEX

AUTHOR INDEX

AUTHOR INDEX

AUTHOR INDEX

AUTHOR INDEX

AUTHOR INDEX

AUTHOR INDEX

AUTHOR INDEX

AUTHOR INDEX

WIP – Renewable Energies
Sylvensteinstr. 2
81369 Munchen
Germany

ISBN 978-1-5108-7155-7

CURRAN ASSOCIATES INC.
proceedings.com

35th European Photovoltaic Solar Energy Conference and Exhibition (EU PVSEC 2018)

Brussels, Belgium
24-28 September 2018

Volume 2 of 3

WIP

35th European Photovoltaic Solar Energy Conference and Exhibition (EU PVSEC 2018)

Brussels, Belgium
24-28 September 2018

Volume 2 of 3

Editors:

P. Verlinden
P. Helm
R. Kenny

ISBN: 978-1-5108-7155-7

Printed from e-media with permission by:

Curran Associates, Inc.
57 Morehouse Lane
Red Hook, NY 12571

Some format issues inherent in the e-media version may also appear in this print version.

Printed by Curran Associates, Inc. (2019)

For permission requests, please contact WIP – Renewable Energies
at the address below.

WIP – Renewable Energies
Sylvensteinstr. 2
81369 Munchen
Germany

Phone: +49 89 72012735
Fax: +49 89 72012791

wip@wip-munich.de

Additional copies of this publication are available from:

Curran Associates, Inc.
57 Morehouse Lane
Red Hook, NY 12571 USA
Phone: 845-758-0400
Fax: 845-758-2633
Email: curran@proceedings.com
Web: www.proceedings.com

TABLE OF CONTENTS
of 35th EU PVSEC 2018 Proceedings Papers

New Materials and Concepts for Photovoltaic Devices

Oral PRESENTATIONS 1CO.1 Advanced Material Development and Analysis for High Performance PV Modules

Oral PRESENTATIONS 1CO.2 Novel Approaches for Special PV Applications

Visual PRESENTATIONS 1CV.4 Fundamental Studies / New Materials and Concepts for Photovoltaic Devices

Silicon Cells

Plenary SESSION 2BP.1 Silicon Photovoltaics

Oral PRESENTATIONS 2AO.4 Characterisation and Modelling of Silicon Cells

Oral PRESENTATIONS 2AO.5 Characterisation and Modelling of Materials and Surfaces for Silicon Photovoltaics

Visual PRESENTATIONS 2AV.3 Heterojunction Solar Cells

Visual PRESENTATIONS 2DV.3 Thin Film and Foil-Based Si Solar Cells/Characterisation & Simulation Methods for Si Cells/Manufacturing & Production of Si Cell

Non Silicon-Based Thin Film Photovoltaics

Visual PRESENTATIONS 3BV.3 Perovskite, Organic and Dye-Sensitised Devices

Concentrator and Space Photovoltaics

Photovoltaic Modules and BoS Components

Visual PRESENTATIONS 5CV.1 PV Module Design, Manufacture, Performance and Reliability

PV Systems - Performance, Applications and Integration

Visual PRESENTATIONS 6CV.2 Design and Installation of PV Systems

PV Economics, Markets and Policies

FUNDAMENTAL AND TECHNOLOGICAL LIMITS TO LOW-LIGHT EFFICIENCY OF CRYSTALLINE SILICON SOLAR CELLS.

B. Conrad[1]*, A. Peter Amalathas[1], and J. Holovsky[1,2]
[1] CTU Faculty of Electrical Engineering Technická 2, 166 27 Prague, Czech Republic
[2] Institute of Physics ASCR, v. v. i., Cukrovarnická 10, 162 00 Prague, Czech Republic
* E-mail: bsmc@alum.mit.edu

ABSTRACT: Low-light applications often bring amorphous silicon and other thin-film cell technologies to mind. While there has been work to improve the low-light efficiency of crystalline silicon solar cells, their potential has not been fully described. The description of fundamental limits, and much of the work to realize those limits, has been limited to one-sun illumination levels and higher. This work aims to establish the fundamental Auger limit for wafer-based silicon solar cells at illuminations lower than one-sun, the additional technological limits imposed by surface recombination, shunt resistance, and recombination in the space charge region, and the degree to which low-light potential of these devices had been realized. High efficiencies demonstrate that in choosing technologies for low illuminations, crystalline silicon solar cells should not be overlooked.
Keywords: c-Si, Fundamentals, Modelling, Silicon Solar Cell, Solar Cell Efficiencies

1 INTRODUCTION

Although solar cell efficiencies are generally reported for standard test conditions under the 1-sun AM1.5G spectrum, cell performance under lower illumination levels is important for a range of cases. These include standard installations in cloudy weather, the early morning and late afternoon, indoor applications, and as subcells in (non-concentrator) tandem devices. It is sometimes claimed that low-light applications should make use of amorphous silicon or other thin film technologies, and crystalline silicon cells. The performance of c-Si modules at low illumination has indeed been shown to suffer greater losses from one-sun values than that of a-Si modules [1]. However, these are observations of achieved performance, rather than potential.

There has been demonstration of c-Si modules [3] and cells [4-6] with improved low-light performance, but these studies have again been limited to empirical measurements of fabricated devices. The vast majority of published work, on the cell level especially, has been focused on improving 1-sun efficiency. With record a-Si solar cells sitting at 16.5% absolute lower efficiency than their c-Si counterparts [2], it is worth considering what the potential efficiencies of wafer-based cells under low light are. However, the limits to the efficiencies of crystalline silicon cells at low illumination levels, while important for the applications described above, have not been described.

Limits to crystalline silicon performance at one sun due to Auger and Radiative recombination are well-known [7], and even the record setting cells are limited by additional technological factors. Of these, recombination in the space charge region (SCR) and shunt resistance have been noted to be particularly damaging at low illumination levels.

In this work, we consider the potential efficiencies of crystalline silicon solar cells under low illumination levels. We first calculate the fundamental limit imposed by Auger and radiative recombination and the Lambertian limit to light-trapping. We next examine non-idealities in device performance, including not only SCR recombination and shunt resistance, but also surface recombination. Finally we will consider some experimental and published values of low-illumination efficiencies and how they compare to the limits described. It is suggested that crystalline silicon solar cells should be considered an attractive choice for illumination levels between 0.01 and 1 sun.

2 METHODS

While there is a standard spectrum for 1-sun efficiency measurements, low illumination conditions occur under a variety of spectra. Therefore, we consider both a scaled version of the AM1.5G spectrum, spectra for different Air Masses as modelled with SMARTS [8], and the diffuse portion of the AM1.5G spectrum (again as modelled with SMARTS).

2.1 Fundamental Limits

The Auger/Lambertian limit is calculated following the formulation of Kerr, Cuevas, and Campbell [7] for Coulomb-Enhanced Auger Recombination, including radiative recombination and photon recycling, with the light induced current determined by the Lambertian limit to light trapping and the index of refraction of silicon.

Calculations and optimisation are performed numerically using a python script. The excess carrier concentration where the Auger Lifetime curve and a line described by Eqn. 1, which determines the lifetime

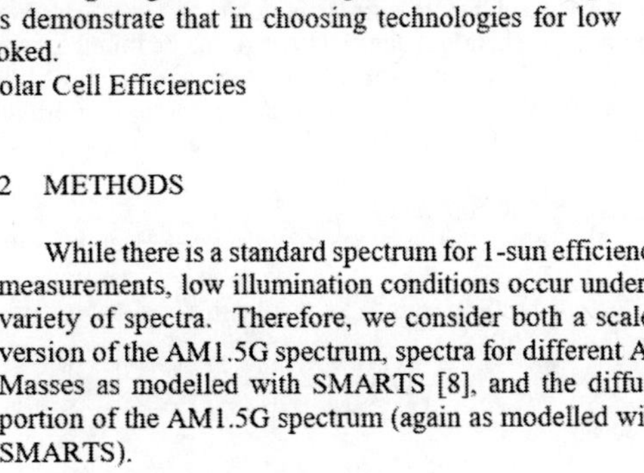

Fig. 1: Carrier lifetime as a function of excess carrier concentration determined by the Auger limit and the values required at certain cell operation points for a 100um n-type base doped at $10^{15}/cm^3$. The black line shows required lifetime for a certain carrier concentration under 1-sun open circuit conditions, with grey dotted lines showing the case for reduced recombination currents, either by reduced illumination or increased carrier extraction. The dashed red line shows the voltage as a function of excess carrier concentration, and the thin lines show the process of determining the operation point and voltage under the Auger limit.

necessary to maintain a certain carrier concentration in the cell at a given recombination rate, expressed as a current density.

$$\tau = \frac{W_b n q}{J_{recombination}} \quad (1)$$

The cell voltage is then calculated from the excess carrier concentration. In the open circuit case, the recombination current is the light-generated current and the Voltage is Voc. As illumination decreases, the line described by Eqn. 1 shifts to the left, resulting in lower carrier concentration and voltage. A similar shift is seen when considering operating points other than open circuit, as the recombination current is decreased due to carrier extraction. Therefore, the cell voltage can be found for any value of extracted current, and the two combined to determine the power. By shifting the extracted current, the maximum power is found. Fig. 1 illustrates this process.

2.2 Surface Recombination

The effect of a non-zero surface recombination velocity, S, taken to be the same on front and rear surfaces, is included by adjusting the effective lifetime downwards, as

$$\frac{1}{\tau} = \frac{1}{\tau_{bulk}} + \frac{1}{\tau_s} \quad (2)$$

where

$$\tau_s = \frac{W_b}{2S} + \left(\frac{1}{D}\right)\left(\frac{W_b}{\pi}\right)^2 \quad (3)$$

with D as the minority carrier diffusivity [9].

2.3 Shunt Resistance

Shunt Resistance is included in a straightforward manner by subtracting V^2/R_{shunt} from the power calculated from the Auger limit, with R_{shunt} expressed in Ωcm^2. Optimisation is carried out using the power generated after shunt resistance losses as the figure of merit.

2.4 Shockley-Reed-Hall Recombination

Shockley-Reed-Hall (SRH) Recombination calculated from Eqns. 4 and 5 is added to the Auger and Radiative recombination based on the carrier concentrations (n, p), lifetime coefficients (τ_{n0}, τ_{p0}), and carrier concentrations for fermi levels at the trap level (n_t, p_t).

$$U = \frac{np}{\tau_{n0}(n+n_t)+\tau_{p0}(p+p_t)} \quad (4)$$

$$\tau_{n/p0} = \frac{1}{N_{defect} v_{th} \sigma_{n/p}} \quad (5)$$

A common simplification when considering SRH recombination is to consider a single trap at midgap, with $\tau_{n0} = \tau_{p0}$. This case is considered for τ_0 of 10^{-3} and 10^{-4} s. We also consider the specific case of iron, as it is one of the major impurities in silicon solar cells. Concentrations of 10^{10} and 10^{11} cm^{-3} are considered, as these are typical of single crystal silicon. FeB is also considered for p-type cells, at 10^{10} cm^{-3}. Trap parameters for these defects are listed in Table 1 from [10,11].

2.4 SRH Recombination in the SCR

SRH recombination in the SCR has been blamed in the past for reduced efficiencies at low light. Because the SCR features more equal numbers of electrons and holes, the

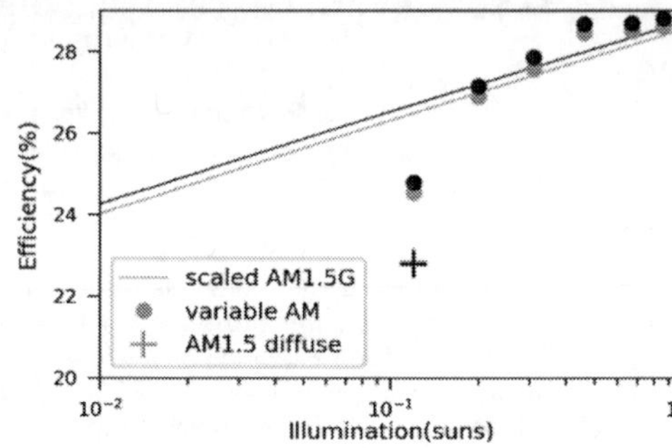

Fig. 2: Auger limit for n-type (light purple) and p-type (dark blue) silicon solar cells with doping of $10^{15}/cm^3$ under illumination levels below one sun and a selection of incident spectra.

Table I: Fe and FeB trap properties

	Fe	FeB
Hole capture cross-section (cm^2)	7e-17	3e-15
Electron capture cross section (cm^2)	5e-14	5e-15
Trap level above valance band (eV)	0.38	0.86

SRH recombination rate is increased over that in the bulk, and is manifested in the I-V curve with a higher ideality factor, though it is not always as high as the commonly cited value of 2 [12]. To accurately calculate the total SCR recombination with spatially varying carrier concentrations, the recombination rate is numerically integrated throughout the SCR. The electron concentrations are determined by:

$$n = n_i \exp\left(\frac{q(\psi - \phi_n)}{kT}\right)$$

with ψ as determined from the depletion approximation, Poisson's equation, and the applied voltage. Hole concentrations are calculated analogously.

The same traps considered for SRH recombination in the bulk are investigated here, and the total recombination in the SCR is multiplied by the fraction of the base that the SCR takes up, to combine it during optimsation with the other recombination rates that occur across the entire base.

3 RESULTS

Limiting Efficiencies have been calculated for devices with both n-type and p-type bases. As low doping levels result in higher Auger Limits, the doping is set to be equal to or greater than 10^{15} cm^{-3}. Lower doping levels provide negligible improvements to the limiting efficiencies. Device thicknesses are limited to 30-600 um.

3.1 Fundamental Limits

Fig. 2 shows the Auger-Lambertian limit to cell efficiency for a scaled AM1.5G spectrum (lines) and spectra generated with SMARTS [8] (circles) for a selection of air masses from AM2.0 (~0.86 suns) to AM18 (~0.12 suns) for both n- and p-type devices.

The decay in scaled AM1.5G efficiency at lower illuminations is essentially due to the reduction in voltage that is the opposite of the effect whereby voltage is increased under concentrated sunlight. The rate of this reduction is set by the ideality factor, which here is approximately 0.8. The differences between this curve and the efficiencies obtainable under different air mass spectra are due to spectral variation. The variable AM spectra benefit the cell between 0.25 and 1.0 suns, but result in a rapid decline in efficiency below 0.2 suns. In standard installations, the spectra with intensity below 0.2 suns will be present only for a very brief time when the sun is close to the horizon, and the system is already particularly susceptible to shading losses regardless of cell efficiency. Non-concentrator multijunction subcells might operate at 0.5 or 0.33 suns (for 2 or 3 junctions), but there will also be additional spectral changes due to absorption of higher bandgap cells.

3.2 Shunt Resistance and Surface Recombination

The effects of shunt resistance are shown in Fig. 3 for n-type cells under a scaled AM1.5G spectrum. P-type cells and other spectrums show similar effects. Bunea et al. showed shunt resistances in excess of 20 kΩcm², indicating that those values which show minimal effect at all but the lowest illumination levels are feasible [4]. As R_{shunt} drops below this, the ideal thicknesses of devices increases, especially at the lowest illumination levels.

Fig. 4 shows that moderate to high surface recombination affects cells similarly at all illumination levels in terms of absolute efficiency. Very low values, such as have been achieved by Bonilla et al. [13] demonstrate less effect on efficiencies approaching one sun than at lower levels. As would be expected, higher

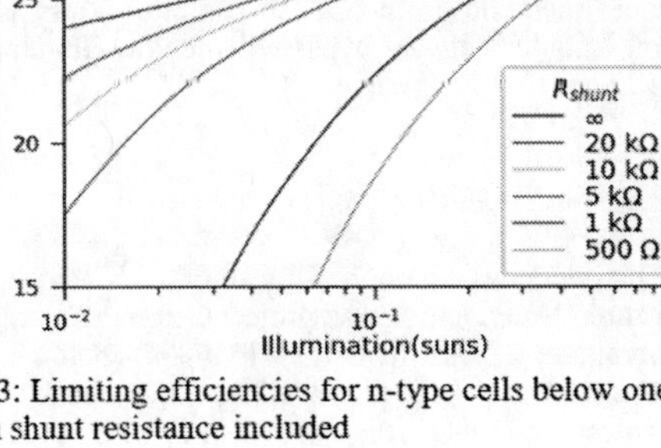

Fig 3: Limiting efficiencies for n-type cells below one sun with shunt resistance included

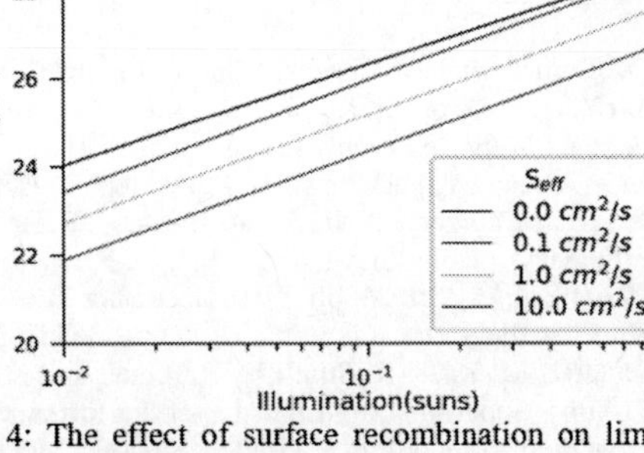

Fig. 4: The effect of surface recombination on limiting efficiencies below one sun.

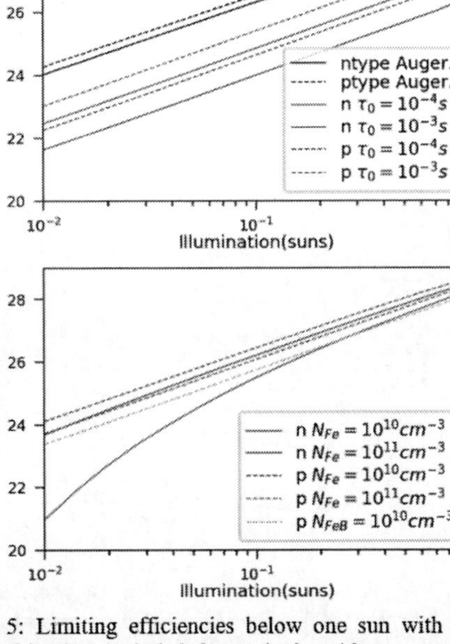

Fig. 5: Limiting efficiencies below one sun with SRH recombination included for a single mid-gap trap with varying lifetime (top) or Fe and FeB in varying concentrations (bottom).

surface recombination results in higher ideal thicknesses for the cells. Higher doping levels are also preferred, as decreasing mobility results in reduced effect of the surfaces on the carrier concentrations in the bulk.

3.3 SRH Recombination in Bulk and SCR

Fig. 5 shows the effect that SRH recombination in the bulk has on the device efficiency. Traps at midgap affect n- and p- type devices similarly, and FeB defects are more harmful to efficiency than a similar concentration of Fe defects. The inclusion of SRH recombination reduces the ideal cell thicknesses. Most of the significant drop in efficiencies at low illumination for n-type cells with higher Fe concentrations is caused directly by the SRH recombination, but a small amount is due to an ideal thickness that would actually be below the 30um limit that has been set in this work. The lower plot in Fig. 5 also indicates that the preferred doping type for the base may vary with illumination level, as the exact dependence efficiency on illumination is different with different types of defect.

Fig. 6 shows that the effect of SRH recombination in the SCR. The effect is small enough that, when simply plotting efficiency, it is overwhelmed by the change in Auger limit over the illumination range, so here it is presented normalized to the p-type Auger limit at each given illumination. Though minimal, the effect does increase for a trap at mid-gap at the lowest illuminations.

Given the minimal impact of the SCR recombination in this analysis, an overarching trend in the effect on ideal thicknesses and doping is not seen.

3.4 Comprehensive Limits and Achieved Values.

Finally, limits are found for when all of the previously described non-idealities are present. This is done with Rshunt set to 20kΩ and S_{surf} to 0.1 cm/s. SRH recombination in both the bulk and the SCR from Fe (for n-type cells) or FeB (for p-type) at a concentration of

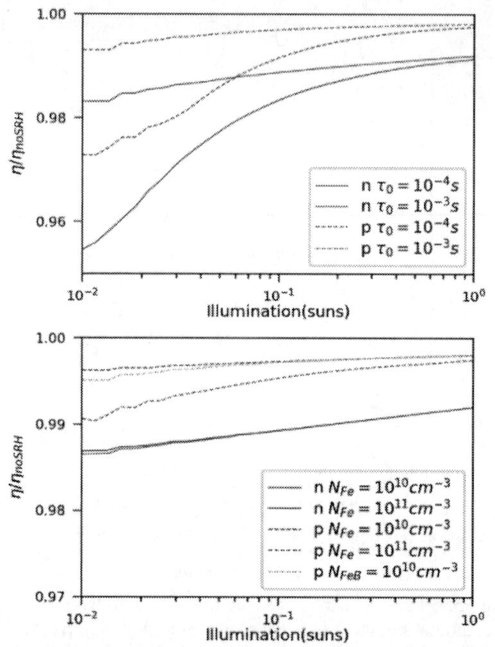

Fig. 6: Effect on limiting efficiencies below one sun of SRH recombination in the depletion region included for a single mid-gap trap with varying lifetime (top) or Fe and FeB in varying concentrations (bottom). Results are shown as a fraction of p-type Auger limit, so that changes are visible despite much larger differences between illumination levels.

$2x10^{10}$ cm-3 is included. As these different factors had opposing effects on the ideal cell thicknesses, including all of them will require a balance in losses between them.

Fig. 8 shows the calculated efficiency limits under these conditions, along with the Auger limits calculated previously. The change in preferred dopant type can be seen here at around .05 suns, with p-type limits being higher than n-type at lower illuminations, and n-type limits being higher at higher illuminations. This is below the relevant level for standard outdoor installations, but may be relevant for indoor or other specialty applications. Furthermore, this crossing point may vary with defect profiles that differ from the case considered here.

Limits here are also calculated for the additional spectra and shown in Fig. 8. Additionally, experimental results from a heterojunction cell measured at varying illuminations are shown, as well as data points representing the few such measurements in the literature [4,5]. None of these are record efficiency devices, so it is the trend with illumination that is of interest. The experimental results show less loss with reduced illumination than the calculated limits. This may be due to series resistance losses, which are not included here, but which are reduced at low illuminations. It indicates that devices approaching limiting efficiencies at one sun should also be capable of doing so at lower illuminations.

4 CONCLUSIONS

Auger limits at low illuminations for n- and p- type wafer-based silicon solar cells are determined, and the

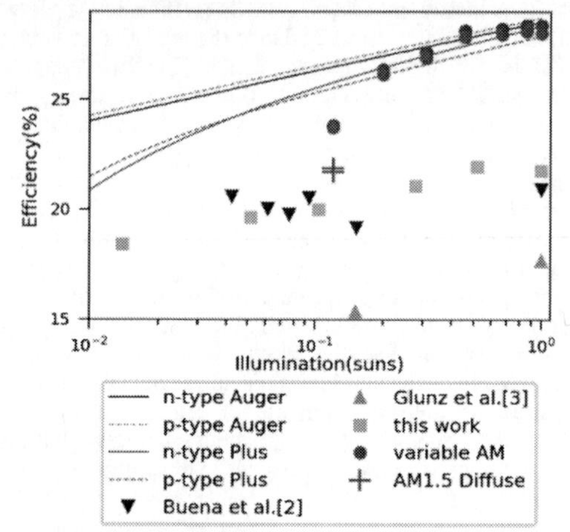

Fig. 7: Limiting efficiencies of n- and p-type wafer based silicon solar cells for a variety of spectra below one sun. Auger limits, and auger limits plus all non-idealities explored in this work are shown, as are experimental results from our work and the literature for comparison.

effects of surface recombination, shunt resistance, and SRH recombination are show to be fairly consistent across illumination levels found in outdoor installations (albeit with better results for very low surface recombination near 1 sun). Losses due to these factors are increased at lower illuminations, but still result in limiting efficiencies that compare favourably with other technologies. It is interesting to note that illumination level may determine the preferred doping type for the base. Experimental results from the literature indicate that cells can approach their limiting efficiencies at low illuminations, potentially even more closely than they do at one sun. Therefore, especially for the levels of illumination encountered in outdoor operation in non-ideal conditions, crystalline silicon should not be bypassed for low-illumination applications.

ACKNOWLEDGEMENTS

This work was supported by the Czech Ministry of Education, Youth and Sports project Centre of Advanced Photovoltaics CZ.02.1.01/0.0/0.0/15_003/0000464

The Si-HJ results presented as "this work" were fabricated and measured at EPFL, with particular assistance from Luca Antognini and Mathieu Boccard.

REFERENCES

[1] N. Reiners and U. Blieske, "Good Modules for Bad Weather," *IEEE 2016 International Energy and Sustainability Conference,* 2016.

[2] M. A. Green,, Y. Hishikawa, E. D. Dunlop, ,D.H. Levi, J. Hohl-Ebinger, A.W.Y Ho-Baillie "Solar cell efficiency tables (version 51)," *Progress in Photovoltaics,* vol. 26, pp. 3-12, December 2017.

[3] K. Ounadjela, P. Rakotoniaina, O. Sidelkheir, O. Laparra, M. Kaes, R. Smith, S. Rummel, and M. Al-Jassim "Superior Low-Light-Level Performance of Upgraded Metallurgical-Grade Silicon Modules" *2012 38th IEEE PVSC* 2012.

[4] G.E. Bunea, K.E. Wilson, Y. Meydbray, M.P.Campbell, and D.M. De Ceuster., "Low Light Performance of Mono-Crystalline Silicon Solar Cells" *2006 IEEE 4th World Conference on PV Energy Conversion,* 2006.
[5] S.W. Glunz, J. Dicker, M.Esterle, M. Hemle, J. Isenberg, F.J. Karnerewerd, J. Knobloch, D. Kray, A. Leimenstoll, F. Lutz, D. Osswald, R. Prau, S. Rein, E. Schaffer, C. Schetter, H. Schmidhuber, H. Schmidt, M. Steuder, C. Vorgrimier and G. Wilenke "High-Efficiency Silicon Solar Cells for Low-Illumination Applications" *2002 29th IEEE PVSC* 2002.
[6] A. Halm, P. Hering, C. Comparotto, L.J. Koduvelikulathu, V.D. Mihailetchi, E. Wefringhaus, and R. Kopecek. "Low Light Intensity Performance of N- and P- type Silicon Solar Cells with Different Architectures" 2011.
[7] M.J. Kerr, A. Cuevas, and P. Campbell "Limiting Efficiency of Crystalline Silicon Solar Cells Due to Coulomb-Enhanced Auger Recombination" *Progress in Photovoltaics.* vol. 11. pp. 97-104. 2003
[8] NREL Simple Model of the Atmospheric Radiative Transfer of Sunshine (SMARTS) 2.9.2 https://www.nrel.gov/rredc/smarts
[9] A.B. Sproul "Dimensionless solution of the equation describing the effect of surface recombination on carrier decay in semiconductors" *J. Applied Physics,* vol. 76, p 2851-4, September 1994.
[10] A.A. Istratov, H. Hieslmair, and E.R. Weber, "Iron and its complexes in silicon", *Applied Physics A (Materials Science & Processing).* vol. 69. July 1999
[11] D.Macdonald and L.J. Geerlings "Recombination activity of interstitial iron and other transition metal point defects in p -and n -type crystalline silicon" *Applied Physics Letters.* Vol. 85 pp. 4061-3. Nov 2004
[12] K.R. McIntosh, P.P. Altermatt and G. Heiser "Depletion-region recombination in silicon solar cells: When does mDR = 2?" *16th EUPVSEC* June 2000
[13] R.S. Bonilla, C. Reichel, M. Hermle, and P.R. Wilshaw "Extremely low surface recombination in 1 Ohm cm n-type monocrystalline silicon" *Phys. Status Solidi RRL.* vol. 11. 1600307. Dec 2016.

SERIES RESISTANCE BREAKDOWN OF SILICON HETEROJUNCTION SOLAR CELLS PRODUCED ON CEA-INES PILOT LINE

L. Basset[1], W. Favre[1], D.Muñoz[1], J.P. Vilcot[2]
[1] Université Grenoble Alpes, CEA, LITEN, INES, F-73375 Le Bourget du Lac, France
[2] CNRS – IEMN, 59650 Villeneuve-d'Ascq, France

Leo.basset@cea.fr
+33479792716

ABSTRACT: We present an analysis of series resistance losses in a 22% efficiency rear emitter bifacial silicon heterojunction (SHJ) solar cell fabricated in the pilot-line of CEA-INES. This analysis is carried out using a well known model [1] but adapted to SHJ cells to take into account considerations of a parallel connection of front TCO and silicon bulk for lateral transport [2], as well as interface resistances between the layers of the SHJ cells [3]. The effect of illumination on the silicon bulk resistivity under working conditions is also taken into account. The model allows for a diagnostic of the cell to give insights on further optimizations.
Our results suggest the biggest part of the resistive losses in SHJ cells stems from the hydrogenated amorphous silicon layers and their interfaces with indium tin oxide and crystalline silicon. This highlights the necessity of interface engineering to improve cell efficiency.
Keywords: Silicon, Heterojunction, Characterisation

1 INTRODUCTION

Silicon heterojunction (SHJ) solar cells are focusing attention as a serious alternative to the mainstream homojunction cell technologies since they demonstrated the best efficiencies for silicon based solar cells, up to 26.7% with back contact design [4].

In the CEA-INES pilot line, bifacial SHJ cells can be fabricated using industrial tools, and record efficiencies of 23.08% and 23.7% were reached in the six busbars and busbarless configurations respectively, on 244.3cm² area.

It is known that several root causes can affect the SHJ cells efficiency: (i) optical losses due to light reflection on the metallizations and parasitic absorption in the thin films layers at the front side [5], (ii) electrical losses due to carrier recombination at both interfaces and in the crystalline silicon bulk and to resistive effects. In this paper, we present a deep analysis of series resistance losses of a 22% rear emitter bifacial SHJ device.

2 SERIES RESISTANCE AND IMPACT ON SHJ CELLS EFFICIENCY

The fill factor (FF) of silicon solar cells is known to be limited by a fair amount by series resistance (Rs) losses. In SHJ cells, the FF losses due to Rs (difference between pseudo-FF (pFF) and FF [6]) is frequently reported to be of several percent [3], [5], [7]. As can be seen on Table 1, this is the case for the studied cell. The cell's total Rs can be extracted using the comparison between the SunsVoc [6] curve and the 1-Sun JV curve [8].

The effect of Rs on conversion efficiency can be approximately assessed by making the assumption that the maximum power point doesn't shift between the J-V and the SunsVoc curves. This gives the expression:

$$\Delta\eta(R_S) = \frac{(pFF - FF) * J_{SC} * V_{OC}}{P_{in}}$$

Where J_{sc} and V_{oc} are the short-circuit current density and open-circuit voltage, and P_{in} is the incident power.

For the studied cell, pFF-FF is approximately 4%, this leads to an approximate 1.1% efficiency loss due to Rs. J-V curves are shown in Figure 1.

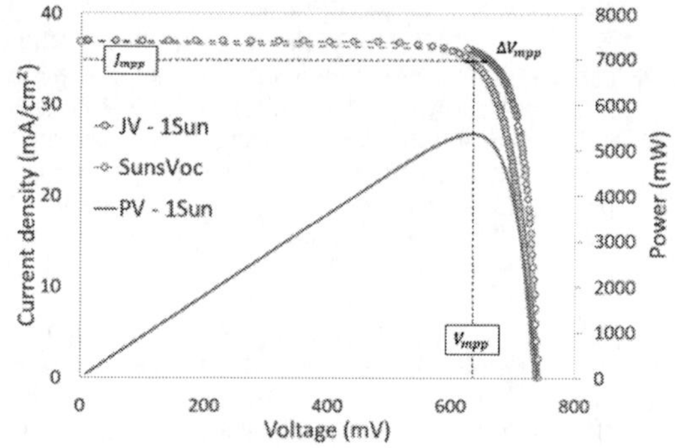

Figure 1 : Current density / Voltage (dotted lines) and Power / Voltage (straigh lines) curves of the studied cell. The blue curves are measured under standard 1 Sun 25°C conditions, and the orange curve reprensents the SunsVoc curve[6].

Voc (mV)	740
Jsc (mA/cm²)	36.96
FF	80.38%
η	21.99%
pFF	84.40%
Rs (Ωcm²)	0.85

Table 1 : Parameters extracted from Figure 1

3 LUMPED SERIES RESISTANCE IN SHJ CELLS

3.1 Origin of series resistance in SHJ cells

Rs has a diverse origin, as it represents the average lumped resistance a charge carrier experiences through its path inside the solar cell through the bulk, the interfaces between layers, the lateral transport to the grid, the contacts with the metallization grid, and the grid itself.

Figure 2 illustrates the typical path of a photo-generated carrier in the cell.

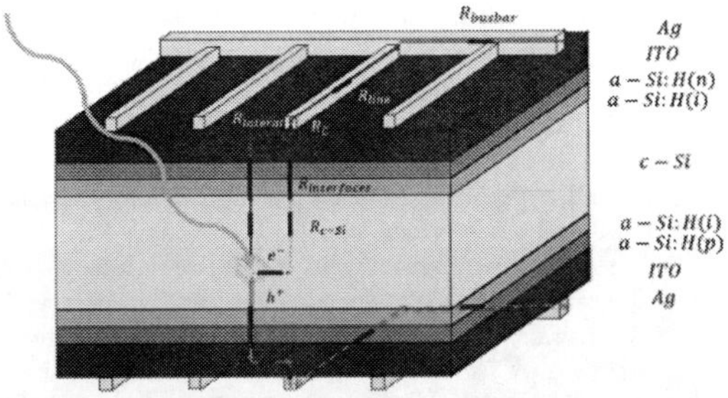

Figure 2 : Schematics of a SHJ solar cell, and path of photo-generated carriers through the different resistive mediums of the cell

3.2 Specificities of SHJ cells

Several models have been proposed in the past to assess the lumped Rs of a cell [5], [6], as an insight for cell diagnostic and optimization. However if used as is, they fail to take into account certain particularities of SHJ cells.

As Bivour et al. pointed out, in rear emitter configuration the low-to-high junction at the crystalline silicon (c-Si(n)) / hydrogenated amorphous silicon (a-Si:H(n)) interface does not prevent a parallel connection between the bulk and the Indium Tin Oxyde (ITO) layer, as is usually the case in front emitter devices [2]. When taking into account the lateral resistive loss to the front grid, one needs to use an effective value of the sheet resistance (Rsh):

$$R_{sh} = \left(\frac{1}{R_{sh,ITO}} + \frac{1}{R_{sh,c-Si}}\right)^{-1} = \left(\frac{1}{R_{sh,ITO}} + \frac{1}{\frac{\rho_{c-Si}}{t_{c-Si}}}\right)^{-1}$$

Where $R_{sh,ITO}$ and $R_{sh,c\text{-}Si}$ are the sheet resistances of the ITO and the c-Si, $\rho_{c\text{-}Si}$ and $t_{c\text{-}Si}$ are the resistivity and thickness of the c-Si.

Because an SHJ cell usually works under high injection level (typically over $2x10^{15}cm^{-3}$), the resistivity of the bulk c-Si(n) under illumination is low enough to play a great part in lateral conductivity, even for lowly doped substrates.

Another particularity of SHJ cells is that the a-Si:H layers as well as the interfaces between a-Si:H layers and ITO or c-Si(n) are substancial as to the total Rs value [3], [5], [9], and their characterization is not straightforward. The models proposed in [1], [10] don't take into account interface resistivities as they are made for homojunction cells designs, where there is usually no interface.

3.3 Lumped series resistance model

The lumped series resistance model used in this paper is mostly based on well known equations from Meier et al: contributions from the grid lines, busbars, ITO to silver screen print contact and transverse path in the absorber are calculated using the equations from [1].

However, adjustments need to be made to adapt to SHJ cells specificities, because of the aforementioned effects and to the bifaciality of the cell.

Bificiality means the metallization grid at the rear is not a full backsheet, but also has localized contacts. Both front and rear grids' contributions to Rs are calculated using the equations for busbar and gridlines contributions. The parallel connection of the front ITO and the c-Si leads to different equations to calculate front (fs) and back surfaces (bs) lateral transport:

$$R_S(lateral\ fs) = \frac{1}{12}p^2 * \left(\frac{1}{R_{Sh.ITO\ fs}} + \frac{1}{R_{Sh,c-Si}}\right)^{-1}$$

$$R_S(lateral\ bs) = \frac{1}{12}p^2 * R_{Sh.ITO\ bs}$$

Where p is the pitch of the metallization gridlines (usually p differs at the front and back surfaces).

Interface resistances at the front and back are taken into consideration directly using the interface resistivity ($\rho_{interface}$) because the interface is all across the cell. More details will be given in the following paragraph.

$$R_S\ (interface)[\Omega cm^2] = \rho_{interface}$$

The calculated lumped Rs corrresponds to the sum of each of the mentionned contributions.

4 EXPERIMENTAL DETAILS

Using the model presented above implies to know various inputs. The c-Si d resistivity is measured in dark condition while its variation with injection level is assessed using the mobility models from PV Lighthouse [11]. Line and busbar resistance per unit length can be extracted easily using commercially available tools (GP Solar's 4-Test for instance).

4.1 Measuring contact resistivity & ITO sheet resistance

Transfer Length Measurement (TLM) [12] can be carried out quite easily, for example using tools such as GP Solar's 4-Test on a cut strip from a finished cell [13]. However, this method applied on the front of an SHJ cell doesn't allow a precise extraction of Rsh and ρ_c as the assumptions of TLM theory are not all valid. Because of the previously mentioned parallel connection of the bulk and front ITO, the sample can't be considered a one-layer thin film, and two-layers models for TLM are not straightforward to carry out [14]. The ITO sheet resistance is likely to be underestimated, leading to a wrong determination of ρ_c.

To correct this effect cells with a thick a-Si:H(i) layer of about 50nm at the front interface were fabricated. The intrinsic layer prevents the current to flow in the c-Si below the ITO during a TLM measurement carried out at the front surface, this test vehicle being quite close to the typical SHJ device under investigation. These samples also allow the extraction of sheet resistance of the ITO by 4-probes measurements.

4.2 Measuring interface resistivity

Interface resistivity is also measured using TLM, but it requires specific test structures such as described in [3] and depicted in Figure 3. These samples are fabricated by selective etching of the ITO, using an inkjet mask to protect ITO locally and form a TLM pattern. An Ag paste is then deposited in alignment with the ITO patterns using screen print.

Because the ITO is etched in between the electrodes, the current flows laterally through the c-Si bulk instead. Indeed, the doped and intrinsic a-Si:H have really low conductivities, so no current flows laterally in them.

When measuring such samples, the injected current goes from the Ag/ITO electrodes to the silicon bulk because conductivity of a-Si:H layers are very low.

The Rsh extracted this way is that of the c-Si and the specific contact resistivity is in fact a sum of interfaces and vertical layers resistances. For instance, the extracted front surface field (FSF) interface resistivity accounts for:

$$\rho_{i,FSF} = \rho_{Ag\backslash ITO} + R(\mathrm{ITO}) + \rho_{\mathrm{a-Si:H(n)/ITO}}$$

$$+R(\mathrm{a-Si:H(n)}) + R(\mathrm{a-Si:H(i)}) + \rho_{\mathrm{a-Si:H(i)/c-Si}}$$

Where R(ITO), R(a-Si:H(n)) and R(a-Si:H(i)) are the transverse resistances through the ITO, a-Si:H(n) and a-Si:H(i) layers, and $\rho_{\mathrm{a-Si:H(n)/ITO}}$ as well as $\rho_{\mathrm{c-Si/a-Si:H(n)}}$ are the interface resistivities of the two successive interfaces at the front surface field.

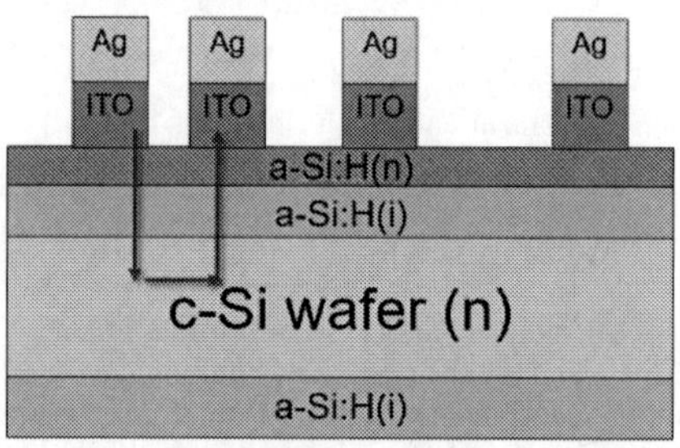

Figure 3 : emitter interface resistivity probing test sample. Red arrows represent the expected current path

For the emitter interface at the back, similar test samples need to be fabricated but with a p-type wafer to avoid the formation of a P-N junction.

5 RESULTS

The full procedure described above was applied to the studied cell. Figure 4 displays the result of the analysis.

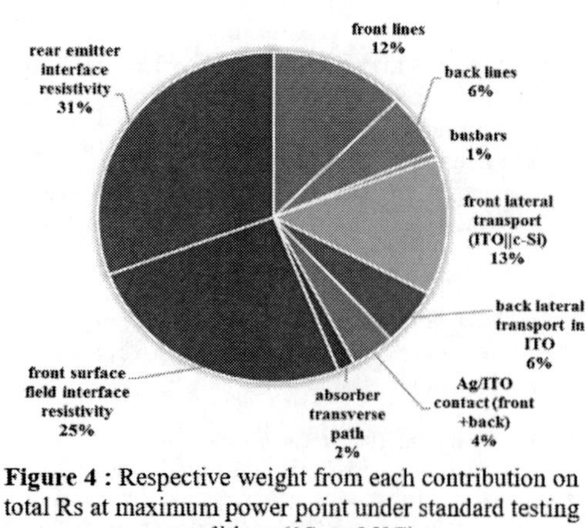

Figure 4 : Respective weight from each contribution on total Rs at maximum power point under standard testing conditions (1Sun, 25°C)

The total lumped Rs determined with the described procedure is 0.95 Ωcm², which is slightly higher than the measured 0.85 total series resistance using J-V SunsVoc comparison.

These results can also be expressed in terms of FF losses due to each contribution ($R_{S,i}$) using the following equation [7]:

$$\Delta FF(Rs_i) = J_{mp}^2 * \frac{R_{S,i}}{V_{OC}J_{SC}}$$

The total calculated FF loss is 4.1%, which again seems quite accurate as it is really close to the measured difference pFF-FF (4.0%).

These results show that the first source of resistive loss is located at the interfaces between the various layers of the cell, weighting 31% for the emitter interface and 25% for the FSF interface.

Losses from the front grid (~12%) as well as front lateral conduction (13%) are also substantial because the metal coverage can't be too high at the front surface to maintain a high Jsc.

The back grid is more restricted in terms of Rsh because it doesn't benefit from c-Si lateral conduction, but metal coverage can be higher without significant impact on Jsc considering illumination coming only from the front side. The small rear pitch of the studied cell then allows for a not too significant weight of the rear lines and lateral ITO transport (6% each).

These results highlight the fundamental role of a-Si:H stacks on overall cell resistance, asserting that interface engineering is essential to further improve the SHJ cells efficiencies. Furthermore, low resistivity substrates could be favorable if bulk lifetime is satisfactory to get less lateral resistance.

The full role of illumination in the series resistance still needs more investigation, notably the influence of the illumination on the determined interface resistivities.

Moreover, a bias is suspected in the TLM, as a significant discrepancy between the value of sheet resistance extracted with TLM and 4-point probe measurement is frequently observed, leading to error in the contact resistivity extraction. This implies a strong uncertainty in the real weight of contact resistance in Rs. This may also be the case for interface resistivities. More experimental data and numerical simulation are finally needed to strengthen our current understanding of the Rs in the SHJ device.

6 CONCLUSIONS

We proposed a method to perform a diagnostic of series resistance in SHJ solar cells, and applied it to a 22% efficiency inverted emitter SHJ cell from CEA-INES. We showed that the biggest part of the losses (56%) stems from a-Si:H layers and their interfaces with ITO and c-Si. Metallizations and lateral transport also have an important weight (19% each). The parallel connection of the bulk and the front ITO layer is of great importance for a correct assessment of lateral conduction resistive losses. Future work will focus on strengthening the model and link it more precisely to experimental data.

7 REFERENCES

[1] D. Meier *et al.*, "Determining Components of Series Resistance from Measurements on a Finished Cell," 2006, pp. 1315–1318.

[2] M. Bivour, S. Schröer, M. Hermle, and S. W. Glunz, "Silicon heterojunction rear emitter solar cells: Less restrictions on the optoelectrical

properties of front side TCOs," *Sol. Energy Mater. Sol. Cells*, vol. 122, pp. 120–129, Mar. 2014.
[3] D. Lachenal *et al.*, "Heterojunction and Passivated Contacts: A Simple Method to Extract Both n/tco and p/tco Contacts Resistivity," *Energy Procedia*, vol. 92, pp. 932–938, Aug. 2016.
[4] K. Yoshikawa *et al.*, "Exceeding conversion efficiency of 26% by heterojunction interdigitated back contact solar cell with thin film Si technology," *Sol. Energy Mater. Sol. Cells*, vol. 173, pp. 37–42, Dec. 2017.
[5] R. Gogolin *et al.*, "Analysis of Series Resistance Losses in a-Si:H/c-Si Heterojunction Solar Cells," *IEEE J. Photovolt.*, vol. 4, no. 5, pp. 1169–1176, Sep. 2014.
[6] R. A. Sinton and A. Cuevas, "A quasi-steady-state open-circuit voltage method for solar cell characterization," 2000.
[7] A. Khanna, T. Mueller, R. A. Stangl, B. Hoex, P. K. Basu, and A. G. Aberle, "A Fill Factor Loss Analysis Method for Silicon Wafer Solar Cells," *IEEE J. Photovolt.*, vol. 3, no. 4, pp. 1170–1177, Oct. 2013.
[8] D. Pysch, A. Mette, and S. W. Glunz, "A review and comparison of different methods to determine the series resistance of solar cells," *Sol. Energy Mater. Sol. Cells*, vol. 91, no. 18, pp. 1698–1706, Nov. 2007.
[9] S.-Y. Lee *et al.*, "Analysis of a-Si:H/TCO contact resistance for the Si heterojunction back-contact solar cell," *Sol. Energy Mater. Sol. Cells*, vol. 120, pp. 412–416, Jan. 2014.
[10] A. Mette, "New Concepts for Front Side Metallization of Industrial Silicon Solar Cells." 2007.
[11] "PV Lighthouse." [Online]. Available: https://www.pvlighthouse.com.au/resistivity. [Accessed: 25-Sep-2018].
[12] H. H. Berger, "Contact resistance and contact resistivity," *J. Electrochem. Soc.*, vol. 119, no. 4, pp. 507–514, 1972.
[13] S. Guo, G. Gregory, A. M. Gabor, W. V. Schoenfeld, and K. O. Davis, "Detailed investigation of TLM contact resistance measurements on crystalline silicon solar cells," *Sol. Energy*, vol. 151, pp. 163–172, Jul. 2017.
[14] R. Halevy and ר. הלוי, *A method for measuring the specific interface resistivity between two semiconductor layers and its application to a heavily doped n-type InP/GaInAs heterostructure.* Technion-Israel Institute of Technology, Faculty of Electrical Engineering, 2011.

EVALUATION OF THE LATERAL UNIFORMITY OF SOLAR SIMULATOR LIGHT FIELDS

Marko Turek*, Stefan Eiternick, Jan Bauer, Christian Hagendorf
Fraunhofer Center for Silicon Photovoltaics CSP
Otto-Eißfeldt-Straße 12, D-06120 Halle (Saale), Germany
*corresponding author: marko.turek@csp.fraunhofer.de

ABSTRACT: Besides the spectral properties and the light intensity, the lateral homogeneity of the light field of solar simulators plays an important role for the accuracy of current-voltage-measurements. We have developed a new measurement approach and corresponding sensor which can determine the lateral inhomogeneity in few milliseconds. Our sensor design relies on a new matrix circuit allowing for a very compact and robust measurement setup. Using our sensor concept, we have evaluated five different commercially available solar simulators of different technologies, e.g. Xenon-based or LED-based, regarding their light field uniformity. We find lateral non-uniformities of up to 9%. However, no correlation between the technology under consideration and the non-uniformity values has been observed. Furthermore, we find that the inhomogeneity pattern can show large differences between various measurement systems even if the overall non-uniformity value is similar. This can be of particular importance when samples of reduced size, e.g. half cells or mini-modules, are measured.
Keywords: solar simulators, uniformity, current-voltage-curves

1 INTRODUCTION

In PV industry, power measurements are performed on all solar cells and modules. On the one side, these measurements serve as an in-line quality control while on the other side, the power data is used to define the price for which each solar cell or module can be sold. Therefore, the power measurements using solar simulators are a crucial step not only in research and development but also in the PV production. A solar simulator classification defined in the IEC 60904-9 standard ensures reliable and comparable measurement conditions. In particular, it requires a quantification of four physical properties of the light field, i.e. the overall intensity, the spectral properties, the lateral uniformity over the test area, and the temporal stability [1].

Quite generally, there are two technological approaches in the design of solar simulator light sources. Solar simulators either include a single point-like light source, e.g. conventional Xenon-based systems, or multiple light sources distributed over a larger area, e.g. LED-based systems [2,3]. The LED-based solar simulators allow for a number of additional measurement applications compared to Xenon-based system, i.e. a detailed series resistance analysis [4] or a rapid quantum efficiency test [5].

In either case, the design of the light source and additional optical components, such as lenses, condensers, and apertures, determine the quality of the light field. In particular, the classification of solar simulator includes a determination of the lateral inhomogeneity of the light field according to IEC 60904-9 [1,6,7]. Thus, each measurement system has to be frequently checked concerning the lateral properties of the light field to ensure accurate current-voltage measurements which are required for a precise determination of a solar cells power. Furthermore, the quality control both in solar cell production as well as in the production of the solar simulators needs to be performed as effortless as possible to reduce maintenance and service costs.

Every power measurement using solar simulator depends sensitively on the employed reference cell or module. In particular, a lateral inhomogeneity of the light field can lead to larger measurement errors when there are design differences between the reference cell and the sample. For example, when the reference cell is much smaller than the sample, the solar cell calibration relies on a small part of the entire light field only. Another example occurs, when half-cells are measured on a system referenced to a full-cell reference. Even if reference cell and sample are of equal size, a difference in the lateral quality between reference and sample can lead to measurement errors.

In our work, we present an innovative measurement approach that includes a new sensor concept together with an adapted measurement electronics which yields the lateral inhomogeneity of a light field in a reliable way. The sensor is based laser separated solar cells [8] integrated into a new matrix concept which allows for a slim and robust sensor design. Based on this measurement approach, we have evaluated five different light sources, i.e. Xenon-based and LED-based solar simulators, regarding their lateral inhomogeneity. We found that the measured non-uniformity was as large as 9%. Furthermore, we found that the inhomogeneity pattern can be very different between two light sources which exhibit a similar overall non-uniformity value. While in some case, the light field can be adjusted, e.g. by shifting the sample or readjustment of the light source, this is not possible in other cases where many localized dark and bright spots were observed. Typical application scenarios of our measurement system include the quality control in the manufacturing process of solar simulators, the installation and maintenance of light sources, as well as the characterization of special cell types and layouts for which no adapted reference exists.

2 EXPERIMENTAL SETUP AND MEASUREMENT APPROACH

2.1 Tested light sources and solar simulators

In this work, we have set the focus on solar simulators for silicon solar cells with an illuminated test area of about 160mm x 160mm. We have analyzed five different commercially available light sources, both Xenon-based and LED-based. The solar simulators are operated in a lab environment according to R&D quality requirements which may differ from industrial standards with regard to a more variable setting of light intensity, spectrum, pulse duration, or measurement time and number of data points. The tested measurement systems are working either in steady-state mode or in flash mode:

1. Xenon technology, steady-state
2. LED (multi-color) technology, steady-state
3. Xenon technology, flash
4. LED (white), steady-state
5. Xenon technology, flash.

All solar simulators are regularly used for R&D applications while some of them can be integrated into a production line as well.

2.2 Measurement approach and sensor technology

In order to implement our measurement approach, we have setup a sensor including 64 individual solar cells obtained by a laser separation process applied to commercially available silicon solar cells. The cells have been tested to be free of any light induced degradation to ensure a stable performance of the sensor. Before integrating the cell into the sensor, each cell has been tested to be within a well-defined parameter range concerning its electrical performance. Thus, it is avoided that the laser process induces any damages on the cells. In a second step, the smaller cells have been connected by an innovative matrix scheme, see Figure 1. Finally, a cover glass is laminated to the front size to obtain a robust measurement setup.

Figure 1: Homogeneity tester "reference cell+" based on laser separated silicon solar cells connected in a matrix circuit.

Having setup the sensor, it is calibrated using a calibration cell. The calibration data set has been included into the data processing which compensates for the differences between the individual solar cells. The light intensity is obtained by measurement of the individual cell currents. The measurement electronics determines simultaneously the current of each row in less than 1 msec. The data acquisition is triggered by the sensor itself. The sensor is placed at the measurement chuck at the position of the solar cell while the measurement electronics sends the measurement data to a graphical user interface, see Figure 2. Due to the short measurement times, a temperature change within the sensor is avoided. In particular, it can be excluded that the different sensors cells operate at different temperatures.

Figure 2: Experimental setup with sensor, measurement electronics and graphical user interface.

3 RESULTS: NON-UNIFORMITY OF FIVE SOLAR SIMULATORS

The results of the non-uniformity measurements are represented a) by an overall uniformity value as defined in the norm, and b) by a 8x8 map of local non-uniformities showing the pattern of the light field, see Figure 3. For these values, we found errors of < 0.1 $\%_{abs}$ for repeated measurements and < 0.5 $\%_{abs}$ of the total error.

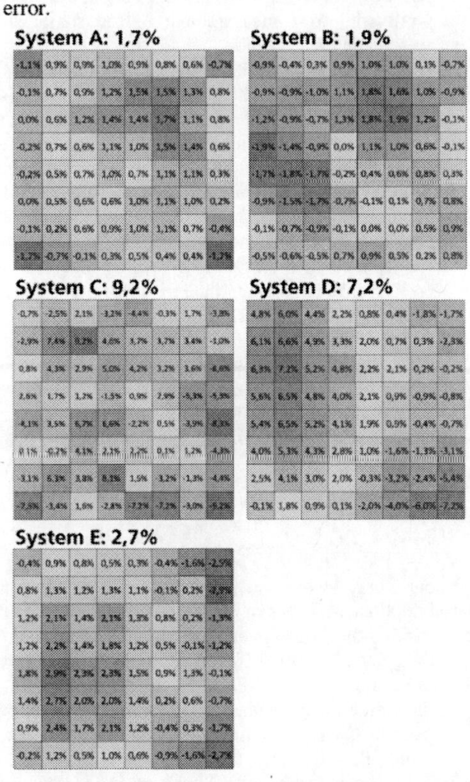

Figure 3: Non-uniformity maps for five solar simulators for solar cells.

From our results, it is evident that there is a wide spread of non-uniformity values among different solar simulator system. In our study, the overall non-uniformity varies between 2 % and 9 %. Furthermore, three different sources of non-uniformities can be classified:

a) Misaligned light field, e.g. system D: The pattern is characterized by a brighter half and a darker half. This could be compensated by better alignment of the light source with the test field.

b) Light field too small, e.g. system A or E: The pattern shows a reduced intensities mostly at the corners. Here, a better optical design might improve the homogeneity, i.e. by increasing the distance between light source and sample.
c) Intrinsic light field non-uniformity, e.g. system C: The pattern shows many local bright spots next to dark areas. Here, an improvement can only be achieved by an optical redesign or improved quality of the included optical components.

Each different non-uniformity class can lead to measurement errors. In order to illustrate possible sources of errors, the following examples are discussed:

a) Misaligned light field: When a solar simulator is calibrated using a full size reference cell, the measurement of a half cell sensitively depends on the positioning of the half cell. For example, placing a half cell on the left side of system D would yield an reduced current of about 5% when compared to a measurement of the same cell placed on the right side.
b) Light field too small: An adjustment of the total intensity of the light field by use of a small reference cell, i.e. of size 2cm x 2cm, sensitively depends on the positioning of the small reference cell. Even an averaging of few measurement points does not represent the relevant overall intensity.
c) Intrinsic light field non-uniformity: An intensity adjustment of the light source using a reference cell of different design than the test samples, e.g. busbar-layout or lateral material inhomogeneities of the cell, leads to a misadjusted intensity resulting in unnecessary errors in the power measurement. Also, the use of a small reference cell placed at one or few position does not give a representative value for the total light field intensity.

4 SUMMARY

In our work, we have analyzed five different types of solar simulators with respect to the quality of their light fields. Among the investigated solar simulators, there are three using the established Xenon technology while two light source are based on LEDs. We found overall non-uniformity values as defined in the IEC 60904-9 standard ranging from 2% to 9%. A correlation between these values and the lighting technology could not be established. Either technology, Xenon or LED, can lead to very uniform light fields. Going beyond the overall quantification of the non-uniformity by a single number, we have studied the light field patterns in more detail. Here, three sources of non-uniformities could be classified: misaligned light fields, light fields that are too small, and lights fields with an intrinsic non-uniformity due to insufficient quality of the optical components.

All measurements presented in this work were performed using a newly developed sensor system "reference cell+". This measurement setup including a compact and robust sensor, the measurement electronics and a graphical user interface provides all relevant data in few milli-seconds measurement time while being easy to use. The sensor can be adjusted to different sizes of light fields and customer specific components, e.g. for the cover glass. Thus, its fields of application lie in PV production facilities to ensure the IEC 60904-9 requirements, in research labs to avoid unnecessary errors in power measurements, and as a quality control tool for solar simulator manufacturers. In further applications, a module-sized sensor will be developed and tested for PV-module solar simulators.

References

[1] IEC 60904-9: 2007 (ed. 2)

[2] S. Kohraku and K. Kurokawa, "New methods for solar cells measurement by led solar simulator," in Photovoltaic Energy Conversion, 2003. Proceedings of 3rd World Conference on, vol. 2, pp. 1977-1980, IEEE, (2003).

[3] M. Bliss, T. R. Betts, and R. Gottschalg, "An led-based photovoltaic measurement system with variable spectrum and flash speed," Solar Energy Materials and Solar Cells, vol. 93, no. 6, pp. 825-830, 2009.

[4] M. Turek, "Current and illumination dependent series resistance of solar cells", Journal of Applied Physics 115, p. 144503, (2014).

[5] T. Luka, S. Eiternick, M. Turek, „Rapid testing of external quantum efficiency using LED solar simulators", Energy Procedia 77, pp. 113-118, (2015).

[6] R.A. Sinton et al., "Critical Evaluation of the Foundations of Solar Simulator Standards", in 44th IEEE Photovoltaic Specialist Conference, Washington D.C., (2018)

[7] H.W. Wilterdink et al; "Practical Assessment of Power Rating Uncertainties for Industrial Silicon Modules"; 35th European Photovoltaic Solar Energy Conference, Brussels, 2018.

[8] S. Eiternick, K. Kaufmann, J. Schneider, M. Turek, "Loss anlaysis for laser separated solar cells", Energy Procedia 55, pp. 326 – 330, (2014).

SIMULATIONS OF OPTIMAL SOLAR CELL ARCHITECTURE AND MATERIAL PARAMETERS FOR SILICON HETEROJUNCTION CELLS ON QUASI-MONO SUBSTRATE: STRATEGIES FOR OBTAINING EFFICIENCIES OVER 20%

Yury Smirnov[1,2*], Vasilii Verbitskii[2], Ilya Nyapshaev[2], Dmitry Andronikov[2], Alexey Abramov[2], Evgeniy Terukov[2]

[1] National Research Nuclear University "MEPhI", Moscow, Russia
[2] R&D Center for Thin Film Technologies in Energetics (TFTE), St.-Petersburg, Russia

* corresponding author's email: y.smirnov@utwente.nl
now with MESA+ Institute for Nanotechnology, University of Twente, 7500 AE Enschede, The Netherlands

ABSTRACT: In a race for a cost reduction, applicability of quasi-mono silicon wafers for fabrication of high-efficiency silicon heterojunction (SHJ) devices is inspected. We present numerical simulations of influence of the bulk lifetime (τ_{eff}) on solar cell output parameters, which are verified by cell results for both front emitter (FE) and rear emitter (RE) configurations. FE outperforms RE configuration in a range of lifetime from 200 µs to 1000 µs mainly due to the enhanced short-circuit current J_{sc}, while at τ_{eff} = 1200 µs efficiencies are getting almost equal for both configurations. We conclude bulk lifetime of 1 ms as a requirement for fabrication of SHJ with efficiencies over 20% which is in good agreement with results of finished cells.

Keywords: quasi-mono silicon, heterojunction, lifetime, simulations

1 INTRODUCTION

Silicon solar cell research and production always implies a trade-off between ultimate cell efficiency and the production cost. In this respect, silicon substrate price and quality play a significant role. Generally, higher quality of bulk material stands for higher efficiencies, although there's a certain dependency on a cell architecture and base resistivity.

Silicon heterojunction (SHJ) cells are known for their high photoconversion efficiencies with recent advances in the 2-side contacted cell architecture exceeding 25% [1]. This is achieved by obtaining high open-circuit voltage (V_{oc}) values provided by low interface defect density which is attributed to the combination of substrate surface cleaning followed by amorphous silicon (a-Si:H) passivating layers deposition. However, high bulk lifetime of the wafers is required to achieve impressive V_{oc}. Thus, a common approach is thinning down initial substrates of high quality Czochralski (Cz) and Float-Zone (FZ) mono-Si wafers. Although, another option of replacing the substrate material with cheaper quasi-mono Si was recently proved to be really promising with impressive achievements of efficiencies above 21,5% for SHJ cells on n-type quasi-mono silicon [2].

For this material, bulk recombination is limiting the performance of the cells. Our previous research has favored rear emitter configuration for high quality Czochralski (Cz) substrates [3]. Applicability of quasi-mono silicon wafers for high-efficiency devices fabrication and preferable device architecture is inspected by numerical simulations of the bulk lifetime's influence on solar cell output parameters. The model is optimized for the layers developed at R&D Center of Thin Film Technologies in Energetics (TFTE).

2 EXPERIMENTAL DETAILS & SIMULATION PARAMETERS

2.1 Experimental details

The SHJ solar cells used in this study had bifacial design. Randomly textured in KOH 6-inch, 180-µm-thick n-type (4 Ω·cm) quasi-mono Si wafers were used for solar cells fabrication. After wafer cleaning, intrinsic and doped a-Si:H layers were deposited on both sides of the substrates by 40.68 MHz (VHF) PECVD. ITO layers were deposited by DC sputtering on top of a-Si layers. An industrial screen printer and a calibrated GridTOUCH system were employed for printing the busbar less contact grid using a low-temperature Ag-based paste with subsequent curing. Current-voltage (I-V) characteristics of produced cells were measured with a solar simulator under standard conditions (AM1.5g, 1000 W m^{-2}, 25 °C). The scheme of the finished device is shown in Figure 1. The doping of front and rear a-Si:H layers depends on a cell configuration, both FE and RE architectures were subjects of investigation.

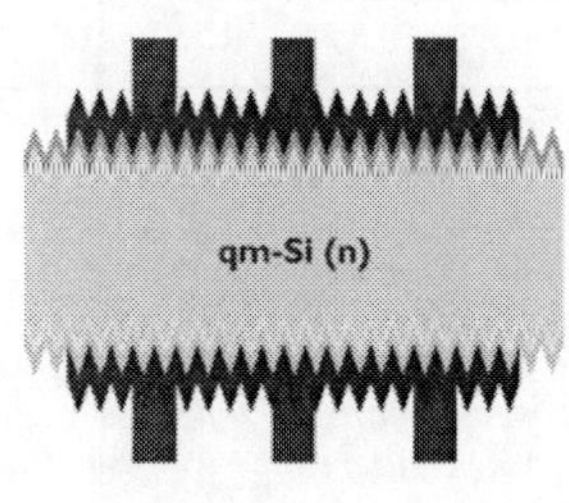

Figure 1: Cross-section of SHJ solar cell

2.2 Simulation parameters

In this study cell performance was simulated using the Quokka extension for MATLAB [4] utilizing quasi-neutrality and conductive boundaries as simplifications to the general semiconductor carrier transport model. Basic physical input parameters such as intrinsic carrier density, bandgap narrowing, Auger and radiative recombination parameters are taken from [5].

Quasi-steady state photoconductance (QSSPC) measurements were used to obtain the effective lifetime of minority carriers and extract the surface recombination velocity. Transfer Length Measurements (TLM) were performed to estimate contact resistances of layers (including a-Si:H/ITO and Ag fingers independently). Minority-carrier effective diffusion lengths have been extracted from a plot of internal quantum efficiency

(IQE) versus photon penetration depth in c-Si in the wavelength regime between 900 and 1000 nm [6]. The described set of values is summarized in Table I.

Table I: Main input parameters used for simulations

Parameter	Value	Estimation
Bulk resistivity (n-type)	4 Ω·cm	QSSPC
SRH lifetime, μs	400	QSSPC
Surface recombination velocity, cm/s	300	QSSPC
Effective diffusion length, mm	0,1	IQE
Lumped R_{series}, Ω·cm2	0,35	TLM
R_{shunt}, Ω·cm2	10^5	IV curve

The developed model of a SHJ solar cell utilizing the measured parameters was used to extract main solar cell characteristics. Comparison of the simulated and measured I-V curve for the fabricated devices is given in Figure 2.

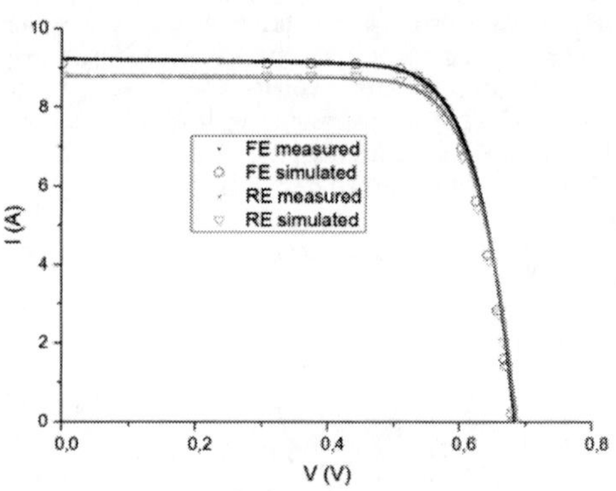

Figure 2: Simulated (open symbols) and measured I-V curves of SHJ cell on quasi-mono substrates

Obtained photovoltaic parameters for front and rear emitter SHJ cells were compared with the model: the results are given in Table II.

Table II: Photovoltaic parameters of simulated and fabricated cells on quasi-mono wafers for rear and front emitter configuration

	Isc, A	Voc, V	FF, %	Eta, %
Measured RE	8,79	0,687	75,85	18,75
Simulated	8,81	0,684	76,10	18,90
Measured FE	9,22	0,683	74,64	19,48
Simulated	9,08	0,685	75,50	19,40

Excellent agreement between experimental data and simulations was achieved proving the relevance of input parameters.

3 RESULTS AND DISCUSSION

For as grown quasi-mono Si wafers with bulk lifetime of 400 μs FE cells outperforms devices in RE architecture. In order to define a critical SRH lifetime, for which the same trend would be observed, bulk lifetime of the substrate was varied from 200 to 1200 μs. Its impact on cell perfomance is given in Figure 3.

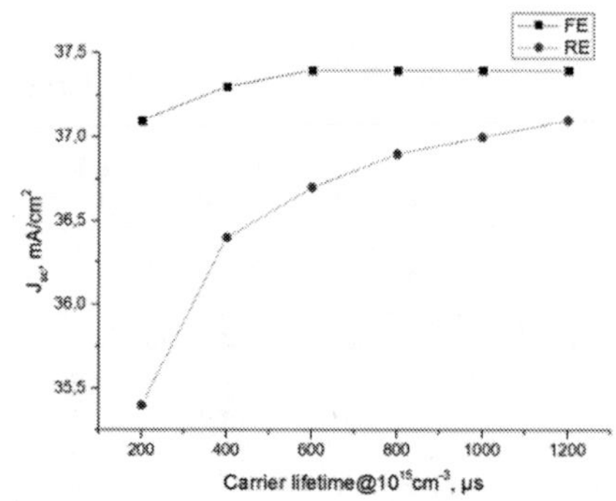

a)

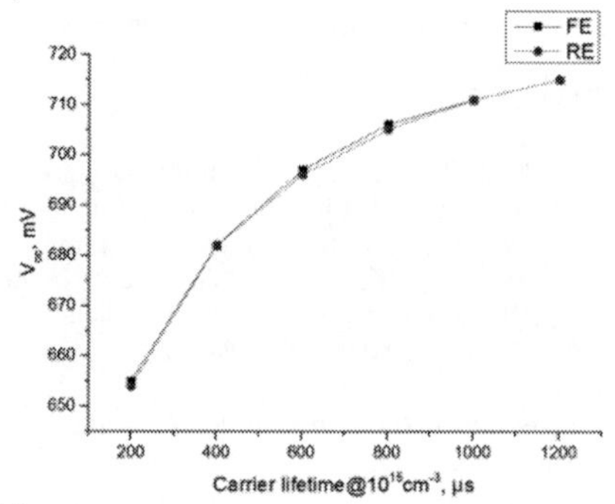

b)

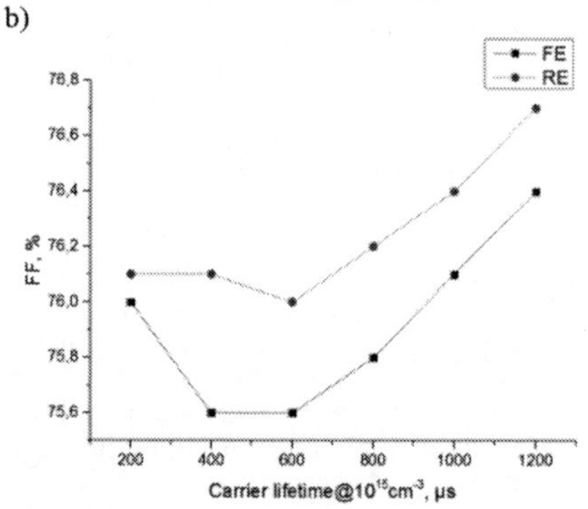

c)

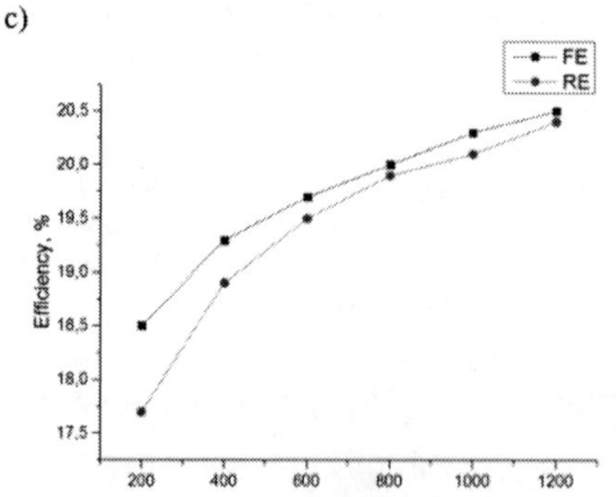

d)

Figure 3: Simulation results for main cell parameters: Jsc (a), Voc (b), FF (c), Efficiency (d) as a function of Shockley–Read–Hall lifetime

As seen from the Figure 2, FE outperforms RE configuration in a range of lifetimes from 200 μs to 1000 μs which may be mainly attributed to the enhanced Jsc, while at τ_{SRH} = 900 μs the efficiencies for both type of cells are reaching almost equal values (~19.8%). Moreover, we observe critical value of lifetime of 1 ms as a requirement for obtaining SHJ cells with efficiency over 20 % (~20,1/20,2 % for RE/FE respectively).

The quality of the substrates was improved by gettering process to obtain lifetime higher than 800 μs. Characteristics of solar cells fabricated on substrates that underwent additional treatment was compared to the predicted values. The simulated and measured cell parameters are given in Table III.

Table III: Photovoltaic parameters of simulated and fabricated cells in FE configuration on quasi-mono wafers with improved lifetime

	τ_{eff}, μs	J_{sc}, mA/cm^2	V_{oc},V	FF, %	Eta, %
Measured	884	37,6	0,705	74,90	19,75
Simulated	800	37,4	0,706	75,80	20,00
Measured	1024	37,6	0,710	75,67	20,12
Simulated	1000	37,3	0,715	76,10	20,30

As expected, improved gettering process providing required passivation quality led to the fabrication of cells with efficiencies over 20% with the best result of 20,12% on full wafer scale.

4 CONLCUSIONS

We have developed the simulation model for defining cell output parameters as a function of bulk lifetime for front and rear emitter configuration. Bulk lifetime of 1 ms is required for fabrication of SHJ solar cells with efficiencies over 20%. Although, the results are relevant for lower bulk lifetimes, since for τ_{SRH} > 1,2 ms, Auger becomes the dominant recombination mechanism and additional simulations with varying wafer resistivity are required. Improved quality of the bulk allowed to obtain high-quality SHJ cells on quasi-mono substrates with efficiencies over 20% on pseudo-square 6" wafers.

5 REFERENCES

[1] D. Adachi, J. L. Hernández, and K. Yamamoto. Applied Physics Letters, 107(23):233506 (2015).

[2] F. Jay et al. Solar Energy Materials and Solar Cells 130 (2014): 690-695.

[3] D. Andronikov et. al. Proceedings of the 33rd EU PVSEC, 732–735 (2017)

[4] A. Fell. IEEE Transactions on Electron Devices 60.2 (2013): 733-738.

[5] A. Fell. IEEE Journal of Photovoltaics 5.4 (2015): 1250-1263

[6] P. Basore , Proccedings of 23rd IEEE Photovoltaic Specialists Conference, pp. 147–152.

INLINE WAFER IDENTIFICATION USING OPTICAL CHARACTER RECOGNITION (OCR)

S. Al-Hajjawi, T. Hammer, J. Haunschild
Fraunhofer Institute for Solar Energy Systems (ISE), Heidenhofstr. 2, 79110 Freiburg, Germany
Contact email: saed.al-hajjawi@ise.fraunhofer.de

ABSTRACT: In order to connect the information from incoming inspection and IV-characterization or even crystallization information, wafers need to be identified and tracked. Identifying each wafer helps to provide a better analysis of the influence or failures of production processing steps on each individual wafer rather than a general statistical analysis approach of broader number of wafers. Nowadays, several wafer identification methods are available to simplify the tracking of wafers. In this work, we present an approach of exploiting the already integrated cameras within the inline characterisation equipment for automatic detection and recognition of wafer IDs. To do so, Optical Character Recognition (OCR) is applied through adapting Tesseract open source software with optimised pre-processed images. Our approach presents an identifier of wafers which don't have BSCs or DMCs but alphanumerical wafer IDs. Before using our algorithm, (alpha)numerical wafer labeling is required. For numerical wafer IDs in best cases, an average ID recognition rate of 99% was achieved. For alphanumerical wafer IDs, automatic software correction is highly relevant to solve the confusion of software when switching from numbers to letters and vice versa. A comparison between wafer ID reading techniques revealed that the best average wafer ID reading accuracy of 98% was obtained using brick slice code reader. The data matrix code reader revealed an average wafer ID accuracy of 93%. Finally, our OCR-based ID reader software had an average wafer ID reading accuracy of 85% for alphanumeric IDs.
Keywords: characterization, software, tracking

1 INTRODUCTION

With the rapid advancement of photovoltaic production, quality control is becoming more significant for research lines as well as for industrial applications. An economical evaluation of capital expenditure (CAPEX) spending for inline metrology and its relative justification with respect to the improvement of cells efficiency is presented in [1]. Single wafer tracking helps to observe the manufacturing steps and hence contributes to technological evaluation and increases the competition.

To date, several wafer labelling techniques for wafer identification are used. We mention here four state of art wafer labeling techniques: Data Matrix Coding (DMC), Alphanumerical wafer labeling, Brick Slice Code (BSC) and wafer ID recognition through grain boundaries. We present a new method by optimizing the utilization of Optical Character Recognition (OCR) to read alphanumerical wafer IDs automatically from the generated images during inline entry characterization measurements; providing probably a way of comparison between this method and the other wafer identification methods.

OCR might not be an ideal alternative for standard identification methods. However, its core advantage lies in the fact that it does not require extra hardware reader but only the already existing cameras along with alphanumerical labeled wafers. Apart from the aforementioned ID matching benefit of using OCR, it functions as a good alternative for industrial applications, where brick slice codes and data matrix codes are not widely used. In this case, OCR functions as wafer ID tracker by detecting and recognizing the wafer ID printed on its surface using the cameras installed within the inline characterization equipment.

2 WAFER LABELING AND ID READING TECHNIQUES

In this work, we mention four wafer labeling and ID reading techniques; pointing on advantages and disadvantages of each and presenting finally our new approach of inline identification of wafers.

2.1 Data Matrix Coding

As shown in Figure 1, IDs can be printed on wafer's surface as Data Matrix Codes (DMCs). Unlike barcode, DMC is a two dimensional code consisting of black and white cells arranged in either a square or rectangular pattern. It is currently the standard wafer identification method in industrial application and it has proved very good detection rates. In multicrystalline wafers, it has the drawback that grain boundaries can form contrasts which interfere with the code and decrease its readability. In addition to this, a loss of active area and laser-induced damage can occur [2].

In order to read DMCs, DMC readers are usually installed within the inline characterization equipment and are relevant for tracking purposes especially in cases where there are no brick slice codes.

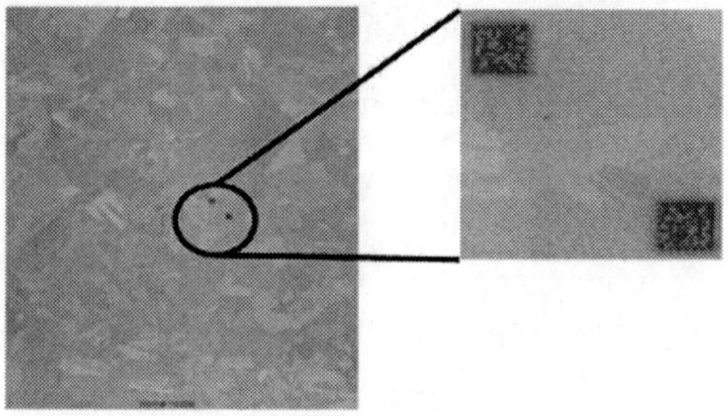

Figure 1: Data matrix coding (DMC) of wafers.

Other current encoding methods are the Quick Response (QR) coding, which differs from DMC in applications as well as in the used pattern. QR coding has a square grid in a white background where a group of black squares are arranged.

2.2 Alphanumerical Wafer Labeling

The alphanumerical labeling format depends on the standards of each manufacturer; see Fig. 2. The ID could contain the year of manufacturing, the experiment number, the wafer ID....etc.). This Labeling technique has the advantage that it is inline, human readable, ideal for research or small experiments. However, it has the same drawbacks as DMC in addition to its lower recognition accuracy.

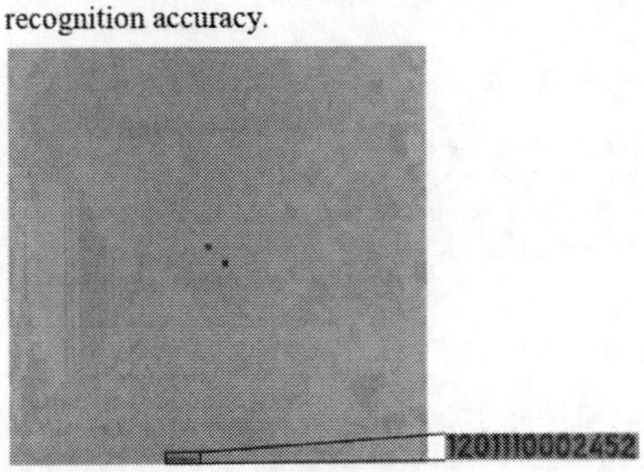

Figure 2: An example of numerical wafer labeling.

2.3 Brick Slice Code

The Brick Slice Code (BSC) grants each wafer a unique ID after slicing. Unlike other wafer labelling methods, BSC is available on a brick level. As shown in figure 3, the first code is a unique code identifying the brick; it contains the producer ID, production year and a free ID. The second code is a V-shape that identifies each wafer individually knowing exactly the position of the wafer inside the brick [3].

One of the main advantages of BSCs is that they are not printed on the surface of wafers. Henceforth, active area loss is avoided. BSCs are read using BSC readers that are usually inline installed and are capable of scanning the BSC on each wafer after slicing. Inaccuracy of detection and recognition can be a reason of dirt or clue residuals.

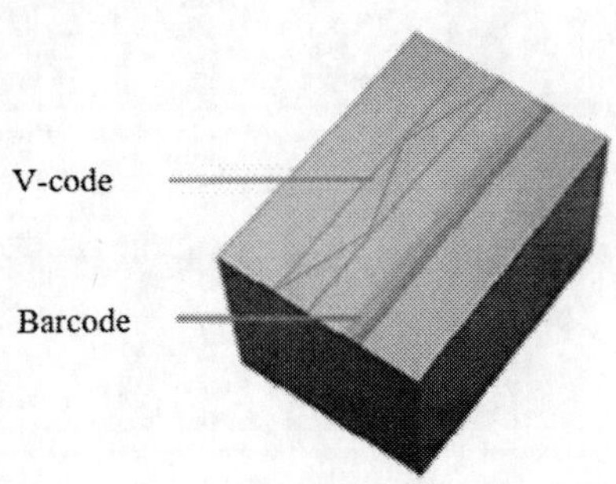

Figure 3: Hennecke Brick Slice Code (BSC) which consists of two different areas marked by laser. Taken from [3].

2.4 Grain Boundaries Detection

In this technique, the grain boundaries are first detected and recognized. As each wafer has its own grain boundary pattern. This property is used as a wafer identification method. Figure 4 shows the grain boundaries of two different multi-crystalline silicon wafers. This technique does not require extra labeling and printing hence there is no loss of active area. To the contrary, it is available only for multi-crystalline wafers, can face recognition problems with neighbor wafers having similar patterns and finally requires complex data analysis and large memory usage.

Figure 4: Grain boundaries of two different multi-crystalline silicon wafers extracted from visual images.

3 APPROACH FOR OCR READING

Our main methodology is the development of a python tool to interact between generated inline images by the Automated Wafer Inspection System HE-WI-06 from Hennecke Systems (AWIS) and the Tesseract software. OCR methods are used to transform images which contain text and are seen in the computer as pixels, into an encoded text. The idea itself belongs to HP between 1984 and 1994. After 2005, HP enabled the software online as an open source tool called Tesseract [4]. In spite of the broad utilisation of OCR in digital libraries, OCR is still far from the desired recognition efficiency. The impact of OCR errors on research in digital libraries is detailed in [5] and [6]. Different approaches are used for OCR accuracy prediction using neural networks [7]. Possible correction methods could be using noisy channel model [8] or merging two different OCR algorithms and applying post correction strategies [9].

Recently, some companies started to implement and integrate OCR within data matrix code readers, allowing by this a feasible data matrix code reading in addition to character recognition for further comparison purposes [10]. The different approaches have not yet reached the perfect accuracy.

In this work, Tesseract software is used. The detection stage starts by extracting Binary Large Objects (Blobs). Afterwards, blobs are filtered out and organized into text lines, curve fitting is done into baselines, descended line, mean line and ascender line; taking an advantage of the software capability of detecting skew lines without even de-skewing them as shown in Figure 5, hence a loss of image quality is avoided.

Figure 5: An example of the Tesseract software detection of skew lines without the necessity of de-skewing them [4].

Following, text lines are separated into characters by estimating the distance between them within words and determining whether to imply a fixed-pitch chopping or a complicated proportional word finding, the latter represents an example of our case where wafer IDs do not have a fix pitch text. Figure 6 shows an example of both cases.

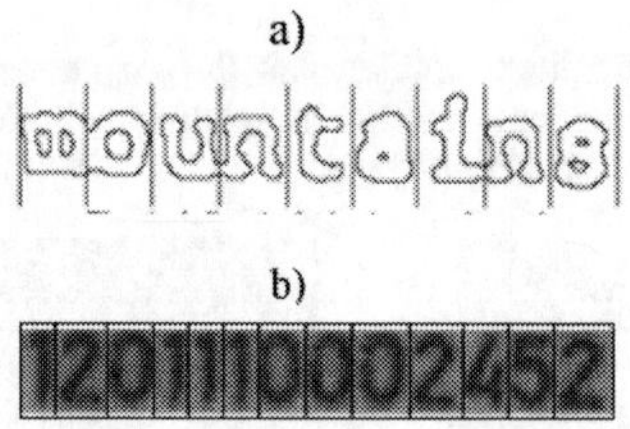

Figure 6: a) Fixed-pitch chopping [4]. b) An example of a wafer ID where complicated proportional word finding instead of fixed-pitch chopping is used.

Unlike fixed pitch text, a decision for non-fixed pitch text can only be made after word recognition using definite and fuzzy space. During the first step of recognition, improvements take place such as chopping the joined characters within a word or associating broken characters as shown in figure 7.

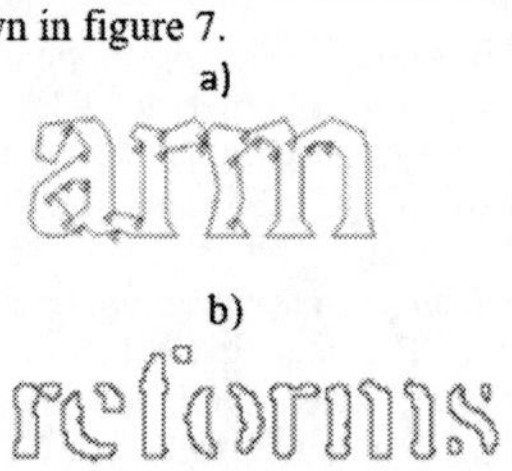

Figure 7: First steps of recognition for two cases a) chopping the joined characters within a word. b) associating broken characters [4].

Words pass then through two consecutive classifiers; the static character classifier and the adaptive classifier. In the static character classifier recognition stage, features are extracted from the outline and a many-to-one matching against the clustered prototype features of the training data is done. In the adaptive learning, a more font-sensitive classifier that is trained by the output of the static classifier and is therefore commonly used to obtain greater discrimination within each document where the number of fonts is limited.

Before implementing OCR, image scaling and smoothing techniques were utilised on good quality images, thanks to the high resolution cameras integrated within AWIS, to attain convenient input images of higher predicted recognition accuracy. However, the inline movement of wafers during image capturing, the small font sizes of wafer IDs (1-3mm) and finally the non-fixed ID pitch complicates the detection and recognition of wafer IDs using the standard Tesseract software. Therefore, we applied further image pre-processing in order to prepare better input images for the Tesseract software; see Fig. 8.

Images of the stain-detection module from the inline measurements by AWIS are used, as the wafer ID can be spotted perfectly due to the high quality cameras and the built in alignment algorithm which preserves good input images for text detection and recognition. In order to reduce the complexity of data to be analysed and focus on the area where the standard wafer ID is printed, the lower side of the wafer is cropped. Each cropped image is then scaled up to provide a better computer vision. Afterwards, image pre-processing techniques are applied including kernel filters, dilation and erosion, ending up with appropriate input images for the Tesseract software. Finally, Tesseract software is used and the results are analysed in term of average character recognition rate and average ID recognition rate.

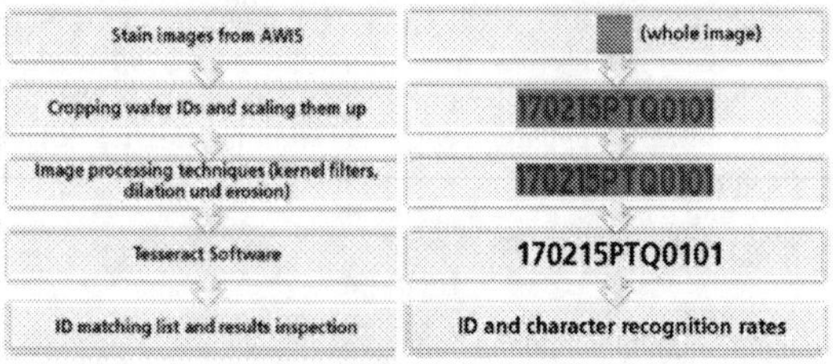

Figure 8: Two schematic diagrams representing the used approach for this work

4 RESULTS

The algorithm is tested on two sorts of wafer labelling: wafers with numerical ID's and wafers with alphanumerical ID's. In both cases, the performance is analysed in terms of processing time (seconds per wafer), average character recognition rate (%) and average ID recognition rate (%). Accordingly, the scaling factor is optimised with high average recognition rates and processing times being enhanced.

4.1 Results of wafers with numerical IDs

Two wafersets containing different characters are used. Each waferset consists of 100 wafers. Different cases were examined as follows: Full image, cropped image, and finally cropped and scaled image. The average character and ID recognition rates are used for evaluation. The results are shown in figure 9.

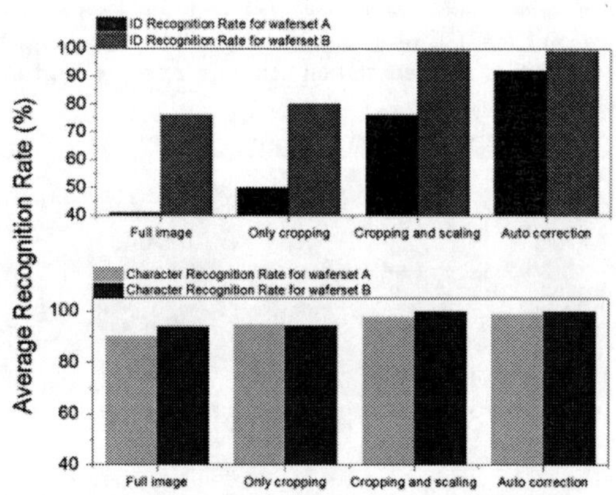

Figure 9: Average ID recognition rate (above) and average character recognition rate (below) for two wafersets with numerical ID's.

The results show a remarkable improvement of the average ID recognition rate from 41% to 92% for waferset A in addition to an improvement of the average character recognition rate from 90% up to 99%. Furthermore, the processing time is also reduced from 2.7s/wafer down to 1s/wafer for whole wafers. Waferset B showed better results where the average character recognition rate increased from 94% to 99% and average ID recognition rate from 76% to 99%. The reason behind the better results of waferset B is the problem the software revealed in detecting number "4", especially that the first part of the wafer ID for the first group is "120344", unlike waferset B which showed fewer problems as the first part of the wafer ID is "120111". In order to explain the selection of the scaling up factor, the impact of it on the average ID recognition error and the processing time is inspected and shown in figure 10. The scaling factors 4 and 5 showed the highest average ID recognition rate. However, the lower processing time when using a scaling factor of 4 instead of 5 explains our selection of this scaling factor for results in figure 9 and for further analysis.

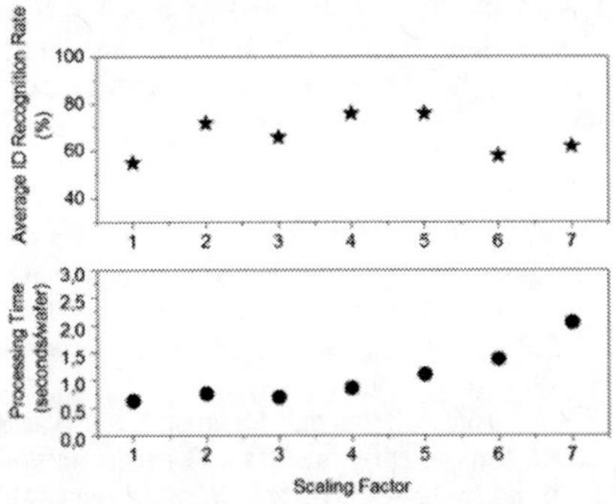

Figure 10: A graph showing the scaling factor against the average ID recognition rate (above) and processing time (below).

4.2 Results of alphanumerical wafers

An example of an alphanumeric wafer ID is "170215PTQ0102". The main problem was the confusion of the software after the detection of letters and switching back to numbers. It has been observed that the software either detects "PTQ" as "PTO" and the following numbers correctly as numbers, or it detects "PTQ" correctly but the following numbers as letters. An example of this would be the false detection of "170215PTQ2503" as "170215PTQZSO3", where the numbers following "PTQ" are detected as letters or "170215PTQ2502" as "170215PT02502"; where "Q" is detected as "0" and so on the following numbers are detected correctly.

To mitigate such a problem, the algorithm is taught to read the three letters correctly and converts automatically the following numbers into the correct corresponding numbers. Table I shows the common false detected characters for a group of wafers with alphanumerical IDs.

Table I: False detected characters and the relative applied correction.

False detected character	Automatic corrected character
PTO	PTQ
o	0
S	5
Z	2
M	13
H	11
I	1
L	1

4.3 Evaluation of wafer ID reading techniques

In order to evaluate each wafer ID reading technique, 1000 wafers were inspected and the average ID reading accuracy; see Fig. 11.

For evaluation purposes, we have taken three factors into account; (i) The average ID accuracy rate (ID was correctly read), (ii) the average wrong ID detection rate (ID was read, but is wrong) and finally (iii) the average no ID detection rate (no ID was read at all).

The brick slice code reader showed the best average wafer ID accuracy rate of 98%. The data matrix code reader revealed a lower detection rate, because some codes could not be read due to a difficult grain contrast. Finally, our ID reader software detected all wafer IDs but it had an average wafer ID accuracy rate 85%, owing to the inaccuracy of the software Tesseract especially for new font types which are not trained.

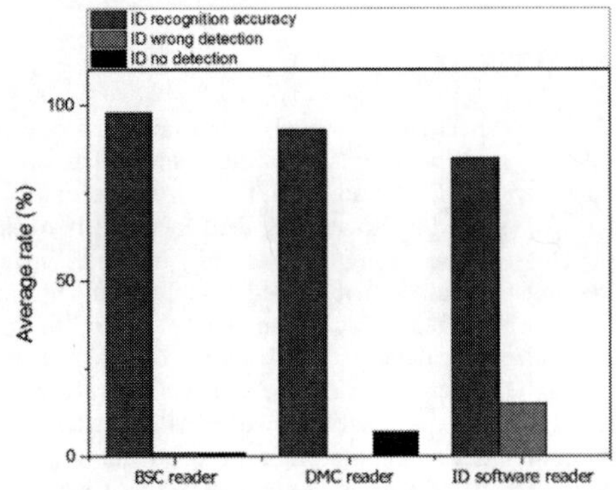

Figure 11: Average ID accuracy rate, wrong ID detection and no ID detection rates of the different wafer ID reading techniques.

4.4 Comparison between wafer ID reading techniques

After introducing the state of art of wafer labeling and reading techniques, it is relevant to sum up the advantages and disadvantages of each technique and compare them with our proposed OCR approach. Table II shows the comparison between the whole wafer labeling and ID reading techniques.

OCR could be a good alternative for the current wafer ID reading techniques; especially that it does not require extra hardware installation for ID detection and recognition. It is human readable and ideal for research or small experiments.

However it still has the drawbacks of active area loss, software false recognition and confusion between numbers and letters, and finally its lowest ID recognition accuracy when compared to DMC and BSC.

On the other hand, utilizing OCR would avoid the problems that could be faced with BSC, such as the inaccuracy of detection and recognition especially for thin wafers as well as the problems of dirt or clue residuals. Finally, taking a closer at the grain boundary wafer identification method, the drawbacks like recognition problem with neighbor wafers, complex data analysis and large memory usage and finally the availability for only multicrystalline wafers would be overcome.

Table II: Comparison between the different wafer Labelling and reading techniques.

Method	Advantages	Disadvantages
Data Matrix Code	• Industrial standard • Character recognition very accurate • readable from wafer to solar cell	• Problem with grain boundary contrast • Loss of active area • Laser induced damage
OCR for alphanumerical/ numerical detection	• It does not require extra hardware install-ation for ID detection and recognition • Human readable • Ideal for research or small experiments	• Same as DMC • Software detection and false recognition of characters • Lowest accuracy
Brick Slice Code	• No loss of active area • Available at brick level	• Inaccuracy of detection and recognition especially for thin wafers • Dirt or clue residuals are problematic
Grain boundary identification	• It Does not require extra labeling and printing • No loss of active area	• Available only for multicrystalline wafers • Recognition problem with neighbor wafers • Complex data analysis and large memory usage

5 CONCLUSION

In this paper, we presented a new methodology of detecting and recognising the numerical and alphanumerical IDs printed on the surface of silicon wafers. We started by illustrating the functionality of the tesseract software along with our implied image processing techniques that helped us to end up with a satisfying approach which can be used to further control/correct the detected BSC's by the BSC reader.

Using OCR tends to enhance an accurate tracking of wafers during manufacturing especially when the position of wafer within the brick is crucial and enables a possible detection and recognition of wafer IDs, using only the already installed cameras within the inline wafer inspection system in the absence of BSCs and data matrix readers.

OCR proved an recognition accuracy up to 85%. Tesseract software error and font size/type are the main contributors to recognition errors but most importantly, the confusion of the software between some numbers and some letters.

ACKNOWLEDGEMENTS

This work was funded by the German Federal Ministry for Economic Affairs and Energy within the project "CUT B" (contract number 0325910A).

REFERENCES

[1] J. Haunschild *et al.*, "Review of Tools and Approaches for Inline Quality Control in High Efficiency Silicon Solar Cell Production: 33rd European Photovoltaic Solar Energy Conference and Exhibition; 870-874," (eng), 2017.

[2] A. Krieg, C. Schmitt, S. Rein, S. Nübling, and P. Schneider, "Wafer Identification Using Laser Marked Data Matrix Codes: 25th European Photovoltaic Solar Energy Conference and Exhibition / 5th World Conference on Photovoltaic Energy Conversion, 6-10 September 2010, Valencia, Spain; 1337-1341," (eng), 2010.

[3] S. Leu, P. Schroll, and R. Peters, "Full Material Tracking - Using BSC (Brick-Slice-Code): 28th European Photovoltaic Solar Energy Conference and Exhibition; 1805-1806," (eng), 2013.

[4] R. Smith, "An Overview of the Tesseract OCR Engine," in *9th International Conference on Document Analysis and Recognition (ICDAR 2007): 23-26 September*, Curitiba, Parana, Brazil, 2007, pp. 629–633.

[5] G. Chiron, A. Doucet, M. Coustaty, M. Visani, and J.-P. Moreux, "Impact of OCR Errors on the Use of Digital Libraries: Towards a Better Access to Information," in *2017 ACM/IEEE Joint Conference on Digital Libraries (JCDL): June 19-23, 2017, Toronto, Ontario, Canada*, Toronto, ON, Canada, 2017, pp. 1–4.

[6] S. Kapidakis, C. Mazurek, and M. Werla, *Research and advanced technology for digital libraries: 19th International Conference on Theory and Practice of Digital Libraries, TPDL 2015, Poznań, Poland, September 14-18, 2015 : proceedings / Sarantos Kapidakis, Cezary Mazurek, Marcin Werla (eds.)*. Cham: Springer, 2015.

[7] J. GONZALEZ, J. KANAI, and T. A. NARTKER, "PREDICTION OF OCR ACCURACY USING A NEURAL NETWORK," in *Series in Machine Perception and Artificial Intelligence*, vol. 29, *Document analysis systems II*, J. J. Hull and S. L. Taylor, Eds., Singapore: WORLD SCIENTIFIC, 1998, pp. 356–370.

[8] O. Kolak and P. Resnik, "OCR error correction using a noisy channel model," in *Proceedings of the second international conference on Human Language Technology Research* -, San Diego, California, 2002, pp. 257–262.

[9] M. Volk, L. Furrer, and R. Sennrich, "Strategies for Reducing and Correcting OCR Errors," in *Theory and Applications of Natural Language Processing, Language Technology for Cultural Heritage: Selected Papers from the LaTeCH Workshop Series*, C. Sporleder, A. Bosch, and K. Zervanou, Eds., Berlin, Heidelberg: Springer Berlin Heidelberg, 2011, pp. 3–22.

[10] SensoPart Industriesensorik GmbH, "VISOR® Code Leser," Feb. 2017. [Online] Available: https://www.sensopart.com/jdownloads/Gesamtkatalog/Visor_Code_Leser.pdf. Accessed on: Aug. 02 2018.

SPECTRAL OPTICAL CHARACTERISTICS OF SILICON NANOWIRE SYSTEM: SIMULATIVE PREDICTION FOLLOWED BY EXPERIMENTS

Mohammad K. Hossain[1,*], Ayman W. Mukhaimer[1, 2] and Billel Salhi[1]
[1]Center of Research Excellence in Renewable Energy (CoRERE), King Fahd University of Petroleum and Minerals (KFUPM), Dhahran 31261, Dhahran (Kingdom of Saudi Arabia)
[2]Department of Mechanical Engineering, King Fahd University of Petroleum and Minerals (KFUPM), Dhahran 31261, Dhahran (Kingdom of Saudi Arabia)
[*]Correspondence to: kamalhossain@kfupm.edu.sa

ABSTRACT: Vertically aligned silicon nanowires (Si-NWs) facilitate efficient light trapping and thus been expected long to improve thin film based solar cell efficiency. A vapor liquid solid technique has been exercised to grow Si-NWs of coverage, *ca.* $6.5 \times 10^8/cm^2$ in chemical vapor deposition reactor. A variation in dimensions (i.e. length and diameter) of as-grown Si-NWs were further characterized through scanning electron microscope. Keeping in mind the geometry of Si-NWs, a model was developed in finite different time domain analysis to extract several optical characteristics (i.e. absorption depth profile, electromagnetic field distribution, Poynting vector, exciton generation rate distribution, etc.) that influence the efficiency of nanowire based thin film solar cell. To the extend, aforementioned optical characteristics were further analyzed spectral wise step by step ranging from near ultraviolet to near infrared. Photon absorption at 700 nm was found to be most effective in this particular Si-NWs model system. Correlated demonstration amongst absorption depth profile, electric field confinement, energy flow distribution and excitation generation rate distribution as investigated in this work opens new avenues and possibilities to improve current nanowires-based optoelectronic devices, particularly thin film solar cell.

Keywords: Photovoltaic; nanowires; absorption; excitation generation; modelling

1 INTRODUCTION

Nanotechnology-based approach such as growing nanometric wires is an emerging field of research and innovation. Efficient light harvesting through photon trapping is a long-expected strategy to implement in a variety of solar cell technologies. In optoelecronic applications, particularly in solar cell, light trapping mechanism contributes at least in three different outlooks such as increased scattering cross section, photon confinement, and incident photon energy propagation through polariton mode [1–4]. Although silicon (Si)-based solar cell dominating current solar cell market the Shockley and Queisser efficiency limit (i.e. ~31% for Si) has become a great challenge. It implies that a small portion of solar spectrum contributes to electricity generation leaving huge amount of photon energy unused. Higher and lower energy phsotons induce thermalization and absorption losses that comprises of nearly 50%. Such energy adds up several problems including heating and degradation in the cell level. Si-NWs, because of its nanometric features possesses great potential improve this efficiency constrain [5-8]. Apart from strong optical absorption, Si-NWs bears other electrical characteristics that lead to reduced production cost [9]. The contribution and vital role played by such Si-NWs can be further understood through predictive analysis of inherent optical characteristics such as absorption depth profile, electromagnetic field distribution, Poynting vector and exciton generation rate distribution. Absorption profile, energy flow, EM field distribution and exciton generation rate at 740 nm (1.8 eV for Si) in addition to spectral analysis of absorption depth profile for Si-NW system has been demonstrated previously [9-11]. However further correlated spectral analysis amongst absorption depth profile, energy flow, electric field localization and exciton generation rate distributions is indispensable.

Here in this work, we have achieved vertically aligned Si-NWs on c-Si wafer followed by further topographic confirmation through scanning electron microscope (SEM). Gold (Au) catalyzed as-fabricated nanowires were found to be in different length and diameter. Based on the observation we investigated and simulated spectral characteristics of Si-NW model, such as absorption depth profile, electromagnetic (EM) field, power flow, exciton generation rate distribution, etc. at different wavelengths of solar spectrum. The results were compared with those obtained in case of flat c-Si slab geometry. Following the understanding learned from predictive simulation, vertically grown Si nanowire was obtained as a first step to realize c-Si nanowire based solar cell for efficient light. A comparative investigation has been carried out to illustrate how an efficient light management can be obtained in Si-NW based solar cell.

2 METHOD and MATERIALS

Si-NWs on c-Si wafer were grown in four steps following the vapor liquid solid route [12-14] as shown in Fig. 1. n-type Si {1-1-1} wafer was received from Siltronix and degreased as per standard procedure. Au nanoparticles as catalyst was obtained by annealing an ultrathin layer of Au at 700 °C. Thereafter the substrate

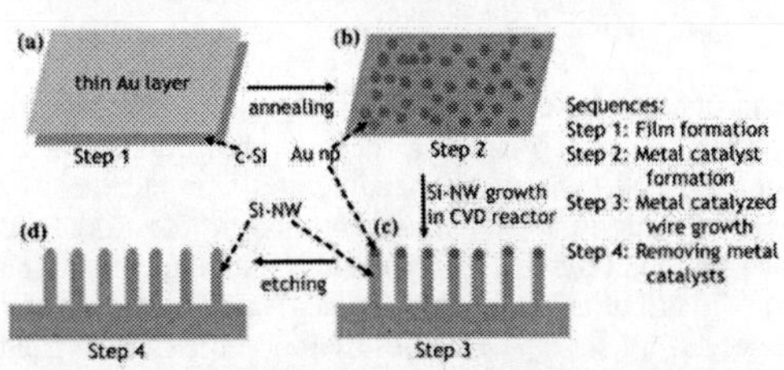

Figure 1: Free-hand schematic to 4 steps VLS process to grow Si-NWs. (a)-(b) indicates film formation, Au formation, nanowire growth and etching respectively.

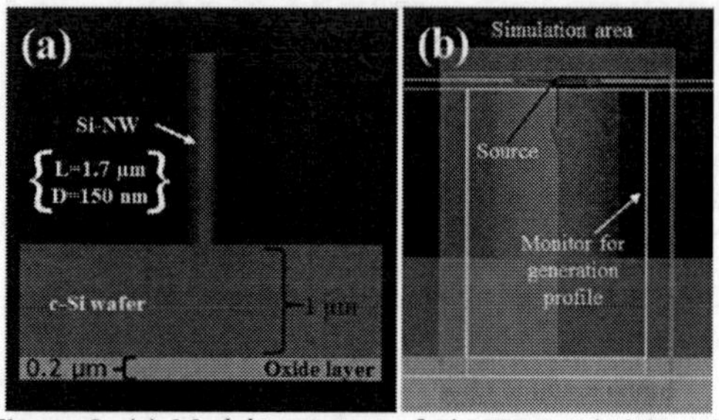

Figure 2: (a) Model geometry of Si-NW on Si slab and oxide layer, and (b) monitors for optical characteristics.

was kept in CVD for 30 mins to grow Si-NWS followed by a standard etching of the Au nanoparticles. The details were mentioned somewhere else [11]. Dimensions including length and diameter were confirmed by SEM. As shown in Fig. 2, a model of Si-NW on c-Si wafer was modelled in FDTD package of Lumerical Solution (ver 8.6). The geometry of model was kept as close as to those of as-grown Si-NWs as reported previously [9-11]. Absorption depth profile, energy transfer, EM field, and exciton generation rate distributions at different wavelengths ranging from near UV to near NIR region were extracted.

3 RESULTS and DISCUSSION

SEM observation of the as-grown Si-NWs as shown in Fig. 3 revealed that nanowires were as long as few microns length. Fig. 3a and Fig. 3b show that Si-NWs were in different length. Depending on Au nanoparticles sizes as indicated in Fig. 3b by white dotted arrow, as-prepared Si-NWs were found to show a wide range of diameters. Efficiency of solar cell is related to the photon

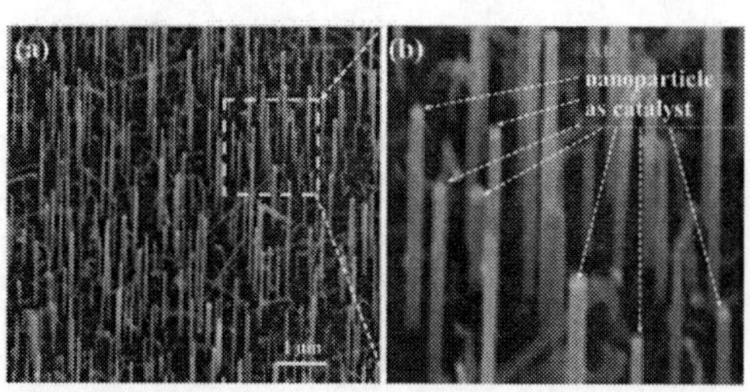

Figure 3: SEM images of (a) Si-NWs grown on c-Si and (b) magnified (2 µm × 2 µm) image of the same as marked by white rectangle showing individual Si-NWs of different length and diameter. Au nanoparticles acted as catalyst are pointed at the top of individual Si-NWs by white dotted arrows.

absorption profile of the absorbing layer. In case of nanowire based solar cell, photon absorption can be confined and enhanced depending and thus increase the light trapping in solar cell. Absorption profile, EM field distribution, Poynting energy and excitation generation rate distribution are inter-related and deterministic factors. Fig. 4a-d shows absorption profile, EM field distribution, Poynting energy and excitation generation rate distribution at solar spectrum of at near UV, 300 nm respectively. Weak distributions were observed near the edge of the nanowire. At 700 nm (i.e. absorption band gap of n-type Si wafer reported to be ca. 1.8 eV (i.e. 700 nm) solar spectrum, absorption profile, EM field

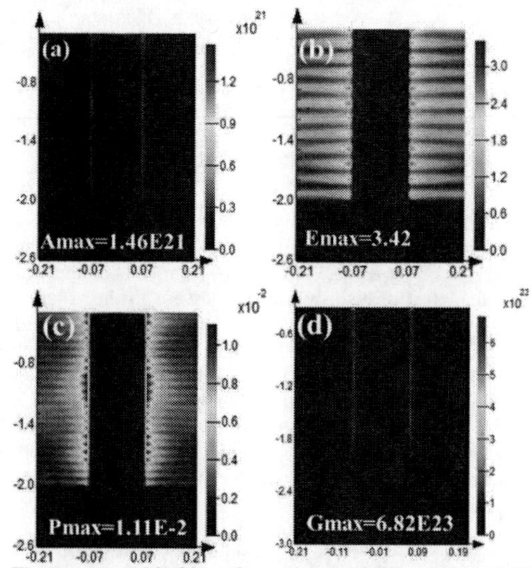

Figure 4: (a)-(b) Absorption profile, EM field distribution, Poynting energy and excitation generation rate distribution at 300 nm solar spectrum.

distribution, Poynting energy and excitation generation rate distribution were found very intense and confined throughout the nanowire as shown in Fig. 5a-d. The plausible reason is the perfect matching of the absorption band gap of n-type Si-NWs.

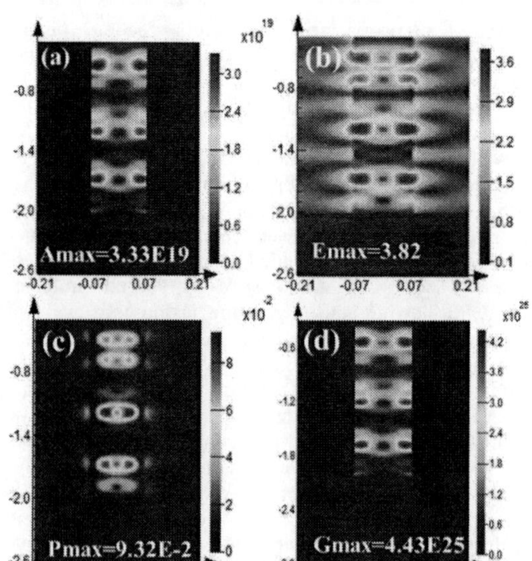

Figure 5: (a)-(b) Absorption profile, EM field distribution, Poynting energy and excitation generation rate distribution at 700 nm solar spectrum.

At near infrared region, 1100 nm absorption profile, EM field distribution, Poynting energy and excitation generation rate distribution started to reduce within the absorbing layer as shown in Fig. 6a-d. Such a correlated investigation amongst absorption profile, EM field distribution, Poynting energy and excitation generation rate distribution support to understand efficient nanometric system to increase photon trapping within the absorbing layer.

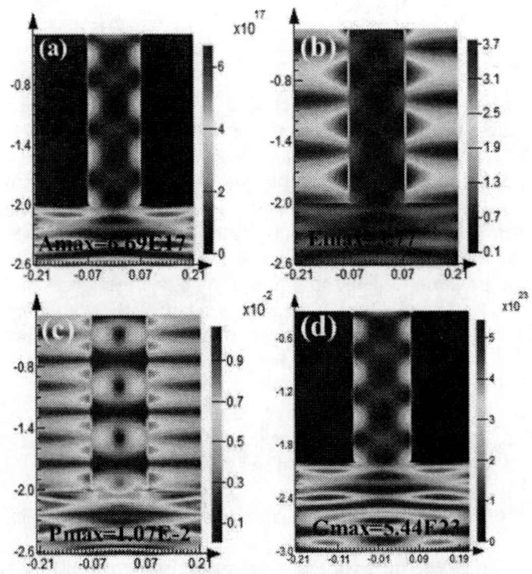

Figure 6: (a)-(b) Absorption profile, EM field distribution, Poynting energy and excitation generation rate distribution at 1100 nm solar spectrum.

4 CONCLUSIONS

A vapor liquid solid technique has been exercised to grow Si-NWs of coverage, *ca.* $6.5 \times 10^8/cm^2$ in chemical vapor deposition reactor. A variation in dimensions (i.e. length and diameter) of as-grown Si-NWs were further characterized through scanning electron microscope. Keeping in mind the geometry of Si-NWs, a model was developed in finite different time domain analysis to extract several optical characteristics (i.e. absorption depth profile, electromagnetic field distribution, Poynting vector, exciton generation rate distribution, etc.) that influence the efficiency of nanowire based thin film solar cell. To the extend, aforementioned optical characteristics were further analyzed spectral wise step by step ranging from near ultraviolet to near infrared. Photon absorption at 700 nm was found to be most effective in this particular Si-NWs model system. Correlated demonstration amongst absorption depth profile, electric field confinement, energy flow distribution and excitation generation rate distribution as investigated in this work opens new avenues and possibilities to improve current nanowires-based optoelectronic devices, particularly thin film solar cell.

5 ACKNOWLEDGEMENT

Authors acknowledge Center of Research Excellence in Renewable Energy (CoRERE), Research Institute, King Fahd University of Petroleum & Minerals, (KFUPM), Dhahran 31261, Saudi Arabia. MKH Acknowledges Deanship of Scientific Research (DSR) at King Fahd University of Petroleum & Minerals (KFUPM) for funding this work through project No. IN151003.

6 REFERENCES

1. Okamoto H, Imura K (2013) J Phys Chem Lett 4:2230–2241.
2. Catchpole KR, Polman A (2008) Opt. Express 16:21793–21800.
3. Hossain MK, Kitahama Y, Biju V et al (2009) J Phys Chem C 113:11689–11694.
4. Hossain MK, Shimada T, Kitajima M et al (2008) Langmuir 24:9241–9244.
5. Kelzenberg, M.D., Boettcher S.W., Petykiewicz J.A., Turner-Evans D.B., Putnam M.C., Warren E.L., Spurgeon J.M., Briggs R.M., Lewis N.S., Atwater H.A. (2010), Nat Mater 9:239–244
6. Polman A., Atwater H.A. (2012), Nat Mater 11:174–177.
7. Yu L.W., Misra S., Wang J.Z., Qian S.Y., Foldyna M., Xu J., Shi Y., Johnson E., Cabarrocas P.R. (2014), Sci Rep 4:4357–4363.
8. Krogstrup P., Jorgensen H.I., Heiss M., Demichel O., Holm J.V., Aagesen M., Nygard J., Morral A.F. (2013).
9. Salhi B., Hossain M.K., Mukhaimer, A.W., AlSulaiman, F.A. (2016), J Electrocer 37: 34–49.
10. Hossain M.K., Salhi B., Mukhaimer A.W. (2016), Plasmonics 12 (5): 1297–1304.
11. Hossain M.K., Salhi B., Mukhaimer A.W., Al-Sulaiman F.A. (2016), Appl Nanosci 6:1031–1036.
12. Schmidt V, Wittemann JV, Senz Gosele U (2009) Adv Mater 21:2681–2702.
13. Suzuki H, Araki H, Tosa M, Noda T (2007), Mater Trans 48:2202–2206.
14. Wagner RS, Ellis WC (1964), Appl Phys Lett 4:89–90.

STUDY OF AMORPHOUS INCUBATION LAYER INHIBITION IN n AND p DOPED μc-Si:H THIN FILMS BY OPTICAL METHODS AND ELECTRON MICROSCOPY

F. Rojas[1], F. Villar[2], J.M. Asensi[2] and J. Bertomeu[2]
[1]Pontificia Universitad Javeriana, Dept. Física – Facultad de Ciencias
Ctra. 7 No. 40-83 (110231) Bogotá, Colombia
[2]Universitat de Barcelona, Dept. Física Aplicada
Av. Diagonal 645. E-08028 Barcelona, Spain
fe.rojas@javeriana.edu.co

ABSTRACT: In the case of doped layers for p-i-n solar cells based μc-Si:H, the control of the thickness of the amorphous incubation zone deserves special attention. Since these being very thin layers, the first stage of amorphous growth can damage the properties of the solar cell, not only electrically, but also can negatively influence through high absorbing losses. In this paper *n* and *p* type doped μc-Si:H thin films were deposited by Hot Wire Chemical Vapor Deposition at low substrate temperature ($T_S \sim 100°C$). Pre-heating of the substrates and cyclically varying hydrogen dilution (CVH) were used to inhibit or reduce the formation of the amorphous phase during the first growth stage of the thin layer. By the implementation of an optical method based on Effective Medium Approximation (EMA) in transmittance and reflectance measurements, it was possible to determine both the optical properties and microstructure in surface, bulk and the initial stage of growth of the films. The results have been correlated with those of transmission electron microscopy. The results showed reduction of incubation amorphous thin films below 10 nm range.
Keywords: *Micro Crystalline Si, Optical Properties, Microstructure*

1 INTRODUCTION

Deposition techniques such as Hot Wire Chemical Vapor Deposition (HWCVD) and Plasma Enhanced Chemical Vapor Deposition (PECVD) can be used to fabricate devices based on α-Si:H or μc-Si:H with different structural characteristics simply by playing with the deposition parameters. The silicon solar cells also permit the possibility of obtain roll-to roll depositions on flexible substrates, such as PEN, thus reduce the production cost. However, the use of flexible substrate involves some restrictions, mainly the substrate temperature, $T_s < 120°C$.

In the case of microcrystalline silicon film growth, it involves two zones that correspond to two growth stages: an incubation zone and a growth zone or "Bulk"

Figure 1 shows for HWCVD the microstructural change in a silicon film, as function of a deposition parameter, such as partial pressure (P_{Part}), distance between filament and susbtrate ($d_{f\text{-}s}$), or hydrogen dilution, among others.

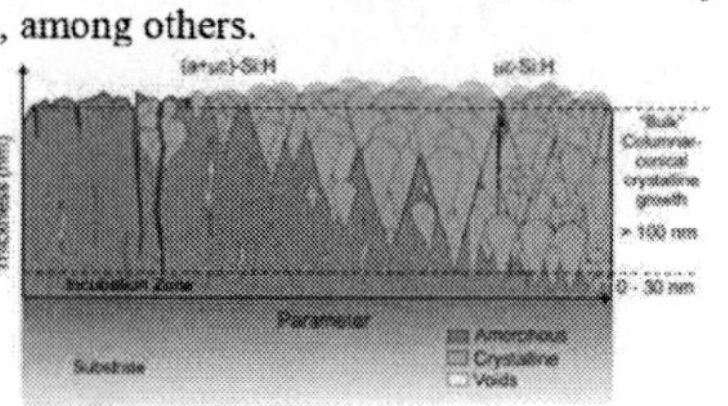

Figure 1: Scheme of transition from the amorphous to microcrystalline phase, in a silicon thin film, as function of a deposition parameter. i.e., the hydrogen dilution (DH).

The growth zone of the thin layer occurs at a certain distance from the interface with the substrate. In this region there coexists a mixture of phases (α+μc)-Si:H that result from the nucleation of nano and/or microcrystals, which increase in size as the thickness of the thin layer increases [1]. In microcrystalline silicon the incubation zone is characterized by the presence of a very thin amorphous zone or by the proto-crystalline silicon formation (p-Si:H) constituted by nanometric crystals (2-5 nm) very dispersed in an amorphous matrix [1].

Two aspects that deserve special attention in the study of p-i-n μc-Si:H solar cells, are:

-The microstructural characteristics of the first layer in contact with the intrinsic layer (p in superstrate structure or n in substrate one), because its structural properties could affect the growth of the intrinsic layer and condition its optoelectronic properties as well.

-The control of the thickness of the amorphous incubation zone, since being very thin layers can be harmful for the properties of the solar cell, not only electrically, but they can also influence negatively by high absorption.

Some authors have used the deposition techniques such as the *layer by layer* (LbL) [2] or the cyclically varying hydrogen dilution (CVH) [3] in order to inhibit the amorphous fraction in the incubation region. In this paper we have selected 4 very thin doped μc-Si:H samples (2 p-type and 2 n-type), grown under different conditions to produce films with different incubation layer. Through measurement of optical transmittance and reflectance and fitting of the spectra by using a three layer model and an effective medium approximation (EMA) [4], microstructure of the films was deduced. Finally, the microstructure deduced from the optical method has been validated with TEM and HRTEM measurements.

2 EXPERIMENTAL

Microcrystalline silicon *n* and *p* doped thin films with thicknesses ranging between 20 to 60 nm were deposited on Corning (1737) glass substrates by HWCVD in a set-up described elsewhere [5]. The deposition of the thin films was carried out by using two Ta filaments (0.5 mm diameter), which were resistively heated up to 1650°C and the substrate was kept at a distance of 4 cm from these filaments.

The deposition conditions presented down in Table I, show that the four samples studied in this work were obtained in deposition conditions, such as, WSH (Without Supplementary Heating), in this process the sample needs 15 min to start growing the crystalline phase, given rise to amorphous phase; CVH (Cyclically Hydrogen Variations), 3 cycles to [(98% D_H, P=6Pa, r_d = 1,1 A/s) to 60s + (D_H=93,3 %, P=1Pa, r_d = 2,0 A/s) to 20 s]; PH (Pre-heating). The substrate temperature is determined by the radiation from the filaments to the substrate (~100 °C).

Table I: Deposition parameters of μc-Si:H n and p doped layers deposited by HWCVD.

Sample	mc-n1 (WPH)	mc-n2 (CVH)	mc-p1 (PH)	mc-p2 (PH)
P (mbar)	1.0×10^{-2}	1.6×10^{-2}	1.0×10^{-2}	1.6×10^{-2}
DH_2 (%)	91.0	90.3	-	93.3/98
% B_2H_6	-	-	4	4
% PH_3	0.8	0.8	-	-

Thickness of the films was measured by a profilometer. The optical characterization was done through T(λ) and R(λ) measurements in the UV/Vis range with a HP8425A spectrophotometer.

In very thin films $10 \leq d \leq 100$ nm, as well as, in films with thickness ≥ 1000 nm, the use of optical methods that take into account only the analysis of the interference fringes and may be insufficient or give wrong results in the calculation of the optical constants of a material. For this reason, we use a model that allows describing the optical properties of materials deposited in thin films, regardless of their thickness. The microstructural study of the thin films was carry out through the optical model based on EMA described in detail in the refs. [4, 6] and using the software Optical 0.1.8. The results were compared with the microstructural characterization obtained by HRTEM measurements and electron diffraction patterns (DP) using a Transmission Electron Microscope (JEOL 2100).

3 OPTICAL MODEL

In the case of thin films, heterogeneities appear in the direction of growth of the film. A suitable option to understand the growth model is to interpret it as multiple structures of thin layers [8, 9, 10]. Each of the layers that make up this system has a defined thickness d_i, a fixed volume fraction f_{ij} and values of n_{ij} and k_{ij} relative to the materials involved in each layer of the multi-structure.

Using the theory of Effective Medium Approximation (EMA) model, each of these values will allow finding a single value of the refractive index n_{EMA_i} and the extinction coefficient k_{EMA_i} for each sublayer of the multi-structure.

In the analysis of the optical system of a multi-structure of sublayers, some of these sublayers can be processed as coherent, that is, when the light through the multiple reflections inside the sublayer keeps the coherence and produces interference [7, 8].The mathematical model used for the processing the optical behavior of the multilayer structure is based on a matrix formalism that takes account of the coherent or incoherent character of each sublayer. [9].

The effective medium approximation allows to estimate the macroscopic properties of a material medium depending on the local properties of the constituent materials, as well as, on the information about the microstructure, such as the concentration of each phase. These methods can be commonly used to predict the optical constants of mixtures of materials of known optical parameters.

Among the EMA models, the Bruggeman model [10] stands out, and is widely used in the analysis of ellipsometry measurements to obtain information about the microstructure of thin layers [11, 12].

The Bruggeman model is based on the following equation:

$$(1-f)\frac{\varepsilon_a - \varepsilon_{EMA}}{\varepsilon_a + 2_{EMA}} + f\frac{\varepsilon_b - \varepsilon_{EMA}}{\varepsilon_b + 2_{EMA}} = 0 \qquad (1)$$

Where ε_{EMA} is the effective dielectric function of the mixture formed by two materials, *a* and *b*, of permittivity ε_a and ε_b respectively, and f is the volume fraction corresponding to material *b*.

We used Bruggeman's EMA model to analyse optical transmittance and reflectance measurements and estimated the microstructure behaviour of the materials in the thin layers. This analysis is a simple and effective alternative compared to other more sophisticated methods of analysis of transmittance and optical reflectance spectra [13, 14].

The use of the EMA optical model in transmittance and reflectance measurements serves as a strategy to identify quantitatively the microstructural parameters such as, relative fractions of deposited material (amorphization or crystallinity), phase changes during the thin film growth, similar to what can be obtained using ellipsometry or through advanced characterisation techniques such as TEM.

4 RESULTS AND DISCUSSION

The figure 2 shows both, the transmittance and reflectance of the μc-Si:H (p and n type doped) films used in this study, as well as the simulated transmittance and reflectance spectra's obtained with the optical model of EMA and the software "Optical 0.1.8" [4].

The simulations correspond to the best fit obtained with 3-layer model (*Incubation, Bulk and Surface Layer*) described in the ref. [6]. For all cases, the thickness obtained by the optical method is similar to the measured by the profilometer. The differences between the spectrum (n and p type doped) are due to changes in the thicknesses of the samples, as well as, to variations in the crystalline fractions.

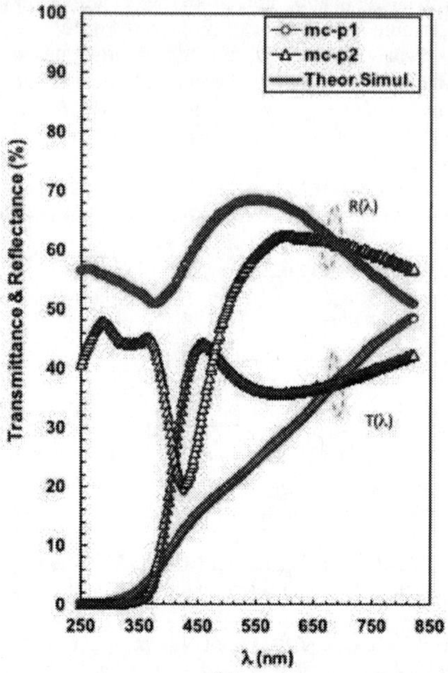

(a)

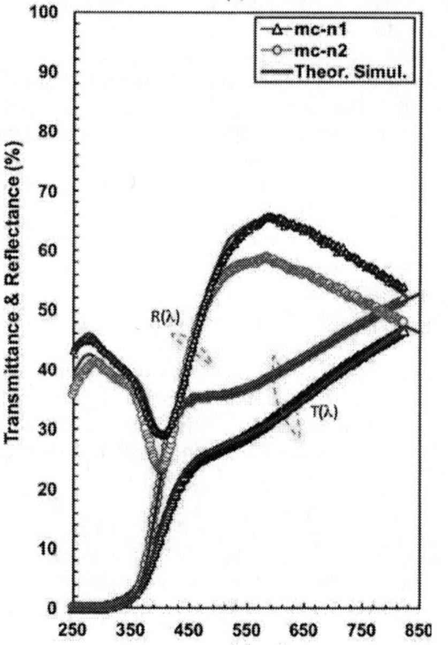

(b)

Figure 2: Transmittance and reflectance spectra and best fittings of *p* (a) and *n* type (b) doped layers of μc-Si:H.

The theoretical simulations of the transmittance and reflectance spectra of the doped layers p and n of μc-Si: H presented in Figure 2, show a good fitting with respect to the experimental values of the same samples.

There are small differences in the UV region of the samples studied. These differences between the theoretical and experimental spectra are due to the fact that the optical model applied to "very thin" layers cannot use the rms roughness as an input parameter in the simulation when it is less than 10 nm. The way to simulate approximately the effect of some of the morphological characteristics on the surface of the layer in the reflectance spectrum is to include in the surface layer a high density of microcavities, as explained in the refs. [15, 12, 16, 17]

The microstructural parameters (thickness, crystalline and amorphous fraction, and voids concentration) corresponding to the best fittings (Fig. 2) obtained from the optical simulation with Optical 0.1.8 [4] are shown in Table II. For each sample, three layers were considered (surface, bulk and incubation).

Table II: Thicknesses, crystalline (f_c), amorphous (f_a), and void (f_v) fractions of the three layers used to fit the transmittance and reflectance spectra of the p and n doped layers of μc-Si:H.

Sample		Surface	Bulk	Incubation
mc-p1 (WSH)	*d*(nm)	2.5	16.4	12.6
	f_c(%)	8	2	0
	f_a(%)	20	84	99
	f_v(%)	72	14	1
mc-p2 (CHV)	*d*(nm)	11.5	23.5	18.2
	f_c(%)	15	87	92
	f_a(%)	25	3	3
	f_v(%)	60	10	5
mc-n1 (PH)	*d*(nm)	14.3	23.5	4.6
	f_c(%)	38	42	88
	f_a(%)	28	52	10
	f_v(%)	43	6	2
mc-n2 (PH)	*d*(nm)	12.4	32.0	4.6
	f_c(%)	38	77	97
	f_a(%)	2	1	2
	f_v(%)	60	21	1

Results in Table II show that all samples have a crystalline phase at all growth stages (incubation, growth phase and surface), except for the incubation phase of mc-p1. Likewise, it is observed that the sample that was obtained by CVH (mc-p2) and the others deposited with previous heating (mc-n1 and mc-n2) showed suppression of the amorphous incubation phase and crystalline fraction in the first stage of growth of the thin film

The best fittings of the spectra were achieved using a 3-layer model described in ref [8]. However, in those cases in which the thickness of the incubation layer is very small or its composition presents a very small difference compared to the growth zone layer, as shown by the sample mc-p2, it could be suggested that a 2-layer model (bulk / surface) would be more appropriate.

By coercing the thickness and relative fractions of the surface layer to keep unchanged and allowing the growth zone (bulk) as the unique variable, the best fitting is achieved with a growth zone thickness equal to 41.2 nm, and a total thickness of 53.1nm (i.e., for the sample mc-p2).

When applying the χ^2 test to correct the thicknesses of both layers, we observed that the thickness of the surface layer was 11.8 nm and that of the Bulk was 41.4nm, that is, the total thickness of the layer was 53.2 nm for the same sample. The results showed a good fit between the experimental and the simulated spectrum; however, in the visible region, the simulated spectrum was lower than the experimental one. This behaviour indicated that the crystalline fraction of the layer should be higher.

To correct the total crystalline fraction, the third layer was added and with it the elements (%) that make it up; so when χ^2 was close to 1, the best fit was achieved and the microstructures as shown in Table II were obtained.

When applying the same procedure to the other layers, it was observed:

a. In surface layer, difference of 0 - 2.6% in the calculus of the thickness layer and maximum variation in the values of the relative fractions in the range of 2 - 12%.
b. In bulk layer, difference of 0.5 - 35% of the thickness layer and the maximum variation in the values of the relative fractions, in the range of 1% - 9%.
c. In incubation layer difference of 0.7 - 23% of the thickness layer and the maximum variation in the values of the relative fractions, in the range of 1% - 9%.

The comparison between the thicknesses obtained by profilometry measurements and the optical method of EMA, is shown in the Table III for the studied samples.

Table III: Comparisons between the thicknesses measured by profilometer and those one obtained by the optical method of EMA with 3-Layers model.

Sample	d_{prof}(nm) ± 5nm	d_{OPT}(nm) ±1%
mc-p1	27.0	31.5
mc-p2	50.5	53.2
mc-n1	49.0	42.4
mc-n2	46.5	49.0

The Figures 3 and 4 show, respectively for p- and n-doped films, the refractive index and extinction coefficient corresponding to the three regions (in incubation, bulk and surface regions). They have been obtained by applying an EMA model and using the thicknesses and fractions of Table II.

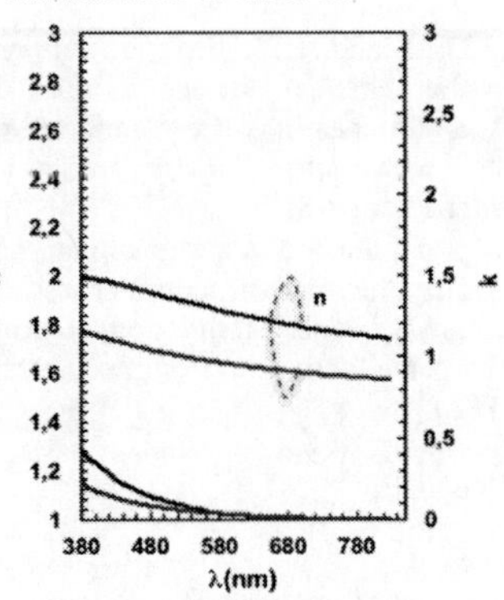

Surface

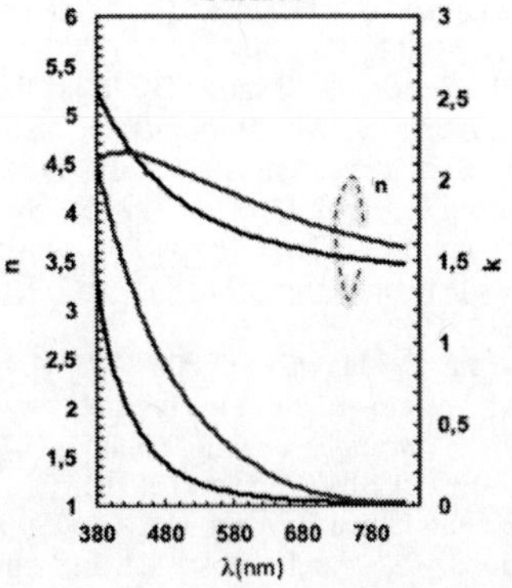

Bulk

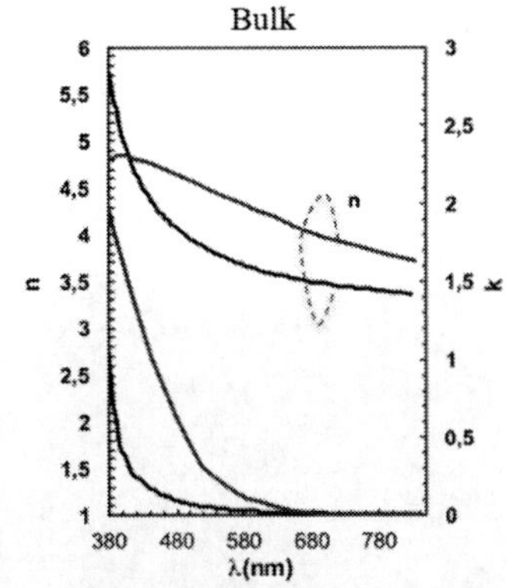

Incubation

Figure 3: Refractive index and extinction coefficient in incubation, bulk and surface region of *p* type doped thin films of μc-Si:H deposited by HWCVD.

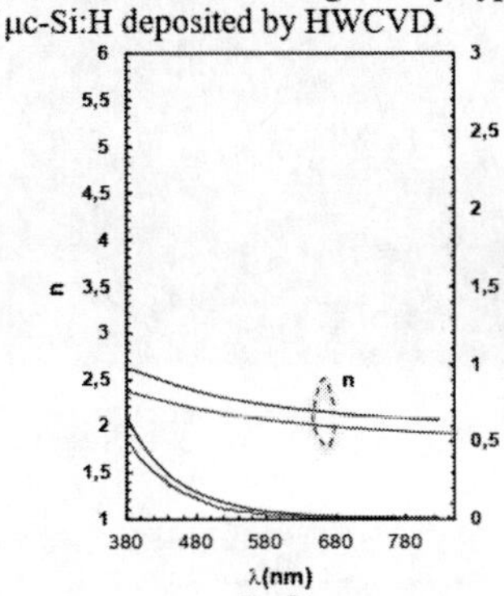

Surface

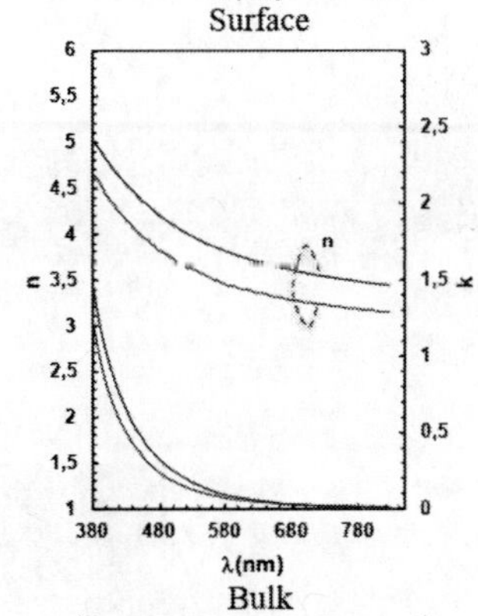

Bulk

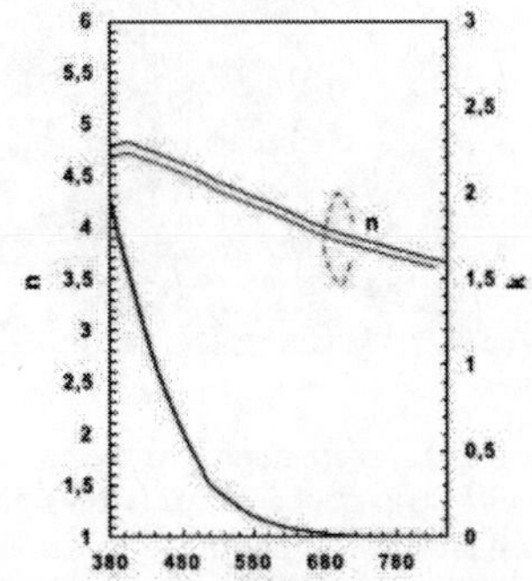

Incubation

Figure 4: Refractive index and extinction coefficient in incubation, bulk and surface region of *n* type doped thin films of μc-Si:H deposited by HWCVD.

The Figure 5 shows HRTEM images to 400000X of the samples (mc-p2 and mc-n1) and the micrograph of the sample mc-p1 (WSH) obtained to 250000X. Only for the pictures (a) and (b) the presence of Si crystalline growth from the interface substrate/film is observed.

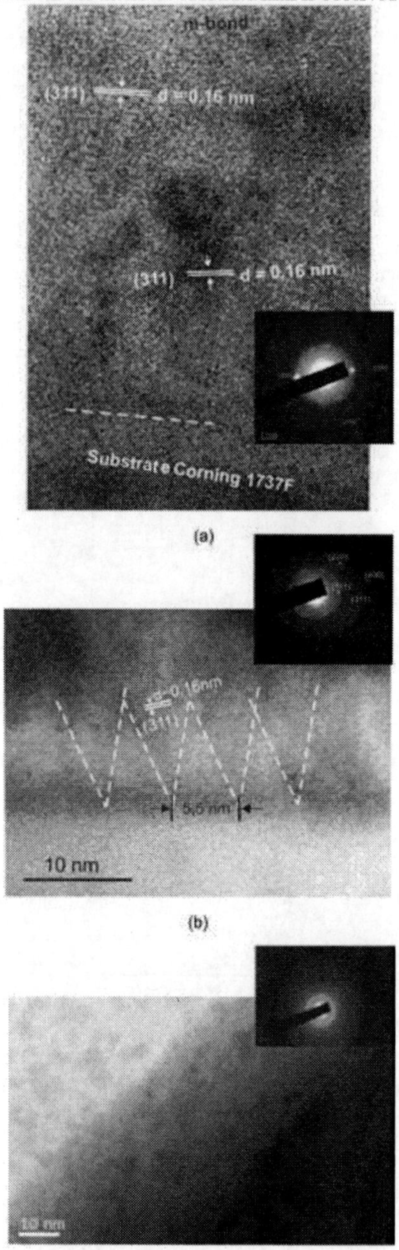

Figure 5: HRTEM micrograph and electron diffraction pattern (DP) of (a) p doped mc-p2 (CHV) and n doped mc-n1 (PH) thin films of μc-Si:H, obtained to 400000X in the incubation region, interface sample/substrate

The micrographs of HRTEM shown the formation of crystals with a conical structure separated by an approximate distance of 5.5 nm (sample 5(a)). Likewise, there are observed fringes of crystalline lattice with approximate atomic spacing of 0.16 nm, corresponding to the orientation (311) characteristic of Si [18].

The low concentration of crystals on the thin film surface contrasts with its high density, either in the first stage of growth of the thin film (incubation), as well as, in the bulk. This feature is consistent with the results obtained through the EMA optical model, whose microstructure was presented in Table II, and confirms the reliability of the optical results.

The electron diffraction patterns (DP) shown in the Figure 5 and, some rings are observed which allows to identify poly-crystallinity in the interface substrate/film. Likewise, some crystals are observed with the orientations (111), (220), (311) and (400) characteristics of c-Si (samples 5(a) and 5(b)). These results are consistent as shown in the HRTEM micrographs (See fig. 5), in which the Si-crystal growth from the substrate is observed.

The figure 5(c) cross-sections and DP of the sample mc-p1 Deposited without supplementary heating shown the amorphous growth of the film from the substrate. This result probes that at low substrate temperature the amorphous fraction dominates the growth process in the film, hence a high porosity in the film and low transport properties.

5 CONCLUSIONS

- The optical model based on the analysis of reflectance and transmittance is reliable as well as suitable to describe with high precision the microstructure of the thin films.

- The TEM and HR-TEM showed crystallization from the substrate, for the samples deposited by CHV and preheating. The mentioned samples have crystals with conical shape, as well as characteristic orientations of c-Si.
- The optical method is a quick diagnostic technique to identify the microstructural characteristics of the thin film as well as the optical properties. The results have demonstrated good correlation with HRTEM.

5 REFERENCES

[1] Collins R. W., Ferlauto A. S., Curr. Opin. Solid-State Mater. Sci.,6 (2002) 425
[2] Van Elzakker G., Tichelaar F. D., Zeman M., Thin Solid Films, 515 (2007) 7460
[3] Villar F., Antony A., Muñoz D., Rojas F., Escarré J., Stella M., Asensi J. M., Bertomeu J., and Andreu J. Mater. Res. Soc. Symp. Vol. 1066 –A06-11 (2008)
[4] Centurioni E., Appl. Phys., 44 (2005) 7532
[5] Villar López F., "Dispositivos fotovoltaicos de capa delgada a baja temperatura", PhD Thesis, , Universitat de Barcelona (2010)
[6] Rojas F., Villar F., Stella M., Bertomeu J., Asensi J.M, "Microstructural analysis of very thin doped μc-si:h films through optical methods" Proceedings (3AV228) -EU-PVSEC Valencia (2010).
[7] Ohta K. and Ishida H, Appl. Opt. 29 (1990) 1952
[8] Katsidis J. C. C and Siapkas D. L., Appl. Opt. 41 (2002) 3978

[9] Prentice J. S. C., J. Phys. D 33, (2000) 3139
[10]Bruggeman D. A. G., Ann. Phys. (Leipzig) 24 (1935) 636
[11]Hamma, S., and Roca i Cabarrocas, P. J., Appl. Phys., 81 (1998) 7282
[12]Summonte C., Mat. Res. Soc. Symp. Proc. Vol. 297 (1993) 395
[13]Kim Y., Appl. Opt., 35, 34 (1996) 6703
[14]Cisneros J. I., Appl. Opt., 37, 22, (1998) 5262
[15]Bosch S., Ferré-Borrull J., Leinfellner N., Canillas A., Surface Science, 453 (2000) 9
[16]Stearns D. G., Gaines D. P., Sweeney D. W., and Guillikson E. M., J. Appl. Phys., 84, (1998) 1003
[17]Nevot L. and Croce P., Rev. Phys. Apl. 15, (1980) 398
[24]Natl. Bur. Stand. (U.S.) Monogr. 25, 13, (1976) 35

APPLICATION OF GENETIC ALGORITHM PARAMETER OPTIMISATION ON CURRENT-VOLTAGE DATA MULTI-CRUSTALLINE SILICON SOLAR CELLS

R.M Dix-Peek, E.E van Dyk, F.J Vorster, C.J Pretorius
Nelson Mandela University
Address: University Way, Port Elizabeth, South Africa
Telephone: +27 41 504 22 33
Email: s212286552@mandela.ac.za

ABSTRACT:

The models relating current and voltage for a diode and other related devices contain many parameters and the relations often contain multiple terms hindering explicit formulae for specific parameters. Optimization procedures are useful for fitting a model to a dataset. In this study a Genetic Algorithm (GA) is proposed as a method to perform Parameter Optimization (PO) of the models used to describe a solar cell. In particular, a Genetic Algorithm for Parameter Optimization (GAPO is applied to the current voltage (I-V) relation of a multi-crystalline (mc) Si (photovoltaic) PV cell. A GA is a type of Evolutionary Algorithm based upon biological processes such as genetic crossover during sexual reproduction, genetic mutation, breeding pair selection, and other similar processes. GAs search the solution space with parallel proposed solutions, increasing the probability of convergence to a global minimum error for the solutions (array of parameters) obtained for the applied model to the data. Other, more classical approaches, such as the Gradient Descent parameter optimization, are prone to acquiring only local minima.
Keywords: Genetic Algorithm, parameter optimization, photovoltaic

1 INTRODUCTION

The models relating current and voltage for the diode and other related systems contain a set of parameters. The relations often contain multiple terms and are often implicit in nature, hindering the determination of explicit formulae for specific parameters. Optimisation procedures can be used to fit the model to the data. In a study by J.A Jervase [1] a Genetic Algorithm (GA) was tested against simulated data sets of varying parameters; the GA used in that study was successful. In this study, a GA is proposed as a valid method to perform the Parameter Optimisation (PO) of the models involved.

GAs are based upon biological processes such as genetic crossover in sexual reproduction. GAs generate a parallel set of random solutions for each iteration, covering a representative portion of the solution space. GAs are therefore more likely to acquire a global minimum error for the parameters obtained relating to the model applied to the data as compared to alternate more classical approaches such as the Gradient Descent method [2]. These approaches are more prone to acquiring only local minima.

In this study, Genetic Algorithm Parameter Optimisation (GAPO) is applied to the dark I-V data of a mc Si PV cell to extract the parameters governing the electrical response of the device in the forward bias, the reverse bias data is included for completeness. The sample understudy was laser cut, a process which induces specific defects. The induced defects negatively affected the performance of the device as indicated by the extracted parameter. Complementary techniques were used to validate the extracted parameters. These techniques include Light Beam Induced Current (LBIC) measurements, Electroluminescence (EL) imaging and Infrared (IR) thermography.

2 THEORY

2.1 Dark I-V Model

There exist multiple models applied to PV cells relating to the I-V characteristics, each model has its own set of assumptions. The models vary in complexity from relatively simple to highly complex. The simpler models are often more practical, and the parameters are easier to extract. The two-diode model was used in the parameter optimisation in this work and is depicted in Figure 1. It provides an adequate description of the forward current-voltage relationship, while still maintaining the ease of two-diode parameter optimisation. In reverse bias this model is not valid, the breakdown mechanisms (and the voltage) are dependent on the defects present in the device.

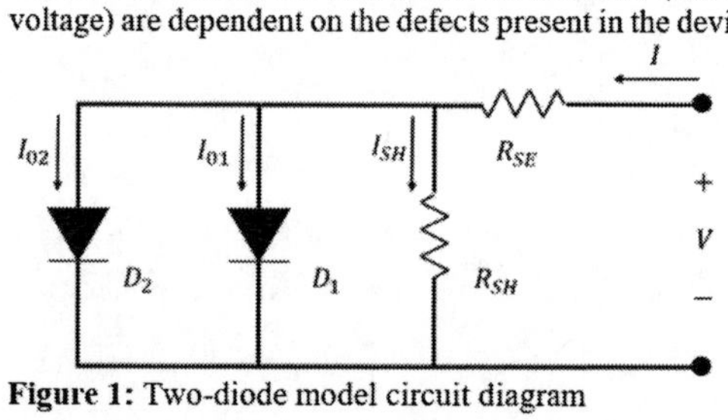

Figure 1: Two-diode model circuit diagram

Eqn .1. is the mathematical model that describes the relationship between current and voltage. Table I lists all the definitions for the symbols within eqn 1. This model does not accommodate the distributed nature of series resistance, nor the lack of homogeneity of cell parameters; however, it does give a fair description of the I-V relation. For the complementary techniques, the independent diode model is assumed, where each local position acts as an independent two-diode model.

$$I = I_{01}\left[\exp\left(\frac{q(V - IR_{SE})}{n_1 k_B T}\right) - 1\right] + I_{02}\left[\exp\left(\frac{q(V - IR_{SE})}{n_2 k_B T}\right) - 1\right] + \frac{(V - IR_{SE})}{R_{SH}} \quad (1)$$

Table I: Symbolic definitions for eqn. 1

Parameter	Value
I_{01}(A)	Dark saturation current of diode 1
I_{02}(A)	Dark saturation current of diode 2
n_1	Ideality factor of diode 1
n_2	Ideality factor of diode 2
R_{SE} (Ω)	Series resistance
R_{SH}(Ω)	Shunt resistance

2.2 Genetic Algorithm Parameter Optimisation (GAPO) [2]

Figure 2 describes the basic structure of GAs. GAs require an initialization step; where the first generation of population of solutions is created. It is an important step, as the closer the initial guesses are to a global minimum, the faster the GA will acquire it. However, if the proposed solutions do not span the solution space, it could reduce the probability of finding a global minimum of error. The next step is the fitness evaluation of the population, or how well the solutions with their parameters fits the data. The fitness evaluation allows for optimal individuals (solutions) to be selected. This leads into the next step, the selection process. In this step, individuals are selected either by random or specific criteria based on their fitness. The individuals selected in the previous step are then used in the crossover process, where the genes (in this case, cell electrical parameters) of a pair are crossed-over. In the case of this algorithm, crossed over is defined as creating a new solution by using a weighted mean of the cell parameters. The following step is the mutation step; this is one of the key factors of GAs. The mutation step is a random perturbation that is applied to random genes within the population. At the termination step, the calculated error is compared to minimum threshold error. If the calculated error is lower, or if the maximum number of iterations are reached the program terminates. This entire process is repeated except the initialization step, until the stopping conditions are met.

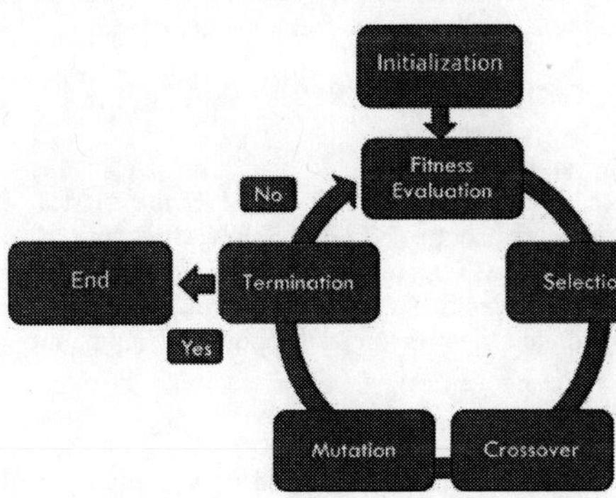

Figure 2: Genetic Algorithm Process Diagram

3 EXPERIMENTAL PROCEDURE

The sample under test was that of a laser-cut scribed mc Si solar cell of size 80 mm x 25 mm. The laser scribing process induces defects into the sample, resulting in very specific optical and electronic features. The I-V data was acquired under dark conditions.

3.1 Dark I-V data

The cell dark I-V data was acquired using a Keithley 2450 source meter unit (SMU). A two-probe method was used, with compensation for the resistance of the probes. The dark I-V data used in the parameter optimization was within the range of -0.1 V to 0.8 V with step size of 20 µV (4500 data points). The dark I-V data the model is then compared to in Figure 3, has I-V data within the range of -1 V and 1 V with steps size of 20 µV.

3.2 GAPO analysis

The GAPO was applied to the corresponding dark I-V data, with population size of 100, max iteration number of 100 and mutation rate of 50%. The iteration number and population size were kept to a minimum so as to limit computational time. The threshold error was not taken into account during this study. The GA went through 100 iterations itself, so as to apply statistical analysis to the parameters extracted from the data.

4 RESULTS AND DISCUSSION

4.1 GAPO

Figure. 3 shows the measured I-V data of the laser cut m-c Si solar cell and the modelled (based upon the extracted parameters) I-V data. The graph shows the contributions of each current mechanism described in eqn. 1. Table II lists the extracted parameters from I-V data using GAPO. The overall fit is depicted by the red curve and labelled "Model" in the graph. The line representing the data is difficult to distinguish from the line indicating the overall model, indicating a good fit. The calculated R^2 value is 0.9998. In the range of (-1 V, 0.4 V) it is difficult to differentiate between the lines representing the Shunt Current, overall model and data. However, upon close inspection it is possible to distinguish between the Shunt line and the I-V Data line in the range of (-1 V, -0.8 V). It is suggested that this observation is a result of a non-linear reverse current through pre-breakdown mechanisms.

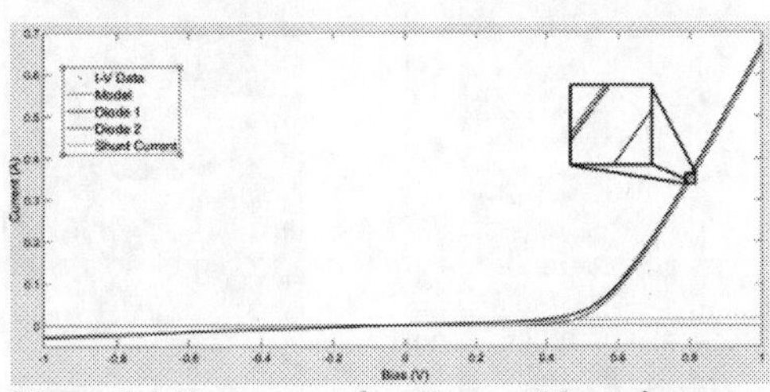

Figure 3: I-V response of an 80 mm x 25 mm laser cut m-c Si solar cell. The optimised model is compared to the measured I-V data as well as the contributions made by Diode 1, Diode 2 and the Shunt are shown.

Table II: Extracted parameters of the laser cut sample.

Parameter	Value
I_{01}(A)	2.920 x 10^{-7}
I_{02}(A)	4.670 x 10^{-6}
n_1	1.836
n_2	6.625
R_{SE} (Ω)	0.148
R_{SH}(Ω)	38.48

Returning to Figure 3 it can be seen that the main contribution of current between -1V and 0.425V is as a result of shunting. This is validated by the low shunt resistance value shown in Table II. The cause of the low shunt value is likely due to damage caused in the laser cutting process of the solar cell. For the remainder of the I-V response in Figure 3, the main current contribution is due to diode 1. The high ideality of the second diode is likely due to laser cutting process inducing defects into the Si energy band gap. [3].

4.2 Complementary Characterisation Techniques

Figure 4 shows the complementary characterization techniques used in this study to validate the results obtained through the GAPO. The breakdown voltage map was generated using reverse bias electroluminescence imaging as described in [4]. The breakdown voltage map (a) indicates sites of low reverse bias breakdown along edges of laser cutting. Due to the fact that breakdown occurs at a low voltage, it is likely that the related defects are due to intra-band thermalization of hot carriers [3]. The carriers are able to tunnel through the band gap due to a number of defect states available within the band.

The reverse bias IR thermal image (b) shows hotspots present at these sites, indicating either breakdown or shunting [5]. Localised heating due high local power density.

The LBIC I_{SC} map (c) shows decreased signal near to the sites of interest. This could be an indication of low material quality and/or shunting. The photogenerated current moves via the alternate current paths.

By using all of the remarks above, the complementary techniques validate the observation of fact low shunt resistance. The shunt was most likely caused by the laser cutting process inducing ohmic like shunts from a continuum of defects in the Si band gap [6].

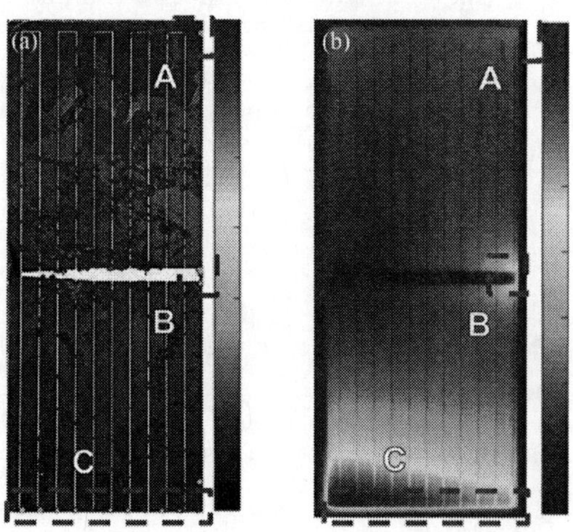

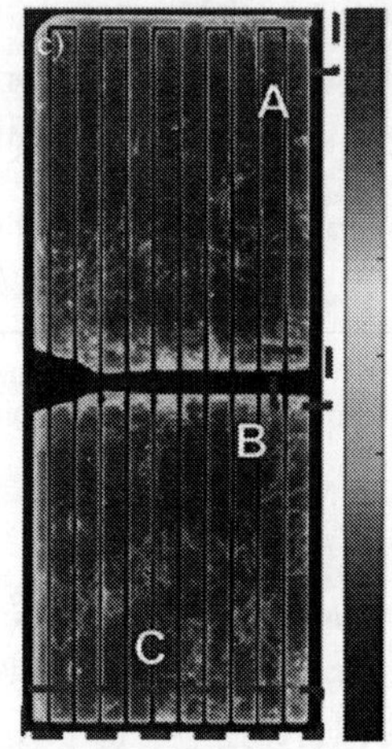

Figure 4: Laser-cut m-c Si solar cell: (a) photograph with Breakdown Voltage map (0..14V), (b) Difference Thermal image, (c) LBIC I_{SC} scan (1.2 .. 1.6 mA)

5 CONCLUSIONS

A Genetic Algorithm for Parameter Optimization was successfully applied to the I-V curve of a mc-Si solar cell. Application of the algorithm allowed for the determination of the current contribution of the various elements described by the two-diode model.

The low shunt resistance determined by the GAPO was validated using complementary techniques. The laser scribing of this particular cell induced defects into the device. It is suggested that proper isolation and an optimized cutting procedure could limit this effect.

6 FUTURE WORK

The GAPO needs to be expanded so as to include temperature dependent analysis. This could then be expanded further to the application of the algorithm to spatially resolved techniques such as LBIC and EL/PL imaging to spatially resolve electrical parameters.

7 ACKNOWLEDGEMENTS

The author would like to thank the Department of Science and Technology (DST) of South Africa, the National Laser Centre (NLC) of South Africa, the African Laser Centre (ALC), Nelson Mandela University, Nelson Mandela University PV Research Group and PVinsight (Pty) Ltd. for their financial support and input into the study.

8 REFERENCES

[1] J. Jervase, H. Bourdoucen, and A. Al-Lawati. (2001). Solar cell parameter extraction using genetic algorithms. Measurement Science and Technology, 12(11), pp.1922-1925.

[2] K. Ming Leung. (2003). GENETIC ALGORITHMS. [online] Available at: http://cis.poly.edu/~mleung/CS4744/f04/ch06/GA3.pdf [Accessed 17 Aug. 2018].

[3] O. Breitenstein (2017). Personal communication. Your article in Physica B. [email].

[4] R. Dix-Peek, E. van Dyk, F. Vorster and C. Pretorius, (2018). Breakdown voltage mapping through voltage

dependent ReBEL intensity imaging of multi-crystalline Si solar cells. Physica B: Condensed Matter, 535, pp.63-66.
[5] O. Breitenstein, J Rakotoniaina, M. Kaes, S. Seren., T. Pernau, G. Hahn, W. Warta, and J. Isenberg. Lock-In Thermography - A Universal Tool for Local Analysis of Solar Cells, Proceedings of 20th European Photovoltaic Solar Energy Conference, (2005), p. 590-593
[6] O. Breitenstein, P. Altermatt, K. Ramspeck, M.A. Green, Jianhua Zhao, A.Schenk, Interpretation of the commonly observed I–V characteristics of c-Si cells having ideality factor larger than two, in: IEEE Proceedings of the Fourth World Photovoltaic Solar Energy Conference, 2006. 879

SILVER NANOPARTICLES ON SUBSTRATE AND SUPERSTRATE: FABRICATION AND NUMERICAL ANALYSIS FOR SOLAR CELL APPLICATIONS

Mohammad K. Hossain[1,*], and Ayman W. Mukhaimer[1,2]
[1]Center of Research Excellence in Renewable Energy (CoRERE), King Fahd University of Petroleum and Minerals (KFUPM), Dhahran 31261, Dhahran (Kingdom of Saudi Arabia)
[2]Department of Mechanical Engineering, King Fahd University of Petroleum and Minerals (KFUPM), Dhahran 31261, Dhahran (Kingdom of Saudi Arabia)
[*]Correspondence to: kamalhossain@kfupm.edu.sa

ABSTRACT: Plasmonics is known to contribute enormously in enhancing solar cell efficiency provided that the challenges how to incorporate plasmonic nanoscatterers has been taken care of effectively. Here in this work, spectral absorption profile for a typical model, silver (Ag) nanoparticles on crystalline silicon along with various interparticle gaps, has been numerically analyzed using finite difference time domain simulation. Parameters (solar spectrum 460, 650 and 1100 nm; Ag nanoparticles diameter of 100 nm; interparticle gaps of 0, 2 and 5 nm; etc.) were chosen in such a way so that the simulative results can be realized with the experimental evidence. Further in-depth lookout has been carried out by extracting excitation generation rate distribution of the same models at different wavelengths of solar spectrum. The predicted understanding was implemented by designing Ag nanoparticles incorporated thin film solar cells with substrate and superstrate configurations. A simple and easy-to-prepare process was introduced to achieve such plasmonic structures on c-Si as well as on transparent conducting oxide suitable for thin film solar cell applications. As-fabricated Ag nanoparticles were found to be 100 nm on average and of wide range of interparticle gap distribution. Such an approach to achieve plasmonic nanoscatterers supported by simulative prediction would be very useful for substrate and superstrate configurations of plasmonic solar cell.

Keywords: Photovoltaic; nanoparticles; absorption; exciton generation; superstrate; substrate; modelling

1 INTRODUCTION

Next generation of thin film solar cells require decreasing the cost, both of materials and processes, significantly, as well as enhanced solar harvesting through efficient photon management [1-2]. Although surface texturing is being used for light trapping so far, but the geometry makes the thin fil vulnerable creating death-zone for precious carriers. Incorporation of plasmonic nanoparticles has been long proposed to efficient solar harvesting. Photon absorption can be increased due to scattering from such nanoparticles and their other inherent properties such as polariton modes and capability to localize incident energy [3]. The proof of concept stays till now low because of the challenge how to embed this plasmonic layer effectively into the absorbing layer. It has been a critical challenge to incorporate plasmonic nanoscatterers in thin film-based plasmonic solar cell design apart from the need of sophisticated and expert hands in this regard. A simple and easy-to-prepare route to fabricate Ag nanoparticles as reported previously along with detailed morphological and structural characterizations [4-5]. The as-fabricated nanoparticles bear a great potential to design thin-film based plasmonic solar cell.

Plasmonic nanoscatterers embedded within cell's individual layers will support the plasmonic effect needed to efficient light management in conventional thin film solar cells. As known, plasmonics is defined as the excitations of conduction electrons at the interface or surface of the nanoscatterers [6–9]. Such layer of nanoscatterers induces light trapping and thus efficiency gets enhanced [10–12]. In addition, polariton modes and energy localizations enhanced exciton generation and collection in solar cell [13–15]. Nevertheless, there is consistence challenges how to embed such nanoscatterers without affecting other associate layers [16–19]. Several techniques have been reported so far, such wet chemical method, laser ablation method, physical vapor method, lithography, etc. However, nanoscatterers size and shape is very critical to control in most of the techniques. Apart from cost reduction and increased efficiency, it is equally important to understand excitation generaton and collection characteristics of such solar cell as well as to devise a way to fabricate effective plasmonic layer in different configuration of solar cell.

Here in this work, an extensive simulation has been carried to deduce spectral absorption profile for such a typical model, Ag nanoparticles on crystalline silicon (c-Si) along with various interparticle gaps. Further investigation has been done to understand excitation generation rate distribution of the same model at different wavelengths of solar spectrum. To implement the understanding gained from simulations, we introduced herewith a simple and effective process to embed Ag nanoparticles in substrate configuration as well as superstrate configuration of thin film solar cell.

2 METHOD and MATERIALS

Ag nanoparticles modeled as 3×3 on c-Si slab was considered as a typical model for FDTD simulations. The choice of nanoparticle size in FDTD simulation, i.e. 100 nm in diameter was based on FESEM confirmation. Interparticle gaps were kept as 0, 2 and 5 nm in entire simulation. Although spectral absorption profile was reported previous, the influence of interparticle gaps as well as correlated spectral exciton generation rate distribution was very important [4]. Therefore, absorption profile of such model was obtained at various incident wavelengths of solar spectrum ranging from near ultra-violet (UV) to near infrared (IR) region. Spectral exciton generation rate distribution, a crucial factor in designing efficient solar cell, was also obtained for the same model at different incident wavelengths. Details on experimental studies will be presented and elaborated in full manuscripts. In short, a layer of ZnO was obtained by sputtering technique on glass substrate followed by another sputtering of Ag ultrathin layer atop and annealing at 600 °C in argon (Ar) (called hereafter Sample A). Sample A was kept for the purpose of superstrate solar cell. Another specimen (called hereafter sample B) was prepared by sputtering an ultrathin layer of Ag on p-type c-Si wafer followed by annealing at 600 °C in Ar. Sample B was kept for the purpose of superstrate solar cell. Topographical and optical characterizations were carried out for both specimens step by step.

3 RESULTS and DISCUSSION

Figure 1a-c shows the absorption depth profile of the proposed model (Fig. 1d, note: gap=0) obtained at incident wavelengths of 460, 650 and 1100 nm respectively. Although a wide range of spectral absorption depth profile ranging from near-UV to near-IR was extracted slice-by slice, only three distributions related to c-Si were mentioned herewith. The interparticle gap play an important role in localizing electromagnetic fields that lead to enhance the absorption profile as well as exciton generation rate distribution. It was evident as observed in Fig. 1a-c. Absorption depth profiles of the proposed model (Fig. 1d, note: gap, g=0 nm) obtained at incident wavelengths of 460, 650 and 1100 nm respectively were represented in Fig. 1a-c. At 460 nm solar spectrum, the distribution was confined near the surface of nanoparticles as well as along the top surface of c-Si. For 650 nm solar spectrum, the distribution was more homogeneous whereas at 1100 nm, very negligible absorption depth profile was noted.

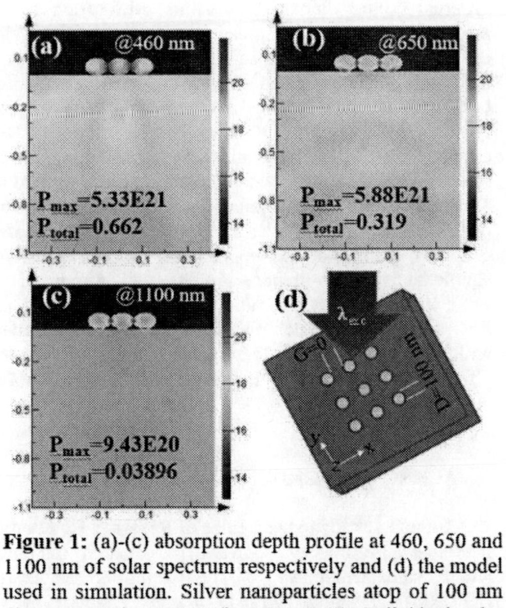

Figure 1: (a)-(c) absorption depth profile at 460, 650 and 1100 nm of solar spectrum respectively and (d) the model used in simulation. Silver nanoparticles atop of 100 nm diameter are in contact (i.e. gap, G=0). Individual color bars represent the corresponding distribution intensity.

As discussed before, interparticle gaps play a crucial role in interpreting plasmonics behavior of such nanoscatterers. Therefore, absorption depth profile for different interparticles gaps were extracted. Figure 2a-c shows the absorption depth profile of the same model with interparticle gap of 2 nm (Fig. 2d, note: gap, G=2) obtained at incident wavelengths of 460, 650 and 1100 nm respectively. Improved and homogeneous distributions were noted in such scenario.

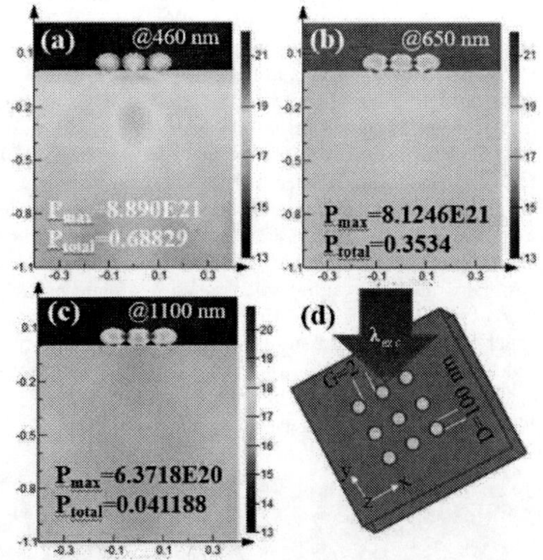

Figure 2: (a)-(c) absorption depth profile at 460, 650 and 1100 nm of solar spectrum respectively and (d) the model used in simulation. Silver nanoparticles atop of 100 nm diameter are 2 nm apart (i.e. gap, G=2). Individual color bars represent the corresponding distribution intensity.

Figure 3a-c shows the absorption depth profile of same model with interparticle gap of 5 nm (Fig. 3d, note: gap, G=5 nm) obtained at incident wavelengths of 460, 650 and 1100 nm respectively. The absorption depth profile was found less distributed compared to those

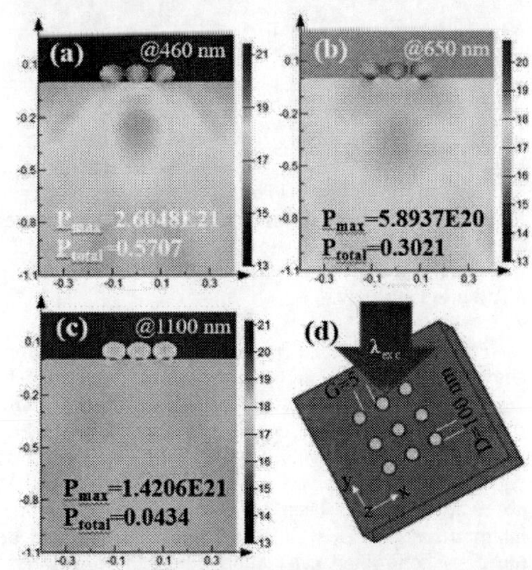

Figure 3: (a)-(c) absorption depth profile at 460, 650 and 1100 nm of solar spectrum respectively and (d) the model used in simulation. Silver nanoparticles atop of 100 nm diameter are 5 nm apart (i.e. gap, G=5). Individual color bars represent the corresponding distribution intensity.

observed in Fig. 2a-c. In addition, irrespective of interparticle gaps, absorption profile was more spread and homogeneous all along the absorbing layer at ~650 nm of indent wavelength of solar spectrum.

The internal quantum yield of a solar cell is directly related to exciton generation rate distribution. Therefore, understanding of spectral generation rate distribution in absorbing layer is indispensable. We have extracted generation rate distribution of one suitable model with the interparticles gap of 2 nm out of the aforementioned three models. Figure 4a-c shows the exciton generation rate distribution of the proposed model (Fig. 2d/4d, note: gap, G=2 nm) obtained at incident wavelengths of 460, 650 and 1100 nm respectively. A range of spectral absorption depth profiles from near-UV to near-IR was extracted slice-by slice. Here in this work, only three distributions related to c-Si were demonstrated to correlate the observation noted in case of interparticle gap of 0, 2 and 5 nm. Homogenous and distributed generation rate distribution was observed at solar spectrum of 650 nm, whereas at the near UV, 460 nm the distribution was confined in the surface and at near IR, 1100 nm, the distribution was less distributed.

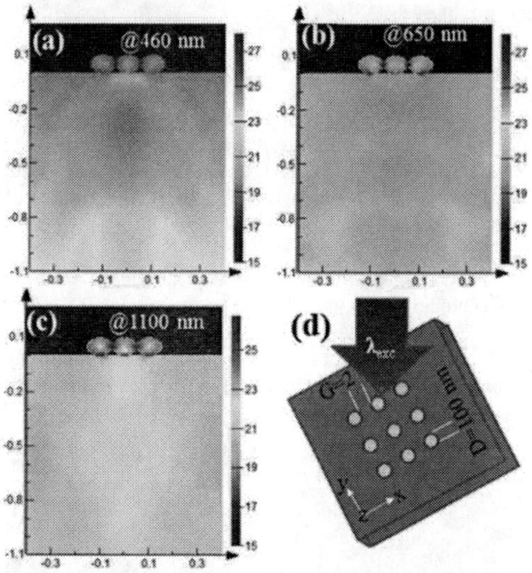

Figure 4: (a)-(c) Exciton generation rate distribution at 460, 650 and 1100 nm of solar spectrum respectively and (d) the model used in simulation. Silver nanoparticles atop of 100 nm diameter are 2 nm apart (i.e. gap, G=2). Individual color bars represent the corresponding distribution intensity.

Following the understanding and experience from abovementioned simulation, an attempt has been taken to design Ag nanoparticle embedded solar cell with substrate and superstrate configuration as shown in Fig. 5a and Fig. 5b respectively. Figure 5a shows as-deposited Ag nanoparticles on p-type c-Si as confirmed by FESEM observations. It has been reported previously that the interparticle gaps as well as nanoparticles size can be tuned by controlling calcination time and temperature. Fig. 5b shows FESEM image of as-deposited Ag nanoparticles on thin layer of ZnO. The size distribution of the as-fabricated Ag nanoparticles was larger compared to those fabricated on c-Si only. In substrate configuration of thin film solar cell, such nanoscatterers are expected to be on c-Si absorber to induce plasmonic effect. Therefore, Ag fabricated on c-Si will improve the process cost as other well as will not affect any associate layer of the solar cell. In case of superstrate configuration, Ag nanoscatterers are expected to be on ZnO layer instead of absorber, c-Si. Hence in both configurations, the proposed fabrication strategy will facilitate to fabricate plasmonic nanoscatterers in thin film based plasmonic solar cell.

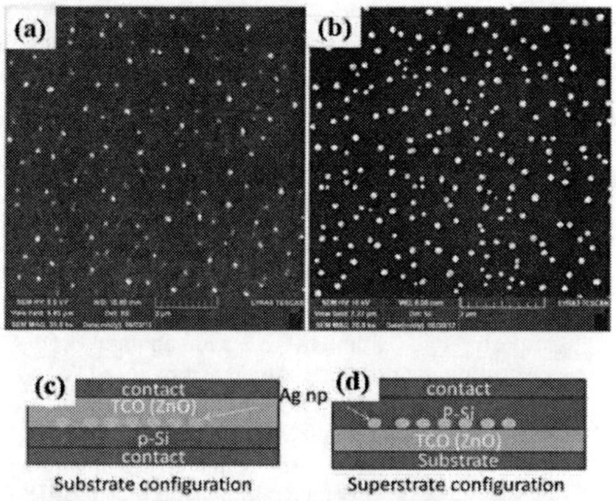

Figure 5: (a)-(b) FESEM micrographs of Ag nanoparticles on c-Si and ZnO respectively, and (c)-(d) substrate and superstrate configuration of the cell respectively.

4 CONCLUSIONS

Interparticle gap dependent spectral absorption profile for a typical model, such as Ag nanoparticles on c-Si, has been numerically analyzed using finite difference time domain simulation. Absorption depth profile was found to be spread homogenously and well-distributed all over the absorbing layer. Interparticle gap of 2 and 5 nm provided higher and well distribution compared to that obtained at zero interparticle gap. Further in-depth lookout has been carried out by extracting excitation generation rate distribution of the same model at different wavelengths of solar spectrum. The predicted understanding was implemented by designing Ag nanoparticles incorporated thin film solar cell with substrate and superstrate configurations. A simple and easy-to-prepare process was introduced to achieve Ag nanoparticles on c-Si substrate as well as on ZnO ultrathin film. Such plasmonic structures embedded within absorbing layer is promising and suitable for efficient light trapping in thin film based plasmonic solar cell applications.

5 ACKNOWLEDGEMENTS

Authors acknowledge Center of Research Excellence in Renewable Energy (CoRERE), Research Institute, King Fahd University of Petroleum & Minerals, (KFUPM), Dhahran 31261, Saudi Arabia. MKH Acknowledges Deanship of Scientific Research (DSR) at King Fahd University of Petroleum & Minerals (KFUPM) for funding this work through project No. IN151003.

6 REFERENCES

1. S. Carretero-Palacios, A. Jiménez-Solano, H. Míguez, ACS Energy. Lett. 1 (2016) 323.
2. C. J. Rhodes , Sci. Prog. 99 (2016) 438.
3. K.-H. Kim, S. Kasouit, E. V. Johnsonc, P. R. Cabarrocas, Sol Energy Mater. 119 (2013) 124.
4. M. K. Hossain, A. W. Mukhaimer, Q. A. Drmoosh, J. Elec Mat. 45 (2016) 5695.
5. M. K. Hossain, Q. A. Drmosh, A. W. Mukhaimer, H. M. Bahaidarah, Appl. Phys. A 117 (2014) 459.
6. B. Rech, H. Wagner, Appl. Phys. A 69 (1999) 155.
7. D. S. Shen, H. Chatham, P. K. Bhat, Sol. Cells 30 (1991) 271.
8. H. A. Atwater, A. Polman, Nat. Mat. 9 (2010) 205.
9. M. K. Hossain, Y. Kitahama, G. G. Huang, T. Kaneko, Y. Ozaki, App. Phys. B 93 (2008) 165.
10. K. Imura, H. Okamoto, M. K. Hossain, M. Kitajima, Nano Lett. 6 (2006) 2173.
11. K.R. Catchpole, A. Polman, Opt. Express 16 (2008) 21793.
12. O. L. Muskens, J. G. Rivas, R. E. Algra, E. P. A. M. Bakkers, A. Lagendijk, Nano Lett. 8 (2008) 2638.
13. M. K. Hossain, Y. Kitahama, V. Biju, T. Itoh, T. Kaneko, Y. Ozaki, J. Phys. Chem. C 113 (2009) 11689.
14. M. K. Hossain, T. Shimada, M. Kitajima, K. Imura, H. Okamoto, Langmuir 24 (2008) 9241.
15. M. K. Hossain, T. Shimada, M. Kitajima, K. Imura, H. Okamoto, J. Microsc. 229 (2008) 327.
16. V. E. Ferry, J. N. Munday, H. A. Atwater, Nano Lett. 8 (2008) 4391.
17. K. Nakayama, K. Tanabe, H. A. Atwater, Appl. Phys. Lett. 93 (2008) 121904.
18. C. F. Bohren, D. R. Huffman, Absorption and Scattering of Light by Small Particles (Chicago: John Wiley & Sons, 2008).
19. D. Derkacs, W.V. Chen, P.M. Matheu, S.H. Lim, P.K.L. Yu, E.T. Yu, Appl. Phys. Lett. 93 (2008) 091107

SPECTRAL CHARACTERIZATION OF TEMPERATURE INCREASE IN CRYSTALLINE SILICON SOLAR CELLS

Ion Urrea, Jaione Bengoechea, Eugenia Zugasti, and Ana Rosa Lagunas
National Renewable Energy Centre (CENER)
Ciudad de la Innovación 7, 31621 Sarriguren (Navarra) Spain
Telephone: +34 948 25 28 00; Fax: +34 948 27 07 74
Email: jbapezteguia@cener.com

ABSTRACT: As it is well known the conversion efficiency of photovoltaic modules decreases as their operating temperature increases. This operating temperature of photovoltaic modules is influenced by the ambient temperature and also by the amount of incident irradiation not converted into electric power, and which is transformed into heat in the photovoltaic module. In this regard, crystalline silicon, with an average decrease of 0.45% per °C above 25 °C, is the most sensitive technology to this effect. Moreover, the heating of photovoltaic modules has also and adverse effect on their durability. This work aims at the spectral characterization of the temperature increase in encapsulated monocrystalline silicon solar cells. More specifically, it is devoted to the quantification of the contribution to the temperature increase of the different spectral regions of incident sunlight. This investigation has been greatly motivated from the question whether it might make sense to develop solutions to reject the near-infrared part of the incident solar spectrum (wavelengths above 1200 nm), which photons are not efficiently converted into electric power by the crystalline silicon solar cells. In this sense, it is mandatory to characterize the contribution to the photovoltaic module temperature increase of this part of the solar spectrum.
Keywords: monocrystalline silicon, thermal performance, near-infrared

1 INTRODUCTION

The increase in the operating temperature of photovoltaic modules has negative effects on both the electricity production and on the service life. Standard crystalline silicon photovoltaic technology shows a relative decrease in the efficiency of 0.45% per °C above 25 °C, and this increase in operating temperature has also an impact on its durability [1]. Due to this double aspect negative impact, several authors have studied this effect, in order to quantify it and also to the aim of developing cooling solutions [2, 3, 4, 5].

The results presented in this work aim at the spectral characterization of the temperature increase in encapsulated crystalline silicon solar cells. More specifically, it is devoted to the measurement of the contribution to the temperature increase of the different spectral regions of the incident light. The objective of this research was to provide experimental data in order to answer the question whether it is essential to develop solutions to reject the near-infrared part of the spectrum (wavelengths above 1200 nm), which sub-band-gap photons are not converted into electric power by the crystalline silicon solar cells. To this aim, it is mandatory to characterize the contribution of this part of the solar spectrum to the total photovoltaic module temperature increase. The quantification of this temperature increase due to the unusable near-infrared absorbed photons would provide solid experimental data in order to discuss about the necessity of designing cooling solutions, such as photonic ones [4]. Besides, one should not forget about the possible increase in the photovoltaic module service time due to the decrease in its operating temperature. Starting from this motivation, the scope of the investigation has been broadened to a systematic measurement of the contribution of the different spectral parts of the incident sunlight to the temperature increase. This investigation has been realized on encapsulated monocrystalline silicon solar cells in open circuit and also at maximum power conditions.

2 EXPERIMENTAL

2.1 Experimental set-up

A schematic representation of the experimental set-up used for spectral characterization of the temperature increase is shown in Figure 1. This indoor experimental set-up is composed by several parts, namely the illumination system (composed of a xenon lamp, several spectral filters and a diaphragm), two thermocouples (T-type; one used to monitor the environmental temperature and the second one attached to the back side of the encapsulated solar cell) and the electronic source and meter, which has been used to obtain the solar cell I/V curve and to set the solar cell at the maximum power point of the I/V curve. Two spectral filters have been used in order to match the spectral distribution of the xenon lamp to a solar simulator of class A.

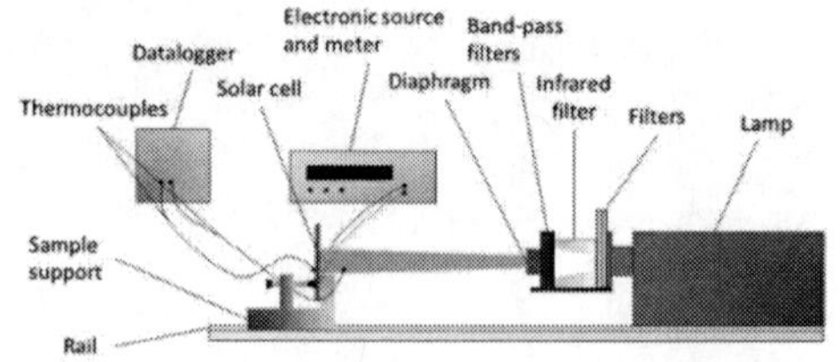

Figure 1: Schematic drawing of the indoor experimental set-up used for the spectral characterization of the temperature increase.

The irradiance on the solar cell was varied changing its distance to the lamp output (which has a divergence semi angle of approximately 14 °), and the diaphragm was used to limit the irradiation only to the solar cell area (see Figure 2). The temperature increase undergone by the illuminated solar cell was monitored by one thermocouple attached to the back side of the solar cell, and its evolution was recorded by a data logger.

Figure 2: Picture of the illuminated solar cell.

The monocrystalline solar cell (with Al-Back Surface Field, and dimensions of 15 mm x 15 mm x 180 μm) was encapsulated using a standard sandwich of materials made of glass-EVA-solar cell-EVA-Tedlar®, with a final larger size of 24 mm x 36 mm.

2.2 Measurement procedure

The temperature increase of a solar cell is the result of an energy balance. On one hand, the amount of the incident sunlight, which is not converted into electric power, is transformed into heat. This heat is dissipated into the environment through conduction, convection and radiative emission, and the environmental temperature is a fundamental parameter in this balance. Therefore, the temperature increase was calculated as the final stabilized solar cell temperature minus the environmental temperature. The error in the temperature increase measurement, obtained statistically, was estimated in approximately 5%, for the range of temperatures measured in this work.

In the procedure applied in the present study, firstly, the solar cell is correctly positioned within the illumination beam at a selected distance from the lamp output. Next, the I/V curve is measured and the relevant electric parameters are obtained from it. After allowing the solar cell to cool down to environmental temperature, it is again illuminated during approximately 10 minutes, until the solar cell temperature reaches an equilibrium with the environment, and the temperature increase is monitored. At the same time, the environmental temperature is also measured. This process is repeated for different irradiance levels obtained varying the distance between the lamp output and the solar cell. The procedure is repeated using different filters in order to spectrally filter the incident light falling upon the solar cell.

The solar cell temperature increase has been measured in open circuit and also at maximum power point situation. It is expected that the solar cell working at its maximum power point will show a minor temperature increase than the solar cell at open circuit condition, since part of the incident light is extracted as electric power and therefore not converted into heat.

3 EXPERIMENTAL RESULTS

In the following sections the results regarding the spectral characterization of the temperature increase are shown. In the first subsection, the temperature increase variation filtering out the near-infrared part of the spectrum incident on the solar cell will be compared to the unfiltered incident light situation. The second subsection is related to the solar cell temperature increase being irradiated with band-pass filters, with central wavelengths ranging from 400 nm to 750 nm.

3.1 Temperature increase comparison filtering out the near-infrared spectrum of incident light

In this case, a 5 cm optical glass cuvette filled with water was used as the near-infrared filter. In Figure 3 the incident xenon lamp spectrum with and without the infrared filtering is shown, together with the spectral transmittance of the 5 cm length cuvette, filled with water.

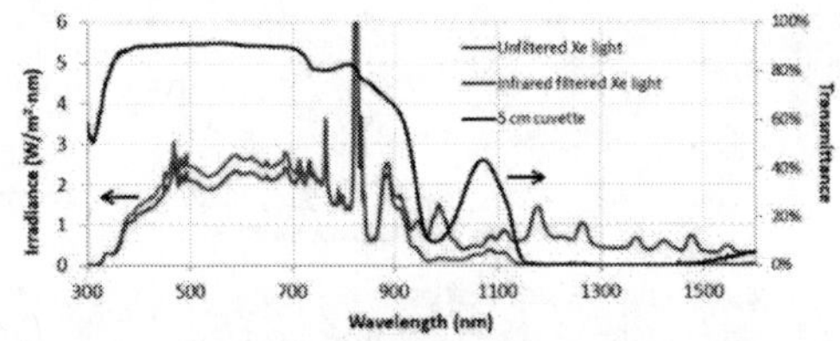

Figure 3: Incident Xe lamp spectrum with and without infrared filtering, and spectral transmittance of the infrared filter.

Following the procedure described before, firstly, the temperature increase has been measured illuminating the encapsulated solar cell with the unfiltered incident light with the solar cell in open circuit condition. Following, the temperature increase has been measured with the infrared filtered incident light, with the solar cell also in open circuit condition. The temperature increase (ΔT) was calculated with respect to the environmental temperature, as $\Delta T = T_f - T_e$, with T_f the solar cell final equilibrium temperature and T_e the environmental temperature. The temperature increase is plotted with respect to the solar cell short circuit current, which is directly related to the usable part of the incident light (see Figure 4), allowing a better comparison of the effect of rejecting the sub-band-gap infrared photons on the final solar cell temperature.

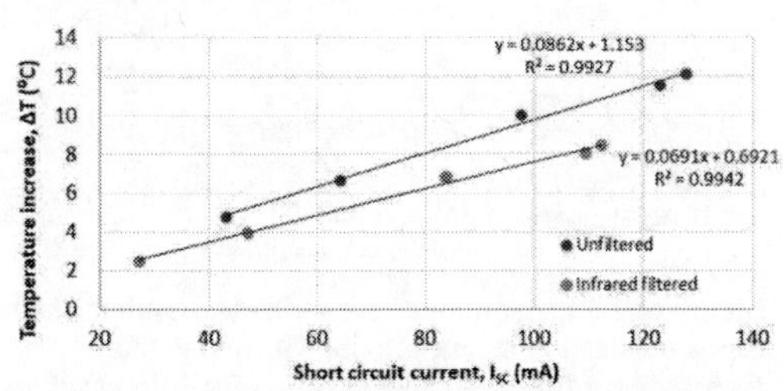

Figure 4: Temperature increase for an encapsulated solar cell in open circuit situation with and without filtering the infrared incident radiation for different short circuit currents.

As can be appreciated, from the previous figure the temperature increase shows a linear dependence with the short circuit current, being lower when the solar cell is illuminated filtering out the infrared light. More specifically, the temperature shows an average decrease of 23.5% when filtering out the infrared part of the incident light. It needs to be pointed out that in this lamp, the infrared light (above 1200 nm) accounted for 13.0% of the total light, while for the sunlight this contribution is of 16.6%. Therefore the decrease in the final temperature due to the infrared filtering could be slightly higher in outdoor conditions.

With a calibrated spectroradiometer, the relation between the solar cell short circuit current and the total incident irradiance can be obtained, allowing the representation of the temperature increase related to the incident irradiance, as can be appreciated from Figure 5.

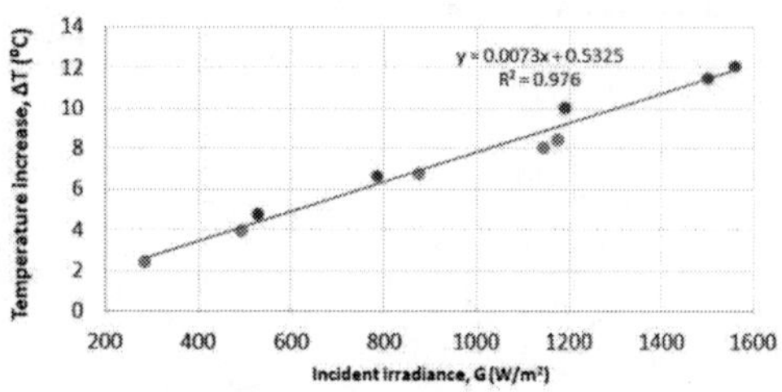

Figure 5: Temperature increase of the encapsulated solar cell related to the total incident irradiance.

As can be appreciated from the previous figure, the temperature increase shows a linear dependence with the incident irradiance, as expected. It needs to be pointed out that in outdoor conditions larger temperature increases of about 30 ºC at an incident irradiance of 1000 W/m^2 have been reported in the literature [6]. The reason for this difference could be due to the less dense packaging of the solar cells used in the present study, which might help the heat dissipation. Thermal performance simulations have been realized, which support this fact.

Next the temperature increase undergone by the encapsulated solar cell filtering out the infrared part of the incident light has been measured with the solar cell now at the maximum power point, and the results are shown in Figure 6.

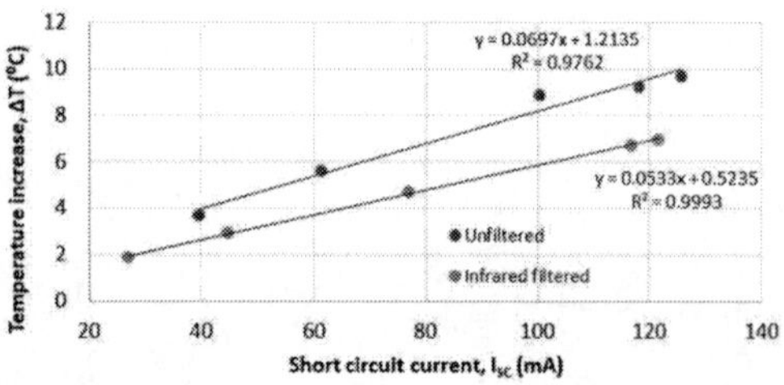

Figure 6: Temperature increase for an encapsulated solar cell at the maximum power point with and without filtering the infrared incident radiation for different short circuit currents.

As can be appreciated from the previous figure, again the temperature increase is lower when filtering the infrared part of the incident light. In average, the temperature increase is 30.9% lower when filtering the infrared light, in comparison to the unfiltered incident light situation. Comparing with the results obtained with the solar cell in open circuit situation, the temperature increase is lower at maximum power point, as expected since in this situation part of the absorbed irradiation is converted into electric power. In average, the temperature increase is 19% lower in the case of the solar cells working at the maximum power point in comparison to the open circuit condition.

3.2 Spectral characterization of the temperature increase using band-pass filters

In this section, the temperature increase experienced by the solar cell when being selectively irradiated at different wavelengths is investigated. In Figure 7 the irradiance of the 8 different spectral filters is shown.

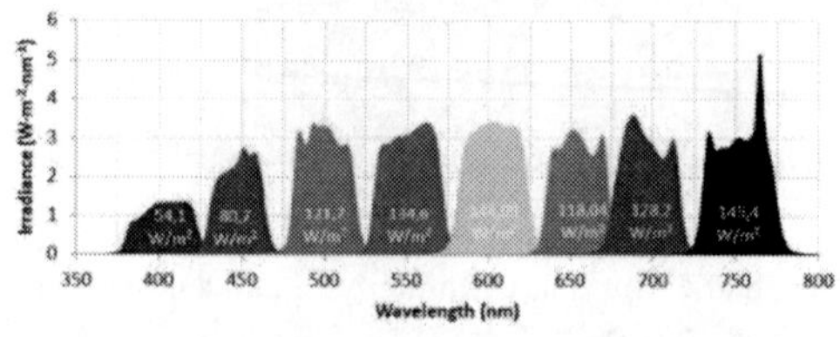

Figure 7: Irradiance of the 8 band-pass filters used in this work.

In this case, the temperature increase could only be measured at the highest irradiance of the xenon lamp, due to the low transmittance of the spectral filters producing low temperature increases. The results of the temperature increase experienced by the solar cell, being irradiated with spectrally selective light, and at the maximum power point are shown in Figure 8.

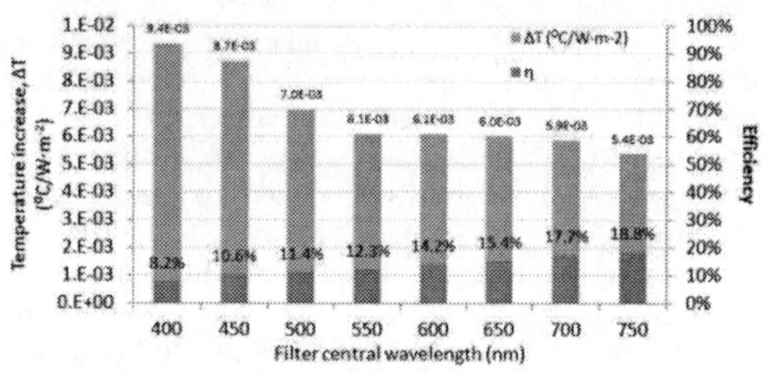

Figure 8: Solar cell temperature increase measured at its maximum power point and solar cell efficiency for different incident wavelengths.

As can be appreciated from Figure 8, the temperature increase declined from 9.4E-3 ºC/(W·m^{-2}) until 5.4E-3 ºC/(W·m^{-2}) for incident wavelengths from 400 nm to 750 nm. The photons with the lowest wavelength show a higher contribution to the temperature through thermalization, since their energy difference, with respect to the silicon band-gap is higher.

4 CONCLUSIONS

In this work the contribution to the temperature increase of different spectral regions of the incident light has been experimentally measured on encapsulated crystalline silicon solar cells. On one hand, the effect of rejecting the infrared part (wavelengths above 1200 nm) of the incident light has been measured and compared to the unfiltered incident light situation, for the encapsulated solar cell at open circuit and maximum power point conditions. In open circuit, the temperature increase showed a reduction of 23.5% when filtering out the infrared light, and leaving the usable part of the incident light intact. This reduction in the temperature increase was of 30.9% when the solar cell was working at its maximum power point.

Besides, the temperature increase undergone by the solar cell when being selectively irradiated at different wavelengths was investigated using 8 band-pass filters, ranging from 400 nm to 750 nm.

In this respect, the temperature increase declined from 9.4 °C until 5.4 °C, per 1000 W/m^2, for incident wavelengths from 400 nm to 750 nm. The photons with the lowest wavelength showed a higher contribution to the temperature through thermalization, since their energy difference, with respect to the silicon band-gap is higher.

ACKNOWLEDGEMENT

This work was cofinanced by the General Directorate of Industry, Energy and Innovation of the Government of Navarre, under the project TEXTUMODU, in its 2017 call.

5 REFERENCES

[1] D. H. Otth, et al., "Assessing photovoltaic module degradation and lifetime from long term environmental tests" Proceedings of the 1983 Institute of Environmental Sciences 29th Annual Technical Meeting, 121-126, (1983).

[2] Olivier Dupré, "Physics of the thermal behavior of photovoltaic devices", Ph. D. Thesis, INSA de Lyon, (2016).

[3] Linxiao Zhu *et al.*, "Radiative cooling of solar absorbers using a visibly transparent photonic crystal thermal blackbody" PNAS, vol. 112, no. 40, 12282-12287, (2015).

[4] W. Li, *et al.*, "A comprehensive Photonic Approach for Solar Cell Cooling", ACS Photonics, 4, 774-782, (2017).

[5] J. M. Russo *et al.*, "Spectral Characterization of the Temperature Performance of Silicon Solar Cells", Renewable Energy and the Environment Technical Digest, (2011).

[6] M. Muller, "Measuring and Modeling Nominal Operating Cell Temperature (NOCT)" PV Performance Modeling Workshop. Albuquerque

ELECTROLUMINISCENCE IMAGING AND LIGHT BEAM INDUCED CURRENT AS CHARACTERIZATION TECHNIQUES OF MULTI-CRYSTALLINE SI SOLAR CELLS

L.A. Sánchez, A. Moretón, M. Jiménez, M. Guada, S. Rodríguez-Conde, M. A. González, O. Martínez and J. Jiménez
GdS-Optronlab, Dpto. Física de la Materia Condensada
Universidad de Valladolid, Edificio LUCIA, Paseo de Belén 19, 47011 Valladolid, Spain.

ABSTRACT: There is an increasing demand for characterizing multicrystalline solar cells at different stages of its service life. Luminescence techniques, e.g. electroluminescence (EL) and photoluminescence (PL), have acquired a paramount interest in the last years. These techniques are used in imaging mode, allowing to take a luminescence picture at a full wafer/cell scale. This imaging approach is fast and sensitive, but has a low spatial resolution, which avoids a detailed analysis of the defect distribution, which can led to misinterpretations about critical parameters as the minority carrier diffusion length, or the internal and external quantum efficiencies. If one complements these techniques with high spatial resolution techniques, such as light beam induced current (LBIC), one can study the electrical activity of the defects at a micrometric scale, providing additional understanding about the role played by the defects in full wafer/cell luminescence images. The combination of the macroscopic and microscopic resolution scales is necessary for the analysis of the full luminescence images in mc-Si solar cells.
KEYWORDS: Solar cells, multicrystalline silicon, UMG silicon, LBIC, photoluminescence.

1. INTRODUCTION:

Photovoltaic (PV) industry is currently dominated by multicrystaline silicon (mc-Si), due its cost-efficiency ratio. The traditional purification method of metalurgical grade silicon is called Siemens process [1]. This process permits to obtain high-quality solar-grade Silicon (SoG-Si), but it is a very energy-intensive process. SoG-Si produced through metallurgical purification processes is called upgraded metallurgical-grade silicon (UMG Si), and it appears as an alternative method to the Siemens procedure due to its reduced production cost and time [2].

Characterization of the electrical activity in UMG Si solar cells is essential to know the distribution of impurities and how they interact with the crystal defects, namely grain boundaries, and intragrain defects, and thereafter how they can affect the cell efficiency [3]. Luminiscence imaging, both electroluminescence (ELi) and photoluminescence (PLi), and light-beam induced current (LBIC) are powerful techniques for the study of the electrical activity of those defects, and the quantum efficiency (QE) [4-7].

ELi is captured on a forward biased solar cell, while PLi is excited by an homogeneously dispersed laser beam. These techniques can provide a full image of the solar cell in a few milliseconds, which promises its in line use. The EL intensity signal is proportional to the excess minority carrier density, so áreas with a high non radiative recombination centers (NRRCs) density appear as dark contrasted. These regions are usually associated with dislocations, grain boundaries (GBs), micro cracks or defective contacts, also impurities can play a relevant role, especially in poorly purified feddstock. In order to relate Eli/Pli images to the cell efficiency the characterization of the contrast in the ELi/PLi must be done .

In LBIC technique, a laser beam is focused on the cell surface using an optical microscope and it is scanned over a surface area of the cell. The photogenerated carriers produce a current which the intensity depends on the trapping of minority carriers. At least two laser wavelengths are necessary to obtain the quantum efficiency of the explored areas of the cell. This technique permits to record high spatial resolution maps (up to 1 μm) at the expense of long acquisition times. We compare herein the results obtained by Eli/PLi and LBIC on UMG-Si solar cells.

2. EXPERIMENTAL:

UMG mc-Si solar cells were characterized by the ELi /PLi and LBIC techniques. The measurements were carried out on a series of UMG mc-Si solar cells with efficiencies varying between 15 and 18.4%.

A homemade LBIC system was used to obtain detailed information about the defects in the solar cells [8]. The scheme of the LBIC setup is shown in Fig. 1. The LBIC apparatus consists of two dual laser diodes providing four excitation wavelengths (639, 830, 853 and 975 nm), allowing different penetration depths in the cell. In this study the 853 nm and 975 nm lines were used. A beam-splitter divides the output beam of the laser (10 mW) into two beams. One of them is used to measure variations of the output power through a photodiode, while the other one is directed into a trinocular microscope, which focuses the laser beam onto the solar cell. Different objectives (20x, 50x, 100x) enable studies at different spatial resolutions, up to 1 μm using the 100x objective, 0,95 numerical aperture (NA). A Si-CCD camera coupled to the microscope ocular is used to collect the light reflected by the sample to drive the auto-focus system. The LBIC scanning is done by moving the sample with a high precision *x-y-z* motorized translation stage (Prior Scientific) over areas as large as 76.7 x 114.5 mm^2 and step sizes as small as 1 μm. The photocurrent is measured by an electrometer (Keithley Instruments). Data acquisition and hardware control were developed in LabVIEW.

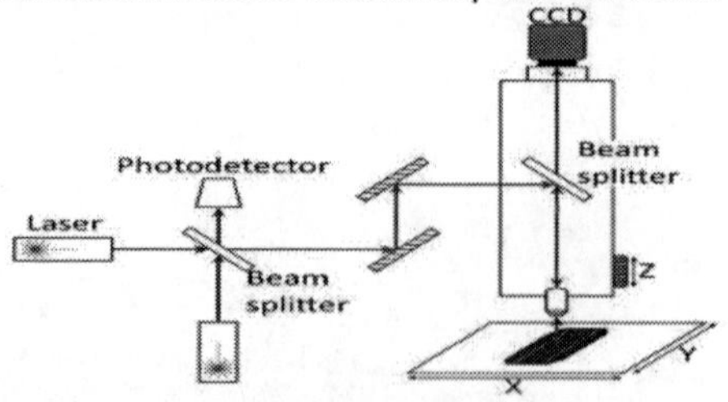

Fig. 1. Scheme of the LBIC system

The Eli/PLi images were acquired using a homemade dual system. A high power 808 nm laser diode was used as the excitation source for PLi measurements. Both PL and El images were acquired with an InGaAs camera.

3. RESULTS AND DISCUSSION:

Figures 2a-c show the EL images obtained in three solar cells, labelled A, B and C, corresponding to different efficiency series, table I. These cells correspond to wafers cut on different positions of the same brick. One can distinguish some patterns common to the three cells, corresponding to grains extending along the brick because of the columnar growth along the solidification direction [9].

The ELi contrast differs for the three cells depending on the position of the wafers in the brick, due to the impurity incorporation gradient, consequence of the directional solidification. At a glance one can establish a correlation between the cell efficiencies, table I, and the darkness of the EL images.

Fig. 2. (a—c) EL images of three solar cells with different efficiencies. The marked areas indicate the zones where LBIC measurements were carried out.

Surprisingly, the grain boundaries (GBs) and intragrain defects in cells A and B present bright contrast, while inside the grains a dark contrast is observed. In order to rule out the possible contribution to this contrast of leakage currents across the GBs [10], photoluminescence images (PLi) were also acquired. The PLi results were similar to ELi; therefore, the bright contrast of the grain boundaries and defects cannot be associated with leakage currents, but with the impurity distribution, and its interaction with the crystal defects. Due to the directional solidification process impurities accumulate in a higher concentration in the upper part of the ingot [9].

The EL intensity calculated as the sum over all the image pixels for the EL image shows a good correlation with the cell efficiency, [11]. A nearly linear correlation between the EL intensity and the cell efficiency was appreciated; evidencing that Eli can be a powerful tool for estimating the cell efficiency

	A	B	C	D
Power (W)	3.85	4.19	4.31	4.46
Average Efficiency(%)	15-16.6	17.2-17.4	17.7-17.8	18.3-18.4

Table. I. Cell efficiencies. A, B and C wafers come from the same mc-Si brick. A is from the top of the brick, B is close to A, and C is closer to the middle of the brick. D is from a different brick.

The LBIC measurements were carried out over the square regions indicated in Figures 1a—c with a 20x objective. 1.8 x 1.8 cm^2 short-circuit current maps obtained with the 853 nm laser are shown in Figures 3a—c. The spatial resolution obtained with the LBIC system allows a more detailed study of the GBs and intragrain defects. In figures 3a and 3b one observes the dark contrast for these defects surrounded by a bright contrasted halo.

Instead of the bright contrast observed in the Eli/PLi images; this discrepancy is due to the different spatial resolutions. The narrow dark contrast lines along the GBs and intragrain defects revealed by LBIC cannot be observed at the full cell scale resolution of the Eli/PLi measurements. The LBIC signal distribution is the consequence of the gettering of impurities by the grain boundaries and the intragrain defects, which leave the surrounding areas depleted of impurities. The impurities are gettered by the strain field of the grain boundaries and dislocations during the solidification. The C series solar cells (figure 3c) do not present this effect and the GBs loss the electrical activity observed in the cells of the upper part of the brick. Note that the GB activity is more marked for cells of type A in which the concentration of impurities is significantly higher as seen by the dark grain bodies as compared to cells B and especially cells C, where the grain cores give bright contrast. This suggests that the electrical activity of the GBs is mainly related to the capture of impurities, and therefore it is governed by the feestock purity[11,12].

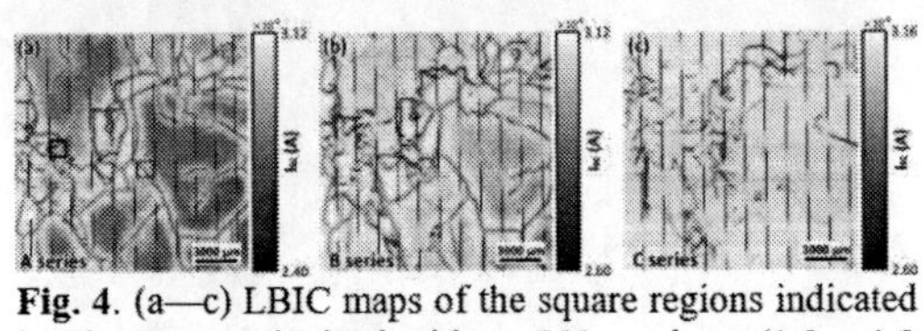

Fig. 4. (a—c) LBIC maps of the square regions indicated in Fig. 2a—c, obtained with an 853 nm laser (1.8 x 1.8 cm2, 20x objective, step size 90 μm).

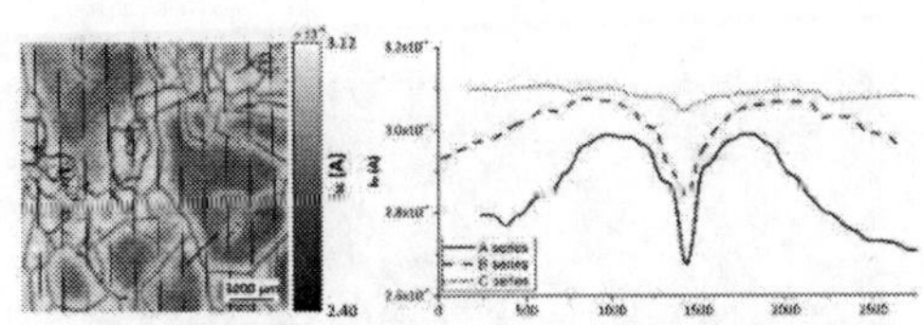

Fig. 5. a) LBIC map of the square region corresponding to A-series solar cell b) LBIC intensity profiles along the red arrow in the LBIC image for A, B and C solar cells series

LBIC intensity profiles across a GB are shown in Fig. 5 for the three cells, A, B and C. A profound carrier depletion is observed for cell A, the depletion is reduced for cell B, and is almost negligible for cell C. This points to the importance of the impurities in the electrical activity of the GBs, the higher the impurity concentration the higher the GB electrical activity. As revealed by the LBIC images, the GBs getter the impurities around them, and the dark contrast at the GBs seems to be related to the presence of impurities, reducing its carrier trapping activity as the material putity improves.

On the other hand, the presence of defects and impurities determines the minority carrier diffusion length (L_{eff}), which can be estimated from the internal quantum efficiencis (IQE) obtained from the LBIC maps for two excitation wavelengths (853 and 975 nm), according to the expression [13]:

$$\frac{1}{IQE(\lambda)} \approx 1 + \frac{1}{\alpha(\lambda)L_{eff}}$$

where $\alpha(\lambda)$ is the Si absorption coefficient at 300 K (534 cm^{-1} for 853 nm and 106 cm^{-1} for 975 nm).
The calculated diffusion length maps for samples A, B and C are shown in Fig. 6

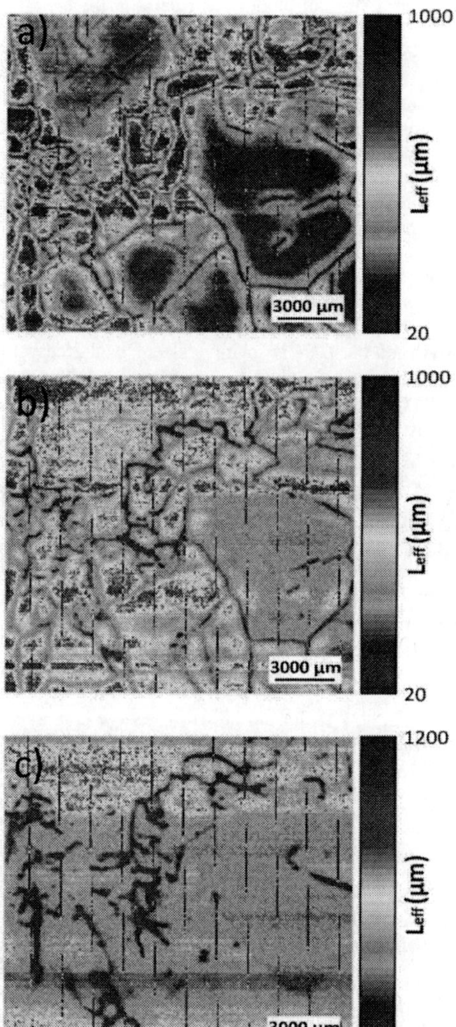

Fig. 6. L_{eff} maps of the A, B and C-series solar cell obtained with 853 nm and 975 nm lasers

The diffusion length increases from A to C. In sample A the diffusion length is severely reduced inside the grains, while it increases around the defects because of the impurity depletion around them. This is the opposite to sample C, where the diffusion length is higher inside the grains, and appears limited at the intragrain defects formed by clusters of dislocations [14]

High resolution LBIC measurements were carried out using a 100x objective with 0.95 numerical aperture (NA). Figure 7 shows 600 x 600 μm² LBIC maps. In this high resolution maps one observes tiny dark spots around the defects. These tiny spots can be associated with metallic clusters without optical and electrical activity, as deduced from the absence of any contrast gradient around them. These tiny precipitates are characteristic of feedstock with low purification as UMG Si. These defects were mainly observed around intragrain defects, and much less frequently around the GBs. This behavior suggests that intragrain defects have a higher impurity gettering efficiency with respect to the GBs.

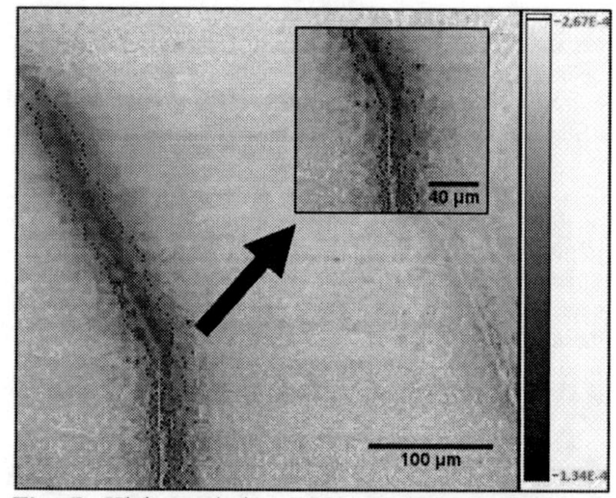

Fig. 7. High resolution LBIC map obtained with an excitation wavelength of 853 nm (600x600 μm2, 100x objective, step size 3 μm) showing an intragrain defect decorated with tiny dark spots

Figure 8 a, b and c shows 400 x 400 μm^2 LBIC maps of A, B and C-series solar cells using a 100x objective. These measurements revealed a slightly bright contrasted line along the core of the defects in all the series. This local bright contrast is due to variations in the light reflectivity.

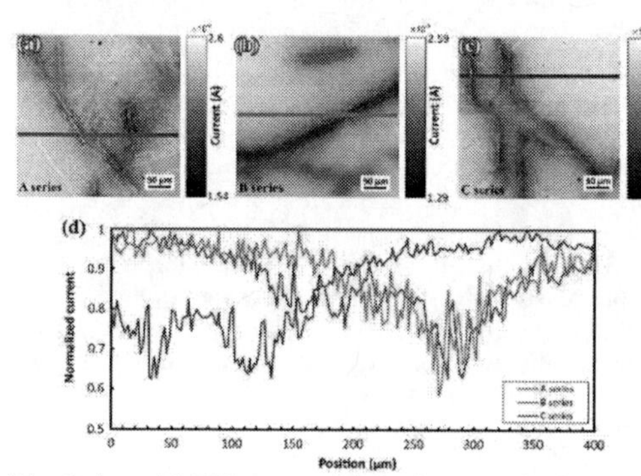

Fig. 8. (a—c) LBIC maps of defective areas in three solar cells of A, B and C series (400x400 μm², 100x objective, step size 2 μm). (d) Normalized photocurrent profiles across the lines marked in Fig. 8a—c.

Figure 8d shows the LBIC contrast profiles across the lines marked in Fig. 8a, b and c. This LBIC contrast is defined by

$$C = \frac{(I_b - I_d)}{I_d}$$

Where I_b is the background photocurrent measured in a defect-free region and I_d is the photocurrent measured in a defective region. I_b (C)>I_b(B)>I_b(A), however, the contrast is similar for the three solar cells near the defects, therefore, the lower efficiencies in the A and B cells is greatly contributed by the presence of impurities inside the grains, which is significantly reduced in cell C.

4. CONCLUSIONS:

UMG Si solar cells with different efficiencies have been characterized through Eli/PLi and LBIC techniques. A full wafer study of the cells with ELi did not reveal the electrical activity of the GBs, but they were blurred by the impurity depleted area around them. High spatial resolution LBIC maps revealed the true trapping activity of the GBs. The LBIC maps of the UMG-Si cells permits to establish the influence of the impurities on the basic parameters of the cells, e.g.L_{eff}, and QE, both external and internal. Also, the interaction between crystal defects in multicrystalline Si and impurities can be analysed. The interpretation of the Eli/PLi full wafer/cell images must be complemented with high spatial resolution measurements.

5. ACKNOWLEDGEMENTS:

This work was supported by the Spanish MINECO project, ref. ENE2014-56069-C4-4-R and "Junta de Castilla y León (Spain)" project number VA081U16. We wish to thank Silicio Ferrosolar for providing the samples studied in this work.

6. REFERENCES:

1. J.O. Odden, H. Halvorsen, H. Rong and R. Gløckner in Silicon for the Chemical and Solar Industry IX (Oslo, Norway, 2008) pp. 75-90.
2. J. Safarian, G. Tranell and M. Tangstad, Energy Procedia **20**, 88-97 (2012).
3. F. Rougieux, C. Samundsett, K.C. Fong, A. Fell, P. Zheng, D. Macdonald, J. Degoulange, R. Einhaus and M. Forster; Progr. Photovolt. Res. Appl. **24**, 725-734 (2016).
4. T. Kirchartz, A. Helbig , W. Reetz , et al.; Progr. Photovolt. Res. Appl. **17**, .394-402 (2009).
5. T. Potthoff, K. Bothe, U. Eitner, et al.; Progr. Photovolt. Res. Appl. **18**, 100-106 (2010)
6. U. Rau; Phys. Rev. B, **76**, 085303 (2007)
7. T. Trupke, B. Mitchell, J. W. Weber, W. McMillan, R. A. Bardos and R. Kroeze, Energy Procedia **15**, 135-146 (2012).
8. B. Moralejo, M. A. González, J. Jiménez, V. Parra, O. Martínez, J. Gutiérrez and O. Charro, J. Electron. Mater. **39**, 663–670 (2010).
9. T. Liu, Z. Dong, Y. Zhao, J. Wang, T. Chen, H. Xie, J. Li, H. Ni and D. Huo, J. Cryst. Growth **355**, 145-150 (2012)
10. O. Breitenstein, J.P. Rakotoniaina, M.H. Al Rifai, M. Werner; Progr. Photovolt. Res. Appl. **12**, 529-538 (2004)
11. L.A.Sánchez, A. Moretón, M. Guada, S. Rodríguez-Conde, O. Martínez, J. Jiménez; MRS Advances 2018 DOI: 10.1557/adv.2018.366
12. T. Buonassisi et al.; Progr. Photovolt. Res. Appl **14**, 513-531 (2006)
13. M.K. Juhl, M.D. Abbott, T. Trupke, IEEE J. Photovolt. **7**, 1074–1080 (2017).
14. B.R. Ryningen, G. Stokkan, M. Kivambe, T.Ervik, O. lohne; Acta Mater. **59**, 7703-7710 (2011)

IV-MEASUREMENTS OF BIFACIAL SOLAR CELLS IN AN INLINE SOLAR SIMULATOR WITH DOUBLE-SIDED ILLUMINATION

A. Krieg[1], J. Greulich[1], K. Ramspeck[2], D. Dzafic[2], N. Wöhrle[1], M. Rauer[1] and S. Rein[1]
[1]Fraunhofer Institute for Solar Energy Systems ISE, Heidenhofstraße 2, 79110 Freiburg, Germany
[2]h.a.l.m. elektronik GmbH, Friesstraße 20, 60388 Frankfurt/Main, Germany
Telephone: +49 761 4588 5488, Fax: +49 761 4588 7621, e-mail: alexander.krieg@ise.fraunhofer.de

ABSTRACT:
Bifacial solar cells allow higher output power densities due to double-sided power generation compared to monofacial cells. An important step is the accurate measurement of the current-voltage (*IV*) characteristics of the cells under realistic conditions. The most accurate option to measure a bifacial solar cell is to illuminate the cells simultaneously from both sides. A new method for irradiance and short-circuit current correction for bifacial solar cells suitable for inline measurements in mass production is presented, where the crosstalk during a single-sided measurement and deviations of the ratio of the nominal to the measured irradiance at double-sided illumination measurements will be corrected by taking the bifaciality coefficient into account. This reduces typical measurement errors for the short-circuit current I_{sc} below 0.1%. The simple approach, adding the front and the rear side irradiance directly to calculate the bifacial irradiance, shows I_{sc} errors of more than 0.5% in typical measuring conditions and should not be used when the solar cell is double-sided illuminated.
Keywords: bifacial solar cell, bifaciality, *IV*-measurements

1 INTRODUCTION

A promising approach to increase energy yield of photovoltaic devices are bifacial solar cells and modules. These devices use the same or even a reduced amount of resources, but allow higher output power densities due to double-sided power generation and thus further reduce levelized cost of electricity compared to monofacial cells and modules. An important step is the accurate measurement of the current-voltage (*IV*) characteristics of the cells under realistic conditions within a setup that is applicable in the photovoltaic (PV) industry. The most accurate and realistic option is to illuminate the cells simultaneously from both sides. Single-sided illumination is easier to be realized, but either the cell has to be measured twice, on the front and on the rear side [1, 2], which is time consuming, or - with equivalent irradiance using the "compensated current" method [3] - the actual variation of the bifaciality coefficient φ of the cells to be measured increases the measurement uncertainty because it has to be assumed constant. With a front side illumination only, variations of the rear side efficiency cannot be measured and higher prices for a claimed power output gain through the rear side are hard to achieve.

However, using a double-sided flash illumination with two independent flash units, the cross talk between the upper and lower part of the measurement chamber has to be taken into account. The measured irradiance under consideration of the bifaciality of the cells has to be corrected to the nominal irradiance by introducing a new advanced method for the evaluation of the short-circuit current. Thus two measurement methods have to be evaluated and compared: a) The simple approach, adding the front and the rear side irradiance, measured by the front and rear monitor cells, to calculate the bifacial irradiance; and b) the new advanced method introduced in this work.

2 CHARACTERISTICS OF THE BIFACIAL CELL TESTER

In order to be able to illuminate and measure bifacial solar cells from the front and the rear side simultaneously in one step, the automated cell tester in the PV-TEC laboratory at Fraunhofer ISE was upgraded in cooperation with the companies h.a.l.m. elektronik and Jonas & Redmann by a bifacial measuring setup with an additional flash box for simultaneous and independent front- and rear-side illumination.

The rear-side flash box of the cell tester, which is synchronized with the front side box and regulated independently, is installed horizontally under the rotating table (Figure 1). A tilted mirror underneath the solar cell directs the light to the rear side of the solar cell. The 45° tilted mirror is mounted on a linear axis and can be moved from the measuring position into a protected rest position under the flash box when it is not in use. In addition to an aluminum-coated mirror, which is designed to reflect the light as unchanged as possible with respect to its spectrum onto the solar cell, other reflectors can be installed on the mount of the linear axis. These allow the electrical parameters of the solar cell to be measured in dependence of specific reflectors, e.g. roofing tiles or floor covers, representing different albedos [4].

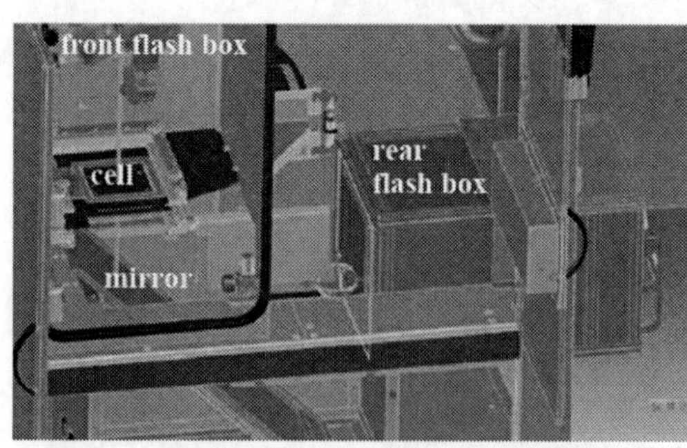

Figure 1: CAD drawing of the inline measuring setup for bifacial cells with the horizontally aligned flash box on the right, the aluminum-coated mirror on the left and the solar cell on the rotating table. Two monitor cells, to measure the front and rear irradiance separately, are mounted on left of the cell.

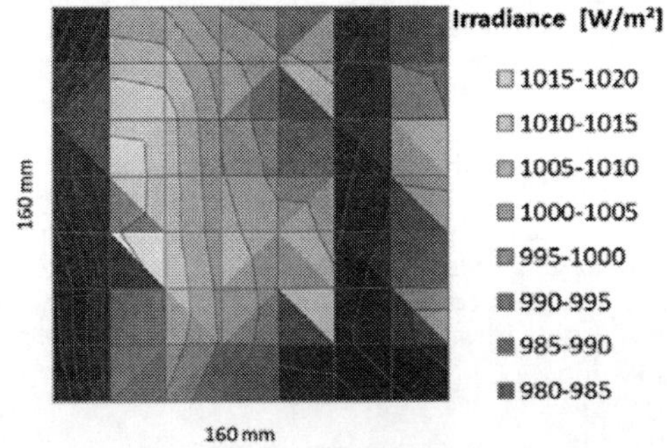

Figure 2: Irradiance measured on cell level at the rear side over an area of 160x160 mm^2 using a mirror to redirect the light to the bifacial solar cell.

The measurements of the irradiance uniformity show that the frame on which the contact bars are mounted affects uniformity of the rear-side illumination. It is also affected as light from the lower part of the flash box is not reflected to the rear side via the mirror but reaches the cell directly. The direct contribution of the irradiance amounts to 13%. However, measurements after an optimization of the uniformity show that a non-uniformity of ± 1.8% on an area of 160 x 160 mm^2 can be achieved, fulfilling the requirements of IEC norm 60904 for single-sided STC measurements.

In order to prevent the front and rear illumination affecting each other, the upper part of the measuring chamber is separated from the lower part by a black plate with a cutout of 166x166 mm^2, which corresponds to the contour of the solar cell to be measured. Due to this passe-partout the cross talk is minimized but not completely prevented. A first monitor cell is mounted next to the cutout on the top half in order to measure the front irradiance. A second monitor cell, which measures the irradiance on the rear side, is mounted face down beside the solar cell in the lower part of the measurement chamber.

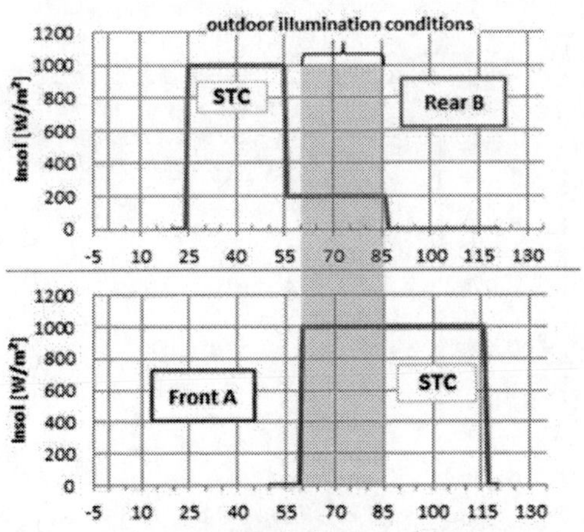

Figure 3: The flash sequences facilitate the STC measurements of the rear and front side and a bifacial measurement under outdoor illumination conditions. Graph taken from [6].

With the configured flash sequence at the cell tester enabling the STC measurements of the front and the rear side and a bifacial measurement within 90 ms (Figure 3), the bifaciality coefficient can be calculated for each cell individually. With the described measuring station, *IV*-measurements with combined bifacial irradiance G_{bi} up to 1700 W/m^2 are possible.

3 IRRADIANCE CORRECTION

3.1 Calibration

The front flash intensity in the inline tool is calibrated to STC using a reference cell calibrated at Fraunhofer ISE CalLab PV cells. The front flash intensity is adapted to achieve the same short-circuit current of the cell's front side in both systems. The rear flash intensity is calibrated the same way matching the short-circuit current of the cell's rear side to STC (1000 W/m^2, AM1.5g, 25°C). A calibration under bifacial illumination is not carried out. How the irradiance is measured and its impact is accounted for, is described in the following.

3.2 Single-sided illumination

In order to be able to compare measurements with each other, the standard test conditions define the nominal front irradiance $G_{0,f}$ = 1000 W/m^2 for the P_{max} measurement.

Assuming that the generated current in the solar cell is linear to the irradiance and a non-reflective chuck is used for the front and the rear measurement, the bifacial short-circuit current $I_{SC,bi}$ is the sum of front short-circuit current $I_{SC,f}$ and the rear short-circuit current $I_{SC,r}$:

$$I_{SC,bi} = I_{SC,f} + I_{SC,r} \quad (1)$$

Since the irradiance G_{meas} at cell level, which is measured by the monitor cell, is not equal to the nominal irradiance G_0 during the measurement of the *IV*-curve, every single measurement point has to be corrected to the nominal irradiance G_0.

The measured uncorrected short-circuit current $I_{SC,uncor}$ using a single-sided illuminated cell, has to be corrected with:

$$I_{SC} = I_{SC,uncor} \cdot \frac{G_0}{G_{meas}} \quad (2)$$

3.3 Double-sided illumination

According to the draft of the IEC norm for measuring bifacial cells, the nominal rear irradiance of $G_{0,r}$ = 200 W/m^2 and $G_{0,r}$ = 100 W/m^2 should be applied in addition to the nominal front irradiance of $G_{0,f}$ = 1000 W/m^2 because of their property to generate additional current when light is impinges on the rear side of the cell [5].

When measuring with double-sided illumination the cells cannot be put on a lowly reflective chuck. Instead they rest on small vacuum holders during inline measurements or directly on the rear contact bars in an offline tester at the PV-TEC and are surrounded by a passe-partout. A tiny light transmitting gap between the cell and the passe-partout is unavoidable, which leads to a light reflection in the measurement chamber and onto the other side of the cell, even if only one flash is active. This light is detected by the monitor cell of the other side of the cell and has to be taken into account for the irradiance correction. A simplified correction of the irradiance $G_{meas,S,bi}$ is the sum of the measured front and rear side irradiance:

$$G_{meas,S,bi} = G_{meas,f} + G_{meas,r} \quad (3)$$

with the subsequent correction of the measured current with equation (2) substituting G_{meas} with $G_{meas,S,bi}$. But this simplified procedure only works 100% correctly when only one side of the cell is illuminated or for cells with bifaciality coefficient $\varphi_{Isc} = 1$. The response of a bifacial solar cell to light reaching the front or the rear side is in most cases not equal, which is described by the bifaciality coefficient φ_{Isc} for the short-circuit current:

$$\varphi_{Isc} = \frac{I_{SC,r}}{I_{SC,f}} \quad (4)$$

Assuming a bifaciality coefficient smaller than 1, the same amount of irradiance causes a lower current at the rear side compared to the front side and thus causes a lower total current. Hence, the usage of equation (3) combined with equation (2) for the irradiance correction leads to an underestimation of the front side current.

A more accurate approach is described in the following. In case of front short-circuit current measurement where no rear side contribution is desired, but not completely avoidable with the present bifacial setups, an equivalent measured front side irradiance $G_{meas,E,f}$ can be calculated by considering the cell's bifaciality coefficient:

$$G_{meas,E,f} = G_{meas,f} + G_{meas,r} \cdot \varphi_{Isc} \quad (5)$$

and accordingly for the rear side irradiance $G_{meas,E,r}$ with no desired front side irradiance:

$$G_{meas,E,r} = G_{meas,f} \cdot \frac{1}{\varphi_{Isc}} + G_{meas,r} \quad (6)$$

The corrected short-circuit current of the front side $I_{SC,f}$ for a nominal front irradiance $G_{0,f}$ with the equivalent measured front side irradiance using equation (2) is:

$$I_{SC,f} = I_{SC,uncor} \cdot \frac{G_{0,f}}{G_{meas,f}+G_{meas,r} \cdot \varphi_{Isc}} \quad (7)$$

and for the rear side:

$$I_{SC,r} = I_{SC,uncor} \cdot \frac{G_{0,r} \cdot \varphi_{Isc}}{G_{meas,f}+G_{meas,r} \cdot \varphi_{Isc}} \quad (8)$$

In principal the single side measurement can be considered as bifacial measurement with a nominal irradiance of $G_0 = 0$ W/m² for the other side. Putting equation (7) and (8) into equation (2) we obtain for the bifacial short-circuit current:

$$I_{SC,bi} = I_{SC,uncor} \cdot \frac{G_{0,f} + G_{0,r} \cdot \varphi_{Isc}}{G_{meas,f}+G_{meas,r} \cdot \varphi_{Isc}} \quad (9)$$

With this the equation the in terms of irradiance uncorrected short-circuit current of double-sided illuminated cells can be corrected to any nominal current for the front side and simultaneously for the rear side. The differences in the response between the front and rear side of a bifacial solar cell are taken into account.

The attentive reader has noticed that we use the short-circuit currents to calculate the bifaciality (equation 4), and then use the bifaciality to calculate the short-circuit currents (equation 9), which is circular reasoning. To resolve this, the bifaciality of a reference cell, which should be similar to the cells to be measured, can be applied in equation 9. Deviations of the bifaciality of the reference and the test cells constitute only a second-order error for the current measurement, compared to a first-order error when using the "compensated current" method.

4 RESULTS

4.1 Cross talk

Even in a measurement where only one flash is used, light reaches the side of the solar cell that is not intended to be illuminated.

If the simple calculation method according to equation (3) is used for the irradiance correction, in which the measured front and rear irradiance are simply added and which only is valid for a bifacial coefficient $\varphi_{Isc} = 1$, the $I_{SC,f}$ for a cell with $\varphi_{Isc} = 0.6$ is 0.4%$_{rel.}$ higher using the advanced method assuming 10 W/m² cross talk (Figure 4).

This error is avoided completely with the more accurate approach (equations 5 and 6) if the bifaciality of the reference and the test cells are the same. In case of a deviation between the bifaciality coefficient of the reference cell which is used for the calibration of the cell tester and the cell to be measured, a second-order error is made: At an assumed deviation of 0.1 of the bifaciality coefficient between the reference cell and the test cell and an irradiance combination of $G_{meas,f}$ = 1000 W/m² and $G_{meas,r}$ = 10 W/m² coming from the crosstalk, this error amounts to an increase in $I_{SC,f}$ of only 0.1%$_{rel}$. A recalculation of the bifaciality coefficient with the corrected short-circuit currents of the front and the rear side from the STC measurements and a subsequent application of equation (9) will further reduce this measurement error.

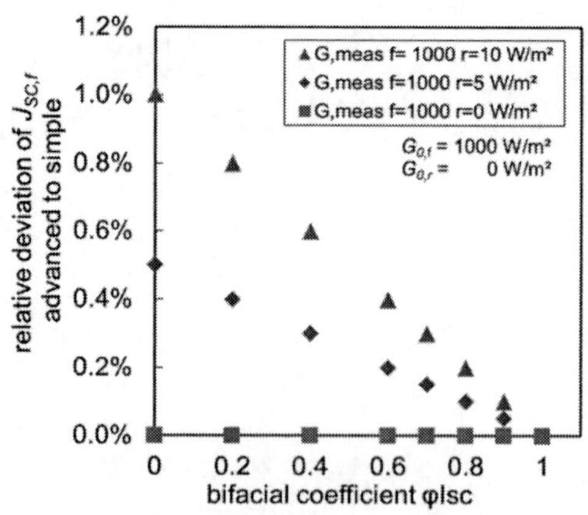

Figure 4: The deviation of $I_{SC,f}$ comparing the simple with the advanced method depends on the bifaciality coefficient and the irradiance from the rear side caused by the cross talk using a single flash.

4.2 Ratio of the nominal to the measured irradiance

For comparability reasons, bifacial measurements should be done with the nominal irradiances of $G_{0,bi} = 1100$ W/m² and $G_{0,bi} = 1200$ W/m². For bifacial measurements with irradiance variations at the front and the rear side of ±10 W/m², which is a typical range around the nominal irradiance $G_{0,f} = 1000$ W/m² and $G_{0,r} = 200$ W/m², the deviation of the $I_{SC,bi}$ also depends on the bifaciality coefficient and the combination of the measured irradiance at the front and the rear side and $I_{SC,bi}$ can be over- or underestimated (Figure 5) if the bifaciality of the cell is not taken into account.

If the ratio of the measured irradiance is the same as the nominal irradiance, for example $G_{meas,f} = 1030$ W/m² and $G_{0,r} = 206$ W/m², the deviation is zero.

The deviation of $I_{SC,bi}$ with $\varphi_{Isc} = 0.6$ and $\varphi_{Isc} = 0.7$, which corresponds to a typical spread of bifaciality coefficients for cells in production, for the irradiance combination $G_{meas,f} = 990$ W/m² and $G_{meas,r} = 210$ W/m² is only 0.08%$_{rel}$. because deviations of the bifaciality constitute only a second-order error. The same deviation in the bifaciality coefficient would lead to a deviation in $I_{SC,bi}$ of 1.75%$_{rel}$ when using the "compensated current" method, due to the to the first-order error characteristics of the bifaciality.

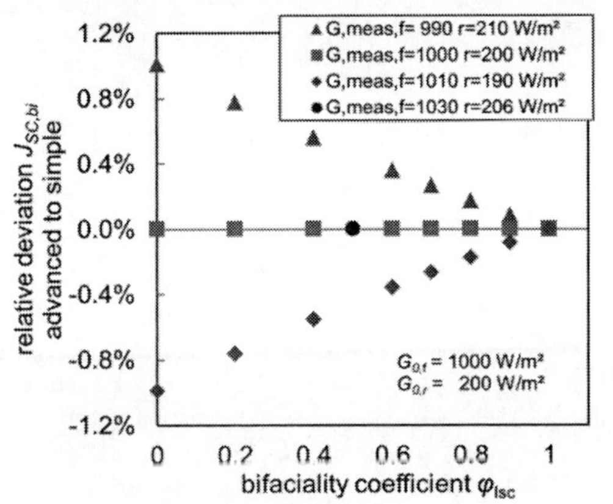

Figure 5: The deviation of $I_{SC,bi}$ comparing the simple with the advanced method depends on the bifaciality coefficient and the ratio of the irradiance from the front and the rear side using two flashes simultaneous.

4.3 *IV*-measurements

To verify the correction method of the bifacial short-circuit current $I_{SC,bi}$ described in section 3, measurements are carried out at the bifacial cell tester at h.a.l.m. elektronik's lab. The tester was calibrated with a reference solar cell which was measured on the front and the rear side under STC in the CalLab PVcells with single-sided illumination on a dark chuck. The bifaciality coefficient of the cell is $\varphi_{Isc} = 0.61$.

Afterwards this cell was measured from the rear and front side under STC and with bifacial illumination with a nominal front irradiance of $G_{0,f} = 1000$ W/m² and a nominal rear side irradiance of $G_{0,r} = 200$ W/m² within one measurement sequence in the bifacial h.a.l.m. tester (Figure 6). The measured irradiance of the front side lies with 1120 W/m² for both measurements clearly above the nominal front irradiance.

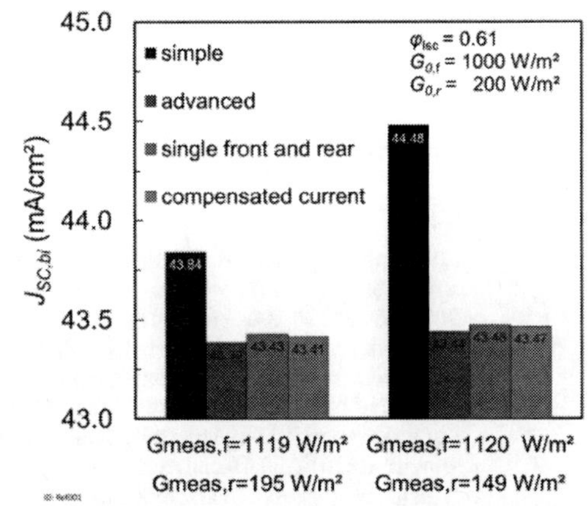

Figure 6: Comparison of the $J_{SC,bi}$ calculated with four different methods. The short-circuit current density $J_{SC,bi}$ calculated where $G_{meas,f}$ and $G_{meas,r}$ are simply added shows in comparison to the others methods an increased $J_{SC,bi}$.

The measured irradiance of the rear side during the first measurement lies with 195 W/m² close to the nominal irradiance and for the second measurement, with 149 W/m², clearly under the nominal irradiance. The short-circuit current density $J_{SC,bi}$ calculated with equation (3) where $G_{meas,f}$ and $G_{meas,r}$ are added, leads to a significantly increased value (0.9-2.3%). The method where the compensated irradiance is used to correct the double-sided measurement agrees well with the $J_{SC,bi}$ coming from the two independent STC measurements and the "compensated current" method for both measurements with deviations below 0.1%$_{rel}$.

5 CONCLUSION

The advanced method proposed in this work leads to a more precise determination of the bifacial short-circuit current $I_{SC,bi}$ by using the bifaciality coefficient φ_{Isc} during simultaneous double-sided illumination measurements. The crosstalk of front and rear illumination during a single-sided measurement and deviations of the ratio of the nominal to the measured irradiance during double-sided illumination measurements are corrected by taking the bifaciality coefficient into account. The error in $I_{SC,bi}$ caused by the deviation of the bifaciality coefficient between the reference cell and the test cell of 0.1 is in typical measuring situations with 0.1% reasonable. The simple approach, adding the front and the rear side irradiance to calculate the bifacial irradiance should not be used for cells with a bifacial coefficient smaller than 1 when the solar cell is double-sided illuminated.

ACKNOWLEDGEMENTS

This work was conducted within the project "BiZePS" (FKZ: 0325909), supported by the German Ministry for Economic Affairs and Energy.

REFERENCES

[1] Hohl-Ebinger, J. and Warta, W. Bifacial solar cells in STC measurement. in Proceedings of the 25th European Photovoltaic Solar Energy Conference and Exhibition. 2010. Valencia, Spain.

[2] Singh, J. P., Walsh, T. M. and Aberle, A. G. (2014), A new method to characterize bifacial solar cells. Prog. Photovolt: Res. Appl., 22: 903–909. doi:10.1002/pip.2341

[3] V. Fakhfouri et al., IEC 60904-1-2: Measurement of current-voltage characteristics of bifacial photovoltaic devices, Presentation at 3rd bifi PV workshop, Miyazaki, Japan, (2016).

[4] Duran, C.(2012). Bifacial Solar Cells : High Efficiency Design, Characterization, Modules and Applications, http://nbn-resolving.de/urn:nbn:de:bsz:352-205361. Dissertation, Uni Konstanz.

[5] draft of IEC 60904 Part 1-2, not published

[6] K. Ramspeck et al., Measurement concepts for bifacial PV-devices- What is really required?, Presentation at 3rd bifi PV workshop, Miyazaki, Japan, (2016).

An examination into the optical coupling between light funnel arrays and underlying substrates

Ashish Prajapati*, Yuval Nissan*, Tamir Gabay* and Gil Shalev**†
*Department of Electrical & Computer Engineering, Ben-Gurion University of the Negev, POB 653, Beer-Sheva 8410501, Israel
**The Ilse-Katz Institute for Nanoscale Science & Technology, Ben-Gurion University of the Negev, POB 653, Beer-Sheva 8410501, Israel.
† Corresponding author: E-mail: glshalev@bgu.ac.il

ABSTRACT: Light trapping and the broadband absorption of the solar radiation are important to various energy harvesting applications. Broadband absorption enhancement was demonstrated with arrays of free-floating silicon light funnels (LF). In the following we use finite-difference time-domain electromagnetic calculations to examine the optical response of the substrate-LF array complexes. We show that the presence the LF arrays on top of an underlying substrates can induce, on top of the strong excitations of the arrays, strong excitations of optical modes in the substrates.
Keywords: *light trapping, broadband absorption, nano photonics, thin film solar cells*

1. Introduction

Light trapping and the broadband absorption of the solar radiation is of interest to various solar energy applications[1–4]. The realization of silicon thin film photovoltaics (PV) technology is challenging due to the indirect nature of silicon bandgap which entails low absorption coefficient. Recently it was demonstrated that the Yablonovitch limit for light trapping can be exceeded using surface arrays of subwavelength structures[5–7]. The excitation of array Mie modes and hybridization of these with substrate guided modes and/or radiation modes, for example, allows light trapping beyond the Yablonovitch limit[8]. Wallentin et al demonstrated InP NP arrays exceeding the ray optic limit with 13.8% photovoltaic efficiency[6]. Yu et al. theoretically demonstrated beyond the Yablonovitch limit with a statistical temporal coupled-mode approach based on wave optics[5], and Callahan et al. demonstrated beyond the Yablonovitch limit provided the absorber hosts a high density of local optical states[7]. Recently, we introduced light funnel (LF) arrays for efficient light trapping bio-inspired by the retinal fovea centralis. We showed an enhancement of the solar radiation broadband absorption of 65% and 36.6% in comparison with continuous thin film and optimized NP arrays, respectively[9]. Also, we show that the coupling of LF arrays to underlying substrates yields ~10% absorption enhancement relative to free floating LF arrays of the same total height[10].
In the current work we numerically examine the optical coupling between LF arrays and underlying substrates. We show that the origin of the strong optical excitations of the array-substrate complexes can be either due to strong excitation of the arrays, strong excitation of the substrate, or excitation of both.

2. Methods and Approach: 3D finite-difference time-domain electromagnetic calculations

The three-dimensional (3D) finite-difference time-domain (FDTD) electromagnetic calculations were performed with Advanced TCAD by Synopsys (Mountain View, CA, USA). We use 50 nm SiO_2 for anti-reflective coating (ARC) and 100 nm of Au bottom reflector. Throughout the work the considered array tiling is cubic and the array period is set to 500 nm[11]. We use plain wave excitation, normally incident with the electric field parallel to the x-axis. We consider periodic boundary conditions along the lateral directions (x-y) and perfectly matched layers along the z-direction. The ultimate absorption efficiency (η_{ult}) is the absorptivity averaged and weighted with standard Air Mass 1.5 Global (AM1.5G) solar spectrum. The materials (silicon, SiO_2) optical constants were taken from the literature[12].

3. Scientific Results

Figure 1a provides an illustration of the LF array-substrate complex. The color code reflects the complex optical absorption of the system. Figure 1b presents the dependency of the total absorptivity spectra, of a 4 μm high LF array on top of a substrate, on substrate thickness (T_s) and their respective η_{abs}^{tot} are presented as well. Figures 1c,d present the absorptivities of the arrays and of the substrates (as well as the η_{abs}^{array} and η_{abs}^{sub}). Noticeably, the absorptivity is very high in the visible spectrum (400 nm – 700 nm) for all substrate thicknesses, and it is obvious that the main contributions to the total absorptivities are due to strong absorption in the arrays. Three points (points 1-3) of strong absorption in the near infrared (NIR) are marked in Figure 1b-d. It is evident that the origin of the strong absorption of point 1 is due to excitations of both the arrays (Mie modes) and the substrate (i.e. e.g. guided modes), the origin of the strong absorption of point 2 is due to strong excitation of the substrate, and the origin of the absorption of point 3 is due to Mie mode excitations (and possible modal hybridizations) in the array. Note, however, that the substrate absorptivity is very low (regardless of substrate thickness) except for a few strong absorption peaks. Interestingly, the η_{abs}^{tot} for a 4 μm LF on the top of 5 μm substrate is twice the η_{abs}^{tot} of a 9 μm thin film and 1.5 times higher than the η_{abs}^{tot} of a 4 μm NP array on the top of a 5 μm substrate, keeping all the design parameters and materials same. Figure 1e presents the absorbed photon density (APD) distribution along the y-z plane for points 1-3 (Figure 1b). In point 1 the excitation and the absorption of both the substrate and the array is evident. Point 2 shows strong excitation of the substrate in which Fabry-Perot oscillations are evident. Point 3 suggests that the absorption is governed by excitation of Mie modes in the array.

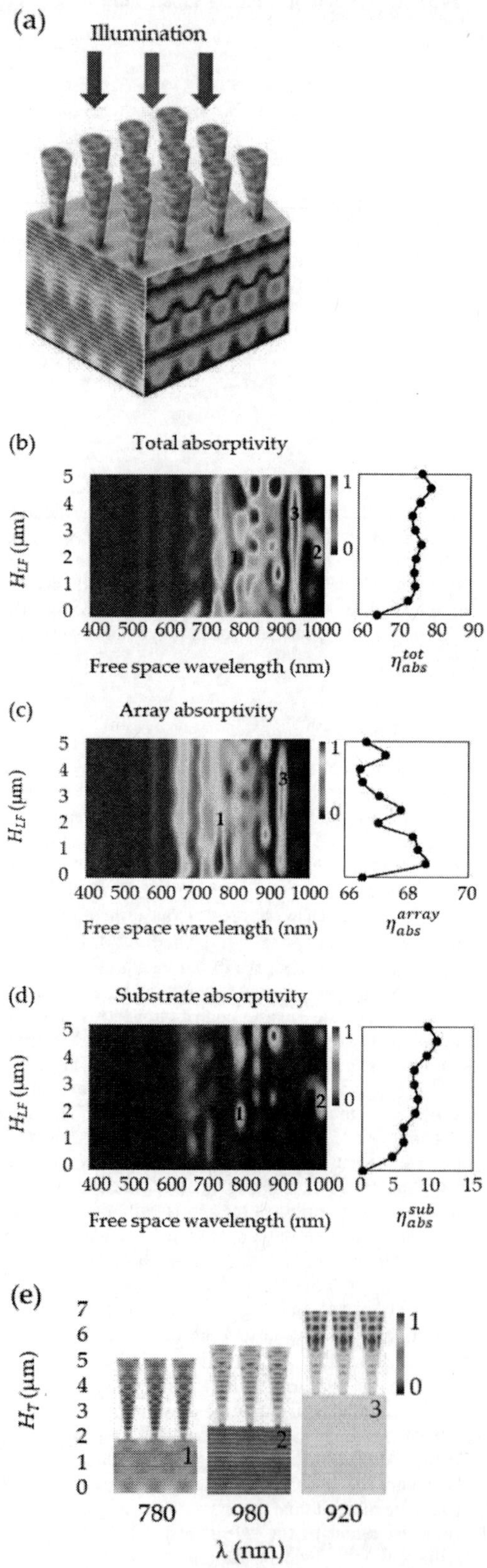

Figure 1. (a) 3D Illustration of a LF array with an underlying substrate (b) The dependency of the total absorptivity (of the LF array-substrate complex) on the thickness of the substrate. The height of array is fixed at 4 μm and the LF bottom diameter is set to Db=100 nm. The respective η_abs^tot is presented as well. (c) Array absorptivity and the respective η_abs^array. (d) Substrate absorptivity and the respective η_abs^sub. (e) APD distribution along the y- plane for the selected points 1-3 (shown in Figure 1b).

4. Conclusions

We demonstrate efficient optical excitation of the underlying substrates using arrays of subwavelength LF, and show that the LF array-substrate complexes exhibit strong absorption peaks at the NIR. The origin of the absorption peaks of the LF array-substrate complexes are traced to either strong optical excitation of the LF arrays, strong excitations of the substrates or excitations of both the arrays and the substrates.

Acknowledgement: We acknowledge the financial support of this work by the Israeli Ministry of Energy.

Conflicts of Interest: The authors declare no conflict of interest.

References

1. James R. Maiolo III, Brendan M. Kayes, Michael A. Filler, M. C. P. & Michael D. Kelzenberg, Harry A. Atwater, and N. S. L. High aspect ratio silicon wire array photoelectrochemical cells. *J. Am. Chem. Soc.* 129, 12346–12347 (2007).
2. Y, N. *et al.* Near-infrared plasmonassisted water oxidation. *J. Phys. Chem. Lett.* 3, 1248–52 (2012).
3. Ingram, D. B. & Linic, S. Water splitting on composite plasmonic-metal/semiconductor photoelectrodes: Evidence for selective plasmon-induced formation of charge carriers near the semiconductor surface. *J. Am. Chem. Soc.* 133, 5202–5205 (2011).
4. Fountaine, K. T., Cheng, W. H., Bukowsky, C. R. & Atwater, H. A. Near-Unity Unselective Absorption in Sparse InP Nanowire Arrays. *ACS Photonics* 3, 1826–1832 (2016).
5. Yu, Z., Raman, A. & Fan, S. Fundamental limit of nanophotonic light trapping in solar cells. *Proc. Natl. Acad. Sci.* 107, 17491–17496 (2010).
6. Jesper Wallentin, Nicklas Anttu, Damir Asoli, Maria Huffman, I. Å., Martin H. Magnusson, Gerald Siefer, Peter Fuss-Kailuweit, F. D. & Bernd Witzigmann, H. Q. Xu, Lars Samuelson, Knut Deppert, M. T. B. InP Nanowire Array Solar Cells Achieving 13.8% Efficiency by Exceeding the Ray Optics Limit. *Sience* 339, 1057–1060 (2013).
7. Callahan, D. M., Munday, J. N. & Atwater, H. A. Solar cell light trapping beyond the ray optic limit. *Nano Lett.* 12, 214–218 (2012).
8. Fonash J. Stephen. *Introduction to Light Trapping in Solar Cell and Photo-detector Devices*. (2012).
9. Shalev, G., Schmitt, S. W., Embrechts, H., Brönstrup, G. & Christiansen, S. Enhanced photovoltaics

inspired by the fovea centralis. *Sci. Rep.* 5, 8570 (2015).

10. Prajapati, A., Nissan, Y., Gabay, T. & Shalev, G. Light trapping with silicon light funnel arrays. *Materials (Basel).* 11, 1–9 (2018).

11. Shalev, G., Schmitt, S., Brönstrup, G. & Christiansen, S. Maximizing the ultimate absorption efficiency of vertically-aligned semiconductor nanowire arrays with wires of a low absorption cross-section. *Nano Energy* 12, 801–809 (2015).

12. Palik, E. D. *Handbook of Optical Constants of Solids. Materials* 11, (2018).

Efficient Light Trapping with Light Funnel Arrays

Ashish Prajapati[1], Yuval Nissan[1], Tamir Gabay[1] and Gil Shalev[1, 2†]

Ashish Prajapati*, Yuval Nissan*, Tamir Gabay* and Gil Shalev**†
*Department of Electrical & Computer Engineering, Ben-Gurion University of the Negev, POB 653, Beer-Sheva 8410501, Israel.
**The Ilse-Katz Institute for Nanoscale Science & Technology, Ben-Gurion University of the Negev, POB 653, Beer-Sheva 8410501, Israel.
† Corresponding author: E-mail: glshalev@bgu.ac.il

ABSTRACT: Light funnel arrays serve as an efficient absorbing layers on account of their light trapping capabilities associated with the excitation of three-dimensional complex Mie modes. Specifically, broadband absorption enhancement of the solar spectrum is noticeable for free floating light funnel arrays. We show that a light funnel (LF) arrays with bottom LF diameter of 100 nm and height of 4 μm exhibit absorption enhancement of ~52.4 % and ~12.5 % relative to thin film and an optimized nanopillar arrays.
Keywords: photovoltaic energy, light trapping, nano photonics, broadband absorption.

1. Introduction

It was demonstrated by others that light trapping exceeding the Yablonovitch limit can be realized with arrays composed of subwavelength features such as nanopillars (NPs), nanocones (NCs), nanoholes, nanospheres etc [1–5]. The Yablonovitch limit assumes "mixing of the light" inside the absorber which allows maximum occupation of the absorber photonic states[6]. Arrays of subwavelength dielectric features introduce additional localized trapped modes (Mie modes), the presence of which, and hybridization of these with other modes, allows light trapping beyond the Yablonovitch limit.

Recently, we introduced vertical arrays of vertical subwavelength LF which is a new efficient light trapping scheme bio-inspired by the retinal fovea centralis[8,9]. In the current work, we numerically examine light trapping in free-floating LF arrays and, specifically, we examine the absorptivity dependence on the LF bottom diameter (D_b).

2. Methods: 3D finite-difference time-domain electromagnetic simulations

The three-dimensional (3D) FDTD electromagnetic calculations are performed with Advanced TCAD by Synopsys (Mountain View, CA, USA). We use 50 nm SiO_2 for anti-reflective coating (ARC). Throughout the work the considered array tiling is cubic and the array period is 500 nm[10]. We use plain wave excitation, normally incident with the electric field parallel to the x-axis. We consider periodic boundary conditions along the lateral directions (x-y) and perfectly matched layer along the z-direction. The ultimate absorption efficiency (η_{ult}) is the absorptivity averaged and weighted with standard Air Mass 1.5 Global (AM1.5G) solar spectrum. The materials (silicon, SiO_2) optical constants were taken from the literature[11].

3. Scientific Results

Figure 1a presents an illustration of a NP and several LFs with various D_b. The color code reflects the complex optical absorption of the system. In the current work the starting point is the geometry of an optimized NP array (NP diameter of 400 nm and array period of 500 nm), and the geometries of the various LFs are obtained by reducing the bottom diameter of the optimized NP. Figure 1b presents the absorptivity spectra of free floating LF arrays with D_b=100 nm, D_b=200 nm and with LF height (H_{LF}) of 4 μm. The absorptivity spectra of thin film, optimized NP array and the Yablonovitch limit of the respective height are presented as well for reference. Evidently, thin film and optimized NP array shows 'smooth' spectra while the absorptivity of the LF arrays is characterized by strong spectral absorption peaks that originate from excitation of various trapped localized modes (Mie modes). Note how the absorptivity of the LF array (D_b=100 nm) reaches the Yablonovitch limit (λ=600 nm, marked in open solid circle), while the absorptivity spectrum of the NP array is far below the limit. Figure 1c shows the respective values of η_{ult}. The LF array with D_b=100 nm exhibits η_{ult} enhancement ~52.4 % and ~12.5 % relative to thin film and an optimized NP array, respectively, whereas LF array with D_b=200 nm exhibits η_{ult} enhancement ~45.6 % and ~7.5 % relative to thin film and an optimized NP array, respectively. Figure 1d presents the dependency of the absorptivity on D_b for H_{LF} of 4 μm and the respective values of η_{ult} are presented as well. Note that LF with D_b=400 nm represents the spectra of optimized NP arrays. Evidently, the η_{ult} of LF arrays for the presented H_{LF} and for all D_b is higher than that of the respective optimized NP array. Note that in the visible range (λ=400-650 nm) the formation of LFs concludes additional absorption. More importantly, for λ>600 nm the LF arrays exhibit absorption peaks at the near infrared (NIR) that are absent in the NP arrays.

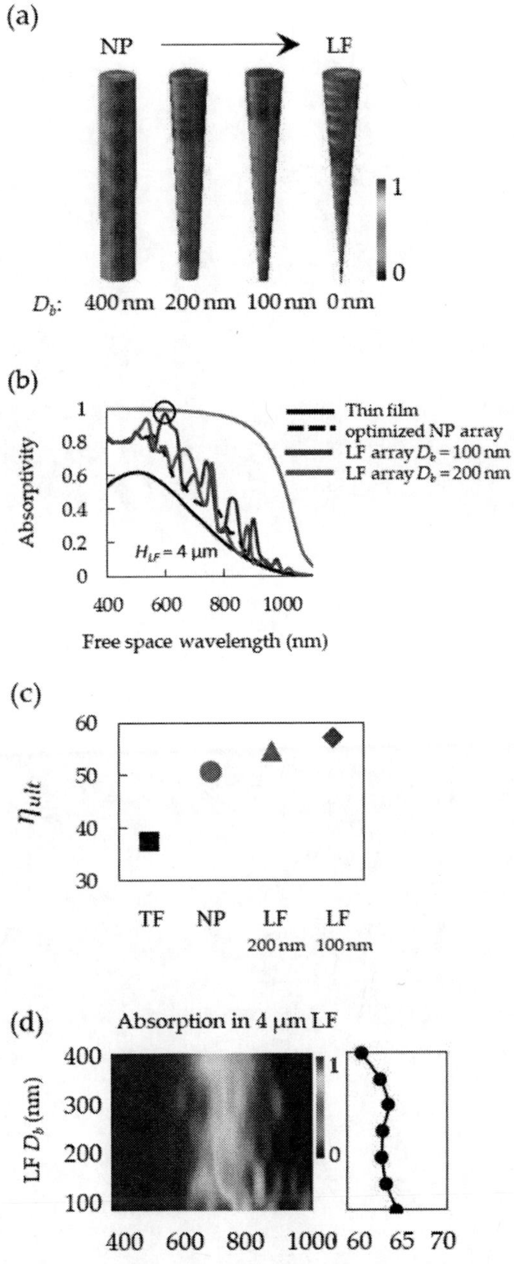

Figure 1. (a) 3D representation of NP and LFs. The colour code reflects the complex optical absorption in the system. **(b)** FDTD calculations of the absorptivity spectra for free-floating LF arrays, optimized NP array and thin film all with a height of 4 μm. **(c)** The respective η_{ult}. **(d)** The dependency of the absorptivity of 4 μm LF arrays on the LF D_b. The respective η_{ult} are presented as well.

4. Conclusions

The broadband absorption of the solar illumination with free floating LF array is presented. We show that the Yablonovitch limit can be reached in free-floating LF arrays.

Acknowledgement: We acknowledge the financial support of this work by the Israeli Ministry of Energy.

Conflicts of Interest: The authors declare no conflict of interest.

References

1. Brongersma, M. L., Cui, Y. & Fan, S. Light management for photovoltaics using high-index nanostructures. *Nat. Mater.* **13**, 451–460 (2014).
2. Yu, Z., Raman, A. & Fan, S. Fundamental limit of nanophotonic light trapping in solar cells. *Proc. Natl. Acad. Sci.* **107,** 17491–17496 (2010).
3. Jesper Wallentin, Nicklas Anttu, Damir Asoli, Maria Huffman, I. Å., Martin H. Magnusson, Gerald Siefer, Peter Fuss-Kailuweit, F. D. & Bernd Witzigmann, H. Q. Xu, Lars Samuelson, Knut Deppert, M. T. B. InP Nanowire Array Solar Cells Achieving 13.8% Efficiency by Exceeding the Ray Optics Limit. *Sience* **339,** 1057–1060 (2013).
4. Wang, K. X., Yu, Z., Liu, V., Cui, Y. & Fan, S. Absorption enhancement in ultrathin crystalline silicon solar cells with antireflection and light-trapping nanocone gratings. *Nano Lett.* **12,** 1616–1619 (2012).
5. Wang, C., Yu, S., Chen, W. & Sun, C. Highly efficient light-trapping structure design inspired by natural evolution. *Sci. Rep.* **3,** (2013).
6. Yablonovitch, E. Statistical ray optics. *J. Opt. Soc. Am.* **72,** 899 (1982).
7. Fonash J. Stephen. *Introduction to Light Trapping in Solar Cell and Photo-detector Devices*. (2012).
8. Shalev, G., Schmitt, S. W., Embrechts, H., Brönstrup, G. & Christiansen, S. Enhanced photovoltaics inspired by the fovea centralis. *Sci. Rep.* **5,** 8570 (2015).
9. Prajapati, A., Nissan, Y., Gabay, T. & Shalev, G. Light trapping with silicon light funnel arrays. *Materials (Basel).* **11,** 1–9 (2018).

10. Shalev, G., Schmitt, S., Brönstrup, G. & Christiansen, S. Maximizing the ultimate absorption efficiency of vertically-aligned semiconductor nanowire arrays with wires of a low absorption cross-section. *Nano Energy* **12,** 801–809 (2015).

11. Palik, E. D. *Handbook of Optical Constants of Solids*. (Academic, 1985).

OVER 22.0% EFFICIENCY FOR THE P-TYPE MONO SILICON PERC SOLAR CELLS BY INDUSTRIAL MASS PRODUCTION TECHNOLOGY

Cheng-Wen Kuo, Ta-Ming Kuan, Wei-Lo Chueh, Li-Guo Wu, Chih-Chiang Huang, Chia-Jung Li, and Cheng-Yeh Yu
TSEC Corporation, No.85, Gaungfu N. Rd., Hsin-Chu 30351, Taiwan.

ABSTRACT: In this paper, we demonstrated that the efficiency of p-type mono PERC cells could be further improved by both front side and backside optimizations. These improved process technologies are including: a) lightly doped emitters (LDE); b) backside passivation layer engineering and laser pattern modification. The lightly doped emitters with a lower surface dopant concentration which helps reduce front surface and emitter recombination; the backside passivation layer and laser pattern optimizations which improve backside recombination and internal reflection of the cells. As compared to the standard p-mono PERC cells, a significant average efficiency gain of 0.3~0.4 % was achieved. We also demonstrated that an average efficiency of 21.51 % with the best efficiency of 22.09 % was achieved with these technologies.
Keywords: PERC, lightly doped emitters, backside passivation

1 INTRODUCTION

Photovoltaic industry had been developed for decades and had been growing rapidly in recent years. It is because of the awareness of global warming effect and shortage of natural resources such as fossil energy. The global photovoltaic market is currently dominated by crystalline silicon (Si) technologies. Recently, there are many research institutes, universities and companies reported on the fabrication of high efficient p-type passivated emitter and rear cell (PERC) structures [1-8]. The PERC applies dielectric passivation layers to the front and rear side in the solar cell [1], which is a significantly better passivation performance in the rear side of solar cell, can be achieved by dielectric layers compared with the full-area aluminum back surface field (Al BSF). The additional dielectric layer also improves the internal rear reflectance that reduces absorption losses in the Al back contact.

On the other hand, the front side emitter is impurity-doping for the silicon wafer thus the cell performance was suppressed due to lifetime reduction [8]. If the emitter is doping lightly, the surface recombination rate is low enough to perform better lifetime. However, the lighter emitter the higher sheet resistance results in the worse metal contact. The paste manufacturers have also continued to improve front contact paste properties to reduce emitter recombination.

In this work, we proposed the improving front side LDE and back side process parameter process parameter applied to screen-printed PERC structure. It could be found that employing optimizing PERC process is better than without optimizing PERC process.

2 EXPERIMENTAL

Figure 1 showed the PERC structure in this work. The high efficiency 6-inch (156.75 mm x 156.75 mm) PERC solar cells were fabricated on 180 μm thick p-type boron doped Czochralski (Cz)-silicon wafers.

The process sequence in this work was shown in Figure 2. Chemical wet etching methods were used for saw damage removal and surface texturing, which include both acidic etching for multi-crystalline silicon wafers and alkaline etching for mono-crystalline silicon wafers. The solution composition of acidic etching was nitric acid (HNO3) : hydrofluoric acid (HF) : deionized water (DI) = 1 : 2 : 1.5, and the solution composition of alkaline etching was potassium hydroxide (KOH) : deionized water (DI) = 1 : 50. Then, the rear surface is polished in order to reduce the roughness and its impact on the effective charge carrier lifetime [5]. There are two conventional phosphorous diffusions processes were employed by furnace to form the n+ emitters. The diffusion process also improved to uniformity for this work.

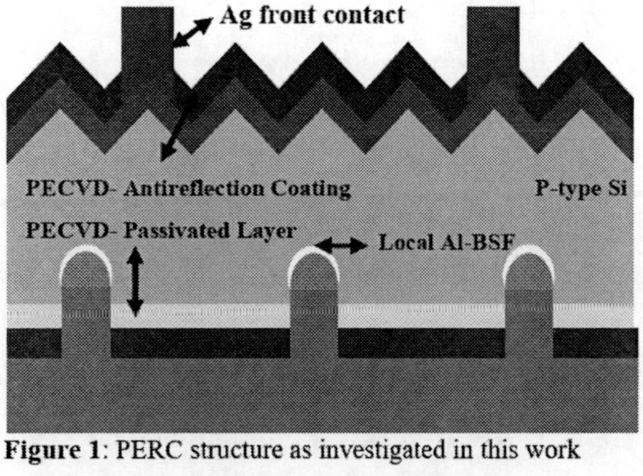

Figure 1: PERC structure as investigated in this work

Directly after the selective emitter formation, the phosphosilicate glass (PSG) was then removed by HF-contained solutions, formed edge isolation in the meantime. In order to decrease PID effect and increase efficiency, we optimized the CVD process parameter to improve. An antireflection coating (ARC) layer of silicon nitride (SiNx: H) and/or SiNx/SiOx dual ARC was deposited by plasma enhanced chemical vapor deposition (PECVD) with thickness of 80 nm. The passivation was carried out on rear side also using a PECVD. Afterward, the rear side laser opening ratio is also optimized [7]. All PERC solar cells have the same laser-ablated opening width in the rear-side passivation stacks which was about 45 μm, as shown in Figure 3. The front side silver and backside silver and grid contact and local back-surface-field. Finally, the current-voltage measurements of solar cells under AM 1.5 illumination were performed.

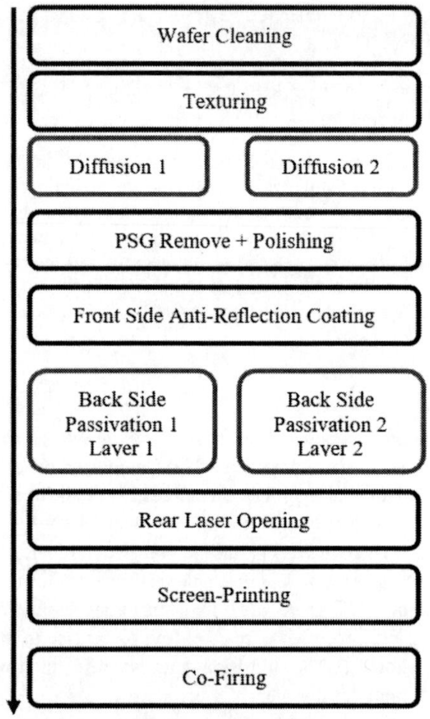

Figure 2 : Process flow for the fabrication of PERC solar cell as developed in this work

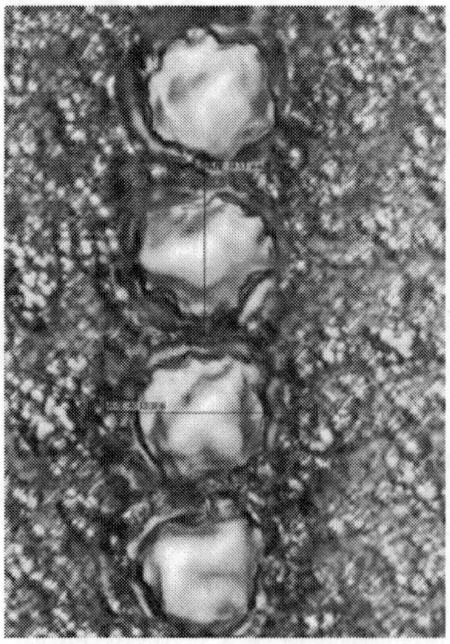

Figure 3 : OM image shown the laser ablated opening width in this work.

3 RESULTS AND DISCUSSION

Table I shows the photovotaic characteristics of Ref., back side passivation 1 (B1), and back side passivation 2 (B2) solar cells. The Ref. solar cell is the TSEC V-cell mass production line at the time, the B1 and B2 is different back side passivation process with the TSEC V-cell mass production line. For B1 solar cell group, the open circuit voltage (Voc), the short circuit current (Isc), and the fill factor (FF) is 0.4mV, 14mA, and 0.34% absolute respectively, which is higher than the Ref. solar cell due to combined contribution of passivation and back reflection of stack layers. For B2 solar cell group, the Voc, Isc, and FF is 1.8mV and 27mA, and 0.21% absolute higher than the Ref. solar cell due to the combined contribution of passivation and back reflection of stack layers, but FF of B2 solar group is less than B1 solar cells group duo to back side passvation pattern design [7].

Table I: The comparison of photovotaic characteristicsfor Ref., B1, and B2 solar cells.

Group	EFF (%)	V_{OC} (V)	I_{SC} (A)	FF (%)
Ref.	21.22	0.6639	9.699	80.50
B1	21.26	0.6652	9.707	80.57
B2	21.37	0.6685	9.744	80.17

In order to further improve the PERC solar cells, the LDE process was employed. Conventional phosphorous diffusion was employed by furnace to form the n^+ emitters with a sheet resistance of 70~80 ohms/sq. The evolution of sheet resistance for the applied laser processes for the different energies is determined by four point probe measurements. The average sheet resistance in two groups is 100 ohm/sq (Diffusion 1) and 120 ohm/sq (Diffusion 2), respectively.

Table II: The comparison of photovotaic characteristicsfor Ref., D1, and D2 solar cells.

Group	EFF (%)	V_{OC} (V)	I_{SC} (A)	FF (%)
Ref.	21.21	0.6615	9.701	80.43
D1	21.26	0.6619	9.715	80.77
D2	21.39	0.6662	9.781	80.20

Table II shows the photovotaic characteristics of Ref., diffusion 1 (D1), and diffusion 2 (D2) solar cells. The Ref. is also TSEC V-cell mass production line, the D1 and D2 is different LDE recipe with the TSEC V-cell mass production line at the time. For the D1 and D2 solar cell group, the Voc and Isc is higher than the Ref. solar cell due to decreased surface recombination velocity. However, FF is 0.23% absolute less because of poor silver contact. In order to improve TSEC V-Cell technology, the optimized front side and back side passivation process were employed. When these two technology were combined, the significant average efficiency gain of 0.3~0.4 % was achieved on 180 μm thick single crystalline silicon with the area of 156.75×156.75 mm^2. Table III shows the The average electrical parameters of PERC cells in R&D mass production line. Figure 4 illustrates the efficiency distribution of our PERC product. It could be found about 0.3% of efficiency gain compared wih Ref. group of Table I.

Table III : The average electrical results of mono PERC cells in mass production line.

	EFF (%)	FF (%)	I_{SC} (A)	V_{OC} (V)	R_S (ohm)
Average	21.51	81.14	9.717	0.6664	0.8519
Best	22.09	81.49	9.810	0.6752	0.8800

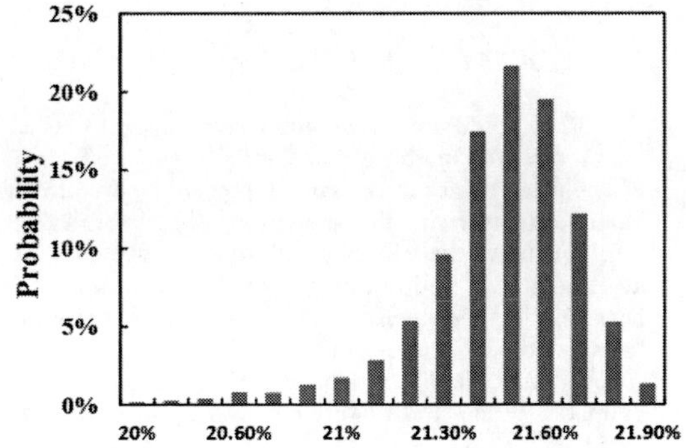

Figure 4: The efficiency distribution of PERC cells line.

4 CONCLUSIONS

In this work, we demonstrated that the efficiency of p-type mono PERC cells could be further improved by both front side and backside optimizations. The lightly doped emitters with a lower surface dopant concentration which helps reduce front surface and emitter recombination; the backside passivation layer and laser pattern optimizations which improve backside recombination and internal reflection of the cells. As compared to the TSEC V-cell mass production line cells, a significant average efficiency gain of 0.3% was achieved. We also demonstrated that an average efficiency of 21.51 % with the best efficiency of 22.09 % was achieved with these technologies.

5 REFERENCES

[1] S. Gatz, T. Dullweber, and R. Brendel, "Evaluation of Series Resistance Losses in Screen-Printed Solar Cells With Local Rear Contacts" IEEE J. Photovolt, vol. 1, pp.37, 2011.

[2] S. Riegel, F. Mutter, T. Lauermann, B. Terheiden, G. Hahn, "Review on Screen Printed Metallization on P-type Silicon" Energy Procedia, vol. 21 pp.14, 2012.

[3] E. Urrejola, K. Peter, H. Plagwitz, and G. Schubert, "Distribution of Silicon in the Aluminum Matrix for Rear Passivated Solar Cells" Energy Procedia, vol. 8, pp.331, 2011.

[4] E. Urrejola, K. Peter, H. Plagwitz, and G. Schubert. "Al–Si Alloy Formation in Narrow P-type Si Contact Areas for Rear Passivated Solar Cells" J. Appl. Phys. vol. 107, pp. 124516, 2010.

[5] E. Urrejola, R.Petres, J. Glatz-Reichenbach, K. Peter, E. Wefringhaus, H. Plagwitz, G. Schubert., "High Efficiency Industrial PERC Solar Cells with all PECVD-Based Rear Surface Passivation" 26th EUPVSEC, pp. 2233, 2011.

[6] S. Mack, A. Wolf, B. Thaidigsmann, E. Lohmüller, U. Jäger, M. Pospischil, F. Clement, D. Eberlein, R. Preu, and D. Biro, "Technology for Mass Production of > 20% Efficient P-type Silicon Solar Cells", 28th EUPVSEC, pp. 761, 2013

[7] C.-W. Kuo, T.-M. Kuan, C.-C. Liu, C.-C. Huang, L.-G. Wu, C.-Y. Yu, "Impact of Laser Opening Ratio for Mass Production High Efficiency Mono-Crystalline Silicon Solar Cells" 29th EUPVSEC, pp. 1392, 2014

[8] S.-Y Chen, Y.-H. Lin, S.-P. Hsu, Y.-Y. Wang, C.-H. Du, "Excellent Commercial Solar Cells with Local Back Surfacefield and Lightly Doped Emitter" 28th EUPVSEC, pp. 1745, 2013

Optimization of Rear Pattern for P-type Mono Bifacial PERC Cells in Mass Production

Hui Li, Minjie Ma, Xiaoqin Yu, Wei Yang, Qintian Ma, Yan Wang
R&D Department, Dongfang Huansheng Photovoltaic (Jiangsu) Co. , Ltd
No. 20, Wenzhuang Road, Economic Development Zone, Yixing, JiangSu, China, 214203. hui.li@dzssolar.com

ABSTRACT: In this work, key factors effect the output power were studied to improve the electrical performance and bifaciality of Bi-PERC cells which were fabricated with thickness of 180um and a bulk resistivity ranging from 0.5Ω · cm to 1.5Ω · cm. Different aluminum paste, weight of aluminum paste, width of finger and the pitch of rear pattern were investigated in this paper. Under industrial mass production condition, it is shown that bifaciality of p-type mono PERC cells is no more than 70%. Moreover, after optimization, usage of aluminum paste reduced from 0.95g to 0.25g, it means 73.7% paste has been saved. And when printed width of aluminum finger is 150um, the pitch of rear pattern is 1.1mm, bifaciality can reached to 72.95%. Most probably, lower and moderate aluminum paste, narrower aluminum finger and wider pitch are the reasons for the higher bifaciality.

Keywords: Bifacial PERC cells, rear pattern, aluminum paste, bifaciality

1 INTRODUCTION

Bifacial silicon solar cells could receive sunlight and produce electricity from both front and back side. These cells are designed with special structure and finger contact, so they can absorb much more light and produce much more power from both sides. In past years, P-type mono bifacial PERC cells have been intensively investigated because of their bifacial power generation, comparable bifaciality, and low cost [1].

The topic of bifacial passivated emitter and rear cells (B-iPERC) is supposed to be the next phase in PERC evolution after PERC technology successfully became mainstream in research and entered industrial mass production [2]. Lots of manufacturers and institutes investigated the industry-related bifacial PERC cells, and bifaciality for p-type mono PERC cells is no more than 70% which will make it hard to push bifacial technology in production[3-5].

In this work, factors of effecting the electrical performance and bifaciality of Bi-PERC cells were studied, such as Al paste, weight of Al paste, width of finger and the pitch of rear pattern. The purpose of this work is to improve the performance and output power of bifacial cells in order to find an economic and efficient approach to fabricate bifacial cells in mass production.

2 EXPRIMENTS

In this experiment, 156.75×156.75 mm^2 p-type CZ Si wafers with thickness of 180um was adopted, and the bulk resistivity is ranging from 0. 5Ω · cm to 1.5Ω · cm. After alkaline texturing and phosphorus diffusion, the wafer was polished about 4-5um of silicon on the rear surface. Afterwards, the rear side is coated with Al2O3/SiNx capping layer and the front side is coated with SiO2/SiNx. Since the Al paste used is not firing through the dielectric stack, this barrier is selectively opened by laser. Later on, aluminum finger was printed on the back and silver finger was printed on the front of the p-type Si wafers. The Bifacial PERC cells were measured I-V performance on the front and rear respectively, the light was shaded from the back during measuring.

3 RESULTS AND DISCUSSIONS

3.1 Effect of different aluminum layer weight on H/W ratio and I-V performance

The performance of the sample prepared with different aluminum layer weight shown in Fig.1 and Table1. Fig.1(a) shows the effects of Al layer weight on width, height and H/W ratio of printed aluminum finger on the back. With the decrease of Al layer weight, the width, height and H/W ratio reduced, but when the Al layer dropped more than 0.25g, the H/W ratio become dropped dramatically, this most likely due to the wider and lower Al finger.

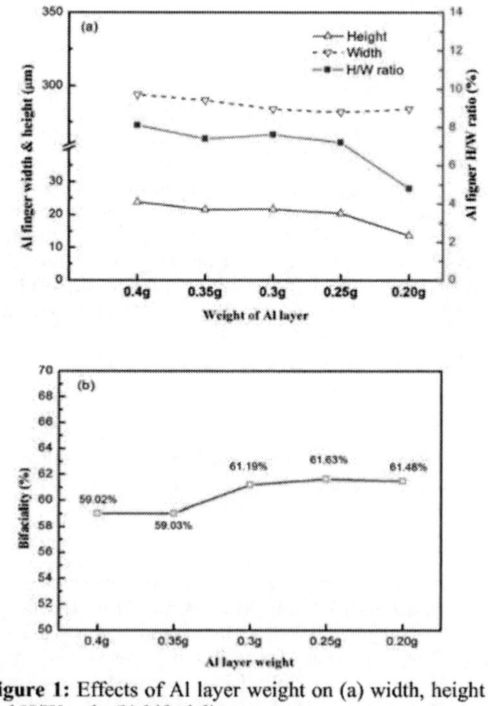

Figure 1: Effects of Al layer weight on (a) width, height and H/W ratio (b) bifaciality

Table1: Electrical performance with different Al weight, the I-V performance were measured on the front and rear respectively, during measuring, the light was shaded from the back.

Group	Weight	F/R	Voc (mV)	Isc (A)	Rs (mΩ)	FF (%)	Efficiency (%)	Bifaciality (%)
G1	0.4g	Front	667.2	9.645	1.7	81.05%	21.35%	59.02%
		Rear	650.4	5.758	98.9	82.20%	12.60%	
G2	0.35g	Front	668.0	9.661	1.8	80.83%	21.35%	59.03%
		Rear	651.1	5.750	102.2	82.25%	12.60%	
G3	0.3g	Front	669.1	9.675	1.9	80.70%	21.38%	61.19%
		Rear	652.7	5.950	95.9	82.30%	13.08%	
G4	0.25g	Front	670.0	9.666	2.0	80.50%	21.34%	61.63%
		Rear	653.8	5.970	97.9	82.30%	13.15%	
G5	0.20g	Front	669.7	9.660	2.0	80.48%	21.31%	61.48%
		Rear	654.0	5.965	99.7	82.24%	13.10%	

The I-V parameter has been measured while illuminated from the front and shaded at the rear side to avoid the influence of rear scattered light. As shown from Fig.1(b), bifaciality has slightly improved when reducing the Al layer weight even as the H/W ratio dropped, it seems that the width of aluminum finger played an important role for Bi-PERC cells.

Table1 lists the I-V performance of bifacial PERC cells with decreasing the aluminum weight, tested under the standard condition of 1000W/m^2, AM1.5 global spectrum and 25°C. All this cells were measured I-V curves on the front side and then been turned over to test the rear parameters. Both open-circuit voltage (Voc) and series resistance (Rs) of front side achived the highest with the value of 670.0mV and 2.0mΩ when the weight is 0.25g. Because less light will penetrate the cells and line resistance of aluminum finger will increase when the rear aluminum finger became wider and lower. Future work will attempt to reduce the series resistance of bifacial cells by designing rear pattern and optimizing aluminum paste.

As shown in Fig.2 about the efficiency trend of the front and rear by using different aluminum layer weight on the back of cells. It seems no obvious difference for the front efficiency, but the rear efficiency become higher when decreasing the Al layer weight from 0.4g to 0.25g, and that value become drop when reduced the weight to 0.2g. In combination with Fig.1, we can confirmed that when rear aluminum weight reduced from 0.4g for local Al-BSF PERC cells to 0.25g, we can get narrower width to make more light penetrate the cell from the back and improve the bifaciality to 61.48%.

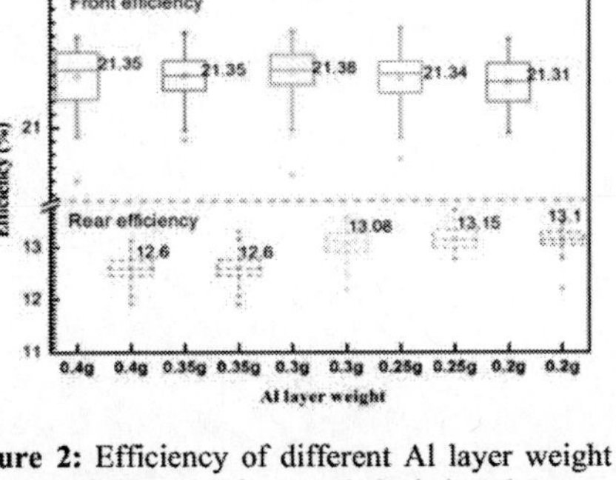

Figure 2: Efficiency of different Al layer weight while illuminated from the front and shaded at the rear side to avoid the influence of rear scattered light (designed width of Al finger is 300um)

3.2 Effect of different aluminum finger width on the front and rear efficiency performance

We tested the aluminum width printed on the rear surface after firing, and found that aluminum finger is nearly 10-15um wider than design. The efficiency performance with different aluminum finger width on the back of p-type Bi-ERC cells is shown in Fig.3, the width is designed 300um, 250um and 150um, respectively. It confirmed that the front cell optical properties are only slightly affected by the width from 300um to 150um. But it is obvious that the rear efficiency is higher when width decreased to 150um due to the reduction of shading area on the back, which can partly explained the higher bifaciality of reaching to 72.09%.

Next work we will attempt to design the aluminum width to 120um,100um, and even narrower that is in the vicinity of finger on the front by adopting aluminum paste with better conductivity and plasticity, and improve the bifaciality to more than 80%.

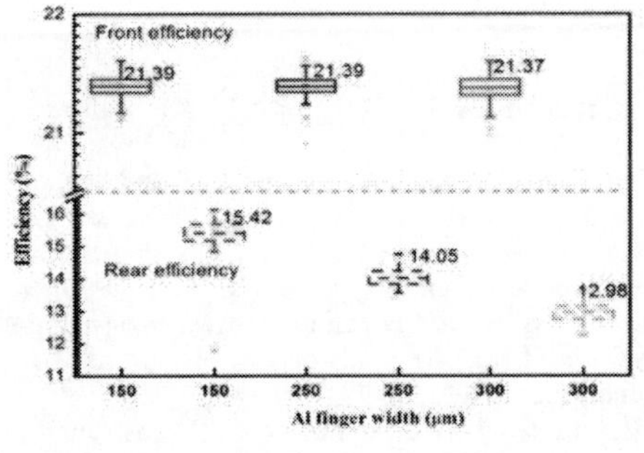

Figure 3: Efficiency with different Al finger width of 300um, 250um and 150um

3.3 Efficiency performance with designing different pitch of rear aluminum finger

Different pitch of aluminum finger on the back of p-type Bi-ERC cells has studied to find suitable pattern, the pitch is designed 0.9mm, 1mm and 1.1um respectively. The electrical performance is as shown in Table1. For back-side irradiance(1000W/m^2, AM1.5, °C25), open-circuit voltage(Voc) and short-circuit current (Isc) of rear side increased obviously when increasing the pitch from 0.9mm to 1.1mm, and series resistance of front side increased at the same time. The rear sides yield Isc=6.989A more current with pitch of 1.1mm than those with pitch of 0.9mm.

Fig.4 shows the efficiency performance, it is normal that increasing the distance between Al fingers on the

Table2: Electrical performance with designing different pitch, the I-V performance were measured on the front and rear respectively, during measuring, the light was shaded from the back.

Group	Pitch	F/R	Voc (mV)	Isc (A)	Rs (mΩ)	FF (%)	Efficiency (%)	Bifaciality (%)
G1	0.9mm	Front	668.7	9.747	2.1	80.45%	21.46%	69.20%
		Rear	661.8	6.659	89.7	82.33%	14.85%	
G2	1mm	Front	669.6	9.760	2.4	80.11%	21.43%	72.39%
		Rear	664.3	6.945	85.9	82.14%	15.51%	
G3	1.1mm	Front	669.6	9.759	2.4	80.08%	21.42%	72.95%
		Rear	664.7	6.989	85.0	82.17%	15.63%	

back can make more light illuminate the cells from the back, so it is easy to explain the relatively better performance of front efficiency 21.42%, rear efficiency 15.63% and bifaciality 72.95%, when the pitch is 1.1mm compared to 1mm and 0.9mm. But the front efficiency has a little drop for that more light has penetrated the cells from the front surface and aluminum passivated area has reduced.

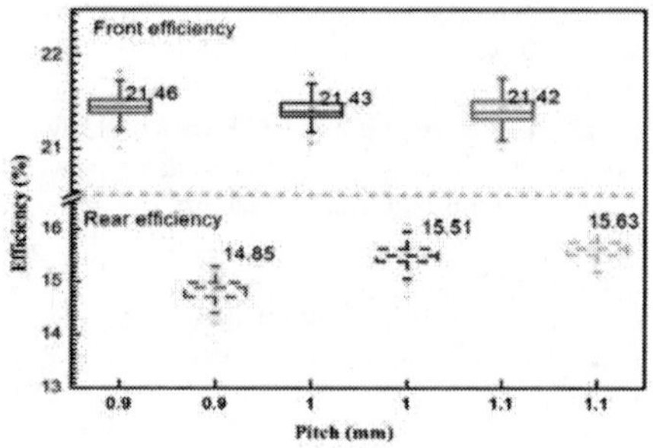

Figure 4: Efficiency with different pitch of rear Al finger pattern (0.9mm, 1mm and 1.1um)

4 CONCLUSIONS

Under industrial mass production condition, it is shown that bifaciality of p-type mono PERC cells is no more than 70%. Moreover, after optimization, usage of Al paste reduced from 0.95g to 0.25g, it means 73.7% Al paste been saved. And when printed width of aluminum finger is 150um, the pitch of rear pattern is 1.1mm, bifaciality can reached to 72.95%.

Most probably, lower and moderate Al paste, narrower Al finger and wider pitch are the reasons for the higher bifaciality. In the future work, we will investigate the much narrower width of 120um, 100um, and even narrower that is nearly the same as finger on the front. Meanwhile, we will also optimize rear pattern and adopt aluminum paste with better conductivity and plasticity to make the bifaciality factor of p-type PERC cells to more than 80%。

5 ACKNOWLEDGEMENTS

The authors would like to thank aluminum paste supplier for fruitful discussion and all members of R&D department from Dongfang Huansheng Photovoltaic Co. .

6 REFERENCES

[1] Fischer G, Müller M, Steckemetz S, etal., Energy Procedia, 2015,515-519

[2] T. Dullweber, C. Kranz, R. Peibst, etal., 31st EU PVSEC, Germany, 2015, p.341-350

[3] Nico Wöhrle, Tobias Fellmeth, Johannes Greulich, etal, Energy Procedia, 2017, 225-234

[4] Yusufoglu UA, Pletzer TM, Koduvelikulathu LJ, etal, IEEE Photovoltaics 2015, 5:320–8

[5] Ilkay Cesar, Petra Manshanden, Gaby Janssen1, etal., IEEE 39th PVSC, USA, June 2014, pp. 1212-1217

82% BIFACIALITY INDUSTRIAL P-TYPE MC-SI BIFACIAL PERC SOLAR CELL

Jianwen Dong, Xianzhao chen, Yibo zhang, Jiaxian Lv, Zhendong Shen, Ji li, Quanhua Ye, Wenwen Wei, Wei Wang, Jian Sheng, Chun Zhang
Solar Energy Research Institute, GCL System Integration Technology Co., Ltd.
5F, GCL Energy Center, No.28 Xinqing Road, SIP Suzhou, Jiangsu, China, dongjianwen@gclsi.com

ABSTRACT: Bifacial PERC is expected to be the next step from mono-facial PERC because of the similar processing between mono-facial PERC and bifacial PERC. Processes relating to the rear side, such as rear surface passivation stacks, laser ablation patterns and rear metallization, were optimized. Industrial mass production p-type mc-Si bifacial PERC solar cells, with average front and rear side efficiency of 20.15% and 15.8% respectively, as measured with a black chuck, are demonstrated in this paper. A detailed optical loss analysis for the best cell with bifaciality factor of 82.5% indicated that the rear metal shading and rear reflection are the dominant J_{SC} loss on the rear side. It is expected that firing through Al paste and multi-busbar can improve rear side efficiency dramatically and mc-Si bifacial PERC solar cell with bifaciality factor of up to 85% can be achieved.
Keywords: multi-crystalline silicon, PERC solar cells, Bifacial PERC

1 INTRODUCTION

Bifacial solar cell and module technology received increasing interest because power generation can be increased by up to 20% in PV systems using bifacial instead of mono-facial modules [1, 2]. Accordingly, the photovoltaic technology roadmap ITRPV predicts a market share of bifacial solar modules of 15% until 2020. Presently, industrial bifacial solar cell concepts are mainly applied on n-type wafers [3-5]. However, a challenge of n-PERT solar cells is that both sides require Ag electrode and hence may increase manufacturing costs.

Bifacial PERC could be fabricated using mono-facial PERC process by changing full area printing to local area printing on the rear surface. Bifacial PERC can reduce Al paste consumption and also generate additional electrical power from rear illumination. P-type Cz-Si bifacial PERC solar cells were widely investigated and bifaciality factor up to 80% was reported by ISFH in 2015 [6]. LONGi solar, Jinko Solar, Trina Solar etc. exhibited their p-type mono-Si bifacial PERC products with bifaciality factor around 75% in SNEC PV POWER EXPO 2018.

The purpose of this paper is to develop and fabricate industrial mass production high efficiency mc-Si bifacial PERC solar cells. Power loss analysis was carried out for rear side of the best mc-Si bifacial PERC solar cell. Industrial mass production p-type mc-Si bifacial PERC solar cells, with average bifaciality factor up to 78.4%, are demonstrated in this paper. Simulation shows that bifaciality factor of 85% in mc-Si bifacial PERC solar cell can be achieved using firing through Al paste and multi-busbar metallization.

2 EXPRIMENTS

156.75×156.75 mm^2 p-type mc-Si wafers with thickness of 180~200μm and bulk resistivity of 1~3 Ω.cm were used in this study. Bifacial PERC solar cells as shown in Fig. 1 were fabricated according to the following process: a) mc-Si wafers were damage-etched in an HNO_3/HF mixture solution, then nano-textures were fabricated by metal-catalysed chemical etching (MCCE). b) Phosphorus diffusion to form PN junction on the front side. c) Rear side etching and front side PSG removal. d) AlO_x/SiN_x film stacks deposited on the rear side and SiN_x film deposited on the front side. e) Laser ablation for rear local contacts. f) Screen printing of rear Al electrode and front Ag electrode. g) Co-firing and anti-LID treatment. Processes c) to g) were optimized to obtain high performance bifacial PERC solar cell. After cells were finished, IV performance for both sides were measured by HALM tester with a black chuck.

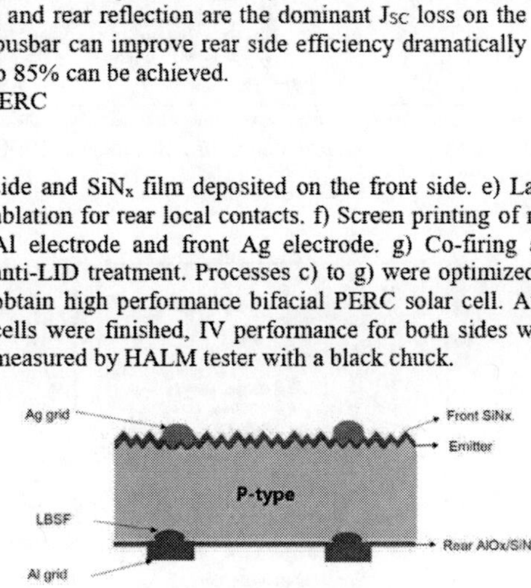

Figure 1: Schematic of bifacial PERC solar cell

3 RESULTS AND DISCUSSIONS

Table 1: Baseline IV performance for both side of mc-Si bifacial PERC solar cells measured by HALM tester with a black chuck.

	Count	Jsc (mA/cm^2)	Voc (V)	FF (%)	Eff (%)
Average-front	19768	39.19	0.650	79.18	20.18
Average-rear	885	27.00	0.640	79.45	13.73
Bifaciality		**69%**			**68.0%**
Best-front	1	39.05	0.652	79.36	20.19
Best-rear	1	27.65	0.646	78.69	14.06
Bifaciality		**71%**			**69.6%**

The processing of mono-facial PERC and bifacial PERC solar cells are similar, but several processes such as rear surface passivation film stacks, local metallization, and rear electrode pattern have been adjusted for the bifacial PERC solar cell in order to optimize the rear efficiency and front efficiency. Based on industrial mass production conditions, mc-Si bifacial PERC solar cells were fabricated, table 1 shows IV performance for both side of mc-Si bifacial PERC solar cells from the baseline process. The average front and rear side efficiency reaches up to 20.18% and 13.73% respectively in mass production, with corresponding bifaciality factor of up to

68%. Fig. 2 shows distribution of front efficiency and bifaciality factor of mass production mc-Si bifacial PERC solar cells.

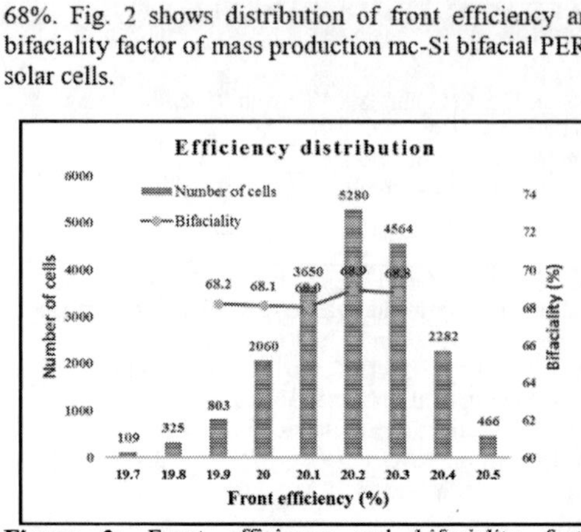

Figure 2: Front efficiency and bifaciality factor distribution of mass production mc-Si bifacial PERC solar cells

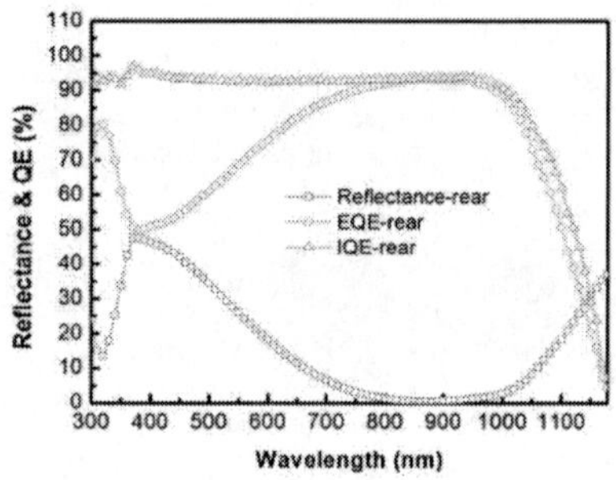

Figure 3: Reflectance and quantum efficiency for the rear side of the best mc-Si bifacial PERC solar cell

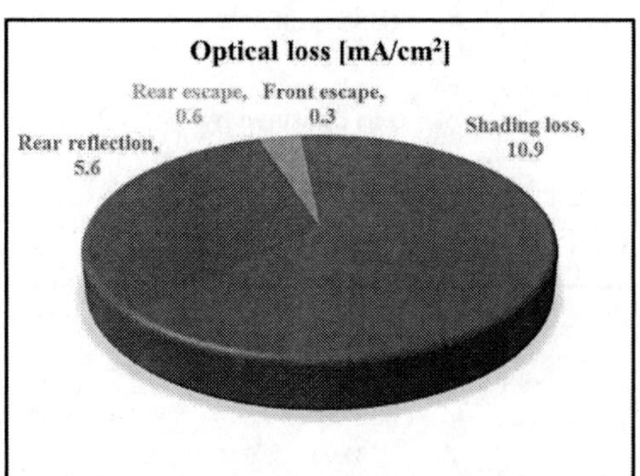

Figure 4: Optical loss analysis for the rear side of the best mc-Si bifacial PERC solar cell

Bifaciality factor of the best cell reached 69.6%, which is similar to the J_{SC} bifaciality factor of 71%. Optical loss analysis was carried out for rear side of the best mc-Si bifacial PERC solar cell. Fig. 3 and Fig. 4 show the quantum efficiency curves and the optical loss analysis result respectively. The biggest component of the rear J_{SC} loss of ~10.9 mA/cm^2 is due to rear metal shading. In this experiment, Al finger width is ~270μm because of bad shape of screen printing Al paste, resulting in rear metal shading of ~30%. Rear reflection is another main reason for the J_{SC} loss, accounting for about 5.6 mA/cm^2. High rear reflection is caused by thick SiN_X/AlO_X stacks which is about 100nm.

Table 2: IV performance for both sides of mc-Si bifacial PERC solar cells measured by HALM tester with a black chuck, after rear metal shading and SiN_X/AlO_X process were adjusted.

	Count	J_{SC} (mA/cm^2)	V_{OC} (V)	FF (%)	Eff (%)
Average-front	236	39.09	0.654	78.79	20.15
Average-rear	236	30.78	0.647	79.35	15.80
Bifaciality		**79%**			**78.4%**
Best-front	1	38.93	0.659	78.43	20.12
Best-rear	1	32.12	0.655	78.95	16.60
Bifaciality		**83%**			**82.5%**

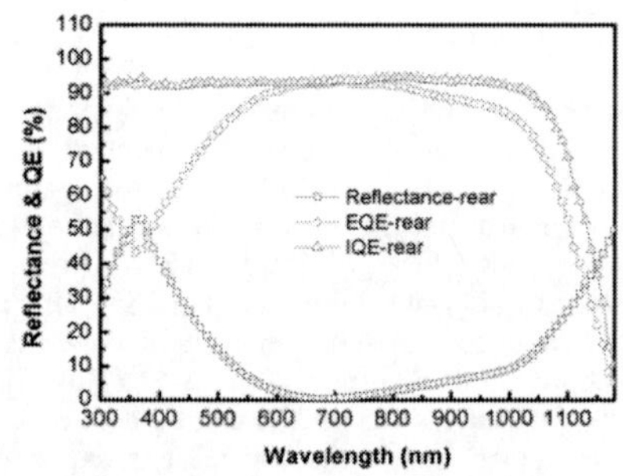

Figure 5: Reflectance and quantum efficiency of the best mc-Si bifacial PERC solar cell after rear SiN_X/AlO_X stacks and metal shading were optimized

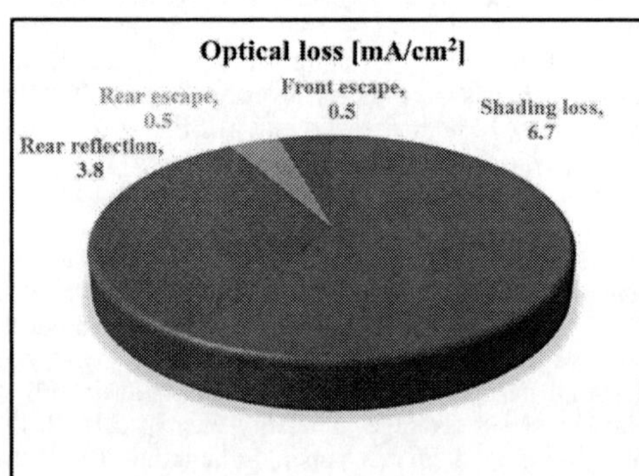

Figure 6: Optical loss analysis for the rear side of the best mc-Si bifacial PERC solar cell after rear SiN_X/AlO_X stacks and metal shading were optimized

After optimizing rear SiN_X/AlO_X stacks and metal shading, SiN_X thickness was adjusted to 80nm and Al finger width was adjusted to 150μm. Rear efficiency and bifaciality factor was increased dramatically. As shown in table 2, rear efficiency increased from 13.7% to 15.8% and bifaciality factor increase from 68% to 78.4%. One of the best cell showed rear efficiency of 16.6% and bifaciality factor of 82.5%. Also, reflectance and quantum efficiency were measured and optical loss analysis was carried out for rear side of the best mc-Si bifacial PERC solar cell. As shown in Fig. 5 and Fig. 6, rear metal shading and rear reflection remain to be the two largest J_{SC} loss at 6.7 mA/cm^2 and 3.8 mA/cm^2 respectively.

In order to further increase rear efficiency and bifaciality factor, further reduction in rear metal shading and rear reflection are necessary. Fig.7 shows the influence of Al finger width on the calculated J_{SC} bifaciality factor. The J_{SC} bifaciality factor increases dramatically with narrower Al finger. J_{SC} bifaciality factor can reach 85% if screen printing of 60 μm Al finger is achieved in mass production. However, it is a challenge to apply 60 μm Al finger in mass production. One reason is that the Al finger would be too narrow to align with laser ablation fingers, the other reason is that R_S of 5 busbar Al grid on the rear is too high as shown in Fig. 8. It is expected that firing through Al paste and multi-busbar are the upcoming technologies that can solve the above problems and mc-Si bifacial PERC solar cell with bifaciality factor of 85% can be achieved.

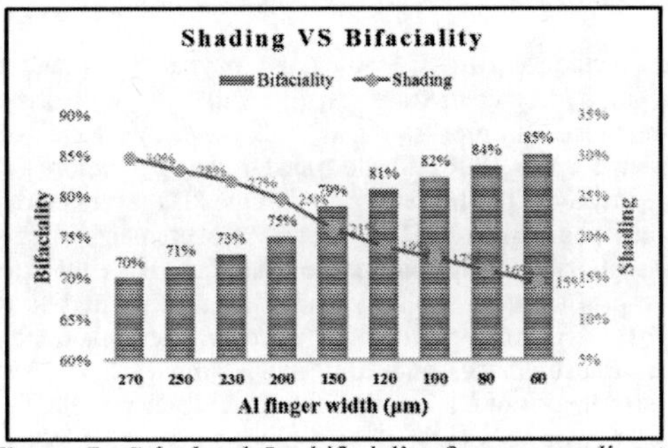

Figure 7: Calculated J_{SC} bifaciality factor according to the specified Al finger width of mc-Si bifacial PERC solar cell

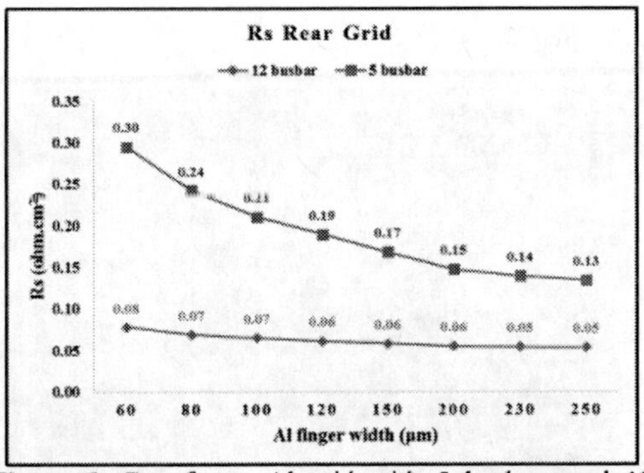

Figure 8: R_S of rear Al grid with 5 busbars and 12 busbars for different Al finger width. Measured Al specific resistivity is about 2.48 x 10^{-5} Ω.cm, Al finger height set to 17μm.

4 CONCLUSIONS

Industrial mass production P-type mc-Si bifacial PERC solar cells, of which average front and rear side efficiency up to 20.15% and 15.8% respectively, as measured by HALM tester with a black chuck, are demonstrated in this paper. A detailed optical loss analysis for the best cell with 82.5% bifaciality factor indicated that the rear metal shading and rear reflection are the dominant J_{SC} loss of rear side. It is expected that firing through Al paste and multi-busbar are the upcoming technologies which can improve rear side efficiency dramatically and mc-Si bifacial PERC solar cell with bifaciality factor of up to 85% can be achieved.

5 ACKNOWLEDGEMENTS

The authors would like to thank Dr. Zhang and all members of R&D department from GCL System Integration Technology Co., Ltd..

6 REFERENCES

[1] Guo S, Walsh TM, Peters M. Vertically mounted bifacial photovoltaic modules: A global analysis. Energy 2013; 61: 447-454.

[2] Janssen GJM, Van Aken BB, Carr AJ, Mewe AA. Outdoor performance of bifacial modules by measurements and modelling. Proceedings of the 5th Silicon PV Conference, Konstanz, Germany, 2015; in press.

[3] I.G. Romijn etc. Industrial cost effective n-Pasha solar cells with> 20% efficiency. In proceedings of the 28th European photovoltaic solar energy conference and exhibition. France: Paris, 2013, 736-740.

[4] Larionova Y, etc. Industrial Ion Implanted Co-annealed and Fully Screen-Printed Bifacial n-PERT Solar Cells with Low-Doped Back-Surface Fields. Oral presentation at the 5th nPV workshop, Konstanz, Germany, 2015.

[5] Guilin Lu, etc. Thin Al_2O_3 passivated boron emitter of n-type bifacial c-Si solar cells with industrial process. Prog. Photovolt: Res. Appl. 2017; 25: 280-290.

[6] T Dullweber, etc. The PERC+ cell: A 21%-efficiency industrial bifacial PERC solar cell. 31st European photovoltaic solar energy conference and exhibition. Hamburg, Germany, 2015.

MAGNETRON SPUTTERED TCO-LAYERS FOR INDUSTRIAL PRODUCTION OF HETEROJUNCTION SILICON SOLAR CELLS

R. Korn, S. Hübner, M. Huber, and P. Wohlfart
Singulus Technologies AG
Hanauer Landstrasse 103, 63796 Kahl am Main, Germany
Phone: +49 6188 440 226; Email: Ronald.Korn@Singulus.de

ABSTRACT: In this contribution we present investigations and developments on sputtering transparent conductive oxides for crystalline heterojunction solar cells using rotary magnetrons. The aim of this work is to define production components and processes for highly transparent and well conducting electrode layers that are both, ideally adapted to the subjacent cell, and cost efficient. In detail, the contribution will deal with the influence of the main growth parameters of indium tin oxide, like gas phase composition, and the magnet bar configuration on layer and cell quality with respect to parameters such as V_{OC}, J_{SC} and efficiency of the final solar cells. Additionally, we will point out the aspects to be considered from the machine builder's point of view to achieve highest possible cell efficiency with cost efficient large scale sputtering systems.
Keywords: Heterojunction, ITO, Magnetron Sputtering, Silicon, Solar Cell

1 INTRODUCTION

Heterojunction (HJT) solar cells nowadays demonstrate the highest conversion efficiency among all kinds of silicon based single junction solar cells. Their huge cell and module efficiency potential, the compact process flow of only four key steps, and the low energy generation cost (LCoE) based on high energy yield have caused tremendous interest in the global PV market. In contrast to the other three key process steps, wet-chemical texturing/cleaning, Plasma Enhanced Chemical Vapor Deposition (PECVD) and screen-printing, the large area deposition of transparent conductive oxide (TCO) layers via magnetron sputtering was not widely applied in mass production of silicon solar cells hitherto, and therefore becomes one of the focuses to master the production of highly efficient HJT solar cells in industrial scale in the future. Besides optimum target material and gas phase composition, the magnetic system of the magnetron sputter source is a key element to achieve well conducting and highly transparent indium tin oxide (ITO) layers without damaging the intrinsic and doped amorphous silicon layers of the HJT cell. Due to the fact, that the customer should never have to change its magnetics to get best results, the processes have to be fully understood by the machine builder and the hardware has to be optimized for these processes in advance.

2 EXPERIMENTAL DETAILS

The samples have been deposited in a horizontal top down sputtering machine equipped with two rotary targets of 550 mm in length and a load lock chamber. As substrates, pure glass, glass with amorphous silicon layer, and 6-inch pseudo-square (M2 size) precursors have been used. After cleaning and texturing the CZ silicon wafers with a thickness of ~150-170 µm, and removing the native oxides by hydrofluoric acid, the amorphous silicon (a-Si:H) was deposited by PECVD to form the a-Si:H(i/n) electron collector on the rear, and the a-Si:H(i/p) hole collector on the front side of the wafer. For deposition of the ITO layers three different ITO target compositions have been used with 90/10, 95/5 and 97/3 relative weight percentage of indium and tin, respectively, and three different magnetic bars. The ITO layers were dynamically sputtered in DC-mode with 4 kW of power, while passing the cathodes twice. Argon (Ar), oxygen (O_2), and an argon/hydrogen mixture ($Ar:H_2$) with 5% of hydrogen were used as process gases. The precursors have been heated up to 180°C in the transfer chamber before ITO deposition. For characterization of the ITO layers on pure glass and glass/a-Si:H substrates, measurements of sheet resistivity by four-point-probe technique, layer thickness by profilometry, mobility and carrier concentration by Hall-effect, and optical absorptance by spectrophotometry in transmittance and reflectance mode have been performed (not all shown here). To characterize the final cells, measurements of the minority carrier lifetimes, temperature corrected J_{SC}, V_{OC}, fill factor, and efficiency have been performed.

3. RESULTS

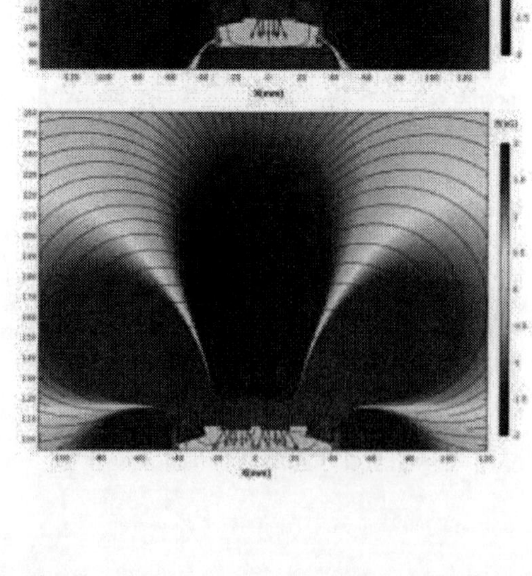

Figure 1: Simulation of unbalanced tri-row magnetics and balanced four-row magnetics.

To achieve the best results for ITO deposition, the used magnetic system is one of the key elements. Since it is impossible for cell manufacturers to have long lasting experiments to find the best magnetic bars for their process, the magnetics have to be optimized at the machine builder's site in advance. Figure 1 exemplarily depicts the simulated magnetic fields for two different magnetic configurations. On the upper end there is the cross sectional view of the magnetic field of a tri-row magnetics. On the lower end, the magnetic field of a four-row magnetics is shown. Within one magnetic system every single row of magnets is made of the same type and strength in the simulation. The resulting difference between the two magnet arrangements is the distance between the field-free point and the magnets or target surface, respectively. By definition, this distance is a measure for the degree of balance of the magnetic system. Therefore, the magnetics on the upper end are quite unbalanced, while the magnetics on the lower end are highly balanced.

The real measurements of two tri-row set-ups (magnetics 1, and 2), and a four-row system (magnetics 3) are shown in figure 2. The distance of the field-free point and the surface of the magnetics No.1 is approximately 60 mm. For magnetics No.2 the distance between the surface and the field-free point is approx. 145 mm. For magnetics No.3 the distance is such large, that the field-free point is no longer visible in the measured area. Therefore, the degree of balance is increasing from magnetics No.1 to magnetics No.3. The maximum measured magnetic field strength $\mathbf{B_{max}}$ is 1636 G for magnetics No.1, 2282 G for magnetics No.2, and 1481 G for magnetics No.3, respectively, each measured in closest distance to the magnets.

Figure 3 presents the resistivity of several ITO-layers, deposited on glass substrates using the 95/5 target without hydrogen, in dependence on the oxygen content in the gas phase for the three different magnetics. The strong dependence of the resistivity of the deposited ITO-layers on the oxygen content and on the used magnetic bar is readily identifiable. The resistivity in dependence on the oxygen content in the gas phase exhibits a minimum at 0.5% to 1.5% of oxygen for all three different magnetic systems. The overall lowest resistivity is reached with the magnetics No.2 at a gas phase composition of 1.25% of oxygen for the 95/5 ITO target.

The unbalanced magnetics No.1 lead to the highest electrical resistivity compared to the ones achieved with the balanced magnetics No.2 and 3. Due to the missing Lorentz force for ions moving along a magnetic field line, a low degree of balance opens up the possibility for charged ions to reach the surface of the substrate. A high degree of balance and high magnetic field improve the confinement of electrons and ions, and therefore lower the probability for sputter damage of the subjacent layers during ITO deposition. When the degree of balance for some magnetics is sufficiently high, the maximum strength of magnetic field can further improve electron and ion confinement and therefore lower the discharge voltage. This is the reason why magnetics No.2 lead to better results than magnetics No.3 with highest degree of balance, but lower maximum strength of magnetic field. Discharge voltages for all three magnetics with 95/5 and 97/3 ITO targets are shown in figure 4.

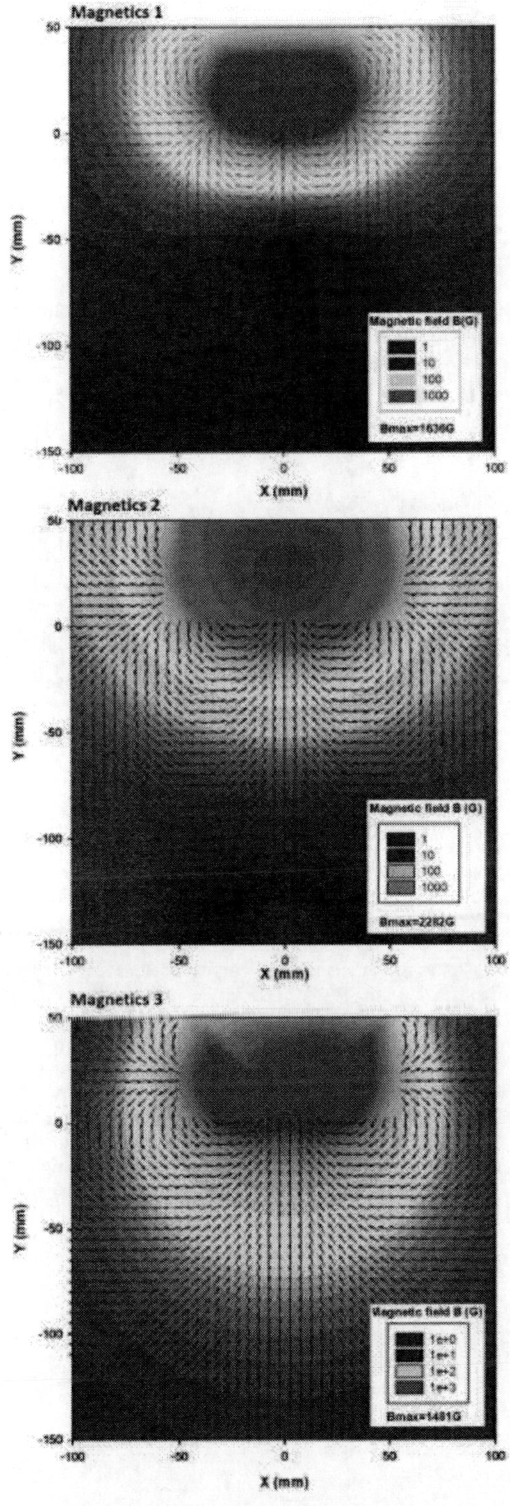

Figure 2: Measurement of the magnetic fields of unbalanced magnetics and balanced magnetics. The strength of the magnetic fields is measured in Gauss (G).

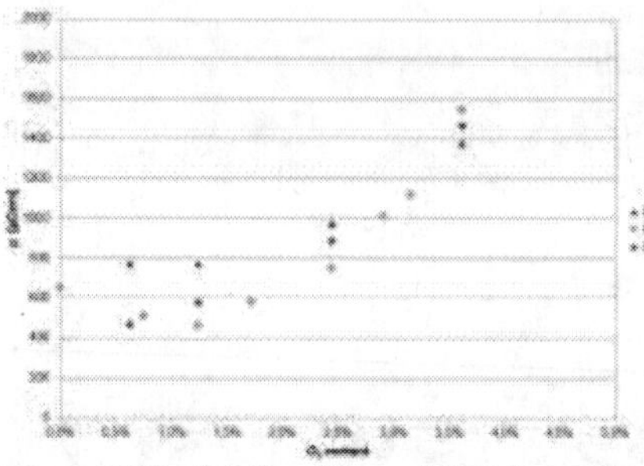

Figure 3: Resistivity of several ITO-layers depending on oxygen content and used magnetics. The layers are deposited without additional hydrogen.

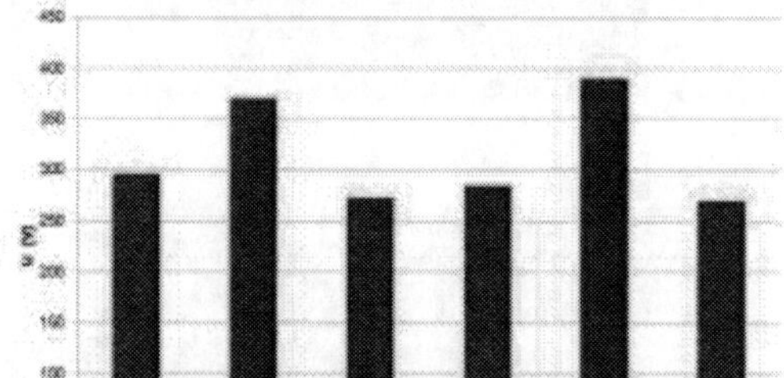

Figure 4: Discharge voltages for all three magnetics with 95/5 and 97/3 ITO targets.

After testing several magnetics with the 95/5 target composition, the best magnetics No.2 have been used to investigate the dependency of the resistivity on oxygen content and target composition. The results are depicted in figure 5. Lowest resistivity is achieved for an oxygen content of 1.25% using the 90/10 ITO target. Figure 6 illustrates the transmission curves of four ITO layers deposited with 95/5 ITO target composition using magnetics No.3, and several oxygen contents, no hydrogen, with layer thicknesses of about 155nm. The maximum transmission is achieved for 3.6% of oxygen in the gas phase. Considering the increasing resistivity with increasing oxygen content in the gas phase (shown in fig. 3 and 5), TCOs for solar cell applications have to be a trade-off between lowest electrical resistivity and highest optical transmission.

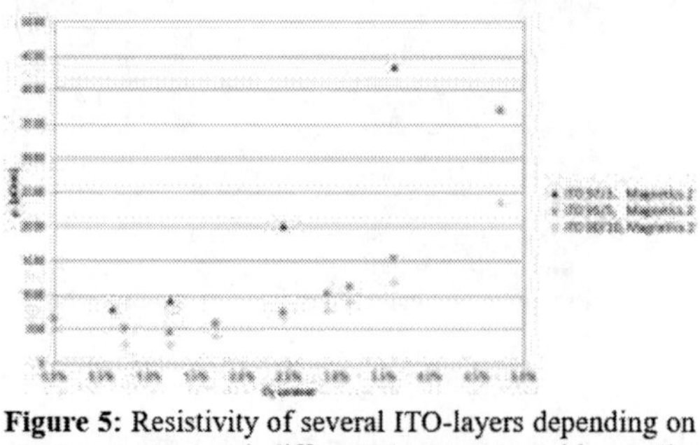

Figure 5: Resistivity of several ITO-layers depending on oxygen content and different target compositions. The layers are deposited without additional hydrogen.

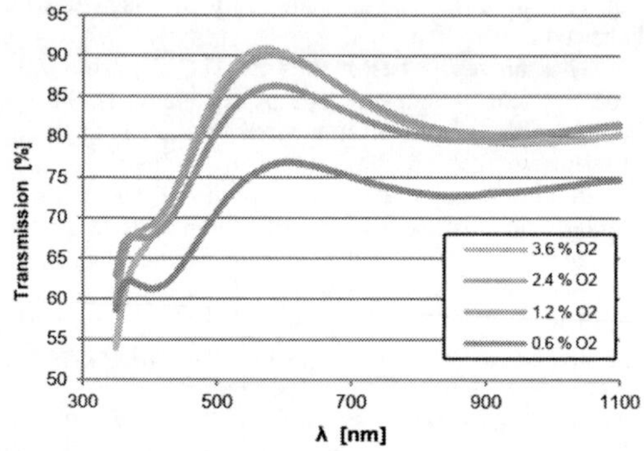

Figure 6: Transmission of four ITO layers with thicknesses of approximately 155nm deposited with several oxygen contents in the gas phase, magnetics No.3, and no hydrogen in the gas phase.

This trade-off between lowest electrical resistivity and highest optical transmission is found to be in optimum with magnetics No. 2, ITO 95/5 target composition, and 1.2% of oxygen in the gas phase. With these settings complete HJT cells have been coated on pre-processed precursors from a cell manufacturer. The reference samples are fully processed by the cell provider with its best known process including the ITO layers on front and rear side achieving efficiencies of up to 22.2%. For the "ITO FS" and "ITO FS & BS" samples (figure 7) the ITO layers were deposited by SINGULUS only on front side, and on front and rear side, respectively. For "ITO FS & BS" samples there is a distinction, whether the silver contact layer was deposited by the cell provider (+Ag ref.) or by SINGULUS (+Ag).

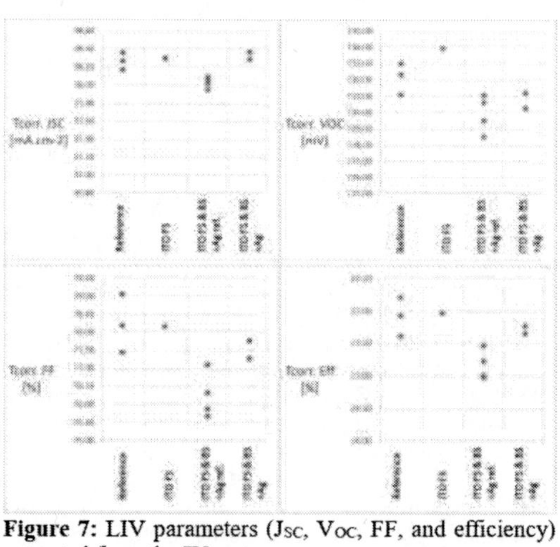

Figure 7: LIV parameters (J_{SC}, V_{OC}, FF, and efficiency) extracted from the IV curves.

All the cells with ITO front side layer and ITO front and rear layer with silver deposited by SINGULUS show very similar values for temperature corrected J_{SC}, V_{OC}, fill factor, and efficiency to the ones of the reference samples. Solely, the cells with front and rear ITO deposited by SINGULUS and with silver layers made by the cell provider reveal lower values for all four parameters. This must be due to the silver deposition process and should not be associated with the corresponding ITO layers. All in all, HJT solar cells have been realized with front and rear side

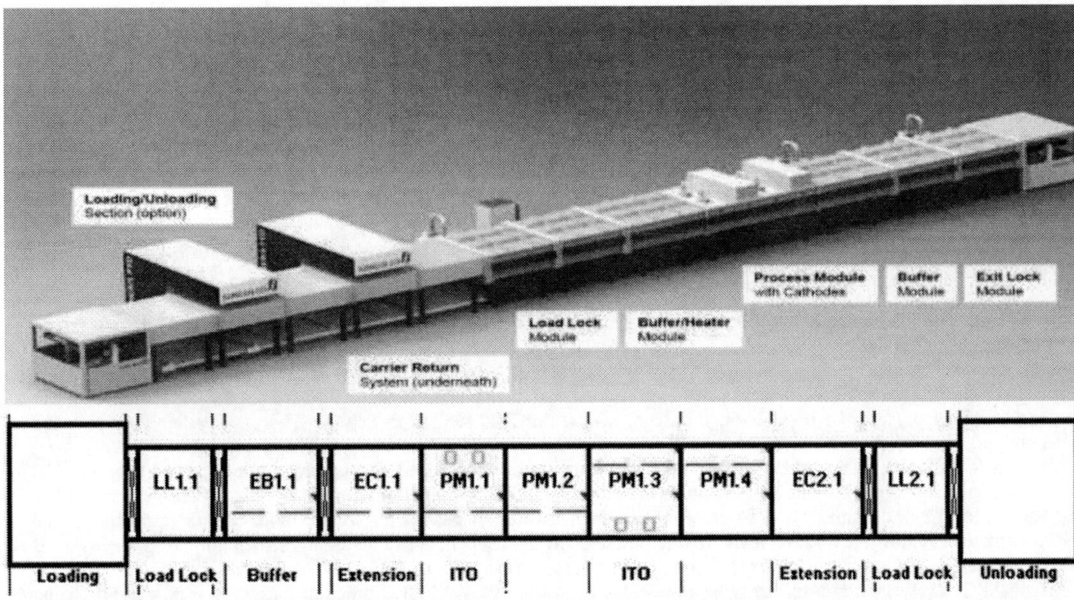

Figure 8: Layout and schematic of an industrial mass production machine for the deposition of transparent conductive ITO layer on crystalline silicon HJT solar cells.

ITO layers deposited by SINGULUS, revealing efficiencies of up to 22% confirming the high quality of the deposited ITO layers.

The design and schematic of an industrial ITO deposition tool is depicted in Figure 8. To achieve best layer properties and lowest sputter damage, the aforementioned optimum in gas flow ratio and magnetic system has to be applied. Beyond that, thermal equilibrium, low background pressure, good gas separation between individual process chambers and overall process stability have to be granted by the tool design. In order to increase machine throughput and to reduce production cost, the deposition process is scaled up to 8 wafers width with a production capacity of 5200 wafers per hour.

4. SUMMARY AND OUTLOOK

In conclusion, HJT solar cells were manufactured with ITO layers deposited by SINGULUS TECHNOLOGIES and were compared to cells manufactured with best known process of the cell supplier. Cells with SINGULUS front- and rear side ITO layers reveal efficiencies of up to 22%, and are on par with the reference samples. The influence of different types of magnetic systems on electrical and optical properties of the deposited ITO layers has been investigated and a fundamental understanding of the requirements for large scale sputtering systems for TCO layers for industrial production of HJT solar cells could be derived.

ATMOSPHERIC PRESSURE CHEMICAL VAPOR DEPOSITION OF IN-SITU DOPED AMORPHOUS SILICON LAYERS FOR PASSIVATING CONTACTS

Agnes Merkle[1], Sven Seren[2], Holger Knauss[2], Byungsul Min[1], Jonathan Steffens[3], Barbara Terheiden[3], Rolf Brendel[1,4], Robby Peibst[1,5]
[1]Institute for Solar Energy Research Hamelin (ISFH), Am Ohrberg 1, D-31860 Emmerthal, Germany
[2]Schmid Group, Robert-Bosch-Str. 32-36, D-72250 Freudenstadt, Germany
[3]University of Konstanz, Universitätsstraße 10, D-78464 Konstanz, Germany
[4]Institute of Solid State Physics, Leibniz Universität Hannover, Appelstraße 2, D-30167 Hannover, Germany
[5]Institute of Electronic Materials and Devices, Leibniz Universität Hannover, Schneiderberg 32, D-30167 Hannover, Germany
Corresponding author: phone: +49-5151-999-306, fax: +49-5151-999-400, email: merkle@isfh.de

ABSTRACT: **A**tmospheric **P**ressure **C**hemical **V**apor **D**eposition (APCVD) is so far used for the deposition of doped and undoped oxides. We upgrade an industrial in-line APCVD tool from Schmid Thermal Systems and develop potential cost-effective APCVD processes for deposition of intrinsic and in-situ-doped amorphous silicon (a-Si) layers. In this work we report for the first time on in-situ phosphorus-doped n^+-type and boron-doped p^+-type APCVD silicon layers. For n^+-type and p^+-type **pol**ysilicon on **o**xide (POLO) junctions based on these layers we demonstrate saturation current densities $J_0 < 6$ fA/cm², implied open circuit voltages iV_{oc} of 721 mV and 730 mV respectively, and simultaneously low junction resistivity ρ_c of 18 mΩcm². These properties enable the implementation of these layers as carrier-selective passivating contacts in solar cells. First industrial proof-of-concept solar cells using boron-doped p^+-type APCVD layers are presented. We show that the p^+-type APCVD POLO junction formation is possible by firing only and present an industrial feasible cell process for its drop-in implementation in current passivated emitter and rear cell (PERC) process flow.
Keywords: Amorphous Silicon, Polycrystalline, Carrier-Selectivity, Passivating Contacts, Silicon Solar Cell

1 INTRODUCTION

The objective to further decrease the levelized costs of photovoltaic energy (LCOE) conversion, as well as the competition on the photovoltaic market demand for high quality products and low production costs. On cell level, this implies a continuous improvement of cell efficiency, realized with lean processes and a minimum investment in new tools sets [1].

Therefore, the solar cell production is currently in a transition phase from the conventional full-area rear aluminum back surface field (Al-BSF) to the passivated emitter and rear cell (PERC) design. As predicted by the International Technology Roadmap for Photovoltaic (ITRPV) [2], the market share of the PERC cell concept will reach 30% in 2018, while the share of cells with Al-BSF is shrinking to 60%.

The PERC technology introduces the passivation of the cell rear side by a locally opened dielectric layer stack and allows a higher cell efficiency due to improved optics and the reduced recombination at the rear-side [3, 4]. A further cell efficiency improvement can be achieved by replacing the local Al-BSF contacts, which still have an increased recombination current, by well passivating carrier selective contacts.

In recent years, the evaluation of different types of carrier selective junctions and their implementation in high efficiency solar cells represent an important topic for many research groups [5].

In contrast to common metal-semiconductor junctions, passivating contacts combine simultaneously low saturation current densities J_0 with low contact resistivities ρ_c resulting in high carrier selectivity

$S_{10} = \log_{10}(V_{th}/(J_0\,\rho_c))$ (V_{th} thermal voltage) [6].

The POLO junctions investigated at ISFH [7] feature a thermally or wet chemically grown thin (1.5-2.2 nm) interfacial oxide layer and a highly phosphorous- or boron-doped polycrystalline silicon layer. The latter was so far deposited by LPCVD (Low Pressure Chemical Vapor Deposition) or PECVD (Plasma Enhanced CVD) technique. Both techniques have their specific advantages.

A further alternative is Atmospheric Pressure Chemical Vapor Deposition (APCVD). The possible advantages of this technology are cost-wise. Working at atmospheric pressure, the industrial-proved APCVD system from Schmid Thermal Systems works without gate chambers (only gas curtains), no vacuum chambers (no pumps) and no plasma sources. This in-line tool enables so far the high-throughput deposition of functional (BSG, PSG, Al_2O_3) or capping (SiO_2, TiO_2) layers in any requested sequence.

However, it was unclear so far whether the residual oxygen concentration in the process zone can be reduced to an extent which allows the deposition of silicon instead of silicon oxide layers.

In this work, we report on in-situ-doped n^+- and p^+-type silicon layers deposited using the APCVD technique. We investigate n^+- and p^+-type POLO junctions based on these APCVD layers and show their suitability for realization of passivating contacts. We implement the p^+-type APCVD POLO junction as a passivating contact at the rear-side of a solar cell with a conventional PERC front-side. These first proof-of-concept solar cells are presented.

We furthermore investigate the formation of p^+-type APCVD POLO junctions by a short high temperature anneal using a fast firing step only. We are able to reproduce the respective results of other groups [8]. Based on the current PERC cell process, we present an industrial relevant process flow for new POLO-PERC solar cells with a p^+-type POLO passivating contact at the rear-side.

2 APCVD DEPOSITION OF SILICON LAYERS

The injector head shown in Figure 1 is a line source with separated supply of SiH_4 and N_2. Doping precursors as PH_3, B_2H_6 etc. can be added to the SiH_4 flow. The directed flow of the process gases assures their very efficient use. The thermally activated reaction occurs directly at the surface of the wafers situated on the transport belt, so that the process is a single-sided coating.

The thickness of the deposited layers and the dopant concentration can be varied by adjusting temperature, gas flows, wafers transport speed or by using multiple injectors in a row. The multiple injectors design also enables the deposition of stacks consisting of different layers.

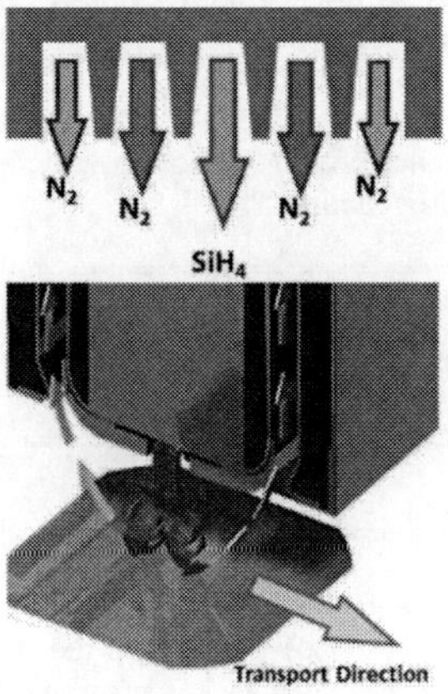

Fig. 1 Injector head of Schmid Thermal Systems APCVD tool

The basic idea for deposition of silicon layers using this tool is the thermal dissociation of SiH_4 in the process zone in an atmosphere with very low oxygen concentration and the precipitation of the silicon atoms as an amorphous silicon layer on the hot substrate.

We demonstrate that gas curtains can assure an atmosphere with sufficiently reduced oxygen amount in the process zone. We redesigned the injector head adapting it to the new specifications needed for deposition of amorphous silicon layers.

We develop a new process with SiH_4 as process gas for intrinsic silicon deposition. By adding PH_3 or B_2H_6 as dopant gases we successfully deposit in-situ n^+-type phosphorus- and p^+-type boron-doped silicon layers.

3 CHARACTERIZATION OF APCVD SILICON LAYERS AND POLO JUNCTIONS

3.1 Optical characterization of intrinsic silicon layers

Using the newly developed deposition process with only SiH_4 as process gas we first deposited undoped APCVD layers on a glass substrate. In order to determine their structure and chemical composition, we used optical methods as spectroscopic ellipsometry and Raman spectroscopy.

We performed spectroscopic ellipsometry measurements in the wavelength range from 240 nm to 1700 nm for five different angles of incidence γ (60°, 65°, 70°, 75° and 80°). The measured data are modeled using a function fit model for amorphous silicon layers of Aspnes *et al.* consisting of a Cody-Lorentz oscillator [9]. Figure 2 shows a representative example of the ellipsometry variables Ψ and Δ as a function of the wavelength.

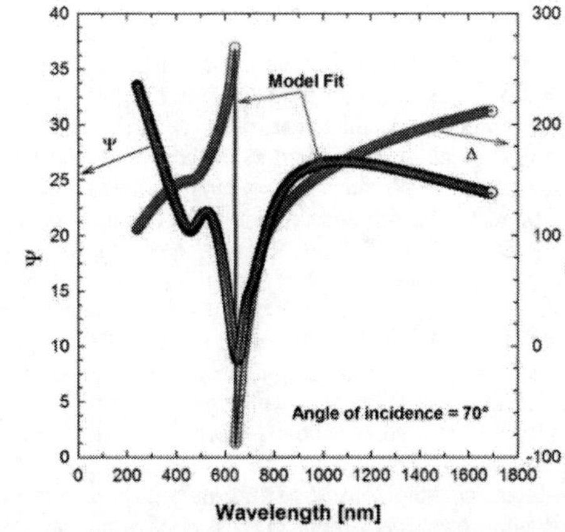

Fig. 2 Measured and fitted spectroscopic ellipsometry data for 70° angle of incidence. Black and green symbols show measured Ψ and Δ data, red lines model fitted data.

The good agreement of measured and model fitted data indicates that the deposited layers are amorphous silicon layers.

We investigate the composition of the deposited layers by Raman spectroscopy. From Raman spectroscopy on partially crystalline silicon layers, one expects a broad peak at around 480 cm^{-1} corresponding to the transversal optical phonon mode in amorphous silicon and a sharp peak at around 520 cm^{-1} corresponding to the longitudinal optical phonon mode in crystalline silicon [10].

Figure 3 exemplarily shows a Raman spectrum of an APCVD layer deposited using the new developed process for intrinsic silicon.

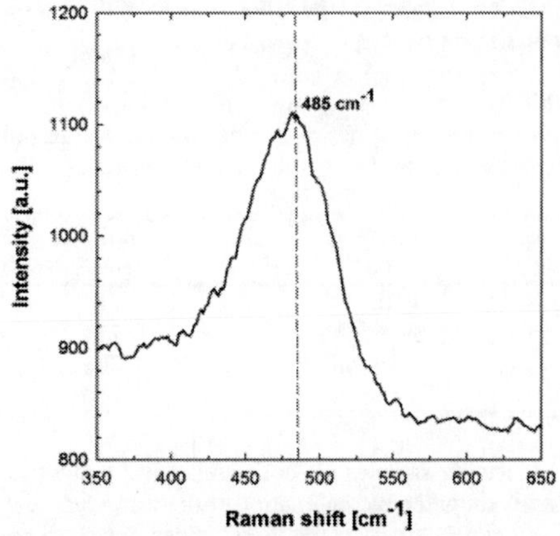

Fig. 3 Raman spectrum of an APCVD silicon layer with a peak at 485 cm^{-1} corresponding to the transversal optical phonon mode of amorphous silicon.

The peak at 485 cm^{-1} is clearly visible while the peak at 520 cm^{-1} is completely missing. Hence, the deposited layer is amorphous without a crystalline phase.

3.2 Electrical characterization of APCVD POLO junctions formed by furnace annealing

All the samples for the characterization of the doped APCVD silicon layers used in our different experiments are prepared using saw damage-etched, (156.75×156.75) mm^2, n-type Czochralski silicon wafers with a base resistivity of 3.6 Ωcm (after high temperature processing) and a thickness of 170 µm. After a wet chemical cleaning of the wafers, a thin interfacial silicon oxide grows in a thermal or wet chemical process, with a thickness of 2.2 nm and 1.9 nm respectively. Using the APCVD tool from Schmid Thermal Systems we deposit phosphorus-doped n^+-type or boron-doped p^+-type silicon layers with a thickness of ~45 nm or ~100 nm on only one side of the wafers. In order to form high quality POLO junctions the samples undergo a high temperature annealing process at 900°C (n^+-type Si layers) or 920°C (p^+-type Si layers). This process crystallizes the amorphous silicon layer, diffuses the dopants, reduces the interface state density and forms pinholes in the thin interfacial oxide. All these processes have an impact on the passivation quality and resistivity of the POLO junction.

An aluminum oxide AlO_x coating deposited by atomic layer deposition (ALD) passivates the other side of the wafers. Subsequently the wafers are annealed at 425° for 30 minutes. We characterize the recombination behavior of these samples performing photoconductance decay (PCD) measurements with a Sinton WCT120 tester.

We characterized four types of POLO junctions: in-situ phosphorus-doped n^+-type POLO junctions with (i) a thermal or (ii) a wet chemical interfacial oxide, and in-situ boron-doped p^+-type POLO junctions with (iii) a thermal or (iv) a wet chemical interfacial oxide. Note that, as the APCVD silicon deposition is one-sided our samples are always not symmetrical. We extract the $J_{0,total}$ values from the measured carrier lifetimes according to the Kane-Swanson method [11] and using the Auger parametrization from Richter [12].

The best values of $J_{0,total}$, implied iV_{oc} and sheet resistivity R_{sheet} for the four types of samples are summarized in Table I.

Table I. Best values of $J_{0,total}$, iV_{oc} and sheet resistivity R_{sheet} for the four groups of samples with 45 nm poly-Si after furnace anneal and without hydrogenation.

Polysilicon layer	Interfacial oxide	$J_{0,total}$ [fA/cm²]	Impl.iV_{oc} [mV]	R_{sheet} [Ω/□]
n^+	thermal	10	718	30
p^+	thermal	15	714	406
n^+	chemical	17	708	27
p^+	chemical	16	714	330

Since our samples are not symmetrical and we do not exactly know the recombination at the wafer side without polysilicon (~2-3 fA/cm²), these values always represent an upper limit for the saturation current densities J_0 of the POLO junctions itself. The low $J_{0,total}$ and high iV_{oc} values, even without hydrogenation, show the good passivation quality of the new APCVD poly-Si layers for both types of interfacial oxide.

In order to investigate the impact of hydrogenation on the passivation quality of POLO junctions we used samples of type (ii) and (iv) with ~100 nm doped polysilicon layers processed as described above until the junction formation. We then passivated only the side without polysilicon layer this time with an AlO_x/SiN_x stack. After PCD measurements, we deposit the same stack also on the side with polysilicon. Subsequently, firing of the samples is performed at 850°C in a belt furnace, as it is usual for PERC solar cells and we perform again PCD measurements. We extract the $J_{0,total}$, iV_{oc} and sheet resistivity R_{sheet} values for the samples with uncoated polysilicon layer (before hydrogenation) and after coating of the polysilicon layer and firing of the samples (after hydrogenation). Table II presents the best values we obtained.

Table II Best values of $J_{0,total}$, iV_{oc} and sheet resistivity R_{sheet} for samples with ~100 nm polysilicon layers before and after hydrogenation.

Polysilicon layer		$J_{0,total}$ [fA/cm²]	Implied iV_{oc} [mV]	R_{sheet} [Ω/□]
Before hydrogenation	n^+	10	712	33
	p^+	17	714	293
After hydrogenation	n^+	6	721	30
	p^+	6	730	185

The $J_{0,total}$ and iV_{oc} values before hydrogenation are similar to the values of Table I, so that the thickness of the doped APCVD layer makes no significant difference in the passivation quality. After coating of the doped polysilicon layer with an AlO_x/SiN_x stack and firing (after hydrogenation), we observe a decrease of the $J_{0,total}$ values to 6 fA/cm² for both groups and an increase of the iV_{oc} values to 721 mV for n^+- and 730 mV for p^+-type junctions.

This improvement is possibly caused by the saturation of dangling silicon bonds with hydrogen atoms released by the hydrogen-containing dielectric stack during the high temperature firing process [13].

Assuming that the high passivation quality of the fired POLO junctions is maintained also in the metallized regions [14], a $J_{0,rear}$ value of 6 fA/cm² is by a factor of 4 smaller than the total $J_{0,rear}$ value of a PERC solar cell with local Al-BSF contacts and AlO_x passivation [15].

Passivating contacts need not only good passivation properties but also low contact resistivity.

In order to evaluate the contact resistivity between an n^+-type polysilicon layer and the n-type wafer we use the transfer length method (TLM) and perform the data analysis according the model of Eidelloth and Brendel [16]. Wafers with n^+-type polysilicon layers on thermal and wet chemical interfacial oxide are used for this purpose. TLM structures are prepared and metalized with evaporated Aluminum as described in Reference [17].

The measurements of the TLM structures for the determination of the combined junction resistivity Al / n^+-type poly-Si / c-Si result in a value of 18 mΩcm². We consider this value as the resistivity between n^+-type poly-Si / c-Si, because the resistivity Al / n^+-type poly-Si is very low [17].

The set of best J_0 of 6 fA/cm² and this ρ_c value for the

n^+-type POLO junctions result in a good selectivity value of S_{10} = 14.4, which would allow a contact limited cell efficiency >25% [6]. Measurements for the p^+-type POLO junctions are in progress.

These results show that POLO junctions with wet chemical oxide have electrical properties comparable to POLO junctions with thermal oxide and are well suited for a cost-effective fabrication of passivating contacts.

3.3 Electrical characterization of APCVD POLO junctions formed by firing only

The stability of the POLO junctions upon firing is a fundamental criterion for their integration in current solar cells with screen-printed metallization. This requirement is fulfilled by the new APCVD POLO junctions (see above). Moreover, it has been shown recently that the firing process itself, which is anyhow necessary in the cell process (formation of the screen-printed contacts, activation of the SiN_x passivation, BO complex de-activation), might be sufficient for the formation of Si-based passivated contacts [8]. This simplifies the process when compared to a furnace-based POLO junction formation. We therefore investigate the formation of APCVD POLO junction during the firing process.

The samples we prepare for this experiment feature one-sided, boron-doped, ~100 nm-thick p^+-type APCVD layer on a wet chemical or an in-situ grown oxide. In this case, the junction formation occurs after coating the wafers with AlO_x/SiN_x stacks on both sides during firing in a belt furnace, as it is typical for industrial PERC solar cells.

Table III shows the best values for $J_{0,total}$, iV_{oc} and sheet resistivity R_{sheet} obtained from PCD measurements as described above.

These values demonstrate that POLO junction formation during firing is possible.

Table III. Best values of J_0, iV_{oc} and sheet resistivity R_{sheet} for samples with APCVD POLO junctions formed during firing.

Polysilicon layer	$J_{0,total}$ [fA/cm²]	Implied iV_{oc} [mV]	R_{sheet} [Ω/□]
p^+ on chemical oxide	18	703	371
p^+ on in-situ oxide	14	697	378

Since our samples are not symmetrical and we do not exactly know the recombination at the wafer side without polysilicon (~2-3 fA/cm²), these values represent an upper limit for the saturation current densities J_0 of the POLO junctions itself. These results are similar to those presented in Reference [8].

4 SOLAR CELLS

4.1 Solar cell preparation

In order to evaluate the boron-doped p^+-type APCVD junction on cell level we implemented it at the rear of a PERC solar cell.

We use (156.75x156.75) mm², boron-doped, 1.6 Ωcm Czochralsky-grown silicon wafers and prepare two types of solar cells:

1. Monofacial PERC cells with full-area rear-side metallization (screen-printing)
2. POLO-PERC with full metallized p^+-type APCVD POLO junction at rear-side (Al evaporation)

The schematic cross sections of the two cell types are shown in Figure 4.

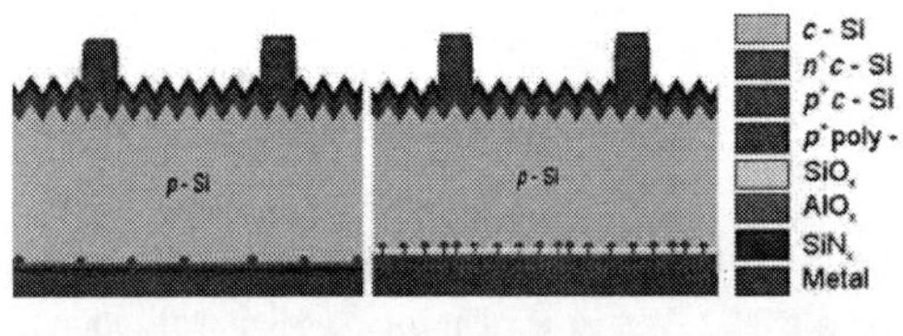

Fig. 4 Cross section of a PERC cell (left) and of a POLO-PERC cell (right)

For the manufacturing of the monofacial PERC solar cells we use a standard process flow described in detail elsewhere [3]. We modify it in order to include the p^+-type POLO junction on the rear-side as a first proof-of-concept.

Figure 5 shows the schematic process flows of the two types of solar cells we fabricated in this study.

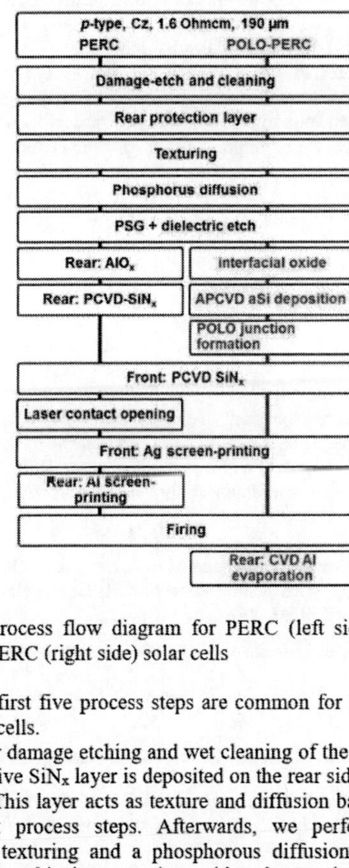

Fig. 5 Process flow diagram for PERC (left side) and POLO-PERC (right side) solar cells

The first five process steps are common for the two types of cells.

After damage etching and wet cleaning of the wafers, a protective SiN_x layer is deposited on the rear side of the wafers. This layer acts as texture and diffusion barrier in the next process steps. Afterwards, we perform an alkaline texturing and a phosphorous diffusion of the front side, achieving an emitter with a sheet resistance of 90 Ω/sq. The rear protection layer and the phosphorus glass are removed by wet chemistry.

The following steps for the PERC solar cells are the passivation of the rear side with an AlO_x/SiN_x stack and of the front side with a SiN_x layer.

In this first proof-of-concept process flow for POLO-PERC solar cells we introduce three new processes related to the formation of the p^+-type POLO junction.

The interfacial 1.9 nm thin oxide grows in a wet

chemical process. Using the APCVD tool from Schmid Thermal Systems we deposit a ~100 nm-thick boron-doped p^+-type silicon layer on the rear side of the wafers. In order to form a high quality POLO junction the wafers undergo a high temperature process at 900°C in N_2 atmosphere. After a short HF-Dip the front side is passivated as described above.

The metallization of the two cell types is only for the front side identical. A finger grid and five busbars are printed using the print on print technique and a commercial silver paste. The rear-side metallization is slightly different for the two cell types.

In case of the PERC solar cell, the dielectric layers on the rear-side are locally ablated to laser contact openings and a full-metallization is realized by screen-printing of an aluminum paste. The contacts formation occurs in a co-firing process in a conventional belt furnace.

Screen-printing on thin polysilicon layers with good contact resistivity and maintaining simultaneously the high passivation quality of these layers is challenging.

In this prove-of-concept POLO-PERC cell process we are focusing on the implementation of the APCVD boron-doped p^+-type layer in a passivating contact at the rear-side of a PERC like solar cell.

Therefore, we choose for the new POLO solar cells a full-area rear-side metallization with 10 µm aluminum, deposited directly on the doped polysilicon layer by chemical vapor deposition (CVD). This technique was successfully applied in the fabrication of record IBC solar cells reported in Reference [18]. In this way, we avoid or reduce a potential degradation of selectivity of the POLO junction by laser ablation and screen-printing.

4.2 Solar cell results

Table IV summarizes the measured one-sun parameters under standard testing conditions (25°C, 100 mW/cm², AM1.5G) of the best PERC and POLO-PERC solar cells.

The measurement of the POLO-PERC solar cell was performed by the ISFH Calibration and Test Center (ISFH CalTeC). The measurement of the PERC cell is an in-house measurement using the LOANA tool from *pvtools*.

Table IV: *I-V* parameters of the best solar cells measured under standard testing conditions (POLO cell calibrated at CalTeC ISFH, PERC cell in-house measurement).

Cell	η [%]	V_{oc} [mV]	J_{sc} [mA/cm²]	*FF* [mV]	*A* [cm²]
POLO	19.5	661	37.7	78.0	92.7
PERC	20.9	662	39.3	80.1	240.7

The cell efficiency of 19.5% of the POLO-PERC solar cell is limited by the low short circuit current J_{sc} and the moderate fill factor *FF*.

This cell has a J_{sc} value of 37.7 mA/cm², which is 1.6 mA/cm² lower than compared to the PERC cell.

In order to explain the low short circuit current we performed spectral response and reflectance measurements on the two cells. Figure 6 presents the external quantum efficiency (EQE) and reflectance (R) curves for the POLO-PERC and the conventional PERC cell. Clear differences can be identified in short and long wavelength regions.

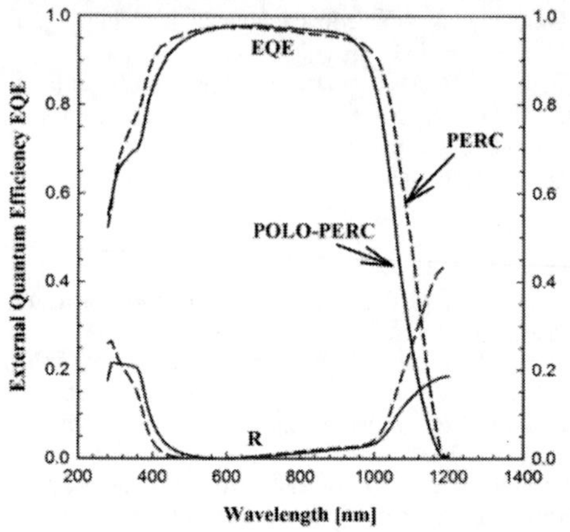

Fig. 6 External quantum efficiency EQE (dark lines) and reflectance R (blue lines) curves for the analyzed POLO-PERC (solid lines) and PERC (dashed lines) solar cells.

As the furnace-based POLO junction formation is an annealing process occurring at 900°C after diffusion of the emitter, phosphorus is deeply driven in the crystalline silicon during this process. Therefore, the emitter of the POLO-PERC cell is deeper as that of the PERC cell, resulting in higher recombination and optical losses at the front of the solar cell. This explains the differences in the EQE and reflectance curves for wavelengths <500 nm.

The reflectance curve of the POLO-PERC cell is shifted towards higher wavelengths, which indicates a thinner front passivation layer.

For wavelengths >900 nm the reflectance and EQE values of the POLO-PERC cell are lower than for the conventional PERC cell, due to parasitic absorption in the ~100 nm-thick polysilicon layer and the absence of a dielectric reflector on the rear-side of the cell.

After processing the POLO-PERC cell was laser cut from the 210 mm diameter wafer. Therefore, high recombination occurs at the strongly laser damaged edges, which further reduces the J_{sc} value.

On the one hand the moderate fill factor of the POLO-PERC cell is due to a higher series resistance (0.8 Ωcm² instead of 0.6 Ωcm² for the PERC cell). One possible reason for the increased series resistance is a higher specific contact resistance of the screen-printed front-side metallization on the deeply driven-in emitter. On the other hand, a lower pseudo fill factor (81.9% instead of 82.9% for the PERC cell) also limits the fill factor. This is possibly a negative effect of the laser cutting. It is obvious, that a process optimization is needed in order to enhance the cell efficiency and that the moderate cell performance does not reflect yet the efficiency potential of the APCVD passivating contacts.

4.3 Simplified POLO-PERC cell process flow utilizing “firing-only” POLO contacts

In Figure 7 we present the process flow for an industrial feasible POLO-PERC process, which is based on a drop-in implementation of the p^+-type APCVD POLO junction as passivating contact in the current PERC cell technology.

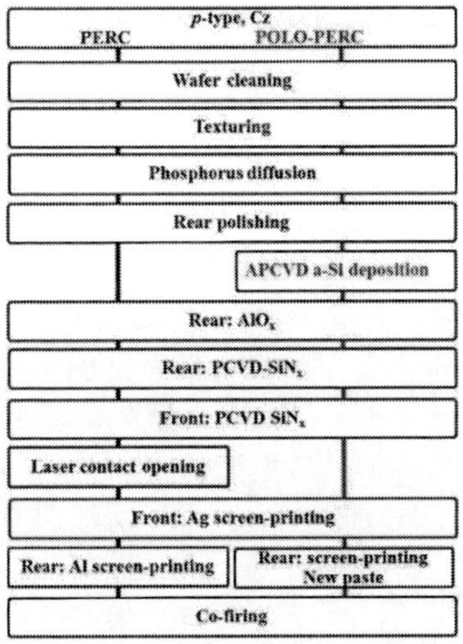

Fig. 7 Comparison of the optimized process flow for POLO-PERC (right) and conventional PERC solar cells

In order to reduce the number of process steps and thus production costs, we target the integration of APCVD POLO junctions with a wet chemical or in-situ grown interfacial oxide and their formation during the firing process.

This process flow has the same number of process steps as the current PERC process, the APCVD deposition being the only new process. It could be a cost-effective way for the implementation of the p^+-type passivated contacts onto the rear-side of mainstream manufacturing of p-type high efficiency solar cells.

5 CONCLUSIONS

For the first time, we demonstrate the deposition of intrinsic and in-situ-doped n^+- and p^+-type silicon layers using an industrial APCVD in-line tool from Schmid Thermal Systems.

Spectral ellipsometry and Raman spectroscopy measurements confirm that these layers consist of amorphous silicon.

The electrical characterization of n^+- and p^+-type furnace-based POLO junctions with thermal and wet chemical interfacial oxide results in similar high $J_{0,total}$ and implied iV_{oc} values.

Before hydrogenation, n^+- and p^+-type POLO junctions with wet chemical oxide achieve best $J_{0,total}$ values of 10 fA/cm² and 17 fA/cm², which correspond to best implied iV_{oc} values of 712 and 714 mV. After hydrogenation of these junctions best $J_{0,total}$ values of 6 fA/cm² and best implied iV_{oc} values of 721 for n^+-type and 730 mV for p^+-type junctions are achieved.

Both n^+-type POLO junctions with thermal and wet chemical interfacial oxide have a low contact resistivity of 18 mΩcm².

For p^+-type APCVD POLO junctions with wet chemical or in-situ grown oxide we demonstrate the possibility of junction formation by firing only, achieving best $J_{0,total}$ values of 18 fA/cm² and 14 fA/cm² respectively, comparable to latest values published in Reference [8].

We successfully implement a boron-doped p^+-type APCVD POLO junction with a chemical interfacial oxide at the rear-side of first proof-of-concept solar cells with conventional (PERC) front-side, achieving an independently certified cell efficiency of 19.5%.

Further process optimization should result in an improvement of the cell efficiency as compared to PERC solar cells with a conventional rear-side.

We target the drop-in integration of p^+-type APCVD POLO junctions with wet chemical or in-situ grown interfacial oxide as passivating contacts on the rear-side of current PERC cells, resulting in an advanced process flow for POLO-PERC solar cells. This cell process includes only one new process and a constant total number of process steps, having the potential for high cell efficiencies.

6 AKNOWLEDGEMENTS

This work was funded by the state of Lower Saxony and the Federal Ministry for Economic Affairs and Energy (BMWi) under grant number 0324042C (KOMPASS).

We thank our ISFH colleagues Thomas Friedrich, Ulrike Sonntag, Sarah Spätlich, David Sylla, Tobias Neubert for the preparation of the samples and solar cells. Furthermore, we thank to Heiko Zunft from Schmid Group for the deposition of the APCVD a-Si layers.

7 REFERENCES

[1] International Technology Roadmap for Photovoltaic (ITRPV), 8th Edition (2017).

[2] International Technology Roadmap for Photovoltaic (ITRPV), 9th Edition (2018).

[3] T. Dullweber, S. Gatz, H. Hannebauer, T. Falcon, R. Hesse, J. Schmidt and R. Brendel, Prog. Photovolt: Res. Appl. **20**, 630–638 (2012).

[4] T. Dullweber and J. Schmid, IEEE Journal of Photovoltaics **6** (5), 1366-1381 (2016).

[5] J. Schmidt und J. Melskens et al., IEEE Journal of Photovoltaics **8** (2), 373 (2018).

[6] R. Brendel, R. Peibst, IEEE Journal of Photovoltaics **6** (6), 1413-1420 (2016).

[7] U. Römer, R. Peibst, T. Ohrdes, B. Lim, J. Krügener, T. Wietler, and R. Brendel, IEEE Journal of Photovoltaics **5** (2), 507-514 (2015).

[8] A. Ingenito, G. Nogay, Q. Jeangros, E. Rucavado, C. Allebé, S. Eswara, N. Valle, T. Wirtz, J. Horzel, T. Koida, M. Morales-Masis, M. Despeisse, F.-J. Haug, P. Löper and C. Ballif, Nature Energy, **3**, 800-808 (2018).

[9] D. E. Aspnes, A. A. Studna, and E. Kinsbron, Physical Revue B, **29** (2), 768-779 (1984).

[10] Zhi Li, Wei Li, Yadong Jiang, Haihong Cai, Yugang Gong, Jian He, Journal of Raman Spectroscopy, **42** (3), 415-421 (2011).

[11] D. E. Kane, R. M. Swanson, Proceedings18th IEEE Photovoltaic Specialists Conference, Las Vegas, USA, 578–583, (1985).

[12] A. Richter, F. Werner, A. Cuevas, J. Schmidt, S. W. Glunz, Energy Procedia, 27, 88-92 (2012).

[13] M. Schnabel, Bas W.H. van de Loo, W. Nemeth, B. Macco, P. Stradins, W.M.M. Kessels, D. L. Young, Appl. Phys. Lett. **112**, 203901 (2018).

[14] Y. Larionova, F. Haase, A. Raugewitz, M. Rienäcker, R. Brendel and R. Peibst, 7th Workshop on Metallization & Interconnection for Crystalline Silicon Solar Cells , Konstanz, (2017).
[15] F. Ye *et al.*, Proceedings of the 43rd IEEE PVSC, Portland, USA, 3360-3365 (2016).
[16] S. Eidelloth, and R.Brendel, IEEE Electron Device Letters, 35 (1), 9-11 (2014).
[17] M. Rienäcker, M. Bossmeyer, A. Merkle, U. Römer, F. Haase, J. Krügener, R. Brendel, and R. Peibst, IEEE Journal of Photovoltaics **7** (1), 11-18 (2017).
[18] F. Haase, C. Hollemann, S. Schäfer, A. Merkle, M. Rienäcker, J. Krügener, R. Brendel, and R. Peibst, Solar Energy Materials and Solar Cells 186, 184-193 (2018).

PRINTED DOPANT SOURCES FOR LOCALLY-DOPED SIOX/POLY-SI PASSIVATING CONTACTS

Z. Kiaee, C. Reichel, F. Feldmann, M. Jahn, J. D. Huyeng, R. Keding, M. Hermle, F. Clement
Fraunhofer Institute for Solar Energy Systems (ISE), Heidenhofstr. 2, 79110 Freiburg, Germany
Phone: +49 761 - 4588 5654; e-mail: zohreh.kiaee@ise.fraunhofer.de

ABSTRACT: This paper presents an industrially relevant solution for the formation of local passivating contacts for back-contact back-junction (BC-BJ) solar cells, based on inkjet- and screen-printing of dopant inks and pastes. The successful realization of passivating contacts utilizing printing technologies is a crucial prerequisite to simplify the fabrication process of BC-BJ cells while approaching theoretical efficiency limit of silicon solar cells. In this work, dopant inks and pastes as well as the printing parameters are tuned in order to establish a high passivation quality within c-Si/SiO_x/poly-Si lifetime samples. The impact of the anneal/diffusion conditions on the surface passivation are studied. Excellent passivation quality is achieved for inkjet-printed *n*-type (phosphorus) poly-Si surfaces with implied open-circuit voltage iV_{OC} of 726 mV and implied fill factor *iFF* of 86.2%. On *p*-type, iV_{OC} of 692 mV and *iFF* of 82.8% and iV_{OC} of 700 mV and *iFF* of 82.4% where measured by utilizing inkjet and screen-printing, respectively. Similar results were realized by Ion-implanted surfaces, herein used as the reference, underlining the high potential of the investigated printing processes.

Keywords: Inkjet, Screen-printing, Passivating contacts, BC-BJ, TOPCon

1 INTRODUCTION

One of the last remaining obstacles to approach the theoretical efficiency limit of silicon solar cells are passivating contacts [1]. Up to now, increasing the efficiency by improving the bulk lifetime and the dielectric surface passivation of silicon wafers has been extensively studied. Thus, in order to realize solar cells with efficiencies approaching the theoretical limit of 29%, passivating contacts are crucial to be developed and practically implemented in photovoltaic devices [2].

Recently, silicon solar cells with tunnel oxide passivated carrier-selective contact (TOPCon) structure [3-10] have demonstrated the great potential and received significant attention. For these types of solar cells, an ultra-thin oxide layer is implemented between polycrystalline silicon (poly-Si) and the crystalline silicon (c-Si) to deliver interface passivation of high quality. Efficiently doped poly-Si layers are used to maintain the quasi-fermi level separation in c-Si (high V_{OC}), and fulfill an efficient majority carrier transport (high *FF*).

As a result, the photo conversion efficiency of TOPCon solar cells has been continuously improved to 25.8% [11, 12], one of the highest efficiencies for crystalline silicon solar cells. A promising route for passivating contacts is their integration into back-contact back-junction (BC-BJ) solar cells indicated by an increase in efficiency of up to 26.1% [9].

As both contacts of BC-BJ solar cells are located at one side of the wafer, the fabrication of such a solar cell requires various patterning steps. Thus, for the formation of local passivating contacts we address printing technology (inkjet- and screen-printing) as an industrially relevant or rather scalable method. The technology of choice would enable a fast and easy fabrication of local passivating contacts and, consequently, an industrially attractive approach for high efficiency and low manufacturing costs.

Inkjet-printing in particular is well suited for BC-BJ solar cells as it allows precise consecutive- or simultaneous local printing of boron- (B) and phosphorus- (P) dopant sources in one printing step [13], which can be combined with one single thermal process step ("co-diffusion") [14]. The appeal of this technology lies in its flexibility, precision, digital printing templates, and the non-contact mode. Because of the latter, wafer breakage and contamination during solar cell manufacturing is reduced, which is important for high efficiency solar cell fabrication.

In this paper, the utilization of printing in order to create local passivating contacts for BC-BJ solar cells is evaluated. The focus is set on inkjet-printing, but in case of B-Doping screen-printing is evaluated, as well. Selected process parameters are tuned in order to achieve adequate quality parameters of recombination and electrical conductivity.

2 EXPERIMENTAL DETAILS

2.1 Lifetime sample preparation

Symmetric lifetime samples were realized on planar shiny-etched (100) oriented phosphorous-doped *n*-type float-zone (FZ) silicon wafers with a specific resistivity of 1 Ωcm and a thickness of 200 µm. To this end, wafers were cleaned according to RCA standard clean procedure and an ultra-thin silicon oxide (SiO_x) layer, approximately 1.3 nm, was thermally grown in a tube furnace at 600°C for 10 minutes on both sides. Subsequently, a 50 nm intrinsic amorphous silicon (a-Si(i)) layer was deposited on both sides of the wafer by LPCVD (deposition temperature of 485°C) (Fig. 1). For reference wafers, B- and P-dopants were introduced into the a-Si(i) layers by ion implantation at ion doses of 1×10^{15} (B) and 2×10^{15} (P) and at ion energy of 2 keV (B) and 5 keV (P) on both sides of the lifetime samples. In order to create dopant sources utilizing printing, B- and P-ink, a *PixDro LP50* inkjet-printer were used. The wafers were dipped in an aqueous solution of 1% hydrofluoric acid (HF) in order to guarantee a clean surface before printing. By a variation of the printing resolution (number of droplets per inch (dpi)) the amount of ink was varied, which results in different thickness values. Each variation was printed on an area of $A = 30 \times 30$ mm². After printing, the wafers were cured on a hotplate in order to evaporate the organic components at temperatures of $T_{HP} = 300$°C for $t_{HP} = 2$ min (B-ink) and $T_{HP} = 200$°C for $t_{HP} = 5$ min (P-ink).

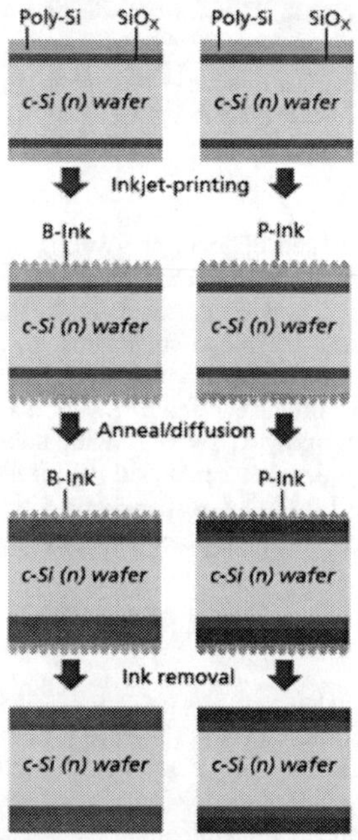

Figure 1: Schematic illustration of the structure of c-Si/SiOx/poly-Si symmetrical lifetime samples. The SiOx layer has been grown thermally in a tube furnace and features a layer thickness of about 1.3 nm. The poly-Si has been deposited by means of LPCVD and features a thickness of 50 nm. The doping of poly-Si layer was realized by inkjet-printing of dopant sources and consequent high temperature anneal.

The curing conditions were determined regarding recommendation of the material suppliers and after performing preliminary experiments. In order to obtain a proper droplet formation and, accordingly, homogeneous layers, the time-dependent piezo-voltage (wave form), firing frequency, or printing resolution were varied. One group of wafers was screen-printed using B-dopant paste. The wafers were subjected to T_{HP} = 400°C on a hotplate for t_{HP} = 3 min.

Afterwards, a high-temperature anneal/diffusion is performed in a tube furnace under nitrogen atmosphere. Four anneal/diffusion profiles were evaluated in these experiments with anneal temperature T_P = 850°C, 900°C or 950°C and annealing time t_P = 10 min or 30 min, as listed in Fig. 2. Following the anneal/diffusion step, residuals were removed in a buffered solution of 1% HF.

2.2 Lifetime sample characterization

After ink/paste removal, the passivation quality on the surface of the symmetrical lifetime samples was measured by means of quasi-steady state photoconductance (QSSPC) measurements through determining implied open-circuit voltage (iV_{OC}) and implied fill factor (iFF). These measurements were performed on a WCT-120 lifetime tester from Sinton Instruments [15]. Thereafter, the wafers were subjected to a hydrogenation process in a remote plasma hydrogen passivation (RPHP) system at 400°C for 30 min. The QSSPC measurements were repeated after RPHP in order to investigate the effect of hydrogenation on passivation quality. The sheet resistance of doped regions was

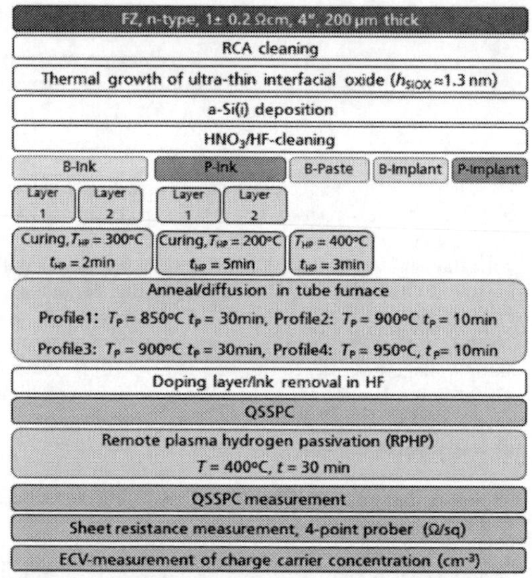

Figure 2: Schematic process sequence for fabrication of lifetime samples. An overview of different groups to investigate influence of the print resolution of the doping layers and the anneal/diffusion plateau time and temperature on passivation quality of layers.

determined by a 4-point prober. Subsequently, the doping profile was measured by electrochemical capacitance voltage (ECV) measurements using a WEP CV21 ECV profiler.

3 RESULTS AND DISCUSSION

3. 1 Inkjet-printing process

In order to optimize the jetting performance for B- and P-inks, the integrated high-speed drop-view camera of the PixDro LP50 was utilized to analyze the droplet formation, while varying the jetting settings. Once an optimal waveform (time-dependent piezo-voltage) for the piezo elements inside the print-head was created, a dynamic droplet analysis (Fig. 3) allows validating that waveform for different nozzles by measuring Volume V, velocity v, and angle of flying droplets in order to achieve precise printing of droplets onto the silicon substrate. It is clearly visible that the selected nozzles (Fig. 3) are performing similar and enable homogeneous processing. Moreover, undesired defects like the formation of so called satellites, which might induce blurry print products can be avoided. Figure 4 shows a microscopic image of droplets on a textured surface printed by using a 10 pl print-head and a print resolution in x- and y-direction of 100 dpi. The droplet diameter is 108 µm. The vertical and horizontal spacing between droplets (pitch) are 140µm and 134 µm, respectively. These dimensions are very feasible for e.g. consideration within BC-BJ architectures. Further improvements are possible by e.g. utilizing print-heads with lower droplet volume, adapting the waveform, and by heating the substrate during printing. Figure 5 shows a printed layer of B-ink on a shiny-etched FZ substrate with a resolution of 300 × 300 dpi. Different printing parameters were evaluated in order to achieve complete wafer coverage of the ink, which resulted in formation of two different sizes of droplets.

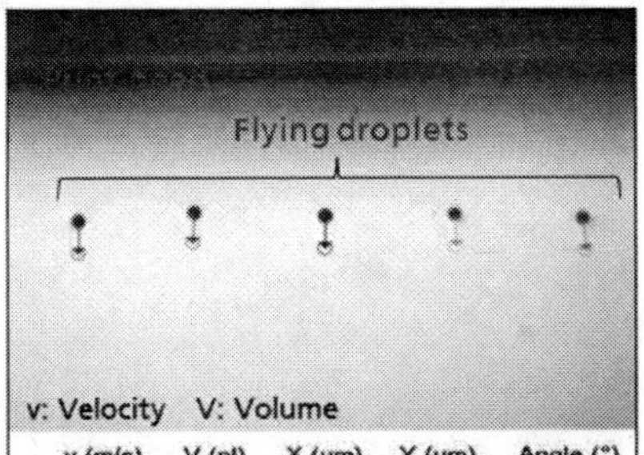

v (m/s)	V (pl)	X (µm)	Y (µm)	Angle (°)
3.09	± 11.0	153	512	-0.56°
2.85	± 11.2	392	492	-2.76°
3.06	± 11.4	661	500	-1.00°
3.05	± 11.2	924	496	1.02°
3.10	± 10.0	1184	505	4.48°

Figure 3: Dynamic analysis of the ink-droplets. Volume *V*, velocity *v*, and angle of flying droplets are measured and controlled in order to achieve precise printing of droplets onto the SiO_x-coated Si substrate.

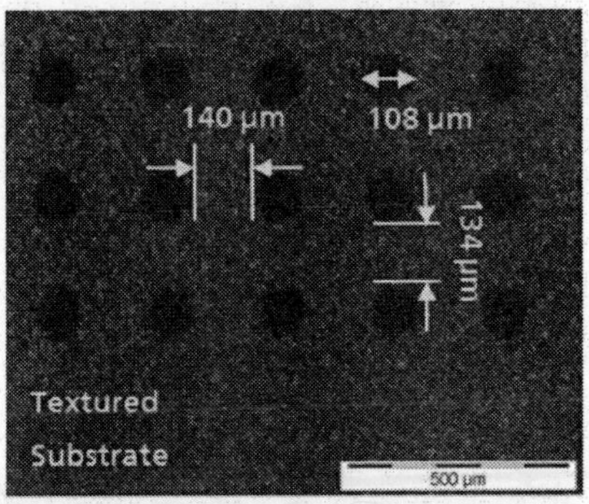

Figure 4: Droplets on a textured surface printed by using a 10 pl print-head and a resolution in x- and y-direction of 100 dpi. Droplet size, vertical and horizontal spacing between lines are illustrated.

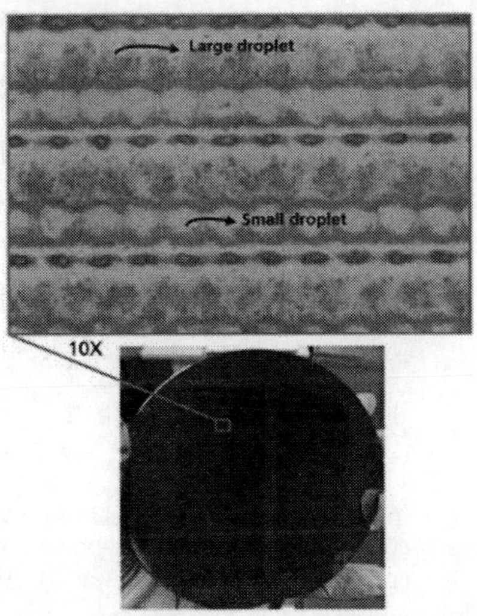

Figure 5: Printed layer on a shiny etched substrate with a resolution of 300 x 300 dpi.

3.2 Passivation quality of lifetime samples

In order to investigate the surface passivation quality, the iV_{OC} and *iFF* values of samples were extracted from injection level dependent lifetime measurements right after the high-temperature anneal/diffusion process (before introducing samples to the RPHP). Considering four different anneal/diffusion processes, the B-ink has an iV_{OC} of around 692 mV being independent of the thermal conditions (Fig. 6(a)), which is similar to the values achieved with the used reference B-ion implantation (Fig. 6 and Table I). In case of B-paste, higher passivation quality of about 700 mV can be achieved at higher diffusion temperatures. The *iFF* of B-ink and B-ion decreases with increasing anneal temperature (Fig. 6(b)). In case of B-paste, a high *iFF* of up to 82.4% can also be achieved at higher diffusion temperatures.

For P-doped samples, the iV_{OC} and *iFF* of P-ink (A) and P-ion improved, respectively, significantly and slightly, by increasing the anneal temperature (Fig. 7(a) and (b)). The improvement resulted in high iV_{OC} and *iFF* for P-ink (A) that is very close to the values of the P-ion reference sample (Table I). After performing RPHP the iV_{OC} of the P-ink (A) improved to 726 mV with an *iFF* = 86.2%, which reveals the excellent passivation quality of the layers. These values are very close to the iV_{OC} = 732 mV and *iFF* = 87.3% obtained from P-ion reference samples after performing RPHP. The P-ink (B) represented a poor passivation quality even at high temperatures. Missing values of iV_{OC} and *iFF* of P-ink (B) at lower temperatures in Fig. 7(a) and (b) are due to the very low carrier's lifetime. In addition of low passivation qualities resulted from P-ink (B), we observed precipitations on the surface of samples after the ink removal step. Table I summarizes the highest iV_{OC} and *iFF* values obtained from each dopant source.

Figure 8(a) shows the phosphorous profiles measured by ECV profiling starting on the poly-Si film and expanded to the c-Si. It can be clearly seen from the profiles that 850°C (30 min) and 900°C (10 min) anneal temperatures are not high enough to enable sufficient diffusion. The abrupt reduction of doping concentration shows the position of the tunnel oxide. The doping profiles were analyzed by EDNA [16]. The sheet resistance of 2.16 kΩ/sq and 1.93 kΩ/sq was calculated, respectively from profiles of samples annealed at 850°C (30 min) and 900°C (10 min). It was observed that by increasing anneal time at 900°C from 10 min to 30 min or increasing the temperature from 900°C to 950°C (10 min), the charge carrier concentration P-ink (A) increases significantly. Specifically, at anneal temperature of 950°C, the surface charge carrier concentration of P-ink (A) is above 10^{20} cm^{-3}, similar to the profile of the P-ion reference sample. The sheet resistance of 272 Ω/sq and 230 Ω/sq were calculated, respectively from P-ink (A) and P-ion profiles. Figure 8(b) illustrates boron profiles measured by ECV profiling. The doping profiles of B-ink samples indicate that an anneal temperature of 950°C leads to excessive diffusion of boron enabling a very deep profile inside c-Si. The calculated sheet resistance of the profile of B-ink samples annealed at 900°C and 950°C are respectively, 228 Ω/sq and 127 Ω/sq. The sheet resistance of the samples measured by a 4-point prober is 306 Ω/sq and 214 Ω/sq, respectively. The calculated sheet resistances are lower than the measured values.

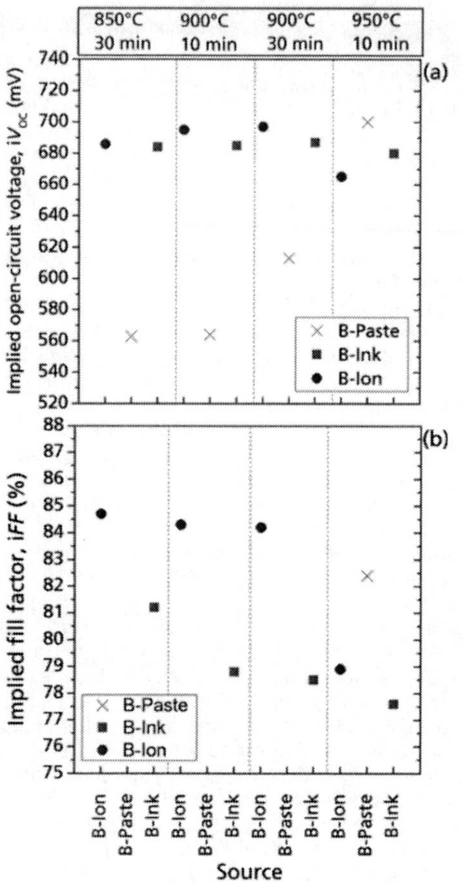

Figure 6: (a) the implied open-circuit voltage (iV_{OC}) and (b) the implied fill factor (*iFF*) vs. boron dopant source for four different diffusion processes, whereas the latter vary in temperature (850°C, 900°C, 950°C) and time (10 min and 30 min).

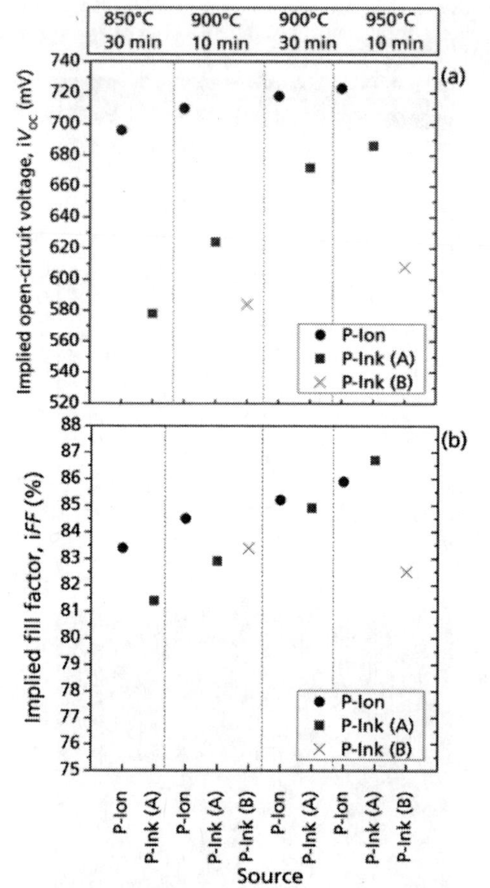

Figure 7: (a) the implied open-circuit voltage (iV_{OC}) and (b) the implied fill factor (*iFF*) vs. phosphorous dopant source for four different diffusion processes, whereas the latter vary in temperature (850°C, 900°C, 950°C) and time (10 min and 30 min).

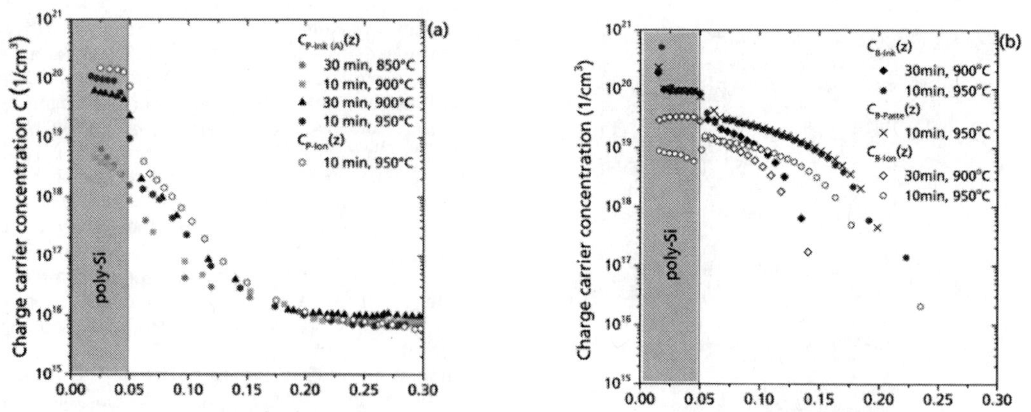

Figure 8: Charge carrier concentration vs. depth for (a) inkjet-printed phosphorus P-Ink (A) and ion-implanted phosphorus P-Ion. (b) inkjet-printed boron B-Ink, ion-implanted boron B-Ion, and screen-printed boron pasteB-Paste.

Table I: Overview of passivating contact's characteristics prepared with inkjet-printed, screen-printed and ion-implanted boron and phosphorous doped poly-Si layers, specifying passivation quality of surfaces before and after applying RPHP (iV_{OC} and iFF), the surface charge carrier concentration (C_s), calculated and measured sheet resistances from ECV profiles and by a 4-point prober.

		After Anneal		After H passivation				
Dopant source	Anneal condition	iV_{OC} (mV)	iFF (%)	iV_{OC} (mV)	iFF (%)	C_s (cm^{-3})	R_{sh} (Ω/sq) from ECV profile	R_{sh} (Ω/sq) from 4pp
P-Ink (A)	950°C, 10 min	686	87.2	726	86.2	10^{20}	272	344
P-Ion (ref)	950°C, 10 min	723	86	732	87.3	10^{20}	230	316
B-Ink	900°C, 30 min	692	82.8	682	80.6	2×10^{20}	228	306
B-Paste	950°C, 10 min	700	82.4	687	83.9	2×10^{20}	170	284
B-Ion (ref)	900°C, 30 min	695	83.2	688	84	2×10^{19}	391	564
B-Ion (ref)	950°C, 10 min	698	84.3	686	83.8	10^{19}	291	515

It is likely because the calculation tool assumes mobility of c-Si for the whole profile, either inside of the poly-Si layer or c-Si. A comparison of the profiles of B-ion and B-ink samples reveals that the amount of surface charge carrier concentration provided by B-ink ($>10^{20}$ cm^{-3}) is much higher than B-ion ($\sim10^{19}$ cm^{-3}) at the same anneal temperatures.

The B-ion sample annealed at 950°C (10 min) exhibited lower surface charge carrier concentration, but a deeper doping profile inside the c-Si, compared to the B-ion sample annealed at 900°C (30 min). The calculated sheet resistance of profiles is 564 Ω/sq and 515 Ω/sq, respectively. The B-paste sample annealed at 950°C (10 min) showed a similar profile as the B-ink annealed at the same condition. It should be noted that the profiles were measured after applying RPHP.

4 SUMMARY AND CONCLUSION

The poly-Si/SiOx/c-Si passivating contacts were realized by utilizing inkjet- and screen-printing of phosphorous and boron dopant sources. Inkjet-printed dopant sources could be homogeneously applied with a feature size of around 100µm and a pitch of around 130µm and 140 µm, respectively in x and y direction. The latter are very suitable for integration on Back-Contact Back-Junction silicon solar cells.

The passivation of both inkjet-printed phosphorous and boron doped contacts were studied as a function of anneal/diffusion condition and it turned out that passivation provided by the former was superior to the latter. It was observed that the excellent passivation quality can be achieved by inkjet-printing of phosphorus ink with iV_{OC} = 726 mV and iFF = 86.2%. Inkjet-printing of boron ink enables the formation of local passivating contacts with iV_{OC} = 692 mV and iFF = 82%. The screen-printed boron paste showed a good passivation quality at high annealing temperatures of 950°C with iV_{OC} = 700mV and iFF = 82.4%. These results pave the way for future works, where BC-BJ solar cells are fabricated with locally doped *p*- and *n*-type areas, simultaneously applied by inkjet-printing.

Ongoing studies focus on further optimizing printing processes towards higher stability, adapting poly-Si layers for even lower recombination, metallization of doped poly-Si layers and, finally, the preparation of solar cells.

5 ACKNOWLEDGEMENTS

The authors would like to thank anyone involved in this work, the research teams at Fraunhofer ISE PV-TEC and clean room facilities.

This work was supported by the German Federal Ministry for Economic Affairs and Energy within the research project "SiBoKo" under contract number 0324075B and the research project "26+" under the contract number 0325827B.

6 REFERENCES

[1] R. M. Swanson, "Approaching the 29% limit efficiency of silicon solar cells," in *31st IEEE Photovolt. Specialists Conference (IEEEPVSC)*, pp. 889-894, 2005.

[2] A. Richter, M. Hermle, S. W. Glunz, "Reassessment of the limiting efficiency for crystalline silicon solar cells," *IEEE J. Photovolt.*, vol.3, no. 4, pp. 1184-1191, 2013.

[3] F. Feldmann, M. Bivour, C. Reichel, M. Hermle, S. W. Glunz, "A passivated rear contact for high-efficiency n-type silicon solar cells enabling high Vocs and FF > 82%," in *28th European PV Solar Energy Conference and Exhibition (EUPVSEC)*, pp. 988-992, 2013.

[4] F. Feldmann, M. Bivour, C. Reichel, M. Hermle, S.W. Glunz, "Passivated rear contacts for high-efficiency n-type Si solar cells providing high interface passivation quality and excellent transport characteristics," *Sol. Energy Mater. Sol. Cells.*, vol. 120, pp. 270-274, 2014.

[5] F. Feldmann, M. Bivour, C. Reichel, H. Steinkemper, M. Hermle, S.W. Glunz, "Tunnel oxide passivated contacts as an alternative to partial

rear contacts," *Sol. Energy Mater. Sol. Cells.*, vol.131, pp. 46-50, 2014.

[6] F. Feldmann, M. Simon, M. Bivour, C. Reichel, M. Hermle, S.W. Glunz, "Efficient carrier-selective p-and n-contacts for Si solar cells," *Sol. Energy Mater. Sol. Cells.*, vol.131, pp. 100-106, 2014.

[7] C. Reichel, F. Feldmann, R. Müller, A. Moldovan, M. Hermle, S. W. Glunz, "Interdigitated back contact silicon solar cells with tunnel oxide passivated contacts formed by ion implantation," in *29th European PV Solar Energy Conference and Exhibition (EUPVSEC)*, pp.487-491, 2014.

[8] D. Yan, A. Cuevas, Y. Wan, J. Bullock, "Passivating contacts for silicon solar cells based on boron-diffused recrystallized amorphous silicon and thin dielectric interlayers," *Sol. Energy Mater. Sol. Cells*, vol. 152, pp. 73-79, 2016.

[9] F. Haase, C. Hollemann, S. Schäfer, A. Merkle, M. Reinäcker, J. Krügener, R. Brendel, R. Peibst, "Laser contact openings for local poly-Si-metal contacts enabling 26.1%-efficient POLO-IBC solar cells," *Sol. Energ. Mat. Sol.*, vol. 186, 2018.

[10] U. Römer, R. Peibst, T. Ohrdes, B. Lim, J. Krügener, T. Wietler, R. Brendel, "Ion implantation for poly-Si passivated back-junction back-contact solar cells," *IEEE J. Photovolt.*, vol.5, pp. 507-514, 2015.

[11] A. Richter, J. Benick, R. Müller, F. Feldmann, C. Reichel, M. Hermle, S. W. Glunz, "Tunnel oxide passivating electron contacts as full-area rear emitter of high-efficiency p-type silicon solar cells," *Progs. Photvoltaics.*, vol. 26, 2017.

[12] M. A. Green, Y. Hishikawa, E. D. Dunlop, D. H. Levi, J. H-Ebinger, A. W. Y. Ho-Baillie, "Solar efficiency tables (version51)," *Prog. Photovolt. Res. Appl.*, vol. 26, 2018.

[13] F. Haase, B. Lim, A. Merkle, T. Dullweber, R. Brendel, C. Günther, M. H. Holthausen, C. Mader, O. Wunnicke, R. Peibst, "Printable liquid silicon for local doping of solar cells," *Sol. Energ. Mat. Sol.*, vol. 179, pp. 129-135, 2018.

[14] R. Keding, D. Stüwe, M. Kamp, C. Reichel, A. Wolf, R. Woehl, D. Borchert, H. Reinecke, and D. Biro, "Co-diffused back-contact back-junction silicon solar cells without gap regions," *IEEE J. Photovolt.*, vol.3, pp. 1236-1242, 2013.

[15] R. A. Sinton, A. Cuevas, "Contactless determination of current-voltage characteristics and minority-carrier lifetime for quasi-steady-state photconductance data," *Appl. Phys. Lett.*, vol. 69, no, 17, pp. 2510-2512, 1996.

[16] K. R. McIntosh, P. P. Altermatt, "A freeware 1D emitter model for silicon solar cells," in *35th IEEE Photovoltaic Specialists Conference (IEEEPVSC)*, pp. 2188, 2010.

MCCE TEXTURED MULTICRYSTALLINE AND MONOCRYSTALLINE SILICON SOLAR CELLS

W. Jooss, I. Melnyk, A. Teppe, S. Madugula, J. Jung-König, O. Voigt, P. Fath
RCT Solutions GmbH, Line-Eid-Str. 1, 78467 Konstanz, Germany
Tel: (+49) 7531-584 70 17, Email: wolfgang.jooss@rct-solutions.com

ABSTRACT: In this work we demonstrate a Metal Catalyst Chemical Etching (MCCE) based inline approach for texturing DWS multicrystalline and monocrystalline wafers. Industrial manufacturing of 450,000 multi PERC solar cells based on the MCCE texturing process is demonstrated with an average efficiency of 20.2% at a customer site. The average V_{oc} and I_{sc} of the manufactured multi PERC solar cells are 653 mV and 9.56 A, respectively. The bath and process are very stable with a bath lifetime of up to 1.4 million wafers. CoO for the MCCE based process is less than 1.25 USDcts/wafer with Ag consumption of less than 0.5 mg/wafer. This is significantly lower than the texturing routes and can be reduced further with recycling of the used Ag. Inverted pyramid structures are obtained for monocrystalline wafers based on the MCCE process with reflectivity in the range of 6% to 10%. Implied V_{oc} of 678 mV is obtained for lifetime samples of mono PERC solar cells, while the IQE and EQE curves are similar to the reference solar cells based on alkaline texturing. Based on the work, an inline tool (i-UniTex) is developed to texture both monocrystalline and multicrystalline wafers.

Keywords: MCCE, texturing, multicrystalline silicon, monocrystalline silicon, inverted pyramids.

1 INTRODUCTION

In recent years, Diamond Wire Sawing (DWS) has been introduced in the manufacturing of crystalline silicon wafer as alternative technology to slurry based sawing. Until now, DWS has nearly replaced slurry based sawing for the slicing of monocrystalline as well as multicrystalline wafers. Since the standard acidic isotexturing process for multicystalline wafers is not as effective due to increased smoothness in the as-cut surface after the slicing process, an effective and economical alternative method is required to texturing the DWS multicrystalline wafers.

The implementation of nano-scale textures have been discussed in recent years as an effective substituting technology, e.g. by Reactive Ion Etching (RIE) [1-2] or Metal Catalyst Chemical Etching (MCCE) [3-4]. Since with those technologies, the reflectivity can be reduced well below 10%, the term "black silicon" has been introduced. Since an RIE based process requires a vacuum chamber along with wet chemical pre- and post-treatments, it is currently seen as a costlier option. MCCE has been introduced as an alternative approach and for its implementation in industry, mainly batch-type tools have been considered, in most cases using Ag metal as the catalyst and H_2O_2 as the oxidant [3]. In a previous study we have introduced an inline approach based on MCCE for texturing multicrystalline wafer [5]. MCCE texturing of monocrystalline wafer offers the opportunity for reduced reflectivity and lower costs compared to conventional alkaline texturing with additives.

2 GENERIC PROCESS FLOW OF MCCE

A generic process flow for Metal Catalyst Chemical Etching (MCCE) of a crystalline silicon wafer is given in Figure 1. It includes all relevant steps necessary to process a wafer from an as-cut wafer to be ready for emitter diffusion. The surface damage after sawing process needs to be removed (about 3 to 5 µm). This can be performed as separate process step or during MCCE. Since MCCE is a process in which a metal is used as a catalyst to promote etching in the vicinity of the metal, the respective metal precipitates need to be deposited, e.g. Ag nanoparticles. Metal nanoparticle deposition is carried out by dipping the wafer into a wet chemical solution containing metal ions, which is in most cases a solution of $AgNO_3$. Additionally HF needs to be present to initiate the deposition. After metal deposition, the texturing process starts, which require an oxidant (e.g. H_2O_2 or HNO_3) and HF.

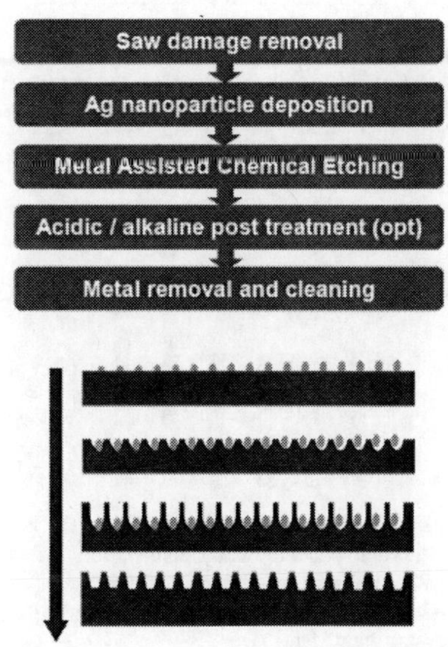

Figure 1: Generic process flow for black silicon based on Metal Assisted Chemical Etching (MCCE).

Depending on the bath composition, a porous Si layer can be formed. Also the surface morphologies can vary based on the bath chemistry. Post-treatments are optionally carried out (either alkaline or acidic) to obtain an optimum structure for high solar cell performance. The optimized structure is a trade-off between low reflectivity (necessary for high J_{sc}) and reduced surface enlargement (necessary for high V_{oc} and FF). In addition, the wafer surface needs to be cleaned from any metal and contaminations prior to emitter diffusion. An overview on MCCE can be found in [6]. In general, all steps of the generic process can be carried out individually, but alternatively, two or more of the steps can be combined in one treatment.

3. PRODUCTION RESULTS FOR MULTI PERC SOLAR CELLS

In a previous paper we have in detail described the process development for MCCE texturing on multicrystalline wafer [5]. This process was then applied for the development of an inline texturing tool (i-BlackTex). This inline tool was installed at a customer and used for the manufacturing of PERC solar cells [7]. The solar cells were processed on DWS wafer (wafer size 156.75 x 156.75 mm²) via the typical industrial manufacturing flow. Table I shows the solar cell parameters of the 450,000 cells subdivided into five batches. It can be seen, that during the processing of the batches only small changes in the solar cell parameters are observed. This demonstrates the stability of the PERC process, including the texturing of the inline iBlackTex tool. The average efficiency of the complete lot was 20.23% which is a very good value for multi PERC solar cells. The average solar cell efficiency of reference PERC solar cells using additives was 19.57% (V_{oc} = 0.642 V, I_{sc} = 9.43 A & FF = 79.9%). The efficiency gain was higher in this case, whereas in general we observe an advantage of about 0.4%abs. Hence MCCE texturing in the iBlackTex tool as compared to additives based texturing leads to significantly higher efficiencies. The gain is mostly due to an increase in I_{sc} by better texturing, but V_{oc} is also significantly higher than the reference solar cells.

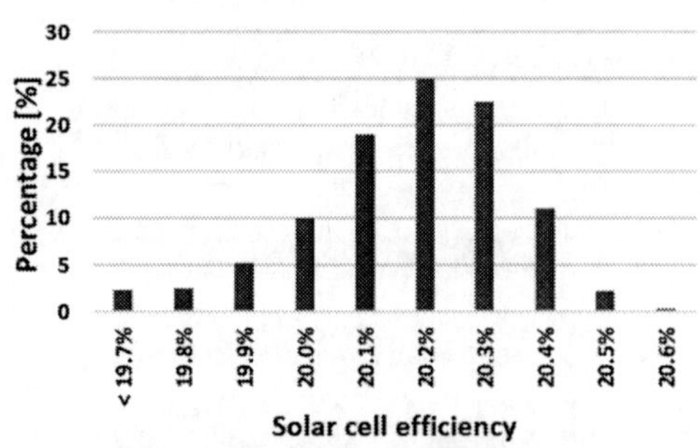

Figure 2: Solar cell efficiency distribution of 450,000 PERC solar cells processed on RCT i-BlackTex inline equipment using MCCE process for texturing of DWS multicrystalline wafer. [7]

Figure 2 shows the solar cell efficiency distribution of the production lot. The distribution is narrow and 80% of the cells have an efficiency of 20.1% or higher. A bath lifetime of more than 1.4 million wafers with consistent etching and texturing demonstrates the stability of MCCE based process in the inline tool.

4. COST-OF-OWNERSHIP

Figure 3 compares the Cost-of-Ownership (CoO) of the new inline approach applying MCCE with the industry standard of acidic isotexturing and an alternative approach of MCCE using batch type processes with additive. The CoO of acidic isotexturing is about 1.0 USDct/wafer, while the MCCE on RCT's inline tool leads to slightly higher costs of about 1.2 USDct/wafer. This is caused by a higher equipment cost as well as slightly higher consumable costs for Ag and cleaning. This moderate increase in CoO results from a very low Ag consumption, which is below 0.5 mg/wafer, leading to a cost contribution of less than 10% in the consumable costs.

Ag recycling, foreseen to be implemented in the near future, will lead to a further reduction of Ag consumption and therefore CoO. Nonetheless, the costs are substantially lower than batch-type MCCE processes, since the equipment cost and the consumable costs for additives are higher. Also conventional acidic texturing plus additives leads to a higher CoO due to the contribution of the additives. Hence, MCCE is also attractive from a cost perspective, when compared to alternative approaches like RIE, batch-type MCCE and isotexturing with additives.

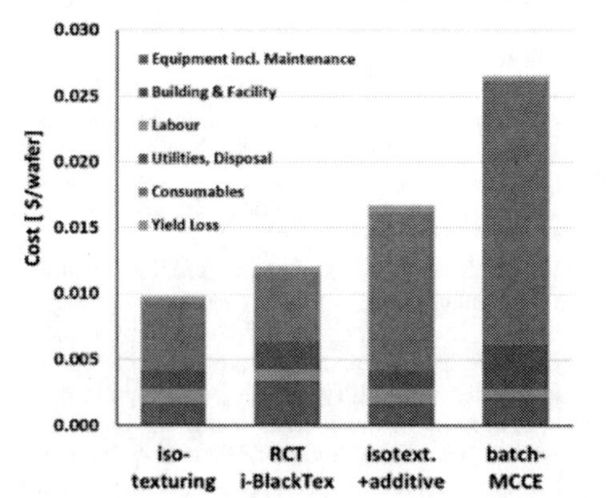

Figure 3: Cost of Ownership analysis of different texturing processes. Acidic isotexturing as applied as standard for texturing of MWSS wafer (left), MCCE in RCT's inline tool (middle), Acidic Isotexturing + additives (center) and batch-type MCCE (right).

Table I: Solar cell parameters of a production lot of 450,000 solar cells processed on i-BlackTex machine of RCT using MCCE process for texturing of DWS multicrystalline wafer [7].

iBlackTex 5BB PERC Production	V_{oc} [V]	I_{sc} [A]	FF [%]	η [%]	R_s [mOhm]	R_{sh} [Ohm]
1-100,000	0.654	9.57	79.6	20.3	2.04	198
100,000-200,000	0.653	9.55	79.6	20.2	2.00	224
200,000-300,000	0.653	9.57	79.5	20.2	2.08	239
300,000-400,000	0.653	9.56	79.6	20.2	2.10	231
400,000-450,000	0.654	9.56	79.6	20.3	2.11	232
Average of 450,000	**0.653**	**9.56**	**79.6**	**20.2**	**2.06**	**224**

4 MCCE RESULTS ON MONO C-SI WAFERS FOR THE FORMATION OF INVERTED PYRAMIDS

The aim of process and equipment development for mono-Si is basically two-fold. First, the random upright pyramid texture obtained by alkaline solutions leads already to reasonable reflectivity, so it is expected that the solar cell efficiency can increase by about 1%rel. by improved texturing. Second, the random pyramid texture of mono-Si is a quite costly process due to the necessity of additives. In the previous section, we already showed an attractive CoO by using MCCE. Therefore we expect a cost advantage for this alternative texturing method.

Several publications have demonstrated that inverted pyramid structures can be obtained by MCCE. Besides pyramids, alternative structures can be obtained with even lower reflectivity. The process development needs to come to an optimum between low reflectivity and low surface recombination (turning into good J_{0e} values).

In the initial texturing experiments, the post treatment after MCCE texturing is varied. Figure 4 shows the inverted pyramid structure obtained after the texturing process. In Figure 5 different reflectivity curves are shown with average reflectivity (400 nm – 1,100 nm) in the range of 6.0% and 10.0%, whereas the reflectivity of the random pyramid textured wafer is about 12.0%. This demonstrates that lower reflectivity values can be obtained using the MCCE based process. Besides low reflectivity values, it is important that those structures can be used for a good emitter with low J_{0e} values (good diffusion and passivation), which is important for high solar cell efficiencies.

The texturing process is applied to PERC solar cells and the lifetime is measured for asymmetric samples. The front surface has an emitter which is passivated by PECVD SiN_x, while the rear surface is passivated by a

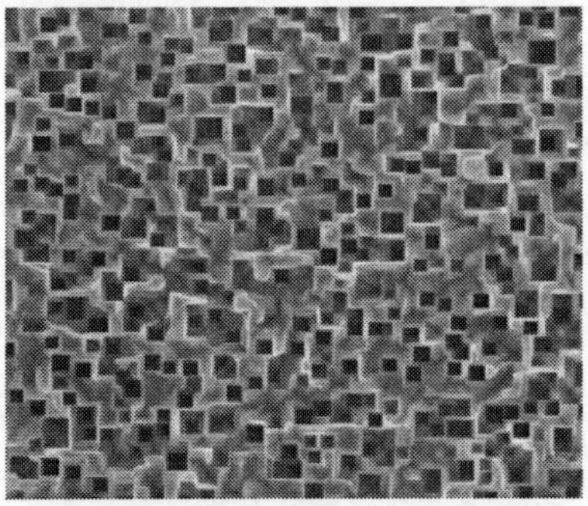

Figure 4: Inverted pyramid structure formation on mono c-Si wafer by MCCE based texturing process showing an inverted pyramid structure.

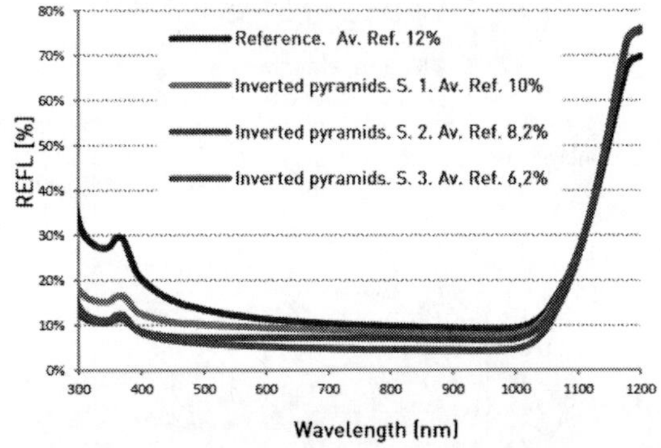

Figure 5: Measured reflectivity curves for four different MCCE textured mono-Si wafer. For reference, a standard random pyramid textured wafer is shown.

$SiON_x/SiN_x$ stack (as for RCT's Centaurus solar cells). Solar cell precursors were manufactured on CZ-Si M2 wafer with a base resistivity of 2 Ω-cm. After firing of the samples, PL measurements were carried out, which were then transferred into an implied V_{oc} (iV_{oc}) image. The iV_{oc} image is shown in Figure 6. The average value of this sample is 678 mV. In general, such voltages are good enough to process PERC solar cells with a V_{oc} of > 660 mV. This already represents a quite good value at this stage of process development. Figure 7 shows the QE curves for an alkaline textured and MCCE textured PERC solar cell. It can be observed that the IQE and EQE curves are almost identical. The higher recombination losses on the front and rear for MCCE textured cell is due to a slight difference in diffusion, rear-side polish and printing process.

5 SUMMARY

In this work we have demonstrated an MCCE based inline texturing approach for DWS multicrystalline and monocrystalline wafers. The developed MCCE process uses a single step for saw damage etching, Ag nanoparticle deposition and metal catalyst chemical etching. Industrial manufacturing of multi PERC solar cells based on the MCCE texturing process with an average efficiency of 20.2% is demonstrated. The bath is very stable with a lifetime of more than 1.4 million wafers. The CoO for the MCCE based process is less than 1.25 USDcts/wafer, which is significantly lower than the other process routes. Encouraging results were obtained for texturing monocrystalline wafers based on MCCE process with reflectivity lower than the reference alkaline textured wafers. Based on the same, an inline tool (i-UniTex) is developed to texture both monocrystalline and multicrystalline wafers.

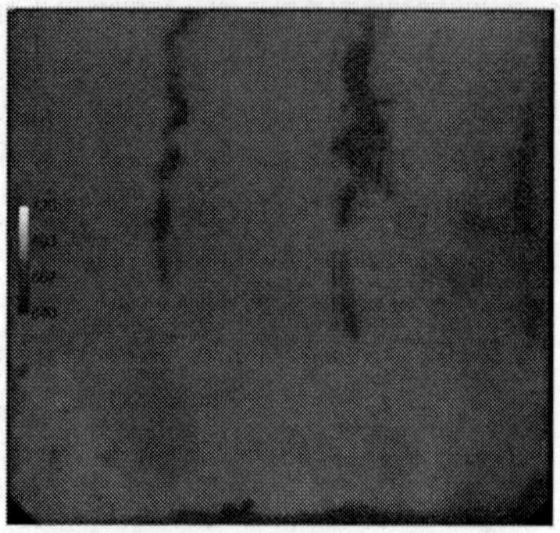

Figure 6: iV_{oc} image of a solar cell precursor. The image is made by PL, which is then transferred into an iV_{oc} image. The average iV_{oc} value corresponds to 678 mV.

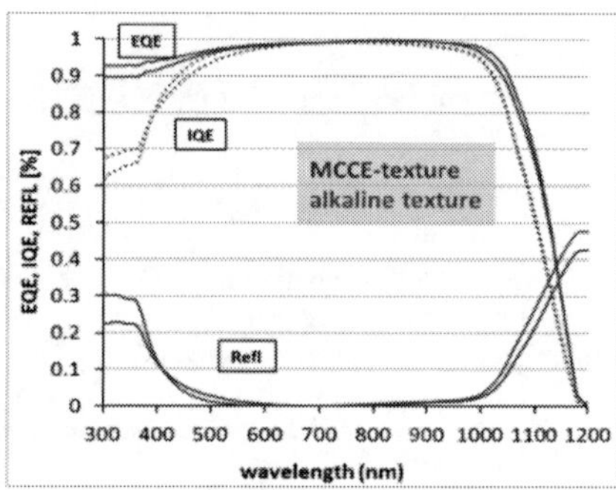

Figure 7: Comparison of QE curves for alkaline textured and MCCE textured PERC solar cells.

ACKNOWLEDGEMENTS

This work is financially supported by the German Ministry of Economic Affairs and Energy under the contract number 0324194A (FlexFab).

REFERENCES

[1] Y. Inomata, K. Fukui, K. Shirasawa, “Surface texturing of large area multicrystalline silicon solar cells using reactive ion etching method”, Solar Energy Materials and Solar Cells, 48, 1997, 237-242.
[2] J. Yoo, G. Yu, J. Yi, “Large-area multicrystalline silicon solar cell fabrication using reactive ion etching (RIE)”, Solar Energy Materials and Solar Cells, 95, 2011, 2-6.
[3] A. Kumaai, “Texturization using metal catalyst wet chemical etching for multicrystalline diamond wire sawn wafer”, Solar Energy Materials and Solar Cells 133, 2015, 216–222.
[4] F. Cao, K. Chen, J. Zhang, X. Ye, J. Li, S. Zou, X. Su, “Next-generation multi-crystalline silicon solar cells: Diamond-wire sawing, nano-texture and high efficiency”, Solar Energy Materials and Solar Cells, 141, 2015, 132-138.
[5] W. Jooss, I. Melnyk, J. Jung König, A. Teppe, T. Werling, O. Voigt, F. Binaie Masouleh, P. Fath, B. Hu, Q. Zhang, I. Wang, P. Tian, J. Huang, J. Yu, J. Qiu, S. Zhou, Q. Huang, X. Gou, “Development and Optimization of a Novel Inline Black Silicon Texturing Process for Increased Solar Cell Performance”, Proc. 33rd European Photovoltaic Solar Energy Conference and Exhibition, Amsterdam, 2017, pp. 368
[6] Z. Huang, N. Geyer, P. Werner, J. de Boor, U. Gösele, “Metal-Assisted Chemical Etching of Silicon: A Review”, Adv. Mater. 23 (2011), 285–308
[7] W. Jooss, I. Melnyk, C. Lai, J. Jung-Koenig, A. Teppe, S. Reddy Madugula, O. Voigt, P. Fath, B. Hu, I. Wang, P. Tian, A. Yang, T. Wang, “Process development for MCCE textured solar cells on multicrystalline DWS wafer”, to be published in Proc. World Conference on Photovoltaic Energy Conversion (WCPEC-7), Hawaii, 2018

GENTLE AND DAMAGE FREE ABLATION OF DIELECTRIC LAYERS USING A FEMTOSECOND LASER SOURCE FOR HIGH EFFICIENCY SILICON WAFER SOLAR CELLS

Jaffar Moideen Yacob Ali[1, 2], Vinodh Shanmugam[1], Armin G. Aberle[1, 2], Thomas Mueller[1]
[1] Solar Energy Research Institute of Singapore, National University of Singapore, Singapore 117574
[2] Department of Electrical and Computer Engineering, National University of Singapore, Singapore 117583

ABSTRACT: Laser opening of dielectric layers without inducing damage to the underlying substrate is an attractive process simplification for ablation intensive silicon solar cell architectures. In this work, ablation of dielectric layers such as SiN_x, AlO_x/SiN_x stack and SiO_2/SiN_x stack is carried out using a femtosecond (fs) laser source at two different wavelengths: 343 nm (UV) and 515 mm (green). At 515 nm wavelength, the laser pulses are absorbed in the underlying silicon substrate, causing defects. However, the defects can be minimised by carrying out a gentle ablation process wherein the laser induced damage can be etched off in less than 5 s of chemical etching. At 343 nm wavelength, the laser pulses are mostly absorbed in the dielectric. Hence, it is possible to selectively ablate the dielectric layers by choosing the right pulse fluence and pulse overlap, thus completely avoiding the requirement of post laser chemical etching process. Operating the fs (ultrashort pulse) laser at high pulse fluence induces severe bulk damage that typically requires a prolonged (15 min) chemical etching and therefore should be avoided. Hence, gentle and damage free fs laser ablation process will result in reduced fabrication time, reduced process costs for PV manufacturers applying laser ablation in their processing sequence, and higher wafer/hour throughput.

Keywords: Laser processing, femtosecond, c-Si, damage-free ablation

1 INTRODUCTION

Several approaches were proposed to reduce the cost of PV electricity [1], such as moving towards high-efficiency solar cells, increasing the fabrication throughput and reducing the silicon usage (thinner wafers). To achieve this, laser processing has the potential to simplify solar cell fabrication sequences and facilitate novel processing techniques that enable high throughput manufacturing. For example, in various high-efficiency solar cell architectures, such as interdigitated back contact (IBC) solar cells, the dielectric layers are selectively ablated and patterned for local contact formation [2]. In addition to that, lasers are also used for local doping for emitter and back surface field (BSF) formation in the fabrication of silicon IBC solar cells [3]. However, a successful laser process is the one that imparts negligible damage to the substrate such that there is no degradation in the solar cell's electrical performance.

In this paper, the ablation of several commonly used dielectric layers in silicon solar cell fabrication are carried out using a femtosecond (fs) laser source with two different wavelengths: 515 and 343 nm. The selection of the laser parameters was based on the single-pulse ablation behaviour using the fs green laser source. The laser induced damage is characterized using photoluminescence (PL) analysis after etching and re-passivation using 70 nm SiN_x on both sides.

2 EXPERIMENTAL DETAILS

The fs laser system produces pulses in the duration of 480 fs and can be tuned to operate at 515 nm and 343 nm wavelengths. The beam scanner has a maximum scanning speed of 20 m/s and the pulse repetition rate can be varied from 200 to 2000 kHz, thus enabling ultra-fast processing for ablation intensive solar cell architectures.

2.1 Sample specifications

Planar *n*-type crystalline silicon wafers with a base resistivity of 3-5 Ωcm and passivated with dielectric layers on both sides were used in this work. The dielectric layers were deposited with the plasma-enhanced chemical vapour deposition (PECVD) technique. The sample schematic is shown in Fig. 1. The dielectric layers and their thicknesses are shown in Table I.

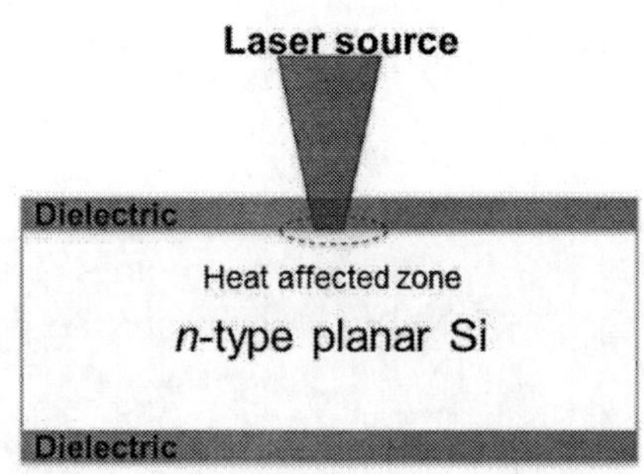

Figure 1: Schematic of the test sample illustrating the process of indirect laser ablation.

Table I: Dielectric layers and their thicknesses.

Dielectric layer	Thickness (nm)
SiN_x	70
AlO_x/SiN_x stack	20 / 50
SiO_2/SiN_x stack	20 / 50

2.2 Post-laser processing and characterisation

Post laser ablation, the samples were dipped in 10% HF solution to remove the remaining dielectric layers on both sides. For fs 515-nm ablation, the samples were etched in KOH solution from 5 s to 15 min for laser damage removal. For fs 343-nm ablation, the samples were directly re-passivated after the HF dip. The re-passivation is via PECVD SiN_x at a thickness of 70 nm on both sides. The samples were then fired in an industrial firing furnace and then characterized using the PL imaging technique. The process sequence for post laser processing and characterization is shown in Fig. 2.

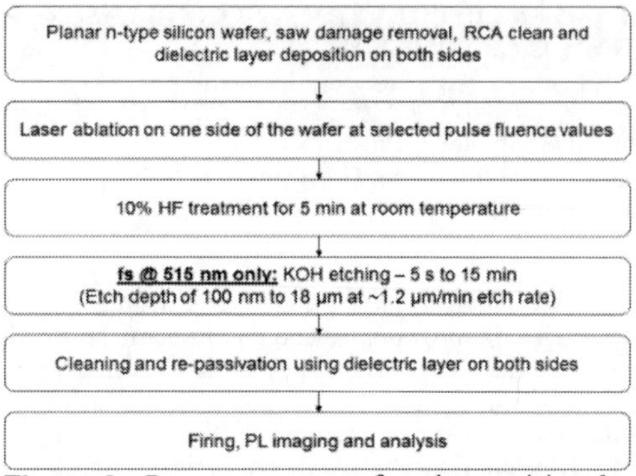

Figure 2: Process sequence for characterizing laser induced damage. KOH etching is used only for fs ablation at 515 nm.

3 RESULTS AND DISCUSSION

Initially, the single-pulse ablation behaviour was analysed using the fs laser in the green wavelength. Single spots were ablated at different pulse fluence (F_P) values. The diameter (d) and depth (h) of the spots were plotted against the logarithm of F_P, in order to estimate the threshold fluence (F_{TH}), spot radius (r) and energy penetration depth (h_0) using Eqs. 1 and 2 [4]. The plot of d^2 and h versus Ln (F_P) for the ablation of 70 nm SiN_x is shown in Fig. 3.

$$d^2 = 2r^2 \ln\left(\frac{F_P}{F_{TH}}\right) \quad \text{Eq. 1}$$

$$h = h_0 \ln\left(\frac{F_P}{F_{TH}}\right) \quad \text{Eq. 2}$$

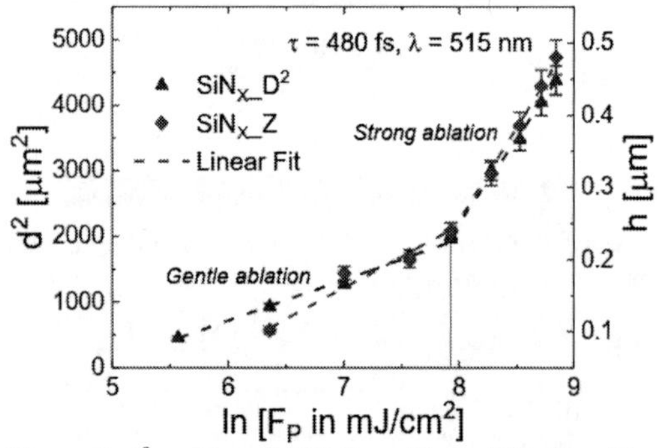

Figure 3: d^2 and h vs ln(F_P) for a 70-nm SiN_x film. The gentle and strong ablation regimes have different single-pulse ablation properties.

As can be seen from Fig. 3, the fs laser ablation has two distinct regimes that are characterized by different slopes and indicated by a vertical drop line. In ultra-short pulse ablation at very high pulse fluence, the rapid electron-lattice collisions result in bulk heating and enhanced ablation [5]. Hence, the laser induced damage experiments were conducted at F_P values from both the gentle and strong ablation regimes.

3.1 Laser ablation at 515 nm

Laser ablation of 70 nm SiN_x was carried out at selected pulse fluence values from both the regimes as denoted in the test structure in Fig. 4. Figure 5 shows PL imaging after re-passivation at different KOH etch durations.

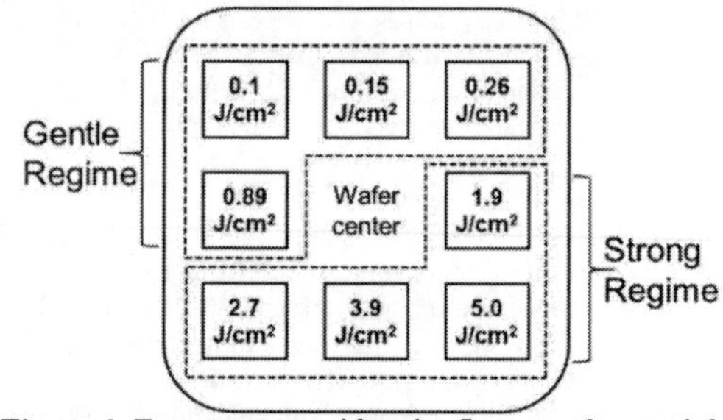

Figure 4: Test structure with pulse fluence values and the corresponding ablation regime. The area in the centre of the wafers acts as a reference.

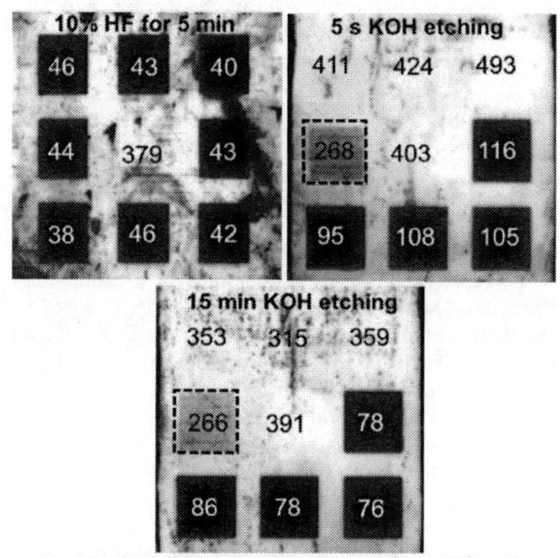

Figure 5: PL imaging after re-passivation at HF dip and two different KOH etch durations. The boxes highlighted by dashed lines indicate the effect of pulse overlap. The value in each box is the effective minority carrier lifetime in µs. Laser damage induced in the gentle regime can be etched off in 5s, whereas, the strong regime needs more than 15 min of etching time.

The damage induced by the gentle regime could be etched off in 5 s, whereas the strong regime needed more than 15 min of KOH etching that corresponds to an etch depth of 18 µm (~1.2 µm/min etch rate). The dashed boxes were ablated with F_P in the gentle regime, however, the effect of pulse overlap resulted in strong ablation. Hence, the selection of pulse fluence and the overlap is important in dielectric ablation using ultra-short pulse lasers.

3.2 Laser ablation at 343 nm

The ablation at 343 nm was carried out at F_P only in the gentle regime so as to optimize the ablation process for a damage-free dielectric removal. The test structure used for fs UV laser ablation and optimization is shown in Fig. 6. The pulse overlap was varied from 6 to 18 µm at a constant line pitch of 2 mm. A nanosecond (ns) laser providing 38-ns laser pulses was also used for comparison. The corresponding PL images after HF dip and re-passivation are shown in Fig. 7.

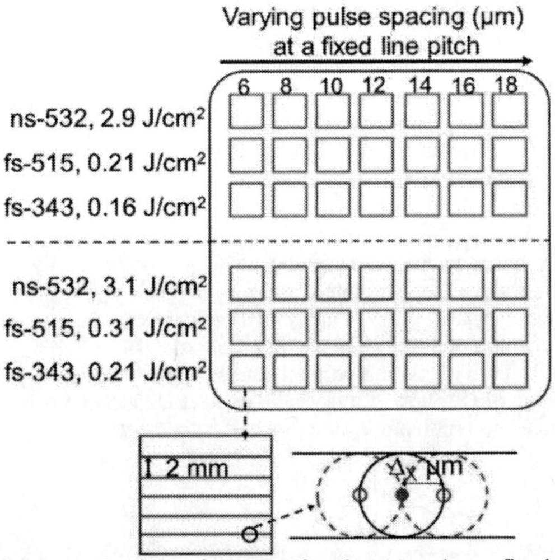

Figure 6: Varying pulse separation (6 - 18 µm) at a fixed line pitch of 2 mm. The numbers on top of the boxes indicate pulse spacing in µm.

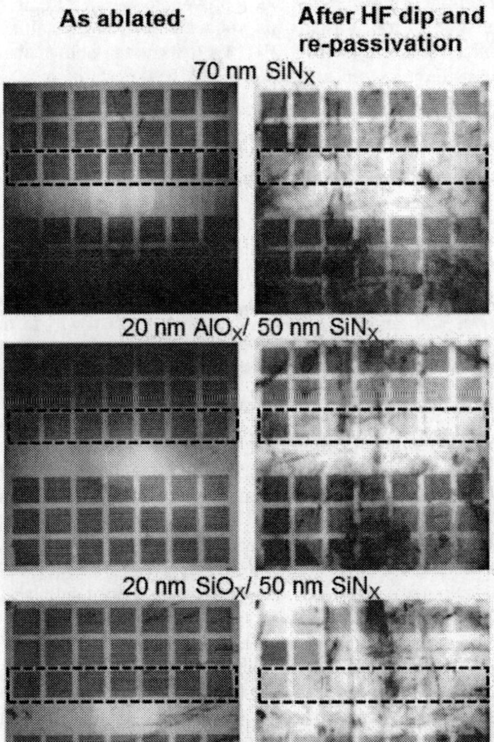

Figure 7: PL imaging after HF dip and re-passivation. The dotted boxes indicate fs UV laser ablation.

The laser damage induced by the ns laser source could not be recovered using HF treatment and re-passivation due to considerable heat diffusion. Therefore, we could see visible dark lines in the PL images of the re-passivated samples. For fs laser ablation at 515 nm, the re-passivated samples show a brighter PL image compared to ns ablation, however, we could still see fine dark lines that might indicate some damage to the underlying silicon. For fs laser ablation at 343 nm, the 'as-ablated' samples at F_P = 0.16 J/cm² (highlighted using dotted boxes) appear dark, which indicates a complete removal of the dielectric layers. However, after re-passivation the boxes appear brighter (highlighted using dotted boxes), implying that the underlying silicon was not damaged during the ablation process.

4 CONCLUSION

Damage free dielectric ablation that does not require a post laser chemical etching step, and a gentle ablation process where the laser induced damage can be etched off in a short time (5 s), were successfully performed using a femtosecond laser source at 343 and 515 nm, respectively. It was also shown that in the strong ablation regime the laser damage cannot be completely removed even during 15 min of KOH etching, indicating the presence of bulk damage. Thus, optimising the fs laser parameters for gentle and damage free ablation will result in considerable savings in processing time and fabrication cost. Hence, fs lasers are highly suitable for industrial mass production of ablation intensive and high efficiency silicon wafer solar cells.

5 REFERENCES

[1] D. M. Powell, M. T. Winkler, H. J. Choi, C. B. Simmons, D. B. Needleman, and T. Buonassisi, "Crystalline silicon photovoltaics: a cost analysis framework for determining technology pathways to reach baseload electricity costs," *Energy & Environmental Science,* vol. 5, pp. 5874-5883, 2012.

[2] M. Dahlinger, K. Carstens, E. Hoffmann, R. Zapf-Gottwick, and J. H. Werner, "23.2% laser processed back contact solar cell: fabrication, characterization and modeling," *Progress in Photovoltaics: Research and Applications,* vol. 25, pp. 192-200, 2017.

[3] M. Ernst, E. Franklin, A. Fell, K. Fong, D. Walter, E.-C. Wang, *et al.*, "Fabrication of a 22.8% Efficient Back Contact Solar Cell With Localized Laser-Doping," *physica status solidi (a),* vol. 214, p. 1700318, 2017.

[4] J. M. Liu, "Simple technique for measurements of pulsed Gaussian-beam spot sizes," *Optics Letters,* vol. 7, pp. 196-198, 1982/05/01 1982.

[5] P. Mannion, J. Magee, E. Coyne, and G. M. O'Connor, "Ablation thresholds in ultrafast laser micromachining of common metals in air," in *OPTO Ireland*, 2003, p. 9.

IMPROVEMENT OF SOLAR CELL EFFICIENCIES FOR ULTRASHORT-PULSE LASER CONTACT OPENING WITH NI-CU PLATED CONTACTS BY OPTIMIZED LCO – FFO PROCESSING ORDER

V. Arya[1], S. Gutscher[1], S. Kluska[1], F.Meyer[1], G. Cimiotti[1], J. Nekarda[1], A. A. Brand[1]
[1]Fraunhofer Institute for Solar Energy Systems ISE
Heidenhofstr. 2, 79110 Freiburg, Germany

ABSTRACT: We demonstrate that the standard fast firing (FFO) process at temperatures around 850°C can significantly reduce the damage caused by a UV picosecond pulse laser ablation for nickel-copper plated silicon solar cells. Although, the metallization via Ni/Cu/Ag plated contacts are a promising technology, disruption in the crystalline structure of silicon due to laser pulse damage pertains, which increases recombination and suppresses the efficiency of solar cell. We realize laser openings of linewidths ~12.7 µm to ~16.9 µm for metallization by plating on semi-finished industrial PERC solar cells and demonstrate that by flipping the chain of processes from FFO followed by laser ablation (LCO) to LCO followed by FFO, significant reduction is seen in laser-damage induced recombination which increases the power conversion efficiency (η) by 0.5% $_{abs}$.
Keywords: laser ablation, recombination, annealing, c-Si, PERC, solar cell, plating

1 INTRODUCTION

Although, the technology of laser contact opening (LCO) and Ni/Cu/Ag light-induced plating for forming front side contacts could offer economic and technological advantages over the industries' standard silver screen printing, it often falls short of expectation. One of the suggested advantages of using Ni-Cu plating would be reduced production costs [1] (studies suggest that 40% of cell processing cost constitutes of silver paste costs [2]). Also, narrower metallization lines (10-30 µm) leading to smaller shading effects, lower specific contact resistivity [3]. Sufficient adhesion of the plated contacts has been demonstrated to be achievable by the surface morphology created by typical ultrafast pulse laser ablation process [4] .

Despite the progress of laser ablated and plated contacts, the resulting solar cell efficiencies (η) are often inferior to those achieved by screen printed contacts on the same solar cell structures. Crystal damage caused by the laser ablation increases recombination, lowering open circuit voltage (V_{oc}) and pseudo-fill-factor (*pFF*) keeping the LCO+Plated contacts from reaching their full efficiency potential.

Damage due to LCO of anti-reflection coatings on silicon solar cells is well researched and published. Drops in V_{oc} [5] and decrease in minority carrier lifetime [6], [7] have been reported. Primarily, amorphized silicon, dislocations cause increased recombination [8]–[10].

LCO for Ni-Cu plating on solar cell emitters demands for the residual damaged zone to be as shallow as possible. Ultrashort pulse lasers with a typical pulse duration of less than 15 ps offer a small thermal diffusion length (40-80nm). Also, optical absorption lengths of the laser radiation of less than 40nm can be achieved by frequency conversion to ultraviolet, typically 355 nm or 343 nm wavelength. Laser ablation with such parameters creates steep pressure and temperature gradients leave an amorphous silicon surface due to rapid cooling in the heat affected zone (HAZ) [9]–[11].

1.1 Recrystallization of Laser Amorphized Silicon

Attempts to improve recombination by employing KOH/NaOH etching after laser ablation has been successfully demonstrated for buried contacts [5]. However, etching away the laser-damaged silicon layer would increase production costs of LCO + Ni-Cu plated contacts. Additionally etching away the top part of the emitter with the highest doping concentration would most likely decrease the emitters' ability to shield the metal contacts and increase contact resistance.

Thermal annealing has been proven to be an effective method of transforming amorphous silicon to crystalline silicon as reported in [12]–[14]. It has been shown that annealing a-Si up to temperatures of 900°C initiates the crystallization of the a-Si layer [12]. In another study, Raman scattering spectra reveals amorphous silicon transforming to nearly fully crystallized poly-Si film after subjecting to temperatures of 700-1100°C [14].

1.2 Fast Firing Oven (FFO)

Fast Firing process (FFO) is a standard process in manufacturing crystalline silicon solar cells. Its primary purpose in the process chain is activating the passivation layer and sintering silver and aluminum pastes to form front and rear side contacts. Typical peak temperatures of the solar cells are around 850°C for several seconds. In the preparation of PERC solar cells specifically, FFO is typically the last processing step following screen printing of metal paste on both sides.

However for manufacturing PERC cells with laser ablated and plated front contacts, the wafers are often processed to the point where they have already received full area aluminum screen printing on the rear side and subsequent FFO. These PERC precursors without front side contacts are then usually subjected to front side laser LCO and plating.

In our work, we change the processing order. The wafers are subjected to front side LCO before FFO is applied in order to utilize the thermal budget of the FFO to anneal the laser damage along with rear side contact formation.

2 EXPERIMENT DESIGN

Semi-finished p-type Cz-Si monocrystalline PERC solar cells were procured for the experiment which were processed on the rear side (LCO + Al paste) but not yet fired. The front side carried KOH textured surfaces with random pyramids. A total of 200 cells were equally divided into two groups namely Group A and Group B. The Group A cells were exposed to front side laser ablation and then subjected to FFO and the process was reversed in the Group B. The whole process flow is shown in Figure 1. It was made sure that all the cells from both the groups receive the FFO process simultaneously so both the groups receive exactly the same thermal annealing treatment.

PERC Precursors - Rear side passivation opened & Al paste screen printed NOT FIRED YET	
Group A: Front side LCO → FFO	Group B: FFO → Front side LCO
Front LCO: UV(355nm) pulse-duration ≈ 15ps (*H-Grid with 100 fingers, 5 Busbars*)	
FFO at 830°C set temperature	
	Front LCO: UV(355nm) pulse-duration ≈ 15ps (*H-Grid with 100 fingers, 5 Busbars*)
Front side PL measurement at 1 Suns	
Nickel + Copper electroplating with HF Dip	
Solar Cell IV Characterization and Analysis	

Figure 1: Process flow chart illustrating the two groups with altered process order of front side LCO before FFO (A) and FFO before front side LCO (B)

2.1 Laser Ablation of SiN_x

The laser employed for front side laser ablation is an ultrashort pulsed UV (355nm) laser with pulse-duration of < 15ps. All the cells were lasered with a complete H-Grid geometry (see Figure 2) comprised of 100 fingers, 5 busbars (each 0.5mm wide).

Both groups were divided into 5 sub-groups and they were lasered upon with five different pulse energies E_{pulse}- 0.5µJ, 0.6µJ, 0.8µJ, 0.9µJ and 1.0µJ while keeping all other laser beam characteristics constant throughout the experiment. Different pulse energies generated different opening sizes in the dielectric layer on the solar cell. The opening widths varied from ~12.5µm at 0.5µJ to ~16.9µm at 1.0µJ as shown in Figure 3. As expected, the openings at a higher pulse energy leave behind much less residue of the dielectric layer hence appear shinier. The residual dielectric layer plays its part in increasing the overall series resistance of the Ni-Cu plated solar cells as will be visible in the IV results in the subsequent sections.

After subjecting both the groups to both LCO and FFO, PL measurements were performed at one sun illumination intensity to characterize the pre-metallization lifetime performance of the cells and the effect of laser pulse energy variation on it.

2.2 Light induced Ni-Cu plating (LIP)

Following the PL characterization, they were subjected to inline Ni/Cu/Ag light-induced plating processes. The cells were dipped in 1 wt% HF solution for 30s to strip off native oxide followed by growing of nickel seed layer in the ablated areas and further received light induced copper and silver plating process. This completes the process of forming the contacts by electroplating. No anneal was performed after plating. The cells followed IV characterization.

Figure 2: Schematic of an H-Grid front metallization grid geometry with 100 finger and 5 Busbars.

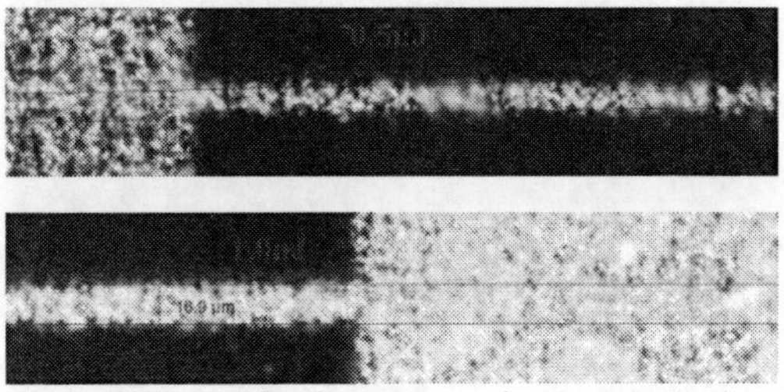

Figure 3: Laser ablated contact openings at 0.5µJ and 1.0 µJ having opening widths of 12.7µm and 16.9µm respectively. The blue part is the SiN_x dielectric film and the shining silver part is the silicon exposed after laser ablation. Openings at 0.5µJ appear cloudy and patchy due to incomplete ablation compared to opening at 1.0µJ.

3 RESULTS

3.1 Photoluminescence Measurements (PL)

PL measurements were performed on all the wafers from both the groups after they received both laser opening and once the rear side passivation was activated after FFO. Figure 4 below shows the mean PL counts of Group A cells (LCO → FFO) juxtaposed with Group B (FFO → LCO) cells at various pulse energies.

Two things can be immediately concluded from this – (i) there is general and intuitive upward trends in the PL counts with decreasing pulse energy, (ii) Group A (LCO → FFO) shows a significant improvement in the PL counts over Group B which do not receive the FFO anneal treatment after laser ablation.

3.2 IV Characterization

The non-degraded cells were subjected to standard IV characterization. A quick look at the Figure 5 directly indicates, keeping in accordance with the PL results, a significant jump in the efficiency of solar cells that received the FFO after LCO at all different laser pulse energies. An improvement of ~0.25%$_{abs}$ at higher pulse energies to >0.50%$_{abs}$ at lower pulse energy is seen in the efficiency (η) when cells are thermally annealed after LCO.

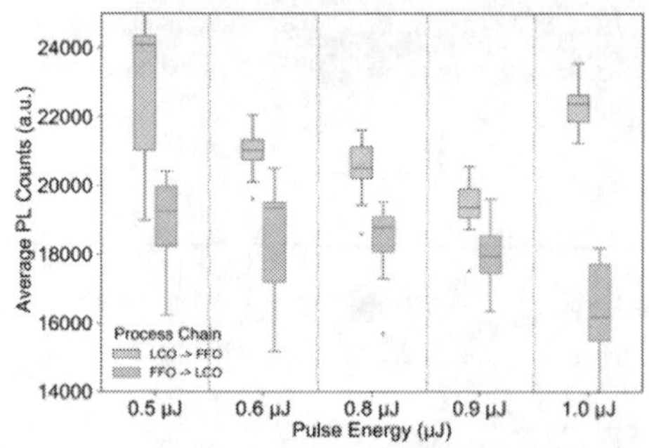

Figure 4: Mean PL counts vs. ablation pulse energy for Group A (LCO → FFO) and Group B (FFO → LCO) cells.

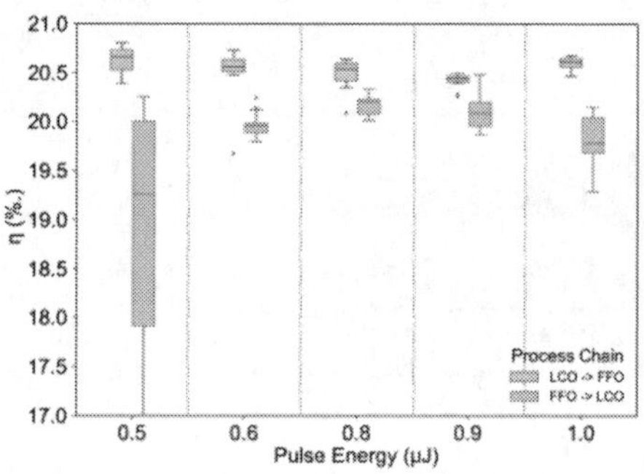

Figure 5: Solar cell efficiency η vs. laser ablation pulse energy (µJ).

3.3 FFO→LCO Process Sequence

Deeper look reveals some interesting insights into the trade-offs that are at work in case of laser + plated solar cells. Looking at the FFO→LCO order separately, we observe that the cells lasered with pulse energy of 0.8µJ demonstrate the best efficiency results, both in terms of median and the scatter of data. The efficiency of cells lasered upon with the minimum pulse energy of 0.5µJ shows a large spread on the efficiency scale indicating higher uncertainty. Even though the V_{oc} trends indicate continuous drop with increasing pulse energies, (as expected – increasing pulse energy should result in higher dislocations and recombination sites, ergo lower V_{oc}) and the j_{sc} shows hardly any significant dependence on pulse energies (see Figure 6), the efficiencies however do not slide off with increasing pulse energies, as one would have expected.

Here the quality of the contact opening, in terms of the residual dielectric layer, and the resistance it offers play a significant role. This will become much clear looking at the Fill Factor (*FF*) and pseudo-Fill Factor (*pFF*) trends of the cells. The pFF describes the fill factor of the solar cell which doesn't take into account, the drop in maximum output power (P_{mpp}) due to the series resistance (R_s) of the solar cell.

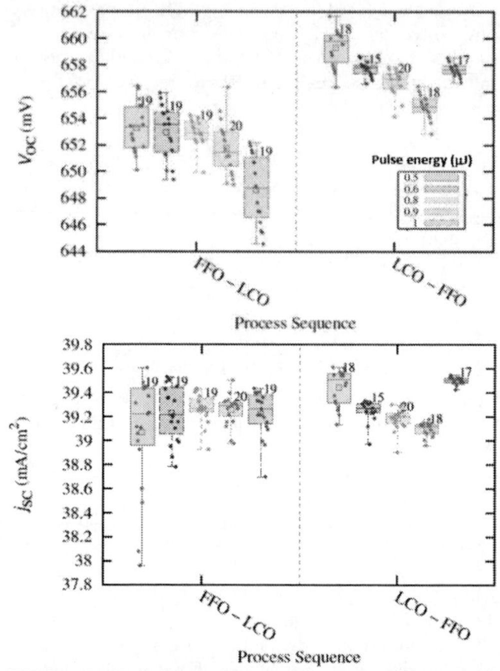

Figure 6: V_{oc} (mV) and j_{sc} (mA/cm²) at different pulse energies for the LCO → FFO and FFO → LCO process sequences.

In Figure 7, the *pFF* of the cells for FFO→LCO process is showing a downward trend with increasing pulse energies, implying higher damage (just as in case of V_{oc}). Interestingly, the downward trend of *pFF* is not seen for the LCO→FFO process and stabilization is seen with various pulse energies.

The R_s trends, as in Figure 7, show decrease in total series resistance with increasing pulse energies. This decrease in R_s is attributed to the quality of contact that the plated metal makes with the emitter in the ablated area. Presence of SiN_x residue in ablated area due to insufficient/lower pulse energies (as seen in Figure 3), indicate to be resulting in a poor or rather 'obstructed' ohmic contact which is indicated in the higher values of R_s for smaller pulse energies.

As a consequence of increasing R_s with decreasing pulse energies, the actual Fill Factor (*FF*) (see Figure 8), in disagreement with *pFF*, comes out to be highest at middle range pulse energies (0.8 & 0.9 µJ) where the amount of drop in V_{oc} is balanced out by lower series resistances. Consequently, the trade-offs between V_{oc}, R_s and *FF* churn out the highest efficiency η at 0.8µJ.

At 1µJ pulse energy, although the R_s is lowest but also the drop in V_{oc} is highest, which results in lower *FF* and hence lower overall efficiency. Contrastingly, at 0.5µJ pulse energy, the V_{oc} is the highest (ergo: high *pFF*) but so is R_s, resulting in lower *FF*. It is also notable that the larger spread of R_s at 0.5µJ pulse energy is also further transposed to a larger spread in *FF* and the end efficiency η. The large scatter in R_s values at lower pulse energies indicates inhomogeneous laser openings, with sporadic spread of residual SiN_x and irregular contacts.

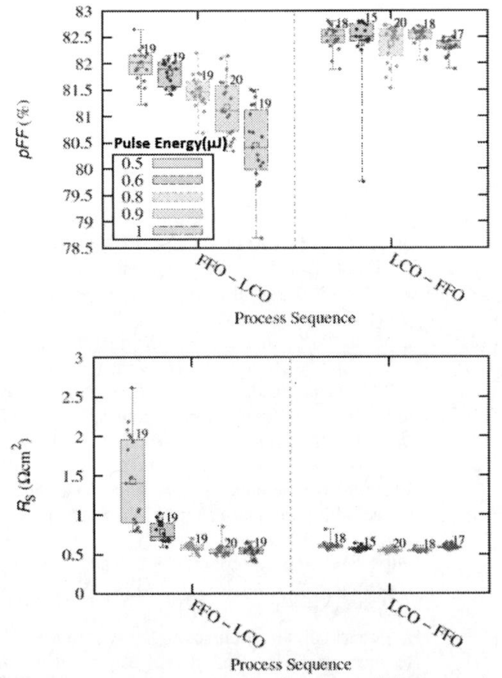

Figure 7: *pFF* (%) and R_s (Ωcm²) at different pulse energies for the LCO → FFO and FFO → LCO process sequences.

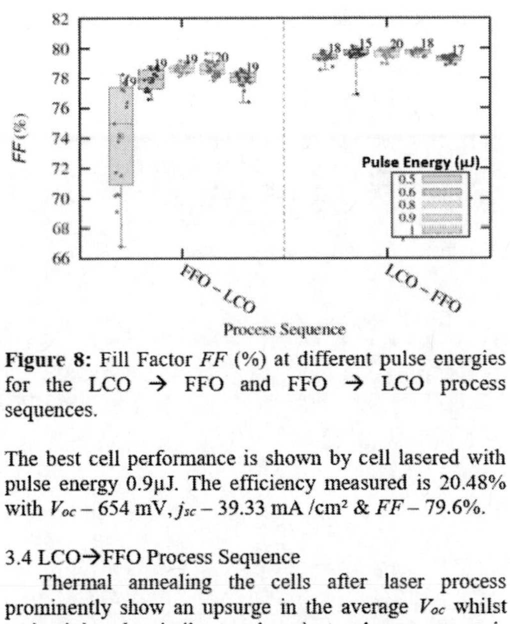

Figure 8: Fill Factor *FF* (%) at different pulse energies for the LCO → FFO and FFO → LCO process sequences.

The best cell performance is shown by cell lasered with pulse energy 0.9μJ. The efficiency measured is 20.48% with V_{oc} – 654 mV, j_{sc} – 39.33 mA /cm² & *FF* – 79.6%.

3.4 LCO→FFO Process Sequence

Thermal annealing the cells after laser process prominently show an upsurge in the average V_{oc} whilst maintaining the similar trend against pulse energy, as is clearly visible in Figure 5, compared to the reversed processing sequence. Also the *FF* stays more constant at all pulse energies (see Figure 8). Both of these factors contribute to higher solar cell efficiencies at all pulse energies and also less sensitivity of η to pulse energy.

The best cell showed an efficiency of 20.8% with V_{oc} – 660 mV, j_{sc} – 39.6mA/cm² and a *FF* of 79.6% and was lasered with the pulse energy of 0.5μJ.

3.5 High Efficiency Laser Ablated and Ni-Cu Plated Solar Cell

The new processing sequence of FFO after LCO was applied to semi-finished PERC precursors from our industrial partner Hanwha QCells GmbH.

The 156mm x 156mm pseudo square wafers featured two different emitter doping profiles namely Group B2 and Group B3. Group B2 has 100 Ω/sq. emitter profile and Group B3 has 150 Ω/sq. emitter profile with a lower surface dopant concentration. Both the groups were subjected to exactly similar laser, plating and firing processes and parameters.

The precursors were received with both sides passivated and the rear side passivation was already ablated. The front side passivation was ablated at ISE and then sent back to the industrial partner for further rear side processing. The PERCs received full area Aluminum paste screen printing on the rear side and were subsequently fired (FFO). The front side Ni-Cu plating was further performed at ISE by Kluska et al [15].

Front side screen printed (with Ag paste) metal grid PERCs were taken as reference cells for performance comparison. The screen printing of the front metal fingers was performed using 30μm screen opening. In both the cases (screen printing and laser ablation), 114 fingers and 6 busbars were applied on the front side of cells.

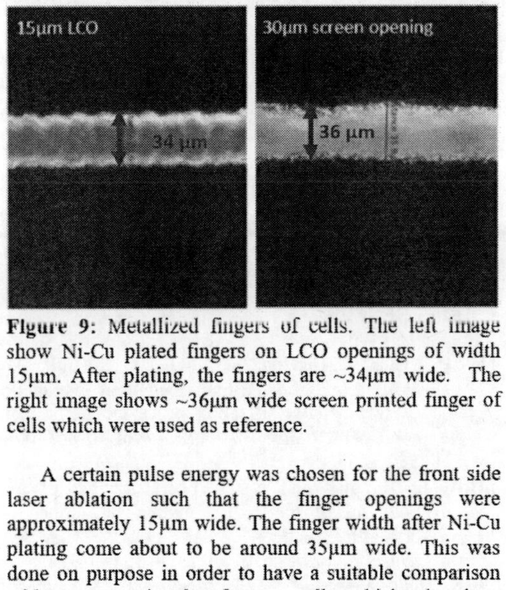

Figure 9: Metallized fingers of cells. The left image show Ni-Cu plated fingers on LCO openings of width 15μm. After plating, the fingers are ~34μm wide. The right image shows ~36μm wide screen printed finger of cells which were used as reference.

A certain pulse energy was chosen for the front side laser ablation such that the finger openings were approximately 15μm wide. The finger width after Ni-Cu plating come about to be around 35μm wide. This was done on purpose in order to have a suitable comparison with screen printed reference cells which also bear ~35μm wide fingers (see Figure 9). A highlight of the best cell results can be seen in the Table 1.

Table 1: Best cell IV results using two different front side metallization methods i.e. LCO+Plating and screen printing juxtaposed.

Emitter Type	Front Metal	η (%)	V_{oc} (mV)	J_{sc} (mA/cm²)	FF (%)
B3	LCO	22.2	680	40.2	81.3
B3	Screen Print	22.1	679	40.2	81.1
B2	LCO	22.2	675	40.0	82.0
B2	Screen Print	22.2	676	40.1	81.7

On an average, the Ni-Cu plating showed, for a screen printing optimized emitter (B2), a similar cell efficiency level compared to the screen printing reference group. The optimized processing sequence of LCO-FFO featured no laser-induced V_{oc} loss compared to the screen printing reference. Similar finger width and low contact resistances for both metallization technologies resulted in equal cell efficiencies for this emitter. The 150 Ω/sq. emitter of group B3 showed also similar V_{oc} level compared to screen printing. The plated cells, in some singular cases, show 0.1%$_{abs}$ efficiency gain in combination with B3 emitter profile in comparison to screen printed reference cells. Overall, no loss was seen in V_{oc} and an equal level of V_{oc}, Eta (η) and *FF* are achieved with laser and plating technology in comparison to screen printing.

4 CONCLUSIONS

We demonstrate the an efficiency advantage of 0,5%$_{abs}$ for laser ablated and Ni-Cu plated PERC solar cells caused by flipping the processing order from performing FFO before front side LCO to performing FFO after the front side LCO. The increased efficiency is due to higher V_{oc} and *pFF*. Therefore, we allocate the improvement in solar cells performance to recrystallization of laser damage after LCO and it warrants further deeper investigation into this phenomena.

5 ACKNOWLEDGEMENTS

This work has been performed under the aegis of project "Groschen" (FKZ: 0324012B), funded by German Federal Ministry for Economic Affairs and Energy.

The authors would also like to extend a word of thanks to HQC Cells GmbH for sharing their technology and precursors for testing at our facility and also for sharing the results.

6 REFERENCES

[1] M. Kamp, J. Bartsch, S. Nold, M. Retzlaff, M. Hörteis, and S. W. Glunz, "Economic evaluation of two-step metallization processes for silicon solar cells," *Energy Procedia*, vol. 8, pp. 558–564, 2011.

[2] A. ur Rehman and S. H. Lee, "Review of the potential of the Ni/Cu plating technique for crystalline silicon solar cells," *Materials (Basel).*, vol. 7, no. 2, pp. 1318–1341, 2014.

[3] S. Kluska *et al.*, "Electrical and Mechanical Properties of Plated Ni/Cu Contacts for Si Solar Cells," *Energy Procedia*, vol. 77, pp. 733–743, 2015.

[4] A. Büchler *et al.*, "Interface oxides in femtosecond laser structured plated Ni-Cu-Ag contacts for silicon solar cells," *Sol. Energy Mater. Sol. Cells*, vol. 166, no. March, pp. 197–203, 2017.

[5] M. Abbott, P. Cousins, F. Chen, and J. Cotter, "Laser-induced defects in crystalline silicon solar cells," *Conf. Rec. IEEE Photovolt. Spec. Conf.*, vol. 61, no. 0, pp. 1241–1244, 2005.

[6] G. Poulain *et al.*, "Characterization of laser-induced damage in silicon solar cells during selective ablation processes," *Mater. Sci. Eng. B Solid-State Mater. Adv. Technol.*, vol. 178, no. 9, pp. 682–685, 2013.

[7] C. Dang *et al.*, "Investigation of laser ablation induced defects in crystalline silicon solar cells," *Energy Procedia*, vol. 55, pp. 649–655, 2014.

[8] S. Steffens *et al.*, "Impact of dislocations and dangling bond defects on the electrical performance of crystalline silicon thin films," *Appl. Phys. Lett.*, vol. 105, no. 2, 2014.

[9] A. A. Brand, F. Meyer, J. F. Nekarda, and R. Preu, "Reduction of picosecond laser ablation threshold and damage via nanosecond pre-pulse for removal of dielectric layers on silicon solar cells," *Appl. Phys. A Mater. Sci. Process.*, vol. 117, no. 1, pp. 237–241, 2014.

[10] S. Rapp, G. Heinrich, M. Wollgarten, H. P. Huber, and M. Schmidt, "Physical mechanisms of SiNx layer structuring with ultrafast lasers by direct and confined laser ablation," *J. Appl. Phys.*, vol. 117, no. 10, pp. 1–9, 2015.

[11] P. Lorazo, L. J. Lewis, and M. Meunier, "Molecular-dynamics thermal annealing model of laser ablation of silicon," pp. 1–18, 2001.

[12] J. Koc, S.; Zavetova, M.; Zemek, "Physical Properties of Amorphous Si: The Role of Annealing," *Czechoslov. J. Phys.*, vol. 25, pp. 83–90, 1975.

[13] H. Bouridah, F. Mansour, R. Mahamdi, N. Bounar, and P. Temple-Boyer, "Effect of thermal annealing and nitrogen content on amorphous silicon thin-film crystallization," *Phys. Status Solidi Appl. Mater.*, vol. 204, no. 7, pp. 2347–2354, 2007.

[14] G. Marcins, J. Butikova, I. Tale, B. Polyakov, R. Kalendarjov, and A. Muhin, "Crystallization processes of amorphous Si by thermal annealing and pulsed laser processing," *IOP Conf. Ser. Mater. Sci. Eng.*, vol. 23, no. 1, 2011.

[15] S. Kluska et al. (2018) "Plated Fine Line Metallization for PERC Solar Cells". *35th EU-PVSEC*, to be published.

PVD METALLIZATION FOR HIGH-EFFICIENCY C-SI SOLAR CELLS: SCENARIO FOR IMPLEMENTATION IN PRODUCTION

A. Hain[1,*], H. Nagel[2], M. Doerr[1], T. Buck[3], Z. Peng[3], R. Lorenz[4], T. Fischer[4], S. Gledhill[2], T. Kroyer[2], M. Huber[1] and P. Wohlfart[1]

[1] SINGULUS TECHNOLOGIES AG, Hanauer Landstrasse 103, 63796 Kahl am Main, Germany
[2] Fraunhofer Institute for Solar Energy Systems ISE, Heidenhofstr. 2, 79110 Freiburg, Germany
[3] International Solar Energy Research Center Konstanz e.V., Rudolf-Diesel-Straße 15, 78467 Konstanz, Germany
[4] teamtechnik Automation GmbH, Heinrich-Hertz-Straße 1, 71642 Ludwigsburg, Germany
*Author for correspondence: Achim Hain, Phone: +49 6188-440-312, e-mail: achim.hain@singulus.de

ABSTRACT: In comparison with the present industry standard for crystalline Si solar cell which is screen printing of large-area Al and local Ag solder pads on the rear, a physical vapor deposited (PVD) Al rear contact has several advantages that makes it ideally suited for new generation high-efficiency solar cells. In this work we describe the boundary conditions for a cost-effective introduction of PVD Al in production. To begin with the cell design a contact resistance below 10 mΩ cm² can be assured by a Si doping concentration higher than 10^{17} cm^{-3} for *p*-type Si and higher than $2x10^{19}$ cm^{-3} for *n*-type Si. In case of 5 busbars an Al thickness of only 1 µm is needed to reduce the lateral resistance loss to less than 0.04 % absolute for cells featuring an efficiency of 23.4 %. By sputtering a thin NiV adhesion layer on the Al and preferably another thin metal layer for good solder wetting, normalised interconnector peel forces exceeding 2 N/mm were obtained. This way standard interconnection technology based on soldering can be utilized for module production. The cost of ownership calculation reveals a cost reduction of 2.6 % on module level if the module efficiency can be increased from 21 % for modules equipped with screen-printed cells to 21.7 % for high-efficiency solar cell modules.

Keywords: metallization, back contact, module integration

1 INTRODUCTION

In industrial production of crystalline Si solar cells the most widespread technology for rear side metallization is screen-printing of large-area Al in combination with screen-printing of local Ag solder pads. The market share of this well-developed process far surpasses 90 %. Because engineers and scientists have utilized this technique for decades now, it is not surprising that it was quickly adapted to passivated emitter and rear contacted (PERC) solar cells, the increasingly important cell type in production which yields efficiencies of up to about 22 %. However, screen-printed Al contacts suffer from shortcomings like possible reduction of the cells' surface passivation due to the applied high sintering temperature, increased effective surface recombination velocity in comparison with boron back surface field (BSF), damage of the *n*-doped BSF of *n*-type cells and damage of thin carrier-selective contacts. Hence, for future production of new generations of crystalline Si solar cells featuring efficiencies in the 23 % range, physical vapor deposited (evaporated or sputtered, PVD) Al is a suitable choice. It makes good ohmic contact to *p*- and *n*-type Si, provides excellent back reflection of infrared light passing through the cells and features very high electrical conductivity.

A promising candidate for the very next generation high-efficiency crystalline Si solar cells in production is characterized by a carrier-selective back-contact. The outstanding surface passivation and carrier transport properties allow for lab efficiencies of 25.7 % in case of a both sides contacted cell design [1] and 26.1 % in case of rear side contacted cell [2]. The latter refers to the present efficiency world record for crystalline Si solar cells and it is not surprising that this cell features a PVD Al back contact.

For a broad industrial utilization feasible technological solutions for making PVD Al solderable have to be considered and an eye has to be kept on the processing costs. In this contribution we identify the technological and financial benchmarks and its market potential.

2 APPROACH

Starting point is the development of an evaporation setup that uses an anti-sticking agent coated on the walls of the deposition chamber [3]. By preventing the evaporated metal from adhering to the walls, the machine's uptime is increased. In combination with a high-throughput vacuum system the costs of metal deposition are significantly reduced. We investigated the optical and electrical performance of PVD Al as a full area rear contact on high-efficiency cells and deposited different metal layer stacks on the Al in order to make the rear contact solderable. The metal layer stacks can be deposited in the same vacuum system at low incremental cost. The peel force of tinned copper ribbons soldered to the metal stacks was measured and high-efficiency mini-modules were prepared. In a thorough cost assessment PVD Al as a full area rear contact is compared with screen-printed rear contacts consisting of large-area Al and local Ag solder pads.

3 RESULTS AND DISCUSSION

3.1 High-throughput high-uptime metal PVD system

As long as the walls of the vacuum deposition system are covered with anti-sticking agent, absolutely no metal is deposited upon. During system operation the anti-sticking agent evaporates with a low rate which is compensated by reconditioning the walls after an adjustable number of coating cycles. Al layers deposited on crystalline Si wafers were investigated by energy dispersive x-ray analysis in a secondary electron microscope (SEM). No residual anti-sticking agent was found in or on the deposited Al or at the interface to the Si substrate. Comparison of Al layers deposited with and without anti-sticking agent revealed the same microstructure and electrical resistivity. An SEM picture of the microstructure is shown in Fig. 1. The measured electrical resistivity was 3.2 μΩ cm. Within the measurement error this is similar to that of bulk Al (2.8 μΩ cm [4]).

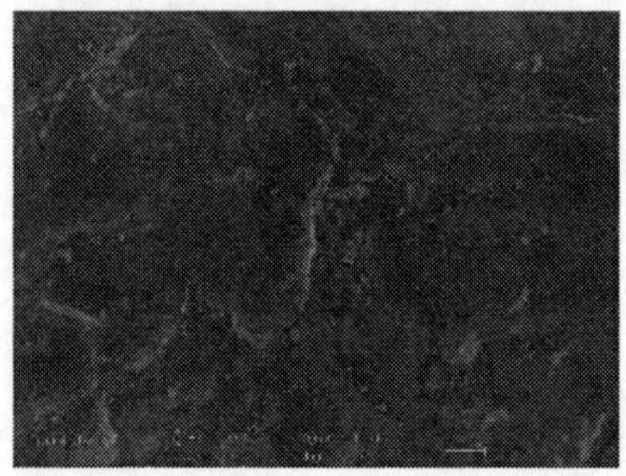

Fig. 1: High magnification SEM picture of thermally evaporated Al back contact on a CZ Si PERC solar cell using anti-sticking agent. The cell features rounded random pyramids on the rear.

The anti-sticking agent is suited for integration in a commercial vacuum-deposition system as schematically depicted in Fig. 2. It operates at a throughput rate of 6000 wafers/h and is characterized by an overall equipment effectiveness of 82.1 %. A detailed cost assessment of this system is presented in paragraph 3.5.

Fig. 2: Layout of a high-throughput high-uptime inline system for vacuum deposition of Al contacts on c-Si solar cells.

3.2 Electrical performance of full-area PVD Al rear contact

Utilizing the measured electrical resistivity mentioned above we analytically calculated the efficiency loss due to the lateral series resistance of a full area PVD Al contact for different numbers of busbars as a function of Al thickness. The maximum power point current density (38.4 mA/cm²) and voltage (610 mV) assumed in the calculations was measured before on a 23.4 % efficient large-area carrier selective rear contact solar cell featuring a PVD Al back contact [5]. Fig. 3 shows that the efficiency loss is below 0.04 % absolute in case of 1 μm thickness of the Al and 5 busbars. This acceptable value can be further reduced by increasing the number of busbars, e. g. below 0.03 % absolute for 6 busbars. In case of 16 busbars in a multi-wire approach the efficiency loss is below 0.04 % absolute for a PVD Al rear contact that is only 100 nm thick.

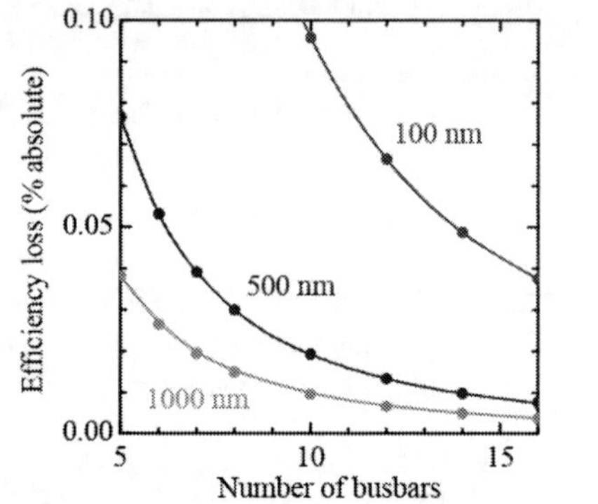

Fig. 3: Calculated efficiency loss of a 23.4 % efficient carrier selective contact Si solar cell as a function of the number of busbars for different PVD Al rear contact thicknesses.

In principal, the contact resistivity between Al and Si can contribute significantly to the cell's series resistance. In practice, however, the contact resistivity can be very low, e. g. on the world record c-Si solar cell mentioned in the introduction the sum of the contact resistivities of PVD Al on *p*- and *n*-type poly Si is less than 3 mΩ cm² [2]. For full area contacts considered in this work 10 mΩcm² are tolerable. In publication [6] the contact resistivity of evaporated Al on n^+-diffused crystalline Si was reported as a function of donor concentration. The relevant graph is reproduced in Fig. 4. It can be seen from the figure, that donor concentrations higher than about $2x10^{19}$ cm^{-3} yield contact resistivities below 10 mΩ cm² as desired. Between *p*-type c-Si and PVD Al contact resistivities below 0.4 mOhm cm² were obtained if the acceptor concentration is higher than 10^{17} cm^{-3} [7]. Hence, in an appropriate cell design it is possible to ensure low contact resistivity between PVD Al and Si.

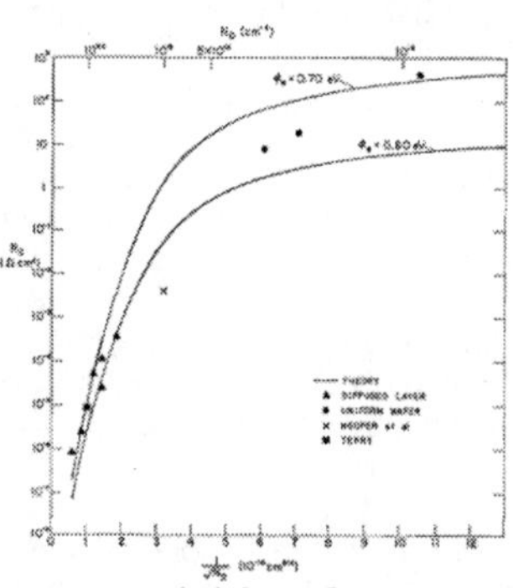

Fig. 4: Contact resistivity of evaporated Al on n^+-diffused Si as a function of donor concentration. The graph was taken from [6].

3.3 PVD of solderable metal stacks

For solder tests we used *p*-type CZ Si PERC cells. They were alkaline textured on both sides, whereby the pyramids on the rear side were leveled to some extend during subsequent horizontal *pn*-junction isolation in a mixture of HNO_3 and HF. The rear surface was passivated by an Al_2O_3/SiN stack, a 1 μm thick Al contact was thermally evaporated and a 100 nm NiV layer was sputtered onto it. Al and NiV were deposited in the same inline high-throughput system without breaking the vacuum. NiV is a good adhesion layer on Al and we found that tinned Cu ribbons can directly be soldered onto it. Local ohmic contacts between Al/NiV and the Si were formed by laser (so-called 'laser-fired contacts'). For one cell batch an additional 50 nm thick Ag layer was thermally evaporated on the NiV layer in order to ensure good solder wetting. 2 mm wide SnPbAg-coated Cu ribbons were manually soldered on the rear of the cells after applying Kester 952s flux on them. Finally, the 90 degrees peel force of the ribbons was measured with an automatic system. The results are shown in Figs. 5 and 6. The peel forces normalised to the ribbon width are higher than 1 N/mm at every point. With respect to solder wetting obviously there were no cavities that locally reduce the peel force. The average values are 2.4 N/mm in case of NiV and 2.1 N/mm in case of Ag top layer. Taking into account the natural spread of measured peel forces the values can be considered equal. They comply with the standard DIN EN 50461 (> 1 N/mm). We observed that mechanical fracture most often took place between NiV and Al for both metal layer stacks. Hence, we conclude that solder wetting of the NiV was not limiting the peel force. This result was confirmed in another experiment where a different inexpensive metal capping layer was deposited on NiV to ensure good solder wetting.

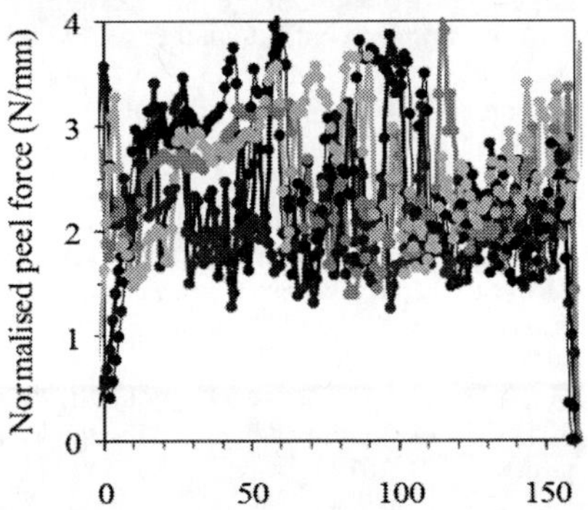

Fig. 5: Measured 90 degrees peel force of tinned Cu ribbons soldered onto 1 µm PVD Al / 100 nm NiV metal layer stack on the rear of PERC CZ Si solar cells. The peel force is normalised to the ribbon width.

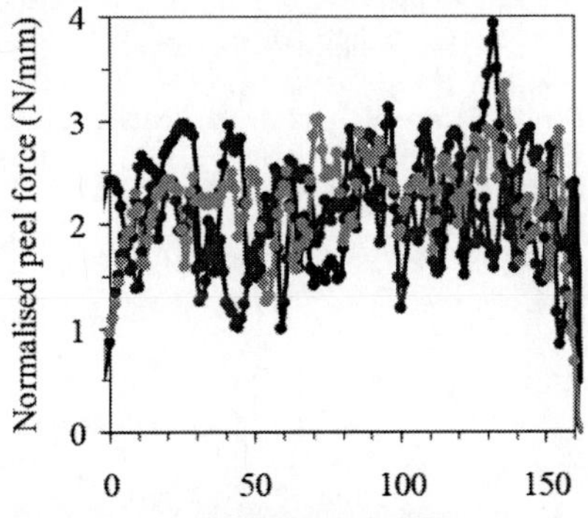

Fig. 6: Measured normalised 90 degrees peel force of tinned Cu ribbons soldered onto 1 µm PVD Al / 100 nm NiV / 50 nm Ag metal layer stack on the rear of PERC CZ Si solar cells.

We investigated if the Al/NiV metal stack can be soldered in an automatic tabber-stringer. The stringer and a completed 3-cell string are shown in Fig. 7. We found that solder wetting of the metal stack is more critical compared with manual soldering because of short cycle time. Hence, on the NiV a thin low-cost metal layer improving solder wetting is recommended for high-throughput cell string production.

Fig. 7: Photo of a tabber-stringer used for soldering CZ Si PERC solar cells featuring a 1 µm thick full area PVD Al rear contact and a 100 nm thick NiV layer on top. The inset shows a photo of the rear of a three-cells string.

3.4 Module manufacturing

As a proof of concept we prepared one-cell modules from both sides contacted 100 cm² large solar cells featuring carrier selective rear contacts covered by PVD Al plus solderable metal stack. The measured aperture area electrical performance parameters of the best module are listed in Table 1. An efficiency (*eta*) of 21.7 % was obtained. Because of accidental handling issues the second diode saturation current density increased by a factor of 10 to 16.5 nA/cm² and hence the pseudo fill factor (*pFF*) reduced by 4 % absolute due to module making. Correcting for it the projected efficiency potential of the module is 23.0 %.

Table 1: Measured illuminated *IV* parameters of a carrier-selective one-cell module featuring PVD Al rear metal contact.

V_{oc} (mV)	J_{sc} (mA/cm²)	*FF* (%)	*pFF* (%)	*Eta* (%)	R_{ser} (mΩ cm²)
699.7	40.7	76.3	79.8	21.7	0.74

3.5 Cost of ownership calculation

Industrial solar cell production is cost and cell efficiency driven. The Mechanical Engineering Industry Association (VDMA) provides a cost of ownership (COO) modeling tool [8] which we used to shed light on the economic potential of full-area PVD Al rear contact in a high-efficiency solar cell production. The cost break up shown in Fig. 8 was obtained for the inline vacuum deposition system depicted in Fig. 2. It is characterized by a throughput of 237 MWp/a at a foreseen average cell efficiency of 22.8 %, which corresponds to 43.4 million 6 inch wafers/a. The calculation was performed for a 1 µm thick full-area Al contact that guarantees less than 0.04 % absolute efficiency loss caused by lateral series resistance in case of 5 busbars as described in paragraph 3.2. It can be seen in Fig. 8, that with the assumed depreciation life of 7 years this cost element accounts for 1.7 € cents/wafer. The expenses for consumables are only

1 € cent/wafer due to the low price of bulk Al. Furthermore, the expenses for utilities are in sum 0.6 € cents/wafer wherein electricity is the main component. Altogether, the costs for full-area PVD Al are 4.0 € cents/wafer.

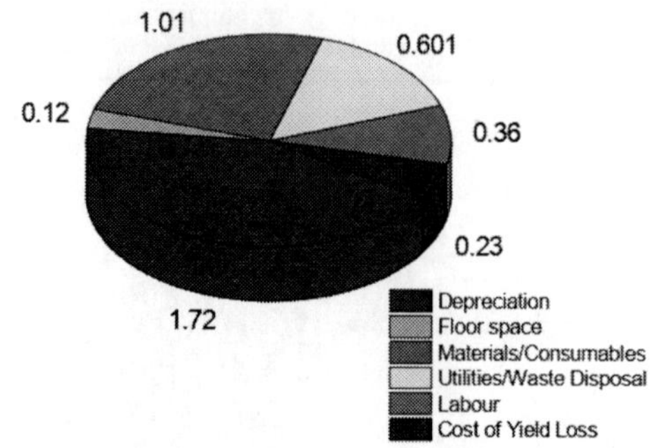

Fig. 8: Calculated add-on process costs in € cents/wafer for full-area 1 µm thick PVD Al by element.

Depositing 25 nm NiV adhesion layer on the Al plus 100 nm inexpensive solder wetting metal layer on the NiV in the same PVD systems results in costs of 5.4 € cents/wafer, see cost distribution in Fig. 9. In comparison with Fig. 8 the share of consumables is higher due to additional metal sputter targets.

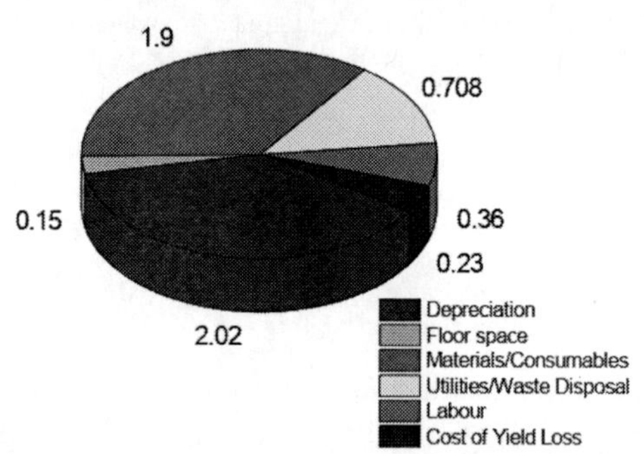

Fig. 9: Calculated add-on process costs in € cents/wafer for a full-area rear contact metal layer stack consisting of 1 µm PVD Al / 25 nm NiV / 100 nm inexpensive solder wetting metal layer.

These results have to be compared to standard screen-printed large-area Al plus screen-printed local Ag solder pads. Utilizing the same base data we obtained 4.6 € cents/wafer. In comparison with PVD Al depreciation costs are lower and consumable costs are higher. The latter is due to the Al paste being more expensive than bulk Al.

In order to identify the most favorable approach in cell production, we look at two scenarios: first we assume a module efficiency of 21 % that could be obtained with screen-printed PERC solar cells in future. This corresponds to a power of 336 W_p for a module being 1.6 m^2 in size. Second we assume 21.7 % efficient modules equipped with carrier-selective contact solar cells featuring 1 µm PVD Al / 25 nm NiV / 100 nm inexpensive metal layer stack. Note that the efficiency potential of carrier-selective contact solar cell modules is higher but here we restrict the assessment on the efficiency we obtained for the prepared one-cell module, knowing that we measured the aperture efficiency and now we assume that this value can be obtained as total area efficiency. It corresponds to a power of 347 W_p if the module would be 1.6 m^2 in size. Adopting a present PERC module selling price of 25 € cents/Wp, the high-efficiency carrier-selective contact solar cell modules can reduce the selling price by 2.6 % if the other costs for cell production are the same.

4 SUMMARY AND OUTLOOK

We have shown that PVD Al plus solderable metal layer stack is an attractive option for full-area rear contact on high-efficiency c-Si solar cells. Because of its favorable electrical and optical properties PVD Al might be a door opener for carrier-selective rear contacts in cell production. In an assessment we found a possible 2.6 % short-term cost advantage of modules equipped with high-efficiency carrier-selective contacted solar cells in comparison with screen-printed PERC cells. If the projected 23.0 % efficiency potential of the carrier-selective modules can be realized, the cost advantage on module level can increase to 8.1 %.

ACKNOWLEDGEMENT

The authors thank the German Federal Ministry for Economic Affairs and Energy for funding within the project 'IdeAl' under contract number 0325889.

5 REFERENCES

[1] A. Richter, J. Benick, F. Feldmann, A. Fell, M. Hermle and S. W. Glunz, Solar Energy Materials and Solar Cells 173(2017) 96-105.
[2] F. Haase, C. Hollemann, S. Schäfera, A. Merkle, M. Rienäcker, J. Krügener, R. Brendel and R. Peibst, Solar Energy Materials and Solar Cells 186 (2018) 184-193.
[3] M. R. Huber, M. Dörr and P. Wohlfart, Proceedings of the 27th EU Photovoltaic Solar Energy Conference, p. 1185-1187, 2012.
[4] https://en.wikipedia.org/wiki/Aluminium.
[5] H. Nagel, D. Eberlein, S. Hoffmann, M. Graf, B. Steinhauser, F. Feldmann, A. Kraft, U. Eitner, M. Glatthaar, M. Hermle, S. W. Glunz, H. Haverkamp, T. Fischer, A. Hain, P. Wohlfart, V. Mertens, J. M. Müller and T. Buck, Proceedings of the 33rd European Photovoltaic Solar Energy Conference, p. 295-299, 2017.
[6] A. Y. C. Yu, Solid-State Electronics 13 (1970) 239-247.
[7] D. K. Schroeder and D. L. Meier, IEEE Trans. Electron Devices 31 (1984) 637-647.
[8] http://www.itrpv.net/Reports/Downloads/Related-documents/.

DEEP LEARNING APPROACH TO INLINE QUALITY RATING AND MAPPING OF MULTI-CRYSTALLINE SI-WAFERS

Matthias Demant[1,3], Patrick Virtue[2,3], Aditya S. Kovvali[1], Stella X. Yu[2,3], Stefan Rein[1]
[1]Fraunhofer ISE, Heidenhofstraße 2, D-79110 Freiburg, Germany
[2]University of California, Berkeley, CA 94720, USA
[3]International Computer Science Institute, 1947 Centre Street, Berkeley, CA 94704, USA

ABSTRACT:
This work shows the first successful application of convolutional neural networks (CNN) for material characterization and process control in solar cell production. We present a fully data-driven machine learning approach for inline quality rating and quality mapping of as-cut multi-crystalline Silicon (mc-Si) wafers. We use Photoluminescence (PL) images to image crystallization related defects in the wafers. We show that we can learn how to quantify these defect patterns based on empirical data and derive a meaningful wafer representation directly from the high-resolution input images by means of deep CNNs. This end-to-end regression model predicts solar cell efficiencies with mean errors of 0.12% for materials of bricks not presented in the training set, which is $25\%_{rel}$ better than our classical methods based on feature engineering. Moreover, we visualize the expected quality distribution for each sample within a spatially resolved activation map. The mapping procedure gives an insight into the "black box" neural network and shows that the quality distribution is in accordance to the expectations of domain experts and similar to spatially resolved quality data like the image of the dark saturation current density (j_0). More details on learning and mapping will be reported within two studies elsewhere.

Keywords: Multi-Crystalline Silicon, Convolutional Neural Networks, CNN, Regression Activation Mapping, Material Rating, Semantic Representation, PERC

1 INTRODUCTION

The rating of mc-Si wafers with regard to the expected solar cell quality directly after wafering allows a fast and cost efficient material control for crystal growers and solar cell manufacturers.

This paper presents results from two comprehensive studies on learning quality rating [1] and quality mapping [2] of as-cut multi-crystalline Si-wafers via convolutional neural networks (CNN) for regression.

We use Photoluminescence (PL) images [3] as input for a convolutional neural network to predict the current-voltage (IV) parameters of the solar cells. PL-images are suitable for an inline analysis of as-cut samples since they reveal crystallization-related defects in mc-Si materials, which correlate with solar cell quality [4].

Nevertheless, for as-cut wafers a determination of the bulk lifetime of excess charge carriers is limited due to the high surface recombination. This impacts the rating of high quality material for high quality solar cell processes. This study investigates high performance materials [5] and passivated emitter and rear cells (PERC) [6] to address this matter.

In previous works [7-13], feature engineering methods have been applied to extract features according to domain knowledge. The design of domain-relevant features is challenging as PL-intensities and PL-contrasts vary in the continuous image space. The defect patterns superimpose in the images and interact during the process [14-16] for example within high-temperature or surface structuring processes.

A meaningful representation of the PL-images with regard to solar cell quality can be learned within deep convolutional neural networks (Section 2). The training of the model requires a representative data set which is acquired within a huge experiment (Section 3). The proof of generalizability for complex models needs to be evaluated on "unknown" test data not presented in training data (Section 4). For a deeper understanding and acceptance of our approach in the PV community we visualize what has been learned in the PL-image (Section 5).

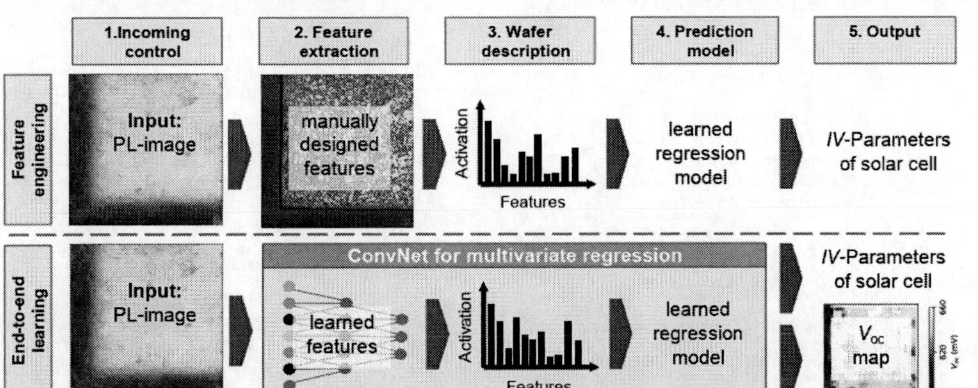

Figure 1: (Top row) Overview of the pattern recognition pipeline for material rating based on as-cut mc-Si wafers. In the middle row the classical approach uses features engineered by human experts and in the bottom row the end-to-end rating learns feature extraction and prediction in the same convolutional neural network. In addition to quality prediction, the regression activation map (RAM) reveals the spatial distribution of open-circuit voltage (V_{oc} map) learned by the network

2 APPROACH

2.1 Pattern recognition pipeline

An overview of the pattern recognition pipeline for material rating based on as-cut mc-Si wafers is shown in Figure 1. Wafers from different materials are measured in an incoming control and PL-images are analyzed. The PL-features are extracted and combined in a histogram which forms the basis of a rating model to predict the IV-parameters of the solar cells.

Following to the classical feature engineering approach [17], the extraction of features is based on a combination of human defined filters. The optimization of the regression model for V_{oc} prediction is based on empirical data.

Within the end-to-end rating model the feature representation and the regression parameters are learned within the same optimization step and are not based on a manually designed feature definition.

2.1 Regression network and activation map

Recently, innovations in deep learning have led to significant breakthroughs in computer vision tasks from classification [18] to localization of objects and activities within an image [19]. Convolutional neural networks with dense connections have been particularly effective in these tasks, primarily due to its additional connections between layers that improve accuracy by allowing data to propagate more efficiently through the very deep network. We convert densely connected network architecture from an object recognition network to a multivariate regression network that predicts solar cell efficiency, open-circuit voltage, short-circuit current and fill factor within one model.

We want to analyze, what has been learned in the model. An activation map [19] with spatial coordinates can be computed for a given PL-image as input to reveal the learned quality distribution of our regression model. It can show us how the network rates dislocations and regions of reduced lifetime due to contaminations from the crystallization crucible. The map is qualitatively compared to j_0-images of the solar cell [20] to confirm the meaningful interpretation of the patterns.

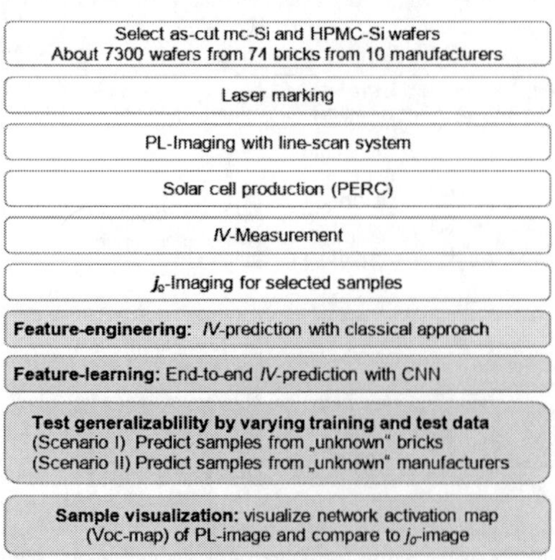

Figure 2: Schemata of the experiment

3 EXPERIMENTAL

As deep neural networks require large training datasets for successful optimization, we collect a wide variety of experimental data.

Figure 2 gives an overview of the experimental approach for the collection and evaluation of this data. The comprehensive data set allows a proof of generalizability by testing the model with completely unknown data.

The experimental process steps are as following: the wafers are sampled from different positions of various bricks from ten manufacturers. The data set contains mc-Si and HPMC-Si material. The samples are laser-marked and PL-images are measured within an incoming-control with a line-scan PL-system [21]. The wafers are processed to PERC solar cells within an industrial production line. *IV*-parameter and j_0-images [20] are measured after solar cell production.

We evaluate the measured images based on the feature-engineering approach [12] and the proposed feature-learning approach. The models are trained with approximately 3000 wafers and tested with the remaining samples. This work presents the results of Scenario I where no wafers of the same brick are in test and training set. In Scenario II all wafers of the tested manufacturer are removed from the training set, which will be presented in [1].

The presented regression activation maps (RAM) are based on a network trained for V_{oc} prediction with PL-images as input. For the computation of the RAM the PL-image is passed through the network. No additional information about V_{oc} from current-voltage measurement is required. The RAM is compared to the j_0-images to visualize what has been learned in the network.

4 RESULTS ON CELL QUALITY PREDICTION

The presented prediction model is capable to quantify complex defect structures and predict the solar cell efficiency (Figure 3) and open-circuit voltage (Figure 4) with low mean absolute error for the prediction of unknown bricks (Table I). The prediction of short-circuit current density (J_{sc}) and fill factor are less accurate. The network improves the rating with manually designed features by about 25%rel for efficiency prediction.

Table I: Prediction of unknown bricks

Method		Eta [%]	V_{oc} [mV]	Jsc [mA/cm²]	FF [%]
Designed features	MAE	0.16	2.73	0.17	0.32
	Corr	0.88	0.89	0.83	0.53
Learned features	MAE	0.12	2.15	0.15	0.28
	Corr	0.93	0.93	0.87	0.67

MAE = Mean absolute error in unit of parameter
Corr = Pearson correlation coefficient in [-1, 1]

The network converges rapidly during training and overcomes the bias-variance trade-off during testing: on the one hand, the network is capable to quantify defect patterns within a complex network structure and on the other hand, the model is generalizable to predict unknown data. The prediction can be computed within only a few milliseconds.

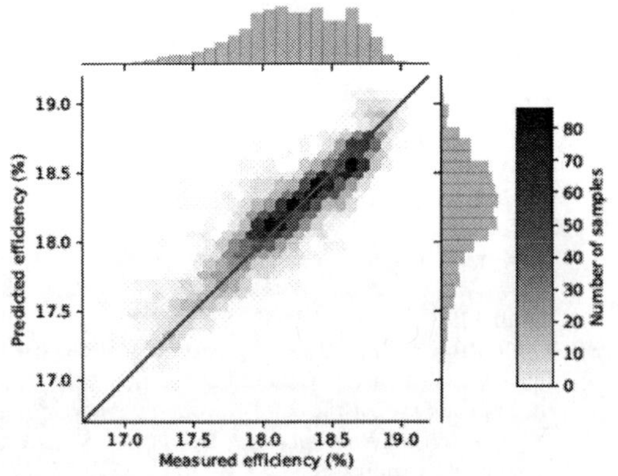

Figure 3: Result on efficiency prediction

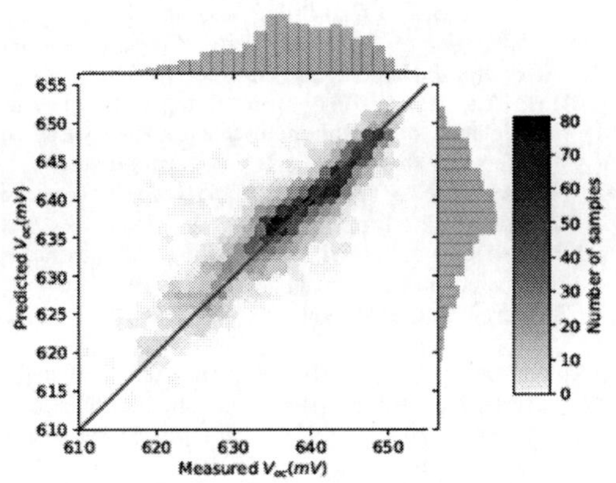

Figure 4: Result on open-circuit voltage prediction

Details on model and rating, as well as the prediction results for materials from unknown manufacturers will be available in the report on our full study [1].

5 RESULTS ON VISUALIZING WHAT HAS BEEN LEARNED IN THE NETWORK

The network learns a prediction model based on the PL-image as input and the solar cell parameter as output, here the open-circuit voltage. What has been learned in the network? The activation map (AM) is a technique, which reveals how different image regions are rated in the network purely based on the PL-image and the trained model. Note that no human input has been given to the network, except the PL-input image and the V_{oc} during model training.

The average of the activation map for regression (RAM) is the prediction result. Thus, the map can be scaled to Volts.

The comparison of PL-inputs, the RAM of the network and the measured j_0-images in Figure 5a) show strong similarities between predicted and measured quality. The network assigns low quality values to dislocation clusters and contaminated regions.

Even for samples with inverted contrast in Figure 5b) the network distinguishes the effect of different structures which are comparable to the j_0-images. The distributed V_{oc} values result in quality estimation with small error.

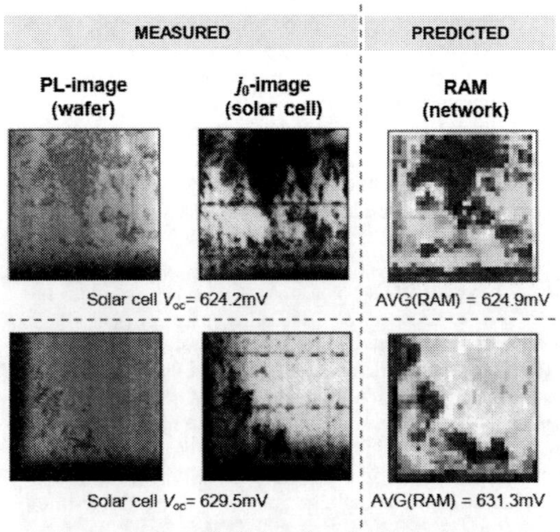

a) Two samples with dislocation clusters and contaminated regions at the edge of the wafer.

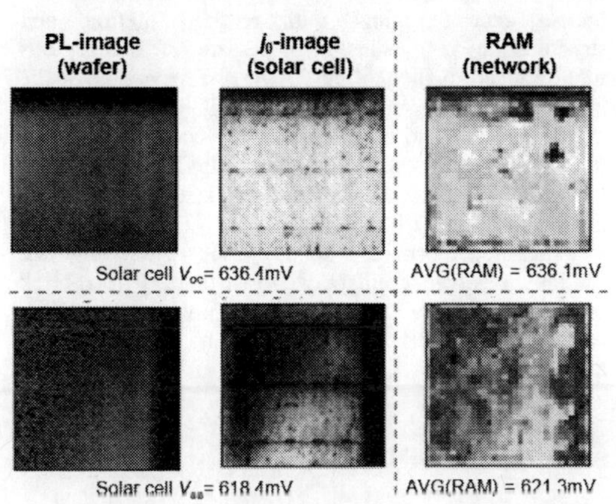

b) Samples with dislocations showing inverted contrasts with (top row) low material quality and (bottom row) high material quality.

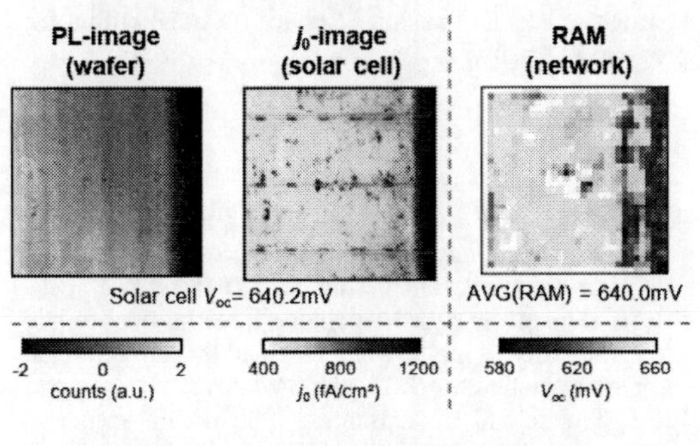

c) Failure case shows activations, which are not directly connected to crystallization defects.

Figure 5: Triplets of (left) PL-image of as-cut wafer, (center) the j_0-image measured after solar cell production and (right) the regression activation map learned by the network for V_{oc}-prediction. The measured and predicted V_{oc} values are annotated below each sample.

Only few examples show artefacts not being related to crystallization defects. For example the line like structure in the RAM in Figure 5c) might be connected to a strong saw-mark.

Further errors might be connected to the low resolution of the RAM with 32px × 32px. Also the exact defect position in the RAM may vary due to the large receptive field of each neuron in PL-image space.

6 CONCLUSION

This work showed the first successful application of CNNs for material characterization and process control in solar cell production. We identified a promising neural network architecture, which requires an intensive training procedure but allows for a generalizable and inline applicable prediction of the material quality. Further details on the learning algorithm will be available in our report in [1].

We applied an activation mapping technique to reveal what has been learned by the network. The regression activation map is in accordance with the expectations of a human expert: regions with reduced lifetime and structural defects in the PL-images are rated as regions with reduced quality in the RAM. A comparison of the network activation map shows the high similarity of the predicted quality distribution in wafer coordinates to the measured j_0-images. A detailed introduction to this mapping technique and further visualization results will be given in [2].

The model can be extended easily by adding e.g. spatially resolved grain boundary data [22] or a brick lifetime parameter [23] to the prediction model, which can offer additional information for material characterization.

ACKNOWLEDGEMENTS

This work was supported by German Federal Ministry of Economic Affairs and Energy (BMWi) in the project "Q-Crystal" under the contract number 0324103A.

M. Demant was supported by a scholarship of the German Academic Exchange Service (DAAD) within the program FITweltweit.

REFERENCES

[1] M. Demant, *et al.*, "Learning quality rating of multicrystalline Si-wafers via convolutional regression networks," submitted 2018.

[2] M. Demant, *et al.*, "Visualizing material quality and similarity of mc-Si wafers learned by convolutional regression networks," submitted 2018.

[3] T. Trupke and R. A. Bardos, "Photoluminescence: a surprisingly sensitive lifetime technique," in *Proceedings of the 31st IEEE Photovoltaic Specialists Conference*, Orlando, Florida, USA, 2005, pp. 903-906.

[4] J. Haunschild, *et al.*, "Quality control using luminescence imaging in production of mc- solar cells from UMG feedstock," in *Proceedings of the 35th IEEE PVSC*, Honolulu, Hawaii, USA, 2010, p. in print.

[5] C. W. Lan, *et al.*, "Grain control in directional solidification of photovoltaic silicon," *Journal of Crystal Growth*, vol. 360, pp. 68-75, 2012.

[6] M. A. Green, "The Passivated Emitter and Rear Cell (PERC): From conception to mass production," *Solar Energy Materials and Solar Cells*, vol. 143, pp. 190-197, 2015.

[7] M. Demant, *et al.*, "Analysis of luminescence images applying pattern recognition techniques," in *Proceedings of the 25th EU-PVSEC*, Valencia, Spain, 2010, pp. 1078-1082.

[8] W. McMillan, *et al.*, "In-line monitoring of electrical wafer quality using photoluminescence imaging," in *Proceedings of the 25th European Photovoltaic Solar Energy Conference and Exhibition*, Valencia, Spain, 2010, pp. 1346-51.

[9] B. Birkmann, *et al.*, "Analysis of Multicrystalline Wafers Originating from Corner and Edge Bricks and Forecast of Cell Properties," in *26th European Photovoltaic Solar Energy Conference and Exhibition*, Hamburg, 2011, pp. 937 - 940.

[10] B. True, *et al.*, "Image processing techniques for correlation of photoluminescence images of as-cut wafers with final cell IV parameters," in *Proceeding of the 26th EU-PVSEC*, Hamburg, Germany, 2011, pp. 1-4.

[11] M. Demant, *et al.*, "Modelling of physically relevant features in photoluminescence images," *Energy Procedia*, vol. 27, pp. 247-52, 3-5 Apr 2012.

[12] M. Demant, *et al.*, "Inline quality rating of multi-crystalline wafers based on photoluminescence images," *Progress in Photovoltaics: Research and Applications*, pp. 1533-46, 2015.

[13] T. Trupke and R. A. Bardos, "Wafer imaging and processing method and apparatus," ed: Google Patents, 2017.

[14] S. McHugo, *et al.*, "Gettering of metallic impurities in photovoltaic silicon," *Applied Physics A*, vol. 64, pp. 127-137, 1997.

[15] A. Bentzen and A. Holt, "Overview of phosphorus diffusion and gettering in multicrystalline silicon," *Materials Science and Engineering B*, vol. 159-60, pp. 228-34, 2008.

[16] S. Castellanos, *et al.*, "High-performance and traditional multicrystalline silicon: Comparing gettering responses and lifetime-limiting defects," *IEEE Journal of Photovoltaics*, vol. 6, pp. 632-640, 2016.

[17] M. Demant, *et al.*, "Inline quality rating of multi-crystalline wafers based on photoluminescence images," *Progress in Photovoltaics: Research and Applications*, 2015.

[18] A. Krizhevsky, *et al.*, "Imagenet classification with deep convolutional neural networks," in *Advances in Neural Information Processing Systems*, 2012, pp. 1097-1105.

[19] B. Zhou, *et al.*, "Learning deep features for discriminative localization," in *Computer Vision and Pattern Recognition (CVPR), 2016 IEEE Conference on*, 2016, pp. 2921-2929.

[20] M. Glatthaar, *et al.*, "Evaluating luminescence based voltage images of silicon solar cells," *Journal of Applied Physics*, vol. 108, pp. 014501/1-5, 2010.

[21] H. Höffler, *et al.*, "Comparison of line-wise pl-imaging and area-wise pl-imaging," *Energy Procedia*, vol. 124, pp. 66-75, 2017/09/01/ 2017.

[22] A. S. Kovvali, *et al.*, "About the Relevance of Defect Features in As-Cut Multicrystalline Silicon Wafers on Solar Cell Performance," in *Silicon PV*, Lausanne, Switzerland, 2018.

[23] B. Mitchell, *et al.*, "PERC solar cell performance predictions from multicrystalline silicon ingot metrology data," *IEEE Journal of Photovoltaics*, vol. 7, pp. 1619-1626, 2017.

SCREEN PRINTED THICK FILM METALLIZATION OF SILICON SOLAR CELLS - RECENT DEVELOPMENTS AND FUTURE PERSPECTIVES

Andreas Lorenz[1], Michael Linse[1], Herbert Frintrup[2], Martin Jeitler[3], A. Mette[4], M. Lehner[5], R. Greutmann[6], H. Brocker[6], M. König[7], Denis Erath[1], Florian Clement[1]
[1]Fraunhofer Institute for Solar Energy Systems (ISE), Heidenhofstraße 2, 79110 Freiburg, Germany
[2]Hans Frintrup GmbH, Röhfeldstr. 20, 53227 Bonn, Germany
[3]G. Bopp + Co. AG, Bachmannweg 21, 8046 Zürich, Switzerland, Phone +41 44/377 66 66, mjeitler@bopp.ch
[4]Hanwha Q Cells GmbH, Sonnenallee 17-21, 06766 Bitterfeld-Wolfen, Germany
[5]Lehner Engineering GmbH, Ebnetstrasse 18, 9032 Engelburg, Switzerland
[6]Gallus Ferd. Rüesch AG, Harzbüchelstrasse 34, 9016 St. Gallen, Switzerland
[7]Heraeus Deutschland GmbH & Co. KG, Heraeusstr. 12-14, 63450 Hanau, Germany

ABSTRACT: Within this work, we present actual results in the field of screen printing front side metallization for Silicon solar cells using standard and knotless screen technology. We show that knotless screens have the potential to improve the printed finger geometry and reduce silver consumption. A fine line front side metallization for PERC (passivated emitter and rear) with a mean finger width of $w_f = 27$ µm is realized using a fine line screen printing process at Fraunhofer ISE. The PERC solar cells obtain a mean conversion efficiency of $\eta = 21.6$ %. Furthermore, we present actual results regarding rotary screen printed front side metallization – a highly promising approach to increase throughput significantly. Using this technology, PERC solar cells are metallized at a printing speed of $v = 333$ mm/s. A mean finger width $w_f = 47$ µm and a mean finger height of $h_f = 13$ µm is obtained.
Keywords: Solar cell metallization, flatbed screen printing, knotless screens, rotary screen printing

1 INTRODUCTION

1.1 Flatbed screen printing - status and current challenges

Since the first published application for solar cells in 1975 [1], flatbed screen printing (FSP) has been established as the leading production technology for front and rear side metallization of crystalline Silicon (c-Si) solar cells. Today, more than 95 % of all c-Si solar cells worldwide are metallized using FSP [2]. The ongoing success of FSP is based on a number of factors: FSP is comparatively easy to handle, reliable and cost-efficient compared to much more elaborate alternatives like photolithography, plating or other printing technologies. On the other hand, increasing the throughput of the flatbed screen printing process further (currently up to 4000 wafers/h on a single line) is one major challenge which makes alternative technologies with significantly higher throughput more and more interesting. If one considers the technological challenge of solar cell metallization from a printer's point of view, the main challenge is the front side metallization of the solar cell. On the one hand, the mean finger width w_f of the grid should be as low as possible to minimize shading losses. Fig. 1 illustrates the progress in reducing the screen printed finger width w_f since 2005 on the basis of published results.

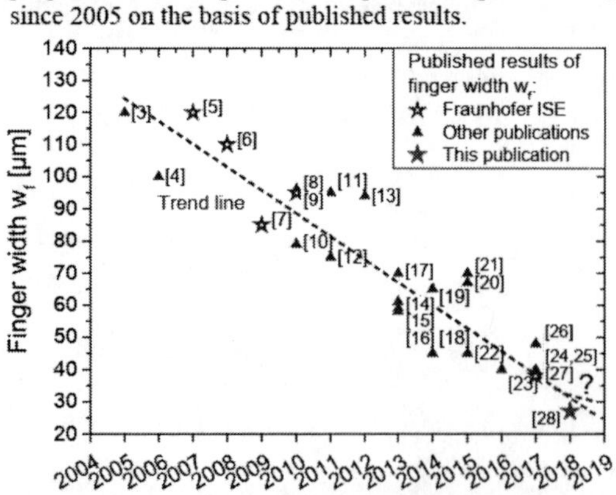

Figure 1: Learning curve of the flatbed screen printing process (2005-2017) regarding finger width w_f based on published results [3–28].

The steep learning curve is the result of intense industrial efforts to optimize paste, screen and machine technology. This impressive development underlines the ongoing potential of screen printing technology for the metallization of solar cells. Beside a narrow finger width w_f, the mean lateral finger resistance R_L or respectively the so-called Grid Resistance R_{grid} has to be sufficiently low to ensure a low series resistance contribution and in consequence a high Fill factor FF of the solar cell. Thus, the ideal contact finger geometry is characterized by a narrow mean finger width w_f and a large contact finger cross-section area A_f (provided that the specific contact resistance ρ_f of the sintered silver paste is sufficiently low). In other words, the electrical aspect ratio AR_{el} as introduced by Pospischil et al. [29] has to be maximized:

$$AR_{el} = \frac{A_{f,\emptyset}}{w_{f,\emptyset}^2} \quad (1)$$

Furthermore, significant variations of the finger height, width and cross-section area along the finger have to be minimized. These variations (commonly known as "mesh marks") are caused by the deformation of the printed finger profile by wire intersections within the screen mesh (Fig. 2) on the one hand and an insufficient sealing (known as "gasket seal effect") between the emulsion along the finger channels in the screen and the rough surface texture of the silicon wafer [30]. Both effects lead to a deformation of the finger profile.

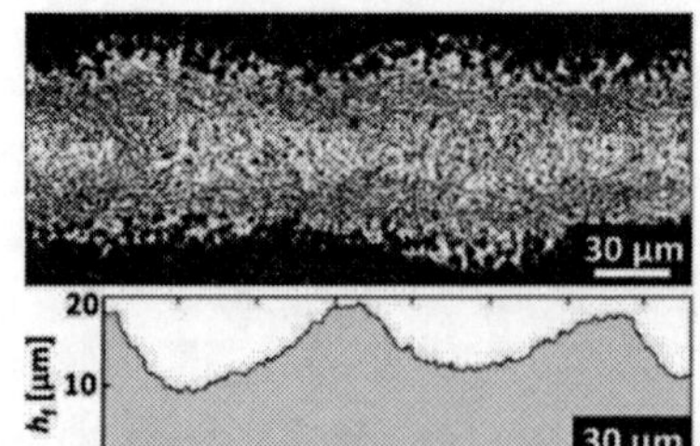

Figure 2: Microscopic image and height profile of a screen-printed contact finger with typical width and height variations ("mesh marks").

1.2 Knotless screens

In recent years, several approaches have been introduced to address the challenge to optimize the finger geometry with respect to uniformity and high AR_{el}. Examples of such approaches which have found their way to an industrial application in the past are stencil printing [31], parallel dispensing [32], screen printing of hotmelt pastes [33] and the application of a double printing sequence [34]. Recently, so-called knotless screens (also called knot-free or super screens) have gained increasing attention [35]. This type of screen is characterized by a 0° mesh angle (orientation of the mesh to the screen frame). Compared to conventional screen meshes with a 22.5° mesh angle, knotless screens have no wire intersections in the open channels for contact fingers. This can be realized by omitting selected threads within the production process of the mesh (Fig. 3). Compared to standard screens, knotless screens have an increased open area ratio of the mesh (open area in proportion to covered area by the wires) and should thus enable a higher amount of paste transfer within the finger channels. Furthermore, one can expect less mesh marks along the contact fingers and thus a better uniformity of the contact fingers. In consequence, a better lateral conductivity of the printed and fired contact fingers as well as a reduction of paste consumption due to the reduction of finger height and width variations can be expected.

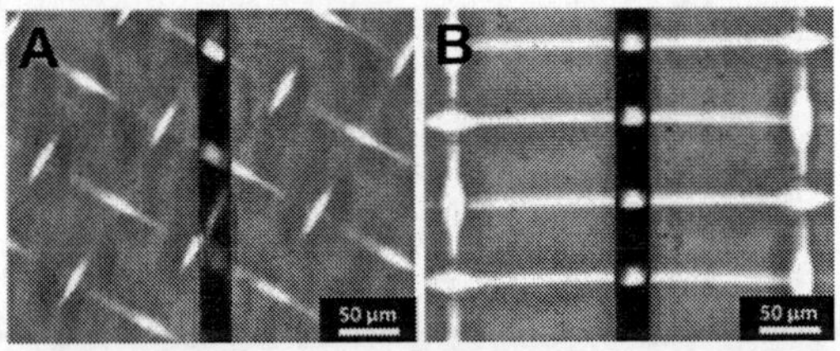

Figure 3: Microscopic images of contact finger channels within the standard screen with mesh angle 22.5° (A) and a Frintrup knotless screen with mesh angle 0° (B). The knotless screen is fabricated by omitting selected threads in the intended finger opening areas during the production process. As a consequence, wire intersections within the finger openings can be avoided.

1.3 Boosting throughput with rotary screen printing

Rotary screen printing (RSP) – a closely related technology to flatbed screen printing – has the potential to combine the technological advantages of thick film metallization with the potential high throughput of rotary printing methods. To date, this technology is primarily used on web-based materials (i.e. label or textile printing) with a printing speed of up to 100 m/min. First attempts to use this technology for solar cell metallization date back to the year 1999 [36]. In the last years, this promising technology has been intensively investigated [37–39] and gained increasing attention within the metallization community [35,40].

Similar to flatbed screen printing, a woven screen mesh covered with a partly open emulsion layer is used as printing form. However, in contrary to the flat screen used in FSP, cylinder-shaped screens are used for RSP. The meshes of such screens can consist of up to 400 threads per inch (mesh count). However, due to stability reasons of the cylinder screen, thread thickness is increased (minimum 25 to 30 µm) compared to flat screens. FSP requires two printing sequences. Firstly, the open areas of the screen are filled using a metal flood bar. Secondly, the paste is pressed through the openings of the flat screen using a flexible squeegee. Using RSP, the paste is constantly pressed through the screen openings by a fixed squeegee within the rotating screen cylinder (Fig. 4). A pre-filling of the screen is thus not required. RSP requires a lower paste viscosity compared to FSP pastes to ensure a good paste transfer through the screen openings. The ability to transfer thick film metallization patterns makes RSP interesting for both – front and rear side metallization of c-Si solar cells.

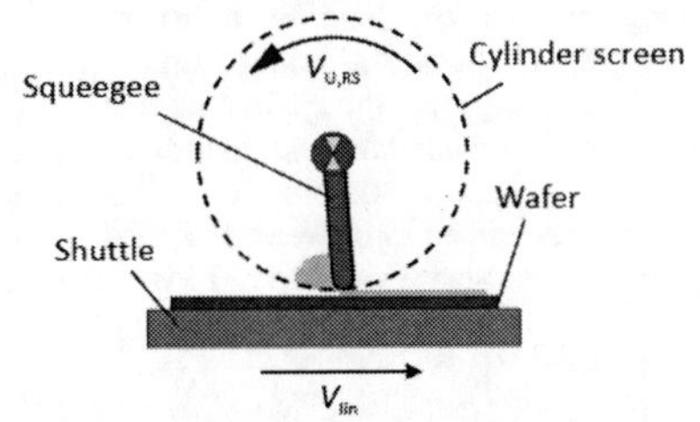

Figure 4: Schematic drawing of a rotary screen printing unit for the metallization of c-Si solar cells. The paste is pressed through the openings of the rotating cylinder screen onto the wafer substrate. The wafer can be transported i.e. by a shuttle system through the printing unit.

Within this work, we compare the results of a state-of-the-art fine line screen and a knotless screen with respect to finger geometry, electric results of the front side metallization and silver consumption. Furthermore, we present recent achievements in the field of rotary screen printing and give an outlook to further research activities.

2 EXPERIMENTAL SETUP

2.1 Outline and parameters of the experiment

Fig. 5 shows the process plan of the experiment. We use state-of-the-art industrially pre-fabricated p-type Cz-Si PERC solar cells (precursors) with SiN_x anti reflection coating and rear-side LCO without metallization on the front and rear side.

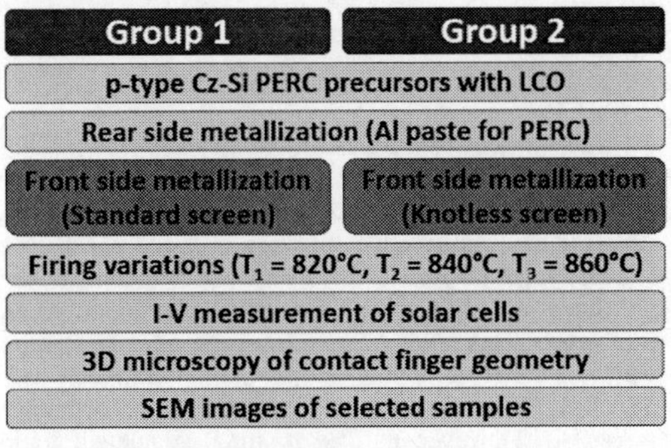

Figure 5: Experimental setup

The front and rear side metallization is carried out on an industrial ASYS screen printing line with Ekra XH-2 screen printing units using a newly developed Heraeus Ag paste and a commercially available Al paste for PERC solar cells. Two experimental groups are metallized using two types of screens. Group 1 is metallized on the front side using a state-of-the-art reference screen A

(380 mesh, mesh angle: 22.5°). Screen A has a grid layout with 5 segmented busbars and 120 contact fingers (nominal finger width w_n = 30 μm, measured finger channel width w_{ch} = 31 μm). Screen B is a knotless screen fabricated by Hans Frintrup GmbH. This screen has a 400 mesh and a grid layout with 5 segmented busbars and 111 contact fingers (nominal finger width w_n = 25 μm, measured finger channel width w_{ch} = 28 μm). For organizational reasons, the number of contact fingers within the grid layout of both screens is different. This leads to different results with respect to j_{sc}, V_{oc} and FF of the fabricated solar cells and has to be considered when comparing the I-V-results of both groups.

During the printing step, the applied mass of Ag paste for the front side metallization is measured inline using precision inline scales. 30 solar cells of each group are metallized, cured in an inline drier and subsequently fired using an industrial fast firing oven (FFO) with 3 peak set temperatures á 10 cells per temperature (T_1 = 820°C, T_2 = 840°C, T_3 = 860°C).

2.2 Characterization of the experimental results

I-V-results of the solar cells are measured using an industrial HALM cell tester with standard conditions. The measurement data includes the total busbar-to-busbar resistance (Grid resistance) $R_{BB\text{-}BB}$ from which the mean lateral finger resistance $R_{L\varnothing}$ can be calculated with known number of contact fingers n_f and finger segment length l_f:

$$R_{L\varnothing} = \frac{n_f \cdot R_{BB-BB}}{l_f} \quad \text{for } r_{BB} \ll r_f \qquad (2)$$

To analyze the three-dimensional geometry of the contact fingers with an adequate statistic, 25-30 microscopic measurements per group are carried out (5 measuring positions on 5 cells per group) using an Olympus LEXT confocal laser-scanning microscope (CLSM) with 500x magnification. The finger geometry (mean finger width $w_{f\varnothing}$, mean finger height $h_{f\varnothing}$ and mean finger cross-section area $A_{f\varnothing}$) are analyzed using an image analyze algorithm (FineUp) developed at Fraunhofer ISE [39]. This algorithm allows a reproducible and statistically sufficient analysis of the mean finger geometry.

3 RESULTS AND DISCUSSION

3.1 Thee-dimensional contact finger geometry

Group 1 using the reference screen A could be printed using typical printing parameters. However, realizing the front side metallization of group 2 using the knotless screen B turned out to be more challenging. The printing quality reacted sensitively to the printing parameters. This might be possibly caused by the smaller width of the finger channels and another emulsion with different surface properties compared to the reference screen A. However, optimal printing parameters could be identified to print the front side metallization with good quality.

Table 1 shows the statistically analyzed results of the three-dimensional finger geometry for both groups.

Table I: Finger geometry (mean values) of group 1 (screen A) and 2 (screen B).

Group	Finger geometry parameter Ø Width $w_{f\varnothing}$ [μm]	Ø Height $h_{f\varnothing}$ [μm]	Ø Cross-section area $A_{f\varnothing}$ [μm²]
1	30 (σ = 1.2)	16 (σ = 1.0)	324 (σ = 22)
2	27 (σ = 1.2)	17 (σ = 0.9)	320 (σ = 18)

The results show a slightly smaller mean finger width of group 2 (screen B) compared to group 1 (screen A) while the mean finger height h_f and finger cross section area A_f of both groups are closely comparable. The smaller mean finger width of group 2 can be most likely traced back to the narrower nominal finger width w_n (respectively finger channel width w_{ch}) of screen B and is thus no clear indication for a benefit of the knotless screen. However, if one calculates the electric aspect ratio AR_{el} based on equation (1), contact fingers of group 1 obtained a smaller $AR_{el,1}$ = 0.36 compared to group 2 with $AR_{el,2}$ = 0.44. This can be demonstrated by a superimposed illustration of an averaged finger profile of selected, representative contact fingers from group 1 and 2 (Fig. 6). The results indicate a benefit for the knotless screen B with respect to the printed finger geometry (aspect ratio). In other words it is possible to print slightly narrower but higher contact fingers with the knotless screen B compared to the standard screen A. The uniformity of the fingers of both groups seems to be comparable according to the measurement data (standard deviation σ of w_f and h_f).

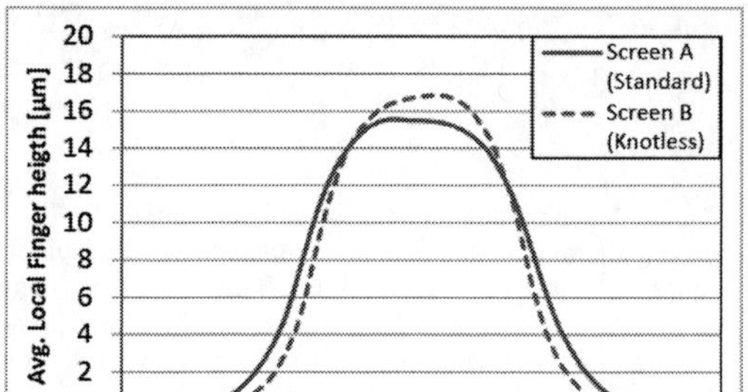

Figure 6: Superimposed mean cross-section profiles of selected, representative contact fingers printed with standard screen A (group 1) and knotless screen B (group 2). An advantage of the knotless screen with respect to the cross-section area A_f, finger aspect ratio and steepness of the flanks is visible.

3.3 Silver consumption

The inline weighing of the applied front side paste revealed that an average mass of m_1 = (88 ± 1.8) mg (group 1) and m_2 = (69 ± 1.8) mg (group 2) wet paste has been applied on the solar cells. However, the total mass of wet paste consumption cannot be directly compared due to the different number of contact fingers in the layout. Yet, if one substracts the known mass of silver paste applied for the busbars (m_{BB} = 15 mg) it is possible to calculate the mean paste consumption per finger with the known number of contact fingers in the grid. This calculation gives a mean mass of $m_{f,1}$ = 0.61 mg per finger for group 1 and $m_{f,2}$ = 0.49 mg per finger for group 2. In consequence, this is an effective wet paste reduction of approx. 20 % of the knotless screen compared to the standard screen. This confirms the ability of knotless screens to reduce the silver consumption und thus generate an economic benefit.

3.2 SEM-Analysis of finger geometry

Fig. 7 shows SEM images of selected, representative contact fingers of both groups. The SEM images also underline the more optimal shape of the contact fingers printed with the knotless screen B. A higher aspect ratio and better uniformity along the finger is visible.

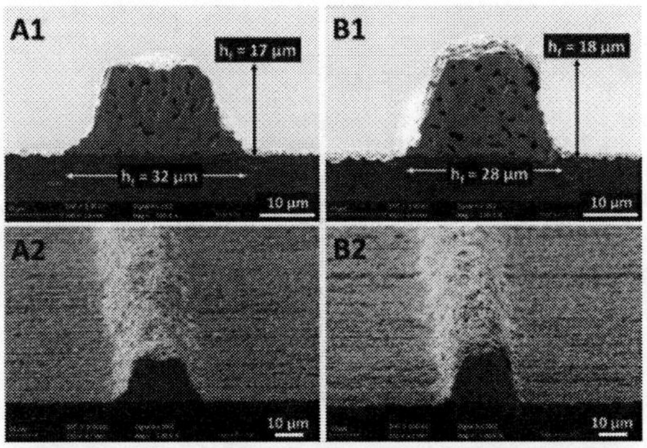

Figure 7: SEM images of selected, representative contact fingers of group 1 (A1 and A2) and group 2 (B1 and B2). This improved finger geometry (higher aspect ratio AR_{el}) and a better uniformity (less mesh marks) of the finger printed width the knotless screen B (group 2) is clearly visible.

3.4 Lateral resistance of the front side grid

The mean lateral finger resistance $R_{L\varnothing}$ of the contact fingers is calculated from the I-V-date with the known number of contact fingers n_f, the segment length of the contact fingers between two busbars l_f and the average busbar-to-busbar resistance $R_{BB\text{-}BB}$ using equation (2). For group 1 (screen A), a mean lateral finger resistance of $R_{L\varnothing}$ = 0.88 Ω/cm (87.9 Ω/m) is calculated. This result corresponds with typical lateral finger resistance values of contact fingers with similar geometry. Group 2 (screen B) obtained a mean lateral finger resistance of $R_{L\varnothing}$ = 1.13 Ω/cm (112.6 Ω/m) and thus a slightly higher lateral finger resistance. This can be explained by the smaller mean finger cross-section area $A_{f\varnothing}$ of group 2 which mainly results from the narrower nominal finger width w_n respectively the actual finger channel width w_{ch} in the screen. In a second step, the layout of the front side grid (number of contact fingers) can be optimized with known average R_L in order to realize an optimal tradeoff between j_{sc} (shading losses) and FF (series resistance losses) and thus maximize the conversion efficiency η.

3.5 I-V results of the solar cells

A direct comparison of the I-V results of both groups is due to the different number of contact fingers within the grid layout of both screens not reasonable. The difference regarding the number and nominal finger width w_n of the contact fingers has a significant impact on j_{sc} (shading) and FF (series resistance) and thus prevents a direct comparison of the resulting conversion efficiency η. However, I-V results of both groups are stated for the sake of completeness (table 2 and Fig. 8). In summary, the I-V results of both groups are promising and slightly better than the predicted conversion efficiency of industrial p-type Cz-Si PERC solar cells for 2018 [2]. Group 2 achieved a mean conversion efficiency of $\eta_\varnothing$ = 21.6 % (best individual cell η_{max} = 21.8 %).

Table II: I-V results of the best solar cells of both groups (please note: a direct comparison is not reasonable due to a different number of contact fingers).

I-V Parameter	Unit	Group 1 (Best cell)	Group 2 (Best cell)
j_{sc}	[mA/cm²]	39.6	40.0
V_{oc}	[mV]	668.2	669.0
FF	[%]	81.8	81.3
η	[%]	21.7	21.8

j_{sc} = Short-circuit current density, V_{oc} = Open circuit voltage, FF = Fill factor, η = Conversion efficiency

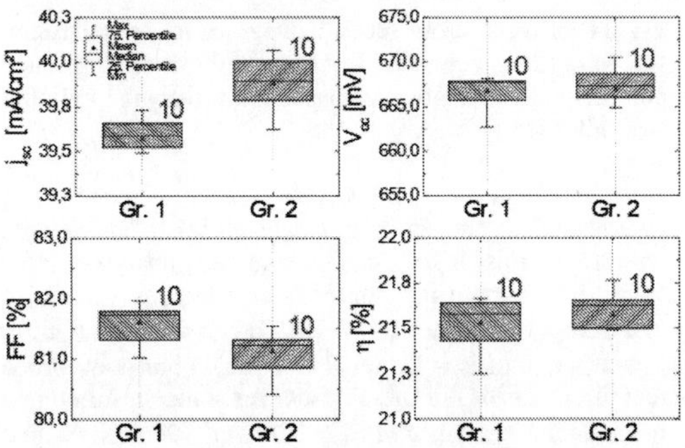

Figure 8: Comparison of the I-V-results of both groups.

4 CURRENT RESULTS OF ROTARY SCREEN PRINTING

The general feasibility of a high throughput front and rear side metallization using rotary screen printing (RSP) has been recently demonstrated in several publications [37,38,40]. While it has been shown that the rear side metallization of Al BSF (aluminium back surface field) solar cells can be realized without FF and η losses compared to FSP metallization [38], front side metallization still represents a bigger challenge due to thicker wires of the cylinder screen mesh and a slightly lower viscosity and yield point of the paste.

Recently, very promising results have been achieved by applying RSP for the front side metallization of p-type Cz-Si PERC precursors (format 125 mm x 125 mm). The experimental setup is similar to previous test series [38] with the excpetion of the Ag paste. For the latest test, a state-of-the-art Ag paste for front side metallization has been slightly diluted with a sufficient thinner and used for the test. A fine line Gallus Screeny© cylinder screen with a mesh count of 400 wires/inch and a 4" grid layout (3 busbars with w_b = 1.0 mm, 85 fingers with w_n = 35 µm) has been fabricated and used for the experiment. The metallized precursors have been subsequently fired and the finger geometry has been analyzed by confocal microscopy and SEM.

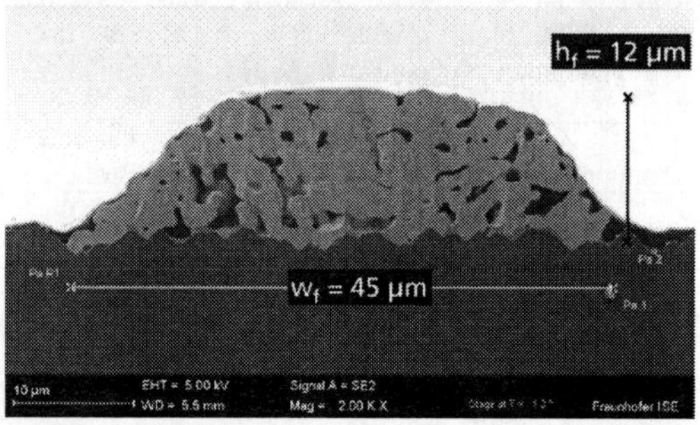

Figure 9: SEM cross section image of an actual rotary screen printed contact finger using slightly diluted commercially available Ag paste and a fine line cylinder screen for rotary screen printing.

Fig. 9 shows a SEM cross section image of a representative contact finger which has been printed within the latest RSP test. The statistical analysis of the finger geometry using confocal microscopy revealed a mean finger width of $w_f = 47$ µm ($\sigma = 2$ µm) and a mean finger height of $h_f = 13$ µm ($\sigma = 2$ µm). Four point probe measurements of the lateral finger resistance resulted in a mean lateral resistance of $R_L = 0.9$ Ω/cm (90 Ω/m) with a standard deviation of $\sigma = 0.07$ µm. In summary, RSP front side metallization has not yet reached the same finger quality as FSP. However, further optimization of screens and pastes will give a realistic chance to achieve closely similar results to the state-of-the-art FSP process. A promising path to achieve this goal might be the development of knotless cylinder screens for rotary screen printing.

5 CONCLUSION

Within this work, a fine line front side metallization using state-of-the-art industrial pastes, screen printing machinery and two types of fine line screens has been realized. Using a standard screen (380-0.014 mesh, mesh angle 22.5°, $w_n = 30$ µm), a mean finger width of $w_f = 30$ µm with an electric aspect ratio of $AR_{el,1} = 0.36$ could be obtained. Using a specially developed Frintrup knotless screen (400-0.016 mesh, mesh angle 0°, $w_n = 25$ µm), an even narrower mean finger width of $w_f = 27$ µm and a higher electric aspect ratio of $AR_{el,2} = 0.44$ could be achieved. Image analysis results of the contact finger geometry and SEM images confirmed that the knotless screen is able to realize narrower contact fingers with a comparable cross section area A_f and a better uniformity along the finger (less mesh marks). This underlines the potential of knotless screen technology to realize a more optimal finger geometry and thus increase the overall solar cell efficiency. The best group of p-type Cz-Si PERC solar cells (group 2) which has been metallized with the knotless screen obtained a mean conversion efficiency of $\eta = 21.6$ %. However, I-V results of both groups cannot be directly compared due the varying number of contact fingers and another nominal finger width w_n of the two screens.

Furthermore, we demonstrated actual research and development results in the field of high-throughput rotary screen printing metallization for Si solar cells. Using slightly diluted commercially available Ag paste and a fine line cylinder screen, a fine line front side metallization could be successfully printed on p-type Cz-Si PERC precursors. A mean finger width of $w_f = 47$ µm and a mean lateral finger resistance of $R_L = 0.9$ Ω/cm could be achieved. This results show the fast progress and the potential of this technology with respect to front side metallization.

6 ACKNOWLEDGEMENTS

This work was partly supported by the German Federal Ministry of Education and Research (BMBF) within the funding program Photonics Research Germany under the contract number 13N13512 (Rock-Star) and by the German Federal Ministry of Economic Affairs and Energy under the contract numbers 0325775E (AdmMo) and 0325823 (CUT-A).

7 REFERENCES

[1] Ralph EL, *Recent advancements in low cost solar cell processing. Proc. of the 11th PVSC*, pp. 315–316 (1975).
[2] ITRPV, *International Technology Roadmap for Photovoltaic (ITRPV): 2017 Results.* 9. Ed. (2018).
[3] Raabe B, Huster F, McCann M, Fath P, *High aspect ratio screen printed fingers. Proc. of the 20th EUPVSEC*, pp. 930–933 (2005).
[4] Hilali MM, Nakayashiki K, Khadilkar C, Reedy RC, Rohatgi A, Shaikh A, Kim S, Sridharan S, *Effect of Ag Particle Size in Thick-Film Ag Paste on the Electrical and Physical Properties of Screen Printed Contacts and Silicon Solar Cells. J. Electrochem. Soc.* 153, S. A5, doi: 10.1149/1.2126579 (2006).
[5] Mette A, *New concepts for front side metallization of industrial silicon solar cells.* PhD Thesis, Albert-Ludwigs-Universität, Freiburg (2007).
[6] Hörteis M, *Aerosol-Printed Silicon Solar Cell Exceeding 20% Efficiency.* Crystal Clear - Workshop on Metallization, Utrecht, 01.10.2008.
[7] Clement F, Menkoe M, Erath D, Retzlaff M, Goetz AK, Hoerteis M, Bartsch J, Hoenig R, Belledin U, Biro D, Preu R, *Industrially feasible mc-Si solar cells with fine line printed front contacts on high emitter sheet resistance towards 17% efficiency. Proc. of the 24th EUPVSEC*, pp. 2172–2176 (2009).
[8] Galiazzo M, Furin V, Tonini D, Cellere G, Baccini A, *Double Printing of Front Contact Ag in c-Si Solar Cells. Proc. of the 25th EUPVSEC*, pp. 2338–2340 (2010).
[9] Erath D, Filipović A, Retzlaff M, Goetz AK, Clement F, Biro D, Preu R, *Advanced screen printing technique for high definition front side metallization of crystalline silicon solar cells. Solar Energy Materials and Solar Cells* 94, pp. 57–61, doi: 10.1016/j.solmat.2009.05.018 (2010).
[10] Falcon T, *High Accuracy, High Aspect Ratio Metallization on Silicon Solar Cells Using a Print on Print Process. Proc. of the 25th EUPVSEC*, pp. 1651–1655 (2010).
[11] Magnone P, Rose R de, Zanuccoli M, Tonini D, Galiazzo M, Cellere G, Guo H-W, Frei M, Sangiorgi E, Fiegna C, *Understanding the impact of double screen-printing on silicon solar cells by 2-D numerical simulations. Proc. of the 37th IEEE Photovoltaic Specialists Conference*, pp. 2177–2180 (2011).

[12] Tavares R, Dobie A, Buzby D, Zhang W, *Optimal Screen Mesh, Emulsion Chemistry, and Emulsion Thickness for Fine-Line Front Metallization Pastes on Crystalline Silicon Solar Cells. Proc. of the 26th EUPVSEC*, pp. 2040–2043 (2011).

[13] Pesce M, Maugeri M, Marsili M, Zarcone M, Tonini D, Bottosso C, Galiazzo M, Tomasi A, *Technological and economic assessment of two-steps printing processes in a mc-Si solar cells production environment. Energy Procedia* 21, S. pp. 24–31, doi: 10.1016/j.egypro.2012.05.004 (2012).

[14] Hannebauer H, Dullweber T, Falcon T, Brendel R, *Fineline Printing Options for High Efficiencies and Low Ag Paste Consumption. Energy Procedia* 38, pp. 725–731, (2013).

[15] Hannebauer H, Dullweber T, Falcon T, Chen X, Brendel R, *Record Low Ag Paste Consumption of 67.7 mg with Dual Print. Energy Procedia* 43, pp. 66–71, doi: 10.1016/j.egypro.2013.11.089 (2013).

[16] Chang Y-H, Su W-M, Huang P-S, Cheng L-W, *Improvement of the solar cell efficiency by fine line print on print technology. 2013 Proc. of the 39th IEEE PVSC* pp. 2176–2178 (2013).

[17] Bottosso C, Tao W, Wang X, Ma L, Galiazzo M, *Reliable Metallization Process for Ultra Fine Line Printing. Energy Procedia* 43, S. pp. 80–85, doi: 10.1016/j.egypro.2013.11.091 (2013).

[18] Voltan A, Bortoletto E, Zamuner M, Martire M, Bertazzo M, Galiazzo M, *Double Printing Feasibility of 35um Printed Ag Finger Width. Proc. of the 29th EUPVSEC*, pp. 1387–1391 (2014).

[19] Thibert S, Jourdan J, Bechevet B, Chaussy D, Reverdy-Bruas N, Beneventi D, *Influence of silver paste rheology and screen parameters on the front side metallization of silicon solar cell. Materials Science in Semiconductor Processing* 27, pp. 790–799, doi: 10.1016/j.mssp.2014.08.023 (2014).

[20] Voltan A, Bortoletto E, Borsato O, Galiazzo M, *Toward 30μm Printed Ag Finger Width by Ultra Fine Line Double Printing. Proc. of the 31st EUPVSEC*, pp. 704–708 (2015).

[21] Shanmugam V, Wong J, Peters IM, Cunnusamy J, Zahn M, Zhou A, Yang R, Chen X, Aberle AG, Mueller T, *Analysis of Fine-Line Screen and Stencil-Printed Metal Contacts for Silicon Wafer Solar Cells. IEEE J. Photovoltaics*, pp. 1–9

[22] Galiazzo M, Voltan A, Bortoletto E, Zamuner M, Martire M, Borsato O, Bertazzo M, Tonini D, *Fine Line Double Printing and Advanced Process Control for Cell Manufacturing. Energy Procedia* 67, pp. 116–125, doi: 10.1016/j.egypro.2015.03.295 (2015).

[23] Galiazzo M, Borsato O, Bortoletto E, *Fine Line Double Printing + for Today and Tomorrow Cell Metallization and Interconnection. Proc. of the 32nd EUPVSEC*, pp. 958–960 (2016).

[24] Galiazzo M, *Status and Perspectives of Metallization Technologies for HVM.* 7th Metallization Workshop, Konstanz, 22.-23. Oct. 2017.

[25] Zhang Y, *Knotless screen printing for crystalline Silicon solar cells.* 7th Metallization Workshop, Constance, 22.-23. Oct 2017.

[26] Xu C, Fies M, Willenbacher N, *Impact of Wall Slip on Screen Printing of Front-Side Silver Pastes for Silicon Solar Cells. IEEE J. Photovoltaics* 7, pp. 129–135, doi: 10.1109/JPHOTOV.2016.2626147 (2017).

[27] Werner S, Lohmüller E, Saint-Cast P, Greulich J, Weber J, Schmidt S, Moldovan A, Brand A, Dannenberg T, Mack S, Wasmer S, Demant M, Linse M, Ackermann R, Wolf A, Preu R, *Key aspects for fabrication of p-type Cz-Si PERC solar cells exceeding 22% conversion efficiency. Proc. of 33rd EUPVSEC* (2017).

[28] Lorenz A, Linse M, Frintrup H, Jeitler M, Mette A, Lehner M, Greutmann R, Brocker H, König M, Erath D, Clement F, *Screen Printed Thick Film Metallization of Silicon Solar Cells - Developments and Future Perspectives.* [This paper - to be published]

[29] Pospischil M, Fellmeth T, Brand A, Nold S, Kuchler M, Klawitter M, Gentischer H, König M, Hörteis M, Wende L, Doll O, Zengerle R, Clement F, Biro D, *Optimizing Fine Line Dispensed Contact Grids. Energy Procedia* 55, S. pp. 693–701 (2014).

[30] Aoki M, Nakamura K, Tachibana T, Sumita I, Hayashi H, Asada H, Ohshita Y, *30 μm fine-line printing for solar cells. Proc. of the 39th IEEE PVSC.* IEEE, pp. 2162–2166 (2013).

[31] de Moor HHC, *Fine-Line Screen Printing For Silicon Solar Cells. 6th workshop on the Role of Impurities and Defects in Silicon Device Processing*, pp. 154–170 (1996).

[32] Pospischil M, Klawitter M, Kuchler M, Jahn M, Efinger R, Schwarz R, Wende L, König M, Clement F, Biro D, *High Speed Dispensing with Novel 6" Print Head. Energy Procedia* 98, pp. 61–65, doi: 10.1016/j.egypro.2016.10.081 (2016).

[33] Olaisen BR, Holt A, Marstein ES, Sauar E, Shaikh A, McVicker K, Salami J, Miranda H, Kim SS, *Hot-melt screen-printing of front contacts on crystalline silicon solar cells. Conference Record of the Thirty-first IEEE Photovoltaic Specialists Conference, 2005.* IEEE, pp. 1084–1087 (2005).

[34] Hoornstra J, de Moor HHC, Eeeber A, Wyers P, *Improved front side metallization on silicon solar cells with stencil printing. Proc. of the 16th EUPVSEC* (2000).

[35] Chunduri SK, Schmela M, *Market Survey Screen Printers 2018* (2018).

[36] Hahne P, *Innovative Druck- und Metallisierungsverfahren für die Solarzellentechnologie.* PhD Thesis, Fernuniversität, Hagen (2000).

[37] Lorenz A, Gredy C, Lehner M, Greutmann R, Brocker H, Rohde J, Senne A, Strauch T, Reinecke H, Clement F, *Progress with Rotational Printing for the Front Side Metallization of Silicon Solar Cells. Proc. of the 32nd EUPVSEC*, pp. 413–419 (2016).

[38] Lorenz A, Münzer A, Lehner M, Greutmann R, Brocker H, Reinecke H, Clement F, *High-throughput front and rear side metallization of silicon solar cells using rotary screen printing. Energy Procedia* 124, S. pp. 680–690, doi: 10.1016/j.egypro.2017.09.343 (2017).

[39] Strauch T, Demant M, Lorenz A, Haunschild J, Rein S, *Two Image Processing Tools to Analyse Alkaline Texture and Finger Geometry in Microscope Images. Proc. of the 29th EUPVSEC* (2014).

[40] Lorenz A, Münzer A, Lehner M, Senne A, Greutmann R, Brocker H, Ott T, Hage F, Clement F, *Towards a high-throughput metallization for silicon solar cells using rotary-printing methods. Photovoltaics International* 37, pp. 48–55 (2017).

RESULTS OF THE PROJECT "ADMMO" – CELL AND MODULE DEVELOPMENT TOWARDS A 318 W MODULE

J. Nekarda[1], A. Brand[1], M. Linse[1], F. Clement[1], J. Schneider[2], M. Turek[2], S. Großer[2], S. Schönfelder[2], T. Urban[3], M. Ehrl[3], M. Müller[3], J. Heitmann[3], C. Yüce[4], N. Willenbacher[4], T. Große[5], R. Böhme[6], M. König[7], H. Frintrup[8], P. Fuß-Kailuweit[9], F. Kersten[10], A. Mette[10], J. Müller[10], R. Preu[1]
[1]Fraunhofer Institute for Solar Energy Systems ISE, Heidenhofstraße 2, 79110 Freiburg, Germany
[2]Fraunhofer-Center for Silicon Photovoltaics, Otto-Eissfeldt-Strasse 12, 06120 Halle (Saale), Germany
[3]TU Bergakademie Freiberg, Leipziger Str. 23, 09599 Freiberg/Sachsen, Germany
[4]Karlsruher Institute for Technology (KIT), Gotthard-Franz-Straße 3, 76131 Karlsruhe, Germany
[5]Meyer Burger Germany AG, An der Baumschule 6-8, 09337 Hohenstein-Ernstthal, Germany
[6]InnoLas Solutions GmbH, Pionierstraße 6, 82152 Krailling, Germany
[7]Heraeus Deutschland GmbH & Co. KG, Heraeusstr. 12-14, 63450 Hanau, Germany
[8]Hans Frintrup GmbH, Röhfeldstr. 20, 53227 Bonn, Germany
[9]WAVELABS Solar Metrology Systems GmbH, Markranstädter Str. 1, 04229 Leipzig, Germany
[10]Hanwha Q CELLS GmbH, Sonnenallee 17 – 21, 06766 Bitterfeld-Wolfen OT Thalheim, Germany
Telephone: +49 761 4588 5563, Fax: +49 761 4588 7621, e-mail: jan.nekarda@ise.fraunhofer.de

ABSTRACT:

In December 2014, the publicly funded project "AdmMo: Advanced multi-crystalline Modules" initiated by Hanwha Q-Cells started with the partners Heraeus, Frintrup, Innolas-Solutions, Meyer Burger Germany, Wavelabs, Karlsruhe Institute of Technology, Bergakademie Freiberg, Fraunhofer CSP and Fraunhofer ISE. The content of the project, which ended on April 30th, 2018, is a comprehensive and consistent further development of PERC cell and module technology. The overall project objective is a reduction of the "Levelized Cost of Electricity" (LCoE) to < 6 € ct / kWh in relation to the southern German region. For this purpose, a substantial cost reduction with simultaneous increase in performance of the cells and modules as well as a service life of the modules extended to 30 years was aspired. Our conference paper gives an overview of the most important results. Among other things, Cz-PERC solar cells with efficiencies > 22 % and based on this a module of 120 half-cells with a power output of 318 watts is demonstrated.

Keywords: PERC, half-cell module, metallization, passivation, laser processing, characterization

1 INTRODUCTION

The main objective of the consortium is the development of a high-profit quality solar module based on PERC cells. The aim was to achieve levelized cost of electricity (LCOE) of 6 cents / kWh in southern Germany. The three most important factors influencing the LCOE were addressed. A) The cost reduction in cell production. B) An increase in efficiency of cells and modules and C) an increased yield.

To reduce costs during cell production, the silver consumption on the front side was reduced to < 60 mg / cell. Furthermore, the investment costs for passivation and laser systems was decreased through substantial increases in throughput, and new solar simulators based on LED technology for cell and module measurement were developed.

To increase the efficiency of the solar cells, special silver pastes were developed in combination with innovative, knotless printing screens. Furthermore, the front side passivation was improved by the deposition of an intermediate oxide. On the back side, the local contact structures and fabrication processes were further optimized. The module efficiency was increased by using half-cells. The yield of the modules was increased by analyzing and reducing various electrical and charge carrier lifetime degradation mechanisms.

2 EXPERIMENT&RESULTS

Metallization:

In the metallization work package, the paste manufacturer Heraeus, the printing screen manufacturer Frintrup, as well as the Karlsruhe Institute of Technology in the field of rheology and the Fraunhofer ISE worked together to optimize the front-side contacting.

Heraeus investigated various glass frits and paste systems with regard to their contacting properties for lowly and very lowly doped emitters based on different paste systems. New glass frits and constantly modified formulations led to contact resistances of < 0.6 mΩ*cm^2 on emitters with a surface doping concentration of 8×10^{20} cm^{-3} at sheet resistances of > 100 Ω/sq.

Likewise, the glass frits were modified with the goal of achieving a higher open circuit voltage after opening the passivation layer. This resulted in the paste SOL 9641B, which demonstrated on average 2.6 mV higher open-circuit voltage compared to previous pastes.

The structures printed with the new pastes show an electrically line resistivity of down to 1.6×10^{-8} Ω*cm depending on the Fast Firing Furnace (FFO) process, which is only marginally lower than that of bulk silver. Thus, the contribution of series resistance resulting from the finger shape inhomogeneity is more significant.

Homogeneous and at the same time very narrow fingers should be made possible by the development of a new generation of printing screens in combination with an Ag paste adapted in their flow properties. As a printing screen, a so-called "knotless" screen was developed in order to avoid wire intersections in the printing channels and therefore improve paste transfer and printed finger uniformity. The novel tissue also prevents scattered radiation, so that more precisely defined print-channels can be produced as a result of the exposure process. Using such screens in combination with an adapted printing process, finger widths of < 30 µm could be demonstrated for the first time (see figure 1 right).

Additionally such screens allowed for continuously increasing aspect ratios of the printed fingers to the range of > 0.6 over the project time (see figure 2). Consequently the silver consumption could be decreased to < 60 mg/cell.

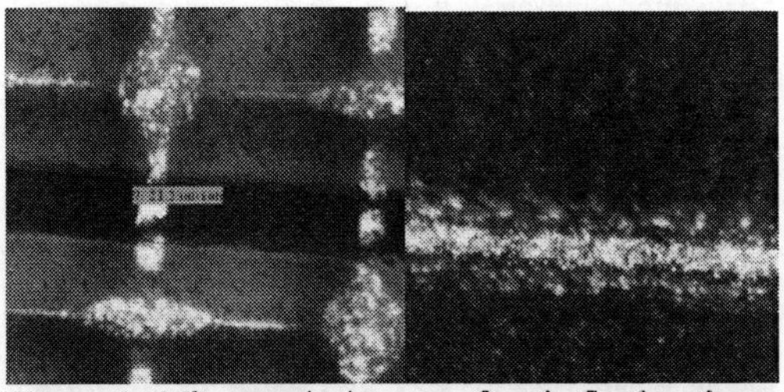

Figure 1: Microscopic images of antireflective tissue with finger opening parallel to the wires (left) and Ag-fingers printed with this tissue.

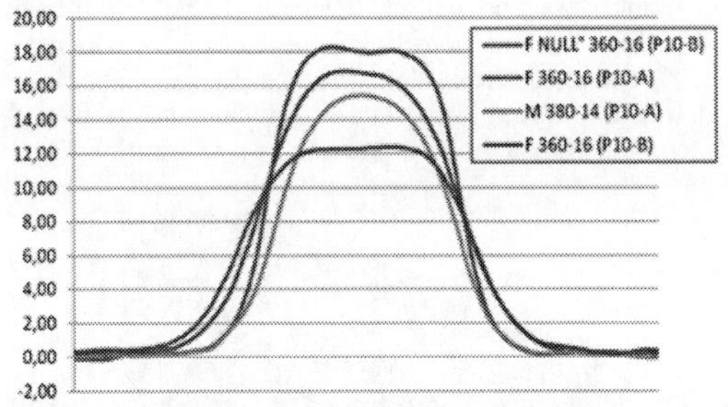

Figure 2: Average finger profiles of various screen / paste combinations (all screens 30 μm target finger width).

Passivation

The aim of the company Meyer Burger was a further development of its FABiA 4.1 platform for the reduction of investment and operating costs as well as an improvement of the passivation quality. The approach is based on combining the layers usually produced in separate systems for front side passivation by means of SiN_x and rear side passivation by means of an aluminum oxide (AlO_x) and silicon nitride (SiN_x) stack in one unit. The savings in the corresponding vacuum chambers and automation have a significant impact on the cost-of-ownership (CoO). This also further reduces yield losses and achieves a higher degree of automation, which additionally reduces the production costs. Figure 3 shows a graphic representation of the entire system and schematically its modular structure.

Core components of the FABiA 4.1 demonstrator system are the process modules for the deposition of AlO_x from trimethylaluminum (TMAL) and nitrous oxide (N_2O) as well as SiN_x from silane (SiH_4) and ammonia (NH_3). In the course of the project, the preferred direction was applied to the coating of the solar cell - backside from above, since the SiN protective layer must be deposited on the AlO_x layer before a front side coating occurs in order to prevent negative influence on the electrical characteristics of the produced solar cells.

Another major concern for reducing costs is to reduce the layer thickness of the AlO_x layer on the PERC solar cell rear side. The TMAL used for this process step is one of the main cost factors on the operating cost side in the two-sided surface passivation. The usual 22 nm in continuous coating systems have already been clearly undercut by competing processes such as atomic layer deposition (ALD) with target layer thicknesses of 6-8 nm. However, as ALD methods have a lower conversion rate of the TMAL material, the TMAL savings amounted to a maximum of 50 %. In order to demonstrate an equivalent low consumption, an AlO_x layer of 8-12 nm had to be developed, which does not show any negative effects on solar cell efficiency. Following preliminary tests to optimize the process parameters with regard to TMAL consumption and homogeneity in the expected throughputs, the electrical characteristics of solar cells with different AlO_x layer thicknesses were compared. It could be clearly shown, that a sufficient decrease of TMAL consumption from 400 mg/min down to 250 mg/min is possible without any impact on the cell efficiency (see Figure 4).

Likewise, a special pretreatment for the pre-oxidation of the front side before SiN_x deposition was investigated. For this purpose pure N_2O plasmas for the solar cell front side were realized on additional plasma sources, whereby reactive oxygen is produced and by oxidation of the silicon surface a ~ 1-1.5 nm thick SiO_2 layer is formed. Due to the slow, targeted growth of open bonds, an improvement in the passivation properties of the cell front was suspected. Most of the experiments were able to confirm an additional improvement of approx. 0.1-0.15 % efficiency.

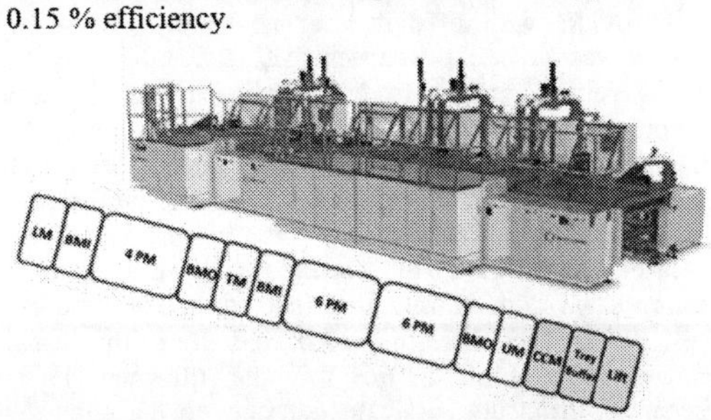

Figure 3: Graphic representation of a FABiA 4.1 for double-sided coating; Pass direction is from left to right

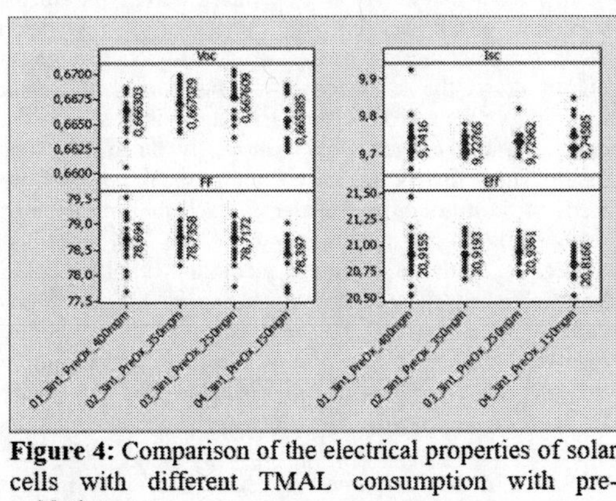

Figure 4: Comparison of the electrical properties of solar cells with different TMAL consumption with pre-oxidation.

Laser processing

There were two objectives regarding local back side contacts.

a) Optimization of the process chain consisting of laser contact opening (LCO), printing of aluminum paste and firing process
b) Reduction of the CoO during laser structuring by increasing the system throughput.

Optimization of the process chain:
The optimization of the backend process chain is extensive. Common process parameters that have a great influence on the quality of the backside contact range from the geometry of the laser aperture, diameter, points, lines, dashed lines and the classical laser parameters such as wavelength, pulse duration, fluence over the condition and the application of the paste to the peak-temperature and the speed of the firing process, to name but a few. In order to be able to evaluate the large parameter space with a manageable sample size, a method based on photoluminescence measurements and solar cell precursors with 33 test fields per cell was developed for the variation of laser parameters. The front of the samples is not metallized, but is otherwise made identically to a PERC solar cell. The passivation on the back is also like a real solar cell. At this "cell precursor", the minority carrier lifetime is measured by means of calibrated photoluminescence with spatial resolution, and the supposed open-circuit voltage (implied V_{oc}) is calculated therefrom (see Figure 5, left). Subsequently, 33 test fields are provided with different laser parameters and contact hole coverage levels. Between these fields 16 reference fields remain unprocessed. Then the reverse side is completely printed with paste and fired, and then the charge carrier lifetime is determined again. Based on the measured values in the reference fields, the influence of the processes on the unopened areas can be determined (see Figure 5, right). This "background" can be spatially interpolated from the reference fields on the intermediate test fields and there be subtracted from the actual measurement result. In this way the influence can be separated from the passivated surfaces to the contacted ones, and each one analyzed separately. Furthermore, the mere consideration of the contacted surfaces and the "before-after" measurement of the recombination current $J_{0,met}$ caused by the contacts can be determined exactly.
With this method a total of 2085 different parameter variations in the backend with a total of 6811 test fields were examined and compared in the project. The graph below shows the influence of the belt speed and peak temperature on a fixed laser parameter and paste on the dark saturation current density caused by the contact. The results show that in this case a higher peak temperature leads to the formation of better contact, the influence of firing time as a function of the chain belt speed is negligible. At the same time, a methodic technique could be established and verified by means of which the so-called "silicon inflow" can be determined in a spatially resolved manner in the porous electrode by a simple imaging of the finished cell backside and an image recognition software. The measured values of the "Silicon Inflow" correlated very precisely with the thickness of the aluminum back surface fields formed at this point, which was measured by means of several dozens of raster electron microscope measurements on cross sections [1].
Furthermore, the project developed a new method for the exact determination of the electrical resistances of the contacts and thus enabled a holistic characterization of the different formed contact structures by means of which the backend process chain could be successfully optimized towards the end of the project [2].

Reduction of CoO:
In addition to process optimization, Innolas-Solutions has continuously improved its rear surface structuring system platform and added other components (see Fig. 7), which now achieve a throughput of 5400 wafers/hour.

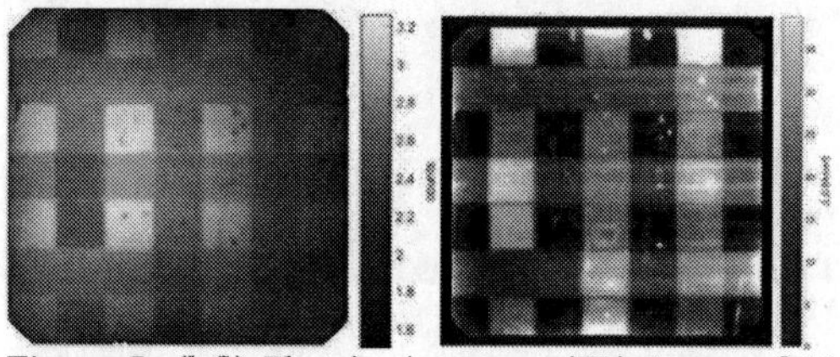

Figure 5: (left) Photoluminescence (PL) Image of 33 rectangular LCO test structures and 16 blank reference fields on the back of a semi-finished PERC solar cell without silver contacts on the front. (right) Additional saturation current density of the LCO test structures ($\Delta j0$) calculated from "left". The empty reference fields define the wafer background and have no additional recombination activity to themselves by definition.

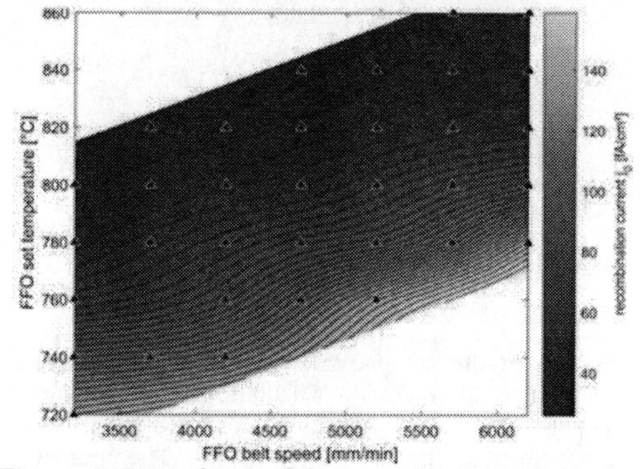

Figure 6: Dependence of the saturation current density j_0, which additionally results from the LCO contacts, as a function of the FFO belt speed and the set peak temperature.

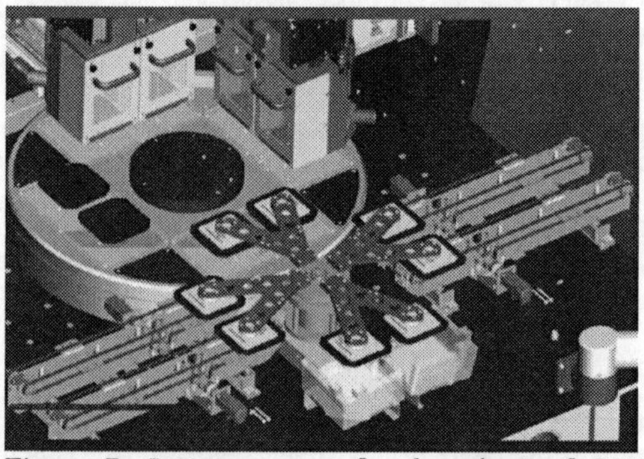

Figure 7: System concept for throughput of 5400 wafer/hour.

Module assembly
The interconnection of solar cells continues to offer great potential for optimization. The project aimed for a module with sub-cells. For this purpose, the influence of cell size, cell cutting processes and connector ribbon

widths was investigated at Fraunhofer CSP. Figure 8 shows the efficiency for full, half, and 1/3 cell modules with narrow (0.8 mm) and wide (1.5 mm) connectors. The fill factor is greater for each cell size due to the lower series resistance for the wide connectors and increases from 75.5 % for full cells to 81.5 % for one-third cells. In terms of efficiency, however, the wide connectors are better only for the whole-cell modules. With half and one-third cells, narrower connectors achieve the higher values. As expected, the smaller currents shift the optimum between electrical and optical losses towards narrower ribbon widths and less shading [3]. The efficiency increases from 18.4 % for whole cells with wide connectors to 19.8 % for third cells with narrow connectors. An increase of 1.4 $\%_{abs}$. or 7.6 $\%_{rel}$.

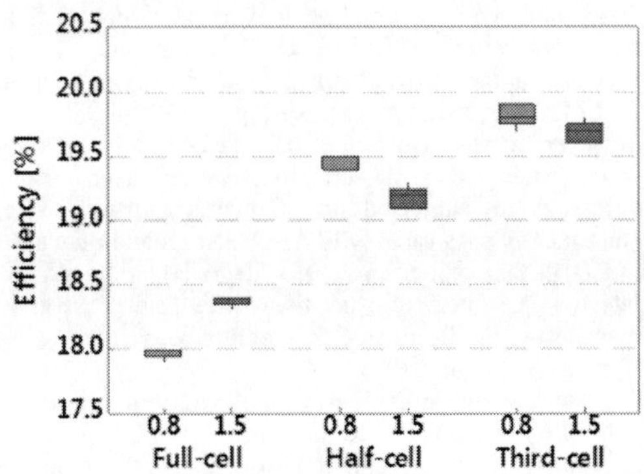

Figure 8: Cell efficiency for full, half, and 1/3 cell modules with 0.8 mm narrow (orange) and 1.5 mm wide (green) connectors.

Figure 9: Half-cell module of the best solar cells of a 650-cell final batch with the most relevant technology improvements of the project partners from the project AdmMo.

Characterization

As part of the project, the company Wavelabs developed a solar simulator based on LED lighting technology, completely assembled, installed and tested at the Fraunhofer CSP. The specifications are:

• Spectral range 350 -1100 nm, intensity and spectrum adjustable

• Illuminance of 1.2 suns (equivalent to 946 W/m² in the target spectrum)

• Illuminated area 2 m × 1 m

• Exposure time at least 250 ms

• homogeneity better than 2 %

• Stability according to IEC 60904-9

The device was designed to be modular in order to ensure optimal maintainability and fault tolerance. The main components of the module tester are used up to 72 individual LED light sources (LED boxes), which are directed in a special arrangement on the module (see Figure 10). Each light source contains a complete set of LED channels, so a single box is capable of displaying the full target spectrum with greatly reduced intensity

In the course of the project, it has been shown that the radiated light output of 60 LED boxes is sufficient to achieve the desired spectral performance. The boxes are controlled simultaneously and the radiated light is measured continuously in the module level with a spectrometer. The measured spectrum is then used to control the performance of individual LED channels with fast control electronics to keep the target spectrum constant during the entire exposure time.

The complete unit was successfully put into operation in december 2017.

Figure 11 shows a comparison of the desired AM1.5g spectrum and the superimposed spectrum of the simulator achieved after adaptation of the individual LED segments, and below this the percentage deviation within the individual spectral ranges.

Figure 10: LED light units of the LED sun simulator.

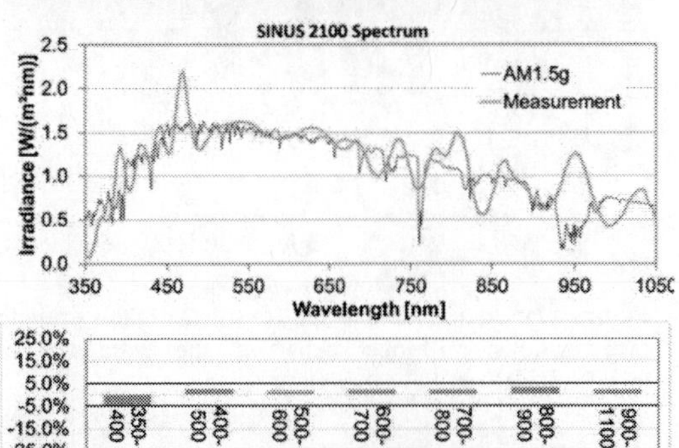

Figure 11: Comparison of the LED spectrum of the SINUS2100p with the AM1.5g reference spectrum. The bar graph below shows the wavelength intervals specified in IEC60904-9 and the summed deviation of the simulator spectrum from the reference spectrum.

Yield: Mechanical degradation

In order to increase the yield of the modules an attempt was made to increase their lifetime to > 30 years and to reduce the effects of known degradation mechanisms. At Fraunhofer CSP, for example, the influence of different backside metallizations on the mechanical stability of the cells was investigated.

For this purpose, solar cells with different backside metallization were produced and tested in the 4-point bending test, then the breaking stress was determined with the help of a finite element model and finally evaluated statistically using the Weibull distribution. The influence of three different pad pastes, different laser processes as well as an un-lasered reference from textured wafers with and without metallization and a foil metallization was investigated [4].

Of the three different pad pastes, Paste 1 and 3 showed a significant reduction in strength of approximately 25 % and Paste 2 a significant reduction of 41 % from the reference.

Compared to reference it could be shown that all lasers lead to damage. This reduces the characteristic fracture stress between 50 % and 70 % and increases the Weibull modulus by 42 % to 119 %.

By contrast, the used FolMet cells showed no change from the reference and are preferable to the classic screen printing PERC cells in terms of strength. The results are shown in the following figure with the following symbol description:

black diamond: textured non metallized reference;

Red dots: LCO Laser with Al paste;

Blue squares: FolMet;

Black squares: Laser Fired contacts with metallization;

Brown circles: Laser without metallization

Accordingly, the influence of the laser process has a subordinate importance for the stability of the cell. Primarily responsible for cell failure were defects associated with different metallization processes.

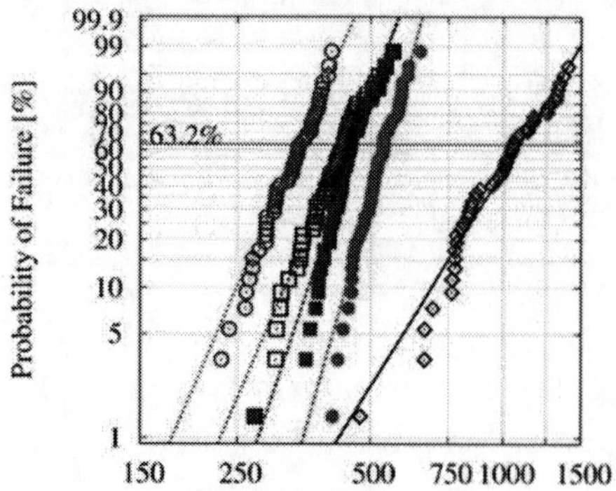

Figure 12: Weibull parameters: Results from the strength investigations on solar cells on the influence of metallization (symbol description see text).

Yield: electrical degradation

At the University of Freiberg, it was investigated how adaptations in the cell production process can reduce the effect of "Light and Elevated Temperature Induced Degradation" (LeTID).

Storage of mc-PERC cells at temperatures of 50-100 °C for > 48 h showed a V_{oc} reduction of up to 10 % [5, 6, 7]. These conditions must be considered relevant for real operation, which requires an analysis of the cause of the degradation and its correlations.

By comparative studies on lifetime samples, a fundamental characterization of the LeTID effect was first performed. By examining wafers with different passivation layer, it could be shown that Al_2O_3 passivation, which was integrated into the manufacturing process with the PERC process, is not solely responsible for the decrease in carrier lifetime (τ_{eff}). Even a pure Si_xN_y layer shows a comparable behavior with degradation at 75 °C and an irradiation power of 300 W/m² [8].

Comparative capacity voltage (CV) measurements were taken before and after storage at 75 °C & 300 W/m² to determine the density of local oxide charges of the passivation layers. It could be demonstrated that the field effect passivation of the layers remains, even if the effective charge carrier lifetime significantly decreases in the LeTID-sensitive process, see Figure 13 [9].

However, it was also found that the LeTID effect only occurs when using dielectric passivation layers. Thus, wafers were subjected to a classic furnace step, subsequently passivated with Al_2O_3 / Si_xN_y and then aged under LeTID conditions. No degradation could be detected. An interaction between wafer and passivation layer during the firing step is therefore seen as the cause of the degradation [9].

The analysis of individual wafers, uniformly selected from the total height of a mc-Si crystal, showed a correlation of the degradation from the position in the crystal. So that interaction with impurities in the crystal is obvious. The concentration of interstitial oxygen O_i was determined on these wafers, but no correlation was found with the LeTID effect size [7].

For the characterization of solar cells during degradation, a semi-automatic measuring station with lighting and temperature control and continuous V_{oc} measurement was set up. This enabled the extensive analysis of various temperature & lighting variations as well as operating modes (V_{oc}/I_{sc}).

As shown in Figure 13, the project developed processes that suppress the LeTID effect and thus reduce the degradation to a level below 1.5 %. Test wafers with adapted process flow showed a stable behavior and thus provided evidence that the FABiA 4.1 platform of the project partner Meyer Burger Germany with adapted process flow is able to produce LeTID stable cells.

The "stabilized" samples showed a constant lifetime over the degradation process.

In parallel to LeTID effect, the reliability of the metallization was investigated in detail on single module level. Therefore, a special interconnection scheme was developed and evaluated. The 6-inch PERC cells are separated in smaller sub-cells, interconnected with two copper ribbons and then encapsulated. The 4-wire interconnection lies partially outside of the module and therefore the interconnection of the rear ribbons can be disconnected by soldering, see Figure 13 a). This enables the measurement of the resistance R_{rear} between both ribbons on cell rear side and the IV measurement of an encapsulated cell. These measurements are performed during thermal cycle (TC) treatment, following the IEC 61215 [10].

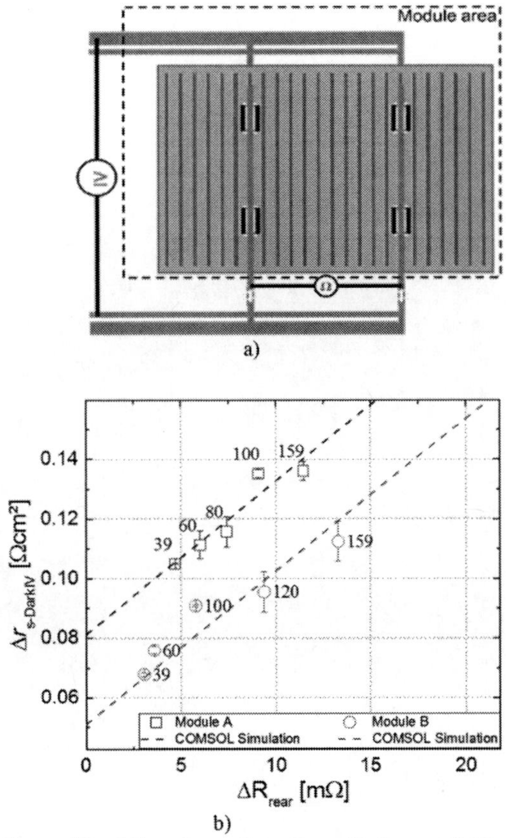

Figure 13: a) Sample configuration with the possibility to measure the resistance (R_{rear}) between rear side ribbons and the dark IV-curve during degradation and b) the dependency of the series resistance ($r_{s\text{-DarkIV}}$, extracted from dark IV-curve) and rear side resistance (each related to the initial value), labels denote the thermal cycle count.

The resistance R_{rear} measured between the ribbons is increasing during TC and correlates with the extracted series resistance ($r_{s\text{-DarkIV}}$) from the dark IV-curve, see Figure 13 b). Simulating the rear resistance increase with COMSOL [11] confirms its influence on $r_{s\text{-DarkIV}}$, considering the typical initial degradation offset. Thus the series resistance, and so efficiency, degradation is driven by the degradation of the rear resistance which is caused by Aluminum layer degradation during TC [12].

Cell & module processing

At the end of the project, the individually developed technologies were merged to produce several industrial solar cell modules. At Fraunhofer CSP several half-cell modules consisting of 120 half-cells and common modules made of 60 whole-cells as references were made. For this purpose, 650 PERC solar cells based on industrial 156×156 mm Cz wafers were first manufactured in the presence of an ISE employee at the leased pilot line of Hanwha Q-Cells. The Al_2O_3/SiN_x passivation was deposited on a Meyer Burger tool, the laser structuring on the developed Innolas platform. For the printing process, partner Frintrup's knotless printing screens in combination with a Heraeus SOL 9641BX Ag paste were selected and a target finger width of 30 µm was sought. It was printed on a half-cell adapted five busbar design with 110 fingers and solder pads on the rear side. The cells were then measured in the cell tester of the pilot line and sorted.

Subsequently, half of the cells on the CSP were singulated using a "scribe & break" laser process, and in each case three industrial modules were constructed and measured from 60 whole cells or 120 half-cells. The power increases on average from 311.7 W for the modules with full cells to 317.4 W for the modules with half-cells. This corresponds to a power increase of 5.7 W or 1.8 %. As expected, the fill factor increases due to lower series resistance losses. The current increases slightly due to narrower connecting ribbons and the larger number of gaps. The voltage decreases slightly due to damage to the cell edges.

The cell-to-module (CTM) power value indicates the percentage of the total cell output the module has. The full cells have an average power of 5.38W. The divided cells still have 2.66W, which corresponds to an efficiency of 22.01 % and 21.76 %. This reflects the losses due to the cell division process. Nevertheless the fillfactor increase leads to half-cell modules with an average of 317.4 W, compared to these with full cells and 311.7 W. This results in CTM values from the full cell to the full cell module of 96.6 % and from the full cell to the half-cell module of 98.4 %. If the lower power value of half-cells is used for the CTM value of the half-cell modules, then a CTM value of 99.5% is achieved.

3 SUMMARY

Within the AdmMo project, five industrial partners and four research institutes have been researching the development of PERC cell and module technology for over three years, addressing the key cost drivers of this new cell technology. Overall, the front side silver consumption could be reduced to less than 60 mg/cell and the finger width to less than 30 µm. In passivation and laser processing, significantly higher throughputs have been achieved through improved platform technology. A new LED-based solar simulator for cell and one for module measurement has been developed and constructed. The degradation of the cells was suppressed by an adapted overall process flow and average efficiencies > 22 % were achieved in the last cell batch. Half-cell modules were produced from these cells and an output of > 318 W was achieved. The half-cell modules show an average of 5.7 watts higher output power than conventional modules of whole cells from the same cell batch.

ACKNOWLEDGEMENTS

This work was conducted within the project “AdmMo” (FKZ: 0325775), supported by the German Ministry for Economic Affairs and Energy.

The project partners also want to thank Hanwha Q-Cells for initiating and planning the project, as well as for providing their pilot line and support in making our final cell batch.

REFERENCES

[1] A. Brand, Dissertation, to be published

[2] J. Hofmann, “Aufbau und Charakterisierung eines Messplatzes zur Untersuchung des Serienwiederstandes an PERC-Solarzellen“, Albert Ludwig Universität Freiburg, 2017.

[3] H. Hanifi, et al., "Optimized Tab Width in Half-cell Modules", Energy Procedia, Volume 92, 2016, Pages 52-59, ISSN 1876-6102, https://doi.org/10.1016/j.egypro.2016.07.009.
[4] Kaule, F., Meyer, S., Schoenfelder, S., "Benchmarking Mechanical Strength Data for New Solar Cell Concepts", EU PVSEC 2017 33rd European Photovoltaic Solar Energy Conference and Exhibition, Amsterdam (Netherlands), 2017
[5] Rampsbeck, K. et al. (2012): Light Induced Degradation of Rear Passivated mc-Si Solar Cells. In: 27th European Photovoltaic Solar Energy Conference and Exhibition, S. 861–865. DOI: 10.4229/27thEUPVSEC2012-2DO.3.4.
[6] Fertig, Fabian; Krauß, Karin; Rein, Stefan (2015): Light-induced degradation of PECVD aluminium oxide passivated silicon solar cells. In: Phys. Status Solidi RRL 9 (1), S. 41–46. DOI: 10.1002/pssr.201409424.
[7] Kersten, Friederike; Engelhart, Peter; Ploigt, Hans-Christoph; Stekolnikov, Andrey; Lindner, Thomas; Stenzel, Florian et al. (Hg.) (2015): A new mc-Si degradation effect called LeTID. Photovoltaic Specialist Conference (PVSC), 2015 IEEE 42nd, 14-19 June: IEEE.
[8] F. Kersten; P. Engelhart; H.-C. Ploigt; F. Stenzel; K. Petter; T. Lindner et al.: A new light induced volume degradation effect of mc-Si solar cells an modules. In: 31st European Photovoltaic Solar Energy Conference and Exhibition, S. 1830–1834. DOI: 10.4229/EUPVSEC20152015-5CO.16.5.
[9] Kersten, Friederike; Heitmann, Johannes; Müller, Jörg W. (2016): Influence of Al2O3 and SiNx Passivation Layers on LeTID. In: Energy Procedia 92, S. 828–832. DOI: 10.1016/j.egypro.2016.07.079.
[10] IEC 61215-Crystalline silicon terrestrial photovoltaic (PV) modules - Design qualification and type approval
[11] COMSOL Multiphysics® v. 5.2a. www.comsol.com. COMSOL AB, Stockholm, Sweden.
[12] Urban, T.; Turek, M., Großer, S., Hagendorf, C., Heitmann, J.: "Reliability Investigation of the Rear Side Metallization of PERC Cells", 33rd European Photovoltaic Solar Energy Conference and Exhibition, 1705-1708 (2017). DOI: 10.4229/EUPVSEC20172017-5BV.4.71

TOWARDS THE MASS PRODUCTION OF HIGH EFFICIENCY PASSIVATED CONTACTS N-TYPE PERT SOLAR CELLS

B. Martel[1], J. Yang[2], J.C Loretz[2], S. Tran[2], T. Blevin[1], C. Bouet[1]
[1]CEA, LITEN, National Institute of Solar Energy, F-73375 Le Bourget du Lac, France
[2]SEMCO Technologies, 395 Rue Louis Lépine, 34000 Montpellier- France

ABSTRACT: In this work, we intent to elaborate a rear side passivated contact n-PERT solar cell structure using high throughput industrial equipment in a cost-effective manner. The passivation of the metal contact, and the charge transport on the rear side are ensured by a thin silicon oxide layer (<2 nm) covered by a poly-silicon (poly-Si) phosphorus doped layer. This structure features a growing interest in the photovoltaic community. The paper presents our latest developments in terms of *tunnel* oxide growth, poly-Si layer elaboration using an LPCVD reactor and phosphorus doping (both ex situ & in situ processes). After growing a thin oxide using combined wet and thermal oxidations, depositing, doping and annealing a phosphorus-doped poly-Si layer, we managed to reach high levels of surface passivation with implied open circuit voltages (iVoc) over 740 mV. Solar cells precursors were prepared including a front lightly doped 120 ohm/sqr boron diffused emitter (homogeneous emitter contactable with standard screen-printing pastes) and a rear passivated contact structure as depicted above. These precursors feature promising iVoc up to 707 mV and saturation recombination current densities (J_0) down to 20 fA/cm^2.
Keywords: Polycrystalline, Back contact, Boron

1 INTRODUCTION

Nowadays, passivated contacts integration using doped polycrystalline silicon (poly-Si) on ultra-thin oxide layers sparks strong interest for solar cells fabrication [1]. Ultra-thin oxide layers are essentially elaborated by chemical or thermal ways, and poly-Si layers are usually obtained by PECVD [2] (Plasma Enhanced Chemical Vapor Deposition) or LPCVD [3] (Low Pressure CVD) deposition of amorphous (a-Si) or microcrystalline Si (µc-Si) layers, usually before crystallization anneal steps. Doping of the poly-Si layer can be done during either layer deposition itself or using ex situ processes. The various combinations of those technological choices lead to different oxide/poly-Si bilayer structures, featuring various properties in terms of surface passivation, thermal budgets and reliabilities regarding solar cell integration. With such passivated contact structures, excellent PV conversion efficiencies were obtained with small surface area devices and lab-scale equipment. However, few studies have been presented about their deployment at the industrial level. The goal of this work is to demonstrate that the integration of these advanced collecting structures in screen-printed high efficiency devices can be done with high throughput industrial equipment in a cost-effective manner. More precisely this work presents our latest developments in terms of tunnel oxide growth, poly-Si layer elaboration using SEMCO Technologies LPCVD reactor followed by ex situ phosphorus (P) diffusion doping, as well as studies about the compatibility of the elaborated structure with thermal budgets associated with a high quality BCl3 boron (B) emitter diffusion conducted with the SEMCO LYDOP technology (up to 1400 wafers furnace capacity). Our final goal is the fabrication of high efficiency bifacial n-PERT solar cells integrating both rear poly-Si passivated contacts and a front boron diffused emitter. In parallel, first results about the in situ doping of the poly-Si layers are presented, such layers being particularly interesting for simplifying the cell fabrication process.

2 EXPERIMENTAL

The wafers used for the study are Czochralski (Cz) 2 ohm.cm n-type P-doped wafers. There oxygen concentration were intentionally chosen <12 ppma to limit the risk of oxygen-induced defects formation during the relatively consequent thermal budget applied to the wafers during the processing steps. A wet saw damage removal and polishing step using 20% KOH solution is completed to initiate the wafers surfaces. At the end of this step, the wafers are rinsed in deionized (DI) water, which could contain or not ozone (O_3). The use of O_3 during the rinsing step enables the formation of a thin chemical oxide. The deposition of a double side intrinsic (i) poly-Si layer is done after a short thermal oxidation, at various temperatures from 630 to 670°C in oxygen atmosphere at low pressure (this oxidation can be done directly in the LPCVD furnace). The i-poly-Si layer is deposited at 630°C, under silane flow using a horizontal LPCVD furnace. In this configuration, the deposited i-poly-Si layer is already microcrystalline. The deposition time is varied to obtained i-poly-Si layers thicknesses from 100 to 250 nm.

Figure 1: HORTUS platform for Poly-Si deposition & B diffusion.

The doping of the i-poly-Si layer is conducted using a conventional phosphorus diffusion step. After removing the Phosphorus Silicate Glass (PSG) layer with an HF solution, a silicon nitride (SiN) capping layer is deposited on the rear side of the wafers using a PECVD reactor.
A KOH texturation is realized to optimize optical absorption and remove the P-poly-Si layer on the front side. A 120 ohm/sqr boron diffusion doping of the front side is done in a horizontal furnace, at 950°C, using BCl_3 gas precursor. An in situ oxidation of the wafers is realized to avoid passivation issues linked to the presence of a remaining boron rich layer [4]. After removing the capping layer and the boron silicate glass (BSG) into HF solution,

a front Aluminum Oxide (AlOx) and SiN layers are deposited to passivate the boron emitter, and a rear SiN layer is deposited to ensure the solar cell bifaciality and provide an additional hydrogen source for the Si bulk and interfaces. Solar cells precursors' quality is evaluated after firing using IC-PCD (Inductively-Coupled Photo Conductance Decay) technique. Figure 2 summarizes the process flow described, as the corresponding solar cell structure.

In parallel, to allow process simplifications, first elaborations of n-type poly-Si layers, in situ doped during the LPCVD deposition treatment, were conducted. Their passivation properties were assessed by IC-PCD.

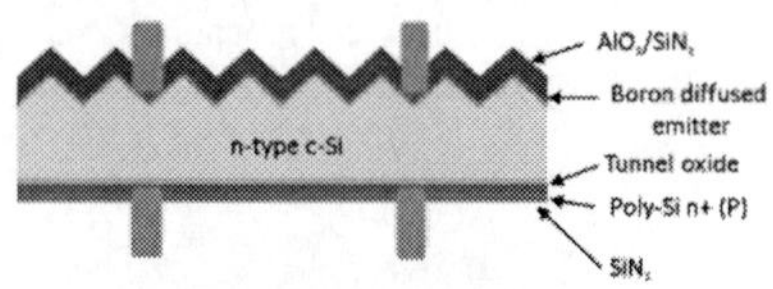

(a)

KOH polishing and chemical oxidation
Thermal oxidation + i-poly SEMCO LPCVD deposition
SEMCO LYDOP Phosphorus diffusion
HF PSG removal
Rear SiN:H capping layer
Short front KOH texturation
SEMCO LYDOP Boron diffusion
HF BSG and capping layer removal
Front AlOx and SiN:H deposition
Rear SiN:H deposition

(b)

Figure 2: Solar cell structure (a), process flow (b)

3 RESULTS AND DISCUSSIONS

In this section, results will be first depicted separately for symmetric structures for the front boron diffusion and then the rear-passivated contact. Finally, solar cell precursors' results will be shown and discussed.

3.1 Front boron diffusion

Sheet resistance of the boron diffused emitter was measured at 120 ohm/sqr after BSG removal with an intra-wafer uniformity <2% (uniformity defined by *(max – min)/(2×average)*) using 4 point probes technique, 9 points mapping. Electrochemical capacitance voltage (ECV) measurements were done at the center of the diffused wafer after BSG removal. Figure 3 presents the resulting electrically active boron concentration ([B]) profile.

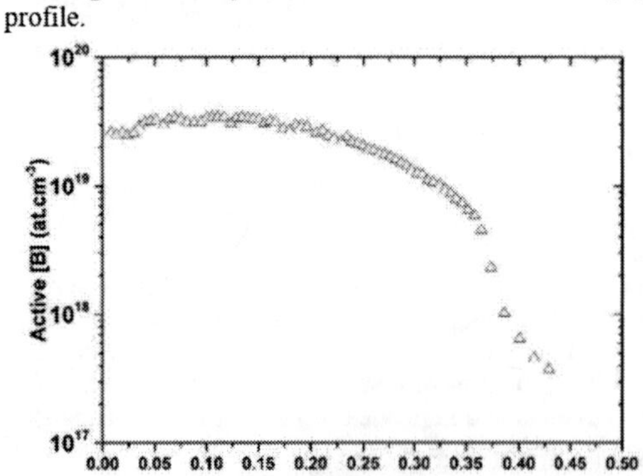

Figure 3: ECV measurement of the active [B] profile for the diffused emitter (KOH polished wafers).

From figure 3, we can extract the surface [B] close to 2.5×10^{19} cm^{-3}, which is a low value reducing surface recombination, while being compatible with state of the art metallization pastes (internally confirmed by determining contact resistivity values). Symmetric boron diffused samples passivated with an AlOx / SiN bilayer featured high implied open circuit voltage (iVoc) up to 702 mV, and low saturation current density (J_0) down to 22 fA/cm^2 after firing.

3.2 Rear poly-Si on thin oxide

We explored two different ways to elaborate the thin interface oxide on the rear side of the solar cell structure. The first one uses a chemical oxidation during a DI-O_3 rinsing sequence only. We noticed that the use of a chemical oxide alone was not compatible with high iVoc and thus high passivation levels after applying a 950°C thermal budget (boron diffusion peak temperature) with symmetric n-poly-Si on oxide structures. Thermal oxidation was shown to be more compatible with elevated thermal budgets. We investigated the growth of the thin oxide (compatible with an efficient charge collection) with a short thermal oxidation at different temperatures with and without the preliminary wet oxidation. Figure 4 presents iVoc results on symmetric structures (KOH polished, without SiN layers), with 250 nm n-doped (ex situ) poly-Si on oxide, after applying the boron diffusion thermal budget. We can notice that very high passivation levels are reached for both oxide elaboration paths: i-Voc up to 740 mV and J_0 down to 5 fA/cm^2.

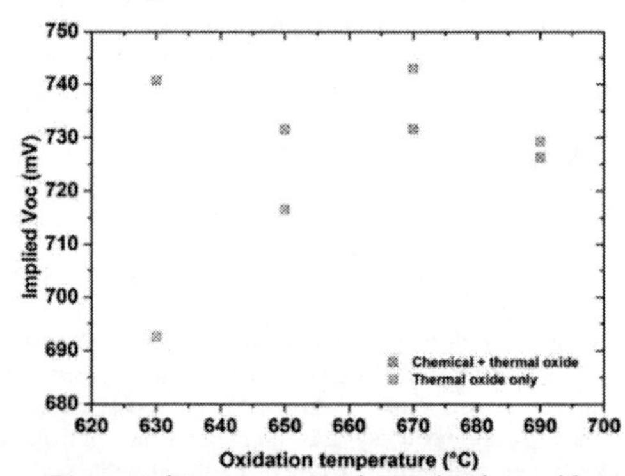

Figure 4: iVoc on symmetric n-poly-Si on oxide structure as a function of the thermal oxidation temperature. Both with and without preliminary chemical oxidation is shown.

From figure 4, we can see that the use of a preliminary wet oxide reduces the oxidation temperature (down to 630°C) needed to achieve a high passivation level. The thermal oxidation only requires a higher oxidation temperature (670°C), to compensate the lack of oxide on the wafer surface before this step.

In addition, we can notice that high oxidation temperatures affect the passivation quality (with & without the wet oxide). This could be explained by thicker oxides that would thus limit the diffusion of P atoms at the surface of the c-Si substrate, reducing the field effect passivation. We also decreased the n-poly-Si layer thickness down to 100 nm and we managed to reach the same high iVoc results on symmetric structures. Figure 5 depicts the ECV active P concentration ([P]) profiles obtained (for both, the 100 nm and 250 nm thick poly layers).

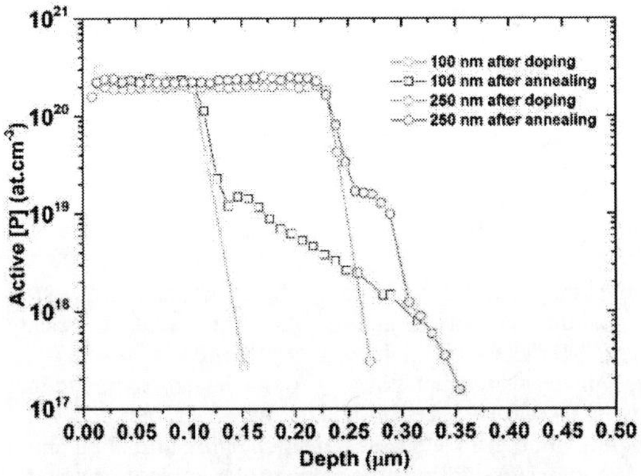

Figure 5: ECV measurements of active [P] for poly-Si/SiO_2/c-Si structures (different poly thicknesses, 250 nm and 100 nm), after diffusion doping and after annealing at 950°C.

From figure 5 we can notice an active [P] between 2 and 3×10^{20} cm^{-3} within the poly-Si layer, and a sharp decrease of the [P] at the oxide/bulk Si interface. A short P diffusion tail is present in the bulk Si after doping. This diffusion tail is enlarged after the 950°C annealing, due to the P diffusion through the oxide.

Regarding the in situ P doping of the poly-Si layers, similar passivation properties were obtained, with i-Voc values of 740 mV (KOH polished surfaces, without SiN layers).

3.3 Solar cells precursors

After studying and developing both boron emitters and n-poly-Si on oxide structures, we integrated both charge collectors in bifacial solar cells, as depicted in figure 2b. On the rear side, the 250 nm thick poly-Si layer was integrated. After passivating the front boron emitter using AlOx and SiN stack and the rear poly-Si using SiN layer, we measured iVoc values up to 707 mV and J_0 down to 20 fA/cm² after firing. These results validate the compatibility between the rear poly-Si layer on oxide elaboration and the front B-diffused emitter.

4 CONCLUSION

This study deals with the integration of rear passivated contacts using poly-Si on oxide, into n-type bifacial PERT solar cells. We demonstrated here that the production of high quality solar cells integrating such advanced structures is compatible with industrial high throughput equipment. The front side light boron emitter is achieved using a BCl_3 diffusion step. The obtained emitters feature low J_0 after passivation and are compatible with actual screen-printing pastes. The rear thin oxide is elaborated using thermal oxidation. A preliminary wet oxide helps to reduce the thermal budget associated with this step. The poly-Si layer as deposited is intrinsic and subsequently doped using a conventional P diffusion. The doping of the poly-Si layer is tuned to be compatible with a subsequent boron diffusion step. Promising solar cells precursors were elaborated, with iVoc up to 707mV. Complete solar cells results are not presented here, as metallization studies are still ongoing.

In parallel, in situ P doped poly-Si layers were elaborated, such an approach being particularly interesting for process simplifications. These layers feature high passivation levels, with i-Voc values of 740 mV (symmetric samples).

5 REFERENCES

[1] F. Feldmann et al., Energy Procedia 124, 31-37, 7th Silicon PV conference 2017

[3] S. Glunz et al., proceeding of the 31st EU PVSEC 2015, Hamburg

[3] M. K. Stodolny et al., proceeding of the 33rd EU PVSEC 2017, Amsterdam

[4] V. D. Mihailetchi et al., IEEE Journal of Photovoltaics, Vol. 8, N°2, March 2018

INDUSTRIAL SOLUTIONS TO MITIGATE LIGHT-INDUCED DEGRADATION IN MC-PERC CELLS AND MODULES

Jian Wu*, Zheng Yao, Daqi Zhang, Guangyong Xiong, Fangdan Jiang, Jing Sun, Jean-Nicolas Jaubert, Guoqiang Xing
Canadian Solar Inc. No. 199 Lushan Rd. SND, Suzhou, China
*Corresponding author: Phone: +86 512 68966889; Email: jian.wu@canadiansolar.com

ABSTRACT: The successful mitigation of light induce degradation (LID) is the key to volume production of mc-silicon PERC solar cells. In this paper, we will report our breakthroughs in mitigating LID of mc-Si PERC cells, named as P4 (Poly Gen4) technology at Canadian Solar Inc. (CSI). These technology advancements include: 1) A unique ingot-casting process to control the impurities within multi-Si wafer materials; 2) An advanced regeneration process to deactivate the defects that cause LID, as well as to increase P4 cell efficiency by ~0.1% abs.; 3) An optimized cell process, especially with regard to the metallization, in order to suppress defect complex formation and to enhance hydrogen passivation of the multi-Si bulk; and 4) An intensified inline process control to create a fast and reliable LID-controlled multi-Si PERC cells and modules. In outdoor test, multi-PERC modules presented an enhanced performance of only 1.0% degradation after 40~45kWh/m^2 light soaking. We also performed indoor accelerated CID testing (0.5A current injection at 85 °C), and the resultant degradation is less than 1.5% after 500 hours, equivalent to two years of hot climate outdoor exposure. CSI's multi-PERC cell capacity was ramped to 1GW by the end of 2017 and will be 5GW by the end of 2018, with cell efficiency of more than 20% on average. In combination with advanced module technologies, CSI's multi-PERC module wattage exceeded 300W.
Keywords: light induced degrade, regeneration, solar cell

1 INTRODUCTION

For the past few years, multi-Si has been the mainstream technology, holding a majority market share. However, multi-Si's leadership of the market is facing a serious challenge from mono-Si, which is demonstrating inherently higher efficiencies and whose wafer costs are rapidly decreasing. It is therefore critically important to implement promising technologies such as diamond-wire sawing, black-silicon texturing and PERC in the mass production of multi-Si cells in order to improve efficiencies and further reduce costs.

By Q3 of 2017, CSI had successfully implemented diamond-wire sawing and state-of-the-art black silicon texturing in all its multi-Si cell production lines, with a total capacity of 4.5GW and a mass production efficiency exceeding 19.2% [1]. The integration of PERC technology in multi-Si cells, however, is much more challenging, particularly because of light-induced degradation (LID) and light- and elevated-temperature-induced degradation (LeTID) issues [2–6]. It has been reported by UNSW and other institutes that two degradation modes exist in multi-PERC: 1) the fast degradation mode, which occurs within 100 hours and is caused by Type 1 defects; and 2) the slow degradation mode, which occurs over a period of up to one thousand hours and is caused by Type 2 defects [3]. Nevertheless, the LeTID effect reported by Hanwha Q-CELLS in multi-PERC is a cause of great concern for this technology [5, 6].

By combining technical innovations throughout ingot material control, cell process optimization, advanced regeneration and intensified inline process control, CSI realized a mass production of LID-controlled multi-PERC cells and modules with a total capacity of over 1GW by the end of 2017; this capacity will be ramped up to more than 4GW by the end of 2018.

In this paper, the performance of CSI's multi-PERC cells and modules will be presented, and the solutions to realizing LID controlled multi-PERC cells and modules will be introduced.

2 MULTI-PERC CELL PROCESS

The starting p-type wafer resistivity is 1.0~2.5 Ω·cm and the wafer thickness is (190 ± 20) μm. First, the front-textured surface was formed by CSI's proprietary state-of-the-art black-silicon texturing (metal-catalysed chemical etching – MCCE). After MCCE texture, an n$^+$-Si emitter is formed by low pressure $POCl_3$ diffusion in the tube furnace, with the sheet resistance of around 110Ω/. The next step is phosphosilicate glass (PSG) removal and rear-side polishing.Al_2O_3 layers deposited by the atomic layer deposition (ALD) technique are used to passivate the rear surfaces. The post-deposition annealing of as-deposited Al2O3 layers is integrated in the subsequent rear-side SiNx anti-reflection coating (ARC) deposition by a tube plasma-enhanced chemical vapour deposition (PECVD) process. The front-side SiNx ARC is also deposited by the tube PECVD process.

Following laser contact opening, the metallization is performed by screen printing and a co-firing process. All the as-fired multi-PERC cells then go through a regeneration process, before the final testing and sorting. Process flow is shown in Fig. 1.

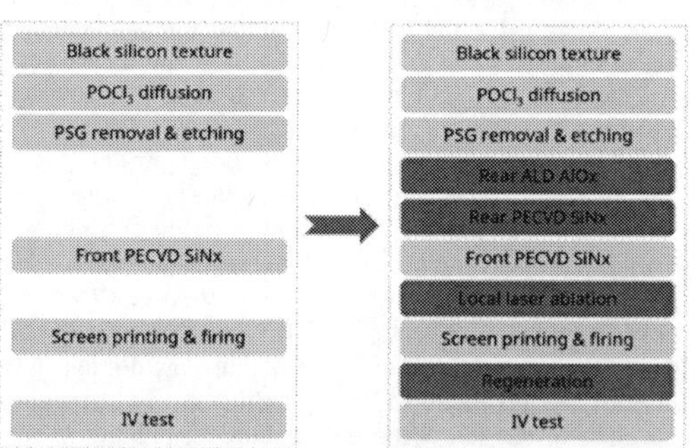

Fig.1 Process flows for conventional multi-Si cells (left), and multi-PERC cells (right).

The efficiency distribution of CSI's multi-PERC cells is shown in Fig. 2 (top); the average efficiency of the cells exceeds 20%, which equates to a 0.9% efficiency gain compared with conventional multi-Si cells based on diamond-wire-sawn (DWS) wafers.

As regards module performance, as shown in Fig. 2 (bottom), the average wattage of standard 60-cell multi-PERC modules exceeds 287W, which is comparable to that achieved by mono-Si modules. In combination with the use of module technologies such as half-cut and multi-busbar, the average wattage of 120-cell multi-PERC modules exceeds 300W, as shown in Fig. 2(bottom). This is again comparable to that for mono-PERC modules. Clearly, the implementation of multi-PERC enhances the competitiveness in terms of performance of multi-Si compared with mono-Si, but at a lower cost.

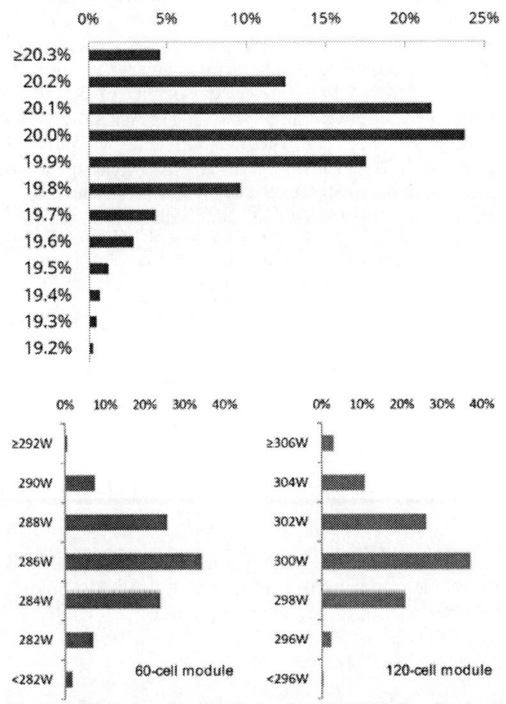

Fig. 2 Efficiency distribution of multi-PERC cells (top) and wattage distribution of 60-cell and 120-cell multi-PERC modules (bottom).

3 SOLUTION TO CONTROLL LIGHT INDUCED DEGRATATION

LID issues are much more challenging for multi-PERC than for mono-PERC. Direct evidence of this is that there are many manufacturers who can produce quality reliable mono-PERC modules, but only a few who are able to produce multi-PERC modules [7]. UNSW and other research institutes have proposed the existence of both a fast degradation mode, occurring within 100 hours and caused by so-called Type 1 defects, and a slow degradation mode, lasting up to one thousand hours and caused by so-called Type 2 defects. The Type 1 defects have been identified as being B–O defect complexes, whereas the Type 2 defects are not yet fully explained. Hydrogen [8] or metal impurities such as Fe, Co and Ni [9] are the most likely suspects for Type 2 defects. UNSW recently reported that these Type 1 and Type 2 defects are also present in p-type mono-Si and n-type mono-Si [10]. To solve the most challenging LID issues for multi-PERC, CSI has implemented several technical innovations:

1) An intensified inline process control to create reliable LID-controlled multi-Si PERC cells and modules.
2) A unique ingot-casting process to control the impurity content within multi-Si wafer materials.
3) An optimized cell process, especially with regard to the metallization, in order to suppress defect complex formation and to enhance hydrogen passivation of the multi-Si bulk.
4) An advanced regeneration process to deactivate the defects that cause LID.

3.1 Accelerating degradation by current injection instead of light soaking

The reason why it is much more challenging to control the LID with multi-PERC than with mono-PERC is primarily the wide variation in quality of multi-Si wafers [6]. Even though innovative steps have been taken to control the impurity concentration of multi-Si ingots, it is still essential to reinforce the inline process control; in particular, a more exhaustive monitoring of degradation rate at the cell level is necessary, in addition to prompt process optimization in response to fluctuations.

Normally the LID of Si solar cells is tested by light-soaking; however, this technique has some drawbacks, such as a test time that is too short (typically 24–72 hours), imprecise control of the wafer temperature, and limited sample volume.

To test the LID performance of multi-PERC cells, CSI uses the current injection method (called "CID"), which offers several advantages, as listed in Table 2. The parameters for the CID method are forward-biased injection current, cell temperature and duration. These parameters are carefully selected after extensive experimental studies to reflect the degradation rate of multi-PERC cells as much as possible. In fact, if setting the injection current and wafer temperature parameters to certain values results in a regeneration-dominant effect, it is a regeneration process; on the other hand, if lower values for the injection current and wafer temperature result in a degradation-dominant effect, it is a degradation process. The chosen CID parameters for cells are 3.5A, 105℃ and 4 hours, which is equivalent to testing LID by light-soaking with the parameters 1,000W/m2, 65℃ and 24 hours, which is shown in Fig.3.

Table I: Comparison of CID over LID in testing the degradation behavior of multi-PERC cells

	CID	LID
Injection intensity	Controlled by forward-biased current (0–10A)	Controlled by light intensity (0–1 sun)
Temperature control	50–150°C adjustable, accuracy <5°C	Not easy to control, sensitive to light intensity
Batch capacity	Up to 80 cells per stack, with small footprint	10–100 cells per tiled array, with large footprint
LeTID correlation	Higher temperature higher correlation	Not easy to control temperature above 75°C

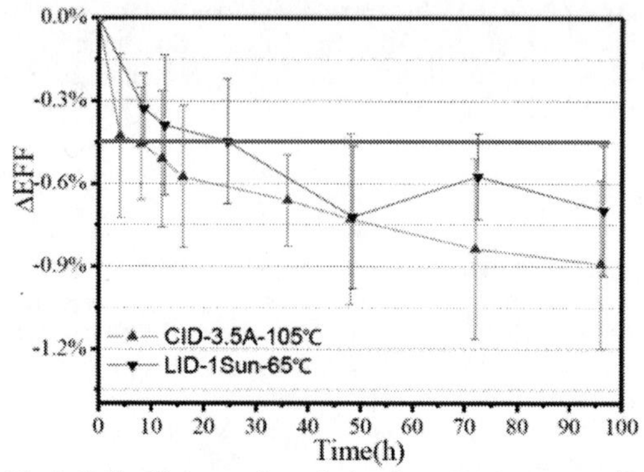

Fig.3 Cell efficiency degradation by CID (3.5A, 105℃) vs light-soaking 1,000W/m2, 65 ℃. CID-4 hours is equivalent to light-soaking 24 hours.

3.2 Unique ingot-casting process

The degradation rate for multi-PERC is dependent on the ingot and wafer material quality. For multi-Si ingots, the trend generally seen is that the degradation rate increases from the top part of the ingot to the bottom part. Additionally, there are various effects on the degradation rate arising from resistivity or doping concentration, oxygen content and structural defect density. The degradation rate is also sensitive to the dopant – B or Ga or a mixture of B and Ga; the beneficial effect of Ga doping, or partial Ga doping, is widely accepted. Fig. 4 shows the notable differences in multi-PERC module degradation, tested by current-induced degradation (CID), between quality-controlled and normal multi-Si wafers.

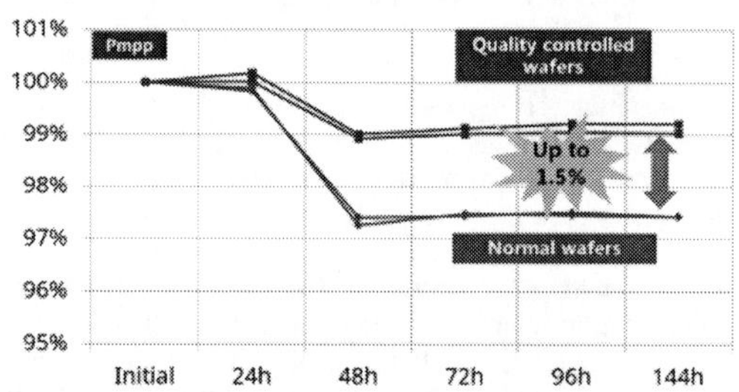

Fig.4 Effect of wafer quality on CID (0.5A 85°C) degradation performance of multi-PERC modules.

3.3 Optimized cell metallization process and regeneration

The significant impact of the firing temperature on degradation has been widely investigated [11, 12]. Either lowering the peak temperature or lowering the cooling rate in the firing process will help to reduce the degradation rate considerably. Many theories to explain these findings have been proposed [11–14]; possible explanations are that a reduced firing temperature will suppress defect formation and/or change the hydrogen content in the multi-Si bulk.

The use of this strategy results in excellent control of the CID of multi-PERC cells, to reduce the average CID by 0.44% abs.

Another key to reducing the degradation rate of multi-PERC is the advanced regeneration process. The regeneration process to address the defect complexes causing LID consists of excess carrier injection, and adequate temperature and duration time [15]. Generally, for mono-PERC a light-induced regeneration (LIR) process using a halogen lamp, LED or laser is used [15]; however, the industrial LIR process is not adequate for use with multi-PERC.

CSI uses a proprietary CIR process; compared with LIR, there are many advantages of CIR, such as a broader process window, higher throughput, lower electricity consumption and lower cost. By using the CIR process, the degradation rate can be reduced by up to 80%.

Fig. 5 shows the correlation between the reduced degradation rate after CIR treatment and the degradation rate without CIR treatment, tested by CID. It can be seen that the higher the degradation rate for multi-PERC cells without CIR treatment CIDw/o CIR, the greater the mitigation coefficient CIDw/o CIR-CIDCIR reflected in the reduced degradation rate. Interestingly, there is a quasi-linear correlation between CIDw/o CIR-CIDCIR and CIDw/o CIR, indicating that the CIR process effectively passivates the defect complexes causing LID. Moreover, it can be seen that the quality of the wafers from some suppliers (suppliers 1 to 4) is not satisfactory, showing a very high degradation rate without CIR treatment. This again reflects the importance of controlling the quality of the ingot and wafer materials in order to produce multi-PERC cells with controlled LID.

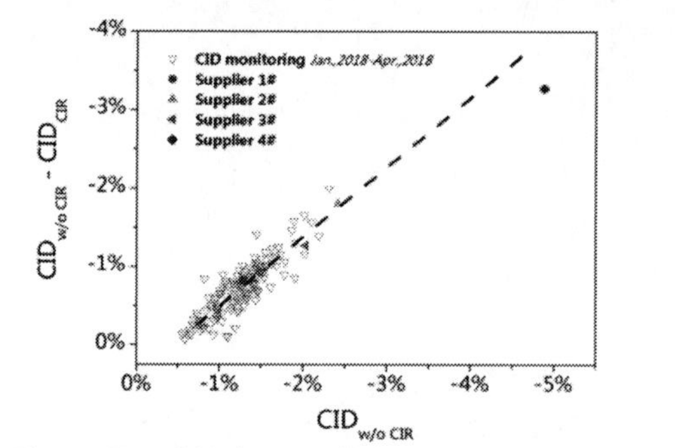

Fig.5 LID mitigation coefficient of the CIR process, reflected in the reduced degradation rate.

3.4 Module LID performances

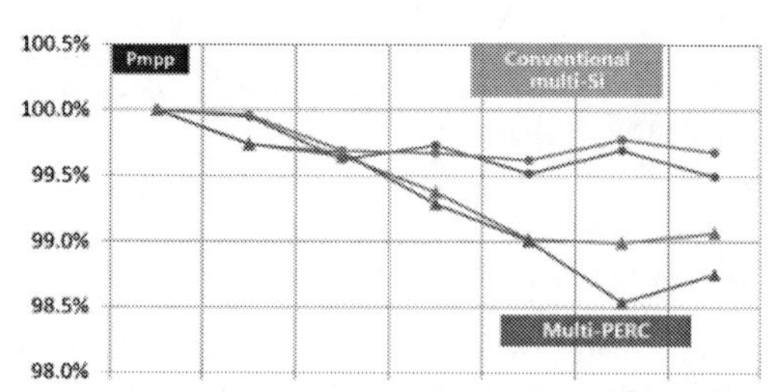

Fig.6 Module performance of CSI's multi-PERC modules undergoing CID testing (0.5A 85°C).

For the module degradation test, the CID method is also used instead of indoor or outdoor light soaking methods, with the set-up and parameters suggested by Hanwha Q-CELLS. Fig. 6 shows the comparison of CID degradation between multi-PERC modules and conventional multi-Si modules. The figure highlights that there is an increase in the module degradation rate of multi-PERC modules, as compared to conventional multi-Si modules. After 300 hours, however, the degradation rate of multi-PERC modules stabilizes, and proves to be less than 1.5% for up to 500 hours of testing; this is equivalent to two years of hot climate outdoor exposure, as proposed by Hanwha Q-CELLS [16].

In outdoor test, CSI's multi-PERC modules presented an enhanced performance of only 1.0% degradation after 40~45kWh/m^2 light soaking, which is shown in Fig.7.

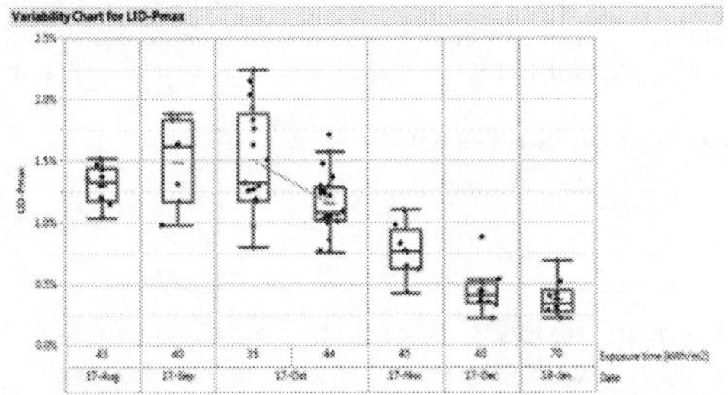

Fig.7 Outdoor test for CSI's multi-PERC modules degradation after 40~45kWh/m^2 light soaking.

All the cell and module degradation results demonstrate that, after taking several technical innovative steps, the LID of CSI's multi-PERC cells and modules can be successfully controlled.

4 CONCLUSIONS

Through innovations in materials, cell processing, advanced regeneration and enhanced inline process control, CSI has successfully demonstrated LID controlled multi-PERC cells and modules in GW capacity. Multi-PERC technology is a must for high-efficiency multi-Si products to compete with upcoming high-volume mono-Si products. In order to further improve competitiveness, the fabrication of bifacial multi-PERC and an integration of advanced technologies to push efficiencies even higher will be essential.

5 REFERENCE

[1] Wang, X. et al. 2017, Photovoltaics International,35th edn, pp. 67–72.

[2] Ramspeck, K. et al. 2012, Proc. 27th EU PVSEC,Frankfurt, Germany, pp. 861–865.

[3] Hallam, B. et al. 2016, physica status solidi (RRL),Vol. 10, No. 7, pp. 520–524.

[4] Luka, T. et al. 2016, Photovoltaics International, 32nd edn, pp. 43–48.

[5] Fertig, F. et al. 2015, physica status solidi (RRL), Vol.9, No. 1, pp. 41–46.

[6] Kersten, F. et al. 2015, Sol. Energy Mater. Sol. Cells,Vol. 142, pp. 83–86.

[7] Fertig, F. et al. 2017, Energy Procedia, Vol. 124, pp. 338–345.

[8] Wilking, S. et al. 2014, Sol. Energy Mater. Sol. Cells, Vol. 131, pp. 2–8.

[9] Luka, T. et al. 2017, Proc. 33rd EU PVSEC,Amsterdam, The Netherlands, pp. 413–417.

[10] Niewelt, T. et al. 2017, J. Appl. Phys., Vol. 121, p. 185702.

[11] Glunz, S. et al. 2001, J. Appl. Phys., Vol. 90, pp.2397–2404.

[12] Bothe, K. et al. 2002, Proc. 29th IEEE PVSC, New Orleans, Louisiana, USA, pp. 194–197.

[13] Unsur, V. et al. 2016, Proc. 43rd IEEE PVSC, Portland, Oregon, USA, pp. 717–719.

[14] Kouhlane, Y. et al. 2016, J. Electron. Mater., Vol. 45,pp. 5621–5625.

[15] Herguth, A. et al. 2009, Proc. 24th EU PVSEC, Hamburg, Germany, pp. 974–976.

[16] Kersten, F. et al. 2017, Proc. 33rd EU PVSEC, Amsterdam, The Netherlands, pp. 1418–1421.

HIGH-EFFICIENCY DIAMOND WIRE-SAWN MC-SI-BASED PERC SOLAR CELLS TEXTURED BY ATMOSPHERIC PRESSURE DRY ETCHING

B. Kafle[1], A. I. Ridoy[1], P. Saint-Cast[1], L. Clochard[2], E. Duffy[2], K. Duncker[3], K. Petter[3], S. Peters[3], M. Hofmann[1]
[1]Fraunhofer Institute for Solar Energy Systems ISE
[2]Nines Photovoltaics
[3]Hanwha Q Cells GmbH
[1]Heidenhofstr. 2, 79110 Freiburg, Germany
[2]IT Tallaght, Dublin 24, Ireland
[3]Sonnenallee 17-21, 06766 Bitterfeld-Wolfen, Germany

Correspondence: bishal.kafle@ise.fraunhofer.de, +49 (0) 761 4588-5499

ABSTRACT

In this paper, we present further results on the integration of a "black-silicon" texturing process on diamond-wire sawn multicrystalline silicon wafer, using atmospheric pressure dry texturing (ADE) in a passivated emitter and rear cell (PERC) solar cell architecture. Following on from our previous work on high-efficiency nanotextured solar cells, we present further optimization of our texturing process, resulting in the uniform formation of spherical caps-like surface morphology with low reflectivity and comparable reflection distribution to the wet-chemical acidic texture based on additives. First cell results show up to 19.8% conversion efficiency. Characterisation and analysis of the cells highlight current limitations and identify potential for improvement: optimization of the firing process and minimization the inherent bulk-related recombination losses in the mc-Si material, which is expected to lead to further gain in conversion efficiencies.

Keywords: black silicon, nanotexture, atmospheric pressure process, dry etching, chemical etching, PERC solar cells

1. INTRODUCTION

1.1 Context:

A significant cost reduction in wafering is possible by the adoption of diamond-wire-sawing (DWS). For mc-Si wafers, it also requires a change in the texturing process as the incumbent wet acidic process is not suitable. The use of additives allows the adoption of the DWS mc-Si but results in a high reflectivity texture. In order to stay competitive against mono-Si cells, mc-Si based cells also need to further improve their efficiency and industries are looking at introducing advanced texturing processes into solar cell production, which promises J_{SC} improvement in comparison to the state-of-the-art techniques. Introduction of such novel textures, however, demands successive optimizations of the cell-processing steps in order to deliver improved electrical performance at cell and module levels.

In this study, a dry chemical etching process (atmospheric dry etching, ADE) is used to form low reflectivity surface texture, as it provides economical, technological and ecological advantages compared to the competing wet-chemical technologies such as metal-assisted chemical etching (MCCE) and other dry-etching alternatives (reactive ion etching - RIE). The advantages of using ADE are summarized as: a) high etching rate and inline modular nature of the etching tool allowing high volume production, b) low cost of ownership (COO) due to no vacuum nor plasma, c) easy abatement of waste gases through standard wet scrubber systems, d) use of environmental friendly F_2 gas with zero global warming potential (GWP), and e) chemical etching without any ion-induced damage in Si.

Formation of nano-scale structures on mc-Si surfaces by using ADE process and their successful integration in a standard Al-BSF mc-Si solar cells [1,2] and on PERC mc-Si solar cells [3] has already been demonstrated, achieving > 20.0% conversion efficiency when using HP grade mc-Si wafers. PERC solar cell is now set to become the standard cell architecture in industrial production [4].

1.2 Approach:

Each etching process step induces a degree of crystal orientation dependency on the mc-Si wafer. The result is a grain dependant texture, with difference in aspect ratio of nanostructures formed in different grains. This non-uniformity makes it more difficult to optimize the emitter formation as well as the PECVD deposition processes, which eventually can limit the internal quantum efficiency (*IQE*) of the solar cell. Meanwhile, a large distribution of reflection limits the external quantum efficiency (*EQE*) and therefore the short circuit current density (J_{SC}) of the solar cell.

In order to further improve our PERC cell results; this work is focussing on narrowing the distribution of reflection within the wafer surface by optimizing the sequence of etching steps toward a more isotropic process. Narrowing this distribution will also improve the overall appearance of the cell. In this study, we present our first results of optimization using ADE-texture on industrial standard diamond-wire sawn *p*-type mc-Si wafer-based PERC solar cells.

2. EXPERIMENTAL

2.1 Sample preparation

The wafers used in this process are large area (156 mm edge length) *p*-type mc-Si. First generation (Gen.1) PERC solar cells were fabricated on high performance (HP) slurry-cut mc-Si wafers, whereas third generation (Gen.3) are fabricated on DW-sawn industrial type mc-Si wafers. The process plan for both groups is shown in Figure 1.

Both group of wafers are first saw-damage etched and then textured using the ADE process, during which the wafers are dynamically transported in an inline mode through the reaction chamber of the ADE tool (Nines ADE-100) with the process described elsewhere in detail [2], followed by a short post-etching step to enlarge the surface structures and facilitate conformal deposition of SiN_x by PECVD. Gen.1 received an alkaline post-treatment step after ADE black silicon texturing. In case of Gen.3, the ADE-texture and the post-etching steps are collectively optimized to achieve a more isotropic result. Afterwards, Q.ANTUM process [5] of Hanwha Q-Cells is applied to prepare PERC solar cells.

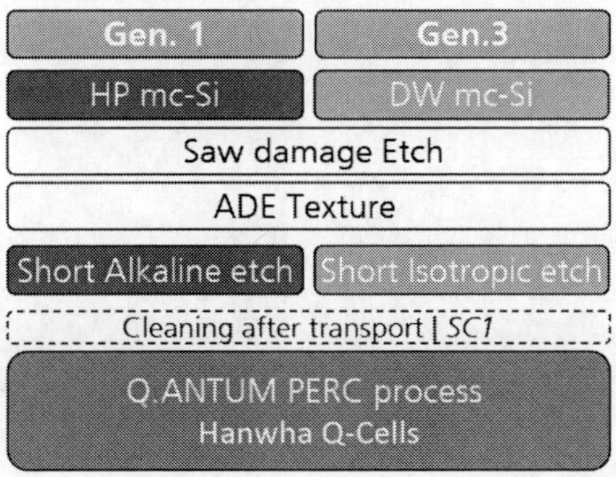

Figure 1: Process plan for ADE-textured mc-Si PERC solar cells of first (Gen.1) and third (Gen.3) generations.

2.2 Characterisation

Illuminated *I-V* measurements are performed under the standard test conditions using an in-house solar simulator that is calibrated with the Fraunhofer ISE CalLab reference. Spatial mapping of quantum efficiency and reflectivity of solar cells is performed at 406 and 960 nm wavelength using light beam induced current (LBIC) method using a PV Tools Loana system with a resolution of 200 µm.

3. RESULTS

3.1 Texturing homogeneity

A more isotropic post-ADE process is used for this work and compared to the results obtain previously [3] with an anisotropic (alkaline) post-etching process. Further increase of the J_{SC} should focus both on lowering of the overall reflection as well as narrowing the spatial distribution of reflectivity. Collective optimization of ADE and post-etching processes is performed on DW-sawn mc-Si wafers in order to achieve best optical and electrical performances in the cell-level. Figure 2 a) shows the scan image of Gen.1 ADE-texture after post-treatment in the alkaline solution, showing a large distribution of reflection in the full-wafer area. SEM images of the darkest and the lightest grains of the mc-Si wafer, respectively Figure 2 b) and c), show the formation of pseudo-pyramid like structures with substantial differences in the aspect ratios, and hence the surface reflection values.

In contrast, Gen.3 ADE-textured DW-sawn mc-Si wafer (scan image shown in Figure 2 d)), has a narrow distribution of reflection in the whole wafer-area. The etching process developed results in spherical cap-like structures with dimensions of 1 µm (see Figure 2 e) and f)) homogeneously in all crystal orientations of the mc-Si wafer. Such characteristic dimensions are expected to cause high scattering of the middle and long wavelengths of visible light, leading to an improved light trapping in comparison to sub-micron wavelength structures. In this particular wafer, the weighted surface reflection measured along the full-wafer area is in the range of 14-17% after the texturing process.

Figure 3 compares the distribution of reflection values at 405 nm from the mapping of the full wafer area of ADE-textured samples after PECVD SiN_x anti-reflective coating. For comparison, reflection data of additive-based texture after SiN_x coating is also included. One should note a significantly smaller reflection distribution for Gen.3 ADE texture. In comparison to the additive-based wet-chemical texture on DW-sawn wafers, Gen.3 ADE texture shows both lower average reflection values as well as a comparable reflection distribution in the full wafer area. Using Gen.3 texture on industrial standard diamond wire mc-Si wafers, PERC solar cells were fabricated.

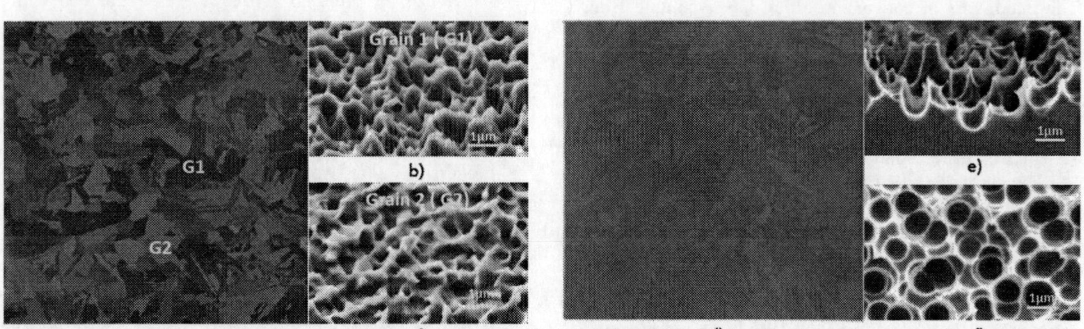

Figure 2: a) Scanned image of Gen.1 ADE-textured DW-sawn wafer and corresponding SEM images of pseudo-pyramid like structures formed in two grains – b) Grain 1 (G1) with lowest reflection and c) Grain 2 (G2) with highest reflection properties; d) Scanned image of Gen.3 ADE-textured DW-sawn wafer and corresponding SEM images showing e) cross-section view and f) top-view of the etched surface. In case of Gen.1 ADE texture, pseudo-pyramid structures with difference in aspect ratios are formed in various mc-Si grains. The etching process developed for Gen.3 ADE texture leads to the formation of spherical cap-like structures homogeneously distributed in all crystal orientations of the mc-Si wafer.

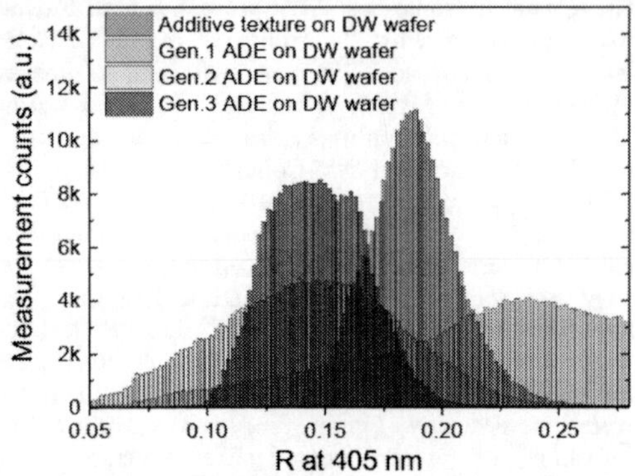

Figure 3: Histograms showing distribution of reflection at 405 nm in the full wafer area for different generations of ADE texture, in comparison to the additive-based acidic texture on DW-sawn wafer.

3.2 *I-V* results

Table 1 lists the *I-V* characteristics of Gen.3 ADE-textured solar cells on industrially used DWS mc-Si wafers. For comparison with our previous results in [3], *I-V* characteristics of Gen.1 ADE textured solar cells on HP mc-Si substrates are also listed.

Table 1: *I-V* characteristics of Gen.1 and Gen.3 ADE textured large area mc-Si wafer-based PERC cells. Please note that Gen.1 ADE texture is fabricated previously on high performance (HP) mc-Si substrates [3]. Gen.3 optimized ADE texture, featuring spherical caps-like surface topography, is used to fabricate PERC-solar cells on industrial type DWS mc-Si wafers.

Texture	Avg./ Best	V_{OC} (mV)	J_{SC} (mA/cm^2)	*FF* (%)	η (%)
Gen.1	Avg. (12)	659	38.2	79.5	20.0
Gen.1	Best [3]	660	38.4	79.4	20.1
Gen.3	Avg. (12)	650	38.1	78.8	19.5
Gen.3	Best	655	38.2	79.7	19.8

ADE-based solar cells of Gen.3 reached an average efficiency of 19.5%, with the best cell achieving 19.8%. The major loss in conversion efficiency is coming from *FF*, which is mainly due to a high series resistance caused by an un-optimized firing process for this batch as no firing variation was performed. Improved *FF* that is comparable to Gen.1 ADE texture would allow an immediate gain of 0.2% absolute in η of Gen.3 ADE-based PERC solar cells.

The short circuit current density (J_{SC}) of this batch is identical to the Gen.1 ADE-textured solar cells. Comparing to the Gen. 1 solar cells with ADE-texture, Gen.3 solar cells have lower V_{OC} values, which is expected to be linked to: a) higher internal quantum efficiency (*IQE*)-related losses in short wavelengths, b) difference in bulk quality of the HP mc-Si and standard mc-Si substrates. These *IQE* maps are compared to Gen.1 solar cells at short wavelength (405 nm) and at long wavelength (940 nm) in Figure 4.

Comparing the short wavelength spectral response in Figure 4 a) and b), Gen.3 texture exhibits slightly more *IQE* losses and would suggest recombination losses at the front surface/emitter. Meanwhile, a much more homogeneous distribution of *IQE* is observed for Gen.3 as a result of improved uniformity of the texturing process.

The *IQE* mapping at 940 nm provides mostly the information coming from the bulk material and the rear-side passivation. Since the rear-side passivation is unchanged, the differences are expected to be mostly due to the bulk. It can be observed that, in comparison to the HP mc-Si used in Gen.1, standard mc-Si wafer used in Gen.3 consists of higher percentage of wafer area exhibiting low bulk lifetime possibly associated to higher percentage of dislocations/impurities. This suggests that a fraction of V_{OC} and J_{SC} is lost due to the inherent lower bulk quality of the standard mc-Si wafer in comparison to HP mc-Si wafer.

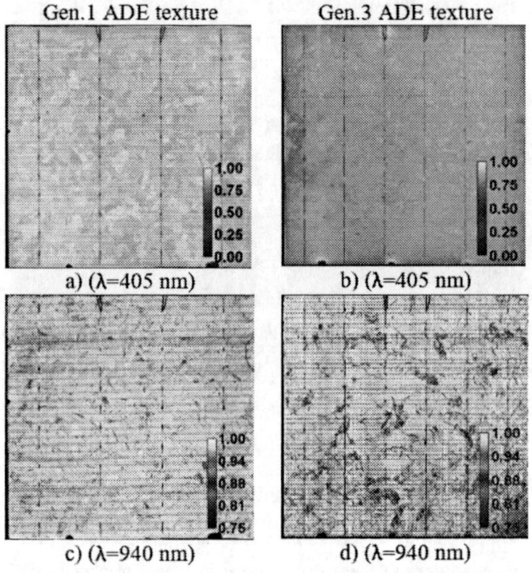

Figure 4: *IQE* mapping at 405 nm for representative ADE-textured solar cells of a) Gen.1 on HP mc-Si and b) Gen.3 on standard DWS mc-Si substrates respectively. In c) and d), *IQE* mapping at 940 nm is plotted for Gen.1 and Gen.3 respectively.

Hence, in order to further increase the conversion efficiency of Gen.3 texture and fully take advantage of the more uniform texture process, further work should focus on the combined optimization of texture and emitter diffusion to further improve the blue response. Using mc-Si of better bulk quality or further improving the inherent bulk quality of mc-Si material during the cell processing steps would also lead to higher efficiency values, directly comparable to previous results [3].

4. SUMMARY

ADE method of black silicon texturing on DW-sawn mc-Si is an alternative to MCCE and RIE due to its technological and ecological advantages. Based on high efficiencies of >20.0% on Gen.1 ADE-texture using high performance mc-Si, we presented a process showing improved spatial homogeneity of the texture in order to increase the J_{SC} gain and enhances the aesthetic appearance of the solar cell. This process is applied to industrial-type DWS mc-Si wafer-based PERC solar cells and reach mean efficiency η_{mean} = 19.5%, with champion efficiency η_{max} = 19.8%. The major loss in conversion efficiency is coming from *FF* and a firing variation is expected to increase the η by +0.2% absolute. We expect to further improve the electrical performance by boosting both V_{OC} and J_{SC} values by: a) minimizing *IQE*-losses in short wavelengths by co-optimization of texture and emitter diffusion processes, and b) avoiding the recombination associated with the material quality of standard DWS mc-Si substrates.

5. ACKNOWLEDGEMENTS

The authors would like to thank all co-workers at Photovoltaics department of Fraunhofer ISE, especially Elisabeth Schaeffer and Felix Martin for their help in solar cell characterization. The work is supported by KIC InnoEnergy within the research project ADE-GLOBAL.

6. REFERENCES

[1] B. Kafle, A. Mannan, T. Freund, L. Clochard, E. Duffy, M. Hofmann, J. Rentsch, and R. Preu, Nanotextured multicrystalline Al-BSF solar cells reaching 18% conversion efficiency using industrially viable solar cell processes, Phys. Status Solidi RRL 9 (8) (2015) 448–452.

[2] B. Kafle, J. Seiffe, M. Hofmann, L. Clochard, E. Duffy, J. Rentsch, Nanostructuring of c-Si surface by F_2 -based atmospheric pressure dry texturing process, Phys. Status Solidi A 212 (2) (2015) 307–311.

[3] B. Kafle, A. I. Ridoy, P. Saint-Cast, L. Clochard, E. Duffy, K. Duncker, K. Petter, M. Hofmann, J. Rentsch, Atmospheric pressure dry texturing enabling 20% conversion efficiency on multicrystalline silicon PERC solar cells, AIP Conference Proceedings 1999 (2018) 50003.

[4] International Technology Roadmap for Photovoltaic (ITRPV): 2017 Results including maturity report: Eighth edition, 2018.

[5] P. Engelhart, D. Manger, B. Klöter, S. Hermann, A.A. Stekolnikov, S. Peters, H.-C. Ploigt, A. Eifler, C. Klenke, A. Mohr, G. Zimmermann, B. Barkenfelt, K. Suva, J. Wendt, T. Kaden, S. Rupp, D. Rychtarik, M. Fischer, J.W. Müller, P. Wawer, Q.ANTUM – Q-CELLS NEXT GENERATION HIGH-POWER SILICON CELL & MODULE CONCEPT, in 26th European PV Solar Energy Conference and Exhibition, Hamburg, Germany (2012).

INCREASED EFFICIENCY WITH CdSeTe LAYER IN FRONT OF CdTe

James Sites, Alexandra Huss, Tao Song
Colorado State University, Fort Collins, CO 80525 USA
Martina Lingg, EMPA, Dübendorf, Switzerland

ABSTRACT: The highest-efficiency CdTe cells at both First Solar and Colorado State University have been made with a layer of the lower band-gap CdSeTe alloy between the emitter and the CdTe. Simulations which include the CdSeTe layer are in agreement with experimental results. The CdSeTe thickness has generally been the order of 1 µm. Most of the solar spectrum is therefore absorbed in the CdSeTe layer, which means that the cell structure is best described as a CdSeTe absorber with CdTe behind to form a back surface field. For Se/Te ratios between 0.1 and 0.4, the band gap decreases from 1.50 eV for CdTe to the neighborhood of 1.4 eV for the alloy. This decrease in band gap implies an increase in current density of approximately 3 mA/cm^2, which is quite close to the amount seen experimentally. The voltage in both simulation and experiment for a single-layer CdSeTe absorber decreases by an amount similar to the 100-meV band-gap difference, but recovers to very nearly the same voltage as the CdTe cell when the CdSeTe/CdTe combination is employed. This recovery is strongly suggestive of a conduction-band offset near 100 meV that creates a back-surface field which reflects electrons at the CdSeTe/CdTe transition. The result demonstrates that the CdTe voltage deficit compared to the ideal can be reduced by approximately the 100-meV difference between the CdSeTe and CdTe band gaps and explains why an increase in absolute efficiency of approximately 2% has been observed.
Keywords: Solar Cells, Cadmium Telluride, Band Structure

1 INTRODUCTION

The record efficiency for thin-film CdTe cells has been pushed considerably higher during the past several years by GE and more recently by First Solar. The current CdTe efficiency record of 22.2% [1] is very similar to that of the competing multicrystalline silicon and thin-film CIGS technologies. The use of a graded CdSeTe layer to assist with the record efficiency has been reported at various conferences and its lower band gap is apparent in First Solar's quantum-efficiency curves. The highest reported efficiency at Colorado State University (CSU) is 19.2% [2] is smaller, but as with First Solar, it incorporated a CdSeTe layer in front of the CdTe. The purpose of the work reported here is to provide likely explanations for the benefit resulting from the inclusion of the CdSeTe layer.

2 DEVICE STRUCTURE AND FABRICATION

The solar-cell structure discussed here is shown in Fig. 1. Several of the key features were discussed at EU-PVSEC-34 [3]. The materials are deposited in successive layers on TEC 10 soda-lime glass from NSG, which is furnished with fluorine-doped tin oxide to provide a transparent conductive-oxide contact. The thin MgZnO (MZO) layer is rf-sputter-deposited and serves as the emitter in place of the more commonly used CdS [4]. Its larger band gap allows greater transmission of short wavelength photons, and its conduction-band offset with CdTe or CdSeTe (shown in Fig. 2) decreases the hole concentration at the front interface.

The absorber layers (CdSeTe and CdTe) are deposited in various thickness combinations by close-space sublimation in a common vacuum system with several deposition stages [5,6]. With the thicknesses of the CdSeTe shown, photons are primarily absorbed in the CdSeTe layer (98% of the photons in the CdSeTe for 1-µm thickness and 90% for 0.5 µm). Hence, the structure might best be described as a CdSeTe cell with a rear layer of CdTe.

Passivation of the CdTe and CdSeTe absorber layers is done in the same vacuum chamber with a single deposition of $CdCl_2$ followed by annealing. A very small amount of copper is then applied to increase the hole density at the back surface of the CdTe. The thin layer of tellurium that follows is used to divide the hole barrier between the CdTe and metal contact and thus to effectively reduce it [7]. A thicker layer of nickel completes the structure, and individual 8-mm square cells are delineated for the experimental cell measurements.

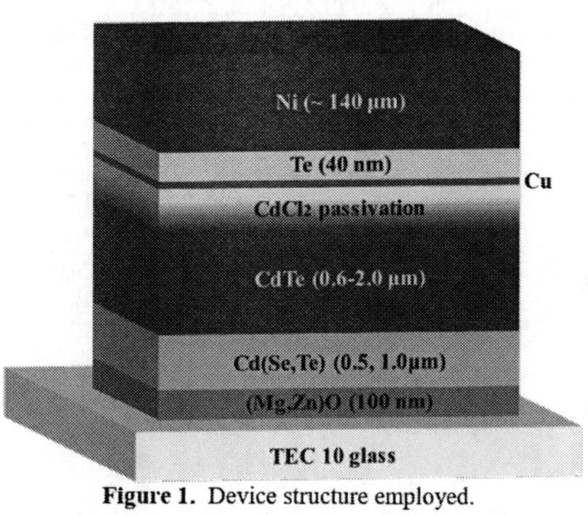

Figure 1. Device structure employed.

3 DEVICE SIMULATION

Figure 2 shows the band diagram for the standard CSU CdTe solar cell in the top panel and for the two-layer CdSeTe/CdTe cell in the lower one. It is assumed here that at least most of the band-gap difference appears in the conduction band. The band gaps for both cells are shown with a forward bias of 0.8 V, which would be approaching the open-circuit voltage of a good-quality cell. At this voltage, there is an approximate balance of current in both directions and therefore little or no net current.

The band diagram for the CdTe absorber in the upper panel shows two of the key features of the CdTe cells fabricated at CSU: the MZO emitter layer that produces the beneficial spike in the conduction band and the Te layer that reduces the band bending at the rear of the absorber. The Te layer, however, is only partially effective, since it still allows some band-bending at the back and thus the forward electron current shown that limits voltage.

The lower panel of Fig. 2 shows the thin CdSeTe layer that is now added in front of the CdTe. By itself, the smaller band-gap absorber (1.4 eV compared to 1.5) would be expected to increase the cell current, but reduce its voltage by approximately the 100-mV band-gap difference. If the band-gap difference in the bilayer occurs primarily in the conduction band, however, there will be a small electron barrier between the CdSeTe and CdTe. This difference would allow the bands to flatten more before the opposite currents are equal, and hence one would expect a voltage closer to that of a CdTe cell. In fact, the conduction-band offset is likely to be somewhat greater than 100 mV with a small negative offset in the valence band [8] so that the sum would be 100 mV.

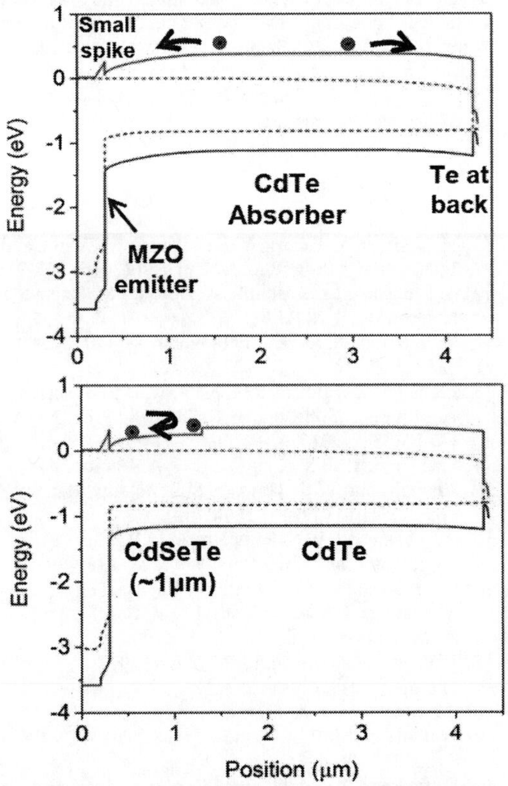

Figure 2. Calculated band diagrams of cells with CdTe-absorber only (above) and two-layer CdSeTe/CdTe absorber (below). Bias in both cases is +0.8 V.

Current-voltage (J-V) curves calculated using SCAPS software [9] are shown in Fig. 3 for three cases which incorporate the same device parameters apart from the absorber band gaps. The blue curve is for a standard 1.5-eV CdTe cell assumed to have a carrier density of 10^{14} cm^{-3}, a bulk recombination lifetime of 5 ns, and a rear-interface recombination velocity of 10^{5} cm/s. The resulting 844-mV voltage is typical of good-quality CdTe cells. When the band gap in the calculation is narrowed to 1.4 eV for a single-layer CdSeTe cell (green curve), the voltage decreases by 100 mV and the current increases by approximately 3 mA/cm^2. The current increase is what would be expected from the increased solar spectrum below the respective band gaps.

Calculation of J-V for the combination CdSeTe/CdTe absorber, again with the same cell parameters except for the band gaps, results in the red curve, which maintains most of the higher current and restores the CdTe voltage to near its original value. Some current is lost because the longer wavelength photons are not fully absorbed by the CdSeTe, and the voltage slightly exceeds the original value because of the larger current. Two variations in the simulation were also tested (1) division of the band-gap difference equally between the conduction and valence bands, where the voltage is only partially restored and (2) a decrease of the CdSeTe carrier density to 10^{13} cm^{-3}. In the latter case, was there was no significant change in the curves.

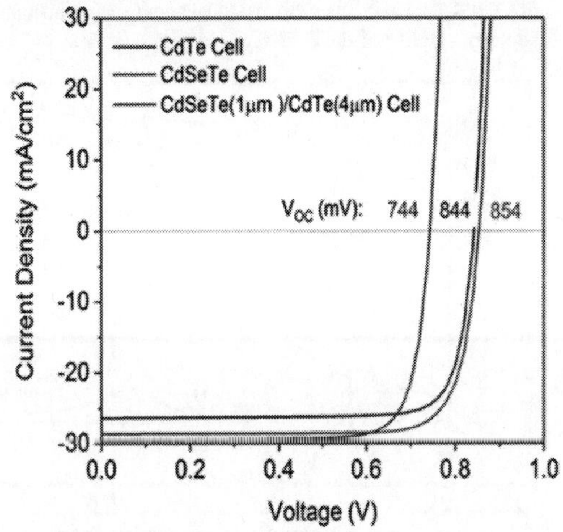

Figure 3. Calculated J-V curves for three cases.

4 EXPERIMENTAL RESULTS

Experimental results from the three types of cells discussed above gave results closely aligned with the simulations. Experimental variation of the Se concentration from 0 to 20% in cells with CdSeTe-only absorbers showed that the cell voltage and band-gap, which was deduced from quantum efficiency and photoluminescence measurements, showed a continuous decrease to the lower Se concentrations. The voltages and currents are compared from the simulations and measurements for the three cases in Table I.

Table I: Simulation comparison with experiment

Absorber	V_{OC} calc	V_{OC} meas	J_{SC} calc	J_{SC} meas
CdTe	844 mV	830 mV	25.9	23.5
CdSeTe	744	720	29.2	26.3
CdSeTe/CdTe	854	829	28.6	26.6

Because the experimental conditions were less ideal than those used in the simulations, the experimental values were smaller for both voltage and current, but the differences in the voltage and current in each of the three cases were similar. The experimental efficiency increased from 14.1% for the CdTe cell to 16.2% for the CdSeTe/CdTe combination, whereas the simulation increase was from 18.0 to 19.5%. The efficiency for the CdSeTe-only cell was slightly less than that for CdTe-only cell [10], which may likely result from a larger hole barrier when the alloy is used by itself.

A second reason for the superiority of the CdSeTe/CdTe combination is the quality of the CdSeTe layer. This difference is apparent in the top panel of Fig. 4, which shows the much larger photoluminescence (PL) intensity from the CdSeTe in the bilayer cell compared to that from a similar CdTe cell. It also shows the 100-meV shift of the peak. A similar large increase has also been seen in the intensity of cathode luminescence (CL) from CdSeTe in the bilayer [11]. The lower panel of Fig. 4 shows the comparison of time-resolved PL between a bilayer cell and one with a CdTe-only absorber. In this case the electron-hole recombination decay time is seen to be significantly longer when the emission comes from the CdSeTe layer than from CdTe, again indicating the superior quality of the CdSeTe.

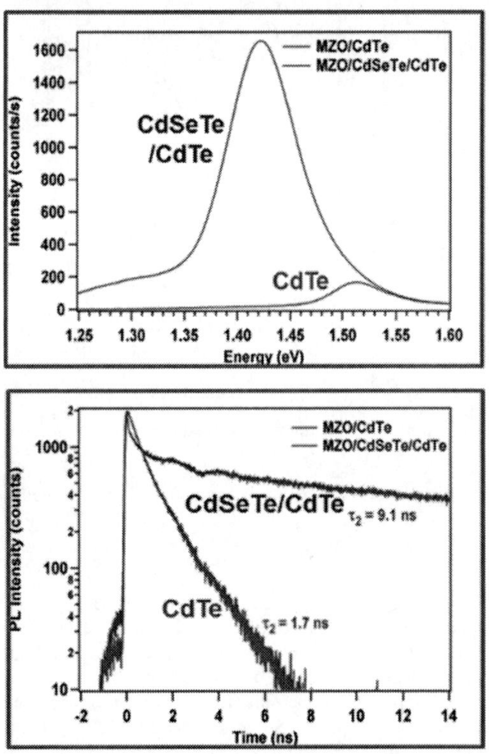

Figure 4. Comparison of spectral PL(top) and time-resolved PL (bottom) between CdTe and CdSeTe/CdTe absorbers.

5 DISCUSSION

There is a clear advantage to the CdSeTe/CdTe absorber combination and a reasonable argument that it primarily results from a back-surface field at the transition. The current increase for CdSeTe compared to CdTe is significant, and it is maintained in the two-layer structure. The voltage restoration very plausibly results from the conduction-band barrier at the transition, since the CdSeTe layer is thin and has low carrier density, and is therefore fully depleted. The voltage in this model can therefore be restored to near that of a CdTe cell if one assumes there are no other changes to the cell structure. To attain still higher voltage, one possibility is an additional back-surface field at the rear of the CdTe layer. A wider band-gap alloy of CdTe at the back of the bilayer could plausibly enhance the voltage further and reduce the voltage deficit to values comparable with crystalline cells.

6 ACKNOWLEDGMENTS

The material presented here is based upon work supported by the U.S. Department of Energy's Office of Energy Efficiency and Renewable Energy Technologies Office (SETO) Agreement Number DE-EE-0007543. Additional funding for equipment development and maintenance came from the NSF Accelerating Innovation Research program, and for Martina Lingg's support from the Swiss National Science Foundation. The authors are particularly grateful to their colleagues Jennifer Drayton, Amit Munshi, W.S. Sampath, Anna Wojtowicz, and Pascal Jundt at Colorado State and Darius Kuciauskas at the U.S. National Renewable Energy Laboratory for assistance with the research.

7 REFERENCES

[1] M.A. Green K. Emery, Y. Hishikawa, W. Warta, and E.D. Dunlop, Prog. Photovoltaics 24 (2016) 905-916.
[2] A. Munshi, J.M. Kephardt, A. Abbas, J.M. Raguse, J-N. Beaudry, K.L. Barth, J.R. Sites, J.M. Walls, and W.S. Sampath, IEEE J. Photovoltaics 8 (2018) 310-314.
[3] J.R. Sites, A. Munshi, J.M. Kephart, D.E. Swanson, A. Moore, T. Song, and W.S. Sampath, Proc. 34th EU-PVSEC (2017).
[4] J.M. Kephart, J. McCamy, Z. Ma, A. Ganjoo, F. Alamgir, and W.S. Sampath, Solar Energy Mat. and Solar Cells 157 (2016) 266-275.
[5] D.E. Swanson, W.S. Sampath, and J.R. Sites, Solar Energy Mat. and Solar Cells 159 (2017) 389-392.
[6] P.S. Kobyakov, A. Moore, J.M. Raguse, D.E. Swanson and W.S. Sampath, J. Vac. Sci, Technol. A32 (2014) 021511.
[7] T. Song, A. Moore, and J.R. Sites, IEEE J. Photovoltaics 8 (2018) 293-298.
[8] B.I. MacDonald, A. Martucci, S. Rubanov, S.E. Watkins, P. Mulvaney, and J.J. Jasieniak, ACS Nano 6 (2012) 5995-6004.
[9] Burgelman, P. Nollet, and S Degrave, Thin Solid Films 361 (2000) 527-532.
[10] M. Lingg, A. Munshi, W.S. Sampath, J.R. Sites, S. Buecheler, and A.N. Tiwari, Proc. E-MRS Spring Meeting (2018).
[11] T.A.M. Fiducia, A. Munshi, K.L. Barth, D. Proprentner, G. West, W.S. Sampath, and J.M. Walls, Proc. 7th WCPEC (2018).

INTRODUCTION OF COPPER BY WET DEPOSITION IN CDTE SOLAR CELLS

Elisa Artegiani, Daniele Menossi, Marco Barbato*, Matteo Meneghini*, Gaudenzio Meneghesso*, Mauro Leoncini, Marco Cavallini and Alessandro Romeo.

LAPS -Laboratory for Photovoltaics and Solid State Physics, Department of Computer Science, University of Verona, Ca' Vignal 1, Strada Le Grazie 15, 37134 Verona, Italy, e-mail address: alessandro.romeo@univr.it

*Department of Information Engineering, University of Padova, Via Gradenigo 6/B, 35131, Padova, Italy

ABSTRACT: CdTe thin film solar cells have reached an efficiency of 22.1%. These solar cells are generally fabricated adding copper to the absorber. Despite different copper-free back contacts have been introduced, the presence of copper has demonstrated to deliver the highest efficiency. On the other hand, Cu is a fast diffuser in CdTe, leading to the formation of carrier traps and resulting in degradation of the devices. For this reason a suitable Cu concentration and the formation of a stable back contact are very important to reach high efficiency and stability. In this paper, a copper chloride wet deposition method has been developed, and a comparison between the addiction of copper through vacuum evaporation and wet deposition is presented. Through a solution, the deposition uniformity of a very small amount of copper can be easily controlled. Moreover, with this method the obtained efficiencies are higher than with same quantity of evaporated copper, and are very similar to the device with larger quantity of evaporated copper but with a dramatic stability improvement.
Keywords: CdTe, back contact, stability.

1 INTRODUCTION

Last year First Solar has achieved a new record efficiency of 22,1% [1] for CdTe solar cells, obtaining efficiencies very close to crystalline silicon. In addiction CdTe modules are the most successful thin film photovoltaic devices in terms of industrial production [2].

One of the main critical technical issue of this technology is the formation of the ohmic contact on the CdTe absorber layer, because of its high work function and the difficulty to dope it. In order to fabricate good performing back contact, copper has been found to be a crucial element which is rarely missing for devices above 16%.. It dopes the CdTe surface as a p^+ region and forms a back contact lowering the barrier height; moreover, in certain conditions, it dopes the CdTe bulk increasing the carrier concentration. On the other hand, Cu diffuses into the CdTe bulk and CdS film shunting the device and in some cases forming carrier traps degrading the device performance [3].

As a consequence, it is very important to optimize the amount of copper which influences the degradation of the cells. Even with methods that limit copper diffusion, such as the application of buffer layers (e.g As_2Te_3 [4] and Bi_2Te_3 [5]), the deposition of a ZnTe:Cu layer [6], copper quantity needs to be calibrated. Also the formation of a Te-rich layer on the CdTe surface through chemical etching, in order to favor the formation of Cu_xTe compounds [3,7].

We have studied a way to minimize and optimize the amount of copper applied without having efficiency reduction. This is done by developing a wet deposition method where copper chloride is used as Cu carrier into the CdTe bulk. Depending on the copper concentration and on the amount of solution used, any very small amount of copper can be uniformly deposited all over the sample area.

Moreover this method optimizes the positive effect of copper on CdTe. Depositing an amount of solution equivalent to a 0.1 nm thick copper layer, the samples have a nice ohmic back contact and efficiencies comparable to those of samples containing a 2.0 nm thick layer deposited by vacuum evaporation. Finally, accelerated aging tests have been performed on both types of samples, revealing that a significant stability improvement has been achieved by the copper chloride wet deposition method.

2 EXPERIMENTAL

2.1 Preparation of the devices

CdTe thin film solar cells are made in superstrate configuration by a low-temperature fabrication process based on vacuum evaporation (VE). On an ITO/ZnO stack as front contact, 150 nm thick CdS thin film is deposited at 150 °C and then annealed at 450 °C in vacuum. Subsequently 7 microns of CdTe thick layer are deposited in the same vacuum chamber and subsequently activated by $CdCl_2$ treatment, which is applied by wet deposition, and annealed in air at 380 °C.

Prior to back contact deposition, the CdTe surface is etched by a bromine/methanol solution in order to remove $CdCl_2$ residues and to generate with subsequent deposited copper the beneficial Cu_xTe compound.

In our standard process, the back contact is made by vacuum evaporation of copper and gold. A subsequent annealing in air at 200°C of the finished device is needed to reach good efficiencies. With this method high efficiencies are obtained with large copper quantities: 2 nm thick layer delivers the best efficiency values.

For the wet deposition method, the copper chloride is diluted in methanol, mixed and deposited on top of the CdTe, the stack is then annealed in air at 200°C. A final gold coating finishes the cells.

The amount of solution applied is equivalent to a deposition of 0,1 nm thick copper layer.

2.2 Characterization

Current density-voltage (JV) measurements are performed using a Keithley Source Meter 2420 at room temperature, using a halogen lamp calibrated with a silicon solar cell under an irradiation of 100 mW/cm^2.

The external quantum efficiency (EQE) was obtained using a commercial LOANA solar cell analysis system, calibrated with a silicon reference sample with known EQE using an incident spotlight of 1 mm x 2 mm area.

Accelerated stress tests (AST) were performed in a specific metal chamber, where a rack of halogen lamps and a temperature system control allows to keep the cells under one sun and 80 °C.

3 RESULTS

3.1 Performance

Reducing copper is one of the possible paths in order to minimize the degradation effects due to its diffusion. In a previous work we have shown that it is possible to successfully reduce the thickness of the evaporated copper layer [8], with a good efficiency but however with a pronounced roll-over.. In fact this might lead to a possible lower CdTe doping of both the bulk and the surface region, and consequently to a lower net carrier concentration and to a higher barrier height at the contact.

In our samples reducing the thickness of copper layer from 2.0 nm to 0.1 nm, the average efficiency drops from 16% to 14%. In these samples the appearance of a marked roll over, suggesting the presence of a high series resistance at the back contact, causes a strong fill factor decrease.

On the other hand, introducing the equivalent to 1 Å thick copper layer at the back contact of the samples, but this time by the introduction of $CuCl_2$, the samples do not show any efficiency drop,. In fig.1 two J-V curves are depicted: one is typical of a sample containing a 2.0 nm thick copper layer by vacuum evaporation, that from now on will be called Ve-Cu, and the other of a sample containing the equivalent to a 1 Å thick copper layer by wet deposition, that will be called CCWD. The two JV characteristics are almost perfectly overlying, as also confirmed by the efficiency parameters of the cells shown in table I, suggesting that the wet deposition system delivers high efficiency and a good ohmic back contact, despite the smaller amount of copper used.

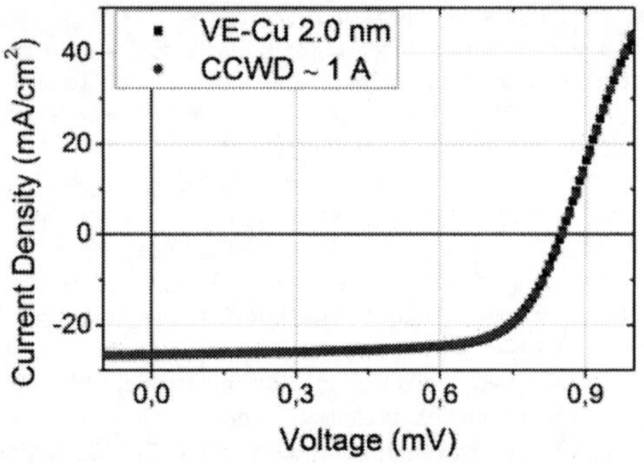

Figure 1: comparison between J-V characteristics of devices with a 2.0 nm thick copper layer by evaporation (black curve), and with the equivalent amount of a 1 Å thick Cu layer by wet deposition (red curve).

Table I: efficiency parameters of devices of Fig. 1: conversion efficiency (η), fill factor (FF), open circuit voltage (Voc) and short circuit current density (Jsc).

sample	Eff (%)	FF (%)	Voc (mV)	Jsc (mA/cm^2)
VE-Cu	15.9	70.6	852	26.4
CCWD	15.8	70.0	852	26.5

3.2 External quantum efficiency

Introduction of copper by copper chloride wet deposition method was developed on the basis of the activation treatment knowledge. The $CdCl_2$ treatment is a key step for the improvement of solar cells efficiency, and it has been known and studied since 1976 [9]. In particular, chlorine reacts with CdTe enhancing the recrystallization of the structure. Moreover it passivates the grain boundaries and it favours the formation of the CdS_xTe_{1-x} intermixed layer at the interface, which reduces the reticular mismatch between CdTe and CdS [10]. As seen from the JV characteristics of fig.1, the chlorine compound optimizes the role of copper, as it favour the formation of an ohmic back contact with a minimum amount of Cu. Moreover it can not be excluded that it also might change the interface and the bulk structure of the absorber.

To analyse this, External Quantum Efficiency (EQE) measurements have been performed on VE-Cu and CCWD samples (see figure 2).

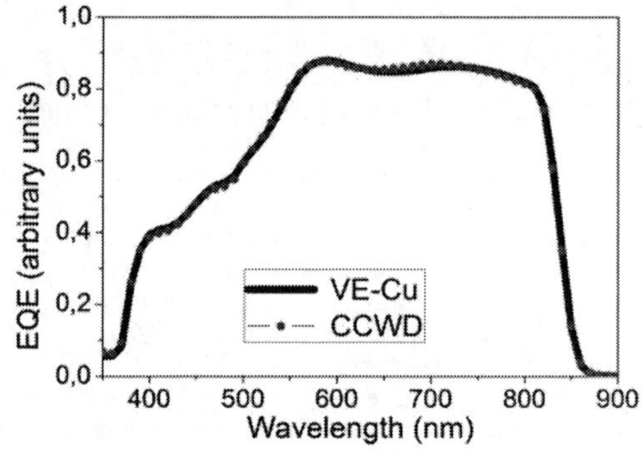

Figure 2: comparison between EQE curves of a VE-Cu sample (black curve) and a CCWD sample (red dots).

The EQE curve of the CCWD sample overlaps almost perfectly the one of the VE-Cu sample. This means that the alternative method used to add copper does not influence the main features of the absorber layer, such as the band gap energy and the optical absorption coefficient.

Moreover, the similar response at long-wavelength region shows that the CdS_xTe_{1-x} intermixed layer has not been modified by the additional chlorine treatment. In fact, it is known that this intermixed layer has a lower band gap than CdTe, and thus its formation allows the absorption of photons with slightly less energy, shifting the EQE response [11]. This means that the $CuCl_2$ treatment does not affect the inter-diffusion of tellurium and sulphur at the interface.

3.3 Aging

In order to investigate the stability of the CCWD samples compared to the one of the VE-Cu samples, accelerated stress tests (AST) have been conducted.

AST were performed in a specific metal box, where a rack of halogen lamps and a temperature-controlled system allows to keep the cells under an illumination of one sun and at a temperature of 80°C. Two set of samples have been stored in the metal box for weeks, and their performance degradation has been analyzed by measuring their efficiency at different time steps. A set was composed by CCWD samples, and the other by VE-Cu samples; both the sets had an initial average efficiency slightly below 14%.

Table II: Average efficiency parameters of CCWD devices at different aging steps.

CCWD				
AST time (h)	Eff (%)	FF (%)	Voc (mV)	Jsc (mA/cm^2)
0	13.9 ± 0.4	65 ± 2	818 ± 8	26.2 ± 0.6
336	13.0 ± 0.3	62.6 ± 0.6	814 ± 4	25.5 ± 0.5
672	12.2 ± 0.4	60 ± 1	800 ± 10	25 ± 1
1008	11.7 ± 0.3	60 ± 1	782 ± 6	25.0 ± 0.8

Table III: Average efficiency parameters of VE-Cu devices at different aging steps.

VE-Cu				
AST time (h)	Eff (%)	FF (%)	Voc (mV)	Jsc (mA/cm^2)
0	13.9 ± 0.8	69 ± 4	849 ± 5	23.6 ± 0.1
336	8.1 ± 0.3	50 ± 1	708 ± 5	22.8 ± 0.7
672	7.2 ± 0.5	49 ± 1	698 ± 5	21 ± 2
1008	7.3 ± 0.4	48 ± 1	694 ± 5	22 ± 1

Table IV: Relative performance parameters (in percentage) of samples after 1008 h of AST.

	Percentage of initial value			
samples	Eff	FF	Voc	Jsc
CCWD	84 %	92 %	96 %	95 %
VE-Cu	53 %	70 %	82 %	93 %

As it can be see comparing table II and table III, the CCWD samples show reduced degradation. In these tables it is possible to read the average performance parameters of the samples at different time steps during the AST. In particular, after 1008 hours of AST the CCWD devices have an average efficiency of (11.7 ± 0.3) % compared to the (7.3 ± 0.4) % of the VE-Cu ones. However VE-Cu cells degrade very fast at the beginning: the larger efficiency drop is in the first 336 h, with an efficiency of (8.1 ± 0.3) %.

In table IV the average performance parameters of the samples after 1008 h of AST as a percentage of their initial value are presented. The data in this table suggest that the fill factor decrease is the main responsible of VE-Cu samples degradation: after 1008 h of AST the VE-Cu samples have reduced their average FF to 70 % of the initial value, while the CCWD samples to 92 %. Moreover, at the end of the test the CCWD samples still have a good average efficiency (84 % of the initial value) while at the same time VE-Cu samples have reduced their average efficiency to 53%.

The higher degradation of VE-Cu samples can be explained considering the larger amount of copper at the back contact, thus the higher copper diffusion. And as already mentioned, by diffusing Cu can form recombination centers and shunt pathways [12], which result in a reduction in performance of the solar cells.

Anyway, accelerated stress tests clearly confirm that the addiction of copper through wet deposition method allows the fabrication of much more stable devices.

4 CONCLUSION

A copper chloride wet deposition method has been developed to add Cu at the back contact of CdTe solar cells. Using a solution allows to simply deposit any ideal small amount of copper evenly over the entire area of the sample, favoring the deposition of a minimum amount of copper. Samples containing an amount of copper equivalent to a 1 Angstrom thick copper layer deposited by wet deposition reach similar efficiency than ones containing a 2.0 nm thick copper layer by vacuum evaporation. This alternative method to introduce copper brings to completely different mechanisms of inclusion of copper inside the CdTe, optimizing the positive effects of Cu. Moreover the samples achieve a dramatic stability improvement. Further details on this work, and further in-depth analysis will be soon published.

5 REFERENCES

[1] S. Krum, S. Haymore, First Solar Achieves Yet Another Cell Conversion Efficiency World Record, First Sol. (2016).

[2] S. Philipps, F. Ise, W. Warmuth, P. Conferences, C. GmbH, Photovoltaics Report, (2018). doi:20.10.2016.

[3] J. Zhou, X. Wu, A. Duda, G. Teeter, S.H. Demtsu, The formation of different phases of CuxTe and their effects on CdTe/CdS solar cells, Thin Solid Films. 515 (2007) 7364–7369. doi:10.1016/j.tsf.2007.03.032.

[4] N. Romeo, a. Bosio, a. Romeo, An innovative process suitable to produce high-efficiency CdTe/CdS thin-film modules, Sol. Energy Mater. Sol. Cells.` 94 (2010) 2–7. doi:10.1016/j.solmat.2009.06.001.

[5] A. Romeo, A. Salavei, I. Rimmaudo, A. Bosio, D. Menossi, F. Piccinelli, et al., Electrical Characterization and Aging of CdTe Thin Film Solar Cells with Bi 2 Te 3 Back Contact, (2013) 1178–1182.

[6] T.A. Gessert, A.R. Mason, P. Sheldon, A.B. Swartzlander, D. Niles, T.J. Coutts, Development of Cu-doped ZnTe as a back-contact interface layer for thin-film CdS/CdTe solar cells, J. Vac. Sci. Technol. A Vacuum, Surfaces, Film. 14 (1996) 806–812. doi:10.1116/1.580394.

[7] I. Rimmaudo, A. Salavei, E. Artegiani, D. Menossi, M. Giarola, G. Mariotto, et al., Improved stability of CdTe solar cells by absorber surface etching, Sol. Energy Mater. Sol. Cells. 162 (2017) 127–133. doi:10.1016/j.solmat.2016.12.044.

[8] E. Artegiani, D. Menossi, C. Ferrari, A. Romeo,

Effects of Ultra-Thin Copper Layer on the Performance and Stability of CdTe/CdS Solar Cells, in: 33rd Eur. Photovolt. Sol. Energy Conf. Exhib., 2017: pp. 1001–1003. doi:10.4229/EUPVSEC20172017-3AO.7.4.

[9] S.I. and Y.H. Nobuo Nakayama, Hitoshi Matsumoto, Kazufumi Yamaguchi, Ceramic Thin Film CdTe Solar Cell, (1976). doi:10.1143/JJAP.15.2281.

[10] I. Dharmadasa, Review of the CdCl2 Treatment Used in CdS/CdTe Thin Film Solar Cell Development and New Evidence towards Improved Understanding, Coatings. 4 (2014) 282–307. doi:10.3390/coatings4020282.

[11] Y. Yan, R.G. Dhere, K.M. Jones, M.M. Al-Jassim, Influence of substrate structure on the growth of CdTe thin films, J. Appl. Phys. 89 (2001) 5944–5948. doi:10.1063/1.1367406.

[12] K.D. Dobson, I. Visoly-fisher, G. Hodes, D. Cahen, Stability of CdTe / CdS thin-film solar cells, Sol. Energy Mater. Sol. Cells. 62 (2000) 295–325.

IN-LINE MOCVD OF ALUMINIUM ALLOYED ZINC SULFIDE: A PATH TO HIGH PERFORMANCE CDTE SOLAR CELLS

A.J. Clayton, O. Oklobia, P.J. Siderfin, S. Jones, G. Kartopu, A.C. Teloeken, D.A. Lamb and S.J.C. Irvine
Centre for Solar Energy Research, Swansea University
OpTIC Centre, Ffordd William Morgan, St. Asaph Business Park, Denbighshire, North Wales, UK

ABSTRACT: An aluminium alloyed zinc sulfide (Zn(Al)S) layer produced on NSG TEC™ C10 substrates using in-line metal organic chemical vapour deposition (MOCVD) was incorporated into thin film cadmium telluride (CdTe) devices as a high resistance transparent (HRT) layer. Inclusion of the Zn(Al)S improved the open circuit voltage (V_{oc}) of the CdTe solar cells. The low electron affinity of the Zn(Al)S was shown by SCAPS to introduce a positive conduction band discontinuity, attributing to the enhanced V_{oc}. This improvement was preserved as the cadmium zinc sulfide (Cd(Zn)S) buffer thickness was reduced. The Zn(Al)S was well adhered to the NSG TEC™ C10 substrate and acted to prevent delamination of the device structure, previously observed, during more aggressive cadmium chloride ($CdCl_2$) activation treatment (CHT), which occurred for devices without the HRT layer. Aggressive CHT parameters improved short circuit current-density for devices with 0.05 μm thick Cd(Zn)S and 0.1 μm Zn(Al)S HRT layer, with some loss to V_{oc} for the thinner Cd(Zn)S. Scanning electron microscopy revealed that grain enlargement increased after more aggressive CHT, reducing the grain boundary density where defect states typically reside.
Keywords: CdTe; Thin Film; CVD Based Deposition; Interfaces; Solar Cell Efficiencies

1 INTRODUCTION

Introduction of a highly resistant transparent (HRT) layer into thin film solar cell devices is a proven method for allowing thinning of the n-type buffer layer [1]. It has been applied to thin film cadmium telluride (CdTe) reducing parasitic absorption losses in the buffer layer [2]. It also limits micro-shorts created through pin-holes between the CdTe and transparent conducting oxide (TCO) and reduces lateral inhomogeneities across the solar cell [3]. This helps to preserve both fill factor (*FF*) and open circuit voltage (V_{oc}).

CdTe solar cells are typically produced using close spaced sublimation (CSS), which is carried out at high temperatures (~600°C) and leads to large polycrystalline grains with low density of grain boundaries [4-6]. Metal organic chemical vapour deposition (MOCVD), on the other hand, is a lower temperature process producing smaller grains, which relies more on the cadmium chloride ($CdCl_2$) activation treatment (CHT) for grain enlargement to occur during the recrystallisation process [6].

The advantage of MOCVD is the ability to introduce extrinsic dopants during deposition to control the p-type character and the acceptor levels in the CdTe [7]. CdTe solar cells are limited by low acceptor concentrations due to the self-compensating nature of CdTe [8]. If the acceptor concentration can be increased via extrinsic doping, through processes such as MOCVD, improvements to V_{oc} may be possible. For this to be achieved, improved grain enlargement of MOCVD-grown CdTe is one of the parameters that needs to be addressed. CHT at high annealing temperatures is necessary for large grains to be achieved in this case.

This paper demonstrates improvement to V_{oc} for CdTe solar cells produced by MOCVD, with values consistently above 0.8 V. The origins of this enhancement have been explored using SCAPS modelling [9] to determine the effect an aluminium alloyed zinc sulfide (Zn(Al)S) HRT has on the energy bands within the device. The CHT parameters were also investigated to increase grain size with correlation to solar cell performances.

2 EXPERIMENTAL

2.1 Large area Zn(Al)S film deposition

Thin film CdTe solar cells were produced using an all-MOCVD process using commercial fluorine doped tin oxide (FTO) coated soda-lime glass substrates (NSG TEC™ C10). A 0.1 μm thick Zn(Al)S HRT layer was deposited onto substrates cut to 10 x 10 cm^2 using an in-line MOCVD process passing the substrate through the reaction zone at a rate of 5.4 cm/min. and growth temperature of 400°C. Diethylzinc (DEZn), ditertiarybutylsulfide (DtBS) and trimethylaluminium (TMAl) were used as precursors.

2.2 Device Fabrication

The Zn(Al)S/FTO samples were then cut to 5 x 5 cm^2 and transferred to a conventional static horizontally configured MOCVD system to produce the rest of the CdTe device structure, including the n-type cadmium zinc sulfide ($Cd_{0.3}Zn_{0.7}S$) buffer layer deposited at thicknesses of 0.05, 0.1 and 0.15 μm and at a growth temperature of 360°C. The CdTe absorber (3.3 μm) was then deposited at 390°C followed by a $CdCl_2$ activation layer deposited at 200°C. In situ annealing treatment [10] was carried out between 420 – 450°C. The organometallic precursors for this MOCVD equipment included dimethyl cadmium (DMCd), diisopropyltellurium (DiPTe), tris(dimethyl-amino)arsine (TDMAAs) as p-type dopant [7], tertiarybutylchloride (tBuCl), DEZn and DtBS.

Excess $CdCl_2$ was rinsed off with deionized water and the devices dried with nitrogen. A post-growth air anneal at 170°C was then carried out for 90 minutes [11]. Gold back contacts were then defined through a shadow mask using thermal evaporation under vacuum at 4 x 10^{-5} mbar.

2.2 Characterisation

Transmittance (T%) measurements of the Zn(Al)S layers were obtained using an Ocean Optics dual-beam light source and HR4000 high resolution spectrometer. Current density – voltage (*J-V*) measurements of completed CdTe solar cells were carried out with a single pass illumination at air mass (AM)1.5 using an Abet

Technologies Ltd. solar simulator employing a light power density output of 100 mW/cm^2 calibrated using a Gallium Arsenide reference cell. External quantum efficiency (EQE) measurements were performed using a Bentham spectral response spectrometer under non-biased conditions over the spectral range 0.3 – 1.0 μm. High resolution scanning electron microscopy (HR-SEM) was carried out on the devices with a Hitachi S-4800 instrument equipped with field emission gun (FEG) using beam settings at 2 - 10 keV and 10 μA. Capacitance – Voltage (*C-V*) measurements of individual solar cells were performed using an Ametek Scientific Instruments ModuLab XM ECS Electrochemical test System.

3 RESULTS AND DISCUSSION

3.1 Zn(Al)S properties

ZnS is a wide band gap (E_g) material, which has a high resistance [12], making it a good candidate for use as an HRT layer in thin film solar devices. Fig. 1 shows the T% spectra of a ZnS layer compared to Zn(Al)S layers alloyed with less than 1 at.% of Al and another film alloyed with > 1 at.% of Al. It shows that alloying with Al slightly widens the E_g of ZnS. Tauc plots of these films gave optical E_g values of 3.6 - 3.7 eV for the ZnS and Zn(Al)S films respectively. Alloying ZnS with small concentrations of Al has been reported [12] to increase conductivity of ZnS, with resistance then increasing for higher concentrations of Al. It was not possible to measure conductivity of the Zn(Al)S films reported here as the resistances of the films were beyond the limits of our 4-point probe instrument (>5 × 10^8 Ω/□). It is likely that the concentrations of Al within these films are beyond the solubility limit and that alloying is likely to occur.

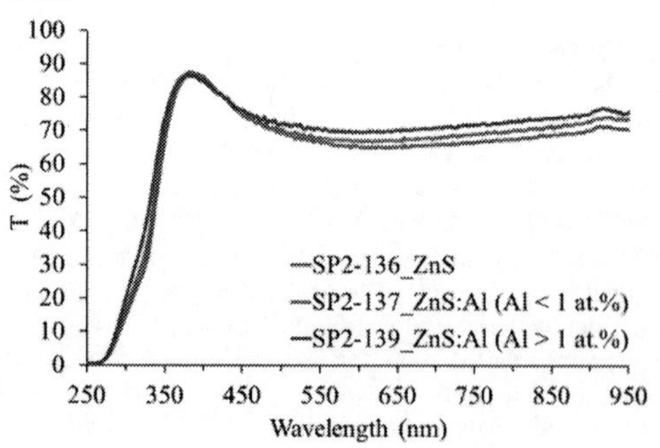

Fig.1: T% spectra of ZnS and Al doped ZnS films

ZnS was also considered as a potential candidate for the HRT layer in thin film CdTe solar cells due to its chemical resistance to etching in hydrochloric acid (HCl). This was relative to zinc oxide (ZnO), typically used as the HRT layer [13], which could be easily removed from the substrate using 1% solution of concentrated HCl. It was also noted that alloying the ZnS with Al increased the chemical resistance to HCl etching. This observation was considered significant for applying a more aggressive CHT process for MOCVD-grown CdTe solar devices, which are treated with $CdCl_2$ under a hydrogen atmosphere, to enable further grain enlargement during the recrystallisation process. This is essential for producing films with low density of grain boundaries and reducing related defect states for enhancing V_{oc}. Fig. 2 shows the etching times for removal of different 0.1 μm HRT layers using 1% HCl solution. It shows ZnS to be significantly more chemically resistant to HCl than ZnO HRT films.

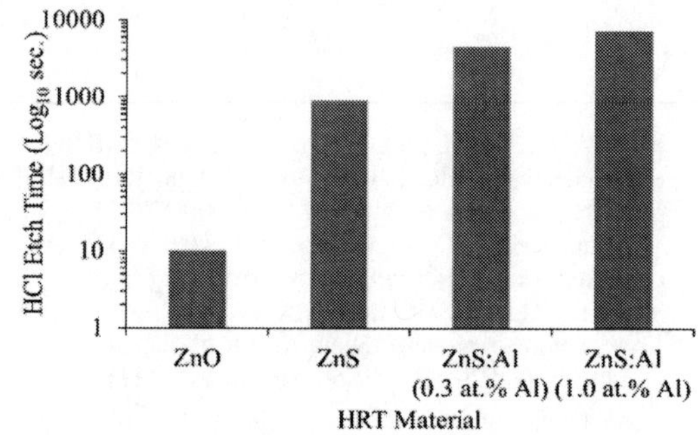

Fig. 2: 1% HCl etching of ZnO with comparison to ZnS and Zn(Al)S.

3.2 Device properties

CdTe solar cell devices were produced under the same conditions, with one device deposited onto an NSG TEC™ C10 substrate with a 0.1 μm Zn(Al)S HRT layer and the other device on a NSG TEC™ C10 substrate without any additional layer. The devices were subjected to an anneal at 440°C during the CHT process (typical anneal carried out at 420°C). The device without the Zn(Al)S HRT layer delaminated at the FTO interface, the effects of which could be seen from the front side of the device through the glass. This is shown in the right-hand side (RHS) image in Fig.3. The equivalent device with the 0.1 μm Zn(Al)S HRT layer (LHS image in Fig. 3) showed no sign of delamination after the more aggressive CHT process.

Fig.3: CHT carried out at 440°C for 10 min. on CdTe solar devices; LHS photo is a device with a Zn(Al)S HRT; RHS photo is a device with no Zn(Al)S HRT.

Devices produced with different CHT parameters were cleaved and HR-SEM was carried out on cross-sections to investigate the effect of CHT anneal temperature on grain size and interdiffusion characteristics. Fig. 4 shows a cross-section of three devices consisting of a 3.3 μm CdTe absorber layer, 0.05 μm Cd(Zn)S buffer layer and a 0.1 μm Zn(Al)S HRT layer with Al ~ 1 at.%. Each device was subjected to a different CHT anneal temperature with annealing time of 10 min. Device 1064 was annealed at 420°C (Fig. 4 (a)), 1163 at 440°C (Fig. 4 (b)) and 1169 at 450°C (Fig. 4 (c)). The thick CdTe layer can be clearly seen in all images. Below the CdTe layer, the Cd(Zn)S and Zn(Al)S cannot be distinguished from one another, but the images do show the FTO layer on the glass substrate.

HR-SEM has shown that the grains for the MOCVD-grown CdTe have gone from a random polycrystalline structure to more columnar with fewer grain boundaries, which nearly extend through the entire thickness of the CdTe layer with the higher temperature CHT. Further optimization of the CHT process will be required to produce large columnar grains that interface with the Cd(Zn)S buffer layer.

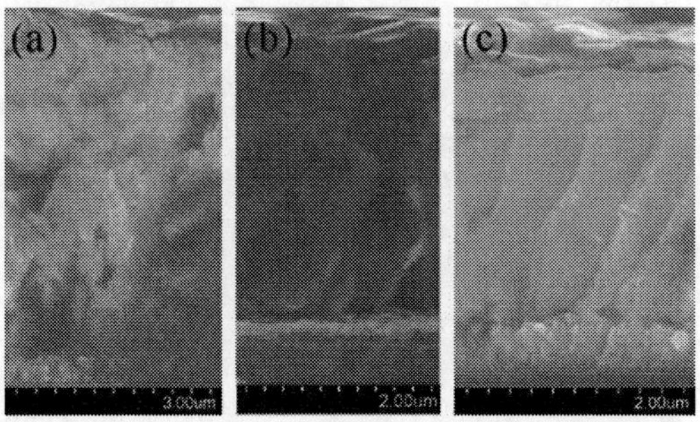

Fig. 4: HR-SEM cross-section of devices with 0.05 μm thick Cd(Zn)S and different CHT anneal set point temperatures; (a) 1064 at 420°C; (b) 1163 at 440°C; (c) 1169 at 450°C. Note, 3 μm scale for (a) and 2 μm scale for (b) & (c).

3.3 Solar cell performances with different HRT layers

CdTe solar cells were produced with 0.1 μm thick ZnS and Zn(Al)S (Al ~ 1 at.%) HRT layers, with comparison to equivalent devices with ZnO HRT layers. The devices with the ZnS HRT were produced in the horizontally configured MOCVD system by sequential growth directly onto the NSG TEC^{TM} C10 substrates. These devices did not show any rectification but very high series resistance and in the case for some cells shunt resistances were as low as 150 Ω cm^2. A previous study on Cd(Zn)S composition for CdTe solar cells [14] also had difficulty in producing solar cells with ZnS.

Devices were fabricated, initially, with 0.15 μm $Cd_{0.3}Zn_{0.7}S$ buffer layer. These devices either had the ZnO HRT, Zn(Al)S HRT or no HRT layer. Other devices were produced omitting the Cd(Zn)S buffer layer altogether to produce a CdTe/HRT device to investigate the effects on V_{oc}. Fig. 5 shows the *J-V* curves for selected cells from each device configuration.

Comparison of the devices with the same buffer thickness (1057, 899 and 1155) revealed that incorporation of a ZnO HRT layer (device 899) resulted in similar V_{oc}, with a slight drop in short-circuit current density (J_{sc}). Inclusion of a Zn(Al)S HRT layer to this same buffer layer thickness device configuration (1155) also resulted in a drop in J_{sc}, but V_{oc} improved nearing 0.8 V. *FF* exceeded 70 % for devices 1057 and 1155, with mean conversion efficiencies (η) of 14.1 and 13.7 %, respectively. Device 899 with the ZnO HRT layer was more variable with *FF* below 70 % and V_{oc} down to 0.71 V in some cases, giving a mean η of 11.9 %.

The difference in device performance became more significant when the Cd(Zn)S buffer layer was omitted from the devices. The device with ZnO (887) had poor V_{oc} with values < 0.6 V. This signified a weak junction formed by the CdTe/ZnO interface. Device 1068, CdTe/Zn(Al)S, had an increase in J_{sc} due to the lack of parasitic absorption in the Cd(Zn)S buffer layer. It also maintained V_{oc} with mean values of 0.79 V. However, *FF* dropped due to a double rectification at forward bias producing a *J-V* curve with an S-shape. Mean *FF* was as low as 42.5 %. The origins of the second rectification will be discussed in a later section.

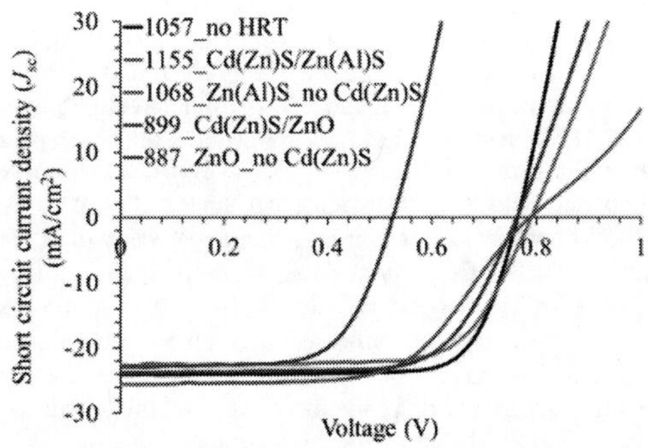

Fig. 5: *J-V* curves for equivalent CdTe solar cells with either a ZnO or Zn(Al)S HRT layer. Comparison is made for devices with a 0.15 μm thick Cd(Zn)S buffer layer with and without a Zn(Al)S HRT layer and with no Cd(Zn)S buffer layer producing a CdTe/HRT interface.

EQE curves for these devices, shown in Fig. 6, highlight the improvement to the short wavelength photocurrent generation when incorporating a ZnO or Zn(Al)S HRT layer with omission of the Cd(Zn)S buffer layer (devices 887 and 1068). The ZnO HRT layer reduces long wavelength photocurrent generation, particularly when there is no Cd(Zn)S buffer layer (device 887). Device 899 shows two short wavelength absorption edges due to the ZnO HRT and the Cd(Zn)S buffer layer. The red-shift of the buffer layer absorption edge compared to devices 1057 and 1155 may be due to a difference in the concentration of Zn in the Cd(Zn)S, with less Zn narrowing the E_g. Except for device 1068, inclusion of ZnO or Zn(Al)S HRT layers reduced photocurrent generation over the red region of the solar spectrum approaching to the CdTe absorption edge. This may imply poorer carrier lifetimes for these devices.

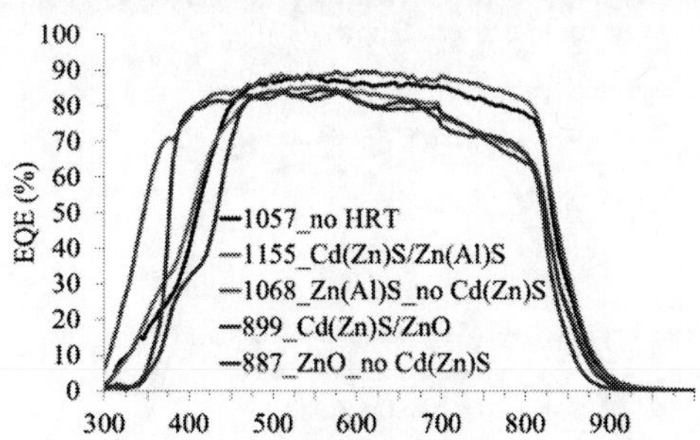

Fig. 6: EQE curves for equivalent CdTe solar cells with either a ZnO or Zn(Al)S HRT layer. Comparison is made for devices with a 0.15 μm thick Cd(Zn)S buffer layer with and without a Zn(Al)S HRT layer and with no Cd(Zn)S buffer layer producing a CdTe/HRT interface.

3.4 Cell performances with varying Cd(Zn)S thickness

Devices were produced with reduced Cd(Zn)S

thicknesses of 0.1 μm (1058) and 0.05 μm (1064), using the 0.1 μm Zn(Al)S HRT layer. Solar cell performance was compared to devices 1155 and 1068, having a 0.15 μm Cd(Zn)S buffer layer and no Cd(Zn)S buffer layer, respectively. *J-V* curves for these devices are represented in Fig. 7 and show the general trend of J_{sc} enhancement as the Cd(Zn)S layer thickness is reduced, until a significant improvement when the Cd(Zn)S buffer layer is omitted from the device (1068). V_{oc} exceeds 0.8 V for devices 1058 and 1064, both having thinner Cd(Zn)S than device 1155. Without the Cd(Zn)S buffer layer, V_{oc} drops to 0.79 V. Fig. 7 also shows a *J-V* curve simulated with SCAPS to represent device 1068 with no Cd(Zn)S buffer. The electron affinity (χ) was varied for the Zn(Al)S HRT layer, which resulted in varying degrees of the second rectification at forward bias. A lower χ value for Zn(Al)S resulted in a more pronounced S-shaped *J-V* curve. The low χ of Zn(Al)S creates a large positive conduction band discontinuity and resulting energy barrier, the cause for the second rectification.

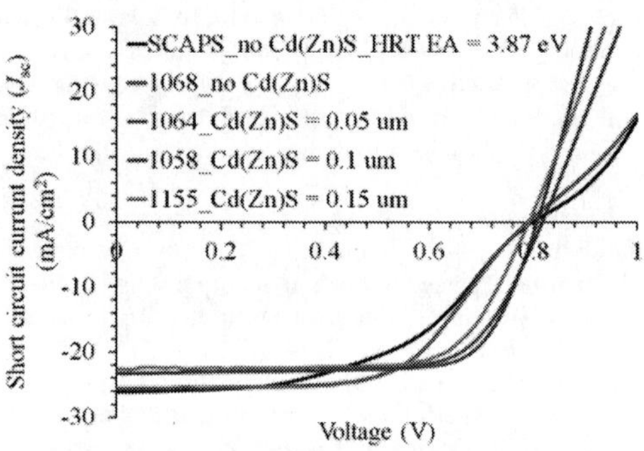

Fig. 7: *J-V* curves of different individual solar cells with variation in Cd(Zn)S buffer layer thickness.

The trend of the J_{sc} is also represented in the EQE curves (Fig. 8), with improvement to short wavelength photocurrent generation with thinning of the Cd(Zn)S buffer. The reduced presence of Cd(Zn)S benefits the EQE in the red region of the solar spectrum towards the CdTe absorption edge. This may be due to a possible improvement in carrier lifetime.

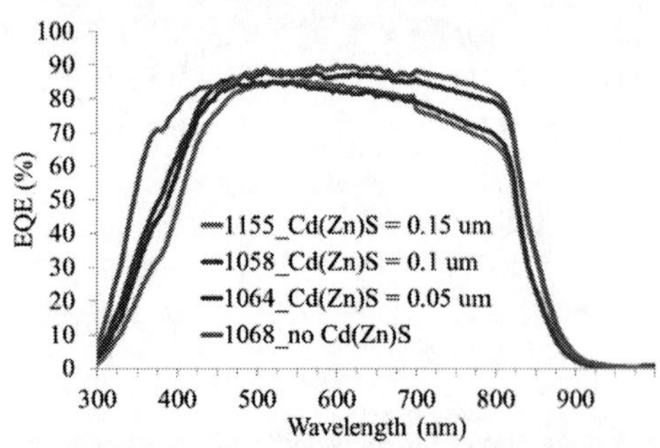

Fig. 8: EQE curves of different individual solar cells with variation in Cd(Zn)S buffer layer thickness.

Mean V_{oc} has been plotted against Cd(Zn)S thickness for each solar cell device (Fig. 9) . V_{oc} enhancement with inclusion of the Zn(Al)S HRT layer can be clearly seen for devices with a Cd(Zn)S buffer layer. Further V_{oc} improvement was observed for Cd(Zn)S thicknesses 0.05 – 0.1 μm, with mean values of 0.817 V (0.05 μm) and 0.813 V (0.1 μm) for devices 1064 and 1058, respectively. Device 1068, which had no Cd(Zn)S buffer layer, showed large variability with a lower mean value of 0.773 V. This was not far from the mean value obtained for devices with 0.15 μm thick Cd(Zn)S buffer without the Zn(Al)S HRT layer. For devices without the Zn(Al)S HRT layer mean V_{oc} generally improved with increasing Cd(Zn)S buffer thickness, being optimal for a thick 0.15 μm layer. This restricted the potential for optimizing the J_{sc}, demonstrating the need for the Zn(Al)S HRT for high solar cell performance.

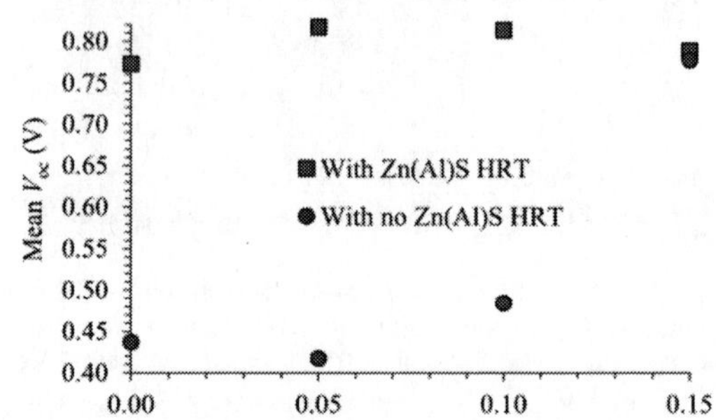

Fig. 9: Mean device V_{oc} plotted against device Cd(Zn)S buffer layer thickness, with and without a Zn(Al)S HRT.

C-V measurements of cells in each device were carried out to investigate the influence of Cd(Zn)S buffer layer thickness on acceptor levels (N_a). It also gave information on the depletion width (W_d) for that cell. Fig. 10 shows that acceptor concentrations are very high ($> 1 \times 10^{16}$ cm^{-3}) for devices with a Cd(Zn)S buffer layer, all having similar depletion widths of 0.3 μm. V_{oc} is enhanced with acceptor concentration and minority carrier lifetime, but lifetimes can reduce for high acceptor concentrations by introduction of point defects and self-compensation [15].

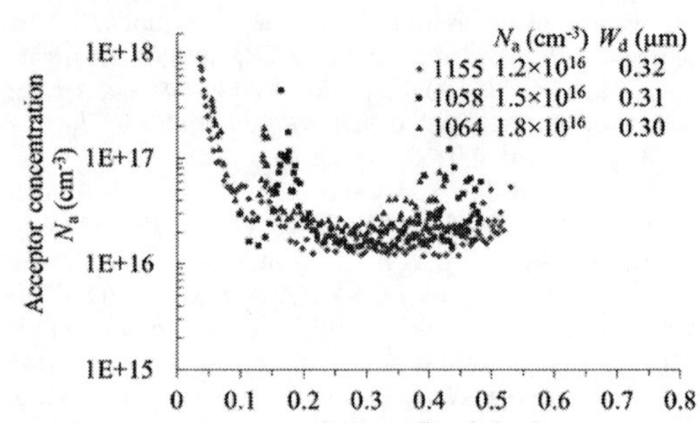

Fig. 10: Acceptor concentrations taken from *C-V* measurements for CdTe devices with different Cd(Zn)S buffer layer thicknesses.

3.5 Effect of CHT parameters on solar cell performances

CHT anneal temperatures of 420 °C (1155, 1058 and 1064) and 440°C (1030, 1154 and 1163) were used for devices with Cd(Zn)S buffer layer thickness of 0.15, 0.1 and 0.05 μm, with an additional CHT at 450°C (1169) used for a device with 0.05 μm Cd(Zn)S thickness. Increasing the CHT anneal temperature to 440°C for the devices with Cd(Zn)S thickness of 0.1 μm benefitted V_{oc}, increasing it to a mean value of 0.825 V over 12 × 0.248 cm^2 cells. This is shown in Fig. 11. The CHT process had a different effect on the devices with 0.15 and 0.05 μm Cd(Zn)S buffer by producing a small drop in V_{oc} (Fig. 11). The higher annealing temperature was detrimental to the device with no Zn(Al)S HRT layer and 0.15 μm Cd(Zn)S buffer.

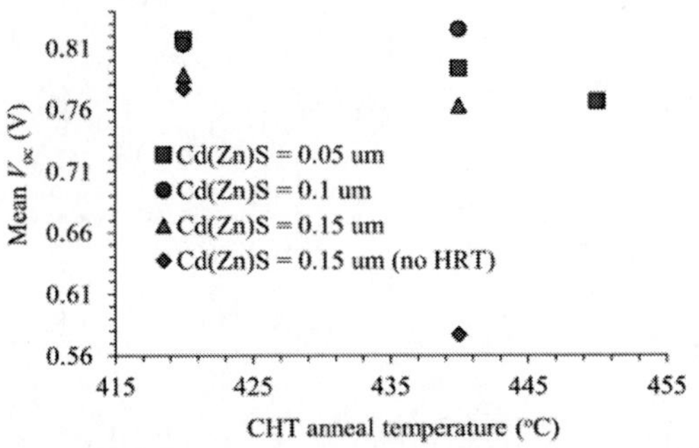

Fig. 11: Mean V_{oc} for solar cell devices with varied Cd(Zn)S buffer thickness, receiving different CHT anneal temperatures.

J_{sc} was observed to increase slightly for the device with 0.05 μm Cd(Zn)S buffer after the higher temperature CHT process (Fig. 12). For some cells, the second rectification at forward bias was exacerbated with the higher anneal temperature. Consumption of the 0.05 μm thick Cd(Zn)S buffer layer, likely at the higher CHT temperature, would lead to a thinning of the Cd(Zn)S layer, trending towards no buffer layer [16]. This is consistent with the small decrease in V_{oc} compared with the lower CHT temperature and the increase in roll over in forward bias.

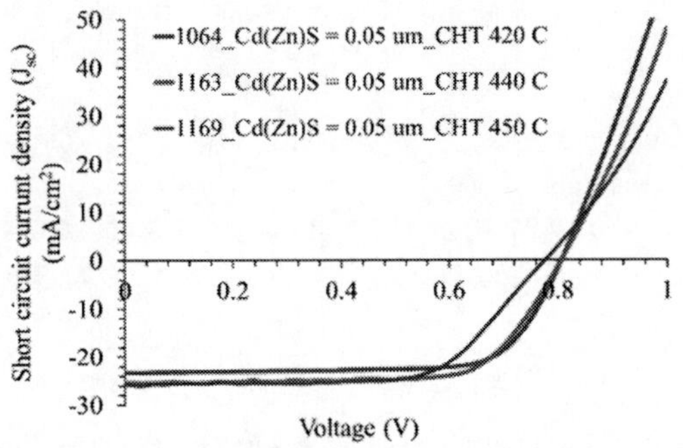

Fig. 12: *J-V* curves of different individual solar cells with variation in CHT anneal temperature for Cd(Zn)S buffer layer thickness 0.05 μm.

The increase in roll-over in forward bias in *J-V* curves, for devices having more aggressive CHT, was attributed to more prominent interdiffusion and the increased influence of the Zn(Al)S HRT at the interface with CdTe. SCAPS modelling was used to simulate the influence of the Zn(Al)S HRT layer. It was found that due to the low χ of 3.9 eV for ZnS [17], energy barriers to minority carriers are created because of large conduction band discontinuities. This is represented in Fig. 13 and shows the band energies for a device with HRT χ of 3.9 eV with comparison to an equivalent device with HRT χ of 4.2 eV. These energy barriers induced by a low χ of the HRT are causing the second rectification in the *J-V* curves. This was also simulated in SCAPS (Fig. 7). However, the energy barrier at the CdZnS/Zn(Al)S interface also has the effect of maintaining high V_{oc} values, and particularly when the Cd(Zn)S buffer layer thickness is reduced. The V_{oc} in these devices, where the acceptor concentration is above 1×10^{16} cm^{-3}, is limited by the minority carrier lifetime and particularly with the minority carrier recombination in the relatively thin depletion region (0.3 μm). The barrier in the conduction band is helping to separate the high electron concentration in the TCO from the high acceptor concentration in the CdTe layer, thus reducing the probability for trap assisted recombination.

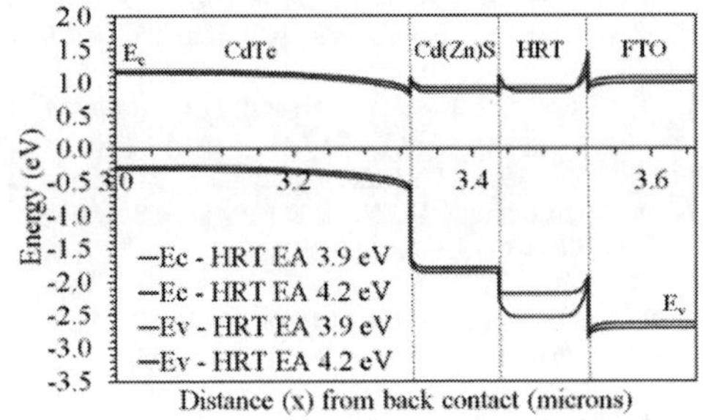

Fig. 13: Energy band diagrams showing conduction band and valance band energies for equivalent CdTe devices with HRT of low χ of 3.9 eV and higher χ of 4.2 eV.

4 CONCLUSIONS

An in-line MOCVD process was successfully used to produce large area HRT Zn(Al)S films on to NSG TEC™ C10 FTO substrates. The Zn(Al)S coated FTO resulted in a significant improvement to solar cell performance of CdTe devices produced by MOCVD. This was achieved through V_{oc} enhancement and was attributed by SCAPS to be due to the Zn(Al)S creating a positive conduction band discontinuity inducing significant energy barriers to recombination. This contributed to the V_{oc} being preserved during thinning of the Cd(Zn)S buffer layer. Removal of the Cd(Zn)S buffer layer altogether, to produce a CdTe/Zn(Al)S junction, did not result in a significant drop in V_{oc} but did affect the *FF* by inducing a second rectification at forward bias. The presence of a Cd(Zn)S buffer layer reduced discontinuity in the conduction band and helped to soften the second rectification observed in *J-V* curves. Increased CHT temperature exacerbated the second rectification at forward bias when 0.05 μm thick Cd(Zn)S layers were used. This was attributed to the increased influence of the Zn(Al)S HRT at the p-n junction. HR-SEM showed that using the Zn(Al)S HRT buffer layer allowed a more aggressive CHT to be used; achieving grain enlargement in the CdTe layer without delamination. Optimization of the buffer layer and the Al concentration in the HRT layer will be necessary to further increase V_{oc}.

5 ACKNOWLEDGEMENTS

The authors would like to acknowledge the European Regional Development Fund (ERDF) and the Welsh European Funding Office (WEFO) for funding the second Solar Photovoltaic Academic Research Consortium (SPARC II) which supported this research.

The authors are also grateful for the free access to the SCAPS software from Marc Burgelman and his team at Gent University, Belgium.

6 REFERENCES

[1] F. Bittau, A. Abbas, K.I. Barth, J.W. Bowers, J.M. Walls, Thin Solid Films 633 (2017) 92.
[2] E. Hernández-Rodríguez, V. Rejón, R. Mis-Fernández, J.L. Peña, Sol. Energy 132 (2016) 64.
[3] H. Mahabaduge, K. Wieland, C. Carter, V. Plotnikov, D. Giolando, Proceedings 37th IEEE Photovoltaic Specialists Conference (2011) 001302.
[4] K. Durose, P.R. Edwards and D.P. Halliday, J. Cryst. Growth 197 (1999) 733.
[5] C.S. Ferekides, U. Balasubramanian, R. Mamazza, V. Viswanathan, H. Zhao, D.L. Morel, Sol. Energy 77 (2004) 823.
[6] K. Durose, M.A. Cousins, D.S. Boyle, J. Beier, D. Bonnet, Thin Solid Films 403-404 (2002) 396.
[7] Y.Y. Proskuryakov, K. Durose, J.D. Major, M.K. Al Turkestani, V. Barrioz, S.J.C. Irvine, E.W. Jones, Sol. Energy Mater. Sol. Cells 93 (2009) 1572.
[8] J. Sites, J. Pan, Thin Solid Films 515 (2007) 6099.
[9] M. Burgelman, J. Verschraegen, S. Degrave, P. Nollet, Thin Solid Films 480-481 (2005) 392.
[10] V. Barrioz, S.J.C. Irvine, E.W. Jones, R.L. Rowlands, D.A. Lamb, Thin Solid Films 515 (2007) 5808.
[11] S. L. Rugen-Hankey et al., Sol. Energy Mater. Sol. Cells, vol. 136, pp. 213–217, 2015.
[12] K. Nagamani, O. Prathap, Y. Lingappa, R.W. Miles, K.T.R. Reddy, Phys. Proc. 25 (2012) 137.
[13] D.A. Lamb, S.J.C. Irvine, A.J. Clayton, G. Kartopu, V. Barrioz, S.D. Hodgson, M.A. Baker, R. Grilli, J. Hall, C.I. Underwood, R. Kimber, J. Photovolt. 6 (2016) 557.
[14] G. Kartopu, A.J. Clayton, W.S.M. Brooks, S.D. Hodgson, V. Barrioz, A. Maertens, D.A. Lamb, S.J.C. Irvine, Prog. Photovolt. Res. Appl. 22 (2012) 18.
[15] A. Nagaoka, D. Kuciauskas, J. McCoy, M.A. Scarpulla, Appl. Phys. Lett. 112 (2018) 192101.
[16] Y. Yan, K.M. Jones, M.M. Al-Jassim, R. Dhere, X. Wu, Thin Solid Films 519 (2011) 7168. V.
[17] H. Ramli, S. Kamal, A. Rahim, T.A. Rahim, M.M. Aminuddin, Chalc. Lett. 6 (2013) 189.

PROBING PHOTOINDUCED DEGRADATION OF $CH_3NH_3PbI_3$ PEROVSKITE FILMS BY KELVIN PROBE AND PHOTOLUMINESCENCE TECHNIQUES

Amalraj Peter Amalathas*[a], Lucie Abelová [ab], Brianna Conrad [a], Branislav Dzurňák [a],
Martin Ledinský[b] and Jakub Holovský [ab]
[a]Centre for Advanced Photovoltaics, Faculty of Electrical Engineering, Czech Technical University in Prague, Technická 2, 166 27 Prague, Czech Republic
[b]Institute of Physics, Czech Academy of Sciences, v. v. i., Cukrovarnická 10, 162 00 Prague, Czech Republic
*E-mail: peterama@fel.cvut.cz

ABSTRACT: In this work, the photoinduced degradation of $CH_3NH_3PbI_3$ perovskite films under illumination in ambient conditions (relative humidity 30-50%) was studied by Kelvin Probe and Photoluminescence techniques. Using Kelvin Probe techniques, we investigated the effects on the work function of the $CH_3NH_3PbI_3$ film on fluorine-doped SnO_2 (FTO) at various photoinduced degradation states in the dark and under illumination. It was found that the work function of $CH_3NH_3PbI_3$ film on FTO was gradually increased in the dark after every 10 min illumination step until the film was completely degraded. The gradual increase in work function due to the degradation can be ascribed to modulation doping of the $CH_3NH_3PbI_3$ by PbI_2 phase. It was also found that the contact potential difference (CPD) of $CH_3NH_3PbI_3$ on FTO was increased under illumination as a result of a positive surface photovoltage relative to the FTO. The steady-state photoluminescence (PL) and time-resolved PL measurements show that photoinduced degradation of $CH_3NH_3PbI_3$ leads to different charge carrier dynamics, as the localized defects and vacancies are increased during the degradation.
Keywords: perovskite, photoluminescence, degradation, Kelvin probe, Fermi level

1 INTRODUCTION

In the past few years, organic-inorganic halide perovskite solar cells have attracted enormous research interest due to their well-suited optical properties as photovoltaics absorber, large carrier mobility, low cost, and low fabrication temperature, with certified power conversion efficiencies in solar cells already exceeding 23%[1]. However, $CH_3NH_3PbI_3$ perovskite may contain inclusions of PbI_2 either as a result of the two-step preparation process or as a result of degradation in the presence of moisture, light illumination, and external voltage bias. Recently, we have investigated the coexistence of the two phases by absorptance spectroscopy [2], Raman spectroscopy [3], photothermal deflection spectroscopy (PDS) [4], and Fourier-transform photocurrent spectroscopy (FTPS) [4]. In this study, we broaden our investigation using photoluminescence and Kelvin Probe techniques.

2 EXPERIMENTAL METHODS

The $CH_3NH_3PbI_3$ layers on glass, FTO and quartz were prepared by two-step sequential deposition method. The substrates were cleaned by sonicating in a 2% solution of Hellmanex detergent diluted in deionized water, acetone and then ethanol, and dried with nitrogen gas. 1M PbI_2 solution in DMF and DMSO (9.5/0.5 v/v) was dissolved under constant stirring at 70 °C for overnight. The PbI_2 solution was first spin-coated on the substrate to form a layered PbI_2 film at 1000 rpm for 40 s. The spin-coated film was dried on a hot plate at 70 °C for 10 min. After drying the PbI_2, 0.22M CH_3NH_3I dissolved in 2-propanol solution was spin-coated on top of the PbI_2 layer using two-step protocol that includes 60 s of loading time followed by 40 s of spinning 4000 rpm. The films, which changed to brown in color during spin-coating were transferred onto a hotplate after spin-coating and was dried at 100 °C for 10 min to crystallize the perovskite. All the solutions and layers were prepared inside the nitrogen glove box.

The contact potential difference (CPD) measurements of the $CH_3NH_3PbI_3$ film on FTO at different photoinduced degradation stages were taken using KP Technology Ltd. scanning Kelvin probe (SKP) system. Steady-state and time-resolved PL measurements were acquired using Cary Eclipse fluorescence spectrophotometer and a time-correlated single photon counting (TCSPC) setup (FluoTime 200, PicoQuant GmbH), respectively.

3 RESULTS AND DISCUSSION

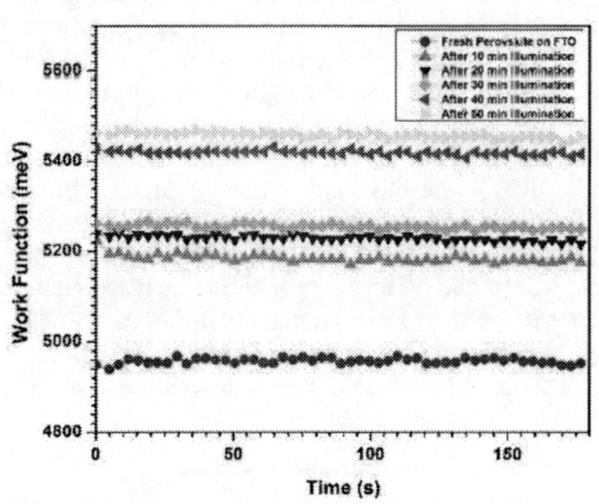

Figure 1: Work Function of $CH_3NH_3PbI_3$ film on FTO in the dark at various state of photoinduced degradation.

In order to investigate the Fermi level changes during the photoinduced degradation steps, CPD measurements were taken after exposing the film to 10 min illumination steps at 1 Sun under ambient conditions of 30% relative humidity, see Fig. 1. The value of the work function of fresh $CH_3NH_3PbI_3$ film on FTO was stable at 4958 meV (Fermi level -4958 meV) as shown in Fig. 1. The (apparent) electron affinity and the bandgap of the $CH_3NH_3PbI_3$ film are 3.9 eV [5] and 1.57 eV [4], respectively, and it seems to have p-like doping as its

Fermi level (-4.96 eV) is offset significantly from the middle of the bandgap located at -4.68 eV. It has previously been shown that the doping in $CH_3NH_3PbI_3$ can be achieved by changing the ratio of MAI and PbI_2 precursor solutions where I^- vacancies lead to n-like doping while $CH_3NH_3^+$ and Pb^{2+} vacancies lead to p-like doping [6].

As shown in Fig. 1, the CPD of $CH_3NH_3PbI_3$ film on FTO gradually increases due to the photoinduced degradation of the $CH_3NH_3PbI_3$ film under ambient conditions. This results in material decomposition into PbI_2 and other components [7] and the slow increase in work function be attributed to modulation doping of $CH_3NH_3PbI_3$ by PbI_2 phase as we have proposed recently [4].

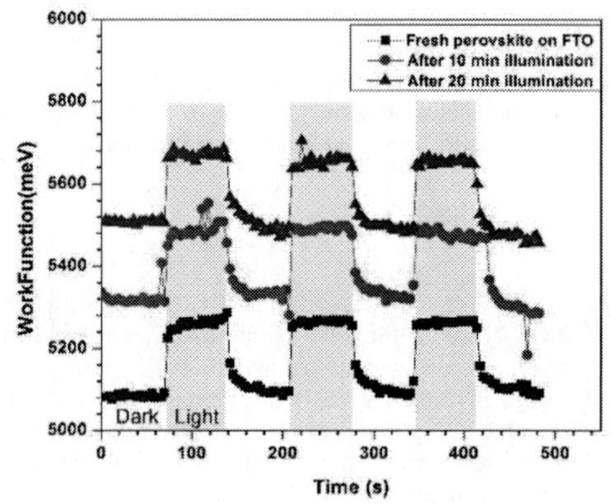

Figure 2: Work function of $CH_3NH_3PbI_3$ film on FTO as a function of time in dark and under light at various state of photoinduced degradation.

After measuring the Fermi level of $CH_3NH_3PbI_3$ film on FTO in dark, the influence of light on Fermi level was investigated. Therefore, the sample was illuminated during CPD measurement by white light in order to see the effect of modulation doping. Fig. 2 shows the work function of $CH_3NH_3PbI_3$ film on FTO as a function of time in dark and under light at various state of photoinduced degradation. In the dark, the work function of the fresh perovskite on FTO glass was stable at around 5100 meV but when the light was turned on the work function increased rapidly to 5270 meV with a change of more than 170 meV. When the light was turned off, the Fermi level returned to its initial value with a slow decay. The observed increase of CPD might be an indication of the effect of modulation doping by photogenerated carriers as proposed in [4] and shown in Fig. 3. After the photogeneration holes are injected to $CH_3NH_3PbI_3$ phase increasing band banding and providing effectively increase of work function. This effect is self-limiting while the band bending causes later also negative charge extraction and charge balancing. This effect may explain slow decay of observed increased CPD after switching off the illumination.

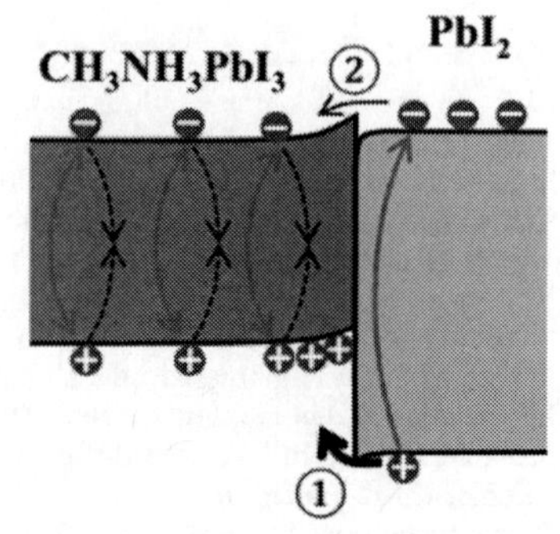

Figure 3: Model of the CH_3NH_3/PbI_2 interface. CBM and VBM refer to conduction band minimum and valence band maximum, respectively. Holes generated in the PbI_2 are injected into the $CH_3NH_3PbI_3$ (process ①). Consequent charging and band bending combined with thermal excitation also allow delayed electron injection (process ②)

Alternatively, the reversible photoresponse can be induced by generated photovoltage at the free surface or at the $CH_3NH_3PbI_3$ /FTO interface, as shown in Fig 4. Fig. 4(a) and (b) shows the band diagram $CH_3NH_3PbI_3$ and FTO in equilibrium and under illumination, respectively. The fresh $CH_3NH_3PbI_3$ layers and FTO are p-doped with work function (φ_{MAPI}) 5.1 eV as shown in Fig. 2 and the metallic-like heavily fluorine-doped SnO_2. When $CH_3NH_3PbI_3$ and FTO are brought together into contact the Fermi level must line up in equilibrium. In this case, the $CH_3NH_3PbI_3$ bands bend down towards the interface as shown in Fig. 4(a), presenting a barrier to majority carriers which are holes. Some fixed positive charge must be created to compensate the depletion. Under illumination, E_f is split into two quasi-Fermi levels of the electron (E_{fn}) and hole (E_{fp}), separately due to the light absorption. Under illumination, electrons are driven into

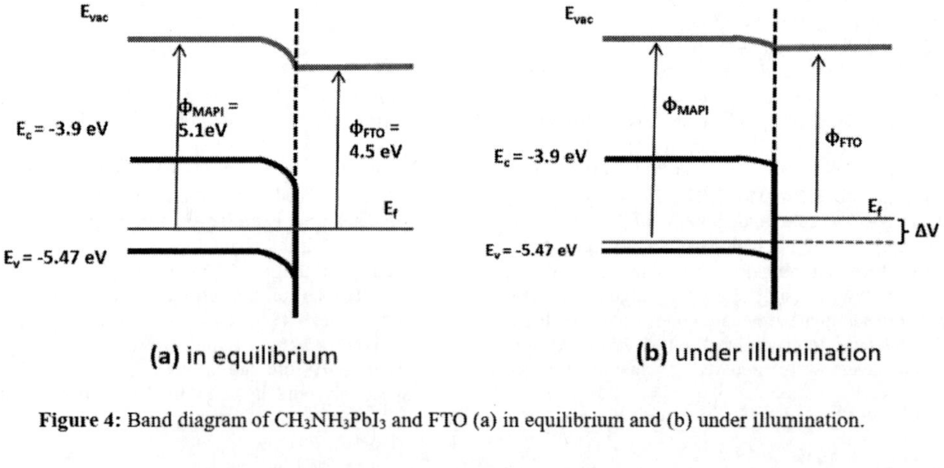

Figure 4: Band diagram of $CH_3NH_3PbI_3$ and FTO (a) in equilibrium and (b) under illumination.

the FTO and holes into the $CH_3NH_3PbI_3$, and the $CH_3NH_3PbI_3$ creates a photovoltage which is positive relative to the FTO. Fermi level of $CH_3NH_3PbI_3$ far from the junction is lower than it was in dark and lower than the Fermi level in the FTO, as shown in Fig. 4(b). As a result, the work function of the $CH_3NH_3PbI_3$ on FTO was increased under illumination.

Further experiments should be made to distinguish between the two proposed mechanisms of CPD increase upon illumination. It was observed that there was no evidence of degradation in the samples during this measurement and that the degradation effect was only observed under illumination from high-intensity light, and not during the Kelvin probe measurement in dark or under low-intensity light in short time.

For steady-state PL and time-resolved PL measurements, $CH_3NH_3PbI_3$ films were prepared either on glass or quartz substrates to avoid quenching at the interfaces. As shown in Fig. 5(b), gradual degradation was observed by exposing the $CH_3NH_3PbI_3$ film to 10 min illumination step at 1 Sun as the yellow color indicates that the $CH_3NH_3PbI_3$ materials decompose into PbI_2.

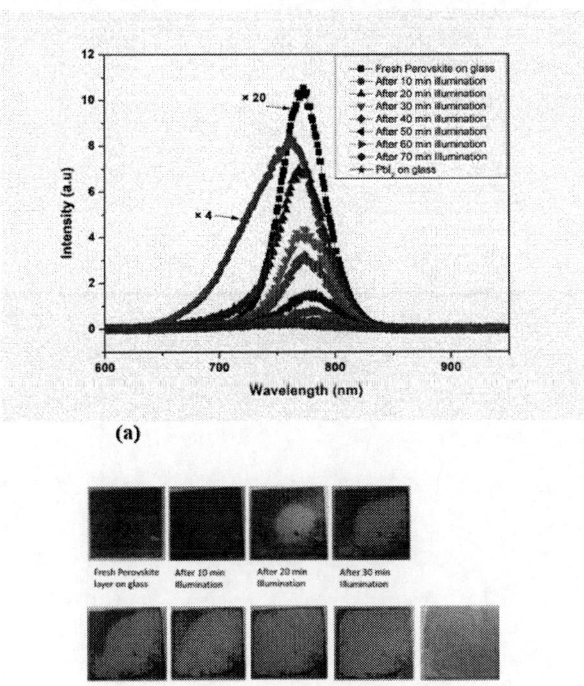

(a)

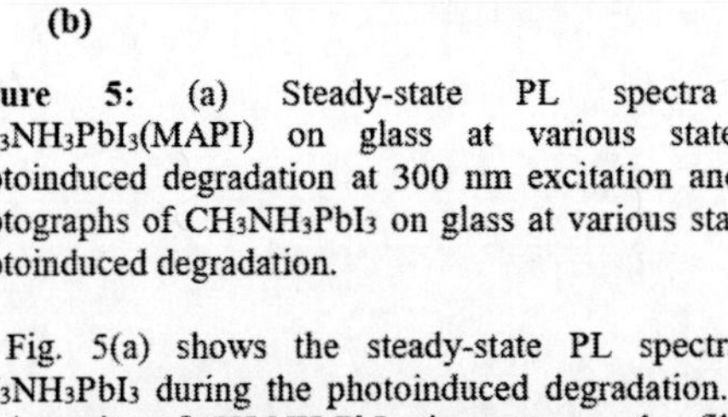

(b)

Figure 5: (a) Steady-state PL spectra of $CH_3NH_3PbI_3$(MAPI) on glass at various state of photoinduced degradation at 300 nm excitation and (b) photographs of $CH_3NH_3PbI_3$ on glass at various state of photoinduced degradation.

Fig. 5(a) shows the steady-state PL spectra of $CH_3NH_3PbI_3$ during the photoinduced degradation. The PL intensity of $CH_3NH_3PbI_3$ decreases as the film is degraded. Prolonged light illumination results in CH_3NH_3 ions decomposition into CH_3NH_2 and H_2. The resulting CH_3NH_2 and H_2 molecules can be diffused out of the $CH_3NH_3PbI_3$ layer as they are highly mobile at room temperature. As a result, localized defects and vacancies are increased and decreases the PL intensity. In addition, photoinduced degradation causes in material decomposition to scattered particles and shows little PL spectral shifts as shown in Fig. 4(a).

Fig. 6 shows the time-resolved photoluminescence measurements of $CH_3NH_3PbI_3$ on quartz at 780 nm at various state of photoinduced degradation. It was observed that photoinduced degradation leads to different charge carrier dynamics.

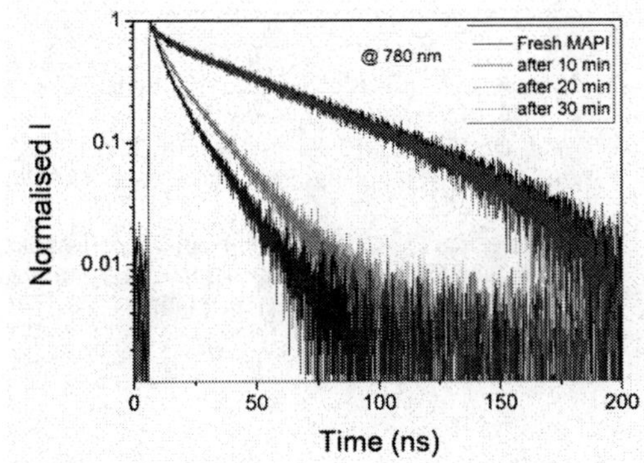

Figure 6: Time-resolved PL spectra of $CH_3NH_3PbI_3$ on quartz at various state of photoinduced degradation

4 CONCLUSION

In summary, the photoinduced degradation of $CH_3NH_3PbI_3$ perovskite films under illumination at 1 Sun in ambient conditions (relative humidity 30%) was studied at various stages of degradation by Kelvin Probe and Photoluminescence techniques. The work function of $CH_3NH_3PbI_3$ film on FTO was gradually increased at dark after every 10 min illumination step until the film was completely degraded. The gradual increase in work function due to the degradation can be ascribed to modulation doping of the $CH_3NH_3PbI_3$ by PbI_2 phase. It was also found that the work function of $CH_3NH_3PbI_3$ on FTO was further increased under illumination as a result of either the same effect of modulation doping or alternatively as a surface photovoltage which is positive relative to the FTO. The steady-state PL and time-resolved PL measurements show that photoinduced degradation of $CH_3NH_3PbI_3$ leads to different charge carrier dynamics, as the localized defects and vacancies are increased during the degradation.

5 ACKNOWLEDGEMENT

The authors acknowledge support via the Centre for Advanced Photovoltaics supported by the Czech Ministry of Education, Youth and Sport (CZ.02.1.01/0.0/0.0/15_003/0000464).

REFERENCES

[1] Jeon, N.J., H. Na, E.H. Jung, T.-Y. Yang, Y.G. Lee, G. Kim, H.-W. Shin, S. Il Seok, J. Lee, and J. Seo, Nature Energy, 2018. **3**(8): p. 682-689.
[2] De Wolf, S., J. Holovsky, S.-J. Moon, P. Löper, B. Niesen, M. Ledinsky, F.-J. Haug, J.-H. Yum, and C. Ballif, The journal of physical chemistry letters, 2014. **5**(6): p. 1035-1039.
[3] Ledinský, M., P. Löper, B. Niesen, J. Holovský, S.-J. Moon, J.-H. Yum, S. De Wolf, A. Fejfar, and C. Ballif, The journal of physical chemistry letters, 2015. **6**(3): p. 401-406.
[4] Holovsky, J., S. De Wolf, J. Werner, Z. Remes, M. Muller, N. Neykova, M. Ledinsky, L. Cerna, P. Hrzina, P. Loper, B. Niesen, and C. Ballif, J Phys Chem Lett, 2017. **8**(4): p. 838-843.
[5] Xing, G., N. Mathews, S. Sun, S.S. Lim, Y.M. Lam, M. Grätzel, S. Mhaisalkar, and T.C. Sum, Science, 2013. **342**(6156): p. 344-347.
[6] Wang, Q., Y. Shao, H. Xie, L. Lyu, X. Liu, Y. Gao, and J. Huang, Applied Physics Letters, 2014. **105**(16): p. 163508.
[7] Abdelmageed, G., L. Jewell, K. Hellier, L. Seymour, B. Luo, F. Bridges, J.Z. Zhang, and S. Carter, Applied Physics Letters, 2016. **109**(23): p. 233905.

EFFECT OF ELECTROLYTIC BATH STIRRING ON OPTICAL, MORPHOLOGICAL AND COMPOSITIONAL PROPERTIES OF CdS THIN FILM

Solomon Werta*, Obi Echendu and Francis Dejene
Department of Physics, University of the Free State, Qwaqwa Campus, Private bag X13 Phuthaditjhaba, 9866, South Africa.
*Corresponding author's email: solomonwerta@gmail.com Phone: +27 78 68 05 771

ABSTRACT: Electrodeposition of cadmium sulphide (CdS) films has been carried out from an electrolytic bath containing cadmium acetate $Cd(CH_3COO)_2$ and sodium thiosulphate ($Na_2S_2O_3$) using simple two electrode configuration. The deposition electrolytic bath stirring rate was varied at 0, 60, 120, 180 and 240 rotations per minute (rpm). The corresponding optical, morphological, surface roughness, compositional and structural properties of the deposited films were characterized using UV-Vis spectroscopy, scanning electron microscopy (SEM), scanning probe microscopy (SPM), energy dispersive spectroscopy (EDS) and glancing incidence X-ray diffraction (GIXRD) respectively. The results revealed that, the estimated energy band gaps were in the range of 2.24 to 2.29 eV as stirring rate decreases, for as-deposited films and 2.28 to 2.33 eV as stirring rate decreases, for annealed films. SEM images show very compact and agglomerated spherical grains. From SPM measurements it is observed that, the surface roughness was 2.50 to 2.70 nm for as-deposited samples and 1.76 - 2.50 nm for annealed ones. EDS shows the films are cadmium rich for both as-deposited and annealed conditions at different stirring rates.
Key words: CdS, stirring, cadmium acetate, two electrode, roughness

1 INTRODUCTION

The Sun offers the most abundant, reliable, sustainable, and environmentally friendly source of energy for the world. To use this energy, solar cells play a vital role in the conversion of solar energy to electrical energy. However, in utilization of this power there are a number of limitations, namely; solar panels use expensive semiconductors, the conventional growth equipment are expensive to install, practically the efficiency of solar cells is not enough yet to convert all sunlight that strikes the panel to energy and due to this, much sunlight that strikes the panel is wasted as heat [1,2]. To address such issues, over the past few decades, research on solar cells has grown significantly, and cost effective semiconductor based photovoltaic devices with simple fabrication process have been introduced and practised as alternative technology [3,4,5]. CdS is one of the most attractive semiconductors for researchers for solar cell application. It is a compound of group II and VI elements having a wide energy band gap of 2.42 eV and widely used as n-type semiconductor material in CdS/CdTe, CdS/CuInSe and CdS/CuInGaSe solar cells [6,7]. Other than for photovoltaic cells as a window material, CdS compound has also been used for different applications such as in gas sensors, field effect transistors, optical filters and multilayer light emitting diodes, and other optoelectronic devices [8,9]. There are different methods applied for the deposition of CdS thin films. Commonly used methods are chemical bath deposition (CBD)[10,11] , electrodeposition [12,13], molecular beam epitaxy (MBE) [14], metal organic chemical vapour deposition (MOCVD) [15], close space sublimation (CSS) [16], successive ionic layer adsorption and reaction (SILAR) [17], screen printing [18], Sol – gel [19], RF sputtering [20], pulsed laser deposition (PLD) [21], and physical vapour deposition (PVD) [22]. Each method has its advantages and disadvantages in device applications. Among these methods electrodeposition method is applied to grow CdS thin film due to the fact that electrodeposition method is one of the convenient methods to grow CdS thin films with the advantages of large area deposition, easy process control, minimum waste generation and self-purification of electrolytes. It is observed that, in the deposition of CdS thin films using electrodeposition method, parameters like concentration of ions in the deposition electrolyte, deposition time, bath temperature, electrolyte stirring, pH, electrodes used and applied voltage play a vital role in the quality of the film [23,24]. If these parameters are optimized, efficient CdS thin film which can be applied for different applications will be produced. Among these deposition parameters, stirring of the electrolytic bath has not been investigated adequately and there is limited reference on this in the literature [25]. Hence, in this paper, the effect of stirring rate of an electrolytic bath solution on the optical, morphological and compositional properties of electroplated CdS thin films using cadmium acetate $Cd(CH_3COO)_2$ as Cd source, is investigated. This has not yet been reported using electrodeposition method with two electrode configuration.

2 EXPERIMENTAL DETAILS

CdS thin films were deposited by using FTO (fluorine-doped tin oxide) substrates. Five pre-cleaned 2 cm × 3 cm FTO substrates were prepared for deposition of CdS thin film at five different stirring rates (0, 60, 120, 180 and 240 rotations per minute). Before the deposition was carried out, the substrates were immersed in de-ionised water with laboratory soap solution and cleaned ultrasonically for 30 min. Then the substrates were cleaned with de-ionized water and finally with ethanol, acetone, methanol respectively using de-ionized water to rinse in between washings, and then dried in air. The deposition electrolyte solution was prepared from 0.3 M of cadmium acetate ($Cd(CH_3COO)_2$) as cadmium source and 0.03 M sodium thiosulphate ($Na_2S_2O_3.5H_2O$) as sulphur source in a 400 ml of deionized water. Both chemicals are laboratory reagent grade purchased from Emsure, Germany with purity of 99.0 %. The electrolytic solution pH was maintained at 1.8 ± 0.02 and dilute sulphuric acid (H_2SO_4) and ammonium hydroxide (NH_4OH) were used to control the pH. The solution was stirred for 6 hr at room temperature until chemicals dissolved well. To accomplish the electrodeposition of CdS thin films, the glass/FTO substrate was attached to carbon plate using polytetrafluoroethylene (PTFE) which serves as working electrode. The space between electrodes was set at 5.5 cm. To determine the possible potential to grow CdS thin films at different stirring rates, a cyclic voltammetry measurement was taken using a computerized Gill AC potentiostat (ACM Instruments, United Kingdom) in two- electrode configuration, with a clean fluorine-doped tin oxide-coated glass (glass/FTO) as the working electrode and a high-purity graphite rod as the counter electrode. In electrodeposition, an electric current is generated due to electrode reactions that occur at electrodes and directional movements of ions in electrolytes as a result of electrode reactions [24,26-28]. Due to electrode reaction, reduction reaction of $S_2O_3^{2-}$ takes place earlier at lower cathodic potential than reduction of Cd^{2+} ions to Cd atom with relatively higher reduction potential of –0.4030 mV at the working electrode which is key process that leads to the formation of CdS. As $Cd(CH_3COO)_2$ split in to Cd^{2+} and CH_3COO^-, those Cd^{2+} ions from the electrolytic solution are then brought to the surface of the electrode by three different processes; migration, convection and diffusion [24,28]. Migration is the movement of ions due to the influence of an electric filed which is generated from the anode and cathode electrodes. Convection is due to agitation and heat elevation of the aqueous solution and diffusion is a process which happens because of concentration difference from higher concentration to lower concentration of atoms. As seen from Figure 1a, cyclic voltammetry measurement was taken in range of 0 to -2000 mV cathodic potential. Due to the fact that $S_2O_3^{2-}$ dissociates at relatively lower cathodic potential, sulphur comes to deposit first on the substrate from around -400 mV (or cathodic potential of 400 mV) and the current density is low. But as the cathodic potential increases in the region between -1300 and -1525 mV both sulphur and cadmium start to deposit and a gradual increase in current density is observed. As the cathodic potential goes up above -1525 mV, more Cd comes to deposit than S and this leads to sharp increase in current density which is observed in the cyclic voltammetry graph. So, the possible range of cathodic deposition voltages for CdS was obtained from the voltammogram measurement in between 1300 – 1525 mV. In between these two voltages different preliminary depositions were done and -1450 mV was settled as the best deposition potential. Five CdS films were then deposited for 30 minutes each at this potential and at five different electrolytic solution stirring rate. The samples were each divided into two and one part annealed at 400 °C for 15 minutes

Optical transmission and absorbance data were obtained from a Shimadzu UV-1700PC spectrophotometer. Scanning electron microscopy (SEM) was used to study the surface morphologies of the samples and scanning probe microscopy (SPM) was used to study the surface roughness of the samples. Energy dispersive spectroscopy (EDS) was used to determine the chemical composition and glancing incidence X-ray diffraction (GIXRD) was used to characterize the structure of the CdS thin films with a Cu-Kα wavelength of 1.54 Å.

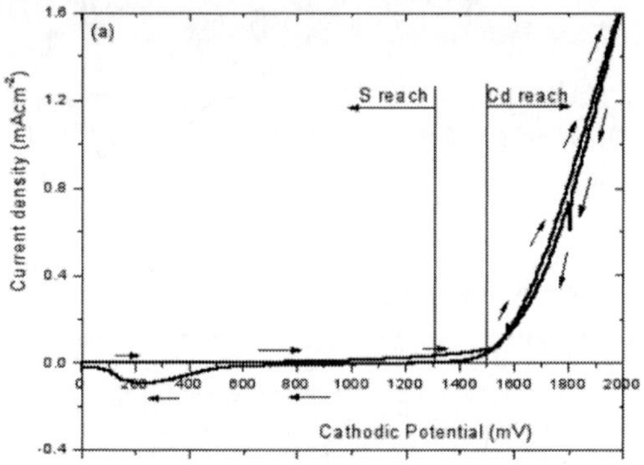

Figure 1. (a) Two-electrode cyclic voltammogram of CdS deposition electrolyte bath with aqueous solution of 0.3 M $Cd(CH_3COO)_2$ and 0.03 M $Na_2S_2O_3$. $5H_2O$ at a pH of 1.80±0.02

3 RESULT AND DISCUSSION

3.1 Optical properties

As shown in figure 2a,b for both as-deposited and annealed samples, at non-stirring state (0 rpm), due to the fact that deposition rate of ED-CdS thin film is influenced by migration of Cd^{2+} and S^{2-} ions, deposition rate is very low, and results the film being thin in thickness. It is obvious that film thickness significantly affects optical properties of semiconductor thin films. Thicker film results in high absorbance and low transmittance, and vice versa. So, at non-stirring rate the absorbance of samples is low. But as stirring rate increases to 60, 120 and 180 rpm the migration of Cd^{2+} and S^{2-} ions towards the substrate increases gradually with increase in deposition rate resulting in thicker films, thicker than at non-stirring. However, when the stirring rate increased to 240 rpm, the absorbance of both as-deposited and annealed films comes down. This is because high turbulence will happen and this affects the electric field between the working and counter electrodes. In addition, the high stirring rate tends to move the ions away from the working electrode at similar rate at which they arrive at the electrode due to high circular speed of rotation it imposes on the ions. As a result, fewer ions get discharged at the working electrode to form the required material. Therefore, in the deposition of thin films using electrodeposition method, the stirring rate and the electric field which is created between the electrodes inside the solution should be balanced. After annealing a decrease in absorbance is observed in the visible to near infrared regions due to thermally induced re-crystallization and complete reaction between un-reacted Cd and S atoms to form CdS thin films. In addition, there could be loss of excess Cd and S depending on their vapour pressure and annealing temperature involved. Figure 3a,b shows transmittance of CdS thin films for both as-deposited and annealed conditions. It is observed that at low stirring rate the transmittance of electrodeposited CdS thin film is high for both as-deposited and annealed conditions. When the stirring rate increases, transmittance of the ED-CdS thin films decreases gradually because the deposition rate of the film is high and which results to increase in film thickness. This continues up to stirring rate of 180 rpm. When the stirring reaches 240, deposition rate decreases as explained earlier and transmittance of ED-CdS starts to increase. Generally, optical property of ED-CdS thin film is dependent on thickness, and thickness of the sample is significantly affected by deposition rate, which is in turn affected by stirring rate.

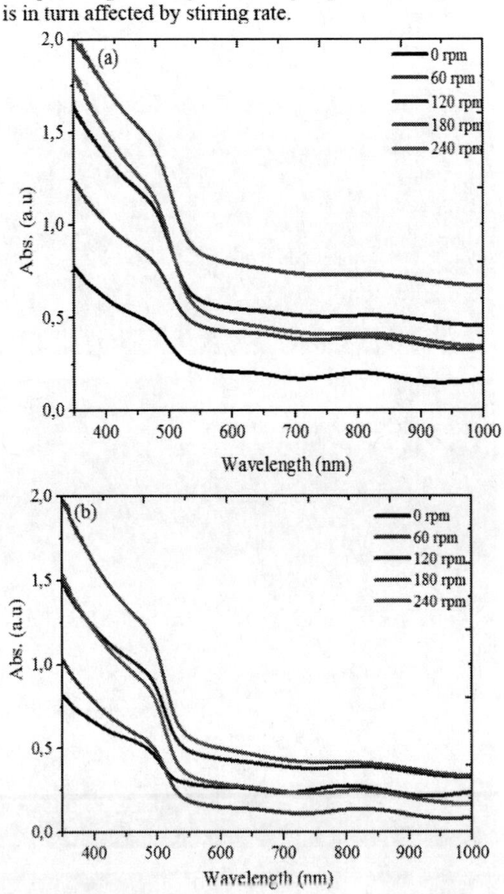

Figure 2. Absorbance of CdS thin films as a function of stirring rate for (a) as deposited and (b) annealed samples

To determine the energy band gap, the square of absorbance (A^2) as a function of photon energy has been plotted in figure 4, and the band gap energy estimated by extrapolating the straight line proportion of the graph to the photon energy (i.e to $A^2 = 0$). The band gap ranges from 2.24 eV to 2.29 eV for as-deposited films and from 2.28 eV to 2.33 eV for annealed films. Figure 4a,b show the graphs of Abs^2 versus photon energy of both as deposited and annealed samples. There is a slight decrease in energy band gap as stirring rate increases. So, at non-stirring rate, the deposition rate is lowest which yields the thinnest CdS sample. Low thickness results in low absorbance and higher energy band gap. But as stirring increases from 60 rpm to 180 rpm the deposition rate increases which leads to increase in thickness of film and decrease in energy band gap. In contradiction, as stirring reaches 240 rpm due to high rate of fluid and ion movement there is decrease in rate of deposition, resulting in low thickness and therefore increase in the energy band gap.

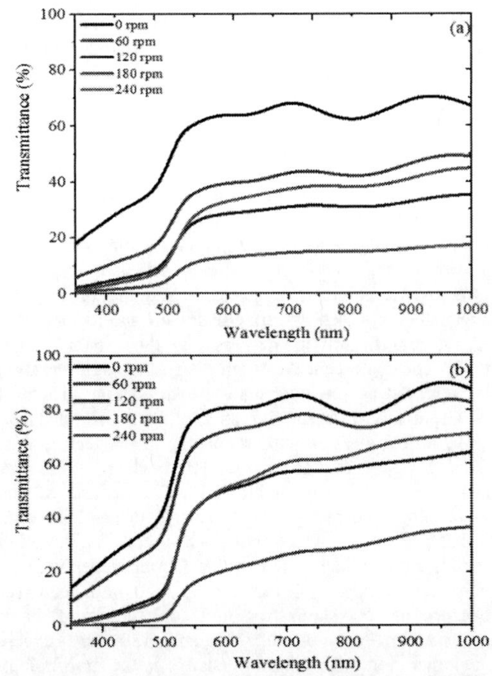

Figure 3. Transmittance of CdS thin films as a function of stirring rate for (a) as deposited and (b) annealed samples.

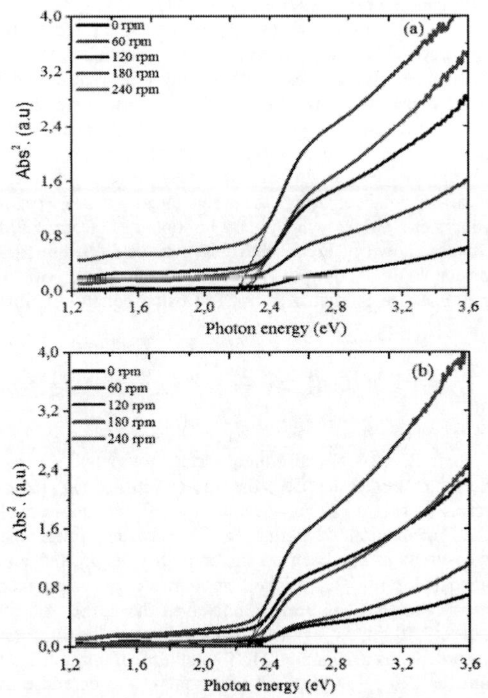

Figure 4. Abs^2 versus photon energy for (a) as-deposited and (b) annealed

3.2 Scanning electron microscopy (SEM)

Scanning electron microscopy (SEM) technique was used to study the morphology of CdS thin films deposited at five different stirring rate of electrolytic bath. Figure 5 shows, the surface morphology of electrodeposited CdS thin films which is covered by spherical grains for both as-deposited and annealed samples. These films grown at 60, 120 and 180 stirring rates have the same morphology. So, ED-CdS thin film grown at 120 stirring rate is reported in addition to 0 and 240 stirring rates. The obtained scanning electron microscopy images of ED-CdS thin film reveal that, at 0 rpm due to low deposition rate there exist pin-holes in the film. Due to the low thickness of the film size of the grains is small compared with other stirring rates. When stirring of the electrolytic solution increase, up to 180 rpm, the deposition rate increases and the substrate is adequately covered by spherical CdS grains and this results significantly in decrease, in the existence pin-holes. The coalescence of small grains and agglomeration is increased as stirring increases. The sample grown at 240 rpm has different morphology with others. There exist large grain size with wider gap in between grains, these happens due to the growth of large grains by coalescence of small grains[24]. The nature of the morphology and the pin-holes exist as a result of high turbulence in the electrolyte occasioned by the high stirring rate, as mentioned earlier. During annealing, the existence of these holes at stirring of 0 rpm decreases because of recrystallization and coalescence of small grains. However, at 60, 120 and 180 rpm except agglomeration there is no big change in the morphology of the sample. Similarly for sample grown at 240 rpm, coalescing of small grains increased.

3.3 Scanning probe microscopy (SPM)

Scanning probe microscopy measurement technique was applied to investigate the surface roughness of ED-CdS thin films. For solar cell or any other engineering applications, measurement or study of surface nature of materials is an important issue [29,30]. SEM image tells us only the shape and magnitude of grain size. The magnitudes of the surface roughness are quantitatively verified by comparing the root mean square (rms)-roughness values which can be obtained from SPM images shown in Figure 6. To save space CdS thin film grown only 120 rpm reported. Root mean square roughness is estimated from SPM using equation below [30,31].

$$R_q = \sqrt{\frac{1}{L}\int_0^L Z^2(x)\,\mathrm{d}x}$$

Where R_q is root mean square surface roughness, Z(x) is the function that describes the surface profile analyzed in terms of height (Z) and position (x) of the sample over the evaluation length "L". Surface roughness measurement has been taken for both as-deposited and annealed films. The highest root mean square surface roughness (R_q) has been estimated in the range of 2.50 nm to 2.70 nm for as-deposited conditions. The highest recorded was 2.70 nm at 240 rpm and the minimum 2.50 nm for 120 rpm. After annealing there is a decrease in roughness of these films and it was in between 1.76 nm and 2.50 nm: 1.76 nm for 120, and 180 rpm and 2.50 nm for 0 rpm. As confirmed by SPM images in figure 8, surface roughness is significantly influenced by stirring rate of the electrolytic solution. Fluorine-doped tin oxide substrate has rough surface properties. Due to roughness of the substrate and deposition rate is low at non-stirring rate, ED-CdS thin film has high surface roughness compared to other stirring rates (i.e between 60 and 180 rpm). Similarly, as stirring rate reaches 240, due to high movement of fluid, deposition rate decreases and leads to increase in surface roughness.

Figure 5. SEM images of CdS thin films deposited at different electrolytic bath stirring rate

However, when the stirring rate was in between 60 and 180, there is balanced migration of ions to the substrate which leads to the film being sufficiently covered by CdS grains. This results in decrease in surface roughness of the samples. Generally surface roughness of CdS thin film is significantly affected by stirring rate of an electrolytic bath. Table 1 summarizes stirring rate versus root mean square surface roughness values for both as-deposited and annealed ED-CdS thin films.

Table 1. Summary of stirring rate and surface roughness of ED-CdS thin films

Rotation per minute	as-deposited roughness (nm)	annealed roughness (nm)
0	2.64	2.50
60	2.52	1.77
120	2.50	1.76
180	2.51	1.76
240	2.70	2.42

3.4 Compositional Characterization

The EDS measurement has been used to confirm the composition of CdS thin films grown at different stirring rates. Figure 7 shows the spectra CdS thin films. From the spectra it is observed that both Cd and S exist in the films. The peaks belonging to Sn, O and Si come from the glass/FTO substrate. It is observed that the ratio S/Cd is 0.90 for non-stirring rate and similarly for 60, 120 and 180 stirring rate in as-deposited films. But as the stirring rate reaches 240, there is a slight improvement in S/Cd ratio to 0.92. After annealing there is decrease in stoichiometry for 0 and 240 stirring rates. It is observed that at zero stirring state, the stoichiometry decreased from S/Cd of 0.90 to 0.87 and for 240 stirring rate it decrease from 0.92 to 0.90 after annealing. However, at 120 stirring rate there is an improvement of S/Cd ratio from 0.90 to 0.91 after annealing and similar observation is made at 60 and 180 rpm. Generally all ED-CdS thin films grown at 1450 mV for 30 min with different stirring were cadmium-rich. Even though, EDS is not perfect technique to quantitatively measure the composition of these materials, but it gives information on the chemical nature of the films. So, the results support the conclusion that 60-180 rpm is recommended stirring to grow ED-CdS thin film for better stoichiometry.

3.5 Structural properties

From optical, morphological and compositional properties of ED-CdS thin film results, the sample grown at 120 rpm has better optical, morphological and compositional property. For this reason, the sample grown at 120 rpm was taken to study the structural properties of the film using glancing Incidence X-ray diffraction measurement. CdS thin film can grow in cubic, hexagonal, orthorhombic or mixed phases depending on deposition conditions [3, 10,32]. According to the results presented in Figure 8 cubic structural ED-CdS thin film was observed. The identified diffraction peaks match those of Joint Committee on Powder Diffraction and Standards (JCPDS) reference files no: 00-075-1546. As can be seen from the XRD spectrum there is the appearance of XRD peaks at 2θ = 26.55, 30.60, 44.00, 51.98 and 54.62 degrees which correspond to diffractions from C(111), C(200), C(220), C(311) and C(222) crystal planes respectively.

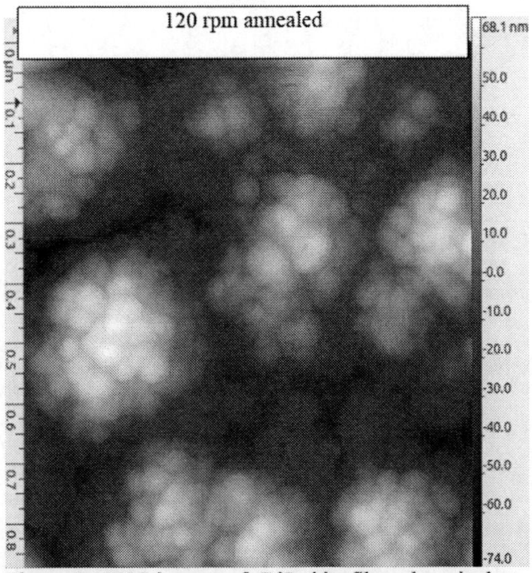

Figure 6. SPM images of CdS thin films deposited at different electrolytic bath stirring rate.

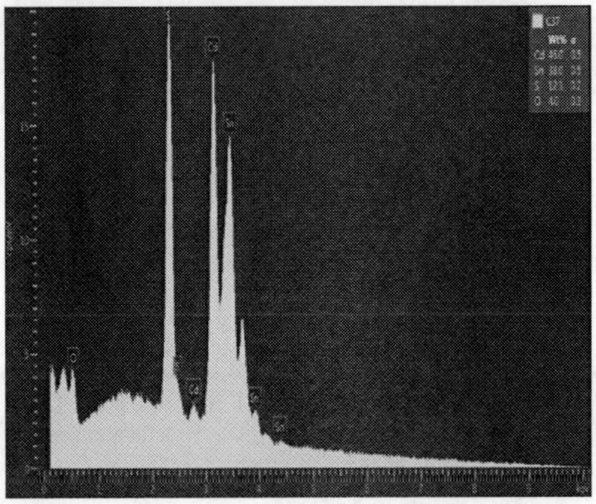

Figure 7. EDS spectra of the CdS thin films grown at different electrolytic bath stirring rate

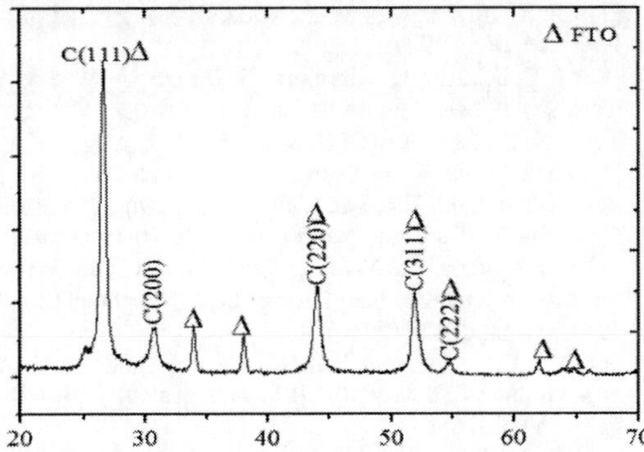

Figure 8. Glancing incidence XRD patterns of ED-CdS thin film grown at 120 rpm

4 CONCLUSIONS

CdS thin films have been electrodeposited successfully on glass/FTO substrates from an electrolytic bath containing cadmium acetate and sodium thiosulphate as cadmium and sulphur precursors respectively using electrodeposition system with 2-electrode configuration. The deposition was carried out at cathodic potential of 1450 mV with five different stirring rates 0, 60, 120, 180 and 240 rpm for 30 min deposition time. Optical, morphological, surface roughness and compositional measurements show that, stirring rate of an electrolytic bath solution has significant influence on physical and chemical properties of electrodeposited CdS thin films.

Acknowledgement
The authors would like to thank the University of the Free State, Research Directorate for funding the research that led to this paper.

Conflict of interest
Authors declare no conflict of interest

REFERENCES

[1]https://www.nature.com/articles/nenergy201732, January 2018
[2] https://www.weforum.org/agenda/2017/08/, January 2018
[3] Ru Zhou, Qifeng Zhang, Jianjun Tian, Daniel Myers, Min Yin, Guozhong Cao , J. Phys. Chem. C **117**,2013 26948–26956
[4] Karuppuchamy, Subbian and Nonomura, K and Yoshida, T and Sugiura, T and Minoura, H, Solid State Ionics, **151,** (2002) 19-27
[5] Jackson, Philip and Hariskos, Dimitrios and Lotter, Erwin and Paetel, Stefan and Wuerz, Roland and Menner, Richard and Wischmann, Wiltraud and Powalla, Michael, Progress in Photovoltaics: Research and Applications, **19**, (2011) 894-897
[6] O. K Echendu and I.M Dharmadasa, Materials chemistry and physics **157** (2015) 39–44
[7] L.L Kazmerski, F.R White and G.K Morgan, Applied Physics Letters **29** (1976) 268–270
[8] M. Kobayashi, K. Kitamura, H. Umeya, A. W. Jia, A. Yoshikawa, M. Shimotomai, and Y. Kato, J. Vac. Sci. Technol. B 18 (2000) 1684-1687
[9] Kuan-Ju Wu, Kung-Ching Chu, Chih-Yu Chao, Yang-Fang Chen, Chih-Wei Lai, Chia-Cheng Kang, Chun-Yen Chen, and Pi-Tai Chou, Nano Lett., **7** (2007) 1908-1913
[10] Fangyang Liu,Yanqing Lai, Jun Liu, Bo Wang, Sanshuang Kuang, Zhian Zhang, Jie Li, Yexiang Liu, J. Alloys Compd. **436** (2010) 305
[11]W.G.C. Kumarage,L.B.D.R.P. Wijesundara, V.A. Seneviratne, C.P. Jayalath, B.S. Dassanayake, Procedia Eng. **139** (2016) 64
[12] O. K. Echendu,U. S. Mbamara, K. B. Okeoma, C. Iroegbu, C. A. Madu, I. C. Ndukwe I. M. Dharmadasa, J. Mater. Sci. Mater. Electron. **27** (2016) 10182
[13] N. A. Abdul-Manaf,A. R. Weerasinghe O. K. Echendu I. M. Dharmadasa, J. Mater. Sci. Mater. Electron. **26** (2015) 2429
[14] P. Boieriu,R. Sporken, A. Adriaens, Yan Xin, N.D. Browning, S. Sivananthan, Nucl. Instrum. Methods Phys. Res. Sect. B: Beam Interact. Mater. Atoms **161–163** (2000) 975
[15] E.W. Jones,V. Barrioz, S.J.C. Irvine, D. Lamb, Thin Solid Films **517** (2009) 2226
[16]V. Krishnakumar, K. Ramamurthi, A. Klein, W. Jaegermann, Band alignment of differently treated TCO/CdS interface, Thin Solid Films **517** (2009) 2558
[17]M. Saglam,A. Ates, B. Guzeldir, A. Astam, M.A. Yildirim, J. Alloys Compd. **484** (2009) 570
[18]P.J. Sebastian, M.E. Calixto, Porous CdS: CdO, Thin Solid Films **360** (2002) 128-132
[19]A.A. Ziabari,F.E. Ghodsi, Sol. Energy Mater. Sol. Cells **105**, (2012) 249-262
[20] Gupta, Akhlesh and Compaan, Alvin D, Applied Physics Letters, **85** 684-686, 2004
[21] B. Ullrich , H. Sakai , Y. Segawa, Applied physics letters **80** (2002) 356-358
[22] Rusu, M and Glatzel, Th and Neisser, A and Kaufmann, CA and Sadewasser, S and Lux-Steiner, M Ch , Applied physics letters. **88** (2006) 143510
[23] I.M Dharmadasa and O.K Echendu, Encyclopedia of Applied Electrochemistry (2014) 680–691
[24] Joel Pantoja Enriquez, Xavier Mathew, Solar Energy Materials & Solar Cells **76** (2003) 313–322
[25] Atapattu, HYR and De Silva, DSM and Pathiratne, KAS and Dharmadasa, IM, Journal of Materials Science: Materials in Electronics, **27** 2016 5415—5421
[26] K. Zarębska, M. Skompska, Electrochim. Acta **56**, 5731–5739 (2011)
[27] F. Kadirgan, D. Mao, W. Song, T. Ohno, B. Mccandless, Turk. J. Chem. **24**, 21–33 (2000)
[28] S. Dennison, Electrochim. Acta **38**, 2395–2403 (1993)
[29]Maeng Jun Kim, Hong Tak Kim, Jin Kyu Kang, Dae Hwan Kim, Dong Ha Lee, Sung Ho Lee, and Sang Ho Sohn, Mol. Cryst. Liq. Cryst., **532** (2010) 21
[30] R.R.L. De Oliveira, D.A.C. Albuquerque, T.G.S. Cruz, F.M. Yamaji and F.L. Leite, www.intechopen.com InTech (2012)
[31] S. Z. Werta, O. K. Echendu, F. B. Dejene, Z. N. Urgessa, J. R. Botha, Applied Physics A: Materials science & processing, **124**(2018)576
[32] K. Gowrish Rao , V.K. Ashith, J. Phys. Chem. **77** (2015) 14–22

Molybdenum bilayer thin film on large area by cylindrical rotating DC magnetron sputtering for CIGS solar cell application

Amol C. Badgujar[a,b], Brijesh S. Yadav[a,c], Ramakrishna M.[a], Suhash R. Dey[c], Rajiv O. Dusane[b] and Sanjay R. Dhage*[a]

[a]International Advanced Research Center for Powder Metallurgy and New Materials (ARCI) PO Balapur, Hyderabad 500005, India
[b]Department of Metallurgical Engineering and Materials Science, Indian Institute of Technology Bombay, Powai, 400076, India
[c]Department of Materials Science and Metallurgical Engineering, Indian Institute of Technology Hyderabad, Kandi, Sangareddy 502285, India
*Email: dhage@arci.res.in

ABSTRACT: Highly conductive and adhesive Molybdenum (Mo) thin film back contact on soda lime glass (SLG) substrate popularly deposited by magnetron sputtering is essential for fabrication of $Cu(In,Ga)(Se,S)_2$, $Cu(Zn,Sn)(S,Se)_2$, CdTe and Sb_2Se_3 based thin film solar cells. Electrical and optical properties of the sputtered Mo thin films are predominantly governed by sputtering process conditions such as sputter power, deposition pressure and substrate temperature. In this work, we investigated influence of deposition pressure on phase constitution, adhesion strength on SLG, electrical and optical properties of bilayer Mo thin films coated on 300 mm X 300 mm X 3 mm Soda lime glass substrate using cylindrical rotating DC magnetron sputtering. The prepared bilayer Mo thin films using optimized sputter process conditions was further used as back contact for $Cu(In,Ga)Se_2$ solar cell device, demonstrated a photo-conversion efficiency of 5.0% on an area of 0.16 cm^2 under 100 mW/cm^2 illuminations.

Key words: DC magnetron sputtering, Molybdenum thin film, deposition pressure, CIGS.

1 INTRODUCTION

Thin film solar cell technologies are expected to provide a cost-effective flexible alternative to conventional Silicon-based solar cells addressing widespread energy demands of mankind. In this regard, $Cu(In,Ga)(Se,S)_2$ (CIGS) thin film solar cells are researched extensively owing to its excellent optoelectronic properties and longer thermochemical stability [1][2]. Molybdenum (Mo) based thin films are preferred back contact material for CIGS solar cells compared to Pt, Al, Ni, and Ag, due to its better thermochemical stability during high temperature processing of CIGS absorber layer apart from good electrical conductivity [3]. Mo also makes ohmic contact with CIGS in the form of $MoSe_2$ at CIGS –Mo interface [4]. Apart from CIGS solar cells, Mo thin film contacts are also popular for CdTe, $Cu(Zn,Sn)(S,Se)_2$ [5] and Sb_2Se_3 [6] based thin film solar cells. Moreover, Mo thin films have a wide range of applications as sensor or actuator component of various microelectromechanical devices such as capacitive pressure sensors [7], carbon-nanotube gate transistors [8], gas sensing devices [9], biosensors [10], acoustic resonators [11] and mechanical switches [12].

Magnetron sputtering is a popular technique to coat high-quality Mo thin films on a large area with superior thickness uniformity. DC magnetron systems with cylindrical rotating targets have advantages of possessing high plasma density, faster deposition rate and enables better control over the sputter deposition process, required for high thickness uniformity over the large area. Apart from that cylindrical rotating targets are known for better target utilization compared to planer targets. Properties of Mo thin film are largely dependent upon various sputtering parameters such as sputtering power, pressure and substrate temperature [13]. Electrical conductivity and adhesion of sputtered Mo thin films to the substrate are often governed by deposition pressure during the sputter process. A thin seed layer of Mo sputter coated at low sputter power and high deposition pressure is reported to enhance adhesion of Mo thin film to the substrate [14,15]. The prime objective of this work is to optimize the sputtering deposition pressure to obtain conductive, well adherent and uniform Mo bilayer thin films on substrates of size 300 mm x 300 mm for CIGS thin film solar cell applications.

2 EXPERIMENTAL

Mo thin films as back contact were coated on soda lime glass (SLG) substrate samples of size 300 mm × 300 mm × 3 mm using a sputter coater equipped with a cylindrical rotating DC magnetron (VISTARIS 600, Singulus Technologies, Germany). The working distance between the target and the substrate was maintained at 80 mm. The sputter, designed to coat substrates placed in a vertical orientation, was tooled to move the substrate carrier at a linear velocity of 0.5 m/min. Mo target was hollow cylindrical (inner diameter 12.5 cm & length 60 cm) with thickness of 5 mm. Sputter system was capable to apply power density as high as 4.25 W/cm^2 to target while Ar gas flow rate can be controlled up to 1000 sccm in the deposition chamber. Before sputtering, the SLG was cleaned in a glass washing machine (Make: Miele, Germany). The cleaning procedure adopted included rinsing with DI water, cleaning with alkaline and neutralizing agent (Extran, Merck) followed by hot DI water rinsing and, finally, drying of the cleaned glass at 90 °C described in detail in our earlier report [16]. The air side surface of the SLG was selected for coating the Mo thin film. A base vacuum of 7.5×10^{-7} Torr was maintained in the deposition chamber prior to initiating the coating process. Sputtering was carried out without intentional heating of substrates. All samples were coated with a thin seed layer of Mo (about 50 nm) deposited at a power density of 0.42 W/cm^2 and Ar gas flow rate of 1000 sccm to ensure good adhesion of bilayer Mo thin films. In the present work, we varied deposition pressure of bulk layer of Mo thin films by varying Ar flow rates from (100 sccm to 1000 sccm) at a constant sputter power density of 3.05 W/cm^2. Further, CIGS solar cell device fabrication was carried out on optimized sputter coated Mo bilayer thin films using following steps; sputtering of CuGa-In precursors and rapid thermal selenization to make CIGS thin film absorber (1400 nm), chemical bath deposited CdS (80 nm) [17], sputtered un-doped ZnO (50 nm) and aluminium doped Al:ZnO (900 nm) [18] followed by mechanical scribing for isolation of cell area of 4 mm X 4 mm.

The thickness of Mo thin film measured at 36 different locations on 300 X 300 mm glass substrate by XDD, SDD XRF system (Helmut Fisher, Switzerland). Electrical properties were measured by four probe equipment (Loresta GP, Mitsubishi, Japan), Phase identification of the all Mo film was carried out by XRD system (D8 Brukar with Cu-Kα source). Morphological studied was done by FESEM (ZEISS 500, Gemini, Denmark), UV-visible-NIR (Varian, Cary 5000) was used to measure optical properties and adhesion was checked by scotch tape test. Current-Voltage characteristics were measured using a solar simulator (94123A; Oriel Instruments) under 100 mW/cm^2 irradiation.

3 RESULTS AND DISCUSSIONS

Sputter process parameters optimization was done to produce high quality Mo thin film for contact application in thin film solar cells. Fig 1 shows XRD patterns of the Mo films prepared at various Ar gas flow rates. All films showed a typical body centered cubic (BCC) Mo structure (ICDD reference No. 004-0809) and no peaks corresponding to FCC phase were observed. Peak corresponding to (110) orientation was found dominant in all samples indicating a columnar growth. The crystallite size of Mo thin films as calculated using Scherrer's equation increased from about 19 nm to 28 nm with Ar flow rate reduced from 1000 sccm to 100 sccm which is consistent with previous reports, as crystal ordering of sputtered thin films increases with reduction in deposition pressure [13]. As compared with reference, the negative shift in peak corresponding to (110) plane was observed for films sputtered at low Ar flow rates indicating the presence of strain in the lattice.

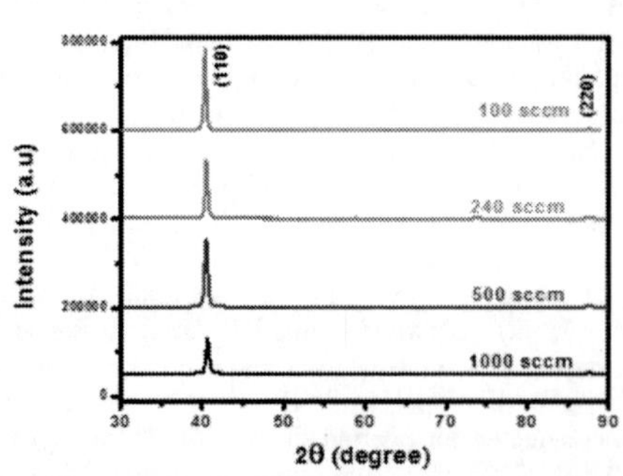

Figure 1: XRD patterns of bilayer Mo thin films sputtered at varied Ar gas flow rates

Fig 2 shows the evolution of the morphology of sputtered bilayer Mo thin films as Ar flow rate is varied from 100 to 1000 sccm. All the films revealed non-porous dead fish shape like microstructure and surface becomes coarser as deposition pressure is increased. A bilayer structure with highly columnar growth consistent with phase analysis of sputtered thin films can be confirmed from the cross section FESEM image in Fig 3.

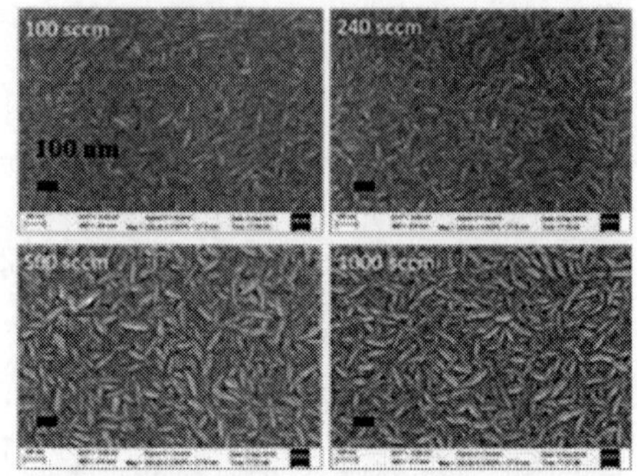

Figure 2: Surface morphology of bilayer Mo thin films sputtered at variable Ar flow rates

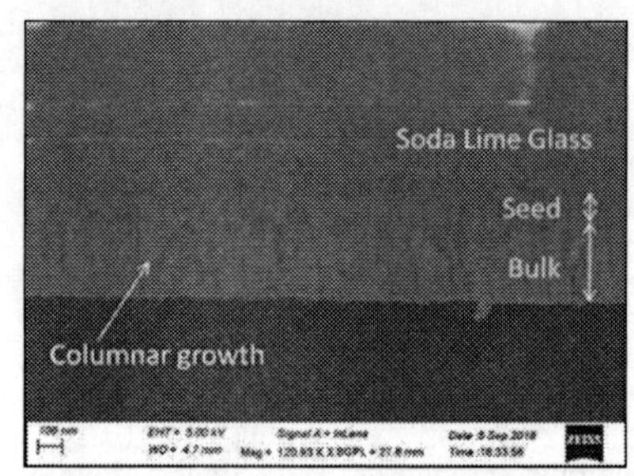

Figure 3: Cross-sectional morphology of bilayer Mo film sputtered at 240 sccm & power density of 3.05 W/cm^2.

Table I summarizes data related to the variation of thickness and uniformity of bilayer Mo films deposited at different deposition conditions. All the films possessed a high degree of thickness uniformity over a large area. A minor variation in thickness was observed towards central region as plasma is expected to be denser at the centre than at the edges of substrate. Overall thickness and deposition rate was found to increase with the increase in deposition pressure and then remained stable thereafter.

Table I: XRF results of Mo seed and bulk layer deposited at variable Ar flow rates

Sputter Process Parameters			XRF Measurements		
Power Density (W/cm^2)	Ar flow rate (sccm)	Deposition Pressure (mTorr)	Average thickness (nm)	Standard Deviation ((Max-Min)/(Min+Max)) %	Deposition rate (nm/min)
Seed					
0.42	1000	6.07	49	2.48	98
Bulk					
3.05	100	0.74	433	2.66	213
	240	1.72	456	3.17	226
	500	3.38	509	3.38	255
	1000	6.07	511	3.47	256

Fig 4 illustrates a variation of the average electrical resistivity of Mo thin films with varied Ar flow rate. As Ar flow rates were reduced from 1000 sccm to 100 sccm, the average electrical sheet resistance and resistivity both decreased consistently to the minimum value of 0.30 Ω/□ and 13.39 μΩ.cm respectively. The reduction in resistivity is consistent with improvement in crystallite size as confirmed from XRD analysis discussed earlier. Larger crystallite size improves the carrier mobility in thin films during electrical transport as charge carrier has to cross a small number of grain boundaries [19]. Measurements on multiple locations on 300 mm X 300 mm samples revealed a high degree of uniformity in electrical resistivity which is complementing the thickness uniformity.

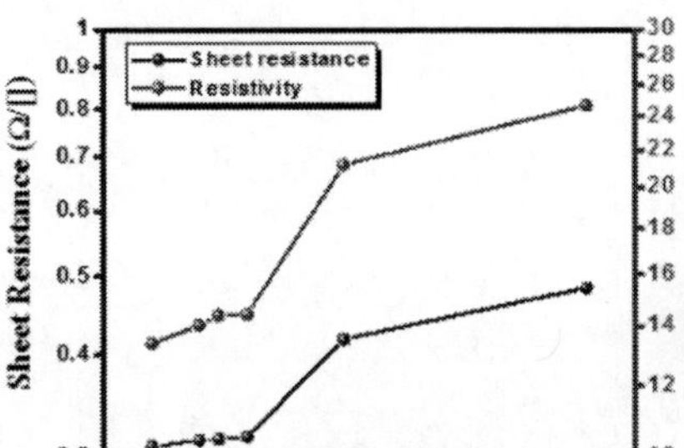

Figure 4: Average Sheet resistance and resistivity of bilayer Mo thin films sputtered at variable Ar flow rates

Reflective back contact enhances absorption of light in thinner absorber which is crucial to improve device performance of solar cells. Fig 5 presents reflectance of sputtered bilayer Mo thin films deposited at various Ar gas flow rates. The reflectance of Mo thin films improved consistently as the Ar flow rate was reduced which is consistent with observed surface morphology of Mo thin films. Reflectance is expected to be poorer for thin films with coarser surface which scatters the incident photons.

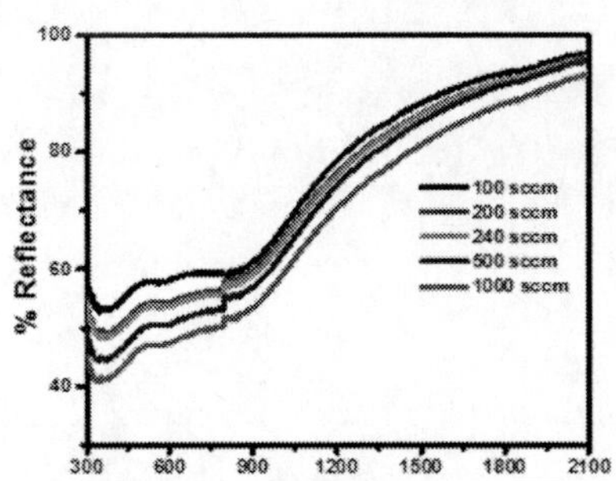

Figure 5: Reflectance variation of bilayer Mo thin films sputtered at variable Ar flow rates

Adhesion of sputtered bilayer Mo thin films is critical for CIGS solar cell fabrication and device performance. We employed a scotch tape test following ASTM D3359-09 standard to test adhesion strength of sputtered Mo bilayer thin films. Although all bilayer Mo films passed the adhesion test, films sputtered at low Ar flow rates up to 200 sccm found to peel partially in the progression of rapid thermal processing for making CIGS absorber layer. However, films sputtered at higher flow rate sustained thermal shock during the CIGS absorber preparation without delamination. Mo bilayer thin films sputtered at 240 sccm found to be optimum as possessed enough adhesion during absorber processing with least resistivity. Fig 6(a) presents camera image of 300 mm X 300 mm Mo thin films on SLG while Fig 6(b) represents camera image of adherent bilayer Mo thin films sputtered at 240 sccm after the peel test.

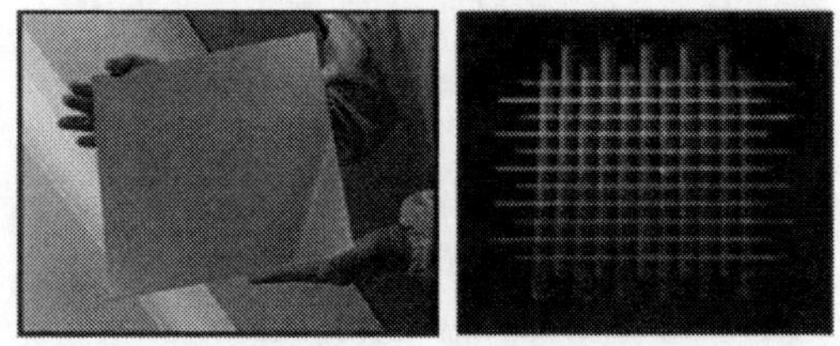

Figure 6: (a) Snapshot of 300mm X 300mm Mo thin film film on SLG substarte; (b) Scotch tape test on Mo film as per ASTM D3359-09 standard

Devices were fabricated in substrate configuration Al:ZnO/i:ZnO/CdS/CIGS/Mo/SLG to unveil the application of bilayer Mo thin film. Current-Voltage (IV) testing of devices was measured under dark and AM1.5 using the solar simulator. Devices fabricated with sputtered bilayer Mo thin film contacts using optimized deposition conditions of 3.05 W/cm^2 & 240 sccm confirmed photo-conversion efficiency of 5% on an area of 0.16 cm^2 as illustrated in Fig 7. As various factors such as quality of other active layers and interfaces are mainly responsible for the performace of the solar cell, all those aspects were not focused and discussed in the present work except the bilayer Mo thin film and demonstration of its application as back contact for CIGS thin film solar cells.

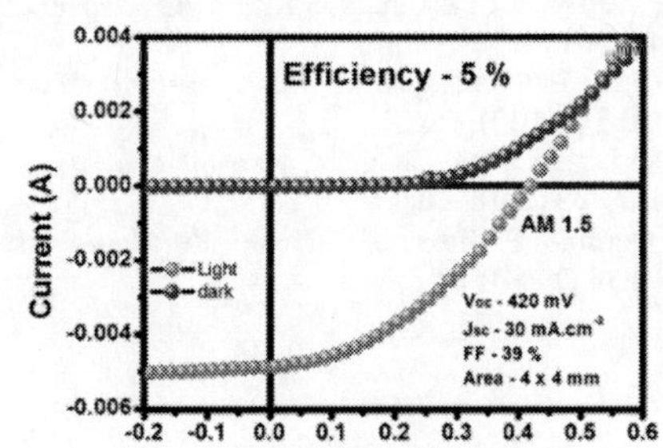

Figure 7: Current-Voltage characteristics of CIGS thin film solar cell under dark and illumination

4 SUMMARY AND CONCLUSION

In summary, bilayer Mo thin films with more than 96 % of thickness uniformity were sputter coated on 300 mm X 300 mm Soda lime glass substrate employing rotating cylindrical direct current magnetron sputtering. Deposition pressure was optimized to obtain well adherent, reflective, highly oriented columnar thin films with a low electrical sheet resistance of 0.31 Ω/□. CIGS thin film solar cell fabricated using optimized bilayer Mo thin film as back contact demonstrated 5% photo-conversion efficiency on the area of 0.16 cm^2 under AM1.5.

5 ACKNOWLEDGEMENTS

Acknowledge the funding from ARCI-Technical Research Center{Ref.No.AI/1/65/ARCI/2014(c)} through Department of Science and Technology (DST), India.

6 REFERENCES

[1] K.L. Chopra, P.D. Paulson, V. Dutta, *Progress in Photovoltaics: Research and Applications12* (2004) 69.
[2] A.C. Badgujar *et al.*, in *2015 IEEE 42nd Photovoltaic Specialist Conference (PVSC)*, IEEE (2015) 1.
[3] S. Raud, M.-A. Nicolet, *Thin Solid Films* 201 (1991) 361.
[4] E. Moons, T. Engelhard, D. Cahen, *Journal of Electronic Materials* 22 (1993) 275.
[5] E. Garcia-Llamas *et al.*, *Solar Energy* 141 (2017) 236.
[6] L. Wang *et al.*, *Nature Energy* 2 (2017) 17046.
[7] R.B. Brown, M.-L. Ger, T. Nguyen, in *IEEE Proceedings on Micro Electro Mechanical Systems, An Investigation of Micro Structures, Sensors, Actuators, Machines and Robots.*, IEEE, 77.
[8] M.M. Aqil *et al.*, *Journal of Nanotechnology* 2017 (2017) 1.
[9] L. Mele *et al.*, *Sensors and Actuators A: Physical*, 188 (2012) 173.
[10] V. Kamakoti *et al.*, *Biosensors* 6 (2016) 36.
[11] F. Martin, P. Muralt, M.-A. Dubois, *Journal of Vacuum Science & Technology A: Vacuum, Surfaces, and Films* 24 (2006) 946.
[12] C. Goldsmith *et al.*, in *2009 IEEE MTT-S International Microwave Symposium Digest*, IEEE (2009) 1229.
[13] S.A. Pethe *et al.*, *Solar Energy Materials and Solar Cells* 100 (2012) 1.
[14] J.H. Scofield *et al.*, *Thin Solid Films* 260 (1995) 26.
[15] A.C. Badgujar, S.R. Dhage, S. V. Joshi, *Thin Solid Films 589* (2015) 79.
[16] B.S. Yadav, A.C. Badgujar, S.R. Dhage, *Solar Energy* 157 (2017) 507.
[17] P. Uday Bhaskarand, S.R. Dhage, *Materials Today: Proceedings 4* (2017) 12525.
[18] S.R. Dhage, A.C. Badgujar, *Journal of Alloys and Compounds763* (2018) 504.
[19] G. Gordillo, F. Mesa, C. Calderón, *Brazilian Journal of Physics* 36 (2006) 982.

P2 NANOSECOND LASER PATTERNING OF CIGSE SOLAR CELL ABSORBER LAYERS: REVEALING ENHANCED CHARGE CARRIER RECOMBINATION WITHIN LASER-AFFECTED ZONES BY TIME-RESOLVED PHOTOLUMINESCENCE SPECTROSCOPY

Christof Schultz[a,*], Andreas Bartelt[a], Cornelia Junghans[b], Sebastian Schmidt[c], Rutger Schlatmann[a,c], Bert Stegemann[a]

[a] *University of Applied Sciences - HTW Berlin, Wilhelminenhofstr. 75a, D-12459 Berlin, Germany*
[b] *Becker & Hickl GmbH, Nunsdorfer Ring 7-9, D-12277 Berlin, Germany*
[c] *PVcomB / Helmholtz-Zentrum Berlin für Materialien und Energie GmbH, Schwarzschildstr. 3, D-12489 Berlin, Germany*

*Corresponding author: schultz@htw-berlin.de, Phone: +49 30 5019-3618

ABSTRACT: Laser-based patterning of CIGSe solar cells for monolithic serial interconnection is still challenging. Reasons are laser-induced damages resulting in alterations of the CIGSe material either due to material inhomogeneities or improper laser patterning parameters. These alterations affect the CIGSe composition as well as its electrical properties. In this study we focus on the electrical properties in the vicinity of the P2 lines, scribed either by laser or needle patterning. The CIGSe films patterned by nanosecond lasers are compared to needle-patterned CIGSe samples serving as a reference. By comparing optical microscopy images with photoluminescence intensity images significantly wider laser-scribed lines were found in the photoluminescence images. Time-resolved photoluminescence decay curves were analyzed for varying distances to the scribe line. For the laser-scribed samples a significant reduction of the charge carrier lifetime is found when approaching the scribe from untreated material areas. It is shown that the shorter charge carrier lifetimes can be related to known CIGSe lattice defect states such as Cu vacancies and Cu anti-sites acting as recombination centers. These results provide further understanding of the laser-material interaction and help optimizing the patterning processes for monolithic serial interconnection of CIGSe solar cells.
Keywords: Ablation; Charge Carrier Lifetime; Cu(In,Ga)Se_2 Solar Cells; Defects; Laser; Photoluminescence; Recombination, Serial Interconnection

1 INTRODUCTION

Laser patterning for serial interconnection of CIGSe (i.e., $CuIn_xGa_{1-x}Se_2$) thin film solar cells is a viable and advanced alternative to mechanical patterning for fabricating monolithic serial interconnected modules. It requires complete and selective layer ablation and is typically done using pulsed lasers of different wavelengths with a Gaussian beam profile. In Fig. 1 a schematic illustration of the monolithic serial interconnection scheme involving three patterning steps, known as P1, P2, and P3, is shown, which are required to form the serial interconnections between the front and back contact while electrically isolating the individual solar cells.

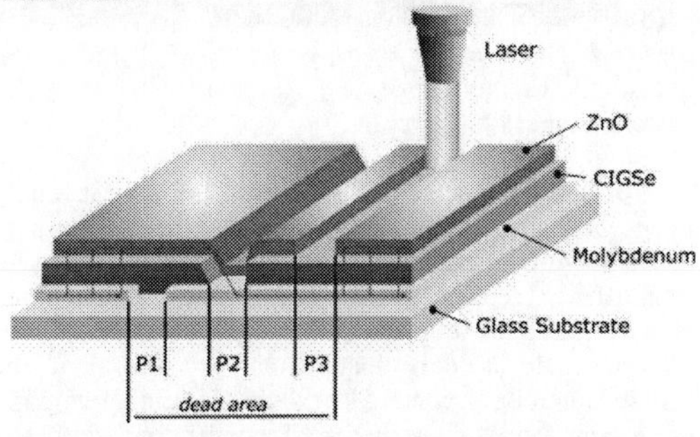

Figure 1: Schematic illustration of the monolithic serial interconnection scheme and the three patterning steps P1, P2 and P3, which can be done by laser or by needle.

In industrial production the P1 and increasingly the P2 patterning step are typically done by picosecond (ps) lasers, whereas P3 laser patterning is still challenging. Thus, it is often still carried out by needle, due to the lack of a reliable laser processes. Though it was recently shown that P3 patterning using ultrashort laser pulses might provide reliable results, slight absorber degradation was still observed [1, 2]. Main challenges for laser-based P2 patterning are material inhomogeneities which lead – when constant laser parameters are applied – to local structural and compositional material changes due to the thermal sensitivity of the CIGSe layer itself [3-5].

An alternative approach of P2 patterning for CIGSe solar cells does not rely on laser ablation but rather makes use of the thermal sensitivity of the absorber material to transform the CIGSe locally into a Cu-rich conductive phase, which forms the interconnection between front- and back contact layer, by applying nanosecond (ns) laser pulses. This approach works even after transparent conductive oxide (TCO) layer deposition, as was initially introduced by Westin et al. [6] and further elaborated by Schultz et al. [7]. Moreover, it was shown that this approach enables also simultaneous patterning of P2 and P3 which could further simplify the process chain [5-7].

Irrespective of the utilized laser patterning method, the energy in the beam tails is below the ablation or phase transformation threshold and, thus, promotes heating of the surrounding material leading to structural damages. Thus, due to the impact of the laser pulses modifications of the electronic and physical properties of the CIGSe layer were found altered close to the P2 scribe lines.

The aim of this contribution is a better understanding of the material degradation following laser patterning. To this end, we analyze the optoelectronic material properties of the laser patterns with different photoluminescence (PL) methods.PL spectroscopy has proven in the past to be a versatile technique to detect material modifications and to help evaluating material qualities by determining optical transitions and quantifying recombination processes [8-10]. To this end, we aim to relate the PL intensity decays measured by time-resolved spectroscopy with defect states revealed by spectrally resolved PL in the laser-affected areas. From this data we determine the extent of the heat-affected zones, the nature of the material modifications and the influence on charge carrier recombination, and compare these results with needle patterned CIGSe layers.

2 EXPERIMENTAL

2.1 Sample preparation and laser patterning

The investigated solar cells were prepared using the baseline process of the Competence Center for Thin-Film and Nanotechnology for Photovoltaics in Berlin (PVcomB) by sequential growth of the functional cell layers on soda lime glass [11]. The samples consist of a SiO_xN_y barrier layer, a 800 nm back contact Mo layer, a 1.6 µm CIGSe absorber layer, and a 40-50 nm CdS buffer layer.

For laser patterning an industrial-near patterning tool (Rofin Baasel Lasertech) is used that allows high precision (< 5 µm) and high speed (up to 1.2 m/s) sample translation. This patterning tool is equipped with several pulsed laser sources, enabling the selection of the proper laser parameters to match the specific properties of the material being processed. Here, a tungsten-carbide needle for mechanical patterning and a nanosecond laser source (λ = 532 nm, τ_{pulse} = 35 ns) were utilized for patterning. The laser beam possesses a Gaussian spatial intensity distribution and a diameter auf $2\omega_0$ = 51.5 µm. The tungsten carbide needle has a nominal tip width of 40 µm.

2.2 Photoluminescence

The optical properties of the patterned and modified absorber layer were derived from spectrally resolved PL measurements at different positions of the sample. The PL spectra were acquired at room temperature with an S&I TriVista spectrometer setup using a laser excitation wavelength of 532 nm. To avoid measurement-induced compositional changes the laser power incident on the sample surface was attenuated by means of optical filters to 0.15 mW.

A commercially available scanning time-correlated single-photon counting (TCSPC, Becker& Hickl, Simple Tau) setup was utilized to measure the time-resolved decay of the photoluminescence (PL), similar to our previous measurements on laser-patterned perovskite samples [12]. The sample was excited by 473 nm diode laser pulses with 50 ps pulse lengths. The PL decay was detected by an InGaS (IGQ-220) detector with 1024 time channels. A long pass filter was used to block all light below 830 nm and hence separate the luminescence light of the CIGSe from the laser light.

3 RESULTS AND DISCUSSION

The CIGSe absorber layer was patterned with multiple lines by nanosecond laser pulses and for the sake of comparison by conventional mechanical needle scratching. Thereby, the laser fluence was systematically varied, covering a range of slight surface deterioration to the ablation of the full layer stack. To determine the extent of the affected areas, which are altered in terms of their composition, laser scribed and needle-patterned lines and their vicinity were inspected by optical microscopy, PL imaging, spectrally resolved and time-resolved PL.

3.1 PL intensity imaging

Recently, we have determined the optimal ns pulse laser fluence for achieving precise and selective CIGSe layer phase transformation, and evaluated the electrical functionality of the P2 interconnect [7]. As a typical example Fig. 2 shows a homogeneous P2 trench obtained at a laser fluence of 0.42 J/cm² and a rather high laser spot overlap of > 99%, imaged by optical microscopy (OM) (Fig. 2a) and by PL (Fig. 2b). In the OM image the P2 features a constant width of about 30 µm and slight melt rims at the scribed line edges. In the PL image the width of the trench appears significantly wider, indicating a depletion of the PL intensity in the vicinity of the scribed line.

(a)

(b)

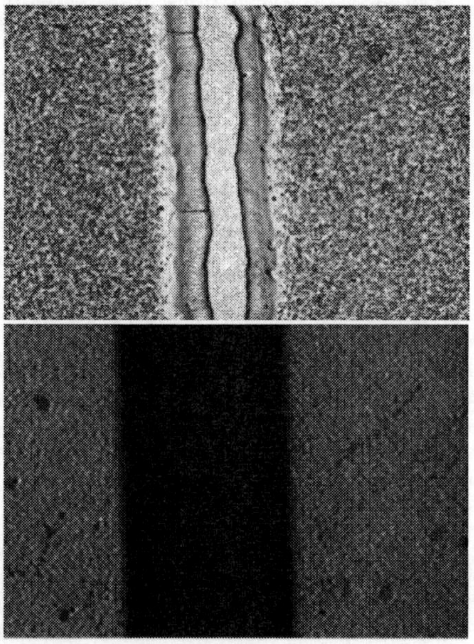

Figure 2: Nanosecond laser patterned P2 line. (a): optical microscopy image, (b): PL intensity image. Width of both images: 200 µm.

For comparison, Fig. 3 shows the corresponding images of needle patterned P2 line. Due to the pure mechanical, heat-free patterning this process is a well-established reference process for CIGSe P2 patterning. However, as shown by the OM image in Fig. 3a, various effects, such as delamination and corrosion of the Molybdenum back contact layer and also chipping might occur and lead to an increased connection width. As shown in Fig. 3b, a depletion of the PL signal was found in the affected range, which nearly coincides with the affected range visible by OM.

A quantitative analysis of these results is given in Fig. 4, comparing the widths of the processing tool and of the P2 trenches as detected by OM and PL for both patterning approaches. It can be concluded, that ns laser-based patterning is advantageous vs. P2 needle pattering with respect to small affected areas. Moreover, OM apparently underestimates the extent of the affected zones.

(a)
(b)
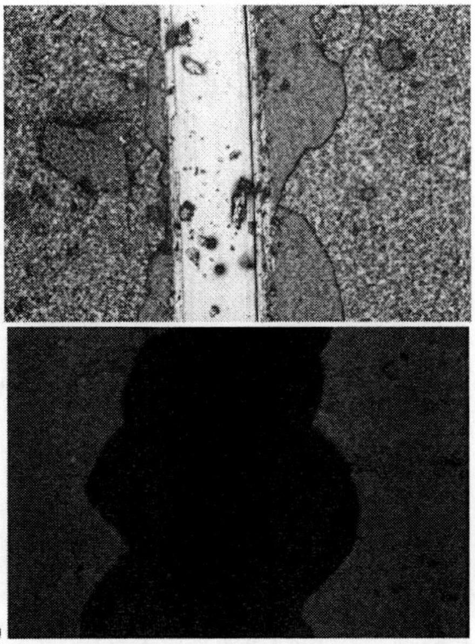

Figure 3: Needle patterned line. (a): optical microscopy image, (b): PL intensity image. Width of both images: 200 µm.

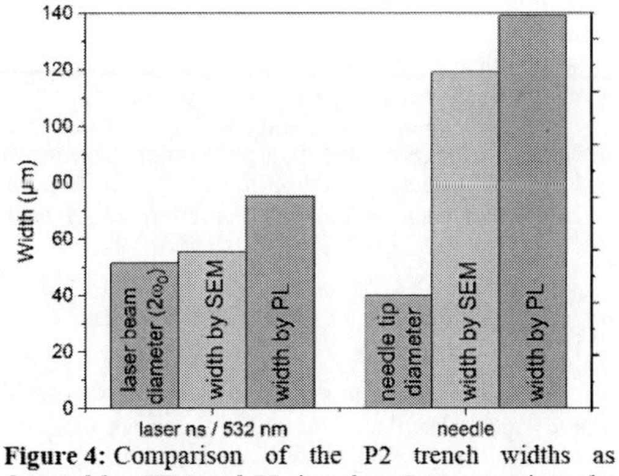

Figure 4: Comparison of the P2 trench widths as detected by SEM and PL imaging. For comparison the size of the patterning toll is given.

3.2 Photoluminescence decay

The time-resolved PL scanning microscopy resulted in images with 512 × 512 pixels and a pixel separation of 850 nm. Each pixel contains the PL decay curve for the corresponding position on the sample. From this 3D-data, the PL decay as a function of the separation from the scribe center was extracted. In order to enhance the signal-to-noise ratio, pixels with the same distance to the scribe center were averaged by vertical binning. The resulting decay curves were best fitted by bi-exponential functions. The bi-exponential nature of the PL decay is most probably dominated by combined contributions from bulk and surface recombination [9, 13], with the latter due to possible surface contaminations and degradations as well as the surface sensitivity of the PL-inducing UV laser light. The bi-exponential fit procedure included the convolution of the experimental response function. In Figs. 5 and 6 the fitting functions are shown, representing the PL decays for different positions in respect to the scribe center.

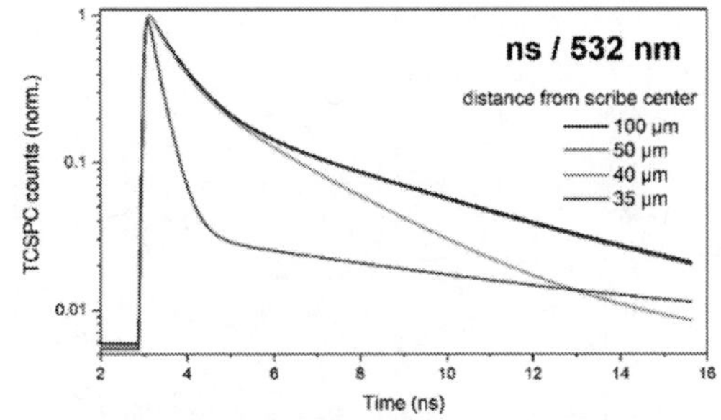

Figure 5: Measured decay curves as function of the distance to the nanosecond laser scribed line center.

In Fig. 5 the PL decay is shown for the case of the ns laser patterning. It can be seen that up to a distance to the scribe line center of approximately 50 µm the PL decay is identical to the decay of untreated areas (e.g. distance of 100 µm). Thus, the CIGSe absorber appears unaffected by the laser scribing up to a 50 µm separation. Approaching the center of the scribed line, first the slow component of the curve becomes weaker and faster (see green 40 µm line) and then the fast component also becomes faster (see red 35 µm decay curve). As the integrated PL intensity decreases at the scribe line edges, these enhanced decays can be related to additional non-radiant recombination processes.

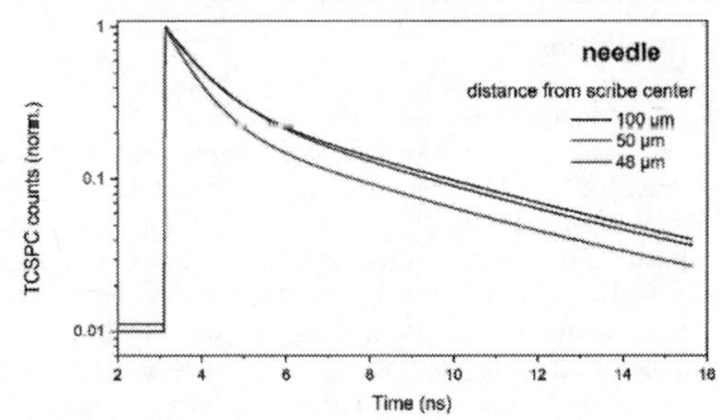

Figure 6: Measured decay curves as function of the distance to the needle scribed line center.

In Fig. 6 the fits of the PL decay for needle scribed lines are shown. In contrast to the decay curves shown in Fig. 5 for ns laser patterning, here the PL decays remain mostly constant as we approach the scribe from the untreated CIGSe material side. Only within a small 2 µm slab of the scribe edge the decay curves become slightly faster (see Fig. 6). Beyond 48 µm no PL signal could be detected. This finding indicates that the needle patterning process induced hardly any additional damages despite chipping and delamination of the absorber layer.

The fitting procedure resulted in the bi-exponential time-constants τ_1 and τ_2 and their respective amplitudes a_1 and a_2. From these parameters the average lifetime

$$\tau_m = \frac{a_1\tau_1 + a_2\tau_2}{a_1 + a_2}$$

can be calculated which is a good measure of the lifetimes in our samples [9, 13, 14]. An average lifetime of $\tau_m \cong 2$ ns is found which remains mostly constant when approaching the needle scribe edge. However, a significant drop down to $\tau_m \cong 400$ ps is observed when approaching the laser scribe edge. The lifetimes of ~2 ns are short compared to lifetimes found on clean CIGSe samples with pristine surfaces, reflecting contaminations and degradations on the surfaces of our samples [15].

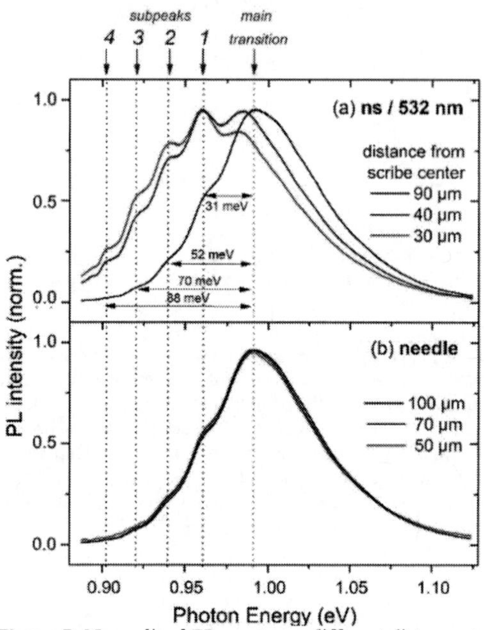

Figure 7: Normalized PL spectra at different distances to the scribe line center: (a) P2 ns laser patterning, (b) P2 needle patterning.

3.3 Correlation to electronic defect states

To understand the enhanced PL signal decay in the vicinity of the laser-based P2 scribes and to identify material modifications induced by the P2 step, we compare our results with spectrally resolved PL data obtained at discrete distances up to 100 µm away from the scribe line center. Typical PL spectra acquired from samples with well working P2 scribes performed by ns laser and by needle scribing are shown in Fig. 7. For both approaches three characteristic, normalized PL spectra are shown. Apparently, in the case of ns laser patterning (Fig. 7a) the shape of the spectrum changes when approaching the center of the scribed line since several subpeaks evolve at the low energy tail, whereas the spectra of needle patterned sample remain nearly constant down to a distance of 50 µm (Fig. 7b). Dotted lines are added to the graphs to indicate the spectral positions of the main transition (at around 0.99 eV) and of the subpeaks, which occur consistently at distances of (31 ± 3) meV, (52 ± 4) meV, (70 ± 6) meV and (88 ± 9) meV from the main transition.

The energetic positions of these subpeaks are plotted in Fig. 8 together with positions of typical defects in CIGSe, i.e. the acceptor levels of Cu anti-sites replacing In (Cu_{In}) and Cu vacancies (V_{Cu}), located in the bandgap (for a detailed description see ref. [10]). It turns out that the positions of the subpeaks agree well with these flat acceptor levels. Apparently, due to the laser impact lattice defects are created which generate flat defect states in the bandgap. These defect states facilitate non-radiative recombination. Thus, within the affected range PL intensity is depleted and charge carrier lifetime is reduced (cf. Fig. 5).

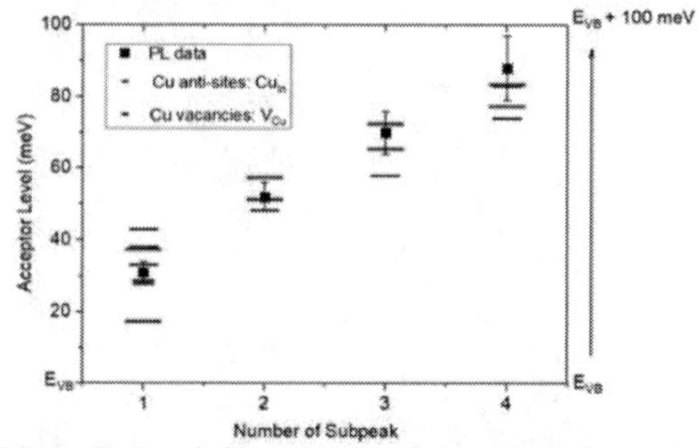

Figure 8: Correlation of subpeak positions from PL measurements and energy levels of Cu-related defect states.

4 CONCLUSIONS

We presented a comprehensive time-resolved and spectrally resolved PL analysis of CIGSe absorber material modifications upon P2 ns laser and needle patterning. We found that both patterning procedures cause modifications of the CIGSe layers near the actual scribing lines. As revealed by PL imaging, the width of the affected area is significantly larger than expected from visual microscopy inspections. While needle patterning is dominated by chipping effects leading to a widening of the trenches, laser patterning creates Cu-related lattice defects up to 50 µm away from the scribe center, which generate shallow acceptor states and facilitate recombination. Consequently, charge carrier lifetimes are significantly reduced within the affected areas. Nevertheless, with repect to the total scribe width, needle based patterning is still outperformed by laser-based patterning.

ACKNOWLEDGEMENTS

The authors gratefully acknowledge the support of the PVcomB team and the HTW laser lab team for experimental support and fruitful discussions. Part of this work was financially supported by the "Kommission für Forschung und wissenschaftlichen Nachwuchs (FNK)" of the HTW Berlin within the ZeLuMi project.

REFERENCES

[1] P. Gečys, E. Markauskas, S. Nishiwaki, S. Buecheler, R. De Loor, A. Burn, V. Romano, G. Račiukaitis, Scientific Reports 7 (2017) 40502.

[2] G. Heise, A. Börner, M. Dickmann, M. Englmaier, A. Heiss, M. Kemnitzer, J. Konrad, R. Moser, J. Palm, H. Vogt, H.P. Huber, Progress in Photovoltaics: Research and Applications 23 (2015) 1291.

[3] M.G. Flammini, N. Debernardi, M. Le Ster, B. Dunne, J. Bosman, M. Theelen, International Journal of Photoenergy 2016 (2016) 7.

[4] J. Hermann, M. Benfarah, S. Bruneau, E. Axente, G. Coustillier, T. Itina, J. Guillemoles, P. Alloncle, Journal of Physics D: Applied Physics 39 (2006) 453.

[5] C. Schultz, M. Schüle, K. Stelmaszczyk, M. Weizman, O. Gref, F. Friedrich, C. Wolf, C.A. Kaufmann, B. Rau, R. Schlatmann, F. Fink, B. Stegemann, in: (Eds.), 2015 IEEE 42nd Photovoltaic Specialist Conference (PVSC), 2015.

[6] P.-O. Westin, U. Zimmermann, M. Ruth, M. Edoff, Solar Energy Materials and Solar Cells 95 (2011) 1062.

[7] C. Schultz, M. Schuele, K. Stelmaszczyk, M. Weizman, O. Gref, F. Friedrich, C. Wolf, N. Papathanasiou, C. Kaufmann, B. Rau, R. Schlatmann, V. Quaschning, F. Fink, B. Stegemann, Solar Energy Materials and Solar Cells 157 (2016) 636.

[8] D. Hebert, J.A.N.T. Soares, A.A. Rocket, Materials Research Sociiety Symposium Proceedings 1195 (2009) 1165.

[9] A. Redinger, S. Levcenko, C.J. Hages, D. Greiner, C.A. Kaufmann, T. Unold, Applied Physics Letters 110 (2017) 122104.

[10] C. Schultz, G.A.F. Basulto, S. Ring, C. Wolf, R. Schlatmann, B. Stegemann, IEEE Journal of Photovoltaics 7 (2017) 1442.

[11] B. Rau, F. Friedrich, N. Papathanasiou, C. Schultz, B. Stannowski, B. Szyzska, R. Schlatmann, Photovoltaic International 17 (2012) 182012.

[12] C. Schultz, F. Schneider, A. Neubauer, A. Bartelt, M. Jošt, B. Rech, R. Schlatmann, S. Albrecht, B. Stegemann, IEEE Journal of Photovoltaics 8 (2018) 1244.

[13] M. Maiberg, T. Hölscher, S. Zahedi-Azad, R. Scheer, Journal of Applied Physics 118 (2015) 105701.

[14] M. Maiberg, F. Bertram, M. Müller, R. Scheer, Journal of Applied Physics 121 (2017) 085703.

[15] W.K. Metzger, I.L. Repins, M.A. Contreras, Applied Physics Letters 93 (2008) 022110.

Photovoltaic glazing from bottom-up electrodeposition of CIGS on patterned Mo/Glass Substrates

Tarik Sidali[1], Adrien Bou[1], Stéphanie Leyder[1], Damien Coutancier[2], Elisabeth Chassaing[2], Bertrand Theys[2], Richard Garuz[3], Damien Barake[3], Pierre-Yves Thoulon[1], Daniel Lincot[2]

1. Crosslux, 1200 Avenue Olivier Perroy 13790 Rousset
2. CNRS, IPVF (Institut Photovoltaïque d'Ile de France), 30 route départementale 128 – 91120 Palaiseau
3. IM2NP, Université Aix Marseille, Avenue Escadrille Normandie Niemen, 13397 Marseille Cedex 20

ABSTRACT: In this paper, semi-transparent solar cells using Cu(In1-xGax)Se2 (CIGS) semiconductors as absorber for BIPV applications is presented. The elaboration process consists in the electrodeposition of a metallic stack onto a sub millimetric hole-patterned molybdenum substrate, followed by thermal selenization. This method allows the selective deposition of the metallic stack and the CIGS growth only on Mo-covered areas. In addition, series interconnection following the P1P2P3 method was carried out to produce minimodules. Several large area devices have been manufactured with a transparency of 50%, leading to an efficiency of 3.5% (V_{OC} of 7.6 V and I_{SC} of 50 mA). On isolated cells, a maximum efficiency of 7.2%, corresponding to a V_{OC} of 365 mV, a J_{SC} of 36 mA/cm² and a fill factor of 55%, was measured.

Keywords: BIPV, CIGS thin films, semi-transparent solar cell, electrodeposition, sub millimetric pattern, bottom-up solar cell.

1 OBJECTIVES

The aim of this work is to develop semi-transparent photovoltaic glazing for building-integrated photovoltaics (BIPV) applications. These devices are based upon the standard thin film CIGS architecture, which consist of a stacking Glass substrate/Mo/CIGS/Buffer layer/ZnO/AZO. To achieve transparency suitable for glazing utilisation, the easiest approach is the structuration of every layer in the device to create small areas free of the opaque photovoltaic stack. This process can be achieved on the complete device, but it exhibits several drawbacks: the difficulty to achieve fine structuration (sub millimetric patterns) without degrading the cell performance, time consuming processes and a waste of Indium and Gallium. In addition, the appearance is impacted by the need to use mm patterns.

To overcome these issues, electrodeposition has unique advantages over classical methods like coevaporation or sputtering since it allows selective deposition on conducting zones in fine patterned areas. This process has the advantage to limit the Indium and Gallium consumption, and to allow reducing the pattern size well below the mm level. Furthermore, electrodeposition has been recognized to be a valid industrial method by the start-up NEXCIS, leading to high efficiencies cells (17.3%) and modules (up to 14%) [1]. The objective of this work is to transfer this method to semi-transparent glazing made on laser patterned substrates and to test the monolithic integration.

2 STATE OF THE ART

The proof of concept of the method has been recently demonstrated by our group by using the electrodeposition of CIGS by the oxide route developed in our laboratory [2,3]. Promising results have been obtained [4], however it is not, up to now, and industrially proven method as that developed by NEXCIS which uses the electrodeposition of stacked metallic layers directly. The present study aims to transfer the NEXCIS process to semi-transparent glazing and presents first results. A preliminary study on lines has been published [5]

3 METHOD

Laser patterning on the Mo/Cu substrate was used to achieve the desired aperture level:

holes with 700 µm diameter with a 900 µm center-to-center spacing distributed across the surface of the modules (Fig. 1). To allow monolithic integration, the P1 step was carried with the laser engraving of the patterned Mo layer with 50 µm wide lines.

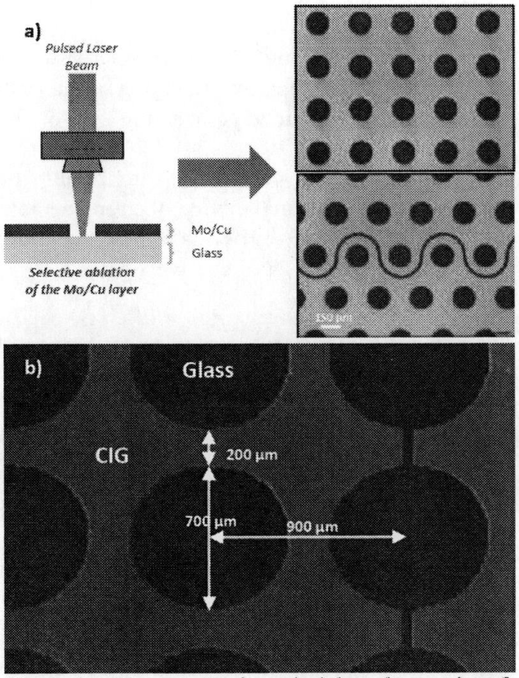

Figure 1. a) Laser patterning principle and examples of various patterns; **b)** SEM image of electrodeposited Cu/In/Ga metallic stack

A stack of Cu/In/Ga layers was deposited by sequential electrodeposition of In and Ga individual layers from separated baths on patterned Mo/Cu substrates. The relative proportions of Cu, In and Ga, are controlled by adjusting the thicknesses of the In and Ga layers. Subsequently, annealing under selenium pressure lead to the formation of the CIGS absorber with the desired composition and a good homogeneity across the surface. The completion of the module was performed by Chemical Bath Deposition of the CdS buffer layer and sputtering of i-ZnO, P2 engraving, ZnO:Al sputtering and P3 engraving.

4 RESULTS

Several 10x10 and 15x15 cm² working minimodules were obtained. A 10x10 cm² plain minimodule led to a V_{OC} of 7.6 V (21 cells by module), I_{SC} of 100 mA (corresponding to 32 mA/cm² of active area) and a fill factor of 44.5%. This corresponds to an efficiency of about 3,5%. For the corresponding 50% semi-transparent minimodule, the photocurrent was reduced to 50 mA while keeping about the values for the other parameters (Figure 2). This means that the efficiency is proportional to the area of the active surface, of 1.7 % in the present conditions. However, the I-V curves indicate the presence of losses associated to large series resistances.

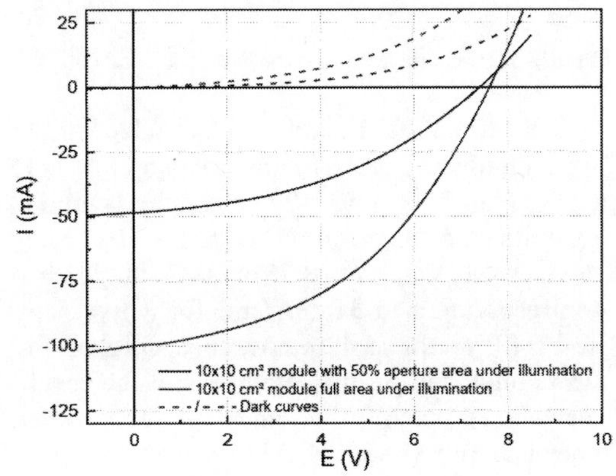

Figure 2. IV characteristics comparison between semi-transparent (**black**) and plain (**blue**) minimodule.

To study this effect, small individual cells (~0.1 cm²) were isolated in patterned zones within the CIGS stripes. The cell parameters reached a V_{OC} of 365 mV, an I_{SC} of 10 mA (corresponding to 36 mA/cm² for active area) and fill factor of 54.5%, leading to an efficiency of 3,6% with 50% transparency. The active area efficiency is thus about 7.2% (Figure 3). The origin of the improvement is the reduction of series resistance losses observed in the minimodules. This means that the P1/P2/P3 interconnections must be improved.

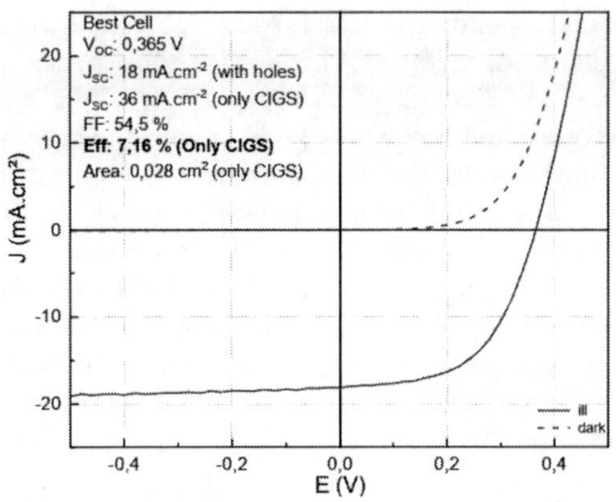

Figure 3. IV characteristic of the best individual cell

Furthermore, In and Ga electrodepositions were conducted on larger modules up to 15x15 cm². The best one (29 cells by module) exhibits similar performances to the 10x10 cm² devices: a VOC of 8.78V, ISC of 61 mA (corresponding to 31 mA/cm² for active area) and a fill factor of 42%. An encapsulation test was conducted on this module and the results were very small V_{OC} and I_{SC} losses after encapsulation (Figure 4).

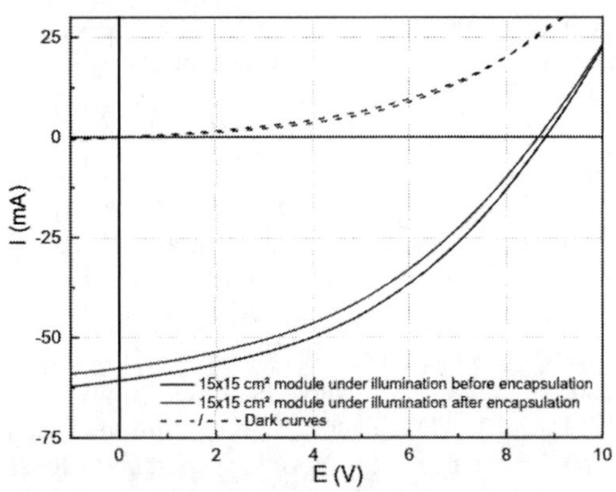

Figure 4. IV characteristics comparison before (**black**) and after (**blue**) encapsulation on a 15x15 cm² device.

5 CONCLUSIONS

In this paper, demonstrated a successful way to produce CIGS semi-transparent modules for BIPV applications. The electrodeposition process exploited to form the absorber allowed to use sub millimetric patterned Mo substrates.

We obtained several working devices with 50% transparency and encouraging efficiency of about 4%. However, performances are still impeded by large series resistance caused by the ZnO:Al window layer. Indeed, after isolating small individual cells within the device, 7.2% efficiency was achieved.

To improve the homogeneity of the electrodeposited layers, optimization of the deposition conditions will be conducted, for example by pulsed cathodic deposition.

Concerning the efficiency, further experiments are under way to increase the open circuit voltage by including a thermal treatment in sulphur vapour at the end of the selenization step.

Finally, in order to improve the appearance of the final device, the replacement of the yellow CdS buffer layer by the more transparent Zn(O,S) will be investigated

6 REFERENCES:

[1] C. Broussillou et al., IEEE Conference Publications DOI: 10.1109/PVSC.2015.7356224.

[2] A. Duchatelet, T. Sidali, N. Loones, G. Savidand, E. Chassaing and D. Lincot, Solar Energy Materials and Solar Cells, Vol 119, p241-245, Dec 2013.

[3] A. Duchatelet, G. Savidand, N. Loones, E. Chassaing and D. Lincot, J. Electrochem. Soc., vol. 161, no 7, p. D3120-D3129, Jan. 2014.

[4] T. Sidali, A. Bou, D. Coutancier, E. Chassaing, B. Theys, D. Barakel, R. Garuz and D. Lincot, Semitransparent photovoltaic glazing based on electrodeposited CIGS solar cells on patterned molybdenum/glass substrates, EPJ PV in press.

[5] A. Duchatelet, K. Nguyen, P.-P. Grand, D. Lincot and M. Paire, Applied Physics Letters, vol 109, no 25, Dec 2016, DOI: 10.1063/1.4971975

Alkali reactivity on aged CIGS absorber deposited on flexible substrates studied by XPS

S. Béchu[1,2], V. Achard[1,3], A. Loubat[1,2], M. Balestrieri[1], T. Hildebrandt[1,3], M. Frégnaux[1,2], J. Vigneron[1,2], D. Lincot[1,4], M. Jubault[1,3], M. Bouttemy[1,2], F. Donsanti[1,3], A. Etcheberry[1,2]

1. Institut Photovoltaïque d'Ile-de-France (IPVF), 30 RD 128, 91120 Palaiseau, France.

2. Institut Lavoisier de Versailles (ILV), Université de Versailles Saint-Quentin en Yvelines, Université Paris-Saclay, 78035 Versailles, France.

3. EDF R&D, IPVF, 30 RD 128, 91120 Palaiseau, France.

4. CNRS, IPVF, UMR 9006, 30 RD 128, 91120, Palaiseau, France

Abstract **— CIGS based solar cells are part of the most popular solar cells developed. Recently, CIGS technology was transferred on flexible substrates with interesting performances. As it was proved that presence of alkali elements in the bulk materials leads to better efficiencies in soda-lime glass substrates, a post-deposition alkali treatment is necessary to achieve high performances when CIGS is deposited on other substrates with no internal alkali source. The control of introduced quantities, localization and distribution of alkali elements (and in certain case, their counter-ions) after post-deposition treatment and ageing is then a crucial information in order to optimize the cells. This study is related to this context and is mainly oriented on alkali reactivity, by considering their evolutions at the surface but also at the sub-surface with Ar^+ profiling, thanks to XPS measurements.**

I. Introduction

Photovoltaic cells based on CIGS ($Cu(In,Ga)Se_2$) absorbers are among the most efficient. In the competitive context of photovoltaic market, the cell architectures are in constant evolution to gain efficiency while decreasing the cost production. In addition, new developments are considered to address new emerging market orientation, such as flexible solar cells dominated at the moment by the organic and nanomaterial sectors [1]. Among the thin films technology, CIGS has proven to be relevant and recent evolutions are oriented towards its adaptation on flexible substrates. In this particular case, unlike the classical soda lime glass substrates on which was developed the initial CIGS cell design, flexible ones do not have an internal source of alkali elements. Past studies [2]-[4] have clearly demonstrated that alkali presence and localization in the CIGS network (mainly at grain boundaries) are key parameters to increase efficiency devices. Therefore, different routes are considered to provide alkali in the absorber from an external source. This additional but essential step is usually performed from the front side of the absorber using post-deposition treatment, *i.e.* after the growth of the layer. It raises the issue of the alkali quantity and depth sank in the material, as the counter-ion behavior. Therefore, a specific attention has to be brought on the CIGS top surface composition possible evolution after treatment, as it will condition the absorber/buffer layers interfaces and their properties in the final stack. In fact, since the buffer layers can be deposited by wet or dry processes, the surface study is then a key point to optimize this step.

Indeed, regardless the substrate on which CIGS is deposited, it is well known that the performances of these solar cells are governed by the bulk absorber properties but its interfaces with the front and back contact layers. Thzy are also crucial for the performances and have to be finely optimized. X-ray Photoelectron Spectroscopy (XPS) enables to obtain overall physico-chemical information by the determination of the composition and chemical environments at the surface, bulk and interfaces of CIGS layers. Specific signatures of deposition process and storage conditions can be highlighted and brought insight about the quality of the absorber considering different surface issues such as alkali elements distribution or absorbers and alkali ageing evolution…

Among the different issues raised by the development of CIGS deposition on flexible substrates (adherence [5], metallurgy [6]…), present work is mainly centered on the question dealing with alkali elements incorporation and distribution evolution in time. In a first part, we investigate the in-depth position of alkali elements (and counter-ions) in the CIGS sub-surface by XPS profiling. The second part addresses the question of the absorber surface state prior to the buffer layer deposition when wet chemistry is employed [7]. We discuss here the critical time of nucleation of the buffer layer and correlate it with the quantities of alkali elements present on the surface during this step. The third part of this work deals with the alkali ageing in the CIGS layers on flexible substrates, with respect to different post-deposition treatments and the traditional behavior observed on classical soda-lime configuration. All these combined approaches evidence complex alkali behavior at the front interface.

II. Results and Discussions

$Cu(In,Ga)Se_2$ layers were deposited on 800 nm DC sputtered molybdenum layer on polyimide foils (Upilex 50S). CIGS films were grown with a three-stage co evaporation process developed at IPVF at low temperature (<400°C). Different post-treatments are studied, such as NaF and NaF/KF, evaporated under Se excess flux.

Samples were transferred to the XPS vessel without specific care from air interactions before XPS analysis. Consequently, chemical modifications of the initial film could be expected on CIGS surface during ageing. Moreover, possible alkali incorporation and migration during ageing must be considered.

Different ratios will be considered to qualify the major elements of CIGS surface: GGI, CGI and 2*VI/(I+3*III) ([Ga]/([Ga]+[In]), [Cu]/([Ga]+[In]) and 2*[Se]/([Cu]+3*([Ga] +[In])), respectively).

A. Alkali elements distribution

We first investigate the position of alkali elements within a CIGS sample deposited on polyimide, with NaF and KF post-deposition treatments, from the top side.

In a first time, XPS profiling using Ar^+ ion gun (approximately 200 -300 nm sputtered) was performed on raw surface of aged-CIGS (4 months) deposited on polyimide (Fig 1) with a NaF/KF post-deposition treatment. The atomic percentages evolution shows a coherent behavior with a decrease of the alkali elements over the etching depth.

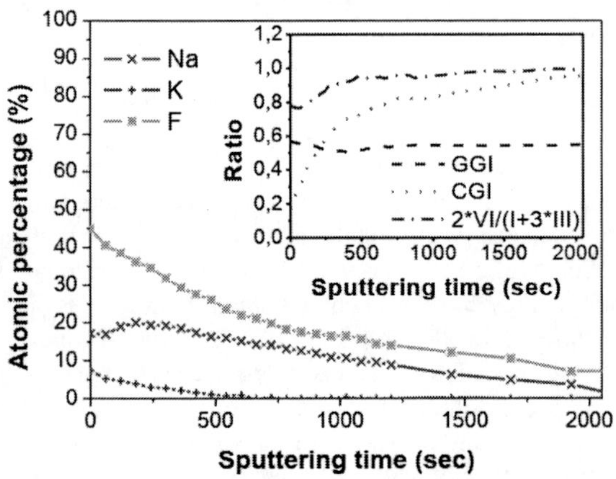

Fig 1: XPS atomic percentage profile of Na, K and F and ratio evolution (GGI, CGI, and 2*VI/(I+3*III)) over sputtering time.

Na and K alkali elements present two different incorporation profiles: the K element is indeed more superficial than the Na alkali element, as expected with the literature [4]. Moreover, it is possible to see in the first seconds of sputtering (up to 250 sec) a small depletion in Na, which can be due to the fact that K is occupying the interstitial sites instead of Na. The counter-ion F is also present throughout the full profile performed, questioning about its specific position and interaction within the absorber.

From these profiles, considering the XPS atomic composition evolution, the presence of an external layer with major alkali elements contribution is established. This layer is Cu-poor as observed in the ratio evolution in Fig 1. The question of the behavior of this external layer during the initial stages of the buffer deposition step is asked. To investigate this crucial issue in the cell elaboration process, samples were rinsed in deionized water (18mΩ) for 90 seconds (3 successive baths of 30 seconds each) to simulate the initial stage of the buffer bath interaction. After this treatment, no more alkali elements were detected on the surface with XPS analysis. XPS profiling was then performed (also over approximately 200-300 nm) and no alkali elements or counter-ions could be detected at the detection limit of the technique (0.1% for Na1s and F1s and 0.5% for K2s photopeaks in atomic percentages). Thanks to this information, the following hypothesis can be stated: at the surface of the aged absorber, a consequent external layer of alkali elements appears.

B. Evolution of alkali surface content over rinsing time

From the work above, an issue is raised about the latency of alkali evacuation during the sample immersion in water and so, in more extent in ammonia medium employed for buffer deposition (CdS or ZnS). Indeed, a certain latency time is expected before the buffer layer nucleation begins.

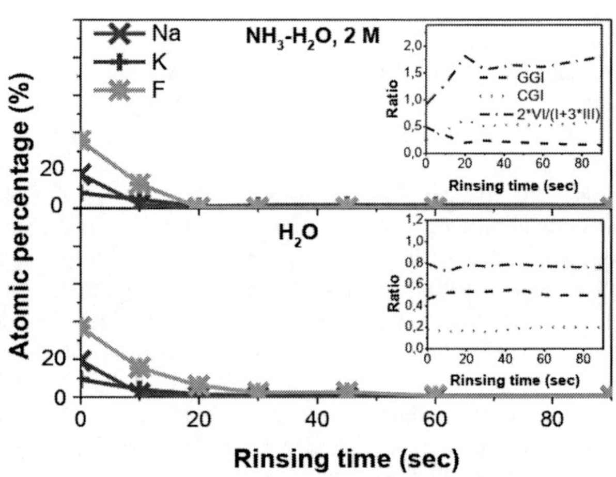

Fig 2: XPS atomic percentages of Na, K and F and ratios evolutions (GGI, CGI, and 2*VI/(I+3*III)) over dipping time in deionized water (bottom) and aqueous ammonia (top).

To investigate the disappearance time of alkali rich layer on the surface, several shorter rinsing times (0 to 90 seconds) were performed on a same batch of aged-CIGS samples on flexible substrate, with NaF and KF post-deposition treatments, in two aqueous environments: deionized water and ammonia solution (to reproduce the CBD process).

Over rinsing time, an impressive decrease of alkali elements is observed, regardless the aqueous environment used (Fig 2).

Indeed, after rinsing for 20 seconds in deionized water, the XPS signal of Na1s photopeak disappears, leading to the assumption that less than 0.1% (in atomic ratio) of Na is present at the surface. K signal remains all over the different rinsing times but in really low quantities (less than 1% in atomic percentage) after 20 seconds of rinsing.

As observed in Fig 2, the ratios evolutions differ from one aqueous environment to another one. Indeed, while ratios remain constant in deionized water over rinsing time, a different behavior is observed in liquid ammonia. The GGI decrease over time, leading to evolution of CGI and 2*VI/(I+3*III). This difference is due to critic a Ga depletion over dipping in liquid ammonia as observed in Fig 3.

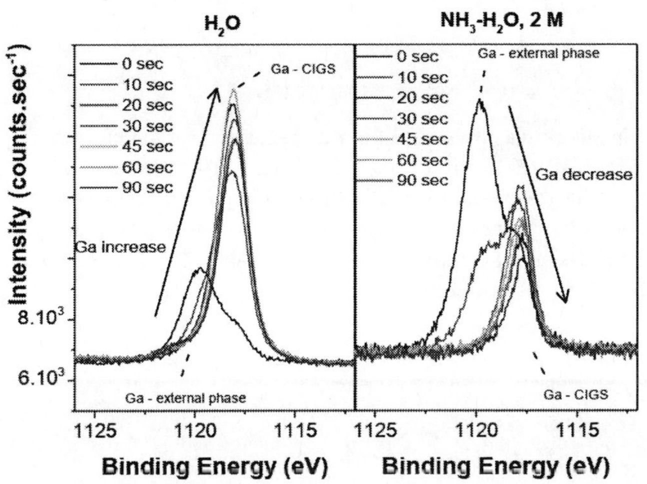

Fig 3: $Ga2p_{3/2}$ XPS spectra for different dipping times in deionized water (left) and aqueous ammonia (right).

This difference emphasizes the issue of the latency time before the buffer layer nucleation. Each buffer material presents specific deposition time, as 5 minutes for 50 nm of CdS and 8 to 20 minutes for ZnS. As this layer deposition is not instantaneous, a certain amount of alkali elements disappears from the surface during the first seconds of rinsing, leading to an important question about the quantity of alkali remaining in the material once the buffer layer starts to grow. This spectacular evolution is then of great importance to determine the correspondence between the surface state and the *in situ* delay time to optimize the absorber/buffer interface properties.

Moreover, a specific contribution in the Ga spectra (Fig 3) is observed at 1119.8 ± 0.1 eV on raw surface (0 sec) after post deposition treatment. By calculation of Auger parameters (not shown here), it is clear that these contributions are not related to oxides but to an external phase with the alkali and counter-ions, as observed by Maticiuc and al [8]. These observations reinforce the presence of the external layer with alkali element already presumed in A Section of the proceeding.

C. Ambient air ageing of the CIGS layer after rinsing

A comparison ageing was performed between CIGS samples deposited on polyimide substrates with a NaF post-deposition treatment and CIGS samples grown on glass. The interest here is to compare the evolution of Na migration when the source of Na is external (post-deposition treatment on polyimide substrate) and when the source is internal (soda lime glass) Specific ageing studies were performed on CIGS with different post-deposition treatments (NaF and NaF/KF).

In the case of the absorber with only NaF post-deposition, we aged a freshly rinsed sample of CIGS for 6 months, to reproduce the process stage before the buffer layer deposition. Regarding the alkali behavior (Fig 4), the Na_{1s} photopeak increase in time evidences its migration toward the surface, as for soda-lime system [9] with a final ratio of NaF/CIGS of 0.09. However two different kinetics are observed for the reemergence of Na. In the case of polyimide substrate (Fig 4, left), the alkali element reemerges only after one month while in the case of glass substrate, the Na reappears after 5 months of air-ageing. This kinetic difference raises the issue about the diffusion of the Na in the absorber. It is however necessary to remind here that another major difference between the two CIGS absorbers is also the deposition temperature. Indeed, the process on polyimide substrate is performed at low temperature, leading to morphology evolution and possibly responsible for a faster Na migration.

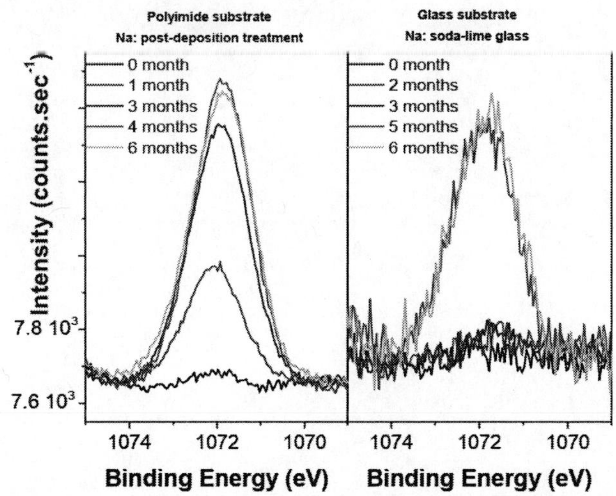

Fig. 4: Na_{1s} photopeaks evolution over air ageing for CIGS sample deposited on flexible substrate with NaF post-deposition treatment after a rinsing step with deionized water (left) and CIGS sample deposited on glass after a deoxidation step with HCl (right).

During air ageing, unusual surface chemical evolutions are also evidenced from XPS spectra observation, with the absence of oxidized Se contribution at 59.0 eV on the Se3d photopeak even after 6 months air ageing on contrary to what is observed on conventional CIGS/soda-lime [10]. This feature is concomitant with a surprising reemergence of the spin-orbit splitting and an increase of the Se branching ratio, leading to the assumption of an improvement of the crystal quality.

III. CONCLUSION

Surface and sub-surface evolutions after post-deposition treatment, rinse and ageing were studied, with a specific focus on alkali behavior. This treatment raised the issue about the importance of the alkali quantity incorporated in the CIGS absorber, as well as its capacity to assimilate it. By comparing those results with similar one on soda-lime substrates previously obtained, it is possible to underline similar behavior with a migration of Na toward the air exposed surface.

This key point leads to the interface optimization, obtained with the following step of the process, the buffer deposition, either by wet chemistry or by physical deposition with ALD process.

Perspectives to this work will be focused on the specific CIGS metallurgy on flexible substrates at low temperature, as it was observed that the XPS signatures differ from the ones obtained on soda-lime substrates. The possibility of an interaction between low temperature deposited CIGS and alkali elements will be studied.

REFERENCES

[1] M. Kaltenbrunner et al., *Nature Communications*, vol. 3, p 770, 2012.
[2] L. Kronik et al., *Advanced Materials*, vol. 10, pp 31-36, 1998.
[3] K. Granath. et al., *Solar Energy Materials and Solar Cells*, vol. 60, pp. 279-293, 2000.
[4] A. Chirila et al., *Nature Materials*, vol. 12, pp. 1107–1111, 2013.
[5] P. Blösh et al., *Thin Solid Films*, vol.31, pp 7453-7457, 2011
[6] V. Achard et al., *IEEE Journal of Photovoltaics*, vol 99, 2018.
[7] T. Hildebrandt et al., *Thin Solid Films*, vol 619, pp 25–32, 2016.
[8] N. Maticiuc et al., *Thin Solid Films*, vol 665, pp 143-147, 2018.
[9] D. Hauschild et al.; *Journal of Applied Physics*, vol 11, p 183707, 2014.
[10] A. Loubat et al., *Thin Solid Films*, vol. 633, pp. 87–91, 2017.

ACKNOWLEDGEMENTS

This work was carried out in the framework of the projects I and B of IPVF (Institut Photovoltaïque d'Ile-de-France). This project has been supported by the French Government in the frame of the program of investment for the future (Programme d'Investissement d'Avenir – ANR-IEED-002-01). The authors thank ZSW for providing CIGS on glass substrates.

MOLYBDENUM BACK CONTACT ENGINEERING USING ULTRATHIN INTERMEDIATE LAYERS FOR SOLUTION PROCESSED $Cu_2(Zn,Cd)SnS_4$ SOLAR CELLS

Siarhei Zhuk[1,2,3], Terence Kin Shun Wong[1], Venkatesh Tunuguntla[3], Sudhiranjan Tripathy[2], Ten It Wong[2], Andrei Stsiapanau[4], Aliaksandr Smirnov[4], Lydia Helen Wong[3], Goutam Kumar Dalapati[2]
[1]School of Electrical and Electronic Engineering, NOVITAS, Nanyang Technological University, Block S2, 50 Nanyang Avenue, Singapore 639798, phone: +65-67906401, fax: +65-67933318, E-mail: ekswong@ntu.edu.sg
[2]Institute of Materials Research and Engineering, A*STAR (Agency for Science, Technology and Research), 2 Fusionopolis Way, #08-05, Innovis, Singapore 138634
[3]Energy Research Institute @ NTU, Nanyang Technological University, 50 Nanyang Drive, Research Techno Plaza, X-Frontier Block, Level 5, Singapore 637553
[4]Department of Micro- and Nanoelectronics, Faculty of Radioengineering and Electronics, Belarusian State University of Informatics and Radioelectronics, Ulitsa Petrusya Brovki 6, Minsk 22013, Belarus

ABSTRACT: Sputter deposited ultrathin (~10nm) CuO and ZnS intermediate layers have been separately incorporated into solution processed Cd-doped $Cu_2(Zn,Cd)SnS_4$ (CZCTS) thin film solar cells to suppress the diffusion of S during sulfurization and the formation of MoS_2 between the absorber and the Mo back contact. X-ray diffraction of sputtered Mo films with CuO and ZnS overlayers showed reduced MoS_2 formation after sulfurization at 600°C. However, CuO results in the lowest sheet resistance for the CuO/Mo bilayer after sulfurization. When introduced into a CZCTS/CdS device structure, the ultrathin CuO and ZnS layers can maintain good adhesion with the absorber layer due to an absence of voids in CZCTS near the back interface. Deposition of CZCTS onto CuO results in a reduction of the crystallite size while ZnS has the opposite effect. Elemental profiling of fabricated CZCTS/CdS solar cells showed reduced diffusion of S into Mo. As a result, CZCTS/CdS device with ZnS intermediate layer have higher power conversion efficiencies of 9.49% compared with the reference device of 8.84%.
Keywords: Kesterite, Thin Film Solar Cell, Characterization

1 INTRODUCTION

The Cd-doped kesterite CZCTS is a promising semiconductor material for high efficiency thin film solar cells. The current highest reported power conversion efficiency η of 11.5% for sputter-deposited pure sulfide kesterite solar cells was achieved when Cu_2ZnSnS_4 (CZTS) film alloyed with Cd was utilized as the absorber layer [1]. CZCTS based photovoltaic devices have demonstrated significant enhancement of device performance compared to conventional CZTS based devices [2,3]. For further efficiency improvement, one potentially fruitful approach is to use solution processing to deposit the CZCTS absorber layer. In a recent report [4], a simple spin coating process for depositing doped CZTS that does not require the use of hazardous hydrazine has been demonstrated.

Since the kesterite semiconductors such as CZTS are prone to decomposition and solid state reaction with the Mo back contact metal, incorporation of ultrathin intermediate layers (IL) can improve the back contact parasitic series resistance. The IL materials reported thus far include: TiN, TiB_2, C, Ag, Al_2O_3 and ZnO [5-8]. CZTS solar cells with Al_2O_3 IL was reported by Liu et al. to have a high η of 9.26% [9]. The improvement in device performance was associated with reduction of MoS_2 thickness along with improvement in CZTS absorber crystalline quality. Incorporation of ZnO ultrathin IL at the back interface of CZTS solar cells similarly resulted in substantial reduction of MoS_2 interfacial layer and inhibition of void formation [10]. In these and other studies, the absorber layer was typically deposited by sputtering. By contrast, the effect of IL on solution processed kesterite absorbers have not been reported thus far. In this paper, we used the spin coating process mentioned above to deposit CZCTS and applied interface engineering to improve the back interface of CZCTS/CdS thin film solar cells. The IL materials incorporated into the Mo/CZCTS contact are cupric oxide (CuO) and zinc sulfide (ZnS) due to presence of Cu and Zn in CZCTS. The main objectives are to investigate the effect of these two ILs on the inhibition of MoS_2 formation during sulfurization and the crystallinity of the absorber layer deposited subsequently above the IL.

2 EXPERIMENTAL DETAILS

2.1 IL deposition and device fabrication

ZnS and CuO IL films (~10nm) were deposited over ~900nm molybdenum (Mo) coated soda lime glass (SLG) by radio frequency (RF) magnetron sputtering of 99.99% purity ZnS and CuO targets respectively. The base pressure and working pressure were $\sim 1\cdot 10^{-6}$ Torr and $4.2\cdot 10^{-3}$ Torr respectively whereas the applied RF sputtering power was 100W. Some samples were used for solar cell fabrication to study the impact of the ILs on structural properties of the CZCTS absorber layer. The remaining samples were subjected to sulfurization in a tube furnace to study the ability of the IL layers to inhibit MoS_2 formation.

The precursor solution for CZCTS thin film deposition was prepared as described previously [4]. After sputter deposition of IL films, ~1μm thick CZCTS absorber layer was deposited by spin-coating of the precursor solution at 4000 rpm for 20s followed by sulfurization at 600°C for 40 min in elemental sulfur vapor. After sulfurization, the tube furnace was allowed to cool naturally to room temperature in ~4 hrs. A CdS buffer layer (~50nm) was deposited over the CZCTS by chemical bath deposition (CBD) method. Finally, indium tin oxide (ITO) film was deposited by RF sputtering to complete the device.

2.2 Characterization methods

The effect of ZnS and CuO IL on the formation of

MoS_2 layer and crystalline structure of the solution deposited CZCTS layer has been investigated by X-ray diffraction (XRD). A θ-2θ scan was performed using a Bruker D8 diffractometer with a Cu K_α source (0.15418nm). Visible Raman spectroscopy with a 514nm laser source at normal incidence was used to study the structural properties of Mo thin films coated with IL materials after the sulfurization. Sheet resistance was measured by the four point probe method with a MicroXact A4P-200 system. Cross sectional scanning electron microscopy (SEM) was carried out using a JEOL JSM6700F field emission SEM. In addition, the elemental out-diffusion of Mo and S after sulfurization was studied using time of flight secondary ion mass spectrometry (TOF-SIMS) for the CZCTS/Mo structure with and without ZnS and CuO IL. A TOF.SIMS-4 system was used for the elemental profiling. The illuminated current density-voltage (J-V) characteristics of the CZCTS/CdS solar cells with and without IL were measured by using a source meter and AM1.5 radiation from a solar simulator with irradiance of 1 Sun.

3 RESULTS AND DISCUSSION

3.1 Mo with ZnS and CuO IL

X-ray diffraction (XRD) was used to analyze the crystalline structure of the sulfurized samples with and without IL (Fig. 1). The diffraction scan of the control sample revealed the presence of MoS_2 phase owing to the peak located at 29.07° (Fig. 1a). This peak is attributed to the (004) diffraction planes of MoS_2 according to JCPDS #00-037-1492. In addition, there is a prominent peak at 40.52° due to the (110) planes of Mo and a minor peak at 73.69° due to the (211) planes of Mo. A weak but discernible MoS_2 peak is observed in the diffraction scan for the Mo film coated with CuO. However, no diffraction peak attributed to the CuO phase was detected for this sample (Fig. 1b). In Fig. 1c, a peak associated with (002) ZnS diffractive plane was observed at 28.62° (JCPDS #00-036-1450). In addition, the XRD scan revealed a peak located at 21.7° which may be due to the presence of Zn(O,S) phase as a result of oxidation. The XRD data of Fig. 1 suggests that the CuO IL is more effective in mitigating the formation of MoS_2 during sulfurization than ZnS IL.

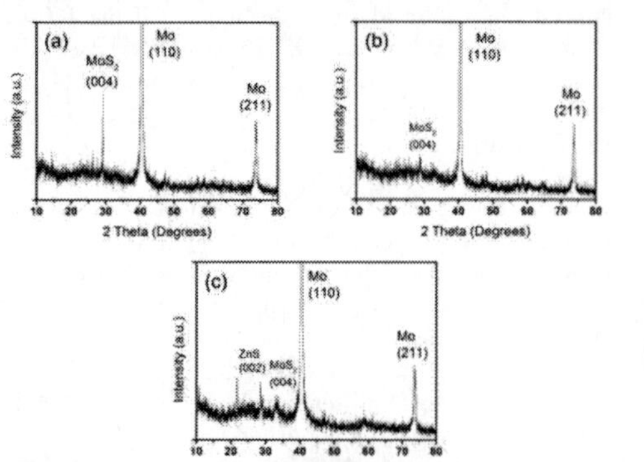

Figure 1: XRD patterns of sulfurized Mo films (a) without IL, with (b) CuO and (c) ZnS IL.

Fig. 2 shows the Raman spectra of the sulfurized Mo samples with CuO and ZnS ILs in comparison with the control sample. Raman peaks located at 380cm^{-1} and 406cm^{-1} attributable to the E^1_{2g} and A_{1g} vibrational modes of MoS_2 phase are observed for all samples [11,12]. This indicates that MoS_2 formation has occurred and is consistent with the XRD data. For the CuO/Mo sample (Fig. 2a), the intensity of these two Raman peaks was lower than that of the control sample indicating the CuO IL has a better S diffusion barrier property. In contrast, the intensity of the MoS_2 peaks for the ZnS/Mo sample (Fig. 2b) was higher due to the enhanced S diffusion from ZnS IL.

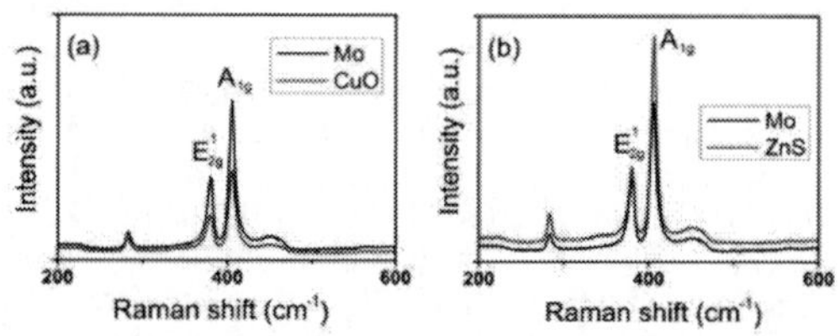

Figure 2: Raman spectra of sulfurized Mo thin films coated with (a) CuO, (b) ZnS ILs in comparison to the control sample.

In order to further evaluate the ability of the ultrathin IL materials to inhibit S diffusion into Mo films, the sheet resistance (R_s) of the samples was measured before and after sulfurization. The measured data is presented in Table 1. The R_s of CuO/Mo and ZnS/Mo are both higher than that of Mo (0.76 Ω/□). However, the R_s of ZnS/Mo (1.17 Ω/□) is higher than the R_s of CuO/Mo (0.88 Ω/□). This is due to the larger energy band gap of ZnS (3.7eV) than CuO (1.5eV). After sulfurization, R_s of all samples increased owing to the formation of the resistive MoS_2 layer. The measured R_s of 1.42 Ω/sq of the sulfurized CuO/Mo sample was lower than the control sample with R_s of 1.62 Ω/sq. On the other hand, R_s of the sulfurized ZnS/Mo sample (1.75 Ω/□) became higher than both the control and the CuO/Mo sample. The R_s data in Table I demonstrates that CuO is a more promising IL for mitigating formation of MoS_2. However, additional series resistance will be introduced to the solar cell.

Table I: Effect of the ILs on sheet resistance of Mo thin films before and after the sulfurization.

	Sheet resistance (R_s), Ω/□		
	Without IL	CuO	ZnS
Before sulfurization	0.76	0.88	1.17
After sulfurization at 600°C for 40 min	1.62	1.42	1.75

3.2 Effect of CuO and ZnS IL on CZCTS solar cells

SEM was used to examine the cross section morphology of CZCTS/CdS solar cells as shown in Fig. 3. A MoS_2 layer can be identified in Fig. 3a as indicated but such a layer could not be observed for Fig. 3b and c. The cross sectional SEM images also show an absence of voids at the interfaces between Mo and CZCTS for all three samples indicating adequate adhesion of the absorber layer to the back contact. However, increased number of voids within CZCTS film was observed for the control sample. This can be explained by the evaporation of volatile secondary phases during the sulfurization. Thus, it can be concluded that deployment of ZnS and CuO ILs can be used to slow down void formation in CZCTS absorber layer.

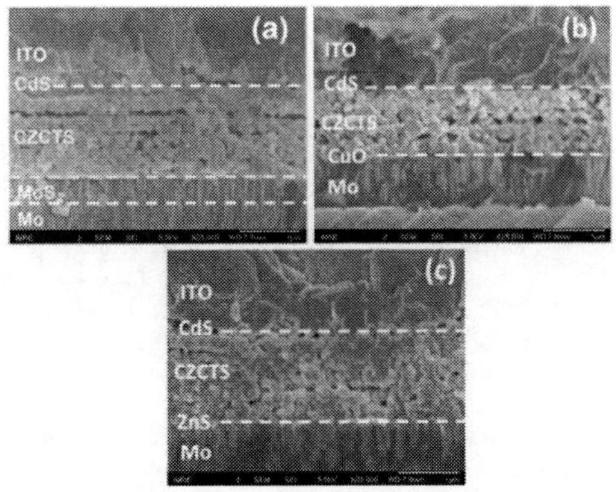

Figure 3: Cross sectional SEM images of fabricated CZCTS/CdS solar cells: a) control sample without IL, b) CuO IL, c) ZnS IL.

The impact of ILs on the crystalline structure of CZCTS absorber was studied using XRD. Fig. 4 shows that for the reference sample and the samples with ILs, the diffraction pattern consists of peaks corresponding to the (112), (200), (220) and (312) planes of CZCTS and the (110) planes of Mo. The average crystallite size (D) was calculated using the Scherrer formula [13], $D = 0.9\lambda/\beta\cos\theta$ where λ is the X-ray wavelength of 0.15148nm, β is the full width at half maximum in radians, θ is the angle of diffraction in degrees. Fig. 5 revealed that the crystallite size of CZCTS absorber layer degrades with incorporation of CuO IL owing to the reduction of crystallite size from 44.1nm to 35.8nm and from 30.8nm to 17.6nm for (112) and (220) diffractive planes respectively. In contrast, it has been observed that crystallinity of the absorber layer improves with ZnS IL incorporation. The study showed increase of crystallite size to 51.8nm and 64.2nm for both (112) and (220) planes respectively.

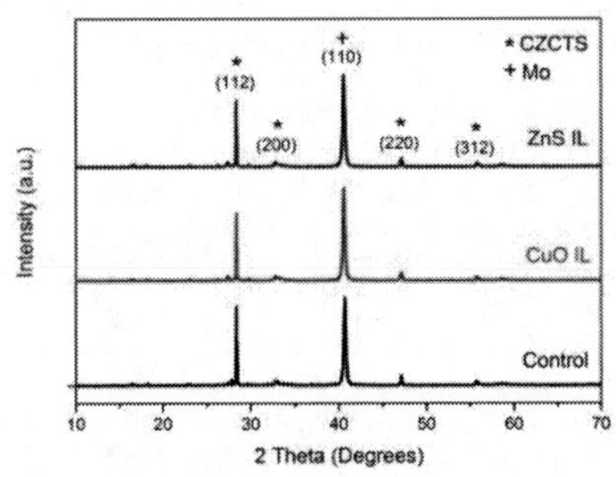

Figure 4: XRD patterns of CZCTS deposited on ZnS IL, CuO IL and on Mo after sulfurization.

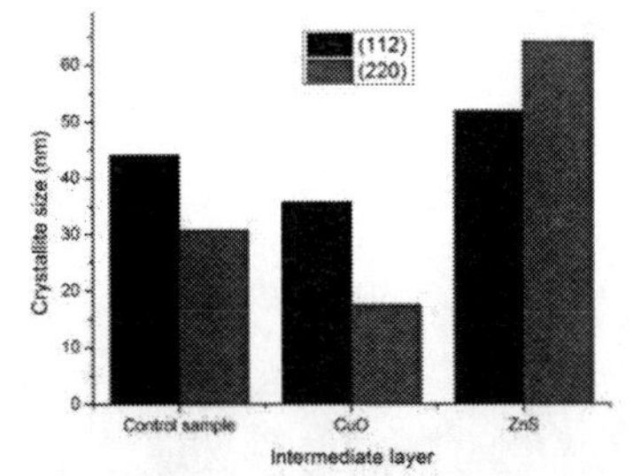

Figure 5: Calculated average size of crystallites related to (112) and (220) diffractive planes of CZCTS.

Fig. 6 shows the TOF-SIMS elemental profile for fabricated CZCTS/CdS solar cells with and without IL. For ease of reference, the approximate location of the various interfaces are annotated at the top of each profile. It can be seen that for Fig. 5a-c, the profile for the metallic elements Cu, Zn, and Sn is quite uniform within the CZCTS layer. For Cd, there is an increase in the SIMS signal intensity near the heterojunction due to the CdS buffer layer. These profiles show that the incorporation of CuO and ZnS IL does not have an adverse effect on the composition profile of the absorber. By superposing the extracted Mo and S profiles from Fig. 5 (not shown), it is found that the peak of the Mo profile is the highest for the reference device and the Mo peaks for the devices with CuO and ZnS IL are similar. However, Mo penetrates further into the CZCTS when CuO is used as IL. For the S diffusion profile, there is more S in the Mo for the reference device than devices with the CuO and ZnS IL. When the two devices with IL are compared, the ZnS IL is slightly more effective in blocking S diffusion.

3.4 CZCTS/CdS thin film solar cell measurements

Fig. 7 shows representative J-V characteristics of solution processed CZCTS/CdS solar cells with and without ZnS IL under AM1.5 illumination. The extracted open circuit voltage (V_{oc}), short circuit current density (J_{sc}), fill factor (FF), series resistance (R_s), shunt resistance (R_{sh}) and η are tabulated in Table II. The reference device has a baseline efficiency of 8.84% with V_{oc} = 615mV; J_{sc} = 23.44mAcm^{-2} and FF = 61.31%. It can be seen that the incorporation of ZnS IL results in enhancement of device power conversion efficiency. The CZCTS/CdS device efficiency η increased to 9.49% due to improvements in V_{oc}, J_{sc} and R_{sh}. However, the fill factor decreased due to increase in R_s. This increase is consistent with the higher sheet resistance found for ZnS/Mo films after sulfurization in section 3.1.

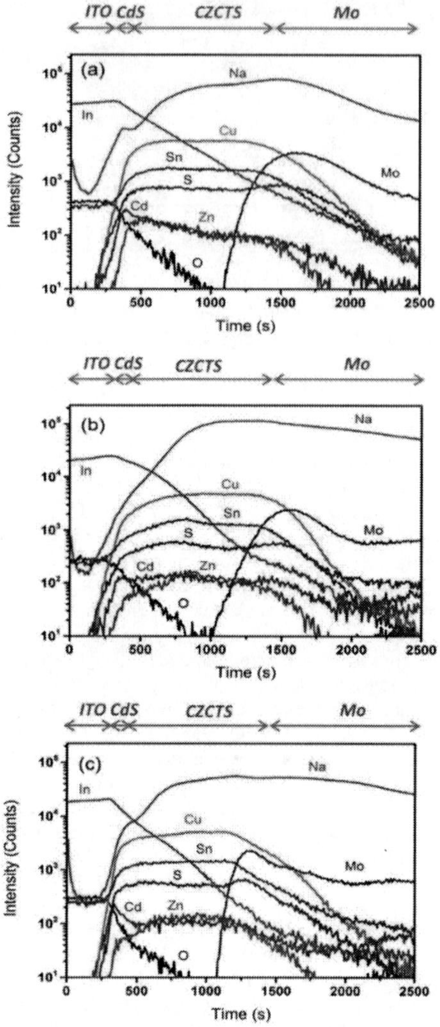

Figure 6: TOF SIMS depth elemental profile for (a) control device, solar cell with (b) CuO IL and (c) ZnS IL.

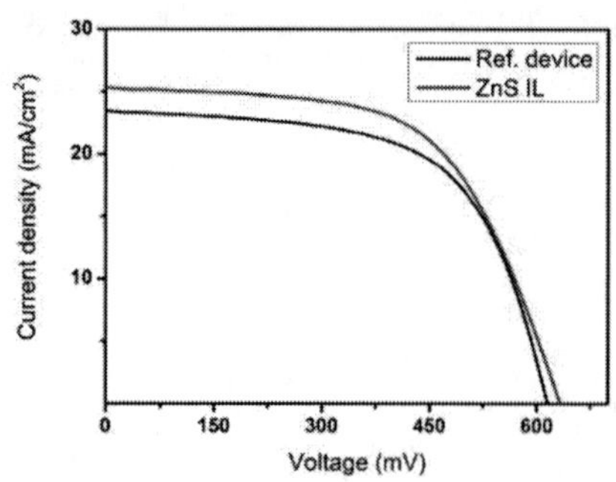

Figure 7: J-V curves for (a) control device, solar cell with (b) ZnS IL.

Table II: Photovoltaic parameters of the fabricated solar cells.

	V_{oc}, mV	J_{sc}, $mAcm^{-2}$	FF, %	R_s, Ωcm^2	R_{sh}, Ωcm^2	η, %
Ref.	615	23.44	61.31	3.36	438.9	8.84
ZnS	633	25.29	59.32	6.26	604.3	9.49

4 CONCLUSION

Back interface engineering by incorporation of CuO and ZnS ultrathin IL has been demonstrated for CZCTS/CdS thin film solar cells. CuO deposited on Mo films resulted in the lowest sheet resistance after sulfurization at 600°C compared with ZnS and Mo sample. Both CuO and ZnS were found to affect the crystallite size of CZCTS absorber layer. Sulfurization of CZCTS/CuO/Mo resulted in reduction of MoS_2 formation. However, CuO resulted in reduction of crystallite size of the CZCTS layer. In contrast, incorporation of ZnS IL led to an increase of MoS_2 formation but a significant improvement of the crystallite size of CZCTS film was observed. Solution processed CZCTS/CdS solar cells with ZnS IL have demonstrated high η of 9.49%.

5 REFERENCES

[1] C. Yan, K. Sun, J. Huang, S. Johnston, F. Liu, B.P. Veettil, K. Sun, A. Pu, F. Zhou, J.A. Stride, M.A. Green, X. Hao, ACS Energy Lett. 2 (2017) 930-936.

[2] Z. Zu, J.M.R. Tan, X. Li, X. Zeng, S.K. Batabyal, L.H. Wong, Adv. Energy Mater. 5 (2015) 1500682.

[3] Y.F. Tay, H. Kaneko, S.Y. Chiam, S. Lie, Q. Zheng, B. Wu, S.S. Hadke, Z. Su, P.S. Bassi, D. Bishop, T.C. Sum, T. Minegishi, J. Barber, K. Domen, L.H. Wong, Joule, 2 (2018) 537-548.

[4] Z. Su, W. Li, Y.F. Tay, G. Asim, L.H. Wong, Proceedings 43rd IEEE Photovoltaic Specialists Conference (2016) 534.

[5] J.J. Scragg, T. Kubart, J.T. Watjen, T. Ericson, M.K. Linnarsson, C. Platzer-Bjorkman, Chem. Mater. 25 (2013) 3162-3171.

[6] F. Liu, K. Sun, W. Li, C. Yan, H. Cui, L. Jiang, X. Hao, M.A. Green, Appl. Phys. Lett. 104 (2014) 051105.

[7] F. Zhou, F. Zeng, X. Liu, F. Liu, N. Song, C. Yan, A. Pu, J. Park, K. Sun, X. Hao, ACS Appl. Mater. Interfaces 7 (2015) 22868-22873.

[8] H. Cui, X. Liu, F. Liu, X. Hao, N. Song, C. Yan, Appl. Phys. Lett. 104 (2014) 041115.

[9] F. Liu, J. Huang, K. Sun, C. Yan, Y. Shen, J. Park, A. Pu, F. Zhou, X. Liu, J.A. Stride, M.A. Green, X. Hao, NPG Asia Materials 9 (2017) e401.

[10] X. Liu, H. Cui, W. Li, N. Song, F. Liu, G. Conibeer, X. Hao, Phys. Status Solidi 8 (2014) 966-970.

[11] T. Hong, C-L. Wu, H-C. Chang, Y-J. Lin, J. Mater. Sci. Mater. Electron. 29 (2018) 351-356.

[12] K. Haynes, R. Murray, Z. Weinrich, X. Zhao, D. Chiappe, S. Sutar, I. Radu, C. Hatem, S.S. Perry, K.S. Jones, Appl. Phys. Lett. 110 (2017) 262102.

[13] B.T. Jheng, K.M. Huang, S.F. Chen, M.C. Wu, Thin Solid Films, 564 (2014) 345-350.

CHARACTERIZATION OF SPUTTERED CDSE$_X$TE$_{1-X}$ FILMS AND THEIR APPLICATION IN CDTE SOLAR CELLS

Chunxiu Li[1], Lili Wu[1*], Fogen Wang[1], Yunlu Chen[1] and Lianghuan Feng[1]
[1] Sichuan University
No.24 South Section 1,Yihuan Road, Chengdu 610064, China

ABSTRACT: $CdSe_xTe_{1-x}$ film has high absorption coefficient and optimum band gap and chemical stability. This makes it attractive for photovoltaic application. $CdSe_xTe_{1-x}$ films with variable concentrations (x = 0 to 1) have been deposited onto quartz glass substrates by radio frequency (RF) sputtering deposition. The crystal structure, surface profile and optical properties of $CdSe_xTe_{1-x}$ films are investigated by XRD, SEM, AFM and an UV-VIS spectrometer. $CdSe_xTe_{1-x}$ films (0<x<1) are of polycrystalline texture over the whole range and exhibit wurtzite structure. The band gap reaches a minimum (1.23 eV) for $CdSe_{0.45}Te_{0.55}$ film. We add $CdSe_{0.45}Te_{0.55}$ films between CdS and CdTe layer for CdTe solar cells. Due to the interdiffusion of $CdSe_{0.45}Te_{0.55}$ and CdTe films, the smallest effective band gaps of absorber $CdSe_xTe_{1-x}$ is 1.425 eV enhancing the photocurrents in long wavelength regions.
Keywords: $CdSe_xTe_{1-x}$ films, RF sputtering, CdTe solar cells.

1 INTRODUCTION

CdTe solar cell has reached a high efficiency of 22.1%, fabricated by the First Solar Corporation and confirmed by National Renewable Energy Laboratory (NREL) [1]. High efficiency CdTe solar cells typically use n-type CdS thin films as window layers to form PN junctions [2]. Due to heat treatments in device fabrication, $CdTe_{1-x}S_x$ in CdTe absorbers will be formed because of the interdiffusion at CdS/CdTe interfaces. Since the solubility of CdS in CdTe has found to be very low [3], only a few amount S can be incorporated into CdTe. So photons can't be absorbed effectively in long wavelength regions. $CdSe_xTe_{1-x}$ films with proper X values have smaller bang gaps than CdTe. It can absorb photons with longer wavelength.

In this paper, we fabricated $CdSe_xTe_{1-x}$ films with different X values on quartz glass substrates by radio frequency (RF) sputtering deposition. Transmittance spectra and X-Ray diffraction (XRD) were performed. We calculated the materials band gaps by transmittance spectra. $CdSe_{0.45}Te_{0.55}$ film has the smallest band gap. So we add $CdSe_{0.45}Te_{0.55}$ film between CdS and CdTe layer to enhance the photo-currents in long wavelength regions.

2 EXPERIMENTAL

2.1 The fabrication of $CdSe_xTe_{1-x}$ films and CdTe Solar cells

$CdSe_xTe_{1-x}$ films were co-sputtered with CdSe and CdTe at room temperature. The purity of CdSe and CdTe targets (Chengdu Ultra-pure Application Materials Limited Liabilities Company) is greater than 99.99%. The chamber is pumped down to a base pressure of 6.0×10^{-4} Pa using mechanical and turbo-molecular pumps. Then Ar gas flowed into the chamber. The pressure was kept stable at 2.0 Pa by adjusting the high vacuum flapper valve. We can get different X of $CdSe_xTe_{1-x}$ films by changing the RF power of CdSe and CdTe. CdTe solar cells were fabricated on commercial SnO_2: F coated soda lime glass substrates supplied by NSG North America. Firstly, a 60 nm thick CdS layer was deposited by chemical bathing deposition (CBD). Then different thickness $CdSe_{0.45}Te_{0.55}$ films were fabricated on it. CdTe was deposited by close space sublimation (CSS) substrate at 550°C. CdTe thickness was about 4μm.The samples were exposed to $CdCl_2$ for 120 s and then annealed at 382 °C for 35 minutes. The back surface of CdTe was etched in a 0.2% Br_2/CH_3OH solution for 4 seconds. At last, a 100 nm Au was evaporated on the etched CdTe surface. The area of each device was 0.24 cm^2.

2.2 Characterization Methods

The film thickness was measured by Ambios Technology XP-2 step profiler. The crystal structure of films was carried out by analyzing the XRD patterns obtained by Dandong Fangyuan DX-2600 X-ray diffractometer with Cu-Kα radiation and the 2θ range between 10° and 70°. A UV-VIS spectrophotometer (Perkin Elmer Lambda-950) was used to record the optical transmittance of the samples in the wavelength range of 300~1000 nm. SEM (HITACHI S4800) was used to get surface images of the films. AFM was carried out by Bruker Nano Inc. DI Multimode 8 to obtain the surface roughness of $CdSe_xTe_{1-x}$ films. The solar cell performance was measured under simulated AM 1.5 radiation using a commercial tool calibrated by a certified silicon standard (Gsolar). QE was measured from 300 nm to 950 nm by QX10-PV Measurements at room temperature.

3 RESULTS AND DISCUSSION

3.1 Structural Characteristics of $CdSe_xTe_{1-x}$ Films

X-ray diffraction patterns of $CdSe_xTe_{1-x}$ films are shown in Fig. 1. CdSe (x=1) film is amorphous. The XRD patterns of $CdSe_xTe_{1-x}$ (x=0.25, 0.45, 0.54) films show one prominent peak at 2θ value of 24.2°, which corresponds to the hexagonal phases of the plane (002). With the increase of X, the intensity of the XRD peaks increase. CdTe (x=0) film has face-centered cubic (FCC) structure. The XRD pattern of CdTe shows two peaks at 2θ value of 23.8° and 39.6°, which corresponds to the FCC phases of the plane (111) and (220).

3.2 SEM images of $CdSe_xTe_{1-x}$ films deposited at different substrate temperatures

We can easily figure out the quality and grain size of $CdSe_xTe_{1-x}$ films through the surface profile. Fig.2 shows the SEM images of $CdSe_xTe_{1-x}$ films with different compositions. It can be observed that $CdSe_xTe_{1-x}$ films are uniform and homogenously spread on the substrate. The film surface is composed of densely packed, nanosized pellets.

The grain size of $CdSe_xTe_{1-x}$ films is one of the factors that have great effects on film properties. We use Nano Measurer-SEM software to measure the particle size of $CdSe_xTe_{1-x}$ films with different compositions. We selected 16 positions on each sample. The statistical results are presented in Fig. 3. CdTe film has the smallest average particle size of about 25 nm. CdSe film has the largest average particle size of about 54 nm.

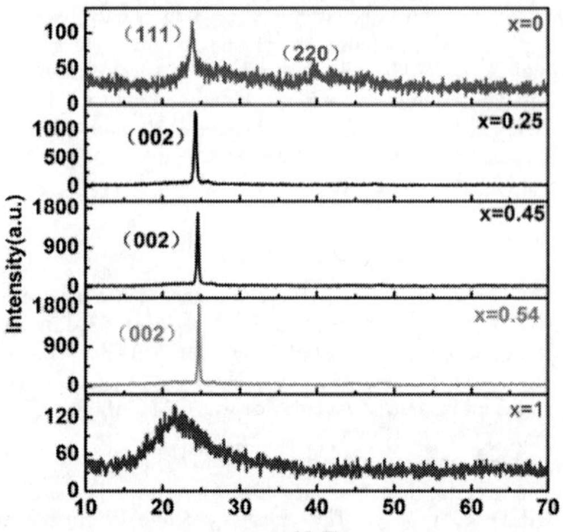

Figure 1: X-ray diffraction patterns of $CdSe_xTe_{1-x}$ films of different composition

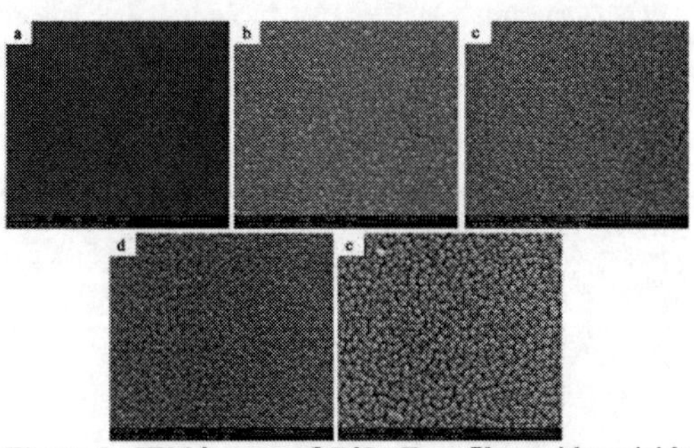

Figure 2: SEM images of $CdSe_xTe_{1-x}$ films with variable concentrations.

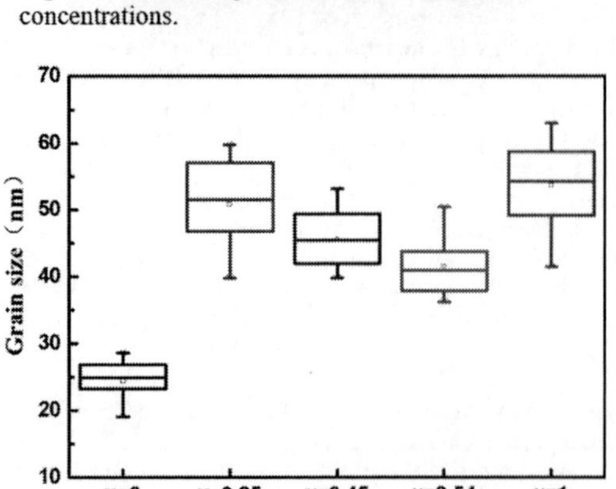

Figure 3: Grain size of $CdSe_xTe_{1-x}$ films with variable concentrations.

3.3 Three-dimentional AFM images of $CdSe_xTe_{1-x}$ films with different compositions.

The surface roughness (RMS) of the films is calculated using the equipment's software routine. The calculated results are shown in Fig. 4. From Fig. 4, we can find that the RMSR values have a certain correlation with X values. We can find that the RMS of CdTe film is the smallest. The RMS of $CdSe_xTe_{1-x}$ films firstly increases then decreases with the increase of X values. The RMS change of $CdSe_xTe_{1-x}$ films is relative to the X values.

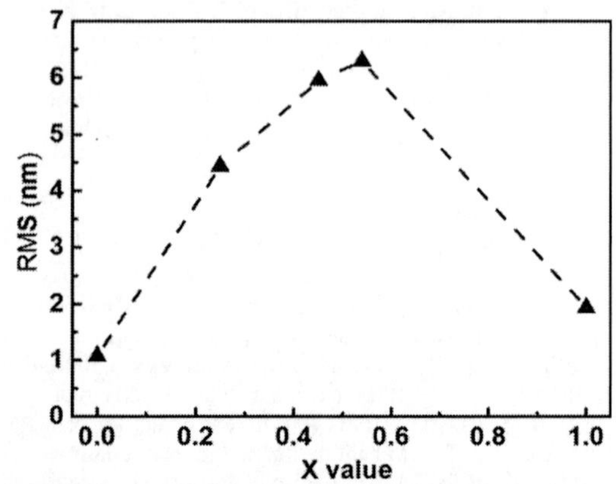

Figure 4: The RMS of $CdSe_xTe_{1-x}$ films with variable concentrations.

3.4 Optical Characteristics of $CdSe_xTe_{1-x}$ Films

The transmittance spectra of $CdSe_xTe_{1-x}$ films of different composition are shown in Fig. 5. The transmittance spectra reveals that $CdSe_xTe_{1-x}$ films can almost absorb all photons when the photon wavelength changes from 200 nm to 500 nm. Optical band gaps (Eg) of films were determined by Tanc relationship. The Eg of $CdSe_xTe_{1-x}$ film firstly de-creases and then increases when x changes from 0 to 1. $CdSe_{0.45}Te_{0.55}$ film has the smallest band gap of 1.23 eV. So $CdSe_xTe_{1-x}$ films can absorb photons with longer wave-length.

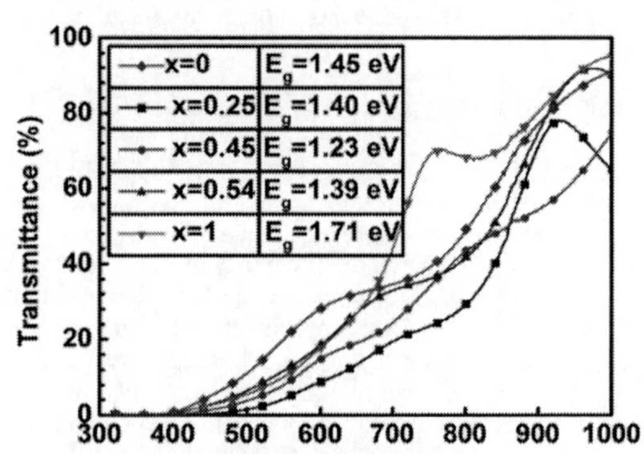

Figure 5: Transmittance spectra of $CdSe_xTe_{1-x}$ films of different composition

3.5 Results of Applying $CdSe_{0.45}Te_{0.55}$ Films to CdTe Solar Cells

After studying the properties of $CdSe_xTe_{1-x}$ films, we add a layer of $CdSe_{0.45}Te_{0.55}$ film between CdS and CdTe layer to enhance the absorption of the long-wavelength photons. The performance parameters for CdTe solar cells are shown in Table I. Short current density (J_{SC}) was calculated according to the quantum efficiency (QE). These results show that adding a layer of $CdSe_{0.45}Te_{0.55}$ can improve the J_{SC} of the device. The measured band gaps decreased from 1.471 eV to 1.425 eV as the thickness of the initial $CdSe_{0.45}Te_{0.55}$ increased from 0 nm to 267 nm. It means that devices with $CdSe_{0.45}Te_{0.55}$ films can absorb photons with longer wavelength. This is confirmed by measured QE curves, as shown in Fig. 6. The V_{OC}'s are generally low as compared to the V_{OC}'s for CdTe devices without $CdSe_{0.45}Te_{0.55}$ layers. The lower V_{OC}'s are mainly due to the reduced effective band gaps of $CdSe_xTe_{1-x}$ alloys.

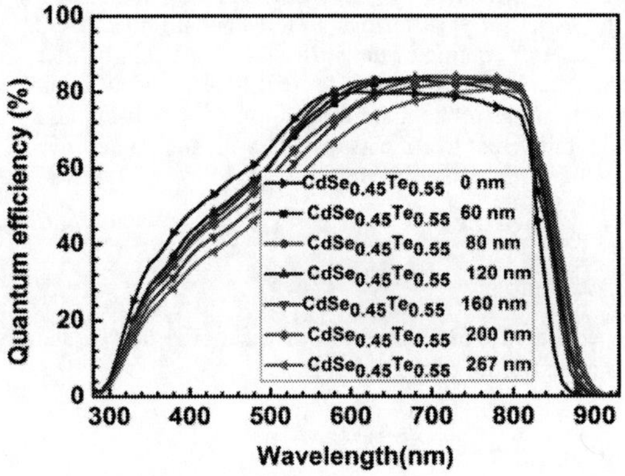

Figure 6: Measured QE spectra of CdTe solar cells with different thicknesses $CdSe_{0.45}Te_{0.55}$ films

Table I: Performance parameters for CdTe solar cells with different thickness of $CdSe_{0.45}Te_{0.55}$ films

$CdSe_{0.45}Te_{0.55}$ thickness (nm)	E_{ff} (%)	V_{OC} (V)	J_{SC} (mA/cm^2)
0	9.87	0.770	21.31
60	8.11	0.701	21.70
80	9.24	0.724	22.05
120	9.32	0.727	22.27
160	8.83	0.732	21.30
200	9.55	0.753	21.71
267	8.08	0.708	20.38

4 CONCLUSIONS

In conclusion, we studied the structural, morphological and optical properties of $CdSe_xTe_{1-x}$ films. The structure of $CdSe_xTe_{1-x}$ films is related to X values. $CdSe_xTe_{1-x}$ films (0<X<1) are hexagonal. $CdSe_xTe_{1-x}$ films are uniform, and grain size is affected by composition. RMSR was affected by both the substrate temperature and the crystal structure. The RMS of $CdSe_xTe_{1-x}$ films firstly increases then decreases with the increase of X values. Due to the bowing effect, band gaps of $CdSe_xTe_{1-x}$ films show parabolic curves. When X is equal to 0.45, the band gap of $CdSe_xTe_{1-x}$ film is 1.23 eV. $CdSe_{0.45}Te_{0.55}$ film can absorb photons with wavelength of 1000 nm. The absorption of photons with long wavelength was enhanced by applying $CdSe_{0.45}Te_{0.55}$ films in CdTe solar cells. Due to the interdiffusion of $CdSe_{0.45}Te_{0.55}$ and CdTe, the effective band gap of absorber $CdSe_xTe_{1-x}$ changes from 1.425 eV to 1.471 eV.

ACKNOWLEDGEMENTS

This work is financially supported by the National High Technology Research and Development Program of China No. 2015AA050610.

REFERENCE

[1] http://investor.firstsolar.com/releasedetail.cfm?ReleaseID=956479
[2] J. J. Li, D. R. Diercks, T. R. Ohno, C. W. Warren, M. C. Lonergan, J. D. Beach and C. A. Wolden, Sol. Energy Mater. Sol. Cells 133 (2015) 208.
[3] K. Ohata, J. Saraie and T. Tanaka, Jpn. J. Appl. Phys. 12 (1973) 1641.

EFFICIENCY IMPROVEMENT OF CDTE SOLAR CELLS WITH ULTRA-THIN CDS LAYER

Mauro Leoncini, Elisa Artegiani, Marco Barbato*, Matteo Meneghini*, Gaudenzio Meneghesso*, Marco Cavallini and Alessandro Romeo

LAPS-Laboratory for Applied Physics, Department of Computer Science, University of Verona, Ca' Vignal 1, Strada Le Grazie 15, 37134 Verona, Italy, tel.:+39-045-8027974, e-mail address: alessandro.romeo@univr.it

*Department of Information Engineering, University of Padova, Via Gradenigo 6/B, 35131, Padova, Italy

ABSTRACT: CdTe solar cells have performed excellent efficiencies up to 22,1%. In order to reach these remarkable values it is very important to reduce or completely remove the parasitic absorbance of the CdS buffer layer. One path is to reduce the thickness and by doing so to allow the light transmittance in the low wavelength region. In this scope, ultra-thin CdS has been grown in order to enhance the photocurrent generated in CdS/CdTe thin film solar cells. However reducing CdS thickness is generally detrimental for the solar cell performance with a standard transparent conductive oxide, due to the formation of shunts for a non homogeneous coating of the buffer layer particularly after CdCl2 treatment that enhances sulphur interdiffusion into the CdTe.. In this paper we present a study of different treatments of the CdS buffer layer, which are introduced in order to stabilize the material and limit the consumption of CdS under the CdTe activation treatment. A study of the effect of different treatments on CdS with goal to improve the layer's structural stability and electrical performance is presented.

Keywords: CdTe, CdS,

1 INTRODUCTION

CdTe solar cells are very competitive thin film photovoltaic devices in terms of production yield, as attested by the remarkable market performance[1]. Nevertheless this technology has also shown to be competitive in terms of efficiency with the latest results from First Solar, where a record efficiency of 22,1% has been obtained[2].

One of the major limitations to high efficiency is the parasitic absorption of CdS layer, which is typically used as n-part of the solar cell junction. So in order to reach high efficiency, it is necessary to increase the current density, which means improving the transparency of the buffer layer. This can be obtained in two different ways: either .by substituting the typical CdS layer with a higher band gap semiconductor or by reducing the CdS thickness.

However in the second case, it is difficult to obtain a deposited thin layer free from pin-holes, without empty spaces among the grains and with a homogeneous grain size distribution to favor nucleation and particle growth kinetics for CdTe [3]. Also because of the extensive inter-diffusion of tellurium and sulfur, which favors of a CdS_xTe_{1-x} intermixed layer (important for the reduction of the reticular mismatch at the interface [4]).

Several works have been made in this scope where even if a significant improvement of the current density is obtained a reduction in V_{oc} and no improvement in J_{sc} below 85 nm is also observed [3].

In this work, we have applied different treatments on 80 nm thick CdS layers in order to recrystallize the layer and to stabilize it to the subsequent CdTe and $CdCl_2$ deposition and annealing.

We have observed that by annealing the CdS in a chlorine containing gas, it is possible to obtain an efficiency enhancement compared to a sample without CdS preparation. An improvement of all the performance parameters (current density, open circuit voltage and fill factor) is gained. Moreover the accelerated stress tests, performed under condition of 1 sun and 80 °C, show a good stability of the samples.

2 EXPERIMENTAL

2.1 Preparation of the devices

CdTe thin film solar cells are made in superstrate configuration by low-temperature fabrication process based on vacuum evaporation (VE). On an ITO/ZnO stack as front contact, CdS thin films with thickness of 80 nm are deposited at a substrate temperature of 150 °C. Then the samples are treated in a furnace tube with temperatures in the range of 500-550 °C under argon and chlorine atmosphere. Subsequently CdTe is evaporated on top of the stack in a vacuum chamber, typically 5 μm are deposited and later activated by $CdCl_2$, applied by a methanol solution, and annealed in air at 380 °C.

Prior to back contact deposition, the CdTe surface is etched by a bromine/methanol solution in order to remove $CdCl_2$ residues but most important to make with the subsequent deposited copper a Cu-Te compound, important for the good performance of the cell [5,6].

Back contact is made by Cu/Au by vacuum evaporation with a successive annealing in air at 200°C.

2.2 Characterization

Completed devices have been studied by means of current density-voltage (JV). JV measurements are performed using a Keithley Source Meter 2420 at room temperature, using a halogen lamp calibrated with a silicon solar cell under an irradiation of 100 mW/cm^2.

The external quantum efficiency (EQE) was obtained using a commercial LOANA solar cell analysis system, calibrated with a silicon reference sample with known EQE using an incident spotlight of 1 mm x 2 mm area.

Accelerated stress tests (AST) were performed by leaving the finished devices in a black box under a rack of halogen lamps (at one sun illumination) and with temperature controlled and kept fixed at 80°C.

3 RESULTS

A layer of 80 nm thick CdS was deposited at 150 °C and subsequently the samples have been treated with three different approaches:

- Annealing at 450 °C in vacuum with a pressure of 10^{-5} mbar for 30 minutes (reference sample).
- Annealing at 550 °C with 100 mbar argon pressure for 30 minutes.
- Annealing at 550 °C with 100 mbar argon pressure for 30 minutes and 5 mbar chlorine-containing gas for 5 minutes.

In order to analyse if the various treatments are changing the structure of the window layer, influencing features such as the band gap energy, transmittance measurements have been performed. In particular, transmittance spectra of the CdS layers have been measured before and after the three above-mentioned treatments. From these measurements the band gap of the layers have been estimated through the Tauc plot method [7] (see figure 1).

In table I the calculated band-gap values are reported: the differently treated CdS layers show a small increase. However the band gap values of the treated samples are similar, and not very much differentiated from the initial ones. So this small band gap increase would not be enough to justify an increase in current density..

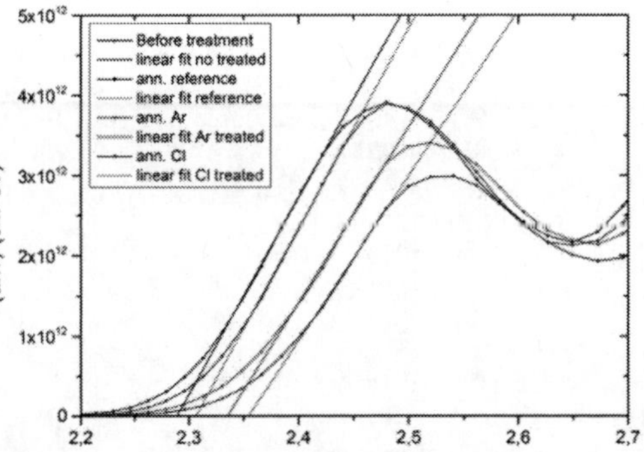

Figure 1: Tauc plot from the CdS transmittance measurements before and after treatment, and linear fit in order to estimate the energy gap.

Table I: Value of the CdS band gap estimated with Tauc plot before and after the various treatments.

Sample	Before treatment (eV)	After treatment (eV)
V883 (Reference)	2.29	2.3
V884 (Ar)	2.28	2.325
V882 (Ar + Cl)	2.26	2.34

Table II: Average efficiency parameters of devices with differently treated CdS.

Annealing	V_{oc} (mV)	J_{sc}(mA/ cm^2)	FF(%)	Eff (%)
in vacuum	733±60	24±4	56 ±8	10 ±2.5
in Ar	732±43	26.3±1	48 ±7	9.4±2
in Ar+Cl	780± 24	27±1	64 ±2	13.5±0.4

The average performance parameters of the cells fabricated with different CdS treated layers are reported in Table II. Devices made with CdS treated by chlorine show a higher average efficiency than the reference samples: (13.5 ± 0.4) % versus (10 ± 2.5) %. Moreover, the data clearly suggest that the chlorine treatment is responsible for the efficiency increase, because a simple annealing in argon even causes a reduction in efficiency to (9.4± 2) %. In particular, chloride treated samples show a superior response in terms of open circuit voltage and fill factor; while argon treated samples maintain the same Voc as the reference devices, but with a fill factor reduction, both samples treated in Ar and in Ar+Cl show a current density increase.

This effect can be clarified with the help of quantum efficiency spectra, shown in figure 2. The spectral response is strongly improved at low wavelength regions, probably due to a re-evaporation of CdS at the high treatment temperature.

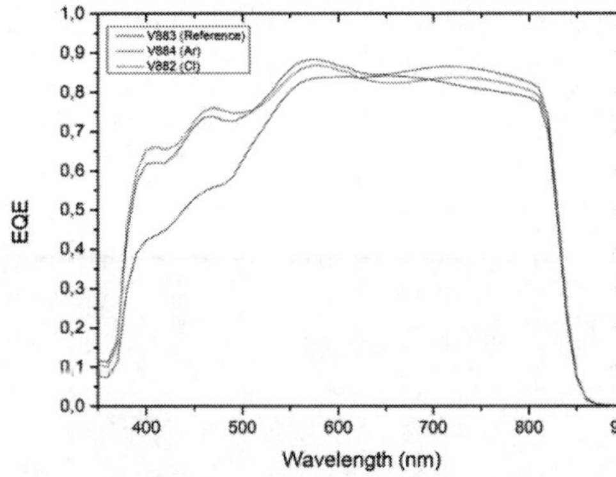

Figure 2: Comparison between External Quantum Efficiency curves of the best cells with differently treated CdS.

The stronger stability of chlorine treated CdS layer is surely attested by the remarkable difference in efficiency. Most probably a different intermixing and a higher stability of the CdS grains have developed after this treatment, demonstrating that also with an ultra-thin CdS layer (less than 80 nm, because of the re-evaporation of the layer) it is possible to obtain higher Voc and FF values.

The annealing process of CdS has been also studied by changing the treatment temperature. In particular, in order to understand if a lower CdS re-evaporation during the treatment could influence the performance of the samples, we have reduced the temperature to 500 °C with the same atmosphere conditions. Results are shown in Table III, in comparison with samples annealed at 550°C, the reference (annealing in vacuum) shows a higher conversion efficiency, given by a general increase of all the parameters. Probably the reduced temperature causes a lower CdS re-evaporation, leading to a more performing junction. On the other hand, the Cl treated samples shows a lower efficiency compared to the 550°C annealed devices: mainly due to a fill factor and a Jsc drop, while the Voc is increased. We can conclude that performing the treatment at lower temperature does not lead to the same improvement from our standard vacuum annealing to the Cl treated samples as observed for treatments at 550°C; the chlorine treatment is less effective at 500°C. For this reason high temperature seems to play an important role, especially in terms of fill factor and current density.

Table III: Average efficiency parameters of devices with differently treated CdS.

Samples	V_{oc} (mV)	J_{sc} (mA/cm^2)	FF (%)	Eff (%)
V851 (reference)	781.9	25.5	61.9	12.37
V852 (Cl)	803	25.5	61.12	12.51

Finally, in order to evaluate the stability of the samples with differently treated CdS, we have performed AST on the finished devices. In particular samples with CdS treated in Ar and Ar + Cl at 550°C, and annealed in vacuum at 450°C (reference) are discussed. In figure 3 the JV curves of the different samples before and after 900 hours of AST are presented. The reference sample shows a good stability, while the treated CdS samples have both a reduction in Voc and FF. After AST cells treated in argon atmosphere at 550 °C and cells treated in vacuum at 450 °C have a little improvement in roll-over; in any case also the chlorine treated sample keep a good performing back contact without rollover appearance. In conclusion the CdS chlorine treatment does not affect the stability of the samples.

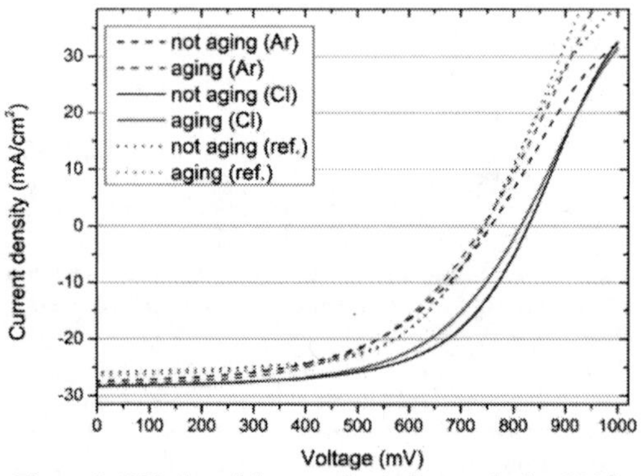

Figure 3: J-V plot of the samples before and after 900 h of AST.

4 CONCLUSION

We have applied different treatments on 80 nm thick CdS layers. Annealing the CdS at 550°C in a chlorine containing gas, an efficiency enhancement is obtained if compared to samples annealed in vacuum or just in argon-atmosphere. A general performance parameters improvement occurs; the open circuit voltage and the fill factor increase, as well as the current density, as confirmed by external quantum efficiency measurements. Thus a different stability of CdS with chlorine treatment is surely attested; probably a different intermixing and a higher stability of the CdS grains have developed after this treatment. Moreover the accelerated stress tests, performed under condition of 1 sun and 80 °C, show a good stability of the samples. More investigations will be presented in a separated paper.

5 REFERENCES

[1] S. Philipps, F. Ise, W. Warmuth, P. Conferences, C. GmbH, Photovoltaics Report, (2018). doi:20.10.2016.

[2] S. Krum, S. Haymore, First Solar Achieves Yet Another Cell Conversion Efficiency World Record, First Sol. (2016).

[3] S.G. Kumar, K.S.R.K. Rao, Physics and chemistry of CdTe/CdS thin film heterojunction photovoltaic devices: fundamental and critical aspects, Energy Environ. Sci. 7 (2014) 45. doi:10.1039/c3ee41981a.

[4] Y. Yan, R.G. Dhere, K.M. Jones, M.M. Al-Jassim, Influence of substrate structure on the growth of CdTe thin films, J. Appl. Phys. 89 (2001) 5944–5948. doi:10.1063/1.1367406.

[5] X. Wu, J. Zhou, a. Duda, Y. Yan, G. Teeter, S. Asher, et al., Phase control of CuxTe film and its effects on CdS/CdTe solar cell, Thin Solid Films. 515 (2007) 5798–5803. doi:10.1016/j.tsf.2006.12.151.

[6] I. Rimmaudo, A. Salavei, E. Artegiani, D. Menossi, M. Giarola, G. Mariotto, et al., Improved stability of CdTe solar cells by absorber surface etching, Sol. Energy Mater. Sol. Cells. 162 (2017) 127–133. doi:10.1016/j.solmat.2016.12.044.

[7] J. Tauc, Optical properties and electronic structure of amorphous Ge and Si, Mater. Res. Bull. 3 (1968) 37–46. doi:https://doi.org/10.1016/0025-5408(68)90023-8.

HIGH PERFORMANCE CU-ZN-SN-SE THIN FILM SOLAR CELLS FABRICATED BY SPUTTERING QUATERNARY COMPOUND TARGET METHOD

Xinchen Li[1,2], Daming Zhuang*[1,3], Ming Zhao[1,3], Ning Zhang[2], Yuandong Liu[2], Peng Liu[2], Yaowei Wei[1], Guoan Ren[1]
1. School of Materials Science and Engineering, Tsinghua University, Haidian District, Beijing, 100084, P. R. China
2. Beijing Sifang Crenergy Optoelectronics Technology Co., Ltd., No. 9, Shangdi 4th Street, Haidian District, Beijing, 100085, P. R. China
3. Key Laboratory for Advanced Materials Processing Technology, Ministry of Education, Tsinghua University, Haidian District, Beijing, 100084, P. R. China

ABSTRACT: This paper presents a method of fabricating pure selenized kesterite thin film solar cells by magnetron sputtering with Cu-Zn-Sn-Se (CZTSe) quaternary compound target. High quality CZTSe absorbers can be fabricated by selenization with annealing process. Due to the existence of adequate Se within the as-sputtered CZTSe precursors, a very low concentration of 2% (v/v) H_2Se in Ar atmosphere was required during the annealing process. Smooth and uniform CZTSe absorbers can be obtained. Thin film CZTSe solar cells have been fabricated by using the absorbers and achieved peak efficiency of 11.74%. The results imply the method of sputtering quaternary compound target can be a simple and effective way to fabricate high performance kesterite thin film solar cells.
Keywords: kesterite solar cell, magnetic sputtering, quaternary compound target

1 INTRODUCTION

Kesterite solar cell has been one of the most promising candidates for next generation thin film photovoltaic technologies. It is satisfied for the demands of earth abundant, non-toxic, stable and high theoretical efficient for high performance solar cell absorber materials. Systematically researches have been made to investigate the fabrication process,[1, 2] the defects,[3-5] the energy level alignment,[6, 7] and new perspective methods to improve the efficiency.[8, 9] The kesterite solar cell with the record efficiency of 12.6% was achieved by using hydrazine-based solution process method with the Cu-Zn-Sn-S-Se (CZTSSe) absorber.[10] The research for boosting efficiency and simplifying fabrication process of kesterite solar cells should be taken efforts to become an alternative commercialized photovoltaic technique in the future.

The kesterite materials include pure Cu-Zn-Sn-Se (CZTSe) and Cu-Zn-Sn-S (CZTS). The bandgap of kesterite materials can be tuned from ~1.0 eV to ~1.5 eV by different Se/S ratios.[11, 12] However, among the high performance kesterite solar cells, the absorbers were mostly achieved with the non-metal elements of high Se/S ratio or even pure selenized material. The possible reasons for the high performance solar cells with such high Se content are listed as follows. (1) The higher formation energy of Se induced defects leads less isolated deep donor defects, which can be recombination centers in the material.[3] (2) The electron trapping effect due to conduction band edge downshift in CZTSe is weaker compared to CZTS.[3] (3) A smaller bandgap allows the fully utilization of photons in longer wavelength. To sum up, with the advantages of less open circuit voltage (V_{OC})-deficit and higher photon-conversion ability, pure selenized CZTSe should be investigated to achieve high performance kesterite solar cells.

In this paper, CZTSe solar cells with the peak efficiency of 11.74% have been achieved. The method of fabricating CZTSe thin film solar cell by magnetron sputtering of quaternary compound target is presented. With the existence of Se element within the precursors, annealing process of the precursors was performed under only 2% (v/v) of H_2Se in Ar atmosphere. Compared to the films annealed under pure Ar atmosphere, the scanning electron microscopy (SEM) images reveal uniform and smooth absorbers can be obtained by using this method. The X-ray diffraction (XRD) and Raman scattering spectroscopy results indicate almost the pure phase trasformation was occurred during selenization. These results suggest that the method of sputtering quaternary compound target can be a simple and effctive way to fabricate high performance kesterite solar cells.

2 EXPERIMENTAL

2.1 Film preparation

The CZTSe precursors were deposited by DC magnetron sputtering with Cu-Zn-Sn-Se quaternary compound target on Mo-coated soda lime glass (SLG) substrate. The targeting composition of the precursor was Cu-poor and Zn-rich stoichiometry of Cu/(Zn+Sn)≈0.80 and Zn/Sn≈1.05. The precursor was deposited at room temperature with a film thickness of 1500 nm.

After the precursor deposition, the samples were transferred into a cold wall vacuum chamber to perform annealing process. The annealing temperature was controlled to 590°C with the heating and cooling temperature rate of 5°C/min. A total flow rate of 200 sccm of 2% (v/v) H_2Se mixed in Ar gas was continuously filling in and pumping out of the chamber to keep the pressure of 50000 Pa. After maintaining at 590°C for 5 min, uniform and smooth CZTSe absorbers can be obtained after the samples cooling to room temperature. As a control group, CZTSe precursors were annealed without any Se atmosphere, which was performed in another cleaned chamber with the same structure and operation procedure to avoid Se contamination.

2.2 Device fabrication

After obtaining the CZTSe absorbers, the samples were deposited a 50 nm-thick of CdS layer by chemical bath process. Layers of 100 nm i-ZnO and 400 nm AZO were deposited by sputtering method. The area of each solar cell was 0.54 cm^2, which was defined by mechanical scribing. No anti-reflectance layer was used in the device fabrication process. The details of device fabrication were discribed in our previous work.[13]

2.3 Characterization

The thickness of different films was measured by KLA-Tencor Alpha-Step D-500 stylus profiler. The SEM surface and cross-sectional images were taken by SEM, SIGMA HV & PV Series, Carl Zeiss. The XRD results were performed by using D8-Advance (Bruker) under Cu Kα (λ=0.15418 nm) irradiation. Raman scattering spectroscopy (Raman, LabRAM HR Evolution) results were measured with 532 nm laser excitation. The composition of CZTSe layer on Mo-coated SLG was measured by X-ray fluorescence (XRF, SPECTRO XEPOS), which was calibrated by inductively coupled plasma atomic emission spectroscopy (ICP-AES, SPECTRO ARCOS EOP). The test of solar cell was performed under standard test condition (STC). The measurement of J-V characteristics of solar cell was performed by using a 1000 W m^{-2} light intensity AM 1.5G AAA solar simulator (Photo Emission Tech) and a Keithley 2420 source meter. The external quantum efficiency (EQE) of solar cell was measured by using a SCS-100 EQE testing system from Zolix with a wight light bias of 0.1 sun intensity.

3 DISCUSSION

3.1 The characterization of CZTSe precursors

The CZTSe precursors were deposited by DC-sputtering with Cu-Zn-Sn-Se quaternary compound target at room temperature. As shown in Figure 1(a) and (b) for the SEM surface and cross-sectional images, a uniform and flat precursor with the thickness of 1500 nm was formed on the Mo-coated SLG substrate. It is shown that there is no obvious particle or void within this layer. The analysis of XRD pattern of CZTSe precursors in Figure 1(c) reveals that three major peaks are at 27.38°, 45.24° and 53.18°, which are correspond to (112), (220) and (312) peaks of CZTSe diffraction patterns.[14] However, due to the overlapping of peaks of the CZTSe and selenized secondary phases such as ZnSe or Cu_2Se, Raman spectroscopy was used to determine the phase purity of the precursor. As the result in Figure 1(d), the Raman peaks at 174 cm^{-1}, 195 cm^{-1}, 224 cm^{-1} and 245 cm^{-1} correlates well with the dominated Raman peaks of CZTSe quaternary compound.[15, 16] The XRD and Raman results indicate that a pure layer of CZTSe precursor can be formed on the substrate after the sputtering deposition process.

However, as the precursors were tried to fabricate solar cells, no photovoltaic characteristics were observed. The reason may be that the sputtering process can only form a CZTSe layer with crystal nucleus on the substrate. Thus, the kesterite crystal structure of CZTSe was not determined. Another reason is that the existence numbers of crystal boundaries may strongly recombine the photon generated carriers. As a result, solar cells based on the precursors showed no photocurrent. The crystallization of the precursors should be conducted to obtain high quality CZTSe absorbers, and to fabricate high performance solar cells.

3.2 The annealing process of CZTSe absorbers

The CZTSe precursors were annealed in vacuum chamber to induce crystallization of this layer. As the above results, it has shown that the CZTSe crystal nucleus formed within film. Firstly, the precursor was transferred into a vacuum chamber for annealing process

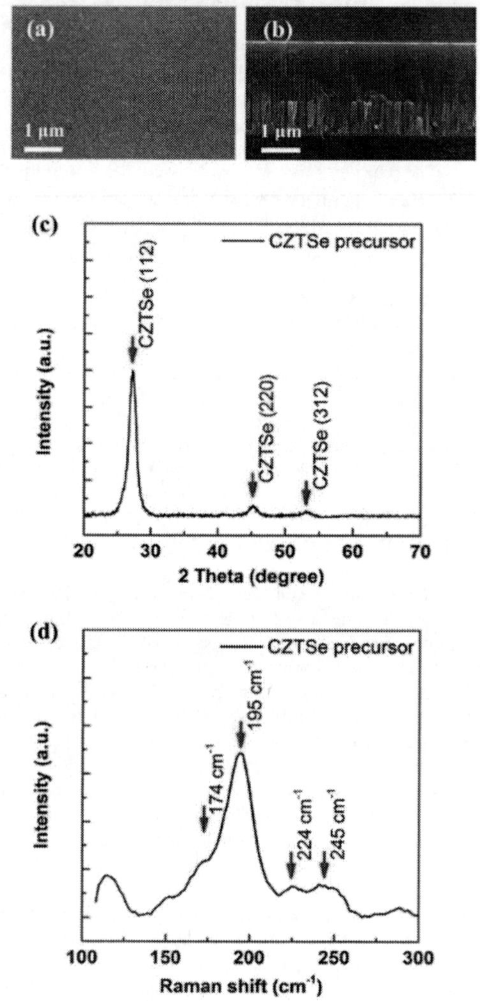

Figure 1: (a) Surface and (b) cross-sectional SEM images of CZTSe precursor on the Mo-coated SLG substrate. (The wight bar in the corner illustrates the length of 1 μm.) (c) XRD pattern of CZTSe precursor. (d) Raman spectrum of CZTSe precursor.

without any additional Se content. The annealing process was performed in pure Ar of 50000 Pa with heating and cooling rate of 5°C/min, and maintaining at 590°C for 5 min. However, as shown in Figure 2(a) and (b), the obtained film showed vast vacancies on the surface and cross-sectional of the CZTSe layer. Mo-coated SLG substrate was directly exposed to the surface. Obviously, based on the absorbers annealed in pure Ar, solar cells performed no photovoltaic capability.

Due to the volatile nature of Sn-Se phases, the annealing temperature of the CZTSe precursor should not be too high. A proper annealing temperature can promote the crystallization of the absorbers. However, the lack of Sn element may lead to the decomposition of the CZTSe absorber. To maintain a proper chemical composition, excess Se content such as Se powder or H_2Se gas is always applied in annealing process to crystallize the CZTSe absorbers. In this work, a cold wall vacuum chamber with a controllable of H_2Se flow rate was applied to anneal the CZTSe precursors. With the same procedure above, the heating and cooling rate were 5°C/min and maintained at 590°C for 5 min. A total flow rate of 200 sccm of 2% (v/v) H_2Se mixed in Ar gas was applied to keep the pressure of 50000 Pa.

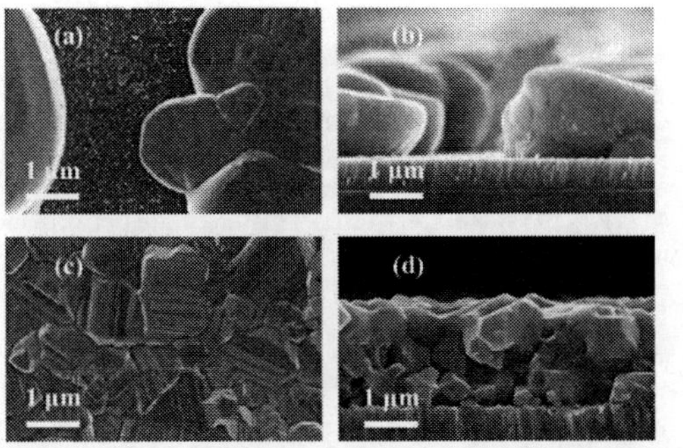

Figure 2: (a) Surface and (b) cross-sectional SEM images of CZTSe film annealed in vacuum chamber without additional Se. (c) Surface and (d) cross-sectional SEM images of CZTSe absorber with the temperature annealing process under H_2Se mixed in Ar atmosphere in the vacuum chamber. (The wight bar in the corner illustrates the length of 1 μm.)

Compared to the absorbers annealed in pure Ar, a smooth and compact absorber layer can be obtained after the CZTSe precursors selenized in H_2Se mixed with Ar atmosphere. As shown in Figure 2(c) and (d), the surface and cross-sectional images illustrate the crystallized CZTSe absorbers. With the annealing process, the size of CZTSe grains was grown up over 1 μm with only a few voids between the grain boundaries. In the corss-sectional image for the interface between the Mo layer and CZTSe absorber, a thin $MoSe_2$ layer with the thickness of ~100 nm can be observed. On the contrary to no obvious $MoSe_2$ in the samples with pure Ar annealing process, this layer was formed during the selenization process. Previous report indicated that the thickness of the $MoSe_2$ layer performed a strong reverse correlation to the CZTSe solar cell performance.[17] The results show that our apporach can form a desirable thickness of $MoSe_2$ layer, leading to a high quality CZTSe absorber on Mo-coated SLG substrate.

To investigate the chemical composition of the selenized CZTSe absorbers, XRF results reveal that the absorbers maintained almost the same stoichiometry of Cu/(Zn+Sn) ≈0.80 and Zn/Sn≈1.05 compared to the precursors. This indicates that the selenization process can surpress decomposition of the CZTSe absorbers. It implies the great advantage of out approach that the CZTSe absorber can be crystallized with proper grain size and chemical composition under a very low concentration of H_2Se in Ar atmosphere.

The XRD measurement pattern of the selenized CZTSe absorbers is plotted in Figure 3(a). From the result, it clearly shows four dominated peaks in the pattern. All the major peaks, namely, the (112) peak=27.38°, (220) peak=45.24°, (312) peak=53.18° and (008) peak=65.90°, are corresponding well to the previous report.[14] Compared to the XRD pattern of CZTSe precursor, the intensity of the major peaks was increased. And the peaks of the absorber became sharper compared to the precursor. The (008) peak was appeared after the selenization of CZTSe absorbers. These results indicate that the crystal grows within the CZTSe absorbers. A few minor peaks with very low intensities were not identified within the pattern, which may be corresponding to the secondary or ternary phases of the absorber. However, as mentioned above, the XRD pattern may not distinguish some of the secondary phases due to peaks overlapping.

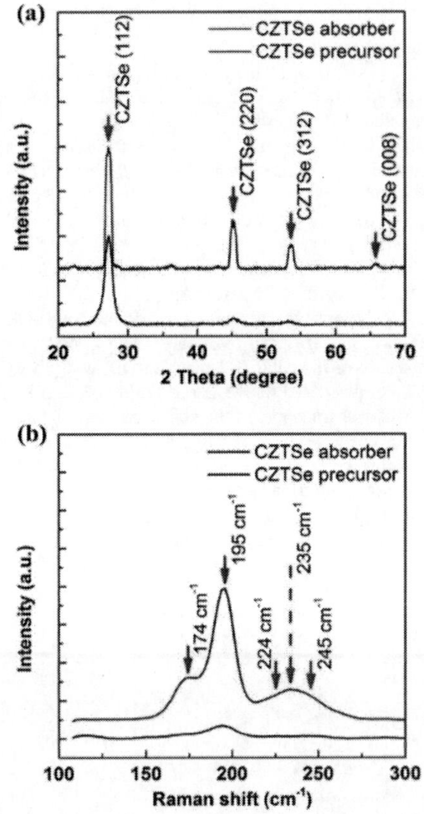

Figure 3: (a) XRD patterns of CZTSe precursor and selenized absorber. (b) Raman spectra of CZTSe precursor and selenized absorber. The XRD and Raman results of the CZTSe precursor were plotted in the same intensity scales with the results of selenized CZTSe absorber.

As shown in Figure 3(b), the Raman spectrum of the selenized CZTSe absorbers is illustrated to determine the phase purity of this layer. Major peaks at 174 cm^{-1} and 195 cm^{-1} are dominated Raman active A mode of kesterite structure. The peaks at 224 cm^{-1}, 235 cm^{-1} and 245 cm^{-1} are corresponding to the E and B mode of CZTSe.[15] As a comparison between the precursors and the absorbers, not only peaks intensity is substantially intensified, but also a peak at 235 cm^{-1} is also observed. The 235 cm^{-1} peak can be attributed to the B mode of kesterite structure, which is distinguished from other crystal structures of CZTSe material.[15] Moreover, Raman peaks of secondary phases such as ZnSe (205 cm^{-1} and 251 cm^{-1}), SnSe (130 cm^{-1} and 150 cm^{-1}), $SnSe_2$ (185 cm^{-1}) or Cu_2Se (263 cm^{-1}) are not identified,[18-20] which indicates homogeneous CZTSe crystal formed within the absorbers. The results give strong evidences that the sputtered CZTSe precursors can be transformed into stable kesterite crystal structure absorbers by the selenization process.

Table 1: Device characteristics of peak efficiency CZTSe solar cell.

	V_{OC}	J_{SC}	FF	PCE
CZTSe solar cell	0.431 V	36.33 mA cm^{-2}	74.99%	11.74%

3.3 The performance of CZTSe solar cells

CZTSe solar cells have been fabricated by using selenized absorbers. As summarized in Table 1, the high performance CZTSe solar cell with peak efficiency performed a V_{OC} of 0.431 V, a short-circuit current density (J_{SC}) of 36.33 mA cm^{-2}, a fill factor (FF) of 74.99% and a power conversion efficiency (PCE) of 11.74%. The J-V characteristics of the solar cell is illustrated in Figure 4(a). The V_{OC} of 0.431 V is almost the same value of reported pure selenized CZTSe solar cells. As illustrated in the insert diagram of J_{dark}-V curve, the CZTSe solar cell showed a remarkable low leakage current at the reverse voltage. The good diode characteristic of the solar cell may lead to the high FF of 74.99%. As discussed above, large grain size can benifit to better carrier transport in the absorbers. And few voids across the absorbers would suppress the leakage current of the solar cells. Thus, the solar cell characteristics illustrate that sputtering quaternary compound target with selenization process can be suitable for the fabrication of high performance CZTSe solar cells.

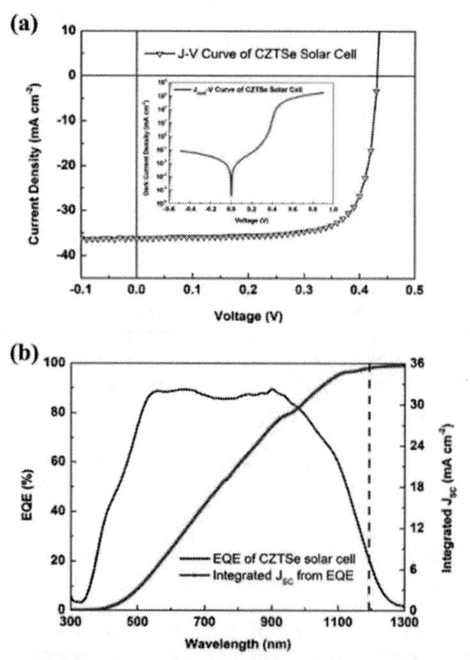

Figure 4: (a) The J-V characteristics (black curve) of the peak efficiency CZTSe solar cell. (insert: the J_{dark}-V characteristics) (b) The EQE spectrum (black curve) of the solar cell, and the integrated J_{SC} from the EQE (red curve) at each wavelength. The blue dash line indicates the extracted band gap from the EQE spectrum.

The EQE spectrum of the CZTSe solar cell is illustrated in Figure 4(b). In the main absorption region wavelength from 500 nm to 1000 nm, the diagram shows a high EQE of 80% to 90%, which indicates high photon conversion ability of the CZTSe solar cells. As the blue dash line plotted in the figure, the extracted band gap from the EQE spectrum is 1.04 eV, which is correlated well with the same band gap of pure selenized CZTSe solar cells. The integrated J_{SC} from the EQE spectrum of CZTSe solar cell is 35.66 mA cm^{-2} , which agrees well with the J-V characteristics. Only a 1.8% difference between the EQE intergrated J_{SC} and J-V characteristics suggests the small spectrum mismatch of the performance measurement.

4 CONCLUSION

In conclusion, high performance CZTSe solar cells with peak efficiency of 11.74% heve been fabricated by using sputtering quarternary compound target with selenization method. Due to the existence of adequate Se within the precursors, a low concentration of H_2Se mixed with Ar atmosphere was applied during the selenization process. Results indicate that the CZTSe absorbers can be crystallized with proper grain size and chemical composition. The analyses of XRD results and Raman spectra reveal that the CZTSe precursors with crystal nucleus can transformed into kesterite crystal structure absorbers homogenously. Solar cells based on the selenized CZTSe absorbers can achieve a peak efficiency of 11.74%. Our demonstration suggests that sputtering with CZTSe quaternary compound target can be a simple and effective method to fabricate high performance kesterite solar cells.

5 ACKNOWLEDGMENT

This work was supported by the Project of Beijing Municipal Science and Technology Commission (Z171100000317002). The authors are also grateful for the analysis support of State Key Laboratory of New Ceramics and Fine Processing.

REFERENCES

[1] Bag S, Gunawan O, Gokmen T, Zhu Y, Todorov TK, Mitzi DB. Low band gap liquid-processed CZTSe solar cell with 10.1% efficiency. Energy & Environmental Science. 2012;5(5).

[2] Lee YS, Gershon T, Gunawan O, Todorov TK, Gokmen T, Virgus Y, et al. Cu2ZnSnSe4Thin-Film Solar Cells by Thermal Co-evaporation with 11.6% Efficiency and Improved Minority Carrier Diffusion Length. Advanced Energy Materials. 2015;5(7).

[3] Chen S, Walsh A, Gong XG, Wei SH. Classification of lattice defects in the kesterite Cu2ZnSnS4 and Cu2ZnSnSe4 earth-abundant solar cell absorbers. Adv Mater. 2013;25(11):1522-39.

[4] Shin D, Saparov B, Mitzi DB. Defect Engineering in Multinary Earth-Abundant Chalcogenide Photovoltaic Materials. Advanced Energy Materials. 2017;7(11).

[5] Bourdais S, Choné C, Delatouche B, Jacob A, Larramona G, Moisan C, et al. Is the Cu/Zn Disorder the Main Culprit for the Voltage Deficit in Kesterite Solar Cells? Advanced Energy Materials. 2016;6(12).

[6] Haight R, Barkhouse A, Gunawan O, Shin B, Copel M, Hopstaken M, et al. Band alignment at the

Cu2ZnSn(SxSe1−x)4/CdS interface. Applied Physics Letters. 2011;98(25).
[7] Crovetto A, Hansen O. What is the band alignment of Cu 2 ZnSn(S,Se) 4 solar cells? Solar Energy Materials and Solar Cells. 2017;169:177-94.
[8] Giraldo S, Saucedo E, Neuschitzer M, Oliva F, Placidi M, Alcobé X, et al. How small amounts of Ge modify the formation pathways and crystallization of kesterites. Energy & Environmental Science. 2018;11(3):582-93.
[9] Xin H, Vorpahl SM, Collord AD, Braly IL, Uhl AR, Krueger BW, et al. Lithium-doping inverts the nanoscale electric field at the grain boundaries in Cu2ZnSn(S,Se)4 and increases photovoltaic efficiency. Phys Chem Chem Phys. 2015;17(37):23859-66.
[10] Wang W, Winkler MT, Gunawan O, Gokmen T, Todorov TK, Zhu Y, et al. Device Characteristics of CZTSSe Thin-Film Solar Cells with 12.6% Efficiency. Advanced Energy Materials. 2014;4(7).
[11] Chen S, Walsh A, Yang J-H, Gong XG, Sun L, Yang P-X, et al. Compositional dependence of structural and electronic properties of Cu2ZnSn(S,Se)4alloys for thin film solar cells. Physical Review B. 2011;83(12).
[12] Fan F-J, Wu L, Gong M, Liu G, Wang Y-X, Yu S-H, et al. Composition- and Band-Gap-Tunable Synthesis of Wurtzite-Derived Cu2ZnSn(S1−xSex)4 Nanocrystals: Theoretical and Experimental Insights. ACS Nano. 2013;7(2):1454-63.
[13] Sun R, Zhuang D, Zhao M, Gong Q, Scarpulla M, Wei Y, et al. Beyond 11% efficient Cu 2 ZnSn(Se,S) 4 thin film solar cells by cadmium alloying. Solar Energy Materials and Solar Cells. 2018;174:494-8.
[14] Nagaoka A, Yoshino K, Taniguchi H, Taniyama T, Miyake H. Growth of Cu2ZnSnSe4 single crystals from Sn solutions. Journal of Crystal Growth. 2012;354(1):147-51.
[15] Khare A, Himmetoglu B, Johnson M, Norris DJ, Cococcioni M, Aydil ES. Calculation of the lattice dynamics and Raman spectra of copper zinc tin chalcogenides and comparison to experiments. Journal of Applied Physics. 2012;111(8).
[16] Guc M, Levcenko S, Izquierdo-Roca V, Fontané X, Arushanov E, Pérez-Rodríguez A. Polarized Raman scattering analysis of Cu2ZnSnSe4 and Cu2ZnGeSe4 single crystals. Journal of Applied Physics. 2013;114(19).
[17] Shin B, Zhu Y, Bojarczuk NA, Jay Chey S, Guha S. Control of an interfacial MoSe2 layer in Cu2ZnSnSe4 thin film solar cells: 8.9% power conversion efficiency with a TiN diffusion barrier. Applied Physics Letters. 2012;101(5).
[18] Xue C, Papadimitriou D, Raptis YS, Richter W, Esser N, Siebentritt S, et al. Micro-Raman Study of Orientation Effects of CuxSe-Crystallites on Cu-rich CuGaSe2 Thin Films. Journal of Applied Physics. 2004;96(4):1963-6.
[19] Perna G, Lastella M, Ambrico M, Capozzi V. Temperature dependence of the optical properties of ZnSe films deposited on quartz substrate. Applied Physics A. 2006;83(1):127-30.
[20] Fernandes PA, Sousa MG, Salomé PMP, Leitão JP, da Cunha AF. Thermodynamic pathway for the formation of SnSe and SnSe2 polycrystalline thin films by selenization of metal precursors. CrystEngComm. 2013;15(47).

IMPEDANCE SPECTROSCOPY OF CDTE PV MODULES – COMPARATIVE STUDY

T. Finsterle[1], L. Cerna[1], P. Hrzina[1,2] and V. Benda[1]

[1]Department of Electrotechnology, Faculty of Electrical Engineering, Czech Technical University in Prague, Technicka 2, 166 27 Prague, Czech Republic.

[2] University Centre for Energy Efficient Buildings, Czech Technical University in Prague, Trinecka 1024, 273 43 Bustehrad, Czech Republic.

Phone: +420 725656969; E-mail: finsttom@fel.cvut.cz

ladislava.cerna@fel.cvut.cz, hrzinap@fel.cvut.cz, benda@fel.cvut.cz

ABSTRACT: One of interesting technology of photovoltaic modules which can be used within an industrial scale is CdTe. These modules are indeed problematic due to Cd content, but still represent interesting alternative to standard crystalline photovoltaic modules, mainly thanks to their significantly higher energy yield. Often discussed problems connected with thin-film modules are their stability, variability of production generations and also the degradation. One of possible method which can reveal early state of degradation connected with shunts is described in this article.
Keywords: Photovoltaics; CdTe; Impedance Spectroscopy; Degradation

1 INTRODUCTION

There exist many diagnostic methods which are commonly used for photovoltaic (PV) modules diagnostics. One of less common, but currently discussed in some papers (e.g. [1]), is Impedance Spectroscopy. Unlike the standard measurement methods, this method measures AC parameters of PV cell, module respectively, and thus provides additional informations about the internal structure properties. In the following paragraphs, utilizing of Impedance Spectroscopy at CdTe PV modules is described.

2 IMPEDANCE SPECTROSCOPY

The AC parameters can be measured using frequency domain technique – impedance spectroscopy. The Impedance Spectroscopy (IS) is characterized by the measurement and analysis of some or all impedance related functions of an electronic device. In IS, the complex impedance $Z(\omega) = R(\omega) + jX(\omega)$ of a device is measured directly within a large range of frequencies using LCR meter. A purely sinusoidal voltage with varying frequency is applied to the terminals of the device under test and the phase shift and amplitude of the voltage and current signals are measured. The ratio between the applied voltage and the resultant current is calculated and this gives the impedance $Z(\omega)$ of the device under test. The plotting of $R(\omega)$ and $X(\omega)$ on a complex plane, in function of the varying frequency, gives the impedance spectrum of the device. This graph is the special case of Nyquist diagram called Cole-Cole diagram (Fig. 1). Method, its possibilities and character is discussed in details e.g. in [2.

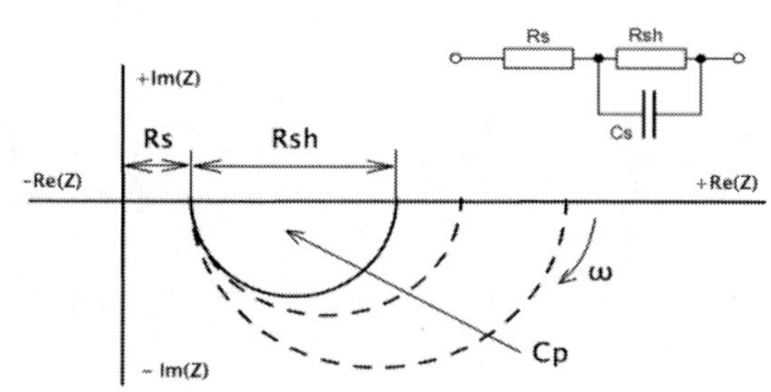

Fig. 1: Cole-Cole diagram of PV cell (module) [3]

3 MEASURED RESULTF OF CDTE PV MODULES

Previously published works proved that the defects and degradation which are not visible at the measured I-V curves have entirely different spectrum measured by IS [4]. This work was focused on crystalline Si PV modules, but there is no limitations regarding the technology, so this method was applied also at the series of CdTe modules.

Measured modules showed degradation from 25 % to almost 50 % at flash tests, but the differences measured by IS are much more detectable – Fig. 2.

The results can be divided into two groups regarding the measured shunt resistance RSH by flash test and by the shunt resistance measured by IS (the resistance estimation is apparent from Fig. 1). Shunt resistance is key parameter when evaluating degradation of thin-film PV modules, because it is strongly dependent on shunt creation within the module inner structure.

It is apparent, that the difference showed by both measurement methods between the modules is of one magnitude higher and thus better detectable in the case of IS method – Table I.

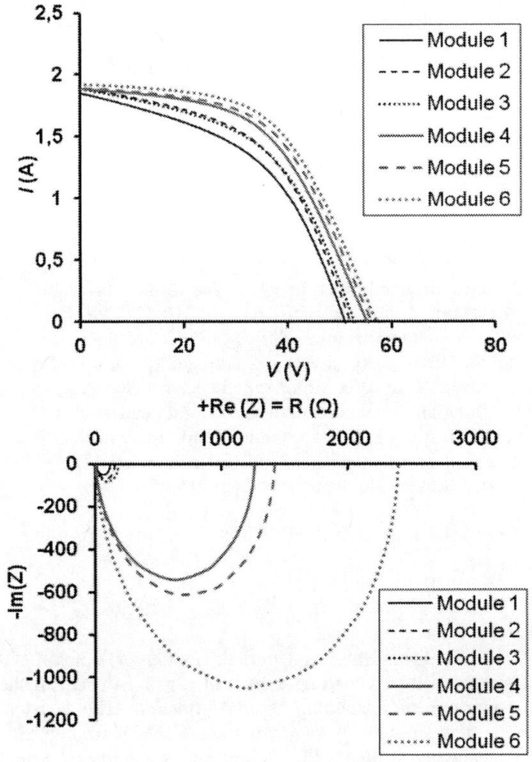

Fig. 2: Comparison of I-V curves and Cole-Cole diagram of measured CdTe PV modules

Table I: Comparison of values of RSH obtained by Flash Test method and IS

Module No.	R_{SH} – FT (Ω)	P_{MAX} – FT (W)	R_{SH} – IS (Ω)
6	365	59,443	2389
5	294	57,035	1262
4	283	56,234	1423
3	122	49,557	186
2	112	49,446	158
1	89	44,636	119

4 CONCLUSION

Although differences between MPP and shunt resistance measured by flash tests showed differences about tens of percents, difference between the modules resistance measured by IS is one order of magnitude higher. Impedance Spectroscopy then represent powerful tool for revealing the degradation of CdTe modules, because it is more sensitive on resistance changes than standard methods.

ACKNOWLEDGMENT

This work has been supported by project of University Centre for Energy Efficient Buildings, SGS SGS16/074/OHK3/1T/13, European Regional Development Fund-Project "Centre of Advanced Photovoltaics" (No.CZ.02.1.01/0.0/0.0/15_003/0000464) and Laboratory of Photovoltaic Systems Diagnostics.

5 REFERENCES

[1] SHAO, Shuyan, Mustapha ABDU-AGUYE, Li QIU, et al. Elimination of the light soaking effect and performance enhancement in perovskite solar cells using a fullerene derivative. In Energy & Environmental Science [online]. 2016, 9(7), 2444-2452 [cit. 2018-03-08]. DOI: 10.1039/C6EE01337F. ISSN 1754-5692. Available at: http://xlink.rsc.org/?DOI=C6EE01337F

[2] CHEVNIDHYA, D., K. KIRTIKARA a C. JIVACATE Dynamic Impedance Characterization of Solar Cells and PV Modules Based on Frequency and Time Domain Analyses. *Trends In Solar Energy Research*. Hough, T.P. New York: Nova Science Publishers, 2006, s. 21-45. ISBN 1-59454-866-8; LCCN: 2005034740.

[3] ČERNÁ, L., FINSTERLE, T., HRZINA, P. and V. BENDA. Note on Cole-Cole Diagrams of Photovoltaic Modules Evaluation. In *Proceedings of EUPVSEC'17*. September 2017.

[4] ČERNÁ, L., FINSTERLE, T., HRZINA, P. and V. BENDA. Impedance Spectroscopy and Its Possible Use for Defects Detection. In Proceedings of EUPVSEC'16. 2016

EFFECT OF TEMPERATURE ON CZTS BASED THIN FILM SOLAR CELL PERFORMANCE

Arbab Chowdhury, Md Mamunur Rahaman, Mohaimenul Islam and Md. Mosaddequr Rahman
Department of Electrical & Electronic Engineering, BRAC University
66-Mohakhali, Dhaka-1212, Bangladesh
*Corresponding author: Md. Mosaddequr Rahman (mosaddeq@bracu.ac.bd)

ABSTRACT: In recent days CZTS based thin film solar cells have attracted much attention due to their favorable electrical and optical properties, low cost and non-toxicity. A maximum laboratory efficiency of 10-12% has been reported for these types of cells. As the cells are subjected to much higher temperatures during outdoor operation, it is important to understand how CZTS cells will perform outdoor. This work investigates the effect of varying temperature on the different components of the dark current, as well as their effects on the cell performance. Among the different components of the dark current, trap-assisted tunneling recombination, thermionic emission and interface recombination are found to have significant contribution to the dark current while thermionic emission is found to be the dominating loss mechanism at high temperature. A calculation of cell efficiency at different temperatures show that the cell efficiency decreases from about 16.3% at 0°C to about only 7.3% at 75 °C.
Keywords: CZTS, Modeling, Thin Film Solar Cell

1 BACKGROUND

Copper zinc tin sulfide (CZTS) as an active solar cell material has drawn interest in the PV community as it has advantages over certain other thin film materials-namely it is made of non-toxic and inexpensive elements [1]. CZTS has a band gap of 1.4 to 1.6 eV, which is close to the ideal band gap for photovoltaic materials (around 1.5 eV), and a high absorption coefficient of 10^4 cm^{-1} [2]. However, the maximum solar conversion efficiency of CZTS is much lower, around 12%, than that of other materials such as CIGS which is around 20-22% [3, 4].

Ito and Nakazawa first reported the photovoltaic effect of CZTS material [5]. In 2010, Todorov et al. reported an efficiency of 9.6%, a world record at that time [6]. However, the best CZTS solar cell laboratory efficiency achieved at the time was 8.4% [7]. Courel Maykel et al. carried out a detailed theoretical analysis of different loss mechanisms and their contribution to the dark current density to understand what limits performance of the CZTS based solar cells [2]. As the solar cells experience high temperature during outdoor operation, it is also important to see how the cell temperature affects the different loss mechanisms and the overall cell performance. The main purpose of this work is to investigate the effect of cell temperature on the different dark current components that limits the performance of the CZTS based solar cells.

In this work, A comprehensive model of an n-CdS/p-CZTS junction solar cell has been developed first, assuming abrupt pn junction. Effect of doping densities, donor and acceptor, on the various components of the dark current and thereby on the performance of the solar cell is then investigated. The cell efficiency is estimated under varying donor and acceptor concentrations as well as the thicknesses of the CdS and CZTS layers in order to find the optimum parameter values. Using the optimum cell parameters, the effect of cell temperature on the different components of the dark current is then investigated.

The remainder of this paper is organized as follows: section 2 describes the model details, section 3 highlights the analysis procedure, section 4 shows the results and discussion part, finally section 5 provides the conclusion obtained from the totality of this work.

2 MODEL DETAILS

2.1 Solar Cell Model

The simplified single-diode model of a solar cell connected to a load, as shown in Fig. 1, is used for the purpose of examining the J-V relation. The constant current source J_{ph} represents the light-induced current generated in the cell, J_D represents the dark current through the cell, R_S is the series resistance that the current encounters when it flows through the cell, R_{sh} is the shunt resistance and is the result of impurities, defects and precipitates in the crystal structure, V is the voltage across load, and J is the current flowing through the load, given by,

$$J = J_D + J_{Rsh} - J_{ph} \tag{1}$$

$$= J_{D0}\left[exp\left(\frac{q(V - JR_S)}{2kT}\right) - 1\right] + \frac{V - JR_S}{R_{Sh}} - J_{PH}$$

Figure 2 shows the energy band diagram of the n-CdS/p-CZTS junction solar cell under illumination. Bandgap values for CdS ($E_{g\text{-}CdS}$) and CZTS ($E_{g\text{-}CZTS}$) are taken as 2.4 eV and 1.5 eV, respectively. The depletion region in the p-CZTS region is ahown to be larger than that in the n-CdS layer, as the later is much more heavily than the former. A heavily doped CdS layer increases the depletion width in the absorber side, thus allowing more electron-hole pairs to be generated in the depletion region which are then quickly separated by the depletion region built-in electric field, thus increasing the photocurrent and cell efficiency. ΔE_C and ΔE_V are the conduction and valence band discontinuities, respectively.

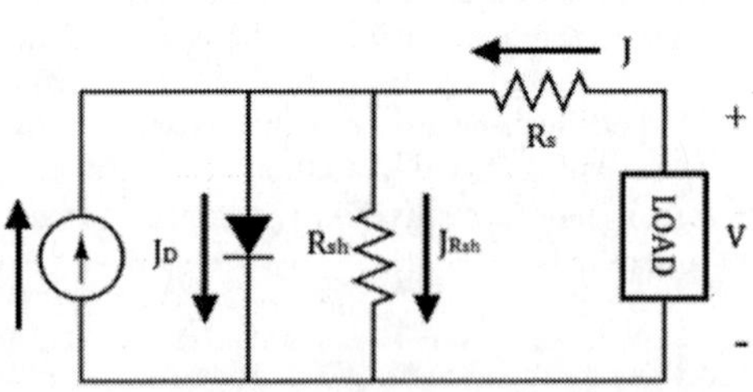

Figure 1: Simplified single-diode model of a solar cell.

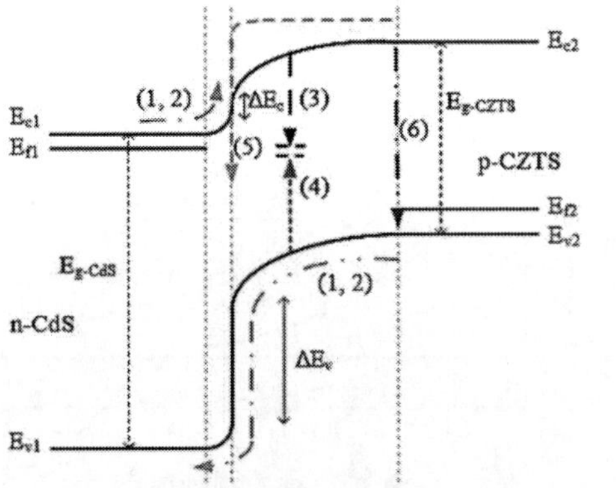

Figure 2: Energy band diagram n-CdS/p-CZTS junction solar cell under illumination showing different loss mechanisms: (1,2) Diffusion and Thermionic Emission, (3) Non-Radiative (4) trap assisted tunneling (5) interface and (6) Radiative losses.

2.2 Dark Current Density

The dark current is the current that flows through the system when it is not under illumination. Figure 1(b) shows the energy band diagram of the n-CdS/p-CZTS junction solar cell under illumination with different components of the dark current. Different components of the dark current shown are diffusion (J_{Diff}), thermionic emission (J_{TE}), non-radiative recombination (J_{NR}) trap-assisted tunneling recombination (J_{TATR}), CdS/CZTS interface recombination (J_{IR}) and radiative recombination (J_{RR}). In Eq. (1), J_{D0} is the summation of all these dark currents, given as,

$$J_{D0} = J_{Diff} + J_{TATR} + J_{IR} + J_{TE} + J_{RR} \quad (2)$$

Electron diffusion current in p-CZTS and hole diffusion current (n-CdS) can be written as [8]:

$$J_{diff\text{-}CZTS} = qD_n \frac{d\Delta n_p}{dx} = \frac{q*Dn*n_{po}}{Ln} \quad (3)$$

$$J_{diff\text{-}CdS} = -qD_p \frac{d\Delta p_n}{dx} = -\frac{q*Dp*p_{no}}{Lp} \quad (4)$$

Here D_n and D_p are the diffusion coefficients of electrons and holes, n_{p0} and p_{n0} are the minority carrier electron and hole concentrations, and L_n and L_p are the minority electron and hole diffusion lengths, respectively.

Trap-assisted tunneling is modeled as [2]:

$$J_{TATR} = q\left[\frac{x_n n_{i\text{-}CdS}(1+F_{CdS})}{\tau_n} + \frac{x_p n_{i\text{-}CZTS}(1+F_{CZTS})}{\tau_p}\right] \quad (5)$$

Here x_n and x_p are depletion widths of CdS and CZTS, $n_{i\text{-}CdS}$ and $n_{i\text{-}CZTS}$ are the intrinsic carrier concentration of CdS and CZTS, τ_n is minority carrier (hole) lifetime in n-CdS, τ_p is minority carrier (electron) lifetime in p-CZTS, F_{Cds} and F_{CZTS} model the effects of tunneling on CdS and CZTS, respectively.

Interface recombination is modeled by [2]:

$$J_{IR} = qSn_{i\text{-}CZTS}\left(1 + \gamma DOSe^{-(\Delta E_C+\Delta E_V)/2kT}\right) \quad (6)$$

Here γDOS is the density of states enhancement factor and S is interface recombination speed.

Total thermionic emission current due to electrons is given by [8]:

$$J_{TE} = A^*T^2 exp\left(-\frac{q\phi_B}{kT}\right) \quad (7)$$

Here Φ_B is barrier height and A* is effective Richardson constant. Radiative recombination losses at the depletion region can be modeled by [2]:

$$J_{RR} = q\left(x_n * B_n * n_{i\text{-}CdS}^2 + x_p * B_p * n_{i\text{-}CZTS}^2\right) \quad (8)$$

Here B_n and B_p are the radiative recombination coefficients for CdS and CZTS respectively.

2.3 Photocurrent Density

The complete equation for photocurrent density J_{PH} is derived assuming an abrupt p-n junction and a uniform carrier concentration. The total photocurrent at a given wavelength can be written as:

$$J_{PH}(\lambda) = J_p(\lambda) + J_n(\lambda) + J_{dr}(\lambda) \quad (9)$$

Here J_p is the photocurrent of n-side (CdS) and J_n is the photocurrent of p-side (CZTS) of the junction; J_{dr} is the depletion region photocurrent. Hole photocurrent density at the depletion edge is derived as:

$$J_p = -q.D_p\left(\frac{dp_n}{dx}\right)_{x_j} = \left[\frac{q\emptyset(1-R)\alpha_1 L_p}{(\alpha_1{}^2 L_p{}^2 - 1)}\right] * \left[\frac{\left(\frac{SpLp}{Dp}+\alpha_1 L_P\right) - \exp(-\alpha_1 x_j)\left(\left(\frac{SpLp}{Dp}\right)cosh\left(\frac{x_j}{Lp}\right) + sinh\left(\frac{x_j}{Lp}\right)\right)}{\left(\frac{SpLp}{Dp}\right)sinh\left(\frac{x_j}{Lp}\right) + cosh\left(\frac{x_j}{Lp}\right)} - \alpha_1 L_P \exp(-\alpha_1 x_j)\right] \quad (10)$$

The photocurrent due to electrons at the depletion edge is derived as:

$$J_n = q.D_n\left(\frac{dn_p}{dx}\right)_{x_j+wd_1+wd_2} = \frac{q\emptyset[1-\mathrm{R}]\alpha_2 Ln}{\alpha_2{}^2 Ln^2 - 1}\exp\left[-\alpha_1(x_j + wd_1)\right] * \exp(-\alpha_2 x_p)\left[\alpha_2 Ln - \frac{\left(\frac{S_n L_n}{Dn}\right)\left[\cosh\left(\frac{H'}{Ln}\right) - \exp(-\alpha_2 H')\right] + \sinh\left(\frac{H'}{Ln}\right) + \alpha_2 Ln\exp(-\alpha_2 H')}{\left(\frac{S_n L_n}{Dn}\right)\sinh\left(\frac{H'}{Ln}\right) + \cosh\left(\frac{H'}{Ln}\right)}\right] \quad (11)$$

Here $\alpha_1(\lambda)$ and $\alpha_2(\lambda)$ are the absorption coefficients of n-CdS and p-CZTS respectively, $\emptyset(\lambda)$ is the number of incident photons per unit area per unit time per unit bandwidth and $R(\lambda)$ is the percent of photons reflected from the surface.

Depletion region photocurrents J_{dr_1}, on the p-side, and J_{dr_2}, on the n-side of the junction, are calculated as:

$$J_{dr_1} = q\emptyset(1-R)\exp(-\alpha x)\Big|_{x_j+wd_1}^{x_j} = q\emptyset(1-R)\exp(-\alpha_1 x_j)[1 - \exp(-\alpha_1 wd_1)] \quad (12)$$

$$J_{dr_2} = q\emptyset(1-R)\exp(-\alpha x)\Big|_{x_j+wd_1+wd_2}^{x_j+wd_1} = q\emptyset(1-R)\ \exp\ [-\alpha_1(x_j + wd_1)][1 - \exp(-\alpha_2 wd_2)] \quad (13)$$

3 ANALYSIS

First, different components of the dark current density and the photocurrent current density are calculated for varying carrier concentration using the cell parameters given in Table I. Further, following additional parameter values are used: interface recombination speed, $S=10^4$ cm/s, series resistance, $R_S=5.8\ \Omega cm^2$, and shunt resistance $R_{Sh}=2.2*10^4\ \Omega cm^2$. An illumination level of 1 sun under AM1.5 condition has been assumed for the photocurrent calculation. The dark and the photocurrent current densities obtained is used in Eq. (1) to obtain the J-V characteristic of the cell for varying doping concentrations. Thus, the donor (N_D) and acceptor (N_A) concentrations are optimized for maximum cell efficiency. The thicknesses of CdS and CZTS films are then optimized to maximize the cell efficiency. The optimized parameters are as follows: $N_D=10^{18}$ cm^{-3}, $N_A=5*10^{14}$ cm^{-3}, CdS thickness=50 nm and CZTS thickness=5 μm. A thinner n-CdS window layer allows more photons to penetrate to the p-CZTS absorber layer, reduces the recombination loss at the surface and contributes to enhanced photocurrent.

Table I: n-CdS and p-CZTS parameters

Parameter	Description	Value
	n-CdS	
τ_n	Minority hole lifetime	$9.1*10^{-9}$s
μ_n	Mobility of hole	50 cm^{2}V^{-1}s^{-1}
ε_n	Dielectric permittivity	$7.969*10^{-13}$ Fcm^{-1}
Eg_{CdS}	Energy band gap	2.4 eV
$m^*_{e(CdS)}$	Electron effective mass	$2.2775*10^{-31}$kg
$m^*_{h\ (CdS)}$	Hole effective mass	$6.377*10^{-31}$kg
N_{c-CdS}	Effective density of states of conduction band	$1.80*10^{19}$cm^{-3}
N_{V-CdS}	Effective density of states of Valence Band	$2.40*10^{18}$cm^{-3}
χ_{CdS}	Electron affinity	4.4 eV
σ	Capture cross section of holes	10^{-13} cm^2
v_{th}	Hole thermal velocity	10^7cms^{-1}
	p-CZTS	
τ_p	Minority electron lifetime	$1.7*10^{-9}$s
μ_p	Mobility of electron	620 cm^{2}V^{-1}s^{-1}
ε_p	Dielectric permittivity	$8.854*10^{-13}$ Fcm^{-1}
Eg_{CZTS}	Energy band gap	1.5 eV
$m^*_{e(CZTS)}$	Electron effective mass	$1.6398*10^{-31}$kg
$m^*_{h\ (CZTS)}$	Hole effective mass	$6.4681*10^{-31}$kg
N_{c-CZTS}	Effective density of states of conduction band	$2.39*10^{18}$cm^{-3}
N_{V-CdS}	Effective density of states of Valence Band	$1.87*10^{19}$cm^{-3}
χ_{CZTS}	Electron affinity	4.1 eV

4 RESULTS AND DISCUSSION

Figure 3 shows the variation in the different components of the dark current with the acceptor doping density calculated for a fixed donor doping density $N_D=10^{18}$ cm^{-3}. From the figure, it can be seen that trap-assisted tunneling recombination is the most significant component of the dark current followed by thermionic emission and interface recombination while radiative recombination and diffusion currents are several orders of magnitude lower, and therefore will have negligible effect on the overall cell performance.

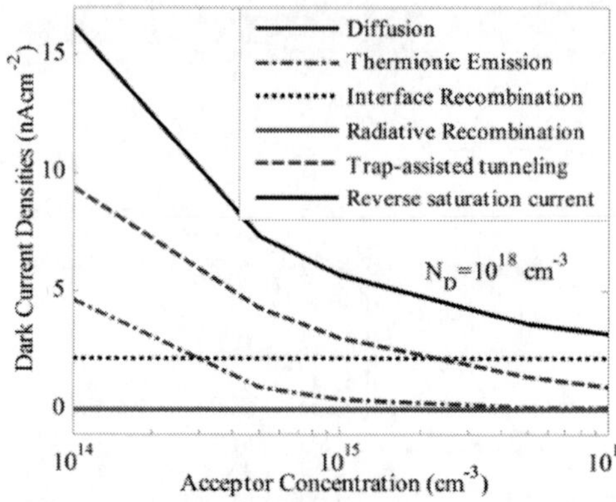

Figure 3: Plots of different dark current components with respect to the acceptor doping density (N_A) of the CZTS film for a fixed donordensity $N_D=10^{18}$ cm^{-3}.

Figure 4 shows the variation in the three most significant components of the dark current along with the total reverse saturation current density with temperature calculated using the optimized values of doping densities. From Fig. 3, it can be observed that the different dark current components and thereby, the total reverse saturation current density, increase almost exponentially with the increase in the cell temperature. At lower temperatures, trap assisted tunneling recombination is the largest contributor to the reverse saturation current density followed by CdS/CZTS interface recombination. However, thermionic emission current density increases at a much greater rate with temperature and becomes the dominating loss mechanism at temperatures above 40°C. Rapid increase in dark current with temperature considerably restricts the efficiency at higher temperature.

The effect of the different components of the reverse saturation current on the current density-voltage

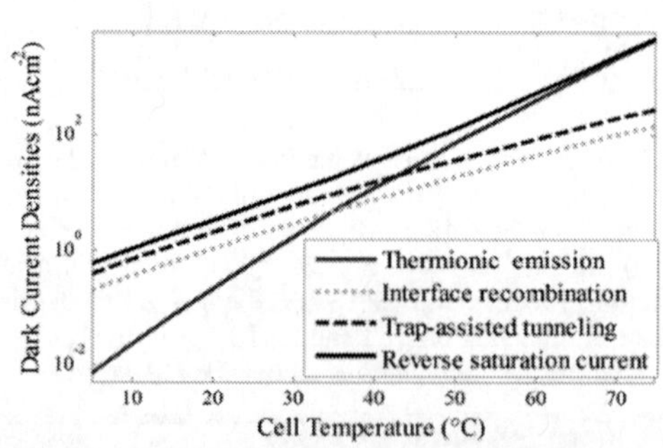

Figure 4: Plots of different dark current components along with the total reverse saturation current density with respect to the cell temperature.

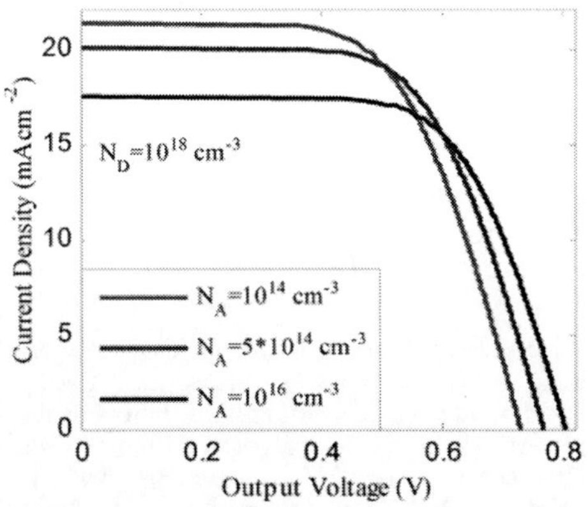

Figure 5: Plots of output current density vs. output voltage with J_{D0} consisting of part of different loss mechanisms.

characteristics of the solar cell is shown in Fig. 4. As the radiative recombination and diffusion currents are much lower compared to the others, they have negligible effect on the current density-voltage graph, and the J-V plots with and without them completely overlap. The J-V plot that includes the trap assisted tunneling and thermionic recombination slightly deviates from that which also includes interface recombination along with these two, indicating only marginal contribution of the interface loss mechanism. The contribution of thermionic emission loss on the J-V plots is more pronounced than that of interface loss mechanism.

From efficiency calculations, it has been observed that ignoring radiative recombination and diffusion losses does not make any noticeable difference in cell efficiency. Thus, the reverse saturation current density can be effectively expressed as the summation of three loss mechanisms; trap assisted tunneling, interface recombination and thermionic emission:

$$J_{D0} = J_{TATR} + J_{IR} + J_{TE} \quad (14)$$

Efficiency of the n-CdS/p-CZTS has been calculated for different temperatures using the optimized cell parameters: $N_D=10^{18}$ cm^{-3}, $N_A=5*10^{14}$ cm^{-3}, CdS thickness=50 nm and CZTS thickness=5 μm for an illumination level of 1 sun under AM1.5 condition. Figure 6 shows how efficiency of the cell decrease with increasing temperatures. Over the 5°C to 75°C range, efficiency decreases from 16.3% to 7.3%, decreasing on average at a rate of 0.13%/°C. Fill factor and the open circuit voltage of the cell are also observed to fall with temperature over the same temperature range. As the operating temperature of a cell is much higher than its surroundings during outdoor operation, this will severely limit the performance of the cell at higher temperatures.

5 CONCLUSIONS

In this work, effect of the different components of the dark current on the n-CdS/p-CZTS cell efficiency at different temperatures has been investigated. The different dark current components considered are the diffusion, trap-assisted tunneling recombination, thermionic emission, radiative recombination and CdS/CZTS interface recombination. A comprehensive model for the n-CdS/p-CZTS junction solar cell has been

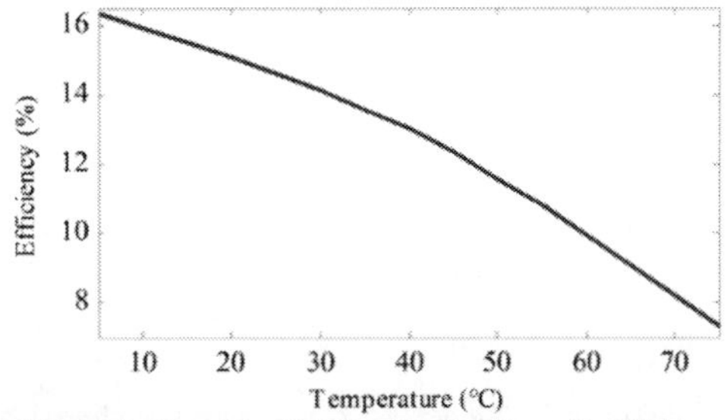

Figure 6: Plot of n-CdS/p-CZTS solar cell efficiency with respect to temperature under 1 sun AM1.5 illumination.

developed first assuming an abrupt pn junction. Effect of doping densities on the various components of the dark current and thereby on the cell performance is then investigated. Using the optimum doping densities, the effect of temperature on the different components of the dark current is then studied.

This study shows that while at low temperatures, thermionic emission has the least contribution, at high temperatures, above 40 °C, it becomes the most dominating loss mechanism, thus severely limiting the efficiency of the cell at high temperatures. A calculation of cell efficiency at different temperatures show that the cell efficiency decreases from about 16.3% at 0°C to about only 7.3% at 75 °C. Since the surface temperature of solar cell during outdoor operation is much higher than that of its surrounding environment, this will severely limit the cell performance in real life operation.

REFERENCES

[1] Suryawanshi, Mahesh & Agawane, Ganesh & Bhosale, S.M. & Shin, Seung Wook & Patil, P & Kim, Jeong-Hoon & Moholkar, Annasaheb, CZTS based thin film solar cells: A status review, Materials Technology,Vol.28 (2013) 98.

[2] Courel Maykel, J.A. Andrade-Arvizu, and O. Vigil-Galán, Loss mechanisms influence on Cu_2ZnSnS_4/CdS based thin film solar cell performance, Solid-State Electronics, Vol.111 (2015) 243.

[3] Flammersberger, Hendrik, Experimental study of Cu_2ZnSnS_4 thin films for solar cells, Solid State Electronics, Vol.2 (2010) 117.

[4] Morales-Acevedo, A. A simple model of graded band-gap solar cells,Energy Procedia, vol-2(1) (2010) 169.

[5] Ito and T. Nakazawa, Electrical and optical properties of stannite-type quaternary semiconductor thin films, Japanese Journal of Applied Physics, Vol. 27(11) (1988) 2094.

[6] K. Todorov, K. B. Reuter, and D. B. Mitzi, High-efficiency solar cell with earth-abundant liquid-processed absorber, Advanced Materials, Vol. 22(20) (2010) E156.

[7] B. Shin, O. Gunawan, Y. Zhu, N. A. Bojarczuk, S. J. Chey, and S. Guha, Prog. Photovoltaics Res. Appl. Vol.21 (2013) 72.

[8] Sze, S.M, & Ng, Kwok. k., Physics of Semiconductor Devices. John Wiley & Sons (2006).

Fabrication of p-type Na doped SrCuSeF and n-type ITO bilayer ohmic tunnel junction and its application to the back contact of CdS/CdTe solar cells

T. Wada[1], K. Miki[1], D. Tamai[1], K. Takaya, Y. Shiina[2], S. Okamoto[2], and T. Okamoto[2]
[1] Department of Materials Chemistry, Ryukoku University, Otsu 520-2194, Japan
[2] National Institute of Technology, Kisarazu College, Kisarazu, Chiba 292-0041, Japan

ABSTRACT: For applications to polycrystalline thin-film tandem-type solar cells, we studied p-type conductive Na doped SrCuSeF and n-type conductive In_2O_3:Sn (ITO) bilayer films. The Na doped SrCuSeF layers were prepared by pulsed laser deposition (PLD), and the ITO layers were prepared by RF sputtering. The bilayer films showed ohmic current-voltage characteristics. A tunnel junction between these two layers was successfully fabricated, becuse p-type Na doped SrCuSeF and n-type ITO layers had sufficiently high carrier concentrations. The Na doped SrCuSeF/ITO bilayer films were applied to the back electrodes of the CdS/CdTe solar cells. A CdTe solar cell with a back electrode of a 10 nm thick $(Sr_{0.9}Na_{0.1})$CuSeF layer and a 200 nm thick ITO layer showed a high conversion efficiency of 14.7% (V_{OC} = 806 mV, J_{SC} = 27.5 mA/cm^2, and FF = 0.661). The efficiency was higher than the efficiency of CdTe solar cell with a SrCuSeF single-layer (864 nm) back contact, 11.6%, and that with a SrCuSeF (34nm)/ITO (200nm) bilayer back contact, 14.3%.
Keywords: p-type, TCO Transparent Conducting Oxide, CdTe, Back Contact, Tandem,

1 INTRODUCTION

Several research groups have fabricated polycrystalline thin-film tandem solar cells. The top cells were fabricated with n-type transparent conductive oxide (TCO) back contacts [1-3]. However, the n-type TCOs were not suitable for the back contact of the polycrystalline thin-film solar cells because their absorber layers, such as CdTe and Cu(In,Ga)Se_2 showed p-type electrical conductivity. Therefore, p-type TCOs have been developed. Some p-type TCOs have been reported. However, the conductivities of the reported p-type TCOs were much lower than those of the n-type TCOs [4]. Tate's group have reported that BaCuSeF and related compounds are promising candidates for p-type TCO [5-7].

We prepared polycrystalline BaCuSeF films with a transmittance exceeding 50% in the visible light region and a p-type conductivity of about 19.4 S/cm by pulsed laser deposition (PLD) [8]. We applied the BaCuSeF film as the back contact of CdS/CdTe solar cells. In the early stage, the efficiency of the CdS/CdTe solar cells was as low as 2.82% [9]. Then, a higher conversion efficiency of 9.91% was obtained by the etching of CdTe surface in a bromine-bromate solution [10]. Recently, we have studied the related compound such as SrCuSeF and obtained a higher conversion efficiency of 11.6% for a CdTe solar cell with a SrCuSeF back contact [11]. The efficiency is comparable to that of a CdS/CdTe solar cell fabricated using a conventional carbon electrode (12.0%) [12]. However, the conductivities of BaCuSeF (1.1 S/cm) [9] and SrCuSeF (0.11 S/cm) [11] films were much lower than that of n-type TCOs such as ITO films (1.3×10^4 S/cm). Therefore, we studied a bilayer film of transparent p-type conductive SrCuSeF and transparent n-type conductive In_2O_3:Sn. We obtained a higher conversion efficiency of 14.3% for the CdTe solar cell with a back contact of the SrCuSeF and ITO bilayer [13]. We also obtained high efficiency of 13.9 % for the CdTe solar cell with BaCuSF/ITO bilayer back contact [14].

In this study, we studied the Na doping in SrCuSeF film in order to increase p-type conductivity of SrCuSeF film and applied Na doped SrCuSeF/ITO bilayer to the back contact of CdTe solar cells. The CdTe solar cell with a back electrode of $(Sr_{0.9}Na_{0.1})$CuSeF layer (10 nm) and ITO (200 nm) bilayer showed a high conversion efficiency of 14.7% (Voc = 806 mV, Jsc = 27.5 mA/cm^2, and FF = 0.661).

2 EXPERIMENTAL PROCEDURE

2.1 Preparation and characterization of Na doped SrCuSeF and ITO films

$(Sr_{1-x}Na_x)$CuSeF ceramic targets with x=0.00, 0.02, 0.05 and 0.10 were fabricated by a solid-state reaction method. The mixed CuSe, SrF_2, SrSe, and NaF powders were heated in N_2 gas atmosphere at 800°C for 10 h. The obtained $(Sr_{1-x}Na_x)$CuSeF powders were pressed into a disk and sintered in N_2 gas atmosphere at 850°C for 10 h. The obtained ceramic target had a relative density of 83-93%.

We fabricated $(Sr_{1-x}Na_x)$CuSeF films by PLD. A KrF excimer laser with a 248 nm wavelength was focused on the $(Sr_{1-x}Na_x)$CuSeF ceramic target. The fabrication conditions were as follows: a laser repetition rate of 10 Hz, a distance of 90 mm between the target and the substrate, an energy density of 1.5 J/cm^2, an Ar gas pressure of 1.3×10^{-1} Pa, and a substrate temperature (Ts) of 200°C, which was the best substrate temperature for CdTe solar cells with BaCuSeF back contact [10].

We fabricated ITO films by RF sputtering at room temperature. A commercially available ITO (SnO_2 10 wt%) ceramic target was used. The thickness of the ITO layers was fixed at about 200 nm.

The crystallographic structures were examined by X-ray diffraction (XRD) analysis. Optical properties were characterized by ultraviolet–visible–near infrared (UV–vis–NIR) spectroscopy. Electric properties were evaluated by the van der Pauw method. Electrical conductivity and the Hall effect were measured using an Ecopia HMS-3000 Hall measurement system.

2.2 Fabrication and characterization of CdTe solar cells with $(Sr_{0.9}Na_{0.1})$CuSeF/ITO bilayer back contact

The CdTe solar cells have a structure of glass substrate/ITO(front electrode)/n-CdS layer/p-CdTe absorber layer/back electrode/Pt electrode. The CdTe layer was deposited by close-spaced sublimation (CSS). Before the deposition of the back contact, we performed $CdCl_2$ treatment and Cu diffusion. After the $CdCl_2$ treatment, the CdTe surface was etched in a bromide-bromate solution at room temperature [10]. Then, Cu diffusion was performed. Finally, we fabricated Na doped SrCuSeF/ITO bilayer back contact. The Na doped SrCuSeF layer was fabricated by PLD on the CdTe surface of CdTe solar cells. To characterize the photovoltaic performance of the CdTe solar cells, the ITO surface was coated with Pt paste and dried at 110°C for 1 h in air. The photovoltaic performance of the CdTe solar cells was characterized by photocurrent density-voltage (J-V) and external quantum efficiency (EQE) measurements.

3 RESULTS

3.1 $(Sr_{1-x}Na_x)$SrCuSeF single-layer, films

Figure 1 shows the transmittance spectra of $(Sr_{1-x}Na_x)$SrCuSeF single-layer films with x=0.00, 0.02, 0.05 and 0.10. Thickness of these films are 570, 560, 670, and 720 nm, respectively. These films do not show a high transmittance in the visible light region ($400 \leq \lambda \leq 800$nm) because these films were deposited at low substrate temperature of 200°C. The substrate temperature of 200°C is too low to obtain transparent (Sr,Na)SrCuSeF films. In order to increase transmittance of (Sr,Na)SrCuSeF films, we fabricated thinner films.

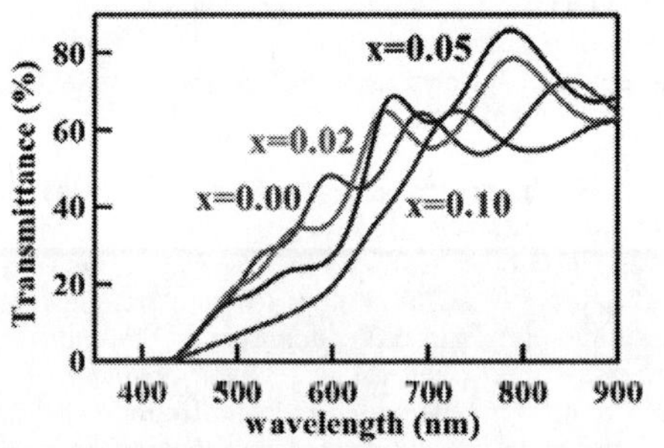

Figure 1: Transmittance spectra of $(Sr_{1-x}Na_x)$SrCuSeF films with x=0.00, 0.02, 0.05 and 0.10.

Table I: Electrical properties of $(Sr_{1-x}Na_x)$SrCuSeF single-layer films with x=0.00, 0.02, 0.05 and 0.10.

x	type	σ (S/cm)	n (/cm³)	μ (cm²/Vs)
0.10	p	33	3.9×10^{21}	5.2×10^{-2}
0.05	p	1.8	3.2×10^{20}	3.5×10^{-2}
0.02	p	0.36	3.1×10^{18}	7.1×10^{-1}
0.00	p	0.34	1.6×10^{19}	1.3×10^{-1}

Table I shows electrical properties of $(Sr_{1-x}Na_x)$SrCuSeF films with x=0.00, 0.02, 0.05 and 0.10. All the films show p-type conductivity. The conductivities, σ, of all the films are more than 0.30 S/cm and increase with increasing Na concentration, x. The carrier concentrations, n, of all the films are more than $1\times10^{18}/cm^3$ and almost increase with increasing Na concentration, x. The $(Sr_{0.9}Na_{0.1})$SrCuSeF (720 nm) shows p-type conductivity of 33 S/cm with carrier concentration of $3.9\times10^{21}/cm^3$ and hole mobility of 5.2×10^{-2} cm²/Vs. Therefore, we deposited bilayer film of thin $(Sr_{0.9}Na_{0.1})$SrCuSeF (10 nm) and ITO (200 nm) bi-layer film.

3.2 $(Sr_{0.9}Na_{0.1})$SrCuSeF/ITO bi-layer films

The $(Sr_{0.9}Na_{0.1})$CuSeF single-layer film (10 nm) showed a high transmittance of more than 85% in the visible light region ($400 \leq \lambda \leq 800$nm). The ITO single-layer film (200 nm) showed a higher transmittance of more than 90% in the visible light region. **Figure 2** shows transmittance spectrum of $(Sr_{0.9}Na_{0.1})$CuSeF (10 nm)/ITO (200 nm) bilayer film. The transmittance corresponds to the superimposed transmittance spectra of the $(Sr_{0.9}Na_{0.1})$CuSeF single-layer (10 nm) and ITO single-layer (200 nm) films. The $(Sr_{0.9}Na_{0.1})$CuSeF (10 nm)/ITO (200 nm) bilayer film shows a high transmittance of 87% in the visible light region.

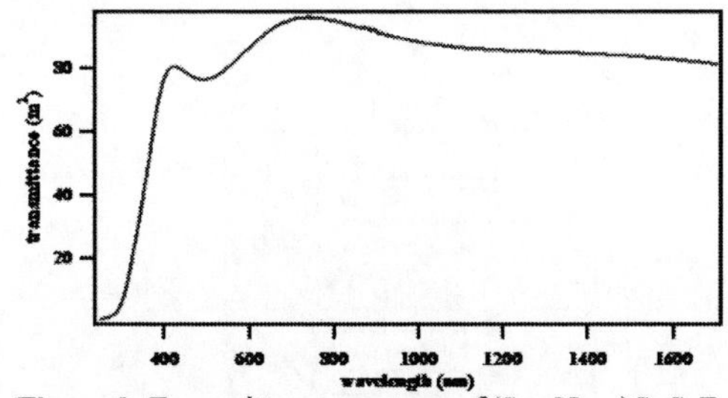

Figure 2: Transmittance spectrum of $(Sr_{0.9}Na_{0.1})$CuSeF (10 nm)/ITO (200 nm) bilayer film.

The conductivity of the $(Sr_{0.9}Na_{0.1})$CuSeF single-layer film (10 nm) is 1.8 S/cm with a carrier concentration of 3.2×10^{20} /cm^{-3} and a hole mobility of 3.5×10^{-2} cm^{2}V^{-1}s^{-1}. The conductivity of the single-layer ITO film (200 nm) is 8×10^2 S/cm with a carrier concentration of 4.1×10^{20} cm^{-3} and a electron mobility of 13 cm^{2}V^{-1}s^{-1}.

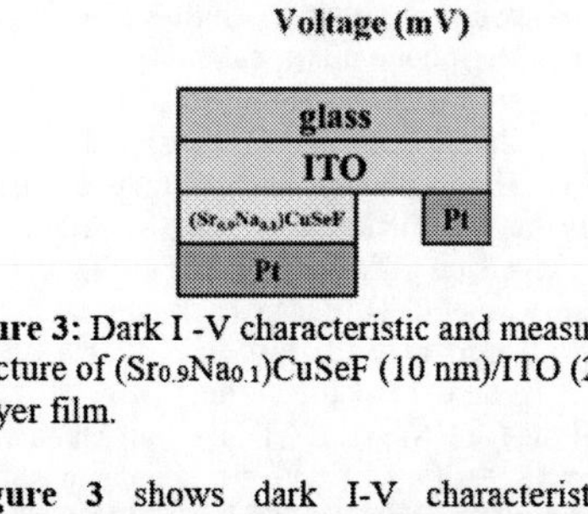

Figure 3: Dark I -V characteristic and measured structure of $(Sr_{0.9}Na_{0.1})$CuSeF (10 nm)/ITO (200 nm) bilayer film.

Figure 3 shows dark I-V characteristic of the $(Sr_{0.9}Na_{0.1})$CuSeF(10 nm)/ITO (200nm) bilayer film. The measured structure is also shown in **Fig. 3**. The $(Sr_{0.9}Na_{0.1})$CuSeF and ITO bilayer film shows an ohmic I-V characteristic. A tunnel junction between p-type conductive $(Sr_{0.9}Na_{0.1})$CuSeF layer and n-type conductive ITO layer was successfully formed because the p-type

$(Sr_{0.9}Na_{0.1})CuSeF$ and n-type ITO layers had sufficiently high carrier concentrations. A tunnel p-n junction can form when p- and n-type semiconductors have carrier concentrations higher than 1×10^{18} cm^{-3} [15].

3.3 CdTe solar cells with Na doped SrCuSeF/ITO bilayer back contact

Figure 4 shows a schematic device structure of CdTe solar cells with $(Sr_{0.9}Na_{0.1})CuSeF$/ITO bilayer back contact. **Table II** shows the photovoltaic performances of various CdTe solar cells with various $(Sr_{0.9}Na_{0.1})CuSeF$/ITO bilayer back contacts. The thicknesses of the $(Sr_{0.9}Na_{0.1})CuSeF$ layers are 40, 30, 20, 15, 10, and 5 nm. The thickness of the ITO layer is fixed value of 200 nm. The device area was in the range of 0.14-0.27 cm^2.

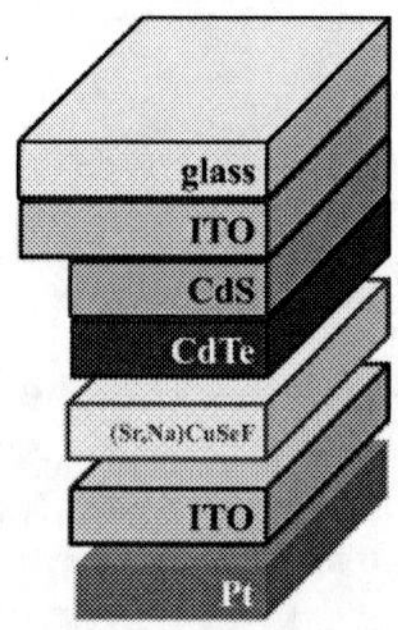

Figure 4: Schematic device structure of $(Sr_{0.9}Na_{0.1})CuSeF$/ITO bilayer film. The thickness of the $(Sr_{0.9}Na_{0.1})CuSeF$ layer was 40, 30, 20, 15, 10, and 5 nm. The thickness of the ITO layer was 200 nm.

Table II: Photovoltaic performance of various CdTe solar cells with $(Sr_{0.9}Na_{0.1})CuSeF$/ITO bi-layer back contact.

$(Sr_{0.9}Na_{0.1})CuSeF$ Thickness (nm)	Voc (mV)	Jsc (mA/cm^2)	FF	η (%)
40	762	25.1	0.394	7.53
30	764	27.6	0.447	9.43
20	785	26.8	0.611	12.9
15	812	28.1	0.602	13.7
10	806	27.5	0.661	14.7
5	813	26.1	0.655	13.9

Table II shows the $(Sr_{0.9}Na_{0.1})CuSeF$ layer thickness dependence of the photovoltaic parameters of the CdTe solar cells, namely, conversion efficiency (η), open-circuit voltage (Voc), short-circuit current density (Jsc), and fill factor (FF). Their photovoltaic parameters depend considerably on the thickness of the $(Sr_{0.9}Na_{0.1})CuSeF$ layer. The conversion efficiency of CdTe solar cells with the $(Sr_{0.9}Na_{0.1})CuSeF$/ITO bilayer back contact increases from 7.5% (40 nm) to 9.4% (30 nm), 12.9% (20 nm), 13.7%(15 nm) and 14.7% (10 nm) with decreasing thickness of the $(Sr_{0.9}Na_{0.1})CuSeF$ layer, and then slightly decreases to 13.9% (5 nm) with further decrease in the thickness of the $(Sr_{0.9}Na_{0.1})CuSeF$ layer. The CdTe solar cell with the $(Sr_{0.9}Na_{0.1})CuSeF$ (10 nm) and ITO (200 nm) bilayer back contact showed the highest conversion efficiency of 14.7%. The FF changes from 0.394 (40 nm) to 0.447 (30 nm), 0.611(20 nm), 0.602(15 nm), and 0.661 (10 nm) with decreasing thickness of the $(Sr_{0.9}Na_{0.1})CuSeF$ layer, and then slightly decreases to 0.655 (5 nm) with further decrease in the thickness of the SrCuSeF layer. The main factor for the increase in the conversion efficiency with a thinner $(Sr_{0.9}Na_{0.1})CuSeF$ layer is increasing FF. We consider that the increase in the FF of the CdTe solar cells with a thinner $(Sr_{0.9}Na_{0.1})CuSeF$ layer is due to the decrease in the series resistance of the CdTe solar cells, because the conductivity of the $(Sr_{0.9}Na_{0.1})CuSeF$ layer is considerably lower than that of the ITO layer.

4 SUMMARY

The photovoltaic performances of some CdS/CdTe solar cells are summarized in **Table III**. The characteristics of the CdS/CdTe solar cells with a carbon electrode and a Cu-doped carbon electrode are shown for reference [12, 16]. The CdTe solar cell with the SrCuSeF single-layer back contact showed an efficiency of 11.6% (Voc = 771 mV, Jsc = 26.4 mA/cm2, and FF=0.57) [11] and that with the ITO single-layer back contact showed an efficiency of 2.75% (Voc = 588 mV, Jsc = 20.4 mA/cm^2, and FF = 0.23) [13]. The efficiency was improved by applying the SrCuSeF/ITO bilayer back contact. The CdTe solar cell with the SrCuSeF (34 nm)/ITO (200 nm) bilayer back contact showed an efficiency of 14.3% (Voc = 804 mV, Jsc = 27.5 mA/cm^2, and FF = 0.65) [14]. The CdTe solar cell with a bilayer back contact of $(Sr_{0.9}Na_{0.1})CuSeF$ and ITO bilayers showed higher efficiency of 14.7% (Voc=806 mV, Jsc=27.5 mA/cm^2, and FF=0.661). We considered that the parameters of photovoltaic performance such as FF were greatly increased by the increase in the electrical conductivity of the back contact fabricated using the $(Sr_{0.9}Na_{0.1})CuSeF$(10nm) /ITO (200 nm) bilayer.

Table III: Photovoltaic characteristics of CdS/CdTe solar cells with $(Sr_{0.9}Na_{0.1})CuSeF$(10 nm)/ITO (200 nm) [SNCSF/ITO bilayer(a)] and SrCuSeF (34 nm)/ITO (200 nm) bilayers [SCSF/ITO bilayer (b)] and SrCuSeF single-layer (860 nm) and ITO single-layer (200 nm) back contacts. The photovoltaic characteristics of the CdS/CdTe solar cells with a carbon electrode [12] and a Cu-doped carbon electrode [16] are shown for reference.

Back contact	Voc (mV)	Jsc (mA/cm^2)	FF	η (%)	Ref.
SNCSF/ITO bilayer (a)	806	27.5	0.661	14.7	This study
SCSF/ITO bilayer (b)	803	27.5	0.647	14.3	13
SrCuSeF (860 nm)	771	26,4	0.572	11.6	11
ITO (200 nm)	588	20.4	0.23	2.75	13
Carbon	821	26.0	0.560	12.0	12
Cu-doped carbon	829	26.5	0.692	15.2	16

SNCSF/ITO bilayer (a): $(Sr_{0.9}Na_{0.1})CuSeF$(10 nm)/ITO (200 nm)
SCSF/ITO bilayer (b): SrCuSeF (34 nm)/ITO (200 nm)

5 References

[1] S. Nishiwaki, S. Siebentritt, P. Walk, and M. C. Lux-Steiner, Prog. Photovoltaics **11**, 243 (2003).

[2] R. Kaigawa, K. Funahashi, R. Fujie, T. Wada, S. Merdes, R. Caballero, and R. Klenk, Sol. Energy Mater. Sol. Cells **94**, 1880 (2010).

[3] A. N. Tiwari, G. Khrypunov, F. Kurdzesau, D. L. Batzner, A. Romeo, and H. Zogg, Prog. Photovoltaics **12**,

33 (2004).
[4] H. Kawazoe, M. Yasukawa, H. Hyodo, M. Kurita, H. Yanagi, and H. Hosono, Nature **389**, 939 (1997).
[5] R. Nagarajan, N. Duan, M. K. Jayaraj, J. Li, K. A. Vanaja, J. Tate, and A. W. Sleight, Int. J. Inorg. Mater. **3**, 265 (2001).
[6] H. Yanagi, J. Tate, S. Park, C.-H. Park, D. Keszler, Appl. Phys. Lett. **82**, 2814 (2003).
[7] J. Tate, P. F. Newhouse, R. Kykyneshia, P. A. Hersh, J. Kinney, D. H. McIntyre, and D. A. Keszler, Thin Solid Films **516**, 5795 (2008).
[8] S. Yamazoe, M. Yoshikawa, and T. Wada, Jpn. J. Appl. Phys. **51**, 10NC40 (2012).
[9] K. Yamamoto, H. Sakakima, Y. Ogawa, A. Hosono, T. Okamoto, and T. Wada, Jpn. J. Appl. Phys. **53**, 05FX02 (2014).
[10] K. Yamamoto, H. Sakakima, Y. Ogawa, A. Hosono, T. Okamoto, and T. Wada, Jpn. J. Appl. Phys. **54**, 08KC01 (2015).
[11] T. Wada, S. Kitabayashi, K. Yamamoto, H. Sakakima, Y. Ogawa, A. Hosono, and T. Okamoto, Proc. **42th** IEEE Photovoltaic Specialist Conf. [DOI: 10.1109/PVSC.2015.7355888].
[12] T. Okamoto, R. Hayashi, S. Hara, and Y. Ogawa, Jpn. J. Appl. Phys. **52**, 102301 (2013).
[13] S. Kitabayashi, Y. Shiina, A. Murata,T. Okamoto and T. Wada, Jpn. J. Appl. Phys. **56**, 08MC18 (2017).
[14] K. Miki, T. Kawabe, Y. Shiina, S. Okamoto, T. Okamoto, and T. Wada, Jpn. J. Appl. Phys. **57**, 08RC19 (2018).
[15] T. Takamoto, M. Yumaguchi, E. Ikeda, T. Agui, H. Kurita, and M. Al-Jassim, J. Appy. Phys. **85**, 1481 (1999).
[16] T. Okamoto,Y. Harada, A. Yamada, and M. Konagai, Sol. Energy Mater. Sol. Cells **67**, 187 (2001).

EVALUATION EMERGING PV PERFORMANCE UNDER ENERGY HARVESTING FOR IOT (INTERNET OF THINGS) APPLICATIONS

Yean-San Long[1*], En-Yun Wang[1], Cho-Fan Hsieh[1], Teng-Chun Wu[1]
[1]Center for Measurement Standards, Industrial Technology Research Institute, Taiwan
*Corresponding author (mickeylong88@itri.org.tw, Phone: +(886) 3 5911026, Fax: +(886) 3 5725540)

ABSTRACT: The measured I-V hysteresis is complicated by the vast array of different materials, lighting sources, and device architectures. In our study, we used RTOS method for I-V then the results showed better accuracy by eliminating in real time the acceptance effect. We also used this method to compare emerging PV cells with the latter cells exhibited promising power conversion efficiency and no hysteresis behavior under indoor lighting simulator. In this study, we defined a performance ratio to compare between indoor lighting and solar and the results will be show different performance trend for the different type PV in indoor/outdoor, which depends on device. That's will be shown what's type PV have the application of energy harvesting for indoor.

Keywords: Organic Photovoltaic, Real-Time Removing Capacitance Effect, One-Sweep Method.

1 INTRODUCTION

Energy harvesting for IoT (Internet of things) application presents a straightforward solution for easily powering those remote devices using clean energy. The power-generating element must be selected after considering the type of energy to be collected from the surrounding environment, whether vibration, light, or heat. The most common elements used are solar, piezoelectric, or thermoelectric. Therefore, the standard of the emerging PV applied in IoT follow

(1) For Outdoor environment, we meet IEC 60904 series, SEMI PV57, SEMI PV69, SEMI PV76 and IEC 61215.

(2) For Indoor environment, we meet SEMI PV57, SEMI PV76 and SEMI PV80.

In general indoor lighting, a lot of different light sources exist. The International Commission on Illumination (CIE) is responsible for publishing binding standards for different types of light sources, called illuminants. A standard illuminant represents a mathematical table of relative energy versus wavelength, used for colorimetric calculations. It is a (theoretical) source of visible light with a set spectrum, determined by convention and therefore provides a worldwide basis for comparing images under different lighting. Which illuminants are useful standards for indoor illumination ? We consider some typical, widely used, light sources for indoor environments (residential and commercial) and relate them to an appropriate illuminant (standard). First, we will propose a standard for the common incandescent light bulb. An incandescent lamp consists of a balloon of (usually) glass in which a filament is placed that is heated to high temperatures by the Joule's effect. To a first approximation the filament can be considered as a black body radiator. In incandescent lamps, the filament usually consists of tungsten heated to a temperature of 2000 to 3000 K. The CIE has agreed upon a standard for the incandescent light bulb, called "illuminant A". This illuminant is intended to represent typical, domestic, tungsten-filament lighting. Its relative spectral power distribution is that of a Planck radiator at a temperature of approximately 2856 K. In other words, a typical light bulb can be characterized by a black body at temperature $T = 2856$ K. Technically, illuminant A is only defined over the spectral region from 300 nm to 830 nm. However, solar cells often absorb longer wavelengths. Therefore, we will not use illuminant A itself to represent the spectrum of the incandescent light bulb, but we will use Planck's law for a black body at a temperature of 2856 K. In the 300–830 nm range, the difference between Planck's law at this temperature and illuminant A never exceeds 1%. However, In indoor lighting, the International Commission on Illumination is an organization devoted to international cooperation and exchange of information between its member countries on all matters relating to the science and art of lighting. In accordance with an agreement between ISO and CIE, standards are published as double logo standards by ISO. Standards produced by the CIE are a concise documentation of data defining aspects of light and lighting, for which international harmony requires a unique definition. Therefore, the standard follow that the CIE has defined some lighting spectrums for different indoor colorful applications and situations. However, the requirements for color testing do not apply to related indoor energy harvesting under different lighting uniformity and stability, for performance measurement of emerging PV and device. Furthermore, countless types of indoor lighting patterns are required to do performance characteristics measurement for different application contexts. The most common performance indicator is the PV efficiency under standard test conditions (STC). It is the maximum electrical power divided by the total irradiance. For indoor lighting illuminance values are measured used a lux meter or spectradiometer.

(a) Lux is a typical unit for measuring indoor lighting.

(b) For outdoors, irradiance is measured in W/m^2 and there is an approximate conversion of 0.0079 W/m^2/ Lux for the sun.

For indoor lighting, fundamental difference between illuminance and irradiance is the weighting of the spectral response. Irradiance includes the power from all wavelengths weighted equally, whereas illuminance weights the power from each wavelength in proportion to the sensitivity of the human (Fig. 1). Although these conditions are seldom achieved in practice (except in the lab), this characterization provides a reasonable guideline for comparing different solar cell types under outdoor conditions. However, the STC are not relevant for indoor applications. Accurate characterization of emerging PV requires level lighting consideration on each very slow temporal response in the I-V curves of the emerging PV are clearly dependent on the voltage sweep direction, even when the sweep time is the order of seconds [1]. Furthermore, the temporal response is dependent on different level lighting consideration.

In addition, we developed the Indoor lighting simulator, meet SEMI PV80, for energy harvesting under Indoor environment for IoT applications. In our study, RTOS (real-time one-sweep) method can real-time remove the capacity effect under solar simulator for I-V successfully and proposes the forward/backward schematic reaction mechanism for DSC [2]. In this study, RTOS method [3, 4] practices for measuring perovskite solar cell performance showed having better accuracy results by eliminating acceptance effect instantly. Additionally, RTOS method will be useful to measuring cell performance more accurately and rapidly when evaluating emerging PV performance in indoor lighting condition. Therefore, we also defined a performance ratio to compare between indoor lighting and solar simulator.

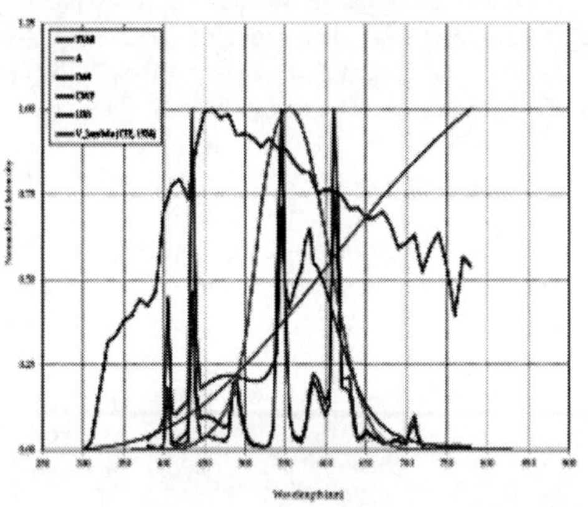

Figure 1: Illumination spectra of CIE Standard Lighting [11, 12]

2 EXPERIMENTS

The samples Sample size ranges from 0.1 cm^2 to 1 cm × 5 cm. These I-V curves were done on a Keithley Source Meter (Model 2460), AM1.5G solar simulator, and Indoor Multi-Light Source System (see Fig. 4). We developed the Indoor lighting simulator, meet SEMI PV80 and follow as:

1) Spectral match, available methods are the use of spectroradiometer. Standard lighting spectrum is defined by the deviation from CIE[12]. For three wavelength intervals of interest, the percentage of total irradiance is specified in Table I. Calculate the spectral match for each wavelength interval, which is the ratio of calculated percentage for the CIE spectrum and the lighting simulator spectrum (Table 1). The data comparison with the solar spectrum shall indicate the spectral match classification as per the following as Table 2.

2) Non-uniformity of irradiance in the test plane, the lux meter is recommended to be used as uniformity detector for determining the non-uniformity of irradiance in the test area of the lighting simulator. The uniformity detector shall have a spectral response appropriate for the simulator. The linearity and time response of the uniformity detector shall conform to the characteristics of the simulator being measured. And the designated test area divided by 25. The area covered by the detector measurements should be 100% of the designated test area. The measurement positions should be distributed uniformly over the designated test area. Therefore, the maximum and minimum irradiance are those measured with the detector(s) over the designated test area, likely in Fig. 2. The calculation of non-uniformity is defined as Eq. (1).

3) Temporal instability of irradiance, the lux meter is recommended to be used as temporal instability detector for determining the temporal instability of irradiance in the test area of the lighting simulator. For lighting simulator the temporal instability is related to the irradiance change of measured data sets during the time of data acquisition (Fig. 3 and Table 3).

Table 1: The ratio of calculated percentage for the CIE spectrum

Wavelength	Bandwidth	Percentage of total irradiance (300 nm – 900 nm)				
Band		A	D65	U30	CWF	TL84
1	300 - 450	3.78	20.31	11.40	17.99	15.18
2	450 - 650	40.15	45.35	81.72	76.73	79.13
3	650 - 900	56.06	34.34	6.88	5.29	5.69

Table 2: Spectral definition of lighting simulator classifications

Classification	Percentage
A	1 %
B	3 %
C	5 %

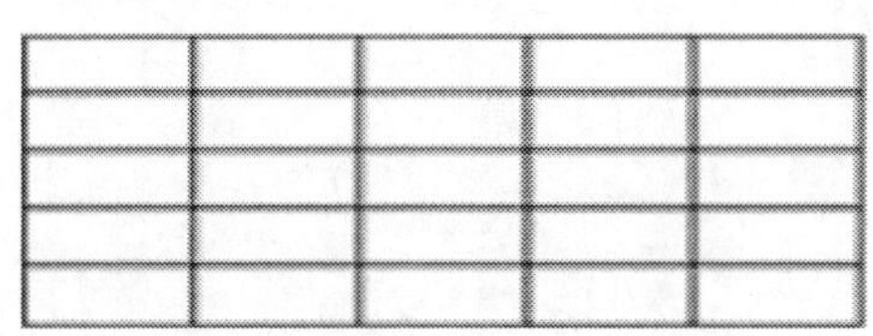

Figure 2: Non-uniformity put a position of detector.

$$Non-uniformity(\%) \equiv \frac{(max-min)}{(max+min)} \times 100\% \quad (1)$$

The Indoor Multi-Light Source included D65 (6500 K, Average North USA sky daylight) / TL84 (4100 K, European shop fluorescent) / CWF (4150 K, cool white fluorescent, shop lighting) / U30 (3000 K, shop lighting) / A (2856 K, typical home lighting), meeting CIE Standard illuminants. The adjustable light intensity (ranging from 0 to 2500 lux) is supplied with an ultraviolet light source for aging testing, with non-uniformity of less than 2% (at 20 cm × 20 cm) and temporal instability of less than 2%. During the I-V measurement in the dark and under Indoor Multi-Light Source System, we followed the testing flowchart in Fig. 5; the scan direction was forward and backward, the sample temperature should be stable at 25

°C with a fluctuation of less than 1 °C, and the irradiance and luminance were determined using a NIST-traceable spectroradiometer (StellarNet) and an NML (National Measurement Laboratory, Taiwan)-traceable lux meter (HIOKI, model FT3424).

Table 3: Temporal instability of indoor lighting simulator.

Item	Average (lx)	Short-term (lx)	Long-term (%)
U30	2709.33	6.35	0.23
	1492.13	1.64	0.11
TL84	2772.08	10.43	0.38
	1507.82	5.61	0.37
D65	1480.16	5.75	0.39
	663.42	2.35	0.35
CWF	2536.80	0.00	0.00
	1199.10	1.91	0.16
A	1902.85	9.01	0.47
	282.92	3.19	1.13

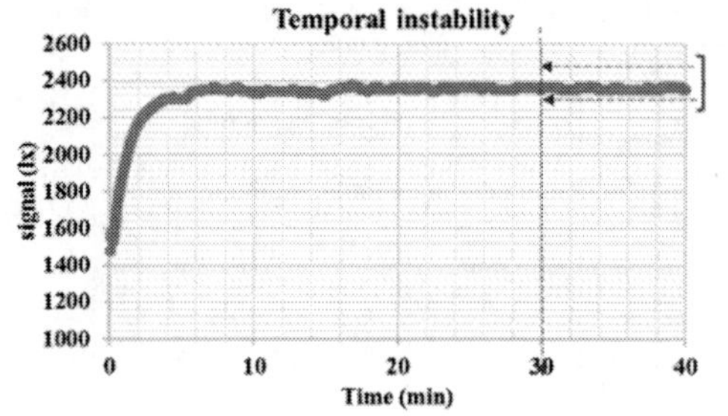

Figure 3: Evaluation of temporal instability.

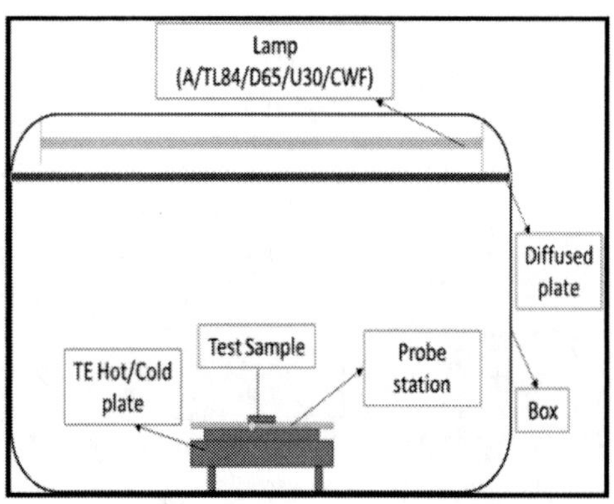

Figure 4: Indoor multi-light source system custom-made by CMS/ITRI for solar cell measurement.

3 ANALYSIS AND RESULTS

In this study, we follow the testing flow [3,4] in the I-V measurement. We used real-time removal of capacity effect; the RTOS method shows the best repeatability (see Fig. 5). A proposed list of steps [2,5] for the measurement of a sample is given in Fig. 8. For most traditional semiconductor PV devices, the current will respond to small changes in bias voltage in less than a millisecond. Sweeping the voltage from open-circuit voltage (Voc) to short-circuit current (Isc) in a few seconds at a constant illumination for such a device will yield accurate currents. Many emerging PV cells respond far more slowly, and consequently, the sweep time must be increased. Fig. 9 shows the I-V for a slowly responding emerging PV cell with different sweep rates.

Therefore, we used the RTOS Method [2] following the testing flowchart (see Fig. 8) in the I-V measurement. applied bias voltage is changed from Isc to Voc (forward scan) or in the reverse direction (backward scan, Voc to Isc). When a capacitance effect occurs, this measurement process produces the phenomenon. In the forward scan, an overshot current appears immediately after an abrupt increase of applied voltage and then the current gradually decreases to an equilibrium state. Sweeping to determine maximum power (Pmax) may not be valid if the device is metastable because of the long times that it sits near Isc or Voc before getting to Pmax shown in Fig. 7(b) to confirm the "near-true" Pmax of emerging PV, especially for samples that depend strongly on sweep direction or rate. This also shows that the asymptotic I-V method may be the only reliable way to obtain efficiency.

In the light intensity under artificial lighting conditions found in offices and factories is less than 5 W/m2 as compared to 100–1000 W/m2 under outdoor conditions, depending on the type of light source and its distance. Moreover, the spectrum can be totally different from the outdoor solar reference spectrum (e.g. AM1.5G). The spectrum depends not only on the type of light source, but also on the presence of reflected and diffused light. Therefore, we shown a definition of performance (η) between indoor lighting and solar simulator below as

$$\eta \equiv (P_{max}/Area)/P_{in}. \qquad (2)$$

Where (a) P_{max} means maximum power of device, (b) Area means area of device, (c) P_{in} means Illuminant power of lighting source. Therefore, we also defined a normalized power density (performance ratio, η_n) to compare between indoor lighting and solar simulator below as

$$\eta_n \equiv \eta_{Indoor}/\eta_{AM1.5G} \qquad (3)$$

Where (a) η_{Indoor} means performance of device at each indoor lighting (A/TL84/D65/U30/CWF), (b) $\eta_{AM1.5G}$ means performance of device at solar simulator.

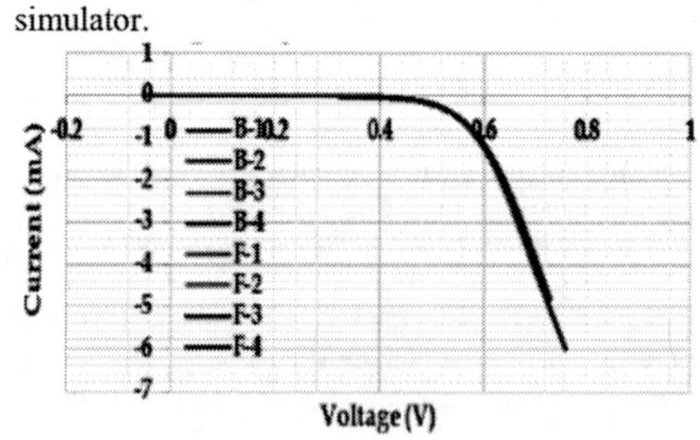

Figure 5: I-V curves measured by RTOS method at dark condition.

4 CONCLUSIONS

The characterization after testing showed that one must measure performance under all typical artificial light sources. Therefore, we propose that each solar cell should be measured varied indoor and outdoor lighting condition for its characterization by RTOS method. In Fig. 8a, we will be compare the Pmax of the different solar cells in light levels in indoor environments to show linearity under near range lighting level, which depends on device. In Fig. 8b shown nearly a constant from 400 lx to 1000 lx, it will be get an average of normalized power density and follow the defined of equation (3). Therefore, the results will be show the different solar cells in light levels in indoor environments to show linearity under near range lighting level, which depends on device. From Fig. 10b, it is also shown the different solar cells in indoor lighting and solar environments to show different performance, ex. DSSC have more 0.26 times for TL84 than D65. We are also known STC under outdoor application, therefore, we also define a indoor lighting-STC for indoor application, that is " A/ 25 °C/ 1000 lx ".

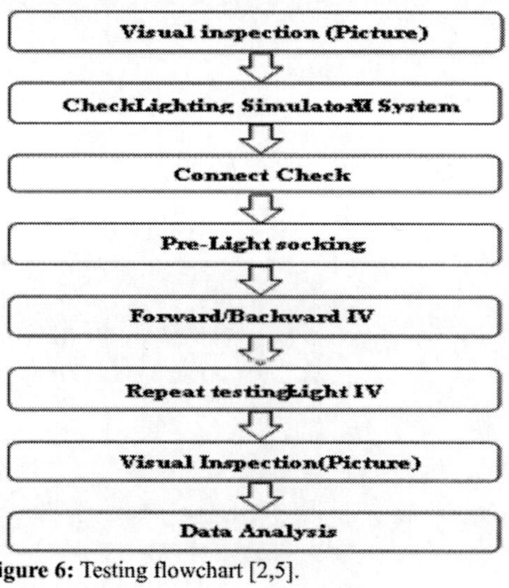

Figure 6: Testing flowchart [2,5].

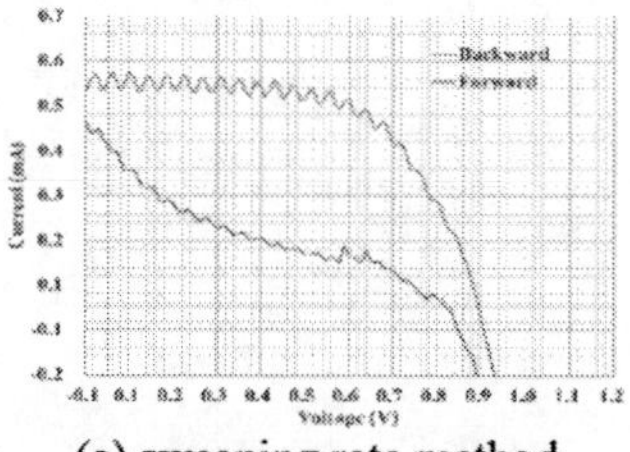

(a) sweeping rate method

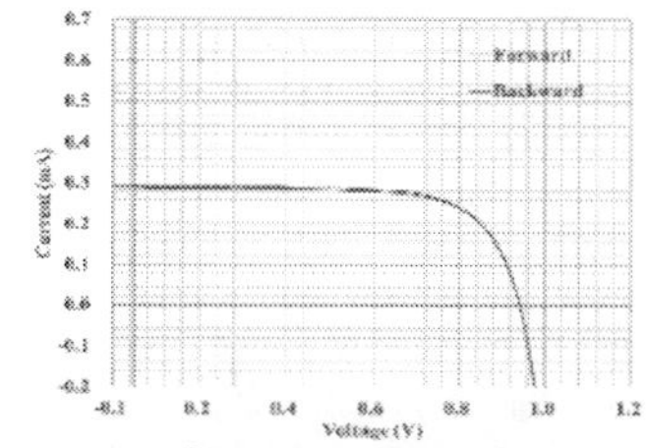

(b) RTOS method

Figure 7: I-V measurement of PSC (at STC) using the (a) sweeping rate method and (b) RTOS method.

In this work, we defined performance and normalized power density, is called performance ratio, shown a comparison for different emerging PV in indoor lighting and solar environments (see Fig. 10). Therefore, the results also find DSSC have more 0.26 times for TL84 than D65, but OPV/a-Si have less 0.5 times for TL84 than D65. That's shown DSSC have more indoor application than others in this case study.

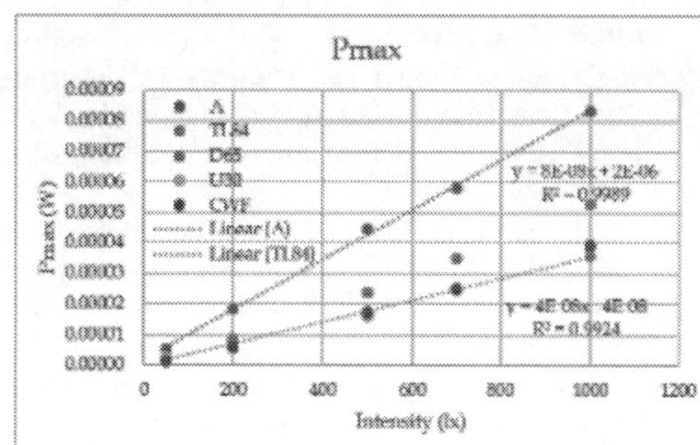

(a) P_{max}

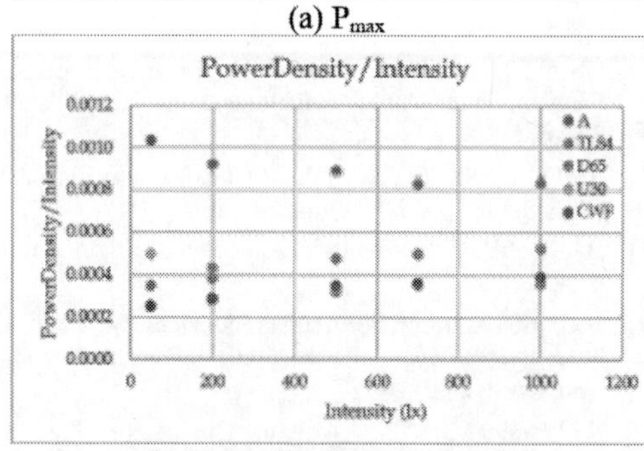

(b) Normalized Power density

Figure 8: Light linearity of DSSC measured by indoor multi-light source system under varied lighting condition (I-V using RTOS method).

When we will be follow through CIE of the defined some lighting spectrums for different indoor colorful applications and situations. However, the requirements for color testing do not apply to related PV indoor performance measurement under different lighting uniformity and stability, for performance measurement of emerging PV and device, and Its characteristics will be having a classification. Furthermore, countless types of indoor lighting patterns are required to do performance characteristics measurement for different application contexts. We should be define a different STC than STC in AM1.5G, which we will be define a STC in indoor lighting as below as "I-V measurement for indoor lighting-STC condition: 1000 lx/25 °C/Illuminant A". Furthermore, our goal is to determine which light source is relevant for a new international standard test method. Upon the indoor characterization of solar cells, it is the spectrum range of the light source that decides the lighting performance, not the light intensity, which is present and affects the cell equally through different light sources.

REFERENCES

[1] T. C. Wu, S. T. Hsu, and Y. S. Long, "New Set-up Procedures and Integrated Measurement System for Organic Photovoltaic (OPV) Module," *International Photovoltaic Science and Engineering Conference (PVSEC23)*, 2013.

[2] Y. S. Long, S. T. Hsu, and T. C. Wu, "Induction of Internal Capacitance Effect in Performance Measurement of OPV (Organic Photovoltaic) Device by RTOSM (Real-Time One-Sweep Method)," *Journal of Energy and Power Engineering*, Vol 8, pp1059-1066, 2014.

[3] Y. S. Long, S. T. Hsu, and T. C. Wu, "Capacitance Effect from High to Low Level Lighting by Real-Time One-Sweep Method for OPV Device," *IEEE PVSC42*, 2015.

[4] Y. S. Long, S. T. Hsu, and T. C. Wu, "A Study of Capacitance Effect in DSC from High to Low Level Lighting by Real-Time One-Sweep Method," *OPTIC2014*.

[5] SEMI PV57-1214, "Test Method for Current-Voltage (I-V) Performance Measurement of Organic Photovoltaic (OPV) and Dye-Sensitized Solar Cell (DSSC)."

[6] US PTO. 8224598, Method for Forming Optimal Characteristic Curves of Solar Cell and System Thereof.

[7] ASTM E490, "Standard Solar Constant and Zero Air Mass Solar Spectral Irradiance Tables."

[8] Handbook of Photovoltaic Science and Engineering, Chapter 16

[9] IEC 60904-1, "Photovoltaic Devices – Part 1: Measurement of Photovoltaic Current-Voltage Characteristics," 2006.

[10] IEC 60904-3, "Photovoltaic Devices – Part 3: Measurement Principles for Terrestrial Photovoltaic (PV) Solar Devices with Reference Spectral Irradiance Data," 2008.

[11] ISO 10526, "CIE Standard Illuminants for Colorimetry," 2007.

[12] CIE S 014-2, "CIE Standard Illuminants for Colorimetry," 2006.

[13] K. Emery, S. Winter, S. Pinegar, and D. Nalley, "Linearity Testing of Photovoltaic Cells," *4th World PV Conf.*, 2006.

[14] CIE S 014-2, "CIE Colorimetry - Part 2: Standard Illuminants for Colorimetry "

[15] ISO 11664-2:2007, "Colorimetry -- Part 2: CIE standard illuminants"

[16] ASTM D 1729, "Standard Practice for Visual Appraisal of Colors and Color Differences of Diffusely-Illuminated Opaque Materials";

[17] SEMI Doc. 5979, "Specification of indoor lighting simulator requirements for emerging PV "

[18] Handbook of Photovoltaic Science and Engineering, Chapter 16

INDOOR CALIBRATION OF LARGE AREA ORGANIC PV MODULES

G. Bardizza*, E. Salis, D. Pavanello, T. Sample, H. Müllejans, E.D. Dunlop
European Commission, Joint Research Centre, Directorate C - Energy, Transport & Climate,
Energy Efficiency and Renewables Unit, European Solar Test Installation
Via E. Fermi 2749, I-21027 Ispra (VA) - Italy
* E-mail: giorgio.bardizza@ec.europa.eu - Tel: +39 0332785529

ABSTRACT: The PV market is currently dominated by crystalline silicon based technologies. However, the emerging thin-film technologies show the potential to compete with them in the future, due to the improved efficiencies and increased lifetime reached. Examples of printed large-area OPV modules have been produced by different companies. There is a clear need of guidelines and protocols on how to perform a reliable calibration of these devices. In this work we present a practical example of a calibration of a large-area OPV module performed indoor under a solar simulator following the procedures described in IEC 60891 and 60904 standards. Preliminary results about a power matrix measurement study performed indoor with the same OPV module under a steady-state large-area solar simulator are also presented.

1 INTRODUCTION

Printed organic PV (OPV) has been demonstrated to have overcome the laboratory discovery phase and to have reached scaled manufacture: examples of solar cell production on an industrial scale for energy production have been shown in the last five years [1] [2]. There are companies capable of producing large-area OPV modules to be installed outdoor. As a consequence their reliable calibration and characterization is a fundamental task to achieve.

In this work we present a procedure for the reliable calibration of a large-area OPV module which allows for an appropriate evaluation of their response time and, consequently, for a reliable calibration fulfilling the requirements specified in the standards (IEC 60891 and IEC 60904 series).

An indoor power matrix of this device has been measured under a continuous large-area solar simulator and the preliminary results are reported. Interesting results have been found confirming what was already reported by our group in a previous work about the power matrix of other OPV devices provided by a different manufacturer [3].

2 DEVICE UNDER TEST

The device measured in this study is a large-area OPV module provided by an external party/company (ESTI code TK801. The device has an active area of 30x200 cm^2 and is formed by 24 cells serially connected. It arrived to the European Solar Test Installation (ESTI) mounted by the company on a rigid metallic frame in order to avoid bending and facilitate the mounting and handling during the measurements. Standard MC4 type connectors are available for electrical connections. A picture of TK801 is shown in Figure 1.

3 CHARACTERIZATION AT STANDARD TEST CONDITIONS

In order to characterize the module at standard test conditions (STC) as defined in IEC TS 61836 [4] we followed the procedure implemented previously in our laboratory and described in [5] [6].

3.1 Time response analysis

The setup of the time response analysis is described in our previous work [5]. Due to space constraints of the setup a smaller mini-module taken from the same printing batch of TK801 (ESTI code TK804) has been used in this case. TK804 has exactly the same type and number of cells of TK801; it is a shorter piece of the same roll (dimensions 30x30 cm^2).

The electrical signals generated by the device under test TK804 (DUT) and by the calibrated c-Si reference cell PX302C (REF) illuminated by the chopped white light from the Xenon lamp were simultaneously acquired while the chopping frequency was scanned from 1.4 Hz to 80 Hz. The irradiance of the incident white light measured with REF on the test plane was 30 W/m^2.

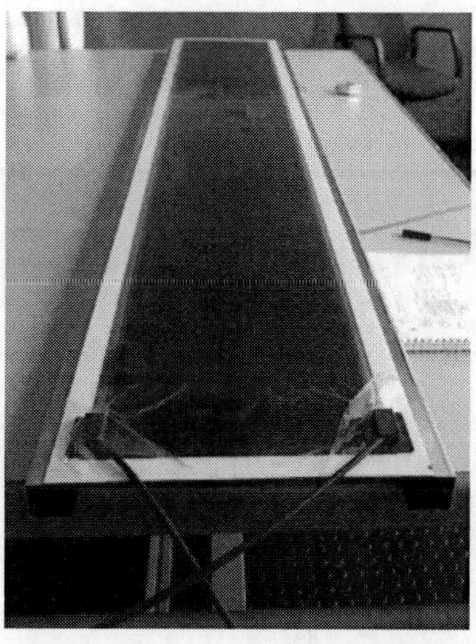

Figure 1 Picture of the device under test TK801. The plastic protection cover visible in the picture has been removed before measurements.

The peak-to-peak variation of the signals of the DUT and REF was analyzed and the results are shown in Figure 2. The ratio of the peak-to-peak signal of DUT and REF stays well constant (within +/- 1% variation) in the scanned frequency range and no distortion of the signals was observe even at high chopping frequencies. These results allowed us to conclude that the current response to the light pulse of TK804 is fast and comparable with crystalline silicon devices used in this case as reference. As a consequence a chopping frequency of approximate (65-70) Hz can be appropriate also for these devices.

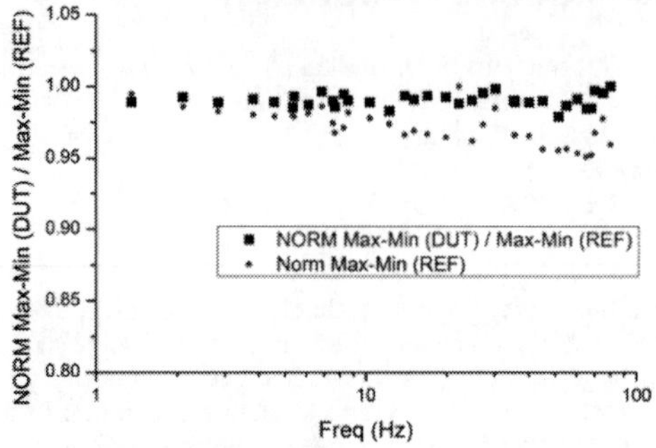

Figure 2 Results of the time response analysis of TK801.

3.2 Spectral responsivity
The spectral responsivity (SR) measurements of TK804 were performed on both Oriel and Pasan systems available at the ESTI for SR measurements. A comparison of the two measurements performed on TK804 with the two apparatus is shown below. SR of TK801 was measured only with Pasan system. A complete description of the two setups is available in [7] and [8] respectively.

3.2.1 Oriel
The Oriel setup for SR measurements basically consists of a continuous illumination from a Xenon lamp, band-pass filters covering the interested wavelength range, a mechanical chopper and lock-in amplifiers. The illuminated area is about 30x30 cm^2. Following the suggestions of the manufacturer the devices under study were initially treated as multi-junction devices. Different LED bias lights were used to saturate separately the junctions during measurements as required in the IEC 60904-8-1 [9]. The SR data for TK804 measured using different bias lights in the Oriel setup are reported in Figure 3 In all these cases a bias voltage was applied in series to DUT to compensate for the voltage generated by the bias light (see V_{bias} in the table). Dark measurements without any bias light and voltage were also performed and reported in the graph. Looking at the results of Figure 3 we could not notice any separate junction, the device behaved as a single junction. This has been confirmed also by the dark measurements which report a SR curve comparable to the ones measured with bias illumination.

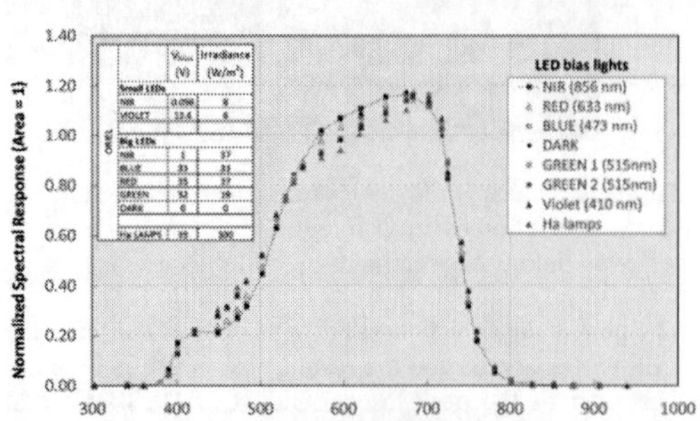

Figure 3 Spectral responsivity data of TK804 measured with Oriel setup using different LED bias lights.

Few differences were noticed in the shape of the SR curves: in particular data measured with NIR and blue LEDs and in the dark resulted in almost identical curves with a pronounced shoulder at about 420 nm and a higher response between (570-670) nm. Data measured with green and violet LEDs and Ha lamps showed at the contrary a less pronounced shoulder at 420 nm and a lower and constantly increasing response between (570-670) nm. The case with red LEDs lays in between these two cases. At the moment there is no clear explanation of this behaviour. Further investigation is foreseen.

3.2.2 Pasan
The SR measurements of TK804 and TK801 were performed also using the Pasan system. This consists of a large-area flash solar simulator (illuminated area 2x2 m^2), band-pass filters covering the interested wavelength range, wide bandwidth amplifiers for the amplification of the electrical signals. The measurements were performed without any bias light within the wavelength range (300-900) nm. The SR data measured with Pasan for TK804 and TK801 are reported in Figure 4. The measured points overlap perfectly and the devices showed an almost identical curve.

3.2.3 Comparison of Oriel and Pasan setups
In Figure 5 the data measured with Pasan and Oriel systems are compared. In particular the curves measured in the dark without bias light in both systems are here compared. The measurements performed with Oriel and Pasan are in perfect agreement. The small differences noticeable in the graph are within the typical accuracies of the systems.

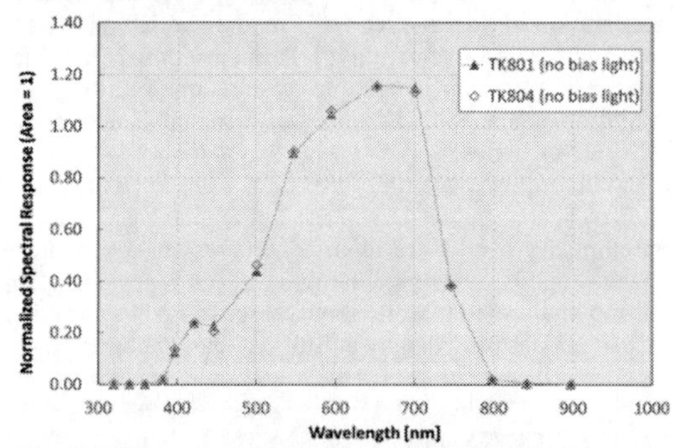

Figure 4 Spectral responsivity data of TK804 and TK801 measured with Pasan setup without bias light.

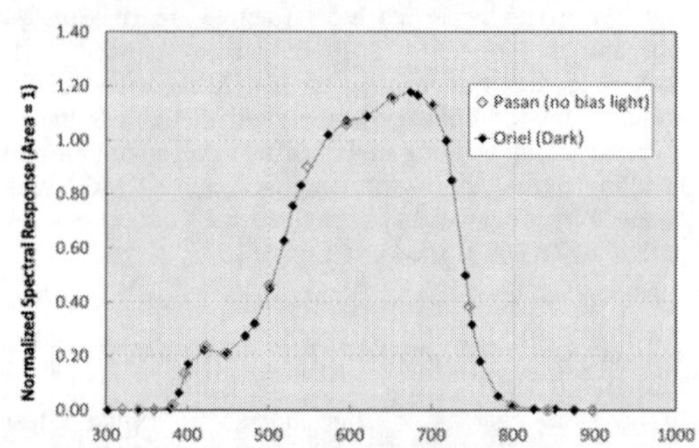

Figure 5 Comparison of the SR measurements of TK804 performed with Oriel and Pasan setups.

A calculation of the spectral mismatch factor (MMF) for TK804 and TK801 was done following the IEC 60904-7 [10]. All the different configurations (bias lights and dark) and the two setups (Oriel and Pasan) utilized in this work were considered. The results are reported in Table 1.

Table 1 Spectral mismatch factor calculation for TK801 and TK804 for the different bias lights and setups (Oriel and Pasan).

TK804 vs PX301A and Pasan IIIB		MMF
ORIEL (PX302C)	NIR	0.9884
	RED	0.9888
	BLUE	0.9882
	DARK	0.9885
	GREEN 1	0.9896
	GREEN 2	0.9890
	Ha LAMPS	0.9894
PASAN (ASP10 wog)	DARK	0.9881

TK801 vs PX301A and Pasan IIIB		MMF
PASAN (ASP10 wog)	DARK	0.9880

3.3 I-V measurements

The I-V characterization of TK801 was performed using a large-area flash solar simulator following IEC 60904-1 [11]. The measurements were performed at an irradiance close to 1000 W/m^2 and a temperature of 25 °C. The I-V curves were corrected to the nominal irradiance 1000 W/m^2 following the IEC 60891 [12] standard. The device was mounted at the test plane with the cells aligned vertically. I-V curves were measured in both single flash (SF) and multiflash (MF) modes and compared. In **Error! Reference source not found.** the comparison of I-V curves measured using both SF and MF methods is shown. The electrical parameters extracted from the measured I-V curves for TK801 are reported in Table 2

A significant difference between the curves measured in MF and SF methods has been observed for both TK804 and TK801. The main difference is observable in the P_{max} values (and consequently in the FF values) reaching a relative difference between SF and MF of about 1.6% in the case of TK801. This difference is not dramatically large, but is systematic and above the stated uncertainty of the systems. We believe this could be due to the sweep time, which is probably too short in I-V curves measured in single flash mode (sweep time ca. 10 ms). This assumption has been confirmed by a set of measurements performed with TK804 at increasing sweep time from 0.2 ms to 120 ms. As shown in Figure 7 the FF (and consequently the P_{max}) increases with the sweep time. Looking in more details above 5 ms (see plot on the bottom of Figure 7) the values do not stabilize even up to about 100 ms. The relative variation of P_{max} going from 10 ms to 100 ms is about 0.5%. Considering the fact that the measurements done with MF method are performed at almost constant voltage which is manually set before each flash, the so obtained 0.5% agrees well with the 1.6% difference calculated before between curves measured in SF and MF methods. The sweep time of 100 ms lies indeed in between. A quick extrapolation of the measured values at higher sweep times can qualitatively explain the 1.6% relative increase of P_{max} going from 10 ms to 1000 ms or higher. In order to clearly and quantitatively prove this a new experiment using illumination from a continuous large-area solar simulator would be required. In this case it would be possible to measure I-V curves at even longer sweep times. Other issues like temperature stability and uniformity could affect the measurements. The results here presented suggest that MF method is recommended for accurate I-V characterization of this OPV module.

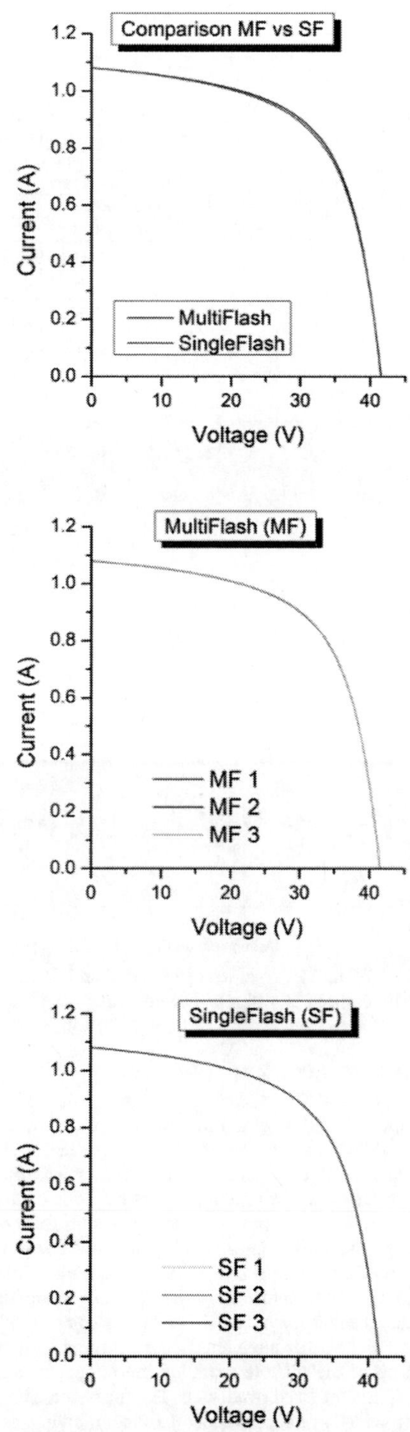

Figure 6 Comparison of I-V curves measured with single flash and multiflash methods for TK801.

Table 2 Electrical parameters extracted from the I-V curves of TK801 measured with both methods. The average values, maximum-minimum absolute and relative variation are calculated. The relative difference between the average values obtained using the SF and MF methods is calculated and showed (Diff1).

TK801	Filename	Date	Temp (°C)	Isc (A)	Voc (V)	Pmax(W)	Fill Factor (%)	Imp (A)	Vmp (V)
Multiflash	TK801_MF_PIIIb_01.stf	Wed Aug 02 2017	25.13	1.0810	41.5405	27.5006	61.2434	0.8504	32.3373
	TK801_MF_PIIIb_02.stf	Wed Aug 02 2017	25.22	1.0798	41.5350	27.4732	61.2553	0.8502	32.3142
	TK801_MF_PIIIb_03.stf	Thu Aug 03 2017	25.23	1.0802	41.5346	27.5202	61.3367	0.8539	32.2274
			average	1.0803	41.5367	27.4980	61.2785	0.8515	32.2930
			max-min	0.0011	0.0059	0.0470	0.0932	0.0037	0.1099
			max-min/avg	0.11%	0.01%	0.17%	0.15%	0.44%	0.34%
Single Sweep 1	TK801_PIIIb_01.stf	Wed Aug 02 2017	24.98	1.0804	41.5758	27.0714	60.2686	0.8464	31.9834
	TK801_PIIIb_02.stf	Wed Aug 02 2017	24.97	1.0809	41.5822	27.0739	60.2367	0.8473	31.9517
	TK801_PIIIb_03.stf	Wed Aug 02 2017	24.97	1.0807	41.5799	27.0727	60.2479	0.8464	31.9850
			average	1.0807	41.5793	27.0726	60.2510	0.8467	31.9734
			max-min	0.0005	0.0064	0.0025	0.0319	0.0009	0.0333
			max-min/avg	0.05%	0.02%	0.01%	0.05%	0.11%	0.10%
			Diff 1	**-0.03%**	**-0.10%**	**1.55%**	**1.68%**	**0.56%**	**0.99%**

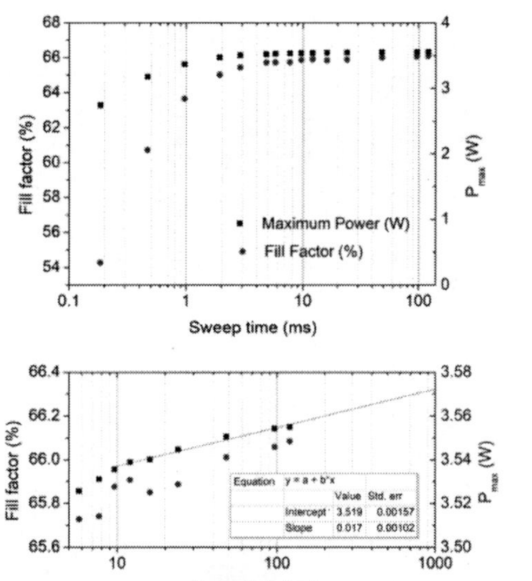

Figure 7 FF and P_{max} values extracted from I-V curves of TK804 measured at different sweep times

4 POWER MATRIX MEASUREMENTS

I-V curves of TK801 were measured, under indoor controlled conditions, at different levels of irradiance, and module temperatures. A class AAA (IEC 60904-9 [13]) large-area continuous solar simulator was used as light source [14]. The irradiance level has been controlled by opening and closing the shutters placed in front of the 11 lamps of the simulator. The module has been cooled down to about 10°C prior to electrical measurements with the use of a temperature controlled climatic chamber and naturally heated up by the light of the solar simulator. Periodical I-V measurements were acquired during the warm up of the module spanning the temperature range 15°C - 55°C at fixed irradiance. The measurements were performed at irradiances close to the following values: (1000, 800, 600, 400, 200, 100) W/m^2 and corrected afterwards to the nominal irradiances. The main electrical parameters V_{oc}, I_{sc}, FF and P_{max} were extracted from the I-V curves and analyzed. In this work we present the variation of P_{max} in the measured range of temperature and irradiance.

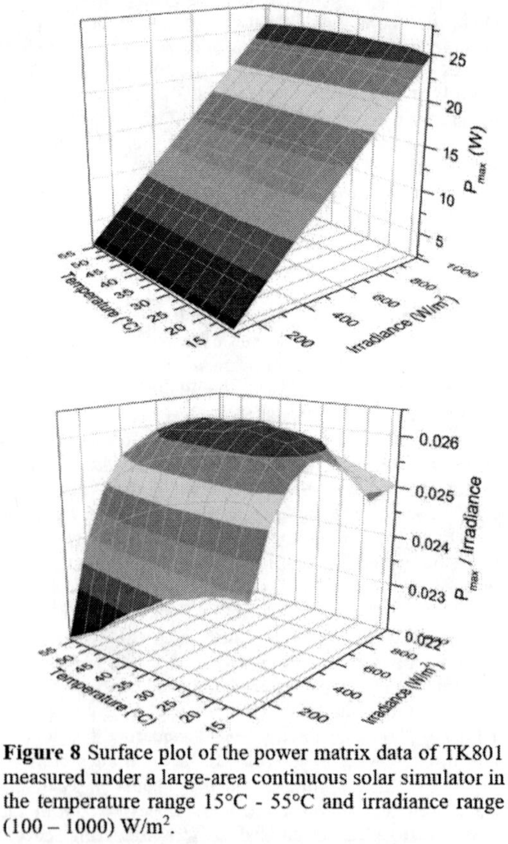

Figure 8 Surface plot of the power matrix data of TK801 measured under a large-area continuous solar simulator in the temperature range 15°C - 55°C and irradiance range (100 – 1000) W/m^2.

The surface plots of Figure 8 show the P_{max} data and P_{max}/Irradiance data as a function of temperature and irradiance. From these plots it is noticeable a linear dependence of P_{max} on the irradiance and a more complex one on the temperature. In the second plot of Figure 8 P_{max}/Irradiance shows different behaviors depending on the irradiance level. At low irradiance levels below

200 W/m^2 P_{max}/Irradiance decreases non-linearly with T. At higher irradiance levels above 200 W/m^2 it shows an initial increase, reaches a maximum at a certain T and going to higher temperatures starts to decrease. This behavior has been observed at all irradiances above 200 W/m^2, but the temperature corresponding to the maximum of P_{max}/Irradiance constantly shifts to higher values going from low to high irradiance levels. A similar behavior was observed already by us in a previous study about power matrix measurements of other OPV mini-modules provided to ESTI by a different company. More details about this are reported in [3]. A more detailed analysis of the power matrix data of TK801 will be included in a separate work focused on this topic.

6 CONCLUSIONS

In this work a general procedure for the accurate and reliable calibration of large-area OPV devices was described and discussed. It allows for an evaluation of their response time and, consequently, for a reliable calibration fulfilling all requirements specified in the standards (IEC 60891 and IEC 60904 series). Multiple I-V measurements were performed at various sweep times and in both single flash and multi flash modes in order to check for the most appropriate configuration to use. The procedure has been applied here to the case of a large-area roll-to-roll printed OPV module provided to our laboratory by an external company. However, the method presented is not limited to this type of device and can be generally applied, in particular to PV devices which are known to have a long response time or whose response is unknown.

An energy rating study based on indoor power matrix measured data of different OPV mini-modules has been presented previously by our group [7]. Interesting results have been found showing potential advantages of OPV technology in specific locations (low latitude and hot climate) if compared with more established PV technologies. In this study we presented the preliminary results of power matrix measurements performed on the large-area OPV module under study. The analysis of the P_{max} variation in the temperature and irradiance ranges under investigation 15 °C -55 °C and (100-1000) W/m^2 confirmed the behavior observed already in a previous study. Considering that these results are from an advanced prototype of an OPV module, this increases the confidence in our energy rating study and in the conclusions raised there.

REFERENCES

[1] M. A. Green, Y. Hishikawa, E. Dunlop, D. Levi, J. Hohl-Ebinger e A. Ho-Baillie, «Solar cell efficiency tables (version 51),» *Prog Photovolt Res Appl.*, vol. 26, pp. 3-12, 2018.

[2] J. Carle, M. Helgesen, O. Hagemann, M. Hoesel, I. M. Heckler, E. Bundgaard, S. A. Gevorgyan, R. R. Søndergaard, M. Jørgensen, R. Garcıa-Valverde, S. Chaouki-Almagro, J. A. Villarejo e F. C. and Krebs, «Overcoming the Scaling Lag for Polymer Solar Cells,» *Joule,* vol. 1, n. 2, pp. 274-289, 2017.

[3] G. Bardizza, E. Salis, A. M. Gracia Amillo, T. Huld e E. D. Dunlop, «Power matrix measurements and energy rating analysis of organic PV mini-modules,» in *Proceedings of the 33th EU PVSEC conference*, Amsterdam (Netherlands), 2017.

[4] IEC TS 61836 Solar photovoltaic energy systems - Terms, definitions and symbols, 2016.

[5] G. Bardizza, D. Pavanello, H. Müllejans e T. Sample, «Spectral responsivity measurements of DSSC devices at low chopping frequency (1Hz),» *Progress in Photovoltaics: Research and Applications*, n. 24, pp. 428-435, 2016.

[6] G. Bardizza, D. Pavanello, R. Galleano, T. Sample e H. Müllejans, «Calibration procedure for solar cells exhibiting slow response and application to a dye-sensitized photovoltaic device,» *Sol. Energy Mater. Sol. Cells*, vol. 160, pp. 418-424, 2017.

[7] M. Pravettoni, A. Komlan, R. Galleano, H. Müllejans e E. and Dunlop, «An alternative method for spectral response measurements of large-area thin-film photovoltaic modules,» *Progress in Photovoltaics: Research and Applications*, vol. 20, pp. 416-422, 2012.

[8] U. Bahrs, W. Zaaiman, M. Mertens e H. Ossenbrink, «Automatic large area spectral response facility,» in *14th European Photovoltaic Solar Energy Conference and Exhibition*, Barcelona (Spain), 1997.

[9] IEC 60904-8 (Ed. 3) Photovoltaic devices – Part 8: Measurement of spectral responsivity of a photovoltaic (PV) device, 2014.

[10] IEC 60904-7 (Ed. 3) Photovoltaic devices – Part 7: Computation of the spectral mismatch correction for measurements of photovoltaic devices, 2008.

[11] «IEC 60904-1 (Ed. 2) Photovoltaic devices – Part 1: Measurement of photovoltaic current-voltage characteristics,» 2006.

[12] IEC 60891 (Ed.2) Photovoltaic devices - Procedures for temperature and irradiance corrections to measured I-V characteristics, 2009.

[13] IEC 60904-9 (Ed.2) Photovoltaic devices - Part 9: Solar simulator performance requirements, 2009.

[14] E. Salis, I. Sharlandzhiev e M. Field, «Feasibility Study for PV Measurements at Varying Irradiances on a Large-Area Steady-State Solar Simulator,» in *Proceedings of the 33th EU PVSEC conference*, Amsterdam (Netherlands), 2017.

[15] F. C. Krebs, M. Hösel, M. Corazza, B. Roth, M. V. Madsen, S. A. Gevorgyan, R. R. Søndergaard, D. Karg e M. Jørgensen, «Freely available OPV - The fast way to progress,» *Energy Technology,* vol. 1, pp. 378-381, 2013.

[16] IEC 61853 (Ed. 1) Photovoltaic (PV) module performance testing and energy rating - Part 1: Irradiance and temperature performance measurements and power rating, 2011.

[17] IEC 60904-10 (Ed.2) Photovoltaic devices – Part 10: Methods of linearity measurement, 2009.

Lead sulfide nanocrystal co-sensitized dye-densitized solar cell: scalable deposition process and significant improvement in power conversion efficacy

Umer Mehmood[1], Amir Al-Ahmed[1]*, Mohammad Afzaal[2], Abbas S. Hakeem[3], Shamsuddeen A. Haladu[4], and Fahad A. Al-Sulaiman[1]

[1]Center of Research Excellence in Renewable Energy, King Fahd University of Petroleum and Minerals, Dhahran-31261, Saudi Arabia
[2]Maths and Natural Sciences Division, Higher Colleges of Technology, Sharjah Campus, 7947, Sharjah, United Arab Emirates
Centre of Research Excellence in Nanotechnology, King Fahd University of Petroleum and Minerals, Dhahran 31261, Saudi Arabia
[4] Department of Basic Sciences and Humanities, College of Engineering, Imam Abdulrahman Bin Faisal University, Dammam 34212, Saudi Arabia
*Presenting Author's E-mail: aalahmed@kfupm.edu.sa

ABSTRACT: Commercial titanium dioxide (TiO_2) paste was tape casted on a pre-cleaned FTO glass substrate and baked at 450 °C for 30 mins. Different concentrations of lead xanthate precursor, $Pb(S_2COEt)_2$ were sprayed on the TiO_2 layer and backed again in air at 120 °C for 2 hr to generate lead sulfide (PbS) nanocrystals. As-deposited PbS nanocrystals deposited on TiO_2 photoanodes were dipped in a dye solution (N719 in methanol) for 24 hr. After sensitization, all the samples were washed with ethanol to remove any unanchored dye molecules. Dye-sensitized solar cells (DSSCs) were completed by depositing a platinum back electrode. 0.22 mM $Pb(S_2COEt)_2$ sprayed cell showed improved power conversion efficiency of 7.9%, which is an improvement of 29% compared to the cell without any co-sensitizer. It is expected that most of the absorbed light was involved in the photocurrent generation and was confirmed by an incident photon-to-current conversion efficiency analysis. This is also in-line with the ultraviolet–visible absorption spectra of the photoanode. However, an ***increase in precursor concentration*** negatively affected the power conversion efficiency, primarily due to the blocking of incident light reaching the prime photosensitizer.

1 BACKGROUND

Ease of fabrication, low manufacturing costs, and competitive efficiency have provided much impetuous toward research on dye-sensitized solar cells (DSSCs). However, nowadays, they are facing extensive competition from other thin film solar cells. Therefore, research are trying improve its performance by various approaches, like, co-sensitization, better hole transport material, doped TiO_2 as anode, developing cheaper and more effective counter electrodes, and so on. Here, we use a simple but scalable spray coating technique to deposit uncapped lead sulfide (PbS) nanocrystals on the TiO_2/FTO surfaces, in another case on the counter electrode i.e. Pt/FTO, from the lead xanthate precursor. Resulting PbS nanocrystals acted as a co-sensitizer and minimize recombination lose. Both cells showed higher power conversion efficiency than the pure in the dye-sensitized solar cells demonstrated marked efficiencies up to 7.9%, which was an improvement of 29% compared cell with absence of a co-sensitizer [1-5].

2 PREPARATION OF LEAD SULFIDE (PbS) NANOCRYSTALS

- Lead potassium ethylxanthogenate was synthesized using a reported procedure and its purity was confirmed by ^{1}H NMR[6].
- To optimize the deposition conditions, $Pb(S_2COEt)_2$ solution was sprayed on pure FTO substrates and baked in air at different temperatures and times. It was observed that baking for 2 hr at 120 °C in atmospheric conditions generated pure cubic PbS nanocrystals (Figure-1).
- Scanning electron images and X-ray powder diffraction were used to confirm the resulting morphologies and phase, respectively.

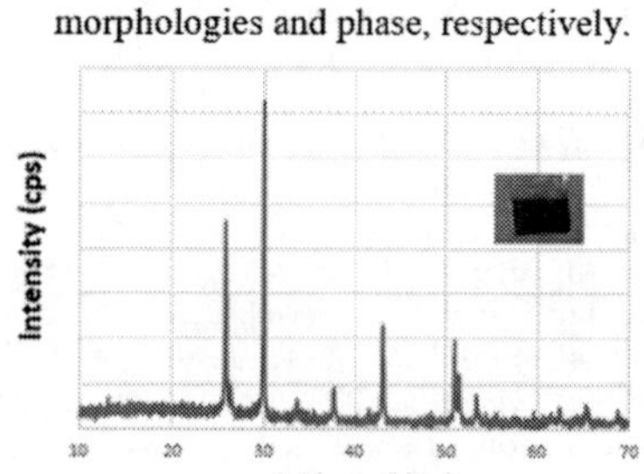

Figure-1. XRD pattern and picture of the backed PbS on pure FTO

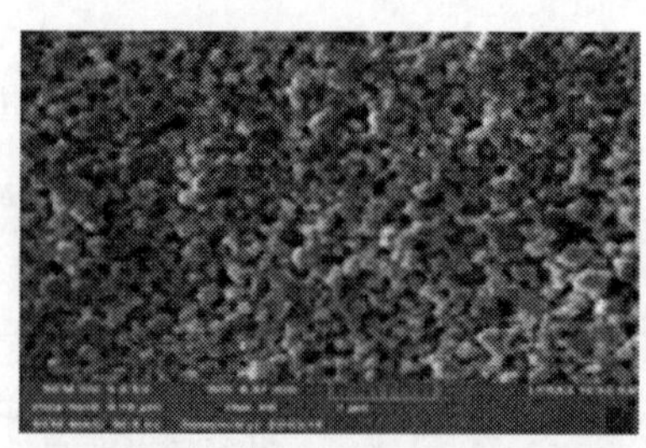

Figure-2. SEM of pure PbS nanocrystals

3 DSSC FABRICATION

- DSSCs were prepared using a conventional tape casting technique as prepared in literature [1].
- DSSC structure without PbS nanocrystals: FTO/TiO_2/Dye(N719)/Pt/FTO
- DSSC structure with PbS nanocrystals: FTO/TiO_2/PbS/Dye(N719)/Pt/FTO
- Different concentration of $Pb(S_2COEt)_2$ solutions sprayed to deposit different amounts of PbS nanocrystals on the TiO_2 and marked as Samples, B, C, D and E.

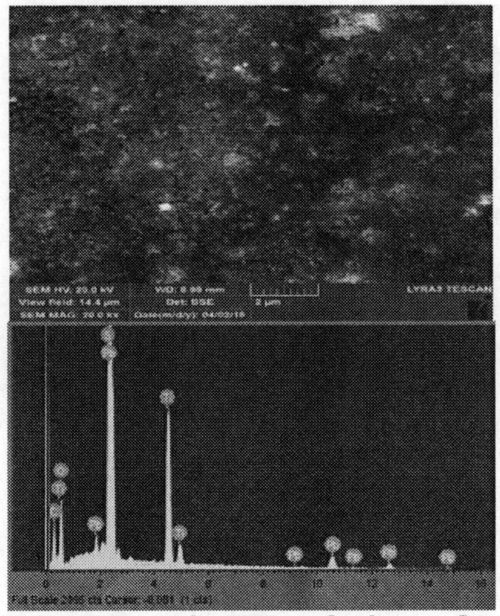

Figure-3 SEM and EXD of the surface of FTO/TiO_2/PbS/Dye

4 RESULTAS AND COMMENTS

- Cell A showed power conversion efficiency of $\eta = 6.11\%$
- Cell B with sprayed 22 mM of $Pb(S_2COEt)_2$ solution showed power conversion efficiency of 7.90% which is nearly 29% increase in efficiency

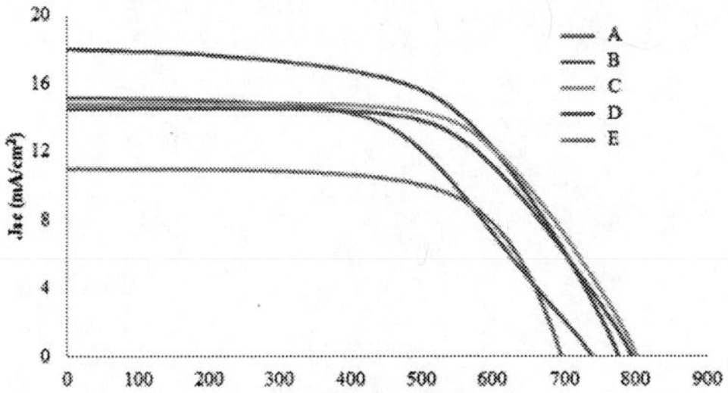

Figure-4. Current–voltage characteristics of DSSCs fabricated using different amounts of PbS precursor.

Table I: Photovoltaic Properties of the DSSCs

DSSCs	jsc (mA/cm2)	Voc (mV)	FF (%)	η (%)
A: (No PbS nanocrystals)	15.14	739	54.66	6.11
B: Sprayed with 22 mM of $Pb(S_2COEt)_2$	17.98	778	56.68	7.90
C: Sprayed with 44 mM of $Pb(S_2COEt)_2$	14.47	800	63.41	7.39
D:Sprayed with 66 mM of $Pb(S_2COEt)_2$	14.48	794	61.42	7.06
E:Sprayed with 110 mM of $Pb(S_2COEt)_2$	11.00	695	67.17	5.12

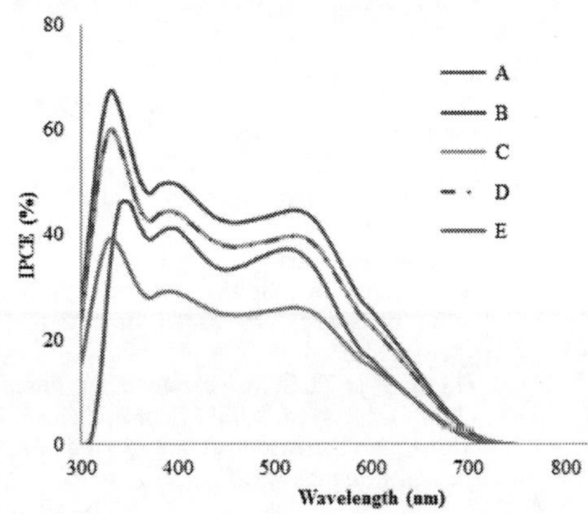

Figure-5. IPCE spectra of the DSSCs

- Increasing PbS concentration negatively impacted the power conversion efficiency of the cells.
- This was attributed to the decrease of transparency (reducing incident light) and agglomeration (can hamper charge collection) of the nanocrystals.
- Incident photo to current conversion efficiency (IPCE) was measured as shown in Figure-5. Higher IPCE value is related to the higher light harvesting by the anode of the cell.

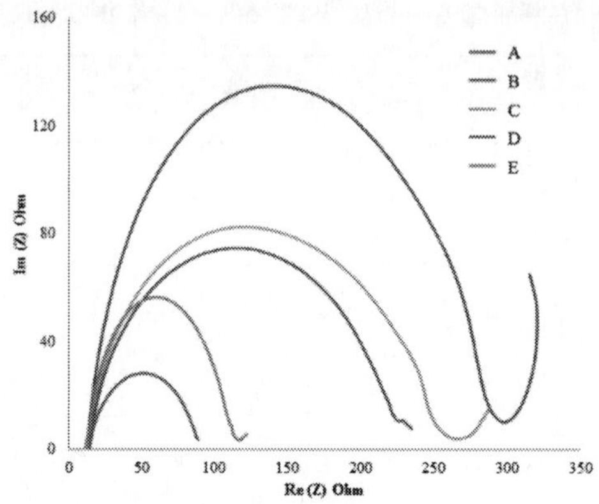

Figure-6. EIS investigation of the DSSCs

- Electrochemical impedance spectroscopy (EIS) of the cells was measured and present in Figure-6.
- All the cells showed large arcs of charge transport resistance (R_3) at the photoanode/electrolyte interface.
- This indicates lower charge recombination and better cell performance. Here, cell 'B' gave highest R_3 value of 288 Ω.

4.1. Conclusion

- A simple air spray technique has been successfully used to deposit PbS crystals on the FTO/TiO_2 surfaces.
- This method is successfully used for the first time to co-sensitize DSSC cells for improving their overall PCEs.
- It is observed that the absorption was decreasing with increasing precursor concentration on photoanodes.
- The highest PCE was obtained for sample B (7.90%) (sprayed with 22-mM $Pb(S_2COEt)_2$, with a 29% increase than that of sample A, i.e., the cell without any PbS nanocrystals

5. AKNOWLEDGEMNT

Amir Al-Ahmed acknowledge the King Abdulaziz City for Science and Technology (KACST) for funding through project No. 12-ENE3204-04 via the Science & Technology Unit at King Fahd University of Petroleum & Minerals.

6. REFERECNES

[1] Umer Mehmood, Amir Al-Ahmed, Mohammad Afzaal, Abbas S. Hakeem, Shamsuddeen A. Haladu, and Fahad A. Al-Sulaiman, IEEE JOURNAL OF PHOTOVOLTAICS, 8 (2) (2018) 512-516

[2][1] A. Hagfeldt, G. Boschloo, L. Sun, L. Kloo, H. Pettersson, Chem. Rev., 110 (2010) 6595–663.

[3][4] H. S. ElBatal, S. Aghazada, S. A. Al-Muhtaseb, M. Gr¨atzel, M. K. Nazeeruddin, Energy Technol., 4 (2016) 380–384.

[4][6] U. Mehmood, S. Ahmed, I. A. Hussein, K. Harrabi, Electrochim. Acta, 173 (2015) 607–612.

[5][7] C. Wang et al., Nano Energy, 1 (2012) 440–447.

[6] [21] J. M. Clark et al., Dalton Trans., 40 (2011) 6893–6900.

[7][15] Y. Liu, J. Wang, Thin Solid Films., 518 (2010) e54–e56.

[8][18] J. B. Sambur, T. Novet, B. A. Parkinson, Science, 330 (2010) 63–66.

[9] [10] P. J. Thomas, D. Fan, P. O'Brien, J. Colloid Interface Sci., 354 (2011) 210–218.

[10][25] Y.-L. Xie, Z.-X. Li, Z.-G. Xu, H.-L. Zhang, Electrochem. Commun., 13 (2011) 788–791.

[11][26] S. Li et al., Int. J. Minerals,Metall.,Mater., 17 (2010) 92–97.

THE EFFECT OF POTASSIUM DOPING ON PEROVSKITE SOLAR CELLS PERFORMANCE AND STABILITY

Marina Vildanova[1], Anna Nikolskaia[1], Sergey Kozlov[1], Oleg Shevaleevskiy[1], Liudmila Larina[1,2]
[1]Solar Photovoltaic Laboratory, Institute of Biochemical Physics, Russian Academy of Sciences
119334, Kosygin St., 4, Moscow, Russia
[2]Department of Materials Science and Engineering, Korea Advanced Institute of Science and Technology
291 Daehak-ro, Yuseong-gu, Daejeon 34141, Republic of Korea

ABSTRACT: We have fabricated a series of air-processed potassium-doped (K-doped) perovskite solar cells (PSCs). The structural and optoelectronic parameters of K-doped perovskite layers were investigated. A comparative study of PV characteristics for the PSCs based on pristine and K-doped perovskite layers was provided under varying illumination conditions. The long-term and thermal stability testing of the PSCs fabricated were performed. As a result, we have shown that the incorporation of the additive metal ions into perovskite structure improved the PV performance and operational stability of the PSCs.
Keywords: perovskite, doping, performance, degradation

1 INTRODUCTION

Organic-inorganic lead halide perovskites are of utmost interest for development of next generation cost-effective high-efficiency solar cells. Perovskite family is generally adopting the ABX_3 structure, where X is an anion (Br^-, I^-, Cl^-), A ($CH_3NH_3^+$, $HC(NH_2)_2^+$, Cs^+) and B (Pb^{2+}, Sn^{2+}) are cations of different sizes. So far, thin-film perovskite solar cells (PSCs) demonstrate significant progress in the photoconversion efficiency and also have numerous of advantages compared to conventional silicon-based solar cells [1].

The PSC's performance is determined by strong optical absorption of the perovskite layer in a visible range of solar spectrum. Worth noting, that absorption characteristics strongly depend on the deposition technique and the quality of the crystal structure [2]. On the other hand, high sensitivity of the perovskite layer to atmospheric humidity, UV radiation and high temperatures results in low stability and rapid degradation of PSCs [3]. Thus, due to the atmospheric influence the hydrogen iodide (HI) and methylamine (CH_3NH_2) is gradually removed from the perovskite lattice converting them back into PbI_2 (see Fig. 1).

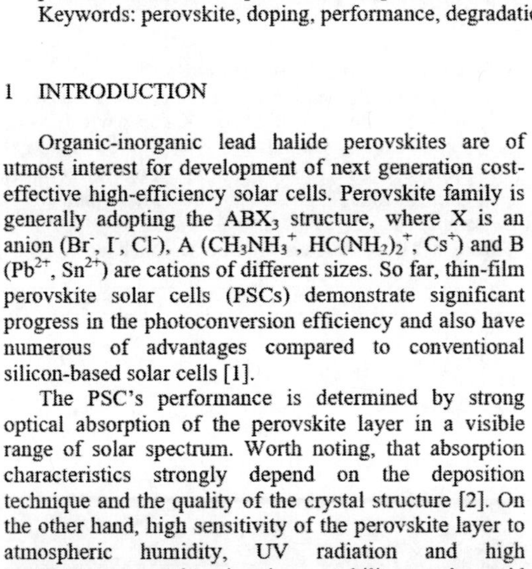

Figure 1: Moisture disruption of the perovskite structure [4]

In order to optimize the perovskite layer, a number of studies have been carried out aimed at increasing the crystallinity, uniformity and quality of the films. The structure, morphology and the optoelectronic properties of perovskite layers can be adjusted. Thus, the combinations of new solvents and dopants have been studied to improve the performance and stability of PSCs: introduction of various metal ions, use of mixed solvents, lead salt precursors, ionic liquids and various additives [5-6]. There are many studies reported on the ion doping in A-, B- and X-sites of the system, particularly for Pb^{2+} replacement in perovskite films [7-8]. It is known so far that the introduction of alkali metals results in the increase of the oxidation resistance in perovskite layer. It also reduces grain boundary recombination and trap states density in the crystal structure [9-10] leading to further improvement of the solar cell performance.

2 EXPERIMENTAL

PSC fabrication process is described in details elsewhere [11]. All the procedures were carried out under ambient conditions and at a relative humidity of 40-50%. K-doping of $CH_3NH_3PbI_3$ was provided by adding 0.02M KI to 1.2M perovskite precursor solution. The image of the PSC lab samples fabricated is shown in Fig. 2.

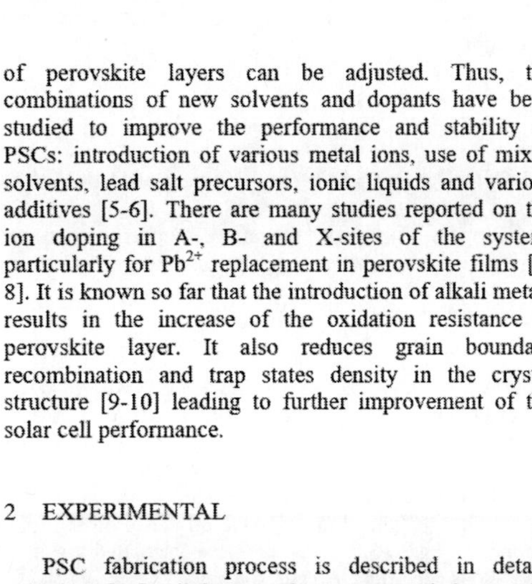

Figure 2: The images of K-doped PSCs

Optical characteristics of the perovskite layers were studied using UV-vis spectrophotometer Shimadzu 3600 with the ISR-3100 integrating sphere (Japan). The structure and morphology was obtained using field-emission scanning electron microscopy (FE-SEM) using Hitachi SU8000. IPCE spectra (incident photon-to-current conversion efficiency) were obtained using QEX10 Solar Cell Quantum Efficiency Measurement System (PV Measurements, USA) in a wavelength range 300-850 nm. PSCs performance was studied under simulated sunlight illumination (AM1.5G, 1000 W/m^2) using the Abet Technologies Solar Simulator (USA). J-V curves were recorded using Keithley 4200-SCS Semiconductor Characterization System (USA) in both forward and reverse scan modes. Neutral-density filters were used to perform the characterization of the PSCs at varying illumination intensities ranging from 10 to 1000 W/m^2.

Power conversion efficiency η (%) was calculated according to the known formula:

$$\eta = \frac{J_{SC} \cdot V_{OC} \cdot FF}{P_{IN}} \cdot 100\%$$

where J_{SC} is a short circuit current density, V_{OC} is an open circuit voltage, FF is a fill factor and P_{IN} is an incident light intensity.

Stability tests of PSCs were conducted in air by measuring the J-V characteristics, while the samples were stored in the dark for 45 days. The thermal stability studies were carried out at 60°C by keeping the samples in the oven (Binder, Germany) for 24 hours under dark conditions.

3 RESULTS AND DISCUSSION

3.1 Optoelectronic and structural characterization of K-doped perovskite absorber layers

A major structural advantage appears when the partial substitution at the B-site affects the Goldschmidt's tolerance factor [12].

$$t = \frac{RA + RX}{\sqrt{2}(R_B + R_X)}$$

Here A is $CH_3NH_3^+$ (R_A = 2.71 Å), B is K^+ (R_B = 1.38 Å) or Pb^{2+} (R_B = 1.19 Å), X is I^- (R_X = 2.20 Å). While K^+ is a larger cation compared to Pb^{2+}, it leads to the decrease of t and octahedral crystal structure.

The optoelectronic characterization of the perovskite layers was performed to study the influence of K-doping on the band gap (E_g). E_g values were calculated for direct transitions according to Tauc's law.

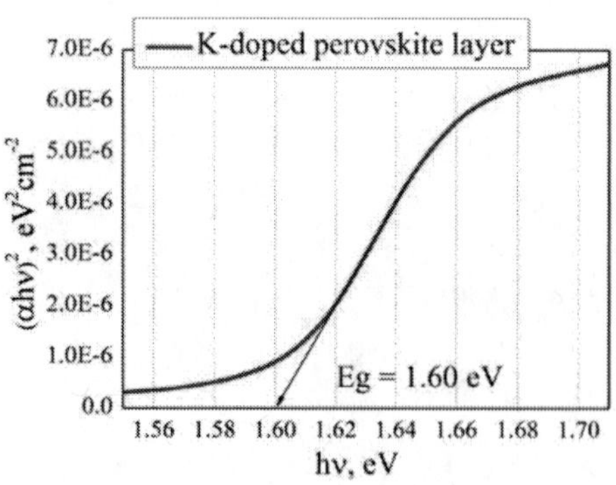

Figure 3: Plot of $(\alpha h\nu)^2$ versus photon energy for K-doped perovskite layer

The band gap for K-doped $CH_3NH_3PbI_3$ was found to be 1.60 eV. Due to a low K-doping level there were negligible changes observed in the absorption coefficient spectra compared to pristine $CH_3NH_3PbI_3$ (E_g = 1.59 eV).

Morphological characteristics based of the K-doped perovskite layer were provided to estimate the film quality. As a result, K^+ additives have shown to make a positive effect on perovskite crystal growth and layer uniformity. The K-doped perovskite film of high coverage is observed in the SEM image given at Fig. 4. The average grain size was estimated to be in a range up to 400 nm.

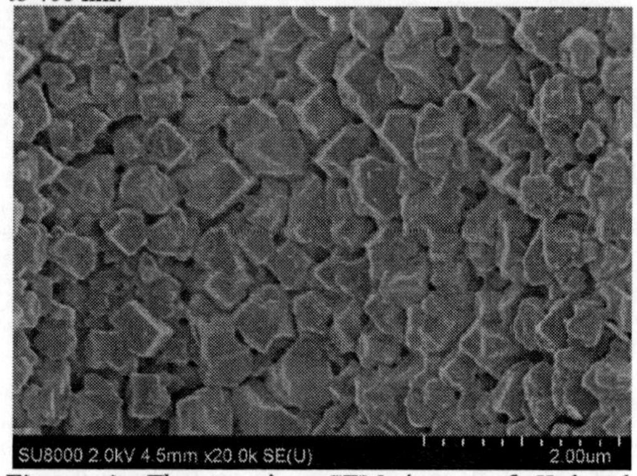

Figure 4: The top-view SEM image of K-doped perovskite film

3.2 Characterization of K-doped PSCs

Comparative studies of J-V characteristics for completed PV devices based on K-doped and pristine perovskite (control sample) are displayed in Fig. 5a. It was observed that K-doping has a positive effect on the PSC performance. The key photovoltaic parameters of the samples fabricated are listed in Table 1.

Table I: PV parameters for the K-doped and pristine PSCs

	PSC	K-doped PSC
V_{OC} [V]	0.920	0.922
J_{SC} [mAcm^{-2}]	17.39	19.03
FF [a.u.]	0.648	0.685
R_{sh} [Ωcm^2]	468.08	682.23
PCE [%]	10.37	12.02

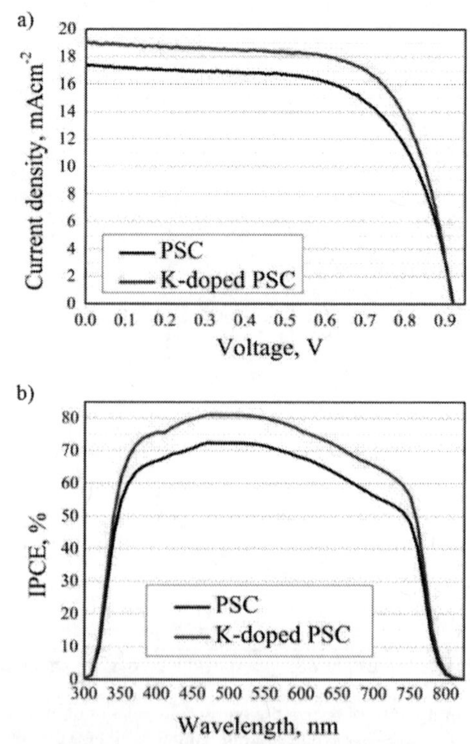

Figure 5: (a) J-V curves and (b) IPCE spectra for pristine (black) and K-doped (red) PSCs

Introduction of K^+ significantly increases the FF value from 0.648 to 0.685. This result is caused by changes in the layer morphology, which increased the shunt resistance and retarded the charge recombination. As a result in K-doped samples J_{SC} increases from 17.39 mA/cm^2 to 19.03 mA/cm^2 thus increasing the photoconversion efficiency by up to 15% in comparison with undoped samples. The photovoltaic parameters obtained are in agreement with the data presented elsewhere [9-10].

A positive effect of K^+ incorporation is also confirmed by the results of IPCE measurements of PSCs. The appropriate spectra are shown in Fig. 5b. It is seen that K-doping enhances IPCE of the PV device at the mostly important range of wavelengths between 550 and 800 nm.

3.2 Low light performance of K-doped PSCs

The performance of both K-doped PSCs and control series was studied under varying illumination conditions from 10 to 1000 W/m^2. Comparative PCE trends for both pristine and K-doped PSCs are plotted in Fig. 6a. As a result, K-doped PSCs have shown better performance at high illumination intensities, while the decrease of intensity level to 10 W/m^2 resulted in a slight drop in PCE.

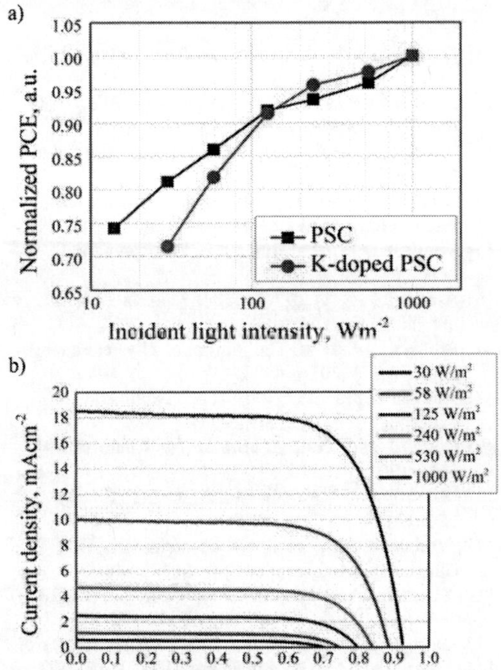

Figure 6: (a) Comparative PCE trends for both pristine and K-doped PSCs and (b) J-V curves of K-doped PSCs obtained under varying illumination conditions from 10 to 1000 W/m^2

J_{SC} of K-doped PSCs shows a linear dependence on light intensity as seen in J-V curves (Fig. 6b); V_{OC} decrease can be explained by the decrease of the Fermi level due to the fewer photoelectrones being generated and injected; FF increase is related to the retarded charge recombination under low-light conditions.

3.3 K-doped PSCs operational stability tests

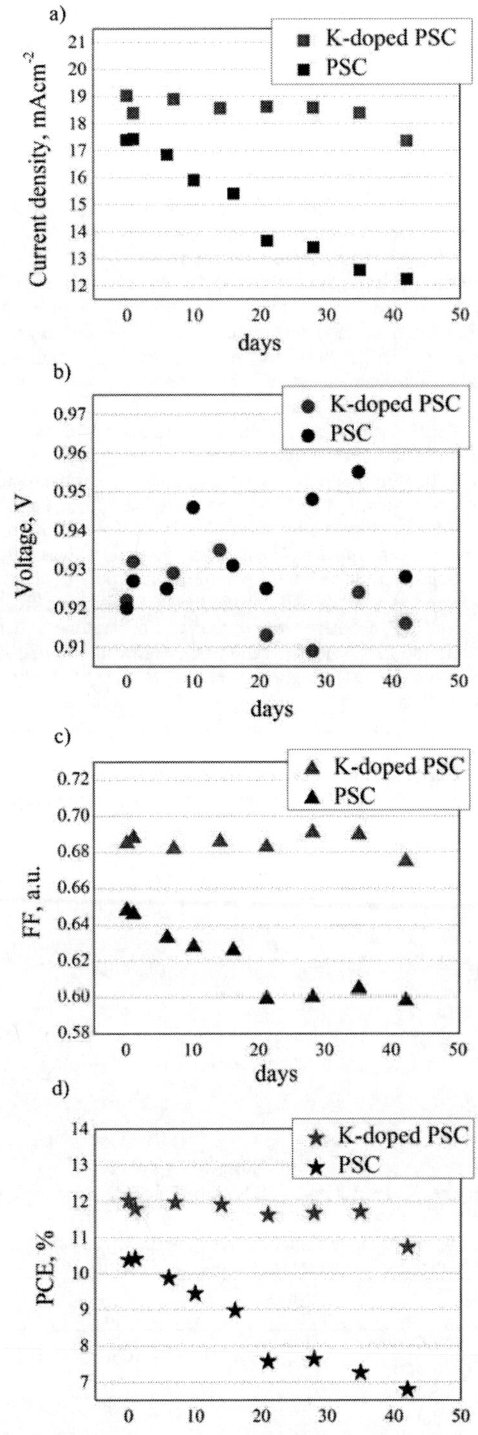

Figure 7: Trends of the photovoltaic parameters of K-doped and pristine PSCs while storing in the air: J_{SC} (a), V_{OC} (b), FF (c), PCE (d)

To observe the degradation of PSCs and the effect of K-doping on the device stability, the stability tests were performed. The first series of both K-doped and pristine PSC samples was exposed to normal temperature and pressure conditions (NTP) in the dark with the average humidity of 50%. The trends of photovoltaic performance of that series are summarized in Fig. 7.

It is observed that K-doping significantly increases the stability of PSCs. The average values of FF and J_{SC} for the exposed K-doped PSCs remain stable in comparison to the pristine PSCs. At the same time the V_{OC} value was slightly decreased. Thus, the drop in the PCE of K-doped PSCs was found to be about 10% compared to 30% for control samples.

Discussion has ensued on the complex effect of the external factors. Thus, the sample exposure under low humidity conditions doesn't result in a significant degradation of PSC performance. The effect of oxygen on perovskite structural properties of perovskite appeared under illumination only, thus initiating the photo-oxidation processes in PSCs [13].

On the other hand, temperature is known to have a critical effect on the stability of PSCs [14]. In order to study thermal degradation mechanisms, an exposure to higher temperature of 60°C was performed for a period of 24 hours. Fig. 3 shows the influence of temperature on the PSCs performance. The degradation at 60°C was very rapid probably due to the well-known problem of Au diffusion. A sharp initial drop in the PCE (resulting from J_{SC} and FF decrease) was observed for pristine PSCs during the first 6 hours of the test. Meanwhile, K-doped PSEs demonstrated higher operation stability and only a slight decrease in the main photovoltaic parameters. The observed result confirmed the data of our measurements provided earlier.

The PCE trends of the thermal exposed series are shown in Fig. 8. The appropriate data are listed in Table 2.

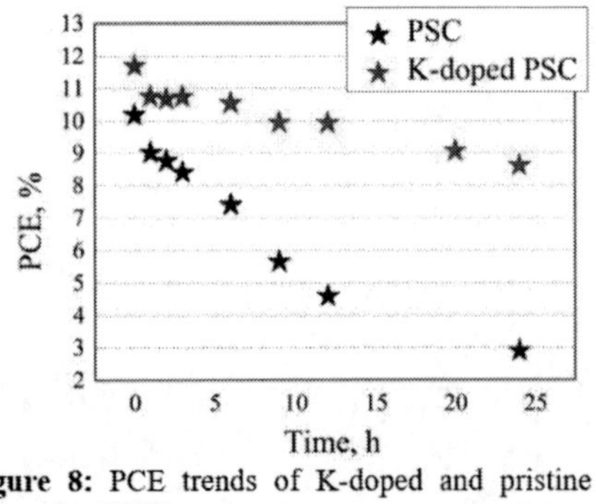

Figure 8: PCE trends of K-doped and pristine PSCs exposed to 60°C

Table II: Average initial and aged (exposed to 60°C for 24 hours) PV parameters of K-doped and pristine PSCs

	J_{SC}, mA/cm^2	V_{OC}, V	FF, a.u.	η, %
PSC (0 hours)	17.37	0.90	0.65	10.17
PSC (24 hours)	7.76	0.83	0.44	2.84
K-doped PSC (0 hours)	18.83	0.92	0.67	11.64
K-doped PSC (24 hours)	16.51	0.90	0.58	8.62

4 CONCLUSIONS

We have fabricated a series of perovskite solar cells with a low level of potassium-doping and studied the effects of K-doping on the structure of $CH_3NH_3PbI_3$ film and PV performance of the solar cells. The structure investigations have shown that the alkai doping results in the rearrangement of perovskite layer morphology, reducing grain boundaries and trap states, thus, retarding surface recombination processes. A study of PV performance of K-doped PSCs has shown that K-doped samples exhibited higher J_{SC}, FF and PCE compared to the control series of undoped PSCs. The improvement of PSC parameters was also confirmed by IPCE measurements and the studies of photovoltaic parameters behavior under varying illumination conditions. The results of the aging tests have shown that the incorporation of the additive metal ions into perovskite structure has a positive effect in terms of PSCs operation stability. We have shown that K-doped PSCs exhibit improved stability both in air and when exposed to higher temperatures.

5 ACKNOWLEDGMENTS

The work is supported by the Russian Science Foundation under grant no. 17-19-01776.

6 REFERENCES

[1] Green M. A. et al. Progress in Photovoltaics 26(7) (2018)
[2] Song Z. et al. Journal of Photonics for Energy 6(2) (2016) 022001.
[3] Berhe T. A. et al. Energy & Environmental Science 9(2) (2016) 323.
[4] Strachala D. et al. Proceedings 33rd European Photovoltaic Solar Energy Conference (2017) 1151.
[5] Wang J. T. W. et al. Energy & Environmental Science 9(9) (2016) 2892
[6] Aldibaja F. K. et al. Journal of Materials Chemistry A 3(17) (2015) 9194.
[7] Frolova L. A. et al. The journal of physical chemistry letters 7(21) (2016) 4353.
[8] Abdi-Jalebi M. et al. Journal of visualized experiments: JoVE 121 (2017)
[9] Chang J. et al. Journal of Materials Chemistry A 4(42) (2016) 16546.
[10] Zhao W. et al. Advanced Science 5(2) (2018) 1700131.
[11] Vildanova M. F. et al. Nanosystems: Physics, Chemistry, Mathematics 8(4) (2017) 540.
[12] Swarnkar A. et al. ASC Energy Letters 3(2) (2018) 286.
[13] Domanski K. et al. Nature Energy 3(1) (2018) 61.
[14] Divitini G. et al. Nature Energy 1 (2016) 15012.

SIMULATION STUDY OF TANDEMS OF PEROVSKITE AND IBC c-SI SOLAR CELLS

S. Silvestre[1*], E. Mas-Marzá[2], J. Puigdollers[1], F. Fabregat-Santiago[2], P. Ortega[1], D. Serra[1].
[1] MNT Group, Electronic Engineering Department, Universitat Politécnica de Catalunya (UPC) BarcelonaTech
C/Jordi Girona 1-3,Campus Nord UPC, 08034 Barcelona, Spain.
Tel. +34 93 4017491, email : Santiago.silvestre@upc.edu
[2] Institute of Advanced Materials (INAM), Universitat Jaume I
12006 Castelló, Spain

ABSTRACT: In this work, the simulation results obtained for a tandem configuration of perovskite, as top cell, and Interdigitated Back-Contacted (IBC) c-Si as bottom cell are presented. Two different models were used for the modelling and simulation of the perovskite and the IBC c-Si solar cells. The fabrication and characterization processes are also summarized in the experimental section. The models were validated by comparison of simulated current-voltage characteristics and real data measured in two fabricated devices. The best efficiency obtained in this study is 25% for a tandem device with the top and bottom solar cells connected in series.
Keywords: Perovskite, IBC c-Si, tandems, simulation.

1 INTRODUCTION

Perovskite solar cells have shown great potential to achieve high-efficiency solar energy conversion at low cost. Moreover, tandem junctions fabricated by placing a wide–band-gap top cell above a silicon (Si) bottom cell have shown very high efficiencies in recent works reported in the literature [1-2]. Perovskite solar cells are promising high-efficiency top cell candidates in perovskite/silicon and perovskite/CIGS tandem cells [1-4].

Accurate models simulating the electrical characteristics of perovskite solar cells help to achieve optimal designs and to improve their characteristics and performance. In addition, simulation studies of tandem configurations based on these models allow obtaining the maximum values of expected efficiencies in several scenarios.

2 EXPERIMENTAL

2.1 Fabricacion Process of the Perovskite and IBC c-Si solar cells.

The details about the fabrication process of the perovskite solar cell are provided in a work previously reported in the literature [5].

IBC c-Si solar cells having an active area of 9 cm^2, were manufactured using high quality FZ <100> c-Si(p) 4" wafers with resistivity and thickness of 2.5±0.3 Ωcm and 280±20 μm respectively.

Cells were fabricated considering next main technological features, namely:

1) Boron and phosphorous diffusions to form p+ (base contacts) and n+ (emitter regions) regions respectively, which are patterned using standard photolithography.

2) A front surface textured with random pyramids and passivated with 90nm ALD Al_2O_3 films, resulting reflectance values below 0.5% at wavelengths λ~600 nm with effective surface recombination velocities below 3 cm/s (@ 1Sun).

3) A back reflector scheme consisting of a thermal SiO_2 (110 nm)/Al (~2 μm) stack was included at the rear sideScanning Electron Microscopy was ~200 nm.

2.2. Solar cells characterization.

Table I shows main electrical parameters of the perovskite and IBC c-Si solar cells measured at STC, including the conversion efficiencies, η, and the coordinates of the maximum power points: *Vm* and *Jm*.

Figures 1 and 2 show the EQE and the I-V characteristics of the perovskite solar cell respectively.

Table I: Solar cells main parameters

Parameter	Perovskite	IBCc-Si
Jsc (mA/cm^2)	17.97	40.95
Voc (V)	1.034	0.646
Vm (V)	0.8	0.532
Jm (mA/cm^2)	15.24	38.11
η (%)	12.15	20.27

The External Quantum Efficiency (EQE) of the perovskite solar cell was performed using a Xenon lamp power source coupled with a monochromator controlled by a computer; the photocurrent was measured using an optical power meter 70310 by Oriel Instruments.

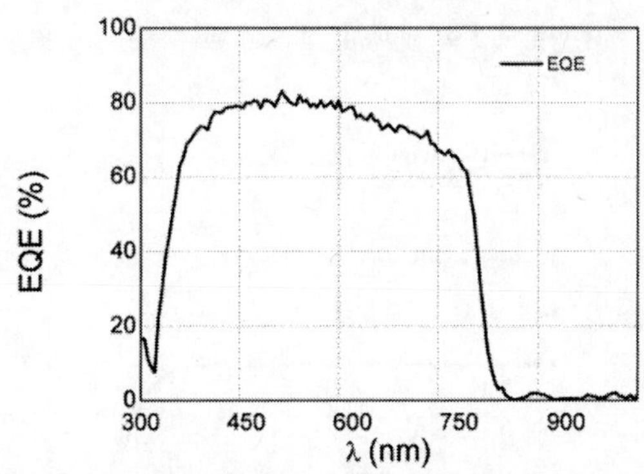

Figure 1: External Quantum Efficiency.

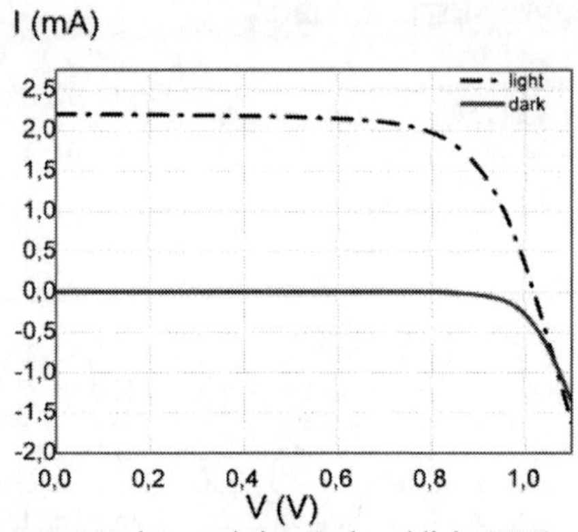

Figure 2: I-V characteristics: Dark and light (STC).

Current-voltage (I-V) curves were performed under 1 sun illumination (100 $mWcm^{-2}$) using a xenon arc lamp simulator (Sun 2000, ABET Technologies) with an AM 1.5G spectral filter and a Keithley 2400, previously calibrated. All measurements were performed with an opaque mask of 0.112 cm^2 and without encapsulation.

The external quantum efficiency obtained for the IBSC c-Si solar cell is given in Fig. 3. The J-V characteristics of the cell in dark and light (STC) conditions are shown in Fig. 4.

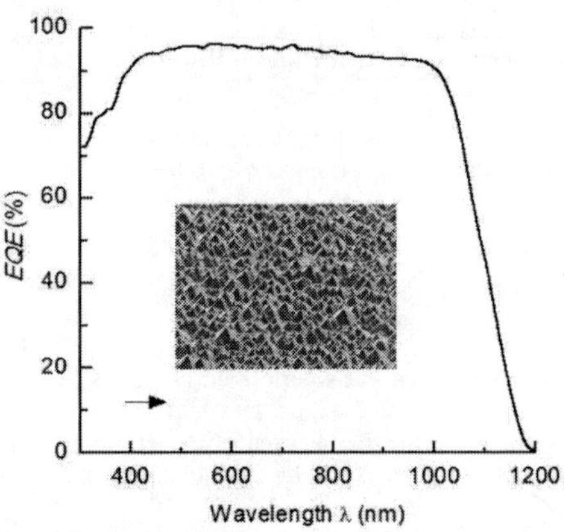

Figure 3: EQE of the IBSC c_Si solar cell.

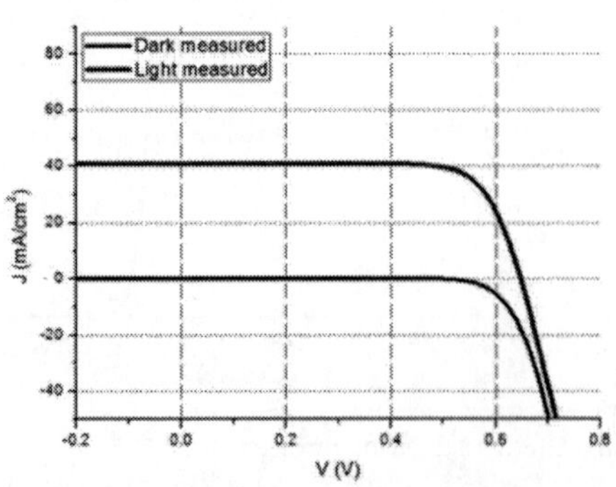

Figure 4: J-V characteristics: Dark and light (STC).

3 SIMULATION MODELS

The model considered for the simulations of the perovskite solar cell is the analytical model introduced by Sun et al. [6]. The equations used in the model to calculate the I-V characteristics of the perovskite solar cell are the following:

$$J_{dark} = (\alpha_f J_{of} + \alpha_b J_{ob})\left[e^{\left(\frac{V}{Vt}\right)} - 1\right] \quad (1)$$

$$J_{photo} = q\, G_{max}[A - B\ e^{-m}] \quad (2)$$

$$J_{light} = J_{dark} + J_{photo} \quad (3)$$

where G_{max} is the maximum absorption, for the perovskite solar cell included in this study qG_{max} = 19.5 mA/cm^2, Vt is the thermal voltage, J_{of} and J_{ob} are the dark diode current densities recombining at the front and back transport layers respectively, J_{photo} is the photocurrent density, J_{dark} is the dark current and J_{light} the solar cell output current density.

Table II: Model parameters

Parameter	Equations		
α_f	$\frac{V'}{e^{V'}-1}+\frac{1}{\beta_f}$	$V' = \frac{(V-V_{bi})}{Vt}$	$\beta_f = \frac{D}{to\, s_f}$
α_b	$\frac{V'}{e^{V'}-1}+\frac{1}{\beta_b}$		$\beta_b = \frac{D}{to\, s_b}$
A	$\alpha_f\left(\frac{1-e^{V'-m}}{V'-m}-\beta_f\right)$		$m = \frac{t_o}{\lambda_{ave}}$
B	$\alpha_b\left(\frac{1-e^{V'+m}}{V'+m}-\beta_b\right)$		$\lambda_{ave} = 100\ nm$

The model parameters: A, B, α_f, α_b and m, are functions of the physical parameters of the solar cell as the build-in potential, V_{bi}, the diffusion coefficient, D, the effective surface recombination velocities at the front, S_f, and back, S_b, interfaces respectively, and the thickness of the absorber layer, t_o, [6-7]. These model parameteres are described in Table II.

The model selected for the simulations of the IBC c-Si solar cell is the well-known: Five-parameter or one diode model, where the relation between the output current and voltage is defined as follows:

$$J = J_{ph} - Jo\left[exp\left(\frac{V+R_sAJ}{nVt}\right) - 1\right] - \frac{V+R_sAJ}{R_{sh}} \quad (4)$$

where: J_{ph} is the photocurrent density, Jo is the diode reverse saturation current density, n is the diode ideality factor, R_s and R_{sh} are the series and shunt resistances respectively, Vt is the thermal voltage, J and V are the output current density and voltage and A is the cell area.

Table III: Model parameters used in simulations.

Perovskite solar cell Sun Model		IBC c_Si solar cell Five Parameter model	
Vbi	*1.01 V*	*n*	*1.12*
to	*250 nm*	*Rs*	*1.1Ω*
Sf	*1.65 10⁴ cm/s*	*Rsh*	*4 103 Ω*
Sb	*67.49 cm/s*	*Jo*	*3.1 10⁻¹⁰ mA/cm²*
J_{of}	*3.44 10⁻¹⁷ mA/cm²*	J_{ph}	*40 mA/cm²*
J_{ob}	*3.42 10⁻¹⁷ mA/cm²*		

Table III shows the values of the solar cells model parameters used in the simulations of the J-V characteristics of the perovskite and IBC c-Si solar cells. The model parameters of the IBC c-Si solar cell were directly evaluated from the J-V characteristics measured in dark and STC irradiance conditions, whereas model parameters of the perovskite solar cell were calculated by using parameter extraction techniques [8].

4 RESULTS

Both simulation models offer a good agreement between the measured J-V characteristics of the solar cells and the simulation results as can be seen in Figs. 5 and 6, where the J-V characteristics in dark and light conditions (STC) are presented for both devices.

The root-mean-square error (RMSE) between the measured and simulated current densities is given in Table IV.

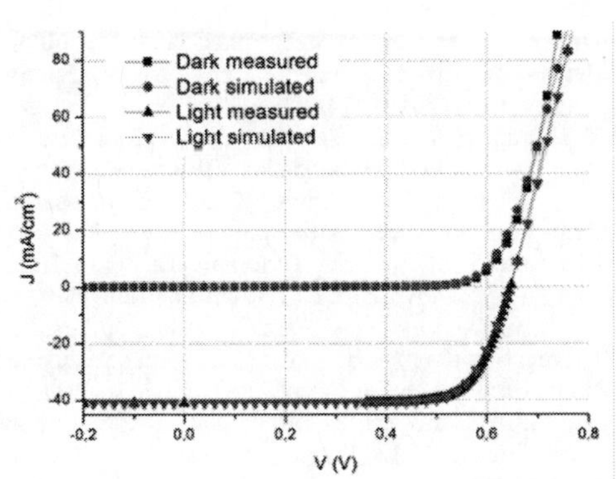

Figure 5: J-V characteristics of the IBC c-Si solar cell.

Figure 7 shows the expected short circuit current density of the perovskite cell, Jsc, as a function of the thickness of the absorber layer, to, and the recombination velocity at the front, Sf. As it can be seen in Fig. 6, the highest values of Jsc are obtained for thickness of the absorber layer in the range of 500nm < to < 700nm and front recombination velocities below 2 10^3 cm/s.

The electrical model of the tandem having the perovskite as top cell and the IBSC c-Si as bottom cell, used in the simulations is given by Fig. 8.

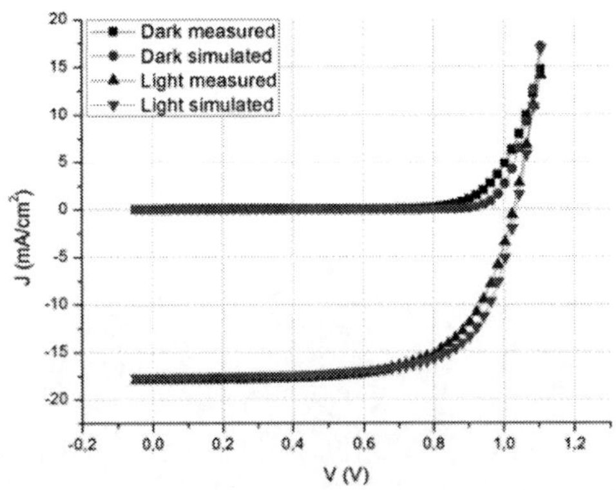

Figure 6: J–V characteristics of the Perovskite solar cell.

Table IV: RMSE obtained for the currents

Solar Cell	RMSE (mA/cm²) dark conditions	RMSE (mA/cm²) light conditions
Perovskite	0.73	0.32
IBC c-Si	2.5	0.58

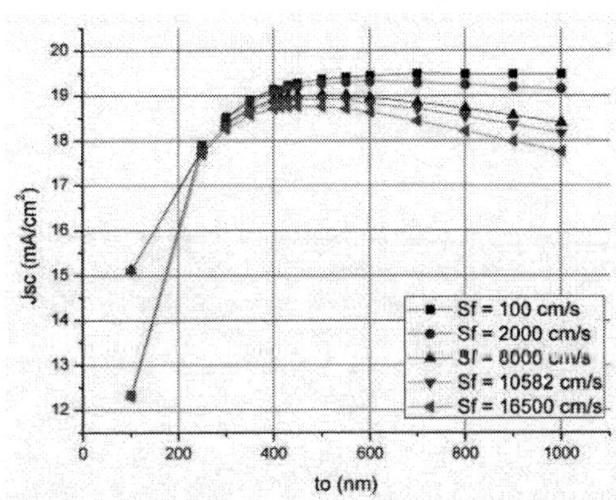

Figure 7: Short circuit current density of the perovskite cell.

Taking into account that the performance of the tandem device is limited by the cell with the lowest current because of the current matching, several values of short-circuit currents were considered for the IBC c-Si bottom cell in order to cover different absorption conditions for the perovskite solar cell. Fig. 9 shows results obtained for the efficiency in the simulations of the tandem configuration.

As it can be seen in Fig. 9, the tandem in series connection shows efficiencies up to 25% for a short-circuit current of the IBC c-Si solar cell of 25 mA. This value corresponds to a total absorption of the perovskite top cell for wavelengths below 600nm. Moreover, the efficiency of the tandem configuration shows a linear behaviour as a function of the short circuit current of the IBC c-Si bottom cell in the range of currents where the current limitation of the tandem is fixed by the IBC c-Si bottom cell. When the current limitation pass to be imposed by the top cell, for Jsc values of the bottom cell above 18 mA, the trend of the tandem efficiency becomes logarithmic. The bottom cell can no longer work at constant voltage conditions and the behaviour of voltage is logarithmic. This fact influences the power and efficiency trends, making them logarithm-shaped as well.

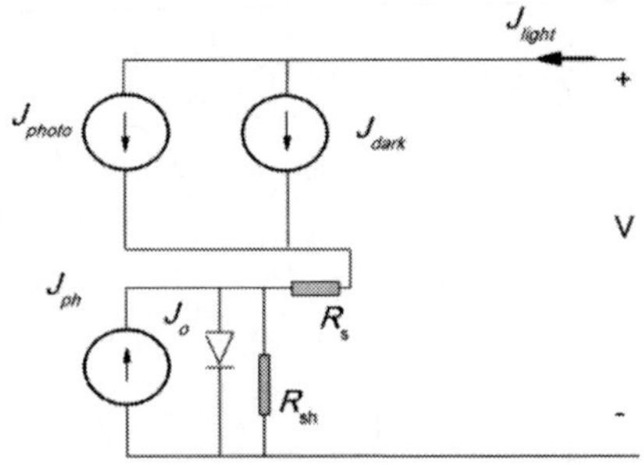

Figure 8: Tandem configuration in series connection.

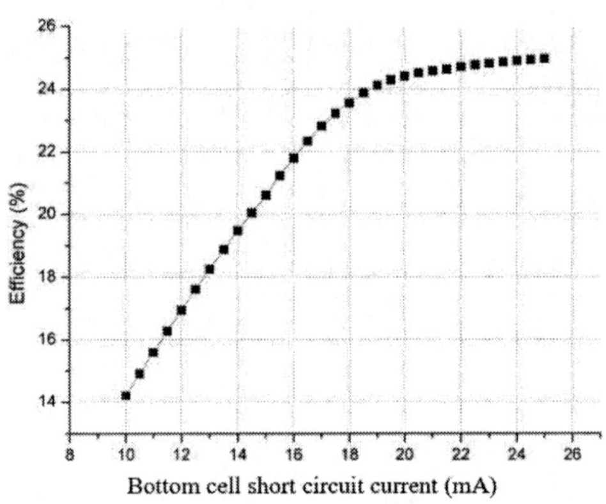

Figure 9: Efficiencies for the tandem configuration obtained in the simulations.

The current of the bottom cell at the maximum power point, I_{MPP}, neglecting the effects of the shunt and series resistance, can be written as follows:

$$I_{MPP} = I_{ph} - I_o\left(e^{\frac{V_{MPP}}{nV_T}} - 1\right) \quad (5)$$

From the above equation, the voltage at the maximum power point, V_{MMP}, is given by the following equation:

$$V_{MPP} = nV_T \cdot \ln\left(1 + \frac{I_{ph} - I_{MPP}}{I_o}\right) \quad (6)$$

As I_{ph} increases linearly and I_{MPP} in a similar way, while all the other parameters remain constant, the behaviour of the maximum power point voltage is logarithmic.

5 CONCLUSIONS

Simulations of the I-V characteristics at STC of a tandem configuration with an active area of 1 cm^2 having the perovskite as the top cell with the IBC c-Si bottom cell, were carried out. Then the efficiencies were evaluated. The top cell and the bottom cell were electrically connected in series. Results obtained in the simulations show efficiencies up to 25% for a short-circuit current of the IBC c-Si solar cell of 25 mA. This value corresponds to a total absorption of the perovskite top cell for wavelengths below 600nm and a bandgap value of the perovskite cell in the range of 1.7-1.8 eV in order to maximise the absorption of the bottom cell.

ACKNOWLEDGMENT

This work has been funded by the Spanish Ministry of Economy and Competitiveness under projects number: ENE2014-56237-C4-2-R.

6 REFERENCES

[1] D.P. McMeekin, G.Sadoughi, W. Rehman, G.E. Eperon, M. Saliba, M.T. Hörantner, A. Haghighirad, N. Sakai, L. Korte, B. Rech, M. B. Johnston, L. M. Herz and H.J. Snaith. A mixed-cation lead mixed-halide perovskite absorber for tandem solar cells. Science 351, no. 6269 (2016), 151-155.

[2] C.D. Bailie, M. G. Christoforo, J. P. Mailoa, A.R. Bowring, E.L. Unger, W.H. Nguyen, J. Burschka, N. Pellet, J. Z. Lee, M. Gratzel, R. Noufi, T. Buonassisi, A. Salleo and M.D. McGehee. Semi-transparent perovskite solar cells for tandems with silicon and CIGS. Energy Environ. Sci. 8 (2015) 956-963.

[3] J. Werner, L. Barraud, A. Arnaud , M. Bräuninger, F. Sahli, D. Sacchetto, N. Tétreault, B. Paviet-Salomon , S.J. Moon, C. Allebé, M. Despeisse, S. Nicolay, S. De Wolf, B. Niesen and C. Ballif. Efficient Near-Infrared-Transparent Perovskite Solar Cells Enabling Direct Comparison of 4-Terminal and Monolithic Perovskite/Silicon Tandem Cells. ACS Energy Letters 1 (2) (2016) 474-480.

[4] T. Duong, N. Lal, D. Grant, D. Jacobs, P. Zheng, S. Rahman, H. Shen, M. Stocks, A. Blakers, K. Weber , T.P. White and Catchpole K.R. Semitransparent Perovskite Solar Cell With Sputtered Front and Rear Electrodes for a Four-Terminal Tandem. IEEE Journal of Photovoltaics 6 (3) (2016) 679-687.

[5] S. Silvestre, E. Mas-Marza, J. Puigdollers, F. Fabregat-Santiago, V. Alfonso. Comparison of simulation models for Perovskite solar cells. Proc. of the 33rd European Photovoltaic Solar Energy Conference and Exhibition, Amsterdam 2017. pp. 1141-1143.

[6] X. Sun, R. Asadpour, W. Nie, A.D. Mohite and M.A. Alam. A Physics-Based Analytical Model for Perovskite Solar Cells. IEEE Journal of Photovoltaics, 5 (5) (2015) 1389-1394.

[7] S. D. Stranks, G. E. Eperon, G. Grancini, C. Menelaou, M.J. Alcocer, T. Leijtens, L.M. Herz, A. Petrozza, and H.J. Snaith. Electron-hole diffusion lengths exceeding 1 micrometer in an organometal trihalide perovskite absorber. Science, 342 (6156) (2013) 341-344.

[8] S.Kichou, S. Silvestre, L. Guglielminotti, LL. Mora-López and E. Muñoz-Cerón. Comparison of two PV array models for the simulation of PV systems using five different algorithms for the parameters identification. Renewable Energy 99 (2016) 270-279.

EXPLORING THE USE OF COLLODION AS A BINDER IN SCREEN PRINTING VEHICLES FOR PEROVSKITE SOLAR CELLS METALLIZATION

C. Montes[1,2], L. Ocaña[1], L. De Sousa-Vieira[1], J.S. Moreno-Ramírez[1], S. González-Pérez[2], B. González-Díaz[2], M. Friend[1] & M. Cendagorta[1].
[1]Instituto Tecnológico y de Energías Renovables, S. A. (ITER)
Pol. Industrial de Granadilla, s/n, E 38600 Granadilla de Abona, Spain. Ph. +34 922 747 700. Fax +34 922 747 701. E-mail cmontes@iter.es
[2]Departamento de Ingeniería Industrial.
Avd. Astrofísico Francisco Sánchez, s/n, 38206 S/C de Tenerife. Spain. Ph. +34 922 316 502.

ABSTRACT: Industrial grade nitrocellulose and different kinds of collodion recipes have been produced in order to find a suitable perovskite-friendly ink. In this line, we fabricated a collodion complex with graphite (as the conductive powder) and using both Chlorobenzene and Toluene (as the solvents). The obtained collodion-based ink (or c-ink) was tested for depositing contacts over a glass substrate via screen printing. These contacts showed a high adhesion to the glass substrate with an average volume resistivity (ρ) of 0.061 Ω•cm after a curing process of 24 hours at room temperature. In order to evaluate whether the c-ink could be used to produce back contacts for perovskite solar cells, it was hand-spread over thin layers of perovskite, resulting in their degradation. This questions the usefulness of collodion as a stand-alone binder agent for producing perovskite-friendly inks. It is still pending to evaluate its use together with perovskite-friendly binders, such as epoxy resins.
Keywords: Back Contact, Screen Printing, Metallization, Electrical Properties, Perovskites.

1 INTRODUCTION

Screen printing is a mature, cost-effective and easily scalable metallization technique, widely used nowadays for photovoltaic (PV) cell manufacturing purposes. Thus, from the point of view of its industrial application, based upon its relative simplicity and low energy consumption requirements in order to produce metal contacts, this technique could become a better alternative to the thermal evaporation processes today in use, as long as inks compatible with perovskite substrates are available.

With this purpose in mind, the Instituto Tecnológico y de Energías Renovables (ITER), has been working on screen printing solutions for perovskite solar cell metallization since 2015, by initially testing different commercial pastes and inks. Although these tests were carried out with perovskite-friendly fabrication methods, the aqueous nature of the vehicles used in their formulation, produced a quick degradation in the finished devices [1].

In order to overcome this matter, the laboratory started developing several conductive ink formulas composed by epoxy resins and anhydrous solvents, as vehicles, along with different kinds of graphite powders, as the conductive material of choice. Optimal proportions of these three components were studied in order to obtain inks suitable for printing purposes. In particular, different solvents were analyzed with the aim to increase the quantity of graphite within the mixture and, subsequently, improve the conductivity of the printed contacts [2].

A step further was undertaken by modifying the ink formulations so as to solely use non-polar solvents, thus preventing the degradation of the perovskite substrates, together with metallic powders, such as aluminium, nickel and copper, as alternatives to graphite, due to their higher bulk conductivities and densities. However, the results obtained indicated that, although visually the integrity of the perovskite layers were indeed preserved, by simply replacing graphite powders with metallic ones, the printed contacts did not have lower volume resistivity, an issue that was attributed to an excess of epoxy resin in the ink formulations used [3].

Therefore, alternative strategies had to be undertaken in order to develop ink recipes that either require less epoxy resin, or use a different kind of binder altogether.

Collodion is a viscous solution that has been around since 1846, which is commonly used as vehicle in paints and varnishes for producing coatings with a high level of adhesion that dry quickly [4-5]. Normal or plain collodion can be obtained by dissolving cellulose nitrate in a mix of ethers and alcohols in approximately a 3:1 volume ratio, respectively [6-7]. On the other hand, ethers, in particular ethyl ether, are being used as an anti-solvent, that is, as a reactive that is applied during the washing treatment stage (one-step deposition method achieved via spin coating) in order to produce thin layers of perovskite [8-10]. Therefore, among the two kind of solvents required in the collodion recipe, the use of ethers (even in greater quantities) along with collodion as binder seem worth exploring, for producing perovskite-friendly screen printing inks.

2 EXPERIMENTAL

2.1 Cellulose nitration

Many factors determine the nitration reaction required to produce nitrocellulose from cellulose, ranging from the temperature at which the reaction occurs to the concentration of the acids involved, through the water content in the nitrating acids used [11].

The chemical reactions involved in the synthesis of nitrocellulose are:

Synthesis of the nitrating agent $NO2^+$

$$\mathbf{HNO_3 + 2\ H_2SO_4 \rightarrow NO_2^+ + H_3O^+ + 2\ HSO4^-}$$

Nitration of Cellulose

$$\mathbf{C_6H_7O_2(OH)_3 + 3\ NO_2^+ \rightarrow C_6H_7O_2(NO)_3 + 3H_2O}$$

The reactions were carried out in a beaker containing a mixture of concentrated Sulfuric Acid, (H2SO4, 98%) and concentrated Nitric Acid (HNO3, 70%), called "Nitrating Mixture" (see Table I), surrounded by an ice-water bath, in order to control the temperature of the reaction. When the acid mixture reached 21°C [12], a sample of 5 grams of commercial cotton, as Cellulose source, was immersed and left to soak into the reaction beaker. Transcurred 20 minutes, sample temperature dropped to 12 °C and the sample was carefully removed from the beaker. The acid residues (HSO4-) bonded to the nitrocellulose were neutralized by employing a washing sequenced procedure that alternated seven beakers, five of which had 1M sodium carbonate (Na2CO3) solution and two with deionized water (18MΩ·cm), while monitoring the pH after each washing. This phase finished when the pH of the solution was neutral. Finally, after removing the nitrocellulose from the beaker, the sample was tightly pressed down, spread as thinly as possible and introduced in a desiccator during 24 hours, in order to obtain a highly dehydrated product.

Table I: Mix of nitrating acids used for the synthesis of nitrocellulose

Sample	$\%H_2SO_4$	$\%HNO_3$	$\%H_2O$
50/50 (vol. %)	56.44	30.49	13.07

2.2 Preparation of the collodion

A systematical procedure was carried out following the original "Collodion" recipe [7], which uses cellulose nitrate (NC) (3-7 wt. %) in Ethyl Ether (DE) (65-75 wt. %) and ethanol (EtOH) (20-30 wt. %), many attempts were carried out in order to decrease the volume percent (Vol. %) of the DE and of the EtOH by employing two well-known perovskite anti-solvents: Chlorobenzene (CB) and Toluene [8,13], as well as Butyl Diglycol (BDG), which is an Glycol Ether widely used for producing inks where a low evaporation rate is required in order to balance the one of the DE [14].

Each collodion sample was produced by mixing 0.5 grams of nitrocellulose from a batch previously synthesized, together with the solvent mixture of choice, by vigorously shaking the mix in a covered beaker during 2 minutes.

Table II: Modified collodion recipes carried out in function of their Vol. %

Trial	DE	BDG	EtOH	CB	Toluene
Original[7]	75.50		24.5	-	-
P2	74.36	-	25.64	-	-
P3	37.66	24.68	24.68	-	-
P4	-	75	25.00	-	-
P6	-	37.66	24.68	-	37.66
P7		37.66	24.68	-	37.66
P8	-	75.00	-	25.00	-
P9	-	75	-	-	25
P10	-	74.03	12.99	12.99	-
P11	-	74.03	12.99	-	12.99
P12	37.66	37.66	-	24.68	-
P13	37.66	37.66	-	-	24.68
P14	14.72	14.72	9.64	-	60.91
P15	21.83	7.61	9.64	60.91	-
P16	19.91	19.91	4.63	-	55.56
P17	17.68	19.91	4.63	-	73.17
P18	19.91	19.91	4.63	-	55.56

Figure 1: Set of produced collodion samples in their corresponding covered beakers.

2.3 Preparation of the inks

Although different collodion-based inks (or c-inks) recipes were tried out, the one that yielded the best results at the time of writing this communication was prepared in four mixing stages: In the first stage, 21.6 g of graphite powders [15] were mixed with 59.14 ml of collodion, prepared as per recipe P3 (see Table II), by using an overhead stirrer [16] equipped with a kneading stainless steel beater, until obtaining a conveniently distributed mixture. In the second stage, a three roll mill [17] was used by giving the mix four passes, in order to homogenize the blend. In the third stage, the overhead stirrer was used again, albeit this time with a steel beater, which works better for lower density mixtures, 47.04 ml of Chlorobenzene and 63.93 ml of toluene were added to the mix. In the final stage, four more passes were given to the mix in the three roll mill.

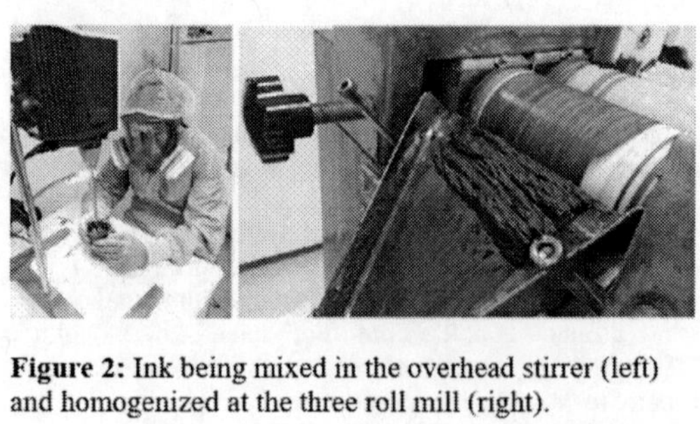

Figure 2: Ink being mixed in the overhead stirrer (left) and homogenized at the three roll mill (right).

2.4 Contact printing and curing

The printing itself was performed using screens [18] capable to handle inks with up to 67 micron particle sizes. The screens were mounted on a screen printer [19], provided with a 70 durometer squeegee blade and a dial indicator for fine adjustment of the screen and the squeegee rod to the tool plate, to finely control the pressure exerted and the thickness of the inks and pastes deposited with an accuracy of micrometers.

The contacts were printed over glass substrates of 175mm x 175mm x 2mm, in order to carry out their electrical characterization. In addition, the inks were also spread by hand on glass substrates of 25mm x 25mm x 2mm with perovskite layer pre-deposited on them, in order to evaluate their effect on these layers.

Figure 3: Contacts being deposited by screen printing.

Finally, the printed contacts were cured in air and at room temperature during 24 hours, while the hand-spread samples were cured during the same period but and in a vacuum desiccator [20], in order to preserve the perovskite layer from any damage due to ambient exposure [21].

3 CHARACTERIZATION

In order to evaluate the percentage of Nitrogen in the produced nitrocellulose, the samples were weighted before the nitration reaction took place and after the drying process was finished. Also, a combustion test was performed.

The printed contacts were characterized following the Standard F 1896 Test Method for Determining the Electrical Resistivity of a Printed Conductive Material [22], according to which, the sheet resistance (Rs) can be derived from printed contacts that comply with a length (L) to width (W) geometric ratio of at least 50:1.

$$\mathbf{R_S = R \cdot W/L} \qquad (\Omega / \square)$$

Also, when the thickness (t) of the contact is known, it is possible to obtain its volume resistivity (ρ) by the following formula:

$$\boldsymbol{\rho = R_S \cdot t} \qquad (\Omega \cdot \mathrm{cm})$$

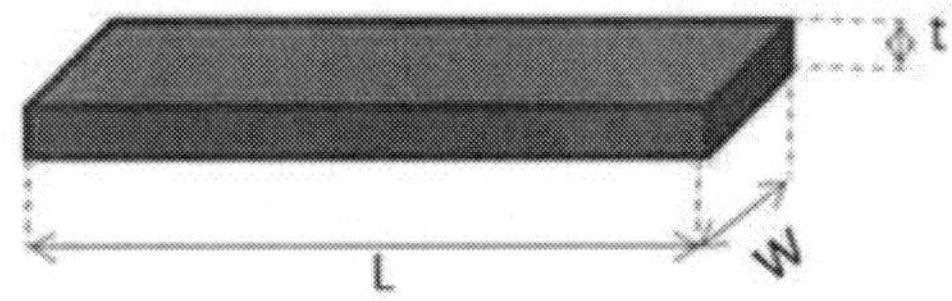

Figure 4: Geometry of a printed contact.

Therefore, screens were designed and produced so as to provide printing patterns that comply with the aforementioned Standard, while being conveniently distributed over the printing area in order to also allow uniformity studies.

The actual measurements were carried out using a semiconductor characterization system [23], configured to operate as an ohmmeter.

Finally, visual inspections of all the printed samples were carried out by using a digital microscope [24], which allowed us to evaluate the morphology, adhesion level and, for the ones hand-spread over perovskite layers, the apparent degree of interaction between the ink and the perovskite.

4 RESULTS AND DISCUSSION

The difference of weight measured before and after the nitration process gave a nitrocellulose with a percentage of Nitrogen of 11.3%. Also, the combustion test performed with a small sample produced a flash deflagration that hardly left any residues. Both results indicate that the obtained nitrocellulose was, in terms of its nitrification, of the industrial kind, that is, of the type used as binder in printing inks and coatings [13].

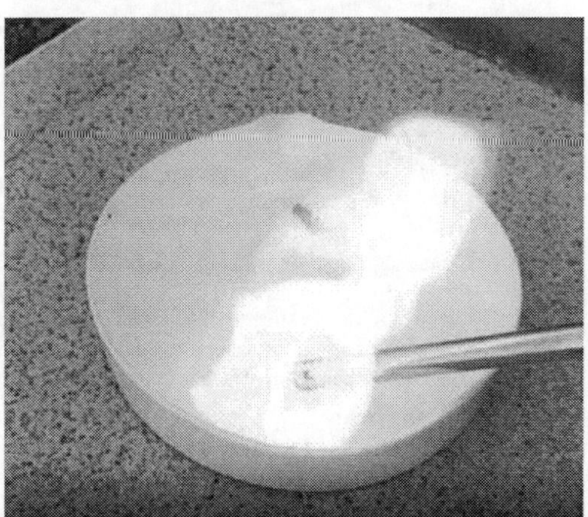

Figure 5: Image of the combustion test performed to a sample of nitrocellulose, showing a flash deflagration.

Although all the recipes of collodion, summarized in Table II, resulted in viscous solutions, with the nitrocellulose properly jellified, the ones that presented a more homogeneous appearance were the mix P3 and P4. All the collodion samples were tested over thin layers of perovskite via drop-casting. As expected, they glued perfectly, producing smooth coatings that quickly dried.

Unfortunately, for all the cases, the layer of perovskite below these coatings became utterly damaged.

The contacts produced with the c-ink via screen printing adhered perfectly to the glass substrate, and the electric measurements performed after 24 hours of curing at room temperature gave an average volume resistivity (ρ) of 0.061 Ω • cm, with a standard deviation of 0.006 Ω • cm. In comparison with the results published in previous communications [2, 3], this is a remarkable result, as it can be seen in Table III. Also, such low values suggest that the c-ink was highly homogeneous, with a vehicle that mostly evaporated during the curing process.

Table III: Summary with the best results obtained with their corresponding ink recipes, expressed in weigh percentaje (wt %).

Powder	ρ (Ω·cm)	Binder	Recipe (wt%)		
			Conductor	Binder	Solvent
Graphite	3.31	Epoxy	36.2	54.1	9.7
Aluminium	3.40	Epoxy	70.4	25.6	4.0
Nickel	6.26	Epoxy	68.4	26.9	4.7
Graphite	**0.061**	**Collodion**	**13.0**	**25.4**	**61.6**

Figure 6: Contacts printed on a 175x17x2mm glass.

However, when the produced ink was hand-spread and cured over the substrates covered with perovskite thin layers, it resulted in their degradation, as it can be seen in Figure 7.

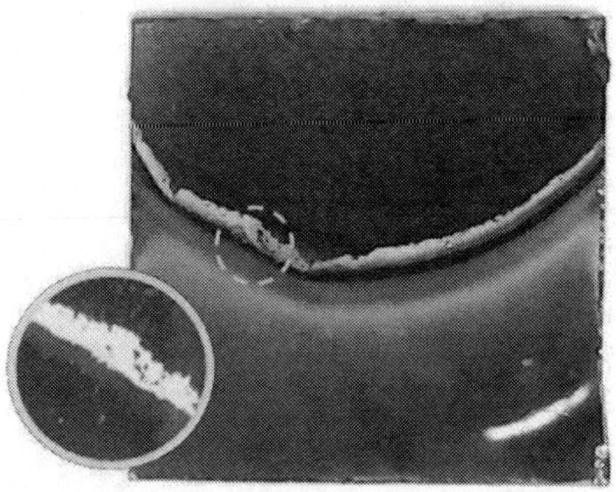

Figure 7: A 25x25x2mm perovskite covered glass with microscopic detail showing a damaged area at the borderline where the ink was hand-spread.

5 CONCLUSIONS

Although the volume resistivity obtained for the contacts printed with the c-ink recipe was very promising; alas its chemical nature was not compatible with perovskite substrates.

It remains to explore the possibility of combining collodion with epoxy resin, in order to produce a mixed binder which, theoretically, could combine the strengths of collodion, which allows to produce inks whose vehicle mostly evaporate when cured at room temperature, leaving a highly conductive layer fixed to the substrates, with the epoxy resin, which accommodates larger amounts of conductive powders and don't seem to damage the perovskite layers.

6 ACKNOWLEDGEMENTS

The authors would like to thank Luisa Quintanilla of JLQ ISMAF, S.L., and Javier A. López-Caballero of Metalor Technologies (Ibérica), S. A., for their continued support throughout this project.

7 REFERENCES

[1] C. Quinto, A. Linares, E. Llarena, C. Montes, O. González, D. Molina, A. Pío, L. Ocaña, M. Friend and M. Cendagorta, 31st European Photovoltaic Solar Energy Conference and Exhibition (2015) 1144.
[2] C. Montes, A. Linares, E. Llarena, O. González, D. Molina, A. Pío, L. Ocaña, C. Quinto, M. Friend and M. Cendagorta, 32nd European Photovoltaic Solar Energy Conference and Exhibition (2016) 1303.
[3] C. Montes, L. Ocaña, C. Quinto, S. González-Pérez. B. González-Díaz, M. Friend and M. Cendagorta, 33rd European Photovoltaic Solar Energy Conference and Exhibition (2017) 1158.
[4] M.R. Fisch, Liquid Crystals, Laptops and Life, World Scientific Publishing, Singapore, 2004.
[5] J.J. Meister, Polymer Modification: Principles, Techniques, and Applications, CRC Press, Boca Raton, FL, 2000.

[6] Kohli, R. Stripable Coatings for Removal of Surface Contaminants. Developments in surface contamination and cleaning. Volume 2, Particle deposition, control and removal [1-4377-7830-5; 1-4377-7831-3]. 2010. 177 -224.

[7] Collodion USP, Material Safety Data Sheet (2001).

[8] J. Troughton, K. Hooper, and T. M. Watson, Nano Energy, vol. 39, pp. 60–68, Sep. 2017.

[9] Clara Aranda,Cesar Cristobal, Leyla Shooshtari, Cheng Li,Sven Huettner and Antonio Guerrero, Sustainable Energy Fuels, 2017,1, 540.

[10]Lidia Contreras-Bernal, Clara Aranda, Marta Valles-Pelarda, Thi Tuyen Ngo, Susana Ramos-Terrón, Juan Jesús Gallardo, Javier Navas, Antonio Guerrero, Ivan Mora-Sero, Jesus Idigoras, and Juan Antonio Anta. J. Phys. Chem. C, 2018, 122 (10), pp 5341–5348.

[11] W. B. Robertson, "Factors Governing the Nitration of Cellulose," Cornell University, 1946.

[12] Francis Sullivan, Laurent Simon , Nikolaos Ioannidis, Subhash Patel, Zohar Ophir, Costas Gogos, Michael Jaffe, Shakeel Tirmizi, Peter Bonnett, and Philip Abbate, Ind. Eng. Chem. Res., 2018, 57 (6), pp 1883–1893.

[13]"WALSRODER™ Nitrocellulose" technical leaflet by The Dow chemical Company (http://msdssearch.dow.com/PublishedLiteratureDOWCOM/dh_08a6/0901b803808a680d.pdf?filepath=/822)

[14] "Butyldiglycol" technical information by BASF Petrochemicals (http://www.solvents.basf.com/portal/streamer?fid=280335)

[15]Graphite powder SC 20 OS from Graphite Kropfmühl GmbH Langheinrichstraße 1, 94051 Hauzenberg Germany.

[16]Overhead stirrer Hei-TORQUE 400 made by Heidolph.

[17]ESSG65 Laboratory Three roll mill, made by Shanghai Espread Industry Co. Ltd.

[18] Stainless steel fabric, 200 mesh, 0.040 D-Ø x 22.5 high tension screens made by SEFAR MAISSA, S.A.U. C/ Alcorcon, 13, 28850 Torrejon Ardoz (Madrid).

[19] MSP-485, Precision Screen Printer. Hary Manufacturing, Inc. Precision Screen Printers and automation equipment (HMI), 24 Cokesbury Road, Lebanon, NJ 08833 USA

[20] Vacuum desiccator model DE02-251-001 by Labbox.

[21] J.Yang, B.D. Siempelkamp, D. Liu, T.L. Kelly, ACS Nano. 24 February 2015, Vol.9(2), pp.1955-1963.

[22] F 1896 – 98 (Preaproved 2004). Test Method for Determining the Electrical Resistivity of a Printed Conductive Material. Test method © by the American Society for Testing Materials (ASTM) International, 100 Barr Harbor Drive, PO Box C700, West Conshohocken, PA 19428-2959, United States.

[23] Semiconductor Characterization System model SCS-4200 by Keithley.

[24] Digital Microscope U500X, focus range 15mm - 40mm, range 50X-500X by Unimake.

PRODUCING UNIFORM AND SMOOTH THIN LAYERS OF PEROVSKITE UNDER AMBIENT CONDITIONS BY ADJUSTING TO THE EXISTING LEVELS OF MOISTURE

C. Montes[1,2], L. Ocaña[1], L. De Sousa-Vieira[1], J.S. Moreno-Ramírez[1], S. González-Pérez[2], B. González-Díaz[2], C. Hernández-Rodríguez[3], M. Friend[1] & M. Cendagorta[1].
[1]Instituto Tecnológico y de Energías Renovables, S. A. (ITER)
Pol. Industrial de Granadilla, s/n, E 38600 Granadilla de Abona, Spain. Ph. +34 922 747 700. Fax +34 922 747 701. E-mail cmontes@iter.es
[2]Departamento de Ingeniería Industrial. [3]Departamento de Física. Universidad de La Laguna.
Avd. Astrofísico Francisco Sánchez, s/n, 38206 S/C de Tenerife. Spain. Ph. +34 922 316 502.

ABSTRACT: Uniform and smooth thin layers of perovskite have been produced under ambient conditions and at different humidity levels. Their optical properties, general morphology, chemical composition, abundance and submicroscopic appearance have been analyzed. Spectral ellipsometry, X-ray crystallography and Scanning Electron Microscope imaging techniques have been used to analyze samples under different relative humidity values.
In the deposition process via spin coating, a variation in the air flow within the process chamber can be achieved through changing the applied clean dry air pressure. This modifies the evaporation rate of the solvents present in the deposition process and, therefore, it could be used as a tool for adjusting the evaporation rate to the existing humidity levels. The obtained results suggest that the efficiency of chlorobenzene as anti-solvent critically depends upon the water:DMF ratio existing in the precursor solution which, in turn, depends on these humidity levels.
Keywords: Perovskite, absorption, thin film, moisture.

1 INTRODUCTION

The use of organometal halide perovskite thin films as the light-harvesting layer on solar cell devices, has led to the development of technologies that have made it possible to achieve unprecedented increases on their power conversion efficiencies (PCE) in a very short period of time [1,2]

Although there are different methods for producing perovskite layers [3-5], since the optoelectronic properties of these films strongly depend on their corresponding processing parameters [6,7], the most extended ones take place under totally dry conditions and make use of wet chemical formulations of its precursor components by dissolving them in polar organic solvents, which then can be deposited by spin coating techniques and subsequently annealed [8,9].

Usually, the issue of exposure to water has been considered over finished devices and in terms of their operational degradation [10]. However, well-controlled studies, performed on the effect of moisture exposure during film formation, have reported a significant improvement on the overall performance in devices fabricated in the presence of water [11], as well as a degree of degradation during operation similar to counterpart ones, manufactured in dry N2-filled glove boxes [12]. Continuing with this approach, recent studies performed on the issue of producing perovskite thin-films in ambient conditions conclude that an intermediate precursor, present after the spinning but previous to the annealing process, is required in order to have an adequate coordination [13, 14].

The Instituto Tecnológico y de Energías Renovables (ITER), with the support of the University of La Laguna (ULL), has been researching extensively on the one-step deposition method, assisted with non polar washing treatment, in order to produce thin films of perovskite in ambient conditions [15,16]. A spectral ellipsometer and a digital microscope have been successfully used for evaluating their quality and further improve the fabrication methodology [17].

Recently, the use of a dehumidification system in our laboratory has made it possible to gain a considerable control over the ambient humidity and, as a result, to acquire the capability of studying the effect that moisture has on the different steps that comprises the production of thin layers of perovskite.

In this work, we evaluated the relative ambient humidity and its influence in the hydration of the perovskite synthesis during the deposition phase. The effect of the DMF washing at different relative humidities was studied, as well as the volume of anti-solvent necessary to obtain uniform and smooth perovskite layers.

2 EXPERIMENTAL

2.1 Fabrication methodology

The whole fabrication process was carried out inside an ISO 7 cleanroom in ambient air, operating at a set temperature of 23 °C and at a relative humidity (RH) set from below 30% to 60%. These two parameters were constantly monitored in order to know the absolute humidity present during the operations. The glass substrates (2.5 x 2.5 mm2) were cleaned sequentially by ultrasonication in a 2% soap solution, deionized water, acetone and finally in 2-propanol. Finally, the substrates were dried with nitrogen gas flow.

The synthesis of the perovskite precursor solution was performed by stirring PbI2 and CH3NH3I in an equimolar ratio of 1:1 (45% wt. in DMF) at 75 °C for 50 min and at room temperature overnight. Finally, the solution was passed through a 0.22 µm PTFE filter prior to its deposition. Then, the perovskite precursor solution (100 µL) was spin-coated at 5000 rpm for 30 seconds and, after 6 seconds, a non-polar solvent, in this case chlorobenzene, was dropped. During the spin-coating process the RH inside the chamber was maintained at a RH range of 6-8 %, depending on the existing ambient humidity, by a constant flow of clean dry air. After deposition process, the sample was annealed at 100 °C for 10 minutes inside a stove and stored in a vacuum desiccator to avoid their degradation [10,11].

In order to diminish and increase the evaporation rate inside the spin coater chamber, the dry clean air flow pressure was varied between 0.2 and 0.4 bar, respectively. This allowed us to study the influence of the evaporation rate of the solvent used in the synthesis (in this case, DFM) and of the anti-solvent used in the "washing" phase (in this case, chlorobenzene), at different relative humidity values, ranging from 30 to 60%. Once the optimum evaporation rate as a function of the relative humidity was determined, the influence of the anti-solvent in the process was evaluated. With this aim in mind, we fabricated perovskite layers using different chlorobenzene volume ratios with respect to the precursor solution, specifically: 3:1, 2:1, 1:1 and 0.5:1.

2.2 Characterization methodology

The characterization process was performed on freshly deposited samples using a digital microscope, in order to identify those showing mirror-like appearance, which denotes their underlying crystalline nature. The optical-absorption characteristics of the samples thus selected were analyzed using spectroscopic ellipsometry, which allows studying the samples in a wide spectral range (245 to 1000 nm). Several reflection measurements were carried out, varying the angle of incidence on the sample, at 65°, 70° and 75°, in order to determine the absorption coefficient and the thickness of the deposited layers [17]. Also, the absorbance values and the band gap were determined from transmittance measurements. The results so obtained were compared with those published in the literature.

The chemical composition, abundance and submicroscopic appearance of the deposited samples were also analyzed using X-ray crystallography techniques and a Scanning Electron Microscope imaging.

3 RESULTS AND DISCUSSION

First, we studied the samples fabricated at different relative humidity values, ranging from 30 to 60 %, varying the dry air flow entering the chamber of the spin coater from 0.2 to 0.4 bar. Figure 1 shows digital microscope images of the samples fabricated in this part of the work. As it can be seen, the majority of the layers present a mirror-like appearance, which is the consequence of a correct $MAPbI_3$ crystallization. The only exception was the sample 18043, which showed a cloudy appearance. Which indicated that the use of a dry air flow of 0.2 bar was not adequate when the relative humidity is low.

Figure 1: Microscope images of the perovskite samples fabricated by varying the dry air flow from 0.2 to 0.4 bar depending on the ambient relative humidity.

From the experimental part of this work, it was concluded that, in order to obtain mirror-like layers, it was necessary that the relation between humidity and dry air flow pressure was adequate, so that when the anti-solvent (chlorobenzene) was deposited, the layer maintained a dark brown appearance (see Fig. 2 (B)), indicating that the crystallization process was maintained during the rest of the fabrication procedure. On the other hand, when the relation between humidity and dry air flow pressure was not adequate, the crystallization process would start when the anti-solvent was deposited, but it would be reversed during the rest of the spinning, leading to a transparent layer before the annealing (see Fig. 2 (A)), and to a sample with a cloudy appearance.

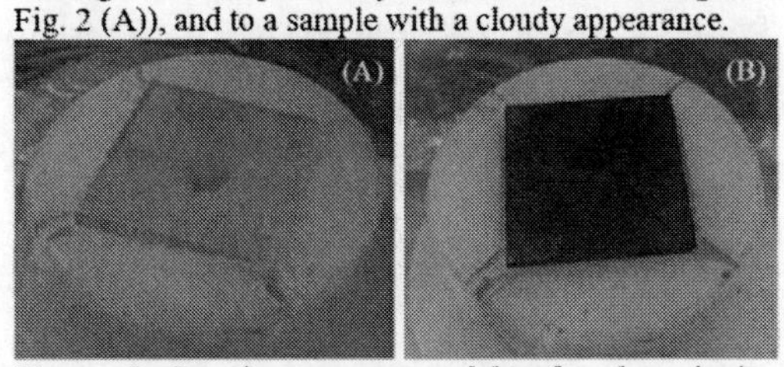

Figure 2: Samples appearance right after the spinning process is finished. (A) Sample 18043, layer deposited with an incorrect relation between humidity and dry air flow. (B) Sample 18101, layer deposited with the correct relation between humidity and dry air flow.

The absorption coefficient results are presented in Figure 3. It is observed that at low relative humidity (32.8%RH), when the evaporation rate was reduced (0.2 bar), the absorption coefficient diminished. For the rest of the cases analyzed, the absorption coefficient values, as well as the tendency of the curves, were similar to each other and coincident with the referenced values [18-19].

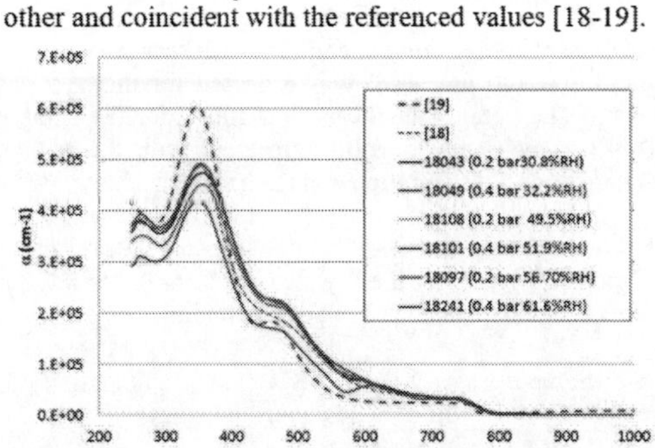

Figure 3: Absorption coefficient obtained for the perovskite samples fabricated by varying the dry air flow from 0.2 to 0.4 bar depending on the ambient relative humidity.

Figures 4 and 5 show the transmittance and absorbance obtained for the perovskite layers. The sample 18043, fabricated at the lowest relative humidity (30.8%) and with the smallest dry air flow (0.2 bar), shows the lowest transmittance, which leads to the greatest absorbance. Nevertheless, this cannot be attributed to an increase in the absorption ability of the layer, but to a defect in the layer uniformity, which results in a cloudy appearance, as it can be seen in the digital microscope images (see Figure 1). Furthermore, taking into account figures 3, 4 and 5, it can be seen that the band gap was adequately positioned, approximately at 780nm (1.6eV).

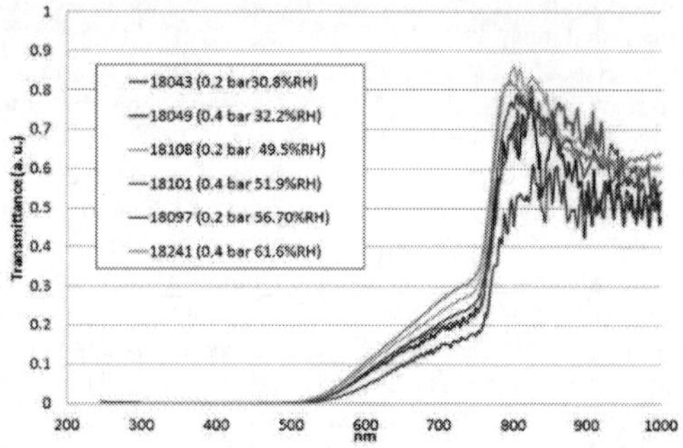

Figure 4: Transmittance obtained for the perovskite samples fabricated by varying the dry air flow from 0.2 to 0.4 bar depending on the ambient relative humidity.

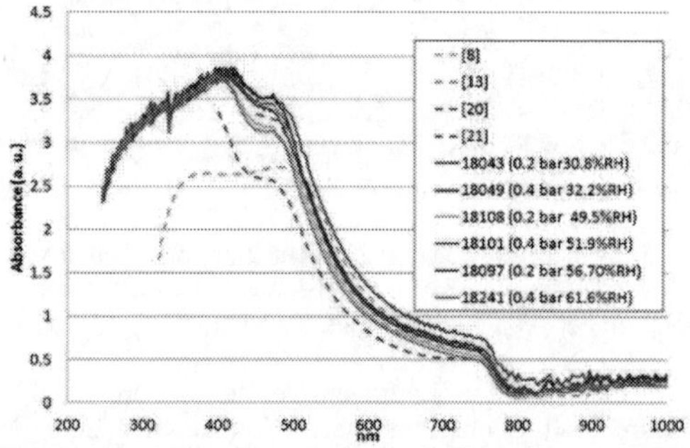

Figure 5: Absorbance obtained for the perovskite samples fabricated by varying the dry air flow from 0.2 to 0.4 bar depending on the ambient relative humidity.

Table I shows the obtained thicknesses of the different samples. In general, the thicknesses were at about 300-350 nm, as it was expected for this type of layers. The greatest thickness was found for the sample 18043, which was in good agreement with the results obtained for the absorption and the transmittance.

Table I: Obtained thickness for the perovskite samples fabricated by varying the dry air flow from 0.2 to 0.4 bar depending on the ambient relative humidity.

Sample ID	Dry air flow pressure (bar)	RH (%)	Thickness (nm)
18043	0.2	30.8	352.70
18049	0.4	32.2	323.43
18108	0.2	49.5	311.36
18101	0.4	51.4	341.02
18097	0.2	56.7	311.86
18241	0.4	61.6	294.80

Taking into account these results, we came to the conclusion that the samples fabricated with low relative humidity (~30%RH) required a high dry air flow pressure (0.4 bar) in order to present a mirror-like appearance, while samples fabricated at high relative humidity (50-60%RH), did not require a specific dry air pressure to crystallize correctly, since they present a good appearance for both pressures: 0.2 and 0.4 bar.

Once the optimum balance between dry air flow pressure and the relative humidity was found, the effect of the anti-solvent used during the washing phase of the layer deposition process was analyzed. For this purpose, the relative humidity and the dry air flow pressure entering the spin coater chamber were maintained constant (50-55%RH and 0.4 bar), while the volume ratios of chlorobenzene deposited were changed with respect to the precursor solution. Thus, the selected chlorobenzene-precursor solution ratios were: 3:1, 2:1, 1:1 and 0.5:1.

Figure 6 shows digital microscope images of the samples fabricated in this part of the work. It can be seen that all the layers presented a mirror-like appearance. Nevertheless, the sample with the smallest volume of chlorobenzene showed noticeable defects.

Figure 6: Microscope images of the perovskite samples fabricated varying the chlorobenzene from 50 to 200 µl at a constant dry air flow pressure of 0.4 bar.

In Figure 7, the absorption coefficient values obtained are presented. As it can be seen, they were similar to each other and to the referenced ones, being slightly higher the ones obtained for the samples fabricated with less volume of anti-solvent.

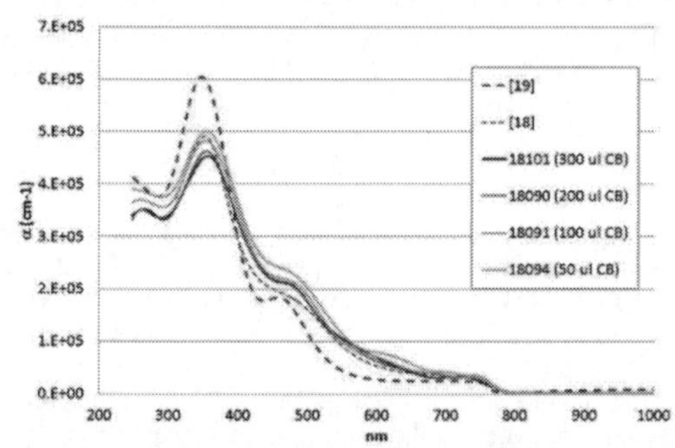

Figure 7: Absorption coefficient obtained for the perovskite samples fabricated by varying the chlorobenzene from 50 to 300 µl at a constant dry air flow pressure of 0.4 bar.

In relation to figures 8 and 9, which show the transmittance and the absorbance obtained for these samples, respectively, it can be seen that the results were quite similar regardless the volume of chlorobenzene deposited. They were also similar to the referenced ones [8, 13, 20-21]. Furthermore, taking into account figures 7, 8 and 9, it can be seen that the band gap was adequately positioned, approximately at 780nm (1.6eV).

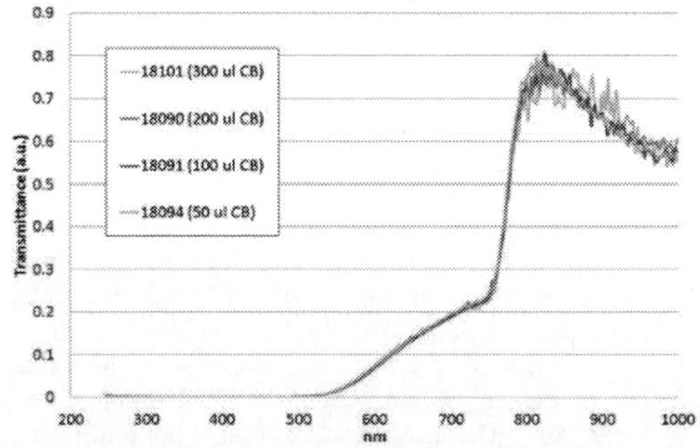

Figure 8: Transmittance obtained for the perovskite samples fabricated by varying the chlorobenzene from 50 to 300 µl at a constant dry air flow pressure of 0.4 bar.

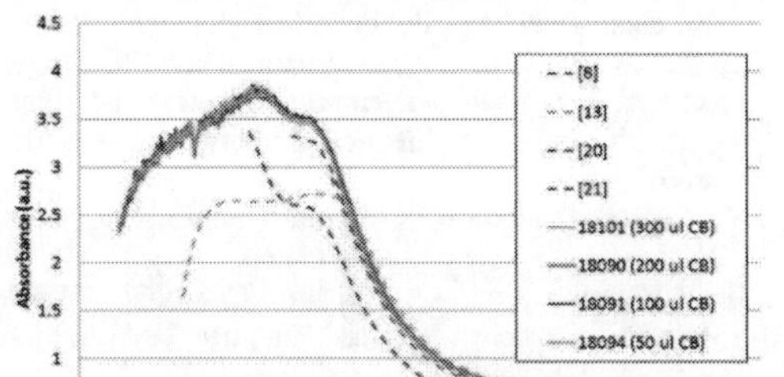

Figure 9: Absorbance obtained for the perovskite samples fabricated by varying the chlorobenzene from 50 to 300 µl at a constant dry air flow pressure of 0.4 bar

On the other hand, by comparing the obtained thicknesses (see Table 2), it can be seen that the greatest thickness was found for the layer fabricated with the biggest volume of anti-solvent, and viceversa, the thinnest layer was the one fabricated with 50µl of chlorobenzene. This relationship between the thickness and the anti-solvent quantity coincides with the results obtained in the absorption coefficient study. Nonetheless, the use of reduced volumes of chlorobenzene led to the formation of defects in the layer, as it can be seen in the central area of sample 18094 (see Figure 6).

Table II: Obtained thickness for the perovskite samples fabricated by varying the chlorobenzene quantity at a constant dry air flow pressure of 0.4 bar.

Sample ID	Chlorobenzene quantity (µl)	RH (%)	Thickness (nm)
18101	300	51.9	341.02
18090	200	54.6	337.53
18091	100	55.1	323.80
18094	50	56.8	320.55

Excluding sample 18094, fabricated with 50 µl of chlorobenzene, due to its central defect, the rest of the samples were analyzed with a scanning electron microscope (SEM). A representative point of every sample was selected. The obtained images are shown in Figure 10. It can be seen that the grain size was similar for the three cases, between 200 and 350 nm. From the microscopic point of view, the layers are uniform and do not present pinholes. Nevertheless, from a macroscopic point of view (see figures 1 and 6), the samples fabricated with lower quantities of chlorobenzene presented superficial defects.

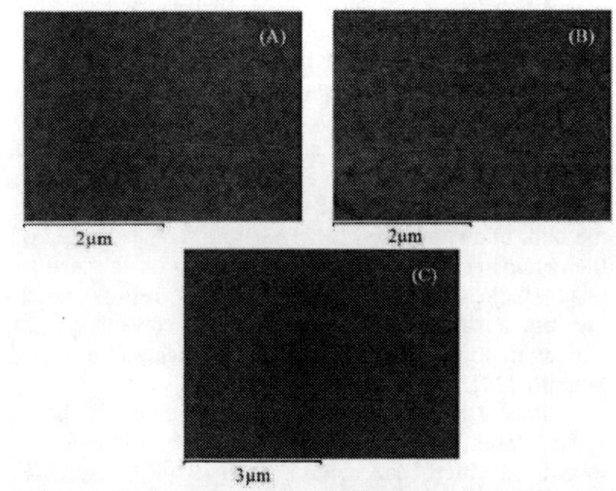

Figure 10: Scanning electron microscopy (SEM) images of samples 18091 (A), 18090 (B) and 18097 (C), fabricated with 100, 200 and 300µl of chlorobenzene, respectively and at a constant dry air flow pressure of 0.4 bar.

Finally, we proceeded to study the best samples fabricated with 300 µl of chlorobenzene at a relative humidity of approximately 50% with the X-ray diffraction technique (XRD). The samples were the 18101 (0.4 bar) and the 18108 (0.2 bar). The $MAPbI_3$ phase was confirmed, as it is shown in Figure 11.

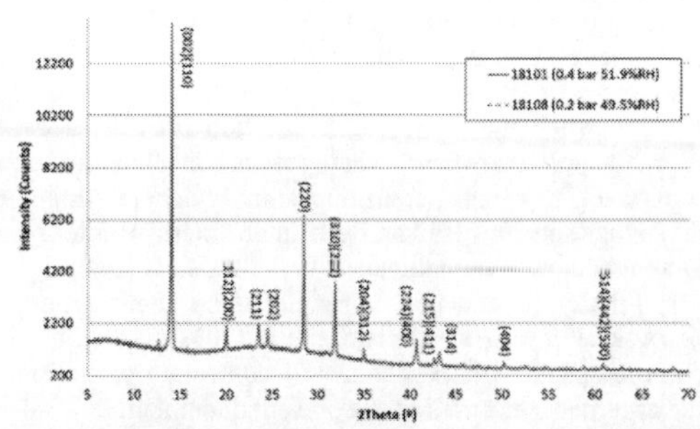

Figure 11: X-ray diffraction pattern of fabricated $MAPbI_3$ layers of samples 18101 and 18108, fabricated with a dry air flow pressure of 0.2 and 0.4 bar, respectively and with 300µl of chlorobenzene.

4 CONCLUSIONS

Uniform and smooth thin layers of perovskite have been successfully obtained under ambient conditions with relative humidity ranging from below 30% to 60%, by way of modifying the airflow applied inside the chamber of the spin-coater, so that at lower moisture levels an increase in the airflow was required, while at higher moisture levels a reduction in the airflow was not necessary. The ellipsometric study performed over those layers indicates that their optical parameters are similar to those published in the literature [8, 13, 18-21].

Mirror-like perovskite layers were also studied using Scanning Electron Microscope and X-ray crystallography techniques. The results showed that the most uniform layers were those deposited with a 3:1

volume rate for the chlorobenzene and the perovskite solution, regardless of the dry clean air flow pressure applied, at relative humidity values of around 50%.

The evaporation stage in a typical spin coating process is the final one, and happens when most of the centrifugal outflow of the deposited liquid virtually stops, so that any further shrinkage is mostly due to solvent loss. In this stage, the rate of evaporation depends on two factors, both happening in the free surface at the remaining liquid layer: the difference in partial pressure (actually chemical potential) of each solvent species, and the bulk of air flowing [22]. Therefore, being the rest of the parameters the same, a change solely on the airflow inside the chamber of the spin coater, results on an increase in the evaporation rates of the solvent species which, in turn, depends on their differences in vapor pressure [23].

Since DMF miscibility in water is total [24], the water present in the precursor solution will be highly mixed, so chlorobenzene can wash it with the DMF. When the ratio water:DMF is high (50-60% RH), the washing mechanism produced by chlorobenzene is very efficient at dragging the water together with the DMF. However, when the ratio water:DMF in the precursor solution is lower (30% RH), the mechanism is less efficient, and an increment in the evaporation rate of the solvents is necessary.

In this work, we have demonstrated that, in order to obtain mirror-like perovskite layers, a right balance of water and DMF is also required in the intermediate precursor solution, that is, the liquid that remains in the chuck at the end of the spinning process, before the annealing takes place. Such right balance is achieved when the crystallization process, that starts when the chlorobenzene is deposited, remains until the end of the spinning process. Therefore, a proper deposition process finishes when the intermediate solution presents a dark brown aspect. However, when the intermediate solution turns back to a transparent appearance, denotes that the crystallization process has been interrupted, which leads to an inadequate crystallization in the finished layer.

Finally, it remains to explore the fabrication of perovskite layers at low humidity (30%), varying the chlorobenzene volume. This will contribute to validate the arguments exposed in the present document.

5 REFERENCES

[1] Akihiro Kojima, Kenjiro Teshima, Yasuo Shirai, and Tsutomu Miyasaka, Journal of the American Chemical Society 17 (2009) 6050.

[2] Research Cell Efficiency Records Chart (webpage). National Center for Photovoltaics, NREL. Acess: February 2017. https://www.nrel.gov/pv/assets/images/efficiency_chart.jpg

[3] Mehran Habibi, Fatemeh Zabihi, Mohammad Reza Ahmadian-Yazdi, Morteza Eslamian. Renewable and Sustainable Energy Reviews 62 (2016) 1012.

[4] Jiandong Fan, Baohua Jia, and Min Gu, Photon. Res. 2 (2014) 111.

[5] Zhaoning Song, Suneth C. Watthage, Adam B. Phillips and Michael J. Heben, J Photon Energy 6 (2016) 022001.

[6] Q. Wang, Q. Dong, Z. Xiao, Y. Yuan and J. Huang, Energy Environ. Sci. 7 (2014) 2359.

[7] H. Zhou, Q. Chen, G. Li, S. Luo, T.-B. Song, H.-S. Duan, Z. Hong, J. You, Y. Liu and Y. Yang, Photovoltaics. Science 345 (2014) 542.

[8] Xiao, M., Huang, F., Huang, W., Dkhissi, Y., Zhu, Y., Etheridge, J., Gray-Weale, A., Bach, U., Cheng, Y.-B. and Spiccia, L. Chem. Int. Ed. 53 (2014) 9898.

[9] Kuen-Feng Lin, Sheng Hsiung Chang, Kai-Hung Wang, Hsin-Ming Cheng, Kuo Yuan Chiu, Kun-Mu Lee, Sheng-Hui Chen, Chun-Guey Wu, Solar Energy Materials and Solar Cells 141 (2015) 309.

[10] Taame Abraha Berhe, Wei-Nien Su, Ching-Hsiang Chen, Chun-Jern Pan, Ju-Hsiang Cheng, Hung-Ming Chen, Meng-Che Tsai, Liang-Yih Chen, Amare Aregahegn Dubale and Bing-Joe Hwang, Energy & Environmental Science 9 (2015) 323.

[11] Giles E. Eperon, Severin N. Habisreutinger, Tomas Leijtens, Bardo J. Bruijnaers, Jacobus J. van Franeker, Dane W. de Quilettes, Sandeep Pathak, Rebecca J. Sutton, Giulia Grancini, David S. Ginger, Rene A. J. Janssen, Annamaria Petrozza, and Henry J. Snaith, American Chemical Society Nano 9 (2015) 9380.

[12] Joel Troughton, Katherine Hooper, Trystan M. Watson, Nano Energy 39 (2017) 60.

[13] Clara Aranda, Cesar Cristobal, Leyla Shooshtari, Cheng Li,Sven Huettner and Antonio Guerrero, Sustainable Energy Fuels 1 (2017) 540.

[14] Lidia Contreras-Bernal, Clara Aranda, Marta Valles-Pelarda, Thi Tuyen Ngo, Susana Ramos-Terrón, Juan Jesús Gallardo, Javier Navas, Antonio Guerrero, Ivan Mora-Sero, Jesus Idigoras, and Juan Antonio Anta, J. Phys. Chem. C 122 (2018) 5341.

[15] L.Ocaña, A. Linares, E. Llarena, C. Montes, O. González, D. Molina, A. Pío, C. Quinto, M. Friend and M. Cendagorta, 31st European Photovoltaic Solar Energy Conference and Exhibition (2015) 1138.

[16] L. Ocaña, C. Quinto, C. Montes, A. Linares, E. Llarena, O. González, D. Molina, A. Pío, C. Hernández-Rodríguez, S. González-Pérez, R. Guerrero-Lemus, M. Friend and M. Cendagorta, 32nd European Photovoltaic Solar Energy Conference and Exhibition (2016) 1316.

[17] C. Montes, L. Ocaña, C. Quinto, S. González-Pérez. B. González-Díaz. M. Friend and M. Cendagorta, 33rd European Photovoltaic Solar Energy Conference and Exhibition (2017) 1181.

[18] Löper, P., Stuckelberger M., Niesen, B., Werner J., Filipič W., Moon, S., Yum, J., Topič M., De Wolf, S., Ballif, The Journal of Physical Chemistry Letters 6(1) (2015) 66.

[19] Leguy A., Hu Y., Campoy-Quiles M., Alonso M., Weber O., Azarhoosh P., Schilfgaarde M., Weller M., Bein T., Nelson J., Docampo P., Barnes P., Chemistry of Materials 27(9) (2015) 3397.

[20] Cheng-Chiang Chen, Sheng Hsiung Chang, Lung-Chien Chen, Hsin-Ming Cheng, Zong-Liang Tseng, Chun-Guey Wu, Solar Energy 139 (2016) 518.

[21] Lung-Chien Chen, Yu-Shiang Lin, Zong-Liang Tseng, Chiale Wu, Feng-Sheng Kao and Sheng-Hui Chen, Nanomaterials 7(7) (2017) 166.

[22] Niranjan Sahu, B. Parija, S. Panigrahi, Indian Journal of Physics 83 (2009) 493.

[23] Bing Cai, Wen-Hua Zhang, Jieshan Qiu, Chinese Journal of Catalysis 36 (2015) 1183.

[24] Ian M. Smallwood, Butterworth-Heinemann, Handbook of Organic Solvent Properties (1996).

HIGHLY EFFICIENT AND ITO-FREE FLEXIBLE COUNTER ELECTRODES EMPLOYING NOVEL COPPER BASED REDOX SHUTTLES IN DYE-SENSITIZED SOLAR CELLS

Authors: Haider Iftikhar [a], Gabriela Gava Sonai [b], Syed Ghufran Hashmi [c*], Peter Lund [a] and Ana Flávia Nogueira [b]

* Corresponding author's email address: ghufran.hashmi@aalto.fi, sghufranh28@gmail.com

[a] New Energy Technologies Group, Department of Applied Physics, Aalto University, P.O. BOX 15100, FI-00076 Aalto (Espoo), Finland.

[b] Laboratório de Nanotecnologia e Energia Solar (LNLS), Chemistry Institute, University of Campinas (UNICAMP), P.O. Box 6154, 13083-970, Campinas, São Paulo, Brazil.

[c] Department of Applied Physics, Puumiehenkuja 2, 02150, Espoo-Finland.

ABSTRACT: We present results on highly efficient and indium tin oxide (ITO) free counter electrodes. These were fabricated via low-temperature based highly conductive single walled carbon nanotubes (SWCNTs), which are printed over a flexible polymer substrate. The printed nanotubes served as an alternative to both the traditional conducting (ITO) and catalyst (Pt) layers of the counter electrode. The SWCNT or Pt based flexible counter electrodes were assembled with the photoelectrodes to produce dye-sensitized solar cells (DSSCs) employing electrolytes based on novel copper redox shuttles. The performance of the fabricated solar cells were investigated through J-V curves and electrochemical impedance spectroscopy (EIS). The SWCNT-based flexible counter electrodes exhibited excellent compatibility and superior catalytic activity with copper electrolytes compared to flexible Pt catalyst based counter electrodes fabricated on flexible ITO-PEN substrates. Our results suggest that SWCNT can be a potential alternative for realizing efficient flexible dye-sensitized solar cells when fabricated with electrolytes based on copper redox shuttles.

Keywords: Dye-sensitized solar cells, copper redox shuttles, single-walled carbon nanotubes, platinum, catalyst, counter electrodes.

1. INTRODUCTION

Dye sensitized solar cells (DSSCs) have been considered as one of the potential alternatives in realizing low cost solar-to-electrical energy production [1-3]. One of their promising features include their efficient photovoltaic performance under low light intensity, which offers great motivation to develop novel materials in order to produce efficient DSSCs which could exhibit high photovoltaic performance under such considerably lower stressful conditions. In this regard, novel copper redox shuttles (as a potential alternative to traditional iodide/triiodide based electrolytes for DSSCs) are recently introduced and have received tremendous attention in contributing high performance DSSCs when measured under both the full sun and low light intensities [4-6]. Interestingly, these high performance based DSSC configurations employing copper electrolytes are mostly reported or compared with either traditional platinum (Pt) and PEDOT catalysts based counter electrodes (CEs) or very recently graphene catalyst layers based CEs which have been deposited over rigid FTO-Glass substrates [4-8]. Here, we report low temperature printable single-walled carbon nanotubes (SWCNTs) films as another potential alternative catalyst material for these novel copper redox shuttles based electrolytes. They provide an opportunity to realize an efficient catalytic material for glass based CEs and may also be realized as a potential fully ITO free flexible counter electrode for their possible application in fully flexible DSSCs employing copper redox based electrolyte solution. These low temperature SWCNTs based catalyst layers were deposited via screen printing on Fluorine-doped tin oxide (FTO) Glass, indium doped tin oxide polyethylenenaphtalate (ITO-PEN) and ITO free polymer polyethylenenaphtalate (PEN) based substrates, which were utilized to assemble the different DSSC configurations. The performance of these configurations are compared with both the high and low temperature processed traditional Pt catalyst layers, which are also deposited on aforementioned FTO-Glass and ITO-PEN substrates. The SWCNTs based CEs demonstrated superior catalytic performance and directly contributed in significantly improved photovoltaic performance over traditional Pt CE based DSSCs. This results suggests that in near future, efficient and fully flexible DSSCs may also be produced via employing these SWCNTs based flexible CEs in combination with these novel copper redox shuttles and may potentially give high performance under low light intensities.

2. EXPERIMENTAL SETUP

2.1 Measurements setup

The photovoltaic performance of the cells was measured under various light intensities with a source meter (Keithley 2420) in the artificial solar simulator (Peccell-Japan). The electrochemical impedance spectra of both the complete DSSCs and symmetrical CE-CE were measured with a potentiostat (Zahner-Elektrik IM6-Germany). Incident photon to collected electron (IPCE) measurements of the fabricated DSSCs was performed with Spectral Response/QE/IPCE measurement system (PV measurements Inc-USA). The measured data is not presented for this study. The sheet resistances of each type of electrodes were measured with four-probe measurements system (HP).

2.2 Fabrication of the test cells

Six different types of CEs (three on rigid FTO Glass and three on different flexible substrates) were fabricated for this study. The details of their fabrication are given below.

2.2.1 Preparation of counter electrodes (CEs)

Thermal Platinization on FTO Glass

For thermal platinization 6 μL of 5 mM chloroplatinic acid hydrate ($H_2PtCl_4.6H_2O$) solution in 2-Propanol was spread evenly on the entire pre-cleaned (15 minutes each in 1% diluted Helmanex solution, Acetone and 2-propanol solvents) TEC-7 FTO-Glass substrates, followed by their sintering in a pre-heated (410 °C) programmable oven (Nabertherm-Germany) for 30 minutes. The substrates were then removed off from the oven and were cooled down to room temperature. These cooled Pt counter electrodes were kept in a vacuum box before the final cell assembly.

Screen-printing of SWCNTs on FTO-Glass

The screen printing of a low temperature based printable SWCNT paste (Brewer Science-USA) was performed over pre-cleaned TEC-7 Glass substrate via semi-automated screen printer (Atma-Taiwan). After printing, the printed electrodes were placed over a pre-heated (120 °C) hot-plate for 20 minutes and were cooled down to room temperature. Similar to thermally platinized CEs, the cooled Glass/SWCNT-CEs were also kept in a vacuum box before the final cell assembly.

Screen-printing of SWCNTs on ITO-PEN and PEN polymer foil for flexible CEs

The flexible ITO-PEN and PEN based counter electrodes were obtained via utilizing same SWCNT printable paste which was again screen-printed on the pre-cleaned sheets (10x10 cm^2) of these substrates. After printing, these flexible substrates were placed on the pre-heated hotplate (120 °C) for 30 minutes to ensure the complete removal of the solvent. After that, these flexible substrates were cooled down to room temperature and were kept in the vacuum box before the final assembly.

Low temperature chemical platinization on FTO-Glass and ITO-PEN substrates

Low temperature chemical platinization over both the rigid FTO Glass and flexible ITO PEN electrodes was performed by first spreading 40 mM Pt solution (in 2-propanol, 6 µl) over each type of substrate. After which the electrodes were dried over a hotplate at 120 °C for 30 minutes. The dried $H_2PtCl_4.6H_2O$ covered CEs (both ITO PEN and FTO Glass) were then chemically reduced via immersing them in 50mM $NaBH_4$ solution containing DI water and ethanol (8:2), for 5 minutes. These chemically reduced platinized electrodes were first washed with DI water and then with 2-propanol and were then dried by compressed air and were kept on a hot plate at 120 °C for 30 minutes. These electrodes were also kept in vacuum box before their final assembly.

2.2.2 Fabrication of photoelectrodes

The photo electrodes for this experiment were kept common to all the complete DSSCs in the study and were prepared as follows:

In the beginning, 10x10 cm^2 pre drilled FTO-Glass plate (TEC-15) was cleaned by the same cleaning protocol as mentioned for glass CEs in the previous section. After that, the substrates were UV-O_3 treated in the same manner as the CEs, for 30 minutes to ensure removal of any remaining organic contaminants. Then the cleaned FTO-Glass plate was subject to the first $TiCl_4$ treatment, by immersing it in 53 mM $TiCl_4$ solution in sealed plastic container which was then placed in a preheated (70 °C) oven (Memmert-Germany) for 30 minutes. After that, the sealed container was removed off from the oven and the FTO-Glass plate was removed from the sealed container and was sequentially washed with DI water and ethanol. After washing, this FTO-Glass plate was dried with compressed air. After this first $TiCl_4$ treatment, the FTO-Glass plate was sintered at 500 °C for 2 hours in an oven (Nabertherm-Germany) and was allowed to cool down to room temperature.

The FTO-Glass plate was then subjected to screen printing of the TiO_2 paste. 3 µm semi-transparent thick layer of nanocrystalline TiO_2 particles was deposited via screen printing by employing 30 NR-D TiO_2 paste (Dyesol) and was dried on a preheated (120 °C) hotplate for 5 minutes. After that, a layer (3 µm) of scattering TiO_2 particles (400 nm) was stacked on semi-transparent nano-crystalline TiO_2 particles based layer, which was again dried at 120 °C for 5 minutes over the pre-heated hot plate. Then the FTO-Glass plate was again placed in the oven for sintering the printed TiO_2 layers at 500 °C for 2 hours and was again cooled down to room temperature. After that, the second $TiCl_4$ treatment (40 mM) was performed as mentioned above and the FTO-Glass plate was again sintered for 30 minutes at 500 °C and was again cooled down to room temperature to complete the fabrication process. The plate was broken into 30 individual photoelectrodes, which were kept in a vacuum box before dye-sensitization process. The area of the printed TiO_2 layers was 0.4 cm^2.

2.2.3 Dye-sensitization of the PEs

0.2 mM concentrated solution of dye Y123 (Dyenamo) was first prepared by dissolving the dye-powder in a mixed solution of Acetonitrile and t-butanol (1:1). The prepared photoelectrodes were then immersed into this prepared dye solution for a period of 16 hours to ensure the complete sensitization of the TiO_2 photoelectrodes.

2.2.4 Electrolyte composition

The electrolyte for this study was prepared (in a glass vial) by dissolving 60.6 mg of $Cu(dmp)_2TFSI$ together with 13 mg of $Cu(dmp)_2(TFSI)Cl$, 12.6 mg of LiTFSI and 32 mg of 4-tertbutylpyradine (4-TBP) in acetonitrile (0.4 ml). The vial of electrolyte was kept in closed plastic tube before final cell assembly.

2.2.5 Complete DSSCs assembly

The complete DSSCs assembly was performed by separating both the photoelectrodes and counter electrodes through a 25 μm Surlyn frame foil (Dupont). A sandwich type configuration was obtained by melting the Surlyn frame foil, which was attached to both the photoelectrode and counter electrode. After that, the copper electrolyte was introduced into the cell channel via drilled holes at the photoelectrode side. The holes on the photoelectrodes were then sealed with a Surlyn cover foil and thin glass cover. After that, the contacts were fabricated by attaching copper tape on the non-active areas of the DSSCs followed by painting the quickly dried silver ink over the copper tapes to complete the cell assembly.

3. RESULTS AND DISCUSSIONS

3.1 Photovoltaic performance

Table 1 represents the photovoltaic performance of the fabricated champion DSSCs (which were assembled with 6 different types of CEs as mentioned in earlier section) whereas the J-V curves of these champion DSSCs are depicted in Figure 1.

Table 1. Photovoltaic parameters of champion DSSCs measured under full sun light intensity with each type of CEs fabricated for this study.

CE Types	J_{sc} (mA/cm²)	V_{oc} (V)	FF	η (%)	R_{ctt} (Ω)
Flexible PEN/ITO-PEN CEs					
PEN ITO SWCNT	7.84	0.93	0.66	4.81	135.38
ITO free PEN SWCNT	7.27	0.90	0.64	4.22	142.31
PEN ITO Chemical Pt	6.22	1.01	0.46	2.89	185.71
Reference FTO Glass CEs					
Glass FTO SWCNT	8.91	1.00	0.65	5.79	117.39
Glass FTO Chemical Pt	7.50	1.02	0.66	5.06	124.25
Glass FTO Thermal Pt	8.45	0.96	0.56	4.52	222.35

Impressively, for both the rigid FTO-Glass and flexible substrates, the SWCNT CE based DSSCs exhibited higher photovoltaic performance than the traditional Pt CE based cells (Table 1). In the case of completely rigid cells, the ones with SWCNT CEs exhibited up to 14% higher solar-to-electrical conversion efficiency compared to the low temperature chemically platinized CE DSSCs, which utilized very high concentration of Pt solution (40 mM). The same SWCNT CEs also outperformed the traditional high temperature processed thermally platinized CEs (by exhibiting significantly higher PCE up to 28% solar-to-electrical conversion efficiency, Table 1). It is worth noticing that the thermally platinized CEs consume significantly lower concentrations (5 mM) of Pt solution and have contributed in demonstrating very high efficiency DSSCs when used with traditional iodide based electrolyte solutions [10-11].

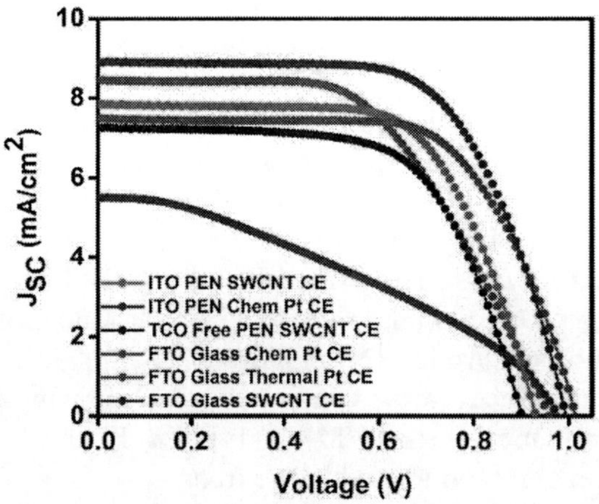

Figure 1. J-V curves of champion DSSCs with each type of CEs fabricated for this study.

On the other hand, the same trend was observed with the DSSCs, which were assembled with rigid FTO-Glass based photoelectrodes (PEs) and flexible counter electrodes. Both the SWCNT-ITO-PEN CE and ITO free SWCNT CE based DSSCs again showed impressive higher photovoltaic performance (i.e. up to 66% higher efficiency with SWCNT-ITO-PEN CE and 46% higher conversion efficiency with ITO free SWCNT-PEN CE).

The lower J_{SC} values obtained with each type of DSSCs fabricated for this study, Table 1, suggest a diffusion based current limitations even for thinner (6-7 μm) TiO_2 layers, which was confirmed in Figure 2. This behavior have been frequently reported for both the bulky cobalt and copper redox shuttles [12-14]. The SWCNT CEs based DSSCs also exhibited impressively higher fill factors (FF) as well as lower overall cell resistance (R_{CELL}) compared to Pt CEs based DSSCs except for chemically platinized CE on rigid FTO glass (Table 1), which is an expected result due to its significantly high concentration than the traditional thermally platinized CE.

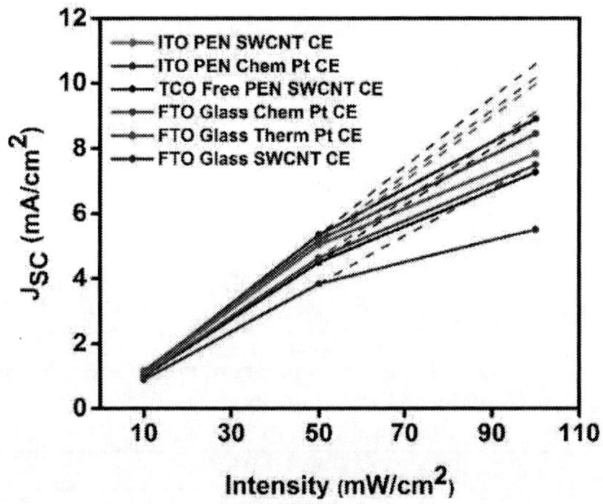

Figure 2. Illustration of the current limitation with different light intensities. The solid lines represents the measured values whereas the dashed line represents the ideal projection.

Nevertheless, the flexible chemically platinized CE based DSSCs exhibited significantly lower (43%) FF, thus raises the concerns regarding not only the chemical compatibility of the complete configuration but also the poor catalytic performance of this highly concentrated catalyst layer when tested with copper electrolytes for this study. On the other hand, the high FF and lower R_{CELL} values obtained with SWCNTs based CE (both on rigid and flexible substrates) indicate superior catalytic performance since it directly affects over these two important photovoltaic parameter of a DSSC [15-17].

The similar trends are observed under low light intensities (Table 2 and 3) where the SWCNT CEs based DSSCs exhibited outstanding performance by revealing higher conversion efficiencies than the traditional Pt catalyst CE based DSSCs on both the rigid and flexible CEs based DSSCs (Tables 2 and 3).

Table 2. Photovoltaic parameters of champion DSSCs measured under 0.5 sun light intensity with each type of CEs fabricated for this study.

CE Types	J_{SC} (mA/cm²)	V_{OC} (V)	FF	η (%)	R_{cell} (Ω)
Flexible PEN/ITO-PEN CEs					
PEN ITO SWCNT	4.88	0.91	0.68	6.09	165.70
ITO free PEN SWCNT	4.50	0.87	0.66	5.17	175.18
PEN ITO Chemical Pt	3.84	1.01	0.49	3.79	262.15
Reference FTO Glass CEs					
Glass FTO SWCNT	5.34	0.98	0.68	7.11	168.02
Glass FTO Chemical Pt	4.63	1.01	0.68	6.36	170.43
Glass FTO Thermal Pt	5.17	0.96	0.58	5.73	306.15

Table 3. Photovoltaic parameters of champion DSSCs measured under 0.1 sun light intensity with each type of CEs fabricated for this study.

CE Types	J_{SC} (mA/cm²)	V_{OC} (V)	FF	η (%)	R_{cell} (Ω)
Flexible PEN/ITO-PEN CEs					
PEN ITO SWCNT	1.08	0.86	0.72	6.63	474.38
ITO free PEN SWCNT	0.99	0.79	0.71	5.58	492.49
PEN ITO Chemical Pt	0.90	0.95	0.57	4.87	1129.30
Reference FTO Glass CEs					
Glass FTO SWCNT	1.13	0.92	0.73	7.51	443.49
Glass FTO Chemical Pt	1.04	0.94	0.72	7.07	558.73
Glass FTO Thermal Pt	1.16	0.90	0.67	7.10	762.06

Interestingly, the photovoltaic performance under low light intensities also suggested that the remarkable performance of SWCNT CEs based DSSC was again obtained as a result of higher FFs and lower R_{CELL}, which offers great motivation to investigate the electrochemical performance in detail. Thus a set of electrochemical data was collected via employing electrochemical impedance spectroscopy (EIS) measurements to evaluate further the performance of these DSSCs and is discussed in the following section.

3.2 Electrochemical Impedance Spectroscopy (EIS) Analysis

The EIS measurements are performed for both the complete DSSCs configurations. Figure 3 represents the EIS spectra of complete DSSCs which are fabricated with SWCNTs and Pt catalyst layers on both the rigid FTO-Glass and flexible substrates. All these measurements were performed underfull sun light illumination and at open circuit voltage. Moreover, the frequency range was kept from 100 mHz to 500 KHz. In this frequency range, the Pt catalyst CE based DSSC typically reveal three impedance arcs at different frequency peaks (Figure 3 a-b). Out of three, the two impedance arcs appear in the low frequency range and corresponds to the diffusion resistance (R_D, 100 mHz – 1 Hz, Figure 3) and the recombination resistance (R_{REC}, 1-100 Hz) which is also known as the resistance of photoelectrode [15-17]. Moreover, the third impedance arc which appears at high frequency peak (500 Hz-1 KHz, Figure 3) and corresponds to charge transfer resistance (R_{CT}) at the counter electrode [15-16].

On the other hand, the case of having a carbon based catalyst materials is slightly different where a very small semicircle appears at very high frequency range which corresponds to the in-pore diffusion resistance (R_{PORE}) [9, 15-17]. Another slight difference in case of having carbon based catalyst layer is the low frequency response (10-100 Hz) which does not belong to only R_{REC} but to a combined R_{REC} and R_{CT} [15-18] and can be distinguished via fabricating CE-CE cells [15-17].

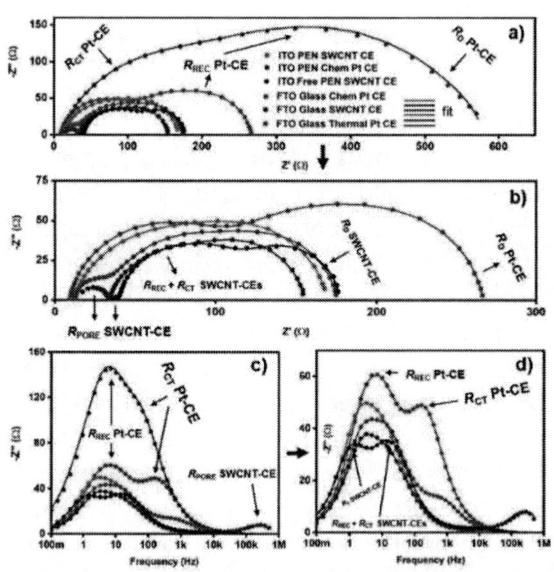

Figure 3. Typical EIS spectra of complete DSSCs fabricated with SWCNT and Pt catalyst layers on rigid and flexible substrates. a-b) Nyquist plots c-d)

frequency vs Imaginary impedance. Figures 3 b and d are plotted for better clarity.

Table 4 and 5 represents the EIS parameters via performing the equivalent circuit modeling [9, 16-17] on the measured EIS data which revealed remarkable electrochemical performance of the SWCNTs CE based DSSCs over traditional Pts CE based DSSCs.

Table 4. EIS parameters of the complete DSSCs using SWCNT as catalytic layer.

CE SWCNT	R_S Ω	R_D Ω	R_{PORE} Ω	$R_{REC}+R_{CT}$ Ω	R_{CT} Actual Ω	R_{CT} Total Ω
ITO-PEN SWCNT	11.9	58.2	0.0004	99.7	5.1	5.1
ITO free PEN SWCNT	36.4	58.4	3.0	79.0		8.0
Glass FTO SWCNT	17	42.5	16.4	80.1		21.4

Table 5. EIS parameters of the complete DSSCs using Pt as catalytic layer.

CE Pt	R_S Ω	R_D Ω	R_{REC} Ω	R_{CT} Ω
ITO-PEN Chemical Pt	11.7	15.98	373.7	180
FTO-Glass Chemical Pt	11.5	46.3	94.6	23.1
FTO Glass Thermal Pt	9.1	34.8	123.1	99.5

Since all the DSSCs are fabricated with same photoelectrodes (PEs), therefore for this study, we took the same approximation (as mentioned in our previous research work [9,17]). In short, the charge transfer resistance (R_{CT}) at the counter electrode for SWCNT CEs based DSSC was determined by first subtracting the R_{REC} of the most efficient counter electrodes among different Pt catalyst (i.e. FTO-Glass chemically platinized CE, which exhibited the lowest R_{CT} = 23.1 Ω, Table 5) from its closest combined response of R_{REC} and R_{CT} of SWCNT CE based DSSC (i.e. ITO-PEN-SWCNT CE based DSSC which exhibited 99.7 Ω as a combined response for $R_{CT}+R_{REC}$ resistance). This R_{CT} value for SWCNT CE based DSSC) was used as a reference value for the rest of SWCNT CEs based DSSCs (by adding in it the R_{PORE} resistance values in order to calculate the total charge transfer resistance ($R_{CT\text{-}TOTAL}$) of these SWCNTs CEs since their combined R_{REC} and R_{CT} response was lower than the aforementioned most efficient chemically platinized CE based DSSC (Figure 3, Table 4). With these approximations, the SWCNT CE based DSSCs revealed remarkable lower R_{CT} values in these complete DSSCs configurations than the traditional Pt catalysts based DSSCs. Among them, the flexible ITO-PEN/SWCNT CE based DSSCs exhibited the lowest R_{CT} (5.1 Ω). Moreover, the ITO free SWCNT CE also exhibited 62% lower R_{CT} (8.0 Ω) than the FTO-Glass/SWCNT CE (21.4 Ω) thus reveals its potential to be utilized with these novel copper redox shuttles as an alternative catalyst and conductive material in the futuristic fully flexible DSSCs.

On the other hand, the Pt catalysts based CEs in these complete DSSC configurations showed poor catalytic performance (Table 5) and suggest that the traditional Pt may not be considered as an efficient catalyst materials for developing both the rigid and flexible DSSCs. Among them, the most efficient CE (i.e. FTO-Glass/Chem Pt CE) revealed 21.4 Ω R_{CT} (Table 5) with very high concentrations of Pt loadings (40 mM Pt solution was utilized) which raises the concerns of high costs of fabrication. Contrary to this, a very thin (~ 500 nm-1 μm) SWCNTs with impressive catalytic activity offers great possibilities to produce efficient DSSC with this promising alternative catalytic material when fabricated with these copper based new redox shuttles. Despite of showing a reasonable R_{CT} on rigid FTO-Glass CE, the flexible Pt CE produced for this study also unexpectedly exhibited poor catalytic performance with very high R_{CT} (180 Ω, Table 5) and raises concerns for their integration in flexible DSSCs with copper electrolytes.

4. CONCLUSIONS

We successfully fabricated DSSCs for the first time with flexible CEs employing low temperature Pt and SWCNT catalysts layers and novel copper redox shuttles based electrolytes. The low temperature SWCNT catalyst layer exhibited remarkable electrochemical performance when deposited over both the rigid and flexible substrates and assembled as CEs with rigid FTO-Glass based photoelectrodes. These CEs demonstrated impressive and higher photovoltaic performance in complete DSSC configuration when compared the traditional Pt catalyst CE based DSSCs. Moreover, these efficient SWCNT CEs also showed significantly lower R_{CT} than the traditional Pt catalyst layers. Our results suggest that SWCNT can also be realized as another alternative in addition to PEDOT and Graphene to the traditional Pt catalyst, which does not seem compatible to these newly introduced copper redox based electrolytes. Moreover, these results indicate that highly efficient and fully flexible DSSC may be potentially produced via employing copper redox based electrolytes and ITO free flexible electrodes loaded with SWCNT as both the conductive and catalyst material which may decrease the overall production cost compared to the traditional and expensive Pt catalyst layers based CEs.

5. ACKNOWLEDGMENTS

This work was financed by Fortum Foundation and Academy of Finland. Haider Iftikhar thanks Fortum Foundation for Doctoral Fellowship. Ghufran Hashmi is grateful to Academy of Finland for the Post-doctoral fellowship (Decision # 287641).

6. REFERENCES

[1] M. Grätzel, Inorg. Chem. 44, 20 (2005), 6841-6851.

[2] T W. Hamann, R. A. Jensen, A. B. F. Martinson, H. V. Ryswyk, J. T. Hupp, Energy Environ. Sci., 1 (2008), 66-78.

[3] G. Hashmi, K. Miettunen, T. Peltola, J. Halme, I. Asghar, K. Aitola, M. Toivola, and P. Lund, Renew. Sust. Energ. Rev 15 (2011) 3717-3732.

[4] Y. Saygili, M. Soderberg, N. Pellet, F. Giordano, Y. Cao, A.B. Munoz-Garcia, S.M. Zakeeruddin, N. Vlachopoulos, M. Pavone, G. Boschloo, L. Kavan, J.E. Moser, M. Grätzel, A. Hagfeldt, M. Freitag, J. Am. Chem. Soc., 138 (2016), 15087-15096.

[5] M. Freitag, F. Giordano, W. Yang, M. Pazoki, Y. Hao, B. Zietz, M. Grätzel, A. Hagfeldt, G. Boschloo, J. Phys. Chem. C, 120 (2016), 9595-9603.

[6] M. Freitag, J. Teuscher, Y. Saygili, X. Zhang, F. Giordano, P. Liska, J. Hua, S.M. Zakeeruddin, J.E. Moser, M. Grätzel, A. Hagfeldt, Nat. Photonics, 11 (2017), 372-378.

[7] L. Kavan, H. Krysova, P. Janda, H. Tarabkova, Y. Saygili, M. Freitag, S.M. Zakeeruddin, A. Hagfeldt, M. Grätzel, Electrochimia Acta, 251 (2017), 167-175.

[8] L. Kavan, Y. Saygili, M. Freitag, S.M. Zakeeruddin, A. Hagfeldt, M. Grätzel, Electrochimica Acta, 227 (2017), 194-202.

[9] G. Hashmi, Janne Halme, Tapio Saukkonen, E.-Leena Rautama, P. Lund, Phys. Chem. Chem. Phys., 2013, 15, 17689-17695.

[10] Q. Wang, S. Ito, M. Grätzel, F. Fabregat-Santiago, I. Mora-Seró, J. Bisquert, T. Bessho, H. Imai, J. Phys. Chem. B, 110, 50 (2010), 25210–25221.

[11] Y. Chiba, A. Islam, Y. Watanabe, R. Komiya, N. Koide. L. Han, Jpn. J. Appl. Phys., 45, 25 (2006). 638-640.

[12] H. N. Tsao, Chenyi Yi, Thomas Moehl, J.-Ho Yum Shaik M. Zakeeruddin, M.K. Nazeeruddin, Michael Grätzel, ChemSusChem 2011, 4, 591 – 594.

[13] R. García-Rodríguez, R. Jiang, E. J. Canto-Aguilar, G. Oskam, G. Boschloo, Phys. Chem. Chem. Phys., 19 (2017), 32132-32142

[14] H. N. Tsao, P. Comte, C. Yi, M. Grätzel, ChemPhysChem 2012, 13, 2976 – 2981.

[15] K. Miettunen, M. Toivola, G. Hashmi, J. Salpakari, M. I. Asghar, P. Lund, CARBON 49 (2011) 528 – 532.

[16] J. Halme, P. Vahermaa, K. Miettunen, , P. Lund, Device Physics of Dye Solar Cells. Adv. Mater., 22 (2010), E210-E234.

[17] G. Hashmi, K. Miettunen, J. Halme, M. I. Asghar, H. Vahlman, T. Saukkonen, Zhu Huaijin, P. Lund, J. Electrochem. Soc., 159, 7 (2012), H656-H661.

SEMITRANSPARENT ORGANIC SOLAR CELLS FROM A TRANSPARENT CONDUCTIVE POLYMER ELECTRODE FOR THE 35TH EU PVSEC 2018

Moon Hee Kang*, Dong Kyo Heo, Da Hyun Kim, Dong Jae Lee
Department of Electrical Energy Engineering, Keimyung University
1095 Dalgubeol-daero, Daegu, South Korea, 42601 (*e-mail: moonhee@kmu.ac.kr)

ABSTRACT: We have fabricated semitransparent organic solar cells (OPV) owing from a transparent conductive polymer electrode. Single layer of conductive polymer can act as not only an anode electrode but hole transport layer (HTL). The figure of merit to using conductive polymer in OPV is that OPV can be fabricated from all-solution process at plastic-compatible low temperature (<200°C). In addition, due to the absence of opaque metal electrode, final OPVs exhibit semitransparency. Sheet resistance (R_{sh}) of developed conductive polymer was 260 Ω/sq. Measured from calibrated light (AM1.5G, 1000W/m^2) I-V, conductive polymer induced semitransparent OPV provides photo-conversion efficiency of 2.0% which is from open circuit voltage (V_{oc}) of 576mV, short circuit current (J_{sc}) of 6.9mA/cm2, and fill factor (FF) of 51.2%.
Keywords: conductive polymer, organic solar cell, semitransparent

1 INTRODUCTION

OPV can have either conventional or inverted structure [1]. Sequentially, fabrication of conventional OPVs follow order of ① substrate, ② anode, ③ HTL, ④ active layer, ⑤ electron transport layer (ETL), and ⑥ cathode while inverted OPVs follow order of ① substrate, ② cathod, ③ ETL, ④ active layer, ⑤ HTL, and ⑥ anode [2], [3]. As the main idea of this research is to eliminate anode electrode step by merging it into HTL, inverted OPV structure was adopted. This is because HTL and anode were adjacent in inverted structure.

Figure 1 shows the geometrical structure of the developed semitransparent inverted OPV with conductive polymer electrode [4]. The main difference compared to traditional inverted OPV is one single layer of conductive polymer (mixture of AI4083 and PH1000 poly(3,4-ethylenedioxythiophene)-poly(styrenesulfonate) (PEDOT:PSS)) works as a HTL and anode electrode together. This eliminates the need of opaque metal anode which is deposited from the high-cost vacuumed evaporator. In addition, due to the use of transparent conductive polymer electrode, final OPV was semitransparent.

Note, no vacuum process or equipment was used for OPV fabrication, and all process were performed in normal air condition (not in nitrogen filled glove box) with plastic-compatible low temperature (<200°C). This will significantly reduce the manufacturing cost and time.

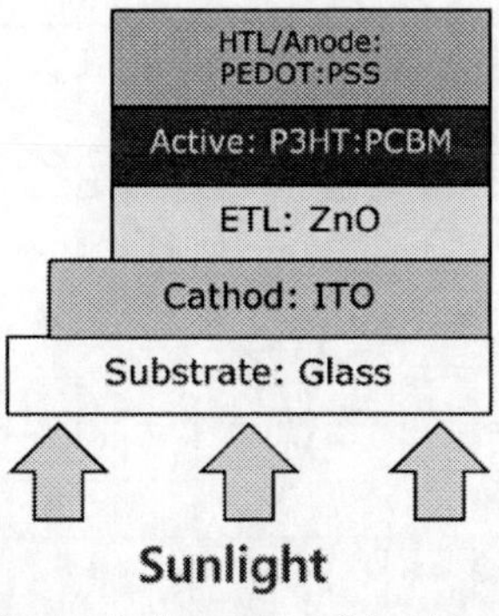

Figure 1: Geometrical structure of developed all-solution-processed conductive polymer electrode inverted OPV

2 EXPERIMENTAL

Fabrication of an all-solution-processed inverted OPV starts from an indium-tin-oxide (ITO) deposited glass substrate. ITO works as a cathode electrode in inverted OPV, and R_{sh} was 10 Ω/sq. On top of the ITO layer, zinc acetate solution was spin-coated followed by thermal annealing at 200°C to form zinc oxide (ZnO) ETL [5], [6] followed by ethoxylated-polyethylenimine (PEIE) cathode buffer layer for work function adjustment. Bulk heterojunction (BHJ) active layer was formed from a mixed solution of 1 to 1 wt% poly(3-hexylthiophene-2,5-diyl) (P3HT) and (6,6)-Phenyl C61 butyric acid methyl ester (PCBM) in 1,2-dichlorobenzene. Different P3HT and PCBM concentration solutions (Table I) were prepared to evaluate the P3HT:PCBM concentration induced cell performance. Finally, conductive polymer HTL/anode single layer was fabricated from the mixed solution of AI4083 and PH1000 grade PEDOT:PSS (Heraeus Clevios^{TM}) [4]. AI4083 PEDOT:PSS is used for interfacial layer due to its high work function, and PH1000 is used for electrode due to its high conductivity. Therefore, in this research, AI4083 was used for HTL and PH1000 was used for anode electrode. In addition, for further R_{sh} enhancement, high boiling point solvents (Ethylene glycol: EG or 2-Ethoxyethanol: 2-ee) were added [7]–[10].

Table I: Chemical composition of various P3HT:PCBM solutions

No.	P3HT	PCBM	1,2-dichlorobenzene
10mg	10mg	10mg	1ml
15mg	15mg	15mg	1ml
20mg	20mg	20mg	1ml

3 RESULTS AND DISCUSSION

3.1 All-solution-processed semitransparent OPV

Figure 2 shows the fabricated final all-solution-processed semitransparent OPVs. Four OPVs were fabricated together in one glass substrate. Active area of each OPV was 2mm by 2mm.

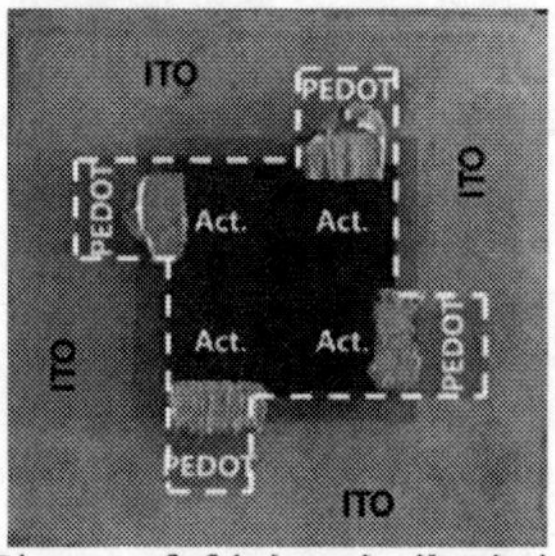

Figure 2: Picture of fabricated all-solution-processed conductive polymer electrode OPV

Figure 3 shows the field-emission scanning electron microscope (FE-SEM) pictures of the fabricated all-solution-processed OPVs with conductive polymer. Vol% concentration of P3HT:PCBM in 1,2-dichlorobenzene was varied from 10mg (0.8%:0.8%), 15mg (1.2%:1.2%), and 20mg (1.6%:1.6%). Thickness of conductive polymer (mixed solution of 1:1= AI4083:PH1000 with 12 vol% EG) spin-coated at 1000rpm was about 89nm. Note, thickness of active layer is strongly dependent on the concentration of P3HT and PCBM though all those layers were spin-coated at the same speed (400rpm).

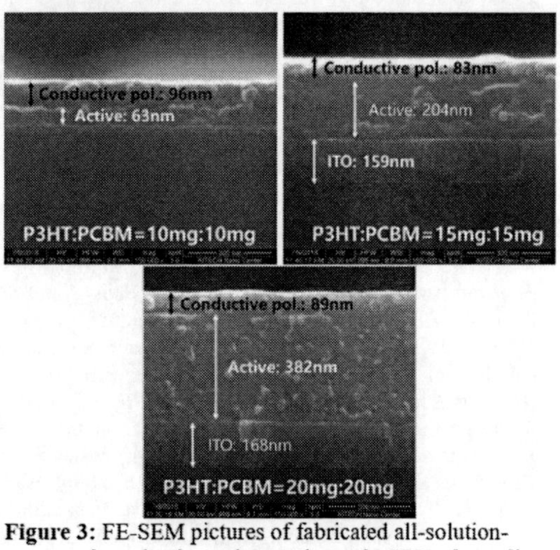

Figure 3: FE-SEM pictures of fabricated all-solution-processed conductive polymer electrode OPVs depending on the concentration of the P3HT:PCBM

3.2 Conductivity of conductive polymer electrodes

To evaluate the performance as an anode electrode, various conductive polymers were deposited. All conductive polymers were spin-coated at 1000rpm for 30sec. Figure 4 plots the R_{sh} of various conductive polymer depending on its additive solvents. PH1000 PEDOT:PSS alone provides R_{sh} of 0.14 MΩ/sq. When we add same volume of AI4083 on the PH1000, R_{sh} dramatically increased to 7.7 MΩ/sq. Therefore, strategies for significant R_{sh} reduction have to be considered. As a first step, we added 12 vol% EG into the PH1000 and AI4083 mixture, and could obtain R_{sh} of 386 Ω/sq. Finally, for further reduction, PH1000 was added more to make vol% of PH1000 and AI4083 to 2:1. As a result, we could obtain conductive polymer's R_{sh} of 260 Ω/sq which is enough to be used as an electrode.

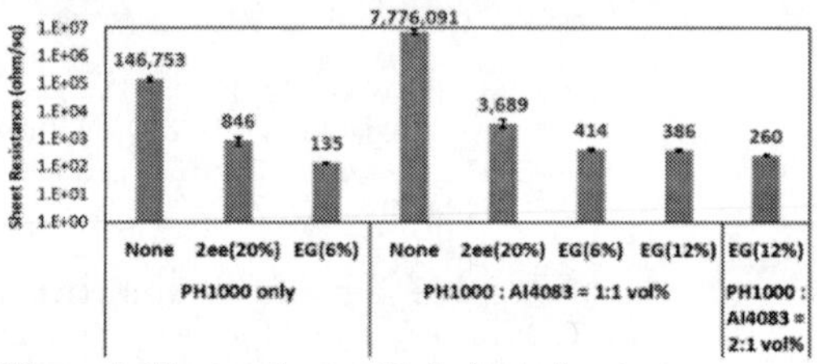

Figure 4: Sheet resistance (R_{sh}) chart of various conductive polymers

3.3 Transparency of fabricated OPVs

This section is to evaluates the performance of the final OPV in terms of the semitransparency. Figure 5 Shows the transmittance curve of the all-solution-processed OPVs as a function of the concentration of the P3HT:PCBM in 1ml 1,2-dichlorobenzene. Obviously, higher concentration of P3HT:PCBM provides lower transmittance. Note, double the concentration reduced the transmittance more than half. This is due to the fact of concentration-induced thickness increase shown in Figure 3.

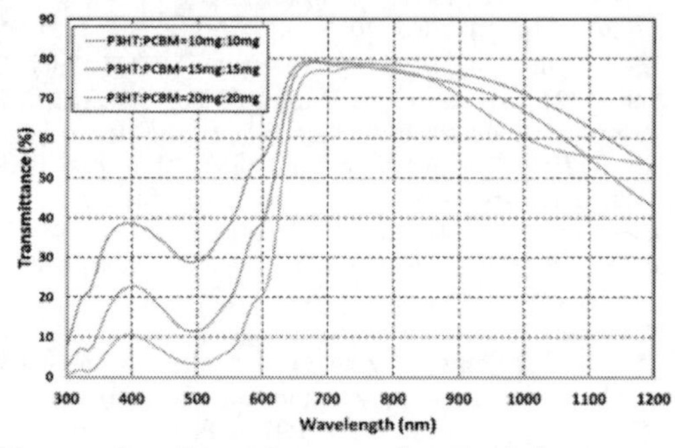

Figure 5: Transmittance of all-solution-process conductive polymer electrode inverted OPV as a function of P3HT:PCBM concentration

3.4 Dark and light I-V of fabricated all-solution OPVs

In this section, I-V characteristics of final devices were measured. Figure 6 plots the dark I-V characteristic of the final all-solution-processed conductive polymer electrode inverted OPVs. As P3HT:PCBM concentration decreased, reverse saturation current (J_o) increased.

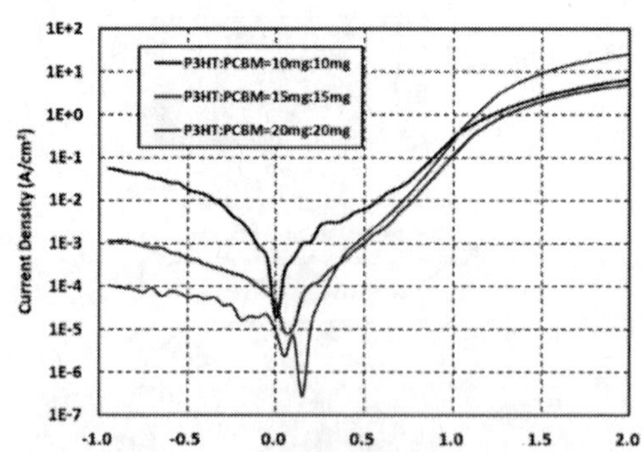

Figure 6: Dark I-V curve of the final inverted OPV

However, in terms of the forward-biased current, P3HT:PCBM=15mg:15mg in 1ml 1,2-dichlorobenzene provides lowest current density, which may provide potential highest open circuit voltage (V_{oc}) as photo-generated currents are opposite to forward-biased current.

Therefore, Figure 7 shows the light I-V characteristic of final all-solution-processed conductive polymer electrode semitransparent OPV when the active layer was made from the solution of P3HT:PCBM=15mg:15mg in 1ml 1,2-dichlorobenzene. AM1.5G spectrum calibrated 1000W/m^2 simulated light was used for light I-V measurement. Fabricated all-solution-processed conductive polymer electrode OPV provided photo-conversion efficiency of 2.0% accounted from V_{oc} of 576mV, J_{sc} of 6.9mA/cm2, and FF of 51.2%.

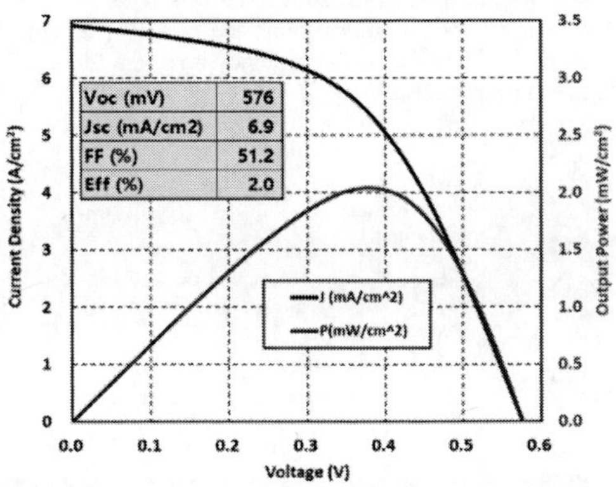

Figure 7: Light I-V curve of the all-solution-processed conductive polymer electrode semitransparent inverted OPV

4 CONCLUSION

In this research, all-solution-processed conductive polymer electrode inverted OPV was fabricated. Conductive polymer (mixture of AI4083 and PH1000 PEDOT:PSS with additive solvent) act as both HTL and anode electrode, together. Achieved best R_{sh} of conductive polymer electrode was 260 Ω/sq. When we apply this conductive polymer on the inverted structure OPV, photo-conversion efficiency of 2.0% was achieved with V_{oc} of 576mV, J_{sc} of 6.9mA/cm2, and FF of 51.2%.

5 REFERENCES

[1] J. J. Van Franeker *et al.*, "All-solution-processed organic solar cells with conventional architecture," *Sol. Energy Mater. Sol. Cells*, vol. 117, pp. 267–272, 2013.

[2] H.-L. Yip and A. K.-Y. Jen, "Recent advances in solution-processed interfacial materials for efficient and stable polymer solar cells," *Energy Environ. Sci.*, vol. 5, no. 3, p. 5994, 2012.

[3] F. C. Krebs, "Fabrication and processing of polymer solar cells : A review of printing and coating techniques," *Sol. Energy Mater. Sol. Cells*, vol. 93, pp. 394–412, 2009.

[4] Y. Zhou, H. Cheun, S. Choi, C. Fuentes-Hernandez, and B. Kippelen, "Optimization of a polymer top electrode for inverted semitransparent organic solar cells," *Org. Electron. physics, Mater. Appl.*, vol. 12, no. 5, pp. 827–831, 2011.

[5] B. Niesen and B. P. Rand, "Thin film metal nanocluster light-emitting devices," *Adv. Mater.*, vol. 26, no. 9, pp. 1446–1449, 2014.

[6] T.-W. Koh *et al.*, "Metal nanocluster light-emitting devices with suppressed parasitic emission and improved efficiency: exploring the impact of photophysical properties," *Nanoscale*, vol. 7, no. 20, pp. 9140–9146, 2015.

[7] L. Hu, H. S. Kim, J. Lee, P. Peumans, and Y. Cui, "Scalable Coating and Properties of transparent Ag nanowire," *ACS Nano*, vol. 4, no. 5, pp. 2955–2963, 2010.

[8] D.-S. Leem, A. Edwards, M. Faist, J. Nelson, D. D. C. Bradley, and J. C. de Mello, "Efficient Organic Solar Cells with Solution-Processed Silver Nanowire Electrodes," *Adv. Mater.*, vol. 23, no. 38, pp. 4371–4375, 2011.

[9] Y. J. Noh, S. S. Kim, T. W. Kim, and S. I. Na, "Cost-effective ITO-free organic solar cells with silver nanowire-PEDOT:PSS composite electrodes via a one-step spray deposition method," *Sol. Energy Mater. Sol. Cells*, vol. 120, no. PART A, pp. 226–230, 2014.

[10] Y. K. Seo *et al.*, "Efficient ITO-free organic light-emitting diodes comprising PEDOT:PSS transparent electrodes optimized with 2-ethoxyethanol and post treatment," *Org. Electron.*, vol. 42, pp. 348–354, 2017.

THE POTENTIAL AND DESIGN PRINCIPLE FOR NEXT-GENERATION SPECTRUM-SPLITTING PHOTOVOLTAICS: TARGETING 50% EFFICIENCY THROUGH BUILT-IN FILTERS AND GENERALIZATION OF CONCEPT

Dongchen Lan and Martin A.Green
Australian Centre for Advanced Photovoltaics
University of New South Wales
Sydney NSW 2052, Australia

ABSTRACT: Increasing the sunlight to electricity conversion efficiency is a key driver for on-going photovoltaic cost reduction. One of the most successful approaches for improving the efficiency is to split sunlight into suitable spectral bands for a set of solar cells with appropriate energy band gaps. Using this concept, in 2014, our group demonstrated the first photovoltaic conversion efficiency above 40%, where an external filter directed a portion of the sunlight normally wasted by commercial InGaP/InGaAs/Ge triple-junction cells to a silicon cell. Here, we analyze the primary power losses in state-of-the-art spectrum-splitting photovoltaic modules and evaluate the potential for further efficiency increase. Two further developments, applying built-in filters and generalizing the present concept, are proposed for next-generation spectrum-splitting photovoltaics, which can make 50% efficiency a realistic target. The design principal is briefed as is the comparison with monolithic multijunction cells.
Keywords: Spectrum-splitting, luminescent coupling, Bragg reflector, mismatch, efficiency

1 INTRODUCTION: BACKGROUND AND MOTIVATION

The multijunction approach promises to reduce the cost of solar photovoltaic electricity production significantly through high energy conversion efficiencies. However, commercial monolithic multijunction solar cells often have power losses due to current-mismatching tolerated in the interests of lattice matching. A typical example is that, in commercial InGaP/InGaAs/Ge triple-junction cells, the Ge cell generates much more photocurrent than the top two III-V cells under sunlight. This results in not only the extra Ge current being wasted but some power dissipated as heat within the cell, increasing challenges in maintaining low operating temperatures where the cells are most efficient. This shortfall significantly limits the photovoltaic energy conversion efficiency in practice and becomes more severe for cells using concentrated sunlight.

To tackle this issue, in 2014, Green et al. [1] used a band-pass filter to divert some light wasted by the Ge cell to a separate silicon cell, demonstrating the first conversion of sunlight to electricity with an efficiency above 40% (Fig. 1). Besides this being a significant milestone in solar cell efficiency, the approach is readily accessible to the industry in that all cells are commercially available. Here, we analyse existing power loss mechanisms in state-of-the-art spectrum-splitting photovoltaic modules and evaluates the potential for further efficiency enhancement. Two further developments are proposed in light of the analysis. One development is to build filters into the rear of the cells themselves that can drive the efficiency towards 45%. The other development is to generalize the concept by having two electrically separate tandem cells with one absorbing light directed from the other which can open the possibility of 50% efficiencies. The design principal and mutual connection between the two developments are briefly discussed as is the comparison with monolithic multijunction cells.

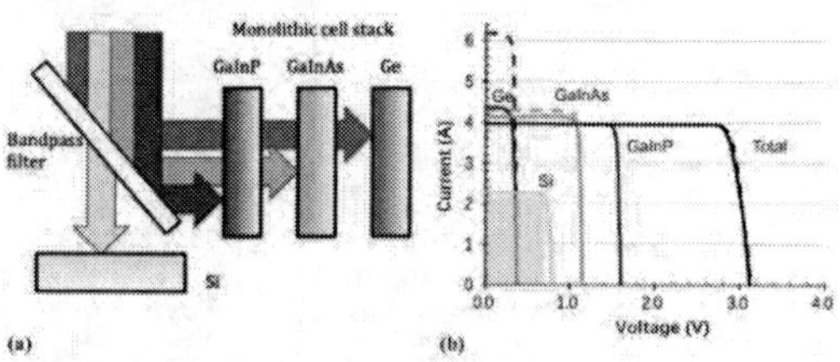

Figure 1: Schematics of (a) an over 40% efficient state-of-the-art spectrum-splitting concentrator photovoltaic module and (b) the current voltage curves (with the dashed lines being the characteristics without the band-pass filter). The external band-pass filter directs a portion of sunlight wasted by the tandem cell to a separate silicon cell. The data are from a most recent publication [2]. Colors are indicative only.

2 EFFICIENCY LOSS MECHANISMS AND POTENTIAL FOR FURTHER EFFICIENCY ENHANCEMENT

Two primary efficiency loss mechanisms can be identified for the above system. First, emissions from high- to low-band-gap cells result in a reduction in the high-band-gap cell's voltage as well as a subsequent power loss due to carrier thermalisation in the low-band-gap cell. Second, there is still some light wasted by the Ge cell that could be diverted to another cell for an increase in efficiency.

The first power loss mechanism, as a result of luminescent coupling [3-11] from high- to low-band-gap cells, is detrimental to the emitting cell's voltage. This is due to the cell's rear emissions increasing its dark saturation current [12]. Presuming rear-emission does not alter other material properties significantly, it can be shown that, when a cell operates at a voltage $V \gg kT/q$, the drop in its operating voltage at a fixed current due to its rear-emission is approximately:

$$\Delta V \cong \frac{kT}{q}\ln(1+\eta n^2), \quad (1)$$

where η is the radiative efficiency of the cell if it has no rear-emissions or other rear surface optical losses, n is the refractive index of the medium underlying the rear of the cell taken to be 3.5 and kT/q is the thermal voltage. If such rear emissions are reflected back to the emitting cells for reabsorption, in the radiative limit where only radiative recombination occurs, i.e., $\eta = 1$, the top two cells will have an increase of 0.067V in each its voltage with a total increase of 0.134V for the tandem device. This would give an absolute increase of ~1.8% (~35.7% × 0.134V/2.7V) in the tandem efficiency. For good real cells with radiative efficiencies ~30%, the tandem voltage will have an increase of ~0.08V, corresponding to an absolute increase of ~1.1% in the tandem efficiency part. Noticeably, the data are for cells using concentrated sunlight and the relative improvements will be larger for 1-sun modules since each cell operates at a lower voltage but with the same absolute improvement.

For the second power loss mechanism, the previous [1, 2] band-pass filter directs out light from 880 to 1030 nm (1.41 to 1.206 eV) only (Fig.2), suggesting that the Ge cell is still generating more photocurrent than the top two cells hence still dissipating some power. In fact, the wavelength range for the directed light can be extended further to ~1190 nm (1.04 eV) where the Ge cell can generate just matched photocurrent, unless it has significant defects resulting in too much bend in its I-V curve before reaching its maximum power point voltage. Noting that silicon has a band gap of 1.11 eV, extending the directed long-wavelength edge to 1120 nm (1.11 eV) can increase the silicon current by ~58%, corresponding to an absolute increase of ~2.6% (~4.4% × 58%) in the silicon efficiency part. However, as silicon can absorb light effectively up to 1120 nm only, light from 1120 to 1190 nm cannot be harnessed by a silicon cell. Hence, replacing the silicon cell by another high-quality cell with a 1.04 eV or slightly lower band gap, such as an InPAs cell, may improve the efficiency further. Compared with the old scheme where only light from 880 to 1030 nm is directed to the silicon cell, the current will be increased by ~85.7% and the voltage will be decreased by ~6.3%. This can boost the detached single cell efficiency by ~74%, corresponding to an absolute increase of ~3.3% in the detached cell efficiency part.

Hence, addressing the two power losses identified above can increase the overall photovoltaic energy conversion efficiency to 45% (40.6%+1.1%+3.3%). Besides tackling these two power loss issues, the efficiency of this system may be further improved by optimizing each cell's band gap or, if needed, increasing the cell number to better harness the solar spectrum.

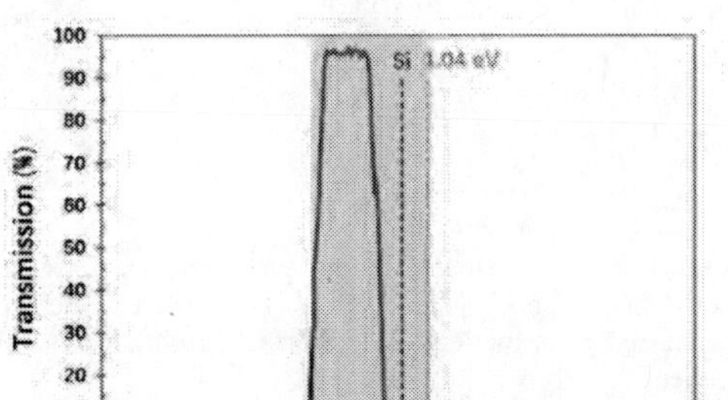

Figure 2: The actual (solid line) and ideal (shaded) transmission spectra of the band-pass filter in the spectrum-splitting photovoltaic module shown in Fig. 1. Note that the transmitted light here is the light diverted to the silicon cell.

3 BUILT-IN FILTERS FOR IMPROVED LIGHT MANAGEMENT

Interestingly, filters with suitable optical windows built into the rear of the cells themselves can address the above two power loss mechanisms simultaneously. For instance, as shown in Fig. 3, a filter capable to reflect light of 750–1190 nm built into the rear of the 1.41 eV GaInAs cell can both reflect its rear emissions (750–900 nm) back for reabsorption and direct the extra light (900–1190 nm) to another cell for additional photovoltaic power production.

Such built-in filters may be realized through Bragg-type structures that are essentially stacked thin-film layers of alternating high and low refractive indices. They use light interference to reflect certain wavelengths but transmit others. Key requirements of these thin-film optical layers include transparency to the wavelengths of interest and good conductivity of electricity. Notably, such filters can be easily built into the rear of a solar cell using current industrial facilities. There have already been works [13, 14] reporting on the use of distributed Bragg mirrors to enhance light reabsorption in monolithic multijunction cells. The method can be adapted, such as by tuning the optical windows, to find its new applications in next-generation spectrum-splitting photovoltaics.

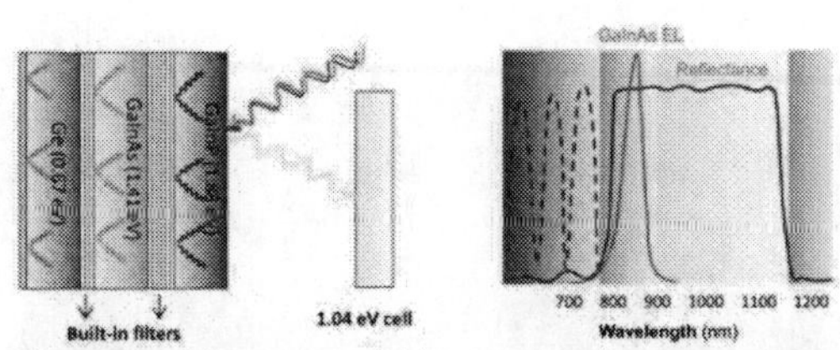

Figure 3: Schematics of (a) a proposed new spectrum-splitting system with built-in filters and the silicon cell replaced by a slightly lower gap (1.04 eV) cell; (b) the GaInAs electroluminescence and an effective rear filter's reflectance. Colors are indicative only.

4 GENERALISATION OF CONCEPT

The concept may be further generalized by having two monolithic tandem cells with built-in filters, where a second tandem absorbs light directed out from the first tandem (Fig. 4). This can improve the overall efficiency on two accounts. First, sunlight may be better harnessed with an increased number of cells of appropriate band gaps. Second, splitting a long cell stack into two short stacks can reduce a range of mismatch issues that limit the performance of series-connected cells. Note that these include more than photocurrent mismatching. For instance, the lowest-band-gap cell normally has the largest dark saturation current which can move the operating points of the other series-connected cells away from their maximum power points. This can have a most severe impact on the performance of the highest-band-gap cell that normally has the smallest dark saturation current. Having two electrically separate tandems ensures that the lowest-band-gap cell can only lightly impact its series-connected medium-band-gap cells instead of all other cells with higher band gaps.

Fig. 4 shows an optimal design using the dual tandem concept where two monolithic cell stacks are lattice matched to the GaAs and InP wafer substrates. We predict the module performance by realistic calculations based on the following constraints that were universally applied for cells of ~1-cm^2 area:

(i) No photovoltaic response at wavelengths below 350 nm and the response at long wavelengths cuts off at the band gap energy (E_g);

(ii) 90% of incident photons in the above range contribute to the photocurrent;

(iii) Each cell's open-circuit voltage is empirically given by $V_{OC} = E_g/q - 0.4 + 0.03\ln(X) - 0.03\ln(Y)$ in volts, where X is the sun concentration and Y is the ratio of the current in the cell if it were the top cell to the reduced current if it were not (approximately equals the cell number N if cells are current-matched);

(iv) The fill factor of a cell stack is taken to be 85% at all suns concentrations.

The robustness of this evaluation method is verified by its ability to accurately predict the performance of a large range of world record devices listed in Ref. [15], though small errors may occur due to different experimental fill factors or cell areas. (A significantly smaller cell area may boost the actual cell efficiency.) Using this realistic evaluation method, the optimal design shown in Fig. 4 is predicted to have a practical efficiency of 51.0% at 300 suns and 51.7% at 500 suns, with an absolute further improvement of ~1% possible using built-in filters.

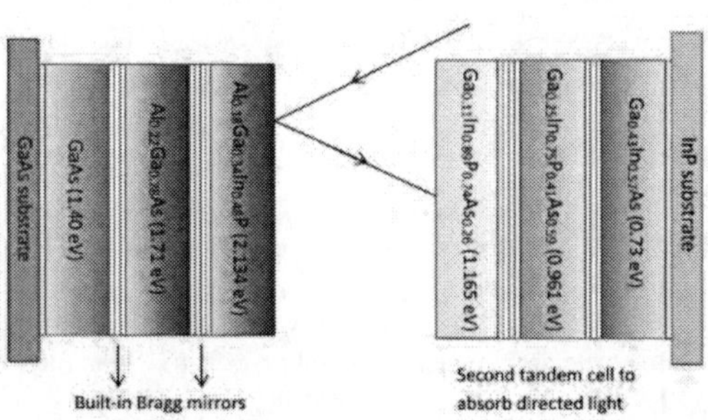

Figure 4: Conceptual design of a next-generation spectrum-splitting module consisting of two monolithic cell stacks lattice-matched to GaAs and InP wafers. Each cell stack has built-in filters and is current-matched (but not necessarily to the same value). Colors are indicative only.

5 THE DESIGN PRINCIPAL AND MUTUAL CONNECTION BETWEEN THE TWO DEVELOPMENTS

For the design of rear Bragg-type mirrors, current industrial facilities can deposit at least 32 optical layers onto the rear of a cell, which is sufficient to make an effective built-in filter. However, provided the filters can divide light appropriately, minimal number of optical layers is desired as this would improve the electrical conductivity and lower the potential fabrication cost. This can be realized by noting two points. First, the built-in feature allows the requirements of filter properties over some wavelengths to be relaxed compared with external ones. Taking the 1.41 eV GaInAs cell in Fig. 3 as an example, its rear filter's reflectivity for wavelengths below 800 nm would not matter because incident light of these wavelengths is largely absorbed before reaching its rear and the cell does not have appreciable self-emissions in this wavelength range. Second, instead of using repeated periods of optical layers, we may design chirped Bragg reflectors by tuning the thicknesses and refractive indices of certain optical layers for more desired reflection patterns.

For the implementation of dual tandem concept, each cell stack is designed to be current-matched, though not necessarily to the same value. The first cell stack is lattice-matched to a GaAs wafer substrate and the second is lattice-matched to an InP wafer substrate. For the cells on the GaAs wafer, band gaps between 1.40 and 2.30 eV is accessible using lattice-matched solid solutions of GaAs, GaP, InP, AlP and AlAs [16]. For the cells on the InP wafer, band gaps between 0.73 and 1.40 eV is accessible using lattice-matched solid solutions of GaAs, InAs, InP, AlAs and GaSb [16]. Fig. 4 shows a rather sophisticated 6-cell module that can take photovoltaic conversion of sunlight to efficiency levels beyond present worldwide expectations. But a much simpler 4-cell module consisting of 1.89/1.40 eV cells on the GaAs wafer and 1.041/0.73 eV cells on the InP wafer, is predicted to deliver an over 48% efficiency under 500 suns, which can already beat the current efficiency record for solar cells of any type.

Notably, the dual tandem approach will not only make better use of the solar spectrum but allow more flexibility in the filter design since the requirement on the spectral shape of the directed light can be significantly relaxed. In particular, the spectrum of the light directed out from the first tandem no longer needs to be a continuous band. It can even be two discrete parts with irregular shapes, which can still be well utilized by a second tandem cell with suitable band gaps (Fig. 5). A more flexible requirement on the reflectance spectrum is likely to lead to even fewer optical layers required to make an effective filter, which can result in both improved efficiencies and reduced fabrication costs.

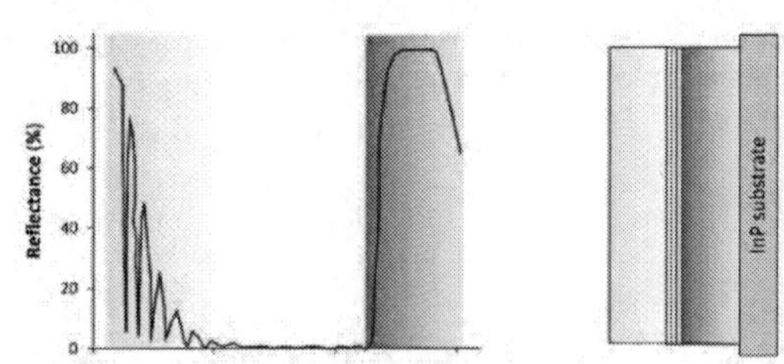

Figure 5: A schematic example of reflected spectrum consisting of two distinct parts (left) that can be well harnessed by a two-cell tandem stack (right). Colors are indicative only.

6 COMPARISON WITH MONOLITHIC TANDEM DEVICES

Compared with the monolithic multijunction approach, the present spectrum-splitting approach has advantages in both device performance and technical implementation.

In terms of potential device performance, the spectrum-splitting approach is obviously useful when the tandem current-matching is difficult due to failure to identify lattice-matched materials of appropriate band gap and good material properties. It is expected to exhibit stronger spectral robustness when the actual spectral condition departs from the designed spectrum since there are two electrically separate devices absorbing high- and low-energy light respectively. Besides tackling the photocurrent mismatch, it can cope with a range of other mismatch issues that can be severe for a series connection of multiple cells. For instance, an often-neglected issue in monolithic quadruple or quintuple junction cells is that the lowest-band-gap cell normally has a much larger dark saturation current which moves the operating points of the other cells away from their maximum power points. Splitting it into two electrically separate devices can obviously reduce the issue. As mentioned previously, optical losses such as luminescent coupling and reflection related losses are minimized too using the presently proposed method of built-in filters. These are all expected to improve the overall photovoltaic energy conversion efficiency further.

In terms of technical implementation, the spectrum-splitting approach can reduce the challenges in wafer bonding because junctions can be physically separate using this method. This is particularly advantageous when there are more than three junctions in pursuit of ultrahigh efficiencies. The approach is compatible with industrial facilities in that Bragg-type filters can be easily built into cells and the current 40.6% efficient spectrum-splitting photovoltaic system [2] used all commercial grade cells. These suggest the approach is technically easier to implement and is likely cheaper than alternatives being investigated internationally. Although spectrum-splitting solar cells may take more room than monolithic tandem cells, the proposed built-in filters save space significantly compared with the current demonstration using external filters.

The proposed new method is also more compatible with future generation photovoltaics in the sense that low-cost photovoltaics often involve combining materials of vastly different properties (e.g., silicon and perovskites), where the traditional monolithic approach may be more restricted in some aspects.

7 SIGNIFICANCE AND CONCLUSIONS

The significance of the proposed next-generation spectrum-splitting approach lies in two aspects, namely its capability to take photovoltaic energy conversion efficiencies to levels beyond present worldwide expectations and accessibility to the industry. It uses not only built-in filters to minimize luminescent coupling related losses and divide sunlight more appropriately, but also two cell stacks to better harness the entire solar spectrum and reduce some mismatch issues. The method is readily accessible to the industry in that a recent landmark result was demonstrated by commercial cells [1] and the proposed Bragg-type filters for the next-generation version can be easily built into cells by industrial facilities.

Provided these filters can split light appropriately, minimal number of optical layers is desired to ensure good electrical conductivity and low fabrication costs. This is attainable by taking advantage of relaxed requirements of filters' optical properties over some wavelengths and tuning the thickness and refractive indices of certain layers. As tandem cells can well harness various spectrums, the dual tandem configuration allows increased flexibility in the filter design too.

Compared with the monolithic tandem approach, the spectrum-splitting approach can save wafer-bonding and various mismatch issues and is likely to show better spectral robustness since different junctions can now be separate. Hence, it is technically easier to implement and can reach higher efficiencies. Compared with currently used external filters, the proposed built-in filters can save space significantly and tackle optical losses better. This technology is also more compatible with future generation photovoltaics where vastly different absorbers are often combined together to produce the highest efficiencies at the lowest costs.

ACKNOWLEDGEMENTS

The Australian Centre for Advanced Photovoltaics commenced operation in February 2013 with support from the Australian Government through the Australian Renewable Energy Agency. Responsibility for the views, information or advice expressed herein is not accepted by the Australian government.

REFERENCES

[1] M.A. Green, M.J. Keevers, I. Thomas, J.B. Lasich, K. Emery, R.R. King, 40% efficient sunlight to electricity conversion, Progress in Photovoltaics: Research and Applications, 23 (2015) 685-691.

[2] M.A. Green, M.J. Keevers, B. Concha-Ramon, J. Jiang, Verlinden, P.J., Y. Yang, X. Zhang, 37% Efficient One-Sun Minimodule and over 40% Efficient Concentrator Submodules, 32nd EU PVSEC, Munich, June, (2016).

[3] A.S. Brown, M.A. Green, Radiative coupling as a means to reduce spectral mismatch in monolithic tandem solar cell stacks theoretical considerations, in: Photovoltaic Specialists Conference, 2002. Conference Record of the Twenty-Ninth IEEE, IEEE, 2002, pp. 868-871.

[4] M.A. Steiner, J.F. Geisz, T.E. Moriarty, R.M. France, W.E. McMahon, J.M. Olson, S.R. Kurtz, D.J. Friedman, Measuring IV curves and subcell photocurrents in the presence of luminescent coupling, IEEE Journal of Photovoltaics, 3 (2013) 879-887.

[5] D. Lan, J.F. Geisz, M.A. Steiner, I. Garcia, D.J. Friedman, M.A. Green, Improved modeling of photoluminescent and electroluminescent coupling in multijunction solar cells, Solar Energy Materials and Solar Cells, 143 (2015) 48-51.

[6] Z. Ren, J.P. Mailoa, Z. Liu, H. Liu, S.C. Siah, T. Buonassisi, I.M. Peters, Numerical analysis of radiative recombination and reabsorption in GaAs/Si tandem, IEEE Journal of Photovoltaics, 5 (2015) 1079-1086.

[7] D. Lan, M.A. Green, Equivalent circuit analysis of radiative coupling in monolithic tandem solar cells, Applied Physics Letters, 106 (2015) 263902.

[8] J.F. Geisz, M.A. Steiner, I. Garcia, R.M. France, W.E. McMahon, C.R. Osterwald, D.J. Friedman, Generalized optoelectronic model of series-connected multijunction solar cells, IEEE Journal of Photovoltaics, 5 (2015) 1827-1839.
[9] D. Lan, M.A. Green, Photoluminescent and electroluminescent couplings in monolithic tandem solar cells, Progress in Photovoltaics: Research and Applications, 24 (2016) 1566-1576.
[10] D. Lan, M.A. Green, Limiting efficiencies of GaInP/GaAs/Ge up-conversion systems: Addressing the issue of radiative coupling, Applied Physics Letters, 109 (2016) 123508.
[11] A. Karani, L. Yang, S. Bai, M.H. Futscher, H.J. Snaith, B. Ehrler, N.C. Greenham, D. Di, Perovskite/Colloidal Quantum Dot Tandem Solar Cells: Theoretical Modeling and Monolithic Structure, ACS Energy Letters, 3 (2018) 869-874.
[12] D. Lan, M.A. Green, Ideal solar cell equation in the presence of photon recycling, Journal of Applied Physics, 116 (2014) 174511.
[13] M. Shvarts, O. Chosta, I. Kochnev, V. Lantratov, V. Andreev, Radiation resistant AlGaAs/GaAs concentrator solar cells with internal Bragg reflector, Solar energy materials and solar cells, 68 (2001) 105-122.
[14] M.A. Stan, A. Cornfeld, Inverted Metamorphic Multijunction Solar Cells with Distributed Bragg Reflector, in, Google Patents, 2008.
[15] M.A. Green, Y. Hishikawa, W. Warta, E.D. Dunlop, D.H. Levi, J. Hohl-Ebinger, A.W. Ho-Baillie, Solar cell efficiency tables (version 50), Progress in photovoltaics: Research and applications, 25 (2017) 668-676.
[16] I. Vurgaftman, J. Meyer, L. Ram-Mohan, Band parameters for III–V compound semiconductors and their alloys, Journal of applied physics, 89 (2001) 5815-5875.

GaAs P-ON-N SOLAR CELLS WITH MOVPE GROWTH RATE OF 120 μm/h

Hassanet Sodabanlu,[1,*] Akinori Ubukata,[2] Kentaroh Watanabe,[1] Takeyoshi Sugaya,[3] Yoshiaki Nakano,[4] Masakazu Sugiyama[1,4]
[1] Research Center for Advanced Science and Technology, the University of Tokyo
[2] Tsukuba Laboratories, Taiyo Nippon Sanso
[3] National Institute of Advanced Industrial Science and Technology (AIST)
[4] Department of Electrical Engineering and Information systems, School of Engineering, the University of Tokyo
*Corresponding author: sodabanlu@hotaka.t.u-tokyo.ac.jp

ABSTRACT: The growth rate of GaAs in our MOVPE reactor has been boosted to 120 μm/h which is limited by the current mass flow controllers of trimethylgallium bubblers. Effects of growth rate on the quality of GaAs and the efficiency of p-on-n solar cells were investigated. The minority hole lifetime in n-GaAs was clearly shortened with an increase in growth rate from 20 to 90 and 120 μm/h using the same growth conditions. The GaAs p-on-n solar cells comprising of 2-μm-thick n-GaAs base grown at 20, 90 and 120 μm/h using a V/III ratio of 20 were evaluated under 100 mW/cm^2 AM1.5G illumination. Because of short hole lifetime in base layer, the J_{sc} as well as conversion efficiency obviously decreased with an increase in growth rate. A degradation of carrier extraction from n-GaAs base layer was revealed by external quantum efficiency spectra at long wavelength range. The average efficiency was diminished from 24.51 to 23.77% when the growth rate of n-GaAs was increased from 20 to 120 μm/h.
Keywords: III-V Semiconductors, Gallium Arsenide Based cells, Photovoltaic

1 INTRODUCTION

During last few years, III-V multiple junction solar cells (MJSCs) have been a focus of research attentions because of their extra high conversion efficiency. The recent record conversion efficiency of 46% has been achieved by 4-junction structure under a concentration light of 508 sun [1]. However, the cost of III-V semiconductor solar cells is relatively high as compared to that of other materials including silicon and CIGS. This limits the usages of III-V semiconductor solar cells to some special tasks, including concentrating photovoltaic (CPV) [2-3] and space applications [4-5]. Much effort has been devoted to reduce the cost of GaAs based solar cells including the epitaxy cost, generally by metalorganic vapor phase epitaxy (MOVPE). This can be realized by either improving material (source) utilization or increasing the growth rate (decrease machine overhead time) [6-8]. Our research group has demonstrated that the growth rate of GaAs in our MOVPE system could be accelerated to 90 μm/h [9-10], by shrinking the boundary layer thickness for the mass transport of a Ga precursor [11]. GaAs p-on-n solar cells exhibited insignificant degradation of cell efficiency which their n-base layers were deposited at 80 μm/h using a V/III ratio of 40 as compared to the growth at 20 and 60 μm/h [9]. A high V/III ratio can plausibly be a reason for no degradation in the cell performance up to 80 μm/h. In addition, several reports have shown that the hole lifetime is less sensitive to the increase in the threading dislocation density in GaAs than the electron lifetime [12-13]. This may make a p-on-n device more robust against an increase in defect density. In other words, the defect-tolerant structure of p-on-n GaAs solar cells may allow the growths of GaAs with a lower V/III ratio with negligible performance degradation [14]. In this work, the growth rate of GaAs was further enhanced to 120 μm/h which was limited by the mass flow controllers of trimethylgallium (TMGa) bubblers in our MOVPE system. The qualities including surface morphology and minority carrier lifetime of GaAs epitaxial layer at this extremely fast growth rate were evaluated and compared with those of GaAs grown at a lower growth rate. Later on, influences of GaAs growth rate on conversion efficiency of p-on-n solar cells were investigated.

2 EXPERIMENTAIL DETAILS

The crystal growth in this study were carried out in a MOVPE reactor (Taiyo Nippon Sanso, HR3335) using standard precursors, including TMGa, trimethylindium (TMIn), trimethylaluminium (TMAl), arsine (AsH_3), phosphine (PH_3), diethylzinc (DEZn) and disilane (Si_2H_6). Si-doped (001) GaAs (2-3×10^{18} cm^{-3}) wafers with 2° misorientation toward <110>, were used as substrates. First, 4-μm-thick unintentionally doped (ud) GaAs layers were deposited with various growth rates of 20, 90 and 120 μm/h using a V/III ratio of 20. Atomic force microscopy (AFM) was utilized to observe the surface morphology of epitaxial layers. In this work, a GaAs p-on-n solar cell structure will be later investigated, therefore, the hole minority lifetime in n-doped GaAs is one of the important parameters to be evaluated. The time-resolved photoluminescence (TRPL) was emplyed in order to characterize the impacts of growth rate on the properties of minority carriers in Si-doped GaAs (electron concentration: ~4×10^{17} cm^{-3}) in the n-AlGaAs/n-GaAs/n-AlGaAs doublehetero (DH) structure. Finally, GaAs p-on-n solar cells were fabricated, in which their 2-μm-thick n-GaAs base layer was grown at a speed of 20, 90 and 120 μm/h, respectively, under a V/III ratio of 20. The current-voltage (I-V) characteristic under AM1.5G illumination and the external quantum efficiency (EQE) spectrum of each solar cell with ZnS/SiO_2 anti-reflection coating (ARC) were evaluated and compared.

3 SURFACE MORPHOLOGIES

AFM images, with a scanning area of 2x2 μm^2, of ud-GaAs grown with various growth rates of 20, 90 and 120 μm/h are shown in Figs. 1(a), 1(b) and 1(c), respectively. The reactor temperature, total pressure and V/III ratio during the growth were 680 °C, 6 kPa and 20, respectively. Ridge-like textures along [001] direction in parallel with the step edges of the vicinal substrate were observed for all the samples. A faster growth rate results in a rougher surface morphology presumably due to a relatively slower surface migration of Ga adatoms compared with the growth rate. The background hole concentrations in ud-GaAs were 1.6x10^{15}, 2.0x10^{16}, and 3.0x10^{16} cm^{-3} for the growth rate of 20, 90, and 120 µm/h, respectively. The 20-µm/h grown sample exhibited notably different surface morphology. A reason is that this particular sample was fabricated with a slightly different reactor configuration.

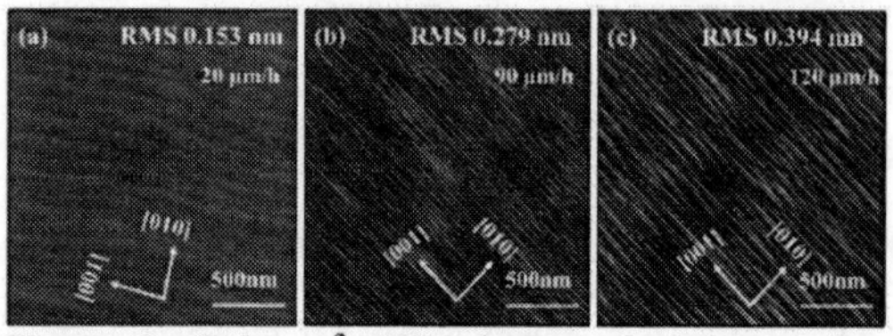

Figure 1: 2×2 µm^2 AFM images depicting surface morphologies of (a) 20, (b) 90, and (c) 120 µm/h grown 4-µm-thick undoped GaAs layers

4 MINORITY HOLE LIFETIME

Figure 2 shows (a) the AlGaAs/GaAs double hetero (DH) structure employed for the characterization of minority carrier lifetime. The tested GaAs layers were Si doped and their active carrier concentrations were evaluated by electro-chemical voltage (ECV) measurement, shown in Fig. 2(b), to be approximately 4-5x10^{17} cm^{-3}. This value is similar to the electron concentration in n-GaAs base layer of GaAs p-on-n solar cells investigated in this work. The reactor temperature, total pressure and V/III ratio during the growth were 680 °C, 6 kPa and 20, respectively. The decay transients measured by time-resolved photoluminescence (TRPL) under 780-nm excitation for the GaAs DH structures grown with various growth rates as described in Fig. 2 is shown in Fig. 3(a). The decay time of each transient was extracted by re-convoluting with an instrument respond function using a single exponential model. The extracted values are given in Fig. 3(a). With a faster deposition speed, the minority hole lifetime was remarkably shortened, most likely due to an increase of non-radiative carrier trap density in n-GaAs. The radiative recombination was plausibly not responsible for this result considering the active carrier concentration shown in Fig. 2(b). The dependency of carrier lifetime on the growth rate is plotted in Fig. 3(b).

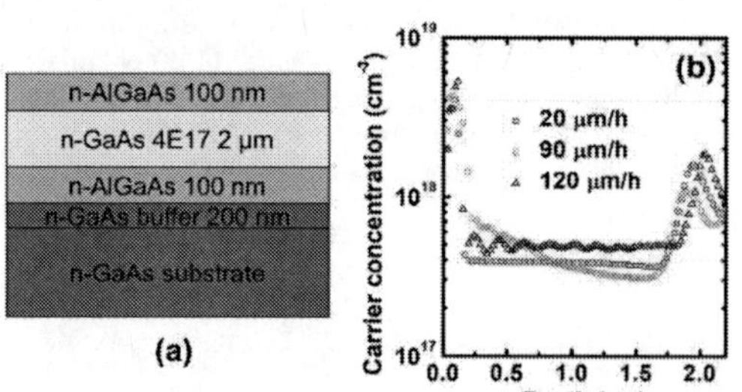

Figure 2: (a) Double-hetero (DH) layer structure for transient PL measurement, and (b) carrier profiles in DH samples measured by ECV

5 PERFORMANCE OF GaAs SOLAR CELLS

Figure 4 shows the schematic drawing of the GaAs p-on-n solar cell investigated in this work. The structure was similar to that reported in our previous study, excepting that AlGaAs back surface field (BSF) layers was employed as a rear hetero barrier instead of a highly doped GaAs layer. The reactor pressure was maintained at 6 kPa, while the growth temperature was decreased from 680 during growth of p-on-n GaAs to 600 °C for growth of InGaP window and GaAs contact layers. The reason is to enhance the doping efficiency in these layers. The InGaP window was slightly Ga-rich in order to decrease light absorption and increase barrier height. Effects of growth rate on the performance of GaAs solar cell were investigated by the 2-µm-thick n-GaAs base layer grown with various growth rates, 20, 90 and 120 µm/h. The area of GaAs solar cell was about 0.043 cm^2, determined by mesa area. This included 31% of metal coverage, resulting in an effective area of about 0.032 cm^2.

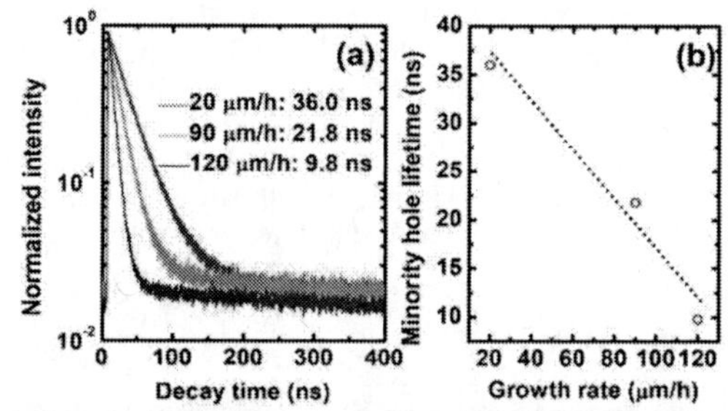

Figure 3: (a) PL decay curves for the DH samples with the core n-GaAs grown at growth rates of 20, 90 and 120 µm/h, and (b) minority hole lifetime in n-GaAs as a function of GaAs growth rate.

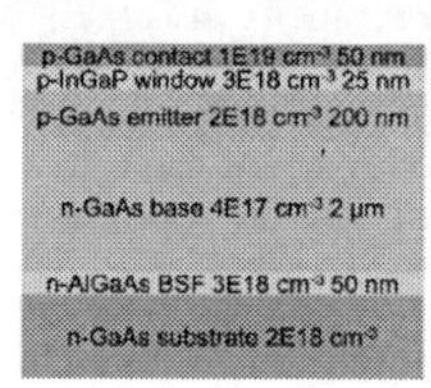

Layer	GR. (µm/h)	V/III	T_g (°C)	P_g (kPa)
Contact	5	20	600	6
Window	1.9	100	600	6
Emitter	20	20	680	6
Base	20/90/120	20	680	6
BSF	5	40	680	6
Buffer	5	20	680	6

Figure 4: The layer structure of GaAs p-on-n solar cells and the growth conditions of each layer.

The current-voltage (I-V) characteristics under AM1.5G (100 mW/cm^2) and external quantum efficiency (EQE) spectra of GaAs p-on-n solar cells of which the n-GaAs base layers were grown at various growth rates with ZnS/SiO_2 anti-reflection coating (ARC) are plotted in Figs. 5(a) and 5(b), respectively. The results here were evaluated using the effective area of 0.032 cm^2. The 90-µm/h grown cell had the lowest open circuit voltage (V_{oc}), most likely due to the fabrication failure, i.e. about 8.7 mV decrease in V_{oc} was observed in this sample after ARC deposition. A decrease in EQE at long wavelength range with an increase in growth rate evidenced the degradation of photo-carrier correction, possibly because of defects in n-GaAs base. This EQE degradation therefore was responsible for the reduced short circuit current density (J_{sc}) with increasing growth rate.

Figure 6 shows distributions of (a) J_{sc}, (b) V_{oc} and (c) conversion efficiency from one fabrication batch of the GaAs p-on-n solar cells grown with various growth rates. The measurement was done using the effective area of 0.032 cm^2. The tendencies, short-dot lines, were plotted using the average values. The J_{sc} was clearly degraded when the growth rate was increased from 90 to 120 µm/h. This result could be well explained by the reduction in minority carrier lifetime in Fig. 3(a) and poor carrier extraction in Fig. 5(b). As aforementioned, the 90 µm/h cells suffered from the reduction in V_{oc} after ARC deposition. Hence, the tendency of V_{oc} in Fig. 6(b) could not be well correlated with the GaAs growth rate. Fig. 6(c) clearly shows the dependence of solar cell efficiency which decreased with an increase in growth rate. Note that, the conversion efficiency of 90-µm/h grown sample was also reduced by V_{oc} degradation. The efficiency of GaAs p-on-n solar cells, averaged over 9 cells, decreased slightly from 24.51 to 23.77% when the growth rate of the n-GaAs base layer was increased from 20 to 120 µm/h.

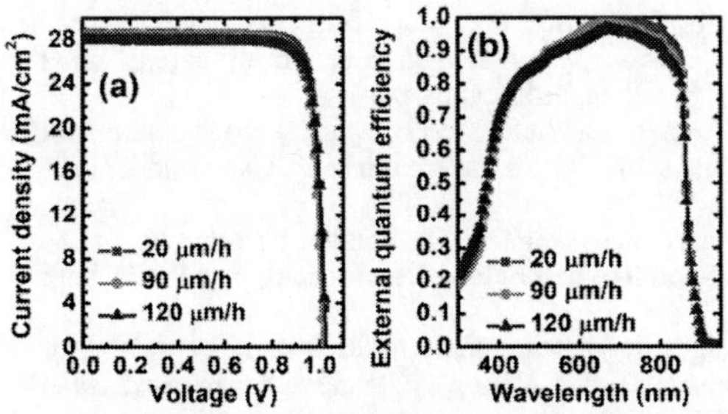

Figure 5: (a) I-V characteristics under AM 1.5G illumination and (b) EQE spectra of GaAs solar cells with the n-GaAs base layer grown at 20, 90, and 120 µm/h. The devices had the effective area of 0.032 cm^2 and evaluated with ZnS/SiO_2 anti-reflection coating.

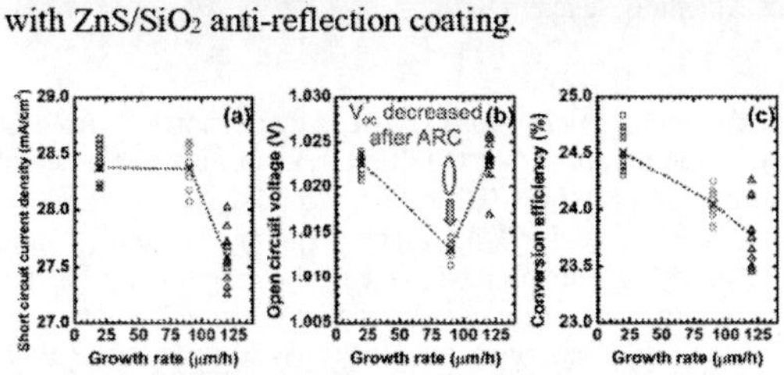

Figure 6: Distributions of (a) J_{sc}, (b) V_{oc} and (c) efficiency of GaAs solar cells grown with various growth rates. The devices had the effective area of 0.032 cm^2 and evaluated with ZnS/SiO_2 anti-reflection coating.

6 CONCLUSIONS

The GaAs growth rate could be increased to 120 µm/h and showed almost no saturation with an increase in TMGa supply. Such fast growth induced lattice imperfection and deteriorated carrier lifetime. Its impact was, however, limited in the performance of GaAs single-junction solar cells and approximately 24-25% efficiency could be realized for a GaAs solar cell, in with the 2-µm-thick n-type base layer was grown in only 1 min at a growth rate of 120 µm/h. This achievement in high-speed epitaxial growth will open a way to the drastic cost reduction in the production cost of III-V photovoltaic cells.

7 ACKNOWLEDGEMENT

Part of this study was supported by the NEDO project: the research and development of ultrahigh-efficiency and low-cost III-V compound semiconductor solar cell modules.

8 REFERENCES

[1] F. Dimroth, T. N. D. Tibbits, M. Niemeyer, F. Predan, P. Beutel, C. Karcher, E. Oliva, G. Siefer, D. Lackner, P. Fuß-Kailuweit, A. W. Bett, R. Krause, C. Drazek, E. Guiot, J. Wasselin, A. Tauzin, T. Signamarcheix, J. Photovoltaics 6, 343 (2016).

[2] S. P. Philipps, G. Peharz, R. Hoheisel, T. Hornung, N. M. Al-Abbadi, F. Dimroth, and A. W. Bett, Sol. Energy Mater. Sol. Cells 94, 869 (2010).

[3] M. Meusel, C. Baur, G. Siefer, F. Dimroth, A. W. Bett, and W. Warta, Sol. Energy Mater. Sol. Cells 90, 3268 (2006).

[4] M. Yamaguchi, Sol. Energy Mater. Sol. Cells 68, 31 (2001).

[5] M. Yamaguchi, Sol. Energy Mater. Sol. Cells 75, 261 (2003).

[6] C. Ebert, A. Parekh, Z. Pulwin, W. Zhang, D. Lee, and D. Byrnes, in Proc. IEEE PVSC-35, Honolulu, HI, USA, 2010, pp. 002007.

[7] K. J. Schmieder, M. K. Yakes, C. G. Bailey, Z. Pulwin, M. P. Lumb, L. C. Hirst, M. González, S. M. Hubbard, C. Ebert, and R. J. Walters, in Proc. IEEE PVSC-40, Denver, CO, USA, 2014, pp. 2130.

[8] K. J. Schmieder, E. A. Armour, M. P. Lumb, M. K. Yakes, Z. Pulwin, J. Frantz, and R. J. Walters, J. Photovoltaics 7, 340 (2017).

[9] H. Sodabanlu, A. Ubukata, K. Watanabe, T. Sugaya, Y. Nakano and M. Sugiyama, J. Photovoltaics 8, 887 (2018).

[10] A. Ubukata, H. Sodabanlu, K. Watanabe, S. Koseki, Y. Yano, T. Tabuchi, T. Sugaya, K. Matsumoto, Y. Nakano, M. Sugiyama, J. Cryst. Growth 489, 63 (2018).

[11] S. Mazumder and S. A. Lowry, J. Cryst. Growth 224, 165 (2001).

[12] A. Jorio, L. Sellami, M. Aubin, and C. Carlone, J. Appl. Phys. 91, 9887 (2002).

[13] C. L. Andre, D. M. Wilt, A. J. Pitera, M. L. Lee, E. A. Fitzgerald, S. A. Ringe, J. Appl. Phys. 98, 014502 (2005).

[14] H. Xu, K. Toprasertpong, A. Delamarre, H. Sodabanlu, K. Watanabe, Y. Nakano, and M. Sugiyama, Jpn. J. of Appl. Phys. 56, 08MC06-1 (2017).

EFFECTS OF IRRADIATION ON TRIPLE AND SINGLE JUNCTION INGAP/GAAS/GE SOLAR CELLS

Roberta Campesato1, Carsten Baur5, Mariacristina Casale1, Massimo Gervasi 2,3, Enos Gombia 4, Erminio Greco1, Aldo Kingma4, P.G. Rancoita2, Davide Rozza2, Mauro Tacconi2,3
1CESI, 20134 Milan, via Rubattino 54 , Italy
2 INFN Milano-Bicocca, Italy
3 and Univ. Milano Bicocca, Italy
4 IMEM-CNR Institute, Parco Area delle Scienze 37/A, 43124 Parma, Italy
5 ESA/ ESTEC, Keplerlaan 1, 2201 AZ Noordwijk, The Netherlands

ABSTRACT: The investigation of the degradation effects on triple-junction (TJ) solar cells, operating in space environment, is of primary importance in view of future space missions towards harsh radiation orbits (e.g. MEO with high particle irradiation intensity) and for the new spacecraft based on electrical propulsion.
In the present work, we report the experimental results obtained from the irradiation test campaign carried out both on (InGaP/GaAs/Ge) TJ solar cells and on single junction (SJ) isotype sub-cells (i.e. InGaP top cells, GaAs middle cells and Ge bottom cells).
In particular, the electrical performances degradation of the irradiated samples was analyzed by means of the Displacement Damage Dose (DDD) method based on the NIEL (Non Ionizing Energy Loss) scaling hypothesis, with DDD doses computed by the SR-NIEL approach.
By plotting the remaining factors as a function of the DDD, a single degradation curve for each sample type and for each electrical parameter has been found. The collapse of the remaining factors on a single curve has been achieved by selecting suitable displacement threshold energies, Ed, for NIEL calculation.
Moreover, DLTS (Deep level Transient Spectroscopy) technique was used to better understand the nature of the defects produced by irradiation. DLTS measurements reveal a good correlation between defects introduced by irradiation inside the GaAs sub-cell and the calculated Displacement Damage Doses.
Keywords: Radiation Damage, Multijunction Solar Cell, Gallium Arsenide Based cells

1 INTRODUCTION

The prediction of the solar cell degradation due to particle irradiation is one of the key elements to properly size a solar array for a given mission.

It is known that the degradation of the solar cells electrical performances is a function of the energy, the fluence and the type of the incident particles (electron or proton).

Since the space environment can be approximated by an omnidirectional spectrum of particles incident on solar cells, a method is needed to predict the on-orbit solar cell performance.

There are two methodologies currently available to perform on-orbit solar cell performance predictions. The first one is the Equivalent Fluence Method developed by JPL (Jet Propulsion Laboratory) [1] and the second one is the Displacement Damage Dose (DDD) Method developed by NRL (Naval Research Laboratory) [2].

The JPL approach is the first implemented and used by the space actors. To employ this method, it is necessary to determine the degradation curves for each electrical parameter (e.g Voc, Isc, Pmax etc.) over a wide range of energies both for protons and electrons. Each degradation curve is obtained by plotting the remaining factor (the ratio of the end-of-life EOL value to the beginning-of-life value) of a given electrical parameter as a function of the fluence. Therefore, this method requires a significant number of irradiation tests with electrons and protons with various energies.

NRL approach is based on the hypothesis that the permanent displacement damage produced by nonionizing events is the main phenomenon that degrades the device electrical performance. With this approach a single characteristic degradation curve, for each electrical parameter, can be obtained. This occurs if the remaining factors are plotted against the displacement damage dose, rather than fluence. The displacement damage dose (DDD) is determined from a calculation of the nonionizing energy loss (NIEL).; for electrons, protons, ions and neutrons passing through an absorber (elemental or compound) the NIEL values can be obtained by means SR-NIEL framework approach ([3] and Chapters 2, 7 and 11 in [4]). The main advantage of the latter method is that it requires less irradiation tests (e,g. only 1 energy for proton and 2 energies for electrons) to predict the EOL behavior of solar cells.

In this paper, the results of electron and proton irradiation tests, performed on triple junction solar cells and related isotype sub-cells manufactured by CESI, are presented. In particular, the radiation hardness analysis has been conducted by the DDD method.

The dependence of the NIEL calculation on the threshold energies for displacement, E_d, is the key point to obtain a single degradation curve as a function of the DDD. The value of the energy threshold has been chosen to make collapse the experimental data in a single degradation curve for each solar cell electrical parameter.

In parallel to this analysis, DLTS (Deep level Transient Spectroscopy) technique has been used on InGaP (top sub-cell) and GaAs (middle sub-cell) samples manufactured as diodes and irradiated together with the solar cells. The diodes were manufactured with exactly the same epitaxial structure of the sub-cells composing the triple junction device.

From the DLTS measurements, it turned out that the concentration of defects introduced by irradiation on GaAs samples is a linear function of the DDD. In particular, the expected linear behavior is obtained if the same value *Ed*, used to make collapse the degradation curves of the GaAs sub-cell, is selected.

2 IRRADIATION TEST CAMPAIGN ON TJ SOLAR CELLS, COMPONENT CELLS AND DIODES

InGaP/InGaAs/Ge TJ solar cells and related component cells with AM0 efficiency class 30% (CTJ30) [5] have been manufactured as 2x2 cm^2 solar cells.

The basic structure of the solar cells is reported in Figure 1. The TJ solar cell is composed by a Ge bottom junction obtained by diffusion into the germanium P-type substrate, a middle junction of (In)GaAs, whose energy gap is around 1.38 eV and a top junction of InGaP with an energy gap of 1.85 eV. Component cells are single-junction cells which are an electrical and optical representation of the sub-cells inside the TJ cell. Therefore, to manufacture them, special attention was put to reproduce the optical thicknesses of all the upper layers present in the TJ structure.

Top and middle sub-cells were also manufactured as diodes with 0.5 mm diameter for DLTS analysis before and after irradiation. In Figure 2 is shown how the solar cells and the diodes were arranged for irradiation tests.

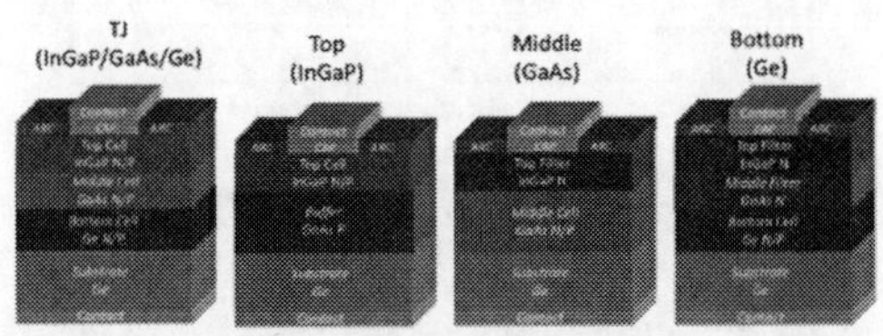

Figure 1: schema of triple junction (TJ) and single junction (SJ) isotype sub-cells (Top, Middle and Bottom)

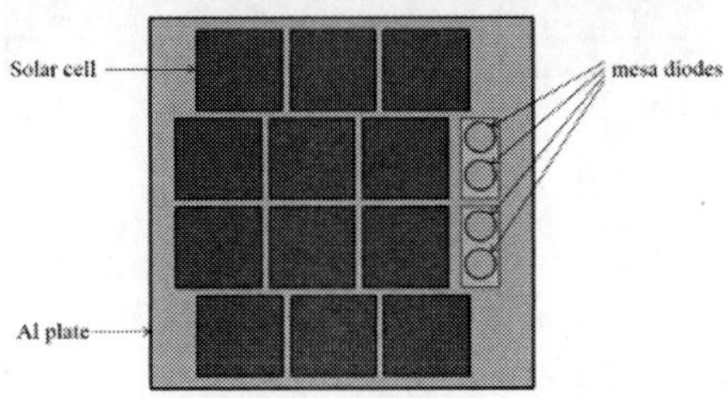

Figure 2: Arrangement of solar cells and diodes on Al plate for proton irradiation

In Table 1 the energy-fluence pairs used for protons and electrons irradiation tests are reported. For each condition, the following (2x2 cm^2) cells were irradiated: 3 TJ, 3 Top, 3 Middle and 3 Bottom. InGaP and GaAs diodes, for DLTS analysis, were irradiated together with the solar cells.

Table 1: Electron and proton irradiation test campaign dedicated to establish a degradation data set.

Protons		Electrons	
Energy [MeV]	**Fluence [p+/cm-2]**	**Energy [MeV]**	**Fluence [e-/cm-2]**
0.7	1.7E+11	1	1.0E+14
	3.3E+11		5.0E+14
	4.5E+11		1.0E+15
1	4.5E+10	1.5	5.0E+14
	2.3E+11		1.0E+15
2	4.2E+11	3	2.0E+14
	8.4E+11		4.0E+14

To calculate the remaining factors of each electrical parameter, all the solar cells have been measured both in BOL and EOL conditions. In particular, two measurements have been performed after the irradiation tests: after self-annealing (i.e. solar cells kept in dry box for 1 month before measurement) and after photon and thermal annealing (i.e. 8 hours at 1 AM0 illumination followed by 48 h at 60°C according to ECSS standards).

In Figure 3 the P_{max} degradation curves after self-annealing of TJ, top, middle and bottom cells are shown. The following semi-empirical equation (Eq.1) is used to fit the experimental P_{max} Remaining Factors (for each particle/energy separately):

$$RPF = \frac{P_{max}(n,E,\phi)}{P_{max}(BOL)} = 1 - C\log_{10}\left(1 + \frac{\phi}{\phi_{crit}(n,E)}\right) \quad (1)$$

with Φ being the particle fluence and with C, Φ_{crit} (critical fluence) being fitting parameters. The parameters n, E refer to the particle type and the particle energy indicating that the degradation and the critical fluence are particle and energy dependent. Equivalent curves can be plotted for each electrical parameter (i.e. Voc, Isc).

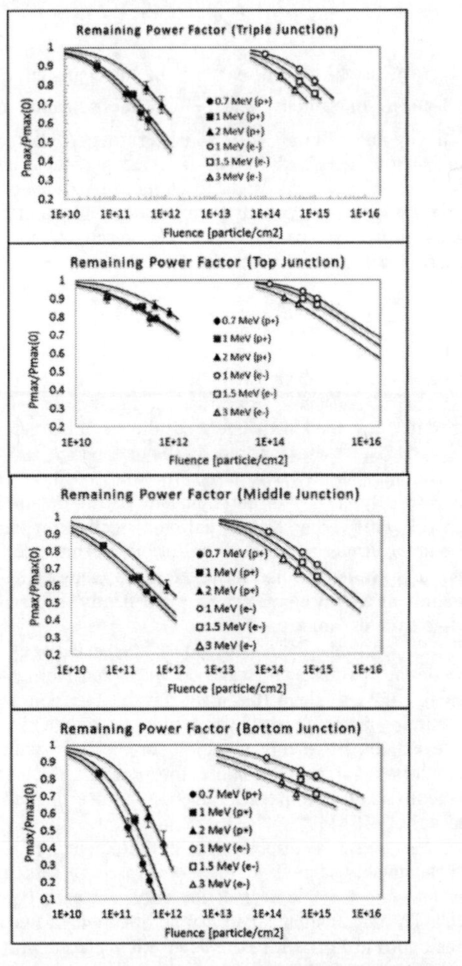

Figure 3: Pmax degradation curves of TJ, top, middle and bottom solar cells as a function of the particle fluence

Analyzing the curves in Figure 3, it turned out that the samples irradiated with electrons degrade as expected, whereas a higher spread in the remaining power factors was observed for the samples irradiated with protons, especially for the TJ and Ge bottom cells. It can also be noticed that Ge bottom junction is high resistant against electrons but it shows a bad radiation resistance against proton irradiation, especially if low energy protons are concerned [6]. Moreover, top junction (InGaP) confirms its strong behavior against irradiation. The analysis of remaining factors against DDD is reported in the next chapter.

3 NIEL ANALYSIS

In this section, the degradation of the photovoltaic parameters of TJ solar cells and single junction sub-cells is investigated using the DDD method.
According to this approach, the remaining factors of each electrical parameter are plotted as a function of the displacement damage dose (DDD). This quantity, expressed in units of MeV/g, can be obtained from (Eq.2):

$$DDD = \Phi \frac{dE_{de}}{d\chi} \quad (2)$$

where Φ is the fluence in cm^{-2} of traversing particles (electrons or protons) and $\frac{dE_{de}}{d\chi}$ is the so-called non-ionizing energy loss NIEL which expresses the amount of energy deposited by an incident particle passing through a material and resulting in displacement processes. This quantity, expressed in units of $MeVcm^2/g$, can be calculated by means of the SR (Screened Relativist) treatment (Eq.3) ([3,4]):

$$\frac{dE_{de}}{d\chi} = \frac{N}{A}\int_{E_d}^{E_R^{max}} E_R L(E_R) \frac{d\sigma(E,E_R)}{dE_R} dE_R \quad (3)$$

where χ is the absorber thickness in g/cm^2; N is the Avogadro constant, A is the atomic weight of the medium; E is the kinetic energy of the incoming particle; E_R and E_R^{max} are the recoil kinetic energy and the maximum energy transferred to the recoil nucleus respectively; $L(E_R)$ is the Lindhard partition function; $d\sigma(E,E_R)/dE_R$ is the differential cross section for elastic Coulomb scattering for electrons or protons on nuclei; E_d is the so-called displacement threshold energy, i.e. the minimum energy necessary to permanently displace an atom from its lattice position.

To perform the calculations, the energy of the incoming particles E was modified considering the energy losses along the path inside the non-active absorbing layers. In particular, it turns out that there is no relevant kinetic energy variation for electrons whereas the losses are not negligible for protons. The actual kinetic energy E, in each junction, was estimated by means of SRIM [7].

Moreover, the displacement threshold energy E_d has been suitably chosen to make collapse the remaining factors into a single degradation curve for each type of cell. The degradation curves can be fitted with a function analogous to Eq.1 where the fluence is replaced with the calculated DDD (Eq.4):

$$\frac{P_{max}(EOL)}{P_{max}(BOL)} = A - C \cdot log_{10}\left[1 + \frac{DDD(E_d)}{DDD_x}\right] \quad (4)$$

being A, C and DDD_x fitting parameters.

The P_{max} degradation curves as a function of the DDD for TJ, top, middle and bottom cells are shown In Figure 4.

The optimal fit for each cell type, was obtained using the E_d values reported in Table 2
In this work, the TJ solar cell was approximated to a GaAs cell single junction cell since the middle cell is the one that mainly affects the overall TJ performances.

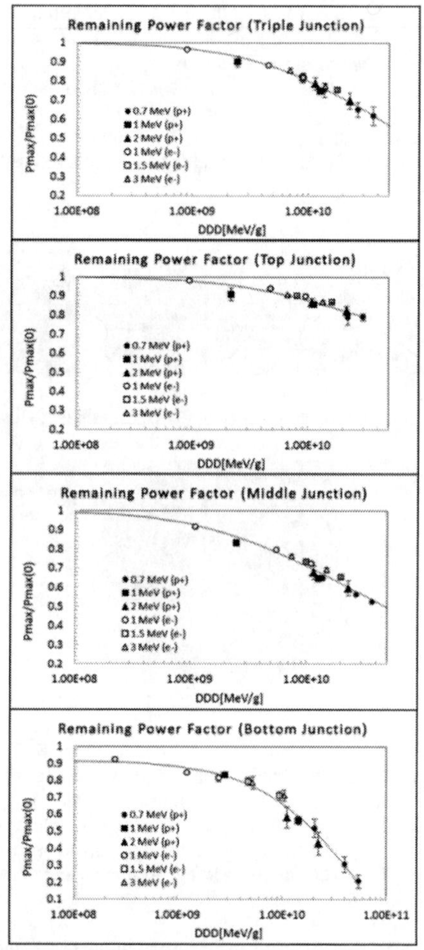

Figure 4: P_{max} degradation curves of TJ, top, middle and bottom solar cells as a function of the DDD.

Table 2: Optimum E_d values found to obtain the best-fit curves shown in Fig.4

Cell type	Ed
InGaP	Ed(In) ≈ 43eV
	Ed(Ga) ≈ 21eV
	Ed(P) ≈ 21eV
GaAs	Ed(Ga) ≈ 21eV
	Ed(As) ≈ 21eV
Ge	Ed(Ge) ≈ 40eV
TJ (GaAs)	Ed(Ga) ≈ 24eV
	Ed(As) ≈ 24eV

A single degradation curve has been obtained also for V_{oc} and I_{sc} remaining factors using the same E_d values reported in Table 2.
In the next chapter, we will see how the E_d value for GaAs has been confirmed from the DLTS analysis performed on GaAs middle junction.

4 DLTS analysis

In order to perform DLTS (Deep Level Transient Spectroscopy) investigations of deep levels induced by electron and proton irradiation, diodes of 0.5 mm in diameter have been prepared using the same epitaxial structure of top and middle sub-cells. The samples were manufactured by means of optical lithography, evaporation of metal contacts and mesa etch to remove the edge defects. Figure 5 shows a picture of a diodes rack used for DLTS analysis.

Figure 5: 0.5 mm dia diodes for DLTS and irradiation analysis

Concerning the InGaP diode (Top sub-cell), due to the relatively high doping concentration and the low reverse bias that can be applied to it, it was only possible to investigate regions near the junction interface dominated by high densities of interface states. These restrictions prevented a detailed study of bulk deep levels on both irradiated and non-irradiated samples.
For this reason, the DLTS study has been mainly focused on GaAs diodes (Middle sub-cell).

Figure 6 shows the DLTS spectra of a GaAs diode irradiated with electrons using different reverse voltages V_r (-0.4 V, -1.5 V, and -2.9 V) at fixed pulse amplitude (1.4 V).

With increasing reverse voltage V_r the spectra show the characteristics of regions at deeper distance from the n+/p interface. The high temperature peak (activation energy = 0.7 eV) labeled E3, which is also present in the non-irradiated samples, is likely to correspond to a defect at the junction interface. For higher reverse biases the DLTS spectra show the presence of at least two levels, labeled E1 (0.21 eV) and E2 (0.45 eV).

These levels have been identified as majority carrier traps in the p-type bulk bases since majority carrier pulses have been used in the DLTS measurements. As a consequence their energy must be referred to the top of the valence band.

These defect states are not observed in non-irradiated samples and show a density which increases with the absorbed dose, hence they are attributed to irradiation induced defects.

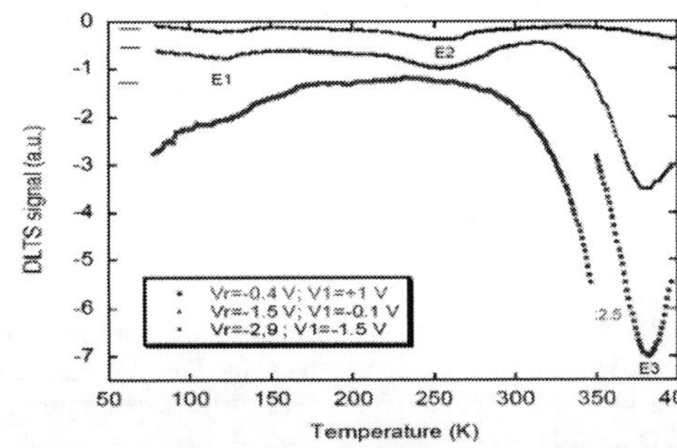

Figure 6: DLTS spectra obtained on a middle junction irradiated with electrons (1 MeV at a fluence of 10^{15} cm-2) using different reverse voltage V_r at fixed pulse amplitude of 1.4 V. Emission rate =46 s^{-1}, pulse width=500 ms, V1=pulse voltage.

In Figure 7 the DLTS spectrum of a middle cell diode irradiated with protons (energy 0.7 MeV and fluence $4.5x10^{11} cm^{-2}$ corresponding to NIEL dose of $3x10^{10}$ MeV g^{-1}) is compared to that of a middle cell diode irradiated with electrons (energy 1 MeV and fluence $1x10^{15}$ cm^{-2} corresponding to NIEL dose of $1.1x10^{10}$ MeV g^{-1}).

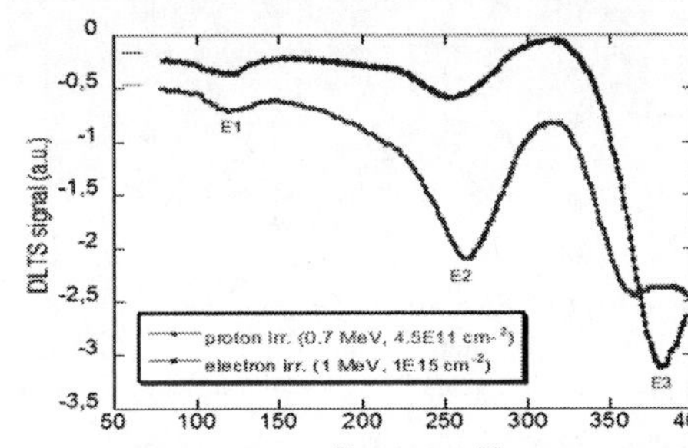

Figure 7: Comparison of the DLTS spectra of two middle junctions irradiated by protons and electrons respectively. Emission rate =46 s^{-1}, pulse width=500 μs, reverse voltage Vr = -1.5 V, pulse voltage V1= -0.1 V

From the analysis of the DLTS graphs obtained at different combinations of particle-energy, it can be observed that:

- The peak E1 is detected at all doses for samples irradiated with electrons whereas at only highest doses for those irradiated with protons.
- The peak E2 is detected at all doses both for electron and proton irradiated samples;

From the DLTS peak height of the different irradiated diodes, it is possible to estimate the defects concentration as a function of the DDD.

In particular, the concentration of E1 traps was correlated to the DDD of the GaAs diodes irradiated with electrons (Figure 8) and the concentration of E2 traps was correlated to the DDD of the GaAs diodes irradiated both with electrons and protons (Figure 9). In both cases, the expected linear behavior is obtained using the same threshold energies E_d, selected for the NIEL analysis of the previous paragraph (i.e. E_d = 21 eV for GaAs). Concerning the peak E2, two independent linear fit with slightly different slopes have been found for protons and electrons.

The good correlation found between the DLTS

measurements and the NIEL analysis can be appreciated considering the Trap Introduction Rate (TIR) of the defect states E1 and E2 as a function of the energy E of the incident particle (Eq.5):

$$TIR(E) = \frac{Trap\ concentration\ [cm^{-3}]}{\Phi(E)[cm^{-2}]} \quad (5)$$

It turns out that the trap introduction rate TIR(E) is proportional to NIEL(E) (in units of MeV cm^2/g) being the proportionality constant given by the slope of the lines in Figure 8 and in Figure 9 (Traps Concentration vs DDD).

The conversion factors between trap introduction rate and NIEL are reported in Table 3

By using these conversion factors, it can be seen from the Figures 10-11 that the NIEL curves well represent the introduction rates of the traps as a function of the energy E of the incident particle.

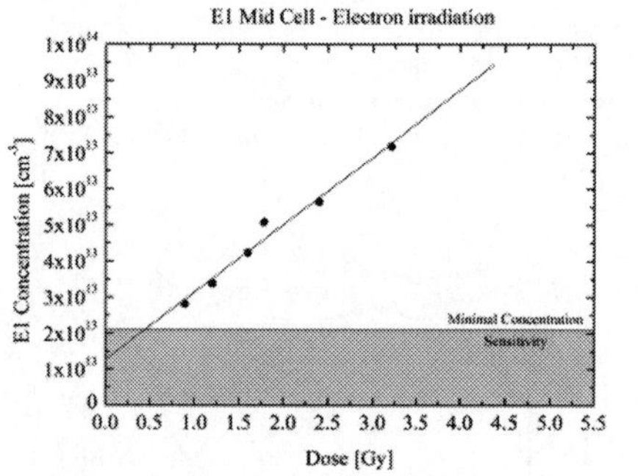

Figure 8: Concentration of E1 traps induced by electron irradiation in middle sub cell as a function of Displacement Damage Dose (obtained with Ed = 21 eV).

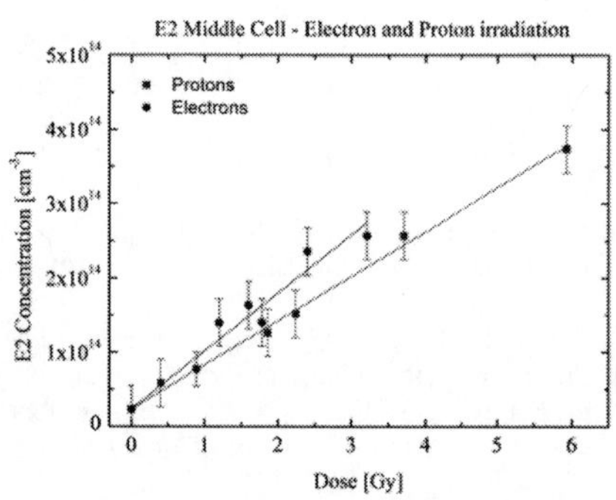

Figure 9: Concentration of E2 traps induced by electron and proton irradiation in middle sub cell as a function of Displacement Damage Dose (obtained with Ed = 21 eV).

Table 3: Conversion factors between Trap Introduction Rate TIR(E) and NIEL

Traps	Particle	TIR/NIEL ($g/MeV/cm^3$)
E1	electron	3.0 X 10^3
E2	electron	1.2 X 10^4
E2	proton	9.3 X 10^3

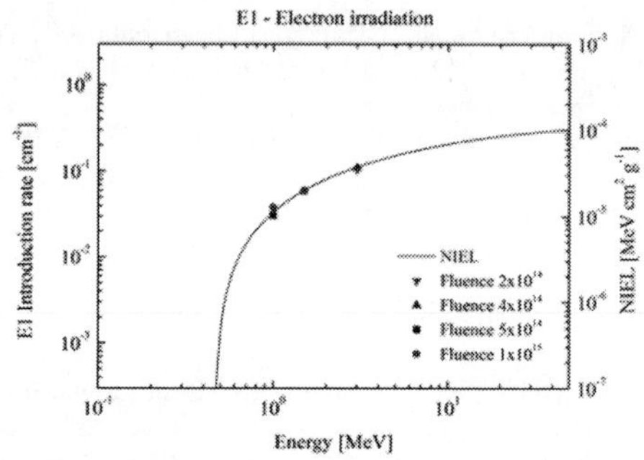

Figure 10: E1 traps introduction rate as a function of incoming electron energy: right scale NIEL values in GaAs sub cell for electrons.

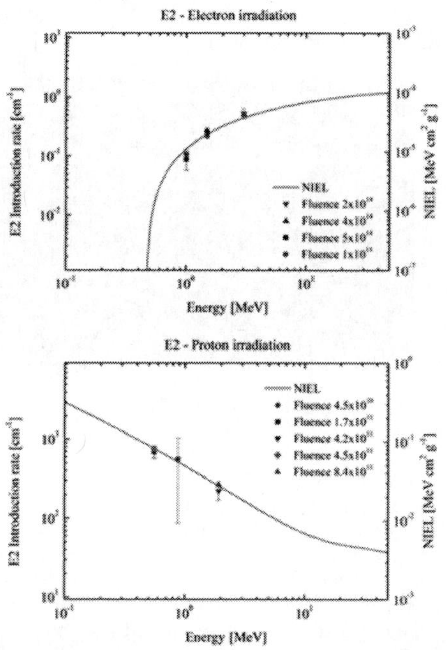

Figure 11: E2 traps introduction rate as a function of incoming proton (top) and electron (bottom) energy: right scale NIEL values in GaAs sub cell for protons and electrons.

5 CONCLUSION

Triple junction (InGaP/GaAs/Ge) solar cells and related single junction (SJ) isotype sub-cells were irradiated with protons and electrons at different energies.

The data were analyzed by means of the Displacement Damage Dose (DDD) method based on the NIEL hypothesis.

It is demonstrated that by choosing an appropriate threshold energy, E_d, for atomic displacement a single degradation curve for each solar cell type and for each electrical parameter can be found.

To better understand the nature of the defects produced by irradiation, DLTS analysis is conducted on the real sub-cells manufactured as diodes and not on "ad hoc" p/n junctions typically used for this kind of measurements.

The DLTS analysis, carried out on middle junctions, indicates two main traps are introduced, E1 having an energy of 0.21 eV and E2 having an energy of 0.45 eV above the top of the valence band. These traps are for majority carriers because the high doping associated with the middle cell junction prevented the detection of minority carrier traps.

From the DLTS measurements it has been possible to correlate the defects concentration and the calculated Displacement Damage Doses. This analysis reveals that complex defects are likely to be introduced at a slightly different rate for electron and proton irradiations.

The set of measurements performed on this work experimentally support the validity of SR-NIEL treatment approach for obtaining the displacement damage doses.

Acknowledgments

Part of this work was supported by ESA contract 4000116146/16/NL/HK with title "Non-Ionizing Energy Loss (NIEL) Calculation and Verification".

6 REFERENCES

[1] Anspaugh "GaAs Solar Cell Radiation Handbook" NASA 1996

[2] S.R. Messengers, et al.: "Modeling solar cell degradation in space: A comparison of the NRL displacement damage dose and the JPL equivalent fluence approaches" PIP 2001, 10.1002/pip.357View

[3] M.J. Boschini, P.G. Rancoita and M. Tacconi (2014), SR-NIEL Calculator: Screened Relativistic (SR) Treatment for Calculating the Displacement Damage and Nuclear Stopping Powers for Electrons, Protons, Light- and Heavy- Ions in Materials (version 3.9.8, September 2018); available at: http://www.sr-niel.org/.

[4] Leroy, C. and Rancoita, P.-G., Principles of Radiation Interaction in Matter and Detection 4th Edition, World Scientific (Singapore) 2016, ISBN 9789814603188

[5] G. Gori, R. Campesato : Photovoltaic Cell Having a high Conversion Efficiency PCT I09111-WO (2009)

[6] C. Baur et al.,"Investigation of Ge component cells", Photovoltaic Specialists Conference, Conference Record of the Thirty-first IEEE, 2005

[7] James F. Ziegler , SRIM – The stopping range and range of ions in matter. Code online at www.srim.org

PROGRESS TOWARDS HIGH EFFICIENCY THIN-FILM III-V QUANTUM DOT SOLAR CELLS FOR SPACE

T. Aho[(1)], A. Tukiainen[(1)], F. Elsehrawy[(2)], A. Khalili[(2)], J. Lyytikäinen[(1)], E. Halonen[(1)], T. Niemi[(1)], M. Guina[(1)], F. Cappelluti[(2)]
[(1)] Optoelectronics Research Centre, Laboratory of Photonics, Tampere University of Technology, P.O. Box 527, FI-33101 Tampere, Finland. Email: mircea.guina@tut.fi
[(2)] Department of Electronics and Telecommunications, Politecnico di Torino
Corso Duca degli Abruzzi 24,10129, Torino, Italy. Email: federica.cappelluti@polito.it

ABSTRACT: This work summarizes our results in the development of high efficiency III-V quantum dot (QD) solar cells, aimed at tackling with two of the most relevant issues posed by QD solar cells (QDSCs), namely the degradation of open circuit voltage and the weak photon harvesting by QDs. In particular, we report our latest achievements in: i) The molecular beam epitaxy growth of high-quality QDSCs, demonstrating V_{oc} as high as 0.94 V and low penalty (~ 40 mV) with respect to the single-junction reference cell. ii) The development by nanoimprint lithography of metal/polymer back reflectors with high diffraction efficiency, enabling four times increase of the QD photogenerated current. Experimental results are discussed with the support of numerical simulations.
Keywords: III-V semiconductors; Quantum Dots; Light Trapping; Thin Film

1 INTRODUCTION

Future applications in space and aeronautics demand for very high efficiency, lightweight and flexible solar cells. To achieve these goals, several approaches are investigated worldwide to bring new concepts in the consolidated III-V single- and multi-junction solar cells, such as thin-film architectures, novel materials, nanostructures, and nanophotonics. The successful exploitation of these concepts may bring cost-effective and attractive options for next generation space solar cells. In particular, QDs have potential impact on diverse aspects of space photovoltaics, such as improved current matching in multijunction cells by bandgap tuning [1] and enhanced radiation hardness [2]. However, a number of issues need to be solved in order to achieve high efficiency QD-based solar cells.

In [3] we have shown through electrical simulations that the bandgap-voltage offset of QD solar cells (QDSC) operating in the thermal limit [4], i.e. with carriers undergoing full thermalization, is expected to be comparable to that one of III-V single-junction cells. In fact, the large degradation of the open circuit voltage (V_{oc}) often observed in experiments is due to extrinsic loss caused by material quality issues. Of the same signature were the conclusions drawn in [5, 6] based on detailed balance arguments. Therefore, material optimization is a key task to demonstrate high-efficiency QDSCs. Moreover, the absorptivity of the solar cell must be substantially increased by suitable photon management approaches. In [3] we demonstrated doubled QD photocurrent by substrate removal through epitaxial lift-off and integration of a rear planar reflector. Much higher enhancement of the QD current is expected by implementing diffraction gratings within such thin-film architecture.

2 METHODS

2.1 Solar cell fabrication and characterization

The InAs/GaAs QD samples and solar cells were grown on p-GaAs(100) substrates using a V90 molecular beam epitaxy (MBE) system. The GaAs-based layers were grown at 580 °C. Growth temperatures for InAs QDs and AlInP were 465-475 °C and ~490 °C, respectively. Thin GaAsP strain-balancing layers were also grown within the undoped GaAs barrier layers between the QD sheets.

6×6 mm^2 solar cells were fabricated by dicing and using a shadow mask process for the front-side grid metal deposition in an electron beam evaporator. No anti-reflection coating was deposited onto the cells. The electrical parameters of the solar cells were determined using light-biased current-voltage (I-V) and external quantum efficiency (EQE) measurements.

2.2 Gratings fabrication and characterization

Grating prototypes were fabricated on double-side polished 350 μm semi-insulating GaAs wafer by nanoimprint lithography (NIL). Being the substrate almost transparent at the QD wavelengths it provides a suitable platform to test the diffraction properties of the grating. The grating is patterned onto a commercial OrmoComp NIL photoresist by using an in-house fabricated Si master. In order to achieve the desired grating dimensions, the master was realized by a sequential optimized processing involving a first patterning step by electron beam lithography into a sacrificial Si wafer and a second one by NIL and dry etching onto the final Si master. From the master, the grating was then transferred to polydimethylsiloxane stamp and finally imprinted to the NIL photoresist. 200 nm Ag was deposited on top of the polymer layer using electron beam evaporator. The fabricated grating has height of 700 nm and period of 2.5 μm, according to numerical optimization aimed at maximizing the diffraction efficiency at the QD wavelengths [7].

The grating morphology was characterized with a scanning electron microscope (SEM). Specular and total reflectance were measured with a PerkinElmer Lambda 1050 spectrophotometer which includes an integrating sphere module. To quantify the amount of the diffracted light, a variable angle measurement technique has been used, where the sample is illuminated with incident angle of 8°, and the diffracted light is collected at variable angles from 8° to 48° with an optical fiber connected to a spectrometer. This allows to quantify the grating diffraction efficiency as the ratio between the collected diffracted power and the incident power. For the grating under study this allowed to analyze diffraction orders of $m=0$ and $m=\pm 1$.

2.3 Modeling

The solar cells were simulated using a drift-diffusion

Poisson model coupled with a rate-equation model for the carrier processes involving the QD states (see the QD electronic structure in Fig.1) and charge transfer between QD states and the continuum bands. Such an approach allows for an accurate and detailed analysis of the solar cell physics and it is developed to be thermodynamic consistent [3, 8].

The grating structures were simulated using the rigorous coupled-wave analysis (RCWA) method with

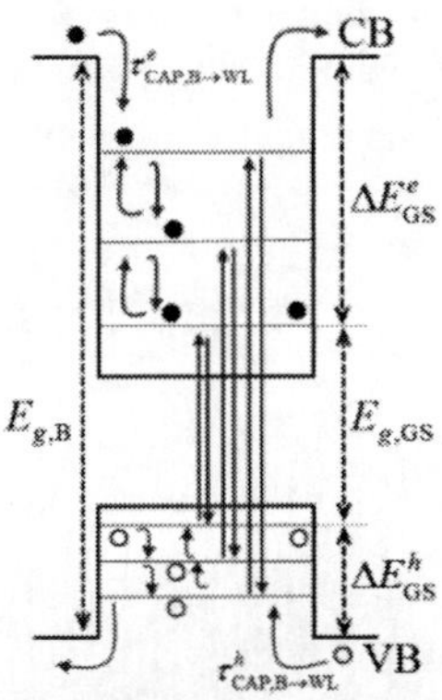

Figure 1: QD electronic structure with the indication of the carrier processes involved in thermally-limited QDSCs. $E_{g,B}$, $E_{g,GS}$ are the GaAs and GS bandgap energy, respectively.

the RSoft DiffractMOD software (Synopsys). Material optical models are derived from the literature or –in the case of the QD stack – from optical or electro-optical characterization. In particular, the reported case study assumes a QD absorption coefficient estimated from the EQE measurement of a previous generation of fabricated QDSCs [3]. The absorbance spectrum was estimated from the absorbed photon density in the photoactive region only, and thus does not include absorption in the contact and metal layers. The corresponding QD photocurrent density can be estimated by integrating the absorbance over the solar spectrum of interest in the 895 nm – 1200 nm range.

3 RESULTS AND DISCUSSION

QDSCs with deep-junction (DJ) and shallow-junction (SJ) configuration were designed and fabricated. As shown in Fig.2, in both cases, the multi-layer QD stack is placed in an intrinsic GaAs region next to the emitter-base junction. The QD placement in the high field region ensures full carrier collection at short circuit. Fig.3 shows the calculated V_{oc} for QD cells with nominal ground state (GS) emission at 1033 nm ($E_{g,GS}$ =1.2 eV) and different arrangements of the QD stack as a function of the QD in-plane density. Assuming realistic non radiative recombination loss in the GaAs interdot layers (SRH lifetime ~ 50 ns), V_{oc} higher than 0.9 V is predicted. Non radiative recombination is minimized by shrinking the intrinsic region, provided that the material quality remains high. The achievable open circuit voltage decreases as the number of layers increase due to the increased radiative recombination through the QD states.

Several solar cells with 10 and 20 QD layers– both with DJ and SJ configuration - were grown by MBE exploiting thin GaAsP strain balancing layers in the GaAs interdot layers. The QD in-plane density was estimated to be about 7×10^{10} cm^{-2}. Depending on the device configuration, QDs showed photoluminescence (PL) peak between 1015 nm and 1040 nm, and wetting layer emission around 920 nm. More details about material characterization and External Quantum Efficiency measurements were reported in [9].

Both DJ and SJ QDSCs showed only 40 mV V_{oc} penalty with respect to their GaAs reference, comparable to that one reported by Bailey et al. [1] with metalorganic vapour phase epitaxy (MOCVD) growth. The highest V_{oc} (0.94 V) was obtained for the SJ structure, which exhibited a fill factor of about 84 %. On the other hand, DJ cells suffered from slightly higher recombination in the thick emitter.

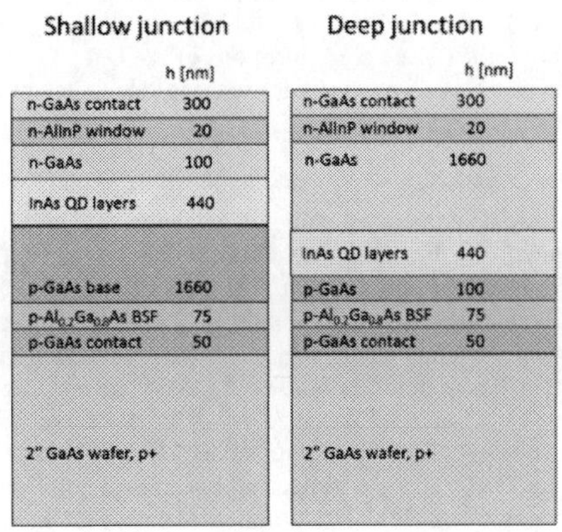

(a)

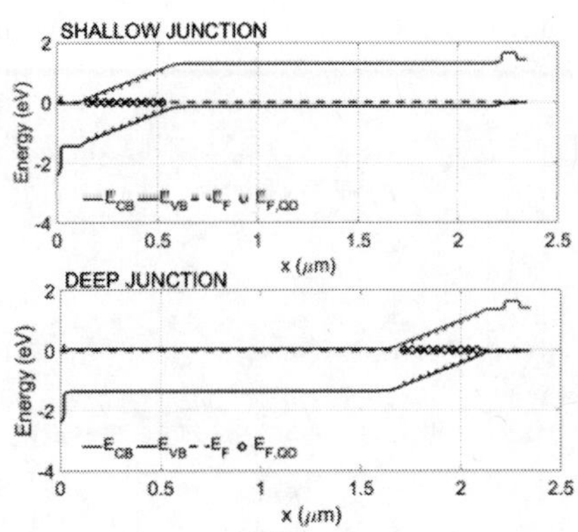

(b)

Figure 2: (a) Schematic structure of the shallow and deep junction QD cells. (b) Corresponding energy band diagram at thermal equilibrium.

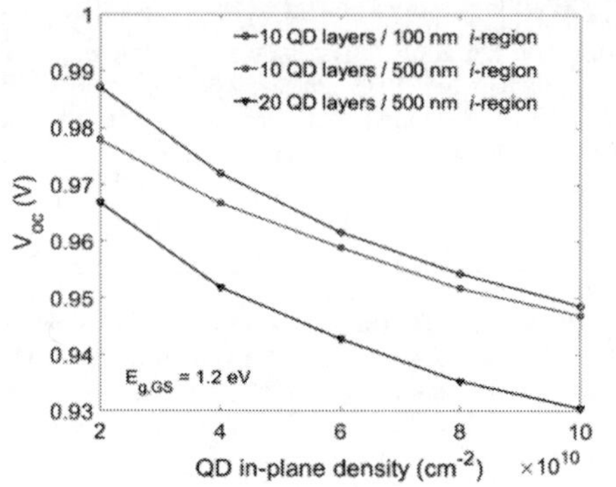

Figure 3: Calculated V_{oc} as a function of the QD in-plane density for different configurations of the QD stack (number of layers and interdot spacing) for QDs with PL peak emission at 1033 nm.

The measured V_{oc} vs. PL emission wavelength for a wide set of fabricated samples is reported in Fig.4: as expected, the V_{oc} decreases with the PL wavelength, i.e. with larger QD confinement. The measured values are in line with those expected from the GS transition energy ($V_{oc}=E_{g,GS}/q-W_{oc}$) [10], using for the bandgap-voltage offset (W_{oc}) reference values for state-of-art thin-film (W_{oc}=0.3 V) and wafer-based (W_{oc}=0.4 V) GaAs single-gap cells. A V_{oc} - bandgap offset of 0.3 V was also reported in [11] for a collection of experimental data of high-quality QDSCs. The graph also shows the bandgap-voltage offset as predicted under radiative limited conditions by the QD-corrected drift-diffusion model for a representative case of a 20 layers QDSC. Overall, the presented results demonstrate a high material quality of the fabricated devices.

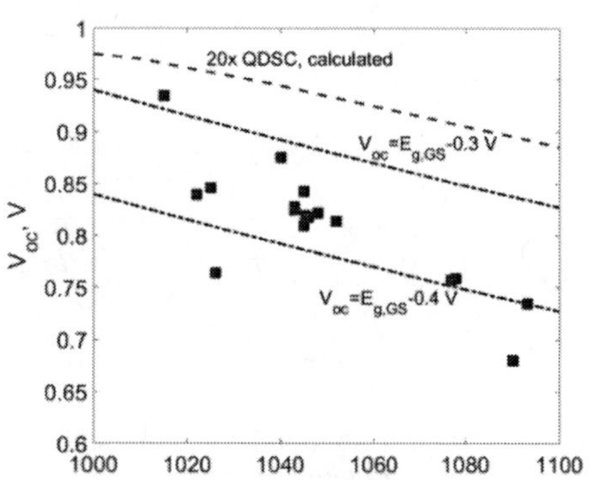

Figure 4: Experimental V_{oc} values of several InAs QDSCs as a function of the PL peak emission wavelength. The black dashed lines indicate reference trends in terms of bandgap voltage offset in state-of-art III-V solar cells [10].The blue dashed line indicates the scaling law predicted by the QDSC physics-based model.

Light-trapping enhancement of the QD photocurrent may be pursued through a thin-film structure implementing a textured reflector in the rear side of the cell. The absorbance spectra reported in Fig.6 for different configurations of the backside textured reflector highlight the remarkable absorption increase at the QD wavelengths: the achievable increase of the QD current with respect to a wafer-based configuration ranges from 4 to 7 depending on the material used for the grating fabrication: polymer (low index case study, n= 1.5) or AlInP (high index case study).

Different geometries and materials have been investigated and described in [12]. Fig.6 summarizes the results of the fabrication of pyramidal metal/polymer back reflectors. Experiments and RCWA simulations demonstrate high diffuse reflectance (Fig. 6 (b)) and efficient diffraction of light (Fig.6 (c)) in the QD wavelength range.

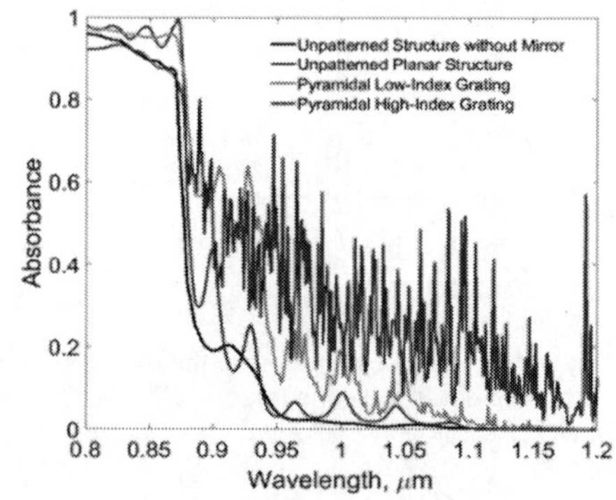

Figure 5: Calculated absorbance spectra for a representative 20 layer QDSC with different photonic configurations. The corresponding enhancement of the short circuit current contribution from the QDs with respect to the wafer-based configuration is 1.6, 4, and 7 for the planar reflector, low index pyramid grating and high index pyramid grating, respectively.

4 CONCLUSION

In conclusion, QDSCs with the highest V_{oc} ever reported for MBE grown QDs and V_{oc} penalty in line with the state-of-art [2, 3] are demonstrated. Both DJ and SJ configurations show promising performance, enabling cell optimization for highest efficiency at beginning of life or for highest radiation hardness. Fabricated metal/polymer photonic gratings demonstrate significant potential for enhancing the QD current contribution. Numerical simulations [4] show that the presented advancements in the technology of QD growth and structures for photon management pave the way to the realization of thin InAs/GaAs QDSCs with target efficiency of 30% - AM1.5G (26% - AM0). The present study – regardless of the specific material system – is also of interest for intermediate band solar cells in concentrator photovoltaics [13]. Finally, the photonic gratings proposed in this work show promising performance and would be suitable for other photovoltaic technologies by proper optimization of the grating geometrical features to match the wavelength range of interest.

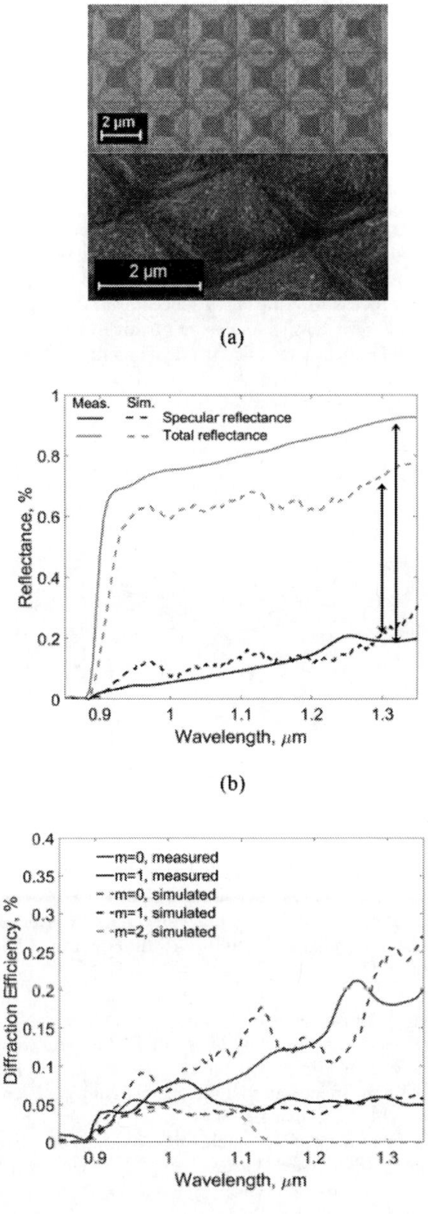

Figure 6: (a) SEM image of the pyramid grating. (b) Measured and simulated reflectance. The difference between the total and the specular reflectance represents the amount of diffuse light (the black double-headed arrows). In contrast, planar structures – not shown here – demonstrated marginal difference between total and specular reflectance. (c) Measured and simulated diffraction efficiency.

Acknowledgements
The work was partly funded by European Union Horizon 2020 project TFQD (Grant Agreement No. 687253) and ERC AdG project AMETIST (Grant Agreement No. ERC-2015-AdG 695116).

4.3 References

[1] C.G. Bailey, D.V. Forbes, R.P. Raffaelle, S.M. Hubbard, "Near 1 V open circuit voltage InAs/GaAs quantum dot solar cells," Appl. Phys. Lett. 98 (16) (2011) 163105.

[2] C.D. Cress et al, "Quantum dot solar cell tolerance to alpha-particle irradiation," Appl. Phys. Lett., 91.18 (2007): 183108.

[3] F. Cappelluti, D. Kim, M. van Eerden, A. P. Cédola, T. Aho, G. Bissels, F. Elsehrawy, J. Wu, H. Liu, P. Mulder, G. J. Bauhuis, J. J. Schermer, T. Niemi, and M. Guina, "Light-trapping enhanced thin-film III-V quantum dot solar cells fabricated by epitaxial lift-off," Sol. Energy Mater. Sol. Cells, https://doi.org/10.1016/j.solmat.2017.12.014 (2018).

[4] V. Aroutiounian, S. Petrosyan, A. Khachatryan, K. Touryan, "Quantum dot solar cells," J. Appl. Phys. 89 (4) (2001) 2268–2271.

[5] Yamaguchi, M., Lee, K. H., Araki, K., Kojima, N., Yamada, H., & Katsumata, Y. (2018). Analysis for efficiency potential of high-efficiency and next-generation solar cells. *Progress in Photovoltaics: Research and Applications*, 26(8), 543-552.

[6] Zhu, L., Akiyama, H., & Kanemitsu, Y. (2018). Intrinsic and extrinsic drops in open-circuit voltage and conversion efficiency in solar cells with quantum dots embedded in host materials. *Scientific reports*, 8(1), 11704.

[7] A. Musu, F. Cappelluti, T. Aho, V. Polojärvi, T. Niemi, and M. Guina, "Nanostructures for light management in thin-film GaAs quantum dot solar cells," in Light, Energy and the Environment, (Optical Society of America, 2016), paper JW4A-45

[8] A. P. Cédola, D. Kim, A. Tibaldi, M. Tang, A. Khalili, J. Wu, H. Liu, F. Cappelluti, "Physics-Based Modeling and Experimental Study of Si-Doped InAs/GaAs Quantum Dot Solar Cells." International Journal of Photoenergy (2018).

[9] A. Tukiainen, J. Lyytikainen, T. Aho, E. Halonen, M. Raappana, F. Cappelluti, M. Guina, Comparison of 'shallow' and 'deep' junction architectures for MBE grown InAs/GaAs quantum dot solar cells. 2018 WCPEC-7, JUNE 10-15, 2018 Waikoloa, Hawaii

[10] R. King, D. Bhusari, A. Boca, D. Larrabee, X.-Q. Liu, W. Hong, C. Fetzer, D. Law,N. Karam, "Band gap-voltage offset and energy production in next-generation multijunction solar cells," Progress. Photovolt: Res. Appl. 19 (7) (2011) 797–812.

[11] K. Tanabe, D. Guimard, D. Bordel, Y. Arakawa, "High-efficiency InAs/GaAs quantum dot solar cells by metalorganic chemical vapor deposition," Appl. Phys. Lett. 100 (2012) (1293905-193905-3).

[12] Aho, T., Guina, M., Elsehrawy, F., Cappelluti, F., Raappana, M., Tukiainen, A., ... & Niemi, T. (2018). Comparison of metal/polymer back reflectors with half-sphere, blazed, and pyramid gratings for light trapping in III-V solar cells. Optics express, 26(6), A331-A340.

[13] A. Luque and A. Martí, "Increasing the efficiency of ideal solar cells by photon induced transitions at intermediate levels," Phys. Rev. Lett. 78, 5014-5017 (1997).

DICHROIC PENTAPRISM FOR THE SPECTRAL SPLITTING OF CONCENTRATED SOLAR RADIATION

Antonio Parretta[1]*, Francesca Menchini[2] and Mario Tucci[2]
[1]University of Ferrara, Physics and Earth Science Department, Via Saragat 1, 44122 Ferrara (FE), Italy
*Phone: +39 3401509953; E-mail: parretta@fe.infn.it
[2]ENEA Casaccia Research Center, Via Anguillarese 301, 00123 Rome (RM), Italy
E-mails: francesca.menchini@enea.it; mario.tucci@enea.it

ABSTRACT: An innovative method for the spectral splitting of concentrated solar radiation is introduced, as an alternative to the well-established monolithic separation of light in a tandem cell. The PV receiver operates a parallel spectral splitting of solar radiation and use a pentaprism coupled with two dichroic filters. The use of a pentaprism allows to reduce the optical losses associated to the spectral splitting process offering: i) absence of air/dielectric interfaces; ii) parallel radiation travelling inside the pentaprism, incident at near normal angles on the dichroic filters; iii) pentaprism operating as a light trap, with reflective internal surfaces and radiation recycled inside it among the three conversion devices.
Keywords: Concentrators, Spectral Splitting, Dichroic Filters, Optical Losses, Simulation.

1 INTRODUCTION

The combined use of solar radiation concentration and spectral splitting makes it possible to enhance significantly the conversion efficiency of a PV system [1, 2]. In PV systems, as an alternative to the well-established monolithic separation of light in a tandem cell, many other designs have been explored for the spectral splitting receivers [3-7]. The scheme currently adopted to realize a parallel spectral splitting system uses a series of dichroic filters, made with periodic sequences of thin and non-absorbent dielectric layers, with a strong contrast on the refractive index. Despite the theory suggests splitting the solar spectrum into many bands to enhance the conversion efficiency, in practice, the added complexity of increasing the number of bands can result in a lower total practical efficiency. Splitting losses are the most responsible causes of efficiency reduction, and include T/R sloped transitions, non-ideal cut-off wavelengths and reflection losses of the dichroic filters. Non-ideal cut-off can occur as an effect of the angular dispersion of light in concentrating configurations, or due to a non-perpendicular incidence of light on the dichroic filter; reflection losses, on the other hand, generally increase at increasing of the number of optical interfaces.

Aim of the present work is to present a receiver to be used in a concentrating photovoltaics system, operating under conditions of parallel spectral splitting and characterized by avoiding the main criticality of this configuration, i.e. the presence of air/dielectric interfaces causing unwanted optical losses and ultimately the reduction of the overall optical efficiency. The proposed photovoltaic receiver (see Fig. 1) represents a technological innovation in that it uses, instead of a series of separate dichroic filters, a pentaprism (PP) free of air/dielectric interfaces, coupled to two or three dichroic filters (DF) and three solar cells, inside which the solar radiation is conveyed and recycled, thanks to the passive internal surfaces made reflective, thus constituting a real optical trap reducing to a minimum the optical losses. The radiation from the primary optical element is made parallel by using a planoconvex lens applied to the entrance face of the PP. To minimize reflection losses, the lens is provided with an ARC. The photovoltaic receiver can be realized with low cost optical components. It is an innovation compared to the existing technology, as it is the first using a pentaprism for the spectral splitting of concentrated solar radiation.

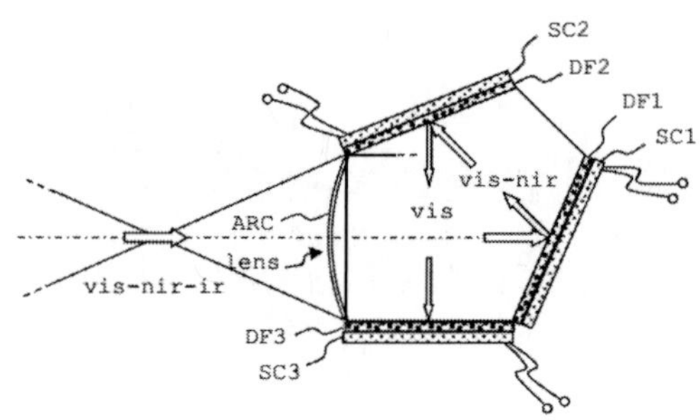

Figure 1: Example of the scheme of a PV spectral splitting receiver based on the pentaprism (PP), operating with concentrated light and provided with three dichroic filters (DF) and three solar cells (SC). The third dichroic filter is optional.

Fig. 1 shows an example of PV spectral splitting receiver based on the pentaprism (PP). The input light is schematized as the superposition of the three spectral bands of e.m. spectrum: vis, nir and ir. The sequence by which light is converted into electricity, whether at increasing or decreasing wavelengths, depends on the type of dichroic filters used. In the example of Fig. 1, the PV conversion occurs on spectral bands with decreasing wavelengths, and the dichroic filters are of the "cold mirror" type. If the PV conversion occurs on spectral bands with increasing wavelengths, the dichroic filters will be of the "hot mirror" type.

The PP-based spectral-splitting system can be used as PV receiver or PV-T receiver, depending on the type of the conversion devices used. In this paper we present the results of the optical simulations carried out on a PV receiver provided with ideal absorbers and operating at 1 sun or at 18x concentration. The operation at 1 sun requires the removal of the lens at the input of the PP. Besides the optical simulations, we present the experimental results relative to measurements of the optical transmittance of the DFs as they are (out of PP) and when glued to the PP. Similarly, we present the results of measurements of the quantum yield of homojunction solar cells of the Ge and c-Si type when they are out of the PP and when are glued to the PP.

2 OPTICAL SIMULATIONS

2.1 Ray-tracing with parallel light

The first optical simulations were carried out with parallel light by removing the planoconvex lens from face (1) (see Fig. 2a) and by using the TracePro optical code for ray tracing [8]. The PP from Edmund Optics, made from fused silica, has a square input face with 25x25mm^2 surface area, and is equipped with two dichroic filters, DF1 and DF2 (the third dichroic filter is not present in these simulations) and three ideal absorbers, AB1, AB2 and AB3 in place of the three solar cells (see Fig. 2a).

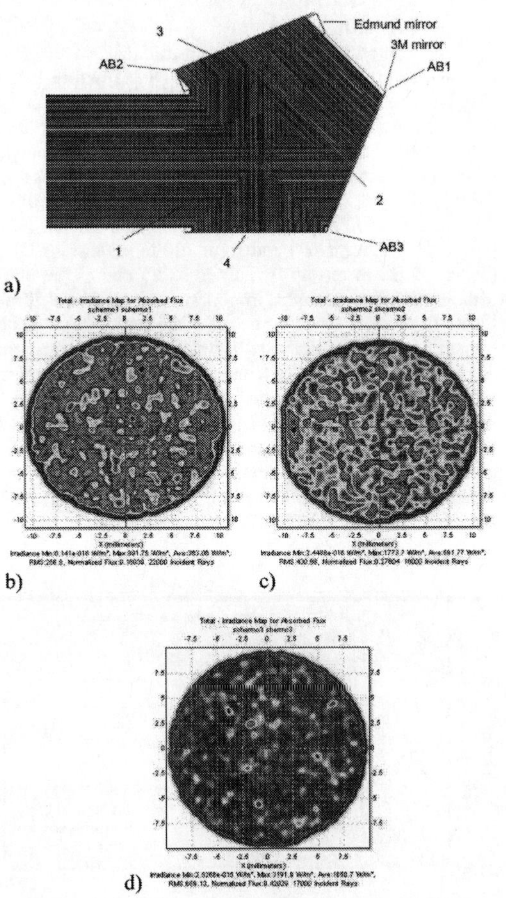

Figure 2: Ray-tracing of the pentaprism operating without concentration (a). Highly homogeneous maps of flux density on AB1 (b), AB2 (c) and AB3 (d) ideal receivers.

The two dichroic filters DF1 and DF2 were selected to simulate a PV spectral splitting receiver operating in the vis-nir-ir spectral range (λ = 400-1800nm). This spectral range can be covered by different types of solar cells. In the present work, we will refer, by way of example, to three types of homojunction solar cells: Ge as the first PV receiver in the nir-ir spectral range (λ = 1000-1800nm), c-Si as the second PV receiver in the red-nir spectral range, (λ = 675-1000nm) and InGaP as the third PV receiver in the vis spectral range (λ = 400-675nm). The dichroic Radiant Mirror Film VM2002 of 3M was chosen as DF1. It is a polymeric film showing a sharp cut-off wavelength, with a reflectivity >90% in the 400-1050nm range and a transmissivity >90% for λ > 1050nm. Due to the strong absorption for λ < 400nm, with hindsight, we recognize that its use as first dichroic filter, as done in this paper, is not suitable. The cold mirror 0°AOI of Edmund Optics, made of borosilicate glass, has a nominal cut-off wavelength of 675nm, and was chosen as DF2. Both dichroic filters are optimized to operate at 0° incidence. The parallel beam from input reaches surface (2) of the PP with an incidence angle of 22.5°; here the nir-ir radiation is transmitted by DF1 and absorbed by AB1. The rest of radiation, vis-nir, is reflected by DF1 and reaches surface (3), again with an incidence angle of 22.5°, where it is transmitted by DF2 and absorbed by AB2. The remaining vis radiation is reflected by DF2 and reaches surface (4) of the PP with an incidence angle of 0°, where it is absorbed by AB3. Fig. 2a shows the ray-tracing of the PP-based receiver with parallel light of AM1.5D spectrum, and Figs. 2b-d show the simulated maps of flux density on the three ideal receivers coupled to the PP. The passive surfaces of the PP were not coated with mirror films. The total flux measured for each map, without AR coating on face (1), was: 17% on AB1, 28% on AB2 and 42% on AB3, with a total optical efficiency of 87%. With AR coating the measured total flux was: 17% on AB1, 29% on AB2 and 44% on AB3, with a total optical efficiency of 90%.

2.2 Ray-tracing with concentrated light

The optical simulations were repeated by operating with concentrated light. The AM1.5D direct solar radiation is collected by a round Fresnel lens (d_1 = 85 mm, f_1 = 110 mm) and transformed into a parallel beam by applying the planoconvex lens (d_2 = 25 mm, f_2 = 25 mm) to face (1) (see Fig. 3a). A concentrated beam on the entrance face of the PP is then produced, projecting a circle of diameter d = 20 mm. The geometric concentration ratio is: $C_{geo} = (d_1/d)^2 = (85/20)^2 = 18x$. Fig. 3b shows the ray-tracing of the receiver schematized in Fig. 3a.

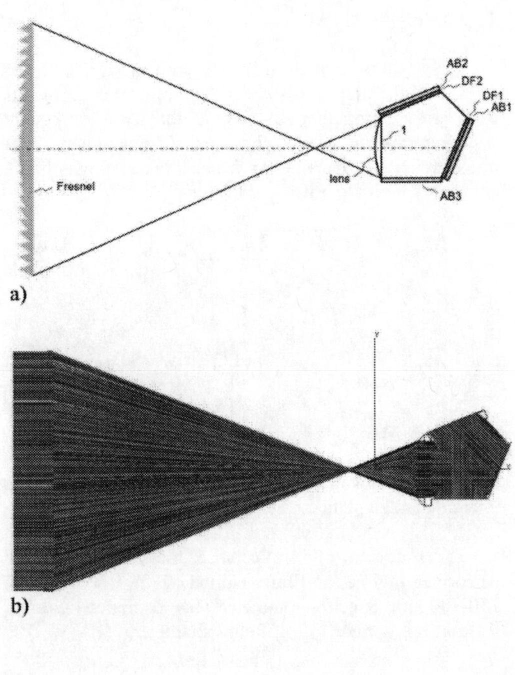

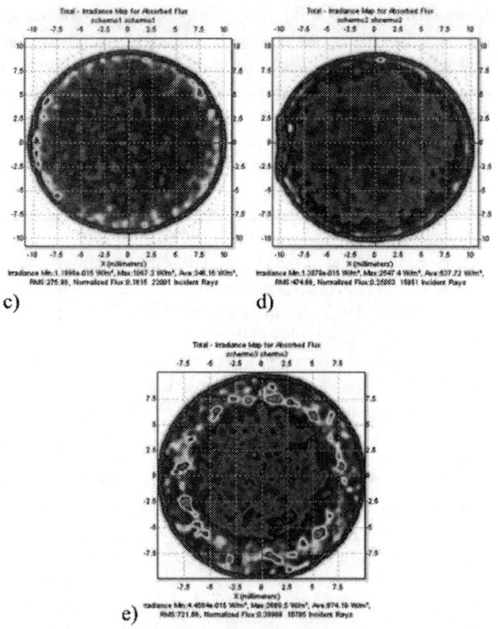

Figure 3: Scheme of the pentaprism (a) and ray-tracing with concentration (b). Maps of flux density on AB1 (c), AB2 (d) and AB3 (e) receivers.

The simulated flux maps on the three ideal receivers, in absence of AR coating on both lenses and of mirror films on the passive surfaces of the PP, are shown in Figs. 3c-e. The percentage of flux absorbed by the receivers was: 16% on AB1, 25% on AB2 and 39% on AB3, with a total optical efficiency of 80%. Unlike the maps of Fig. 2, we notice now a slight unevenness of the flux density, to be attributed to the chromatic aberrations produced by the two lenses.

3 EXPERIMENTAL

Fig. 4 shows photos of the PP with the two dichroic filters and the planoconvex lens. The optical components were sealed together by using the two-component "Wacker SilGel® 612" glue, which guarantees high transparency comparable to fused silica, of which are composed both the prism and the lens.

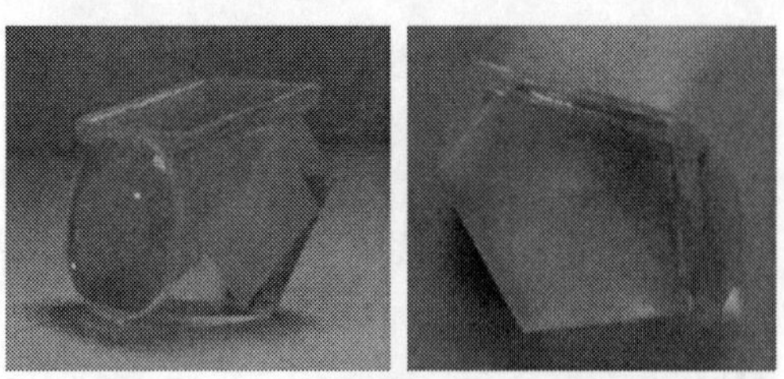

Figure 4: Photos of the pentaprism with the two dichroic mirrors and the planoconvex lens.

Transmittance and reflectance measurements were performed in a Perkin Elmer Lambda 950S UV-VIS-NIR 350-2000nm Spectrophotometer also equipped with an integrating sphere of 150mm diameter. (UV / Vis resolution: 0.05 nm - 5.00 nm; NIR resolution: 0.20 nm - 20.00 nm).

Two different solar cells were used to characterize the prism. The first is a c-Si based high efficiency solar cell fabricated on a pyramid textured c-Si substrate; the second is a Ge based solar cell fabricated on a flat substrate. Their Quantum Yield were measured in the spectral range from 350nm to 1700nm, through a monochromator and lock-in amplifier system. Measurements were commonly performed in alternate 30Hz.

4 OPTICAL MEASUREMENTS

4.1 Spectral transmittance

Fig. 5 shows the spectral transmittance of the two dichroic filters (curves "DF1" and "DF2") before being glued to face (1) and (2) of the pentaprism. We note a steep rise of transmittance for both filters at 1060nm and 720nm, respectively. The two curves undergo a significant change when the filters are glued to the pentaprism. We observe for both, in fact, a smoother rise and a downward shift of the cut-off wavelength, from 1060 to 980nm for DF1 (curve "T prism nir-ir") and from 720 to 690nm for DF2 (curve "T prism red-nir"). This is likely due to the non-orthogonal flux of light impinging the dichroic films located at the exit ports of the pentaprism. The transmittance shift, therefore, is dependent on the incident angle of the impinging light and in the Fig.5 the maximum shift is observed for the optical transmittance of the DF1 dichroic filter. Moreover, a glue undesired absorption at ~1380nm is evident from the transmittance curve of the glued DF1 dichroic filters. Fig. 5 shows also the transmittance curve measured on the exit port of the PP (face (4)), with a cut-off wavelength of 680nm.

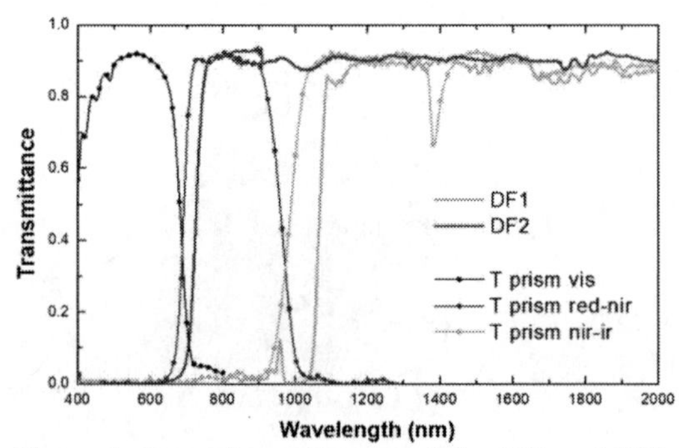

Figure 5: Spectral transmittance of the DF1 and DF2 dichroic filters at different operating conditions.

4.2 Quantum Yield

Fig. 6 shows the quantum yields of a c-Si and a Ge solar cell measured at different operating conditions. The QY of reference are those of the two cells out of the prism: curve "QY Si out-prism" (dashed black) and curve "QY Ge out-prism" (solid black). The quantum yield of the c-Si cell was then measured by moving it along the three exit ports of the pentaprism: (1) in presence of the DF1 filter, (2) in presence of the DF2 filter, and (3) with no filter. The respective results are the curves "QY Si nir" (dashed green), "QY Si red" (solid red), and "QY Si vis" (solid blue). The quantum yield of the Ge cell was measured only at the exit port (1) of the pentaprism, in presence of the DF1 dichroic filter, as shown by the curve "QY Ge nir-ir" (solid green). An interesting effect due to different kind of substrate in the two solar cells is evident looking at the wavelength cut-off of both solar cells. Comparing the QY of the two cells in the wavelength range between 900nm and 1100nm, we notice, indeed, that the presence of the texture in the c-Si based solar cell allows a photon collection very close to the pentaprism transmission ("T prism nir-ir"), whereas the Ge based cell shows a collection cut-off wavelength centered at around 1000nm, that is in between the two transmittance spectra reported in Fig.5 for the nir-ir region. This effect is essentially due to a difference in the solar cell substrate. The substrate texture in the c-Si cell allows a larger angle acceptance of the impinging light with respect to the flat one in the Ge cell, and this affects the transmittance of port (2) covered by DF1 dichroic filter.

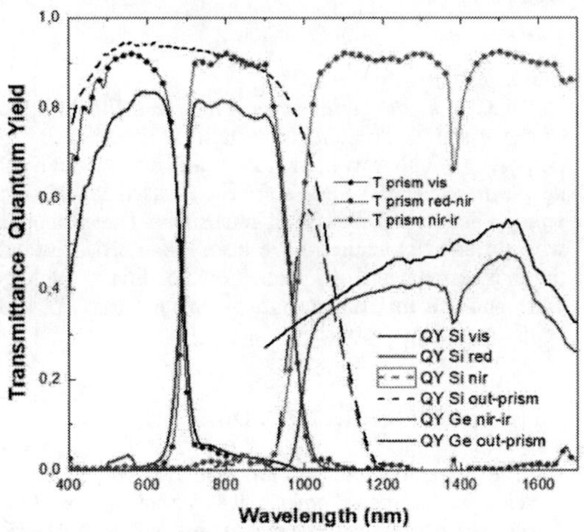

Figure 6: Quantum yield of the c-Si and Ge cells at different operating conditions.

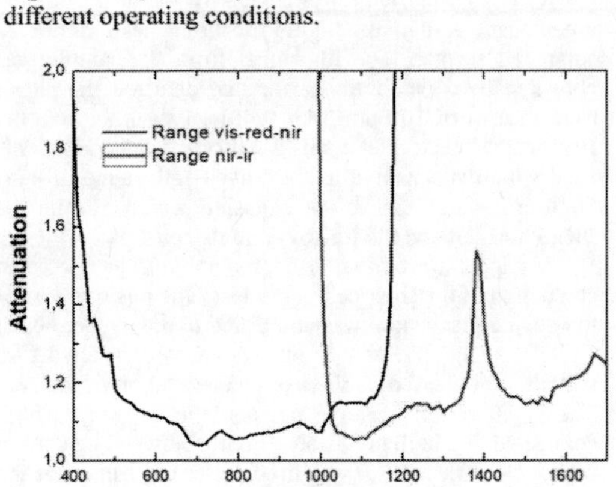

Figure 7: Attenuation factor operated by pentaprism with respect to the photonic flux, as function of wavelength.

By comparing the quantum yield QY of the cells out of the prism with that QY* of the cells exposed to each face of the prism, it is possible to evaluate the prism attenuation, in terms of flux of photons impinging the prism at each wavelength, as the ratio between QY and QY*, as reported in Fig.7. The attenuation is around 10% in the spectral range between 500nm and 1600nm. The undesired effects of transmittance in the blue region could be compensated by providing an appropriate ARC on face (1).

5 CONCLUSIONS

An innovative method for the spectral splitting of concentrated solar radiation is investigated through simulations and experiments, as an alternative approach to the well-established monolithic separation of light in a tandem cell. In this study the PV receiver operates a parallel spectral splitting of solar radiation using a pentaprism coupled with two dichroic filters. This configuration allows to reduce the optical losses associated to the spectral splitting process offering: i) absence of air/dielectric interfaces; ii) parallel radiation travelling inside the pentaprism, incident at near normal angles on the dichroic filters; iii) pentaprism operating as a light trap, with reflective internal surfaces and radiation recycled inside it among the three conversion devices. A 10% attenuation, in terms of flux of photons, has been demonstrated even without appropriate antireflection coating at input and coating by mirrors of the PP passive surfaces, which indeed could be adopted to enhance the effectiveness of the pentaprism.

REFERENCES

[1] C. Stanley, A. Mojiri, G. Rosengarten, "Spectral light management for solar energy conversion systems", Nanophotonics 5 (2016) 161–179.

[2] A. Mojiri, R. Taylor, E. Thomsen, G. Rosengarten, "Spectral beam splitting for efficient conversion of solar energy - A review", Renewable and Sustainable Energy Reviews 28 (2013) 654–663.

[3] A. Antonini, M. Butturi, P. Zurru, M. Norton and A. Parretta, "Development of a high/low concentration photovoltaic module with dichroic spectrum splitting", Progress in Photovoltaics, 23 (2015) 1190–1201.

[4] A. Antonini, M.A. Butturi, P. Di Benedetto, E. Milan, D. Uderzo, P. Zurru, D. Sartore, A. Parretta, "Mirror based spectral splitting CPV system", Proceedings 25th European Photovoltaic SolarEnergy Conference, Vol. I (2010) 172–175.

[5] D. Vincenzi, A. Busato, M. Stefancich, G. Martinelli, "Concentrating PV system based on spectral separation of solar radiation", Physica Status Solidi (a), 206 (2009) 375–378.

[6] C. Maragliano, M. Chiesa, and M. Stefancich, "Point-focus spectral splitting solar concentrator for multiple cells concentrating photovoltaic system", Journal of Optics, 17 (2015) 105901.

[7] C. Maragliano, A. Zayan, and M. Stefancich, "Three-Dimensional Point-Focus Spectral Splitting Solar Concentrator System", Int. J. of Optics and Applications, 4 (2014) 6–11.

[8] http://lambdares.com/

COUPLING EFFECTS IN INGAP/INGAAS/GE TRIPLE JUNCTION PV CELLS OF DIFFERENT STRUCTURES

Paraskeva Vasiliki[1], Armani Nicola[2], Malchiodi Alessandro[2], Trespidi Franco[2], Timo Gianluca[2], Livera Andreas[1], Georghiou George[1]
[1]FOSS Research Centre for Sustainable Energy, ECE, University of Cyprus, 75 Kallipoleos St., Nicosia, 1678, Cyprus
[2]RSE, Ricerca Sistema Energetico, Casino Mandelli, Loc. Le Mose, 29122 Piacenza, Italy

ABSTRACT: Photoluminescence spectroscopy with contemporary two excitation lasers of different wavelength was used to demonstrate the impact of the structure of the solar cells on the coupling effects. Cells with AlInGaAs/InGaAs heterojunction and higher spacer thickness in the middle junction were found to enhance coupling between the top and middle junction. In the presence of varying NIR power and fixed green laser power both the top junction integrated luminescence and the short-circuit current of the device start to plateau at the same NIR power, thus demonstrating the transition of the current limiting junction in the device and particularly the transition from the middle to the top current limiting junction.
Keywords: multi-junction solar cell, photoluminescence, concentrator cells

1 INTRODUCTION

Multi-junction (MJ) devices used in CPV applications utilize the solar spectrum more efficiently than single junction ones via lower thermalization losses. In recent years several MJ designs have achieved efficiencies way higher than 40% [1,2]. As such designs continue to set conversion efficiency records there is a growing interest in the fundamental phenomena associated with these devices.

Due to the series connection of junctions, coupling effects are apparent in those devices. Coupling effects are attributed to optical and/or electrical interaction of junctions. Due to coupling effects, the operating point of one junction is greatly affected by the operating point of the remaining junctions in the tandem in the presence of a specific bias voltage at the terminals of the device. Recently, luminescent coupling effects attracted significant attention [3,4]. Luminescent coupling effect is a specific type of coupling effect and refers to the phenomenon where photons emitted from a junction are reabsorbed by another junction with a smaller band-gap thus changing the operating voltage of the junctions in the tandem [5]. Therefore it is attributed to optical interactions between junctions. Luminescent coupling effects are strongly dependent on the intensity of the light bias applied to the device since recombination increases under strong light bias conditions when the operating point of the irradiated junction is closer to open-circuit [6]. Furthermore, the spectrum of the applied light bias determines the presence and the intensity of the luminescent coupling current [7]. While coupling effects are possible from higher band-gap junctions to lower ones and vice versa, luminescent coupling effects are possible only from higher band-gap junctions to lower band-gap ones. In this paper, we investigate coupling effects where the operating voltage of the middle junction changes (with the application of light) and thus due to the series connection of junctions this determines the operating voltage of the top junction in the tandem device. In that way coupling effects have been examined from lower band-gap junctions to higher band-gap ones and this was achieved by using double wavelength Photoluminescence (PL). Previously, investigation of coupling effects from lower band-gap junctions to higher band-gap ones has been undertaken using spatially-resolved Electroluminescence (EL) [8]. While a significant amount of work has been undertaken in the area of luminescent coupling effects, general coupling effects have yet been studied extensively.

The major goal of this paper is to investigate the effect of the material structure on coupling effects. For this purpose, a double wavelength PL set-up with a green and near infrared (NIR) laser source was used at different voltage biases and excitation intensities. Three samples with different structures have been examined. Through those measurements the excitation conditions (voltage bias) and the material design parameters that enhance coupling will be established.

2 EXPERIMENTAL PROCEDURE

The experimental set-up used for the multiple wavelength PL measurements was based on two laser diodes with emission at 520 nm (green) and 785 nm (NIR) that were focused with a circular spot of 60 μm for the green beam, and about 100 μm for the IR beam on the TJ solar cell surface. The PL signal from the sample was collected by a parabolic mirror that detected the signal over an area of 500 μm^2. An additional lens focused the light to the spectrometer slit. A Silicon spectrometer was used which is sensitive in the wavelength range of 475-1030 nm, which covers the emission region of both the InGaP top cell and the InGaAs middle cell.

A copper bench was used for appropriate electrical contacting of the solar cell under test. The positive part of the device (back side) was connected to the copper bench while the negative part (front side) was connected to a metallic pin pressed on the front metallization of the cell. The positive and negative parts of the cells were both connected to a high precision current source. The current source (Keithley 2430) was used to set the voltage at the terminals of the cell under examination and also to control the injected current. The voltage was changed at each light bias intensity. Initially, measurements in the absence of voltage (open-circuit conditions) were carried out and afterwards they were repeated in the presence of voltage. The voltage applied on the device was selected to be in the range 0 - 2 V which is lower than the open-circuit of the cells which is around 3 V.

PL measurements were carried out in the presence of two excitation light sources, the green and the NIR one. The green light was set at a fixed power (9.2 mW), while the NIR source power varied in the range 0-15 mW. The top junction emission was collected each time in order to establish the amount of coupling in each sample.

Three samples were examined (labeled A, B, C). The samples have identical sizes of 1 x 1 cm^2. The detailed structure of the three samples is reported in Table I. The samples under investigation were InGaP/InGaAs/Ge lattice matched TJs, grown in the n-on-p configuration on (001) 6° off towards the (110) direction Ge substrates [9]. The cells have different window, emitter and base layer thicknesses in the top InGaP junction. The In content of the InGaP compound is very similar for all the samples and close to the lattice matched value (49.5%) as measured by high resolution X-Ray Diffraction. The top layer in all the samples under investigation has a band-gap at 1.8 eV and therefore absorbs in the wavelength region 300-690 nm. The doping has been changed only in the emitters of the top and middle cells, keeping all the other doping values in the remaining structure the same. Furthermore, the thickness of the middle junction emitter and base layers are identical for all the three samples. The emitters of the middle cells were deposited with 1% of Indium content. However, in the case of samples B and C, 10 % of aluminium was added in the emitter layer in order to get a quaternary material (AlInGaAs). This is indicated in Table I by adding (Al) in the middle emitter layer. Also, the spacer thickness in sample A is less than that in samples B and C. The middle layer absorbs in the wavelength region 600-880 nm.

Table I: Structure of the triple junction device under examination.

Sample	A		B		C	
Structure	T*(μm)	D*	T (μm)	D	T(μm)	D
			Top cell (InGaP)			
window	0.025		0.02		0.02	
emitter	0.15	n_2	0.085	n_2	0.085	$n_2/2$
spacer	0		0.05		0.05	
base	0.8		0.84		1.48	
			Middle cell (InGaAs)			
window	0.03		0.03		0.03	
emitter	0.085	n_1	0.085(Al)	n_1	0.085(Al)	$n_1/2$
spacer	0.0038		0.03		0.03	
base	3		3		3	

T*:thickness, D*: Doping

The three samples have been characterized with a solar simulator under ASTM G173 1.5D spectrum with a concentration of 500 suns. The electrical results are reported in Table II.

Table II: Electrical performances of the samples under examination

Sample	A	B	C
Isc (A)	5.23	6.70	5.48
Voc (V)	2.97	2.94	2.97
FF	76.7	87.7	87.0
Efficiency(%)	23.8	34.5	28.3

3 RESULTS AND DISCUSSION

The intensity and spectrum of the light bias applied on the MJ solar cells determine the current limiting junction and provide access to each junction in the cell separately [10]. Since the amount of radiative recombination is related with the amount of coupling between the top and middle junctions in the device, the integrated PL intensity of the radiative recombination from the band-gap edges of the top and middle junction was measured at different excitation intensities of the NIR light source. The top junction emission was collected each time. The decrease of the top junction emission by increasing the NIR power and by keeping the visible excitation fixed is attributed to coupling [8] and thus will demonstrate the amount of coupling in each sample. Specifically, the increase of the NIR power increases the operating voltage of the middle junction resulting in a transition of the junction from reverse to forward bias conditions. This consequently reduces the operating voltage of the top junction since the total voltage of the tandem is constant due to the series connection of junctions and the coupling effects. The reduction of the operating voltage of the top junction results in top junction emission reduction and this is an indication of the amount of coupling in the tandem. It is worth noting that the bottom junction was not taken into consideration in our analysis. In the presence of just green light or green and NIR light, the photocurrent of the Germanium bottom junction is negligible. The very small photocurrent in the junction at all light biases under investigation in combination with the low shunt resistance of the junction make that junction shunted. Thus, the Germanium bottom junction does not influence the measurement results of the top or the middle junction.

An NIR source of varying intensity and a green light source of fixed intensity (9.2 mW) were used in order to study the coupling from the middle to the top junction. The luminescent signal from the band-edge of the top junction was acquired at each NIR irradiance value. The reduction of the top junction emission will indicate coupling effects and will demonstrate which material structure enhances coupling. The top junction emission from all the samples at a voltage bias of 1 V is depicted in Fig.1 .

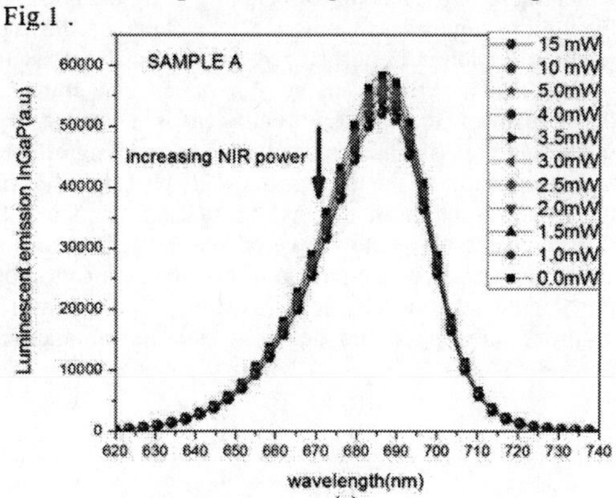

(a)

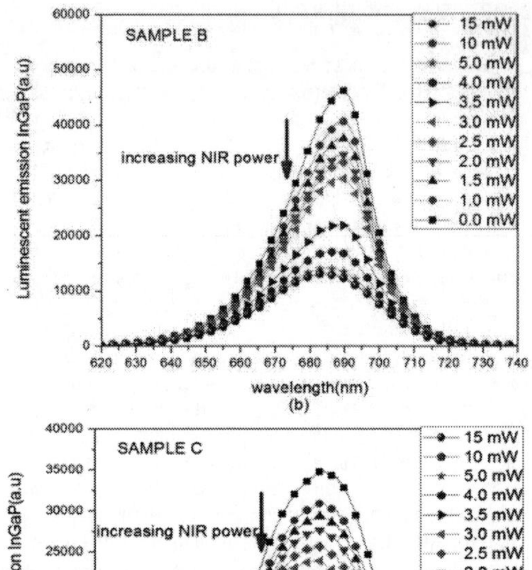

Figure 1: Luminescent emission from the InGaP junction as a function of NIR laser power, by keeping the green laser power constant (9.2 mW) at a voltage bias of 1 V for samples (a) A, (b) B and (c) C.

Comparing the top junction emission reduction with increasing NIR power in the samples in Fig. 1 it can be observed that the reduction of the top junction emission in sample A is significantly lower compared to the reduction measured in the other two samples. In order to better compare the differences in top junction emission reduction which indicate the amount of coupling in the samples, in Fig. 2 the luminescent emission quenching of the top junction is plotted in normalized units for different NIR power. In all the samples under examination, quenching of the integrated top junction emission is observed by increasing the NIR laser intensity due to coupling effects from the middle to the top junction. The NIR light lies in the response region of the middle InGaAs junction and thus increases the photovoltage of that layer. Due to the series connection of the junctions the voltage increase of the middle junction causes a reduction of the forward voltage of the top one and the emission of the top junction decreases.

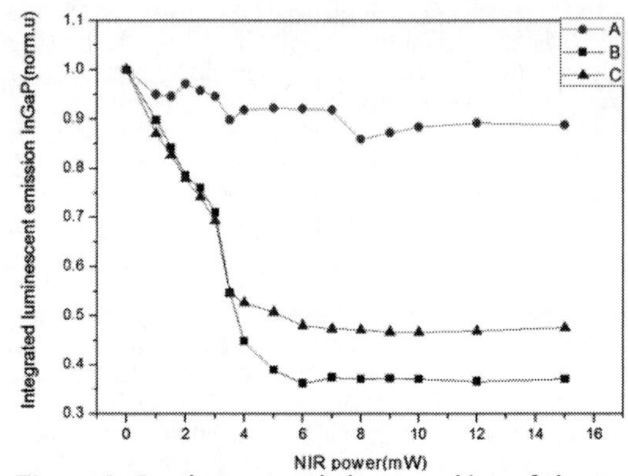

Figure 2: Luminescent emission quenching of the top junction in the presence of fixed green light power (9.2 mW) and varied NIR light power at a voltage bias of 1 V for all samples under investigation.

As depicted in Fig. 2 the luminescent emission reduction of the top junction in sample A is the lowest, indicating lower coupling at all NIR light bias intensities. A possible interpretation is related to the different emitter type and spacer thickness in the middle cell. The presence of a homojunction (n-InGaAs/p-InGaAs) and a thinner spacer in the middle junction of sample A generates a lower increase in the middle junction photovoltage with varying NIR light than that obtained in the n-AlInGaAs/p-InGaAs heterojunction of samples B and C. This results in lower voltage reduction on the top cell and consequently in lower reduction in the top junction emission of sample A.

The luminescent emission reduction of the top junction in samples B and C is much higher compared to that obtained in sample A thus indicating that coupling is much higher in those samples due to the presence of the heterojunction as discussed above. The reduction of the top junction emission in samples B and C as a function of the NIR laser power is divided into two regions. In the first region (0-4 mW) no distinct difference in the top junction emission reduction in the two samples can be observed indicating that similar coupling is present in both samples. However, at NIR intensities in the range 4-15 mW, the top junction emission reduction of sample B becomes notably higher compared to that of sample C. The difference in the top junction emission reduction at higher NIR intensities is mainly attributed to the different base thicknesses in the top junction. The presence of lower thickness in the base of the top junction in sample B results in lower recombination current in the top junction and thus contributes to higher top junction emission reduction in the sample.

The luminescent emission quenching of the top junction with varying NIR light power was also examined at voltage biases of 0 V and 2 V. The results are similar with those depicted in Fig. 2 indicating no significant voltage dependence of the top junction emission reduction. The independence of the top junction emission quenching from the voltage bias indicates that the operating voltage of the top junction remains the same.

One important characteristic of the emission curves of the samples shown in Fig. 1 is the exponentially dropping edges. Energetically lower/higher dropping edges are the exponential curves present just below and above the band-gap of the top junction emission spectrum and represent the top junction emission from the states below and above band-gap [11]. Examination of the slopes of the dropping edges at each light bias intensity showed that the slope of the energetically lower dropping edge changed more with increasing NIR light compared to the slope of the energetically higher dropping edge. This observation can be clearly seen in Fig. 3 for all samples under investigation, indicating that the distribution of excited carriers inside the top junction changes significantly, owing to the reduction of the top junction operating voltage. Another important observation concerns the fact that the change of the energetically lower dropping edge is much higher in sample B compared to the one observed in samples A and C, demonstrating that the number of excited carriers near band-gap changes significantly in sample B with the change of NIR power. This might be related to the coupling effects in the samples. The higher change in carriers just below band-gap with NIR light power in sample B is a result of higher coupling effects between top and middle junction in accordance with Fig. 2. On the other hand, sample A appears to have the lowest change in the slope of the energetically lower dropping edge demonstrating the smaller change in the number of carriers in that region due to lower coupling effects in the sample in good agreement with the experimental results of Fig. 2. The lower coupling effects affect the forward bias of the top junction less and thus the change in the number of band-gap edge carriers.

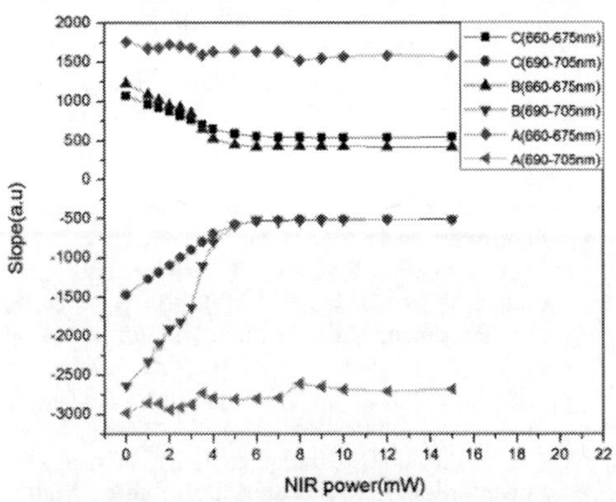

Figure 3: Slope of the lower (690-705 nm) and higher (660-675 nm) dropping edges of the PL emission of the top InGaP junction for samples (a) A (b) B and (c) C. The measurements were taken at a voltage bias of 1 V.

The luminescent emission of the middle junction was also investigated as a function of voltage bias. The InGaAs junction integrated PL intensity of sample A is presented in Fig. 4. A linear behaviour is observed up to a NIR power of 4.5 mW, followed by an exponential behaviour at higher laser powers. This can be explained by the fact that the middle cell becomes forward biased when the NIR light power reaches 4.5 mW. No significant voltage dependence of the integrated middle junction emission was detected in agreement with the top junction emission results.

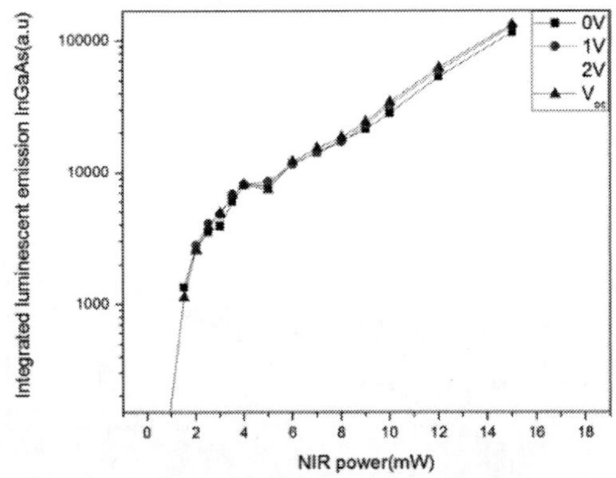

Figure 4: Integrated PL emission of the middle InGaAs junction in the presence of fixed green light power (9.2 mW) and variable NIR light power at different voltage bias in sample A.

Investigation of the PL emission of the cells and its correlation with the short-circuit current was also considered. In previous publications, spectrometric characterization was performed on the cells in the presence of varying total irradiance. This method indicates the current matching point between the junctions through the change in the increase rate of the short-circuit current. In order to investigate the current matching by changing the NIR light intensity and by keeping the green light intensity fixed, the short-circuit current was collected in these conditions alongside the luminescent emission from both middle and top junctions. The results for samples B and C are shown in Fig. 5(a) and 5(b) respectively. A clear change in the increasing rate of the short-circuit current is observed in both samples indicating the current matching point of the junctions. The change in the increasing rate of short-circuit current coincides with the change in the reduction rate of the luminescent emission of the top junction and the change in the increasing rate of the luminescent emission of the middle junction and it is in good agreement with recently published results [12,13]. The transition point of the short-circuit current and the luminescent emission of the top and the middle junctions in both samples is apparent at NIR intensities around 4.5 mW. At NIR intensities lower than 4.5 mW the luminescent emission of the top junction drastically reduces while the emission from the middle junction increases linearly along with the short-circuit current. However, at NIR intensities above 4.5 mW the top junction emission remains constant while the middle junction emission increases significantly. At the same time the short-circuit current remains constant or increases slightly. These observations provide evidence that at NIR intensities below 4.5 mW the tandem device could be middle junction limited while above that value could be top junction limited. The transition from middle to top current limited junction can explain the observations in emission and current. The observations hold if we assume that the bottom junction is shunted. We believe that due to the low photocurrent of the Germanium junction and its low shunt resistance, the junction is shunted at all light bias under examination. This means that the bottom junction is not participating in the interactions between junctions and the triple-junction device behaves like a double junction one.

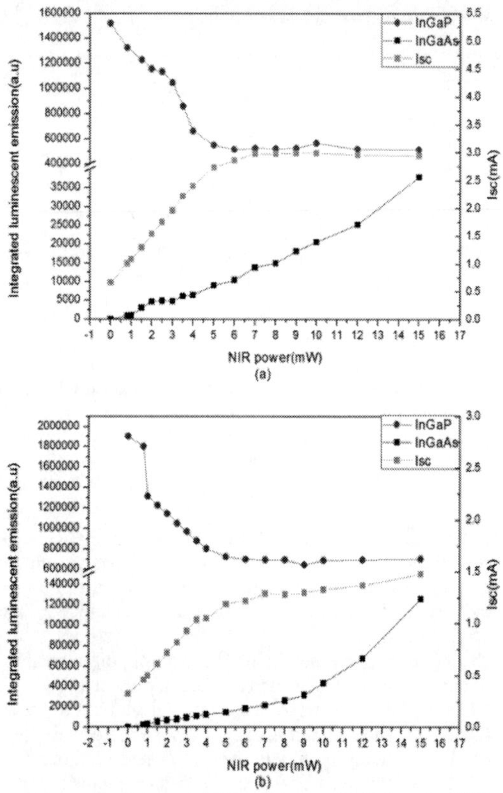

Figure 5: Integrated PL of the top InGaP and middle InGaAs junctions alongside the short-circuit current of the tandem InGaP/InGaAs/Ge in the presence of varying NIR intensity in the samples (a) B and (b) C.

4 CONCLUSIONS

The work presented in this paper provides an insight into radiative recombination processes occurring in triple junction (TJ) InGaP/InGaAs/Ge devices at different light wavelengths and intensities. Double wavelength PL was used to investigate coupling between top and middle junctions. The coupling effects observed with double wavelength PL with fixed visible and variable NIR light sources, at different voltage bias are influenced by the multi-junction structure. In particular, the presence of a heterojunction was found to promote coupling effects in a tandem device. Samples with n-AlInGaAs/p-InGaAs heterojunction in the middle junction and thicker spacer thickness demonstrated higher reduction of the top junction luminescence in the presence of NIR light due to enhanced coupling effects. Furthermore, higher doping levels in the emitter of the middle junction in combination with lower base thickness in the top junction enhance coupling between top and middle junctions. Voltage dependence of coupling effects was not detected in the voltage range under examination since in the presence of light bias used in the experiments and voltage biases in the range 0-2 V the operating voltage of a junction is mainly determined by the light bias. Thus, in the presence of fixed light bias and varying voltage between 0-2 V the operating point of a junction changes only slightly thus resulting in a negligible change in the luminescent emission. Finally, emission from both the top and middle junctions is in good agreement with the short-circuit current behaviour of the tandem. The change in the slope of the short-circuit current coincides with the change in the slope of the top junction luminescent emission thus demonstrating a transition in the current limiting junction in the TJ device.

The paper gives an insight into the structural aspects of multi-junction cells and thus provides a better understanding of the material structure properties of those cells which is essential for the improvement of the design of tandem cells. By taking into account coupling effects we become aware of another parameter that might affect the real performance of such devices and leads to a better understanding of the real behaviour of these devices.

ACKNOWLEDGMENT

This project has received funding from the European Union's Horizon 2020 research and innovation program under the project TWINPV (grand agreement No. 692031)

5 REFERENCES

[1] T. Tibbits, P. Beutel, E. Oliva, C. Karcher, in Proc.29th Eur. Photovolt. Sol. Energy Conf. Exhib. (2014) 1975.

[2] F. Dimroth, M. Grave, P. Beutel, U. Fiedeler, C. Karcher, T.N.D. Tibbits, E. Oliva, G. Siefer, M. Schachtner, A. Wekkeli, A.W. Bett, R. Krause, M. Piccin, N. Blanc, C. Drazek, E. Guiot, B. Ghyselen, T. Salvetat, A. Tauzin, T. Signamarcheix, A. Dobrich, T. Hannappel, K. Schwarzburg, Prog. Photovoltaics Res. Appl. 22 (2014) 277.

[3] D.J. Friedman, J.F. Geisz, M. a. Steiner, IEEE J. Photovoltaics 4 (2014) 986.

[4] M.A. Steiner, J.F. Geisz, T.E. Moriarty, R.M. France, W.E. McMahon, J.M. Olson, S.R. Kurtz, D.J. Friedman, IEEE J. Photovoltaics 3 (2013) 879.

[5] S.H. Lim, J. Li, E.H. Steenbergen, Y. Zhang, Prog. Photovoltaics Res. Appl. 21 (2011) 344.

[6] V. Paraskeva, M. Hadjipanayi, M. Norton, M. Pravettoni, G.E. Georghiou, Sol. Energy Mater. Sol. Cells 116 (2013) 55.

[7] M.A.Steiner, J.F.Geisz, Appl. Phys. Lett. 100 (2012) 25116.

[8] G.W. Shu, J.Y. Lin, H.T. Jian, J.L. Shen, S.C. Wang, C.L. Chou, W.C. Chou, C.H. Wu, C.H. Chiu, H.C. Kuo, Opt. Express 21 (2013) 123.

[9] N. Armani, G. Abagnale, M. Cornelli, G. Timò, E. Malvisi, F. Farina, G. Carbi, F. Rossi, F. Fabbri, L. Nasi, Energy Procedia 84 (2015) 34.

[10] S. Kurtz, K. Emery, J. Olson, Proc.24th IEEE Photovolt. Spec. Conf. (1994) 1733.

[11] C. Karcher, H. Helmers, M. Schachtner, F. Dimroth, A.W. Bett, Prog. Photovoltaics Res. Appl. 22 (2013) 757.

[12] D. Lan, J.F. Geisz, M.A. Steiner, I. Garcia, D.J. Friedman, M.A. Green, Sol. Energy Mater. Sol. Cells 143 (2015) 48.

[13] D. Lan, M.A. Green, Prog. Photovoltaics Res. Appl. 24 (2016) 1566.

ISOTYPE HETEROINTERFACES IN TRIPLE-JUNCTION SOLAR CELLS: INFLUENCE ON IV-CURVE SHAPE AND OPTIMIZATION

M.A. Mintairov[*1,a], V.V. Evstropov[1,b], S.A. Mintairov[1,c], M.Z. Shvarts[1,d] & N.A. Kalyuzhnyy[1,e]
1 Ioffe Institute
26 Politekhnicheskaya, St Petersburg 194021, Russian Federation, tel/fax: +7(812) 2972173
A - mamint@mail.ioffe.ru, B – vvevstropov@gmail.com, C - mintairov@scell.ioffe.ru,
D - shvarts@scell.ioffe.rssi.ru, E - nickk@mail.ioffe.ru

ABSTRACT: Studies of the *GaInP/GaAs/Ge* SC bottom intergenerator part (layers between the *GaAs* and *Ge* subcells) has been carried out. For this purpose the comparison of IV-curves of *GaInP/GaAs/Ge* solar cells with different intergenerator part has been made. It was found that *p-AlGaInP/p-GaInP/p-GaAs* heterobarriers are responsible for previously observed peculiarities on triple-junction solar cells IV-curve. To study the photo-activity of bottom intergenerator part a special sample of double-junction *GaAs/Ge* solar cells has been created. For the sample, a set of IV-curves was recorded at different incident light irradiation and a photovoltaic dependence of the open circuit voltage on photogenerated current V_{oc}-J_g was obtained. The analysis of obtained characteristics shows that, in contrast to the top, the bottom intergenerator part has no counteracting photovoltaic effect.
Keywords: GaInP/GaAs/Ge solar cell, GaAs/Ge solar cell, counteracting photovoltaic effect, isotype heterointerface

1 INTRODUCTION

Triple-junction solar cells (SC) have a complex structure that includes a large (more than 15) number of semiconductor layers. As is known [1-3], "parasitic" heterointerfaces that increase the resistive losses of solar cells can appear on the boundary between layers. In addition, as shown in [2, 3], a counter photovoltaic effect can arise in the multi-junction SC tunnel part. Thus, the study of the properties of heterointerfaces, especially in the regions between subcells (intergenerator parts) is an urgent task. Previously [2,3], we carried out a detailed investigation of the top intergenerator part which include layers between the top (*GaInP*) and middle (*GaAs*) subcells of the triple-junction *GaInP/GaAs/Ge* SC. It has been shown that it can generate counteracting electromotive force and in some critical cases (non-degenerate tunnel junction) it can distort multi-junction SC IV-curve. In [1], the hetrointerfaces influence on IV-curve has been studied. Heterointerfaces IV-curves have been extracted from full device ones, but the location of heterointerfaces in solar call structure was not determined.

In this paper, the study of the bottom intergenerator part (between *GaAs* and *Ge* sub-cells) is carried out. Note that the same sample as in [1] was used as a sample with IV-curve peculiarities.

2 OBJECTS OF STUDY

Three samples were used for studying bottom intergenerator part. Two of them were *GaInP/GaAs/Ge* triple-junction solar cells with only one difference in the bottom intergenerator part structures (Fig. 1). Comparison of sample characteristics allow to locating the heterointerfaces caused to appearance of IV-curve peculiarities and obtaining their IV-curves. In addition to these *GaAs/Ge* duble-junction solar cell has been used. This sample has been created to investigate the photo-activity of bottom intergenerator part. To observe such effect the tunnel junction with low peak current has been integrated in the double-junction structure. All other layers were similar to GaAs/Ge part of the sample 1 structure (Fig 1).

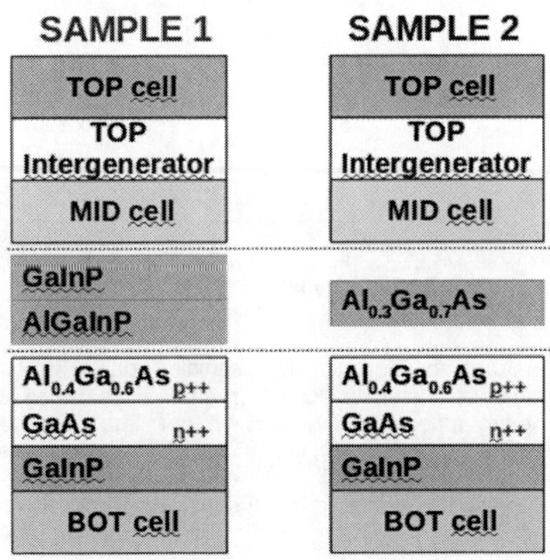

Figure 1: GaInP/GaAs/Ge triple-junction solar cell structures (sample 1 and sample 2) with detailed bottom multi-layer intergenerator parts.

3 INFLUENCE OF NON-OPTIMIZING HETEROBARRIERS ON MULTI-JUNCTION IV-CURVE

IV-curves of two triple-junction SCs (sample 1 and sample 2) having different structure of bottom intergenerator part (Fig. 1) have been compared (Fig. 2).

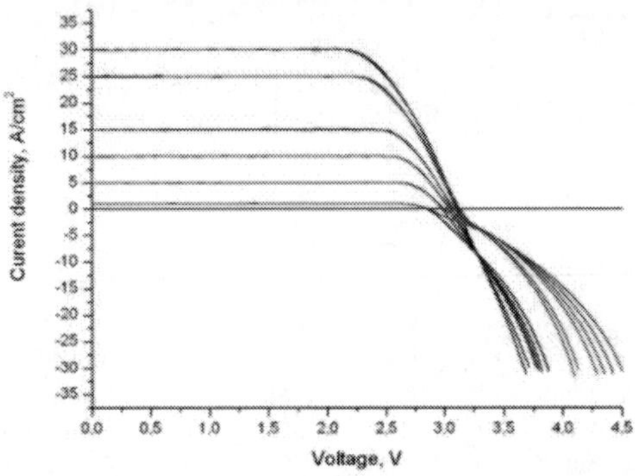

Figure 2: Experimental IV-curves of sample 1 (red lines) and sample 2 (blue lines).

Sample 1 has the same heterointerface peculiarities of IV-curves (with inflection)) as was shown in [1]. The sample 2 IV-curves don't have such inflection. The comparison of sample band diagrams (Fig. 3) shows that sample 2 has no parasitic isotype barriers in valence band.

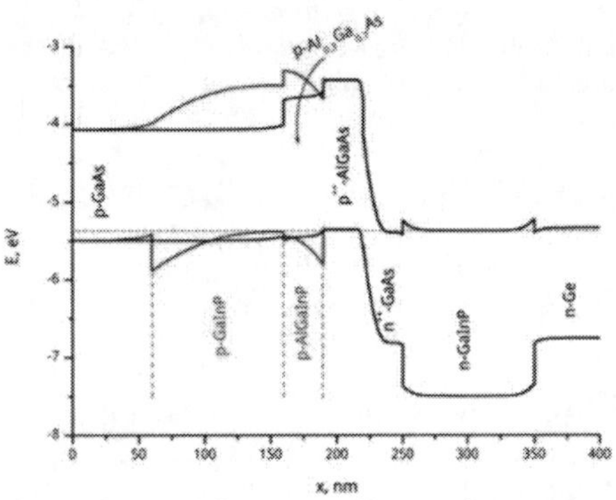

Figure 3: Band diagrams of bottom intergenerator part for sample 1 (red line) and sample 2 (blue line).

Due to the fact that samples have similar structure (except layers that gives an appearance of heterobarriers in valence band) the voltage subtracting of samples IV-curves allows one extracting «parasitic» isotype heterointerfaces IV-curves. The results of such subtracting are shown on Fig. 4.

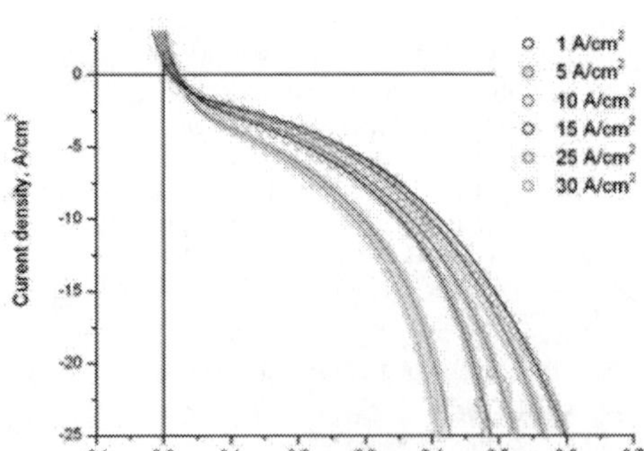

Figure 4: Extracted (symbols) and fitted by eq 1 (lines) heterointerfaces IV-curves. Extracted IV-curves have been obtained by voltage subtraction of experimental IV-curves of sample 2 and sample 1 (Fig. 2).

Three following exponential equation have been used for describing extracted heterointerfaces IV-curves:

$$J = J_{0,B} \cdot \exp\left(-\frac{V_B}{E_B}\right) - J_{0,F1} \cdot \exp\left(\frac{V_{F1}}{E_{F1}}\right) - J_{0,F2} \cdot \exp\left(\frac{V_{F2}}{E_{F2}}\right) \quad (1)$$

The free parameters of eq.1 have been obtained for all heterointerfaces IV-curves (Table 1). Voltage ideality factors (E_b, E_{F1}, E_{F2}) determining current flow mechanisms are constants. So, a current flow mechanism does not depend on a photogenerated current. On the contrary, pre-exponents ($J_{0,B}$, $J_{0,F1}$, $J_{0,F2}$) determining the rate of recombination channels increasing at higher photogenerated currents. Note that the change in the pre-exponent $J_{0,F1}$ is not significant (less than an order of magnitude), this can occur due to a change in the thickness of the barriers. However, its increasing greatly affects IV-curve shape. Other pre-exponents weakly affect only the upper ($J_{0,B}$ affect the IV-curve shape at currents more than 0 A/cm^2) and lower ($J_{0,F2}$ at currents less than -15 A/cm^2) regions of the characteristic.

Table I: Parameters obtained at heterenterfaces IV-curve fitting (Fig. 4) by eq. 1.

J_g A/cm^2	$J_{0,B}$ A/cm^2	$J_{0,F1}$ A/cm^2	$J_{0,F2}$ A/cm^2
1	2.3	1.35	-
5	3.0	1.45	$1.5 \cdot 10^{-6}$
10	4.0	1.65	$8.0 \cdot 10^{-6}$
15	5.0	1.77	$30 \cdot 10^{-6}$
25	5.5	2.39	$150 \cdot 10^{-6}$
30	6.0	2.45	$200 \cdot 10^{-6}$
	E_B V	E_{F1} V	E_{F2} V
for all J_g	0.025	0.205	0.040

4 STUDY OF PHOTO-ACTIVITY OF BOTTOM INTERGENERATOR PART

In [2, 3] we show that tunnel junction of top intergenerator part (between top and middle subcells) is photo-active and producing counter electro-motive force. To study such effect in bottom intergenerator part we use a special double-junction structure which as mentioned above has a similar structure to the sample 1, but with tunnel junction having low peak current. Low peak current allows to easily detecting tunnel junction photo-activity. Fig. 5 shows the experimental V_{oc}-J_g dependence for the sample under study. As is known, such dependence after a voltage shift by a certain value $V_{a,oc}$ [4, 5] coincides with the resistiveless dark IV-curve. As shown in [4], this correction depends on the imbalance of the photogenerated currents (at full balance it becomes zero) and usually does not exceed 0.05 V. The obtained IV-curves were recorded in the voltage range from 0.1 to 0.5 V. The magnitude of these voltages is significantly exceeds $V_{a,oc}$, therefore, to simplify the calculations, it was not taken into account. As can be seen (Fig. 5), the V_{oc}-J_g dependence, and, consequently the dark IV-curve, is well described by a two-exponential model:

$$J_g = J_{02} \cdot \left(\exp\left(-\frac{V_{OC}}{2kT} \right) - 1 \right) + J_{03} \cdot \left(\exp\left(-\frac{V_{OC}}{3kT} \right) - 1 \right) \quad (2)$$

where k – Boltzmann constant, T – absolute temperature. Approximation by formula (2) of the V_{oc}-J_g dependence allows determining the parameters $J_{02}=5.5 \cdot 10^{-13}$ A/cm^2 and $J_{03}=5.5 \cdot 10^{-9}$ A/cm^2. These parameters have been used to obtain a resistiveless light IV-curves (dashed lines in Fig 6).

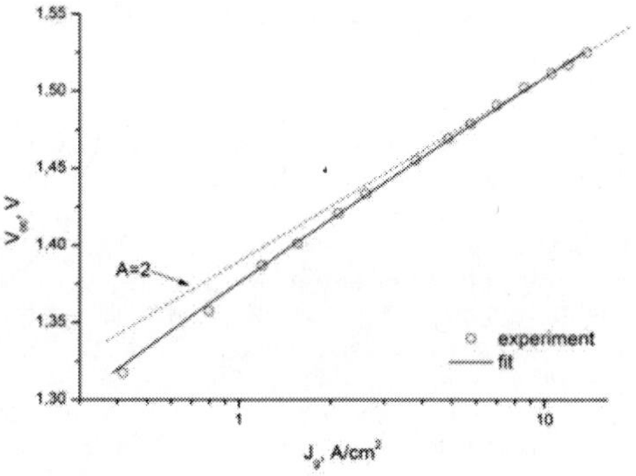

Figure 5: Experimental (symbols) and calculated (solid line) J_g-V_{oc} curves of the GaAs/Ge SC.

As can be seen (Fig. 6), the obtained IV-curves do not coincide with the experimental ones. The voltage subtraction of the resistive-less IV-curves from the experimental ones results in a set of IV-characteristics of the intergenerator part for different J_g (Fig. 6 on the left). Intergenerator part IV-curve does not depend on J_g, therefore the bottom intergenerator part is not photoactive in contrast to the top one [2,3].

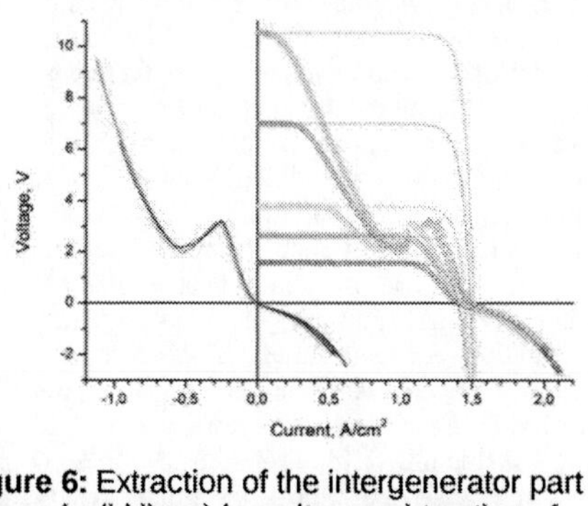

Figure 6: Extraction of the intergenerator part IV-curves (solid lines) by voltage subtracting of experimental IV-curves (symbols) from light resistive-free ones (dash lines).

So there is no photo-activity in bottom tunnel junction of GaInP/GaAs/Ge solar cells, which is most likely connected with full light absorbtion in GaAs sub-cells.

5 CONCLUSION

The intergenerator parts sometimes significantly affect the performance of multi-junction solar cells. Therefore, the definition of the main properties of these parts is important for the purpose of further increasing the number of sub-cells.

The bottom intergenerator part (between GaAs and Ge photovoltaic p-n junctions) has been experimentally analyzed. The comparison of IV-curves of two GaInP/GaAs/Ge solar cells with different intergenerator part has been made. It allows determining the location of non-optimized heterobarriers. The comparison of band diagram shows that such barriers appear between *p-AlGaInP/p-GaInP/p-GaAs* layers. The substitution of *p-AlGaInP/p-GaInP* to $p\text{-}Al_{0.3}Ga_{0.7}As$ in bottom intergenerator part improve IV-curve shape of triple-junction solar cells.

Also the photo-activity of bottom intergenerator part has been investigated. For this purpose, the method of extracting the IV-curve of the intergenerator part (previously developed for top intergenerator part between GaInP and GaAs p-n junctions), has been used. It has been found that, in contrast to the top, the bottom intergenerator part has no counteracting photovoltaic effect.

5 ACKNOWLEDGMENTS

The work has been supported by the Russian Science Foundation (Grant №17-79-30035).

References

[1] M.A.Mintairov, V.V.Evstropov, N. A. Kalyuzhnyy, S. A. Mintairov, N.Kh.Timoshina, M.Z.Shvarts, AIP Conf. Proc., Aix-les-Bains (France) , 2015, v. 1679, p. 050007

[2] M.A. Mintairov, V.V.Evstropov, S. A. Mintairov, M.Z.Shvarts, .A. Kozhukhovskaia, N. A. Kalyuzhnyy, Proc. of 33 PVSEC, Amsterdam (Netherlands) 2017, pp. 1275 – 1277

[3] M.A. Mintairov, V.V.Evstropov, S. A. Mintairov, M.Z.Shvarts, .A. Kozhukhovskaia, N. A. Kalyuzhnyy, JOP Conf. Series, Saint Petersburg (Russia), v. 917, (2017), p 052034

[4] M.A. Mintairov, V.V. Evstropov, N.A. Kalyuzhnyy, S.A. Mintairov, V.D. Rumyantsev, N.K. Timoshina, M.Z. Shvarts, V.M. Lantratov, Semiconductors, v.46, 8, 2012, pp: 1051-1058

[5] M. A. Mintairov, N. A. Kalyuzhnyy, V. V. Evstropov, V. M. Lantratov, S. A. Mintairov, M. Z. Shvarts, V. M. Andreev, A. Luque, IEEE Journal of Photovoltaics, v.5, 4, 2015, pp: 1229 - 1236

INTERNAL SUN TRACKING CPV4ALL MODULE: IMPROVEMENT AND NOVEL CHARACTERIZATION OF THE 4-LOBED PARABOLIC MIRROR WITH TOLERANCE MANUFACTURING

Sarah Bernardis[1], Gerd Ickes[2], Harro Hagedorn[2], Holger Reus[2], Torsten Schmauder[2], Michael Schottner[3], Harmen Rooms[3], Pieter-Jan Bolt[3], Joep Bos-Coenraad[4], Sverre G.D. van Es[4], Erik J. Haverkamp[4], John J. Schermer[4], Julien Gaume[1], and Philippe Voarino [1]

[1] Univ. Grenoble Alpes, French Commission for Atomic and Alternative Energies (CEA), LITEN, Department for Solar Energy, National Institute of Solar Energy, Concentration PV Lab, 73375 Le Bourget du Lac, France

sarah.bernardis@cea.fr

[2] Buhler Leybold Optics, 63755 Alzenau, Germany

[3] TNO, Solar Research Solliance, 5656 AE Eindhoven, The Netherlands

[4] Radboud University, Institute for Molecules and Materials, Applied Materials Science, Heyendaalseweg 135, 6525 AJ Nijmegen, The Netherlands

ABSTRACT: The SunCycle module is a hybrid concentration system, transforming captured solar radiation in electricity and heat. Novelty of this system is that each module composing the system is designed with two rotating optical stages (prism and mirror) to avoid the need of external sun-tracking. In this work, we evaluate the impact on the final module performances of two different coatings of the 4-lobed parabolic mirror, an Aluminum (Al) and a Silver (Ag) stack, and their manufacturing tolerances. To do so, we propose two mirror measurement techniques, named "Pilot-Line" and "R&D" mirror tests, backed-up by solar simulator measurements and simulations. With the pilot-line tests, the measurements from the Ag-coated mirrors showed an 8 % increase in average reflectivity compared to the Al-coated mirrors. The variation within the Al batch was 1,6 % and 1 % for the Ag batch. The variation between repeated measurements of the same mirror was lower than 0,6 %. These values, combined with the fast measurement time, make this method suitable for pilot-line implementation. In both characterization methods, the surface shape of the tested mirrors showed the same features due to molding. All measurement techniques agreed on the Ag-coated mirror reflectivity improvement, compared to the Al-based one, leading to a similar increase in module power production.

Keywords: CPV, mirror, coating, optical characterization, manufacturing

1 INTRODUCTION

The EC FP7 CPV4ALL project aims at bringing the SunCycle module at production level. The SunCycle CPV module (described in the patent referenced in [1] and schematized in **Figure 1**) transforms the captured solar radiation in electricity and heat. Each module produces a geometrical concentration ratio of 870X thanks to the design of its Fresnel prism, a 4-lobed parabolic mirror, a DTIR-like refractive secondary-optical element (SOE) and a III-V triple-junction solar cell.

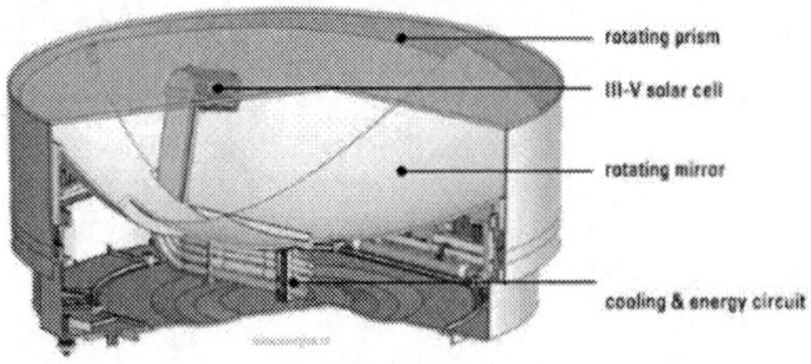

Figure 1: Cross-cut of a SunCycle module where the two rotating optical stages are composed of the Fresnel prism and the 4-lobed parabolic mirror. Image curtesy of *SunCycle*.

To reach the project target, a pilot line has been assembled and modules produced. Current modules have mirrors prepared with Al reflective layers capped with a dielectric protective layer. Within the project the Consortium investigated the possibility of improving the reflective properties of the mirror. Here, an Ag–based mirror layer stack capped with a dielectric layer is proposed, deposited and tested.

Surface imperfections, like variations of mirror curvature and surface roughness, may cause optical losses. Knowing these imperfections and effects on the final performance at an early stage of the manufacturing process can help the manufacturability and thus improve the performance of the whole system. Trespidi *et al.* presented a grazing-angle illumination method to evaluate defects [2]. Given CPV modules are designed to work with a wide range of wavelengths and incidence angles, the selection of the mirror coating plays an important role in the overall efficiency. Production of such an optical part demands, however, in-line control with specific manufacturing tolerances. CPV4ALL mirrors have so far being tested post-production via solar simulator measurements. Here we propose two measurement set-ups, developed specifically to determine the spatially-resolved optical efficiency of this mirror, one referred to as "R&D mirror test" and the other one "Pilot-Line mirror test" (mainly because the characterization time makes the latter more suitable to pilot line takt time restrictions). To establish tolerances a theoretical approach has been undertaken. Results from solar simulator measurements, mirror spatially-resolved optical efficiency measurements and simulations will be matched to validate the proposed characterization methods and to propose manufacturing tolerances.

2 SAMPLE PREPARATION AND METHOD

Within the scope of ramping up to production and improving even further the performances of the module, optical properties of the 4-lobed parabolic mirror are investigated within the project.

2.1 Mirror Substrate and Coating Deposition

The substrate of the mirror is in PolyCarbonate (PC) material; the shape of the substrate is obtained by means of injection compression molding, to minimize internal stresses and to produce a perfect, mirror-like surface. Once the substrate is ready, it is then coated with a reflective and dielectric layer stack. Mirrors with an Al reflective or an enhanced, highly reflective Ag layer are coated with dielectric protective layers.

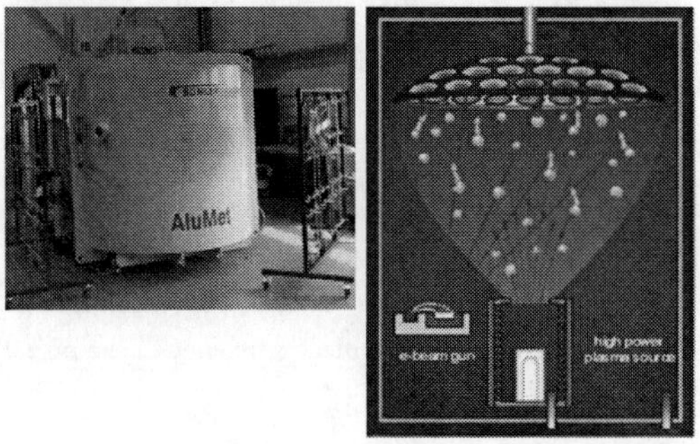

Figure 2: Deposition scheme of the two reflective coatings. On the *left*, the pilot-line equipment used to deposit the Al-based mirror; on the *right*, the deposition system used to deposit the Ag-based coating using a SyrusPro 1350 high vacuum coater.

Ag-mirrors are prepared with a SyrusPro 1350 high vacuum coater, equipped with a planetary system, electron beam evaporator and plasma source (**Figure 2**). After loading and fixing the PC-substrates onto the planetary system, the machine is evacuated down to a pressure of 3E-6 mbar before the coating process can start. This high vacuum pumping serves as a degassing time to achieve a very low background pressure to minimize residual gas contamination. Next, the electron beam evaporator is activated, the materials – one after the other are evaporated while case by case – in parallel the plasma ion source is running.

Within this study, 7 Al-coated and 7 Ag-coated mirrors are produced. The mirrors are produced in two batches; all 7 mirrors of each type are produced in one coating-deposition run. One additional Al-coated mirror is included in the study and is intentionally damaged. First, all mirrors are evaluated with the Pilot-Line mirror test characterization method, based solely on the use of the bare mirrors. Using this quick reference analysis, mirrors from both batches are ranked from the highest to lowest average reflectivity. Second, mirrors of each type showing highest and lowest reflectivity (plus the damaged mirror) are characterized with the R&D mirror test that includes the use of the SOE, but no solar cell. Third, these same five mirrors are assembled with SOE and solar cell for verification characterization (solar simulation measurements).

2.1 Pilot-Line Mirror Test

Simulations are used to predict the influence of mirror manufacturing defects on test results and hence on the final module performance. By using an optical method based on a complete ray tracing modelling already developed within this project [3], it is possible to compare the electrical output of an ideal system composed either of an "Al-based" or a "Ag-based" mirror.

Two approaches are proposed. A *forward (simulation-based) case* allows to calculate energy loss due to surface roughness and uses the SOE and the solar cell. In this case, irradiance maps are generated and they allow predicting energy loss in the direct light configuration. Simulations are performed via the optical software Zemax. An optical ray-tracing model is developed to simulate different types of randomly distributed surface defects. Tactile measurements of these mirrors reveal 0.25mm RMS and 1.50mm Peak-to-Valley (PtV) features [4]; the surface roughness simulation reproduces here these values via a Monte-Carlo approach, with Zernike deformation functions and grid-based perturbations with higher frequency.

A *backward (experimental-based) case* allows projecting the reflected light from the 4-lobed parabolic mirror on a "Hartmann-Screen". In this case, the SOE/solar cell are replaced by a 4 point light source positioned in the focus points of the mirror lobes. Optical analysis of the reflected light of the mirror allows measuring surface defects. This is possible via the movements of the spots in the Hartmann-Screen setup proposed as "Pilot-Line mirror test" set-up (**Figure 3**). For both models, position and tilt errors of the components and surface imperfections of the primary mirror are analysed.

The measurement set-up consists of a wide-angle camera, a white screen, a Hartmann-plate of the same dimensions of the screen and a four-point light source positioned in the mirror focal points. The Hartmann-screen is used to get quantitative information about a large number of discrete points on the mirror surface.

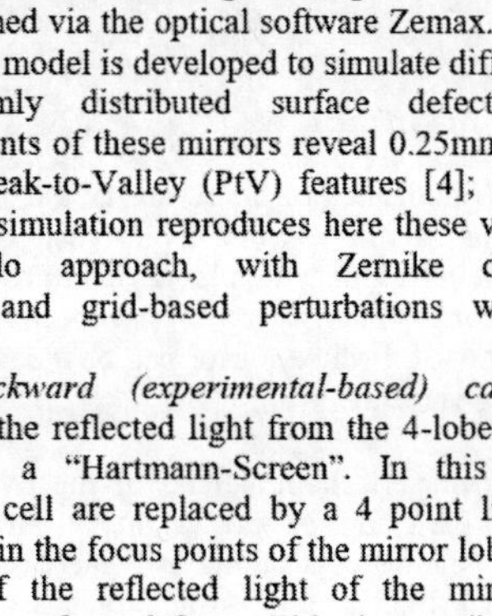

Figure 3: Image of the constructed Pilot-Line set-up with automated analysis of 145 evenely distributed points on the mirror surfaces, including precise assessment of local curvature and reflectivity. When the screen is removed, the set-up allows collecting high resolution images of all 4 mirror lobes with a strong enhancement of local curvatures.

As an example, **Figure 4** shows an unprocessed image on the screen where the smallest curvatures on the mirror surface appear with high contrast. From position, intensity and size of these spots on the screen, the local tilt and the effective reflectivity of the mirror surfaces can be derived. Alignment errors of the whole mirror or the light sources can also be assessed from the positions of these spots.

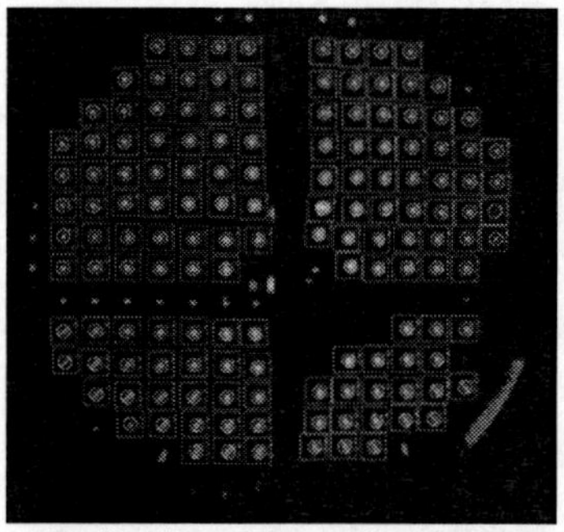

Figure 4: Visualization of the results of a measurement of one of the sample mirrors. Every (magnified) spot displacement corresponds to a local curvature and every brightness corresponds to the local reflectivity. Small misalignments of a whole mirror can be assessed based on overall movements.

This characterization method of the overall mirror surface has two advantages: 1. Local surface defects (linked to coating deposition) and curvature mismatch (linked to substrate manufacturing) are measured together at the real installation tilt angle of module. 2. The set-up is easily assembled and the measurement is performed in a few seconds, qualifying for production-line takt-time requirements.

2.2 R&D Mirror Test

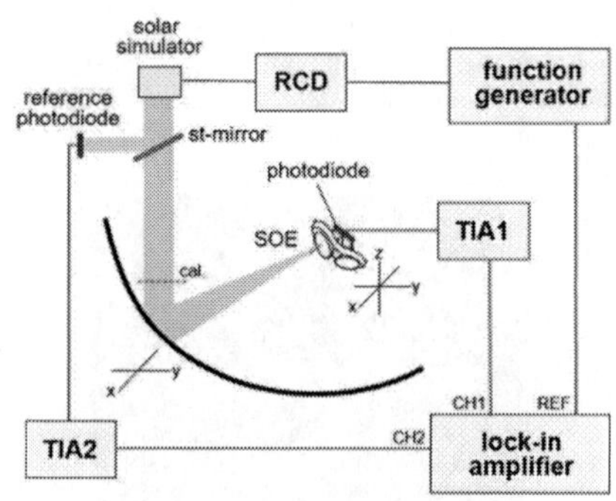

Figure 5: R&D mirror test set-up scheme.

The spatially-resolved analysis of the optical efficiency of the mirror is performed with the set-up depicted in **Figure 5** and pictured in **Figure 6**. The mirror is placed in an inclined support equipped with a bridge that permits holding and sliding a detector or probe that consists of the actual SOE of the module equipped with a Hamamatsu photodiode. In other words, the photodetector replaces the regular triple-junction solar cell in the module in order to avoid possible issues with current matching. The signal of the photodiode is measured over a constant resistor V using a trans-impedance amplifier. The optimal position of the detector with respect to the mirror is determined by maximizing V using four identical light sources illuminating spots of ~ 50 mm in diameter in the center of each of the four-mirror lobes. Each light source consists of a high power LED. Each is mounted behind a 0.39 mm diameter pinhole positioned in the focal point of a lens, with a focal length of 150 mm and an aperture of 50 mm. This guarantees production of parallel bundles of light with a maximum angular aperture of 0.18°.

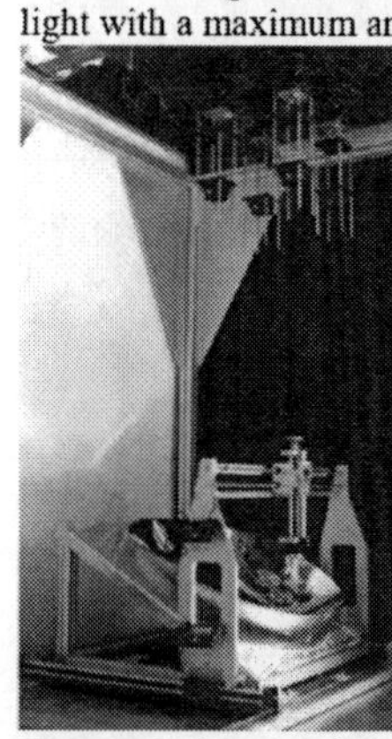

Figure 6: Image of the set-up.

After optimizing the position of the photodiode, its position with respect to the mirror is fixed. Subsequently the spatially resolved optical efficiency of the mirror is mapped by repeatedly translating the mirror mount over 2 cm intervals in the x and y directions and monitoring the detector readout while the mirror is illuminated by a single light source at a fixed position. The optical efficiency η_{opt} at each position is given by the ratio of the detector read-out V and the calibration or reference signal V_{ref} that is obtained by measuring the intensity of the light just before it reaches mirror (*i.e.* at the position indicated with "cal." in **Figure 5**) using the bare photodiode without SOE. Note that therefore η_{opt} represents the combined optical efficiency of the mirror-SOE combination.

2.3 Mirror Characterization Verification

These modules, without Fresnel prism, are used to evaluate the impact of the mirror coating on the electrical performances. Both Al- and Ag- coated mirrors are tested and compared. Angular tolerance to chassis misalignment as well as mirror acceptance angles are investigated through indoor experimental measurements performed with a solar simulator with ±0.4° collimated light (**Figure 7**). IV performances are then recorded for both types of mirrors under CSTC conditions (AM1.5D; 1000 W/m2; T = 25°C).

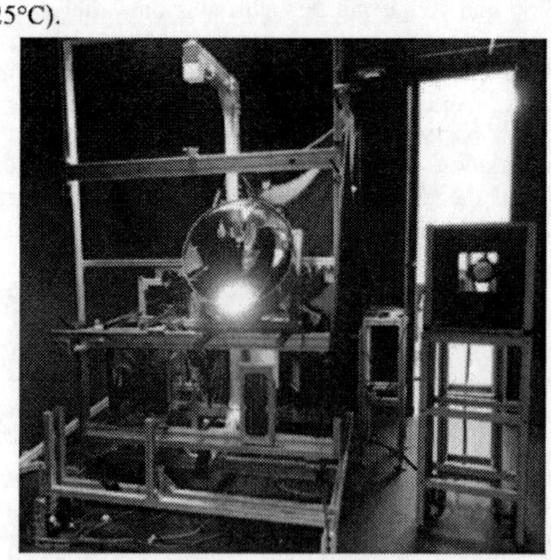

Figure 7: Helios SAV Solar simulator used to verify mirror characterization techniques developed. Unlike the previous two methods, where the mirror was characterized alone, here the secondary optical element with the solar cell are used.

3 RESULTS AND DISCUSSION

Reflectivity measurements are shown in **Figure 8** for the Al-coated mirrors and in **Figure 9** for the Ag-coated mirrors. This latter shows 97% reflectivity within the visible range (380-780 nm). These measurements, performed with a reflectance spectrometer, are angle of incidence (AOI) dependent.

Simulation results on the impact of surface roughness of the Al-based mirror are presented in **Figure 10** where a perfect system is compared to a system with surface defects. These are the output of the *forward case* cited in the Method section. Surface imperfections show significant change in the irradiance maps calculated at the focal plane of the cell (asymmetry). An energy loss lower than <1% is deduced.

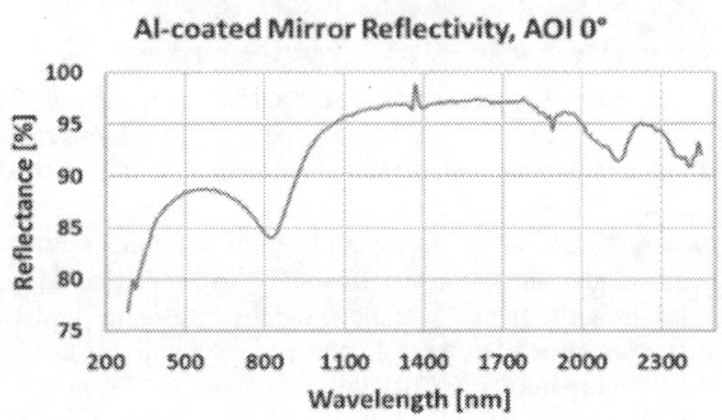

Figure 8: Reflectivity measures of Al-coated mirrors (please note the peak below 1400 nm is a measurement artifact).

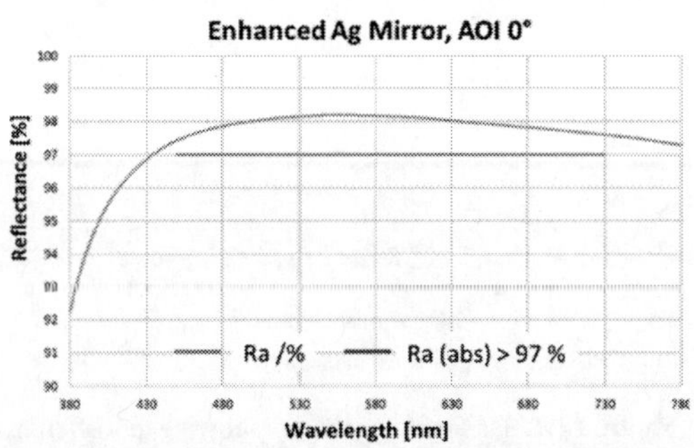

Figure 9: Reflectivity measures of Ag-coated mirrors over the visible range.

Measurements from the Ag- and Al-coated mirror batches are collected. They showed small features with high contrast mainly caused by the molding process of the mirror PC substrate. The mirror coatings showed neglegible variations as they were produced in the same batch. For each mirror the average reflectivity is calculated. For the Al-coated mirror batch, the maximum value is 86.7% (mirror L) while the minimum is 82.5% (mirror J). For the Ag-coated mirror batch, 92.7% (mirror A, shown in **Figure 11**) and 90.7% (mirror D). respectively.

The observed deviation of the spot positions in the Hartmann-Screen setup had a standard deviation of about 0.4 mm and PtV of about 1 mm. This corresponds in the used setup to line of sight error from the direction of the sun of 0.1(0.24) degree. These values are acceptable within the angular size of the sun.

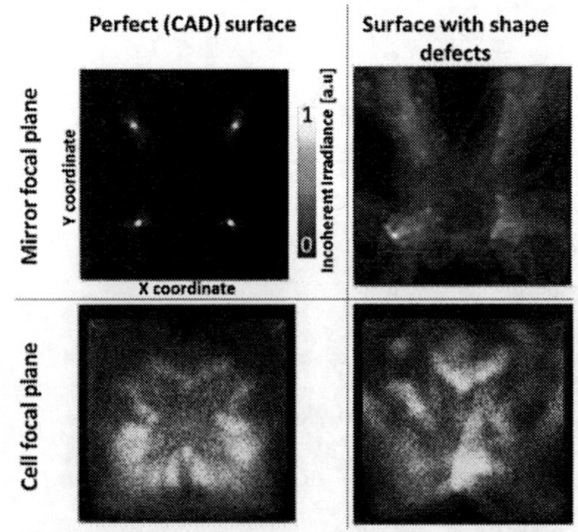

Figure 10: Simulated energy distributions in the focal plane of the parabolic mirror and the solar cell behind the secondary optics: a) In the focal plane of the nominal parabolic mirror. b) In the focal plane of a parabolic mirror with randomly distributed surface irregularities. c) On the solar cell from the *perfect* parabolic mirror. d) On the solar cell from the surface *with defects*.

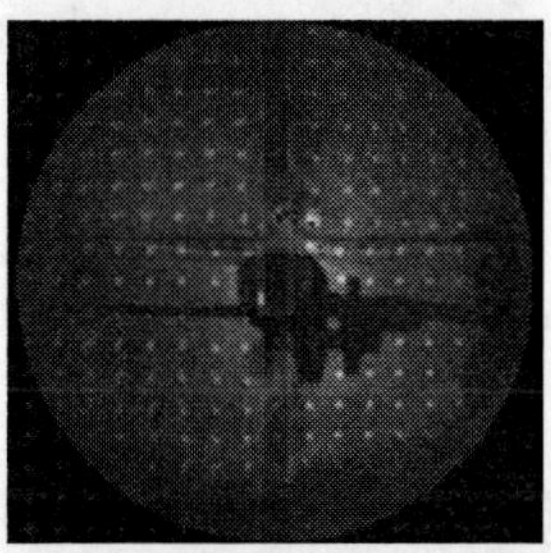

Figure 11 : Ag-coated mirror (A): mirror recorded without Hartmann-screen (green), the ideal reference pattern (blue), the measures spot pattern (red) for the maximum average reflectivity. The features in the green image and the red spot positions are very similar. The deviation from the red spots to the blue pattern might be caused by a slight misalignment of the light sources and can be removed from the derived local surface curvatures.

A different visualization of the optical image presented in **Figure 11** is presented in **Figure 12**. Here all collected points are plotted. When the average reflectivity is higher, the dispersion of the relative reflectivity is tighter (green dots). The opposite is seen in the case of lower average reflectivity (red dots).

Figure 13 shows the results obtained using the R&D mirror test on Al-coated mirrors (top row) and Ag-coated mirrors (bottom row). The location of the horizontal and vertical ridges separating the lobes of the mirror are indicated with the two thin crossing lines in the figure. The shadow casted over the two bottom lobes by the bridge holding the detector is clearly visible and excludes the corresponding area of the mirror from analysis. The mapping clarifies that the vertical ridge dividing the left from the right lobes has a severe impact on the local η_{opt} while the horizontal ridge does not.

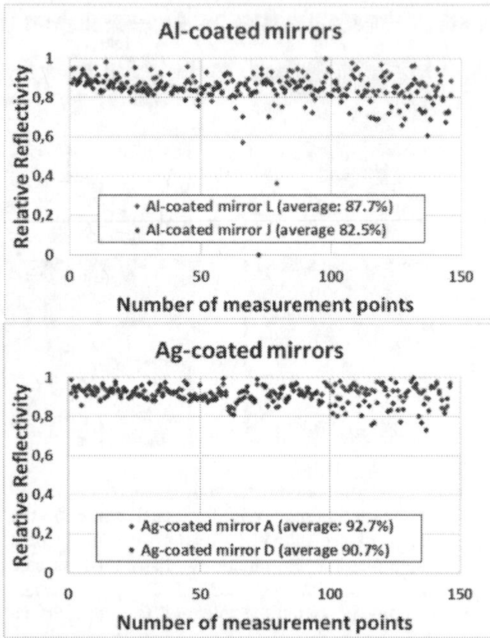

Figure 12: Relative reflectivity of the mirrors with maximum (green) and minimum (red) average reflectivity. Top: Al-coated and, bottom, Ag-coated mirrors.

The average optical efficiency of the Al-based mirrors including SOE ranges from 0,609 to 0,616 while this efficiency increases to range from 0,670 to 0,680 for the Ag based mirrors. The relative improvement of 10% is in line with the simulations and quick pilot-line mirror test. In addition the spread in the measured reflectivity of the worst and best performing mirror within each set is quite small, indicating a good reproducibility in the production of the mirrors.

By recording mappings obtained while the SOE-photodiode detector is deliberately translated or rotated with respect to its optimal position, the 10% efficiency-loss tolerances for the optimal alignment of the SOE with respect to the mirror were demonstrated to be 2-3 mm, ~1 mm, ~1 mm and ~5° for respectively the x-, y-, z- and φ-misalignment.

In all characterization methods the Al-coated mirrors have lower reflectivity and they result in lower power output with respect to the Ag-coated mirrors. Differences between techniques appear to be related to light source alignment.

R&D test results are confirmed via solar simulator measurements (**Figure 14**) in the case of the Ag-coated mirrors, while the opposite trend is found for the Al-coated mirrors. While the R&D mirror test has to first locate the optimal position for the SOE and photodiode (position that yields the highest performance during the measurement, *i.e.* misalignment with respect to the mirror), the solar simulator test performs measurements on the SOE and solar cell mounted and fixed in the module as they would be in outdoor operation conditions (*i.e.* no optical adjustment possible – only module alignment with respect to the "sun", the lamp of the solar simulator). Differences are thus easily obtained by a slight adjustment of the SOE with respect to the mirror in the R&D test. It is estimated that if the SOE had been fixed in the same way as in the fixed module configuration, the measured optical efficiency might have been 1-2% lower for each mirror.

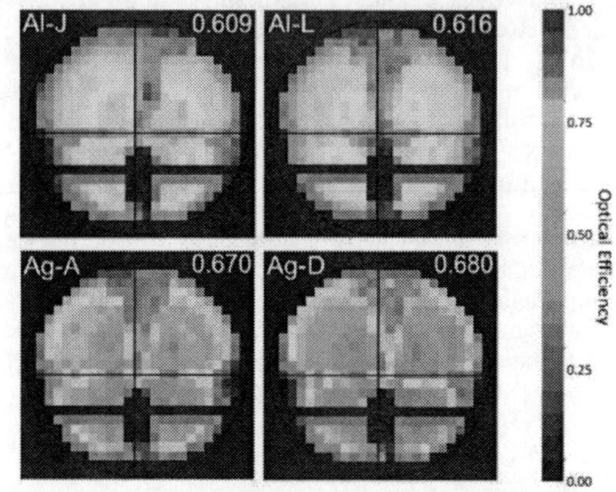

Figure 13: 2D optical efficiency scan obtained using the R&D mirror test of the worst and best performing mirrors from the two sets. One set with Al- coatings (top row) and the other with Ag-coatings (bottom row). For each measurement the SOE probe is placed in the optimal position and the average optical efficiency (η_{avg}) over all valid measurements is indicated in white. The applied solar simulator aperture is 25 mm and the x and y scan step sizes for the mirror are 20 mm.

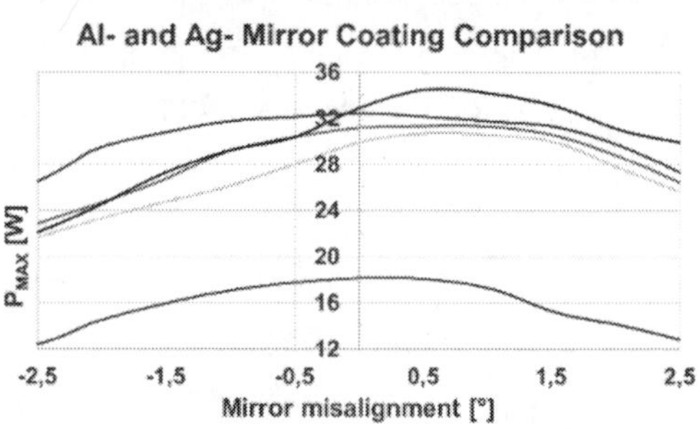

Figure 14: Solar simulator results, showing the difference between the Al- (green lines) and Ag- (red lines) coatings and the damaged Al-coated mirror (blue line).

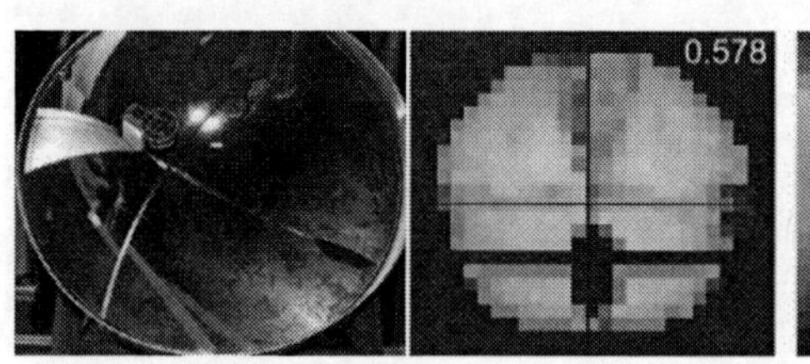

Figure 15: Damaged Al-coated mirror. Left: image of the damaged surface. Right: the R&D mirror test gives an optical efficiency of 0.578.

As a matter of comparison, an intentionally damaged mirror is investigated. The R&D mirror test shows 20% decrease in optical efficiency (**Figure 15**). This mirror, installed in the final module configuration, shows 40% power loss as measured with the solar simulator (**Figure 14**, blue line).

3 CONCLUSIONS

In conclusion, Al- and Ag- mirror coatings for the SunCycle CPV module are produced and tested. Characterization is performed via two methods, implemented specifically for this 4-lobed parabolic mirror design. The Pilot-Line mirror test delivers detailed and high contrast information of the mirror surfaces (especially on local surface curvatures) within seconds, making this technique compatible with requirements of in-line testing. The analysis of repeated measurements on the same sample as well as between different samples allows to propose mirror manufacturing tolerances. Absolute comparisons would benefit from an improved alignment of the light sources and better mounting points on the mirror substrates. Furthermore, the number of measurement points can be increased by increasing the number of holes in the Hartmann screen. Whereas the Pilot-line mirror test provides information on the mirror shape for a discrete number of points, the R&D mirror test provides information on the spatially distributed light capturing efficiency of the combined mirror and SOE. This is a highly relevant parameter because also in the final module the light from the mirror will be collected via the SOE. Small imperfections may not be detrimental if the reflections can still be captured by the SOE while it was found that slightly larger imperfections can be partly compensated for by repositioning the SOE. The time required to perform a mirror scan and the spatial resolution of the obtained data can be balanced. That is, decreasing the exit aperture of the solar simulator and the applied step intervals in the x and y directions, increases the resolution of the measurement but then it also becomes increasingly time consuming to scan the mirror. In practice an aperture diameter of 25 mm and step sizes of 20 mm was found to provide sufficient resolution in an acceptable time (3-4 hours) for off-line research on the system performance. Results of both tests are verified via solar simulator measurements; power output ranges between 35 W for the highest reflectivity Ag-coated mirror and 31 W for the Al-coated mirror. Measurement methods all find the same trends and measurements are within uncertainties validating their use to characterize CPV4ALL mirrors.

ACKNOWLEDGMENTS

This work is supported by the EU within the FP7 project CPV4ALL under grant agreement N° 296014.6.1. The authors wish to thank Consortium Partners for support in substrate preparation and technical discussions.

4 References

[1] R. Bijl and P. Penning. Device for converting solar energy, WIPO Patent No. 2007117136. 2007

[2] F. Trespidi, E. Malvisi, and R. Parmesani, *Shape characterization of the primary mirror of a mirror based solar concentrator*, AIP Conference Proceedings **1881**, 030009 (2017)

[3] P. Voarino, C. Dominguez, E. Bijl, P. Penning, *Optical performance analysis of a novel tracking-integrated concentrator through ray tracing*, 10th International Conference on Concentrator Photovoltaic Systems, AIP Conf. Proc. **1616**, 215-219 (2014)

[4] S. Bernardis, P. Voarino, J. Gouffa-Folliet, H. Rooms, M. van de Zande, P. Penning, and Mathieu Baudrit, *Tracking-Integrated CPV4ALL System Installation and Analysis of Mirror Tolerance Manufacturing*, Oral Communication, 14th International Conference on Concentrator Photovoltaics, April 16-18th, 2018, Puertollano, Spain

OPEN AND CLOSED LOOP CONTROL OF SOLAR TRACKER APPLIED TO SMALL SIZE HCPV SYSTEM

Luz Elena Peñaranda Chenche[1], Marcelo Barros de Almeida[2], Roberto Mendes Finzi Neto[1], Oscar Saul Hernandez Mendoza[1], Enio Pedone Bandarra Filho[1]
[1]School of Mechanical Engineering, [2]School of Electrical Engineering, Federal University of Uberlandia, Av. Joao Naves de Avila, 2121, 38400-902 Uberlandia, MG, Brazil

ABSTRACT: In last years, the solar energy research field has experienced a rapid increase in a wide range of applications. In this sense, High Concentration Photovoltaic (HCPV) systems appears as one of the most promising technologies. However, this kind of systems are highly sensitive to errors in solar tracking control or mechanical clearances in the structural movement of the tracker. Thus, an accurate solar tracking system is necessary aiming to maintain the amount of produced energy in favorable conditions. This paper presents the implementation and evaluation of two control strategies for a dual axes sun tracking system in a small size HCPV system connected to a dummy load.
Keywords: Tracking, Multijunction Solar Cell, Concentrators, Energy Performance

1 INTRODUCTION

Due to his highest photoelectric conversion efficiency the High concentrating photovoltaic (HCPV) devices has becoming the most promising technology in this field. However, the small tolerance in the acceptance angle of the incoming solar light received by the cells demands an accurate solar tracker in order to maintain its high performance [1]. Though, there are two factors that can strongly influence on maintaining the deviation below the acceptance angle. The first one is related to assembly, mounting, alignment and stiffness of the concentrator system. The second factor corresponds to the precision of the tracking system that is related to the quality of the sun sensor and to mechanical factors including backlash and friction [2]. Algorithms for sun tracking can be classified in two groups, open loop and closed loop [3]. An open-loop algorithm uses information as date, time and geolocation to find the angles for the tracking mechanism. Otherwise, the closed loop approaches are able to track the apparent movement of the sun without regard on the geographical location but depends on a feedback mechanism that uses an optical sensor to measure sunlight and requires tuning of the controller closing the loop [4].

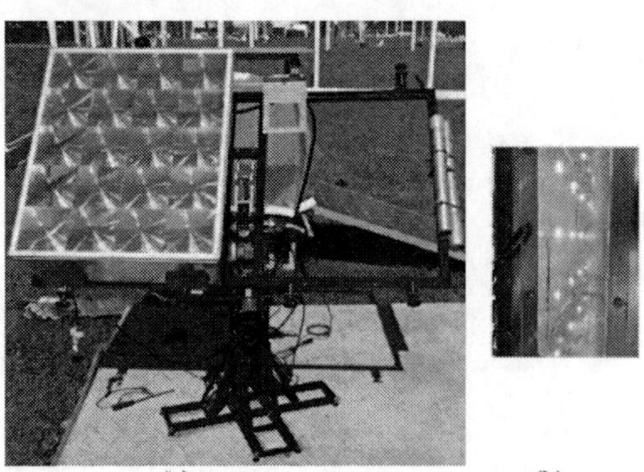

(a) (b)

Figure 1: Sun tracker HCPV: a) General view b) Alignment detail

In this work, an open and a closed loop control strategies are implemented and evaluated in order to determine the main difference between them in a small size HCPV system connected to a resistive dummy load. The solar tracker device uses a rotary and linear actuator for orientation of the solar modules in azimuth and zenith solar direction. Control system uses two strategies to find the solar position, the open and the closed loop. The open loop uses global position sensor (GPS) to find the true north and applies this value into the solar orientation equations in order to determine the position of the sun, regardless the sunlight.

Closed loop, on the other hand uses a sun sensor to refine the sun position.

2 METHODOLOGY

The proposed solar tracker was designed with two different approches for positioning. The first one is based on open loop strategy, using only GPS data for determining azimuth and zenith angles. The second approach combines GPS data with information from a solar sensor in a closed loop for obtain feedback, providing better results when tracking. Both strategies are described below and represented in Fig. 2.

Initially, the tracker must be positioned for a maximum horizontal excursion. The mechanical lower limits are fed into controller (minimum angles for azimuth and zenith in the current position). When the tracker is turned on, it moves the solar panel to the opposite direction in both axes. Using encoders information, the solar tracker can calculate the maximum excursion angles for azimuth and zenith.

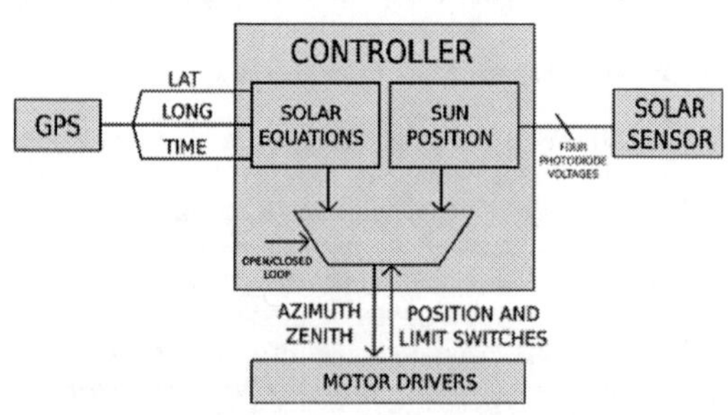

Figure 2: Controller structure

When GPS data is available, latitude, longitude and time are determined and the optimal solar position is calculated according to PSA Algorithm [5].

Although initial positioning is always accurate when the solar tracker starts, mechanical clearances are accumulated during the operation and the difference between calculated angles and real position becomes gradually larger. In such situation it is required a new calibration process, restarting the controller. As this operation is time consuming and represents a loss in converted energy, a new strategy was developed, using an ISS-A5 solar sensor from Solar MEMS Technologies. This sensor has a field of view with ten degrees of aperture, allowing a fine-tuning operation.

When using the solar sensor, the solar tracker uses the GPS only for the initial positioning. After moving the solar panel to the calculated position, the solar sensor is used to keep the solar panel at maximum incidence angles of sun rays. The solar sensor is repeatedly evaluated in a closed control loop, fixing the panel position at each 30 seconds and allowing the tracking. As the aperture angle is narrow (ten degrees), cloudy situations or any other conditions, where some low sun radiation are detected by weak voltage values from the solar sensor. In this case, the solar tracker will use only GPS information for proper positioning, waiting for better sun ray's conditions for reestablish the solar sensor usage.

The controller was implemented in a STM32F407VG microcontroller, an ARM Cortex M4 processor with float pointing from ST Microeletronics. For GPS positioning, it was used the Maestro module A2200-A and an active antenna

3 RESULTS

Outdoor tests were carried out with a total duration of 22 days in order to verify the adequate operation of the implemented sun tracking control algorithms. Measurements of the electrical properties (IV curve) of the HCPV module and the direct normal irradiance (DNI) were performed in a continuous way each hour while the temperatures of the HCPV module and the ambient were measuring with a sample rate of 3 seconds

The characterization IV curves (as shown in Fig. 4) allow to determine the maximum power point generated in the module for a specific operating condition.

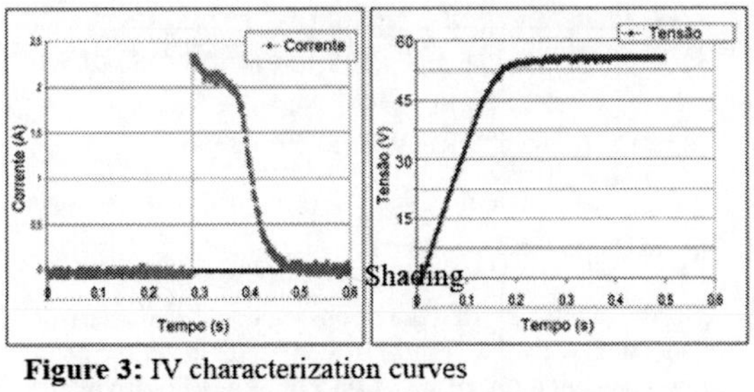

Figure 3: IV characterization curves

Hence, with this curve is possible calculate the amount of power every hour in the PV module and also determine the main parameters of the model that describes the electrical behavior of PV devices [6]. Furthermore, the short circuit current in the module was measured every minute, this together with the known power data allowed to calculate the resulting power data for each known current data by applying the five-parameter model.

3.1 Open Loop Tracker Results

The results obtained in the third test day of open loop control solar tracker operation, are shown in Fig. 5, here is possible to observe the generated power of the HCPV system among all the day. It can be noted that the control algorithm worked relatively fine, but there exist some power drops relate to mismatches in the acceptance angle of the incidence irradiance over the solar cells. This behavior that can be attributed to mechanical clearances in the solar tracker structure and to computed errors in the controller.

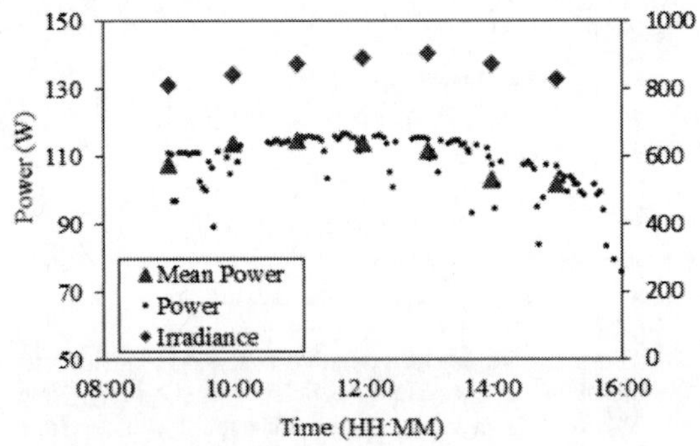

Figure 4: Power produced by the HCPV system with open loop control solar tracking

The photoelectric conversion efficiency of the HCPV system was calculated at each hour of the test from the DNI data and the mean of the power in Fig. 4. The average of the efficiency and the module temperature are summarized in Table I.

3.2 Closed Loop Tracker Results

Similar, with Fig. 4, the amount of power produced by the HCPV systems when the closed loop control was implemented is showed in Fig. 5. Thus, when compared to Fig. 5 (open loop control), it is possible to note that the profile of electric power generation during the day is higher stable when the closed loop control operates, and power generation drop ensue only due events of shading caused by clouds.

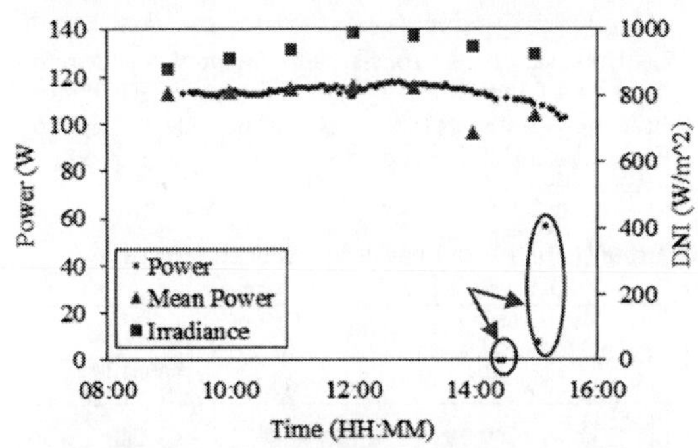

Figure 5: Power produced by the HCPV system with closed loop control solar tracking

Also, it can be noted that the mean of power in the point at 14h is highly affected by the power drops due to shading.

In Fig 6. it's shown a detailed view of the power produced by the HCPV system operating with open loop solar tracking contrasting with the power produced when closed loop control solar tracking is implementing. Here is evident that for most of the time the control strategy that allows to generated the highest power is the closed loop approach.

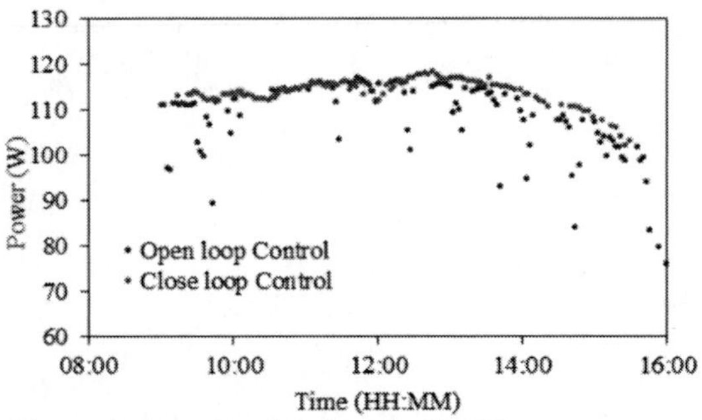

Figure 6: Power produced by the HCPV system: open loop vs closed loop control solar tracking

Furthermore, a more detailed analysis is carried out emphasizing in the system's efficiency as shows in Table I. Where, it's possible to see that the lower average efficiency among all the test day was obtained for the closed loop control approach, but this low efficiency may be due to different reasons. Firstly, Table I shows that the higher module temperature value of the test days also corresponds to the test day with closed loop implementation, and as it was proved in many works, high module temperature has a negatively influence on PV efficiency. Shading, also can affect in some way the efficiency, because as was see earlier this phenomenon has a strongly effect in the mean power generated.

Table I: HCPV efficiency and temperature for different tests days

Control	Efficiency (%)	Module Temp. (°C)	Outdoor Temp. (°C)
Open loop	27,31	59,63	28,1
Closed loop	26,77	65,07	32,8

However, when considering the power generation capacity of the HCPV system for each of the control strategies tested (Table II), it is clear that by implementing closed loop control, the system produces more energy.

Table II: HCPV rate power for different tests days

Control	Mean Power (Wh)	Day Power (Wh)	Power capacity (%)
Open loop	80,79	646,34	65,58
Closed loop	103,30	877,98	88,23

In Table 2, the first column represents the average energy generated during the test day. The second is the total amount of energy generated during the day. Finally, the "Power capacity" is the ratio between the energy that could have been generated if the system was worked in optimum conditions and the amount of energy that actually was produced throughout all the test day. Thus, it is possible to verify that the solar generation system achieves the highest production capacity when the closed loop control is used. However, the calculated values in Table 2 for the closed-loop test also considers the losses obtained in power by shading (Fig. 7) which correspond to 5.49% of the total losses, if these losses are unconsidered, the value of the power capacity of the system may increase to 93.74%.

Finally, the average of the HCPV system efficiency for all of the 22 test days, at each hour are showed in Fig. 7 and 8. Where it can be observed that the average efficiency of the system is between 23% and 32%, however, these efficiency values correspond to the generated power when the HCPV system is connected to a resistive dummy load (off grid), so it is expected that the overall efficiency will decrease if the system is connecting to the grid.

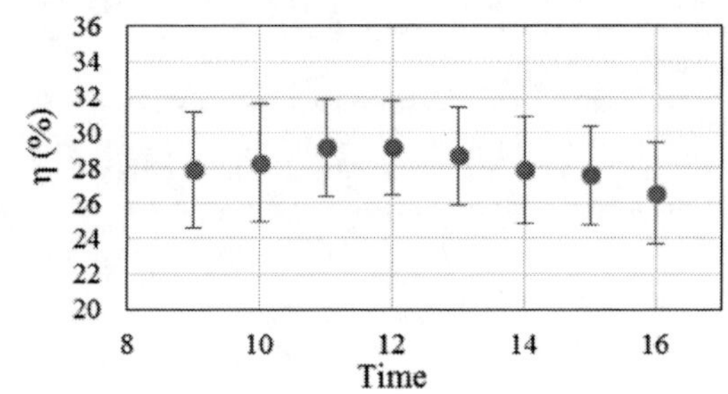

Figure 7: Mean efficiency through the day

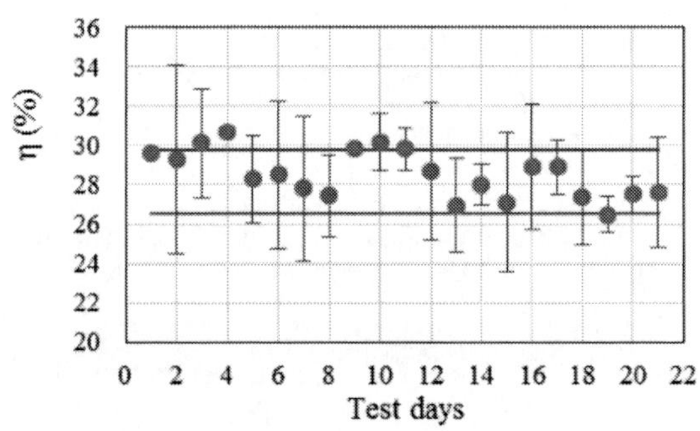

Figure 8: Mean efficiency for each tests day.

4 CONCLUSION

An analysis of the constructive characteristics associated with the control system solar position of the power generation High Concentration Photovoltaics, demonstrates the necessity of the development of a robust mechanical system, without clearances.

Despites the high module temperatures that can be achieve with the closed loop control, it was proved that this is the best approach to obtain the highest power capacity production of the HCPV system and to ensure the correct incident of the irradiance in the receiver cells avoiding mechanical clearances.

The results shown an average efficiency in power

generated between 23% and 32% for the implemented HCPV system.

5 ACKNOWLEDGEMENTS

The authors acknowledge the CAPES, CNPq, FAPEMIG and CTG-Brasil for financial support.

6 REFERENCES

[1] Z. Mi, J. Chen, N. Chen, Y. Bai, R. Fu e H. Liu, "Open-loop solar tracking strategy for high concentrating photovoltaic systems using variable tracking frequency," Energy Conversion and Management, (2016) 142-149.
[2] R. Garrido e A. Díaz, "Cascade closed-loop control of solar trackers applied to HCPV systems," Renewable Energy, vol. 97, (2016) 689-696.
[3] R. Singh, S. Kumar, A. Gehlot e R. Pachauri, "An imperative role of sun trackers in photovoltaic technology: A review," Renewable and Sustainable Energy Reviews, vol. 82, (2018). 3263-3278.
[4] G. Liu, A. O. Baba e L. and Zhu, "Hybrid controller for dual axes solar tracking system," in 36th Chinese Control Conference, Dalian, (2017).
[5] M. Blanco-Muriel, D. C. Alarcón-Padilla, T. López-Moratalla e M. Lara-Coira, "Computing the solar vector," Solar Energy, vol. 70, (2001) 431-441.
[6] L. E. Peñaranda, O. S. Hernandez e E. P. B. Filho, "Comparison of four methods for parameter estimation of mono- and multi-junction photovoltaic devices using experimental data,," Renewable and Sustainable Energy Reviews, vol. 81, (2018) 2823-2838.

STANDARDS FOR PV - OVERVIEW OF IEC RELATED PV STANDARDS AND HOW THEY CONTRIBUTE TO REDUCED COSTS OF ENERGY

Tony Sample
European Commission, DG JRC, Directorate C – Energy, Transport and Climate, Energy Efficiency and Renewables Unit, Via E. Fermi, 2749, I-21027, Ispra (VA), Italy
Tel +39 0332 789062, tony.sample@ec.europa.eu,

ABSTRACT: The IEC International Standards and CA (Conformity Assessment) activities facilitate world trade by removing technical barriers between countries and/or regions. Within the area of PV the IEC Technical Committee 82 (TC 82) has developed 113 documents (up to September 2018). Within TC 82 Working Group 2 develops standards for modules (non-concentrating). The paper will give an overview of the current activities for WG2, briefly describing the 44 published documents, the 21 ongoing revisions and 17 new projects. More detail for some projects, such as the revision of the 61215 standard series will be presented.

Keywords: PV standard, measurement, qualification testing and safety

1 INTRODUCTION

Continued growth in the PV industry was reflected in that the newly installed capacity was over 90 GW in 2017 [1]. It is a truly international business from material supply, product manufacturing through to installation of systems. As such it is important to facilitate trade between different countries within the whole chain from materials to deployment.

As stated on the IEC website, www.iec.ch, "The International Electrotechnical Commission (IEC) International Standards and Conformity Assessment (CA) activities facilitate world trade by removing technical barriers between countries and/or regions. This helps to develop new markets and achieve economic growth". There are currently 104 Technical Committees and 100 subcommittees developing standards within the IEC Organisation.

The IEC Technical Committee 82 (Solar photovoltaic energy systems) is tasked with writing all standards in the field of Photovoltaics and as such has developed and published many standards covering the entire field from light input to a photovoltaic cell to and including the interface with the electrical system(s) to which energy is supplied.

Beyond the obvious stated ideals of the IEC of removing technical barriers between countries, the use of standards is usually associated with the reduction of risk. This may be as simple as adopting the same methodology for measurement and labelling of modules maximum power, so that the end user have confidence that different manufacturers are ultimately traceable to the SI unit of power (W). The use of type approval for performance and safety standards have facilitated the expansion of the PV industry over the last decades.

This paper will outline both existing standards and those currently under development within IEC TC 82 WG2 (modules, non-concentrating). It will also briefly cover the aspects of conformity assessment under the IECRE scheme.

2 STATUS OF PV MODULE STANDARDS WITHIN IEC TC 82 WG2

Since its establishment in the early 1980's IEC TC 82 WG2 has developed a number of standards and technical specifications. The technical areas covered by these documents can be grouped together in the following areas;

- *Measurement Principles*
- *Type Approval and Safety Tests*
- *Energy Rating*
- *Specialized Stress Tests*
- *Module Components*
- *Module Materials*
- *Quality assurance*
- *Increased Reliability*

The order also generally reflects the order when they were developed within WG2, with the basic measurement principles being the first, followed by type approval and safety, etc. The development also reflects the growth of the PV industry and the greater appreciation of the need for material assessment and control of manufacturing QA together with the need to demonstrate increased reliability.

This has been reflected in an increase in the total number of documents issued from WG2, which has increased from 16 in 2012 [2] to a total of 44 today, with 38 ongoing projects, either revisions or amendments of existing standards (21) or new documents (17). Each of these areas will now be addressed in greater detail below.

2.1 Measurement principles

The full list of documents covering measurement principles, together with the current status within WG2 is listed in Table I. As can be seen in Table I a number of these documents have been revised a number of times and are now on the third or fourth revision. But you can also see that as PV technologies developed additional standards have been issued, such as the IEC 60904-1-1 and IEC 60904-8-1 to cover the I-V measurement and the measurement of spectral responsivity of multi-junction photovoltaic devices, published in 2017. Continued technological change is also reflected in the IEC TS 60904-1-2 which Measurement of current-voltage characteristics of bifacial photovoltaic (PV) devices which is about to be published.

The last item in Table I is a Technical Report which is being compiled on the state of the art of measurement of emerging PV devices, (OPV, DSC, and Perovskite). The measurements of these devices present technical challenges due to their slow response times, meta-stable characteristics and differing stabilisation procedures required to accurately characterise them.

Table I: IEC TC 82 WG2 documents relating to measurement principles

IEC number	Edition	Status	Title
60891	Ed 3	Under revision	Procedures for temperature and irradiance corrections to measured I-V characteristics
60904-1	Ed 3	Under revision	Measurement of photovoltaic current-voltage characteristics
60904-1-1	Ed 1	Published (2017)	Measurement of current-voltage characteristics of multi-junction photovoltaic devices
60904-1-2	Ed 1	In preparation	Measurement of current-voltage characteristics of bifacial photovoltaic (PV) devices
60904-2	Ed 3	Published (2015)	Requirements for reference solar devices
60904-3	Ed 4	Under revision	Measurement principles for terrestrial photovoltaic (PV) solar devices with reference spectral irradiance data
60904-4	Ed 2	Under revision	Reference solar devices - Procedures for establishing calibration traceability
60904-5	Ed 2	Published (2011)	Determination of the equivalent cell temperature (ECT) of photovoltaic (PV) devices by the open-circuit voltage method
60904-7	Ed 4	Under revision	Computation of the spectral mismatch correction for measurements of photovoltaic devices
60904-8	Ed 3	Published (2014)	Measurement of spectral responsivity of a photovoltaic (PV) device
60904-8-1	Ed 1	Published (2017)	Measurement of spectral responsivity of multi-junction photovoltaic (PV) devices
60904-9	Ed 3	Under revision	Solar simulator performance requirements
60904-10	Ed 3	Under revision	Methods of linearity measurement
60904-13	Ed 1	Published (2018)	Electroluminescence of photovoltaic modules
TR (TBA)	Ed 1	In preparation	Measurement protocols for emerging PV devices (OPV, DSC, and Perovskite)

2:2 Type approval and safety tests

Probably the most important group of standards issued by WG2 relate to type approval and safety tests. The full list of documents covering type approval and safety tests together with the current status within WG2 is listed in Table II.

Given the importance of the type approval tests 61215 and 61730 these are continually being examined to see if modifications are appropriate of is new tests should be included. Currently the IEC 61215 Ed2 series has been circulated as committee drafts with a number of proposed modifications and additions, briefly listed below;

- Procedures for Bifacial Modules
- Use of Representative Samples
- Addition of Dynamic Mechanical Load Test (between UV and TC50)
- Addition of Test for Potential Induced Degradation (new leg)
- Simulator Requirements
- References to Retest Guidelines
- Weight on Junction Boxes during Thermal Cycling
- Correction to Monolithically-Integrated Hot-Spot Endurance Test
- Number of Modules in Sequence A (reduced from 3 to 1)
- Removal of Nominal Module Operating Temperature (NMOT)
- Very Low Currents During Thin Film Tests
- Limit Bypass Diode Testing to Three Diodes
- Revert Insulation Test to 2005 Version
- Bending Test (for flexible modules) (after DAH, Mech Load)
- Addition of Test for Detection of Light and Elevated Temperature Induced Degradation (LeTID)

For the IEC 61730-1 and -2 safety standards amendments have been launched, for the IEC 61730-1 these include;

- Instructions for reporting of I_{sc} for bi-facial modules.
- A requirement for Class 0 modules (for use in restricted access areas), so that they are not required to meet the glass breakage test, MST 32
- Weathering requirements for relied-upon insulation.
- A component level test for assessment of distance through insulation (dti) for use with thermally deformable materials.
- Clarifications on the usage of CTI values.

For the IEC 61730-2 the proposed changes include;

- Clarification to flow chart indicating which tests may be optional
- Complete new procedure for Sharpness of edges test MST 06
- Addition of drawings to clarify procedure for Ignitability test MST 24
- Correction to Module breakage test MST 32
- Correction to Robustness of termination test MST 42
- Modification to sampling requirements in Annex A
- Compatibility with changes in IEC 61730-1 Ed.2 Amd.1

The IEC TS 62915 Retesting for type approval, design and safety qualification was published in 2018. However, as the revisions of the IEC 61215 series and 61730 amendments progress toward publication it will be necessary to amend the IEC TS 62915 to reflect the changes and new tests added to the documents.

One new document in preparation is the IEC 63163 which is aimed at photovoltaic modules suitable for consumer application which can have shorter-term operation in general open-air climates. The approach so far chosen by the project team is the application of similar test from the IEC 61215, but with reduced severities.

Table II: IEC TC 82 WG2 documents relating to type approval and safety

61215-1	Ed 2	Under revision	Terrestrial photovoltaic (PV) modules - Design qualification and type approval - Part 1: Test requirements
61215-1-1	Ed 2	Under revision	Part 1-1: Special requirements for testing of crystalline silicon photovoltaic (PV) modules
61215-1-2	Ed 2	Under revision	Part 1-2: Special requirements for testing of thin-film Cadmium Telluride (CdTe) based photovoltaic (PV) modules
61215-1-3	Ed 2	Under revision	Part 1-3: Special requirements for testing of thin-film amorphous silicon based photovoltaic (PV) modules
61215-1-4	Ed 2	Under revision	Part 1-4: Special requirements for testing of thin-film Cu(In,GA)(S,Se)2 based photovoltaic (PV) modules
61215-2	Ed 2	Under revision	Part 2: Test procedures
61730-1	Ed 2 Amd 1	Under revision	Photovoltaic (PV) module safety qualification - Part 1: Requirements for construction
61730-2	Ed 2 Amd 1	Under revision	Photovoltaic (PV) module safety qualification - Part 2: Requirements for testing
62915	Ed 1	Published (2018)	Photovoltaic (PV) modules – Retesting for type approval, design and safety qualification
63163	Ed 1	In preparation	Terrestrial photovoltaic (PV) modules for consumer products - Design qualification and type approval

2.3 Energy rating

The IEC 61853 standard series was completed with the publication of the IEC 61853-3 and IEC 61853-4 in august 2018, see Table III. An amendment to IEC 61853-2 has been launched with the stated aim of,

- Modification of the data filtering requirements for determination of Nominal Module Operating Temperature (NMOT).

2.4 Specialized Stress Tests

A number of specialised stress tests have been published or are under development, see Table IV. Only the IEC 61701 is currently under revision, the main areas to be updated are;

- The scope has been updated to better reflect the applicability of the standard.
- Test methods and requirements have been condensed and harmonized with the new editions of 61215-1, 61215-2, and 61730-2.
- The salt mist test references have been updated to harmonize with changes to 60068-2-52.
- A normative annex has been added to provide guidance on which of the test methods in 60068-2-52 are applicable to different applications. This includes references to new test methods in the latest edition of 60068-2-52.

Development of new methods for detection of potential-induced degradation of thin-film devices (IEC TS 62804-2) and the production of delamination within crystalline silicon based modules (IEC TS 62804-1-1) are ongoing.

Table III: IEC TC 82 WG2 documents relating to module energy rating

61853-1	Ed 1	Published (2011)	Photovoltaic (PV) module performance testing and energy rating – Part 1: Irradiance and temperature performance measurements and power rating
61853-2	Ed 1 Amd 1	Under revision	Part 2: Spectral responsivity, incidence angle and module operating temperature measurements
61853-3	Ed 1	Published (2018)	Part 3: Energy rating of PV modules
61853-4	Ed 1	Published (2018)	Part 4: Standard reference climatic profiles

Table IV: IEC TC 82 WG2 documents relating to specialized stress tests

61701	Ed 3	Under revision	Salt mist corrosion testing of photovoltaic (PV) modules
62716	Ed 1	Published (2013)	Ammonia corrosion testing
62759-1	Ed 1	Published (2015)	Transportation testing - Part 1: Transportation and shipping of module package units
62782	Ed 1	Published (2016)	Cyclic (dynamic) mechanical load test
62804-1	Ed 1	Published (2015)	Test methods for the detection of potential-induced degradation - Part 1: Crystalline silicon
62804-1-1	Ed 1	In preparation	Test methods for the detection of potential-induced degradation - Part 1-1: Crystalline silicon - Delamination
62804-2	Ed 1	In preparation	Test methods for the detection of potential-induced degradation - Part 2: Thin-film
62892	Ed 1	In preparation	Extended thermal cycling of PV modules - Test procedure
62938	Ed 1	In preparation	Non-uniform snow load testing for photovoltaic (PV) modules
63140	Ed 1	In preparation	Partial shade endurance testing for monolithically integrated products

Tests for the extended thermal cycling of PV modules (IEC 62892) to investigate thermal fatigue failure of tabbing ribbon solder bonds are under development. The test procedure in this standard was developed based on analysis of the stress on tin-lead solder bonds on crystalline silicon solar cells in a glass superstrate type package.

The non-uniform snow load testing for photovoltaic (PV) modules (IEC 62892) is about to be published. Because typical field failures of PV modules caused by snow load show glass breakage and frame bending, the test method aims at reproducing the load under which such failures occur.

The partial shade endurance testing for monolithically integrated products (IEC TS 63140) is looking to develop a test method to assess the impact of partial shade on monolithically integrated PV modules.

2.5 Module Components

A number of module component tests have been published or are under development, see Table V. The safety requirements for J-boxes (IEC 62790) and DC connectors (IEC 62852) are currently under revision as amendments.

The main areas to be updated are;

- Updated references
- Clarification on creepage and clearance distances
- Updated (RTE/RTI or TI) requirements

2.6 Module Materials

Over the past 8 years WG2, in collaboration with the activities of the International PV Quality Assurance Task Force (PVQAT) [3], have develop a range of documents relating to the characterisation of materials used in PV modules, see Table VI. The IEC 62788 series has been established to contain most of these activities with sub-parts dedicated to particular activities, such as;

-1) Encapsulants
-2) Backsheets and frontsheets
-5) Edge seals
-6) General tests
-7) Environmental exposures – Accelerated weathering tests of polymeric materials
-8) Materials and coatings, Abrasion and environmental testing

Although the current activity under -8 may be moved to -7 as it has a closer link to environmental exposure tests.

Activities on encapsulants under the 62788-1-x series are mostly published, with the exception of the 62788-1-7

Table V: IEC TC 82 WG2 documents relating to component testing

62790	Ed 1 Amd 1	Under revision	Junction boxes for photovoltaic modules - Safety requirements and tests
62852	Ed 1 Amd 1	Under revision	Connectors for DC-application in photovoltaic systems – Safety requirements and tests
62916	Ed 1	Published (2017)	Bypass diode electrostatic discharge susceptability testing
62979	Ed 1	Published (2017)	Bypass diode - Thermal runaway test

Table VI: IEC TC 82 WG2 documents relating to module materials

62805-1	Ed 1	Published (2017)	Method for measuring photovoltaic (PV) glass - Part 1: Measurement of total haze and spectral distribution of haze
62805-2	Ed 1	Published (2017)	Method for measuring photovoltaic (PV) glass - Part 2: Measurement of transmittance and reflectance
62788-1-1	Ed 1	In preparation	Measurement procedures for materials used in photovoltaic modules – Part 1-1: Polymeric materials used for encapsulants
62788-1-2	Ed 1	Published (2016)	Part 1-2: Encapsulants - Measurement of volume resistivity of photovoltaic encapsulants and other polymeric materials
62788-1-4	Ed 1	Published (2016)	Part 1-4: Encapsulants - Measurement of optical transmittance and calculation of the solar-weighted photon transmittance, yellowness index, and UV cut-off wavelength
62788-1-5	Ed 1	Published (2016)	Part 1-5: Encapsulants - Measurement of change in linear dimensions of sheet encapsulation material resulting from applied thermal conditions
62788-1-6	Ed 1 Amd 1	Under revision	Part 1-6: Encapsulants - Test methods for determining the degree of cure in Ethylene-Vinyl Acetate
62788-1-7	Ed 1	In preparation	Part 1-7: Test procedure for the optical durability of transparent polymeric PV packaging materials
62788-2	Ed 1	Published (2017)	Part 2: Polymeric materials – Frontsheets and backsheets
62788-5-1	Ed 1	In preparation	Part 5-1: Edge seals - Suggested test methods for use with edge seal materials
62788-5-2	Ed 1	In preparation	Part 5-2: Edge seals - Edge-seal durability evaluation guideline
62788-6-2	Ed 1	In preparation	Part 6-2: General tests - Moisture permeation testing with polymeric materials
62788-6-3	Ed 1	In preparation	Part 6-3: Adhesion testing of interfaces within PV modules
62788-7-2	Ed 1	Published (2017)	Part 7 - 2: Environmental exposures - Accelerated weathering tests of polymeric materials
62788-8-2	Ed 1	In preparation	Part 8-2: Materials and coatings for the irradiant incident surface of photovoltaic modules or similar solar devices: Abrasion and environmental testing

which is looking at the long term weathering of encapsulant materials. This is primarily a material test and would be performed on small coupons.

The other document under development is the IEC 62788-1-1. This is an overarching document describing when and how he other encapsulant documents should be used and what information can be gained e.g. for use in datasheet reporting (aiding module design or material research and development); process- and manufacturing-control etc.

The IEC 62788-2 details the characterisation of polymeric backsheet and frontsheet materials. Although not currently under revision the project team is continuing to develop testing methods and also to define appropriate pass fail criteria, which will form the basis of a future revision.

The development of IEC 62788-5-1, Suggested test methods for use with edge seal materials and the IEC 62788-5-2, Edge-seal durability evaluation guideline look at how to evaluate edge seals used in some PV module constructions.

Under development within the area of general tests are the IEC 62788-6-2, a general test method which could be used for moisture permeation testing of polymeric materials. The IEC 62788-6-3 is a method for testing the adhesion testing of interfaces within PV modules.

The published IEC TS 62788-7-2 outlines various accelerated weathering tests of polymeric materials. This will form the basis of weathering requirements which may then be called upon by other material tests, where pass fail criteria may also be added.

The ongoing development of the IEC 62788-8-2 is looking at evaluating the durability of materials and coatings to abrasion. This mostly concerns sand abrasion. It is possible that this document may be re-allocated under the -6 section dealing with general tests.

In addition to the 62788 series there are also two standards relating to glass, IEC 62805-1, measurement of total haze and spectral distribution of haze and IEC 62805-2, measurement of transmittance and reflectance.

2.7 Quality assurance and increased reliability

WG2 has one document relating to module manufacturing QA, the IEC TS 62491, see Table VII. The technical specification was originally published in 2016 and aims to provide greater confidence in the ongoing consistency, performance and reliability of certified PV modules. Its requirements are defined with the assumption that the quality management system of the module manufacturer has already fulfilled the requirements of ISO 9001 (or equivalent) quality management system. The document is currently under revision and will be re-issued as an international standard. The revision includes;

- A change of title to better reflect the scope
- Feedback from certifying bodies who use the document as a basis for factory inspections.
- Updating to the ISO 9001 (2015)

2.8 Increased reliability

Originally WG2 looked for a way to differentiate performance in multiple climates and mounting configurations under the IEC 62892-x standard series which was started in 2013. Unfortunately, during its development it became apparent that it not achievable to the initial scope of the document, as such the project series was abandoned. However, some very good science was contained with the original work and two parts of these were selected to be developed as stand-alone documents. The extended UV testing of encapsulants became the 62788-1-7 and the extended thermal cycling of PV modules retained the IEC 62892 number. Both of these have been described earlier in this paper.

The work to try and define climate specific testing highlighted the need to deal with higher temperature operations. The current type approval and safety tests assume a maximum module temperature of ~70°C with a 20°C safety factor leading to the requirement for a relative thermal endurance, relative thermal index or thermal index (RTE/RTI or TI) for relied upon insulation of at least 90°C. This temperature, for modules mounted in a traditional open rack system, typical of utility scale installations is not exceeded even in extreme climates such as Phoenix Arizona or Riyadh Saudi Arabia.

However, PV modules may be installed in places where the ambient temperatures exceed those defined in the referenced standards or installed in mounting systems that restrict cooling (for example on roof-tops, or integrated in buildings) so that the operating temperature exceeds that assumed in the referenced standards.

As such WG2 started a project, IEC TS 63136, see Table VII. The objective of the project team is to;

- To determine the maximum operating temperatures that the module and components will experience in particular ambient conditions and mounting systems.
- To modify the testing specified in the reference standards in order to qualify the product (module or component) for operation at temperatures higher than specified in those standards.

The current approach of the project team is to use the 98th percentile of the modules maximum temperature in a particular location as the selection criteria. The 98th percentile has been chosen to reflect a module being at a high temperature for a significant time (~175 hours in a year), as opposed to a maximum temperature reached only for some minutes. It has then defined two additional levels, beyond what is currently covered in the standards as;

- Original Requirement: $T_{98\%}$=70°C or less
- Level 1: 70°C < $T_{98\%} \leq$ 80°C
- Level 2: 80°C < $T_{98\%} \leq$ 90°C

Table VII: IEC TC 82 WG2 documents relating to quality assurance, high temperature and extended stress testing

62941	TS to IS	Under revision	Terrestrial photovoltaic (PV) modules - Quality system for PV module manufacturing
63126	Ed 1	In preparation	Guidelines for qualifying PV modules, components and materials for operation at higher temperatures
63209	Ed 1	In preparation	Extended-stress testing of photovoltaic modules for risk analysis

Another aspect of increased reliability (reduced risk) is the use of extended testing of PV modules. An extensive overview of the subject by Kurtz et al [4] showed that "The number of these tests and the frequency of their use by the community imply that they provide value."

Problem for manufacturers is that a particular customers/clients tend to prefer a particular test, based on previous experience. So as manufacturers deal with multiple clients they have to test to most, if not all of the available commercial offerings. As a result WG2 was asked if it could propose its own extended testing scheme. Therefore, a new project IEC TS 63209 (Table VII), was started in August of 2018 with the aim of defining an extended test sequence for PV modules. As with all of the commercial offerings to date the project will start from IEC 61215 as the basis. The technical specification:

- Is not intended to replace IEC 61215 as a pass-fail test procedure for market entry, but to provide a standardized method for the type of testing that is required by stakeholders wishing to have more information than is provided by IEC 61215.
- Is not intended to identify all failure modes. It is not possible to do highly accelerated stress testing without causing some changes that are irrelevant to field performance and/or masking some changes that are important.

The project team will look at incorporating the key elements from the existing commercial test sequences as they move forward to create the first draft.

3 IECRE conformity assessment

Although many international standards are written by the technical committees of the IEC, how they are assessed to have been applied across the industry is the task of conformity assessment. With this in mind the IEC established a new conformity assessment for renewable energy (IECRE) in 2014. I will briefly describe some of the features of the IECRE scheme, for more details see Kelly et al [5,6] The IECRE currently has three sections for;

- Wind Energy
- Solar Energy
- Marine Energy

The three schemes have some common points and have sections relating to both "factory" and "field" aspects of projects.

In regard to PV projects, factory aspects include certification of hardware (modules, inverters, BOS) and auditing of the manufacturer's quality management system. Field aspects include system design, installation, commissioning, operation and maintenance [5]. Since its establishment in 2014 a great deal of effort has been made by the whole of TC 82 to ensure the required standards are in place for the scheme.

Certification agencies have been approved for participation in the system, and efforts are underway to promote the value of certification as a standard practice in the PV industry. Initial certificates have been issued for some of the points on the project timeline (Fig. 1) e.g. module manufacturing quality assurance to IEC TS 62941.

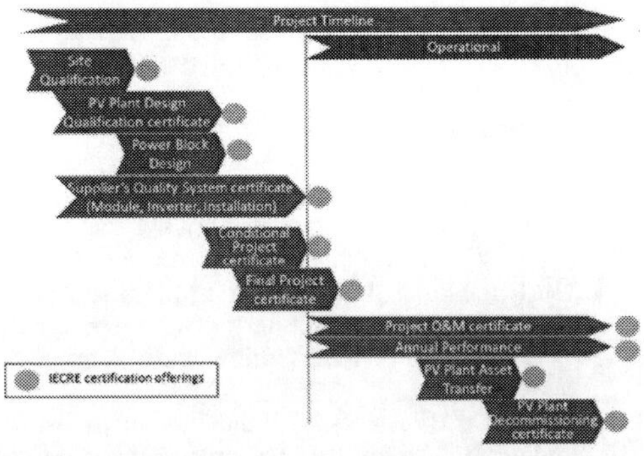

Figure 1: Possible certification offerings for a typical PV system

5 CONCLUSIONS

IEC TC 82 WG2 has been very active in updating its standards as well as developing new standards of interest to the PV industry. The participation of the International PV Quality Assurance Task Force (PVQAT) as a type A Liaison to the IEC TC82 has strengthened the interactions between PVQAT and TC 82 WG2.

Although the membership in WG2 is limited to those appointed by their National Committees the overall participation in WG2 continues to increase with more than 230 current members. Anyone wishing to join any of the IEC technical committees should request their national committee to nominate them.

Membership in the Task Groups (TG) within PVQAT are open to all who wish to participate via requests from the website www.pvqat.org

ACKNOWLEDGEMENTS

The author would like to thank all current and past members of IEC TC 82 WG2, without their dedication the standards described in the paper would not exist.

5 REFERENCES

[1] A. Jäger-Waldau, EPJ Photovoltaics 9, 6 (2018)
[2] J.H. Wohlgemuth, Proceedings 27th EU PVSEC 2976-2980
[3] www.pvqat.org
[4] S. Kurtz, K. Whitfield, N. Philips, T. Sample, C. Monokroussos, E. Hsi, et al. Proceedings 33rd EU PVSEC 1302-1310
[5] G. Kelly, A. Häring, S. Hogan, G. Ball, I. Repins, M. Heinze, et al. "Defining the Value of IECRE Certifications for Providing Confidence in PV System Performance" Proceedings of World Conference on Photovoltaic Energy Conversion (WCPEC-7) (2018)Waikoloa, Hawaii.
[6] G. Kelly, E. Spooner, G. Volberg, G. Ball, J. Bruckner, Proceedings 41st IEEE Photovoltaic Specialist Conference, (2014) 914 - 018.

TYPE APPROVAL AND SAFETY CONSIDERATIONS FOR BIFACIAL PV MODULES: REQUIREMENTS FOR IEC 61215 AND IEC 61730

Bengt Jaeckel[1], Guido Volberg[2], Christos Monokroussos[3], Georg Muelhofer[4] and Arnd Roth[5],

[1] Fraunhofer Center für Silizium-Photovoltaik CSP, Walter-Hülse-Straße 1, 06120 Halle, Germany,
bengt.jaeckel@csp.fraunhofer.de, Tel. +49-177-7503123
[2] TÜV Rheinland LGA Products GmbH, Am Grauen Stein 29, 51105 Köln, Germany
[3] TÜV Rheinland Shanghai Co. Ltd., Lane 777, West Guangzhong Road, Shanghai 200072, P.R. China
[4] Fraunhofer Institute for Solar Energy Systems ISE, Heidenhofstraße 2, 79110 Freiburg
[5] VDE Renewables, Siemensstrasse 30, 63755 Alzenau, Germany

ABSTRACT: Bifacial c-Si PV modules are in fast commercialization and therefore in the need for appropriate testing according to common PV standards, namely IEC 61215 and IEC 61730 series. Both standards were just updated recently but only keeping mono-facial modules in mind. Due to the fast growing market share a project team is working hard on an update to include the needs to correctly qualify bifacial PV modules.
The main concern here is the increased current that is generated from the backside. Unfortunately this is also the most complex part to simulate as the backside contribution strongly depends on module design, mounting structure and ground albedo. Corresponding wording is added to the latest drafts but not all tests are fully covered, yet. This is mainly related to the hot-spot test which is dependent on multiple inputs.
Furthermore bifacial might have a quite different behavior (degradation and recovery) for the resistance against high voltage stress (PID).

Keywords: IEC 61215, IEC 61730, Bifacial, c-Si, Fundamentals, type approval

1 INTRODUCTION

Bifacial c-Si PV modules are in fast commercialization in the last years and will gain a significant market share over the next view years (see FIGURE 1, [1.]).

Why do Bifacial modules /systems gain such interest in the last years and will in the next? Actually the bifacial PV technology concept is not new. As early as 1966, a US patent regarding bifacial solar cell with a pnp junction structure was granted to a Japanese researcher [6.]. But it took almost 50 years for the industry to gain the understanding and production technologies to manufacture such cells cost effectively and reliably in large quantities. Nowadays, passivated emitter rear totally diffused (PERT), passivated emitter rear cell (PERC) and heterojunction (HJT) are the three mainstream technologies for bifacial PV c-Si cells. These new cell concepts are bifacial per design and can be used in mono and bifacial module. This leads directly to the main advantage: If used in a bifacial module design the same solar cell will produce a higher output power without extra cost. This is advantageous in several ways and lead to a reduction in CO2-foot print and LCOE.

For manufactures it is essential to have standards that are up to date and applicable to their product. However, as of today no published standard exists for type and safety certification, nor testing for bifacial PV modules. For PV modules mainly IEC 61215 [2.] and IEC 61730 [3.] series are important. They deal with type approval and safety certification, respectively. While the first one looks into output performance, the second one looks into safety aspects of the product. Here certain new challenges occur for bifacial PV modules as they collect the light from both sides and therefore generate slightly higher voltages but mainly higher currents.

Within this contribution the relevant sections of both standard series will be reviewed and reasoning for change will be given based on scientific evidence collected from the different participating laboratories.

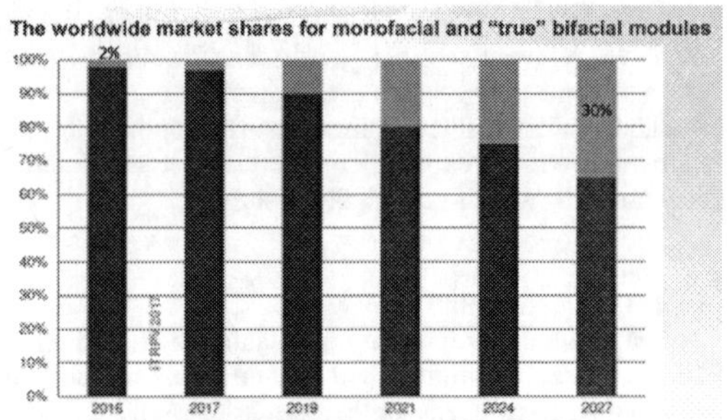

Figure 1: Market outlook for Bifacial c-Si PV modules over the next years [1.].

2 MAIN DIFFERENCES: MONO VS BI-FACIAL

Based on the long-term experience of measuring, testing and modeling parameters of mono-facial modules the results are typically quite reliable.

Although the concept of using both sided of the solar cell for light collection the experience is very limited. Based on the concept it is obvious that the current of the device will increase if both sides are used for photon collection. However, the amount of light shining on the backside of the module is the bit unknown. It strongly depends on the reflection of all surrounding structures and obstacles. The following list highlights some of the issues for bifacial modules:

- Different ground Albedo (see FIGURE 2)
- Albedo is not a constant and can vary on vegetation and weather conditions (summer/ winter/ rain/ snow)
- Backside illumination further depends on module height over ground, row-to-row-spacing in a larger system, surrounding of the system
- Backside shading of cells/modules due to module design (junction box, name plate, back rails…) and mounting structure (rails, foundation pols, Inverters, combiner boxes...)

Based on those challenges it is clear that energy yield prediction is challenging due to the contribution from backside. Furthermore, more detailed measurement and modeling procedures are under development within the community to improve the outcome of predictions for bifiacial modules.

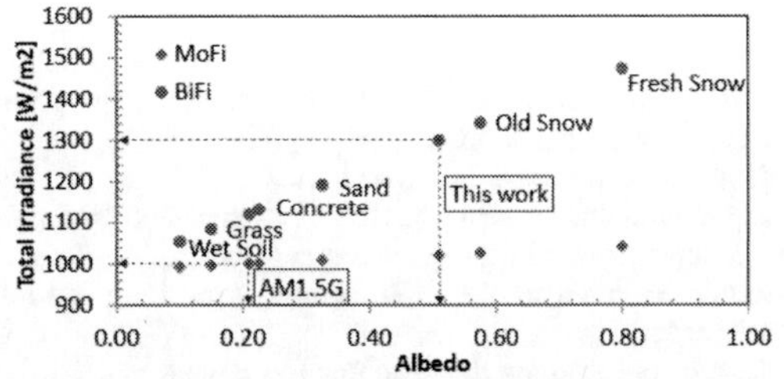

Figure 2: Simulation of total irradiance as function of Albedo for a bifacial module. A 30% increase in total irradiance (+300W/m²) is achieved with a 51% Albedo. As the simulation shows this covers all typical grounds where also high irradiance levels can be expected. Snow is a special case as this may lead to almost no front side illumination. Taken from [5.].

3 IDENTIFIED AND ADDRESSED ISSUES

IEC 61215 and IEC 61730 series were published in 2016. The development of both series was done keeping mono-facial modules in mind. Therefor the specifics for bifacial modules were not considered. The newly formed project team is working hard to include in the next committee draft (CD) the relevant changes. Those are described next.

3.1. PASS-FAIL CRITERIA:

The safety test requirements and pass criteria are clearly set and are current dependent and the current depends on rating of the PV module. So, if the rating is done correctly, including bifacial gains, the pass/fail criteria are fully applicable as set.

Here care must be taken especially for the fuse rating as this will directly influence MST 13 (Continuity test for equipotential bonding) and MST 26 (Reverse current test). The change from a mono to a bifacial module configuration will increase the current by approx. 10-20%. Assuming a typically mono-facial short circuit current of 9.0A typically a 15A fuse is used to include the required safety factors. Increasing the bifacial short circuit current to e.g. 11.0A will than lead to a fuse rating of 20A (next fuse size that is available). So test requirements will change from 15A to 20A in this example. All other bifacial impacted MST tests directly link to MQTs (02, 03, 07, 22, 51, 54) and will discussed for IEC 61215.

Generally the pass-fail are also clear for IEC 61215, but there are some clarifications needed, as the backside was formally treated as "non-power-relevant". For bifacial modules the power degradation can happen on both sides, but will contribute differently to the overall power rating based degradation. Therefor a clarification of the use of Gate #1 und Gate #2 is needed to include bifaciality coefficients.

To further do that reliably it is necessary to clarify the exact determination of bifacial coefficients for STC and low light power determination (MQT 6.1 and 7).

The general procedure to measure bifacial modules is given in IEC 60904-1-2 [4.]. The reference to IEC 60904-1-2 and the text to better describe the apparatus to measure front- and backside performance of the module is included in the upcoming CD.

3.2. STRESS TESTS

UV preconditioning test (MQT 10): The current proposal is to add: "For bifacial modules repeat the procedure of UV irradiation on the rear-side of the modules." As most bifacial modules are currently glass/glass constructions the question is why this extra portion of UV should be added. UV light is known to be more severe for polymeric materials. So for all polymeric backsheet this would be a good addition, independent of a mono- vs. bifacial technology discussion.

There are also known UV induced cell related degradation phenomena, but may not be the dominant part. In the end it would be better to request this for all modules as it is also required for IEC 61730.

Thermal cycling test (MQT 11): For the thermal cycling test no changes were proposed. But also to align this with the requests for change of the bypass diode thermal test (MQT 18.1) the wording for the current should also updated to read: "Test current is set to Front-side (STC) + Backside stated max current (e.g. 200W/m2)".

Bypass diode thermal test (MQT 18.1): As higher currents can appear in bifacial modules also the bypass diodes should be capable of dealing with this increase. There the following text is added: "Test current is set to Front-side (STC) + Backside stated max current (e.g. 200W/m2) times 1.25".

Stabilization (MQT 19): All PV modules must be stabilized prior there initial MQT 6.1 measurement to verify their nameplate rating. The general procedure is given for MQT 19. Also here only mono-facial modules were considered and no special requirements and procedures are given for bifacial modules. This could be particular important as it is allowed to qualify an alternative stabilization technique. This could be for example a current injection. This actually would stress the front and the backside, light soaking on the other hand mostly only the front side. The new cell technologies like PERC and PERT do typically have different LID effects (e.g. LeTID). As for instance LeTID can require long light soaking procedures (several weeks) alternate procedures are under development. A Round Robin Test is currently developed for that to qualify a detailed procedure that can be used without going through the currently required verification procedure [7.], [8.].

3.3. HOT SPOT STRESS TESTS

The first experiments done for hot-spot testing on bifacial modules resulted in different results. The reasons for that are plentiful.

Generally there are two principle options to do the test. Assuming a certain "worst-case" albedo the back side of the module might be illuminated with 300W/m2. Now it is possible to ether

- illuminate the front side with 1300W/m2 and follow the standard procedure of MQT 9 or
- illuminate the front side with the standard 1000W/m2 and add 300W/m2 to the back side.

It has to be noted that the alternative language can be to use the "Backside stated max current" (e.g. 200W/m² - here the test would have to be done at 1200W/m2).

Both procedures have advantages and disadvantages. The first method is relative straight forward to perform but

require a constant sun simulator capable of illuminating the front side with 1300W/m2. Here the debate is about how realistic this approach is.
The front- and backside illumination seems quite logical as both surfaces are supposed to see light and generate charge carriers.
To perform the regular Hot-spot test a minimum BBB-class constant sun simulator is required. The question is what requirements a backside illumination apparatus must have as a minimum. From the first performed test it seems that the classification is of secondary importance. Therefor a CCC-class sun simulator will be recommended, but not required. A simple intensity tunable LED-based array to homogeneously illuminate the backside of the cell string under test should be sufficient. To generate the current the appropriate current the light intensity should be adjusted that way that it matches the currents determined under STC conditions for the backside while determining the bifaciality coefficients.
The next issue is that there will be an "intrinsically" induced higher operating module temperature. The normal limit for 1000W/m2 is set to 50°C +/-10K. With only a little bit of ventilation it is even problematic to stay within this limit for 1000W/m2 illumination as it can be seen in FIGURE 3 (black curve till minute 80). Increasing the illumination ether from front side (first couple of minutes in the graph) and extra backside illumination (from minute 80) will increase the temperature of the full module. Extra cooling may be needed or alternatively a different set point might want to be defined for bifacial modules that operate in high albedo regions.
The definition of the shading situation is quite tricky. There might be constant shades e. g. due to name plates, junction boxes or back rails on the PV module or due to mounting structure shading. These variables together with the difficulty to determine worst-case shading on the front side alone [10.] make the Hot-spot test for bifacial modules quite tricky.

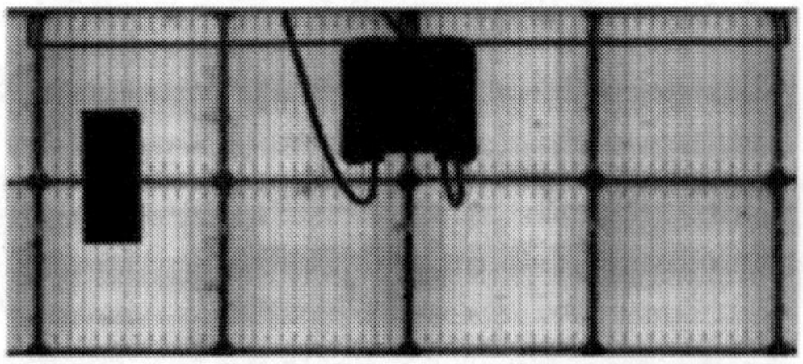

Figure 3: EL image of a bifacial PV module from the backside nicely showing the shading effect of nameplate and junction box [9.].

FIGURE 4 shows two scenarios. Starting with minute 45 the mono-facial configuration is tested. With a 40% shade the highest temperatures were achieved in the illuminated part of the solar cell (purple curve → ≈ 150°C). The area with the shade only heated up to ≈ 100°C.
Starting minute 80 the module was also illuminated from the backside resulting in a temperature increase for the shaded cell (front side only), but also for the reference cell that reached here a temperature of ≈ 68°C that is outside the currently allowed limit. Using an extra 300W/m2 from the backside the shaded cell temperature increased by ≈ 20K (till minute 130). Starting at minute 130 the front-side shaded cell was shaded the same way from the backside. The total current in the module dropped accordingly (see green curve). The temperature in the non-shaded area almost stayed constant (≈+0.5K). The temperature in the shaded area dropped by ≈5K due to the shading also of the sensor now – resulting in a less severe scenario.
As it can be seen on this example more work is needed to fully understand the correlations and to define "worst-case" hot-spot test scenario. However, the new standard should at least address the need for higher irradiation levels and request the test following the mono-facial scenario.
As last note it must be mentioned that the procedure for wafer-based technologies (WBT-4.8.5.2 - MQT 09.1) to determine high and low shunt resistant cells would also require front and back side flash-test illumination. Due to limited availability of double-sided flash testers and the lack of knowlegde it is currently suggested to follow the mono-facial procedure.

4 SUMMARY

Within this contribution the relevant sections of IEC 61215 and IEC 61730 standard series were reviewed and reasoning for change was given to the extend necessary.
As relevant sections that may require modification the following clauses and tests are identified:
For IEC 61215 the section 5 "Marking and documentation" is very critical as it defines the name plate values. Those are then important for Gate 1: "Verification of rated label values" and Gate 2: "Maximum power degradation during type approval testing". Both gates are closely correlated to MQT 6.1: "Measuring at STC". Details are added for the tests to address the gap to determine bifacial's output power and pass-fail criteria.
MQT 07, MQT 10, MQT 19 as well as the relevant MST tests from IEC 61730 were briefly discussed.
MQT 09, the "Hot-spot endurance test" of the modules is still one of the not well-understood tests to clarify for bifacial modules.
The PID test, that is new in the CD proposal as MQT 21 and is based on IEC 62804-1, does not include reference to bifacial specifica. Recent research results show that PID has to also be considered on front- and backside of bifacial modules as the degradation processes will be different [11.].

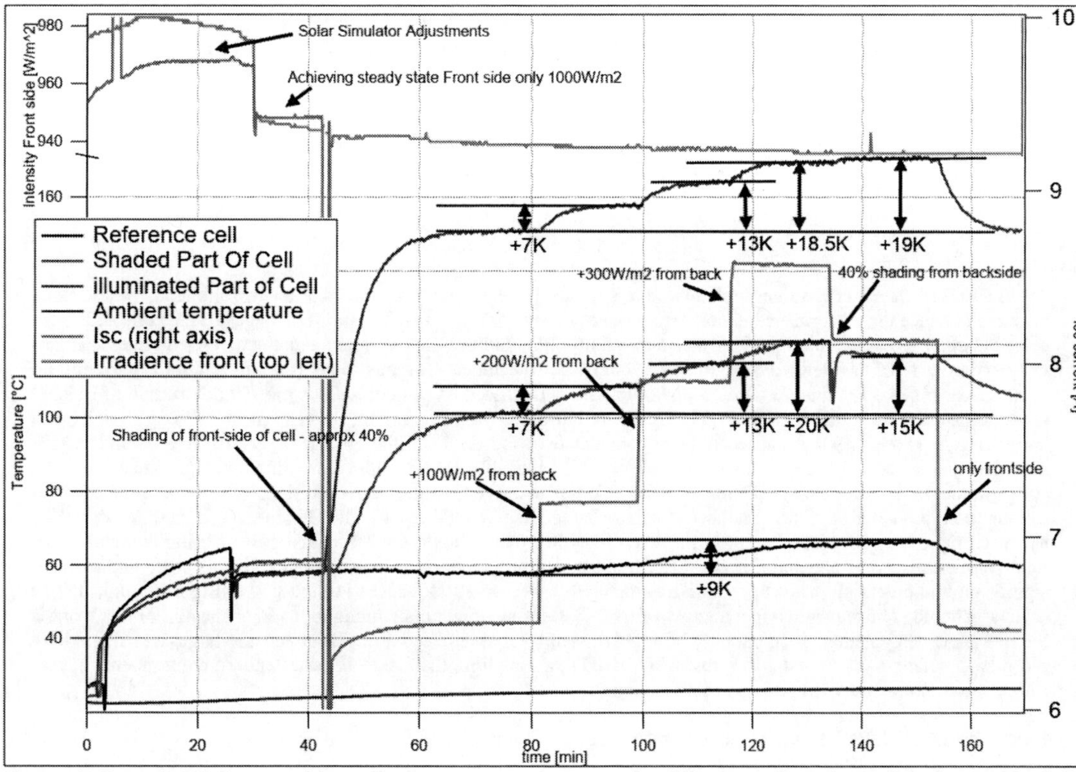

Figure 4: Example HS-Test: resulting cell temperatures and currents for different configurations of shading and backside illumination.

3 LITERATURE

[1.] Solar Power Europe, "Global Market outlook 2018-2022", 2018

[2.] IEC 61215-all parts: "Terrestrial photovoltaic (PV) modules – Design qualification and type approval", 2016.

[3.] IEC 61730 all parts: "Photovoltaic (PV) module safety qualification", 2016

[4.] IEC 60904-1-2: "IEC 60904: Photovoltaic devices - Part 1-2: Measurement of current-voltage characteristics of bifacial photovoltaic (PV) devices", 2018

[5.] C. Monokroussos, X. Y. Zhang, E. Lee, N. Wang, C. Zou, J. Bonilla Castro, M. Schweiger, W. Herrmann, "Rear Face Spectral Irradiance at 1-Sun and Application to Bifacial Module Power Rating", EU PVSEC, 2018, Brussels

[6.] Mori, H. 1966, "Radiation energy transducing device", U.S. Patent 3278811A

[7.] Ingrid Repins, M. Köntopp, "LeTID round robin", 2018/2019

[8.] M. Pander, M. Turek, J. Bauer, T. Lukal, Chr. Hagendorf, M. Ebert, R. Gottschalg, "Benchmarking Light and elevated temperature induced degradation (LeTID)", EU PVSEC, 2018, Brussels

[9.] C. Monokroussos, X.Y. Zhang, E. Lee, Y.H. Wang & C. Zou, „Rear Face Spectral Irradiance at 1-Sun and Application to Bifacial Module Power Rating", EU PVSEC, 2018, Brussels

[10.] D. Philipp, H. Manuel & G. Mülhöfer, "Hot-Spot Endurance Test - Modifications for Bifacial Photovoltaic Modules", EU PVSEC, 2018, Brussels

[11.] K. Sporleder, Chr. Hagendorf, "PID Benchmark on rearside of bifacial solar cells", PV workshop NREL 2018, Denver

DESIGN STUDY OF A DOUBLE-SIDE ILLUMINATION SOLAR SIMULATOR FOR BIFACIAL SILICON PV MODULES CHARACTERISATION BASED ON LOW-COST LED BIAS LIGHT

D. Shaw, J. Lopez-Garcia, R.P. Kenny, L. Pinero-Prieto, E. Ozkalay
European Commission, Joint Research Centre (JRC), Directorate C - Energy, Transport and Climate, C02 - Energy Efficiency and Renewables Unit,
Via E. Fermi 2749, 21027 Ispra (VA), Italy
Tel +39 0332 785896 / Fax:+39 0332 789 992
email: david.shaw@ec.europa.eu

ABSTRACT: Bifacial crystalline Si photovoltaic modules have attracted considerable interest in the last years, since they can enhance the performance in comparison to traditional PV modules because both sides of the cell can absorb solar radiation. To assess their performance and quality, PV modules are measured and characterized under standard test conditions (STC) as defined by the International Electrotechnical Commission (IEC). Since there is currently no standard for bifacial PV modules, one of the main issues that need to be addressed is the standardisation of bifacial measurements for PV devices. The IEC TC82 is currently working on the development of a draft technical specification. Three different approaches are considered in the draft for the electrical performance of bifacial PV modules, that is, natural sunlight, a solar simulator with adjustable irradiance level for single-side illumination and a solar simulator with double-sided illumination. In order to reproduce the real conditions with both front and rear illumination, the latter approach has been used to characterise bifacial solar cells. However, at a module scale, this approach represents a technical and logistical challenge for many due to the associated cost of using two controlled light sources instead of one and it is considered that the current equipment is not suitable for this setup and requires significant and potentially expensive modifications. However, recent advances in high-power Light Emitting Diode (LED) technology have resulted in a number of LED-based sun simulators coming to market. The aim of this work is the design, development and validation of a double-sided solar simulator based on an existing large-area single long pulse flash solar simulator and a low-cost LED-based rear bias light that can fulfil the standard requirements of rear illumination (up to 200 W/m^2) and non-uniformity.

Keywords: Bifacial; PV modules; characterization; LED; standard.

1 INTRODUCTION

Bifacial crystalline Si photovoltaic modules have attracted considerable interest in the last years, since they can enhance the performance in comparison to traditional PV modules because both sides of the cell can absorb solar radiation. To assess their performance and quality, PV modules are measured and characterized under standard test conditions (STC) as defined by the International Electrotechnical Commission (IEC) [1]. Since there is currently no standard for bifacial PV modules, one of the main issues that need to be addressed is the standardisation of bifacial measurements for PV devices. The IEC TC82 is currently working on the development of a draft technical specification IEC TS 60904-1-2 [2]. Three different approaches are considered in the draft for the electrical performance of bifacial PV modules, that is, natural sunlight, a solar simulator with adjustable irradiance level for single-side illumination and a solar simulator with double-sided illumination. In order to reproduce the real conditions with both front and rear illumination, the latter approach has been used to characterise bifacial solar cells. However, at a module scale, this approach represents a technical and logistical challenge for many due to the associated cost of using two controlled light sources instead of one [3] and it is considered that the current equipment is not suitable for this setup and requires significant and potentially expensive modifications [4]. However, recent advances in high-power Light Emitting Diode (LED) technology have resulted in a number of LED-based sun simulators coming to market. The aim of this work is the design, development and validation of a double-sided solar simulator based on an existing Class AAA large-area pulsed solar simulator combined with a low-cost LED-based rear bias light that can fulfil the standard requirements of rear illumination (up to 200 W/m^2) and non-uniformity. A prototype LED simulator with illumination area of 40 cm x 40 cm has been characterised for non-uniformity, temporal stability, and spectral content. The simulator has been validated by comparison of measurements made at 200 W/m^2 on a mini-module with equivalent measurements on a pulsed solar simulator. Measurements using the combined simulators have been compared with those from the equivalent rear irradiance driven power gain yield measurement with single-side illumination.

2 METHODOLOGY

2.1 LED Simulator design and construction.

The technology chosen was an LED integrated array COB (circuit on board) module, where each 30 mm diameter LED light emitting surface has between one to two hundred individual series and parallel connected blue emitters below the orange-red local phosphor covering. A matrix of fourteen of these low-cost COBs was then envisaged to cover the area of the bifacial module under test with sufficient uniformity. The LED production "bin" was chosen to be nearly the highest power and photon-flux available on the market, however for higher reliability and repeatability the 'second-bin' flux level was chosen with at least 13000 lumens at 4000K at 25°C. The next most important element for the simulator design was ensuring that the LED output was stable. Thermal management is critical for this so our system design employs good passive thermal management with thermal interface material and heatsink on the matrix mounting bars and a non-solder spring-click holder for each COB, allowing future upgradeability, keeping the LEDs as close as possible to 25°C.

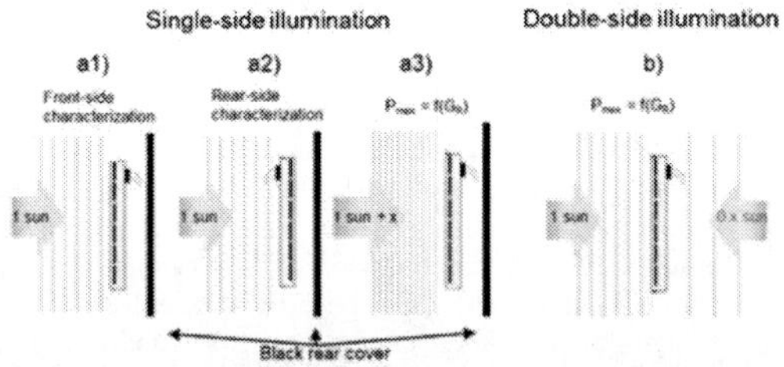

Figure 1 *Single-side illumination approach based on equivalent irradiance method and double-sided illumination method using two light sources.*

We supply all COB LEDs in parallel with a single program voltage from a commercial 1 kW laboratory PSU, at nominal 72 V. An Arduino control system was used to adjust the PSU output into 2000 ms pulses for testing purposes, interface to the ESTI control systems and raise the array power to 350 W/m^2, or to reduce the measured irradiance to well below 100 W/m^2, from the nominal 200 W/m^2. This prototype LED simulator with non-uniformity projected to be < 5 % over an illumination area of 40 cm x 40 cm was designed and constructed and the non-uniformity, temporal stability, and spectral content has been characterised.

Pulsed operation of the LED simulator ensures a very long lifetime of the LEDs. The LED simulator has been programmed for a steady state output over a 2 second time period, which facilitates synchronization with the pulsed simulator it is paired with. The irradiance may be controlled by adjusting the drive voltage and/or the distance to the test device. By adjusting the distance between 25 cm and 60 cm combined with acceptable drive voltages, the irradiance in the test plane may be controlled in a range between 100 W/m^2 and 300 W/m^2.

A prototype LED simulator with non-uniformity projected to be < 5 % over an illumination area of 40 cm x 40 cm has been designed and constructed. The non-uniformity, temporal stability, and spectral content have been characterised.

The test module is a purpose constructed n-type glass-glass frameless 4-cell bifacial mini-module with a size of 40 cm x 40 cm, sufficiently small to be measured within the illumination area of the prototype LED Simulator. The cell area is about 16 cm x 16 cm.

2.2 I-V measurement methods

According to the draft IEC TS 60904-1-2, the contribution from the light incident on the opposite side of the device under test must be eliminated completely during the measurement by creating a non-irradiated background. The background is considered to be non-irradiated if the irradiance does not exceed 3 W/m^2, at any point, on the non-exposed side of the device. A previous electrical characterization of the modules was carried out with an opaque or non-reflective painted wooden panel with irradiance below 3 W/m^2 in order to avoid the illumination of the non-exposed side.

Double-sided and single-sided reference I-V curve measurements (as shown schematically in figure 1) were performed with a Pasan IIIb (PIIIb) pulsed solar simulator (with a nominal sweep time of 10 ms) at ~1000 W/m^2, 25.0 ± 0.1 °C and corrected to STC (1000 W/m^2, 25°C and AM1.5G). For all measurements involving the PIIIb simulator, the so-called multi-flash (MF) method [5] was employed as the 10 ms pulse is too short for accurate measurement of the high capacitance module under test. In this method, the final I-V curve is constructed from typically 35-40 individual flashes. Each point of the I-V curve was obtained by applying a constant voltage to the module which was then subjected to a flash from the solar simulator at the desired irradiance level. The resultant module I-V curve was then obtained by performing an interpolation of the measured points [6, 7].

The spectral responsivity of the front and rear side of the bifacial modules was measured using a differential spectral responsivity technique with a large area pulsed solar simulator equipped with a number of bandpass filters to obtain illumination of modules with quasi-monochromatic light [8] and the spectral responses were used to calculate the Mismatch factors (MMF). No differences between front and rear spectral responses were observed.

As described in the draft IEC TS 60904-1-2, bifaciality refers to the ratios between the main I-V characteristics of the rear and front side of a bifacial device, typically at Standard Test Conditions (STC). This is quantified with reference to bifaciality coefficients, namely the short circuit current bifaciality coefficient φ_{Isc}, the open-circuit voltage bifaciality coefficient φ_{Voc} and the maximum power bifaciality coefficient φ_{Pmax}. This last one is defined as:

$$\varphi_{P_{max}} = \frac{P_{max,rear}}{P_{max,front}} \qquad (1)$$

where $P_{max,rear}$ and $P_{max,front}$ are the module output power measured when illuminating only the rear and the front side at STC, respectively, with a non-irradiated background. These coefficients are required in order to calculate the P_{max} at an equivalent irradiance level when a single-side illumination is used [2].

IEC TS 60904-1-2 states that for the equivalent rear irradiance driven power gain yield measurement with single-side illumination, the P_{max} of the device must be measured at equivalent irradiance levels G_{Ei} corresponding to 1000 W/m^2 on the front side plus different rear side irradiance levels G_{Ri}. The equivalent irradiance levels are determined as functions of the bifaciality coefficient φ according to equation 2 and equation 3:

$$G_{E_i} = G_F + \varphi \times G_R \qquad (2)$$

$$\varphi = Min(\varphi_{I_{sc,STC}}, \varphi_{P_{max,STC}}) \qquad (3)$$

where φ is the minimum of the I_{sc} and the P_{max} bifaciality coefficients φ_{Isc} and φ_{Pmax}. Besides the STC measurement at 1000 W/m^2 (for G_R = 0 W/m^2), at least two different equivalent irradiance levels are required, so two specific P_{max} values must be reported: $P_{max,BiFi10}$ and $P_{max,BiFi20}$, for G_{R1} = 100 W/m^2 and G_{R2} = 200 W/m^2, respectively. For the module under test, φ = 0.935 and the equivalent irradiance levels for G_{R1} = 100 W/m^2 and G_{R2} = 200 W/m^2 are G_{E1} = 1093 W/m^2 and G_{E2} = 1187 W/m^2, respectively.

2.3 Setup for LED simulator measurements

For the double-sided measurements, the LED simulator is mounted vertically at the back of the device under test at an adjustable distance. The distance has been set at 50 cm for these tests. This configuration enables illumination of the rear side by LED and front side by PIIIb simultaneously. Separate single-side measurements are also possible. Identical data acquisition software and hardware system are used for both single- and double-side I-V measurements. In order to eliminate reflections between the two light sources which might cause irradiance gain and non-uniformity, black masking and apertures around the module are used as suggested in the draft IEC TS 60904-1-2. Nevertheless, a small gain of about 0.5% in I_{sc} is observed at STC. In order to ensure the optimal module position with lowest possible non-uniformity within the measurement area, Zemax virtual prototyping ray-tracing software is used (Figure 2).

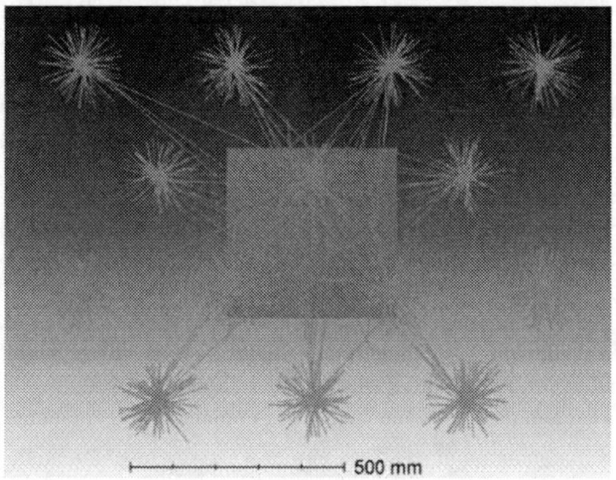

Figure 2 *Position of the 14 LEDs (green) and measurement plane (orange).*

3 RESULTS

3.1 LED Solar Simulator Characterization

The non-uniformity on the measurement plane (40cm x 40 cm) located at the center of the LED solar simulator and has been measured according to Standard IEC 60904-9 [9]. The LED simulator has been found to have Class B irradiance non-uniformity of 4.45%. Temporal stability of the LED simulator has been examined for both short term instability (STI) and long term instability (LTI) as described in IEC 60904-9. Both short term and long term instabilities are Class A, 0.45% and less than 0.25% (for 100 ms sweep time), respectively.

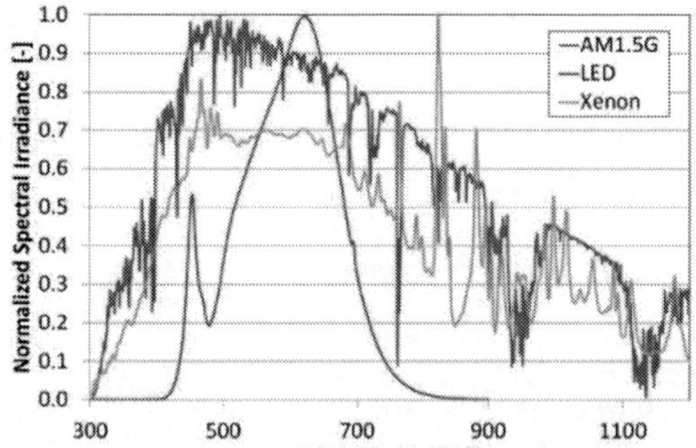

Figure 3 *Solar spectrum AM1.5G, spectrum of the 4000K LED simulator at 350 W/m² and spectrum of a xenon Class AAA solar simulator at 1000 W/m².*

The spectrum of the LED simulator is in the classification range for some wavelength ranges defined in IEC 60904-9, but the overall spectrum is worse than Class C since the spectrum of the LED does not extend beyond 800-900nm (Table I). Figure 3 shows a comparison of the global spectrum AM1.5, the spectrum of a xenon class AAA solar simulator in our lab and the spectrum of the LED based solar simulator.

Pulse-to-pulse stability of the LED is examined even it is not specified in the standard, as it is crucial to have repeatable light intensity on the measurement plane for all measurements when using the MF method. The standard deviation between the light intensity on the measurement plane for different flashes is around 0.5 %.

Table I *The spectral content of LED compared to AM1.5G according to Standard IEC 60904-9.*

Wavelength range nm	Percentage of total irradiance in the wavelength range 400 nm - 1100 nm		Spectral match to all intervals	Class
	IEC 60904-9	LED		
400-500	18%	11%	0.62	B
500-600	20%	38%	1.9	C
600-700	18%	44%	2.39	-
700-800	15%	6%	0.42	C
800-900	13%	0%	0.03	-
900-1100	16%	0%	0.02	-

3.2 IV measurement

IV characteristics of the bifacial mini-module were measured using both the PIIIb and LED solar simulators. Figure 4 shows the I-V curves of the back side of the bifacial 4-cell mini-module at 200 W/m² with black cover by using PIIIb and LED solar simulators separately. The IV curve (blue) measured using the PIIIb is MMF corrected. The IV curve (green) measured at 200 W/m² using the LED simulator has lower values compared to IV curve measured by PIIIb at 200 W/m² mainly due to the poor spectral response of the LEDs which leads to a large MMF. The effective irradiance method, stated in Section 3 of Standard IEC 60904-7, may be applied for the correction of spectrum mismatch [10]. The effective irradiance to generate the I_{sc} in the bifacial mini-module as generated at 200 W/m² is calculated by equation 4.

$$G_{eff} = G_{ref} \times MMF \qquad (4)$$

where G_{eff} is the effective irradiance, G_{ref} is the reference irradiance (here 200 W/m²) and MMF is the mismatch factor of the device under LED illumination. Since a measurement at 200 W/m² is available from the PIIIb simulator, an analogous but improved approach is to use this measurement as a reference. By adjusting the voltage control, and hence intensity of the LEDs, an effective irradiance is applied so that the short circuit current is matched to that obtained using the PIIIb thereby correcting for the MMF. As can be seen from Figure 4, the two IV curves measured by PIIIb and LED 'effective irradiance' (blue and red, respectively) are virtually indistinguishable.

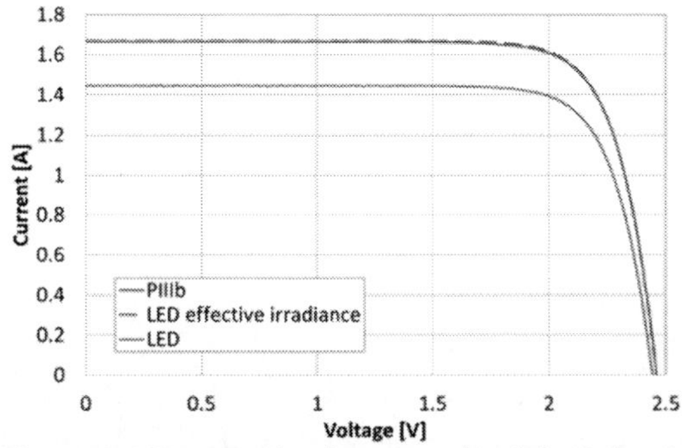

Figure 4 I-V curves for the rear side of the bifacial module using PasanIIIb and LED at 200 W/m².

The equivalent rear irradiance driven power gain yield measurement with single-side illumination described in IEC 60904-1-2 is used as the reference measurement to validate the LED simulator when used for double-sided illumination. Figure 5 shows the IV curves obtained for the two approaches. Front side only at STC with black rear cover and rear side at 200 W/m² with black rear cover IV curves are also shown for reference. The two measurement methods give similar IV characteristics for the module but the double-side illumination has around 0.6% higher I_{sc} compared to single-side equivalent irradiance method. Single-side measurements with and without black rear cover using the PIIIb and LED simulator demonstrates that there is a gain of about 0.6% in I_{sc} when the PIIIb is used without a rear cover. As the LED simulator flashes through to the PIIIb side, there is not significant difference (less than 0.1%) between measurements with and without black rear cover. The difference matches closely the extra I_{sc} of the double-side illumination compared to the single-side equivalent irradiance measurement, demonstrating that the gain is mainly due to stray/parasitic reflection from the back of the mini-module at the LED simulator side.

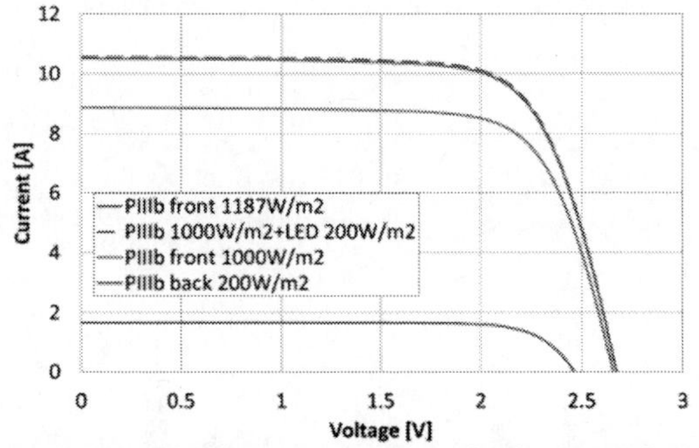

Figure 5 IV curves measured with the equivalent irradiance single-side method and double-side illumination method. Front side only at STC and rear side at 200 W/m² IV curves are also shown for reference.

Table II shows electrical characteristics obtained using the two methods to characterize a bifacial module. It can be seen that main difference between two methods are in the I_{sc} and P_{max}, 0.6% and 0.77% respectively. After correction of the double-side illumination current for the stray/parasitic reflected light, as proposed in IEC 60904-1-2 [2], the difference between the methods is less than 0.3% for I_{sc} and P_{max}.

***Table II** Comparison key parameters measured with the equivalent irradiance single-side method and double-side illumination method using the LED simulator.*

	I_{sc} [A]	V_{oc} [V]	P_{max} [W]	delta I_{sc} [%]	delta V_{oc} [%]	delta P_{max} [%]
PIIIb 1187 W/m²	10.51	2.671	20.50	-	-	-
PIIIb 1000 W/m² + LED 200 W/m²	10.57	2.672	20.65	0.60	0.04	0.77
PIIIb 1000 W/m² + LED 200 W/m² corrected	10.53	2.672	20.55	0.20	0.05	0.29

4 CONCLUSIONS

The prototype LED simulator has demonstrated very good performance, enabling illumination of the rear side of a bifacial module at variable light levels to above 300 W/m². When combined with a commercial Class AAA pulsed solar simulator, double-side illumination produces results very close to those obtained with the Class AAA solar simulator using equivalent irradiance and single-side illumination.

The non-uniformity below 5% means it meets the requirements of draft IEC TS 60904-1-2 for use with bifacial modules. The spectral match to AM1.5G is outside Class C, but this may be compensated by a mismatch correction, or using the effective irradiance method.

The module used for testing purposes had a very good bifaciality coefficient and good rear-side cell current matching, so the IV curves obtained with single-side illumination method were practically identical to those obtained with the double-side method. For modules having poorer rear-side performance the double-side illumination approach may be necessary in order to obtain valid results, confirming the potential utility of this LED simulator. Furthermore, in real outdoor conditions, the rear-side illumination varies in intensity and uniformity according to the albedo, and also with daily and seasonal variations. The uniformity of the LED simulator may also be readily adjusted, by changing the geometry or by varying the powering of individual LEDs, which may enable the performance of bifacial modules to be evaluated over the full range of outdoor conditions.

The modular simulator design means that extension of the area to allow measurement of full size 60-cell modules will be straightforward and at low cost. The full size LED simulator will be tested paired with a flat-bed 100 ms pulsed solar simulator (SPIRE) available at ESTI for double-sided measurements, as this would allow single pulse measurements. This configuration would be convenient for factory testing of bifacial modules.

Acknowledgements

The authors thank Franscesco Noseda, Diego Pavanello, Roberto Galleano and Tony Sample for help in the setup, and measurements, and ECN-TNO for providing the test module. This project has received funding from the EMPIR programme co-financed by the Participating States and from the European Union's Horizon 2020 research and innovation programme.

REFERENCES

[1] IEC 61836, "Solar photovoltaic energy systems - Terms, definitions and symbols" IEC, 2016.
[2] IEC TS 60904-1-2, "Measurement of current-voltage characteristics of bifacial photovoltaic (PV) devices" IEC, 2018 Draft.
[3] S. Roest, W. Nawara, B. B. Van Aken and E. Garcia Goma, "Single side versus double side illumination method IV measurements for several types of bifacial PV modules" Proc. 33rd European Photovoltaic Solar Energy Conference and Exhibition, Amsterdam, 2017.
[4] B. Newman, A. Carr, K. Groot, N. J. J. Dekker, B. B. Van Aken, A. Vlooswijk and A. Van de Loo, "Comparison of bifacial module laboratory testing methods" Proc. 33rd European Photovoltaic Solar Energy Conference and Exhibition, Amsterdam, 2017.
[5] J. Lopez-Garcia, B. Haile, D. Pavanello, A. Pozza and T. Sample, "Characterisation of n-type bifacial silicon PV modules" Proc. 32rd European Photovoltaic Solar Energy Conference and Exhibition, 1724-1729, 2017.
[6] N. Ferretti, Y. Pelet, J. Berghold, V. Fakhfouri and P. Grunow, "Performance testing of high-efficient PV modules using single 10ms flash pulses" Proc. 28th European Photovoltaic Solar Energy Conference and Exhibition, 3184-3187.
[7] A. Virtuani, G. Rigamonti, P. Beljean, G. Friesen, M. Pravettoni and D. Chianese, "A fast and accurate method for the performance testing of high-efficiency C-Si photovoltaic modules using a 10Ms single-pulse solar simulator" Proc. 38th IEEE Photovoltaic Specialists Conference, 496-500.
[8] R. Van Steenwinkel, "Measurements of spectral responsivities of cells and modules" Proc. 7th European Photovoltaic Solar Energy Conference and Exhibition, 325-329, 1986.
[9] IEC 60904-9, "Solar simulator performance requirements" IEC, 2007.
[10] IEC 60904-7, "Computation of the spectral mismatch correction for measurements of photovoltaic devices" IEC, 2008.
[11] "CREE company Ltd.," 2016. [Online]. Available: http://www.cree.com/led-components/media/documents/XLampXML.pdf.

COMPARISON OF LAYOUTS FOR SHINGLED BIFACIAL PV MODULES IN TERMS OF POWER OUTPUT, CELL-TO-MODULE RATIO AND BIFACIALITY

Andrew Mondon, Nils Klasen, Esther Fokuhl, Max Mittag, Martin Heinrich, Harry Wirth
Fraunhofer Institute for Solar Energy Systems (ISE), Heidenhofstraße 2, 79110 Freiburg, Germany
andrew.mondon@ise.fraunhofer.de

ABSTRACT: This contribution analyses two different module topologies for shingled solar cells that increase module power at normal operation and under partial shading conditions. A bifacial shingled parallel string layout and a bifacial shingled matrix layout are compared in terms of module output by experiment and cell-to-module (CTM) analysis. After verification of the CTM model for modules samples (0.475 m²), a calculation for full size modules is performed. Both shingled layouts, excel in output power and $CTM_{efficiency}$ in comparison to a conventional module. The shingled modules are also compared regarding partial shading response, with the matrix layout showing superior power output in most investigated cases. Finally, the newly developed concept of "SlimLine" module fabrication is introduced, which uses structured encapsulants to simplify shingle cell placement and alignment having the potential of increased assembly speed and reliability.
Keywords: Shingled, Bifacial, Electrical Properties, Module Integration, Module Manufacturing, CTM, Shading

1 INTODUCTION

The concept of shingled solar cells is currently undergoing a renaissance. It was first introduced by Dickson *et al.* in 1960 [1]. Conventional solar cells feature busbars on the front surface, which cause shading loss. On solar cells for shingling, the busbars are positioned at the edge of the cell, which is then overlapped by a neighbouring cell; hence this inactive cell area as well as the cell interspace is overlapped by active cell area, significantly increasing the share of active area within the module [1].

This contribution compares two different module layouts (see Figure 1), which are parallel shingled strings and shingled cells in matrix layout initially named "shingle-roof patterned solar cell array" [2]. The two shingled module layouts have been manufactured in the size of 475 x 1000 mm²; subsequently they have been analysed. The modules feature bifacial shingled cells (pSPEER, p-type Shingled Passivated Emitter Edge and Rear) as previously presented by Baliozian *et al.* [3].

The modules are compared regarding power output at STC condition and power output considering bifaciality. Additionally, comparisons based on cell-to-module (CTM) [4, 5] analysis are carried out. The potential module performance is calculated for full size modules and compared to a conventional full size solar cell module.

The response to partial shading is examined for both shingled concepts by applying multiple partial shading scenarios, both with a covered module rear side and under bifacial operation.

Furthermore, we present a new shingled module manufacturing concept named "SlimLine", which is based on structured EVA foils. With this concept a fast and simple placement and alignment of shingled cells is achieved.

2 MATERIALS & METHODS

2.1 Examined module layouts

Two shingled cell module concepts are examined, i.e. parallel string layout (see Figure 1, top) and matrix layout (see Figure 1, bottom). Both shingled modules feature bifacial shingled solar cells (pSPEER) in the size of 148x22 mm²; in the matrix module also "half" pSPEER cells in the size of 78x22 mm² are implemented.

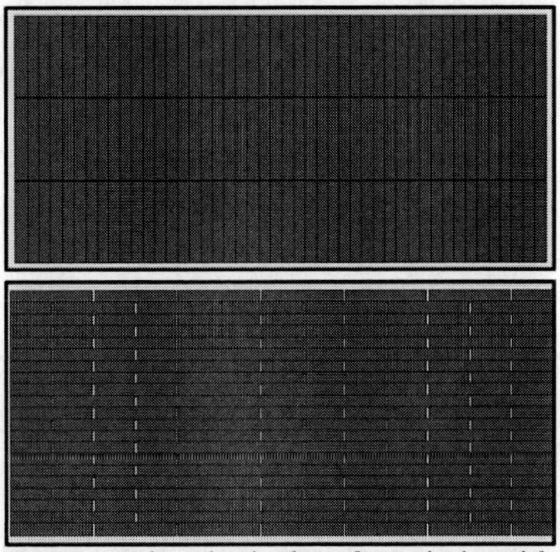

Figure 1: Schematic sketches of examined module layouts; top: parallel shingled strings layout, bottom: matrix layout

The first module (parallel string layout) comprises three strings of 46 cells each. The cells are connected in series within a string and the strings are interconnected in parallel. Since a much higher number of cells is incorporated into the module than for a standard module, substrings can be connected in parallel and only one bypass diode per module is required. The chosen orientation allows maximum module area utilisation and therefore a maximum number of incorporated cells.

The second module (matrix layout) comprises six and a half cells per row in parallel connection and 21 rows connected in series. Again the orientation was chosen for the reason of maximum module area occupancy.

The pSPEER cell's architecture corresponds to the cells presented by Baliozian *et al.* [3]. In the parallel strings setup each front side busbar of one cell is connected directly to a rear side busbar of the covering cell. In the matrix set up the shingle cells are additionally shifted along their long edge and connect two cells of the next row. Within a cell row this results in a parallel connection of all front side busbars via the rear side busbars of the covering row of cells and vice versa. Also one half shingled cell is included into each row to achieve a rectangular cell area in shape of the module glass.

2.3 Fabrication of modules

pSPEER cells are produced at Fraunhofer ISE PV-TEC. The process is presented by Baliozian *et al.* in reference [3].

Electrically Conductive Adhesive (ECA) is printed selectively onto the busbars of the solar cells by a semi-automatic stencil printing tool. The cells are positioned by a six-axis robot at Fraunhofer ISE Module-TEC. ECA curing is performed during lamination.

In this contribution we are presenting "SlimLine" as a new manufacturing technology for simple and fast shingle module fabrication on large scale. The concept is based on the use of structured EVA foils instead of planar foils. The structure provides terraces as a guide for manual or automated positioning of cells with easy cell alignment and a well-defined shingle overlap in terms of length and height. A photo of a string, positioned on a SlimLine foil is presented in Figure 2.

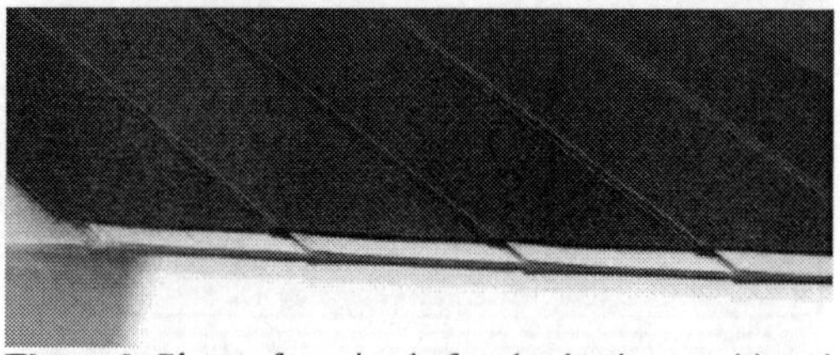

Figure 2: Photo of a string before lamination, positioned on a pre-shaped "SlimLine" encapsulant foil

ECA curing is performed during module lamination. A schematic sketch of the layup cross section is shown in Figure 3.

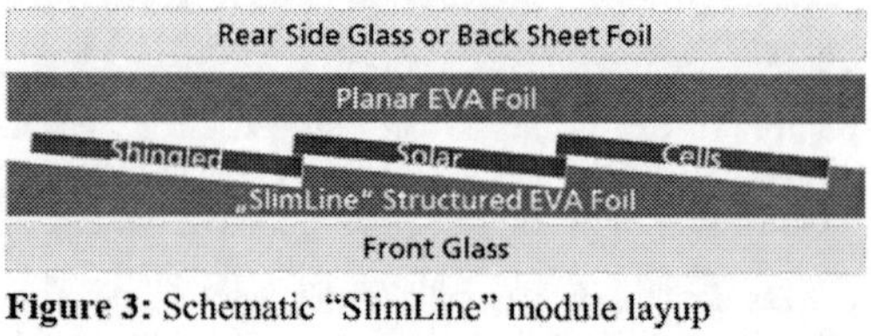

Figure 3: Schematic "SlimLine" module layup

By the use of structured EVA foils, the steps of transport of strings to the module glass, soldering, positioning of strings on the glass and interconnecting are performed in a single step.

2.2 Module efficiency analysis

We analyse the cell-to-module (CTM) ratio and the losses from module integration of each concept based on the fabricated samples of each module. We use the methodology presented by Hädrich *et al.* [4] which was extended for shingled modules [5, 6]. We perform the CTM analysis for standard testing conditions (STC) as well as for bifacial operation using the Ge-method [7] for a bifaciality of 65% and with a rear side irradiance of 10% and 20%. The used materials are characterized at Fraunhofer ISE and the resulting data is used as input for the analysis. We perform the CTM analysis regarding efficiency and module power of each concept.

The electric cell data which is used for calculations is listed in Table I, which are averaged from multiple cell measurements of the used batch of cells for the fabricated modules.

Table I: Cell data input for simulation of module outputs

Type	Area [cm²]	I_{SC} [A]	V_{OC} [V]	P_{MPP} [W]	FF [%]	η [%]
pSPEER cell	32.56	1.22	0.65	0.62	77.8	19.0
½ pSPEER	16.28	0.61	0.65	0.31	77.8	19.0

2.4 Power measurement during partial shading

Both fabricated sample modules are exposed to multiple partial shading scenarios during constant illumination in a climate chamber with integrated class B solar simulator at irradiance of approximately 1000 W/m². An illumination of the rear side of approximately 275 W/m² is achieved by means of a reflective cloth at the walls of the chamber. Shading is carried out by fixing a sheet of black cardboard in the desired shape and position directly on the front side of the module. All scenarios are considered for both, a covered and an open module rear side. The examined scenarios are graphically shown in chapter 3.4.

3 RESULTS & DISCUSSION

3.1 Measurement results on sample modules and CTM analysis

Figure 4 shows electroluminescence (EL) images of the sample modules. These show some variations in grey scale, which mainly indicates cell mismatch. The measured module outputs are listed in Table II, including results for bifacial measurements for 10% and 20% rear side irradiance, following the GE-method.

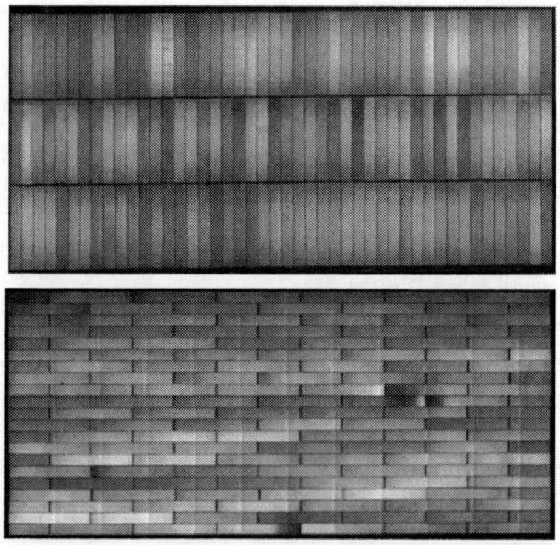

Figure 4: Electroluminescence images of the fabricated modules; top: parallel strings layout, bottom: matrix layout

Table II: Key parameters of IV measurement results of the fabricated modules at STC and with 10% and 20% additional rear side (RS) illumination

Type	I_{SC} [A]	V_{OC} [V]	P_{MPP} [W]	FF [%]
Parallel Strings	3.39	30.0	78.8	77.5
+10% RS	3.62	30.0	84.0	77.3
+20% RS	3.86	30.1	89.6	77.1
Matrix	7.30	13.6	76.2	76.9
+10% RS	7.76	13.6	81.2	76.7
+20% RS	8.21	13.7	86.2	76.6

There are 138 SPEER cells integrated into the parallel string module and only 136.5 cells in the matrix layout module. This explains the lower power output for the matrix module. Additionally, the current of the parallel string layout is lower due to a series connection of 46 cells, while in the matrix layout there are only 21 cell areas connected in series. The lower current corresponds with less resistive losses in the module, which can be observed by the increased fill factor of the parallel string layout in comparison with the matrix module.

The bifaciality calculates to 69.2% for the parallel strings module and 65.3% for the matrix module.

The calculated CTM results for power output under STC and bifacial operation are listed in Table III.

Table III: Calculacted module outputs and CTM factors for the examined module layouts, with CTM_P for power and CTM_η for efficiency and RS for rear side irradiance

Type	P_{Cells} [W]	P_{MPP} [W]	η [%]	CTM_P [%]	CTM_η [%]
Parallel-Strings	85.7	81.9	17.3	95.6	91.5
+10% RS	91.6	87.6		95.6	
+20% RS	97.5	93.4		95.6	
Matrix	84.8	80.2	17.4	94.6	91.1
+10% RS	91.2	85.4		94.6	
+20% RS	97.7	90.6		94.5	

A discrepancy between CTM calculations and the measured modules was expected due to cell mismatch, which is not included in the simulation. The relative deviance lies between 4% for the parallel layout and 5% for the matrix layout.

3.2 Full size module calculations

We have extended the CTM module calculations from the previous section for a module glass size of 1000 x 1670 mm². In this size a number of 468 cells in form of 6 strings of 78 cells may be integrated into a parallel string module. For the matrix module a number of 507 cells fit into this area in form of 6.5 cells per row and 78 rows. These calculations are compared to a conventional module with 60 solar cells. The supposed dimensions are listed in Table IV.

Table IV: Supposed module dimensions for CTM calculations on full size modules

Module type	Top & bottom margin [mm]	Side margin [mm]	String / cell spacing [mm]	Overlap [mm]
Conventional	30.7	18.5	5.4	n.a.
Parallel strings	14.75	14	15	1.25
Matrix	14.75	14	1	1.25

The CTM calculation results are listed in Table V. Here only the resulting maximum power point P_{MPP} and the power density ρ_P are given.

Table V: Calculated module outputs for full size modules

Type	P_{MPP} [W]	ρ_P [W/m²]	CTM_{Power} [%]	$CTM\eta$ [%]
Convent.	270.7	162.1	97.0	83.8
+10% RS	289.0	173.1	97.2	
+20% RS	307.3	184.0	97.0	
Parallel Strings	285.2	170.8	98.1	91.3
+10% RS	305.0	182.6	98.1	
+20% RS	325.2	194.7	98.2	
Matrix	305.3	182.8	97.4	97.8
+10% RS	325.2	194.7	96.9	
+20% RS	345.1	206.6	97.0	

The power output increases in the following order: Conventional, parallel strings and matrix layout. The main cause for this increase is an increased total incorporated cell area, which is accompanied by an increased total cell power. The magnitude of CTM_{Power} ratio is similar to the conventional module, yet the $CTM_{efficiency}$ ratio is outstandingly high. Firstly, this is a result from employing more active cell area and resulting in reduced cell and string spacing. Secondly, the active cell area is further increased by covering the busbar area by active cell area. Thirdly, the reduced current from reduced cell sizes and resistance decrease from parallel connections both lead to reduced resistive losses.

3.3 Power measurement during partial shading

The investigated partial shading scenarios are shown in Figure 5. Here an electroluminescence image of the examined modules is overlain by rectangles in the shape of the desired shading areas. The shading scenarios are numbered and colour coded according to the notes below Figure 5. Scenarios 1, 2 and 3 are single shaded cells and 4, 5, 9 and 10 are two shaded cells. These were picked for reasons of fundamental comprehension. Scenarios 7 and 8 depict module coverage by leaves, dirt or similar. Scenarios 6, 11, 12, 13 and 14 were chosen to resemble systematic covering by buildings, trees and neighbouring modules. Scenario 15 resembles posts and branches.

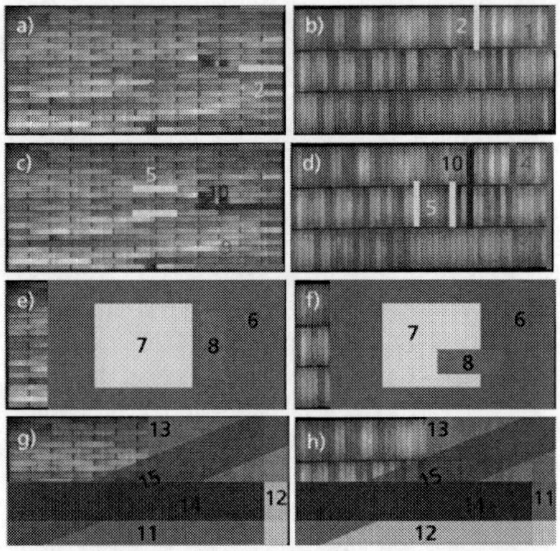

Figure 5: Partial shading scenarios; colour coding: a) & b) orange: scenario 1, yellow: scenario 2, blue: scenario 3; c) & d) orange: scenario 4, yellow: scenario 5, blue: scenario 9, red: scenario 10; e) & f) orange: scenario 6, yellow: scenario 7, blue: scenario 8; g) & h) orange: scenario 11, yellow: scenario 12, blue: scenario 13, red: scenario 14, purple: scenario 15

The resulting maximum power under partial shading is normalised to the respective initial unshaded maximum power for each scenario and plotted in Figure 6. By using normalised values, a direct comparison of power change is simplified.

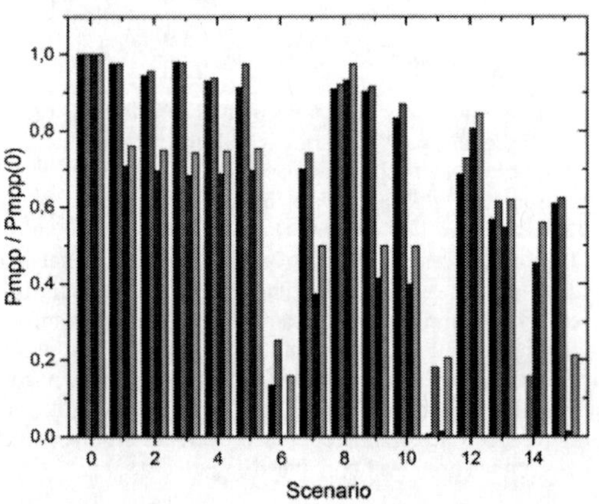

Figure 6: Maximum power at partial shading scenarios from Figure 5, normalised to unshaded case; black for the matrix module under STC, grey for the matrix module under bifacial conditions, dark blue for the parallel strings module under STC and light blue for the parallel strings module under bifacial conditions.

For most partial shading scenarios the matrix layout shows a higher power output than the parallel strings layout. Shading of a single cell leads to 33% reduction for the parallel strings layout and to a power reduction of only 2.5% for the matrix layout (scenarios 1-5). Shading of 25% of the module area leads to 63% reduction for the parallel strings layout and to a power reduction of only 30% for the matrix layout (scenario 7). Some cases were examined, in which an interruption of all series connections was planned for one module and not for the other (scenarios 6, 11, 14 and 15). To avoid performance losses, these scenarios require particular consideration when positioning these module types on the site.

With additional rear side illumination the power reduction by front side partial shading is lower in all cases within measurement uncertainty. This is mainly caused by a reduced ratio of shaded to non-shaded area and by increasing the photocurrent of the shaded cell. This applies even for scenarios which result in series connection interruption for single sided illumination cases.

4 SUMMARY & CONCLUSION

In this work we have compared two different shingled module concepts in terms of potential output power and shading behaviour and presented a new manufacturing method for shingled solar cell module fabrication (SlimLine).

The IV measurements of the module samples has demonstrated a higher output power for the parallel string layout, since more active cell area can be used for this specific configuration. The CTM calculations of both module concepts meet the measured outputs with an accuracy of 4% - 5% for both, STC and bifacial operation even though a high cell mismatch was observed.

In the case of full-size modules the occupancy by active cell area is maximized by the matrix layout. Compared to the conventional module, both examined shingled layouts show an increased power output as a result from increased active area and decreased resistive losses.

During investigation of partial shading response, we observe that the parallel strings layout performs similar to an expected shading response of a conventional layout. Less module power reduction by partial shading is observed for the matrix layout in most examined scenarios, due to the matrix of parallel and series interconnection.

In conclusion, the choice of shingled layout depends on the one hand on the number of cells of a given size which are to be incorporated into a module. On the other hand, potential shading on a site may also influence the choice of layout. Depending on the expected shading, both module layouts show increased performance under partial shading with more advantages for the matrix layout.

While a gain of 12.5 % from a full size conventional module to a matrix module was calculated, the manufacturing and cell costs of shingled modules may be increased due to usage of more wafers per module. Innovative manufacturing methods such as SlimLine may decrease total module fabrication costs by allowing very narrow cell overlap. Shingled module concepts offer main advantages in cases where highest power density and good partial shading response is beneficial, where the latter is satisfied best by the matrix layout.

ACKNOWLEGEMENTS

The authors want to thank Daniel von Kutzleben for fabrication of SlimLine foils, Michel Erlemann and Ädem Minat for ECA printing, Christoph Kutter, Gábor Horváth and Celine Hilger for robot operation, Martin Wiese and Felix Basler for lamination processes,

Alexandra Schmidt for calibrated IV measurements, Jibran Shahid for adaption of SmartClac.CTM software and Hebin Manuel and Daniel Weißer for conducting shading experiments. Also we want to thank the German Federal Ministry for Economic Affairs and Energy (BMWi) for funding the work in scope of the project PV_BAT400 under the funding code 0324145.

References

[1] D. C. [J.] Dickson, "Photovoltaic semiconductor apparartus or the like," US 2938938 A, May 31, 1960.

[2] W. Schmidt and K.-D. Rasch, "New interconnection technology for enhanced module efficiency," *IEEE Trans. Electron Devices*, vol. 37, no. 2, pp. 355–357, 1990.

[3] P. Baliozian *et al.,* "Bifacial p-type silicon shingle solar cells - the pSPEER concept," *Sol. RRL*, vol. 1700171, 2018.

[4] I. Hädrich, "Unified methodology for determining CTM ratios: Systematic prediction of module power," in *Proceedings of the 4th International Conference on Crystalline Silicon Photovoltaics*, 's-Hertogenbosch, Netherlands, 2014.

[5] M. Mittag and M. Ebert, "Systematic PV module optimization with the cell-to-module (CTM) analysis software," *Photovoltaics International*, no. 36, pp. 97–104, 2017.

[6] N. Woehrle *et al.,* "Solar cell demand for bifacial and singulated-cell module architectures," *36th Photovoltaics International*, pp. 48–62, 2017.

[7] A. Schmid, G. Duelger, G. Baarah, and U. Kraeling, "IV Measurement of Bifacial Modules: Bifacial vs. Monofacial Illumination," in *Proceedings of the 33rd European Photovoltaic Solar Energy Conference and Exhibition*, Amsterdam, 2017, pp. 1624–1627.

IMPACT OF USING SPECTRALLY RESOLVED GROUND ALBEDO DATA FOR PERFORMANCE SIMULATIONS OF BIFACIAL MODULES

M. R. Vogt,[1,*] T. Gewohn,[1] K. Bothe,[1] C. Schinke,[2] R. Brendel[1,2]
[1]Institute for Solar Energy Research Hamelin (ISFH), Am Ohrberg 1, 31860 Emmerthal, Germany
[2]Dep. Solar Energy, Leibniz University of Hannover, Appelstr. 2, 30167 Hannover, Germany
*Corresponding author: m.vogt@isfh.de, Tel: +49-511-762-5505

ABSTRACT: We simulate the average annual current gain $f_{bif,spe}$ of bifacial PV modules versus monofacial PV modules for various ground albedos and mounting conditions. Our in-house developed ray tracing framework DAIDALOS is used to compute various spectrally resolved rear and front side photon flux. We demonstrate that the annual current gain $f_{bif,spe}$ and as a consequence the energy yield of a bifacial module may vary by up to 35%$_{abs}$ when considering different ground albedos and by up to 10%$_{abs}$ when considering different mounting heights. Considering the difference between using spectrally resolved albedo data and averaged albedo data with respect to wavelength, we find relative deviations in the bifacial gain of up to 19.5%$_{rel}$. We find that the deviation is reduced to 2.6%$_{rel}$ by weighting the albedo with both the external quantum efficiency of the photovoltaic module and the spectrum.
Keywords: Albedo, Bifacial PV Modules, Ray Tracing.

1 INTRODUCTION

The market share of bifacial PV modules is expected to increase from about 2% in 2016 to about 30% within the next ten years [1]. This prediction is driven by the ability of bifacial PV modules to convert both light incoming from the front and the rear side of the module, which enables higher yearly energy yields for bifacial modules than for monofacial modules [2][3][4][5][6].

The bidirectional hemispherical ground reflection [7][8] (albedo) has major impact on the rear illumination of a bifacial module [9][10][11][12]. Thus, the albedo determines the energy yield of a bifacial module offers compared to a monofacial module with the same front side efficiency in a given environment. The albedo is a wavelength dependent material property [8]. Nevertheless, the vast majority of modelling tools and studies calculating the outdoor performance of bifacial modules is unable to consider spectrally resolved irradiance and spectrally resolved albedos[5][11][13][14]. Instead, the albedo is considered wavelength independent by integrating it over all wavelengths and weighting it by the incoming spectrum. Guerrero-Lemus et al. [4] counted that there are nearly 400 published articles on bifacial PV technology. However, we could only find very few studies [10][15][16] which consider the spectral dependencies of albedo and its impact on bifacial PV.

Andrews and Pearce [10] found a difference of up to 10.5%$_{rel}$ in PV system energy yield calculations between weighting the albedo with the incoming spectrum only and weighting the albedo with the spectrum as well as the external quantum efficiency of the module. In a later study [15], they found that this effect is dependent on the PV technology and the specific albedo. However, in both studies they did not perform the yield calculation using spectrally resolved input data. Russel, Saive et al. [16] calculated IV curves considering spectrally resolved albedo and integrated albedo. They found a difference of 7.5%$_{rel}$. under the AM1.5g spectrum. These studies show the potential impact of neglecting the wavelength dependence of albedo data.

This work uses forward ray tracing with our in-house developed ray tracing framework DAIDALOS [17][18] in order to analyze the current gains of bifacial modules in optical (background) scenarios. Our approach allows for the use of spectrally resolved input data as well as to model all geometries in three dimensions. Hereby, we are able to determine the systematic error when neglecting spectral effects e. g. by the use of different weighting methods. For our simulations, we consider realistic mounting conditions and use mean yearly irradiance data from outdoor measurements.

2 APPROACH

For better understanding, we use the following designations: Averaged albedo values α are given as regular numbers, bifacial gains f_{bif} as %$_{abs}$ and relative deviations between bifacial gains Δf calculated with different methods as %$_{rel}$.

2.1 Spectrally resolved albedo data

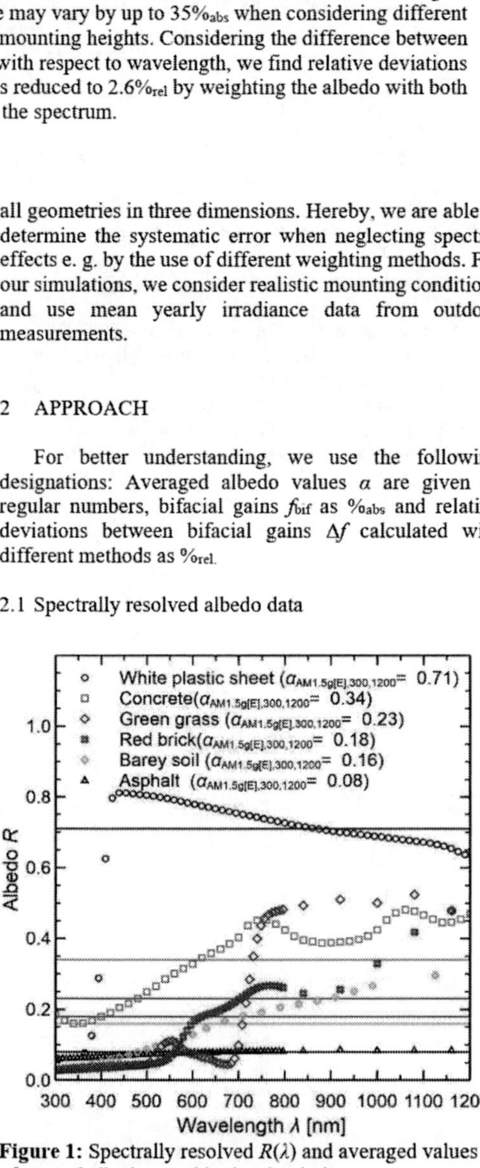

Figure 1: Spectrally resolved $R(\lambda)$ and averaged values α of ground albedos used in the simulation.

Figure 1 shows the albedo data [7][8] of six different ground materials taken from Ref. [19], but originally measured by [20], and the averaged albedo values $\alpha_{AM1.5g[E],300,1200}$ weighted by the spectral irradiance of the Am1.5g spectrum from 300 nm to 1200 nm (see sec. 4.1).

2.2 Simulation Domain

Our simulation domain is a bifacial PV module (length: 1.5 m; width: 1 m; thickness: 5 mm) hovering above the ground, as depicted in Fig. 2. The module mounting height is varied from 0.1 m to 1 m. The module is tilted 35° towards south. The simulation is performed in three dimensions. We detect the amount of incoming light rays on each of the six module surfaces with spectral resolution.

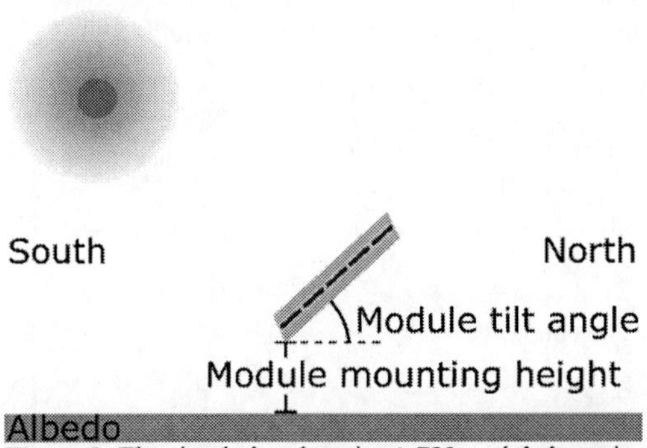

Figure 2: The simulation domain: A PV module hovering above the ground.

2.3 Mean annual light source (MALS)

In order to allow for a realistic representation of daylight within a simulation, its spectral and angular distribution is required. We utilize an in-house developed mean annual daylight model [21] [18] which uses irradiances measured over 14 years (1992 – 2006) in Hamelin, Germany. The measurements were performed using a pair of pyrometers to obtain values of horizontal global and diffuse irradiance with a temporal resolution of 5 minutes. Our model calculates the spectral irradiance from these non-spectrally resolved measurements. The model includes the impact of the sun's position and clouds on the angular and spectral distribution. Scattering of light is treated wavelength independent. A mean spectrally resolved annual daylight distribution results for the full celestial hemisphere, which we partition into solid-angle intervals of 5° azimuth and 5° altitude (Fig. 3). Each of these intervals contains its own spectral distribution for diffuse and direct light.

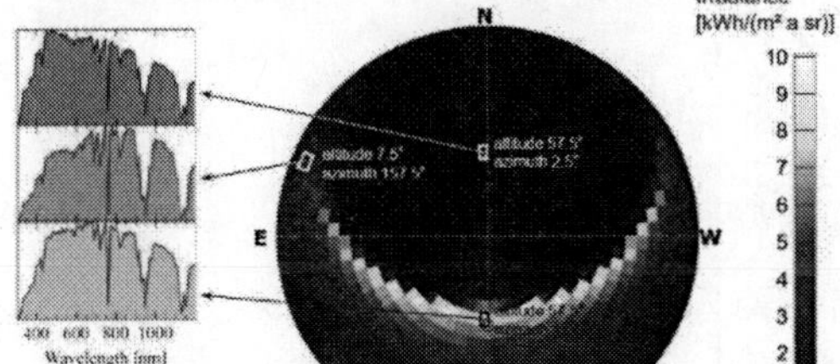

Figure 3: The mean annual daylight distribution partitions the celestial hemisphere into solid-angle intervals of 5° azimuth and 5° altitude. Each of these intervals contains its own spectral distribution of diffuse and direct light.

2.4 Experimental validation of module EQE simulation procedure

We use separate ray tracing simulations to calculate external quantum efficiencies (EQEs) of c-Si PV modules. Our 3D ray tracing [22][23] models the geometries and the spectrally resolved optical properties of the glass cover, the encapsulation, the cell metallization and cell interconnects, the surface texture of the cells, dielectric layers, the Si wafer and the local cell contacts. The resulting simulated EQEs are in good agreement with measured EQEs as we show in Fig. 4.

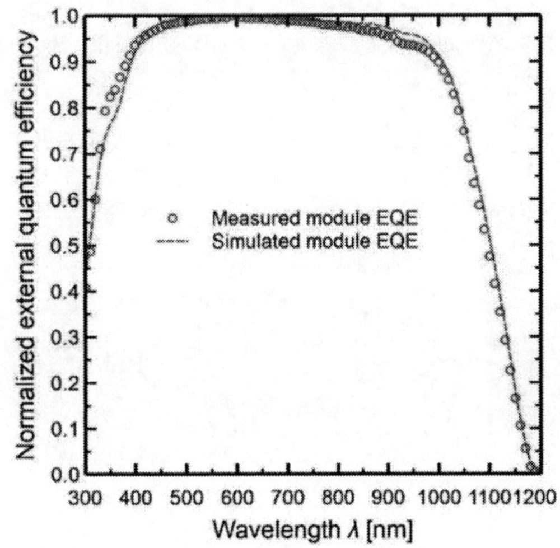

Figure 4: Comparison of measured and simulated external quantum efficiencies.

2.5 Averaged annual current and bifacial gains

We calculate the average annual current of a module

$$I_x(y) = \int_{300nm}^{1200nm} \Phi_{x,y}(\lambda) \times EQE_x(\lambda) \times d\lambda, \qquad (1)$$

where x is either the front or rear side of the module and y can be used to denote the type of albedo (e.g. green grass or red brick) used in the simulation. The photon flux $\Phi_{x,y}(\lambda)$ represents the amount of photons hitting side x of the module in the ray tracing simulation where albedo y is assumed for the ground, $EQE(\lambda)$ indicates the external quantum efficiency of the module and λ denotes the wavelength. Since this study is focused on the impact of albedo we assume the same EQE for the rear and front side of the module, which results in a bifaciallity of the module equal to 1. This is technologically possible, as modules with bifacialities above 0.99 [24] have been fabricated.

This average annual current can serve as a current rating for PV modules. A current rating can be used to evaluate optical performance of PV modules under certain reference irradiance and albedo conditions without the additional work of calculating energy yield. While energy yield calculations simplify input parameters such as non-spectrally resolved irradiances or module degradation in order to preserve a high time resolution, our approach averages over time in order to preserve spectrally resolved irradiance model. Moreover, such a current rating could be proportional to an energy rating for modules with similar thermal behavior and with a linear dependence of the electrical power output on the short circuit current.

We calculate the bifacial gain $f_{bif,spe}(y)$ in terms of average annual current according to

$$f_{bif,spe}(y) = \frac{I_{rear}(y)}{I_{front}(y)}, \quad (2)$$

with the current of each side $I_x(y)$ calculated according to (1). This allows us to focus on the bifacial aspect of the module and evaluate the rear side relative to the front side or a bifacial module relative to a monofacial module. Please note, that the bifacial gain $f_{bif,spe}(y)$ considered over an typical irradiation year in Hamelin and that the assumption of a bifaciality equal to 1 leads to higher $f_{bif,spe}(y)$ values than modules with lower bifacialites would achieve.

3 RESULTS

3.1 Impact of module mounting height and albedos

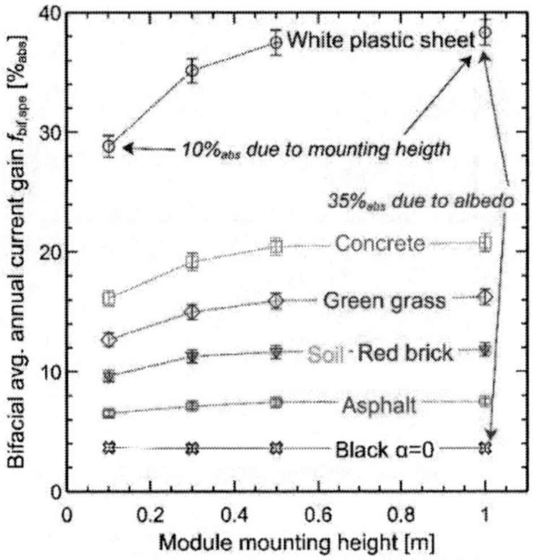

Figure 5: Average annual current gain of bifacial modules compared to monofacial modules.

Figure 5 shows the average annual current gain $f_{bif,spe}$ according to our simulations. The results for a ground albedo of zero (Fig. 5, magenta) show, that a bifacial gain of 3.6%$_{abs}$ is achieved due to rear side irradiance originating from the north half of the hemisphere, as shown in Fig. 3. This light hits the module without being reflected at the ground. Reasonably, this 3.6%$_{abs}$ gain is thus independent from the module mounting height. The bifacial gains due to reflection from the ground are dependent on the ground albedo and on the module mounting height. For the bifacial module simulated above the white underground, a 10%$_{abs}$ increase in average annual current is achieved by mounting it 1 m above the ground instead of 0.1 m. Making the same change in mounting height above green grass or asphalt only leads to a 4%$_{abs}$ or 1%$_{abs}$ increase in average annual current, respectively. This demonstrates that high ground reflectivity increases the current gain caused by higher mounting heights. This is a consequence of the larger ground area "seen" by the module when increasing its mounting height.

3.2 Impact of different spectral albedos

Figure 6 shows the photon flux $\Phi_{front,y}(\lambda)$ hitting the front side of the module for four different albedos, including perfectly black albedo (α=0), green grass, red brick and the white albedo. We use the front side photon flux $\Phi_{front,y}(\lambda)$ from Fig. 6 and the EQE from Fig. 4 for calculating the annual averaged front side current $I_{front}(y)$ according to Eq. 1. For the front side, the largest difference between the different albedos in terms of annual averaged front side current $I_{front}(y)$ is 6%$_{abs}$. This difference is caused by photons hitting the ground and being reflected on to the module.

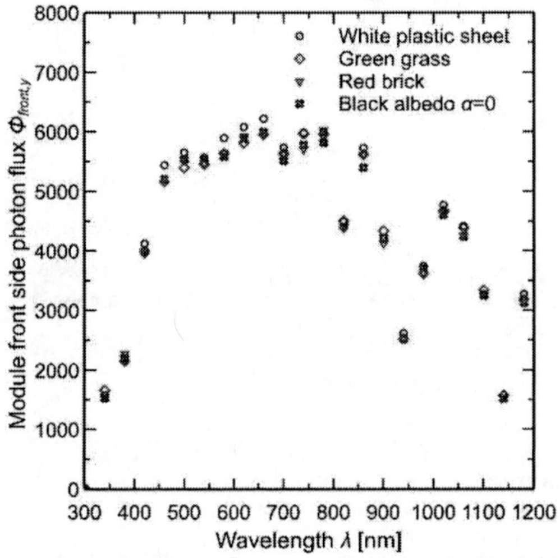

Figure 6: The photon flux $\Phi_{front,y}(\lambda)$ hitting the front side of the module for four different albedos out of 1E8 simulated photons.

Figure 7 shows the photon flux $\Phi_{rear,y}(\lambda)$ hitting the rear side of the module for the same four different albedos. However, in contrast to Fig. 6, the spectral dependencies of the albedos are easily identified, such as the peak for the green grass at 540 nm and the increase caused by the red albedo above 600 nm.

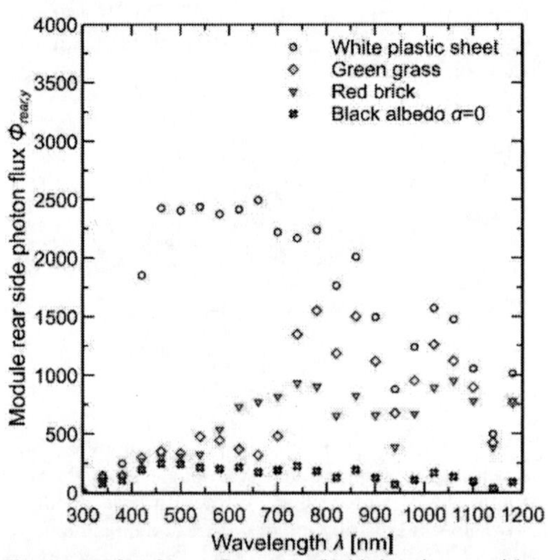

Figure 7: The photon flux $\Phi_{rear,y}(\lambda)$ hitting the rear side of the module for four different albedos. For the rear side the albedo has a larger impact than on the front side.

Table I: The weighted averages of albedos according to Eq. 3-5. Both the AM1.5g and the spectrum of the mean annual light source (MALS) as well as the EQE of a c-Si module are used for weighting.

Row	Albedo type (albedo identifying number in Ref. [19])	Asphalt (42)	Bare soil (8)	Red brick (41)	Green grass (12)	Concrete (18)	White plastic sheet (57)
1	$\alpha_{AM1.5g[E],300,4000}$	0.08	0.19	0.24	0.24	*	*
2	$\alpha_{AM1.5g[E],300,1200}$	0.08	0.16	0.18	0.23	0.34	0.71
3	$\alpha_{AM1.5g,EQE}$	0.08	0.17	0.20	0.27	0.36	0.73
4	$\alpha_{MALS[E],300,1200}$	0.08	0.15	0.17	0.22	0.33	0.70
5	$\alpha_{MALS[\Phi],300,1200}$	0.08	0.18	0.21	0.28	0.36	0.71
6	$\alpha_{MALS,EQE}$	0.08	0.17	0.19	0.26	0.35	0.72

*Albedo data ends at wavelengths shorter than 4000nm

4 IMPACT OF SPECTRALLY RESOLVED AND AVERAGED ALBEDOS ON BIFACIAL GAIN

4.1 Different approaches for averaging albedo data

In order to evaluate the deviations of the bifacial gain factor when using averaged albedo data instead of taking their spectral properties into account, we have to calculate averaged albedo data first. A reasonable option for this is calculating the average of the albedo data weighted by the spectrum [10][15][16]

$$\alpha_{x[E],A,B}(y) = \frac{\int_A^B R_y(\lambda) \times E_x(\lambda) \times d\lambda}{\int_A^B E_x(\lambda) \times d\lambda}, \quad (3)$$

where $R_y(\lambda)$ is the albedo of the ground material y, $E_x(\lambda)$ is the spectral irradiance and A and B define the integration interval. The most commonly used integration interval is 300 nm to 4000 nm [10][15], since this wavelength range contains more than 99 % of the overall irradiance defined by the AM1.5g reference spectrum. However, this neglects the external quantum efficiency $EQE(\lambda)$ of the module.

Thus Andrews and Pearce [10] proposed a different approach for calculating a weighted average of albedo data: They called it spectrally responsive albedo and calculated it according to

$$\alpha_{x,EQE}(y) = \frac{\int R_y(\lambda) \times \Phi_x(\lambda) \times EQE(\lambda) \times d\lambda}{\int \Phi_x(\lambda) \times EQE(\lambda) \times d\lambda}, (4)$$

where $\Phi_x(\lambda)$ is the photon flux of the incoming spectrum x. The integration interval is determined by the condition that the $EQE(\lambda)$ is significantly larger than zero. The inherent disadvantage of using spectrally responsive albedo is that the resulting averaged values depend on a certain module technology and must be recalculated when considering other technologies [15].

Thus, we also test a mixture of both weighting methods according to

$$\alpha_{x[\Phi],A,B}(y) = \frac{\int_A^B R_y(\lambda) \times \Phi_x(\lambda) \times d\lambda}{\int_A^B \Phi_x(\lambda) \times d\lambda}. \quad (5)$$

Using the incident photon flux $\Phi_x(\lambda)$ instead of the spectral irradiance removes the potential thermalization losses from the weighting process. For silicon modules, we propose to select a wavelength range from 300 nm to 1200 nm.

Table I lists the weighted averages of the six albedos used in this study according to Eq. 3-5 for both the AM1.5g spectrum and the spectrum of the mean annual light source. The different averaged albedos of asphalt, concrete and the white plastic sheet show variations smaller than 0.03. In contrast, the different averaged albedos of soil, red brick and green grass show deviations of up to 0.07. Using the photon flux instead of the spectral irradiance for weighting puts less weight on the UV-VIS region, which increases the averaged albedos of soil, red brick and green grass.

Comparing $\alpha_{AM1.5g[E],300,1200}$ (row 2) and $\alpha_{MALS[E],300,1200}$ (row 4) as well as $\alpha_{AM1.5g,EQE}$ (row 3) and $\alpha_{MALS,EQE}$ (row 6) shows that differences between the averaged albedos due to the different spectrums used are below 0.01.

4.2 Methods for evaluating the impact of spectrally resolved and averaged albedo data on the bifacial gain

In order to investigate the impact of spectrally resolved ground albedo data, we compare two simulations for each ground material: One simulation using spectrally resolved albedo data as input and another simulation using albedo data that is averaged according to Eq. 3-5. The averaged albedo values used are listed in Tab. I. Due to the lack of spectral information in the results of the simulation we have to calculate the bifacial gain for the non-spectrally resolved simulations as

$$f_{bif,av}(\alpha) = \frac{\Phi_{rear}(\alpha)}{\Phi_{front}(\alpha)}, \quad (6)$$

where $\Phi_x(\alpha)$ are the number of simulated photons hitting side x of the module hovering above albedo $\alpha(y)$.

We calculate the differences between the spectrally resolved and non-spectrally resolved simulations according to

$$\Delta f(y,\alpha) = \frac{|f_{bif,spe}(y) - f_{bif,av}(\alpha)|}{f_{bif,spe}(y)}, \quad (7)$$

where $f_{bif,spe}(y)$ is the bifacial gain according to Eq. (2) from the simulation with spectral dependencies. Please note that for better comparison across different albedos the are given as relative differences ($\%_{rel}$).

4.3 Impact of averaged albedo data on calculated bifacial gain

Using the data from table I, we perform simulations for a module mounting height of 0.5 m.

Table II: The differences between the simulations with full spectral dependency and the relative changes for the weighted average albedos according to Eq. 3-5 for both the Am1.5g spectrum and the spectrum of the mean annual light source.

Row	Albedo type (albedo identifying number in Ref. [19])	Asphalt (42)	Bare soil (8)	Red brick (41)	Green grass (12)	Concrete (18)	White plastic sheet (57)	Average $\Delta f(y,\alpha)$
0	$f_{bif,spe}(y)[\%_{abs}]$	0.074	0.116	0.128	0.159	0.204	0.374	-
1	$\Delta f(y,\alpha_{AM1.5g[E],300,4000})[\%_{rel}]$	0.1	11.1	19.5	3.6	*	*	8.6
2	$\Delta f(y,\alpha_{AM1.5g[E],300,1200})[\%_{rel}]$	0.1	2.5	5.7	8.2	1.9	1.4	3.3
3	$\Delta f(y,\alpha_{AM1.5g,EQE})[\%_{rel}]$	0.1	2.6	2.8	3.9	2.1	1.0	2.1
4	$\Delta f(y,\alpha_{MALS[E],300,1200})[\%_{rel}]$	0.1	5.9	7.0	10.7	6.1	2.7	5.4
5	$\Delta f(y,\alpha_{MALS[\Phi],300,1200})[\%_{rel}]$	0.1	4.0	5.8	7.5	2.1	1.4	3.5
6	$\Delta f(y,\alpha_{MALS,EQE})[\%_{rel}]$	0.1	2.6	0.8	1.6	0.0	1.0	1.0

*Albedo data ends at wavelengths shorter than 4000nm

Table II lists the simulation results for the six different albedos. In row 0, the bifacial gain from the simulations considering spectrally resolved albedo data (see also Fig. 5) are listed. Since the AM1.5g spectrum is defined as in plane irradiance there is currently no hemispheric mean annual light source for it thus the spectrally resolved results are in terms of other MALS derived from weather data in Hamelin (see sec 2.3).

The other rows list the relative differences when switching to the averaged albedo data $\Delta f(y,\alpha)$. The closest match is the spectrally responsive albedo for the MALS spectrum with an average difference of $1\%_{rel}$ across all investigated albedos. It is followed by the spectrally responsive albedo for the AM1.5g spectrum with $2.1\%_{rel}$. Restricting the integrating interval to 300-1200 nm reduces the average relative difference from $8.6\%_{rel}$ to $3.3\%_{rel}$ if spectral irradiance is used for weighting. For the albedo data considered in this study, using the photon flux instead of the spectral irradiance lowers the average deviation for the MALS spectrum from $5.4\%_{rel}$ to $3.5\%_{rel}$.

However, our results suggest that the calculation accuracy for the yield of bifacial modules can be improved even when using averaged albedo data by limiting the integration interval in Eq. (3) to the wavelength range from 300 nm to 1200 nm. This can be further improved by weighting the albedo with the EQE of the module.

5 CONCLUSIONS

We demonstrate the necessity for the use of spectrally resolved data when calculating the performance of bifacial modules.

Our in-house developed ray tracing framework DAIDALOS is used to compute various spectrally resolved rear and front side photon fluxes considering all geometries in three dimensions as well as the directional irradiance distribution of a typical year in Hamelin. The spectral distribution of the photon flux incoming on the rear side module is influenced by the spectral properties of both the light source and the albedo. In contrast, the front side mainly shows the spectral distribution of the spectrum.

We simulate the average annual current gain $f_{bif,spe}$ of bifacial PV modules versus monofacial PV modules for various ground albedos and mounting conditions. For bifacial module above a albedo of zero, facing southward at a tilt angle of 35°, we demonstrate that the $f_{bif,spe}$ is equal to $3.6\%_{abs}$ due to irradiation directly onto the rear side of the module. When considering different ground albedos the average annual current gain $f_{bif,spe}$ can vary up to $35\%_{abs}$. When considering different mounting heights the average annual current gain $f_{bif,spe}$ can vary by up to $10\%_{abs}$.

Considering the difference between using spectrally resolved albedo data and averaged albedo data with respect to wavelength, we find relative deviations in the bifacial gain of up to $19.5\%_{rel}$. We find that the deviation is limited to $2.6\%_{rel}$ by weighting the albedo with both the external quantum efficiency of the photovoltaic module and the spectrum. Since weighting the albedo with the external quantum efficiency makes the result technology dependent, we show that limiting the integration range from 300 nm-4000 nm to 300 nm-1200 nm can reduce the deviations from $19.5\%_{rel}$ to $8.2\%_{rel}$.

ACKNOWLEDGMENT

The research work leading to this article was carried out within the EMRP 16ENG02 project „Advanced PV energy rating“. The EMRP is jointly founded by the EMRP participating countries within EURAMET and the European Union This work was partly funded by the state of Lower Saxony and the Federal Ministry for Economic Affairs and Energy (BMWi) under grant number 0324171C project "NextStep".

References

[1] "International Technology Roadmap for Photovoltaic (ITRPV)," no. Eighth Edition. 2017.

[2] A. Cuevas, A. Luque, J. Eguren, and J. del Alamo, "50 Per cent more output power from an albedo-collecting flat panel using bifacial solar cells," *Sol. Energy*, vol. 29, no. 5, pp. 419–420, 1982.

[3] L. Kreinin, N. Bordin, A. Karsenty, A. Drori, and N. Eisenberg, "Experimental analysis of the increases in energy generation of bifacial over mono-facial pv modules," in *26th European Photovoltaic Solar Energy Conference and Exhibition, Hambourg, Allemagne*, 2011, pp. 3140–3143.

[4] R. Guerrero-Lemus, R. Vega, T. Kim, A. Kimm, and L. E. Shephard, "Bifacial solar photovoltaics - A technology review," *Renew. Sustain. Energy Rev.*, vol. 60, pp. 1533–1549, 2016.

[5] G. J. M. Janssen, B. B. Van Aken, A. J. Carr, and A. A. Mewe, "Outdoor Performance of Bifacial Modules by Measurements and Modelling," *Energy Procedia*, vol. 77, pp. 364–373, 2015.

[6] T. Dullweber *et al.*, "Present status and future perspectives of bifacial PERC + solar cells and modules," *Jpn. J. Appl. Phys.*, vol. 57, no. 08RA01, 2018.

[7] F. Nicodemus, J. Richmond, and J. Hsia, "Geometrical considerations and nomenclature for reflectance," *Sci. Technol.*, vol. 60, no. October, pp. 1–52, 1977.

[8] G. Schaepman-Strub, M. E. Schaepman, T. H. Painter, S. Dangel, and J. V. Martonchik, "Reflectance quantities in optical remote sensing-definitions and case studies," *Remote Sens. Environ.*, vol. 103, no. 1, pp. 27–42, 2006.

[9] L. Kreinin, N. Bordin, A. Karsenty, A. Drori, D. Grobgeld, and N. Eisenberg, "PV module power gain due to bifacial design. Preliminary experimental and simulation data," in *2010 35th IEEE Photovoltaic Specialists Conference*, 2010, pp. 002171–002175.

[10] R. W. Andrews and J. M. Pearce, "The effect of spectral albedo on amorphous silicon and crystalline silicon solar photovoltaic device performance," *Sol. Energy*, vol. 91, pp. 233–241, 2013.

[11] U. A. Yusufoglu *et al.*, "Simulation of energy production by bifacial modules with revision of ground reflection," *Energy Procedia*, vol. 55, pp. 389–395, 2014.

[12] U. A. Yusufoglu, T. M. Pletzer, L. J. Koduvelikulathu, C. Comparotto, R. Kopecek, and H. Kurz, "Analysis of the annual performance of bifacial modules and optimization methods," *IEEE J. Photovoltaics*, vol. 5, no. 1, pp. 320–328, 2015.

[13] G. J. Ward, "The RADIANCE Lighting Simulation and Rendering System," in *Proceedings of the 21st Annual Conference on Computer Graphics and Interactive Techniques*, 1994.

[14] X. Sun, M. R. Khan, C. Deline, and M. A. Alam, "Optimization and performance of bifacial solar modules: A global perspective," *Appl. Energy*, vol. 212, pp. 1601–1610, 2018.

[15] M. P. Brennan, A. L. Abramase, R. W. Andrews, and J. M. Pearce, "Effects of spectral albedo on solar photovoltaic devices," *Sol. Energy Mater. Sol. Cells*, vol. 124, pp. 111–116, 2014.

[16] T. C. R. Russell, R. Saive, A. Augusto, S. G. Bowden, and H. A. Atwater, "The Influence of Spectral Albedo on Bifacial Solar Cells: A Theoretical and Experimental Study," *IEEE J. Photovoltaics*, vol. 7, no. 6, pp. 1611–1618, 2017.

[17] H. Holst *et al.*, "Application of a new ray tracing framework to the analysis of extended regions in Si solar cell modules," *Energy Procedia*, vol. 38, pp. 86–93, 2013.

[18] C. Schinke, M. R. Vogt, and K. Bothe, "Optical Modeling of Photovoltaic Modules with Ray Tracing Simulations," in *Photovoltaic Modeling Handbook*, 2018, pp. 27–92.

[19] C. Gueymard, "SMARTS2, a simple model of the atmospheric radiative transfer of sunshine: algorithms and performance assessment," p. 84, 1995.

[20] D. E. Bowker, R. E. Davis, D. L. Myrik, K. Stacy, and W. T. Jones, "Spectral reflectances of natural targets for use in remote sensing studies," NASA Ref. Publ. 1139, 1985.

[21] M. Winter, H. Holst, M. R. Vogt, and P. P. Altermatt, "Impact of realistic illumination on optical losses in Si solar cell modules compared to standard testing conditions," in *31st EU PVSEC*, 2015, pp. 1869–1874.

[22] M. Winter, M. R. Vogt, H. Holst, and P. P. Altermatt, "Combining structures on different length scales in ray tracing: analysis of optical losses in solar cell modules," *Opt. Quantum Electron.*, vol. 47, no. 6, pp. 1373–1379, 2015.

[23] M. R. Vogt *et al.*, "PV module current gains due to structured backsheets," *Energy Procedia*, vol. 124, pp. 495–503, Sep. 2017.

[24] H. Schulte-huxel *et al.*, "Flip-Flop Cell Interconnection Enabled by an Extremely High Bifacial Factor of Screen-Printed Ion Implanted n-PERT Si Solar Cells," in *EU PVSEC 32*, 2016, pp. 407–412.

TOWARDS IEC TS 60904-1-2: ASSESSING THE REQUIREMENTS FOR IRRADIANCE ON THE NON-ILLUMINATED SIDE OF BIFACIAL PV MODULES WITH SINGLE LIGHT SOURCE TESTING

Tian Shen Liang, Mauro Pravettoni, Jai Prakash Singh, Yan Wang, Yong Sheng Khoo
Solar Energy Research Institute of Singapore (SERIS), National University of Singapore, Singapore, 117574
Tel.: (+65) 6567 8073, tianshen@nus.edu.sg

ABSTRACT: This paper proposes a mounting structure with low-reflective back-side panel to reduce the irradiance on the non-illuminated side of the bifacial PV module below 3 W/m², as required by draft standard IEC TS 60904-1-2. The paper also demonstrates a method to quantify the irradiance on the non-illuminated side and construct a 2-D irradiance map. The irradiance map shows the effectiveness of mounting structure with back-side panel of two low-reflective materials (one has low diffusive reflectance and another has low specular reflectance), as they fulfil the requirement of draft standard. On the other hand, the characterisation results revealed the relationship between the optical properties of the low-reflective materials (i.e. diffusive versus specular reflectance) and the distance of the test modules from these materials, and their impact on the short-circuit current (I_{SC}) of bifacial PV modules. It shows that the distance of the test modules from the low-reflective materials with high diffusive reflectance should be large enough to prevent the counter effect, in which the diffusive reflectance leads to an increase of I_{SC}.
Keywords: Bifacial, characterization, experimental methods, module, photovoltaic

1 INTRODUCTION

Current-voltage characterization of bifacial PV modules at Standard Test Conditions (STC) is challenging, requiring an extension of the definition of STC that includes the spectral and total irradiance on the rear side of the module and a regulation of spectral and total irradiance on both sides of the module [1]. The Technical Specification IEC TS 60904-1-2 [2,3], based on the existing STC measurements, has been proposed as a first step towards standardization of the indoor and outdoor characterization of bifacial PV cells and modules.

IEC TS 60904-1-2 distinguishes indoor characterization into single-side and double-side illuminations. Single-side illumination involves separate flashing of each side of the bifacial PV module using a single light source (while keeping the other side in the dark), then flashing the front side at an equivalent irradiance, G_{equiv} to simulate the additional contribution of the rear side irradiance on top of the 1000 W/m² front irradiance. The equivalent irradiance, G_{equiv} is given as:

$$G_{equiv} = 1000 + \varphi \cdot G_{rear} \quad (1)$$

where φ and G_{rear} are the bifaciality factor and rear-side irradiance, respectively. Bifaciality factor is determined by taking the minimum value of either the ratio of the rear and front short-circuit current or the ratio of the rear and front maximum power. At least three equivalent irradiance levels are required by the draft standard. As an example, for φ = 80%, the first, second, and third equivalent irradiances are 1080 W/m² (for $G_{rear,1}$ = 100 W/m²), 1160 W/m² (for $G_{rear,2}$ = 200 W/m²), and 1240 W/m² (for $G_{rear,3}$ = 300 W/m²), respectively.

To minimize the unintended current contribution from the non-illuminated side, the draft standard requires the irradiance on the non-illuminated side to be below 3 W/m² (representing 0.3% of 1000 W/m², the front-side total irradiance at STC).

Irradiance on the non-illuminated side is associated with the diffused irradiance from surroundings, reflections, optical properties of the module, the position of the module and geometrical disposition of the cells [4–6]. Therefore, the Technical Specification proposes control measures such as placing non-reflective material at a suitable distance from the non-illuminated side and fixing the module to an aperture to limit the exposure of illumination within the size of the PV module. Despite that, IEC TS 60904-1-2 does not specify the details on how to select the non-reflective materials, the required optical properties and the measures to attain the "suitable distance" from the bifacial PV module. To measure the irradiance on the non-illuminated side, the Technical Specification requires at least five symmetrical points on the non-illuminated side, for instance, P1-P3-P5-P7-P9 (as shown in Figure 1).

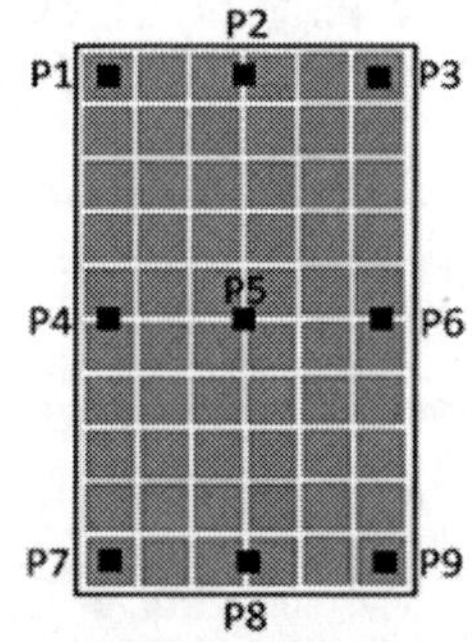

Figure 1 Positions for measuring the irradiance on the non-illuminated side (at least 5 symmetrical positions), as per IEC TS 60904-1-2.

This paper is in conjunction with the first international bifacial PV module proficiency testing campaign coordinated by SERIS (Solar Energy Research Institute of Singapore), which is expected to give more insight on the reproducibility of the procedure described in the Technical Specification and will end in late 2019: this work assesses the part of the Technical Specification concerning the indoor characterization with a single light source. The paper focuses on the study of the choices of low-reflective material behind the module based on the optical properties, proposes a mounting structure with low-reflective back-side panel, and method of quantifying the irradiance on the non-illuminated side of the bifacial PV module. Lastly, the paper investigates the impact of the back-side panel on the characterization results of bifacial PV modules.

2 METHODOLOGY

2.1 Designing a bifacial PV module characterization mounting structure and choosing the low-reflectance materials for back-side panel

A mounting structure aiming to reduce the irradiance on the non-illuminated side below 3 W/m^2 is illustrated in Figure 2. The structure was designed to be easily attached to the existing mounting rack in a Class AAA large-area solar simulator. It was constructed with light weight, black anodized aluminium extrusion (6-series/slot width 10/30x30 mm). The dimension of the mounting structure is 2.2 m wide and 1.2 m high, which is suitable for both 72- and 60-cells bifacial PV modules. The structure also consists of two adjustable aluminum extrusions for securing the test module of different sizes. Following the recommendations of draft standard, a black acrylic sheet with the size of 2.2 m x 1.2 m (5mm thickness) was used as the back-side panel. The reflectance of both a commercially available low reflectance coating (LRC) sample and of the black acrylic sheet itself were investigated to study the effectiveness of different low reflectance materials on achieving low irradiance on the non-illuminated side.

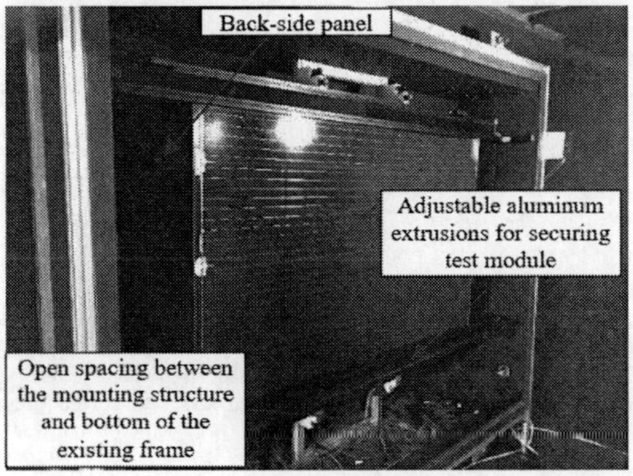

Figure 2: A mounting structure with black acrylic back-side panel and adjustable aluminum extrusions for securing bifacial PV test module of different sizes.

Measured using the Cary 7000 UV/VIS spectrophotometer, the spectral reflectance of: (a) low reflectance coating (LRC), (b) black acrylic sheet, (c) a black cloth, and (d) a white back-sheet are shown in Figure 3, where the wavelength band of spectral responsivity of c-Si is highlighted. Both low reflectance coating and black acrylic sheet have lower than 5 % of reflectance within the spectral responsivity range of c-Si. On the other hand, the spectral reflectance of black cloth is close to 10 %. As a comparison, white back-backsheet has more than 50 % of reflectance at the VIS and NIR, then gradually reduces to less than 40 % at IR and beyond. According to the manufacturer, the black material on the low reflectance coating has 99.9 % of specular absorptance. The manufacturer's claim was verified by measuring the specular and diffusive reflectance of the sample, with results shown in Figure 4 and Figure 5. In contrast, the specular reflectance of black acrylic sheet is much higher, in particular for the wavelength band between 750 nm and 1000 nm.

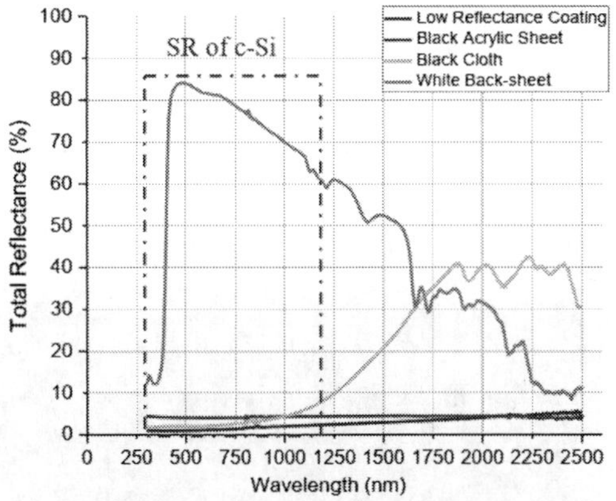

Figure 3: Spectral reflectance of low reflectance coating (LRC), black acrylic sheet, black cloth, and white back-sheet. The band of spectral responsivity of c-Si lies in between 300 nm to 1200 nm.

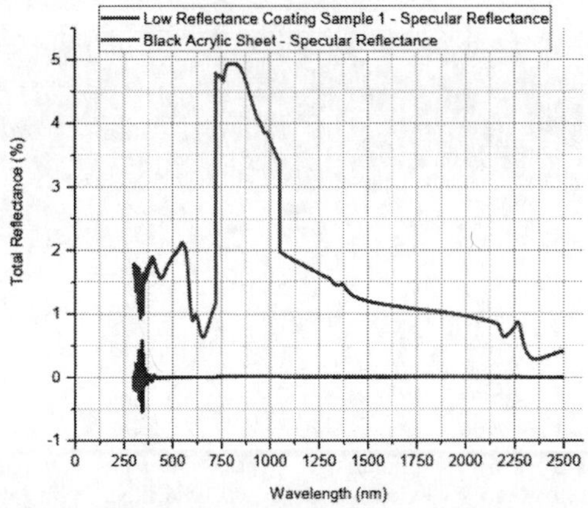

Figure 4 Specular reflectance of the acrylic sheet is higher, close to 5 % for wavelengths between 750 nm and 1000 nm.

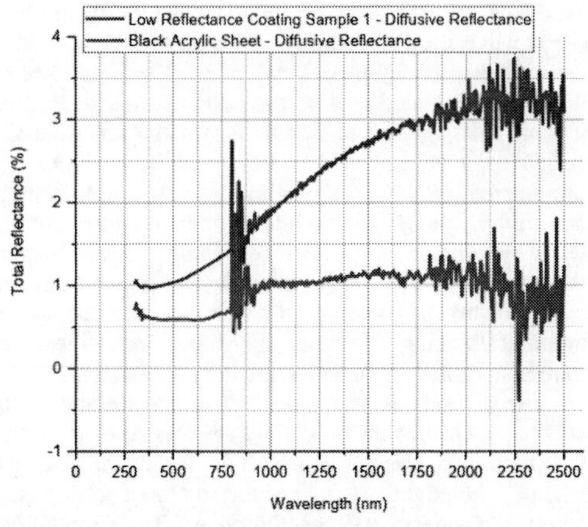

Figure 5 LRC has higher diffusive reflectance than acrylic sheet; the difference is even higher as the wavelength increases.

2.2 Measuring the irradiance intensity on the non-illuminated side of the bifacial PV test module

The experimental investigation was furthered by covering LRC on one side of the acrylic panel, as shown in Figure 7 a. The side of the acrylic panel coated with LRC (referred to as "Configuration 1") and another side of the black acrylic panel without LRC (referred to as "Configuration 2") were tested with the procedures described in the following paragraph. For the purpose of performance comparison, the acrylic panel was dismounted to measure the irradiance on the non-illuminated side under normal (existing) characterization test conditions (referred to as "Reference configuration").

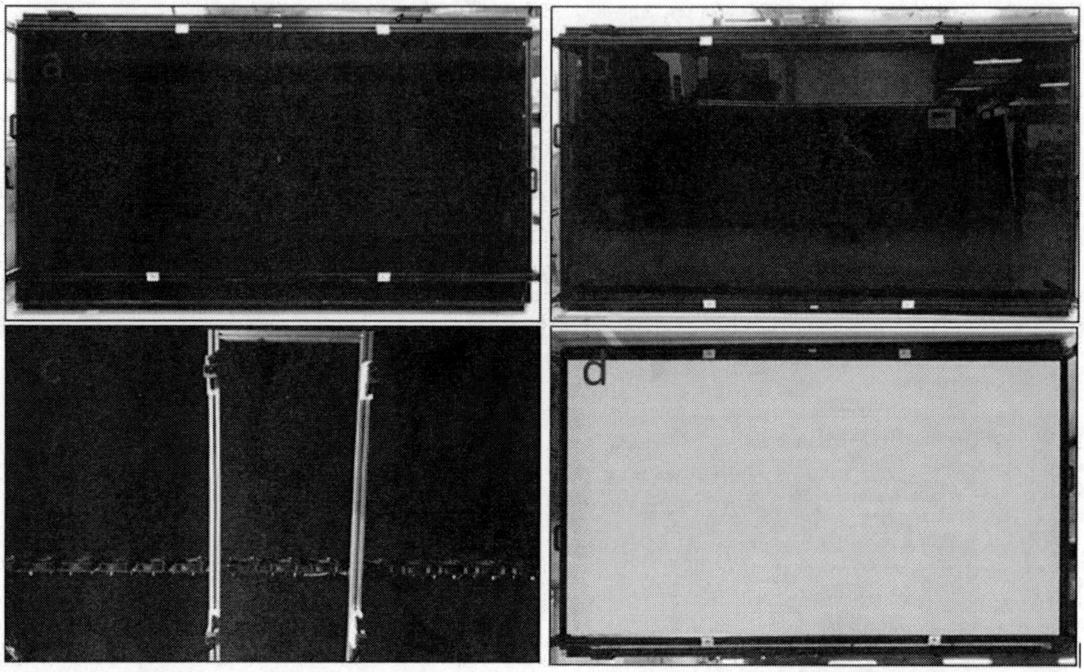

Figure 6 a) Configuration 1, back-side panel covered in low reflectance coating (with high diffusive and low specular reflectance); b) Configuration 2, acrylic sheet (with low diffusive and high specular reflectance) as the back-side panel; c) Reference configuration – without back-side panel; d) Back-side panel covered in white back-sheet (high diffusive reflectance, for test in Section 2.3).

The measurements were carried out at the SERIS PV Module Testing Laboratory darkroom (an ISO/IEC 17025 accredited testing laboratory), with a large Class AAA 100 ms pulsed solar simulators by h.a.l.m. The irradiance on the non-illuminated side of the 60-cells bifacial PV test module (glass/glass and frameless structure; module size is 160 cm x 96 cm) was measured (while the front was illuminated at 1000 W/m^2) using a 100 mm^2, high-detectivity silicon photodiode (Edmund Optics, model DUV Photodiode with Ceramic Housing). The distance of the test module was 10 cm further away from the back-side panel, instead of closing to the panel because of the required spacing between the panel and the non-illuminated side for maneuvering the photodiode.

The photodiode was placed at the centre and at the corners of each of the 60 cells. Care was taken to ensure the photodiode was in plane with the testing module. The silicon photodiode was connected to a photodiode amplifier and digital oscilloscope. The photodiode amplifier (an Edmund Optics current-to-voltage transimpedance amplifier; used scale range: 20 nA to 2 mA, corresponding to levels of irradiance from 0.64 W/m^2 to 64 W/m^2 respectively) converted the low irradiance signal measured by the photodiode to voltage signal, which was recorded and shown by the digital oscilloscope.

The photodiode was calibrated to check the linearity of the response at the low irradiances from 1 to 200 W/m^2 expected at the non-illuminated side of the testing module. To calibrate the photodiode, it was faced towards the light source of the solar simulator to measure the irradiance at 11 intensity levels, from 100 W/m^2 to 1100 W/m^2. The calibration curve, as shown in Figure 6 and the R-square value confirms the linearity of the photodiode below 400 W/m2; linearity was then extrapolated down to lower irradiances.

Referring to Equation (2), the irradiance (in W/m^2) was then converted from the output current (in mA) based on the linear slope of the calibration curve (in mA/W/m^2). The output current was calculated from the voltage output recorded by the digital oscilloscope and the gain (in V/A). A 2-D map for the irradiance incident on the non-illuminated side of the module for each configuration was then constructed using the measured irradiance and via bilinear interpolation.

$$\text{Irradiance}_{non-illuminated\ side}\ (W/m^2) = \frac{\text{Voltage(V)}}{Gain\ (V/A) \times 1000 \times Linear\ slope\ (\frac{\text{mA}}{W/m^2})} \quad (2)$$

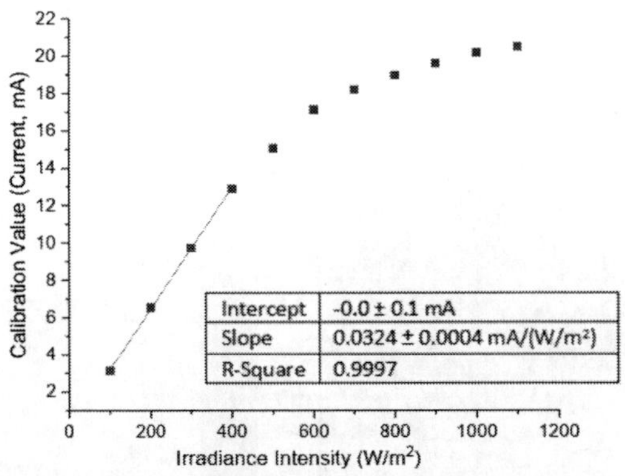

Figure 7 Calibration curve of the silicon photodiode for irradiance measurements under three configurations. The current (in mA) measured by the silicon photodiode is linearly proportional to the irradiance intensity produced by the solar simulator between 100 W/m^2 and 400 W/m^2, linearity was then extrapolated down to lower irradiances.

2.3 Investigating the impact on the short-circuit current, ISC and maximum power, PMPP

Subsequently, three bifacial PV modules from different manufacturers, with different size, number of cells and cell technologies were fixed on the mounting structure with the front side of the module facing the solar simulator light source and flashed at 1000 W/m^2. The details of these test modules are given in Table 1. The test modules were mounted close to the back-side panel, at a distance of 1 cm. The short-circuit current, I_{SC} and maximum power, P_{MPP} were recorded for analysis. The test was conducted for all configurations, with an additional configuration using white back-sheet as the back-side panel (as shown in Figure 7 d). Another set of characterisation tests was also performed by mounting the test modules at 10 cm further from the back-side panel for all configurations.

Table 1 Details of the bifacial PV test modules, including the brand, model, and cell technology

Bifacial PV Module	Model and Technology
Longi	LR6-72BP-365M (72 cells), P-type, with aluminum frame
Linyang	LYGF-QP60TPG-335 (120 cells), N-type half-cut, frameless
Jinergy	JHNM60-355 (60 cells), HJT, frameless

3 RESULTS AND DISCUSSIONS

3.1 Irradiance intensity on the non-illuminated side of bifacial PV test module

Figure 8, Figure 9, and Figure 10 are the irradiance map measured without the back-side panel (Reference configuration), with LRC coated back-side panel (Configuration 1) and with black acrylic back-side panel (Configuration 2), respectively. The grayscale indicates the irradiance measured by the silicon photodiode in W/m^2. Lines of different colour distinguish the regions with various irradiance intensity. For example, in Figure 9, the region encircled by red line has irradiance of 1 W/m^2 and below, while the region bounded between yellow and orange lines has irradiance between 1.5 W/m^2 and 3 W/m^2.

The irradiance on the non-illuminated side, measured under Reference configuration, ranges from 4 W/m^2 to a maximum of 11 W/m^2 at the borders. Reflection from the aluminium mounting rack are evident, as indicated by regions with higher irradiance between the horizontal positions of 50 cm and 70 cm as well as 100 cm and 120 cm. Irradiance at the bottom edge of the module is particularly higher than other edges, due to the reflection of the light entering via the large spacing between the module edge and the bottom of the existing frame (refer to Figure 1).

The irradiance on the non-illuminated side measured under both Configuration 1 and 2 are uniform (based on the requirement of draft standard), as most area measures irradiance at or lower than 3 W/m^2. However, the edges of the test module have higher irradiance (due to reflections from the side of the testing module), with a maximum of 5 W/m^2 for Configuration 1 and 6 W/m^2 for Configuration 2. The average irradiance for Configuration 1 and 2 are 1.54 W/m^2 and 1.50 W/m^2, respectively. Both are below the requirement of draft standard. In contrast, the average irradiance on the non-illuminated side measured under Reference configuration is 6.69 W/m^2, more than double of maximum value specified in the draft standard requirement.

The average irradiance on the non-illuminated side is reduced by 77 % and 78 % with low reflectance coating and black acrylic sheet, respectively. The results highlight the importance and effectiveness of back-side panel on reducing the average irradiance on the non-illuminated side during the single-light source testing.

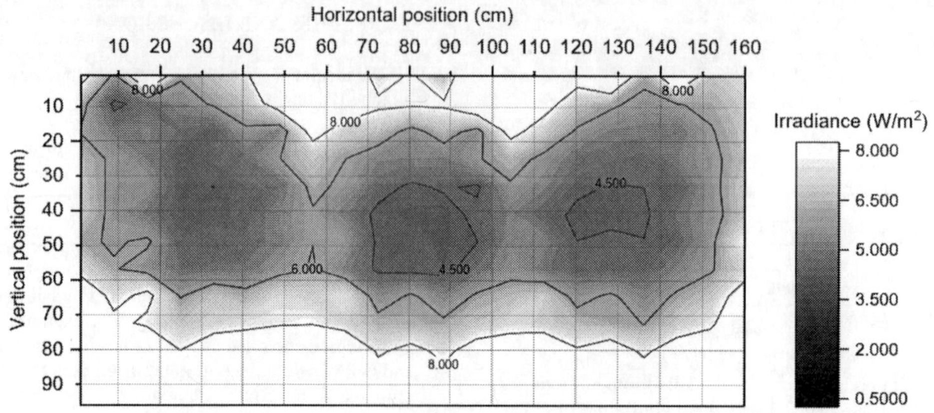

Figure 8 Reference configuration, regions between the horizontal positions of 50 cm and 70 cm as well as 100 cm and 120 cm has higher irradiance, due to the reflections from the aluminium mounting rack. The bottom of the test module is subjected to the reflection of the light entering through the spacing between the bottom edge of the test module and the existing frame. The calculated average irradiance on the rear side of the test module is 6.69 W/m^2.

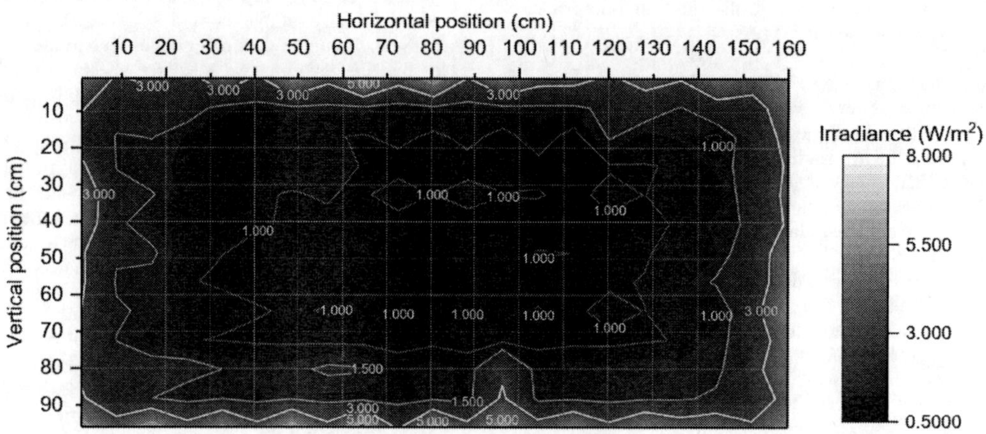

Figure 9 Irradiance map of the non-illuminated side measured under Configuration 1. A large area has irradiance lower than 3 W/m2, with higher irradiance observed at the edges of the test module. The calculated average irradiance is 1.54 W/m^2.

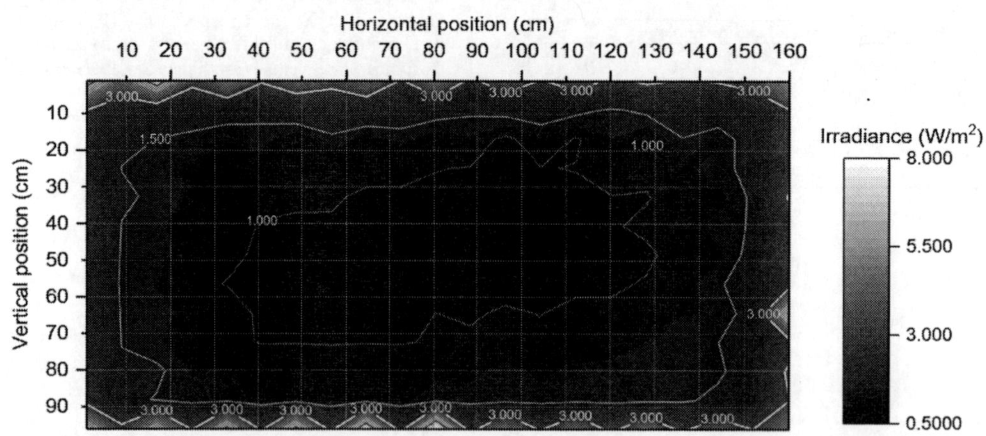

Figure 10 Irradiance map of the non-illuminated side measured under Configuration 2. The irradiance is uniform as almost all area has irradiance lower than 3 W/m^2. Similar to the Configuration 1, higher irradiance was observed at the edges of the test module. The calculated average irradiance is 1.50 W/m^2.

3.2 Impact on the short-circuit current, I_{SC} and maximum power, P_{MPP}

Table 2 summarises the I_{SC} and P_{MPP} of three modules measured under four configuration, at two distances between back-side panel and test modules (1 cm and 10 cm). When the back-side panel and modules are close to each other, it negates the capacity of the LRC to reduce the irradiance on the non-illuminated side. This is mainly due to the surface of the LRC which has a higher diffusive reflectance. Light enters through the gap between cells, incident on the surface of the LRC and reflected diffusively onto the non-illuminated side of the cells, resulting in unwanted current generation. The intensity of the diffusive reflectance is larger when the test module is closer to the surface of LRC, therefore contributing more reflected irradiance on the non-illuminated side.

In contrast, light incident on the acrylic sheet surface is specularly reflected at an angle normal to the acrylic sheet surface. Much of the reflected light escape through the cell gaps, instead of impinging on the cell area. Therefore, I_{SC} is reduced by 0.1 % under Configuration 2 for Longi and Linyang modules. This also explains the average irradiance on the non-illuminated side measured under Configuration 1 is higher than Configuration 2. An experimental method coupled with ray-tracing simulation have to be formulated in order to quantify the differences in light travel mechanism behind the module for Configuration 1 and 2.

Table 2 I_{SC} and P_{MPP} of three modules measured under four configuration at two distances between back-side panel and test modules (1 cm and 10 cm). The percentage in the bracket quantifies the difference between the I_{SC} of test modules measured under Configuration 1, 2, and White back-sheet and the I_{SC} measured under Reference configuration.

Test	Longi				Linyang				Jinergy			
	I_{SC}		P_{MPP}		I_{SC}		P_{MPP}		I_{SC}		P_{MPP}	
Reference configuration	9.73		368.8		9.70		294.5		9.12		316.4	
Distance between back-side panel and test modules	1 cm	10 cm	1 cm	10 cm	1 cm	10 cm	1 cm	10 cm	1 cm	10 cm	1 cm	10 cm
Configuration 1	9.75 (0.2%)	9.7 (-0.3%)	370.3	368.3	9.7 (0.0%)	9.64 (-0.6%)	295.4	293.6	9.16 (0.4%)	9.07 (-0.5%)	317.1	314.5
Configuration 2	9.71 (-0.2%)	9.71 (-0.2%)	368.0	368.6	9.69 (-0.1%)	9.66 (-0.4%)	295.5	294.3	9.12 (0.0%)	9.10 (-0.2%)	316.7	315.8
White back-sheet	9.97 (2.4%)	9.95 (2.2%)	378.5	377.3	10 (3.1%)	10.02 (3.4%)	306.9	307.4	9.36 (2.6%)	9.36 (2.6%)	324.8	326.6

Conversely, Configuration 1 and 2 produce lower I_{SC} for larger distance between the back-side panel and test modules. The intensity of the diffusive reflectance decreases when the test module is further away from the LRC. Therefore, Configuration 1 produces even lower I_{SC} than Configuration 2 when the test module is placed further from the back-side panel. The small magnitude of I_{SC} reduction for Configuration 2 when the distance increases from 1 cm to 10 cm is attributed by the existence of light beam passing through the cell gaps at various angles. As the distance between the back-side panel and test modules increases, the intensity of the reflected light beam from the surface with high specular reflectance diminishes at a lower rate than the reflected light from the surface with high diffusive reflectance. Overall, LRC with high diffusive reflectance, putting at a larger distance from the test modules produces the greatest I_{SC} reduction for all test modules. In addition, larger gap between back-side panel and test modules promotes air ventilation, thus reducing the thermalization time.

The use of white back-sheet behind the modules results in the increase of I_{SC} for all modules. The results are consistent regardless of the distance between the back-side panel and the test modules. This stresses once more the importance of covering highly reflective materials as the back-side material for single light source characterisation test, since it could contribute an uncertainty contribution of 2 % to 3 % to I_{SC} and P_{MPP}.

4 CONCLUSIONS

This paper focused on the recommendations proposed in the draft standard following the requirement of the standard to maintain the irradiance on the non-illuminated side of the bifacial PV modules below 3 W/m^2. The paper proposed a mounting structure with low-reflective back-side panel, which fulfils the draft standard requirement. The paper also demonstrated a rigorous method to quantify the reflections of the background environment to the rear side of bifacial modules, when the front side is illuminated at 1000 W/m^2. The analysis of the characterisation results revealed the relationship between the optical properties of the low-reflective materials and the distance of the test modules from these materials, and their impact on the short-circuit current of the bifacial PV modules. It has showed that the low-reflective materials with high diffusive reflectance requires a larger distance from the test modules to prevent the reflected light from impinging onto the cell areas.

ACKNOWLEDGEMENTS

The Solar Energy Research Institute of Singapore (SERIS) is a research institute at the National University of Singapore (NUS). SERIS is supported by the National University of Singapore (NUS), National Research Foundation Singapore (NRF) and the Singapore Economic Development Board (EDB). The authors would also like to thank Ryan A. Alinsod, Cai Yutian, Srinath Nalluri and Carlos Enrico Clement for their assistance during the experiments; and the kind support of Longi, Linyang and Jinergy for having provided testing modules.

REFERENCES

[1] T. S. Liang, D. Poh, and M. Pravettoni, "Challenges in the pre-normative characterization of bifacial photovoltaic modules", *Energy Procedia*, to be published.
[2] V. Fakhfouri, "IEC 60904-1-2: Measurement of current-voltage characteristics of bifacial photovoltaic devices," presented at the 4th Bifi PV Workshop, Konstanz, Germany, 2017.
[3] IEC TS 60904-1-2 ED1, in draft version at the time of writing. *Photovoltaic devices - Part 1-2: Measurement of current-voltage characteristics of bifacial photovoltaic (PV) devices*.
[4] J. P. Singh, A. G. Aberle, and T. M. Walsh, "Electrical characterization method for bifacial photovoltaic modules," *Sol. Energy Mater. Sol. Cells*, vol. 127, pp. 136–142, 2014.
[5] C. Duran, P. Hering, T. Buck, and K. Peter, "Characterization of Bifacial Silicon Solar Cells and Modules: A New Step," in *Proc. 26th Eur. Photovolt. Sol. Energy Conf.*, 2011, pp. 1550–1554.
[6] B. B. Van Aken, M. J. Jansen, A. J. Carr, G. J. M. Janssen, and A. A. Mewe, "Relation between indoor flash testing and outdoor performance of bifacial modules," in *Proc. 29th Eur. Photovolt. Sol. Energy Conf.*, 2014, no. September, pp. 2399–2402.

CHARACTERIZATION OF BIFACIAL PV MINI-MODULES USING FRONT- AND DOUBLE-SIDE ILLUMINATION

Sebastian Dittmann[1], Stephan Krause[1], Jörg Bagdahn[1], Hyeonwook Park[2], Soo-Young Oh[2], Sascha Esefelder[3], Torsten Brammer[3], Byong Su Kim[4], Sebastian Schindler[5], Hamed Hanifi[5]

[1]Anhalt University of Applied Sciences, Bernburger Str. 55, 06366 Köthen, Germany
E-mail: sebastian.dittmann@hs-anhalt.de, Tel.: +49 3496 67 2354

[2]Yeungnam University, 280 Daehak-ro, 38541 Gyeongsan, Republic of Korea
[3]Wavelabs Solar Metrology Systems GmbH, Markranstaedter Str. 1, 04229 Leipzig, Germany
[4]LG Electronics, 168 Suchul-daero, 730-030 Gumi, Republic of Korea
[5]Fraunhofer-Center for Silicon Photovoltaics CSP, Otto-Eissfeld-Str. 12, 06120 Halle, Germany

ABSTRACT: Our paper reports on the comparison of different measurement methods including indoor front-side and double-side illumination. We present a measurement protocol with different irradiance level for front and the rear-side of bifacial modules, which can be used as input for energy rating approaches such as IEC61853-1. The indoor measurements are performed on a high flexible LED sun simulator with a separate rear-side illumination.
We present a comparison of the double-side measurement and the IEC60904-1-2 approach, which show a good agreement for lower rear-side illumination (±1%) but we observed higher deviation of ±5% for higher rear-side illumination e.g. 1000 W/m² (front) and 1000W/m² (rear). Furthermore, our results show that the used rear-side material influences the bifacial gain and it is wavelengths dependent. A transparent backsheet shows slightly higher performance at low light condition.

Keywords: Bifacial modules, LED Sun Simulator, Characterisation

1 INTRODUCTION

To date, it is well known that bifacial solar cells and modules can utilize incident sun light via front- and rear-side absorption. This allows higher energy production under various installation conditions e.g. large solar farms, roof top installations and building integration (BIPV). However, to date existing photovoltaic (PV) standards do not allow for evaluation of the bifacial gain so far.

The draft standard IEC60904-1-2 [1], which is currently under evaluation, describes characterization methods for bifacial PV modules. This draft standard allows three different options to rate a bifacial PV module. This options are: (1) as known as G_e-method is an indoor front-side measurement with defined irradiance levels based on the bifaciality of the module, (2) is an indoor measurement with double-side illumination, and (3) is an outdoor measurements with different rear-side intensities. With regards to option (1) it only entails front-side measurement it is not clear how the backside material, e.g., glass or transparent backsheet influences the measurement results and if these results represent the real operating case of bifacial modules i.e. double-side illumination with different irradiance and spectral content on rear-side. Moreover, energy relevant measurements for bifacial modules are not discussed yet. The published energy rating standard IEC61853-1,-2 [6,7] describes performance measurement at different irradiances and temperatures levels only for monofacial modules.

Our paper reports on the comparison of different measurement methods including indoor front-side and double-side illumination. We present a measurement protocol with different irradiance level for the front and the rear-side, which can be used as input for energy rating approaches. The indoor measurements are performed on a high flexible LED sun simulator with a separate rear-side illumination.

2 METHODOLOGY

2.1 Experimental setup

The experimental setup consist of two separate light sources based on light emitting diodes (LED) which can be either used simultaneously or as single light sources. Figure 1 shows the principle set-up. As front-side illumination (upper part), we use WAVELABS's SINUS-220 LED sun simulator. This simulator consists of 21 single LED channels and has a Class A spectrum classification at 1000W/m². Figure 1b shows the prototype of WAVELBS's rear-side engine. It consists of 18 LED channels with a Class A rated spectrum. However, for both light sources it is possible to control the single LEDs to create spectra unlike AM1.5g. The front-side irradiance can be varied step-less in the range of 100 to 1200W/m², while the upper limed of the rear-side is limited to 1000W/m².

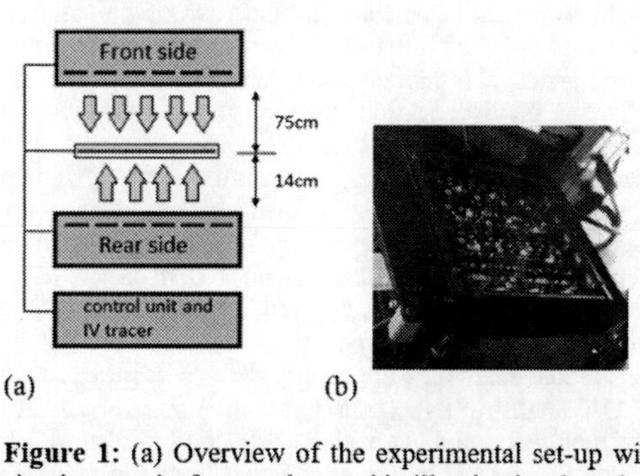

(a) (b)

Figure 1: (a) Overview of the experimental set-up with simultaneously front and rear-side illumination based on LEDs, (b) Prototype of WAVELABS's rear-side light engine based on 18 LEDs

2.2 Samples

The authors used a set of bifacial mini-modules (one encapsulated PV cell) of different manufactures and rear-side covers. The mini-modules can be summarized in three groups. Group I consists of three mini-modules with n-type bifacial PV cells with multi wire busbars, a glass front cover and a transparent polymer backsheet (BS). In the following these samples are labeled as TTx. Group II labeled with GGx consists of n- and p-type bifacial PV cells with different numbers of busbars. For these samples a low iron solar glass is used as front and rear-side cover. Group III represents the same cell technology as Group II. However, instead of the transparent BS we used a white and black BS as rear-side cover, which is indicated with TW and TB respectively. As encapsulation material we used for all mini-modules ethylene vinyl acetate (EVA). In order to investigate the difference between the glass and the transparent BS for sample TB2-R the bifacial PV cell is flipped i.e. the rear-side of the cell faces the glass. Table 1 shows an overview of the used mini-modules and abbreviation.

Table 1: Overview of the used mini-modules and abbreviation. In Group III, *F* and *R* indicate the front or rear-side of the cell facing the glass respectively.

Gr.	ID	Cell type	Busbars	Material front/rear
I	TT1 to TT3	n-type	multi wire	glass/trans.BS
II	GG1a, GG1b	n-type	4 busbars	glass/glass
	GG2	n-type	3 busbars	glass/glass
	GG3	p-type	5 busbars	glass/glass
III	TW1-F	n-type	multi wire	glass/white.BS
	TB1-F, TB2-F	n-type	multi wire	glass/black.BS
	TB2-R	n-type	multi wire	glass/black.BS

2.3 Measurement methods

First, we carried out an initial characterization of the described samples. This includes current-voltage (I-V) curves at standard test condition (STC), performance at different irradiance levels (100, 200, 400, 700, 1000 and 1200 W/m²), and spectral response (SR) measurements. The front-side light source is used for the initial performance measurements. In the case of the bifacial mini-modules both front and rear-side characterization were carried out. Moreover, in order to investigate the influence of the rear-side material of the mini-modules and cell properties. We carried out measurements with spectra unlike AM1.5g. Therefore, three different wavelength ranges were defined and the Isc were measured at a low wavelength range (350 to 650 nm), at a middle range (550-850 nm), and a high range (850-1100 nm). Figure 5 shows the simulator's spectrum at the three defined wavelength ranges.

Double-side measurements were performed with WAVELABS's SINUS-220 LED sun simulator and the designed rear-side light engine as descried in section 2.1. We performed I-V curves at different front and rear-side irradiance according to Table 2. In addition, we performed I-V curves at higher rear-side irradiances such as 1000f/400r, 1000f/700r, and 1000f/1000r (unit W/m²). These measurements were compared to the G_e-method of the IEC60904-1-4 [1].

All described I-V measurements were performed at 25°C ±2°C with traceability to a WPVS reference, which has been calibrated at Fraunhofer ISE. In the case of the single side illumination on the bifacial min-modules reflection from the back side were less than 3 W/m².

Based on outdoor measurement data of our outdoor test facility [3] in Bernburg, Germany (51°N, 11°E) the authors defined a measurement protocol to characterize bifacial cells and modules at different front and rear-side irradiances. Two pyranometer measure the front and the rear-side irradiance in an open-rack configuration. The subsurface is grass of green/yellow color. The open-rack is south oriented and has an inclination of 35°. The pyranometers are mounted 1.4 m above the ground. We selected typical summer days in June and July 2018, indicated as clear, partial cloudy and overcast. Clear sky days are days with a low diffuse irradiance and no clouds, partial cloudy days have direct and high diffuse light due to bright/white clouds and overcast days have a closed cloud cover.

For each single day type we selected different front-side irradiances levels of interest (100, 200, 400, 700, 1000 and 1200W/m²) and defined the corresponding rear-side irradiance.

3 RESULTS AND DISCUSSION

3.1 Measurement protocol for bifacial cells and modules

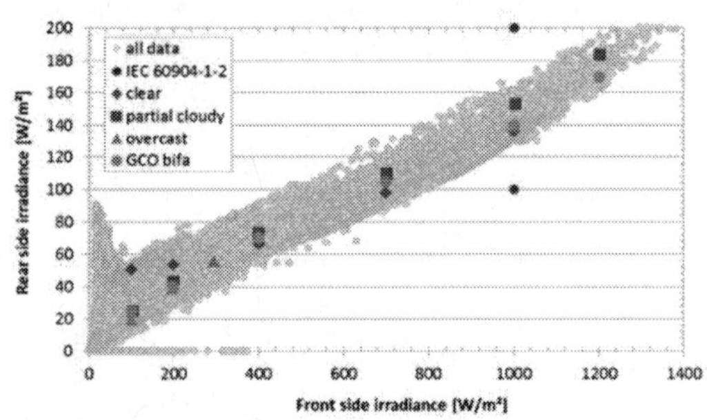

Figure 2: Outdoor front and rear-side measurements (grey) in open-rack configuration with grass subsurface. Blue diamonds represent the defined irradiance levels of interest for the clear day, red rectangle the partial cloudy day, green triangle the cloudy day, and the red dots the IEC60904-1-2 draft. Orange dots are the defined measurements values of Gf and Gr.

Figure 2 shows the rear-side irradiance (Gr) as function of the front-side irradiance (Gf) in an open-rack configuration (grey dots). For each day type defined Gf levels of interest are highlighted. Blue diamonds represent the clear day, red rectangle the partial cloudy day, green triangle the cloudy day, and the red dots the IEC60904-1-2 draft. It is clear visible that the IEC does not represent the rear-side condition at 1000W/m² at this specific site. While analyzing the measured Gf we see an average value of about 145 W/m² at 1000 W/m² for clear and partial cloudy days. At clear days and Gf < 300W/m² Gr is rather stable with values around 50 W/m². However, at days with a higher diffuse irradiance such as partial cloudy and cloudy days Gr is decreasing constantly down to 25 W/m² if Gf < 200 W/m². The ratio of Gr/Gf is 0.14 for clear days, 0.15-0.25 for partial cloudy and 0.2 for overcast days.

Table 2 shows the defined values of Gr and the corresponding Gf values (see also orange dots in Figure 2). In addition, we add the defined values of the IEC60409-1-2. Please note that due to the limitation of the first prototype we set Gr to 100 W/m² if Gf ≤ 400 W/m².

Table 2: Measurement protocol for bifacial cells and modules.

#	Gf [W/m²]	Gr [W/m²]	Comment
1	1200	170	
2	1000	200	IEC60904-1-2
3	1000	140	
4	1000	100	IEC60904-1-2
5	700	100	
6	400	70	*in this study:
7	200	40	set to
8	100	25	100 W/m²

3.2 Initial characterization

Figure 3 shows the Isc of the front and rear-side at STC and low light conditions using only the front-side light source. The mini-modules of Group I (multi wire n-type) show the highest current generation at STC and low light, while the mini-modules of Group II show slightly lower values. The deviation between the samples of the same technology is within ±0.2%. While analysing the bifaciality ($\delta = I_{sc}r / I_{sc}f$) we observed the highest value for Group I with 90% and 85% for GG1a, GG1b, and GG2. Sample GG3 (5 busbars, p-type) shows the lowest bifaciality. It is only 65% at STC. However, comparing the bifacial at low light down to 100W/m² we observed for all samples only a slide decrease of <1%.

Comparing the mini-modules of Group III with Group I we see +6.5% current gain due to the white BS. This is mainly due to the internal reflection [4,5]. The mini-module TB2-F with the black BS shows instead losses of -2.5%. If we change the back side material to a glass instead of the transparent BS we see losses of -1% in the Isc.

Double-side illumination with Gf=1000W/m² and Gr=200W/m² (green dots) show a current gain of ~17% for all bifacial samples except GG3, which shows a gain of 13% due to the lower bifaciality of the p-type PV cell.

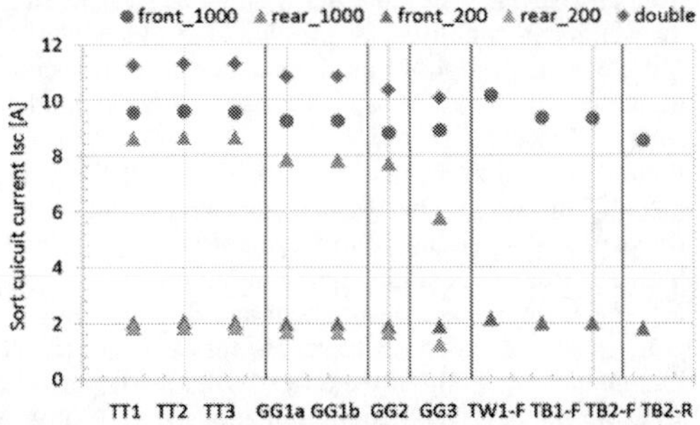

Figure 3: Isc of front-side and rear-side measurements at STC (blue/grey) and low light 200W/m² (orange/yellow). Double-side illumination (green) with Gf=1000W/m² and Gr=200W/m².

Efficiency curves normalized to the value at 1000 W/m² as shown in Figure 4 show significant differences between the front and rear-side of the cell and the used backside material. In fact, the rear-side of sample TT1-R (blue dotted line) shows lower performance like the front-side. The difference between front and rear at 200 W/m² is -10%. This is mainly due to the higher recombination losses and the optical losses in the wavelength range under 650 nm (see Figure 5). However, if we compare the sample with the glass as back cover (TB2-R) it shows even higher losses at low light conditions due to the lower spectral response at wavelengths >950nm. The difference in power to the front-side is -13%. The best performing mini-module is TW1-F with the white BS. This mini-module shows a power gain at low light condition of +4.5% while the mini-module with the black BS shows losses of -2%.

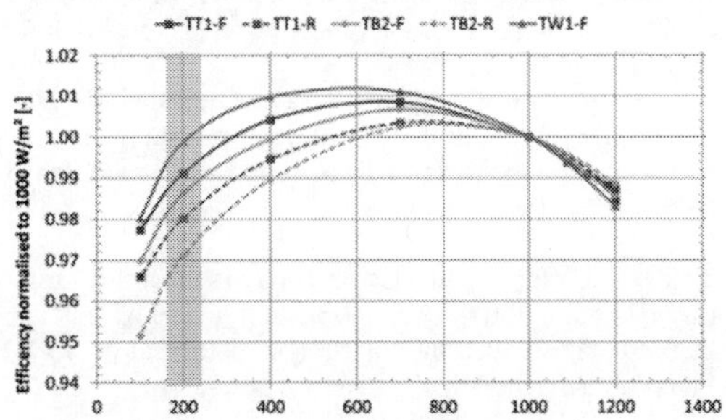

Figure 4: Results on irradiance dependency of mini-modules of Group I and III. I-V curves were taken at 100, 200, 400, 700, 1000 and 1200 W/m² and 25°C. The efficiency is normalised to the value at 1000W/m².

3.3 Spectral response and current generation at spectra unlike AM1.5g

Figure 5 shows the spectral response measurements of the mini-modules in Group I and III together with the simulator's spectrum at three defined wavelength ranges. It is clearly visible that the current generation of the rear-side of the bifacial cells is reduced over the full wavelength range. Moreover, we found that below 650 nm optical losses of the rear-side are driven by the transparent BS properties. However, above 950 nm the response is higher than the one with the glass as rear cover (TB2-R).

While analysing the SR data of the mini-modules with the white and black BS, it shows that TW1-F has the highest performance while TB2-F the lowest even lower than the TT1-R (transparent BS). This can be explained again due to internal reflections caused by the white BS.

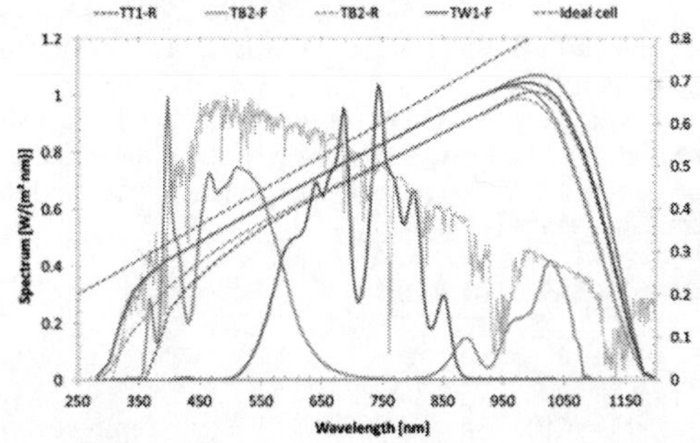

Figure 5: Spectral response of the mini-modules of Group I and III. *F* and *R* indicate the front or rear-side of the cell facing the glass respectively. Simulator's spectrum with different spectra unlike AM1.5g (Low λ-range 350-650nm, Mid λ-range 550-850nm and High λ-range 850-1000nm).

Considering Isc measurements with the defined spectra unlike AM1.5g we see the influences of the lower SR due to the optical and material properties. Table 3 shows the differences in Isc at the low, mid and high wavelengths range compared to TT1-F. Differences in Isc of about 20% in the low wavelengths range could be observed for TT1-R (transparent BS). At the higher wavelengths range the same mini-module shows losses of about -4%, while TB2-R (glass cover) shows -12.5% losses.

The results of the SR and measurements with spectra unlike AM1.5 confirm the different current generation and low light behavior of the tested mini-modules of Group I and III.

Table 3: Differences in Isc at different spectrum ranges (Low λ-range 350-650nm, Mid λ-range 550-850nm and High λ-range 850-1000nm) compared to sample TT1-F (glass/transparent BS).

	TT1-R	TB2-R
Low λ-range	-19.8%	-14.0%
Mid λ-range	-7.7%	-8.8%
High λ-range	-3.9%	-12.5%

3.4 Double-side vs. IEC90409-1-2 draft (G_e-method)

This section presents the comparison of the double-side illumination and the IEC60409-1-2 (G_e-method). Figure 6 shows the Isc measurements of mini-module TT2 as function of irradiance. The yellow diamonds are the front-side measurement with the upper light source, the grey triangles are the rear-side measurements with the rear-side light source, the blue dots are the double-side illumination, the orange triangles are the IEC approach, and the green rectangle are the sum of the single front and rear-side measurements.

The single front and rear-side measurement show a linear trend. As expected, however, the Isc of the rear-side is lower due to the reported rear-side losses. Moreover, the measurement according to the IEC approach and the double-side illumination show the same trend in the irradiance range from 200 to 1200W/m². The deviation between these two approaches is here within ±1%, as shown in Figure 7. While analysing the data beyond 1200 W/m² or less than 200 W/m² the slope of the double-side illumination decreases. The deviation increases up to ±5%.

Considering the sum of the single front and rear-side measurement ($I_{sc}f + I_{sc}r$) we observed the same value as the double-side measurement. The deviation over the full irradiance range is only ±0.8%.

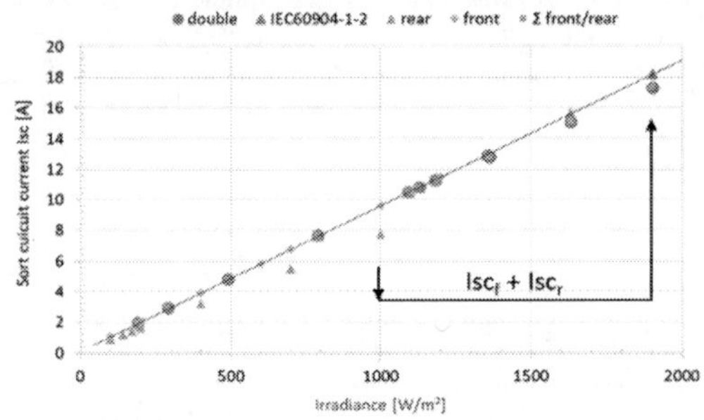

Figure 6: Isc of mini-module TT2 measured with upper light source (yellow diamonds), rear-side light source (grey triangles), double-side illumination (blue dots), IEC approach (orange triangles), and the sum of the single front and rear-side measurements (green rectangle).

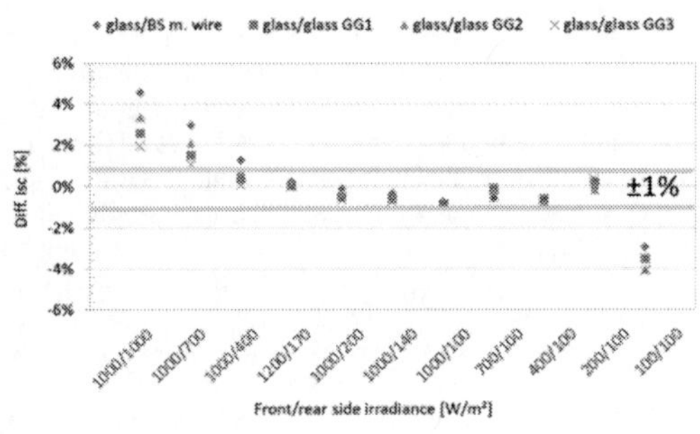

Figure 7: Deviation between double-side illumination and IEC60904-1-2 draft (G_e-method). $\Delta I_{sc} = (I_{sc(Ge)} / I_{sc(double)}) - 1$.

4 CONCLUSION

The paper presents a double-side sun simulator based on LED with a flexible range of irradiance and spectrum to provide a realistic measurement for bifacial cells and modules. We presented a definition of different irradiance levels for the front and the rear-side. This irradiance levels give more information as just one single measurement condition as presented in IEC60904-1-2. Moreover, the proposed irradiance levels can be used as input for energy rating approaches of bifacial modules such as IEC61853-1.

Furthermore, our results show that the used rear-side material influences the bifacial gain and it shows their wavelengths depending. A transparent BS shows slightly higher performance at low light condition.

The double-side measurement and the IEC60904-1-2 draft show a good agreement for lower rear-side illuminations (±1%) but we observed higher deviation of ±5% for higher rear-side illuminations e.g. 1000W/m² (front) and 1000W/m² (rear).

However, current we work on the validation of the double-side measurement with outdoor data at different albedo scenarios and a measurement protocol with different rear-side spectra.

7 REFERENCES

[1] V. Fakhfouri et al., "IEC 60904-1-2 draft: Measurement of current-voltage characteristics of bifacial photovoltaic devices", Presentation at 3rd bifi PV workshop 2016, Miyazaki, Japan

[2] G. Arnoux1 et al., "Toward the standardisation of the power rating of bifacial solar devices", Presentation at 5th bifi PV workshop 2018, Denver, USA

[3] S. Dittmann, et al., "Development of Characterization Techniques and Applications of Bifacial Solar Cells and Modules", 33rd European Photovoltaic Solar Conference and Exhibition, 2017, Amsterdam, The Netherlands

[4] Liu et al., Solar Energy Materials and Solar Cells, Volume 144, January 2016, Pages 523-531

[5] Koentopp et al., IEEE JOURNAL OF PHOTOVOLTAICS, VOL. 3, NO. 1, January 2013

[6] IEC61853-1 "Photovoltaic (PV) module performance testing and energy rating - Part 1: Irradiance and temperature performance measurements and power rating", 2011

[7] IEC61853-2 "Photovoltaic (PV) module performance testing and energy rating - Part 2: Spectral responsivity, incidence angle and module operating temperature measurements", 2016

8 ACKNOWLEDGEMENT

This work was funded by the German Federal Ministry of Economic Affairs and Energy (BMWi) and Zentrales Innovationsprogramm Mittelstand (ZIM) in the project BiFaTest (ZF4020807 LT6). We would like to acknowledge the help Kai Sporleder of Fraunhofer CSP for the support on some of the indoor measurements and Sebastian Schindler and Hamed Hanifi for providing samples from the BiFaLo project (BMWi: 034088B).

HOT-SPOT ENDURANCE TEST
MODIFICATIONS FOR BIFACIAL PHOTOVOLTAIC MODULES

Daniel Philipp, Hebin Manuel, Esther Fokuhl and Georg Muelhoefer
Fraunhofer ISE, Institute for Solar Energy Systems
Heidenhofstr. 2, 79110 Freiburg, Germany; +49 761 4588-5414; daniel.philipp@ise.fraunhofer.de

ABSTRACT: This paper aims at identifying suitable procedures for hot-spot endurance tests on bifacial PV modules. Drawing on the relevant background, i.e., normative requirements and available studies, it considers existing theoretical models and procedures for their application on bifacial modules. The resulting assumptions are subjected to experimental tests on four modules and in three approaches varying in their irradiance of module sides as well as location and degree of cell shading. In a second cycle, the first two approaches were repeated with extreme variations in irradiance level foreseen by the standard IEC 61215-2:2016. As it turns out, worst-case hot-spot temperatures appear in all three approaches, indicating a necessity for comprehensive testing. However, worst-case hot-spot temperatures are highly sensitive to irradiance intensity in general to an extent that may affect their passing or failing a hot-spot test, even if the test is done within the specified range of 900 to1100 W/m². It is therefore advised to reconsider the currently permitted leeway in equipment precision for both bifacial and conventional monofacial modules.

1 MOTIVATION AND BACKGROUND

Bifacial solar modules can operate at higher currents compared to conventional modules as they use irradiance from both their front and rear sides. However, the electrical point of operation of single modules or cells can vary in a wider range, as more factors influence the overall irradiance. Thus, bifaciality presents particular challenges of its own not only for the system layout but also with regard to reliability. This paper addresses the behavior of bifacial PV modules under inhomogeneous illumination, especially under partial shading. The results and findings are based on a series of shading experiments on several commercially available bifacial modules. The major aims of this study are to

- contribute to the adaptation of important testing standards (IEC 61215-2:2016 and IEC 61730-2:2016) for bifacial modules
- enhance the general understanding of the behavior of bifacial modules under inhomogeneous front or rear side illumination.

2 THEORY

2.1 Focus of hot-spot testing acc. to IEC 61215-2:2016

Almost every PV module type commercially available on the world market has been tested according to the hot-spot endurance test described in the above-mentioned quality standard.

The purpose of the hot-spot test acc. to IEC 61215 is to "determine the ability of a module to withstand hot-spot heating effects, e.g. solder melting or deterioration of the encapsulation. While absolute temperature and relative power loss are no criteria of this test, the most severe hot-spot conditions are utilized to ensure safety of the design." [1] In fact, the idea of the IEC hot-spot test is to check whether the module suffers substantial damage under <u>worst-case</u> shading and operation conditions. While the exact temperature of the hot spot is irrelevant, the electrical safety must not be negatively affected.

Experience in hot-spot testing at Fraunhofer ISE showed that critical damages that may result in the module failing the standard requirements are caused by temperatures between 160 °C and 200 °C for typical glass/backsheet PV modules. The exact value is of course strongly depending on the specific materials used.

To obtain the worst-case hot-spot condition, the standard requires operating the module in short circuit at 1000 ±100 W/m² while particular ambient conditions (T_{amb}, wind) assure that the module temperature (T_{mod}) is kept stable at 50 °C ± 10 K [1]. The shading situation leading to the highest temperature is determined and maintained for one hour.

As the amount of dissipated energy grows approximately linearly with the increase in irradiance (see Figure 1), it is important to point out that the permitted ranges for irradiance (±100 W/m²) and module temperature (±10 K) can lead to significant variation in hot-spot temperatures even within the allowed limits for the equipment.

Side note: In real PV module operation such worst-case operation conditions are quite rare, while minor temperature differences between 15 and 25 K relative to the average module temperature are much more common. These situations do normally not lead to fatal module defects, but even minor over-temperatures can be relevant with regard to long-term stability due to the influence of temperature on chemical degradation [2]. Therefore, in practice it is desirable to avoid local overheating.

2.2 Development of local overheating

Hot spots occur if for any reason the short-circuit current of a cell is reduced to a level below the string operation current. This may happen due to shading or as a consequence of cell defects, such as deactivating cell cracks. In this situation, the cell with the reduced short-circuit current operates in reverse bias and dissipates energy in form of heat.

The amount of dissipated energy depends on the reverse characteristic of the shaded cell, the overall irradiance on the module, and the point of operation of the unshaded cells. Figure 1 shows the simplified I-V curve of a cell string of 20 cells with one cell shaded to receive reduced irradiance relative to the other cells (50 and 80 %), and with the shaded cell assumed to have high (upper graphs) and low (bottom graphs) shunt resistance, respectively. The model illustrates that the amount of dissipated energy strongly depends on the reverse characteristic of the shaded cell as well as the current of the shaded and unshaded cells. Furthermore, it can be seen that the likelihood that the shaded cell is pushed into reverse bias is much lower when an external load is connected, which reduces the current.

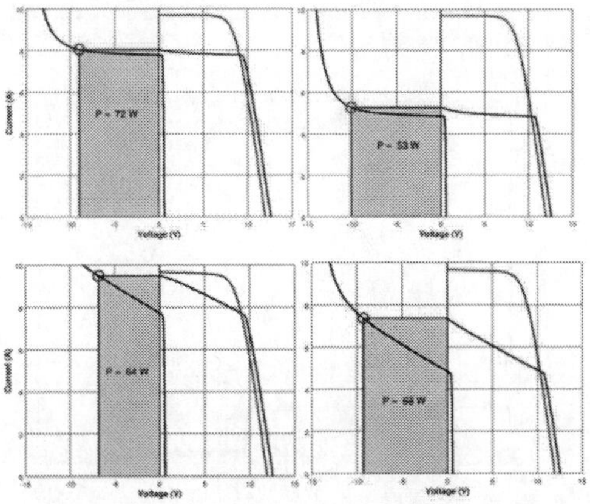

Figure 1: Simplified I-V curve of a cell string of 20 cells with one cell shaded to 80 % (left) and 50 % (right) irradiance compared to the others, with the shaded cell assumed to have high (top) or low (bottom) shunt resistance. Legend: Green: unshaded cells; **Black**: shaded cell; Blue: resulting curve; Gray area: dissipated energy

In addition to the above-mentioned factors, which influence the dissipated energy, local shunts play an essential role for the actual temperature of a hot spot. According to testing experience of the TestLab PV Modules at Fraunhofer ISE, the most critical hot spots typically occur on cells which have areas with relatively low shunt resistances. In these cases, the highest temperatures are not necessarily measured under the shading condition with the highest total amount of dissipated energy, but in conditions where the dissipated energy is most concentrated on a small spot of the shaded cell.

2.3 Considerations for bifacial modules

For bifacial modules, the first question is if and how the additional light from the rear can be factored in to an adapted test procedure. Following the purpose of the current standard, to obtain the worst-case condition the first consideration concerns the irradiance reaching the module.

For instance, it could be assumed that 20 or 30 % of the front irradiance reaches the module's rear side due to a reflecting environment. Typically, the values may be a lot lower in practice. However, in case of an extremely reflecting environment even higher reflection rates may be possible. Adding 20 % to 1100 W/m² (max. allowed irradiance for testing monofacial modules) would result in a total irradiance of 1320 W/m² for a bifacial test. Assuming 30 %, the irradiance would rise to 1430 W/m², respectively.

The second consideration refers to size and position of the shaded area. As shown in Figure 1, both the overall irradiance and the difference in irradiance between shaded and unshaded cells substantially affect the dissipated energy. However, in terms of the resulting hot-spot temperature the size of area dissipating this energy matters strongly, which is exemplarily illustrated in Figure 2 for a bifacial cell shaded either 100 % from the rear side or 20 % from front side. In case of 100 % rear side shading, the whole cell heats up, in case of 20 % front side shading mainly the unshaded area (80 %) heats up. Respectively different temperature values are expected in both cases.

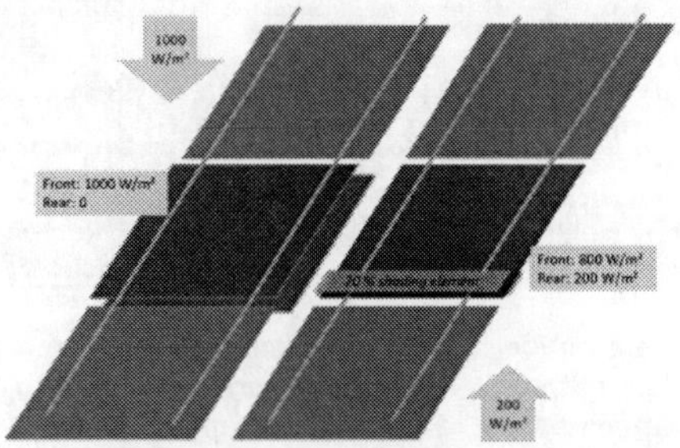

Figure 2: Schematic representation of a shaded bifacial cell. Left string: the rear side of a cell is fully shaded; right string: the front side is 20 % shaded. Assuming a bifacial factor of 1, the amount of dissipated energy is similar in both cases.

Last but not least, it has to be considered if and how design elements, e.g., junction boxes or frames, which may continuously affect the overall irradiance reaching the cells, should be taken into account.

3 EXPERIMENTAL PART

3.1 Testing environment

To analyze the influence of shading on bifacial modules, four samples have been exposed to artificial sunlight in a combined climate chamber and steady state solar simulator (Figure 3). The device guarantees comparable conditions in terms of temperature and illumination throughout the experiments. Furthermore, the intensity of the irradiance can be varied between 80 and 100 % by dimming the lamps.

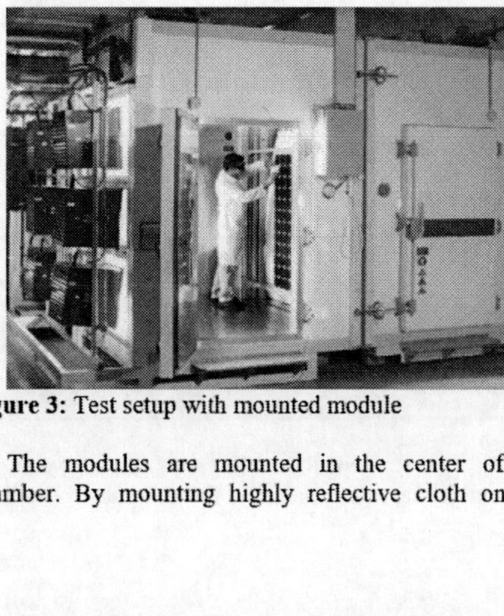

Figure 3: Test setup with mounted module

The modules are mounted in the center of the chamber. By mounting highly reflective cloth on the chamber's rear side, floor, and ceiling, the light passing the module laterally, at the top and the bottom is reflected to the rear side of the module, allowing partial rear-side illumination between 17 and 20 % for full size modules. Table I shows the distribution of rear side illumination over a PV module.

Table I: Irradiance measured on nine positions of the module surface (front and rear sides) after being mounted in the climate chamber

Measurement position	Irradiance [W/m²]	
	front	rear
Top left	1074	200
Top middle	1113	187
Top right	1076	181
Center left	1118	230
Center middle	1113	199
Center right	1109	198
Bottom left	1169	197
Bottom middle	1118	199
Bottom right	1171	200

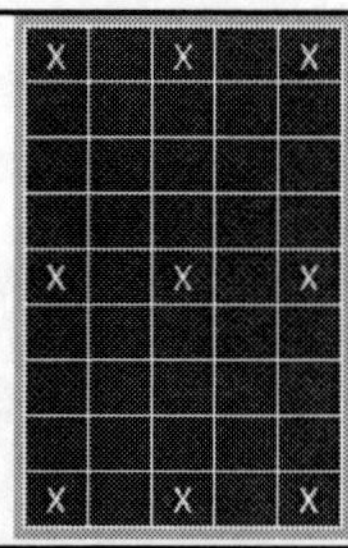

Thin cardboard pieces (thickness 2 mm) were used as shading elements. The same material was used to shade the entire rear side of the module in approach 1.

3.2 Results

Four bifacial PV modules of two different manufacturers were used for the experiments. All modules consist of a 60-cell design of similar size. The following table shows an overview of the samples and their ID.

Table II: Test modules and their characteristics for monofacial front and rear side irradiance

ID	Type	P_{max} [W] front/rear	I_{sc} [A] front/rear	Bifaciality I_{sc}/P_{max}	Remark
M01	A	266/159	9.12/5.28	0.58/0.60	mono
M02	B	268/207	9.07/7.17	0.79/0.78	poly
M03	C	242/186	8.72/6.57	0.75/0.76	mono
M04	D	266/159	9.12/5.28	0.58/0.60	mono

Experiment I – Three approaches to illumination and shading

The four modules were exposed to controlled irradiance and temperature conditions by means of a solar simulator integrated in a climate chamber. Due to modified reflective properties of the inside of the climate chamber they could be illuminated from both sides as shown in Table I. A hot-spot test similar to IEC 61215-2:2016 was conducted under three different shading conditions and irradiance settings. In the following, they are referred to as "approaches" (see Table III).

Table III: Three different irradiance / shading combinations applied on bifacial modules in the conducted hot-spot endurance test

Approach	Irradiance on front side	Irradiance on rear side	Shading of individual cells
1	100%	0%	on front side
2	100%	~20%	on front side
3	100%	~20%	on rear side

In a first step the cells with the lowest shunt resistances were determined by fully shading the cells one by one and measuring the I-V characteristics of the module. The shape of the curve indicates low or high shunt resistance. (Cells with local shunts are able to carry higher current in a fully shaded situation).

As it turned out, none of the modules and cells under review exhibited low shunt resistances. This fact is relevant to the interpretation of the following results.

The following graphic (Figure 4) exemplarily shows the selection of cells of one of the tested modules (M02). As can be seen, none of the shaded cells carry a high current in shaded condition, i.e., none of the shaded cells has low shunt resistance.

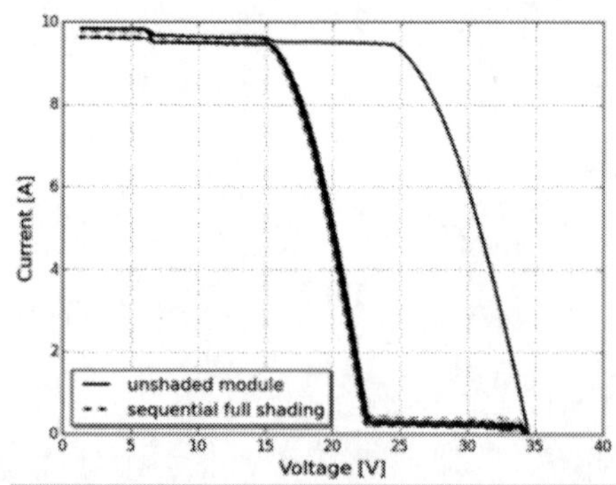

Figure 4: Exemplary I-V curves of a module with fully shaded cells and none of the cells indicating low shunt resistance.

In a second step, the worst-case shading situation was identified for three cells per module. This was done by varying the degree of shading while the module was short-circuited. The maximum temperature was measured with an IR camera.

The following graphics (Figure 5 to Figure 7) show the maximum temperatures vs. the respective degree of shading of the tested cells and modules for each of the testing approaches. Figure 5 represents approach 1, Figure 6 approach 2, and Figure 7 approach 3, respectively. As no low-shunt resistance cells could be detected on any of the modules, the findings correspond to theoretical assumptions that in case of high-shunt resistance cells the worst-case shading situations appear at lower degrees of shading (< 40 % in approach 1; < 60 % in approach 2). In approach 3, an energetically similar situation results from a higher degree of rear side shading.

Table IV shows the highest measured hot-spot temperatures for each module and approach. Unexpectedly, for M01 the highest overall temperature was measured in approach 1 (without rear side illumination). Also, for M02 and M04 the temperature was higher in approach 1 than in approach 2 (with rear side illumination). However, for these modules the highest temperature was measured in approach 3 in which the cell was shaded on the rear side.

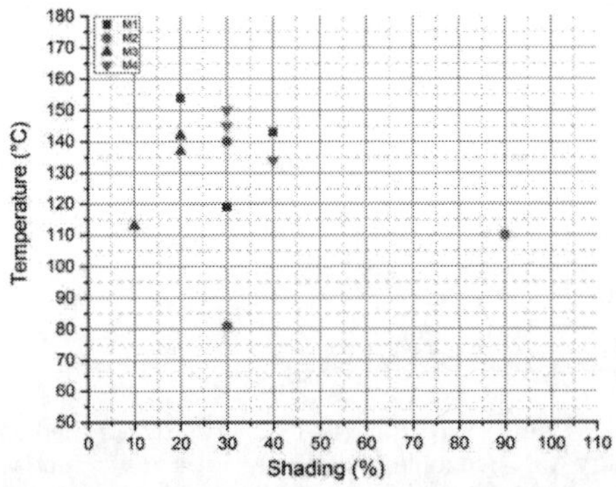

Figure 5: Approach 1 – three cells per module shaded on the front side. For high-shunt resistance cells, the highest temperatures are typically measured at shading levels < 40 %.

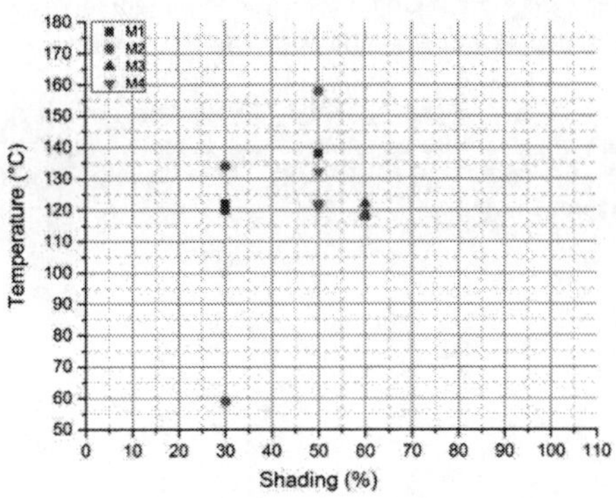

Figure 6: Approach 2 – three cells per module shaded on the front side. For high-shunt resistance cells, the highest temperatures are typically measured at shading levels < 60 %.

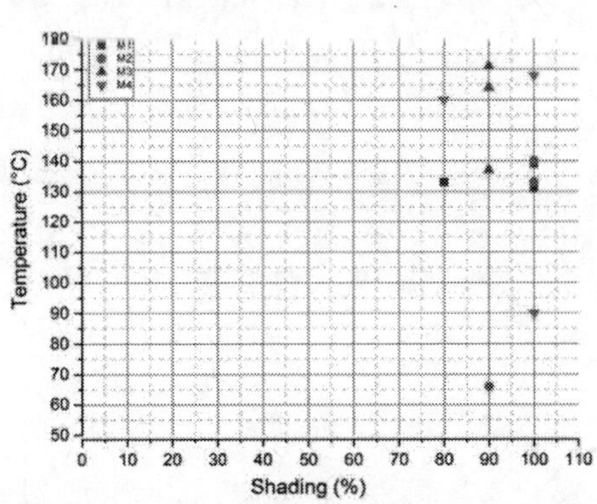

Figure 7: Approach 3 – three cells per module shaded on the rear side. For high-shunt resistance cells, the highest temperatures are typically measured at shading levels > 80 %.

Table IV: Max. measured hot-spot temperatures for each module and approach. **Bold**: highest temperature

Test approach	Max. temperature [°C]			
	M01	M02	M03	M04
1	**152.5**	150.0	148.5	144.8
2	134.0	131.8	**161.4**	121.8
3	139.5	**167.8**	140.1	**171.6**

Experiment II – Influence of irradiance level

Approaches 1 and 2 were repeated on one module from experiment I, but at higher (close to 1100 W/m²) and lower (close to 900 W/m²) irradiance, which represents the upper and lower range of the allowed intensity acc. to IEC 61215-2:2016. In the following they are referred to as approach 1' and 2', respectively. The exact irradiance intensities are given in Table VI.

The results (Table V) show that the changes in irradiance correlate with significant changes in the maximum hot-spot temperature, which range between 8 and 25 K. However, the increase in temperature seems also to be depending on the approach.

Table V: Results of experiment II

Cell ID	Shading [%]	max temp. at high irrad. [°C]	max temp. at low irrad. [°C]	change [K]
		Approach 1'		
19	20	134.5	122.6	11.9
22	20	134.5	118.8	15.7
24	10	124.5	116.6	7.9
		Approach 2'		
24	60	119.3	94.2	25.1
29	60	117.5	97.6	19.9
45	60	121.8	100.9	20.8

Table VI: Irradiance intensity in experiment II (W/m²)

	Approach 1'		Approach 2'	
	high	low	high	low
Front (avg.)	1097	920	1097	920
Rear (avg.)	0	0	200	165

4 SUMMARIZED FINDINGS AND CONCLUSION

In the first experiment, the hot-spot temperatures of bifacial modules with and without 20 % rear side illumination were determined. Considering the degree of shading which leads to the highest measured temperature, the results correspond to theoretical considerations. But although the highest temperatures were measured in situations with rear side illumination (either approach 2 or 3) on three of the four modules, it was not expected that on one module (M01) the highest temperature was measured without rear illumination. Also, hot-spot temperatures were unexpectedly higher for M02 and M04 in approach 1 (without rear illumination) compared to approach 2 (with). However, for these modules the highest temperature was measured in approach 3, where the cell is shaded on the rear side. Although no temperature-induced defects could be observed on any of the tested modules, the absolute differences in hot-spot temperatures for the applied shading approaches range from 20 to 30 K – a range which could be significant in terms of a pass/fail evaluation, depending on the used materials.

Theoretical considerations indicate a direct correlation between overall irradiance and dissipated energy leading to hot spots. Although experiment I partially yielded other worst-case hot-spot temperatures than predicted on the grounds of this assumption, a corresponding relation could be observed in experiment II where hot-spot tests at around 900 and 1100 W/m² were conducted. However, the results also show that the irradiance level affects the maximum hot-spot temperatures differently for different cells. They also depend on whether only the front or both sides are illuminated. In the modified irradiance approach 1', without rear light, the increase in irradiance led to an increase in hot-spot temperatures between 8 and 16 K. In case of approach 2', with rear illumination, the increase in hot-spot temperatures ranges from 20 to 25 K.

Summarizing the results, it is obvious that both the level of front and rear side irradiation and the location and degree of shading are significant for the maximum hot-spot temperature. These factors may well substantially influence whether a module passes or fails the standard hot-spot test, even if the test equipment conforms to the requirements for monofacial modules specified in IEC 61215-2:2016. Measurement results did not always fit the expectations derived from theory. These deviations could be explained by several factors. They could be related to the equipment and test procedure (e.g., homogeneity of the light source – see Table I – or isolating properties of the shading element) or the individual test sample (e.g., bifaciality, presence of permanent shading elements on the rear side such as the j-box, and temperature-dependent changes of the reverse characteristic of shaded cells).

For the hot-spot test on bifacial modules, the following conclusions are drawn from these experiments and considerations:

- Following the idea of the standard to address worst-case situations, it seems necessary to somehow factor in rear side irradiance during the hot-spot test, as it can have significant influence on temperature.
- However, to determine a reasonable level for this additional illumination it is necessary to reconsider the provisions for hot-spot testing of monofacial modules. The current standard allows a wide irradiance range between 900 and 1100 W/m², which is already significant in terms of pass/fail.
- To cover all possible scenarios for worst-case temperatures as shown in Table IV, it seems necessary to both, illuminate the front and rear sides simultaneously and shade different cells along the lines of the three approaches outlined in this paper.

In addition to these conclusions, attention is drawn to the overall level of irradiance in the testing of bifacial modules in adapted standard versions. Although the influence of irradiation on temperature could be shown, fixing a reasonable degree of rear side illumination cannot be settled by scientific considerations alone. The authors of this study propose two possibilities: Either the standard specifies the overall irradiance for testing of bifacial modules (e.g., front irradiance: 1000 W/m² and add. 25 % from rear side). In this case, the end user would have to be made aware that installations allowing a considerably higher degree of rear side irradiance are no longer covered by the standard. This could be done by a provision in the standard specifying the conditions the test is designed for. Or, alternatively, the manufacturer could state design limits for his product, which could then be used as references to configure the irradiance settings in the hot-spot test and other safety-relevant stress tests related to the current generated in a PV module, such as the bypass diode test.

6 ACKNOWLEDGEMENTS

This work was funded by the German Federal Ministry for Economic Affairs and Energy (Contract Number 0324036A).

7 REFERENCES

[1] IEC 61215-2:2016-03, Terrestrial photovoltaic (PV) modules – Design qualification and type approval – Part 2: Test procedures (IEC61215-2:2016).
[2] Kurtz, S. et al. (2011): Evaluation of high-temperature exposure of photovoltaic modules. In: Prog. Photovolt: Res. Appl. 19 (8), S. 954–965. DOI: 10.1002/Pip.1103.

INTERLABORATORY COMPARISON OF METHODOLOGIES FOR MEASURING THE ANGLE OF INCIDENCE DEPENDENCE OF SOLAR CELLS

Nicholas Riedel[1], Adrián A. Santamaría Lancia[1], Mekbib Amdemeskel[1], Sune Thorsteinsson[1], Peter B. Poulsen[1], Fabian Plag[2], Ingo Kröger[2], Lenneke H. Slooff[3], Mark J. Jansen[3], Anna J. Carr[3], Petra Manshanden[3], Martin Bliss[4], Tom Betts[4], Iñigo Petrina Jauregui[5], Mikel Ezquer Mayo[5], Jose L. Balenzategui[6], Ruben Roldan[7], Ulli Kräling[8], Ghassan Baarah[8], Beniamino Iandolo[9], Rasmus S. Davidsen[9], Anders Thorseth[1], Carsten Dam-Hansen[1], Gisele A. dos Reis Benatto[1]

[1]Department of Photonics Engineering, Technical University of Denmark, Frederiksborgvej 399, 4000, Roskilde, Denmark
[2]Physikalisch Technische Bundesanstalt (PTB), Bundesallee 100, D-38116 Braunschweig, Germany
[3]Energy Research Centre of the Netherlands (ECN>TNO), P.O. Box 1, 1755 ZG Petten, The Netherlands
[4]Centre for Renewable Energy Systems Technology (CREST), Loughborough University, Loughborough, Leicestershire, LE11 3TU, UK
[5]Spanish National Renewable Energy Centre (CENER), C/Ciudad de la Innovación 7, 31621 Sarriguren (Navarra), Spain
[6]CIEMAT, Department of Energy, Avda. Complutense 40, 28040 Madrid, Spain
[7]University of Applied Sciences and Arts of Southern Switzerland, Institute for Applied Sustainability to the Built Environment (SUPSI - ISAAC), Campus Trevano, CH-6952 Canobbio, Switzerland
[8]Fraunhofer Institute for Solar Energy Systems ISE, CalLab PV Modules, Heidenhofstrasse 2, 79110 Freiburg, Germany
[9]Department of Micro- and Nanotechnology, Technical University of Denmark, Ørsteds Pl., 2800 Kongens Lyngby, Denmark

ABSTRACT — The aim of this work is to compare angle of incidence (AOI) measurement setups for solar cells between laboratories with such capability. For the first time, we compare relative light transmission measurements among eight laboratories, whose measurement techniques include indoor and outdoor methods. We present the relative transmission measurements on three 156 mm x 156 mm crystalline-Si (c-Si) samples with different surface textures. The measurements are compared using the expanded uncertainties provided by each laboratory. Five of the eight labs showed an agreement better than ±2% to the weighted mean between AOIs from -75° to 70°. At AOIs of ±80° and ±85°, the same five labs showed a worst case deviation to the weighted mean of -3% to 5% and 0% to 18%, respectively. When measurement uncertainty is considered, the results show that measurements at the highest incidence angle of ±85° are problematic, as measurements from four out of the six labs reporting uncertainty were found non-comparable within their stated uncertainties. At 85° AOI a high to low range of up to 75% was observed between all eight laboratories.

1 INTRODUCTION

Intercomparisons between laboratories are important to ensure reliable and accurate measurements to the highest level of confidence. Laboratories have different methods, equipment setups and procedures, and it is of great value to assess the comparability of their results. Moreover, the IEC 61853-2 standard, which provides procedures on how to conduct AOI measurements on PV devices was recently published [1]. Thus, a round-robin on AOI measurements will allow us to validate the newly adopted standard, provide feedback to standardization bodies, and establish a baseline that future round-robin campaigns can improve upon. The results from this intercomparison will provide data which will be used in a future study to investigate the impact of measurement deviation on modelling and energy prediction.

The core objective of this work is to determine the level of agreement in the relative transmissivity measurements performed at eight laboratories for each angle of incidence θ. The relative transmissivity is commonly referred to in other works as the incidence angle modifier (IAM). The range of angles extends from -85° to +85° in steps of 5°. The results are analyzed using the E_n number statistical method as outlined in ISO 17043:2010 [2]. This approach is commonly used in proficiency testing and it provides insight to the equivalence of the participating laboratories' test results. The data obtained from the participating laboratories is analyzed with the following objectives: (i) evaluate whether the measurement deviations from the weighted mean are within each partner laboratory's stated uncertainties; (ii) validate if the procedures stipulated by the existing IEC 61853-2 result in comparable measurements when different measurement techniques are used; and (iii) determine the main sources of uncertainty that contribute to non-agreement of measured values between partners.

Although numerous round-robin programs have been conducted for PV measurements at Standard Test Conditions (STC) [3-5], the literature shows comparatively few works on round-robins or laboratory comparisons of angular dependent measurements. The intercomparisons of AOI measurements to date have compared a limited number of laboratories and methods. For example, the authors in [6] compared outdoor measurements performed in real-time at Sandia National Laboratories (SNL) and CFV Solar Test Lab, which lay roughly 10 km apart from each other. These two labs used unique methods to measure the AOI response of full-sized PV modules and found an acceptable level of agreement between their measurements. The authors in [7] compared AOI measurements again performed at SNL on full-sized PV modules to the IAM data for the same modules in PVsyst's database, or to measurements performed at an unnamed third party lab. The authors found that these three sources often had significant deviations for the same module type, up to a 14% difference in relative transmissivity at large AOIs. In this work, we present the AOI measurements made on three unique single cell (156 x 156 mm) laminates by eight European laboratories. The measurement systems encompass five different light sources including one lab that performed the measurements outdoors.

TABLE I: Description of participating laboratory measurement systems [8-11].

Laboratory	Test condition	Rotation stage	Description of light source
CENER	Outdoor	1-axis, 0° to 90°, Automated	Natural sunlight. Diffuse directly measured
CIEMAT	Indoor	1-axis, +90° to -90°, Automated	Continuous collimated broadband halogen lamp (1kW DXW)
CREST	Indoor	1-axis, +90° to -90°, Automated	Pasan 3b flasher with broadband Xe arc lamp (class AAA)
DTU	Indoor	1-axis, +90° to -90°, Automated	Energetiq (EQ-99FCX) broadband laser driven light source
ECN>TNO	Indoor	1-axis, +90° to -90°, Manual	Pasan flasher with broadband Xe arc lamp (class AAA)
Fraunhofer ISE	Indoor	1-axis, +90° to -90°, Manual	Pasan flasher with broadband Xe arc lamp (class AAA)
PTB	Indoor	2-axis, +90° to -90°, Automated	Tuneable laser system with broadband bias lamps
SUPSI	Indoor	1-axis, +90° to -90°, Automated	Pasan flasher with broadband Xe arc lamp (class AAA)

2 EXPERIMENTAL DETAILS

2.1 *Partner laboratories and procedures*

Eight scientific institutions from six European countries are involved in the measurement comparison. These include the Department of Photonics Engineering at the Technical University of Denmark (DTU Fotonik), Physikalisch Technische Bundesanstalt (PTB), the Energy Research Centre of the Netherlands (ECN>TNO), the Centre for Renewable Energy Systems Technology (CREST) at Loughborough University, the Spanish National Renewable Energy Centre (CENER), the Laboratory of Photovoltaic Solar Energy PVLab at CIEMAT, the University of Applied Sciences and Arts of Southern Switzerland (SUPSI), and Fraunhofer Institute for Solar Energy Systems (ISE). A brief description of each laboratory's measurement setup is described in Table I [8-11].

The measurement systems used at CREST, ECN>TNO, Fraunhofer ISE and SUPSI are based on a flash system used for full-sized modules. However, each one of these labs used a different approach to build the rotation stage. The rotation stage at ECN>TNO, for example is only capable of holding small laminates (roughly 20 x 20 cm), while the CREST, Fraunhofer ISE, and SUPSI rotation stages can accommodate small laminates up to full-sized modules. The CIEMAT and DTU systems are only capable of testing single cell laminates and the PTB system can accommodate mini-modules with up to four 156 x 156 mm cells. DTU was the only laboratory that performed the test with a light source that did not cover the full active cell area (i.e. the diameter of the collimated light beam in the DTU system was 12 mm). The two-axis tracker at CENER can accommodate full-sized module and since single cell laminates were used as the devices under test (DUTs) in this round-robin, they utilized the tracker's area to perform the test on all DUTs at once.

All the participant laboratories were asked to use their standard techniques for AOI measurements. The axis of rotation was predefined in order to allow the results from the different partners to be directly comparable. The definition for which angular direction was positive and negative was made explicitly clear in a memorandum document that all partners received. The document also specified the location of the rotational axis and tolerances for the precise location of the front of the PV cell surface within the laminate. The samples made for this round-robin have an offset of approximately 1.5 mm ±0.5 mm from the rear side (back sheet) to the front surface of the PV cell. The participant laboratories were encouraged to use this tolerance when mounting the samples in their measurement system, although larger deviations from the rotational axis to the center of the PV cell would be acceptable for the outdoor measurement system.

The physical quantity measured by the partner labs is the short circuit current (I_{SC}) of each DUT over the specified angular range. The data of ultimate interest, however, are the relative transmissivity $\tau(\theta)$ values. The relative transmissivity $\tau(\theta)$ - sometimes referred to as the IAM - represents the percentage of in-plane direct beam irradiance available to the PV cell for conversion to electricity. The $\tau(\theta)$ is calculated by normalizing the I_{SC} measured at each angle to the I_{SC} value measured at normal incidence. No procedures were provided to the partner labs as to how to correct for fluctuations in the test environment during the measurement (e.g. spectrum, irradiance or temperature) and instead it was left up to each partner to use their standard correction approach. A common temperature coefficient for I_{SC} was provided to the partner labs in the case temperature corrections were necessary.

2.2 *Devices under test (DUT)*

The round-robin includes measurements of three different PV devices in duplicate (i.e. each lab measures six DUTs in the total). Redundant samples were used so there would always be a backup in the case that any one sample became damaged. Electroluminescence (EL) images were taken before each laboratory started testing to ensure that no damage occurred during transportation. The analysis of EL images taken before the first lab's measurement and after the final lab's measurement showed that no cell damage had occurred during transportation. However, the glass of one sample was damaged about halfway through the campaign, which highlights the necessity of duplicate samples in round-robin campaigns. The measurements made on the damaged sample are not published here.

All DUTs have the following specifications: (i) An active cell area of 156 mm x 156 mm; (ii) full area dimensions of 200 mm x 200 mm; (iii) 3.2 mm thick finely textured PV glass superstrate; (iv) ethylene-vinyl acetate (EVA) encapsulant; (v) two tabs as metal contacts; and (vi) a flat polymeric backsheet with slight curvature around the cell edges. The differences between the DUTs are the cell types and cell texturing. Two DUTs are mono-crystalline standard silicon; two DUTs are multi-crystalline black silicon textured under reactive ion etch (RIE) treatment (referred to as 'Black-Si A' hereafter) [12]; and two DUTs are multi-crystalline black silicon textured under atmospheric pressure dry etching (ADE) treatment (referred to as 'Black-Si B' hereafter) [13]. The edges of samples were covered with non-transparent tape to prevent measurement artifacts at large incident angles.

IEC 61853-2 specifies that three measurements shall be taken at each angle. Six DUTs and 38 angles per sample would mean that each laboratory needs perform 684 measurements. In most cases, this was not possible due to time constraints and therefore only one measurement per angle, per sample was provided by most laboratories.

2.3 *Uncertainty and Analysis of Results*

Six of the eight labs provided the expanded uncertainty ($k = 2$) for the value of relative transmissivity $\tau(\theta)$ at each measured angle θ. Two of these labs provided DUT specific uncertainty while the other four provided an uncertainty estimate that covered all samples in this round-robin.

The uncertainty of each lab's measurement is critical for establishing comparability through the E_n number calculation (1). The two labs that were not able to provide measurement uncertainty were assumed to have the worst case uncertainty. That is for every angle measured, the highest uncertainty submitted by any lab is used as the measurement uncertainty for the labs without uncertainty.

For every sample and every angle, an E_n number is calculated per (1).

$$E_n = \frac{x_i - X_{ref,i}}{\sqrt{UC_i^2 + UC_{ref,i}^2}} \quad (1)$$

Wherein x_i is the individual laboratory's measured relative transmissivity $\tau(\theta)$ and UC_i is the expanded ($k = 2$) uncertainty of the lab's measurement of $\tau(\theta)$. The reference value $X_{ref,i}$ is the weighted mean of all partner's measured $\tau(\theta)$ values for a given sample at a given angle. Here the measurements are weighted by the uncertainty provided by each partner. Weighting the results in this manner has the consequence of shifting the X_{ref} value towards the measured values (x_i) of the laboratories with lower uncertainty. For every sample and every angle, the $X_{ref,i}$ value is calculated using equation (2).

$$X_{ref,i} = \frac{\sum_{i=1}^{N} \frac{x_i}{\sigma_i^2}}{\sum_{i=1}^{N} \frac{1}{\sigma_i^2}} \quad (2)$$

Wherein σ_i is the $k = 1$ uncertainty of the lab's measurement. The value σ_i is squared in (2) to arrive at the variance, which is additive by nature, whereas the standard deviation is not. Finally, $UC_{ref,i}$ is the expanded ($k = 2$) combined uncertainty of $X_{ref,i}$ and is calculated per (3). Calculating UC_{ref} in this way yields a value that is always lower than any of the participating labs' declared uncertainties.

$$UC_{ref,i} = \frac{2}{\sqrt{\sum_{i=1}^{N} \frac{1}{\sigma_i^2}}} \quad (3)$$

The relative transmission measurements from each laboratory are said to be coherently within their declared uncertainty when $-1 \leq E_n \leq 1$. In other words, the condition $-1 \leq E_n \leq 1$ is met when the difference between a lab's measurement (x_i) and the reference value ($X_{ref,i}$) is less than or equal to the root sum of squares (RSS) of the lab's declared uncertainty (UC_i) and the reference uncertainty ($UC_{ref,i}$). The sign of E_n provides a convenient way of discerning whether a lab's measurement is high (E_n = +) or low (E_n = -) relative to the weighted group mean.

3 RESULTS

The expanded uncertainty provided from each partner laboratory is shown as a function of AOI in Figure 1. Since there were three unique DUTs in the campaign, some labs accordingly reported an expanded uncertainty for each of the three DUTs. The uncertainty from partners who did provide DUT specific uncertainty (CREST and PTB) is represented by range bars and a dot symbol. The range bars show the maximum and minimum uncertainty reported by the laboratory while the dot in the middle represents their median uncertainty. The same convention is used to represent UC_{ref}. Note how UC_{ref} is less than the uncertainty of all laboratories at each angle. The uncertainty provided by partners with non-DUT specific uncertainty is represented with 'x' symbols. Figure 1 shows a trend of increasing uncertainty with increasing AOI, wherein a range of 0.1% to 3% at normal incidence and a range of 2.5% to 9.3% at 85° AOI is observed. The specific reasons for this increasing trend will be unique to a given measurement system (e.g. increasing non-uniformity or uncertainty of the measured angle θ). In the DTU measurement system, for example, the increase of uncertainty at large angles is primarily due to an increased contribution of uncertainty from the measured angle θ.

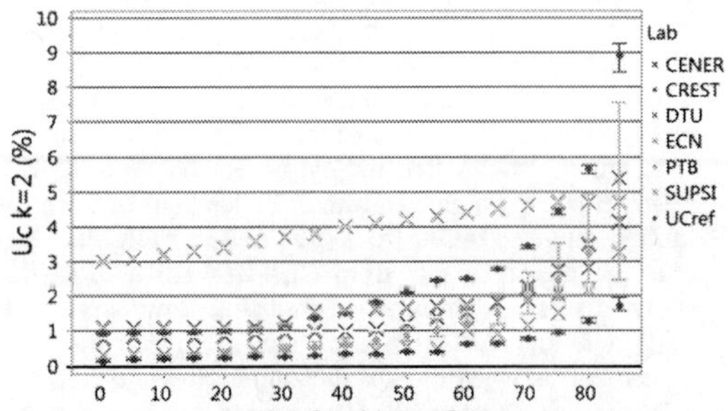

Figure 1: Expanded uncertainty from six laboratories as a function of AOI. The expanded uncertainty of the reference (UC_{ref}) is also shown. Labs providing a single uncertainty value at each angle are noted with an 'x' whereas labs with uncertainty specific to each DUT are noted with dots and range bars.

The measurements on sample Black-Si A performed by the eight partner laboratories are shown in Figure 2. We show the high-level results from this sample only because the measurement differences between the labs showed a consistent trend and because sample Black-Si A showed the worst case range (max-min) among the eight labs. Figure 2 shows an increasing measurement range between the labs with an increasing AOI. This result follows the same trend of increasing uncertainty with increasing AOI purported by the partner laboratories. In the positive direction, the measurement range is less than 0.1 (10%) up to a +60° AOI, while in the negative direction the range is less than 0.1 up to a -70° AOI. We attribute the difference between measurements at positive and negative angles to an artifact in some participant's results, not because of an inherent non-symmetry in the DUT itself. The maximum disagreement of 0.69 (69%) and 0.75 (75%) is observed at +85° and -85° AOI, respectively. This large disagreement is mostly driven by the outdoor measurement performed by CENER. If this measurement is removed, the range at 85° AOI is decreased to 0.23 (23%) in the positive direction and 0.36 (36%) in the negative direction.

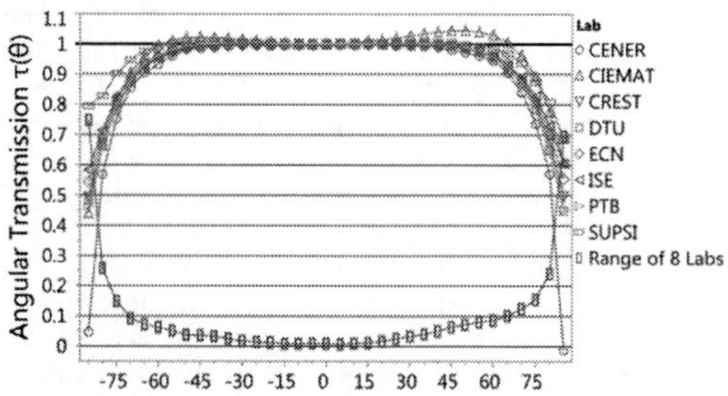

Figure 2: Overlay of transmission as a function of AOI θ as measured by eight laboratories on sample Black-Si A. The range of all measurements is also shown.

In Figure 3 we show the difference between each lab's angular transmission measurement $\tau(\theta)$ and the reference value X_{ref}. Recall that there are three DUTs, and thus three X_{ref} values at each AOI. The error bars drawn around each data point represent the maximum and minimum difference to the reference value; the center of the data point between the bars represents the median difference. Note that the deltas of -33% to -45% at ±85° reported by CENER have been removed from the plot in order to create a finer view of the y-axis.

Figure 3 shows that the agreement between all labs relative to the weighted mean is better than ±2% in the limited range of AOIs from -30° to 25°. However, it is only the measurements from CIEMAT and SUPSI that show dispersion from the group at low angles. If CIEMAT and SUPSI's data are excluded, then the agreement between the six remaining labs is better than ±2% for AOIs -50° to 50°. Beyond -50° to 50°, CENER's measurements become increasingly lower relative to the other labs'. If CENER's measurements are also excluded, five of the eight labs have an agreement of better than ±2% from the weighted mean on all samples between -75° to 70°. At ±80° these same five labs show worst case agreement of -3% to 5% relative to the weighted mean (Black-Si A), and a best case agreement of -1% to 0.5% relative to the weighted mean (Mono-Si). And finally, at ±85° the five labs show worst case agreement of 0% to 18% relative to the weighted mean (Black-Si B) and a best case agreement of 0.5% to 8.5% (Mono-Si). The worse agreement on the Black-Si as compared to Mono-Si sample could be due to the difference in surface structures or differences in low-light behavior.

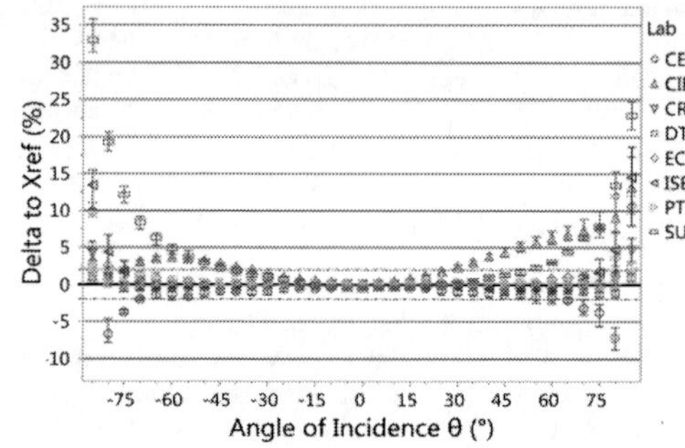

Figure 3: Plot showing the difference between each lab's measurement and the reference value X_{ref}. The error bars show each lab's max and min difference to X_{ref}. The red dashed reference lines are drawn at ±2%.

In Figure 4 we present the results from the E_n number calculation as a function of AOI for the Black-Si A sample. The profile observed in Figure 4 is similar to that in Figure 3 because the data in Figure 3 are the numerator used in the E_n calculation. Figure 4 shows 39 (of 280 total) measurements where the results are not comparable within the stated or assumed uncertainties. In other words, 14% of the time E_n is greater than 1 or less than -1, wherein all instances occur at AOIs $\geq |\pm 40|°$. PTB and CREST are the only two labs whose measurements result is $-1 \leq E_n \leq 1$ for all AOIs. Thus, we can say that their measurements agree to the weighted mean within the stated uncertainties for all measurement angles.

The CIEMAT data show 10 instances of non-comparability. Therefore the worst case uncertainty assumed for the CIEMAT measurements is clearly too conservative to make them comparable to the rest of the partners. Interestingly, CIEMAT shows several instances of non-comparability in the positive angular direction, but none in the negative direction. This is due to the non-symmetrical measurements they obtained (Figure 2). The reason behind the non-symmetry is still uncertain, but we believe it is likely due to an offset in the vertical axis of rotation from the middle of the PV cell. In addition, it is also possible that the optics unique to the CIEMAT measurement system result in the partial polarization of light from the halogen lamp, which in return would alter the reflection and transmission properties.

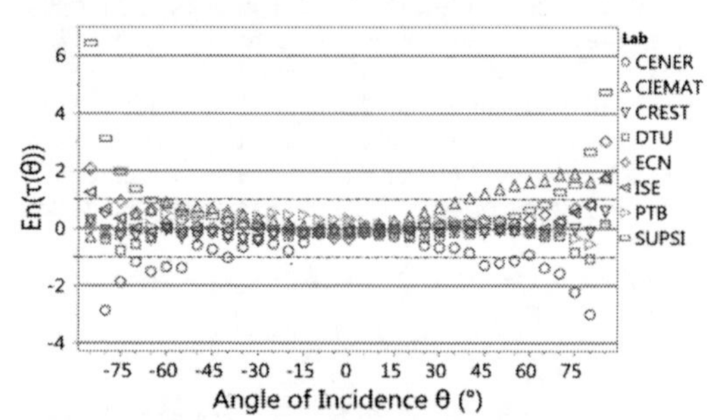

Figure 4: E_n number as a function of AOI θ for the measurements from eight laboratories.

The CENER data show 16 instances of measurements that deviate from the weighted mean beyond their reported uncertainty level. The CENER measurements are consistently lower than the measurements from the other seven partners and this difference increases at higher AOIs. The CENER measurements were the only measurements in this round-robin performed outdoors. However, we do not conclude that the low results are attributable to the outdoor approach itself. To better understand the differences that can occur indoors versus outdoors, we propose to extend the measurement campaign to additional laboratories performing the AOI test outdoors.

The SUPSI data show 8 instances where their measurements do not agree within the stated uncertainty and all 8 instances occur at AOIs $\geq |\pm 70|°$. Although the SUPSI measurements are consistently higher than most of the labs at every AOI (Figure 2 and Figure 3), they are also the lab with the most conservative uncertainty. However, the stated uncertainty of 4.6% to 4.7% at AOI $\geq 70°$ is not conservative enough for comparability. The reason(s) for SUPSI's higher than average angular transmission measurements are still under investigation. At the time of writing the high measurements are believed to be due to unwanted reflections within the test bed. These reflections are more pronounced in the negative direction than in the positive direction. The SUPSI data show that establishing comparability is not dependent on the light source used. CREST, ECN>TNO, ISE and SUPSI all used Xe flash lamps to measure the samples and yet SUPSI shows significant deviations from the other three labs.

DTU, ECN>TNO and ISE all show generally good agreement to the weighted mean; each lab shows one or two instances where their measurements are not comparable within the stated or assumed uncertainties. For DTU this occurs at AOI of 80° while for ECN>TNO and ISE it occurs at the extreme angles of ±85°. In actuality the measurements from ECN>TNO are closer to the weighted mean than ISE's at these high AOIs, since the uncertainties provided by ECN>TNO are more conservative than the maximum uncertainty assigned to the ISE measurements.

To understand the uncertainty level necessary for CIEMAT and ISE to be comparable to the weighted mean, we solve (1) for UC_i while leaving $E_n = 1$ and $X_{ref,i}$ and $UC_{ref,i}$ constant based on results from the other six labs. The results are shown in Figure 5 for the Black-Si A sample. The maximum uncertainty at each AOI reported by any of the other six labs is also shown. Since the expanded uncertainty limit for ISE is less than 1% out to ±65°, we can conclude that their measurements agree well to the weighted mean within that range. However, an uncertainty between 11.7% and 17.7% at the highest angles of ±85° would be required for comparability.

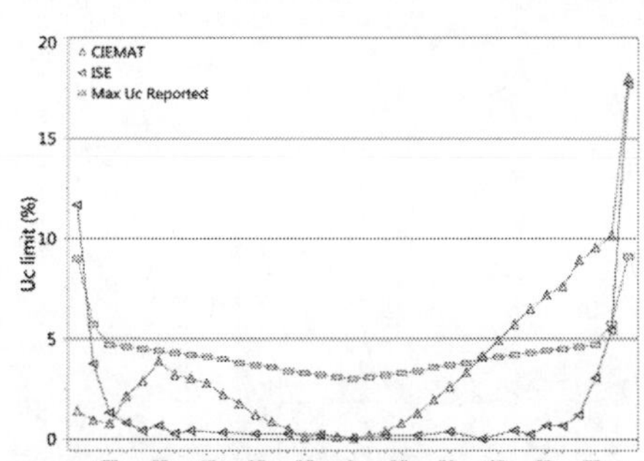

Figure 5: Minimum uncertainties required in order for CIEMAT's and ISE's measurements to be comparable to the weighted mean of the other six labs.

As mentioned, the measurement differences between labs showed the same trend across all three samples. Nevertheless we wish to give a full perspective of how the E_n number varied across samples and angles. In Figure 6, we show the number of labs that had an E_n number outside the limit of comparability. Figure 6 shows that at the extreme AOIs half or more of the labs have deviations that are not fully covered by their uncertainty budgets. This result highlights the difficulty of performing accurate measurements at high AOIs. Figure 6 includes all eight laboratories and assumes the worst case uncertainty for the ISE and CIEMAT data.

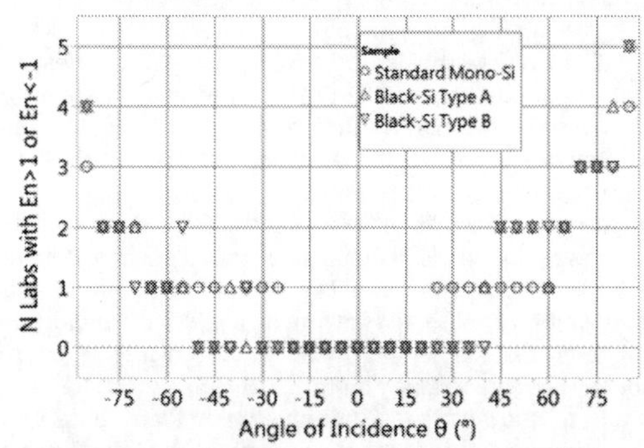

Figure 6: Plot showing the number of labs that do not comply to the condition $-1 \leq E_n \leq 1$ as a function of AOI.

IEC 61853-2 states that the rotational symmetry of the test system shall be verified at -80° and 80° AOI. It further states that the deviation in the relative transmissivity at these two angles shall not deviate by more than 2%. We performed this check for all eight laboratories and the results for the Black-Si A sample are shown in Figure 7. The limits of ±2% are indicated by the dashed red reference lines. It can be seen in Figure 7 that the labs who were the most comparable always meet the IEC requirement for symmetry. On the contrary, the labs with the lowest comparability and largest deviations from the weighted mean, showed symmetry that did not comply with the standard. This result suggests that it is a good practice to verify the symmetrical performance of a test system for measurement of the angular dependency of PV devices.

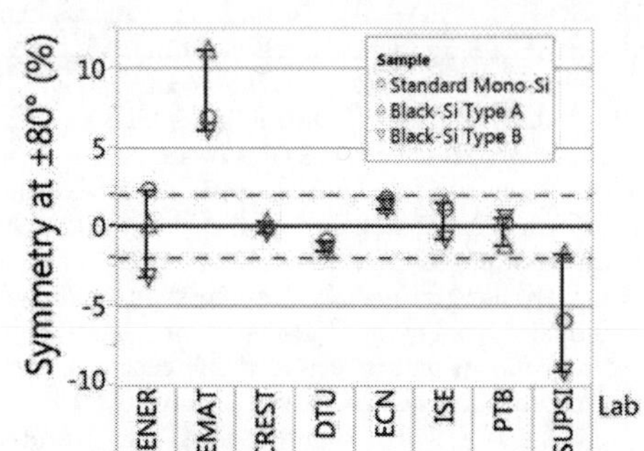

Figure 7: Variability plot showing the symmetry of the relative transmissivity measurements at ±80°. Three data points are shown for each lab, one for each sample.

4 SUMMARY

The results from an international round-robin between eight laboratories performing angular dependent measurements have been presented. Five of the eight labs showed an agreement better than ±2% between AOIs from -75° to 70°. At AOIs of ±80° and ±85°, the same five labs showed a worst case deviation to the weighted mean of -3% to 5% and 0% to 18%, respectively. These worst cases were observed on the 'Black-Si A' and 'Black-Si B' samples. The results showed that measurements at the highest incidence angle of ±85° are problematic, where a high to low range of up to 75% was observed among the eight labs.

The results provided by CREST and PTB agreed to the weighted average within their stated uncertainties for all angles. Similarly, the results from DTU and ECN>TNO agreed to the weighted average within their uncertainties, for all angles except 85°. At high angles of incidence (≥ 70°) the measurements from three to five labs were found not comparable within their stated or assumed uncertainties. The more frequent non-comparability observed at high angles of incidence suggests that labs may need to reconsider a more conservative uncertainty budget at angles ≥ 70°.

In a future work we will investigate the influence of the measurement deviations observed in this round-robin on modelled energy production. Furthermore, we plan to extend the round-robin to obtain more data from laboratories performing the AOI test outdoors.

5 ACKNOWLEDGEMENTS

We would like to thank all the partner laboratories for all their time and work performing the measurements. The facility and part of the work at DTU has been made possible by the grant "LEDMET - Center for LED metrology" from Innovation Fund Denmark.

6 REFERENCES

[1] "IEC 61853-2, Photovoltaic (PV) module performance testing and energy rating - Part 2: Spectral responsivity, incidence angle and module operating temperature measurements." 2016.

[2] "ISO/IEC 17043:2010 Conformity Assessment – General Requirements for Proficiency Testing International Organization for Standardization." 2010

[3] B. Mihaylov *et al.*, "Results of the SOPHIA intercomparison Part-1: STC, low-irradiance conditions and temperature coefficient measurements of C-Si technologies" in *29th Eur. Photovolt. Sol. Energy Conf. Exhib.,2014*

[4] D. Dirnberger *et al.*, "Progress in photovoltaic module calibration: Results of a worldwide intercomparison between four reference laboratories" *Meas. Science and Tech.*, vol. 25. no. 10, 2014.

[5] E. Salis *et al.*, "Improvements in world-wide intercomparison of PV module calibration," *Solar Energy,* Volume 155, October 2017, Pages 1451-1461

[6] B. H. King *et al.*, "Recent advancements in outdoor measurement techniques for angle of incidence effects," in IEEE 42nd *Photovoltaic Specialist Conference (PVSC)*, 2015

[7] S. Boppana *et al.*, "Impact of uncertainty in IAM measurement on energy predictions," *7th World Conference on Photovoltaic Energy Conversion (WCPEC-7) 2018*

[8] M. W. Amdemeskel *et al.*, "Indoor measurement of Angle resolved light absorption by antireflective glass in solar panels", *34th Eur. Photovolt. Sol. Energy Conf. Exhib., 2017*

[9] J.L. Balenzategui and F. Chenlo, "Measurement and analysis of angular response of bare and encapsulated silicon solar cells," *Sol. Energy Materials & Sol. Cells*, vol. 86, pp. 53–83, 2005.

[10] F. Plag , I. Kröger, T. Fey, F. Witt, & S. Winter "Angular-dependent spectral responsivity - Traceable measurements on optical losses in PV devices" *Prog. Photovoltaics Res. Appl.*,Special Issue Paper, pp. 1–14, 2017.

[11] L.H. Sloof *et al.,* "Looking at the annual yield from various angles: Optical model verification for structured glass" in *33rd Eur. Photovolt. Sol. Energy Conf. Exhib., 2016*

[12] R. S. Davidsen *et al.*, "Black silicon laser-doped selective emitter solar cell with 18.1% efficiency," *Sol. Energy Materials & Sol. Cells*, vol. 144, pp. 740–747, 2016.

[13] B. Kafle *et al.*, "Plasma-free dry-chemical Texturing process for high-efficiency multicrystalline silicon solar cells," *Energy Procedia*, vol. 92, no. September, pp. 359–368, 2016.

QUANTITATIVE EVALUATION OF PV DEVICE LINEARITY WITH THE TWO-LAMP METHOD

Harald Müllejans and Elena Salis
European Commission[1], Joint Research Centre, Directorate C - Energy, Transport & Climate,
Energy Efficiency and Renewables Unit, European Solar Test Installation (ESTI), Ispra, Italy
harald.muellejans@ec.europa.eu; elena.salis@ec.europa.eu
[1]The views expressed are purely those of the authors and may not in any circumstances be regarded as stating an official position of the European Commission.

ABSTRACT: The non-linearity of PV reference cells can be determined by the two-lamp method according to IEC 60904-10. However, the data so obtained only give a series of local non-linearities, i.e. between two irradiance levels differing by a factor of two. Here a scheme is presented on how to obtain the global non-linearity from these measurements over a much larger irradiance range. Essentially, the local non-linearities are added for neighbouring irradiance ranges, using suitable interpolation to obtain intermediate data points. The method is illustrated with measurements of a PV reference cell for which all local non-linearities are well below 1% whereas the global non-linearity from 25 W/m^2 to 1000 W/m^2 is almost 3%. It is recommended that this methodology be considered in the next revision of IEC 60904-10.
Keywords: calibration, reference cell, photovoltaic, standards, measurement method

1 INTRODUCTION

The linearity dependence of a PV device short-circuit current versus irradiance (hereafter called simply linearity) as defined in IEC 60904-10 [1] is required in the application of several other IEC standards. As currently defined by the standard, the linearity check is only a pass/fail test with an arbitrary limit of 2% deviation. The two-lamp method is mentioned in the standard as one possible method to obtain data to make the decision on whether a PV device is linear or not. However, inherently the two-lamp method gives information about the linearity between two irradiance levels which differ by a factor of two. Currently, there is nothing stipulated on how to use the information so obtained to deduce the linearity over a larger irradiance range and make a decision on the linearity of a PV device. Therefore, the two-lamp method so far only gives partial information and comparison with results from other methods (also included in IEC 60904-10) on linearity is not directly possible [2].

The two-lamp method is based on the flux addition principle used in photometry to determine the linearity of optical detectors - such as photodiodes - over a broad range of irradiances, typically several orders of magnitude. Several implementations can be found even recently in the literature [3, 4] with sophisticated data analysis methods based on matrix algebra. However, they are deemed too complex for implementation in the field of PV and therefore a simpler scheme is presented below.

In the two-lamp method essentially the detector response to two light sources is measured with both light sources illuminating the detector, first individually and then both together. From these measurements the deviation from linearity, i.e. the deviation of proportionality of the detector output with respect to incident irradiance, is readily determined for each single triplet of measurements. Here a scheme will be presented on how to obtain the non-linearity over a larger irradiance range by combining the results of a series of measurements with varying irradiance.

The two-lamp method can be used to determine the linearity of a PV reference cell's output (normally short-circuit current) against incident irradiance. It can easily be implemented even at low cost. Furthermore, the two-lamp method does not require any reference device to measure the actual irradiance, but rather it can determine the linearity of a PV device just from the latter's own measurement and is therefore suitable as a primary method for the determination of linearity of PV devices. The only required knowledge is the calibration value of the PV device, which is the natural reference point and is normally the single point known from device calibration. Furthermore, for single-cell devices the spatial non-uniformity is irrelevant because anyway averaged on the cell area and the spectral match to the reference spectrum is not very important.

The two-lamp method was introduced in the IEC standard as a qualitative method [5] on the basis of work by Emery, who described a suitable apparatus [2], after it had already previously been proposed for application in PV [6]. Based on the measurements made and presented [2], it was already clear that the data could not be compared as such with the results of the more traditional differential spectral responsivity (DSR) method.

In a first step the two-lamp method provides data points about what we can call the local non-linearity of a PV device, where local means that the information is only about the irradiance range spanned by the measurements, which is always a factor of two. In the literature [7] and in IEC 61315 [8], several schemes have been developed to combine these local non-linearities to obtain the global non-linearity, which is the non-linearity over the entire irradiance range spanned by all measurements. This is relatively straightforward in the case the neighbouring irradiance intervals are continuous, i.e. the lower irradiance level of one interval is identical to the higher irradiance interval of the next [7]. However, this ideal case might not always be achievable for experimental reasons. Therefore, here a method is presented on how to deal with the more general case in which a number of measurements has been made with the two-lamp method and the irradiance intervals show overlap. For this case, the data analysis of the two-lamp method is extended such that it provides quantitative information about the non-linearity of the PV device over the full irradiance range.

2 TWO-LAMP METHOD

2.1 Measurement principle

In the two-lamp method two different light sources A and B are used to illuminate a PV device each singularly as well as together. The irradiances by the two light sources should be roughly equal. The respective short-circuit currents generated in the PV device are measured. After also measuring the short-circuit current generated by possible background illumination (i.e. after excluding both light sources A and B) and subtracting it from the original measurements, the three short-circuit currents generated by the light sources A and B (individually or combined) are obtained:

$$I_A = I_{meas\ lamp\ A} - I_{meas\ room} \quad (1)$$

$$I_B = I_{meas\ lamp\ B} - I_{meas\ room} \quad (2)$$

$$I_{AB} = I_{meas\ lamps\ A\ and\ B} - I_{meas\ room} \quad (3)$$

2.2 Apparatus, device under test and measurements

Here we used the steady-state large area solar simulator APOLLO at the European Solar Test Installation (ESTI) (shown in Figure 1).

Figure 1: Photo of the APOLLO solar simulator with all eleven lamps active.

This simulator consists of 11 steady-state Xenon high pressure lamps, which all illuminate the test area of 2 m by 2 m. Each lamp has a mechanical shutter such that within a few seconds it can be excluded from contributing to the overall irradiance of the device under test [9]. Here, single lamps and group of lamps forming a virtual lamp were used to illuminate the device under test. For example for a measurement by the two-lamp method the device under test was first illuminated by a group of four lamps (virtual lamp A), then by a group of four different lamps (virtual lamp B) and finally by all eight lamps together as well as without any lamp (background). In each case the short-circuit current generated in the device under test was measured with a dedicated transimpedance amplifier connected to a digital storage oscilloscope. For the lower irradiance triplet additional wire mesh filters were used. The total irradiance range spanned was from 25 W/m^2 to 1000 W/m^2.

The device under test was an encapsulated crystalline silicon PV reference cell with an active area of 2 cm by 2 cm (ESTI code: SD74). The calibration value at standard test conditions (STC) had been determined previously as 124.17 mA ± 0.60 mA and was taken as the reference point.

2.3 Data analysis

The non-linearity according to IEC 60904-10 was calculated as

$$D_{lin}(I_{AB}) = I_{AB} / (I_A+I_B) - 1 \quad (4)$$

Traditionally each value of non-linearity D_{lin} provides the non-linearity at the short-circuit current I_{AB} with respect to current $I_A \approx I_B$ or

$$I_{ave} = (I_A+I_B) / 2 \quad (5)$$

In PV we are normally interested in the non-linearity at irradiances lower than the reference point, which is traditionally 1000 W/m^2 (as defined in standard test conditions). Therefore (as will become clearer below) it is more convenient to calculate the non-linearity NL as

$$NL(I_{ave}) = (I_A+I_B) / I_{AB} - 1 \quad (6)$$

which now gives the non-linearity at the current I_{ave} with respect to I_{AB}. For example, if A and B each generate about 500 W/m^2 on the test plane and therefore together 1000 W/m^2, the non-linearity at 500 W/m^2 with respect to 1000 W/m^2 is immediately obtained. Please note that the signs of NL and D_{lin} are opposite.

For further information at lower irradiances additional measurements have to be made by the two-lamp method while varying the irradiance produced by the lamps. If for a subsequent measurement the irradiance is adjusted such that the combined irradiance of the lamps is equal to the average irradiance of the previous measurement, i.e.

$$I_{AB2} = I_{ave1} \quad (7)$$

then the data analysis is straightforward: it just requires simply adding non-linearities (Strictly, two correction factors which are the respective non-linearities plus the value 1 should be multiplied). As the latter condition was not met here, a suitable scheme is developed below, using interpolation between already measured points and successively calculating the non-linearity information over a larger irradiance interval.

3 RESULTS

The series of measurements by the two-lamp method together with the results of the non-linearity determination are summarized in Table I.

Table I: The first three columns contain the measured short-circuit current data triplets of device SD74 for various combinations of lamps, after background (less than 0.01 mA) subtraction (equations 1 to 3). The fourth column is the non-linearity calculated according to equation (4) (following IEC 60904-10 ed.2). The fifth column is the average short-circuit current according to equation (5) and column 6 the overall non-linearity based on equation (6) and the procedure described below.

I_A [mA]	I_B [mA]	I_{AB} [mA]	D_{lin}	I_{ave} [mA]	NL
3.20	3.18	6.41	0.30%	3.19	-2.86%
3.21	3.38	6.61	0.42%	3.29	-2.93%
6.56	6.61	13.23	0.47%	6.58	-2.39%
11.65	11.18	22.91	0.37%	11.41	-1.92%
22.82	22.63	45.70	0.53%	22.73	-1.53%
43.24	44.60	88.41	0.65%	43.92	-1.03%
56.45	43.23	100.34	0.67%	49.84	-0.92%
52.10	68.51	121.40	0.65%	60.31	-0.68%
				121.40	-0.03%
				124.17	0.00%

The highest generated short-circuit current of 121.40 mA is close, but not equal, to the calibration value. This measurement had a raw NL of -0.65 % for the average current of 60.31 mA with respect to the combined current of 121.40 mA, which therefore should be regarded as having 0% non-linearity as it is the reference point for this first measurement. However, this value does not correspond to the calibration value obtained by the usual calibration procedure. So, a linear function through these two points (60.31,-0.65%) and (121.4,0%) was fitted. In the next step the value of this linear function at the calibration value is determined and then used to normalise the linear function. This ensures that the non-linearity at the calibration value is 0%, which should be the case as this is the actual reference point for the non-linearity evaluation. Consequently, to the original 121.40 mA a NL value of -0.03% is assigned and so on. Furthermore, the value of this normalised linear function at the short-circuit currents of 88.41 mA and 100.34 mA are determined, as the next two measurements have these combined short-circuit currents (Table I). The process is summarised in Figure 2.

With these intermediate points calculated and the respective raw NL (not shown) the following two points on the global non-linearity function can be determined. This is shown in Figure 3. Now the following intermediate point at 45.70 mA is calculated based on a linear interpolation between the nearest two data points already present. The process is then repeated until all measured data points have been assigned a value of the global NL.

The final result showing all values of the global NL as well as the values of the non-linearity determined according to equation (4) are shown in Figure 4.

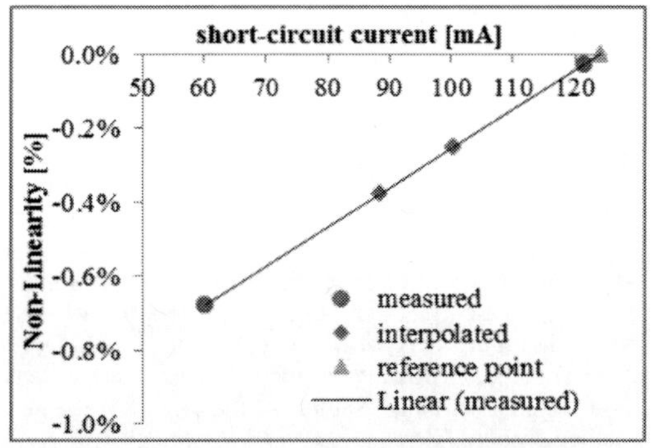

Figure 2: The reference point (green triangle) at the short-circuit current corresponding to the calibration value has by definition a non-linearity of 0%. From the measurement with two lamps together generating nearly this short-circuit current the two red dots can be calculated. Linear extrapolation and normalization is used to obtain 0% at the reference point. Linear interpolation is used to obtain the non-linearity at the two intermediate points (blue diamonds). Go to Figure 3.

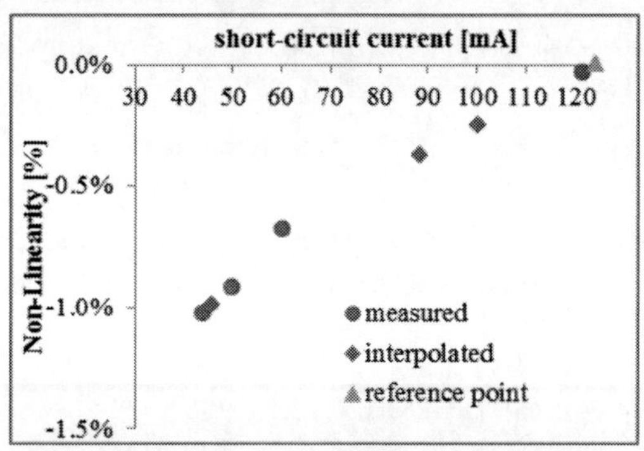

Figure 3: After the points in Figure 2 are determined the next two measurements (Table I) (leftmost two red dots here) can be added to the graph and another interpolated point (leftmost blue diamond) can be calculated. This procedure is repeated until all measurements are added. Go to Figure 4.

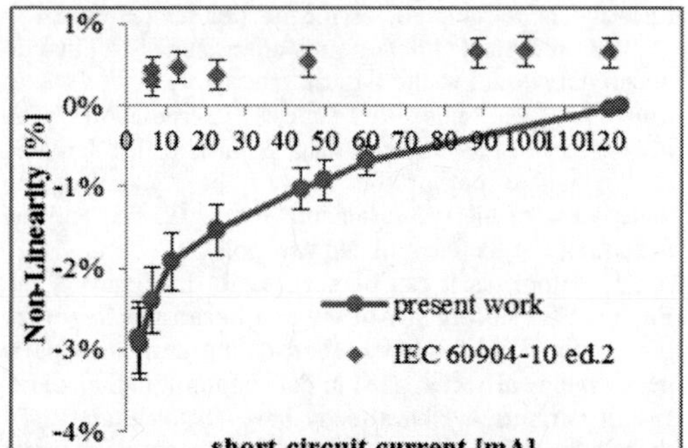

Figure 4: The full non-linearity function is obtained (now only showing the measured points). Here the blue diamonds show the result of the calculation according to IEC 60904-10 ed. 2, i.e. D_{lin} (equation (4)). Note the opposite sign. The local non-linearities given by D_{lin} are all below 1%, whereas the global non-linearity over the full range (25;1000) W/m^2 is close to -3%.

The result of this analysis is global non-linearity information over a large range of short-circuit current (roughly corresponding to an irradiance range of 25 W/m^2 to 1000 W/m^2). It can be seen that over this irradiance range the cell should be deemed non-linear according to IEC 60904-10 ed. 2 as it exceeds the limit of 2% deviation. If we are only interested in the non-linearity down to 100 W/m^2, the cell would just be linear as the non-linearity value is slightly below the 2% limit. However, the current data analysis in the standard (also shown in Figure 4) gives all values around +0.5%, so the device would incorrectly be classified as linear in any case. This is a striking example of the benefit of the new method proposed here.

The non-linearity can then also be plotted as a function of irradiance such as in Figure 5, thanks to the reference point used.

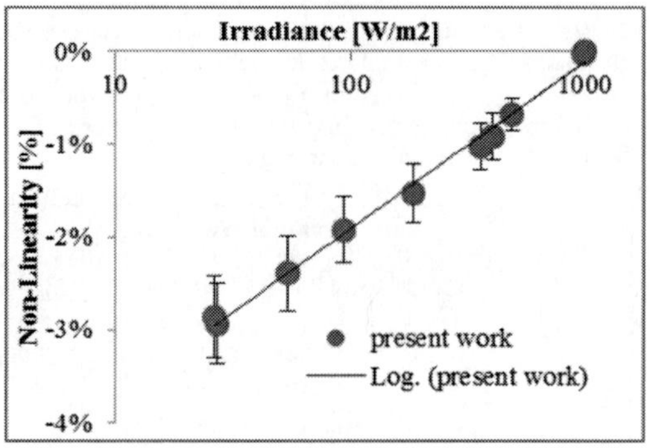

Figure 5: Presenting the data form Figure 4 on a semi-logarithmic irradiance scale.

4 MEASUREMENT UNCERTAINTY

The measurement uncertainty has been estimated as follows: each measurement by the two-lamp method has an uncertainty of 0.18% (k=2) based on a detailed analysis [10]. The main contribution to the uncertainty budget is made by the lamp stability over the time period it takes to measure a data triplet. Other minor contributions are due to data acquisition, amplifier non-linearity and variation of device junction temperature.

The interpolated data points are assumed to have an uncertainty equal to the largest uncertainty of the values used to determine the linear interpolation. The uncertainty of the following data point is then calculated as the square-root of the sum of the squares of the uncertainty of the measurement (always 0.18%) and the uncertainty of the interpolated data point.

Therefore, as it can be seen easily in Figure 4 and Figure 5, the uncertainty of the non-linearity is larger for those data points at lower short-circuit current. This is because they are referenced to data points at higher short-circuit currents, which already have an uncertainty. To the latter then the uncertainty of each separate measurement has to be added.

5 SUMMARY, DISCUSSION AND CONCLUSIONS

The two-lamp method has been applied to determine the non-linearity of a PV reference cell. A new data analysis scheme is proposed, which provides the global non-linearity based on a series of measurements by the two-lamp method. Compared to the calculation prescribed currently in IEC 60904-10 ed.2 the results are significantly different. This is attributed to the fact that the method in the standard only determines the local non-linearity between the irradiance levels of each separate triplet of measurements, whereas the scheme proposed here is able to determine the global non-linearity over the entire irradiance range spanned by the measurements. The result obtained here also compares favourably with the results of linearity determination by other independent methods [11]. For the device under test investigated here it is shown that the global non-linearity can be well above 2% even when all local non-linearities are well below 1%. This should be of concern as a PV device would be deemed linear in one case and non-linear in the other. It is strongly advised to revise the two-lamp method in IEC 60904-10 in the next revision.

Here we deliberately chose a non-linear PV reference cell in order to demonstrate the methodology and problematic aspects. Most modern PV reference cell are linear to a degree that with current methods it is very difficult to detect any non-linearity.

7 ACKNOWLEDGEMENTS

The research work leading to this article was partly carried out within the EMRP ENG55 project "Towards an energy-based parameter for photovoltaic classification" (PhotoClass). The EMRP is jointly funded by the EMRP participating countries within EURAMET and the European Union. We thank the Physikalisch-Technische Bundesanstalt (PTB), Germany, for providing the reference cell for measurements after the end of the project. We acknowledge our colleagues at ESTI, W. Zaaiman for preliminary work on the two-lamp method and D. Pavanello and M. Field for assistance in operating the APOLLO simulator.

8 REFERENCES

[1] IEC 60904-10 ed. 2 (2009-12) "*Photovoltaic devices - Part 10: Methods of linearity measurement*".

[2] K. Emery, S. Winter, S. Pinegar, D. Nalley, IEEE 4th World Conference on Photovoltaic Energy Conference, Vol. 2 (2006) pp. 2177–2180.

[3] D-J. Shin, S. Park, K-.L Jeong, S-N. Park, D-H. Lee, Metrologia 51 (2014) 25–32.

[4] B. H. Hamadani, A. Shore, J. Roller, H. W. Yoon, M. Campanelli, Metrologia 53 (2016) 76–85.

[5] K. Emery, private communication.

[6] L. P. Boivin, Metrologia 30 (1993) 355-360.

[7] C. L. Sanders, Applied Optics 1 (1962) 207-211.

[8] IEC 61315 ed. 2 (2005-10) "*Calibration of fibre-optic power meters*"

[9] E. Salis, I. Sharlandzhiev, M. Field, Proceedings of the 33rd European Photovoltaic Solar Energy Conference and Exhibition. Amsterdam, pp. 1548 - 1553.

[10] H. Müllejans, E. Salis; under preparation.

[11] M. Bliss, T. Betts, R. Gottschalg, E. Salis, H. Müllejans, S. Winter, I. Kröger, K. Bothe, D. Hinken, J. Hohl-Ebinger, under preparation.

PRACTICAL ASSESSMENT OF POWER RATING UNCERTAINTIES FOR INDUSTRIAL SILICON MODULES

Harrison W. Wilterdink, Adrienne L. Blum, Cassidy L. Sainsbury, Ronald A. Sinton
Sinton Instruments
4720 Walnut St., Suite 102, Boulder, CO 80301, U.S.A.

ABSTRACT: Measurement uncertainties in industrial module power ratings are often treated as a topic of mind-numbing scope and complexity, effectively leading to a lack of uncertainty reporting in actual practice. For typical commercial silicon modules in production line settings, we show that power rating uncertainty is primarily due to the reference module(s) and procedure(s) used to calibrate the solar simulator (flash tester) irradiance. The uncertainty "added" at the production line flash test step is only the test reproducibility, which can be quite small (e.g., ± 0.2 %, $k = 1$) and can be assessed by straightforward methods described here.
Keywords: Silicon, Module, Module Manufacturing, Qualification and Testing

1 INTRODUCTION

1.1 Previous work

A full characterization and reporting of production line power rating (P_{MAX}) uncertainties is arduous and likely outside the technical expertise of many module manufacturers. Such a task requires:

- Understanding the many factors that can affect flash test results
- Understanding the traceability chain (Figure 1) used to establish standard test conditions (STC)

Regarding the first point, exemplary work has already been done by Monokroussos et al. [1]; we reproduce here a list from that work indicating several factors that can contribute to flash test uncertainties (Table I).

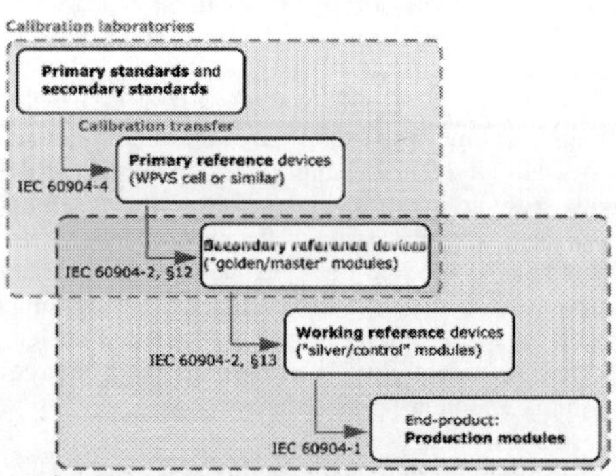

Figure 1. Example traceability chain used to establish STC for module power rating measurements.

Table I. Factors contributing to flash test uncertainty, reproduced from Monokroussos et al. [1]. Uncertainties are reported with a coverage factor (k) equal to one.

Factor	Uncertainty
Reference module P_{MAX}	± 1.03 %
Reproducibility and repeatability	± 0.5 %
Drift of reference module	± 0.3 %
Non-uniformity of irradiance	± 1 %
Measurement uncertainty of non-uniformity of irradiance	± 0.25 %
Temperature accuracy	± 0.5 °C
Temperature non-uniformity	± 1 °C
Temperature gradient	± 0.5 °C
Temperature probe contact quality	± 0.5 °C
Electronic (current, voltage, irradiance)	± 0.2 %
Spectral mismatch (due to production tolerance)	± 0.3 %
Contact resistance	± 20 mΩ
Other influences	< ± 0.3 %
Total (see reference [1])	**± 1.31 %**

1.2 Present work

This work will focus on uncertainties in module P_{MAX} at STC, since that is often the most commercially relevant electrical parameter of a module. We can significantly simplify the uncertainty analysis in Table I by removing uncertainty components that cancel from the final measurement as a result of the flash tester irradiance calibration. Best practice is to adjust the irradiance so that the measured P_{MAX} values of one or more reference modules match their known STC values. P_{MAX} matching is used (rather than I_{SC} matching) because P_{MAX} is much less sensitive to non-uniform irradiance than I_{SC} [1–3].

This procedure obviously corrects for non-STC irradiance levels; however, it can also compensate P_{MAX} for several other systematic effects in Table I, such as those due to temperature, contact resistance, spectral mismatch, non-uniform irradiance, as well as accuracy of the electronics used to measure current, voltage, and irradiance signals. Two examples illustrate how the compensation works:

1. If the measurement electronics cause a systematic under-reporting of module current, then P_{MAX} is also under-reported and irradiance is set > 1000 W/m^2 to compensate
2. If module temperature is < 25 °C (and is not corrected using temperature coefficients), then P_{MAX} is over-reported and irradiance is set < 1000 W/m^2 to compensate

It is easy to understand why the irradiance compensation works for errors in current (example 1), since current is typically proportional to irradiance. However, it also works for errors in voltage (example 2), because relative changes in current and voltage are interchangeable near P_{MAX} ($dV/V = -dI/I$).

This compensation is completely valid, and yields accurate reference module P_{MAX} values at the exact time of the irradiance calibration. However, it is less understood and appreciated that the compensation is *also valid for production module* P_{MAX} measured after irradiance calibration, as long as two conditions hold:

- Reference modules and production modules are identical (size, number of cells, cell technology,

- spectral response, angular response, temperature coefficients, etc.)
- Production line conditions that influence P_{MAX} (temperature, contact resistance, etc.) are constant between irradiance calibrations

Deviations from these ideal conditions because of production line variation affect the P_{MAX} uncertainty by increasing the flash test reproducibility.

In the ideal case, all systematic current and voltage errors are compensated, and therefore removed from the uncertainty budget. Total uncertainty of the production line power rating reduces to only three components, as in Table II. The main uncertainty contribution is from the reference module itself, while uncertainties "added" at the production line are quite small in comparison. Reference module uncertainty is outside the control of module manufacturers, but may decrease as calibration laboratory test procedures improve over time. For example, recent results [4] report measurement of a silicon module with P_{MAX} uncertainty ($k = 1$) of ± 0.57 %, rather than ± 1.03 % as listed in Table I.

Table II. Proposed uncertainty budget for power rating of production modules (simplified version of Table I, after removing systematic uncertainties that cancel as a result of flash tester irradiance calibration). See Section 2 for discussion of reproducibility and repeatability term.

Factor	Uncertainty
Reference module P_{MAX}	± 1.03 %
Reproducibility and repeatability	± 0.2 %
Drift of reference module	± 0.3 %
Total (sum in quadrature)	± 1.09 %

2 EXPERIMENT

We propose a method to test the claims of Section 1.2. The method also quantifies the reproducibility term in Table II, which may differ for various production lines and test apparatus, thus allowing for a straightforward assessment of power rating uncertainties in any given production setting.

2.1 General procedure

The procedure is described below (see also Figure 2):

1. Obtain a sample set of N modules, representative of the module production as a whole
2. Randomly select one module, and use it as a "silver" module to calibrate the flash tester
3. Measure all samples in the set (including the module from step 2) in random order
4. Repeat steps 2 – 3 until each module in the set has been used once as a silver module

The procedure has an implicit requirement to reference the power of each silver module to a single reference device ("golden" module). The details of this device are unimportant, other than that it has stable electrical characteristics for purposes of calibration transfer.

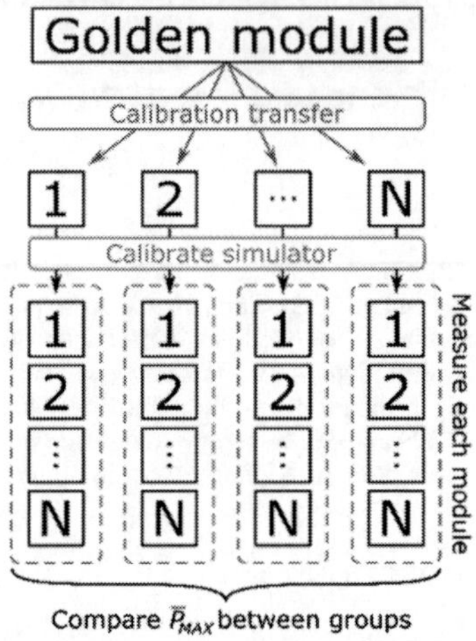

Figure 2. Schematic illustration of the experimental procedure.

The data analysis compares the average module power of each calibration group ($\bar{P}_{max,N}$). If most systematic effects are compensated, the differences in $\bar{P}_{max,N}$ should be dominated only by random effects (i.e., flash test reproducibility); there should be no significant difference in $\bar{P}_{max,N}$ values. The comparison can be framed rigorously using an analysis of variance (ANOVA) approach that tests the null hypothesis $\bar{P}_{max,1} = \bar{P}_{max,2} = \ldots = \bar{P}_{max,N}$. The ANOVA within-groups mean-square value can be used to directly quantify flash test reproducibility [5] for any given production setting.

The basic procedure can be scaled to production settings by using large sample sizes that are representative of overall module variability (e.g., spectral response, angular response, cell cracking, current mismatch due to I_{SC} sorting). The time scale of the experiment can also be selected to include relevant external sources of variability (different flash test shifts/operators, varying ambient temperatures, etc.). With proper experimental design, the reported reproducibility combines all these sources of variability and their possible interactions.

2.2 Specific procedure and experiment

In our case, the sample set is ten commercial modules from Trina Solar, each made from 60 cells of standard technology (multicrystalline silicon with aluminum back-surface-field). All modules were measured using Sinton Instruments' FMT-500, an industrial flash module tester that measures Suns-Voc and lifetime data in addition to standard current-voltage data.

For industrial relevance, we performed the procedure three times using distinctly different values for flash tester irradiance non-uniformity according to IEC 60904-9 [6]. Chosen values (Figure 3) span the range from nearly Class A (2.2 %) to nearly Class C (4.9 %). This is instructive, as irradiance non-uniformity is commonly thought to be a major contributor to production line P_{MAX} uncertainties. This experiment can be understood as simulating two different production scenarios: First, simulating the difference in production P_{MAX} that would be measured by different flash testers calibrated to the same set of reference modules. Second, simulating the difference in production P_{MAX} that would be measured by a single flash tester experiencing gradual changes in irradiance non-uniformity due to regular use over time.

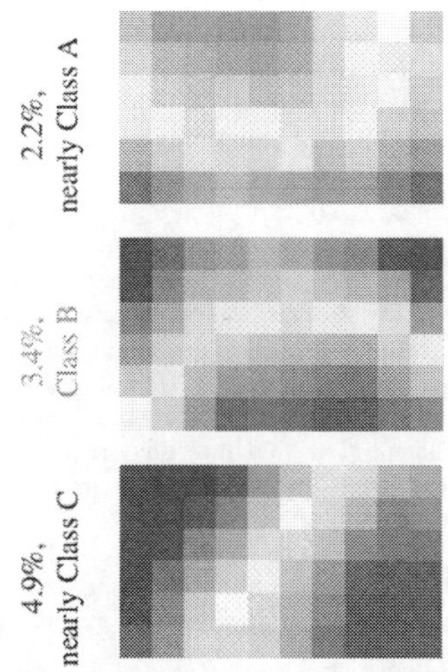

Figure 3. Irradiance measured across the module test plane, color-coded as a relative heat map. Irradiance non-uniformity according to IEC 60904-9 [6] is noted.

2.2 Results and discussion

Raw P_{MAX} data is shown in Figure 4. Visually, there is no obvious indication the measured P_{MAX} values are correlated to either the calibration group or flash tester irradiance non-uniformity. The first point supports the idea from Section 1.2 that systematic errors are accurately compensated when using reference modules that are nominally identical to the production modules under test. This is examined further in Figure 5, where the data is aggregated by calibration group. Visually, we still see no significant difference; this result is confirmed by the ANOVA F-test described in Section 2, which yields p-values ≥ 0.95 (much greater than any standard significance threshold: e.g., 0.05). Note—pre-planning for this study suggests the experimental design should detect $\bar{P}_{max}$ differences ≥ 0.1 % with adequate statistical power $(1 - \beta \geq 0.8)$, so this is unlikely to be a null result due to an underpowered experimental design.

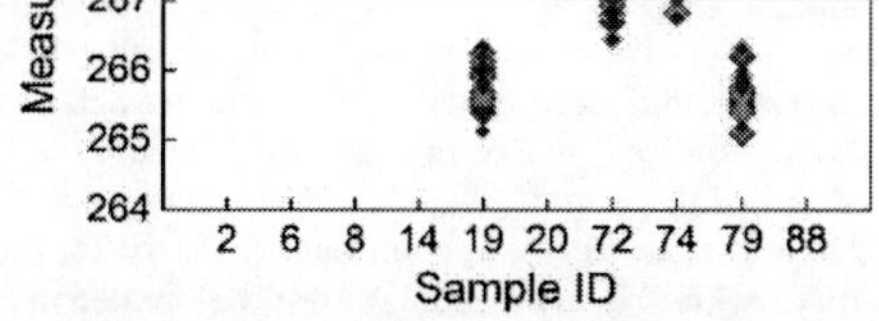

Figure 4. Raw P_{MAX} values measured for each sample. Data includes variation due to calibration group and irradiance non-uniformity.

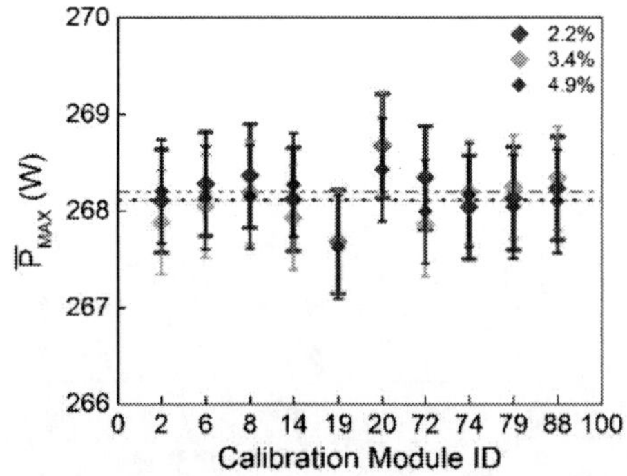

Figure 5. Same data as Figure 4, aggregated for each calibration group. Error bars are the flash test reproducibility (± 0.2 %, $k = 1$).

The most interesting feature of Figure 5 is the lack of difference in average P_{MAX} *even as the flash tester irradiance non-uniformity varies* by several percentage points. This is examined further in Figure 6, which aggregates the data across calibration groups in order to show only irradiance non-uniformity effects. In this case, there is < 0.02 % difference in average P_{MAX}, *independent of the flash tester irradiance non-uniformity*. Contrast this result against previous work [1], which suggests that using a flash tester with irradiance non-uniformity of 5 % (vs. 2 %) would increase the P_{MAX} uncertainty by 0.36 % (absolute, $k = 1$). Closer analysis resolves this apparent discrepancy: The previous work added the systematic P_{MAX} errors due to irradiance non-uniformity during the reference and production measurements as if they are random and uncorrelated. This is incorrect, since the same simulator is used for both measurements. Thus, both reference modules and production modules experience similar P_{MAX} errors due to irradiance non-uniformity, which can then cancel by the logic of Section 1.2. Using the Monte Carlo results of the previous work to analyze the uncertainties in this manner, we expect a flash tester with irradiance non-uniformity of 5 % (vs. 2 %) would increase the P_{MAX} uncertainty by < 0.07 % (absolute, $k = 1$), in general agreement with the experimental results presented here. The revised uncertainty decreases further when considering the interaction of irradiance non-uniformity and cell mismatch (I_{SC} sorting tolerance). The previous work tabulated these effects using assumed cell mismatch values between 2 % and 5 %. In our case, we characterized the cell mismatch of each module using published methods [7], and found the values to be much smaller (≤ 0.5 %), corresponding to much smaller P_{MAX} errors from the interaction of irradiance non-uniformity and cell mismatch.

Previous work [8, 9] has already shown a lack of correlation between flash tester spectrum classification and spectral mismatch uncertainties. The data presented here also call into question the relevance of the flash tester irradiance non-uniformity classification to P_{MAX} uncertainties for production line scenarios. At the very least we can observe that flash test procedures are equally important, if not more important, in determining P_{MAX} uncertainties than the actual flash tester specifications. When using the best practice of calibrating flash tester irradiance to reference module P_{MAX}, the flash tester classification requirements are actually quite minimal.

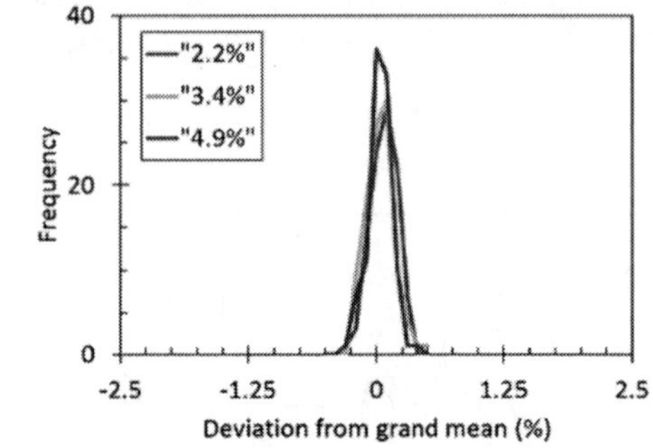

Figure 6. Same data as Figure 5, aggregated across calibration groups and displayed as a histogram in order to compare irradiance non-uniformity effects.

A few more insights appear by reanalyzing the data as if the flash tester irradiance calibrations had been performed using reference module I_{SC} (rather than P_{MAX}). The result (Figure 7) clearly illustrates the superiority of using P_{MAX} for irradiance calibration. Despite the fact that module I_{SC} is insensitive to temperature, contact resistance, and other voltage effects, it is *extremely* sensitive to irradiance non-uniformity. This is because module I_{SC} is not truly a module property, but rather, *determined by a single cell* limiting the module current at the zero-bias point of the module current-voltage curve. Thus, module I_{SC} is not indicative of the average irradiance over the entire module area, it is only indicative of average irradiance over a single cell area (and is also convoluted with a single cell's reverse bias and shunt resistance characteristics). Module P_{MAX}, on the other hand, is effectively a weighted average of the irradiance incident on each cell [8, 10], and so more accurately represents the average irradiance over the entire module area. This desirable property is in addition to the P_{MAX} compensation properties discussed at length throughout this paper. We therefore conclude that production line flash tester irradiance calibration using I_{SC} should be eschewed in favor of P_{MAX} calibration, which will always result in lower production line power rating uncertainties.

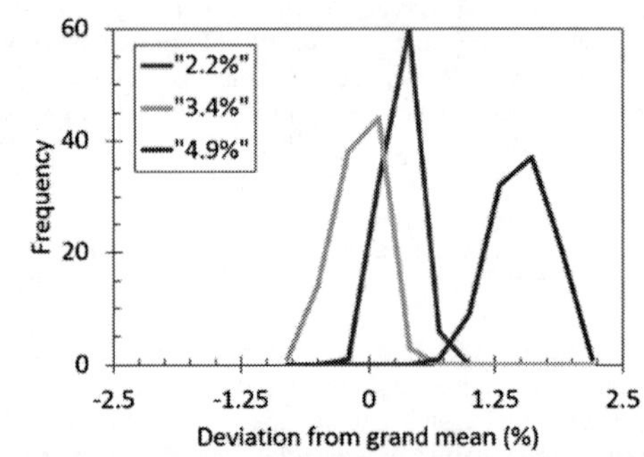

Figure 7. Same data as Figure 6, after reanalyzing the raw data assuming the use of module I_{SC} (rather than P_{MAX}) for flash tester irradiance calibration.

These points are also relevant when I_{SC} and V_{OC} in addition to power rating, as shown in Figure 8. One can view the calibration question—I_{SC} vs. P_{MAX}—as a tradeoff where calibrating to P_{MAX} gives lower P_{MAX} uncertainty at the expense of I_{SC} uncertainty, and vice-versa for I_{SC} calibration [1, 3]. However, even when calibrating flash tester irradiance using module I_{SC}, there is a rather large spread in reported I_{SC} values. As discussed before, this is because I_{SC} is effectively determined by one cell in each module, and so varies quite considerably between samples. We should note that I_{SC} uncertainties can become rather large (> 2 %) in the case of P_{MAX} calibration. This is acceptable given the relatively minor role of I_{SC} in module sale and system design (e.g., fuse sizing) compared to P_{MAX} (e.g., system sale price and electricity pricing). One final observation is the very slightly increased spread of V_{OC} values when using I_{SC} calibration vs. P_{MAX} calibration. Since I_{SC} calibration is less consistent at setting the average irradiance over the entire module area, it is possible that this V_{OC} spread is due to logarithmic variation of V_{OC} with average irradiance. The fact that this small variation is noticeable indicates just how variable the irradiance set by an I_{SC} calibration is.

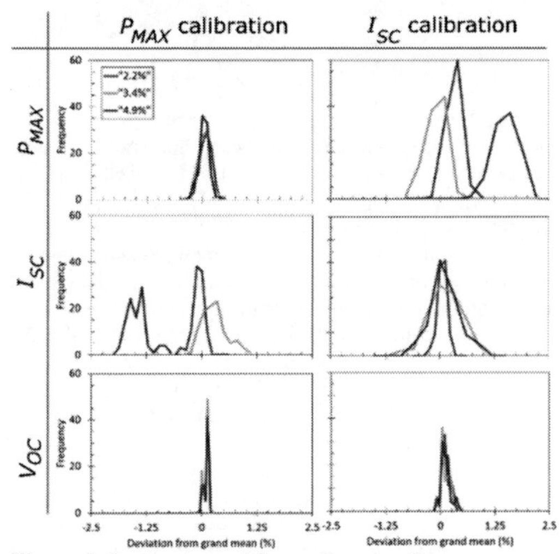

Figure 8. Comparison of I_{SC} vs. P_{MAX} irradiance calibrations for spread in reported I_{SC}, V_{OC}, and P_{MAX}.

3 CONCLUSIONS

Production line P_{MAX} uncertainties can be surprisingly small, and their assessment quite straightforward. Arguably, this is because production line flash test measurements are used to rate P_{MAX} among modules which are nominally identical in their construction, and thus, their electrical properties. Taking advantage of this fact by proper choice of reference modules allows several systematic errors in flash test measurements to be canceled. In the ideal case, the predominant P_{MAX} uncertainty component is then reference module itself, and uncertainties "added" by the flash test are quite small (± 0.2 %, $k = 1$). The exact amount of uncertainty added by flash test reproducibility will depend on the details of each production line, but can be quantified according to the methods of this paper.

Compare this to the conditions of a calibration laboratory, where every module must be treated as a "black box" with unknown physical and electrical characteristics to be diligently measured against existing primary and secondary reference devices. Each reference device carries uncertainties from many sources (for example, see published reports from NREL [11]). In a production line setting, however, the black box approach is overkill. The use of reference devices which are known to be similar to the production modules permits many simplifications, which is the essence of the methods presented here.

A particularly surprising result of our work is that *effects of flash tester irradiance non-uniformity can cancel entirely* from measured P_{MAX} values: Within the range of irradiance non-uniformity values tested here (2.2 % – 4.9 %), we saw virtually no impact (< 0.02 % difference in average P_{MAX}). Since irradiance non-uniformity is generally assumed to be a major contributor to flash test P_{MAX} uncertainties, this result is good news for the PV industry as a whole. In general, it means that robust flash test operating procedures can ensure the accuracy of STC P_{MAX} ratings, largely independent of flash tester classification. In an industry with high capital expenditure, the ability to use simpler and more cost-effective test equipment without sacrificing measurement accuracy is highly beneficial. There are also potential benefits associated with relaxing classification requirements for existing flash testers, thus decreasing the amount of production downtime required for regular maintenance to maintain classification.

One of the positions of this paper is that robust calibration are the most critical factor for achieving stable production line power rating uncertainties. Practical implementation of a robust calibration procedure should be based on reference module P_{MAX}, using modules that are nominally identical to the production modules to be tested. The procedure would ideally include multiple reference modules to mitigate effects of module drift and metastability. Module drift—e.g., due to contact resistance increases of the electrical connector with repeated use—may be monitored by several methods for measuring series resistance. Metastabilities may be best monitored by minority carrier lifetime characterization, such as that enabled by Suns-Voc measurement and analysis.

4 ACKNOWLEDGEMENTS

The authors would like to thank the State Key Laboratory of PV Science and Technology of Trina Solar for supplying the modules used in this study.

5 REFERENCES

[1] C. Monokroussos, D. Etienne, K. Morita, V. Fakhfouri, J. Bai, C. Dreire, U. Therhaag and W. Herrmann, "Impact of Calibration Methodology into the Power Rating of c-Si PV Modules Under Industrial Conditions," *Proc. of 28th EU PVSEC*, Paris, 2013.

[2] H. Wilterdink, A. Blum, C. Sainsbury and R. Sinton, "Spatial non-uniformity of irradiance and uncertainty in power rating for c-Si modules," *Proc. of NREL PVRW*, Denver, 2016.

[3] W. Herrmann and W. Wiesner, "Modelling of PV Modules - The Effects of Non-Uniform Irradiance on Performance Measurements with Solar Simulators," *Proc. of EU PVSEC*, Glasgow, 2000.

[4] D. H. Levi, C. R. Osterwald, S. Rummel, L. Ottoson and A. Anderberg, "Self-reference procedure to reduce uncertainty in module calibration," *Proc. of NREL PVRW*, Denver, 2018.

[5] D. C. Montgomery, Introduction to Statistical Quality Control (7th Edition), John Wiley & Sons, Inc., 2013.

[6] IEC 60904-9 Ed. 2.0, "Solar simulator performance requirements," 2007.

[7] H. Wilterdink, R. Sinton, C. Sainsbury, A. Blum, J. Dinger, and W. Dobson, "Measuring current mismatch in a module," *Proc. Workshop on Crystalline Silicon Solar Cells and Modules: Materials and Processes*, Winter Park, 2018.

[8] R. Sinton, H. Wilterdink, J. Dinger, and A. Blum, "Critical Evaluation of the Foundations of Solar Simulator Standards," *Proc. 44th IEEE PVSC*, Washington D.C., 2017.

[9] H. Wilterdink, R. Sinton, A. Blum, E. Schneller, K. Davis, and H. Field, "Effectiveness of the IEC 60904-9 Spectral Match Classification for Industrially-Relevant Silicon Solar Cells," *Proc. 25th EU PVSEC*, Amsterdam, 2017.

[10] R. Evans, K. H. Kim, X. Wang, A. Sugianto, X. Chen, R. Chen and M. A. Green, "Simplified technique for calculating mismatch loss in mass production," *Solar Energy Materials & Solar Cells*, vol. 134, pp. 236-243, 2015.

[11] K. Emery, "Photovoltaic Calibrations at the National Renewable Energy Laboratory and Uncertainty Analysis Following the ISO 17025 Guidelines," Technical Report TP-5J00-66873, 2016.

1ST INTERNATIONAL ROUND ROBIN ON EL IMAGING: AUTOMATED CAMERA CALIBRATION AND IMAGE NORMALISATION

Karl G. Bedrich*[1], Jing Chai[1], Yan Wang[1], Armin G. Aberle[1], Martin Bliss[2], Matevž Bokalič[3], Bernd Doll[4], Marc Köntges[5], Alexandra Huss[6], Juan Lopez-Garcia[7], Yong Sheng Khoo[1]

[1]Solar Energy Research Institute of Singapore, National University of Singapore, Singapore
[2]Centre for Renewable Energy Systems Technology (CREST), Wolfson School of Mechanical, Electrical and Manufacturing Engineering, Loughborough University, Loughborough, Leicestershire, LE11 3TU, UK
[3]University of Ljubljana, Faculty of Electrical Engineering, Tržaška cesta 25, SI-1000 Ljubljana, Slovenia
[4]Institute Materials for Electronics and Energy Technology (i-MEET), Martensstraße 7, FAU Erlangen-Nürnberg, D-91058 Erlangen, Germany
[5]Institute for Solar Energy Research Hamelin (ISFH), Am Ohrberg 1, 31860 Emmerthal, Germany
[6]Department of Physics at Colorado State University (CSU), Fort Collins, CO 80523, USA
[7]European Commission, DG JRC, Directorate C - Energy, Transport and Climate, Energy efficiency and Renewables Unit, Via Fermi 2749, 21027 Ispra (VA), Italy
*Corresponding Author Email: karl.bedrich@nus.edu.sg

ABSTRACT: Results from the first international Round Robin on electroluminescence (EL) imaging of PV devices are presented. 17 Laboratories across Europe, Asia and the US measured EL images of ten commercially available modules and five single-cell modules. This work presents a novel automated camera calibration and image scaling routine. Its performance is quantified through comparing intensity deviation of corrected images and their cell average. While manual calibration includes additional measurement of lens distortion and flat field, the automated calibration extracts camera calibration parameters (here: lens distortion, and vignetting) exclusively from EL images. Although it is shown that the presented automated calibration outperforms the manual one, the method proposed in this work uses both manual and automated calibration. 501 images from 24 cameras are corrected. Intensity deviation of cell averages of every measured device decreased from 10.3 % (results submitted by contributing labs) to 2.8 % (proposed method), For three images the image correction produced insufficient results and vignetting correction failed for one camera, known of having a non-linear camera sensor. Surprisingly, largest image quality improvements are achieved by spatially precise image alignment of the same device and not by correcting for vignetting and lens distortion. This is due to overall small lens distortion and the circumstance that, although vignetting caused intensity reduction of more than 50%, PV devices are generally positioned in the image centre in which vignetting distortion is lowest.
Keywords: Electroluminescence, Evaluation, PV module, Software

1 FIRST INTERNATIONAL ROUND ROBIN ON EL IMAGING

Electroluminescence (EL) imaging is a fast and non-destructive method for spatial characterisation of photovoltaic (PV) cells and modules [1]. An international inter-lab comparison (Round Robin) was conducted to further harmonise EL imaging methods, to estimate its potential for quantitative analysis and to qualify the respective IEC TS 60904-13:2018, which was at the time a draft standard [2].

17 laboratories across the EU, USA and Asia took part at this first Round Robin (RR). The experiments were conducted between January 2016 and September 2017. An initial evaluation on image quality of the RR data set was presented in [3]. This second work compares results, obtained from two different camera calibration methods, with results submitted by the contributing laboratories.

In this section an overview of the RR is given. Section 2 and 3 present two different methods (manual vs automated) for vignetting and lens calibration. Finally, EL intensity deviation of the RR data set is presented in Section 4.

Five single-cell-mini modules (short 'cells') and 10 commercially available modules are measured (Fig. 1). EL images are captured at two given currents corresponding to 10% and 100% of the devices nominal short circuit current (I_{sc}). Additionally, calibration images to correct for vignetting and lens distortion are taken (manual calibration, Fig. 5a).

Some laboratories submitted results for multiple camera or imaging setups. This resulted in 24 laboratory-camera combinations (short 'camera'). For the sake of anonymity, these sets are assigned single letters for CCD and CMOS cameras sensors (A-U) and double letters, starting with 'X' for InGaAs camera sensors (XA-XC).

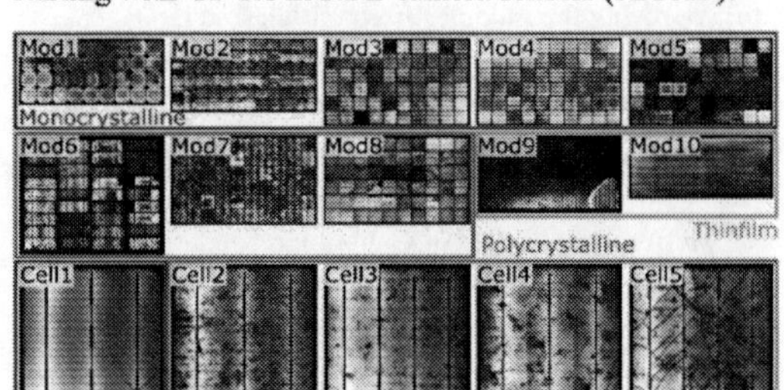

Figure 1: PV devices measured at the Round Robin

Two types of intensity deviation ($E_{I,i}$ and $E_{A,i}$, Fig. 2) are evaluated in this work: Intensity deviation of corrected EL images (I_i) and their cell averages (A_i). For both, the root mean square (RMS) difference between all images or cell averages (placeholder: #) and their temporal average is calculated. The temporal average (mean_z) is defined as element-wise average of multiple images.

$$E_\# = \sqrt{\mathrm{mean}_{x,y}((\# - \mathrm{mean}_z(\#))^2)}\ [\%] \qquad (1)$$

For both parameters, I_i is divided by its bit depth (255, for 8-bit images) and the border of 50 pixels around the device is removed.

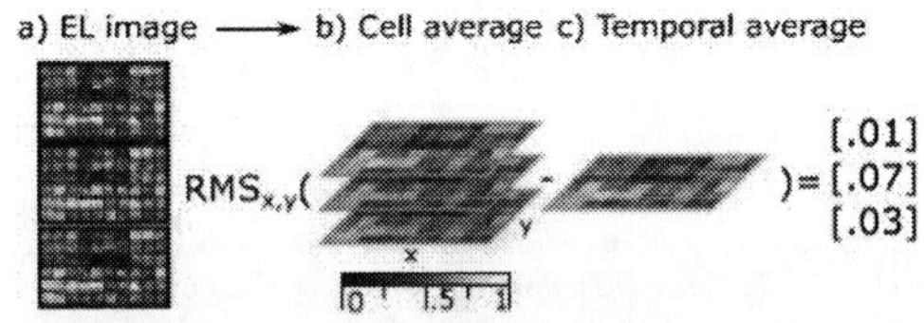

Figure 2: Scheme for calculating the intensity deviation from cell averages ($E_{A,i}$) of corrected EL images (only 3 shown for clarity)

Nine measurement protocols from eight labs containing a quantitative analysis of the taken images are submitted. An elementary part of quantitative analysis is the measurement of cell averages of the EL intensity. Seven labs stated that their cell averages are obtained fully automated. However, a direct comparison of the intensity deviation of submitted cell averages (E_A) of Mod1-10 (Fig. 3) reveals a median over 15% (Fig. 4b). If the cell intensities are chosen randomly in a uniform distribution from 0 to 100%, the deviation would be only twice the size (Fig. 4a).

The measurement plan specified to scale intensities between 0% (darkest cell) and 100% (brightest cell). However, in Fig. 3 cell averages of camera F and K do not contain a 0% or 100% cell. The E_A can be reduced to 10% (Fig. 4c) if all cell averages (A_i) of every module are normalized as follows:

$$\overline{A_i} = \frac{A_i - \min(A_i)}{\max(A_i) - \min(A_i)} [\%] \qquad (2)$$

In this work, intensity deviation of EL images of the same device and imaged by different cameras is reduced to 2.8% (Fig. 4d). For this, a method for image scaling and two methods for removal of lens distortion and vignetting are presented. It is stated that EL signal deviation due to different device temperatures, instabilities and cracks, introduced during transport between the labs, are not analysed. However, the value obtained in this work is close to the EL signal uncertainty of 2.8%, estimated in [3].

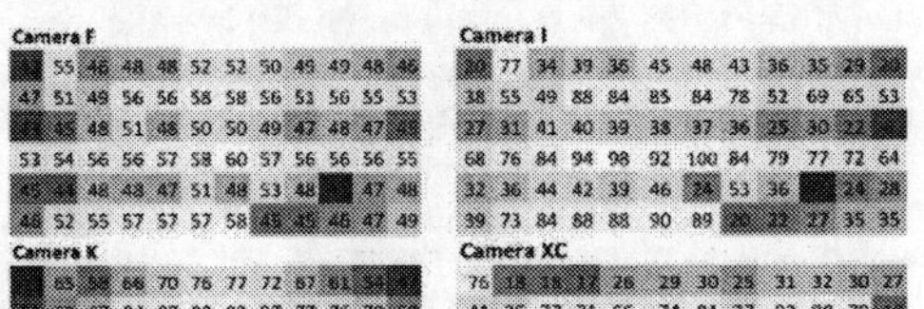

Figure 3: Submitted cell averages [%] of Mod2 (Fig. 1) at 100% I_{sc} from four example cameras (F, I, K, XC)

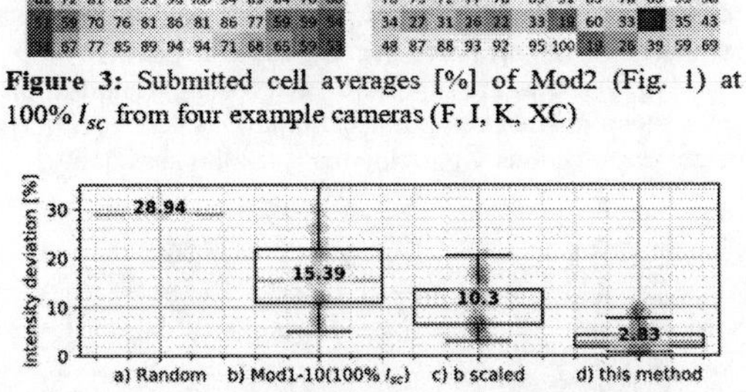

Figure 4: Intensity deviation (E_A, Eq. 1) of a) Random values for comparison; b) submitted cell averages of Mod1-10 at 100 % I_{sc}; c) same, after application of Eq. 2; d) Result for procedure presented in this work (Fig. 17)

Camera calibration and image correction are extensively discussed in [3]. In contract to this, this paper focusses on a novel automated calibration method to extract vignetting and lens distortion from EL images directly. It therefore does not rely on additional images taken (Fig. 5b). Also, a method for image intensity scaling is introduced.

Sensor nonlinearities and spatial variations of pixel sensitivity (especially for InGaAs cameras) can heavily reduce the calibration quality. Their calibration and correction are not investigated in this work. All vignetting maps, lens calibrations and corrected images analysed in this work can be found at [4].

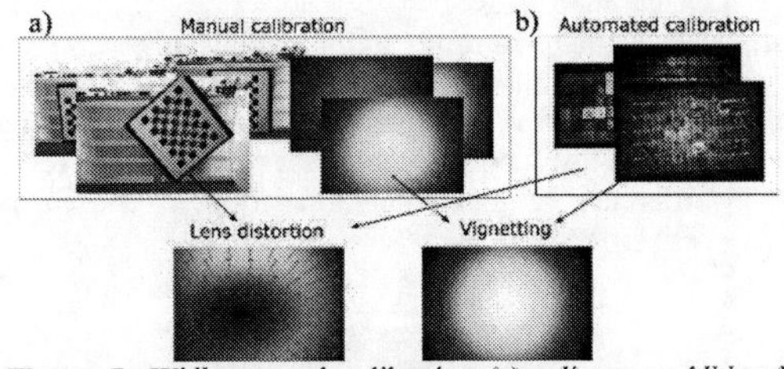

Figure 5: While manual calibration (a) relies on additional calibration images, automated calibration (b) extracts calibration parameters (lens distortion, vignetting) from the EL image itself

2 VIGNETTING MEASUREMENT

Vignetting describes the gradual intensity reduction from the image centre to its corners. The vignetting map (I_V) is scaled 0 to 1 and is of the same size as an EL image (I_{EL}). For linear camera sensors, vignetting correction is done by element-wise division $\frac{I_{EL}}{I_V}$. The effective vignetting map is specific for the camera system used [5]. Its dependency on wavelength, exposure time and aperture are not analysed in this work.

2.1 Manual calibration

All laboratories were asked to submit at least ten illuminated and five dark (background) images, taken with the same imaging setup. To illuminate the camera sensor homogeneously, a mobile phone or tablet displaying a red screen is placed on top of the camera lens at different positions. It is referred as method A in [5]. Using a red screen is originally proposed in [6]. The later publication [5] however discouraged its use due to vignetting overcorrection from using a red-, instead of a near infrared (NIR) light source.

For 22 cameras vignetting images are submitted, while two are missing. Background images, needed for vignetting measurement, are not available for four cameras. For those cases, the background level is set to the maximum position of the first peak of the intensity histogram from one of the background images of the submitted EL image data of that camera.

2.2 Automated calibration

The main challenge for vignetting calibration is the necessity for a homogenous light source to illuminate the camera sensor. The methods discussed in [5] attempt to eliminate this inhomogeneity by spatial or temporal averages. However, all methods rely on additional calibration images. In this work, I_V is extracted from EL images through a mixed spatial and temporal average of all EL images taken at 100% I_{sc}. For cameras measuring both modules and cells, 15 images are taken. For cameras measuring only cells, only five images are available. For a sufficient quality, the author recommends using at least 30 images. The algorithm to extract vignetting from EL images is as follows (Fig. 6):

1. Subtract background images, if available, or remove estimated background level from EL images.
2. Mask background areas through applying a threshold condition, such as Otsus method [7]. Average unmasked areas with a spatial mean filter. Here, a kernel size of 10% of the image width is used.
3. Divide every averaged image by its maximum and apply a temporal average on all images, excluding background areas.
4. To smoothen the result and to remove empty areas a polynomial filter, the radial average and the Kang-Weiss function are applied independently [5]. The output resulting in the respective lowest RMS difference to the temporal average is chosen as calibration map (I_V).

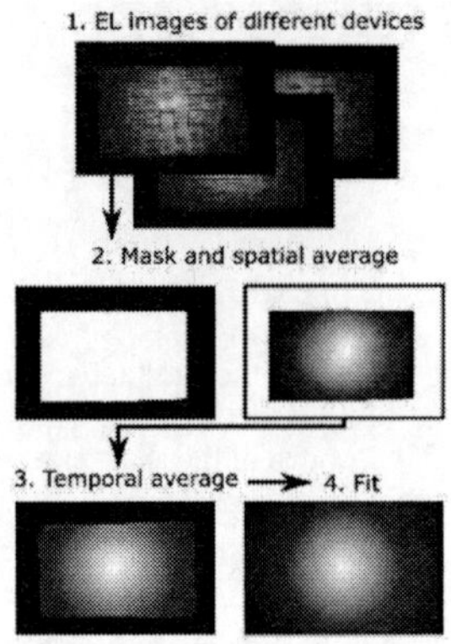

Figure 6: Algorithm to estimate vignetting of one camera from multiple EL images of different PV modules

The differences between manual and automated vignetting measurement are summarized in Table 1.

Automated vignetting calibration strongly depends on the quality and quantity of the available EL images. As Fig. 7 shows, automated calibration can fail when the temporal average is not capable of averaging out the devices spatial inhomogeneity and/or not a sufficiently large area is covered.

Table 1: Differences between both evaluated vignetting measurement methods

Manual	Automated
Waveband	
Visible (red)	NIR (EL)
Distance light source to camera	
Close	Same as EL (image plane)
Evaluated area	
Whole image	Only where devices are imaged
Pixel dependent sensitivity measured?	
Yes	No
Image fit	
Not needed	Either polynomial, Kang-Weiss or radial average

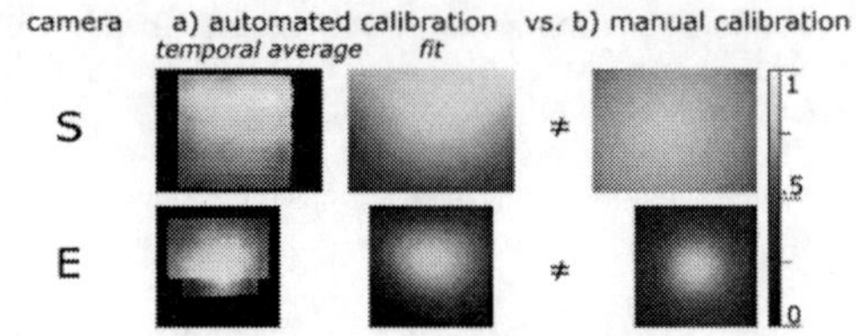

Figure 7: a) Temporal average and derived fit of automated calibration for camera S and E; Strong differences to manual calibration (b) indicate failure of automated calibration

2.3 Vignetting quality

The improvement due to vignetting correction (Q_V, Fig. 8) is quantified through comparing the straightness of an image ($q_V(I)$) before (I_{EL}) and after vignetting correction ($\frac{I_{EL}}{I_V}$):

$$Q_V = 1 - \frac{q_V\left(\frac{I_{EL}}{I_V}\right)}{q_V(I_{EL})} [\%] \quad (3)$$

The image straightness is defined as the standard deviation (σ) of a line wise horizontal and vertical image average (A_x, A_y):

$$q_V(I) = \sigma(A_x(I)) + \sigma(A_y(I)) \quad . \quad (4)$$

Background areas are excluded from A_x and A_y using a threshold condition, such as Otsus method.

Figure 8: Example vignetting map (a), uncorrected- (b) and vignetting corrected image (c); A_x and A_y displayed as green and blue lines; Q_V = 78.7%

It is stated, that Q_V depends on the homogeneity of the imaged PV devices and that a high Q_V is not a guaranty for high calibration quality since the undistorted source image is unknown. For robust quality metrics, refer to [5].

The Q_V distribution of the Round Robin data set is shown in Fig. 9a. Next to it, Fig. 9b displays the average of the individual I_V maps. Since I_V is bound between 0 and 1, an average of 1 (100%) corresponds to zero distortion from vignetting. It can be seen, that manual calibration tends to have a lower average vignetting and with it a higher extend of vignetting distortion. As [5] discusses, vignetting can depend on the waveband of the light source used to measure it. It was found that visible red light causes a more pronounced vignetting and therefore a vignetting overcorrection in comparison to NIR.

Fig. 9 allows the suggestion that for the Round Robin data set automated vignetting calibration generates slightly more homogenous vignetting maps at a higher quality.

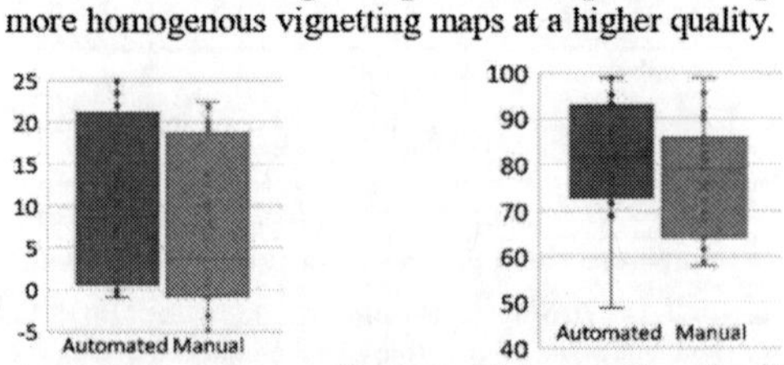

a) Relative improvement (Q_V) [%] b) Average vignetting [%]

Figure 9: Comparison of vignetting quality parameters

3 LENS DISTORTION

3.1 Grid detection

To estimate image lens distortion, reference points are detected. In case of manual calibration, checkerboard patterns (Fig. 5a) are imaged and the corner points of the black squares are detected using the C++ library OpenCV [8], [9]. For the automated calibration the following steps are applied:

1. Detect the corners of a bright quadrilateral object (here: PV device) in the image.
2. Estimate the number of cells and busbars from the horizontal and vertical image average along the detected device edges.
3. Refine positions of all detected lines using a maximum likelihood approach.

Example algorithms for step 1 and 3 are provided by the main author in [3]. Another routine, incorporating machine learning, is detailed in [10].

The quality of the resulting cell grid lines is inspected visually using the image material from the RR. In this work, grids build by PV devices are mostly detected correct (Fig. 10a). Images with low signal-to-noise ratio and images displaying only part of the PV device cause erroneous results. Also, the number of cells of the thin-film modules Mod9-10 is mostly detected wrong. The intensity reduction of their cells is often inferior to the intensity deviation from intrinsic and extrinsic defects and therefore hard to detect using machine vision (Fig. 10c,d).

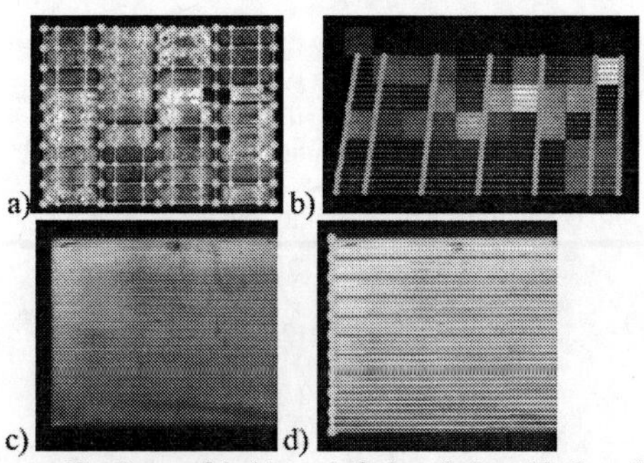

Figure 10: Successful (a) and faulty (b,d) grid detection; c: Cells of thin film modules are often hard to differentiate

The qualitative success rate is listed in Table 2. Devices with visually correct results are counted as 1 (Fig. 10a). Devices with incorrect results are counted as 0. Device position, number and position of cells as well as number and position of busbars are evaluated separately.

Table 2 shows, that the device position is detected correctly in 98% of all cases. Number of cells and busbars are detected correctly in 85% of all cases.

Table 2: Grid detection success rate for 255 analysed images of the Round Robin data set [%]

		Device position	n. cells	n. busbars
1	All	98.4	84.6	90.5
2	Only SNR>5 and fully imaged devices (={2})	98.7	86.5	92.2
3	{2} - only cells	100.0	100.0	100.0
4	{2} - only modules	98.1	79.6	88.9
5	{2} - only thin film	93.1	6.9	N.A.
6	{2} - only InGaAs	94.6	81.1	71.0

3.2 Lens calibration

For both manual and automated calibration, the camera intrinsics and distortion parameters are calculated using the OpenCV function 'calibrateCamera'. Amongst the available standard, rational and thin prism lens model, the model resulting in the lowest reprojection error (usually the standard model) is chosen for every calibration individually. The differences between manual and automated calibration are summarized in Table 3.

Table 3: Differences between both evaluated lens calibration methods

Manual	Automated
Input	
Images of a checkerboard pattern	EL images of PV modules
Waveband	
Mostly visible	Near infrared (EL)
Distance	
Mostly in image plane	Image plane (EL)
Evaluated area	
Limited to checkerboard-pattern positions	Limited to PV-module positions
Factors limiting calibration quality	
• Number of available calibration images • Checkerboard pattern: print quality and pad straightness • Camera focus (need to be adjusted for lenses without IR-correction)	• Number of available EL images of PV modules • EL image quality (sharpness and noise) • Quality of grid detection and precision of edge-alignment algorithm

3.3 Lens calibration quality

To assess the lens calibration quality, four quality numbers are applied to the RR data set: a) detected corner points, b) deflection assessment, c) relative improvement of the line accuracy, and d) point uniformity.

<u>a) Detected corner points.</u> For every camera, an average of 19 images of a checkerboard pattern at different positions within the image are submitted. The provided checkerboard pattern contains 8x6 corners. This results in an average of around ~900 corner points available for every manual calibration (Fig. 11a).

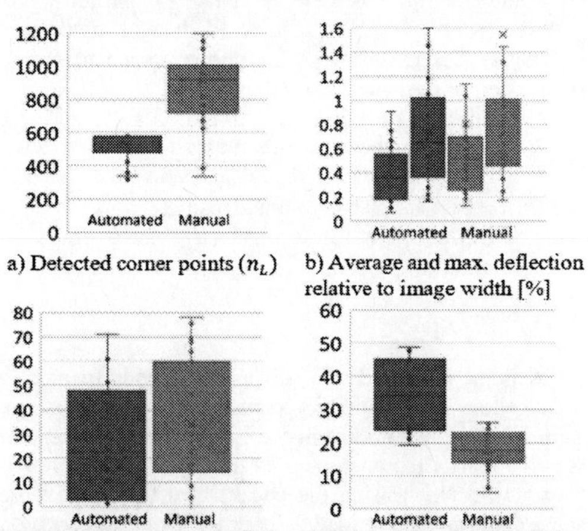

a) Detected corner points (n_L)
b) Average and max. deflection relative to image width [%]
c) Rel. improvement (Q_L) [%]
d) Point uniformity (U_L) [%]

Figure 11: Comparison of lens calibration quality parameters

For the automated calibration, only the crystalline modules Mod1-Mod8 can be used for calibration since the one-cell mini modules and thin-film modules do not provide the minimum number of 4 corner points in x and y dimension.

Only EL images at 100% I_{sc} are used for calibration since these images have a significantly higher image quality. Also, the images devices are generally not moved for 100% and 10% I_{sc} images. With eight available module images and detected grids of 5x9 to 7x13 (dependent on number of cells in the module), an average of ~550 corner points (n_L) is available per camera.

b) Deflection assessment. In this work, deflection describes the spatial shift due to correction from lens distortion. Both automated and manual calibration lead to similar results for average and maximal deflection (Fig. 11b). With 0.5% relative to image width, average deflection is so small, that lens distortion is hardly visible. However, if two images from different cameras are compared, the distance between features can be up to 50 px (given an image width of 5000 px and an opposing deflection of 0.5%).

c) Relative improvement of the line accuracy. The relative improvement (Q_L), relates the straightness of gridlines (G), before ($G_{i,0}$) and after correction from lens distortion ($G_{i,LC}$). The straightness ($q_L(G)$) is defined as average of all vector magnitudes, build by the difference of an ideal straight gridline (G_{ideal}) to the given (G).

$$Q_L = 1 - \text{mean}_i\left(\frac{q_L(G_{i,LC})}{q_L(G_{i,0})}\right)[\%] \quad (5)$$

$$q_L(G) = \text{mean}(|G - G_{ideal}|) \quad (6)$$

As Fig. 11c shows, relative improvement is in most cases positive although the result varies strongly from 5% to almost 80%. Improvement is roughly 8% higher for manual calibration. This is reasonable since the detection of black corners on a white background should have a higher precision than the detection of cell corners in an EL image.

d) Point uniformity: For high quality lens calibration, corner points should be distributed across the whole image [11]. Fig. 12 compares two extreme cases. In Fig. 12a the chosen checkerboard pattern is too small or far away and only moved within a small area. In Fig. 12b corner points detected from Mod1-Mod8 in the EL images are widely distributed. The uniformity of the corner point distribution (U_L) is determined as follows:

1. Split the imaged area in 10x10 cells.
2. Count number of corner points within each cell.
3. Divide all cells by the highest value in the grid to obtain the discrete point density (ρ_{10x10})
4. Define point uniformity (U_L) as average of ρ_{10x10}:

$$U_L = \text{mean}(\rho_{10x10}) \quad (7)$$

A point uniformity of 100% corresponds to a fully homogenous distribution of corner points in an image. Surprisingly, point uniformity of the automated calibration is twice that of the one obtained from manual calibration (Fig. 11d). When using the checkerboard pattern in the same plane as the PV module then PV modules occupy a larger area in the image.

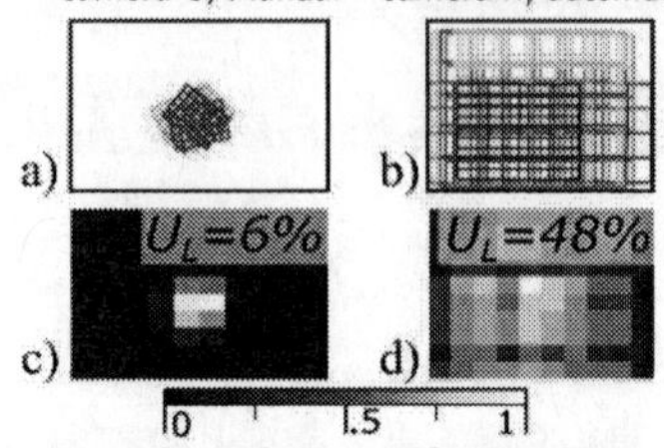

Figure 12: Comparison of a non-uniform (a,c) to a uniform (b,d) corner point distribution; top: gridlines built by detected corner points from imaged checkerboard pattern (a) and PV modules (b); bottom: point density discretized on a 10x10 grid

Therefore, the point uniformity extracted from the EL image of the PV module is higher. Additionally, less care might have been given to the additionally imaged checkerboard patterns, since only one of eight labs stated in their submitted protocol to execute calibration and correction from lens distortion as part of their EL imaging routine.

For the Round Robin data set it is found (Fig. 18) that the quality indicator (Q_{L2}) based on corner point number (n_L) and -uniformity (U_L) is more reliable than Q_{L1}:

$$Q_{L2} = U_L \cdot n_L \quad (8)$$

4 IMAGE COMPARISON

To compare the ability of manual and automated calibration to reduce camera dependent distortion, 15 EL image sets at 100% and 10% I_{sc} from all 24 cameras are corrected according to the following procedure (Fig. 13). A complete image set includes two consecutively captured EL images and two background images.

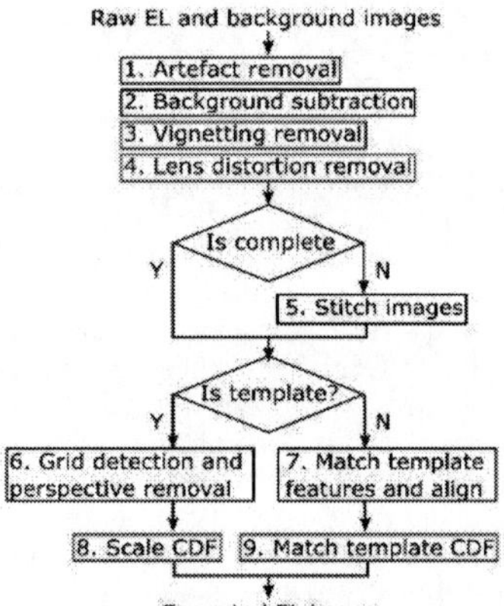

Figure 13: EL image correction procedure; Items in green boxes are discussed in this work

The EL image correction procedure contains the following steps:

1. Removal of image artefacts
   - Single-time-effects: Both taken EL images are averaged. If the image difference exceeds a given threshold, the corresponding minimum is used instead [3].
   - Hot pixels: All image areas where the ratio image vs. median filtered image exceeds a threshold are set to the median filtered result [3].
2. The background image is (after step 1) subtracted from

the EL image.

3. The corrected EL image (I_{EL}) is divided by the vignetting map (I_V).
4. Lens distortion is removed (Section 3.2).
5. For two cameras, larger modules are only available as part images (Fig. 14a). These images are stitched together in two steps:

- Estimate the position of a part image in the stitched image (depending on size of overlap either through key-point- or template matching).
- Average all part-images after perspective warp using 'blurry rectangles' as weights (Fig. 14c).

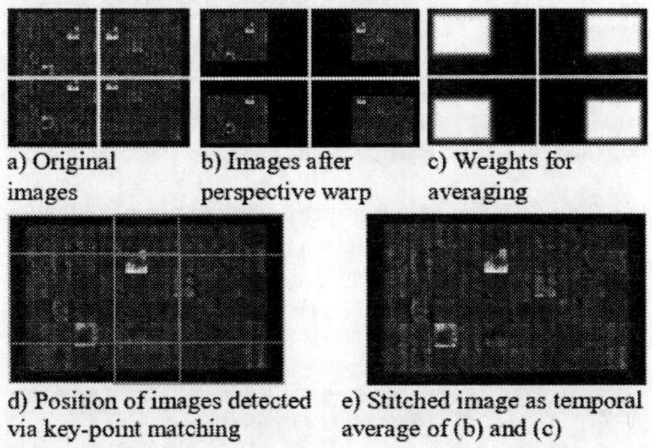

Figure 14: Image stitching steps

6. All images, taken by camera M, are set as template images due to their sufficient quality and no missing images. Perspective distortion is obtained from the result of grid detection (Section 3).
7. For all other cameras, perspective distortion is obtained through matching features between a template image (e.g. Mod3 from camera M and Mod3 from camera P). From the different positions of matched features, a single perspective transformation (homography) matrix is calculated.
   Perspective correction using a single homography matrix is not suitable to perfectly align images, especially if lens distortion remains, the images are of low quality (and only few features match) or the imaged devices are slightly bent. A precise alignment algorithm, similar to the one presented in [3] is used to reduce remaining distortion (only case 'bestPrecise').
8. To equalize the image intensity spectrum a cumulative distribution function (CDF) of the image intensity is calculated. The intensities at the left end of the CDF (1 and 2%) and the right end (98 and 99%) are calculated (Fig. 15). The intensity difference between 1 and 2% respective 98 and 99% is subtracted (left) or added (right) to obtain the representative intensity range.

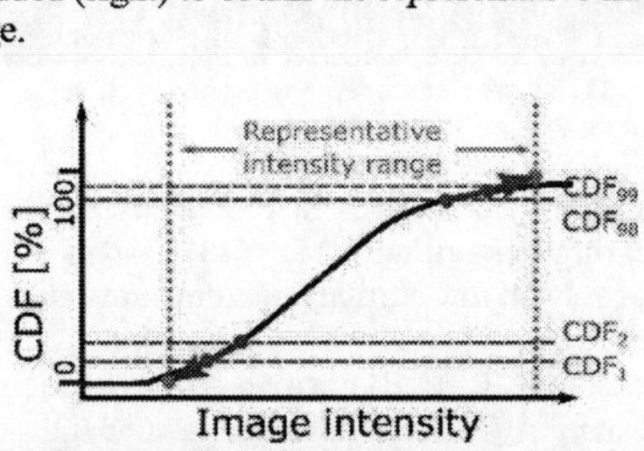

Figure 15: Scheme to obtain the representative intensity range from evaluating the CDF at 1, 2, 98 and 99 %

9. If the image is not a template (I_i) the representative intensity range (i_{mn}, i_{mx}) is determined from a linear fit ($y = mx + n$) of the images CDF to the templates CDF (Fig. 16):

$$i_{mn} = \frac{i_{T,mn}-n}{m}, \; i_{mx} = \frac{i_{T,mx}-n}{m} \qquad (9)$$

$$m,n \rightarrow \operatorname{argmin}\left(\operatorname{abs}\left(\mathrm{CDF}(I_T) - \mathrm{CDF}(I_i \cdot m + n)\right)\right) \qquad (10)$$

Powell's method [12] is used to solve Eq. 10.

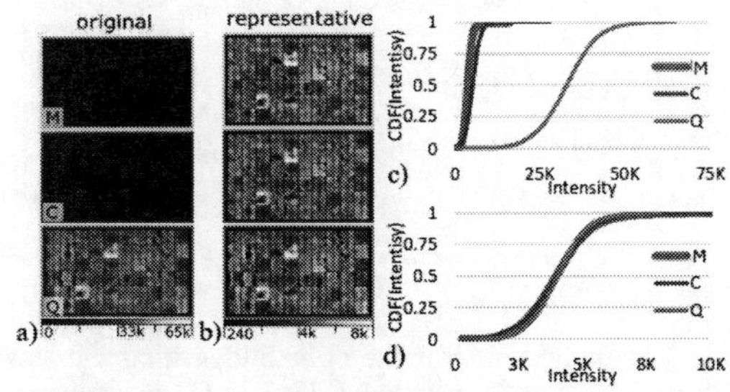

Figure 16: Mod7 at 100% I_{sc}, imaged by cameras M, C and Q; template image is M; a) images at original intensity range; b) images at representative intensity range; c) CDF before fit; d) CDF after fit using Eq. 10

Image similarity is compared for the following cases:

- **base** – Image correction according to Fig. 13, but without vignetting removal {step 3} and lens correction {step 4}.
- **manual** – Vignetting removal and lens correction using only manual calibration (if available, else no correction).
- **automated** – Vignetting removal and lens correction using only automated calibration (if available, else no correction).
- **best** – Vignetting removal and lens correction using either manual or automated calibration, dependent on the highest relative improvement Q_V, Q_{L2} (Eq. 3, 8)
- **bestPrecise** – same as case best using 'precise alignment' {step 7}.
- **+100% I_{sc}** – same as case bestPrecise but using only high current EL images to allow direct comparison to cell averages, submitted by the Round Robin contributors.

Fig. 17 displays the intensity deviation of all corrected EL images (E_I) and their cell averages (E_A) of all compared cases. Both boxplots follow a similar trend. Surprisingly, both E_I and E_A increase significantly for manual calibration. Possible reasons are already discussed in Section 2.3 and 3.3.

For both parameters, automated calibration outperformed case base and case manual. This indicates, that a calibration using only EL images can result in higher quality EL image correction in comparison to no or manual calibration. Case best, which uses either manual or automated calibration, outperforms case automated.

The additional precise alignment of case bestPrecise causes the largest decrease for E_I, did however not influence E_A since image intensities remained mainly unchanged and only deflection changed slightly.

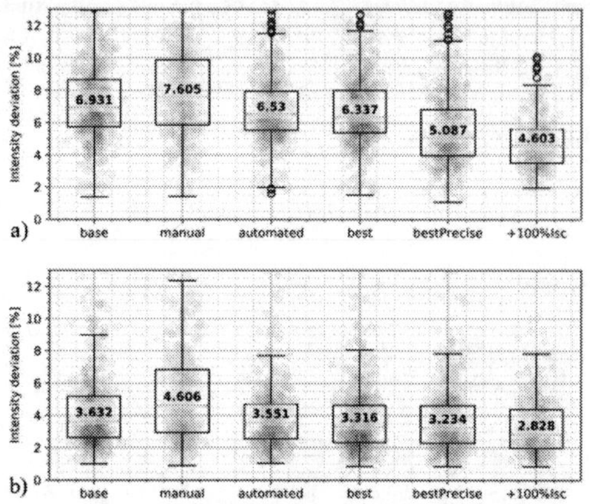

Figure 17: Intensity deviation of corrected EL images (E_I, a) and their cell averages (E_A, b)

The individual influence of both vignetting and lens distortion correction is visualized in Fig. 18. It contains an additional case best2. For vignetting, this case chooses either manual or automated calibration depending on the respective higher average vignetting (Fig. 9b).

For lens distortion, the initial relative improvement Q_L is used as selection criterion. For cell averages the positive effect of lens calibration can be neglected (Fig. 18b, right). However, the relative improvement, based on the distribution of detected corner points (Q_{L2}) used in case 'best' is favourable to Q_L which only considers the grid straightness. For vignetting calibration no difference between case 'best' and 'best2' is visible. This allows the conclusion that statistically more homogenous vignetting maps also generate better image corrections. Again, it must be emphasized that the vignetting calibration method used in the Round Robin is found to cause vignetting overcorrection and is not encouraged any more [5].

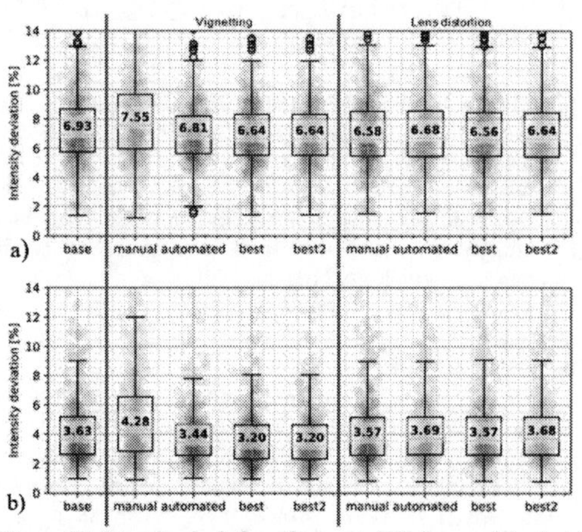

Figure 18: Intensity deviation of corrected EL images (E_I, a) and their cell averages (E_A, b) for exclusive correction from vignetting (middle) or lens distortion (right)

The following figures are an excerpt of [4]. In most cases, images corrected in case bestPrecise are visually more similar than the ones of case base (Fig. 19).

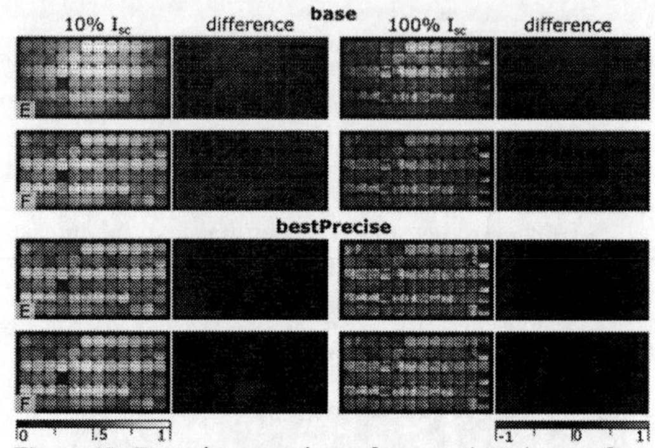

Figure 19: Example comparison of corrected EL images from Mod2 (1st, 3rd column) and difference to EL images from template camera M (2nd, 4th column) for two cameras (E, F) of case base (top) and case bestPrecise (bottom)

Some EL image differences display variable EL signal either due to instabilities (Fig. 20 left) or due to introduction of extrinsic defects (cracks) during the course of the Round Robin (Fig. 20 right).

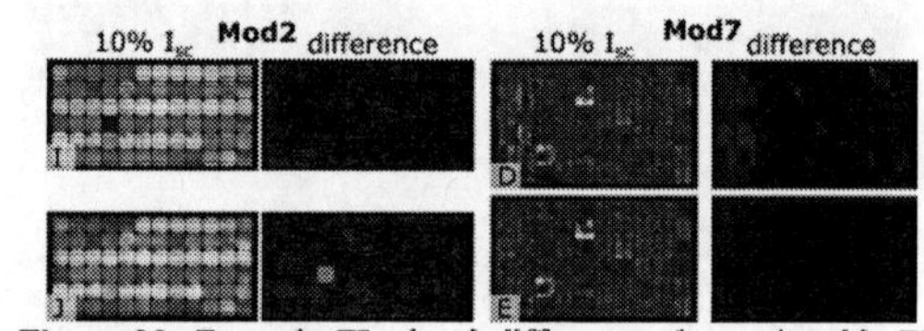

Figure 20: Example EL signal differences due to instable EL signal (left) or cracks (right); case bestPrecise

Corrected EL images from camera P are the only ones that clearly show effects of a nonlinear camera sensor. This nonlinearity causes amplified low image intensities and suppressed higher intensities. This can be seen through a more pronounced vignetting for low current (darker) EL images in comparison to high current images (Fig. 21 left vs right). Since both images are corrected with the same vignetting map, remaining vignetting can be found in the low current image and vignetting overcorrection is visible for the high current image.

It is stated that only K and P are consumers product cameras with a CMOS sensor. All other cameras are based on a silicon CCD or InGaAs sensor.

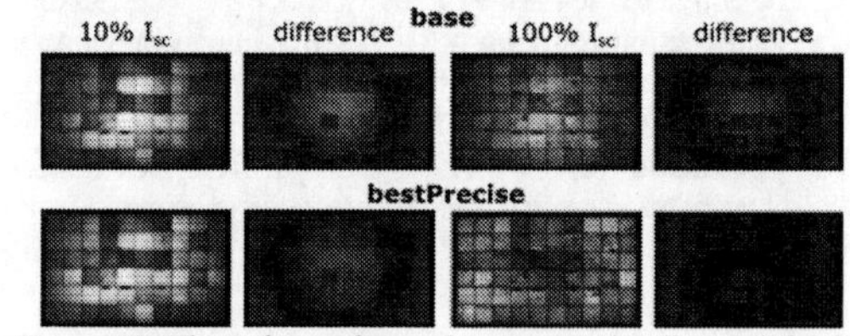

Figure 21: Vignetting under- and overcorrection (bottom) for the non-linear camera P (Mod8)

Amongst all images in case bestPrecise, the image processing algorithm (Fig. 13) failed three times exclusively on low current (and often low quality) 10% I_{sc} images:

- camera K, R (Cell1): wrong device rotation
- camera A (Mod5): failed image stitching

5 CONCLUSION

This paper presents and validates a novel automated camera calibration which does not depend on additional calibration images. In comparison to manual calibration it allowed for more accurate image correction. Even higher accuracy is reached when, dependent on presented quality parameters, either the manual or the automated calibration is chosen.

This work further presents methods to scale image intensities and to measure intensity similarity. For all 501 processed image sets, the proposed image correction case bestPrecise failed on three images. Vignetting correction produced insufficient results for only one camera, known to have a non-linear camera sensor. Non-linearity correction will be covered in future work.

The comparison of intensity deviation shows that contrary to prior assumptions, the largest EL quality improvement comes from precise spatial alignment of images to be compared, followed by vignetting correction. Especially for cell averages, the positive effect of lens calibration is negligible. For 100% I_{sc}, case bestPrecise resulted in a three times lower intensity deviation of cell averages (2.8%) than the one generated from data submitted by the Round Robin contributors (10.3%, Fig. 4).

All key features that enable quantitative EL imaging are included in QELA, a freely accessible software for EL image analysis [13]. It uses the automated calibration for its integral image correction routine. It is planned to enable submission of own manually taken camera calibrations which will only be applied if the quality metrics presented in this work indicate a higher quality.

7 REFERENCES

[1] T. Fuyuki, H. Kondo, T. Yamazaki, Y. Takahashi, and Y. Uraoka, "Photographic surveying of minority carrier diffusion length in polycrystalline silicon solar cells by electroluminescence," *Appl. Phys. Lett.*, vol. 86, no. 26, p. 262108, 2005.

[2] IEC, "IEC TS 60904-13: Photovoltaic Devices - Part 13: Electroluminescence of photovoltaic modules." 2018.

[3] K. G. Bedrich, "Quantitative Electroluminescence Measurement of PV Devices (Thesis)," Loughborough University, 2017.

[4] K. G. Bedrich, "1st International Round Robin on EL imaging: Automated Camera Calibration and Image Normalization - Additional material," 2018. [Online]. Available: https://www.researchgate.net/publication/327767524_35th_EU_PVSEC_2018_5BO111_Additional_material_corrected_images_and_camera_calibration_maps.

[5] K. G. Bedrich, M. Bokalič, M. Bliss, M. Topič, T. R. Betts, and R. Gottschalg, "Electroluminescence Imaging of PV Devices : Advanced Vignetting Calibration," *IEEE J. Photovoltaics*, vol. PP, no. 99, pp. 1–8, 2018.

[6] M. Topič, J. Raguse, K. Zaunbrecher, M. Bokalič, and J. R. Sites, "Electroluminescence of Thin Film Solar Cells and PV Modules - Camera Calibration," in *26th European Photovoltaic Solar Energy Conference and Exhibition*, 2011, pp. 2963–2966.

[7] N. Otsu, "A threshold selection method from gray-level histograms," *IEEE Trans. Syst. Man. Cybern.*, vol. 9, no. 1, pp. 62–66, 1979.

[8] V. Pisarevsky, "Introduction to OpenCV," *Agenda*, vol. 42, pp. 433–434, 2010.

[9] A. Datta, J.-S. Kim, and T. Kanade, "Accurate camera calibration using iterative refinement of control points," *2009 IEEE 12th Int. Conf. Comput. Vis. Work. ICCV Work.*, pp. 1201–1208, Sep. 2009.

[10] S. Deitsch, C. Buerhop-Lutz, A. Maier, F. Gallwitz, and C. Riess, "Segmentation of Photovoltaic Module Cells in Electroluminescence Images," 2018.

[11] P. Galeone, "Camera calibration guidelines," 2018. [Online]. Available: https://pgaleone.eu/computer-vision/2018/03/04/camera-calibration-guidelines/. [Accessed: 23-Aug-2018].

[12] M. J. D. Powell, "An efficient method for finding the minimum of a function of several variables without calculating derivatives," *Comput. J.*, vol. 7, no. 2, pp. 155–162, 1964.

[13] K. G. Bedrich, "QELA," 2018. [Online]. Available: https://qela.seris.sg.

Electroluminescence for Power Loss Prediction of Photovoltaic Modules

Timo Kropp, Markus Schubert, Jürgen H. Werner
University of Stuttgart, Institute for Photovoltaics and Research Center SCoPE, Stuttgart, Germany,
timo.kropp@ipv.uni.stuttgart.de

ABSTRACT: Electroluminescence imaging usually only qualitatively maps the local electrical performance of photovoltaic modules. For example, due to their low luminescence intensity, cell cracks as well as partially and fully disconnected cell areas are visible in case of modules built from crystalline silicon. However, so far, a direct quantitative prediction of the electrical properties of modules just from electroluminescence images has yet to be accomplished. Here, we present a novel two-image approach for calculating a local series resistance mapping of the total module. Using the resistance mapping as simulation input, we straightforwardly predict the electrical performance of the total module. The simulation model is based on a simplified series resistance area-element segmentation. We experimentally validate and confirm the simulation results by the characterization of a photovoltaic module with cracked cells from on-site hail impacts. The deviation of the predicted and directly measured current/voltage-curve is less than 4.3% at the maximum power point.
Keywords: Electroluminescence, PV Module, Defects, Simulation, Reliability, Photovoltaic, Performance

1 INTRODUCTION

Electroluminescence (EL) imaging spatially resolves local and overall electrical properties of photovoltaic (PV) modules. By visual inspection, those images easily reveal mechanical damages/impacts that degrade photovoltaic modules [1]. For example, a mechanically damaged contact grid separates the cell in fully and partially disconnected cell fragments. Via a locally reduced luminescence intensity, those defects are directly visible in EL images. However, in most cases, these images only provide qualitative information about the pure existence of defective cells or the type of defects in PV modules.

First approaches purely using EL images for quantifying the module performance only evaluate the contribution of individual cells to the total module voltage [2]. Furthermore, a simplified modelling approach for describing the electrical performance of a module with cracked cells was developed [3]. However, the presented simulation model has no input from or direct relation to EL images and is not capable of predicting the power loss of a specific damaged module.

The present contribution demonstrates a novel two-image electroluminescence modelling approach for calculating a series resistance map of a damaged PV module. By area-element segmentation of the resulting series resistance mapping, we generate the input parameters of a simulation model. With data sheet information and a self-calibrating approach, our simulation model straightforward predicts the resulting electrical performance in good agreement with experimental results of a module with cracked cells from on-site hail impacts.

2 MODELLING PRINCIPLE

Evaluating two EL images taken under low (<10% I_{sc}) and high (>30% I_{sc}) current injection, the operating cell voltage of individual cells can be calculated as introduced by Potthoff [2]. Figure 1 shows the high current EL image in Fig. 1(a) and the low current EL image in Fig. 1(b).

We extend the approach of Potthoff et.al. for determining the operating voltage of each individual cell: We calculate the local series resistance $r'_{i,s}$ (x,y) cell number i at the location (x,y) by comparing the luminescence intensity of the cell in the high current image to the maximal luminescence intensity $\phi_{i,ref}$ (x_0,y_0) of the very same cell. Then we derive the local series resistance

$$r'_{i,s} = \frac{n_{id}V_{th}\, ln\left(\frac{\phi_{i,ref}}{\phi'_i}\right) + J_0\left(\frac{\phi_{i,ref}}{C}\right) r_{i,ref}}{J_0\left(\frac{\phi'_i}{C}\right)} \quad (1)$$

for each individual cell i. The thermal voltage V_{th}, the ideality factor n_{id}, reference resistance $r_{i,ref}$ as well as the recombination current density J_0 are assumed to be constant for all cells, unrelated to the structural defect. The recombination current density J_0 is directly calculated from the open circuit voltage V_{oc} of the module. The calibration factor C is determined from the low current EL image as proposed by Potthoff [2], neglecting the voltage drop along series resistances. The ideality factor n_{id} remains as only uncertainty and needs to be adjusted or calibrated between $1.0 < n_{id} < 1.4$ as typical value for multi-crystalline silicon solar cells.

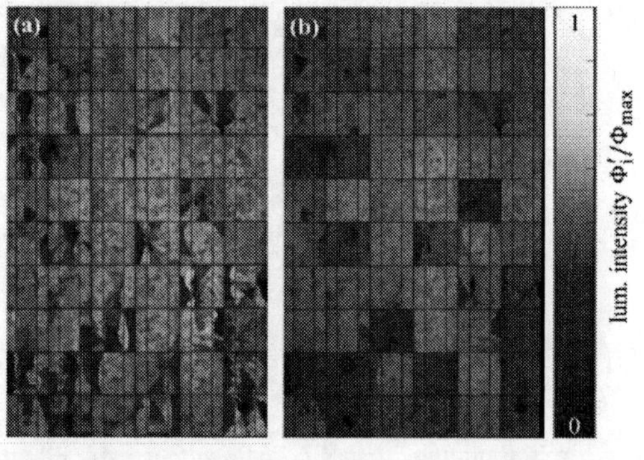

Figure 1: Electroluminescence images of a damaged multi-crystalline silicon photovoltaic module, (a) at supplied current of 37% with 100 s exposure time and (b) with 7% of short circuit current I_{sc} = 8.3 A and 360 s exposure time. Both images are normalized to their maximal luminescence intensity.

In the high current EL image, the reference luminescence intensity $\phi_{i,ref}(x_0, y_0)$ is given by the maximal intensity in each individual cell i.

Via Eq. (1), all local luminescence intensities $\phi_i'(x,y)$ from the high current EL image are transposed into the series resistance mapping $r_{i,s}'(x,y)$ for all (x,y) of each individual cell i. Since all luminescence intensities $\phi_i'(x,y)$ are referenced to the maximal luminescence intensity $\phi_{i,ref}(x_0, y_0)$ of cell i, the local series resistance $r_{i,s}'(x,y)$ is calculated in relation to the reference series resistance $r_{i,ref}(x_0, y_0) = d\,\overline{r_s}$ at the location of the reference luminescence $\phi_{i,ref}(x_0, y_0)$.

The reference series resistance $r_{i,ref}(x_0, y_0)$ is the minimal localized specific series resistance of cell i due to the highest luminescence $\phi_{i,ref}(x_0, y_0)$. We approximate this minimal resistance with the displacement factor d, as deviation from the mean series resistance $\overline{r_s}$ of a defect free cell extracted from the data sheet information via the one diode model. The displacement factor d is calculated iteratively from the mean value of three defect free cells in the series resistance map $r_{i,s}'$ using the bisection method.

As a result, the mean series resistance of the three defect free cells is then identical to the mean series resistance $\overline{r_s}$ of a defect free cell extracted from the data sheet via the one diode model.

Figure 2(a) shows the series resistance mapping $r_{i,s}'$ with colored segmentation categories in Ωcm^2. The space in-between the cells is cut out and excluded for the series resistance mapping. Figure 2(b) qualitatively shows the histogram of categorized series resistances $r_{i,s}'$ of one cell i with the relative area $A_{i,seg}/A_c$ of a segment as share of the total cell area Ac. The mean series resistance $r_{i,seg}$ in each segment is used for the simulation model shown in Fig. 2(c), predefined in ten segments. Each series resistance is weighted by its relative area $A_{i,seg}/A_c$. All constant diode model parameters (green box in Fig. 2(b)) are directly extracted from the data sheet or calculated from a diode fit.

Figure 2(c) shows the resulting simulation model for cell i. All cells $i = 1, 2, \ldots, N$, with the total number of cells N, are connected in series and simulated with the bypass-diode configuration of the module.

By varying the ideality factor n_{id}, the simulation can be further calibrated. The reverse characteristics for describing the re-verse current density J_{rev} are numerically modelled as proposed by Quaschning [4], with the breakthrough voltage V_{br} = -15 V as well as the breakthrough model parameters $n = 2.3\times10\text{-}3$ and $a = 1.9$. By choosing an identical ideality factor for all cells and a typical value between $1.0 < n_{id} < 1.4$ for multicrystalline silicon solar cells, we find very good agreement with measured results.

3 EXPERIMENTAL

For evaluating the simulated results, we prepare a multi-crystalline silicon PV module for measuring the current-voltage curves of each cell individually. The characterized module (Chinalight Solar, CLS-230P, maximum power P_{mpp} = 230 W_p) was already installed prior to the experiments and shows multiple cracked cells after on-site hail impacts (compare Fig. 1(a) and Fig. 1(b)). Small holes drilled through the back sheet encapsulation are placed inbetween the cells dircetly reaching the bus bar interconnections. At the interconnections inbetween the cells, the positive and negative contact of each cell is accessible.

A Keithley 2651a source-meter captures the current/ voltage-curves of each cell as well as of the total module while a halogen-lamp solar simulator irradiates the module with the intensity Φ_{irad} = 600 W/m². All current/voltage-

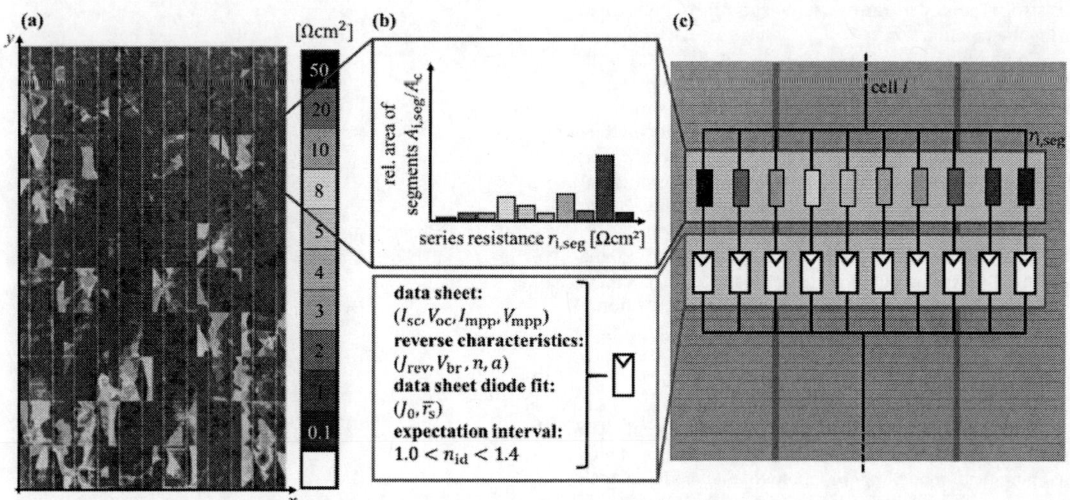

Figure 2: Series resistance mapping (a) with colored segmentation categories (b). Separated into ten area segments with the relative area $A_{i,seg}/A_c$ of segments, the cells i=1,2,..., N with the cell area A_c are modeled in (c) defined by the series resistance segments $r_{i,seg}$ and the constant diode model parameters extracted from the data sheet. The ideality factor n_{id} is identical for all cells and varied between $1.0 < n_{id} < 1.4$. The reverse characteristics for describing the reverse current density J_{rev} depending on the breakthrough voltage V_{br} = -15 V as well as the breakthrough model parameters $n = 2.3\times10^{-3}$ and $a = 1.9$ are numerically modeled as proposed by Quaschning [4]. All other parameters are either extracted from the data sheet or calculated from the series resistance mapping and scaled by the relative area of the segment. All cells i are connected in series and simulated with the bypass-diode configuration of the module.

curves are temperature and irradiation corrected for deriving the standard test condition (STC) electrical characteristics.

Figure 3(a) shows maximal power $P_{i,mpp}$ of the individual cells extracted from the current/voltage-measurements normalized to the maximal measured cell power $P_{mpp,max}$ in the module.

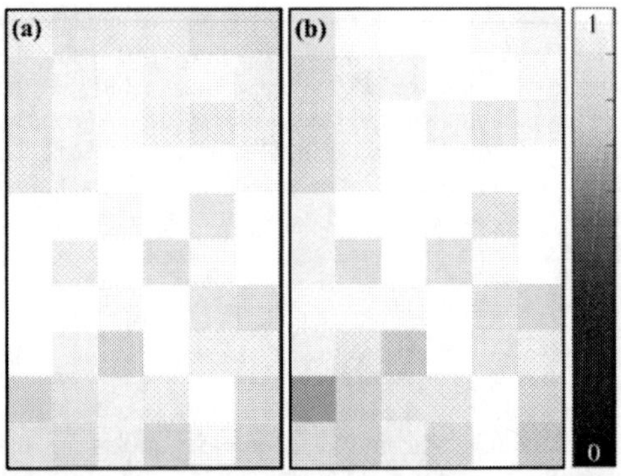

Figure 3: Measured maximal power output (a) and simulated maximal power output (b) for all individual cells. All cell powers are normalized the maximal measured or simulated cell power in the module and colored by the same false color scale. The location of defective cells and the normalized cell power is in good agreement comparing simulation and measurement.

Comparing the EL image in Fig. 1(a) with the measured normalized cell powers, Fig. 3(a) matches the location of strongly damaged cells. The cells located at the top of the module in Fig. 3(a) show the strongest deviation, since there are no damaged cells in the EL images. However, this can be explained by an increased series resistance induced by the measurements (only in this case) using the junction box.

Comparing the measured results in Fig. 3(a) with simulated results in Fig. 3(b), a very good agreement between the individual normalized cell powers is found. The maximal measured power of the total module $P_{mpp,meas}$= 209 W ± 5% (at STC) is also in very good agreement with the simulated power expectation interval 200 W < $P_{mpp,sim}$ < 206 W, depending on the chosen ideality factor n_{id}.

This maximal deviation of 4.3% compared to the measured power output is in the same range of accuracy, resulting from common current/voltage characterization.

3 CONCLUSION

This work has presented the proof-of concept of a novel electroluminescence two-image series resistance mapping and modelling of photovoltaic modules with cracked solar cells. The series resistance mapping is successfully extracted from two luminescence images with low (<10% I_{sc}) and high (>30% I_{sc}) current injection. Using the data sheet information of the photovoltaic module, the simulation model is capable of predicting the expected power output with an error less than 4.3%.

For the application of the proposed method to a variety of defect types, including shunts or potential induced degradation (PID), an approximation of the change in parallel resistance needs to be included.

Furthermore, the impact of image quality on the simulation results should be evaluated.

ACKNOWLEDGEMENT

We gratefully acknowledge funding by the German Federal Ministry for Economic Affairs and Energy (BMWi) under Contract No. 0324069A.

REFERENCES

[1] S. Kajari-Schröder, I. Kunze, U. Eitner, M. Köntges, "Spatial and orientational distribution of cracks in crystalline photovoltaic modules generated by mechanical load tests," Solar Energy Materials and Solar cells, vol. 95, pp. 3054-3059 (2011)

[2] T. Potthoff, K. Bothe, U. Eitner, D. Hinken, M. Köntges, "Detection of the voltage distribution in photovoltaic modules by electroluminescence imaging," Progress in Photovoltaics: Research and Applications, vol. 18, pp. 100-106 (2010)

[3] M. Köntges, I. Kunze, S. Kajari-Schröder, X. Breitenmoser, B. Bjørneklett, "The risk of power loss in crystalline silicon based photovoltaic modules due to micro-cracks," Solar Energy Materials and Solar cells, vol. 99, pp. 1131-1137 (2011)

[4] V. Quaschning, and R. Hanitsch, "Numerical simulation of current-voltage characteristics of photovoltaic systems with shaded solar cells," Solar Energy, vol. 56, pp. 513-520 (1996)

Magnetic Field Imaging (MFI) of Solar Modules.

Dominik Lausch[1,2], Markus Patzold[1,2], Maik Rudolph[1], Chia-Mei Lin[1,2], Jens Fröbel[2], Kai Kaufmann[1,2]

[1]DENKweit GmbH, Marthastraße 13, DE-06108 Halle (Saale), Germany
[2]Fraunhofer Institute CSP, Center for Silicon Photovoltaics, Halle (Saale), Germany
*Dominik.Lausch@DENKweit.de

Solar energy has become increasingly economical, especially in recent years. Defects in solar systems can endanger this economic efficiency. In solar modules electric currents generated by electrical or light induced injection flow distributed over the module area. Each flowing current generates a magnetic field depending on the strength and direction of the electric current. Defects limiting the conversion efficiency of solar modules are changing the quite specific current distribution of a working solar module and hence also the magnetic field distribution. Therefore, by analyzing the magnetic field distribution one can trace back to electrical defects within the solar module. In this work, a novel approach to analyze the direction and strength of the electrical currents by analyzing the magnetic fields is shown by using a new sensor capable of measuring all three spatial directions (x, y and z). The magnetic field is recorded as a function of the position by stroking over the solar module. The method was applied to different electrical defects such as broken cross and inter connectors, defective junction boxes, cracks, solder flux variations and to investigate single solder joints. The influence of the lamination process and thermal cycling on broken connectors is analyzed.

Keywords: Magnetic Field Measurement (MFI), Solar Modules, Imaging Technique, Electric Defects

1 INTRODUCTION

In recent years, the solar industry has been economically successful thanks to high cost savings and increased production capacities and is consistently developing to become the cheapest source of energy. Due to the high learning curve of PV as a semiconductor technology, an end of cost reduction is not in sight. It can be expected that the production of solar modules will increase sharply. On the other hand, however, there are quite high defect rates as a recent study by TüV Rheinland has shown (30 % of the examined solar modules showed defects). However, the reliability of solar modules is essential for a secure energy supply and financial payback. There is a large variety of electrical and other defects limiting the conversion efficiency over time. Solder connections of the solar modules and the junction box in particular seem to cause major problems currently. Established methods such as electroluminescence or thermography are only sensitive to voltage drops at the p-n transition (and effective lifetime according to Fuyuki approximation [1]) or lack spatial resolution and sensitivity for industrial applications [2] respectively. Therefore, a method would be desirable which can analyze current strength and current direction quantitatively, contactlessly and fast.. The MFI method shown in this publication fulfills this needs and can be used in an production environment as well as in the field. The method also opens the opportunity to increase the understanding of the device physics.

Under working conditions or electrical current injection solar modules have a quite typical distribution of the flowing electrical current. An electrical current generates a magnetic field which is described by the fundamental relationship between magnetic fields and electrical current, the Biot-Savart Law,

$$d\vec{B} = \frac{\mu_0}{4\pi}\frac{Id\vec{l}^2 \times \vec{r}}{r^3}$$

where B is the magnetic field, I is the current strength $d\ell$ is an element of the conductor, the constant $\mu 0$ is the permeability of free space, and r is the distance between the conductor and the sensor. If the geometry of the current distribution is well known, the current can be calculated from the magnetic field knowing only the separation distance between the current and the magnetic field sensor. Without knowing the geometries the mathematic solution is quite advanced since the Maxwell equation are nonlinear. As an example, in the case of an infinite straight wire, the resulting magnetic field encloses the conductor circularly. The details of this mathematical calculation can be found in [3, 4]. In solar modules the current flows are way more complex leading to a complex magnetic field surrounding the solar modules. However, a functional solar module has a quite typical magnetic field fingerprint. Electrical defects cause a change of the current distribution in the module and thus also a change of the magnetic field fingerprint compared to defect-free modules. By analyzing the magnetic field distribution in all three spatial directions x, y and z as a function of the position (over the module) one can trace back to the flowing electrical currents and therefore to all electrical defects.

2 MEASUREMENT DEVICE AND ROUTINE

The use of magnetic field sensors to analyze electrical currents has been around for a long time. Some magnetic sensors of today include Hall effect sensors, flux gates, induction coils, magnetic force microscope tips, magnetoresistive sensors and SQUIDs (superconducting quantum interference devices). An overview about the sensitivity can be found in [5]. It is important to choose the appropriate sensors for our application. In semiconductor industry SQUID sensors are often used for defect localization in 3D integrated circuits [6]. In PV single coils combined with a pulsed laser have already been used by Buchwald et al. [7] for solar cell defect localization. However, imaging all three magnetic field components over a larger area by using a magnetic field line sensor has not been performed before. Due to the lack of available sensors we had to develop a line array sensor board including a concept for data read out and transfer (patent pending).The length of the line array is variable, width and height are 30 mm and 1.2 mm respectively. For PV applications we chose a length of 160 mm which equals roughly the width of a solar cell. A gapless interconnection of multiple sensor boards can be realized. The developed sensor board has a lateral resolution along the board of about 0,4 mm and a resolution in movement direction depending on speed, measurement distance and the speed of data acquisition. The magnetic field resolution is about 0.3 µT in all three components. Data acquisition can be performed up to 200 Hz.

The measurement routine is as follows: As long as flowing currents are guaranteed by light or electrically induced current the generated magnetic fields can be measured by simply stroking over the solar module. In order to guarantee reproducible measurements, we have developed a x-y-z robot. The closer the measurement device the stronger the signal since the magnetic field decreases quadratically with distance according to (1). Therefore, i.e. measuring on the back sheet of the solar module gives higher magnetic field values than measuring from the glass side. By analyzing and processing the data, one can characterize quantitatively the current flowing through connectors, cracks, shunts, solder flux variations and even quantify single solder points and so on. Due to the complexity of the data and the needed analyzing speed for inline application, we have developed neuronal networks for a fast and reliable data analysis. Our method to find anomalous regions will be explained in upcoming publication in detail. Currently, various devices such as a lab tool, a handheld device and an inline tool are under development. Please note that both the sensor as well as the measurement routine have a patent pending in all major markets. In the following some measurement examples are shown and their results are discussed.

3 BROKEN CONNECTORS

When using electrical injection the majority of the electrical current is flowing through the ribbons as shown schematically in Fig. 1(a). The arrows indicate the current direction. The solar cell strings within a module are usually connected by a cross connector on the top and bottom of the module. In case of a broken middle cross connector all electrical current has to flow through the left and right busbar (in case of a three busbar cell) as visualized in Fig. 1(a). In this case the overall series resistance Rs of the whole module is increased since the broken cross connector is a bottle neck for the charge carriers. In Figure 1(b) and (c) the x-component of the corresponding magnetic field of one solar cell of the solar module is shown in 2D and 3D respectively. The missing current at the middle busbar caused by the broken cross connector leads to a missing magnetic field signal (marked with a red circle in Fig. 1(b)). In Figure 1(c) the same measurement is shown as a 3D plot. Since the current is increased in the other two busbars the magnetic field is increased as well. The busbars are connected with the small grid fingers on solar cell level. Therefore, the current distribution returns normality which can be seen by the increasing magnetic field signal of the middle busbar and by the decreasing signal at left and right busbar. The waving form of the signal is caused by the single soldering points where the current is changing its direction from the front side to the back side or vice versa which will be discussed later. In Electroluminescence (EL) measurements single broken cross connector can only hardly be classified or are not visible at all, because the EL signal is mainly proportional to the voltage drop at the p-n junction and approximately to the minority carrier life time (Fuyuki approximation) [1]. Here no evidence for a broken connector is visible.

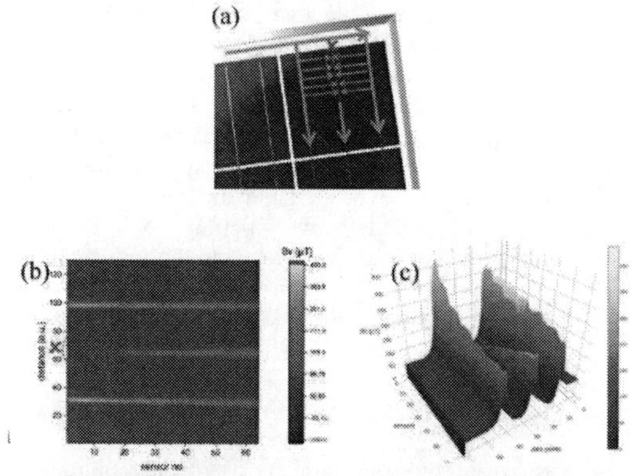

Figure 1: (a) Illustration of a section of a solar module. The red arrows schematically illustrate the current direction under electrical injection. (b) Magnetic field measurement (x-component) of a cell within a solar module. At the red cross the solder joint of the cross connector is defective. (c) 3D image of the same measurement. Due to the broken cross connector a higher current is flowing through working busbars.

In Figure 2(a) a similar measurement of a broken cell connector connecting two solar cells is shown. The current distribution is similar to the case of a broken cross connector but here the cell before the broken cell connector is of course influenced as well. Also here, the current has to flow through the two working cell connectors as shown in Fig. 2(a) which is confirmed by the MFI in Fig. 2(b). The magnetic field signal is low around the broken connector. Note that the lower right connector is also broken. Out of these measurements the quantitative electrical current can be calculated which will be shown elsewhere. In the EL measurement in Fig. 2(c) the broken cell connector can only be seen in the second cell by reduced indirect radiative recombination (dark area) around the middle busbar. However, the real cause cannot be extracted by EL measurement since the lower EL signal can have various reasons, i.e. back surface field defects, p-n junction issues, simply various contact issues or simply measurement issues. The broken busbar in the lower right corner cannot be seen within this measurement. MFI in contrast clearly indicates all broken cell connector since it is sensitive to the flowing electrical currents.

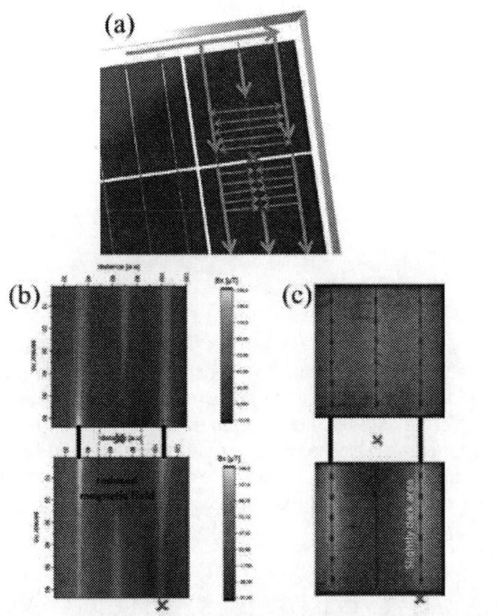

Figure 2: (a) Illustration of a section of a solar module. The red arrows schematically illustrate the current direction under electrical injection. (b) Magnetic field measurement (x-component) of two connected solar cells within a solar module. The red crosses illustrate the broken interconnectors. The low magnetic field indicates the broken connector. (c) Electroluminescence measurement of the same two solar cells. The broken interconnector can only be seen at the second cell. However, the lower EL signal can have various reasons and the broken interconnector cannot be identified clearly.

4 JUNCTION BOX FAILURES

Shunted bypass diode: In Figure 3 the x, y, and z component of the magnetic field of a defect-free junction box (3a, b, c) and a junction box having a shunted bypass diode (3d, e, f) is shown. The direction and the path of the electrical current is visualized by the dotted lines. The shown measurement is performed by simply stroking over the plastic housing of the junction box on the rear side of the solar module. In a defect-free junction box the current is flowing the regular way along the plus and minus pole of the contacted solar module (i.e. see Fig 3(b)). When the diode is shunted, the current flow deviates from normal status. In Figure 3d, e, f the left bypass diode is shunted as marked in Fig. 3(h). Due to the shunted bypass diode the current is flowing through the bypass diode instead of the first string and only through the second and third string of the solar module. A changed magnetic field distribution can be seen easily in all three components in Fig. 3(d,e,f) indicating that the current has changed direction and is flowing through the bypass diode. The measurement was performed on a junction box which was filled with a potting to protect the contacts. After getting rid of the potting we have measured a linear I-V characteristic of the bypass diode which is typical for a shunted diode. The measurements shown can be performed using electrical carrier injection under production conditions as well as light induced carrier injection in field operation.

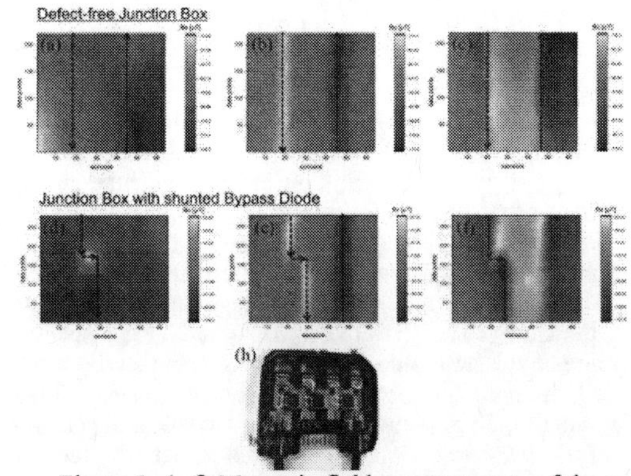

Figure 3: (a-f) Magnetic field measurements of the x, y and z component of a junction box measured by stroking over the plastic housing. The junction box was filled with a so called potting. In (a-c) a defect-free junction box is shown. The current flow is illustrated by the arrows. (d-f) Junction box with shunted bypass diode. The current is flowing through the shunted diode (d). (h) Photography of an open junction box without potting.

Partly soldered junction box: In Figure 4 a line plot of the y-component of the magnetic field across a partly soldered junction box before and after re-soldering is shown. The line scan was extracted out of an area resolved measurement exactly at the position of the soldered contacts as shown in Fig. 4(b). The black line is measured directly after production. The unbalanced peak of the two contacts is striking. Both, the height and the peak position are not symmetric as it can be seen in Fig. 4(a). By investigating the left contact in detail, it was observed that the solder dot was not in the middle of the solder contact but slightly shifted to the right leading to the right shift of the peak and the lowered signal. It was not investigated if this contact issue leads to a significant increase of the series resistance of the solar module but the reliability of the contact is very likely decreased. This module will tend to fail after few years in operation. After prior investigation we have re-soldered this contact. After re-soldering the peak position and height is again symmetric as we usually observe for defect-free junction boxes indicating that we have successfully repaired this module.

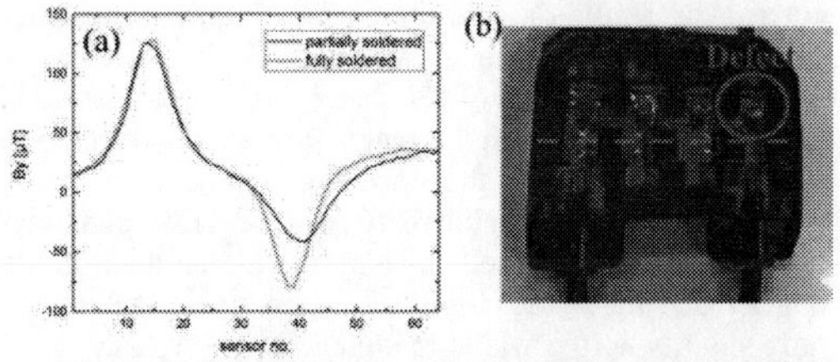

Figure 4: (a) Line scan (y-component) of the magnetic field at a junction box before (black line) and after re-soldering (red line) of the left contact (minus pole). (b) Photography of an open junction box. The line scan in (a) was taken out of a measurement along the dotted line.

5 SINGLE SOLDER POINTS

In Figure 5 (a-c) the derivative of the z-component dBz/dy of the magnetic field of a soldered solar cell within a solar module is shown. The z component is influenced by the current flowing perpendicular to the surface hence it is an indirect quantitative measure for the current flowing through the soldered contacts from the ribbons into the p-n junction. The derivation is sensitive to local maxima and is used to image the contacts with better clarity. The derivative at a maximum has a positive value on the slope side and a negative value on the waste side. Within the ribbon the electrical current flows parallel to the p-n junction. Only at the soldered contacts the electric current can flow perpendicular into the p-n junction region. The current path length is assumed to be of about 1-2 µm in length. In Figure 5 the measurement of three different solar cells are shown where we did not solder 2 (a), 4 (b) and all solder joints (c) of the middle busbar. At the non-soldered joints no magnetic field signal is detected indicating that no current is flowing from the ribbon to the into the solar cell. The influence of the missing solder joint cannot be seen in an EL measurement as shown in Fig. 5 (d-f).

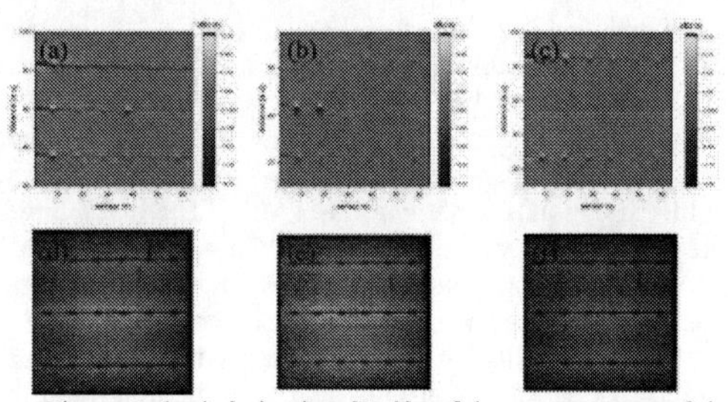

Figure 5: (a-c) derivative dBz/dy of the z-component of the magnetic field measurement of a soldered solar cell. The z component can be used as measure of the current flowing perpendicular to the surface. At the middle bus bar 2 (a), 4 (b) or all (c) solder contacts were left out. In (d-f) the corresponding EL measurements are shown. In EL measurements no influence of the missing contacts can be seen.

In Figure 6 a solar cell was soldered using different solder temperatures in order to produce solder pads having different contact quality. Changing the solder temperatures results in different electrical resistances of the single contacts. As it can be seen in Figure 6 the current distribution across the contacts is not uniform as expected. The peak height depends on the amperage. Therefore, the quality of the individual solder contacts can be determined from the magnetic field measurement. If this correlates to the local resistance has to be investigated in detail. Interestingly, we have observed that in a stable and excellent solar module production the quality of single solder joints is comparable and varies only slightly across whole modules. Only single contacts are missing. This changes dramatically after a thermal cycle (TC) test. In general thermal cycle tests have a huge influence on all soldered contacts of solar modules. A comparison of solar modules before and after TC tests will be done in future works.

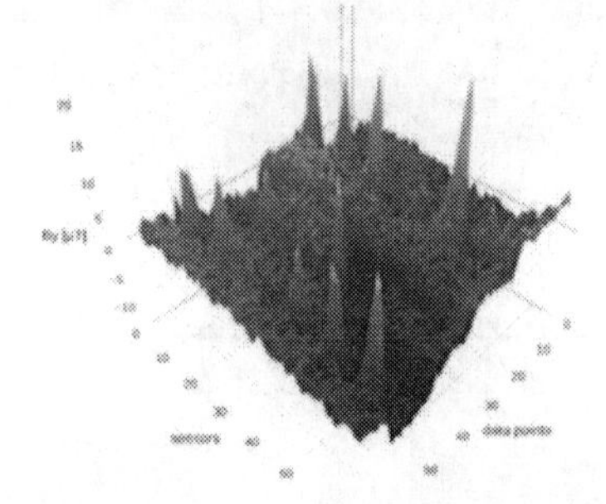

Figure 6: y-component of the magnetic field (indicting the current flowing through the solder contacts into the p-n junction) of a solar cell soldered using different solder iron temperatures. The solder iron temperatures influence the quality of the single soldered contacts. The peak heights are a measure of the electrical current flowing through the contact.

6 INFLUENCE OF LAMINATION AND TC CYCLE TEST ON DEFECTIVE SOLDER JOINTS

There are various discussions about the influence of the lamination process on non- or wrong-soldered contacts. It is from time to time discussed that missing contacts are pressed against the silver finger by the lamination realizing an electrical "press" contact. Figure 7 shows examples of the influence of the lamination process as well as thermal cycles (as they can occur in field operation) on the contact quality. In Figure 7 (a-c) the MFI of a reference sample having 4 defect-free soldered busbars is shown. As one can assume the contact quality isn't changed by the lamination step. In this example the contact quality stays also constant during the thermal cycle (TC) test. It should be noted that this is not always the case since the TCs have a strong impact on the contact quality. The TC test was performed 57 times with temperatures between -40°C and 85°C. In Figure 7(d) a solar cell with one partly soldered busbar (marked with a red rectangle) before lamination is shown. After lamination (Fig. 7(e)) the contact is still faulty. Therefore, the lamination of the solar module has not improved the contact. The reason for this could be that the EVA has flowed under the contact and thus prevented a press contact. After TC test the contact is still defective as expected. A clearly different behavior can be observed at the sample in Fig. 7(g-i). Also this sample has a defective busbar before module lamination marked with a red rectangle in Fig. 7(g). However, after lamination the contact seems to be defect-free since the busbar carries electrical current indicated by the magnetic field as can be seen in Fig. 7(h). Likely the ribbon is simple pressed on the busbar. But after TC test this contact is defective again as can be clearly seen in Fig. 7(i). Likely this happens under field operation as well. In summary, it is therefore important to control the contacts before lamination in order to guarantee defect-free solar modules.

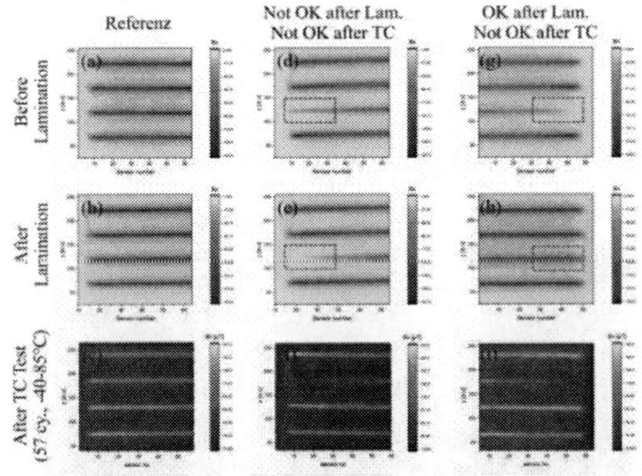

Figure 7: Magnetic field imaging of a reference sample and two samples with a broken busbar before lamination, after lamination and after thermal cycle (TC) tests. After lamination, sample 2 (d-f) shows no improvement. In contrast, the broken busbar in sample 3 is no longer visible after lamination (h). However, this busbar breaks again in operation, as the TC test clearly indicates (i).

7 CONCLUSION AND OUTLOOK

In solar modules electric currents generated by electrical or light induced injection flow distributed over the module area. Each flowing current generates a magnetic field depending on the strength and direction of the electric current. Defects limiting the conversion efficiency of solar modules are changing the quite specific current distribution of a working solar module and hence also the magnetic field distribution. Therefore, by analyzing the magnetic field distribution one can trace back to electrical defects within the solar module.

Within this contribution the analyzation of the direction and strength of the electrical currents by analyzing the magnetic fields is shown. For this a sensor to measure all three spatial directions (x,y and z) as a function of the position by only stroking over the solar modules has been developed. Various applications of the method are currently under development. Here the method was applied to different electrical defects such as broken cross and inter connectors, defective junction boxes, cracks, solder flux variations and single solder joints. Also it was shown that broken cross connectors behave different after lamination and TC cycle testing lamination. Even if an electrical contact is formed by the lamination step a non-soldered contact is very likely to break after TC cycle testing. Hence of full control of the solder contacts over the whole production as well as in field has to be performed.

8 REFERENCE

[1] FUYUKI, Takashi ; KONDO, Hayato ; YAMAZAKI, Tsutomu ; TAKAHASHI, Yu ; URAOKA, Yukiharu: Photographic surveying of minority carrier diffusion length in polycrystalline silicon solar cells by electroluminescence. In: Applied Physics Letters 86 (2005), Nr. 26, S. 262108

[2] EDEN, Anthony: Solar Collector Systems Analysis Using Infrared Scanning Techniques, Bd. 0254. In: Proc. SPIE 0254, Thermal Infrared Sensing Applied to Energy Conservation in Building Envelopes: Thermosense III, S. 208–212

[3] WIKSWO, J. P.: The Magnetic Inverse Problem for NDE. In: Weinstock H. (eds) SQUID Sensors: Fundamentals, Fabrication and Applications. NATO ASI Series (Series E: Applied Sciences), vol 329. Springer, Dordrecht, S. 629–695

[4] CHATRAPHORN, S. ; FLEET, E. F. ; WELLSTOOD, F. C. ; KNAUSS, L. A. ; EILES, T. M.: Scanning SQUID microscopy of integrated circuits. In: Applied Physics Letters 76 (2000), Nr. 16, S. 2304–2306

[5] LENZ, J. E.: A review of magnetic sensors. In: Proceedings of the IEEE 78 (1990), Nr. 6, S. 973–989

[6] KNAUSS, L. A., S. I. WOODS, AND A. OROZCO: Current imaging using magnetic field sensors. In: Microelectronics Failure Analysis Desk Reference 5 (2004), S. 303–311

[7] BUCHWALD, Rajko ; KÖSTNER, Stefan ; DRECKSCHMIDT, Felix ; MÖLLER, H. J.: Investigation of Defects in Solar Cells and Wafers by Means of Magnetic Measurements. In: Solid State Phenomena 178-179 (2011), S. 331–340

UTILIZING DIGITAL LIGHT PROCESSING AND COMPRESSED SENSING FOR PHOTOCURRENT MAPPING OF ENCAPSULATED PHOTOVOLTAIC MODULES

George Koutsourakis[a], Martin Bliss[b], Thomas R Betts[b] and Ralph Gottschalg[c,d]
[a] National Physical Laboratory (NPL), Hampton Road, Teddington, Middlesex, TW11 0LW, UK
[b] Centre for Renewable Energy Systems Technology (CREST), Wolfson School Mechanical, Electronic and Manufacturing Engineering, Loughborough University, Loughborough, Leicestershire, LE11 3TU, UK
[c] Fraunhofer-Center for Silicon-Photovoltaic (CSP), Otto-Eissfeldt-Straße 12, 06120 Halle
[d] Fachbereich Elektrotechnik, Maschinenbau und Wirtschaftsingenieurwesen (EMW), Hochschule Anhalt, Bernburger Str.57 06366 Köthen

ABSTRACT: Photocurrent mapping can provide useful spatial information about the electrical and optical properties of a photovoltaic (PV) device under actual operating conditions. Although it is a well-established technique for PV cells, direct current mapping measurements of PV modules is impractical and time-consuming to be applied. One has to mechanically shade specific cells of the PV module or destructively access the cell to be measured. In this work, non-destructive, automated current mapping of encapsulated PV modules is demonstrated. A commercial Digital Light Processing (DLP) projector is utilised in order to apply compressive sampling for current mapping of PV modules. This method is non-destructive, cost effective and significantly fewer measurements are needed for acquiring a current map compared to raster scanning methods. When applying compressive sampling, a series of patterns is projected on the sample, the current response is measured for each pattern and the current map is acquired using an optimisation algorithm. Specific shading strategies, voltage bias settings and I-V curve details are investigated for optimised compressive sampling.
Keywords: Current mapping, Compressed sensing, Non-destructive testing, PV modules

1 INTRODUCTION

Accessing the electrical parameters of individual cells in encapsulated photovoltaic (PV) modules is not an easy task. Several methodologies have been demonstrated in order to achieve this, most based on mechanical shading of cells. Dark I-V curves of individual cells of a module can be measured by completely shading the cell [1]. Differential I-V analysis and partial shading can be used to extract the shunt resistance and output current of individual cells in encapsulated modules [2]. Partial shading of cells in a module has also been used to extract their series resistance [3]. Shutter techniques have been used to access the response of individual cells in concentrating PV modules [4].

This work demonstrates a methodology for non-destructive automated current mapping of encapsulated PV modules without the need of cumbersome mechanical shading. Current mapping of PV modules is realised even in the presence of by-pass diodes without having to remove them or alter the structure of the PV module. The proposed approach utilises an alternative method for current mapping of PV devices based on the application of the compressed sensing (CS) sampling theory on LBIC measurements [5].

The CS sampling theory suggests that an N element signal can be measured by acquiring K observations, where K<<N [6][7]. With the standard LBIC method a point by point scan is performed by a light beam, reading the current response at every point to acquire the current map. When using CS current mapping a series of predefined random binary illumination patterns is projected on the sample. The current response is measured for each pattern and the final current map is acquired applying an optimisation algorithm. The main advantage of this technique is that fewer measurements are required to reconstruct the final current map than in standard LBIC measurements. This can lead to significantly shorter measurement times for current mapping. The signal to noise ratio is greatly increased, as almost half of the sample is illuminated by each pattern.

This technique has already been successfully applied to small area devices [8]. The initial small area experimental setup utilises a digital micromirror device chip (DMD) [9] in an optical system. In this work, a digital light processing (DLP) projector is utilised for CS current mapping of PV modules. DLP projectors are also based on a DMD chip, which actually creates the projection. Using a DLP based system for CS current mapping of PV modules shows some practical advantages when it comes to PV module current mapping. Such a system provides a perfect means to project the necessary partial shading and to project patterns on the sample under measurement for CS current mapping. Additionally, high signal to noise ratio for every projected pattern is achieved. Automated current mapping of cells in encapsulated modules is realised using this methodology. This is achieved by projecting patterns on the cell under measurement while light biasing the rest of the cells at the same time. This is the most significant advantage of this method, since any shading pattern can be implemented using a DLP projector. Current mapping of a PV module with by-pass diodes is achieved for the first time, without any mechanical shading.

2 COMPRESSED SENSING CURRENT MAPPING

In order to apply CS for acquiring measurements, a series of sampling functions $\{\varphi_m\}_{m=1}^{M}$ have to be projected on the PV device. Random binary matrices of ones and zeroes can be used as patterns, as they are easy to implement and satisfy the requirements for compressive sampling [10]. These matrices of ones and zeroes are translated to patterns of black and white pixels, to be projected on the sample. The patterns are expressed in a one row vector form, meaning that an $\sqrt{N}\cdot\sqrt{N}$ pixel square image is expressed as an N element vector. For every projected pattern, the current response of the PV device is

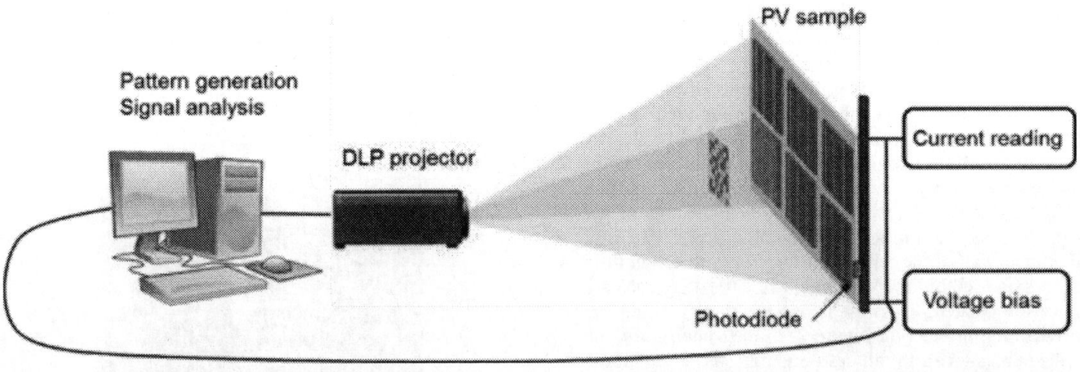

Figure 1: The experimental layout. An image is projected on the PV device. A part of it covering a single cell includes the specific pattern for compressive sampling while the rest of the cells are fully illuminated

measured, populating the measurement vector **y**. Since the projected patterns are known and included in the sensing matrix Φ, a solution to the underdetermined problem can be approximated by the **x** vector with the minimum ℓ1 norm [11], in other words:

$$\hat{\mathbf{x}}=\text{argmin}\|\mathbf{x}\|_1 \text{ subject to } \Phi\mathbf{x}=\mathbf{y}$$

With this method, current maps can be acquired with much fewer measurements than what a raster scan would require. This is because much fewer patterns than the pixels of the current map are projected on the sample. Using random patterns of 100 by 100 pixels, a 10000 pixel current map is acquired. The ℓ1 magic package in MatLab is used for reconstruction [12] and the primal dual ℓ1 minimisation algorithm is used from this toolbox. The time needed for reconstruction of the final current map is less than a minute, which is negligible compared to the data acquisition time. The CS current mapping procedure for PV devices is already analytically described in [8].

3 EXPERIMENTAL SETUP

The projector used for this work is a commercial Acer P7605 DLP projector with a 370W metal halide lamp. It is a reasonably high intensity projector, capable of generating a brightness of 5000 ANSI lumens. An important advantage of DLP technology is the high contrast ratio [13], which allows the system to efficiently generate a black pixel equivalent to a masked shaded spot on the PV module.

The basis of DLP projectors technology is the use of a beam of white light generated by the internal lamp. The beam is collimated and divided into its red, green and blue spectrum by a colour filter wheel. The micro-mirrors of the DMD operate at a frequency much higher than the human eye can perceive. The colour wheel is synchronised with the DMD and the colours are displayed at a very high rate so that eventually only the final combination of colours can be observed. This creates significant spectral and temporal variations of irradiance for PV characterisation. To overcome this limitation, the colour wheel of the projector was removed from the light path. In its final state, the projector only projects in black and white, the spectrum being the stable spectrum of the lamp.

Even without the colour wheel there are still periodic variations of intensity. The projector's internal DMD control behaves as if the colour wheel is still in place and is switching constantly, which leads to intensity variations. Although the projector's projections are controlled through LabVIEW environment, there is no absolute control of the internal DMD chip of the projector. The micromirrors that create the projections still turn on and off at a rate faster than 50 μs, causing temporal variations of irradiance on the sample's signal. These variations are short and periodic, with a period of approximately 4 ms. The signal of each period is averaged and recorded, and each current measurement averages 10 such readings, for reducing measurement noise due to intensity variation. Due to this reason, the sampling rate is very slow, approximately 2 samples per second.

The average absolute value of the irradiance on the sample plane is approximately $30W/m^2$. This is significantly lower than what is used in standard test conditions for PV modules (1000W/m2), but it is adequate for acquiring valid measurement results in this study. The long term stability of the light source is also monitored simultaneously with the measurements, to ensure that any instability does not affect measurements. This is achieved using a photodiode installed next to the PV sample under test. A long term variation of light intensity of approximately 3.5% over a period of 30 minutes has been recorded.

This work uses 6-cell crystalline Silicon (c-Si) mini modules, which are produced in-house at CREST. All terminals of each individual cell were extended to the outside of the encapsulation, allowing direct contact with each cell

4 VOLTAGE BIAS CONSIDERATIONS FOR CURRENT MAPPING OF PV MODULES

When applying CS current mapping to a PV module, some implications of the applied voltage bias have to be considered. Ideally, when all the cells of the module are identical and uniformly illuminated, by setting the PV module to short circuit conditions (0V), each cell should individually be also at short circuit condition. With the projector as the light source, some small voltage differences for each cell arise due to the expected illumination non-uniformities. When CS sampling is applied, the patterns are projected onto one cell of the module while the rest are fully illuminated. If the PV module is biased to 0V, it is expected that the shaded cell under measurement is actually reversed biased, while each of the others will be slightly forward biased, resulting in a sum of 0V for the module.

To verify this behaviour, all the cells of the custom mini-module are individually contacted and their voltage was measured while the PV module was biased at 0V. A sequence of 100 shading patterns are projected on one of the cells, each pattern increasing shading level (proportion of dark pixels randomly distributed over the area of the cell) by 1%. This means that the first projection fully illuminates all of the cells while the 100th projection fully shades one cell of the PV module while the other cells are fully illuminated. This test is applied to cell 1 and the results are presented in Figure 2. In the same figure, the voltage bias levels for all cells are indicated when all the cells are fully illuminated. This shows the effect of the non-uniformity of the projection. Nevertheless, as it will be demonstrated such non-uniformity issues do not prevent measurements to be applied.

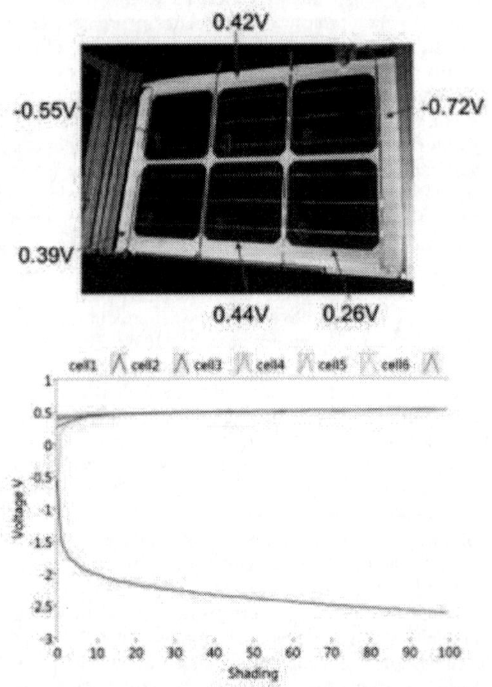

Figure 2: Top, the voltage bias of each cell when all the cells are fully illuminated and the module voltage is set to 0V. Bottom, the voltage levels of individual cells when cell1 is shaded with different shading levels

From Figure 2 it is clear that as shading is increased, the shaded cell becomes more reverse biased while the rest of the cells are forward biased as expected. This also shows that initial small non-uniformities of irradiance have no impact on voltage bias levels if one of the cells is shaded. In the case where one cell is underperforming compared to the others in the same module, it is reversed biased even when unshaded. To confirm this, similar defect-emulating masks to the previous section were used to reduce the output of some of the cells of the module. Three of the cells were covered with different sizes of these masks, as presented in Figure 3. The largest mask was placed on cell 3. When all the cells are illuminated, the cell with the largest mask is reversed biased, while all the others are forward biased at very similar levels. Running the same procedure as before and measuring the voltage of each cell while shading patterns are projected onto cell 5 (not masked), the graph at the bottom of Figure 3 is acquired.

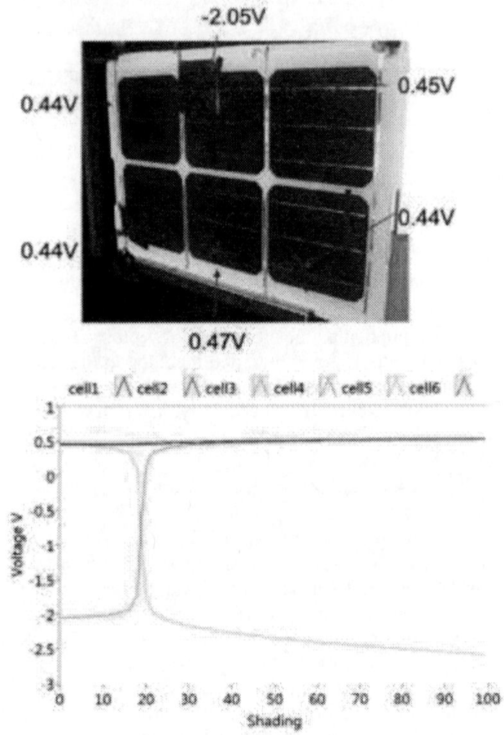

Figure 3: Top, voltage bias of each cell when all the cells are fully illuminated and the module voltage is set to 0V. Masks are applied on some of the cells. Bottom, the voltage levels of each cell when cell1 is shaded to different levels.

These results show that even if several cells of a module are underperforming, only the worst of them will be reversed biased. By increasing the shading levels on any of the rest of the cells, there is a point above which the shaded cell (and not the underperforming one) becomes reversed biased. The cell to be measured has to be the current limiting cell in a module, when applying CS current mapping. A forward voltage has to be applied to the module to bring the cell under test to operate close to short circuit conditions. Similar considerations and approaches have been applied for measuring spectral response of multijunction solar cells [14] and PV modules [15].

The same voltage measurements as above were applied with the the voltage bias set to 2.5V across the module terminals. This was not selected arbitrarily, but is just below the value of 5/6ths of the V_{OC} of the module (V_{OC}=3.15V) when all the cells are illuminated. Since for the patterns used for compressive sampling 40% of the pixels are at the "on" state (60% shading), the cell that is measured will be close to 0V when 2.5V forward bias is applied to the module.

This behaviour is confirmed in Figure 4 for cell1. By applying 2.5V forward bias to the module and for 60% shading, cell1 is close to short-circuit conditions (slightly forward biased ≃0.07V). When applying the patterns for compressive sampling to the other cells, the forward voltage bias of the cell under test is in the region of 0.02V-0.08V for a module bias at 2.5V. Because in a PV module where individual cell contacts are inaccessible these direct measurements are not be possible, a module forward bias of $((N-1)/N)*V_{OC}$ should be applied as a rule of thumb. This ensures that the cell under measurement is always close to short-circuit conditions, or at least only very slightly forward biased. This voltage bias approach is not being introduced in this work, but has been reported and successfully applied in spectral response measurements [15] and angular response measurements of PV modules [16]. In our case, as long as any cell of the module does not perform less than 40% of the cell that is measured, CS current mapping measurements will be valid as the limiting cell will be that under measurement. More precisely, no cell should have similar performance with the measured (limiting) cell when the latter is illuminated with the patterns, as this will influence measurements. In that case, sparser sensing matrices (fewer pixels at the "on" state) can be used for meaningful measurements. This may decrease the SNR ratio of measurements but this approach is necessary for applying this method successfully.

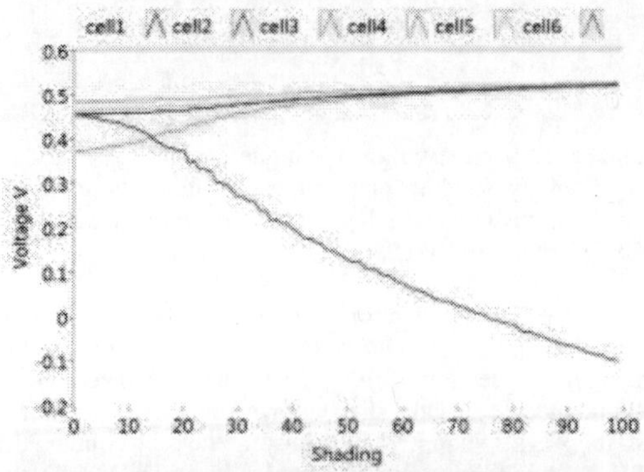

Figure 4: The voltage level of each cell of the custom module, when cell1 is shaded with different shading levels and the module is kept at 2.5V forward bias.

5 CURRENT MAPPING RESULTS FOR PV MODULES OF SERIES INTERCONNECTED CELLS

Following the individual cell test cases and the voltage bias considerations, the next step is to test whether CS current mapping of PV modules is practically feasible. Each cell is sampled with the necessary sensing patterns for compressive sampling consecutively, while the rest of the module is fully illuminated. All the cells are connected in series and the current is measured by contacting only at the PV module terminals.

As 100 by 100 pixel random patterns are used, the reconstructed current map of the entire PV module consists of 60000 pixels. Because of the slow sampling rate due to the previously described signal variations, measurement time was approximately 30 minutes for each cell. The reconstruction process is performed offline, using the $\ell 1$ reconstruction algorithm. This means that as soon as measurement acquisition of one cell has finished, the reconstruction process starts and the patterns move to the next cell at the same instant. The module voltage bias was set at 2.5V, as described in the previous section.

CS current mapping measurements with and without the defect-emulating masks are realised and presented in Figure 5. The current maps are acquired with 30000 measurements, an undersampling level of 50% of what a point by point scan would need. The measurement SNR value for these current measurements is approximately 1000. This is mainly due to the instability of the light source, as there is no direct control of the internal DMD of the projector.

These initial results using the custom mini module show that the projection approach for current mapping is a very convenient way for acquiring current maps of encapsulated PV modules. No mechanical shading is required to bring the cell under test to limiting conditions. The sampling patterns themselves not only realise the measurement but also provide the necessary shading that ensures the current limiting conditions.

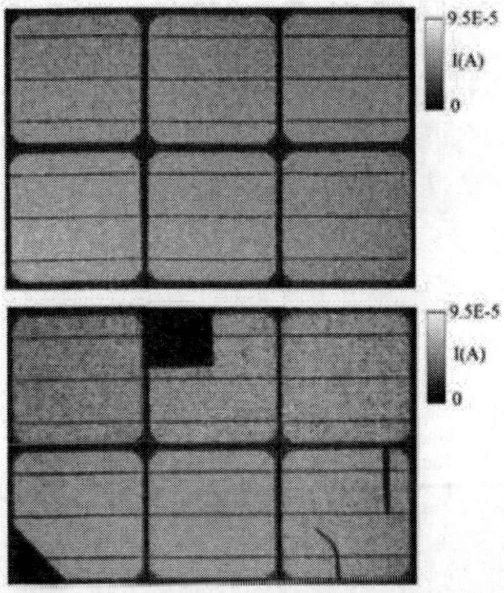

Figure 5: CS current map of the 6 cell PV module used in these experiments. The CS current map is produced having acquired 30000 (50%) measurements

The results show that the method works, even in the case with current mismatches between the cells. In reality, mismatches may exist in commercial PV modules, but such high levels of underperformance of individual cells are more likely to appear in aged or damaged PV modules rather than brand new ones. Some increased noise in the reconstructed maps is visible in the case of the masked cells and it affects the reconstruction process. This is more likely due to the current mismatch influencing measurements. Such issues were also observed in the case of LBIC measurements for PV modules with additional partial shading, where defects of individual cells were masked when contacting the module [17]. In the case of CS current mapping, this increases the noise levels of the reconstruction process.

6 BYPASS DIODES AND I-V CURVE CONSIDERATIONS

Commercial PV modules always have by-pass diodes

installed in order to avoid damage from reverse biased underperforming cells. However, this has implications when applying CS current mapping, since by-pass diodes provide alternate current paths for locally reverse biased cells or cell sub-strings. To establish what modifications to the CS method would be necessary in such a case, two by-pass diodes were added to the custom PV module used here. The PV module now consists of two sub-strings of three cells each with a bypass across. This was straightforward to implement since the contacts of each cell were already extended outside the module. The masks on three of the cells used in the previous section were also applied. The configuration of the PV module with the bypass diodes and defect-emulating masks is presented in Figure 6.

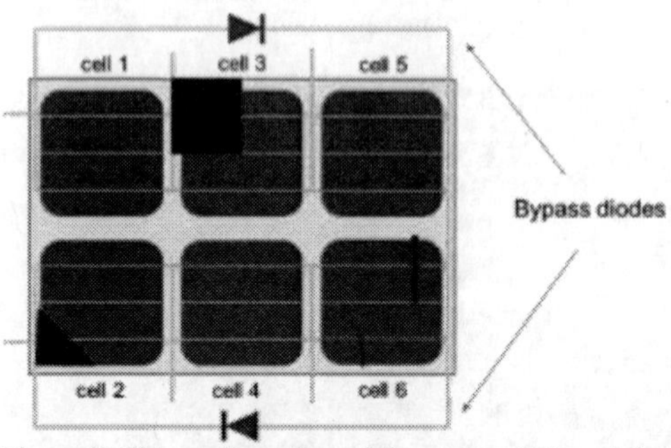

Figure 6: The configuration of the custom PV module of this work after installing two bypass diodes to create two sub-strings of three cells each

As a first step, the by-pass diodes were disconnected and a sequence of random patterns with increasing levels of shading is applied to one of the cells while the rest are fully illuminated, similarly with the voltage measurements of the previous section. I-V curves of the module for each shading level are acquired. This means that for the first I-V curve the cell is completely shaded (all pixels dark for this cell) while the last I-V curve is the same in all cases, as there is no shading at all. 10 of these I-V curves when each of cells 3 (large 'defect'), 4 (no defects) and 6 (two small 'defects') are shaded are presented in Figure 7.

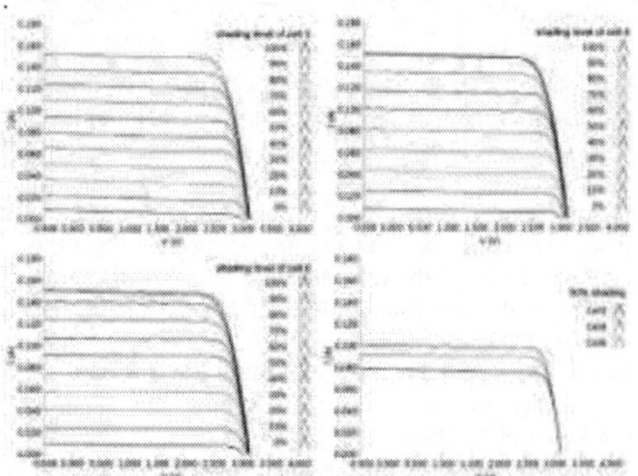

Figure 7: I-V curves of the PV module without by-pass diodes, acquired by shading one cell each time with different levels of shading. One I-V curve for each cell case with 50% shading is also included.

The I-V curves of the module for shading of any of the cells exhibit similar behaviour, while for 50% shading the maximum current depends on the performance of the cell shaded. This is precisely what makes CS current mapping work in the case that all the cells are connected in series. However, when by-pass diodes are installed, the I-V curves differ significantly from the case when all the cells are simply series connected. Following the same procedure as before to acquire the same series of I-V curves for shading the same three cells, the results of Figure 8 are acquired. It is noteworthy that even with no shading from the projector, the bypass diodes are activated at short circuit. This is due to the masks that have been applied on some of the cells, resulting in current mismatch that activates the by-pass diode of the worse string.

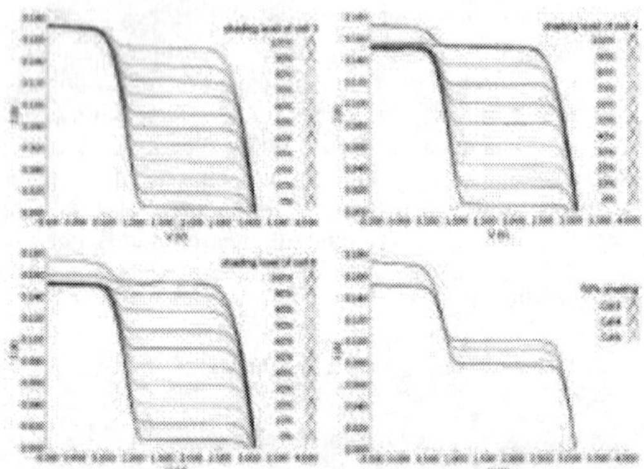

Figure 8: I-V curves of the PV module with bypass diodes, acquired by shading one cell each time with different levels of shading. One I-V curve for each cell case with 50% shading is also included.

For the application of CS current mapping measurements, a forward voltage bias as in the case of the previous section is necessary. Nevertheless, simultaneously with the voltage bias, an additional strategy is also adopted: by adjusting the projection on the PV module, one string is completely shaded, while the sub-string that includes the cell that is measured is properly illuminated. The patterns are projected on the cell under test and the rest of the cells of this specific sub-string are fully illuminated. The procedure is illustrated in Figure 9. As a first step, I-V curves are acquired as before, following this shading strategy. The results are presented in Figure 10. The influence of the shaded sub-string is almost negligible and it is barely visible at the bottom of the graphs, where a very low current exists for voltage values higher than 1.1V. If this voltage region is not considered, the I-V curves resemble those that would be acquired from a 3-cell mini module, with a small voltage drop. This is extremely convenient for the application of CS current mapping, since the current measured at the contacts of the PV module is only influenced by the limiting cell (measured cell) of the illuminated sub-string. This is clear when the I-V curves for 50% shading with this procedure for the three measured cells are displayed in the same graph. It is apparent that the current of the I-V in this instance depends on the performance of the cell of interest.

This above strategy where the sub-string not containing the cell under test is shaded is used for CS current mapping. A forward voltage bias of 0.6V is applied during measurements. This value was selected by considering the I-V curves of Figure 10. The illuminated cells of the module produce a voltage of around 1V, hence, by choosing a forward bias of 0.6V the cell under test is operating very close to short-circuit conditions. This approach also shows that by acquiring the above I-V curves before measurements, one can reveal which cell underperforms. An initial diagnostic test can be set before CS current mapping measurements, where I-V curves are acquired with the same shading level for each cell while illuminating the rest of the string, following the above shading strategy for disabling cell sub-strings. Such tests are implemented in seconds. This procedure can reveal which cells underperform, the cells can be sorted depending on their performance and the correct forward voltage bias levels for the CS current mapping procedure can be chosen. More importantly, with this procedure the correct sparsity levels for the sensing matrices can be determined, to ensure that the patterns will shade the cells to a lower level than the output of the worst performing cell.

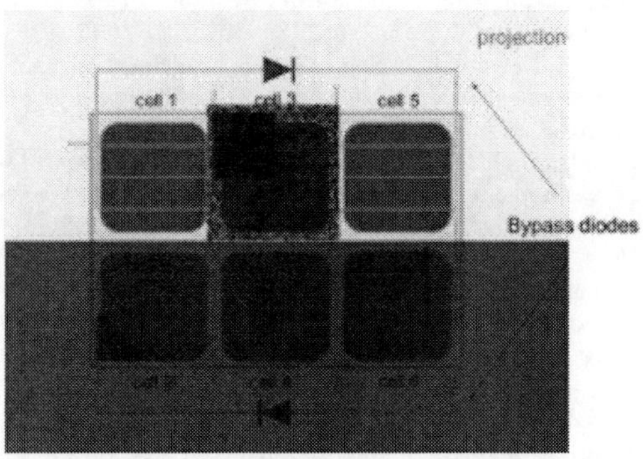

Figure 9: The process for CS current mapping measurements of a cell when by-pass diodes are installed in the module. The patterns are projected on the cell under test and the rest of the cells of this specific sub-string are fully illuminated, while the other sub-string is not illuminated.

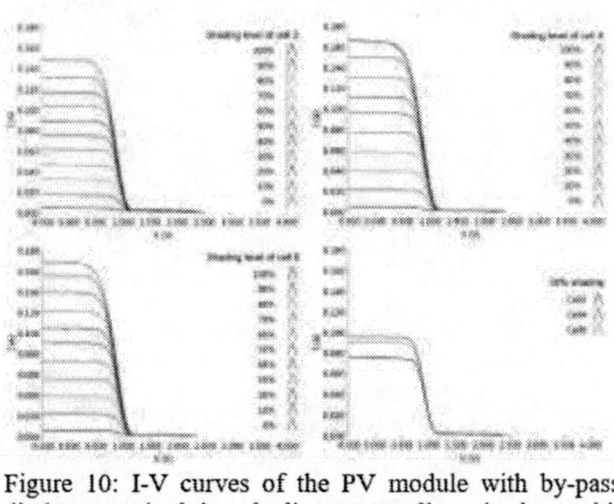

Figure 10: I-V curves of the PV module with by-pass diodes, acquired by shading one cell each time with different levels of shading, while completely shading the other string. One I-V curve for each cell case with 50% shading is also included.

Following the procedure of Figure 9 for all cells of the mini module with the by-pass diodes installed, the current map shown in Figure 11 is acquired. The reconstructed current map for an undersampling level of 50% (30000 measurements, 60000 pixels) is presented. In the same figure, a photograph of the module during measurements is also presented. The results are very similar to the case of the measurements for the mini module before installing by-pass diodes. The noise is slightly increased, since the average measurement SNR is reduced to ~700 after installing the bypass diodes. Nevertheless, meaningful measurements are acquired and this demonstrates for the first time that current mapping of a PV module with by-pass diodes can been achieved.

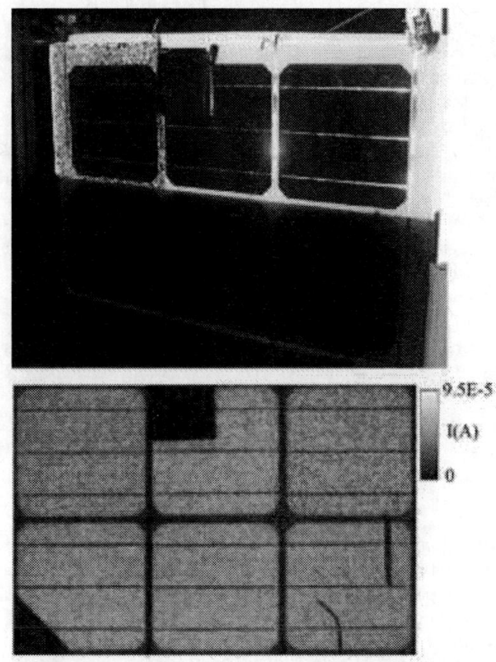

Figure 11: On top, a picture of the PV module during measurement when bypass diodes are included. Bottom, the CS current map of the PV module with by-pass diodes at 50% undersampling.

7 CONCLUSIONS

Non-destructive, automated current mapping of PV modules has been demonstrated in this work. Current maps of individual PV cells in encapsulated modules can be acquired without having to tap the cells, remove bias diodes or use mechanical shading. The approach includes a prototype DLP projection based current mapping system that has been developed at CREST, which utilizes a commercial DLP projector and compressive sampling for current mapping of PV devices.

The demonstrated methodology can become a very useful tool for spatial characterisation of crystalline silicon based or thin film PV modules. The tools utilised to achieve this approach exhibit some very practical properties for spatial characterisation. CS current mapping always requires fewer measurements than what a raster scan would need. The measured signal is significantly amplified due to the projection of patterns instead of a raster scan, resulting in an enhanced measurement SNR.

Experimental results prove that a custom DLP projector based CS current mapping system for PV modules is feasible. DLP projection technology is a perfect means for creating and projecting structured illumination patterns for current mapping of PV modules. Specific shading strategies were developed in order to isolate the cell under test. It is demonstrated that such shading strategies can also be applied to acquire I-V curves of the module to detect if any of its cells is underperforming. This diagnosis process takes seconds using the DLP projector and can sort the cells of a module in order of performance, even in the presence of by-pass diodes. This can also be useful in order to determine the sensing matrix sparsity necessary for CS current mapping measurements.

As discussed already, this prototype system cannot achieve very high accuracy and optical resolution as it uses a commercial DLP projector. This is mainly because of the high noise levels and the low irradiance produced by this system. A custom system will provide more control on the DMD chip and will be able to reduce temporal variations of light intensity. This would decrease noise levels and provide higher accuracy. Latest advances in high power DLP display technology for cinema applications can be utilized for increasing optical power. The proposed measurement method and system aids to the development of a realistic solution for current mapping of PV modules. Moreover, the approaches and strategies demonstrated can also be utilised to develop additional characterisation tools for PV modules based on DLP technology and the CS theory.

ACKNOWLEDGMENTS

This work was funded through the European Metrology Research Programme (EMRP) Project 16ENG02 PV-Enerate. The EMRP is jointly funded by the EMRP participating countries within EURAMET and the European Union. This work is co-funded by the UK National Measurement System.

REFERENCES

[1] L. De Bernardez and R. H. Buitrago, "Dark I–V curve measurement of single cells in a photovoltaic module," *Prog. Photovoltaics Res. Appl.*, vol. 14, no. 4, pp. 321–327, 2006.

[2] G. B. B. Alers, J. Zhou, C. Deline, P. Hacke, and S. R. R. Kurtz, "Degradation of individual cells in a module measured with differential IV analysis," *Prog. Photovoltaics Res. Appl.*, vol. 19, no. 8, pp. 977–982, 2011.

[3] Y. S. Kim, S. M. Kang, B. Johnston, and R. Winston, "A novel method to extract the series resistances of individual cells in a photovoltaic module," *Sol. Energy Mater. Sol. Cells*, vol. 115, pp. 21–28, 2013.

[4] M. D. Yandt, J. P. D. Cook, K. Hinzer, and H. Schriemer, "Shutter technique for noninvasive individual cell characterization in sealed CPV modules," *IEEE J. Photovoltaics*, vol. 5, no. 2, pp. 691–696, 2015.

[5] S. R. G. Hall, M. Cashmore, J. Blackburn, G. Koutsourakis, and R. Gottschalg, "Compressive Current Response Mapping of Photovoltaic Devices Using MEMS Mirror Arrays," *IEEE Trans. Instrum. Meas.*, vol. 65, no. 8, pp. 1945–1950, 2016.

[6] D. Donoho, "Compressed sensing," *IEEE Trans. Inf. Theory*, vol. 52, no. 4, pp. 1289–1306, Apr. 2006.

[7] E. J. Candès, J. K. Romberg, and T. Tao, "Stable signal recovery from incomplete and inaccurate measurements," *Commun. Pure Appl. Math.*, vol. 59, no. 8, pp. 1207–1223, 2006.

[8] G. Koutsourakis, M. Cashmore, S. R. G. Hall, M. Bliss, T. R. Betts, and R. Gottschalg, "Compressed Sensing Current Mapping Spatial Characterization of Photovoltaic Devices," *IEEE J. Photovoltaics*, vol. 7, no. 2, pp. 486–492, Mar. 2017.

[9] L. J. Hornbeck, "The DMDTM Projection Display Chip: A MEMS-Based Technology," *MRS Bull.*, vol. 26, no. 04, pp. 325–327, Apr. 2001.

[10]R. Baraniuk, M. Davenport, R. DeVore, and M. Wakin, "A Simple Proof of the Restricted Isometry Property for Random Matrices," *Constr. Approx.*, vol. 28, no. 3, pp. 253–263, Jan. 2008.

[11]R. G. Baraniuk, "Compressive Sensing," *IEEE Signal Process. Mag.*, vol. 24, no. July, pp. 118–121, 2007.

[12]E. J. Candes, J. Romberg, and T. Tao, "Robust uncertainty principles: exact signal reconstruction from highly incomplete frequency information," *IEEE Trans. Inf. Theory*, vol. 52, no. 2, pp. 489–509, Feb. 2006.

[13]D. S. Dewald, D. J. Segler, and S. M. Penn, "Advances in contrast enhancement for DLP projection displays," *J. Soc. Inf. Disp.*, vol. 11, no. 1, p. 177, 2003.

[14]J. Burdick and T. Glatfelter, "Spectral response and I-V measurements of tandem amorphous-silicon alloy solar cells," *Sol. Cells*, vol. 18, no. 3–4, pp. 301–314, 1986.

[15]Y. Hishikawa, Y. Tsuno, and K. Kurokawa, "Spectral Response Measurements of PV Modlues and Multi-Junction Devices," in *22nd European Photovoltaic Solar Energy Conference (22nd EU PVSEC)*, 2007, pp. 2765–2769.

[16] W. Herrmann, M. Schweiger, and L. Rimmelspacher, "Solar Simulator Measurement Procedures for Determination of the Angular Characteristic of PV Modules," in *29th European Photovoltaic Solar Energy Conference (29th EU PVSEC)*, 2014, vol. 2, pp. 2403–2406.

[17]P. Vorasayan, T. R. Betts, and R. Gottschalg, "Limited laser beam induced current measurements: a tool for analysing integrated photovoltaic modules," *Meas. Sci. Technol.*, vol. 22, no. 8, p. 085702, Aug. 2011.

FIELD ANALYSIS AND DEGRADATION OF MODULES AND COMPONENTS IN DISTRIBUTED PV APPLICATIONS

Hongjie. Hu[1], William Gambogi[2], Kaushik Roy Choudhury[2], Lucie Garreau-Iles[3], Thomas Felder[2], Steven MacMaster[2], Oakland Fu[1] and T.-John Trout[2]
(1) DuPont (China) R&D Center, Shanghai, China; (2) DuPont, Wilmington, DE, USA; (3) Du Pont de Nemours International S.A., Geneva, Switzerland

ABSTRACT: The Chinese Photovoltaic (PV) market grew rapidly in 2017 with 53GW of new installations. Distributed PV (DPV) accounted for 19.4GW growing from 4.2GW in 2016. Examples of DPV include small scale (usually < 20MW) PV installations on commercial and residential rooftops, over-water installations and small size off-grid installations. The DPV segment is expected to keep growing in the next few years. However, along with the high-speed growth and development, field failures and underlying mechanisms/root causes need to be assessed. A field investigation by DuPont of more than 1GW and ~300 installations, ranging from newly installed to 30+ years old with a variety of module types including mono, poly, fixed, tracking, glass/backsheet, glass/glass, and bifacial, in DPV and ground installations. The data shows higher module visual failure rate for DPV (32.5%) compared with large scale ground PV (22.8%), while backsheet failure rate in DPV is > 2.5 times higher than that of conventional PV. Examples are shown of module and component failures including solar cell corrosion, hot spots, EVA browning and glass delamination, and backsheet yellowing, cracking and thickness erosion. To understand failure mechanisms, non-destructive analysis and lab destructive analysis were conducted and defects were analyzed.
Keywords: Distributed PV; module; field analysis; rooftop

1 INTRODUCTION

The PV market grew rapidly in the past decade all over the world. Global new PV installation reached ~100GW in 2017 while module price dropped significantly. Distributed PV (DPV) is the fastest-growing segment in China. DPV installation reached 4.2GW in 2016 and increased to 19.4GW in 2017. Meanwhile, endless cost down pressure caused more and more quality and reliability concerns about PV modules and components. PV modules are expected to maintain their performance over greater than 25 years' service life. Output power generation, safety and appearance are expected to be maintained to meet the financial returns of the installation and the expectations of the field owner [1].

However, indoor reliability evaluation is a systematic, time consuming and expensive undertaking. It requires knowledge of environmental stress levels, with appropriate equipment and well-designed experiments. There is no single test machine that can simulate the full range of environmental stresses including UV, humidity, thermal cycling, dynamic load, chemical corrosion, sand abrasion, etc. together. Single or double applied stresses cannot simulate the synergistic effects of these environmental stresses. Accelerated aging test protocols in current codes & standards also need to be improved according to stress analysis and comparison with field performance. [1,2]

Therefore, long-term field outdoor performance has been recognized as the ultimate test for all module components and manufacturing quality. [3] DuPont has conducted field inspections for the last 7 years of more than 1GW and ~300 installations, ranging from newly installed to 30+ years old, with a variety of module types including mono, poly, fixed, tracking, glass/backsheet, glass/glass, and bifacial, in over-water, rooftop, and field settings all over the world. In this work, statistical analyses for distributed PV (rooftop, overwater, etc.) and conventional large-scale ground mounted PV installations are compared. Typical environment stresses for rooftop applications and the impact to module design are analyzed and discussed.

2 EXPERIMENTS

2.1 Non-Destructive Analysis Methods

The use of non-destructive analysis (NDA) methods on fielded and accelerated test modules allows the direct comparison of performance and tracking of performance through various exposure times in the field and exposure periods under accelerated tests. [2] The NDA methods employed include:

a) Visual assessment
b) Optical properties (color and gloss) measurement
c) IR spectroscopy
d) Thermal imaging
e) Current-voltage (IV) characteristics
f) Electroluminescence (EL) imaging
g) Wet leakage testing
h) Dry insulation testing

These methods have been extensively used in module field or lab analysis. IV measurement provides insights into power loss and the resistive losses that can contribute to overall power loss. EL imaging has been effective in identifying specific areas of the module that have developed problems, and helps identify area for destructive analysis methods. Wet leakage and dry insulation are used to identify overall issues with the electrical insulation of the product and can be used to isolate these problems for further investigation. Thermal imaging provides information on areas of the module that may degrade faster than others due to over-heating from resistive losses. IR spectroscopy is effective in identifying the outer layer of the backsheet and changes to all components during exposure. Optical properties and visual assessment track the appearance, and may be related to materials property changes.

2.2 Destructive Analysis Methods

For modules that warrant further investigation, destructive methods of sampling and measuring the material properties including physical (structure, composition, and layer thicknesses), mechanical (tensile, elongation, adhesion), optical (reflectance, transmission, color, gloss), chemical (water vapor and oxygen transport rate) and electrical (dielectric breakdown) can be applied to samples that are delaminated or cored from the target module. These sampling techniques rely on the physical integrity of the underlying materials, and sometimes multiple attempts are needed to get a clean, physically intact example.

3 RESULTS AND DISCUSSIONS

3.1 Overview of roof and ground mounted PV field failure rates

DuPont has inspected more than 1GW PV installations all over the world through 2018. Visual defects of different module components were analyzed with statistical methods. The results indicate that the total module defect rate is 22.3%, while cell and backsheet account for the top two defect components and constitute most of the observed defects. [4] Figure 1 shows the statistical results comparison between ground mounted and roof mounted PV installations. It is notable that backsheet visual defects (cracking, delamination, yellowing, inner layer cracking) in roof-mounted installations are ~2.5 times more prevalent than in ground mounted installations, 22.3% vs 9.2%.

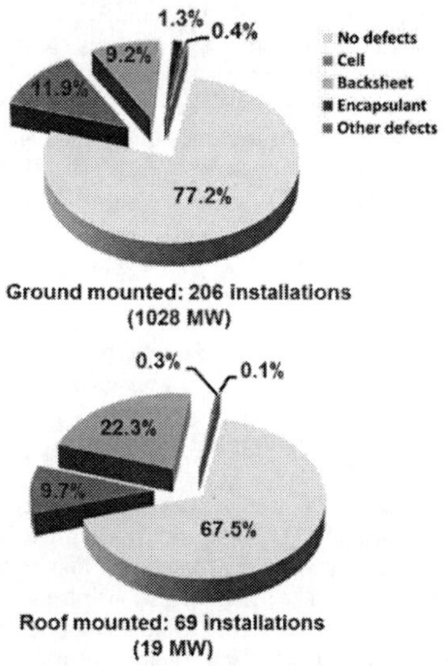

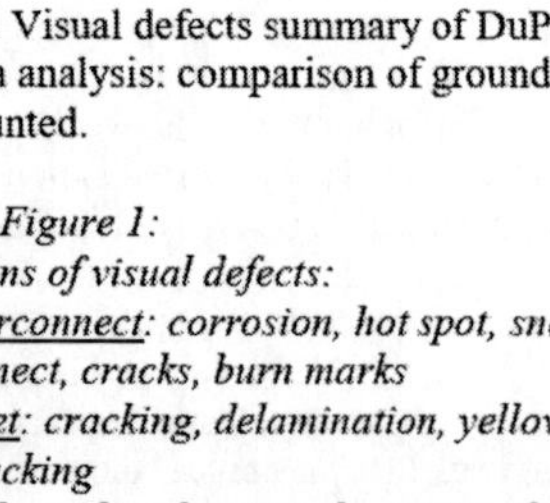

Figure 1 Visual defects summary of DuPont 2018 global field data analysis: comparison of ground mounted vs. roof mounted.

Note for Figure 1:
Definitions of visual defects:
<u>Cell/interconnect</u>: corrosion, hot spot, snail trails, broken interconnect, cracks, burn marks
<u>Backsheet</u>: cracking, delamination, yellowing, inner layer cracking
<u>Encapsulant:</u> discoloration, browning, delamination
<u>Others</u>: glass defects, loss of AR coating, junction box defects

The reason for this difference might be two-fold. First, a module system in a rooftop environment usually has higher operating temperature, which is the main aging factor for polymer materials. Rooftop systems are typically 15°C hotter than ground-mounted systems according to a recent study by NREL. [5] Close-fit installations on rooftops could produce even higher temperature due to poor heat dissipation from the module rear side. This trend with temperature is similar to the effect from high temperature seen in climate data. Hot arid and tropical climates showed higher backsheet and encapsulant failure rate than temperate climates. [4] Second, backsheet materials used for rooftop and ground-mounted systems are often different. A higher ratio of poor quality backsheet materials are used for roof mounted PV due to cost considerations and reduced monitoring by most roof system owners.

Different backsheet materials show different failure rates in the field. Table I shows defect rate by backsheet types. PA, PVDF, PET and FEVE coating-based polymer backsheet and glass backsheets were listed. PA-based backsheets show 45.7% field rate while PVDF-based backsheets show 10.6% defect rate. By comparing with 2017 results, both PA and PVDF-based backsheets show higher failure rate in the field by prolonging field service time beyond four or so years. [2] Glass backsheets (for G/G modules) were first added in 2017, and show 12.8% defect rate which is also very high. By considering the average field service age, Tedlar® PVF-based backsheet is the oldest (9.7 yrs) but has the lowest defect rate, only 0.05%.

Table I: Defect rate by backsheet types

Material	Averaged Age (yrs)	Fields/MW	Defect rate (MW basis)
PA	6.1	19/121	45.7%
Glass	8.3	9/9	12.8%
PVDF	3.4	47/240	10.6%
PET	5.1	46/251	3.4%
FEVE	3.4	26/155	2.3%
Tedlar®	9.7	99/111	0.05%

PA = Polyamide
PVDF = PolyVinylidene DiFluoride
PET = PolyEthylene Terephthalate
FEVE – FluoroEthylene VinylEther

3.2 Environmental stresses and requirements for rooftop PV applications

The higher defect rate of roof-mounted PV suggests higher requirements for components than with ground-mounted PV. Table II on the following page shows roof-mounted module requirements and impact on module and backsheet materials.

Since roof-mounted PV is built on residential or industrial roofs, safety comes first and is the most important requirement. It requires modules and backsheets have good fire resistance and electrical insulation properties. Secondly, durability is important for PV systems to provide 25 years of service life with stable power generation. Roof-mounted, especially close fit roof-mounted modules, will encounter higher operating temperature due to poor heat dissipation conditions. The temperature range (periodic highest and lowest temperature within one day or year) is also higher than that of ground-mounted systems in the same environment. Good thermal stability for a backsheet requires high softening and /melting point, good thermal resistance and low coefficient of thermal expansion. Higher reflected UV to the back surface of the moduleis also expected for open rack module installation due to higher reflectivity of cement or painted steel tiles. [6] Color stability should be considered. Additionally, less weight and good aesthetics are also desired for roof-mounted systems. Double-sided glass modules are less suitable for roof- mounted applications due to the 20-30% higher weight than a glass/polymer backsheet module.

Regarding fire resistance, Tedlar® PVF has been granted RP15 (ASTM E162), the highest safety rating in the radiant panel flame spread test, the best test for PV modules. Tedlar® PVF-based backsheets have been used in PV modules for over 35 years in the f8ull range of climate conditions. Tedlar® PVF meets all Federal Aviation Administration / European Aviation Safety Agency fire, smoke and toxicity (FST) regulatory requirements for aircraft, and has been widely used in aircraft interiors for Boeing and Airbus, etc. since the Boeing 747 launch in 1963. It has been also used as the protective layer for metal and fabric laminates in construction and wallcovering decoration fields which also have high fire resistance requirements.

Table II: Roof mounted PV application requirements and impacts to module and backsheet

Application requirements	*Impact to module*	*Requirement for Backsheet*
Safety	Fire resistance Electrical insulation	• Fire resistance • Flame spread index (FSI) • Electrical insulation
Durability	Ensure 25yrs' lifetime and performance	• Long term field proven record • Superior durability after various aging tests
High Temp.	High Temp. stability	• Softening/Melting point • Thermal resistance (RTI; TI) • Coefficient of thermal expansion (CTE)
Greater Temp. Range	High thermal cyceling resistance	• Strong mechanical properties in MD and TD directions • Better co-efficient of thermal expansion (CTE)
High UV Albedo for Open Rack Installations	High Backsheet UV resistance	• High UV resistance
Aesthetics	Color stability; Different colors; New structure designs	• Color stability • Different colors
Lightweight & Easy Installation	Glass/backsheet structure; flexibility module; new designs	• Avoid to use glass backsheet except in BIPV

Higher module operating temperature and wider temperature range require high backsheet softening temperature and low coefficient of thermal expansion (CTE). Figure 2 gives inner layer softening temperature and maximum CTE of typical backsheet materials and structures. TPT shows the highest inner layer softening temperature due to the high thermal stability of Tedlar® film. Tedlar® PVF film has ~190°C softening temperature while that of PVDF and polyolefin film is only about ~150°C and ~100°C, respectively. High softening temperature ensures Tedlar® film is more resistant to hot spot issues. On the other hand, CTE of TPT is the lowest among tested samples in Figure 2. CTE of TPT is only ~1/3 of PA-based backsheet and ~1/6 of PET-based backsheet (PET/PET/Primer). Low CTE is desirable for high thermal cycle applications such as on a roof and in the desert.

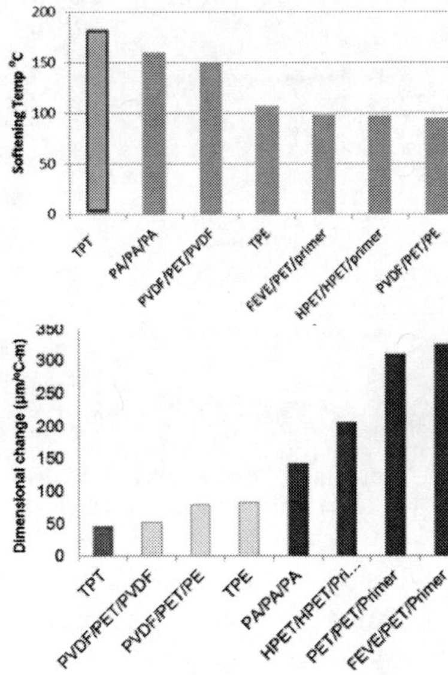

Figure 2 Inner layer softening temperature (Upper) and maximum CTE (Lower) of typical backsheet materials and structure

Elongation is a direct indicator of polymer degradation. Under thermal aging, different backsheet materials show totally different performance. Figure 3 shows elongation of PVDF and Tedlar® PVF films along TD axis after 70°C temperature storage. Elongation of five brands of common commercial PVDF films all dropped to less than 50% after 1000 hours aging. There is no elongation loss for Tedlar® PVF films even after 2000 hours aging under the same test conditions. Significant elongation loss and embrittlement of PVDF films in the TD direction might due to post-crystallization. [7,8] Elongation loss of PVDF films has been observed after UVA exposure, damp heat aging, and pressure cooker testing as well, and is consistent with field and accelerated test cracking of PVDF-based backsheets, which indicate inherent weakness of PVDF films and backsheets. [4]

PVF films were found to be uniformly stable to thermal, UV exposure, damp heat exposure and pressure cooker testing in both MD and TD directions and would be expected to retain their mechanical properties. Retained mechanical properties can be important if there is damage, cracking, or loss of mechanical properties in the PET core of backsheet. Backsheet cracking can result in electrical insulation failure of modules, an important safety requirement.

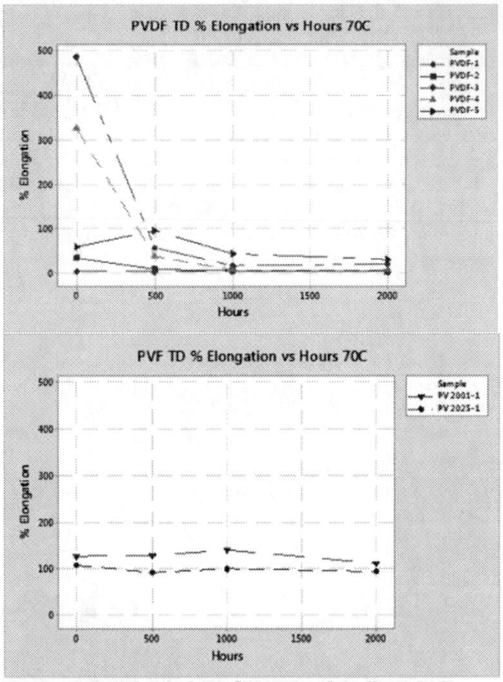

Figure 3 TD Elongation of PVDF and Tedlar® PVF films after 70°C temperature resistance test in thermal chamber

4 CONCLUSIONS

Distributed PV has grown aggressively in recent years. But concerns about module quality have been expressed due to significant price decreases. A field investigation by DuPont has accumulated more than 1GW modules data which covers different age, climate, manufacture, installation type, module type and backsheet material. Backsheet visual defect rate of roof mounted PV system is found to be 2.5 times of that of ground mounted PV system, which indicates more critical requirements for roof-mounted PV applications. Safety and durability were recognized as top two requirements. Fire resistance, and electrical insulation are extremely important. Environmental stresses such as high operating temperature, greater thermal cycling, strong UV reflectivity from roofs were identified as critical stresses. These requirements mandate high flame spread resistance, high softening temperature, and low CTE of backsheets. Comparison test data of typical backsheet material and structure was analyzed. TPT has the highest softening temperature, highest flame spread rating and lowest CTE, which indicate superior fire resistance and excellent thermal stability. Elongation at break of PVDF films dropped after thermal aging, likely due to post-crystallization, and is consistent with field observation and lab testing results after UV, damp heat, pressure cooker, and high temperature exposure. Tedlar® PVF films are stable after these aging tests, maintaining MD and TD mechanical properties.

Roof-mounted PV has higher safety and durability requirement than ground mounted PV. Higher defect rates of roof mounted PV are generated by critical environment stresses in roof application and underscore the importance of reliable materials in PV modules.

5 REFERENCE

[1] W. Gambogi, T. Felder, S. MacMaster, et al, IEEE PVSEC (2016).

[2] H. Hu, W. Gambogi, T. Felder, et al, The 11th SNEC, Shanghai, China (2017).

[3] A. Skoczek, T. Sample, and E. D. Dunlop, Wiley InterScience, 2008.

[4] H. Hu, K. Choudhury, W. Gambogi, et al, Energy Taiwan Forum, Taipei (2018).

[5] D. C. Miller, M. Kempe. 35th IEEE Photovoltaic Specialists Conference (PVSC '10) Honolulu, USA (2010).

[6] R. Chadysiene and A. Girgzdys, J. Env. Eng. And Land. Mgmt., 16 (2008) 83-88.

[7] G. Oreski, G.M. Wallner. Solar Energy 79 (2005) 612-617.

[8] T. Hosoda and T. Yamada, J. Appl. Polym. Sci., 2014.

CLIMATE-SPECIFIC DAMAGE ACCUMULATION OF SOLDER BONDS IN SILICON PHOTOVOLTAIC MODULES

Michael Owen-Bellini[1], Jiang Zhu[2], Daniel Montiel-Chicharro[2], Thomas R. Betts[2], Ralph Gottschalg[3]
[1]National Renewable Energy Laboratory, Golden, CO, USA, 80401
[2]Centre for Renewable Energy Systems Technology, Loughborough University, Loughborough, LE11 3TU, United Kingdom
[3]Fraunhofer Centre for Silicon Photovoltaics, Otto-Eißfeldt Straße 12, 06120 Halle (Saale), Germany

ABSTRACT: Damage accumulation of solder bonds in 6-cell photovoltaic (PV) mini modules is evaluated for 9 locations covering a variety of climate types. Finite-element methods are used to simulate the inelastic deformation of the solder bonds induced by changes in module temperature and mismatched thermal expansion coefficients between materials in the module. It is shown that accumulated damage is a function of mean daily maximum module temperature, temperature fluctuation frequency and mean daily maximum temperature change, with the latter showing the strongest correlation to accumulated damage and being the most influencing factor. Damage accumulation is also calculated for the standard thermal cycling protocol, TC200, as outlined in the IEC61215. The number of standard thermal cycles required to generate the same damage as one year in the field for each climate is then determined.

Keywords: photovoltaics, reliability, solder bonds, power loss, series resistance, finite-element, simulation

1 INTRODUCTION

Solder bond fatigue is a commonly reported failure mechanism responsible for a steady increase in series resistance and subsequent power loss in photovoltaic (PV) modules [1]. Fatigue in solder bonds is induced by cyclic thermomechanical stress from continual fluctuations in module temperature during normal operation. The rate of fatigue is dependent on materials, module design and the operating environment [2]. Standardised certification testing procedures subject modules to the same conditions regardless of intended deployment location, and while these tests are typically used to certify modules for upwards of 20 years of operation, actual solder bond degradation rates cannot be expected to be identical across all environment types. Higher rates of degradation are often reported for hotter environments [3]. In the interest of more accurate calculations for the levelized cost of energy (LCOE) for potential PV installations, a better understanding of the operating environments influence on solder fatigue is required.

The purpose of this work is to identify the characteristics of module temperature profiles which have the most significant influence on solder fatigue and the climates which most exhibit these characteristics. A finite-element model (FEM) is developed to simulate the thermomechanical behaviour and accumulation of solder bond damage of PV mini-modules for 9 different locations across a range of climate types. Damage accumulated for one year in each climate is compared with damage accumulation during standard thermal cycling testing, as outlined by IEC61215 [4] such that a yearly equivalent number of cycles can be determined.

2 METHODS

2.1 Finite-Element Model

A 2-dimensional finite-element model (FEM) has been developed using COMSOL Multiphysics to reflect mini-modules fabricated at Loughborough University. A similar method was previously proposed by Bosco et al [5]. The modules are representative of conventional configurations found in the industry. consist of six mono-crystalline silicon cells connected in series via solder-coated copper-based ribbons. The assembly is encapsulated in ethylene-vinyl acetate (EVA) and laminated together with a float glass front cover and Tedlar/Polyester/Tedlar backsheet. Figure 1 shows a top-down and cross-sectional schematic of the assembly.

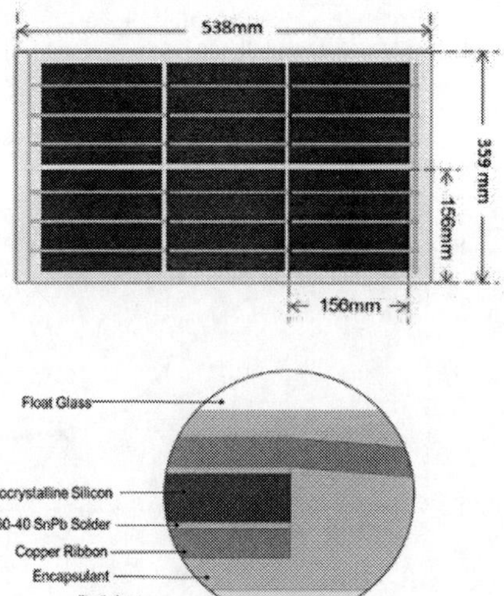

Figure 1: Top-down and cross-sectional schematic of PV mini-modules fabricated at Loughborough University.

Inelastic deformation (damage) of the solder bonds during normal operating conditions is the primary investigation of this work and is modelled using Anand's viscoplastic model [6]. Anand's model calculates inelastic strain according to the applied conditions i.e. stress, temperature and the deformation resistance of the solder. Deformation resistance is governed by the structural characteristics of the solder and is quantified in Anand's model using a single scalar value. The evolution of the structural characteristics of the solder material when subjected to temperature, stress and time is also calculated in Anand's model.

Table 1: Material propeties for each component in the model

Component	CTE (10^{-6} K)	Young's Modulus (GPa)	Poisson's Ratio	Density (g/cm^3)
Cell	2.7	Anisotropic	0.28	2.3
Glass	8	73	0.23	2.5
Backsheet	5.04	3.5	0.29	2.5
EVA		Temperature-dependent	0.25	0.95
Ribbons	17	105	0.35	8.5
Solder	21	Viscoplastic	0.4	8.4

Appropriate simulation of deformation and stress mechanisms in the assembly require accurate characterisation of the materials in use. The values presented for each material have been furnished from manufacturer datasheets or otherwise determined experimentally. Material properties are outlined in Table 1. The necessary parameters to calculate inelastic deformation of 6040PbSn using Anand's model has previously been determined by Wang et al [7]. The elastic modulus of EVA undergoes significant change depending on temperature and has been shown to impact the thermomechanical behaviour of modules and the stress applied to solder bonds [8]. Therefore temperature-dependent storage modulus for EVA is also considered in this work and measured for 5mm diameter post-cure, free standing samples using a TA-Instruments Dynamic Hybrid Rheometer for temperatures ranging -40 to 150°C at 1Hz with a temperature ramping rate of 5°C/min (Figure 2). Anisotropic material properties are also considered for the silicon cells which exhibit a loading direction-dependent Youngs modulus [9].

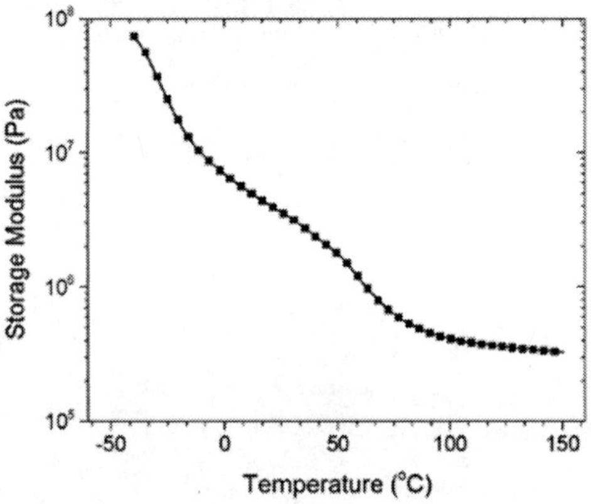

Figure 2: Temperature-dependent storage modules of EVA measured at 1Hz

Mesh quality is an important consideration for FEM simulations. As such, a mesh refinement is conducted to increase the confidence in simulations whilst also attempting to keep computation time to a minimum. A custom mesh was developed using the finest possible mesh as a benchmark. Comparing accumulated solder damage following simulation of 10 standard thermal cycles yielded a result from the custom mesh that fell within 2% of the finest mesh's result. A screen capture of the mesh at the edge of the inner cell is shown in Figure 2.

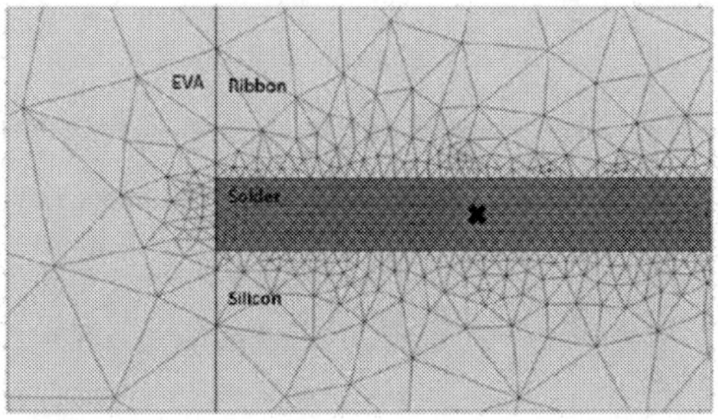

Figure 3: Screen capture of the mesh at the edge of the inner cell

2.2 Weather Data

Weather data is obtained from nine locations from the baseline surface radiation network (BSRN). Details for each location are outlined in Table 2. For each location, one year of minutely data is obtained and each site is categorised by climate according to the All-India survey climate classification system. Module temperature data is estimated from weather data using the thermal model described by Fuentes [10]. Modules are assumed to be in an open-rack mounting configuration. The thermal model parameters are optimised using weather and module temperature data from a PV installation at Loughborough University.

Table 2: Locations used in study

Location	Short Name	Climate Type
Darwin, Australia	DAR	Hot & Humid
Papua New Guinea	MAN	Hot & Humid
Gobabeb, Namibia	GOB	Hot & Dry
Solar Village, Saudi Arabia	SOV	Hot & Dry
Cabauw, Netherlands	CAB	Warm & Humid
CENER, Spain	CNR	Warm & Humid
Oklahoma, USA	E13	Temperate
Carpentras, France	CAR	Temperate
Regina, Canada	REG	Cold

3 RESULTS

3.1 Standard Thermal Cycles

Standard thermal cycles are first simulated to understand the behavior of the solder and is later used for comparison to the outdoor environments. Figure 3 shows the rate of energy dissipation of the solder bonds at the middle cell. It is shown that the bottom bond has the highest rate of energy dissipation (damage accumulation). In the interest of reducing computation time of subsequent simulations, viscoplastic modelling is only considered for this bond.

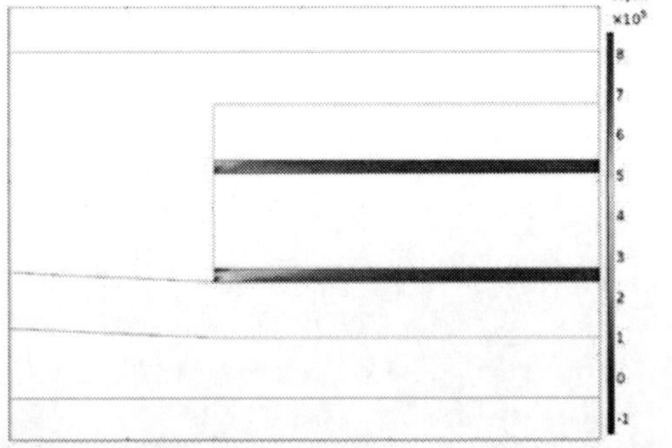

Figure 4: Rate of accumulation of energy dissipation (W/m³) for solder bonds in the middle cell at the onset of the high temperature dwell during a thermal cycle

Creep dissipation density during one standard thermal cycle is shown in Figure 4. The hyperbolic sine term in Anand's model manifests as an equivalent increase in damage accumulation during the temperature ramps. The greatest accumulation of damage occurs during the ramping up period due to the high stress generated in the ramp and the solders sensitivity to elevated temperatures.

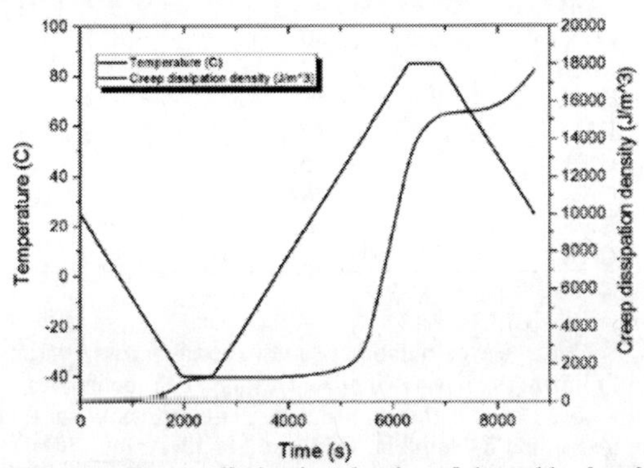

Figure 5: Energy dissipation density of the solder bond during one standard thermal cycle

3.2 Outdoor Climates

Inelastic strain energy density is calculated through one-year of data in each location outlined in Table I. Results from the simulations are presented in Figure 5. The starting point for each climate is at the beginning of the calendar year. It can be seen that the climates which cause higher module temperatures, are also responsible for the highest accumulation of damage on the solder bonds, which is quite unsurprising. However there exists a disparity between the Hot and Dry climates, SOV and GOB, and the Hot and Humid climates, DAR and MAN. The reasons for this disparity will be discussed in the next sub-section. Whilst the hotter climates are in rough agreement with each other, it can be seen that the temperate climates and warm and humid climates exhibit large differences in damage accumulation. E13 is found to have the highest damage accumulation amongst these and comes under the temperate classification, whereas the next climate in terms of damage is CNR, a warm and humid climate. Following this we have the next temperate climate, CAR, which is unexpectedly met with a similar accumulation of damage from the Cold climate, REG. The climate with the least accumulation of damage by a significant margin is CAB, a Warm and Humid climate. Whilst CAB and REG both demonstrate a similar rate of accumulation of damage during the first 150 days, REG then accelerates dramatically as it enters the summer months. During these months it experiences much higher temperatures than CAB, which has milder summers.

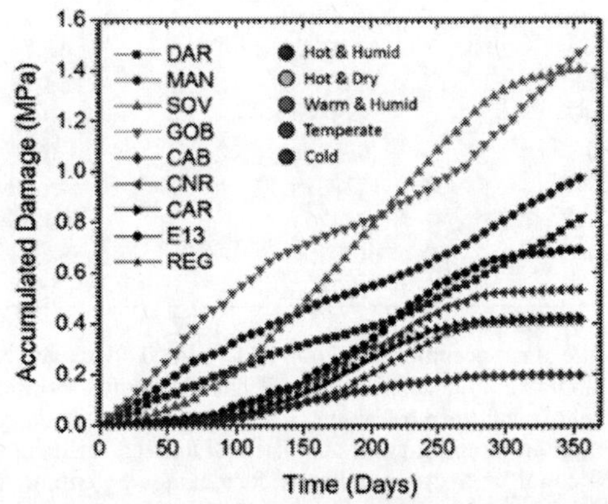

Figure 6: Accumulated solder bond damage for one year in each location

To understand the factors influencing solder fatigue, 3 crucial parameters of module temperature profiles are examined to identify any correlations. These are mean daily maximum module temperature, number of temperature ramping events and the mean daily maximum ramping event magnitude. Table 3 summarises these characteristics for each of the locations in the study. It can be seen that for the two locations which exhibit the highest mean daily maximum module temperatures, SOV and MAN, there is a difference in accumulated damage of approximately 40%. Interestingly, SOV experiences significantly fewer total ramping events than MAN, but does have a higher mean daily maximum ramping event size by 4.58°C, which might suggest that the larger ramping event sizes is what accounts for the increased damage. For the two locations which follow in terms of mean daily maximum module temperature, GOB and DAR, there is a significant disparity in the accumulated damage over one year with a difference of nearly 80%. While GOB does exhibit a slightly higher mean daily maximum module temperature of 0.59°C, it experiences far fewer total ramping events.

The stand-out difference between these two locations is in the mean daily maximum ramping event size, which sees a difference of more than 13.2K, again adding credence to the idea that climates with larger ramping events result in a higher accumulation of damage as opposed to locations with more temperature fluctuations. Three locations, CNR, CAR and REG, have very similar mean daily maximum ramping event sizes. Amongst them, CNR has the highest accumulated damage which is likely attributable to the higher mean daily maximum module temperature and total number of ramping events. The location with the lowest accumulated damage of the three, REG, has a significantly lower mean daily maximum module temperature by 10°C and also has the lowest mean daily maximum ramping event size. The location with the lowest accumulated damage of all sites, CAB, also has the lowest mean daily maximum event size by a wide margin.

Table 3: Summary of module temperature profile characteristics and total accumulated solder bond damage for one year in each location

Location	Accumulated Damage (Pa)	Mean Daily Max. Module Temperature (°C)	Total Ramping Events	Mean Daily Max Ramping Event Size (K)
GOB	1520830	61.57	2111	46.48
SOV	1410173	63.77	3052	41.9
MAN	1003800	63.13	7927	36.5
DAR	849194	60.98	7220	33.3
E13	694768	46.8	5709	34.57
CNR	529819	45.09	6102	32.2
CAR	424794	44.05	3865	32.25
REG	409204	34.44	5812	31.5
CAB	197878	35.9	6372	23.6

While the accumulation of solder bond damage is attributable to a combination of the 3 thermal profile properties outlined in Table 3, it seems that the mean daily maximum ramping event size has a significant influence and could be a strong indicator for damage potential. A strong correlation is observed in Figure 6 between the accumulated damage and mean maximum ramping event size for each location over one year.

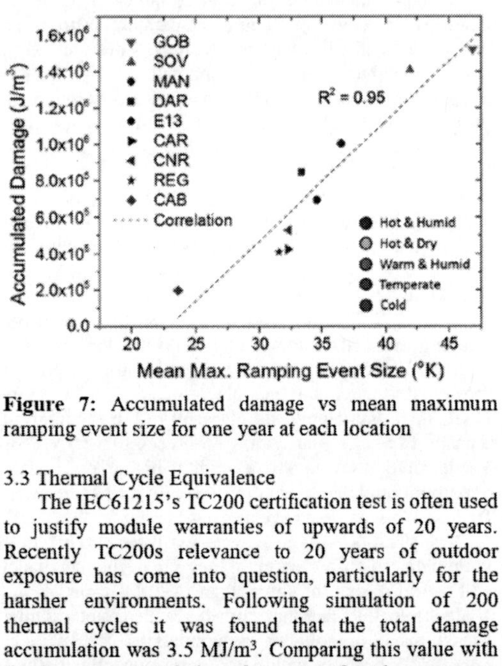

Figure 7: Accumulated damage vs mean maximum ramping event size for one year at each location

3.3 Thermal Cycle Equivalence

The IEC61215's TC200 certification test is often used to justify module warranties of upwards of 20 years. Recently TC200s relevance to 20 years of outdoor exposure has come into question, particularly for the harsher environments. Following simulation of 200 thermal cycles it was found that the total damage accumulation was 3.5 MJ/m³. Comparing this value with the damage accumulation of one year of outdoor exposure in each location a year equivalent number of thermal cycles can be determined, as summarized in Table 2. It is clear that 200 standard thermal cycles falls short of generating an equivalent amount of damage as 20 years of exposure. Even in less harsh environments where solder fatigue is not often a reported field failure due to other mechanisms being more prominent.

Table 4: Yearly equivalent thermal cycles and number of years tested by TC200 for each location

Location	Yearly Equivalent Thermal Cycles	Number of Years Tested by TC200
GOB	86	2.3
SOV	80	2.5
MAN	57	3.5
DAR	48	4.2
E13	39	5.1
CNR	30	6.7
CAR	24	8.3
REG	23	8.7
CAB	11	18.2

4 CONCLUSIONS

Damage accumulation of solder bonds across a variety of climate types was simulated using FEM techniques. It has been shown that hot and dry environments such as those found in Namibia (GOB) and Saudi Arabia (SOV) are most detrimental to the solder bonds. Beyond this, there was no correlation to be found between solder bond damage and climate-type as defined by the classification system used in this work, with disparities in damage shown for the temperate, cold and warm and humid climates. Whilst mean maximum module temperature and total number of ramping events does play a role in the accumulation of solder bond damage, a strong correlation was found between accumulated damage and the mean daily maximum ramping event size. This lends credence to the IEC61215 standard thermal cycling profile to accelerate damage, which uses large ramps.

Comparing the total damage accumulation in both indoor test profiles and outdoor exposure showed that indoor testing grossly under-stresses the solder bonds, with 200 cycles being equivalent to 2.3 years and 18.2 years in the most and least damaging environments, respectively. However, these results come with the caveat that the simulations do not consider changes in properties of the other materials which could be dependent on the applied conditions and result in an increased rate of damage accumulation over time.

5 REFERENCES

[1] D. C. Jordan and S. R. Kurtz, "Photovoltaic Degradation Rates-an Analytical Review," *Prog. Photovoltaics Res. Appl.*, vol. 21, no. 1, pp. 12–29, Jan. 2013.

[2] N. Bosco, T. J. Silverman, and S. Kurtz, "The Influence of PV Module Materials and Design on Solder Joint Thermal Fatigue Durability," *IEEE J. Photovoltaics*, vol. 6, no. 6, pp. 1407–1412, 2016.

[3] D. C. Jordan, J. H. Wohlgemuth, and S. R. Kurtz, "Technology and climate trends in pv module degradation," *27th Eur. Photovolt. Sol. Energy Conf. Exhib.*, no. October, pp. 3118–3124, 2012.

[4] International Electrotechnical Commision, "IEC 61215: Crystalline Silicon Terrestrial Photovoltaic (PV) Modules—Design Qualification and Type Approval," *International Electrotechnical Commission*, vol. 3. 2005.

[5] N. Bosco, T. J. Silverman, and S. Kurtz, "Climate specific thermomechanical fatigue of flat plate photovoltaic module solder joints," *Microelectron. Reliab.*, vol. 62, pp. 124–129, 2016.

[6] L. Anand, "Constitutive equations for hot-working of metals," *Int. J. Plast.*, vol. 1, no. 3, pp. 213–231, 1985.

[7] Z. N. Cheng, G. Z. Wang, L. Chen, J. Wilde, and K. Becker, "Viscoplastic Anand model for solder alloys and its application," *Solder. Surf. Mt. Technol.*, vol. 12, no. 2, pp. 31–36, 2000.

[8] M. Owen-bellini, D. Montiel-chicharro, J. Zhu, T. R. Betts, and R. Gottschalg, "Influence of Viscoelastic Properties of Encapsulation Materials on the Thermomechanical Behavior of Photovoltaic Modules," vol. 8, no. 1, pp. 183–188, 2018.

[9] J. J. Wortman and R. A. Evans, "Young's Modulus, Shear Modulus, and Poisson's Ratio in Silicon and Germanium," *J. Appl. Phys.*, vol. 36, no. 153, 1965.

[10] M K Fuentes, "A Simplified Thermal Model for Flat-Plate Photovoltaic Arrays," *Sandia Report*. Sandia National Laboratories, 1987.

EVALUATION OF TECHNOLOGY-DEPENDENT MAXIMUM POWER POINT CURRENT AND VOLTAGE DEGRADATION IN A TEMPERATE CLIMATE

Sascha Lindig[1,2*], Philip Ingenhoven[1], Giorgio Belluardo[1], David Moser[1], Marko Topič[2]
[1]Institute for Renewable Energy, EURAC Research - Viale Druso 1, 39100 Bolzano, Italy
[2]University of Ljubljana, Faculty of Electrical Engineering, Tržaška c. 25, 1000 Ljubljana, Slovenia
*Phone: +39 0471 055 715; E-mail: sascha.lindig@eurac.edu

ABSTRACT: In this paper the maximum power point current and voltage degradation of eight different photovoltaic technologies (27 systems) installed in a temperate climate is studied. The plant was installed in 2010 and seven years of production are considered. A performance loss over time has been observed for all systems. Most systems show a stronger relative decrease in current compared to voltage. The connection between performance losses broken down into these two parameters can lead to findings of which degradation modes appear in the PV system. Depending on the predominant performance loss parameter, specific degradation modes were suspected and several PV plant inspections, including IV curve measurements and electroluminescence as well as infrared imagery, took place to confirm these hypotheses. Degradation in current was detected in all systems and can be connected to hot cells, fractured cells, soiling as well as other root causes. A decrease in voltage was not observed for crystalline systems but is severe for copper indium (gallium) selenide as well as amorphous silicon systems and is suspected to be caused by a reduction in the conductivity of the metal contacts and a decreasing ability of the photovoltaic active material to separate the charge carriers. The outcomes of this study might contribute to O&M related improvements of failure detection.

Keywords: PV system, performance degradation, system performance

1 INTRODUCTION

Photovoltaic devices convert the energy of the sun directly into electricity. This technology became widely available in the late 1990's and evolved from being a niche product to a widely excepted energy source and business investment. To ensure a safe, reliable and profitable operation of PV systems, possible risks during operation need to be minimized and performance reducing effects fast detected in order to be eliminated. The project "Solar Bankability" was created to contribute to the reduction of investment risks in photovoltaic projects [1]. Further research in this direction is crucial and will eventually contribute to a further increase of confidence in photovoltaics as a mature energy source.

PV systems are subject to continuous cycles of temperature, humidity, irradiation, mechanical stress and soiling. These stresses lead to the occurrence of degradation modes (DM). DM are effects that reduce the output power of a PV system irreversibly and might create safety problems [2]. Each mode has other causes and is triggered differently. It develops either in isolation or in combination with other DM or technical risks and might lead to the failure of a PV module. The term failure is for electrotechnical devices defined as "the termination of the ability of an item to perform a required function" [3].

An important part plays the detection of DM and other performance decreasing occurrences through operation & maintenance (O&M) routines. The ideal case would be to detect DM and other technical risks early and reliable with minimal effort.

On one hand O&M companies perform field inspections where, among other things, IV-curves are recorded or electroluminescence images taken, to detect DM or other performance impairing technical risks. On the other side system degradation is discovered by studying performance rating parameters of a PV system, for example the maximum power or the performance ratio (PR), and calculating the performance loss (PL). Although a great amount of publications focuses on degradation and improvement of degradation models, a deeper understanding of the current and voltage behavior of a PV system could lead to the connection of the aforementioned operations. The correlations between the maximum power point current (I_{MPP}) and irradiation as well as the maximum power point voltage (V_{MPP}) and temperature are indicators, which might contribute to the understanding of a decrease in a PV systems performance. A better understanding of the time dependent behavior of I_{MPP} and V_{MPP} values, which are available from most data-loggers of installed PV systems, might enhance the knowledge of what is happening in the PV module over time. Especially in regards to solar cell parameters, like short-circuit current density, open-circuit voltage, fill factor, series resistance or shunt resistance, this is a promising approach to find root causes for degradation.

This paper presents a study of the time evolution of I_{MPP} and V_{MPP} values of eight different PV module technologies (27 PV systems) installed in a temperate climate for seven years. Each PV system consists of one or two strings. The idea of the investigation is to find dependencies between certain DM/technical failures and patterns in the time series. Therefore, next to the statistical degradation studies, field visits were conducted including IV-curve measurements, visual inspections and capturing electroluminescence (EL) as well as infrared (IR) images.

2 METHODOLOGY

2.1 PV system

The PV installation in question consists of 27 different PV systems of varying size, which cover eight different technologies and have an overall nominal power of 59kW. The test facility is located at the Airport Bolzano Dolomiti in South Tyrol/Italy and installed at a fixed tilt of 30° and an orientation of 8.5° west of south. The prevailing climate is according to Köppen classified as Cfb (temperate, without dry season, warm summer). In close proximity to the test side, a dedicated meteo-station is operated. Here, the direct normal and diffuse horizontal irradiance as well as the global plane-of-array (POA) irradiance, the wind speed and the ambient temperature is measured with a measurement frequency of 1 min. The electrical parameter of the inverter have a resolution of 15 min and because of that, the weather data are averaged on a 15 min interval. The technologies installed are micromorphous silicon (µmorph), amorphous silicon (a-Si), cadmium telluride (CdTe), copper indium (gallium) selenide (CIGS), silicon solar cells made out of a heterojunction with an intrinsic thin layer (HIT), monocrystalline silicon (mc-Si), polycrystalline silicon (pc-Si) and polycrystalline silicon string ribbon (ribbon). Three different kind of a-Si systems are considered, namely with single, double and triple junctions.

2.2 Data preparation

Before any calculation is performed, the data are subject to certain filters, correction and normalization. First, all 15-min data, which have a POA irradiance lower than 500W/m² and higher than 1200W/m² are excluded. At a later stage, the temperature coefficients from the PV module datasheets are used to perform temperature corrections. These temperature coefficients, obtained for V_{OC} and I_{SC}, are retrieved at a POA irradiance (G_{POA}) of 1000W/m². This filter is a compromise of excluding not too many data and reducing the uncertainty of applying the coefficients. A second filter concerns the performance ratio. The PR is calculated by dividing the final (array) yield $Y_{f(a)}$ (depending if AC- or DC-power is evaluated) with the reference yield Y_{ref} [4]. The yields are ratios of measured values of power/irradiance and nominal values.

$$PR = \frac{Y_f}{Y_{ref}} = \frac{P_{AC}/P_{nom}}{G_{POA}/G_{STC}} \quad , \quad PR_{DC} = \frac{Y_a}{Y_{ref}} = \frac{P_{DC}/P_{nom}}{G_{POA}/G_{STC}}$$

Here, P_{nom} is the nominal power of the PV system and G_{STC} is the irradiance at Standard Test Conditions (STC with cell temperature T_{STC} of 25°C, irradiance G_{STC} of 1000W/m² and air mass of 1.5). When studying the PV performance, it is advisable to use DC related performance metrics in order to eliminate eventual influences due to inverter degradation.

The filter excludes all data points whose 15-min PR value is not within a ±2σ-interval of its most likely value (PR_{mode}) within the month. Like this, data corresponding to inhomogeneous irradiance conditions between the weather station and the PV system are eliminated [5]. Approximately 50% of the data (excluding nights) are filtered out.

In a next step, measured values for MPP current and voltage, I_{meas} and V_{meas}, are corrected to lower seasonality. Both parameter are corrected for temperature. Furthermore, I_{meas} is additionally corrected for irradiance because of its strong irradiance dependency. The following formulas are used:

$$V_{T_{corr}} = V_{meas} * \frac{1}{[1+\beta*(T_{cell}-T_{STC})]}$$

$$I_{G,T_{corr}} = I_{meas} * \frac{G_{STC}}{G_{POA}} * \frac{1}{[1+\alpha*(T_{cell}-T_{STC})]}$$

where β is the temperature coefficient in voltage [%/K], α the temperature coefficient in current [%/K] and T_{cell} is the cell temperature, which is calculated by [6]:

$$T_{cell} = T_{amb} + (NOCT - T_{NOCT}) * \frac{G_{POA}}{G_{NOCT}}$$

$NOCT$ is the Nominal Operating Cell Temperature from the datasheet and is determined under the following conditions: T_{NOCT} = 20°C, G_{NOCT} = 800W/m² and a wind speed of 1m/s. Although measured cell temperatures are available, it was decided to use the modelled values because the measured data are incomplete and some temperature sensors lost its connection to the backside of the PV modules for a certain period.

Afterwards, the corrected values are normalized according to [7] in order to achieve better comparability of the behavior among the systems.

$$V_{norm}[\%] = \frac{V_{T_{corr}}}{V_{STC} * n_{modules_{series}}}$$

$$I_{norm}[\%] = \frac{I_{G,T_{corr}}}{I_{STC} * n_{modules_{parallel}}}$$

Finally, the normalized current, voltage and performance ratio is aggregated to monthly values. This was done to remove strong data oscillation due to short term variations.

2.3 Calculation methods

In order to verify the degradation of the different systems, the yearly performance loss (PL) of normalized monthly current and voltage time series as well as PR values is calculated using the software R. Thereby, seasonal-trend decomposition using LOESS (STL) is applied on the prepared data [8], [9]. By using STL, the seasonality and a remainder are separated from a set of measured time series data to receive a clear trend over time, and it is described by:

$$Y_t = T_t + S_t + R_t$$

STL uses locally weighted regression and smoothing to filter out the trend. This statistical method was chosen in order to reduce the performance loss uncertainties by removing seasonal effects and noise from the time series. To receive the PL a simple linear regression of the trend line is performed. The parameter of the regression line are a, which is the gradient of the function, and b, which is the intercept with the y-axis. By applying the following formula:

$$PL = 12 * a$$

the yearly performance loss is calculated. Figure 1 shows I_{norm} of a pc-Si PV system together with its trend and the regression line of the trend.

Finally, the uncertainty of the PL is determined. Therefore, the confidence interval of the regression functions gradient is computed using a confidence level of 95%. The half width of the absolute interval value is multiplied by 12 (months per year).

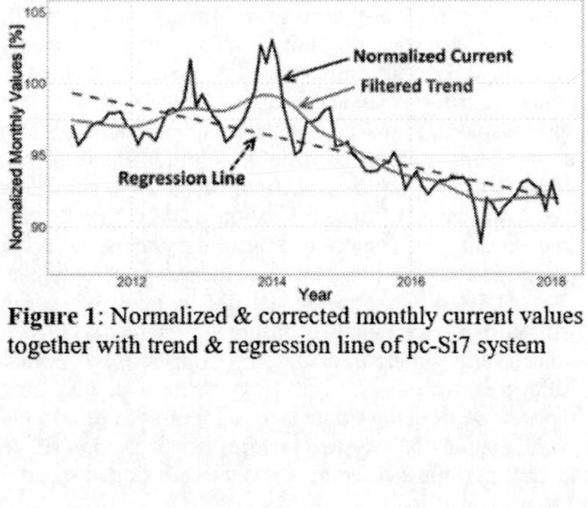

Figure 1: Normalized & corrected monthly current values together with trend & regression line of pc-Si7 system

3 RESULTS

The following figures represent the yearly PL for I_{norm}, PR and V_{norm}, divided in crystalline and thin-film systems. For every PV system the PL values including uncertainties are shown. The horizontal red line indicates a yearly performance loss of 1%, which is a minimal performance guarantee many PV manufactures ensure for the first 10 years. When looking at Figure 2, it is visible that most crystalline systems operate within the guaranteed performance levels. For thin-film systems, more than the half experience a degradation of more than 1%/a.

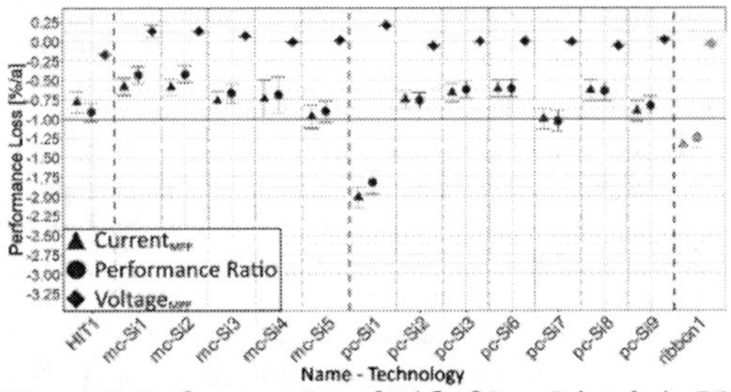

Figure 2: Performance Loss [%/a] of I_{norm} (triangles), *PR* (circles) and V_{norm} (rhombi) of all crystalline PV systems

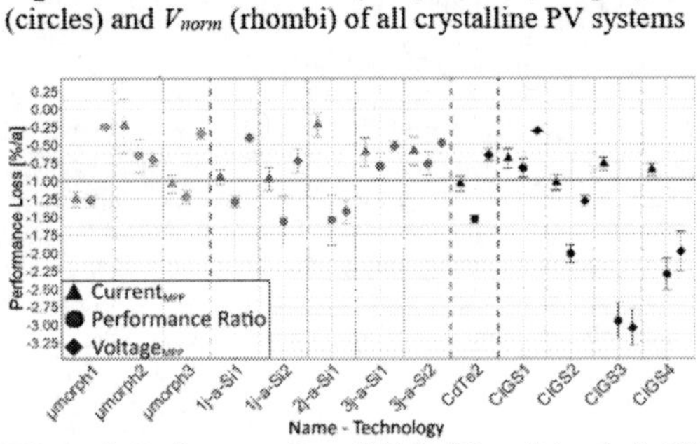

Figure 3: Performance Loss [%/a] of I_{norm} (triangles), *PR* (circles) and V_{norm} (rhombi) of all thin-film PV systems

For most systems under observation, the current degrades more strongly in comparison to the voltage, with loss rate values between 0.25% and 2% per year. The main drivers for this decrease in I_{norm} and a resulting reduced power output are a degradation in the short circuit current combined with a lowered fill factor due to a decreasing shunt resistance. This effect might be caused by soiling and occurring DM. The studied PV installation is not subject to periodical cleaning, and soiling in dry periods can have a noticeable impact. The most commonly reported DM occurring in crystalline systems installed in moderate climates are according to Jordan et al. [10] hot spots, potential induced degradation (PID) and circuity discoloration (due to corrosion). Additionally, systems, which are in operation since more than 10 years, show frequently encapsulant discoloration, a DM with relatively low severity. In respect to thin-film systems, absorber corrosion and glass breakage seem to have a bigger impact [10, 11]. It has to be stressed that DM are rarely appearing isolated but often simultaneously, and interact with each other, since different modes are triggered by similar influences such as PV technology, climate or mounting type. Apart from the simultaneous appearance of DM the consideration of PV system performance, instead of single module performance, complicates the detection of specific DM. Nevertheless, it was decided to study PV system degradation since system data, available from inverter, are more easily accessible and more relevant within the scope of this study and future applications.

Performance losses detected in crystalline silicon systems are exclusively due to a decreasing current. It is visible that monocrystalline (mc-Si) systems degrade less than polycrystalline (pc-Si) PV panels. Furthermore, pc-Si systems seem to show a wider distribution of performance loss rates than mc-Si. Pc-Si1 shows the highest degradation rate of all crystalline silicon systems in consideration, and is exceeding the maximum guaranteed performance loss stated on the datasheet (90% of initial power after 10 years). One reason can be traced back to two cells with considerably higher temperatures within the string, which were detected via infrared (IR) imagery. An IR image of one cell is shown in Figure 4. This effect might be caused by an accumulation and thermal recombination of charged carriers in the bad cells due to hot-spots or corroded/detached connections. The hotspots are having a visible impact on the IV curve of the system. Figure 5 shows the normalized measured IV curve at 872.3W/m² and 51.5°C cell temperature together with the computed initial curve at the measurement conditions and the initial curve at STC (using model according to De Soto et al. implemented in PVLIB [12, 13]). At low voltages the gradient of the curve is higher than in the initial state, which is assumed to be caused by cell mismatch and a reduced shunt resistance. This in turn creates alternate current pathways and ultimately contributes to the detected loss in I_{norm}. Another reason for the high loss in current might be the selected front-side glass of pc-Si1. The glass has a pyramid like pattern, visible in Figure 4, and it is expected to accumulate soiling "more efficiently".

Figure 4: IR image of pc-Si1 system including hot cell, Front-side glass of pc-Si1

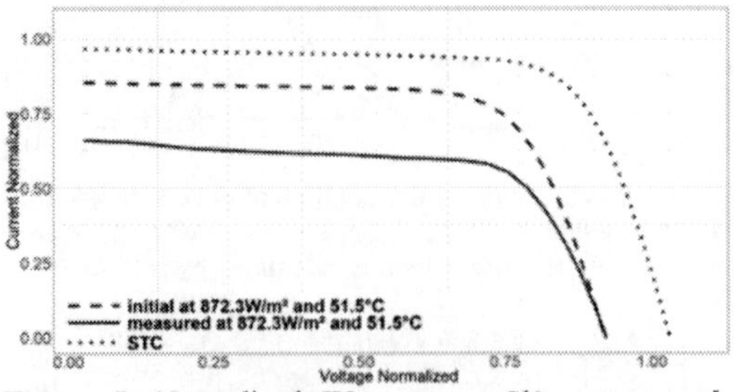

Figure 5: Normalized IV curve pc-Si1 system, red: measured curve, blue dashed: initial curve at measurement conditions, black dotted: initial curve at STC

A similar degradation pattern, although with higher current values, has been detected in the IV curves of pc-Si7 and mc-Si5. This pattern in turn can partly be traced back to fractured cells in some of the modules, visible in Figure 6 and Figure 7, where electroluminescence images of cells and modules are depicted.

Figure 6: EL image of module of pc-Si7

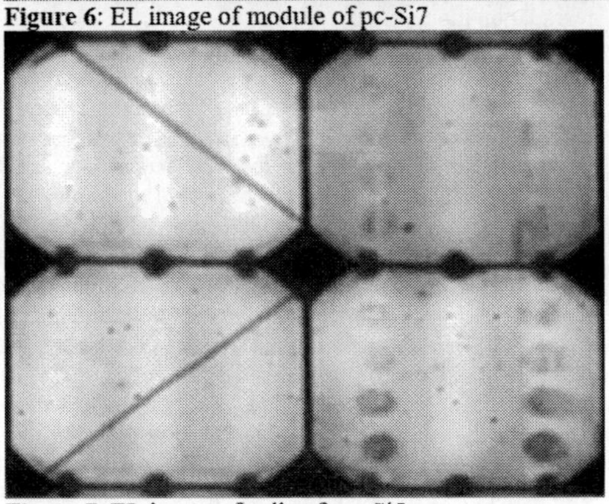

Figure 7: EL image of cells of mc-Si5

Circuitry discoloration was detected in several cells of the pc-Si2 system, which leads to a partial disconnection of the cells affected, connected to an associated shunt resistance loss. Figure 8 shows a cell, which has circuitry discoloration and cracks, possibly stemming from enlarging snail tracks, and an EL-image of the same cell, where a partial cell disconnection can be seen.

Figure 8: EL image and picture of pc-Si2 cell

Apart from the degradation modes shown, other minor degradation limiting effects were detected for crystalline systems, such as minor cell cracks or back-sheet cracking. The ribbon system can be affiliated to pc-Si systems due to their crystal structure and HIT to mc-Si systems. Their degradation behavior is similar to the one of these systems.

It became apparent that the detection of DM within thin-film systems is somewhat more complex. The lack of a crystalline structure and the low operating current makes the usage of EL imagery more difficult. Furthermore, the apparent difference in performance loss behavior between crystalline and thin-film systems (visible in Figure 2 and Figure 3) leads to the conclusion, that different DM develop over time. Especially the CIGS and a-Si systems under observation behave differently. Here, higher performance losses in V_{norm} were observed. The highest overall losses of 2-3% per year in PR, partly due to a high degradation in V_{norm}, are recorded for CIGS 2, 3 and 4. The IV curve for CIGS3 is shown in Figure 9 (initial curves computed using model according to Vergura [14]). It can be seen that a decreasing voltage primarily triggers the overall performance loss. For this system, EL images were done, visible in Figure 10. In the initial state, the EL image looks normal but after field exposure some of the modules emit almost no light anymore. This was a very surprising discovery and until now it was not possible to connect it to a specific root cause.

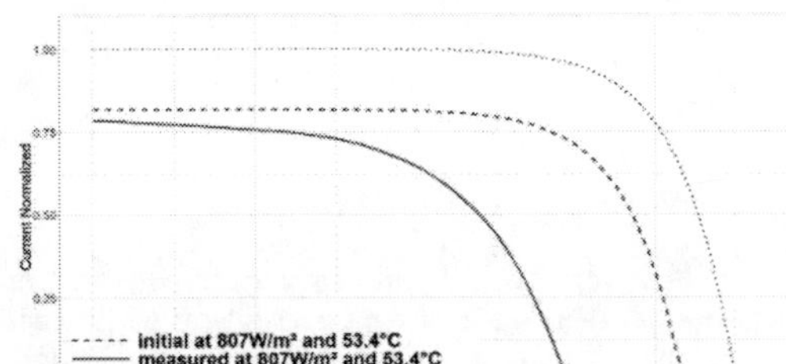

Figure 9: Normalized IV curve CIGS3 system, red: measured curve, blue dashed: initial curve at measurement conditions, black dotted: initial curve at STC

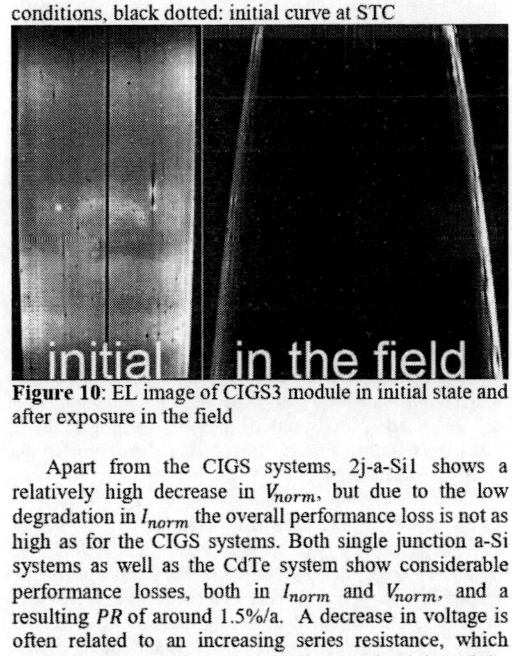

Figure 10: EL image of CIGS3 module in initial state and after exposure in the field

Apart from the CIGS systems, 2j-a-Si1 shows a relatively high decrease in V_{norm}, but due to the low degradation in I_{norm} the overall performance loss is not as high as for the CIGS systems. Both single junction a-Si systems as well as the CdTe system show considerable performance losses, both in I_{norm} and V_{norm}, and a resulting PR of around 1.5%/a. A decrease in voltage is often related to an increasing series resistance, which might be connected to a decreasing conductivity of the metal contacts. Another root cause might be a chemical process within the active material, which affects the efficiency of the charge carrier separation. Micromorph silicon systems consist of a junction between amorphous silicon and microcrystalline silicon. Their temperature dependency is similar to the one of amorphous silicon systems (when comparing the temperature coefficients) and the degradation behavior seems to be closely related as well.

A summary of the discussed results can be found in Table 1.

Table 1: Summary of systems degradation behaviors

Characteristic	Crystalline systems	Thin-film systems
Performance Ratio	Moderate decrease (0.4-1.1%/a)	Higher decrease (0.6-2.4%/a)
Current	Loss in power is primarily due to decrease in current	Power loss can be traced to both, loss in current and in voltage
Voltage	No visible decrease in voltage	
Affected electrical parameter	$I_{SC}, R_{sh}, FF,$	$I_{SC}, R_{sh}, FF, V_{OC}, R_s$
Degradation modes	Hot cells, damaged cells, metallization degradation	Metallization degradation, degradation of active material (assumptions)

4 SUMMARY & CONCLUSION

In this work, a study of the current and voltage behavior of different PV systems over time is performed. The results are linked to the performance ratio evolution, which is proportional to the evolution of the absolute power output over time. While considering different silicon and non-silicon based PV technologies, it became apparent that the technologies, which account for the majority of installed system on the market, poly-crystalline and mono-crystalline silicon, experience the lowest performance loss over time. This observation is attributed to the maturity of this technology due to decades of research, development and field experiences but also to the stability of the crystalline base material. The study is conducted based on the examination of field data, whereby seven years of outdoor exposure are considered. Eight different technologies, divided into 27 different PV systems, are subject of the study. After pre-filtering the raw data, the current and voltage values are corrected, considering their dependency on temperature and irradiance, normalized using nominal data-sheet values and aggregated to receive monthly values. The trend lines of the time series data of I_{norm}, V_{norm} and PR are then extracted using STL. Finally, the trend lines are subject to the application of simple linear regression and the resulting monthly gradient is aggregated to compute the yearly performance loss.

It has been shown that the main driver for the decrease in PV power performance is a declining current output. The majority of systems under surveillance show this trend. Through the execution of EL, IR and visual inspections as well as IV curve recordings a decrease in current can be affiliated to damaged/partly disconnected cells, hot cells and circuitry discoloration. These prevailing degradation modes lower the shunt resistance and directly affect the operating current. The mentioned degradation modes were exclusively detected in crystalline systems.

The degradation of CIGS, and partly also a-Si, systems seem to be rooted from different causes. Three out of four CIGS systems show an unusual high performance loss rate in voltage. A decrease in voltage is often connected to an increasing series resistance. It is assumed that this is connected with a reduction in the conductivity of the metal contacts or a decreasing ability of the photovoltaic active material to separate the charge carriers.

It became apparent that the connection of specific degradation modes to PV system performance is a difficult task. The detection of a single degradation mode formation within a PV system, instead of a single module, is complicated because of the relatively low impact on electrical parameter. Furthermore, the appearance of degradation modes in thin-film systems is more difficult to detect compared to crystalline systems. Yet, we are convinced that further studies to improve the detection time and quality will contribute to early detection methodologies. This will ultimately lead to the development of early-state counteract mechanisms within O&M routines.

For a qualitatively more profound study, measurements on module level are suggested. This type of studies would serve as an advancement of knowledge in degradation mode occurrence and severity.

To better track the changes within the systems in the future, periodical recordings of IV curves will be crucial. The IV curve evolution over time is providing essential information about the change in series and shunt resistances, which in turn are directly affecting the current and voltage behavior of the system in question. Furthermore, visual inspections, IR and EL imaging are viable tools for the connection of the IV curve evolution and occurring degradation modes. A more thorough study of performance degradation of PV installations and its root causes will enrich the common knowledge about the occurrence and severity of degradation modes and other technical risks in specific climates. This expertise will be viable for the improvement of O&M best practices and help to develop reliable insurance models. Apart from that, the accuracy of prediction scenarios of the performance of future PV systems can be improved based on these studies, which in turn will contribute to setting up stronger business cases for PV installations.

ACKNOLEDGEMENTS

The research leading to these results has received funding from the European Union's Horizon 2020 programme under GA. No. 721452 – H2020-MSCA-ITN-2016.

In addition, the authors would like to thank Alexander Astigarraga for his support in carrying out the field tests.

REFERENCES

[1] D. Moser, M. Del Buono, M. Jahn, M. Herz, M. Richter and K. De Brabandere, "Identification of Technical Risks in the PV Value Chain and Quantification of the Economic Impact on the Business Model," *Progress in Photovoltaics: Research and Applications,* vol. 25, no. 7, pp. 592-604, 2017.

[2] M. Köntges, G. Oreski, M. Herz, U. Jahn, P. Hacke and K.-A. Weiß, "Assessment of Photovoltaic Module Failures in the Field," IEA PVPS Task 13, Subtask 3 Report IEA-PVPS T13-09:2017, 2017.

[3] IEC60050-191, "International Electrotechnical Vocabulary: Chapter 191: Dependability and quality of service," International Electrotechnical Comission, Geneva, Switzerland, 1990.

[4] IEC 61724:1998, "Photovoltaic system performance monitoring—Guidelines for measurement, data exchange and anlysis," Geneva, Switzerland, 1998.

[5] D. Moser, M. Pichler and M. Nikolaeva, "Filtering Procedures for Reliable Outdoor Temperature Coefficients in Different Photovoltaic Technologies," *Journal of Solar Energy Engineering,* vol. 136, no. 2, 2013.

[6] E. Skoplaki and J. A. Palyvos, "On the temperature dependence of photovoltaic module electrical performance: a review of efficiency/power correlations," *Solar Energy 83,* pp. 614-624, 2008.

[7] S. Ransome and J. Sutterlueti, "Using the Loss Factors Model to Improve PV Performance Modelling for Industrial Needs," in *29th EU PVSEC,* Amsterdam, 2014.

[8] R. B. Cleveland et al., "STL: A Seasonal-Trend Decomposition Procedure Based on LOESS," *Journal of Official Statistics,* 1990.

[9] A. Phinikarides, N. Kindyni, G. Makrides and G. E. Georghiou, "Review of Photovoltaic Degradation Rate Methodologies," *Renewable and Sustainable Energy Reviews,* vol. 40, pp. 143-152, 2014.

[10] D. C. Jordan, T. J. Silverman, J. H. Wohlgemuth, S. R. Kurtz and K. T. VanSant, "Photovoltaic failure and degradation modes," *Progress in Photovoltaics,* vol. 25, pp. 318-326, 2017.

[11] R. M. Smith, D. C. Jordan and S. R. Kurtz, "Outdoor PV Module Degradation of Current-voltage Parameters: Preprint," in *National Renewable Energy Laboratory,* Golden, CO, 2012.

[12] W. De Soto, S. A. Klein and W. A. Beckman, "Improvement and validation of a model for photovoltaic array performance," *Solar Energy,* vol. 80, pp. 78-88, 2006.

[13] W. F. Holmgren, R. W. Andrews, A. T. Lorenzo and J. S. Stein, "PVLIB Python 2015," in *42nd Photovoltaic Specialists Conference,* New Orleans, 2015.

[14] S. Vergura, "A Complete and Simplified Datasheet-Based Model of PV Cells in Variable Environmental Conditions for Circuit Simulation," *Energies,* vol. 9, no. 5, pp. 1-12, 2016.

DELAMINATION OF CIGS THIN FILM PHOTOVOLTAIC MODULE IN DESERT CLIMATE

A. Abdallah[1], A. Abotaleb[1], M. Buffière[1], G. Stephan[2], S. Richter[2] and C. Hagendorf[2]
[1]Qatar Environment and Energy Research Institute (QEERI), Hamad bin Khalifa University (HBKU), Qatar Foundation, P.O. Box 34110, Doha, Qatar
[2]Fraunhofer Center for Silicon Photovoltaics CSP, Otto-Eißfeldt-Straße 12, 06120 Halle (Saale), Germany

ABSTRACT: In this paper, the reliability of copper indium gallium selenide (CIGS) photovoltaic (PV) technology under high temperature, high soiling rate and high UV exposure conditions of Doha (Qatar) was studied. Thin film delamination was found to be the main dominant failure mechanism that affects the module performance and reliability for CIGS modules installed in desert climate. Using both, non-destructive and destructive techniques, different aspects have been reported to contribute to the initiation and propagation of the interfacial failure. These are; the chemical surface modification of the transparent conductive oxide (TCO) and the ethylene vinyl acetate (EVA) encapsulant, the additives to EVA material and the ingress of moisture into the interface. As a result, a degradation of the module nominal power by 35% was measured after two years from installation.

Keywords: Photovoltaic; CIGS Module; delamination; degradation.

1 INTRODUCTION

Studying the performance and reliability of photovoltaic (PV) modules in a desert environment with high temperature, soiling rate and humidity is very important to the deployment of PV power into the electrical grid in Qatar and more generally in the MENA region. Long-term reliability of photovoltaic modules and systems requires a systematic study on monitoring performance and degradation mechanisms.

Studying the effects of climatic conditions on PV performance and reliability is important, for example, to determine the PV technology that produces the maximum energy yield and stable performance at a specific location. Various failure modes have been observed in PV thin film technologies in desert environment, e.g. dry heat delamination known for copper-indium-gallium-diselenide (CIGS), CdTe and amorphous/micro-crystalline silcion (a-Si/μc-Si) thin film modules [1-2], which resulted in a drop in the performance.

Despite the fact that PV modules are subjected to standard testing and certification, some of these modules may show several failure modes in the field. This has risen the question about the need for additional testing simulating the desert climates, for instance, in order to improve their reliability. This paper focuses on interfacial failure defects of CIGS module. Advanced indoor module characterization techniques will be used to investigate the module failure root causes and related yield losses. These data are useful to the PV cell and module manufacturers since it could be used as a feedback to improve the PV module and system performance and reliability. Additionally, the data will allow to develop extended indoor testing protocolls to ensure a long-term reliable performance of PV module installed in desert climate [3-4]. Furthermore, these results will provide experience to energy providers in the desert climates with respect to on-site testing, performance monitoring and operation & maintenance (O&M).

2 EXPERIMENTAL DETAILS

2.1 The solar test facility

The Solar Test Facility (STF) is approximately 200 kWp of different solar PV technologies ranging from silicon to thin film technologies. The STF is located at the Qatar Science & Technology Park (QSTP) in Qatar (25°17′12″N, 51°32′0″E). A meteorological station monitors the irradiance components, air temperature, wind speed and direction. A data acquisition system measures current-voltage, power for each array at maximum power point (MPP) conditions (Müller+Ziegler sensors). The module temperature was measured using one thermocouple (Pt-100 sensors) attached to the back-side of one module.

Figure 1: The solar test facility, Qatar Science and Technology Park (QSTP), Doha, Qatar. Various silicon and thin film technologies were tested in real outdoor testing conditions

2.2 Module characterization

Various CIGS modules have shown thin film delamination after two years of operation under desert environmental conditions. For root cause analysis, one affected module was selected and investigated at the PV Defect Diagnostics Labs at Fraunhofer CSP and QEERI. In a first instance various nondestructive techniques were applied, such as optical microscopy, electroluminescence (EL), static thermography, current-voltage (IV), dark lock-in thermography (DLIT) and wet leakage testing. Secondly, the interfacial failure in the layer stack was investigated further through target preparation and micro-analytic tools and for local chemical analysis time-of-flight secondary ion mass spectrometry (ToF-SIMS) . The major aim was to understand the root cause of thin film delaminations that lead to power degradation and long-term reliability issues in the field.

3 RESULTS AND DISCUSSION

Firstly, for PV modules operated in desert climate, it is important to consider the high irradiance and high operating temperature typical to a desert climate (Fig. 2). A module temperature T_{mod} of 70 °C was measured during summer months, while module temperatures between 20 and 30 °C were measured at low irradiance conditions below 250 W/m^2 during winter months from November to March. Apart from that, depending on the incident irradiance and wind speed and direction, a variation of the ambient air temperature and module temperature was observed. High module temperature is measured during the noontime where the solar irradiance is maximal.

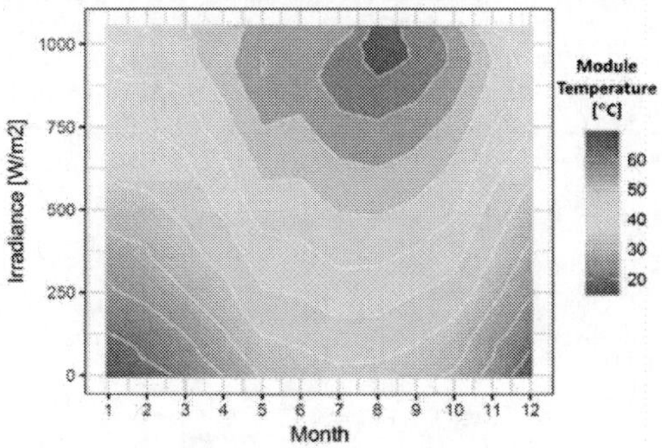

Figure 2: Irradiance and module temperature measured at the Solar Test Facility in Doha, Qatar during one year.

Further detailed optical microscopy measurement at a delamination site shows that these interfacial failures propagate further along the laser scribe and/or perpendicular to it. The driving force for delamination crack propagation in thin films is known to be due to the presence of high biaxial-compressive residual strain in the film during deposition and/or a poor adhesion at the interface [5]. Additionally, the in-diffusion of moisture and reactive species through the module edge sealant could result in excessive delamination and propagation and subsequent corrosion of the Transparent Conductive Oxide (TCO).

Secondly, visual inspection of CIGS modules showed thin film delamination as shown in Fig. 3. Nine areas of the module delamination were selected for further investigations. High-resolution optical microcopy shows that thin film delamination initiates often near the P3 scribe line (Fig. 3b) and from the module edges (Fig. 3c).

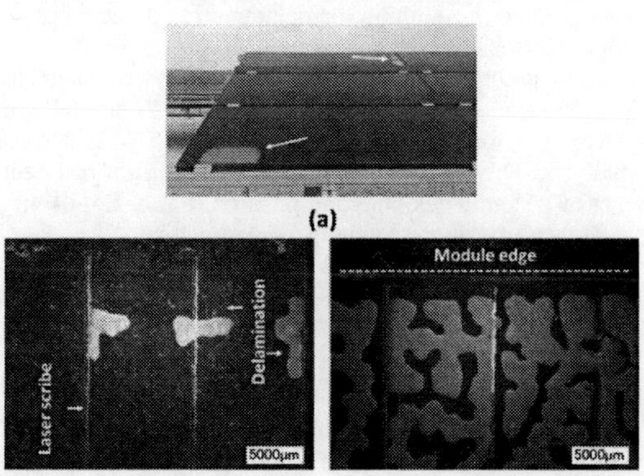

Figure 3: (a) A photograph of CIGS modules showing thin film delamination after two years of installation in desert climate. A large delamination area is visible at the module edge (indicated by arrow). Optical micrographs showing worm-like thin film delamination (b) initiated from the laser scribe and (c) initiated from the module edge.

In order to localize the delamination within the module layer stack solar cell/encapsulant/glass, high-resolution optical microscopy images have been acquired on module level with the module inspection microscopy setup in dark field mode (Fig. 4). Imaging the module surface from the sunny side shows delaminated (bright) and not delaminated (dark) areas extending on the TCO covered regions of the CIGS layer stack. P1 and P2 scribes do not inhibit the spreading of the delamination crack. This indicates an enhanced adhesion of the polymer encapsulant to the bare molybdenum (Mo) surface of the P3 scribe bottom. Furthermore, indications of delaminated absorber material close to P1/P2 scribe have been observed. The delaminated regions stop abruptly at the P3 laser scribes.

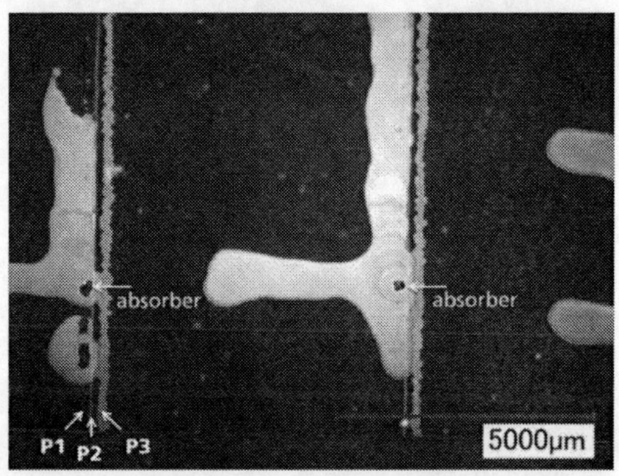

Figure 4: Optical microscopy top view on delaminated region close to laser scribes: the delamination is affecting TCO covered regions of the CIGS layer stack whereas the bare Mo surface at the P3 scribe suppresses the extension of the delamination crack.

Next, IV measurements were performed at Standard Testing Conditions (STC) for the CIGS module with thin film delamination. Table 1 presented the results before and after thin film degradation. A clear drop in the module nominal power P_{max} by 35% was measured due to thin film delamination after two years from installation in desert climate.

Electroluminescence (EL) imaging (Fig. 5a) shows that a large area of the module is inactive (no luminescence was observed), which indicates significant degradation of the module. Static thermography imaging (Fig. 5b) shows an increase in the temperature (weak hot spots) in areas with EL activity. Therefore, shunting of module areas with possible degraded contacts can be concluded.

Table I: Electrical parameters of CIGS thin film module at Standard Test Conditions (STC) before and after degradation

Module	Before	After
Power output, P_{max} [W]	120.0	78.0
Module efficiency, η [%]	11.0	8.0
Voltage at P_{max}, V_{mpp} [W]	42.7	33.1
Current at P_{max}, I_{mpp} [A]	2.5	2.4
Open circuit voltage, V_{oc} [V]	55.2	57.5
Short circuit current, I_{sc} [A]	3.32	3.20

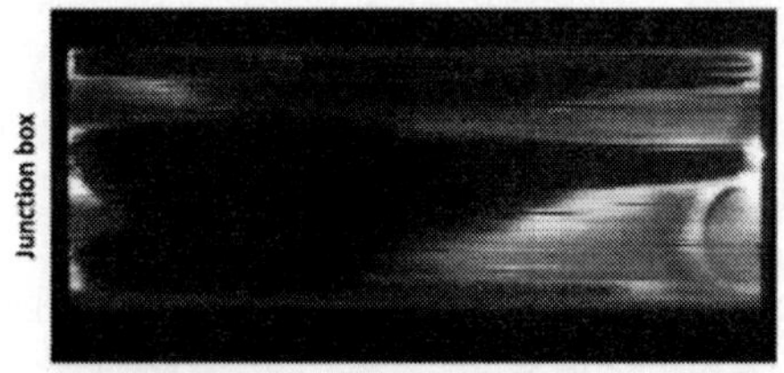

(a)

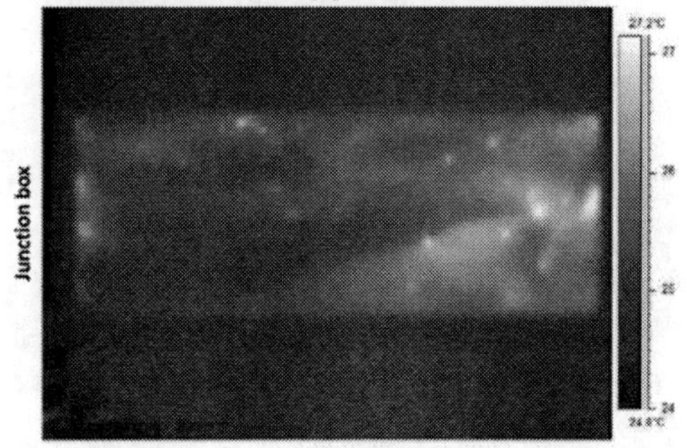

(b)

Figure 5: (a) Electroluminescence EL image (I = 0.25 A, V = 64 V) shows dark areas with no luminescence, (b) Static thermography image (I = 0.35 A, V = 70 V) shows slightly increased temperatures in areas with EL activity.

Shunt positions were investigated further using Dark Lock-In thermography (DLIT). Fig. 6 shows the amplitude image for different points (P1 to P4). Optical microscopy investigation at local shunted positions did not give a clear correlation between visible defects/particles and the local shunt. Nevertheless, the result shows that the local shunts are independent from the observed strong power degradation.

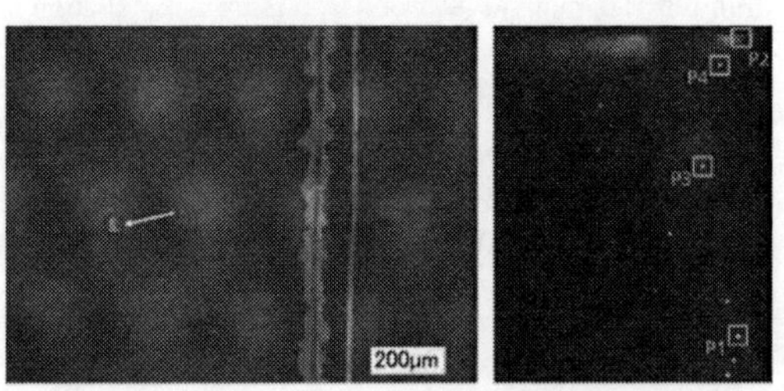

Figure 6 (a) Light optical microscopy image within a region of interest P3 showing a delamination point indicated by an arrow, (b) Dark Lock-In thermography (DLIT): amplitude image (U = 65 V, I = 0.3 A) showing different shunt positions (P1 to P4)

Before applying the destructive testing, the CIGS module has passed the wet leakage test performed under the following condition: Voltage = 997.3 V; Duration = 2 min 39 s, Resistance = 75.28 MΩ.

Moreover, in order to determine the defect microstructure and composition, the regions of interest have been separated from the module, i.e., destructive testing. Afterwards the laminate has been separated at the delamination interface, which exposes the delamination surfaces for further analysis. Fig. 7a shows a delamination cavity in the module. At the same position the cell surface (Fig. 7b) and the encapsulant surface (Fig. 7c) have been investigated. The delamination interface has been found at the cell interface. Nevertheless, visual discolorations at the cell surface indicates whether remaining residuals or a chemical reaction on top of the cell stacks. Furthermore, like in Fig. 6 we observed absorber material (red arrows in b) and c)) at the encapsulation interface, which indicates a poor adhesion of the absorber material near the scribes.

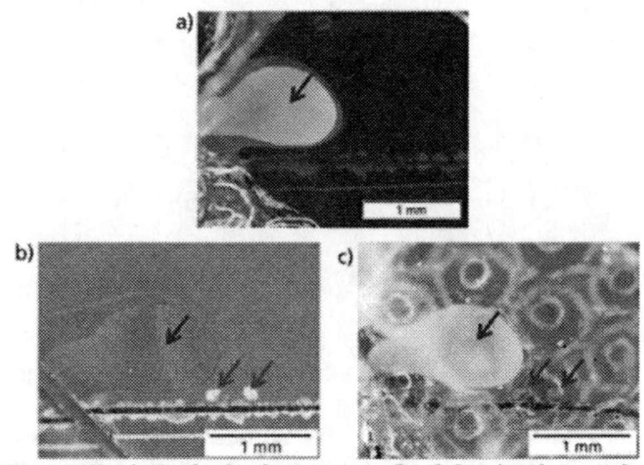

Figure 7: a) Optical microscopy of a delamination cavity (blue arrow) in the laminate; after separation at the same position the b) cell surface and c) encapsulant surface is exposed, showing delaminated absorber material in c) (red arrows).

Based on the microstructural and chemical investigation, we conclude that the delamination is occurring at the TCO/encapsulant interface indicating a chemical degradation at such interface under elevated temperature stress conditions. This stress could pull out absorber material, preferred near the scribes, which maybe support the delamination process (see also Fig. 3b). Regions with a different elemental composition (e.g. Mo in P3 scribes) inhibit the delamination process up to a certain stress level.

In order to understand the surface chemical modification we have performed Time-of-Flight Secondary Ion Mass Spectrometry (ToF-SIMS) surface analysis. The local chemical surface spectrum has been measured at the delamination position. For comparison, a reference position without delamination has been investigated as well. Fig. 8 shows a part of the spectra indicating multiple peaks. Significant differences are the appearance of signals according to PDMS (Polydimethylsiloxane) at around 73 amu. This signal is assigned to $SiC_3H_9^+$, a characteristic PDMS fragment. This residual is possibly originated by a bonding agent in the encapsulant material, which is used to improve the adhesion to the glass surface. This supports the finding that locally absorber material sticks at the encapsulant due to a good adhesion to the TCO top surface. From these results, further chemical reactions like additional TCO degradation cannot be excluded. Nevertheless, the results show that a delamination goes with a chemical change the TCO/encapsulant interface and interacts strongly with the mechanical stresses and the adhesion of the encapsulant to the TCO, absorber and Mo interface.

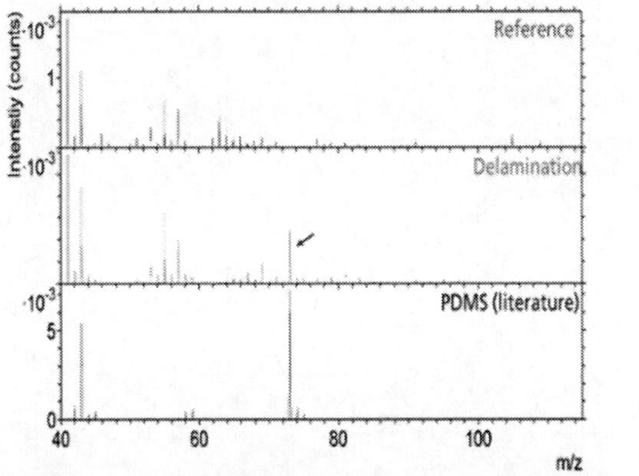

Figure 8: Positive ToF-SIMS surface spectra at the CIGS cell surface in a non-delaminated region (reference) and a delaminated region with a marked significant deviation in the spectrum at 73 amu (indicated by an arrow).

Exemplarily, ToF-SIMS spectra in the range of 40-120 amu are shown in Fig. 8 for the CIGS surface of a delaminated area and a reference position. In addition, a literature mass spectra of a Polydimethylsiloxane (PDMS) is shown for comparison. On the delaminated area some typical peaks are observed (e.g. at 73 amu) which are indicative for siloxane molecular fragments (SiC3H9+). ToF-SIMS measurements at the encapsulant surface have been performed (not shown here). The encapsulation material was identified as EVA with the help of characteristic mass spectra of known EVA material (data not shown here). Accordingly, it is assumed that the siloxane peaks originate from the reaction of a silane-based adhesion promoter [6]. Adhesion is obtained through the formation of siloxane compounds at the interfaces leading to the observed peak at 73 amu.

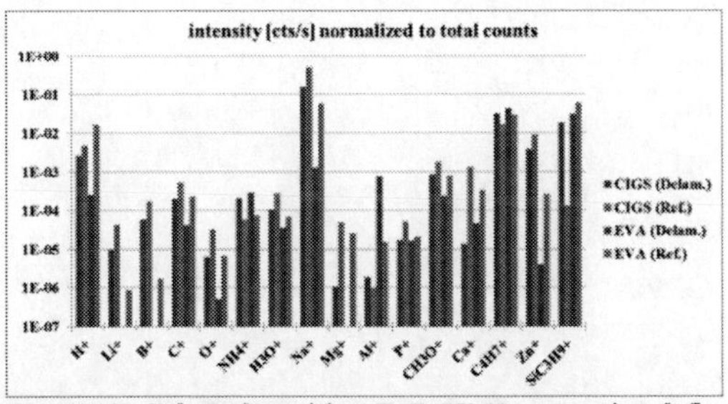

Figure 9: Selected positive ToF-SIMS mass signal for both, EVA and TCO/CIGS surfaces at the delaminated area and at the reference position.

Further selected mass signals are compared for both, EVA and TCO/CIGS surfaces (normalized to total yield of each measurement) and show characteristic differences, depicted in Fig. 9. Ammonium and some organic fragments have a higher intensity at the delaminated position for both - CIGS and EVA compared to the reference positions. In contrast, oxygen fragments and a few metals (Li, Mg) show lower signals for delaminated CIGS and EVA. Siloxane-related fragments (SiC3H9+) have clearly lower intensity at the reference CIGS surface in contrast to the delaminated CIGS as well as both EVA surfaces. The delaminated EVA presents some special distinctions, compared to the other surfaces; it has the lowest intensities for H, Li, B, C, O, Na and Zn, but shows significant highest signal intensities for Al (dopant in TCO layer) and slightly higher intensities for ammonium and some organic components (see C4H7).

Based on the microstructural and chemical investigations of the EVA and TCO/CIGS delamination we conclude that thermal stresses, initial horizontal cracks as well as a chemical degradation of silane based coupling agents play a major role for delamination. However, due to the complex processes occurring during thermal aging of organic/anorganic multilayer stacks with unknown additive composition a further more systematic investigation will be required to unveil the underlying failure mechanisms. At first, the thermo-mechanical stresses and multi-material response has to be analyzed. Secondly, the polymer aging properties and additive properties are required. Finally, the interactions with contaminations and degradation at the interface (e.g. Na) have to be investigated [7-8].

4 CONCLUSION

The paper deals with a reliability issue of CIGS thin film PV modules due to delamination of high operating temperatures, high UV irradiance and high relative humidity. A typical feature of this failure mode has been identified: thin film worm-like delamination defects guided by scribes in several modules. Various characterization techniques were used to study the initiation and propagation of these delamination defects. The delamination observed at the module edges is affecting the TCO/encapsulant interface and could be explained by in-diffusion of moisture and/or other reactive species through the edge sealant. ToF-SIMS has shown that adhesion promoting additives in the EVA encapsulant are involved in a chemical modification and subsequent mechanical weakening of the interface. For individual CIGS modules after two-years of installation, these worm-like delamination defects can be related to a nominal power drop by 35%. Therefore, it is crucial that CIGS thin-film modules are developed to match the needs for extreme climates and are tested with adapted test sequences which adress i.e. UV aging and increased sealing properties. For example, optimization of the laser scribing process, e.g., laser properties and processing speed, could lead to the reduction of the number of initiation points for interfacial failure.

5 REFERENCE

[1] Tzikas C., Gómez G., van den Donker M., Bakker K., Smets A.H.M., Folkerts W. "Do Thin Film PV Modules Offer An Advatge Under Partial Shading Conditions?", European PV Solar Energy Conference and Exhibition EU PVSEC, 25-29 September 2017, Amsterdam, the Netherlands.

[2] Wendlandt S., Berendes S., Weber T., Berghold J., Krauter S., Grunow P. "Shading investigations on thin film modules", European PV Solar Energy Conference and Exhibition EU PVSEC, 20-24 June 2016, Munich, Germany.
[3] Zhul J., Koehl M., Hoffmann S., Berger K.A., Zamini S., Bennett I., Gerritsen E., Malbranche P., Pugliatti P., Di Stefano A., Aleo F., Bertani D., Paletta F., Roca F., Graditi G., Pellegrino M., Zubillaga O., Cano Iranzo F.J., Pozza A., Sample T., Gottschalg R. "Changes of solar cell parameters during damp-heat Exposure", Prog. Photovolt: Res. Appl. 2016; 24:1346–1358.
[4] Lechner P., Schnepf J., Stellbogen D. "Performance Characterization and Extended Reliability Testing of CIGS PV Modules", European PV Solar Energy Conference and Exhibition EU PVSEC, 25-29 September 2017, Amsterdam, the Netherlands.
[5] Hutchinson J.W., Suo Z., "Mixed Mode Cracking in Layered Materials", Advances in Applied Mechanics. Vol. 29 (1992) 29-63.
[6] Plueddemann, E.P., "Adhesion Through Silane Coupling Agents", The Journal of Adhesion, 2:3 (1970) 184-201.
[7] Antinucci, J.M., Dickens, S.H., Fowler, B.O. Hockin, H.K., McDonough, W.G., "Chemistry of Silanes: Interfaces in Dental Polymers and Composites", J. Res. Natl. Inst. Stand. Technol. 110 (2005) 541-558.
[8] Keller, U., Mortelmans, H. "Adhesion in Laminated Safety Glass –What makes it work?" Glass Processing Days, 13–16 June 1999, ISBN 952-91-0885-0.

PV MODULE AND SOLAR GLASS TRICKLING SAND TESTING

G. Mathiak D. Grimm, L. Rimmelspacher, W. Herrmann, F. Reil, J. Althaus, TÜV Rheinland, Germany.
TÜV Rheinland Energy GmbH, Am Grauen Stein, 51105 Cologne, Germany.
A. Morlier, Institut für Solarenergieforschung GmbH Hameln, Am Ohrberg 1, 31860 Emmerthal, Germany.
Email: Gerhard.Mathiak@de.tuv.com

ABSTRACT: More and more PV systems are installed in deserts with hot and dry climates. Higher module temperatures, ultraviolet radiation, sand storms and dust soiling are the main challenges for special PV module designs and the module test requirements. In this work the abrasion effect of blown sand is investigated. Sand trickling tests according to DIN 52348 [1] and ASTM D968-05 [2] were performed. The test setup includes a funnel for the sand and a tube to guide the falling sand to the sample. Parameters as sand particle speed, test duration and the sand type were varied. For comparison blowing sand tests according to IEC 60068-2-68 method Lc2 [3] were conducted with modules and glass samples.
Keywords: Qualification and testing, Module reliability, Desert application, Sand abrasion, Sand trickling test, Blowing sand test, Glass coating, Transmittance and reflectance measurement

1 INTRODUCTION

There are some environmental situations in desert areas that can have a negative effect on the efficiency of PV modules. In addition to dust and sand contamination (soiling), resistance to sand drifts and sandstorms, higher UV radiation, cleaning and higher temperatures are the main challenges for the PV module design and product qualification testing. Sandstorms can have a great influence on the glass surface of the modules and thus strongly influence their performance. In the case of photovoltaic modules that are installed relatively close to the ground, the kinetic energy of the sand particles accelerated by wind gusts and sandstorms is sufficient to cause damage to the glass surface of the modules. In most cases, standardized ageing tests are also available for the evaluation of PV modules especially the front glasses at certain locations.

To date, no accelerated ageing guideline has been developed for the degradation of PV modules due to sandstorm impact, in order to realistically take into account performance losses caused by sand particle abrasion. In general, the type of stress on solid-state abrasion of materials is described by the properties of the erosive particles (size, shape, hardness and density) and the external conditions (impact velocity, angle of incidence and temperature). The numerous parameters that can occur at the same time, which define the sandstorm erosion of materials in desert areas, show the complexity of this observation. The threat of erosion effects on front and rear side of PV modules leads to a growing expectation for a suitable test procedure. With the help of test procedures, plant operators could adapt and develop an optimized risk analysis for performance and cost calculations to the expected service life of PV modules.

The draft standard for PV abrasion NP1365/IEC 62788-8-2 [4] defines procedures for sand trickling tests. In this standard small glass samples are tested. Differences in the hardening of the glass samples and the full-size glasses may lead to differences in the test results. Verification measurements with full-size modules and different glass samples have to be performed.

The diagnostic method for spectral transmittance and reflectance measurements of glass samples with integrating sphere and for PV modules with short current measurements of sanded cells with partially shading or direct contacting was described in [5,6].

2 SAND ABRASION TEST STANDARDS

For sand abrasion testing typically sand trickling tests and blowing sand tests are applied. In this study two different sand trickling abrasion tests and one blowing sand test were studied.

2.1 Trickling Sand (Free Falling Sand)

The sand trickling tests according to DIN 52348 and ASTM D968-05 are usually used in the industry to investigate the abrasion resistance of glass and surface coatings. A typical test with 3 kg sand takes 7 min.

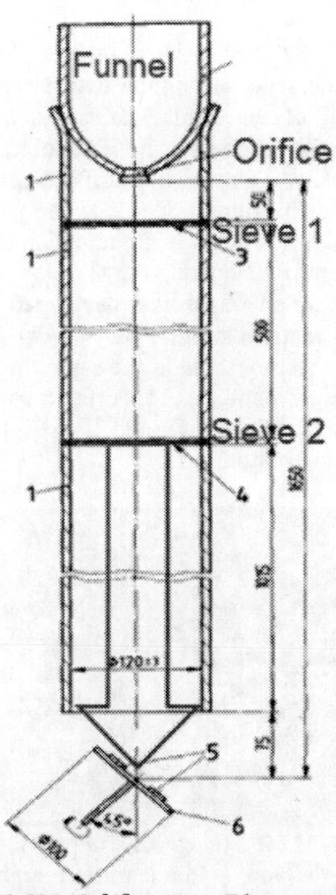

Figure 1: DIN 52348 [1] set up. Blue arrow shows the final free fall of the sand grains towards the rotating sample.

The DIN 52348 set up consists of tubes with 120 mm inner diameter. Sieves are used for homogenization. The free falling height after the last sieve 1.015 m. Standardized natural silica sand with 95 % SiO_2 and grain diameter of 0.5 mm to 0.7 mm is mainly used for this test. Silica sand produces a higher rate of abrasion compared to the most natural desert sands. The impact angle of the sand beam is 45° to the sample. The test is with continuous sample rotation rate of 250 rpm. As a reference sample a special crown glass shall be processed. The adaption to tempered glass sheets and modules in original size is without sample rotation.

The ASTM D968-05 test setup has a thinner tube (25 mm or 50 mm) and no homogenization sieves. The tube height is 0.91 m. No sample rotation is foreseen.

2.2 Blowing Sand (Wind tunnel)

The sand blowing test according to IEC 60068-2-68 method Lc2 uses as abrading sand silica sand with at least 95 % SiO2 particles with grain diameters in the range 150 µm to 850 µm. The test is performed inside a closed wind tunnel chamber. The air temperature can be chosen between ambient and high temperature up to 70°C. During the test, wind speeds of 18 m/s are capable of blowing large particle sand, whilst wind gusts can reach up to 29 m/s. The test sand concentration for natural conditions is 0.18 g/m³. The sand concentration of 2.2 g/m³ is applied for most PV module tests. Typical test duration for PV applications is 2 hours per vulnerable area i.e. on the front glass and the backsheet.

3 TEST SET UP

The DIN 52348 test setup (see Fig. 3) consists of a funnel for the sand and a tube with two sieves to guide the free fall of the sand. The sieves homogenize the particle concentration over the exposition area. The final tube piece is 1.015 m high and has a inner diameter of 120 mm. The maximum speed of the free falling sand particles is 4.4 m/s. Sand with grain diameter of 0.7 to 0.9 mm was used for comparison tests.

The test set up was further developed. Main task was the integration the rotation table (see Fig. 2B). Four 5 cm samples can be processed at the same time with a crown glass sample as reference. This effort was made to reduce inhomogeneity of falling sand caused by sand grains reflected to the bottom side.

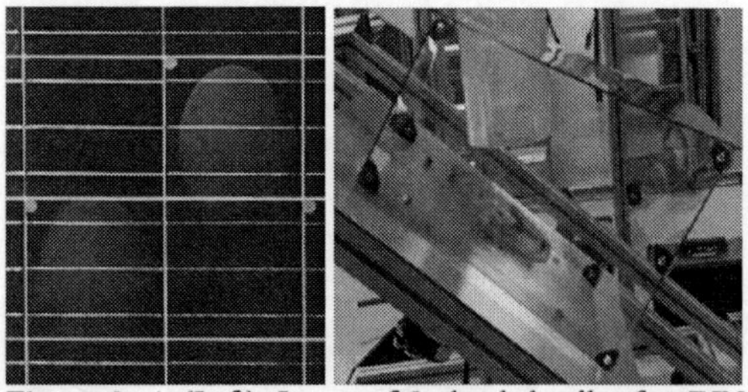

Figure 2: A (Left): Image of 2 abraded cells after DIN 52348. At the top is the abrasion higher due to the reflection of the sand grains to the bottom side. B (Right): Rotating sample table under acrylic glass cover.

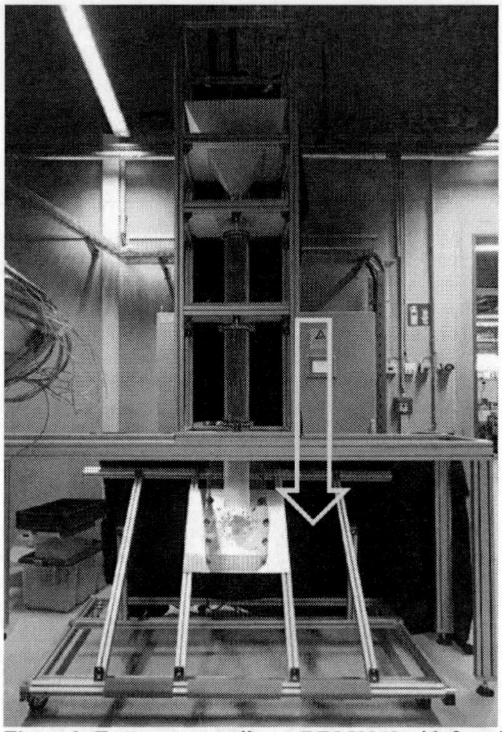

Figure 3: Test setup according to DIN 52348 with funnel and acrylic glass tubes. Under the tube it is possible to position full-size PV modules and glasses on the tilted table. On the image the rotating sample holder is mounted. The yellow arrow shows the falling height from the last sieve to the sample

The speed of the falling sand was measured with two photodiodes. The measured speed of 5.08 m/s confirmed the calculated free falling speed of 4.6 m/s for the 1.015 m long tube of the DIN 52348 test setup.

An extension of the tubes with a change in particle velocity leads with the same sand type to significant transmittance losses, both for PV modules and for solar glass samples. A significant factor with regard to transmittance losses is the sand grain size distribution.

4 SAND TRICKLING TEST

PV modules and solar glass were tested with the sand abrasion test apparatus according to DIN 52348 and ASTM D968-05. In addition, the sand particle speed (variation of tube length), and the grain sizes were varied.

Tests on PV modules, solar glass in original size and small glass samples with and without antireflective coating were performed. PV modules were characterized by the short circuit current, solar glass by spectral transmittance and reflectance with and without light trap.

4.1 Different glass types

Four different glass samples were investigated under same conditions: Float and microstructured glass with and without anti-reflective coating. The reflectance measurements with light trap (see Fig. 4a-b) showed severe differences between the glasses. The glasses with anti-reflective coating showed minor increase of the reflectance, while the effect of the anti-reflective coating was decreased significantly. A microscope was used to detect surface defects as scratches and chipping of the glass (see Fig. 5a-d). The 4 images appear similar. The images in this resolution reflect the destructive force of the sand abrasion test, but not the optical effect.

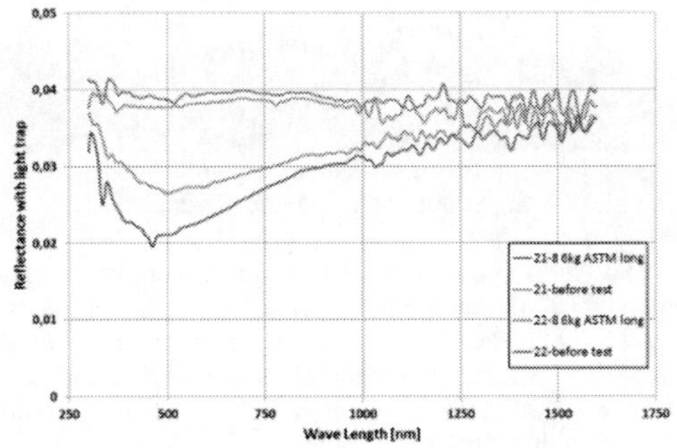

Fig. 4a: Float glass with/without ARC (21, 22)

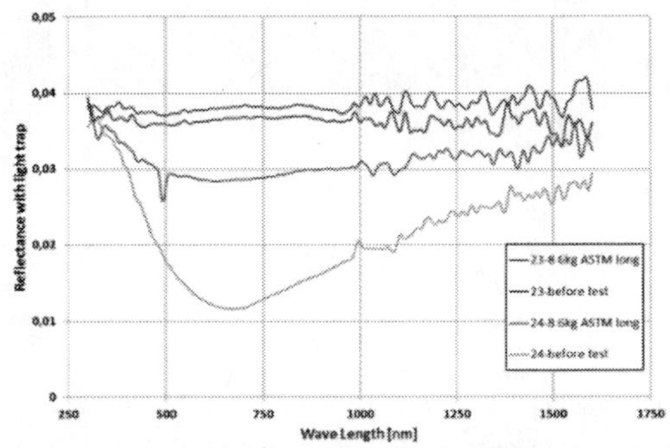

Fig. 4b: Microstructured glass with/without ARC (23, 24)

Figure 4a-b: Spectral reflectance of 4 different glass types under same test conditions before and after 6 kg standard sand load (long tube ASTM test configuration).

Fig. 5a: Float glass without ARC (21)

Fig. 5b: Float glass with ARC (22)

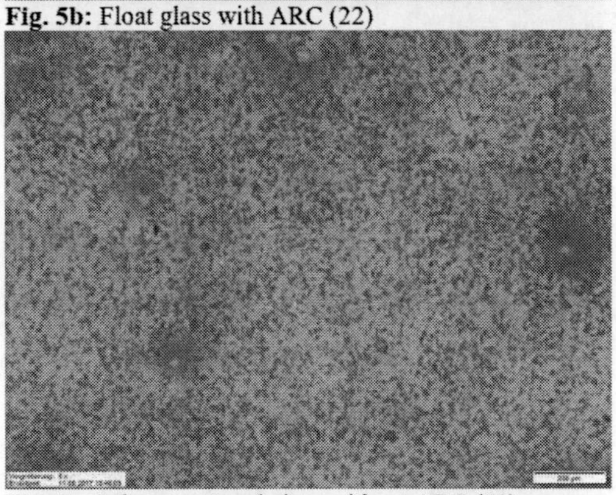

Fig. 5c: Microstructured glass without ARC (23)

Fig. 5d: Microstructured glass with ARC (24)

Figure 5a-d: Microscopy images of 4 different glasses under same test conditions after 6 kg standard sand load (long tube ASTM test configuration).

4.2 Rotating glass samples

Four different glass samples can be processed on the rotating table. Tests were also performed with solar thermal glass (see Fig. 6). The transmittance of rotated and non-rotated samples is very close

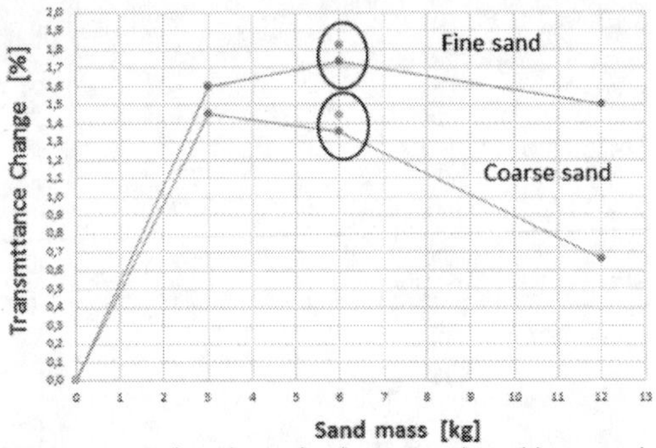

Figure 6: Solar thermal glass sample without anti-reflective coating with transmittance gain due to abrasion with fine and coarse sand after DIN 52348. With 6 kg sand, additional tests with rotating samples were performed. The transmittance of rotated and non-rotated samples is very close (inside red circle).

4.3 Cleaning and drying of glass samples

One glass sample with anti-reflective coating was abraded with 18 kg after DIN 52348. This sample and the non-abraded sample were cleaned and dried with the following 3-step procedure (see Fig. 7):

- Wiped with deionized water and dried with paper.
- Dried at 80°C for 30 min on a heating mat.
- Soaked during night in water and dried with paper

Reflectance of the abraded sample increased by the heating mat step and increased by the soaking step. The non-abraded sample behaved different. The heating mat step reduced the reflectance. The soaking step increased the reflectance but not over the initial values.

Tests with samples without anti-reflective coating showed much minor effects. The significant reflectance increase of the ARC sample can be explained by the abrasive opening of the Anti-reflective coating structure.

This measurement is not only relevant for measurement procedures in the laboratory. In the field condensated water can increase the reflectance of glasses especially of glasses with abraded anti-reflective coatings. Soiling can increase the effect.

Figure 7: Cleaning and drying procedure with abraded and non-abraded glass samples

4.4 PV module: Removal of backsheet

Another method to compare measurements of glass samples and modules is to remove backsheet, encapsulation and cell (see Fig. 8).

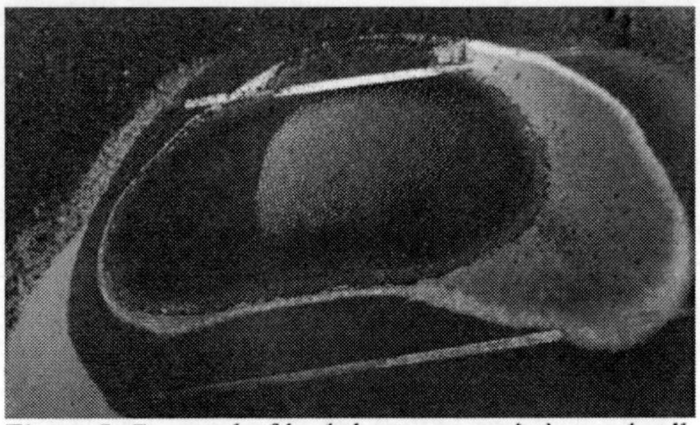

Figure 8: Removal of backsheet, encapsulation and cell. The glass with pyramids on cell side is visible.

So it is possible to measure transmittance and reflectance of the module glass which was measured before. Even when the glass has a pyramid structure it wa possible to measure transmittance and reflectance. In Fig. 9 the spectral reflectance (with light trap) of a module glass after abrasion test is shown.

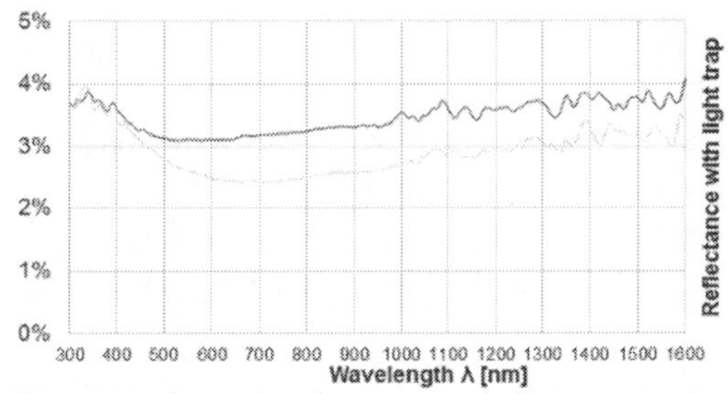

Figure 9: Spectral reflectance (with light trap) of a module glass after abrasion test

5 BLOWING SAND TESTS

Blowing sand test with a module and small glass samples was performed at 63°C with an incidence angle of 0°. The test duration was 2 h.

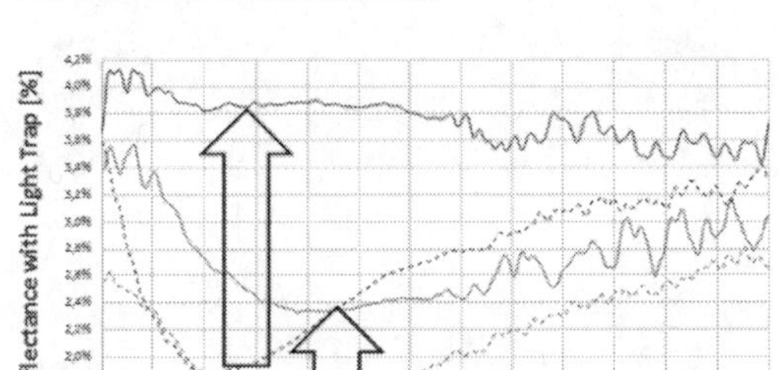

Figure 10: Spectral reflectance (with light trap) of two samples with anti-reflective coating of two different manufacturers before and after blowing sand test.

In Fig. 10 the blowing sand test of two samples with anti-reflective coating of two different manufacturers is shown. One sample lost completely anti-reflective behavior while the other sample remains anti-reflective.

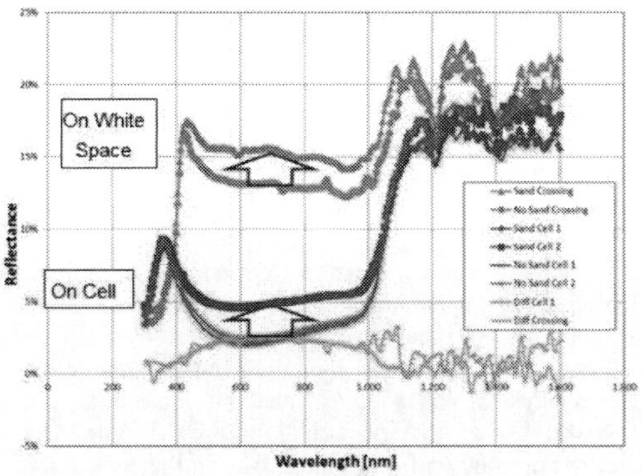

Figure 11: Commercial c-Si module: Spectral reflectance on the cell and between the cells (white space) before and after blowing sand test. The reflectance gain was 2.5 % in the range from 500 nm to 900 nm.

In Fig. 11 the blowing sand test of a commercial c-Si with anti-reflective coating is shown. The spectral reflectance was measured on the cell and between the cells (white space) before and after blowing sand test. The reflectance gain was 2.5% in the range from 500 nm to 900 nm. The short circuit current after the blowing sand test was reduced by 3.75% The results are comparable with the results of trickling sand tests according to DIN 52348 with 18 kg sand.

6 CONCLUSIONS

With the new sand abrasion test setup PV modules and solar glass samples were tested according to DIN 52348 and ASTM D968-05. For PV modules, the DIN 52348 standard is more practical as the ASTM D968-05 test, due the larger test area close to the size of a solar cell and the more homogeneous sand distribution caused by homogenization sieves and sample rotation.

An extension of the tubes with a change in particle velocity leads with the same sand to significant transmittance losses, both for PV modules and for solar glass. The sand speed was measured with two photodiodes and confirmed the calculated free falling speed.

PV modules and solar glass sheets with anti-reflective coating react more sensitive to the abrasion stress. PV modules without coating can even gain transmittance by minor roughening of the surface. The cleaning and drying process can affect the reflectance especially of samples with abraded anti-reflective coatings.

Water condensation (rain, dew or cleaning) of abraded anti-reflective coatings must be considered for performance measurements and yield predictions, especially in combination with soiling.

Blowing tests according to IEC 60068-2-68Lc2 were performed with glass samples and modules and showed comparable results as sand trickling tests.

The development of glass coatings with anti-reflective, anti-soiling and easy-clean behavior is fundamental for desert applications of photovoltaics. Therefore, a standard to qualify PV modules for arid climate conditions requires the definition of a sand abrasion test.

A new standard for the sand abrasion test of solar glass is currently discussed in the PVQAT working group 12 "Soiling and Dust" as a part of NP IEC 62788-8-2. In this standard small glass samples are tested. Differences in the hardening of the glass samples and the full-size glasses may lead to differences in the test results. Verification measurements with full-size modules and a laboratory round robin test with different glass samples shall be performed.

From real weather data and reported PV module damages in the field, appropriate test parameters for standard test procedures have to be deduced. More data from fielded PV modules and from laboratory tests need to be collected to define pass/fail-criteria on a scientific basis. This concerns the transmittance of the front glass with durable coating and the safety aspects for backsheet insulation.

7 REFERENCES

[1] DIN 52348: 1985:02: Testing of glass and plastics, abrasion test, sand trickling method.

[2] ASTM D 968-05:2010: Standard Test Methods for Abrasion Resistance of Organic Coatings by Falling Abrasive.

[3] IEC 60068-2-68:1994 Environmental testing - Part 2-68: Tests - Test L: Dust and sand

[4] IEC 62788-8-2 (NP) Measurement procedures for materials used in photovoltaic modules - Part 8-2: Materials and coatings for the irradiance incident surface of photovoltaic modules or similar solar devices: Abrasion and environmental testing.

[5] G. Mathiak, M. Hansen, M. Schweiger, L. Rimmelspacher, W. Herrmann, J. Althaus, F. Reil: „PV Module Test for Arid Climates including Sand Storm and Dust Testing" Proc. 32nd EU PVSEC, Munich, 2016, 5BO.11.4, 1568-1572.

[6] G. Mathiak, N. Pfeiffer, L. Rimmelspacher, W. Herrmann, J. Althaus, F. Reil, C. Holze, A. Morlier: "PV Module Sand Abrasion Testing", Proc. 33rd EU PVSEC, Amsterdam, 2017, 5CO.5.2, 1383-1390.

8 ACKNOWLEDGEMENT

The work presented has been elaborated in the frame of the project 'Fidelitas' (FKZ: 0325735A) supported by the German Federal Ministry for Economic Affairs and Energy.

BACKSHEET CHALKING – THEORETICAL BACKGROUND AND RELATION TO BACKSHEET CRACKING AND INSULATION FAILURES

Paul Gebhardt[1], Luciana Pitta Bauermann[1], Daniel Philipp[1]
[1]Fraunhofer Institute for Solar Energy Systems, Heidenhofstr. 2, 79110 Freiburg, Germany
paul.gebhardt@ise.fraunhofer.de

ABSTRACT: This paper assesses the phenomenon of PV backsheet chalking and its effect on solar module reliability. We review the current state of literature on the chalking effect in general and discuss the commonly accepted photocatalytic mechanism of polymer degradation in paints and coatings. We then apply this general knowledge to solar backsheets and discuss its relevance for solar modules reliability and performance, including the experience we have gained from investigations on field-exposed modules. After a physical and chemical analysis identifying the powder as titanium dioxide (TiO_2), the important factors for the occurrence of backsheet chalking are determined as (a) the type of polymer layers, (b) the polymer formulation, and (c) the amount and location of TiO_2 in the backsheet. Finally, two cases of field-aged modules exemplify different degradation pathways, that both originate from the same degradation mechanism. Although backsheet chalking can, in some cases, be an indication for further degradation, an analytical investigation of the backsheet is necessary to determine the actually associated risk.

Keywords: Backsheet, Degradation, Durability, Polymer film, PV materials, Reliability, Safety, TiO_2

1 INTRODUCTION

Backsheet chalking is a failure mode of field-exposed modules that has recently been reported more frequently. It is observable as a white powder on the outer backsheet layer, which can be removed by extensive wiping or washing with water. The phenomenon is often observed in conjunction with a rough, matte backsheet surface, which is prone to mold formation and often appears in context with other, more severe module failures – such as backsheet cracking between cells and along busbars, and in some cases complete insulation failures. Furthermore, it seems that some types of backsheet polymer composite combinations are particularly affected by this degradation path or show more severe consequences.

Figure 1: Finger with white powder from chalking backsheet

An in-depth evaluation and classification of this failure mode is therefore needed in order to assess the severity of backsheet chalking and allow estimation of the resulting long-term effects for different backsheet foils. In this paper, we discuss the suitability of backsheet chalking as an early warning sign, as has already been applied in the similar area of TiO_2 pigmented paints[1]. Further, we determine the necessary actions to estimate failure development in the case of appearing backsheet chalking in the field.

2 POWDER ANALYSIS

The white powder observed on chalking backsheets was removed from the backsheet either mechanically with a spatula or with adhesive tape and subsequently analyzed by scanning electron microscopy (SEM), energy-dispersive X-ray spectroscopy (EDX) and Raman spectroscopy.

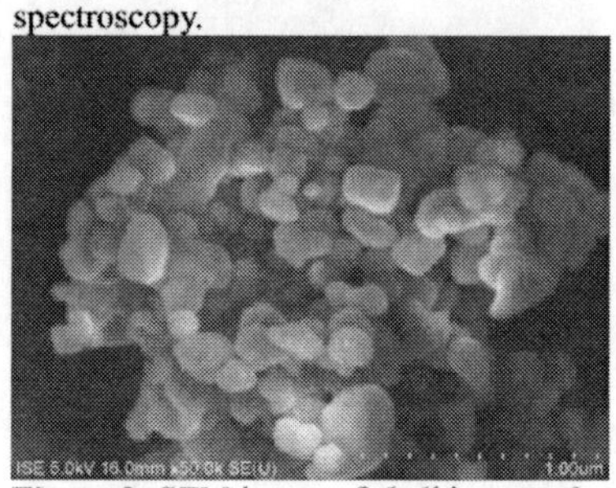

Figure 2: SEM image of chalking powder

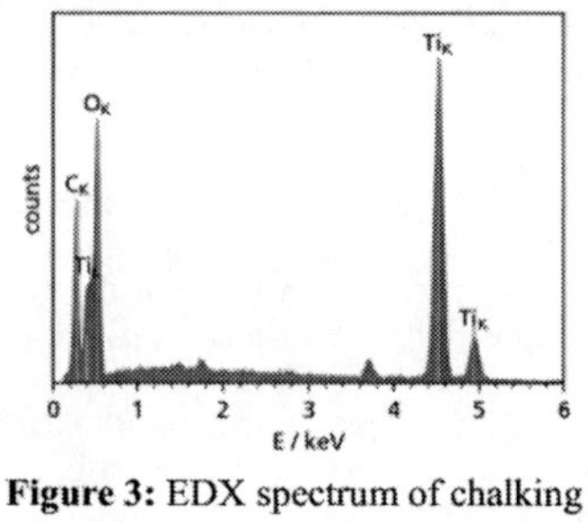

Figure 3: EDX spectrum of chalking powder

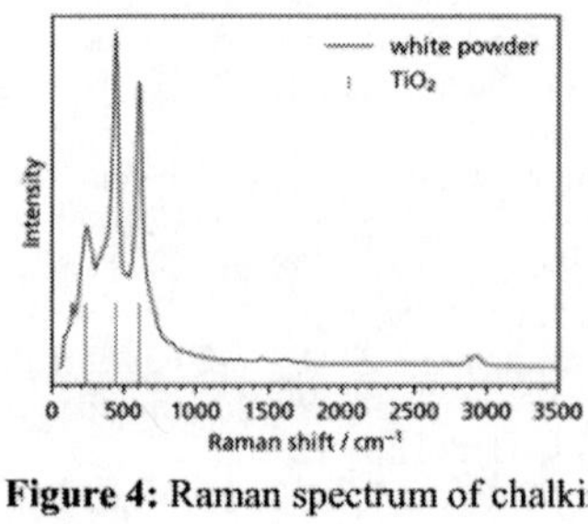

Figure 4: Raman spectrum of chalking powder

The SEM image (Figure 2) shows particles of a 100–400 nm in size, while the EDX (Figure 3) and Raman (Figure 4) identify the particles as titanium dioxide (titania, TiO_2): EDX shows the expected peaks for titanium and oxygen atoms and the signals in the Raman spectrum correspond to TiO_2 rutile phonons. The latter method is also very useful for identifying the organic polymer and presence of TiO_2 of individual backsheet layers, as performed in our lab in conjunction with fourier-transformed infrared spectroscopy (FTIR, not shown).

3 THEORETICAL BACKGROUND: TIO_2'S ROLE AS PIGMENT AND PHOTOCATALYST

TiO_2 pigment particles are added to the backsheet formulation in order to protect it from UV light. TiO_2 is by far the most commonly used inorganic pigment in coatings and paints due to its chemical stability and inertia, non-toxicity, and high refractive index[2]. The chalking phenomenon of TiO_2 has been known since the 1940s in the context of pigmented paints and coatings and could be linked to the photochemical properties of TiO_2[3].

In order to avoid pigment agglomeration and improve the dispersion of TiO_2 in the polymer matrix of the backsheet, pigments are often coated with a thin layer of inorganic oxides such as alumina or silica and used in combination with an organic dispersant[2]. A good dispersion of TiO_2 pigments in the backsheet provides an optimal particle size to efficiently protect the backsheet by reflection of light, or, to a smaller extent, conversion to heat.

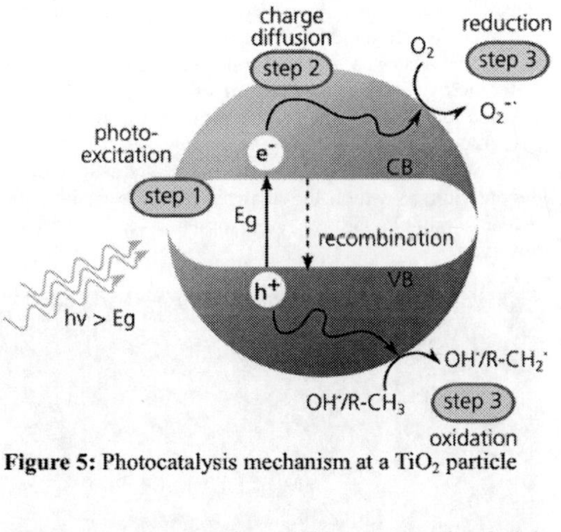

Figure 5: Photocatalysis mechanism at a TiO_2 particle

However, as a semiconductor, TiO_2 can also absorb light and initiate a photocatalytic process (Figure 5), which aggravates the degradation that is caused by UV light [4]: If the energy of incoming photons is larger than the band gap, electrons can be elevated across the band gap and electron hole pairs are created (step 1). In a second step, these charge carriers can be separated and subsequently diffuse to the particle surface if they don't recombine into a lower energy state, e.g., at bulk defects. At the surface, electrons and holes can perform reduction and oxidation reactions, respectively, and thereby form radical species (step 3): Reduction of oxygen (O_2) leads to the formation of the reactive superoxide radical ($O_2^{\cdot-}$), while holes can produce either hydroxyl radicals or form reactive alkyl species by directly abstracting hydrogen atoms from organic molecules, such as the polymer binder or the surrounding polymer matrix of the backsheet.

In any case, the formed radicals accelerate the degradation of the polymeric dispersant and the surrounding polymer. This degradation on the backsheet surface then leads to a detachment of pigment particles from the foil and to the observed chalking effect. The exact properties of TiO_2 pigments thereby have a strong effect on the extent of observed chalking, and are affected by the particle size, crystal structure (rutile/anatase ratio), amount and location (surface or bulk) of defects, TiO_2 surface functionalization, and coverage with inorganic oxides.

It is important to note that both chalking and photocatalysis are not exclusive properties of titania: Chalking has also been observed on polymer films with manganese oxide[5], and other, less common white pigments also exhibit photocatalytic activity, albeit to a smaller extent. A replacement of TiO_2 with an alternative pigment would therefore not necessarily avoid the chalking effect, but instead require the use of a higher pigment concentration in order to match TiO_2's shielding abilities.

3.1 Effects of chalking in solar backsheets

In principle, all polymer types can exhibit some degree of backsheet chalking. Even the well-tried and tested combination of a PET core and PVF protecting layers, such as TPT, can be affected, as could be reproduced in the lab by combined UV damp-heat aging 190 kWh/m², 85 °C, 85 % r. h., 1000 h[6].

However, the severity of failures linked to backsheet chalking, such as backsheet cracking and insulation failures, are strongly dependent on (a) the kind of polymer layers included in the backsheet composite, (b) the polymer formulation and addition of stabilizers, and (c) the amount and location (*i.e.* which layer) of TiO_2 pigments present in the backsheet.

Concerning the polymer material (a), the increase in reported cases of backsheet chalking with severe module failures seems to correlate with the increased use of alternative polymers, in contrast to the conventional combination of PET core layer and fluoropolymer outer and inner layer. Indeed, we have no knowledge of a report of a module failure in which pigment chalking of such a backsheet type could be proven to be the root cause.

On the other hand, there are also many examples of backsheets consisting of "new" polyamide (PA) or polyolefin layers as main polymer components which perform reliably in various outdoor scenarios. Some of these newly developed materials are even more durable than their conventional counterparts[6]. It therefore seems that (b) the exact production parameters and formulation of the polymer, *i.e.* the addition of stabilizers against the effects of UV irradiation, temperature, moisture, and diffusing gases has a large impact on the durability of the backsheet. The right combination and quantities of these components − which have to be determined in extensive reliability tests during the development phase − might be harder to optimize for some polymer materials than for others, which could explain the recent increase in occurrence of backsheet-related module failures.

As mentioned in (c), the use of TiO_2 in different backsheet layers can result in differences of degradation, as illustrated in Figure 6. Most commonly, TiO_2 pigments are used in the outer backsheet protection layer as UV blocker (Figure 6, left). In this case the discussed photocatalysis mechanism can substantially damage the respective layer, via the produced radicals, and release TiO_2 particles, which are then observed as chalking powder. This process does not in itself affect module safety or performance directly; in the long run, however, this can initiate a cascade of further degradation: Without proper protection by the outer backsheet layer, UV light and moisture reach the core backsheet layer, which directly reduces the electric insulation and safety of the module. Furthermore, moisture can diffuse further into the modules and lead to corrosion of cells and connectors, which has a negative impact on module performance.

In some cases, TiO_2 pigments reside in a layer inside the backsheet sandwich, *e.g.* core- (Figure 6, right) or inner/adhesion layer. In principle, this can lead to a very similar degradation behavior and damage the layer that comprises the pigments, with the difference that TiO_2 particles are constrained inside the backsheet and no chalking is observed on the outside. Possible further degradation pathways are independent of the exact pigment location. This case is therefore usually detected only after further degradation occurs, *e.g.* loss of adhesion with the consequence of low insulation properties or cracking of the backsheet.

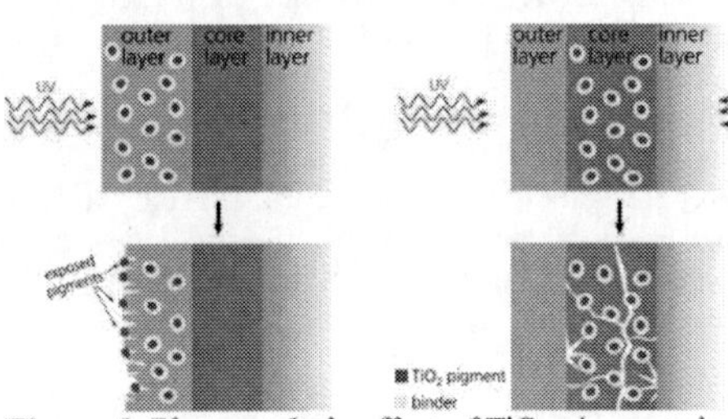

Figure 6: Photocatalytic effect of TiO_2 pigments in the outer (left) and core (right) backsheet layer

4 BACKSHEET CHALKING ON FIELD-EXPOSED MODULES

4.1 Exemplary cases of field-aged modules

In general, weak chalking does not necessarily lead to the severe module degradation modes discussed above; we believe that certain module types affected by backsheet chalking could function without further problems throughout the expected lifetime, although this would be difficult to verify with certainty in a lab.

Backsheet chalking can, however, be an early indication of future degradation. As an example, the microscope analysis of a chalking backsheet cross section shows cracks in the outer PA protection layer, which contains TiO_2 (Figure 7). Although the respective modules showed no other peculiarities in the field, further degradation, including lowered insulation properties and reduced power output, is very likely as discussed above and already indicated by a weak adhesion between the inner PET and PA backsheet layers, leading to delamination. In this case, crack formation could also be observed in the inner PA layer, which also contains TiO_2.

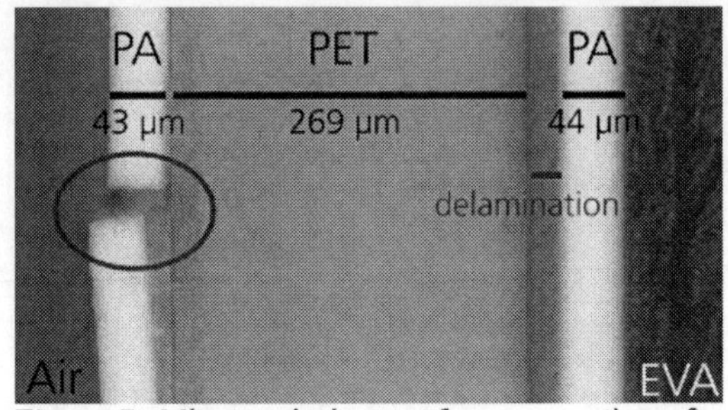

Figure 7: Microscopic image of a cross-section of a backsheet that exhibits chalking shows a crack in the outer protection layer. Both white PA layers contain TiO_2

In another case of field-aged modules, severe drops in insulation properties were observed, in some cases leading to complete failure of modules. A microscopic analysis of the backsheet cross-section of modules with low insulation revealed a severely damaged, TiO_2-containing PP layer (Figure 8, right). This is an example for the same photocatalytic mechanism (Figure 6, right) that does not lead to a visible chalking effect because the TiO_2 particles are still confined inside the backsheet sandwich. The resulting cohesion loss of the white PP layer was observed across the whole module.

Another module from the same type and outdoor site with decent insulation values was analyzed in the same way. The respective backsheet was intact at locations that were covered from front irradiation by solar cells. Between the cells, however, it was also strongly damaged. We conclude that these modules only represent an earlier state of degradation, *e.g.*, due to slightly different exposure parameters in the field, and that the degradation will continue, going down the path of the more damaged modules.

This is another example of TiO_2-related module degradation, in which the analysis of the backsheet can reveal a module failure before other severe effects are observed.

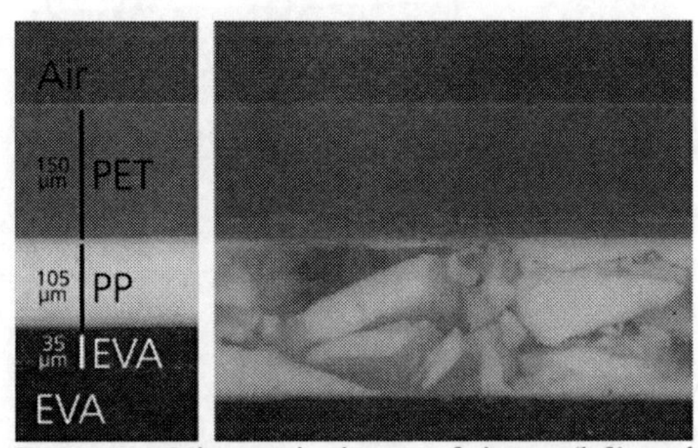

Figure 8: Microscopic image of intact (left) and damaged (right) backsheet cross section. The scale is the same for both images

5 CONCLUSION

In the recent past an increasing number of chalking backsheet findings has coincided with severe cracks of the backsheet and insulation losses. At the same time, we also see an increase of cases with weak insulation without the occurrence of backsheet chalking.

As discussed above, backsheet chalking can, in some cases, be an indication for backsheet cracks and provide an early warning for further degradation. But the exact course of further development depends on various properties of the backsheet. With knowledge of the underlying degradation mechanism, an analytical investigation of the backsheet can help understanding the actually associated risk and future development.

6 EXPERIMENTAL

Cross-sectional samples were prepared by cutting pieces of the backsheet and EVA from locations of interest from the module. These polymer laminates were then embedded in epoxy resin and afterwards ground and polished in order to reveal a flat and even surface at the location of interest. The thickness of individual polymer layers was measured with an optical microscope. The values given in the microscope images are averages of 10 individual measurements.

Electron microscopy and energy-dispersive X-ray spectroscopy (EDX) measurements were carried out using an Environmental Scanning Electron Microscope ESEM (FEI Quanta 400) with integrated EDX spectroscopy system (EDAX Genesis) at an acceleration voltage of 10 kV.

The Raman spectroscopic measurements were performed using a WiTec Alpha 500 Raman microscope with a 532 nm laser, which is obtained using a frequency doubled Nd:YAG laser with a maximum excitation power of around 17 mW. The CCD detector was thermoelectrically cooled to -67 °C.

7 REFERENCES

[1] B.A. van Driel, T.A. Wezendonk, K.J. van den Berg, P.J. Kooyman, J. Gascon, J. Dik, Determination of early warning signs for photocatalytic degradation of titanium white oil paints by means of surface analysis, Spectrochimica acta. Part A, Molecular and biomolecular spectroscopy 172 (2017) 100–108.

[2] S. Farrokhpay, A review of polymeric dispersant stabilisation of titania pigment, Advances in colloid and interface science 151 (2009) 24–32.

[3] A.E. Jacobsen, Titanium Dioxide Pigments: Correlation between Photochemical Reactivity and Chalking, Ind. Eng. Chem. 41 (1949) 523–526.

[4] U.I. Gaya, A.H. Abdullah, Heterogeneous photocatalytic degradation of organic contaminants over titanium dioxide: A review of fundamentals, progress and problems, Journal of Photochemistry and Photobiology C: Photochemistry Reviews 9 (2008) 1–12.

[5] G. Liu, S. Liao, D. Zhu, J. Cui, W. Zhou, Solid-phase photocatalytic degradation of polyethylene film with manganese oxide OMS-2, Solid State Sciences 13 (2011) 88–94.

[6] Philipp Schmid, Marktübersicht, Charakterisierung und Studien zum Alterungsverhalten von polymeren Isolationsfolien zur Anwendung in der Photovoltaik-Industrie. Diplomarbeit, Ulm, 2011.

COMBINED-ACCELERATED STRESS TESTING FOR ADVANCED RELIABILITY ASSESSMENT OF PHOTOVOLTAIC MODULES

Michael Owen-Bellini[1], Peter Hacke[1], Sergiu Spataru[2], David C. Miller[1], Michael Kempe[1]
[1]National Renewable Energy Laboratory, Golden, CO, USA, 80401
[2]Aalborg University, Fredrik Bajers Vej 5, 9100 Aalborg, Denmark

ABSTRACT: A combined-accelerated stress test (C-AST) has been developed to more accurately simulate the natural environment and capture the interdependencies between various stress factors. The stress factors simulated include temperature, humidity, high fidelity ultraviolet (UV) light exposure, cyclic dynamic mechanical loading, rain spray and system voltage bias. A demonstration experiment is established in the form of a backsheet differentiation test. The test includes six 4-cell p-PERC mini-modules with 3 backsheets based on polyvinylidene fluoride, polyamide, and polyvinyl fluoride. As well as two types of polyethylene-co-vinyl-acetate (EVA), one UV blocking and one UV transparent. The experiment demonstrated C-ASTs ability to identify backsheet cracking and other failure mechanisms such as UV-induced degradation, solder bonds breakage, minor corrosion, cell cracking, some of which had been missed by conventional accelerated stress tests.

Keywords: accelerated ageing, combined stress, photovoltaics, reliability

1 INTRODUCTION

Historically, accelerated stress testing protocols for reliability assessment of photovoltaic (PV) modules were developed to address known failures discovered in the field. As a result, design type qualification standards, such as the IEC 61215 [1], consist of a series of single or double stress factor tests designed to determine susceptibility of module designs to failure. While some of the testing is intended to capture unknown failure modes and mechansims, testing is of limited duration and thus mostly targeting specific mechanisms because designing a comprehensive test is cost prohibitive and beyond the scope of a qualification test. As such, undiscovered failure mechanisms are still being found in modules deployed in the field, often resulting in significant financial and energy production losses. Recent examples include backsheet cracking [2], potential-induced degradation [3], grid-finger corrosion [4], light and elevated temperature induced degradation (LeTID) [5] and snail trails [6]. Many of these are a result of a combination of multiple stress factors which were simply not applied in conventional tests or may require much longer duration testing at reasonable acceleration rates to be uncovered. Some of these examples follow from material interactions, which may not have been discovered in component-level tests.

The necessity for combining multiple stress factors into fewer tests has been recognised for some time, as demonstrated by efforts made by various other groups [7][8][9][10]. While important results have been produced, these efforts are typically limited to combining up to 3 stress factors targeting specific mechanisms. In this work, a combined-accelerated stress testing (C-AST) capability has been developed which combines multiple stress factors from the natural environment including light, humidity, temperature, rain, mechanical loads and voltage stress. The test has been developed with the philosophy that it is agnostic to a-priori known failure mechanisms such that mechanisms in new module designs or materials may be identified before deployment. C-AST is applied here to mini-module specimens to allow for possible adverse material interactions.

2 TESTING CAPABILITIES

The purpose of C-AST is to combine multiple stress factors of the natural environment into a single test, which is capable of identifying degradation modes resulting from a complex use environment. These include: backsheet cracking, hydrolysis and photodegradation of polymers, delamination, potential-induced degradation (PID), light-induced degradation (LID), corrosion of metal interconnects and PV cells, thermomechanical fatigue, crack development of cells, etc.

Testing capabilities have been developed in a modified Atlas XR-260 climate chamber which is configured for housing up to six 2x2 cell mini-modules. Modules are seated horizontally in rigid, Teflon-enveloped steel frames (Figure 1). Internal chamber temperature and relative humidity can be controlled between -40°C to 90°C and <5% to >95%, respectively. Four water-cooled Xenon arc lamps are mounted approximately 0.8 m above the sample plane which deliver up to two sun irradiance, spectrally filtered for AM1.5 G at 340 nm. Reflective troughs are installed below the sample plane to simulate ground albedo, reflecting ~10% of UV light onto the rear surface of the samples. The xenon arc lamps are driven by an AC power supply causing the light to fluctuate at ~120Hz. It is well known that the UV output for xenon lamps is subject to change as the lamps age; therefore, it is necessary to monitor and adjust the intensity of the lamps. This is achieved using four spectrally filtered (340 nm) light-pipe photodiodes and an additional two total UV radiometers (TUVRs). Total solar irradiation on the sample plane is monitored using six LICOR pyranometers, each of which are positioned as closely to the module mounts as possible so that irradiance at the surface of the samples can be measured.

Figure 1: Inside view of the Atlas XR-260 chamber

In addition to the Xenon lights, four white-light LED strips have been installed to deliver a minimum of 50W/m^2 light bias to the sample plane such that a voltage bias may be generated over the cell junctions and application of system voltage stress can be maintained in a representative manner, if required.

Hydraulic actuators are used to apply dynamic mechanical loads to the samples via 100 mm diameter steel rings covered with Teflon. Displacement of the ring and subsequent pressure on the front surface of the modules is controlled using a mechanical pressure valve and mechanical stop.

High voltage stress, up to 1500 V, is delivered via two Stanford power supply units. Each unit can be configured to delivered positive or negative bias. Electrically isolated Mooshimeter digital multimeters are used to monitor leakage current on the high-voltage side of each module.

In-situ module performance monitoring is also integrated by utilising individual load resistors which allows for determination of P_{max} of the mini-modules. Current and voltage are measured via Verivolt isolated sensing devices.

Two T-type thermocouples are attached to the rear-surface of each module to monitor module temperatures using NI 9211 ADC cards.

Control and monitoring of all the systems is implemented using a Queued State Machine in NI LabView. Further details for the software implementation is described by Spataru et al [11].

3 DEMONSTRATION EXPERIMENT

Despite being certified for field deployment, some backsheets have undergone significant degradation after deployment in relatively short amounts of time. Two examples of this include Polyamide (PA), composed primarily of nylon-12, and Polyvinyl1-1,1-difluoride (PVDF)-based backsheets, which each suffered significant cracking and delamination following various extents of field exposure, as soon as 4 years in the case of the PA.

A demonstration of C-AST is established in the form of a backsheet differentiation experiment. The test is conducted on six 4-cell p-type passivated emitter rear contact (PERC) mini-modules with various encapsulant/backsheet configurations. Three backsheet types are tested including PVF/Polyethylene terephthalate/PVF (TPT), PA and PVDF/PET/EVA. Additionally, these have been divided into two sample types, one utilizing a 100% UV absorbing EVA and another a 100% UV non-absorbing EVA, as per Table 1. In each case the same EVA type is used on both the front and backsides of the cells.

Table 1: Matrix of materials

	TPT	PA	PVDF
UV Pass EVA	✓	✓	✓
UV Block EVA	✓	✓	✓

The C-AST protocol is based on ASTM D-7869, a weathering test for clear coat automotive paints designed to simulate tropical conditions observed in south Florida, USA. ASTM D-7869 is used here as a baseline from which the C-AST protocol is developed. One important feature of ASTM D-7869 is that the applied stress magnitudes are limited to the extremes of the natural environment and so there is little risk of producing unrealistic failure mechanisms [12]. Adjustments have been made to be more relevant to stress factors experienced by PV modules. These include a moisture spray to the front and backside of the module to ensure saturation of the PV backsheet, high voltage bias (+500 V) replicating system voltage, mechanical loading for snow and wind loads, an increased temperature to achieve 90°C module temperature during periods of illumination. While higher temperatures have been observed [13] [14], 90°C is above typical operating conditions and just enough to provide a reasonable acceleration while not becoming unrealistic and the inclusion of a dry and humid freeze cycle to -20°C. A graphical representation of a single cycle is shown in Figure 2.

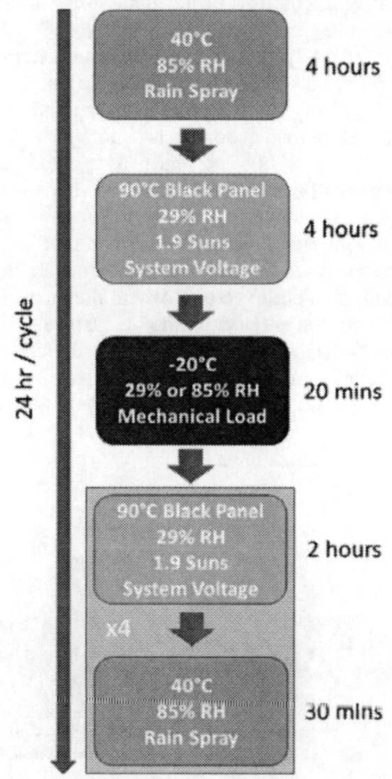

Figure 2: Graphical representation of a single cycle of the C-AST protocol, time spent at each stage does not include ramping time

Appropriate saturation and desiccation of the backsheet is considered an important factor in the test cycle. As such, water vapour transition rate (WVTR) through the backsheet is calculated using diffusivity models described by Kempe [15] to ensure that the test cycle is adequate to meet the criteria. Relative humidity at the backside of the cells during the test cycle shows that saturation and desiccation are achieved at the relevant periods of the cycle.

In addition to appropriate moisture permeation, the use of mini-modules calls for an understanding of the relevant mechanical load to induce the same level of stress as seen in full-sized modules. Larger modules are more susceptible to damage for a prescribed load than smaller modules due the reduced displacement with decreasing module size. A full-sized module was loaded with a 1 kPa static load. Displacement was measured diagonally across the module surface. Calculation of the bending moments at the centre of the full-sized module was used to determine that a mechanical displacement of 1.1mm at the centre of the mini-modules would induce the same bending moments. It is worth noting that the mounting configuration for the mini-modules in chamber are likely not representative of conventional mounting systems for full-sized modules and therefore further work is required to understand the potential differences in stress distribution.

4 TESTING RESULTS

4.1 Backsheet Shrinkage and Cracking

The PA backsheet was immediately seen to pull away from the edge of the module (Figure 5). Although shrinkage of PA backsheet on PV modules has not been widely reported, dimensional stability is a common occurrence in polyamide materials [16] and it is likely that shrinking on fielded modules has not been discovered previously due to the presence of a module frame. Displacement of the backsheet from the module edge was monitored and found to progress linearly throughout testing at a rate of 7 μm/cycle. In addition to the edge displacement, a visibly increased conformality around the cell ribbons was observed and, where the unaged PA had a visibly textured surface, a smooth surface was present after C-AST. Delamination also occurred along the edge of the module, likely a result of the strain induced by shrinking, and was found to be most severe at the corners (Figure 5), however delamination did not propagate away from the specimen corners.

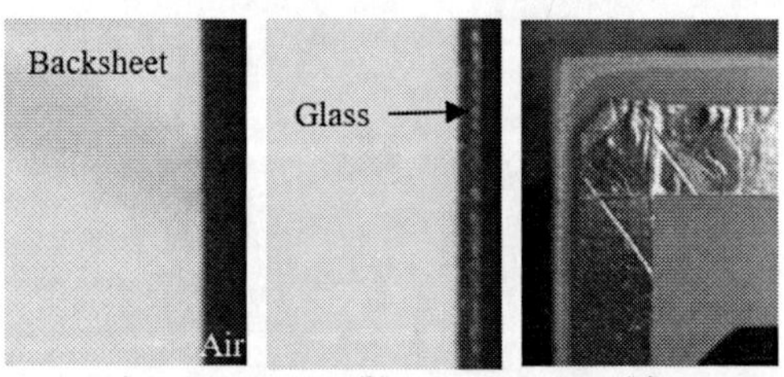

(a) (b) (c)

Figure 3: Image of (a) unaged backsheet, (b) aged backsheet with visible displacement of 0.8 mm from the module edge at 4.5 months of test and (c) delamination from edge and corner of module.

PA backsheet cracking had not previously been identified using established accelerated stress test methods yet was later discovered in the field. It was demonstrated by Rummens [17] that following a sequential test of 1000 hours at 85°C/85% RH and 6 hours at dry 150°C backsheet cracking did occur. Lyu et al [18] showed that PA undergoes embrittlement and a reduction in tensile strength following weathering according to the A3 condition in IEC TS 62788-7-2, which sees exposure conditions of 0.8W/m^2/nm at 340nm, black panel temperature 90°C and relative humidity 20%. However, cracking does not appear until mechanical loading is applied. Cracking of the PA backsheet occurred at 4.5 months (Figure 6) of the C-AST experiment. Cracking occurred predominantly along the cell ribbons where increased conformality had been observed and presumably tensile strain was highest. This pattern was found to be consistent with cracking previously reported in field-aged modules. The occurrence of cracking is considered an important success for C-AST, as it demonstrates the ability to induce the same changes in relevant mechanical properties and cracking behaviour seen in the field. The mode of crack formation and propagation as well as the affected layers remain to be confirmed for this study relative to that in a PV module.

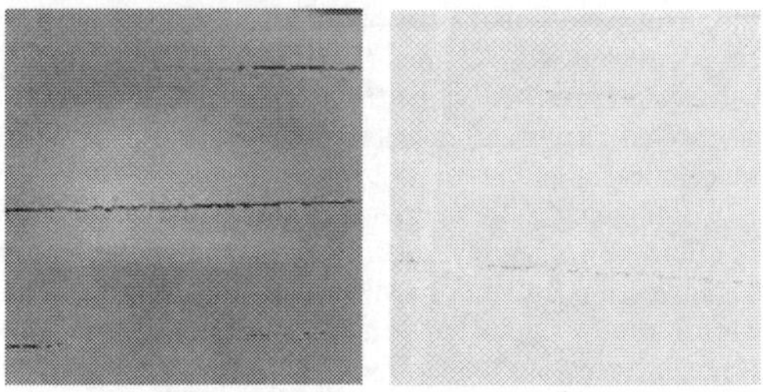

(a) (b)

Figure 4: PA backsheet cracking after (a) 4.5 months in C-AST and (b) 4-5 years of field deployment

4.2. UV-Induced Degradation

UV-transparent encapsulants have re-entered the market in recent years due to the potential for increased efficiencies brought about by an increasing blue response in modern PERC cells [19]. However, this has led to a re-emergence of the issue of UV-induced degradation. Witteck et al demonstrated that degradation of the passivation surface of PERC solar cells occurs when sufficient UV light is allowed to reach the passivation interface. Degradation manifests as a reduction in open-circuit voltage (V_{oc}) and power loss [20].

Figure 7 compares electroluminescence (EL) images for TPT modules with UV-pass and UV-block EVA. A bright ring can be seen in the UV-Pass module which is consistent with the mechanical loading rings which sit on the glass surface and cause shading from the chamber lights. The increased brightness indicates a higher minority carrier lifetime on the parts of the cell shaded by the Teflon ring than in the more exposed areas of the cells. On the modules with UV-block EVA, no such ring is observed.

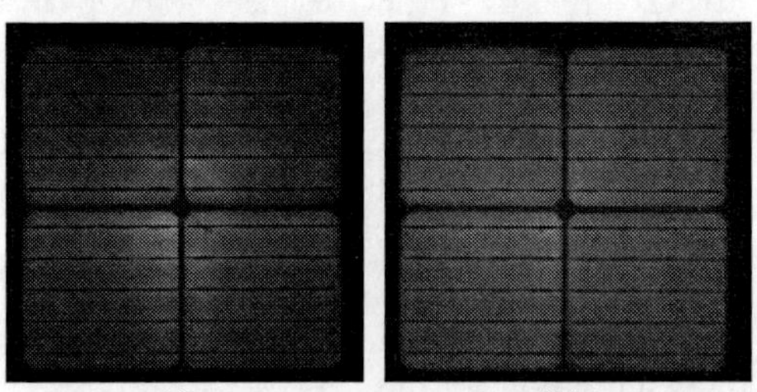

(a) (b)

Figure 5: EL images measured with I_{sc} current injection comparing (a) UV-pass EVA module with (b) UV-block EVA module

In addition, reduction of short-circuit voltage (V_{oc}) was found to be much higher in UV-Pass modules (Figure 8). No significant change was observed in I_{sc}. The absence of significant change in V_{oc} for the UV-block EVA modules suggests that UV-induced degradation of the passivation layer in the p-PERC cells is the primary mechanism by which degradation has occurred.

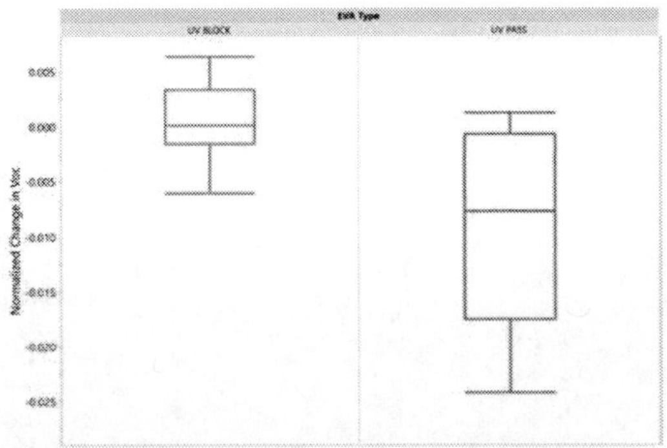

Figure 6: Change in V_{oc} for modules UV-Block and UV-Pass EVA after 6 months in test with total incident irradiance dose of 4484 kWh/m^2

4.3 Solder joint failures

Early in test, one module showed no output power during the light soaking periods. However, IV characterisation and EL imaging continued to yield reasonable results. Figure 9 compares module temperature to output power for the affected module. Reasonable power output is delivered for temperatures < 40°C. Destructive analysis later showed that a critical solder joint connection had degraded (Figure 10), however, at lower temperatures, contraction of the module assembly allowed for the joint to reconnect and thus pass electrical current. Without in-situ monitoring, a problem may not have been detected until much later as intermediate IV and EL measurements indicated that there was no problem.

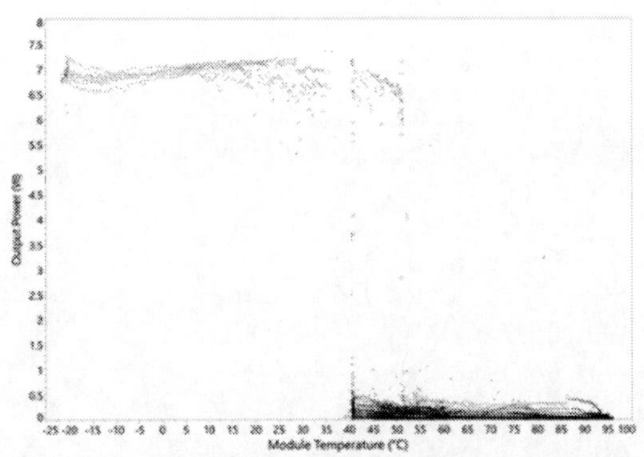

Figure 7: In-situ module temperature and module output power for module damaged by solder fatigue

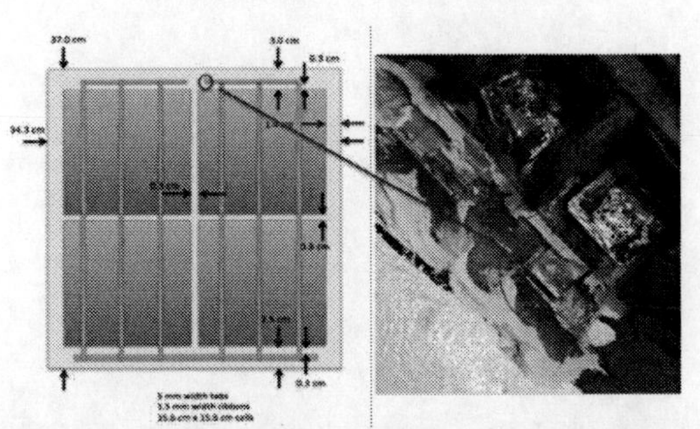

Figure 8: Solder joint disconnection at critical point between junction box and interconnect ribbon

Solder joint failure was determined to be an issue with the quality of the solder assembly, which was performed manually in-house. However, solder failure demonstrated the ability of C-AST to identify manufacturing issues.

5 CONCLUSIONS

A combined-accelerated stress testing capability was developed to more appropriately simulate the complex influence of stress factors of the natural environment. Cracking of a PA backsheet occurred following 4.5 months of testing in a visibly similar manner to what has previously been observed for 4-5 years of field-exposed material. This type of degradation was not detected in the standard qualification testing protocols which apply individual rather than combined stress factors. The usefulness for combined accelerated stress testing of new module designs and materials is therefore demonstrated in this work.

Despite those successes, the PVDF material did not undergo delamination as seen in the field. This is likely due to the lack of consideration for multiple climate types. The testing protocol in this work was designed to simulate tropical conditions in Florida with temperature humidity levels at the upper extremes of the natural environment. While low temperatures were included, the testing environment was warm and wet and largely unrepresentative of colder or drier climates known to affect PVDF.

6 ACKOWLEDGEMENTS

This work was supported by the U.S. Department of energy under Contract No. DE-AC36-08GO28308. with the National Renewable Energy Laboratory. The authors would like to thank Kent Terwilliger and Greg Perrin for their continued efforts towards to the maintenance and development of the experimental systems. We also thank Tadanori Tanahashi, Keiichiro Sakurai (AIST), Bill Gambogi, John Trout, Kaushik Roy Choudhury, and Nancy Phillips (DuPont), Tsuyoshi Shioda (Mitsui Chemical), and Frank Choltco-Devlin (Itek Energy) for materials and helpful discussions.

7 REFERENCES

[1] International Electrotechnical Commision, "IEC 61215-2: Crystalline Silicon Terrestrial Photovoltaic (PV) Modules—Design Qualification and Type Approval," *International Electrotechnical Commission*. Geneva, 2016.

[2] W. Gambogi *et al.*, "A comparison of key PV backsheet and module performance from fielded module exposures and accelerated tests," *IEEE J. Photovoltaics*, vol. 4, no. 3, pp. 935–941, 2014.

[3] H. H. and J. B. S. Pingel, O. Frank, M. Winkler, S. Oaryan, T. Geipel, "Potential Induced Degradation of Solar Cells and Panels," *35th IEEE PVSC.*, pp. 2817–2822, 2010.

[4] P. Hacke *et al.*, "Test-to-Failure of Crystalline Silicon Modules Preprint," in *35th IEEE PVSC.*, 2015, no. July 2010.

[5] F. Kersten *et al.*, "A New mc-Si Degradation Effect called LeTID," 2015.

[6] S. Richter, M. Werner, and C. Hagendorf,

"Understanding the Snail Trail Effect in Silicon Solar Modules on Microstructural Scale," in *27th European Photovoltaic Solar Energy Conference and Exhibition*, 2012, no. January, pp. 1–4.

[7] M. Koehl, S. Hoffmann, and K.-A. Weiss, "Combined Stress Factor Testing of PV Modules," in *Proceedings of 29th European Photovoltaic Solar Energy Conference and Exhibition*, 2014.

[8] A. Masuda *et al.*, "Sequential and combined acceleration tests for crystalline Si photovoltaic modules," 2016.

[9] W. Gambogi, "Sequential Stress Testing to Predict Photovoltaic Module Durability," in *7th World Conference on Photovoltaic Energy Conversion*, 2018.

[10] G. Tamizhmani, "Long-term sequential testing (LST) of PV modules," in *Photovoltaic Module Reliability Workshop*, 2012.

[11] S. Spataru, P. Hacke, and M. Owen-Bellini, "Combined-Accelerated Stress Testing System for Photovoltaic Modules," in *Proceedings of 45th IEEE Photovoltaics Specialists Conference*, 2018.

[12] M. Nichols *et al.*, "An improved accelerated weathering protocol to anticipate Florida exposure behavior of coatings," *J. Coatings Technol. Res.*, vol. 10, no. 2, pp. 153–173, 2013.

[13] M. D. Kempe, D. C. Miller, S. Kurtz, J. M. Moseley, and Q. A. Shah, "Field testing of thermoplastic encapsulants in high-temperature installations," *Energy Sci. Eng.*, vol. 3, pp. 565–580, 2015.

[14] M. D. Kempe, D. C. Miller, A. Zielnik, D. Montiel-chicharro, J. Zhu, and R. Gottschalg, "Survey of Mechanical Durability of PV Backsheets," in *IEEE PVSC*, 2017.

[15] M. D. Kempe, "Modeling of rates of moisture ingress into photovoltaic modules," *Sol. Energy Mater. Sol. Cells*, vol. 90, no. 16, pp. 2720–2738, Oct. 2006.

[16] S. Gerbig, "Polyamide Moisture Absorption | Relative Dimensional Change of Various Nylon Products - Prospector Knowledge Center," *Prospector*, 2014.

[17] F. Rummens, "Combined testing to assess backsheet reliability." .

[18] Y. Lyu, J. H. Kim, A. Fairbrother, and X. Gu, "Degradation and Cracking Behavior of Polyamide-Based Backsheet Subjected to Sequential Fragmentation Test," *IEEE J. Photovoltaics*, vol. PP, pp. 1–6.

[19] M. R. Vogt *et al.*, "Optical constants of UV transparent EVA and the impact on the PV module output power under realistic irradiation," *Energy Procedia*, vol. 92, pp. 523–530, 2016.

[20] R. Witteck, B. Veith-wolf, H. Schulte-huxel, and A. Morlier, "UV-induced degradation of PERC solar modules with UV-transparent encapsulation materials," *Prog. Photovoltaics Res. Appl. Res. Appl.*, no. March, pp. 409–416, 2017.

A MULTIDIMENSIONAL OPTIMIZATION APPROACH TO IMPROVE MODULE EFFICIENCY, POWER AND COSTS

Jibran Shahid, Max Mittag, Martin Heinrich
Fraunhofer Institute for Solar Energy Systems ISE, Heidenhofstraße 2, 79110 Freiburg, Germany
jibran.shahid@ise.fraunhofer.de

ABSTRACT: In the design of common solar modules, one of the main tasks is to achieve maximum output power or efficiency. Due to the complex interaction between different parameters of PV materials and module configuration this is a difficult task. The complexity increases when costs attributed to components and processes have to be considered. We use an approach to optimize module power and costs simultaneously based on the cell-to-module ratio which is usually used to describe the impact of module design and materials on module power. We present optimization routines to identify ideal module configurations. We apply five different types of optimization algorithms to the cell-to-module model, to find the optimal values of input parameters that maximize module power and efficiency and minimize the respective cell-to-module losses as well as the specific module costs [€/Wp]. The algorithms applied within this study are evaluated regarding their accuracy towards optimal output of the PV module and their computing power. We optimize a 285 W reference module and increase power by 5.8% and efficiency by 0.45 $\%_{abs}$. Specific costs are reduced by 0.9 €ct/Wp.

Keywords: Cell-to-module ratio, CTM, Multidimensional optimization, efficiency, power, costs

1. INTRODUCTION

Integrating solar cells into a PV module causes different loss and gain mechanisms leading to a module power that is different from the power of the solar cells before module integration [1–4]. The cell-to-module efficiency ratio $CTM_{efficiency}$ describes the ratio between the module efficiency and the mean efficiency of the solar cells before integration (Figure 1).

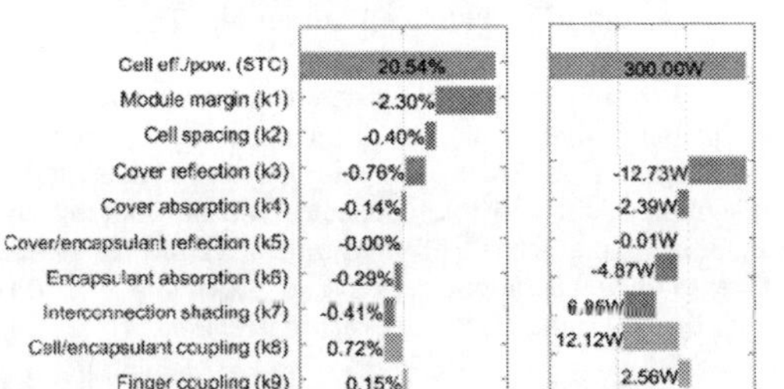

Figure 1: Cell-to-module (CTM) loss and gain factors for an exemplary photovoltaic module.

Module design and materials are chosen to achieve high CTM ratios to maximize the electrical power output or efficiency. In order to reduce the losses and increase the gains related to module integration, a profound understanding of all factors that influence the module output is essential.

Hädrich *et. al.* proposed a methodology to calculate CTM ratios and to analyze PV modules regarding efficiency and power changes resulting from the integration of solar cells into modules [5]. The methodology presented by Hädrich *et. al.* is based on 13 different factors to describe gains and losses attributed to module components, physical effects or important interfaces. In previous work, we added two electrical k-factors, extended the model for additional module concepts and performed detailed analyses of single components and loss factors [6–8].

In this paper, we present an approach based on a combination of CTM analysis, cost modelling and optimization algorithms. We apply different optimization algorithms in order to minimize CTM losses and specific costs [€/Wp] simultaneously. We evaluate optimization algorithms on the basis of their accuracy towards optimal output of the PV module and their computing power.

2. APPROACH

Our work is based on the CTM methodology as proposed by Hädrich *et. al.* and several extensions as presented earlier [2, 6, 8–11]. In this work we describe the optimization using algorithms and a cost model that has been added to the existing set of models.

2.1. Objective function of the system

In order to apply the optimization algorithms, an objective function has to be implemented for the output of the CTM model that defines the targeted value. In this objective function, decision variables and constant variables should be recognized.

$$PV\ output = f(x_1, x_2, \dots, x_n, y_1, y_2, \dots, y_n) \quad (1)$$

In (1) f is a function that describes the optimization target: the CTM ratio, the output of the PV module or the specific cost of the PV module. Decision variables x_n can be changed within the optimization and y_n are constant variables that are fixed within the optimization. Simply speaking, the decisions variables are those input parameters a module manufacturer can influence and modify such as cell distance, ribbon cross section etc. while the constant variables are not directly accessible for optimization.

The user can either define one parameter as a decision variable or a number of parameters for optimization. All other variables will act as a constant y_n for the objective function.

Once the objective function has been implemented, different algorithms can be applied and the results of the optimization algorithms can be compared on the basis of criteria given above. Several optimization algorithms are applied to this CTM model in order to optimize either the output power or efficiency or cost of the PV module. Additionally, optimization algorithms can be applied simultaneously in order to increase efficiency or power and at the same time reducing the cost of the module.

2.2. Optimization algorithms

Five different optimization algorithms are used in this study. The algorithms are chosen on the basis of their different optimization approach and popularity.

The first algorithm is the Nelder-Mead Algorithm (NMA) which is most commonly used in a numerical method which does not require the derivative of the objective function [12]. The second algorithm is the Interior Point algorithm (IPM) which is widely used to solve linear and nonlinear optimization problems [13]. The last three algorithms come in the category of random search method; names are Particle Swarm Optimization (PSO), Crow Search Algorithm (CSA) and Antlion Optimization (ALO) [14–16]. These algorithms are based on a random search method and use search agents in order to converge to the solution. The number of search agents used in this study is 20 for each random search algorithm.

The comparing criteria among these algorithms are computing power and the time taken to converge to the optimum result. If the algorithm is able to reach the optimum result, the computing power is measured and the number of iterations is counted. The last factor is related to convergence time which also depends on the performance of the machine on which the algorithm is running.

2.3. Selected optimization parameters

We demonstrate the optimization using selected parameters as decision variables. The parameters selected for this study are cell and string distance, the width and thickness of the cell ribbon and the thickness of the front glass. Constraints for these parameters are set as shown in Table I.

Table I: Boundaries of selected parameters.

[mm]	Min	Ref.	Max
Cell distance	0.5	2	10
String distance	0.5	2	10
Cell ribbon width	0.5	1.5	2.5
Cell ribbon thickness	0.05	0.2	0.3
Front glass thickness	1	3.2	10

In the given range, some of the values are not practically applicable. For example, 1 mm front glass thickness is not a common thickness. In reality a glass with this thickness is more expensive compared to common thicknesses of 2 to 4 mm. In order to use 1 mm front glass, also other measures have to be taken into account, e.g. the frame type. We extended parameter boundaries i.e. of glass and cell ribbon thickness to check the capabilities of the algorithms. For practical application parameter limits can be adjusted easily.

Cell and string distance play an important role in increasing the efficiency, power and specific cost of the PV module. In order to understand the effects of variations of these parameters on the output of the PV module, we perform a detailed analysis.

When the cell gap increases (and the module size is adapted accordingly), the inactive area share becomes larger, which increases the loss factor attributed to cell spacing (k2, see [5]). Cell spacing losses belong to the geometrical effects which describe the efficiency, but not the power of the PV module. An increase in inactive area reduces module efficiency (Figure 2).

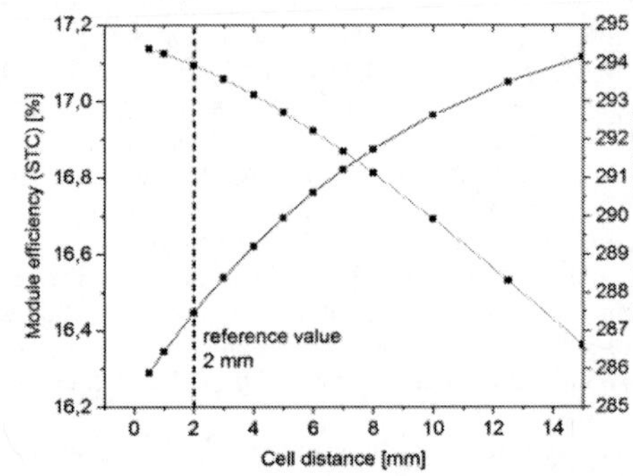

Figure 2: Efficiency and power graph for cell distance.

An increase in module power due to internal reflection gains is attributed to a separate factor k11 [7]. An increase in cell distance increases the electrical losses in ribbons (k12) due to the additional ribbon lengths. The change in string distance affects the losses in string ribbons (k13) accordingly. Gains from k11 show a non-linear, flattening behavior with the increase of cell distance while k2 and k12 (electrical losses in ribbons) increase linearly with cell distance. Therefore an optimum cell spacing exists.

Width and thickness for interconnector ribbons are parameters that can be influenced by a module manufacturer in order to avoid losses. Increasing the thickness of the cell interconnector will not increase the shading of the active cell area but lower the resistance. Due to this the output power and efficiency increases with the cell ribbon's thickness (Figure 3).

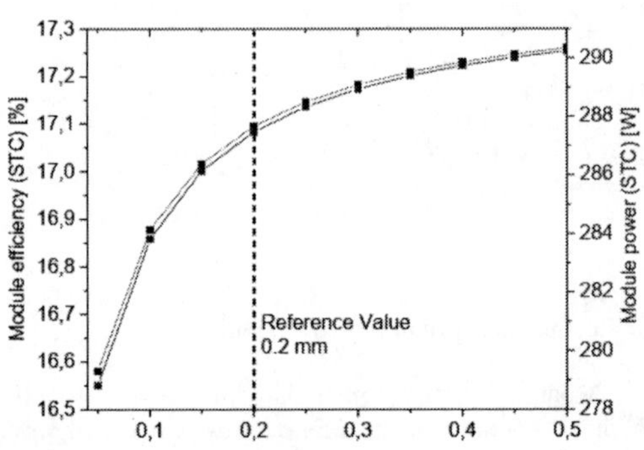

Figure 3: Efficiency and power graph for cell ribbon thickness.

The width of the ribbon affects the shading of active cell area (k7). Assuming a cell with continuous busbars, shading will occur if the width of the ribbon exceeds the busbar width. With the ribbon width, the cross section of the ribbon changes, affecting the resistive losses (k12). Therefore, the cross section area of cell ribbon should be increased without active area shading. Additionally, gains from reflection from the ribbon to the solar cell occur (k10). Depending on cell metallization, ribbon properties and the electrical currents an optimum regarding power and efficiency for interconnector dimensions exists. The optimum width might not be the width of the busbar (Figure 4).

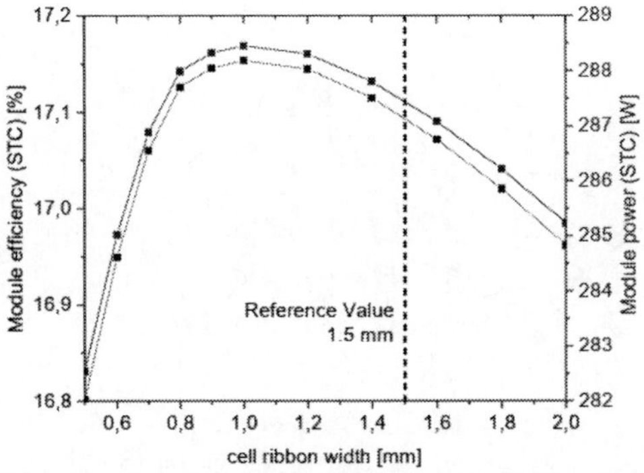

Figure 4: Power and efficiency for a variation of the ribbon width. The continuous busbars on the cell have a width of 0.8 mm.

2.4. Cost modelling

To perform an optimization of specific costs (€/Wp) we create a cost model that considers material costs of relevant components as well as manufacturing costs. Manufacturing costs include equipment (4.5%), facility (1.0%), labor (6.0%), spare parts (1.0%), utilities (5.0%) and waste disposal costs (0%). We use a combined surcharge of 17.5% of the material costs as manufacturing overhead (which are approx. 15% of the total reference module costs as displayed in Figure 5) based on data extracted from the SCost model [17].

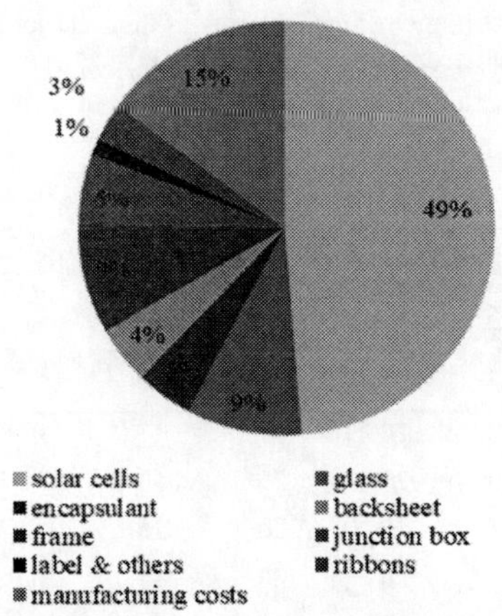

Figure 5: Cost structure of a reference module

We use the module specifications (size etc.) and the module material costs to calculate the costs shares of different module areas as shown in Figure 6. We distinguish costs related to inactive areas (i.e. cell spacing), cost related to active areas (solar cell), cost that are independent of module dimensions (i.e. junction boxes and labels) and costs depending on the circumference (frame). As can be seen in Figure 6, active area (with cells) displays the highest costs within the module due to the large area share. Calculating the cost per area, the inactive and active area cost the same (not considering cells).

Inactive area contributes to a certain extent to module power generation (backsheet gains) [8]. Inactive area power gains increase with larger cell spacing but additional gains become marginal when increasing cell distances. Costs increase linearly with module size. Therefore an optimum exists regarding the size of the inactive areas and the attributed power.

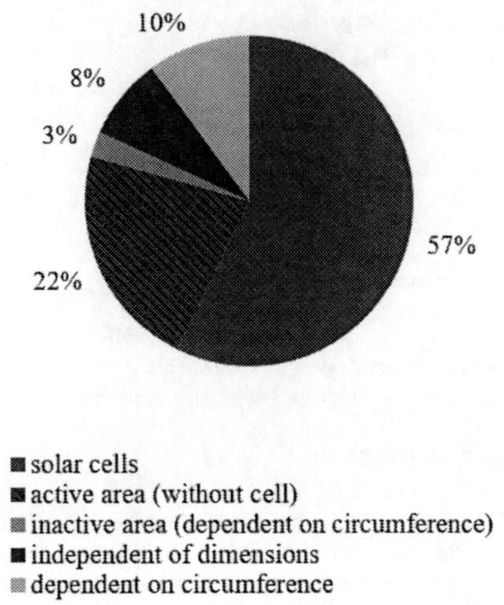

Figure 6: Area cost share structure of a reference module

Specific module costs (€/Wp) use the total module costs as described above and the module power. Since module power results from the combination of initial cell power and CTM effects, module integration and CTM power changes need to have an impact on specific costs.

We calculate the power losses attributed to different CTM factors and combine those results with the cost model. Our results are discussed below.

2.5. Reference module

In order to evaluate the results of the optimization algorithms, a reference study is performed. We assume a module consisting of 60 H-pattern solar cells with 5 continuous busbars (156x156 mm², I_{MPP} = 8.571 A). The initial cell efficiency is 20.54% (5.00 Wp). Additional parameters of commercially available module materials (encapsulant etc.) are used as input. Reference values of the selected materials are given in Table I.

We perform a CTM analysis of the reference module and find the output power to be 286.1 W_p which corresponds to a CTM_{power} ratio of 0.95. The module efficiency is 17.02 % with a $CTM_{efficiency}$ ratio of 0.83. This result from the reference study will be used to compare the results of the optimization algorithms. The cost of ownership for the reference module setup are calculated 72.95 € which gives the specific cost 25.5 €ct/W_p.

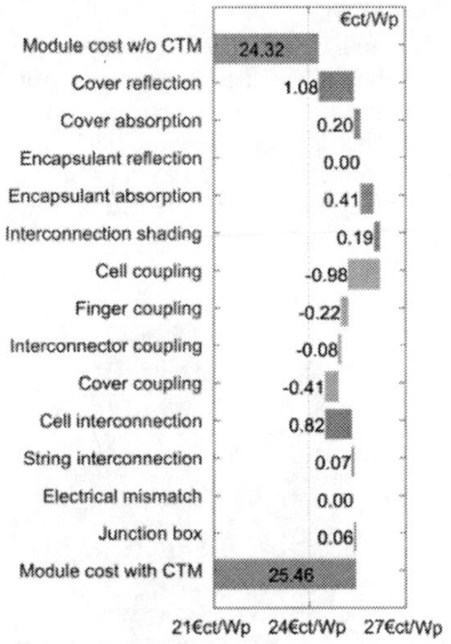

Figure 7: Waterfall diagram of specific cost of the module with and without CTM power changes.

3. CASE STUDIES

A few of the available module configuration parameters are selected as decision variables. The selected parameters for this study are the cell and the string distance, the width and thickness of the cell ribbon and the thickness of the front glass. Other input parameters remain constant. We apply the algorithms and perform a module optimization regarding power, efficiency and specific costs. All calculations are performed assuming standard testing conditions (STC).

3.1. Power of the PV module

The first optimization is related to the output power of the PV module. As displayed in Figure 8 the different algorithms find a different optimum. A maximum module power of 302.9 W_p is calculated. The NMA, PSO and IPM are able to reach the highest possible output power. They set the cell and string distance to their upper boundary which is 10 mm. By increasing the cell gap, the backsheet reflection gain can be increased.

Algorithms also reduce the width of the cell connector to 0.9 mm in order to reduce the ribbon shading losses. Since a reduction in ribbon width increases the electrical losses of the cell connector, the algorithms increase the thickness of the cell connector in order to compensate and to reduce the resistive losses.

The glass thickness is also reduced to 1 mm to its lower boundary from its initial value of 3.2 mm in the reference module.

The efficiency of the module is 16.61%, which is less than our reference study. This is due to fact that the cell spacing and the total module area increased. Therefore the inactive area share increased lowering the efficiency. The power maximization also reduces the specific cost of this module to 24.7 €ct/W_p.

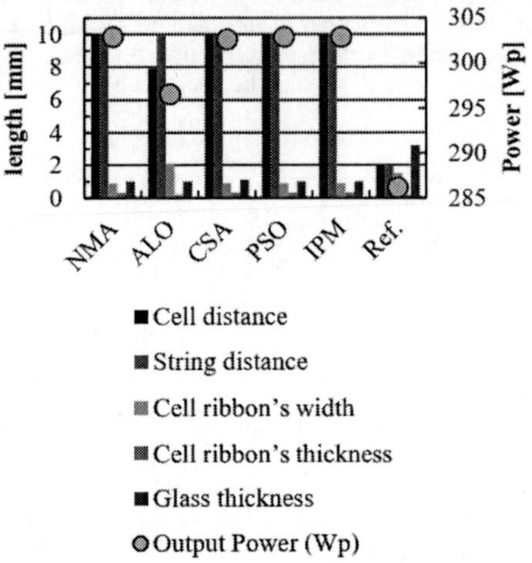

Figure 8: Module power and selected input parameters after application of optimization algorithms on CTM modelling.

Among the algorithms with accurate results, the fastest is achieved by IPM, which only takes 9 iterations and reaches the maximum power in only 4.75 seconds as shown in Table II. All random search algorithms are unable to reach the optimum result. Nonetheless, CSA, a random search algorithm, is ranked second. It takes 20 iterations and approximately 13 seconds to reach the second highest power which is 302.62 W_p. In comparison to the other algorithms, the weakest result is achieved by ALO which takes around 14.8 seconds and 20 iterations and achieved 296.55 W_p .

Table II: Algorithms results in terms of their computing power, time and accuracy.

	Objective f(x) evaluations count	Algorithm Iteration	Search agents	Time Lapse (sec)	Deviation from optimal value (abs)
NMA	242	147	-	17.0	0.0
ALO	500	20	20	14.8	-6.35
CSA	201	20	20	15.5	-0.28
PSO	200	20	20	15.3	0.0
IPM	66	10	-	7.2	0.0

3.2. Efficiency of the PV module

The second study focusses on the efficiency of the PV module. As we mention before, efficiency is affected by changes in inactive module area. Therefore efficiency optimization includes power, active and inactive module areas.

The highest module efficiency calculated is 17.47%, which is 0.45%$_{abs.}$ more than the reference study. In this study, again the NMA, CSA and IPM are able to reach the highest efficiency. The algorithms reduce the cell and string distance to their lower boundary, which is 0.5 mm as shown in Figure 9. The width of the cell ribbon also decreases to 0.9 mm, algorithms predict this result due to fact that it will decrease the shading of the cell and also this width will not lead to severe electrical losses. The small cell gap reduces cell spacing losses. Beside this, cell ribbon losses also decrease. The IPM, CSA and NMA have identical results. The thickness of the cell connector is 0.3 mm and the top glass thickness is 1 mm which are same that we have observed in Section 3.1.

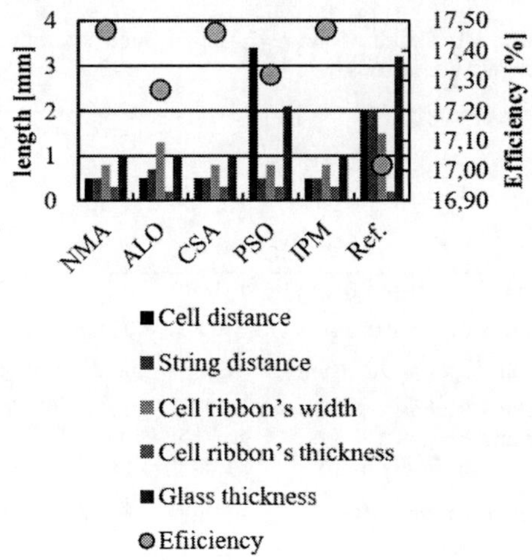

Figure 9: Module efficiency and selected input parameters after application of optimization algorithms on CTM-modelling.

The area of the PV module is reduced because of the reduced cell and string distance, this also results in the lower cost of ownership value which is 74.4 €, but in comparison to the previous study, the specific cost increas to 25.73 €ct/W_P. This is because the output power is reduced to 289.21 W_P by achieving the highest efficiency.

Table III: Algorithms results in terms of their computing power, time and accuracy.

	Objective f(x) evaluations count	Algorithm Iteration	Search agents	Time Lapse (sec)	Deviation from optimal value (abs)
NMA	285	172	-	18.9	0.0
ALO	500	20	20	15.3	-0.2
CSA	201	20	20	17.7	-0.01
PSO	60	6	20	4.7	-0.15
IPM	133	20	-	8.89	0.0

As shown in Table III, once again the IPM achieves the highest efficiency within the given constraints. The IPM takes 20 iterations in 8.89 seconds, whereas the NMA takes 172 iterations in 18.09 seconds to reach the same result as IPM. The second promising result is calculated by the PSO, which calculates 2.1 mm optimal glass thickness. Beside this, the cell distance is 3.4 mm and 0.50 mm string distance. The NMA and IPM predict 0.8 mm to be the optimal width of the cell connector to reach the highest efficiency. In this study, the least optimum result is observed by ALO, which calculates an optimum efficiency of 17.27%. PSO takes around 20 iterations in 14.80 seconds, which is the longest time among all algorithms.

3.3. Specific Cost of the PV module

In the third study, the specific costs of the PV module are optimized which are defined as the cost per watt peak at STC. Calculations now have to include models for module power, active and inactive area and their respective costs.

The same set of input parameters as in the other studies is used. We assume that the costs of materials (i.e. the front glass) remain constant even if material parameters change (i.e. the thickness). Therefore, we only consider changes in module area (absolute and relative) and ribbon lengths to be cost relevant. We are also assuming that the cross-sectional area of the cell ribbon does not influence cost.

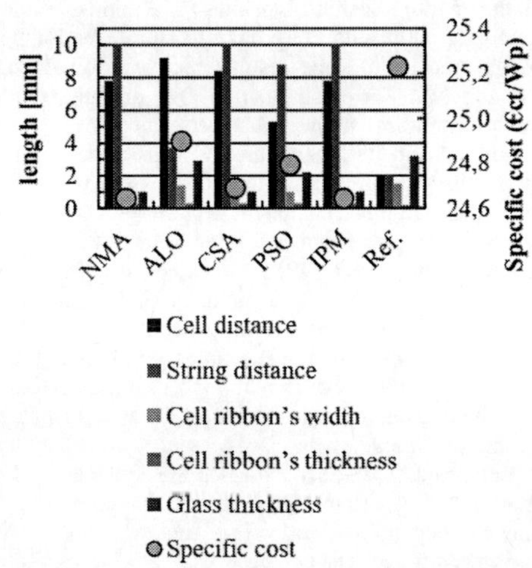

Figure 10: Specific module costs and selected input parameters after application of optimization algorithms on cost and CTM-modelling.

After performing all calculations using the different optimization algorithms we find the results displayed in Figure 10. The cell and string distance are set to their upper boundaries which is 10 mm. The width and thickness of the cell connector are 0.9 mm and 0.3 mm and the thickness of the glass is 1 mm which is the same as we have observed in our last two studies related to power and efficiency. The NMA and IPM find the lowest specific cost of the module with 24.6 €ct/W_P. This specific cost is only 0.1 €ct/W_P less than what we have observed in the first study related to optimization of power of the PV module.

CSA reaches the second minimum specific cost within 15 seconds and takes 20 iterations as shown in Table IV. Among the best three algorithms which reach the least specific cost, IPM is the fastest to achieve the optimum result in 8.5 seconds and NMA is slowest (23.9 seconds, 154 iterations) to reach the same result. The algorithm calculating the highest costs in this study is ALO, which calculates 24.9 €ct/ W_P in 15.3 seconds.

Table IV: Algorithms results in terms of their computing power, time and accuracy.

	Objective f(x) evaluations count	Algorithm Iteration	Search agents	Time Lapse (sec)	Deviation from optimal value ($_{abs}$)
NMA	252	154	-	23.9	0.0
ALO	500	20	20	15.3	-0.26
CSA	201	20	20	15.0	-0.05
PSO	60	6	20	4.2	-0.15
IPM	121	19	-	8.5	0.0

3.4. Extended constraints

Among all studies, IPM and NMA performed well regarding the quality of the result and computation time. All the random search algorithms are unpredictable in terms of their results, they have to run several time in order to reach the same result which was obtained by IPM and NMA in a single run. Out of three random search algorithms only CSA performed well in the Section 3.3 and PSO performed well in Section 3.1 but not in others studies. Other than that, NMA was able to reach the same result as IPM, but it takes more computational power and time. Due to this, we carried all the above studies using IPM and extended constraints of the selected parameters. So far in order to increase the output power and reduce the specific cost, all the algorithms selected the upper boundary of cell and string distance. On the other hand, algorithms selected the lower boundaries of cell and string distance in order to increase the efficiency.

In this study, we have extended the constraints of cell distance, string distance and the thickness of the cell connector beyond practical values as shown in Table V to analyze additional effects.

Table V: Constraints of the selected input parameters

[mm]	**Min**	**Ref.**	**Max**
Cell distance	0.5	2	100
String distance	0.5	2	100
Cell connector width	0.05	0.2	0.35

We perform three optimization runs (module power, module efficiency and specific costs). Results are displayed in Table VI.

Optimizing for maximum power now leads to 310.2 Wp. We observe a significant increase in cell and string distance. The algorithm predicted an optimal cell distance of 31.4 mm and string distance should be 49.8 mm. Differences between string and cell spacing result from the specific resistance of string and cell ribbons. Beside this the width of the cell connector is also increased to 1 mm in order to handle the current increase in the PV module. The remaining parameters (cell ribbon thickness and front glass thickness) reach the same values as we have observed in previous studies.

Optimizing the module efficiency, IPM calculates the same set of values that was predicted in the Section 3.2 except for the width of the cell connector. Here the width of cell connector is decreased to 0.8 mm. As the cell gap (and therefore the backsheet gain) is reduced in this study, the reduced cross-sectional area of the cell connector is sufficient to limit the electrical losses in the module. The efficiency achieved in this study is 17.52% which is 0.05 $\%_{abs}$ higher than what we observe in Section 3.2.

Table VI: Result of the revised first three studies using different boundaries.

	Power	**Efficiency**	**Specific Costs**
Cell distance (mm)	31.50	0.50	7.90
String distance (mm)	49.80	0.50	11.40
Cell ribbon width (mm)	1.00	0.80	0.90
Cell ribbon thickness (mm)	0.30	0.30	0.30
Glass thickness (mm)	1.00	1.00	1.00
Module power (Wp)	310.6	290.0	303.4
Efiiciency (%)	12.87	17.52	16.71
Specific cost (€ct/Wp)	26.2	25.0	24.6
Objective f(x) evaluations count	128	60	197
Algorithm Iteration	18	9	29
Time Lapse (sec)	10.24	4.01	13.00

Optimizing the specific cost aims at increasing the output power and while reducing the cost of ownership. In this study, we observe a change in cell distance to 7.9 mm and a string distance of 11.4 mm. The specific cost achieved in this study is 24.6 $€ct/W_p$ and the output power of the module is 303.4 W_p. Previously in Section 3.3, we have observed that the algorithms have predicted the same set of parameters which was observed in Section 3.1, but in this study, the algorithm has predicted a significantly different result than the study related to the output power of the module.

Regarding the applications of algorithms for module optimization we find the choice of boundaries to be relevant. No "automatic" module optimization appears to be possible without deriving constraints from manufacturing (i.e. technical limits for glass sizes or thicknesses) or final module use. The additional consideration of PV systems and power plants is required for module optimization at this point. An unrestricted use of module parameters as derived from optimization will not be possible without creating conflicts in module use. Especially, the importance of mounting and installation as well as the share of system costs may be mentioned here [18].

CONCLUSION & SUMMARY

We add a cost model and different algorithms for module optimization to existing cell-to-module (CTM) analysis models. The prior enables us to perform an analysis regarding specific module costs (€/Wp) for flexible module layouts. The latter allows the multi-dimensional optimization of modules.

We apply these extensions and optimize a reference module regarding module power, efficiency and specific costs by selecting five input parameters for optimization.

Results of the multidimensional power optimization show an increase in module power of 16.8 Wp (+5.9%) compared to the reference module and assuming the same initial cell power. The CTM_{power} increased from 0.95 to 1.01. An increase in module efficiency by $0.45\%_{abs}$ is observed after optimizing the efficiency of the reference module. As the specific cost of the reference module is 25.5 €ct/W_p, with the optimization 24.6 €ct/W_p is reached.

We observe that increasing the cell and string distance will increase the output power of the module, but show that there is a limit on increasing the cell gap beyond practically relevant boundaries for spacing. Other than this, the algorithms optimize the cross-sectional area of the cell ribbon according to the output power. We observe that the higher power module requires a larger cross-sectional area of the ribbon due to increase of current in the maximum power point. The cross-sectional area is decreased, when the output power of module decreases.

In this study, we have assumed that front glass thickness and the cell ribbon cross-section do not influence cost. In future, we would like to add cost dependencies with input parameters of the PV module integration and try to make the optimization model more robust.

We used the algorithms NMA, ALO, CSA, PSO and IPM in this study. In this paper we have carried out 3 studies using all algorithms. In each study, the random search algorithms i.e. ALO, CSA and PSO delivered varying results. Random search algorithms have to run several time on the optimization problem in order to reach the optimum. Sometimes they got stuck in local minima were not able to reach the global minima. So far only PSO was able to reach the optimum in the study related to output power optimization and CSA was able to reach the optimum in specific cost. Besides this, the most promising algorithm is IPM for this application. NMA is also able to achieve the same result that was obtained by IPM, but it takes more computing power and time. Beside them, all the random search algorithms were not able to achieve the desired results.

ACKNOWLEDGEMENT

We would like to thank the German Ministry of Economic Affairs and Energy (FKZ 0324033) for their funding.

REFERENCES

[1] ITRPV, "International Technology Roadmap for Photovoltaic (ITRPV): Results 2017, Ninth Edition," 2018.

[2] M. Mittag *et al.*, "Systematic PV module optimization with the cell-to-module (CTM) analysis software," *Photovoltaics International*, no. 36, pp. 97–104, 2017.

[3] I. Hädrich *et al.*, "PV Module efficiency analysis and optimization," PV-Rollout, Boston, Massachusetts, USA, 2011.

[4] H. Wirth *et al.*, *Photovoltaic Modules: Technology and Reliability*. Berlin/Boston: De Gruyter, 2016.

[5] I. Hädrich *et al.*, "Unified methodology for determining CTM ratios: Systematic prediction of module power," *Solar Energy Materials and Solar Cells*, vol. 131, pp. 14–23, 2014.

[6] M. Mittag *et al.*, "Electrical and thermal modeling of junction boxes," 33rd EUPVSEC, Amsterdam, 2017, pp. 1501–1506.

[7] M. Mittag *et al.*, "Analysis of backsheet and rear cover reflection gains for bifacial solar cells," *33rd European PV Solar Energy Conference and Exhibition*, vol. 2017.

[8] M. Mittag *et al.*, "Cell-to-Module (CTM) analysis for photovoltaic modules with shingled solar cells," in *44th IEEE PV Specialist Conference PVSC*.

[9] N. G. Dhere, "Comparison of optical transmittance spectra of new tempered glass with that of superstrate glass extracted from a field-aged PV module," Sandia National Laboratories Photovoltaics Program Tech Brief #: 26 MDRC-10/10/99, 1999.

[10] N. Woehrle *et al.*, "Solar cell demand for bifacial and singulated-cell module architectures," *36th Photovoltaics International*, pp. 48–62, 2017.

[11] C. Reise *et al.*, "From bifacial PV cells to bifacial PV power plants: The chain of characterization and performance prediction," *Photovoltaics International*, no. 38, pp. 1–10, 2018.

[12] J. C. Lagarias *et al.*, "Convergence Properties of the Nelder--Mead Simplex Method in Low Dimensions," *SIAM J. Optim.*, vol. 9, no. 1, pp. 112–147, 1998.

[13] J. Gondzio, "Interior point methods 25 years later," *European Journal of Operational Research*, vol. 218, no. 3, pp. 587–601, 2012.

[14] S. Mirjalili, "The Ant Lion Optimizer," *Advances in Engineering Software*, vol. 83, pp. 80–98, 2015.

[15] F. Marini *et al.*, "Particle swarm optimization (PSO). A tutorial," *Chemometrics and Intelligent Laboratory Systems*, vol. 149, pp. 153–165, 2015.

[16] A. Askarzadeh, "A novel metaheuristic method for solving constrained engineering optimization problems: Crow search algorithm," *Computers & Structures*, vol. 169, pp. 1–12, 2016.

[17] S. Nold *et al.*, "Cost modeling of silicon solar cell production innovation along the PV value chain," in *27th EUPVSEC 2012, Frankfurt, Germany, 2012*.

[18] Ran Fu, David Feldman, Robert Margolis, Mike Woodhouse, and Kristen Ardani, "U.S. Solar Photovoltaic System Cost Benchmark: Q1 2017,"

THE COMPLETED IEC 61853 STANDARD SERIES ON PV MODULE ENERGY RATING, OVERVIEW, APPLICATIONS AND OUTLOOK

Thomas Huld, Ana Gracia Amillo*, Tony Sample, Ewan D. Dunlop, Elena Salis, Robert Kenny
European Commission, Joint Research Centre, Viale Fermi, 2749, Ispra, Italy. Tel: +39 0332783989
ana.gracia-amillo@ec.europa.eu

ABSTRACT: We present an overview of the IEC 61853 standard series "*Photovoltaic (PV) module performance testing and energy rating*" paying special attention to the last two parts (3 and 4) which were published in August this year. The standard defines a methodology to estimate the performance of PV modules considering real working conditions defined by six datasets representative of the major climatic conditions likely to be encountered by PV installations. The standard not only provides these climatic datasets (Part 4) and the models to estimate the PV performance (Part 3), but it also describes the methodology to test and measure the relevant characteristics of the PV modules under evaluation (Parts 1 and 2) so as to obtain the input data required by the said models. The output of the standard, the so called "*Climate Specific Energy Rating*" parameter (*CSER*), can be used to compare different PV technologies in terms of their performance under the same climatic conditions, or to compare the behaviour of a specific technology under the different climatic regions included in the standard, helping PV manufacturers, installers and investors to choose the most suitable PV technology for every location.
Keywords: energy rating, energy performance, PV module, modelling.

1 INTRODUCTION

The International Electrotechnical Commission (IEC) Standard series 61853 "*Photovoltaic (PV) module performance testing and energy rating*" has been under development for nearly 20 years but has just been finalized with the publication of the final two parts (3 and 4) in August 2018. Back in 1995 the IEC Technical Committee 82 started working on a standard on energy rating of PV modules that would take into consideration the effects of the solar spectrum, module temperature and reflection on the module's surface on the estimation of the PV performance. The complexity of the whole methodology, both in terms of testing and modelling, resulted in the original attempt to write a standard being abandoned. In 2008 a proposal to divide the standard into four different parts eased the process and led to the publication in 2011 of the first part of the standard series [1], followed by Part 2 in 2016 [2]. The last two parts, 3 and 4 [3, 4], have just been published completing the whole IEC 61853 standard series.

The aim of the standard is to provide PV manufacturers, installers and buyers with a tool to obtain a simple but realistic estimate of the performance of a PV module at different climatic conditions. The *Climate Specific Energy Rating* (*CSER*) parameter calculated with the standard can be used to compare the performance of different PV technologies in a more meaningful way than by just comparing the PV power declared by the manufacturer and measured at Standard Test Conditions (STC), which are hardly ever met under real working conditions. These conditions are a module temperature of 25 °C, 1000 W/m^2 of in-plane irradiance with a spectral content as defined in the IEC 60904-3 [5].

CSER is a dimensionless parameter calculated as the ratio of the estimated energy yield of a PV module under a certain climate conditions to the energy yield that would have been obtained if the module efficiency would have been the one measured under STC. The whole calculation method of the energy rating *CSER* is based on one year of hourly data which, even though it could be representative of the conditions of a certain climatic region, it would not be enough so as to perform an accurate estimation of the energy yield of a PV module installation located in a specific emplacement. For that purpose, data for that specific location and installation would be needed including climatic data for a period of time of several years and information about possible presence of shadows, snow or dust on the modules, as well as information about the installation settings, such as azimuth (orientation) and inclination angles. For an accurate prediction of the total energy yield, the effects of degradation and ageing of the PV modules should be considered as well. On the contrary, the purpose of the IEC 61853 standard series is to provide a realistic estimation of the energy rating of a PV module located in a generic site with certain climatic conditions where the modules are assumed to be installed facing the equator with an inclination angle of 20 degrees. Notwithstanding, the estimation methodology defined in the standard could be applied to calculate the energy yield of a real PV module installation, provided losses due to presence of obstacles, dirt or snow and the effects of degradation are applied to the PV output estimated by the methodology presented in the IEC standard. Besides the above-mentioned losses and degradation, the standard does not consider either transient behaviors such as light induced changes or thermal annealing.

The scope of the standard covers mono-facial and single junction modules without restrictions in the PV technology, although it may not work properly for technologies whose performance changes with time or experience significant non-linearities in their characteristics used for the modelling.

2 IEC 61853 PHOTOVOLTAIC (PV) MODULE PERFORMANCE TESTING AND ENERGY RATING

The IEC 61853 standard series is based on three components: a set of meteorological datasets which correspond to various climatic conditions, a concatenation of various mathematical models to estimate the PV power output taking into consideration various effects to simulate real working conditions different from the STC and a set of measurements and parameters that are the input to the said estimation models. These measurements include a matrix of maximum power output under various levels of irradiance and module's temperature, the spectral responsivity of the PV device, the reflectance of the surface of the module's cover and the module's thermal properties.

All these aspects are detailed distributed in the four distinctive parts that constitute the complete standard series as described in the following:

2.1 IEC 61853-1. Irradiance and temperature performance measurements and power rating

The first part of the standard, published in 2011, defines a methodology to measure the PV module power (W) at maximum power of operation (Pmax) for a defined set of in-plane irradiance and module temperature conditions shown in Table I. These measurements can be obtained following three different procedures which include outdoors measurements under natural sunlight with or without using a solar tracking system and indoor testing with a solar simulator. The former two are generally less expensive than the indoor testing but they are less accurate and may require several months to record all the required measurements. The technical specifications for applying each procedure are detailed in this part of the standard. A sample of three modules of every evaluated PV module type should be tested. Prototypes of new designs can be also tested as the scope of the standard is not limited to commercial products.

Table I: Seven irradiance and four temperature levels applied. The irradiance should correspond to a spectral distribution of AM1.5G (IEC 60904-3 [5]).

Irradiance	Module temperature (°C)			
($W.m^{-2}$)	15	25	50	75
1100	NA	✓	✓	✓
1000	✓	✓	✓	✓
800	✓	✓	✓	✓
600	✓	✓	✓	✓
400	✓	✓	✓	NA
200	✓	✓	NA	NA
100	✓	✓	NA	NA

(NA: not-applicable)

Measurements not need to be recorded at exactly the levels defined in Table I. An interpolation, according to IEC 60891 [6], can be applied either with respect to temperature or irradiance.

The result of applying this testing procedure is a 22-elements matrix of maximum power that will be used as input data in Part 3 of this standard.

2.2 IEC 61853-2. Spectral responsivity, incidence angle and module operating temperature measurements

The methodology applied to estimate the PV module output takes into consideration various effects to model the performance of the PV module under conditions different from the STC used to define the reported Pmax in the module's datasheet.

This part of the standard describes the testing procedures and specific setting conditions for evaluating and quantifying these effects. One of these procedures aims at quantifying the effect of the increased angle of incidence between the light source and the normal to the module in the reflected irradiance that is consequently not absorbed by the active material, not contributing to the power generation. A second effect measured in this part of the standard is related to the spectral responsivity of the module that will define the irradiance effectively absorbed by the active material. The last effect evaluated here is related to the estimation of the temperature of the module which depends on the received irradiance, the ambient temperature and the cooling effect of the wind. For most PV technologies, the temperature reached by the modules, along with the received irradiance, is one of the factors which most significantly affects the power generation.

The output of Part 2 of the standard should include the parameter to model the angle of incidence effect (a_r) and the coefficients required to estimate the PV module's operating temperature under different levels of in-plane irradiance, ambient temperature and wind conditions (u_0 and u_1). In addition to these, the spectral responsivity of the module (*SR*) should be obtained as well from the various measurements performed in this part. All these parameters are required as input data by the models defined in Part 3 in order to estimate the PV power output.

Similarly to Part 1 a sample of three modules has to be analyzed here, one per studied effect.

2.3 IEC 61853-3. Energy rating of PV modules

The third part of the standard, published in August 2018, defines the complete methodology to calculate the PV module energy ratings. The models applied estimate for every hour of a complete reference year the energy output (Wh) used to estimate the *Climate Specific Energy Rating* (*CSER*) parameter (dimensionless) used for comparison purposes. The full methodology is the concatenation of various steps so as to calculate for every hour two sets of data:

1. in-plane irradiance, broadband and spectrally resolved, corrected for angle of incidence effects, AOI and the spectral effects that define the effectively absorbed irradiance (G_{corr}),
2. module's temperature (T_{mod}) estimated using the ambient temperature (T_{amb}) and wind speed (v) data included in the climatic datasets contained in Part 4 and the in-plane irradiance corrected for angle of incidence effects ($G_{corr,\,AOI}$) as input data.

These two sets of hourly values, the effective irradiance and the module temperature (G_{corr}, T_{mod}) are used to estimate the power output by interpolation, and extrapolation when necessary, of the efficiency values derived from the values of the power matrix obtained in Part 1 of the standard.

The model used to consider the AOI effects on the received irradiance applies a different treatment to the diffuse and beam component as shown in Equations 1 and 2 respectively.

$$D_{corr,j} = D_j \cdot \left\{1 - exp\left[-\frac{1}{a_r}\left(\frac{4}{3\pi}\left(\sin\beta + \frac{\pi-\beta-\sin\beta}{1+\cos\beta}\right) + (0.5a_r - 0.154)\left(\sin\beta + \frac{\pi-\beta-\sin\beta}{1+\cos\beta}\right)^2\right)\right]\right\} \quad (1)$$

$$B_{corr,j} = B_j \cdot \left[\frac{1-exp\left(-\frac{\cos\theta_j}{a_r}\right)}{1-exp\left(-\frac{1}{a_r}\right)}\right] \quad (2)$$

Where B_j and D_j are, respectively, the uncorrected in-plane broadband beam and diffuse irradiance of hour j, and θ_j is the incidence angle between the normal to the PV module and the sun for the same hour j. All these values are tabulated in the datasets provided in Part 4 of the standard. β is the inclination angle, expressed here in radians, of the module which is fixed at 20° and a_r is the factor obtained in Part 2 from the analysis of the AOI test.

The sum of the broadband beam and diffuse irradiance corrected for AOI effects is the global in-plane broadband irradiance corrected for AOI effects ($G_{corr, AOI}$), which is used to estimate the correction coefficient to be applied to the global in-plane spectrally resolved irradiance (R) of the climatic datasets so as to correct these values for the AOI effect.

The second model estimates the in-plane irradiance modified to account for the spectral responsivity of the PV module material. A spectral correction factor, C_s is calculated for every hour according to Equation 3. This factor is needed to obtain the corrected irradiance for angle of incidence and spectral effects, G_{corr} (Eq 4).

$$C_{s,j} = \frac{1000 \cdot \int SR(\lambda) \cdot R_{corr,AOI,j} d\lambda}{G_{corr,AOI,j} \cdot \int SR(\lambda) \cdot R_{STC} d\lambda} \tag{3}$$

$$G_{corr,j} = C_{s,j} \cdot G_{corr,AOI,j} \tag{4}$$

Where SR is the spectral responsivity measured at different wavelengths as defined in Part 2 of the standard and in IEC 60904-8 [7], $R_{corr, AOI, j}$ is the global in-plane spectrally resolved irradiance corrected for AOI at the hour j and R_{STC} is the spectral intensity for the standard test condition spectrum AM1.5G (IEC60904-3 [5]).

Besides the corrected in-plane irradiance, the second parameter required to estimate the hourly values of power output is the module temperature (T_{mod}). This variable is estimated, according to Equation 5, using as input data the ambient temperature, wind speed, in-plane AOI corrected irradiance and the thermal parameters obtained in Part 2 of the standard, u_0 and u_1.

$$T_{mod} = T_{amb} + \frac{G_{corr,AOI}}{u_0 + u_1 \cdot v} \tag{5}$$

Where T_{amb} and v are respectively the ambient temperature and wind speed provided in tabulated form in the climatic datasets included in Part 4. u_0 and u_1 are the thermal behavior coefficients defined in Part 2 in the methodology for calculating the module operating temperature.

Once T_{mod} and G_{corr} are defined for every hour, by linear interpolation, or extrapolation, of the efficiency values derived from the power matrix measured in Part 1, it is possible to obtain the PV power output (P_{mod}) for every hour as well as the energy output (E_{mod}). The sum of all these hourly values is the annual module energy output, $E_{mod, year}$. From this value, the dimensionless parameter $CSER$ is calculated according to Equation 6.

$$CSER = \frac{E_{mod,year} \cdot G_{ref}}{P_{max,STC} \cdot H_p} \tag{6}$$

Where G_{ref} is equal to 1000 $W \cdot m^{-2}$ as the irradiance applied at STC to measure the maximum power at STC ($P_{max, STC}$). H_p is the yearly global in-plane irradiation provided in Part 4 for every climatic region.

The outcome of this third part of the standard is the $CSER$ parameter calculated for the PV module under analysis for the six climatic regions defined in Part 4. A possible PV installer or user could use the $CSER$ value for the climatic region that suits best his particular location to obtain a rough estimate of the yearly energy output ($E_{mod, year, s}$) provided the local yearly in-plane irradiation is known (Equation 7).

$$E_{mod,year,s} = \frac{CSERS \cdot P_{max,STC} \cdot H_{p,s}}{G_{ref}} \tag{7}$$

Where $H_{p,s}$ is the yearly in-plane irradiation ($kWh \cdot m^{-2}$) for the user specific location. $CSERs$ is the $CSER$ estimated for the considered PV module obtained using the climatic dataset that best represents the climate of the specific location of the user.

2.4 IEC 61853-4. Standard reference climatic profiles

The last part of the IEC standard has the six standard reference climatic profiles. These datasets contain hourly values for a complete year of environmental data such as ambient temperature (T_{amb}, °C) and wind speed (v, $m \cdot s^{-1}$). In addition to these, they also contain satellite-retrieved irradiance data of horizontal global and beam broadband irradiance (G_h and B_h, $W \cdot m^{-2}$), in-plane global and beam broadband irradiance (G and B, $W \cdot m^{-2}$), and spectrally resolved in-plane global irradiance integrated into 32 spectral bands of different width as specified in the datasets (R, $W \cdot m^{-2}$). The irradiance values are calculated from satellite images using the SPECMAGIC algorithm [8, 9]. Besides these variables, the solar elevation angle (°) and the angle of incidence between the sun and the normal to the surface of the module (θ, °) required to apply the correction for AOI effects are also provided. The modules are defined as being oriented facing the equator with an inclination angle of 20°.

Part 4 also provides the yearly in-plane irradiation value (H_p, $kWh \cdot m^{-2}$) for six locations chosen as representative of the six climatic regions (tropical humid, subtropical arid, subtropical coastal, temperate coastal, high elevation above 3000m and temperate continental) defined as representative of the most common climatic conditions that PV installations may encounter. This climatic classification may not correspond to the traditional climatic regions, such as the ones defined by Köppen-Geiger [10] which are strongly influenced by precipitation and its effect on vegetation. The reference climates used in this part of the standard were defined according to the $CSER$ values obtained for various PV technologies in diverse locations analyzed. As shown by Huld et al [11] the variability in the PV module measurement data may lead to an uncertainty in the module performance ratio, parameter estimated with a similar methodology as the one applied for the $CSER$ calculation, of about 2% for the analysis carried out over Europe. Therefore, the different climatic datasets were defined so that the $CSER$ reached values differed when possible by that threshold, so as to guarantee that the differences in the $CSER$ values are due to climatic differences and not to uncertainties in the input data.

2.5 Application of the IEC 61853 standard series to different technologies

Figure 1 shows the $CSER$ values reached by four different PV technologies under the six reference climatic regions.

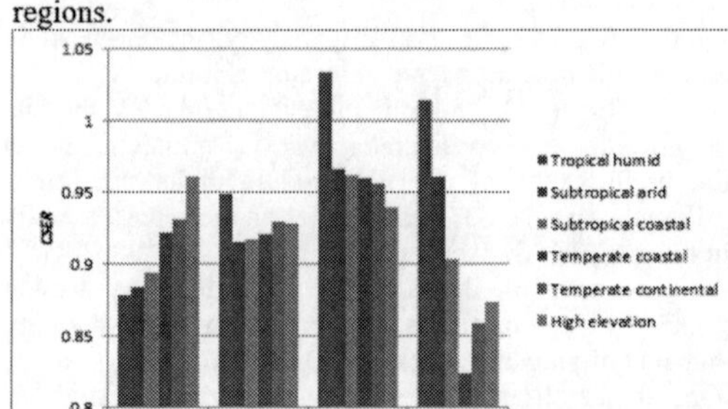

Figure 1: *CSER* calculated at the six reference climatic regions for four PV technologies.

The climatic conditions of the tropical humid region enhances the performance of the amorphous silicon (a-Si) to the extent that it would perform better than expected by the power output measured under STC conditions declared by the manufacturer. In opposition to this, the crystalline silicon technology (c-Si) would see its performance limited by the high temperatures reached under that climate and other arid conditions. However, its performance would considerably improve in high elevation areas as shown by the results obtained in the reference climate "High elevation". As an example of the emerging PV technologies, we have included in this analysis one module of organic PV (OPV) which benefits from its good performance under warm temperatures but gets its performance significantly diminished under low irradiance conditions like those characteristic of the "Temperate coastal" climate.

We highlight that the estimation methodology defined in this standard does not account for transient effects that can affect the performance of technologies such as amorphous silicon [12].

3 QUANTIFICATION OF THE VARIOUS EFFECTS

In order to better understand the impact on the *CSER* values of the different factors accounted for in the IEC standard, several maps have been created to graphically show their impact on the obtained *CSER* values or on the in-plane irradiation or PV production [13].

3.1 Angle of incidence effect

As shown in Figure 2 the AOI effect causes a decrease in the irradiation entering the PV module that varies between -1.5% and -4.5% which would in turn decrease the power output. The effect is generally higher in cloudy areas.

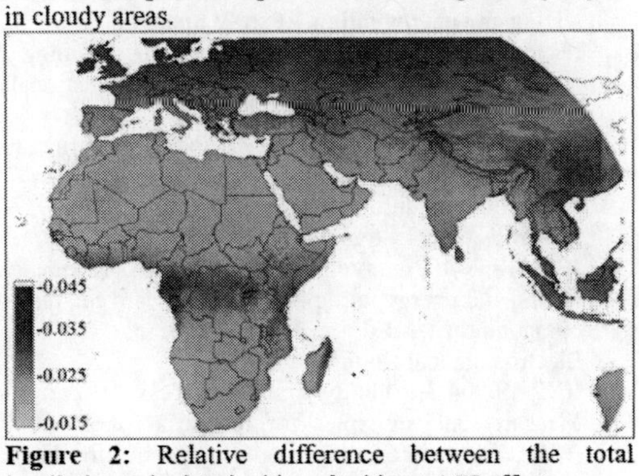

Figure 2: Relative difference between the total irradiation calculated with and without AOI effect.

For the IEC standard modules are assumed to be installed facing the equator with a fixed inclination angle of 20°. The difference in the *CSER* values obtained for c-Si technology with that configuration or from considering an inclination angle of 40° is shown in Figure 3.

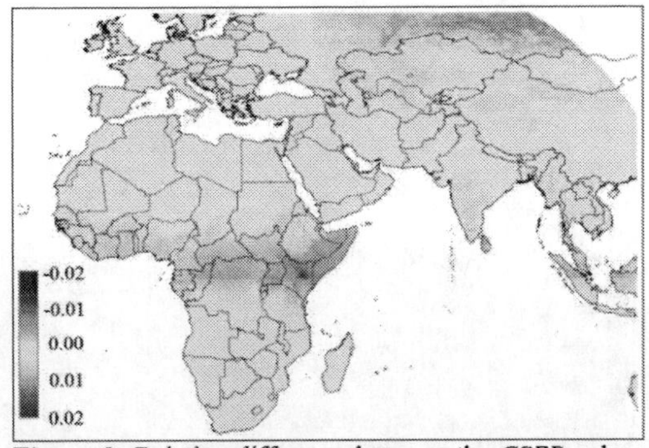

Figure 3: Relative difference between the *CSER* values obtained for c-Si modules when inclined 40° and 20°.

For most regions the absolute difference is below 0.5% (0.005), although at high latitudes, *CSER* at 40° inclination angle could be up to 2% higher than the values at 20° based on standard glass air interface.

3.2 Module temperature

For most PV technologies the temperature reached by the modules under working conditions would significantly limit its performance. The model included in the standard for the estimation of T_{mod} considers the cooling effect of the wind which may increase the module's power output by decreasing its temperature. Figure 4 shows the difference in the obtained *CSER* values for c-Si modules with and without considering effect of the wind. In extreme cases like the Sahara, including the wind effect leads to a significant increase of 6%, while in most regions the impact is lower.

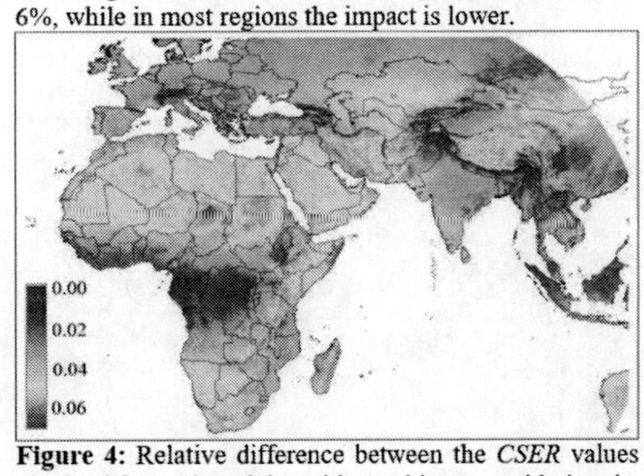

Figure 4: Relative difference between the *CSER* values obtained for c-Si modules with or without considering the wind effect.

3.3 Spectral effects

Figure 5 shows the relative difference in the energy production values obtained for the CdTe technology with and without taking into consideration the spectral effects.

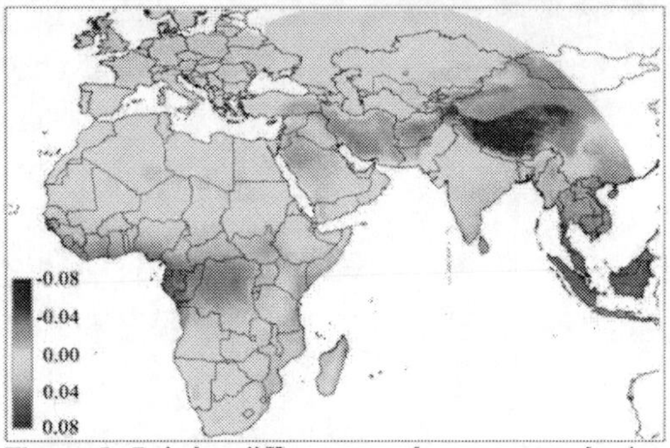

Figure 5: Relative difference on the energy production obtained for CdTe modules with or without considering spectral effect.

The spectral effect for CdTe technology ranges from -6% to 6% while for crystalline silicon technology is lower due to its broader spectral responsivity, varying between -5% to 2% depending on the region (results not shown here). In general, the narrower the spectral responsivity, the higher impact due to considering the spectral effect and a spectral distribution of the received irradiance different from the STC one. An exception is found in Tibet where both technologies show a high negative effect. The high elevation and low temperatures of this region may lead to a very low water vapour content, which in turn results in a high infrared proportion in the received irradiance compared to the AM1.5G spectrum. This part is not effectively used leading to a negative impact on the PV power output.

3.4 World map of *CSER* for c-Si modules

Due to the lack of spectrally resolved irradiance data for some regions, the *CSER* values shown in Figure 6 have been calculated without considering the spectral effects.

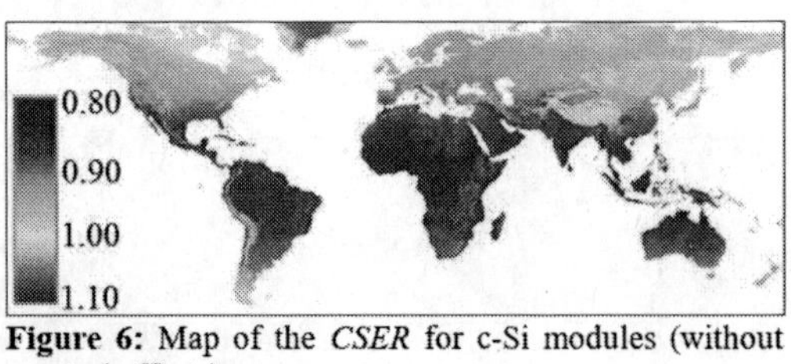

Figure 6: Map of the *CSER* for c-Si modules (without spectral effects).

The distribution of the *CSER* values calculated for C-Si modules show a clear relation with the temperature as it is one of the main factors that affects the module performance.

4 POSSIBLE FUTURE REVISION AND SCOPE EXTENSIONS

The stability date set for Parts 1 and 2 of the IEC 61853 standard are 2019 and 2020 respectively, when both documents will be revised in order to be reconfirmed, withdrawn, amended or replaced by a revised edition. Some possible revisions or changes to the actual content of these first two parts could regard the scope of the whole standard which is limited at the moment to mono-facial and single-junction devices. An energy rating methodology linked to the power matrix, reflectance due to angle of incidence effects and a procedure to determine thermal behavior for multiple-junction devices or bifacial modules could be developed in the future. In fact, bifacial modules are the subject of a new IEC Technical Specification under development [14], which could include in a future revision a methodology for applying an energy rating analysis like the one presented in the IEC 61853 standard.

Another possible improvement could aim at simplifying the procedures described Part 2 to define the module normal operating temperature and the thermal coefficients u_0 and u_1 so that more test laboratories and PV manufacturers could perform the said testing procedures.

Regarding the power matrix defined in Part 1, for some PV technologies it could be beneficial for modelling their performance to have additional power output measured at lower levels of irradiance than those currently required in Part 1. Similarly, it has been observed that for some climatic conditions the temperature reached by the modules is lower than the minimum temperatures applied in the procedure defined in Part 1 which requires that extrapolation is applied in order to estimate the power output at certain hours. However, these changes could make the performance of the measurements for the power matrix more complex.

5 REFERENCES

[1] IEC 61853-1. Photovoltaic (PV) module performance testing and energy rating – Part 1: Irradiance and temperature performance measurements and power rating. Edition 1.0 International Electrotechnical Commission (2011).

[2] IEC 61853-2. Photovoltaic (PV) module performance testing and energy rating – Part 2: Spectral responsivity, incidence angle and module operating temperature measurements. Edition 1.0 International Electrotechnical Commission (2016).

[3] IEC 61853-3. Photovoltaic (PV) module performance testing and energy rating – Part 3: Energy rating of PV modules. Edition 1.0 International Electrotechnical Commission (2018).

[4] IEC 61853-4. Photovoltaic (PV) module performance testing and energy rating – Part 4: Standard reference climatic profiles. Edition 1.0 International Electrotechnical Commission (2018).

[5] IEC 60904-3. Photovoltaic devices – Part 3: Measurement principles for terrestrial photovoltaic (PV) solar devices with reference spectral irradiance data. Edition 3.0 International Electrotechnical Commission (2016).

[6] IEC 60891. Photovoltaic devices – Procedures for temperature and irradiance corrections to measured I-V characteristics. Edition 2.0. International Electrotechnical Commission (2009).

[7] IEC 60904-8. Photovoltaic devices – Measurement of spectral responsivity of a photovoltaic (PV) device. Edition 3.0. International Electrotechnical Commission (2014).

[8] R. Müller, T. Behrendt, A. Hammer, A. Kemper. A new algorithm for the satellite-based retrieval of solar surface irradiance in spectral bands. Remote Sensing 4 (2012) 622-647.

[9] A. Gracia Amillo, T. Huld, R. Müller. A new database of global and direct solar radiation using the

Eastern Meteosat Satellite. Models and validation. Remote sensing 6 (2014) 8165-8189.
[10] W. Köppen. Das geoghaphisca System der Klimate, in Handbuch der Klimatologie, edited by Köppen W and Geiger G. 1.C. Gebr, Borntraeger, (1936) 1-44.
[11] T. Huld, E. Salis, A. Pozza, W. Herrmann, H. Muellejans. Photovoltaic energy rating data sets for Europe. Solar Energy 133 (2016) 349-362.
[12] D. Polverini, R. Kenny, T. Huld, G. Tzamalis. Toward a model for a-Si modules performance behaviour under real outdoor operating conditions. Solar Energy 116 (2015) 388-398.
[13] T. Huld, A.M. Gracia-Amillo. Estimating PV module performance over large geographical regions: the role of irradiance, air temperature, wind speed and solar spectrum. Energies 8(6) (2015) 5159-5181.
[14] IEC TS 60904-1-2. Photovoltaic devices – Part 1-2: Measurement of current-voltage characteristics of bifacial photovoltaic (PV) modules. Draft Technical Specification (2018).

IDENTIFYING HIGH UNCERTAINTIES IN PV SOILING MEASUREMENTS WHEN COMPARING TWO DEVICES

A.T. Alasfour, F.G. Alzubi, and A. Alkandary
Kuwait Institute for Scientific Research
Safat 13109, Kuwait

ABSTRACT: While recently there has been a growing interest in measuring the effect of soiling on photovoltaics (PV) using two similar PV devices, little research has been undertaken to investigate the uncertainties stemmed from the inherent performance asymmetries between the two devices. In this study, we evaluated the accuracy of a direct soiling loss (SL_D) measurement method and another corrective soiling loss (SL_C) method using commercial mc-Si modules. We demonstrate that SL_D errors and tolerances can be significantly different between different pairs of similar PV devices. Although the corrections made by SL_C can significantly improve accuracy, it was unable to improve error tolerances and requires periodic PV characterization. Furthermore, a preliminary investigation of soiling rate (SRate) uncertainties suggests that they are highly sensitive to the standard deviation of soiling loss measurements.
Keywords: PV soiling, soiling loss, measurement uncertainties

1 INTRODUCTION

Dust settlement on the surface of PV systems (soiling) remains a serious problem in dust-intensive environments as it can cause a significant loss in revenue while increasing operation and maintenance costs. Power reductions as high as 20% have been reported in arid climates by many research groups [1-3]. Consequently, there has been a growing interest in quantifying and monitoring such an environmental factor as the deployment of PV systems in arid climates is rapidly increasing.

Early measurements of soiling induced losses (SL) using two PV devices have dated back to the early 1970's [4]. Since then, performance-based measurements have been widely adopted by PV practitioners and researchers for a better evaluation of lost power. This soiling evaluation approach compares the short circuit current and/or maximum power of two coplanar PV devices, where one is used as a clean reference while the other is a test device left to naturally soil. As of today, this technique is typically used in two methods: (1) to directly compare the electrical outputs of the two devices, denoted as SL_D and (2) to make the electrical comparison after applying a correction(s) factor, denoted as SL_C. The latter method is used to improve the accuracy of the measurement by using a correction factor to normalize any weather and/or performance discrepancies between the two devices. Moreover, a few variations of the SL_C method exits as different research groups aimed to make further improvements [5, 6].

Although the two methods have been heavily used, rather less attention has been paid to the errors and uncertainties associated with them. For a true isolation of the SL, the two PV devices must experience identical weather conditions and must have identical electrical characteristics. Such factors, however, are not considered in the SL method and thus are expected to be seen as measurement artifacts. The extent of these artifacts depends on the degree of both weather and PV deviations. SL_C on the other hand, although improves measurement accuracy, its errors remain unclear. One of the correction factors used to compensate complex differences in external quantum efficiency (EQE), angular response (Ar), and parasitic resistances (Pr) between the two devices is typically oversimplified. Therefore, errors will depend on both the degree of PV deviations and the accuracy of the correction factors. Furthermore, since PV characteristics change over time under field conditions, it is expected that the correction factors used in SL_C will eventually yield higher errors than SL_D; in such case, they will need to be re-evaluated to maintain acceptable accuracy. Understanding the extent of errors when measuring SL is of great value as it enhances energy prediction models, optimizes PV plant cleaning frequency, and isolates other weather effects.

In this study, we monitor the soiling loss of six commercially available mc-Si modules, of the same model, in the field to investigate the errors and uncertainties associated with the SL_D and SL_C methods. In addition, we examine the error and uncertainties in SRates extrapolated from SL data using the least squares regression method.

2 EXPERIMENTAL

The experiment was performed on 7 commercial mc-Si modules of the same model that have not been previously exposed to sunlight. Prior to field testing, the samples went through a benchmarking process using an indoor AAA solar simulator to identify the performance deviations between the modules as a result of manufacturing quality, device pre-conditioning, and different temperature coefficients. The Corrective parameters necessary for computing SL_C were obtained in this step. Figure 1 illustrates a flowchart that describes this pre-installation process in greater detail.

After this initial benchmark, the modules were taken out to the field to start measuring SL. One reference sample was used as the clean sample, while the other 6 test samples were used as the soiled samples. Using an outdoor IV tracer, each sample was measured every 10 min throughout the day. Irradiance and module temperature data were collected using a reference cell coplanar to the samples and a thermocouple attached on the back sheet of each sample. SL_D was determined by comparing the Pmax of each soiled test sample to the clean reference sample. SL_C on the other hand, further accommodated for performance and weather discrepancies between the reference and test samples to improve the isolation of SL.

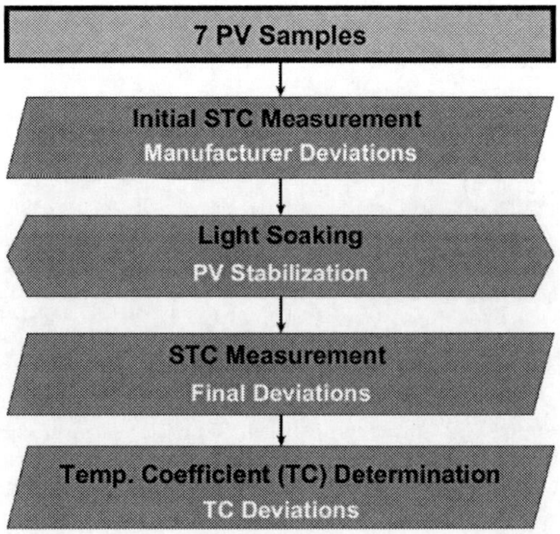

Figure 1: Flowchart showing the pre-installation process for all the PV samples before outdoor testing.

For each test sample, SL_D and SL_C were recorded as daily mean values. In an effort to minimize errors stemming from angular misalignment, high irradiance fluctuations, correction errors, and equipment faults, SL_D and SL_C were only recorded when three filtration criteria were met. The first criterion required SL to be computed using data measured two hours before and after solar noon. The second criterion constrained data to be used only when irradiance was between 900 W/m^2 and 1100 W/m^2. The last criterion necessitated that the first standard deviation of SL_D and SL_C computed every ten minutes throughout the day must be lower than 2%.

For a quantitative SL measurement error analysis, instead of only cleaning the reference sample and leaving the test samples to naturally soil, all of the samples were cleaned daily. This "all-clean" condition was selected because the true value of SL is known to be zero. The SL_D and SL_C values of multiple days were compared to assess their measurement accuracy. Furthermore, the long-term accuracy of the SL_C methods was investigated by re-benchmarking all the samples a year later to update the correction factors. The samples were then re-installed back in the field and the new factors were used to calculate new corrective soiling loss, defined as SL_{C2}. The differences between old and new soiling loss values were analyzed. Finally, we highlight the important relationship between SL measurement repeatability and SRate uncertainty.

3 RESULTS AND DISCUSSION

3.1 Measurement Errors When Using SL_D

Figure 2 presents the SL_D values for each test sample calculated every ten minutes throughout a period of 25 days when both the reference and test samples were clean. Less than 5% of the collected data achieved the true value. Between all the test samples, the soiling loss showed a large range from 3% to -5%. On average, sample two yielded the most accurate soiling values, while sample five was the least accurate with a mean error of -0.5$\%_{abs}$ and -2.4$\%_{abs}$ respectively. On the other hand, sample five showed the highest measurement repeatability, while sample showed the least with a standard deviation (σ) error of $\pm$0.6$\%_{abs}$ and $\pm$1.5$\%_{abs}$ respectively. This experiment clearly demonstrates that unless the electrical characteristics of the reference and test samples are well understood, the extent of measurement error and tolerances are difficult to estimate.

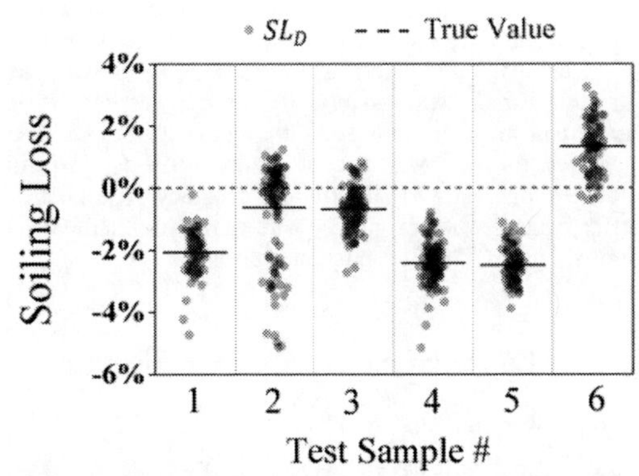

Figure 2: SL_D error deviations when computing soiling loss using different PV samples from the same manufacturer. The plotted values were computed every ten minutes.

High error deviations were also present when the daily mean SL_D was considered as shown in figure 3. Such deviations stem from constantly changing weather conditions and inherent differences between PV characteristics. Consequently, an adequate estimate of measurement errors and tolerances will require multiple days of evaluation.

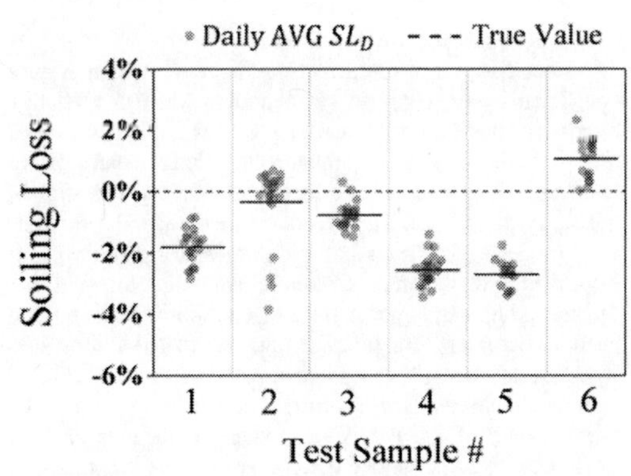

Figure 3: SL_D error deviations when computing soiling loss using different PV samples from the same manufacturer. The plotted data represent daily mean values.

3.2 Measurement Errors When Using SL_C

To improve the accuracy and precision of the SL_D measurement, a different corrective method was investigated where the differences in temperature, irradiance, and Pmax between the reference and test samples were normalized back to STC conditions using measured temperature coefficients and a power correction factor. Figure 4 illustrates a plot that compares the daily mean values of SL_D and SL_C for three test samples. Corrections improved the accuracy of the SL value nearly 95% of the times it was applied. However, a slight increase in the standard deviation of error was observed. It is suspected that this increase was a consequence of possible non-synced data between PV temperature/irradiance and the IV curve where a thirty-second mismatch is possible due to set-up limitation. The implications of SL_C having accuracy improvements, but not repeatability indicates that this corrective technique is more suitable for measuring soiling loss in real time, but not for extracting SRates. Errors in SRates will be discussed more extensively in another section. It is important to note, however, that out of the six test samples, three showed accuracy improvements, two did not, and one showed slightly less accuracy. The lack of effectiveness when using SL_C with those three samples is likely a result of PV mischaracterization.

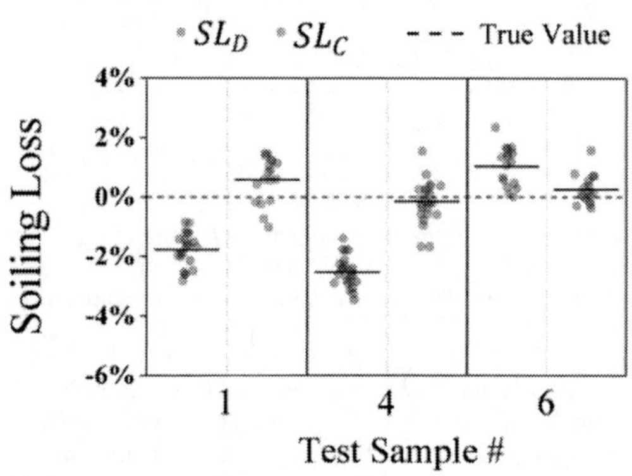

Figure 4: Comparison between SL_D and SL_C errors for three PV samples. The plotted data represent daily mean values.

3.3 Long-term Accuracy of SL_C

Since SL_C requires corrective factors (i.e. temperature coefficients and a power correction factors) that change over time, the long-term accuracy of this method becomes questionable. Figure 5 compares the daily mean soiling loss values using corrective factors obtained at the start of the study period (i.e. SL_{C1}), and re-assessed values a year later (i.e. SL_{C2}). The accuracy of measuring soiling loss improved with all samples when re-evaluating this method. On the other hand, when the old correction factors were used a year later, four samples underestimated soiling loss by 3%, and one sample underestimated as much as 5% on average. These large errors demonstrate that the performance of the PV devices changed differently over time because SL_{C1} yielded better results at the beginning of the study. This highlights the importance of periodically accommodating changes with appropriate corrective factors, otherwise, soiling loss errors could become more significant than those of SL_D.

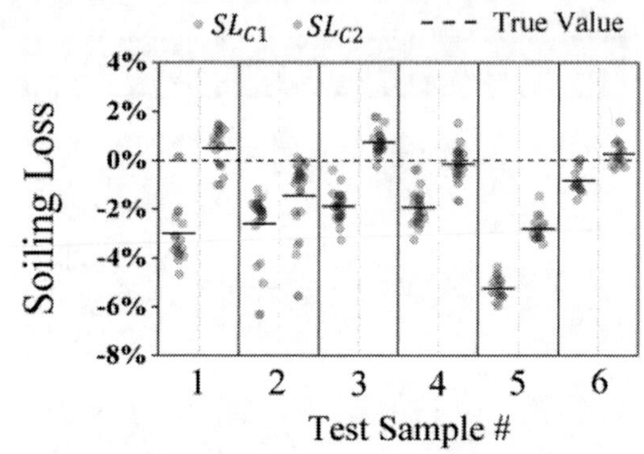

Figure 5: Comparison between SL_{C1} (i.e. using correction factors prior to the start of the study) and SL_{C2} (i.e. using correction factors one year later) errors for all the PV samples. The plotted data represent daily mean values.

3.4 The Uncertainty of Extracted SRates

Many PV practitioners are interested in SRates which are defined as the daily soil-induced power loss. This is of great value for energy prediction models and estimating the next cleaning event for a PV system. The rate is highly dependent on a site's weather characteristics and the PV technology intended to be used. Therefore, it is often obtained from real SL measurements. However, since measurements fluctuation are expected when using two devices, as discussed earlier, uncertainties are thus cascaded down to SRates. Figure 6 presents three possible SRates derived from a test sample's SL_D values during a period of ten days. Because of the sample's standard deviation, the range of possible SRates is from 0.33%/day to 0.59%/day. The predicted SL difference after ten days between the lowest and highest SRates is 2.6% and 4.0% after fifteen days. These significant estimation differences reveal that SRates are highly sensitive to SL measurement deviations. Therefore, to improve SRate accuracy, SL fluctuation must be kept at a minimum.

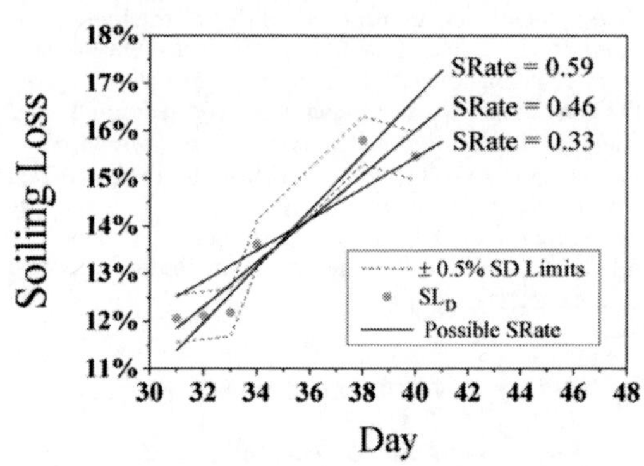

Figure 6: Three possible SRate values were derived from a test sample's soiling loss data and its standard deviation.

4 CONCLUSIONS

Commercially available mc-Si modules of the same model were subjected to prolong dust-intensive in-field conditions where the errors of measuring soiling loss using the SL_D and SL_C methods were studied. Amongst six samples, the mean daily SL_D error ranged between 2.7%$_{abs}$ below the true value and 1.0%$_{abs}$ above. It is rather clear, that such a large deviation in errors between the different samples is primarily a result of performance asymmetry and thus contributes as a source of high uncertainty in SL_D measurements. Furthermore, the standard deviation of error between the six samples ranged between ±0.4%$_{abs}$ and ±1.4%$_{abs}$. This inconsistency further acts as an additional source of uncertainty in SL_D measurements as repeatability is different for each sample, which is caused by both daily weather and performance variations. Caution is advised when using two devices to measure soiling loss, as different devices can yield different errors and different error tolerances.

As for SL_C, the daily mean error for half the samples showed significant improvements while none were seen for the other half. This is indicative that accurate PV characterization is crucial to the effectiveness of this method. With regards to measurement repeatability, however, the standard deviation of error slightly increased from that of SL_D. The increase can be attributed to a slight offset of sampling time between temperature, irradiance, and IV curve data which ultimately affects the accuracy of corrections. Moreover, re-evaluating the correction factors a year later was necessary to maintain acceptable accuracy. For example, using old correction values, the mean error of one sample was as high as 5%$_{abs}$ below the true value. The minimum re-evaluation frequency to maintain a desired level of accuracy needs further investigation. Nonetheless, it is important to note, that even annual indoor characterization can be difficult for some PV operators thus raising practicality issues.

When obtaining SRates for soiling forecasts, a range of possible SRates is to be expected because of measurement deviations when measuring SL. Preliminary results illustrated that SRate uncertainties are highly dependent on the repeatability of SL measurements. Further investigations of this relationship are required, as SRates are widely used without a clear understanding of their accuracy.

5 ACKNOWLEDGMENTS

The work was financially supported by the Kuwait Foundation for the Advancement of Sciences (KFAS). Facility and equipment support was provided by the Kuwait Institute for Scientific Research (KISR). The authors wish to thank A. Al-Qattan for his continuous support and encouragement.

6 REFERENCES

[1] H. A. AlBusairi and H. J. Möller, "Performance Evaluation of CdTe Modules Under Natural Outdoor Conditions in Kuwait," in *European Photovoltaic Solar Energy Conference and Exhibition* Valencia, Spain, 2010: European Photovoltaic Solar Energy Conference and Exhibition

[2] B. Guo, W. Javed, B. W. Figgis, and T. Mirza, "Effect of Dust and Weather Conditions on Photovoltaic Performance in Doha, Qatar," in *IEEE Photovoltaic Specialists Conference* Doha, Qatar, 2015: IEEE.

[3] R. K. Jones; *et al.*, "Optimized Cleaning Cost and Schedule Based on Observed Soiling Conditions for Photovoltaic Plants in Central Saudi Arabia," *IEEE Journal of Photovoltaics,* vol. 6, no. 3, pp. 730-738, 11 March 2016 2016.

[4] S. AAM, "Effect of dust on flat plate collectors," in *International Solar Energy Congress*, New Delhi, 1978, pp. 960-4: Pergamon Press.

[5] M. Gosteinl, T. Düster, and C. Thuman, "Accurately Measuring PV Soiling Losses With Soiling Station Employing Module Power Measurements," in *IEEE Photovoltaic Specialists Conference* New Orleans, LA, USA, 2015 IEEE.

[6] L. Micheli, M. Muller, and S. Kurtz, "Determining the effects of environment and atmospheric parameters on PV field performance," in *IEEE Photovoltaic Specialists Conference* Portland, Or, USC, 2016: IEEE.

DC-DC POWER OPTIMIZERS FOR BUILDING-INTEGRATED PHOTOVOLTAIC APPLICATIONS – A SIMULATION-BASED EVALUATION

J. Eisenlohr, S. Gasparotto, H.R. Wilson, T. E. Kuhn
Fraunhofer Institute for Solar Energy Systems ISE, Heidenhofstraße 2, 79110 Freiburg, Germany
Phone: +49 (0) 761 4588 5745 Email: johannes.eisenlohr@ise.fhg.de

ABSTRACT: DC-DC power optimizers present a promising approach to minimize losses in photovoltaic systems integrated into buildings (BIPV). Compared to free-standing PV plants, BIPV systems are often geometrically very complex, leading to a more complicated electric circuit and effects like partial shading and hot spots. We addressed the complex electrical design with a detailed modeling approach that includes the relevant physical quantities to analyze the time-resolved yield of a BIPV system. This comprises the calculation of the time-resolved irradiance on each individual solar cell by ray-tracing, the temperature-dependent electric performance of each solar cell, the electrical interconnection including bypass and blocking diodes and the DC/AC conversion. As each BIPV system is geometrically unique and requires detailed electrical planning, DC-DC power optimizers, which ensure that individual modules are always operated at their individual maximum power point, could simplify this planning process. However, this requires DC-DC converters that are highly efficient also under the specific operation conditions that can strongly vary between different BIPV systems, during the day or the year. In this work, we extended the Fraunhofer ISE BIPV simulation tool and included DC-DC optimizers to be able to quantify the benefits of optimizers for specific BIPV projects.
Keywords: Building Integrated PV (BIPV), System Performance, Façade, Shading, Simulation, Ray Tracing, DC-DC optimizers.

1 INTRODUCTION

Photovoltaic systems that are integrated into buildings (building-integrated photovoltaics, BIPV) face several challenges. Especially, they often suffer from partial shading leading to significant mismatch between individual cells, modules and/or strings of a BIPV system. One promising option to reduce the detrimental effect of mismatch is the use of DC-DC power optimizers. By DC-DC conversion, these optimizers ensure that each module (or more generally: sub-unit of the system connected to an optimizer) is operated at its individual maximum power point (MPP). By converting current and voltage along an equal-power curve from the individual MPP, it can be ensured that the current values of all optimized units connected in series are equal. Different technological implementations, mainly buck-only or buck-boost converters, are available on the market, as well as different control strategies, e.g. maintaining a constant string voltage or levelling to the highest current in the string. Typical conversion efficiencies stated by manufacturers are quite high (between 98 and 99.5 %). To decide whether additional DC-DC power optimizers are a good option for a BIPV system, the potential gain in yield has to be balanced with the additional effort and cost of having more electronic devices in the system. In this paper, we demonstrate how the potential gain in yield can be quantified for the various operating conditions of a BIPV system. Therefore, we present an extension to the BIPV simulation tool of Fraunhofer ISE and results for examples of BIPV systems with DC-DC power optimizers.

2 APPROACH: DETAILED BIPV SIMULATION TOOL

During recent years, Fraunhofer ISE has developed a very detailed BIPV simulation tool suite [1,2]. So far, DC-DC power optimizers were not included. In the work reported here, we added methods to the tool to simulate such optimizers. Two steps were necessary for that: (i) MPP tracking is done at a lower level for each optimized unit, replacing the central MPP tracking by the inverter and (ii) the DC connection of the optimized units has been adapted according to different control strategies. Buck or buck-boost converters can be chosen and the DC output is calculated either based on the maximum occurring current or for a fixed string voltage. The former approach is realized for example in optimizers by Tigo [3], the latter one for example by Solaredge [4]. The finite efficiency of optimizers, depending on the input voltage and power and the output voltage, is taken into account e.g. by a parametric relationship [5] .

Fig. 1 shows the workflow of the simulations. Compared to previous publications about our BIPV simulation tool, step 4 has been divided into 4a and 4b to integrate DC-DC optimizers. In 4a, for each optimized sub-unit (typically a module), the IV curves at all time steps are calculated considering the internal electrical design including bypass diodes etc. The maximum power point of every optimized unit is extracted. In step 4b, the electrical interconnection between all optimized sub-units is considered. Based on the control strategy of the optimizer system, the individual MPP values are combined to DC values of the whole string or system. The efficiency of the DC-DC converters in dependence on the operating conditions of the converter (input voltage and power, ratio between input and output voltage) is a crucial input for step 4b.

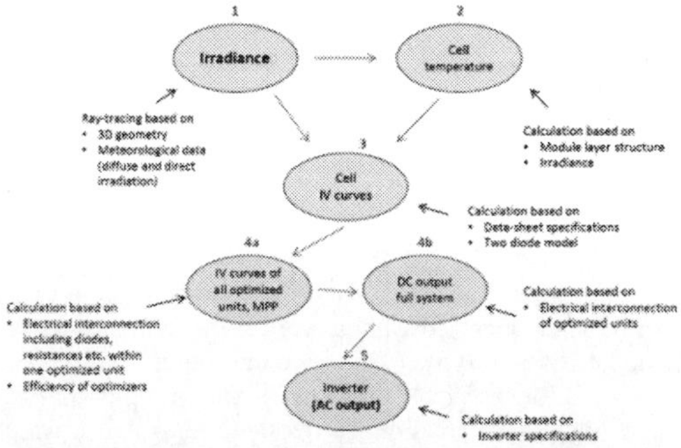

Figure 1: Schematic description of the five main steps for a detailed BIPV simulation. For each time step, the irradiance on each solar cell (1), the temperature of each solar cell (2) and the cell IV characteristic for each temperature/irradiance combination (3) are calculated. The IV curve of every optimized sub-unit is calculated (4a) and then combined with the IV curves of all other units according to a certain control strategy resulting in the DC output of the whole system (4b). Finally, the AC output is calculated considering the inverter specifications (5).

With this add-on to the tool suite, it is now possible to investigate complex and complete BIPV systems with DC-DC optimizers. In a what-if analysis, we can compare systems with and without optimizers and take into account also different options for utilizing bypass diodes. We especially investigate and optimize systems with modules in different orientations and PV slat systems that combine use of daylight within the building, shading to avoid high thermal loads especially in summer, and electric yield. Our simulations are time-resolved and for each time step consider the behavior of each individual solar cell in the system according to the corresponding cell irradiance and temperature. In the following, the five main steps of the simulation tool are explained briefly:

2.1 Irradiance calculation

To calculate the time-dependent irradiance on each solar cell of the system, the ray-tracing tool RADIANCE [6] is used. The main inputs required for this step are: the 3D geometry of the building and its surroundings including optical properties of the materials and meteorological data including diffuse and direct irradiation and temperature. As the 3D geometry is needed in a RADIANCE format, we have developed methods to extract exactly the required information from architectural models [7]. It should be emphasized that the 3D model has to contain also the relevant surroundings, e.g. neighboring buildings or trees that influence the irradiance on the building skin. The meteorological data can be measured onsite or reference data sets for the specific geographical location can be used. Based on diffuse and global irradiance values, we calculate a sky model [8] and an effective irradiance at the cell positions including an incidence angle modifier to account for the varying sensitivity of the PV module at different angles of incidence.

2.2 Temperature calculation

In order to calculate the cell temperature of each solar cell for each time step, differently detailed models can be used: from a simple approach using ambient air temperature and irradiance level (calculated in step 1) with a product-specific linear factor, through node models considering the module layer structure, up to detailed, spatially resolved thermal simulations. A detailed comparison and validation of different models can be found for example in [9].

2.3 Calculation of relevant cell IV characteristics

For all occurring combinations of irradiance and temperature, the cell IV characteristics are required. To obtain them, we usually derive temperature-dependent parameters of the two-diode model from the data sheet specifications; for details see [10]. It is important to mention that the IV characteristic is not only needed in the "useful" quadrant (positive voltage and current, where power can be extracted), but also for negative voltages and negative current, as such operation points can occur in series and parallel connections when parts of the system are shaded. A detailed description of how reverse bias conditions are modelled in the presented tool suite can be found in [10].

2.4 Calculation of the system IV characteristic (DC output)

This step has now been separated into two steps. Knowing the cell IV characteristics at a specific temperature/irradiance combination, the electrical interconnection within one optimized sub-unit and additionally the IV characteristics and location of blocking and bypass diodes within such a sub-unit, the IV curve of this subunit can be calculated based on Kirchhoff's circuit laws in step 4a. By determining the maximum power point of this IV curve, all cell operation points (current and voltage of each cell) can be extracted.

In step 4b, all individual DC outputs of the optimized sub-units are combined according to the actual electrical interconnection and the control strategy and efficiency of the DC-DC optimizers.

2.5 Calculation of the AC output

Based on the data sheet parameters of the inverter, the inverter efficiency is calculated as a function of the input power and the input voltage using the model of Schmidt and Sauer [11]. Considering the DC result of step 4b as inverter input, the final AC output of the system is calculated and can be optimized by varying any of the inputs.

2.6 Post-processing of simulation results

Knowing all the time-resolved values of irradiance, temperature, current (DC of subunits and the full system), voltage (DC of subunits and the full system) and AC power, many quantities of interest can be derived/calculated. By varying steps 4a and 4b, especially the influence of optimizers on individual operating conditions and on the final system yield can be evaluated. Also, typical performance indicators like performance ratio, specific yield in kWh/kW_{peak} or electrical mismatch of the chosen electric circuit plan can be extracted. When the system is divided into subsystems that feed into different phases of the grid, the time-resolved balance between these different systems can be evaluated.

3 SYSTEM EXAMPLE

To demonstrate the versatility of the presented BIPV simulation tool and its extension regarding DC-DC power optimizers, a configuration example combining cells with various effective irradiance levels is investigated. In a BIPV project that is currently in the planning phase, PV slats that are installed in front of the glazed façade to allow electricity generation, daylight use in the building and shading especially during summer, simultaneously, are under investigation. The planned BIPV system has several different orientations, as all façades will be equipped with the PV slats. Basic effects of the optimizers and the electrical circuit in general, including diodes, can be studied already in an abstract geometrical model as presented in Fig. 2. Considering an actual building with more slats, more storeys etc. can be treated in a straightforward way using the developed tools.

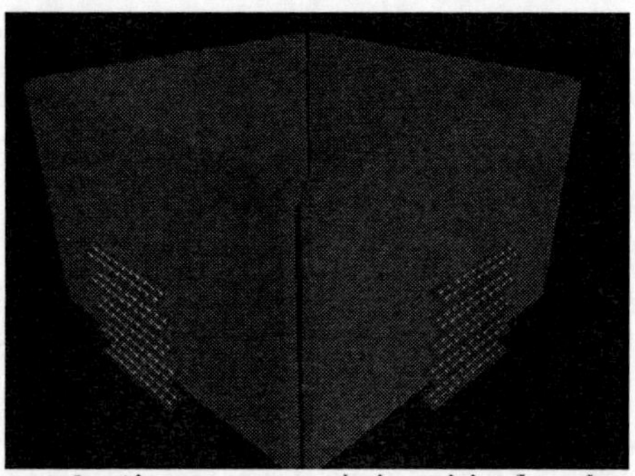

Figure 2: Abstract, geometrical model of a planned BIPV system. All four walls with orientations south-east, south-west, north-east and north-west are equipped with five PV slats each. Each PV slat contains 2 x 10 solar cells, indicated by the green dots and arrows.

In the following, the application of the five main steps of the simulation tool to this specific project is explained briefly.

3.1 Irradiance calculation

The abstract geometrical model was created directly in RADIANCE. Thus, also parameters like dimensions, vertical gap distance or rotation angle of the slats can be easily varied and optimized. A similar model with additional information about the glazing of the façade and the rooms behind the façade has been used for daylight calculations. As meteorological input, data from Solargis for the geographical location have been used. To import a complete building geometry including the surroundings, rather than only a simplified, abstract model, we have developed filters and wrappers to extract the required information from Building Information Modelling data formats like the industry foundation classes (IFC) [7,12].

3.2 Temperature calculation

In this case, a simple linear temperature model has been assumed, as the PV slats are well ventilated from the front and the back.

3.3 Calculation of relevant cell IV characteristics

Based on STC measurements, two diode parameters have been extracted and used for all simulations. In the current planning phase, the exact solar cell type that will be used has not yet been decided, so results are preliminary. Due to the modular design of the complete simulation tool, the IV curves for all temperatures and irradiance levels calculated in this step can be replaced later on and the rest of the simulation can be easily repeated without further changes.

3.4 Calculation of the system IV characteristic (DC output)

To analyze how the cells in the configuration in Fig. 2 can be interconnected without very large mismatch losses, four different types of interconnection have been simulated. In case 1 (4 subsystems), the group of slats on each façade orientation is treated separately, which means they are not electrically connected. This would require four individual inverters (one for each orientation). As the inner cell row and the outer cell row of each slat are illuminated quite differently, case 2 (8 subsystems) considers eight subsystems: two for each of the four orientations, one for the inner cells and one for the outer cells. This would require four cables for each PV slat and a total of eight individual inverters. For case 3 (2 subsystems), all inner cells are connected in series, regardless of their orientation. In the same way, all outer cells are connected in series. This would require two inverters. Clearly, by connecting different orientations in series, the power output strongly decreases, as north-east and north-west oriented modules usually receive significantly less solar radiation. Bypass diodes, which have been added to each PV slat, decrease the loss. Finally, for case 4 (2 subsystems, opt.), two DC-DC-optimizers are used for each orientation, one for the inner and one for the outer cells. In total, eight optimizers and two inverters would be needed.

Fig. 3 shows the time-resolved DC output for the four different cases for a sunny day (meteorological data from March 14th, 2015). The global horizontal irradiance is also plotted in red. It can be seen that the scenario with two subsystems combined with four optimizers is very close to the one with 8 different subsystems and significantly higher than for the other two scenarios. Only minor differences occur due to the efficiency losses of the DC-DC conversion process. For the efficiency, a parametric model was used [5]. Knowledge of the precise efficiency for all operating conditions of the specific optimizers used is crucial to correctly quantify the potential gains. This requires detailed efficiency descriptions in the data sheet of DC-DC optimizers for various operating conditions, which would allow a parametric model to be fitted. However, right now, such detailed data is often not made available by the manufacturers. Calculated for a complete year, the four cases lead to the specific yields given in Table 1.

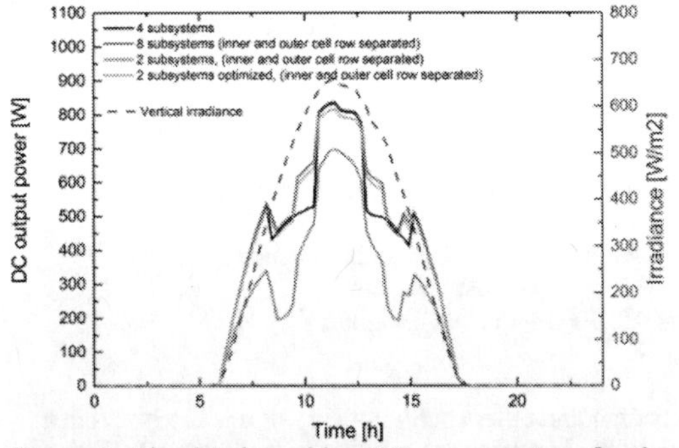

Figure 3: Simulated resulting DC power output for the system in Fig. 2 for a sunny day (March 14th, 2015). The red curve (right axis) shows the global irradiance on a horizontal plane.

Table I: Specific yield for the four different cases

	Spec. Yield [kWh/kWp]
Case 1 (4 subsystems)	649
Case 2 (8 subsystems)	705
Case 3 (2 subsystems)	502
Case 4 (2 subsystems, opt)	680

4 CONCLUSIONS AND OUTLOOK

In this paper, we presented a BIPV simulation tool for detailed simulations of building-integrated photovoltaic systems including DC-DC optimizers. Based on the 3D geometry of the building and its surroundings, the time-resolved irradiance on each solar cell position is calculated via ray-tracing using meteorological data for the relevant geographical location. Subsequently, the temperature and the electric behaviour of each cell are modelled and the system's properties are determined considering parallel and series connections of cells, diodes, modules and optimizers. Finally, the AC output is calculated for each time step based on the parameterized inverter efficiency. All these calculation methods are combined in a tool suite and programmed using Python and basic linux tools.

With this tool, complex BIPV systems can be planned, evaluated and optimized. The potential benefit of applying DC-DC optimizers to a specific system can be quantified. The final decision, whether optimizers are a good solution for a BIPV system, has to take not only technical but also economical aspects into account, which have not been addressed in this paper. However, knowing the yield accurately is mandatory for a sound economic analysis.

The discussion of an example of a complex BIPV system has demonstrated the flexibility and applicability of the presented tool. A detailed understanding of these BIPV systems using our tools is possible and necessary to achieve an optimum yield and fail-safe operation.

ACKNOWLEDGEMENTS

The research leading to these results was partly funded by the German Federal Ministry for Economic Affairs under the project number 03SBE0005B.

BIBLIOGRAPHY

1. W. Sprenger, H. R. Wilson, and T. E. Kuhn, "Electricity yield simulation for the building-integrated photovoltaic system installed in the main building roof of the Fraunhofer Institute for Solar Energy Systems ISE," Solar Energy **135**, 633–643 (2016).
2. J. Eisenlohr, H. R. Wilson, and T. E. Kuhn, "Detailed Modelling of Building-Integrated Photovoltaics - From Component and Environmental Data to the System Output," (2017); 5 pages / 33rd European Photovoltaic Solar Energy Conference and Exhibition; 2118-2122.
3. https://www.tigoenergy.com/ts4/.
4. https://www.solaredge.com/products/power-optimizer#/.
5. C. Deline and S. MacAlpine, "Use conditions and efficiency measurements of DC power optimizers for photovoltaic systems," in *IEEE Energy Conversion Congress and Exposition (ECCE), 2013. 15-19 Sept. 2013, Colorado Convention Center, Denver, CO* (IEEE, 2013), pp. 4801–4807.
6. "Radiance - A validated Lighting Simulation Tool," https://www.radiance-online.org/.
7. J. Eisenlohr, C. Eller, C. Leifgen, M. Boudhaim, C. Maurer, W. Sprenger, T. E. Kuhn, "Flexible filtering of heterogenous data using the example of the design and simulation of building integrated photovoltaics," BauSIM, Karlsruhe (2018).
8. R. Perez, R. Seals, and J. Michalsky, "All-weather model for sky luminance distribution—Preliminary configuration and validation," Solar Energy **50**, 235–245 (1993).
9. Y. B. Assoa, L. Mongibello, A. Carr, B. Kubicek, M. Machado, J. Merten, S. Misara, F. Roca, W. Sprenger, M. Wagner, S. Zamini, T. Baenas, and P. Malbranche, "Thermal analysis of a BIPV system by various modelling approaches," Solar Energy **155**, 1289–1299 (2017).
10. W. Sprenger, *Electricity yield simulation of complex BIPV systems* (Fraunhofer-Verlag, 2013).
11. H. Schmidt and D. U. Sauer, "Praxisgerechte Modellierung und Abschätzung von Wechselrichter-Wirkungsgraden," Sonnenenergie **4**, 43–47 (1996).
12. M. Boudhaim, "A BIM-compatible Framework to Assess the Thermal and Optical Performance of Envelope Systems in Architecturally Special Buildings," in *30th International Conference on Efficiency, Cost, Optimisation, Simulation and Environmental Impact of Energy Systems (ECOS)* (2017).

SWITCHED-CAPACITORS AS LOCAL CONVERTERS FOR SNAKE PV MODULES: A COST/EFFICIENCY EXPLORATION

Pieter Bauwens[*,a], Apostolis Bakovasilis[b,c], Patrizio Manganiello[c], Eszter Voroshazi[c], Jan Doutreloigne[a], Jef Poortmans[c,d], Francky Catthoor[c,d]
[a] CMST – ELIS, imec and Ghent University, Tech Lane Ghent Science Park - Campus A, Technologiepark 15, B-9052 Ghent, Belgium
[b] Aristotle University of Thessaloniki, University Campus, 54124 Thessaloniki, Greece
[c] Imec, EnergyVille Campus, Thor Park 8320, 3600 Genk, Belgium
[d] ESAT, KU Leuven, Kasteelpark Arenberg 10, 3001 Heverlee, Belgium

ABSTRACT: In order to reduce the negative effect of partial shading and other sources of current mismatch within a module, smart reconfigurable modules allow altering the connections between groups of cells (cell-strings). With a proper algorithm managing these connections, we can make sure that the majority of the cells are operating close to their MPP, even when a part of the module is shaded. Such a smart reconfigurable module consists of some extra components. Switches are needed to change the interconnection scheme. Small, local converters collect power from multiple cell-strings. They step-up the voltage to reduce the current on the central bus they are connected to. At the end where we connect to the string-level bus, a module converter further regulates the voltage for the grid or the PV array. This topology was presented before where we showed that a smart reconfigurable module could recover up to 70% of the power lost to partial shading. In this paper we take a closer look at the local DC-DC converter. More precisely, we present a cost-efficiency analysis of different converter topologies. Taking into account practical limitations (economical limitations, number of components, maximum switch currents, maximum capacitance values, etc..) we estimate efficiency and projected cost. We show that Dickson pump (CR3) with 30-35mΩ switches is the best candidate. This would result in a chip cost of about €1.5

Keywords: DC-DC-Converter, Economic Analysis, Energy Performance, Shading, Configurable Topology

1 INTRODUCTION

A standard silicon photovoltaic (PV) module consists of typically 60 or 72 cells connected in series. While this series connection might result in a more practical voltage-current profile, it is also the cause of the large drop in output power when such a PV panel is partially shaded. A local current mismatch will limit the total PV module current to the one of the shaded cell. The current solution, inserting three bypass diodes [1] in anti-parallel between substrings, is an improvement but is ultimately sub-optimal: (a) even if only one cell of a substring is shaded, the generated power of the entire substring is bypassed, (b) bypass diodes themselves have to carry a large current which leads to additional power loss and possibly heating issues [2] and (c) the insertion of bypass diodes reshape the power-voltage curve creating local maxima [3], requiring more complex Maximum Power Point Tracking (MPPT) algorithms able to track the global maximum. Simple MPPT algorithms [4] such as the Open Circuit Voltage method typically have a low MPPT efficiency (< 90%). More complex algorithms, e.g. Perturb & Observe (P&O), can result in higher efficiency, but require sensing of both voltage and current and might still miss the global MPP. In any case, multiple local MPP and especially fast changing shading patterns are difficult to deal with for these MPPTs. These conditions occur especially in domestic PV plants or urban-integrated PV arrays, such as Building Integrated PV (BIPV).

Another approach to mitigating this issue, one that is evidenced at recent Solar Industry exhibitions, is the move towards so-called ‘smart PV modules’. In these smart modules, the power optimizers or micro-inverters are integrated at module level, and not at the array level. The MPPT is performed at module level as well, resolving the inter-module current mismatch problems. Intra-module current mismatch can be addressed by replacing bypass diodes with local optimizers [5]. However, the granularity is still fairly limited and comes at a relatively high cost.

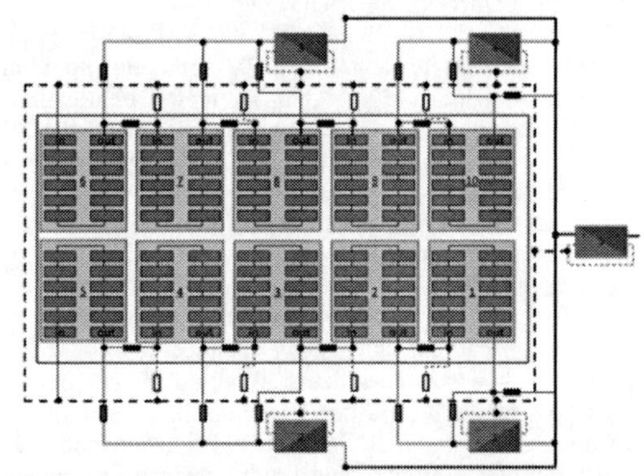

Figure 1: Smart reconfigurable module topology, using switches (small boxes), local converters (boxes 1, 2, 3, 4)

Going a step further, we arrive at the concept of reconfigurable topologies. In such a topology, multiple run-time configurations are enabled, depending on the current operating condition, thus allowing the majority of the cells operation at their MPP. Ideally, every cell can be connected to every other cell, but this would require an unreasonable amount of connections, switches, etc. The most suitable granularity level has to be explored by taking into account the increased manufacturing cost, the increased losses due to resistivity and addition of dynamic elements such as the switches and local DC-DC converters. While some approaches to a reconfigurable smart PV module have been presented in literature [6], the optimal trade-off between realization cost and improved energy output is still not clear. We previously published an approach that takes these aspects into account [7] (Figure 1). These modules are able to establish different interconnection schemes along small groups of cells (cell-strings). These connections can be either in parallel or in series by use of switches. The intra-module converters ensure suitable current and voltage level and allow a direct control of the operating point of the cell-string. They are connected in parallel to the module level converter via a central bus. The module converter can be used for the connection with either the grid or the rest of the PV array. Under non-uniform conditions, this reconfigurable topology can recover up to 70% [8] [9] of the power that was lost in a traditional configuration. It is clear that these topologies are only relevant if we succeed in keeping the cost of the different elements as low as possible. Also here there is a clear trade-off between cost and efficiency. In this paper we take a closer look at the cost-efficiency optimization of the local converter. We examine different converter topologies and estimate their cost and efficiency.

2 TOPOLOGIES

2.1 General

The goal of the local converter is to have a very cheap (~€1) solution that regulates the cell-strings output current to be conformable with the module converters specifications. A lower current on the central bus also implies lower resistive losses. It is important to note that since the outputs of these local converters are connected in parallel, such converters will not integrate any output voltage control algorithm. This voltage will be determined instead by the module converter. The local converters conversion ratio will thus determine the cell-string voltage.

We are aiming at a low-cost fully integrated solution (PowerSoC). As inductors are bulky and integrated inductors are costly, we opt for a switched-capacitor solution. In the following sections, we take a closer look at the investigated topologies. Different topologies have different conversion ratios, number of components (switches, capacitors), require different current levels, generate different switching losses, etc.. Higher conversion ratios typically mean lower bus currents and module converter currents, but higher converter complexity and cost.

2.2 Ladder converter

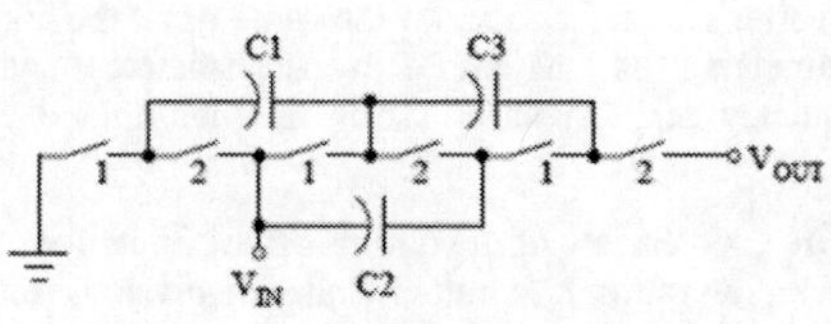

Figure 2: Ladder topology with conversion ratio 3

The ladder topology (Figure 2) achieves a fixed conversion ratio for a specific input node. Changing the input source (Vin) can alter the conversion ratio. This topology is suitable also for non-integer conversion ratios m:n. The odd-numbered switches are turned on during phase 1, the even-numbered switches during phase 2. The number of capacitors and switches needed to achieve a conversion ratio of n is 2n-3 and 2n respectively. All capacitors and switches support a voltage equal to the input. For smaller conversion ratios, this topology uses few switches, but quite a lot of capacitors.

2.3 Dickson charge pump

The Dickson charge pump is mostly known for 1:n conversion ratios. Like the ladder, the odd-numbered switches turn on during phase 1 and the even numbered switches turn on during phase 2. The Dickson topology has the advantage of using fewer capacitors for the same conversion ratio compared to the Ladder topology. In general, for a fixed conversion ratio n, the Dickson pump uses n-1 capacitors and n+4 switches. The capacitor in the lowest stage supports the input voltage, while the rest support twice the input voltage. The blocking voltage of the switches is either the input voltage or twice the input voltage.

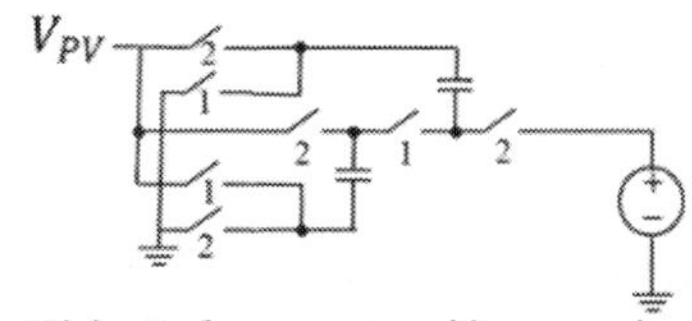

Figure 3: Dickson charge pump with conversion ratio 3

Other topologies such as the series-parallel converter, connecting all capacitors in parallel to the input voltage in the first stage and connecting them in series in the second stage, typically require too many switches to be cost-effective here

3 METHODOLOGY

3.1 Cost

In a switched-capacitor converter, the cost is defined by the cost of the capacitors and switches. As we are aiming for an integrated converter, we assume a HV planar integration chip technology for the MOSFET switches. A very rough estimate puts the cost on €0.1/mm² die area. When designing for low on-resistances (larger area of MOSFET), we can assume that the die area is inversely proportional to the on-resistance (about 69Ωmm²). Therefore we can state that the main drivers of the cost are the MOSFETs' on-resistance and the number of components. The cost of the capacitors will mainly be determined by their number, not so much by their exact value. Note that the switches will dominate the cost, well above the other components. See also Figure 5.

3.2 Efficiency

The different topologies are simulated using realistic models for the capacitors and switches. The main contributors to the power losses are the conduction losses and the switching losses. Eq. (1), with R_{ON} the on-resistance of the MOSFET, ESR_{CAP} the series resistance of the capacitor, N_{SW} and N_{CAP} the number of switches and capacitors respectively and I_{PV} the PV input current, and Eq. (2), with V_{SW} the voltage swing of the switches, I_{SW} the current through the switches and f_s the switching frequency, represent the conduction losses and the switching losses of a ladder with conversion ratio 2, respectively.

$$P_{loss,Cond} \approx \left(N_{SW}\cdot\frac{R_{ON}}{2}+N_{CAP}\cdot ESR_{CAP}\right)\cdot I_{PV}^2 \qquad (1)$$

$$P_{loss,Switch} \approx g(R_{ON},V_{SW},I_{SW})\cdot N_{SW}\cdot f_s \qquad (2)$$

It is clear from these equations that factors that increase the efficiency, will also increase the cost, which is why this cost-efficiency analysis is useful for further development of the local converter. It is worth to note that, for the value of capacitance we need, in the order of tenths of µF, and the RMS current they have to process, low ESR capacitors will be used. Thus, their impact on the converter losses is significantly lower than the one of the MOSFETs' on-resistance.

4 RESULTS

Even though Eq. (1) and Eq. (2) refer specifically to a ladder with conversion ratio 2, they can be used to make some observations that can be extended to any of the switched-capacitors topologies we studied, helping to reduce the solution space. The conductions losses can be lowered by reducing the input current, which is done by using half-cells (See Figure 1). It is also clear that the number of components (N_{SW}, N_{CAP}) should be as low as possible; this not only lowers the power loss, but also reduces the cost of the converter. For this reason, we limit ourselves to the conversion ratios 2, 3, and 4. The on-resistance of the MOSFET affects not only the conduction loss, but also the switching losses. A lower on-resistance seems beneficial, but also increases the cost. On top of that, large MOSFETS with a low on-resistance have a large gate, requiring a larger current to be driven at a certain frequency. The frequency itself also poses limitations on the system: a smaller frequency requires larger capacitors; a higher frequency requires larger gate currents. For practical reasons we want the capacitors to be smaller than 100μF, and a gate current in the 250mA to 500mA range. Figure 4 shows achievable switching frequencies for a certain maixmum gate current. For switches of 30-35mΩ, the switching frequency will be around 200kHz.

For a conversion ratio of 2, the ladder topology reduces to a simple H-bridge, with 1 capacitor and 4 switches. Simulation showed that a conversion ratio of 3 and 4 was more efficient using the Dickson charge pump, using 7 and 8 switches respectively. An extra capacitor is needed at the input to smooth the voltage ripple. The projected cost of these converters is depicted in Figure 5.

For all topologies we assumed an irradiance of 1000W/m². Upper and lower boundaries were defined by connecting the local converter to a single cell-string, or to a double cell-string (2 cell-strings in parallel, doubling the current, see Figure 1). For each topology, we determined a solution point were the balance between MOSFET on-resistance, cost and efficiency is optimal. Table I collects the results.

For each topology, the module converters (MC) input voltage is shown. This is the voltage that the module converter should fix, in order for the cell-string to operate

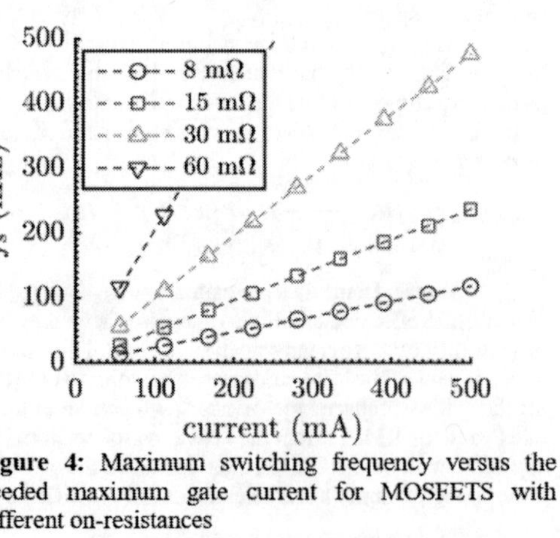

Figure 4: Maximum switching frequency versus the needed maximum gate current for MOSFETS with different on-resistances

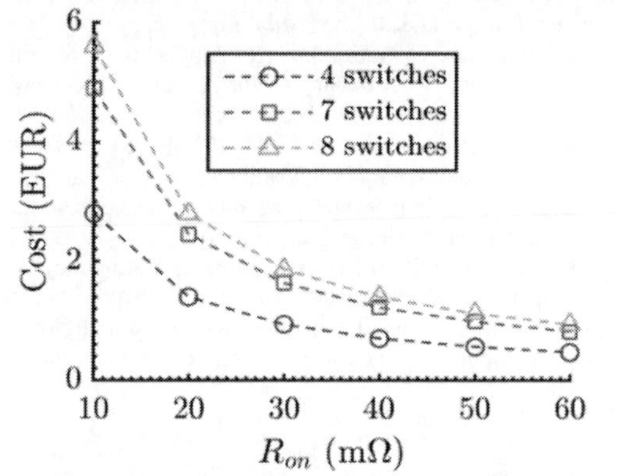

Figure 5: Projected converter cost for converter with different amounts of switches and their on-resistances

at its MPP. Together with this, also the MOSFET maximum current is shown, allowing us to estimate the feasibility of a fully integrated converter.

The results above are for an irradiance of 1000W/m² but they can be easily reprojected for other irradiance levels. The whole point of the reconfigurable module is to only activate the local converters when there is an partial shading. This implies that the local converters will often work in conditions lower than 1000W/m². Simulation shows that the efficiencies in Table 1 increase by 1-2% when the irradiance is decreased due to lower conduction losses. For very low irradiances (<300W/m²) this effect is less obvious mainly due to the larger relative effect of the switching losses (See Figure 6).

5 CONCLUSION

The above analysis effectively uses a worst-case scenario to select the most promising solutions for the local converter in the smart reconfigurable module. The selection is based on chip cost, converter efficiency, requirements and limitation due to cell-string and module converter characteristics. At the same time, the converter parameters as MOSFET on-resistance, switching frequency and capacitor sizing are determined for the

Table I: Overview of the optimization results for conversion ratios 2, 3, and 4. Underlined values indicate corresponding extremes of the range

Topology	Ladder (CR2)	Dickson (CR3)	Dickson (CR4)
Capacitors	1 (2) <70μF	2 (3) <100μF	3 (4) <120μF
Switches	4	7	8
R_on (mΩ)	[20 ÷ 36]	[30 ÷ 35]	[32 ÷ 34]
η_min @MPP (%)	[89 ÷ 93]	[89 ÷ 91]	[89 ÷ 90]
Chip cost (€)	[0.76 ÷ 1.37]	[1.37 ÷ 1.60]	[1.61 ÷ 1.95]
MC volt. @MPP (V)	[10.4 ÷ 11.6]	[15.5 ÷ 17.5]	[20.5 ÷ 23.5]
SW current @MPP (I)	[8.5 ÷ 9]	[11 ÷ 13]	[11 ÷ 13]

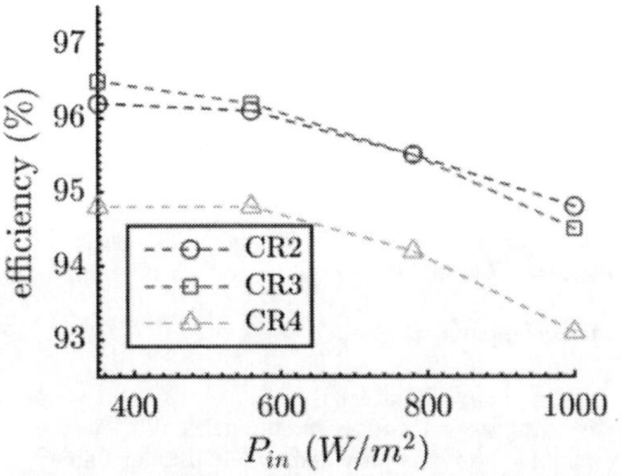

Figure 6: Converter efficiencies as a function of input power

selected topologies. Ladder and Dickson topologies allow for a lower number of components, reducing the cost and related losses. To reduce the constraints posed on the module converter, a conversion ratio of at least 2 should be considered. With this, we show that Dickson pump (CR3) with 30-35mΩ switches is the best candidate. This would result in a chip cost of about €1.5. The downside of this approach is the fixed conversion ratio. In future work, we are looking at topologies for multiple conversion ratios. This may require more components, but the energy trade-off will be larger as well.

5 ACKNOWLEDGEMENTS

The authors gratefully acknowledge imec's SiPV industrial affiliation program and its partners. This project has received funding from the European Union's Horizon 2020 research and innovation programme under the Marie Skłodowska-Curie grant agreement No. 751159.

6 REFERENCES

[1] E. Diaz-Dorado, A. Suárez-García and C. Carrillo, "Influence of the shadows in photovoltaic systems with different configuration of bypass diodes," in *2010 SPEEDAM*, 2010.

[2] K. Kim and P. Krein, "Reexamination of Photovoltaic Hot Spotting to Show Inadequacy of the Bypass Diode," *IEEE J. Photovolt.*, vol. 5, no. 5, pp. 1435-1441, 2015.

[3] B. A. Alsayid, S. Y. Alsadi, J. S. Jallad and M. H. Dradi, "Partial Shading of PV System Simulation with Experimental Results," *Smart Grid and Renewable Energy*, vol. 4, no. 6, pp. 429-435, 2013.

[4] A. Reisi, M. H. Moradi and S. Jamasb, "Classification and comparison of maximum power point tracking techniques for photovoltaic system: A review," *Renewable and Sustainable Energy Reviews*, vol. 19, pp. 433-443, 2013.

[5] Jinko, "http://www.jinko-smart.com/maxim.html".

[6] R. Dorn, "Photovoltaic module utilizing a flex circuit for reconfiguration". Patent WO/2008/076301, 2006.

[7] M. Baka, F. Catthoor, D. Soudris and A. Papanikolaou, "Configurable Module Topology to Recover Power Lost due to Current Mismatch," in *29th EUPVSEC*, Amsterdam, 2014.

[8] M. Baka and F. S. D. Catthoor, "Smart PV Module Topology with a Snake-Like Configuration," in *31st EUPVSEC*, Hamburg, 2015.

[9] P. Bauwens, J. Govaerts, M. Baka, F. Catthoor, K. Baert, G. Vandenbroeck, H. Goverde, D. Anagnostos, J. Doutreloigne and J. Poortmans, "Reconfigurable Topologies for Smarter PV Modules: Simulation, Evaluation and Implementation," in *32nd EUPVSEC*, Munich, 2016.

LIFE CYCLE ASSESSMENT OF CIGS PV MODULES: UPDATE OF CURRENT PRODUCTION CONDITIONS IN GERMANY AND INVESTIGATION OF A PLANNED FACTORY IN CHINA

Ann-Kathrin Briem[1*], Michael Held[2]; Bernhard Dimmler[3]

[1]Department Life Cycle Engineering (GaBi), Institute for Acoustics and Building Physics (IABP), University of Stuttgart, Germany
[2]Department Life Cycle Engineering (GaBi), Fraunhofer Institute for Building Physics (IBP), Stuttgart, Germany
[3]NICE Solar Energy GmbH, Schwäbisch Hall, Germany

[*]Phone: +49 (0)711 970-3164, ann-kathrin.briem@iabp.uni-stuttgart.de

ABSTRACT: This study investigates the potential environmental impacts of copper-indium-gallium-diselenide (CIGS) PV modules using life cycle assessment (LCA). Detailed manufacturer data provided by NICE Solar Energy GmbH for an existing production line in Germany and for a factory under construction in China were available. The results show that the environmental impacts from CIGS module production in China are higher compared to the German reference production. The sensitivities related to the regional conditions of the production location (electricity generation mix and climate-dependent electricity demand) are investigated for the Chinese production. Regarding CIGS electricity generation, irradiance values of 1,100 kWh/(m²a) and 1,700 kWh/(m²a), module efficiencies of 14.6%, 16.0% and prospective 18.0%, a lifetime of 30 years, a performance ratio of 0.75 and a degradation rate of 0.5% per year are taken into account for all installation scenarios. The results show that the higher impacts from a production in China compared to a production in Germany can be compensated during the use phase by an ideal site selection. For a German production, the GWP_{100} results range from 15.2 to 30.0 g CO_2-eq./kWh for the investigated installation scenarios and for a Chinese production this range is 19.0 to 37.3 g CO_2-eq./kWh.
Keywords: Life Cycle Assessment (LCA), CIGS, CO2 footprint, Sustainable, Module Manufacturing

1 INTRODUCTION

Trends in the photovoltaic (PV) industry and market show an ongoing shift to Asia in general and especially to China. In 2017, 70% of the worldwide PV module production took place in China [1]. Furthermore, the Chinese market is the largest globally with 53 GW installed in 2017 and with a total installed capacity of 131 GW equaling 32% globally it is the leading country in this regard as well [2,3]. In Europe, Germany is the country with the highest installed PV capacity (42 GW at the end of 2017) being one of the early pioneer countries of this technology. After a decline in recent years, the German market is growing again and currently the largest European market with 1.8 GW installed in 2017 [3].

At the same time technological advances lead to improvements in production technologies. Not only increased module efficiencies, but also lower energy- and material consumption of production processes are in the focus of producers and researchers [4]. Especially thin-film PV technologies, such as copper-indium-gallium-diselenide (CIGS), are being further developed and show great potential on lab-scale [5,6]. Consequently, the challenge is applying these improvements on an industrial scale [5]. During the last years, the industrial CIGS production processes have been significantly improved by NICE Solar Energy GmbH (former Manz CIGS Technology GmbH) in a 20 MW_p/a production line in Germany [7,8]. Furthermore, a new CIGS factory with a capacity of 306 MW_p/a is under construction in China with the newest available process equipment being employed [9].

To assess the influence of these trends on the environmental impact of PV technologies, life cycle assessment (LCA) studies have to be conducted and regularly updated. Regarding CIGS PV technology, few studies can be found that reflect current production technology and locations. Therefore, the aim of this study is to investigate the potential environmental impacts of CIGS modules first under present production conditions in Germany and second for the mentioned factory under construction in China.

2 STATE OF THE ART OF LCA OF CIGS PV

Numerous LCA studies have been published regarding PV technologies. However, most studies focus on crystalline-silicon (c-Si) technologies. There are fewer studies about thin-film PV and the majority of these focuses on CdTe modules. Furthermore, most studies focus mainly on the analysis of global warming potential (GWP) or other energy-related indicators; other impact categories are rarely evaluated. This is the conclusion of several recent meta-studies regarding LCA in the field of PV, for example Gerbinet et al. (2014) [10], Bhandari et al. (2015) [11] and Peng et al. (2013) [12]. Koppelaar (2016) [13] further concludes that on average seven years lie between data collection and publication of the reviewed studies. In addition, Bracquene et al. (2018) [14] find that, due to outdated life cycle inventory (LCI) data, recent studies most likely overestimate the environmental impact of PV modules. As a consequence, it can be concluded from the literature that more transparency in reporting, methodological consistency and updated LCI data are necessary in order to better understand and compare the results of different studies.

One important public source for LCI data is a report by the International Energy Agency's Photovoltaic Power Systems Programme (IEA PVPS) Task 12 Group [15] including data for commercial PV technologies. Furthermore, the same group published a methodological guideline regarding parameters for conducting LCA on PV systems consistently [16]. Additionally, the "Product Environmental Footprint" (PEF) screening report [17] initiated by the European Commission provides current data, although it is partly based on the IEA PVPS Task 12 LCI report as well.

Regarding LCA of CIGS PV, various studies are shortly described in the following paragraph. The results of these studies are to be compared later to the results of this study. Thus, only recent and sufficiently transparent studies using similar parameters are reviewed that allow a comparison. Results for scenarios taking into account a solar irradiation of 1,700 kWh/(m²a) are considered and summarized in Table I, as this represents a typically evaluated value, e.g. for a Southern European location.

Table I: Assumptions and results of selected LCA studies in the literature for scenarios with 1,700 kWh/(m²a)

Study	Loc	LT	ME	PR	Sys	Results	
						EPBT	GWP
[18]	SE	30	12.0	0.75	R	1.0	30.0
[19] EU	SE	30	11.7	0.77	R	1.02	21.44
[19] CN	SE	30	11.7	0.77	R	1.01	27.64
[20]	USA	30	10.9	0.75	-	-	36.0
[21]	-	30	14.0	0.80	G	1.1	25.4

Loc Installation location, solar irradiation of 1,700 kWh/(m²a)
LT Module lifetime [years]
ME Module efficiency [%]
PR Performance Ratio [-]
Sys System (R = rooftop, G = ground mounted)
EPBT Energy Payback Time [years]
GWP Global Warming Potential [g CO_2-equivalents/kWh]
SE Southern Europe

Our previous study [18] is based on detailed manufacturer data from the year 2008 for a 30 MW_p/a production line by then Würth Solar. However, no balance of system (BOS) components were taken into account. A frequently cited study is de Wild-Scholten (2013) [19], which investigates six PV technologies and two production sites (Europe (EU) and China (CN)). Data for CIGS were provided to the author by the market-leading producers Avancis and Solar Frontier. The results show a higher global warming potential (GWP) for a Chinese production, due to higher emissions from the coal-based Chinese electricity generation mix compared to the European mix. Collier et al. (2014) [20] investigate and compare two novel PV technologies to commercial technologies including CIGS. For this purpose, a model with the "GaBi ts" software was created. The US grid mix is used for the production and BOS components are not considered. The results in the upper part of the range presented here could be due to the comparably low module efficiency of 10.9% that is assumed. Leccisi et al. (2016) [21] use the Ecoinvent database (v3.1) as background system for their investigation of a ground-mounted system. In the foreground, data published by the IEA PVPS Task 12 group are used [15]. Japan is assumed to be the production country for the studied CIGS modules and the corresponding electricity mix is used.

In conclusion, the presented results of the reviewed studies and scenarios for 1,700 kWh/(m²a) range from 21 to 36 g CO_2-eq.kWh.

3 IMPLEMENTATION OF LCA FOR CURRENT CIGS PV MODULE PRODUCTION

3.1 Background of LCA methodology

LCA is an internationally established methodology for evaluating the environmental impacts caused by a product or service during its life cycle. Typically, the investigation of this product system includes mining of raw materials, manufacturing, usage, as well as recycling and disposal at the end of life. The method is standardized in ISO 14040 [22] and 14044 [23] and can be used to support decision-making, e.g. during product development, policy-making and marketing. The implementation of LCA is highly iterative, but generally comprises four interconnected phases: Goal and scope definition, life cycle inventory (LCI) analysis, life cycle impact assessment (LCIA) and interpretation [22,23].

3.2 Goal and Scope of this study

The goal of this study is the investigation of potential environmental impacts of a current CIGS standard module of NICE Solar Energy GmbH with 14.6% total area module efficiency. The scope of this study includes all life cycle phases (resource mining, production of the module, packaging and transport, use phase, i.e. electricity generation, and end of life). The focus is put on the production phase, since the relevant influencing factors are expected to be found there. The functional unit is 1 kWh of produced (AC) electricity and the reference flow is 1 m² of module area.

3.3 Life cycle inventory analysis

The main data basis for this study is detailed manufacturer data provided by NICE Solar Energy (Schwäbisch Hall (SHA), Germany) and turnkey CIGS factory supplier Manz AG (Reutlingen, Germany). This includes the materials used and their quantities, as well as energy consumption and auxiliary materials used in production. Measured values from a two-week measurement in autumn 2015 are available from the CIGStab in SHA. The data quality can be classified as very good, as it is real production data. Based on this measurement data, planning data for a larger factory in China were developed by Manz AG. The upscaling of the production capacity from approx. 20 MW_p/a to 306 MW_p/a results in uncertainties in the planning data, which should be confirmed or adjusted in the future by the actual state after completion and commissioning of the new factory in Chongqing (CQ).

In the event of data gaps, assumptions based on expert interviews and the literature are made in consultation with the manufacturer. Some assumptions regarding the Chinese framework conditions are made in consultation with Shenhua, Manz's Chinese partner. The database system of the GaBi 8.7 software (Service Pack 36, 2018) [24] used to create the LCA model serves as the background system.

Based on this information, a detailed parameterized LCA model is created taking the following process steps into account individually:

- Manufacturing of substrate and cover glass,
- Sputtering of molybdenum back contact,
- Co-evaporation of the semiconductor elements for the CIGS layer,
- Chemical bath deposition of buffer layer (CdS),
- Sputtering of iZnO and ZAO layers,
- and lamination.

Other process steps are included as well, such as washing and structuring steps, contacting and the junction box. In addition to these production-related process steps, the technical building systems (TBS) are taken into account, as they contribute significantly to the overall energy demand of the factory. They include process cooling, air-conditioning, compressed air and process gas supply, water treatment and other systems required for operation at the production site. Thus, the entire production process is accounted for in the model.

The available data regarding the TBS included measurement data for some but not all relevant systems. Therefore, the total electricity demand of the TBS had to be estimated. For climatic conditions similar to those in Germany a TBS share of 30% of the total electricity demand is assumed based on a previous study [18], for which data was measured over the period of a year at the same location, and the current data for the CIGSfab in SHA. This assumption is considered plausible by the manufacturer and an interviewed expert [25]. For other climatic conditions the TBS share is likely higher or lower. For the conditions in CQ, where ambient temperature and humidity are higher compared to Germany [26,27], a TBS share of 40% of the total electricity consumption is assumed and later evaluated in a sensitivity analysis. Furthermore, it is assumed that no additional heating is required, as waste heat can be used within the factory.

3.4 Life cycle impact assessment

The CML 2001 methodology (January, 2016) [28] is chosen for impact assessment and the following impact categories are selected:

- Global warming potential with a time frame of 100 years (GWP_{100})
- Acidification potential (AP)
- Eutrophication potential (EP)
- Photochemical ozone creation potential (POCP)
- Abiotic depletion potential (ADP) divided into "ADP fossil fuels" and "ADP elements"

Additionally, the non-renewable primary energy demand is evaluated ($PED_{non\text{-}ren}$).

POCP was initially considered for the LCIA. However, the results in this category are not discussed in this paper in order to avoid misinterpretations. One reason for this is that nitrogen monoxide (NO) has a negative characterization factor in the CML characterization model. This leads to negative results, especially in glass production and transport processes, which can be justified, but can lead to incorrect conclusions in the interpretation. Another problem is that NO is often considered via the group of nitrogen oxides (NO_x) in the datasets and is not listed separately. The described effect of negative results only occurs in processes with a detailed breakdown of NO_x emissions. Since the datasets are not consistent in this regard, the significance of this impact category is limited.

A similar issue can also be pointed out for ADP. Due to the high sensitivity related to uncertainties of elementary flows as well as due to the applied allocation methods, results of the ADP elements might vary significantly depending on the used background data making conclusions difficult to find. Therefore, ADP results are not included in this paper, again to avoid misinterpretations. Furthermore, the category ADP elements as an environmental impact category considers only the depletion of resources in relation to their natural reserves, but not their criticality, which is also based on economic and geopolitical factors.

4 LCA RESULTS

4.1 Production of 1 m² CIGS module

The relevant assumptions and parameters of the German reference scenario and the Chinese base scenario are summarized in Table II. The main difference lies in the location-dependent electricity demand and supply.

Table II: Main assumptions and parameter settings of the reference scenarios GER and CQ

Assumptions	Base scenario: Germany (GER)	Base scenario: Chongqing, China (CQ)
Material consumption	The same consumption based on measured data	
Data basis of the process electricity demand	Measured data from SHA (ca. 20 MW_p/a)	Planning data for CQ (306 MW_p/a)
Electricity generation mix	GER	CQ
TBS share (of total electricity demand)	30%	40%
Additional heating	Yes	No

Detailed GWP_{100} results of the production of 1 m² CIGS module for both base scenarios are shown in Figure 1. The impacts from glass manufacturing, other materials and auxiliaries, the electricity demand of production processes and the electricity demand of the TBS are separately indicated. All results are summarized in Table VI in the appendix for GWP_{100}, AP, EP and $PED_{non\text{-}ren}$ (cf. section 3.4).

The results show that the environmental impacts from a CIGS module production in China with 91 kg CO_2-equivalents/m² are higher than from the studied reference production in Germany with 67 kg CO_2-eq./m². Taking a closer look at the GWP_{100} of glass manufacturing, the impacts of the Chinese scenario are only slightly higher. The influence of the various upstream processes of the energy sources is low here, since most of the emissions originate from fuel combustion and melting of the raw materials. It is assumed that similar standards apply in Europe and China regarding glass manufacturing [29]. Similarly and due to the model, there are only minor differences in the materials and auxiliaries, since for most substances the same datasets based on global production mixes are used in both scenarios.

The higher environmental impacts of the Chinese production are mainly due to higher emissions from the electricity generation of the Chinese grid mix, but also due to the higher climate-dependent electricity demand of the TBS of the prospective Chinese factory, as can clearly be seen in Figure 1 for the GWP_{100}. This also applies to all other investigated impact categories (see Table VI in the appendix).

To identify the sensitivities related to these parameters, the underlying Chinese regional electricity grid mix and the percentage of the TBS electricity demand are varied with regard to the influence of different climatic conditions at different Chinese locations. The resulting bandwidths are shown in Figure 2.

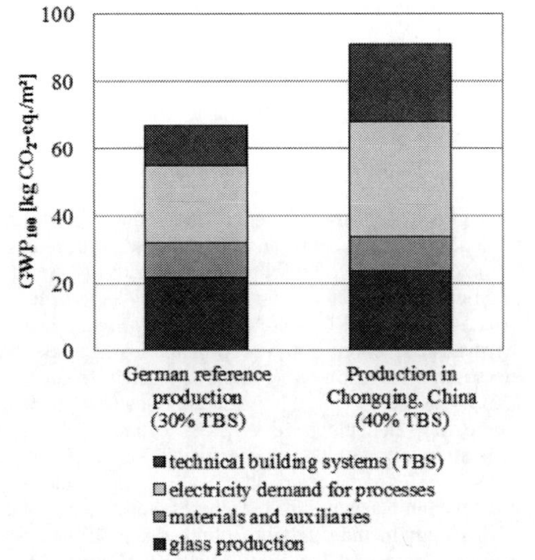

Figure 1: GWP_{100} results for the production of 1 m² CIGS module in Germany and China

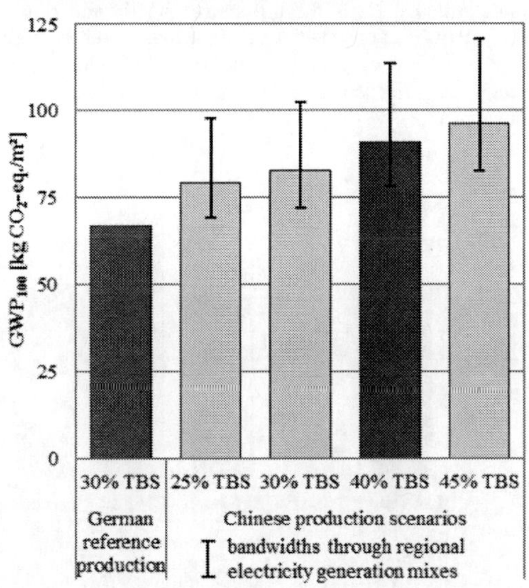

Figure 2: GWP_{100} results and bandwidths of various Chinese production scenarios.

The columns represent the production results for different TBS shares calculated with the grid mix in Chongqing, where the planned CIGS factory is under construction with the dark blue representing the base scenario. The black bars indicate the bandwidths due to different regional Chinese electricity grid mixes with varying shares of coal and renewable energy sources. In total, the results for a Chinese production regarding the GWP_{100} range from 69 kg CO_2-eq./m² (Southern China, 25% TBS) to 121 kg CO_2-eq./m² (Northern China, 45% TBS). Therefore, the lower end of this range for the GWP_{100} is similar to a German production under present conditions.

4.2 Life cycle results of 1 m² CIGS module

The life cycle results including not only the production but also packaging, transport and end of life of the module are summarized in Table VI in the appendix for both the German and Chinese base scenario. While packaging has a negligible influence on all impact category results, transport can have a noticeable effect. Figure 3 illustrates the GWP_{100} and EP of 1 kg of freight for the considered transport routes (within EU: 1,000 km truck transport; sea route CN-EU: 20,000 km container ship, 1,500 km freight train in China and 1,000 km truck in EU; within CN: 1,000 km truck and 1,500 km freight train). For an installation in the production country or region, the transport accounts for a few percent of the overall environmental impacts. However, if modules are transported overseas, e.g. from China to Europe, the emissions from container ships lead to an increase in the results, especially for EP and AP. Regarding the end of life, recycling options based on [17] with credits were assessed. The results show that recycling has minor but positive effects on the overall results.

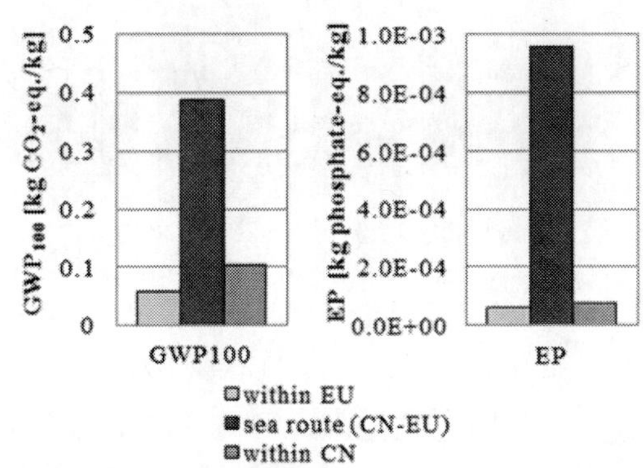

Figure 3: GWP_{100} and EP of 1 kg of freight for the investigated transport routes.

4.3 Electricity generation with CIGS modules

The environmental impacts of CIGS PV power generation are investigated. Installation locations with irradiance values of 1,100 kWh/(m²a) and 1,700 kWh/(m²a) as well as (commercial) module efficiencies of 14.6% and 16.0% are considered for both production scenarios. A lifetime of 30 years, a performance ratio of 0.75 (representing a rooftop installation) and a degradation rate of 0.5% per year are taken into account for all installation scenarios. The assumptions for the use phase are based on the IEA PVPS Task 12 Methodological Guidelines [16] and summarized in Table III. The investigated module efficiencies are based on total area efficiencies including all losses.

In order to calculate the environmental profile of PV power generation per kWh, all environmental impacts from the life cycle of the module (production, packaging, transport and end of life) are related to the total amount of electricity produced during use. It is assumed that both for a German and a Chinese production the installation will take place in the respective country, or for Germany in Europe. Thus, only land transport is considered.

Table III: Use phase assumptions based on [16]

Parameter	Value(s)	Unit
Total area module efficiency	14.6 16.0	%
Performance ratio (roof top installation)	0.75	-
Solar irradiation	1,100 1,700	kWh/(m²a)
Module lifetime	30	years
Degradation	0.5	% / year
Inverter lifetime	15	years

Although the focus of this study is put on the module, the balance of system (BOS) components are taken into account for completeness. Cables and mounting structures are based on LCI data from IEA PVPS Task 12 [15]; the inverter is based on internally available data and its lifetime is assumed to be 15 years. The GWP_{100} results per kWh for eight scenarios are shown in Figure 4.

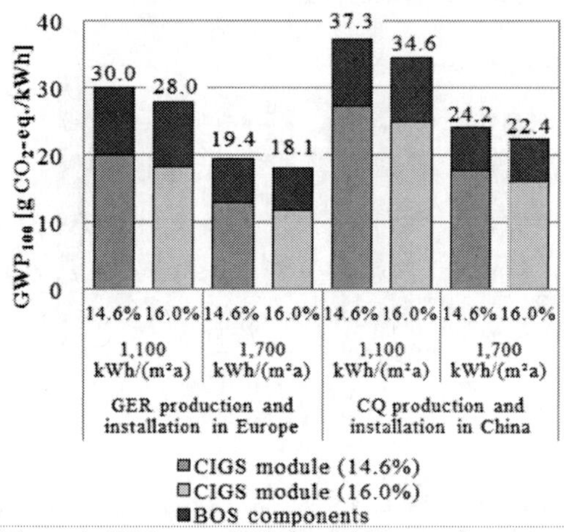

Figure 4: GWP_{100} per kWh for various scenarios

The GWP_{100} results range from 18.1 to 37.3 g CO_2-eq./kWh. Comparing the scenarios with each other reveals the influence of the investigated parameters (module efficiency, production location and solar irradiation). The increased module efficiency from 14.6% to 16.0% leads to a decrease of 1.3 to 2.7 g CO_2-eq./kWh. German production results in 4.3 to 7.3 g CO_2-eq./kWh lower GWP_{100} results compared to the investigated Chinese production. The specific conditions at the installation site, i.e. solar irradiation, have the most significant influence on the results. While the results range from 28.0 to 37.3 g CO_2-eq./kWh for scenarios with 1,100 kWh/(m²a), the range is 18.1 to 24.2 g CO_2-eq./kWh for scenarios with 1,700 kWh/(m²a). Therefore, the higher impacts from a production in China compared to a production in Germany could be compensated during the use phase by an ideal site selection.

Thus, the most relevant influencing factor on the environmental impacts of CIGS electricity generation is the solar irradiation at the installation site followed by the electricity mix used for manufacturing at the production location, as also shown in section 4.1 (cf. Figure 2). The influence of increased module efficiency on these results is lower in comparison. Additionally, the assumed lifetime, performance ratio and degradation are not analyzed further here; however, these parameters also have an influence on the results and should be kept in mind.

4.4 Scenarios 2030

In order to take into account future developments in the energy sector, future electricity generation mixes for the year 2030 (from the GaBi ts database [24]) are considered for the production of 1 m² CIGS module. Technological advancements of the production technology are not considered, as it would go beyond the scope of this study. The following scenarios are meant to point out the general direction of future developments.

Results for GWP_{100} are displayed in Figure 5. For a German production, the 2030 projection of the German electricity generation mix compared to the reference mix (2014) leads to a 13% reduction of the impacts. Comparing the prospective 2030 Chinese electricity generation mix with the reference situation in Chongqing reveals a 10% decrease of the GWP_{100}. The Chongqing mix taken into account for the Chinese reference scenario of this study includes relatively high shares of renewable energies, reflecting the local situation. However, since the "CN: Electricity grid mix (2030) ts" refers to China as a whole, a comparison with the Chinese grid mix today ("CN: Electricity grid mix ts", reference year 2014) is more suitable. Here, the difference is 20% and reflects the energy sector transformation China is undergoing. The reduction of potential impacts is even higher for other categories, such as AP (cf. values in Table VI in the appendix).

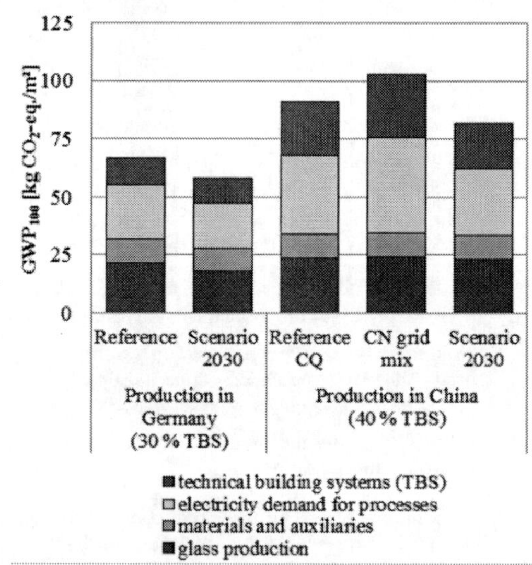

Figure 5: GWP_{100} results for reference and scenarios 2030.

For electricity generation with future CIGS PV modules, both presented production scenarios 2030 are considered and a total area module efficiency of 18.0% is assumed in accordance with NICE Solar Energy GmbH. Otherwise, the same assumptions as in section 4.3 (cf. Table III) apply. The GWP_{100} results per kWh electricity produced are in the range of 15.2 to 29.4 g CO_2-eq./kWh and displayed in Figure 6.

Compared to the reference scenarios (14.6%) a reduction of 4.3 to 7.9 g CO_2-eq./kWh results from the investigated potential future developments. This equals a reduction of approximately 20% for each scenario.

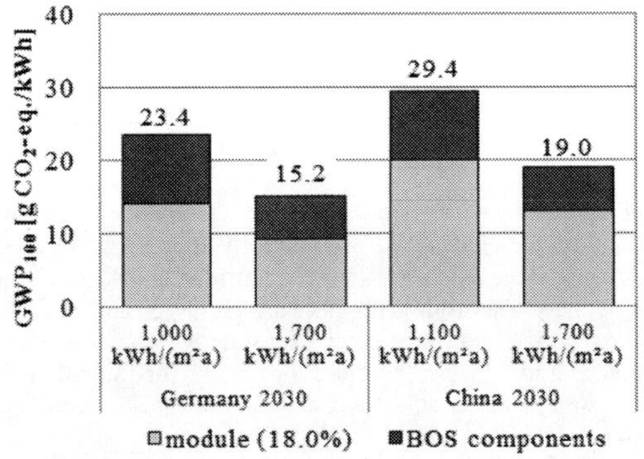

Figure 6: GWP_{100} of future CIGS PV electricity production.

4.5 Non-renewable energy payback time (NREPBT)

In the following, the NREPBT for CIGS PV is presented under the framework conditions of the scenarios examined above including the future scenarios. To determine the NREPBT, the annual electricity yield for the examined efficiencies and irradiation values is calculated. The values resulting from the six possible cases are summarized in Table IV. The performance ratio of a rooftop installation and the degradation over 30 years are included proportionately for one year.

Table IV: Annual yield of 1 m² CIGS PV system

Annual yield [kWh/a]		
Module	Solar irradiation	
efficiency	1,100 kWh/(m²a)	1,700 kWh(m²a)
14.6%	112	173
16.0%	123	190
18.0%	138	214

For the calculation of the NREPBT, the non-renewable PED resulting from the life cycle of a CIGS PV system (module and BOS components) is compared to the amount of PV electricity generated annually by this system. The non-renewable PED of the respective reference electricity mix in Germany (GER), Europe (EU-28) or China (CN) is used for this purpose. Note that for the prospective module efficiency of 18.0%, the future electricity generation mixes are used both as production and reference mixes.

Table V: Non-renewable energy payback time of CIGS PV systems

Non-renewable energy payback time [years] for different production/reference electricity mixes				
Module	Solar irradiation 1,100 kWh/(m²a)			
efficiency	GER/GER	CQ/CN	CN/CN	CN/GER
14.6%	1.5	1.4	1.6	1.9
16.0%	1.4	1.3	1.5	1.8
18.0%	1.4*	-	1.5*	2.2*
Module	Solar irradiation 1,700 kWh/(m²a)			
efficiency	GER/EU-28	CQ/CN	CN/CN	CN/EU-28
14.6%	1.0	0.9	1.0	1.3
16.0%	0.9	0.9	1.0	1.2
18.0%	0.8*	-	0.9*	1.3*

* calculated with future electricity mixes (DE 2030, CN 2030, EU-28 2030)

While for 1,100 kWh/(m²a) a range of 1.3 to 2.2 years results, for 1,700 kWh/(m²a) this is 0.8 to 1.3 years. Locations with higher irradiation values lead to an energy amortization at an earlier point in time. However, this also depends strongly on the reference electricity mix under consideration, e.g. on its share of renewable energies. All calculated values are summarized in Table V.

4.6 Comparison to results from the literature

The results of this study for the global warming potential are compared to the literature values discussed in section 2 regarding CIGS PV modules. Scenarios for 1,700 kWh/(m²a) are shown in Figure 7. The different assumed module efficiencies of the studies are indicated in parentheses (for other assumptions see Table I).

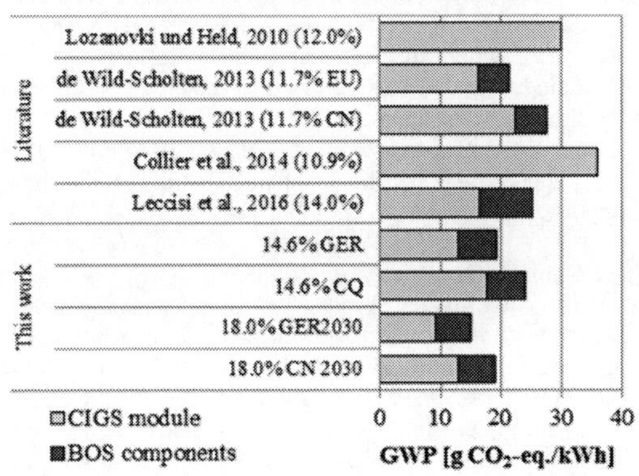

Figure 7: Comparison of GWP literature values with the results of this work for CIGS electricity generation at 1,700 kWh/(m²a)

Compared to our previous study of a German CIGS production line (30 MW_p/a) in 2010, the results of this study show that the environmental impacts have decreased significantly. Compared to results from previous LCA studies found in literature for CIGS, the investigated CIGS modules are lower, but still in the same range. The potential differences in the underlying assumptions should be kept in mind when comparing these results.

5 DISCUSSION AND CONCLUSIONS

5.1 Environmental profile of CIGS module production

The results of this work show that the environmental impacts caused in the production of CIGS modules dominate the life cycle results. Packaging and short transport routes can usually be neglected. However, intercontinental overseas transport can significantly increase the environmental impacts and should not be neglected. Especially with further efficiency improvements of the production, the share of other life cycle phases will increase.

The investigated environmental profile of CIGS modules are considered to be very reliable based on the input data. The quality and robustness of the input data is high, since the data used for the production processes are measurement data of a real production. Uncertainties in the data arise with regard to the electricity consumption for the technical building systems. Here, measurement data were available for some systems, but the total consumption had to be estimated. The estimate for a production in Germany is based on expert opinions and is also plausible with regard to the values from the previous study. This is a well-founded assumption, which should, however, be validated in the future by measured data.

Compared to our previous study in 2010, the results of this study show that the environmental impacts have decreased significantly due to more energy-efficient, less material-intensive production processes.

5.2 Recycling

Recycling at the end of life of the modules has the potential to reduce the environmental impacts. Although the results were not presented in this paper, the elementary resource scarcity potential (ADP elements) is particularly of interest in this context. Material recycling can make a major contribution towards making resources available again for further use. At present, only metals such as copper are recycled, while the recycling of semiconductor materials from PV modules is still not feasible due to the small amount of material to be recycled. By developing a successful commercial recycling process for the semiconductor materials, these important raw materials could also be recovered.

5.3 China as the country of production

The investigated production in Chongqing has a higher environmental impact than the German reference production. This is mainly due to the high influence of the electricity generation mix on the results. The bandwidth resulting from the sensitivity analysis regarding regional electricity mixes confirms this influence. However, these results should not immediately lead to the conclusion that the CIGS modules should not be produced there. China is already the largest PV producer and has the largest PV market worldwide. It can therefore be reasonable to produce locally, even if other production locations still bring advantages in terms of environmental impact at the moment. China is also undergoing an energy system transformation and the Chinese electricity mix will be based on fewer emissions in the future. The analysis of future scenarios 2030 confirmed this expectation. A similar conclusion is found by Stamford and Azapagic (2018) [30] who investigated the production of c-Si systems in Europe and China. The authors emphasize that acidification is much higher for a Chinese production and that future impacts will depend strongly on the development of the Chinese electricity generation mix; a finding that can be confirmed by this study.

At the same time, the results presented are based on planning data and estimations for the technical building systems. The actual climatic conditions at the production site are a significant influencing factor in this context, as shown by the sensitivity analysis. Therefore, updates of this study in the coming years are generally recommended for the Chinese situation in order to validate or supplement the results and to record the developments.

In general, the investigation of sensitivities related to the exact production location within a large country or region, such as China, makes sense especially for thin-film PV (in comparison to the supply-chain of c-Si), since the entire manufacturing of the module takes place in one location.

5.4 CIGS electricity generation

Several conclusions can be drawn with regard to electricity generation. The environmental impacts per kWh depend largely on the parameters of the use phase. Above all, the solar irradiation and orientation of the modules at the site, the actual lifetime and the performance ratio have a major influence. The results show that the higher impacts from a production in China compared to a production in Germany can be compensated during the use phase by an ideal installation site selection. The selection of the production site and the associated environmental impacts from production are thus partly put into perspective. However, the exact consideration of all factors of a specific situation is necessary in order to be able to make concrete statements.

Furthermore, the BOS components can have a significant influence on the results. Their share of the results for electricity generation is approximately 30% or higher in some cases, such as in the 2030 scenarios. Thus, a closer look at the components and their environmental impacts is necessary, as this was not possible within the scope of this study.

The analysis of the energy payback time shows marginal differences for the considered scenarios in relation to a lifetime of 30 years. The important finding at this point is that a CIGS module can generate clean electricity in any scenario for decades after it has already compensated for its production. The results of this work confirm the general consensus in science in this regard.

5.5 Comparison with other technologies

A comparison of CIGS modules to other PV technologies shows that CIGS seems to be at the lower end of the bandwidth based on the available literature, such as [19] and [21]. However, due to the lack of current and publicly available life cycle inventory data or studies that present updated results for all technologies, no well-founded comparison of the PV technologies can be made at the moment. This study contributes to this debate by providing updated results for CIGS PV.

In general, there is no doubt that PV electricity has significantly lower environmental impacts than fossil energy sources and can thus make and continue to make an important contribution to the global energy transition. However, the topic of resource efficiency should continue to be discussed. Efforts must be made to reduce resource consumption and make it more sustainable.

6 REFERENCES

[1] Fraunhofer ISE, Photovoltaics Report, 10.08.2018.

[2] International Energy Agency (IEA) PVPS, 2018 Snapshot of Global Photovoltaic Markets. Report IEA PVPS T1-33:2018, 2018.

[3] M. Schmela, Z. Alsam, T. Döring, R. Rossi et al., Global Market Outlook For Solar Power 2018-2022, 2018.

[4] ITRPV, VDMA, International Technology Roadmap for Photovoltaic (ITRPV): Results 2017. Ninth Edition: March 2018, 2018.

[5] T. Feurer, P. Reinhard, E. Avancini, B. Bissig, et al., Progress in thin film CIGS photovoltaics - Research and development, manufacturing, and applications, Prog. Photovolt: Res. Appl. (2016).

[6] ZSW, ZSW stellt neuen Weltrekord bei Dünnschicht-Solarzellen auf: Wirkungsgrad der CIGS-Photovoltaik steigt immer schneller, Stuttgart, Germany, 15.06.2016.
[7] S. Enkhardt, Manz entwickelt CIGSfab weiter, 17.03.2014, https://www.pv-magazine.de/2014/03/17/manz-entwickelt-cigsfab-weiter/.
[8] Manz AG, Manz und Shanghai Electric arbeiten zusammen an der Weiterentwicklung der Solartechnologie, Shanghai, China, 31.05.2016.
[9] I. Flaig, Manz ergattert Rekordauftrag in der Solarbranche, Stuttgarter Zeitung, 24.01.2017.
[10] S. Gerbinet, S. Belboom, A. Léonard, Life Cycle Analysis (LCA) of photovoltaic panels: A review, Renewable and Sustainable Energy Reviews 38 (2014) 747–753.
[11] K. Bhandari, J. Collier, R. Ellingson, D. Apul, Energy payback time (EPBT) and energy return on energy invested (EROI) of solar photovoltaic systems: A systematic review and meta-analysis, Renewable and Sustainable Energy Reviews 47 (2015) 133–141.
[12] J. Peng, L. Lu, H. Yang, Review on life cycle assessment of energy payback and greenhouse gas emission of solar photovoltaic systems, Renewable and Sustainable Energy Reviews 19 (2013) 255–274.
[13] R.H.E.M. Koppelaar, Solar-PV energy payback and net energy: Meta-assessment of study quality, reproducibility, and results harmonization, Renewable and Sustainable Energy Reviews (2016).
[14] E. Bracquene, J.R. Peeters, W. Dewulf, J.R. Duflou, Taking Evolution into Account in a Parametric LCA Model for PV Panels, Procedia CIRP 69 (2018) 389–394.
[15] R. Frischknecht, R. Itten, P. Sinha, M. de Wild-Scholten et al., Life Cycle Inventories and Life Cycle Assessments of Photovoltaic Systems. Report IEA-PVPS T12-04:2015, 2015.
[16] R. Frischknecht, G. Heath, M. Raugei, P. Sinha, et al., Methodology Guidelines on Life Cycle Assessment of Photovoltaic Electricity: 3rd edition. Report IEA-PVPS T12-06:2016, 2016.
[17] P. Stolz, R. Frischknecht, F. Wyss, M. de Wild-Scholten, PEF screening report of electricity from photovoltaic panels in the context of the EU Product Environmental Footprint Category Rules (PEFRC) Pilots, 2016.
[18] A. Lozanovski, M. Held, Update of the environmental indicators and energy payback time (EPBT) of CIS modules in Europe and scenario analysis of solar cell printing, 2010.
[19] M.J. de Wild-Scholten, Energy payback time and carbon footprint of commercial photovoltaic systems, Solar Energy Materials and Solar Cells 119 (2013) 296–305.
[20] J. Collier, S. Wu, D. Apul, Life cycle environmental impacts from CZTS (copper zinc tin sulfide) and Zn3P2 (zinc phosphide) thin film PV (photovoltaic) cells, Energy 74 (2014) 314–321.
[21] E. Leccisi, M. Raugei, V. Fthenakis, The Energy and Environmental Performance of Ground-Mounted Photovoltaic Systems—A Timely Update, Energies 9 (2016) 622.
[22] EN ISO 14040, Environmental management - Life cycle assessment - Principles and framework, 2006.
[23] EN ISO 14044, Environmental management - life cycle assessment - Requirements and guidelines, 2006.
[24] thinkstep, GaBi ts: Software-System and Databases for Life Cycle Engineering, Stuttgart, Echterdingen, Germany, 1992-2018.
[25] K. Eberhardt, Personal conversation, Stuttgart, Germany, 2017.
[26] climate-data.org, Klima Chongqing https://de.climate-data.org/location/5383/.
[27] climate-data.org, Klima Schwäbisch Hall, https://de.climate-data.org/location/58705/.
[28] Universiteit Leiden - Institute of Environmental Sciences (CML), CML-IA Characterisation Factors, https://www.universiteitleiden.nl/en/research/research-output/science/cml-ia-characterisation-factors.
[29] Italian Trade Commission, China Glass Industry Market Report 2011, 2011.
[30] L. Stamford, A. Azapagic, Environmental Impacts of Photovoltaics: The Effects of Technological Improvements and Transfer of Manufacturing from Europe to China, Energy Technol. 6 (2018) 1148–1160.

APPENDIX

Table VI: Overview over the results for AP, EP, GWP_{100} and $PED_{non\text{-}ren}$ for the evaluated scenarios

	Impact category	AP	EP	GWP_{100}	$PED_{non\text{-}ren}$
	Unit	kg SO_2-eq.	kg phosphate-eq.	kg CO_2-eq.	MJ
1) GERMAN REFERENCE SCENARIO (SHA)					
Total life cycle of 1 m² CIGS module (production, packaging, transport, EoL)		2.30E-01	2.82E-02	6.73E+01	8.61E+02
Production of 1 m² CIGS module in Germany (German reference scenario)		2.41E-01	2.97E-02	6.70E+01	8.76E+02
Packaging (cardboard and wood palette)		4.89E-04	2.31E-04	0.00E+00	-3.55E+00
Transport within the EU (only road)		3.96E-03	1.01E-03	9.60E-01	1.32E+01
End of life with recycling credits		-1.46E-02	-2.72E-03	-6.56E-01	-2.47E+01
Results for 1 kWh CIGS PV electricity generation					
CIGS module					
14.6%	GER 1,100 kWh/(m²a)	6.74E-05	8.39E-06	2.00E-02	2.56E-01
	GER 1,700 kWh/(m²a)	4.36E-05	5.43E-06	1.29E-02	1.66E-01
16.0%	GER 1,100 kWh/(m²a)	6.15E-05	7.65E-06	1.83E-02	2.34E-01
	GER 1,700 kWh/(m²a)	3.98E-05	4.95E-06	1.18E-02	1.51E-01
BOS components					
14.6%	GER 1,100 kWh/(m²a)	4.88E-05	3.00E-06	1.00E-02	1.27E-01
	GER 1,700 kWh/(m²a)	3.16E-05	1.94E-06	6.48E-03	8.22E-02
16.0%	GER 1,100 kWh/(m²a)	4.80E-05	2.93E-06	9.72E-03	1.23E-01
	GER 1,700 kWh/(m²a)	3.11E-05	1.90E-06	6.29E-03	7.96E-02
2) CHINESE BASE SCENARIO (CQ)					
Total life cycle of 1 m² CIGS module (production, packaging, transport, EoL)		5.21E-01	5.19E-02	9.66E+01	1.07E+03
Production of 1 m² CIGS module in China (Chongqing base scenario)		4.41E-01	3.87E-02	9.09E+01	1.02E+03
Packaging (cardboard and wood palette)		4.89E-04	2.31E-04	0.00E+00	-3.55E+00
Transport to the EU (sea route)		9.50E-02	1.57E-02	6.35E+00	7.78E+01
End of life with recycling credits		-1.46E-02	-2.72E-03	-6.56E-01	-2.47E+01
Results for 1 kWh CIGS PV electricity generation					
CIGS module					
14.6%	CQ 1,100 kWh/(m²a)	1.28E-04	1.11E-05	2.73E-02	3.01E-01
	CQ 1,700 kWh/(m²a)	8.29E-05	7.21E-06	1.77E-02	1.95E-01
16.0%	CQ 1,100 kWh/(m²a)	1.17E-04	1.02E-05	2.49E-02	2.75E-01
	CQ 1,700 kWh/(m²a)	7.56E-05	6.58E-06	1.61E-02	1.78E-01
BOS components					
14.6%	CQ 1,100 kWh/(m²a)	4.88E-05	3.00E-06	1.00E-02	1.27E-01
	CQ 1,700 kWh/(m²a)	3.16E-05	1.94E-06	6.48E-03	8.22E-02
16.0%	CQ 1,100 kWh/(m²a)	4.80E-05	2.93E-06	9.72E-03	1.23E-01
	CQ 1,700 kWh/(m²a)	3.11E-05	1.90E-06	6.29E-03	7.96E-02
3) FUTURE SCENARIOS 2030					
Production of 1 m² CIGS module in Germany (German grid mix 2030)		2.27E-01	2.78E-02	5.80E+01	5.57E+02
Production of 1 m² CIGS module in China (Chinese grid mix 2030)		2.92E-01	3.24E-02	8.20E+01	1.06E+03
For comparison: Production of 1 m² CIGS module in China (Chinese grid mix, reference year 2014)		4.91E-01	4.32E-02	1.03E+02	1.17E+03
Results for 1 kWh CIGS PV electricity generation (2030 projection)					
CIGS module					
18.0%	GER 2030 1,100 kWh/(m²a)	5.14E-05	6.34E-06	1.41E-02	1.31E-01
	GER 2030 1,700 kWh/(m²a)	3.33E-05	4.10E-06	9.10E-03	8.47E-02
	CN 2030 1,100 kWh/(m²a)	6.79E-05	7.53E-06	2.00E-02	2.55E-01
	CN 2030 1,700 kWh/(m²a)	4.39E-05	4.88E-06	1.30E-02	1.65E-01
BOS components					
18.0%	GER 2030 1,100 kWh/(m²a)	4.71E-05	2.85E-06	9.37E-03	1.18E-01
	GER 2030 1,700 kWh/(m²a)	3.05E-05	1.85E-06	6.07E-03	7.65E-02
	CN 2030 1,100 kWh/(m²a)	4.71E-05	2.85E-06	9.37E-03	1.18E-01
	CN 2030 1,700 kWh/(m²a)	3.05E-05	1.85E-06	6.07E-03	7.65E-02

LIFE CYCLE ASSESSMENT OF PV-BATTERY SYSTEMS FOR A CLOAKROOM AND CLUB BUILDING IN ZURICH

Philippe Stolz[1*], Rolf Frischknecht[1], Thomas Kessler[2], Yvonne Züger[2]
[1]treeze Ltd., Kanzleistrasse 4, 8610 Uster, Switzerland
[2]City of Zurich, Switzerland
[*] Corresponding author: stolz@treeze.ch, phone: +41 44 940 61 93

ABSTRACT: The Office for Building Engineering of the City of Zurich plans the construction of a cloakroom and club building. The building and the floodlights of the surrounding soccer fields use electricity mainly in the evening. That is why the installation of a photovoltaic (PV) system in combination with a battery storage system is evaluated in the pre-project phase. The environmental footprint of the PV system with multi-crystalline silicon modules and of current, future and second-life lithium-ion batteries is quantified within the life cycle assessment of the building. The self-consumption share of PV electricity increases from 31% to 62% if a 60 kWp PV system is complemented by a 100 kWh battery storage. The complementary grid electricity mix strongly influences the environmental impacts of electricity consumed by the cloakroom and club building. The installation of a PV system and a battery storage leads to a 10-17% reduction in greenhouse gas emissions compared to the full coverage of the electricity demand by the average Swiss supply mix. The addition of a current battery system does not yield any further reduction compared to the "PV only" option. With the renewable electricity mix of the City of Zurich, the installation of a PV system and a battery storage leads to higher environmental impacts of the electricity consumed by the cloakroom and club building, irrespective of the type of battery used. A future increase in energy density, production optimisations and second-life batteries bear a significant potential to reduce the environmental impacts of battery storage systems.

Keywords: photovoltaics, PV system, storage, batteries, lithium-ion, building, self-consumption, LCA, CO_2 footprint, environmental effect

1 INTRODUCTION

The City of Zurich has committed itself to develop towards the 2'000-Watt Society [1]. This vision is based on the premise that the limited worldwide energy resources should be fairly allocated among countries and generations and used in a way that environmental impacts are minimized [2]. To achieve the goals of the 2'000-Watt Society, the City of Zurich strives to reduce the primary energy demand related to energy purposes such as heating, hot water, electricity and fuels from 3'900 watt per capita (average 2012-2016; [3]) to 2'000 watt per capita. In addition, a reduction of the per capita greenhouse gas emissions from an average 4.7 ton CO_2-equivalent (t CO_2-eq) between 2012 and 2016 [4] to 1.0 t CO_2-eq per capita in 2050 is pursued. The measures to realise the targeted reduction pathways include, among others, the promotion of energy efficiency and the increased use of renewable energy [5].

Photovoltaic (PV) systems have one of the highest potentials among renewable electricity generation technologies in Switzerland [6]. The cumulative installed PV capacity has more than doubled between 2013 and 2016 [7]. The vast majority of PV systems in Switzerland is decentralised and either mounted on or integrated in buildings [8]. Buildings with PV systems are increasingly equipped with a battery storage system for various reasons. Batteries may be used to increase the share of self-consumed PV electricity and can contribute to the integration of additional PV electricity into the grid by shaving production peaks. Furthermore, a battery storage system can be used as a backup in case of blackouts and may prevent the need for enlarging the connection for power supply [9].

The combination of a PV system with a battery storage results in an increase of the environmental impacts of building construction. A detailed understanding of the environmental impacts of currently available battery systems is therefore an important prerequisite for the design of buildings with low environmental impacts over their entire life cycle. The rapid development in battery research and in battery applications raises the question as to how the environmental impacts may change in the future. Potential future optimisations in battery manufacturing and the repurposing of traction batteries used in electric vehicles is therefore also taken into consideration in the evaluation of the environmental footprint of batteries used in stationary storage systems.

This study aims to analyse the life cycle environmental impacts of a PV system in combination with a battery storage system using the specific case of a cloakroom and club building that is currently planned by the Office for Building Engineering of the City of Zurich. In the pre-project phase, the installation of a roof-integrated PV system and a battery storage system is evaluated [10]. A battery storage system helps to increase the share of self-consumed PV electricity since the building and the floodlights of the surrounding soccer fields use electricity mainly in the evening. Three types of battery systems are investigated: current lithium-ion (Li-ion) batteries, repurposed (second-life) Li-ion batteries previously used in electric vehicles and future Li-ion batteries with an increased energy density and environmentally optimized production. The environmental impacts of the PV system and the battery storage systems are quantified by means of life cycle assessment (LCA). The main components and processes contributing to the environmental impacts are identified. The results are then used to calculate the average environmental impacts of the electricity supplied to the cloakroom and club building depending on the presence of a battery storage system and the battery type.

2 METHODOLOGY AND SCOPE

2.1 PV system

The functional unit of the LCA of the manufacture and disposal of the PV system is 1 kWp maximum power output. The LCA of the PV system comprises the manufacture of the PV modules, the slanted-roof construction, the electric installation and the inverter. The planned PV system is integrated in the roof of the cloakroom and club building. Since the LCA of the PV system is done as part of a building LCA (not shown in this paper), an allocation of the environmental impacts of the PV system to the functions of electricity generation and weather protection is not needed. 100 % of the environmental impacts of the PV system are attributed to electricity in the assessments presented in this paper.

2.2 Battery storage systems

The functional unit of the LCA of the manufacture and disposal of the battery systems is 1 kWh storage capacity. In the LCA of battery storage systems the manufacture or repurposing (in case of second-life batteries) of Li-ion batteries including the production of the battery management system, the cooling system, the battery cells and the packaging are taken into account. The manufacture of electronics for battery control, the wiring and the system housing are also included.

2.3 Electricity consumption

The electricity consumed by the cloakroom and club building is assessed using the functional unit of 1 year of electricity consumption. The annual electricity demand may be covered from the grid, the PV system and the battery system. According to the technical bulletin SIA 2040 of the Swiss Society of Architects and Engineers that sets the rules for building LCAs, the environmental impacts of PV electricity supplied to the grid are subtracted from the impacts of the construction of the PV system using the same environmental intensity [11]. This means that the environmental impacts of the PV system are only accounted for to the degree the PV electricity is self-consumed. The environmental impacts of the battery storage system are fully allocated to the building based on the assumption that the battery is used only by the building itself and does not provide any services to the grid. The analysis in this paper is limited to the electricity consumption of the cloakroom and club building and does not include other parts of building LCAs such as construction, heating energy and induced mobility.

2.4 Background data and indicators

The life cycle inventories (LCIs) of the PV system and the battery systems are linked to KBOB life cycle inventory data DQRv2:2016 [12], which is an extensively updated version of ecoinvent data v2.2 [13]. The life cycle impact assessment results are shown using the indicator greenhouse gas emissions [14]. The indicators cumulative energy demand [15] and total environmental impacts according to the Swiss ecological scarcity method 2013 [16] were also analysed but are not shown here. In general, the results are similar for the three indicators analysed [17]. The analyses were carried out using SimaPro v8.4.0 [18].

3 LIFE CYCLE INVENTORIES

3.1 PV system

The planned PV system of the cloakroom and club building is integrated in the roof and has a maximum power output of 60 kWp. It consists of 230 multi-crystalline silicon (multi-Si) framed PV modules with a maximum power output of 161 Wp/m^2 [10]. In accordance with the draft Product Environmental Footprint Category Rules (PEFCR) for PV Electricity [19] it is assumed that 1 % of the modules supplied is rejected and that 2 % of the PV modules are defective during the time of operation. The replacement of these modules is accounted for. Generic LCIs are used to model the inverter, the electric installation and the integrated slanted-roof construction [20]. It is assumed that the life time of the PV system is 25 years and that the inverter is replaced once within this time. The annual yield of the PV modules at the location of the cloakroom and club building is 950 kWh/kW$_p$, including degradation [10].

3.2 Battery storage systems

3.2.1 Current Li-ion batteries

The LCIs of battery storage systems are compiled based on information reported in literature and obtained in expert interviews. Li-ion batteries are characterised by a high energy density, a high efficiency and a long life time [9] and are frequently used for mobile and stationary applications. There is a variety of Li-ion batteries with the chemical composition of the cathode as the main difference. Changes in the cathode composition are used to influence important features such as the energy and power density, the life time, the operational safety and the costs [21, 22]. Li-ion batteries for stationary applications often have an iron-phosphate ($FePO_3$) cathode, which ensures a long life time and high operational safety [23].

Very detailed and transparent LCI data of battery production were reported in [24], which are based on primary data from a Norwegian manufacturer of Li-ion batteries with a nickel-cobalt-manganese oxide (NCM) cathode. Due to their high energy density, these batteries are frequently used in electric vehicles. The LCIs based on [24] and published in [25] are used for the storage system with current Li-ion batteries since primary data on the production of $LiFePO_3$ batteries were not available. A previous study also revealed that the contribution of the cathode to the total environmental impacts of Li-ion batteries is small [26].

The analysed LiNCM battery weighs 253 kg and has a storage capacity of 26.6 kWh, which yields an energy density of 105 Wh/kg [24]. The battery cells contribute 60 % to the total battery weight and are composed of a $Li(Ni_{0.33}Co_{0.33}Mn_{0.33})O_2$ cathode and an anode based on graphite. The separator is made of a porous polyolefin film and lithium-hexafluorophosphate is used for the electrolyte. Battery cell production is very energy-intensive with an electricity demand of 22.7 kWh/kg. This electricity is supplied by an Eastern Asian mix since battery cells are mainly produced in that region [24, 25]. Besides the battery cells, the LiNCM battery also requires a battery management system, a battery cooling system and packaging [24]. For Li-ion batteries used in stationary storage systems, the batteries are packed in a housing made of steel, connected to each other with cables and equipped with a control unit [26].

The storage capacity of Li-ion batteries continually decreases with use and ageing. The life time of Li-ion batteries is uncertain, which is partly due to its dependence on temperature and the depth of discharge. For current batteries, the life time is assumed to be 15 years or 5'000 charging cycles [9, 22, 27].

3.2.2 Future Li-ion batteries

The LCI of the future battery system is based on the LCI of current battery systems [24, 25] and modified in relevant aspects. Future Li-ion batteries are projected to reach an energy density, which is approximately twice as high compared to today's batteries [9, 21, 22, 28]. This increase is taken into account by assuming an energy density of future Li-ion batteries of 210 Wh/kg. There is also intensive research on new materials to be used in electrodes and electrolytes [29]. For instance, the graphite based anodes are expected to be increasingly blended with silicon in order to enhance the capacity and some of the cobalt and manganese contained in the cathode is expected to be replaced by nickel (shift from $Ni_{0.33}Co_{0.33}Mn_{0.33}$ to $Ni_{0.6}Co_{0.2}Mn_{0.2}$ or $Ni_{0.8}Co_{0.1}Mn_{0.1}$) [22, 30]. However, LCIs for the future development of these components are not available and a preliminary analysis of current batteries revealed a minor contribution of the anode, the cathode and the electrolyte to the environmental impacts. These potential changes in materials are therefore neglected in the LCA of future Li-ion battery storage systems.

Instead it is assumed that battery manufacturers will focus on the most important materials and processes in order to reduce the environmental impacts of future Li-ion batteries. The electricity demand of battery cell production and the copper used in the anode were found to be particularly relevant for the environmental footprint of the battery. Production optimisations in these areas are taken into account in the LCIs of future Li-ion batteries manufactured by environmentally conscious companies. A reduction in the electricity demand of future battery cell production to 15.0 kWh/kg was reported to be achievable in the medium term [28]. It is assumed that this electricity demand will be covered by large-scale open ground PV systems with multi-Si modules. The improvement of the module conversion efficiency is approximated by a PV installation located in Spain, where the annual yield is about 20 % higher compared to the location of important Asian battery producers [31]. For the production of anodes used in future Li-ion batteries it is assumed that only secondary copper is used, which causes significantly lower environmental impacts compared to the production mix of primary copper. The housing, cabling and electronics required for stationary storage systems with future Li-ion batteries decreases as the energy intensity of the batteries increases. The demand of these components is therefore reduced by a factor of two.

The life time of future Li-ion batteries is projected at approximately 10'000 charging cycles or up to 25 years [9, 27, 32]. In the present study, a life time of 20 years is assumed for storage systems with future Li-ion batteries.

3.2.3 Second-life Li-ion batteries

Stationary storage systems with repurposed Li-ion batteries are an alternative to the recycling of the materials in used traction batteries. Electric vehicles usually have stronger requirements regarding the energy density, so batteries are changed when they reach about 70-80 % of the initial capacity. Some of these batteries may be suited for a second use in stationary storage systems. However, there are several logistic, technical and economic challenges of sorting and repurposing used batteries [33-35]. An ongoing pilot project tried to overcome these difficulties by only using batteries that were previously used in electric scooters of the Swiss Post [36]. Primary data from this project are used to compile an LCI of second-life Li-ion batteries.

The effort of sorting the used batteries from electric scooters is negligible since only one type of batteries is collected. Each battery is equipped with a newly developed electronic print, which allows the individual management of each battery cell. The possibility to bypass cells that are defective or have a low storage capacity substantially facilitates the testing of the used batteries [36]. After their first use in electric scooters, the batteries retain an economic value. The production efforts of Li-ion batteries therefore need to be allocated to the two use phases, which is done according to their economic value based on current prices. According to this allocation key the second use of the Li-ion batteries in stationary storage systems bears one seventh of the environmental impacts of the initial battery production. The price paid for the used batteries from electric scooters may significantly change in the future, which would affect the share of environmental impacts of battery production to be allocated to the second use. The battery production is modelled using the LCIs of current Li-ion batteries [24, 25].

The energy density of the second-life battery system is 84 Wh/kg (approximately 80 % of the capacity of new batteries [33-35]). Performance tests of the second-life Li-ion batteries developed in the pilot project [36] are being carried out, but results were not available to the time this study was carried out. It is assumed that the stationary storage system with second-life Li-ion batteries is operated for 10 years.

3.3 Electricity consumption of the cloakroom and club building

Different options are considered in the LCA of the yearly electricity consumption of the cloakroom and club building: 1. grid supply only (no PV system and no battery storage), 2. 60 kWp PV system without battery storage, 3. 60 kWp PV system and 100 kWh battery storage. The last option is further divided into three alternatives: 3a. current Li-ion battery, 3b. future Li-ion battery, 3c. second-life Li-ion battery. The projected electricity demand of the planned cloakroom and club building is 106 MWh/a. The consumption profile exhibits a distinctive evening peak, which is mainly caused by the electricity demand of the floodlights used for the surrounding soccer fields (see Figure 1).

The annual electricity demand of the planned building is covered from the grid, the PV system (options 2 and 3) and the battery system (option 3). The self-consumption share of PV electricity in option 2 is estimated at 31 % and the self-sufficiency share is 18 %. The self-consumption share and the self-sufficiency share increase to 62 % and 35 % respectively, if the 60 kWp PV system is complemented by a 100 kWh battery storage (option 3). Hence even with a battery storage system, almost two thirds of the yearly electricity demand are covered by the grid. The grid electricity is modelled using two different mixes, namely the Swiss supply mix and the renewable electricity mix of the Zurich City Administration.

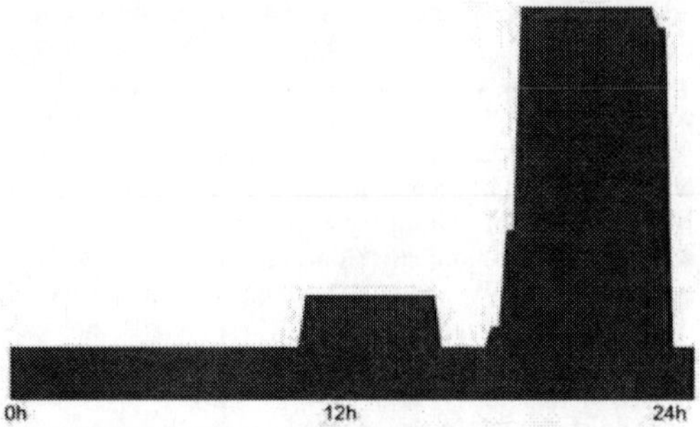

Figure 1: Forecast electricity consumption profile during a typical day of the cloakroom and club building that is currently planned by the City of Zurich.

4 RESULTS AND DISCUSSION

4.1 PV system

The greenhouse gas emissions of the PV system planned for the cloakroom and club building are shown in Figure 2 and compared to a similar roof-integrated PV system with multi-Si modules based on a generic LCI of KBOB life cycle inventory data DQRv2:2016 [12]. The results are normalised to the functional unit of 1 kWp maximum power output. The difference in the greenhouse gas emissions of the two PV systems is very small, which is mainly due to the use of generic LCI data and default assumptions for the PV system of the planned cloakroom and club building. The multi-Si PV modules foreseen for this building have a higher conversion efficiency (16.1 %) compared to the assumed efficiency in the LCI of multi-Si PV modules in KBOB life cycle inventory data DQRv2:2016 (14.7 %) [12, 37]. However, the lower environmental impacts due to the higher module conversion efficiency are largely compensated by the aluminium frame that is foreseen for the modules of the cloakroom and club building but not included in the generic LCI.

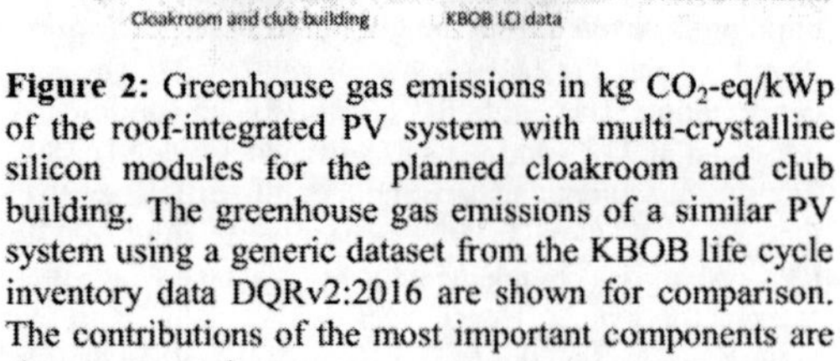

Figure 2: Greenhouse gas emissions in kg CO_2-eq/kWp of the roof-integrated PV system with multi-crystalline silicon modules for the planned cloakroom and club building. The greenhouse gas emissions of a similar PV system using a generic dataset from the KBOB life cycle inventory data DQRv2:2016 are shown for comparison. The contributions of the most important components are shown separately.

The supply of the PV modules contributes approximately 79 % to the total greenhouse gas emissions of the planned PV system of the cloakroom and club building. The shares of the roof-integrated substructure, the inverter and the electric installation are 9 %, 8 % and 3 %, respectively. The greenhouse gas emissions caused by the transport of the PV modules and the BOS component and by the construction of the PV system are negligible.

4.2 Battery storage systems

The greenhouse gas emissions of battery systems with current, future and second-life Li-ion batteries per kWh storage capacity are depicted in Figure 3. The contributions of the most relevant materials and processes are shown separately. The greenhouse gas emissions of the storage system with current LiNCM batteries (185 kgCO_2-eq/kWh) are significantly higher than the greenhouse gas emissions of storage systems with future and second-life LiNCM batteries (49.4 kgCO_2-eq/kWh and 80.6 kgCO_2-eq/kWh, respectively).

The manufacture of the Li-ion batteries contributes 79 % to greenhouse gas emissions of current battery storage systems. The electricity demand of Li-ion battery cell production (48 %) is by far the most relevant process for the greenhouse gas emission of current battery systems, followed by the cathode (9 %) and the anode (5 %). The remaining processes and components of current Li-ion battery manufacturing include for instance the battery cooling system, the packaging and the electricity demand of battery assembly. The housing, electronics and cabling used for the stationary storage system and the transport of the battery system to the place of installation contribute 21 % to the total greenhouse gas emissions of battery systems with current LiNCM batteries.

Future Li-ion batteries are assumed to have an energy density, which is twice as high compared to current Li-ion batteries. In addition, process optimisations in the electricity demand of battery cell production, the electricity mix and in the supply of copper for the anode are taken into account. The magnitude of the change in greenhouse gas emissions due to the higher energy density and the process optimisations is similar. The share of future LiNCM battery production in the total greenhouse gas emissions amounts to 60 %, whereby the cathode (17 %), the anode (9 %) and the remaining battery components including assembly (20 %) are particularly relevant. The electricity demand of future battery cell production contributes 3 % to the total greenhouse gas emissions of the battery storage system analysed. The housing, electronics, cabling and transport cause about 40 % of the greenhouse gas emissions of storage systems with future LiNCM batteries.

The greenhouse gas emissions of the storage system with repurposed Li-ion batteries are approximately 56 % lower than the emissions of the current battery system. The main reason for this difference is that the reused batteries in a stationary storage system bear only one seventh of the environmental impacts of battery manufacturing based on the economic allocation using current prices. The share of battery manufacturing in the total greenhouse gas emissions of storage systems with second-life Li-ion batteries is 32 %. The printed wiring board and the energy required for repurposing of used batteries contribute 19 % to the greenhouse gas emissions. The remaining 49 % of the total greenhouse gas emissions of storage systems with repurposed

LiNCM batteries are caused by the housing, electronics, cabling and transport.

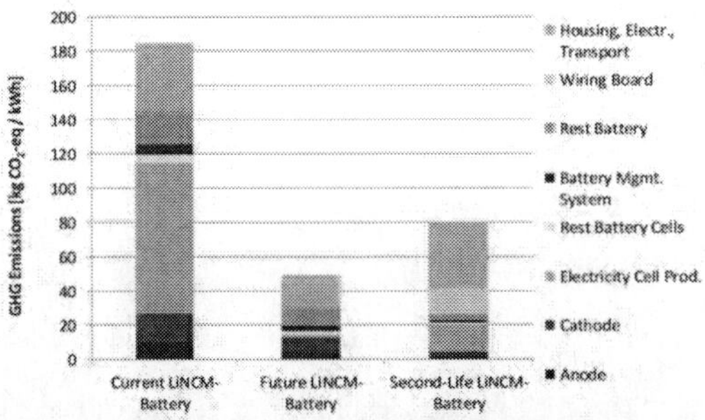

Figure 3: Greenhouse gas emissions in kg CO_2-eq/kWh of storage systems with current, future and second-life Lithium-ion batteries with a Nickel-Cobalt-Manganese (NCM) cathode. The contributions of the most important components and processes are shown separately.

4.3 Electricity consumption of the cloakroom and club building

The greenhouse gas emissions of the different options to cover the yearly electricity demand of the planned cloakroom and club building are shown in Figure 4 and Figure 5. It becomes evident that the results strongly depend on the grid electricity mix used to model the share of electricity demand not supplied by the PV system and the battery storage system.

With the Swiss electricity supply mix, the difference in greenhouse gas emissions of the five options to cover the electricity demand of the cloakroom and club building is small (Figure 4). The greenhouse gas emissions amount to 14.8 tCO_2-eq/a if the electricity demand is completely covered by the average Swiss supply mix. These emissions are reduced by about 10 % to 13.4 tCO_2-eq/a if a 60 kWp PV system is installed on the planned cloakroom and club building. In this case, 31 % of the electricity generated by the PV system are self-consumed by the cloakroom and club building. The electricity demand in this option is covered by 18 % PV electricity (self-sufficiency share) and 82 % grid electricity. The reduction in greenhouse gas emissions occurs because the emission intensity of electricity generated by the planned PV system is lower compared to the Swiss electricity supply mix.

The complementation of the 60 kWp PV system by a 100 kWh battery storage system leads to a significant increase of the self-consumption share (62 %) and the self-sufficiency share (35 %). However, the greenhouse gas emissions of electricity consumption are similar to those of the "PV only" option. The greenhouse gas emissions of battery production outweigh the difference in emission intensity of PV electricity and the Swiss supply mix. If future increases in energy density and production optimisations are taken into account for the storage system, the greenhouse gas emissions of the electricity consumption of the cloakroom and club building decrease to 12.3 tCO_2-eq/a (minus 8 % compared to the option with a PV system and a storage system with current Li-ion batteries). A further reduction may be achieved by future improvements in the PV module conversion efficiency and the PV module production processes. These developments were not considered in the present study. The greenhouse gas emissions of the option with a PV system and a storage system with second-life Li-ion batteries amount to 12.8 tCO_2-eq/a and are in between the emissions of the options with current and with future battery systems.

The Zurich City Administration uses a renewable electricity mix, which is mainly generated in hydropower plants and contains shares of wind power and photovoltaic electricity. If the grid electricity used by the planned cloakroom and club building is supplied by this renewable electricity mix, the absolute greenhouse gas emissions of all options considered are substantially lower (see Figure 5). In this case, the full coverage of the electricity demand by grid electricity is environmentally favourable with greenhouse gas emissions of 1.2 tCO_2-eq/a.

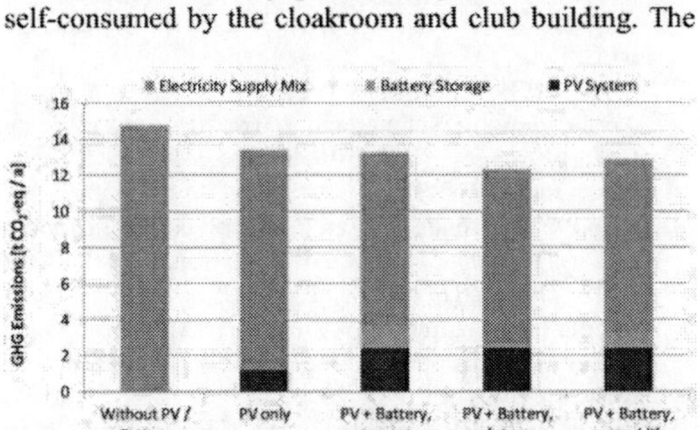

Figure 4: Greenhouse gas emissions in t CO_2-eq/a of the electricity consumed by the planned cloakroom and club building for the options without PV system and without battery system, with a 60 kWp PV system only (self-consumption share: 31 %, self-sufficiency share: 18 %) and with a 60 kWp PV system and 100 kWh storage systems using different types of Lithium-ion batteries (self-consumption share: 62 %, self-sufficiency share: 35 %). The electricity supplied from the grid is modelled with the **Swiss supply mix**. The total electricity consumption is projected at 106 MWh/a.

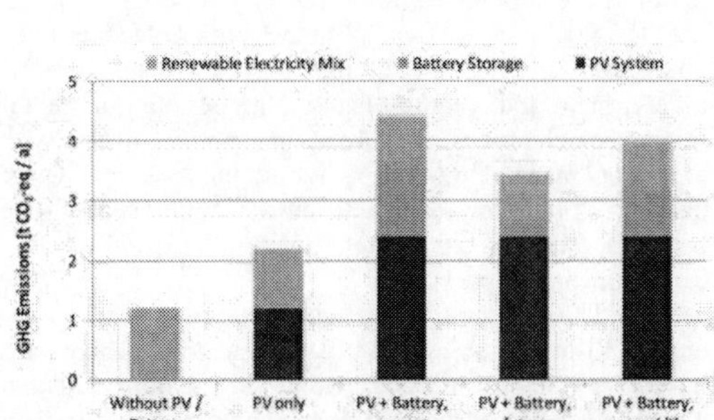

Figure 5: Greenhouse gas emissions in t CO_2-eq/a of the electricity consumed by the planned cloakroom and club building for the options without PV system and without battery system, with a 60 kWp PV system only (self-consumption share: 31 %, self-sufficiency share: 18 %) and with a 60 kWp PV system and 100 kWh storage systems using different types of Lithium-ion batteries (self-consumption share: 62 %, self-sufficiency share: 35 %). The electricity supplied from the grid is modelled with the **renewable electricity mix** used by the Zurich City Administration (ewz Ökopower). The total electricity consumption is projected at 106 MWh/a.

Due to the high share of hydropower, the renewable electricity mix used by the Zurich City Administration causes fewer greenhouse gas emissions per kWh than PV electricity. Hence, the installation of a PV system results in an increase of greenhouse gas emissions of electricity consumption to 2.2 tCO_2-eq/a. A storage system with current Li-ion batteries further doubles these emissions (4.4 tCO_2-eq/a) because of the greenhouse gas emissions caused by battery production on the one hand and the higher self-sufficiency share on the other hand. The greenhouse gas emissions of the electricity consumption of the planned cloakroom and club building for the options with a PV system and a storage system with future or second-life Li-ion batteries are 3.4 tCO_2-eq/a and 4.0 tCO_2-eq/a, respectively.

5 CONCLUSION

The cloakroom and club building that is currently planned by the Office for Building Engineering of the City of Zurich was used as a showcase to analyse the environmental footprint of PV-battery systems. The forecast electricity consumption profile of the building is characterised by a distinctive evening peak, which is caused by the floodlights of the surrounding soccer fields. In this case, the complementation of the planned 60 kWp PV system with a 100 kWh battery storage system leads to a significant increase of the self-consumption share of PV electricity from 31 % to 62 %. However, even with a PV system and a battery storage approximately two thirds of the electricity consumption are covered with grid electricity. The complementary grid electricity mix strongly influences the environmental impacts of the electricity used by the cloakroom and club building. The installation of a PV system and a battery storage system leads to a 10-17 % reduction in greenhouse gas emissions compared to the full coverage of the electricity demand by the average Swiss supply mix. The addition of a storage system with current Li-ion batteries does not yield any further reduction compared to the "PV only" option if the complementary electricity supply is covered with the Swiss supply mix. With the renewable electricity mix of the Zurich City Administration, the installation of a PV system and a battery leads to higher environmental impacts of electricity consumed, irrespective of the type of battery used. However, a future increase in energy density, production optimisations and second-life batteries bear a significant potential to reduce the environmental impacts.

The LCIs of the PV system, the storage system with current Li-ion batteries and the grid electricity mixes are based on reliable and transparent information sources. The data quality of these LCIs is considered as fair. The LCIs of storage systems with future and second-life Li-ion batteries required assumptions on key parameters such as the energy density and the life time. The environmental impacts are very sensitive to changes in these key parameters, which renders the results of the storage systems with future and second-life batteries more uncertain. The LCA of the yearly electricity consumption of the cloakroom and club building is based on a forecast consumption profile and the corresponding self-consumption share and self-sufficiency share. These parameters also have a high influence on the environmental impacts. These uncertainties need to be taken into consideration when interpreting the results but they are not expected to change the overall conclusions drawn in this paper.

6 ACKNOWLEDGMENT

This study was financially supported by the Office for Building Engineering of the City of Zurich.

7 REFERENCES

1. Stadt Zürich, Gemeindeordnung der Stadt Zürich. 2017.
2. Spreng DT & Semadeni M, Energie, Umwelt und die 2000 Watt Gesellschaft. 2001, CEPE working paper Nr. 11, CEPE Centre for Energy Policy and Economics Swiss Federal Institute of Technology: Zürich. p. Online Datei. Retrieved from e-collection.ethbib.ethz.ch/show?type=incoll&nr=420.
3. UGZ, Primärenergiebilanz Stadt Zürich. 2018, Umwelt- und Gesundheitsschutz Zürich (UGZ), Stadt Zürich: Zürich. https://www.stadt-zuerich.ch/content/dam/stzh/gud/Deutsch/UGZ/Umwelt%20%26%20Energie/Energie%20in%20Zahlen/2000-Watt-Indikatoren/%3e%20Dokumente%20und%20Publikationen/Prim%c3%a4renergiebilanz_Daten.pdf.
4. UGZ, Treibhausgasemissionen Stadt Zürich. 2018, Umwelt- und Gesundheitsschutz Zürich (UGZ), Stadt Zürich: Zürich. https://www.stadt-zuerich.ch/content/dam/stzh/gud/Deutsch/UGZ/Umwelt%20%26%20Energie/Energie%20in%20Zahlen/2000-Watt-Indikatoren/%3e%20Dokumente%20und%20Publikationen/Treibhausgasemissionen_Daten.pdf.
5. Gessler R & Volland B, Roadmap 2000-Watt-Gesellschaft. 2016, Stadt Zürich, Umweltdelegation des Stadtrats: Zürich.
6. Prognos, Die Energieperspektiven für die Schweiz bis 2050; Energienachfrage und Elektrizitätsangebot in der Schweiz 2000 - 2050. 2012, Bundesamt für Energie, BFE: Bern.
7. IEA-PVPS, National Survey Report of PV Power Applications in Switzerland 2016. 2017, International Energy Agency (IEA) Photovoltaic Power Systems Programme (PVPS).
8. IEA-PVPS, Trends 2017 in Photovoltaic Applications - Survey Report of Selected IEA Countries between 1992 and 2016. 2017, PVPS T1 - 23 : 2013, International Energy Agency (IEA) Photovoltaic Power Systems Programme (PVPS). Retrieved from www.iea-pvps.org.
9. Swissolar, PV-Anlagen mit Batterien. 2016, Schweizerischer Fachverband für Sonnenenergie, Swissolar: Zürich.
10. Stettler Y, Garderoben- und Clubgebäude, Zürich Höngg. Projektbeschreibung Photovoltaik Vorprojekt. 2018, Basler & Hofmann AG: Zürich.
11. SIA, Merkblatt 2040: SIA-Effizienzpfad Energie. 2017, Schweizerischer Ingenieur- und Architektenverein (SIA): Zürich. p. 28.
12. KBOB, eco-bau, & IPB, KBOB Ökobilanzdatenbestand DQRv2:2016; Grundlage für die KBOB-Empfehlung 2009/1:2016: Ökobilanzdaten im Baubereich, Stand 2016. 2016, Koordinationskonferenz der Bau- und

Liegenschaftsorgane der öffentlichen Bauherren c/o BBL Bundesamt für Bauten und Logistik. Retrieved from www.ecoinvent.org.

13. ecoinvent Centre, ecoinvent data v2.2, ecoinvent reports No. 1-25. 2010, CD-ROM, Swiss Centre for Life Cycle Inventories: Duebendorf, Switzerland. Retrieved from www.ecoinvent.org.
14. IPCC, The IPCC fifth Assessment Report - Climate Change 2013: the Physical Science Basis. 2013, Working Group I, IPCC Secretariat: Geneva, Switzerland.
15. Frischknecht R, Wyss F, Büsser Knöpfel S, Lützkendorf T, et al., Cumulative energy demand in LCA: the energy harvested approach. The International Journal of Life Cycle Assessment, 2015. 20(7): p. 957-969. http://dx.doi.org/10.1007/s11367-015-0897-4.
16. Frischknecht R & Büsser Knöpfel S, Swiss Eco-Factors 2013 according to the Ecological Scarcity Method. Methodological fundamentals and their application in Switzerland. 2013, Environmental studies no. 1330, Federal Office for the Environment: Bern. Retrieved from http://www.bafu.admin.ch/publikationen/publikation/01750/index.html?lang=en.
17. Stolz P, Frischknecht R, Kessler T, & Züger Y, Ökobilanz PV-Anlage und Batterie für das Garderoben- und Clubgebäude in Zürich Höngg. 2018, Amt für Hochbauten, Fachstelle nachhaltiges Bauen, Stadt Zürich: Zürich.
18. PRé Consultants, SimaPro 8.4.0. 2017: Amersfoort, NL.
19. Technical Secretariat of the PEF Pilot "Photovoltaic Electricity Generation", Product Environmental Footprint Category Rules: Production of Photovoltaic Modules used in Photovoltaic Power Systems for Electricity Generation (NACE/CPA class 27.90 "Manufacturing of other electrical equipment"). 2016.
20. Frischknecht R, Itten R, Sinha P, de Wild Scholten M, et al., Life Cycle Inventories and Life Cycle Assessments of Photovoltaic Systems. 2015, International Energy Agency (IEA) PVPS Task 12.
21. Helms H, Julius J, Claudia K, Giegrich J, et al., Weiterentwicklung und vertiefte Analyse der Umweltbilanz von Elektrofahrzeugen. 2016, Umweltbundesamt.
22. de Haan P & Zah R, Chancen und Risiken der Elektromobilität in der Schweiz. 2013, vdf Hochschulverlag AG: Zürich.
23. EnergieSchweiz, Stationäre Batteriespeicher in Gebäuden (Dokument zur Vernehmlassung). 2018, EnergieSchweiz, Bundesamt für Energie BFE: Bern.
24. Ellingsen LA-W, Majeau-Bettez G, Singh B, Kumar Srivastava A, et al., Life Cycle Assessment of a Lithium-Ion Battery Vehicle Pack. Journal of Industrial Ecology, 2014. 18(1): p. 113-124.
25. Stolz P, Messmer A, & Frischknecht R, Life Cycle Inventories of Road and Non-Road Transport Services. 2016, treeze Ltd.: Uster CH.
26. Messmer A & Frischknecht R, Ökobilanz von PV-Batterie-Systemen. 2015, treeze Ltd.: Uster.
27. EASE / EERA, European Energy Storage Technology Roadmap. 2017 Update. 2017, European Association for Storage of Energy and European Energy Research Alliance: Brussels.
28. Cox B & Bauer C, Environmental assessment of current and future passenger vehicles in Switzerland. 2018, Paul Scherrer Institut PSI, commissioned by the Swiss Federal Office for Energy SFOE: Villigen.
29. Fraunhofer ISI, Gesamt-Roadmap Lithium-Ionen-Batterien 2030. 2015, Fraunhofer-Institut für System- und Innovationsforschung ISI: Karlsruhe.
30. Romare M & Dahllöf L, The Life Cycle Energy Consumption and Greenhouse Gas Emissions from Lithium-Ion Batteries. 2017, IVL Swedish Environmental Research Institute: Stockholm.
31. Pérez-López P, Gschwind B, Blanc P, Frischknecht R, et al., ENVI-PV: an interactive Web Client for multi-criteria life cycle assessment of photovoltaic systems worldwide. Progress in Photovoltaics: Research and Applications, 2016: p. 1-15.
32. EASE / EERA, European Energy Storage Technology Roadmap towards 2030. Technical Annex. 2017, European Association for Storage of Energy and European Energy Research Alliance: Brussels.
33. Schaufenster Elektromobilität, Studie: Second-Life-Konzepte für Lithium-Ionen-Batterien aus Elektrofahrzeugen. Analyse von Nachnutzungsanwendungen, ökonomischen und ökologischen Potenzialen. 2016, Begleit- und Wirkungsforschung Schaufenster Elektromobilität (BuW): Frankfurt am Main.
34. Neubauer J, Smith K, Wood E, & Pesaran A, Identifying and Overcoming Critical Barriers to Widespread Second Use of PEV Batteries. 2015, National Renewable Energy Laboratory NREL: Golden, CO.
35. Reid G & Julve J, Second Life Batteries As Flexible Storage For RenewableEnergies. 2016, Bundesverband Erneuerbare Energie e.V. (BEE): Berlin.
36. Sattler M, Hausammann B, & Held M, Stromspeichersystem mit Second-Life Akkumulatoren (SL-Speicher). Jahresbericht vom 26. Juni 2017. 2017, Bundesamt für Energie BFE: Bern.
37. Stolz P, Frischknecht R, Wyss F, & de Wild Scholten M, PEF screening report of electricity from photovoltaic panels in the context of the EU Product Environmental Footprint Category Rules (PEFCR) Pilots, version 2.0. 2016, treeze Ltd. commissioned by the Technical Secretariat of the PEF Pilot "Photovoltaic Electricity Generation": Uster, Switzerland.

INFLUENCE OF THE MODULE TEMPERATURE ON THE PERFORMANCE AND EL-IMAGE OF PRECRACKED PV-MODULES

Claudia Buerhop[1], Marcus Krause[1], Thilo Winkler[1], Jens Hauch[1], Christian Camus[1,2], Christoph J. Brabec[1,2,3]
[1]ZAE Bayern e.V., D-91058 Erlangen, Immerwahrstraße 2,
[2]i-MEET, FAU Erlangen-Nürnberg, D-91058 Erlangen, Martensstraße 7
[3] Helmholtz-Institut Erlangen-Nürnberg (HI ERN), D-91058 Erlangen, Egerlandstraße 3

ABSTRACT: EL-imaging is a valuable tool to analyze PV-modules for the presence of cracked cells. Normally, these inspections are done in air-conditioned labs or at cool nights outdoors. However, the PV-modules are supposed to generate energy outdoors in the field. Therefore, they operate at significantly elevated temperatures which are much higher than the temperatures at the inspection time. In order to investigate the temperature influence on the performance of precracked PV-modules, EL-images and IV-curves were recorded of warm and mechanically loaded modules. The study illustrates that the performance of cracked cells changes with temperature and pressure. There is the tendency that modules at elevated temperatures perform better than those at low temperatures under the same loading situation.

Keywords: Electroluminescence, temperature, cracks, mechanical loading, performance

1 INTRODUCTION

Currently, Electroluminescence (EL)-imaging is the most attractive method used to detect and visualize PV-modules with precracked cells in the field. How these precracked PV-modules perform during their remaining lifetime is of great interest and studied in current research work [1-4]. The focus is on wind and snow loads. There have been effective techniques developed during the last years for carrying out the EL-measurements at night. However, how do these precracked PV-modules perform at the elevated module operating temperatures? Temperature differences of 60 K and even more are reasonable, especially in summer: hot at daytime and cold at night [2].

For the first time we present our performance study of the precracked PV-modules in which the modules are mechanically loaded at elevated temperatures. The temperatures were chosen close to operating temperatures of the modules in the field, up to 85°C. The precracked modules had cracks and fractures of differing severity, "crack-poor" means only a small number of cells with existing cracks and "fracture-rich" describes that many cells with open cracks are present and potentially more cracked cell will be generated. The modules were loaded uniformly up to 1000 Pa which represents wind pressure of strong storms.

2 EXPERIMENTAL PROCEDURE

For this investigation precracked, crystalline PV-modules (230 W) were investigated. According to the data sheet, the temperature coefficient for P_{mpp} is $\alpha = -0.47\%/K$, temperature coefficient for I_{sc} $\beta = 0.045\%/K$, and the temperature coefficient for V_{oc} $\gamma = -0.34\%/K$. The modules were previously loaded up to 5,000 Pa using the mechanical load test-setup, as described elsewhere [2]. They had differing numbers of cells with cracks and fractures (cells with darkened cell areas in the EL-image). The cracks were mostly identified in the second and fifth long cell rows. Three tested modules are discussed in detail, see Table 1 for more information about the module characteristics.

Table 1: Description of the modules investigated in detail at $T = 25°C$ and $p = 0$ Pa, number of cracked cells and fractured cell (dark or black area in EL-image) and characteristics of the IV-curve

	Module A	*Module B*	*Module C*
Cracked cells	4	12	6
fractured cells	0	4	12
IV-curve	normal	normal	with step

The test facility applies the pressure by vacuum pumps and air blowers. Thus, underpressure as well as over-pressure can be used to simulate snow, wind loads and wind suction, respectively. Pressures p between 200 and 1000 Pa simulating rather moderate wind speeds were applied at steps of 200 Pa. The module temperature was increased using a halogen lamp field. Thus, module temperatures of T_{mod} = 25, 35, 45, 55, 65, 75, and 85°C were realized. The temperature was recorded with a PT100 attached to the module's backside. Strain gauges recorded the displacement at several positions on the glass surface.

Simultaneous EL-imaging and IV-measurements during loading of the modules at various temperatures were taken and evaluated.

3 RESULTS AND DISCUSSION

The impact of module temperature and moderate loading on the EL-image and the IV-curve is shown for three modules exemplarily in Figure 1, Figure 2, and Figure 3. Module A, shown in Figure 1, has only a few cells with closed cracks exhibiting no dark or black area. Neither temperature increase nor pressure rise cause significant changes. In contrary, modules B (Figure 2), and module C (Figure 3) exhibit many cells with cracks and fractures with dark areas. Upon loading, the cracks open and the number of dark areas increases, as seen elsewhere [2-4]. Increasing the temperature influences the EL-intensity distribution hardly visible by eye, as will be shown quantitatively below.

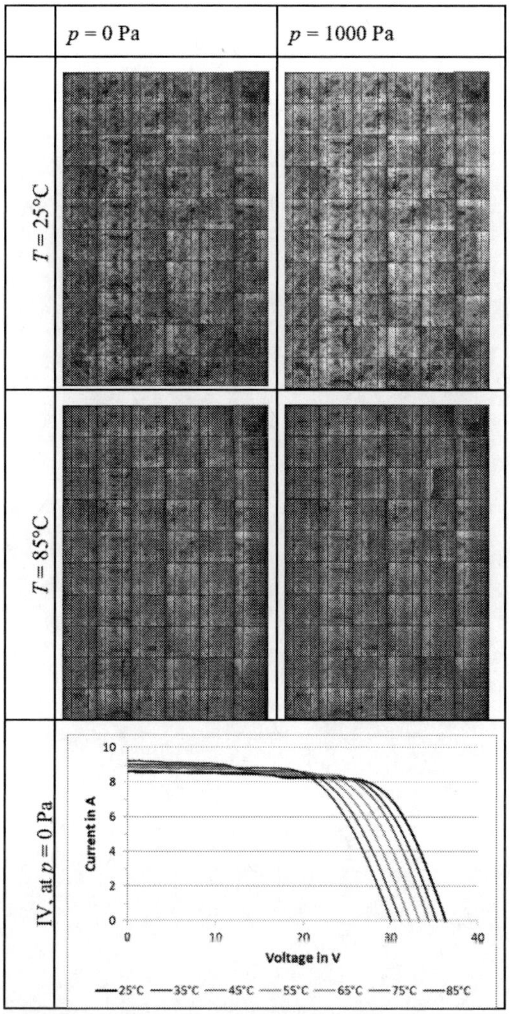

Figure 1: EL-images of PV-module A at different module temperatures T_{mod} and mechanical load pressures p. Frames mark cells with cracks [5].

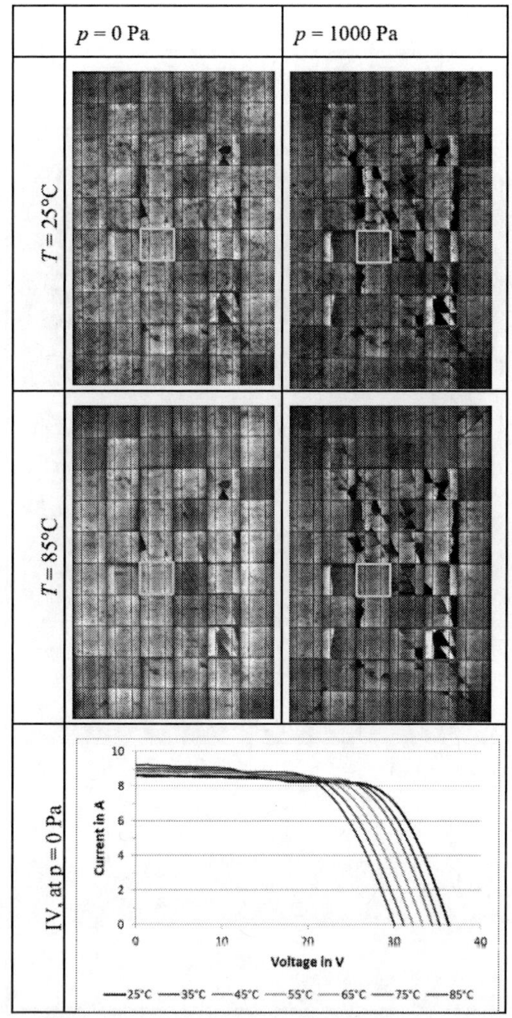

Figure 2: EL-images of PV-module B at different module temperatures T_{mod} and mechanical load pressures p. Green and red frames mark cells with cracks discussed in more detail, the yellow frame indicates the reference cell [5].

Module C with many cracks exhibits intensity changes in the EL-image as well as obvious differences in the IV-curves. While at low temperatures a step-shaped IV-curve is visible, at rather high temperature, above 45°C, the step disappears. The steps are caused by fractured cells in two substrings, whose electrical contacts recover at higher temperatures.

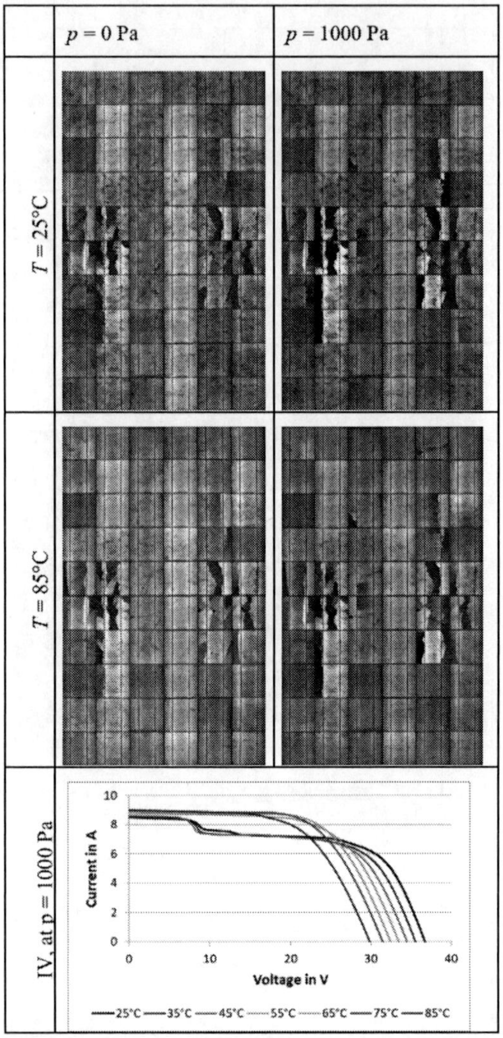

Figure 3: EL-images of PV-module C at different module temperatures T_{mod} and mechanical load pressures p. Green frames mark cells with cracks discussed in more detail, the yellow frame indicates the reference cell [5].

Studying the IV-curves includes observing the changes of open-circuit voltage V_{oc}, short-circuit current I_{sc} and maximum power point power P_{mpp} with temperature for p = 0 Pa and p = 200 Pa, as presented in Figure 4. A low pressure of p = 200 Pa is chosen because such wind loads are more likely. V_{oc} decreases with increasing temperature, as expected. The fairly good module A shows similar behavior for loaded and unloaded conditions, I_{sc} increases and P_{mpp} decreases. In contrast, the measured data for module B and C fluctuate considerably while they decrease. The increase of I_{sc} is stronger than expected from the temperature coefficient stated in the data sheet. The power P_{mpp} however decreases less than expected. The data points are above the calculated values (dashed line) at high temperatures.

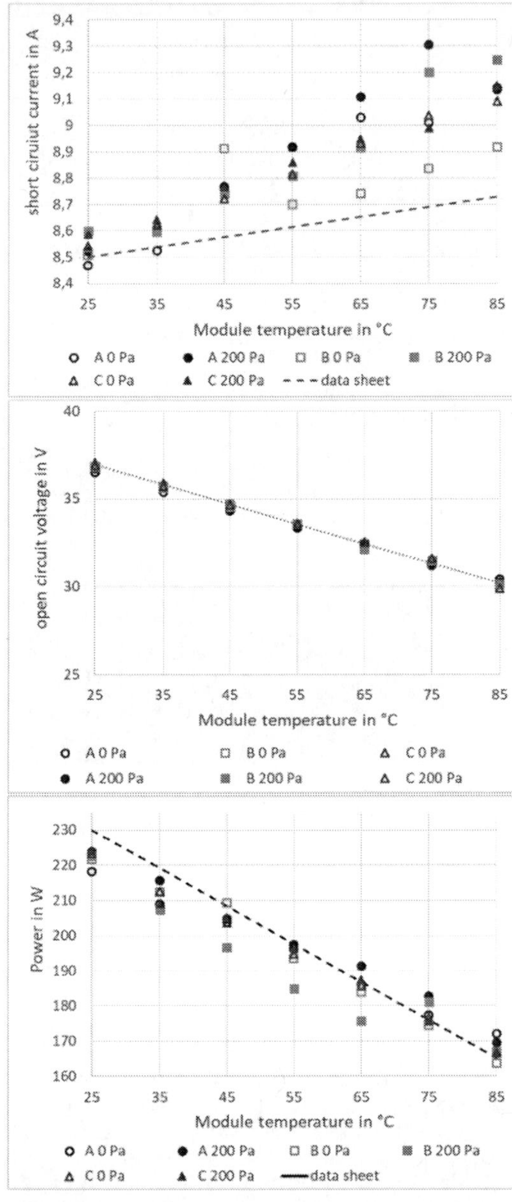

Figure 4: short-ciruit current (top), open-circuit voltage (middle), and P_{mpp} (bottom) as a function of module temperature for 3 modules A, B, and C, unloaded (p = 0 Pa) and loaded (p = 200 Pa)

Measured performance data are listed in Table 2. The good module's (A) performance is hardly affected by pressures up to 1000 Pa, as seen in Table 2, but significantly by temperature increase. The modules with more open cracks, on the other hand, show power reduction with both increasing pressure and increasing temperature. However, the impact on the power due to loading is less severe at elevated temperatures. For example, the power loss of module C at T = 25°C is about 16% whereas at 85°C it is roughly 6%. In other words, increasing the temperature reduces the power of the unloaded module significantly more than that of the loaded module, e.g. 23% at p = 0 Pa versus 13% at p = 1000 Pa for module C.

Note, throughout all measurements the data sheet temperature coefficient seems to be higher than the measured ones. For p = 0 Pa, the temperature coefficient for the power is α = -0.39 ± 0.04%/K, for p = 200 Pa, α = -0.38 ± 0.01%/K. This is mainly due to the strong increase of the temperature coefficient of I_{sc}, at p = 0 Pa results β = 0.10 ± 0.03%/K, at p = 200 Pa results β = 0.13 ± 0.01%/K.

Table 2: Measured power at different loads and module temperatures

	P	p = 0 Pa	p = 1000 Pa
Module A	T = 25°C	217 W	220 W
	T = 85°C	175 W	173 W
Module B	T = 25°C	221 W	192 W
	T = 85°C	166 W	150 W
Module C	T = 25°C	222 W	186 W
	T = 85°C	170 W	161 W

This unexpected phenomenon of a less than expected power loss at high temperatures can be explained by analyzing the EL-images carefully. Figure 5 illustrates a series of EL-images of two cells of module B, E7 and E8, with cracks and fractures for all investigated temperatures at two pressures. The image series reveal that the EL-intensity / the darkness of cell fragments fluctuate. The EL-image visualizes how the electrical contact of the fractured cell parts alter. Electrically unconnected cell parts appear black, fairly well connected fragments grey and bright. A clear trend with increasing temperature or pressure does not exist, according this study. The changing of the cell appearance with temperature was also observed by Silverman [6], the influence of pressure was not analyzed.

However, the intensity distribution of the EL-signal is analyzed more closely. As a measure for the cell quality and failure presence the median and the 25%-quantile of the EL-intensity of the cell are chosen. Figure 6 visualizes the EL-intensity for two bad cells and a good reference cell as a function of module temperature. The EL-signal for the chosen intact reference cell (cell C6 of module B) increases strongly and steadily with temperature.

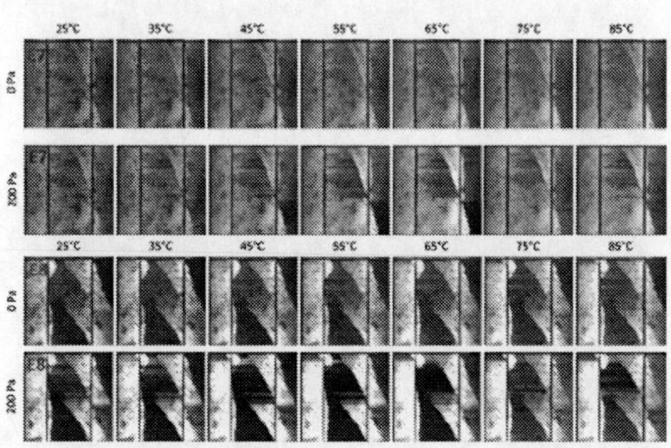

Figure 5: EL-images of cell E7 and E8 of module B for various module temperatures at p = 0 Pa and p = 200 Pa

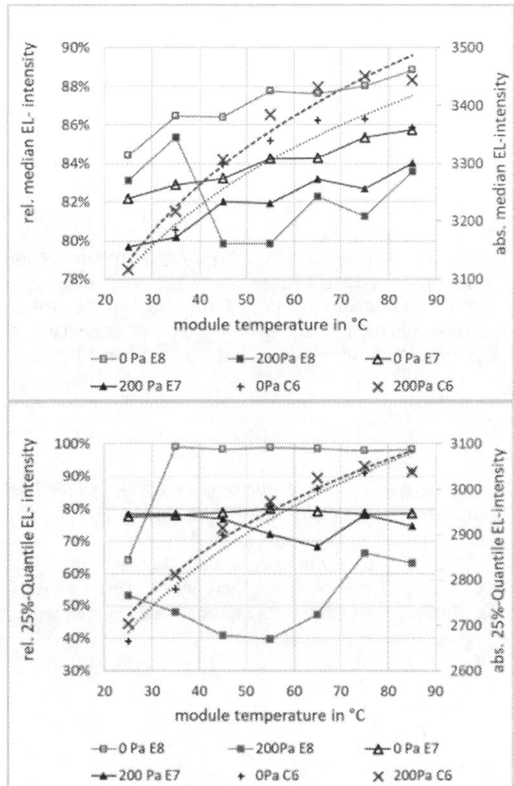

Figure 6: absolute EL-intensity for reference cell C6 (right axis) and relative EL-intensities of two cracked cells E7, E8 (left axis) of module B (see Figure 2, Figure 5)

Evaluating the cracked cells, the median was calculated and related to the median and 25%-quantile of the reference cell in the same image. This ratio, the relative median and quantile of the EL intensity are shown in Figure 6 for the two cells and two pressures. In general, it is obvious that the EL-signal increases with temperature. If nothing changes, the ratio is expected to be constant. That is observed for E7 at p = 0 Pa, as shown the quantile in Figure 6. In case the quantile drops, the dark area should increase or its intensity decrease. Since no new cracks are generated at such low pressures (also visible in the images), the intensity must be reduced. This means physically that the electrical connection between the crack faces becomes worse. If the ratio increases, the contrary takes place. Both effects can be seen in the EL-images, in Figure 5, and the ratios, Figure 6. An electrical impact on the module power is reasonable if the EL-intensity is rather low for the area separated by the cracks [7], which holds only for the black area in cell E8. The evaluation of the 25%-quantile seems to be more promising for analyzing the cell quality of broken cells than the median, since the low EL-signals of the dark areas are described better.

Both, crack-poor and fracture-rich, modules show the expected decrease of V_{oc} with temperature increase, which is due to the temperature-dependent behavior of the semiconductor. However, the behaviour of I_{sc} and P_{mpp} differ from the expected calculated values. Here, the influence of the cell fracture becomes relevant. Disconnected areas reduce the active cell area and therefore the short-circuit current. According to the electrical connection, the current fluctuates with temperature and consequently the module power.

4 CONCLUSION

The investigations demonstrate that the module temperature has an impact on the performance and appearance of cracked cells in EL-images. Up to now, it is not predictable whether the crack structures recover or become worse with increasing temperature. This knowledge is of extreme importance for the interpretation of EL-images taken at cold nights and it is strongly recommended to take this temperature-dependence into account in the future.

5 ACKNOWLEDGEMENTS

ZAE Bayern gratefully thanks the German Federal Ministry for Economic Affairs and Energy (BMWi) for financial funding of the project AQUAM.

The authors sincerely thank the Allianz Risk Consulting GmbH / Allianz Zentrum für Technik (AZT) in Munich, Germany for supporting the project with a large number of PV-modules.

6 REFERENCE

[1] C. Buerhop-Lutz, et al., 44th IEEE (2017)
[2] C. Buerhop, et al., Progress in Photovoltaics (2018), 261-272
[3] E. J. Schneller, et al., 44th IEEE (2017)
[4] C. Buerhop-Lutz, et al., 33rd EU-PVSEC (2017), 1451 - 1456.
[5] M. Krause (2018) Bachelor Thesis; Fachhochschule Lübeck
[6] T. J. Silverman and S. Huang, NREL reliability workshops (2018);
[7] R. Amesöder (2017) Master Thesis; FAU Erlangen-Nürnberg

STRESS TEST METHODS TO PREDICT OUTDOOR PERFORMANCE OF MONOFACIAL AND BIFACIAL MODULE DESIGNS

William Gambogi[1], Thomas Felder[1], Steven MacMaster[1], Kaushik Roy Choudhury[1], Bao-Ling Yu[1], Nancy Phillips[1], Axel Borne[2], Hongjie Hu[3] and T.-John Trout[1]

(1) DuPont, Wilmington, DE, USA; (2) Du Pont de Nemours International S.A., Geneva, Switzerland; (3) DuPont (China) Research & Development and Management Co., Ltd., Shanghai, P.R.C.

Correspondence to: William.J.Gambogi@DuPont.com

ABSTRACT: To ensure long term reliability of PV modules in the field, test methodologies need to be developed to assess the long-term stability of modules and materials. Multistress test protocols have been developed that show materials degradation and changes in physical properties, and demonstrate known failure mechanisms seen in the field. Stress tests using sequential exposures of temperature/humidity, UV and mechanical stress have been developed which correlate to several different PV module degradation modes. Backside cracking was observed with polyamide (PA) and polyvinyldifluoride (PVDF) based backsheets after MAST stress testing including DH followed by repetitive applications of backside UV and TC. The same results were seen utilizing both mini-modules, and full size modules tested in a third-party laboratory. Inner layer cracking observed with PET based backsheets was replicated with a mini-module using a weatherometer cycle incorporating water spray. Delamination in glass/glass modules was replicated using a sequential approach incorporating UV exposure, dynamic mechanical load, thermal cycling, and humidity freeze stress segments.

Keywords: backsheets, materials, module, durability, reliability, testing

1 INTRODUCTION

Accelerated laboratory stress tests which are predictive of field performance are critical to establish reliability and bankability of photovoltaic modules. Much of the PV industry relies on IEC qualification conditions to assess durability of PV module designs [1-3]. Most of the included stress sequences apply a single stress, with one applying short durations of the included stresses of damp heat, UV (front and back), and HF. These are insufficient to predict long term performance. Both longer stress times and multistress protocols are being addressed by several research and standards groups, with a goal to better predict long term reliability in the application.

Sequential testing refers to the use of multiple types of environmentally based stress exposures applied in a sequential process. As the PV module application environment contains many types of stressors acting synergistically, replicating observed field failures in a laboratory setting requires use of sequential stresses.

We have considered UV, temperature, humidity, thermal cycling and mechanical stress as key elements of a sequence representative of the application environment [4]. A number of groups have presented research directed at similar combinations [5-9], including application of simultaneous stress conditions as a research program [10]. We have utilized stress levels and durations based on industry norms, application conditions, and the expected dose in the outdoor environment. The approach has been to replicate the field stresses with combination sequences, and then apply accelerations to achieve the same result in a shorter duration sequence. We have carefully approached test acceleration using high stress regimes, to avoid provoking degradation that will never occur in the application. Work has focused on sequential stress conditions and durations that are compatible with equipment readily available within the industry making these test methods readily used for the evaluation of new materials and module designs. Our goal is to develop sequential test protocols with a minimized test duration, providing results which show the same failure modes as field observations.

2 BACKSIDE SEQUENTIAL STRESS TEST

To assess the durability of the backside of the module to environmental stresses, we have developed a methodology that applies sequential steps of higher temperature/humidity (damp heat), UV and thermal cycling, "MAST", with the original sequence shown in Figure 1A.

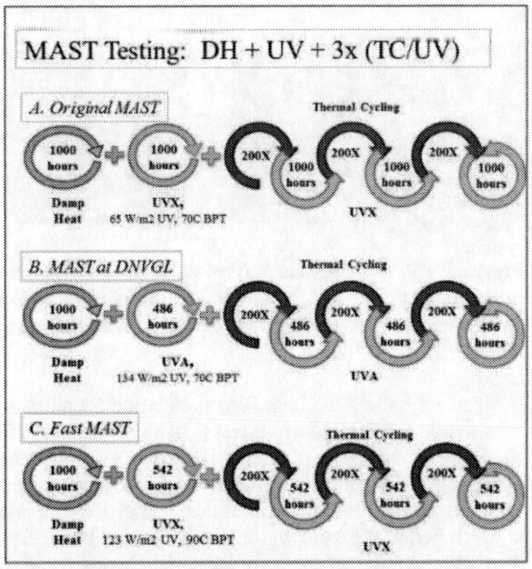

Figure 1. MAST Sequences: A) original sequence; B) full size modules; C) fast MAST sequence with accelerated UV exposure

The 1000h damp heat duration used in this test sequence is consistent with previous work assessing the backsheet degradation of fielded modules [11] and is consistent with work at NREL predicting hydrolysis damage of core PET in backsheets in various outdoor locations [12]. The UV level applied in this sequence is based on a 12% albedo level over >20 years. The thermal cycling levels are also appropriate for long term reliability [13].

We have previously reported on the results of this work as it relates to several backsheet types including PVF (Tedlar®, PVDF, PET and polyamide (PA) based backsheets [14, 15]. In PVDF backsheets, we have observed cracking in the machine direction (MD) of the backsheet consistent with loss in mechanical properties observed in PVDF films. In PA backsheets, we have also found cracking of the backsheet consistent with loss in mechanical properties found in this backsheet. In PET backsheets, yellowing is observed with long term sequential testing.

These results correlate with extensive field studies, where we have observed degradation on the backside (junction box side) of the backsheet which include yellowing and cracking [16]. Backsheets can be identified in the field at a materials level by analysis of data from a portable FTIR, and differences between materials sets are noted. Our field studies have demonstrated cracking of PVDF-based backsheets and PA-based backsheets.

We have applied this testing method to full size commercial modules using a third party testing group, DNV GL. Shown in Figure 2 is the results of MAST testing of commercial modules with PVDF-based backsheet. As has been demonstrated previously in four-cell minimodules, we found cracking in the MD direction directly over busbar ribbons. We believe that the cracking is intially over the busbar ribbons becase these are areas of localized mechanical stress and the failure occurs there first. In the commercial modules tested, the cracking was observed on all busbar ribbons in the module. Results in this testing validated the mini-module results from our laboratories, and are found to correlate with field observations [18].

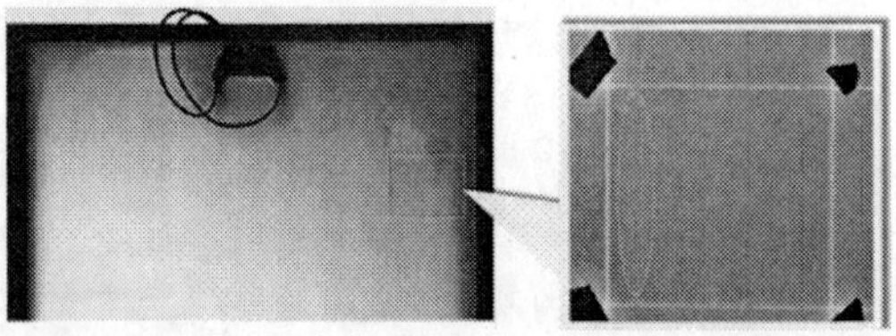

Figure 2. PVDF backsheets after MAST at DNVGL on a full size module. Cracks are in the machine direction, along the bus bar ribbons in each cell over the entire module.

Table 1 shows the improved correlation found using the module accelerated stress test (MAST) approach with field observations in comparison to single stress exposures. By comparing single stress test conditions to sequential multistress test conditions, we see that the later is much better at predicting failures observed in the field over a broad range of backsheet material classes.

Table I. Comparison of single and sequential stress test results to field observations

Stress	PPE	PVDF	Polyamide	TPT or TPE	Comment
Field	Yellowing Mech Prop Loss Cracking	Cracking Front Side Yellowing	Yellowing Mech Prop Loss Cracking	Low Defects	Effects of simultaneous and sequential stresses
Damp Heat (1000 hrs)	Slight Yellowing	No Change	Mech Prop Loss	No Change	Misses UV Degradation
UV (4000hrs)	Yellowing Mech Prop Loss	No Change	Mech Prop Loss	No Change	Misses hydrolysis and moisture
DH/UV/TC (MAST)	Yellowing Mech Prop Loss Cracking	Cracking Front Side Yellowing	Yellowing Mech Prop Loss Cracking	No Change	Combines key stresses, gives best correlation

We are aiming to improve the test, with goals to make the total duration shorter while maintaining compatibility with testing of minimodules to full size modules. The UV exposure was accelerated by increasing the irradiance level (and keeping the same dosage) and temperature. As described in Figure 1C. Mini-module exposures achieved the same results as the original MAST sequence, maintaining correlation to field results. New exposure results in reduction of total test time to just over 5 months.

A xenon device may be preferred because it has its excellent match to the solar UV spectrum and will in general be expected to give the best correlation field results [17], but this is not currently practical for full-size modules with today's UV exposure chambers. However there are activites within the industry to develop higher intensity UVA systems and metal halide systems that are more appropriate for full size modules. Some newer approaches which are showing promising results include replacement of thermal cycles with humidity freeze, or by incorporating water spray into the UVX exposure as a thermal shock.

3 FRONTSIDE SEQUENTIAL STRESS TEST

Our initial approach to frontside sequential test was to utilize the water spray feature of a commerical weathering device, to include heat and UV with repetitive segments of front water spray providing a thermal shock and resulting mechanical stress as in Figure 3. This resulted in front-side cracking on the inner layer of some PET backsheets as shown in Figure 4A. Damage to the front side of the backsheet is increasingly common in fielded modules, in the form of yellowing and cracking. Some PVDF-based backsheets have been found to yellow after less than five years in the field, and we have recently begun to observe cracking of the inner layer of PET backsheets in the field [16]. These cracks are easy to miss, but become apparent when backlighting the module (Figure 4B). Destructive analysis using nanoindentation showed the inner layer of the backsheet was found was embrittled [19]. Analysis also showed that the crack did not extend behind the cells indicating the importance of UV exposure in this failre mechanism.

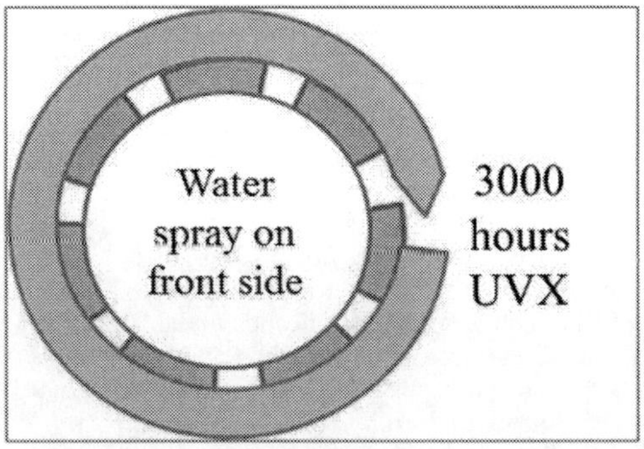

Figure 3. Sequential test of UV exposure with water spray. Exposure conditions: 90 °C BPT, 1.1 W/m2 –nm UV, cycle: 102 min UV followed by 18 min UV + water front spray, 3500h (~ 5y outdoor equivalent).

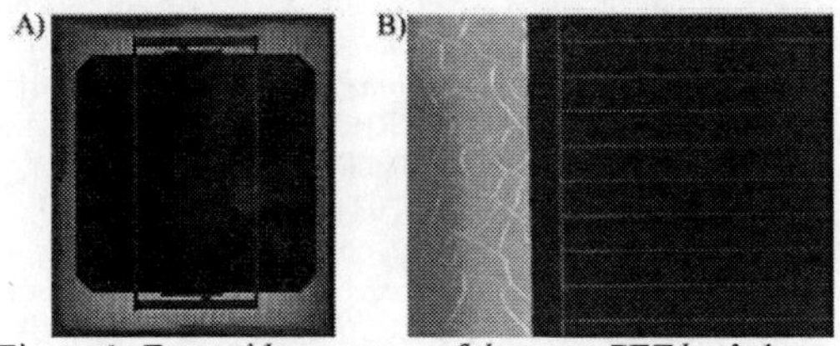

Figure 4. Front side exposure of the same PET backsheet. A) Front side exposure in xenon exposure incorporating waterspray (conditions in Figure 3). B) Fielded module with 6 years service in North America with 30% power loss in 5 years of service (6% linearized power loss per year).

4 SEQUENTIAL STRESS TEST USING DYNAMIC MECHANICAL LOAD

Wind can provide significant stress onto modules, and to replicate that dynamic environment, we incorporated a DML sequence in addition to frontside UVA, TC, and humidity freeze (HF) as shown in Figure 5.

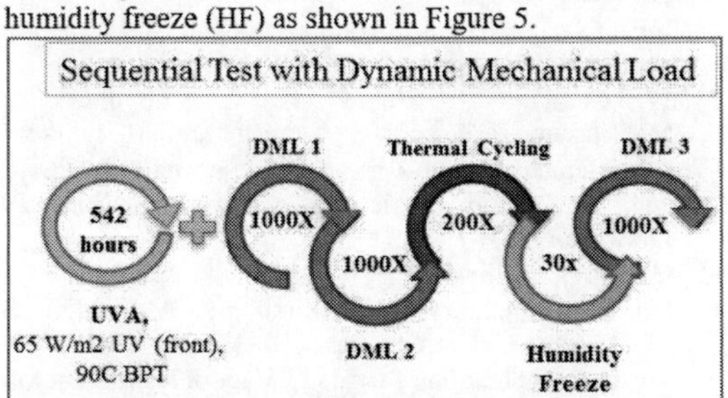

Step	Stress
1	UVA front exposure: 65 W/m2 (300-400 nm)
2	DML1: 1000 cycles, +/- 1500 Pa loading at 1/6 Hz
3	DML2: 1000 cycles, +/- 1500 Pa loading at 1/6 Hz
4	TC200 -40°C to 85°C, 200 cycles
5	HF30: Humidity Freeze, 30 cycles
6	DML3: (opt) 1000 cycles, +/- 1500 Pa loading at 1/6 Hz

Figure 5. Multistress test sequence used to assess glass/glass (GG) and glass/backsheet (GB) module designs

The frameless mounting and the additional rigidity of the glass/glass module can make these module designs more susceptible to wind load compared to glass/backsheet modules. Further, glass/glass module designs are breathable and EVA degradation products are trapped, creating localized internal pressure, with the glass/glass structure stiffer and less able to dissipate the internal stress.

This test sequence described in Figure 5 was applied to commercial GG and GB modules with an inspection at each step of the sequence. Results are summarized in Table II, which shows the progression of delamination in the glass/glass module. Slight delamination was visible after UV exposure and DML; this was then exacerbated by thermal cycling and humidity freeze. While significant delamination resulted on both front and back sides of the glass/glass module, no delamination was observed in the companion glass/backsheet modules.

Table II. Summary of results of multistress testing of glass/glass and glass/backsheet module designs

	G/G modules	G/Backsheet modules
UVA	No change	No change
DML1	No change	No change
DML2	Slight delamination on front	No change
TC200	Delamination on front, encapsulant voids on back	No change
HF30	Severe delamination on front, multiple encapsulant voids on back	Slight yellowing on edge, no delam
DML3 not performed		

The results of the DML sequential test match closely with delamination in glass-glass modules we have observed in the field. Figure 6 below shows examples of this effect in fielded modules.

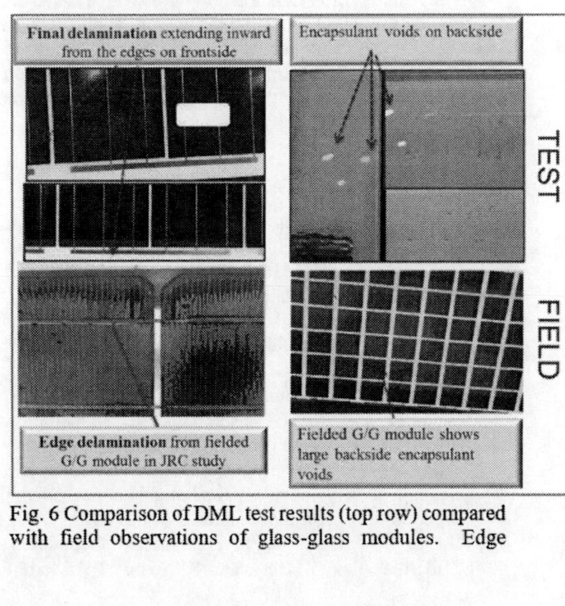

Fig. 6 Comparison of DML test results (top row) compared with field observations of glass-glass modules. Edge

delamination is seen on the frontside, voids are seen on the backside.

4 DURABILITY TESTING OF BIFACIAL MODULES

Bifacial modules present additonal requirements for durability that need to be addressed using the proper test conditions. For bifacial modules, both glass/glass and glass/backsheet stuctures have been introduced commercially.

As indicated in the earlier section, glass/glass structures have been found to show glass/encapsulant delamination when a UVfront/DML/TC/HF test sequence is applied while a glass/backsheet module from the same manufacturer does not show this delamination.

We suspect that UV exposure on the front encapsulant and glass initates the delamination due to loss in adhesion at that interface. In the case of bifacial modules, UV is incident on both the front and the back of the module, leading to the possibility of adhesion loss on both sides of the glass/glass structure. In addition to this consideration, encapsulant degradation products (acetic acid) and gas entrapment in glass/glass structures will apply additional stress. The higher operating cell temperatures found in glass/glass modules [20] may accelerate these effects, and also suggest lower output power for the glass/glass structure.

As in the case of monofacial PV modules, bifacial modules should not be assessed based on extended single stress conditions but should rather be evaluated using sequential stress conditions where the impact of the synergistic effects of these stresses are found to better correlate with outdoor exposure.

5 CONCLUSIONS

Sequential stress protocols are essential to developing PV modules with long term reliability. Several different sequential stress protocols have been identified which replicate field degradation modes not captured in current qualification tests. We are working to develop more aggressive and shorter duration sequences which cause the same effects, without causing degradation not observed in the field. Different sequences may be necessary to replicate different failure modes, with opportunities to find commonalities between the tests as more information becomes available.

6 REFERENCES

[1] IEC 61215 "Crystalline silicon terrestrial photovoltaic modules- design qualification and type approval" http://www.iec.ch/ (2016)

[2] IEC 61730 "Photovoltaic Module Safety Qualification" http://www.iec.ch/ (2016)

[3] S. Kurtz, K. Whitfield, N. Phillips, T. Sample, C. Monokroussos, E Hsi, I. Repins, P. Hacke, D. Jordan, J. Wohlgemuth, P. Seidel, U. Jahn, M. Kempe, T. Tanahashi, Y. Chen, B. Jaeckel, and M. Yamamichi , "Qualification Testing Versus Quantitative Reliability Testing of PV - Gaining Confidence in a Rapidly Changing Technology," *European PV Solar Energy Conference and Exhibition*, Amsterdam (2017).

[4] T.J. Trout, W. Gambogi, T. Felder, K. R. Choudhury, L . Garreau-Iles, Y. Heta, and K. Stika; "PV Module Durability – connecting field results, accelerated testing, and materials," *IEEE PVSEC*, Washington,D.C. (2017)

[5] C. R. Osterwald, T. J. McMahon. "History of Accelerated and Qualification Testing of Terrestrial Photovoltaic Modules: A Literature Review," *Progress in Photovoltaics: Research and Applications*, Vol. 17 (2009): pp. 11-33. https://dx.doi.org/10.1002/pip.861

[6] P. Hacke, D. Trudell, K. Terwilliger, N. Bosco, E. Gelak, and S, Kurtz, "Application of the NREL Test-to-Failure Protocol for PV Modules," *19th Workshop on Crystalline Silicon Solar Cells and Modules: Materials and Processes*, Vail, CO (2009)

[7] M. Kohl, "Accelerated Service Life Testing of Photovoltaic Modules", in *Service life prediction of exterior plastics: Vision for the future* (pp.199-213), C. White, et. al. ed. (2015) DOI 10.1007/978-3-319-06034-7

[8] K. Masuda, "Improvement of reliability and development of acceleration test methods for photovoltaic module," *Japan-Korea Joint Symposium on Advanced Solar Cells*, (2015) http://rrpdb.db.aist.go.jp/RRPDB/system/Koukai_e.List?Person==Collaborative%20Module-Reliability%20Research%20Team

[9] T. Ngo, "Effects of UV on power degradation of photovoltaic modules in combined acceleration tests," *Japanese Journal of Applied Physics* (55), 5, 52301-1 to 8 (2016)

[10] S. Spataru, P. Hacke, M. Owen-Bellini, "Combined-accelerated stress testing system for photovoltaic modules, *WCPEC-VII*, Waikoloa, Hawaii, (2018).

[11] Gambogi, O. Fu, Y. Heta, K. Hashimoto, J. Kopchick, T. Felder, S. MacMaster, A. Bradley, B. Hamzavytehrany, V. Felix, T. Aoki, T. J. Trout, Tony Sample, "A comparison of key PV backsheet properties from fielded module exposures and accelerated test conditions", *NREL PV Module Reliability Workshop* (2013)

[12] M. Kempe, "Understanding the temperature and humidity environment inside a PV module," *NREL PV Workshop* (2013)

[13] IEC 62892 "Test procedure for extended thermal cycling of PV modules" (CDV 2016)

[14] W. Gambogi, T. Felder, K. R. Choudhury, S. MacMaster, B. L. Yu, H. J. Hu, and T. J. Trout, "Accelerated testing using multiple sequential stresses and comparison to field performance," *NREL PV Workshop* (2018)

[15] Gambogi, Phillips WCPEC7 2018

[16] K. R. Choudhury, W. Gambogi, T. Felder, S. MacMaster, L. Garreau-Iles, H.J. Hu, R. Khatri, T. J. Trout, "Degradation of fielded PV modules from across the globe," *WCPEC-VII*, Waikoloa, HW (2018)

[17] IEC 62788-7-2 "Measurement Procedures for Materials Used in Photovoltaic Modules – Part 7-2: Environmental exposures – accelerated weathering tests of polymeric materials." http://www.iec.ch/ (2017)

[18] K. R. Choudhury, W. Gambogi, T. J. Trout, R. Sharnais, "Backsheet degradation in accelerated sequential testing of full-size commercial modules," *WCPEC-VII*, Waikoloa, HW (2018)

[19] K. Stika, C Westphal, J. Kapur, R. G. Raty, J. Li, J. Kopchick, W. Gambogi, B. Hamzavytehrany, A. Bradley, J. Marsh, B. Foltz, "Mapping Chemical and

Mechanical Property Degradation in Photovoltaic Modules, *IEEE PVSC,* (2014)
[20] K. Arihara, R. Koyoshi, Y. Ishii, M. Kadowaki, A. Nakahara, H. Nishikawa, T. Takayama, T. Nishimura, K. Ogawa, Y. Chiba, "Reliability and long term durability of bifacial photovoltaic modules using transparent backsheet", *Japanese Journal of Applied Physics*, (2018)

EARLY POTENTIAL INDUCED DEGRADATION (PID) DETECTION IN THE FIELD: SOLAR CELL SHUNT RESISTANCE MEASUREMENT AT DIFFERENT BIAS CONDITIONS

Michalis Florides*, George Makrides, George E. Georghiou
Photovoltaic Technology Laboratory, FOSS Research Centre for Sustainable Energy,
Department of Electrical and Computer Engineering, University of Cyprus
Panepistimiou 1 Avenue, 1678, Nicosia, Cyprus
* florides.michalis@ucy.ac.cy

ABSTRACT: Potential Induced Degradation (PID) is still a serious threat for the PV industry and it is expected to get worse with the trend to increase the operating voltage of PV systems. Therefore, a detection method which can detect PID at an infant stage is necessary in order to increase the reliability of PV and preserve their lifetime. In this paper a PID detection method based on the solar cell shunt resistance is presented. A simulation shows at which bias conditions the shunt resistance has to be measured in order to detect PID before 1% power is lost. The simulation results are then verified by experimental measurements.

1 INTRODUCTION

Potential induced degradation (PID) has been considered an important degradation mechanism in the last decade, especially for high voltage PV systems. Several methods to prevent it or slow its progression have been proposed, however, their practical implementation is subject to cost, performance and long term reliability [1], [2]. The performance of a solar cell degrades fast when tested in the laboratory for PID under accelerated conditions. On the other hand, when the solar cells (PV module) operate in the field and are subject to PID, their performance can degrade slowly and over a long time. In this way, this failure mechanism can stay undetected for a long time by conventional power loss detection routines due to measurement uncertainties and hence a new detection technique is required [3].

Electroluminescence (EL), thermal imaging (IR) and outdoor luminescence (OL) are detection methods that have been proposed for potential induced degradation (PID), however, none of them are cost effective to detect PID in the field, i.e. while the PV system is operating. Their high cost in equipment and labour make then unattractive. Electroluminescence and outdoor luminescence require extra cost due to PV module removal and PV string wiring modification respectively. Therefore, a detection mechanism is required in the field to warn users for PID development. Hacke and Luo have shown a power loss estimation technique due to PID based on dark IV curves [4], [5]. Our work [6]–[8] is in the same direction by examining the shunt resistance (R_{sh}) of the solar cell.

The shunt resistance (R_{sh}) is a parameter that changes steeply with PID [9], [10] and it could be used for the early PID detection in the field before any significant power loss occurs. In this work, in order to evaluate its appropriateness, the shunt resistance due to PID is measured at different bias conditions. Initially, a simulation was performed to demonstrate at which bias conditions the shunt resistance is best to be measured in order to detect PID before 1% of power is lost. The simulation results were then verified with experimental measurements.

2 EXPERIMENTAL SETUP

For this experiment, single p-type polycrystalline silicon solar cells encapsulated as mini-modules were used. The mini-modules are PID prone with a cell area of 243.4 cm2, 3 mm thick glass, x-23-105 EVA and TPT Akasol PV1000V back-sheet.

To keep the cell temperature constant, an indoor test setup (Fig. 1) has been built which allows the temperature of the cell under test to be controlled while a high voltage bias is applied at the same time to cause PID. A high voltage power supply (HV PSU) was used for this purpose.

The glass facing surface of the mini-module was placed on the metallic temperature controlled plate which is connected to the positive (+ve) terminal of the HV PSU. The positive terminal of the mini-module was connected to the negative (-ve) terminal of the HV PSU. The terminals of the mini-module were connected to a device which was monitoring the shunt resistance (R_{sh}) of the cell while the high voltage bias was applied.

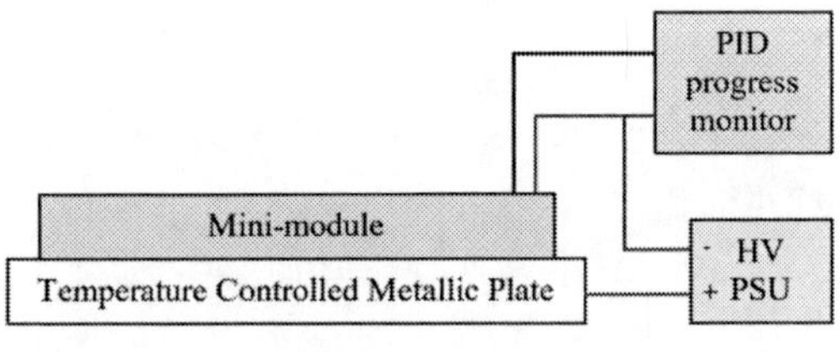

Figure 1: Indoor PID test setup for inducing PID and measuring the shunt resistance (R_{sh}) of a solar cell (single cell PV mini-module).

The setup was placed in a dark room in order not to affect the measurement of the shunt resistance (R_{sh}) from any photocurrent due to light. The measuring error of the equipment used is shown in Table I.

Table I: Equipment Measurement Error

Equipment	Measurement Error
HV PSU	< 1%
Flasher (Pasan A+++)	< 3%
Cell temperature sensor	< 0.5%
PID progress (shunt resistance) monitoring	< 0.5%

3 METHODOLOGY

The following methodology was followed in order to obtain the results, any deviation from this will be noted in the results section.

3.1 Simulation

In order to identify the best bias conditions for the assessment of the occurrence of PID, a model was developed based on the one-diode model for c-Si PV cells. The model was used to calculate the estimated shunt resistance (R_{sh-e}) of the solar cell under dark conditions as its actual shunt resistance (R_{sh-a}) changes due to PID. The shunt resistance was estimated by applying different forward bias conditions while varying the actual shunt resistance. Since the actual shunt resistance value decreases with increasing PID [9], [10], the PID progression was modelled by different values of shunt resistance. The value of the shunt resistance was varied exponentially from 1 mΩ to 1 kΩ. For the simulation, the reverse saturation current (I_{rs}) at a cell temperature of 25 °C was set equal to 5.82 nA and the ideality factor (n) equal to 1.15 [11]. Since the temperature coefficient of the cell shunt resistance when it varies with PID progression is unknown, the cell temperature (T_{cell}) was set to 25 °C in order to demonstrate the concept.

In addition, the power loss (P_{loss}) was estimated based on the ratio of the output power change with shunt resistance (R_{sh}). The output power was calculated by finding the maximum power operating point of the cell. This was undertaken in order to assess whether it is possible to detect PID before any significant power loss occurs by using the shunt resistance method. The cell temperature (T_{cell}) was set constant at 25 °C in order to be able to directly compare the calculated power loss with the measured one from the Flash test which is also performed at 25 °C.

The series resistance (R_s) of the cell during the simulation was set to 18 mΩ for all cases.

3.2 PID Progression

In order to induce PID in the solar cell, a voltage of -1000 V DC was applied. During the HV bias, the cell temperature (T_{cell}) was kept constant at 60 °C (+/- 2 °C) to accelerate the test time. The PID progression was monitored by measuring the shunt resistance (R_{sh}) at different bias conditions. The measurements were performed at a forward bias voltage of 150 mV and 500 mV and at a reverse bias voltage of 1 V. The voltage was applied to the electrical terminals of the cell and the shunt resistance was calculated based on the measured current. The shunt resistance was measured semi-continuously with a sampling rate of 10 seconds, i.e. every 10 seconds each of the three measurements was performed for a duration of 75 ms. This was done in order not to bias the cell continuously and possibly affect its PID progression.

Although the experiment was done to verify the simulation results at forward bias conditions, the shunt resistance was also measured at -1 V in order to verify that the measured shunt resistance at low forward bias conditions is close to the shunt resistance measured at low reverse bias conditions.

3.3 Power Loss Measurement

The output power of the cell was measured with a Flash test at standard test conditions (STC) before the PID test and every time the measured shunt resistance (R_{sh-m}) reached a value of 10, 5, 2.5, 1, 0.5 and 0.25 Ω. The measured shunt resistance values were obtained from the measurements at the -1 V condition since the reverse dark IV is linear up to -4 V [6] and gives a more accurate measurement of the actual shunt resistance (R_{sh-a}). The Flash test was performed within less than 15 mins every time the measured shunt resistance was reaching the desired value to reduce any recovery effects.

4 RESULTS

4.1 Simulation

Fig. 2 shows the estimated shunt resistance (R_{sh-e}) and cell power loss (P_{loss}) with the actual shunt resistance (R_{sh-a}) for various values of forward bias current. The PID progress is modelled by decreasing the actual shunt resistance as described in the methodology. In order to assess the best bias conditions to detect PID occurrence, the estimated shunt resistance was calculated at different forward bias currents.

From the results, it can be seen that as PID progresses the estimated shunt resistance (R_{sh-e}) is constant and does not change with the varying actual shunt resistance (R_{sh-a}) until it decreases below a value which depends on the forward bias current. In addition, the constant R_{sh-e} region is wider with increased bias current, i.e. with increased bias current the R_{sh-a} has to reduce more (i.e. PID has to progress more) in order to observe a change in R_{sh-e}. Our target is to detect PID before 1% power loss occurs. From the results in Fig. 2, the bias current has to be less than 100 mA (solid line with x marks) in order for PID to be detected before the output power decreases by 1%. This is deduced by looking at the highest bias current value, which crosses the shunt resistance line (dashed line), on the right side of the point at which the power loss (solid line) equals to 1%. Therefore, for the early detection of PID the R_{sh-e} has to be measured at low bias conditions.

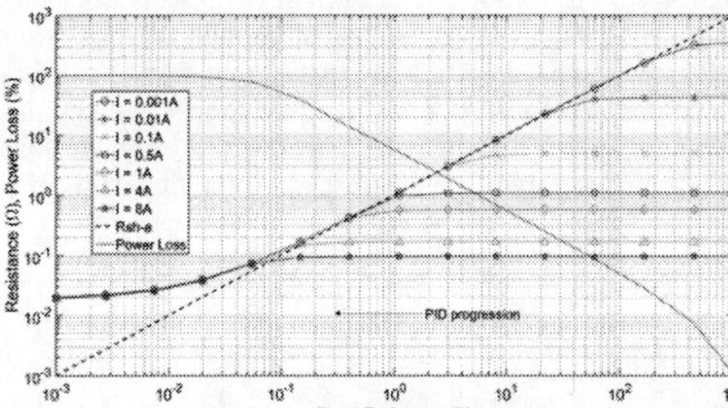

Figure 2: Estimated shunt resistance (R_{sh-e}) and percentage cell power loss (P_{loss}) at STC vs actual shunt resistance (R_{sh-a}) for various values of forward bias current. The cell output power before any power loss occurs is 3.02 W. The more the PID progresses the lower the shunt resistance is. R_{sh-a} is shown as well in order to see at which points the estimated R_{sh-e} matches the actual R_{sh-a}.

4.2 Experimental Measurements

The measured shunt resistance (R_{sh-m}) data and the power measurements at STC are shown in Fig. 3. It can be observed that the measured shunt resistance of the cell decreases fast at the beginning as PID progresses while the output power of the cell (P_{cell}) does not change significantly. The measured shunt resistance at -1 V reduces from 33 Ω to 5 Ω (75% reduction) while the power output of the cell reduces by less than 1%. A similar observation can be done for the measured shunt resistance at 150 mV which decreases from 20 Ω to 3 Ω (85% reduction) before 1% power loss is observed. On the other hand, the measured shunt resistance at 500 mV started to change when the cell output power loss was higher than 7%. This indicates that the shunt resistance is a good parameter to be measured for the early PID detection (< 1% power loss) at low bias conditions as predicted by the simulation. It is also observed that for significant power loss (> 1%), the shunt resistance has to be lower than 5 Ω. The power loss of the cell reached 50 % of its initial power at a shunt resistance value of 0.25 Ω.

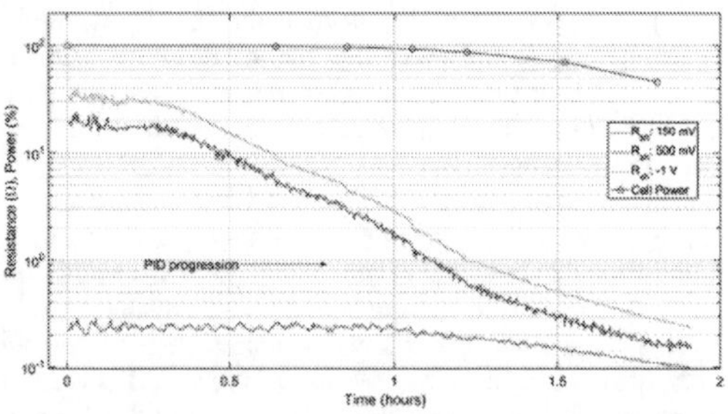

Figure 3: Measured shunt resistance ($R_{sh\text{-}m}$) at different bias conditions and cell output power (P_{cell}) vs PID progression. $R_{sh\text{-}m}$ was measured at a cell temperature (T_{cell}) of 60 °C and the initial cell power was 3.2 W at STC.

The IV curves at STC used to obtain the maximum output power of the cell (P_{cell}) in Fig. 3 are shown in Fig. 4 in order to show the cell power reduction with PID progression. The first parameter that changes with PID progression is the fill factor (FF) and then the open circuit voltage (V_{oc}) confirming once again this well-known behaviour. The short circuit current (I_{sc}) seems to change quite similarly to the open circuit voltage but this is not expected and can be attributed to the input resistance of the IV tracer. For a 60-cell PV module the IV tracer input resistance is not that important, however, for a single solar cell it makes a significant difference as its shunt resistance becomes less and less with PID.

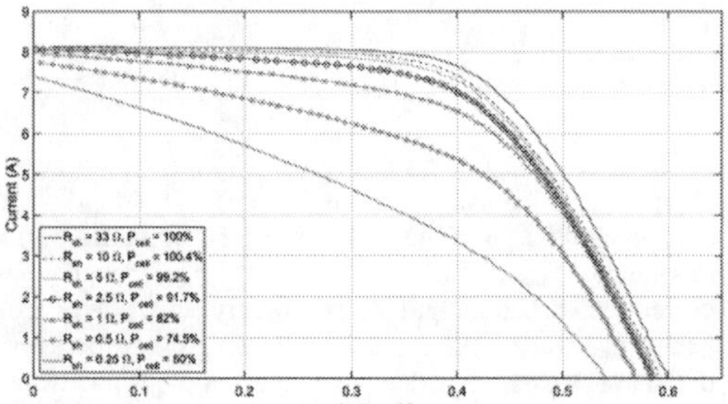

Figure 4: IV curves at STC for various measured shunt resistance ($R_{sh\text{-}m}$) values at -1 V with PID progression. $P_{cell\text{-}100\%}$ = 3.2 W at STC.

5 CONCLUSIONS

The work in this paper demonstrated by a simulation analysis that the shunt resistance (R_{sh}) of a solar cell can be used to detect PID before 1% power loss occurs. The simulation predicted that the shunt resistance has to be measured at a bias current lower than 100 mA in order to detect PID before 1% power is lost. The simulation results were confirmed with experimental measurements which demonstrated that at low bias conditions the shunt resistance changes by more than 50% before the cell power decreases by 1%.

6 ACKNOWLEDGEMETS

This work was funded through the EdGeWIse project which is co-financed under the ERA-NET-MED Transnational Calls of the European Union under the grand agreement number KOINA/ERANETMED/1114/01.

The Austrian Institute of Technology (AIT) is gratefully acknowledged for the help with the PID prone cells.

7 REFERENCES

[1] S. Pingel, O. Frank, M. Winkler, T. Geipel, H. Hoehne, and J. Berghold, "Potential Induced Degradation of Solar Cells and Panels," in *35th IEEE Photovoltaic Specialists Conference (PVSC)*, 2010, pp. 002817–002822.

[2] J. Berghold, O. Frank, H. Hoehne, S. Pingel, B. Richardson, and M. Winkler, "Potential Induced Degradation of solar cells and panels," in *25th European Photovoltaic Solar Energy Conference and Exhibition*, 2010, pp. 3753 – 3759.

[3] W. Luo, Y. S. Khoo, P. Hacke, V. Naumann, D. Lausch, S. P. Harvey, J. P. Singh, J. Chai, Y. Wang, A. G. Aberle, and S. Ramakrishna, "Potential-induced degradation in photovoltaic modules: a critical review," *Energy & Environmental Science*, vol. 10, pp. 43–68, 2017.

[4] P. Hacke, K. Terwilliger, and S. Kurtz, "In-Situ Measurement of Crystalline Silicon Modules Undergoing Potential-Induced Degradation in Damp Heat Stress Testing for Estimation of Low-Light Power Performance," Breckenridge, Colorado, 2013.

[5] W. Luo, P. Hacke, J. P. Singh, J. Chai, Y. Wang, S. Ramakrishna, A. G. Aberle, and Y. S. Khoo, "In-Situ Characterization of Potential-Induced Degradation in Crystalline Silicon Photovoltaic Modules Through Dark I-V Measurements," *IEEE Journal of Photovoltaics*, vol. 7, no. 1, pp. 104–109, 2017.

[6] M. Florides, G. Konstantinou, V. Venizelou, G. Makrides, and G. E. Georghiou, "Potential Induced Degradation (PID) Correlation to Leakage and Reverse Bias Currents," in *44th IEEE Photovoltaic Specialits Conference (PVSC)*, 2017.

[7] M. Florides, G. Makrides, and G. E. Georghiou, "Early potential induced degradation (PID) detection in the field: Voltage Measurement Methods," in *33rd European Photovoltaic Solar Energy Conference and Exhibition (EUPVSEC)*,

2017.

[8] M. Florides, G. Makrides, and G. E. Georghiou, "Characterisation of the Shunt Resistance due to Potential Induced Degradation (PID) in Crystalline Solar Cells," in *45th IEEE Photovoltaic Specialits Conference (PVSC)*, 2018.

[9] M. Bahr and K. Lauer, "Analysis of Activation Energies and Decay-time Constants of Potential-induced Degraded Crystalline Silicon Solar Cells," *Energy Procedia*, vol. 77, pp. 2–7, 2015.

[10] A. Masuda, Y. Hara, and S. Jonai, "Consideration on Na Diffusion and Recovery Phenomena in Potential-Induced Degradation for Crystalline Si Photovoltaic Modules," *Japanese Journal of Applied Physics*, vol. 55, no. 2S, 2016.

[11] E. L. Meyer, "Extraction of Saturation Current and Ideality Factor from Measuring Voc and Isc of Photovoltaic Modules," *International Journal of Photoenergy*, vol. 2017, 2017.

IMPORTANCE OF POWER STABILIZATION OF CRYSTALLINE PV MODULES

[1]R. Ebner, [1]G. Újvári, [2]W. Mühleisen, [2]Ch. Hirschl, [3]G.C. Eder, [3]Y. Voronko, [4]F. Vollmaier,
[1]AIT Austrian Institute of Technology, Center for Energy, www.ait.ac.at, Giefinggasse 2, 1210 Vienna, Austria, T +43 50550-6628, F +43 50550-6390, rita.ebner@ait.ac.at
[2]CTR Carinthian Tech Research AG, Europastr.12, 9524 Villach/St. Magdalen, Austria, www.ctr.at
[3]OFI Austrian Research Institute for Chemistry and Technology, Vienna, Austria, www.ofi.at
[4]PVP Photovoltaik GmbH, Wies, Austria, www.pvp.co.at

ABSTRACT: Power output results of module measurements performed in independent laboratories often appear lower than those indicated on the label. The reason for this effect is that several factors that may affect a PV module power output are not taken into account. PV modules made of crystalline silicon solar cells suffer an initial slight loss in peak power upon irradiation, an effect known as "light induced power stabilization" [1,2]. As many module producers know this effect, this light induced degradation (LID) is included in their labeled output power value and, as a consequence also in the product guarantee. When modules are subject to IEC reliability testing, the modules need to show a stable output before starting the tests, thus, they must be exposed to natural sunlight or an artificial light source in a solar simulator over $20kWh/m^2$. This power stabilization procedure is compulsory according to the standard IEC 61215 for certifications of PV modules [3]. When different combinations of module materials are tested for durability in e.g. damp heat (DH) or thermal cycle (TC) testing, a stable condition of the solar cells is needed as reference state. This allows distinguishing between the various ageing effects caused by physical or chemical changes of the polymeric materials and/or solar cells. In the performed experiments, durability testing results with and without light soaking (40kWh) were compared.

KEYWORDS: LID, c-Si, DH

1 INTRODUCTION

Crystalline p-type boron doped silicon solar cells generally exhibit a degradation of conversion efficiency during the first hours of exposure to the sun light. This light induced degradation (LID) is associated with the formation of the well-known boron oxygen complex which acts as a harmful defect and reduces the minority carrier diffusion length accordingly [4-10]. LID is therefore related to both, boron and oxygen concentration. For wafers with bulk resistivities in the range of 1-2 Ωcm, LID can reduce the efficiency by up to 5% relative in CzSi due to the relatively high oxygen concentration, while LID is typically about 1% in case of multicrystalline Si solar cells. Multicrystalline silicon wafers are partly compensated by phosphorous and contain a higher boron concentration as compared to non-compensated reference wafers with same resistivity. In this paper the influence of durability tests on mono- and multicrystalline PV modules with and without light soaking will be investigated.

2 EXPERIMENTS

The purpose of this analysis is to evaluate the importance of the initial stabilization process of mono- and multicrystalline photovoltaic mini-modules.

1-cell and 6-cell test modules with crystalline Si cells (4 busbar cells, mono and multi) with PET-based backsheets and glass on the front side were produced for use in climate specific accelerated ageing tests. The modules were characterized by means of power output determination, electroluminescence (see FIG.1), photoluminescence, UV fluorescence and dark-lock-in thermography.

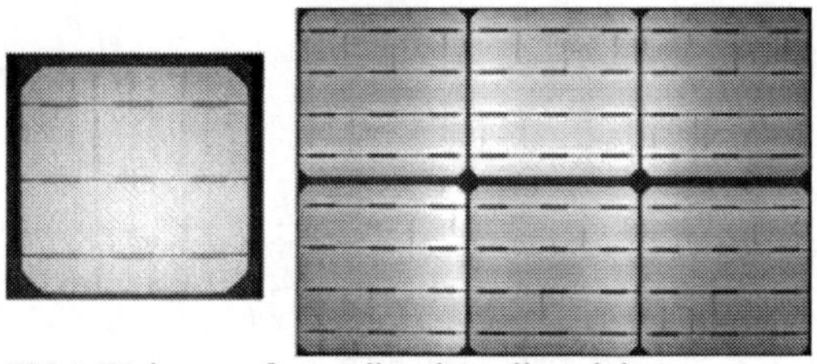

FIG.1 EL image of a 1-cell and 6-cell module.

Respectively the half of the 1-cell and the 6-cell modules were stabilized by light soaking at 40kWh before undergoing artificial ageing in climatic chambers, damp heat test (DH). The other half of the modules were not stabilized before testing under DH conditions (85°C, 85% RH, for 500h and 1000h).

2.1 Monocrystalline PV modules

2.1.1 One-cell-monocrystalline-modules

FIG.2, FIG.4 and FIG.6 show the maximum power output (PMPP) of 1-cell monocrystalline modules before and after preaging. FIG.3, FIG.5 and FIG7 show the change of PMPP in percent after pre-aging at 40kWh.

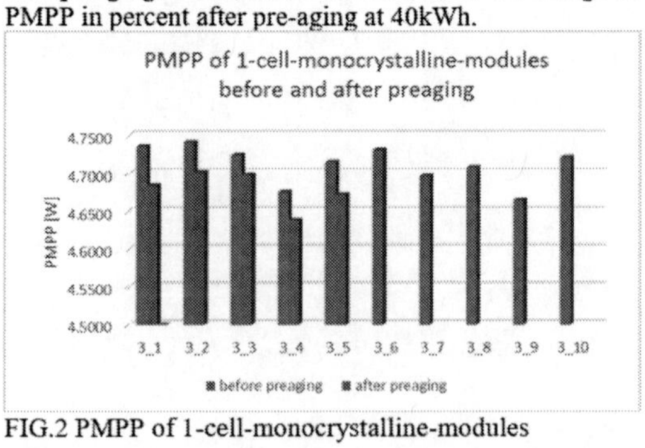

FIG.2 PMPP of 1-cell-monocrystalline-modules

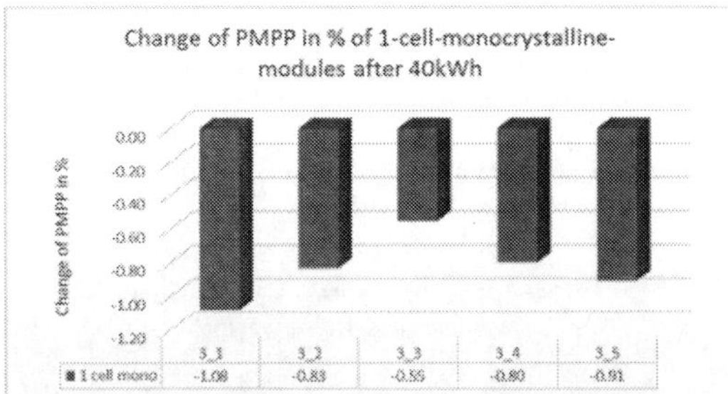

FIG.3 Change of PMPP of 1-cell-monocrystalline-modules

FIG.3 shows a reduction of the PMPP after preaging between 0.55% and 1.08%

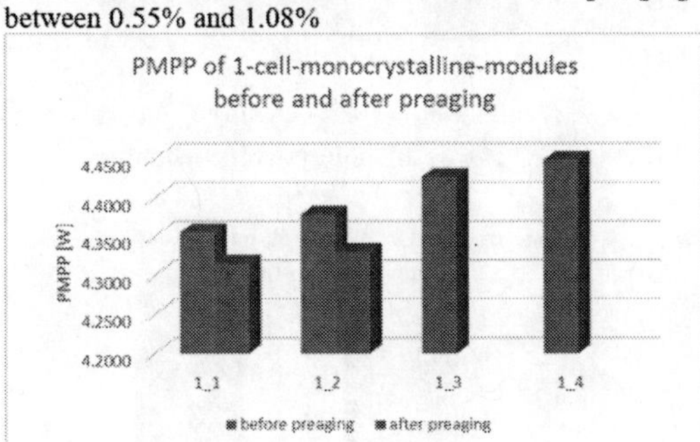

FIG.4 PMPP of 1-cell-monocrystalline-modules

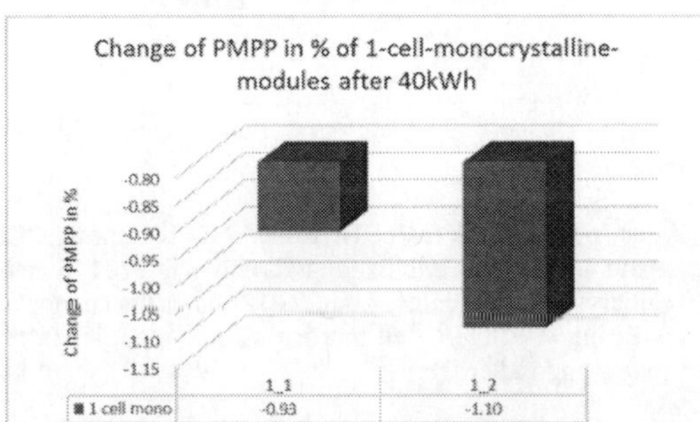

FIG.5 Change of PMPP of 1-cell-monocrystalline-modules

In FIG.5 a reduction of the PMPP after preaging of approximately 1% could be identified.

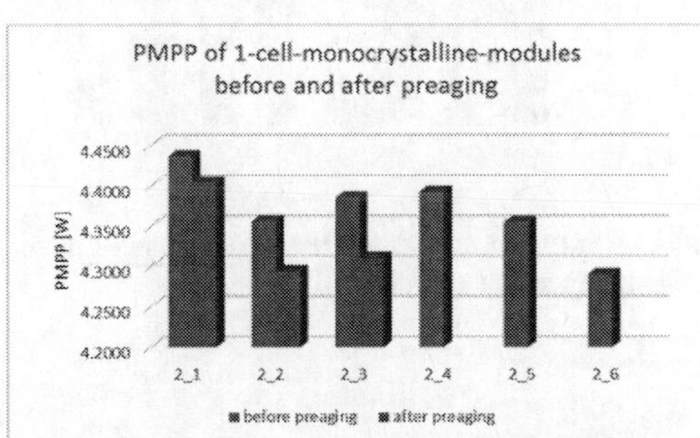

FIG.6 4 PMPP of 1-cell-monocrystalline-modules

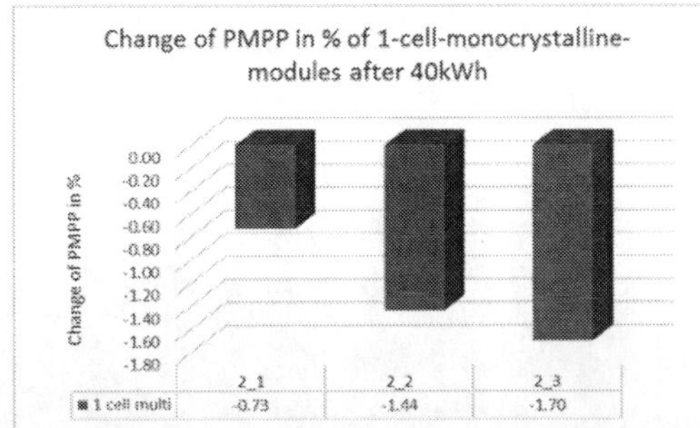

FIG.7 Change of PMPP of 1-cell-monocrystalline-modules

In FIG.7 a reduction of the PMPP between 0.7% and 1.7% could be identified.

After performing the DH tests, FIG.8 shows the behaviour of all stabilized and non-stabilized 1-cell-monocrystalline modules. And FIG.9 shows the change of the PMPP in % of the 1-cell-monocrystalline-modules (after preaging and after DH). The change of % is always referenced to the initial measurements.

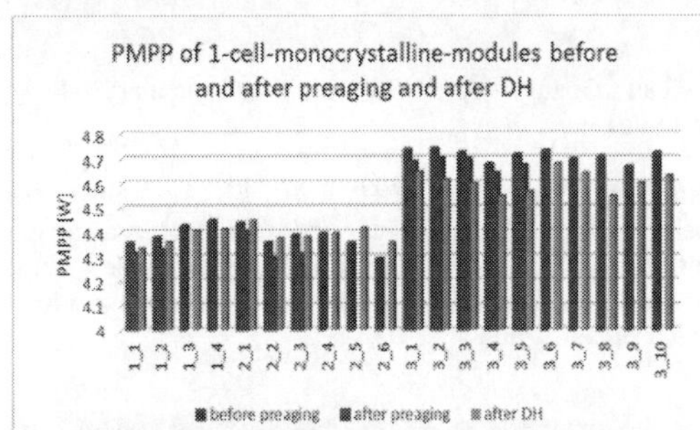

FIG.8 PMPP of 1-cell-monocrystalline-modules (before and after preaging and after DH)

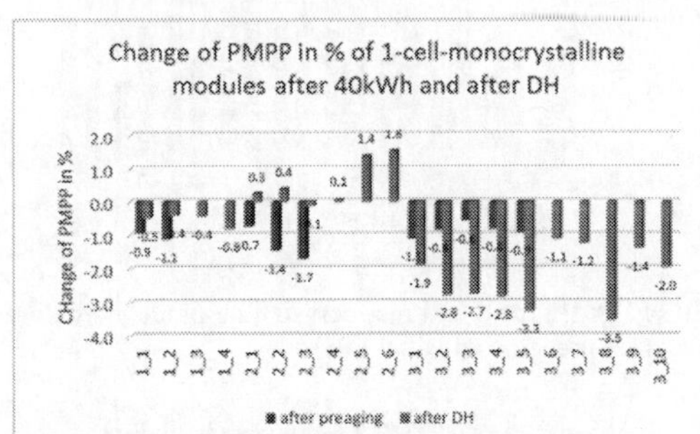

FIG.9 Change of PMPP in % of 1-cell-monocrystalline-modules (after preaging and after DH)

2.1.2 Six-cell-monocrystalline-modules

FIG.10 and FIG.11 indicates the results of the 6-cell-monocrystalline- modules before and after preaging. A reduction of the PMPP between 0.35% and 0.8% could be identified.

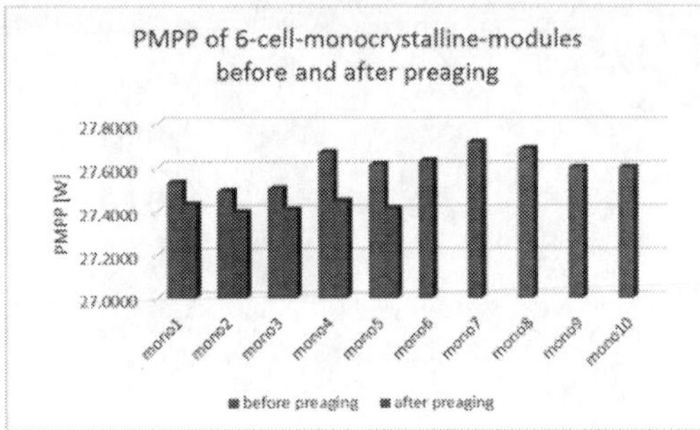

FIG.10 PMPP of 6-cell-monocrystalline-modules

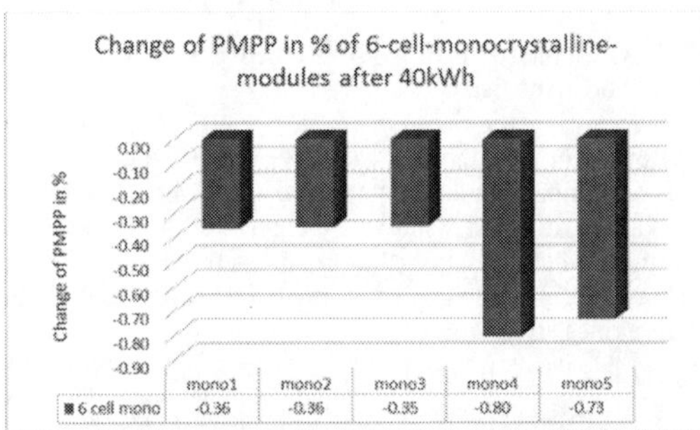

FIG.11 Change of PMPP of 6-cell-monocrystalline-modules

After performing the DH tests, FIG.12 shows the behaviour of the stabilized and non-stabilized 6-cell-monocrystalline modules. And FIG.13 shows the change of the PMPP in % of the 6-cell-monocrystalline-modules (after preaging and after DH).

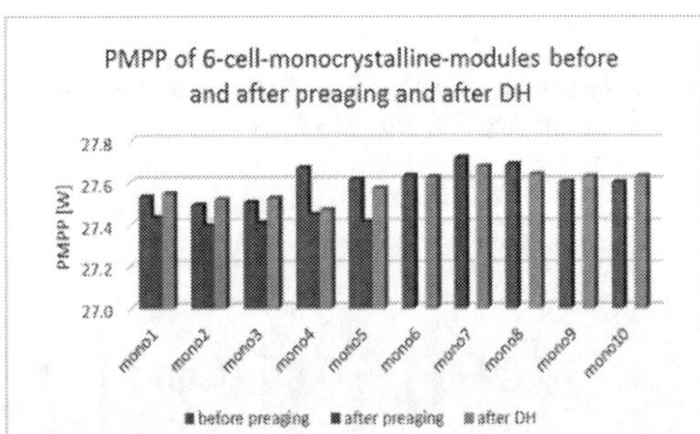

FIG.12 PMPP of 6-cell-monocrystalline-modules (before and after preaging and after DH)

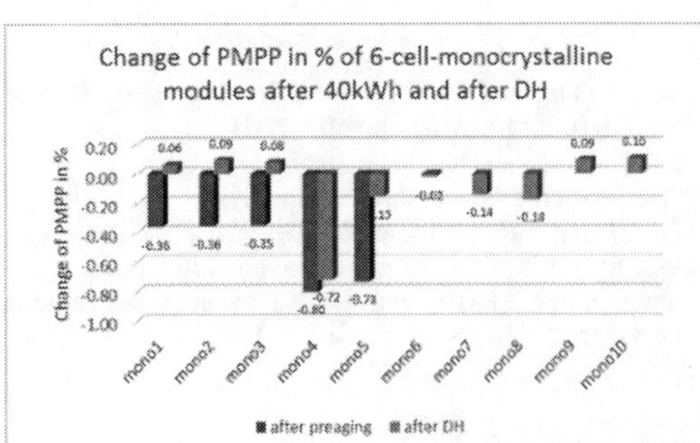

FIG.13 Change of PMPP in % of 6-cell-monocrystalline-modules (after preaging and after DH)

2.2 Multicrystalline PV modules

FIG.14 shows the maximum power output (PMPP) of the 6-cell multicrystalline modules before and after preaging. And FIG.15 shows the change of PMPP in percent after pre-aging at 40kWh. A reduction of the PMPP between 0.7% and 1% could be identified.

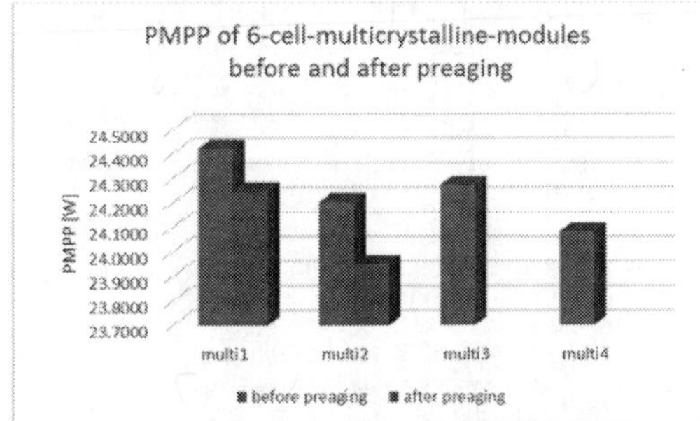

FIG.14 PMPP of 6-cell-multicrystalline-modules

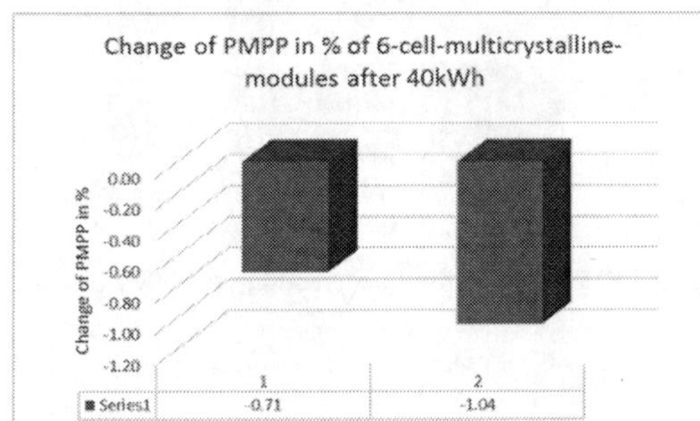

FIG.15 Change of PMPP of 6-cell-multicrystalline-modules

After performing the DH tests, FIG.16 shows the behaviour of the stabilized and non-stabilized 6-cell-multicrystallin modules. And FIG.17 shows the change of PMPP in % of the 6-cell-multicrystalline-modules (after preaging and after DH).

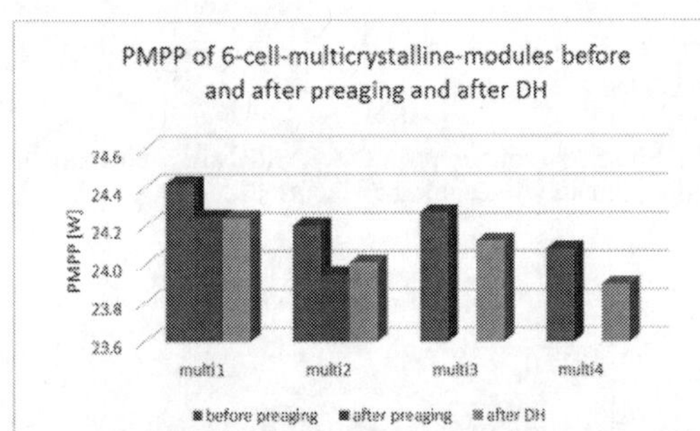

FIG.16 PMPP of 6-cell-multicrystalline-modules (before and after preaging and after DH)

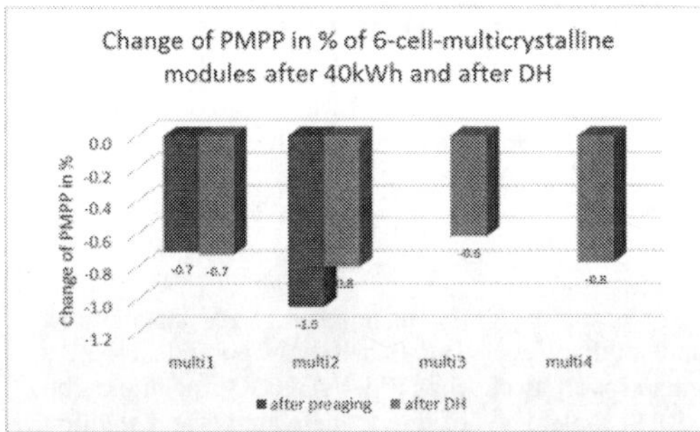

FIG.17 Change of PMPP in % of 6-cell-multicrystalline-modules (after preaging and after DH)

3 RESULTS

The power of crystalline PV modules typically decreases by around 1%. It can be concluded from our experiments that stabilized and non-stabilized modules are difficult to compare. A different behaviour of mono- and multicrystalline modules and of 1-cell and 6-cell-modules also could be identified. The degradation of 1-cell-modules is stronger than that of 6-cell-modules.

4 CONCLUSIONS

Since the 1970s, boron-doped Czochralski-grown silicon (Cz-Si) has been known to degraded in performance when exposed to light or when minority carriers are injected in the dark, until a stable efficiency is reached. This is due to the formation of a specific metastable boron oxygen complex in the Cz-Si material. It was then less commonly known that the same effect may occur on multi-crystalline material. Today, it is well known that power of crystalline PV modules typically decreases by around 1%, and losses of over 4% have also been reported. With PERC-cells even power losses of up to 6% are expected.

Further experiments are still ongoing in order to confirm the necessity that all cells/modules need a stable output before durability tests could be performed.

5 REFERENCES

[1] J. Schmidt, "Light-induced degradation in crystalline silicon solar cells", Solid State Phenomena Vols. 95-96 (2004) pp. 187-196.

[2] M. Gostein, L. Dunn, "Light soaking effects on PV modules: overview and literature review", 37th IEEE PVSC 2011.

[3] IEC61215-1 and IEC61215-2, International Standard, Terrestrial PV modules – Design qualification and type approval – Part 1: Test requirements, Part 2: Text procedures, Edition 1.0, 2016.

[4] W. Vaaßen, "Qualität von PV-Kraftwerken-Felderfahrung, Technische Risiken und Reduktionsmaßnahmen", PV-Symposium 2016, Kloster Banz, Bad Staffelstein, Deutschland.

[5] Photovoltaics Report, Fraunhofer Institute for Solar Energy Systems, Freiburg, 12 July 2017.

[6] K.M. Broek, I.J. Bennett, M.J. Jansen, N.J.C.M. van der Borg, W. Eerenstein, "Light and current induced degradation in p-type multi-crystalline cells and development of an inspection method and a stabilization method", 27th EU PVSEC, Frankfurt, Germany, 2012.

[7] M.A. Munoz, "Influence of initial power stabilization over crystalline Si PV modules maximum power", Progress in Photovoltaics, volume 19, issue 4, pp. 417-422, 2011.

[8] A. Colli, J. Vasquez, W.J. Zaaiman, "Initial stabilization of a statistical sample of forty-four monocrystalline PV modules", Renewable Energy 75, pp. 326-334, 2015.

[9] H. Wang, A. Wang, H. Yang, J. Zhang, J. Huang, „Performance degradation of crystalline silicon solar module in various ultraviolet radiation area", IEEE, 978-1-5090-2724-8/16, 2016.

[10] F. Kersten, P. Engelhart, H.Ch. Ploigt, A. Stekolnikow, T. Lindner, F. Stenzel, M. Bartzsch, A. Szpeth, K. Petter, J. Heitmann, J.W. Müller, „Degradation of multicrystalline silicon solar cells and modules after illumination at elevated temperature", Solar Energy Materials & Solar Cells 142 (2015) 83-86.

Acknowledgement: This work was conducted as part of the Austrian "Energy Research Program" project INFINITY. Funded by the Austrian Climate and Energy Fund and the Austrian Research Promotion Agency (FFG), both are gratefully acknowledged.

RESISTANCE OF PID SENSITIVE PV MODULES TO ALTERNATING HIGH VOLTAGE STRESS (A-HVS)

Bengt Jaeckel[1], Jürgen Arp[2]
[1] Fraunhofer Center für Silizium-Photovoltaik CSP, Walter-Hülse-Straße 1, 06120 Halle, Germany,
bengt.jaeckel@csp.fraunhofer.de, Tel. +49-177-7503123
[2] PV Lab Germany GmbH, Gartenstrasse 36, 14482 Potsdam, Germany, j.arp@pv-lab.de

ABSTRACT: Potential induced degradation (PID) is a severe problem in a significant number of PV installations globally. Ultimately the problem is reliably resolved on cell and module level. But this is only possible for new installations and with good pre-qualification checks of components as well as checks on delivered PV modules. The presented method targets all existing PV installations having the PID problem. As of today, there are several solution providers in the market that offer PID mitigation devices. Those operate differently and will be differently effective to recover or stabilize the output power. In most cases the recovery works, at least to a certain extend. The long term impact of such devices has not been investigated yet.
The proposed test and test sequence offers a fast initial check to see the effect of alternating potentials on the PID sensitive PV module.
Based on the findings PV modules, that have been recovered successfully can grouped into two types: Modules that are stable and those that degrade in the cycling A-HVS test. The test basically indicates a potential long term risk for a PV module exposed to high alternating bias even after a first successful recovery step utilizing constant recovery potential in the field.

Keywords: c-Si, Durability, Reliability, Degradation, Energy Performance, PID

1 BACKGROUND AND MOTIVATION

Potential induced degradation (PID) is a well-known, not resolved degradation phenomenon, yet [1.]. Most dominant is PID in not grounded (negative) floating potential PV systems where negative potentials of up to -500V (p-type c-Si) can exist. Module power losses of almost 100% were observed. This translates into significant system power losses (>>10%) and therefor to critical financial (up to 100%) losses.

Over the past years several PID mitigation devices were developed to recover the PV modules and therefor to protect the long-term investment of the PV system. To recover the system performance different approaches exist:

a) System Grounding
b) Anti-PID boxes
   - Type A: Potential shifting towards positive (for p-type c-Si) potentials vs ground
   - Type B: Application of a recovering potential overnight/non-operating hours

In any case the applicability depends on the system wiring design. All types have to be proven to work in general and have different advantages and disadvantages, if applicable.

Grounding is in principle simple and a very reliable solution, but due to safety measures not always possible. Also the recovery may take quite long as shown by Pingel [2.].

The Anti-PID boxes can be divided into two types and require different levels of system modifications and might be therefore not applicable. Generally Type A should be the best, as it totally avoids potentials that lead to the PID-shunting of p-type c-Si cells [6.][10.][11.]. But it has to be noted that the absolute potential versus ground on the positive side will go up to maximum system voltage allowed by the components. That might trigger other and/or not known high voltage induced defects [3.]. Type B allows normal operation of the system during day and only applies a potential for recovery during non-operating hours of the system (typically during the night). The impact to the system wiring scheme is here very limited and the involved cost for the new hardware is limited (neglecting labor for installation and maintenance of this extra hardware, including spare parts). Therefor the focus in this study is on Type B.

Previous studies have shown that not all modules react the same and expected way to the recovery during non-operating hours [4.]. The information that is needed to decide on a certain PID mitigation device depends is how good and reliable the device will work – both to recover the PV modules and to operate reliably till the end of the system life (typically assumed 25 years). From a financial point of view the recovery should happen as quickly as possible and the modules should regain 100% of their initial power output. The possible changes in relative output power are illustrated in FIGURE 1. Example 1 is a PID resistant module and Example 2 the trend what would happen without any actions taken. The area between the blue curve of Example 1 and the other Example scenario is equivalent to a loss in energy production, neglecting the even worse impact of PID for low light performance [12.].

From FIGURE 1 it is obvious that to minimize the low in energy production the PID mitigation device must work as efficient as possible (Example 5 is preferred). The scenarios described in FIGURE 1 can be generally observed in PV installations.

To achieve a strong regain in output power the regeneration rate during PV non-operating hours and temperature induced daytime regeneration must be significantly higher compared to the daily degradation when high negative potentials exist. It has been shown that both, degradation and regeneration are strong function of temperature [5.][6.][7.][8.][9.]. Local weather, soiling and season of the year will further strongly impact the rates of degradation and recovery.

The target of this work is to identify potential long-term risks (power recovery and safety concerns) caused by alternating potentials of several hundred volts as they occur with the use of PID mitigation boxes that apply high potentials versus ground.

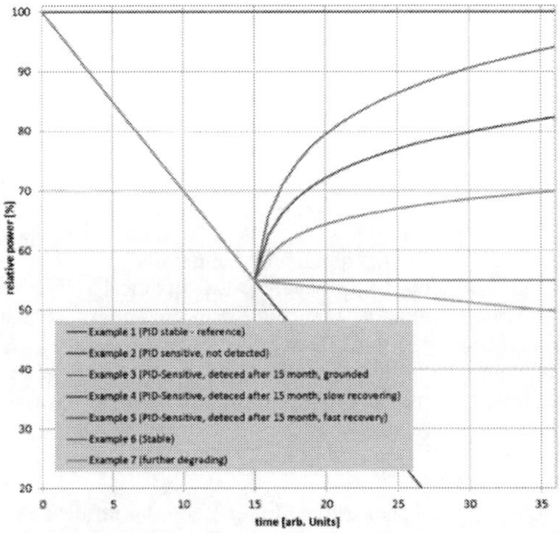

Figure 1: Different PID scenarios with different effective PID mitigation devices.

2 EXPERIMENTAL SETUP AND A-HVS-TEST BACKGROUND

The experiment is designed to investigate the influence of alternating application of high voltages as they will occur while using PID mitigation devices that apply high recovery potentials during non-operating hours. To first check for the recovery potential of the fielded, PID affected modules, each module was exposed to a constant recovery step. New modules were first run through an IEC 62804-1 Aluminum test to check for general PID susceptibility.

For the test the following environmental assumptions were made:

- Maximum potential during operation: -500V
- Maximum potential applied for recovery: +1000V
- Day and night are equally long – 12hours
- Day and night have similar/same temperatures

For the stress level acceleration an Arrhenius behavior was assumed considering typical reported activation energies of 0.7eV [16.] [17.] [18.].

The temperature during the alternating high voltage stress test was set to 85°C for acceleration and 85% rel. humidity to act as a front-side electrode. This is the same setup as defined in IEC 62804-1 [15.]. To better control the applied surface potential the front side of the modules was additionally covered with an Aluminum foil during high voltage stress (additional stress due to homogeneity, see [13.][14.]). Therefor the setup significantly accelerates degradation and recovery with a factor in the range for 100-1000.

As of now the test is designed as a susceptibility test. Due to different climate conditions it cannot be transferred directly into a defined number of years operating time, yet.

Some of the modules were taken from PV power plants and already showed initial PID degradation. That means that the modules already saw some outdoor stress. The detailed impact of this stress is outside the scope of this investigation. For this study the different PID pre-aging (outdoor in the field vs accelerated indoor test) are assumed to cause the same behavior.

3 RESULTS

FIGURE 2 to 4 show the EL imaging results of the A-HVS-Test for three different modules. The results for Figure 2 and 3 are not very different except the module in FIGURE 2 shows a slight improvement during the A-HVS test whereas the module in FIGURE 3 a small degradation (yellow markings). Those modules are classified as "stable" – no further degradation-during the test. The STC flash test performance data is shown in FIGURE 5 (left side). Such modules most likely will follow the Example 6 curve in FIGURE 1.

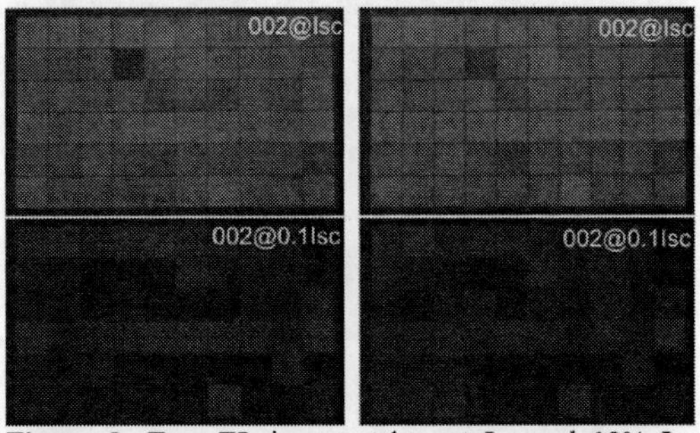

Figure 2: Top: EL images taken at Isc and 10% Isc before and after A-HVS. No significant changes are visible in the EL images. This is confirmed with the performance tests.

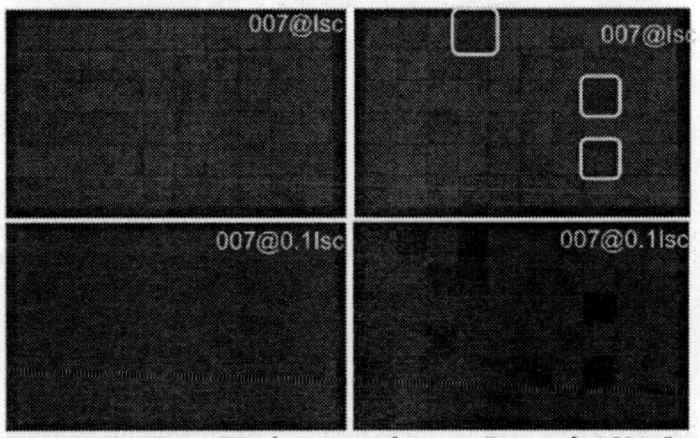

Figure 3: Top: EL images taken at Isc and 10% Isc before and after A-HVS. A slight degradation especially in the 10% Isc images is visible.

The second group of modules is represented by the one shown in FIGURE 4. A strong degradation during the A-HVS-Test is clearly visible and also confirmed with performance measurements (see FIGURE 5 right side). The drop in power is above the set 5% mark where some showed a degradation of more than 50% (see FIGURE 4 and module no. 005 in FIGURE 5). Such a module most likely will follow the Example 7 curve in FIGURE 1.

The STC flash test results are given in FIGURE 5. They confirm the visual impression of the qualitative EL-images in FIGURE 2 to 4. The graphic also shows the measurements after an additional recovery step (yellow bars) and after a long-term ware house storage (blue bar). Both extra steps did not show a significant change in module performance. That means that those modules can hardly be recovered anymore.

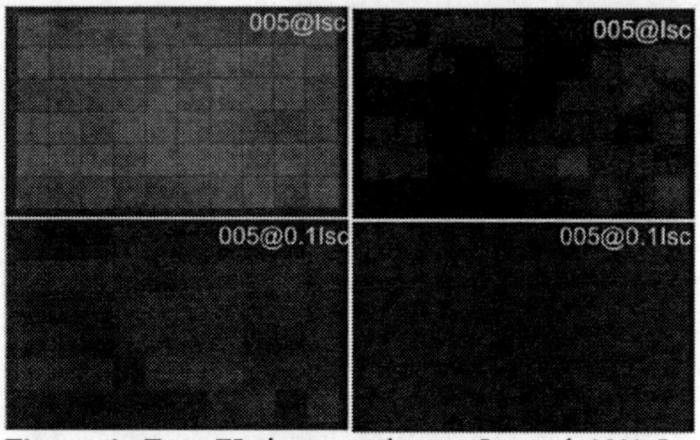

Figure 4: Top: EL images taken at Isc and 10% Isc before and after A-HVS. A significant degradation after the 200 cycles is clearly visible. The EL image taken at low current shows almost no signal any more.

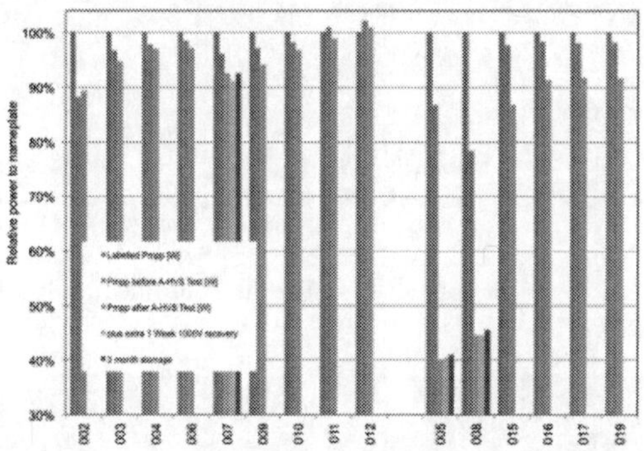

Figure 5: Normalized PV module performance data collected under STC conditions. The two module types ("stable" left and "degrading" right) are easily distinguishable. For the degraded modules an extra regeneration test was performed and furthermore a test after 3 month of ware house storage.

4 SUMMARY AND CONCLUSIONS

Analyzing the results, the modules can be grouped into two types of PID sensitive modules after initial recovery:

1. Stable (FIGURE 2 and 3) and
2. Degrading (FIGURE 4) under A-HVS

Earlier investigations [4.] lead to the assumption, that there might exist modules of a third group where the modules output power improves during the A-HVS-Test. But this expected behavior was not found in this experiment but may depend on stage of recovery/degradation. The modules herein all showed relatively strong PID susceptibility.

The results clearly show that the regeneration and degradation rates of PID susceptible modules can be quite different. Due to the higher potential for recovery (1000V vs. -500V) all modules should recover, but they do not. Based on these observations the „blind" installation of Anti-PID boxes that apply a recovery potential during non-operating hours seems to include a risk.

The presented results are a first step to further understand the recovery of PID affected modules. But also show that a constant potential test as defined in IEC 62804-1 or in the long-term probably also in IEC 61215 might not be suitable to understand and verify PV modules' susceptibility towards high voltage stresses. The constant application of a positive or negative potential is good for screening the general sensitivity but may lead to a fail of certain module types. The A-HVS-Test can look into the risk of using a A-HV-box within the constraints of test setup complexity. This is especially true for the real versus laboratory module temperatures during day and night. Also the durations of degradation are typically not the whole day – more only when there is morning dew or rain. Salt or other soiling will definitively more harm the modules as it applies degrading potentials more effectively and also for longer periods of time, depending on their conductivity and ambient humidity.

Using the proposed procedure, the technical risk for a false decision can be reduced due to the better technical understanding of the PID degradation and recovery rates. Reducing the technical risk also reduces the financial risk to invest in the not-appropriate PID counter measure device.

5. OUTLOOK

More tests are planned to further study the influence for the temperature (here 85°C) as well as the set duration of the cycles (here 15min). Those studies will help to better understand the recovery vs degradation rates. Herein the activation energy was assumed to be the similar to argue for the same durations even this is not correct as the stress during "night" (+1000V) was twice the "day" stress (-500V). That could mean that the observed A-HVS-Test behavior of the modules could change significantly if e.g. the test is performed at 50°C or 95°C. Further the influence of different environmental conditions on the acceleration factor will be analyzed and included in the acceleration model in future work.

6. LITERATURE

[1.] S. Pingel, O. Frank, M. Winkler, S. Daryan, T. Geipel, H. Hoehne and J. Berghold, Potential Induced Degradation of solar cells and panels, 35th IEEE Photovoltaic Specialists Conference, 2010

[2.] S. Pingel et al, „Recovery methods for modules affected by Potential induced Degradation", 27th European Photovoltaic Solar Energy Conference, 2012

[3.] N. Bosco, P. Hacke, S, Kurtz, Adhesion degradation of the metallization-encapsulant interface, 33rd European Photovoltaic Solar Energy Conference, 2017

[4.] B. Jaeckel, M. Cosic and J. Arp, Investigation of c-Si Modules degradation and recovery effect under potentials, 29th European Photovoltaic Solar Energy Conference, 2014

[5.] H. Nagel, R. Pfeiffer, A. Raykov and K. Wangemann, Lifetime warranty testing of crystalline silicon modules for potential induced degradation, 27th European Photovoltaic Solar Energy Conference, 2012

[6.] Chr. Taubitz, M. Schütze, M. B. Koentopp, Towards a kinetic model for Potential induced degradation, 27th European Photovoltaic Solar Energy Conference, 2012

[7.] J. Hattendorf et al, „Potential-Induced Degradation and Temperature-driven Regeneration: A Realistic Simulation", 28th EU Photovoltaic Solar Energy Conference, 2013

[8.] S. Hoffmann et al, "Influence of Humidity and Temperature on the potential induced degradation", 27th European Photovoltaic Solar Energy Conference, 2012

[9.] A. Raykov et al, "Climate model for Potential Induced Degradation of crystalline Silicon", 27th European Photovoltaic Solar Energy Conference, 2012

[10.] V. Naumann et al, "Micro Structural Root Cause Analysis of Potential Induced Degradation in c-Si Solar Cells", Energy Procedia 2012

[11.] V. Naumann et al, "The role of stacking faults for the formation of shunts during potential-induced degradation of crystalline Si solar cells", Physic status Solidi, 2013, DOI 10.1002/pssr.201307090

[12.] Arp, J. Jaeckel, B. Behrschmidt, J., "Yield losses of PID-affected PV systems - simulation of yield losses beyond power loss", 32nd European Photovoltaic Solar Energy Conference, 2014

[13.] Hoffmann, S. Koehl, M., "Influence of Humidity and Temperature on the potential induced degradation", 27th EU Photovoltaic Solar Energy Conference, 2012

[14.] Hoffmann, S. Koehl, M., "Modelling the Leakage Current for Potential Induced Degradation", 28th EU Photovoltaic Solar Energy Conference, 2013

[15.] IEC TS 62804-1, Photovoltaic (PV) modules – Test methods for the detection of potential-induced degradation – Part 1: Crystalline silicon, 2015

[16.] P. Hacke, K. Terwilliger, R. Smith, S. Glick, J. Pankow, M. Kempe, S. Kurtz, I. Bennett, M. Kloos., System voltage potential-induced degradation mechanisms in PV modules and methods for test, Conference Record of the IEEE Photovoltaic Specialists Conference, 2011

[17.] J. Berghold, P. Grunow, P. Hacke, W. Hermann, S. Hoffmann, S. Janke, B. Jaeckel, S. Koch, M. Koehl, G. Mathiak, S. Pingel, L. Poehlman, P. Reinig, A. Ukar, „PID test round robins and outdoor correlation", 28th EU Photovoltaic Solar Energy Conference, 2013

[18.] B. Jaeckel, M. Cosic, Chr. Flueckiger, „PV module degradation under high potentials – a comparative study between test setups", 28th EU Photovoltaic Solar Energy Conference, 2013

Outdoor performance characterization of a novel shadow tolerant module

K. Sinapis[1]*, R. Pacheco Bubi[1], L.H. Slooff[2], W. Folkerts[2], L.A.G. Okel[2], M.J. Jansen[2]
[1]*Solar Energy Application Centre (SEAC), High Tech Campus 21, Eindhoven, The Netherlands*
[2]*Energy research Centre of the Netherlands (ECN), Westerduinweg3, Petten, The Netherlands*
Email: sinapis@seac.cc Phone:+31 612339374
* Corresponding author

ABSTRACT: Large scale implementation of PV in urban areas asks for more shadow tolerant modules and more size flexibility. Power optimizers and micro inverters can be used to improve shadow tolerance of standard c-Si modules but the electrical interconnection and the size of cell groups in the standard module design is a limiting factor when it comes to further increase shadow tolerance. In this work another approach is used in which the cells in the module are smaller and series connected in building blocks. The resulting lower currents allow the use of small in laminate diodes and the building blocks give access to more size flexibility and improved shadow robustness.
Keywords: TESSERA, micro inverter , BIPV

1 Introduction

Solar photovoltaic (PV) systems today play a significant role in the energy mix and their penetration is forecasted to keep growing in the future. Particularly the application of building integrated PV (BIPV) and building added PV (BAPV) systems are projected to thrive in the following years as a result of increasing electricity prices for the residential sector and decreasing PV component costs. Residential and small commercial PV systems are typically installed in an urban environment. Roofs and terraces are often affected by shade coming from the close proximity of buildings, poles, antennas, dormers, etc. introducing electrical and thermal mismatch losses between cells and modules.
Conventional solar module designs consist of 60 or 72 cells connected in series, forming 3 groups of 20 or 24 cells which are protected via a by-pass diode. When partial shading is present on a standard solar module, shading half a cell of a group can cause the current to flow through the by-pass diode. As a result 1/3 of the output is lost. Additionally the voltage will be 1/3 less causing operational issues to the inverter by reaching the lower end of the Maximum Power Point (MPP) voltage. Parallel interconnection has been proved to be more shadow tolerant than series connection but the high current of 6 inch cells reaching up to 9A is not desirable to be connected in parallel in the standard solar module.

2 The TESSERA module concept

The TESSERA solar module concept [1] makes use of smaller cells. It includes 6 inch Metal Wrap Through (MWT) c-Si cells which have been cut to produce 16 mini cells. The mini cells are then connected in series. In total 64 mini cells are connected in series forming a Vmpp of around 33V at standard test conditions (STC). Blocks of 64 mini cells are then connected in parallel forming modules with custom size and thus custom current levels. In this work a full size module with 15 blocks of 64 mini cells are connected in parallel forming a shadow tolerant module with 960 mini cells.
The advantage of such a solution is that the TESSERA module can be customized in terms of size by keeping a lower amount of building blocks which will always have a similar Voc. Even under partial shading conditions the voltage will remain constant and only the current will be reduced. In this way the inverter will operate at high efficiency regardless of the amount of partial shading.

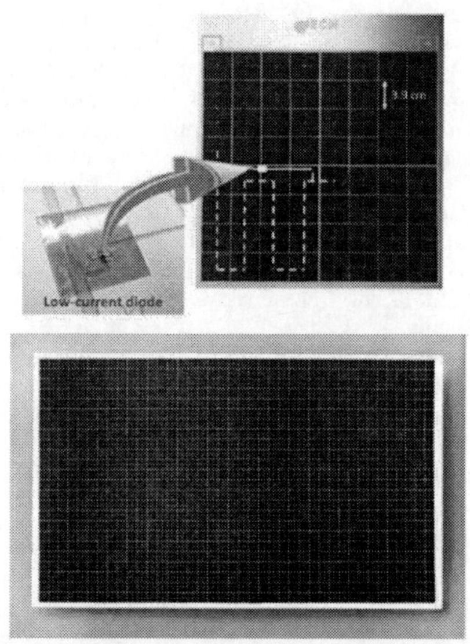

Figure 1. Building block of TESSERA module concept consisting of 4 full size MWT cells (upper side). Full module of 1.6m2 consisting of 960 mini cells (lower side)

3 Field test

Six full size TESSERA modules (around 1.7m^2) have been produced and have been installed outdoors for more than nine months. The TESSERA modules have been equipped with micro inverters from HELIOX. An additional system serving as a reference has been installed next to the TESSERA system with standard 60 cell (three substrings of 20 cells). The same micro inverters have been used for the reference system.
The systems are fully monitored in terms of electricity production for the DC and the AC part with high accurate power analysers. Moreover temperature sensors are deployed and a secondary standard pyranometer in the plane of the solar modules to measure the irradiance. A schematic of the field test is provided in figure 2.

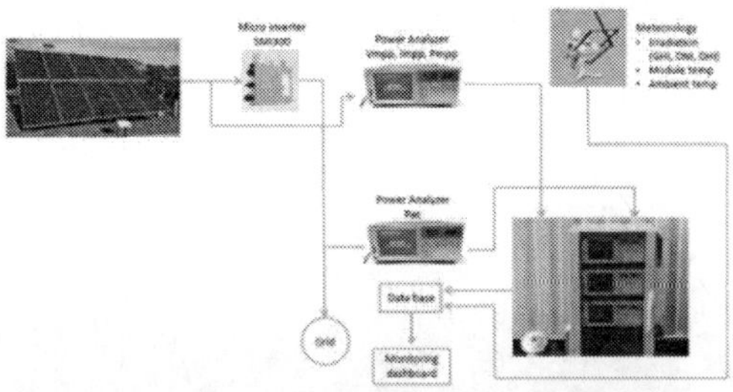

Figure 2. Schematic of the monitoring system

During the field testing artificial shadowing [2] was casted on the two systems in the same way. Specifically a pole has been positioned to the south side of the systems (fig. 3) at the exact same place and thus casting the same shade extension throughout the field testing period.

Figure 3. The six TESSERA full size modules are shadowed by a pole positioned in the center.

4 Results

The field test provided a wealth of data during winter and summer months. In figure 4, the production of a TESSERA and a reference module can be seen. While most of the day the production is almost equal, during the partial shading period the TESSERA module produced 15% more energy.

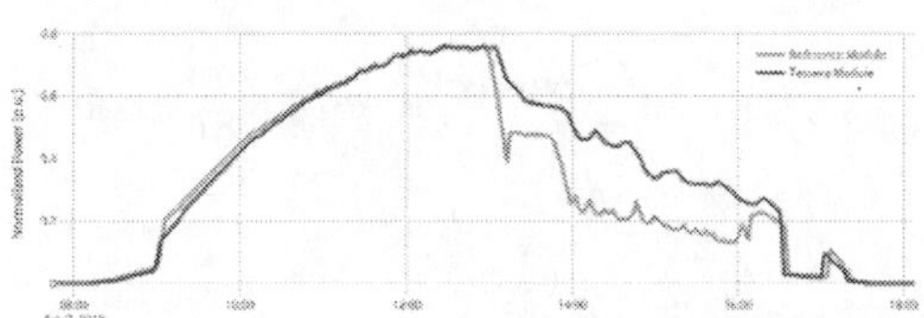

Figure 4. Power production under artificial shade during a sunny day. The TESSERA module produced 15% more DC energy.

The TESSERA modules have a steady voltage output regardless of the partial shading conditions. When partial shading occurs only a portion of the current is reduced. In this way the power electronics can work more efficiently. In figure 5 the relation of voltage, power output and inverter efficiency can be seen. The data set includes more than six months of data. It can be seen that indeed the voltage output stays in a certain range which is only affected by the module temperature and not by partial shading.

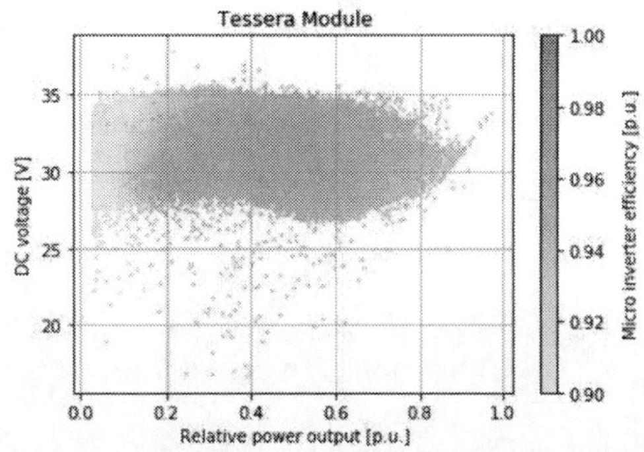

Figure 5. Correlation of Vmpp, power output and micro inverter efficiency of the TESSERA module..

For the sake of comparison, figure 6 shows the correlation under the same operating conditions for the reference module connected to the same micro inverter. It can be observed that the micro inverter of the TESSERA modules is operating in higher efficiency during partial shading operation due to the fact that the voltage output remains steady. In this way an additional AC power bonus can be achieved when using the TESSERA module.

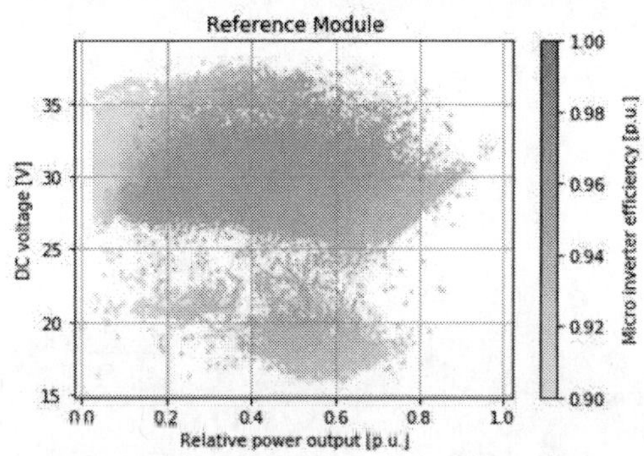

Figure 6. Correlation of Vmpp, power output and micro inverter efficiency of the reference module.

5 Conclusions

In this work, the superior behavior of the TESSERA technology has been demonstrated in an outdoor test set-up under realistic partial shadow conditions. During a sunny day the TESSERA module obtained 15% more DC energy than the reference module. Moreover the steady voltage output of the TESSERA module is boosting the energy output further by operating the power electronic devices in an optimal way.

6 Acknowledgments
This work is supported by "Netherlands Enterprise Agency" (RVO) via the project: Scalable Shadow Tolerant Module with grant number TEID215030

7 References

[1] A.J. Carr, K. de Groot, M.J. Jansen, E.E. Bende, J.A.M. van Roosmalen, L.Okels, W.Eerenstein, "Tessera: Scalable, Shade Robust Module" *42th IEEE Photovoltaic Specialists Conference*, 2015.

[2] K. Sinapis, G. Litjens, M. van den Donker, W. Folkerts, W.G.J.H.M. van Sark, "Outdoor characterization and comparison of string and module level power electronics under clear and partially shaded conditions," *Energy Sci. Eng.*, vol. 3, pp. 510–519, 2015 http://dx.doi.org/10.1002/ese3.97.

COMPARATIVE ASSESSMENT OF ANTI-SOILING NANOCOATED PHOTOVOLTAIC MODULES IN AN ARID DESERT ENVIRONMENT

Abdullah Alkandary, Abdulwahab Al-Asfour, Feras G. Alzubi
Energy and Building Research Center, Kuwait Institute for Scientific Research
Safat 13109, Kuwait City, Kuwait

ABSTRACT: In this work, we assess the performance of three commercial anti-soiling PV nanocoatings under a harsh arid desert climate. The two main parameters monitored under this study were optical transmission losses and PV power output losses. The study began mid-March, with anticipation to experience Kuwait's turbulent sand and dust storm season. Hence by the end of August, sand and dust storms, rain storms, scattered rains, and muddy rains have permeated the experiment's weather conditions. Two of the coatings were hydrophobic and showed slight enhancement of PV performance under dry dusty weather. With a maximum power loss enhancement value (as compared to the uncoated reference PV module) reaching 4.37% and 4.74%. While the third coating, with a hydrophilic nature, performed almost identically to the uncoated PV glass surface (which is also commonly hydrophilic).
Keywords: Photovoltaic, Soiling loss, Optical losses, Power Production, Anti-soiling, Nanocoatings, Wettability, Contact angle.

1 INTRODUCTION

Soiling loss due to dust accumulation on photovoltaic (PV) modules' surfaces is a critical factor of PV arrays or power plants performances. Dust settlement on PV glass surface is primarily dependent on local environment, dust properties, and surface characteristics [1]. One of the possible solutions to deter the aforementioned factors is the treatment of PV glass surfaces with anti-soiling nanocoatings. Despite their extensive employment in the field, anti-soiling nanocoatings' effectiveness in harsh arid environments is still not fully determined. Of the main parameters investigated to determine nanocoatings effectiveness on PV modules are electrical characterization of coated PV's (i.e. energy production and power loss), optical behavior of the surface, mass of settled dust, and thermoelectric analyses (PV surface temperature) [2-6].

Kuwait lies in one of the world's driest and highest dust-intensive regions [7,8]; an arid environment ideal for testing anti-soiling nanocoatings to their limits in harsh natural conditions. Soiling loss has severely reduced PV energy production by up to 24% in Kuwait[1] [9].

In our current study, we assess nanocoatings' power and optical losses in Kuwait's climate for a period of five-and-a-half months; starting from mid-March to the end of August 2018—the time of the year when Kuwait experiences the highest recorded days of sand and dust storms [10].

2 EXPERIMENTAL

2.1 Outdoor Setup

Four polycrystalline silicon PV modules were mounted at the Kuwait Institute for Scientific Research's outdoor testing facility (shown in Fig. 1). The PV modules were oriented facing south at a 20° tilt angle. One of these modules was uncoated and used as a reference, the other three modules were coated with three different commercial PV anti-soiling nanocoatings.

[1] The published value was for a study of a period of three months. The same research team has recently detected an energy yield reduction due to soiling loss of around 56% in Kuwait.

Fig. 1. Kuwait Institute for Scientific Research outdoor PV testing facility (Alshuwaikh Campus, Kuwait City).

Fixated adjacent to the PV modules was a manufacture-provided tempered glass sheet (identical to that used to seal the PV modules under test, with dimensions of 195cm x 98.5cm x 3mm). The glass sheet was sub-divided into four sections of nearly equal areas, as can be seen in Fig. 2: one was kept uncoated as a reference area and the other three were coated with the three commercial coatings.

Fig. 2. Tempered PV glass under test with four segments: uncoated, C1, C2, C3.

Throughout the study, all the tested PV modules in addition to the PV glass sheet were left to soil naturally without any intervention (i.e. cleaning or artificial soiling). The coatings were selected according to the following criteria:

1. Company explicitly claims anti-soiling PV coatings,
2. Product could be applied after PV installation (in-field),
3. Ease of manual application and uniform surface distribution.

Table 1. lists the selected coatings' specifications—as claimed by each manufacturer.

Table 1. Coatings C1, C2, and C3 specifications as claimed by their manufacturers.

Coating	Wettability	Power gain	Anti-dust claims
C1	Hydrophobic - up to 118°	-	Repels bugs, dirt, and debris.
C2	Hydrophobic - up to 105°	More than 10%	Less dirt and dust adherence.
C3	Hydrophilic	Up to 8%	Does not allow dust, mud, oil, and dirt adherence.

2.2 Methodology

Right after mounting the PV modules at the testing facility, the manual coating process took place according to the following procedure:

1. Sanitizing the surface with ethanol and wiping it off with a microfiber cloth, followed by water rinsing, mopping using a squeegee, and drying off with a clean microfiber cloth.
2. Dispensing *double* the amount recommended by each manufacturer[2] of the nanocoating on the PV glass (using a scaled dispenser), followed by manual forceful rubbing of the coating in a circular motion using a clean microfiber cloth.
3. Checking the coating's distribution using a water spray test (which also serves as a wash-up step for the excess product).

The tempered PV glass sheet followed the same process, after dividing the areas and recalculating the amount of coating to be applied on each area.

The characterized parameters were contact angle (CA), transmission loss due to soiling SL_T, and power loss due to soiling SL_P. Contact angle measurements were performed using a DataPhysics goniometer for benchmarking the coatings. As for SL_T, a handheld transmission meter by Solar Spectrum was deployed on a weekly basis. SL_P was acquired by fixed I-V tracers on each PV module which then sends data to our indoor servers on a daily basis[3].

3 RESULTS AND DISCUSSION

3.1 Contact Angle

In order to benchmark the commercial nanocoatings' contact angles, deionized water droplets were dispensed on the surface of coated glass slides and then contact angles were identified. It was observed that the measured contact angles had variant values at different periods of time, which led to taking two contact angle measurements for each coating–one instantly after the coating has been applied, and another after 24 hours. The results in Fig. 3 show that C1 and C2 approach their manufacturers promised hydrophobicity at the 24-hour test. While C3 has deviated from being nearly super-hydrophilic, when freshly coated, to a 44.9° after 24 hours; which exceeds the reference glass slide's contact angle that was found to be 32.5°.

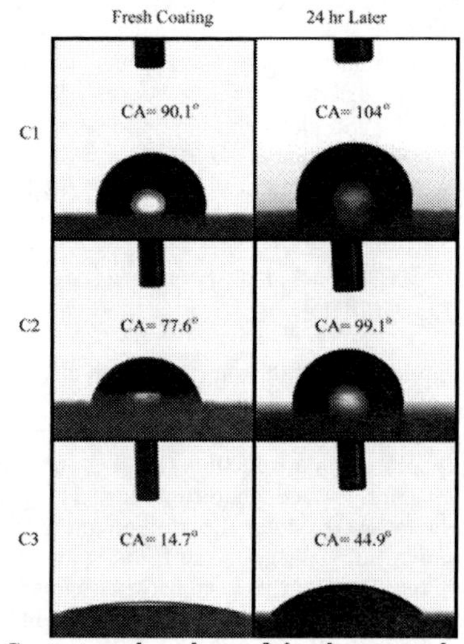

Fig. 3. Contact angle values of the three tested coatings C1, C2, C3 on glass slides, right after the coating process and 24 hour later.

3.2 Transmission Soiling Loss (SL_T)

Transmission loss percentages of the three nanocoated areas of the tempered PV glass, in addition to the uncoated reference area are depicted in Fig. 4. On this figure, five different zones (labeled in red circles) were specified as the regions with steep soiling losses (in terms of both transmission and power).

The first day transmission measurement showed no transmission losses for the coated regions, implying highly transparent coatings. A higher transparency value would indicate lower dust-induced shading. It is clear that along the whole experiment, the reference uncoated area has generally undergone less SL_T than the three coatings. Meaning that, in terms of transmission loss due to soiling, the untreated surface showed better light transmittance than the coated surfaces. In the first three weeks, labeled in Fig. 4 as Zone 1, there seems to be a steep incline in SL_T, i.e. high transmission loss, for all samples which is due to the turbulent weather conditions. Muddy rains followed by sand storms and scattered rain in that period of time has affected transmission levels dramatically, increasing SL_T up to 93%, 87%, 90%, 90% for uncoated, C1, C2, and C3 respectively. Those weeks were followed by a week of scattered rain and a sand storm in between which resulted in a remarkable behavior of the tested surfaces. The uncoated and C3 samples (which are both hydrophilic) faced less SL_T, in fact from that point on, they both outperformed the hydrophobic C1 and C2 in terms of

[2] Our preliminary testing followed the amount specified by the companies and weak surface coverage was observed.

[3] Data points of some days were ruled out due to machinery errors.

SL_T. This behavior is likely the result of a cementation process which consequently eliminates the coating's effectivity, making the surface as if it were uncoated [5,11]. Also, allowing for better dust accumulation by creating a strong dust-to-dust bond on the surface; which explains the soiling build up that led to transmission loss in Zone 4–when no sand or dust storms were present. Table 2 shows the SL_T per day during the five red zones in Fig. 4 taken from regression curves over each zone's time period.

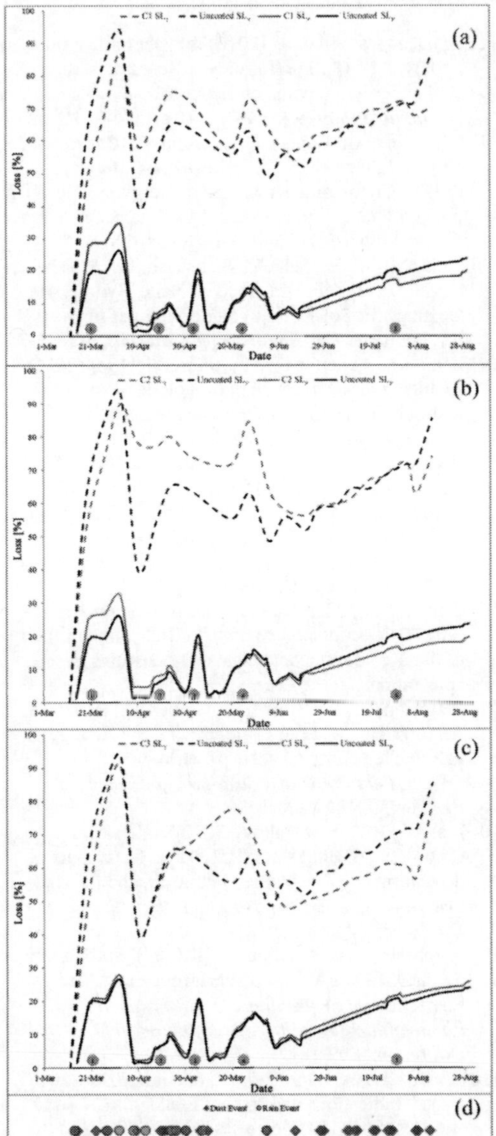

Note: Red circles and lines represent the five Zones of interest, specified based on significant steep losses.

Fig. 4. Transmission and power loss percentages due to soiling of coatings (a) C1, (b) C2, and (c) C3 as compared to the reference uncoated samples (represented in black lines). In addition to (d), major dust and rain events over the testing period [12].

Table 2. Soiling transmission loss per day of the four PV glass areas under test (uncoated, C1, C2, C3) during five specified periods of time.

Zone	Date	Transmission Loss Per Day (%)			
		Uncoated	C1	C2	C3
1	March 12 - April 3	4.08	3.82	3.95	4.11
2	April 10 - April 24	1.76	1.16	0.10	1.57
3	April 24 - May 3	-0.15	-0.33	-0.67	0.52
4	May 20 - May 28	0.96	1.86	1.58	-0.83
5	June 21 - August 14	0.47	0.21	0.27	0.36

It is worth mentioning that Fig. 4(d) shows dust *events*; which are sand and dust storms officially recorded by the Metrological Department of the Directorate General of Civil Aviation of Kuwait [12]. Nevertheless, statistically, the number of days with dust and sand fallout per year in Kuwait is 255.4 days (70% of the year) [13]. That explains the soiling power and transmission losses around Zone 4.

3.3 Power Soiling Loss (SL_P)

Also plotted in Fig. 4 are the power losses due to soiling (SL_P) of the four PV panels under test staring from mid-March to the end of August 2018. It is quite obvious that in Zone 1, under intensive muddy rain events, the two hydrophobic nanocoatings C1 and C2 underperformed the uncoated module by reaching a severe 34.3% and 32.1% power max loss respectively; while the uncoated module has peaked at 25.7% loss in that same period. Furthermore, and in spite of the clean showers that occurred on April 4th and the decrease of SL_P in all modules under test, the uncoated module outperformed the coated modules up until the end of Zone 2 (when rain events were mostly accompanied by sand storms and mud). From Zone 3 and on, C1 and C2 showed a minimal enhancement of power loss due to soiling, reaching a maximum difference of SL_P by the end of the August with 4.37% and 4.74% respectively. Throughout the study, C3 (hydrophilic coating) has underperformed the uncoated module in terms of SL_P. Table 3 shows daily SL_P during the five zones of interest.

Table 3. Soiling power loss per day of the four PV glass areas under test (uncoated, C1, C2, C3) during five specified periods of time.

Zone	Date	Power Loss Per Day (%)			
		Uncoated	C1	C2	C3
1	March 15 - April 3	0.96	1.31	1.26	1.03
2	April 10 - April 24	0.57	0.46	0.45	0.59
3	May 1 - May 6	3.43	2.90	2.89	3.03
4	May 21 - May 30	0.65	0.57	0.57	0.67
5	June 18 - August 30	0.22	0.18	0.17	0.22

From the presented results, an overview of the study suggests weak preventative properties of hydrophobic nanocoatings under muddy rains or even separate intensive rain and sand storm cycles. This agrees with the functionality limitations on hydrophobic-lotus-like surfaces. A hydrophobic surface requires water drops to exceed the size of dust particles in order for the particle pickup and slide to occur [14]. That would be difficult with the presence of agglomerated mud stains on the surface–let alone the cementation buildup due to high heat, enough humidity, and airborne dust. Figure 5 shows the tested PV glass right after the first muddy rain event. It is very obvious that hydrophobic coatings (C1 and C2) have enhanced soiling agglomeration, while a more spread soiling dispersion is noticed on the hydrophilic surfaces (uncoated and C3). The two hydrophilic surfaces also show traces of rain drops, distinguishable by the relatively clean spots bordered by soil.

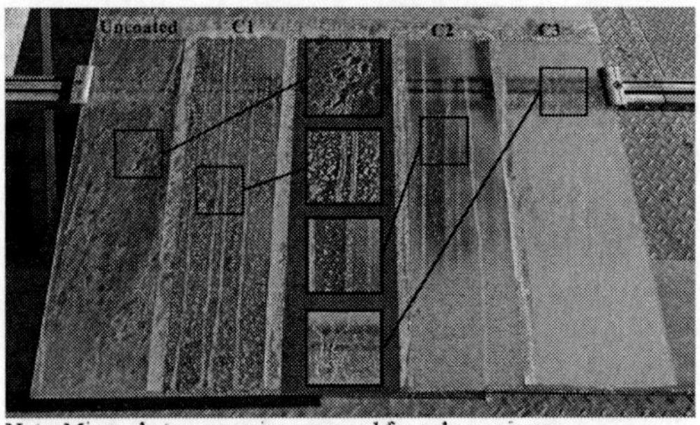

Note: Minor photo-processing was used for a sharper image.

Fig. 5. The tested PV glass sheet with selected areas of each tested region magnified. (photo taken on March 19th, 2018).

During the last three months of the study, the weather was dry and dusty which reflected on the coatings' performance, giving the hydrophobic (C1 and C2) a slight gain in soiling power over uncoated and C3. The results suggest an event-specific conclusion that needs further comprehensive investigation which is: in extreme weather conditions, hydrophilic surfaces showed better performance under muddy rains while hydrophobic surfaces performed better under dry dusty climates

4 CONCLUSIONS

The three nanocoatings were tested in Kuwait climate's harshest times of the year. The results found that commercial anti-soiling nanocoatings show minimal – to null – effectivity in minimizing PV power loss due to soiling in arid desert climates. Even with the maximum enhancement of +4% SL_p that the hydrophilic nanocoatings introduced on a head-to-head comparison to uncoated PV; the severe power losses at certain times associated with soiling would compensate for the power gains. That is why seasonal energy production of coated and uncoated modules will be investigated in the future. Although, in certain times of the year, frequent rainfall events occur at arid regions, they often carry airborne dust (or are accompanied by it due to the ease of exciting dust particles in the region) which enhances cementation build up on glass surfaces.

In spite the fact that SL_T and SL_P seem to follow similar trends under changing weather conditions, an absolute relationship between optical losses and power losses due to soiling is hard to establish. Our findings produced some PV glass surface-specific SL_T and SL_P correlation, which would be further investigated and validated. As for the present data, the suspected correlations would exclusively reflect on that single season of the year with difficulties to deduce a generalized soiling power and transmission loss relationship.

The production of customized site-specific anti-soiling nanocoatings is needed for better utilization in arid parts of the world, with consideration of natural dust fallout in those regions. As previously mentioned for instance, Kuwait's atmosphere carries airborne dust 70% of the year. Thus, designing anti-soiling nanocoatings for arid region is ought to consider dusty atmospheres as the norm.

As for additional future considerations, in-lab testing and simulation of extreme weather cycles would provide a better understanding of anti-soiling nanocoatings limits and ways to naturally and artificially revive the coatings' functionality.

5 REFERENCES

[1] Mani, M., & Pillai, R. (2010). Impact of dust on solar photovoltaic (PV) performance: Research status, challenges and recommendations. *Renewable and Sustainable Energy Reviews,14*(9), 3124-3131.

[2] Piliougine, M., Cañete, C., Moreno, R., Carretero, J., Hirose, J., Ogawa, S., & Sidrach-De-Cardona, M. (2013). Comparative analysis of energy produced by photovoltaic modules with anti-soiling coated surface in arid climates. *Applied Energy,112*, 626-634.

[3] Fathi, M., Abderrezek, M., Friedrich, M., & Tabani, K. (2017). Beneficial Effects of Nano Hydrophobic Coatings for Solar Photovoltaic Modules in Dusty Environment. *Key Engineering Materials,733*, 32-35.

[4] Fathi, M., Abderrezek, M., & Friedrich, M. (2016). Reducing dust effects on photovoltaic panels by hydrophobic coating. *Clean Technologies and Environmental Policy,19*(2), 577-585.

[5] Ilse, K. K., Rabanal, J., Schonleber, L., Khan, M. Z., Naumann, V., Hagendorf, C., & Bagdahn, J. (2018). Comparing Indoor and Outdoor Soiling Experiments for Different Glass Coatings and Microstructural Analysis of Particle Caking Processes. *IEEE Journal of Photovoltaics,8*(1), 203-209.

[6] Gholami, A., Alemrajabi, A. A., & Saboonchi, A. (2017). Experimental study of self-cleaning property of titanium dioxide and nanospray coatings in solar applications. *Solar Energy,157*, 559-565.

[7] Maghami, M. R., Hizam, H., Gomes, C., Radzi, M. A., Rezadad, M. I., & Hajighorbani, S. (2016). Power loss due to soiling on solar panel: A review. *Renewable and Sustainable Energy Reviews,59*, 1307-1316.

[8] Prakash, P. J., Stenchikov, G., Kalenderski, S., Osipov, S., & Bangalath, H. (2014). The impact of dust storms on the Arabian Peninsula and the Red Sea. *Atmospheric Chemistry and Physics Discussions,14*(13), 19181-19245.

[9] Alzubi, F. G., & Alasfour, A. T. (207). Soiling Effect on Photovoltaic Modules' Performance in Arid Environment. *Proceedings 33rd European Photovoltaic Solar Energy Conference and Exhibition*, 1809-1812.

[10] Al-Dousari, A., Doronzo, D., & Ahmed, M. (2017). Types, Indications and Impact Evaluation of Sand and Dust Storms Trajectories in the Arabian Gulf. *Sustainability,9*(9), 1526.

[11] Ilse, K. K., Figgis, B. W., Werner, M., Naumann, V., Hagendorf, C., Pöllmann, H., & Bagdahn, J. (2018). Comprehensive analysis of soiling and cementation processes on PV modules in Qatar. *Solar Energy Materials and Solar Cells,186*, 309-323.

[12] Kuwait Meteorological Department's official website: http://www.met.gov.kw/

[13] Al-Dousari, A. M., Al-Dousari, N., Ramadhan, A., Ahmed, M., & Aba, A. (2014). Assessments of Dust

Fallout Within Kuwait. *Proceedings 1st International Conference on Atmospheric Dust,*149-155.
[14]Yu, M., Chen, S., Zhang, B., Qiu, D., & Cui, S. (2014). Why a Lotus-like Superhydrophobic Surface Is Self-Cleaning? An Explanation from Surface Force Measurements and Analysis. *Langmuir,30*(45), 13615-13621.

6 ACKNOWLEDGEMENTS

The presented work is part of project EA048C that is financially supported by the Kuwait Foundation for the Advancement of Sciences (KFAS). Facility and equipment support was provided by Kuwait Institute for Scientific Research (KISR) under Renewable Energy program.

ASSESSMENT OF TECHNOLOGY- AND WEATHER-SPECIFIC TEMPERATURE LOSSES OF VARIOUS PHOTOVOLTAIC TECHNOLOGIES

Christian Camus[1]*, Adrian Adrian[1,2], Josef Bogenrieder[3], Jens Hauch[1] and Christoph J. Brabec[1,4,5]
[1] Bavarian Center for Applied Energy Research (ZAE Bayern), Immerwahrstr. 2, 91058 Erlangen, Germany; [2] International Solar Energy Research Center Konstanz e.V., Rudolf-Diesel-Straße 15, D-78467 Konstanz, Germany; [3] Bavarian Center for Applied Energy Research (ZAE Bayern), Unterkotzauer Weg 25, 95028 Hof, Germany; [4] Materials for Electronics and Energy Technology (i-MEET), Friedrich-Alexander-Universität Erlangen-Nürnberg, Martensstraße 7, 91058 Erlangen, Germany; [5] Helmholtz Institute Erlangen-Nürnberg for Renewable Energy (HI ERN), Egerlandstraße 3, 91058 Erlangen, Germany
*Corresponding Author: Christian Camus; christian.camus@zae-bayern.de

ABSTRACT: In this paper, the temperature-induced performance losses of eight different photovoltaic (PV) module technologies under various real weather conditions are analyzed. Therefore, the electrical performance data of one year gathered at the ZAE Bayern's PV test plant in Arzberg, Germany was correlated to on-site weather data and module temperature measurements. By classifying the measurement days into distinct weather categories an in-depth analysis of the weather- and technology-dependent performance is realized in this study. For each weather category, the different PV technologies are ranked with respect to their electrical yield. Subsequently, the temperature behavior of the technologies is analyzed for each of the categories in order to investigate, in how far the temperature-related yield losses are responsible for the weather-dependent performance differences that are observed for the different technologies. This analysis shows that the performance behavior is one of the main causes for the performance differences that are observed. However, it also becomes clear that further losses due to differences in spectral and / or angular sensitivity or weak light performance also influence the performance significantly. In the future, the results of this study will support system designers in choosing the appropriate PV technology for a given site based on the local climate conditions.
Keywords: Outdoor performance, yield analysis, PV technologies, temperature behavior

1 INTRODUCTION

Usually, studies comparing the performance of different PV technologies assess the cumulative performance of the modules at a single [1] or a limited number [2, 3] of sites. Thereby, the results of these studies are highly dependent on the site of investigation. Hence it is unclear to what extent they can be transferred to other sites In contrast, it would be highly desirable to obtain direct insights into the weather- and climate-dependent performance of a given PV module technology from the monitoring data obtained from ideally just one site. Furthermore, the root-causes determining this weather- and climate-dependent performance need to be understood in order to improve the performance for specific target climatic conditions. Amongst potential root-causes, spectral response, weak light performance, angel dependence and temperature coefficients can be expected to be most significant.

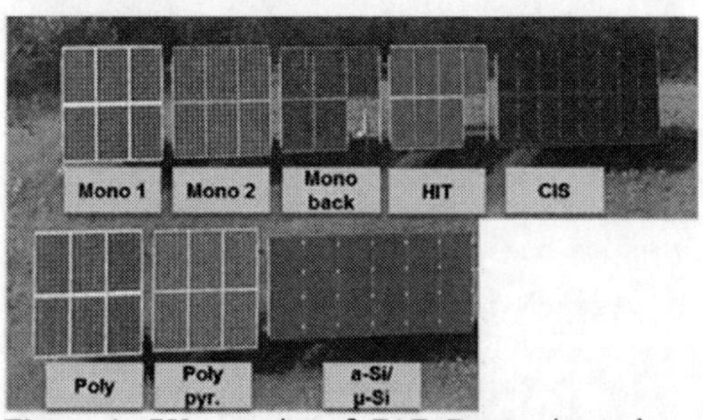

Figure 1: PV test site of ZAE Bayern in Arzberg, Germany. All PV technologies are mounted under identical conditions (from upper left to lower right): *Mono 1, Mono 2, Mono back, HIT, CIS, Poly, Poly pyr., a-Si/µ-Si.*

Table I: List of all module technologies analyzed in this work and their relevant properties.

Abbreviation	Technology	Power [Wp]	Temp.-Coeff. [%/K]	Total Power [kWp]
aSi-µSi	Micromorph tandem module (frameless)	130	-0.24	1.56
CIS	$Cu(In,Ga)(S,Se)_2$	125	-0.39	2.00
HIT	Heterojunction with intrinsic thin layer	240	-0.30	1.68
Mono 1	Monocrystalline silicon (vendor 1)	280	-0.41	1.68
Mono 2	Monocrystalline silicon (vendor 2)	300	-0.41	1.80
Mono back	Monocrystalline silicon with backside contacts	345	-0.30	1.73
Poly	Polycrystalline silicon	255	-0.45	1.53
Poly pyr	Polycrystalline silicon with pyramid-textured glass	260	-0.40	1.56

Table II: PV Technologies with the highest and lowest electrical yield for each weather category as obtained from a previous study

	Highest Electrical Yield			Lowest Electrical Yield		
Weather category	High quantity	Medium quantity	Low quantity	High quantity	Medium quantity	Low quantity
Tropical day	HIT	Mono 2	-	CIS	CIS	-
Summer day	Mono 2 / HIT	Mono 2	-	CIS	CIS	-
Sticky day	HIT	Mono 2	Mono 2	CIS	CIS	CIS
Normal day*	Mono 2	Mono 2	Mono 2	CIS	CIS	aSi-µSi
Rainy day	Poly	Mono 2	Mono 2	CIS	CIS	CIS
Frost day	Mono 2	Mono 2	Mono 2	aSi-µSi	CIS	Poly Pyr
Ice day	CIS	aSi-µSi	aSi-µSi	Poly Pyr	Poly Pyr	Poly Pyr

Therefore, in a previous study [4], we introduced a novel method for classifying the monitoring data of different PV technologies based on the weather conditions present during monitoring. By this approach we were able to deduce a performance matrix describing the electrical yield performance of eight different PV technologies (**Table I**, **Figure 1**) for 21 weather categories (**Table II**). Following this methodology, the weather conditions at the site of interest merely need to be broken down into the weather categories of this study so that one can easily assess the relevance of each category result (i.e. loss) for the site of interest. Thus, designers of PV systems are enabled to choose the most appropriate technology for a given site by estimating the amount of temperature-related losses, which are particularly important for highly relevant PV markets such as the Middle East or California.

By analyzing the temperature-related losses of PV modules of different technologies for each weather category independently, this study, for the first time, allows to transfer the learnings about the severity of temperature-related losses to virtually any site in the world whilst spectral and weak light effects have been discussed elsewhere [8].

2 APPROACH

This paper analyzes the root-causes of weather-dependent performance differences of eight different photovoltaic (PV) module technologies (**Table I**, **Figure. 1**) that were observed previously [4]. The weather dependence of the PV technologies was investigated by correlating the monitoring data to the weather data for a one-year period in Arzberg, Upper Franconia, Germany. Therefore, the measurement days were categorized into 21 meteorological weather categories based on a classification scheme of the German Weather Service (DWD) [5] and the cloud coverage situation [6]. For these categories, the module technologies with the highest and lowest electrical yield for days of a given category were identified (**Table. II**). However, so far this ranking was purely empirical and the reasons for the performance differences remained unknown. The temperature behavior (i.e. temperature coefficient), spectral and angular sensitivities as well as the weak light performance can all affect the performance of PV modules at a given installation site [1,7]. Therefore, we analyze the influence of one of the most significant impacts on PV performance – the temperature dependence of the electrical yield – by quantifying the temperature-related losses for each weather category based on electrical yield and module temperature data. Thereby, the weather categorization also allows for transferring the learnings to virtually any other site of interest.

3 EXPERIMENTAL DETAILS

The PV installations of the eight PV technologies (*Mono 1*, *Mono 2*, *Mono back*, *HIT*, *CIS*, *Poly*, *Poly pyr*. *aSi/µSi*) were installed at the ZAE Bayern PV test site. And monitored during the period from March 1 to 2016 - February 28, 2017 for the present study. Details on the layout of the different PV systems at the test site as well as on the instrumentation for weather monitoring are described in [4] in greater detail. For the present study, module temperature data was obtained by attaching thermocouples to the back side of the modules. The temperature obtained from the thermocouples was then corrected by 2°C for glass/backsheet and 3°C for glass/glass PV modules in order to obtain the cell temperature according to [9]. Adhesion of the thermocouples was monitored closely. Days for which it was found that one or more thermocouple lost contact to the modules were excluded from the temperature data evaluation. As a result the matrix depicted in **Figure 2** was obtained. Here, the entries in each matrix field state the numbers of days that were evaluated for the respective category.

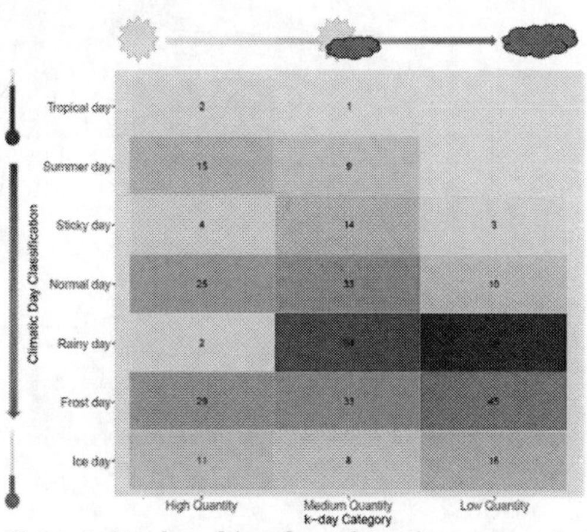

Figure 1: Number of days for each weather category that were included in the temperature analysis.

4 RESULTS

Figure 3 displays the module temperature as a function of irradiance. Obviously, the actual operating temperatures for the different module technologies differ significantly for almost all weather conditions. For a given irradiance, differences in the average temperature can be as high as 8°C. Thereby, *CIS* always exhibits the highest temperature whereas *Poly pyr.* exhibits the lowest. This already partially explains the inferior performance of the *CIS* modules under almost all weather categories since they also exhibit one of the highest temperature coefficients (**Table I**).

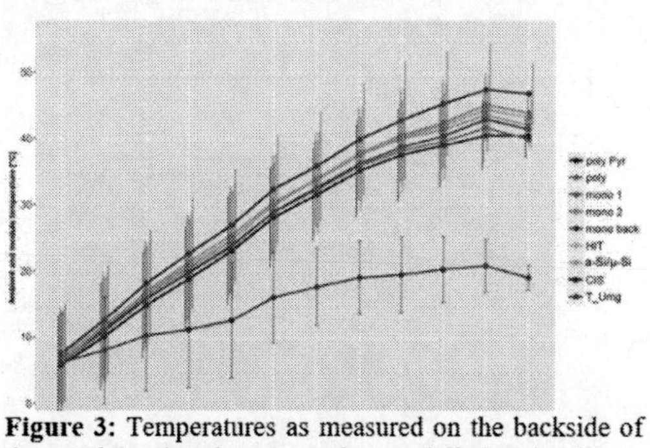

Figure 3: Temperatures as measured on the backside of the modules by thermocouples at differing irradiance levels. The ambient temperature is also shown (dark blue trace).

Analogously, **Figure 4** shows the temperature profiles during the course of an average day of each category. Note that due to the deviations in the occurrence of days of the various categories the averageing may or may not be an adequate representation of a "true" average profile. It should however, be an appropriate approximation to first order.

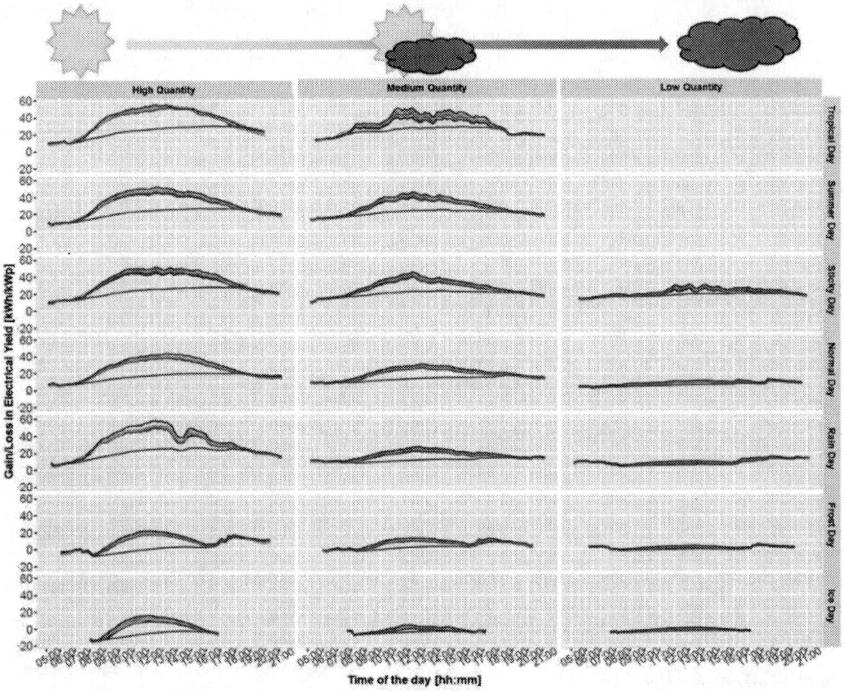

Figure 4: Temperature profiles of the evolution of module temperature for the various PV technologies during the course of an averaged day of the respective weather category.

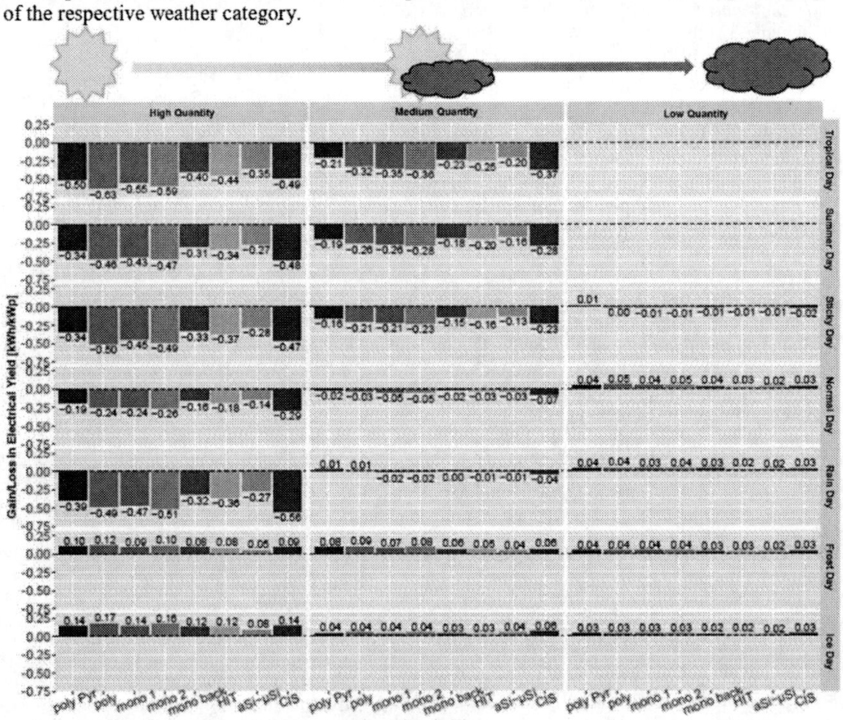

Figure 5: Calculated daily absolute losses in specific electrical yield due to deviations from STC for each module technology for all weather categories.

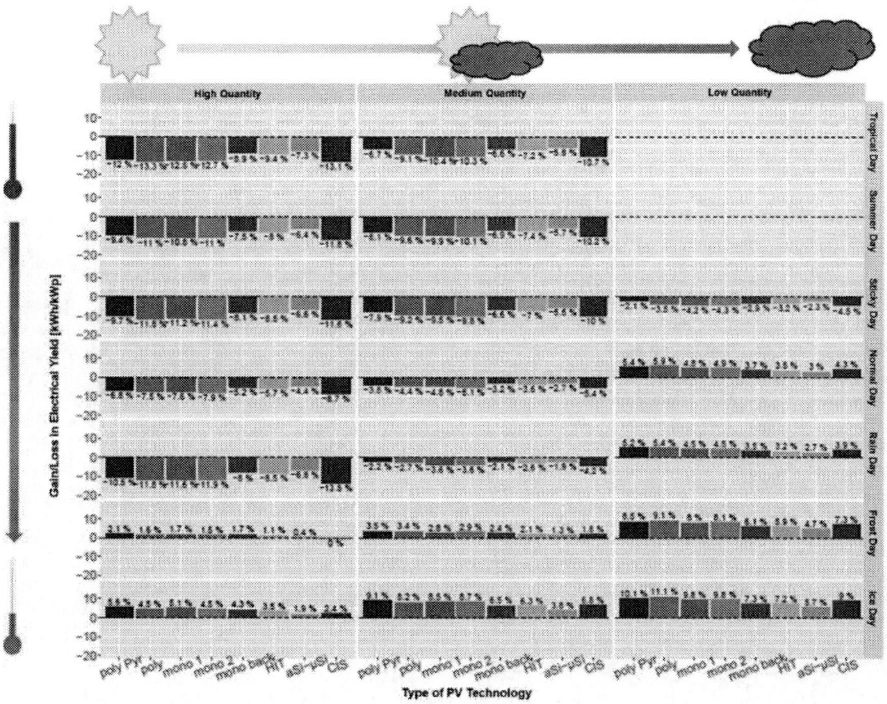

Figure 6: Daily relative electrical yield loss or gain which is attributed to the thermal behavior of the various PV module technologies for all weather categories that occurred during the course of this study.

Based on the module temperatures, the losses $\Delta Yield$ (for T_m >25°C) and gains (for T_m <25°C) resulting from the deviation dT of the module temperature from 25°C were calculated by applying the temperature coefficients γ of the modules (see **Equations (I)-(IV)**), which were also verified experimentally in the course of this study, to the measured electrical yields.

$$dT = T_{mod} - 25°C \qquad \text{(I)}$$

$$\%dev = dT \cdot \gamma \qquad \text{(II)}$$

$$P_{STC} = \frac{P_{measured}}{(1 + \%dev)} \qquad \text{(III)}$$

$$\Delta P = P_{STC} - P_{measured} \qquad \text{(IV)}$$

$$\Delta Yield = \frac{\int \Delta P \; dT}{installed\ nominal\ power} \frac{5}{60} \qquad \text{(V)}$$

where, P_{STC} and $P_{measured}$ represent the expected power output at STC and the measured one, respectively. The resulting losses and gains per kWp are shown in **Figure 4** in absolute values and in **Figure 5** in relative values. As expected, the losses are most significant for the categories of highest irradiance. The losses also vary significantly within one category. Generally, *HIT* and *Mono back* exhibit the lowest losses due to their superior temperature coefficients. Both technologies are also among the most efficient technologies for warm and sunny categories (tropical day, high quantity; summer day, high quantity; sticky day, high quantity) with *HIT* being the most efficient for two of them, e.g., temperature-related relative losses of 20% with respect to the best-performing module were observed for summer days and sunny conditions. Strikingly, in spite of its rather poor temperature coefficient, *Mono 2* reaches the same electrical yield as *HIT* for summer days, which shows that also other factors like weak light or spectral behavior, which will be analyzed separately, contribute to the module performance. Nevertheless, this study clearly reveals the importance of the temperature-related losses for sunny regions, which are minimized for *HIT* and *Mono back* modules.

5 DISCUSSION

The results of this work clearly reveal the importance of the temperature behavior of the various PV technologies. It has been shown that temperature-related losses in electrical yield can amount to 12-13% for sunny conditions, whilst other technologies only exhibit losses of around 6%. Hence, for sunny regions module technologies like *HIT* or *Mono back* appear to be the most attractive choices. Even though the Levelized cost of energy calculation is out of the scope of the present paper it can be assumed that in spite of their higher price these modules are likely to compensate for this initial investment by significantly higher electrical yield.

However, it has also been shown, that the temperature behavior is not the only important factors, which becomes clear by the superior performance of the *Mono 2* modules. It was shown elsewhere that the weak light performance is equally important even under sunny conditions [8]. In contrast, modules with high temperature coefficients may be an appropriate choice for colder regions. In particular, the high temperature of the *CIS* modules may become beneficial for snowy conditions as it appears to enhance snow sliding, which was shown in [10].

6 ACKNOWLEDGMENTS

The research project “Smart Grid Solar” was co-financed by the European Union through the European Regional Development Fund and by the Free State of Bavaria.

7 REFERENCES

[1] A. Louwen, A. d. Waal, R.E.I. Schropp, A. Faaij, W. v. Sark, Prog. Photovolt: Res. Appl., 25 (2017) 218
[2] M. Schweiger, J. Bonilla, W. Herrmann, A. Gerber, U. Rau, Prog. Photovolt: Res. Appl., 25 (2017) 968
[3] B. Zinßer, PhD Thesis, Universität Stuttgart (2010)
[4] J. Bogenrieder, M. Hüttner, P. Luchscheider, J. Hauch, C. J. Brabec., Prog. Photovolt: Res. Appl., 26, (2018) 74
[5] German Weather Service (DWD) website; visited on Nov. 29th 2017:https://www.dwd.de/DE/service/lexikon/Functions/glossar.html?lv2=101334&lv3=101452
[6] B. O. Kang, K.-S. Tam., Solar Energy 94 (2013) 102
[7] A. Louwen, R.E.I. Schropp, W. v. Sark, A. Faaij, Solar Energy 155 (2017) 1339
[8] C. Camus, A. Adrian, J. Bogenrieder, J. Hauch, C. J. Brabec, i-PVTC, Cassis, France (2018), DOI: 10.13140/RG.2.2.24789.88802
[9] G. Tina, R. Abate; Proc. 14th IEEE Mediterranean Electrotechnical Conf. (2008) 579
[10] J. Bogenrieder, C. Camus, M. Hüttner, P. Offermann, J. Hauch, C. J.Brabec, J. Ren. Sust. En. 10, 021005 (2018); doi: 10.1063/1.5001556

NOVEL COMPUTATIONAL FLUID DYNAMICS MODELING OF SPATIAL CONVECTIVE HEAT TRANSFER OVER PV-MODULES MOUNTED ON AN INCLINED SURFACE WITH AN UNDERLYING AIR GAP

Mohammed Gofran Chowdhury[1, 2, 3], Lucas Somma[2, 5], Dirk Goossens[1, 4], Hans Goverde[2, 3], Imre T. Horváth[2, 3], Eszter Voroshazi[2, 3], Jef Poortmans[1, 2, 3, 6], Francky Catthoor[1, 2]
[1]KU Leuven ELECTA/ ESAT - MICAS
[2]IMEC
[3]IMEC (EnergyVille campus)
[4]KU Leuven Department of Earth and Environmental Sciences, Geo-Institute
[5]University of Applied Sciences Upper Austria
[6]University of Hasselt Physics Department
[1]Kasteelpark Arenberg 10, 3001 Leuven, Belgium (name.surname@kuleuven.be)
[2]Kapeldreef 75, 3001 Heverlee, Belgium (name.surname@imec.be)
[3]Thor Park 8320, 3600 Genk. Belgium
[4]Celestijnenlaan 200E, 3001 Leuven, Belgium
[5]Wels, Austria (lucas_somma@hotmail.com)
[6]Campus Diepenbeek, Belgium (name.surname@uhasselt.be)

The energy yield of PV modules depends on solar insolation, wind speed, wind direction, ambient temperature, the mounting structure, the building structure, and the local topography. The interaction of all these factors results in either a heating up or a cooling down of the module. Variations in temperature result in variations in the module's energy yield. Hence it is imperative to have accurate information on cell and module temperature. State of the art studies already reports on the dependency of module temperature on wind velocity. However, the current literature does not consider fine grain details or assumes a constant heat transfer or an average temperature over the module surface. In this paper, we work on the fine grain details and include volumetric heat generation from the cell and correlate it with the forced convection over PV module surfaces. A test PV setup was mounted on an inclined roof with a 3.5 and 5.5 cm air gap underneath the module. Inlet freestream wind velocities of 3, 4 and 5 ms-1 were considered We only considered forced convection because, for 3, 4 and 5 ms-1 freestream velocities natural convection was negligible. We proposed a feasible method to include the fine details with the thin layer feature of COMSOL inside the module. By doing so, the computational effort was significantly reduced. Results were compared to experimental data collected in a wind tunnel. Overall, there was good agreement between the simulated and the measured results. The mismatches that were observed arose mostly from overestimations in the simulation. The modeling technique can be used in future studies on the effects of wind velocity and wind direction on temperature distributions over PV setups of different configurations.
Keywords: Façade, Rooftop, Module, Performance, PV System, PV Module, Simulation, Wind Modelling, Computational fluid dynamics

1 BACKGROUND

Solar photovoltaic (PV) module energy yield depends on various ambient conditions and module components. The dependency on such dynamic ambient conditions makes it difficult to optimize non-flexible modeling approaches. Solar insolation, wind speed, wind direction, ambient temperature, the mounting structure, the building structure, and the local topography have a combined effect on energy yield. Before developing a method to describe all these environmental effects together, it is imperative to have a good understanding of these elements separately. In this paper, we focus on simulating the effect wind has on the surface temperature of a PV module mounted on an inclined surface with air gaps underneath.

Typically a commercial PV module converts 13-20% of the incident solar irradiation into electricity [1]. The remaining part is converted into heat. This heat increases the cell's temperature, which is transferred to the module surface both on the front and on the back. Depending on the technology, every increase of one degree in temperature of a silicon solar cell results in a decrease of approximately 0.45% relative efficiency [2] [3]. Moreover, the surface temperature is not constant but varies over the module when the latter is subjected to wind [4] [5]. The temperature difference between the front and the back of a module can mount to 20°C [6].

Analytical models have been developed to correlate the wind speed and the heat transfer coefficient based on thermal boundary layer theory. These models usually consider smooth, horizontal, isothermal flat plates with parallel wind flow [7]. A growing body of literature recognizes that there are limitations in using such analytical formulations for PV applications [8] [9]. This paper reports progress in the fine-grain analysis of coupled computational fluid dynamics (CFD) and heat transfer models to simulate the surface temperature of a PV module on an inclined surface with a 3.5 and 5.5 cm air gap underneath.

2 STATE OF ART

Extensive work has been done on module setups that go beyond the basic horizontal configuration. Wilson et al. studied the heat transfer inside air gaps under a PV module in which they considered average module temperature[10]. Xing et al. followed the assumptions of Wilson et al. and studied the effects of the tilt angle, mounting structure, and air gap size on module temperature and power output [11]. In their study, they considered the variation of solar irradiation due to changes in the tilt angle. They validated 2D simulations using experimental data from Trinuruk et al. [12]. Roeleveld et al. analyzed the heat transfer coefficient of a PV module, focusing on the air channel underneath the module. In their 2D simulation model, they specified the convective heat flux created by the wind and the natural convective heat flux as an input boundary condition [13]. Jubayer et al. and Karava et al. analyzed the forced convective heat transfer of PV modules in various setups of ground and roof-mounted systems in 3D. These studies focused on the array/system level, kept PV module surface temperature constant, and analyzed the convective heat transfer characteristics [14] [15].

Mirzaei et al. experimentally investigated the impact air gaps underneath a building integrated (BIPV) module have on cooling. They reported that air gaps and panel arrangements significantly change the air flow characteristics [16]. Zhang et al. used Mirzaei et al.'s experiments to validate a 3D CFD heat transfer model. Their model showed the highest deviation of 23.1% with measurement for a 0.5 m/s inlet wind speed. The overall average error was 9.7% for isothermal and 8.0% for non-isothermal cases. They observed large deviations on the top surface's windward edge of the PV surface [4]. These papers are the closest related work to our approach today.

3 RESEARCH CHALLENGES

However, the assumption of constant heat transfer or an average temperature over a module's surface will translate into non-negligible overestimations of the power output. Overestimations of up to 4.3% have been reported [5]. Studying the spatial variations of the heat transfer coefficient through experiments is useful but has some limitations. This is because the number of measurement points on a module's surface is limited. Thermocouples surely cannot be placed on solar cells because they will create shading. One way to overcome this is to use infrared camera measurements. However, infrared images also capture unwanted reflections from the module surface. Another solution is to develop a validated model by placing thermocouples on the surface as near to the cell region as possible without interfering with the electricity output. This model can be used for fine-grain temperature variations. Such models were not considered in the literature so far. A validated model could further be used to explore different configurations of PV module setups for other novel applications of building applied PV

4 RESEARCH SCOPE

Here, we propose a feasible path towards the development of a bottom-up physical model. This model can simulate (1) the spatial temperature variations over an inclined PV module that (2) is subjected to different wind velocities and contains an air gap underneath it, and (3) considers a variable volumetric heat transfer from the cells. In this context, a coupled CFD and heat transfer model is the best option to simulate the temperature variations, caused by wind flow, in the thermal boundary layer near the PV surface. We wanted to integrate the front glass, encapsulant, cell, and black polythene back sheet and use these as heat generation sources. The primary challenge is to integrate the effects of the heat transfer from thin 0.2 mm solar cells and apply these to a 54.5×34.6×0.4 cm, six-cell module. We have worked to reduce complexities, memory requirements and the time required for convergence. In the simulation, we have used a resistive thin layer instead of resolving all thin layers as a 3D domain

5 TEST SETUP

The tested setup is a 2×3 cell module of 54.5×34.6×0.4 cm mounted on an inclined roof surface (angle of inclination: 10°). The roof rests on a building structure 54.5 by 52.7 cm. An air gap was left between the building and the roof. Wind speed and surface temperature were measured at four points along the centerline of the module, both on the front and the back surface, at 4.0, 19.6, 35.4, and 50.9 cm from the leading edge (Fig. 2a). Wind speed was measured at 1.5 mm above the surface with a digital Furness FC016 manometer (precision: 0.01 ms-1). The surface temperature was measured using thermocouples (precision: 0.1°C). Illumination was 300 Wm^{-2}. A Keithley 2602A instrument was used to measure the electrical performance of 6 cells independently. Ambient temperatures for 3, 4 and 5 ms^{-1} freestream wind velocities were 21.7, 23.0 and 23.4°C respectively, and atmospheric pressure, 1 atm.

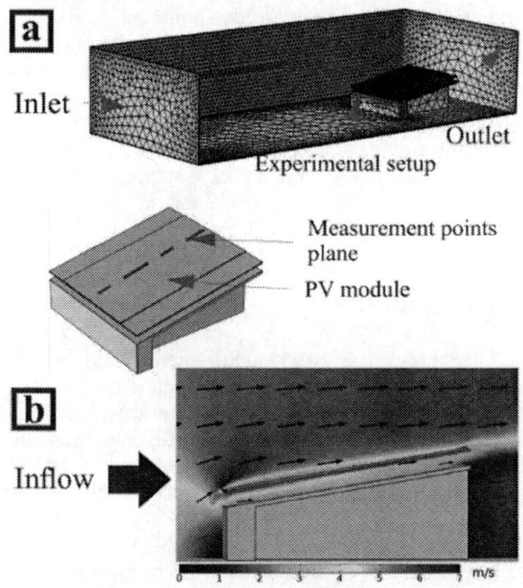

Figure 1: (a) 3D surface mesh of the wind tunnel test section (b) Wind velocity vector and magnitude for a 3.5 cm air gap and 5 ms^{-1} inlet wind speed.

6 FORCED CONVECTION

Three aerodynamic conditions were simulated, with freestream wind speeds of 3, 4 and 5 ms^{-1}. Fig.2 shows the local Richardson number (Ri), calculated from the experimental results. Ri is the ratio of a buoyancy term and a flow shear term [7]. For this calculation, we chose the third measuring point on the module because it was located well downstream, where the boundary layer is well established. For the three wind velocities selected, $Ri \leq 0.01$. Natural convection could, therefore, be neglected [7] [17].

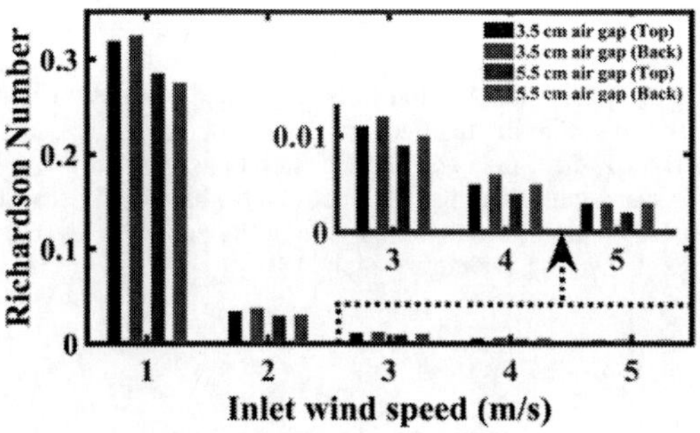

Figure 2: Richardson numbers for different inlet wind speed conditions, for 3.5 and 5.5 cm air gaps.

7 CFD MODEL SETUP

The CFD model is a 3D domain built in Comsol 5.3. The bulk domain of the fluid (air in this case) was an unstructured tetrahedral mesh to capture turbulence and vorticity properly. A dense mesh was applied near the PV module. For the gradual development of the fluid and thermal boundary layers in the near-wall region, prismatic cells were used. A total of 7.388×105 mesh elements were used. Wall function is used to approximate the boundary layer region. Fig 1(a) shows the surface mesh configuration for the bulk CFD domain with inlet and outlet. Within the PV module, we considered glass (3.2 mm), EVA (0.6 mm) and a black back sheet (0.2 mm). The thin layer feature of Comsol 5.3 was used to simulate six full-sized cells within the EVA layer. The thickness of the cells was 0.2 mm. Heat generation from the cell was calculated from the short-circuit current of the cell separately. For the airflow, a standard $k-\varepsilon$ model was used with wall functions on the wind tunnel floor and the module's front and back surfaces. CFD and heat transfer were coupled using the Kays-Crawford method. Heat transfer within the module took place via conduction, and from the module towards the surrounding area, via convection.

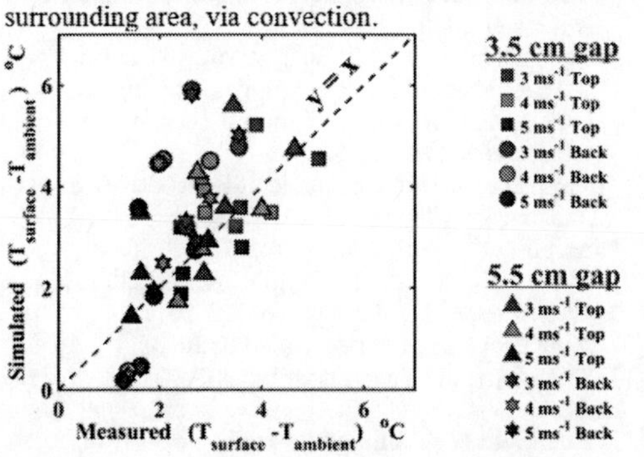

Figure 3: Simulated versus measured surface temperature, for a 3.5 and a 5.5 cm air gap

8 RESULTS AND DISCUSSION

The patterns for the wind velocity vector and magnitude were very similar for the 3, 4 and 5 ms-1 freestream inlet wind conditions and resemble the pattern in Fig. 1(b). Hence only the case for a 3.5 cm air gap and a 5 ms-1 freestream inlet case is presented here.

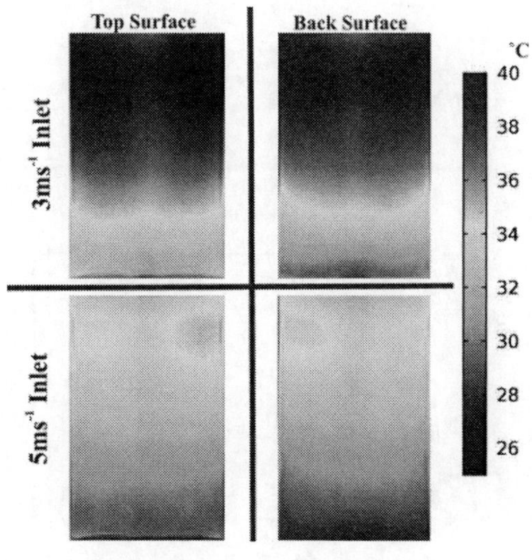

Figure 4: Surface temperature on the module's top surface, for a 5.5 cm air gap.

To validate the model, wind tunnel experiments were performed in the environmental wind tunnel of the Geography and Tourism Division, KU Leuven, Belgium. In Fig. 3 the simulated result is compare Toto the experimentally measured result. Overall there is good agreement between the simulated and the measured result. Most of the discrepancies are because of overestimations of the simulated results, except for the first point on the front surface, where there is an underestimation. This underestimation can be explained

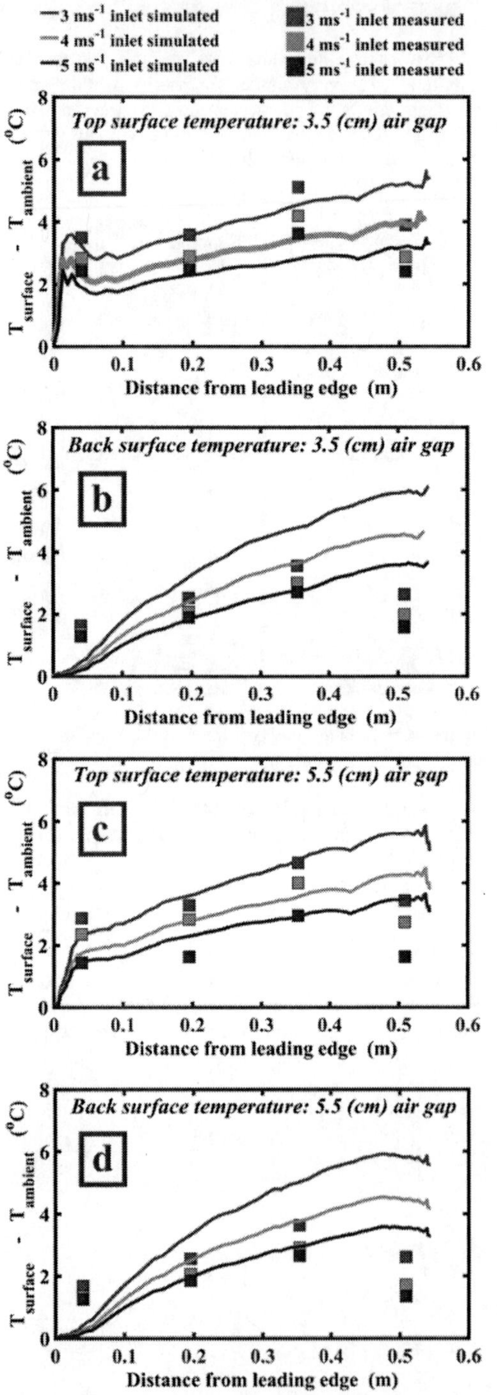

Figure 5: Temperature profile along the central line of the module

by the flow separation that occurred near the module's surface. Zhang et al. reported a similar effect [4]. Less heat is carried away from the module because of the sudden drop in velocity. This suggests that the thermal boundary layer region should be improved in the CFD model.

From Fig. 4 it can be seen that the upwind part of the module shows a fairly uniformly distributed temperature. The central region of the module shows a gradual change. Near the trailing edge high-temperature regions start to develop in the corners. As the air is already heated up, less heat is carried away from the module in these zones.

Fig. 5 shows the surface temperature from the leading edge towards the trailing edge of the module. The top surface is situated in the fully developed boundary layer, and therefore the temperature data show good agreement. The trend for the back surface is comparable, but there is a mismatch. This mismatch could have happened because of the region of transitional flow in the boundary layer, as explained in Chowdhury et al. [18].

9 CONCLUSIONS

In this paper, we report on the progress in developing a bottom-up physical model to simulate the spatial temperature distribution over a PV module subjected to wind. The- model includes the fine grain details of heat transfer from the cells towards the module surface. We proposed a feasible method to integrate all these fine details with a thin layer feature inside the module. The model allows a better understanding of the effect of air gaps underneath a module have on the module's temperature. These insights are necessary for accurate energy yield estimations. Another conclusion is that, for 3, 4 and 5 ms-1 free stream wind velocities, it sufficed to consider only forced convection in the coupled CFD-heat transfer model.

10 ACKNOWLEDGMENTS

The authors gratefully acknowledge IMEC's SiPV industrial affiliation programme and its partners. The work in this paper was partially funded by Kuwait Foundation for the advancement of Sciences under project number P115-15EE-01.

11 REFERENCES

[1] Tonui et al., Sol. Energy, vol. 81, 2007.
[2] Evans Tonui et al., Sol. Energy, vol. 27, 1981.
[3] Tonui et al., 25th Eur. Photovolt. Sol. Energy Conf. September 2010.
[4] Zhang Tonui et al., Sol. Energy, vol. 147, 2017.
[5] H. Goverde et al., Sol. Energy, vol. 147, 2017.
[6] H. Goverde et al., Sustain. Energy Technol. Assessments, vol. 11, 2015.
[7] Bergman et al., Fundamentals of Heat and Mass Transfer, 7th Edition.
[8] Sharples et al., Sol. Energy, vol. 62, 1998.
[9] Armstrong et al., Appl. Therm. Eng., vol. 30, 2010.
[10] Wilson et al., Sol. Energy, vol. 85, 2011.
[11] Xing et al., Sol. Energy, vol. 103, 2014.
[12] Trinuruk et al., Thammasat Int. J. Sc. Tech., vol. 14, 2009.
[13] Roeleveld et al., Energy Procedia, vol. 78, 2015.
[14] Jubayer et al., Sol. Energy, vol. 133, 2016.
[15] Karava et al., Appl. Therm. Eng., vol. 31, 2011.
[16] Mirzaei et al., Sol. Energy, vol. 107, 2014.
[17] Bird et al., Transport Phenomena, 2nd Edition
[18] Chowdhury et al., Sustain. Energy Technol. Assessments, 2017 (under review)

THE STUDY ON ANTI-PID PERFORMANCE OF HIGH EFFICIENCY BIFACIAL CELL MODULE

Xia Cai, Zhichun Ni*, Chengjin Chen, Po Ke, Henglei Chen, Haibo Cao
Suzhou Talesun Solar Technologies Co., Ltd. Changshu, Jiangsu, 215542, P. R. China
* Email: Paul@talesun.com

ABSTRACT: This paper mainly studied the correlation between the PID attenuation of P/N type bifacial cell and the adhesive film. At present, with the development of glass-glass module technology and bifacial cell technology, glass-glass bifacial cell module is an important direction of industry development. Currently P type monocrystal PERC bifacial cell and N-type PERT bifacial cell are typical representatives of the bifacial technology. However, the PID problem of bifacial cell is more serious than single side cell. How to solve the bifacial cell PID problem has become a hot topic. We mainly studied the correlation between two different adhesive film (EVA&POE), and the anti-pid attenuation ability of P/N type bifacial cell. The influence of the water vapor transmission rate and resistivity of the adhesive film on PID attenuation of bifacial cell was studied. After PID 192h test, For P type, power decay of the front side is lower than rear side, while for N type cell, power decay of the rear side is lower than front side. Both P&N type bifacial cell module with POE are much better than with EVA. For POE, recipe adjustment affects PID performance obviously.
Keywords: bifacial cell, PID, water vapor transmission rate, resistivity

1 INTRODUCTION

Potential induced degradation (PID) in photovoltaic (PV) module presents a great challenge for the solar industry, as it could lead to catastrophic failure of the devices [1]. It has previously been reported that certain PV modules suffered a 50% loss in maximum power after a six-month exposure to the outdoor environment due to PID [2]. Addressing PID is critical for ensuring 25-year lifetime of PV modules, and the development of PV [3]. Extensive research has been conducted in recent years to examine, analyze, and prevent PID in PV modules [4]–[14].The PID phenomenon exists not only in the conventional silicon module, but also in the bifacial cell double glass module. The main external factors affected the PID phenomenon of the module are the environment temperature, the environment humidity and the voltage potential. The internal factors are the cells, encapsulation film, and glass and so on.

The encapsulation film is the key material for protecting the module, which isolate cells with the atmosphere environment. The encapsulation film need tolerance all sorts of environment stress, and need have reliable insulation, water resistance and aging resistance. The encapsulation film is crucial to the outdoor reliability, power degradation and life time of the module.

The water-resistance of the encapsulation film determines the effect of the environment humidity on the module. And the resistivity of the encapsulated film determines the electron transfer rate in the module. We studied the correlation between the two different adhesive film (EVA&POE), and the anti-pid attenuation ability of P/N type bifacial cell. At the same time, the influence of the water vapor transmission rate and resistivity of the encapsulation film on the PID attenuation of bifacial cell was also studied.

2 Experimental method

Double glass module samples are choosed and PID tested in the conditions of 85℃, 85% relative humidity (RH) and -1500V bias voltage. The schematic diagram of the PID test is shown in Figure 1. The materials of the samples are same except the encapsulation film and cells. Schematic structure of N type bifacial silicon cell and P type bifacial silicon cell, as shown in Figure 2.

The effect of different adhesive film on the P/N type bifacial cell anti-pid performance was obtained by means of the crossover experiment of P/N type bifacial cell with EVA and POE. At the same time, the correlation of water vapor transmission rate and resistivity of the EVA/POE film with the PID attenuation of module samples is studied.

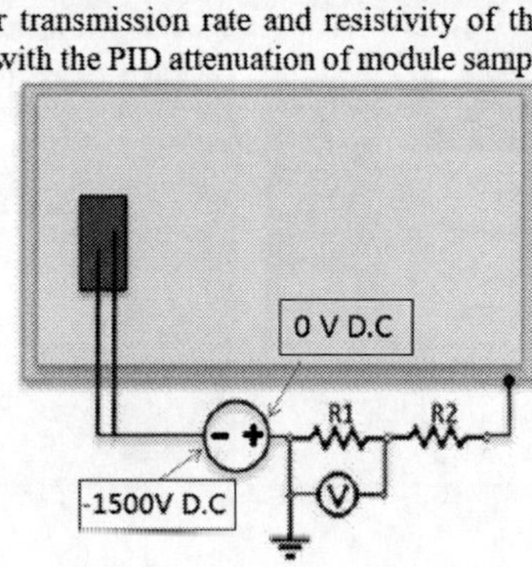

Figure 1: The schematic diagram of the PID test

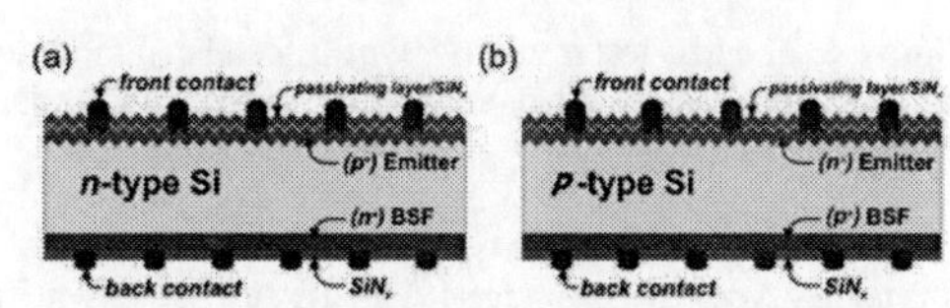

Figure 2: Schematic structure of N type bifacial silicon cell (a) and P type bifacial silicon cell (b)

3 EXPERIMENTAL RESULTS

In order to quickly analyse the anti-PID performance of P/N type cell bifacial with super anti-PID EVA and POE. We prepared module with one bifacial cell in the sample, and then PID tested in the conditions of 85℃, 85% relative humidity (RH) and -1500V bias voltage for 96h, 192h and 280h. The overall EL image after PID was shown in Figure 3. The EL test current remained at 4A. As the Figure 3 shown the N type bifacial cell with POE under PID tested 280h, the power attenuation is more obvious than P type bifacial cell.

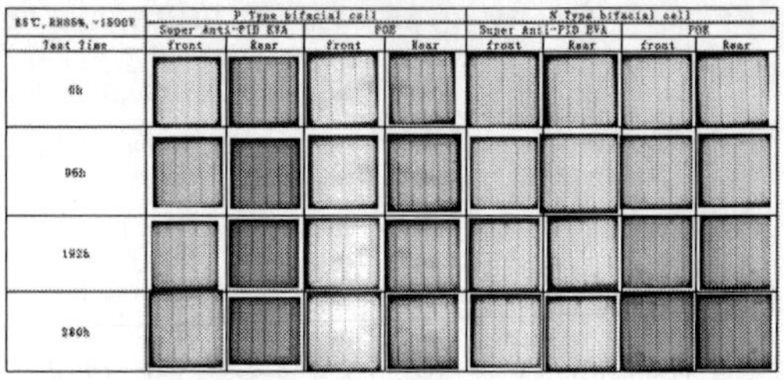

Figure 3: The EL image of P/N type bifacial cell with super anti-PID EVA/POE after PID test

The PID test result of P type bifacial cell module and N type bifacial cell module with normal EVA under 192h is shown in Figure 4. The front side power degradation of N type bifacial cell module is much more serious than the P type cell module. And the front side power attenuation ratio gap between two type modules is 16.29%. The rear side power degradation of P type bifacial cell module is much more serious than the N type cell module. And the rear side power attenuation ratio gap between two type modules is 21.08%.

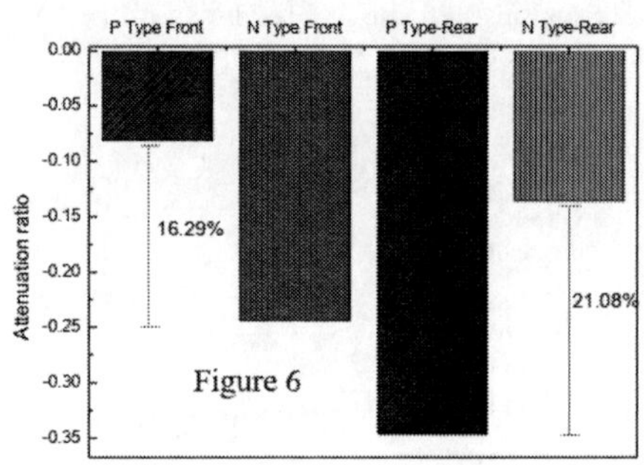

Figure 4: The PID test result of P type bifacial cell module and N type bifacial cell module with normal eva under 192h

The PID test result of P type bifacial cell module with normal EVA and POE under 192h PID test is shown in Figure 5. The power degradation of the module with POE is much better than with EVA, it can be seen that the POE exhibits better protection against PID than the PID-resistant EVA. The front side power attenuation ratio of module with EVA is larger than with POE, and the gap is 4.91%. At the same time, the rear side power attenuation ratio of module with EVA is also larger than with POE, but the gap as high as 28.7%.

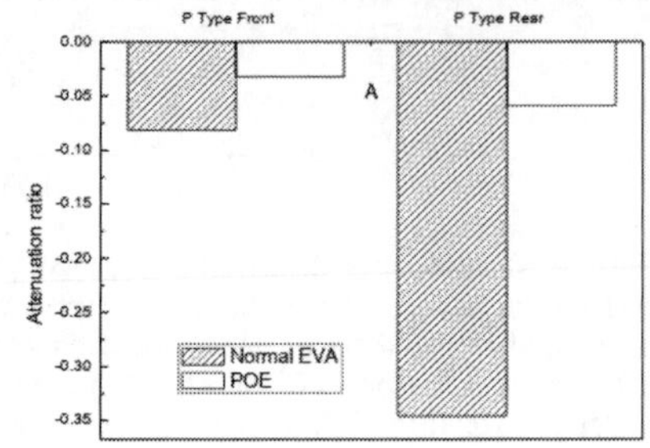

Figure 5: The P type bifacial cell module PID tested 192h with EVA and POE

The PID test result of N type bifacial cell module with normal EVA and POE under 192h is shown in Figure 6. The front side power attenuation ratio of module with EVA is larger than with POE, and the gap is 6.1%. At the same time, the rear side power attenuation ratio of module with EVA is also larger than with POE, and the gap as high as 5.36%. And the performance of the front side is much better than the rear side, whether it's with EVA or with POE.

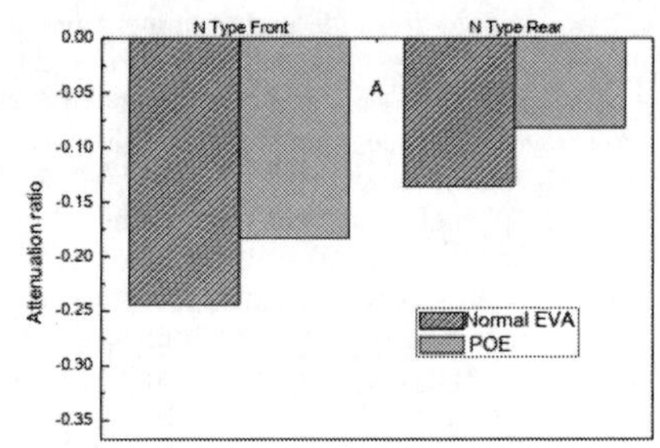

Figure 6: The N type bifacial cell module PID tested 192h with EVA and POE

According to Figure 5 and Figure 6, the power degradation of two types module with normal POE are greater than 5%. The next work, the P/N type bifacial cell modules with super anti-pid POE were further tested for 480h under the conditions of 85℃, 85% relative humidity (RH) and -1500V bias voltage, the power degradation of two types module less than 5%.As shown in Figure 7. When the PID test time is 96h, the Anti-PID performance of front side of N type bifacial cell module is better than P type bifacial cell module. However, the curve of N type bifacial module appears the turning point when the test time is 192h. The performance of N type bifacial module is poorer than P type bifacial module after 192h. And PID test process, the gap of the power degradation of this two modules grown larger.

Furthermore, the rear side power degradation of two types bifacial cell module after PID tested, as shown in Figure 8. When the PID test time is 192h, the Anti-PID performance of N type bifacial cell module is better than P type bifacial cell module. However, the curve of N type bifacial module drop seriously when the test time is 288h. The performance of N type bifacial module is poorer than

P type bifacial module after 288h.

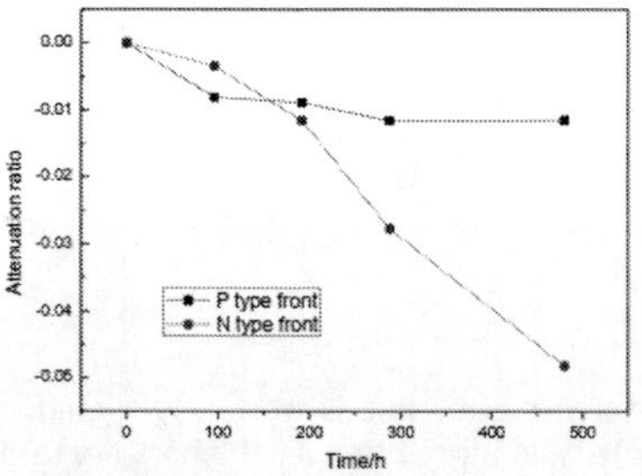

Figure 7: The front side of P/N type bifacial cell module performance after PID test

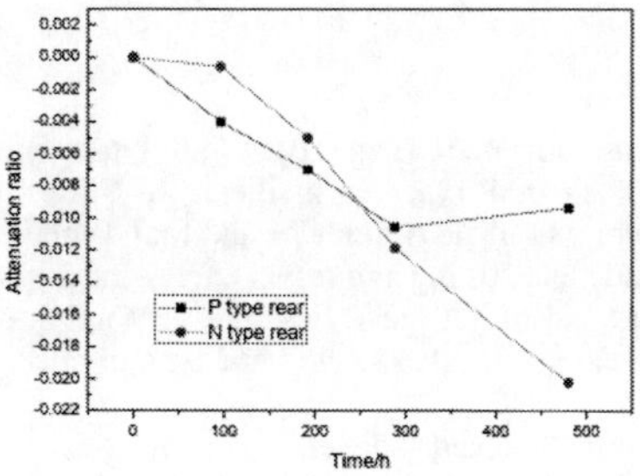

Figure 8: The front side of P/N type bifacial cell module' performance after PID test

4 EXPERIMENTAL ANALYSIS

Therefore, according to the results of previous experiments, the Anti-PID performance of P/N type bifacial cell module with POE is better than EVA. The relationship between PID performance of module with resistivity and water vapor transmission of encapsulated film was analyzed.

The water vapor transmission rate and resistivity of the EVA and super anti-PID POE film were shown in Table 1. The water vapor transmission rate of super anti-PID POE is much smaller than EVA. The resistivity of super anti-PID POE is at least one order of magnitude higher than EVA. Therefore, the higher the water vapor transmission rate, the worse the anti-PID performance. The higher resistivity, the better anti-PID performance, whether it's P type bifacial cell module or N type bifacial cell module.

Table I: The water vapor transmission rate and resistivity of the EVA and super anti-PID POE film

	Unit	**EVA**	**Super anti-PID POE**	**Reference Standard**
Water Vapor Transmission Rate	g/m²/day	28.3	3.2	ASTM F1249
Resistivity	Ω.cm	1.20E+16	1.50E+17	IEC 62788-1-2

5 CONCLUSIONS

In this study, PID in P/N type bifacial silicon PV modules with the double glass structure was investigated. The front side power degradation of N type bifacial cell module is much more serious than P type after 192h PID test, but the rear side power degradation of P type is much poorer. Among the two investigated materials, POE provides the best protection against PID with the slower degradation rates. The performance of N type bifacial cell module is much more serious than P type when under long time PID test. Finally, we analysed the relationship between PID performance of module with resistivity and water vapor transmission of encapsulated film. The higher the water vapor transmission rate, the worse the anti-PID performance. The higher resistivity, the better anti-PID performance, whether it's P type bifacial cell module or N type bifacial cell module.

6 REFERENCE

[1] W.Luo et al., Energy Environ. Sci., (2017) vol. 10, no. 1
[2] E. Schneller, N. S. Shiradkar, and N. G. Dhere, Proc. 40th IEEE Photovolt. Spec. Conf., Denver,CO, USA, (2014) pp. 3216–3219
[3] W. Luo et al., Sol. Energy Mater. Sol. Cells, (2018) vol. 174
[4] S. Pingel et al., Proc. 35th IEEE Photovolt. Spec. Conf.,, Honolulu, HI, USA, (2010), pp. 2817–2822
[5] P. Hacke et al., , Proc. 37th IEEE Photovolt. Spec. Conf., Seattle, WA, USA, (2011) pp. 000814–000820
[6] M. B. Koentopp, M. Krober, and C. Taubitz,. IEEE J. Photovolt., (2016) vol. 6, no. 1, pp. 252–257
[7] P. Hacke et al., IEEE J. Photovolt., (2015) vol. 5, no. 6, pp. 1549–1553,
[8] V. Naumann, C. Hagendorf, S. Grosser, M. Werner, and J. Bagdahn, Energy Procedia, (2012) vol. 27, pp. 1–6
[9] W. Luo et al., IEEE J. Photovolt., (2017) vol. 7, no. 1, pp. 104–109
[10] K. Hara, S. Jonai, and A. Masuda, Sol. Energy Mater. Sol. Cells, (2015) vol. 140, pp. 361–365
[11] H. Nagel, P. Saint-Cast, M. Glatthaar, and S. Glunz, Proc. 29th Eur. Photovolt. Sol. Energy Conf. (2014) pp. 2351–2355
[12] D. Lausch et al., IEEE J. Photovolt., vol. 4, no. 3, (2014) pp. 834–840
[13] S. Hoffmann and M. Koehl, Prog. Photovolt., Res. Appl., vol. 22, no. 2, (2014) pp. 173–179
[14] D. Lausch et al., Energy Procedia, vol. 55, (2014) pp. 486–493

ANNUAL DEGRADATION RATES OF CRYSTALLINE SILICON PHOTOVOLTAIC MODULES IN DIFFERENT CLIMATIC ZONES IN CHINA

Hao Song[1*], Ping Xu[1,2], Zhao Wu[1], Yuan Xia[1], Min Yun[1]
[1]National Center of Supervision and Inspection on Solar Photovoltaic Product Quality(CPVT), Wuxi Institution of Supervision & Testing on Product Quality, 214028, Wuxi, China
[2]College of Mechanical and Electrical Engineering, Hohai University, 213022, Changzhou, China
[*]Email: cpvtsonghao@163.com

ABSTRACT: Module degradation rates in different climatic zones are investigated in this work. Three outdoor PV systems are installed in different places in China. Modules are uninstalled and moved to a local laboratory to do the STC I-V measurements every three months. Degradation rates differ due to the different climatic features, and also the module types. The environment factors affect degradation rate in varying degrees. Glass-backsheet module is sensitive to UV, while dual-glass module is sensitive to a damp heat environment. Dual-glass module has a minor degradation rate than glass-backsheet module in the three sites. Energy output of module is affected by the power degradation. Electricity/power gain is related to the less degradation rate.
Keywords: Multi-Crystalline, Module, Degradation, Reliability

1 INTRODUCTION

Photovoltaic (PV) modules are supposed to serve outdoor decades of years [1]. Understanding the degradation behavior of module is an important part of the investigation of long-term reliability and durability [2-5]. The degradation mechanism of PV module is still not clear now, which is affected by multiple factors such as installation site, mounting conditions, manufacturing process, module types and material [6, 7].

In this work, two types of modules are installed and in operation in three different climatic zones in China. Indoor STC measurements are tested every three months, module degradation modes are investigated and compared in the period of about 2 years.

2 EXPERIMENTAL AND METHODS

2.1 Test sites

Three outdoor test fields are selected, which are in different climatic zones, namely Shunde, Wuxi and Xining. According to the Köppen-Geiger climate classification system, the above three sites belong to three different climatic zones (see Table I).

Table I: Test site specifications

	Shunde	Wuxi	Xining
latitude	22.90 °N	31.51 °N	36.65 °N
longitude	113.20 °E	120.39 °E	101.71 °E
altitude	12 m	24 m	2280 m
tilt angle	22.5°	30°	35°
climatic zone⁺	Cwa	Cfa	Dwb
climatic feature⁺	Subtropical Dry Winter	Humid subtropical climate	Warm-summer humid continental climate

⁺According to Köppen–Geiger climate classification system.

2.2 Modules

Two types of modules are installed in the above three sites, traditional backsheet (KPf) modules (Type A) and dual-glass modules (Type B). Both types employ the same batch of P-type multi-silicon Al-BSF solar cells. There are four Type A modules and four Type B modules at Shunde and Xining site respectively, and eight Type A modules and four Type B modules at Wuxi site. All the modules are connected to the same type of micro-inverter.

2.3 Test method and schedule

The field test started at August 2016. All the modules were installed after an initial indoor STC(AM 1.5G, 1000 W/m^2, 25°C) current-voltage curve measurement. Every three months, the modules are uninstalled and moved to a local laboratory to do the STC I-V and EL measurements. There is also a control module for each type at the three sites, which makes sure the measurement in a stable level.

3 RESULTS

Outdoor field test is ongoing from August 2016. The degradation data is collected for about two years. With the module maximum power output(P_{max}) changes, module degradation mode and rate are investigated.

3.1 Changes of electrical parameters

Module power decreases with time. To find out what parameters affect power degradation, module electrical parameters are tracked. The changes with time of module type A installed in Wuxi is illustrated as Fig.1. The degradation has a positive correlation with changes of short-circuit current, which suggests the light induced degradation may be the main cause of module degradation in the early period.

3.2 Degradation rates in different climatic zones

To quantify the module degradation, definite power degradation rate:

degradation rate = (initial power - current power)/initial power

Degradation rates differ due to the different climatic features (see Table II), and also the module types.

The type A module has a first-year degradation rate of 1.68% at Shunde (Subtropical climate, damp heat), while 2.95% at Xining (Warm-summer humid continental

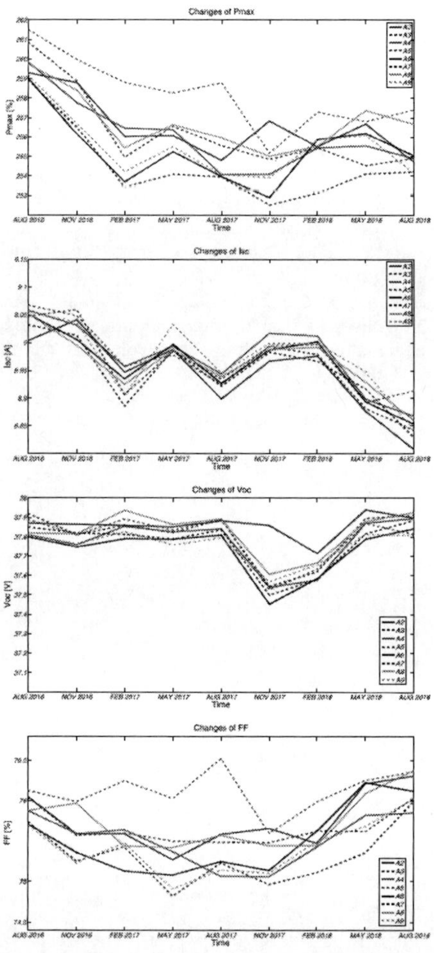

Figure 1: Changes of module electrical parameters of Type A at Wuxi

climate, with high dose level of UV), and annual degradation rate of 1.27% and 1.90%, which suggests that glass-backsheet module is sensitive to UV irradiation.

By contrast, type B module has a first-year degradation rate of 1.58% in Shunde and 0.53% in Xining. Damp heat environment may be an important factor to dual-glass module.

Comparing type A and B, dual-glass module has better performance than traditional backsheet module. Especially at Xining, dual glass module has a much less annual degradation rate than backsheet module, 0.69% vs. 1.90%.

3.3 Degradation trend

In different climatic zones, the same type modules have the similar degradation trends, but the degradation rates differ (see Fig. 2). Module power decreases faster at the beginning (about 6 months), following a relatively smooth decline trend.

Table II: Module degradation rate at the three sites

	Shunde	Wuxi	Xining
type A	*glass-backsheet module*		
Degradation rate at 1st year	1.68%	1.79%	2.95%
Degradation rate at 2nd year	0.85%	-0.03%	0.85%
Annual degradation rate at first two years	1.27%	0.88%	1.90%
type B	*dual-glass module*		
Degradation rate at 1st year	1.58%	1.08%	0.53%
Degradation rate at 2nd year	-0.04%	0.18%	0.84%
Annual degradation rate at first two years	0.77%	0.63%	0.69%

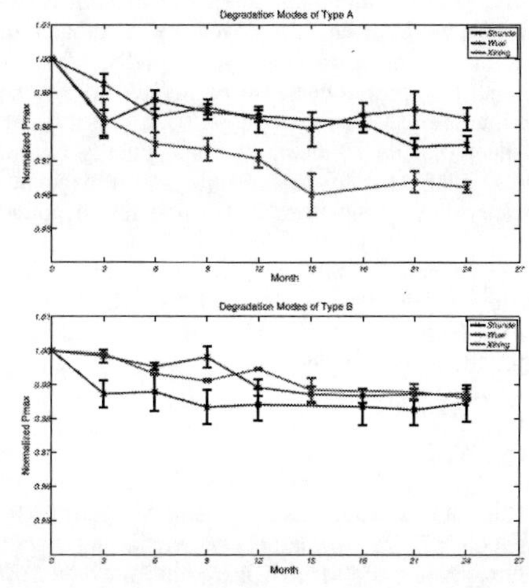

Figure 2: Module degradation trends

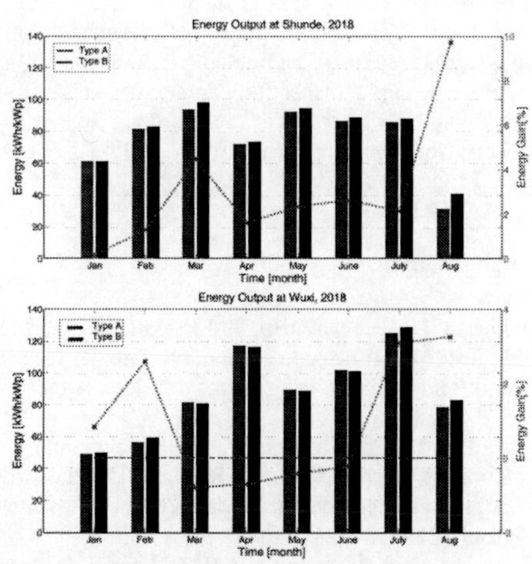

Figure 3: Module energy output

3.4 Energy output

All the modules are connected to the same type of micro-inverter, which ensure maximum power point tracking for every module. The daily energy output of each module is recorded by the micro-inverter. The energy output per watt of year 2018 is showed in Fig. 3, Type B modules produce a little more electricity than Type A, both in Shunde and Wuxi, corresponding to less degradation rates than traditional backsheet module.

4 CONCLUSIONS

Module degradation rates in different climatic zones are investigated in this work. Three outdoor field tests are ongoing in different places in China. Degradation rates differ due to the different climatic features, and also the module types. The environment factors affect degradation rate in varying degrees. Glass-backsheet module is sensitive to UV, while dual-glass module is sensitive to a damp heat environment. Employed the same batch of P-type multi-silicon Al-BSF solar cells, there are differences in power decrease of the two types. Dual-glass module has a smaller degradation rate than glass-backsheet module in the three sites. Energy output of module is affected by the module power degradation, electricity/power gain is related to the less degradation rate.

The degradation rate is also an important parameter to estimate field performance and reliability of the PV module. What is more, field experience is required for module lifetime prediction.

ACKNOWLEDGMENTS

The authors would like to thank the entire MRAT team at CPVT. We especially acknowledge the support of Dr Haitao Xiang and Mr Jie Zhao from Jolywood(Suzhou) Sunwatt Co., Ltd. We also thank Ms Xian Dong, Ms Jianmei Xu, Mr Zhouhua Wu and Dr Qiangzhong Zhu for the helpful discussions. We also thank all the partners for their help: ShunDe SYSU Institute for Solar Energy, Guangdong Testing Institute of Product Quality Supervision(Shunde), Qinghai Product Quality Supervision and Inspection Institute, and Qinghai Tianchuang New Energy Technology Co., Ltd.

This work was supported by the National High-tech R&D Program of China (863 Program) under Grant 2015AA050303.

REFERENCES

[1] Jordan, D. C., Silverman, T. J., Sekulic, B., and Kurtz, S. R. (2017) PV degradation curves: non-linearities and failure modes. Prog. Photovolt: Res. Appl., 25: 583-591.

[2] Jordan, D. C., Silverman, T. J., Wohlgemuth, J. H., Kurtz, S. R., and VanSant, K. T. (2017) Photovoltaic failure and degradation modes. Prog. Photovolt: Res. Appl., 25: 318–326.

[3] Ishii, T., and Masuda, A. (2017) Annual degradation rates of recent crystalline silicon photovoltaic modules. Prog. Photovolt: Res. Appl., 25: 953–967.

[4] Schweiger, M., Bonilla, J., Herrmann, W., Gerber, A., and Rau, U. (2017) Performance stability of photovoltaic modules in different climates. Prog. Photovolt: Res. Appl., 25: 968–981.

[5] Bogenrieder J, Hüttner M, Luchscheider P, Hauch J, Camus C, Brabec CJ. (2018) Technology - specific yield analysis of various photovoltaic module technologies under specific real weather conditions. Prog Photovolt: Res Appl., 26:74 - 85.

[6] Jordan, D. C., Wohlgemuth, J. H. and Kurtz, S. R. (2012) Technology and Climate Trends in PV Module Degradation. Proceedings of the 27th European Photovoltaic Solar Energy Conference and Exhibition.

[7] Jordan, D. C., Kurtz, S. R., VanSant, K., and Newmiller, J. (2016) Compendium of photovoltaic degradation rates. Prog. Photovolt: Res. Appl., 24: 978–989.

NEW DEVELOPMENTS IN MODELING AND NUMERICAL SIMULATIONS OF TEMPERATURE AND YIELD OF COMMERCIAL PHOTOVOLTAICS PANELS UNDER DESERT ENVIRONMENT

Said Ahzi, Shahzada Pamir Aly, Nicolas Barth, Benjamin W. Figgis, Amir Abdallah
Qatar Environment and Energy Research Institute, Hamad Bin Khalifa University
Qatar Foundation, P.O. Box 5825, Doha, Qatar.

ABSTRACT: Due to intrinsic nature of photovoltaic (PV) cells, they can utilize only a small portion of the incident solar spectrum for photoelectric conversion. The unutilized energy builds up inside the PV cells, which in turn raises their working temperature and reduces their electrical conversion efficiency. This indicates that higher temperatures have an adverse effect on the electrical performance of PV devices. A PV system designer therefore needs an accurate temperature prediction of his PV modules under operation, which is not an easy task, owing to the continuously fluctuating field conditions. This review highlights some common practices in the literature for developing thermal models to estimate PV cells temperature and using it to find the electrical yield. A variety of self-developed in-house thermal codes from the authors are also presented and compared with those from the literature.
Keywords: Photovoltaic (PV); Thermal behavior; PV cell temperature; Electrical yield; Modeling.

1 INTRODUCTION

Unlike other solar technologies, photovoltaic (PV) cells have the ability to directly convert sunlight into electricity. On the contrary, solar thermal technologies use optical concentrators to heat a fluid, which in turn drives a turbine to generate electricity. This makes PV technology very easy to be implemented both on small-scale and large-scale.

The most common PV cells technologies are: Mono-crystalline, poly-crystalline and thin-films. Both mono- and poly- are made from silicon, but their manufacturing processes are different. Whereas, thin-films are made by depositing a substrate on a solid surface like glass. Examples of such substrates are amorphous silicon, cadmium telluride (CdTe), copper indium gallium selenide (CGIS) and dye-sensitized solar cell (DSC).

Although the electrical conversion efficiencies of these PV cells have increased significantly in last decades, they still cannot perform to their full potential in field operating conditions. One of the most influential environmental factors, adversely affecting the PV system's efficiency, is its working temperature. There is almost a linear drop in the silicon PV efficiency, for every degree rise in the PV cell temperature [1]. However, thin-film PV cells are less affected by temperature, as compared to their crystalline silicon counterparts. Reducing the temperature of the PV device using passive or active cooling techniques can enhance its electrical efficiency [2].

Due to this interdependence of PV temperature and the electrical efficiency, this review deals with thermal and electrical performance metrics of a PV systems. The term PV system can refer to an individual PV cell, a module, a panel or an array. But in this study, it refers to a PV panel (notified otherwise).

Thermal performance for PV systems is measured through their PV cells temperature. Several methods have been developed to approximate the PV cell temperature. These methods range from simple empirically derived correlations [3], to detailed numerical models [4].

Electrical performance is measured via maximum electrical power and/or electrical efficiency of the PV system. Similar to thermal models, electrical models also vary from simple expressions (measuring only the maximum power) [5] to complex set of equations (giving complete *I-V* characteristics of a PV system) [6,7].

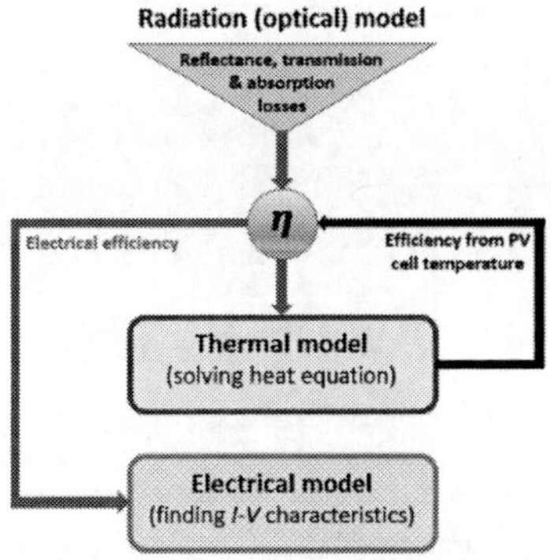

Fig. 1 Multi-physics modeling scheme including the radiation (optical), thermal and electrical models applied to the simulation of the performance of a photovoltaic device.

2 MULTI-PHYSICS MODELING

For estimating the electrical performance of any PV system via an electrical model, the knowledge of PV cells temperature is required, which is obtained via a thermal model. Similarly, the thermal model needs an estimate of total absorbed solar irradiation, which is achieved via radiation model (also known as an optical model). Thus, all these models working together make the complete PV system model a multi-physics model [8]. Summary of this multi-physics approach is shown in Fig. 1.

2.1 Radiation (or optical) model

Not all the solar irradiance incident on the front surface of a PV panel reaches the PV cells. As a typical PV panel comprises of different layers, stacked upon each other, there are reflection, transmission and absorption losses, through each layer, until the light finally reaches the PV cells. These losses are estimated through an optical model.

There are different ways to account for such losses. The simple and straightforward approach is to assume a cumulative absorption and reflectance coefficients of each surface, and use them to remove that percentage of irradiance from each layer, as the light passes through it, from top to bottom.

In reality, the reflection, transmission and absorption of a light wave, inside a layer, is highly dependent on its wavelength. As the entire solar spectrum consists of a wide range of these wavelengths, thus each photon of light reacts differently to different layers. Therefore, the second, yet more accurate approach is to calculate the actual absorption of the incident irradiance, after all the optical losses. This actual absorption is usually found by Markov process, which calculates the probability of each photon being either reflected, transmitted or absorbed [9].

The authors have implemented both the simple (cumulative absorption coefficient) [10–13] and the detailed (Markov process) approaches in their previous works [14]. The more elaborative detailed work, although slightly better, was found to be in agreement with the simple approaches. Fig. 2 shows a cross-sectional side view of a PV panel with stacks of the different main layers. The figure also illustrates the possibility of each photon being reflected, transmitted or absorbed by each layer, depending on its wavelength.

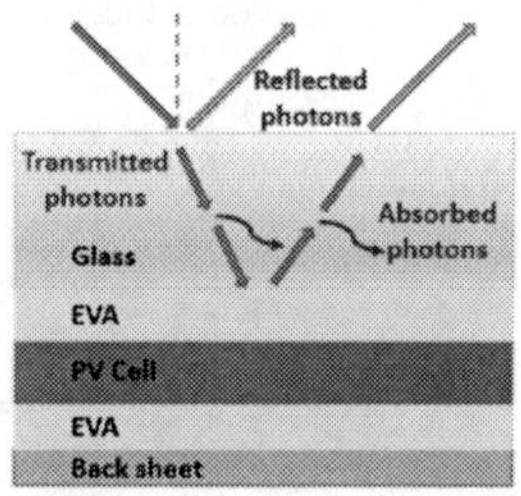

Fig. 2 Different layers of a PV panel illustrating possible reflectance, transmittance and absorption of a single photon of light, for each layer

2.2 Thermal model

As stated earlier, the thermal model needs the estimate of the total amount of absorbed solar radiation in each layer, especially the PV cells layer, which absorbs most of the solar irradiance. This is because layers like glass and EVA (encapsulant) are designed such that they minimize reflectance and absorption losses, to ensure that maximum sunlight reaches the PV cells. On the other hand, the back-sheet layer is designed such that it reflects all the irradiance reaching it, back to the PV cell.

The main purpose of any thermal model is to estimate the PV cell temperature. As empirical approaches are very case specific [5] and the simplified approaches loose accuracy [15], the authors have opted for numerical methods to develop their own thermal models. The models have made use of realistic boundary conditions, with simplifications that reduce the complexity of the model, without compromising its accuracy.

The most common numerical approaches are the finite difference (FD) method and the finite element (FE) method. The basic idea behind implementing the numerical approaches is to find the temperature distribution through the physical domain of the PV panel, from which PV cell temperature can be obtained. This is achieved by solving the heat equation for each layer:

$$\rho C \frac{\partial T(x,y,z,t)}{\partial t} = \nabla \cdot k \nabla T(x,y,z,t) + Q \qquad (1)$$

Here, ρ is the density, C is the material's specific heat capacity, $T(x,y,z,t)$ is the instantaneous spatial temperature, Q is the internal heat generation and x, y, z are the spatial coordinates.

Using the FD method, both 1-D and 2-D models were developed [10–13]. Whereas with FE method, 2-D and 3-D models were developed [4,16,17]. It was further shown in [16], that excluding the side frame of a PV panel does not have any noticeable effect on the prediction accuracy of each thermal model. Furthermore, including the effect of thermal mass of each layer (via C) and using more realistic boundary conditions for radiative heat loss, our models outperformed other models from the literature. The FD models used in-house self-developed codes for solving the heat equation (1), while commercial software (ANSYS and COMSOL) were used for the FE models.

Fig. 3 shows a FD nodal gird. If only Y-direction (thickness direction) is considered, then we get 1-D FD and if both X- and Y- directions are considered, we get 2-D FD.

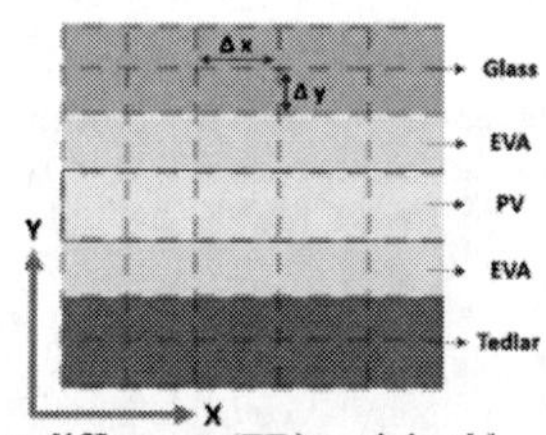

Fig. 3 A finite difference (FD) nodal grid

Similarly, Fig. 4 shows the top view of the developed 3-D FE model, with temperature contours showing at a random operating condition. It was found that that the in-house self-developed FD codes were much faster than FE methods, in terms of computational speed and have almost the same prediction accuracy. Thus, for simulating longer periods (days, weeks, months or even years) of temperature predictions, FD codes are a prudent choice.

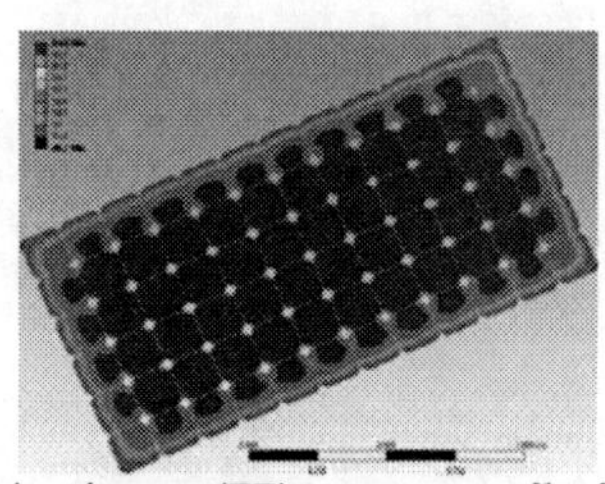

Fig. 4 A finite element (FE) temperature distribution contours at a random operating condition

To find the temperature distribution inside the PV panel, both FD and FE developed methods only require the knowledge of incident global irradiance, ambient temperature and the wind speed. Furthermore, both developed methods are capable of performing steady-state as well as transient analysis. The models have been rigorously tested against different field conditions, typically present in a desert environment like Qatar.

To demonstrate the prediction accuracy of the developed thermal models, a rainy day with extreme weather conditions in the desert environment was considered. The experimental data for this day was taken from our solar test facility (STF), Doha, Qatar. The experimental data to be used by the thermal models consists of incident total global plane of array (POA) irradiance (Fig. 5), ambient temperature (Fig. 5) and wind speed (Fig. 6). The chosen PV panel is of poly-crystalline silicon technology and has a total of 60 cells in series, with a rated power of 220 W at standard test conditions (STC).

From among the different developed thermal models by the authors, the temperature prediction results from 2-D FD model for the PV backsheet, along with experimentally measured temperature are shown in Fig. 7. It can be noticed that even with these extreme weather conditions, the developed model maintains its prediction accuracy. For more calm weather conditions, the results were even better [11–13]. It should be noted that all the validated results assume that there is no dust deposition on the surface of the PV panel, as soiling causes shading and which is not taken care of in the proposed thermal models [18,19].

2.3 Electrical model

The most important performance metric of any PV system is its electrical performance and is measured by output power. The electrical behavior of any PV device is specified by its complete current-voltage (*I-V*) curve. This *I-V* curve depends on the solar irradiance reaching the PV cells (estimated by radiation/optical model) and the PV cells temperature (estimated by thermal model).

One of the most conventional method to find this *I-V* curve is by representing the PV device with an equivalent electrical circuit, with diodes and resistance in parallel as well as series. This approach is preferred because of its good accuracy and satisfactory complexity. The *I-V* characteristics of the PV device at any operating condition are then found by solving the electrical equation representing the complete circuit. The authors have implemented both one-diode and two-diode models in their previous works [6,7]. A schematic of this circuit is presented in Fig. 8.

The electrical circuit equation representing the PV device consists of some unknown PV parameters, which need to be identified before the equation can be utilized for finding the *I-V* characteristics. There are various ways to find these parameters. The authors have opted for different types of optimization techniques, such as particle swarm optimization (PSO), cuckoo search algorithm and hybrid metaheuristic optimization methods [6,7].

Using these optimization techniques, though the electrical circuit is solved correctly, the values of the PV parameters is sometimes not physically correct. Furthermore, the authors mainly focused on the influence of temperature on these PV parameters and ignored the irradiance effect.

The authors are currently working on more improved electrical models [20], which considers the effect of both the irradiance and temperature on the *I-V* characteristics of the PV device. Some preliminary results from this new electrical model [20] are shown in Fig. 9. The results show that the proposed model fits well with the datasheet *I-V* curves, especially at the maximum power point (the knee of each curve), which is the most important point of interest for a PV system designer.

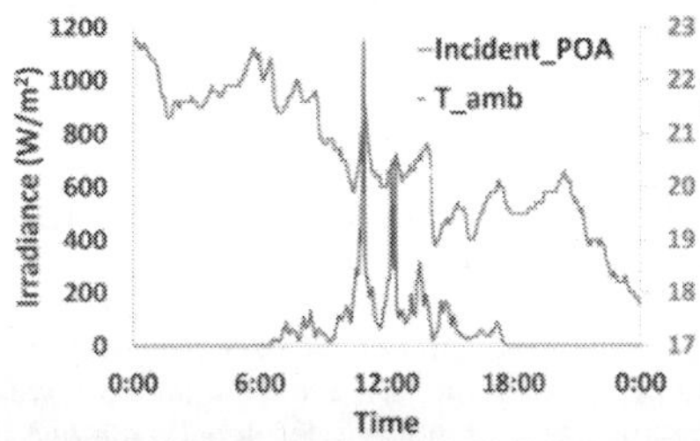

Fig. 5 Field data for incident irradiance and ambient temperature (from STF, Doha, Qatar on 03-26-2014)

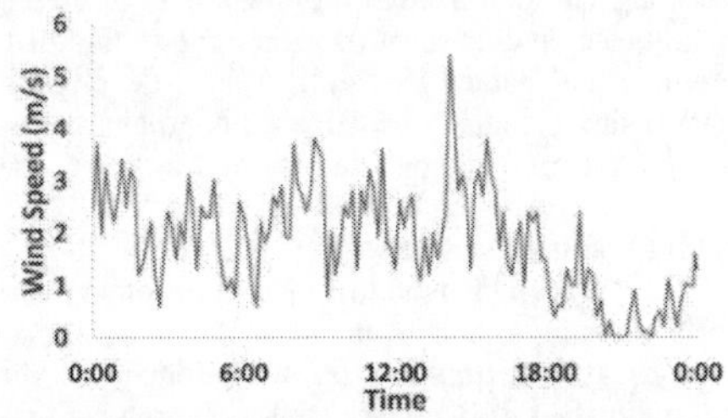

Fig. 6 Field data for wind speed (from STF, Doha, Qatar on 03-26-2014)

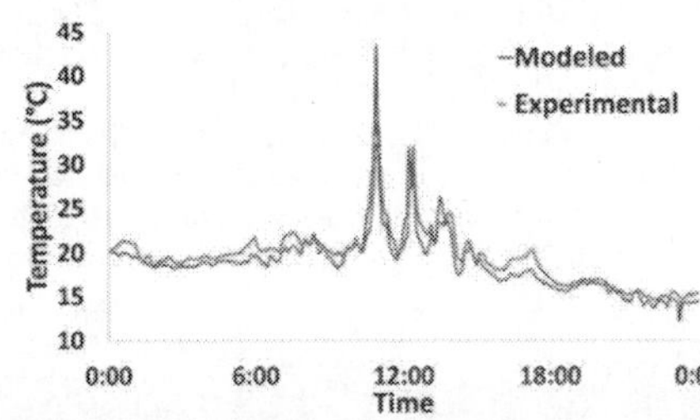

Fig. 7 Modeled vs. experimental backsheet temperatures (from STF, Doha, Qatar on 03-26-2014)

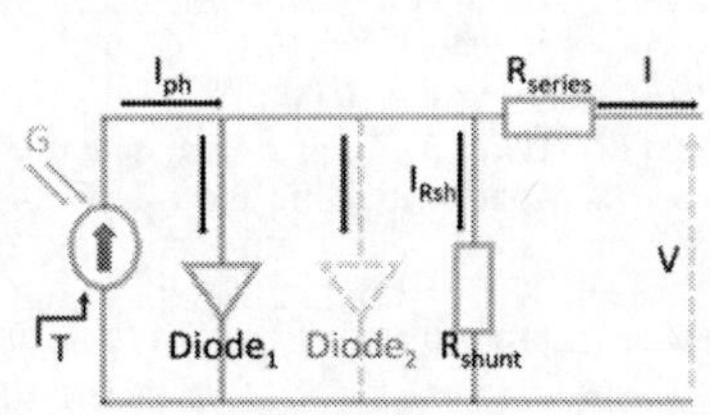

Fig. 8 One and two diode electrical circuit representing a PV device

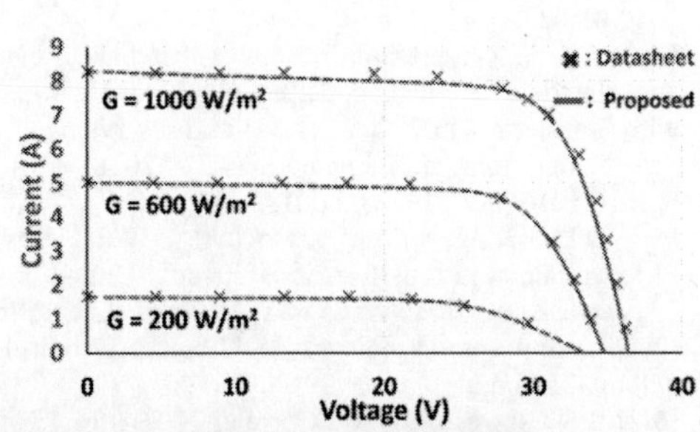

Fig. 9 *I-V* characteristics curves of the PV panel: modeled vs. datasheet provided information

3 CONCLUSIONS

This review briefly highlights the most important aspects of the multi-physics modeling involved in a PV system design. It is shown that the most important performance metric of any PV system is its electrical power output, which is estimated through an electrical model. This electrical model needs inputs such as total available irradiation to the PV cell and its working temperature. The authors have developed both one-diode and two-diode electrical models.

The total available irradiance is approximated through a radiation model, which makes this estimate by calculating the total irradiation losses due to reflectance, transmittance and absorption, throughout the different layers of a PV panel. The authors have developed both simple (using cumulative absorption coefficients for each layer of PV panel) and detailed (using Markov process to calculate wavelength dependent absorption of solar spectrum) radiation models.

The PV cell temperature on the other hand is calculated using a thermal model, which basically solves a heat equation within the physical domain of the PV panel, with a set of boundary conditions subject to a real field operating PV panel. The authors developed a variety of finite difference (FD) and finite element (FE) based numerical thermal models. These thermal models are capable of performing steady-state as well as transient analyses, for a variety of weather conditions experienced in a desert environment like Qatar.

As all these models directly/indirectly influence each other, therefore one of the main aims of the authors is to reduce the prediction error in each model, so that the overall error in the entire PV system model is reduced.

REFERENCES

[1] E. Skoplaki, J.A. Palyvos, On the temperature dependence of photovoltaic module electrical performance: A review of efficiency / power correlations, Sol. Energy. 83 (2009) 614–624. doi:10.1016/j.solener.2008.10.008.

[2] S.P. Aly, N. Barth, B. Figgis, E. Essadiqi, M. Faqir, A. Ennaoui, S. Ahzi, Mitigating the effect of heat and dust to enhance solar panels efficiency, in: 2016 Int. Renew. Sustain. Energy Conf., IEEExplore. (2016): pp. 835–841. doi:10.1109/IRSEC.2016.7983870.

[3] M. Koehl, M. Heck, S. Wiesmeier, J. Wirth, Modeling of the nominal operating cell temperature based on outdoor weathering, Sol. Energy Mater. Sol. Cells. 95 (2011) 1638–1646. doi:10.1016/j.solmat.2011.01.020.

[4] N. Barth, Z.S. Al Otaibi, S. Ahzi, Irradiance, thermal and electrical coupled modeling of photovoltaic panels with long-term simulation periods under service in harsh desert conditions, J. Comput. Sci. 27 (2018) 118–129. doi: 10.1016/j.jocs.2018.05.001.

[5] D.L. King, J.A. Kratochvil, W.E. Boyson, Photovoltaic array performance model, United States, Department of Energy, Sandia National Laboratories, (2004). http://prod.sandia.gov/techlib/access-control.cgi/2004/043535.pdf.

[6] J.C.S. Barroso, J.P.M. Correia, N. Barth, S. Ahzi, M.A. Khaleel, A PSO algorithm for the calculation of the series and shunt resistances of the PV panel one-diode model, in: 2014 Int. Renew. Sustain. Energy Conf., IEEExplore. (2014): pp. 1–6.

[7] N. Barth, R. Jovanovic, S. Ahzi, M.A. Khaleel, PV panel single and double diode models: Optimization of the parameters and temperature dependence, Sol. Energy Mater. Sol. Cells. 148 (2016) 87–98.

[8] S. Ahzi, S.P. Aly, N. Barth, New Developments in the Modeling and Simulations of the Thermal Behavior and Electrical Yield of Photovoltaics Panels With the Consideration of Desert Environmental Conditions, in: 2017 5th Int. Renew. Sustain. Energy Conf. (IRSEC), IEEExplore, (2017). doi:In press.

[9] S.Z. Aljoaba, A.M. Cramer, S.A. Rawashdeh, B.L. Walcott, Thermoelectrical Modeling of Wavelength Effects on Photovoltaic Module Performance—Part II: Parameterization, IEEE J. Photovoltaics. 3 (2013) 1034–1037. doi:10.1109/JPHOTOV.2013.2261117.

[10] J.C. Sánchez Barroso, N. Barth, J.P.M. Correia, S. Ahzi, M.A. Khaleel, A computational analysis of coupled thermal and electrical behavior of PV panels, Sol. Energy Mater. Sol. Cells. 148 (2016) 73–86. doi:10.1016/j.solmat.2015.09.004.

[11] S.P. Aly, S. Ahzi, N. Barth, B.W. Figgis, Two-dimensional finite difference-based model for coupled irradiation and heat transfer in photovoltaic modules, Sol. Energy Mater. Sol. Cells. 180 (2018) 289–302. doi:10.1016/ j.solmat.2017.06.055.

[12] S.P. Aly, N. Barth, B.W. Figgis, S. Ahzi, A fully transient novel thermal model for in-field photovoltaic modules using developed explicit and implicit finite difference schemes, J. Comput. Sci. (2018). doi:10.1016/j.jocs.2017.12.013.

[13] S.P. Aly, S. Ahzi, N. Barth, A. Abdallah, Using energy balance method to study the thermal behavior of PV panels under time-varying field conditions, Energy Convers. Manag. (2018). doi:In press.

[14] M. Sahli, J.P.M. Correia, S. Ahzi, S. Touchal, Multi-physics modeling and simulation of heat and electrical yield generation in photovoltaics, Sol. Energy Mater. Sol. Cells. 180, (2018) 358–372.

[15] G.M. Tina, A. Gagliano, An improved multi-layer thermal model for photovoltaic modules, in: Comput. Energy Sci. (SpliTech), Int. Multidiscip. Conf., IEEE, (2016): pp. 1–6. doi:10.1109/SpliTech.2016.7555927.

[16] S.P. Aly, N. Barth, S. Ahzi, A numerical study for approximating cells temperature inside a PV module, in: Proceeding 3rd Therm. Fluids Eng. Conf. (2018): pp. 1575–1584. doi:10.1615/tfec2018.rce.022023.

[17] S.P. Aly, N. Barth, S. Ahzi, A Three-Dimensional Finite Element Based Dynamic Thermal Model of PV Modules with an Improved Thermal Network, in: 2017 5th Int. Renew. Sustain. Energy Conf. (IRSEC), IEEExplore. (2017). doi:In Press.

[18] S.P. Aly, P. Gandhidasan, Photovoltaic panel cleaning machine, US 9130502 B1, (2015). https://www.google.com/patents/US9130502.

[19] S.P. Aly, P. Gandhidasan, N. Barth, S. Ahzi, Novel dry cleaning machine for photovoltaic and solar panels, in: Proc. 2015 IEEE Int. Renew. Sustain. Energy Conf. IRSEC 2015, IEEExplore. (2015). doi:10.1109/IRSEC.2015.7455112.

[20] S.P. Aly, N. Barth, S. Ahzi, Effect of both irradiance and temperature on the electrical performance of PV devices, in: ms in preparation (2018).

REDUCING UNCERTAINTY IN OUTDOORS PV MODULE CHARACTERISATION

C. H. Rossa, F. Martínez-Moreno*, E. Lorenzo

Instituto de Energía Solar – Universidad Politécnica de Madrid. Grupo de Sistemas Fotovoltaicos (IES-UPM).
Campus Sur UPM. Ctra. Valencia km. 7. ETSI Sistemas Telecomunicación. 28031 Madrid, Spain.
*Corresponding author: francisco.martinez@ies.upm.es

ABSTRACT: Quality control of PV modules in large grid-connected PV plants is typically accomplished by means of indoors testing at qualified laboratories which use solar-simulators. Nevertheless, outdoors on-site measurement of PV modules has several advantages: the devices required to measure the PV modules are much cheaper than solar simulators; the costs and effort related to disassemble and to transport the PV modules to the laboratories are entirely avoided; and the safety of the modules against undesirable cracks and/or module breakage during their manipulation is drastically improved (manipulation is just reduced to unplug and to plug the connectors). Besides, the uncertainties related to outdoors PV module measurements are also strongly reduced when reference devices to measure effective incident irradiance and solar cell temperature are properly selected. These uncertainties are even lower when considering the different response of each PV module to the same real weather conditions is taken into account, especially those related to cell temperature. So uncertainties of outdoors measurements can be reduced close to those ones obtained with the indoors characterisation. This works deals with the temperature dispersion inside PV modules and between different PV modules, as well as how to manage this dispersion when the PV modules are characterised outdoors to obtain accurate results with a reduced uncertainty.

Keywords: outdoors module characterization, uncertainty, temperature dispersion, PV system

1 INTRODUCTION

Quality control of PV modules in large grid-connected PV plants uses to be based on indoors testing at qualified laboratories which use solar-simulators [1] [2] [3]. These devices allow to characterise PV modules just at Standard Test Conditions (STC, $G^* = 1000$ W/m^2 and $T_C^* = 25$ ºC) without weather restrictions. Besides, as operation conditions are totally controlled, the dispersions due to temperature are minimized. Nevertheless, the solar-simulators are very expensive. Additionally, costs related to transport the PV modules from the PV installations to the laboratory should be taken into account, which is not negligible at all because laboratories use to be far away from the PV plants.

Outdoors on-site measurement of PV modules has several advantages if compared to the use of solar-simulators: the devices required to measure the PV modules are much cheaper than solar simulators; the costs and effort related to disassemble and to transport the PV modules to the laboratories are entirely avoided; and safety of the modules against undesirable cracks and/or module breakage during their manipulation is drastically improved (manipulation is just reduced to unplug and to plug the connectors). Besides, the uncertainties related from outdoors PV module measurements are also markedly reduced when reference devices to measure effective incident irradiance and solar cell temperature are properly selected [4] [5]. In fact, other authors have concluded that the natural sunlight is preferred for PV measurements because it has a higher uniformity than solar-simulators [2] [6] [7] [8]. These uncertainties are even lower if the different response of each PV module to the same real weather conditions is taken into account, especially those related to cell temperature. In fact, once a good reference device calibrated to measure effective incident irradiance has been selected, the temperature dispersion becomes in the major contribution to the overall uncertainty in the module characterisation [5]. So uncertainties of outdoors measurements can be reduced close to those obtained with indoors characterisation.

In this work it is shown an experimental campaign of measurements to evaluate the dispersion of cell temperature at three different levels: at cell level, at module level and at array level. Later, different models to extrapolate to STC the measurements performed over 16 PV modules for 9 months have been analyzed, including the effect of this temperature dispersion to obtain results with the lower uncertainty.

2 METHODOLOGY AND EXPERIMENTAL ASSEMBLY.

In order to evaluate the dispersion of temperature at different levels inside a PV array and to check which is the best model to extrapolate to STC in terms of maximum power uncertainty, some tests have been performed over the modules of two c-Si PV arrays of 5kW$_p$ installed by the Instituto de Energía Solar (hereafter IES-UPM) at the facilities of the Universidad Politécnica de Madrid in the South Campus (Fig. 1).

Figure 1: Two PV arrays of 5 kWp grid-connected at the IES-UPM terrace where the PV modules used in the experiments are installed.

2.1 Temperature dispersion.

On the one hand, the cell temperature from the 21 PV modules of one of the arrays has been observed continuously during 6 months. These temperatures have been measured with PT1000 sensors class B (±0.3°C of tolerance) glued on the back of each module, storing the values with an interval of 5 minutes. These sensor have been glued in three different configurations:

- 21 PV modules have a single PT1000 glued just at the center cell of the module. These sensors allow to analyze the dispersion inside the PV array (Fig. 2).
- 6 PV modules have two additional PT1000 sensors glued to other two cells according to international standards recommendations [9] [10]. This configuration allows to study the dispersion intra-module (Fig. 3).
- 3 PV modules have another two PT1000 sensors glued to the central cell which already has a sensor. So, the dispersion intra-cell can also be evaluated (Fig. 4).

Figure 2: View of the PT1000 glued just at the center cell of the 21 PV modules to analyze the temperature dispersion inside the PV array.

Figure 3: View of one of the six PV modules with three PT1000 glued at different cells according international standard to analyze the temperature dispersion inside a single module.

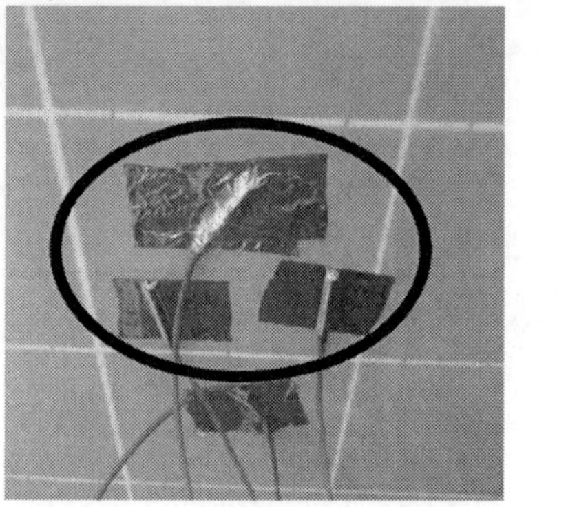

Figure 4: View of one of the three PV modules with three PT1000 glued just at the center cell to analyze the temperature dispersion inside a single cell.

2.2 Measure of I-V curves.

On the other hand, 16 out of the 45 modules distributed on the two PV generators have been characterised, measuring 46 times the I-V curves of each single module during these 7 months. These measurements were done with a twin capacitive load [11] which allows to obtain both the I-V curve of the module under test and the one related to a reference module previously calibrated, under real operation conditions. This way, uncertainties which could arise from measuring each module separately are reduced [12] [13]. Furthermore, as the reference module and the test module are exactly same, uncertainty due to angular and spectral response is avoided [5] [14]. Then, the values of the short-circuit current, I_{SC}, and the open-circuit voltage, V_{OC}, obtained from this reference module allow to calculate the effective incident irradiance [15] and the equivalent cell temperature [16]. Simultaneously to each single I-V curve obtained, the temperatures measured by the PT1000 glued at the center cell of both modules are stored with a datalogger.

2.3 Extrapolation to Standard Test Conditions.

Once the I-V curves of the PV modules have been measured, their maximum power point has to be extrapolated to STC. Now the objective is to determine which one of the different methods of extrapolation leads to the lowest dispersion of the maximum power point at STC. So, the parameter which is going to be evaluated is the standard deviation of the maximum power point after its extrapolation: the lower the standard deviation, the lower the uncertainty

In this work several extrapolation methods have been selected to perform this translation. These methods can be divided in two main groups. The first one is the group in which every single point of the I-V curve is extrapolated to STC (Fig. 5). This kind of methods is based on the temperature coefficients of current and voltage (usually known as alpha and beta, α and β) [9]. The second one is the group in which only the maximum power point of the I-V curve is extrapolated to STC (Fig. 6). In this case the extrapolation can only depend on the maximum power point at STC of the reference module (proportional correction considering the calibrated power of the reference module). But it also can rely on the temperature coefficient of power (usually known as gamma, γ) [17] [18].

The different alternatives in each group depend on the cell temperature used to extrapolate these points, according the conclusions from the temperature dispersion.

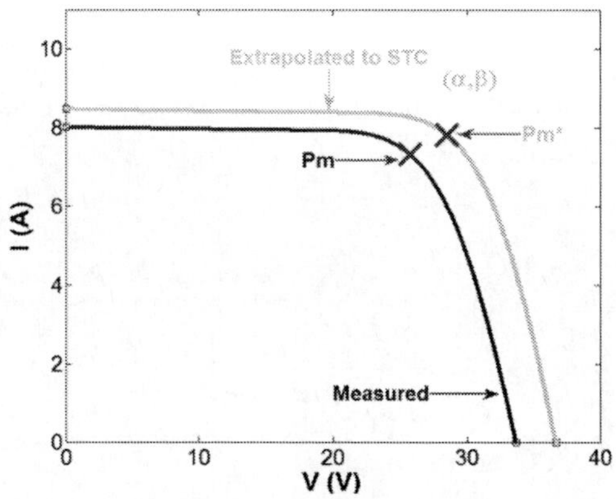

Figure 5: Extrapolation method based on the translation of every single point of the curve I-V.

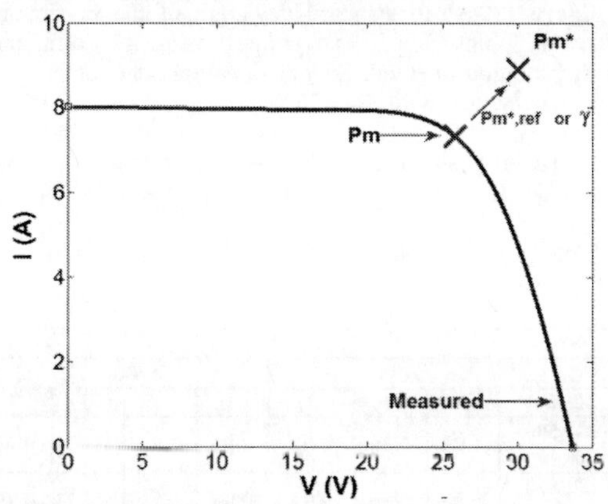

Figure 6: Extrapolation method based on the translation of just the maximum power point.

3 RESULTS.

3.1 Temperature dispersion.

Fig. 7, Fig. 8 and Fig. 9 show the temperature dispersion inside a single cell, inside a PV module and inside an array, respectively. It is possible to appreciate the same behaviour on the three graphics: the temperature dispersion increases at higher irradiances. In the case of a single cell, the dispersion reaches up to 3°C for irradiances around 1000 W/m^2. When considering a single PV module, differences up to 7°C are obtained. Finally, the temperature dispersion reaches up to 14°C. It should be mention that in these graphics, the higher dispersions related to irradiances below 700 W/m^2 (mainly between 200 W/m^2 and 400 W/m^2) are due to the shadow cast over the PV modules from nearby trees.

So, it is clear that different modules located at the same structure, receiving the same irradiance and at the same room temperature have not exactly the same cell temperature. In fact, the temperature difference can be higher than 10°C when the PV module under test and the reference module are not close to each other (separation between modules higher than 10 m).

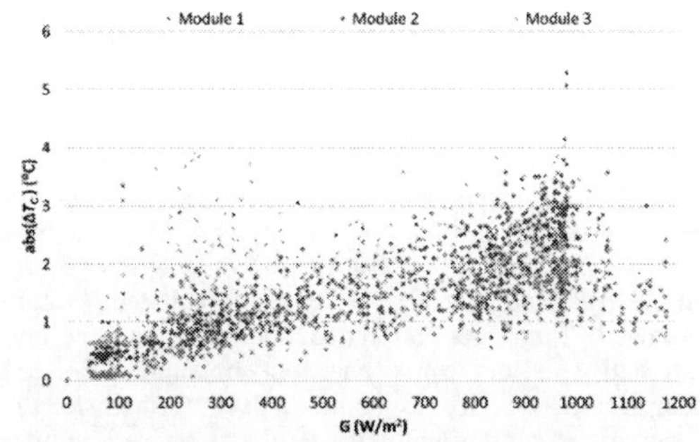

Figure 7: Difference of temperatures between three PT1000 glued at the same cell located at the center of a PV module. The differences observed on 3 PV modules are presented.

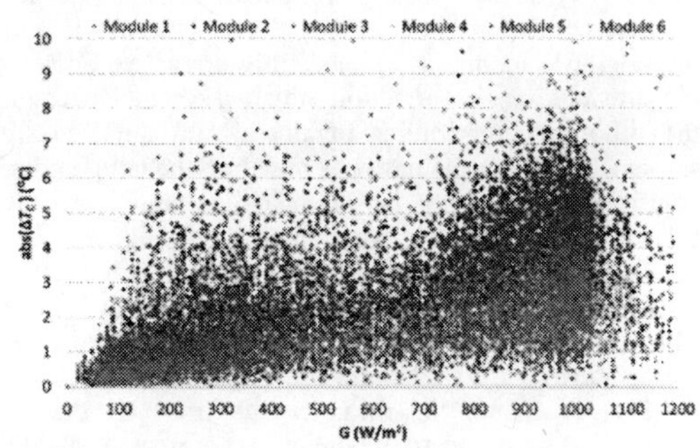

Figure 8: Difference of temperatures between three PT1000 glued at three different cells of the same PV modules according the international standards. The differences observed on 6 PV modules are presented.

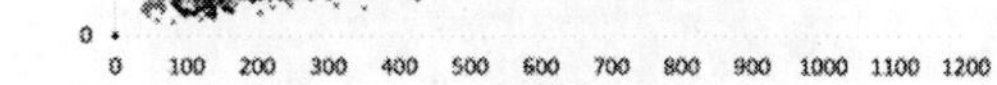

Figure 9: Difference of temperatures between the PT1000 glued at the central cell of 21 PV modules of the same array. The light points represent all the measurements, while the dark ones represent only the measurements related to clear days and once the points affected by shadows from trees have been removed.

3.2 Measure of I-V curves and extrapolation to STC.

The previous analysis of the temperature dispersion reports that different PV modules have also different temperatures. So, we are making a mistake when the I-V measurement of a module under test is extrapolated to STC using the cell temperature measured from a reference module, although this one is installed in the same structure close to the module under test. This fact is usually ignored and leads to an increase of the uncertainty in the determination of the maximum power point.

This is why we have done some corrections to the classical extrapolation methods to include the dispersion of temperature we have observed. We have studied seven methods of extrapolation, three related to the extrapolation of the whole I-V curve (Group 1) and four related to the extrapolation of just the maximum power point (Group 2).

In the Group 1, the first alternative is based just on the international standard [9] with the cell temperature obtained from the reference module. The second alternative is the same one, but obtaining the cell temperature directly from the PT1000 glued to the module under test. Finally, the third alternative is similar to the first one but including a correction related to the difference between the temperatures measured at the central cell of both modules with their corresponding PT1000.

Already in the Group 2, the fourth alternative relies on a direct relationship according the calibration value of the reference module [11]. The fifth alternative adds to the previous one a correction which takes into account the different temperatures obtained from the PT1000 sensors. The sixth alternative is based on the temperature coefficient of power, γ, [17] [18] and assuming that the temperature of the PV module under test is equal to the temperature obtained from the open circuit voltage of the reference PV module. Finally, the seventh and last one is like the previous one, but including the difference of temperature measured with the PT1000 sensors.

Fig. 10 shows the mean results from all the 16 modules. The standard deviation is presented of the maximum power point after applying each one of these extrapolation alternatives to the 46 measurement done in the last 7 months. The seven different methods of extrapolation are represented in the x-axis. In the y-axis is represented the standard deviation in %. As can it be noticed, methods 2 and 5, which consider the difference of temperature between modules, lead to lower values of standard deviation (uncertainty). It is worth to comment that these results include all the measurements done in all types of climate conditions: sunny days, days with certainly amount of high clouds, clouds passing, wind gust, etc.

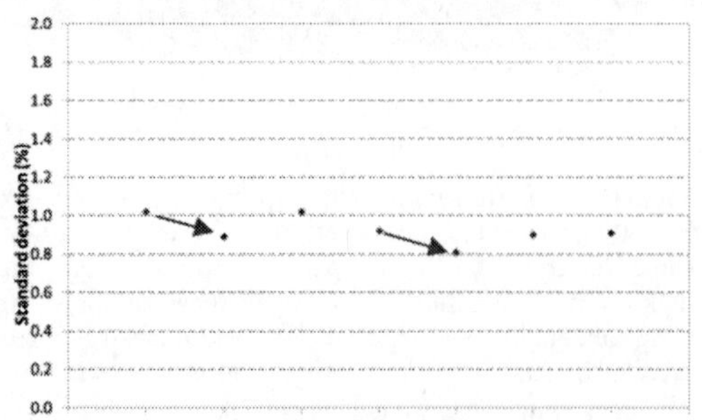

Figure 10. Mean standard deviation of the maximum power point after extrapolate all the measurement using 7 different extrapolations methods. Alternatives 1 to 3 are related to the extrapolation of the whole I-V curve. Alternatives 4 to 7 are related to the extrapolation of just the maximum power point.

Fig. 11 shows the same results, but now taking into account just the measurements obtained in sunny days and with irradiances higher than 850 W/m^2. As it can be noticed, the standard deviation related to the methods which take into account the temperature difference (2 and especially 5) have not only a lower value but also a lower variation regardless the weather conditions, especially the one related to the extrapolation just based on the maximum power point.

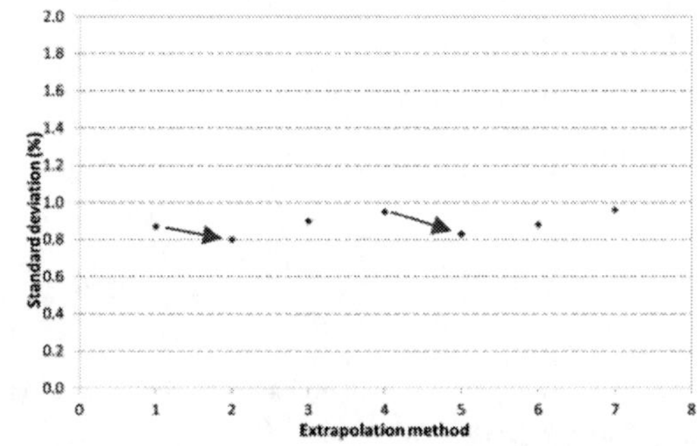

Figure 11. Mean standard deviation of the maximum power point after extrapolate using 7 different extrapolations methods only the measurement obtained in sunny days and with G > 850 W/m^2

Table I summarize the results presented in the previous figures.

Table I: mean standard deviation of the maximum power point related to Fig. 10 and Fig. 11.

	σ (%)						
	IV			Pm			
Method	**1**	**2**	**3**	**4**	**5**	**6**	**7**
G≥ 500 W/m^2	1.02	0.89	1.02	0.92	0.81	0.90	0.91
G≥ 850 W/m^2	0.87	0.80	0.90	0.95	0.83	0.88	0.96

4 CONCLUSIONS.

Once the sensor of irradiance used in outdoors characterisation of PV modules has been selected to reduce its contribution to the overall uncertainty, the main responsible of mistakes and increase of dispersion in the maximum power extrapolated to STC is the cell temperature.

This work reports that the difference of temperature inside a single cell, inside a single module and inside an array increase with incident irradiance and reach up to 3°C, 7°C and 14°C at 1000 W/m^2, respectively.

So, in order to obtain a low uncertainty in the results of the power characterisation it is crucial to considerer the differences in the temperature between different modules, even when they belong to the same manufacturer and model and they are installed in the same structure, receiving the same irradiance and with the same room temperature. When this fact is considered and the proper correction is applied, the standard deviation of the extrapolated power can be reduced between 0.1% and 0.2%. Besides, extrapolation methods which just rely on the maximum power point have a lower variation regardless the weather conditions. This allows to perform the characterisation with good results at a wider range of irradiances, room temperatures and wind speeds.

5 ACKNOWLEDGEMENTS

Authors would like to thank to CNPq (Conselho Nacional de Desenvolvimento Científico e Tecnológico), for the financing which made possible this research.

6 REFERENCES

[1] Alonso-Abella, M., Silva J.P., Fabero F., Vela N., Chenlo F. *PV module peak power: comparison between manufacturer data and laboratory.* 23rd European Photovoltaic Solar Energy Conference: 2837-2840. (2008). doi:10.4229/23rdEUPVSEC2008-4AV.3.21

[2] Kang G-H., Kim K-S., Yu G-J., Ahn H., Han D-Y. *Comparison test of performance at STC and uncertainty for PV modules.* 24th European Photovoltaic Solar Energy Conference: 3356-3358. (2009). doi:10.4229/24thEUPVSEC2009-4AV.3.19

[3] Coello J., Carreño R., Carames M., Galindo J.L. *Results of module testing in Enertis solar during the quality control of 19 MW and 13 MW PV power plants in Spain.* 24th European Photovoltaic Solar Energy Conference: 3494-3497. (2009). doi:10.4229/24thEUPVSEC2009-4AV.3.58

[4] King D.L., Boyson W.E., Kratochvil J.A. *Photovoltaic array performance model.* Sandia National Laboratories. Report SAND2004-3535 (2004). doi:10.2172/919131.

[5] Carrillo J.M., Martínez-Moreno F., Lorenzo C., Lorenzo E. *Uncertainties on the outdoor characterisation of PV modules and the calibration of reference modules.* Solar Energy 155, 880-892 (2017). doi:10.1016/j.solener.2017.07.028

[6] Keogh W.M., Blakers A.W. *Accurate measurement, using natural sunlight, of silicon solar cells.* Prog. Photovoltaics Re. Appl. 12, 1-19 (2004). doi:10.1002/pip.517

[7] Müllejans H, Zaaiman W., Merli F., Dunlop E.D. *Comparison of traceable calibration methods for primary photovoltaic reference cells.* Prog. Photovoltaics Res. Appl. 13, 661-671 (2005) doi:10.1002/pip.625

[8] Müllejans H., Zaaiman W., Galleano R. *Analysis and mitigation of measurement uncertainties in the traceability chain for the calibration of photovoltaic devices.* IOP Publ. Meas. Sci. Technol. 20 (2009). doi:10.1088/0957-0233/20/7/074101

[9] IEC 60891:2009. *Photovoltaic devices- Procedures for temperature and irradiance corrections to measured I-V characteristics.*

[10] IEC 61853-1:2011. *Photovoltaic (PV) module performance testing and energy rating – Part 1: Irradiance and temperature performance measurements and power rating.*

[11] Muñoz J., Lorenzo E., Carrillo J.M., Moretón R. *Design of a twin capacitive load and its application to the outdoor rating of photovoltaic modules.* Prog. Photovoltaics Res. Appl. 23, 247-252 (2015) doi:10.1002/pip.2425

[12] Caamaño E., Lorenzo E., Zilles R. *Quality control of wide collections of PV modules: lessons learned from the IES experience.* Prog. Photovoltaics Res. Appl. 7, 137-149 (1999) doi: 10.1002/(SICI)1099-159X(199903/04)7:2<137::AID-PIP249>3.0.CO;2-C

[13] Nofuentes G., Aguilera J., Santiago R.L., De la Casa J., Hontoria L. *A reference-module-based procedure for outdoor estimation of crystalline silicon PV module peak power.* Prog. Photovoltaics Res. Appl. 14, 77-87 (2006) doi: 10.1002/pip.636

[14] Hishikawa Y., Doi T., Higa M., Takenouchi T., Ohshima H., Yamagoe K. *Reproducible outdoor I-V curve measurement by the use of PV module irradiance sensors and comparison with indoor results.* 33rd European Photovoltaic Solar Energy Conference: 1370-1373. (2017). doi: 10.4229/EUPVSEC20172017-5BO.6.5

[15] IEC 60904-2: 2015. *Photovoltaic devices – Part 2: Requirements for photovoltaic reference devices. corrections to measured I-V characteristics* (2009).

[16] IEC 60904-5: 2011. *Photovoltaic devices – Part 5: Determination of the equivalent cell temperature (ECT) of photovoltaic (PV) devices by the open-circuit voltage method.*

[17] Osterwald, C.R.. *Translation of device performance measurements to reference conditions.* Solar Cells 18, 269–279 (1986).

[18] Fuentes M., Nofuentes G., Aguilera J., Talavera D.L., Castro M. *Application and validation of algebraic method to predict the behaviour of crystalline silicon PV modules in Mediterranean climates.* Solar Energy 81, 1396-1408 (2007). doi: 10.1016/j.solener.2006.12.008

BIPV MODULES WITH PLATED BIFACIAL N-PERT CELLS AND SMART WIRE INTERCONNECTION: MANUFACTURING, MONITORING AND ENERGY YIELD ANALYSIS

Arvid van der Heide[1], Hans Goverde[1], Konstantinos Spiliotis[2], Jonathan Lehmann[2], Loic Tous[1], Richard Russell[1], Dirk Saelens[2], Johan Driesen[2], Eszter Voroshazi[1], Jef Poortmans[1,2,3]
[1]: imec (Energyville campus), Thor Park 8320, 3600 Genk, Belgium
[2]: KU Leuven, departments ESAT – ELECTA/Building Physics, Kasteelpark Arenberg, 3001 Leuven, Belgium
[3]: UHasselt, Martelarenlaan 42, 3500 Hasselt, Belgium
arvid.vanderheide@imec.be

ABSTRACT: This contribution handles about BIPV modules made with plated bifacial n-PERT cells from imec. For the cell interconnection the advanced smart wire connection technology (SWCT) was used that can reduce silver consumption at cell level considerably. These modules were made for installation in a facade installation of the KU Leuven, with monitoring of module performance, weather conditions and module temperatures. The modules are all glass-glass, both transparent and white. The modules were inspected with EL, and *I-V* measurements at standard test conditions were performed (also from the back side for the transparent modules). The modules were installed in September 2017, monitoring started one month later. The most important finding was that the bifacial gain when using a white surface behind the module was only considerable when the modules were ventilated. The probable reason is that without ventilation, the extra heat generated by double-sided illumination results in a module temperature that is so high, that the decrease in V_{oc} and *FF* almost completely compensates for the gain in I_{sc}. The monitoring data still have to be analysed in more detail, including a comparison with energy yield calculations with the imec energy yield software that has been adapted for bifacial modules.
Keywords: Bifacial, Building Integrated PV (BIPV), SWCT, monitoring, ventilation

1 INTRODUCTION

Bifacial cells and modules have gained in interest because of their potential to obtain a higher energy yield than for the usual monofacial devices. The bifacial market share is still at most 5 %, but is expected to increase to 20 % in 2024 according to the International Technology Roadmap for Photovoltaic 2017 [1]. Another important point is to minimise the silver consumption in solar cell and module production, since the PV sector is already consuming 5-10 % of global silver production. When using multi-wire interconnection like the smart wire connection technology (SWCT), the required conductivity of the fingers can be reduced drastically. The best way to do this is by silver plating, since it allows a very low finger height, in contrast to screen printing technology. An aesthetic advantage of SWCT is the uniform outlook of the modules, making them particularly suitable for BIPV application.

Although not optimal for silver reduction, the SWCT technology is usually applied for cells that have been metallised by screen printing, since this is the current industrial technology. This work, is aimed at the combination of plated bifacial cells and SWCT interconnection to make modules. The goals of this work are to assess the electrical performance of these innovative modules under real outdoor conditions, and, on the longer term, to quantify the accuracy of the imec energy yield model and to study the reliability of the technology in outdoor conditions.

So first glass-glass mini-modules were prepared using plated bifacial cells from imec, applying SWCT interconnection. The modules were made in transparent and white (using white encapsulant). After mounting edge junction boxes, these frameless modules were integrated in a BIPV monitoring setup as curtain wall facade elements. Within this building facade, slots were created with and without rear side ventilation, and also the background surface behind the modules has been varied (black, white and aluminium). Both the performance and environmental conditions in the vicinity of the different modules have been monitored for almost a year now. Measured energy production data are presented and discussed, taking into account the different module types and conditions chosen in this experiment.

2 MODULE FABRICATION AND INSTALLATION

2.1 Module fabrication

In total 12 glass-glass modules were made, 7 transparent and 5 white ones. Each module contained 9 series connected cells, the white modules were made using white encapsulation material behind the cells.

The bifacial n-PERT cells used for the modules were made at imec, and had plated Ag metallisation on both front and rear side. The total amount of Ag per cell was 8 mg (~10 times less than the 90 mg for standard screen printed cells in 2017). The average efficiency of these cells was 20.7 %, with a FF of 78.8 %. Using a more complex process with selective BSF, n-PERT cell efficiency at imec was > 22 % [2], but in this case a process with uniform BSF [3] has been applied to reduce complexity, but leading to some efficiency decrease.

The solar glass plates were all 3.2 mm thick and did not have an anti-reflective coating (ARC), the size of the plates was 58 cm x 58 cm.

For encapsulation, a hybrid material from the company Novopolymers was used, made of both EVA and polyolefin. The transparent version used, HFT01, contains an additive to absorb the UV radiation, which was necessary to protect the SWCT foil. Meyer Burger is currently developing a UV-stable foil to solve this issue, so that a higher power output of the module can be reached. At the rear side, HFT01 was used for the transparent modules, while the (pigmented) white version HFW01 was used at the rear to make the white modules. Along the edges, the modules were sealed with black butyl tape of 10 mm width and 1.2 mm thickness, to reduce moisture ingress from the sides to a minimum. An additional effect of butyl sealing is that the encapsulant does not leak out at the edges, reducing contamination of the laminator. This is a clear process advantage, especially in the case of coloured encapsulants.

The SWCT foil used to interconnect the cells was received from Meyer Burger, and was made with 18 Cu wires of 300 μm diameter that were coated with an In52Sn48 solder alloy, having a melting point of 118 °C. For cost reasons, the indium in the solder alloy for SWCT is nowadays being replaced by another metal, with good reliability results. However, at the time of module fabrication this foil was not yet available.

The lay-up of cells and SWCT foil was done manually, the connection between the strings could be made in a simplified way since the cells had a diagonal finger pattern at the rear side. This enabled going directly from the last cell of the first string to the first one of the next string by choosing the SWCT foil direction to be perpendicular to the string direction at the last cell. Since no solder strips are needed to connect string ends, the process is not only simpler, but resistance losses are reduced at the same time. In Figure 1, a picture of one of the modules is shown just before lamination, placed on the 60 cm x 60 cm heating plate of the small laminator. In 2018, imec has also installed a larger laminator that is capable to laminate standard sized 60 cell modules.

Figure 1: Module on heating plate of the small laminator just before lamination. A frame is applied around the edges to avoid pinching of the module edges.

A picture of the resulting modules, but still without junction boxes, is shown in Figure 2.

Figure 2: The bifacial glass-glass modules, both white and transparent

The SWCT wires were connected to solder strips with a lead free solder alloy, and the solder strips finally exit the module for connection to edge junction boxes. Since the strips exit the modules on opposite module edges, two edge junction boxes per module were used, one per polarity. Since the junction boxes each have two connection sockets, it enabled measurement of voltage and current via separate cables.

2.2 Installation of the modules

Eight modules have been incorporated in the BIPV setup of the KU Leuven. The modules have been divided in groups with and without natural ventilation. The way the ventilation is created is shown in Figure 3.

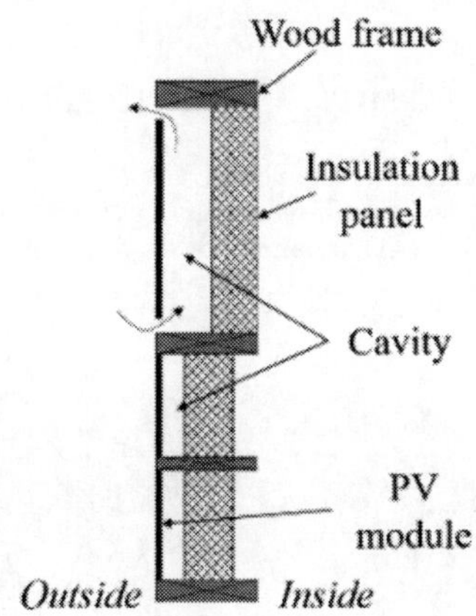

Figure 3: Facade cross section showing construction and ventilation flow (curved arrows)

On the insulation panel of the non-ventilated transparent modules, a reflective surface was placed to reflect radiation transmitted through the module back to the rear side of it. The surface was white for 8555-5, black for 8555-3 and aluminium for 8555-7. The setup and an explanatory schematic are shown in Figure 4.

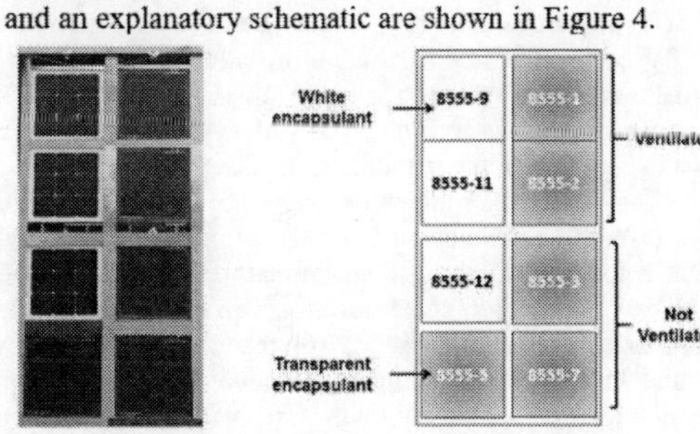

Figure 4: Modules mounted in BIPV monitoring setup and explanatory schematic

The module monitoring started late September 2017, using maximum power point tracking (MPPT) calculated from *I-V* measurements for each individual module by applying programmable electronic loads. A full *I-V* curve is swept every 15 seconds to ensure operation at MPP even at dynamic, fast-changing weather conditions. A novel sweeping algorithm is used which screens mostly the area of the expected MPP, based on the I_{sc} reading [4]. To avoid heat dissipation, the PV module is operated at the maximum power point in between the *I-V* sweeps. The temperature of the back of the modules is monitored with sensors attached to it and logged in 1-second intervals. Finally, the output of the meteorological measurements such as ambient temperature, wind speed and irradiance are logged in minute-long intervals. Using these datasets, the imec model for PV energy yield calculation [5] will be tested for its accuracy for these bifacial modules later on.

The bottom up energy yield model of imec has originally been developed for monofacial modules. The energy yield model has been adapted for the correct modelling of these modules by coupling it to the ray-tracing program Radiance. Using this program, the exact spatially resolved irradiance on each cell can be simulated and used as input for the energy yield simulations tool.

3 MODULE RESULTS AND MONITORING

3.1 Module characterisation by EL and *I-V*

All modules were inspected by electroluminescence (EL) to check for the uniformity of the contact formation, cell cracks or other defects. A typical picture is shown below.

Figure 5: EL picture of a transparent module

In all 12 modules, only one cell was found to have a crack, but since only 8 modules could be fit into the BIPV setup, this module was not installed anyway. Some small non-uniformities can be observed in the EL pictures, but these are related to wafer or cell production issues, not to module manufacturing issues.

The *I-V* measurements of the modules were performed on a flatbed LED solar simulator at Soltech (since the new module solar simulator at imec, with 18 LED types for high spectrum accuracy and with module temperature control was not yet taken into operation). Since there was no calibrated module of similar type available, the intensity was set with a calibrated monofacial module with industrial cells, having a different dimension. Being not optimal for a right calibration of the intensity, this could have caused some error (either up or down) in the determination of I_{sc} and P_{mpp} of the modules. The *I-V* measurements on the transparent modules were done from both sides, all performed with black back sheet behind the module, to avoid illumination from behind as much as possible. The total reflection of this back sheet was around 5 %, being almost constant over the entire wavelength range of 350 nm – 1200 nm).

The *I-V* measurements of the plated bifacial cells used inside the modules were also performed on a black background, with the GridTouch chuck from Pasan. Since this chuck has no temperature regulation, it should ideally be used with a flash illumination to avoid cell heating. Since a constant illumination is applied in the cell solar simulator at imec, the temperature control is difficult. Although an infrared sensor is placed in the chuck to measure the cell temperature, in practice an error of several mV in V_{oc} can occur.

The *I-V* results of the modules (and also the cell *I-V* results for comparison) are shown below.

Condition/ variable	Bifacial cells	Trans-parent	Trans-parent	White
Illumination	1 sun front	1 sun front	back	1 sun front
# modules	-	7	7	5
I_{sc} (A)	9.82	9.23	8.88	9.48
V_{oc}/cell (mV)	652	657	656	658
FF (%)	78.8	75.0	75.2	74.5
P_{mod} (W)	-	40.9	39.4	41.8
P_{cells} (W)	-	45.4	45.4	45.4
P_{mod}/P_{cells}	-	0.90	0.87	0.92

Table I: *I-V* results for the bifacial modules and those for the cells (averaged) for comparison

The I_{sc} of the modules is considerably lower than that for the cells, even for white encapsulant. Apart from calibration issues that could have caused errors up or down, this can be attributed to the absence of an ARC on the module glass, the UV absorption in the encapsulant and finally the fact that the I_{sc} of busbarless cells is corrected to the situation without any wire (so after encapsulation the wire shadowing reduces I_{sc}, in contrast to the situation for cells with busbars). The bifaciality factor of the module I_{sc} is quite high (96%), similar to the value of these imec bifacial cells. The module V_{oc} per cell is 5-6 mV higher than for the cells, this is probably related to the temperature control issues of the cell chuck mentioned before. The *FF* is somewhat reduced compared to what you would expect for SWCT, because of the fact that the SWCT foil did not extend up to the edge of the cells, resulting in no contact to the last 3 fingers on the cells. As a result of the somewhat lower I_{sc} and *FF*, the ratio between module power and power of the cells is reasonable but not very high: around 0.90 for transparent modules and 0.92 for the white ones.

3.2 Monitoring results

When interpreting the energy production data for these modules being used at different conditions, it is of course important to know that the individual modules of the same type are almost equal in power when measured at STC. For the transparent ones, the module power at STC varied in a small range from 41.1 W to 41.5 W, except module 8555-7 that had 39.7 W due to a lower *FF* of 73.5 %. For the white ones, the range was even smaller, namely 41.7 W to 41.9 W. So fortunately, it is quite reasonable to consider modules of the same type to be equal in performance, except that 8555-7 could have some power loss at high light intensities due to a higher series resistance in this module.

In Figure 6 one can find the monthly energy generation per module from October 2017 until August 2018.

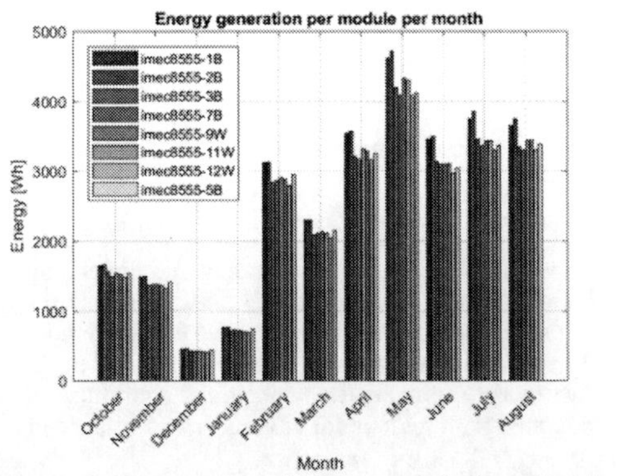

Figure 6: Energy generation per module from October 2017 to August 2018. Modules with the suffix "B" are bifacial while those with the suffix "W" are monofacial (but with bifacial cells) with white encapsulant.

Looking at the graph in Figure 6 and referring to the module positions/conditions in Figure 4, it is clear that the most important difference in energy generation is related to the difference between ventilated and non-ventilated modules. During the winter months there is almost no effect of course, but in the months with the highest energy generation, the difference is around 10 % for the transparent modules. For the white modules, there is also a positive influence of the ventilation, but this is limited to only 3-4 %.

Another observation is that the transparent ventilated modules are performing around 5 % better than the white ventilated modules, while there is almost no difference between the module types for the non-ventilated situation. It is clear that in the ventilated case there is a beneficial effect from the bifacial modules, since the white modules even had a higher power at STC but a lower energy production.

Concerning the background comparison, somewhat surprisingly there is a rather limited positive influence (1-2 %) of the white background (8555-5) compared to the black background (8555-3). However, this was also in the part without ventilation, where the improvement by using bifacial modules is low anyway.

So it seems that there can be an extra energy production for bifacial modules indeed when using a reflecting surface behind it, but for an optimal result in a facade it seems really necessary to provide enough ventilation to remove the extra heat generated in the module because of the double-sided illumination.

In the future a detailed analysis of the results will be made, including a comparison with energy yield calculations preformed with the imec energy yield modeling software, adapted for bifacial modules.

4 CONCLUSIONS

Glass-glass modules were successfully manufactured using plated bifacial cells made at imec. The modules were made in two types, transparent and white (using white encapsulant). The *I-V* results at STC for the modules were satisfying. The cell to module power ratio was 0.9 (transparent) - 0.92 (white), but taking into account that the glass had no ARC coating, that the encapsulant had to be UV absorbing, that the wire shadowing lowers the I_{sc} compared to cell level and that the SWCT foils were a bit short (somewhat reducing the *FF*), this is an acceptable result. EL pictures showed only one crack in a module that did not have to be installed anyway, since a few extra ones had been made. The monitoring in the BIPV facade installation on the VLIET building from KU Leuven were started in October 2017.

The energy production results for the modules at different conditions showed that the difference between ventilation and no ventilation played a very important role, even more so for the transparent bifacial modules (around 10 %). Also a gain was observed for transparent modules compared to the white ones, but only for the ventilated modules (around 5 %), without ventilation the gain is almost reduced to zero. A probable explanation for this is that without ventilation, double-sided illumination increases the module temperature so much that the current gain is almost completely compensated by a reduction of V_{oc} and *FF*.

5 ACKNOWLEDGEMENTS

The authors gratefully acknowledge imec's SiPV industrial affiliation programme and its partners. This work has been conducted within the EFRO SALK project, which receives the support of the European Union, the European Regional Development Fund ERDF, Flanders Innovation & Entrepreneurship and the Province of Limburg. We also thank the company Meyer Burger for their support on the SWCT, especially the foil supply.

6 REFERENCES

[1] ITRPV 2017

[2] Loic Tous, Richard Russell, Emanuele Cornagliotti, Angel Uruena, Patrick Choulat, Michael Haslinger, Joachim John, Filip Duerinckx, Jozef Szlufcik, "22.4% bifacial n-PERT cells with Ni/Ag co-plated contacts and Voc ~691 mV", Energy Procedia, Vol. 124, September 2017, pp. 922-929

[3] Loic Tous, Richard Russell, Emanuele Cornagliotti, Filip Duerinckx, Jozef Szlufcik, "Impact of selective BSF on performance of bifacial nPERT cells with Ni/Ag co-plated contacts", Metallization workshop, Konstanz, October 23-24,2017

[4] Spiliotis K., Yordanov G., Van den Broeck G., Goverde H., Baert K. and Driesen J., "Towards Accurate, High-Frequency I-V Curve Measurements of Photovoltaic Modules Applying Electronic Loads". 33rd EUPVSEC, Amsterdam, The Netherlands, 25-29 September 2017, pp. 1561-1565

[5] Goverde H., Goossens D., Govaerts J., Catthoor F., Baert K., Poortmans J. and Driesen J., "Spatial and temporal analysis of wind effects on PV modules: Consequences for electrical power evaluation", Solar Energy, Volume 147, 2017, pp. 292-299

ADAPTABLE PV PERFORMANCE MODELLING FOR INDUSTRIAL NEEDS

Steve Ransome[1] and Juergen Sutterlueti[2]
[1]Steve Ransome Consulting Ltd, Kingston upon Thames #99 KT2 6AF, U.K.
[2]Gantner Instruments GmbH, 6780 Schruns / Austria

ABSTRACT: Precise PV measurements and accurate performance models are vital in understanding and optimizing the system energy yield. An adaptable combination of features from the "Loss Factors Model" (used to define the normalised parameters) and the "Mechanistic Performance Model" (which determines the optimized fitting method) has been used on performance measurements such as individual module full IV curves (8 derived parameters in Arizona) and monitored large arrays (I_{MP} and V_{MP}) from Gantner Instruments at other sites with similar algorithms. It allows PV performance to be predicted and measurements to be validated with reasons for any underperformance or degradation to be identified and quantified.
Keywords: Energy Rating; Energy Performance; Modelling

1 INTRODUCTION

Electrical measurements of individual modules or string connected arrays on site are made with differing numbers of parameters depending mainly on whether the monitoring system sweeps between I_{SC} and V_{OC} (IV curves) or just measures at maximum power point tracking (MPPT).

As it is easier to measure IV curves of single modules than for strings of arrays, more parameters (6 or more) are available for individual modules.

For multi MW_P solar plants either one (just P_{MP}) or two (I_{MP} and V_{MP}) measured values are usually available for strings.

Performance modelling needs to be adaptable to make the best use of whichever data parameters are available.

Figure 1 and table I give the optimum parameters to measure vs. the number of parameters. They are as independent as they can be from each other, are meaningful and they cover the full IV curve.

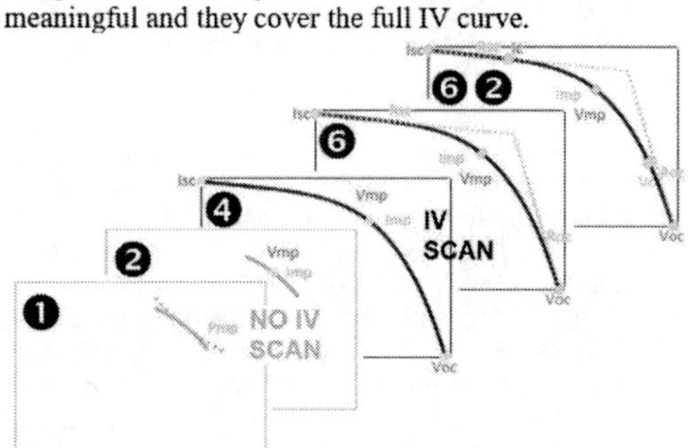

Figure 1: Optimum parameters to measure at increasing levels of detail from MPPT (❶ and ❷) to IV curve (❹ to ❻+❷).

Table I: Optimum parameters to measure (as in figure1)

#Params	Parameter names
(just MPPT)	
❶	P_{MP}
❷	I_{MP}, V_{MP}
(full IV curves)	
❹	I_{SC}, I_{MP}, V_{MP}, V_{OC}
❻	I_{SC}, R_{SC}, I_{MP}, V_{MP}, R_{OC}, V_{OC}
❻+❷	Add I@V_{MP}/2 (check mismatch), and V@I_{MP}/2 (Schottky contact rollover)

An explanation for the choice of parameters is as follows – (see Python Code 1 for glossary).

- ❶ The simplest PV performance measurements (1 parameter P_{MP}) can give PR_{DC} (=$mP_{MP}/rP_{MP}/G_I$).
- ❷ 2 parameters (I_{MP} and V_{MP}) can be analysed to give more information on I and V separately.
- ❹ If IV curves are taken, then I_{SC} and V_{OC} can be also be found and analysed (to make 4 parameters)
- ❻ If IV curves are good quality (i.e. smooth and high precision) then R_{SC} and R_{OC} can also be fitted (6 parameters).
- ❻+❷ For the highest quality IV measurements check "curvature parameters" can be taken such as I@V_{MP}/2 and V@I_{MP}/2 (and comparing these with expected values from extrapolating I_{SC} and R_{SC} or V_{OC} and R_{OC}) to make 8 in total.

These parameters are illustrated in Table I and figure 2.

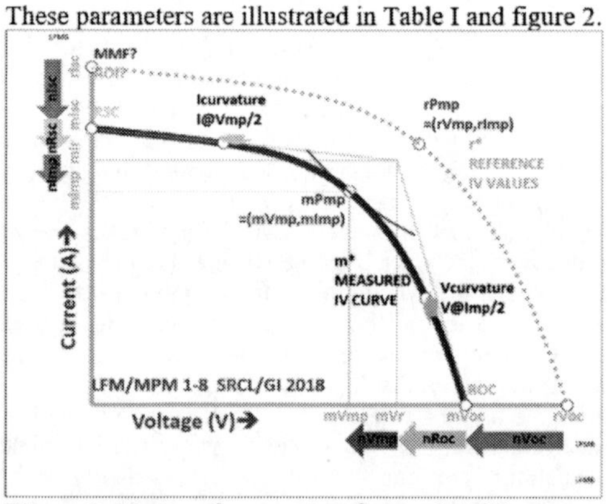

Figure 2: Measured parameters from a 6+2 IV curve used for LFM/MPM analysis

The LFM [1-5] had previously only been used to analyse 6 independent parameters from an IV curve, but in this study it is shown how it can analyse fewer values as in figure 1 depending on what data is available.

The following python code shows how to normalise the measured parameters in figure 1 such that their product equals PR_{DC}.

Python Code 1:

```
# CALCULATIONS IN PYTHON SRCL and GI Sep 2018
# Key to naming convention
# m* = measured value
# r* = reference value at STC
# d* = difference measured - reference
# *_U = not temperature compensated
# *_T = temperature compensated
# alpha, beta, gamma = temperature coeffs
# Gi = POA irradiance kW/m^2
Tstc = 25 # STC temperature celsius
# 1 parameter
PRDC = mPmp / rPmp / Gi
# 2 parameters named to distinguish dc or ac
nIdc = mImp/rImp / Gi
nVdc = mVmp/rVmp
# 4 parameters (note nFFI * nFFV = FF)
nFFI = mImp/mIsc
nFFV = mVmp/mVoc
# temporary calcs for easier maths
# (= intercept of RSC and ROC tangents)
mIr = (mIsc*mRsc-mVoc)/(mRsc-mRoc)
mVr = mRsc*(mVoc-mIsc*mRoc)/(mRsc-mRoc)
# 6 parameters default for LFM
dTmod = mTmod - Tstc
nIsc = mIsc/rIsc/Gi
nRsc = mIr/mIsc
nImp = mImp/mIr*rIsc/rImp
nVmp = mVmp/mVr*rVoc/rVmp
nRoc = mVr/mVoc
nVoc_U = mVoc/rVoc
# correct V by beta_Voc and (temp - 25C)
nVoc_T = nVoc_U * (1-beta_Voc * dTmod)
# curvature parameters for 6+2 check
nIc = mI(@Vmp/2)/(mIsc-mVmp/2/mRsc)
nVc = mV(@Imp/2)/(mVoc-mImp/2*mRoc)
```

The LFM PR_{DC} definitions for 2, 4 and 6 parameters respectively are given in equations below (where PR_{DCx} means the equation is for x parameters).

$$PR_{DC2} = [\quad nI_{DC} \quad] \times [\quad nV_{DC} \quad] \text{❷}$$
$$PR_{DC4} = [nI_{SC} \times nFF_I \quad] \times [\quad nFF_V \times nV_{OC}] \text{❹}$$
$$PR_{DC6} = [nI_{SC} \times nR_{SC} \times nI_{MP}] \times [nV_{MP} \times nR_{OC} \times nV_{OC}] \text{❻}$$

Figure 3 illustrates visually how these values all multiply to give the usual PR_{DC}.

(Note the 6+2 version has two extra parameters to quantify the curvature from cell mismatch or rollover.)

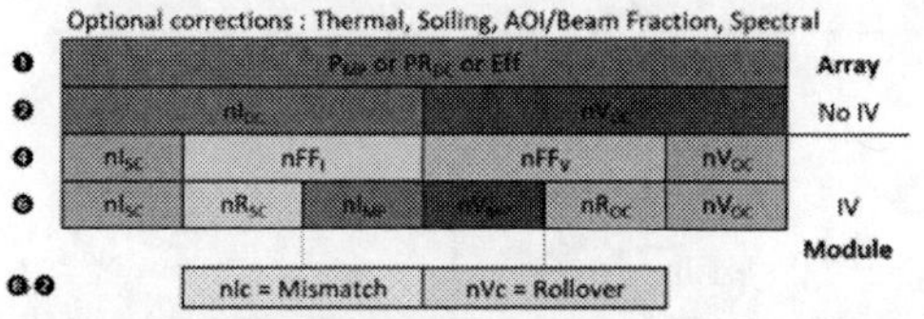

Figure 3: Showing how the PR_{DC} is the product of different numbers of parameters from 2 to 6.

The MPM was originally developed to characterise only the PR_{DC} but has been generalised to be able to fit any of the normalised LFM parameters in code 1 e.g. nR_{SC}, nV_{OC} as in equations for PR_{DCx} and table II where

$dT_{MOD} = (T_{MOD} - 25)$;

G_I = plane of array irradiance (kW/m^2),

WS = wind speed (ms^{-1}).

The MPM equation to fit any normalised LFM parameter is -

$$nLFM = C_1 + C_2 \times dT_{MOD} + C_3 \times Log_{10}(G_I) + C_4 \times G_I$$

Table II: Explanation of MPM coefficients

Coefficient	Dependency	Comment	Unit
C_1	Performance Tolerance	Actual/ Nominal	%
C_2	Delta T_{MOD}	Temperature Coefficient	%/K
C_3	log_{10}	low light fall	%
C_4	G_I	high light fall	%

The fitting modelling of the MPM has therefore been combined with the LFM parameters [3].

Figure 4 shows the effect of monitoring and modelling a large array in the UK at 1-minute instantaneous intervals using just the measured parameter nP_{AC} and then modelling the fits with the MPM vs time, irradiance and module temperature then calculating the residual error "Difference" which is very low usually <±2.5%. The only discrepancies are during quickly changing irradiance periods as the array is large and the irradiance sensor is small, so cloud fronts do not pass over the sensor and modules at the same time.

The same methods can be used to fit individual modules with full measured IV curves (so up to 8 parameters can be extracted, the usual 6 LFM plus the curvature coefficients).

Figure 5 has plots of measured vs. fitted values for four parameters nI_{SC}, nFF_I , nFF_V and nV_{OC} (left) for a CdTe module (top) vs. a c-Si (bottom) showing good accurate fits but different behaviour e.g. the CdTe has a worse low light nV_{DC}.

The other columns show the greater detail from analysing more parameters centre (measured) and right (fitted) for the same two modules. Here we can see the nR_{OC} is worse for the CdTe at high light levels (caused by the higher $I^2.R_{SERIES}$ loss) and the CdTe also has a poorer low light nR_{SC} (caused by a collapse in R_{SHUNT}). This is partially offset by a better nV_{MP}

Figure 6 shows a module with both poor low light and high light PR_{DC} behaviour.

Adding more coefficients to the analysis the reasons for this can be found, here the causes are found to be nR_{SC} (~R_{SHUNT}) and nV_{OC_T} at low light whereas it is due to nR_{OC} (~R_{SERIES}) at high light. These values can be identified and quantified.

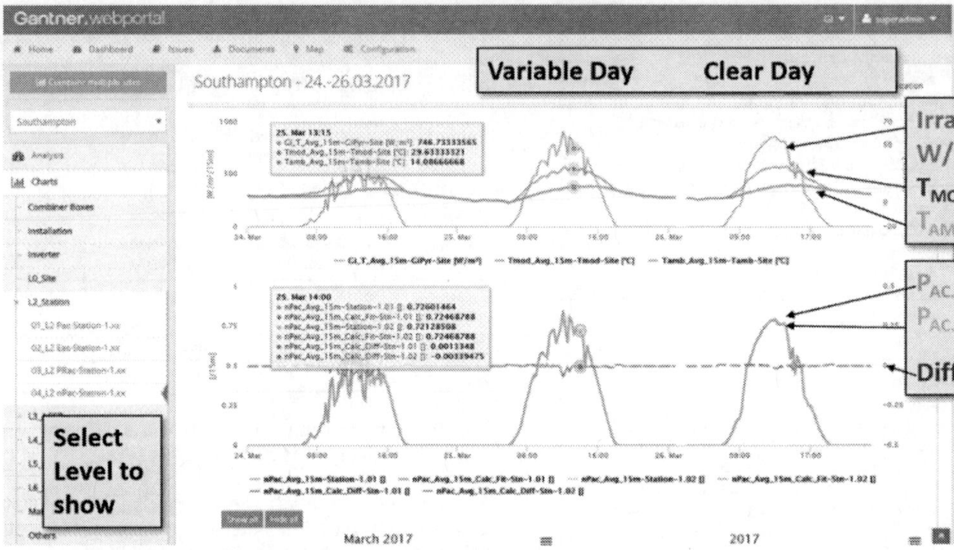

Figure 4: Analysis of a large array measured in the UK for (left to right) poor, variable weather and clear days showing weather (top) and predicted vs. measured nP_{AC} (bottom) with good agreement < ±2.5%.

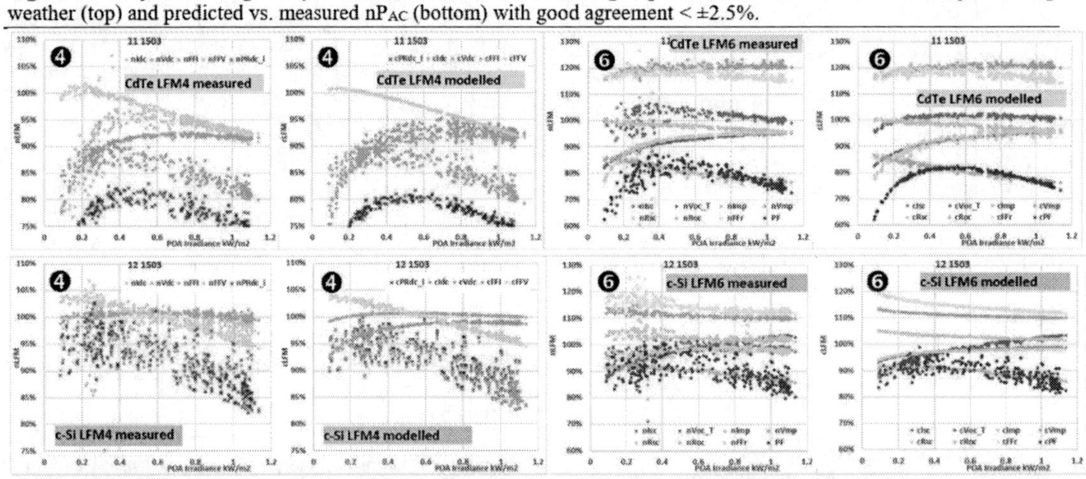

Figure 5: Extended LFM parameters fitted by MPM methods at different irradiances (left to right: LFM 4 measured vs. modelled; LFM 6 measured vs. modelled) for a CdTe module (#11 top) and vs. a c-Si module (#12 bottom) for one-month March 2015 at Gantner Instruments' data at their OTF in AZ.

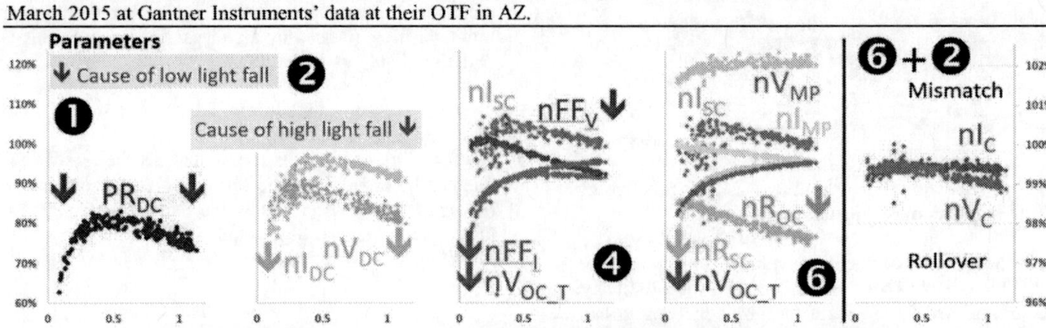

Figure 6: As more parameters are analysed the causes and rates of performance changes can be identified more closely.

	Cause of Low light drop	Cause of High light drop
❶	PR_{DC} no cause known	PR_{DC} no cause known
❷	nI_{DC} could be I_{SC} or R_{SHUNT}	nV_{DC} could be V_{OC} or R_{SERIES}
❹	nFF_I (possibly R_{SHUNT}?) and nV_{OC_T} (V_{OC})	nFF_V (possibly R_{SERIES}?)
❻	nR_{SC} (definitely R_{SHUNT}) and nV_{OC_T} (V_{OC})	nR_{OC} (definitely R_{SERIES})
❻+❷	None Both are flat so this module has no problems with cell mismatch (nI_C) or Schottky back contact (nV_C)	

Another advantage of using normalised coefficients for the MPM is that it is easy to calculate expected performance at the standard PV module test conditions as in Table III.

Table III: Definitions of test conditions

	standard test	PVUSA	nom. operating cell temp	low temp.	low irrad	high temp
	STC	PTC	NOCT	LTC	LIC	HTC
G_I (kW/m^2)	1	1	0.8	0.5	0.2	1
T_{AMB} (C)	-	20	20	-	-	-
T_{MOD} (C)	25	-	-	15	25	75
WS (ms^{-1})	0	1	1	0	0	0
Tilt (degrees)	-	-	45	-	-	-
AM (#)	1.5	1.5	1.5	1.5	1.5	1.5

The MPM fit can be used to estimate the module's performance under specific conditions such as STC (1kW/m^2, Tmod=25C) and Low Irradiance (0.2kW/m^2, Tmod=25C).

The following tables IV and V show the differences between the some of the parameters for the cSi and CdTe modules analysed for each 6 months 2014 to 2016 to evaluate degradation. (Column "yymm" is 2 numbers for the year and 2 numbers for the month").

Table V shows the CdTe module has a degrading nR_{SC} at low light caused by a drop in low light R_{SHUNT}

Table IV: A slight fall in nR_{SC} for CdTe only at high light levels (STC)

STC	yymm	nVoc	nRsc	nRoc
12 c-Si	1403	100.0%	96.5%	88.6%
12 c-Si	1409	99.8%	96.5%	88.7%
12 c-Si	1503	100.0%	98.0%	88.9%
12 c-Si	1509	99.7%	97.1%	88.5%
12 c-Si	1603	99.9%	97.6%	88.9%
12 c-Si	1609	100.2%	95.8%	88.5%
11 CdTe	1403	95.8%	95.8%	75.2%
11 CdTe	1409	96.0%	95.0%	75.8%
11 CdTe	1503	95.0%	94.9%	74.4%
11 CdTe	1509	94.9%	93.8%	76.0%
11 CdTe	1603	95.1%	94.9%	75.0%
11 CdTe	1609	95.1%	93.4%	78.3%

Table V: A worse fall in nR_{SC} for CdTe only at low light levels (LIC)

LIC	yymm	nVoc	nRsc	nRoc
12 c-Si	1403	92.4%	96.1%	91.7%
12 c-Si	1409	92.3%	95.9%	91.7%
12 c-Si	1503	92.3%	95.9%	92.0%
12 c-Si	1509	92.2%	96.2%	91.9%
12 c-Si	1603	92.2%	95.7%	91.6%
12 c-Si	1609	92.4%	95.7%	91.9%
11 CdTe	1403	89.1%	91.1%	85.3%
11 CdTe	1409	88.7%	91.0%	85.3%
11 CdTe	1503	87.6%	86.9%	84.7%
11 CdTe	1509	87.4%	86.4%	85.2%
11 CdTe	1603	87.7%	86.7%	85.0%
11 CdTe	1609	87.5%	86.2%	86.3%

Figures 7 and 8 give the differences between the STC and LIC values of nR_{SC} and nR_{OC} for c-Si (7) and CdTe modules (8)

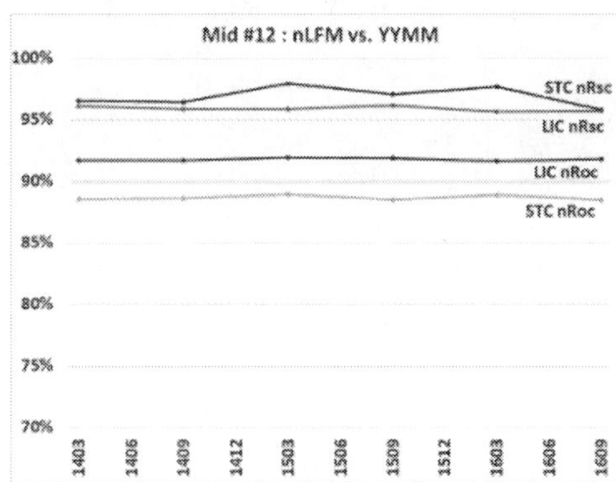

Figure 7: LFM parameters for a c-Si module every 6 months with good(high) and stable(flat) performance.

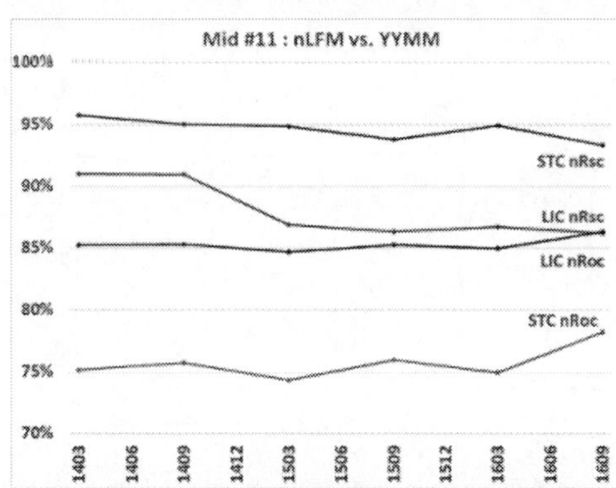

Figure 8: LFM parameters for a CdTe module every 6 months with poorer(lower) and unstable(sloped) performance with worsening nR_{SC}.

3 CONCLUSIONS

- The Loss Factors Model (LFM) and Mechanistic Performance Model (MPM) have been combined to give an advanced analysis of MPPT or IV data using the same procedure.
-

- Optimised coefficients used are "Meaningful, Orthogonal, Robust and Normalised" (MORN).

- Instantaneous measurements can be validated in real time.

- The reasons and magnitudes for any faults/underperformance can be easily found.

- Seasonal effects and degradation can be identified and quantified (e.g. %/year V_{OC}).

- This procedure confirms optimum output behavior or else identifies faults which should be quickly rectified to minimise any downtime or errors.

- The LFM/MPM method has been added to Gantner Instruments' SaaS platform for advanced analysis and fault detection.

- For further data see www.gantner-instruments.com, www.gi-cloud.io, www.otf.gantner-instruments.com.

References

[1] Ransome et al, 44th PVSC (2017) Washington DC "How to Choose the best Empirical Model for Optimum Energy Yield Predictions" https://de.slideshare.net/sandiaecis/21-gi-8thpvpmctalkabqfinalext170508t19new

[2] Ransome et al, PVSEC27 (2017), "7TuPo Quantifying and analysing the variability of PV module resistances R_{SC} and R_{OC} to understand and optimise kWh/kWp modelling"

[3] Ransome et al, PVSEC27 (2017), "7MoO.5.4 "Optimised fitting of indoor (e.g. IEC 61853 matrix) and outdoor PV measurements for diagnostics and energy yield predictions"

[4] http://www.gantner-webportal.com www.gi-cloud.io www.otf.gantner-instruments.com

[5] Sutterlueti, PVPMC8 (2017), "Optimized PV Performance using State of the Art Monitoring for Increased Asset Value"

INVESTIGATION ON THE QUALITY OF ADHESIVE JOINTS OF SHINGLED SOLAR CELLS BY ACCELERATED LIFETIME TESTING

Ingo Ullmann, Dominik Rudolph, Jorge Rabanal-Arabach*, Andreas Schneider, Andreas Halm
International Solar Energy Research Center (ISC), Rudolf-Diesel-Str. 15, D-78467 Konstanz, Germany
*Universidad de Antofagasta, Av Angamos 601, Antofagasta, Chile
Corresponding author: Phone: +49 7531 36183302; Fax: +49 7531 3618311; Email: ingo.ullmann@isc-konstanz.de

ABSTRACT: Electrically conductive adhesive (ECA) interconnections between solar cells are gaining more and more relevance since they provide a solution to realize lead free, elastic bonds, which inflict low mechanical stress on the solar cells. Here we investigate the long term stability of such interconnection joints by accelerated degradation testing in a climate chamber. For that purpose, mini-modules with shingled cells using different ECA formulations and different module materials are assembled and their electrical properties are recorded via IV and EL measurements before and after climatic chamber tests. We found varying power losses for the different ECA´s after damp heat testing (DH2000) and thermal cycling (TC800) testing. Modules of one ECA are stable after both tests, one configuration shows average power degradations around 1.4 to 3 % after both tests and the third one tested revels an average power drop of 25 % after DH2000 testing while it is stable after TC800.
Keywords: PV Module, shingling, reliability

1 INTRODUCTION

Shingled modules are readily available on the market [1-3]. For interconnecting shingled cells ECA gluing is investigated by many groups and found to be a promising technology [4-6]. Long term outdoor performance data proving their stability against environmental influences is not yet available. The interconnection of the shingle pieces to a shingled string, so called supercell, is either realized by ECA gluing or soldering. The behaviour of mini-modules of encapsulated supercells connected by ECA gluing is investigated in this work. The aim is to identify ECA bonds which are mechanically, chemically and electrically stable against accelerated aging tests to compensate the lack of data on the long term outdoor performance of shingled modules.

2 APPROACH AND SAMPLE PREPARATION

6" mc-solar cells were metallized with a special shingle design as shown in Figure 1, which allows a splitting of the cells in five individual shingle cell stripes by laser. This design was found out to be the optimum for the case of a multicrystalline 6" Al-BSF solar cell [7]. ECA is dispensed by a x-y-z dispenser stage, shown in Figure 2, onto the busbar of each stripe. Then the stripes are shingled with a fixed overlap of 1.5 mm in a manual process by the help of stair shaped aluminium plate (step height 200 µm). One ribbon at the beginning and one ribbon at the end of the string are ECA attached for contacting the module. Then the shingle string is transferred to a hot plate to cure the adhesive connections. Curing is done at 150°C or 180°C for between 20 seconds and 1 minute. Afterwards the materials for the module, listed in Table 1, are stapled on each other and then processed in the laminator. For that purpose we assembled mini-modules using four different kind of ECA, which are listed in Table 2. ECA 1, 2 and 4 were used in TC and ECA 1, 2 and 3 were used in DH. The ECA´s differ in the adhesive material (epoxy and acrylic based) and amount of conductive filler material (between 45% and 55%). Furthermore, the mini modules are assembled with commercially available EVA or Polyolefin (POE) as encapsulation and white or transparent back sheet as rear cover in order to quantify their impact on module reliability. Modules are characterized by IV and EL after lamination and then after each following test to monitor changes in the module performance.

2.1 Test and equipment description

The climate tests described in the standard IEC 61215, are thermal cycling -40°C / +85°C (TC) and damp heat 85°C / 85% r.H. (DH) test. The mini module samples with a size of 200mm x 200mm dwell for 800 cycles in TC and 2000 hours in DH.

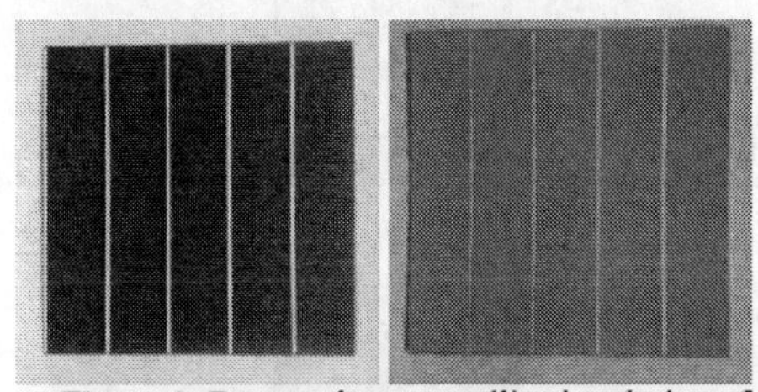

Figure 1: Front and rear metallization design of shingle printed 6"-cells.

Figure 2: Equipment of the mini module line with x-y-z dispenser and hot plate for curing of ECA.

Table 1: Material combinations used for the modules presented in this work.

Encapsulant	Backsheet	Structure
POE	White	ECA 2 1.5 mm overlap Ribbon: - size: 2.5x0.15 mm - material: SnPbAg
POE	Transparent	
EVA	Transparent	
EVA	White	

Table 2: Different ECA's used in this work

Name	glue basics	cure temp.	cure time	resistivity	ECA amount per shingle
ECA 1	Epoxy	150°C	30 sec	5×10^{-4}	21 mg
ECA 2	Epoxy	170°C	60 sec	1×10^{-2}	7 mg
ECA 3	Acrylic	150°C	10 sec	1.1×10^{-3}	5 mg
ECA 4	Epoxy	180°C	20 sec	1×10^{-3}	8 mg

Using a laser scanning microscope (LSM) the profiles of dispensed ECA lines were measured. Typically they reach a width between 250 µm – 480 µm and height of 55 µm to 116 µm. As shown in Figure 3 a) the shape can be round or as in Figure 3 b) trapezoid depending from the used glue. For comparison a screen printed ECA line is shown in Figure 3 c). The screen printed lines have a width of 350 µm and a height of 37 µm and show the typical marks of the screen mesh.

a)

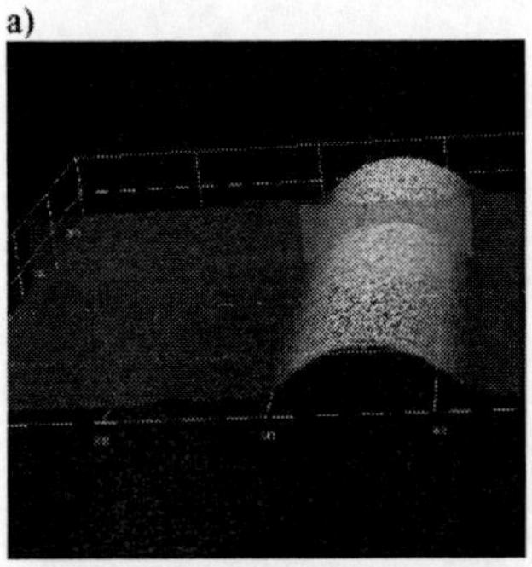

b)

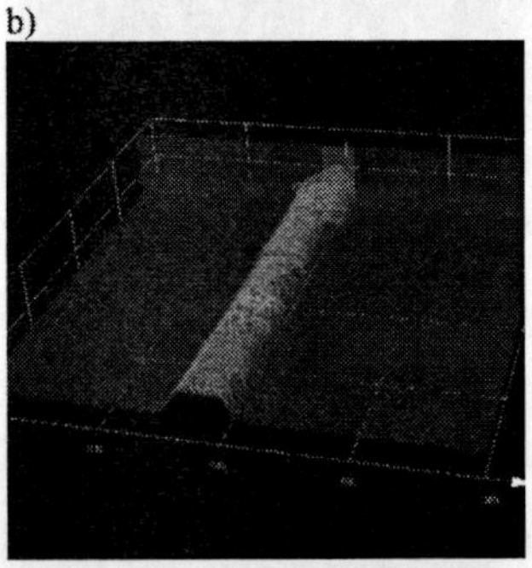

c)

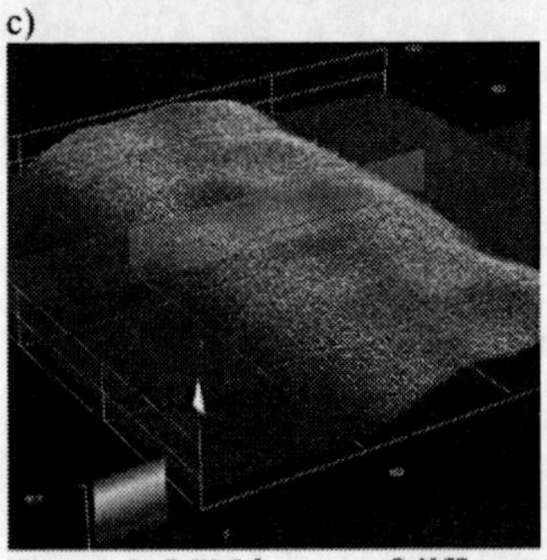

Figure 3: LSM images of different ECA lines
a) dispensed with 480 µm width, 55 µm height
b) dispensed with 250 µm width, 70 µm height
c) screen printed with 350 µm width and 37µm height.

3 RESULTS

3.1 Thermal cycling testing, 800 cycles

During TC testing IV measurement were done every 100 cycles. TC 800 results, as shown in Figure 4 and Figure 5, reveal a power degradation lower than 5% in average for all ECA and BOM combinations tested. ECA 1 shows a stronger drop in Power loss after TC800, in which one module dropped on 6% loss. One module in three different module systems dropped down after TC400 and fall out after TC500. All others stayed under 3% power loss and show very stable conditions.

In all cases, the power loss is comparable to the loss in FF and based on an increase in series resistance of the modules which allows to conclude that the ECA bond quality is decreasing. One reason for bad bond quality is a poor connection to the coatings of the ribbons glued on cell as electrical output. Some ECA are not able to glue a tin coated surface. Therefore debonding under climate stress is probably the reason for the FF drop.

Figure 5 shows the power losses of several shingle mini-modules containing five shingle pieces assembled to a supercell, ECA 2 and same ribbons and overlap after 100 to 800 temperature cycles between -40°C and +85°C. All the material combinations reveal stable connections although one module lose 7% after TC400 and fails after TC500 (EVA-transparent back sheet group).

In Figure 5, ECA 1 which is epoxy based has marginally higher relative losses of power and FF then the others. Modules containing ECA 2 and ECA 4 show very stable performance. All of the mini-modules are assembled with EVA and white back sheet.

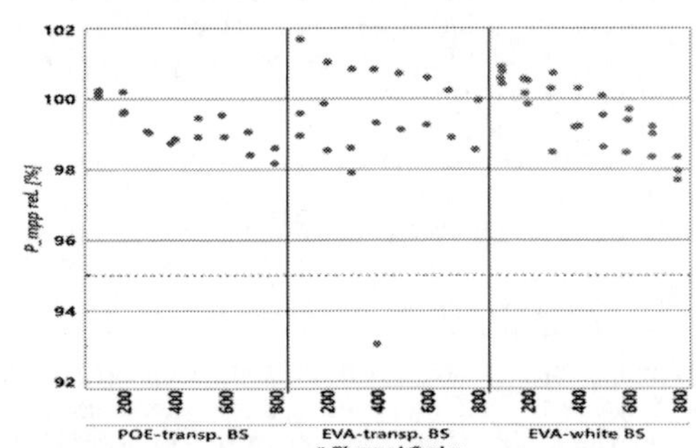

Figure 4: BOM in TC

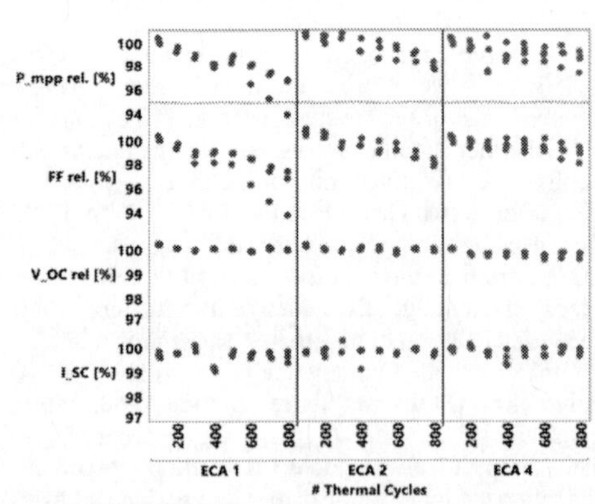

Figure 5: ECA in TC

3.2 Damp heat testing, 2000 hours

During DH, the samples were measured every 1000 hours. After damp heat testing one ECA reveals distinctly power loss while two other ECA´s were stable inside 5% (see figure 6). A significant degradation for ECA2 with 7.2% in average is to observe while 2 modules dropped on more than 70% power loss. Only 2 from 8 modules with ECA2 are inside the 5% limit. ECA1 and ECA3 show stable results after DH2000 by an average drop of 0.5 %. Generally all modules in figure 6 were made with EVA as encapsulant and white back sheet.

The different material combinations after 2000 hours in 85°C/85% r.H. (Figure 7) show a clear drop in power for the combination EVA/transparent back sheet and a further power loss of combination EVA/white back sheet of several modules. POE based modules seem stable in both TC and DH.

The clear increasing of power loss and loss in FF and therefor an increasing serial resistant seem to be only in modules with EVA. Indication for periodical failure during mounting or laminating could not be identified.

An acceptable reason could be the three times higher water vapour permeability of EVA compared to POE. This leads to a higher risk of corrosion and therefore to instable bonds.

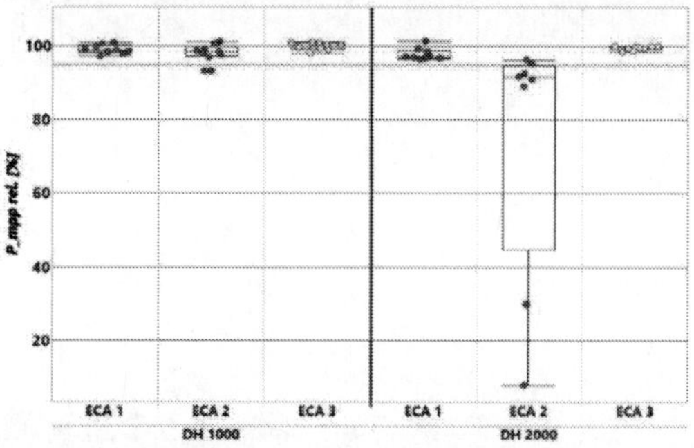

Figure 6: ECA in DH

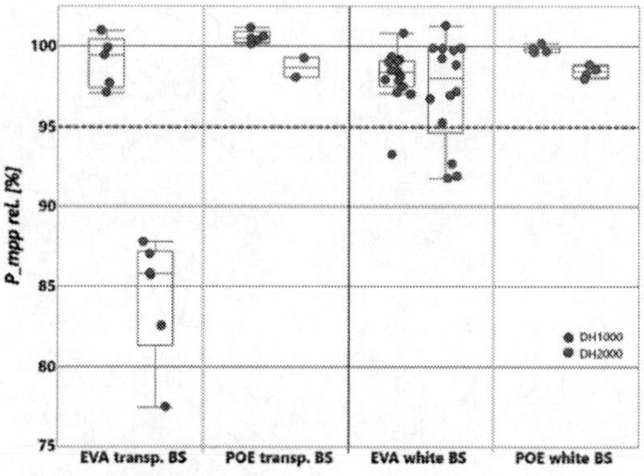

Figure 7: BOM in DH

3.3 EL and IR images

The IR image in figure 8 shows a mini-module with local heating in an area were the ribbon is connected. An incomplete glue line or a rising adhesive bond is the reason. Serial resistant increase and power as well FF loose. The line heats up by less current feed. Figure 9 shows normal and planar heating with fixed interconnections.

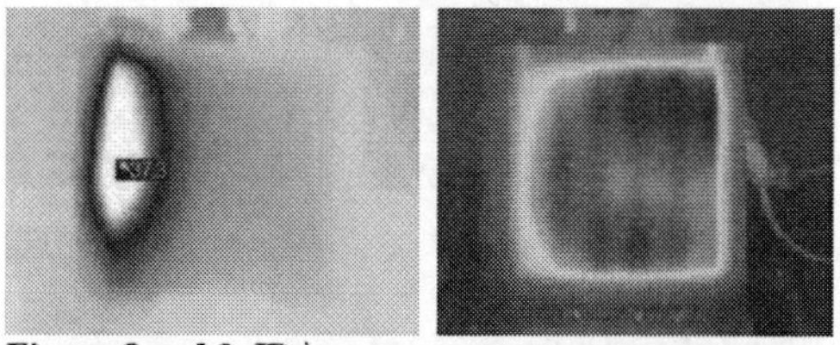

Figure 8 and 9: IR image

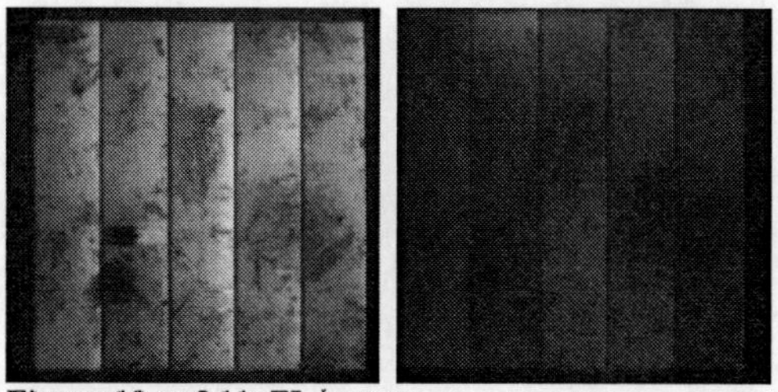

Figure 10 and 11: EL image

This EL images 10 and 11 show the same mini-module before and after 800 cycles TC. Under the same measurement conditions we can see darker areas which point on inactive zones were the power dropped.

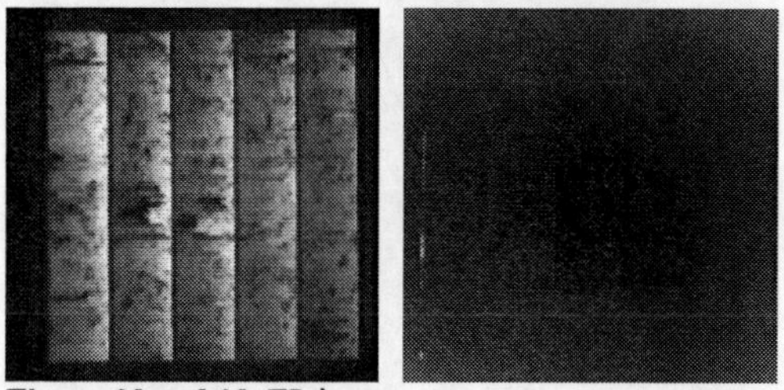

Figure 12 and 13: EL image

The EL pictures 12 and 13 compare the mini-module after lamination and after 2000 hours 85°C/85% r.h. The module dropped 70% in power after DH 2000 and show a significant increase in series resistance. Only a small area on the left side show a bright stripe while the shingle cells in the middle are very poor shining. The corresponding IR image of figure 13 is shown figure 8.

4. CONCLUSIONS

We have presented power degradation results after TC and DH treatment for shingled mini-modules assembled using different ECA for cell interconnection and different module materials. Stable configuration are identified. As main degradation cause a deterioration of the bond quality and thus an increase in module series resistance is identified. IR pictures often show a local heating at the glued ribbon by insufficient adhesion. The strong power drop of mini-modules with transparent back sheet is not due to the structure of the back sheet material as the white back sheet has higher water vapour permeability and therefore a higher influence on corrosion of the glue. The bond quality plays a major role for the formation of durable mechanical bonding and reliable electrical connections in shingling of solar cells.

5. ACKNOWLEDGEMENT

This work was supported by the Federal Ministry for Economic Affairs and Energy under FKZ: 0324109B.

6. REFERENCES

[1] http://www.seraphim-energy.com
[2] https://us.sunpower.com
[3] http://www.solaria.com/

[4] Guy Beaucarne, "*Materials challenge for shingled cells interconnection*"

[5] Max Mittag, Tobias Zech, Martin Wiese, David Bläsi, Matthieu Ebert, Harry Wirth, "*Cell-to-module (CTM) Analysis for Photovoltaik Modules with Shingled Solar Cells*"

[6] N. Klasen, A. Mondon, A. Kraft, U.Eitner, "*Shingled Cell Interconnection: Aiming for a new generation of bifacial PV-modules*"

IV MEASUREMENT OF BIFACIAL PV MODULE USING MONOFACIAL ILLUMINATION BY APPLYING DYNAMIC BIFACIALITY COEFFICIENTS

Alamera N. Alquennah and Abdulwahab T. Alasfour
Energy and Building Research Centre
Kuwait Institute for Scientific Research, Safat 13109, Kuwait
Tel.: +965 99236171, E-mail: aquennah@gmail.com

ABSTRACT: Bifacial (BiFi) photovoltaic (PV) modules increase the overall yield as they generate power from both sides. There has been a growing interest in developing new testing procedures to characterize BiFi devices using monofacial solar simulators. One proposed method to estimate the power generated from both sides measures the P_{max} of only one side with an equivalent irradiance intensity (G_E) as that of the incident irradiance on both sides. This P_{max} determination technique is known as the G_E method. In this study, we investigate the accuracy of the G_E method in estimating P_{max} by comparing the results with the sum of P_{max} measured from both sides separately. Using the G_E method, errors in power determination were found to be as high as 4.8%. Furthermore, using a dynamic fill factor bifaciality coefficient, instead of the proposed short-circuit current or power bifaciality coefficients, when applying the G_E method improved accuracy by more than two folds.
Keywords: Bifacial, standard, power estimation, single side

1 INTRODUCTION

Bifacial BiFi PV modules can capture incident light from the front and rear sides of the module and thus have more power output compared to monofacial PV modules. Under optimal site design, bifacial modules can increase the energy yield up to 25% compared to monofacial modules, if an albedo coefficient of 0.5 is provided [1]. Hence, accurate power ratings of BiFi modules are required for designing and simulating BiFi power plants. For characterizing monofacial PV modules and finding their maximum power P_{max}, the indoor and outdoor measurements proposed by IEC 60904 [2] are followed by laboratories and industries. These measurements do not consider the rear side power output of bifacial devices. Hence, V. Fakhfouri et al. proposed a new standard for indoor and outdoor IV measurements of BiFi PV devices which is still under development [3]. The proposed standard includes methods for determining the power generation gain yielded by the bifaciality. The indoor methods consist of measurements using single-side illumination solar simulators and double-side illumination solar simulators. In the single-side illumination method, P_{max} of the BiFi PV device is found by measuring the front side under elevated irradiance intensity that accounts for front irradiance intensity G_F and rear irradiance intensity G_R. The elevated front irradiance, namely the equivalent irradiance G_E, is defined as a function of the STC bifaciality coefficient of the device under test. Including the bifaciality coefficient accommodates the difference between the two sides of the BiFi device. The BiFi device under test should be measured under at least three different values of G_E to report its power rating. The current and power bifaciality coefficients deviate with respect to the incident irradiance which are affected by the technology, module's design and the position of the junction box [4].

Different research works were carried out to assess the G_E method and evaluate its accuracy in estimating P_{max} of BiFi PV devices [4-7]. Furthermore, calculation methods were proposed to find P_{max} under bifacial illumination based on monofacial illumination measurements [5] and [8]. Schmid, Alexandra, et al. [5] proposed a comparison study between IV measurements using monofacial illumination and bifacial illumination. The results showed that the mean deviation in P_{max} obtained by the G_E method and bifacial illumination measurement is 0.5% for G_F=1000W/m^2 and G_R=200W/m^2. The comparison results in S. Roest et al. work [4] showed that P_{max} deviation reached up to +1.8% and down to -1.2%. Another assessment for the G_E method was proposed by Zhang, Yating, et al. [6] in which the comparison test was done on three different types of BiFi single cells laminates. The results showed that applying the G_E method for characterizing BiFi laminates with low bifaciality coefficient requires series resistance R_s correction. The proposed study by Deline et al. [7] assessed the accuracy of the G_E method in estimating P_{max} of outdoor field measurements. The comparison results showed that the G_E method could accurately estimate P_{max} with 1-2% difference for multiple rear irradiance conditions.

In this study, we evaluated the accuracy of the G_E method in estimating the total power output of a BiFi module with high bifaciality coefficients using a monofacial solar simulator. Furthermore, we investigated the change of different bifaciality coefficients with respect to irradiance at 25°C and the effectiveness of using dynamic bifaciality coefficients instead of STC bifaciality coefficients in estimating the total power output using the G_E method.

2 EXPERIMENTAL

The PV module used in this study is a frameless monocrystalline silicon (mc-Si) BiFi technology consisting of 9 cells with no junction box shading. The IV-characteristic measurements were carried out in accordance to IEC 60904 by using an indoor monofacial pulsed solar simulator of class AAA. The module's temperature was maintained at 25°C throughout the measurements. All measurements were done for each side separately, by covering one side with an opaque black sheet and the simulator's black curtain. Each measurement was repeated three times and their average was used in the study. Irradiance correction in accordance to IEC 60891 [9] was applied on the measured IV characteristics to eliminate the effect of any slight irradiance shift on the measurements.

First, we wanted to investigate the accuracy of the G_E method in estimating the total power output generated from both sides of the utilized BiFi module. Thus, the total power output P_T which is the sum of the single-sided measured P_{max} from the front and rear sides were found, as shown in equation (1). Hence, the front side of the PV module was measured under G_F=1000W/m^2 and the rear side was measured under a range of irradiance between 100W/m^2 and 400W/m^2.

$$P_T = P_{max\,Front} + P_{max\,Rear} \qquad (1)$$

The approximated P_T of the BiFi module was found by measuring the front side only under G_E. As shown in equation (4), G_E is a function of the desired G_R and the module's bifaciality coefficients at STC. The power bifaciality coefficient φ_{Pmax} is defined as the ratio between P_{max} of the rear and the front sides and the current bifaciality coefficient φ_{Isc} is defined as the ratio between short-circuit current I_{sc} of the rear and the front sides, as shown in equations (2) and (3), respectively. Therefore, both the front and the rear sides were measured separately at STC. It was found that the value of φ_{Isc_STC} is lower than the value of φ_{Pmax_STC}. As a result, the front side was measured under the obtained G_E by using φ_{Isc_STC} and the measured P_{max} is named $P_{\varphi Isc\_STC}$. This measurement is named static G_E measurement.

$$\varphi_{Pmax} = \frac{P_{max\,Rear}}{P_{max\,Front}} \qquad (2)$$

$$\varphi_{Isc} = \frac{I_{sc\,Rear}}{I_{sc\,Front}} \qquad (3)$$

$$G_E = 1000\ \mathrm{W/m^2} + \varphi_{min(P_{max}, I_{sc})@STC} \cdot G_R \qquad (4)$$

In order to have a quantitative assessment for the accuracy of the G_E method in estimating P_T, we defined the error of estimation, named e, as shown in equation (5).

$$e(\%) = \left(1 - \frac{P_{max\,G_E}}{P_T}\right) \times 100 \qquad (5)$$

It is found that the error increases as G_R increases and as G_E deviates from the irradiance at which the bifaciality coefficient is calculated. Therefore, the change of φ_{Isc} and φ_{Pmax} with respect to irradiance was investigated. As follows, each side of the module was measured separately under a range of irradiance between 100W/m^2 and 1400W/m^2. The results showed that φ_{Pmax} and φ_{Isc} deviate with respect to irradiance, thus this deviation may affect P_T estimation using the G_E method.

Therefore, the next step was to investigate the effectiveness of using dynamic bifaciality coefficient to obtain G_E in estimating P_T. The dynamic bifaciality coefficient is the bifaciality coefficient at G equals the sum of G_F and G_R. The results showed that the difference between using φ_{Isc_STC} and dynamic φ_{Isc} in calculating G_E is minor. As a result, the effect of using the dynamic φ_{Pmax} was only examined and the measured P_{max} is defined as $P_{\varphi Pmax}$. However, using φ_{Pmax} did not enhance the estimation compared to using φ_{Isc_STC}.

In order to understand the deviation in φ_{Pmax} with respect to irradiance, the fill factor of the front side FF_f and the fill factor of the rear side FF_r were compared. Moreover, the investigation of the change in bifaciality coefficients with respect to irradiance was extended to include FF and V_{oc} bifaciality coefficients. The results showed that φ_{Voc} has very minor variations with respect to irradiance. On the other hand, φ_{FF} which comprises the changes in parasitic resistances showed a higher deviation. Hence, the effectiveness of using the dynamic φ_{FF} in estimating P_T was investigated. The IV measurements of the front side were carried out under the obtained G_E and the measured P_{max} is defined as $P_{\varphi FF}$.

3 RESULTS AND DISCUSSION

Fig. 1 shows the IV curves of the front and rear sides at STC which indicate that there is no distortion in the IV curve of the rear side due to the shade of the junction box. The obtained bifaciality coefficients at STC are illustrated in Table I. Since φ_{Isc} has the minimum value, it was used to calculate G_E in the static G_E measurements.

Table I: Bifaciality coefficients at STC of the utilized BiFi module

φ_{Isc}	φ_{Pmax}	φ_{Voc}
0.912	0.929	0.997

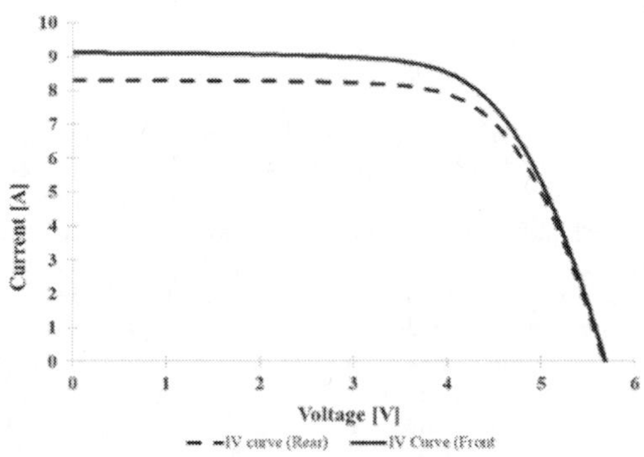

Figure 1: The IV curves of the front and the rear sides of the BiFi module at STC.

Fig.2 shows the total power output P_T and the estimated P_T using the static G_E method. The results showed that errors increase as G_R increase. The error reaches up to 4.8% when estimating P_T under G_F=1000W/m^2 and G_R=400W/m^2.

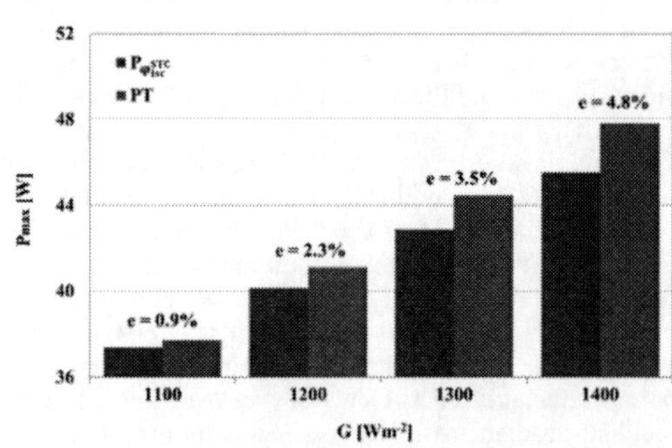

Figure 2: Total power output P_T and estimated P_T using the static G_E method for 1000 W/m^2 front irradiance and four desired rear irradiance. The error of estimation is shown.

The changes in φ_{Pmax} and φ_{Isc} with respect to irradiance are shown in Fig. 3. The results showed that φ_{Isc} deviates slightly with 0.3% standard deviation among the measured range and 0.1% standard deviation in the irradiance range above 1000W/m^2. When G_R=400W/m^2, the difference between using dynamic φ_{Isc} and using φ_{Isc_STC} is equal to 1.4 W/m^2 only. On the other hand, the difference between using φ_{Isc_STC} and using φ_{Pmax}, for the same G_R, is 11.2W/m^2. Fig. 4 shows P_T, $P_{\varphi Isc\_STC}$, and $P_{\varphi Pmax}$ for the four desired rear irradiance. The presented error of estimation is of the obtained $P_{\varphi Pmax}$.

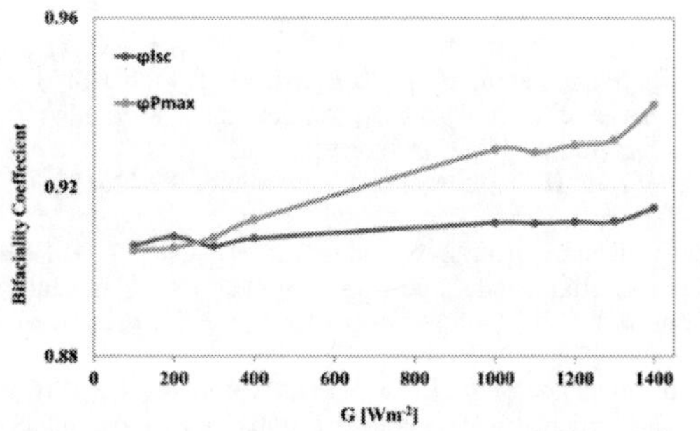

Figure 3: Current and Power bifaciality coefficients with respect to irradiance

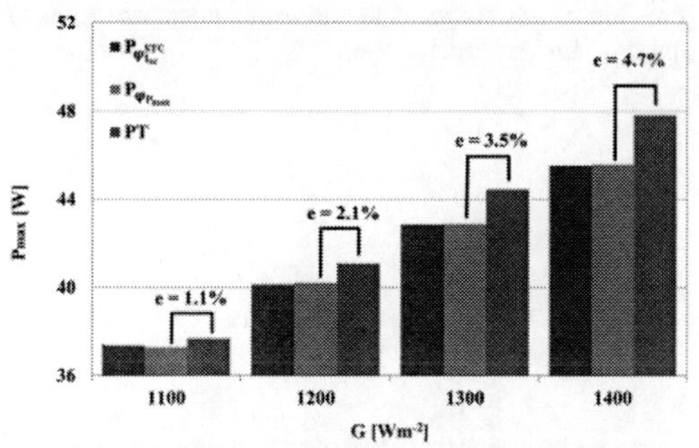

Figure 4: Total power output P_T, estimated P_T using the static G_E method and estimated P_T using dynamic power bifaciality coefficient in the G_E method for 1000W/m^2 front irradiance and four desired rear irradiance. The error of estimation of the last is presented.

The static G_E method yielded a better estimation for G_R=100W/m^2 compared to the results of using dynamic φ_{Pmax}. However, the results of the two measurements were equal for higher G_R.

Fig. 5 illustrates FF_f and FF_r with respect to irradiance. The results showed that FF_r is greater than FF_f in all measurements. Moreover, the difference between the FF of the two sides increases with respect to irradiance which may be due to the disparity in the series resistance and shunt resistance between the two sides. This resistive difference could be a source of error for the static G_E method.

The change of φ_{Voc} and φ_{FF} with respect to irradiance is illustrated in Fig. 6. The standard deviation of φ_{Voc} values in the measured range is 0.04%; V_{oc} bifaciality coefficient does not change with respect to irradiance. Since φ_{FF} is greater than 1, the obtained G_E by using φ_{FF} is greater than the sum of G_F and G_R.

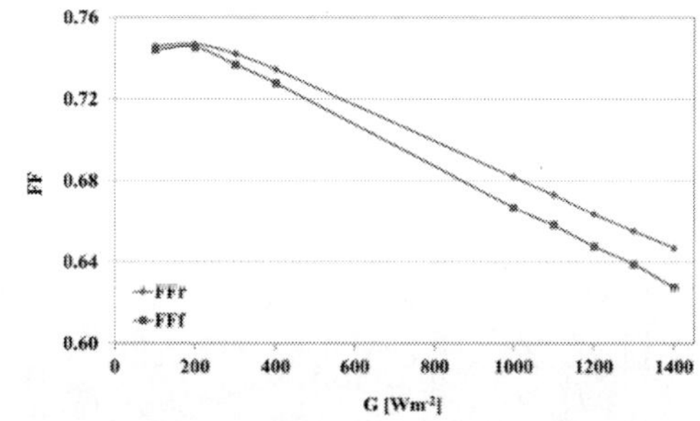

Figure 5: Fill Factor of the front side and Fill Factor of the rear side with respect to irradiance

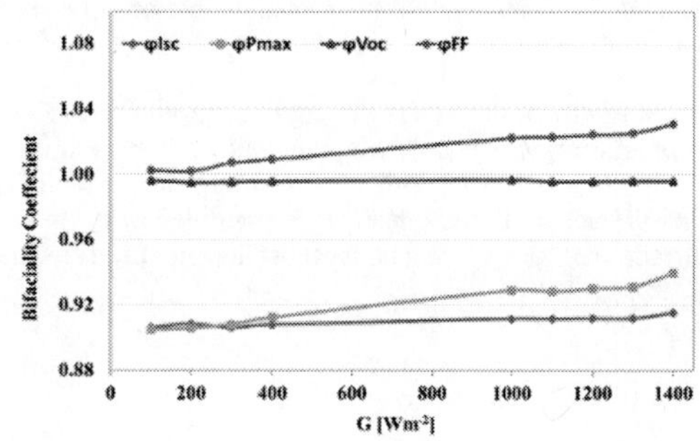

Figure 6: Current, power, voltage and fill factor bifaciality coefficients with respect to irradiance

The estimated P_T by using dynamic φ_{FF} for the four desired G_R and the error of estimation are illustrated in Fig.7. Using φ_{FF} in the G_E method enhanced the estimation compared to using φ_{Pmax} and φ_{Isc_STC}. The results show that the error when G_R=100W/m^2 and G_R=400W/m^2 were 0.3% and 2%, respectively. The errors of estimation using STC bifaciality coefficient and dynamic bifaciality coefficients in the G_E method are shown in Fig. 8.

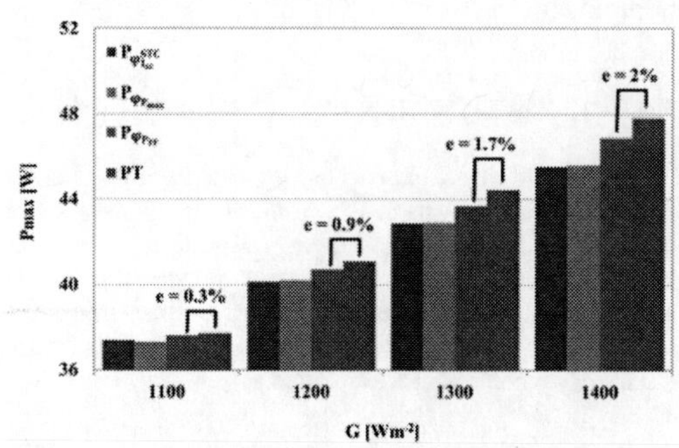

Figure 7: Total power output P_T, estimated P_T using the static G_E method, estimated P_T using dynamic power bifaciality coefficient in the G_E method and estimated P_T using dynamic fill factor bifaciality coefficient in the G_E method for 1000W/m^2 front irradiance and four desired rear irradiance. The error of estimation of the last is presented.

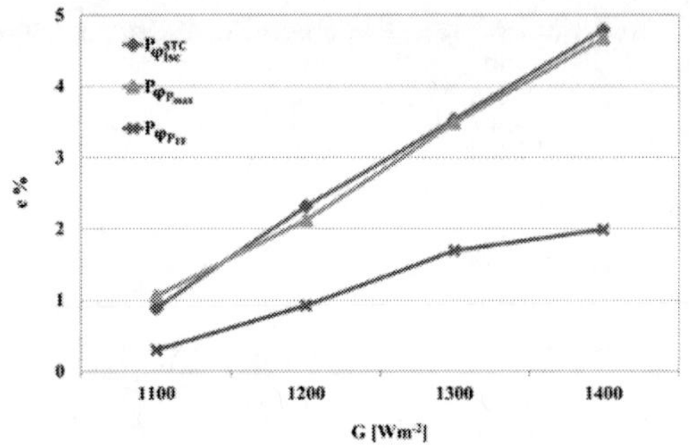

Figure 8: The error of power estimation using short-circuit current bifaciality at STC, dynamic power bifaciality and dynamic fill factor bifaciality in the G_E method for $1000W/m^2$ front irradiance and four desired rear irradiance.

In future work, the effectiveness of using φ_{FF} at STC in estimating P_{max} using the G_E method can be examined to investigate the possibility of reducing the number of measurements. Furthermore, BiFi modules with lower current and power bifaciality coefficients need to be measured.

4 CONCLUSION

The estimated total power output by using the G_E method was off by 0.9% and 4.8% for $100W/m^2$ and $400W/m^2$ rear irradiance, respectively. Hence, the G_E method is effective in estimating the total power output when the desired rear irradiance is low.

Moreover, This study revealed that the resistive differences between the front and rear sides could be a strong source of error for the G_E method. However, further investigations are required.

Dynamic FF bifaciality showed the highest estimation accuracy. The error of estimation was 0.3% and 2% when the rear irradiance was $100W/m^2$ and $400W/m^2$, respectively.

5 ACKNOWLEDGMENTS

We would like to express our profound gratitude to Dr. Bader Aldalali for generously granting us his permission to use the bifacial PV module as a sample for our work. We would also like to express our sincere thanks and appreciation to Dr. Ayman Alqattan for his continuous support and for allowing us to use the PV characterization laboratory whenever it was needed.

6 REFERENCES

[1] Yusufoglu, Ufuk A., et al. "Analysis of the annual performance of bifacial modules and optimization methods." IEEE Journal of Photovoltaics 5.1 (2015): 320-328.

[2] IEC 60904-1-Edition 2 2006, Photovoltaic Devices – Part 1: Measurement of Photovoltaic Current-Voltage Characteristics

[3] V. Fakhfouri et al., IEC 60904-1-2: Measurement of current-voltage characteristics of bifacial photovoltaic devices, Presentation at 3rd bifi PV workshop, Miyazaki, Japan, (2016).

[4] S. Roest et al., "Single side versus double side illumination method IV measurements for several types of bifacial PV modules", Proceedings 33rd European Photovoltaic Solar Energy Conference, Vol. I (2017): 1427-143

[5] Schmid, Alexandra, et al. "IV measurement of bifacial modules: Bifacial vs. Monofacial illumination." Presented at the 33rd European PV Solar Energy Conference, (2017)

[6] Zhang, Yating, et al. "Comparison of Double-Side and Equivalent Single-Side Illumination Methods for Measuring the I–V Characteristics of Bifacial Photovoltaic Devices." IEEE Journal of Photovoltaics 8.2 (2018): 397-403.

[7] Deline, Chris, et al. "Assessment of Bifacial photovoltaic module power rating methodologies—Inside and out." IEEE Journal of photovoltaics 7.2 (2017): 575-580.

[8] Singh, Jai Prakash, Armin G. Aberle, and Timothy M. Walsh. "Electrical characterization method for bifacial photovoltaic modules." Solar Energy Materials and Solar Cells 127 (2014): 136-142.

[9] IEC 60891-Edition 2 2009, Photovoltaic devices - Procedures for temperature and irradiance corrections to measured I-V characteristics.

PERFORMANCE SIMULATIONS OF A 72-CELL, A-SI HET MODULE WITH DIFFERENT TAB-INTERCONNECTION GEOMETRIES

Julien Eymard[1], Benjamin Commault[1], Felix Gerenton[1], Thomas Guérin[1], Raphael Clerc[2], Mathieu Hebert[2]

1 - Univ. Grenoble Alpes, CEA, LITEN, INES, LMPV, F-38000 Grenoble, France
2 - Univ Lyon, UJM-Saint-Etienne, CNRS, Institut d Optique Graduate School, Laboratoire Hubert Curien UMR 5516, F-42023, SAINT-ETIENNE, France

ABSTRACT: The use of half-cells is known to be an effective way to increase cell-to-module (CTM) ratio, thanks to lower resistive losses. Multi-wire interconnection allows reducing silver paste consumption; offering better effective shading compared to flat ribbons. The number and the diameter of wires influence the series resistance and photo-generated current values in opposite ways, leading to an optimal value in CTM. In this paper, the impact of grid resistance and effective shading of wires on the optimum number and diameter of wires is evaluated. Two module architectures are considered, half- and full-cells. Decreasing grid resistance by a factor two, even if it allows using fewer wires, has only a small impact on CTM gain (+ 0.3 %). A large range of possible values of effective shading is found in the literature. The sensitivity analysis shows that it has a major impact on CTM (± 1.3 %), at constant consumption in amount of wires. The use of half-cells allows to employ less wires with a smaller diameter than in full-cell modules. It is a double gain, in CTM performance (+ 2.2 %) and in consumption of raw material (- 40 %).

Keywords: cell-to-module ratio, CTM, multi-wire, half-cell, effective shading; optimization;

1 INTRODUCTION

The photovoltaic (PV) solar energy industry has grown with an average of + 50% per year in the last years. In order to strengthen their position in the market, PV companies seek cost reductions and an increase of technical performance. A standard photovoltaic module is composed of photovoltaic cells electrically connected by ribbons and encapsulated in polymers and glass. Cell-to-module (CTM) analysis is useful to quantify losses and gains in the module. It related the performance of the module with the performance of the cells composing the module, in order to identify opportunities for improvement. In this work, the CTM classification proposed by Haedrich *et al.* [1] is used as it includes optical effects: reflection, absorption, shading and cell / module coupling, and electrical losses like resistance and mismatch.

Reduction of power losses by optimization of the PV module manufacturing process is a way to reduce costs. A popular way is cutting cells into two half parts. The result is a current two times lower, thus dividing resistive power losses in the module by a factor four. Another way that may induce lower power losses is larger wires diameter. This latter results in higher cell shading but smaller series resistance. Increasing the number of wires has both previous effects, but also decreases the finger resistance of the solar cell [2]. Thus, it exists a compromise resulting in an optimum in the number and diameter of the wires. Such optimization is a common approach [3]. The purpose of this work is to analyse the variation of this optimum when varying the effective shading of wires and the grid resistance. A specific case is here presented, a monofacial PV module with 72 heterojunction (HJT) Si-solar full- and half-cells with a multi-wire interconnection.

First, we present modelling considerations about the electrical resistance of the wires and their shading impact on cells. Second, parameters of this case are characterized and explained, such as the effective diameter and line resistance. Third, results of the optimization in terms of number and diameter of wires are detailed. To conclude, a sensitivity analysis on the optimum of effective shading and grid resistance is presented.

2 MODELLING CONSIDERATIONS

2.1 Modelling of the solar cell and module

The 72 full-cells are supposed to be identical and described by a single diode model with five parameters: the photo generated current I_{ph_c}, the diode saturation current I_{0_c}, the ideality factor of the dioden_{id_c}, the series resistance of the cell without finger contribution R_{s_c} and the shunt resistance R_{sh_c}. The current I and voltage V are thus given by the following equation:

$$I = I_{ph_c} - I_{0_c}\left(e^{\frac{V+IR_{s_c}}{n_{id_c}V_{th}}} - 1\right) - \frac{V + IR_{s_c}}{R_{sh_c}}$$

The parameters are extracted from curve fitting of the IV characteristics of each cells. The mean values of the parameters are given in Table I. The half-cell case is supposed to be an ideal case: no additional losses occur at the cutting step. In the half-cell case, the photo-generated current I_{ph_c} and the diode saturation current I_{0_c} are divided by a factor two, whereas the series resistance R_{s_c} and the shunt resistance R_{sh_c} are multiplied by a factor two.

Table I: Mean of the five parameters of the single diode model used in simulation for the full-cell case

$I_{ph_c}\,(A)$	$I_{0_c}(nA)$	n_{id_n}	$R_{s_c}(m\Omega)$	$R_{sh_c}(\Omega)$
9.187	0.7607	1.234	3.999	2.417

When the 72 full-cells or 144 half-cells are connected in series and encapsulated to make the module, only the photo-generated current $I_{ph_{cell}}$ and the series resistance R_{s_c} are affected. The three other parameters I_{0_c}, n_{id_c} and R_{sh_c} are kept constant.

2.2 Change in photo-generated current I_{ph} from cell to module

Several effects due to module architecture can reduce or increase the photo-generated current of the solar cell measured in air:

- Reflection and absorption in glass and encapsulant, shading of interconnection ribbons, cause losses in incident flux
- Cell coupling (index matching between cell and encapsulant), glass/air interface coupling with fingers, wires and backsheet causes additional gains compared to the non-encapsulated case

Varying the number and diameter of the wires mainly affects: (i) the shading of the solar cell, (ii) the coupling between wires and glass/air interface. The wire shading is defined as the ratio of the projected wire area on the solar cell area. The shading ratio k_s is defined as the area of the wire n_w with a diameter d_w and a length on the cell L_w on the cell area A_c, corrected by the area of the finger laying under the wires:

$$k_s = d_w\, n_w \left(L_w - n_f l_f\right)\frac{1}{A_c}$$

Coupling between glass and wires is explained by the reflection to the cell of a certain percent of the incoming light falling on wires [4]. This results in an effective shading diameter of wire smaller than the physical diameter. Shading of wires and coupling between glass and wire affect the intensities of light reaching the solar cell, and thus directly affect the photo-generated current. The photo-generated current of the module I_{ph_m} is thus linked to the photo-generated current of the cell I_{ph_c} by the following equation, where k_{stack} represent the change in I_{ph} which is independent of the number and diameter of the wires:

$$I_{ph_{module}} = I_{ph_c}(1 - k_s)\, k_{stack}$$

The value k_{stack} encompasses losses effects like absorption and reflection by the optical layers, but also gain effects like index matching between cell and encapsulant, coupling of fingers with glass/air interface and coupling of cell and backsheet. Thus, the factor k_{stack} can be smaller or greater than one.

2.3 Additional series resistance R_s from cell to module

The cells used are busbar-less and have been measured with GridTouch method [5]. The additional series resistance of the module is the sum of the contributions of fingers R_{s_f}, wires R_{s_w} and string ribbons R_{s_r}. These parameters depend on the number of wires n_w:

$$R_{s_m} = R_{s_c} + R_{s_f} + R_{s_w} + R_{s_r}$$

The finger contribution in the series resistance of the module is proportional to the grid resistance $R_{s_{grid}}$, to the number of cell n_c, and inversely proportional to the square of the number of wires n_w. Both contributions of the front and back side must be added. :

$$R_{s_f} = \frac{R_{s_{grid}}}{12\, n_w^2}\, n_c$$

The wire contribution can be divided in two parts: one due to length of the wire on the solar cell $R_{s_{wc}}$ and one due to distance d_{ic} between cells:

$$R_{s_{wc}} = \frac{4\, \rho_w}{\pi\, d_w^2\, n_w}\left(\frac{2L_w\, n_c}{3} + d_{ic}\, n_s\left(n_{c_s} + 1\right)\right)$$

3 CHARACTERIZATION OF THE INPUT PARAMETERS OF THE MODEL

In this part, the method is detailed and results of the measurement for the parameters of the case model are presented.

3.1 Characterization of line resistance of wires

The line resistance of the wire is characterized with the transmission line measurement method. A ramp of voltage is applied to two wires at selected distances between them. At a particular distance, applying a voltage ramp and measuring the slope of the IV curve gives the resistance between the wires. Repeating the procedure at various distances between voltage probes leads to a linear relation. The line resistance is the slope of the curve; the contact resistance is the value for a zero distance (Figure 2):

$$R_{line} = 3.419 \pm 0.003\ m\Omega/cm$$

The wire resistivity can be deduced from this value and the line resistance of other wire diameters can be inferred: $\rho_w = 1.81 \pm 0.02\ 10^{-8}\ \Omega.m$. Similarly, the line resistance of the string interconnection ribbons is $R_{s_{line}} = 0.12\ m\Omega/cm$.

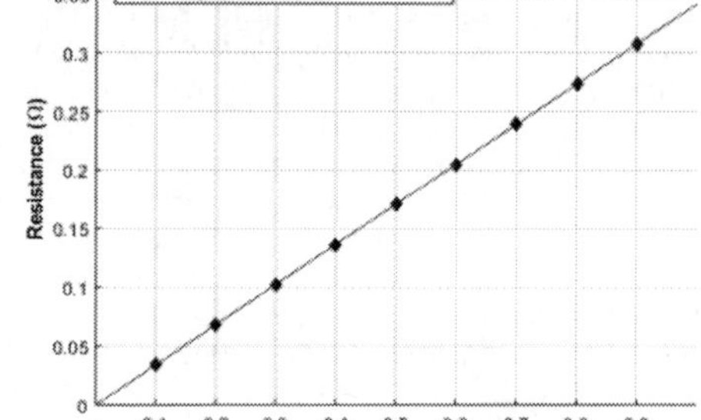

Figure 2: Measurement of the line resistance R_{line_w}. The results show a linear relation, with a relative error on the value around 0.1%. The contact resistance between probes and wire is negligible (smaller than $2.10^{-5}\ \Omega$).

The grid resistance can be characterized with a similar method. A foil of 18 wires, with wire interspacing of $l_w = 8.1\ mm$, is applied on the front side, with wires extending on both sides of the module, as depicted in the Figure 3.

The contact between the Ag coating of the wires and the metallization paste is small enough to be indistinguishable in the measure (Figure 4). The value obtained for the front grid resistance is $R_{s_{grid}} = 32{,}5 \pm 1{,}0\ m\Omega/cm$. During this measure, the current can also flow through the TCO even if its resistivity is several orders of magnitude higher. The resulting value is an equivalent resistance of finger and TCO.

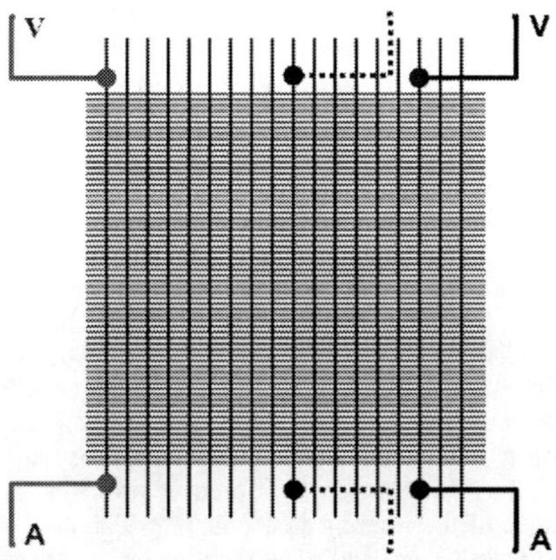

Figure 3: Measurement principle of the line resistance of the grid. Red probes are fixed to the first wire. Black probes are moved from one wire to another. The measurement is independent of the wire resistance.

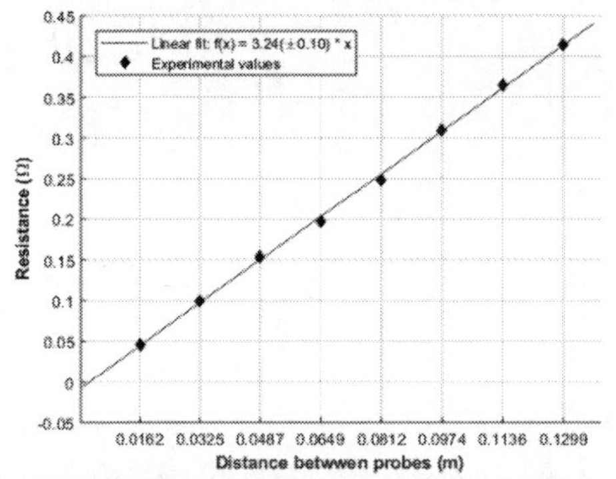

Figure 4: Measurement of the line resistance of the grid R_{line_g}. The deviation from linearity may be explained by the loss of contact between fingers and wires and the non-uniformity of the section of fingers.

3.2 Characterization of the absorption of the stack and effective diameter of the wires

The effective optical diameter of the module can be measured by varying the number of interconnecting wires, using the same cells, encapsulant and glass. The ratio of the photo-generated current of the module to the one of the cell has an affine dependency with the number of wires n_w. The y-axis intersection gives the value of k_{stack}, allowing to deduce the effective diameter from the slope of the curve:

$$\frac{I_{ph_{module}}}{I_{ph_c}} = \left(1 - d_{eff_w} n_w \left(L_w - n_f l_f\right)\frac{1}{A_c}\right) k_{stack}$$

Six mini-modules have been made, with 3, 6, 9, 12, 18, 24 wires of the same measured diameter of $d_w = 260 \pm 2\ \mu m$ and the results of the measure of their short circuit current is summarized in Table II.

From the linear fit of the curve in Figure 5, we get a value of $k_{stack} = 0.970 \pm 0.002$ and a value of effective diameter of $d_{eff} = 243 \pm 2\ \mu m$, with 95% confidence bounds. It corresponds to an effective wire shading equal to $l_{eff_w} = d_{eff_w}/d_w = 0.94 \pm 0.02$. This is different from the expected theoretical value of at least 0.75, mentioned in [4] and [7]. The theoretical effective shading should be around 0.7 when the cell is not encapsulated, due to the amount of light hitting the wire and directly reflected to the cell due to the round wire shape. When the cell is encapsulated, a portion of the impinging light is totally reflected at the air/glass interface while another portion is partially reflected. In this case, the effective shading can be as small as 0.35.

Table II: Cell and module short circuit currents for the six mini module made with varying number of wires.

n_w	$I_{sc_c}(A)$	$I_{sc_m}(A)$
3	9.320	9.001
6	9.243	8.892
9	9.262	8.873
12	9.283	8.837
18	9.292	8.764
24	9.251	8.657

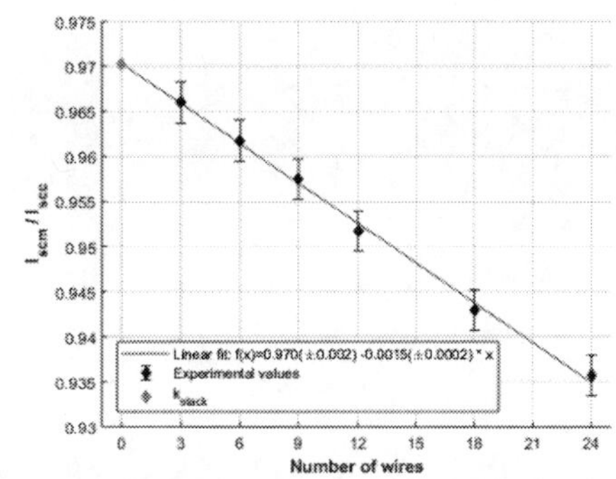

Figure 5: Short circuit current I_{sc} ratio between module and cell as a function of the number of wires n_w. The experimental values are the mean of three successive measurements, and the error is the standard deviation of the six modules.

No defects, such as trapped air around the encapsulated wires, could be seen via optical microscopy. Such defects could have explained the difference given the additional air/encapsulant interface. In the paper by McIntosh it is claimed that a small wire reflectance (R<0.5) can explain such a high value [8]. The difference in effective shading could be explained by the surface roughness of either the wire after melting and/or the glass. An increasing number of wires means also smaller coupling between cell and encapsulant, higher optical path of the incoming light in glass and encapsulant due to reflections. This different explanation needs further investigation.

4 RESULTS OF SENSITIVITY ANALYSIS

Resistivity of wires and string ribbons are determined with a good accuracy and the technological limit is almost reached: from a manufacturing stand point (wire production and handling during module manufacturing), it is difficult to decrease these values. On the other hand, there is a large range of possible values for effective shading of wires: from 35% to almost 100% as discussed in the previous part. Moreover, as the series resistance due to fingers has a quadratic dependence on the number of wires, a small change in grid resistance can lead to huge changes in series resistance.

4.1 Reference case

The reference case is taken with the following parameters:

- Cells parameters as mentioned in 2.1
- Wire line resistance: 3.419 $m\Omega/cm$
- Distance inter-cell / inter-string: $3/5\ mm$
- Measured grid resistance of $32.5\ m\Omega/cm$
- Measured effective shading of 94%
- Measured stack absorption of 97%

We investigate optimums in the range of 200 to 300 μm in core diameter, values commonly found in the industry. The results of the optimization are given in Figure 6. The optimum in CTM power is arbitrarily set at zero and we show contour lines from this reference level.

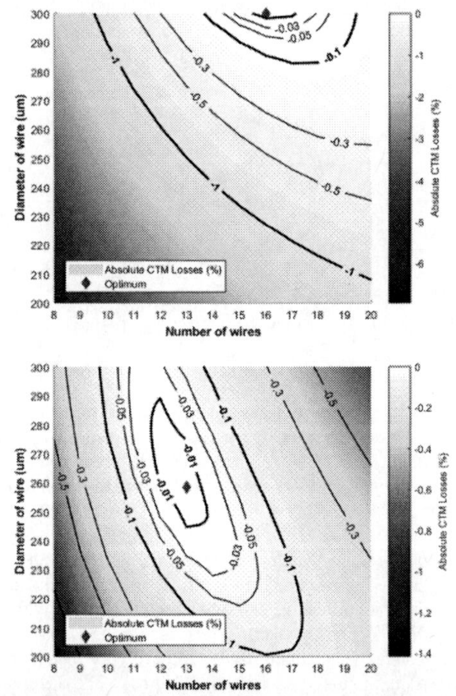

Figure 6: Absolute CTM Losses in percent compared to optimum case. The optimum is arbitrary put at zero level. Bold black contour lines are shown for 0.01%, 0.1% and 1% losses from optimum case. (Top) Full-cell design. The corresponding CTM_P of the optimal point is 91.3% (Bottom) Half-cell design. The corresponding CTM_P of the optimum is 93.5%.

The optimal number of wires in the full-cell case is 16 wires with a diameter of 300 μm. This optimum is found on the boundaries of the domain. This optimum corresponds to a CTM power of 91.3%. The optimal number of wires in the half-cell case is 19 with a diameter of 258 μm, corresponding to a CTM power of 93.5%. Going from full- to half-cell case presents a gain of 2.2%, which is the value expected [9]. Moreover, using half-cells decreases the volume of wires by 40% lower with respect to full-cells.

The optimal zone is flat, allowing choice flexibility, *e.g.* in the full-cell case (Figure 6, top graph), 9 wires with a 290 μm diameter produce the same level of losses than 20 wires with a diameter of 210 μm; *e.g.* fixing the diameter at 260 μm leads to an optimal wire number of 19.

4.2 Impact of grid resistance on optimal number and diameter of wires

In this part, we look at how the optimal number and diameter of wires varies when the grid resistance changes. All other parameters have been kept constant and particularly the cell resistance. Modifying the grid resistance without changing the cell series resistance is possible in two ways: changing either the shape factor of the finger, with an increasing height and a constant width, or the silver paste (with a different resistivity). The grid resistance of the reference case is 173 Ohm.cm²; a range between two times smaller and two times greater than this value is investigated.

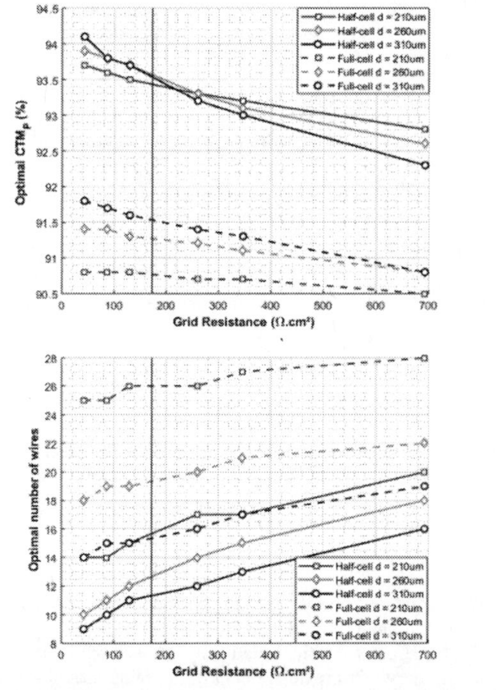

Figure 7: (Top) Results of the optimization in terms of CTM power for three diameters as a function of grid resistance. (Bottom) Corresponding optimal number of wires. Full-cell results are plotted in dashed line and half-cell in solid line. The vertical black line correspond to grid resistance value of the reference case in part 4.1

Results are shown in the Figure 7. The smaller the grid resistance, the greater the CTM power, with a corresponding optimal number of wires decreasing. The increase in CTM power is more pronounced for large wire diameters in both full- and half-cell case. The optimal number of wires is more dependent on the wire diameter in the full-cell case than in the half-cell one. Dividing the grid resistance by a factor two for halved-cells implies a gain in CTM of only +0.3% for diameter of 300 and 260 μm. However, the optimal number of wires falls from 13 (with 260 μm in diameter) to 11: this results in a possible decrease of the volume of wires of 15%.

4.3 Impact of effective shading on optimal number and diameter of wires

Effective shading ratio directly affects the photo-generated current and has a first order influence on power and CTM ratio. We consider the optimal number of wires varies and the CTM for three diameters. The effective shading ratio ranges from 40% to 100%.

Results are presented in Figure 8. Reducing the effective shading of the wires increases the CTM power and the optimal number of wires at the same time. The diameter of wires has almost no impact on the optimal CTM power in the half-cell case, as shown by the overlapping curves. However, the corresponding optimal number of wires is changing: 15 wires with a diameter of 200 µm for a 100% effective shading and 11 wires with a diameter of 300 µm. By reducing the effective shading ratio, the gain in CTM power is smaller in the half-cell case than in the full-cell case.

As the CTM gain is not affected by the wire diameter in half-cell, it is possible to choose the configuration of the lowest raw material consumption, *i.e.* 200 µm in diameter. Thus, going from the reference case in half-cell to a 50% effective shading with 200 µm (and thus 20 wires) implies same material consumption and a 1% gain in CTM power.

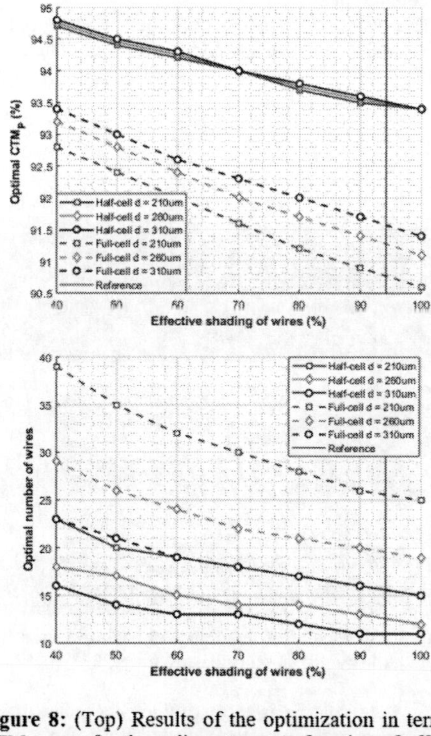

Figure 8: (Top) Results of the optimization in terms of CTM power for three diameters as a function of effective shading. (Bottom) Corresponding optimal number of wires. Full-cell results are plotted in dashed lines and half-cell in solid lines. The vertical black line corresponds to effective shading ratio of 94%, the reference case in part 4.1

5 CONCLUSION

A simple one-diode based model has been used to predict CTM ratio of a 72 equivalent hetorojunction Si-solar cell module with multi-wire interconnection technology. The number and the diameter of wires influences the series resistance and photo-generated current values in opposite ways, leading to an optimal value in CTM. We have studied how the optimum design in terms of number and diameter of wires is influenced by a change in grid resistance and effective shading of wires, in full-cell and half-cell architectures. Using half-cell allows to employ less wires with a smaller diameter than full-cell modules. As a result, both CTM performance and consumption of raw materials are improved.

Decreasing grid resistance, even if it allows using fewer wires, has only a small impact on CTM gain. On the other hand, the sensitivity analysis shows that effective shading value has a major impact on CTM, for constant raw material consumption. Thus, to make the optimization approach relevant, an accurate characterization of the effective shading of wires is needed.

As one of the major benefits of multi-wire interconnection is to reduce silver paste consumption, gain in the volume of wires should be balanced with volume of silver paste. The joint cost optimization should be considered as the next research advancement.

REFERENCES

[1] I. Haedrich, U. Eitner, M. Wiese, and H. Wirth, "Unified methodology for determining CTM ratios: Systematic prediction of module power," *Sol. Energy Mater. Sol. Cells*, vol. 131, pp. 14–23, Dec. 2014.

[2] J. Walter, M. Tranitz, M. Volk, C. Ebert, and U. Eitner, "Multi-wire Interconnection of Busbar-free Solar Cells," *Energy Procedia*, vol. 55, pp. 380–388, Jan. 2014.

[3] J. Roeth, A. Facchini, and N. Bernhard, "Optimized Size and Tab Width in Partial Solar Cell Modules including Shingled Designs," *International Journal of Photoenergy*, 2017. [Online]. Available: https://www.hindawi.com/journals/ijp/2017/3609109/. [Accessed: 17-Apr-2018].

[4] S. Braun, G. Hahn, R. Nissler, C. Pönisch, and D. Habermann, "The Multi-busbar Design: An Overview," *Energy Procedia*, vol. 43, pp. 86–92, Jan. 2013.

[5] "The GridTouch contacting system." 2014.

[7] J. Ufheil, P. Papet, and T. Söderström, "Smart Wire Connection Technology," *28th Eur. Photovolt. Sol. Energy Conf. Exhib.*, pp. 495–499, Nov. 2013.

[8] K. R. McIntosh, M. D. Abbott, M. B. Edwards, R. Evans, and Y. Yao, "Optical Evaluation of Multi-Wire Modules," *32nd Eur. Photovolt. Sol. Energy Conf. Exhib.*, pp. 15–19, Jul. 2016.

[9] S. Guo, J. P. Singh, I. M. Peters, A. G. Aberle, and T. M. Walsh, "A Quantitative Analysis of Photovoltaic Modules Using Halved Cells," *International Journal of Photoenergy*, 2013. [Online]. Available: https://www.hindawi.com/journals/ijp/2013/739374/. [Accessed: 25-May-2018].

NOVEL POWER-ENHANCING RIBBON FOR SOLAR CELL INTERCONNECTION

Tong Zhou[1], Markus König[1], Andreas Henning[1], Daniel Hanselmann[1], Stephan Hoffmann[2], Luciana Pitta Bauermann[2], Esther Fokuhl[2], Dirk Eberlein[2], Achim Kraft[2], Ulrich Eitner[2], Wolfgang Pranger[3], Angelika Schütz[3]
1) Heraeus Deutschland GmbH & Co. KG, Heraeus Photovoltaics, Heraeusstrasse 12-14, 63450 Hanau, Germany
2) Fraunhofer Institute for Solar Energy Systems (ISE), Heidenhofstrasse 2, 79110 Freiburg, Germany
3) Ulbrich of Austria GmbH, Industriestrasse 1, 7052 Müllendorf, Austria

ABSTRACT:
Crystalline silicon solar cells are usually interconnected by solder-coated copper ribbons which are soldered onto the busbars of the front side metallization and the rear side solder pads. Commonly ribbons used for five busbar solar cell interconnection are 0.8-1 mm wide, 0.25 mm thick and reflect ~84-95% of the incident light out of the module. As a consequence, up to 10 W power loss may occur in typical 60-cell modules due to shading of the ribbons. This work presents the development of a novel selectively coated power-enhancing ribbon (SCR™) which yields a module power gain of 1.88 W in average. An effective width (EW) of 70% was obtained for SCR™ and verified by light beam induced current (LBIC) measurements compared to standard ribbons with an EW of 90%. The 20%abs. lower value for the SCR™ is a result of improved internal reflection and demonstrates the higher current generation capability. The plug-and-play application allows simple substitution of standard ribbons without equipment modification which was proven in industrial process lines. Reliability tests demonstrate high reliability exceeding IEC 61215/61730 requirements.

Keywords: solar module, interconnection, solder ribbon, CTM, internal reflection, power gain

1 INTRODCUTION AND SCOPE

Crystalline solar cells are usually interconnected in series using solder-coated copper ribbons. Shading of the incident light in the ribbon area typically results in power loss of up to 10 Watt per 60-cell module with five busbar solar cells when standard solder ribbons are applied. There are various products available on the market which reduce shading through light reflecting structures on the front side of the ribbon [1–3] or through a continuous reflective coating. In both cases the industrial tabber-stringer equipment has to be modified with e.g. pattern recognition to align the front side parts of the ribbon to the front side of the cells. Additional waste is unavoidable due the need for alignment and adds to already high material costs compared to low-cost standard ribbons. Therefore, most of these products are being used for high-efficiency niche products and cannot be found in standard application. The scope of the presented work was to overcome the common issues of power-enhancing interconnection ribbons and provide a market-ready solution.

Three major requirements were identified for the development of a novel and innovative solderable interconnector:

a) Plug-and play application on all common stringer types,
b) Increased module power,
c) Reliability and stability of module performance.

This paper presents the results of the development: A novel selectively coated ribbon SCR™ featuring reliable power-enhancing properties at fully industrial process capability.

2 PERFORMANCE EVALUATION

In order to allow conventional processing in all common types of stringers without the need for modification a special ribbon design was developed. The design features improved reflection of the incident light and thus improves the module power output significantly.

2.1 Internal reflection

The working principle for the usage of back reflected light in a solar module is shown in Figure 1 for a standard solder ribbon and for the SCR™, respectively.

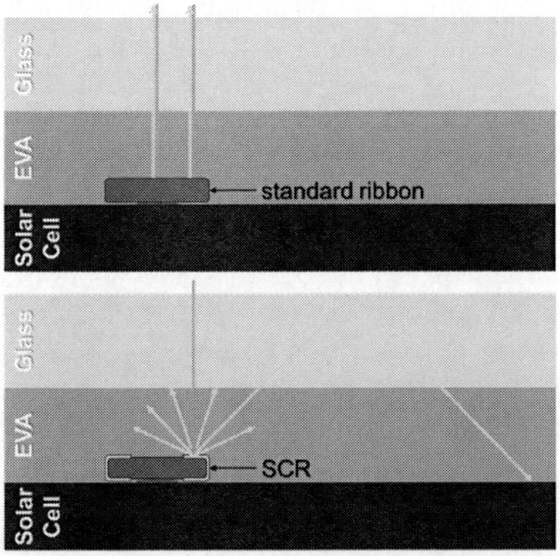

Figure 1: Schematic of incident and back reflected light for standard solder ribbon (top) and for SCR™ (bottom).

A white coating is located along the edges of the SCR™ to improve the internal reflection in the module. Thus, a higher fraction of the incident light is reflected back under low angles. This light is then reflected back onto the solar cell when hitting the glass surface at an angle below the critical angle for total reflection. In contrast, standard solder coated ribbons mostly reflect the incident light at the similar angles and less internal reflection occurs.

2.2 LBIC measurements

The effect of back reflected light on the current generation of the solar cell was investigated using light beam induced current measurements (LBIC). An area of 20 by 10 mm² of interconnected and laminated industrial solar cells was scanned by lasers of different wavelengths (405nm, 532nm, 658nm 780nm, 940nm and 1064nm). The generated currents were recorded, analyzed and compiled to heat maps as shown in Figure 2. The legend for those maps represents the external quantum efficiency and thus, areas where the incident light generated a local current. The results show bright areas along the edges of SCR™ indicating that the reflected light contributes to the current generation. In contrast, almost the entire area of the standard ribbon appears dark indicating that only a very small no current generation is attributed to the ribbon area.

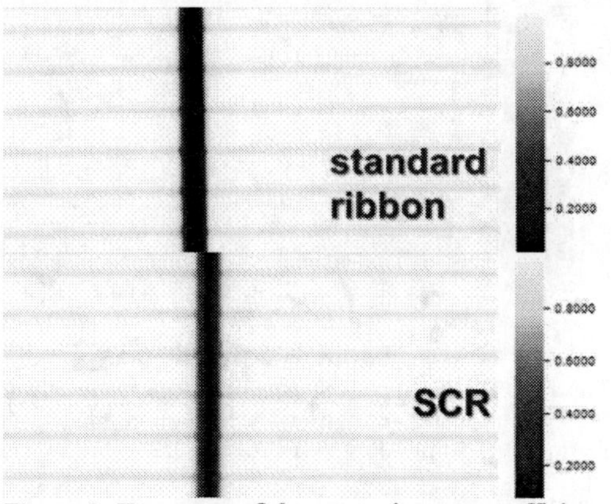

Figure 2: Heat maps of the external quantum efficiency of laminated solar cells interconnected with standard ribbon (top) and SCR™ (bottom). Bright areas represent local current generation due to laser irradiation, dark areas represent shaded areas.

Integration of the locally generated current distribution over the busbar areas gives the parameter of "effective optical width" (EW) [4], i.e. the relative area of the busbars where no current is generated from back-reflection of the incident light. An EW of 70% was obtained for SCR™ compared to standard ribbons with an EW of 90%. The 20%abs. lower value for the SCR™ is thus obtained due to improved internal reflection and demonstrates the higher current generation capability.

2.3 Cell-to-module (CTM) analysis

In order to understand the effect of current gain from internal reflection on the power output of common solar modules, a calculation of the cell-to-module losses was performed using Fraunhofer ISE SmartCalc.CTM [4, 5]. The input data was based on five busbar Cz-PERC cells with 5.19 W and common industrial module components. The ribbon dimensions were identical (1.0 mm wide and 0.2 mm thick) for standard and SCR™. As shown in the waterfall diagram in Figure 3 the SCR™ yields a higher gain from optical coupling ("k10") resulting in a calculated total power gain of 1.85 W.

3. INDUSTRIAL APPLICATION

The industrial application as well as high performance of the final module are crucial requirements for new materials especially in solar industry. The design of the SCR™ targets plug-and-play applicability in all common stringer types along with a power gain through improved internal reflection.

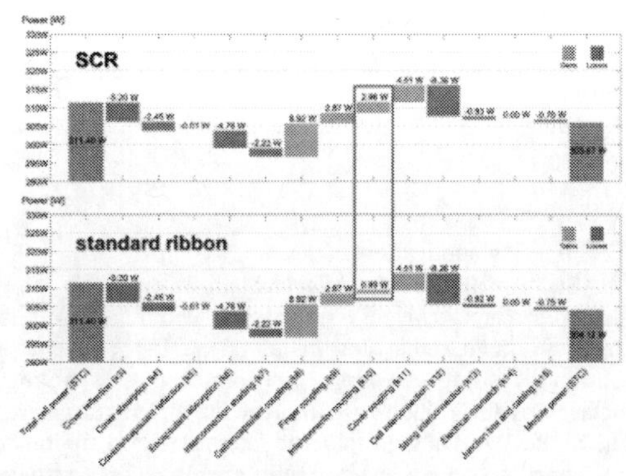

Figure 3: Cell-to-module analyses for SCR™ and standard ribbons. The power gain of 1.85 W is mainly due to improved light coupling.

3.1 Statistical proof of power gain

To prove the CTM analysis results two groups of 60-cell modules with five busbar Cz-PERC cells were fabricated in an industrial process line. Each group consisted of 42 modules to ensure the statistical significance of the expected power gain. One group was equipped with standard ribbons and the other group was equipped with SCR™, both of identical width of 1.0 mm and thickness of 0.2 mm. All Cz-PERC solar cells were selected from the identical power class and underwent a pre-conditioning procedure to minimize degradation or regeneration effects of the module performance. The modules were fabricated according to the manufacturer's product specification.

The power distribution for both groups is displayed in the histogram in Figure 4 and shows a significant power gain of 1.88 W in average.

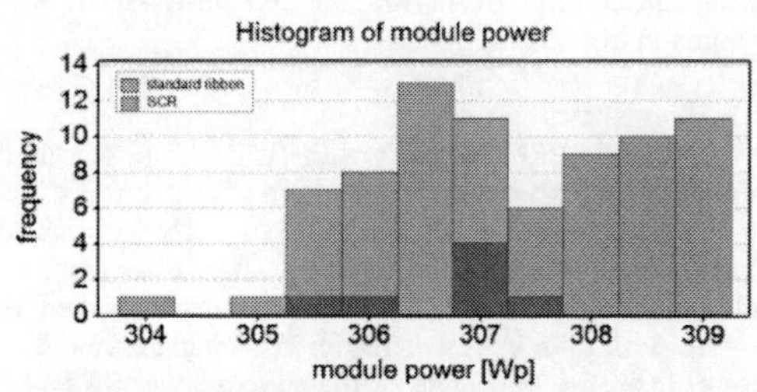

Figure 4: Distribution of module power of 60-cell Cz-PERC modules with standard ribbons and SCR™ produced in an industrial process line.

Table 1 shows the minima, maxima and average values of module power for both groups, standard ribbon and SCR™, respectively. The values are in excellent agreement with the CTM calculations and demonstrate the validity of the SCR™ concept.

Table I: Module power of 60-cell Cz-PERC modules with standard ribbons and SCR™ produced in an industrial process line.

Module power	Standard ribbon	SCR™	delta
Minimum [W_p]	304.21	305.65	1.44
Maximum [W_p]	307.26	309.20	1.94
Average [W_p]	306.27	308.15	1.88

3.2 Soldering application

Successful and cost-effective application of a new ribbon product demands processing without cuts and material waste. The design of the SCR™ features continuous solder coatings on both, top and bottom surfaces of the ribbon to comply with this requirement. The SCR™ thus can be soldered like a standard ribbon at any position onto every cell metallization surface independent from the busbar layout. The applicability and the processing ability was tested and proven on all industry-relevant processes such as

- contact soldering,
- infrared soldering,
- inductive soldering.

The rework ability was demonstrated in case of cell breakage or defective solder joints.

The adhesion force of soldered SCR™ was determined by 180° peel tests. The average adhesion force is ≥ 1 N/mm on both, front and backside metallization of silicon solar cells. Thus, the common industry requirements for adhesion are met.

4. RELIABILITY

Product warranties for silicon PV modules usually cover 10 years while product performance is certified for up to 30 years. This sets high requirements for the introduction of new material components. In this work the most relevant endurance tests for interconnectors were carried out to analyze the reliability of SCR™ compared to standard ribbons.

4.1 Thermal cycling

60 cell PERC modules with SCR™ and standard ribbons were exposed to temperature cycles from -40°C to +85°C according to IEC 6121-2:2016. The IV data of the modules were measured initially, after 200 cycles and after 400 cycles, respectively. The results displayed in Figure 5 show identical behavior for both groups. The power loss of < 2% can be attributed to a lower fill factor (FF) for both, SCR™ and standard ribbon. The open circuit voltage (V_{oc}) and short circuit current (I_{SC}) remain unaffected by thermal cycling.

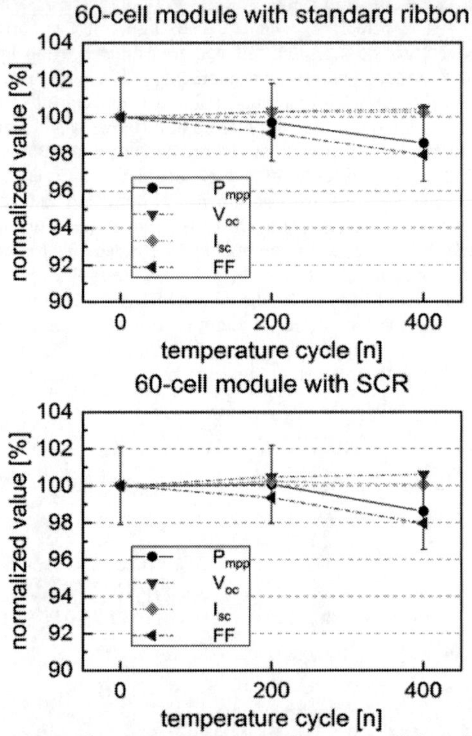

Figure 5: Electrical performance of 60-cell modules with SCR™ (top) and standard ribbons (bottom) after 400 thermal cycles according to IEC 61215-2:2016.

4.2 UV irradiation at increased temperature and humidity

A combined treatment of UV irradiation at elevated temperature and humidity was carried out to simulate a demanding environment for accelerated ageing. The combination of UV irradiation, high temperature and the presence of water promote chemical reactions and thus accelerate possible degradation effects. The setup was defined in order to focus on the durability of the white coating under harsh conditions as well as the solder interconnection.

3-cell PERC modules with standard ribbons and SCR™ were exposed to UV irradiation of 170 W/m² at high relative humidity of 85% and increased temperature. The temperature was set to 60°C for the incoming air in the climate chamber. However, the temperature at the samples was higher and more approximately 80°C due to additional heating from the irradiation source. The IV performance of the modules was measured after 60, 120, and 180 kWh/m², respectively. The duration of each segment was 500 h, adding up to a total of 1500 h combined UV and damp heat exposure. The results in Figure 6 demonstrate that the electrical performance for both groups is not changed after the maximum UV dose of 180 kWh/m² and 1500 h damp heat treatment. The modules with SCR™ as well as the standard ribbon show a constant performance of all relevant IV parameters within the measurement range.

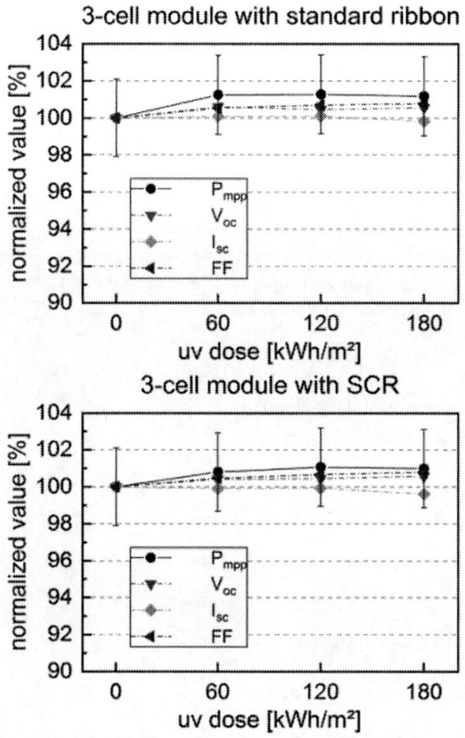

Figure 6: Electrical performance of 3-cell modules with standard ribbon (top) and SCR™ (bottom) after a combined UV and damp heat treatment (1500 h at 180 kWh/m²).

5 CONCLUSIONS

A novel power-enhancing solder ribbon was developed and analyzed for application, performance and reliability. The design allows for plug-and-play application in all relevant stringing and soldering equipments without modification or material cuts. Sufficient mechanical adhesion as well as a power gain through improved usage of back reflected light were demonstrated on industrial scale. The reliability of the SCR™ was proven. No significant degradation was obtained after 400 temperature cycles and after a treatment of 170 kWh/m² UV irradiation and elevated temperature and humidity.

Further investigations of the product reliability are in progress. Design and process optimizations are planned to further increase the product performance as well as reliability.

6 ACKNOWLEDGEMENTS

The authors gratefully acknowledge the support from Ute Mareck, Jörg Walther, Robert Pohl of Solarworld Industries GmbH.

This work was funded by the German Ministry for Economic Affairs and Energy under project "CONNECT" FKZ 0324052.

REFERENCES

[1] E. M. Sachs *et al.*, "Light-capturing interconnect wire for 2% module power gain," 24th EUPVSEC, Hamburg, Germany, 2009, pp. 3222–3225.

[2] W. Muehleisen *et al.*, "Comparison of output power for solar cells with standard and structured ribbons," *EPJ Photovolt.*, vol. 7, p. 70701, 2016.

[3] J. Zemen, M. Strasser, K. Turlapati, "Performance an yield evaluation of C-Si modules buildt with reflective interconnectors," 29th EUPVSEC, Amsterdam, Netherlands, 2014.

[4] *cell-to-module.com.* [Online] Available: http://www.cell-to-module.com. Accessed on: Feb. 28 2018.

[5] M. Mittag *et al.*, "Systematic PV module optimization with the cell-to-module (CTM) analysis software," *Photovoltaics International*, no. 36, pp. 97–104, 2017.

SUPER LIGHTWEIGHT FLEXIBLE HJT SOLAR PANELS

A.Abramov[1], D.Andronikov[1], K.Emtsev[1], D.Orekhov[1], I.Shakhray[2], E.Terukov[1], E.Terukova[1], S.Yakovlev[1]

[1] R&D Center of Thin Film Technologies in Energetics, St.-Petersburg, Russia
[2] Avelar Solar Technology, Moscow, Russia

ABSTRACT: Lightweight flexible solar panels market is rapidly developing. The main goal in that field is to derive high power at low weight and on limited area. In this work we report on development HJT lightweight solar panels with high efficiencies. Our approach in this development is reduction of solar cell thickness, utilization of different available on market plastics, minimizing interconnection shadowing and getting gain from bifaciality of the cells as also of the panel.

Keywords: lightweight solar panels, flexible solar panels, HJT solar panels, HJT technology

1 INTRODUCTION

Attention to the lightweight solar panels is mainly caused by their application in a wide range of mobile power sources as for example wearable panels designed as a power supply for gadgets in hiking tours, panels used as a basis of energy sources for solar powered drones and boats as also panels integrated in buildings and other constructions. Such type of applications usually require flexible panels to provide their placement on non-flat surface. Most of the lightweight flexible panels, nowadays available on a market, are based on different thin film technologies suffering from low efficiency, encapsulation problem or using of relatively massive metallic foil substrates required for high temperature technology processes. HJT solar cells application in lightweight flexible solar panels can enable high efficiency and low temperature production processes.

2 APPROACHES TO HIGH POWER AT LOW WEIGHT

HJT solar cells used for this study were manufactured on HJT production line, which has been successfully converted from micromorph one by Hevel company [1]. Hevel company has successfully implemented of 150 um thin wafers without modules quality losses for HJT solar cells production [2] that was possible due inherent advantage of HJT technology, which does not require diffusion or other high temperature production steps.

Thinner wafers usage in lightweight solar panels can lower the weight by more than 100 g/m^2. Furthermore, our recent study with using of R&D pilot line has been shown that developed HJT technology allows one to reduce cells thickness down to 90 um with only a small efficiency decrease about 3 rel.% (see Fig.1). Additionally it can benefit to panel's flexiblilty (Fig. 2).

Flexibility of the c-Si solar cells strongly increases with thickness reduction, which merges with a general trend in c-Si based PV technologies caused by continuous efforts in production costs reduction. It is possible to achieve very low silicon utilization of about 150 g/m^2 or 1,1 g/W.

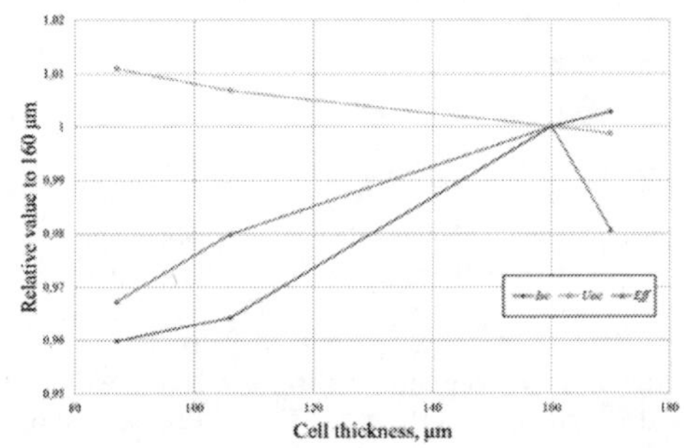

Figure 1 – Dependence of cell parameters on cell thickness

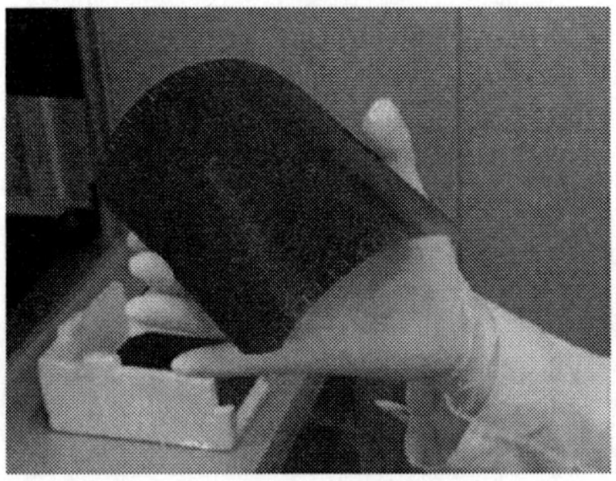

Figure 2 – Photo of thin cell bowing

Another approach to reduce the weight is using of different types of plastics specially developed by different companies. This approach enabled us to get specific power above 130 W/kg for front side illuminated panels at standard test conditions (STC) which corresponds to panel efficiency of about 20 % for the solar panels made of 4x4 M2 wafers (Fig. 3). Total thickness of the panels is less than 1 mm with allowed curvature radius as low as 70 mm per solar cell with M2 size.

	Isc=8,98 A; Voc=11,5 V; Impp=8,43 A; Vmpp=9,4 V; FF=76,7 %; Pmpp=79,25 W Width: 660 mm; Length: 880 mm Thickness: <1 mm; Weight: 0,582 kg

Figure 3. Photo, parameters and technical data of 16 M2 size cells panel

Thinner encapsulant, front sheet and back sheet material usage in lightweight solar panels can lower the weight by more than 300 g/m^2. Additionally it can benefit to panel's efficiency due to higher transparency of thinner encapsulants.

Further improvement of the super lightweight solar panels require optimization of both cells and panel technology. According to our estimations, optimization of wafers and plastics thickness will allow us to get specific power value about 250 W/kg.

Both busbar (BB) and SmartWire Connection Technology (SWCT) [3] from Meyer-Burger Company have been successfully used for the assembly of the panels. The latter has a clear advantage because of lower sensitivity to cells breakage, which can become more significant with shift to super lightweight design, as also SWCT provides lower shadowing of the cells that can give gain in efficiency.

Bifacial design providing two-side sensitivity to incident light has been used for both HJT cells and panels. Using of the panels bifaciality, especially in case of using solar panels for solar powered planes, is also big advantage of our panels. Preliminary measurements have shown that using of our HJT cells with SmartWire interconnection provide typical value of bifaciality factor > 0.92%. It means that taking albedo at 20 % (average value for crops) one can get about 250 W/kg at our current status of development with using of cells from mass production line and 300 W/kg on panels with optimized design.

3 APPLICATIONS

During our project we have implemented different types of customized modules in real application.

1. Drone boat that can be used for guarding, patrolling of borders, goods delivery, intelligence service, sea bottom scanning, monitoring of environment. Drone boat can reach maximum speed of 20 knots, can autonomously sail during 90 days and has low capacity 100 kg (Fig.4).

Figure 4. Photo of drone boat

2. Tent with solar panels for touristic camping. Weight of tent is 50 kg and it can produce up to 2 kW of power (Fig. 5).

Figure 5. Photo of camping tent with lightweight flexible solar panels.

3. Catamaran «Ecovolna»

The boat has crossed Russia from Saint-Petersburg to Astrakhan in frames of regatta "Ecovolna" ("Ecowave") that took 90 days. It is equipped with solar panels of 9,5 kW capacity with area of 57 m^2 and Li-ion accumulators with power of 70 kW solar panels. It can fit up to 12 people and can reach maximum speed is 15 km/hour (Fig. 6).

Figure 6. Photo of Catamaran «Ecovolna»

4. Project: on rowing boat across the South Ocean

Famous Russian traveler Fyodor Konyukhov will go alone on the rowing boat equipped with lightweight flexible solar panels (Fig. 7) from port Dunedin (New Zealand) to cape Horn (Chili) in 2018 and than to cape Luin (WA, Albany) in 2019 and in 2020 to Australia and back to New Zealand. Distance to be covered: 27 000 km.

Figure 7. Photo of rowing boat with solar panels and Fyodor Konyukhov

All above mentioned projects have been successfully implemented and we intend to monitor modules in terms of their degradation and functionality in long term.

4 CONCLUSIONS AND OUTLOOK

Lightweight flexible solar panels is an emerging market that develops rapidly. Vast application areas are already tested and many more are expected emerge

In our work we have proven that use of thin wafers and thin encapsulation materials enables to increase power per kg in flexible solar panels.

One of the important question is reliability of lightweight flexible panels. Testing according to PV standards is going to be conducted in future in order to understand how reliable and safe can flexible panels be and in order to optimize their structure in accordance with intended type of installation.

Acknowledgments: Hevell LLC, Saint-Petersburg State Polytechnical University, Fyodor Konyukhov and Oscar Komyukhov.

REFERENCES

[1] D. Andronikov, A. Abramov, S. Abolmasov, K. Emtsev, G. Ivanov, A. Kukin, I. Nyapshaev, D. Orekhov, A.Semenov, G. Shelopin, E. Terukova, E. Terukov, N. Belkova, A. Dubrovskiy, P. Ishmuratov, A. Ivanov, D. Saykin, I. Shakhray, A. Smirnov, V. Tarasov, V.Timakov, A. Tomchinskiy, G. Kekelidze. A SUCCESSFUL CONVERSION OF SILICON THIN-FILM SOLAR MODULE PRODUCTION TO HIGH EFFICIENCY HETEROJUNCTION TECHNOLOGY, 33rd European Photovoltaic Solar Energy Conference and Exhibition, p.732(2017).

[2] D. Andronikov, A. Abramov, S. Abolmasov, K. Emtsev, G. Ivanov, I. Nyapshaev, D. Orekhov, A. Semenov, G. Shelopin, E. Terukova, E. Terukov, N. Belkova, A. Dubrovskiy, P. Ishmuratov, A. Ivanov, D. Saykin, I. Shakhray, A. Smirnov, E. Sokolov, V. Tarasov. A TRANSITION TO THINNER SI WAFERS AT HJT MASS PRODUCTION: AHEAD OF ITPRV SCHEDULE, 35th European Photovoltaic Solar Energy Conference and Exhibition (2018).

[3] A. Faes, M. Despeisse, J. Levrat, J. Champliaud, N. Badel, M. Kiaee, T. Söderström, Y. Yao, R. Grischke, M. Gragert, J. Ufheil, P. Papet, B. Strahm, B. Cattaneo, J. Cattin, Y. Baumgartner, A. Hessler-Wyser, C. Ballif. Smartwire solar cell interconnection technology, 29th European Photovoltaic Solar Energy Conference, Amsterdam 2014; pp 2555-2561

INFLUENCE OF NEAR FIELD SHADOWING ON THE PERFORMANCE RATIO OF THIN FILM MODULES

Stefan Wendlandt[1,2], Lars Podlowski[1]
[1] PI Photovoltaik-Institut Berlin AG, Wrangelstr. 100, D-10997 Berlin, Germany
Phone: +49 (30) 814 52 64 0, Fax: +49 (30) 814 52 64 101, e-mail: wendlandt@pi-berlin.com
[2] University of Paderborn, Faculty of Electrical Engineering, Informatics and Mathematics, Department for Electrical Energy Technology & Sustainable Energy Concepts, Pohlweg 55, D-33098 Paderborn, Germany

ABSTRACT:

When planning a photovoltaic system or being responsible for its maintenance, many issues have to be considered. In order to minimize any kind of risks the properties of the module technology must be well understood. One of the major potential risks is damage due to partly shading of PV modules. In such conditions the shaded solar cells are operating in reverse bias and thus can heat up significantly. In extreme situations such a high temperature gradient may occur that it can cause glass breakage - especially at thin-film modules where not fully-tempered glass can be used. In all module testing standards only stationary shadowing situations are investigated. However, during daytime also short-term shadowing situations can occur, e.g. through cleaning or inspection of modules, and therefore such situations must also be considered as certain module designs or technologies can be sensitive for this. This paper presents measurements before and after near field shadowing of two different types of thin-film modules. The effect of short-term shading on the IV curves at different irradiance levels as well as image-based electroluminescence analysis were carried out. Finally, the initial and after near-field shadowing IV curves where used to simulate the performance ratio for different locations. The paper will end up in a discussion about the influence of near-field shadowing on the performance ratio at thin film modules

Keywords: shadowing, hot spot, performance ratio

1 INTRODUCTION

Although the market share of crystalline silicon panels has even increased in the past years still a few large manufacturers of CdTe and CIGS technology are present, and also new CIGS factories are currently under construction. Since solar panel design of thin-film PV is very different to c-Si panels which also affects the behaviour under partial shading, in this paper we study the effect of short-term near-field shadowing on different thin-film technologies and the influence on the performance and the energy yield. The motivation for this work is the observation of lower energy yield after daylight cleaning of thin-film PV generators.

Some fundamental research results on this subject from various institutions were reported. It was published that worm-like defects were visible after shadowing of CIGS modules [1]. The analysis of such panels with SEM cross sectional views identified holes/voids within the absorber layer with a typical defect size of ~5 µm [2]. By applying different imaging techniques like ReBEL, ILIT, DLITfor and DLITrev the hot spot formation of the defects can by observed in real-time [2][3]. Richter *et al.* [4][5] and Vasko *et al.* [6] developed also electrical-thermal models to simulate the shadowing effect. Silverman *et al.* [7] and Wendlandt *et al.* [8] reported about a power loss at STC after shading. Silverman pointed also out that the parallel resistance R_p increases after shading. Combining these insights with the results of Grunow *et al.* [9] we expect that the performance ratio and energy yield of thin-film modules will become worse after even short-term shadowing - especially at locations with a high rate of weak light conditions.

For the proof of our assertion we have experimentally investigated the impact of short-term shadowing on the performance and the IV curves of two different types of commercial thin-film products, and we have used these results to calculate the impact on the annual energy yield at different locations.

2 EXPERIMENTAL SETUP

The experiments were carried out for the two market dominating thin film technologies CdTe and CIGS. The test samples have been commercial products, and two different test sequences were applied.

Test Sequence #1

This test sequence #1was applied onto two CdTe modules. First the modules were stabilized (following the IEC 61215: 2016) and then initially characterized (STC, weak light behavior and EL). Both modules then were mounted on our outdoor test side in Berlin / Germany, facing south at an inclination angle of 40°. For 10 days the panels were kept at MPP and P_{max} was logged ($dt = 1$ min). Then the shadowing experiment has be applied to one module (test module) while the other module (reference module) was completely shadowed. The shadowing element has the shape like a roll of a cleaning robot. The roll has been moved with a speed of $v = 17$ m/min along the module surface at daylight while the module was kept in MPP. The physical dimension of the roll lead to a complete shadowing of a couple of solar cells for a short time – exactly what the operation of a cleaning robot at daytime would mean. After this short-term shadowing the power of both modules were logged again for 10 days outdoor, and then a final STC/EL characterization in the lab were conducted.

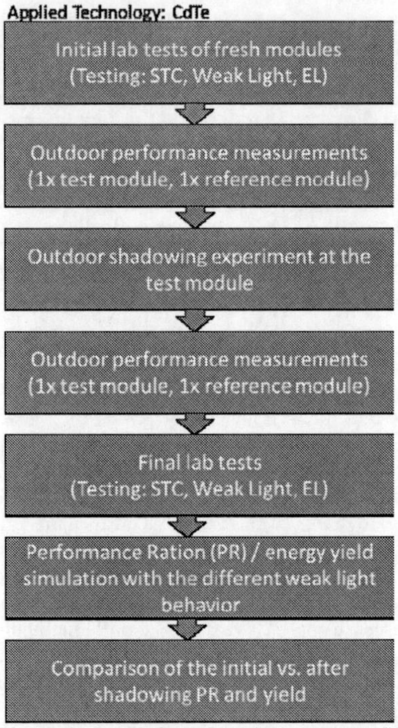

Figure 1: Applied outdoor test sequence #1.

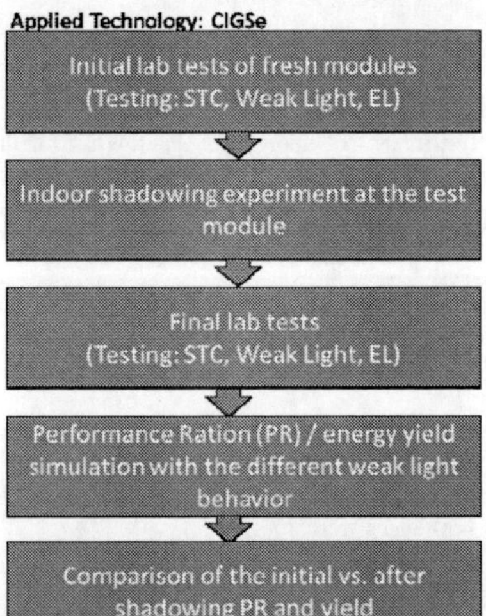

Figure 2: Applied indoor test sequence #2

Based on the initial and final lab data a PVsyst analysis was run to simulate energy yields and performance ratios of panels which have suffered from shading, compared to the reference panel. The entire test sequence is schematically shown in Figure 1.

Test Sequence #2

The second test is a pure indoor test, and even the shadowing experiment has been carried out in a flash solar simulator (class A) at STC. During a test approx. 10% of the module surface was covered when a 10 ms flash was applied. Then the shadowing element was moved along the panel to a different position and the panel got flashed again. This got repeated until every cell of the panel was covered one time. This procedure was applied to a stabilized CIGS module.

Like for the CdTe panel in test sequence #1, also the experimental results from test #2 have been used later for performance simulations.

3 RESULTS

3.1 EXPERIMENTAL RESULTS

Test Sequence #1:

Table I shows the IR images of the CdTe test module before, during (2x) and after the shadowing test.

Table I: IR images of test module at different states

<table>
<tr><th></th><th>IR image of test module</th></tr>
<tr><td>Before shadowing</td><td>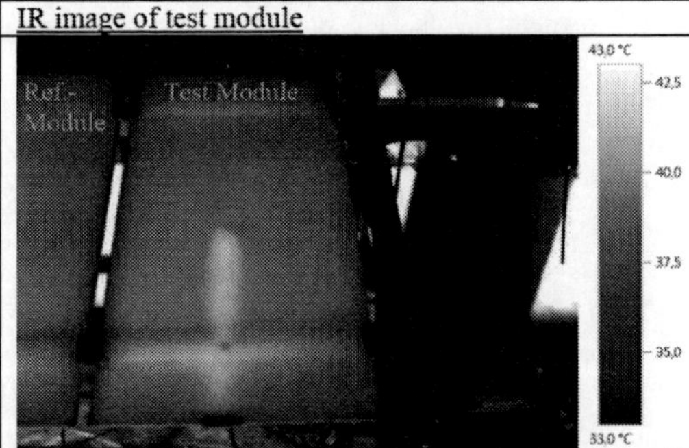
</td></tr>
<tr><td>During shadowing #1 (shading bar in the module center)</td><td>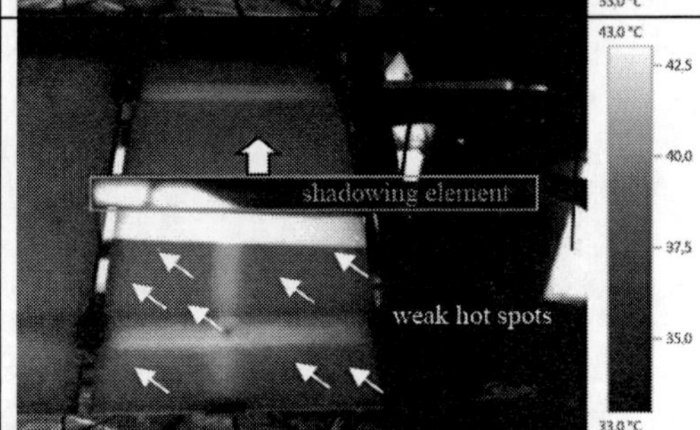
</td></tr>
<tr><td>During shadowing #2 (shading bar at the module top)</td><td>
</td></tr>
<tr><td>After shadowing</td><td>
</td></tr>
</table>

The images show that small hot spots become visible in IR after moving the shadowing element along the module surface. Their temperature difference amounts to $dT = 1\ldots3$ K compared to the normal operating module positions. The small spots are still present after a certain time of unshaded operation. This result proves that the effect is irreversible. The EL images of the reference and test module before and after shadowing are presented in Table II (Please note: the reference module wasn't never shadowed).

Table II: EL results of test sequence 1 at CdTe module

	Initial EL image	EL image after outdoor testing
Reference module		
Test module		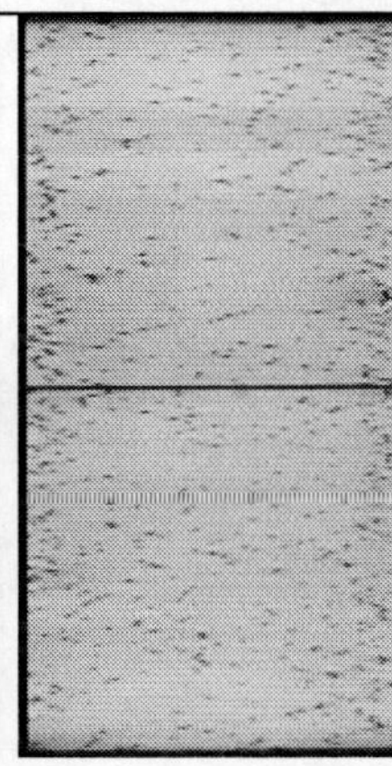

Compared to the initial image the test module shows many dark spots in the EL image after shadowing. This is a clear proof of the irreversible damage of the module after even a short-term shading of complete cells.

Very interesting is how the IV characteristic becomes affect by the shadow – especially when measuring them at different intensities. The IV curves of the test module are shown in Figure 3.

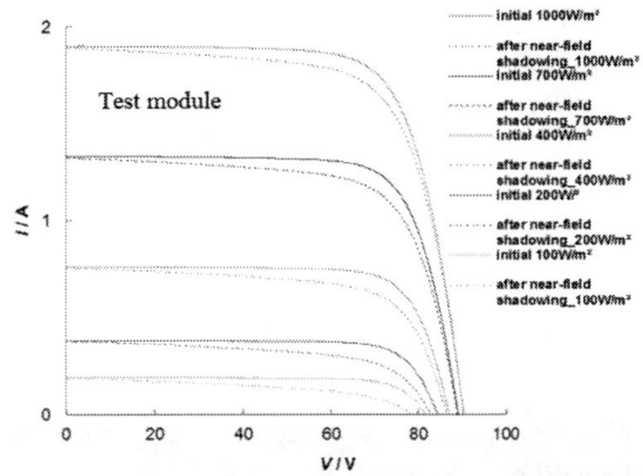

	Irr / (W/m²)	$\frac{P_{max,final}}{P_{max,initial}}$	$\frac{P_{max,final}}{P_{max,initial}}$	$\frac{V_{max,final}}{V_{max,initial}}$	$\frac{I_{max,final}}{I_{max,initial}}$	$\frac{V_{oc,final}}{V_{oc,initial}}$	$\frac{I_{sc,final}}{I_{sc,initial}}$	$\frac{FF_{final}}{FF_{initial}}$
test module	1000	-5.3%	-0.5%	-4.8%	-0.5%	-0.2%	-4.5%	-0.5%
	700	-7.2%	-1.3%	-6.0%	-0.6%	-0.3%	-6.4%	-0.6%
	400	-12.7%	-2.0%	-10.9%	-1.0%	-0.4%	-11.5%	-0.9%
	200	-23.4%	-6.1%	-18.4%	-1.9%	-0.4%	-21.6%	-1.9%
	100	-39.7%	-15.6%	-28.5%	-4.0%	-0.5%	-36.8%	-4.0%
reference	1000	0.0%	-0.2%	0.2%	-0.4%	0.0%	0.3%	-0.3%
	700	0.2%	-0.2%	0.4%	-0.1%	-0.2%	0.5%	-0.1%
	400	0.1%	0.6%	-0.5%	0.0%	-0.2%	0.3%	0.0%
	200	0.3%	-0.4%	0.7%	0.1%	-0.1%	0.3%	0.1%
	100	0.6%	1.1%	-0.4%	0.4%	-0.1%	0.3%	0.4%

Figure 3: IV curves before and after near-field shadowing (Plot: test module, Table: test and reference module).

The IV curve set shows the following properties after shadowing:

1. The slope in the range of I_{sc} is increased -> reduced R_p
2. The V_{oc} becomes more reduced with lower intensities -> as a result of reduced R_p. The V_{oc} loss at 1,000 W/m² is an indication degraded diode factor.
3. FF is reduced
4. P_{max} is reduced, too

The shadow's influence on the weak light behavior is presented in Figure 4.

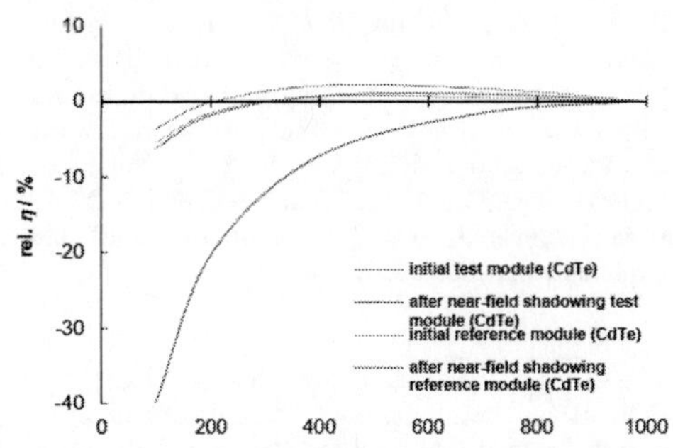

Figure 4: Weak light behavior of the CdTe test and reference module, both initially and after near-field shading (Please note: the reference module wasn't never shadowed).

From Figure 4 a strong impact of the shading on the weak light behaviour is evident. While initially the module efficiency is 3.5% less at 100 W/m² compared to STC, the under-performance sharply increases to 38.9% after shading.

The results of the outdoor energy yield measurements are presented in Figure 5

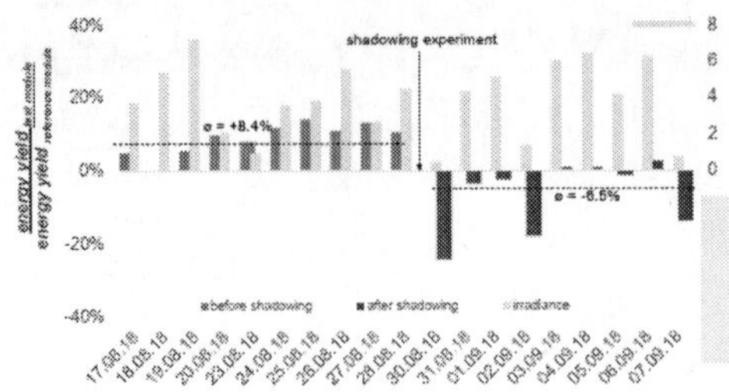

Figure 5: Daily energy yield before and after shadowing

It shows the daily energy yield of the test sample relative to the reference module before and after shading, as well as the daily solar irradiation on these particular days. Before shading the test module has a significantly higher energy yield (+8.4% in average). After having been affected by the shadow it has a lower yield of around –6.5% in average.

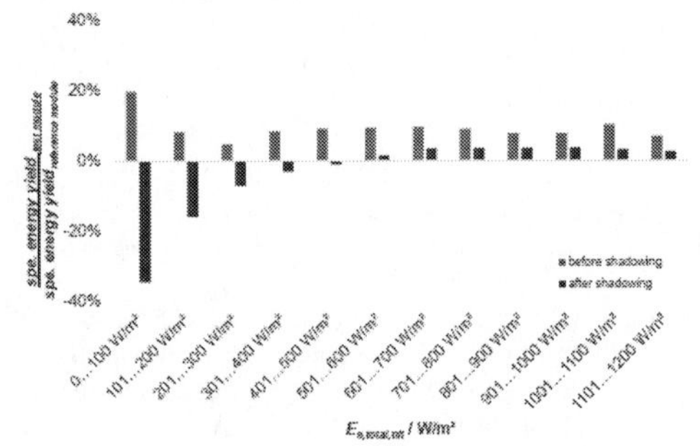

Figure 6: Relative specific energy yield in kWh/(kWh/m²) for different irradiation levels. Outdoor measurements before and after shadowing.

Figure 6 presents a more detailed analysis, showing the yield losses in dependence on the irradiation level. Before shading the test module has a fairly uniform higher yield at all irradiance levels. After the experiment the test module shows a strong decrease in performance which clearly scales with the irradiation level below 600 W/m². Above 600 W/m² the yield is still a little higher compared to the reference module but it is clearly worse compared to yield measurements of the test module before shading.

Test Sequence #2:

Although the shading condition has been very short (only 10 ms during the flash), a similar result as for test #1 can be found. Many dark spots can be observed for the CIGS module. Table III shows the initial and final EL images after indoor shadowing.

Table III: EL results of test sequence #2 for the CIGS module

	Initial EL image	Final EL image after indoor shadowing
Test module	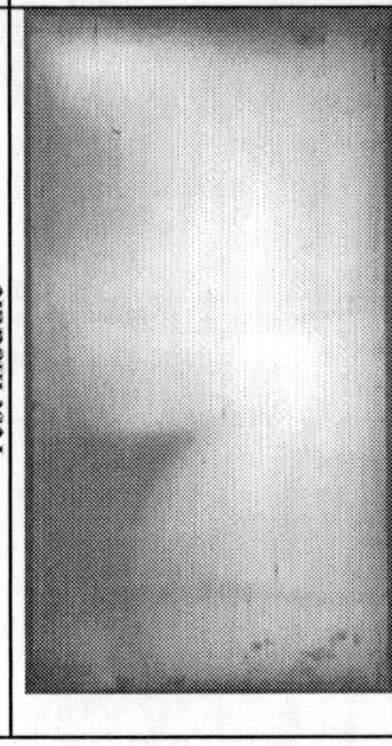	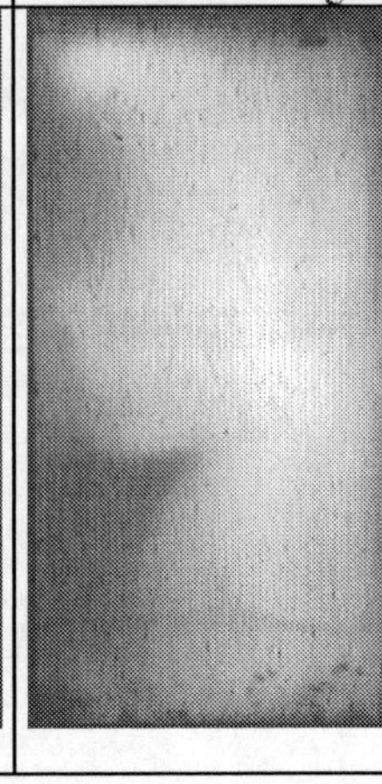

The dark spots are widespread over the module after shadowing during flashing had been applied. A systematic pattern of the spots is not observable and their origin has to be studied on the material level.

The shadowing influence on the IV curves is displayed in Figure 7.

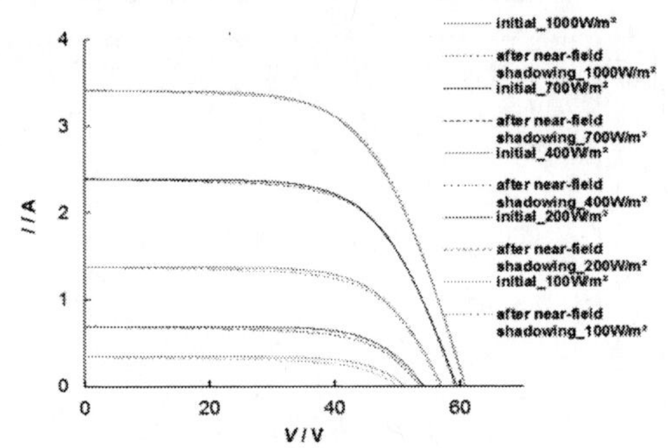

Irr / (W/m²)	$\frac{P_{max,final}}{P_{max,initial}}$	$\frac{V_{max,final}}{V_{max,initial}}$	$\frac{I_{max,final}}{I_{max,initial}}$	$\frac{V_{oc,final}}{V_{oc,initial}}$	$\frac{I_{sc,final}}{I_{sc,initial}}$	$\frac{FF_{final}}{FF_{initial}}$
1000	0.7%	1.1%	-0.4%	-0.5%	0.2%	1.0%
700	-0.6%	0.7%	-1.3%	-0.4%	0.2%	-0.5%
400	-3.4%	-1.5%	-2.0%	-0.6%	0.1%	-2.9%
200	-7.8%	-2.0%	-5.9%	-1.0%	0.1%	-7.0%
100	-15.9%	-3.2%	-13.1%	-2.3%	0.0%	-14.0%

Figure 7: IV curves before and after near-field shadowing at different irradiation levels

Similar to the CdTe module also the CIGS module shows a reduced V_{oc} at lower intensities. The reduction could be described by a lower R_p. P_{max} and FF and is also strongly dependent on the irradiance level.

The weak light behaviour of the CIGS module is displayed in Figure 8.

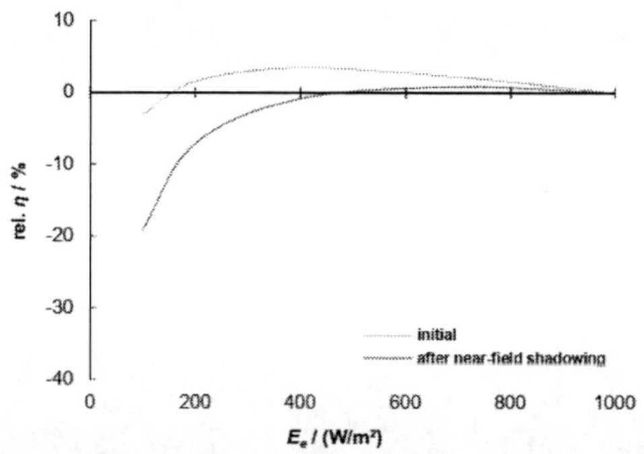

Figure 8: Weak light behavior of the CIGS module initially and after near-field shadowing

Before the CIGS module got shaded it had a stable relative efficiency for intensities down to approx. 150W/m². After shading the efficiency drops like previously examined for CdTe module.
How this impacts the performance for different location is shown in the next section.

3.2 YIELD SIMULATION RESULTS

The module power output of course is an important product parameter - but finally only the energy yield is the crucial aspect for evaluating the performance of a PV power plant. Therefor the determined weak light data of the panels got imported into PVsys in order to run simulations for various location.

Table IV: Locations and radiation of the simulated places

Location	diffuse irradiance rate
Berlin / D	0.54
Freiburg / D	0.55
Loughborough / UK	0.56
Sevilla / ESP	0.28

For all locations the same solar system configuration has been taken and the PV generator was orientated to achieve the highest annual energy yield, respectively. Table V presents the simulation results for the different locations by using data for the test samples which got short-term shaded.

Table V: Annually simulated energy yield and performance ratio losses

Location	Energy yield losses	PR losses
Berlin / D (CIGS)	-0.70%	-0.74%
Freiburg / D (CIGS)	-0.71%	-0.74%
Loughborough / UK (CIGS)	-0.72%	-0.61%
Sevilla / ESP (CIGS)	-0.45%	-0.38%
Berlin / D (CdTe)	-10.0%	-10.3%
Freiburg / D (CdTe)	-10.0%	-10.3%
Loughborough / UK (CdTe)	-10.1%	-10.4%
Sevilla / ESP (CdTe)	-8.85%	-8.97%

The simulation results show that the system at all locations would suffer from decreased yield and a lower PR per year. Figure 9 displays more detailed the dependency on the cloudiness factor.

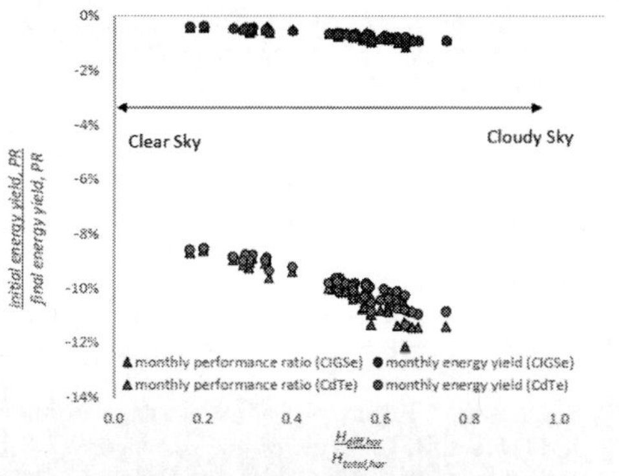

Figure 9: Dependency of the performance ratio and energy yield loss on the cloudiness factor

The simulation data also allow to analyze the cloudiness factor dependency. An agreement to the date from Fig 5 and Fig. 6 the yield and performance losses are more detectable at cloudy days or month (e.g. January or February).

4 SUMMERY

Thin film PV modules are a high-tech investment good which have to resist many challenges to convert sunlight into electrical energy for 20 years reliably at low failure rates. We studied what impact short-term near-field shadowing can have on the long-term performance of CdTe and CIGS modules.

We could determine that even a 1-time short-term shadowing can cause non-reversable power degradation for both module technologies. This results specifically in a much worse weak-light behavior after shadowing. Also in the EL images many dark spots were visible after shading the modules, demonstrating point-defect caused by local breakthrough of the semiconductor material.

Our outdoor performance analysis shows that the energy yield gets reduced after shadowing. Since the module power is reduced at low irradiation levels, this has an impact especially for locations with a high diffuse light ratio. These findings could be confirmed by simulations

Our advice to O&M operators is to turn off the inverter at every situation which may cause even short-term shadowing (e.g. cleaning or walking in front of the modules). Regarding product standards we recommend to carry out a weak light measurement at E = 100 W/m² before and after the hot spot test in order to detect module damages.

5 ACKNOWLEDGEMENTS

This study was supported by a grant from the Federal Ministry for Economy Affairs and Energy on the basis of a decision by the German government under contract number 0324193B.

6 REFERENCES

[1] P-O. Westin, U. Zimmermann, L. Stolt, M. Edoff:

"*Reverse Bias Damage in CIGS Modules*", 24th European Photovoltaic Solar Energy Conference, 21-25 September 2009, Hamburg, Germany

[2] S. Johnston, D. Sulas, H. Guthrey, J. Liu, L. Mansfield, T. J. Silverman, M. Al-Jassim, E. Palmiotti, A. Rockett, A. Gerber: "*Thin-Film Module Reverse-Bias Breakdown Sites Identified by Thermal Imaging*", 2018 World Conference on Photovoltaic Energy Conversion (WCPEC-7) Waikoloa, Hawaii, June 10–15, 2018

[3] S. Johnston, I. Repins, N. Call, R. Sundaramoorthy, K. M. Jones, B. To: "*Applications of imaging techniques to Si, $Cu(In,Ga)Se_2$, and CdTe and correlation to solar cell parameters*", Photovoltaic Specialists Conference (PVSC), 2010 35th IEEE

[4] M. Richter, J. Neerken, J. Parisi: "*Reverse-Voltage Bias Induced Shunt Formation in $Cu(In,Ga)Se_2$ Thin Film Solar Cells: an Approach with Three-Dimensional Electro-Thermal Simulations*", 32nd European Photovoltaic Solar Energy Conference and Exhibition

[5] M. Richter, F.-S. Babbe, A. Vetter, S.J. Heise, J. Ohland, I. Riedel, J. Parisi: "*Origins of Shunt Signals of $Cu(In,Ga)Se_2$ Solar Cells Investigated by Two-Dimensional Simulations*", 31st European Photovoltaic Solar Energy Conference and Exhibition

[6] A. Vasko, A. Vijh, V. G. Karpov: "*Hot spots spontaneously emerging in thin film photovoltaics*", Solar Energy, Volume 108, October 2014, Pages 264–273

[7] T. J. Silverman, M. G. Deceglie, R. L. Garris, C. D., Sarah Kurtz, X. Sun, M. d A. Alam: "*Thermal and Electrical Effects of Partial Shade in Monolithic Thin - Film Photovoltaic Modules*", 42nd IEEE Photovoltaic Specialists Conference, New Orleans, Louisiana, June 14…19, 2015

[8] S. Wendlandt, S. Berendes, T. Weber, J. Berghold, S. Krauter, P. Grunow: "*Shadowing Investigations on Thin Film Modules*", 32nd European PV Solar Energy Conference and Exhibition, 20th – 24th June 2016, Munich, Germany

[9] P. Grunow, S. Lust, D. Sauter, V. Hoffmann, C. Beneking, B. Litzenburger, L. Podlowski: "*Weak light performance and annual yields of PV modules and systems as a result of the basic parameter set of industrial solar cells*", 19th European Photovoltaic Solar Energy Conference, 7-11 June 2004, Paris, France

ENERGY YIELD COMPARISON BETWEEN BIFACIAL AND MONOFACIAL PV MODULES: REAL WORLD MEASUREMENTS AND VALIDATION WITH BIFACIAL SIMULATIONS

Johanna Bonilla Castro, Magnus Herz; Christos Monokroussos, Markus Schweiger
TÜV Rheinland
TÜV Rheinland Energy GmbH, Am Grauen Stein, 51105 Cologne, Germany
Johanna.Bonilla@de.tuv.com

ABSTRACT: The photovoltaic (PV) market has seen the entry of different PV module technologies during the past years. The success of these PV technologies as key market players is linked to their long-term reliability and energy yield (EY) performance. Bifacial PV modules are rapidly increasing their market share, leading to changes in the existing PV standards and modifications in the current measurement practices. With this, debates about conventional EY simulations and their deviations from real-world performance have arisen. To cope with these questions, we investigated the EY performance under real-operating conditions of four commercial bifacial PV samples,12 monofacial c-Si and two CdTe thin-film PV modules during one year at TÜV Rheinland test site in Cologne, Germany. The PV samples were characterized in the laboratory before and after the outdoor exposure to determine final degradation rates. Together with these measurements, the collected climate data were used for EY simulations with PVsyst. The uncertainties of the results were quantified focusing on performance and rear side irradiation.
Keywords: Bifacial, performance, bifacial simulation, energy yield, PVsyst

1 INTRODUCTION

In the PV industry, the watt-peak thinking has been the most significant criteria followed by all market players to make PV business decisions, by either manufactures, systems installers, investors or final users. As part of the traditional way of operating in the PV market, laboratory tests are required in order to commercialize PV technologies. These measurements are performed under standard test conditions (STC; 1000 w/m^2, AM=1.5, 25°C). Additionally, simulations are used to support decision-making.

However, the concept of energy instead of power is becoming more and more important and the PV industry is moving away from the Watt-peak to the kWh thinking. As a result, the share of *new* technologies, as the case of bifacial PV modules has increased. Similar to what happened with other PV technologies; the introduction of bifacial PV modules has led to and is still causing modifications in the existing measurements procedures. In this context, uncertainties regarding their real-world energy performance and doubts about the implementation of conventional EY simulations have arisen.

For a better understanding of the real energy performance of bifacial PV modules and the improvement of simulation tools, we investigated four commercial bifacial PV modules (2 PV types). The research was conducted from three different perspectives: In the laboratory, under real-world conditions and by using simulation tools. For technology intercomparison, 12-monofacial c-Si (6 PV types) and two thin-film PV samples underwent the same assessment. For each PV type, a spare sample(s) remained in laboratory for reference purpose. **Figure 1** shows the research approach and **Table I** describes the investigated PV samples.

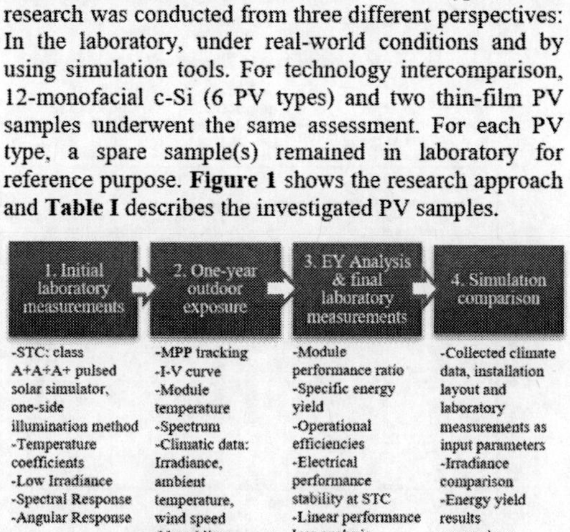

Figure 1: Research approach.

Table I: Investigated PV specimens.

PV module	Cell type	N° of exposed PV samples	N° of spare PV samples (reference)
mofi 1&2	poly c-Si	2	1
mofi 3&4	mono c-Si	2	1
mofi 5&6	mono c-Si	2	1
mofi 7&8	mono c-Si	2	1
mofi 9&10	mono c-Si	2	1
mofi 11&12	mono c-Si	2	1
bifi 1&2	n-pert	2	1
bifi 3&4	n-pert	2	1
thin-film 1&2	CdTe	2	3
Total		18	11

Note: Monofacial c-Si PV modules were designated as mofi and bifacial PV samples, as bifi.

2 LABORATORY INITIAL CHARACTERIZATION

Both the PV modules assigned for outdoor exposure and those stored for reference were submitted to initial measurements and non-destructive laboratory tests to analyze their electrical characteristics and determine their indoor performance.

The samples were characterized in the laboratory using a class A+A+A+ pulsed solar simulator according to IEC60904-9 [1]. For the bifacial PV samples, the one-side illumination method [2] was applied. A non-reflective background was used in order to avoid the illumination of the non-exposed side. The results of this characterization were used as reference for STC corrections.

The measured bifaciality coefficients at STC, defined as the efficiency ratio rear/front of the selected bifacial PV samples for EY testing, are shown in the following table.

Table II: Bifaciality coefficients at STC, values included spectral mismatch (SMM) corrections.

Bifacial PV Module	φP_{MPP}	φV_{OC}	φI_{SC}
bifi 1	0.892	0.994	0.892
bifi 2	0.871	0.993	0.848
bifi 3	0.860	0.993	0.853
bifi 4	0.889	0.995	0.869

The temperature coefficients of current (α), voltage (β) and power (γ) were measured on one sample for each PV module type. **Table III** shows these results.

Table III: Temperature coefficients per module type measured at 1000 w/m^2

PV module type	Current Temp. Coeff. α [1/K]	Voltage Temp. Coeff. β [1/K]	Power Temp. Coeff. γ [1/K]
mofi 1&2	0.0005	-0.0029	-0.0041
mofi 3&4	0.0004	-0.0028	-0.0039
mofi 5&6	0.0004	-0.0029	-0.0040
mofi 7&8	0.0004	-0.0030	-0.0042
mofi 9&10	0.0005	-0.0029	-0.0040
mofi 11&12	0.0005	-0.0032	-0.0042
bifi 1&2	0.0004	-0.0030	-0.0042
bifi 3&4	0.0005	-0.0030	-0.0041
thin-film 1&2	0.0004	-0.0028	-0.0028

The performance of the PV samples at different irradiance was assessed in the laboratory in accordance with IEC 61853-1 [3], as shown in **Figure 2.** The efficiency loss @200W/m2 was higher for bifacial compared to monofacial PV modules, being the rear side less efficient than the front (See **Table IV**)

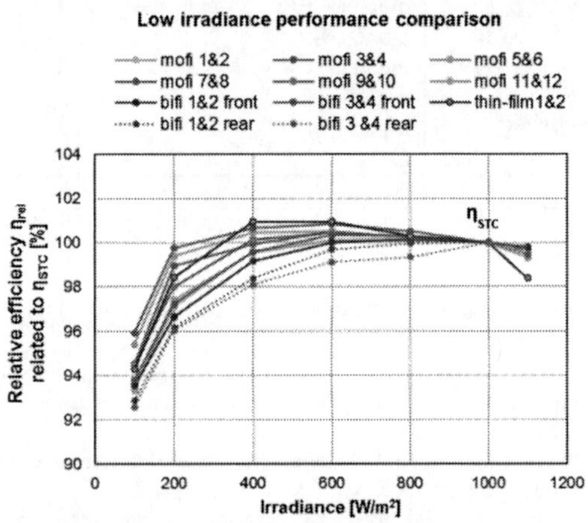

Figure 2: Low irradiance comparison of the PV modules at 25°C, grouped per type (9 PV types).

Table IV: Efficiency loss per module type at 200 W/m^2 and 25°C

PV Module type	Efficiency loss @200 W/m^2	
	Front	Rear
mofi 1&2	-2.9	-
mofi 3&4	-1.0	-
mofi 5&6	-0.6	-
mofi 7&8	-0.3	-
mofi 9&10	-2.0	-
mofi 11&12	-2.6	-
bifi 1&2	-3.3	-3.8
bifi 3&4	-2.8	-4.0
thin-film 1&2	-1.6	-

The spectral response (SR) of all PV modules was measured in accordance with IEC 61853-2 [4] within a wavelength of 300 to 1250 nm and a resolution of 10 nm. **Figure 3** shows the results of these measurements.

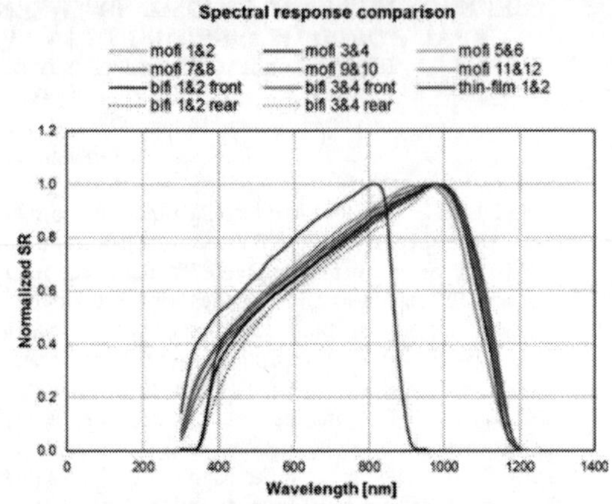

Figure 3: Normalized spectral response (SR) of the PV modules per type (9 PV types).

The spectral mismatch for output power characteriszation with the pulsed solar simulator (SMM) was computed in accordance to the IEC 60904-7 [5], considering the SR results (**Figure 3**) and a primary calibrated crystalline WPVS reference cell used as primary reference of the solar simulator. The obtained values were used for correcting the STC measurements. **Table V** shows the computed SMM factors.

Table V: Spectral mismatch per type (9 PV types)

PV Module type	SMM	
	Front	Rear
mofi 1&2	1.0044	-
mofi 3&4	1.0093	-
mofi 5&6	1.0031	-
mofi 7&8	1.0044	-
mofi 9&10	1.005	-
mofi 11&12	1.0014	-
bifi 1&2	1.0082	1.0108
bifi 3&4	1.0022	1.0061
thin-film 1&2	0.9677	-

The angular response of the PV modules was also measured in the laboratory according to IEC 61853-2 [4], as shown in **Figure 4.**

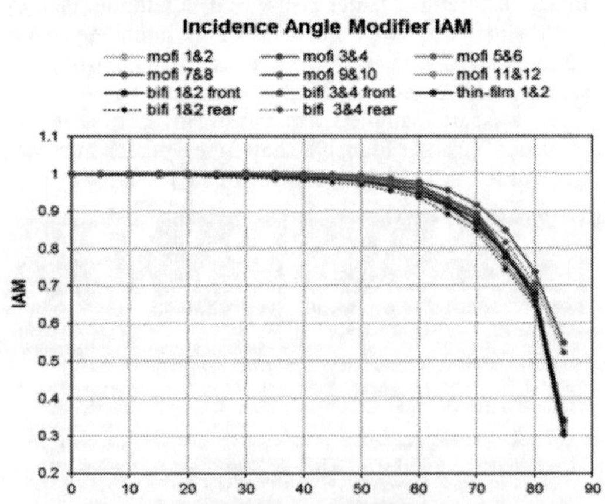

Figure 4: Angular response of the PV modules.

2 COMPARISON OF OUTDOOR ENERGY YIELD MEASUREMENTS

The outdoor energy yield assessment took place at the test site of TÜV Rheinland in Cologne, Germany. After the initial measurement sequence, the PV samples assigned for outdoor exposure (2 per module type) were mounted on an open rack installation. **Table VI** summarizes the main climatic factors and mounting conditions of the test site.

The energy yield performance of the PV modules as well as the meteorological conditions were monitored and recorded during 12 months of field deployment. The test samples had identical measuring systems with synchronous data recording interval of 30 s. Each PV module was equipped with two Pt100 sensors to measure the back of module temperature (T_{BoM}) with a sample rate of 30 seconds. Individual electronic DC loads were installed, so that all test samples operated continuously in maximum power point (MPP-Tracking) with a data acquisition frequency of 30 seconds. Current-Voltage (I-V) curves were recorded in regular time intervals of 10 minutes. The in-plane global irradiance (G_{PoA}), horizontal (G_{Hor}), diffuse (G_d) and rear irradiance (G_{rear}) were measured with pyranometers. The ambient temperature (Tamb), spectral irradiance, precipitation, wind speed and humidity were measured as well. The in-site data acquisition is shown in **Table VII.**

Table VI: Description of Cologne outdoor test site

Location	Cologne, Germany
Köppen-Geiger climate classification	Cfb (temperate)
Mounting conditions	2 racks, fixed-facing south
Tilt angle	35°
Annual in-plane global solar irradiation, H_{POA}	1231 kWh/m²
Low irradiance fraction (<200 W/m²)	16.8 %
Average ambient temperature (G_{PoA}> 15 W/m²)	16.1 °C
Height of PV modules above ground	1.5m
Ground albedo	30% (gravel)
Pitch	11m

Climatic factors are based on measurements from August 2017 to July 2018.

Figure 5: Outdoor mounting of bifacial PV modules at test site Cologne (Germany).

Table VII: Equipment and data acquisition frequency of field measurements

Parameter	Equipment/ Technique	Frequency
T_{BoM}	2 PT100 temperature sensors	30 sec
I-V curve measurements	Electronic loads	10 min
P_{MPP}	MPP tracker	30 sec
G_{PoA}, G_{Hor}, G_d, G_{rear}	Pyranometer	30 sec
T_{amb}, *Precipitation and humidity*	Sensors (instantaneous measurements)	30 sec
Spectral irradiance	Spectrometer (300-1600nm)	1 minute

2.1 Module performance ratio

The module performance ratio (MPR) is a relation of the actual energy production in the monitoring period and the expected yield when continually operating at STC efficiency.

$$MPR = \frac{\left(\sum_{months} P_{MPP}\right) / P_{MPP,STC}}{\left(\sum_{months} G_{PoA}\right) / 1000 Wm^{-2}} \cdot \% \qquad [1].$$

Equation [1] was applied for a period of one year: From August 2017 to July 2018. Here, the measured energy, thus the sum of the 30-second maximum-tracked power output ($\sum P_{MPP}$) values is divided by the maximum power output at standard test conditions ($P_{MPP,STC}$). In the denominator, the in-plane global irradiation ($\sum G_{PoA}$) is divided by the standard irradiance. The values of $P_{MPP,STC}$ in equation [1] were selected according to the following criteria:

- For monofacial PV modules, the results of indoor characterization with the SMM correction
- For bifacial PV technologies, the measurements of the front side ($P_{MPP,STC\ front}$) with SMM correction.
- For thin-film technologies, the nominal power output ($P_{MPP,Nom}$), as the electrical stabilization process was incomplete.

The annual results of MPR are shown in **Figure 6.**

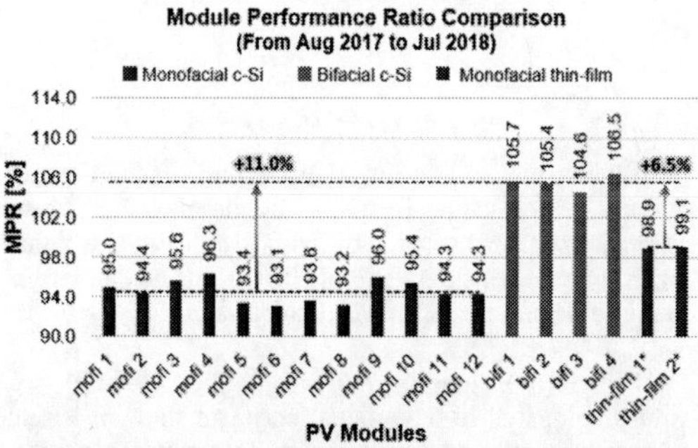

** Values referred to label P_{MPP}. All others PV modules are referred to indoor P_{MPP} at STC with SMM correction and were electrically stabilized*

Figure 6: Comparison of module performance ratio of the 18 PV samples (Baseline one year: From August 2017 to July 2018).

The MPRs of bifacial PV modules were on average 11% higher than c-Si monofacial samples and 6.5% higher compared to thin-film specimens. Differences in MPR amongst pairs of PV modules were also found for the one year of field measurements.

2.2 Area-normalized operational efficiency

The computation is based on

$$\eta_{operational} = \frac{\sum_{months} P_{MPP}}{\left(\sum_{months} G_{PoA}\right) * A_m} \cdot \% \qquad [2].$$

Where the sum of the 30-second tracked power output ($\sum P_{MPP}$) values is divided by the sum of the measured in-plane global irradiance (G_{PoA}) and the solar module area (A_m).

Bifacial PV modules showed higher operational efficiencies compared to monofacial ones, as show in **Figure 7**. Differences amongst PV modules of the same type were also relevant.

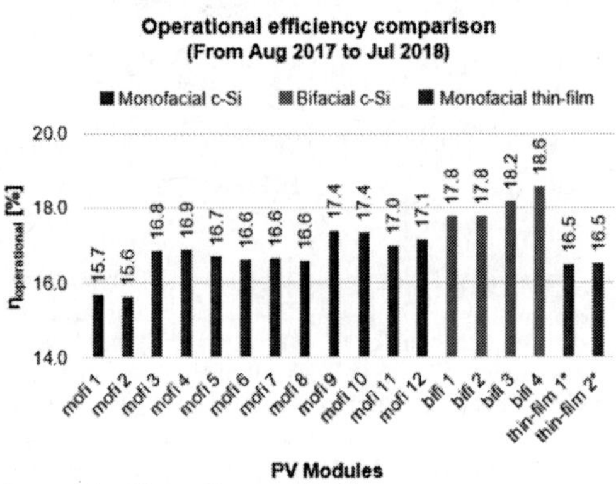

**PV modules with possible metastable effects*

Figure 7: Comparison of operational efficiencies of the 18 PV samples (Baseline one year: From August 2017 to July 2018)

2.3 Specific Energy Yield

The total energy generated per watt-peak (EYspec) of each PV sample during the one-year outdoor exposure was also calculated.

The computation is based on

$$EY_{spec} = \frac{\sum_{months} P_{MPP}}{P_{MPP,STC}} \cdot \frac{Wh}{W_{peak}} \qquad [3].$$

Where the sum of the 30-second tracked power output ($\sum P_{MPP}$) values is divided by the maximum power output at standard test conditions ($P_{MPP,STC}$). The values taken by $P_{MPP,STC}$ in equation [3] were selected according to the same criteria as for MPR calculation.

Bifacial PV modules showed higher specific yield values compared to monofacial ones. As show in **Figure 8**, the specific yield of bifacial PV modules was on average 10.4% higher than c-Si monofacial samples and 6.2% higher compared to thin-film specimens. Differences amongst pairs of PV modules were also found for the one year of field measurements.

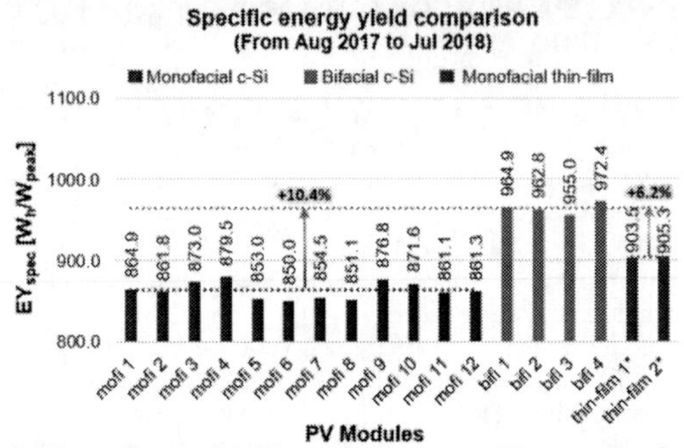

** Values referred to label P_{MPP}. All others PV modules are referred to indoor P_{MPP} at STC with SMM correction and were electrically stabilized*

Figure 8: Comparison of specfic energy yield of the 18 PV samples (Baseline one year: From August 2017 to July 2018)

2.4 Quantification of individual Loss/Gain Mechanisms

The individual loss/gain mechanisms of EY performance were quantified using a linear performance *loss* model [6], resulting from a comprehensive combination of laboratory measurements and the climate dataset of the test site.

Table VIII: Description of loss/gain mechanism due to climatic conditions and solar module characteristics

MPR Variation Mechanism	PV module	Test site
Temperature, ΔMPR_{TEMP}	Temperature coefficients	Irradiance weighted module temperatures
Low irradiance, ΔMPR_{LIRR}	Efficiency curve	Frequency distribution of solar irradiation in irradiance classes
Spectral response, ΔMPR_{SMM}	Spectral responsivity	Average sunlight spectrum
Angle of incidence, ΔMPR_{AoI}	Angular response	Frequency distribution of solar irradiation in angle of incidence classes
Soiling effects, ΔMPR_{SOIL}	surface properties	Environmental conditions
Bifacial gain, ΔMPR_{BIFI}	Bifaciality coefficients (φ_{bifi})	Ground albedo and mounting conditions (shading effects)

The combined loss mechanisms resulted in a calculated PV module performance ratio (MPR_{CAL}), as follows:

$$MPR_{CAL} = 100\% - \Delta MPR_{TEMP} \pm \Delta MPR_{LIRR} \qquad [4].$$
$$- \Delta MPR_{SOIL} \pm \Delta MPR_{SMM} - \Delta MPR_{AOI} + \Delta MPR_{bifi\,A}$$

Each variation mechanism was independently analyzed. For all PV modules, an annual value of ΔMPR_{SOIL}= -0.5% was considered, based on the measurements of two reference cells, one soiled and one regularly cleaned. For bifacial PV modules the total irradiance (front + rear) was considered. Therefore, for the given period the integration of G_{PoA}+ φ_{bifi} •G_{rear} The bifacial gain φ_{bifi} defined, as the minimum value amongst the bifaciality coefficients of current and the output power.

The annual results are shown in **Figure 9**. and **Table IX**. The losses in MPR for bifacial PV samples were comparable to the ones found in monofacial c-Si PV technologies.

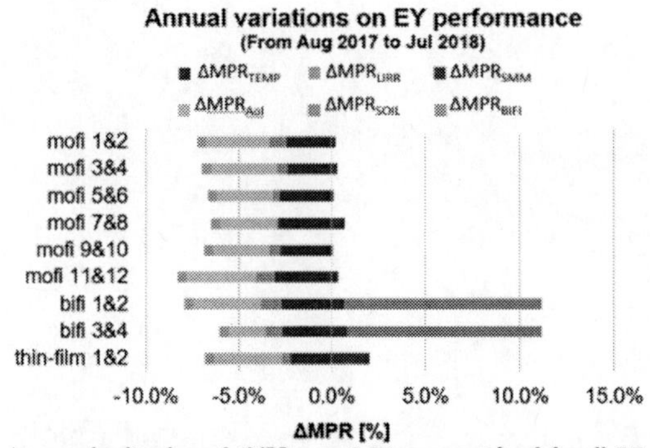

An annual soling loss of ΔMPR_{SOIL}= -0.5% was considered for all PV samples based on the measurements of two reference cells, one soiled and one regularly cleaned

Figure 9: Annual losses and gains of energy yield of the 18 PV modules (grouped per PV type

Table IX: Description of the source information for the two simulated scenarios

PV Module	Variations in ΔMPR [%]				
	TEMP	LIRR	SMM	AoI	BIFI
mofi 1&2	-2.4	-1.0	0.2	-3.3	0.0
mofi 3&4	-2.4	-0.5	0.3	-3.6	0.0
mofi 5&6	-2.8	-0.3	0.1	-3.0	0.0
mofi 7&8	-2.9	-0.1	0.7	-3.0	0.0
mofi 9&10	-2.8	-0.6	0.0%	-3.0	0.0
mofi 11&12	-3.1	-1.0	0.3	-3.7	0.0
bifi 1&2	-2.7	-1.1	0.7	-3.6	10.5
bifi 3&4	-2.7	-0.9	0.8	-2.0	10.4
thin-film 1&2	-2.2	-0.4	2.0	-3.7	0.0

An annual soling loss of ΔMPRSOIL= -0.5% was considered for all PV samples based on the measurements of two reference cells, one soiled and one regularly cleaned

3 SIMULATIONS RESULTS

Simulations were carried out with PVsyst Version 6.7.4. The best and worst case scenarios were simulated to compare amongst results. The optimum scenario (Case A), using the meteorological measurements obtained during the one-year outdoor exposure and results of initial laboratory measurements. The worst-case scenario (Case B) was computer-generated using the climatic files of PVsyst database and the PV modules label information provided by the manufactures. Simulations were performed for one sample of each bifacial PV type. **Table X** summarizes the two scenarios.

The View Factor Model in PVsyst showed for the rear side irradiance (G_{rear}) a strong linear correlation between measured and simulated values, as show in **Figure 10**. However, when the values per month were analyzed in detail, it showed an annual overestimation of G_{rear} of 16%. Since, for the simulation a constant albedo value of 30% (gravel) was used, the model failed to simulate extreme conditions as snow covering (leading to a higher albedo) during winter (See **Figure 11**).

For the best-case scenario, an annual deviation within ±1.3% was found (See **Table XI**). The deviation between the MPR values increased during the winter, were differences of -8% were found (**Figure 12**). This correlates with snow coverage conditions. By using only software data and label information as input parameters, mayor deviations were found.

Table X: Description of the source information for the two simulated scenarios

Variable	Case A (best case)	Case B (worst case)
PV module: (bifi 1&bifi 3)	Laboratory measurements	Label information (datasheets)
Mounting conditions	as real (see table VI)	as real (see table VI)
Meteorological	Real-world measurements	Database from PVsyst

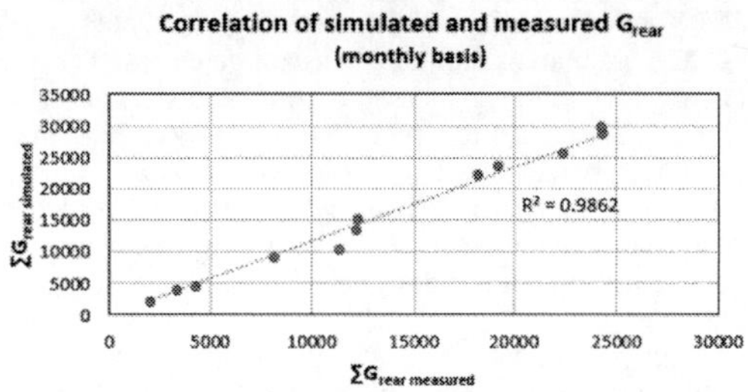

Figure 10: Correlation of monthly simulated and measured G_{rear}

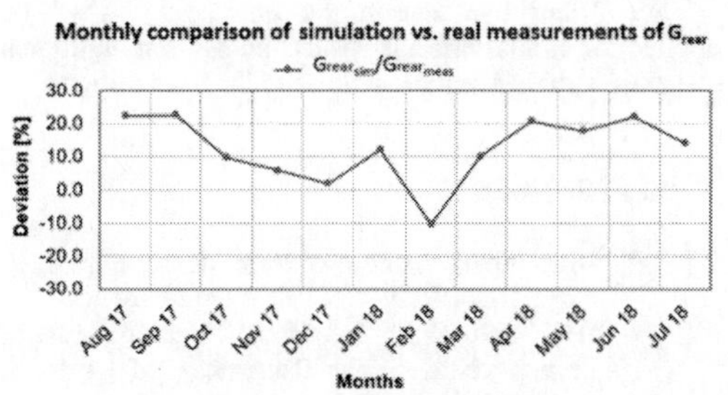

Figure 11: Monthly deviation of simulated values vs. real measurements: Rear Irradiance (Grear)

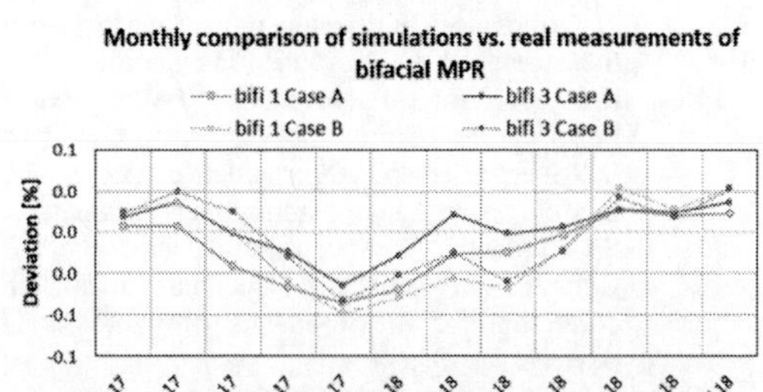

Figure 12: Monthly deviation of simulated values vs. real measurements for different scnearios: MPR of bifacial PV samples (bifi 1 and bifi 3)

Table XI: Annual results for the two simulated scenarios

	Case A		Case B	
	bifi 1	bifi 3	bifi 1	bifi 3
Annual MPR simulated [%]	105.63	105.99	107.22	106.39
Mean Bias Error, MBE [%]	-0.04	1.34	1.46	1.72

5 CONCLUSIONS AND OUTLOOK

The results of outdoor measurements revealed clear advantages in energy yield performance of bifacial over monofacial PV modules. Annual gains in module performance ratio of 11% were found when compared to c-Si monofacial PV technologies and 6.5% compared to thin-film samples. On daily basis, energy yield variations of up to 20% were observed, identifying the sky clearness index and air mass as driving factors behind specific energy yield gains of bifacial PV modules. The individual loss analysis revealed that losses are comparable between monofacial and bifacial PV samples. For instance, the observed operating temperatures were consistent amongst all PV modules, independently of their PV technology.

The simulation showed consistent outcomes. The use of extensive laboratory measurements and high accurate climate datasets led to a reduction of the mean bias error to ±1.3% for the View Factor Model. The obtained mean bias errors for each bifacial MPR were higher, when using the climatic database of the PVsyst software and the nominal output power of the PV modules as stated on the label.

Electrical stabilization effects, however, need to be further analyzed to enhance the precision of computer simulation tools and consequently reduce investments risks.

For further investigation, the same approach will be applied for similar bifacial PV technologies at additional test sites with different climatic and albedo conditions.

7 REFERENCES

[1] IEC 60904-9: Photovoltaic devices; part 9: Solar simulator performance requirements

[2] IEC 60904-Part 1-2 draft: Measurement of current-voltage characteristics of bifacial photovoltaic (PV) devices.

[3] IEC 61853-1: Photovoltaic (PV) Module Performance Testing and Energy Rating; Part 1: Irradiance and temperature performance measurements and Power Rating

[4] IEC 61853-2: Photovoltaic (PV) Module Performance Testing and Energy Rating; Part 2: Spectral response, incidence angle and module operating temperature measurements

[5] IEC 60904-7: Photovoltaic devices; part 7: Computation of the spectral mismatch correction for measurements of photovoltaic devices

[6] Schweiger, M. et al. 2017 "Understanding the energy yield of photovoltaic modules in different climates by linear performance loss analysis", IET Renew. Power Gen. [DOI:10.1049/iet-rpg.2016.0682].

[7] Schweiger, M. et al. 2017, "Performance stability of photovoltaic modules in different climates", Progress in Photovoltaics: Res. Appl. [DOI: 10.1002/pip.2904].

[8] Mermoud, A. et al 2017, "Bifacial shed simulations with PVsyt", Bifacial Workshop 2017 Konstanz, Germany.

A NOVEL HEAT DISSIPATING MATERIAL FOR ENHANCING THE PERFORMANCE OF PHOTOVOLTAIC PANELS

Min-An Tsai[1], Kuan-Wu Lu[1,], Wei-Lun Yang[1,], Sih-Hong Chen[1,], Hsieh-Cho Fan[1], Hung-Sen Wu[1] , Teng-Chun Wu[1]
[1]Center for Measurement Standards, Industrial Technology Research Institute, Taiwan

ABSTRACT: From a long-term/short-term perspective, the solar energy is a 20 to 25 years investment industry. The reliability of solar modules should be as important as the power output. Therefore, the high reliability and high regional weatherability product is more popular by engineering procurement construction (EPC) company. Here we designed a series of rigorous testing procedures for the different module included damp heat, thermal cycle, potential induced degradation, slat mist and a composite test, to analysis the performance and reliability.

Keywords: TEPV, reliability

1 Introduction

According to the BOE/MOEA, Taiwan PV industry is facing both price and trade competition. As a result, the industry needs to adopt competitive strategies like product differentiation and market segmentation in order to create blue ocean markets. In addition to maintaining a high level of capacity utilization, raising prices, increasing product value and maximizing profitability must also be the high priority for Taiwan PV industry development. Therefore, strict international testing standards have been adopted for the Taiwan Excellent PV Award assessment. Products which pass the strict review process and product testing are granted awards to recognize this achievement. This makes the selection criteria much severer than current requirements in other overseas markets.

The Taiwan Excellent PV Awards are based on the latest international IEC photovoltaic standards. For PV module category products, in addition to safety qualification test items and performance test items, the efficiency requirements were raised (monocrystalline ≧ 18.3 %, multicrystalline ≧ 17.4 %, Thin-Film ≧ 13.8%) and salt mist corrosion test for the island climate, electro attenuation test, and warm damp zone of climate specific test. For the solar cell test items, in addition to solar cell efficiency (monocrystalline ≧ 20.6 %; multicrystalline ≧ 19.5 %), power degradation testing was also added. Only products which pass all of the test items and maintain maximum power at 95 % of pre-testing levels (see attachment) can win the award.

2 Experiments and details

In order to improve the quality and competitiveness of Taiwan photovoltaic (PV) products and encourage companies to focus on high-value development, Bureau of Energy, the Ministry of Economic Affairs (BOE/MOEA) has held the fourth Excellent PV Product Assessment Program (Taiwan Excellent PV, TEPV) to select and award high-quality PV products from two major categories "silicon photovoltaic module" and "silicon solar cell". For "silicon photovoltaic module", assessments include not only safety and performance, but also extra reliability tests including salt mist corrosion test for the island climate, potential induced degradation test, and warm damp zone of climate specific test. For "silicon solar cell", assessments cover measuring efficiency of solar cells, as well as the new item of potential induced degradation test.

Assessment items of PV module: Comply with all required safety, performance, efficiency and reliability regulations.

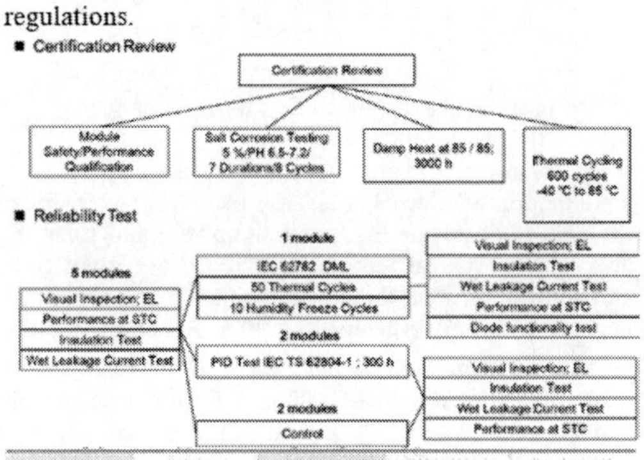

Selection project	Guideline	Request content	Qualified threshold	Mode of review
Safety	IEC 61730-2:2004 or the latest version	Safety test	Meet the reference standard	Book review
performance	IEC 61215:2005 IEC 61646:2008 or the latest version	Performance Testing	Meet the reference standard	Book review
effectiveness	IEC 60904-1:2006	STC maximum power test, efficiency calculation based on module frame area	Single crystal germanium ≧18.3 %	Measured
			Polycrystalline germanium ≧17.4%	
			Film ≧ 13.8%	
Reliability (salt spray test)	IEC 61701:2012 Severity 6	Salt spray test for 56 days	6 qualified	Book review
			Power attenuation ≦ 5%	
Reliability (wet heat test)	IEC 61215:2005 IEC 61646:2008	DH 3000	4 qualified	Book review
			Power attenuation ≦ 15 %	
Reliability (thermal cycle test)	IEC 61215:2005 IEC 61646:2008	TC 600	3 qualified	Book review
			Power attenuation ≦ 15 %	
Reliability (electro-attenuation test)	IEC TS 62804-1:2015	PID 85 °C 85 %RH ±1000 V 300 hours	4 qualified	Measured
			Power attenuation ≦ 5%	
Reliability (composite test)	IEC TS 62782:2016	DML (2000 Pa) - TC50 - HF10	4 qualified	Measured
			Power attenuation ≦ 5% (single item) power attenuation ≦ 8 % (full range)	

Figure 1: the test flow chats for the module.

In accordance with IEC 61730-2:2004 and IEC 61215:2005 or the updated standard, the applicant is required to provide a copy of the test report and verification certificate obtained by the participating product in the certification body for review.

According to the IEC 60904-1 standard, five solar photovoltaic modules that have been pre-treated (Preconditioning Test) provided by the applicants measure the maximum power under the STC and calculate the efficiency according to the measured module frame area. Confirm whether all meet the qualification threshold: the efficiency of single crystal solar photovoltaic module is 18.3 %; the efficiency of polycrystalline solar photovoltaic module is 17.4%.

Assessment items of solar cells : comply with all

required performance and reliability regulations

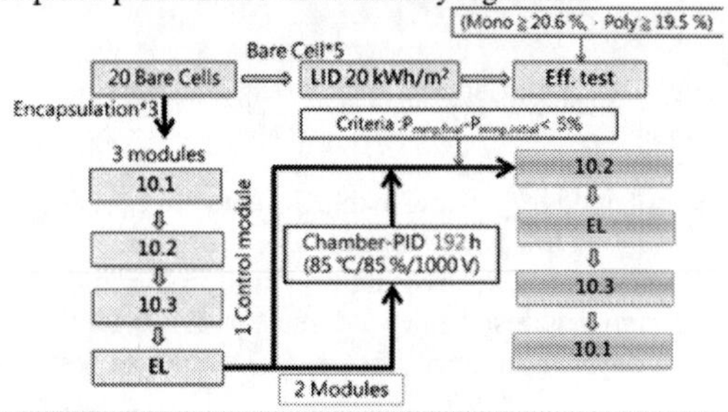

Selection project	Guideline	Request content	Qualified threshold	Mode of review
effectiveness	IEC 60904-1:2006	Efficiency after exposure of 20 kWh m-2	Single crystal germanium 20.6 %	Measured
			Polycrystalline germanium 19.5 %	
Reliability (electro-attenuation test)	IEC TS 62804-1:2015	PID 85 °C 85 %RH ±1000 V 192 hours	Power attenuation ≤ 5%	Measured

Figure 2: the test flow chats for the solar cells.

20 solar cell samples provided by the applicants, 5 were randomly selected as test samples. After exposure to 20 kWh m^2, measure the maximum power and solar cell area under STC, calculate the efficiency of 5 solar cells, and confirm whether they meet the pass threshold: single crystal solar cell efficiency ≧ 20.6 %; polycrystalline solar cell efficiency≧ 19.5%. The efficiency calculation takes the efficiency measurement of the execution unit and uses the rule of "unconditional rounding off" to take the second decimal place, and the third place is unconditionally discarded. For reliability test item, Among the 20 solar cell samples provided by the applicants, 3 samples were randomly selected as test samples. Three small modules are tested according to step 1 in the following table. After completion, take two small modules and test in step 2.

table 1: the test item for the solar cells.

step	1	2
Test items	Pre-test (10.1/10.2/10.7/10.15/EL/MST 15)	1. Test according to the following test environment and conditions: (1) Environment: constant temperature and humidity machine (2) Temperature: (85 ± 2) °C (3) Relative humidity: (85 ± 3) %RH (4) Voltage: ±1000 V (one module for positive and negative voltage) (5) Test time: 500 hours 2. Post test (10.1/10.2/10.7/10.15/EL)
determination	10.1/ 10.2/ 10.15/ MST 13	10.1/10.2/10.3/10.15

3 Results

For the module, the test results as show as the figure 3. The product code A, B, E, F were relatively sensitive to PID(+1000 V), and the degradation ratio was greater than 2.5 %. And the product code E and F were sensitive to the composite test.

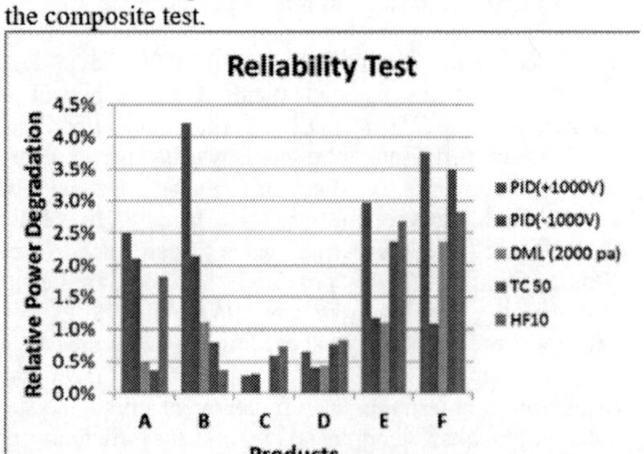

Figure 3: the relative power degradation for the modules.

For the solar cell, the test results as show as the figure 4. The product code A, B, were relatively sensitive to PID(+1000 V) more than for PID(-1000 V).

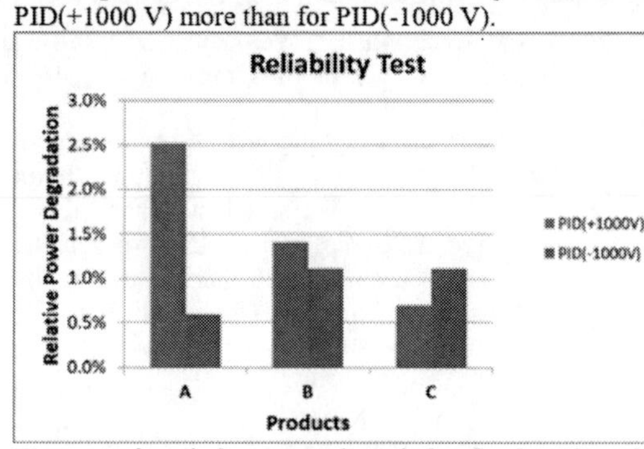

Figure 4: the relative power degradation for the solar cell.

4 Conclusions

Through this assessment program, the BOE/MOEA aims to encourage the industry to adopt tough standards which can achieve the synergy of simultaneous improvement of technology and quality, enhance industry image and boost consumer trust to spur on PV industry development in Taiwan. By holding this Excellent PV Product Assessment Program, the industry is given further incentive to manufacture higher quality products.

5 References

[1] IEC 61215-1:2016 Terrestrial photovoltaic (PV) modules - Design qualification and type approval - Part 1: Test requirements.
[2] IEC 61215-2:2016 Terrestrial photovoltaic (PV) modules - Design qualification and type approval - Part 2: Test procedures
[3] IEC 61701:2011 Salt mist corrosion testing of photovoltaic (PV) modules
[4] IEC TS 62804-1:2015 Photovoltaic (PV) modules - Test methods for the detection of potential-induced degradation - Part 1: Crystalline silicon
[5] IEC TS 62782:2016 Photovoltaic (PV) modules- Cyclic (dynamic) mechanical load testing
[6] IEC 60904-1-1:2017 Photovoltaic devices - Part 1-1: Measurement of current-voltage characteristics of multi-junction photovoltaic (PV) devices
[7] IEC 61730-2:2016 RLV Photovoltaic (PV) module safety qualification - Part 2: Requirements for testing

MULTISCALE ANALYSIS OF SILICON-BASED PHOTOVOLTAIC MODULE PERFORMANCE IN A 19 YEARS-OLD POWER PLANT

LI Min[1], LE DINH Khoa[1], Hideya OCHIAI[2], Ikusaburo KURIMOTO[3], Atsushi FUJITA[4], Yuhsuke TODA[4]
[1]Girasol Energy Inc., Hongo 7-3-1 Bunkyo-ward Tokyo Japan, li.min@pplc.co
[2]The University of Tokyo, Japan
[3]National Institution of Technology, Kisarazu College, Japan
[4]ITES Co.,Ltd.(Japan)

ABSTRACT: A typical photovoltaic power plant is composed by photovoltaic modules connected both serially and parallelly. This special architecture makes it extremely difficult to inspect performance at module-level. We are developing an intelligent, remote and automatic inspection solution to solve this problem by adopting both Internet of Things technologies and big-data analysis. In this research, we deployed a novel power line communication technology on a vintage power plant about 19 years to collect operating data on module-level. Meanwhile, performance of each module was measured through professional equipment. Through statistical analysis of dynamic operating data of each module and comparison with static performance data, we found that there was a correlation. This result indicates that we may be able to develop a big-data approach to supersede the traditional way of inspecting module performance. In addition, the performance data we obtained from this old power plant showed varying degrees of performance degradation. This difference not only displays the importance of module-level data to power plant performance analysis, but also helps to improve the traditional model of photovoltaic power plant performance prediction under the premise of uniform degradation rate.
Keywords: PV Module, Degradation, IoT, Smart Inspection, Old PV Power Plant.

1 RESEARCH PURPOSE

A typical photovoltaic power plant is composed by photovoltaic modules connected both serially and parallelly. This special architecture makes it extremely difficult to inspect performance at module-level. Currently, the module performance of a power plant is mainly detected through a combination of remote monitoring technologies such as string monitoring and field detection instruments[1][2]. However, this method is not only subject to environmental constraints, but also has high requirements for manpower and professionalism. With the development of module-level communication technologies, we have realized that there was a possibility to develop intelligent detection technologies combining the Internet of Things and big data analysis together in order to significantly improve the performance monitoring of photovoltaic power plants. In this paper, we will introduce our preliminary work on it.

2 APPROACH

We used a novel power line communication technology (PPLC-PV) to monitor a 19 years-old silicon-based photovoltaic power plant in Japan and collected operating data from some of its modules. At the same time, we used professional instrument to measure performance data such as the I-V curve of monitored module and compared it with operating data.

2.1 Field photovoltaic plant

The field photovoltaic power plant (***fig 1***) is located in Chiba Prefecture, Japan. It uses crystalline silicon modules produced by SHARP and has a nominal power of 120W. The power plant has two arrays. Onehas 15 strings and the other has 13. Each string consists of 12 modules. These 2 arrays share one inverter. The power plant began its operation since 1999 and is still in service.

Figure 1: A 19 years-old PV power plant

2.2 PPLC-PV to collect operating data

We adopted a pulse power line communication technology called PPLC-PV[3] to monitor the PV modules (***fig 2,3***) in the field. We attached sensors to every one of 48 modules. The sensors measured the module operating data and modulated the current signal into the power cable to send the data. The signals were read through a clamp current meter at the primary junction box.

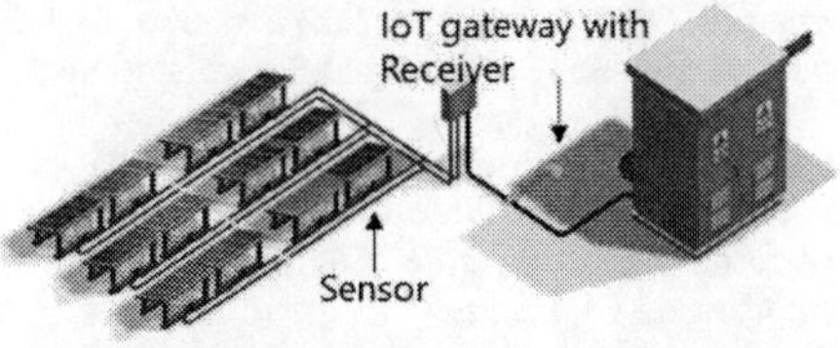

Figure 2: Image of the IoT System Architecture

Figure 3: Sensors and clamp meter

Compare with the existed module-level monitoring solution[4], PPLC-PV has less layers and is easier to be deployed.

2.3 Module performance measurement

Meanwhile, we used an I-V tracer (PV Analyzer PVA14300, Nippon Kernel) to measure the performance of monitored modules (***fig 4***).

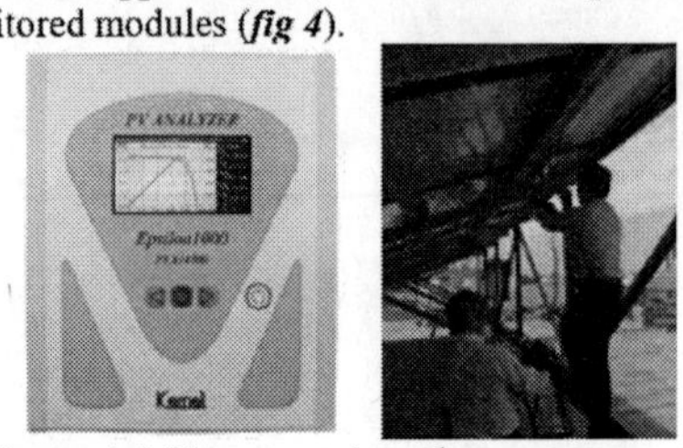

Figure 4: I-V tracer and on-site measurement

3 SCIENTIFIC INNOVATION AND RELEVANCE

3.1 Analysis of operating data on module-level

We collected the operating data such as the operating voltage, temperature etc. on each of 48 modules through PPLC-PV solution during the period from March, 2017 to May, 2018. Analysis on these data showed that modules worked at different voltage levels though in the same string. The difference of operating voltage between modules will enlarge during the daytime (***fig 5***).

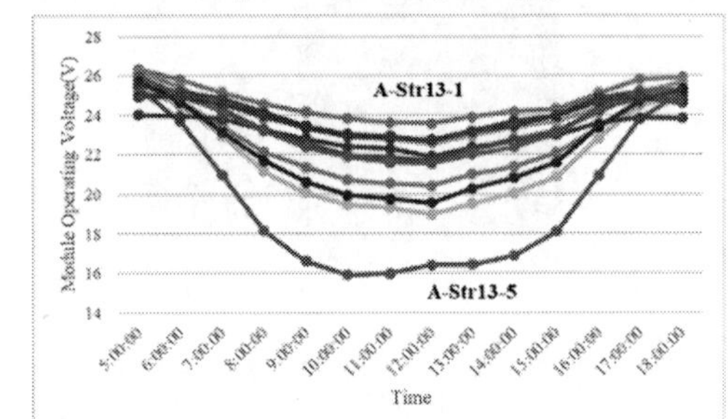

Figure 5 One of the 4 strings in tested field. Each line represents 1 module in this string.

Since the operating current in the same string should be the same, the difference in operating voltage can reflect the performance. Thus, it is easy to make assumptions that the best module in A-Str13 is No.1 and the worst is No.5.

3.2 Analysis of performance data of each module

In order to verify the results obtained in 3.1, we measured the performance data (***table I,II***) of each monitored module using an I-V tracker.

Table I: Pmax for each monitored module

Pmax(W)	#1	#2	#3	#4	#5	#6	#7	#8	#9	#10	#11	#12
A-Str15	74	70	67	71	71	75	54	94	67	84	77	86
A-Str14	67	60	62	70	64	60	78	75	82	71	70	79
A-Str13	91	67	79	72	59	75	72	75	80	68	60	72
A-Str12	83	69	72	70	80	81	72	72	51	69	78	78

Table II: Measurement results on monitored modules

	Pmax[W]	Vpm[V]	Ipm[A]	Voc[V]	Isc[A]	F.F.
Mean	72.4	21.1	3.4	29.4	4.2	0.6
Max	93.8	23.3	4.2	30.7	4.8	0.7
Min	51.3	19.6	2.5	28.8	3.2	0.4
Variation	8.7	0.8	0.4	0.4	0.4	0.1

We measured these data between 12:21 and 12:50, on May 11th, 2018, which was a sunny day. Normally a modification on the measured I-V data, such as IEC 60891[5] etc., should be made, however, we did not modify our data here. Therefore, we focused on the deviations of data rather than the changes on percentage. Considering all modules in this field plant have similar nominal performance (120W) at the very beginning, the deviations between the data indicate that the degradation of module performance is not uniform.

3.3 Comparison between operating data and performance data

By comparing both of operating voltage data and Pmax data of modules in A-Str13, we found that our assumption made in 3.1 is supported by I-V measurement result. The I-V curve of A-Str13-1 and A-Str13-5 are as of figure 6. (***fig 6***). Both of Vmp and Imp of A-Str13-5 are degraded heavily.

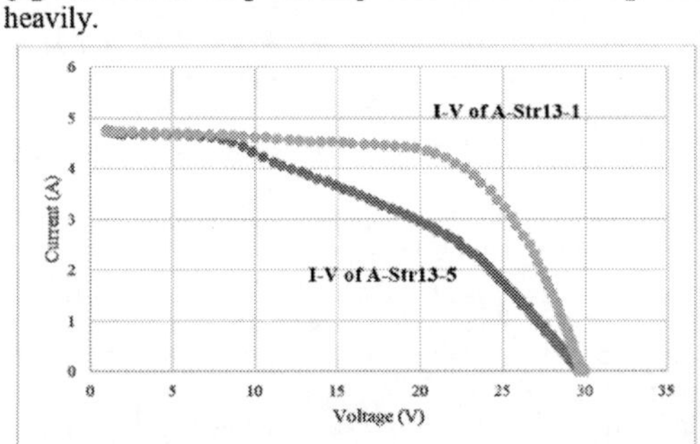

Figure 6: I-V curves of A-Str13-1 and A-Str13-5

Furthermore, we performed a simple comparison between the data obtained in 3.1 and 3.2 to verify the validity of the operating data. Correlation between the operating data and the performance data was found here (***fig 7***). This result indicates that we may be able to develop a big-data approach to supersede the traditional way of inspecting module performance.

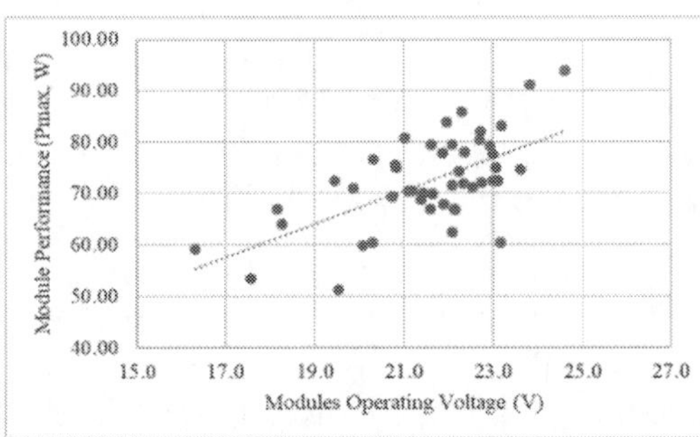

Figure 7: Correlation between operating voltage and performance

3.4 Further Considerations

In Section 3.3, we presented a way to identify module performance by analyzing operating data on modules, which demonstrated the value of module-level monitoring to the development of smart inspection applications. The current identification method is not only subject to environmental constraints, but also has high requirements for manpower and professionalism. An intelligent, remote and automatic method of module performance identification can benefit the performance maintenance of photovoltaic power plants especially in countries like

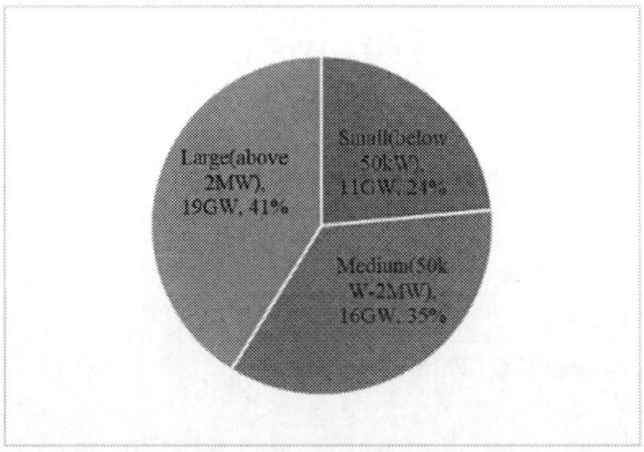

Figure 8: Amount of certificated PV plant in Japan, as in Feb 2018[6]

Japan with a great number of small-medium sized photovoltaic plants (***fig 8***) yet expensive labor.

In addition, the performance data we obtained from this old power plant is also quite valuable. Firstly, the deviations between the data indicate that the degradation of module performance is not uniform. This result may help to improve the traditionally-adopted performance prediction model of PV plants which is mainly under the premise of uniform degradation rate. Moreover, by applying the environment parameter collected from local weather station and irradiance database provided by government agency, it is possible to make a rough revaluation of module performance. We estimated the mean Pmax should be around 80W, which means modules may have lost nearly one-third of their performance in the past 19 years.

4 PRELIMINARY RESULTS

Module performance inspection is important to operation and maintenance of photovoltaic power plants. In this paper, we displayed an approach to identify module performance by analyzing its operating data. The correlation between the operating voltage data and performance data on the same module indicates that we may be able to develop a big-data approach to supersede the traditional way of inspecting module performance and improve the maintenance efficiency on PV power plants.

Besides, the performance data we obtained from this old power plant showed varying degrees of performance degradation. This result may help to improve the traditional model of photovoltaic power plant performance prediction under the premise of uniform degradation rate.

5 REFERENCES

[1]. Shashwata Chattopadhyay et al., All India Survey of Photo voltaic Module Degradation 2014: Survey Methodology and Statistics, 2015 IEEE 42nd Photovoltaic Specialist Conference (PVSC)

[2]. David A. Quansah et al., Reliability and Degradation of Solar PV Modules—Case Study of 19-Year-Old Polycrystalline Modules in Ghana, Technologies, 2017.

[3]. Hideya Ochiai and Hiroyuki Ikegami, "PPLC-PV: A Pulse Power Line Communication for Series-Connected PV Monitoring", IEEE SmartGridComm, Australia, 2016.

[4]. F Di Napoli et al., A power line communication on DC bus with photovoltaic strings, 2014, IET.

[5]. IEC. IEC 60891:2009 Photovoltaic devices—Procedures for Temperature and Irradiance Corrections to Measured I-V Characteristics; IEC: Geneva, Switzerland, 2009.

[6]. Certification List Publication, METI, Japan: (https://www.fit-portal.go.jp/PublicInfo)

[7]. Solar Radiation Database, NEDO, Japan: (http://app0.infoc.nedo.go.jp/)

[8]. Weather Database, Japan Meteorological Agency: (https://www.data.jma.go.jp/)

6 ACKNOWLEDGEMENTS

This research was partially supported by Green University of Tokyo Project (GUTP). We thank our GUTP members who provided fields and insight that greatly assisted the research.

COMPREHENSIVE STUDY OF RELIABILITY OF PHOTOVOLTAIC MODULES OF VARIOUS CONFIGURATION UNDER STATIC AND DYNAMIC MECHANICAL LOADING CONDITIONS USING FINITE ELEMENT ANALYSIS

Bharath Masetty[1, 2], Narendra Shiradkar[2], S. Patwardhan[2]
[1]Department of Mechanical Engineering, Indian Institute of Technology Kharagpur, email: bharathmasetty@iitkgp.ac.in
[2]National Centre for Photovoltaic Research and Education, Indian Institute of Technology Bombay, Powai, Mumbai – 400076, India. Ph : +91 022 2576 9420, email: naren@ee.iitb.ac.in

Keywords: PV Module Reliability, Cell Cracking, Mechanical Loading, IEC 61215.

1 INTRODUCTION

The performance and durability of photovoltaic modules are strongly affected by the physical condition of their component layers, geometry, and mounting conditions. A proper evaluation of sensitivities of each design variable would help us strengthen the key aspects that affect their mechanical durability in various end use environments. Application of finite element analysis to study the stress and temperature distributions under simple static and dynamic conditions have been suggested previously [1]-[4]. This paper presents a comprehensive study of a broad spectrum of module configurations under practical static and dynamic loading conditions. The objective is to provide comparative information about stress distributions and component failure of the modules of Glass-EVA-Cells-EVA-Backsheet and Glass-EVA-Cells-EVA-Glass configurations with varying glass and frame thickness, loaded in various mounting conditions. In case of Glass - Glass configurations, the variation in behaviour with and without frame has also been studied. Finite element analysis software ANSYS has been used for this purpose. A 72 cells (6 x 12) module with approximate dimensions of 0.9 m x 1.9 m has been chosen for the analysis. In order to reduce the computational complexity, a simplified model is made by eliminating screen-printed layers of the cells. Since our geometry is symmetrical both along and across the geometry, a quarter module - as shown in figure 1(b) - has been used by incorporating the symmetry boundary conditions on the cut surfaces. This allows us to use a finer mesh and obtain more accurate results in a given time. The geometry consists of cells embedded in a thin layer of EVA with Glass and Backsheet (or glass layer in case of glass-glass modules) bonded on front and back respectively as illustrated in Figure 1(a). The material parameters of the all the component layers are presented in Table I.

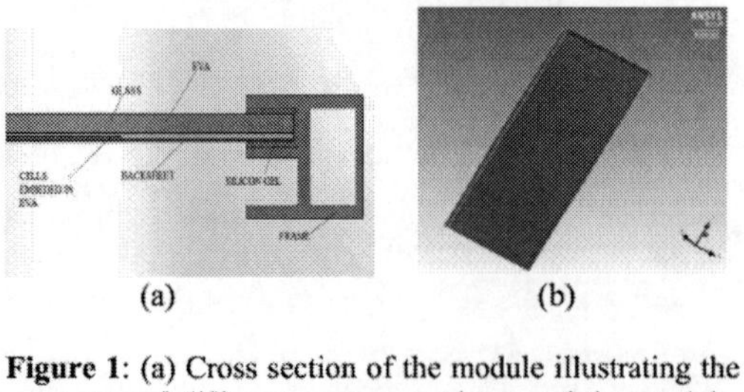

(a) (b)

Figure 1: (a) Cross section of the module illustrating the geometry of different component layers of the module. (b) Simplified Quarter Assembly.

2 LOADS AND GEOMETRY

2.1 Loading Conditions

1. IEC 61215 Mechanical Load test
2. Static load equivalent to a person standing on the module.
3. Modal analysis and a harmonic uniform load test (1000 Pa and 2400 Pa).

Table I: Physical properties of the materials.

Material	Elastic Modulus (Pa)	Poisson Ratio	Density (g.cm^-3)
Glass	7.20E+10	0.23	2.50
Silicon	1.60E+11	0.28	2.33
EVA	6.55E+07	0.15	0.95
Frame(Al)	7.00E+10	0.33	2.70
Silicon Sealant	1.00E+09	0.33	1.28
Backsheet	2.14E+09	0.40	1.50

2.2 Geometry

A total of five different module configurations have been selected which consists of three Glass-Backsheet type geometries with varying glass thickness (3.2 mm (3.2G-B), 2.6 mm (2.6G-B) and 2.0 mm (2G-B)), and two Glass-Glass type geometries with varying front glass thickness (1 mm (1G-2G), 2.0 mm (2G-2G) were considered. The rear glass thickness in both the 1G-2G and 2G-2G is maintained a constant 2.0 mm. Figure 2 illustrates all the mounting configurations used for various studies presented in this paper. The locations that highlighted in the blue on the rear side of the quarter geometry shown are the clamping point used to simulate the corresponding mounting conditions.

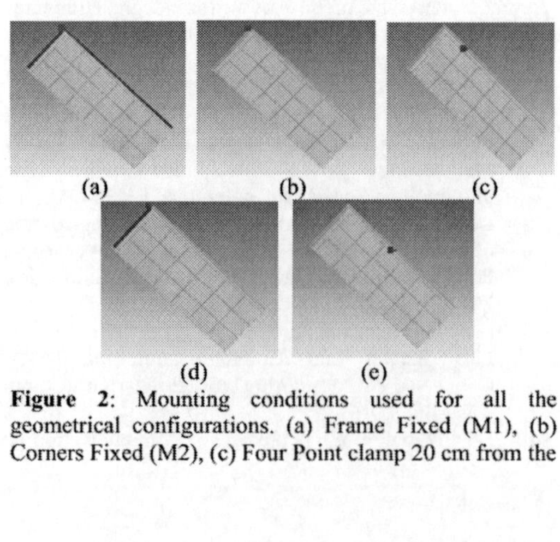

Figure 2: Mounting conditions used for all the geometrical configurations. (a) Frame Fixed (M1), (b) Corners Fixed (M2), (c) Four Point clamp 20 cm from the

top (M3), (d) Top and Bottom Frame fixed (M5), (e) Four Point clamp 40 cm from the top (M4).

3 SOLVER METHODOLOGY

Finite element analysis (FEA) is a popular numerical method used to obtain solutions for the boundary value problems corresponding to complicated domains (geometries) which are formulated by using the governing equations and boundary conditions. The basic principle of FEA is to initially divide the entire domain into finite elements of very small size and reduce the problem to the determination of solution variables like special displacements, stress states etc. only at the nodes (corners) of these small elements. Then, for a given element, solutions at all the points are obtained through polynomial extrapolation of the nodal solutions of that particular element. So necessarily, we are solving the problem assuming that the solution function is a piecewise polynomial function defined separately for each element in the geometry.

In the current study, two types of analyses were conducted, Static structural analysis and Dynamic analysis. In the static analysis, the governing equations are three-dimensional equilibrium equations. The boundary value problems are formulated by using these equations along with three-dimensional elasticity equations and strain-displacement relations. By using the piece-wise polynomial assumption of solution and integration by parts, the governing equations are converted into a set of simple algebraic equations which can be solved by using techniques like Matrix inversion method and Newton Raphson method depending on the nature of the equations obtained. Once the values of solutions at nodal points are evaluated, results at all other points and contours of stress and displacements can be easily obtained.

In the Dynamic analysis, though the governing equations and steps involved are different, the concept of discretization of the domain into small elements and piece-wise polynomial approximation of the solution remains the same. Initially, a modal analysis was performed which involves solving of frequency equations obtained by assuming that nodal displacements are harmonic in free vibration. The purpose of this was to evaluate the undamped natural frequencies of the system whose information is required to conduct the Dynamic study. The governing equation for the dynamic analysis can be obtained by using Newton's second law of motion. More detailed explanation of the methods used for the current study can be found in [5] and [6].

4 RESULTS OF STATIC ANALYSIS

4.1 IEC 61215 Uniform Static Load

A static uniform load of 2400 Pa and 5400 Pa has been applied on the front side of the module geometries. Figure 3 illustrates the stress distributions in cells and glass are shown for the 3.2G-B module in M4 mounting condition. Figure 3(a) is the rear side of the quarter module model, the region in red where the stresses are higher is the central region of the module. Since it would be cumbersome and difficult to compare the contours of all the combinations analyzed, the plot showing the variation of peak values of maximum principal stress in cells is presented in Figure 4. It was observed that for a given module, the distributions of maximum principal stress in the component layers are significantly different for different mounting conditions. However, an intuitive pattern that could be observed from these contours was that maximum stresses in cells are observed only in the central region of the module irrespective of the mounting configuration and stresses increased with increasing deformation. This can be confirmed by correlating the deformation values presented in Table II with the peak stresses presented in Figure 4(a). So, it can be verified that with stresses in cells are directly dependent on deformation of the module. The mounting conditions that result in smaller deformations are preferable for avoiding the failure of cells. Since it is a uniform loading test, maximum deformations were observed at the center of the module for all the combinations. In the glass, greater stress concentration is observed at the mounting locations. Silicon cell generally has its tensile strength around 200 MPa and glass has a value of around 180 MPa. It can be seen from Fig.4(a) that stresses in cells are below their tensile strength and do not fail under 2400 Pa load in any of the mounting configurations

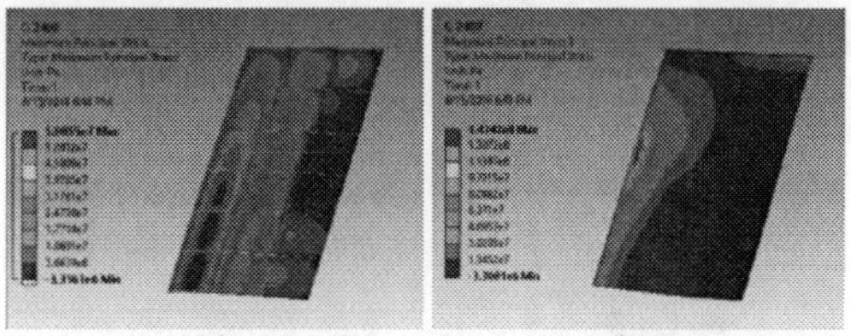

(a) (b)

Figure 3: Contours of Maximum principle stress in (a) Cells and (b) Glass under 2400 Pa uniform static load in M4 Mounting condition.

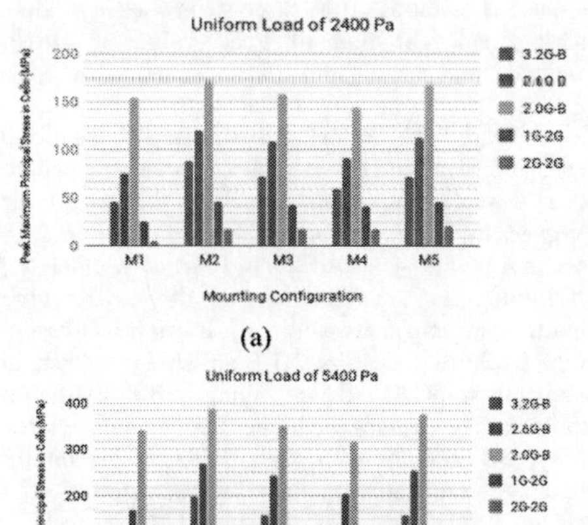

(a)

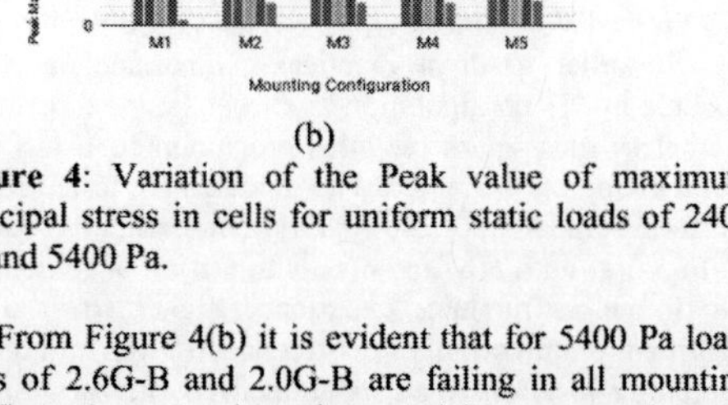

(b)

Figure 4: Variation of the Peak value of maximum principal stress in cells for uniform static loads of 2400 Pa and 5400 Pa.

From Figure 4(b) it is evident that for 5400 Pa load, cells of 2.6G-B and 2.0G-B are failing in all mounting configurations. In glass-glass modules, stresses in cells are well below the failure limit and very less as compared to the glass-backsheet modules for both the loads. Stresses in M1 and M4 are lower compared to other mounting configurations analyzed. The glass-glass modules considered for this analysis are framed, analysis of frameless modules is presented in further sections.

Table II: Maximum deformation(cm) of the module observed for all the tested combinations under 2400 Pa uniform static load.

Mounting Condition	3.2G-B	2.6G-B	2.0G-B	2G-2G	2G-1G
M1	2.61	4.2	7.1	0.92	1.73
M2	10.47	13.76	19.54	6.53	8.92
M3	7.34	9.77	14.64	3.65	5.63
M4	4.78	6.69	11.35	2.14	3.49
M5	9.91	12.83	18.79	5.4	7.84

4.2 Static load equivalent to a person standing on the module.

In order to simulate the condition of a person walking on the module, a localized load equivalent to 80 Kgf has been applied at three different locations, not at the same time but one after the other at regions highlighted in Figure 5. Only the M4 mounting configuration has been selected for the current analysis since it is most commonly employed.

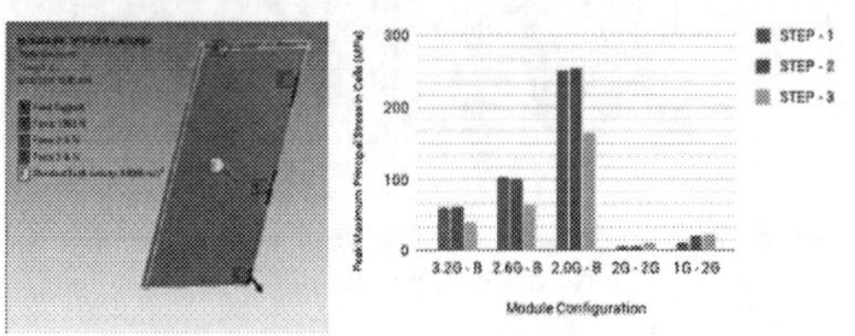

(a) (b)

Figure 5: Locations of the three steps chosen to simulate walking and Variation of Peak value of maximum principal stress in cells for different steps.

In Figure 5(a), let location B is step-1, location C is step-2 and location D is step 3. The area over which the force is applied is 75cm^2 which is the area typically occupied by an adult foot. The application points are separated by a distance of 25cm from each other. Figure 5(b) illustrates the variation of the peak value of maximum principal stress in cells. It can be observed that all the modules except for 2G-B are able to withstand the load. Cells in 2G-B fails are failing in the first two steps itself since the stresses are higher than its tensile strength. So, typical modules with glass thickness greater than 2 mm are theoretically invulnerable to a person walking over it.

4.3 Framed Vs Frameless Glass - Glass Modules

In order to do a practical comparison, a 2G-2G module in M3 configuration is chosen, since most of the frameless glass-glass modules are mounted using four-point clamping technique. The modules are tested for the uniform static load of 2400 Pa. By comparing the contour in Fig. 6(a) with 6(b) and 6(c) with 6(d), it is evident that the frameless modules experience higher stress and a significant amount of stress concentration around the mounting locations than its framed counterpart. Such a behaviour is observed in frameless modules because the mounting points are the only locations available to transfer the load to the roof or ground, whereas in framed modules, the laminate is bonded on all four sides to the frame which provides a greater area to transfer the load to the frame which in turn transfers it to the ground or roof. For the current choice of load, the cells are safe but the glass in frameless 2G-2G module fails.

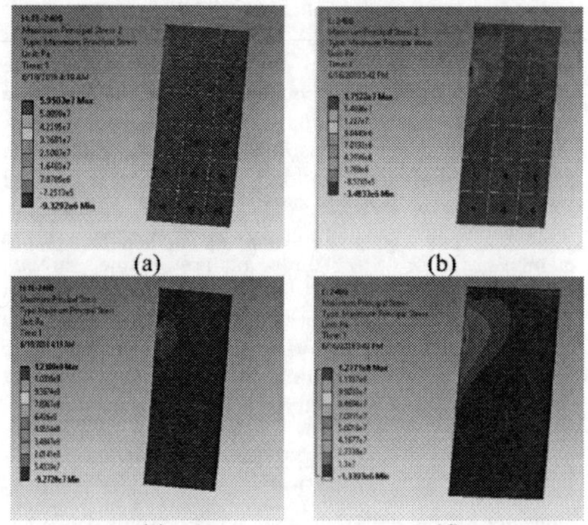

(a) (b) (c) (d)

Figure 6: (a) & (c) maximum principal stress contours in cells and glasses of frameless 2G-2G module respectively. (b) & (d) are maximum principal stress contours in cells and glasses of framed 2G-2G module respectively.

4.4 Effect of Frame Thickness

In this section 3.2G-B module in M4 mounting condition under 2400 Pa uniform, static load is analyzed for three different values of frame thickness (2 cm, 3.0 cm, and 3.5 cm). Figure 7 shows the contours of maximum principal stresses in glass for the three combinations. By looking at the contours, it can be easily understood that as the frame thickness increases, the stresses become more distributed which results in a decrease of the peak value of maximum principal stresses from 7(a) to 7(c).

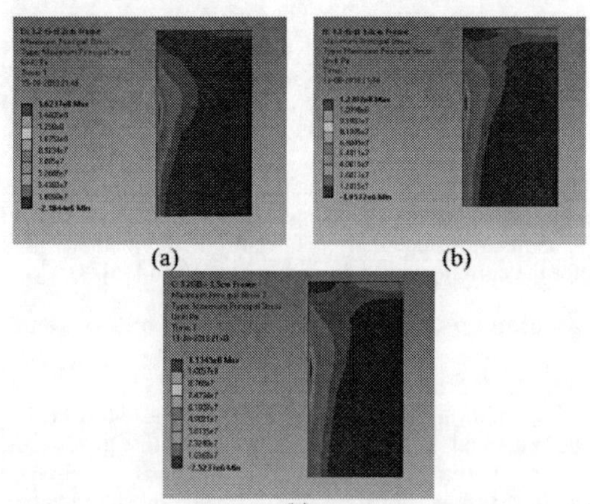

(a) (b) (c)

Figure 7: Maximum principal stress distribution in the glass of 3.2 G – B with a frame thickness of (a) 2 cm, (b) 3 cm and (c) 3.5 cm under 2400 Pa uniform static load.

The reason for this pattern is that with an increase in frame thickness, more amount of elastic matter is available between the mounting point and the module stack which could bear the load which results in the decrease of stress concentration at the mounting points. Upon further extrapolation of this phenomenon, it can be intuitively hypothesized that after increasing the frame thickness to a certain point, the stresses in all the component layers should become independent of mounting point and the results should be close to the results obtained in M1 (frame fixed) mounting condition. To validate this, the results of 4 different combinations of 3.2 G-B are compared in Table III.

Table III: The peak stresses in cells and glass along with the maximum deformation values of four different combinations of 3.2 G-B.

Frame Thickness (cm)	Mounting Condition	Max. Stress in Glass (MPa)	Peak Stress in Cells (MPa)	Maximum Deformation (cm)
3.0	M4	123.97	111.59	3.59
3.5	M4	113.45	106.13	3.19
3.5	M1	104.48	86.78	2.12
2.0	M1	107.16	85.183	2.08

The comparison presented in Table III supports our extrapolation because the stresses and deformation decreased with the frame thickness increased from 3.0 cm to 3.5 cm and can be seen approaching towards the values obtained for 3.5 mm framed module in M1 configuration. Furthermore, if our hypothesis is true, the effect of variation of frame thickness should be minimal in case of M1 mounting condition. It can be confirmed by the comparison of results of modules with 2.0 cm and 3.5 cm in the M1 condition which are almost the same.

5 RESULTS OF DYNAMIC ANALYSIS

5.1 Modal Analysis

The values of natural frequencies shown in Table IV are obtained by conducting a modal analysis which is the first step to conduct any type of dynamic load test. It involves solving of frequency equations obtained from the equations of motion for free vibration.

Table IV: Undamped natural frequencies of principal mode of vibration for various module and mounting configurations.

Mounting Condition	3.2 G - B	2.6 G - B	2.0 G - B	2G - 2G	1G -2G
M1	18.955	15.923	13.715	29.53	24.15
M2	9.112	8.7601	8.225	10.543	10.406
M3	11.391	10.558	9.576	14.701	13.531
M4	14.091	12.733	10.976	19.226	17.217
M5	9.593	9.093	8.404	11.583	11.114
M3 (Frameless)	-	-	-	6.595	5.416
M4 (Frameless)	-	-	-	9.576	7.881

For a particular module configuration, M1 has the highest natural frequency and M2 has the least. Natural frequency also depends on the mass of the geometry and increases and the module mass increases. For example, for a particular mounting configuration, the frequency increases with increasing glass thickness. It also explains the reason why glass-glass modules have relatively higher natural frequency than glass-backsheet modules for a given combination. It is because of the same fact that the frequencies of frameless modules are less compared to others. Higher is the value of natural frequency, safer is the module because the probability of occurrence of resonance due to external dynamic load in practical conditions decreases as the natural frequency of the module increases.

5.2 Harmonic Load at Principle Frequencies

In this section, the modules are subjected to uniform harmonic load of 1000 Pa and 2400 Pa at the corresponding principal (1st mode) natural frequencies of all the combinations. The effects of damping were also considered. From Figure 5(a), it is evident that cells in all the module and mounting configurations do not show any failure for 1000 Pa Harmonic load since the stresses are well below the failure limit. In case of 2400 Pa Harmonic load, from figure 5(b), it can be observed that stresses in cells of all the combinations except 2G-B in M2 configuration have their stresses well below the failure limit. When we compare these results to of static analysis, higher stresses and failure is mainly in combinations with higher natural frequency. This dependence can be identified from the mathematical model related to dynamic analysis, as described in [5]. The data shown in fig.5 does not follow a similar trend as that of corresponding static analysis because the analysis is carried out for each configuration at its natural frequency, whose values are illustrated in Table IV.

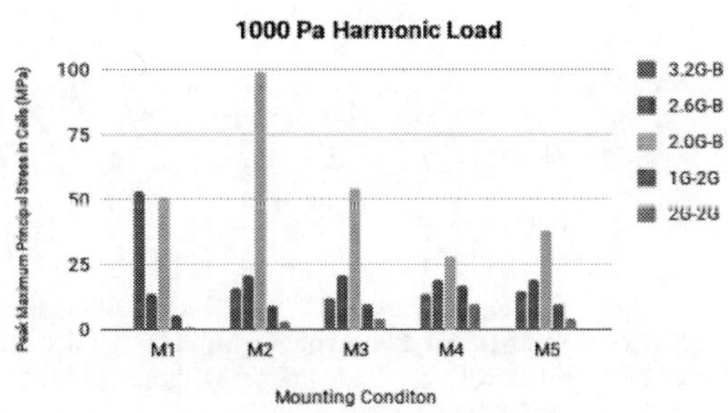

(a)

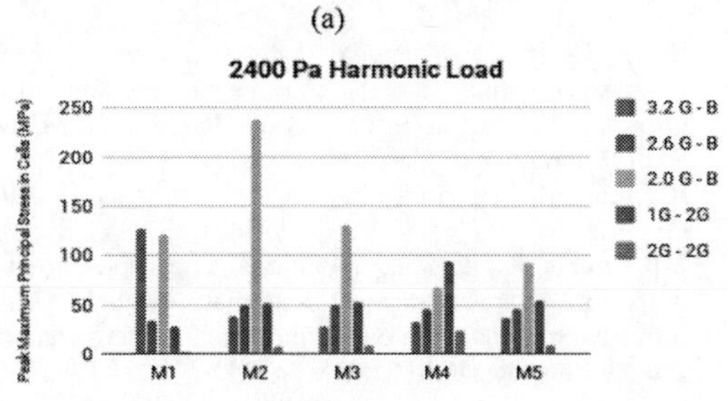

(b)

Figure 8: Variation of the Peak value of maximum principal stress in cells and glass under harmonic loads of 1000 Pa and 2400 Pa at corresponding principal modes of vibration considering the effects of damping.

6 CONCLUSION

Firstly, in the studies related to static loading conditions, 3.2G-B and 2.6G-B are able to sustain most of the combinations related to uniform 2400 Pa load and load caused by a person standing. 2G-B can be seen failing in significantly more number of combinations when compared to other module configurations. It can also be observed that modules in M1 and M4 mounting conditions exhibit greater reliability and M2 configuration exhibit failure in more number of combinations. Secondly, for a given module type, the amount of glass that is used to manufacture 3.2G-B modules is greater than the amount used to manufacture 1G-2G module but it can be observed from the above results that 1G-2G clearly performs better than 3.2G-B type module. Even though minimum thickness of commercially available tempered glass currently is limited to 2 mm, this result provides an insight and incentive to develop 1 mm thick tempered front glass for PV modules. So In order to increase the mechanical reliability of the module, a glass-glass module of 1G-2G type would be a better choice over thicker Glass-Backsheet type module like 3.2G-B. In the results related to dynamic load tests, combinations having higher natural frequency exhibit greater stresses than others. It is evident from the fact that 3.2G-B module is able to withstand most of the static conditions but shows completely opposite behaviour in dynamic situations because of its high natural frequency. Finally, a particular module is expected to exhibit greater reliability in M1 and M4 mounting configurations and modules relatively thinner glass on both the sides could be preferred over Glass-Backsheet type with thicker glass. Also, framed Glass - Glass modules outperform all other combinations in almost all the studies. So it can be said that these modules would offer higher mechanical durability than others, along with low moisture ingress due to Glass-Glass packaging.

REFERENCES

[1] Yixian Lee, Andrew A.O. Tay, Stress Analysis of Silicon Wafer-Based Photovoltaic Modules Under IEC 61215 Mechanical Load Test, Energy Procedia, Vol. 33 2013, pp. 265-271.

[2] J. Dong, H. Yang, X. Lu, H. Zhang, J. Peng, Comparative study on static and dynamic analyses of an ultra-thin double-glazing PV module based on FEM, The 7 th International Conference on Applied Energy – ICAE2015, Energy Procedia, Vol. 75 2015, pp. 343–348.

[3] Artūras Kilikevičius , Audrius Čereška , Kristina Kilikevičienė, Analysis of external dynamic loads influence to photovoltaic module structural performance, Energy Procedia, Vol.75 2015, pp. 343-348.

[4] Yixian Lee, Andrew A.O. Tay, Finite Element Thermal Analysis of a Solar Photovoltaic Module, Energy Procedia, Vol 15 2012, pp. 413-420.

[5] Rao, S. S. Mechanical vibrations / Singiresu S. Rao. 5th ed. ISBN 978-0-13-212819-3.

[6] An Introduction to the Finite Element Method, Reddy, J. N. ISBN 0-07-051355-4.

FROM BIFACIALITY TO YIELD: DIFFERENT BIFACIAL CELL TECHNOLOGIES MAY DIFFER EVEN MORE IN ANNUAL OUTDOOR PERFORMANCE

Christian Reise, Ghassan Baarah, Erdmut Schnabel, Ulli Kräling, Björn Müller
Fraunhofer Institute for Solar Energy Systems ISE
Heidenhofstraße 2, 79110 Freiburg, Germany
Phone +49 761 4588 5282 christian.reise@ise.fraunhofer.de

Sungho Chang, Youngho Choe, Hyunshik Cho
LG Electronics, 168, Suchul-daero, Gumi-si 730-903, Korea
Phone +82 54 469 0712 jason.cho@lge.com

ABSTRACT: A comparative outdoor test of different bifacial P-PERC and N-PERT PV modules was designed and operated in Freiburg, Germany. According to LG Electronic's specifications, the chosen mounting geometry and albedo deliver a high bifacial gain. While this is not comparable to typical commercial PV systems, it emphasizes differences in the bifaciality and the bifacial gain of the module samples.
A greater bifacial gain was expected already for the N-PERT modules. However, the measurements revealed more differences in annual and in high resolution performance data. Laboratory measurements do confirm the observed outdoor behavior, given they are performed under bifacial conditions as well.
Keywords: Bifacial, Energy Performance, Monitoring

INTRODUCTION

Bifacial PV modules are able to convert solar irradiation from both surfaces. Therefore, a larger energy production—compared to standard monofacial modules—is expected. The additional energy delivered by bifacial PV modules is commonly called bifacial gain (BG). It is influenced both by PV system properties and by properties of the PV module itself, especially by the solar cell technology.

End users may distinguish different cell technologies by different conversion efficiencies and bifaciality factors, but the difference in annual energy production might be bigger than expected from these two figures. In this contribution, we compare the performance of different bifacial modules made from P-PERC and N-PERT cells. Several module properties like temperature dependency or LID are affected by the choice of cell technology. Here, we concentrate on differences in efficiency and bifaciality and their dependence on the irradiance level.

A comparative outdoor test of different PV modules was designed and operated for one full year in Freiburg, Germany. According to LG Electronic's specifications, the chosen mounting geometry and albedo deliver a high bifacial gain [1]. While this is not comparable to typical commercial systems, it emphasizes differences in the bifaciality and the bifacial gain of the module samples.

1 BIFACIAL GAIN & BIFACIALITY

The fraction of irradiance received at the rear surface of the module G_{back} may be expressed as optical bifacial gain BG_{opt} (which is a system or geometry property):

$$BG_{opt} = \frac{G_{back}}{G_{front}}$$

In opposite to the front side of the module, the availability of rear surface irradiance depends on a larger number of factors as

- o the mounting geometry (module height, module tilt angle, row-to-row distance)
- o the ground albedo and its homogeneity
- o the mounting structure

Beside this, the rear side irradiance may vary during the day and the year in a manner quite different from the front side irradiance.

The rear side electricity production of a bifacial module is roughly proportional to the optical gain, but will be reduced by the bifaciality factor φ. This factor is a module property and usually defined as ratio of STC power or efficiency values:

$$\varphi = \frac{P_{back,stc}}{P_{front,stc}} = \frac{\eta_{back,stc}}{\eta_{front,stc}}$$

So the bifacial gain of the module may be estimated as:

$$BG_{mod} = \frac{P_{back}}{P_{front}} = \frac{G_{back}\,\eta_{back}}{G_{front}\,\eta_{front}}$$

$$= BG_{opt}\,\varphi$$

Values of φ differ remarkably between p- and n-type cells and can be measured quite accurate in the laboratory. For this purpose, new standards for bifacial PV module characterization are under development. However, questions remain: Is this simple relation sufficient? Is φ a constant?

This is the main topic of our contribution, based on a full year of measured outdoor module performance data and a complementary indoor characterization, performed for different products on two different undergrounds.

From measured values of outdoor module performance, φ may be derived from values of BG_{mod} and BG_{opt}:

$$\varphi = \frac{BG_{mod}}{BG_{opt}}$$

BG_{opt} may be measured by a pair of irradiance sensors, mounted back to back in plane of array. BG_{mod} may only derived using an additional monofacial reference module:

$$BG_{mod} = \frac{P_{back}}{P_{front}} = \frac{P_{bifacial} - P_{monofacial}}{P_{monofacial}}$$

2 OUTDOOR MEASUREMENTS

The outdoor test setup with different bifacial and monofacial module samples was installed on the roof of one of Fraunhofer ISE's buildings in Freiburg, Germany, see Figure 1. The symmetrical setup was operated for one full year. IV curve and power measurements were performed on all modules in regular intervals of 5 min and 1 min. The albedo of the membrane was measured several times and reached values between 75% and 80%. For the given geometry, this lead to an average irradiation gain of 35%.

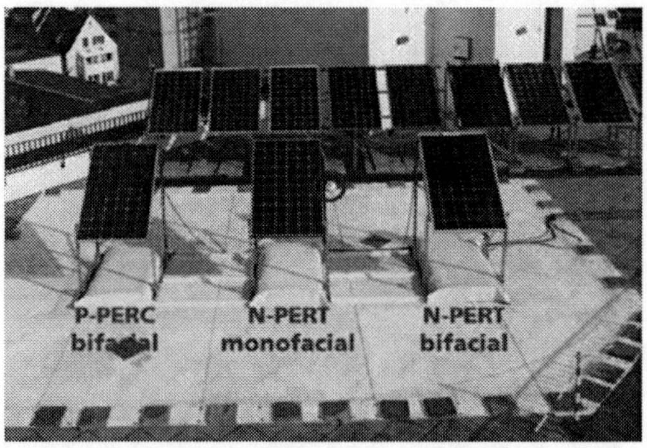

Figure 1: The 3 modules under consideration are mounted in the front row above a bright roofing membrane. For symmetry reasons, the two bifacial modules are placed on the outer positions, while the monofacial reference module is placed in the center. On the right edge of the reference module, two pyranometers are mounted back to back.

2.1 Annual outdoor results

The average annual optical gain BG_{opt} measured by the two pyranometers is close to 35%. From this optical gain and values of φ from laboratory tests, a module bifacial gain $BG_{mod,est}$ may be estimated to values of some 30% for the N-PERT module and some 20% for the P-PERC module. Table 1 compares these estimates to observed $BG_{mod,obs}$. It is seen that both modules miss the expected level of bifacial gain. The N-PERT module shows a higher bifaciality and is closer to the expected bifacial gain than the P-PERC module.

Table 1: Annual outdoor results

	BG_{opt} %	Φ %	BG_{mod} est. %	BG_{mod} obs. %	**obs. / est. %**
P-PERC	34.7	58.2	20.2	16.5	**81.7**
N-PERT	34.7	87.4	30.3	26.5	**87.2**

2.2 High resolution outdoor results

To investigate these deviations in more depth, the measured data is used in its original time step. Figures 3 and 5 depict the dependency of BG values on front side irradiance. While annual average BG_{opt} is close to 35%, instantaneous values vary from 20% to 65%. BG_{opt} clearly depends on the mounting geometry and on the actual position of the sun for a given point in time. Consequently, also BG_{mod} varies between 10% and 35% for the P-PERC module and between 15% and 45% for the N-PERT module.

According to the formulae presented in Section 1, φ may be derived from individual values of BG_{mod} and BG_{opt}. Despite φ is assumed to be a module property, it is obviously not constant. The plots in Figures 3 and 5 show values of φ in a range from 35% to 60% for the P-PERC module and from 60% to 90% for the N-PERT module.

With the P-PERC module, a clear trend is visible. Stable values of φ are seen above some 100 W/m², and a decrease in φ is observed from about 500 W/m² onwards. The situation is less clear with the N-PERT module, however, φ is on a remarkable higher level here.

3 LABORATORY MEASUREMENTS

A simple approach to determine φ in the laboratory utilizes two single sided measurements in a solar simulator. For a given irradiance level G, single sided power or efficiency values are determined while the other module surface is covered. Figures 4 and 6 present such measurements on the two bifacial modules. The upper graphs depict the efficiencies, the red curve in the lower plots gives the resulting φ(G) as ratio of both efficiencies:

$$\varphi(G) = \frac{P_{back}(G)}{P_{front}(G)} = \frac{\eta_{back}(G)}{\eta_{front}(G)}$$

Compared to the trend visible in the measured data (bifaciality decreases with irradiance), the trend seen in the single sided φ(G) curves (bifaciality increases with irradiance) follows the wrong direction. Obviously, the singled sided measurements do not represent the real operating conditions under bifacial irradiance.

This problem may be solved through measurements of φ(G) under bifacial irradiance conditions. Fraunhofer ISE operates a solar simulator which is able to measure full size bifacial modules under bifacial irradiance [2]. The light of a large pulsed solar simulator is reflected by a pair of mirrors to both sides of the test sample. The overall irradiance level may be controlled at the light source, while the back irradiance may be reduced using filter meshes. Figure 2 presents a number of combinations of front and back irradiance used for this investigation.

The results of such bifacial measurements are presented in Figures 4 and 6 as well. In contrast to the single sided measurements, now bifaciality decreases with irradiance. This is consistent with the trend seen in the outdoor data, however, deviations remain regarding the absolute level.

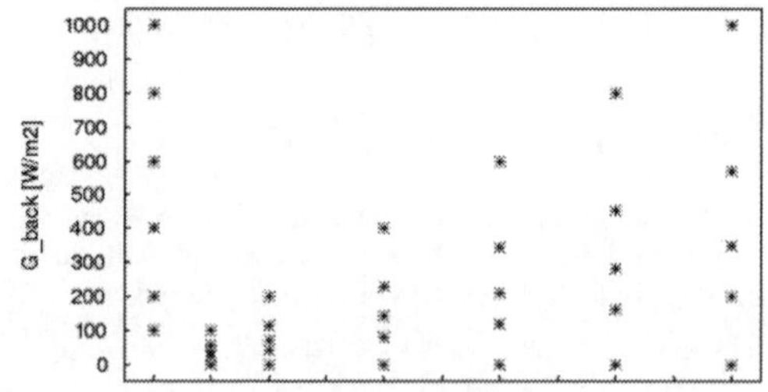

Figure 2: Possible combinations of G_{front} and G_{back} in Fraunhofer ISE's bifacial pulsed solar simulator.

P-PERC

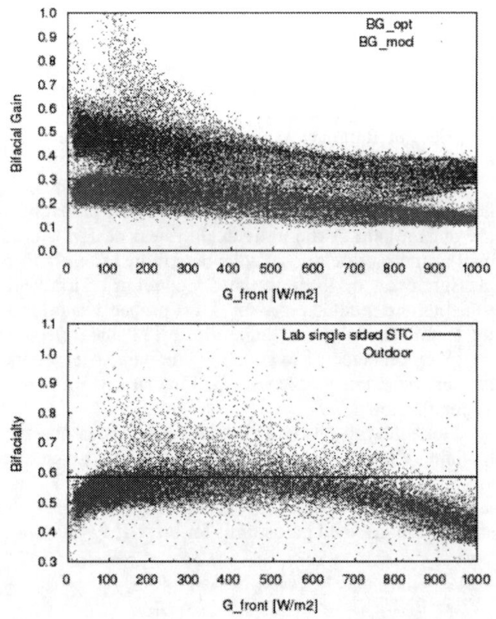

Figure 3: Top: bifacial gain values, red dots show BG_{opt}, green dots show BG_{mod}; bottom: resulting bifacialty φ for the P-PERC module.

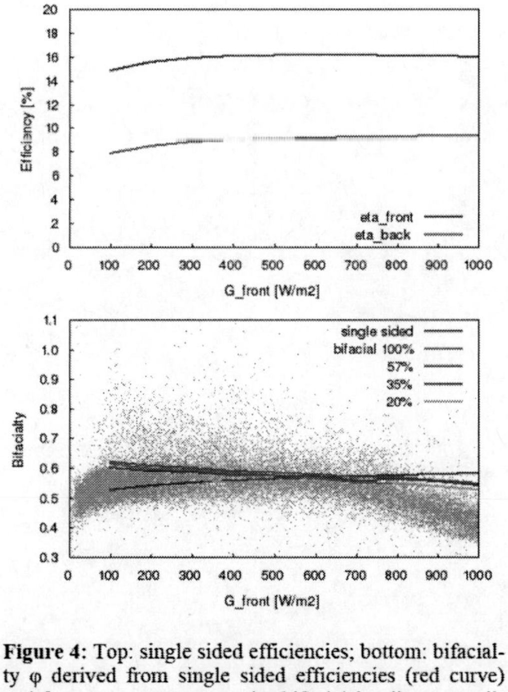

Figure 4: Top: single sided efficiencies; bottom: bifacialty φ derived from single sided efficiencies (red curve) and from measurements under bifacial irradiance conditions (all other curves) for the P-PERC module.

N-PERT

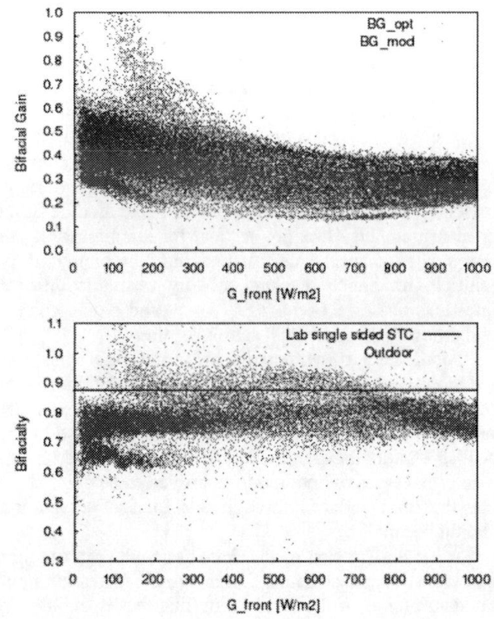

Figure 5: Top: bifacial gain values, red dots show BG_{opt}, green dots show BG_{mod}; bottom: resulting bifacialty φ for the N-PERT module.

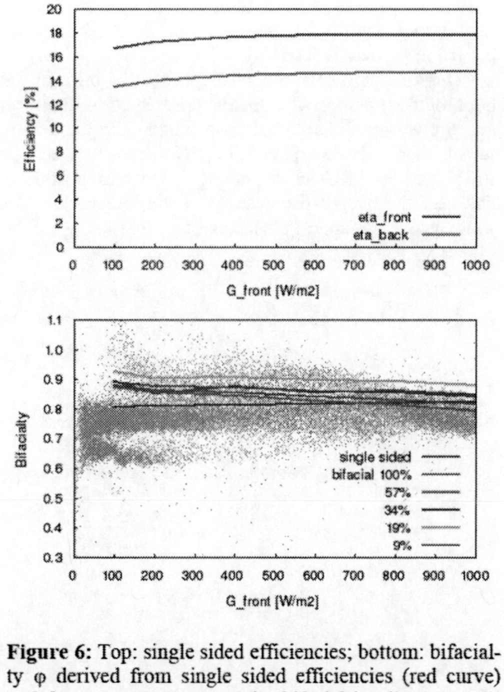

Figure 6: Top: single sided efficiencies; bottom: bifacialty φ derived from single sided efficiencies (red curve) and from measurements under bifacial irradiance conditions (all other curves) for the N-PERT module.

4 DISCUSSION

In the straight forward indoor–outdoor comparison shown in Figures 4 and 6, some differences remain between the gray dots (φ from test site) and the purple lines (φ from bifacial solar simulator measurement with the rear side irradiance set to some 35% of the front side level). There are several contributions to these deviations.

The comparison of optical bifacial gain, determined by two single spot pyranometer measurements, to module bifacial gain, determined from two full size module power measurements, does not account for inhomogeneity in rear side irradiance. This situation could be improved by multiple irradiance sensors, possibly raising additional questions on correct sensor placement and on the correct weighting of the individual sensor readings.

Also, the test setup suffered from some space restrictions and, in consequence, from the installation of only one monofacial reference module. Therefore, the bifacial output of the P-PERC module was referred to the N-PERT monofacial module. As the normalized front side efficiencies of both modules proved to be quite similar, the error induced through this kind of evaluation should be small.

A certain fraction of the deviations between outdoor and laboratory measurements is due to the different angular distribution of the incident irradiance. Especially on the rear side, the outdoor light comes in at rather large angles of incidence, while the light incidence in the simulator is close to normal – leading to lower reflection losses.

Finally, the comparison of bifacial module output power to a monofacial reference module power might need a precise correction for the module temperatures. Differing operating temperatures caused by different optical properties of the modules may distort the simple approach described in Section 1.

Despite all mentioned challenges of a bifacial indoor–outdoor comparison, the data presented in Figures 4 and 6 give some confidence in understanding the behavior of bifacial PV modules. The differences between N-PERT and P-PERC cell technology are well reproduced, and also the overall dependency of the bifaciality φ on irradiance levels appears in a consistent way.

CONCLUSION

A greater bifacial gain was expected already for the N-PERT modules. Both indoor and outdoor measurements confirmed the higher absolute efficiency and the higher bifaciality of the N-PERT product. However, the measurements revealed some issues which are of importance for future yield predictions of bifacial PV systems.

The bifaciality factor φ is not a constant value, but depends on the overall irradiance level and probably on the optical gain of the individual system design. An annual average value of φ may be determined from outdoor measurements or from accurate modelling of irradiance statistics and module behavior. Most probably, a bifaciality factor derived from single sided STC measurements (as given on some of today's module data sheets) is not the best estimator for the realistic bifacial performance of a specific module.

Laboratory measurements do confirm the observed module behavior under outdoor conditions, given they are performed under bifacial conditions as well. Such measurements provide the information necessary for state-of-the-art yield predictions for bifacial PV systems.

REFERENCES

[1] S. Chang, Y. Choe, H. Cho, H.J. Park, Ch. Reise, G. Baarah, E. Schnabel, U. Kräling, B, Müller: Annual outdoor performance of LG bifacial modules. 5th bifiPV Workshop 2018, Denver, CO.

[2] A. Schmid, G. Dülger, G. Baraah, U. Kräling: IV measurement of bifacial modules: bifacial vs. monofacial illumination. Proc. 33rd EU PVSEC 2017, Amsterdam.

WATER DIFFUSION SIMULATION OF THE PHOTOVOLTAIC LAMINATE BASED ON GRAVIMETRIC MEASUREMENT OF PACKAGING MATERIALS

Akhilesh Dadaniya and Naresh Varma Datla
Department of Mechanical Engineering, Indian Institute of Technology Delhi -110016, India
a.dadaniya@ mech.iitd.ac.in, datla@mech.iitd.ac.in

ABSTRACT: Photovoltaic modules (PV) are expected to be durable against exposure to outdoor environments for prolonged time durations. The dynamic outdoor environment is in direct contact with the polymeric back sheet of the module that leads to frequent moisture absorption and desorption within the module. Considering these varying environmental conditions as well as the multilayer nature of the packaging materials, it is challenging to predict the water concentration accurately and thereby mechanical durability of the module (back sheet and encapsulation). Therefore, there is a need for systematic multilayer experimental and computational approach for water diffusion prediction at each layer of the PV module laminate. The composite back sheet is modeled as both single and two-layered material for characterizing the water diffusion using permeation and gravimetric techniques. The encapsulation material is characterized using the gravimetric technique. The gravimetric test was performed for water uptake of PV module laminate and compared the results with the simulated water mass uptake. Water concentration within PV laminate (primer layer) was simulated under the exposure to outdoor environment condition for single and two-layer back sheet.
Keywords: PV module, Water absorption, Back sheet, Gravimetric, Primer layer.

1. INTRODUCTION

Full fillment of end users and manufacturer expectation for consistent power output from the photovoltaic module (PV) is still a big challenge. PV solar cells are exposed to the outdoor environment for the prolonged duration of time. This continuously reduces the electrical and mechanical reliability due to synergistic effects of humidity, temperature, and ultraviolet (UV) irradiation. Interface degradation of back sheet and encapsulation leads debonding which introduce corrosion and oxidation in PV module [1].
Different types of back sheet laminate configuration are using in silicon-based commercial PV Module. Tedlar based symmetrical back sheet consist with Tedlar, Polyethylene Terephthalate (PET), Tedlar are using most. Tedlar based unsymmetrical configuration consists of Tedlar, PET, and primer also using because of the good affinity of primer with encapsulation. Laminate configuration of the back sheet leads low reliability because of lower interfacial strength at the interface compared to bulk. PET primer interface shows relatively low interface strength compare to the other interface of the laminate [2]. It is very difficult to replace multilayer back sheet laminate with the homogeneous material because of numerous advantages over it. Ethylene vinyl acetate (EVA) is used as an encapsulation material which works as adhesive as well as a protective layer.
To develop the proper understanding for interface degradation of multilayer back sheet under variable environment exposure needs systematic experimental and computational approach for water diffusion prediction.
Present research work mainly concerns with the characterization of the packaging material of the PV module through gravimetric and permeation techniques. Water diffusion parameter of the backsheet determined for the single and two-layer model. Water absorption study of the EVA was performed using the gravimetric technique. Water absorption study was also performed for PV laminate without cell layer.
Experimentally obtained parameter of the back sheet and EVA layer used in PV laminate simulation for water mass uptake and compared with experimentally obtained water mass uptake. Water concentration prediction was also performed under the exposure of the realistic outdoor condition by considering back sheet as single (homogeneous) and double layer configuration.
Multilayer back sheet approach provides a more realistic water concentration at the different interface in the PV module compare to single layer homogeneous approach.

2 MATERIAL AND METHOD

2.1 Material

The three PV module commercial materials considered in this study are the encapsulation layer, the back sheet, and low iron content float glass. Encapsulation material used was fast cure ethylene vinyl acetate (EVA) with a thickness of 700 μm. Back sheet laminate consists of polyvinyl fluoride (PVF, 25μm), polyethylene terephthalate (PET, 225μm) and primer layer (100 μm) glued together by adhesives with total stacking thickness 360μm. A laminate consists of polyvinyl fluoride (PVF, 25μm) and polyethylene terephthalate (PET, 225μm) also used in this study. A PV module laminate (60×30 mm^2) was prepared by laminating back sheet (360 μm), and EVA (325 μm) with the low iron contained float glass (3.8 mm) using a vacuum bag procedure (see Fig. 1).

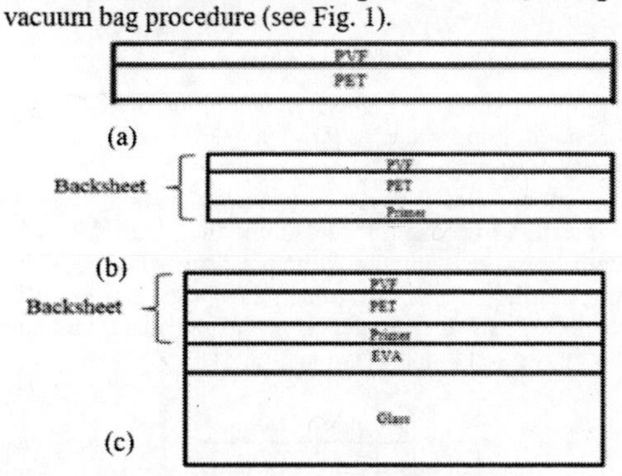

Figure 1: Schematic of the back sheet laminate (a) without primer layer (b) with a primer layer and (c) PV laminate.

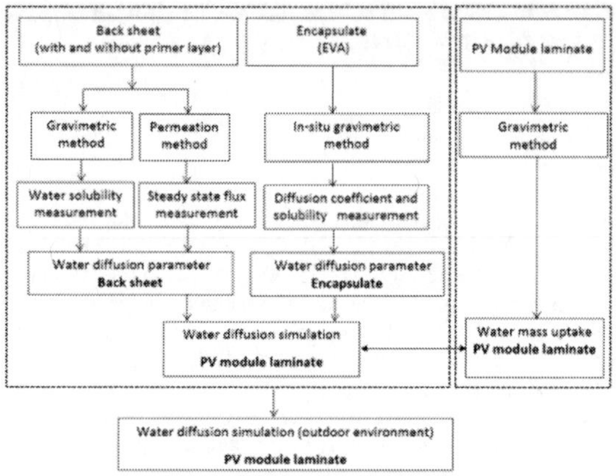

Figure 2: Flowchart of the methodology followed to determine the water diffusion properties of back sheet and EVA that is further used to simulate water diffusion in the PV module.

2.2 Experiments

Figure 2 shows the methodology followed for water diffusion characterization of the back sheet, encapsulate and PV module laminate. Ethylene vinyl acetate (EVA) encapsulant was characterized using the gravimetric technique. Water absorption tests were done to determine the water diffusion coefficient and solubility of EVA. Fickian diffusion model was used to determine the water diffusion coefficient of the EVA [3, 4].

$$\frac{M_t}{M_\infty} = 1 - \frac{8}{\pi^2}\sum_{n=0}^{n=\infty}\frac{1}{(2n+1)^2}e^{-D(2n+1)^2\pi^t t/4l^2} \quad 1$$

where M_t is the total mass uptake of water at time t, M_∞ is the saturated mass uptake and D is the water diffusion coefficient.

Water absorption tests were also used to determine the solubility coefficient of back sheet layer configuration (both with and without the primer layer). The mathematical relation for equivalent solubility is given in the equation (2) used to determine the solubility coefficient of the primer layer.

$$l_{total} \times S_{average} = l_{PVF-PET} \times S_{PVF-PET} + l_{primer} \times S_{primer} \quad 2$$

where S represents the water solubility of the layers measured from water absorption test, l represents the thickness of the layer.

Permeation tests were conducted using Mocon permeation instrument to determine the permeability for the back sheet (with and without a primer layer), which was further used to determine the permeability coefficient of the primer layer using ideal laminate theory is given in the equation 3 [3, 5].

$$\frac{l_T}{P_T} = \frac{l_{PVF\_PET}}{P_{PVF\_PET}} + \frac{l_{Primer}}{P_{Primer}} \quad 3$$

where P represents the permeability.

Gravimetric test of the PV module laminate for water mass uptake was performed for the validation of the measured water diffusion parameter of the back sheet and EVA.

3 WATER DIFFUSION SIMULATIONS

Fick's second law of diffusion was implemented in a 2-D finite element model by using Abacus software (figure 3). Diffusion parameters from gravimetric and permeation tests were used. Normalized concentration approach was used to eliminate concentration discontinuity at the interface. Single (homogenous) and double layer back sheet model considered in this study. Total water mass uptake of the PV module laminate estimated for a given boundary condition (24, 38 and 50^{0}C and 82% RH) and compared with the experimental data. Variation of water mass concentration over time was estimated at the primer layer of the PV module under the exposure outdoor environment condition (Delhi, India) and compared the results for single and double layer back sheet model.

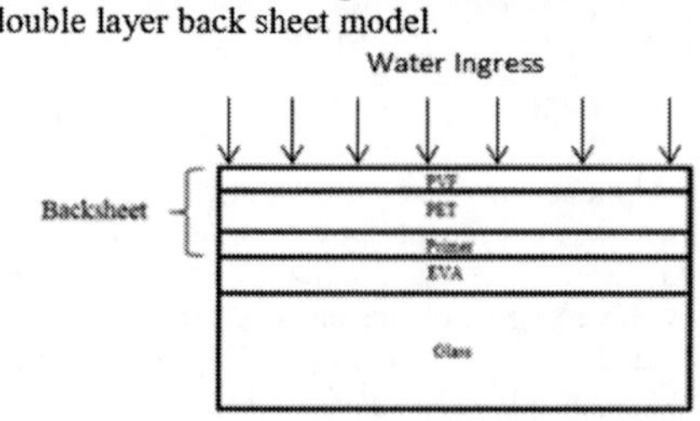

Figure 3: 2-D finite element model for determining water concentration.

4 RESULTS

4.1 Gravimetric Tests

Fickian diffusion model (equation 1) used to determine the water diffusion coefficient of The EVA. Diffusion coefficients obtained through the gravimetric test of EVA shows a good agreement with the Arrhenius equation, as shown in figure 4.

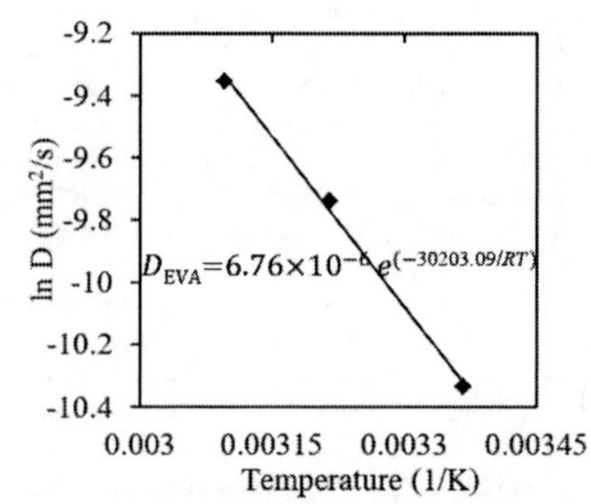

Figure 4: Temperature-dependent diffusion coefficient for EVA.

Figure 5 shows the Arrhenius plot of the water solubility measured at three different temperature. It follows the Arrhenius equation with the good agreement.

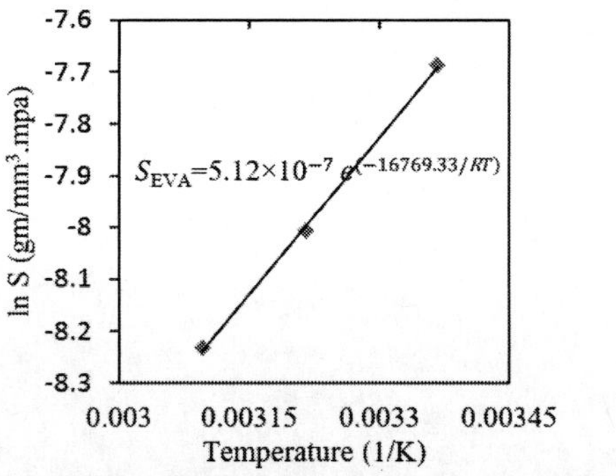

Figure 5: Temperature-dependent water solubility coefficient for EVA.

Water absorption tests of the back sheet (with and without primer layer) were also performed for solubility measurement. Figure 6 shows the solubility coefficient for a back sheet with and without a primer layer and a primer layer at a different temperature and follows a good agreement with the Arrhenius equation. Water solubility coefficient decreases with the inclusion of the primer layer. Primer layer shows the low water solubility compare to the PVF_PET stack. So the water solubility of the PVF_PET_Primer laminate shows the lower value compared to the PVF_PET laminate.

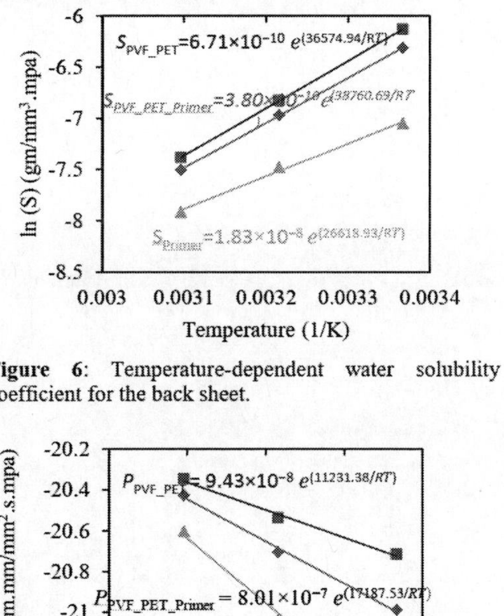

Figure 6: Temperature-dependent water solubility coefficient for the back sheet.

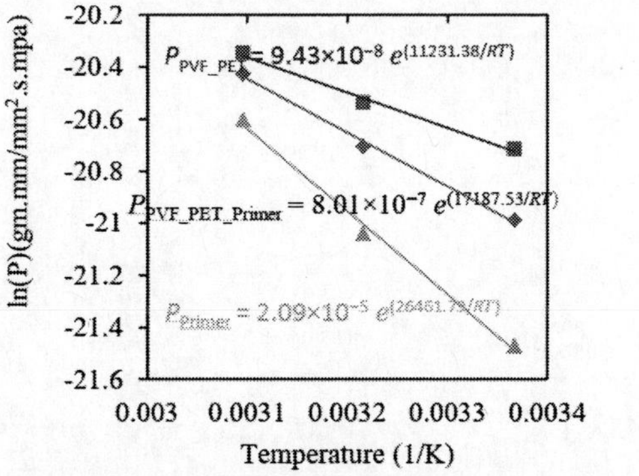

Figure 7: Temperature-dependent permeability coefficient for back-sheet.

Water permeability tests of the back sheet (with and without primer layer) were also performed for permeability coefficient measurement using Mocon permeation instrument. Figure 7 shows the permeability coefficient for a back sheet with and without a primer layer at a different temperature and follows a good agreement with the Arrhenius equation. Water permeability coefficient decreases with the inclusion of the primer layer. Primer layer shows the relatively low water permeability compare to the PVF_PET stack. So the water permeability of the PVF_PET_Primer laminate decreases with the inclusion of the primer layer.

4.2 Water Diffusion Simulation

Figure 8 shows the comparison of the mass water uptake of the PV modulate laminate between simulation and experimental. Water mass uptake simulation (continuous plot) using the water diffusion parameter of the back sheet and EVA shows the good agreement with experimentally obtained water mass uptake of the PV module laminate.

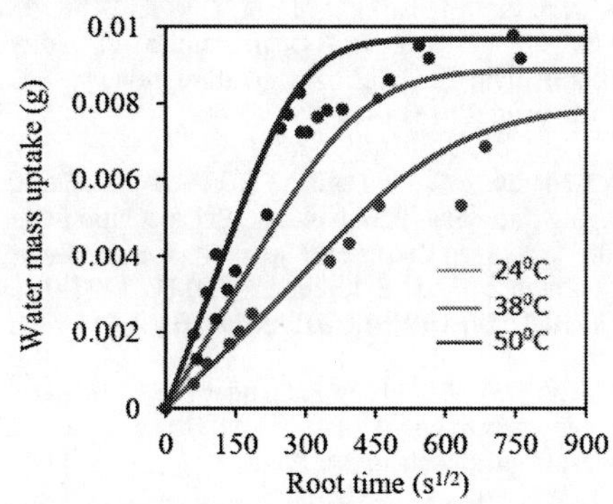

Figure 8: Water mass uptake for PV module laminate at different temperature and fixed relative humidity 82%.

Water diffusion simulation was also performed under the exposure of the outdoor environment condition of Delhi, India with the consideration of the single and double layer back-sheet model. Water concentration prediction was performed inside the primer layer for both the model. Predicted water concentration (figure 9) shows the considerable difference. Consideration of the homogenous back-sheet model overpredicts the water concentration inside the primer layer compare to the two-layer back-sheet model.

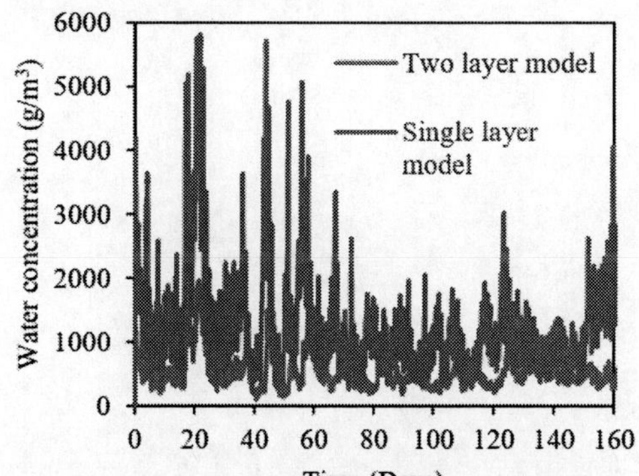

Figure 9: Water concentration prediction inside the PV module laminate (primer layer) under the exposure of outdoor environment.

5. CONCLUSION

Water diffusion, solubility and permeability coefficient increases with the temperature and also follow the good agreement with the Arrhenius equation. The significant differences were seen for solubility and the permeability coefficient of the homogenous back sheet and an individual layer of primer and PVF_PET layer.
Homogeneous backsheet model over predicts the water concentration inside the primer layer compare to the two-layer backsheet model. Two-layer back sheet model gives the realistic concentration prediction compare to the homogenous back sheet model under the exposure of the outdoor environment condition.

6. REFERENCES

1) A. Ndiaye, A. Charki, A. Kobi, C.M.F. Kébé, P.A. Ndiaye, V. Sambou, Degradations of silicon photovoltaic modules: A literature review, Solar Energy. 96 (2013) 140–151.

2) A.Dadaniya, N.V. Datla, Effect of Accelerated Aging on the Photovoltaic Encapsulation/Glass Interface, 33rd European Photovoltaic Solar Energy Conference and Exhibition. (2017) 1727–1729. doi:10.4229/EUPVSEC20172017-5BV.4.81.

3) ASTM D570-98(2010)e1, Standard test method for water absorption of plastics, ASTM International, West Conshohocken, PA, 2010.

4) Crank, J. (1979). The mathematics of diffusion. Oxford University Press.

5) ASTM International, Standard Test Method for Water Vapor Transmission Rate Through Plastic Film and Sheeting Using a Modulated Infrared Sensor, Astm. 90 (2015) 1–6. www.astm.org.

INVESTIGATION AND ANALYSIS OF BIFACIAL PHOTOVOLTAICS MODULES WITH REFLECTIVE LAYER

Ernest SNG[1,2], Chong Xian ANG[1], Idris Li Hong LIM[1]
[1] University of Glasgow, University Avenue, Glasgow G12 8QQ, UK
[2] REC Solar Pte Ltd, 20 Tuas South Ave 14, Singapore 637312, Singapore
Contact: ernest.sng@recgroup.com

ABSTRACT: While the analysis of optical gains and power transfer of monofacial photovoltaics module are relatively well established, there are many nuanced effects that contribute to bifacial cell to module power transfer. To improve the bifacial module power during indoor flash test, reflective layers had been included in the inter-cell spacing. This reflective layer improves bifacial module STC maximum power point (Pmpp) and current. This paper presents the investigation and analysis of bifacial PV modules with reflective layer at the inter-cell spacing. Three main bifacial module configurations with reflective layer are studied where the reflective surface was inserted at different positions, (1) behind the rear glass or transparent back sheet, (2) in between the glass and encapsulant, and (3) in between the encapsulant. A numerical model is created to simulate the current gain from each configuration with varying inter-cell spacing with varying illumination positions.
Keywords: Bifacial, PV Module, Ray Tracing

1 INTRODUCTION

Photovoltaic (PV) energy sources could achieve grid parity with traditional energy sources by either cost reduction or increase in efficiency with new technology. Currently the cost of PV modules had exponentially reduced due to improvement in manufacturability, economic of scale, and improved technology. Traditionally, the implementation of new technology had always been associated with an increase in cost. For PV manufacturers to achieve the grid parity goal, new technology implemented had to be cost effective and reliable. One such potential technology enabler to lower the cost of PV is the bifacial PV cells.

In a typical solar cell, light is only collected from the front side of the cell. Bifacial PV cells enable the collection of light from the both side of the cell, thus generating additional power in the same photovoltaics module [1][2]. By generating more power at lower cost, bifacial design enables PV energy sources to be attractive to consumers. Through the changing of module materials in the construction of bifacial PV modules, an increase of energy yield is achieved without a large increase in cost [3].

Bifacial PV cells can be encapsulated into the typical monofacial solar module structure, which has cells sandwiched between the encapsulant with glass on the front side and white back sheet on the rear, as shown in Figure 1 Bifacial cells in monofacial moduleFigure 1. While bifacial modules come in a few different structures, the standard bifacial module in Figure 2 of glass/glass or glass/transparent back sheet could capture additional sunlight to the rear from ground reflectance [4][5][6]. However, without the white back sheet like of monofacial solar module that is contributing to module internal reflectance [7], the transmission of light during STC results in the power loss as compared to the monofacial modules with the white back sheet on the rear [8].

Recent developments of bifacial modules with a reflective layer inserted between the cell increase the module front side current without totally removing the bifacial capability of the module [8][9]. In this structure, various configurations of varying locations of the reflective surface were proposed. The three main configurations differs where the reflective surface was inserted: (1) behind the rear glass or transparent back sheet in Figure 3, (2) in between the glass and encapsulant in Figure 4, and (3) in between the encapsulant in Figure 5. In the monofacial module structure and bifacial module structure with reflective layer of configuration 1 and 2, the incoming light from the front of the module is reflected onto both the front and rear of the cell. The light ray traces are shown in Figure 1, Figure 3, and Figure 4 with arrows marked "A". The standard bifacial module structure and all three bifacial module structures with reflective layer allows the ground reflectance to be capture by the rear of the active cell, as shown in Figures 2 to 5. The location of the reflective layer determines the amount of ground reflectance reaching the cell.

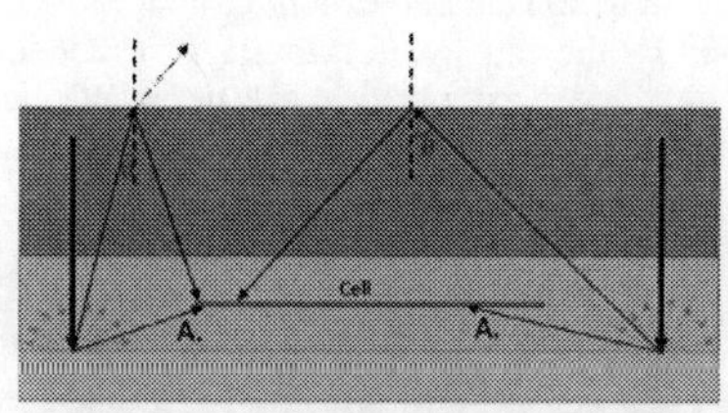

Figure 1 Bifacial cells in monofacial module

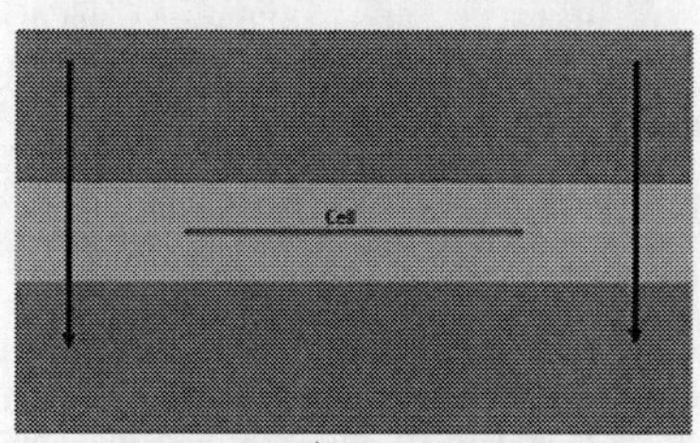

Figure 2 Standard bifacial module

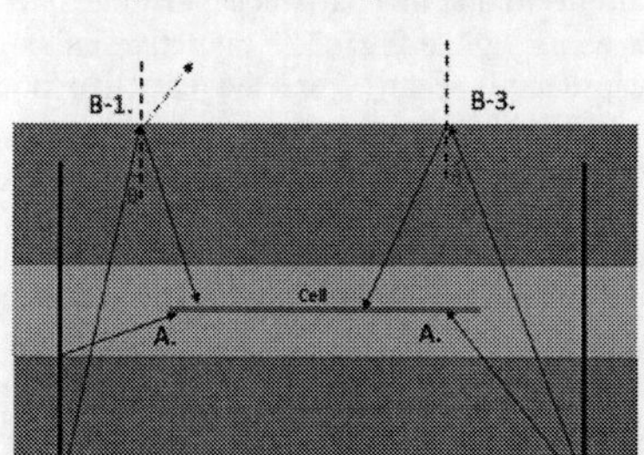

Figure 3 Bifacial module reflective layer configuration 1

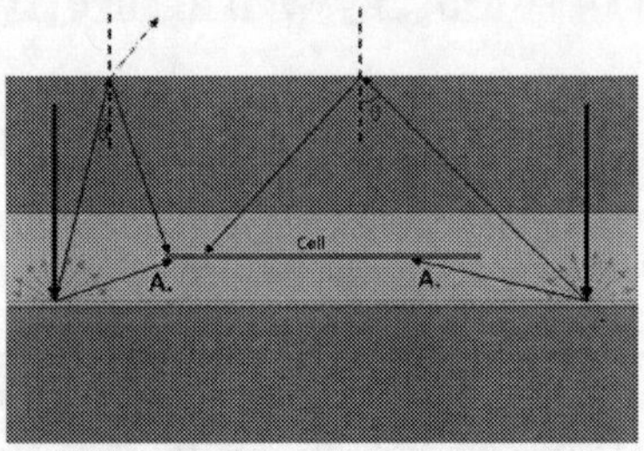

Figure 4 Bifacial module reflective layer configuration 2

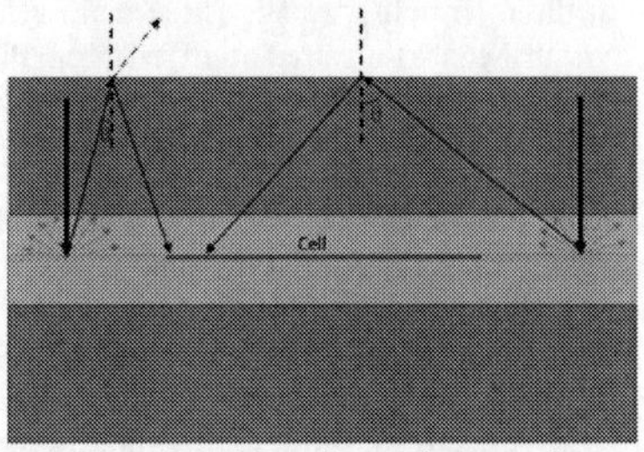

Figure 5 Bifacial module reflective layer configuration 3

The influence of the cell gap to monofacial cells in monofacial modules were widely reported. With wider cell gap, the reflectivity area increases and in turn the cell current [7][10][11]. For bifacial cell and module, strategic placement of the reflective surface affects the module performance during STC and outdoor monitoring.

In this paper, the front and rear current of the bifacial module structures highlighted above will first be analyzed. The equations on light scattering and probability of collection by the active cell area are first modelled with optical ray trace. Next, physical measurements performed on the materials are provided as inputs into the numerical model.

The rest of this paper is organized as follows: Section 2 provides the development of the numerical model with the optical ray trace. Simulation results of the different configurations will be discussed in Section 3. A parametric study on the varying inter-cell gap with reflective layer will be presented. This will be followed by the discussion and analysis of the performance of each configuration. Conclusions are presented in Section 5.

2 METHODOLOGY OF SIMULATION

2.1 Assumptions and experimental inputs

Firstly, some assumptions of the refractive index of the materials were made. The glass and encapsulant refractive index are assumed to be the same, as the optical loss through reflection at this particular junction is very small and can be negligible [12]. The refractive index of glass and encapsulant is set at 1.5 and the refractive index of air is set at 1 [7][12].

Secondly, it is considered that the reflective layer will scatter the light that falls on it into different directions. In order to cover all possible directions of the scattered light, the directions are divided into two segments which is the azimuth and polar directions. The assumption of the direction of light was made the same such that it is scattered uniformly at interval of 3 degrees in the polar coordinates and at intervals of 6 degrees in the azimuth coordinates, resulting in a total of 3600 rays. The reflected rays were also assumed to be unpolarized, thus consisting of parallel and perpendicular light.

Lastly, the white area is divided into unit areas of $1mm^2$ which reflects an individual ray into each unit area of the module. Two experiment inputs were essential for the numerical model were measured by spectrometer, the normalized reflected radiant intensity and the reflectance of the reflective layer. The wavelength of 632nm was chosen to represent the angular scattering of the spectrum AM1.5G on the reflective layer. Its weighted average reflectance is also calculated with respect the AM1.5G spectrum.

2.2 Equations

When light falls on the white area, it will be scatted into different directions. However, the intensity of the scattered light will vary according to the reflectivity of the white inter-cell gap. The calculation of the irradiance at certain direction can be calculated using Equation (1) [7][8]. where irradiance at a certain direction, S(θ) is reflected by the back-sheet where aue is the unit area of the back sheet, Φp is the Power of incident light per unit area, WARbs is the back sheet weighted Average Reflectance, Sn(θ) is the measured normalized reflected radiant intensity of the back sheet, θ is the polar angle, and φ is the azimuth angle.

$$S(\theta) = aue.\Phi p\,.(WARbs)\frac{Sn(\theta)}{\iint Sn(\theta)\sin\theta d\varphi d\theta} \quad (1)$$

The white inter-cell gap will reflect off the ray that incident on it into different directions. Rays will either transmit out of the module or reflected internally inside. A matrix consisting of the coordinated of the module had been simulated to determine the possible locations of the reflected rays. Some assumptions made are the uniform separation of scatted rays across the polar and azimuth angle, and the non-contribution of cells irradiance after two bounces. The contribution of reflected light that hits the rear side of the cell the position vector of the ray are computed by the Equations (2) and (3) [7][8], where t_1 is the thickness of rear encapsulant .For those rays that were not scattered to the rear of the cell and which were reflected by front glass back to the cell, Equations (4) and (5) were applied, where t_2 and g are the thickness of front encapsulant and glass respectively. For incident angle more than critical angle of 41.8°, total internal reflection is considered. If the angle of incident is less than the critical angle of air/glass (41.8°) the reflectance and transmittance radiant intensity are calculated with Fresnel equation.

$$x2 = x1 + (t_1) * \tan(\theta)\cos(\varphi) \quad (2)$$
$$y2 = y2 + (t_1) * \tan(\theta)\sin(\varphi) \quad (3)$$
$$x3 = x1 + (2g + t_1 + 2t_2) * \tan(\theta)\cos(\varphi) \quad (4)$$
$$y3 = y1 + (2g + t_1 + 2t_2) * \tan(\theta)\sin(\varphi) \quad (5)$$

As discussed earlier, the irradiance reflected by the back sheet and incident on the rear of the cell could be calculated if x2 and y2 is situated on the cell with Equations (6) and (7) for the ray reflected to the rear and front of the cell respectively [7, 8, 13].

$$P_r = \int_{q1}^{q2}\int_{p1}^{p2}\int_{0}^{\pi}\int_{\alpha}^{\frac{\pi}{2}} S(\theta).T1(x2, y2).R_g.\,d\theta\, d\varphi\, dx\, dy \quad (6)$$

$$P_f = \int_{q1}^{q2}\int_{p1}^{p2}\int_{0}^{\pi}\int_{0}^{\alpha} S(\theta).T1(x3, y3).R_g\, d\theta d\phi dx dy \quad (7)$$

The current gain calculated by the numerical model includes the reflectance of the back sheet and the internal scattering. However, only two bounces were considered for the calculation as it was assumed that radiant intensity would be negligible after two bounces. This assumption could be further investigated by calculating the absorption in glass and EVA. The EVA encapsulant are typically of the same refractive index as glass, in order to reduce losses. Hence, the absorption losses of light, $\alpha_{medium}(\lambda)$, traversing in the glass and EVA could be calculated using Equation (8), where the power of the incident light is Φp (λ). The absorption coefficient of the medium is described in Equation (9), where t is the layer thickness, and θ is the angle of propagation. This equation could be applied to both glass and encapsulant collectively.

$$\alpha_{medium}(\lambda) = \frac{reflected\ (\lambda) + transmitted\ (\lambda)}{incident\ light\ (\lambda)} \quad (8)$$

$$Absorption_{medium}(\lambda) = \Phi p\ (\lambda)[1 - e^{-\alpha_{medium}(\lambda)t}\ ] \quad (9)$$

Equation (10) calculates the current gain from the reflective area to the rear of the bifacial cell for the monofacial module structure and bifacial module structure with reflective layer of Configurations 1 and 2. Equation (11) calculates the current gain from the reflective area to the front of the cell and Equation (12) calculates the total current gain to the cell from the reflective area.

$$Current_{\text{gain rear}} = \frac{\left(\frac{I_{sc.r}}{I_{sc.f}}\right) * P_r}{\Phi p\ .\text{cell area}} \quad (10)$$

$$Current_{\text{gain front}} = \frac{P_f}{\Phi p\ .\text{cell area}} \quad (11)$$

$$Current_{\text{gain}} = \frac{P_{f+\left(\frac{I_{sc.r}}{I_{sc.f}}\right)*P_{rear}}}{\Phi p\ .\text{cell area}} \quad (12)$$

3 RESULT AND DISCUSSIONS

3.1 Front illumination current gain comparison

The simulated current gains with and without consideration of absorption with varying inter-cell gap of the three different bifacial module structure with reflective layer configuration as illustrated in Figure 3, Figure 4, and Figure 5 where discussed in this section. From Figure 6 the current gain reduces from Configuration 1 to Configuration 3. Difference between the three configurations is the distance between the reflective layer to the cell.

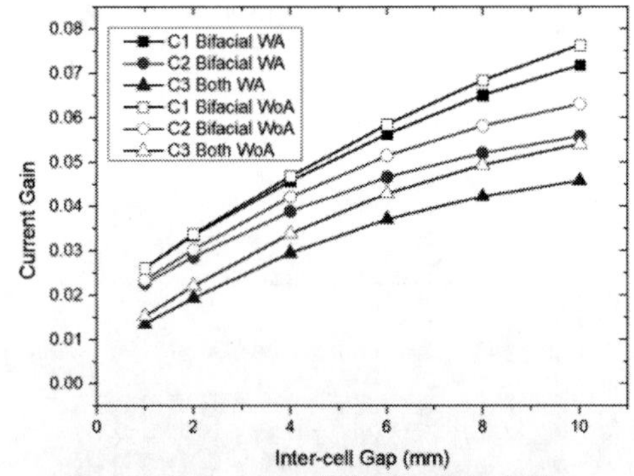

Figure 6 Current gain of bifacial modules for three configurations (C1, C2, C3) with (WA) & without (WoA) absorption

The additional distance between the reflective layer and the cell increases the number of rays being reflected onto the cell. In the illustration of all the bifacial modules configuration cross section, the distance travelled by the rays absorbed on the front side is twice as compared to those absorbed by the rear. In Figure 6 the difference between the current gain with and without consideration of absorption loss could be used to explain the contributions of front and rear to the current gain. Similarly, this is seen in Figure 7 where the current gain is breakdown into front and rear. Configuration 3 has no contributions from the rear as the reflective layer is on the same plane as the cell. Whereas Configuration 2 has minimum rear contribution with increasing cell gap as most light rays are reflected towards the front glass as compared to Configuration 1. At inter-cell gap of 1mm to 10mm, Configuration 1 has the highest current gain of the three. Configuration 1 has the highest rear current gain from the rear surpassing Configuration 2 rear current across 1mm to 10mm. With front illumination, Configuration 1 rear current gain exceeds configuration 2 and 3 front current gain. A higher current gain could be achieved for Configuration 1 if the bifaciality factor of the cell improves. Currently, bifaciality factor of 0.8 was used for Equation (10) for this model.

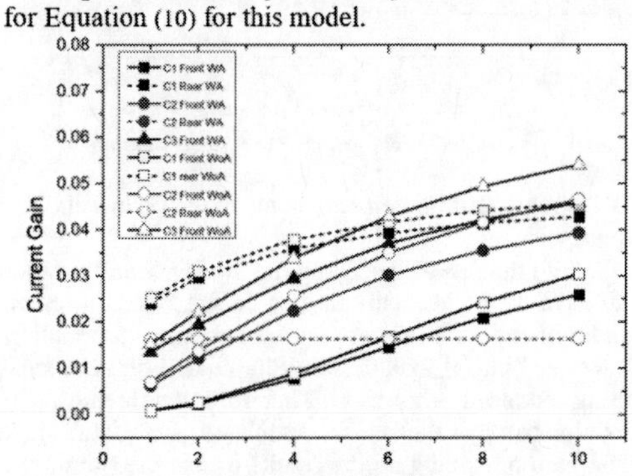

Figure 7 Rear and front current gain of bifacial modules for three configurations

3.2 Rear illumination current gain comparison

While Configuration 1 has highest current gain for front side illumination, there would be little current gain from rear illumination as all the light rays that are incident on the inter-cell gap are reflected out of the module. Configuration 2 and 3, light rays would reflected to the rear of the cell as illustrated in Figure 8 and Figure 9.

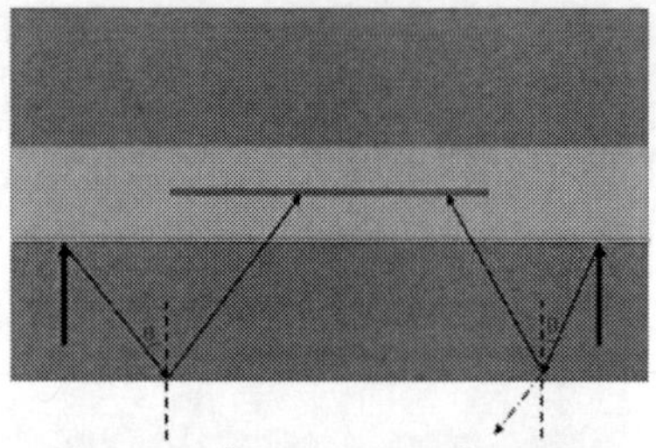

Figure 8 Configuration 2 rear illumination

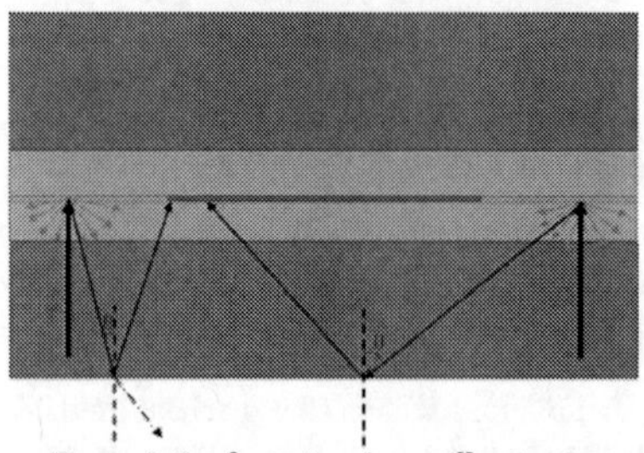

Figure 9 Configuration 3 rear illumination

Figure 10 plots the simulated current gain with rear illumination with Configuration 2 and 3. As the reflective layer distance from the cell between the two configurations are not significant, the resulted current gain is comparative close.

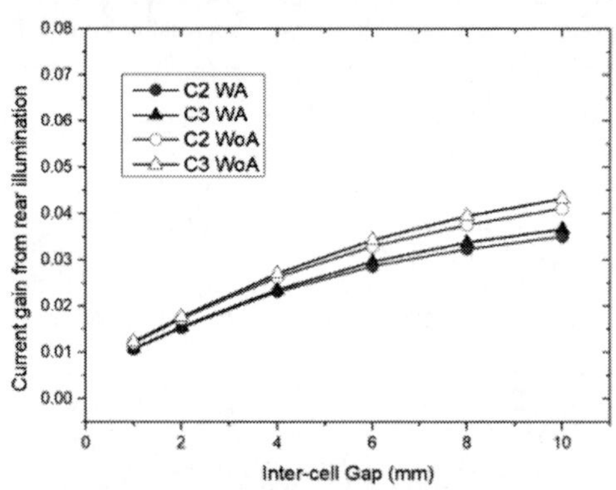

Figure 10 Current gain front rear illumination

3.3 Double sided illumination comparison on bifacial and monofacial cell

While the results discussed previously, Configuration 1 to have the highest current gain during standard testing condition indoor flash test with front side illumination. However, bifacial modules are illuminated on both sides during outdoor energy yield. Therefore, the three bifacial module configurations are simulated for both sided illumination. Adding on the results of front side and rear side illumination current gain, Configuration 2 and 3 surpass Configuration 1 current gain. The front and rear current gain for Configuration 2 results in the highest current gain across 1mm to 10mm inter-cell gap distance. This leads to the disparity between indoor and outdoor performance for the different module configurations

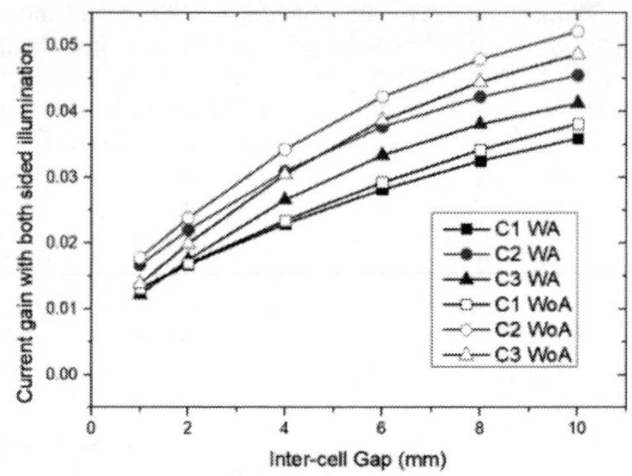

Figure 11 Current gain with both sided illumination

In monofacial modules, reflective rear encapsulants that are on the same plane as the cells are used to reduced absorption losses as the light rays are reflected on the same plane as the cell. Additionally, this remove the possibility of rays being reflected onto the inactive rear of a monofacial cell. In Figure 12, the three configurations were compared

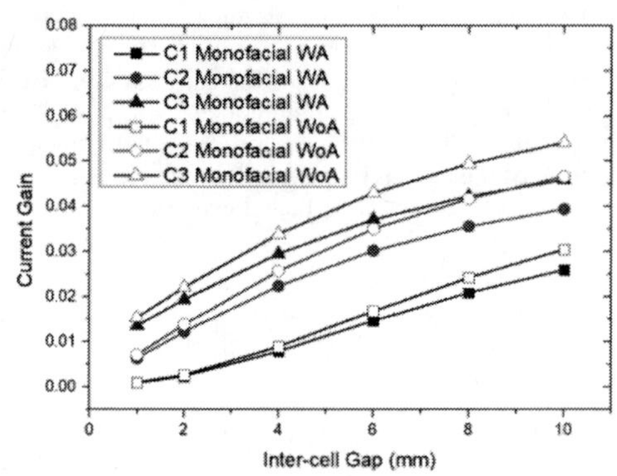

Figure 12 Current gain of monofacial modules for three configurations from front illumination

To compare the reflective layers effect on both monofacial and bifacial modules during pseudo outdoor conditions, a both sided illumination was used to simulate the current gain for monofacial modules shown in Figure 13. While Configuration 2 has the highest current gain across 1mm to 10mm inter-cell gap, Configuration 1 and 3 has little different at a typical 2mm inter-cell gap. The outdoor energy yield of bifacial module would exceed monofacial modules, however the outdoor performance is still not as distant between the bifacial modules.

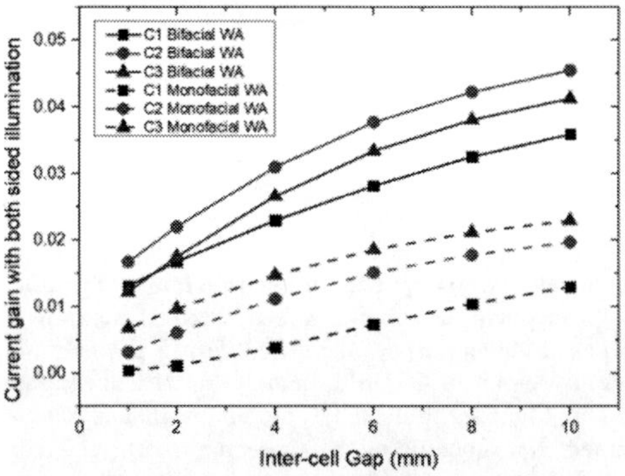

Figure 13 Current gain for monofacial and bifacial modules with both sided illumination

For front side illumination shown in Figure 6, at 2mm inter-cell gap it has been shown that Configuration 1 has the highest current gain of 3.4%. In comparison, for the monofacial module group Configuration 3 has the highest the current gain at 1.9% with front side illumination which is half of the bifacial modules. For the rear illumination, Configuration 2 and 3 has comparable gain while Configuration 1 has no additional gain as the reflective layer reflects the rays out of the module. In outdoor energy yield setting the modules are illuminated on both sides, this results the change in current gain for bifacial modules Configuration 1 to 1.7%, Configuration 2 to have 2.2%, and Configuration 3 to 1.8%. In addition, the simulation results show most of the current gain for bifacial modules with 2mm inter-cell gap comes from the rear of the cell from front side illumination.

4 CONCLUSION

This paper discussed three main bifacial module configurations with reflective layer at different positions are studied in detail; namely Configuration 1: Glass/EVA/ Bifacial Solar Cell/EVA/Glass/Reflective Layer, Configuration 2: Glass/EVA/ Bifacial Solar Cell/EVA/Reflective Layer /Glass, and Configuration 3: Glass/EVA/ Reflective Layer/ Bifacial Solar Cell/EVA /Glass. Optical ray trace model for the three configurations was created with inputs from test measurements. The simulation on the three numerical models created takes into consideration of absorption loss to simulate the current gain from each configuration with varying inter-cell spacing. The corresponding performance gain was evaluated for models with and without absorption loss From the simulated results, the optimal configuration for bifacial cell considering absorption losses in the module materials is proposed.

7 REFERENCES

[1] A. Cuevas, A. Luque, J. Eguren, and J. del Alamo, "50 Per cent more output power from an albedo-collecting flat panel using bifacial solar cells," *Sol. Energy*, vol. 29, no. 5, pp. 419–420, Jan. 1982.

[2] G. J. M. Janssen, B. B. Van Aken, A. J. Carr, and A. A. Mewe, "Outdoor Performance of Bifacial Modules by Measurements and Modelling," *Energy Procedia*, vol. 77, pp. 364–373, Aug. 2015.

[3] R. Guerrero-Lemus, R. Vega, T. Kim, A. Kimm, and L. E. Shephard, "Bifacial solar photovoltaics – A technology review," *Renew. Sustain. Energy Rev.*, vol. 60, pp. 1533–1549, Jul. 2016.

[4] S. Guo, T. M. Walsh, and M. Peters, "Vertically mounted bifacial photovoltaic modules: A global analysis," *Energy*, vol. 61, pp. 447–454, Nov. 2013.

[5] U. A. Yusufoglu *et al.*, "Simulation of Energy Production by Bifacial Modules with Revision of Ground Reflection," *Energy Procedia*, vol. 55, pp. 389–395, 2014.

[6] S. Wang *et al.*, "Bifacial Photovoltaic Systems Energy Yield Modelling," *Energy Procedia*, vol. 77, pp. 428–433, 2015.

[7] S. Guo *et al.*, "Investigation of the short-circuit current increase for PV modules using halved silicon wafer solar cells," *Sol. Energy Mater. Sol. Cells*, vol. 133, pp. 240–247, 2015.

[8] J. P. Singh, S. Guo, I. M. Peters, A. G. Aberle, and T. M. Walsh, "Comparison of Glass/Glass and Glass/ Backsheet PV Modules Using Bifacial Silicon Solar Cells," *IEEE J. Photovoltaics*, vol. 5, no. 3, pp. 783–791, May 2015.

[9] J. A. M. van Roosmalen, S. L. Luxembourg, J. Liu, L. A. G. Okel, and B. B. Van Aken, "White Bifacial Modules – Improved STC Performance Combined with Bifacial Energy Yield," *32nd Eur. Photovolt. Sol. Energy Conf. Exhib.*, pp. 42–47, Jul. 2016.

[10] W.-S. Su *et al.*, "Optimization of the output power by effect of backsheet reflectance and spacing between cell strings," in *2011 37th IEEE Photovoltaic Specialists Conference*, 2011, pp. 003218–003220.

[11] M. B. Koentopp, M. Schu?tze, D. Buss, and R. Seguin, "Optimized Module Design: A Study of Encapsulation Losses and the Influence of Design Parameters on Module Performance," *IEEE J. Photovoltaics*, vol. 3, no. 1, pp. 138–142, Jan. 2013.

[12] I. M. Peters, Y. S. Khoo, and T. M. Walsh, "Detailed Current Loss Analysis for a PV Module Made With Textured Multicrystalline Silicon Wafer Solar Cells," *IEEE J. Photovoltaics*, vol. 4, no. 2, pp. 585–593, Mar. 2014.

[13] K. R. McIntosh, R. M. Swanson, and J. E. Cotter, "A simple ray tracer to compute the optical concentration of photovoltaic modules," *Prog. Photovoltaics Res. Appl.*, vol. 14, no. 2, pp. 167–177, Mar. 2006.

BENCHMARKING LIGHT AND ELEVATED TEMPERATURE INDUCED DEGRADATION (LETID)

Matthias Pander*[1], Marko Turek[1], Jan Bauer[1,2], Tabea Luka[1], Christian Hagendorf[1],
Matthias Ebert[1], Ralph Gottschalg[1,2]
[1]Fraunhofer Center for Silicon Photovoltaics, Otto-Eissfeldt-Str. 12, 06120 Halle (Saale), Germany
phone: +49-345-5589-5215, fax: +49-345-5589-5999, matthias.pander@csp.fraunhofer.de
[2]Anhalt University of Applied Sciences, Bernburger Str. 55, D-06366 Köthen, Germany

ABSTRACT: Light Induced Degradation (LID) is a well-known cell effect, impacting in the first hours of testing or first months of operation. All c-Si device types are affected, typically with less than 5% degradation. The known mechanisms (Boron Oxygen complex formation and Iron Boron pair dissociation) are identified during certification by IEC 61215:2016 standard stabilization procedures carried out at around 50°C. Unfortunately, there is a second degradation mode which is only apparent at higher temperatures. This Light and Elevated Temperature Induced Degradation (LeTID) is not identified during certification. Typically, there is no specific information commonly available for system planners to estimate the magnitude of the losses due to this effect.

Testing at higher temperatures of PERC devices resulted in much higher degradation levels (up to 10 %), but slower degradation rates than BO. Therefore, this new type of degradation was termed Light and elevated Temperature Induced Degradation (LeTID). Initially LeTID was thought to only affect multi-PERC but recent studies have shown that mono-PERC may be affected equally. LeTID is not fully understood but can virtually be eliminated in production.

Currently there is no generally accepted standard to test for LeTID degradation on cell and module level. Thus it is not possible to identify modules susceptible to LeTID. The question arises to what extent currently available modules contain LeTID-susceptible cells, and the magnitude of long-term yield losses. For this purpose, nine module types from different manufacturers were procured from different suppliers and subjected to a LeTID-specific test procedure. Power losses up to 6 % were found during the LeTID-Test. The three multi-Si modules tested show moderate degradation or are even stable. Some of the commercially available mono-Si PERC modules show high LeTID.

Keywords: LeTID, LID, PERC, PV Module, Degradation

1 INTRODUCTION

Passivated Emitter Rear Contact (PERC) Solar Cells are rapidly gaining market share [1]. This new technology does have many advantages over conventional Aluminum Back Surface Field (Al-BSF) devices, but it may have additional reliability issues. Reliability is typically being tested through the certification [reference IEC61215] at independent test houses. This standard includes tests for most known failure mechanisms and sets appropriate conditions to test for these. Most cells show a small amount of Light Induced Degradation (LID) in the initial weeks of operation. A test at 50°C and an irradiance of 1000W/m^2 for a minimum of 2 times 5 kWh/m^2 is included in the certification cycle to identify LID sensitive devices.

It was found recently that PERC devices may be susceptible to another type of LID, which is apparent only at elevated temperatures. This is termed Light and elevated Temperature Induced Degradation (LeTID) [2]-[4]. PERC technology is more affected than Al-BSF [6]. Initially, LeTID was thought to only affect multi-PERC, as the grain boundaries in the multi-Si material may contribute to degradation ([4],[9]). However, it has been shown recently that mono-PERC can be affected equally [5].

LeTID is a cell effect typically associated with the bulk of the device. It correlates well with the current passing through the cell. The traditional LID mechanisms express themselves in the first hours of testing or first months of operation with power losses around 3 % (year 1 degradation). These known mechanisms are:

- Iron Boron (FeB) pair dissociation (Fe-B-LID)
- Boron Oxygen complex formation (B-O-LID)

Susceptibility to these mechanisms is normally identified in type approval. Testing at higher temperatures for PERC devices has been shown to result in much higher degradation levels in some cases with significantly slower degradation rates than B-O-LID but potentially of higher magnitude. Therefore, this new type was termed Light and elevated Temperature Induced Degradation (LeTID) [2].

Hanwha QCells linked LeTID to degradation seen in the field and thus demonstrated that LeTID is a relevant degradation mechanism in outdoor operation [7]. The observed degradation of >7% in one year in Nicosia (Cyprus) is critical for PV systems. As a temperature activated failure mode, LeTID develops faster in hot climates due to the increased operation temperatures. Results of outdoor data from Hanwha QCells have shown that lab-degradation tests with 290 h at 75°C CID (Mpp mode) are comparable to one year outdoor exposure in Nicosia (Cyprus) [7]. The effect is nevertheless relevant even in temperate climates due to the long operation times power degradation might occur more slowly within 5 to 10 years.

The root cause analysis of LeTID is inconclusive and ongoing. Some authors suggest the degradation to be caused by defect complexes of mobile hydrogen and an intrinsic crystal defect [10]. During these studies adequate testing procedures have been developed on cell level which supports testing conditions on module level. Since LeTID might lead to substantial yield loss the following questions shall be investigated and answered in this publication:

1. What are the factors influencing LID/LeTID degradation?
2. What are reliable conditions for testing LeTID on module level?
3. How susceptible are commercially available PERC modules to LeTID?

2 INFLUENING FACTORS FOR LID/LETID

A recombination active defect is formed especially in PERC solar cells at elevated temperatures ($\geq 50°C$) under illumination. This reduces Voc, Isc, FF and efficiency η. The maximum degradation is typically reached after 300 to 400 h at 75°C. Going to even higher test temperatures accelerates the degradation [10]. However, new recombination active defect types might also appear that are unrelated to LeTID [8].

Light beam induced current (LBIC) images exhibit nearly homogeneous degradation of the whole cell area, hence it is assumed that the underlying LeTID defect is a volume defect [9],[10]. It is assumed that impurities in the Si material influences LeTID, however, at the present time no final mechanism is known. In mc-Si PERC cells non-active grain boundaries show a slower or weaker degradation [9]. There is even a thickness dependence of the LeTID effect shown by Bredemeier et al. [11]. The two latter findings are only two pieces of the puzzle to reveal the origins of the LeTID effect. It seems that hydrogens plays a role as well as other elements [10],[11],[12]. Pretreatment of PERC solar cells in the dark with different heat budgets, e.g. during firing processes or intentionally dark anneal, changes the LeTID effect significantly for mono-Si and mc-Si as well [8],[10],[13],[14]. The PERC cells investigated for instance in [12] exhibit a stronger degradation towards a dark anneal of about 175°C, and a lower LeTID effect was seen by increasing the dark anneal temperature further. Screening the literature above it turned out that common LeTID tests conditions are 75°C and illumination intensities from approx. 0.5 to 1 sun respectively and operating conditions with equivalent currents in the dark (so called current induced degradation - CID, see e.g. [15]) trigger LeTID reliable if the cells are prone to this effect.

B-O-LID as well as LeTID is driven by injected charge carriers. Thus, the degradation can either be activated by illumination or application of an external current (CID). Different equivalent operational modes are described in literature for testing LeTID:

- LID Voc mode ≅ carrier injection with an Isc-equivalent injection level
- LID MPP mode ≅ carrier injection with a reduced injection level of around (Isc – Impp) ~ 0.5 … 1 A

Higher injection accelerates degradation, but the maximum degradation is only reached in MPP mode [2]. After reaching the maximum degradation regeneration is observed. This can be close to the initial level.

Temperature sensor positioning

1 – Module in center, centred on back of the module
2 – Module in center, on bottom on back of module
3 – Module right, front on back of module
4 – Module right, centered on back of module

Figure 1: Climate chamber setup

3 MODULE BENCHMARK

3.1 Sampling

Module samples were obtained anonymously from various wholesalers. An initial quality check was performed for all modules and includes visual inspection, EL imaging and STC measurements. The aim was to ensure that no modules are tested with defects potentially causing additional performance reduction due to thermal stress (e.g. 'normal cell cracking'). STC power had to be within specifications. The final test batch included nine module types with two modules each. For rough classification we divide the test batch into:

- 6 mono-Si module types
- 3 multi-Si module types

3.2 Testing Procedure

Based on the review in section 2, the test conditions for the modules have been chosen. Since the known LID defects may impact the results of the study a preconditioning was defined to distinguish between B-O-LID, FeB-LID and LeTID. For B-O-LID the module temperature is controlled at 25°C and current near I_{SC} is injected. These conditions ensure that only B-O-LID is activated. To control the condition of FeB, a dark storage of 10 h was carried out before each intermediate power measurement. For temperature control, the module temperature was measured at 4 representative locations (see Figure 1), which were identified in a pre-test.

After preconditioning the actual LeTID test procedure is carried out. The conditions had to consider that too low temperature may address different dominant processes or won't activate LeTID. Similarly, slow degradation rates lead to long test times and consequently high costs. The lower temperature limit is roughly 50°C in this case. Too high temperatures may shift annealing and degradation processes, resulting in unrealistically low degradation. This may lead to conditions where maximum degradation is not observed. Additionally, LeTID-unrelated defects may be caused when module materials do not withstand the temperatures. This leads to an upper temperature limit of 85°C for standard modules. The injection level has to be close to field conditions, what normally means MPP-Tracking. Based on these considerations the following conditions are common in the community:

- Internal carrier injection: 1000 W/m², MPP-Tracking, 75°C in Solar Simulator (LID – MPP mode)
- External carrier injection: I = 1 A, 75°C in dark climate chamber (CID – MPP mode)

These tests have been shown to be equivalent and trigger the same degradation mechanism. In this work, external carrier injection is used as suggested by [5]. The testing scheme is summarized in Figure 2.

4 RESULTS AND DISCUSSION

The preconditioning was ended after 168 h, with retesting after 24 and 72 h, since the power change was below 0.3 % for the last interval. Temperature control was quite challenging due to joule heating with 9 A current injection during the B-O-LID testing. It was possible to realize a mean temperature of 25 °C with a fluctuation range of ±6 K. Total power loss for preconditioning was less than 1 %, meaning that the tested modules contain stable or stabilized cells regarding B-O-LID.

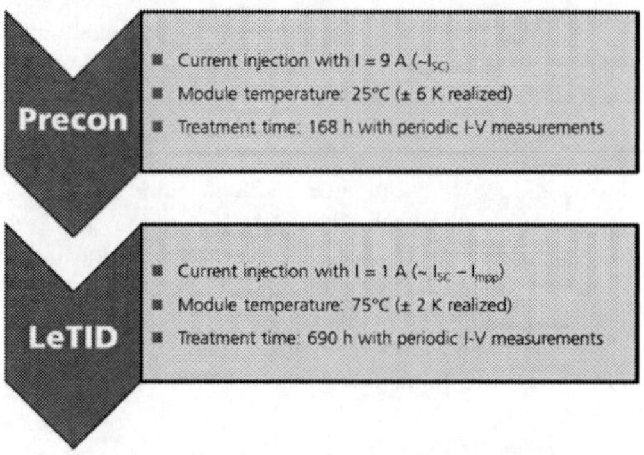

Figure 2: Testing scheme for LeTID module benchmark

In the LeTID test two module types had negligible power loss, one multi-Si as well as one mono-Si (Figure 3). The multi-Si devices in the current sample lost less than 2 % of power. Three mono-Si modules had substantial losses of more than 3 % up to 6 %. Another finding is that devices of the same manufacturer can scatter, but most are close to each other. This may happen if cell quality is not sufficiently controlled and LeTID susceptible and stable cells are present in the cell batch.

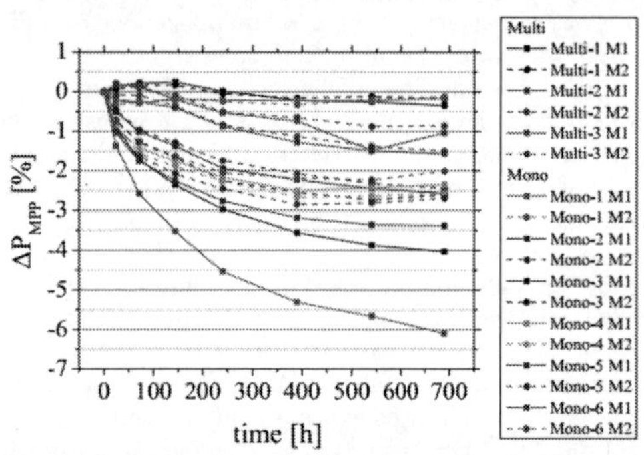

Figure 3: Measured power degradation during LeTID test

EL imaging is useful to qualitatively (in first order approximation) identify the degradation state of cells in the modules. As can be seen from Figure 4, there are also degraded cells in the stable modules, but in total the individual degradation is not enough to affect module power. For the module with the highest degradation almost every cell is affected by LeTID (Figure 5) and leads to the significant power loss. In contrast, module 2 from the same manufacturer has fewer degraded cells and consequently less power loss.

Due to the dependency on individual cell degradation the module parameters degrade in different ways (Figure 7). Some modules show reduction mainly in short circuit current I_{SC} (~ 1 - 2.7 %). Modules with highest power losses have also substantial losses in V_{OC}. The FF varies from improving to degrading, which can be an artifact of interconnection and changing mismatch. Larger FF degradation was only observed for mono-Si.

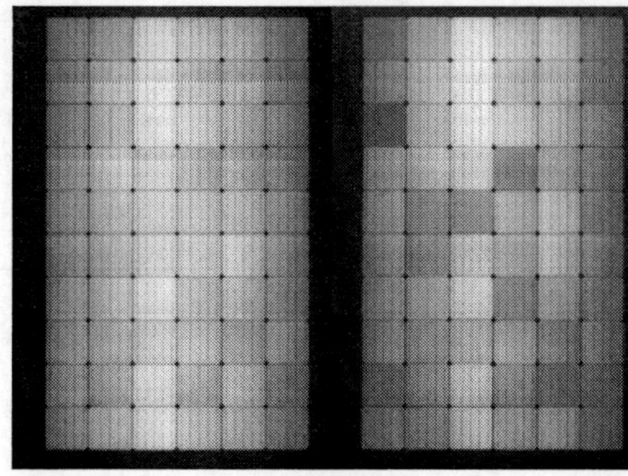

Figure 4: EL images (0.85 A, 15s) of manufacturer Mono-1, M1 (left: initial, right: after LeTID)

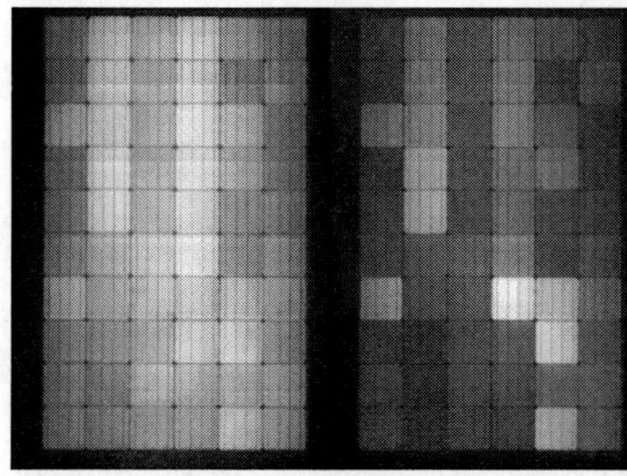

Figure 5: EL images (0.85 A, 15s) of manufacturer Mono-6, M1 (left: initial, right: after LeTID)

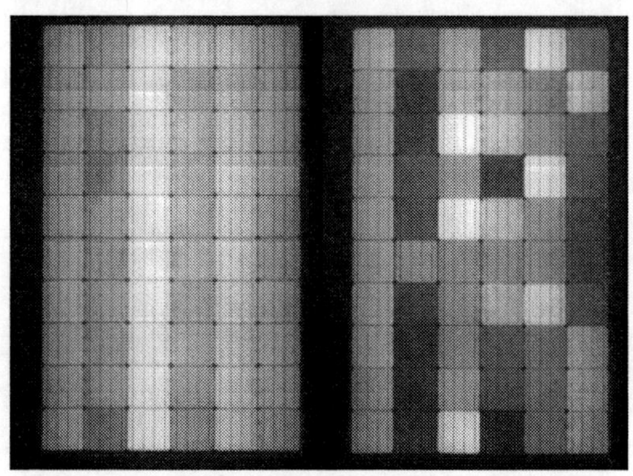

Figure 6: EL images (0.85 A, 15s) of manufacturer Mono-6, M2 (left: initial, right: after LeTID)

Figure 8 shows the comparison of the measured I-V-curves at the end of the different test stages. The module shows only slight changes between after preconditioning. After the LeTID test I_{SC} as well as V_{OC} degraded substantially and also the FF decreases.

5 CONCLUSION

It was verified that the chosen LeTID test conditions could reliably differentiate modules prone to this degradation mechanism. The preconditioning for B-O-LID resulted in minor degradation (<1 % power loss). For product qualification the type of degradation is usually of minor interest. Therefore, it might be sufficient to directly start with the LeTID test procedure. In this case one should be aware that the LeTID test conditions significantly accelerate possible B-O-LID. Therefore additional intermediate measurements in the first week are useful to detect any related degradation before a possible regeneration starts.

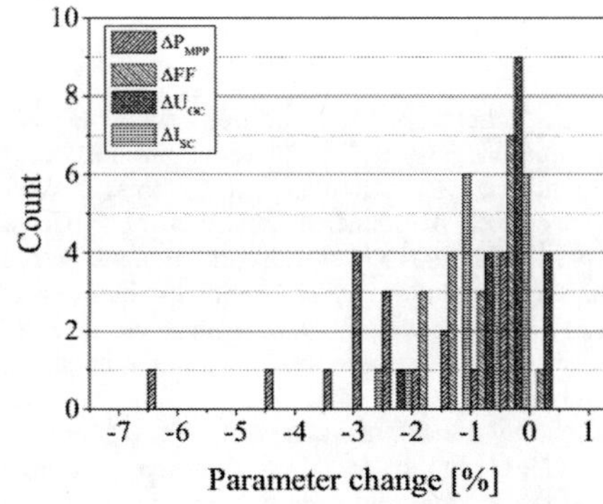

Figure 7: Histogram of module parameter variation due to LeTID

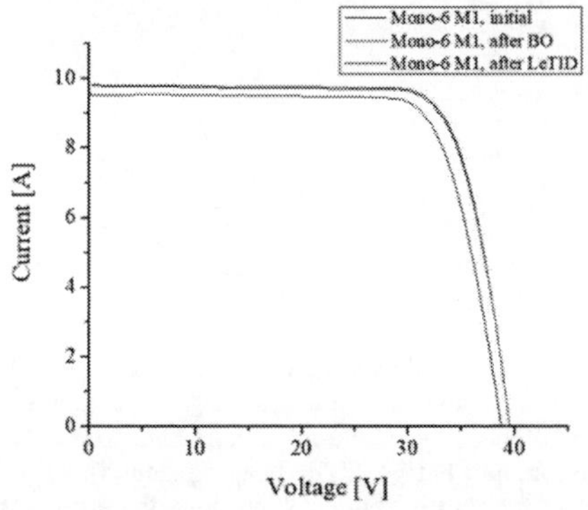

Figure 8: I-V-curve of worst module in LeTID test

In this benchmark, some commercially available mono-Si PERC modules show high LeTID. Currently the number of tested module types is too small to allow generalization about what is more affected (mono or multi) and how many modules in the market might suffer from LeTID. This benchmark goes some way to quantify the potential magnitude of the problem. The modules with moderate LeTID tended not to scatter as much as the worse ones. This indicates that some manufactures might need additional control of their cell quality or that in some production processes there is a wide range of LeTID sensitivity of the individual cells. This clearly indicates that the statistic in this work is not yet robust enough to make final judgements between technologies. The variability makes it more difficult to define a sensible number to identify problematic types, two modules certainly seems to be on the small side.

LeTID is not identified during the IEC certification cycle but potentially affects system performance significantly. An adequate testing procedure should be included in the standard. It is critical for everybody with an interest in the long term performance of a PV system (e.g. installers, financiers or owners) to verify with the manufacturer what yield losses might be expected if LeTID susceptible PERC modules are installed. For manufacturers, especially ones buying cells on stock market, diagnostic and non-destructive tests for incoming cells are needed as a quality assessment.

LeTID can apparently be controlled in the manufacturing process, as some devices are only marginally affected. Tight manufacturing control seems to play a part as susceptible modules appear to have a significant scatter in their susceptibility.

6 ACKNOWLEDGEMENTS

Jens Froebel, Frank Wenger and Ulli Zeller are gratefully acknowledged for testing and sample characterization. Michael Lohmeier is acknowledged for procurement of samples and negotiation with procurement department.

7 REFERENCES

[1] International Technology Roadmap for Photovoltaic (ITRPV), 8th Edition, 2017
[2] F. Kersten et al., Degradation of multicrystalline silicon solar cells and modules after illumination at elevated temperature, Sol. Mat. Ener. Sol. Cells 142 (2015) 83-86
[3] Fung et al, Ener. Proc. 124 (2017)
[4] T. Luka et al., "Intra-grain versus grain boundary degradation due to illumination and annealing behavior of multi-crystalline solar cells", Sol. Ener. Mat. & Sol. Cells 158 (2016)
[5] F. Fertig et al., Ener. Proc. 124 (2017) 338-345
[6] K. Ramspeck, Light Induced Degradation of rear passivated mc-Si solar cells, 27th PVSEC, 2012
[7] F. Kersten et al., System performance loss due to LeTID, Silicon PV 2017
[8] T. Luka et al., "Defect formation under high temperature dark-annealing compared to elevated temperature light soaking", Sol. Ener. Mat. & Sol. Cells 187 (2018)
[9] T. Luka et al., "Microstructural investigation of LID sensitive mc-PERC solar cells", Ener. Proc. 124 (2017)
[10] T. Niewelt et al., "Understanding The Light-Induced Degradation At Elevated Temperatures: Similarities Between Multicrystalline and Floatzone P-Type Silicon", Prog Photovolt Res Appl. 26 (2018), 533-542
[11] D. Bredemeier et al., "Possible Candidates for Impurities in mc-Si Wafers Responsible for Light-Induced Lifetime Degradation and Regeneration", Sol. RRL 2 (2018), 1700159
[12] M.A. Jensen, "Evaluating root cause: The distinct roles of hydrogen and firing in activating light- and elevated temperature-induced degradation", J. Appl. Phys. 124 (2018), 085701
[13] D. Chen et al., "Evidence of an identical firing-activated carrier-induced defect inmonocrystalline and multicrystalline silicon", Sol. En. Mat. Sol. Cells 172 (2017), 293-300
[14] C. Chan et al., "Modulation of Carrier-Induced Defect Kinetics in Multi-Crystalline Silicon PERC Cells Through Dark Annealing", Sol. RRL 1 (2017), 1600028
[15] F. Kersten et al., "New light induced volume degradation effect of mc-Si solar cells and modules", in Proc. 31st EU PVSEC, Hamburg, Germany , 1830-1834

MULTI-IMAGING OF PV MODULE INHOMOGENEITIES IN 17 KW PV POWER PLANT AND MUTUAL CORRELATIONS

Matevž Bokalič, Kristijan Brecl, Marko Topič
University of Ljubljana, Faculty of Electrical Engineering
Tržaška 25, SI-1000 Ljubljana, Slovenia
E-mail: Matevz.Bokalic@fe.uni-lj.si

ABSTRACT: We have performed a detailed spatial conversion efficiency inhomogeneity inspection of a 17 kW PV power plant using an in-house developed multi-imaging system that comprises visual, UV fluorescence, EL near infrared and EL short infrared spatial imaging methods applied to full-size PV modules. All the images were comparatively analysed using an in-house developed software package. A statistical analysis of detected inhomogeneities and correlations between them is performed on a per-module basis. Several correlations between inhomogeneities were detected. *Disconnected areas* have high correlation with *Cracks*, which is obvious since cracks cause them. High correlation is also present between different types of cracks, as well as between different types of *Contact finger interruptions*, because these inhomogeneities were classified into types based on their intensity. Negative correlation between *Bad Edges – Perpendicular* and *Compression cracks at random positions* was not expected, but may provide a clue that *Compression cracks* are predefined by the cell manufacturing process. A 90% correlation between *Compression cracks at random positions* and *UV – Dots* suggests that UV – fluorescence can be used to detected *Compression cracks*. The correlation between *UV – Along Ribbons* and *Contact finger interruptions* may explain that the non-UV-fluorescent cell area along the ribbons is caused by the micro cracks below ribbons, which also cause contact finger interruptions.
Keywords: Electroluminescence, Inhomogeneities, Imaging, Multicrystalline Silicon, PV Module

1 INTRODUCTION

Inhomogeneities in general have negative impact on performance of PV modules, reveal manufacturing deficiencies and indicate possible further degradation. Various inhomogeneities can be detected differently using different spatially resolved characterization methods like visual inspection, UV fluorescence, electro-luminescence (EL), lock-in thermography etc. [1,2]. A combination of techniques allows identifying more inhomogeneities more precisely and provides insight into causes and physical mechanisms behind inhomogeneities.

Inhomogeneity inspection is usually performed on per-inhomogeneity basis and brings detailed knowledge about single inhomogeneities, with emphasis on cracks [3,4], contact finger interruptions [5,6], inhomogeneities detected by UV fluorescence [7], potential induced degradation [8], etc.

However, the way in which correlations interact with each other is less researched. For that reason, we will statistically evaluate the presence and magnitude of inhomogeneities present in a PV power plant. Then we will correlate these data between the modules in the power plant. While correlation is not evidence of causation, it still provides important information on how inhomogeneities appear and, based on the knowledge about inhomogeneities, it allows to predict eventual causes. Knowing how inhomogeneities are correlated makes their detection easier and provides clues on their origin.

In this paper, we carefully examine all the modules of the 17 kW PV power plant visually, by electro-luminescence and by UV fluorescence. The power plant was disassembled for testing after 5 years of operation. The presence of inhomogeneities in the acquired images is evaluated on the per-module basis. Correlation between inhomogeneities is evaluated by calculation of Pearson coefficients between all classified inhomogeneities. Results of inhomogeneity presence and their cross-correlations are presented and discussed.

2 METHODOLOGY

2.1 Spatial characterisation of PV-power plant

The PV power plant on the roof of the Faculty of Electrical Engineering within University of Ljubljana (N46°02.69', E14°29.38') consists of 74 polycrystalline silicon modules and 2 test modules. It was in operation for five years before we disassembled it to perform, among other tests, precise spatial characterization of all the modules. We have used the imaging inspection system built in-house at Laboratory of Photovoltaics and Optoelectronics (LPVO). The imaging setup is positioned indoor in the dark room and comprises three cameras in different spectral ranges:

- a consumer camera for high-resolution visual and UV fluorescence image,
- a 16-megapixel silicon CCD camera for near infrared (NIR) part of the EL spectrum and
- an InGaAs plane-array camera for short-wave infrared (SWIR) part of the EL spectrum.

The detailed description of the imaging system is given in [9].

2.2 Image preparation and inspection

After acquisition, background was subtracted from EL images and in case of EL-NIR images vignetting was corrected [10]. Visual and UV fluorescence images were not corrected. The inspection of the images for inhomogeneities was performed manually, but using a previously presented inspection software [9] that displayed the three images (visual, UV fluorescence and EL) side-by-side, with option of swapping EL image between NIR and SWIR. The software also guided the inspection process through the list of inhomogeneities and stored the results to a database.

The list was constructed from our previous work and literature [11–14], and includes (only detected shown):

- *Disconnected areas*
- *Cracks*
    - *active, semi active, inactive, compression*

- *Contact finger interruptions*
  - *compete, strong, weak*
- Manufacturing marks (*Dark dots*)
  - *perpendicular, parallel, shading*
- Silicon brick mark (*Bad edges*)
  - *parallel, perpendicular, frame, corner*
- *Ribbon* related failures
  - *corrosion, improper interconnect*
- *UV* exhibited inhomogeneities
  - *average area, cracks, along ribbon, dots*

For each inhomogeneity, two parameters were determined:

- Presence – the number of cells per module, where the inhomogeneity is present,
- Magnitude – the magnitude of the inhomogeneity:
  - linear inhomogeneities (1-D: e.g. regular cracks) in average cumulative inhomogeneity length per cell,
  - areal inhomogeneities (2-D: e.g. disconnected areas, contact finger interruptions) in average percentage of cell area affected,
  - point inhomogeneities (0-D: e.g. compression cracks, dark dots, UV-dots) in average number of inhomogeneities present per cell.

For evaluation, inhomogeneity's presence was multiplied by magnitude to reflect the cumulative length/area/number of inhomogeneity in the module. To identify possible correlations, Pearson correlation coefficient was calculated between all inhomogeneities for each module:

$$Correl(X,Y) = \frac{\sum(x-\bar{x})(y-\bar{y})}{\sqrt{\sum(x-\bar{x})^2 \sum(y-\bar{y})^2}} \quad (1)$$

3 RESULTS

3.1 PV power plant performance

In order to put the results in context, it is important to know how the inspected PV power plant was performing before and after the test. Figure 1 presents the energy performance ratio (PR) [13] of the power plant one and a half year before and after inspection. In light grey there is a raw daily PR, while the light red presents a temperature corrected daily PR. The linear regression line of temperature corrected PR shows no degradation.

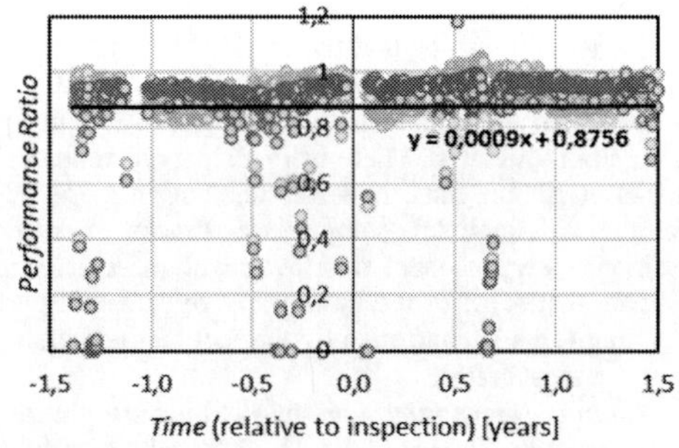

Figure 1: Performance ratio of the power plant

3.2 Inhomogeneities

Visual inspection did not reveal any damages or inhomogeneities in any module. In UV fluorescence, the standard pattern of UV fluorescent rectangles in the middle of the cell and non UV fluorescent ribbons along the edges of the cells were always present, however there were additional non UV fluorescent features present in the images. Electroluminescence images were, as expected, the most revealing showing multitude of inhomogeneities. We mostly used NIR EL images for inhomogeneity detection due to a higher resolution of the sensor.

In general, the EL images of the modules showed reasonable number of inhomogeneities, however there were some exceptions where inhomogeneities like cracks or contact finger interruptions were very frequent. This is also indicated by high standard deviation of cumulative inhomogeneity presence, which is often greater than its average presence. The results are summarized in the left part of the Figure 2.

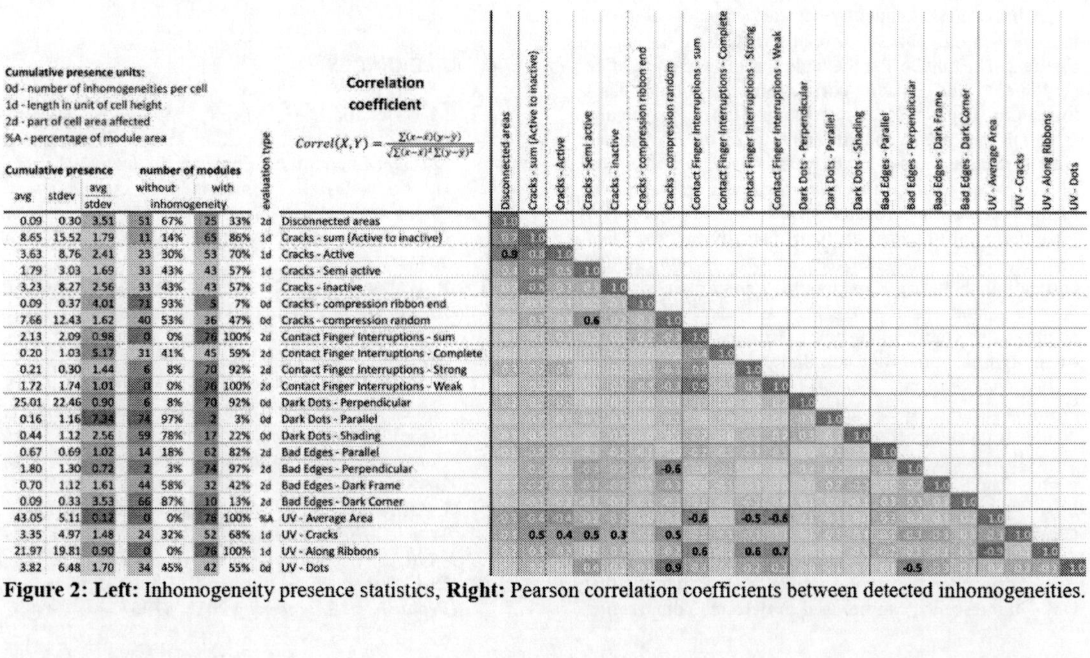

Cumulative presence avg	stdev	avg/stdev	number of modules without inhomogeneity		with inhomogeneity		evaluation type	
0.09	0.30	3.51	51	67%	25	33%	2d	Disconnected areas
8.65	15.52	1.79	11	14%	65	86%	1d	Cracks - sum (Active to inactive)
3.63	8.76	2.41	23	30%	53	70%	1d	Cracks - Active
1.79	3.03	1.69	33	43%	43	57%	1d	Cracks - Semi active
3.23	8.27	2.56	33	43%	43	57%	1d	Cracks - inactive
0.09	0.37	4.01	71	93%	5	7%	0d	Cracks - compression ribbon end
7.66	12.43	1.62	40	53%	36	47%	0d	Cracks - compression random
2.13	2.09	0.98	0	0%	76	100%	2d	Contact Finger Interruptions - sum
0.20	1.03	5.17	31	41%	45	59%	2d	Contact Finger Interruptions - Complete
0.21	0.30	1.44	6	8%	70	92%	2d	Contact Finger Interruptions - Strong
1.72	1.74	1.01	0	0%	76	100%	2d	Contact Finger Interruptions - Weak
25.01	22.46	0.90	6	8%	70	92%	0d	Dark Dots - Perpendicular
0.16	1.16	7.24	74	97%	2	3%	0d	Dark Dots - Parallel
0.44	1.12	2.56	59	78%	17	22%	0d	Dark Dots - Shading
0.67	0.69	1.02	14	18%	62	82%	2d	Bad Edges - Parallel
1.80	1.30	0.72	2	3%	74	97%	2d	Bad Edges - Perpendicular
0.70	1.12	1.61	44	58%	32	42%	2d	Bad Edges - Dark Frame
0.09	0.33	3.53	66	87%	10	13%	2d	Bad Edges - Dark Corner
43.05	5.11	0.12	0	0%	76	100%	%A	UV - Average Area
3.35	4.97	1.48	24	32%	52	68%	1d	UV - Cracks
21.97	19.81	0.90	0	0%	76	100%	1d	UV - Along Ribbons
3.82	6.48	1.70	34	45%	42	55%	0d	UV - Dots

Figure 2: Left: Inhomogeneity presence statistics, **Right:** Pearson correlation coefficients between detected inhomogeneities.

Disconnected areas were present in one third of the modules, but their average cumulative presence was below 8% of cell area per module, with high standard deviations.

Cracks were quite common, with 86% of the modules having at least one occurrence of crack. The average cumulative crack length was 9 cell lengths, out of which 4 were active – meaning that the crack had an influence on EL intensity in the surrounding areas. The standard deviation of 16 cell lengths indicates very unequal distribution of cracks. There were almost no compression cracks at ribbon ends, however there were many (8 in average per module) compression cracks at random positions. Very unequal distribution of the cracks may indicate that some of the modules were damaged during handling. This is confirmed by the fact that test modules have the most cracks.

Contact finger interruptions (CFIs) were present in all the modules, but mainly due to the weak type of CFIs. Strong and complete CFIs in average affect only 40% of one cell area in the whole module in contrast to 1.7 cell areas for the weak CFIs. The distribution of CFIs is quite homogeneous for weak occurrences, but highly inhomogeneous for complete occurrences.

Dark dots were mainly observed in perpendicular orientation to the busbar. They were present in 92% of the modules with quite homogenous distribution.

Bad edges were present in all but two modules, mostly in orientation perpendicular to the busbars but also in parallel orientation.

UV fluorescence was present in all modules. The maximum part of UV fluorescent area was reaching 50% of the module area with average at 43%. The 50% limit should represent the limit for these PV modules in this power plant at this point in time. Anything below 50% indicates a presence of certain inhomogeneities in the module.

Comparison between EL NIR and EL SWIR images reveals very little differences. The exceptions are some grains that appear darker in SWIR images. This change is most likely due to varying diffusion lengths [16].

3.3 Correlations

A list of most frequently detected inhomogeneities is presented in the table of correlations in the right part of Figure 2. The numbers in the table are Pearson correlation coefficients and notable correlations have bold black formatting. In general, the correlation is the highest among *Disconnected areas* and different types of *cracks*, and among different types of *Contact finger interruptions*. This is expected since *Disconnected areas* are caused by multiple cracks, and because *Cracks* and *Contact finger interruption* are divided by their intensity (*active, semi-active, inactive* and *complete, strong, weak*). A 0.6 correlation between *Compression cracks at random positions* and *Semi-active cracks* may show that compression cracks often cause further cracking of the cells. Significant negative correlation between *Bad Edges – Perpendicular* and *Compression cracks at random positions* is not expected, but may provide a clue that compression cracks at random positions are predefined by the cell manufacturing process. This connection is also visible in 0.5 correlation between *UV – Dots* and *Bad Edges – Perpendicular*. Strong correlation between *UV – Cracks* and cracks was reported before [7] and is confirmed herein. A 0.9 correlation between *Compression cracks at random positions* and *UV – Dots* suggests that UV – fluorescence can be used to detected compression cracks. The correlation between *UV – Along Ribbons* and *Contact finger interruptions* was not expected, but it may explain that the non-UV-fluorescent cell area along ribbons is caused by the micro cracks below ribbons, which also cause contact finger interruptions. Furthermore, the lack of correlation between *UV – Along Ribbons* and *Contact finger interruptions – complete* may reveal that complete finger interruptions have different origin than strong and weak ones.

4 CONCLUSION

We have presented the results of the first multi-imaging inspection of the 17 kW PV power plant. Multi-imaging includes visual, UV fluorescence, NIR EL and SWIR EL. The image analysis was done manually but using a software program that systematically lead the analysis and stored the results. The presence of inhomogeneities was statistically analysed based on per-module cumulative presence. The correlations were analysed based on the Pearson's coefficient calculated between different inhomogeneities in the module.

The correlations exist between different types of inhomogeneities. Correlation between inhomogeneities detected by UV-fluorescence method and cracks and contact finger interruptions allows the use of UV-fluorescence for indication of these inhomogeneities. Different correlations for compression cracks at random position suggest that they may be predefined by cell manufacturing process.

The analysed correlations provide opportunities for further research, and require correlation analysis on cell, as well as on module performance level, which is our future research focus.

5 ACKNOWLEDGMENTS

The authors acknowledge the financial support from the Slovenian Research Agency (Research Programme P2-0197, Project J2-7105).

6 REFERENCES

[1] M. Bokalič, M. Topič, *Spatially Resolved Characterization in Thin-Film Photovoltaics*, Springer International Publishing, Cham, **2015**.

[2] D. Abou-Ras, T. Kirchartz, U. Rau, *Advanced Characterization Techniques for Thin Film Solar Cells*, Wiley-VCH Verlag GmbH & Co. KGaA, **2011**.

[3] P. B. Poulsen, D. Sera, G. A. dos Reis Benatto, P. Hacke, H. R. Parikh, S. V. Spataru, *33rd European Photovoltaic Solar Energy Conference and Exhibition* **2017**, 1466.

[4] H. R. Parikh, S. Spataru, D. Sera, C. Mantel, S. Forchhammer, G. Benatto, Alves dos Reis, N. Riedel, P. B. Poulsen, in *Proceedings of 7th World Conference on Photovoltaic Energy Conversion*, **2018**.

[5] I. Zafirovska, M. K. Juhl, J. W. Weber, J. Wong, T. Trupke, *IEEE Journal of Photovoltaics* **2017**, *7*, 1496.

[6] D.-C. Tseng, Y.-S. Liu, C.-M. Chou, *Mathematical Problems in Engineering* **2015**, 879675.

[7] M. Arnaud, M. Siebert, I. Kunze, G. Mathiak, M. Köntges, in *44th IEEE Photovoltaic Specialist Conference (PVSC)*, IEEE, Washington, DC, USA, **2017**, pp. 1710–1716.
[8] K. Brecl, M. Bokalič, M. Topič, *J. Sol. Energy Eng* **2017**, *139*, 031011.
[9] M. Bokalič, K. Brecl, M. Topič, *33rd European Photovoltaic Solar Energy Conference and Exhibition* **2017**, 1682.
[10] K. Bedrich, M. Bokalič, M. Bliss, M. Topič, T. R. Betts, R. Gottschalg, *IEEE Journal of Photovoltaics* **2018**, 1.
[11] M. Bokalič, Conversion Efficiency Inhomogeneities of Photovoltaic Devices, University of Ljubljana, **2013**.
[12] M. Köntges, S. Kurtz, C. Packard, U. Jahn, K. A. Berger, K. Kato, T. Friesen, H. Liu, M. Van Iseghem, *Review of Failures of Photovoltaic Modules*, IEA, **2014**.
[13] *IEC NP 60904-13 Electroluminescence of Photovoltaic Modules (82/901/NP)*, **2014**.
[14] K. Bedrich, et. al, *EL RR (Electroluminescence Round Robin) Project Draft*, **2016**.
[15] K. Brecl, M. Topič, *Journal of Renewable and Sustainable Energy* **2016**, *8*, 043501.
[16] P. Würfel, T. Trupke, T. Puzzer, E. Schäffer, W. Warta, S. W. Glunz, *J Appl Phys* **2007**, *101*, 123110.

ENHANCEMENT OF RELIABILITY FOR PHOTOVOLTAIC MODULES BY MORE SEVERE TEST IN ACCORDANCE WITH PARTICULAR ENVIRONMENT OF TAIWAN

Chia-Yuan Gao[1], Chun-Hong Lin[1*], Bo-Cheng Kuo[1], Chih-Wen Huang[2], Chen-Hsiung Chen[2], Qi-Rong Li[2]
Taiwan electric research & testing center, Taouan, Taiwan[1]
Bureau of Standard, Metrology & Inspection, Taipei, Taiwan[2]
[*]Corresponding author. Tel.: +886 3 4839090; Fax: +886 3 4838119
E-mail: hank@ms.tertec.org.tw

ABSTRACT: Taiwan is located at a relatively low latitude area, resulting that the temperature during the summer season is usually high. Therefore, photovoltaic (PV) modules shall have excellent capability to resist high temperature and high humidity in order to ensure the reliability and output performance of the PV module. In addition, the PV modules must be able to resist the salt corrosion because Taiwan is a sea-island country, which is not only increasing the humidity but also increasing the salt-mist corrosion. The certificate program of PV Taiwan plus is a more severe for high temperature and high humidity test, as well as for salt-mist test. The certificate program of PV Taiwan plus test can enhance the reliability of the PV module in accordance with the particular environment of Taiwan. Moreover, according to the actual market surveys of various types of failure statistics and different experimental proofs, the causes of failure of PV modules have 5 affecting factors after PV modules are used for a long-term time, including time, temperature, humidity, sunlight exposure, and voltage. In this test, the above-mentioned 5 affecting factors are adjusted in order to comply with particular environment of Taiwan.
Keywords: Reliability, PV modules, Salt mist, High temperature, High humidity

1 INTRODUCTION

Present qualifications and certification standards of photovoltaic (PV) modules are mainly focused on the design confirmation, type approval and safety inspection, but lack the relevant methods to assure reliability and life time of the PV modules. It is vital how to assure the reliability and performance of the PV modules within 25 years of usage, which is usually guaranteed by the PV module manufacturers. Therefore, the certificate program of PV Taiwan plus has been implemented in order to assure the reliability of the PV modules. In addition, the reliability of the PV modules is significantly correlated to the surrounding environment such as temperature, humidity, and salt mist.

The region of Taiwan has an average relative humidity above 80% because Taiwan is surrounded by seas. In addition, Taiwan is located at a relatively low latitude area, resulting that the temperature during the summer season is usually high. Therefore, PV modules shall have excellent capability to resist high temperature and high humidity in order to ensure the reliability and output performance of the PV module. Moreover, through the high temperature and high humidity test, defects will be detected, including cell peeling off, ethylene vinyl acetate (EVA) peeling off, discoloration, surface fog, browning, darkening of connecting wires, transparent conductive oxide corrosion, soldered point corrosion, thin-film yellowing, junction box degumming falling off, etc. Taiwan is a sea-island country, which is not only increasing the humidity but also increasing the salt-mist corrosion. Therefore, the PV modules must be able to resist the salt corrosion. Particularly, the PV modules, which is installed on salt beach lands or marine culture zones shall have greater ability to resist salt corrosion than the others zone.

According to the actual market surveys of various types of failure statistics and different experimental proofs, the causes of failure of PV modules have 5 affecting factors after PV modules are used for a long-term time, including time, temperature, humidity, sunlight exposure, and voltage. In this test, the above-mentioned 5 affecting factors are adjusted in order to comply with particular environment in Taiwan. Moreover, the range of score points for affecting factors from 1-5 (extremely week to extremely high) is provided and presented in radar maps. In accordance with the particular environment of Taiwan, the temperature and humidity of the high temperature and high humidity test are increased to 85°C and 85% respectively, and the testing time is extended to 3000 hours. In addition, the testing level of the salt-mist test is enhanced to the highest level (Level 6), which based on adjusted affecting factors.

In this paper, the certificate program of PV Taiwan plus is both addressed and investigated. It is a more severe for high temperature and high humidity test, as well as for salt-mist test. Furthermore, a more severe test can not only enhance the reliability of the PV module in accordance with the particular environment of Taiwan but also promote the development of the PV industry all over the world.

2 METHOD

The method and requirement are established in this specification in order to determine the high performance and durable reliability characteristics of PV modules.

2.1 Safety requirements

PV modules shall have excellent structural design, which is basic performance and safety. Design qualification and type approval of crystalline silicon and thin-film terrestrial PV modules comply with the requirements of CNS 15114 [1] and CNS 15115 [2] respectively. The structure and safety of PV modules comply with the testing requirements of CNS 15118-1 [3] and CNS 15118-2 [4].

2.2 Power generation performance

PV modules shall have the high performance with

maximum output power in order to ensure the quality of electricity supply after the installation of the power plant. Crystalline silicon and thin-film terrestrial PV modules are further tested for the maximum output performance, which is according to 10.2 in CNS 15114 and 10.2 in CNS 15115. The test for crystalline silicon PV modules and the thin-film PV modules are based on the quantity of the battery composition and dimension, respectively. Furthermore, the testing results must fulfill PV Taiwan plus technical specification as mentioned in Table 1 and Table 2.

2.3 High temperature and high humidity test

According to the local weather conditions in Taiwan, PV modules shall have excellent capability to resist high temperature and high humidity in order to ensure the reliability and output performance of the PV module during its useful lifetime. Moreover, through the high temperature and high humidity test, defects will be detected, including cell peeling off, EVA peeling off, discoloration, surface fog, browning, darkening of connecting wires, transparent conductive oxide corrosion, soldered point corrosion, thin-film yellowing, junction box degumming and falling off, etc. Therefore, the test for PV Taiwan plus technical specification increases the testing time for high temperature and high humidity test to 3000 hours.

The crystalline silicon PV module and the thin-film PV module after the test shall comply with the requirements of CNS 15114 and CNS 15115, respectively. During the testing process, the module shall be no occurrence of visually inspected defects. Moreover, the maximum output degradation value shall not exceed 8% of the value for initial measured after test. In addition, the results for insulation resistance and the leakage current tests must be the same as previously testing results.

2.4 Salt-mist test

Because Taiwan is a sea-island country, the PV modules must be able to resist the salt corrosion. Particularly, the PV modules is installed on salt beach lands or marine culture zones, which shall have better ability to resist salt corrosion. Therefore, the highest level (Level 6) is required for the PV Taiwan plus technical specification. Fig. 1 shows the cycle level for salt-mist test. In addition, during the testing process, the module shall be no occurrence of visually inspected defects. Moreover, the maximum output degradation value shall not exceed 8% of the value for initial measured after test. [5].

3 RESULTS AND DISCUSSION

IEC 61215 [6-8] and IEC 61730 [9-10] are still adopted as basic standards of the international PV industry until now. IEC 61730 defines the basic structural requirements for PV modules and specify the electrical and mechanical safety during operation. IEC 61215 defines the output performance, stability, and reliability of PV modules during the operating period and specify the functionality of PV modules. However, researches indicate that the modules pass the most basic functionality test of specification for IEC 61215, which only means that the modules are no major defects in the design of PV modules and able to pass the test in the initial field environmental. Nevertheless, the determination for mid and long-term operating functions as well as various failure modes are relatively weak.

The statistical data of failure, which is conducted by the German system operator is shown Fig. 2. According to the statistical data, the cause of failure for mid and long-term PV modules are mainly due to potential induced degradation (PID), Schottky diode, cell sheets and wire corrosion, EVA discoloration, sealing decomposition and failure, and cell sheet cracks [11]. In addition, according to various experiments, the results indicate that the time, temperature, humidity, sunlight exposure, and voltage are positive correlation with the factors of the failure for PV modules. The radar maps are used to illustrate the relationship between the test items used in this specification and factors of various types in order to demonstrate the point of the test for each item and the reasons for using these items.

According to the actual market surveys of various types of failure statistics and various experimental proofs, the cause of failure of PV modules after a long-term use is correlated to time, temperature, humidity, sunlight exposure, and voltage. In this specification, the test items including above mentioned 5 elements are adjusted in accordance with particular environment of Taiwan. Moreover, according to the comparison between each of the adjusted elements and the original IEC test criteria, test criteria severity, and consideration on the particular environment in Taiwan, the score points range from 1~5 (extremely week to extremely high) are provided and presented in radar maps. In addition, the analysis is conducted on the feasibility of the specification. The affecting factors of feasibility for each test item include cost of test, time for testing, complexity, universality and energy consumption. The range of score points for affecting factors from 1-5 (extremely week to extremely high) is provided and presented in radar maps.

Because Taiwan is surrounded by seas and located at a relatively low latitude area, the region in Taiwan has an average relative humidity above 80% and high temperature during summer season. Therefore, the temperature, humidity, and time are increased to 85℃, 85% RH, and 3000 hours, respectively. Fig. 3 shows the radar map of relevant elements for high temperature and high humidity test. The score points of relevant elements for time, temperature, humidity, sunlight exposure, and voltage are 5, 5, 5, 1, and 1 respectively. Fig. 4 shows the radar map of feasibility for high temperature and high humidity test. The score points of feasibility for cost, time, complexity, universality, and energy consumption are 4, 5, 3, 5, and 3, respectively.

The content for salt mist and humidity are specially high because Taiwan is a sea-island country; what more important is that the PV modules installed on salt beach lands or marine culture zones. Fig. 5 shows the radar map of relevant elements for salt-mist test. The score points of relevant elements for time, temperature, humidity, sunlight exposure, and voltage are 5, 3, 5, 1, and 1 respectively. Fig. 6 shows the radar map of feasibility for salt-mist test. The score points of feasibility for cost, time, complexity, universality, and energy consumption are 3, 4, 3, 4, and 3, respectively.

The decay rate of maximum power for different PV module with different testing time is shown in Fig. 7. It is observed that the decay rates of maximum power for all of PV modules are not exceed 8% of the value for initial measured before testing time of 2000 hours. Nevertheless, the decay rate of maximum for the PV module 1 is substantially increased after 2000 hours, which suggests that the humidity have penetrated considerably into PV module from back sheet or front sheet, resulting in the PID effect [12]. In addition, it can be clearly seen that the decay of maximum power for PV module 1 and 2 are trending slow after testing time of 2000 hours. Moreover, the degree of maximum power decay for the PV module 2 is slower than the others. A slower decay in PV module 2 suggests that the PV module 2 have a better ability for resisting the high temperature and high humidity.

Table 3 shows the results of the sault-mist test for different PV modules. It can be clearly seen that the decay rates of maximum power for all of PV modules are not exceed 3% of the value for initial measured, which suggest that the influence of salt mist is slight for the PV modules. In addition, according to the results of salt-mist test and high temperature and high humidity test, the salt mist, temperature, and humidity for influences of PV modules are not obvious connection.

4 CONCLUSION

In summary, the decay rate of maximum for the PV module 1 is substantially increased after testing time of 2000 hours, which suggests that the humidity have penetrated considerably into PV module from back sheet or front sheet, resulting in the PID effect. Furthermore, according to the results of high temperature and high humidity test, it is necessary that the testing time of high temperature and high humidity test is extended to 3000 hours in accordance with particular environment in Taiwan. Judging by the results of salt-mist test and the high temperature and high humidity test, the influence of high temperature and humidity test for PV modules is more serious than salt-mist test.

Table I: The output performance requirement for Crystalline silicon PV module

Module type	Module output maximum power (W)				Remarks
	2018	2019	2020	2021	
60 pieces of 6" single crystal silicon cell PV modules	295	300	305	310	Module dimension smaller than 1.7m x 1.1m
60 pieces of 6" poly crystal silicon cell PV modules	280	285	290	295	Module dimension smaller than 1.7m x 1.1m
72 pieces of 6" single crystal silicon cell PV modules	356	362	368	374	Module dimension smaller than 2m x 1.1m
72 pieces of 6" poly crystal silicon cell PV modules	336	342	348	354	Module dimension smaller than 2m x 1.1m
96 pieces of 5" single crystal silicon cell PV modules	325	330	335	340	Module dimension smaller than 1.7m x 1.1m

Table II: The output performance requirement for Thin-film PV module

Thin-film PV module dimensional area (A)	Module output maximum power (W)				Remarks
	2018	2019	2020	2021	
A≦1.0m^2	103	106	109	112	
1.0 m^2 <A ≦ 1.6m^2	155	160	165	170	
1.6 m^2 <A ≦ 2.0m^2	205	210	215	220	
A>2.0m^2	311	317	323	329	

Table III: The results of the sault-mist test for PV modules

PV module \ Maximum power	Initial value (W)	After salt-mist test (W)	Decay rate (%)
Module 1	276.6	275.6	0.36
Module 2	281.4	280.7	0.25
Module 3	286.6	279.7	2.41

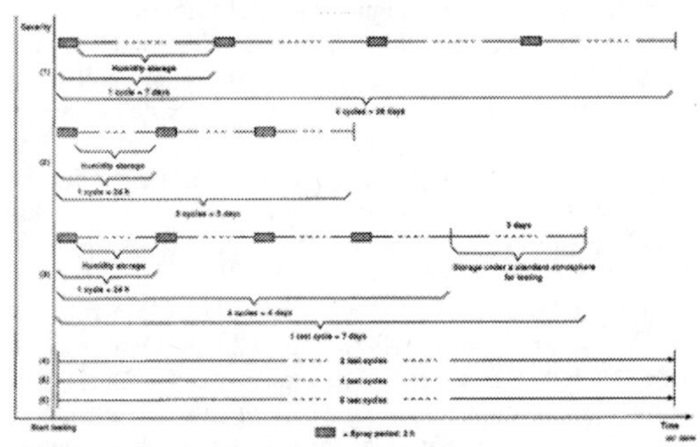

Figure 1: Cycle level of salt-mist test

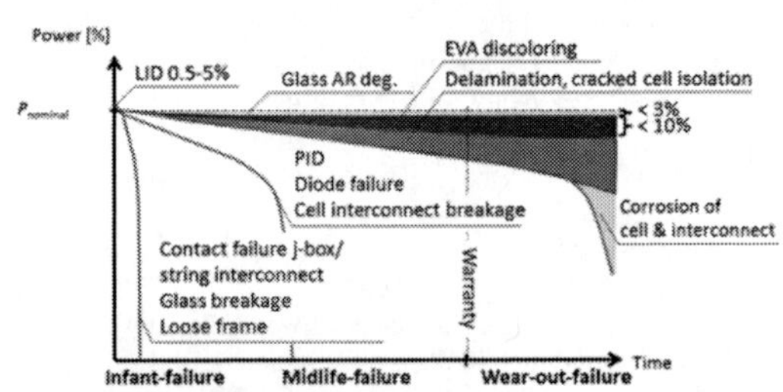

Figure 2: The statistical data of failure

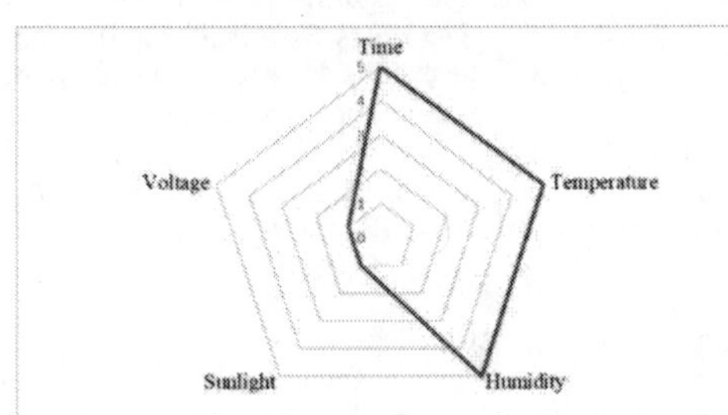

Figure 3: The radar maps of relevant elements for high temperature and high humidity test

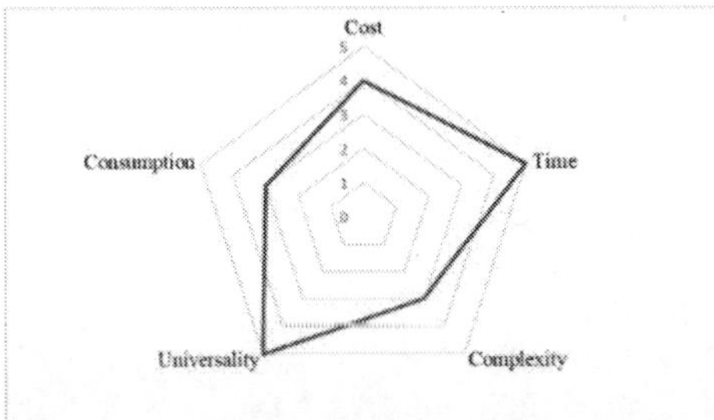

Figure 4: The radar maps of feasibility for high temperature and high humidity test

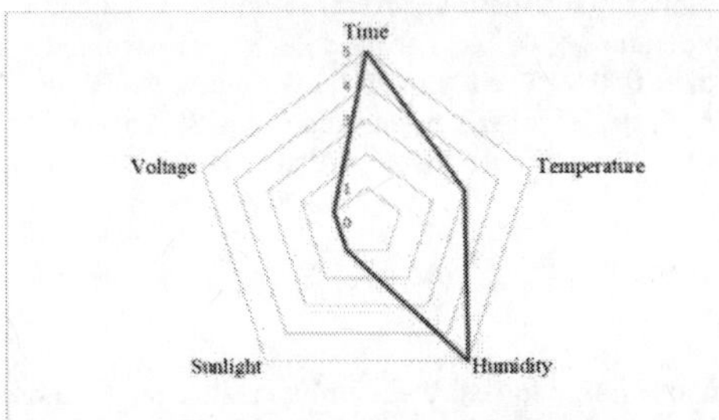

Figure 5: The radar maps of relevant elements for salt mist

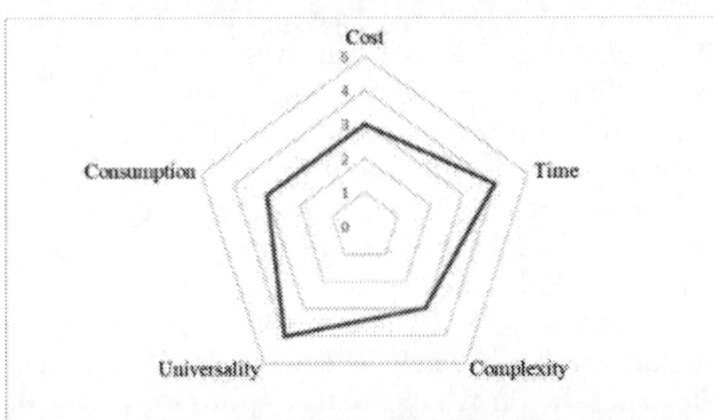

Figure 6: The radar maps of feasibility for salt mist

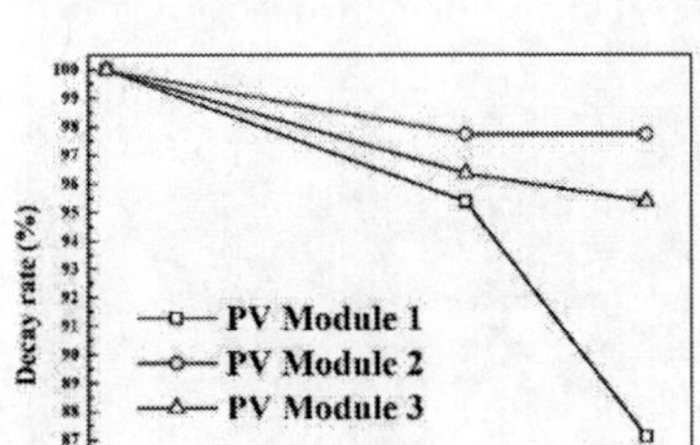

Figure 7: The decay rate of maximum power for different PV module with different testing time

References

[1] CNS 15114 Crystalline silicon terrestrial photovoltaic (PV) modules - Design qualifications and type approval.

[2] CNS 15115 Thin-film terrestrial photovoltaic (PV) modules - Design qualifications and type approval.

[3] CNS 15118-1 Photovoltaic (PV) module safety qualifications - Part 1: Requirements for construction.

[4] CNS 15118-2 Photovoltaic (PV) module safety qualifications - Part 2: Requirements for testing.

[5] CNS 15196 Salt mist corrosion testing of photovoltaic (PV) modules

[6] IEC 61215-1 Terrestrial photovoltaic (PV) modules – Design qualifications and type approval – Part 1: Test requirements (2016).

[7] IEC 61215-1-1 Terrestrial photovoltaic (PV) modules – Design qualifications and type approval – Part 1-1: Special requirements for testing of crystalline silicone photovoltaic (PV) modules (2016).

[8] IEC 61215-2 Terrestrial photovoltaic (PV) modules – Design qualification and type approval – Part 2: Test procedures (2016).

[9] IEC 61730-1 Photovoltaic (PV) module safety qualifications – Part 1: Requirements for construction (2016).

[10] IEC 61730-2 Photovoltaic (PV) module safety qualifications – Part 2: Requirements for testing (2016).

[11] P. Hacke and K. Terwilliger et al., Journal of photonics for energy, 5(1) 053083 (2015).

[12] M. Köntges and S. Kurtz et al., International energy agency-PVPS, T13-01 (2014).

THE INFLUENCE OF THE EVA FILM QUALITY ON THE DEGRADATION OF PV MODULES UNDER HARSH TEST CONDITIONS

Kristijan Brecl[1], Matevž Bokalič[1], Chiara Barretta[2], Gernot Oreski[2], Barbara Malič[3], Marko Topič[1]

[1]University of Ljubljana, Faculty of Electrical Engineering, Ljubljana, Slovenia
[2]Polymer Competence Center Leoben GmbH, Leoben, Austria
[3]Jožef Stefan Institute, Ljubljana, Slovenia

kristijan.brecl@fe.uni-lj.si

ABSTRACT: Photovoltaic modules are normally connected in series and therefore exposed to a high system voltage between the cells and the grounded frame. To study the potentially induced degradation, several in-house developed PV modules were exposed to a damp-heat test under RH 85% and temperature of 60 °C. The modules were assembled with EVA films of different ages and exposed to a high voltage bias of +1000 V. The PV modules with the aged EVA films exhibit very early and severe degradation of solar cells and a high degree of silver corrosion. A DSC analysis reveals a lower crosslinking degree of aged EVA films which leads into the peroxide leftovers that react with silver conduction lines.

Keywords: PID, silver corrosion, PV modules, encapsulant

1 INTRODUCTION

In the last decade, photovoltaics (PV) became an important renewable energy source. Crystalline silicon PV modules share over 90% of the global PV market [1] with a total global installation capacity of 400 GW in 2017 [2]. Many PV systems from small size of several kW up to large PV plants with several MW are in operation. To reduce the joule losses in PV systems, PV modules have to be connected in series. Currently, the PV system voltages are up to 1000 V in Europe. Thus, the PV modules are exposed to a high voltage bias stress. Since strings of PV modules are usually not grounded a PV module has to withstand a high voltage stress of up to a half of the system voltage between the cell and the grounded frame. The high voltage stress can cause a potentially induced degradation (PID). The potential induced degradation mechanisms differ on the solar cell type used in the PV module. P-type crystalline silicon PV modules are usually more sensitive to high negative voltage bias [3-5], while the n-type solar cells are more sensitive to the positive bias. However, the high positive bias can also deteriorate a p-type PV module performance [6-8], especially if the materials used in the module are not of the highest quality level.

In the paper we will present the influence of the EVA film age on the degradation of a positive voltage biased PV module.

2 METHODOLOGY

The degradation behaviour of positively voltage biased PV modules was studied on in-house developed small area PV modules (Figure 1). The module fabrication procedure followed a typical commercial manufacturing process, only on a small scale and manually. We used commercially available pseudo-square high efficient p-type monocrystalline solar cells with no special treatment or production steps to make the solar cells PID resistant. The cells were cut to form small-area solar cells with a dimension of 78 x 39 mm2 by a laser. The laser cuts the cells up to the half of the total depth from the back side; afterwards the cells are broken apart. Each module consists of 12 small-area solar cells connected in series. Due to the small dimension of the module, we used a thinner aluminum frame to seal them and a smaller junction box usually used in thin-film technologies. In all modules we used the same textured solar white glass and a backsheet film. The only difference was the age of the EVA film used for the lamination. The first film was fresh (denoted as "fresh" in the paper), the second was less than half a year (the limit of the usability of the film) stored in our Laboratory (denoted as "aged" film in the paper) and the third film (denoted as "expired" film in the paper) was kept in our Laboratory for more than five years. The laminates consisting of "fresh" and "aged" films were visually good with an excellent adhesion. The "expired" film could not be laminated well. The laminate could be easily peeled off the glass substrate. Anyhow, all three films were optically comparable with no signs of material yellowing.

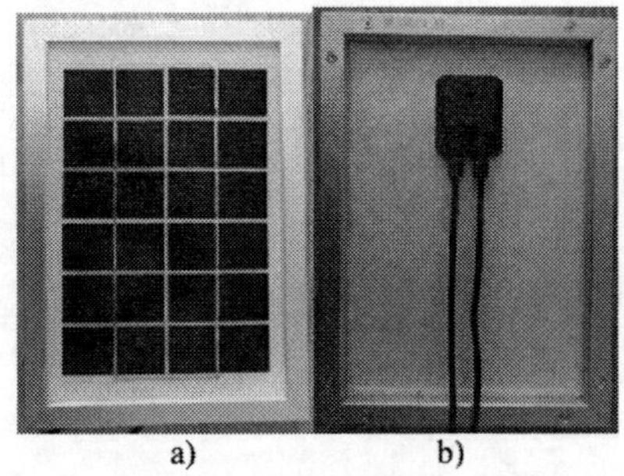

a) b)

Figure 1: Photo of in-house developed test PV module: a) front site, b) back side.

All modules were exposed to an accelerated aging procedure DH80/60 in a climate chamber with a relative humidity of 80% and temperature of 60 °C. During the test, the modules were connected to a high voltage bias of +1000 V. The voltage was applied between short-circuited module contacts and the aluminium frame. During the over 2400-hour test, the modules where frequently taken out for electrical characterisation and electroluminescence (EL) inspection.

3 RESULTS

The leakage current that flows between the solar cells and the grounded aluminium frame was constantly monitored. In all modules we could see a slow increase of the leakage from the beginning of the test to a saturated value of 8 μA, and 6 μA for modules assembled with "aged" and "expired" EVA film, respectively. The PV module with the "fresh" EVA film exhibits much lower leakage current of 1 μA.

The performance of the module was measured under a sun simulator where we could observe a really fast and severe degradation of modules assembled with aged EVA films (Figure 2).

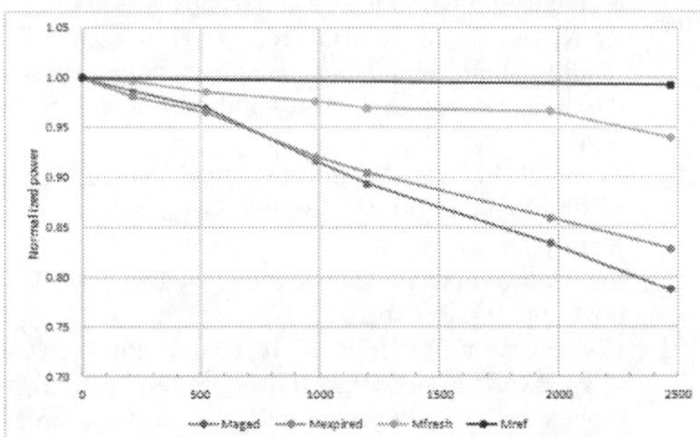

Figure 2: Normalized power output of tested modules.

As it can be seen from Fig. 2, the degradation of the power output starts from the beginning of the test with a linear rate of around 0.2 %/day. The 5 % power loss that is a usual pass criterion in qualification tests [9] was reached after approximately 700 hours of DH85/60 test. Similar results we got from the electroluminescence imaging. Figure 3 shows the EL image of the PV module with the "aged" EVA film before and after the DH85/60 test. The average EL response of the module after the DH85/60 test was 62% lower than before the test. As it can be seen from the Figure 3 the degradation of solar cells in the module is really severe. The cells with a little weaker EL response before the teste were completely degraded.

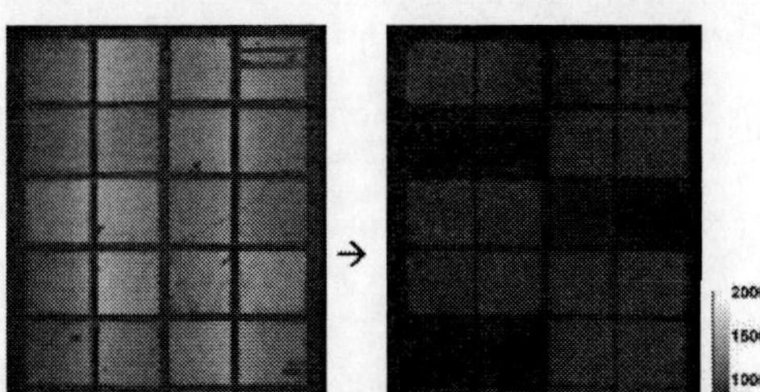

Figure 3: EL image of the PV module with the "aged" EVA film before and after the DH85/60 test.

The EL response of the PV module assembled with the "fresh" EVA film changes very slowly over time and is only 10% lower after the 2400-hour DH85/60 test. The EL image before and after the test is presented in Fig. 4.

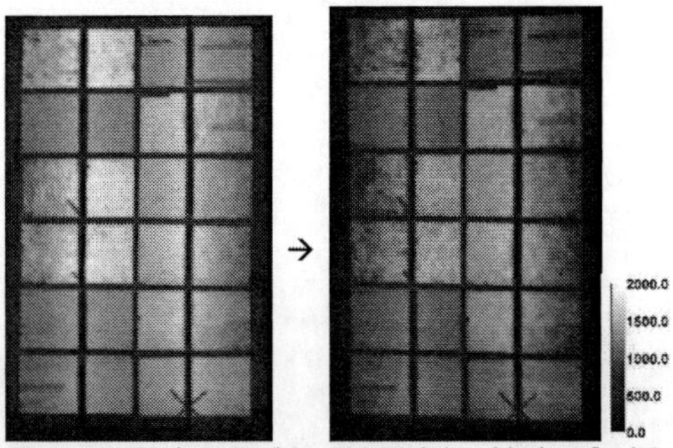

Figure 4: EL image of the PV module with the "fresh" EVA film before and after the DH85/60 test.

Beside the power loss, the modules with "aged" and "expired" EVA film exhibit also a corrosion of silver lines and formation of bubbles above them as it is shown in Fig. 5.

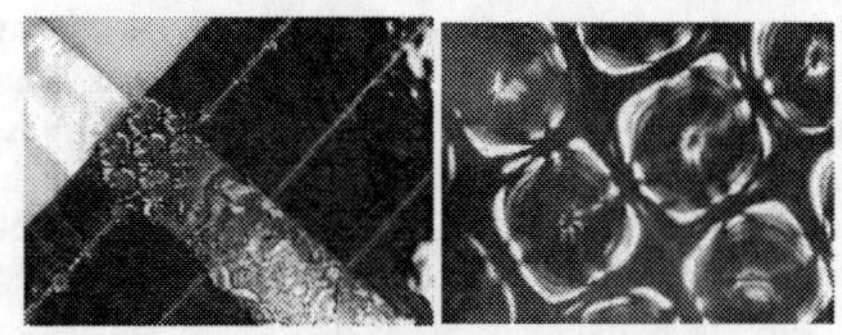

a) b)

Figure 5: Photos of corroded fingers (a) and bubbles on the busbar (b).

To confirm the influence of the EVA film age on PID, we calculated the crosslinking degree of the EVA films by the help of differential scanning calorimetry (DSC). DSC is a thermos-analytical technique to detect endothermic/exothermic processes by measuring absorbed or released amount of energy. The DSC results of the "aged" and "fresh" EVA film in an unprocessed and laminated stage are presented in Figs 6 and 7.

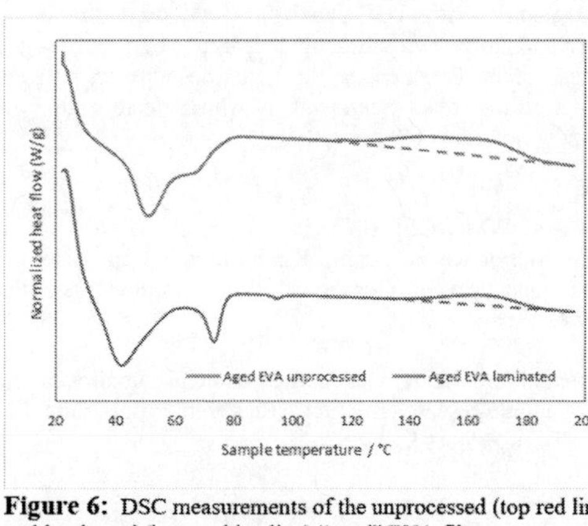

Figure 6: DSC measurements of the unprocessed (top red line) and laminated (bottom blue line) "aged" EVA film.

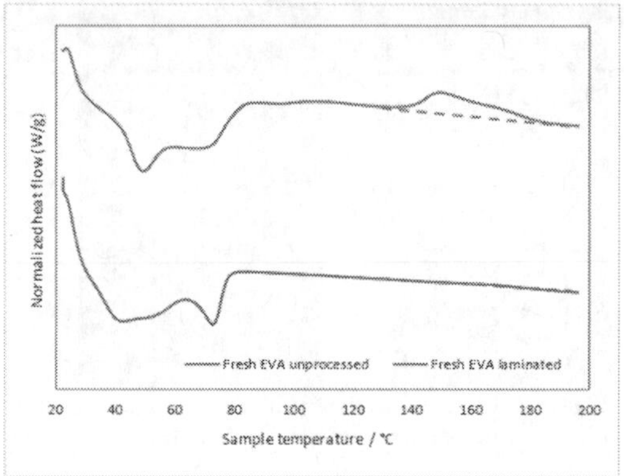

Fig. 7. DSC measurements of the unprocessed (top red line) and laminated (bottom blue line) "fresh" EVA film.

The DSC analysis reveals that almost 40% of the peroxides in the "aged" EVA film did not react. Presumably, these leftovers of peroxides react with silver which leads into the corrosion of finger and busbar conduction lines as observed in modules with "aged" and "expired" EVA film. This is in good agreement with other studies published in literature, where several authors reported a correlation between poorly crosslinked EVA and discoloration of soldering ribbons as well as corrosion of the silver gridlines [10]–[13].

4 CONCLUSION

Small-scale PV modules were exposed to a high voltage bias of +1000 V and tested under damp-heat DH85/60 conditions for more than 2400 hours. The modules were assembled with different qualities of the EVA encapsulant with regard to their age. Three different EVA films were used: "fresh" EVA film, less than half a year old "aged" EVA film and more than five years old "expired" EVA film. The modules with aged EVA films exhibit a severe degradation of solar cells and silver line corrosion. With DSC analysis we could prove an unsatisfactory crosslinking degree of the aged encapsulant. Presumably, the peroxide leftovers in aged EVA films react with silver which leads into the conduction lines corrosion.

ACKNOWLEDGEMENT

The Authors thank Brigita Kmet (Jožef Stefan Institute, Ljubljana, Slovenia) for the valuable measurements of the EVA films.

The Authors acknowledge the financial support of the Slovenian Research Agency (Research Programme P2-0197 and Project J2-7105).

5 REFERENCES

[1] "Photovoltaics Report - Fraunhofer ISE," Fraunhofer Institute for Solar Energy Systems ISE. [Online]. Available: http://www.ise.fraunhofer.de/en/publications/studies/photovoltaics-report.html. [Accessed: 13-Dec-2017].

[2] A. Jäger-Waldau, "Snapshot of photovoltaics – February 2018," EPJ Photovolt., vol. 9, p. 6, 2018.

[3] S. Pingel et al., "Potential Induced Degradation of solar cells and panels," in 2010 35th IEEE Photovoltaic Specialists Conference, 2010, pp. 002817–002822.

[4] H.-C. Liu, C.-T. Huang, W.-K. Lee, and M.-H. Lin, "High Voltage Stress Impact on P Type Crystalline Silicon PV Module," Energy Power Eng., vol. 05, no. 07, p. 455, Sep. 2013.

[5] N. G. Dhere, N. S. Shiradkar, and E. Schneller, "Evolution of Leakage Current Paths in MC-Si PV Modules From Leading Manufacturers Undergoing High-Voltage Bias Testing," IEEE J. Photovolt., vol. 4, no. 2, pp. 654–658, Mar. 2014.

[6] M. K. P. Hacke, "Characterization of Multicrystalline Silicon Modules with System Bias Voltage Applied in Damp Heat." 2011.

[7] Z. Xiong, T. M. Walsh, and A. G. Aberle, "PV module durability testing under high voltage biased damp heat conditions," Energy Procedia, vol. 8, pp. 384–389, Jan. 2011.

[8] K. Brecl, M. Bokalič, and M. Topič, "Examination of Photovoltaic Silicon Module Degradation Under High-Voltage Bias and Damp Heat by Electroluminescence," J. Sol. Energy Eng., vol. 139, no. 3, pp. 031011-031011–6, Mar. 2017.

[9] IEC, "Crystalline Silicon Terrestrial Photovoltaic (PV) Modules—Design Qualification and Type Approval." International Electrotechnical Commission, Geneva, Switzerland, Standard No. IEC-61215 Ed. 2, 2005.

[10] G. Oreski, A. Rauschenbach, C. Hirschl, M. Kraft, G. C. Eder, and G. Pinter, "Crosslinking and post-crosslinking of ethylene vinyl acetate in photovoltaic modules," J. Appl. Polym. Sci., vol. 134, no. 23.

[11] G. Oreski, A. Eder, L. Neumaier, C. Hirschl, W. Krumlacher, and R. Ebner, "Concepts for PV modules optimized for different climatic conditions: Backsheets and encapsulants," presented at the 27th International Photovoltaic Science and Engineering Conference, Otsu, 2014.

[12] S. Jonai, K. Hara, Y. Tsutsui, H. Nakahama, and A. Masuda, "Relationship between cross-linking conditions of ethylene vinyl acetate and potential induced degradation for crystalline silicon photovoltaic modules," Jpn. J. Appl. Phys., vol. 54, no. 8S1, p. 08KG01, Jul. 2015.

[13] M. Köntges et al., "Influence of the Curing State of Ethylene-Vinyl Acetate on Photovoltaic Modules Aging," 28th Eur. Photovolt. Sol. Energy Conf. Exhib., pp. 2832–2837, Nov. 2013.

PV MODULE GLASS STRESS TESTING

Gerhard Mathiak, Lars Falk, Stefan Felden, Lorenz Rimmelspacher, Werner Herrmann, Florian Reil
TÜV Rheinland Energy GmbH, Am Grauen Stein, 51105 Cologne, Germany
Tel.: +49 221 806-3165, Fax: +49 221 806-1350, E-Mail: Gerhard.Mathiak@de.tuv.com

ABSTRACT: Glass breakage is one of the main failure modes found in the field. Mechanical stress caused by hail impact, wind or snow load or localized heating due to hot spots or light arcing are leading to forces which can break the front glass. Understanding the cause for glass breakage is important for future failure avoidance.

Elastic deformations or strains lead to the birefringence of glass so that stress can be made visible. A further developed instrument uses the scattered light method to determine through-the-thickness stress distribution in flat glass products. The instrument was used to measure glass stress in tempered and non-tempered solar glass and to perform rootcause analysis to glass breakage in the field. If the surface stress of one side turns to a higher tensile stress (+50 MPa) glass breakage can occur.
Keywords: PV module reliability, Photoelasticity, Glass stress, Glass breakage

1 INTRODUCTION

Glass stress can be analyzed by using the photoelastic effect [1,2]. Elastic deformations or strains lead to the birefringence of glass (Fig. 1). Stress vertical to the light direction can be made visible with a polariscope. Any stress in the test material will create structures or fringes.

A further developed instrument uses the scattered light method to determine through-the-thickness stress distribution (residual and loading stress) in annealed, heat-strengthened and fully tempered flat glass products. The stress parallel and vertical to the main direction can be separated from two stress measurements.

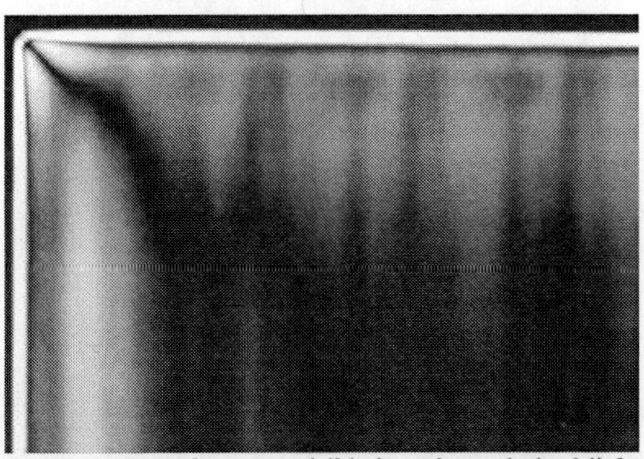

Figure 1: Stress becomes visible by using polarized light in combination with a polarization filter (Corner of hardened solar glass). Especially close to the corner, the glass is susceptible to breakage.

2 SCATTERED LIGHT POLARISCOPE

The commercial 5 cm long instrument emits a polarized laser beam, which is passed obliquely through the glass panel which is optically coupled by immersion oil (see Fig. 2). During the measurement the laser beam's polarization is modulated. The laser light scatters on small particles or bubbles inside the glass (elastic Rayleigh scattering) and the intensity of the scattered light depends on the polarization state of the laser beam (see Fig. 3). During the measurement, the scattered light intensity variations along the laser beam are recorded. From that data absolute optical retardation (see Fig. 4) at each point along the laser beam is calculated. The stress (see Fig. 5) vertical to the beam is calculated from the gradient of optical retardation. With two measurements by turning the instrument in 90° step, the separated stresses in x and y direction can be calculated.

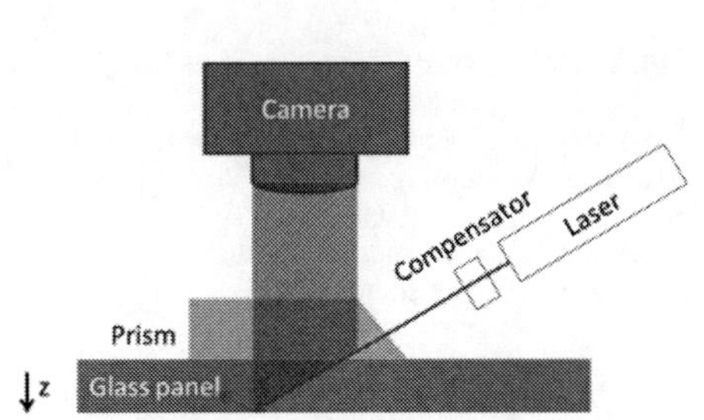

Figure 2: Principle of a scattered light polariscope (side view)

Figure 3: Negative image of the scattered and reflected laser beam by the camera. The beam enters the glass left at the top surface and is reflected/transmitted at the bottom surface

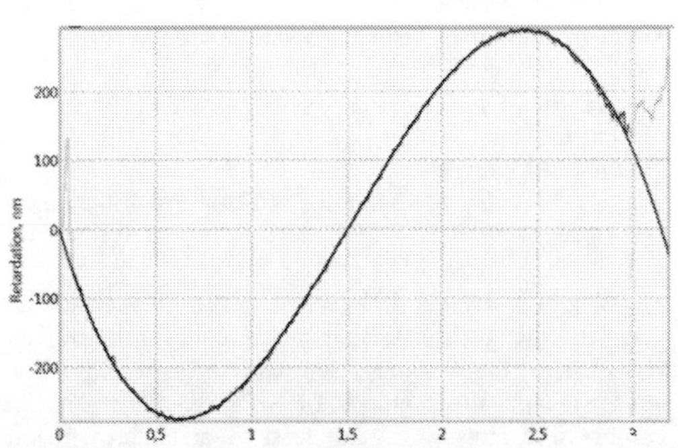

Figure 4: Retardation through the thickness of the glass sheet. A tempered 3.2 mm thick float glass with anti-reflective coating was measured. A parabola of 3rd order was used for fitting, while yellow pixels were excluded.

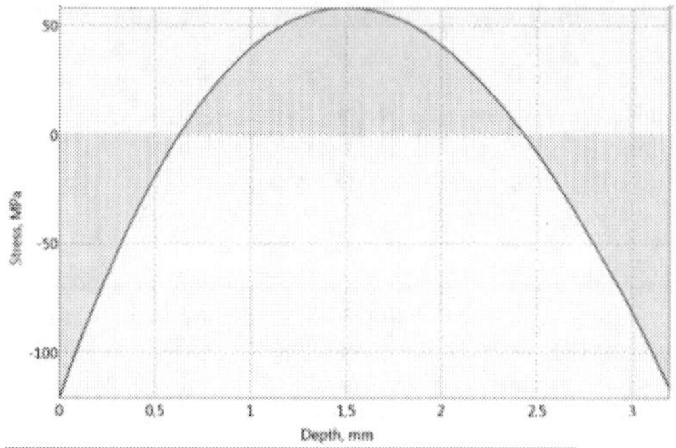

Figure 5: Due to the fast cooling process the surface stress is compressive stress (below zero). The compression zones at both sides are 20 % of the glass thickness. The tension zone in the middle is 60 % of the glass thickness.

3 ELASTIC BENDING

The first application of the instrument is tl measurement of uniform module loading by snow or wind. Beam or plate bending is characterize neutral axis in the middle of the beam wnere no compressive or tensile stresses exist (see Fig. 6).

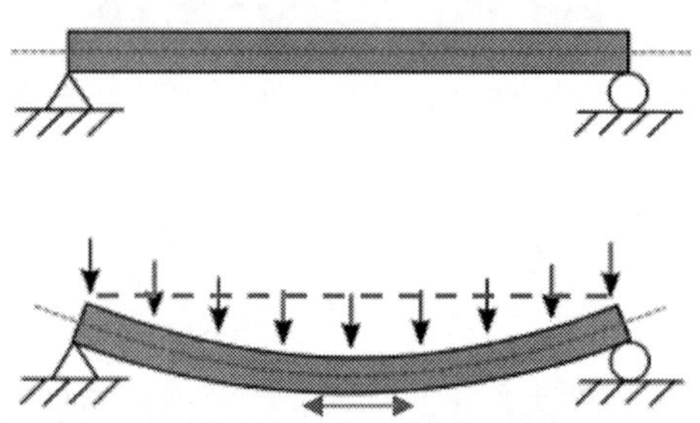

Figure 6: Uniform loading of a plate: Tensile glass stress at the bottom surface can lead to glass breakage.

The main parameters of this configuration can be calculated analytically:

- Maximum displacement on the middle line is:
  $\Delta = 5/32 \; p/E \; L^4/H^3$
- Maximum surface stress (at any point of the bottom surface) is:
  $\sigma = 3/4 \; p \; L^2/H^2$

with p = Pressure, E = Modulus of Elasticity, L = Length, H = Height.

Weight by weight the increase of the stress inside the glass and the surface stress is measured until the module breaks. For the breakage of glass only high positive stress, tensile stress above 40 MPa, is relevant while compressive stress does not lead to crack growth and glass breakage.

Another configuration for testing and calibrating of the instrument is the four point bending test after ASTM C1279-13 [3].

3.1 Hardened Solar Glass

A hardened solar glass (50 cm x 50 cm) was loaded with weights as in the configuration of Fig. 6. The stress is measured for loads in the range from 0 kg to 96 kg (see Fig. 7). The surface stress of the bottom side was extrapolated from the maximum measurement depth due to strong reflection to the glass thickness. The surface stresses at both sides were plotted in Fig. 8. The bottom surface values were corrected as in Fig. 7.

A linear extrapolation of this curve leads to the critical stress of +50 MPa at 300 kg. Fig. 8 shows the set-up with a load of 227 kg, without glass breakage.

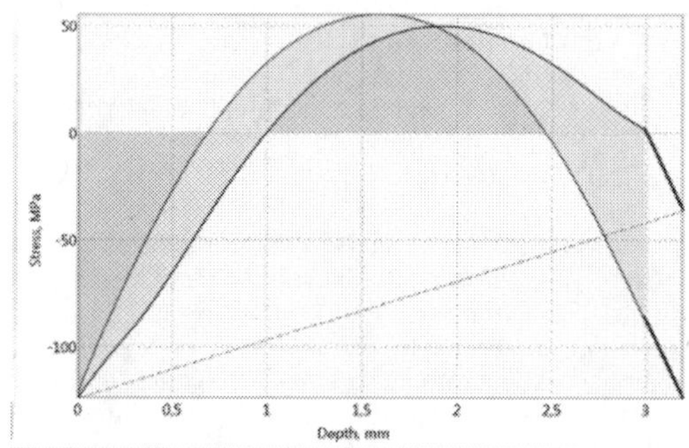

Figure 7: Glass stress: Example of a tempered solar glass (3.2 mm thick float glass with anti-reflective coating). (Red curve: 0 kg; Blue curve: 96 kg load). The stress was extrapolated to 3.2 mm thickness.

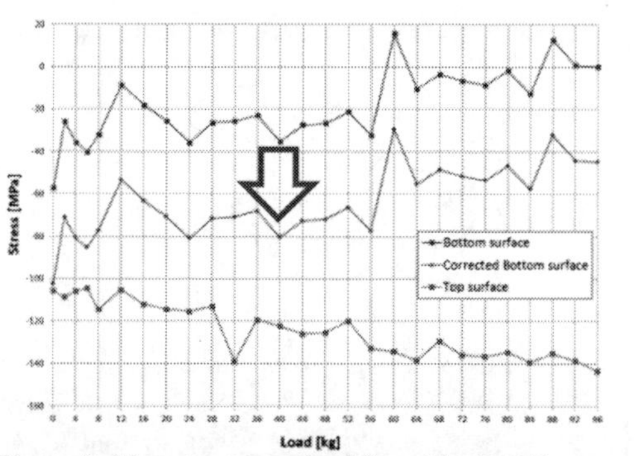

Figure 8: Glass stress over load from 0 kg to 96 kg. (Red: bottom surface stress; violet: corrected bottom surface stress; blue: top surface stress).

Figure 9: Solar glass with a load of 227 kg without glass breakage. Stress measurements were performed up to a load of 96 kg.

3.2 Double glass module with non-tempered glass

This test method was used for the rootcause analysis for glass breakage of thin film modules with non-tempered glass. The test method was verified by loading of steel weights on the central axis of the module with two back rails (see Fig. 10). Weight by weight the increase of the stress inside the glass and the surface stress were measured until the module broke (see Fig. 11). For the breakage of glass only high positive stress, tensile stress above 40 MPa, is relevant while compressive stress does not lead to crack growth and glass breakage (see Fig. 12 and 13).

Figure 10: High mechanical forces were applied to the module by steel weights placed in two rows on hard foam parallel to the backrails.

Figure 11: After 330 kg loading both glass sheets of the module break.

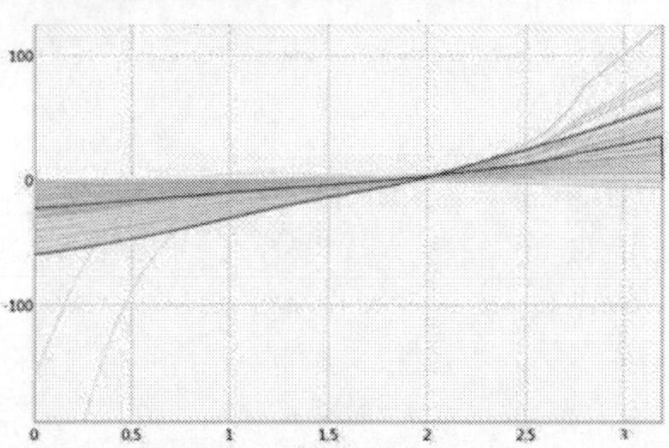

Figure 12: Surface stress vertical to the rail during mechanical load test. Front glass stress is plotted as a function of the depth of all loads. The stress of the non-hardened glass grows linearly with the depth and with the load with the maximum at the bottom side of the front glass.

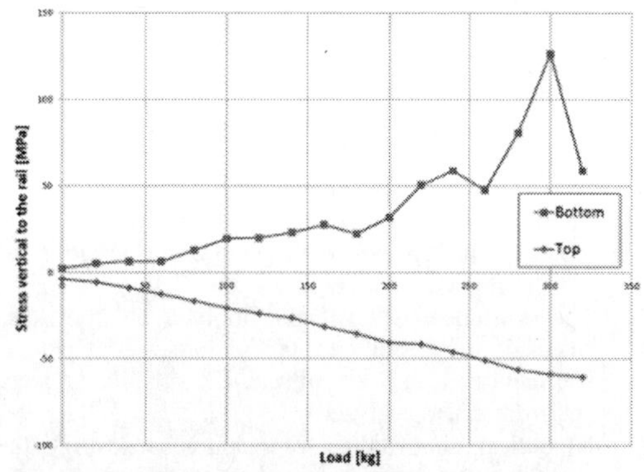

Figure 13: Bottom and top surface glass stress as a function of load. The test ended when the glass broke at load of 330 kg with extrapolated stress of +60 MPa at the bottom side.

4 NUMERICAL SIMULATION

FEM simulations were performed with SOLID WORKS. As example a module glass sheet (2m x 1m, 3.2 mm thick) was loaded with 125 Pa with fixed edges at the top and bottom side. The stress (Fig. 14) and the displacement (Fig. 15) are shown.

Further Simulation shall investigate the effect of the frame, second glass sheet and the laminate (backsheet and encapsulation).

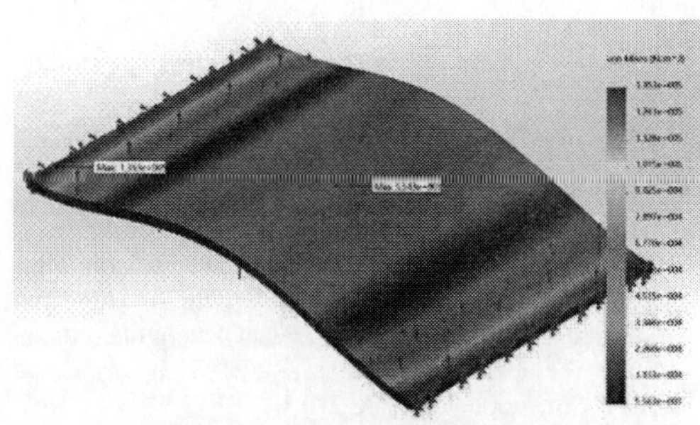

Figure 14: Von-Mises stress of a 3.2 mm thick glass sheet bended with 125 Pa.

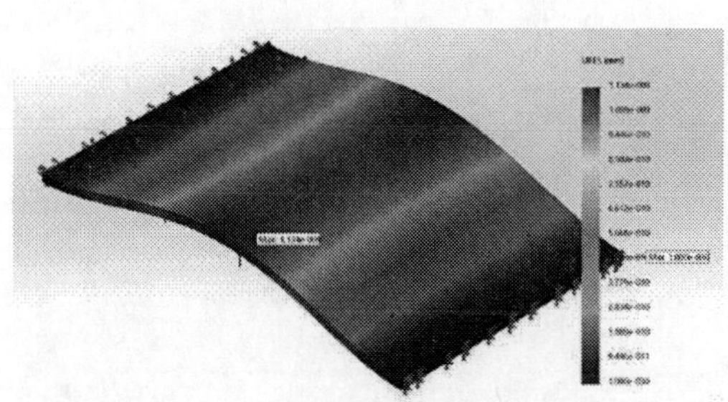

Figure 15: Displacement of a 3.2 mm thick glass sheet bended with 125 Pa.

5 CONCLUSIONS

The scattered light polariscope was applied for investigation of solar glasses and PV modules.

- Different solar glasses (toughened and non-toughened) with and without antireflective coating were measured from both sides.
- The measured stress of the top surface was reproducible and linear. The bottom surface stress of the front glass shows higher scatter.
- The instrument is a valuable tool for the rootcause analysis and the failure reproduction.
- Mechanical load tests of the IEC standards were performed and analyzed.
- Numerical simulations were started to verify the results due to different boundary conditions.
- The instrument is well suited for field inspections due the low weight and low power consumption.
- The instrument was also used to measure surface stresses of the glass by inhomogeneous heating of the glass generated with heat mats, reverse current injection, bypass diode current and hot air.
- The effect of the module fixation by clamps can be analyzed in detail by using static mechanical load and environmental chamber tests of IEC standards.
- The instrument can also be used for quality tests of tempered and non-tempered solar glass.

6 ACKNOWLEDGEMENT

The work presented has been elaborated in the frame of the project "Fidelitas" (FKZ: 0325735A) supported by the German Federal Ministry for Economic Affairs and Energy.

7 REFERENCES

[1] H. Aben, C. Guillemet: Photoelasticity of Glass, Springer-Verlag (1993).
[2] H. Scholze: Glas ed.3, Springer-Verlag (1988).
[3] ASTM C1279-13, Standard Test Method for Non-Destructive Photoelastic Measurement of Edge and Surface Stresses in Annealed, Heat-Strengthened, and Fully Tempered Flat Glass, ASTM International, West Conshohocken, PA, 2013.

DETERMINATION OF MOISTURE INGRESS AND DIFFUSION PATH IN ENCAPSULATION LAYERS OF STANDARD PV MODULES

L. Neumaier[1], D. Holzmann[1], W. Mühleisen[1], J. Zikulnig[1], C. Hirschl[1]
CTR Carinthian Tech Research AG, Europastr. 12, 9524 Villach/St. Magdalen, Austria
(Author and Contact: lukas.neumaier@ctr.at; T: +43 4242 56300 206)

ABSTRACT: Within this work miniature digital temperature sensors were utilized to monitor the moisture ingress in a specially prepared photovoltaic mini module during accelerated aging. This work was conducted within the Austrian flagship project INFINTIY, which deals with the development of climate-adapted PV solutions in order to achieve yield-optimized PV systems with extended system lifetimes. As a further step in conducting research related to moisture ingress in PV modules, within this work a mini PV module was manufactured, comprising eight digital temperature and humidity sensors in front and behind a solar cell (substituted by an aluminium sheet). The goal here was to determine the diffusion path of the humidity diffusing through the backsheet to the front side (sunny side) of the laminate. Here it was essential to gain knowledge about the duration of the moisture ingress to the front part of the test module during accelerated aging as elevated humidity causes corrosion effects on e.g. the grid fingers (reduced electrical conductivity).

Keywords: Moisture ingress, monitoring, accelerated aging

1 INTRODUCTION

Intensive research activities continuously increased the quality of PV modules over the last years in order to achieve long-term stable and efficient products. Nevertheless environmental, climate-specific stress factors such as UV radiation, temperature and humidity are still considered as important degradation factors which need to be addressed in terms of product development and design. Detailed analysis and research efforts are crucial in order to minimize the impact of these effects. Within the cooperative research project INFINITY specific challenges according to aging and degradation effects of PV modules as well as single components are investigated. Within this work the influence of humidity/moisture ingress is being simulated and analysed in detail. Previous work already proved the suitability of the experimental setup and showed very practical results for material and component optimization. This work is based on an approach for monitoring humidity ingress in PV modules using miniaturized digital humidity and temperature sensors [1,2,3]. Here the objective was to determine the diffusion path of humidity inside the encapsulation layers of PV modules. Particular interest lies in the moisture ingress behind but especially also in front (sunny side) of the encapsulated solar cells.

2 APPROACH

For monitoring the humidity ingress into the specially produced PV modules, an Arduino Due based analysis unit, capable of logging and evaluating up to 25 miniature digital humidity and temperature sensors (Sensirion SHT21), was utilized [2]. The device was developed in order to achieve high-resolution long term monitoring data and is able to be applied in an isolated operation mode using an USB power bank. This enables an utilization of the devise also on-site for monitoring real-life data.

2.1 Technical Realisation

Via a simple graphical user interface the logging interval as well as other presets can be adjusted. For the determination of the moisture ingress (diffusion path), a specially prepared mini PV module (22 x 22 cm²) was built up (see Figure 1). One sensor was positioned in in the encapsulation layer pointing towards the backsheet. Additionally to the previously conducted work, here the moisture content was also monitored in the encapsulation layer on the front side of the module (towards glass, see Figure 1). Here 7 sensors were positioned in a stepwise arrangement starting close to the edge and ending in the middle of the sample (see Figure 2). Because of the fragility of a standard silicon solar cell, which would break due to the sensors imprint, a sheet of thin, flexible aluminium was utilized for replacing the cell. As encapsulation material standard PV grade ethylene-vinyl acetate (EVA) foil was utilized. The backsheet is also commonly utilized for PV modules and is based on polyethylene terephthalate (PET). For manufacturing the module, standard lamination conditions were used (150°C, 850 mbar, total lamination time: 17 min).

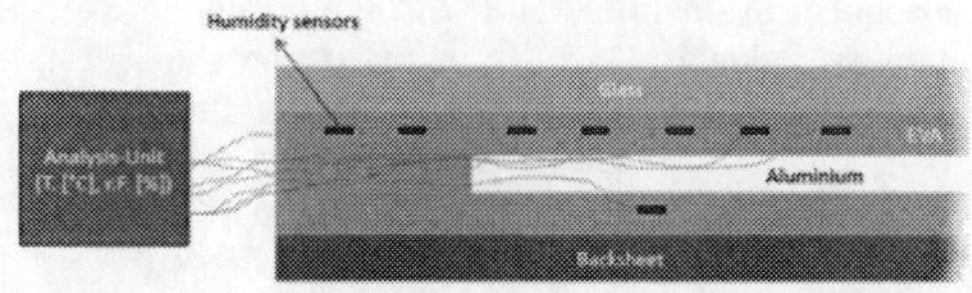

Figure 1: Schematic representation of the experimental setup

Figure 2: Test Module equipped with a set of 8 humidity and temperature sensors (sensor 8 located behind aluminium inlay)

2.2 Accelerated Aging

Before investigating the moisture ingress and furthermore the diffusion path in an accelerated aging test (damp heat), the initial status (after lamination) of the sample was determined. Next, the modules were stored in a climate-chamber (Vötsch VC3 4018) and exposed to elevated humidity and temperature (85°C, 85 % r.H., damp heat testing; see Figure 3). In order to achieve sufficient data, the moisture ingress into the test module was continuously monitored for 350 h of damp-heat testing. To ensure sufficient accuracy, the measuring interval was set to 3 minutes.

Figure 3: Experimental setup in climate chamber; accelerated aging (damp-heat conditions)

3 RESULTS

3.1 Initial Status

The initial humidity content of the encapsulation material of the test module was determined directly after the manufacturing procedure on-site (see Figure 4). Due to a malfunction of Sensor 4, it was excluded from subsequent evaluations. The sensors indicate an average humidity value of 35 % at room temperature (25°C), homogeneously distributed over the whole module area.

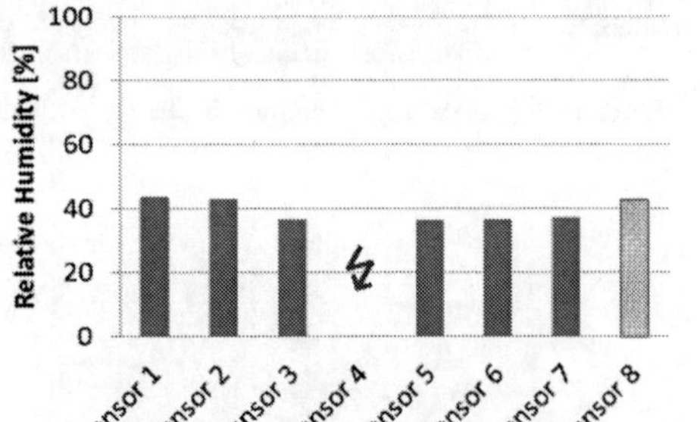

Figure 4: Relative humidity (25°C) of all integrated sensors inside the PV module directly after the manufacturing procedure (malfunction of Sensor 4)

3.2 Accelerated Aging

For determining the moisture ingress into the laminate during accelerated aging, the module was stored in a climate chamber (damp heat conditions) for 300 hours. Figure 5 displays the humidity progression in the encapsulation foil as function of time on both sides of the barrier layer (Al layer). At this point it should be mentioned that after 20 h of testing, sensor 3 delaminated for not explainable reasons (see Figure 6). It is assumed that contaminations on the aluminium sheet or the sensor itself caused this circumstance (e.g.: soldering flux, oily residues).

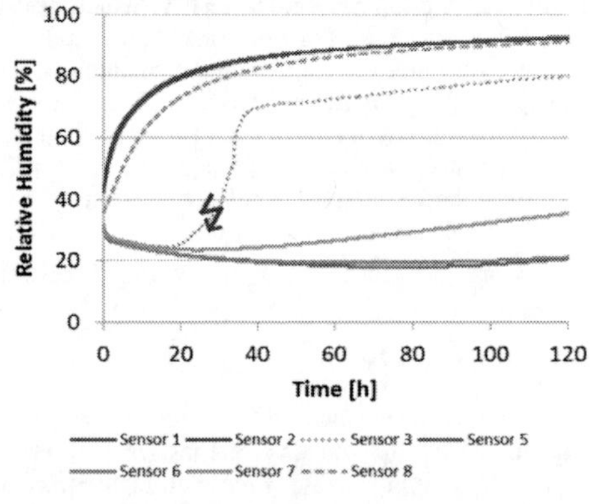

Figure 5: Moisture ingress as function of time during accelerated aging (DH conditions); Sensor 3: delamination issue after 20 h of testing

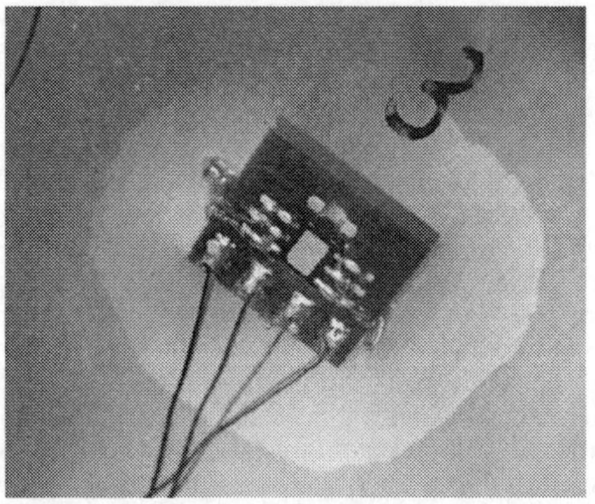

Figure 6: Delamination of sensor 3 after 20 h of accelerated aging

For the sensors located in the region outside of the aluminium sheet (sensor 1 & sensor 2) as well as for the sensor located at backside between the backsheet and the aluminium layer the humidity values exponentially increase almost simultaneously and saturate close to the adjusted value of the climate chamber. Sensor 1 and sensor two show a slightly faster increase, which can be explained by the fact that they are located close to the modules' edge, where also lateral humidity ingress contributes to the raise of the sensor value. Sensor 8 is positioned in the centre of the module. Nevertheless most of the humidity enters through the backsheet and is limited to the moisture vapour transmission rate (MVTR) of the used polymer(s).

In a next step, a spatial analysis of the sensors was conducted by correlating the relative humidity values with the distance of the sensors to the modules' edge (see

Figure 7). Additionally different storage times were plotted in one diagram (see Figure 8).

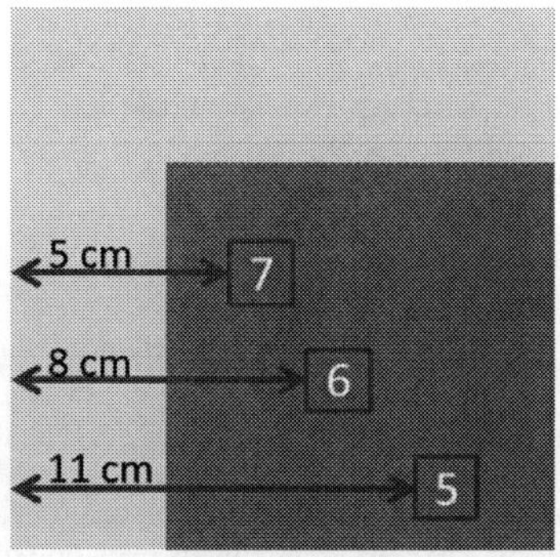

Figure 7: Schematic representation of the sensor's distance to the module edge

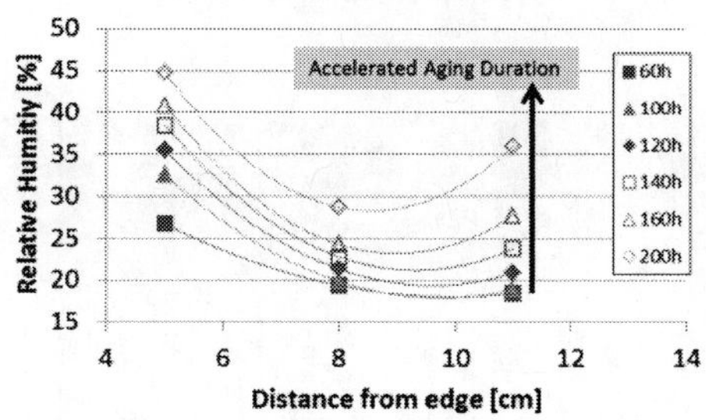

Figure 8: Relative humidity values of 3 selected sensors at different accelerate aging times

It can be seen that there is a symmetric curve shape of the relative humidity value around the centre of the module, indicating an evenly distributed humidity ingress.

4 SUMMARY & OUTLOOK

The moisture ingress analyses as well as the investigation of humidity diffusion properties described in this work were an important step for evaluating and analysing material properties of encapsulation foils and backsheets within the research project INFINITY. Here one key aspect was to determine the behaviour of a certain material composition during accelerated aging (damp heat conditions). It can be stated, that with the current PV module setup the already after about 20 hours of testing, the EVA is moisture saturated resulting from the humidity diffusing mainly through the backsheet. The EVA region in front of the solar cell is not affected by this circumstance, because it serves as water barrier layer. Here the diffusion can only take place through the cross section of the EVA layer(s), which as a matter of course takes much longer compared to the big surface of the modules' backside. After 250 h of accelerated aging the relative humidity value of the centred sensor still does not reach 40 % of relative humidity (see Figure 9). The current setup works well besides some minor drawbacks (delamination issue & one inoperable sensor). Future work will include a revise of the sensor design, comprising a thinner layout for avoiding possible delamination problems. Here the limiting factor is the utilized sensor, which will be replaced by a thinner alternative. Additionally the interconnection approach as well as the wiring of the senor needs to be redesigned in order to achieve a sufficiently thin device.

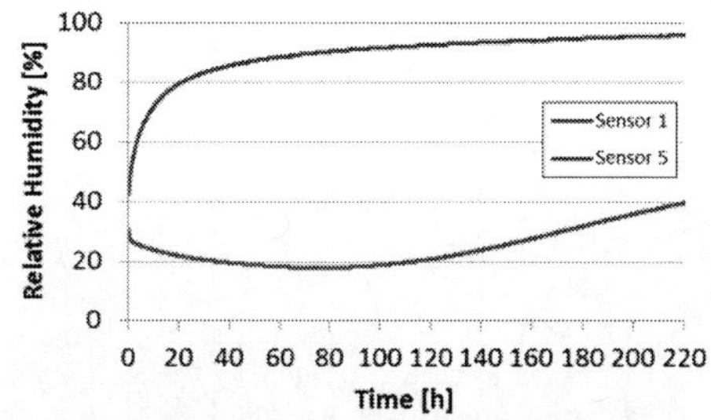

Figure 9: Comparison of the relative humidity results of sensor 1 (located close to the edge; besides barrier layer) and sensor 5 (centred sensor; sunny side of the module, between glass and barrier layer)

As another future step, the temperature and relative humidity of the manufacturing procedure i.e. the lamination process will be monitored with the presented equipment. The aim is here to investigate the stability of the sensors during a standard lamination procedure.

5 ACKNOWLEDGEMENTS

This work was conducted as part of the Austrian "Energy Research Program" project INFINITY. Funded by the Austrian Climate and Energy Fond and the Austrian Research Promotion Agency (FFG), both are gratefully acknowledged.

6 REFERENCES

[1] M.D. Kempe, "Control of moisture ingress into photovoltaic modules", 31st IEEE Photovoltaics Specialists Conference and Exhibition, Lake Buena Vista, Florida, 2005 doi:10.1109/PVSC.2005.1488180

[2] M. Jankovec et al., "In-SituMonitoring of Moisture Ingress in PV Modules Using Digital Humidity Sensors," in IEEE Journal of Photovoltaics, vol. 6, no. 5, pp. 1152-1159, 2016
doi: 10.1109/JPHOTOV.2016.2583779

[3] L. Neumaier, W. Mühleisen, C. Hirschl, G.C. Eder, M. Aichinger, L. Plessing, A. Martin, A. Zimmerman, "Monitoring of moisture ingress in PV module laminates during accelerated aging tests", 33th European Photovoltaic Solar Energy Conference and Exhibition, Munich, 2017

A Benchmark for Visual Identification of Defective Solar Cells in Electroluminescence Imagery

Claudia Buerhop[1], Sergiu Deitsch[2,3,4], Andreas Maier[4], Florian Gallwitz[2,3], Stephan Berger[1], Bernd Doll[1,5], Jens Hauch[1], Christian Camus[1,5], Christoph J. Brabec[1,5,6]

[1]ZAE Bayern e.V., D-91058 Erlangen, Immerwahrstraße 2,
[2]Technische Hochschule Nürnberg Georg Simon Ohm, 90489 Nürnberg, Hohfederstraße 40,
[3]Energie Campus Nürnberg, 90429 Nürnberg, Fürther Straße 250,
[4]Lehrstuhl für Mustererkennung, FAU Erlangen-Nürnberg, 91058 Erlangen, Martensstraße 3,
[5]i-MEET, FAU Erlangen-Nürnberg, D-91058 Erlangen, Martensstraße 7,
[6]Helmholtz-Institut Erlangen-Nürnberg (HI ERN), D-91058 Erlangen, Egerlandstraße 3

ABSTRACT: In this paper a dataset consisting of 2,426 solar cells extracted from high-resolution electroluminescence (EL) images is used for automated defect probability recognition. The collected images contain both functional and defective solar cells with varying degrees of degradation both in monocrystalline and polycrystalline solar modules. The images were labeled by expert who categorized the solar cells by the likelihood of a defect within each image. The labeled images can be used for development of computer vision and machine learning methods for automatic detection of different defects, like cracks, fracture interconnects, PID, and cell quality and for the purpose of predicting the power efficiency loss.

Keywords: EL-imaging, visual inspection, machine learning

INTRODUCTION

Electroluminescence (EL) imaging is an established technique for the visual inspection of photovoltaic modules [1, 2]. EL imaging allows to capture high-resolution images of photovoltaic modules and enables identification of defects in individual solar cells [2, 3]. However, manual inspection of EL images is extremely tedious and requires expert knowledge. Therefore, an automated visual inspection process is more desirable.

The development of automatic visual inspection methods typically first involves the collection of data. The data includes images of both functional and defective solar modules. Additionally, to identify defects at the solar cell level, the EL images of solar modules must be segmented into individual solar cells first [4].

The data in EL imaging that captures a variety of degradation types in solar modules is, however, not cheap. Solar modules do not break on a regular basis but rather over a long period of time. Artificial aging as an alternative is, however, economically not a viable solution. The data collection can therefore be cumbersome or even a hindering factor for the development and quantitative evaluation of visual inspection algorithms.

For these reasons, the datasets used for the evaluation of visual inspection methods in the literature are rather small, capture only specific defect types, and are not made public to gain advantage of possessing such data. As most authors work with different datasets, the comparison between different inspection methods is therefore practically not possible.

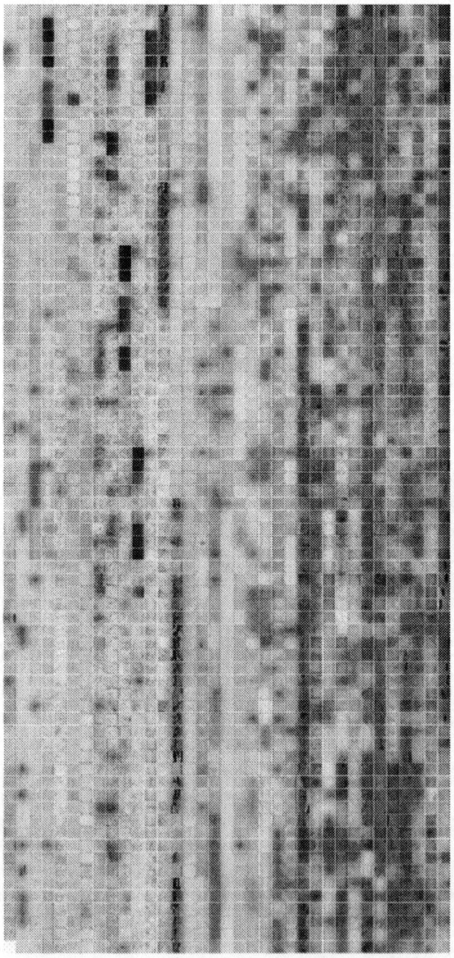

Fig. 1: An overview of all 2,426 solar cell images in the dataset. The images are overlayed by a color that specifies the increased defect likelihood of the underlying solar cell.

In order to support good scientific practices and to encourage the development of visual inspection methods, we devised a dataset consisting of labeled solar cell images, which were extracted from high-resolution EL images of photovoltaic modules. All images were labeled with the help of expert who classified the solar cell images into different defect likelihood categories. Using the provided data, we trained a classifier that can predict defect probabilities in solar cells at an accuracy of 88.42% using images only. This allows us to quickly spot breakage trends in solar modules and decide whether additional measurement is useful to validate the potential loss of power efficiency.

Our dataset is publicly available at https://github.com/zae-bayern/elpv-dataset or with QR code in **Fig. 2**.

Fig. 2: QR code link for https://github.com/zae-bayern/elpv-dataset.

EXPERIMENTAL PROCEDURE

Our dataset consists of 2,426 different images of solar cells, see

Fig. 1. The images were extracted from 44 different solar modules of which 26 are polycrystalline and 18 monocrystalline. In **Fig. 3** one of the polycrystalline PV modules is shown with two exemplary cells and their calculated defect probability.

Our labeling interface then presented one solar cell image at a time to the experts who had to answer two questions:

(1) is the solar cell defective or functional?
(2) are you sure or unsure?

From the corresponding answers, we inferred a set of conditional probabilities (0%, 33%, 67%, and 100% defect likelihood) that define the labels of each solar cell image [5]. An overview of the amount of cells in each category of defect likelihood is shown in Tab. 1.

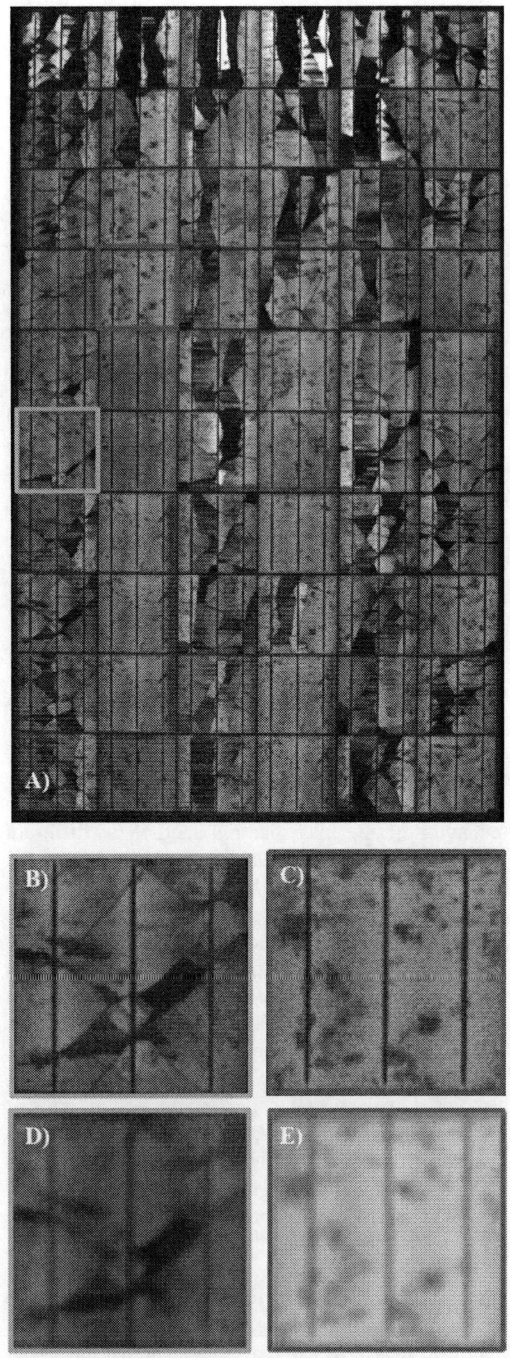

Fig. 3: A) EL image of one of the used 44 PV modules. B) Magnification of cell A8 (yellow) with defects. C) Magnification of cell B4 (blue) without defects. D-E) corresponding magnification of the cells with defect likelihood color grading.

Tab. 1: Amount of the corresponding labelled EL images of each category of defect likelihood.

defect likelihood	mono-crystalline	poly-crystalline
0%	588	920
33%	117	178
67%	56	50
100%	313	402

When training an automatic classifier on this dataset, the class imbalance should be taken into consideration. This can be done, for instance, by down-weighting the categories with large numbers of samples.

For the quantitative evaluation of the visual inspection methods on our dataset we suggest several metrics. The accuracy (in percent) can be used to determine how well the method in question identifies various categories in the dataset. The accuracy is the fraction of true positives (TPs) and true negatives (TNs) with respect to the total number of samples.

Given the imbalance of categories in the dataset, an alternative is the F1-measure which combines both the precision and recall in terms of their harmonic mean [6].

A receiver operator characteristic (ROC) curve created by plotting the true positive rate against the false positive rate can be used to visualize the behavior of the classification method at different thresholds. The area under the curve (AUC) of the ROC curve aggregates classifier's performance into a single scalar and specifies its overall discriminative power (in percent) [7].

Finally, a 2-dimensional confusion (or error) matrix can provide insights into understanding in which categories errors are mostly made and how systematic are these errors. For a confusion matrix, the actual labels are plotted against the labels predicted by the visual inspection method in terms of number of samples corresponding to this combination. Ideally, only the diagonal elements of the matrix would contain non-zero entries.

We also suggest to compute the metrics per type of solar modules, i.e., individually for monocrystalline and polycrystalline. These annotations are also included in our dataset. The discrimination of described metrics between the two module types allows to determine whether the visual inspection method works more accurately for specific type of solar wafers.

RESULTS AND DISCUSSION

We used the dataset to train a deep regression network end-to-end directly on image pixels using the provided labels. We used 75% of the images (1,968 cells) for training and 25% (656 cells) of the images for testing. We used a pretrained VGG [8] architecture consisting of 19 layers and further refined it using the training data.

The resulting network is able to discriminate between various defect probabilities at accuracy of 88.42%. The F1-score is 88.39% and the ROC AUC is 94.7%. This indicates that the trained deep regression network is able to differentiate various defects with a high accuracy.

CONCLUSION

The dataset of the 2426 labeled EL images of single PV cells supports the development of fast and effective evaluation algorithms for the detection of defective, performance relevant solar cells. The presented network is able to detect various defect probabilities for cracks, fracture interconnects, PID and cell quality with an accuracy of 88.42%.

ACKNOWLEDGEMENTS

ZAE Bayern gratefully thanks the German Federal Ministry for Economic Affairs and Energy (BMWi) for financial funding of this project. This work was additionally funded by Energie Campus Nürnberg (EnCN) and partially supported by the Research Training Group 1773 "Heterogeneous Image Systems" funded by the German Research Foundation (DFG).

REFERENCES

[1] M. Köntges (2014) IEA.
[2] K. Bothe, et al., 21st European Photovoltaic Energy Conference (2006) 597.
[3] Y. C. Chiou, et al.; Sens. Rev 31 (2011).
[4] S. Deitsch, et al.; Solar Energy (2018) to be submitted.
[5] S. Deitsch, et al.; Solar Energy (2018) to be submitted.
[6] D. D. Lewis, HLT '91 Proceedings of the Workshop on Speech and Natural Language (1991) 312.
[7] T. Fawcett; Pattern Recognition Letters 27 (2006) 861.
[8] K. Simonyan, et al.; CoRR abs!1409.1556 (2014).

SINGLE DIODE MODEL APPLIED TO PV MODULE AGING

Nikola HRELJA[1], Mike VAN ISEGHEM[1], Eric LAJOIE-MAZENC[1], Eric MOULINES[2], Elodie VERNET[2]
[1]EDF R&D, site des Renardières, 77818 Moret-sur-Loing, France
[2] CMAP École Polytechnique, Route de Saclay, 91128 Palaiseau
Telephone number: 0033763722019 E-Mail: nikola.hrelja@edf.fr

ABSTRACT: The outdoor exposure of PV modules to different climatic variables and physical stresses reduces the power delivery capacity of a PV module over time. The degradation of the PV performance over lifetime is usually determined within the variation of Power at maximum power point (P_{MPP}) and Fill Factor (FF) where the same magnitude of losses of the P_{MPP} can have different causes with different evolution over time. The physic based understanding and modeling of the performance, using the IV curve measurements, can be used to estimate important device parameters and evaluate the health state of the system. In this context the physical model of PV could provide more granulated information about degradation and underlying mechanisms to make better predictions about service life. Single diode model parameter evolution was estimated from 8 years of outdoor IV curve measurements under a range of operating conditions without translating to STC. The results show that the main ageing factors are closely related to the decrease in short-circuit current due to the optical degradation and increase in series resistance regardless of outdoor operating conditions.
Keywords: Degradation, Single Diode Model, CMA-ES, parameter estimation

1 INTRODUCTION

To estimate the profitability of the photovoltaic investment it is essential to accurately quantify and define all the relevant risk factors of long term performance losses. The photovoltaic performance output is known to decline over time and to quantify it, the degradation rate is commonly used.
The plant operating lifetime is mainly determined by the stability and resistance of PV modules to different internal and mechanical loads. These loads are closely related to differences in meteorological exposure depending on the location of the installation and are affecting various module technologies differently. Depending on the technology and its geographical location the degradation rates will be different, specific climatic conditions provoke certain degradation modes more severe. Exemplarily, modules deployed in hot and humid climates were reported with a considerably wider variety of degradation modes. [1]

There are several degradation rate calculation methodologies which are based on various statistical approaches [2] but they fail to provide any information on the mechanisms behind degradation. The degradation of the PV module is closely related to the variation of the internal parameters of the solar cell, namely the short circuit current, ideality factor, saturation current and series and shunt resistances. Observing their evolution in a range of different operating conditions could provide clues about degradation mechanisms.
The commonly observed degradation mechanism include material corrosion, delamination, discoloration of encapsulant over cells, and breakage or cracks of modules. These degradation modes affect the PV characteristics in three ways: decreasing transmittance τ, increasing series resistance R_s, and decreasing shunt resistance R_{sh} [3]. There are several factors such as corrosion of glass, discoloration of encapsulant and degradation of anti-reflecting coating which can cause optical degradation.

The series and shunt resistance evolution represents the summation of several loss mechanisms in a solar cell. Losses due to resistance introduced in cell solder bonds, emitter and base regions, cell metallization and cell-interconnect bus bars all contribute to the value of R_s [4]. Shunt resistance can arise from imperfections on the device surface as well as from leakage currents across the edge of the cell [5]. The R_{sh} value represents any parallel high-conductivity paths across the solar cell p-n junction where an increase in R_s results in a drop in voltage output while a decrease in R_{sh} will reduce current output [6]. Work by Dyk and Meyer demonstrated that by increasing R_s from 0.36 to 1.8 Ω both the P_{MPP} and FF were reduced by 25%, making it clear that the performance is strongly influenced by loss resistance values. Therefore, both R_s and R_{sh} needs to be recognized and their evolution understood in order to improve the module service life forecast. According to the International Energy Agency Photovoltaic Power System Program report: Review of Failures of Photovoltaic Modules [7] if the shunt resistance decreases due to shunt paths in the PV cell and/or interconnections, the I-V curve near ISC becomes sloped. A lower slope of the I-V curve near V_{oc} indicates an increase of the series resistance in the PV module. The series resistance in the module could increase due to an increase in interconnection resistance, corrosion in the junction box or interconnects and slacks joints.

The focus of the approach is to track the evolution of IV curves and SDM parameters over time in stable irradiation and temperature conditions to gain insights into the degradation of PV devices. The evolution of single diode parameters, namely the series resistance (R_s) and shunt resistance (R_{sh}) could at least partially reveal the fundamental mechanisms of degradation and their influence on the open circuit voltage V_{oc}, short circuit I_{sc} and PMPP. Therefore, both Rs and R_{sh} need to be recognized and we need to understand their evolution with changing irradiance (see figure 2) and temperature conditions in order to improve the module performance forecast.

2 METHOD

To evaluate the parameter values in different operating conditions the narrow window method is used (figure 1) where parameters are extracted at certain outdoor temperature and irradiance conditions, initially chosen from standard IEC 61853-1.Without translating to STC conditions the uncertainty of the measurements is reduced. We defined narrow windows with a range of G ± 2W/m2 and T ±1°C. The size of this window is a compromise between sufficient number of data points and acceptable spread between IV curves measured within the same year. Data from high irradiation conditions are used since they show less spread. The single diode model (SDM) is used for comparing the evolution of parameters. It is confirmed that this model provides an accurate prediction of outdoor measurements [8], at the same time allows a good compromise between accuracy and simplicity while avoiding the additional complexity of the double diode model.

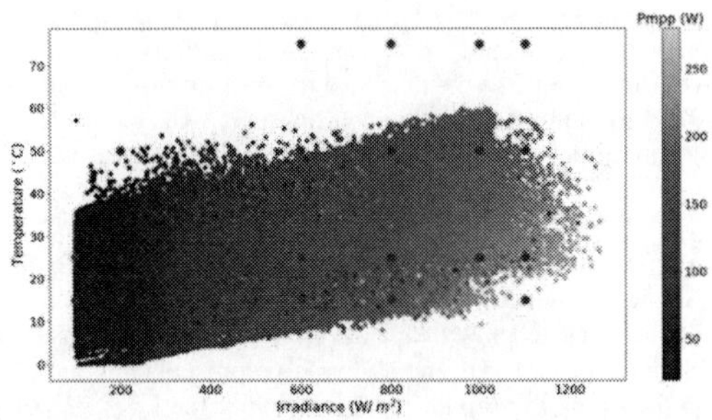

Figure 1: Data points where IV curves are available shown over the Irradiance and temperature domain. The initial reference points are shown in red.

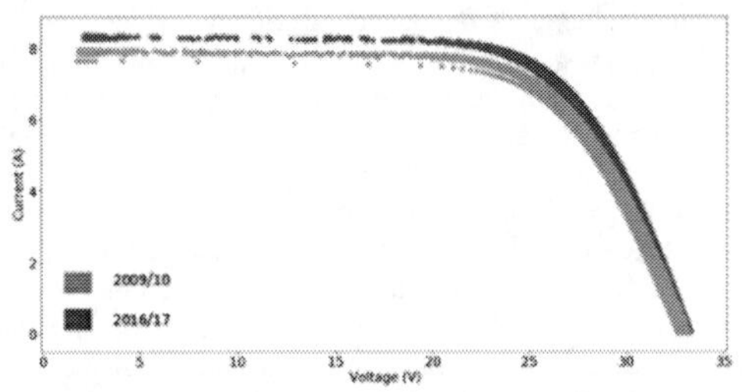

Figure 2: Example of selected IV curves at 1000 ± 2W/m² Irradiance and T =50 ±1 °C

The single diode model (SDM) is used for comparing the evolution of parameters.

$$I = I_{ph} - I_0\left(exp\left(\frac{V+R_s I}{nV_t}\right) - 1\right) - \frac{V+R_s I}{R_{sh}} \qquad (1)$$

Where:

$I_{ph} =$ Photo-generated current (A)

$I_0 =$ Dark saturation current (A)

$n =$ Diode ideality factor (unit less)

$V_{th} = \frac{N_s k T_c}{q}$ Thermal voltage (V) for the module, which is determined from cell temperature $T_c\,(K)$ Boltzmann constant $k\,\left(\frac{J}{K}\right)$ and the elementary charge q (Coulomb)

$k =$ Boltzmann constant ($1.38066 \times 10^{-23}\,J/K$)

$q =$ Elementary charge (1.60218×10^{-19} Coulomb)

$R_s =$ Series resistance (Ω)

R_{sh}=Shunt resistance (Ω)

The values of R_s are considered for the module level. Parameter estimation for a single diode model is challenging and non-trivial due to the implicit equation describing the relationship between current and voltage. Equation 1 cannot be solved for current (or voltage) explicitly using elementary functions. It can be expressed as a function of voltage $I = I(V)\ (or\ V = V(I))$ by using the transcendental Lambert's W function [9] as used in [10]. Lambert's W function is the solution $W(x)$ of the equation $x = W(x)\exp[W(x)]$ using this function we can write $I = I(V)$ as shown in equation 2.

$$I = \frac{R_{SH}}{R_{SH}+R_S}(I_L + I_O) - \frac{V}{R_{SH}+R_S} - \frac{nV_{th}}{R_S} W\left(\frac{R_S I_O}{nV_{th}}\frac{R_{SH}}{R_{SH}R_S}\exp\left(\frac{R_{SH}}{R_{SH}+R_S}\frac{R_S(I_L+I_O)+V}{nV_{th}}\right)\right) \qquad (2)$$

The values of the parameter extracted by optimization techniques depend on the objective function used in the algorithm. In this study the root mean squared error (RMSE) was used and the difference between measured current and the one obtained using the estimated parameters was calculated as shown in equation 3. Here, N is the number of samples, $i_{estimated}$ and $i_{measured}$ are the estimated and measured values for current respectively, the set of parameters, which produce the smallest RMSE are used as extracted. The ideal value of the objective function should be 0, however because of the electrical noise and measurement errors it should be minimized as much as possible.

$$RMSE = \sqrt{\frac{1}{N}\sum_{i=1}^{N}(i_{estimated} - i_{measured})^2} \qquad (3)$$

For the nonlinear optimization problem the Covariance Matrix Adaptation Evolution Strategy was used (CMA-ES) [11]. CMA-ES is a stochastic numerical optimization algorithm for difficult (non-convex) optimization problems in continuous search spaces. Root finding and optimization methods require initial estimates of parameter values, selection of an error metrics and setting of conditions to determine convergence. Different choices for these aspects may results in different parameter values being obtained from the same data. The initial parameters were estimated from the first IV-curve measurement at 4 different operating conditions and were not changed after 8 years. The absolute error between iteration that is acceptable for convergence was set on 1.e-5 and the optimization procedures converged for every IV curve parameter estimation in around 400 iterations with the RMSE between 0.005-0.008.

The parameter estimation for a single I-V curve takes approximately 3 minutes calculation time on an i5 8GB RAM PC.

3 DATA USED

From an EDF monitoring test site located in a "Csa" dry-summer or Mediterranean climate (according to the Köppen-Geiger climate classification) large quantities of data were collected in 1 minute interval. 8 years of IV curves measurements from Poly Si 60 cell module were used to estimate the difference in parameter values at the first year of the installation and in one year after 8 years of outdoor operation. The number of data points in each period were comparable. Maximum power point and the fill factor were extrapolated from the I-V curves.

Weather data includes irradiance, ambient temperature and module temperature. Missing values, outliers representing measurement or sensor error were omitted from analysis.

4 RESULTS

IV curve characteristics and SDM parameters (R_S, R_{SH}, I_{ph}) were estimated in 4 different narrow windows namely: 1000W/m^2-50°C, 1000W/m^2-45°C, 950W/m^2-45°C and 900W/m^2-40°C. High irradiation conditions were chosen because the IV curves variation is reduced. Additional filtering of 2 standard deviation range around the mean values was used to additionally control the spread of measured and estimated values.

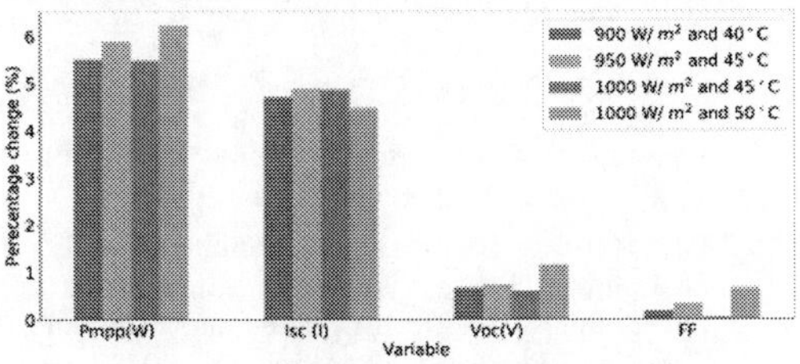

Fig 3: Percentage change of performance indicators after 8 years of outdoor installation, at 4 different G, T conditions

Percentage change of IV mean values before and after 8 years of outdoor operation in different conditions (fig 3) shows that the largest decrease regardless of the operating conditions (figure 3) is in Pmpp, followed by the Isc, 5.8% and 4.7% respectively. The percentage change in Voc is approximately 0.5% where the change in FF is around 0.3%.

The change in photocurrent averaged over all operating conditions (figure 4) is -4.7% where the values of photocurrent don't show any temperature dependencies.

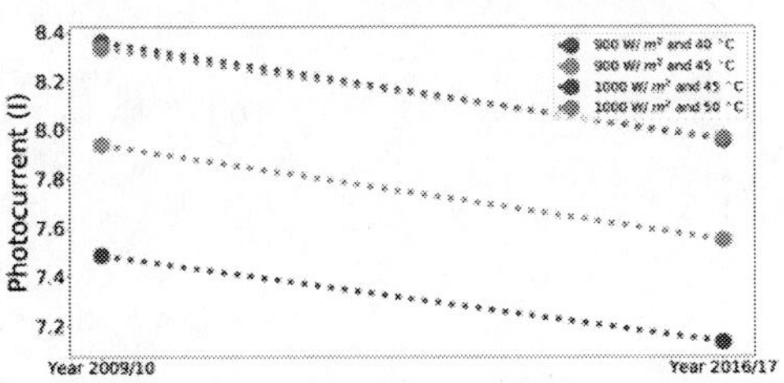

Figure 4: Difference in photocurrent after 8 years of outdoor exposure in 4 different irradiance and temperature combination

The dependence of photocurrent I_{ph} on incident solar irradiation G and cell temperature T is expressed as:

$$I = I_{ph} = \left[I_{pv,ref} + k_1\left(T - T_{ref}\right)\right]\frac{G}{G_{ref}}\tau \quad (4)$$

Where $I_{pv,ref}$ is photocurrent at the reference conditions, usually the standard test conditions (STC) of irradiance 1000 W/m^2, cell temperature 25°C, and air mass 1.5; k_I (A/K) is the temperature coefficient of short-circuit current, T and T_{ref} are the actual and reference cell temperature, G and G_{ref} are the actual irradiation and reference irradiation, the dimensionless variable τ measures the transmittance of PV module, which decreases over time due to the optical degradation. The initial value of transmittance is 1 [12].

τ	Isc (A)	Voc (V)	PMPP	FF
1	8.3	33.44	198	0.69
0.75	6.2	33.2	156	0.68
0.5	4.3	32.8	112	0.69

Table 1: Effect of decreasing transmittance on PV characteristic parameters.

In table1 it is shown how the decreasing transmittance τ reduces both Isc and Voc. When the transmittance is reduced for 25% both the Isc and Pmpp are reduced approximately for the same percentage where FF and Voc show slight variation.

Comparing the series resistance in different narrow windows (figure 5) don't show any changes in the slope. Averaging over all operating conditions the change in R_s is 9.4% after 8 years.

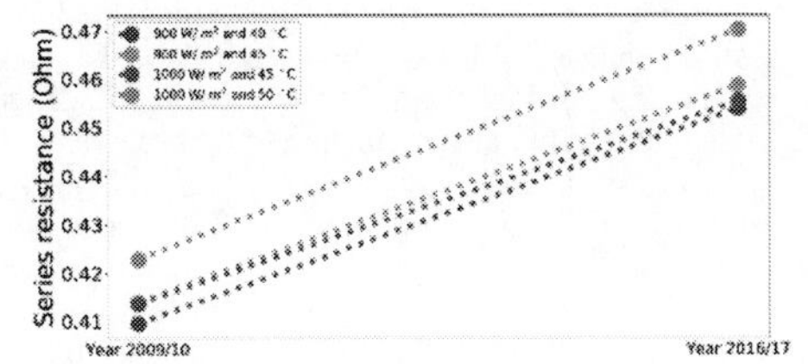

Figure 5: Difference in series resistance after 8 years of outdoor exposure in 4 different irradiance and temperature combinations

The values of shunt resistance don't show considerable degradation in 8 years of outdoor operation. At the same time the values of shunt resistance is difficult to estimate from IV-curves and SDM.

In order to improve the module performance forecast and service life prediction the temporal evolution of the parameters and their dependency on changing irradiance and temperature conditions needs to be understood.

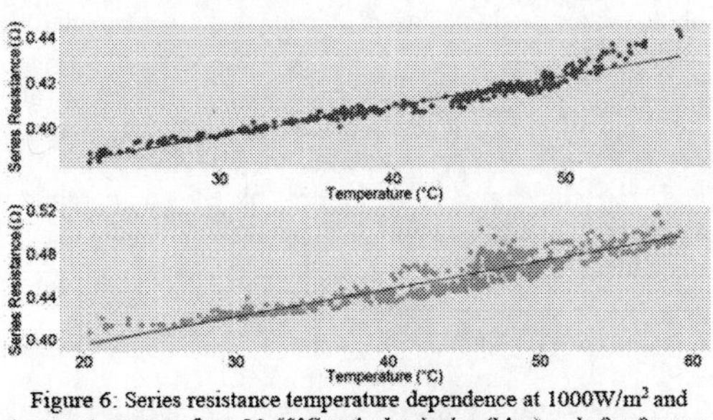

Figure 6: Series resistance temperature dependence at 1000W/m^2 and temperature range from 20-60°C at the beginning (blue) and after 8 years of outdoor exposure (orange).

Comparing the series resistance temperature dependence show no changes in the slope in time (table 2). Values of R^2, α, β show no significant change in time.

	First Year	Last Year
R^2	0.942	0.874
α	0.356	0.344
β	0.0013	0.0025

Table 2: coefficients of the linear models of series resistance temperature dependence in first year of operation and after 8 years.

5 CONCLUSION

In this study an outdoor IV curves parameter estimation methodology was applied to study the degradation process of PV. The method relies on single diode circuit based model of PV electrical characteristics. The approach is based on a narrow window method of different environmental condition (irradiation and temperature) combinations and allows to analyze the outdoor data without conversion to STC.
The results show that the main aging factor of fielded PV module is the decrease of short-circuit current where the main power loss is mostly caused by the optical degradation of the PV modules followed by the increase of series resistance. Possible degradation modes affecting the analyzed module are yellowing and browning of EVA, delamination and appearance of bubbles or anti-reflecting coating degradation. Increase in series resistance suggest that there is a possible corrosion on metal bars or interconnectors in the PV module.
Here, the transmittance τ (t), R_{sh} and the series resistance R_s (t) are time dependent variables presenting the aging process of materials over time, thereby the I-V characteristics are not only a function of irradiation G and cell temperature T, but also a function of time t.

The results of the study should be confirmed with the visual inspection of the fielded module and electrical characterization in the laboratory. The ageing patterns are highly dependent on PV technology and climate that the modules are exposed to. Results should be compared with the IV curves gathered from different climatic conditions or/and different technologies to be able to make specific climate and technology related aging studies.

More than 8 years of IV curves measurements in outdoor should be used to detect degradation modes that are appearing after that period. The shunt resistance values are not showing any degradation in 8 years of outdoor operation.
One of the weaknesses of using a single diode model is not considering the possible mismatch among the cell of photovoltaic module, caused by cell cracks, partial shading or hotspots.

To improve the yield and service life prediction forecasts of the PV plants it is essential to include the degradation and related degradation modes based on parameter evolution which should be technology and climate dependent.

6 ACKNOWLEDGMENT

This study was supported by the European Union's Horizon 2020 programme Solar-Train under GA. No. 721452.

7 REFERENCES

[1] D. C. Jordan, T. J. Silverman, J. H. Wohlgemuth and S. R. Kurtz, "Photovoltaic failure and degradation modes," *Progres in Photovoltaic,* pp. 3018-326, 2017.

[2] D. C. Jordan and S. R. Kurtz, "Photovoltaic Degradation Rates-An Analytical Review," NREL, Denver, 2015.

[3] D. C. Miller, E. Annigoni, A. Ballion, J. G. Bokria, L. S. Bruckman and D. M. Burns, "Degradation in PV encapsulation transmittence: an interlaboratory study towards a climate-specifi test.," in *Photovoltaic Specialis Conference (PVSC) IEEE,* 2015.

[4] M. A. Green, "Solar Cells Operating Principles, Technology and System Applications," University of New South Wales, Australia, 1998.

[5] T. McMahon, T. Basso and S. Rummel, "Cell shunt resistance and photovoltaic module performance," in *Proceedings of the 25th PVSC,* Washington DC, 1996.

[6] v. E. Dyk and E. Meyer, "Analysis of the effect of parasitic resistances on the performance of photovoltaic modules," *Renewable Energy,* pp. 333-344, 2004.

[7] M. Kontes, S. Kurtz, U. Jan, K. A. Berger, K. Kato, T. Friesen, H. Liu and M. V. Iseghem, "Review of Failures of Photovoltaic Modules; Photovoltaic Power System Programme," Paris, 2014.

[8] G. Mekrides, B. Zinsser, M. Shubert and E. G. Georghiou, "Energy yield predicition errors and uncertainties of different photovoltaioc modules," *Progres in Photovoltaic,* pp. 500-516, 2013.

[9] R. e. a. Corless, "On the Lambert W function," *Advances in Computational Mathematics,* pp. 329-359, 1996.

[10] A. Jain and A. Kapoor, "Exact analytical solutions of the parameters of real solar cells using Lambert W-function," *Solar Energy Materials & Solar Cells,* pp. 269-277, 2004.

[11] N. Hansen, "The CMA eolution strategy: a comparing review," *Towards a New Evolutionary Computation,* vol. 192, pp. 75-102, 2006.

[12] Y. Mahmoud, W. Xiao and Z. H.H., "A simple approach to modeling and simulation of photovoltaic modules," *IEEE Trans Power Electron,* vol. 24, pp. 1198-208, 2009.

ELECTROLUMINESCENCE SYSTEM FOR ON-SITE CHARACTERIZATION OF PV MODULES

Ezquer M., Barrenetxea A., Moracho J, Diaz J., Legarrea A., Lagunas A.R.
National Renewable Energy Centre (CENER)
Ciudad de la Innovación 7, 31621 Sarriguren (Navarra) Spain
Telephone: +34 948 25 28 00; Fax: +34 948 27 07 74
Email:mezquer@cener.com

ABSTRACT: The characterization of photovoltaic modules at the installation itself by using the electroluminescence (EL) technique is becoming an essential and increasingly more sought-after tool in the PV sector. CENER (National Renewable Energy Centre of Spain) has developed a customized proprietary system CELSOS (CENER'S ELECTROLUMINESCENCE SYSTEM ON-SITE) to undertake this type of high quality characterization on the field regardless of the existing level of irradiance. In parallel to this measurement set-up, a new analytical software has been also implemented in order to allow the automatic detection, identification and classification of potential problems on the modules under test (i.e. presence of cracks, PID pattern, corrosion, metallization problems, etc) by the implementation of advanced functionalities to analyze of the EL images captured on-site.

1 INTRODUCTION

Although electroluminescence (EL) measurements of PV modules in the laboratory remain as a very useful characterization service in order to identify the potential presence of defects or problems in PV modules, the PV market is increasingly seeking options to undertake this type of characterization at the PV plant itself, due to the wide-ranging advantages involved on it:

- No need to disassemble the modules from their mounting structures
- Transportation of the modules is avoided
- No risk of potential damage in the modules during the entire characterization process (disassembling, transport, unpacking, re-mounting…)
- No need for the installation to stop electricity production over long periods
- No need of disconnecting the PV modules individually
- Higher speed in the measurements because several modules can be captured in the same EL image
- Measurements can take place alongside other types of measurements in the field (e.g.: IV curve tracing, thermal imaging measurements, etc.).

In short, the possibility of taking EL measurements in situ represents a considerable reduction in the economic cost of the test, as well as an avoidance of risk of damage to the samples throughout the whole of the corresponding process.

After having set out the significant advantages offered by being able to take EL measurements at the PV installation itself, there is still a technological challenge in the form of obtaining quality images outside the controlled conditions of the laboratory. One of the most complicated issues when performing this test in the field is the lack of a controlled darkness environment. This is the reason why most companies offer these characterization services at night or during sunset/sunrise. This is not a bad solution, but only undertaking testing at night involves some inherent problems like fewer working hours for taking measurements, increased physical/electrical risk due to worsening visibility, impossibility of running in parallel with other types of measurements, etc.

Nevertheless, in recent years some companies are starting to develop advanced systems to perform this kind of EL measurements at daylight conditions. In this way, some of the systems are still in testing period [1], and the ones that are already in commercial phase present a final price that is normally quite high [2].

In order to minimize the cost of the final system and, at the same time, to improve the technical limitations found in some of these commercial systems, CENER's PV department has worked in developing a fully customized EL system (**CELSOS:** ***CENER'S ELECTROLUMINESCENCE SYSTEM ON-SITE***) that integrates a special controlling software that allows the collection of high quality EL images 24/7, regardless of the existing level of irradiance.

In parallel, making use of CENER's wide experience performing EL measurements of different types of PV modules at laboratory level [3], an advanced analytical program to process the EL images captured on the field has been also implemented. This new software includes diverse functionalities for automatic detection and classification of different potential defects [4] (i.e. presence of cracks, PID pattern, corrosion and soldering problems, metallization failures, etc).

2 CELSOS EXPERIMENTAL SET-UP

The set-up of the system designed by CENER to carry out the EL measurements on-site is based on the integration of these individual components:

- InGaAs camera (resolution: 640pixels x 512pixels)
- Multi-range power supplies up to 15 kW and 1500 V
- Band-pass optical filters
- Wide-angle lenses with high transmittance in the near infrared region.
- All-rounder laptop with control software

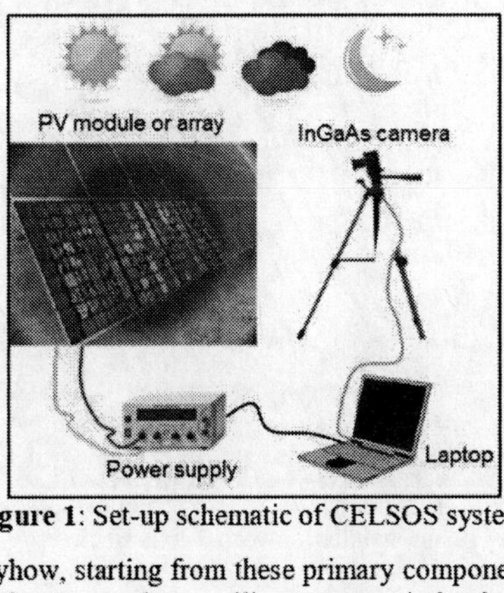

Figure 1: Set-up schematic of CELSOS system

Anyhow, starting from these primary components the designed set-up and controlling program is implemented to provide great versatility to the final systems. In this way, different equipment of several suppliers can be easily integrated in the set-up. The possibility of exchange any of the components in the system implies a great advantage, because in the future any of the parts can be replaced by another one newer, cheaper or with better technical features. Moreover, depending on the particular conditions of the PV installation to be tested and the concrete goal of the characterization procedure, specific components (lenses, optical filter, kind of camera, etc) can be exchanged to improve the results in this concrete application.

Moreover, the high output power supply has been integrated to bias electrically one or several entire strings of PV modules at the same time, which considerably increases the speed of the measurements, and avoids the need of disconnecting the modules during the measuring process. The power ranges of these supplies extend up to 15 kW and 1500 V in order to be able to bias all kinds of PV strings.

Figure 2: CENER's technician performing EL measurements on the field

As it has been mentioned before, in addition to integrating high performance measuring equipment, a customized control software has been also implemented. This program includes tools based on advanced digital processing (i.e. lock-in detection) in order to obtain high quality EL images even in high irradiance conditions. By using this software, any configuration parameter of the test can be conveniently varied to obtain the best EL image depending on the specific environmental conditions at any given moment.

Figure 3: EL image of a PV string of cracked c-Si modules on an installation with 2-axis trackers

3 ANALYTICAL SOFTWARE

In parallel to the development of the system to carry out the EL measurement on the field, CENER has implemented a complete analytical program in order to process the captured images and to obtain an automatic detection and identification of different kinds of potential defects on each of the modules measured on the same EL image (typically 4-6 modules/image): presence of cracks, PID patterns, metallization problems, corrosion, etc [5] [6].

This analytical software has been entirely programmed by using Python that is an open source software, very versatile and easy to handle. The first version of the program includes several functionalities implemented for the particular case of c-Si modules, such as:

1) Processing of the raw EL image: automatic noise reduction, removing of possible erroneous pixels, and sharpness improvement
2) Automatic detection of the borders of the modules included in the EL image in order to examine them in an individual manner latter on.
3) Automatic correction of the original image's perspective to obtain a perfectly squared image of the module.

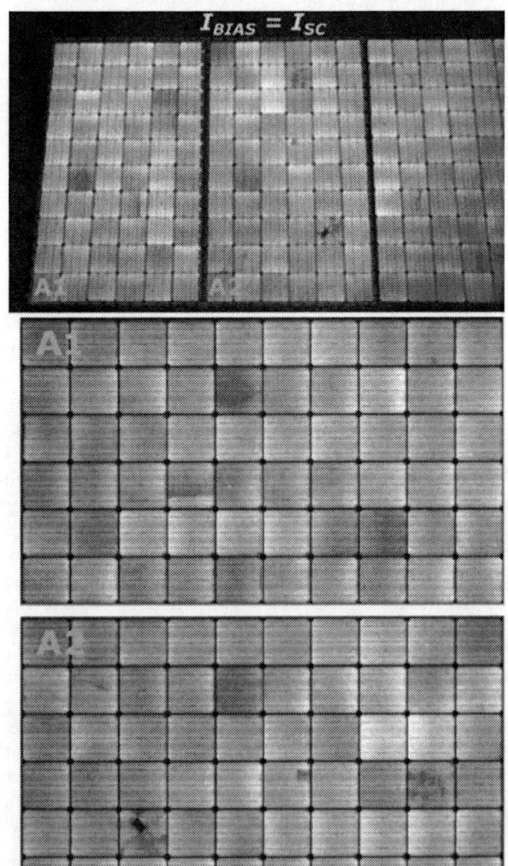

Figure 4: (top) EL image of PV modules mounted in 1-axis tracker; (bottom) Individual images of 2 modules after first processing steps: border module's detection; perspective correction; cell's limits detection.

4) Direct overlay of the visible and EL images, both captured at the same time by the system.

Figure 5: Overlay of visible and EL images of a PV array in an installation integrated in a building façade

5) Automatic detection of the exact limits of the cells in the module under analysis.
6) Multiple statistics obtained from the detailed analysis of EL values on each cell (i.e. area in pixels, average, median, standard deviation, histogram, etc).
7) Analysis of spatial distribution of EL values on both separated axes (X and Y) by using algorithms based on kernel matrixes.
8) Automatic detection, location, and classification of the main defects found in the module under analysis: cracks, isolated areas, PID pattern, corrosion problems, etc. Specific functionalities have been programmed in order to enable the identification of each kind of defect.
9) Generation of a complete report with all the data related to the capture of the original image (date, hour, location, configuration parameters, equipment, etc) and the main results obtained during the analysis, including all the generated images.

4 EXAMPLES OF APPLICATION

This chapter shows some examples of real cases that can illustrate the interesting results obtained from the analysis carried out by using different functionalities of the developed software, starting from EL images captured on-site by CENER's system (CELSOS).

4.1 – CRACK'S DETECTION

One of the most interesting tools provided by the developed analytical program is the automatic location and detection of possible B and C-type cracks [7]. Type-C cracks are the ones that provoke a complete electrical isolation of a certain region of the cell, leading to a decrease in the I_{SC} value of the affected cells. Type-B cracks provoke a significant increase in the series resistance of the affected area, but this region is not still disconnected. It is important to mention that type-B cracks can easily evolve to type-C over time, if the module suffers from any mechanical stress. However, by analyzing only the EL image captured at high biasing current level (typically around I_{SC} level), which normally is the only one captured on the field at daylight conditions, it is frequently difficult to distinguish between B-type and C-type cracks. In this way, starting only from the high biasing EL image, the program is able to detect and identify all the C-type cracks, although it is possible that sometimes these cracks can correspond to B-type cracks with very high series resistance.

Next figure shows the EL image of 4 modules captured on the field at sunlight conditions (Irradiance ≈ 700W/m^2), and the automatic results obtained from the analysis of one of these 4 modules.

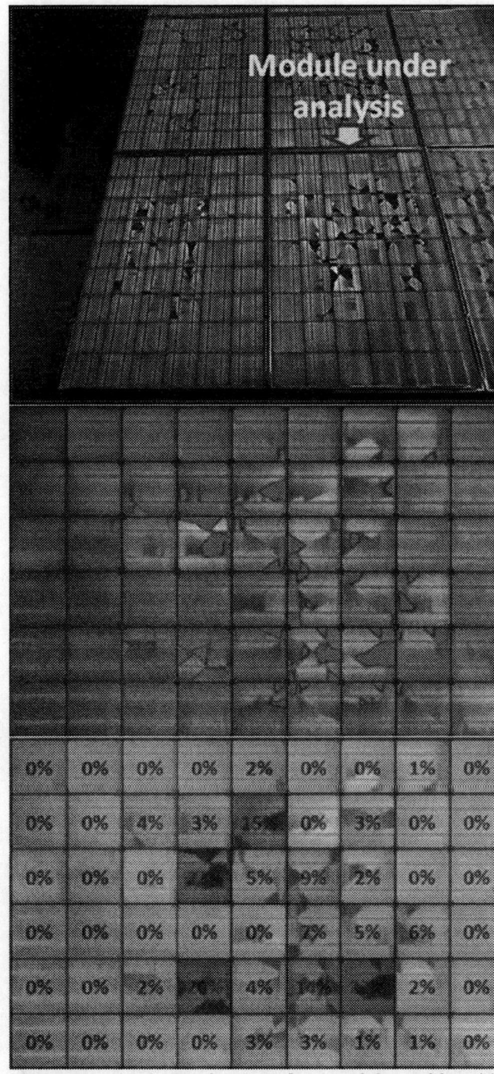

Figure 6: (top) EL image of 4 modules with wide presence of cracks, captured with a biasing level equal to I_{SC} and an irradiance level of 700W/m^2 approx.; (middle) automatic detection of regions isolated by cracks in the analyzed module (in green); (bottom) calculation of the percentage of inactive areas of every cell in the module.

As it can be seen in Figure 6, the program automatically detects the regions that are electrically isolated because of the wide presence of cracks (C-type presumably) and estimates the percentage of inactive area of every cell in the module. This is a very important result because normally the highest value of percentage of inactive cell's area is the one that is going to determine the decrease of the maximum power of the module because of the cracks [7].

4.2 –DETECTION OF PID

EL tests carried out on-site to discover the possible presence of PID (Potential Induced Degradation) problems are typically performed at night. The optimized detection capacity of the CELSOS system allows the possibility of detecting this kind of problem by measuring also during the day. Moreover, if the on-site EL measurements are carried with a low level of irradiance (< 200W/m^2 approx.) the system can also obtain a high-quality low biasing image, which is quite more affected by the possible presence of PID.

In this way, to illustrate the capabilities of the CELSOS system, Figure 7 shows the EL images of some modules located just at the end of the string, captured during twilight with an irradiance level of approximately 100W/m^2, and at 2 different levels of current's biasing. The EL images show 2 modules (C1 and C2) that are heavily affected by PID phenomenon, and a third module (C3) that presents slighter PID symptoms, that can be distinguished in a better way in the lower biasing image.

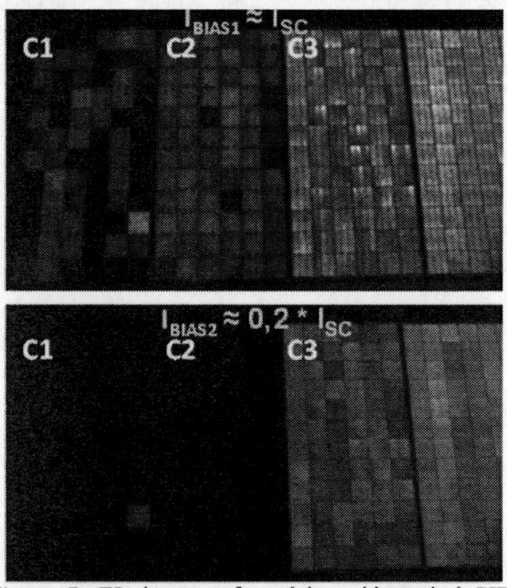

Figure 7: EL images of modules with typical PID pattern, captured at irradiance level of 100W/m^2 approx. and different biasing levels: (top) biasing current similar to module's I_{SC} value; (bottom) biasing current equal to 20% of module's I_{SC} value.

For cases wherein the presence of PID is not as notable as in the previous example, the analytical software includes an additional functionality that allow the estimation of the operating voltage of every cell in the module in order to quantify the percentage of deviation among standard cells and cells affected by PID. [8][9][10].

In order to exemplify the possibilities of this analytical tool, Figure 8 shows in a first step the on-site EL image captured at high biasing. After the perspective corrections and cell detection, the figure shows the calculated values of percentage deviation in the operating voltage of every cell compared to the value of a cell without PID problems, for each of the 3 modules analysed.

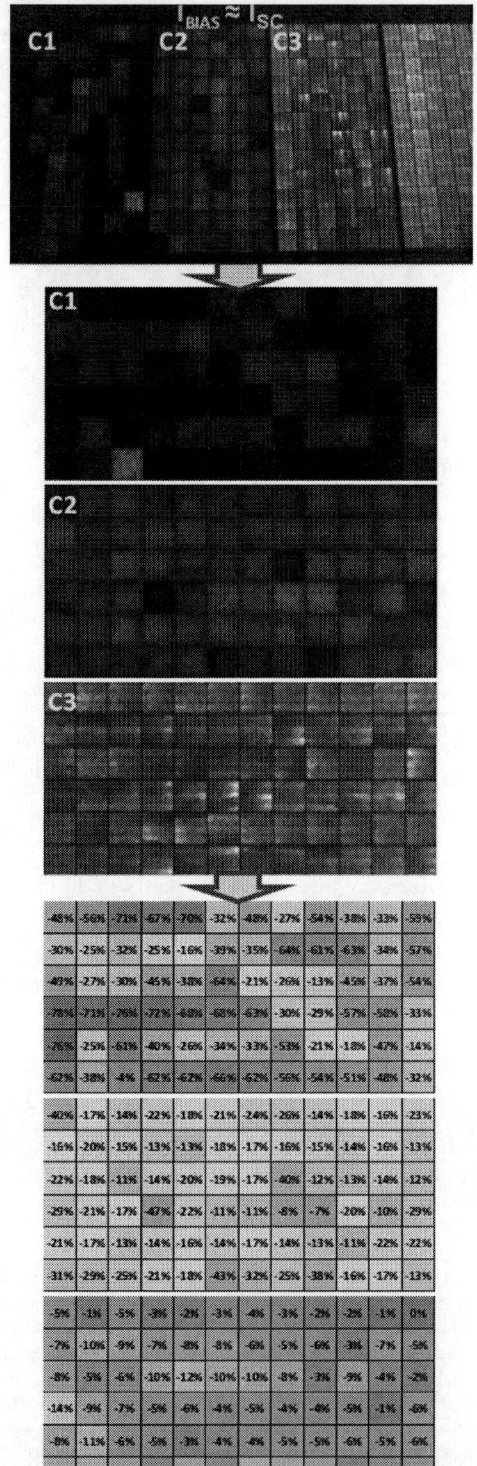

Figure 8: (top) EL image of modules with typical PID pattern, captured with a biasing level equal to I_{SC} and an irradiance level of 100W/m^2 approx.; (middle) Individual EL images of the 3 modules after perspective transformation & cell's limits detection; (bottom) calculation of the decrease in the operating voltage of every cell compared to the value of a cell without PID.

4.3 –SOLDERING FAILURES

Similarly to previous cases, the analytical program also includes a specific functionality to automatically detect and identify the potential presence of tabbing interconnection problems, due to soldering failures between the metallic tabbing and the cell or between two interconnection tabs. By using differential analysis on both axes of the image, the program is able to distinguish between a dark area in the EL image of the cell originated by a crack, and a dark region caused by a soldering problem.

Thus, Figure 9 shows an illustrative example of two modules with soldering problems in one of their cells. In this concrete case, the on-site EL image has been captured with a biasing level equal to I_{SC} and an irradiance level of 650 W/m^2. The result obtained from the analysis of one of the modules (D1) shows the perfect detection of the cell affected by soldering problems (red area), and also the clear differentiation regarding to the inactive area in the cell affected by cracks (green area).

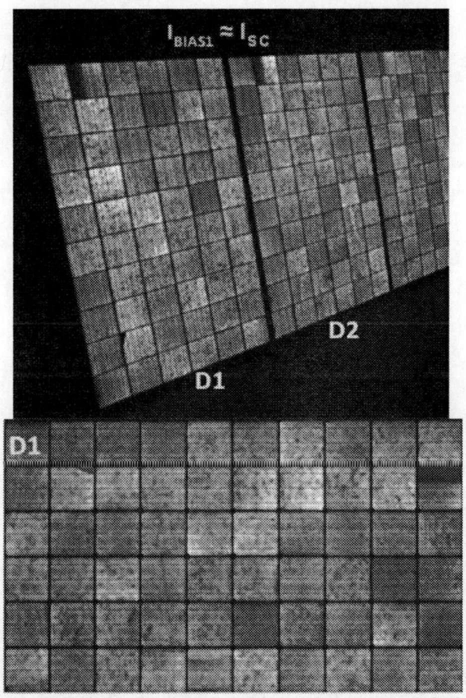

Figure 9: (top) EL image of modules with soldering failure in one of the interconnections tabs, captured with a biasing level equal to I_{SC} and an irradiance level of 650 W/m^2 approx.; (bottom) automatic detection of the cell area affected by the interconnection problem (in red) and the isolated region in other cell originated by a type-C crack (in green).

Moreover, as it can be seen in this example, an accurate analysis of the modules can be performed even starting from an EL image captured with a focusing angle of the camera quite far from the perpendicular condition.

4.4 – CORROSION PROBLEMS

The final example of functionality included in the analytical program is focused on the detection of possible problems of corrosion, normally originated for the penetration of moisture into the inner layers of the module [11]. This way, next example shows the on-site EL image of several modules, one of them (E1) clearly affected by corrosion around the busbars. The EL image has been captured with a biasing level equal to ISC and an irradiance level of 600W/m^2.

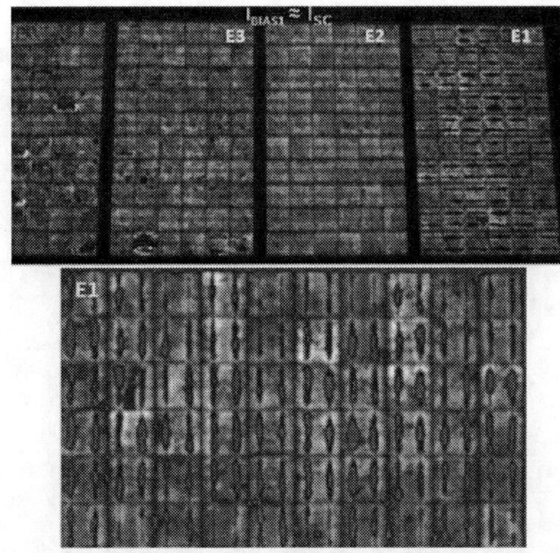

Figure 10: (top) On-site EL image of 3 modules, one of them (E1) with humidity corrosion problem, captured with a biasing level equal to I_{SC} and an irradiance level of 600W/m^2 approx.; (bottom) automatic detection of the areas affected by the corrosion problem (in green).

As it can be seen in the example of Figure 10, the program is able to locate and identify the main regions affected by corrosion problems. In this concrete case, almost all the cells in the module are affected by this corrosion issue that is centred along the busbars of the cells.

5 CONCLUSIONS

CENER has developed a customized proprietary system, named as CELSOS, in order to undertake high quality characterization on-site by using the EL technique. The designed system presents high versatility and it can work 24/7, regardless of the existing level of irradiance, because of the designed set-up and the special algorithms of detection implemented in the control program.

In parallel, an advanced analytical software has been also implemented in order to treat the EL images captured on the field to allow the automatic detection and identification of potential problems on the modules under analysis. In this moment, specific functionalities for the detection of some of the most important kinds of defects have been already included: isolated areas by cracks, PID patterns, soldering failures, corrosion problems, etc.

6 FURTHER WORK

The analytical software is in a constant process of improvement, by upgrading the algorithms already implemented and by adding new functionalities to the program for the detection of other kinds of defects.

Moreover, in this moment some interesting researches are ongoing in order to implement a new tool that estimates, in an accurate way, the possible decrease in the maximum power of the module with the analysis of the EL images captured on-site.

Finally, the implementation of an extension of this analytical application to other PV technologies different from c-Si (i.e. CIGS, CdTe, a-Si) is expected in the near future.

7 REFERENCES

[1] O. Martínez, et Al., "Implementation of a friendly daylight electroluminescence system for the inspection of solar PV panels", 33rd European PV Solar Energy Conference, 2017.

[2] Stoicescu L., et Al., "DAYSY: Luminescence imaging of PV modules in daylight", 31st European PV Solar Energy Conference, 2014.

[3] Cuadra J.M., et Al., "Comparison of different PV modules evolution after being submitted to special test-to-failure protocol beyond the IEC61215:2005 standard", 29h European PV Solar Energy Conference, 2014.

[4] M. Köntges, et Al., "Non-visual failures found in PV modules", IEA PVPS Task 13 Meeting, March 19-20 2013.

[5] Sergiu Spataru, et Al., "Quantification of solar cell failure signatures based on statistical analysis of electroluminescence images", 33rd European PV Solar Energy Conference, 2017.

[6] G. Behrens, et Al., "Digital image processing algorithms for quality-enhancement of electroluminescence picturing in PV-fields", 33rd European PV Solar Energy Conference, 2017.

[7] M. Köntges, et Al., "Quantifying the risk of power loss in PV modules due to micro cracks", *25th European PV Solar Energy Conference*, 2010.

[8] M. Köntges et Al. "Quantitative analysis of PV-modules by electroluminescence images for quality control", 24th European PV Solar Energy Conference, 2009.

[9] T. Potthoff et Al., "Detection of the voltage distribution in photovoltaic modules by electroluminescence imaging", *Progress in Photovoltaics: Reseach and applications*. 2010; 18:100-106.

[10] Ezquer M., et Al., "New analytical software for electroluminescence results out of patterns identified after IEC-61215 standard tests", 27th European PV Solar Energy Conference, 2012.

[11] Kempe MD, Jordan DC., "Evaluation and modeling of the potential effects of a module manufacturing anomaly", Prog. Photovolt Res Appl. 2017;25: 982–988.

CHARACTERIZATION OF FRAMELESS DOUBLE GLASS SILICON CELLS MODULES WITH A NEW ACCELERATED AGING TEST PROCEDURE

J. Dupuis[1*], E. Lajoie-Mazenc[1], F. Sicard[1], D. Binesti[1], S. Mousel[2], K. Radouane[2]
[1]EDF R&D, EDF Lab Les Renardières, Avenue des Renardières, 77250 Moret Loing et Orvanne, France
[2]EDF Renouvelables, 100 Esplanade du Général de Gaulle, 92932 Paris La Défense Cedex, France
*Corresponding author: Julien.dupuis@edf.fr

ABSTRACT: A new accelerated aging test procedure for photovoltaic modules developed by EDF R&D is presented. This test sequence focus on activating specific failure mode mechanisms. The different sequences are then applied to two sets of unframed double glass modules to evaluate its robustness in detecting weaknesses. The results show very robust module conception with degradation inferior to 3% despite aging test sequences well beyond classical IEC standards. Using the procedure and extended aging tests, evidence of slight yellowing during damp heat test were detected. The robustness of the module tested suggests that the aging test sequence need to be completed with more precise characterization method and adapted to the different module technologies available (half cut cell, thin films…).
Keywords: Reliability, Experimental Methods, PV Module, Qualification and testing

1 INTRODUCTION

Nowadays, photovoltaic modules are becoming more and more robust regarding accelerated ageing test based on IEC or UL standards [1-2]. It is not unusual to find module ongoing two, three or even more times each major tests of the different standards without degrading more than several percent of efficiency. In the meantime the need of robust evaluation of new modules for PV projects is crucial regarding new modules technologies like PERC-cell module, module with new encapsulant, frameless module with thin glass, half-cell module design, thin connecting wires… This is why current research institutes and companies are developing alternative aging tests by changing or adding sequences in the IEC or UL standard procedures or combining some accelerated tests in order to highlight particular types of degradation [3-4].

EDF R&D, involved in the quality system developed by EDF Renewables for their photovoltaic projects, has considered this problem by developing a new aging test procedure based on failure mode mechanisms, its experience in testing photovoltaic modules since years and others labs practices and procedures. This procedure is then tested on two sets of frameless double glass 72-cells silicon modules from tier one manufacturers in order to validate its robustness in detecting specific module weaknesses.

2 NEW AGING TEST PROCEDURE

The EDF R&D procedure is composed of five parallel sequences (and an additional reference control sequence). The logic behind each sequence is explained below. All sequences begin with a preconditioning step of at least 20 kWh (40 kWh for PERC cell modules). The full procedure required 10 modules.

2.1 Potential Induced Degradation (PID) sequence – 1 module

By comparing indoor measurement and outdoor exposition, EDF R&D has identified that IEC criteria were not always sufficient to guarantee absence of PID for a long period of operation. This is why we have proposed an extended PID test with a slightly higher temperature as shown in Figure I. We also add an intermediate characterization (I(V) test and electroluminescence) to better follow a potential degradation.

2.2 Mechanical Loading (ML) sequence – 2 modules

If the IEC standard recommends to make a damp heat sequence prior to the mechanical loading [1], we identified that adding a thermal cycling test after mechanical loading is more relevant for the failure modes investigated (see Figure I). Thermal cycling will help in revealing micro-cracks that may appear during mechanical tests.

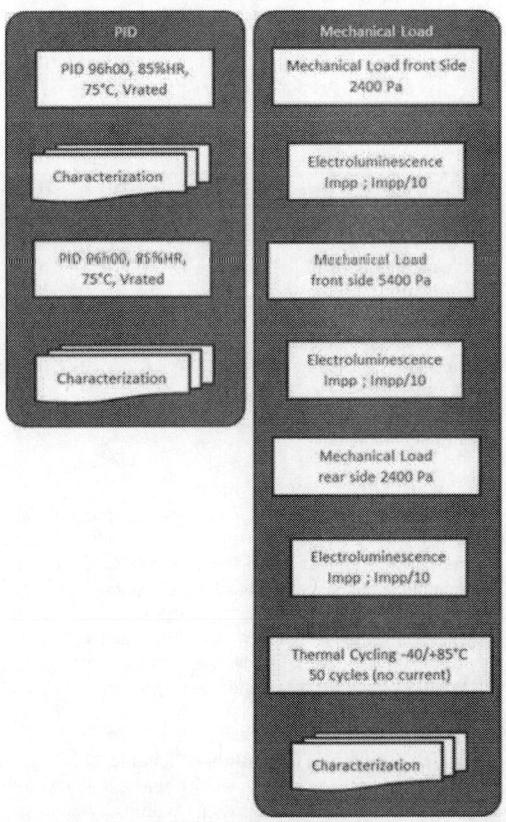

Figure 1: Detailed PID (left) and Mechanical Loading (right) sequence from EDF R&D. Vrated is the maximum voltage allowed by the module (1000 V or 1500 V usually)

Table I: Failing modes tested for PID and ML sequences

Failure modes tested	PID	ML
Cell cracks		X
Electrical connections		X
Mechanical integrity		X
Delamination		X
Corrosion	X	
Encapsulant chemical degradation		
Encapsulant optical degradation		
Joint integrity	X	
PID	X	
Electric Insulation	X	

2.3 Damp Heat (DH) and Potential Induced Degradation (PID) sequence – 2 modules

1000 hours damp heat seem too low to cover real conditions in tropical environment during several decades. Moreover, this duration is also not sufficient to reveals defect in the majority of modern modules. We suggested to extend the duration to 3000 hours and add a PID test to verify if the encapsulant has kept its PID free properties (if any). The detailed sequence is presented in Figure II.

2.4 Thermal Cycling sequence – 2 modules

We decided to add a humidity freezing test between two classical sequences of thermal cycling test instead of just extending the amount of cycles or the thermal amplitude (see Figure II). The objective is to use freezing in order to reveal potential cell cracks initiate during the first thermal cycling test.

Table II: Failing modes tested for DH & PID and TC sequences

Failure modes tested	DH & PID	TC
Cell cracks		X
Electrical connections		X
Mechanical integrity		X
Delamination	X	X
Corrosion	X	
Encapsulant chemical degradation	X	
Encapsulant optical degradation		
Joint integrity	X	X
PID	X	X
Electrical Insulation	X	X

2.5 UV and DH sequence – 2 modules

For this sequence presented in Figure III, our approach is to increase UV exposure in order to focus on edge sealant, encapsulant and backsheet behaviors (yellowing, delamination…). We associate UV with damp heat test, knowing that these two stress are recognized to accelerate degradation on the mentioned module elements. The rest of the sequence (humidity freeze, thermal cycling and PID) will validate the whole module integrity after UV and humidity exposition.

2.6 Reference module for control – 1 module

In order to keep robust results through the whole sequences, each characterization within ageing test sequence is carried out using a reference module of the same characteristics than the tested ones (see Figure III).

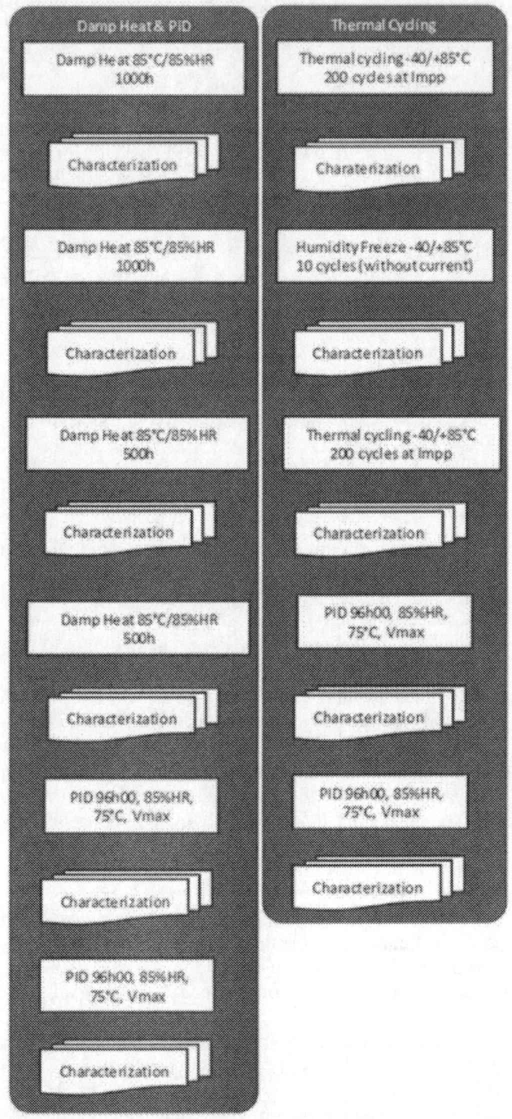

Figure II: Detailed DH-PID (left) and Thermal Cycling (right) sequences from EDF R&D

Table III: Failing modes tested for UV & DH sequence

Failure modes tested	UV & DH
Cell cracks	
Electrical connections	
Mechanical integrity	
Delamination	X
Corrosion	X
Encapsulant chemical degradation	X
Encapsulant optical degradation	X
Joint integrity	X
PID	X
Electrical Insulation	X

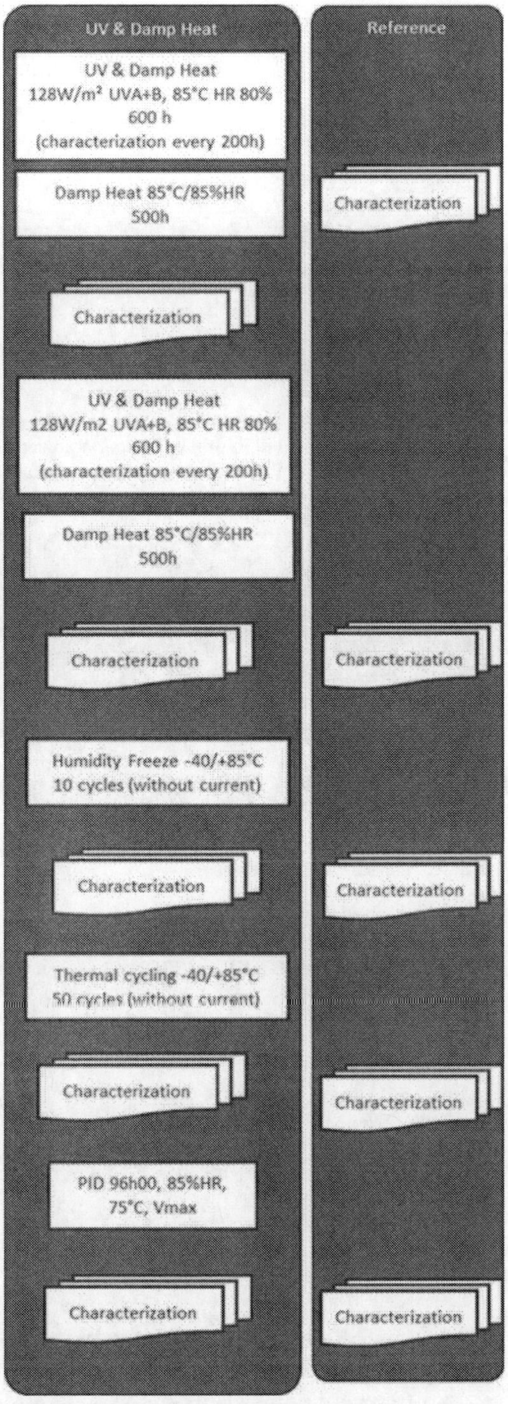

Figure III: Detailed UV-DH (left) and Reference module control (right) sequences from EDF R&D

3 APPLICATION ON DOUBLE GLASS MODULES

In order to evaluate the effectiveness of the procedure in detecting PV modules failing mode mechanisms, we have tested two sets of double glass photovoltaic modules coming from two different PV manufacturers (Man 1 and Man 2). Modules characteristics are very similar and are detailed in the Table IV.

Table IV: PV modules characteristics

PV module characteristics	Man 1	Man 2
Cell Technology	Al-BSF	Al-BSF
Silicon type	mc-Si	mc-Si
Cell number	72	72
Front/rear side	Glass	Glass
Glass thickness (mm)	2.5/2.5	2.5/2.5
Front Encapsulant	EVA	EVA
Rear encapsulant	White EVA	White EVA
Frame	No	No
Module power	320 Wp	315 Wp

I(V) measurements are done using a PASAN flash tester AAA+ SunSim 3b, Electroluminescence (EL) measurement are done on a Greateye LumiSolarProfessional system. Preconditioning is done using OSRAM HMI lamp with 1000W/m² power at 50°C during 20 hours. UV exposure is done in a CTS climatic chamber using 6 lamps able to deliver 128W/m² UV and temperature between 25°C and 100°C. TC, PID and DH tests are done in CLIMATS climatic chamber able to adjust temperature between -50°C and 120°C and 10% to 95% humidity.

3.1 Preconditioning

Before ongoing accelerated ageing tests, all modules underwent a 20kWh preconditioning step. Power variations are detailed in Table V

Table V: Relative variation after preconditioning. Degradation values are overestimated due to temperature probe problem.

Preconditioning impact (%)	Man 1	Man 2
Pmax	-1.76	-1.73
Voc	-0.64	-0.54
Isc	-0.75	-0.80
Fill Factor	-0.38	-0.40

Both manufacturers' modules show similar degradation after preconditioning with nearly 1.7% degradation. These results are slightly below what is observed usually the Light Induced Degradation (LID [5]) phenomenon on this kind of modules. The reason was identified to be cause by a mal-functioning temperature probe: a measurement confirmation was done on spare modules and has given a power degradation near 1.3% only, more in accordance with literature. It has to be noted that only the preconditioning values are affected in the following results presented in this paper.

3.2 PID sequence results

We present in Figure IV results of the two manufacturers for the PID sequence.

No degradation is observed with the PID sequence. A small performance improvement seems to occur but the reference module comparison highlights that the gap is staying in the uncertainty of the measurement. Moreover, Electroluminescence results are not showing any visual change during PID which confirm no significant change at the module level.

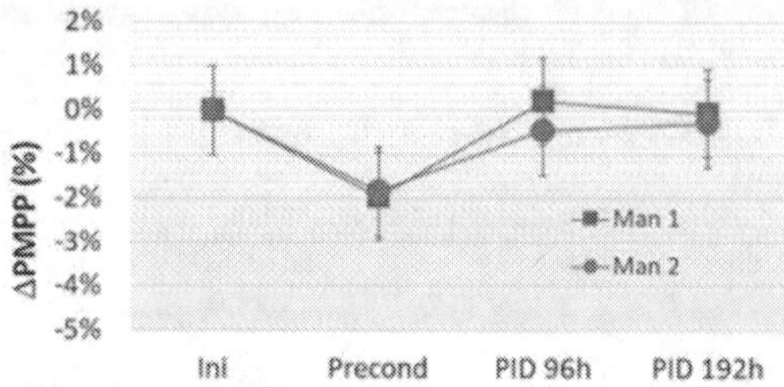

Figure IV: Relative evolution of the maximum power of the modules with PID test sequence. Measurement uncertainty is ±1%.

These results are in accordance with the manufacturer datasheet (PID free) and the fact that unframed double glass module are usually very resistant to PID due to high waterproofing capability and less sensitive to polarization (limited to the clamps) [6-7].

3.3 ML sequence results

The power variation results before and after ML and TC50 are shown in Figure V.

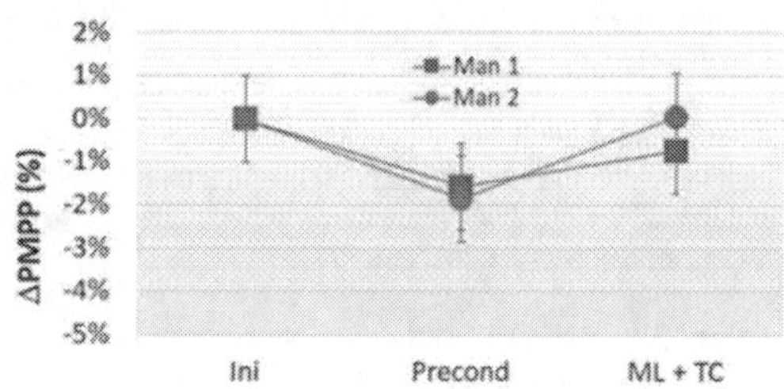

Figure V: Relative evolution of the maximum power of the modules with ML test sequence. Measurement uncertainty is ±1%.

Despite their high mechanical flexibilities, no power degradation is observed after ML and TC50 for both modules. Electroluminescence pictures are not showing any cracks highlighting the robustness of both modules.
Double glass module are more robust to ML than Glass Backsheet thanks to the limited stress on cell during module deformation. This is explained by the mechanical neutral fibre of the module which is located at the cells level contrary to the glass backsheet structure (if both front and rear glasses have the same thickness) [8]. This is why TC50 is added to the sequence to reveal cells' mechanical weaknesses but even this sequence did not reveal any cell cracks.

3.4 DH & PID sequence results

As seen in Figure VI, The degradation stays limited for both sets of modules. Modules of "Man 2" are performing slightly better than "Man 1" (0.8% power degradation versus 1.7%).
This difference may be explained by the apperance of some yellowing of the edge of the modules after the first 2000h of DH (Figure VII).
EL picture is not showing any evident sign of cell degradation at the edge of the module (see Figure XI a) and b)) despite 3000h DH and 192h PID 1500V.

As no edge protection is implemented on these modules, moisture ingress barrier and PID resistances are mostly provided by the EVA encapsulant highlighting its quality.

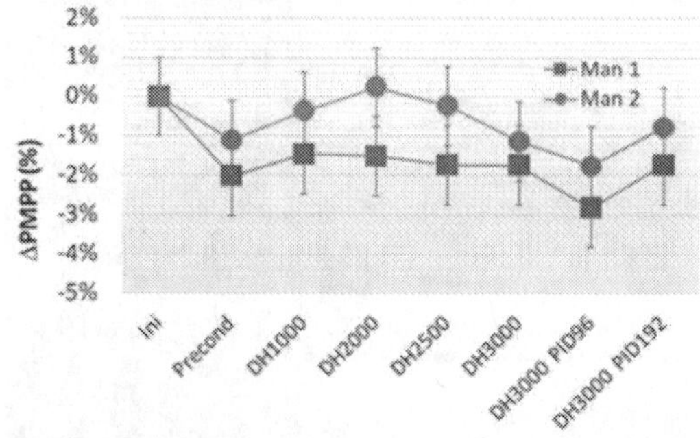

Figure VI: Relative evolution of the maximum power of the modules with DH & PID test sequence. Measurement uncertainty is ±1%.

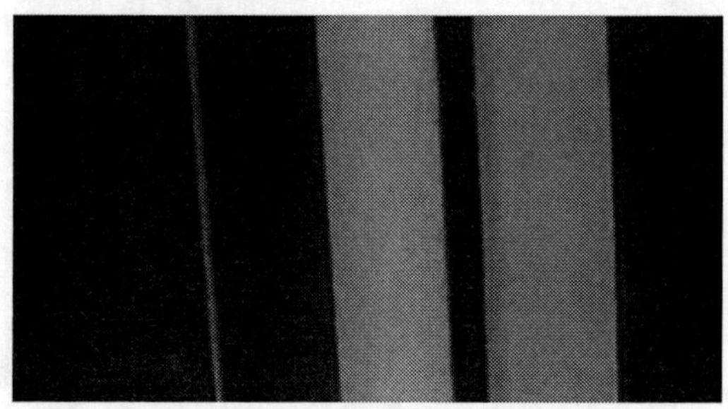

Figure VII: Evidence of yellowing at the edges of a tested module compared to the reference.

3.5 TC sequence results

As seen in Figure VIII, Both types of modules show excellent behavior with TC, HF and PID stress as none of them show a significant decrease in efficiency (<1%). Cell soldering and thermal stress are well controled in both modules.

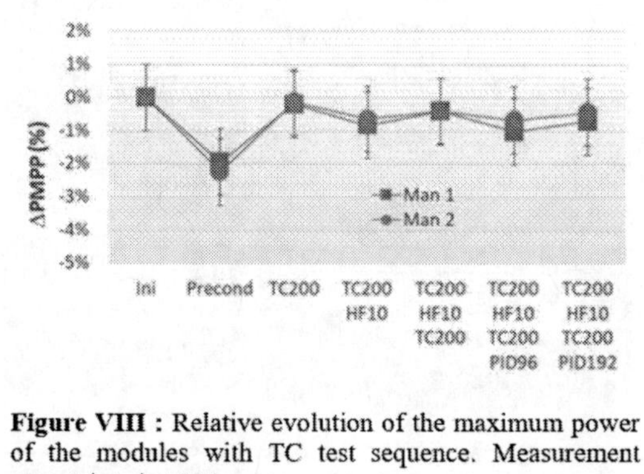

Figure VIII : Relative evolution of the maximum power of the modules with TC test sequence. Measurement uncertainty is ±1%.

3.6 UV & DH sequence

Figure IX shows that UV exposure seems to have limited effect on the performance of both sets of module despite the high UV dose exposure (total of 1200h of UV at 128W/m^2). Surprisingly, the combination of UV and DH did not entail some EVA yellowing as it was observed in the DH & PID sequence. More investigations are in progress to understand the phenomenon.

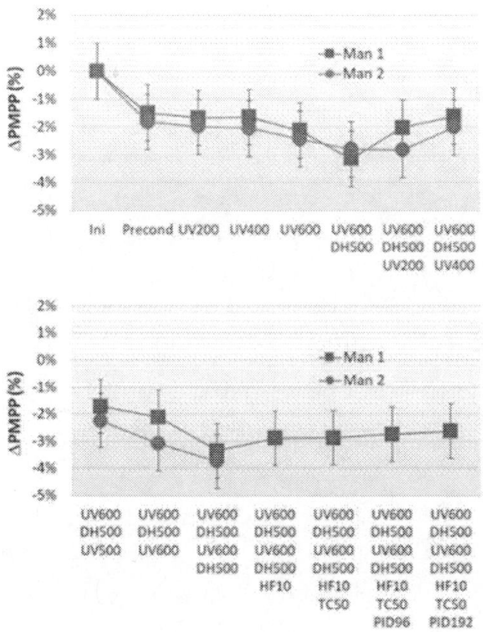

Figure IX: Relative evolution of the maximum power of the modules with UV & DH test sequence. Measurement uncertainty is ±1%. Data are presented on two tables for more clarity.

4 DISCUSSION

The application of this new aging tests sequence on double glass modules results in degradation (power loss) less than 3% after each sequences.
This mean that on one hand these modules are really robust compared to state of the art results [2] and are very well suited for our photovoltaic projects. On the other hand, this also mean that our aging test procedures is still not severe enough or not adapted at these types of module to put in evidence the major failure mode mechanisms.

As the degradation stays limited, we choose to add steps of DH after the PID step of the DH & PID sequence in order to achieve a total of 4500h of DH as EVA is often degrading above 3000h of DH [9-10]. The degradation stay limited with less than 3% degradation (see Figure X) but this time some evidence of corrosion is visible on EL especially on the cell placed at the corner of the module as shown in Figure XI c) and d) which could announce a more severe drop in power. Work are in progress to improve EL picture treatment to be able to detect picture changes earlier and to avoid longer aging test sequence.

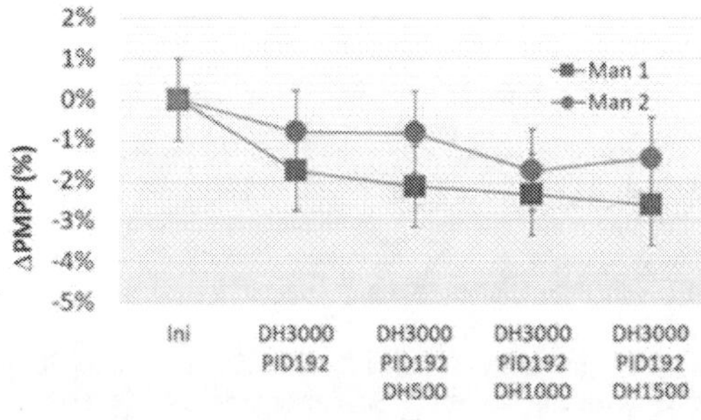

Figure X: Relative evolution of the maximum power of the modules with extended DH & PID test sequence. Measurement uncertainty is ±1%.

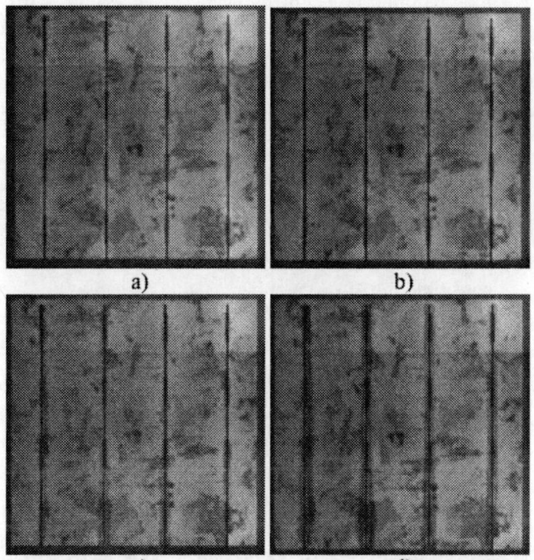

a) b)

c) d)

Figure XI: Electroluminescence pictures of a cells from a module corner in different aging state: a) initial; b) after the DH + PID sequence; c) after the DH + PID sequence + 1000h DH; d) after the DH + PID sequence + 1500h DH

5 CONCLUSION AND OUTLOOK

A new accelerated aging test procedure developed by EDF R&D is presented and tested. This procedure goes well beyond classical IEC standard and has been developed with the objective in detecting specific failure mode mechanisms. All the sequences of this program have been tested on two sets of unframed double glass modules. Despite the absence of frame, both sets of samples show very robust results after each sequence. Only few defect were visible: the DH & PID sequence have put in evidence edge yellowing but limited to the very edge (no evident yellowing of the encapsulant on the solar cells) and without any high impact on the modules performance (less than 3% degradation). If the classical IEC and UL standards would not have been sufficient to highlight these weaknesses, our procedure has not put in evidence sharp degradation that may have been expected. On one hand, we can first conclude on the robustness of the tested modules but on the other hand the aging test procedure has yet to be improved to be able to reveal modules failure mode mechanisms in lower time. In addition to this evolution, work will be carried out to get more precise characterization in order to detect first preliminary degradation effects to avoid time consuming aging test sequences. Lastly, the test sequences need to be adapt to the technology used within the modules (thin films, half cut cell…) and the climate aimed for better relevance.

6 ACKNOLEDGMENTS

The authors want to thanks the EDF R&D laboratory team for the help provided in performing module testing.

7 REFERENCES

[1] International Electrotechnical Commission, IEC 61215 standard - Crystalline silicon terrestrial photovoltaic (PV) modules – Design qualification and type approval (2016)
[2] T. Doyle et al., 2018 PV module reliability scorecard, DNV GL (2018)
[3] A. Funcell, Proceedings NREL PV Module reliability Workshop (2012)
[4] S. Kurtz et al., NREL Technical Report, Photovoltaic Module Qualification Plus Testing (2013), https://www.nrel.gov/docs/fy14osti/60950.pdf, consulted the 4th September 2018
[5] D. H. Macdonald et al., Proceedings 13th Workshop on Crystalline Silicon Solar Cell Materials and Processes (2003)
[6] Y. Zhang et al., RSC Adv., 5 (2015), 65768-65774
[7] B. Braisaz et al., Proceedings 42th IEEE PVSC (2015)
[8] C. Buerhop et al., Prog. Photovolt Res Appl. Vol 26, 4 (2018) 261-272
[9] E. Cabrera et al., Proceedings 31st EUPVSEC (2015), 1935-1941
[10] E. Cabrera et al., Proceedings 32nd EUPVSEC (2016), 1573-1578

ANTI-REFLECTIVE PROPERTIES OF REACTIVE ION ETCHED GLASSES

E. Zugasti, J. Bengoechea, A. Turumbay, M. J. Rodríguez and A. R. Lagunas
NATIONAL RENEWABLE ENERGY CENTRE (CENER)
C/Ciudad de la Innovación 7, 31621 Sarriguren (Navarra) Spain
Telephone: + 34 948 25 28 00; email: ezugasti@cener.com

ABSTRACT: Currently, most of the glasses used in the PV industry include anti-reflective coatings (ARC) in order to reduce the reflection at the air-glass interface and increasing, in this way, the absorption of light within the solar cells. However, the reliability of these coatings for a service life of more than 20 years remains still a challenge. This work investigates the possibility of texturing the front glass surface in order to reduce the reflection in a broad spectral band by a single RIE (Reactive Ion Etching) process, using CF_4 as the active species. The processed glasses showed a decrease in global spectral reflectance compared with non-textured glasses. This observation has been correlated with their surface morphology, through the measurement of roughness parameters with a contact surface profiler and AFM (Atomic Force Microscope). In addition, the dependence of the spectral transmittance with the incident light angle has been characterized.
Keywords: etching, texturisation, glass, optical properties

1 INTRODUCTION

Glass has a natural reflectance in the solar spectrum region that should be decreased in order to maximize performance of PV modules. To reduce this natural reflectance, two alternative approaches, antireflective coatings (ARC) and texturing processes, are usually applied to the glass surface.

Regarding ARC approach, glass substrates with a refractive index of around 1.5 require a coating with a refractive index near 1.2 to reduce their reflectance. This type of materials does not exist in nature and this issue is solved by tuning the refractive index of the coating to be applied through modified porosity. These ARC are commonly deposited by sol-gel processes consisting of a porous silica layer or, to a lesser extent, by vacuum-based sputtering. However, their durability during the module service life is still a challenge [1].

Therefore, the possibility of texturing the front glass surface in order to reduce the reflection in a broadband spectral region by a single RIE (Reactive Ion Etching) process has been investigated in recent years [2-6]. The RIE process is able to reduce reflection at the front surface of the glass, but with the clear advantage that the resultant texturized glass has the potential to be more durable compared to the ARC currently used in the PV industry, since adhesion would not be a concern. Additionally, regarding thin film production, other benefit is that it could withstand high temperature processes without degrading.

Last, optical reflectance of coatings or textured surfaces varies depending on the angular incidence of incoming solar radiation [7]. Thus, it is worth accounting their behavior under different angles of incidence [8].

In this work, RIE has been used to texture glass as a way to reduce its reflectance and its morphological and optical properties have been characterized. Besides, in order to understand its behavior under field conditions, where larger incidence angles occur, transmittance of these textured glasses has been measured at different angles of incidence.

2 EXPERIMENTAL

2.1 Processing

Borosilicate glasses (dimensions of 24 mm x 36 mm and a thickness of 2 mm) have been texturized by the technique of reactive ion etching (RIE) on one surface. After cleaning the glasses in acetone and alcohol, a texturing process through a RIE system AK 400 of Meyer Burger has been applied. Through variation of bias voltage, gas flow, pressure, gases ratio and time, a set of experiments has been designed in order to obtain texturized glasses with different surface morphologies and optical properties. The process parameters have been modified in these ranges:

- Time: from 10 to 30 min
- Total gas flow: 80 sccm
- Ratio CF_4/Total flow ($Ar+CF_4$): from 0.5 to 1
- Pressure: from $5.0x10^{-2}$ mbar to $8.0x10^{-2}$ mbar
- RF Bias Voltage: from 200 V to 500 V

In Table I, the process parameters of RIE textured glasses are shown, remarking in bold letters some experiments representative of the obtained results and which will be analyzed in detail.

Table I: Process parameters of selected RIE experiments

Sample	Time (min)	CF_4 flow (sccm)	Ar flow (sccm)	Pressure (mbar)	RF Bias voltage (V)
GDR_01	20	60	20	$6.5x10^{-2}$	250
GDR_02	30	80	0	$8.0x10^{-2}$	200
GDR_03	10	40	40	$5.0x10^{-2}$	300
GDR_04	30	80	0	$5.0x10^{-2}$	300
GDR_06	**30**	**40**	**40**	**$8.0x10^{-2}$**	**300**
GDR_07	10	80	0	$5.0x10^{-2}$	200
GDR_08	20	60	20	$6.5x10^{-2}$	250
GDR_09	30	40	40	$5.0x10^{-2}$	200
GDR_10	10	80	0	$8.0x10^{-2}$	300
GDR_11	**10**	**40**	**40**	**$8.0x10^{-2}$**	**200**
GDR-12	10	80	0	$5.0x10^{-2}$	500
GDR-13	10	80	0	$5.0x10^{-2}$	400
GDR-14	**10**	**80**	**0**	$5.0x10^{-2}$	**300**
GDR-15	10	40	40	$5.0x10^{-2}$	500
GDR-16	**10**	**40**	**40**	$5.0x10^{-2}$	**400**
GDR-17	10	40	40	$5.0x10^{-2}$	300
GDR-18	10	60	20	$5.0x10^{-2}$	400

2.2 Characterization

The spectral reflectance, between 250 nm and 1640 nm, of the textured glasses is measured with an integrating sphere and two spectroradiometers (one for the wavelength range from 250 nm to 900 nm, and the second from 800 nm to 1640 nm). The lamps used to cover all the above mentioned spectrum are also two: a deuterium lamp (for UV and visible range) and a halogen lamp (for visible and IR range).

The light is coupled to the measurement system via optical fiber from the lamps to the sphere, and then, from the integrating sphere to the spectroradiometers (see Figure 1). Reflectance is measured at an incidence angle of 8°. Global spectral reflectance has two components: diffuse and specular. With the set-up described in Figure 1, global and diffuse spectral reflectance are measured and specular reflectance is obtained by subtracting the diffuse component to the global reflectance. To that end a variable surface is needed which can be: totally reflecting, to measure global reflectance, or a trap, to quantify only diffuse reflectance

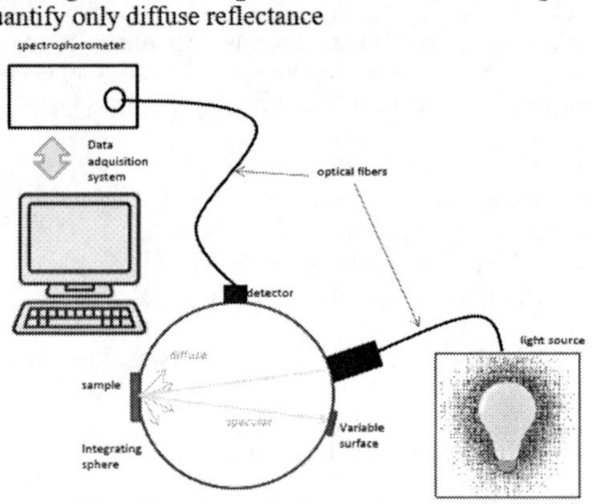

Figure 1: Drawing of the set-up used for the characterization of reflectance.

The global spectral transmittance at different light incident angles (0°, 10°, 20°, 30°, 40°, 45° and 50°) has also been measured with the same set-up but in a different configuration (Figure 2).

Etching depth has been characterized with a Dektak surface contact profiler, placing a physical mask over the glasses before texturisation. Moreover, for the analysis of roughness, several sweeps have been performed on the etched area in order to have statistical information.

At last, roughness and morphology of the glasses have been examined by an AFM (Nanotech, model AFM43) in dynamic mode.

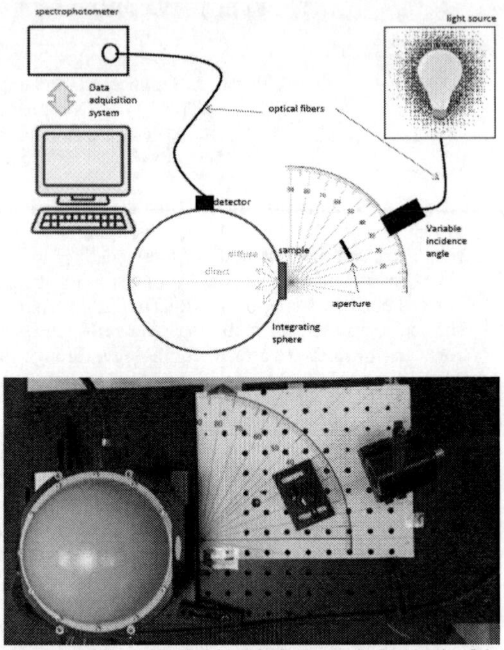

Figure 2: Drawing (top) and photograph (bottom) of the set-up used for the characterization of the global transmittance at different light incidence angles.

3 RESULTS AND DISCUSSION

3.1 Etching rate

In this case, the etching rate of the RIE process depends mainly on the RF power led by the RF bias voltage, being less significant the variations of pressure or gases ratio, as shown in Figure 3, where all the experiments have been depicted. The etching rate is almost linearly proportional to the bias voltage, varying from 7 nm/min to 52 nm/min for RF bias voltage from 200 V to 500 V.

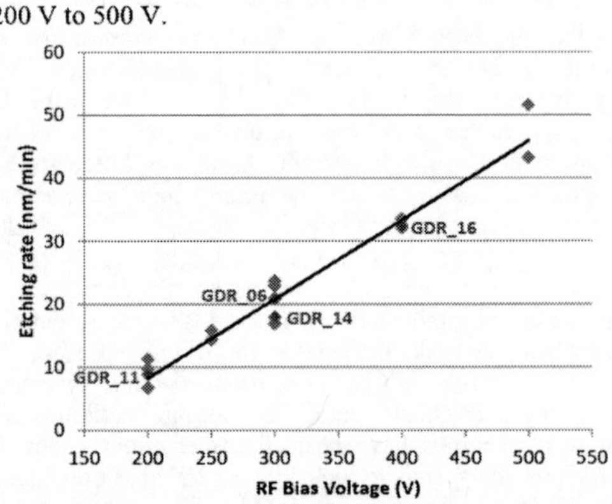

Figure 3: Etching rate as a function of RIE RF bias voltage depicted for all textured glasses. GDR_11, GDR_06, GDR_14 and GDR_16 have been highlighted.

3.2 Reflectance

Global, diffuse and specular spectral reflectances were measured as described in section 2.2. With the aim of comparing the textured glasses, the spectral reflectance has been averaged in the wavelength range between 300 nm and 1150 nm, where crystalline silicon solar cells show absorption (Figure 4). As it can be observed, a reduction of reflectance compared with flat glass has been obtained by texturing borosilicate glasses by means of RIE.

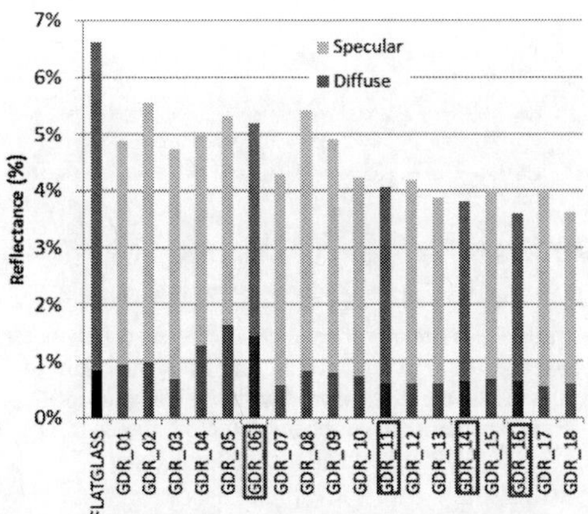

Figure 4: Average reflectance in the wavelength range between 300 nm and 1150 nm.

The global reflectance of the RIE texturized glasses is lower, compared to a flat glass, for all the process parameters tested in this work. The texturized glass which shows the best result is GDR_16 that has a global reflectance of 3.6% (averaged in the range of 300 nm to 1150 nm). If compared with flat glass whose global reflectance in the same range was 6.6%, it accounts for an absolute decrease of 3.0%.

Regarding the ratio between diffuse and specular components of every glass, the diffuse component is minimum at reference flat glass since the surface is not texturized, as expected. Depending on the texture grade, this proportion changes among samples. However, it can be observed that, for all textured glasses, the specular component dominates, having the diffuse component values between 14% and 28% of the total reflectance.

3.3 Roughness

Roughness has been characterized for every textured glass through the parameter "Roughness average", Ra, which is the arithmetic average of the absolute values of the structures height along the sampling length. Ra varied from 14 nm to 147 nm when characterized with a contact surface profiler. Diffuse reflectance component is dependent on this roughness, as shown in Figure 5.

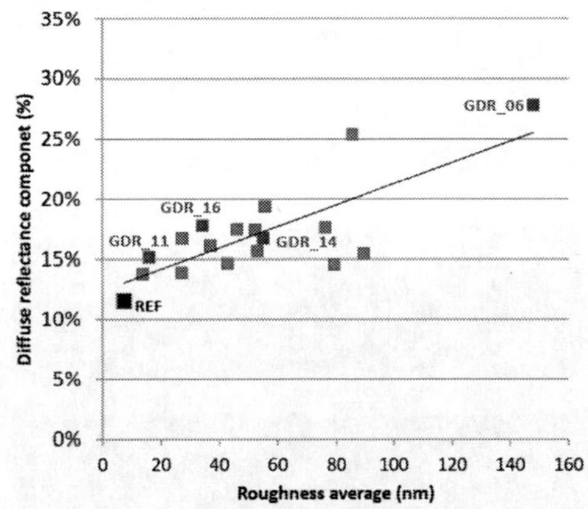

Figure 5: Average diffuse reflectance component in the wavelength range between 300 nm and 1150 nm as a function of the roughness average

As it is discerned in the figure above, the higher the surface roughness, the higher is the diffuse reflectance component of the global reflection

Based on their characteristics, four textured glasses with different average roughness have been selected. Their values of Ra, global reflectance and the reduction in reflectance due to texturisation are summarized in Table II.

Table II: Reflectance and profiler roughness of selected RIE textured glasses

	Average roughness (nm)	Averaged global reflectance (%)	Absolute decrease in reflectance from reference (%)
GDR_11	16	4.1	2.5
GDR_16	34	3.6	3.0
GDR_14	79	3.8	2.8
GDR_06	147	5.2	1.4

In order to study the morphology of these glasses, topology AFM images, depicted in Figure 6, have been captured.

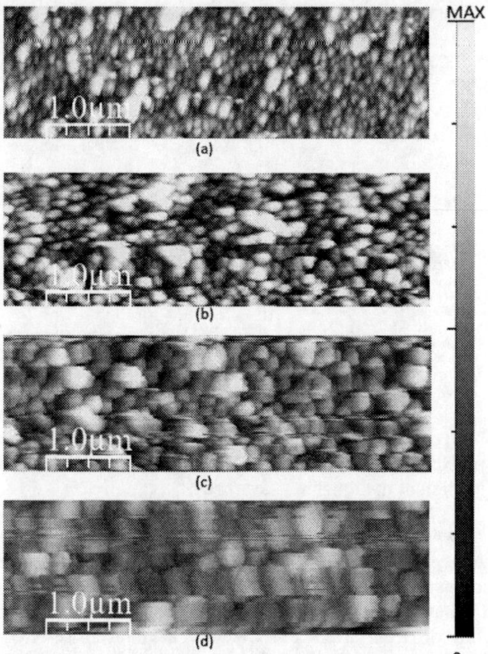

Figure 6: Topology AFM images of different RIE textured glasses: (a) GDR_11, (b) GDR_16, (c) GDR_14 and (d) GDR_06. The scale bar maximum limit (MAX) is different for each image, being 24 nm, 50 nm, 96 nm and 195 nm, respectively

The topology images show that the morphology of the textured glasses is different. The glasses with higher peaks have also broader grains. In Table III, Ra results, quantified with the AFM software [9], together with the ones obtained previously with the mechanical profiler are presented for selected samples.

Table III: Roughness analysis of selected RIE textured glasses carried out by AFM and profiler

Profiler	GDR_11	GDR_16	GDR_14	GDR_06
Ra (nm)	16	34	55	147
AFM				
Ra (nm)	4	11	15	20

As expected, the values of Ra obtained with the AFM are not equal to the ones obtained with the profiler. This is caused by the intrinsic difference in vertical and horizontal resolution of the two techniques joint with the differences in the areal swept. However, although the values of roughness measured with both techniques are different, they follow the same rising trend, from GDR_11, GDR_16, GDR_14 to GDR_06, so the same analysis could be made with the two techniques.

Last, to find out if there is a connection between the roughnesses of the RIE textured glasses and the resultant global reflectance, these two parameters have been represented in Figure 7, for all RIE textured glasses.

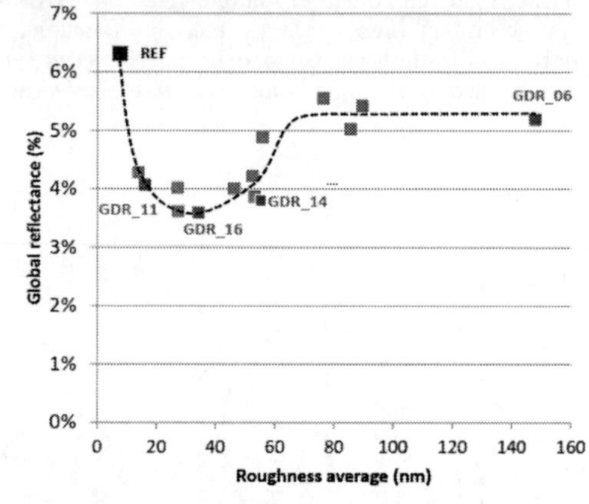

Figure 7: Global reflectance in the wavelength range between 300 nm and 1150 nm as a function of the roughness average measured with the contact profiler. All the experiments have been depicted, being GDR_11, GDR_16, GDR_14 and GDR_06 highlighted

As shown in the figure above, there is a minimum of reflectance for a roughness average measured with the profiler of 40 nm, approximately. The lower and higher roughness values lead to higher reflectance results.

3.4 Dependence of spectral transmittance with the incident light angle

Related to the decrease in reflectance, an improvement of global spectral transmittance has been also obtained by texturing borosilicate glasses with RIE, using CF_4 as the active species compared with flat glass.

The resultant spectral hemispherical transmittances of the more significant texturized glasses are shown in the figure below.

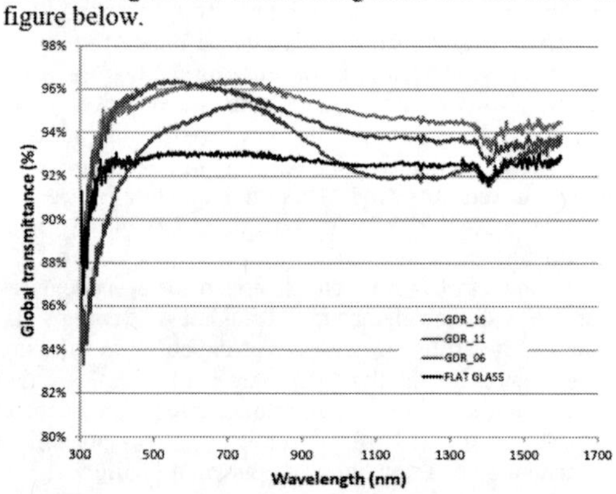

Figure 8: Global spectral transmittance measured on the RIE texturized glasses with different process parameters in comparison with flat glass at normal incidence

As can be valued from Figure 8, the global transmittance of the RIE texturized glasses is higher compared to a flat glass in the whole spectral range and it results, for GDR_16, in an improvement of average transmittance of more than 2.5% (absolute).

When incident light hits glass at an angle higher than the normal, an increased percentage of the irradiance is reflected from the surface of a PV module reducing the in-plane irradiance reaching the PV cells [10].

Some of the textured glasses have been measured at

different incident angles, more representative of real outdoors operation of PV modules, in order to quantify potential IAM (Incident Angle Modifier) factor variation. The selected textured glasses for an in-depth characterization of the incidence angle dependence are:

- Reference: a flat glass that has not undergone any treatment to use it as a reference glass
- GDR_06: which has the lowest improvement in transmittance compared with flat glass,
- GDR_11: with average global transmittance results
- GDR_16: which has attained the best transmittance results in these experiments

The spectral transmittance results have been integrated in the wavelength range, from 350 nm to 1100 nm, significant for silicon solar cells in order to quantify its benefit, see Figure 9.

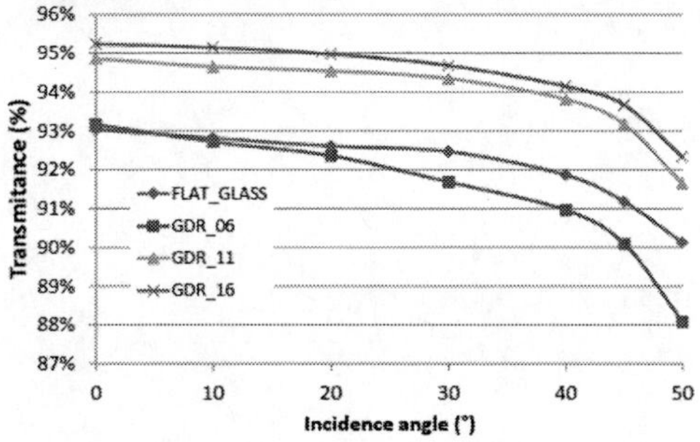

Figure 9: Hemispherical transmittance measured on RIE texturized glasses at different incidence light angles integrated between 350 nm and 1100 nm

A minor variation of incidence angles around the normal incidence will not mean a huge variation of transmittance. However, transmittance losses compared to normal incidence are highly significant for incidence angles exceeding 40 °.

Performance modeling tools, such as PVSys, usually employ a simplified model to determine incidence angle losses and the ASHRAE model is one of the most common methods. To simplify calculations, this model uses a modifier—the incidence angle modifier (IAM), b_0, based on the angle of incidence, according with the following equation:

$$IAM = 1 - b_0 \left(1/\cos(AOI) - 1\right) \quad (1)$$

This effect can be also predicted using a model developed by Martin and Ruiz [11]. Its formulation is used by the International Electrochemical Commission (IEC) standard 61853-2 [12].

$$F(AOI) = [1-\exp(-\cos(AOI)/a_r)]/ [1-\exp(-1/a_r)] \quad (2)$$

With a_r the value which best fits the measured data.

These two parameterizations are presented for each texturized glass in the following table:

Table IV: Fit parameters, b_0 for ASHRAE model and a_r for IEC 61853-2, found to minimize error root-square-mean

	b_0	a_r
FLATGLASS	0,0510	0,19380
GDR_06	0,0903	0,23762
GDR_11	0,0509	0,19382
GDR_16	0,0474	0,18951

The textured glass GDR_06 has higher values of a_r and b_0, which translated into a higher dependence with the AOI, even if this glass has worse behavior than the flat glass reference. The best texturized glass in terms of transmittance, GDR_16, has, in addition, the lowest dependence with the AOI although fit parameters are very similar among samples.

4 CONCLUSIONS

Enhanced antireflective characteristics have been achieved in borosilicate glasses by applying a single RIE process involving CF_4 as reactive species. The absolute reduction in reflectance compared with flat glass is close to 3 %.

Full characterization of roughness of the textured glasses has been done through AFM and DekTak surface contact profiler, having found the minimum reflectance at an average roughness of 40 nm, measured with the mechanical profiler, or 11 nm if it is measured by AFM.

Dependence of the transmittance of the textured samples on the incident angle is similar to the one obtained by flat glass but with a slight improvement for two of the texturized glasses, being the optimum the corresponding to sample GDR_16.

ACKNOWLEDGEMENT

This work was financed by General Directorate of Industry, Energy and Innovation of the Government of Navarre, under the project acronym DESAFIO, in its 2018 call. The authors also want to thank Cristina Pinto for her contribution on optical measurements.

REFERENCES

[1] S. C. Pop, V. Abbaraju, B. Brophy, Y. S.Yang, S. Maghsoodi and P. Gonsalves, "A Highly Abrasive-Resistant, Long-Lasting Anti-Reflective Coating for PV Module Glass", IEEE 40th Photovoltaic Specialist Conference (2014)

[2] X. Li, T. Abe and M. Esashi, "Deep reactive ion etching of Pyrex glass using SF6 p`lasma", Sensors and Actuators A87, pp 139-145 (2001)

[3] J. H. Park, N-E. Lee, J- Lee, J.S. Park and H. D. Park, "Deep dry etching of borosilicate glass using SF6 and SF6/Ar inductively coupled plasmas", Microelectronic Engineering 82, pp 119-128 (2005)

[4] E. Hein, D. Fox and H. Fouckhardt, "Glass surface modification by lithography-free reactive ion etching in an Ar/CF4 plasma for controlled diffuse optical scattering", Surface & Coatings Technology 205, S419-S424 (2011)

[5] Y. M. Song, G. C. Park, E. K. Kang, C. I. Yeo and Y. T. Lee, "Antireflective grassy surface on glass substrates with self-masked dry etching", Nanoscale Research Letters, 8-505 (2013)

[6] N. V. Toan, M. Toda and T. Ono, "An investigation of processes for glass micromachining", Micromachinges, 7, 51 (2016)

[7] C. Armenta-Deu, "A correlation model to compute the incidence angle modifier and to estimate its effect on collectible solar radiation", Renewable Energy Vol. 1, N° 5/6, pp 803-809 (1991)

[8] E. A. Sjerps-Koomen, E. A Alsema and C. Turkenburg, "A simple model for PV module reflection losses under field condtions", Solar Energy Vol. 57, N° 6, pp 421-432 (1996)
[9] I. Horcas, R. Fernández, J.M. Gómez-Rodríguez, J. Colchero, J. Gómez-Herrero and A.M. Baró "WSxM: A software for scanning probe microscopy and a tool for nanotechnology", Review of Scientific Instruments 78, 013705 (2007)
[10] B. Marion, "Numerical method for angle of incidence correction factors for diffuse radiation incident photovoltaic modules", Solar Energy Vol 147, pp 344-348 (2017)
[11] N. Marin and J. Ruiz, "Calculation of the PV modular angular losses under field conditions by means of an analytical model", Solar Energy Materials and Solar Cells, 70, pp 25-38 (2001)
[12] IEC 61853-2

DIGITALIZATION IN PV – VIRTUAL APPLICATION OF REAL WEATHER DATA ON PV MODULES FOR LIFETIME PREDICTION

Ulli Zeller[1,*], Matthias Pander[1], David Dassler[1,2], Bengt Jaeckel[1]
[1] Fraunhofer-Center for Silicon Photovoltaics, Otto-Eissfeldt-Strasse 12, 06120 Halle (Saale), Germany
[2] Anhalt University of Applied Sciences, Faculty EMW, Bernburger Str. 55, 06366 Koethen, Germany
* = corresponding author: phone: +49-345-5589-5219, fax: +49-345-5589-5999, ulli.zeller@csp.fraunhofer.de

ABSTRACT: Lifecycles of products used outdoors are primarily influenced by given operating conditions. These conditions mainly vary depending on the location. Hence one, possibly, can determine one primary driver of all environmental parameters but in general a number of climatic parameters (e.g. temperature, relative humidity, wind, precipitation, UV exposure and others) have to be considered relevant. Working with real meteorological data in lifetime simulation and durability tests leads to large quantities of data. This work introduces an approach to reduce these large quantities to ensure feasible implementation in lifetime simulation by the example of finite element analysis on PV modules.
Keywords: digitalization, lifetime simulation, data management

1 AIM

Lifetime of outdoor products under operating conditions is influenced by a great variety of parameters. Especially in lifetime simulation, these parameters are implemented to represent real environmental loads in abstracted and simplified simulation models of real products of virtual prototypes. In general, aim of lifetime simulation is to show one products behavior under operating conditions for whole life cycles.

By the use of virtual simulation models, just an idea or a virtual prototype of the product is necessary and with state of the art computing hardware simulation time of complete life cycles, for example several decades, can be reduced down to fractions of hours. Within such simulations processes of fatigue, degradation and damage as well as functional capability of a product can be analyzed and evaluated over its expected lifetime.

"Hence, the new degradation science examines degradation of a material or system, guided by real-world or realistic outcomes, whose fundamentals include modeling, monitoring, and prediction of a degradation process, as well as intervention, feature selection, and optimization aimed at improvement of materials and reduction of system failures.

A new interdisciplinary approach to degradation science calls for the involvement of materials science, physics, chemistry, statistics, computer science, engineering, and energy industries in the investigation of real-world degradation of energy material over their lifetimes." [1]

Using real weather data for appropriate load testing of PV modules requires vast quantities of time and costs, respectively. As a result of digitalization and computer aided engineering such comparable testing with real weather data are performed by simulation software with corresponding material and load data implemented.

The PV qualification standard IEC 61215 is known to work for temperate climates (e.g. Central Europe) but in case of e.g. desert climate module loadings are significantly higher over the years of operation. Module manufacturers tackle this problem by extending existing tests without systematically addressing the specifics of individual regions to proof product reliable.

Current PV standards for PV performance testing and energy rating (IEC 61853) use some sort of classification for standard reference climate profiles. Similar research avoided these standard reference climate profiles and analyzed spatiotemporal variability of atmospheric variables to assess solar energy potential using remote sensing and weather forecast data on the example of the UAE [2]. This work goes one step further by introducing a concept for using real weather data for simulation of lifetime behavior.

2 APPROACH

2.1 Environmental influences on PV modules

PV modules in operation are exposed to all naturally occurring environmental conditions. These are temperature, wind, irradiation, precipitation and humidity. Depending on their location of operation, they are subjected to typically environmental conditions of the well-known climate classification by Köppen-Geiger (Figure 1).

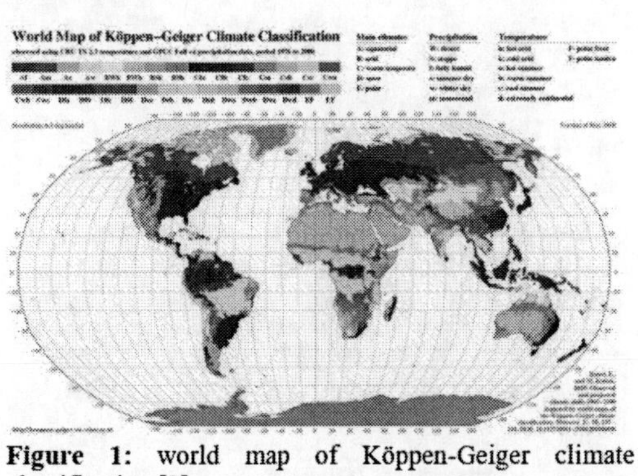

Figure 1: world map of Köppen-Geiger climate classification [3]

Representing the global distribution of each environmental parameter individually, Figure 2 to Figure 6 illustrate the inhomogeneity very clearly.

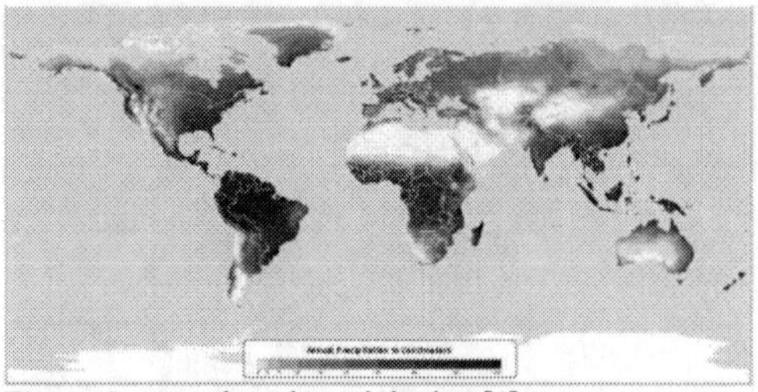

Figure 2: annual total precipitation [4]

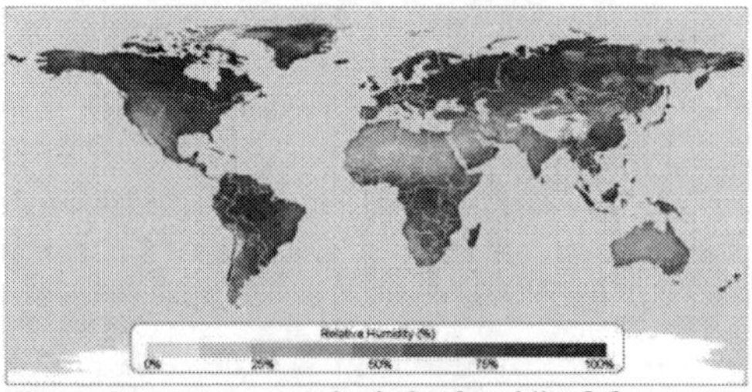

Figure 3: average annual relative humidity [4]

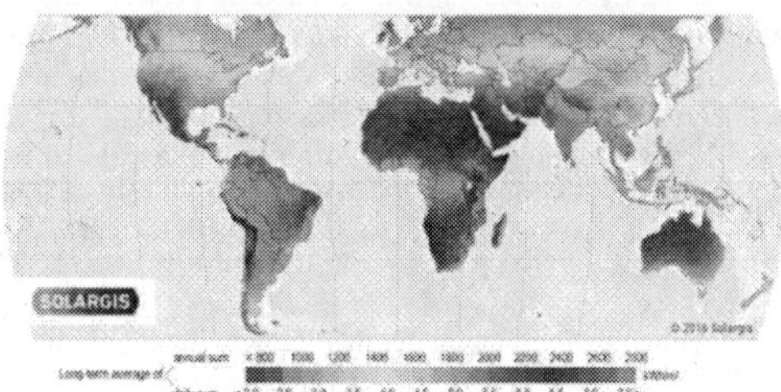

Figure 4: global horizontal irradiation (GHI Solar Map © 2017 Solargis [7]

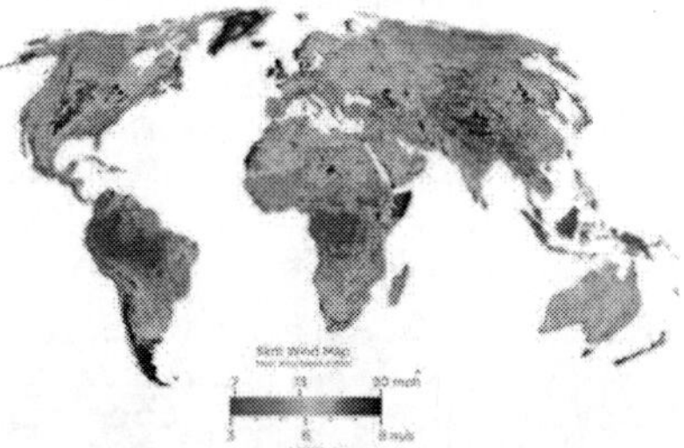

Figure 5: global wind speed (© 2015 Vaisala) [8]

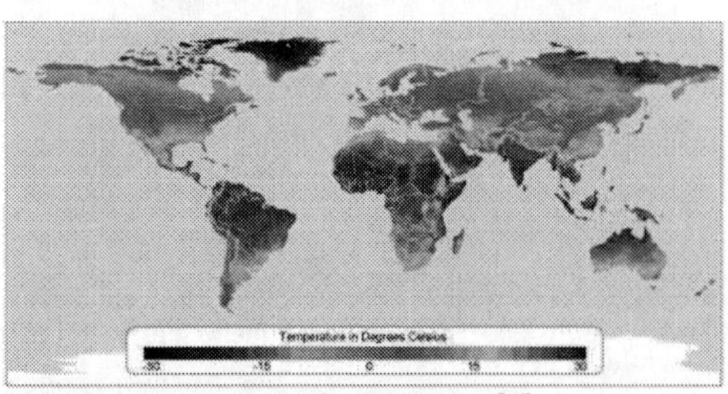

Figure 6: average annual temperature [5]

Table 1 shows the global mean/average values of these five parameters.

Table 1: global mean/average values

	mean/avg.	Ref.
precipitation	1156 mm	[5]
relative humidity (ground +10m)	73.3 %	[6]
global horizontal irradiation	1439 kWh/m²/a	[6]
wind speed (ground +10m)	5.7 m/s	[6]
ambient temperature (land regions)	9.6°C	[9]

For the working example three locations have been selected representing typical conditions regarding their climate classification (see Figure 1) with current commercial potential for PV plants. The selection of locations was based on climatic categorization – moderate, tropical and desert climate – and further filtered in order to get economically relevant regions and metropolises. The considered locations were Berlin (Germany) as reference for moderate climate, Singapore for tropical and Doha (Qatar) for desert climate. Their average weather conditions compared to the global mean values (from Table 1) are shown in Figure 7.

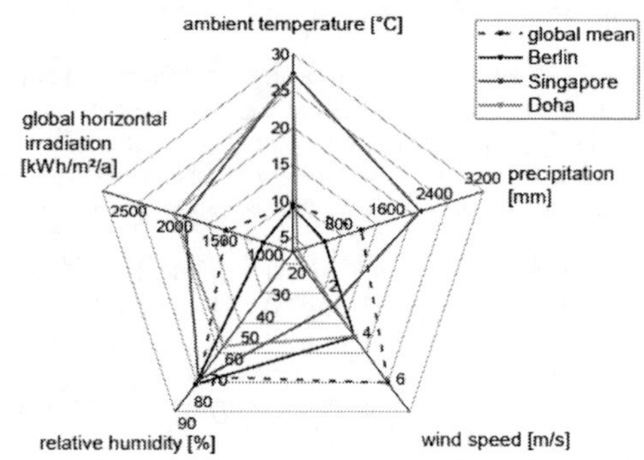

Figure 7: global mean values of environmental parameters and site-specific values for three examples of use (Berlin, Singapore and Doha)

Depending on the climate region, the environmental parameters can promote different degradations with amplifying or decreasing superposition. Moisture diffusion in the module is dependent on external relative humidity as well as module temperature. Module temperature depends on irradiation, ambient temperature and wind speed (convective heat transfer). Increasing temperature is always a way to accelerate aging processes in the polymer components. In contrast high temperatures are softening the polymer and reduce load on the solar cells resulting in lower micro-cracks. The down-side might be additional loads on the cell interconnectors or creep in the solder connects.

Therefore module temperature is a crucial parameter for simulation. In this contribution we limit ourselves to thermo-mechanical loading of the interconnector ribbons and the related fatigue failure. With respect to seasonal variations, day-night-shifts and variable cloudiness with resulting module shading, for this work temperature is set as input parameter. Based on weather data from 2014 module temperatures were calculated (see 3.1) as input parameter for interconnector related PV module's fatigue simulation.

2.2 Data management

Common weather and climate databases provide hourly mean values of environmental parameters as highest data resolution for analysis periods of one year or more. To assess climatic influences regarding PV degradation and fatigue depending on its location, resolution per minute would be necessary for more realistic simulation of temperature changes, for example due to clouding. Considering other parameters like wind, even higher resolutions lead to a sensible data basis.

With the provided resolution of hourly mean data one full calendar year could be divided into 8760 time steps (of thermal loading). For each of the three locations the corresponding weather data has been acquired as described in 3.1.

2.3 Target value

For fatigue assessment of interconnectors there are different concepts using plastic strain of the ribbons or plastic strain energy density as introduced in [10]. In this work we use strain energy density evolution as indicator for ribbon fatigue. The evolution of plastic deformation is path dependent and thus dependent on the load history. Nevertheless individual load steps can be analyzed for an increment of plastic strain energy. This is related to specific one-stage experiments data and gives the particular fatigue progress. The fatigue progress of individual steps can be simply added in first order approximation according to Miner's rule. There are definitely limitations and more dedicated models may be used but it is sufficient for superposition of e.g. thermal and mechanical results.

Keeping this in mind the simulation in this work will deal with interconnector fatigue due to only thermal-mechanical loads due to temperature changes.

2.4 Simulation model

For fatigue simulation of solar cell interconnector due to thermal loading the general setup of a PV module was reduced to a two-dimensional model describing the cross-sectional area of the PV laminate neglecting the frame. Further simplification in order to model reduction was done by consideration of five solar cells within a laminate, like previous investigations on interconnector fatigue [10-13]. The scheme of simulation model – divided into global and submodel – is shown in Figure 8.

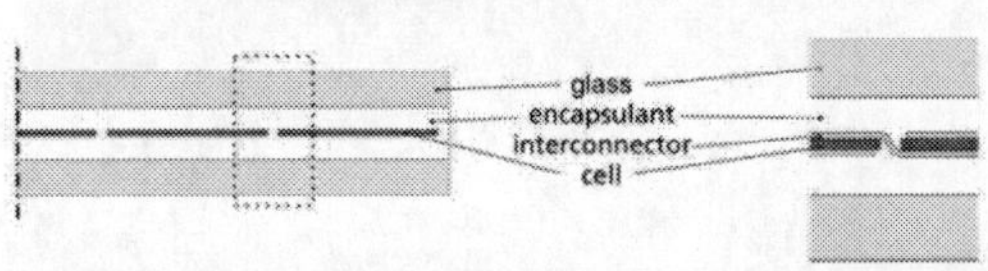

Figure 8: scheme of simulation model - global (left) and dashed section of submodel (right) with each of its components considered

The global model is simplified by use of symmetry while neglecting the interconnector which means considering glass, encapsulant and silicon components only. That is justified by the representative share through the thickness of the modeled cross-section, meaning the gap between interconnectors of one cell is much wider than the interconnector itself.

The finite element mesh shown in Figure 9 displays each models components colored by corresponding materials.Figure 9

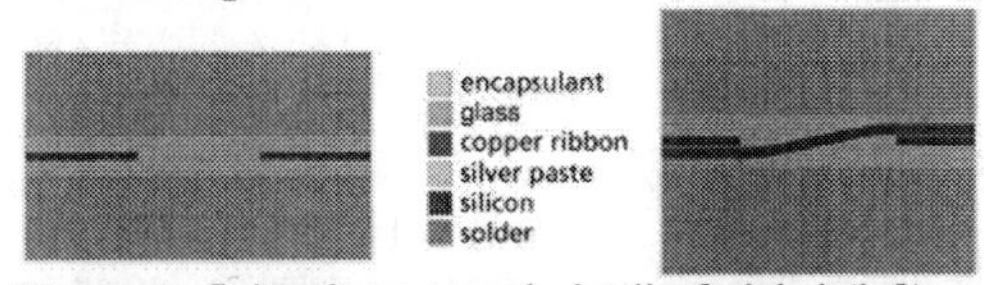

Figure 9: finite element mesh detail of global (left) and submodel (right) with corresponding materials

Boundary conditions for thermal loading simulation were set as horizontal displacement restriction of global models symmetry axis as well vertical displacement restriction of the bottom right corner (mesh node) of the model. The thermal loading was applied homogeneously to the global model. The resulting deformation was transferred to the corresponding interfaces (left and right) of the submodel. This mechanical loading was also superimposed by thermal loading applied to the submodel. The plastic strain energy density accumulated by temperature loading was set as objective criterion (and target value, see 2.3) for evaluation of interconnector fatigue like done in [13].

2.5 Workflow

The working example was processed according to the scheme shown in Figure 10 with temperature as only input parameter. The application to other physics and environmental parameters should be possible.

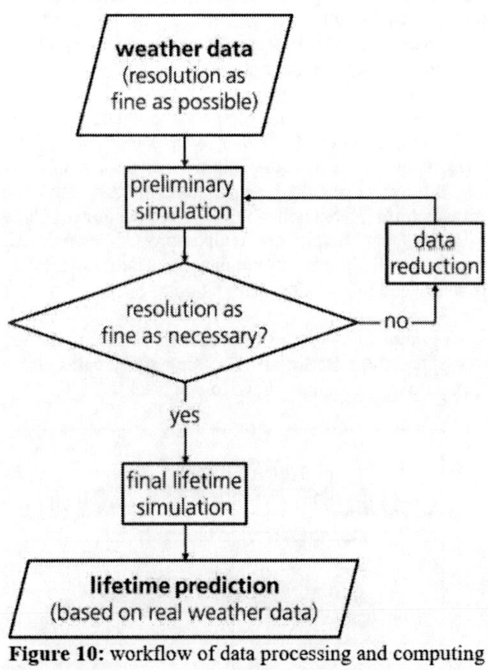

Figure 10: workflow of data processing and computing

3 RESULTS

3.1 Data acquisition

The foundation for further statistical evaluations and simulation is the collection of weather data for several locations in the climates to be investigated. A complete annual profile is to be provided for all five named main influencing variables, with at least an hourly resolution.

For this purpose, different databases (e.g. national weather services) and simulation program were considered and compared in terms of data availability and quality. Based on this comparison the simulation environment INSEL® was chosen. It is a graphical simulation system for programming applications from the entire renewable energy sector [14] and has a weather simulation tool of all necessary parameters, with varying annual values. The model calculates hourly average values for irradiance (GHI & POA), ambient temperature, humidity and wind speed for each location.

Furthermore, for a specific PV module (240 Wp) and installation (30°, South) the module temperature was simulated for each location. For each hour the module was held in its maximum power point and based on its nominal operating cell temperature, irradiance and ambient temperature the module temperature was calculated [15].

3.2 Data analysis & processing

The hourly data gives a starting data set of 8760 time steps for one year. The first step is the reduction of the temperature data by identifying turning points (sign changes) since only a change in load direction represents a part of cyclic loading. This already reduces the data set by dropping time steps of monotonous temperature increase or decrease. The next step introduces a rainflow counting algorithm that identifies full cycles based on the four-point criterion [16] which simply formulated drops one small temperature change between two big ones. It is also possible to add a temperature change threshold to even more reduce the data set.

These steps affect the temperature change velocity which may have an effect due to viscoelasticity of the encapsulant. Since the temperature changes are initially slow this effect should be negligible in this case. The ultimate form of reduction is the identification of daily minimum and maximum temperatures. Clearly this procedure will significantly reduce the database and may probably lead to a lack of information.

In Figure 11 to Figure 13 the levels of data reduction are displayed for Berlin at decreasing time scales (from one year over one month down to one week).

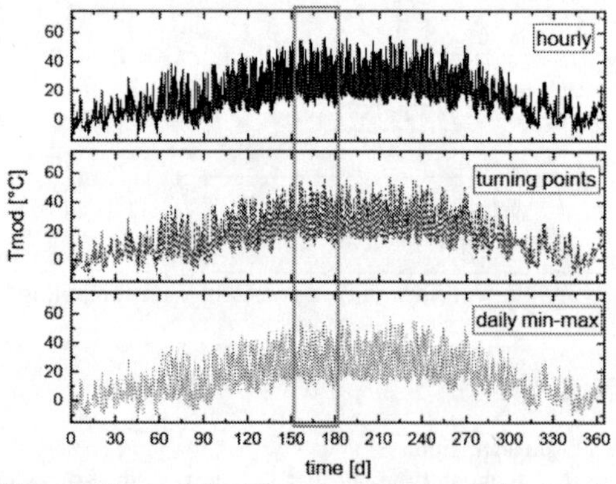

Figure 11: module temperature for resolution modes @Berlin (full year 2014, magnification of orange box in Figure 12)

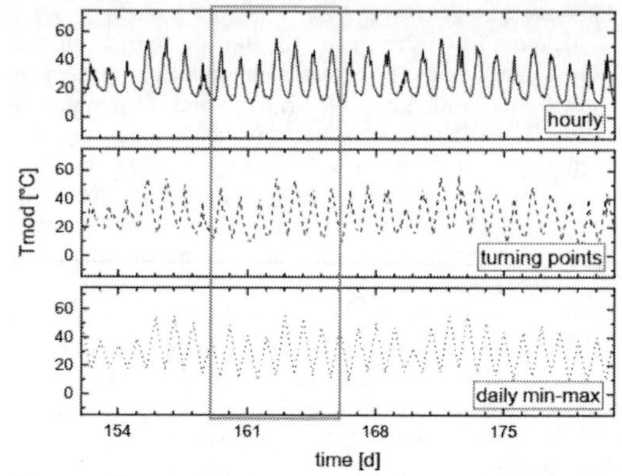

Figure 12: module temperature for resolution modes @Berlin (June 2014, magnification of orange box in Figure 13)

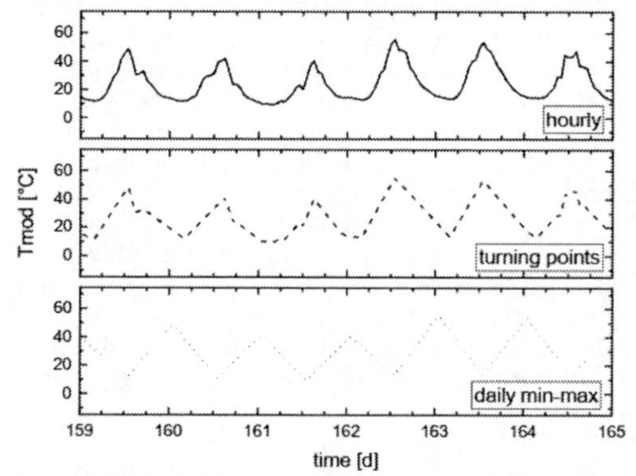

Figure 13: module temperature for resolution modes @Berlin (June 8th – 14th 2014)

In Figure 14 to Figure 16 the resulting data points are displayed for all three locations with increasing level of reduction.

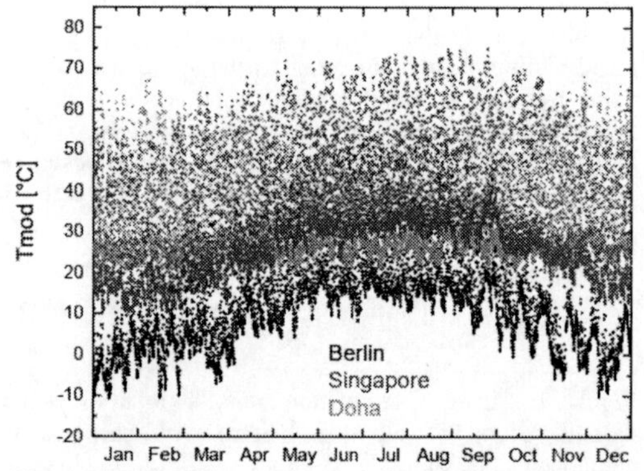

Figure 14: module temperature for hourly resolution at Berlin, Singapore and Doha

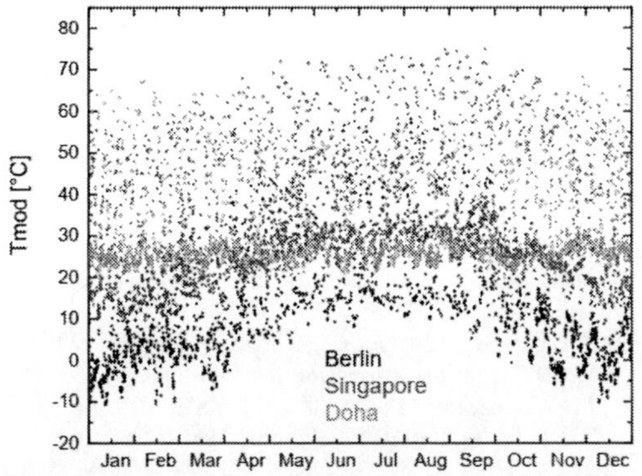

Figure 15: module temperature for turning points resolution at Berlin, Singapore and Doha

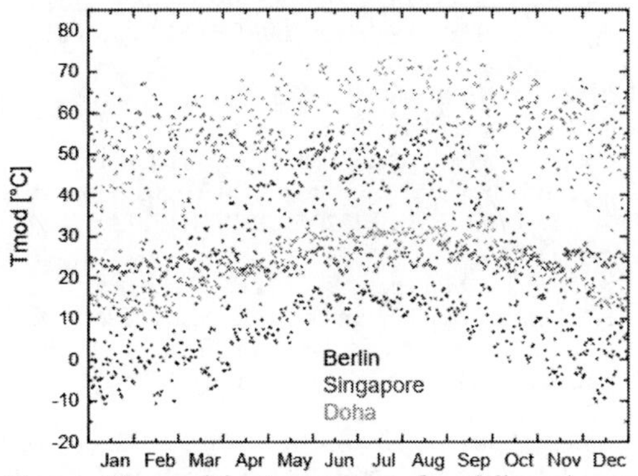

Figure 16: module temperature for daily min-max resolution at Berlin, Singapore and Doha

3.3 Simulation

In implementing the preliminary simulation with the highest loading data resolution (hourly), the resulting computational time (~1 day) and file size (~60 GB) was huge. And these facts would make future sensitivity analysis, parameter studies and optimization infeasible or uneconomic. Thus, the data reduction iteration according to the workflow (Figure 10) resulted in two more load data sets with lower resolution (turning points, daily min-max). The resulting target value of interconnector fatigue for all three sets of data resolution as well as the computational performance data is shown in Figure 17.

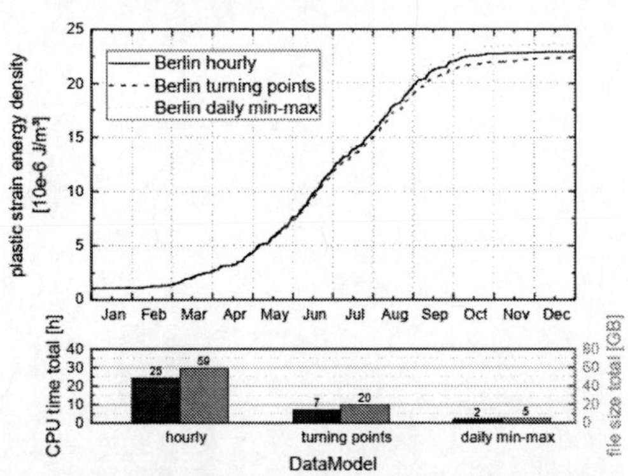

Figure 17: comparison of data resolution with respect to interconnector fatigue (plastic strain energy density as target value), CPU time and file size

The reduction of data resolution had minor influence (~3%) on the resulting target value of interconnector fatigue, but huge impact (up to 90%) on performance, which means significantly reduced computational time (~2 hours) and reduced file size (~5 GB).

3.4 Application

With the established workflow for data reduction and implementation of real weather data to lifetime and fatigue simulation we are enabled to virtual location assessment and comparison (regarding interconnector fatigue) just by weather data. As example, the target value for the selected locations Berlin, Singapore and Doha was computed as described (Figure 18).

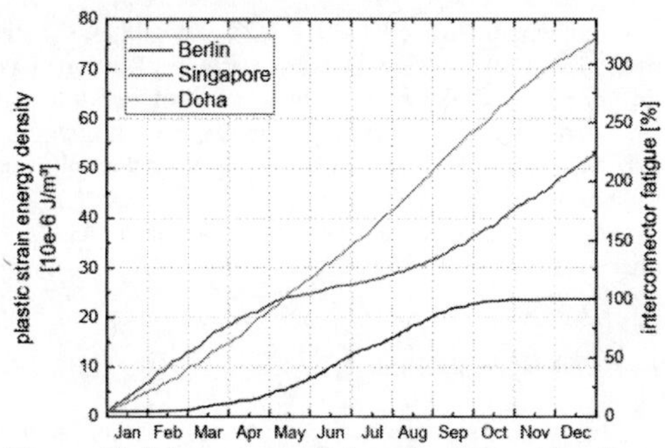

Figure 18: interconnector fatigue for different locations

The digital comparison shows temperature dependent relation of interconnector fatigue between the locations. This means that identical PV modules in Singapore (and Doha) degrade approximately 220% (320%) as fast as in Berlin regarding interconnector fatigue due to temperature loading.

Thus, by having access to site-specific environmental data basis, it is possible to just estimate PV modules operational lifetime compared to other locations. Furthermore, by enabling a digital twin to implement current weather data to the simulation, predictive maintenance can reduce effort and costs due to unscheduled downtime by fatigue and failure of PV modules.

Another application of this work can be digital material design, which means optimizing each material parameters to meet the desired requirements of target values, e.g. interconnector fatigue due to temperature loading. Such design studies can include variations of cell and module layout, too.

4 SUMMARY

Lifetime of outdoor products - like PV modules - under operating conditions is influenced by a great variety of parameters. Especially in lifetime simulation, these parameters are implemented to represent real environmental loads in abstracted and simplified simulation models of real products. In general, aim of lifetime simulation is to show one products behavior under operating conditions for whole life cycles. By the use of simulation models complete life cycles, for PV modules several decades, can be reduced down to fractions of hours.

Within such simulations processes of fatigue, degradation and damage as well as functional capability of a PV module can be analyzed and evaluated over its expected lifetime. Design studies can be evaluated very quickly with virtual prototypes and thus the product innovation can be improved.

In this work we demonstrate that the implementation of real weather data could be realized within a feasible simulation model and reasonable effort of computation time. Different geographic locations could be simulated and evaluated comparing PV modules lifetime estimation under local prevailing weather conditions, respectively.

A comprehensive model that combines both mechanical and temperature-related effects of fatigue has not yet been developed. The interaction of cell loading and interconnector loading is a question of optimization and was not addressed in this paper. Although the presented model was only used for mechanical investigations of module components, virtual models for other loads can also be developed.

5 ACKNOWLEDGEMENT

This work was financially supported by the Federal Government of Saxony-Anhalt within the project "PV Extrem" (No. 1604/0009).

6 REFERENCES

[1] R. H. French, R. Podgornik, T. J. Peshek, L. S. Bruckman, Y. Xu, N. R. Wheeler, A. Gok, Y. Hu, M. A. Hossain, D. A. Gordon, P. Zhao, J. Sun, G.-Q. Zhang, "Degradation science: Mesoscopic evolution and temporal analytics of photovoltaic energy materials", J.COSSMS, Vol. 19 no. 4 (August 2015), pp. 212-226, DOI:10.1016/j.cossms.2014.12.008.

[2] I. Gherboudj, H. Ghedira, "Assessment of solar energy potential over the United Arab Emirates using remote sensing and weather forecast data", RSER, Vol. 55 (2016), pp. 1210-1224, DOI:10.1016/j.rser.2015.03.099.

[3] M. Kottek, J. Grieser, C. Beck, B. Rudolf and F. Rubel, "World Map of Köppen-Geiger Climate Classification updated", Meteorol. Z., pp. 259-263, 2006. DOI:10.1127/0941-2948/2006/0130.

[4] Used by permission of The Center for Sustainability and the Global Environment, Nelson Institute for Environmental Studies, University of Wisconsin-Madison, "atlas of the biosphere", [Online]. Available: nelson.wisc.edu/sage/data-and-models/atlas/maps.php. [Accessed 09. 03. 2017].

[5] "The World Bank: Average precipitation in depth", [Online]. Available: http://data.worldbank.org/indicator/AG.LND.PRCP.MM. [Accessed 10. 03. 2017].

[6] These data were obtained from the NASA Langley Research Center Atmospheric Science Data Center, "NASA Surface meteorology and Solar Energy: Global Data Sets", [Online]. Available: https://eosweb.larc.nasa.gov/cgi-bin/sse/global.cgi. [Accessed 10. 03. 2017].

[7] SOLARGIS, [Online]. Available: http://solargis.com/assets/graphic/free-map/GHI/Solargis-World-GHI-solar-resource-map-en.png. [Accessed 09. 03. 2017].

[8] VAISALA, [Online]. Available: http://www.vaisala.com/Vaisala%20Documents/Scientific%20papers/Vaisala_global_wind_map.pdf. [Accessed 09. 03. 2017].

[9] "BerkeleyEarth.org", [Online]. Available: http://berkeleyearth.lbl.gov/regions/global-land. [Accessed 10. 03. 2017].

[10] M. Pander, R. Meier, M. Sander, S. Dietrich, M. Ebert, "Lifetime estimation for solar cell interconnectors" Proc. 28th EU-PVSEC (2013), pp. 2851-2857, DOI:10.4229/28thEUPVSEC2013-4CO.10.3

[11M. Pander, S. Dietrich, R. Meier, M. Ebert, "Fatigue analysis of solar cell interconnectors due to cyclic mechanical loading", Proc. 32nd EU-PVSEC (2016), pp. 1589-1597, DOI:10.4229/EUPVSEC20162016-5CO.13.2

[12] M. Pander, R. Meier, S. Schindler, U. Zeller, M. Ebert, "In-laminate fatigue of solar cell interconnectors: experimental investigation of sensitivities", Proc. 33rd EU-PVSEC (2017), pp. 1457-1461, DOI:10.4229/EUPVSEC20172017-5CO.8.3

[13] R. Meier, M. Pander, S. Grosser, S. Dietrich, "Microstructural optimization approach of solar cell interconnectors fatigue behavior for enhanced module lifetime in extreme climates", Energy Procedia 92 (2016), pp. 580-568, DOI:10.1016/j.egypro.2016.07.020

[14] doppelintegral GmbH, Webpage "What is INSEL?". URL: http://www.insel.eu/en/what-is-insel.html [12.09.2018].

[15] INSEL® manual, INSEL block reference for Solar Electricity, p. 466-469.

[16] G. Marsh et al, "Review and application of Rainflow residue processing techniques for accurate fatigue damage estimation", International Journal of Fatigue 82, 2016

SCANNING ACOUSTIC MICROSCOPY AS A NON-DESTRUCTIVE METHOD FOR THE INVESTIGATION OF PV MODULE COMPONENTS

Laila Veríssimo Mesquita, Djamel E. Mansour, Daniel Philipp, Luciana Pitta Bauermann
Fraunhofer Institute for Solar Energy Systems ISE
Heidenhofstrasse 2, 79110 Freiburg im Breisgau, Germany
T: +49 761 4588-5879
laila.verissimo.mesquita@ise.fraunhofer.de; luciana.pitta.bauermann@ise.fraunhofer.de
https://www.ise.fraunhofer.de

ABSTRACT: The application of new non-destructive testing for the inspection of photovoltaic modules is needed to advance the assessment of module quality and, thereby, improve their long-term performance. Scanning Acoustic Microscopy (SAM) has proved to be a powerful technique for the investigation of failures and defects in a great range of materials, including solar cells and modules. In this work, defects in PV modules are analyzed by SAM and the images are compared with well-established non-destructive methods, such as Dark Lock-In Thermography (DLIT) and Electroluminescence (EL). Acoustic micrographs were collected from different layers of the module and allowed the visualization and assessment of depth profiles of the embedded PV module components individually. SAM can detect irregularities on the encapsulant, backsheet and the solar cell at high lateral resolution and help elucidate defects detected with the other techniques. Moreover, the acoustic beam can be focused at the encapsulant/cell interface to visualize eventual failures occurring on these two layers separately. Therefore, SAM delivers complementary information to EL and DLIT on the quality of PV modules.
Keywords: PV module, failure analysis, scanning acoustic microscopy, electroluminescence, dark lock-in thermography

1 INTRODUCTION

The comprehensive characterization of PV modules is fundamental to assess the quality of the components, the interactions between them, and the likely causes leading to the development of failures and defects throughout a module's lifetime. Therefore, the need for cutting-edge non-destructive testing (NDT) is crucial to collect reliable information on PV module components and simultaneously reduce the need for destructive testing. Particularly challenging, however, is the non-destructive characterization of PV components which are encapsulated within a module because of the restricted number of techniques available to fulfill this purpose [1].

Therefore, the objective of this paper is to demonstrate the application of Scanning Acoustic Microscopy (SAM) for the inspection of PV modules and some defects. SAM is a NDT and has been recently applied to failure analysis of PV modules and cells [1,2,3,4]. A comparison with Electroluminescence (EL) and Dark Lock-In Thermography (DLIT) is also discussed. Because EL is one of the major imaging techniques used to assess PV module failures [1,5,6], it was applied as a reference technique. On the other hand, DLIT has just recently emerged as a new powerful technique for failure analysis of PV modules because it provides information on the thermal behavior across the whole module at high spatial resolution, besides complementing and overcoming some of the limitations of EL and conventional passive IR cameras [6]. Its basic principle consists in stimulating the module by applying a periodic current signal, while a highly sensitive IR camera detects and measures the surface temperature stimulated by the current signal [6,7].

The basic principle of Acoustic Microscopy consists in the propagation of focused waves through a coupling medium directed at the studied specimen. Once the waves encounter the specimen surface, part of the sound energy is reflected and the waves can then be monitored on an oscilloscope. Essentially, sound waves are reflected upon encountering an interface between materials of different acoustic impedances. The acoustic impedance of a given material is the product of the sound speed and the material density. Thus, the acoustic reflectance at the interface between two materials of differing properties is a measure of their impedance mismatch [8]. This is known as the pulse-echo technique [9]. In other words, because the sound waves respond to the physical properties [10], acoustic microscopy makes a very robust NDT to measure not only mechanical properties, but also to study interfaces, defects, material homogeneity, and anomalies in a range of materials, spanning from biological tissue to metals, polymers and others [8,9].

In SAM, the entire object under investigation is scanned pixel-by-pixel to generate an acoustic micrograph from the desired structure. The versatility of SAM lies in the different images it can develop through different imaging techniques. Acoustic micrographs can be generated based on the maximum peak amplitude of returning echoes at a given depth across the sample thickness. The amplitude of the echoes determines the contrast on the micrograph [11]. This imaging mode thus generates a projection of the structure across the selected thickness (i.e., depth). In contrast, the time-of-flight (TOF) imaging mode yields micrographs based on the relative travel time of echoes within a specified depth range, thus generating a topographic view of the studied structure. An additional imaging mode is the acoustic cross-section, which depicts the time domain signals, thereby revealing the depth profile along any vertical plane of the studied specimen [11,12].

Particularly, SAM has been extensively applied to inspect interfaces and material defects, such as delamination in carbon-fiber reinforced polymer matrix composites [12] and in polymer coatings [13], wire bond interconnects and metal interfaces [14], microelectronic components, bonded wafers, solder bumps in flip-chip contacts, soldered interconnects in silicon solar cells

[10,11], and voids in rear contacts of PERC solar cells [2]. For failure analysis of PV modules, SAM has been used to investigate cell cracks, blisters, and delamination within the encapsulant [1], faults, delamination, and soldering defects on cell interconnects, encapsulation voids and homogeneity [3], and internal defects and grain structure in polycrystalline silicon panels [4].

2 EXPERIMENTAL

2.1 Samples

PV modules (20 x 20 cm²) were prepared by laminating a layer of polymer-based backsheet foil (BS) and 2 layers of ethylene vinyl acetate (EVA) sandwiching a monocrystalline silicon cell and a transparent glass cover (Figure 1).

2.2 Scanning Acoustic Microscopy

The PV modules were placed flat into a water tank and scanned by a SAM HD² (PVA TePla Systems GmbH, Westhausen, Germany) using a 30 MHz transducer at ambient conditions as depicted in Figure 1. Acoustic micrographs from different depths were acquired to allow inspection of the PV components. Acoustic cross-sections of the PV modules were extracted to visualize the depth profile of the PV components. The modules were measured from the backsheet side. An analysis from the glass side at the frequency applied presented limitations compared to the backside of the module, such as high attenuation of sound and insufficient focus at the interfaces of interest.

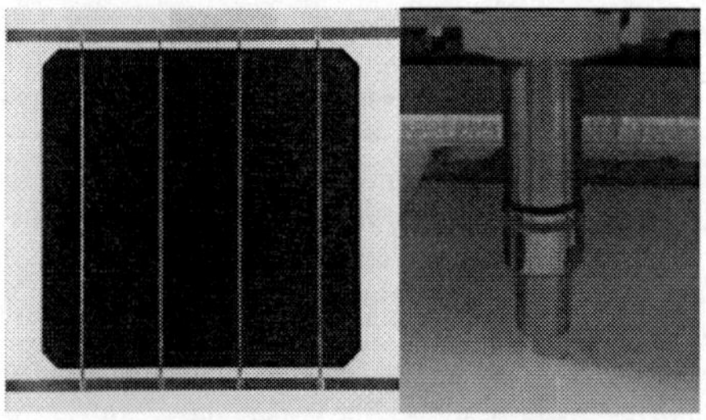

Figure 1. Left: photograph of a PV module sample seen from the glass side after the cell is encapsulated with backsheet, EVA and glass. Right: setup for a SAM investigation of a PV module. The transducer (center) and the module are immersed in distilled water.

2.3 Dark Lock-In Thermography

The modules were stimulated by applying a 9 A current signal at 0.5 Hz. The thermal images were acquired using an IR thermography camera (IRCAM Quadriga 1310k SM pro), sensor size and type, respectively: 1280 x 1024 pixels, InSb FPA 1.5 to 5 μm spectral response, and 107 Hz full frame rate. The modules were measured from the backsheet side.

2.4 Electroluminescence

The PV modules were measured using the LOANA Solar Cell Analysis System (pv-tools GmbH, Germany) at Fraunhofer ISE. The modules are measured from the glass side.

3 RESULTS AND DISCUSSION

The acoustic micrographs from the EVA bulk and EVA/solar cell interface within the PV module (Figure 2) shown in Figure 1 (module 1) were recorded at amplitude mode. Figure 2 (left) reveals details on the EVA texture and the arrangement of other module components, such as the cross-interconnects at the top and bottom edges of the module, the rear side ribbons, and the cell boundaries imprinted on the EVA. Additionally, parts of the cell are already detected on the right-hand side (bright spots).

Deeper down into the module, the main features of the cell can be visualized in Figure 2 (right). The silver pads equally spaced apart are visible along the ribbons.

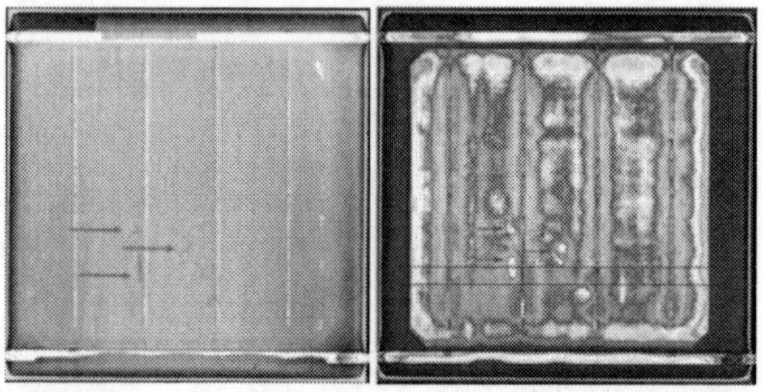

Figure 2. Acoustic micrographs taken from the bulk EVA (left) and cell (right) in module 1. Dimples are indicated by arrows.

The module was damaged during handling which caused some visible shallow dimples on the backsheet foil. These dimples are indicated by arrows in Figure 2. Interestingly, these dimples detected in the EVA layer exactly match to the bright contrast features of irregular shape which appear on the cell. Because the dimples are spots where the encapsulation is locally thinner, we hypothesized that the bright features detected on the cell likely indicate a thinner EVA.

Additional bright areas arise mostly at the cell edges and at the upper region, and extend somewhat towards the length of the cell in between the ribbons. Like the dimples, this leads again to the hypothesis that these bright areas indicate the EVA is locally thinner. In contrast, the areas immediately surrounding the ribbons are depicted as darker patches.

In order to visualize the depth profile of layers involved in these findings, acoustic cross-sections were extracted along vertical sections of the module indicated by red lines in Figure 2 (right), where line 1 intersects the largest dimple and line 2 intersects a defect-free region. The images are presented in Figure 3.

The acoustic cross-sections in the time domain reveal the internal arrangement of the PV components. Since the SAM measurement direction was done from the BS towards the glass, the signals correspond to the existing interfaces within the module, such that the top signal corresponds to the water/BS interface, then BS/EVA interface, cell and finally the glass signal (see the glass cover at the lower right-hand side in Figure 3).

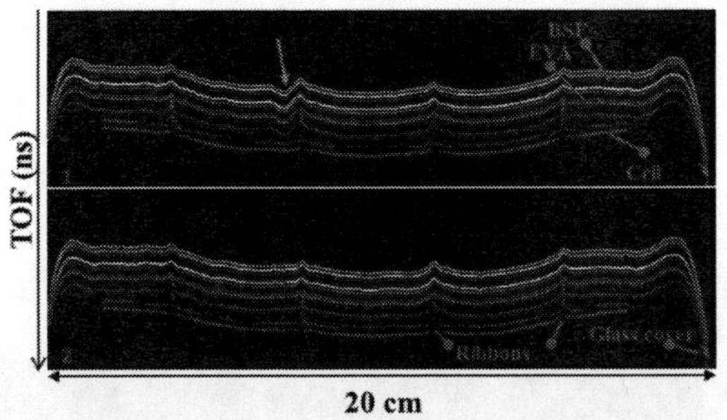

Figure 3. Acoustic cross-sections of module 1. The section lines within the module are shown in Figure 2 (right).

The arrow shown in Figure 3 (1) indicates that the dimple occurs on the BS and reaches the EVA layer. The solar cell is not affected (see comparison in Figure 3 (2)). Acoustically speaking, this means that the BS and EVA located on top of the dimple are closer to the cell (note the TOF axis), i.e., it takes a shorter amount of time for the sound to reach the cell at the dimple compared to a defect-free area depicted on the bottom image. This confirms the EVA is locally thinner as a result of the squeezing of the BS (and consequently the EVA).

Additionally, note that the cell boundaries appear much closer to the BS and EVA interfaces. The same effect is visualized at the glass edges, where the BS and EVA interfaces show a steep drop towards the front glass. This indicates a relatively high bending of the top layers and a sensitive EVA thickness reduction towards the glass edges [3]. This observation is supported by the fact that there is a high stress at the module edges caused by the pressing of a protective membrane against the module during lamination, also known as lamination pinch-out [15]. This ultimately leads to the bending of the layers and encapsulant thickness reduction along the module edges. Here, the same effect seems to happen at the cell edges as well.

In contrast to the cell edges, the EVA and BS signals located immediately along the ribbons take longer to reach the cell when compared to the regions equidistant to the ribbons. That means that the encapsulant at those locations is slightly thicker, providing that its mechanical properties are the same and the lamination process is homogeneous. Unlike the bright areas, these thicker EVA areas are displayed as darker patches in Figure 2 (right).

Figure 4 shows the corresponding DLIT and EL images. The DLIT image shows the temperature distribution after a quarter period of the electrical stimulation. Bright areas indicate surface areas with higher temperatures. The DLIT captures the surface on top of the dimples (indicated by arrows) as warmer areas. This is because the EVA is locally thinner. Therefore, it propagates the heat generated more quickly to the backsheet surface. Similarly, it appears that the edges of the cell are brighter as well, possibly due to the same reason, which may suggest a correlation between SAM and DLIT in detecting EVA thickness variations.

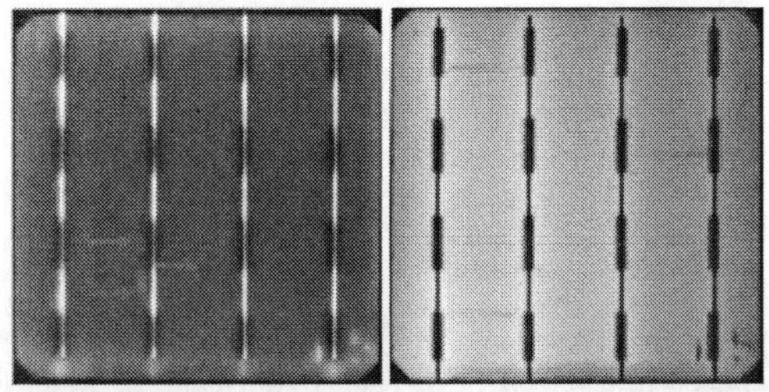

Figure 4. DLIT (left) and EL (right) images of the module.

A sharp change in the heat distribution, however, does not happen and this implies that the encapsulant thickness in those areas does not vary sharply, as it was also suggested with SAM. The latter, however, is able to display the heterogeneity more clearly. The EL image (Figure 4) confirms that the cell is electrically active and undamaged. Besides, shunts and some minor finger detachments were detected. These, however, were not detected by SAM.

Another module (module 2) presented a crack in the solar cell, which is depicted at high resolution in Figure 5. A Prewitt filter was applied to a portion of the crack to enhance the edges (Fig. 5, right) and the details of the crack. It is important to mention that the crack is detected from the backsheet side. The corresponding DLIT and EL images are presented in Figure 6. The findings with SAM are consistent with the DLIT and EL [1]. The crack length is about 50.7 mm long.

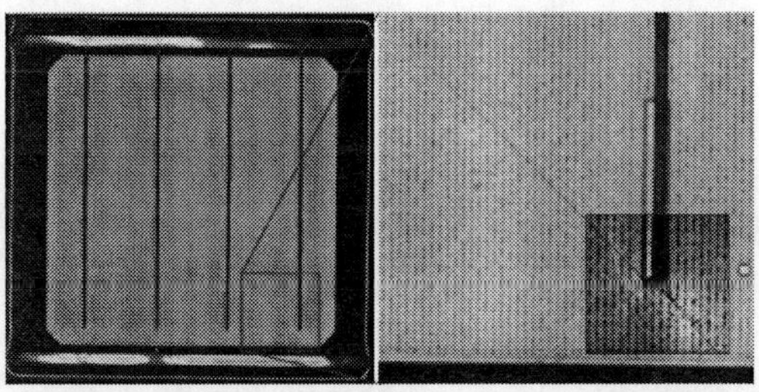

Figure 5. Acoustic micrographs of module 2 displaying the crack detected. To the right, the crack is visualized in detail and after applying a Prewitt filter.

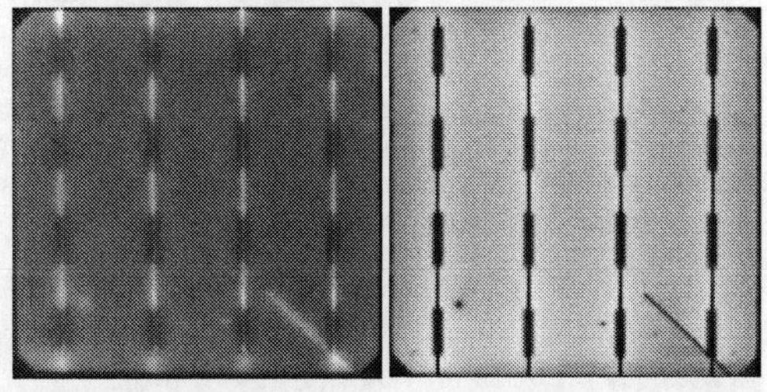

Figure 6. DLIT (left) and EL (right) images displaying the crack in module 2. The DLIT image shows the temperature distribution after a quarter period of the electrical stimulation.

A third exemplary module (module 3) presented lots of visible air bubbles after the lamination and these can be seen in the photograph in Figure 7 (left). The corresponding acoustic micrograph at the EVA/cell interface is presented in Figure 7 (right). It displays the large air bubbles on the left cross-connector (note the bubble as a bright area on the image in detail), which formed within the BS and EVA layers, but smaller air pockets were detected along the rear side ribbons and these are indicated by arrows on the image in detail.

The very fine resolution achieved with SAM allowed a clear visualization of the location and shape of the air pockets detected on the rear side of the module. These air pockets seem to form only on the silver pads next to a ribbon, which may indicate that the EVA did not adhere completely to the silver pads, thus giving rise to air gaps. In acoustic micrographs, air pockets and delaminated areas are easily detected as very bright features because air-filled interfaces reflect almost all incoming sound [10-12].

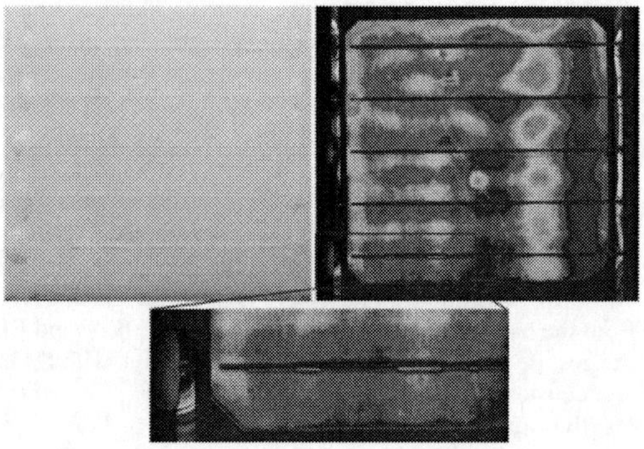

Figure 7. Left: photograph of the backside of module 3. Right: acoustic micrographs from module 3. In detail, note the air pockets indicated by arrows.

The corresponding DLIT and EL images are shown in Figure 8. The DLIT image shows that the uppermost ribbon and parts of the lower ribbon appear darker than the other ribbons. At the lower ribbon, the module surface where the detected air pockets are located (Fig. 7, image in detail) has a lower temperature compared to a defect-free area and, therefore, appeared darker on the DLIT image (see dark parts along the ribbon). At the air pockets, the propagation of heat generated by the constant current on the solar cell is disturbed by the presence of air-filled gaps underneath the backsheet surface, similarly to that observed in delaminated areas in PV modules [16]. Because air-filled pockets have a lower thermal conductivity, their presence leads to a poor dissipation of heat towards the backsheet, which ultimately results in a lower temperature contrast at the module surface (BS) over the air pockets.

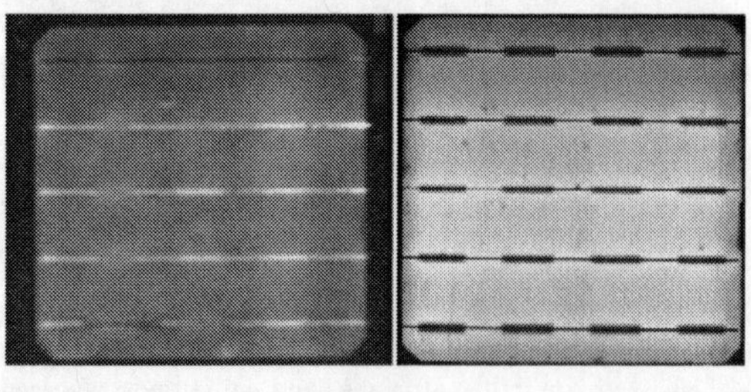

Figure 8. Amplitude DLIT (left) and EL (right) images of module 3.

The EL image shows that the cell is electrically active at the lower part of the cell and that the lower ribbon is intact. Therefore, this suggests that the dark areas along the lower ribbon on the DLIT image likely arise due to the air pockets detected. Because the uppermost ribbon also appears dark, we obtained an acoustic micrograph from that ribbon to check for the presence of similar air pockets. The corresponding image is shown in Figure 9.

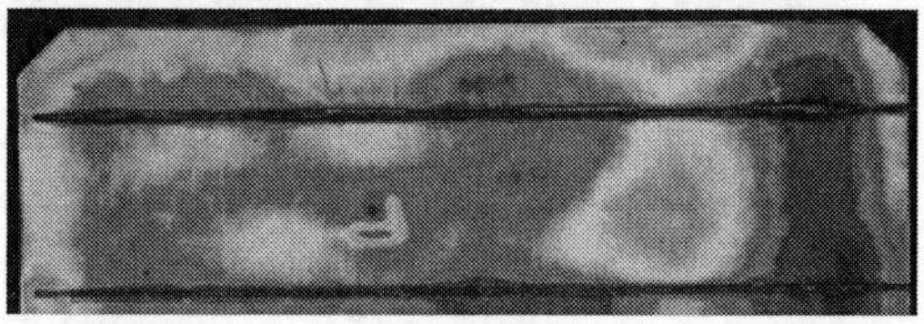

Figure 9. Acoustic micrograph with a detailed view of the uppermost ribbon in module 3 displaying bright features along the silver pads (bright thin areas indicated by arrows).

Like in the lower ribbon, high-reflectance features were also detected on the silver pads (indicated by arrows), which seems to indicate the presence of air gaps, although their shape looks much different compared to the air pockets detected at the lower ribbon (Fig. 7). However, no other bright features were detected along the ribbon, which suggests that other factors may be involved in the heat distribution in that particular ribbon. A closer look at the EL and DLIT images (Fig. 8) reveals that the upper part of the cell indeed looks slightly darker and this indicates that a lower current density, most likely due to a higher series resistance, may be the major reason for the darker appearance of the ribbon on the DLIT image.

4 CONCLUSIONS

This work demonstrates that the SAM allows the investigation of PV modules at any depth range selected by the user, thus enabling a detailed analysis of the backsheet, encapsulant, and the cell individually without carrying out destructive tests. The acoustic micrographs were acquired at high spatial resolution and enabled a clear visualization of the vertical profile of the PV module components, and defects within the BS and EVA, and the cell. The depth profile analysis indicated that SAM can be applied to detect thickness variations within the BS and encapsulant layers. SAM has proved to be a very sensitive technique to detect interfacial faults, such as cracks on the solar cell, bubbles and air pockets at different interfaces. A correlation between SAM and DLIT images with regard to the encapsulation homogeneity and adhesion defects could be verified, suggesting both techniques are complementary to study defects of this nature. Therefore, SAM is a reliable and complementary technique to DLIT and EL in investigating PV modules. The results shown here were all performed in 1-cell modules, but full size modules can also be analyzed.

ACKNOWLEDGEMENTS

This work was funded by the German Federal Ministry of Economic Affairs and Energy (BMWi) (Contract Number 0324145, acronym PV-BAT400). We thank Sandor Stecklum, Esther Fokuhl and Hebin Manuel for assisting in the interpretation of the DLIT and EL images.

References

[1] Y. Voronko, G. Eder, M. Weiss, M. Knausz, G. Oreski, T. Koch, K. Berger, Proceedings of the 27th European Photovolatic Solar Energy Conference and Exhibition (2012) 3530.

[2] K. Dressler, M. Rauer, M. Kaloudis, S. Dauwe, A. Herguth, G. Hahn, IEEE Journal of Photovoltaics, vol. 5, no. 1 (2015) 70.

[3] E. Gerritsen, T. Djuric, A. Schneider, S. Schröder, P. Czurratis, 29th European Photovolatic Solar Energy Conference and Exhibition (2014).

[4] Z. Xu, S. Gao, J. Huang, Y. Wang, Advanced Materials Research, vol. 764 (2013) 9.

[5] J. A. Tsanakas, L. Ha, C. Bauerhop, Renewable and Sustainable Energy Reviews, vol. 62 (2016) 695.

[6] S. Stecklum, D. Philipp, C. Völker, 32th European Photovolatic Solar Energy Conference and Exhibition (2016).

[7] E. Fokuhl, D. Philipp, L. P. Bauermann, S. Stecklum. RET.Con 2018: 1. Regenerative Energietechnik-Konferenz in Nordhausen (2018) 87.

[8] A. Briggs, Reports on Progress in Physics, vol. 55, no. 7 (1992) 851.

[9] M. Sinha, D. J. Buckley, In J. E. Mark (Ed.), Physical Properties of Polymers Handbook, (2007) 1021.

[10] S. Brand, G. Vogg, M. Petzold, Microsystem Technologies, vol. 24, no. 1 (2017) 779.

[11] S. Brand, P. Czurratis, P. Hoffrogge, D. Temple, D. Malta, J. Reed, M. Petzold, Journal of Materials Science: Materials in Electronics, vol. 22, no. 10 (2011) 1580.

[12] S.-L. Gao, J.-K. Kim, Composites Science and Technology, vol. 59, no. 3 (1999) 345.

[13] I. Alig, M. Bargmann, H. Oehler, D. Lellinger, M. Wanner, et al. Journal of Physics D: Applied Physics, vol. 44 (2011).

[14] G. Vogg, T. Heidmann, S. Brand, Microelectron Reliab, vol. 55 (2015) 1554.

[15] M. Köntges, G. Oreski, U. Jahn, M. Herz, P. Hacke, K.-A. Weiss, et al., IEA PVPS Task 13, Subtask 3 Report IEA-PVPS T13-09:2017 (2017).

[16] A. Sinha, O.S. Sastry, R. Gupta, Nondestructive Testing and Evaluation, vol. 31, no. 1 (2016) 1.

USING UV LEDS FOR PV MODULE AGING AND DEGRADATION STUDY

Stefan Mitterhofer, Marko Jankovec, Marko Topič

University of Ljubljana, Faculty of Electrical Engineering, Laboratory of Photovoltaics and Optoelectronics
Tržaška cesta 25, SI-1000 Ljubljana, Slovenia
+386 1 4768 848, Stefan.Mitterhofer@fe.uni-lj.si

ABSTRACT: Two experimental setups using ultraviolet (UV) light-emitting diodes (LEDs) for the purpose of accelerated photovoltaic (PV) module and material aging are built.
The first setup consists of 8 small chambers, each equipped with one UV LED. Radiance, operation mode (pulsed or constant), duty cycle and frequency can be controlled. It is used to degrade glass-EVA-glass samples at room temperature (T). After more than 6500 hours, a loss of UV absorptance and to a lesser extent a reduced transmittance in the low-wavelength visible spectrum can be observed.
The second setup is to be used inside a climatic chamber and consists of an aluminum plate, able to hold an array with up to 20 LEDs. The LEDs are water cooled and sealed with UV-transparent glass to protect them from high humidity. The radiance of the LEDs is individually adjusted to reach a homogeneous irradiance distribution at the target area. The plate can be used as a building block for larger arrays.
Keywords: degradation, LED, UV light

1 INTRODUCTION

Various environmental factors influence the degradation of PV modules mounted in the field. The most crucial are T, relative humidity (RH) and UV light [1]. The latter has been shown to degrade the polymeric materials used in PV modules, causing embrittlement, changes in transmissivity of the encapsulant and delamination [2].

Conventional setups with UV light sources use fluorescence tubes [3], mercury lamps [4] or xenon arc lamps with corresponding wavelength filters [2]. However, LEDs are gaining an increased market share in applications requiring light in the UV spectrum [5]. This trend is driven by big technological advancements and various advantages of LEDs compared to the conventional UV light sources. These advantages include a higher efficiency, a longer lifetime, more constant radiance and less heat generation [6].

Current IEC standards include UV light only as a UV preconditioning test, subjecting the module to at least 15 kWh/m^2 at (60 ± 5) °C without RH control [7]. There are no combined cycles defined in these standards, including elevated T, RH and UV light at the same time.

New combined cycles are being investigated for this purpose [8-10]. The goal is to duplicate failure and degradation modes observed in the field in a shorter amount of time, while not producing any others by using overly harsh parameters. The advantages of the combined tests include the possibility of using fewer modules and fewer parallel tests, reducing costs and time needed [10].

Thus, it is imperative to evaluate the option of using UV LEDs in PV module and component degradation. For this purpose, two experimental setups including these devices are built.

In this paper, the design of both setups is presented. Section 2 shows the design of a small chamber with eight UV LEDs. Furthermore, it presents the results of the degradation experiment of glass-EVA-glass samples in this setup, and the measurements of the LEDs' degradation. Section 4 shows the design of a water-cooled UV LED array, able to be used inside a climatic chamber.

2 UV LED CHAMBER

2.1 Experimental setup

The first setup consists of 8 chambers, each equipped with a single LED as a light source. A custom driver [11] for every LED and the corresponding software enable their pulsed operation with control of radiance, frequency and duty cycle, or their constant operation with control of radiance, shown in Fig. 1.

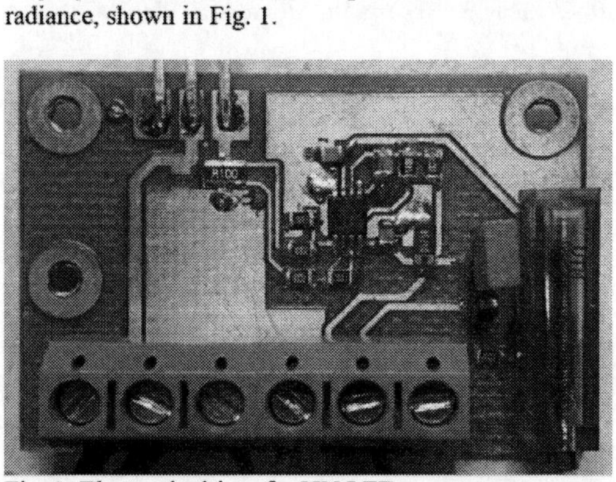

Fig. 1: Electronic driver for UV LEDs

Two different types of LEDs are used: Four LED Engin LZ1-00UV00-0000 with a peak wavelength of 365 nm, and four Vishay Semiconductors VLMU3500-385-060 with a peak wavelength of 385 nm. Each is mounted on a heatsink and air-cooled by a fan.

6 × 6 cm big glass-EVA-glass samples are being degraded in the chambers. Their transmittance between 280 nm and 1200 nm is measured in regular intervals with a spectrophotometer.

The radiance of the LEDs is evaluated with an integrating sphere and a Hamamatsu S1226-8BQ silicon photodiode in the same intervals, enabling the analysis of the LEDs' degradation over time. The input current to all LEDs is readjusted to keep their radiance constant over time.

The impact of two factors is analysed: the irradiance, controlled by varying the current input to the LEDs, and the frequency during pulsed operation.

The four 365 nm LEDs are operated at various currents. The highest irradiance on a sample is (190 ± 15) W/m². The other three are dimmed, achieving an irradiance of 75%, 50% and 25% of this value, respectively. The error is estimated. The largest source of uncertainty is the measurement with the photodiode.

Three of the four 385 nm LEDs are pulsed with a duty cycle of 50%. The peak irradiance on the sample is (162 ± 15) W/m². The frequencies are 1 Hz, 10 Hz and 100 Hz. The fourth LED is operated constantly, but dimmed to 50% of the peak radiance of the other LEDs. Thus, the integrated radiance of those four LEDs is the same. Table I lists the parameters of all LEDs used in the experiment.

Wavelength [nm]	Operation mode	Frequency [Hz]	Peak power [%]
365	constant	n/a	100
365	constant	n/a	75
365	constant	n/a	50
365	constant	n/a	25
385	constant	n/a	50
385	pulsed	1	100
385	pulsed	10	100
385	pulsed	100	100

Table I: LED parameters

The LEDs degrade over time. To avoid using them at currents above their nominal current according to the datasheet, it is reduced by the same factor for all LEDs after 3900 hours of operation. The highest irradiance on the samples of the 365 nm LEDs used afterwards is (174 ± 15) W/m², the peak irradiance on the samples of the pulsed 385 nm LEDs is (149 ± 15) W/m².

2.2 EVA degradation

The experiment is ongoing for more than 6500 hours at room T. After this time, the measurements of the samples show two types of degradation, most pronounced at the highest used irradiance.

- The spectral transmittance in the UV A spectrum of the light decreases, especially below the cutoff wavelength of approximately 370 nm, shown in Fig. 2.
- The transmittance in the visible spectrum decreases slightly in the early stages of degradation, most pronounced at low wavelengths, depicted in Fig. 3. However, it does not change after approximately 1000 hours of degradation.

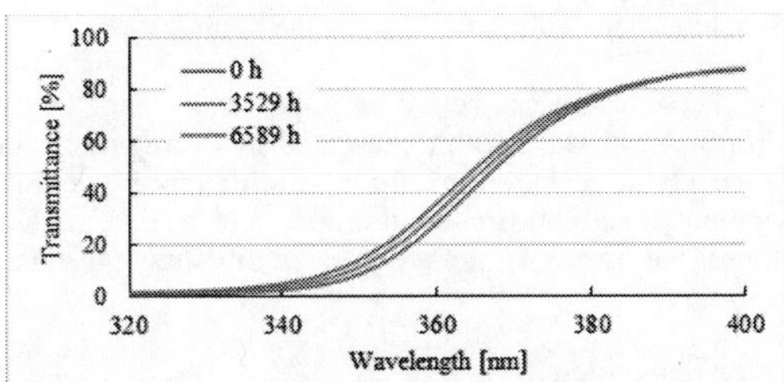

Fig. 2: Increase of UV light transmittance

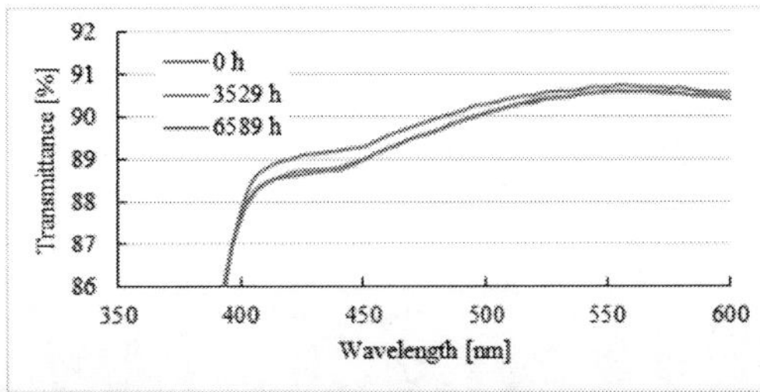

Fig. 3: Loss of visible light transmittance

The largest loss of UV light absorptance can be observed under high intensities of light. The changes to the transmittance in the visible spectrum are too small to yield conclusive results about the influence of the irradiance.

Table II lists the incline of a linear regression of the transmittance between the first measurement after 168 hours, and the last measurement after 6589 hours i_{lr}. The samples taken are degraded under 365 nm light.

Average radiance [W/m²]	i_{lr} at 350 nm [% / 1000 h]	i_{lr} at 365 nm [% / 1000 h]
180 ± 15	0.96	1.30
135 ± 15	0.45	0.69
90 ± 10	0.23	0.65
45 ± 10	0.31	0.49

Table II: Change in transmittance in of the samples at two specific wavelengths

There are no changes in the spectral transmittance of EVA measureable after the degradation with 385 nm light. The experiment is ongoing.

2.3 LED degradation

The LEDs are degrading over time. Operating them under pulsed conditions at high frequencies (10 Hz, 100 Hz) do not change the degradation compared to constant operation. However, no degradation could be measured yet at the LED operated pulsing with a frequency of 1 Hz. The results are shown in Fig. 4.

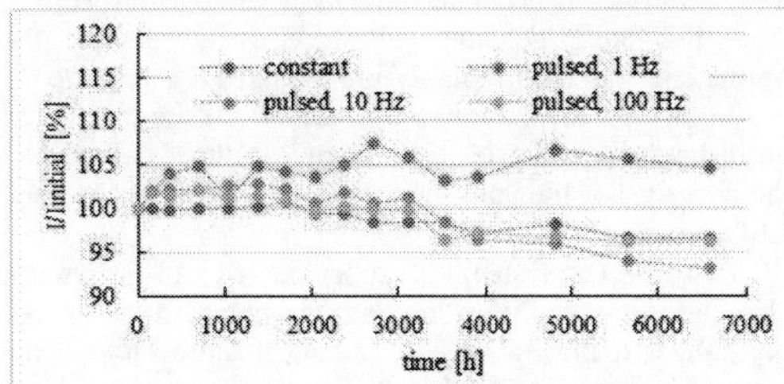

Fig. 4: Degradation of the radiance of UV LEDs at room temperature

The degradation is very slow, demonstrating the long lifetime of LEDs compared to other UV light sources. This result emphasizes their applicability in PV module and component aging.

3 UV LED ARRAY

3.1 Irradiance calculation

The LEDs are approximated as nonperfect Lambertian emitters [12].

$$E(r,\theta) = E_0(r) \cdot \cos^m \theta \qquad (1)$$

θ is the viewing angle, $E_0(r)$ the irradiance at distance r in front of the LED and m a parameter describing the radiance distribution.

$$m = \frac{-\ln 2}{\ln(\cos \theta_{1/2})} \qquad (2)$$

$\theta_{1/2}$ is a LED-specific parameter, describing the viewing angle at half irradiance. The irradiance at an arbitrary point (x, y) of a flat surface at distance z from the LED, with an LED at the position (x_0, y_0), is then:

$$E(x,y,z) = \frac{I_0 \cdot z^{m+1}}{[(x-x_0)^2 + (y-y_0)^2 + z^2]^{(m+3)/2}} \qquad (3)$$

$$I_0 = E_0(1\ \text{m}) \qquad (4)$$

To obtain the irradiance of an array of n LEDs, the irradiances from the single LED sources are summed up.

$$E_{total}(x,y,z) = \sum_{i=1}^{n} E_i(x,y,z) \qquad (5)$$

3.2 Irradiance optimization

Operating all LEDs at their nominal current results in an inhomogeneous irradiance at the target area. It is stronger in front of the center of the array, and weaker in the corners. There are various possibilities to homogenize the irradiance:

- Optimizing the positions of the LEDs
- Adding optical components
- Dimming the central LEDs

The choice is made to optimize position and radiance of the LEDs. An iterative optimization method is devised [13] and improved.

An array of optimization points (OPs) is inhomogeneosly spread over the target area. Their positions depend on the placement of the LEDs on the array. Every LED is connected to a set of OPs closest to it. With this basic geometry in place, two steps for the optimization of the irradiance are repeated.

In the first step, the irradiance at all OPs is calculated according to equation (5). The average of those values, and the averages of the values for the OPs connected to each LED are calculated.

In the second step, the radiances of the LEDs are changed individually. In the end, the obtained values are normalized to the highest used radiance. A flowchart of the method is shown in Fig. 5.

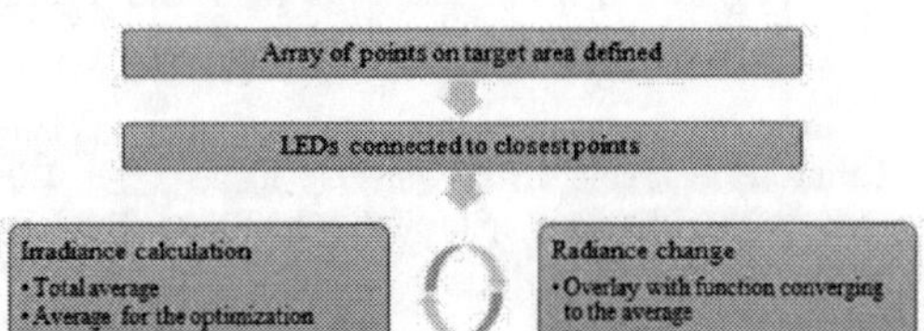

Fig. 5: Flowchart of the irradiance optimization method

To obtain the ideal distance between the LEDs, the optimization is repeated, the irradiance over the entire target area calculated, and the results compared.

Fig. 6 shows the standard deviation and the difference between minimal and maximal irradiance during the first 50 steps of the iteration method. The target area is a 50 × 50 cm² square in a distance of 10 cm to the LED array.

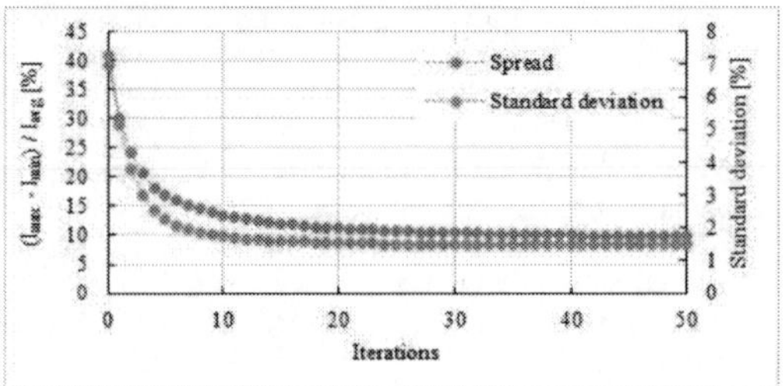

Fig. 6: Convergence of the optimization method

The LED array for this setup is a 9×9 checkerboard style array, using 40 LEDs with $\theta_{1/2} = 55°$. The result after 100 steps of the iteration is shown in Fig. 7. The blue bars indicate the radiance of the LEDs, relative to the maximum. The irradiance calculated is shown in Fig. 8. The standard deviation is 1.5%, the difference between minimum and maximum 9.8%.

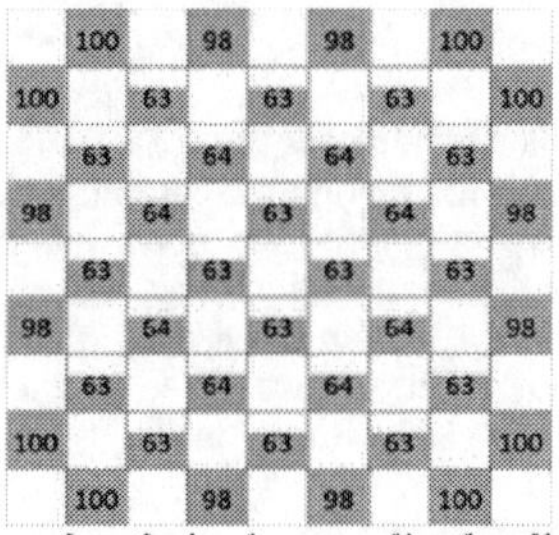

Fig. 7: Array and optimized, normalized radiance in [%]

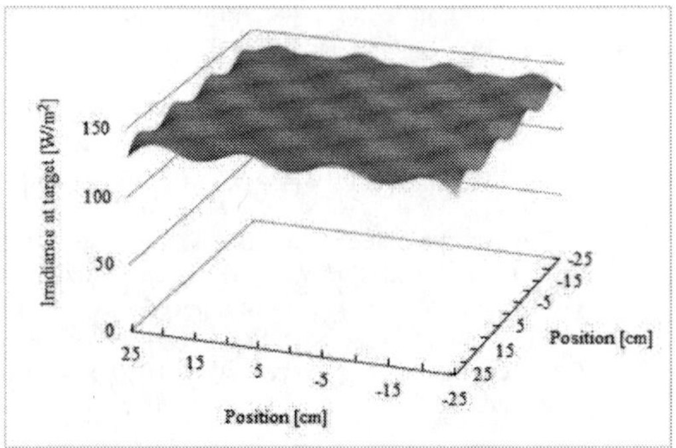

Fig. 8: Calculated irradiance after the optimization

These results can be further improved by using larger arrays. An 11×11 LED checkerboard style array results in a standard deviation of 0.5%, a difference between minimum and maximum of 2.7%, and a 30% higher irradiance. However, it uses 60 LEDs, 50% more than the 9×9 array.

3.3 Array design

Each LED is mounted on a custom designed PCB [14], shown in Fig. 9. A single master controller can set the LED power individually, connected in a daisy-chain topology. A universal LED pattern is designed, able to fit various types of LEDs.

These PCBs are mounted on an aluminum plate. Channels in the back of the plate are used for water-cooling. A plate of highly temperature resistant acrylic glass is used to seal these channels. A groove inside the aluminum plate filled with a rubber gasket, as well as a special isolating tape, placed between glass and aluminum plate, provide the necessary isolation to avoid water egress.

Fig. 9: LED driver used on the array

The front side is covered with a UV transparent glass. Because the array will be used inside a climatic chamber at elevated temperatures and humidity, a groove inside the aluminum plate filled with a gasket provide isolation against humidity ingress to the LEDs. The front side of the design is shown in Fig. 10.

Fig. 10: Design of a plate for the LED array

A 26 × 33 cm^2 plate can hold up to 20 LEDs. It can be used as a building block to upscale the setup for larger areas. For a 50 × 50 cm^2 target area, 4 plates are used, with 10 LEDs on each plate.

Preliminary testing of the prototype provides a proof of concept for the drivers and the water cooling design. Further steps will characterize the array working in a climatic chamber at elevated temperatures and humidities, and the degradation of the LEDs in these conditions.

4 CONCLUSION

The UV LEDs have proven their long life-time and slow degradation at room T and RH. However, also polymeric materials only degrade slowly in those conditions. This emphasizes the need to apply UV light and other stress factors at the same time in accelerated aging and degradation testing.

The LED array design and calculations for the irradiance optimization have been presented. Preliminary tests show a proof of concept for the electronic drivers and the cooling system. Future work will include testing the array at elevated T and RH in the climatic chamber and analyzing the degradation of the radiance.

5 ACKNOWLEDGEMENT

This project has received funding from the European Union's Horizon 2020 programme under GA No. 721452.

6 REFERENCES

[1] A. Ndiaye, A. Charki, A. Kobi, C.M.F. Kebe, P.A. Ndiaye and V. Sambou, "Degradations of silicon photovoltaic modules: A literature review", Solar Energy, vol. 96, pp. 140-151, 2013.

[2] M.D. Kempe, "Ultraviolet light test and evaluation methods for encapsulants of photovoltaic modules", Sol. Energy Materials and Sol. Cells, vol. 94, no. 2, pp. 246-263, 2010.

[3] M. Koehl, D. Philipp, N. Lenck and M. Zundel, "Development and application of a UV light source for PV-module testing," Reliability of Photovoltaic Cells, Modules, Components, and Systems 2, pp. 741202, 2009.

[4] X. Gu, Y. Lyu, C. Yu, C.-C. Lin, J. H. Kim and T. Nguyen, "Reciprocity study and spectral UV wavelength effect on PV components degradation," in *3rd Atlas-NIST workshop on PV materials durability*, Gaithersburg, MD, 2015.

[5] Yole Developpement, "UV LEDs – Technology, manufacturing and application trends report, 2018.

[6] Y. Muramoto, M. Kimura and S. Nouda, "Development and future of ultraviolet light-emitting diodes: UV-LED will replace the UV lamp", Semiconductor Science and Technology, vol. 29, no. 8, pp. 0848004, 2018.

[7] International Electrotechnical Commission IEC, "IEC 61215-1:2016, Terrestrial photovoltaic (PV) modules – Design qualification and type approval – Part 1: Test requirements", 2016.

[8] A. Borne, "New combined cycles provide new options and challenges for durability assessment", 7th SOPHIA workshop, Freiburg, 2017.

[9] T. Sample, "New combined cycles for future test standards in IEC TC82 WG2", 8th SOPHIA workshop, Ljubljana, 2018.

[10] P. Hacke, "New combined cycles provide new options and challenges for durability assessment – Combined-accelerated stress testing (C-AST)", 7th SOPHIA workshop, Freiburg, 2017.

[11] J. Slapšak, B. Strojin, A. Štuhec, Ž. Rojec and M. Jankovec, "Development of a small area LED solar simulator", Proc. of the 52nd Int. Conf. on Microelectronics, Devices and Materials MIDEM, pp. 163-167, Ankaran, 2016.

[12] D. Wood, "Optoelectronic semiconductor devices", Prentice-Hall International, 1994.

[13] S. Mitterhofer, M. Jankovec and M. Topič, "A setup for in-situ measurements of potential and UV induced degradation of PV modules inside a climatic chamber," Proc. of the 53rd Int. Conf. on Microelectronics, Devices and Materials MIDEM, Ljubljana, pp. 65-70, 2017.

[14] Ž. Korošak, "Modularno svetilo iz svetlečih diod za osvetljevanje fotonapetostnih modulov", Master thesis, University of Ljubljana, 2018.

MULTI-WIRE INTERCONNECTION: THE IMPACT OF THE LAMINATION PROCESS AND THE ENCAPSULANT PROPERTIES ON THE SOLDER JOINT FORMATION

Rik Van Dyck[1,2] rik.vandyckt@imec.be, Tom Borgers[2] tom.borgers@imec.be, Jonathan Govaerts[2] jonathan.govaerts@imec.be, Philippe Nivelle[2,3] Philippe.Nivelle@imec.be, Arvid van der Heide[2] arvid.vanderheide@imec.be, Stijn De Jonge[1] stijn.dejonge@kuleuven.be, Eszter Voroshazi[2] eszter.voroshazi@imec.be, Jozef Szlufcik[2] jozef.szlufcik@imec.be, Aart Willem Van Vuure[1] aartwillem.vanvuure@kuleuven.be, Jef Poortmans[1,2,3] jef.poortmans@imec.be

[1]KU Leuven, Oude Markt 13, 3000 Leuven. Belgium, [2]Imec/Energyville, [3]UHasselt, [4]Imomec

ABSTRACT: Multi-wire designs for photovoltaic modules already proved its advantages in the past. In previous contributions, several multi-wire interconnection approaches were presented, both for back-contact [4] and 2-side contacted cells [4,5]. This paper reports the progress in the woven multi-wire interconnection which enables a combined lamination and soldering process using low-temperature melting point solder alloys. This work focuses on the solder joint formation and aims to correlate the viscoelastic properties of the encapsulant material during lamination with the electrical and visual characterisation of the solder joint. To do so, three analysis techniques are used: electroluminescence (EL) imaging, current-voltage (IV) measurements and a cross-section analysis. The aim is to optimize the lamination process and material selection for optimal solder joint formation. Rheological measurements on the encapsulation polymers revealed the viscosities at various temperatures. Small modules were fabricated and tested. According to the results of these tests, the optimal lamination temperature is around 165 °C and at which the viscosity of the encapsulant should be at least 7500 Pa·s. One of the tested encapsulant materials met these requirements and had the best electrical properties according to the EL-imaging and IV-measurements.
Keywords: Characterisation, Encapsulation, Module Manufacturing, Multi-Wire Interconnection

1 INTRODUCTION

While most of the c-Si PV modules today are made using two or three busbars, the advantages of multi-wire technologies have already been demonstrated [1,2,3]. In 2013 Braun et al. proved a 50% silver consumption reduction and a 0.32% absolute efficiency increase using a multi-wire design [1]. As a result, there is a growing interest towards these multi-wire approaches for two-side contacted bifacial and busbar-free cells [4,5]. This trend is in line with the predictions put forward in the ITRPV roadmap [6] and is attracting attention due to the potential it holds for a significantly improved trade-off between optical (finger shading) and electrical (resistive transport) losses, while the potential reduction in cell metallisation is promising from a cost perspective. These benefits are valid at the cell-level, but also materialize into the module-level performance. In this context, two such technologies are already in an advanced stage of development. Although both these techniques implement a multi-wire interconnection, there are fundamental differences.

One approach effectively mimics the standard technology by soldering wires on finger solder pads, replacing the busbars. As in standard production, this step is then followed by a separate encapsulation process [7]. Another approach applies a contact foil incorporating the interconnect wires directly onto the metallised cell. In this case, the wires of the contact foil are soldered directly to the metal fingers of the cell. The contact foil does not act as encapsulant, so extra polymer material needs to be provided before the lamination process [3].

As a next-generation development, imec proposes an approach in which the wires are immediately incorporated into the encapsulant material before lamination [5]. Moreover, the soldering and lamination are conducted in one process, hence there is an interplay between the different materials. This paper will focus on the influence of the encapsulant characteristics and lamination temperature on the solder joint formation during lamination.

2 EXPERIMENTAL SET-UP

2.1 Sample Preparation

The cells used are commercial c-Si bifacial solar cells with three busbars and a screen-printed front- and backside metallisation. To make samples without busbars from three-busbar cells, a laser cutter is used to make strips of 156 mm x 7.8 mm, which are cut between the busbars, as illustrated in figure 1. Each strip represents 1/20th of a 156-mm cell, and one wire is sufficient for each strip to represent the multi-wire interconnection.

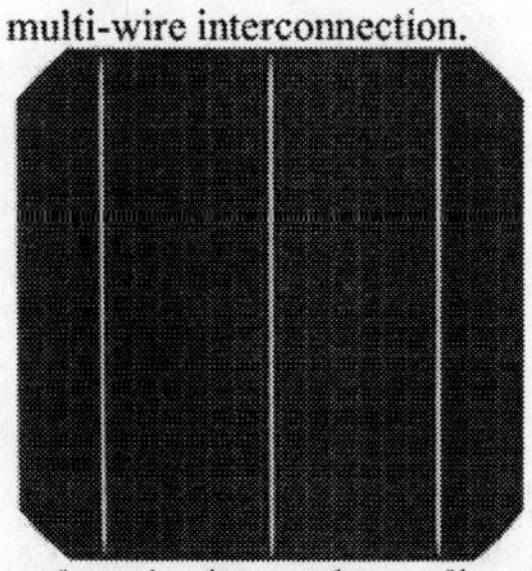

Figure 1: A three-busbar solar cell with red lines indicating the path of the laser.

Next, a layup is prepared of which an overview is given in figure 2. Three of the cell strips are placed between an encapsulant polymer foil and thin glass sheet with a thickness of 0.7 mm. This thin glass is used to facilitate the cross-section analysis. On both sides of the strip, a copper wire with a diameter of 200 µm is placed between the encapsulant foils and the strips. The wires are coated with a SnBiAg-based solder alloy with a low melting point temperature of 139 °C.

The glass has an important function during cool down after lamination and during thermocycling reliability testing. The polymer encapsulant has a larger coefficient of thermal expansion (CTE) than the Si cell, but the CTE of the glass is similar to that of the Si cell. When the cell cools down, and no glass is used, the polymer can shrink freely, pulling on the conductive wire and loosening all the electrical contacts.

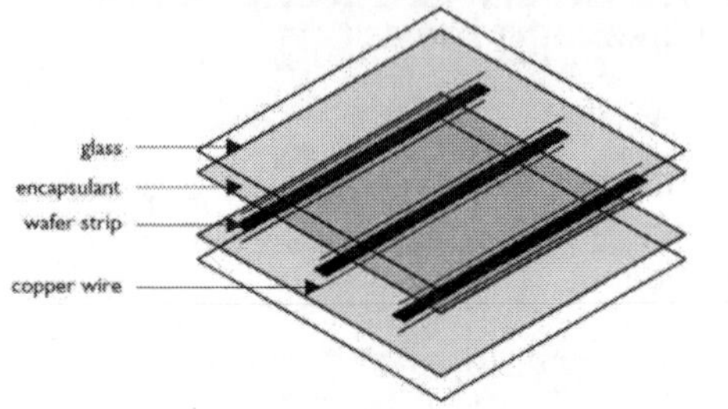

Figure 2: An exploded view of the sample lay-up.

The lamination cycle for the lay-ups consists of different steps, each with a different purpose. The laminator temperature is programmed at a constant value, but the effective temperature may differ due to an offset between the programmed and effective temperature. Different steps during the lamination cycle are as follows:

- When the lay-up is put in the laminator, a vacuum is applied to eliminate air bubbles in the laminate. At this point, the lay-up lays on the laminator pins and does not contact the hot plate of the laminator. The ongoing radiation heating is too small to start melting the polymer and the solder.
- Then, the pins are lowered so the lay-up makes physical contact with the hot plate. The polymer encapsulant and the solder will start to heat up through conduction, but no pressure is applied yet. Previous experiments have shown that the timespan between the lowering of the pins and the moment the membrane starts applying pressure on the lay-up has no significant effect on the solder joint formation.
- In the next step, pressure is applied through a membrane, pushing from the top of the laminate. The solder alloy flows and forms a solder joint while the polymer melts and embeds the interconnection and the cells. The lay-up is now called a laminate.
- The final step is to remove the formed laminate from the laminator and let it cool down to room temperature.

2.2 Encapsulant physical properties

Three different encapsulant materials are used in this work and are further referred to as type A1, A2 and B. By determining the viscosity of the three different encapsulants, it is possible to predict which viscoelastic properties the encapsulant should have to optimize the lamination and soldering process. When the encapsulant viscosity is too low at the used lamination temperature, the polymer could flow between the fingers and the conductive wires, preventing a solder joint to form. On the other hand, when the viscosity is too high, the impregnation of the wires could be incomplete, leading to the formation of voids in the laminate which lowers the mechanical performance of the module

Viscosity measurements were performed on an ARES 2K strain-controlled rheometer under nitrogen gas. The temperature was controlled by a forced convection oven and viscosity was measured within a temperature interval from 130°C to 170°C. Figure 3 shows the viscosity of the three encapsulant polymers, which were all thermoplastic polyolefins.

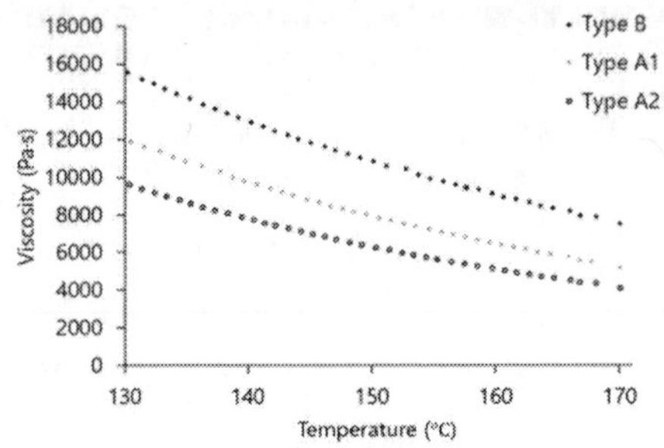

Figure 3: The viscosity as a function of the temperature for the three encapsulant materials.

2.3 Analysing Methodology

Two parameters are varied in this work: the type of encapsulant (type A1, A2 and B) and the programmed lamination temperature. The actual temperature can differ up to one degree due to a temperature offset. Table 1 gives an overview of the combinations which were tested. This matrix allows comparing different encapsulation materials at similar viscosities (but at a different temperature) and at a similar temperature (but with a different viscosity).

Table 1: the different combinations of encapsulant and lamination temperatures and the associated viscosities.

Viscosity (Pa·s)	Temperature (°C)		
	Type A1	Type A2	Type B
10250	-	-	153
9350	142	-	159
7500	153	142	170
5800	-	153	-
5200	170	159	-
4100	180	170	-

All set-ups are tested and characterised using three analysing methods: electroluminescence (EL)-imaging, current-voltage (IV)-measurements and a cross-section analysis. EL-imaging and IV-measurements are both fast and non-destructive electrical characterisation techniques. EL-imaging gives a general idea whether an electrical contact between the wires and the fingers is made. IV-measurements give a quantitative representation of the quality of the laminate by providing both the fill factor (FF) and the pseudo fill factor (pFF). This latter one is similar to the FF but without the losses from the series resistance. Since this series resistances can mainly be attributed to the interconnection and should be as low as possible, the difference between the pFF and FF – the FF-drop – should be minimized. While IV- and EL-analyses give a first impression of the quality of the laminate, a cross-section analysis is needed to further characterise the solder joints. For example, an EL-image cannot distinguish between a wire laying on the fingers giving a pressure contact and a real solder joint contact. Based on specific defects or effects seen in the EL-images and in combination with the IV-measurement of the sample, regions of interest are selected. These regions are cut out of the laminate. The cutting operation leaves behind rough surfaces and local damage, and therefore the cut-out area is slightly over-dimensioned to avoid damaging the region of interest. It is a combination of science and art to achieve a surface which is smooth enough to give trustworthy results. After a sequence of grinding steps with decreasing coarseness, followed by a polishing step with a diamond suspension, the surface is then analysed using a scanning electron microscope (SEM) in combination with an energy-dispersive x-ray spectroscopy (EDX).

3 RESULTS

3.1 Electrical characterisation

After lamination, EL-images were captured. Figure 4 shows an example of an EL-image of one cell strip, where each finger is visible as a vertical line. From these EL-images, it is clear that the edge of the laminate is the most critical, and most of the missing contacts are in this region, which is also concluded in reference [8]. These finger defects are visible as not illuminated fingers. Since the number of missing contacts can be visualised, EL-imaging can be used to monitor the degradation of the laminate during reliability tests. Such reliability testing is discussed more in detail by J. Govaerts et al. [8].

Figure 4: Example of an EL-image.

Each cell strip has around 73 fingers and the number of fingers which did not illuminate are counted. The outer two fingers are always excluded to avoid side effects. The average number of unconnected fingers from the three strips of each sample is plotted as a function of the temperature at which the sample was laminated. Next, these values are plotted against the viscosity at the used temperature, which is calculated by linearly interpolating the measured viscosity values as discussed in section 2.2. Figure 5.a and 5.b give the number of unconnected fingers as a function of temperature and viscosity respectively.

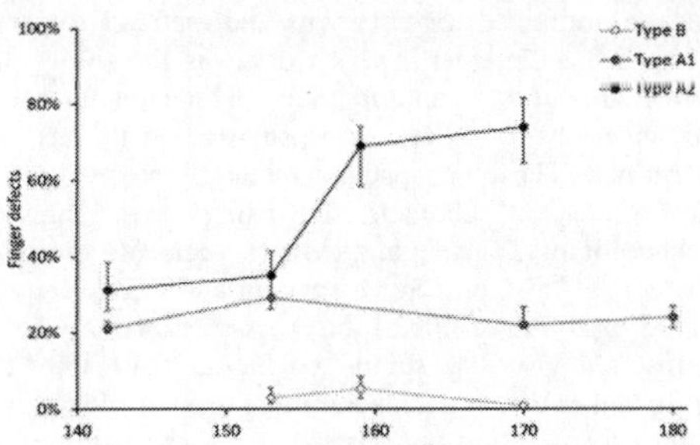

(a)

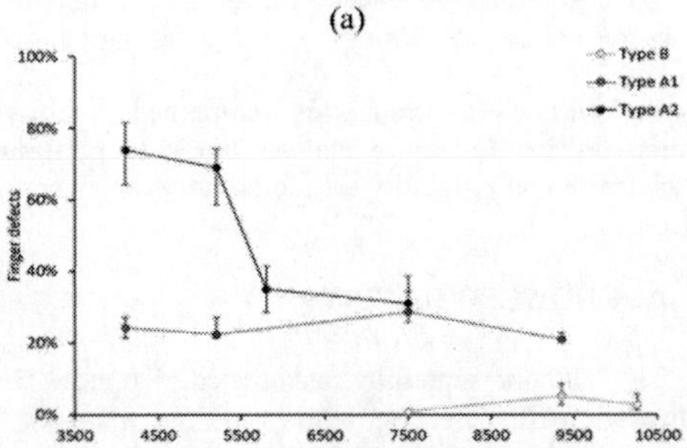

(b)

Figure 5: Number of unconnected fingers as a function of (a) the lamination temperature and (b) the complex viscosity.

The IV-measurements provides the FF and the pFF and for each sample, the FF-drop is calculated as the difference between these two values. The FF-drop is plotted against the temperature and the viscosity and given in figure 6.a and 6.b respectively. Note that all FFs measured are significantly lower than the FF of the cells before sample preparation (which is around 80%). This is attributed to damage induced by the laser scribing process.

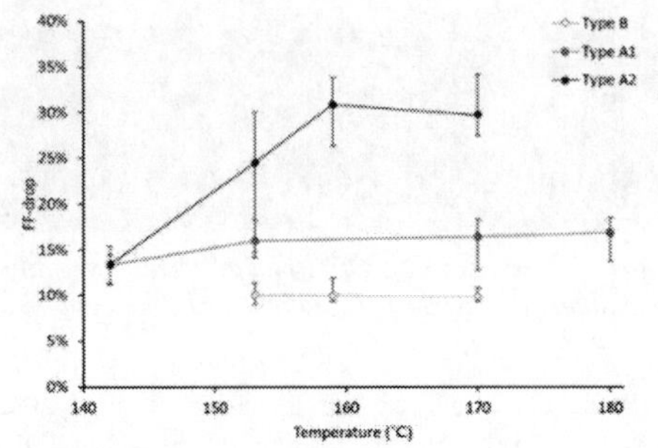

(a)

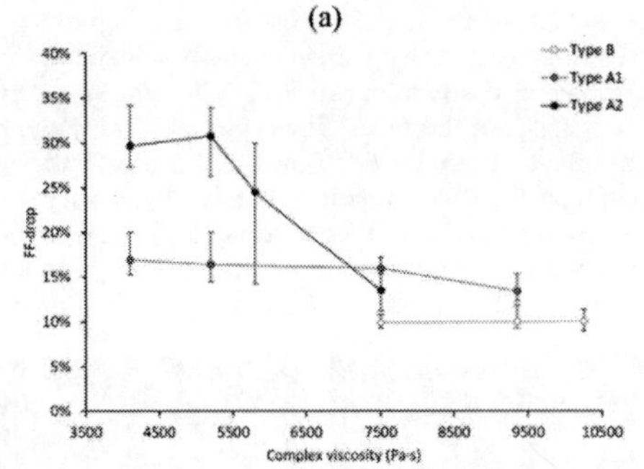

(b)

Figure 6: The FF-drop as a function of (a) the lamination temperature and (b) the complex viscosity.

The results from the IV-measurements are in line with the results from the EL-imaging and from them, three things can be stated:

- At all temperatures, both the number of unconnected fingers and the FF-drop was the lowest for type B encapsulant.
- At higher temperatures (159°C and above), there is a serious decrease in the performance of type A2.
- From these results, there is no clear relationship between the temperature and the performance of the different encapsulant materials, hence no ideal situation can be pointed.
- In the range of viscosities evaluated, the higher viscosities tend to be favourable.

By performing the cross-section analysis, the aim is to find the phenomena explaining the observations.

3.2 Cross-section analysis

The cross-section analysis with SEM-imaging and EDX-scanning reveals both the flow behaviour of the encapsulant and solder material and the phase composition of the solder joint. An example of a SEM-image is given in figure 7, where the different components are indicated. The dark- and light-grey zones within the solder are Sn-rich and Bi-rich phases respectively.

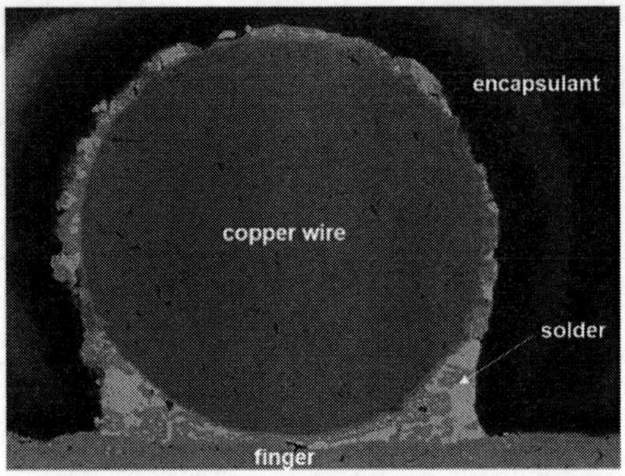

Figure 7: Example of a SEM-image where the different components are marked. The diameter of the copper core is 200 μm.

From the result of the cross-section analysis, an influence of the viscosity on the flow behaviour of the encapsulant and the solder can be observed. Figure 8 gives a comparison between different joints from laminates with decreasing viscosity from (a) to (d). When the viscosity of the encapsulant becomes lower than 7500 Pa·s, the encapsulant material starts flowing underneath the wire where it pushes away the solder. In this way, it can prevent a solder joint to form. This unwanted phenomenon was only observed with samples of which the encapsulant viscosity was lower than 7500 Pa·s during the lamination.

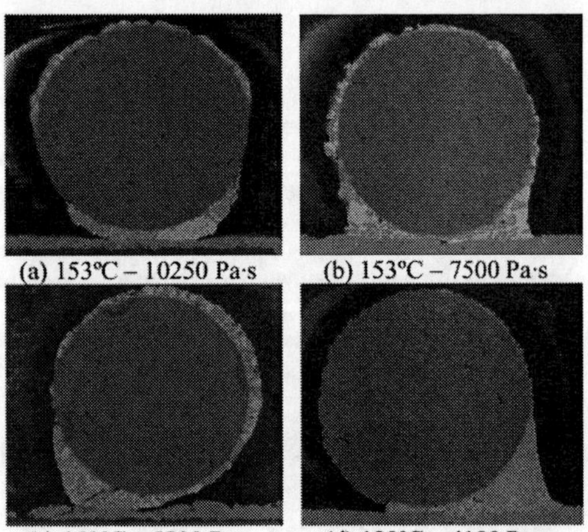

(a) 153°C – 10250 Pa·s (b) 153°C – 7500 Pa·s

(c) 153°C – 5800 Pa·s (d) 180°C – 4100 Pa·s

Figure 8: SEM-images of different solder joints with variable viscosity. The types of encapsulant used were (a) type B, (b) type A1, (c) type A2 and (d) type A1.

Also, an influence of the temperature on the flow behaviour of the solder and on the microstructure of the solder joint is visible. In figure 9.a, no solder has flown towards the cell's metallisation. The amount of material left on the wire decreases when the temperature is increased. At 170 °C, almost all solder material has flown towards the cell's metallisation to form a solder joint.

Secondly, the mixture of the different metallic phases of the solder is affected by the lamination temperature. At a higher temperature, the alloy forms thin, long and continued phases which are well- dispersed. This is ideal to stop crack propagation, which makes the joints better resistant to fatigue.

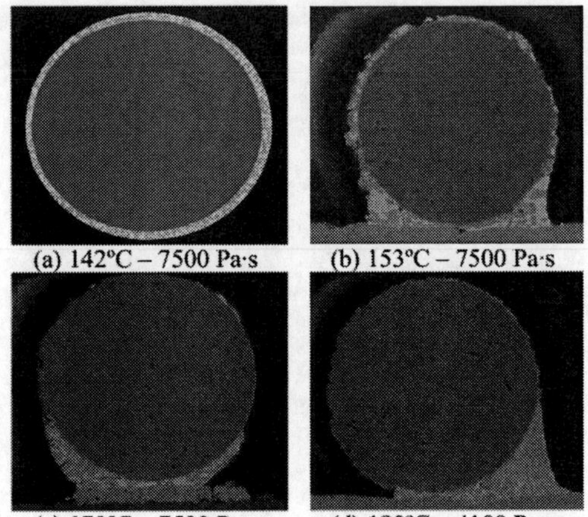

(a) 142°C – 7500 Pa·s (b) 153°C – 7500 Pa·s

(c) 170°C – 7500 Pa·s (d) 180°C – 4100 Pa·s

Figure 9: SEM-images of different solder joints with increasing lamination temperature. The types of encapsulant used were (a) type A2, (b) type A1, (c) type B and (d) type A1.

Higher temperatures would increase the flowability of the solder material, but lower temperatures are interesting from an industrial point-of-view. According to these results, the lamination temperature should be around 165°C at which the encapsulant should have a viscosity of at least 7500 Pa·s. From the different types which were tested in this work, type B corresponds the best to these requirements with a viscosity of 8000 Pa·s at 165°C. This can explain the better results of the electrical characterisation compared to the other encapsulants.

4 CONCLUSIONS & OUTLOOK

To develop a novel module technology based on a multi-wire interconnection design, the lamination cycle must be optimized together with the material selection. The region of interest in this work was the solder joint where the goal is to optimize the solder joint formation. The viscosity of three encapsulation materials was determined. These encapsulant materials were tested on small scale and characterised using three analysing methodologies: EL-imaging, IV-measurements and cross-section analysis using SEM-imaging and EDX-scanning. According to the results of the cross-section analysis, the encapsulant viscosity should be higher than 7500 Pa·s during lamination at a temperature of around 165 °C. One of the encapsulation materials meets these requirements and gave significantly lower FF-drop as a result of the series resistance and showed fewer defects in the EL-imaging.

To further develop this combined multi-wire interconnection, reliability analysis should be performed. Some results on reliability testing are already reported in [8]

5 ACKNOWLEDGEMENTS

The authors gratefully acknowledge imec's SiPV industrial affiliation programme and its partners, the support of the Flemish government through the funded project SolsThore (IWT.155025) and the Soft Matter, Rheology and Technology department from KULeuven for performing viscosity measurements. This project receives the support of the European Union, the European

Regional Development Fund ERDF, Flanders Innovation & Entrepreneurship and the Province of Limburg.

6 REFERENCES

[1] S. Braun et al., *Multi-busbar solar cells and modules: high efficiencies and low silver consumption*, Energy Procedia, vol. 38, 2013 pages 334-339

[2] S. Braun et al., *Solar Cell improvement by using a Multi Busbar Design as Front Electrode*, Energy Procedia vol. 27 pages 227-233

[3] T. Söderström. et al., *Smart Wire Connection Technology*, EUPVSEC, 2013

[4] T. Borgers et al., *Multi-wire interconnection technologies weaving the way for back contact and bifacial solar cells*, IEEE PVSC, 2016

[5] T. Borgers et al., *Combined interconnection and lamination of bifacial busbarless cells trough woven wiring*, EUPVSEC, 2017

[6] G. Cellere et al., *International Technology Roadmap for Photovoltaics*, ninth edition, 2018

[7] J. Walter et al., *Multi-wire interconnection of busbar-free solar cells*, Energy Procedia, vol. 55, 2014, pages 380-388

[8] J. Govaerts et al., *Progress in Encapsulant-Integrated Multi-Wire Interconnection*, EUPVSEC, 2018

NANOINDENTATION ANALYSIS OF PV MODULE POLYMERIC COMPONENTS AFTER ACCELERATED AGING

Djamel E. Mansour[1], Fabian Swientek[2], Ismail Kaaya[1], Daniel Philipp[1], Luciana Pitta Bauermann[1]
1 Fraunhofer Institute for Solar Energy Systems ISE
Heidenhofstrasse 2, 79110 Freiburg im Breisgau, Germany
+49-761-4588-5880
2 Anton Paar Germany GmbH
Helmuth-Hirth-Str. 6, 73760 Ostfildern, Germany
Djamel.eddine.mansour@ise.fraunhofer.de
https://www.ise.fraunhofer.de

ABSTRACT: The degradation and failure of a PV module can be caused by mechanical property changes in the polymeric components during its lifetime. Previously most of the investigation of the mechanical changes, for example hardness, as well as the aging has been carried out with the polymeric materials as bulk. With nanoindentation the change in hardness of encapsulant and backsheet can be spatially resolved with a high lateral resolution. The nanoindentation tests were carried out on the cross-section of glass/encapsulant/backsheet laminates after sequential doses of 500 h damp-heat (DH) at 85 °C and 85 % r.h. Backsheets with different water vapor permeation rates were used. The results show an increase in hardness (up to 40 MPa from 500 h to 2000 h DH) for the PET-based backsheet (BS1) with the highest water vapor permeation rate, which suggests a high degree of hydrolysis of the PET core layer. The hardness values of the encapsulant in the PV laminate showed a uniform increase with a slightly depth-dependence after DH aging. Especially after 500 h DH exposure, the hardness values were slightly higher near the BSF/encapsulant interface.

Keywords: encapsulant, backsheet, damp-heat (DH) exposure, degradation, nanoindentation, hardness, water vapor permeation, hydrolysis.

1 INTRODUCTION

The lifetime of a PV module is generally limited by the degradation of the constituent parts. The polymeric materials - encapsulants and backsheets - are particularly susceptible to degradation influencing the PV module's long-term reliability. These components are essential for mechanical protection and electrical insulation. When PV modules are exposed to environmental conditions like heat, humidity and UV irradiance, they suffer chemical degradation, triggered by a complex interaction of different aging mechanisms [1]. An increased level of moisture content in polymers leads to loss of molecular weight, loss of adhesion, loss of electrical insulation properties, and higher vulnerability to mechanical failure. Moisture can diffuse into PV modules through the permeable backsheets and the encapsulant layers towards the solar cell weakening their interfacial adhesive bonds, resulting in delamination and an increased number of ingress paths [2]. A deeper understanding of this process is of high importance for a better reliability and durability of PV modules [3].

Numerous attempts have been made to estimate water concentration in encapsulants using gravimetric methods [4] and water vapor transmission rate measurements [5]. However, the influence of the mechanical properties of encapsulants inside the PV module has not been thoroughly investigated. There is still a missing link between the aging process and its consequences in the context of mechanical properties [6]. The stability of the encapsulant aged in damp-heat (DH) has been investigated by using this material as granules [7] and free standing films [4]. The hydrolytic stability of polyethylene terephthalate (PET) bulk, a common core layer of PV backsheets (BS), has been studied at different temperatures and different relative humidities (RH) [8]. The degradation of adhesion strength between the backsheet and encapsulant due to moisture penetration was investigated for crystalline silicon photovoltaic mini-modules. The adhesion strength was measured by 90° peel tests, carried out at specified time intervals during accelerated aging tests [9]. The adhesion decreases quickly at the beginning and stabilizes after a certain time. An increased humidity accelerates the loss of adhesion. The thermo-viscoelastic properties of the encapsulant play an important role for adhesion, since the Young's modulus expressing the adhesive stiffness is strongly temperature-dependent [10].

This paper summarizes the research result on the change in hardness occurred in the encapsulant and PET-based BSs under DH conditions. The encapsulant was laminated between glass and different backsheets. The use of a backsheet allows entrance of humidity in an order of magnitude comparable to state of the art PV modules. The aging of the encapsulant in a PV module-like design is very important, because the water vapor diffusion, solubility and accumulation may differ on its bulk or on the interfaces. The object of this research is to identify the mechanical/hardness properties changes along the thickness of the BS and encapsulant. For this purpose, the chosen methodology is nanoindentation, as it allows measuring the hardness of the materials with a high spatial resolution.

To the best of our knowledge, the nanoindentation is a new destructive technique for mechanical testing of the PV module components after accelerated aging. In this way the effect of the hydrolysis as well as the oxidation of the polymers can be measured and localized from the external backsheet surface throughout the encapsulant to the glass.

2 EXPERIMENTAL PART

2.1 Lamination of the samples

PV laminates (18 x 18 cm^2) were prepared by laminating a layer of backsheets (BS), 2 layers of ethylene-vinyl acetate (EVA) as encapsulant and a non-tempered transparent glass cover (Fig. 1). The backsheet foils both have PET as core layer and water vapor permeation values reported for different temperatures on to their datasheets: BS1 = 3.9 g/m²·d at 38°C / 90% r.h, BS2 = 0.7 g/m²·d at 23 °C / 85% r.h.

2.2 Accelerated aging

The laminates were exposed to high humidity and temperature. The conditions of the damp-heat (DH) tests were 85 °C and 85 % r.h. during time intervals of 500 h. The accelerated aging test was based on the International Electrotechnical Commission (IEC) standard "IEC61215: Terrestrial photovoltaic (PV) modules–design qualification and type approval" [11].

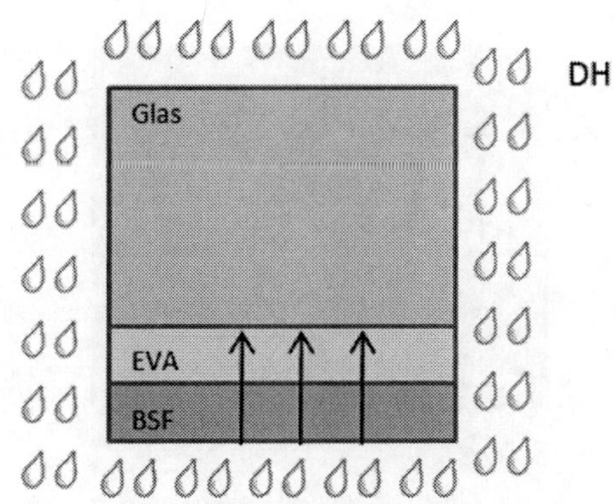

Figure 1: Schematic illustration of the laminate during accelerated aging

2.3 Cross-section sample preparation

The cross-section of the laminate was prepared after the accelerated aging by using a glass cutter to cut the glass followed by a sharp scalpel to cut the polymer layers. The cross-sections were extracted at a distance of at least 3 cm from the edge, to avoid effects from lateral humidity transmission. The piece of laminate was embedded in epoxy resin and polished with polycrystalline diamond suspension (Fig.2).

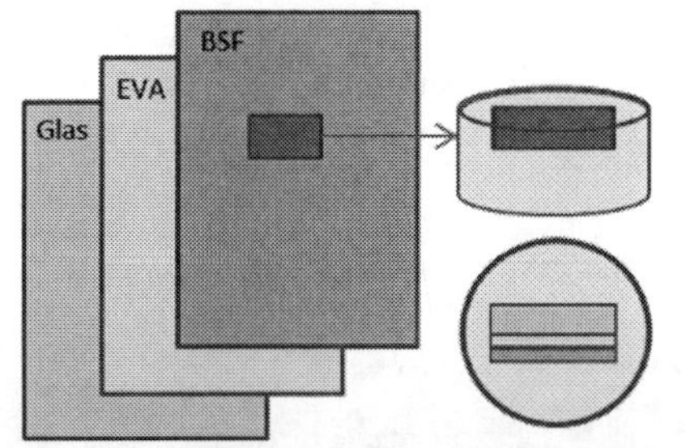

Figure 2: Schematic illustration of the cross-section preparation for the nanoindentation

2.4 Nanoindentation testing

All experiments were performed using the Nano Hardness Tester NHT[3] from Anton Paar Company. The system has load and displacement resolution of up to 500 mN and 200 µm respectively. The three-side pyramid with an area-to-depth function similar to a Vickers diamond indenter geometry, was used in all experiments as shown in Fig. 3. Based on Oliver and Pharr [12] the hardness (H) is defined as the pressure the material will support under load and is calculated from the peak indentation load (P_{max}) divided by the projected area (A) according to :

$$H = \frac{P_{max}}{A} \qquad \text{eq 1.}$$

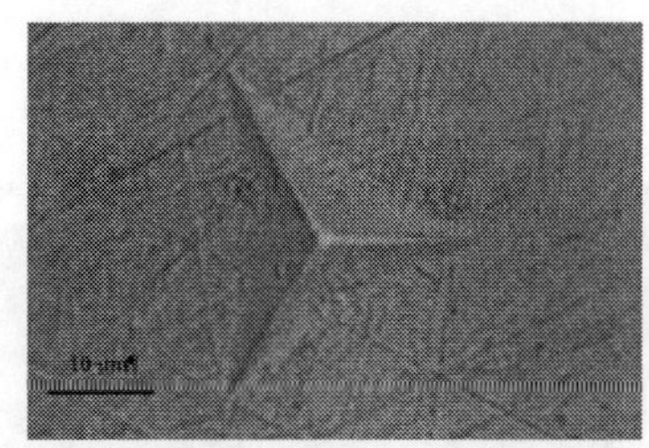

Figure 3: Microscopic image of the three-side pyramid indented on the PET-based backsheet

The indentations were performed at a distance of 44 µm between each point. The possible distance between the indentations is inversely proportional to its depth. A deeper indentation affects a larger area of the polymer and the distance between the measurements have to be large enough to avoid an overlapping.

3 RESULTS AND DISCUSSION

3.1. Ideal depth of nanoindentation

The hardness of the polished cross-sectional material might be affected by the sample preparation. In order to have a coherent value of hardness of the polymers and remove the influence of the exposed surface, a pre-measurement was conducted to find the ideal depth of the nanoindentation tip. A load is applied until the hardness value stays constant. The values of maximum load were fixed at 50 mN for BS and 100 mN for EVA.

A typical graph of load versus indenter displacement used for the calculation of hardness of BS and EVA is shown in Fig.4. For similar load (ca. 50 mN) the indenter in the BS (3,500 nm) is much shallower than in the EVA (30,000 nm). The hardness is calculated from eq. 1 by maintaining the load constant and measuring the indenter displacement or the indented area left after the unloading of the indenter.

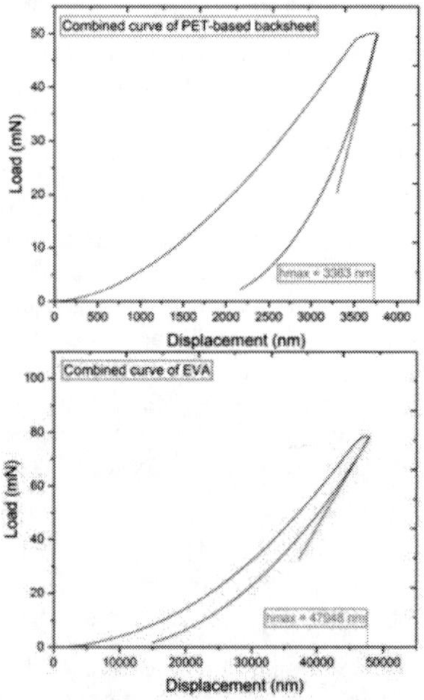

Figure 4: Combined curves of load versus indenter displacement used for the calculation of hardness of BS (top) and EVA (bottom)

3.2 Changes in hardness of BS during DH aging

The laminates were aged for 4 intervals of 500 h DH each. The hardness was measured after each interval. Fig.5 shows the deformations left on the BS surface after the indentations.

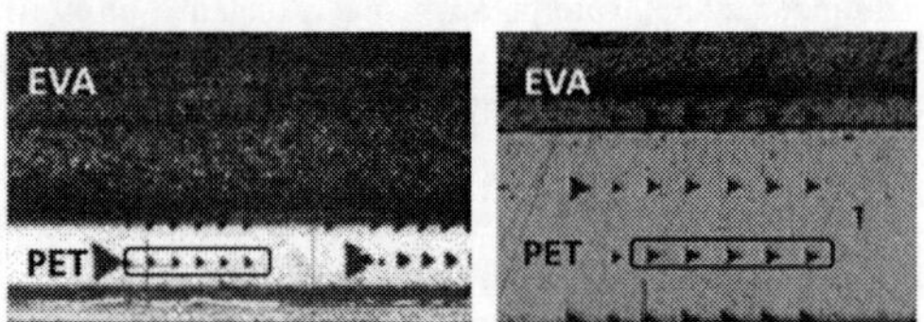

Figure 5: Microscopic images of the cross-section of the laminates after indentations on the BS1 (left) and BS2 (right) as laminated in glass/EVA/BS stack and aged under DH conditions. The values of the five indents on the same plane are averaged.

In this work the hardness of the core layer of the BSs was calculated on the middle of the PET layer. The measurement of the hardness of two BSs at different aging steps is shown in Fig.6. Five indentations were performed on the each line parallel to the interfaces and an average value was calculated.

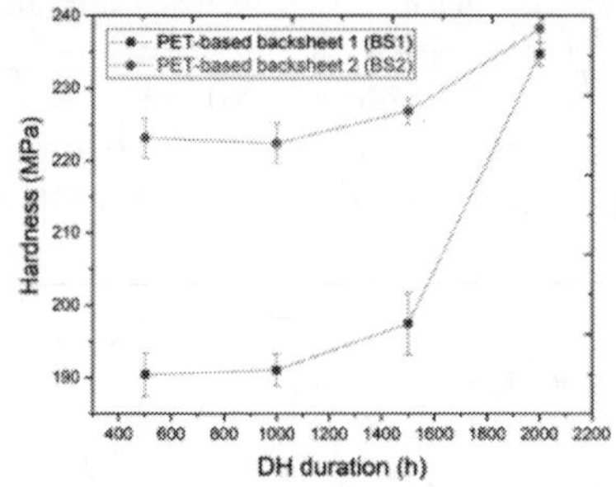

Figure 6: Changes in hardness of BSs after different intervals of DH aging

The initial hardness values of BS1 are lower than those of BS2, due to the difference of PETs. The change in hardness during the DH aging is much higher for the BS1 than for the BS2. This could be explained by the higher water vapor permeation rate of BS1, due to which the hydrolysis could happen to a higher extent. Or the PET-based BS2 is more stable.

3.3 Changes in hardness of EVA during DH aging

As the EVA is more elastic than BS, its deformation after the indentation is barely visible. The positions of the indentations are drawn as red triangles in Fig.7. The depth 1 is close to the BS and it is in contact with water vapor in the early stages of DH. The depths 2 and 3 are further in the interior of the EVA layer. Depth 4 is the furthest away from the BS and closest to the glass.

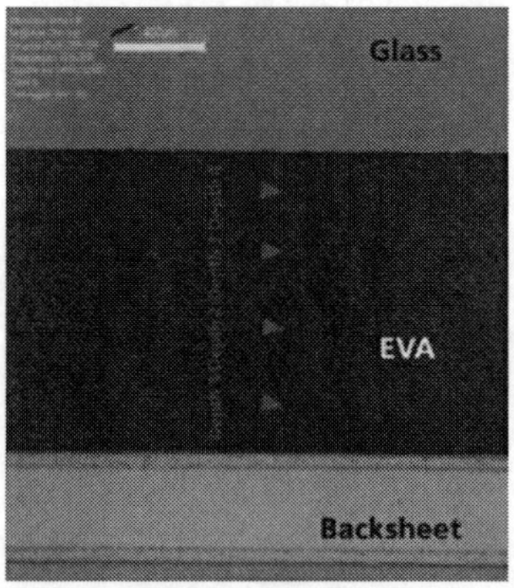

Figure 7: Microscopic images of the cross-section of the glass/EVA/BS laminate. Red triangles indicate the measurement points (depth) lateral from 1 (closest to backsheet) to 4 (closest to glass).

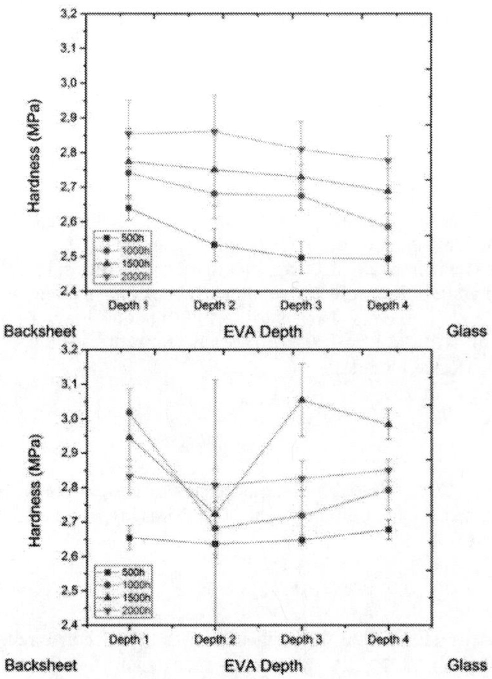

Figure 8: Changes in hardness along the thickness of the EVA laminated with BS1 (top) and BS2 (bottom) after different intervals of DH aging

According to Fig.8 (top) after 500 h DH, the EVA laminated with BS1 shows a stronger increase in hardness close to the BS than in the middle of the EVA or close to glass. After each DH interval, the EVA gets more homogeneous along its thickness and the hardness increases in general. The hardness of the EVA is slightly higher close to the BSF interface in all aged stages.

The laminate with BS2 (Fig.8, bottom) shows a homogeneous hardness along the EVA layer after 500 h and 2000 h DH. This hardness increases with the aging from 500 h DH to 2000 h DH. The hardness for the EVA aged after 1000 h DH and 1500 h DH show unexpected values. We suppose that the displaced EVA area through the indenter was very large and an overlapping between adjacent indenters happened invalidating the result. It might also be that the first point (at depth 1) was too close to the BS and the value was not correct due to the proximity of the interface.

4 CONCLUSION

In this work, nanoindentation was used as a suitable destructive method to investigate the changes in hardness occurred during DH aging of components in the PV modules with a high spatial resolution. The hardness of two different PET-based backsheets increased by prolonged DH exposure time, especially after 1500 h. The increase in hardness of BS1 (high water vapor permeation) was as expected higher than in BS2 (low water vapor permeation) due to a higher moisture ingress. The DH exposure induced a slight increase in the hardness values of the encapsulant. Using the BS1 in the PV laminate, the encapsulant shows an increase hardness values close to the interface BS/Encapsulant. The increase in hardness of BS and encapsulant may indicate an early stage of the PV degradation. Additional research is planned to identify a level of hardness where delamination is about to happen. These are the first results showing the possibility of measuring the change in hardness of the polymeric material of a PV module due to aging. Further work is being done to improve the quality of the measurements by better defining the position and depth of the indents. Also other characterization techniques are going to be used to complement and give stronger arguments for our conclusions. In the future, degradation models will be used to extract the kinetic parameters and correlate them with the different failure modes induced by the accelerated aging tests.

5 ACKNOWLEDGEMENTS

We would like to express our gratitude to Paul Pavlov and his team in Anton Paar Germany GmbH Company for providing much expertise on the Nanoindentation. This work has been funded by the European Union's Horizon 2020 programme under the Marie Sklodowska Curie Actions GA No. 721452, acronym SOLAR-TRAIN.

References

[1] A. Badiee, I.A. Ashcroft, R.D. Wildman; International Journal of Adhesion and Adhesives 68 (2016) 212

[2] N. Park, C. Han, and D. Kim, Microelectron. Reliab., 53 (2013) 1922

[3] R. Meitzner and S.-H. Schulze, Sol. Energy Mater. Sol. Cells, 144 (2016) 23

[4] T. Swonke, and R. Auer, Proceedings .SPIE, (2009) 74120A-1

[5] S. Rashtchi, P. D. Ruiz, R. Wildman, and I. Ashcroft, Proceedings SPIE 8472 Reliability of Photovoltaic Cells, Modules, Components, and Systems V, (2012), 847200)

[6] D. Wu, J. Zhu, T.R. Betts, R. Gottschalg. Prog Photovoltaics Res Appl, 22 (2014) 796

[7] M. Çopuroğlu, M. Şen Polym. Adv. Tech. 15 (2004) 393

[8] James E. Pickett, Dennis J. Coyle. 98 (2013) 1311

[9] D. Wu, J. Zhu, T.R. Betts, R. Gottschalg. Prog Photovoltaics Res Appl; 22 (2014) 796

[10] M. Köntges, G. Oreski, U. Jahn, M. Herz, P. Hacke, K. A Weiss IEA PVPS (2017)

[11] ICE 61215, Terrestrial photovoltaic (PV) modules-Design qualification and type approval, International Electrotechnical Commission, (2016)

[12] W. C. Oliver, G. M. Pharr, J. Mater. Res., 6 (1992) 1564

DAYSY CONTACTLESS MEASUREMENTS OF PID SHUNT RESISTANCE IN INSTALLED PV MODULES

Liviu Stoicescu, Michael Reuter;
Solarzentrum Stuttgart GmbH
Rotebühlstr. 145, 70197 Stuttgart

ABSTRACT: This paper describes a new method to measure the shunt resistance R_p of each solar cell in installed PV modules. The shunt resistance R_p determines the power output of the PV module especially at low irradiance Ψ and is therefore crucial for calculating a reliable power forecast as well as for evaluating the effects of degradation and ageing particularly if the module is affected by Potential Induced Degradation (PID). Our method uses two daylight photoluminescence (dPL) images captured with the DaySy Tool under daylight conditions at two different solar irradiances levels Ψ_1 and Ψ_2. The method is applied to a healthy and a PID affected module: The healthy module shows a mean R_p = 5710 Ω/cm² with negligible power loss at low irradiances. However the PID module shows a heavily reduced mean R_p = 588 Ω/cm² leading to a power loss of -42 % at Ψ = 100 W/m² compared to the healthy module. Hence our method reflects the expected behavior of PID affected modules.
Keywords: field inspection, electroluminescence, photoluminescence, PID

1 DAYLIGHT LUMINESCENCE

As of 2018 more than 400 GWp solar power has been installed worldwide and utility scale solar parks with more than 5 MWp have become critical to renewable energy production. However, in the meantime experience shifts the 20 year promise of faultless operation into the mythical realm. Environmental factors such as humidity, heat, strong winds, floods, hailstorms, hurricanes and tornados, sometimes combined with inadequate structural design and poor workmanship impose high stress on the PV panels themselves which will lead to failure and ultimately power loss. In order to replace damaged modules, claim insurance policies and mitigate losses sensitive and meaningful early detection methods are required to be applicable on utility scale installations.

The patented [1, 2] DaySy daylight luminescence method and tool was introduced in 2014 [3] and has since been used by Solarzentrum Stuttgart GmbH and its customers for qualitative daylight EL & PL inspection of PV installations with a total of more than 1 GWp worldwide. We now introduce a quantitative scientific evaluation method for daylight photoluminescence (dPL) imaging, which works at string level and can be implemented into UAV aerial inspections. Our method is therefore applicable to MW scale PV installations for fast evaluation of degradation and ageing particularly in case of shunt deteriorating PID defects. The method simplifies due diligence inspections since it can be used to simulate and forecast the expected power output of the inspected PV installation.

2 APPROACH

The DaySy daylight luminescence method and tool applies a digital lock-in filtering technique and therefore requires modulation of the luminescence of the imaged solar cells between the two working points short circuit (sc) and open circuit (oc) in order to create difference PL (dPL) images. This modulation can be easily performed at string level using a simple IGBT switch [1-3].

The luminescence signal of a solar cell

$$\Phi = C\, J_d + \Phi_{sc,0} \qquad (1)$$

is proportional to the current density J_d flowing through the pn junction plus an almost negligible true short circuit component $\Phi_{sc,0}$ [4].

The DaySy Daylight Luminescence technique produces difference Photoluminescence (dPL) images with a signal strength

$$S = \Phi(\mathrm{oc}) - \Phi(\mathrm{sc})\,, \qquad (2)$$

while eliminating the influence of the true short circuit component $\Phi_{sc,0}$, as well as any reflected daylight [1-3]. However, the series resistance R_s, influences the luminescence $\Phi(\mathrm{sc})$ in short circuit, while the parallel resistance R_p, affects the luminescence $\Phi(\mathrm{oc})$ under open circuit conditions.

In the following we will use the characteristic resistance $R_c = V_{oc}/J_{ph}$ to show that for series resistances $R_s < 1/2\, R_c$ the luminescence $\Phi(\mathrm{sc}) = 0$ in short circuit. Then we use the effect of the parallel resistance R_p, on the luminescence $\Phi(\mathrm{oc})$ under open circuit conditions in order to calculate the exact value of R_p itself.

2.1 Short Circuit Conditions

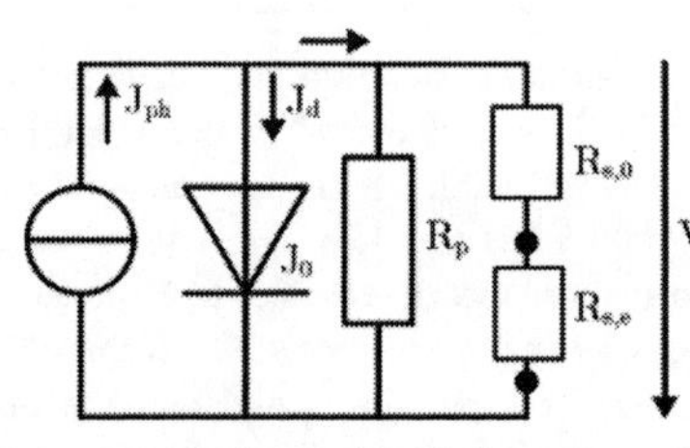

Image 1: One diode model of the solar cell in short circuit over the (external) series resistance R_s

Image 1 shows the one diode model of the illuminated solar cell under short circuit conditions. Note that the series resistance $R_s = R_{s,0} + R_{s,e}$, where $R_{s,0}$ is the cells own internal series resistance, while $R_{s,e}$ distributes the effects of the external circuit $R_{s,\mathrm{circuit}}$ containing cables, contacts and IGBTs on each cell of the inspected PV string as $R_{s,e} = R_{s,\mathrm{circuit}}/(\text{number of cells})$.

Under short circuit the photo generated current

$$J_{ph} = J_d(V_{sc}) + V_{sc}\,\frac{R_s + R_p}{R_s\, R_p} \qquad (3)$$

splits between the diode and the resistances R_s and R_p, and in turn generates a short circuit voltage $V_{sc} \neq 0$. With the safe assumption of $J_d(V_{sc}) \approx 0$ and $R_s \ll R_p$ the short circuit voltage calculates as

$$V_{sc} \leq J_{ph} \frac{R_s}{1 + R_s/R_p} \leq J_{ph} R_s \,. \tag{4}$$

If we furthermore write $R_s = x\,R_c$ in relation to the characteristic resistance R_c, the above equation further simplifies to

$$V_{sc} \leq x\,J_{ph} R_c \leq x\,V_{oc}. \tag{5}$$

The last approximation for V_{sc} calculates together with the saturation current density

$$J_0 \approx J_{ph} \exp(-V_{oc}/V_T)\,. \tag{6}$$

the current through the pn junction as

$$\begin{aligned} J_d &= J_0(exp(V_{sc}/V_T) - 1) \\ &\approx J_{ph} \exp((x-1)V_{oc}/V_T). \end{aligned} \tag{7}$$

If we use a thermal voltage $V_T \approx 26$ mV and open circuit voltages $V_{oc} > 600$ mV the current J_d through the pn junction in short circuit becomes negligible for series resistances R_s which are a fraction $x < 1/2$ of the characteristic resistance R_c.

Ultimately we can now write the DaySy photoluminescence signal

$$S = \Phi(\mathrm{oc}) = C\,J_d(V_{oc}). \tag{8}$$

as the open circuit luminescence of the solar cell which is generally described as the cell's photoluminescence.

2.2 Open Circuit Conditions

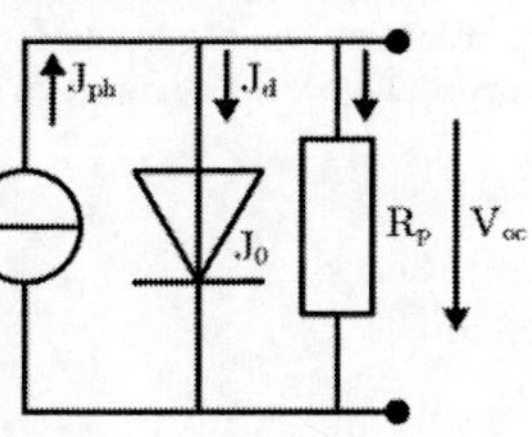

Image 2: One diode model of the solar cell in open circuit conditions

Image 2 shows the one diode model of the illuminated solar cell under open circuit conditions where the photo generated current is split between the pn junction and the cell's parallel resistance R_p as

$$J_{ph}(\Psi) = \Psi J_{ph,0} = J_d(\Psi) + V_{oc}(\Psi)/R_p\,. \tag{9}$$

Here Ψ denominates the present irradiance as a fraction of the standard test conditions (STC) irradiance of 1 kW/m^2 which is used to determine the photo generated current $J_{ph,0}$ under STC.

Parallel resistances R_p therefore lower the luminescence signal according to

$$S(\Psi) = C\left[\Psi J_{ph,0} - V_{oc}(\Psi)/R_p\right]. \tag{10}$$

In order to get rid of the proportionality constant C, we use two images taken at similar irradiance intensities Ψ_1 and Ψ_2 such that $V_{oc}(\Psi_1) \approx V_{oc}(\Psi_2)$. Hence the normalized difference $\langle\Phi\rangle'$ in signals yields

$$\langle\Phi\rangle' = \frac{S(\Psi_1) - S(\Psi_2)}{S(\Psi_1)(\Psi_1 - \Psi_2)} \tag{11}$$

$$\langle\Phi\rangle' = \frac{J_{ph,0}}{\Psi_1 J_{ph,0} - V_{oc}(\Psi_1)/R_p}, \tag{12}$$

and allows the calculation of the parallel resistance as

$$R_p = \frac{V_{oc}(\Psi_1)}{J_{ph,0}} / (\Psi_1 - 1/\langle\Phi\rangle')\,. \tag{13}$$

While we can easily substitute $J_{ph,0}$ with the short circuit current density determined under STC, we need to consider the effect of the parallel resistance R_p on the open circuit voltage V_{oc}: Form eq. 9 it follows with $\Psi = 1$ and

$$R_p = R_c/x = V_{oc,0}/(x\,J_{ph,0}) \tag{14}$$

that

$$J_d(V_{oc})/J_{ph,0}(V_{oc,0}) = 1 - \mathrm{x}\,. \tag{15}$$

Using the approximation

$$J_d \approx J_0 \exp(V_{oc}/V_T)\,. \tag{16}$$

and

$$J_d \approx J_0 \exp(V_{oc}/V_T)\,. \tag{17}$$

we can write the influence of the parallel resistance $R_p = R_c/x$ on the open circuit voltage as

$$V_{oc,0} - V_{oc} \approx -V_T \ln(1 - \mathrm{x})\,, \tag{18}$$

which becomes negligible for $\mathrm{x} < 1/2$ compared to open circuit voltages $V_{oc} > 600$ mV. Therefore we can neglect the effect of parallel resistances $R_p > 2\,R_c$ and approximate the irradiance specific open circuit voltage as

$$V_{oc}(\Psi_1) = V_{oc,0} + V_T \ln \Psi_1 \tag{19}$$

3 APPLICATION TO PID

The formulations of the previous section calculate the shunt resistance R_p from two dPL images $S(\Psi_1)$ and $S(\Psi_2)$ taken at two similar, but different Irradiance intensities Ψ_1 and Ψ_2

Image 3 displays dPL measurements of two PV modules at two irradiance levels: dPL image of a healthy PV module (a) at Ψ_1 = 400 W/m2 and (b) at Ψ_2 = 50 W/m2. Both images show a uniform luminescence pattern. In contrast image 1 (c) shows a PID affected module at Ψ_1 where some cells at the border already show a typical loss in luminescence. This effect is even stronger in image (d) at Ψ_2, where the module is already thoroughly checkered, which is very typical for PID.

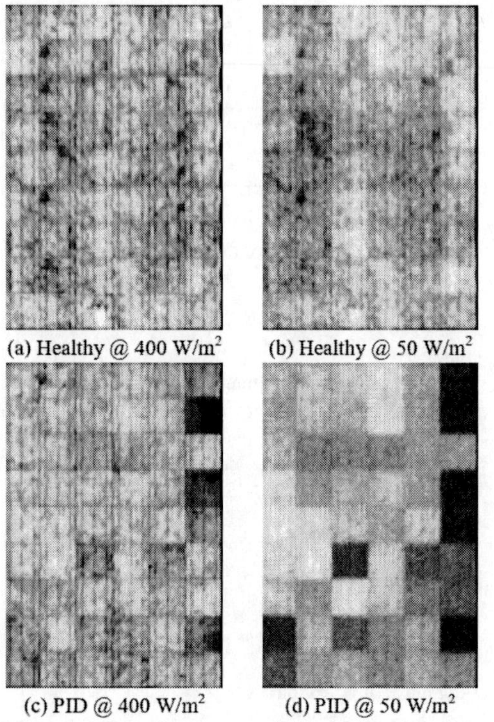

(a) Healthy @ 400 W/m^2 (b) Healthy @ 50 W/m^2

(c) PID @ 400 W/m^2 (d) PID @ 50 W/m^2

Image 3: DaySy dPL images of a healthy module show in (a) at Ψ_1 = 400 W/m^2 and in (b) at Ψ_2 = 50 W/m^2 a homogeneous luminescence distribution. The PID affected module displays in (c) at Ψ_1 already dark cells at the border and with decreasing irradiance more dark cells appear in (d) at Ψ_2. This effect is typical for low shunt resistances which are generally attributed to PID. All images use a differently scaled false color map.

Image 4 uses eq. 11 to calculate the normalized difference $\langle\Phi\rangle'$ from images captured at Ψ_1 = 400 W/m2 and Ψ_2 = 350 W/m^2 and plots it against the calculated shunt resistance R_p from eq. 13. For the healthy module we measure a mean R_p = 5 710 Ω/cm2 while the PID module shows a mean R_p = 588 Ω/cm2. From the extracted values we calculate for the PID module a power loss of 42% at Ψ = 100 W/m^2 compared to the healthy module.

4 CONCLUSION

We have extracted the shunt resistance R_p of solar cells in installed PV modules for identifying early stage PID, measuring the effects of aging and degradation, as well as calculating reliable power forecasts under all irradiance conditions. The method is evaluated in the German research project PARK, which aims at detecting faults and defects of photovoltaic (PV) modules in solar parks by means of luminescence measurements in an efficient and cost-effective way. The project further strives to develop prognostic tools for estimating the effect of faults and defects on the electrical performance of PV modules and PV parks.

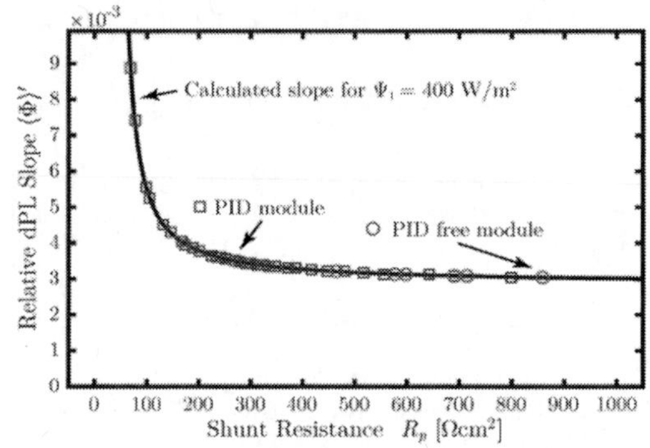

Image 4: The calculated normalized difference $\langle\Phi\rangle'$ creates a clear correlation between image data and shunt resistance R_p. The normalized difference extracted from two images at Ψ_1= 400 W/m^2 and Ψ_2= 350 W/m^2 is marked and correlated to shunt resistance values. Here the PID affected module displays shunt resistances R_p. mostly below 500 Ω/cm^2, while the healthy module shows much higher R_p.

5 REFERENCES

[1] L. Stoicescu, M. Reuter und J. H. Werner, „Method and apparatus for testing photovoltaic modules", US9680412 B2 (2017).

[2] L. Stoicescu, M. Reuter und J. H. Werner, „Method and apparatus for testing photovoltaic modules", CN104685349 B (2017).

[3] L. Stoicescu, M. Reuter, J.H. Werner, „DaySy: Luminescence Imaging of PV Modules in Daylight", *in Proc. 29th European PV Solar Energy Conference and Exhibition*, hrsg. von T.P. Bokhoven (WIP, München, 2014), S. 2553 - 2554.

[4] U. Rau, „Reciprocity relation between photovoltaic quantum efficiency and electroluminescent emission of solar cells", Physical Review B **76** (2007).

INVESTIGATING THE EFFECT OF SOILING ON THE POWER PRODUCTION OF PV PANELS EXPOSED TO WIND: WIND TUNNEL APPROACH

Lundholm Rickard[a]
rickard.lundholm@esat.kuleuven.be
T: +46 76 425 94 29

Goossens Dirk[a,b]
dirk.goossens@kuleuven.be
T: +32 476 06 67 65

Goverde Hans[c,d]
hans.goverde@imec.be
T: +32 468 11 51 40

Poortmans Jef[a,c,d,e]
jef.poortmans@imec.be
T: +32 474 94 00 83

Chowdhury Mohammed Gofran [a,b,c]
gofran.ghowdhury@imec.be
T: +32 16 28 39 75

Govaerts Jonathan[c,d]
jonathan.govaerts@imec.be
T: +32 16 28 86 75

a. KU Leuven ELECTA/ ESAT, Kasteelpark Arenberg 10, 3001 Leuven, Belgium
b. KU Leuven Department of Earth and Environmental Sciences, Geo-Institute, Celestijnenlaan 200E, 3001 Leuven, Belgium
c. IMEC, Kapeldreef 75, 3001 Heverlee, Belgium
d. IMEC (EnergyVille campus), Thor Park 8320, 3600 Genk, Belgium
e. University of Hasselt Physics Department, Campus Diepenbeek, Belgium

ABSTRACT: Studies that have incorporated both dust and wind have been either review papers [1] or were experimentally performed in outdoor conditions [2][3]. This research paper aims to systematically study the effect that soiling has on the power production of solar cells while exposed to wind in a controlled environment. Dust was homogeneously displaced across a PV module in a dust wind tunnel. Using a dust wind tunnel, the experiments were done in an indoor controlled environment so that the variables tested are isolated and their trends become more visible. Dust densities spanned from 0 to 40 gm^{-2} in increments of $5gm^{-2}$. Wind speeds spanned from 0 to 4 ms^{-1} in increments of $1ms^{-1}$. Measurements were taken at thermal equilibriums and maximum output power was computed. The results show for 1 ms^{-1} the gain in Pmax caused by wind is negatively affected by the amount of soil on the module. Results also show that the gain caused by wind in cells located upwind are negatively affected by soil but that the trend decreases more downwind. This suggests that cooling by wind on a longer module would see a positive cooling effect due to soil.
Keywords: Soiling, Dust, Wind, PV, Power

1 INTRODUCTION

The electrical performance of photovoltaic cells is primarily dependent on the irradiance and the temperature of the cell [4]. It is known that the cooling effect from wind has a positive effect on the electrical performance. It is also known that a layer of dust on a PV module reduces the irradiance, and therefore the maximum power output. The effects of dust and wind have been studied individually both experimentally and theoretically, but to get a complete understanding of how dust affects the power output in the presence of wind (a common situation in field conditions) their combined effect needs to be known.

2 SETUP

The experiments were performed on 10 diced mono crystalline silicon cells connected in parallel so that the performance of each cell can be calculated. The cells are aligned in a straight line and laminated between two borosilicate glass panels. Using Kelvin sensing, the electrical characteristics of each individual cell can be measured without factoring in the power lost via the resistance in the wire. To record the I-V measurements a Keithley 2602a source meter was chosen due to its ability to perform dual measurements. This was necessary to ensure measurements of all the cells in an as short as possible time frame as the measurement of only one cell would have taken about 30 seconds. The module itself had to be removed from the experimental location in the wind tunnel at the end of each test for an additional dust layer to be added, and therefore all wires were made to become disconnected with ease.

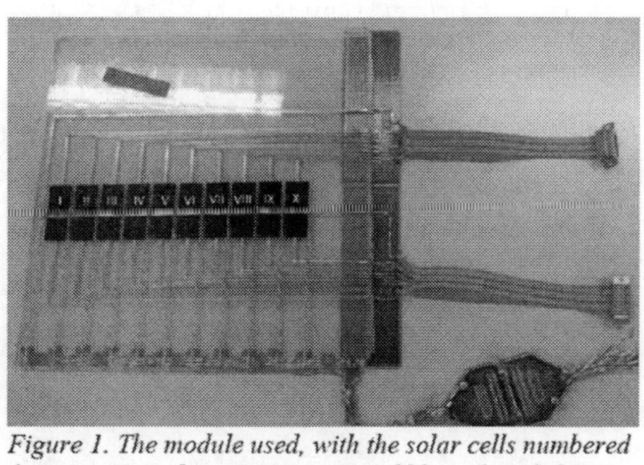

Figure 1. The module used, with the solar cells numbered from I to X. In the experiments wind blows from left to right.

The experiments were conducted in the environmental dust wind tunnel of the Division of Geography and Tourism, KU Leuven, Belgium [5]. A LED light fixture was assembled and placed as close as possible to the wind tunnel test section's glass roof for maximum light intensity. In the testing environment it was calculated that the light intensity reached approximately 190 Wm^{-2}. Clean, dust-free air was used during the experiments. After each experiment, dust was added to the module by feeding the airflow with dust particles and letting them settle spontaneously on the module. Using this technique, the dust is homogenously displaced across the entire module to ensure that all cells are exposed to the same starting conditions. The maximum difference in dust density between the 10 cells was only 0.84%. By having the cells aligned in a straight line along the wind tunnel's test section, the effect of the wind blowing over the soiled module becomes apparent. To be able to detect the cooling effect of the wind, 11 thermocouples were incorporated in the top glass of the module, immediately below the surface of the glass.

Figure 2. Solar panel with dust density layer of 40 gm^{-2}

3 PROCEDURE

Nine experiments of four runs each (one for each wind speed) were performed, with dust densities progressively increasing from 0 gm^{-2} to 40 gm^{-2} in increments of 5 gm^{-2}. Once the dust layer had been created on the module, the wind tunnel was carefully cleaned, and the module was placed at a location where the light intensity is strongest and the most consistent. The lights were turned on and the temperature measurements started. I-V measurements were done at two different times for each wind speed. The first measurement was done when the surface of the module had warmed up and had reached thermal equilibrium. After that the wind was turned on, causing the surface of the module to cool down towards a second equilibrium. The second I-V measurement was performed when that second thermal equilibrium had been reached. Once done, the wind was turned off and a new run could start once the original equilibrium had again been reached. This was repeated in increments of 1 ms^{-1} for wind speeds up to 4 ms^{-1}.

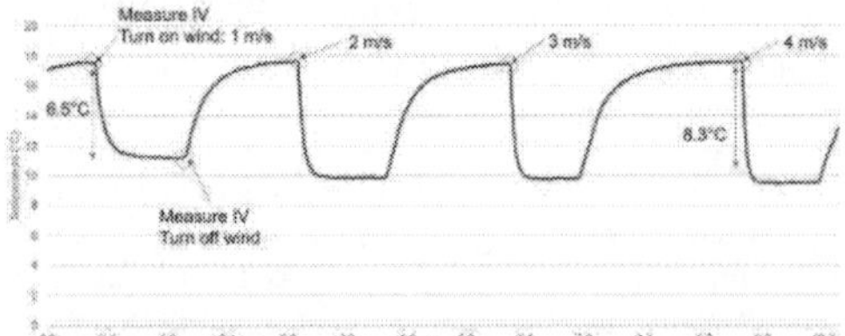

Figure 3 The measurement procedure, including all wind speeds. Times of taking I-V measurements are shown for the first cycle.

4 RESULTS

The cooling effect of the wind is shown in Fig. 4. Cooling was substantial, up to 9 °C for individual cells, but a plateau was reached at a wind speed of approximately 3 ms^{-1} after which higher wind speeds no longer resulted into an increased cooling.

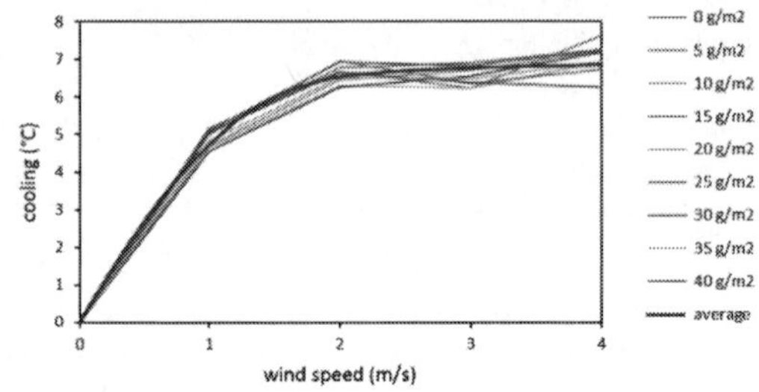

Figure 4 Cooling of the module's surface by wind. Data refer to the average of the 11 measuring points on the module.

The output power of the different cells is displayed in Figure. 5. From this data several trends can be observed:

- The power output decreases exponentially as a function of dust deposition density
- Differences in power output resulting from an increase in wind speed decrease with dust deposition density
- The differences in power output become smaller the more a cell is located downwind on the module
- At 1 ms^{-1} the effect of soiling becomes most apparent having a negative effect on the cooling of the module by the wind
- The trend across the module shows that the effect of soil has a negative effect more upwind on the module. The more downwind on the module, the less the effect of soiling is on the percent gain by the wind, suggesting that a positive gain could be possible for a longer module.

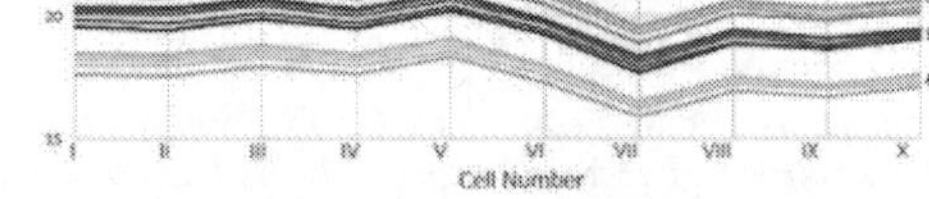

Figure 5 Output power produced by each cell, for all dust densities and wind speeds investigated in this study

For a homogeneous layer of dust across a module the soil has a negative effect on the Pmax of the module due to the reduction in irradiation that reaches the cells. The wind has a positive effect on the Pmax due to the cooling across the module. For a wind speed of 1ms^{-1} the more soil on the module the less effective the cooling effect of wind is. Closer inspection also shows that the cells located more upwind of the module are negatively affected by soil but as the wind travels downwind this effect decreases. This trend suggests that for a long enough module, the soil could have a positive effect on the cooling. A closer look at how dust and wind affect the short circuit current, open circuit voltage, and efficiency of the panel will be discussed in a later paper.

5 REFERENCES

[1] T. Sarver, A. Al-Qaraghuli, and L. L. Kazmerski, "*A comprehensive review of the impact of dust on the use of solar energy: History, investigations, results, literature, and mitigation approaches*," Renewable and Sustainable Energy Reviews, vol. 22, pp. 698–733, 06 2013.

[2] M. Abderrezek and M. Fathi, "*Experimental study of the dust effect on photovoltaic panels energy yield*" Solar Energy, vol. 142, pp. 308 – 320, 2017.

[3] P. Singh, S. Singh, M. Lal, and M. Husain, "*Temperature dependence of iv characteristics and performance parameters of silicon solar cell*," Solar Energy Materials and Solar Cells, vol. 92, no. 12, pp. 1611 – 1616, 2008.

[4] R. Messenger, D. Y. Goswami, H. M. Upadhyaya, T. M. Razykov, A. N. Tiwari, R. Winston, and R. McConnell, *Photovoltaics Fundamentals, Technology and Application*. Reading, Massachusetts: Taylor and Francis Group, LLC, 2007.

[5] D. Goossens and E. V. Kerschaever, "*Aeolian dust deposition on photovoltaic solar cells: the effects of wind velocity and airborne dust concentration on cell performance*," Solar Energy, vol. 66, no. 4, pp. 277 – 289, 1999.

PROPOSING AN ELECTRO-THERMAL SPICE MODEL TO INVESTIGATE THE EFFECT OF PARTIAL SHADING ON CIGS PV MODULES

J. Carolus[1,2], Z. Purohit[1,3], T. Vandenbergh[1], M. Meuris[1,2], B. Tripathi[3], and M. Daenen[1,2]
[1]Hasselt University, Martelarenlaan 42, 3500 Hasselt, Belgium
[2]imec, Kapeldreef 75, 3000 Leuven, Belgium
[3]Pandit Deendayal Petroleum University, Raisan, Gandhinagar, Gujarat 382421, India

ABSTRACT: Copper indium gallium (di)selenide (CIGS) photovoltaic (PV) modules are promising for building integrated PV (BIPV) applications because of their low weight to power production ratio. However, previous research shows that (partial) shading of a monolithically interconnected CIGS PV module can cause permanent damage to the shaded region of the module, even when the shading exposure time is very limited. [1, 2] This should be avoided in order to guarantee a 20 year life-span of the BIPV module. In this proceeding, we are investigating the impact of partial shading on CIGS PV modules. For this purpose, we are developing an electro-thermal SPICE model which maps the current density, voltage and temperature distribution of the CIGS PV module in case of shading. First simulations show a significant reverse bias over the partially shaded cell. Also interesting is the fact that the peak current density located in the non-shaded region of the partially shaded solar cell can be visualized. The model allows us to predict when a CIGS PV module may be permanently damaged in shading conditions and whether bypass diodes should be included or not.

1 INTRODUCTION

Since last decade, thin-film photovoltaic (TFPV) modules have been proven a highly efficient and cost competitive PV technology compared to most dominant PV technologies. [3] Due to its low weight to power production ratio, TFPV is interesting for building integrated PV (BIPV) applications. With the increasing demand of TFPV technologies, its long-term reliability and predictive modelling on its performance has gained a lot of attention. However, among various reliability issues, partial shading has drawn attention due to the permanent damage and its effect on the electrical performance of the PV module, even if the shading exposure time is very limited. [1, 4] In case one or more cells are shaded in a series connected configuration, the shaded cells will operate under a reverse bias, the shaded region now acts as a load to the rest of the PV module. This will force the shaded cells to go into breakdown if the voltage, and thus the number of in series connected cells, is high enough. Consequently, the solar cells material will be impacted and can cause immediate efficiency losses. In the present work, we briefly report the working principle of the electro-thermal model together with the first results. The experimental results show a significant deviation in electrical performance under partial shading conditions, whereas electro-thermal simulations allow us to visualize the distribution in electrical and thermal behaviour to predict the onset of failure of a CIGS PV module operating under partial shading conditions. The experimental results together with the simulation results will give us a better picture of the effect of shading on CIGS PV modules

2 EXPERIMENTAL

In this study, commercially available CIGS semi-fabricates on a glass substrate were encapsulated onto a glass front cover. The cells stack was composed out of a molybdenum (Mo) back contact, a CIGS absorber layer, a cadmium sulfide (CdS) buffer layer, an intrinsic zinc oxide (i-ZnO) passivation layer and an aluminium doped zinc oxide (ZnO:Al) front contact. The CIGS module contained 31 in series-connected solar cells of 14.5 x 0.4 cm2. In order to mimic (partial) shading of the CIGS PV module in the lab, opaque tape was used to cover one or more solar cells. To evaluate the electrical performance of the (partially) shaded module, dark and illuminated current-voltage (IV) measurements were performed at standard test conditions (STC). While conducting the dark IV measurements, in-situ infrared thermography was performed in order to investigate the reverse breakdown voltage of the solar cell.

3 MODEL DEVELOPMENT

The model is based on the single-diode model as described in:

$$I = I_L - I_0\left[exp\left(\frac{V + IR_S}{nV_T}\right) - 1\right] - \frac{V + IR_S}{R_{SH}}$$

with I the current, I_L the photocurrent, I_0 the reverse saturation current, R_S the series resistance, R_{SH} the shunt resistance, n the ideality factor, V the voltage and V_T the thermal voltage. However, the integration of the monolithic interconnection, which can be simplified to a series and shunt resistance, respectively R_{Smod} and R_{SHmod}, is not included in this equation but are included in the model. Also the parameters for the reverse characteristics are not described in this proceeding.

Since the superposition principle with voltage-independent photocurrent does not comply with the experiments for CIGS PV modules as shown by Sun et al. [5], we limited our model to two states which can occur in

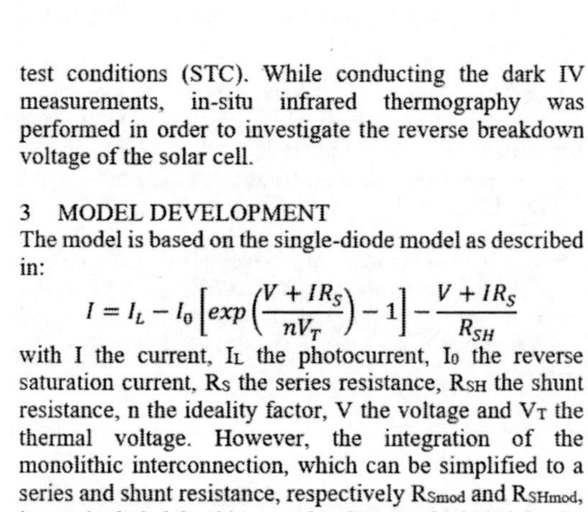

Figure 1 Visual representation of the topology of the pre-defined elements as included in the model together with the resistance of the back and front contact respectively shown as blue and red resistances.

one TFPV module: 1) fully shaded regions (0 W/m2) and 2) fully irradiated regions (1000 W/m2). For both the shaded and illuminated solar cells. The parameters are extracted as included in the single-diode model. For simplicity, only the reverse behaviour was taken into account for the shaded cells and the forward behaviour for the illuminated cells. The pre-defined element approach we included in the model allows us to obtain a voltage, current density and temperature distribution over the PV module. For this purpose, each in series connected solar cell of the TFPV module is divided into a predefined number of finite elements. Each element is composed of a current source, a diode and a shunt resistance and connected to four adjacent elements by the series resistance of the top grid (TCO) and the series resistance of the back contact (Mo). This is shown in Figure 1, in which the elements and their electrical connections to their adjacent elements is shown. Next to the topology of the pre-defined elements, Figure 1 also shows the integration of the monolithic interconnection. Note that this figure represents only one cell divided in six pre-defined elements including the monolithic interconnection (shown as green resistances).

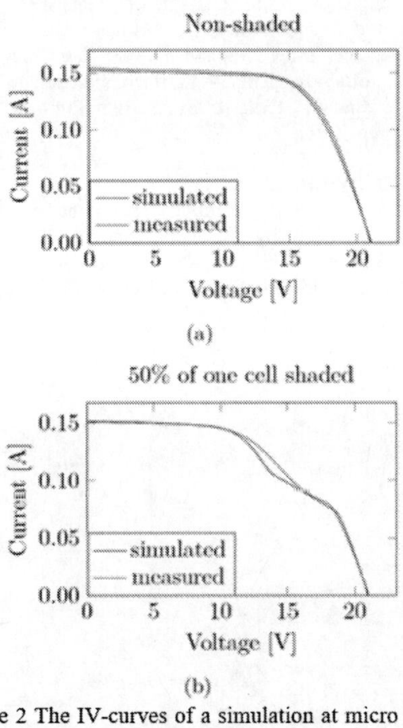

Figure 2 The IV-curves of a simulation at micro scale in comparison with the measured IV-curves with (a) an unshaded cell and (b) a 50% shaded cell.

4 MODEL VALIDATION

Figure 2 shows the IV curves of both the measured PV modules and the simulated PV modules. From here, it is clear that the model is accurate in case of no shading (a). When 50% of one cell is shaded, the IV curve deviates slightly in the region of the bending point (b). This might be solved by using the double-diode model instead of the one-diode model. Nevertheless, since the effect of this small deviation is neglectable at the operation point of the PV modules, the one diode model will be used.

5 RESULTS

In what follows, the simulation results of a 8 x 8 cm^2 CIGS mini-module composed out of 20 in series connected CIGS cells in which the tenth solar cell is 100% shaded for 50% of its area will be briefly discussed. Each cell was divided into 25 square shaped finite elements of equal size per cm^2. Figure 3 represents the mini-module as included in this simulation together with the opacity level for each cell.

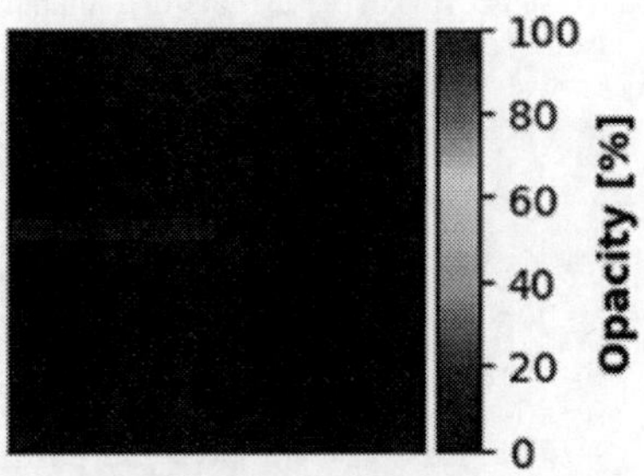

Figure 3 Representation of the 8 x 8 cm^2 CIGS mini-module with 50% of the area of the tenth cell totally shaded.

Figure 4 shows the voltage distribution of the CIGS mini-module. It is clear that the shaded part is subjected to a significant reverse bias of up to 3.5 V. Even the non-shaded part of the third cell is subjected to a (lower) reverse bias. From this plot, it is clear that the third cell is fully acting as a load while only 50% of its area is shaded. Furthermore, a gradient can be observed around the nonshaded region of the third cell.

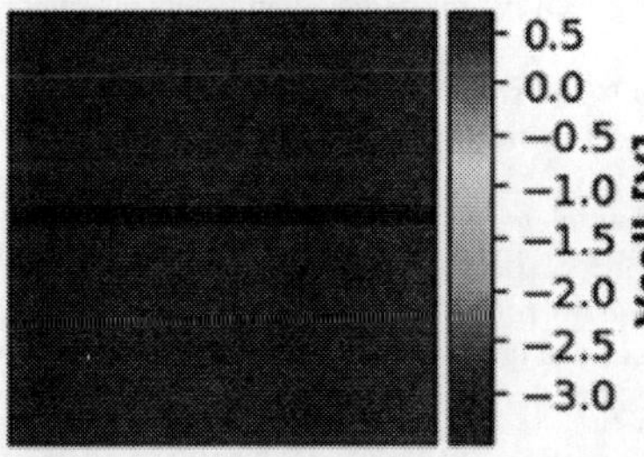

Figure 4 The voltage distribution of the partially shaded minimodule.

In Figure 5, the distribution of the current density of the shaded PV modules is shown. It is clear that the shaded region acts as a bottle neck for the current while the current density of the non-shaded region of the shaded cell exceeds the nominal current density of the mini-module. Also interesting is the gradient of current density at the

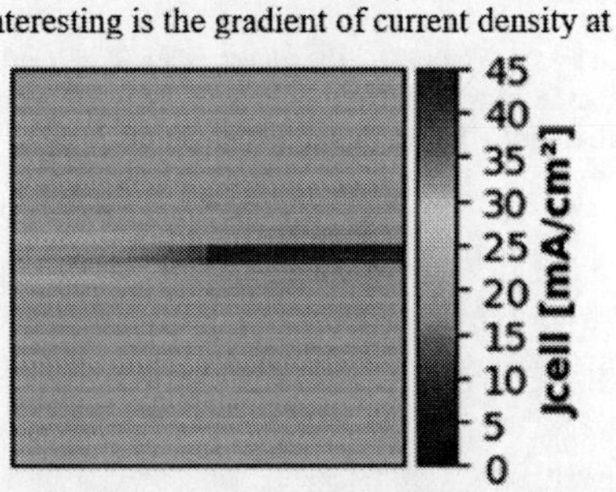

Figure 5 The current density distribution of the partially shaded mini-module.

interface between the shaded and non-shaded part of the third cell.

The temperature distribution of the partially shaded mini-module is shown in Figure 6. A slight decrease in temperature at the shaded region is observed while a temperature increase is observed at the non shaded region of the partially shaded solar cell. This is in accordance with previous results. However, the thermal output of our model is not yet validated and the temperature scale is therefore not absolute.

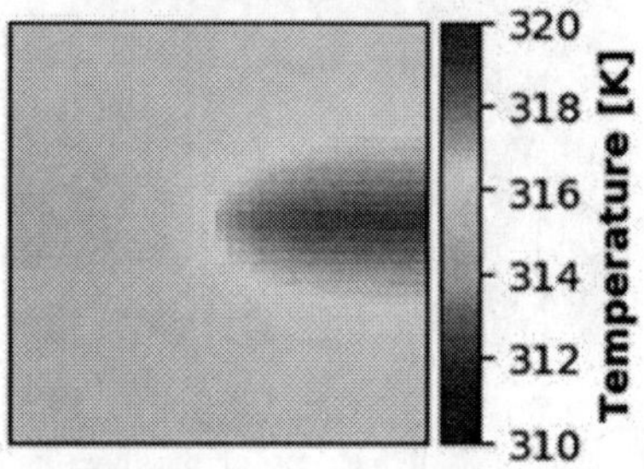

Figure 6 The temperature distribution of the partially shaded mini-module.

4 CONCLUSIONS

In this proceeding we proposed a pre-defined element electro-thermal SPICE model in order to investigate the impact of shading on monolithically interconnected CIGS PV modules. The model was simplified by limiting the possible irradiation levels (0 W/m2 for the shaded part and 1000 W/m2 for the illuminated part). This allowed us to use the single-diode model even when the superposition principle with voltage-independent photocurrent does not hold for CIGS solar cells.

The model is electrically validated, and the IV curves of the measurements and simulations are in accordance with each other. Nevertheless, a slight deviation of the IV curve around the bending point can be observed in case of 50% shading while in case of no shading, no significant difference can be observed. The results might be improved by using the double diode model instead of the one diode model.

This tool outputs the distribution of the current density and voltage of the PV module in case of a shaded cell. Due to a significant reverse bias over the shaded cell, i.e. 3.5 V in this case, a reverse current is enables to flow through the shaded region of the solar cell, which now acts as a load and dissipates power. Note that the reverse bias stretches over the full length of the partially shaded solar cell and thus is the partially shaded solar cell fully acting as a load. The thermal distribution is also carried out by SPICE. This is done by the Foster RC thermal model. The results are in accordance with other simulations, but still needs to be validated in the future by lab experiments.

5 REFERENCES

[1] I. Repins and T. Silverman, "Shadows from People and Tools Can Cause Permanent Damage in Monolithic Thin-Film Photovoltaic Modules," 33rd European Photovoltaic Solar Energy Conference and Exhibition, pp. 1422–1426, nov 2017.

[2] J. Bai, Y. Cao, Y. Hao, Z. Zhang, S. Liu, and F. Cao, "Characteristic output of PV systems under partial shading or mismatch conditions," Solar Energy, vol. 112, pp. 41–54, feb 2015.

[3] M. A. Green, K. Emery, Y. Hishikawa, W. Warta, and E. D. Dunlop, "Solar cell efficiency tables (Version 45)," Progress in Photovoltaics: Research and Applications, vol. 23, pp. 1–9, jan 2015.

[4] J. Parisi, M. Vrenegor, and M. Richter, "Reverse-Breakdown Stability of Cu(In,Ga)Se2 Thin-Film Solar Cells," 33rd European Photovoltaic Solar Energy Conference and Exhibition, pp. 1017–1019, nov 2017.

[5] X. Sun, M. A. Alam, J. Raguse, R. Garris, C. Deline, and T. Silverman, "A Physics-Based Compact Model for CIGS and CdTe Solar Cells: From Voltage-Dependent Carrier Collection to Light-Enhanced Reverse Breakdown Preprint," tech. rep., 2015.

6 ACKNOWLEDGEMENTS

This study has received funding from the project PV OpMaat, financed by the cross border collaboration program Interreg V Flanders-Netherlands with financial support of the European Funds for Regional Development.

ADVANCED TESTING OF PV MODULE ENCAPSULANTS

S. Pingel[1-2*], S. Janke[2], B. Stannowski[2], S. Fechner[1] and L. Podlowski[1]

[1]: PI Photovoltaik-Institut Berlin AG (PIB), Wrangelstr. 100, 10997 Berlin, Germany
[2]: PVComB, HZB Helmholtz Zentrum Berlin
[*]: now with Fraunhofer ISE

ABSTRACT: Encapsulation materials are of critical importance for long-term reliability and safety of PV modules. Only months in the field may lead to drastic power output degradation for example due to PID and also in the longer run adhesion and discoloration issues (Hot Spot) can reduce power output or even may lead to critical electrical safety issues. These findings are not sufficiently covered by IEC 61215 and other standard testing. Additional and different types of tests are required for an in-depth understanding of the encapsulation material's impact on PV module performance. PI Berlin has developed a test sequence that covers many field relevant aspects for PV encapsulations. For example: i) Potential Induced Degradation (PID), ii) Hot Spot Durability and iii) High Temperature Adhesion (delamination). Additionally, material properties which impact the processability in module manufacturing are tested. Also, very new aspects like the effect of encapsulation aging on PID are evaluated.
Keywords: Reliability, PID, PV Module

1 INTRODUCTION

Close to 100% of the modules in the PV market are laminated using encapsulation material which is based on a certain chemical raisin. The manufacturing companies have different recipes for different products. These vary in the base material and a product specific composition of different additives which are essential e.g. for the chemical cross linking, UV stabilization, glass adhesion, etc.

There are many different encapsulation materials used in the PV industry. According to ITRPV [1] the most dominant material will remain to be ethylene-vinyl-acetate (EVA) with more than 70% predicted market share in 2025 but also polyolefin (POE) materials will become more relevant, here more than 20% share in 2025 is predicted.

Ten industry relevant encapsulation materials are compared in the test procedure presented in this paper. The material types range from EVA to POE, but also more exotic materials like polyvinyl butyral (PVB) and ionomer are tested.

The test procedure covers different aspects relevant for module operation in harsh field conditions. In the Fig.1 below the test sequence is schematically shown.

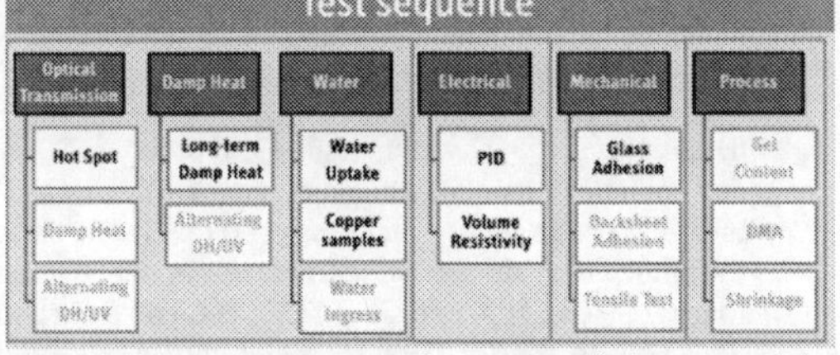

Fig.1: The test sequence for encapsulation materials. Highlighted in black are the tests addressed in this work while the tests in grey are not shown within this paper.

Central are the following four aspects:

- Potential Induced Degradation (PID) [2]
- Hot Spot Durability [3,4]
- Water interaction
- High Temperature Adhesion (Delamination) [5]

In the last few years many module manufacturers have diversified their product range from glass-backsheet (GB) modules to glass-glass (GG) modules. Glass-glass modules are expected to be more stable in the long run, product warranty is typically extended for this type of modules. Since the two types of modules are relevant and the share of GG modules is expected to rise in this study both module configurations are tested in parallel.

2 MATERIALS AND TEST SEQUENCE

In the following table (Tab.1) the ten encapsulation materials compared in this project are listed. Besides EVA and POE also an ionomer and a PVB material is included. This makes the benchmark testing even more interesting since the variability of material's characteristics is much broader and also extreme cases are covered. Most of the tested materials are chemically crosslinking, besides the ionomer that is crosslinked via ionic bonds and the non-cross-linking PVB.

The materials vary in their UV transmission, where two types are separated:

i) materials with high transmission in the UV range (so called UV-transmissive - UVT) which is usually used in front of the solar cells to allow also current generation by UV photons.

ii) materials with low UV transmission which is achieved by adding UV absorbers that are meant to protect the layers below from UV aging. These materials are typically called UV-close (UVC) and are used behind the solar cells to reduce the UV transmitted to the backsheet which might degrade from the UV stress.

No.	Chemical base	Cross-linking	UV-transparency
1	EVA (c)	Yes	UVT
2	EVA (c)	Yes	UVC
3	EVA (s)	Yes	UVC
4	Ionomer (s)	Yes	UVC
5	POE (s)	Yes	UVT
6	POE (c)	Yes	UVT
7	POE (c)	Yes	UVT
8	POE (c)	Yes	UVC
9	POE (c)	Yes	UVC
10	PVB (s)	No	UVC

Tab.1: Encapsulation materials compared in the test sequence. Materials marked with (c) are combined UVT/UVC materials and (s) stands for single materials.

The test samples designed for the different tests are described at the beginning of the particular chapters. Typically, the test samples were made from all ten types of encapsulation materials. Only in case of single solar cell coupons only seven variations were laminated and tested. These consisted of four single materials (Tab.1: (s)) and three UVT/UVC combinations (c).

All samples were laminated with a 20min lamination process at 150°C. The process was qualified by the Soxhelt method: all chemically cross-linking materials had a gel content level in the target range given by the manufacturers. Only the PVB material needed an adaption of the process due to bubble formation.

For characterization IV curves were recorded with a Halm flasher, EL images were taken with a high-resolution Si camera from Sensovation. For visual inspection photos were taken and a Lambda 1050 spectrometer from Perkin Elmer was utilized for transmission measurements.

Before the results are presented here an important side note: for all results the type of material is named but to protect the intellectual property of the participating encapsulation producers the companies and product name are not disclosed. Additionally, the results are randomly mixed. This means that for example EVA.1 in DH must not correspond to EVA.1 in PID.

3 DAMP HEAT TESTING

3.1 Long-term Damp Heat test

Damp Heat (DH 85°C / 85% RH) testing is one of the frequently applied tests in PV industry for certification and quality control. According to IEC61215 [6] a DH test for 1000 hours is necessary for module qualification. The modules are inspected visually, electrical insulation is measured and output power degradation is determined (power degradation criterium <5%). In PV industry DH tests are typically extended to 2000 or even 3000 hours [7]. In this project testing was further extended to DH5000 to identify quality differences between the materials. For this test single cell coupons were made from p-type multi crystalline Al-BSF cells with SiN_x AR coating and screen printed & fired contacts. For module making a standard PV ribbon was soldered on the busbars of the cells. For lamination encapsulation material was placed on the front and rear side of the cells. As mentioned above a total of seven combinations were used, four single materials and three encapsulation combination. In the latter case the UVT material was placed in front of the solar cell and the UVC material on the rear side. Per combination two module designs GG and GB were laminated. For each variation two coupons were made. The materials are compared concerning the long-time reliability of the output power. The power degradation versus DH duration is shown in Fig. 2 below. In most cases up to DH3000 degradation is negligible small.

In GG configuration modules are stable for 3000 h since water ingress is limited by the front and rear glass to the edges of the module. The samples made with the two types of EVA material are among the first to degrade in the interval DH3000-5000, most likely the formed acetic acid leads to corrosion of the cell metallization. POE.2 also shows a borderline result for DH5000 with about 4% degradation while the other two POE, the PVB and ionomer samples are with <2% power degradation stable in DH5000.

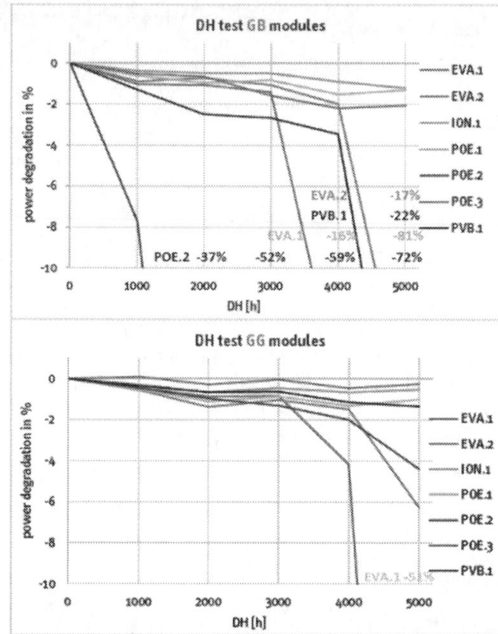

Fig. 2: Power degradation in DH5000 for GB (top) and GB (bottom) single cell modules.

For GB modules the result changes due to faster and stronger water ingress via the backsheet. POE.2 seems to lead to DH susceptible modules since the samples degrade by 8% already after DH1000, a clear fail. The EVA materials again show earlier and even stronger degradation compared to GG around DH3000-4000. For GB also PVB is showing degradation after DH4000. As found already for GG the ionomer and two POE materials achieve stable results in DH5000. Water indicator tests (not shown in this paper) inspired by [8] show that GB samples are influenced by humidity already before being placed in the climate chamber (color change at 8% after storage in the laboratory). For GG samples in a distance of 4 cm from the sample edge a level of about ~8% RH is reached after ~DH500 (latest DH1000). This comparison shows that DH stability varies with the encapsulation material and also the module layup has a significant influence. It is important to test all relevant combinations.

3.2 DH comparison SHJ versus multi BSF cells

Also for DH some GG coupons were made from n-type bifacial silicon hetero-junction (SHJ) cells. In these cells the amorphous silicon passivation layers are covered by a transparent conductive oxide (TCO: ITO) on front and rear side. Due to the temperature sensitivity of the thin passivation layers a low temperature Ag-paste screen printing metallization (dried and cured at low temperature) was applied. The PV ribbons were glued with a conductive adhesive (ECA) to the busbars of the cell. These type of modules or cells is expected to be more susceptible to DH degradation due to the low temperature metallization process and the TCO that is known to degrade when exposed to humidity [9].

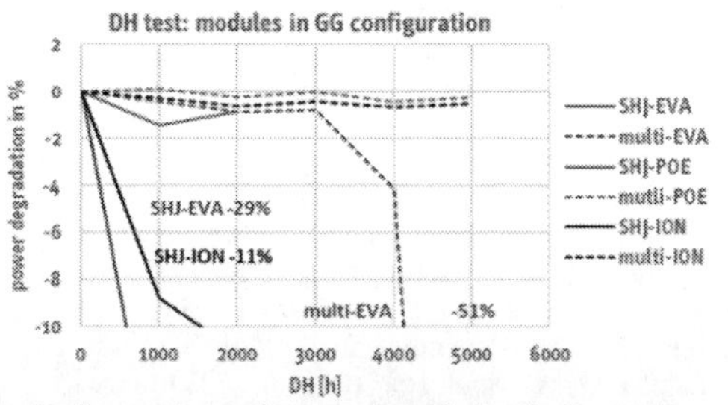

Fig. 3: Comparison of power degradation in Damp Heat for GG modules made from SHJ and multi solar cells.

The results of the comparison SHJ versus standard multi BSF cells is shown above in Fig. 3. In combination with EVA, SHJ cell degrades much faster (SHJ DH2000 -29% vs multi -4% DH4000). EL images in Fig. 4 below show the degradation of the SHJ cell (corrosion of TCO and or metallization) after DH2000 while the multi cell is stable.

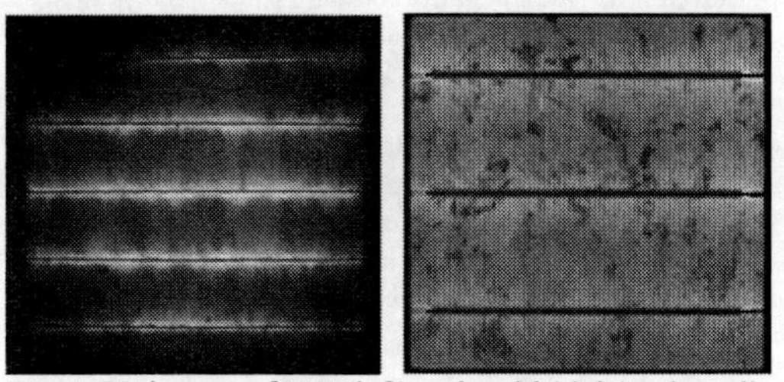

Fig. 4: EL images of SHJ (left) and multi (right) solar cells in GG-EVA coupons after DH2000.

In this comparison the combination of ionomer with this type of SHJ solar cell degraded quite fast, maybe the root cause of this strong degradation is the susceptible cell metallization combined with the stiffness of the ionomer or the chemistry of the ionomer leads to accelerated cell degradation. Ionomer materials are known to have a low water vapor transmission rate (WVTR) what makes the material group interesting for solar cells that are susceptible to humidity, fast water ingress should not be the problem. Laminated with one type of POE material both type of cells pass DH2000 (SHJ) DH5000 (multi) with power degradation rate <2%. This comparison shows that a material may perform good with one type of cell but in case of another cell technology results may be different.

4 WATER INTERACTION

4.1 Water Uptake

Samples for water uptake were made by laminating two or three sheets from one type of material to achieve roughly one mm thick samples. This allows comparable samples for all materials that vary in delivered sheet thickness (350µm – 750µm). The laminated samples were dried and weighted, then exposed to 60°C warm water.
The water uptake was determined by the weight difference before and after final drying (storage at 50°C for 24 hours) after the finished water bath test.
In Fig. 5 below the results of the water bath test are shown. After 63 hours in the water bath the EVA materials gained about 0.2% in weight. Surprisingly the ionomer on a similar level, while the POE materials gained with about 0.05% substantially less. The PVB showed the highest water uptake in this test, >4%.

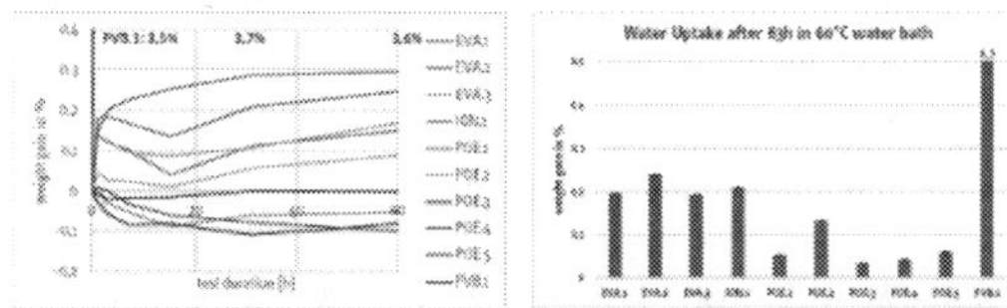

Fig. 5: Weight change during 60°C water bath exposure (left) and relative weight gain due to water uptake (right).

This test shows that by the selection of encapsulation material the amount of water that might diffuse into the module can be influenced. How fast water actually diffuses into a GG module is a different question since the water ingress is limited to the module edge. This could be shown in another test with water indicator [8], here the ionomer showed the best water barrier characteristic while in the water uptake test the material gained significant in weight by water absorption.

4.2 Copper samples in Damp Heat

For this test 0.1 mm thin sheets of bare copper were laminated between encapsulation material in GG and GB samples. In four alternating DH500 UV 25 kWh/m² test steps an image of the front and rear side of the samples was taken.
Depending on encapsulation chemistry and increasing water ingress chemical reactions with the copper occurred. Clearest color change was detected for the ionomer samples as shown in Fig. 6 below for the GB sample.

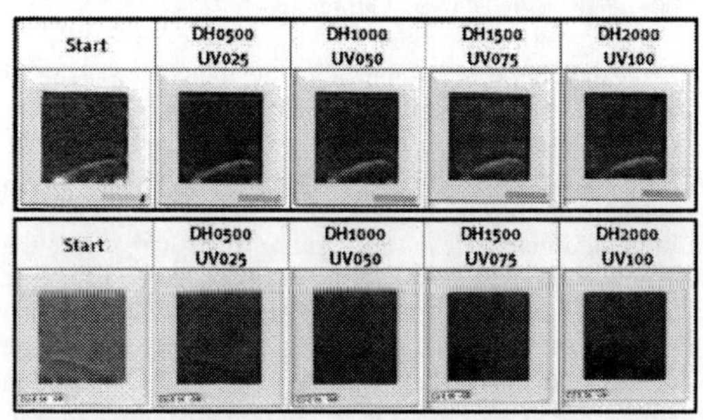

Fig. 6: Images of the GB copper sample in alternating Damp Heat UV test.

For the other encapsulation materials water ingress is typically clearly visible and also color change due to chemical reaction e.g. the formation of copper oxides was clearly visible. The ionomer shown above produced the most colorful result.

5 HIGH TEMPERATURE PERFORMANCE

5.1 Hot Spot testing

Solar cells in a solar module that generate less current compared to the system current turn from power producer to consumer and dissipate power. The reason for lower current might be shadowing (partial or full) or cell degradation (cell cracking, PID or others). If the solar cell current under reverse bias flows locally (e.g. due to shunts or breakdown areas) high temperatures may occur in laboratory hot spot testing or in the field [3,4] where temperatures in the range of 180°C and above are reported.

Focus of this investigation is degradation of the

transmission in a GG samples with an area of 10 x 10 cm² exposed to high temperatures. Gas diffusion (e.g. oxygen) into the sample is limited to the edges by the glass-glass configuration. This is to some extend comparable to a GB solar module where the front glass and the solar cell act as barrier for diffusion. The area in front of the solar cell is accessed only via module backsheet and the edges of the silicon wafer. Before starting the test, initial state was recorded by images and transmission measurements.
Per encapsulation one of two samples was exposed to i.) high temperature in the dark and the other to ii.) high temperature with additional irradiation by an HRI lamp.
i) Dark test: 500 h @ 180°C + 1000 h @ 190°C
ii) Light test (0.6 suns): 500 h @ 180°C + 500 h @ 190°C

The test temperature in the second test step was increased from 180°C to 190°C since degradation was small after the first step. The spectral transmission was measured over a wide spectral range. A change in the transmission spectrum was found especially in the short wavelength range. From this part of the spectrum the yellowness index (YI) was determined. In Fig.7 below is shown the change of the YI versus test time.

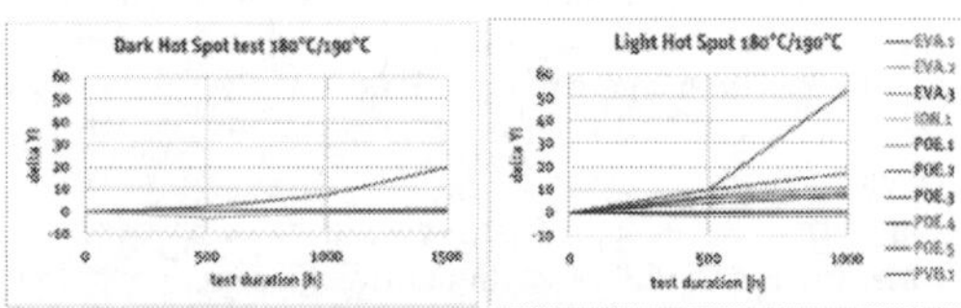

Fig. 7: Change of YI in the dark hot spot test (left) and in the irradiated hot spot test (right). UVC materials (marked in orange) showed more yellowing.

In the dark hot spot test the only thing to mention is the YI increase for PVB.1. All other encapsulation materials are stable even when exposed to 180 or 190°C for 1000 h. The ionomer even increases in transmission what results in a decrease of the YI.
With additional light (HRI lamp with ~0.6 suns) the situation changes, the YI increased faster. Especially some (not all) samples with UV absorber (UVC) degrade (marked in orange in Fig.7). For these YI increased in the range of 5 to 20. Again PVB.1 shows the strongest reaction with ΔYI>50 after 1000 h in the irradiated hot spot test. The color change in the 1000 h shorter irradiated test is more severe compared to the 1500 h dark test where ΔYI<20 was determined for PVB.1.

Below in Fig. 8 the images of an EVA sample in the dark hot spot test are shown. On one side the glass of this sample cracked and due to the increased oxygen ingress the browning due to oxidation is happing in the crack area and is not limited to the edge of the samples.

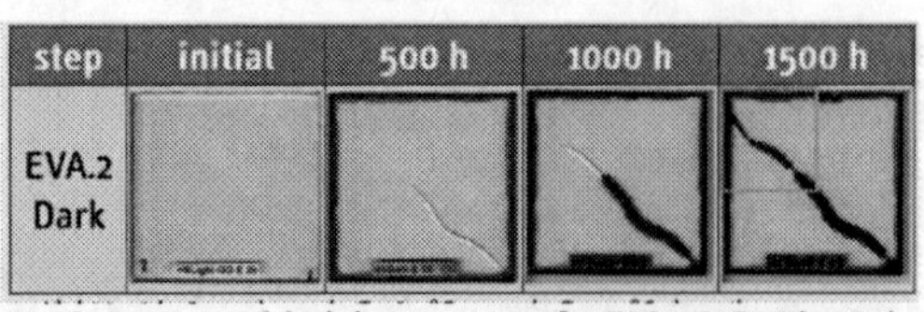

Fig 8: Images of dark hot spot test for EVA.2. In Fig. 9 the upper left quarter indicated by the blue square is shown.

The final images at the end of the two tests are shown in the Fig. 9 below. In the hot spot test with light for five of six UVC materials the yellowing is clearly visible and also the non-uniform distribution of the UV absorbers (EVA.1&2 and PVB.1) is eye catching. The PVB material shows in both tests a strong degradation starting from the sample edge, potentially the material is in comparison more open for oxygen diffusion.

Fig. 9: Final images of one quarter (2.5 cm x 2.5 cm) of the samples at the end of the dark hot spot test (top) and irradiated hot spot test (bottom). UVC materials are marked in orange.

IEC 61215 [6] includes only UV preconditioning with 15 kWh/m². This is to be seen in contrast to measurements in [10] where already within one year in the desert Negev in Israel about 120 kWh/m² are collected in the UV range. This motivated additional long-term testing at 60°C with 100 W/m² UV radiation (alternating DH500 + UV 25 kWh/m²) and a total dose of 100 kWh/m² (+ DH2000). Although the total duration is with 1000 h the same as in the irradiated hot spot test this resulted in negligible transmission change. As discussed in [11] UV aging should be accelerated by temperature increase and if at all only to a small extend by increase of the UV irradiation.
The results show that testing hot spot performance of encapsulation materials comparable to field conditions needs to be carried out at elevated temperatures and light with sufficient UV component is regarded to be crucial.

5.2 Mechanical Glass adhesion

Temperature dependent delamination was tested by recording peel curves at three to four different temperatures for different materials in [5]. Adhesion decreased with increasing temperature. Since delamination at elevated temperature occurring in the field might lead to degradation or even safety issues a simplified test for glass adhesion was developed for the encapsulation test sequence. For this test glass adhesion GB samples were laminated from glass, two layers of encapsulation and a white backsheet. Half of the samples were aged in alternating DH 2000 UV 100 kWh/m². Before the adhesion test 1 cm wide stripes were cut. The samples were fixed in a climatic chamber and each GB sample was loaded with 1,2,3,4 and 5 kg weights in parallel. Then the temperature in the climatic chamber was slowly ramped from about 20°C to 120°C with a rate of approx. ~0.8K/min. For all loaded stripes the temperature of failure meaning the beginning of delamination was recorded. Images of the test chamber and a sample after test is shown in Fig. 10 below. For the shown sample four stripes are completely peel while the stripe with the lowest force of about 1 N/mm is not completely peeled.

Fig.10: Photo of the glass adhesion test in the climatic chamber (left) with different weights per samples. On the right a photo of a sample after the peel test.

The result of the test is shown in Fig. 11, as expected the

temperature of failure decreased with increasing load. On the left side the data of unaged samples is shown while the data of the measurements after aging are shown on the right.

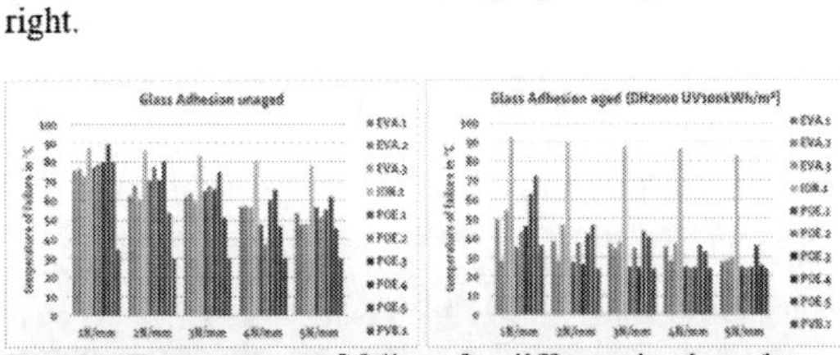

Fig. 11: Temperature of failure for different loads and ten encapsulation materials for unaged samples (left) and samples after aging in DH2000 UV 100 kWh/m² (right).

For unaged material all EVA and POE materials fail at typical operating temperatures of 60°C and a load of 5 N/mm. For a reduced load of 1 N/mm the temperature of failure increases to 70-80°C. The ionomer performed best (80-90°C) for all loads while PVB showed the earliest adhesion failure for all five loads (30°C).
Besides the ionomer that showed improved adhesion after aging, all material failed earlier. Typical failure for a load of 2 N/mm was found in a temperature range of 20-50 °C. Even for a load of only 1N/mm the temperature of failure is well below typical operating temperatures (50-60°C). This result indicates that adhesion test at operating temperature or above are important to qualify a material for a hot climate (highest reported module operating temperatures are above 90°C). Further testing is needed to identify if the combined DH/UV aging is causing the drastic degradation of the glass adhesion or if DH or UV stress alone leads also to earlier failure.

6 PID testing

6.1 Ultra-harsh PID test

For PID the same set of samples as for long term DH testing was laminated for the test (the layup is described in chapter 3.1). Half of the GG and GB coupons with PID sensitive multi solar cells were aged in DH500 followed by a UV25 kWh/m² step. Per variation two samples were tested in PID. The total of 56 PID samples were tested in a DH chamber operated at 85°C and 85% RH. The front glass of the coupons was covered by a grounded aluminum plate and the cells were connected to -1500V. Compared to standard PID testing these conditions are much more severe. This makes the test highly accelerated because of the high temperature of 85°C, complete grounding of the glass surface and the humidity in the chamber. Last but not least the test voltage of 1500V is the upper limited that might occur in systems with high system voltages and worst-case grounding/inverter layout.
The samples were exposed to PID stress with readouts after a total of 12, 48, 168, 500 and 1000 h. Due to the severe nature of the test it is expected to be selective concerning the PID performance of the encapsulation. The result of the test is shown in Fig.12 (a & b) below.
As expected a strong variability in the long term PID test is found. Ageing (DH 500 UV 25 kWh/m²) leads to earlier occurrence of PID, but not in all cases. Comparing the two tested module layups GB and GG, the second tend to degrade faster, potentially due to the glass on the rear side of the module that acts as diffusion barrier for PID supporting species. For example, EVA forms acetic acid in DH that might accelerate PID.

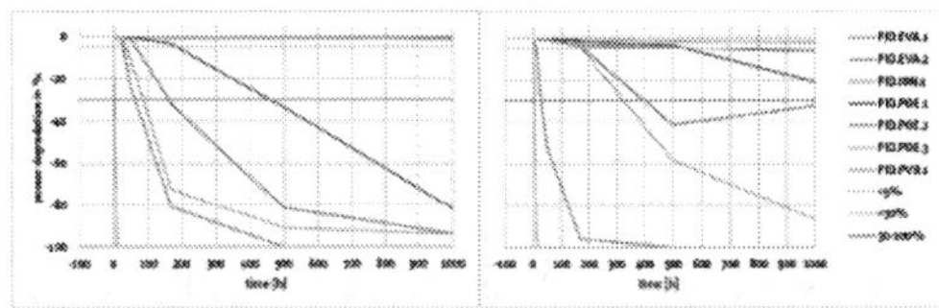

Fig. 12 a: Power degradation in the PID test for unaged GG (left) and GB (right) samples in 1000 h PID test.

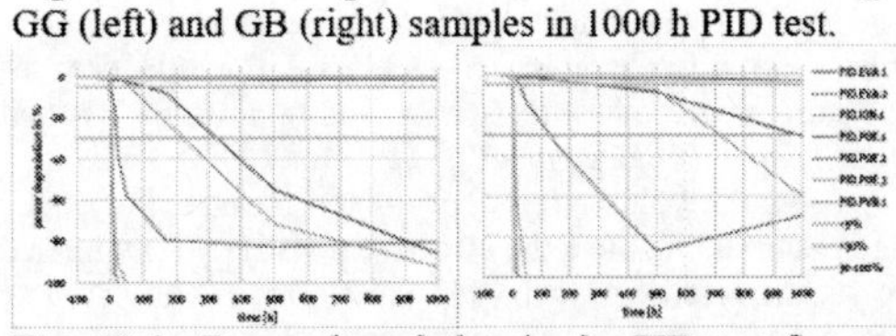

Fig. 12 b: Power degradation in the PID test for aged GG (left) and GB (right) samples in 1000 h PID test.

Looking the different encapsulation materials, samples with PVB generally showed extremely fast degradation. For EVA results are extremely mixed, samples made with EVA.2 degraded fast while EVA.1 showed PID only after prolonged testing. Only EVA.1 showed a tendency for recovery in the long-term test. Samples with POE encapsulation typically showed improved PID stability. On one side POE.3 performed similar compared to EVA.1, POE.1 shows good stability. Top runners are POE.2 and the ionomer that had extraordinary PID stability and passed in almost all cases the 1000h test with substantially less than 5% power degradation. The result shows the importance of PID testing before an encapsulation material is selected for a product (GG or GB) that might be exposed to high system voltages or other field conditions that lead to potential induced degradation.

6.2 Volume Resistivity

Since PID is accompanied by a leakage current flowing from the grounded glass to the solar cell one of basic parameters that is obvious to test is the volume resistivity of the encapsulation material. To measure volume resistivity single layers of encapsulation were laminated flat between two sheets of fluorinated ethylene-propylene copolymer (FEP). Half of the samples were aged in the FEP sandwich in a combined step of DH 500 and UV 25 kWh/m². The thickness of the samples was measured, then the samples were placed between two electrodes in a Faraday cage and 1500 V was applied from one electrode. The leakage current was recorded for 20 min. For evaluation the median value in the range of 15-20 min was used to calculate the volume resistivity. In Fig. 13 below the resistivity for 25°C and also 60°C for fresh and aged materials are presented.

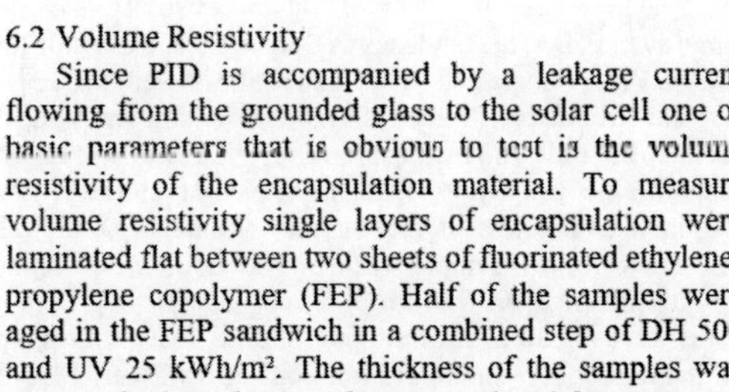

Fig. 13: Measured resistivity unaged/aged for 25°C/60°C.

Since generally volume resistivity measurements of high resistance material (>10^{16} Ωcm) showed poor reproducibility the result shown in the left graph is capped at this resistivity level. At 25°C additional aging leads to an increased or similar resistivity. Going to 60°C leads to a drastic drop in resistivity (factor of 5 to 25). At 60°C aging seems to lead to lower resistivities. Results for the EVAs goes from >10^{15} Ωcm for 25°C to >10^{14} Ωcm for 60°C, while ionomer and POEs start with >10^{16} Ωcm and end at 60°C typical in the range >10^{15} Ωcm or >10^{16} Ωcm, where no difference is accessible by the this type of measurement. Lowest resistivity was found for PVB that goes from ~10^{12} Ωcm to ~10^{11} Ωcm for 25°C/60°C.

From the volume resistivity (aged measured at 60°C) and material thickness the effective resistivity is estimated for the covered cell area (243 cm²). This rough estimate for each front encapsulation material is compared to average PID results after 168h test (unaged / aged GG and GB). The result is also shown in Fig. 14 below.

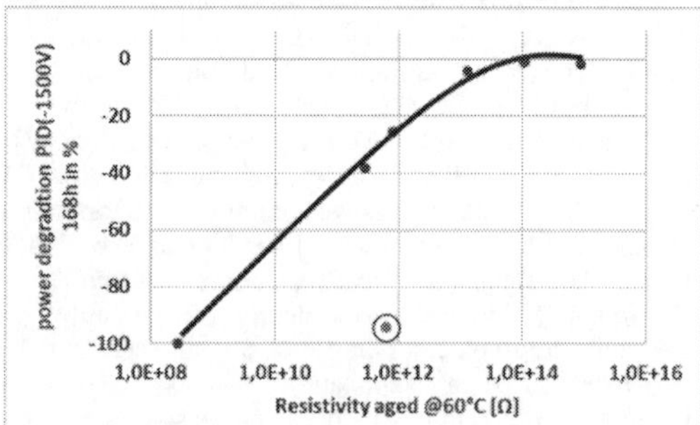

Fig. 14: Correlation of calculated effective resistivity (aged 60°C) with degradation after 168h PID test.

A tendency is clearly visible. Higher resistivity leads to improved PID. This tendency is only valid if the result for one EVA is neglected. This material showed for EVA typical volume resistivity but PID degradation was almost 100% after only 168 h of testing. One parameter that might explain the run-away result is the comparatively low laminated thickness of the sample. Around 350µm was measured for this sample while all the other encapsulants lay in the range 400-750µm.

This investigation shows that volume resistivity is a suitable quality parameter to control PID, effects of temperature and aging must be considered.

7 SUMMARY

In this paper an excerpt of the results produced in the first round of PI Berlin's advanced test procedure for encapsulation material is presented. Most relevant findings in this benchmark test are the following:

- Standard cells are well protected by most current encapsulation materials in long-term DH. Results vary especially by module design (GG or GB). Other cell concepts e.g. SHJ cells or cells made with non-standard (e.g. copper plating) need a careful selection of a suitable encapsulation material. The investigated materials vary significantly in their water uptake, water ingress and also in the interaction with copper.
- Realistic transmission stability testing of an encapsulation in extreme temperature environments (Hot Spot) is only possible if the samples are exposed to high temperatures and additionally exposed by light with significant UV contribution. Here some materials with UV absorber showed significant yellowing.
- Temperature dependent glass adhesion testing revealed delamination at rather low loads and typical module operating temperature (or even below). Preaging lead to even earlier adhesion failure. These results need to be considered especially for hot (and humid) climates.
- When does PID occurrence depends on the type of encapsulation material, module design and aging history. Encapsulation volume resistivity is a suitable parameter for material selection to prevent PID.

8 ACKNOLEDGEMENTS

We would like to thank all companies participating in this project, also for their patience. Also we thank Mr. Lackner from the Solinex GmbH for providing the white backsheet material used for different tests.

9 LITERATURE

[1] International Technology Roadmap for Photovoltaics (ITRPV) – 2017 Results
[2] Berghold et al.: "PID: From material properties to outdoor performance & quality control counter measures" SPIE Optics 2015
[3] Wendlandt et al.: "Thermal stress analysis at encapsulation and backsheet materials for PV modules" EUPVSEC 2015
[4] Janke et al.: "Comparison of hot spot endurance tests: Temperature behavior of bare vs. encapsulated crystalline silicon cells", EUPVSEC 2014
[5] Oreski et al.: "Comparative study of temperature dependent delamination behavior of four solar cell encapsulants to glass and backsheet-laminate", EU PVSEC 2011
[6] IEC 61215 Terrestrial PV modules - Design qualification and type approval Edition 3 (2016)
[7] Pingel et al.: "Investigation of DH degradation mechanism and correlation to an accelerated test procedure (HAST) ", EUPVSEC 2012
[8] Miyashita et al.: "Measuring Method of Moisture Ingress into Photovoltaic Modules", Japanese Journal of Applied Physics 2012
[9] Morales-Vilches A. et al. „ITO-free silicon heterojunction solar cells with ZnO:Al/SiO2 front electrodes reaching a conversion efficiency of 23 %", WCPEC-7 (2018)
[10] Köhl "The challenges of testing the UV-impact on PV-modules", PVMRW (2012)
[11] Shioda et al.: "UV accelerated test based on analysis of field exposed modules" EUPVSEC 2011

SCALING UP LASER LINE PHOTOLUMINESCENCE IMAGING FOR OUTDOOR INSPECTIONS

Gisele A. dos Reis Benatto[1], Mingjin Chi[1], Adrian A. Santamaria Lancia[1], Nicholas Riedel[1], Ole B. Jensen[1], Anders Thorseth[1], Carsten Dam-Hansen[1], Sune Thorsteinsson[1], Peter B. Poulsen[1]

[1]Department of Photonics Engineering, Technical University of Denmark, Frederiksborgvej 399, 4000, Roskilde, Denmark

ABSTRACT – As outdoor electroluminescence is implemented in the field, with several developments for automated inspection, an increasing concern rises upon the limitation that the electrical connections will bring to the inspection speed. In the effort to avoid both electrical contact and time limitations for PV inspections, here we present the progress in scaling up the well-known cell level photoluminescence imaging to a module level. The final goal is to build up a system that can perform PL excitation and image acquisition from the distance on PV modules and brought easily to the field by unmanned aerial vehicles (UAV).
Keywords: Photoluminescence imaging, PV inspections, Fault detection, Laser line.

1 INTRODUCTION

In order to ensure the expected return on investment (ROI) of both small and large-scale PV installations, regular fault detection is highly desirable. At the same time, frequency and level of detail in large power plant inspections is often limited by manpower and cost. Luminescence based imaging can be used to rapidly and accurately detect a large range of major and minor faults in PV modules such as cell cracks, broken interconnections, potential induced degradation (PID), among others differently from current field inspection methods such as infrared thermography and electrical characterization [1]–[3]. As outdoor electroluminescence is implemented in the field, with several developments for automated inspection, an increasing concern rises upon the time limitation that that the appropriate electrical connections to the PV system will bring to the inspection speed.

Luminescence can be induced not only by electricity as performed in electroluminescence (EL), but can also be driven by light therefore called photoluminescence (PL). The sun would be the straightforward way to induce PL, however, as the sunlight is orders of magnitude more intense than the luminescence signal, the need for modulation between V_{OC} and I_{SC} to obtain the PL image require again electrical connections. Another possibility would be to acquire images while the PV panels are at different temperatures (i.e. times of the day), what maximize the time demand of the inspection. There are valuable strategies in the literature to avoid electrical contracting, such as contactless EL and outdoor contactless PL [4], [5]. However, these strategies require the images to be acquired at a very short distance or masks touching the surface of the panels, what represent a speed limitation to a fast UAV-based inspection scenario.

In order to avoid both electrical contact and time limitations for PV inspections, the development of a line shaped laser light source for inducing PL on a module level is the most promising technical solution and it is under development. The final goal of the setup is to be able to induce PL with a single laser light source in a homogeneous line shape over a PV module, where the light source can be switched on and off synchronized with the camera acquisition for the purpose of ambient light subtraction.

Here we describe the prototype and perform a laser line PL image of a 36 cells module with luminescence inducing light source and camera at one meter distance from the PV sample, corresponding to fully contactless PL scanning tool, developed until the moment for low natural light ambient condition. A fast scan concept is presented, taking 2 seconds to scan 2 cells and the final module PL montage is compared with an EL image of the same module. The results from the PL laser line scan demonstrate the possibility of a fast mirrored scanning to be constructed in the future.

2 EXPERIMENTAL

In the following described experiment, an InGaAs camera from Raptor Photonics, a 25 mm sapphire lens and a long-pass filter to cut out the laser emission were used. The sample is a module with 36 multi-crystalline silicon 15.6 cm cells arranged in a 6 x 6 matrix, mechanically stressed showing several cracked cells.

PL images under low natural light were acquired using as PL excitation source an 800 nm laser diode at two different optical power intensities at open circuit (Fig. 1). The camera

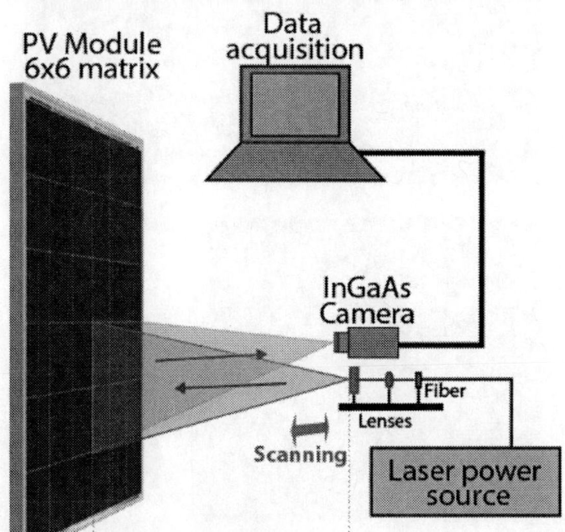

Figure 1: PL laser line scan setup.

was kept fixed in its position at 1.2 meter from the sample, comprising 4 cells in its field view. The laser beam coming from a fiber was shaped into a line using lenses with <5% of optical power loss. The laser fiber was placed at 1 m from the sample, with the line presenting 10 mm width and 360 mm length, with intensity of ~ 4 suns and covering 2 cells each scan. The line beam was mechanically scanned over the region of interest with the use of a hand crank, allowing linear scanning speeds. The scan over the field of view of the camera was performed in 4 seconds, and the images were acquired with a 28 ms exposure time. The image processing was performed using the software ImageJ 1.47t. The setup was built as a freely movable setup, since the goal is to have a line scan not attached to the installation structure, but mounted in movable source, as an UAV.

3 RESULTS

The main challenge towards the field application of PL generated by a line shaped light source without enclosures is the low signal to noise ratio in relation to ambient light, even lower than an regular EL signal. This challenge needs to be carefully considered, since the amount of information presented by PL images change with the excitation light source intensity [6]. The PL signal induced by low light intensities (in the order of few suns) presents higher PL signal in isolated or enhanced series resistance regions. In parallel, at such (relatively) low inducing light intensities, not only PL is observed by also contactless EL [4], [6], due to the solar cell's very efficient lateral current spreading. Consequently, PL imaging has the ability of not only provide data concerning cracks and crystallographic defects, but also broken contact fingers and enhanced resistance areas. These results have been shown previously both experimentally and by simulations [7], [8]. Good quality PL images reported are performed under 2.5 suns [8], [9].

The resulting module PL image shown in Fig. 2 and the corresponding EL image shown in Fig. 3. The inset in Fig. 2 shows a section acquired in a single scan, and the histogram of the PL and EL insets is also shown. The PL and EL are comparable in their level of detail in detecting mode A (micro-cracks), B and C cracks. Some artifacts in the PL image can be observed without a clearly correspondence with the EL image, possibly are due the scanning procedure still to be fully optimized.

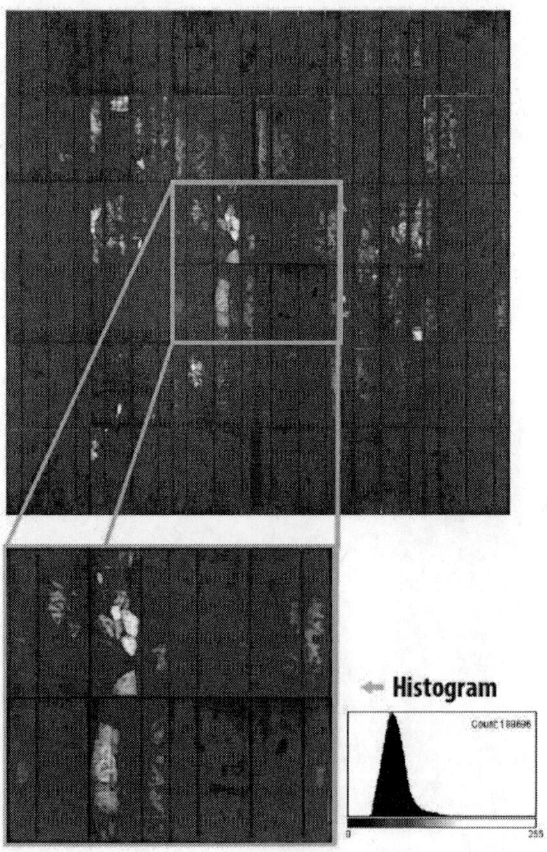

Figure 2: Laser line PL image. Inset shows zoomed in an area obtained with a single laser scan and its correspondent histogram.

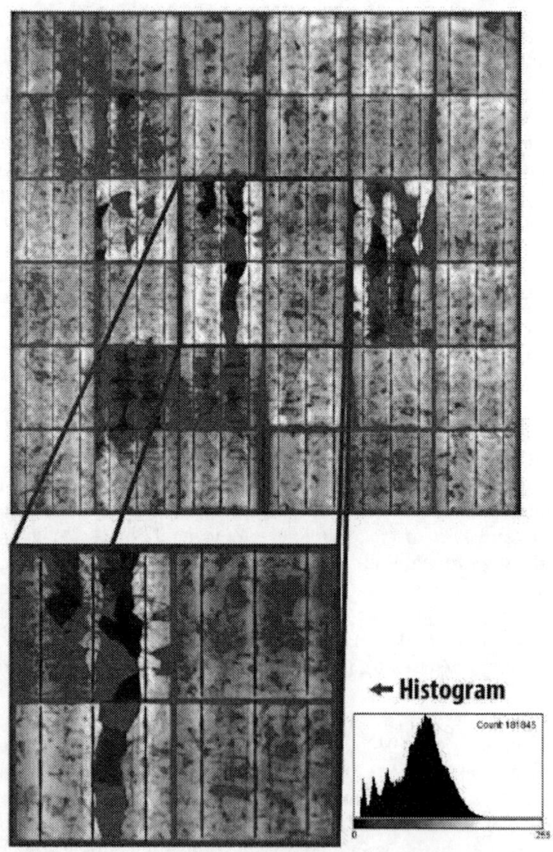

Figure 3: EL image. Inset shows zoomed in of the correspondent EL image from the laser line scan PL obtained with a single laser scan and its correspondent histogram.

The combination of the laser beam shape and the scanning speed provided good quality of images in relation to the laser intensity and exposure time of the image acquisition, showing cracks clearly identifiable and isolated regions appear highlighted. Compared with previous works [10], we have improved the scanning uniformity with a mechanic tool moved by a hand crank and defined the focal distance for the given line shape determined by the optics.

When the histograms of the inset luminescence images are compared, the higher intensity of the EL image allow us to observe the high contrast of the dark areas, evidenced in the high counts of low pixel numbers (see the 3 small peaks to the left of the EL histogram – inset Fig.3). Even though the highlighted isolated regions (cracks B and C) are clearly visible in the PL image, the low PL intensity does not provide enough contrast for efficient histogram analysis. Still, these results suggest important guidelines for the next steps of the development of this technology.

4 CONCLUSIONS AND OUTLOOK

In this work, laser beam shape and scanning uniformity (relative to the scanning speed) were defined to obtain good qualitative comparison of the PL image of a PV module in relation to its EL image. We presented PL images with laser line scan of a 6x6 matrix multi-crystalline PV module with the laser beam at a distance of 1 meter from the PV sample. Acquisition of PL images was feasible at such distance presenting well defined cracks and highlighted enhanced series resistance isolated regions (B and C cracks).

The lower luminescence intensity of the PL signal compared with EL still represents a challenge for outdoor applications. The continuation of this work consists of further development of the light source to overcome the low PL intensity.

At the same time, engineering efforts for a compact automatic scanning tool also need to be addressed for the final outdoor PL scan application for a full contactless luminescence imaging independent of sun irradiance conditions.

5 REFERENCES

[1] M. Köntges *et al.*, "Review of Failures of Photovoltaic Modules," *IEA-Photovoltaic Power Syst. Program.*, pp. 1–140, 2014.

[2] J. B. S. Koch, T. Weber, C. Sobottka, A. Fladung, P. Clemens, "Outdoor Electroluminescence Imaging of Crystalline Photovoltaic Modules: Comparative Study between Manual Ground-Level Inspections and Drone-Based Aerial Surveys," *32nd Eur. Photovolt. Sol. Energy Conf. Exhib.*, pp. 1736–1740, 2016.

[3] M. Bliss, X. Wu, K. G. Bedrich, J. W. Bowers, T. R. Betts, and R. Gottschalg, "Spatially and spectrally resolved electroluminescence measurement system for photovoltaic characterisation," *IET Renew. Power Gener.*, vol. 9, no. 5, pp. 446–452, 2015.

[4] S. Johnston, "Contactless electroluminescence imaging for cell and module characterization," *2015 IEEE 42nd Photovolt. Spec. Conf. PVSC 2015*, 2015.

[5] R. Bhoopathy, O. Kunz, M. Juhl, T. Trupke, and Z. Hameiri, "Outdoor photoluminescence imaging of photovoltaic modules with sunlight excitation," *Prog. Photovoltaics Res. Appl.*, no. July, pp. 14–16, 2017.

[6] J. Wong, "Griddler: Intelligent computer aided design of complex solar cell metallization patterns," *IEEE Photovolt. Spec. Conf.*, pp. 933–938, 2013.

[7] W. W. M. Kasemann, D. Grote, B. Walter, W. Kwapil, T. Trupke, Y. Augarten, R.A. Bardos, E. Pink, M.D. Abbott, "Luminescence Imaging for the Detection of Shunts on Silicon Solar Cells," *Prog. Photovolt Res. Appl.*, vol. 16, no. January 2008, pp. 297–305, 2008.

[8] I. Zafirovska, M. K. Juhl, J. W. Weber, O. Kunz, and T. Trupke, "Module inspection using line scanning photoluminescence imaging," *32nd Eur. Photovolt. Sol. Energy Conf. Exhib.*, 2016.

[9] I. Zafirovska, M. K. Juhl, and T. Trupke, "Efficient Detection of Finger Interruptions from Photoluminescence Images," *33th EUPVSEC*, 2017.

[10] G. A. dos Reis Benatto *et al.*, "Photoluminescence Imaging Induced by Laser Line Scan : Study for Outdoor Field Inspections," *IEEE 7th World Conf. Photovolt. Energy Convers.*, pp. 1–5, 2018.

KEY PERFORMANCE INDICATORS AND PV MODULE RELIABILITY

Gerhard Kleiss
KLEISS CONSULTING
Post Box 190184, 53037 Bonn, Germany info@kleiss-consulting.com

ABSTRACT: This work investigates typical Key Performance Indicators used in solar industry for assessment of the performance of solar modules and Balance-of-Systems (BOS) components. It is most important from the point-of-view of an investor in PV systems to have a comprehensive insight into the reliability of solar modules and inverters as those form a major capital expenditure of the project. Since reliability and longevity of such components is crucial for the long term financial success of the project, corresponding key performance indicators (KPI) are desired. To this end solar industry has traditionally provided KPI's for both solar modules as wells as inverters on the base of relating the number of failed (claimed) product to the number of installed (or sold) product. The denominator of this KPIs may be referred to as either related to the same period as the numerator or to a cumulated number of installations. In the frame of this study it is shown, that currently used KPIs are not sufficient when comparing the ability of solar manufacturers to produce reliable solar modules or inverters. Reference is made to a failure model based on a Weibull distribution. Analytical models based on the Weibull distribution function are given.

KEYWORDS: Failure Rates, Failure Model, Reliability, Solar Modules, Solar Inverters, Weibull Distribution

1 INTRODUCTION

The need for assessment of solar manufacturers' product quality is obvious, since both PV modules and inverters are representing the major part of a PV system's CAPEX. All manufacturers are advertising with high end quality statements. For the investor it is difficult to select the most reliable product. Once a manufacturer (or a distinguished product of the manufacturer) has been shortlisted by the customer an in-depth study may start: Often, manufacturers are audited and their product is scrutinized by numerous tests during production, prior to shipment or during field acceptance procedures. Prior to these intimate quality investigations general selection processes for manufacturer and/or products creates a need to assess and compare Key Performance Indicators KPI (often referred to as "failure rates"), Likewise, there are also many cases, when purchasing choices are made on such KPI, only. This work investigates three different KPI and their applicability to solar industry by taking into account the comparatively long operation times of PV products (typically 25 years for solar modules and 10 - 15 years for inverters):

(a) KPI_1 is defined as: fraction of "failed product within a distinct period Δ" (counter) and "modules shipped within this distinct period Δ" (denominator):

$$KPI_1(t) = \frac{\sum_\Delta product\ failed\ \ at\ t}{\sum_\Delta product\ shipped\ at\ t} \qquad \text{(eqn. 1)}$$

It may be noted that the sum during the time period Δ per (eqn. 1) could be expressed for PV modules both as "pieces modules shipped" and as "wattage of modules shipped". The difference is minor when larger time periods Δ are considered. Typically, Δ is chosen to be one year or one month.

(b) KPI_2 is defined by replacing the denominator of (eqn. 1) with the total historically shipped product:

$$KPI_2(t) = \frac{\sum_\Delta product\ failed\ at\ t}{\sum product\ shipped\ until\ t} \qquad \text{(eqn. 2)}$$

It may be noted that KPI_2 (as per (eqn. 2)) is today more frequently encountered in failure statistics of BOS components.

(c) KPI_3 replaces also the counter of (eqn. 2) with the sum over all claimed products until time t:

$$KPI_3(t) = \frac{\sum product\ failed\ until\ t}{\sum product\ shipped\ until\ t} \qquad \text{(eqn. 3)}$$

We want to note that generally the term "failed product" could mean "defective product" or "claimed product" or "returned product" (with each definition leading to different numbers). It is obvious that manufacturers do not use standardized nomenclature when publishing relevant statistics. In the context of this report we are using the idealized assumption that all manufacturers would use the same language to describe "failed product".

2 METHOD

2.1 Review of the General Failure Mode Model

A Weibull distribution function is chosen as general model for product failures: Solar products shall fail after a time x according to a parameter density function f:

$$f(x) = \frac{\alpha}{\beta} \cdot \left(\frac{x}{\beta}\right)^{\alpha-1} \cdot e^{-\left(\frac{x}{\beta}\right)^{\alpha}} \qquad \text{(eqn. 4)}$$

The two parameters contained in (eqn. 4) are called lifetime β and form factor α.

The cumulated failure function $F(x)$ for modules ("product" in general) is the integral of $f(x)$:

$$F(x) = \int_0^x f(x)dx = 1 - e^{-\left(\frac{x}{\beta}\right)^{\alpha}} \qquad \text{(eqn. 5)}$$

$F(x)$ approaches 1 for $x \to \infty$, reflecting that every product shall fail over time. The lifetime parameter β as defined above (eqn. 4) reveals per (eqn. 5) at time $x = \beta$:

$$F(\beta) = 1 - e^{-1} = 62{,}34\% \qquad \text{(eqn. 6)}$$

The hazard function $h(x)$ describes the general probability for PV modules to fail at time x under condition that there was no fail until time x:

$$h(x) = \frac{f(x)}{1-F(x)} \qquad \text{(eqn. 7)}$$

The schematic time evolution of $h(x)$ is also known as the so-called bathtub curve, where several sections can be distinguished with respect to the form factor α:

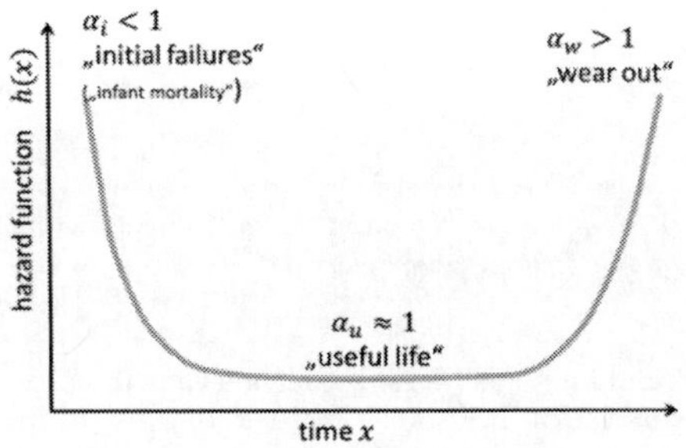

Fig. 1: Bathtub curve, schematically

The general model anticipates that every section of the bathtub curve can be expressed as to follow Weibull characteristics with a particular set of parameters lifetime β and form factor α. During the first phase ("initial failures") the model is treated in a particular way. Details are given in [1] and further expansion in [2]. This leads to the following formula:

$$\mathrm{F}(x) = 1 - q \cdot e^{-\left(\frac{x}{\beta_i}\right)^{\alpha_i} - \left(\frac{x}{\beta_u}\right)^{\alpha_u} - \left(\frac{x}{\beta_w}\right)^{\alpha_w}} - (1-q) \cdot e^{-\left(\frac{x}{\beta_u}\right)^{\alpha_u} - \left(\frac{x}{\beta_w}\right)^{\alpha_w}} \qquad \text{(eqn. 8)}$$

In (eqn. 8) the subscripts i, u, w refer to the 3 sections of the bathtub curve (Fig. 1): "initial (infant) mortality", "useful life" and "wear-out".

The first term of (eqn. 8), corresponding to the 3 exponential functions, models the impact of the failure modes responsible for "useful life" and "wear out" during the first "infant mortality" phase. In the context of this work, this is omitted and short-term failures are not considered.

Likewise, we adopt a further simplification, namely to neglect the "wear out phase". This means that all contributions to the failures are deducted from to (eqn. 4) and (eqn. 5) only. We are confident that this approach is appropriate as long as the wear-out failures are not yet looked at. Manufacturers are granting warranty terms only, if there is evidence that the product will not create massive claims during the typical 25 years term.

In conclusion, the model reduces to (eqn. 4) and (eqn. 5) as given above.

An alternative possibility to include the "wear-out-phase" has been suggested elsewhere [2]. In this case the cumulated parameter density reads:

$$F(x) = \begin{cases} x \leq \mathrm{T}: & 1 - e^{-\frac{x}{\beta_u}} \\ x > \mathrm{T}: & 1 - e^{-\left(\frac{x-\tau}{\beta_{w-\tau}}\right)^{\alpha_w}} \end{cases} \qquad \text{(eqn. 9)}$$

2.2 Generation Rate and "Useful Life" Period

Having stated the two approximations (as per (eqn. 5) and (eqn. 9)) it must furthermore be considered that the production rate of solar modules may vary over time. Let

$$g(t) \qquad \text{(eqn. 10)}$$

be the number of modules produced per time at time t (the "generation" rate). $g(t)$ is denoted in terms of "pieces produced per time interval" and different types of $g(t)$ are discussed in section 3.

A further assumption is chosen in the context of this study, namely that the production date relates to the sales date and – more important – to the date of installation ("fielding"), which can be considered as the first triggering point for ageing / degradation since the exposure to environmental conditions starts then. If statistics is looked at on a year-by-year resolution this is probably a valid statement as neither manufacturers nor retailers or installers intend to operate with a large working capital.

The restriction to the "Useful Life" phase of the complete model is given through (eqn. 4) and (eqn. 5) and can be expressed as to restrict the form factor α to unity:

$$\alpha = 1 \qquad \text{(eqn. 11)}$$

It is important to consider the history of failed modules: Product is reported to fail at time x and was installed in the past at time t. Consequently, (eqn. 4) can be expressed as:

$$f(x,t) = \begin{cases} \frac{1}{\beta} \cdot \mathrm{e}^{-\frac{x-t}{\beta}}, & x > t \\ 0\ , & x \leq t \end{cases} \qquad \text{(eqn. 12)}$$

At time x, the infinitesimal number δN of modules shall fail (these have been installed at time t, with $t < x$):

$$\delta N(x,t) = \delta x \cdot \frac{1}{\beta} \cdot \mathrm{e}^{-\frac{x-t}{\beta}} \cdot g(t) \qquad \text{(eqn. 13)}$$

δx in (eqn. 13) is an infinitesimal time interval and corresponds to the "distinct time interval" Δ introduced in (eqn. 1) and (eqn. 2) of section 1.

In addition, (eqn. 13) takes the possibly varying production rate $g(t)$ into account.

For the full number $\delta N(x)$ of failed modules between x and $x + \delta x$ all historic installation dates $t < x$ shall contribute:

$$\frac{\delta N(x)}{\delta x} = \int_{t=0}^{x} \left(\frac{1}{\beta} \cdot \mathrm{e}^{-\frac{x-t}{\beta}} \cdot g(t) \right) dt \qquad \text{(eqn.14)}$$

(eqn.14) is the numerator of KPI_1 and KPI_2. In consequence, (eqn. 1) through (eqn. 3) transform into:

$$\mathrm{KPI}_1 = \frac{\int_{t=0}^{x} \left(\frac{1}{\beta} \cdot \mathrm{e}^{-\frac{x-t}{\beta}} \cdot g(t) \right) dt}{g(x)} \qquad \text{(eqn. 15)}$$

$$\mathrm{KPI}_2 = \frac{\int_{t=0}^{x} \left(\frac{1}{\beta} \cdot \mathrm{e}^{-\frac{x-t}{\beta}} \cdot g(t) \right) dt}{\int_0^x g(t)\, dt} \qquad \text{(eqn. 16)}$$

$$\mathrm{KPI}_3 = \frac{\int_0^x \int_{t=0}^{x} \left(\frac{1}{\beta} \cdot \mathrm{e}^{-\frac{x-t}{\beta}} \cdot g(t) \right) dt\, dx}{\int_0^x g(t)\, dt} \qquad \text{(eqn. 17)}$$

3 RESULTS

3.1. "Useful Life" and Constant Production

The first approximation discussed here is the restriction to "useful life" and the choice of a constant production rate

$$g(t) = c = const. \qquad \text{(eqn. 18)}$$

This leads to:

$$\mathrm{KPI}_1 = 1 - e^{-\frac{x}{\beta}} \qquad \text{(eqn. 19)}$$

$$\mathrm{KPI}_2 = \frac{1 - e^{-\frac{x}{\beta}}}{x} \qquad \text{(eqn. 20)}$$

$$\mathrm{KPI}_3 = \frac{1}{x} \cdot \left(\beta \cdot e^{-\frac{x}{\beta}} + x - \beta\right) \qquad \text{(eqn. 21)}$$

Under these conditions, the KPIs depend on the ratio of the observation time x and the lifetime parameter β, only. Typically, when $x < \beta$ the KPIs deliver small numbers. However, it can already be seen here, that KPI_1 delivers less favorable (i.e. higher) numbers as x increases (prior to saturation). In other words: a manufacturer with a longer history will always have higher KPI_1 compared to a newcomer company, even if both of them have the same product quality level (expressed by the lifetime parameter β).

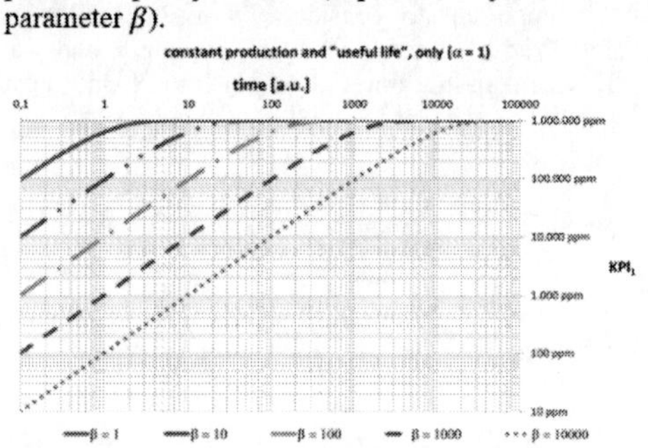

Fig. 2: KPI₁ in the "constant" case after elapsed time x for different lifetime parameters β

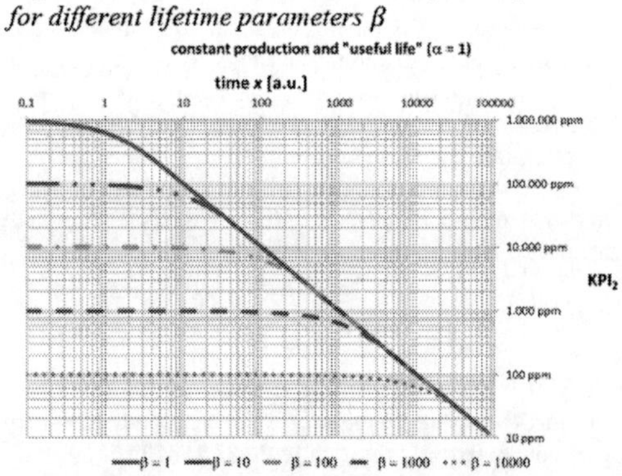

Fig. 3: KPI₂ in t in the "constant" case after elapsed time x for different lifetime parameters β

As can be easily the definition of the KPI can be difficult for the comparative quality assessment of manufacturers with constant production rate. Over time all KPIs will deliver equal results (1 or 0) and without knowledge of the time period since when a manufacturer is active shipping product to the market it is impossible to draw conclusions from these values.

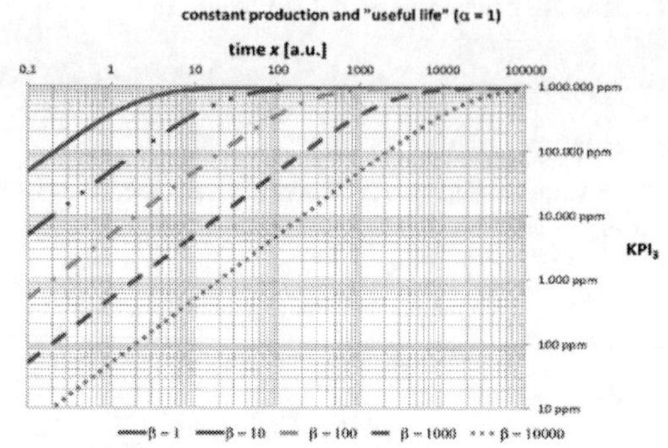

Fig. 4: KPI₃ in t in the "constant" case after elapsed time x for different lifetime parameters β

3.2. "Useful Life" and Linear Production Growth

In this variant it is anticipated that the "useful life" assumption ($\alpha = 0$) is still valid, but now it is assumed that the production rate is constantly growing:

$$g(t) = c \cdot t \qquad \text{(eqn. 22)}$$

In this case (eqn. 15) through (eqn. 17) convert to:

$$\mathrm{KPI}_1 = 1 - \frac{\beta}{x}\left(1 - e^{-\frac{x}{\beta}}\right) \qquad \text{(eqn. 23)}$$

$$\mathrm{KPI}_2 = \frac{2}{x} \cdot \left[1 - \frac{\beta}{x}\left(1 - e^{-\frac{x}{\beta}}\right)\right] \qquad \text{(eqn. 24)}$$

$$\mathrm{KPI}_3 = \frac{2}{x^2}\left(\frac{x^2 - 2\beta x + 2\beta^2}{2} - \beta^2 \cdot e^{-\frac{x}{\beta}}\right) \qquad \text{(eqn. 25)}$$

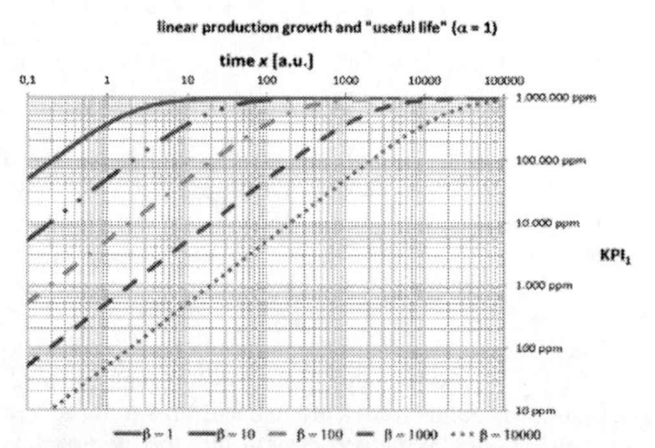

Fig. 5: KPI₁ in the "linear" case after elapsed time x for different lifetime parameters β

The general performance of the KPI is found to be very similar for the "constant" and the "linear" production rate case, whereas the analytical expressions are somewhat different.

In order to estimate typical values for the "useful life" rate, a magnitude of $\beta = 1000\,years$ can be considered ([3], [2], [4]) as a possible value. Please note, that in this case the units of x in the graphs are consequently also years. Again, the strong variation of all KPI's in Fig. 2 through Fig. 7 during the first 20 years shows the fact that no clear information can be drawn from the KPIs under linear or constant production.

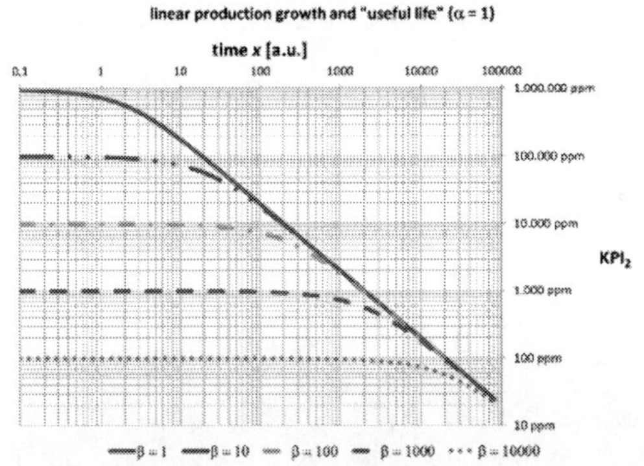

Fig. 6: KPI_2 in the "linear" case after elapsed time x for different lifetime parameters β

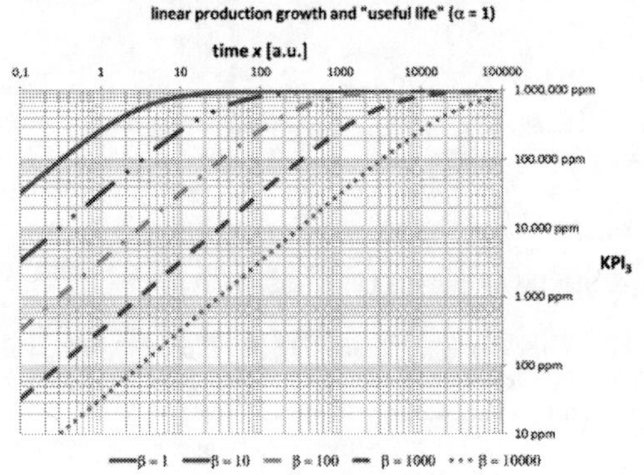

Fig. 7: KPI_3 in the "linear" case after elapsed time x for different lifetime parameters β

3.3. "Useful Life" and Exponential Growth of Production

The assumption of constant or linearly rising production rate is not a realistic scenario for PV industry. Therefore, an exponential growth production rate is considered:

$$g(t) = a \cdot e^{b \cdot t} \qquad \text{(eqn. 26)}$$

From world market data [5] for PV module shipments the factors a and c can be deducted as shown in Fig. 8:

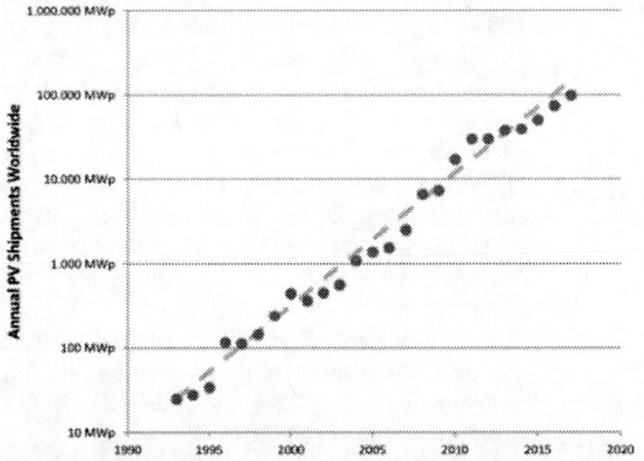

Fig. 8: Exponential growth rate for world wide PV shipments. The factor b=0.3617/year (as of (eqn. 26) was extracted.

Solving the core integral of (eqn. 14) delivers the numerator of the integral:

$$\frac{\delta N(x)}{\delta x} = \frac{a}{1+\beta \cdot b} \cdot \left[e^{x \cdot b} - e^{-\frac{x}{\beta}} \right] \qquad \text{(eqn. 27)}$$

In this scenario the KPI's time evolution is given by the following equations, both of them independent of a. Only the lifetime parameter β and the exponential growth rate b remain relevant.

$$KPI_1 = \frac{1}{1+\beta \cdot b} \cdot \left[1 - e^{-x \cdot \frac{1+\beta \cdot b}{\beta}} \right] \qquad \text{(eqn. 28)}$$

$$KPI_2 = \frac{b}{1+\beta \cdot b} \cdot \frac{e^{b \cdot x} - e^{-\frac{x}{\beta}}}{e^{b \cdot x} - 1} \qquad \text{(eqn. 29)}$$

$$KPI_3 = \frac{\frac{1}{\beta \cdot b+1}\left(e^{b x} + \beta\, b\, e^{-\frac{x}{\beta}} \right) - 1}{e^{b x} - 1} \qquad \text{(eqn. 30)}$$

When $b=0.3617/year$ as extracted is taken to draw the graphs for the KPIs we obtain Fig. 9 through Fig. 11:

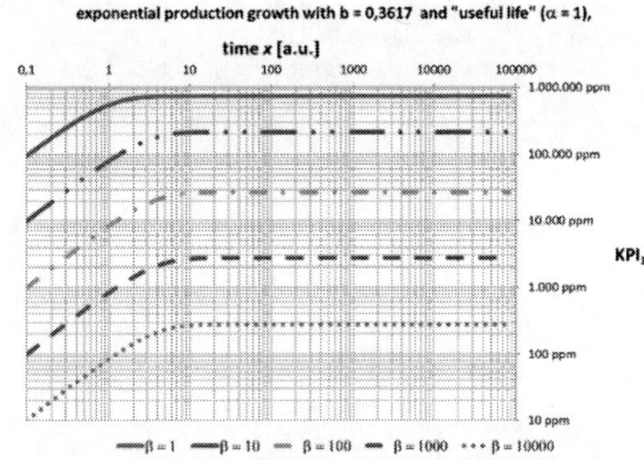

Fig. 9: KPI_1 in the "exponential" case with b=0.3617 after elapsed time x for different lifetime parameters β

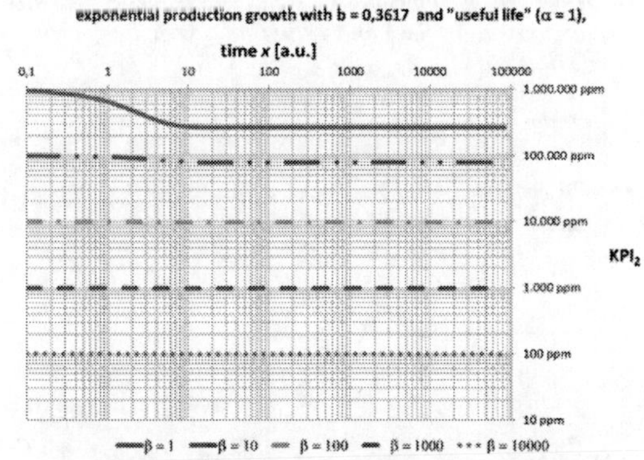

Fig. 10: KPI_2 in the "exponential" case with b=0.3617 after elapsed time x for different lifetime parameters β

As a first important result it is found that KPI_2 shows a flat time evolution for most quality levels (lifetimes β) in the case of the exponential production growth.

The levelling-off for KPI_1 and KPI_3 occurs earlier than in the other cases studied, but would still be not satisfying enough to draw conclusions for the quality of a manufacturer.

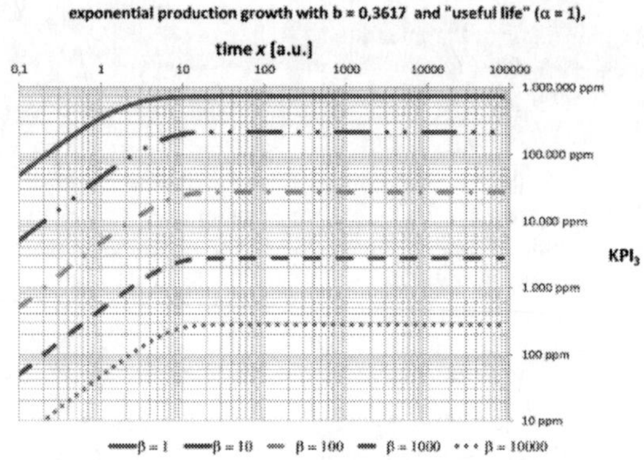

Fig. 11: KPI_3 in the "exponential" case with b=0.3617 after elapsed time x for different lifetime parameters β

3.4. Test of the Exponential Model Against Actuals

It is instructive to study the KPI's of a hypothetical company that existed during the time period given in Fig. 8 and experienced exactly the same growth as the world market. The theoretical (analytical) models are shown in the following graphs in comparison to the actual data based on a year-by-year calculation.

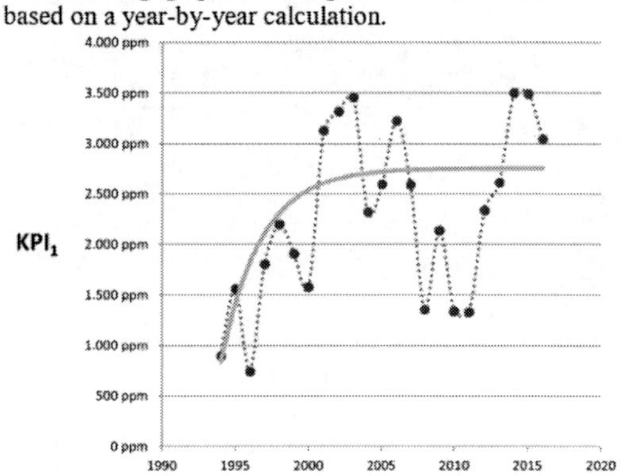

Fig. 12: KPI_1 in the "exponential" case with b=0.3617 with lifetime parameters β=1000 years: theoretical line (orange, straight line) and values on a year-by-year base (blue dots).

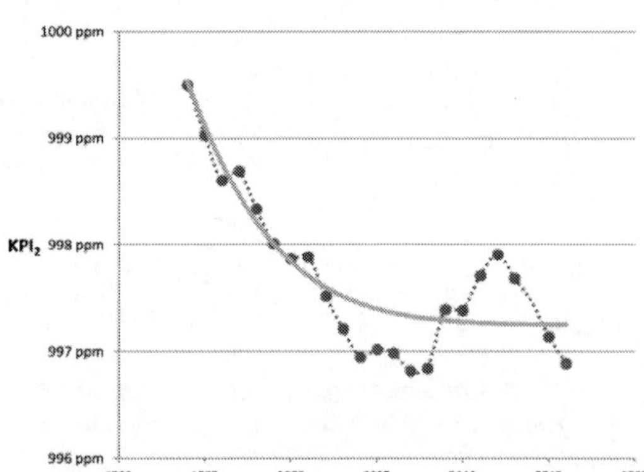

Fig. 13: KPI_2 in the "exponential" case with b=0.3617 with lifetime parameters β=1000 years: theoretical line (orange, straight line) and values on a year-by-year base (blue dots).

Again, it is found that the variability on production output in KPI_2 is extremely low in this exponential scenario, which makes KPI_2 a suitable tool to asses production quality: true differences in KPI_2 could be found when the quality level (i.e. β) will change.

KPI_1 and KPI_3 vary considerably, due to production quantities, only. Such scatter would mask any change in the quality level β.

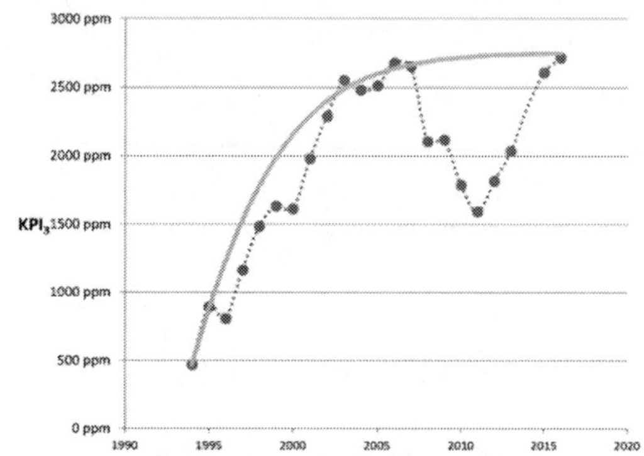

Fig. 14: KPI_3 in the "exponential" case with b=0.3617 with lifetime parameters β=1000 years: theoretical line (orange, straight line) and values on a year-by-year base (blue dots).

4 SUMMARY AND DISCUSSION

On the base of modelling the "useful" life with Weibull parameters several KPI have been evaluated in different scenarios of production growth.

- Only KPI_2 can deliver satisfying statements in the case of exponential growth
- KPI_1 and KPI_3 show high year-to-year variability
- Unlike KPI_2 in the exponential growth scenario, all other scenarios and KPI show little comparability for manufacturers with history less than 10 years

It should be added, that for a comprehensive evaluation of the quality level of a manufacturer also KPI_2 will represent only a limited viewpoint. It is important to survey on-going production [6] and not to look at historic data or one-time tests.

5 REFERENCES

[1] C. D. Tarum, *Classification and Analysis of Weibull Mixtures*, no. 724. Warrendale, PA 15096-0001 U.S.A.: The Engineering Society for Advancing Mobility Land Sea Air and Space, 1999.

[2] G. Kleiss, "Estimating Future Recycling Quantities of Pv Modules in the European Union," *32nd Eur. Photovolt. Sol. Energy Conf. Exhib.*, p. 53175, 2016.

[3] J. Wohlgemuth and D. Cunningham, "Using accelerated tests and field data to predict module reliability and lifetime," *23rd Eur. Photovolt. Sol. Energy Conf. Exhib.*, no. September, pp. 2663–2669, 2008.

[4] N. R. Wheeler *et al.*, "Degradation pathway models for photovoltaics module lifetime performance," *Conf. Rec. IEEE Photovolt. Spec. Conf.*, pp. 3185–3190, 2013.

[5] Solar Power Europe, *Global Market Outlook For Solar Power / 2015 - 2019*. 2015.

[6] B. Jaeckel *et al.*, "A New Standard for Holistic Quality Assurance," *Proc. 26th Eur. PV Sol. Energy Conf.*, pp. 3484–3490, 2011.

Assessments for distance through insulation (DTI) of PV encapsulant/backsheet materials

Hsin-Hsin Hsieh[1], Wen-Hsien Wang[2], Min-Tsung Kuan[2], Cheng-Chu Wang[2], Cho-Fan Hsieh[1]
[1] Center for Measurement Standards, Industrial Technology Research Institute, Hsinchu, 310, Taiwan ROC.
[2] Materials and Chemistry Laboratories, Industrial Technology Research Institute, Hsinchu, 310, Taiwan ROC.

The amended safety requirements of PV modules is proposed in the latest edition IEC 61730-1:2016. It is noted that the distance through insulation (DTI) test replaces partial discharge test due to its authenticity to the practical applications. The insulation layers containing encapsulants and backsheet is especially concerned for its safety characteristics in the increasing system voltage of > 1000 V. According to the IEC standard, the minimum thickness of the insulation should not be less than 150 μm for a PV system of less than 1000 V, while a thickness of > 300 μm is required for a PV system of < 1500 V. To study the actual thickness of insulation layer, encapsulant/backsheet laminates are carried out to evaluate the thickness. In the current research, we report the DTI values of 2 encapsulants and 3 backsheets laminates. A polyolefin-based encapuslant is also contained for its highly demanding in the glass-to-glass PV systems. Furthermore, the triple-layer backsheets are deployed for this study. This may facilitate the solar manufacturers to control or screening the quality as IQC. This paper delivers the preparation method of DTI samples, and their testing results are also revealed for discussion.
keywords：Distance through insulation (DTI); backsheet; encapsulant; IEC 61730-1:2016

1. Introduction

The rise of PV system voltage has drawn much attention due to the lower Levelised Cost of Electricity (LCoE) [1]. According to ITRPV report, the LCoE will generally decrease from (32.9~33.9) %; for example, the estimation for LCoEs are 0.073 USD/kWh and 0.049 USD/kWh for 1000 V and 1500 V, respectively. It is also reported that the system with max. voltage of 1500 V will be around 20 % in 2019, and significantly increase to 50 % in 2021. It is evident that the high system voltage of PV will become a mainstream in the near future. System voltage safety is related to the insulation properties of the package materials [2-3]. Partial discharge test of the backsheet material was the most common method to verify the insulation property of PV modules. It is supposed to help predict wear out of the insulation materials after long term exposure to electric fields [4]. However, the material test may not be consistence to the real application of PV module for two reasons: 1) there very few cases where the geometry in PV module resembles the test; 2) the AC test shows only few information to predict DC performance [5]. Thus, the partial discharge test may not fulfill the requirements of real PV installation.

Recently, the partial discharge test has been removed and the insulation thickness measurement was incorporated in the new edition of IEC 61730 [6]. The distance through insulation (DTI) is defined as the shortest distance between conductive components such as solder ribbon and the outer layer of the backsheet. According to the document, the thickness of thin layers is regulated; for example, the thickness of 150 μm and 300 μm for 1000 V and 1500 V system voltage, relatively.

Concerning the variety of the encapsulants and backsheets, the insulation properties of the polymeric materials are highly focused and presented [3,7]. In this paper, the DTI results of the commercial are presented to evaluate the feasibility of polymeric materials for solar module. And potential encapsulant Polyolefin (PO) is also compared in the current study.

2. Testing methods

The DTI measurement is processed as flow chart in Fig. 1. Sample is prepared as semi-products as shown in Fig. 2. The sandwich contains the most components except solar cell and upper EVA. And a conductive component illustrated as Tin strip is 800 μm diameter. The sample should be laminated with instructed process and then remove the release film and glass after the lamination. The top view of the DTI sample is shown in Fig. 3.

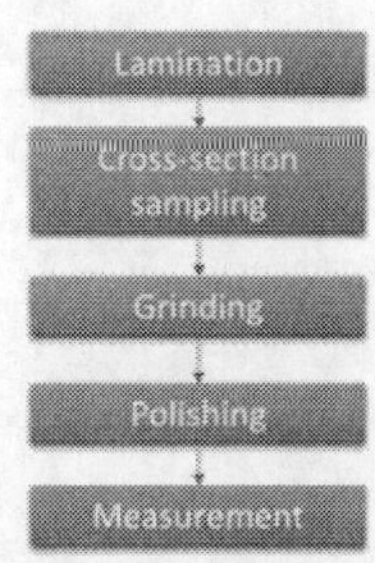

Fig. 1 Flow chart of the DTI measurement.

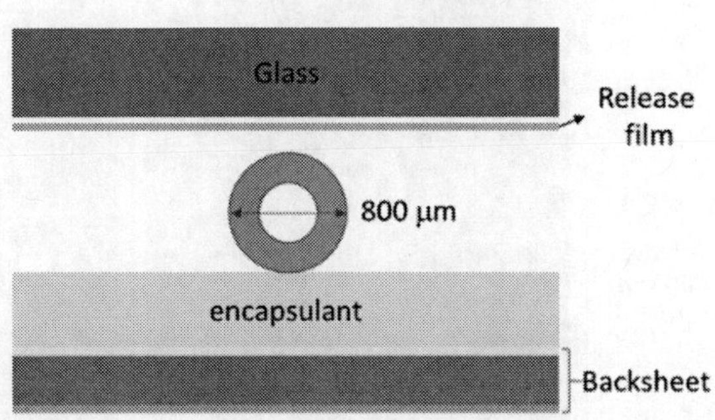

Fig. 2 Schematic of the encapsulant/backsheet laminate structure.

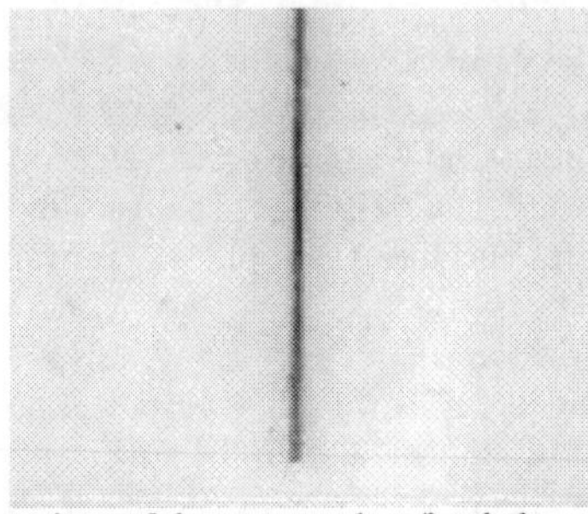

Fig. 3 Top view of the encapsulant/backsheet laminate.

3. Results

The original thickness of raw materials is revealed in Table 1. Commercial EVA is measured of 590 μm, and the polyolefin (POE) are developed in non-curing/curing types, in which the thickness are (300~750) μm and (300~620) μm, relatively. By applying the metallurgical process, the typical cross-sectional images of encapsulant/backsheet laminate are depicted in Fig. 4. According to the images, the tin strip were observed generally distorted due to the high pressure when lamination. The hole in the center of the tin strip could be neglected for its original manufacturing. And the Sn strip was embedded in the encapsulant and the thickness reduced after lamination. It is because encapsulants have a feature of rheology when heating up to the melting point of around 120°C. The thickness of the laminate sample reduced due to the lamination process. It is suggested that the more rheological the encapsulant (for example, POE), the thinner the insulation layer after lamination. On the other hand, POE of curing type (POE-C) showing a thicker layer after lamination should be due to the low rheology property of POE-C.

For the backsheet, the variety of the layer composition affects the results of the thickness of backsheet. For example, the backsheet containing EVA (E-layer) is measured thinner after lamination. It is because EVA encapsulant has a cross-linking reaction with E-layer of backsheet, thus reducing the thickness of backsheet. On the other hand, Tedlar (T-layer) and

Table 1. Thickness of raw materials.

Raw materials	Thickness (μm)
Encapsulant	
EVA	590
Polyolefin-non curing (POE)	300/400/500/750
Polyolefin-curing (POE-C)	300/400/500/620
Backsheet	
TPT	350
TPE	350
Acrylic	400

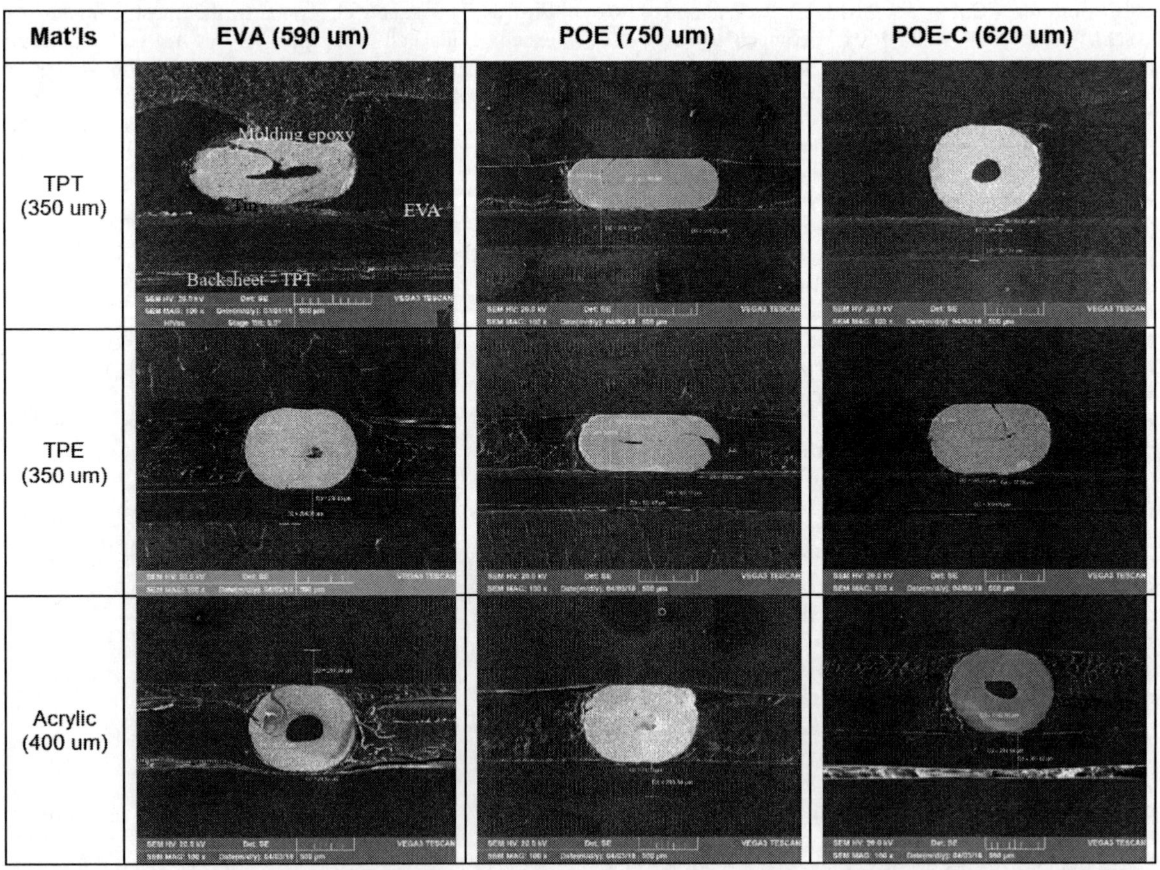

Fig 4. Cross-sectional images of polymeric encapsulant and backsheet laminate.

Table 2. DTI values of polymeric encapsulant and backsheet laminate.

Mat'ls (thickness)	EVA 590 um	POE				POE-C			
		300 um	400 um	500 um	750 um	300 um	400 um	500 um	620 um
TPT (350 um)	437.1	349.6	342.4	360.4	354.1	347.8	342.4	355.0	364.9
TPE (350 um)	294.1	223.5	267.6	221.7	321.7	173.0	188.2	196.0	345.2
Acrylic (400 um)	297.3	245.1	227.1	277.5	293.3	252.3	266.7	291.9	294.6

Polyester (P-layer) are thermodynamically stable < 150°C, thus remaining intact during the lamination process. And the acrylic backsheet contains an adherent paste which cross-links with encapsulant, thus a reducing thickness is detected. More DTI results of commercial and POE laminates are delivered in Table 2. In general, the DTI values are independent from the thickness of encapsulants. It is shown that the DTI values of TPT laminates were measured around 350 μm (±10 μm), which is close to its original thickness before lamination. And TPE laminates have thinnest DTI values when comparing to the other types of backsheet. As described above, it is due to the cross-link reaction between the E-layer and encapsulant. The acrylic laminates have a low variety of DTI values. It is mainly due to the stability of chemistry, yet the low DTI values less than 300 μm may fail the requirement of a 1500 V PV system. According to the measuring results, it is clear that the DTI values are determined by the layer content and interaction of materials, rather than the thickness of encapsulants and backsheets.

3. Conclusions

This paper illustrates the detailed DTI preparation of PV polymers in accordance with IEC 61730 standard. And the testing results of the DTI measurement are delivered to reveal the availability for high voltage application. According to the testing results, the DTI values are independent to the thickness of encapsulants, but related to the interaction between the materials and the layer content. In this research, the variety of the backsheet composition determines the thickness of insulation layer. The backsheet containing E-layer is measured thinner after lamination. It is because EVA encapsulant has a cross-linking reaction with E-layer of backsheet, thus reducing the thickness of backsheet. On the other hand, T-layer and P-layer are thermodynamically intact < 150°C of lamination temperature.

References

[1] International Technology Roadmap for Photovoltaic Eight Edition. 2017.

[2] B. Jaeckel, G. Volberg, J. Althaus, G. Kleiss, P. Seidel, M. Beck, A. Roth, "Safety of Photovoltaic Modules-Focus on Insulation Coordination," 33rd European Photovoltaic Solar Energy Conference, Amsterdam (2017).

[3] Jaeckel, B., et al, "Safety of Photovoltaic Modules –an Overview of the Significant Changes Resultingfrom Maintenance of IEC 61730 Series", 29th European Photovoltaic Solar Energy Conference, Amsterdam (2014)

[4] IEC 61730-1 Ed. 1 PV Module Safety Qualification - Part 1: Requirements for Construction (2004).

[5] John Pern, Module encapsulation, materials, processing and testing, APP International PV reliability workshop (2008).

[6] IEC 61730-1 Ed. 1 PV Module Safety Qualification - Part 1: Requirements for Construction (2016).

[7] Bengt Jaeckel, Guido Volberg, Joerg Althaus, Gerhard Kleiss, Peter Seidel, Markus Beck, and Arnd Roth, "Safety of photovoltaic modules- focus on insulation coordination," 32nd European Photovoltaic Solar Energy Conference, Amsterdam (2017).

[8] Marc Brandenburg and Marc Evertz, "Distance through insulation (DTI) – A must for the safety of backsheets" (2016)

Transient effects and internal series resistance calculation on bifacial silicon PV modules.

Juan Lopez-Garcia*, Diego Pavanello and Tony Sample

European Commission, DG JRC, Directorate C - Energy, Transport and Climate, Energy efficiency and Renewables Unit, Via Fermi 2749, 21027 Ispra (VA), Italy

*Tel: +39 0332783720, Fax: +39 0332789268, e-mail: juan.lopez-garcia@ec.europa.eu

ABSTRACT: The interest on bifacial c-Si PV modules has increased in the last years since they can increase the performance in comparison to traditional PV modules. However, one of the main issues to correctly measure bifacial modules is still a lack of standardisation. High-efficiency PV modules such as bifacial ones may have a significant internal capacitance which can result in I-V measurement artefacts due to transient effects when measured with short pulse durations in conventional pulsed solar simulators and lead to inaccurate output power measurements. Furthermore, the calibration of PV modules requires the measurement of their I-V characteristics at Standard Test Conditions (STC), which involves a curve translation, requiring among others the internal series resistance R_s of the module. R_s influences the module maximum power P_{max}, indicating in some way the quality of the device. This work analyses the capacitive effects on commercially available bifacial PV modules measured with conventional pulsed solar simulator, their influence on the output power accuracy and the variation of internal series resistance of the front and rear side of the module.

Keywords: Bifacial; crystalline Silicon; PV Modules; Capacitance.

1. INTRODUCTION

Bifacial crystalline Si photovoltaic (PV) modules market has grown significantly over the last years since they offer the opportunity for increased performance, in comparison to conventional monofacial PV modules, because both sides of the cells can absorb solar radiation, utilizing the scattered light from the ground and surroundings. To assess their performance and quality, PV modules are measured and characterized in a controlled environment under standard test conditions (STC) as defined by the International Electrotechnical Commission (IEC) [1]. Although the IEC Technical Committee 82 is currently working on the development of a draft technical specification [2], there is currently no standard for bifacial PV modules. One of the main issues that need to be addressed is how to measure bifacial PV modules correctly. Conventional pulsed light simulators are generally used in the industry and testing laboratories for the measurement of module performance due to the possibility to have very controlled environmental conditions and their ease of use. Moreover, class AAA solar simulators nowadays lead to a spectral mismatch factor close to unity for standard c-Si modules. However, high-efficiency PV modules such as bifacial ones may have a significant internal capacitance which can result in I-V measurement artefacts due to transient effects when measured with short pulse durations and lead to inaccurate output power measurements [3-7]. This capacitance produces hysteresis in the response of the modules under varying applied voltage or light intensity, and introduces charge storage effects which prevent rapid changes of voltage. As a consequence, the I-V measurement of high capacitance modules is influenced by the sweep-time and the sweep-direction effect [8]. Several methods have been proposed to measure high capacitance modules including steady-state simulators or simulator with very long pulse durations (>100-200 ms). These are offered by a few manufacturers only and are likely very expensive solution [8]. Conventional flash simulator (with pulse durations in the range 2 to 100 ms) may be used by applying particular approaches as the multi-flash (MF) [9], Dragon Back [4], Kopel method, hysteresis measurements or in general, sectional ones that require a larger number of flashes and are time-consuming imposing some constraints for module rating in the industry [5, 8, 10].

As mentioned before and in parallel to that, the calibration of PV modules requires the measurement of their I-V characteristics at STC. The correction requires among others the internal series resistance of the PV module. The series resistance R_s influences the maximum available power of a PV module, indicating in some way the quality of the device [11]. This work analyses the capacitive effects on commercially available bifacial PV modules measured with conventional pulsed solar simulators, their influence on the output power accuracy and the variation of internal series resistance according to IEC 60891[12]. One of the objectives of this work is to analyse, for a representative set of commercial modules, weather the minimum sweep durations to measure correctly the modules within the measurement uncertainty can be achieved by pulsed solar simulators.

2. EXPERIMENTAL APPROACH

2.1. Module characteristics

Commercially available bifacial crystalline Si modules (5 glass/glass and 1 glass/foil, both framed and frameless) from 5 different manufacturers representative of the market were characterized indoor at the European Solar Test Installation (ESTI). Modules consisted on 60 or 72 cells in three sub-strings and included several cell technologies such as PERC, p-PERC, n-PERT and HIT. In addition, two monofacial c-Si modules of 60 and 72 cells, respectively, were also measured as reference. The characteristics of the modules are listed in Table I. The modules were light soaked following the standard IEC 61215-2 [13] to eliminate possible light induced degradation (LID) effects.

2.2. Electrical performance.

I-V curve measurements were performed with a large-area long pulsed flash solar simulator (Spire SLP4600 with two Xenon linear lamps and pulse duration up to 100ms at 1000W/m²). In order to analyse the transient effects, the pulse duration was varied from 2 ms to 100ms at 1000W/m² as well as the sweep-direction (forward from I_{sc} to V_{oc} and backward from V_{oc} to I_{sc}) with stable temperature (within

±0.5°C during all the measurements). The electrical characterization of the modules was carried out with a non-reflective black wooden rear panel with irradiance on the non-exposed side <3W/m^2.

Table I PV module characteristics

Module	Frame	Partial self-shading on the rear	φ_{Pmax} (%)	N° Cells	Technology
A	Yes	Yes	58	60	PERC
B	Yes	Yes	66	60	n-PERT
C	Yes	No	86	72	HIT
D	No	No	97	60	PERC
E	No	Yes	76	60	n-PERT
F	No	No	68	60	p-PERC
G	Yes	-	-	60	Monofacial
H	Yes	-	-	72	Monofacial

As described in the IEC technical specification draft for bifacial modules IEC TS 60904-1-2 [2], bifaciality refers to the ratios between the main I-V characteristics of the rear and front side of a bifacial device, typically at Standard Test Conditions (STC). This is quantified with reference to bifaciality coefficients for the short-circuit current, φ_{Isc}, the open-circuit voltage, φ_{Voc} and the maximum power, φ_{Pmax}. This last one is defined as:

$$\varphi_{Pmax}(\%) = \frac{P_{\max Rear}}{P_{\max Front}} \times 100 \qquad (1)$$

where $P_{\max Rear}$ and $P_{\max Front}$ are the module output power measured when illuminating only the rear and the front side at STC, respectively, with a black rear panel. Prior to the I-V measurements at different sweep-times, the front and rear I-V curves under STC conditions were measured according to the technical specification IEC TS 60904-1-2 in order to obtain the bifaciality coefficients listed in Table I. For the purpose of this work, we focused on the P_{max} bifaciality coefficient.

2.3. Series resistance calculation

IEC 60891 ed. 2 (correction procedure 1) defines two equations for I–V curve translation from irradiance G_1 and temperature T_1 to irradiance G_2 and temperature T_2. For the determination of the series resistance R_S a number N of I–V curves is measured at different irradiances. Then (N − 1) curves are translated to the I–V curve with the highest irradiance (corresponding to the highest short circuit current Isc) using the translation equations and an initial value for the series resistance $R_S = 0$. By changing R_S in steps of 10 mΩ, the proper value of R_S is found when the deviation of the maximum output power of all translated I–V curves coincides within ±0.5%. The temperature of the PV device should be the same for all N irradiances at which the I–V curves are acquired; however, IEC 60891 allows a variation of ±2 °C. Since the series resistance is strongly dependant on the temperature [11], it was maintained stable within ±0.5°C. A detailed explanation of the R_S calculation method is described in [11].

The front and rear series resistances of the bifacial modules were calculated according to the standard IEC 60891[12], varying the irradiance levels in steps of 100W/m^2 from 200W/m^2 up to 1000W/m^2, in order to calculate the series resistance for different irradiance ranges.

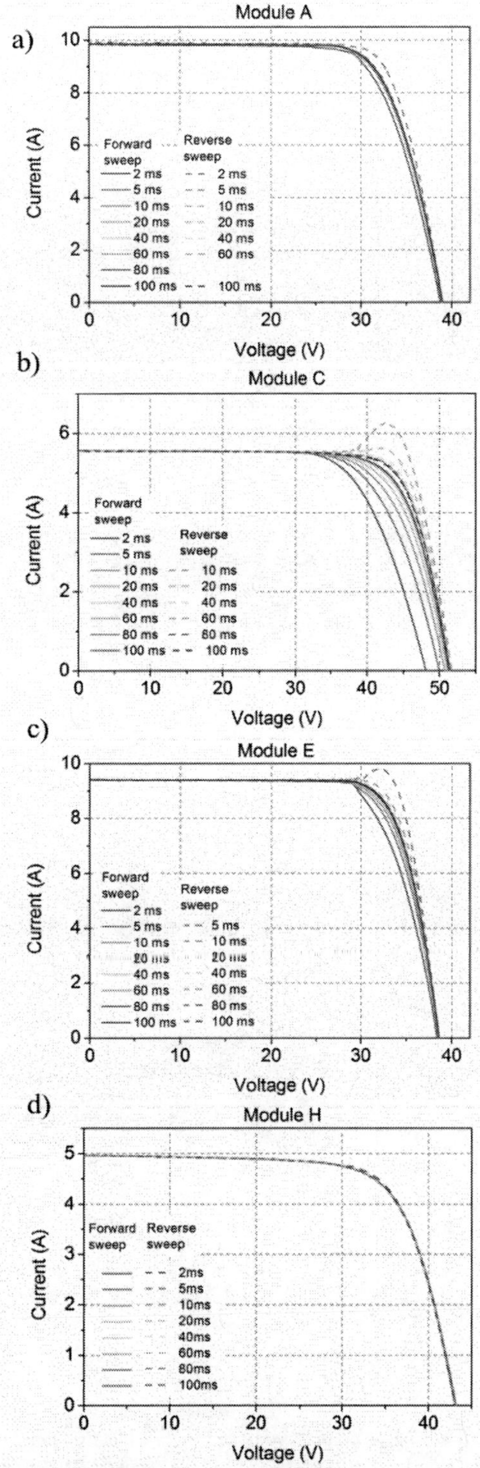

Figure 1 I-V curves measured in forward and reverse sweep for several sweep-times of representative a-c) bifacial and d) monofacial modules.

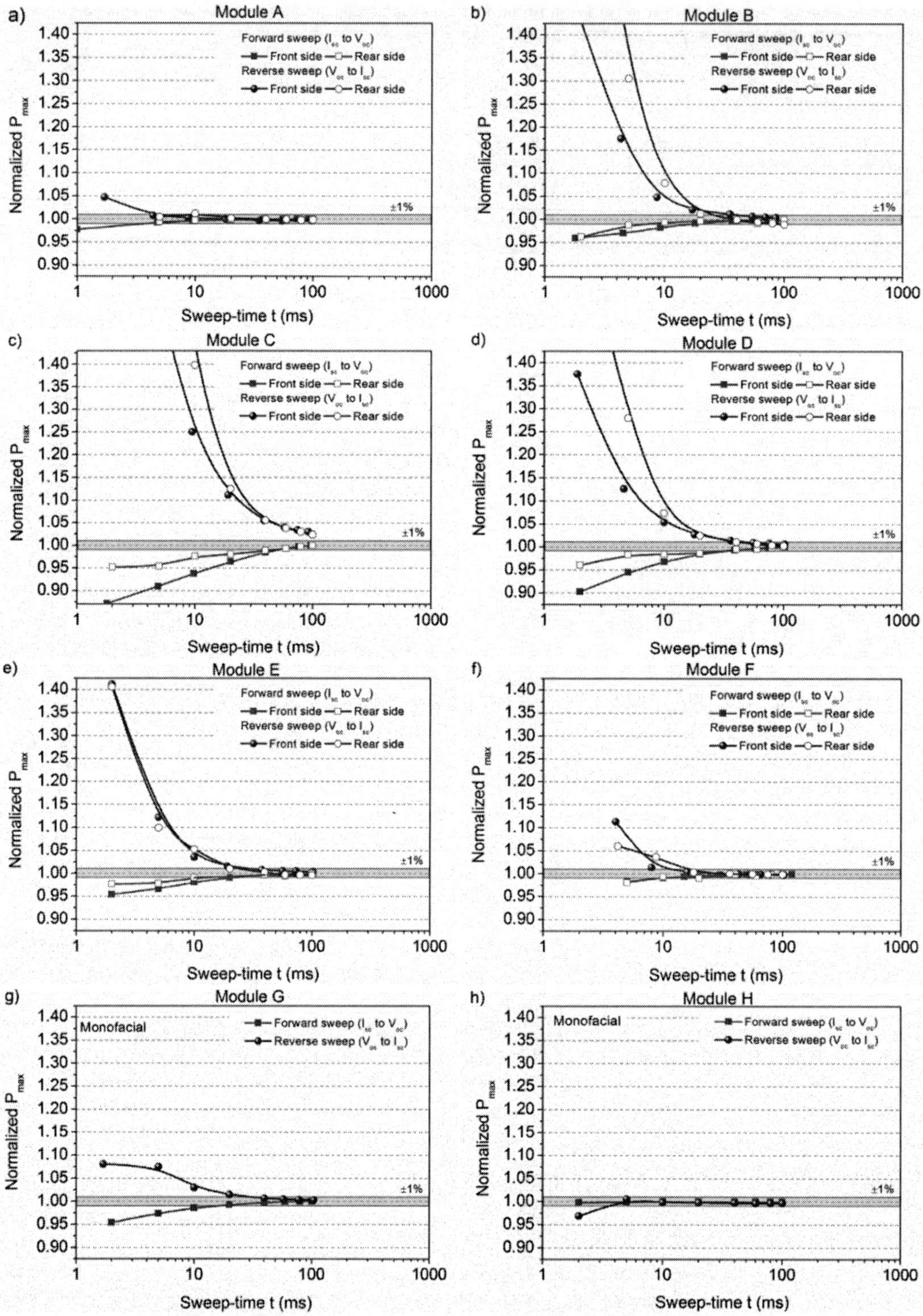

Figure 2 P_{max} normalized to maximum value as a function of the sweep-time of different a-f) bifacial and g-h) monofacial PV modules measured in forward and reverse sweep.

3. RESULTS AND DISCUSSION

3.1. Capacitive effects.

Figures 1a-c and figure 1d show the I-V curves of representative bifacial and monofacial c-Si PV modules, respectively, measured with a pulsed solar simulator with a single flash from I_{sc} to V_{oc} (Forward sweep) and from V_{oc} to I_{sc} (Reverse sweep) for several sweep-times. I-V curves were corrected to 1000W/m^2 and 25°C. A slight difference between forward and reverse sweep is observed for module A (figure 1a). However, module C, displayed in figure 1b, shows strong capacitive effects with very different performance for forward and reverse sweep. The reverse I-V curves present measurement artefacts in the maximum power point region (mpp) for short pulse times leading to an incorrect determination of the P_{max}. It is also observed the increase in the FF and V_{oc} (and then in the P_{max}) with the increase of the sweep-time for forward sweeps. Module E (figure 1c) represents an intermediate case between module A and module C, showing less extreme capacitive effects than in module C for intermediate sweep-times. This means that a longer sweep-time (conventional pulsed simulators are in the order of 10 ms) is needed for loading the cell capacitance and correctly evaluating the module maximum power of these high-efficient high-capacitive modules. The suitable sweep-time is achieved when both forward and reverse measurements have the same value within the repeatability error. Figure 1d shows the I-V curves in forward and reverse sweep for monofacial c-Si modules (module H) used as reference modules at ESTI. No capacitive effects due to sweep-time and sweep direction are observed. Most of the bifacial modules tested showed capacitive effects and a marked difference in the P_{max} and FF between the forward and reverse sweep for different pulse times. This mismatch decreased as the sweep-time increased from 2 ms up to 100ms.

Figures 2 a-h display the output power P_{max} normalized to the maximum as a function of the sweep-time measured in forward and reverse sweep for the front and rear side of the bifacial PV modules under test. The monofacial module measurements are also included as reference. We have set a ±1% deviation as threshold (grey area in the figure 2) for P_{max} based on the best ESTI P_{max} expanded uncertainty (U95% k=2) for full side modules in a pulsed solar simulator. Considering that both front and rear P_{max} are normalized to their respective maximum values, it should be noticed the differences between the front and rear side for the forward (and reverse) sweep in figures 1b to 1e. This indicates that the sweep-time needed for the convergence of forward and reverse curves can be different for the measurement of the front and the rear side of a bifacial PV module. From figure 1a and 1f (module A and module F, respectively) is observed that these two relatively high-efficiency modules are an exception and do not show capacitive effects for pulse duration of typical solar simulators of 10 ms or higher. Module E, B, G and D shows a convergence of the forward and reverse sweep curves for pulse durations of 20, 30, 30 and 40, respectively, within the threshold set. However, module C (HIT) shows a strong transient effect, so longer than 100 ms pulse duration is needed in order to measure the electrical characteristics corretly. For the monofacial module H, shorter pulses such as 4 ms are enough to the correct evaluation of the module maximum power.

Depending on the PV module technology and case by case, a shorter pulse solar simulator can be enough for the I-V measurement of the front and rear side. In general, it has been observed for the tested PV modules that a sweep-time below 40 ms could be sufficient to obtain similar values in forward and reverse sweep within the threshold set. For a specific module type, to achieve a high accuracy measurement, a preliminary analysis of the minimum sweep time needs to be done. That is, the difference from the real or

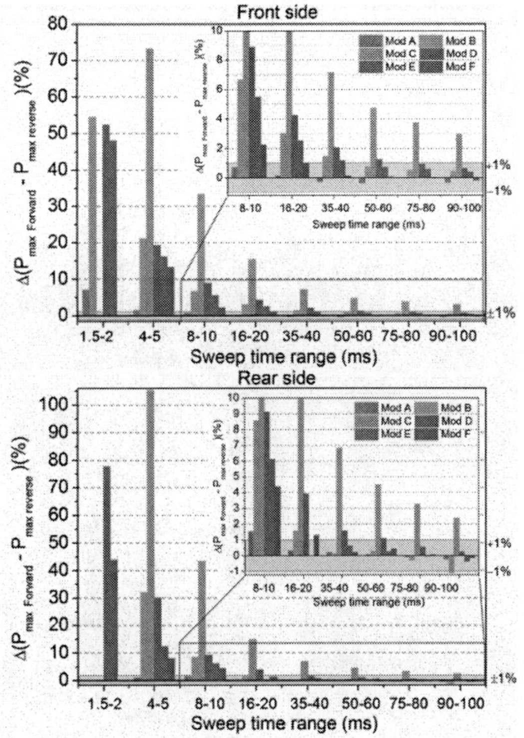

Figure 3 Variation between P_{max} measured from I_{sc} to V_{oc} (forward) and P_{max} from V_{oc} to I_{sc} (reverse) for the front side and rear side of bifacial modules under test. The P_{max} threshold of ±1% is included as grey area.

more accurate value within the uncertainty should be evaluated before using expensive or time-consuming characterisation method. For clarification, figure 3 summarizes the differences between the P_{max} measured under forward and reverse sweep for measurement of the front side and rear side of the bifacial modules.

3.2. Series resistance R_S measurements.

The series resistance values have been object of interest in the present work because it enters into play in the translation of measured I-V curves to different irradiance and temperature values. Although curves measured in indoor controlled environment in most cases do not need R_S for an accurate correction (because laboratory conditions of irradiance and temperature are close to target conditions), this is not often the case for outdoor measurements. The question that arises is whether a single set of equations (used for traditional monofacial modules) can be used also for bifacial modules, using the same parameters for both faces. The results have shown than the bifacial PV module design plays an important role in the R_S calculation and then in the procedures for temperature and irradiance correction since different R_S values for the front and rear side would imply a double set of correction parameters. However, it was recently reported that a same set of temperature coefficients for the front and rear can be used in bifacial modules without a relevant difference between sides [14].

Figure 4 shows the optimum value of the series resistances as a function of the irradiance range for relevant irradiance intervals and for the front and rear side of the modules under test. It has been determined previously that the uncertainty in the R_S calculation is ±10% at ESTI [11].

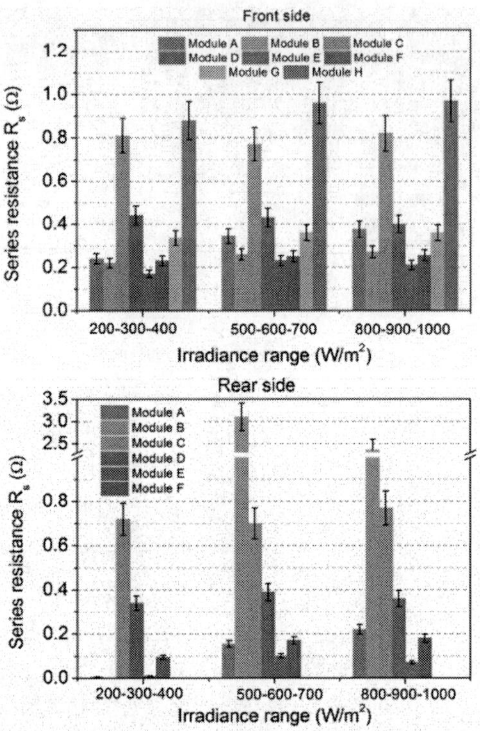

Figure 4 Series resistances R_S for the front and rear side of the modules under test for different irradiance intervals.

For the front side measurements, it is observed that, in general, the R_S is not dependant on the irradiance range. However, for specific cases such as modules A, B and E, the difference between R_S at low irradiance and at higher irradiance intervals was significantly out of the measurement uncertainty. Most of bifacial modules tested (and the monofacial module G) showed low R_S values between 0.18Ω and 0.42Ω, with the exception of module C (HIT technology). Module C and module H (monofacial) showed marked higher R_S around 0.8Ω and 0.88-0.98Ω, respectively. It is observed that the lower the R_S, the higher the module efficiency. It should be also observed the differences in R_S between cell technologies and within the same cell technology.

Figure 4 also shows the optimum series resistance measured on the rear side of bifacial PV modules under test. It should be noted that the design of module, in particular, the junction box location and frame "self-shading" have a strong influence on the I-V curves of the rear side. Partial shading of the rear cells due to covering by the lips of the frame or junction boxes lead to non-uniform rear irradiance and kinks in the rear I-V curves leading to an incorrect or in same case to the impossibility to calculate the series resistance on the rear side. Modules A, B and E show up to three marked kinks in the I-V curves of the rear side due to the partial shading of the cells by the junction box (all three cases) and the frame (modules A and B) which reduce the rear I_{mpp} and P_{max} [14] and lead to an incorrect fitting of the curves for the calculation of the series resistance. The R_S of the rear side of partially shaded modules did not fulfil the ±0.5% deviation of P_{max} according to the standard and disagree with the front side value of R_S. This is the case of modules A, B and E, where it was not possible to calculate the R_S value or not reliable results were obtained. In the other side, in general, a difference Δ R_S for the front and rear side R_S below 10 % was obtained for modules with a good design without partial rear shading (modules C, D and F).

4. CONCLUSIONS

Bifacial crystalline silicon PV modules from different manufacturers were characterised indoor with a pulse solar simulator at different sweep times in forward and reverse sweep. Care should be taken when labelling high-efficiency bifacial PV module since the characterisation method and simulator plays an important role in the accuracy of the maximum output power. Even if most of the current bifacial PV modules present capacitive effects at very short pulse durations, depending on the PV module technology and case by case, a relatively short sweep time (<40 ms) can be enough for the I-V measurement of the front and rear side within a 1% difference between forward and reverse sweep. If this threshold can be relaxed to 2% for example, most of the bifacial modules under test can be tested with a conventional pulsed solar simulator.

The series resistance of the bifacial modules were calculated from measurements of the front and rear side. R_S is not, in general, dependant on the irradiance range. There are not significant differences between the R_S of the front and rear side for a particular bifacial technology and it is, however, strongly dependant on the module design and rear irradiance mismatch. This implies that different set of R_S parameters are not necessary for the front and rear side if the modules design is appropriated without partial self-shading on the rear.

REFERENCES

[1] IEC 61836, "Solar photovoltaic energy systems – Terms, definitions and symbols", 2016.

[2] IEC DTS 60904-1-2, "Measurement of current-voltage characteristics of bifacial photovoltaic (PV) devices", 2018.

[3] G. Friesen, H.A. Ossenbrink, Sol. Energy Mater. Sol. Cells, 48 (1997) 77-83.

[4] N. Ferretti, Y. Pelet, J. Berghold, V. Fakhfouri, P. Grunow, 28th EUPVSEC, Paris, 2013, pp. 3184 - 3187.

[5] A. Virtuani, G. Rigamonti, P. Beljean, G. Friesen, M. Pravettoni, D. Chianese, 38th IEEE Photovoltaic Specialists Conference, Austin, TX, 2012, pp. 496-500.

[6] H.A. Ossenbrink, W. Zaaiman, J. Bishop, 23rd IEEE Photovoltaic Specialists Conference, Louisville, KY, 1993, pp. 1194-1196.

[7] A.S. Eeles, R. Gottschalg, T.R. Betts, 43rd IEEE Photovoltaic Specialists Conference, Portland, OR, 2016, pp. 3693-3697.

[8] C. Monokroussos, D. Etienne, K. Morita, C. Dreier, U. Therhaag, W. Hermann, 27th EUPVSEC, Frankfurt, 2012, pp. 3687-3692.

[9] J. Lopez-Garcia, B. Haile, D. Pavanello, A. Pozza, T. Sample, 32nd EUPVSEC, Munich, 2016, pp. 1724 - 1729.

[10] K. Ramspeck, S. Schenk, L. Komp, A. Metz, M. Meixner, 29th EUPVSEC, Amsterdam, 2014, pp. 1253-1256.

[11] G. Trentadue, D. Pavanello, E. Salis, M. Field, H. Müllejans, Meas. Sci. Technol., 27 (2016).

[12] IEC 60891 ed. 2, "Photovoltaic devices: Procedures for temperature and irradiance corrections to measured I-V characteristics", 2009.

[13]IEC 61215-2, "Crystalline silicon terrestrial photovoltaic (PV) modules: design qualification and type approval", 2016.

[14] J. Lopez-Garcia, D. Pavanello, T. Sample, IEEE J. Photovoltaics, 8 (2018) 960-968.

PERFORMANCE OF POLYMERIC PV BACKSHEETS IN DIFFERENT UV TESTS

Jürgen Jung[1] and Shigeo Suga[2]
[1]Agfa-Gevaert NV, Septestraat 27, 2640 Mortsel, Belgium – juergen.jung@agfa.com
[2]Suga Test Instruments Co., 6-10-2, Shinjuku, Shinjuku-ku, Tokyo, 160-0022, Japan

ABSTRACT: The safety requirements of polymeric PV backsheet include long term durability under environmental conditions, for which the daylight filtered Xenon based UV weathering test (IEC TS 62788-7-2 A3) has been defined as new dedicated test methods, that also considers elevated chamber temperatures and moisture levels regarded typical for application. Pass-/fail criteria are still under discussion as of IEC TC82 WG2 and in this transition period, legacy and ad-hoc UV tests are still in use all over the marketplace, including UVA340, UVB313 and high intensity metal halide (HIUV-M) exposure. This study benchmarks those four UV weathering tests using a selection of 6 polymeric backsheet films with different design. Typical post evaluation analysis, including retention of mechanical properties, discoloration (yellowing) and retained solar reflectivity are conducted and results are set into perspective in view of different backsheet designs.

Keywords: Qualification and Testing, Durability, Polymeric Backsheet, UV Weathering

1 INTRODUCTION

One of the important safety requirements for polymeric PV backsheet is the long term durability under environmental conditions, including UV and elevated levels of temperature and moisture. For laboratory testing, a daylight filtered Xenon UV weathering test has been defined by IEC TS 62788-7-2:2017 that consequently also considers elevated chamber temperatures and moisture levels regarded typical for backsheet in PV modules. However, legacy and ad-hoc UV tests are still in use in the marketplace, including UVA340, UVB313 and high intensity metal halide exposure. This study aims at benchmarking those three tests in comparison to the new reference UV Xenon test. For that benchmark a selection of 6 polymeric backsheet films with different design were exposed under these four different conditions and results of the typical post evaluation analysis, including retention of mechanical properties, discoloration (yellowing) and retained solar reflectivity are discussed.

2 EXPERIMENTAL

2.1 Materials

A set of six marketable polymeric backsheets for backside lamination of PV modules has been selected for a benchmark of different UV tests (see 2.2), including laminated, coextruded and/or coated backsheet designs TPE, KPE, PPE, PO, FCC as well as the single-layer PET based backsheet UniQoat of Agfa-Gevaert NV. This study is focused on fresh films and the change of their intrinsic properties in UV weathering, acknowledging that other failure modes of backsheets in context of encapsulation materials, e.g. adhesion, can only be observed in laminated engineering samples, so-called coupons [1, 2].

2.2 UV weathering tests

Replicates of the BS films were exposed from either side, (outer) air-facing layer and (inner) cell-facing layer towards light source, to a cumulative UV dose of 162 kWh/m^2. This dose is defined by IEC TS 62788-7-2 as representative for ½ of the lifetime dose that a backsheet would typically undergo in 10% Albedo condition in Arizona desert. In case of UVB313 with extreme content of artificial UVB, the exposure was limited to 33 kWh/m^2.

i. UV Xenon: The reference UV weathering test for polymeric components in PV modules has been defined by IEC TS 62788-7-2 [1] based on daylight filtered Xenon exposure, with the A3 condition being recommended for testing of white PV backsheets (see table I).

ii. UVA340: This still is the standard UV test on PV module level as of IEC 61215-1:2016, and in lack of a specific test has been used for component testing, too. IEC TS 62788-7-2 and -2 provide more background [1, 2].

iii. UVB313: This represents an outdated legacy test from the time before ed 1 of IEC 61215:2004: extreme harsh test conditions are based on artificial UVB radiation, which is not observed on earth surface.

iv. HIUV-M: A high intensity UV test using filtered Metal Halide with > 10x UV intensity compared to UV Xenon is occasionally used as fast screening test of polymeric materials.

Details of the UV test methods are given in Table I and in figure 1, the spectral power distributions (spd) are presented in absolute irradiance units [W/m^2/nm].

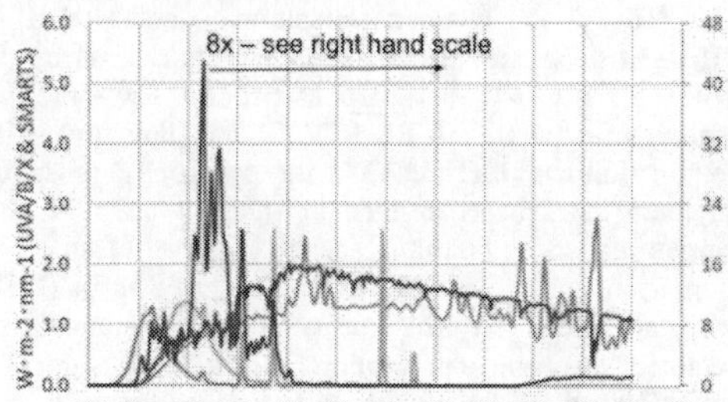

Figure 1: spectral power distribution of the 4 UV weathering test in comparison to standardized natural daylight (SMARTS 2.9.5, CIE Publ. 85 Table 4).

For discussion of the photolytic action of UV radiation on polymeric material, a plot of these spectra, normalized to their UV content in the range 300 nm – 400 nm is useful (270 nm – 400 nm in case of UVB313): Figure 2. clarifies that from the four test methods in this study, daylight filtered Xenon provides by far the best spectral match to natural daylight (represented by SMARTS 2.9.5), whereas the other three contain relative portions of UVB non-representative of natural daylight, i.e. potentially causing "artificial" degradation (e.g. extreme yellowing).

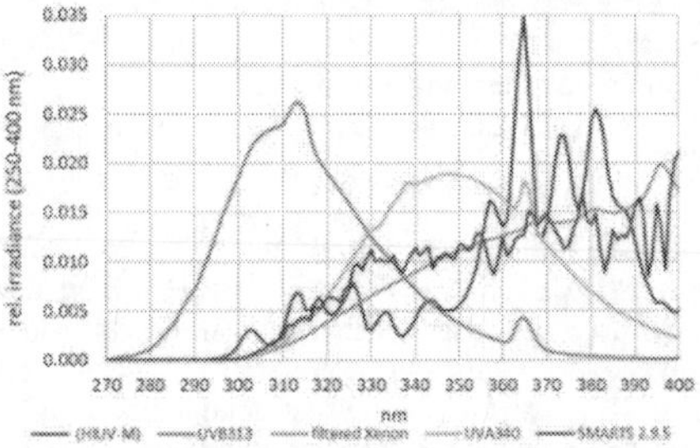

Figure 2: spectral power distribution (spd) of the 4 UV weathering test in comparison to standardized natural daylight (SMARTS 2.9.5, CIE Publ. 85 Table 4).

In addition to spd, also irradiance level, chamber air temperature and humidity are important parameters, as these determine the balance of UV degradation to additional failure pathways, such as (photo-)hydrolysis.

Test	UV Xenon	UVA340	UVB313	HIUV-M
Reference	IEC 62788-2	IEC 61215-2 ed 2	IEC 61215-2 ed 1	
Apparatus	Suga GX-75	Atlas UV tester	Suga FUV	Suga MST
Lamp	7.5 kW Xenon	UVA340 Fluorescent	UVB313 Fluorescent	6kW Metal Halide lamp
Inner filter	Quartz	None	None	Quartz
Outer filter	#295	None	None	#500
Irradiance level	81 $W \cdot m^{-2}$	1.31 $W \cdot m^{-2} \cdot nm^{-1}$	0.71 $W \cdot m^{-2} \cdot nm^{-1}$	1240 $W \cdot m^{-2}$ (> 20 SUN)
wavelength range control	300 - 400 nm	340 nm	310 nm	300 - 400 nm
Sampling	500 / 1000 / 1500 / 2000 h	275 / 550 / 1110 / 2220 h	100 / 200 / 400 / 800 h	16 / 32 / 66 / 132 h
Total radiant exposure	162 $kWh \cdot m^{-2}$	162 $kWh \cdot m^{-2}$	33 $kWh \cdot m^{-2}$	162 $kWh \cdot m^{-2}$
CHT	65 °C	not controlled (est. 63 °C)	not controlled (est. 63 °C)	65 °C
BPT	85 °C	60 °C	60 °C	not controlled (est. 90 °C)
Humidity	20 %RH	no control ("dry", ca 6%)	no control ("dry", ca 6%)	20 %RH

Table I: details of the test conditions of the four UV weathering tests in this study.

2.3 Post evaluation analysis

After UV test the backsheets were analyzed as follows:

1. Visual inspection for eventual cracking, blistering, chalking, delamination is performed under 1000 lux at 30 cm viewing distance.

2. Retention of mechanical properties was determined by tensile testing: (a) retained tensile strength ($rTS = TS_{aged}/TS_{fresh}$) and retained elongation to break ($rEtB = EtB_{aged}/EtB_{fresh}$) are discussed as pass/fail criterion in context of IEC 61730-1:2016 as proof the weathering stability of a polymeric BS (UV, T and moisture). The method follows ISO 527-3 with specific details for polymeric backsheets defined in IEC TS 62788-2 [2]: gripper distance 50 mm, speed of load frame of 50 mm/min and average results from test of 5 strips (MD) were analyzed. Typical measurement tolerance for "retained" mechanical properties, i.e. ratio of measured TS and EtB aged vs. fresh slightly depend on material, but typically in the order of $\sigma_{rTS} \approx 10\%$, and $\sigma_{rEtB} \approx 20\%$.

3. Solar reflectivity of the cell side of the backsheet contributes to the power output of the PV module. Retained total hemispherical reflectivity rSolarRefl = R_{aged}/R_{fresh} is calculated from the photon-weighted reflectivity 300 nm – 1250 nm IEC TS 62788-2 [2], with R measured before and after UV ageing, using a double beam spectro-photometer (Perkin Elmer, Lambda 950) in 8°/d measurement geometry.

4. Discoloration is characterized by measured change of the Yellowing index YI before and after ageing, using a spectro-colorimeter (Konica-Minolta CM-5) with d/8° geometry and following the definitions for $\Delta YI = \Delta YI_{aged} - \Delta YI_{fresh}$ in IEC TS 62788-2 [2].

3 RESULTS

Fig. 3 and 4 show the results after exposure of (inner) cell side and (outer) air side, respectively.

3.1 Cell side exposure

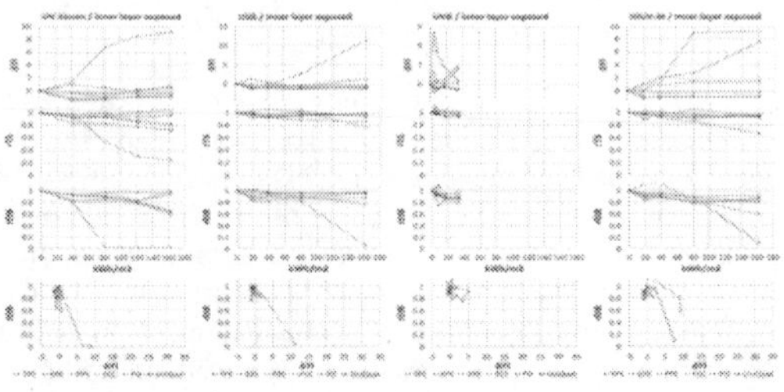

Figure 3: [row 1 – 3 from top to bottom] ΔYI, rTS and rEtB as function of exposure dose from (inner) cell-side for [from left to right] UV Xenon, UVA340, UVB313 and HIUV-M. [4th row] "correlation plot" rEtB vs ΔYI.

3.2 Air side exposure

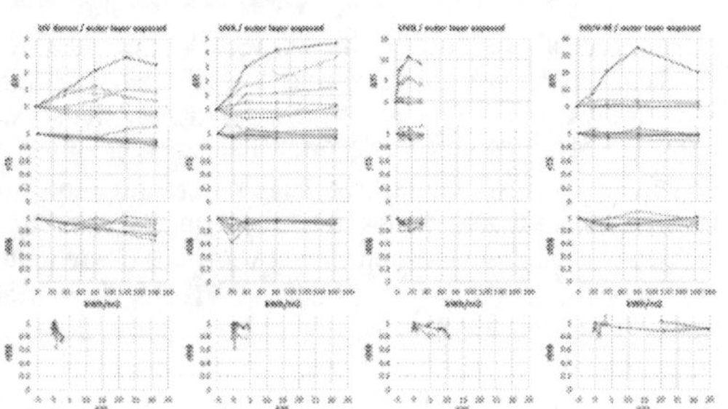

Figure 4: [row 1 – 3 from top to bottom] ΔYI, rTS and rEtB as function of exposure dose from (outer) air-side for [from left to right] UV Xenon, UVA340, UVB313 and HIUV-M. [4th row] "correlation plot" rEtB vs ΔYI.

3.3 Pooled "correlation-plots" across all UV exposures

In fig. 5 the results from fig. 3 and fig. 4 have been pooled across UV exposure method and inner / outer layer exposure. For some of the materials tested, correlations of varying degree are found between yellowing and change of mechanical properties (fig. 5), that however differ for the various UV exposure methods tested. In several cases, yellowing provides a false negative (especially for PET based materials in UVB313 and HIUV-M) or does not flag an eventual change in mechanical properties (UV Xenon).

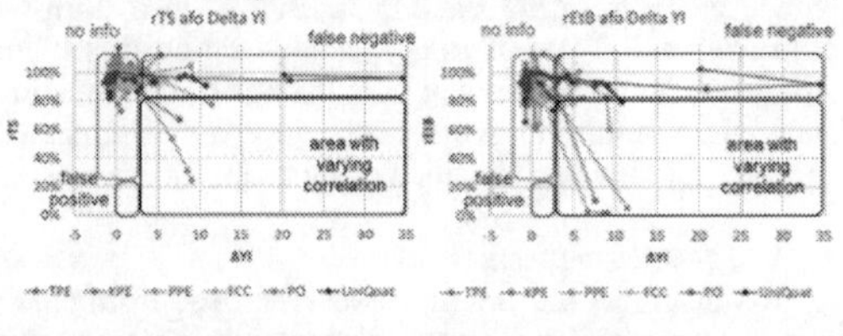

Figure 5: pooled plots (cell-side and air-side exposure UV Xenon, UVA340, UVB313 and HIUV-M) for a) [top] rTS vs. ΔYI and b) [bottom] rEtB vs. ΔYI.

3.4 Visual inspection

Visual inspection of discoloration is in line with measurement. In addition, cracking of E-layer is observed for PPE in direct exposure of cell side in UV-Xenon and HIUV-M, both from 81 kWh/m^2 on. No cracking of E-layer is observed in the "dry" UVA and UVB tests. Moisture may thus be a contributing factor. Cracking of E-layer is even observed in indirect exposure (i.e. exposure of air side with E-layer away from the light source), but it is less pronounced and starts later (at 162 kWh/m^2). E-layer cracking may thus be (photo-)hydrolytic and/or oxidative in nature and occur in combination with thermal stress. Cracking of E-layer has recently been reported in field studies in context of loss of power production due to inverter tripping in wet periods.

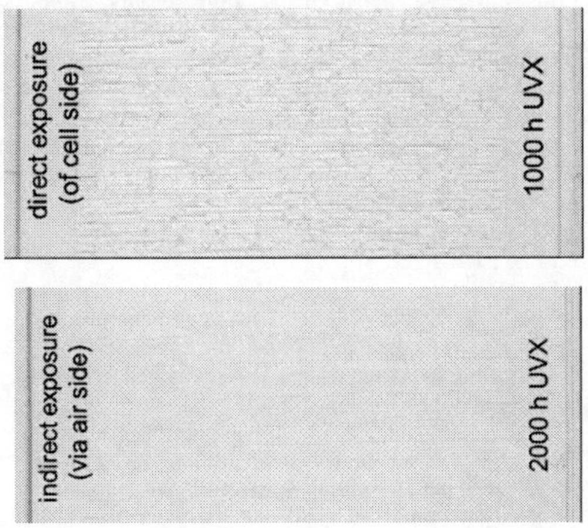

Figure 6: Cracking of E-layer on cell-side of PPE observed in UV Xenon for a) [top] direct exposure (of cell-side layer) at 81 kWh/m^2, b) [bottom] indirect exposure (via air-side) after 162 kWh/m2.

3.4 Retained Solar reflectivity

For several materials there is a correlation of yellowing to the change in retained solar reflectivity rSolarRefl, namely a loss of ca. 0.2-0.25% per unit ΔYI. It was shown, that 10% loss in BS reflectivity reduces PV module output by 0.3-0.5%, depending on cell distance and other factors of module design [3]. Thus, 1 ΔYI yellowing of inner layer is related to ≈ 0.5% ÷ 50 = 0.01% of PV module output. PO shows a spectrally "neutral" (not yellowish) change of reflectivity during tests: the retained reflectivity rSolarRefl reduces by 3% and recovers, indicating some structural changes in the polymer. UniQoat keeps its solar reflectivity in all tests for all times.

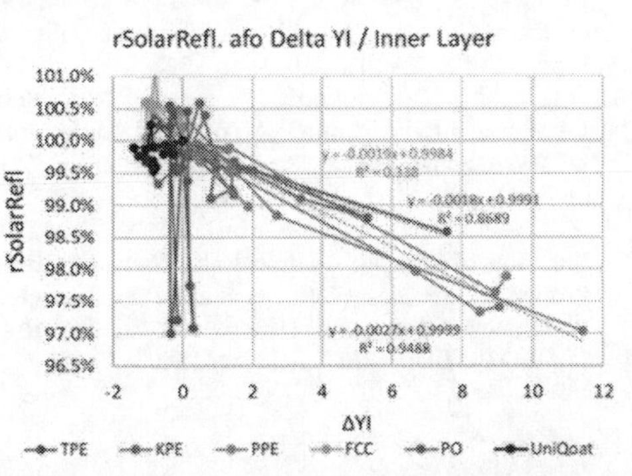

Figure 7: pooled plots (cell-side exposure UV Xenon, UVA340, UVB313 and HIUV-M) for rSolarRefl vs. ΔYI.

4 DISCUSSION

4.1 UV test performance

Obviously, the four different UV tests in this study are not equivalent. The daylight filtered Xenon test IEC TS 62788-7-2 [1] provides the best spectral match to natural daylight and is conducted at a level of T and %RH, that IEC TC82 WG2 regards as most representative for simulation of field performance of polymeric BS in PV modules. UVA, UVB and filtered metal halide are not suitable as agnostic tests, both due to artificially high content of UVB and the lack of moisture (in UVA and UVB).

4.2 Material performance

It is known from literature, that standard grade PET degrades under UV exposure without a suitable UV protection [4]. Therefore, T, K, F or UV stabilized PET functions are used as protective air side layers in laminated or coated backsheet designs with a PET core. These layers and in addition E-layers and/or some primer function may also be used on the cell side. In UniQoat, UV and hydrolytically stabilized PV grade PET is used over the full thickness of the single layer backsheet, whereas in PPE designs eventually only a 50 micron outer layer is PV-grade and UV protective [5].

Fig. 8 gives a schematic overview over backsheet constructions currently encountered. Most of current backsheet designs are laminated, coated or co-extruded. Occasionally, backsheets are classified by their air side protection as T, K, P etc., but in each of the "class" various grades of polymers, formulation and design differences are encountered. In respect to PET, often standard (electrical) grade PET is used as core material for cost competiveness, whereas PV grade PET represents hydrolytically and UV stabilized PET (P^{hUV}). The market observes a cost driven trend to reduce the thickness of UV protective air side layers (K, T, P^{hUV}). Thickness of backsheets is designed to match the requirements for distance through insulation (dti) for 1000V and 1500V system Voltage. It is worth mentioning, that E-layers, but also thermoplastic PO (with melting points in the 160°C range) are at least partially displaced under practical lamination conditions (e.g. 145°C, 12 min) and therefore their fresh thickness does not fully contribute to dti (see IEC TS 62788-2:2017 [2]).

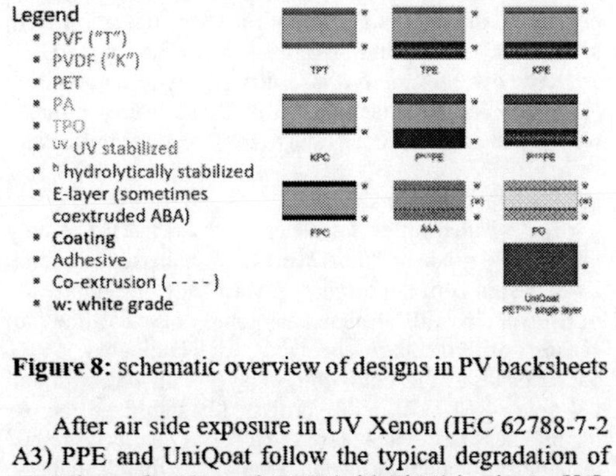

Figure 8: schematic overview of designs in PV backsheets

After air side exposure in UV Xenon (IEC 62788-7-2 A3) PPE and UniQoat follow the typical degradation of rEtB shown for PV-grade PET (with TiO2) in Fig. 3 of [6]. That material has been proven to sufficiently retains its mechanical properties after 324 kWh2 (equiv. 4000 h UV

Xenon IEC 62788-7-2 A3). This UV dose represent the cumulative lifetime-dose of 25 years backside exposure via 10% Albedo in Arizona (see IEC TS 62788-7-2 [1]). In the same paper [6] the authors have indicated, that for TiO2 loaded PET, the A3 test condition may actually be harsher then direct equivalent natural weathering of PET (TiO2) because of artificially high erosion of PET around TiO2 pigments, which is not observed even in direct exposure of air side layers in Florida (i.e. not Albedo limited and even exposed to direct precipitation).

In cell side exposure, however, PPE undergoes fatal degradation of its inner E-layer (cracking), which leaves the core PET unprotected ("UV leakage"). In UniQoat, UV and hydrolytically stabilized PV grade PET is present over its full thickness and in addition, the inner surface is modified to introduce additional UV protection together with the adhesive function. Any "UV leakage" through some laminated, thin protective T-, K-, P-, F-, E-layers on either cell-side or air side is thus excluded by design.

4.3 Post evaluation performance

There is no generic correlation between yellowing and retention of mechanical properties, that holds for different formulations and designs of polymeric backsheet films across different UV exposure tests. This has been pointed out before [7]. Yellowing can be induced by small changes in a small number of molecules and discoloration is therefore not necessarily correlated with degradation of mechanical or electrical properties. For the same amount of yellow molecules, the optically measured level of yellowing also depends on white pigment concentration (e.g. TiO2) and layer stack design of the backsheet. One shall assess this correlation individually for every sheet type [2], also considering the layer stack design.

Yellowing and mechanical degradation are typically two different modes of material change, each with its own sensitivity to influencing factors, given by the test conditions and the material properties.

5 CONCLUSION

5.1 UV test performance

UVA, UVB and HIUV-M are no equal alternatives to the new daylight filtered Xenon (IEC 62788-7-2, A3) that has been designed for representative safety tests of polymeric material in PV modules. Large deviation from spectral power distribution of natural daylight in the UVB range below 320 nm, but also differences in temperature (chamber and specimen) as well as moisture, produce results that are not representative for intended application.

5.2 Backsheet performance

For weathering performance of backsheets, not only the material grade or "backsheet class", but also the overall layer design is of importance. UV leakage or degradation of E-layers in 'xPE' designs may represent one of the root causes of cracking observed for laminated "PET backsheets" in the field. Single-layer UniQoat with full PV-grade PET (UV and hydrolytically stabilized) across its thickness represents therefore a "class" on its own: UV leakage is ruled out by design. Even more: that design also provides extended protection against partial erosion, which for thin air side layers (T, K, P, F) results in fatal loss of protective function in case of a core made from standard grade PET.

For better understanding of degradation, micro-mechanical and analytical analysis of cross sectioned backsheet films seem more suitable [9] to understand the actual failure modes in view of the design of the actual PV module, rather than "oversimplified" binning of backsheet films into material classes (PET, PVDF, PVF, PA, ...).

5.3 Post evaluation performance

Yellowing is not sufficient to prove mechanical degradation of a polymeric backsheet. Studies based on YI, as only measure of change, provide only results on discoloration, but do not necessarily provide insights on functional degradation. [8 and many others]. Use of Yellowing is tempting due to its simplicity and non-destructive character. Yellowing is however not suitable as agnostic pass-/fail criterion, especially in the context of IEC 61730-1 (safety assessment, component / material qualification & certification).

6 REFERENCES

[1] IEC TS 62788—7-2:2017 Measurement procedures for materials used in photovoltaic modules - Part 7-2: Environmental exposures - Accelerated weathering tests of polymeric materials

[2] IEC TS 62788-2:2017 Measurement procedures for materials used in photovoltaic modules – Part 2: Accelerated weathering tests of polymeric materials

[3] K. R. McIntosh, J. Jung, M. D. Abbott and B. A. Sudbury Determination and Evaluation of a Backsheet's Intrinsic Reflectance, AIP Conference Proceedings 1999, 020018 (2018); https://doi.org/10.1063/1.5049257

[4] A. Gok, D. K. Ngendahimana, C. L. Fagerholm, R.H. French, J. Sun, L. S. Bruckman, Predictive models of poly(ethyleneterephthalate) film degradation under multifactor accelerated weathering exposures, PLoS ONE 12(5): e0177614, 2017,

[5] J. Jung, The art of standardizing the testing of polymers in PV-modules, Proc. 8th SOPHIA workshop, Ljubljana, 2018

[6] E. Parnham, A. Whitehead, S. Pain, W. Brennan, Comparison of Accelerated UV Test Methods with Florida Exposure for Photovoltaic Backsheet Materials, Proc. 33rd EUPVSEC, 2017, pp 1345 - 1348

[7] E. Parnham, A. Seaman, A. Whitehead, W. Brennan, E. Ashford, Yellowing of PV Backsheets in Accelerated Tests Can Be Used as a Realistic Indication of Possible Field Failures – Fact or Fiction?, Proc. 32nd EUPVSEC, 2016, pp 1528-1531

[8] M. Köhl, D. Philipp, N. Lenck, M. Zundel, Development and Application of a UV Light Source for PV-Module Testing, Proc. 24th EUPVSEC, 2009, pp 3402-3405

[9] Y. Lyu, J.H. Kim, A. Fairbrother, X. Gu, Depth Profiling of Optical, Chemical and Nano-mechanical Properties of Glass / Encapsulant / Backsheet PV Laminates Aged under Different Intensities of UV Light, 33rd EUPVSEC, 2017, pp 1340 – 1344

INVESTIGATING THE DEGRADATION OF FRONT AND REAR SIDES OF C-SI PV CELLS EXPOSED TO ACETIC ACID

Nikoleta Kyranaki[1]*, Jiang Zhu[1], Ralph Gottschalg[2], Thomas R. Betts[1]
Centre for Renewable Energy Systems Technology (CREST), Wolfson School of Mechanical, Electrical and Manufacturing Engineering, Loughborough University, LE11 3TU, United Kingdom
Fraunhofer Center for Silicon Photovoltaics CSP, Halle, Germany
*N.Kyranaki@lboro.ac.uk, +44 (0)1509 227161

ABSTRACT: The goal of this paper is to investigate the reactions responsible for the degradation of the front and rear side of c-Si solar cells that are immersed in acetic acid solutions. The types of degradation are characterised by electrical, optical and chemical methods. The identified degradation modes are compared to those observed in PV cells undergoing damp-heat aging. The degradation of the power output of a PV cell exposed under damp-heat conditions typically is observed after 3000-5000 hours of exposure. The purpose of this work is to research if acetic acid immersion of PV cells can result in the same types of degradation observed for damp-heat aging and how these degradation modes affect their power output. It is observed that acetic acid immersion and damp-heat exposure show a similar mechanism of degradation. This degradation is identified to occur because of reaction of acetic acid with lead, tin and aluminium. Lead and tin migration causes separation of the ribbons and the silver contacts from the cells. Aluminium corrodes severely because of presence of moisture, causing a series resistance increase of the cells.
Keywords: Back contact, degradation, metallization

1 INTRODUCTION

PV modules that are encapsulated with ethylene-vinyl acetate (EVA), typically show degradation of their power output after Damp-Heat (DH) exposure for 3000-5000 hrs [1]. The most relevant degradation path is that acetic acid is produced because of moisture penetration and high temperature and attacks the metallization, the ribbons and the aluminium (Al) back contact of the PV cell [2, 3]. The exact degradation mechanisms and the impact of the different mechanisms on the power output of a solar cell, are not fully understood yet. The purpose of this work is the simulation of the effects of DH aging by acetic acid immersion, the identification of the responsible aging mechanisms and the separation of the impact of corrosion on the different components of a PV cell.

2 EXPERIMENTAL SETUP

Commercially available c-Si PV cells soldered with Pb-Sn-Cu ribbons are immersed into acetic acid solutions of different concentrations at room temperature. The different concentrations are 30%, 10% and 0.3% v/v. The two first concentrations are used for accelerating aging and better understanding of the reactions. The third concentration is that detected in PV modules aged under damp-heat conditions [4]. The acidity and the temperature of the solution are monitored with a pH meter and kept at pH 1.9+/-0.2, 2.3+/-0.2 and 3.4+/-0.2 respectively. The PV cells are soldered manually using a commercial standard flux, and a standard 2-mm-wide Pb-Sn-Cu ribbon. For better mechanical stability and for the ability to age only one side of the cell, the cells are half encapsulated in a glass-EVA-cell structure (Fig. 1). For the investigation of corrosion of the back contact of the cells, the laminate structure is the same as for the front side aging but the front side of the cell faces the encapsulant and the back side is exposed.(Fig. 2).

Figure 1: The structure of the half-encapsulated PV cells that are used for the investigation of the front contact corrosion.

Figure 2: The structure of the half-encapsulated PV cells used for the investigation of the Al back contact corrosion.

3 CHARACTERISATION METHODS

The process of the experiment and the characterisation methods is presented on the flowchart (Figure 3) below:

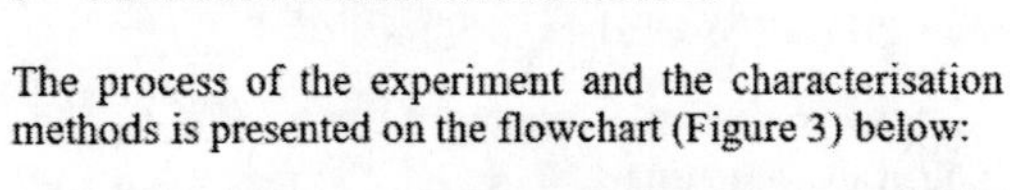

Figure 3: The flowchart of the experimental process.

3.1 I-V curve measurements

The samples were removed from the acetic acid solution every 2 days for Light I-V curve measurements at STC by Pasan IIIb solar (flash) simulator.

3.2 Electroluminescence

Electroluminescence (EL) images are captured every 2 days for the observation of the impact of the acetic acid on the contacts and the ribbons (increased series resistance, shunting).

3.3 Material/chemical analysis (SEM-EDS analysis, optical microscopy)

Optical microscopy is used as a non-destructive method for the observation of material changes during the experiment, due to acetic acid exposure. SEM-EDS analysis is used as the final destructive characterisation method for the identification of the migration of the different elements, due to material reaction with acetic acid.

4 THEORY OF POSSIBLE REACTIONS

In this work the reactions that are mainly reported in literature between acetic acid and the different components of a PV cell, will be verified.

4.1 Silver front contact and metallisation corrosion

For the manufacturing of the solar cell front contacts, normally a silver paste is applied by screen printing, which includes glass frits containing lead oxide. Acetic acid attacks the lead oxide, according to the following reaction [5]:

$$PbO+2CH_3COOH \rightarrow (CH_3COO)_2Pb+H_2O. \quad (1)$$

Because of this reaction, the silver contact is separated from the cell.

4.2 Back contact corrosion

During the acetic acid immersion of the cells, most likely the Al from the Al-back contact reacts with acetic acid. The reaction is presented below [6]:

$$Al + 3H^+ \rightarrow Al^{3+} + 3/2H_2\uparrow \quad (2)$$

This process is responsible for reducing electrical conductivity and an increase of the series resistance [7].

4.3 Ribbon corrosion

For the construction of the ribbons of a PV cell, normally alloys of tin (Sn) and lead (Pb) are used as a solder. Acetic acid attacks the solder, and the reaction is:

$$Sn + 2H^+ \rightarrow Sn^{2+} + 2H_2\uparrow \quad (3)$$

$$Pb + 2H^+ \rightarrow Pb_{2+} + 2H_2\uparrow \quad (4)$$

Because of the solder leaching, the silver contact and the ribbon are separated [6]

5 RESULTS

5.1 Light I-V curves

In Figure 4 the degradation of the performance parameters of the cells immersed in different concentrations of acetic acid solution is presented. For 30% concentration the degradation of short circuit current (I_{sc}), fill factor (FF) and maximum power (P_{max}) (-8%) are bigger than those for 10% concentration (-2%). The increase of series resistance is most likely the reason for the FF degradation, as change of the slope of the I-V curve close to open circuit voltage (V_{oc}) is observed.

Figure 5 illustrates the performance parameters for the cells with the Al-back side exposed to the acetic acid. The cells immersed in higher concentrations (30% and 10% v/v) show a slower degradation than the cells immersed in 0.3% concentration. The power output degradation is totally driven by FF reduction, and the I_{sc} remains almost constant. The reduction of the P_{max} is around -25% for all of the samples, but for the higher concentrations, this level of reduction is reached after 900 hrs, in contrast to the samples immersed in 0.3% concentration, which reach this reduction in only 200 hrs of immersion. This likely happens because the reaction of Al with acetic acid produces aluminium acetate which inhibits further reaction. The Al-back contact corrosion seems to be more serious than the ribbon and silver contact corrosion because the degradation of the Al-back contact causes a bigger reduction on the power output of a PV cell (-25%) than the front side degradation (-8% worst case).

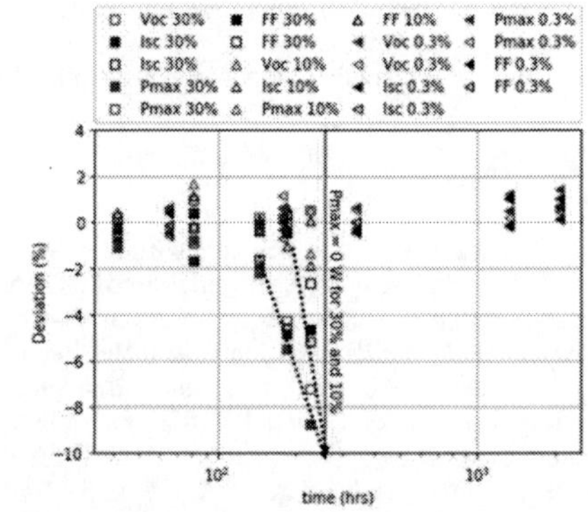

Figure 4: Degradation of the performance parameters of the cells immersed in different concentrations of acetic acid solution.

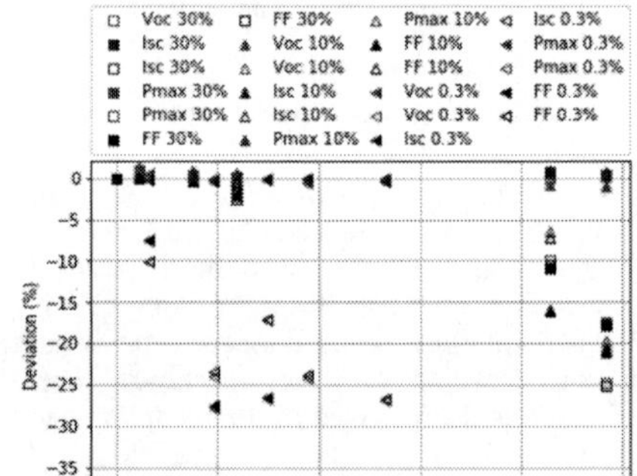

Figure 5: Degradation of the performance parameters of the cells immersed in different concentrations of acetic acid solution for Al-back contact aging.

5.2 Electroluminescence

Figures 6 a and b represent the EL images of an unaged and an aged cell. The bright areas along the busbars of the immersed into acetic acid PV cell, are an indication of series resistance increase.

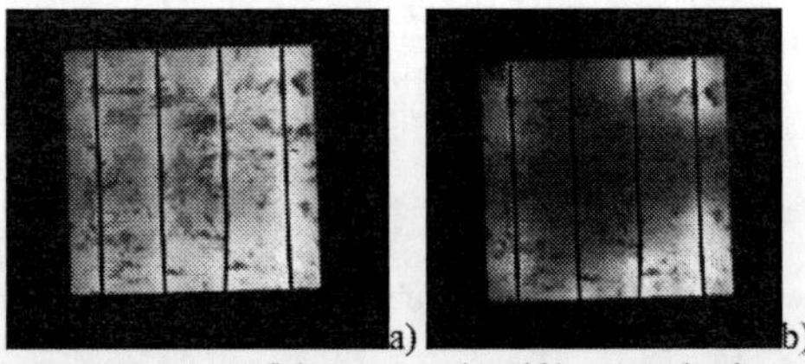

Figure 6: Images of a) an unaged and b) an aged solar cell.

5.3 Material analysis

Figures 7 a and b show the optical microscopy images of the metallization and the ribbon of an unaged and an aged in 30% concentration of acetic acid solution cell. Brown patterns on the ribbon and the metallization are visible which are identified by EDS analysis as copper (Cu) from the ribbons (Figure 10). The colour of the cell appears to be brighter blue for the aged cell. This is an indication of corrosion of the anti-reflection coating. The optical microscopy image of the cell immersed in 10% concentration of acetic acid solution (Figure 8 a and b) shows some white patterns on the metallization. On the EDS analysis of the samples (Figure 11) the white patterns are recognized that are caused because of Sn and Pb migration from the ribbons or the contacts and the metallization. According to the measurement of the ribbon composition, Pb is observed to dissolve more (Figure 12). This phenomenon can be explained, because for 30% concentration of acetic acid, the dissolution of Sn and Pb is very fast, and the core Cu of the ribbons is more exposed into acetic acid, so, it also starts to dissolve. For 10% concentration, the reaction is slower and the most observable migration is this of Sn and Pb.

Figures 9 a and b illustrate the optical microscopy images of an unaged and an aged back contact of a silicon solar cell. The leaching of the Al is visible, as some white and brown patterns as well. The Al leaching can be confirmed also by the EDS analysis of the sample (Figure 13) and the white and brown patterns are identified as Sn/Pb and Cu, respectively.

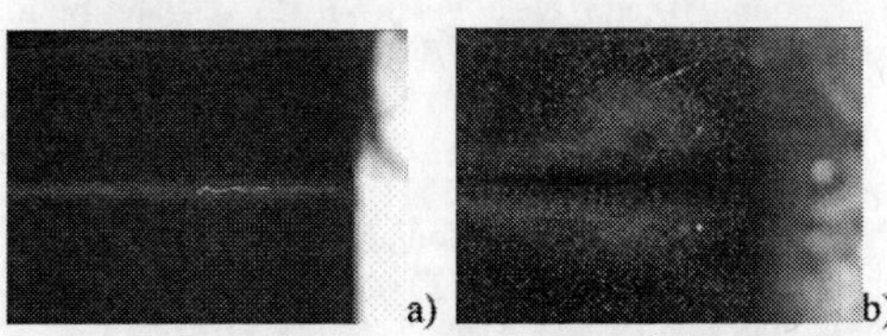

Figure 7: Optical microscopy images of the metallization and the ribbon of a) an unaged and b) an aged cell in 30% concentration of acetic acid solution.

Figure 8: Optical microscopy images of the metallization and the ribbon of a) an unaged and b) an aged cell in 10% concentration of acetic acid solution.

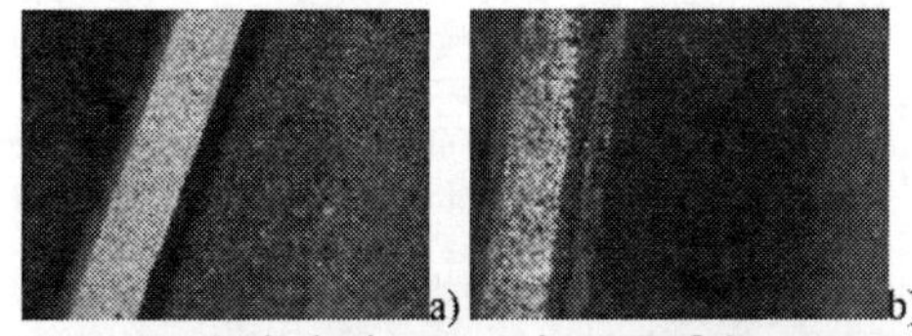

Figure 9: Optical microscopy images of a) an unaged and b) an aged back contact of a silicon solar cell.

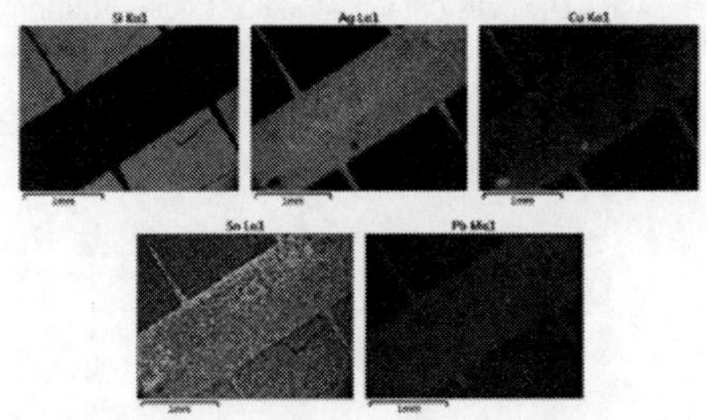

Figure 10: SEM-EDS analysis of the silver contact and the metallization of an aged cell in 30% concentration of acetic acid solution.

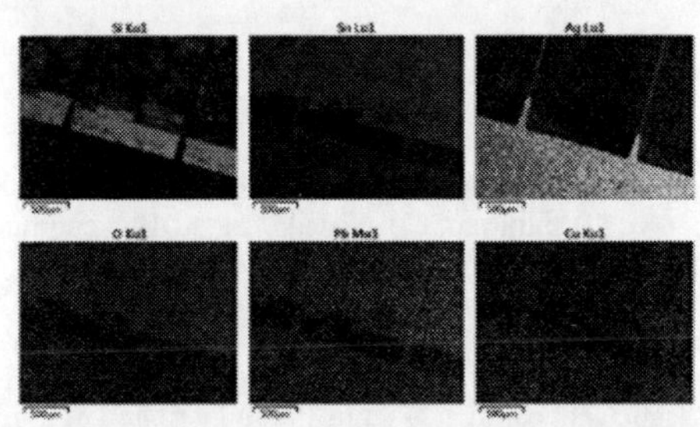

Figure 11: SEM-EDS analysis of the silver contact and the metallization of an aged cell in 10% concentration of acetic acid solution.

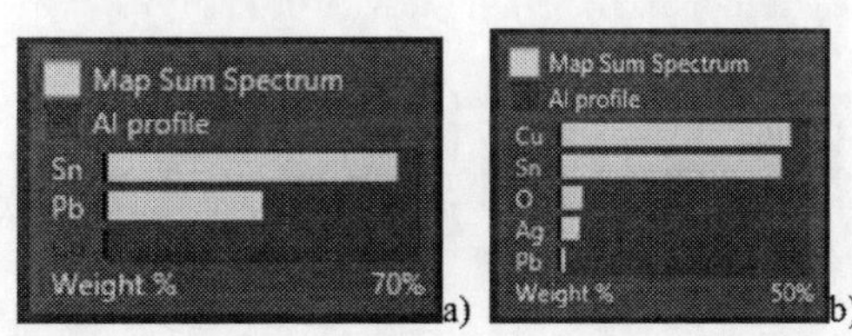

Figure 12: Material composition of a) an unaged and b) an aged ribbon.

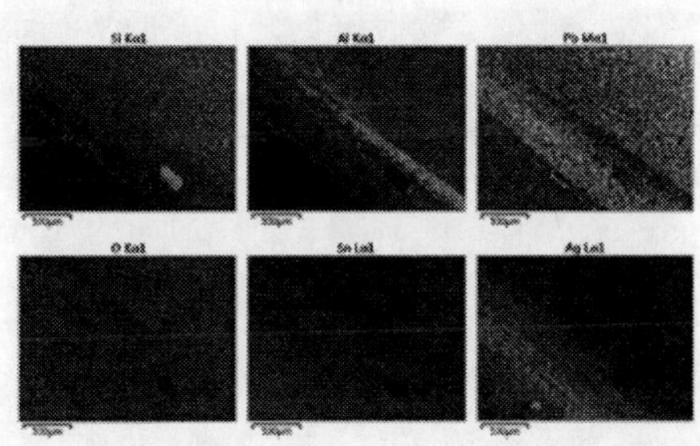

Figure 13: SEM-EDS analysis of an aged back contact of a silicon solar cell.

6 COMPARISON WITH DAMP-HEAT

6.1 Light I-V curves

In Figure 14 the degradation of a 6-cell mini-module aged in DH conditions (85°C/85%RH), is visualized. By combining the impact of front side acetic acid immersion (I_{sc} and FF reduction) and of Al-back contact acetic acid immersion (FF reduction), a similar degradation pattern to that caused by DH aging is observed. The degradation is more accelerated and sudden for high concentrations of acetic acid. The cells immersed in 0.3% concentration, for duration of immersion of 2000 hrs, do not show any degradation. This behaviour is similar to that of the modules aged under DH conditions, as, for 2000 hrs of DH aging, degradation is also not observed.

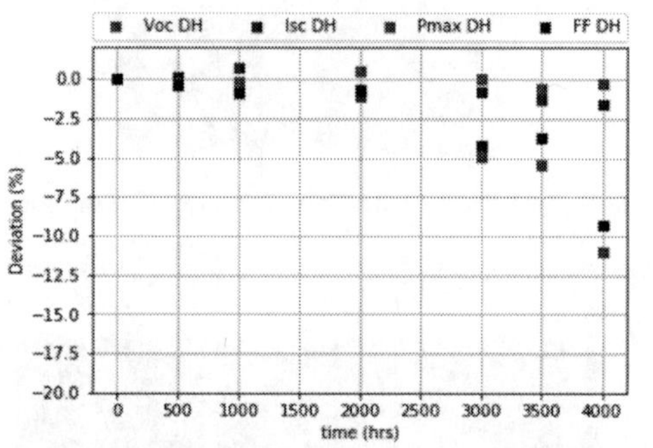

Figure 14: Degradation of the performance parameters of a 6-cell mini-module aged in DH conditions (85°C/85%RH).

6.2 EDS analysis

From the EDS analysis of a ribbon of a cell immersed in acetic acid solution (Figure 15) and a ribbon of a mini-module aged in DH conditions (Figure 16) it is clear that the material reactions are similar to acetic acid immersion, as the alloy of Sn and Pb is visibly corroded, and the Cu is detectable, because of the dissolution of Sn and Pb.

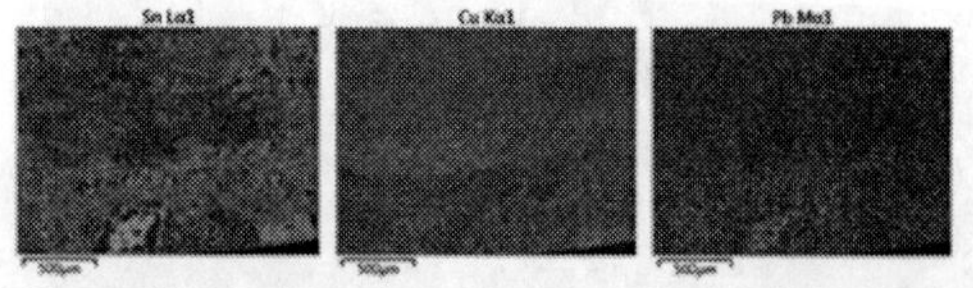

Figure 15: SEM-EDS analysis of a ribbon of a cell immersed in acetic acid solution.

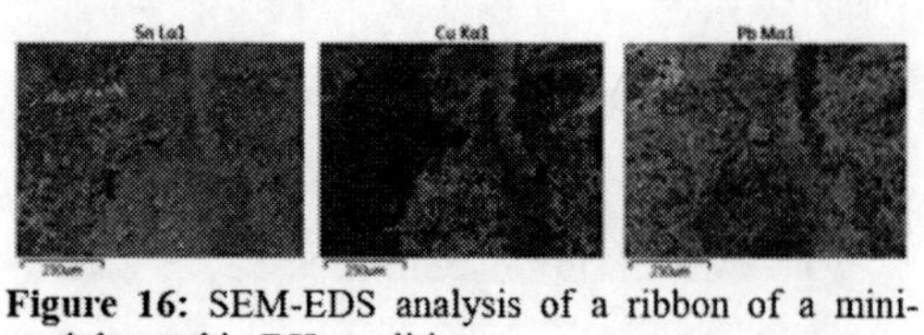

Figure 16: SEM-EDS analysis of a ribbon of a mini-module aged in DH conditions.

7 CONCLUSIONS

For front side immersion in 30% concentration of acetic acid the degradation of P_{max} is bigger and driven by FF degradation. The I_{sc} degradation is smaller and probably caused by the antireflection coating degradation. For 10% the degradation of the performance parameters is smaller and more sudden. The degradation for the cells immersed in 0.3% concentration of acetic acid is negligible. When the Al-back contact is immersed into acetic acid of high concentration does not corrode severely, because of the aluminium acetate (inhibitive complex) that is produced by the reaction of Al with acetic acid. On the contrary, when the acetic acid concentration is low, water is highly corroding for the Al. From these findings we can conclude that the most serious power output reduction occurs because of moisture corrosion of the Al-back contact.

From the material analysis it can be concluded that for the highest concentration of acetic acid, copper has also migrated from the ribbons towards the cell, as the dissolution of the alloy of Sn and Pb is faster. Pb is observed to dissolve even more. For a lower concentration, the Sn and the Pb are mostly detected. In addition, the leaching of the Al from the Al-back contact is visible.

From the comparison of acetic acid immersion of the cells with damp heat aging of modules, we can conclude that acetic acid immersion and DH aging show a similar degradation pattern of the electrical parameters of the modules/cells, FF reduction mainly due to Al-back contact degradation (smaller FF reduction for the front side immersion), and I_{sc} reduction due to front side degradation (no I_{sc} reduction is observed for the Al-back side immersion). In addition, from the EDS analysis, it is observed that the mechanism responsible for the degradation that occurs under DH condition is similar to the mechanism reproduced by acetic acid immersion, as Sn and Pb migration from the ribbons are observed for both acetic acid immersion and DH aging.

8 ACKNOWLEDGEMENTS

This study has received funding from the European Union's H2020 programme SOLAR- TRAIN under grant agreement No 721452.

9 REFERENCES

[1] J. Zhu et al., "Changes of solar cell parameters during damp-heat exposure," Progress in Photovoltaics, vol 24, no. 10, pp. 1346–1358, 2016.

[2] Iea-pvps, Assessment of PV Module Failures in the Field Assessment of PV Module Failures in the Field. 2017.

[3] K. Whitfield, A. Salomon, S. Yang, I. Suez, and S. Corporation, "Damp Heat versus Field Reliability for Crystalline Silicon,", 28th IEEE PVSC, 2012.

[4] B. Ketola, A. Norris, "Degradation Mechanism Investigation of Extended Damp Heat Aged PV Modules.", 26th EU PVSEC, 2011.

[5] A. Kraft et al., "Investigation of Acetic Acid Corrosion Impact on Printed Solar Cell Contacts," IEEE Journal of Photovoltaics, vol. 5, no. 3, pp. 736–743, 2015.

[6] H. Xiong et al., "Corrosion behavior of crystalline silicon solar cells," Microelectron. Reliab., vol. 70, pp. 49–58, 2017.

[7] A. Masuda et al., "Degradation by acetic acid for crystalline Si photovoltaic modules," Jpn. J. Appl. Phys., vol. 54, no. 4, 2015.

[8] A. I. Onuchukwu and F. K. Oppong-Boachie, "The corrosion behaviour of aluminum alloy AA1060 in p-quinone and acetic acid media," Corros. Sci., vol. 26, no. 11, pp. 919–926, 1986.

PV MODULES INSPECTION THROUGH PHOTOLUMINESCENCE IN DAYLIGHT CONDITIONS

Luis San José[1], Rebeca Herrero[1], Ignacio Antón[1] and G. Sala[1]
[1]Instituto de Energía Solar, Universidad Politécnica de Madrid
luis.sanjose@ies.upm.es
Madrid, Spain

ABSTRACT: As PV modules are exposed to outdoor conditions, inspection proceedings are needed in order to evaluate the possible degradation that modules can suffer. The main inspection methods look for non-invasive procedures that allow obtaining inspection data while modules are operating on-field. Photoluminescence (PL) uses recombination phenomena that take place while the module is exposed at light to get information of the cell performance. A method for inspecting PV modules at adverse (low irradiance) light conditions based on outdoor PL is proposed in this work. The outdoor PL measurements obtained are compared against electroluminescence measurements and indoor PL in order to contrast the accuracy of the measurements.
Keywords: Electroluminescence, Photoluminescence, Image processing, PV plant, Inspection.

1 INTRODUCTION: PHOTOLUMINESCENCE VS ELECTROLUMINESCENCE

Because of the configuration of solar cells, its physical structure permits to absorb light (photons) and frees electrons. Due to the reversibility of this process and the imperfections at atomic level of the materials, solar cells are able to emit low energy light when an electron is recombined with a hole and recovers its position in the semiconductor atomic structure. Depending on the bandgap of the semiconductor material used, this process needs higher or lower luminous energy to take place and the re-emitted light spectrum from the recombination may vary.

Based on these physical principles, electroluminescence and photoluminescence are two different techniques in which the cell emission is captured through an image to evaluate the cell health and performance. These images provide spatial high-resolution information of the cells and because they are quite fast, both methods are suitable for being used as inline tool for spatially resolved characterization.

A brief review of these techniques in several works previously done is given below.

Electroluminescence (EL) is a technique that requires supplying current through the cell terminals below the short-circuit current. The electrons from the injected current facilitate electron-hole pair recombination phenomenon, which emits photons. In the case of Silicon cells, this emission can be detected with a CCD camera with a Si-based sensor. Due to the low energy of the emission, these measurements are usually performed indoor under no light. With the images acquired while biasing the cell, different parameters of the cell can be observed or deduced.

Based on this technique, some important previous work has been carried out. For example, Glatthaar et al. [1] use LBIC (Light Beam Induced Current), another inspection technique which consists on mapping cells through a laser point light or monochromatic light in order to determine the local carrier's collection efficiency of each point, and compare their EL measurements with it. One of the main cons to avoid inspection of the solar cells with the LBIC technique is the long time that takes to perform a cell map with a punctual light. Then, EL images are used to determine the dark saturation current of the cell, taking two images at different applied voltages, and to inspect and measure the magnitude of lateral voltages variation and lateral current flow.

In [2] and [3], Bothe et al. compares the information given by the dark lock-in thermography (DLIT), the LBIC and the EL images when characterizing cells in production lines.

As they explain, with a DLIT image of the cell they are able to see areas of increased thermal emission and therefore, determine areas where the recombination effects of light-generated carriers are enhanced; which indicates shunt defects, grain boundaries and dislocations mostly.

Additionally, with LBIC images, they inspect the carrier recombination effects at grain boundaries and dislocations. Moreover, through EL images voltage profile between busbars are determined.

Finally, they conclude that with EL images they are able to reduce inspection times and are able to get the information that both previous methods provide.

Fuyuki et al. [4] conclude also that the EL measurement local signal is directly proportional to the collection efficiency measured with LBIC.

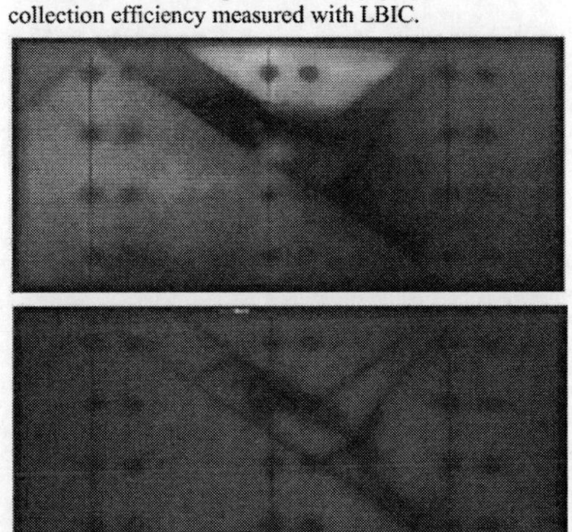

Fig. 1 Electroluminescence (up) and photoluminescence (down) images of the same solar cell.

On the other hand, photoluminescence (PL) is a technique that does not require manipulating the device contacts. When a solar cell is illuminated with a light source, some of the light received is re-emitted in higher wavelengths (lower energy) due to electron-hole pairs recombination and also because the carriers generation process is not perfect. The bandgap existence in the union of the semiconductor implies that the less-energetic photons will not split an electron from its position in the structure, and then, the photon will be radiated by the cell. This implies that despite this method is contactless, the PL signal will be weak and difficult to detect because of the debility of the light emission compared to the light that cell receives. Mostly, PL measures are taken indoors and in complete darkness to boost the signal-noise ratio and avoid other light sources than the needed for the cells excitation.

In this context, T. Lavrenko et al. [5] works use EL and PL images to characterize CIGS cells. With the EL emission spectrum analysis, the external quantum efficiency (EQE) of the device under test is deduced, while with the PL emission intensity, a relationship with the open circuit voltage (V_{OC}) and the performance of the cell is established. Similar works have been carried out [6], relating the PL intensity signal with the efficiency of the solar cell.

On this topic, T. Trupke et al. [7] contrast the PL measurement with calibrated density image measurements (CDI) and establish a direct relationship between the relative pixel count and the local effective minority carrier lifetime, showing the utility of the information given by PL measurements too.

In summary, both methods represent fast ways for PV devices characterization and allow revealing parameters related to defects such as:

- Surface recombination velocity.
- Lateral voltages variation & lateral current flow.
- Local minority carrier diffusion length.
- Grain boundaries and dislocations.
- Shunts and metal contact defects.

2 EL & PL IN PV PLANTS

Performing EL and PL measurements in field-exposed modules is a challenge, due the conditions that outdoors have to be faced. For EL, the module connections must be manipulated in order to bias the device, and in both, daylight conditions represent an amount of noise that has to be avoided in order to obtain a clear image.

In the case of outdoor PL measurements without daylight, a light source along with an adequate spectral sensor is needed for the cell emission

However, performing these measurements on outdoor operating modules would be interesting in order to visualize its degradation over time. The main causes of module performance degradation outdoors are:

- The encapsulant degradation.
- The shunt resistance decrease.
- The light induced degradation (LID).
- The potential induced degradation (PID) due to insulation failures in the electrical circuit.
- The humidity and temperature conditions at which modules are exposed.

EL and PL measurements allow quantifying the degradation and serving as quick inspection tools for diagnosing modules performance.

Some previous works, such as [8] T. Kropp et al. use the EL to detect cell defects in function of the local emission reduction. They use a capacitor bank (charged with the modules production) in order to supply power at voltages higher than V_{OC} of the modules. With a NIR camera synchronized with the extracting and injection process of the capacitor bank, they capture a set of images, use averaging algorithms for processing them, and obtain a clear EL measurement. Because the power injection is produced at a voltage higher than the V_{OC}, the photoluminescent effect is enhanced and therefore is added to the measure.

Following this, M. I. Oprea et al. [9] uses a SWIR camera for EL imaging and an Impedance Spectroscopy (IS) for measuring the capacitance of modules and detect defects of shunting type due to the measurement given by IS.

Other authors like M. Barbato et al. [10] performs PID tests at different voltages and temperatures, monitoring the module parameters (V_{OC}, I_{SC}, FF, P_{MAX}, R_S and R_{SHUNT}) measuring the variation of EL (with an InGaAs camera) emission and EQE. After PID stress, the surface recombination velocity increases and therefore a decrease in the EL emission is observed corresponding to a reduction in the EQE at the UV region. Surface polarization effects are also observed in the EL images, as well as the appearance of dark regions in the module cells related to micro shunting defects.

Further, A. Pozza et al. [11] performs an analysis of modules that have been working on field through EL and LBIC characterization. With this, scratches, finger interruptions, cell cracks and dark areas can be detected. Because all the EL measurements were performed under the same test conditions, they have reached to relate their EL signal with the module performance, comparing the histogram integral of each image against the maximum power of the corresponding module.

In the PL topic, R.Bhoopathy et al. [12] perform PL measurements of a module by changing the optical generation rate of one cell of a module string. By using a high power white light source, they shade a single cell and control its biasing conditions. The shaded cell is illuminated up to one sun using the light source, forcing the biasing of the rest of the string cells to operate between the operating point (MPP) and the open circuit. Taking to images, the PL emission can be separated from the strong background.

Finally, in regard of drone applications, G. Benatto et al. [13] performs PL outdoor measurements with a laser scan line, obtaining images better than EL for failures and defects inspection.

In the next paragraph, a novel method is proposed for acquiring PL measurements of Silicon PV modules outdoor in adverse light conditions, that is, with low irradiance profiles.

3 METHODOLOGY

The proposed method uses the properties of the Silicon emissivity (centered on 1150nm) along with the solar spectrum shape, where the H_2O vapor absorption valley at 1120nm reduces significantly the amount of light received. Because of this attenuation of the solar light, with an adequate sensor, PL emission can be strongly detected outdoors. However, the encountered problem is that for low irradiance profiles, the emissivity of the Silicon is reduced too, which makes the PL signal to be poor. Fig.2 helps to illustrate the searched spectral region where the PL measurement in these conditions would be optimized, It also shows the Silicon emissivity and the SR of two different camera sensors.

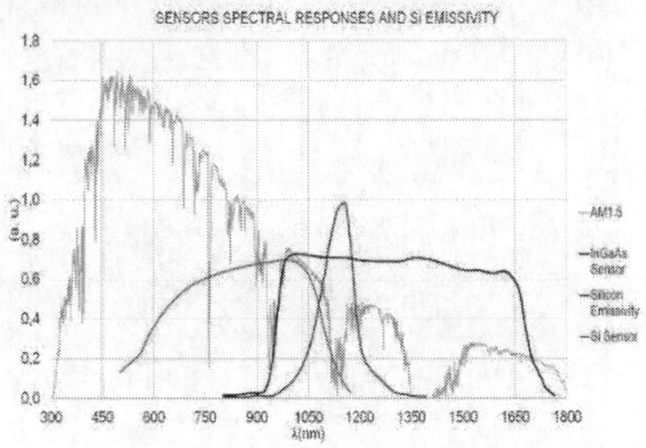

Fig. 2 AM 1.5 spectrum, emissivity of the Silicon, and espectral responses of InGaAs sensors and Si sensors

Aiming to perform PL characterization of multicrystalline Silicon modules at daylight adverdse conditions, a PL set up has been developed; consisting on an InGaAs sensor-based camera and a source light (a scheme is shown in Fig.3). Because of the low emission that it is expected to have, due to the low amount of irradiation, a light source is needed. The light source has been used to excite the device under test obtaining at least two images: one of the device under the light source excitation and other without the light source excitation (the darkness image).

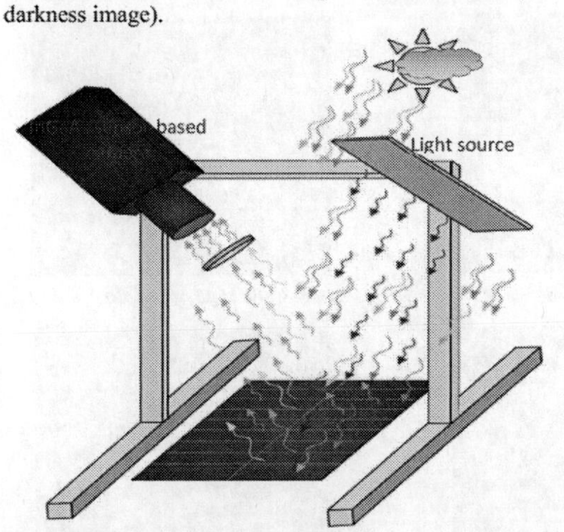

Fig. 3 Measurement set up, the daylight and the light source excites the solar cell, which emits light in larger wavelenghts that can be captured by an InGaAs sensor. Filters allow improvind the images by rejecting the non-useful wavelengths.

Using some image processing algorithms, the subtraction of the darkness image from the image of the illuminated device will eliminate the background spectral noise and return the PL signal of the excited device.

In order to improve the signal-noise ratio different spectral filters are used and tested and electroluminescence and photoluminescence images have been acquired indoors in order to previsualize the desired outdoors measurement result and evaluate the intensity of the PL signal. These preliminary measurements have been performed on a single cell, where different conditions have been tested.

Indoors, different intensities of the light source with the same exposition time for the images have been proved in order to observe the PL intensity variation as shown in Fig.4. As can be deduced, the PL signal intensity is displaced and spread through the histogram values linearly with the intensity of the light source.

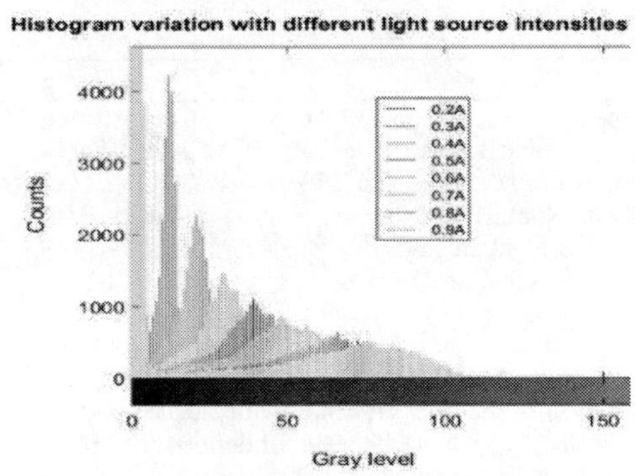

Fig. 4 Gray levels in PL get brighter linearly with higher light source intensity.

Filters also play an important role in the PL measurement, specially, when the noise is very high, like in outdoors. Four different filters have been tested in the preliminary measurements: one long pass filter under the Silicon emissivity central wavelength (filter 1), another long pass filter near the central Si emissivity central wavelength (filter 2) and two band pass filters over the bell shape of the spectral emissivity (filters 3 and 4).

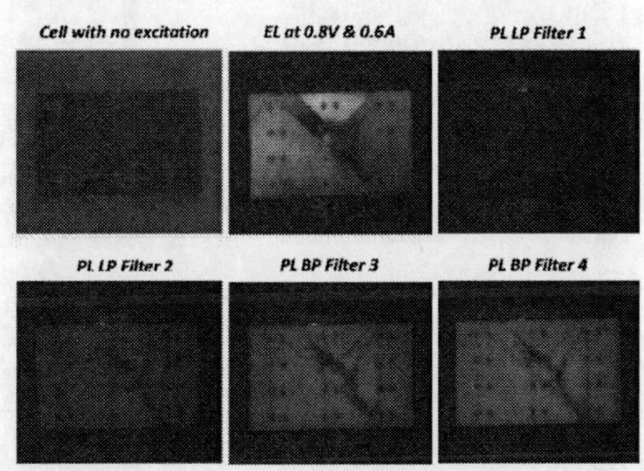

Fig. 5 Indoor EL and PL measurements of a Silicon solar cell . Images are the result of subtracting the excited image from the darkness image. Despite different exposition times have been needed for each of the PL images, all images graylevels are scaled from 0 to 1, showing the better performance for bandpass filters.

As can be seen in Fig.5, band pass filters, such as filters 3 and 4 reduce significantly the noise and therefore where more effective than the long pass filters (1 and 2) that let more wavelengths of the spectrum pass. However, for the bandpass filters, due to their spectral narrowness, longer exposition times and/or more intenser light is needed. The longer the exposition time, the higher the PL signal captured.

4 RESULTS: CASE STUDY

A PV module has been used for testing the proposed method indoors and outdoors. The outdoors measurements have been carried out in different daylight conditions, mainly at sunrise and rainy days, while exciting the module with the light source .

The module used is a SLK60P6L model from Siliken with the following parameters:

Table I: Parameters from the PV module

		Units
Pmpp (Power at maximum PP)	240	Wp
Vmpp (Voltage at maximum PP)	29.6	VDC
Impp (Current at maximum PP)	8.12	A
NOCT (Nominal Op. Cell Tº)	46	ºC
Voc (Open circuit voltage)	37	VDC
Isc (Short circuit current)	8.61	A
FF (Fill Factor)	75.34	%

This studied module presents a severe degradation with most of the cells broken or with isolated parts from the busbars, which results ideal for deffects identification and PL and EL measurements performance.

Two cells (marked on Fig.6, C1 and C2 onwards) that presents different breakage types have been selected to show the PL results using filter 3.

C1 is a cell that presents numerous cracks and that has only the parts near to the busbars connected to the active circuit. On the other hand, C2 has 2/3 of its surface in good conditions and only 1/3 is disconnected from the active circuit.

In order to obtain a complete cell PL image, the camera has been fixed at a given distance from the module and a set of images has been acquired by placing the light source at different positions (this is not an indispensable proceeding if a high power light source is used). With image processing algorithms and subtracting the darkness image, PL images of both cells have been obtained as shown in Fig.7 (all the measurements were performed with the filter 3 mounted on the camera).

Fig. 6 Indoors mounted set up and selected cells C1 and C2 for the case study.

As can be observed in Fig.7, damaged areas of the cell appear brighter in the PL images, because recombination phenomenon is enhanced in the isolated areas.

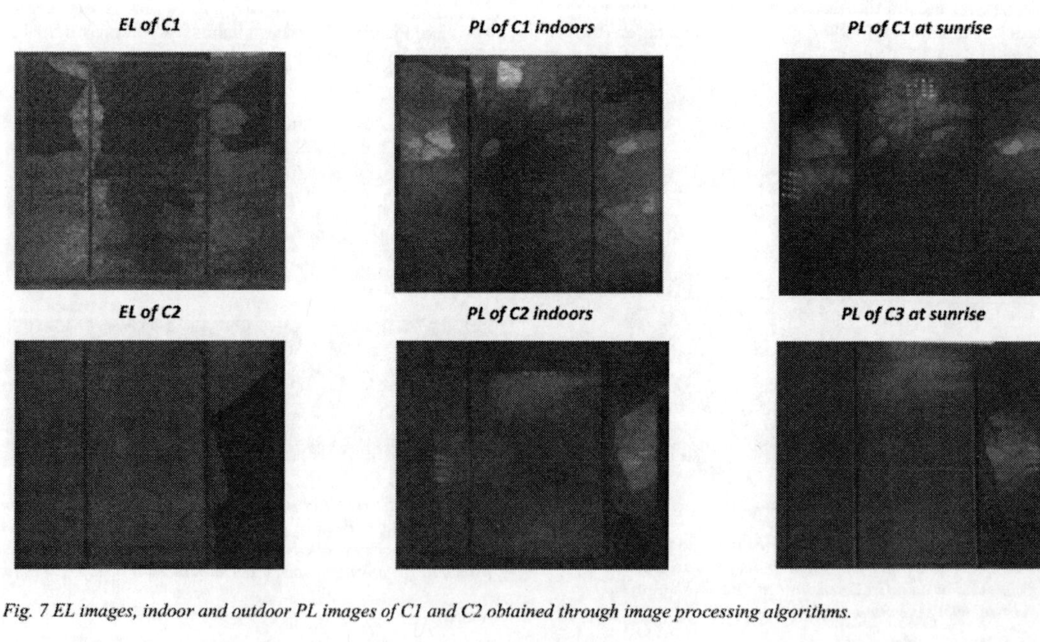

Fig. 7 EL images, indoor and outdoor PL images of C1 and C2 obtained through image processing algorithms.

5 CONCLUSSIONS AND FUTURE WORK

With the presented method, PL measurements have been performed on module with low irradiance outdoors. Through the test of different filters indoors, a set of desirable parameters for the indoor PL measurements and for a good image acquisition (a clear PL image) have been studied. Despite having a low power light source, thanks to the image processing, similar results to the indoors measurements have been reached and defects visible at the EL images have been correlated with the PL images.

Because the main aim of this research is to be able to inspect complete PV modules, as future work, the complete set of filters have to be tested outdoors in order to prove their performance at daylight conditions. Finally, different measurements during different weather conditions have to be performed outdoor.

6 ACKNOWLEDGMENTS

This work has been supported under the program RETOS-COLABORACIÓN (RTC-2015-3982-3 MINECO/FEDER, UE). Moreover, we acknowledge the financial support provided by the Comunidad de Madrid through the program MADRID-PV-CM (S2013/ MAE2780)

7 REFERENCES

[1] Glatthaar, M. et al., "Spatially resolved determination of the dark saturation current of silicon solar cells from electroluminescence images". JOURNAL OF APPLIED PHYSICS 105, 113110 (June 2009)

[2] Bothe, K. et al., "Electroluminescence imaging as an in-line characerisation tool for solar cell production". 21st European Photovoltaic Solar Energy Conference, 4-8 September 2006, Dresden, Germany.

[3] Kendig, G. et al., "Characterization of defects in photovoltaics using thermoreflectance and electroluminescence imaging". 35th IEEE Photovoltaic Specialists Conference, 20-25 June 2010 Honolulu, HI, USA.

[4] Takashi, F. et al., "Photographic surveying of minority carrier diffusion length in polycrystalline silicon solar cells by electroluminescence". Applied Physics Letters, volume 86, 262108, (2005).

[5] Lavrenko, T. et al., "Advanced Luminescence Imaging of CIGS Solar Cells". 10.4229/27th EUPVSEC 2012-3BO.4.3. (2012).

[6] Martínez, O. Et al., "Photoluminescence imaging of solar grade mc-Si wafers and solar cells as a tool for efficiency qualification" EXMATEC 13 (Expert Evaluation and Control of Compound Semiconductor Materials and Technologies) 8-10 June 2016.

[7] Trupke, T. et al., "Photoluminescence imaging of silicon wafers". Applied Physics Letters, volume 89, 044107, (2006).

[8] Kropp, T. et al., "Self-Sourced Daylight Electroluminescence From Photovoltaic Modules". IEEE Journal of Photovoltaics, Vol. 7, No. 5, September 2017.

[9] Oprea, M. I. et al., "Detection of potential induced degradation in c-Si PV panels using electrical impedance spectroscopy," IEEE 43rd Photovoltaic Specialists Conference (PVSC), Portland, OR, 2016, pp. 1575-1579.

[10] Barbato, M. Et al., "Potential induced degradation of N-type bifacial silicon solar cells: An investigation based on electrical and optical measurements", Solar Energy Materials and Solar Cells, Volume 168, Pages 51-61, ISSN 0927-0248, (2017).

[11] Pozza, A., and Sample, T., "Crystalline silicon PV module degradation after 20 years of field exposure studied by electrical tests, electroluminescence, and LBIC". Prog. Photovolt: Res. Appl., 24: 368–378. doi: 10.1002/pip.2717 (2016).

[12] Bhoopathy, R. et al., "Outdoor photoluminescence imaging of photovoltaic modules with sunlight excitation". Prog Photovolt Res Appl. 2018;26:69–73. https://doi.org/10.1002/pip.2946 (2017).

[13] Benatto, Gisele Alves dos Reis et al. "Photoluminescence Imaging Induced by Laser Line Scan: Study for Outdoor Field Inspections". Proceedings of 7th World Conference on Photovoltaic Energy Conversion. IEEE. 2018.

ACCREDITATION FOR THE PV INVERTER TEST BENCH OF THE PV LABORATORY

Urs Muntwyler, André Werder, Luciano Borgna, Daniel Gfeller, Eva Schüpbach

Bern University of Applied Sciences BFH, Engineering and Information Technology, Photovoltaic Laboratory
Jlcoweg 1, CH-3400 Burgdorf, Switzerland
Tel. +41 34 426 68 37, urs.muntwyler@bfh.ch

ABSTRACT: The Photovoltaic Laboratory (PV Lab) at BFH is involved in the measurement of PV inverters since nearly 30 years. In the first years after 1988, stationary PV installations were used. On the new laboratory building in Burgdorf, a flexible 60 kWp PV plant allowed a wide range of measurements [1]. However, this is not very practical. For this purpose, four linear, high-speed PV array simulators with high stability (12 kW, 20kW, 25kW & 100kW) have been developed. In 2015, a new multistring-simulator with 3 x 11,5 kW was built. These projects were partly supported by the Swiss Federal Office of Energy SFOE.

In 2014 we started a project for the accreditation of the PV inverter test bench. Based on the quality management system QM-Pilot, which is standard at BFH, we could establish the processes. In September 2017 we got the accreditation for PV inverters. It gives the industry and the researchers qualified test processes for their measurements.

The project efforts were mainly supported by the Swiss Federal Office of Energy SFOE. Further support was given by the Swiss Competence Center of Energy Research for future grids (SCCER-Furies) and the Berner Fachhochschule Technik und Informatik BFH-TI.

Keywords: PV inverter tests, accreditation, PV-inverter, testing, PV-simulator, EMC

1 THE ACCREDITED MEASUREMENTS

The accreditation closes the effort of the researcher of the PV LAB in Burgdorf of the BFH-TI in the measurement of PV inverter. This started 30 years ago and finalized in a very unique test surrounding we use for student courses and projects. The new PV array simulators for 100 kW and 3x11,5 kW are mainly used for research and projects with industry. The goal of the project was to get the accreditation to give the industry more confidence into our measurements. Additionally, more and more certificates are needed to distribute the products worldwide. We can award such certificates now.

The accredited measurements are:

EN 50530 (IEC 50530)

Overall efficiency of grid connected photovoltaic inverters

EN 61000-3-2 (IEC 61000-3-2)

Electromagnetic compatibility (EMC): Part 3-2: Limits - Limits for harmonic current emissions (equipment input current <= 16 A per phase)

EN 61000-3-12 (IEC 61000-3-12)

Electromagnetic compatibility (EMC): Part 3-12: Limits - Limits for harmonic current produced by equipment connected to public low-voltage systems with input currents > 16 A and >= 75A per phase

CISPR 11 (EN 55011)

Radio interference caused by electromagnetic waves

Industrial scientific and medical equipment – Radiofrequency disturbance Characteristics – Limits and methods of measurement

1.1 QM – Pilot quality system

There were already accredited measurements processes established at Bern University of Applied Sciences. One of them is the "exhaust treatment" measurement of the "exhaust treatment facility" in Nidau/ Biel. Like the PV LAB it is part of the Institute "Energy + mobility". After the start of the project we could use the QM-Pilot quality management system which was introduced at BFH. It allows the administration of all the different processes and their descriptions.

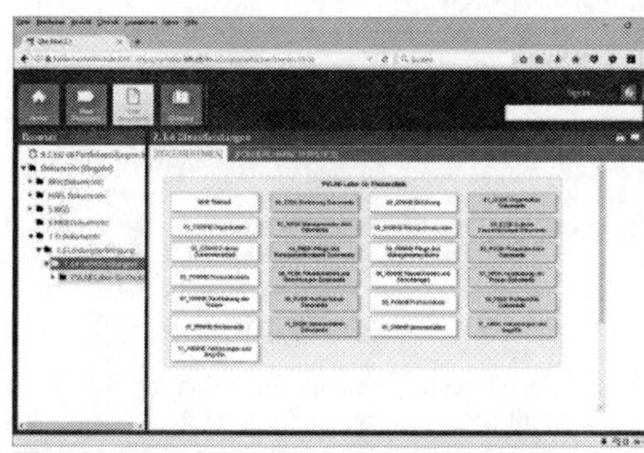

Figure 1: Structure of the QM-Pilot quality management system for the PV inverter test laboratory

1.2 QM – Pilot quality system

To have efficient and comfortable test surroundings, we moved the inverter test bench area to the first floor.

Figure 2: The accredited PV inverter test LAB with the "golden sample" PV inverter (Photo: PV LAB).

The main inverter test benches of 100kW and 3x11,5kW are in the same room. Some of the other inverter test benches are outside the closed area. This allows us to work in parallel and to offer researchers and students a simple access to the test equipment which is not accredited.

2 THE ACCREDITED MEASUREMENTS IN DETAIL

2.1 PV inverter test standard EN50530

The development of the PV inverter test standard was supported by the PV LAB. It allows a characterization of the performance and efficiency of PV inverters. PV inverters developed well in the last 30 years. They reach high efficiencies, are precise and answer quick on changes in the insolation. The test configuration:

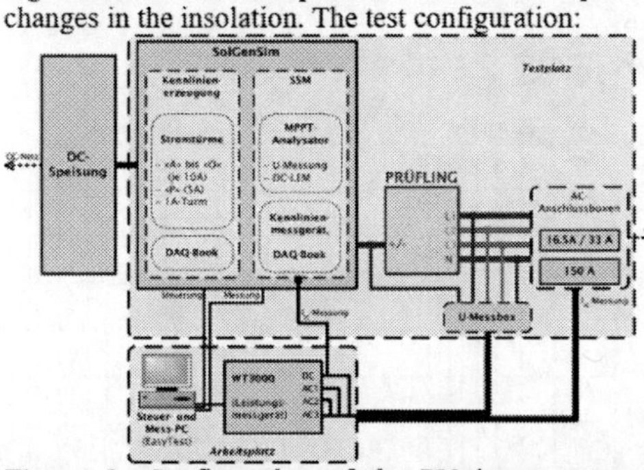

Figure 3: Configuration of the PV inverter test (EN 50530)

Figure 4: "Golden sample" – the reference PV inverter for the test bench (Photo: PV LAB).

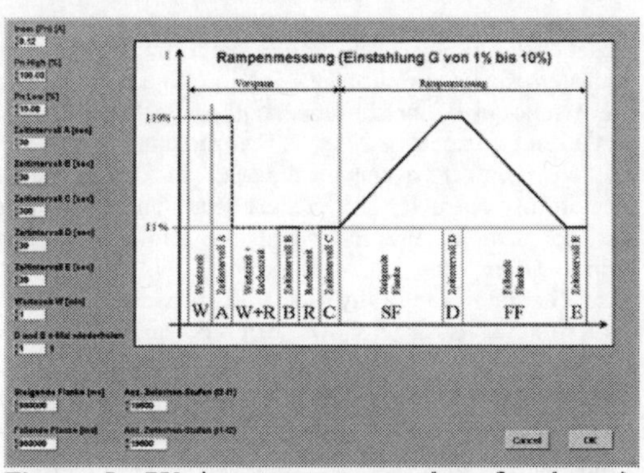

Figure 5: PV inverter test procedure for dynamic measurement

EN 61000-3-2 (IEC 61000-3-2) - Electromagnetic compatibility (EMC)

Part 3-2: Limits - Limits for harmonic current emissions (equipment input current <= 16 A per phase)

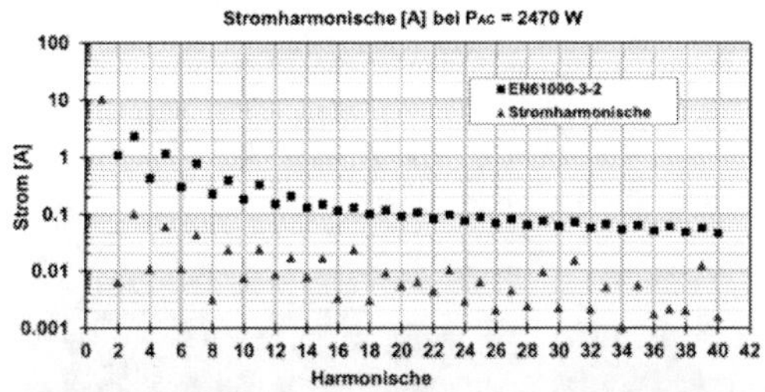

Figure 6: Comparison on the harmonic currents with the Norm (EN 61000-3-2

EN 61000-3-12 (IEC 61000-3-12) Electromagnetic compatibility (EMC):

Part 3-12: Limits - Limits for harmonic current produced by equipment connected to public low-voltage systems with input currents > 16 A and >= 75A per phase

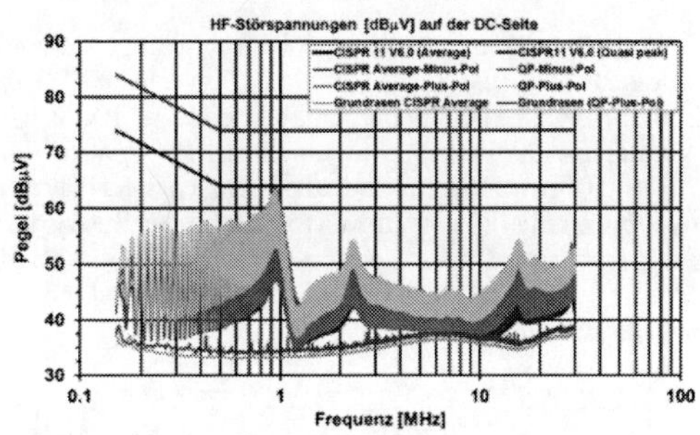

Figure 7: HF-interference voltage within the limits

CISPR 11 (EN 55011)

This standard deals with the radio interferences:

Radio interference caused by electromagnetic waves

Industrial scientific and medical equipment – Radio-frequency disturbance Characteristics – Limits and methods of measurement

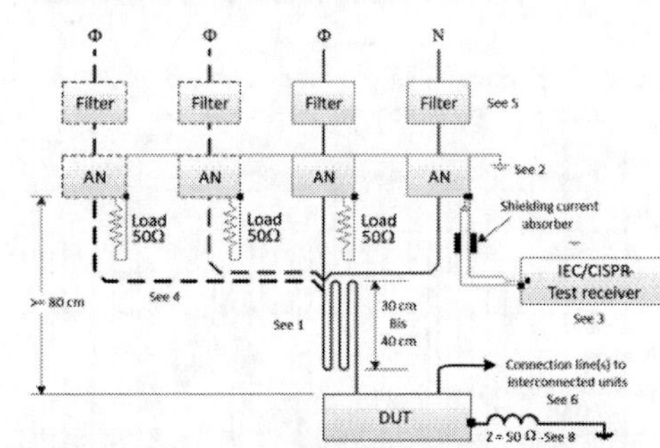

Figure 8: Basic measurement setup for the CISPR 11 Ed.6.0 2015-06 test

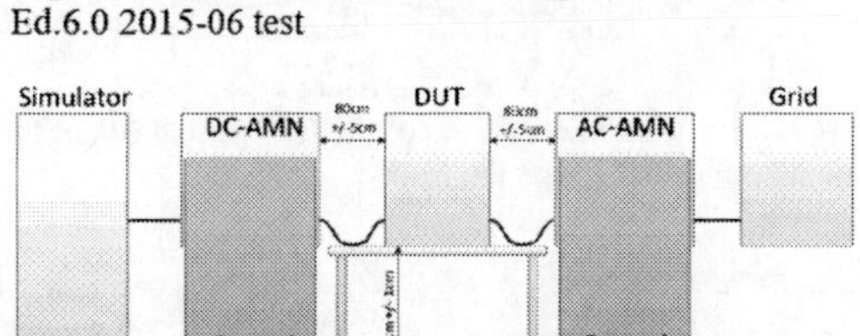

Figure 9: Measurement setup for small devices

CISPR measurement on a customer example:

Figure 10: CISPR measurement of a customer unit – the Power-Blox PBX-200 in the PV inverter test lab (Photo: PV LAB)

The Powerblox is a special unit for isolated microgrids from the Swiss Company Power-Blox AG in 5200 Brugg/ CH. It contains batteries, a PV solar regulator for the battery charge, a small PV inverter for 230 V/ 50Hz (or 60Hz) and allows to connect several Powerblox units in a swarm to form a bigger microgrid. In this scenario, all these connectors have to be measured according the CISPR standard. The following charts show these results:

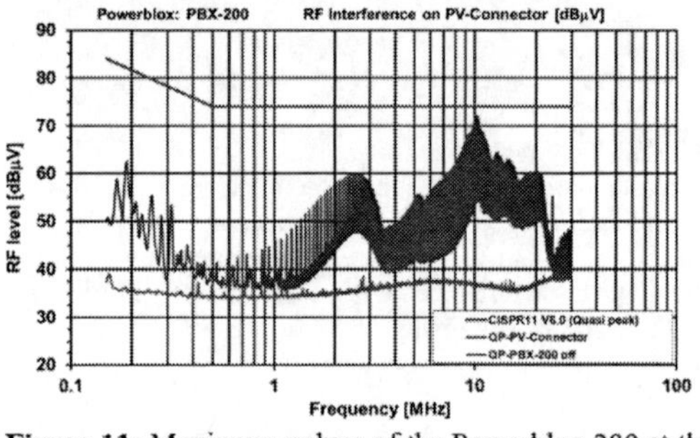

Figure 11: Maximum values of the Powerblox 200 at the PV connectors – pass the test!

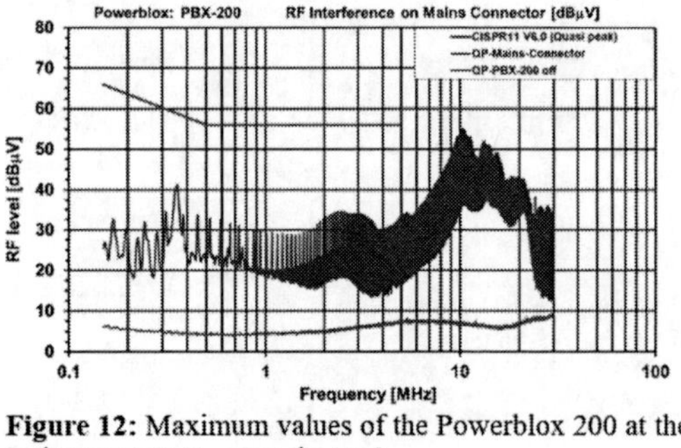

Figure 12: Maximum values of the Powerblox 200 at the main connectors – pass the test!

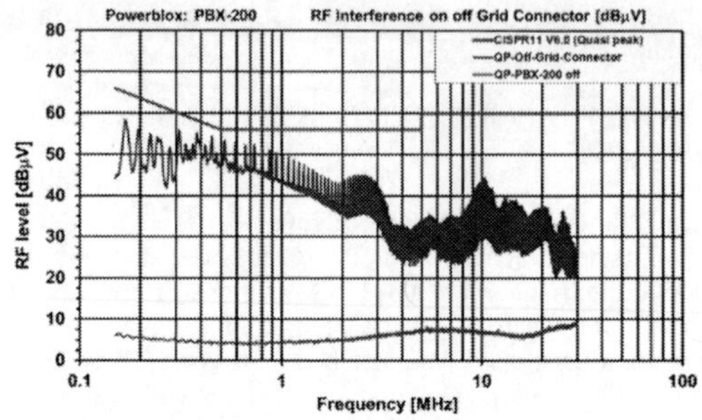

Figure 13: Maximum values of the Powerblox 200 at the island network connectors – pass the test!

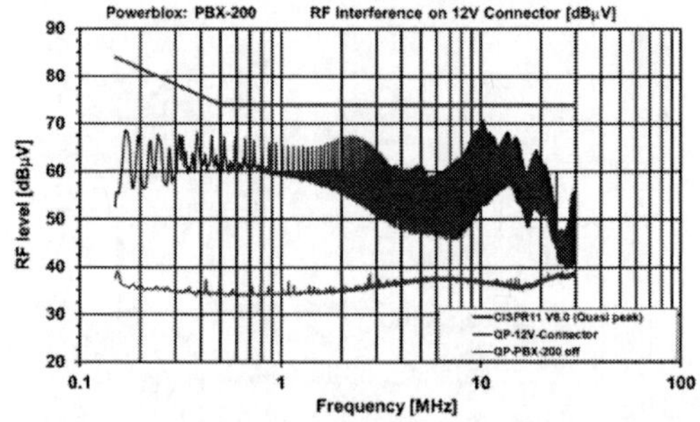

Figure 14: Maximum values of the Powerblox 200 at the 12 V connector!

The customer gets a full report of test results with all the technical measuring conditions. This can be needed for the export of his product in foreign countries. It's possible to get the test reports in several languages as German, French, Italian and English.

2.2 Use of the accredited PV inverter test bench

Since the accreditation, the request for measurements and measurement support went up. There are companies from industry and business which want to check and improve their devices. The service goes often over the real measurements. Our specialists have a broad experience in power electronics and can support the industry. This could also be done by projects supported from Innosuisse, which supports the efforts of the PV LAB if the company is based in Switzerland.

We often get systems for the test, which are not at all or not fully operable. This creates delays and extra costs for the customer. Proper manuals are also an important topic. Often, the software access for the system accreditation is not fully operable. Especially the PV battery storage systems shows such problems.

2.3 Measurement of PV battery storage systems

PV storage systems are a hot topic in the PV market. There are hundreds of systems in the market in mid Europe. However, the test standard for these systems has yet to be established similar to the PV inverters 30 years ago. The PV LAB supports the test standard from HTW Berlin according to the "Effizienzleitfaden". This complex measurement gives a clear picture about the static and dynamic behavior of such systems. The measurement is much more complicated as a PV inverter. However, the process is similar. The PV LAB established an automatic test bench for PV battery systems within the "accredited PV test laboratory".

Figure 15: PV LAB measures PV storage systems (Photo: Muntwyler)

The PV-LAB testbench for the automatic measurement according the “Effizienzleitfaden” [2] is controlled and regulated through the self-developed software “PV BATTTEST”[3] . It allows a 24hour/day and seven days a week working procedure. This needed to speed up the measurement process. It takes about one week to test a system.

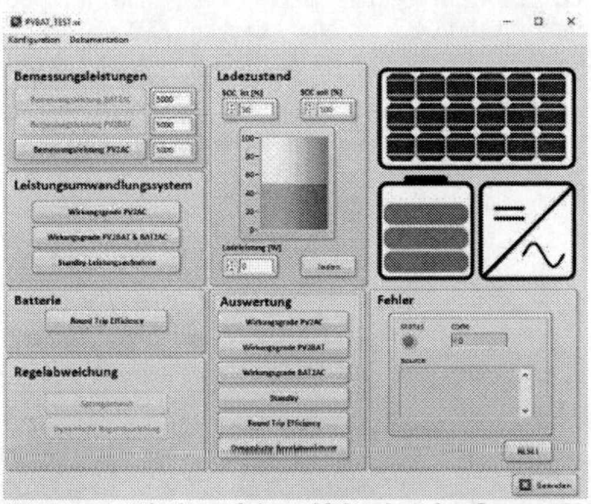

Figure 16: Display of the self-developed software “PV BATTTEST”, it allows us a fully automated test of PV battery systems in the PV LAB of the BFH-TI in Burgdorf.

3 ADDITIONAL COMPONENTS

The PV Lab at BFH Switzerland got his certification by the Swiss Association for Quality and Management Systems (SQS) in September 2017. The effort is supported by the Swiss Federal Office of Energy (SFOE) and the SCCER (Swiss Competence Centers for Energy Research) initiative. The test bench offers the researchers and the industry tests according to the relevant standards in a proven and checked manner.

Figure 17: The researchers from the PV LAB BFH in Burgdorf [4]

4 ACKNOWLEDGMENTS

This research is part of the activities of the Swiss Centre for Competence in Energy Research on the Future Swiss Electrical Infrastructure (SCCER-FURIES), which is financially supported by the Swiss Innovation Agency (Innosuisse - SCCER program). We also gratefully acknowledge funding from Bern University of Applied Sciences BFH, Burgdorf, Switzerland.

5 REFERENCES

[1] Photovoltaik, Strom aus Sonnenlicht für Verbundnetz und Inselanlagen, Heinrich Häberlin, VDE Verlag/ electrosuisse Verlag, 2. Ausgabe 2010
[2] Energieeffizienzleitfaden, HTW Berlin, Version 1.0; August 03/2017
[3] EffiBat Software, Jonas Wälten, Master Science Engineering MSE diplom report, 2018
[4] Jahresbericht PV Labor, BFH-TI, Januar 2018, www.pvtest.ch

DEVELOPMENT OF A MULTICHANNEL DATA LOGGER FOR THE MEASUREMENT OF MICROCLIMATIC PARAMETERS WITHIN ENCLOSURES OF PV INVERTERS

J. Zikulnig[1], L. Neumaier[1], W. Mühleisen[1], Ch. Hirschl[1], D. Holzmann[1], H. Heigl[2]
[1] CTR Carinthian Tech Research AG, Europastr. 12, 9524 Villach/St. Magdalen, Austria
(Author and Contact: johanna.zikulnig@ctr.at; T: +43 4242 56300 276)
[2] Fronius International GmbH, Guenter Fronius Straße 1, 4600 Thalheim bei Wels (heigl.hannes@fronius.com)

ABSTRACT: Microclimatic conditions inside of the enclosures of photovoltaic (PV) inverters are suspected to have a tremendous impact on degradation and failure modes of the devices. To address this issue, a custom multi-channel data logger has been designed which is capable of continuously measuring temperature, relative humidity and wetness in confined space. The collected data is written to a flash memory and additionally sent to a server via GSM to enable remote real-time monitoring. Consequently, the instrument is designed to be integrated in a PV inverter and to be utilised for the characterisation of the respective microclimatic conditions and the identification of error sources, where required. Preliminary field test results showed that the ambient conditions inside of an inverter are rather dependent on its duty than on environmental situations. The findings can be useful to optimise the design of PV inverters and their enclosures and hence contribute to an increased durability and a prolonged life-time.

Keywords: PV inverter, data logger, wireless monitoring, microclimatic conditions

1 INTRODUCTION

With the increasing standard in photovoltaic (PV) technologies the demands for long-term stability and reliability of complete PV-systems have risen as well. Due to the establishment of maximum power point (MPP) tracking systems and the increasing amount of off-grid system installations the corresponding technologies have been advancing making inverters the centrepieces of PV-systems. However, this development is also being accompanied by longer operating times and higher technical requirements for inverters. While there has been a scientific focus on examining the influence of environmental parameters on the failure and degradation modes of PV modules [1-3], the monitoring of PV inverters has widely been neglected. Within the scope of the Austrian flagship project INFINITY climate specific degradation factors have been observed extensively. The findings indicate that for PV inverters not the ambient climate, but rather the microclimate within the inverter's enclosure, which can contain extreme conditions, might be responsible for observable degradation effects. Hence, as part of this research project a multichannel data logger for the continuous measurement of temperature, humidity and condensed water within the enclosure of spatially distant PV inverters during operation has been designed. The aim of this development is to investigate potential relations between device failures and locally focused variances in temperature and humidity, as well as possible condensation effects on specific parts. Furthermore, the data logger is utilised to observe and understand possible sources of noticeable changes in the monitored parameters and, consequently, could help to sustainably improve the quality of the design of inverters and their enclosures.

2 APPROACH

To meet the objectives, the design of the data logger has to fulfil a variety of highly specific requirements. As it should be integrated in a PV inverter's enclosure with limited space it has to have certain geometric dimensions. While a high measurement resolution is rather secondary, the instrument has to be reliable, long-term-stable, robust and as independent of any infrastructure (e.g. wireless network or local area network) as possible.

2.1 Technical Realisation

The designed data logger most importantly consists of a central micro controller unit (MCU, *Arduino MKRZero*) and a GSM/GPS module (*Adafruit FONA 808*), as schematically visualised in the block diagram in Figure 1. The power is supplied via a 5 V USB connection (inherent port inside of the inverter), which is complemented with a 1000 mAh 3.7 V Li-Ion battery providing enough power to cover possible 2 A peaks occurring during data transmission.

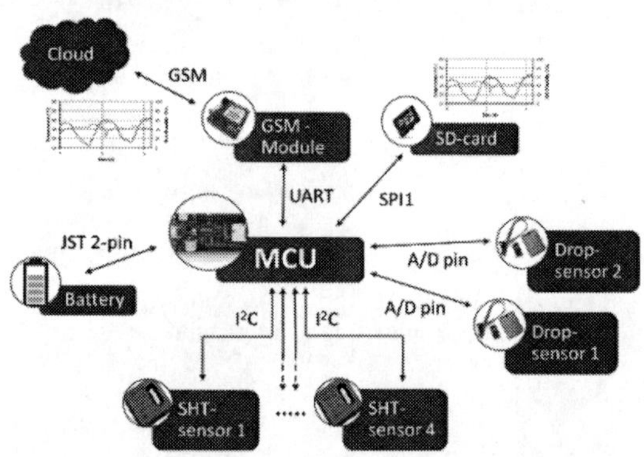

Figure 1: Schematic representation of the hardware design

The setup has been equipped with four sensors for a combined measurement of temperature and relative air humidity (*Sensirion SHT21*, referred to as TH-sensor in the following) and two capacitive sensors to detect actual wetness based on changes in the dielectric constant when contacting liquids (referred to as drop-sensors in the following). Despite the inverter's comparatively small inner volume it is necessary to install at least four temperature-humidity sensors to cover all corners and two wetness sensors in bottom areas where condensed water would collect. The measurement results are written to a non-volatile memory (SDHC card) on a regular basis and additionally sent to a server via GSM communication to enable remote monitoring and data export. The measurement values are queried with a frequency of 1 Hz and a statistical evaluation of the sensor data is implemented. Due to the communication via GSM, the data logger is independent of local area networks, which allows for installations in infrastructurally restricted areas. In addition, device parameters, such as the data-logging and the server updating interval, can be adjusted from afar, which ensures a maximum of flexibility while implementing a highly specialised device.

2.2 Characterisation Test Setup

In the first instance the data logger module's basic functionality and suitability for the given task were evaluated simulating extreme ambient conditions and particular changes in temperature and humidity in a climatic chamber (*Vötsch VC³ 4018*). The tests shall confirm that the module works reliably within the predefined specifications. This means that it is supposed to measure ambient temperature as well as the relative humidity and evaluate the respective maximum, minimum and average measurement value of each sensor for the given logging interval. Furthermore, it has to be capable of detecting the formation of water droplets. The data shall then be logged to a non-volatile memory every five minutes, while the GSM logging interval is ought to be 20 minutes. The overall hardware setup is supposed to work reliably in a temperature range between -30°C and 80°C, which is limited by the tolerances of its individual components, and a relative humidity between 0 % and 100 % in the given temperature range. Hence, the tests have primarily been specified to confirm the device's functionality, as well as its capability to withstand possible harsher environmental conditions. Figure 2 shows an example for one of the specified temperature and humidity profiles. Further tests were inspired by meteorological data of locations with rather extreme climate conditions, whereby a one-year cycle was projected to a testing runtime of 24 hours. The reference climate data was obtained from DWD (Deutscher Wetterdienst).

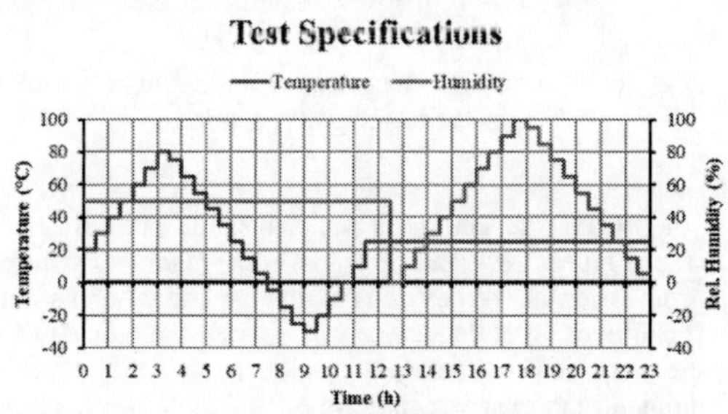

Figure 2: Exemplary specified temperature and relative humidity profile for one of the tests in the climate chamber

To simulate operation under the conditions of intended use, the module was placed inside a plastic enclosure with metal parts mounted on the inside to provoke condensation of air humidity due to their low specific thermal capacity when varying the temperature.

2.3 Field Test Setup

For the field test two PV inverters have been equipped with data loggers, as illustrated in Figure 3. Inverter 1 was then operated under laboratory conditions, while inverter 2 was mounted outdoors on a roof. The ambient conditions of inverter 1 can be considered as controlled laboratory environment with a surrounding temperature between 21°C and 24°C. Inverter 2 has been exposed to outdoor conditions, whereat the ambient temperature was simultaneously recorded with a separate instrument. The outdoor location is A-4600 Thalheim bei Wels in Upper Austria (Köppen-Geiger climate classification: Dfb - temperate/humid continental climate) and the tests presented in this paper were conducted in July 2018.

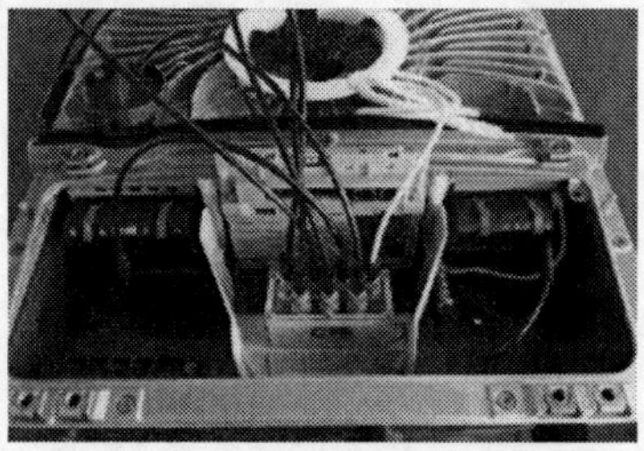

Figure 3: Assembly situation inside the PV inverter

The sensors were placed at dedicated identical positions inside of the two inverters. TH-sensor 1 was placed at the upper power section, TH-sensor 2 at the DC connection area, TH-sensor 3 at the AC connection area and TH-sensor 4 at the lower power section. The drop sensor 1 was positioned at the connection area and the drop sensor 2 at the power section where condensed water would be expected to collect.

3 RESULTS AND DISCUSSION

3.1 Characterisation Test Results

The climate chamber test results confirmed the hardware's suitability for the given application with respect to the required temporal and measurement resolution, as well as its durability and stability. All sensors gained reasonable data and were able to reproduce the specified profiles, as illustrated in Figure 4 and Figure 5. The minimum, the maximum and the average measurement values for each sensor lie close to one another, hence the minimum and maximum values are difficult to see in the plots. Figure 5 features a magnified curve between the hours 2 and 3 to illustrate how the progression of the average measurement values is framed by the corresponding minimum and maximum values.

The waviness of the relative humidity values observable in Figure 5 is reasonable and can be explained due to the slow adaption of the relative humidity by the climate chamber during temperature changes. The simultaneously recorded reference curve of the climate chamber during the test run show a quasi-identical shape. During several test runs condensation could provoke the formation of droplets which were perfectly recorded by the drop sensors validating the device's suitability for the given task.

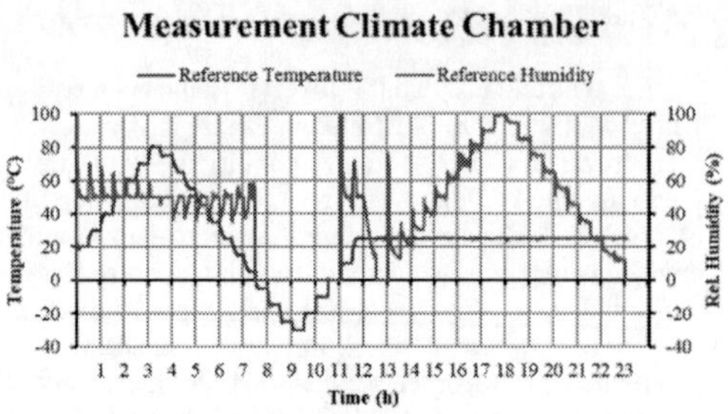

Figure 4: Climate chamber reference measurement during the test presented in Figure 2

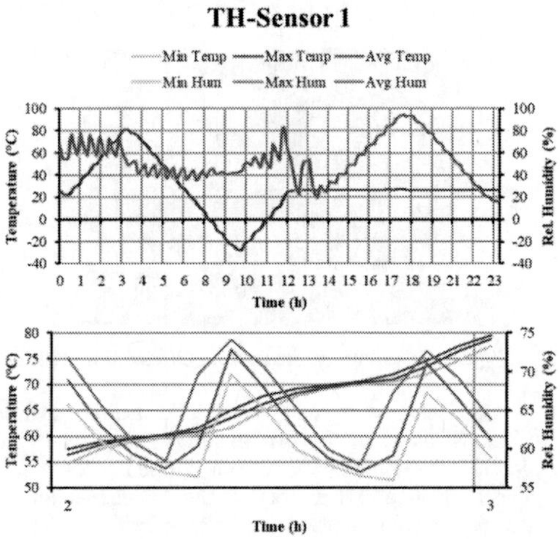

Figure 5: Measurement results (minimum, maximum and average values) of the TH-sensor 1 during the test presented in Figure 2

Subsequently to the tests, a visual observation and a technical confirmation of full functionality indicated that the device is capable of withstanding harsh environmental conditions.

3.2 Field Test Results

As described in 2.3, inverter 1 and inverter 2 have been operating under laboratory conditions and under outdoor conditions, respectively. The measurements were started at noon, hence the time scaling of the subsequent figures does not refer to a certain date but to an absolute duration since the beginning of the records. The presented test run for inverter 1 took 19 days, while inverter 2 was tested for 22 days.

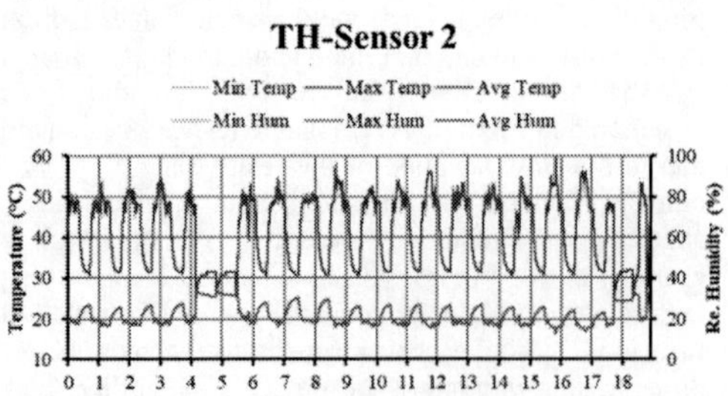

Figure 6: Inverter 1 - temperature and relative humidity in the upper power section

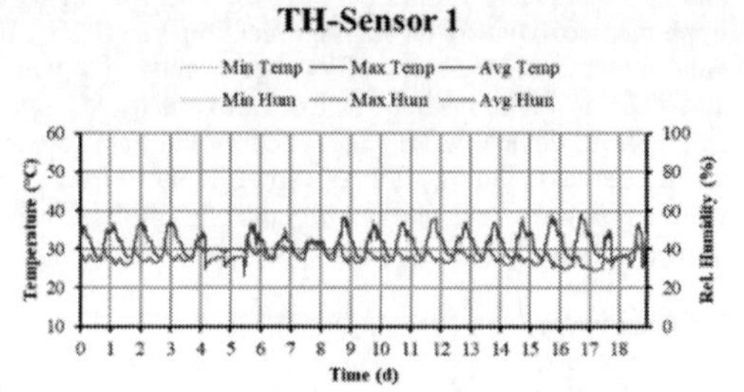

Figure 7: Inverter 1 - temperature and relative humidity in the DC connection area

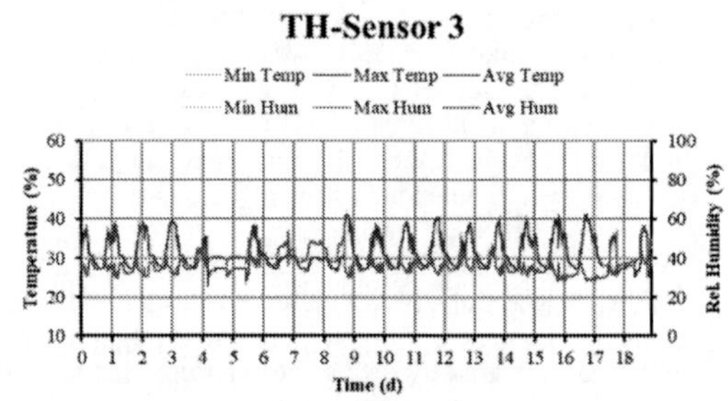

Figure 8: Inverter 1 - temperature and relative humidity in the AC connection area

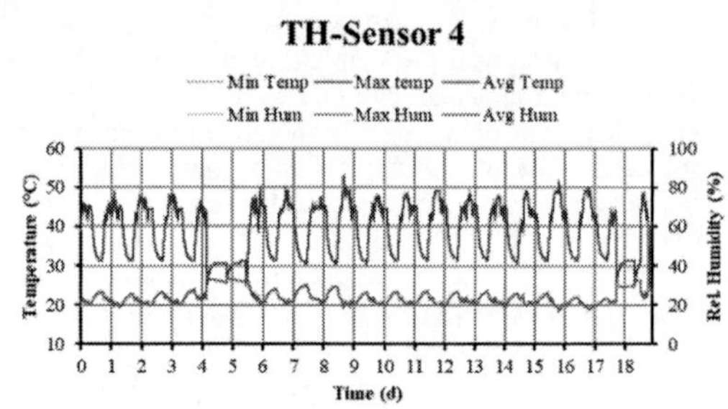

Figure 9: Inverter 1 - temperature and relative humidity in the lower power section

Inverter 1: The measurement results from inverter 1 (see Figure 5-8) indicate that the inner temperature progression of the device is highly dependent on its duty. It is obvious that the device was not operating on day 5 as the recorded temperature values match the expected ambient laboratory conditions. The local temperatures under load vary depending on their actual position. The DC connection area reaches temperatures of up 59.5°C (see
Figure 7), while the maximum temperature in the upper power section does not exceed 39°C (see Figure 6). The temperature and humidity profiles of TH-sensor 1 (upper power section) and TH-sensor 3 (AC connection) are very similar, which presumably attributes to the physical proximity of those two sensors.

The relative air humidity in the upper power section and the AC connection area (mean values 36 % and 37 %) is higher than in the DC connection area and the lower power section (mean values 22 % and 24 %) (see Figure 5-8). While the relative humidity profile of TH-sensor 2 and TH-sensor 4 follows the principle of indirect correlation with the temperature, a less plausible profile is observed with TH-sensor 1 and TH-sensor 3. Here the relative humidity initially increases with increasing temperature. This might be due to the active cooling with a fan which causes an exchange of air in this area.

Neither of the drop sensors recorded any condensation or wetness inside of the inverter during this test run.

Inverter 2: The outside temperature profile on the rooftop has been recorded using a separate instrument and the profile is illustrated in Figure 10.

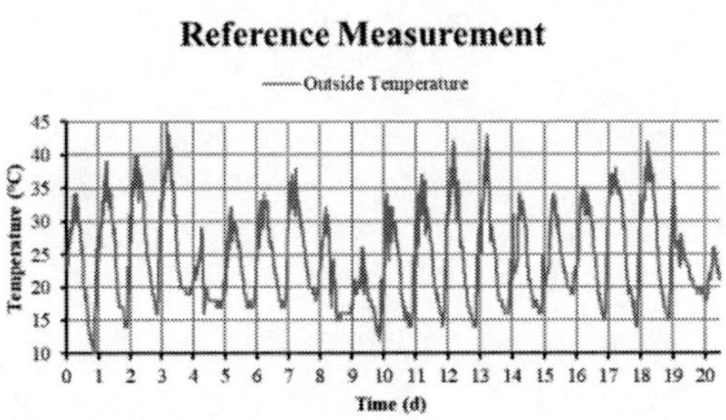

Figure 10: Reference Temperature on the rooftop for the outdoor test of inverter 2

Again, it can be observed that the recorded temperature profile at different locations inside of the device is rather dependent on the inverter's duty than on the outside temperature (see Figures 10-13). However, even though the ambient conditions do not seem to have a drastic influence, it can be observed that the maximum temperature peaks of TH-sensor 1 (Figure 11) and TH-sensor 3 (Figure 13) on day 4 and day 9 are lower than the other peaks. This perfectly agrees with the moderate outside temperature on the respective days, as illustrated in Figure 10. The maximum temperature in the DC connection area reaches 61.5°C (see Figure 7), which is just marginally higher than under laboratory conditions (59,6°C), which emphasises the low influence of ambient conditions. The maximum temperature values recorded by the other sensors are slightly higher than under laboratory conditions. Again, a similarity between the temperature and humidity profiles of TH-sensor 1 (upper power section) and TH-sensor 3 (AC connection) due to their physical proximity can be observed. Besides that, for the externally mounted PV inverter only the humidity profile in the DC connection area looks as expected. TH-sensor 3 and TH-sensor 4 recorded comparatively fast fluctuations following temperature fluctuations during load. This can again be attributed to the air exchange due to active cooling with the fan. The relative humidity profiles measured by TH-sensor 1 and TH-sensor 3 are again higher (mean values 45 % and 46 %) than the values recorded by TH-sensor 2 and TH-sensor 4 (mean values 25 % and 32 %).

The drop sensors did not record any condensation or formation of droplets inside of the inverter during this test either.

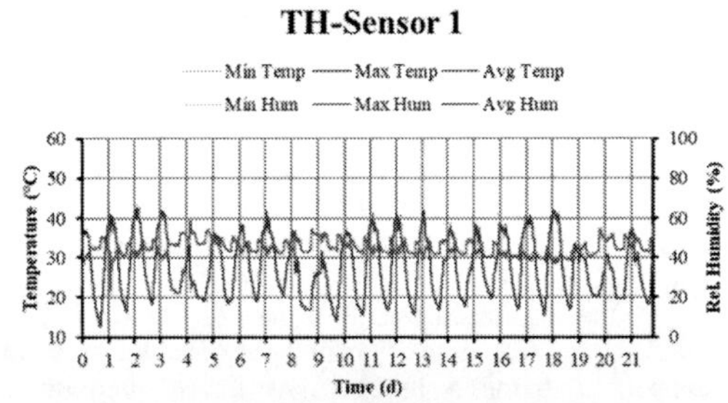

Figure 11: Inverter 2 - temperature and relative humidity in the upper power section

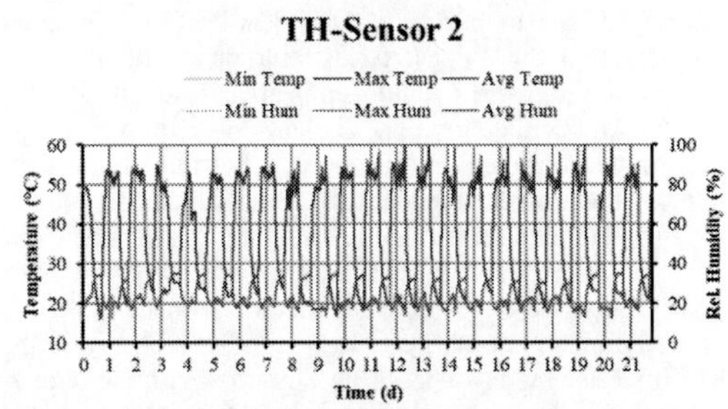

Figure 12: Inverter 2 - temperature and relative humidity in the DC connection area

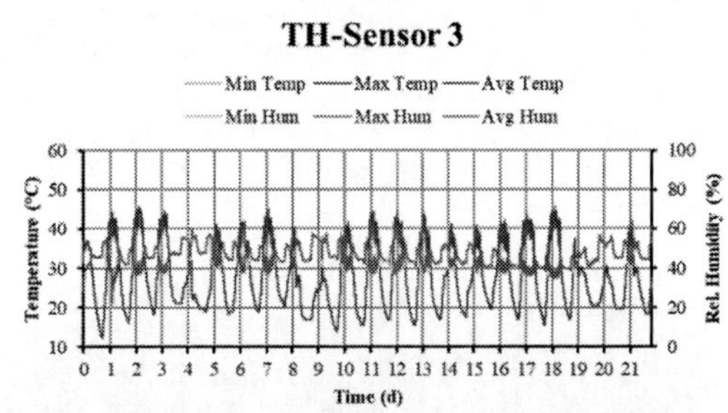

Figure 13: Inverter 2 - temperature and relative humidity in the AC connection area

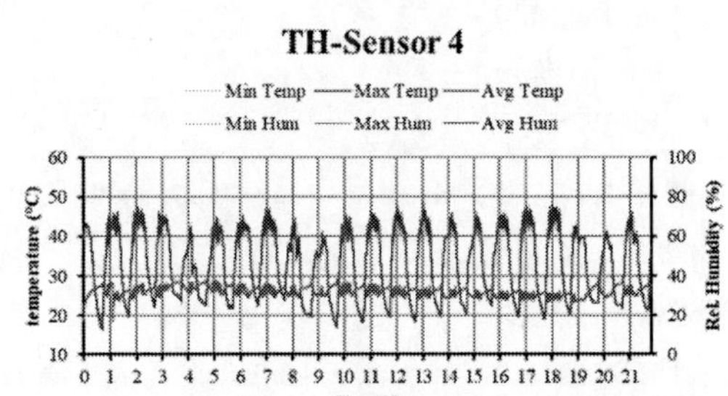

Figure 14: Inverter 2 - temperature and relative humidity in the lower power section

4 SUMMARY & OUTLOOK

The observations as part of this work indicate that the microclimatic conditions inside the enclosure of PV inverters are highly dependent on their duty, while the influence of the ambient conditions seems to be marginal. The measurement profiles of the sensors integrated in the inverter operated under laboratory conditions and the externally mounted inverter on a rooftop do not differ drastically. However, in periods without load the inner temperatures adapt to the outer conditions. Generally, in the DC connection area, a lot of heat is generated, still the other device areas are hardly influenced by this condition. The active cooling with a fan seems to be particularly effective in the upper power section and the AC connection area as the relative humidity profile indicates an exchange of air in the respective areas.

Neither of the droplet sensors recorded any wetness during the test run, which is expectable, as the temperature-humidity profiles have never approximated their dew point. Furthermore, this indicates that there has not been any ingress of liquids from outside.

However, the measurement results collected by the designed data logger are powerful indicators to evaluate the design of the PV inverter. It could be identified that the active cooling works well in some areas, while the cooling in the DC connection area and the lower power section might be improved. Furthermore, it became obvious that during the test run no condensed water collected inside the PV inverter and that no liquid could ingress the enclosure.

The first trials were conducted using new PV inverters with moderate environmental factors. In a next step the microclimatic conditions inside of aged or retired PV inverter models should be observed as the enclosure and the seals might degrade over time. Further studies might reveal if there is a connection between the degradation and corrosion of used materials and the ambient conditions, such as high humidity or salty air and could help to adapt the design for particular climate zones. Generally, the influence of more extreme seasonal conditions in Austria as well as in other climate zones is still to be observed and long term measurements will be conducted.

5 ACKNOWLEDGEMENTS

This work was conducted as part of the Austrian "Energy Research Program" project INFINITY. Funded by the Austrian Climate and Energy Fond and the Austrian Research Promotion Agency (FFG), both are gratefully acknowledged.

6 REFERENCES

[1] Jordan, D. C., Silverman, T. J., Wohlgemuth, J. H., Kurtz, S. R., and VanSant, K. T. (2017) Photovoltaic failure and degradation modes. *Prog. Photovolt: Res. Appl.*, 25: 318–326. doi: 10.1002/pip.2866.

[2] K. Yedidi, S. Tatapudi, J. Mallineni, B. Knisely, J. Kutiche and G. TamizhMani, "Failure and degradation modes and rates of PV modules in a hot-dry climate: Results after 16 years of field exposure," *2014 IEEE 40th Photovoltaic Specialist Conference (PVSC)*, Denver, CO, 2014, pp. 3245-3247.

[3] Dobaria B.V., Sharma V., Adeshara A. (2018) Investigation of Failure and Degradation Types of Solar PV Plants in a Composite Climate: Abstract After 4–6 Years of Field Operation. In: SenGupta S., Zobaa A., Sherpa K., Bhoi A. (eds) Advances in Smart Grid and Renewable Energy. Lecture Notes in Electrical Engineering, vol 435. Springer, Singapore

RELIABLE-THERMAL DESIGN OF SMART CONVERTERS FOR NZEB APPLICATION

G. Graditi*, G. Adinolfi, R. Ciavarella, A. Scognamiglio, V. Palladino
ENEA - Italian national agency for new technologies, energy and sustainable economic development
Portici Research Centre - 80055 Portici (Naples) - Italy
*Corresponding author: 081-7723400 giorgio.graditi@enea.it

ABSTRACT: In recent years, renewable energy sources are undergoing a great increase. In particular, photovoltaic systems are assuming a new role. They are employed in traditional plants, but they have also been integrated in new contexts, such as DC micro-grids and Net Zero Energy Buildings. In these scenarios, it is fundamental to optimize the generators behavior and their reliability. The key idea of this paper is the design of efficient Smart Converters able to maximize the energy with suitable performance indices and also to provide information and ancillary services. A software tool has been developed to support NZEB design by optimized Smart Converters employment. The software permits to choose the energetic scenario and to obtain the more suitable converter design through a graphic user interface.
The evaluation of thermal stresses and reliability indices influencing the converters behavior is carried out by dedicated subroutines. The developed tool provides non dominated Smart converter solutions for specific application contexts. The tool is produced by means of MathWorks MATLAB environment. A case study is presented.
Keywords: Photovoltaics, Net Zero Energy Building, Smart Converters.

1 INTRODUCTION

Actual energetic scenarios seem to take energetic and reliability advantages by Smart converters introduction [1]. Generation and absorption units characterizing NZEB (Net Zero Energy Building), and DC microgrids can be equipped with Smart converters. In this paper the attention is focused on NZEB context.

In detail, NZEB represents a highly performing building that autonomously provides to its energetic generation by RES, such as PV, geothermal energy, Fuel Cells, etc. (Fig. 1). The figure shows different RES, generation and consumption units. In typical NZEBs, PV plants, Fuel Cells and storage systems can be adopted and suitably matched with absorbed energy.

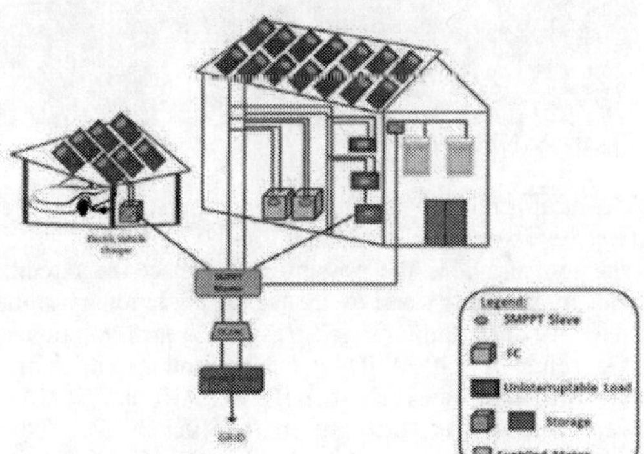

Figure 1: NZEB Scenario

The main goal of a NZEB concerns the balance of energy consumption and energy supply also by RES.

To obtain this result also the energetic system is changing from a centralized type to a distributed and web energy generation system. The development of appropriate devices is required to measure and control the energetic fluxes and data information [2].

In Section II a presentation of the architecture system is reported, in Section III the developed software architecture is reported with its electro-thermal and reliable- thermal subroutines description. In Section IV a case study and its results are proposed.

2 SMPPT CONVERTER

Nowadays, SMPPT (Smart Maximum Power Point Tracking) converters represent promising solutions for NZEB e NZEC (Net Zero Energy Cluster) contexts, as presented in Fig. 1. The introduction of this Smart converter can improve the energy management in complex scenarios. In fact, every generation/ consumption device can be equipped with a Smart Converter, this expedient assures the tracking of the maximum power point by PV and the data acquisition to connect with other SMPPT devices.
For this reason, authors provide a tool to support their optimized design.
The proposal architecture is based on a Master/Slave converter system (Fig. 2), where the master operates like a central controller. Each slave is placed on a single source/load to data and energy control.

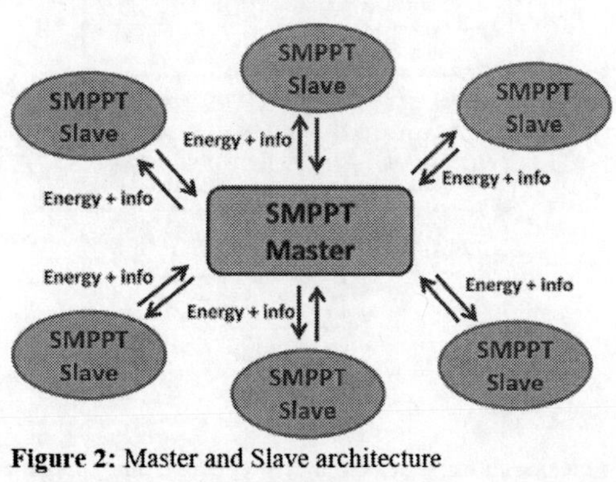

Figure 2: Master and Slave architecture

The master continuously exchanges information with slaves about energy flows and it tries to balance the energy production and consumption; in addition, it manages loads schedule. For example, if the master receives an energy demand relative to a particular slave, it can "decide" to require supplementary power to balance the energy equilibrium or to decrease the energy bill scheduling it in a more advantageous time slot.
The controller can also manage uninterruptable loads and

it can reschedule not urgent consumptions requests.

3 RET TOOL STRUCTURE

In recent times many energetic tools were established to define NZEB design and configuration and to implement RES in residential context [3-4].
The proposed software deeply differs from the other ones and it cannot be classified as proposed in [2-3]. The aim of the Reliable Electro–Thermal Design tool development consists in supporting designers to identify optima switching converters able to control energetic and information fluxes.
The functional schematic of the tool, called RET (Reliable Electro – Thermal Design), is represented in Fig 3 [5]. The structure provides a step by step design to select an optimum Smart converter. The software tool allows many configurations by a flexible architecture.

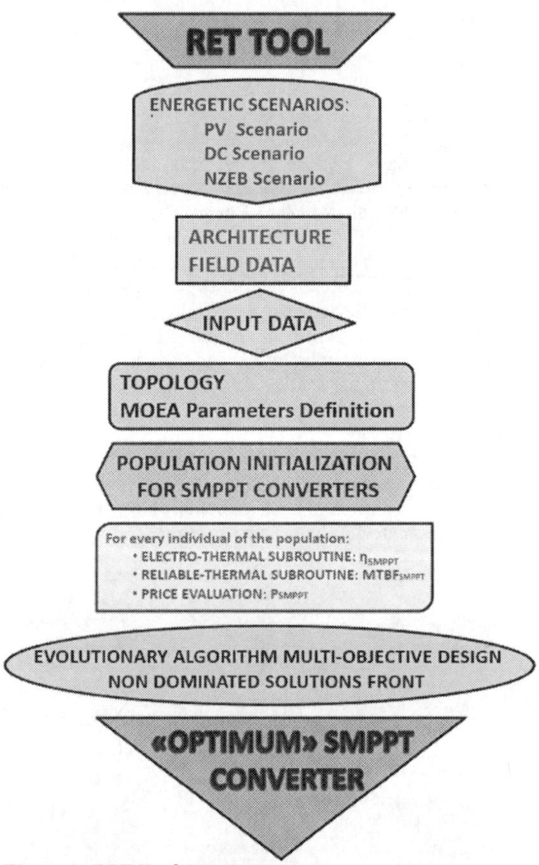

Figure 3: RET Tool Structure

In details, the first selection step concerns the energetic scenario; there are three different options (Fig. 4): PV installations, DC microgrids and NZEB scenarios. In the PV one, the user can design a converter to the maximization of the generated power and to the online tracking of the PV generator Maximum Power Point. In the second case it will achieve a smart device for the contribution to energetic control strategies of DC microgrids. Finally, the converter can be also designed to domestic NZEB contexts.

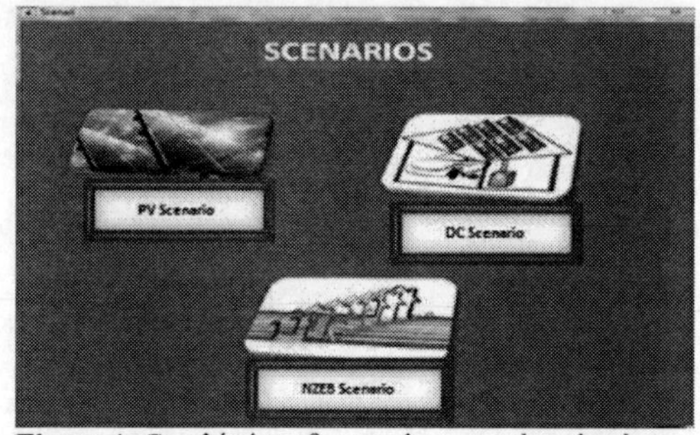

Figure 4: Graphic interface to the scenario selection

After the energetic scenario and the architecture selections, the designer can introduce field data for the specific application. They include buildings coordinates, available roof and nearby surfaces for renewables installations and family units number (Fig. 5). In addition, operating data have to be selected by proper file containing real meteo-climatic data.

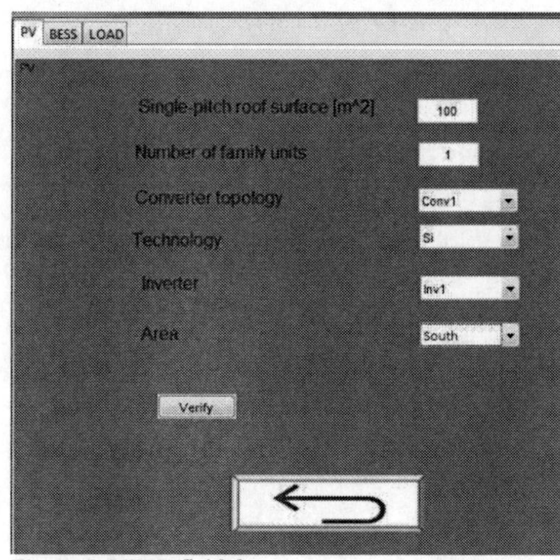

Figure 5: NZEB field data

The user can also import custom operating data files relative to a specific installation site.
The tool provides the possibility to select the circuital converter topology and to choose the evolutionary multi-objective algorithm among Strength Pareto Evolutionary Algorithm II (SPEA-II) and Non dominated Sorting Genetic Algorithm II (NSGA-II). SPEA-II and NSGA-II were selected as they are more efficient than other algorithms to solve multi-objective optimization problems with great amount of individuals [6-7]. After the evolutionary method selection, the first individual generation is randomly created by choosing into commercial electronic component databases to set up the initial population. Every converter device must respect technical requirements verified by a checking phase. In case of uncompliant device, another one is randomly selected in the database. The chromosome constituting the single converter solution is obtained only after devices selection and constraints validation.

3.1 Subroutines

An electro-thermal and reliable-thermal subroutines execution allows evaluating the single chromosome

solution performance in real operating conditions characterized by electric and thermal stresses. The mentioned subroutines permit to evaluate the need and to suitably size heatsink and thermal dissipation systems to the proper converter functioning operations.
In details, the electro-thermal subroutine calculates the back PV generator temperature T_{bk_PV}, for every working condition and it determinates the maximum stressing temperature among all the considered ones. Then the software attests design constraints also taking into account thermal effects on components assets. After this analysis, if necessary, the tool RET designs a heatsinks for needy components.
Finally the electro-thermal subroutine returns the efficiency fitness function η_{SMPPT}, for every working condition.

The reliability evaluation is a critical aspect for every electronic system, which operates in continuous functioning scenarios. In this case the military reliability prediction model (Military Handbook 217F) [8] is applied to measure the Smart Converters Mean Time Between Failure ($MTBF_{SMPPT}$) on the basis of the topology and the thermal information.

The last step of the RET tool, consists on the Smart Converter price evaluation.

4 CASE STUDY

In this section a NZEB case study is described. At the tool execution start, the energetic scenario is selected.
This study investigates generation/consumption devices concerning a single-family home. A 100 sqm rooftop area is available for RES installation. Energetic needs are typical of a family made up of four people who come home at about 6 p.m. Their house is equipped with main domestic appliances. On the basis of the energy absorption a PV plant constituted by twenty PV HIP-200NHE1 generators is designed. Each PV module is equipped with its own Smart converter placed on the generator back side.
The HIP-200NHE1 electrical data are reported below:

- Maximum power (Pmax): 200W
- Max. power voltage (Vpm): 40.0V
- Max. power current (Ipm): 5.00A
- Open circuit voltage (Voc): 49.6V
- Short circuit current (Isc): 5.50A

The residential NZEB is placed in South Italy. The Smart converter is designed taking into account the site operating conditions.
Moreover a 5kWh storage system is employed. It contributes to the energetic requirements fulfillment in high consumption time slots.

In the following figure (Fig. 6) RET results for the NZEB under investigation are represented.

3D and 2D graphs report dominant (red) and not dominant solutions (black) obtained for the PV Smart converter design. The 2D graphs offer details about MTBF and Efficiency versus costs. The designer can choose the best trade off solution thanks to a drop down list.

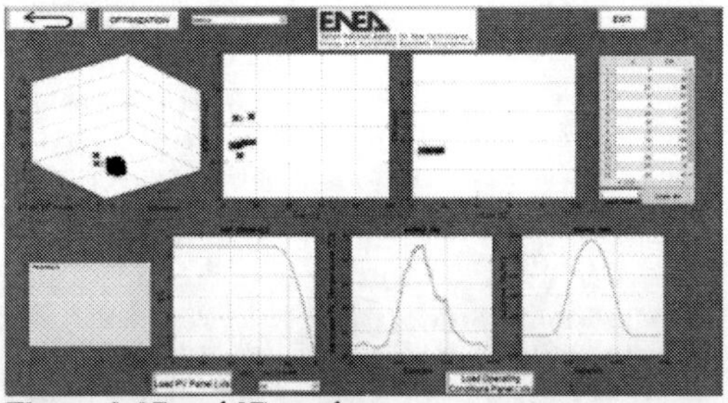

Figure 6: 3D and 2D graphs

5 CONCLUSION

This paper focus on a software tool developed to support the NZEB designer in generation/ consumption match by Smart converters employment. A preliminary description on NZEB context is reported and then SMPPT converter and the architecture are presented. Smart converters are able to maximize the generated energy by continuously impedance matching operations.
A case study is carried out about a single-family NZEB placed in South Italy whose energetic balance is guaranteed by a PV installation and a storage system suitably matched to the domestic needs and requirements

6 REFERENCES

[1] M. A. Guerrero-Martinez, M. I. Milanes-Montero, F. Barrero-Gonzalez, V. M. Miñambres-Marcos, E. Romero-Cadaval and E. Gonzalez-Romera "A Smart Power Electronic Multiconverter for the Residential Sector" Sensors, vol.17 (2017).

[2] A. Scognamiglio, G. Adinolfi, G. Graditi, E. Saretta "Photovoltaics in Net Zero Energy Buildings and Clusters: enabling the smart city operation", The 6th International Conference on Applied Energy, 61, 1171 – 1174, (2014).

[3] S. G. Attia, A. De Herde "Early design simulation tools for net zero energy buildings: a comparison of ten tools", 12th Conference of International Building Performance Simulation Association, Sydney, 94-101 (2011).

[4] M. Iqbal, M. Azam, M. Naeem, A.S. Khwaja, A. Anpalagan, "Optimization classification, algorithms and tools for renewable energy: A review", Renewable and Sustainable Energy Reviews, vol.39, 640–654, (2014).

[5] G. Adinolfi, R. Ciavarella, V. Palladino, M. Valenti, G. Graditi "A multi-objective optimization design tool for Smart Converters in photovoltaic applications" International Symposium on Power Electronics, Electrical Drives, Automation and Motion, 794-799, (2018).

[6] H. Mejbri, K. Ammous , H.Morel , A. Ammous,"Optimal design of power converter using multi-objective genetic algorithm", International Review on Modelling and Simulations (I.RE.MO.S.), vol. 5, no.2, 793-801, (2012).

[7] H. Visairo, M.A. Medina, J.M Ramirez, "Use of Evolutionary Algorithms for Design Optimization of Power Converters", 22nd International Conference on Electrical Communications and Computers, vol.1, 268-272, (2012).

[8] Handbook of 217 "PlusTM Reliability Prediction Models", Quanterion Solutions, Dec. 2014

SUBSTRING-MPPT FOR 4-TERMINAL 3-SUBSTRING MODULES

R. Brace[1)], A. Neumann[2)], T. Czarnecki[1)], R. Merz[2)]
1) BRC Solar GmbH, Robert-Bosch-Str. 49, 69190 Walldorf, Germany
2) University of Applied Science Karlsruhe, Moltkestraße 30, 76133 Karlsruhe, Germany
*corresponding author: richard.brace@brc-solar.de

ABSTRACT: State of the art residential photovoltaic (PV) systems are sensitive to partial shading. Shaded modules subjected to lower solar irradiance result in lower current in a PV-string. To minimize power losses string inverters activate bypass diodes in a shaded module resulting in unused power potential. Common module-based power optimizers (MPO) or module inverters lead to mismatch losses in shaded PV-modules. Individual substring maximum power point tracking minimizes mismatch and shading losses for a partially shaded module, resulting in a higher shaded module power output compared to MPOs and bypass diodes. A first unoptimized Substring-MPPT prototype adapted to one substring shaded by 25 % and another one shaded by 30 % raises the output power from 34 % of the theoretical maximum to 90 % whilst avoiding hotspot generation.

Keywords: maximum power point tracker, partial shading, photovoltaic string, bypass diode, substring power loss,

1 INTRODUCTION

State of the art residential photovoltaic systems connect a number i of photovoltaic (PV) -modules M_Y ($Y = 1...i$) consisting of j substrings S_x ($x = 1...j$) in series to form a PV-string. Figure 1a) depicts such a PV-module M_Y consisting of $j = 3$ substrings S_1, S_2 and S_3, with bypass

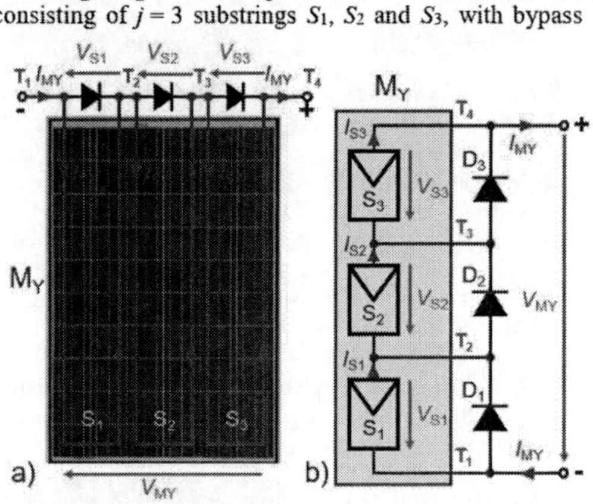

Figure 1: a) Serial connection of substrings S_x in a standard 4-terminal PV-module b) M_Y with protection diodes D_x result in a module voltage $V_{MY}=V_{S1}+V_{S2}+V_{S3}$ and a common module current $I_{MY} = I_{S1} + I_{S3} + I_{S3}$ for homogeneous irradiance.

diodes D_x anti-parallel to each individual substring terminal T_x and T_{x+1} to prevent power losses of unshaded substrings. The serial connection of substrings S_x with substring voltage V_{Sx} results in a total module voltage $V_{MY} = \sum V_{Sx}$ and a common module current I_{MY}. Figure 1b) shows the PV-modules electrical schematic with inactive diodes D_x and the current through each substring I_{Sx} results in $I_{S1} = I_{S2} = I_{S3} = I_{MY}$. String inverters with efficiency $\eta_{ST\_MPPT} \leq 99$ % operate strings at maximum power point (MPP). The serial connection of i PV-modules in a string results in a maximum power $P_{ST,MPP} = \sum V_{MY} I_{ST,MPP}$ for all i modules.

Activated bypass diodes D_z short circuit shaded substrings S_z ($z = 1...j$) and allow the string current I_{ST} to match MPP current of unshaded modules in the string. Depending on the severity of the shading string power may increase. Unfortunately, short circuiting of partially shaded substrings via activated bypass diodes results in $V_{Sz} < 0$ and thereby $P_{Sz} < 0$. Consequently, bypassed substrings do not contribute to the string power. Module-based power optimizers (MPO) provide individual operation for each module M_Y at its MPP_{MY} and reduce mismatch losses of the PV-system in case of partially shaded PV-strings. Common MPOs limit the current I_{Sx} of all substrings S_x to the lowest current I_{Sz} of a shaded substring S_z in a module M_Y and translate the limited module current $I_{MY} = I_{Sz} < I_{ST}$ to the string current I_{ST}. Therefore, MPOs result in mismatch losses within partially shaded PV-modules [1]. Available substring MPPT need a number $u = 2x$ Terminals and are incompatible to state of the art Modules with a number of $w = x + 1$ terminals, consequently leading to time consuming and expensive changes in module production. Here we present a substring maximum power point tracker compatible to state-of-the-art modules with w terminals [2].

2 PURPOSE

PV-cells in a substring S_x allow the generation of a current I_{Sx} proportional to solar irradiance $E_{e,x}$. Each substring S_x is composed of k PV-cells connected in series producing a substring current I_{Sx} which can be physically described using an approximated single diode model [3] with $I_{Sx} \approx I_{SC,x} - I_0 (\exp(V_{Sx} / (k\, V_T)) - 1)$ with $I_{SC,x}$ being the short circuit current of the substring, I_0 the diode reverse saturation current and, V_T the temperature voltage. Therefore, a substring S_x provides an output power $P_{Sx} = V_{Sx} I_{Sx} = V_{Sx} (I_{SC,x} - I_0(\exp(V_{Sx} / (k\, V_T)) - 1))$ with an MPP of $P_{Sx,MPP} = V_{Sx,MPP}\, I_{Sx,MPP} = \max(P_{Sx})$.

Figure 2 shows a mismatch in currents $I_{S1} \neq I_{S2} \neq I_{S3}$ of each of $j = 3$ substrings, with $n = 20$ cells each, individually proportional to inhomogeneous irradiances $E_{e,1} \neq E_{e,2} \neq E_{e,3}$ due to dirt or shading. This results in individual power P_{Sx} with each an individual substring MPP $P_{S1,MPP} = V_{S1,MPP}\, I_{S1,MPP}$, $P_{S2,MPP} = V_{S2,MPP}\, I_{S2,MPP}$ and $P_{S3,MPP} = V_{S3,MPP}\, I_{S3,MPP}$, with the power mismatch $P_{S1,MPP} \neq P_{S2,MPP} \neq P_{S3,MPP}$. The differences to the power points $P_{Sx,MY}$ which each substring S_x contributes to the maximum module power $P_{MY,MPP} = \sum V_{Sx,MY}(I_{Sx}) I_{MY,MPP}$ at $I_{Sx} = I_{MY,MPP}$ show the potential power increase $\Delta P_{Sx} = P_{Sx,MPP} - P_{Sx,MY}$ using individual substring MPP tracking. When considering a standard PV-module M_Y slight differences in irradiance or substring mismatch result in a common module current I_{MY} with a maximum power $P_{MY,MPP} = V_{MY,MPP}\, I_{MY,MPP} < \sum P_{Sx,MPP}$. Larger

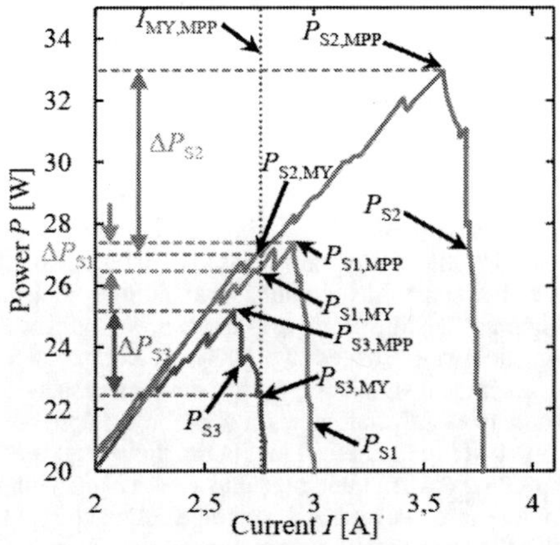

Figure 2: Mismatch of substrings S_x in a standard 4-terminal module M_Y due to unequal irradiance show the potential power output increase $\Delta P_{Sx} = P_{Sx,MPP} - P_{Sx,MY}$ by tracking each substrings MPP $P_{Sx,MPP}$ individually instead of the module MPP with substring power contribution $P_{Sx,MY}$.

differences in irradiance ΔE_e in a PV-string cause the activation of diodes D_z bypassing the shaded substrings S_z with $P_{Sz} < 0$, resulting in a larger module current I_{MY} to achieve I_{ST}, but with a greater power loss of an affected module M_Y. Furthermore, these differences in irradiance on individual modules in a PV-string due to partial shading can result in a complete bypass and zero power generation $P_{MZ,bypass} = 0$ of the affected modules M_Z to achieve I_{ST}, although the shaded modules power $P_{MZ} > 0$ with MPP_Z $P_{MZ,MPP} = V_{MZ,MPP}\, I_{MZ,MPP}$, $I_{MZ,MPP} < I_{ST}$ can still be harvested. To harvest maximum energy from shaded substrings S_z and unshaded substrings S_x ($x \neq z$) in a PV-module M_Y, MPP trackers (MPPT) attached to terminals T_x and T_{x+1} of each substring S_x operate all substrings independently from each other. A first prototype of substring-MPPTs $MPPT_x$, designed for standard 4-terminal PV-modules, prove independent operation of series connected substrings and harvests shaded substrings power P_{Sz}.

3 APPROACH

Figure 3 replaces the protection diodes D_x connected between terminals T_x and T_{x+1} with individual substring maximum power point trackers $MPPT_x$ in parallel to each serially connected substring S_x in a junction box of a standard 4-terminal PV-module M_Y with $j = 3$ substrings. For a shaded substring S_z, $MPPT_z$ adjusts the voltage V_{Sz} to the MPP voltage $V_{Sz,MPP} \approx 1/3\ V_{MY}$ and the current I_{Sz} to the MPP current $I_{Sz,MPP} < I_{MY}$. The power balancing and MPP tracking (PBM) provides the power $P_{z,MPP} = V_{Sz,MPP} I_{Bz,MPP}$ to $MPPT_z$ attached to the shaded substring S_z, allowing $I_{Sz,MPP} + I_{Bz,MPP} = I'_{MY}$ and the power $P_{PBM} = V_{MY} I_Q$ lowers the current $I_{MY} = (I_{Sz} + I_{Bz}) - I_Q < I'_{MY}$ if one or more substrings S_z are shaded. Therefore, in case of shaded substrings S_z, a current interface fits the current I_{MY} to the current I_{ST}.

Homogeneous irradiance on the PV-string with $I_{Sx} = I_{MY} = I_{ST}$ results in $I_{Bx} = I_Q = 0$ A, leading to a true serial connection solely composed of the substrings S_x, and the PBM sets $MPPT_x$ and current interface to standby for maximum energy yield. Standby operation with

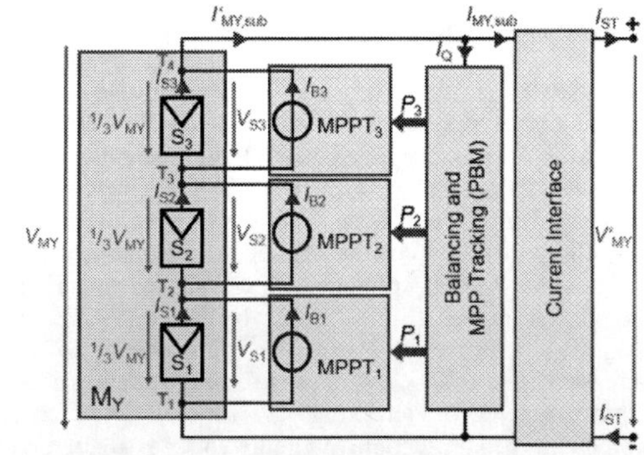

Figure 3: Substring-MPPT for independent I_{Sx} and V_{Sx} of serial substrings S_x. Power balancing sets $V_{Sx}=1/3\ V_{MY} \approx V_{Sx,MPP}$ and current interface fits I_{MY} to I_{ST} and V_{MY} to $V_{MY'}$.

$V_{MY} = V'_{MY}$ prevents switching losses in same way the Low Cost MPPT (LCMPPT) increases its efficiency η_{LCMPPT} [4]. The target efficiency $\eta_{SUB} = \eta_{LCMPPT} > 99.5$ % of the inactive $MPPT_x$ without shading represents DC losses $P_{VDC} = (R_{DSon} + R_{LDC})\,(I_{ST})^2$ due to the resistance R_{DSon} of non-switching metal-oxide-semiconductor field-effect transistors and R_{LDC} of inductors integrated in the current interface [5]. In case of partial shading, global MPP tracking of the string inverter activates the current interface to adjust the module current I_{MY} to the string current I_{ST}, resulting in an operation of all modules M_Y of the PV-string at MPP conditions.

4 EXPERIMENTAL RESULTS

First conducted laboratory measurements of a shaded module M_Y fitted with the prototype of the Substring-MPPT regulate $V_{Sx} = 1/3\ V_{MY}$ and provide proof-of-concept curves with a comparison to the same shaded module M_Y with the standard bypass diodes D_x.

Figure 4 shows the current $I_{MY,sub}$ and the power $P_{MY,sub}$ depending on the voltage V_{MY} of a measured module M_Y with substring S_2 shaded by roughly 60 %. The equipped prototype of the Substring-MPPT harvests the individual substring powers P_{S1}, P_{S2} and P_{S3} of the unshaded substrings S_1 and S_3 and shaded substring S_2 with voltages $V_{Sx} > 0$. The individual $MPPT_x$ adapts each

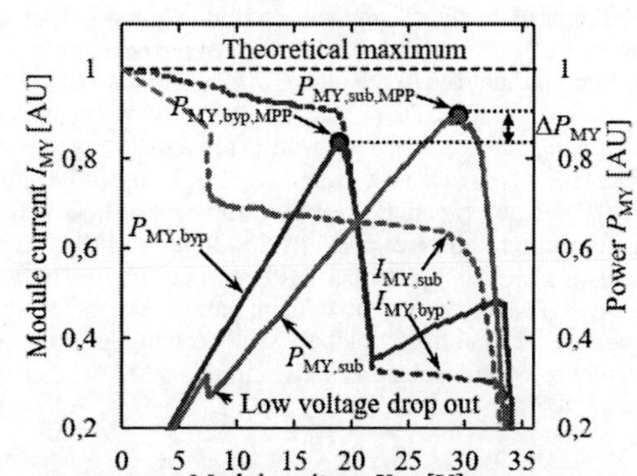

Figure 4: Module M_Y using the Substring-MPPT prototype results in a higher MPP output $P_{MY,sub,MPP}$ than $P_{MY,byp,MPP}$ which uses diode D_2 to bypass the shaded substring S_2.

substring current I_{Sx} to fit the current $I_{MY,sub} = I_{S1} - I_Q = I_{S2} + I_{B2} = I_{S3} - I_Q$ to achieve $P_{MY,sub,MPP} = V_{MY,sub,MPP} I_{MY,sub,MPP} = \sum V_{Sx} I_{MY,sub,MPP}$. Diode D_2 bypasses S_2 with $P_{S2} < 0$ to achieve $P_{MY,byp,MPP} = V_{MY,byp,MPP} I_{MY,byp,MPP}$. Experimental results prove $P_{MY,sub,MPP} > P_{MY,byp,MPP}$ and show an increase in maximum power point from 79.8 % using bypass diode technology to 89.2 % of the theoretical maximum using the Substring-MPPT prototype with one shaded substring. The output power P_{MY} increases by $\Delta P_{MY} \approx 12$ % to $P_{MY,sub,MPP}$ compared to $P_{MY,byp,MPP}$, with potential of an increase of $\Delta P_{MY,max} \approx 25$ % through implementation of an active substring MPP tracking. Use of power electronic components needing certain minimum voltages to function correctly leads to a low voltage drop out of the Substring-MPPT at lower module voltages.

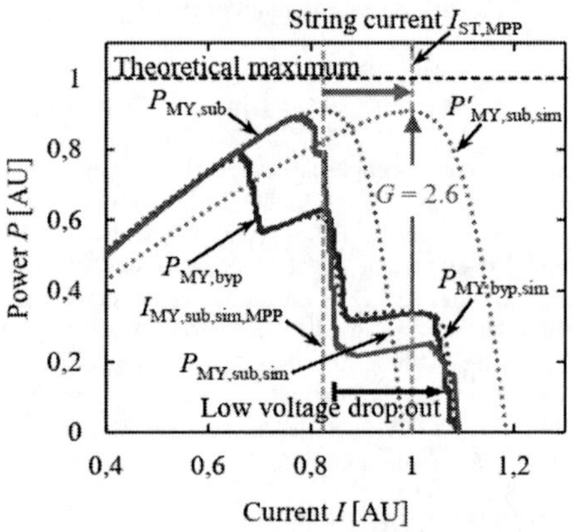

Figure 5: Substring-MPPT reaches a higher module power output MPP $P_{MY,sub}$ by harvesting shaded substring power and the current interface translates $P_{MY,sub}$ to $P'_{MY,sub}$ to adapt to global string MPP current $I_{ST,MPP}$.

Figure 5 shows the module output power $P_{MY,sub}$ depending on I_{MY} of a measured module M_Y with substrings S_2 and S_3 shaded by roughly 25 % and 30 % respectively with the equipped Substring-MPPT prototype. Comparison to $P_{MY,byp}$ of the same module identically shaded with bypass diode technology shows an increase in maximum power output. The power is normalized to the theoretical maximum possible power output of the module with S_2 shaded as above over the current normalized to the global MPP current $I_{ST,MPP}$. The simulated power $P_{MY,sub,sim}$ of the prototype of the substring balancer PBM with ideal components reaches an increase of power by a factor $G = 2.6$ compared to the simulated power output $P_{MY,byp,sim}$ with bypass diodes. The active current interface converts the modules MPP current $I_{MY,sub,sim,MPP}$ to the global MPP current $I_{ST,MPP}$ of the string. The theoretical maximum power output of the module at PV-string global MPP conditions can be achieved by implementing active substring MPP tracking.

5 DISCUSSION AND CONCLUSION

The first prototype of the Substring-MPPT shows proof-of-concept of increasing power performance by regulating the substring voltages $V_{Sx} = 1/3\ V_{MY}$ in respect to the voltage V_{MY} over the three serial connected substrings. Substring balancing optimizes the power of a shaded or mismatched 4-terminal PV-module M_Y compared to standard bypass diodes. The current interface adapts the MPP current $I_{MY,MPP}$ to the strings global MPP current $I_{ST,MPP}$ allowing a forward biased operation of the individual substrings whilst avoiding hotspots. Experimental shading scenarios show a power gain of 8 % for one substring shaded by 60 %. A comparison of measurement and simulation for two substrings shaded by 25 % and 30 % allow an increase in module output power from 34 % of the theoretical maximum power to 90 % by a factor $G = 2.6$ when integrated into a PV-system with the current interface converting the module output current to a global MPP current. The implementation of an active substring MPP tracking and optimization of the Substring-MPPT can provide an additional potential power output increase to the theoretical maximum. For unshaded modules the Substring-MPPT reduces all power losses to minimal DC losses over the integrated non-switching metal-oxide-semiconductor field-effect transistor and inductors. In case of shading, global MPP tracking of the string inverter activates the Substring-MPPTs current interface translating the lower module MPP current $I_{MY,MPP}$ to the higher PV-string MPP current $I_{ST,MPP}$.

6 REFERENCES

[1] maxim integrated, solar cells optimization-1.pdf, https://www.maximintegrated.com/en/markets/solar-energy.html (20.09.2018)

[2] R. Merz, T. Czarnecki, A. Neumann, EP2018/054159, to be published

[3] John A. Duffie et al., *Solar Engineering of Thermal Processes*, Fourth Edition, Wiley and Sons, Hoboken, NJ, 2013

[4] A. Neumann, T. Czarnecki, R. Merz, in proc. 32. Symposium Photovoltaische Solarenergie (OTTI, Regensburg, 2017), D13

[5] R. Brace, A. Neumann, R. Merz, in proc. 33. Symposium Photovoltaische Solarenergie (Conexio, Pforzheim, 2018), D3

NEW PV SYSTEM CONCEPT – INDUCTIVE POWER TRANSFER FOR PV MODULES

Fabian Carigiet[1,2]*, Raphael Knecht[1], Thomas Baumann[1], Christoph J. Brabec[2], Franz P. Baumgartner[1]
[1] ZHAW, Zurich University of Applied Sciences, School of Engineering, Institute of Energy Systems and Fluid Engineering, Technikumstrasse 9, CH-8401 Winterthur, Switzerland, www.zhaw.ch\~bauf
*phone: +41 (0) 58 934 7292; e-mail: fabian.carigiet@zhaw.ch, www.zhaw.ch/=cari
[2] FAU, Friedrich-Alexander University Erlangen-Nürnberg, Institute Materials for Electronics and Energy Technology, Martensstraße 7, 91058 Erlangen, Germany

ABSTRACT: The proposed new PV system concept is based on several AC modules that are connected in series using inductive power transfer. These modules include a cell matrix that is connected to a module integrated DC/AC inverter. The high frequency AC current flows through the primary side planar coil generating a magnetic flux. Outside of the PV module, there is a clamp including ferromagnetic material for the magnetic circuit that caries the magnetic flux to the secondary winding. The magnetic flux induces an AC current in the secondary winding, which is formed by the common cable. An AC/AC converter is placed at the end of the PV module strings to generate the 50 Hz and to connect the PV power plant to the electricity grid. This new PV system concept is a fundamentally new approach of the electricity transmission in the field of PV system technology. It is not restricted to the replacement or optimisation of an individual system component, but it requires the continuing development of the PV module construction and the contactless connection technology to the common cable. The proposed inductive power transfer per each PV module opens up a complete new field for the PV system technology.
Keywords: PV system, inductive power transfer, AC modules, module integration,

1 INTRODUCTION

In 2016, the PV system related costs amounted to 1200 $/kWp for commercial applications and system sizes of 1 MW according to the NREL benchmark report. In that case, the soft costs are disregarded including the land acquisition, sales tax, overhead, net profit, etc. The cost share for the PV modules is 51% whereby half of that falls to the share of the cell production. The inverter cost amount to 11%. The costs for structural and electrical balance of system (BOS) are 14% and 12%, respectively. The labour costs including the necessary equipment for the installations correspond to 12% [1].

The levelized cost of electricity depends not exclusively on the initial costs. There are other factors such as lifetime of PV system components as well as maintenance cost due to system component defects. The IEA PVPS task 13 analysed frequently occurring PV module and system defects. In the first year after commissioning, the main defects represent potential induced degradation, connector failures, junction box defects or defective bypass diodes [2].

The proposed new PV system concept affects three quarter of the present PV system in terms of costs with potential cost reductions regarding BOS, installation and maintenance. Additionally, it should eliminate common system component failure.

2 NEW PV SYSTEM CONCEPT

The goal is to develop a new PV system that uses an integrated inductive power transfer (IPT) technology for the power feed-in between the cell matrix and the string wiring. The DC current produced by solar cell matrix is converted into high frequency AC current within the laminated area of the PV module. The output of this inverter is connected to a planar coil that generates a magnetic flux.

A connection element (magnetic clamp in Figure 2) has to be developed that includes the magnetic coupling between the primary winding formed by the planar coil and secondary winding formed by the common cable. The clamp contains the magnetic circuit based on high frequency magnetic material. Optionally, this clamp could serve as module retainer by attaching it to the module in an interlocking and force-fitting manner. The magnetic flux generated by the planar coil is transported over a magnetic circuit and it induces a small voltage in the high voltage common cable (the number of turns of the primary side is greater than the number of turns of the secondary side). The high voltage can be achieved by connecting multiple modules to the common cable, in similar manner. A high frequency power cable will be used. At the end of a module string, there is an AC/AC converter because the used transmission frequency within the string is higher than 50 Hz.

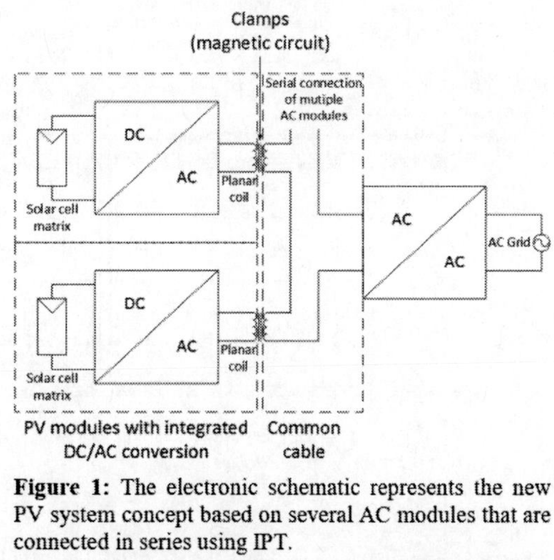

Figure 1: The electronic schematic represents the new PV system concept based on several AC modules that are connected in series using IPT.

A control and communication system has to be developed that synchronises all AC PV modules and allows an upscaling of the PV power plant (>100 kWp).

The whole system can be described as a high

frequency transformer based grid-connected inverter where the high frequency AC stage is inside the PV module. Figure 1 shows the electrical schematic of the proposed new PV system concept and Figure 2 and 3 illustrates its mechanical setup.

In a first step, a DC/AC/DC system will be developed where the common cable carries DC current. This provides a functional prototype at an early stage without the need for the complex synchronization and communication system.

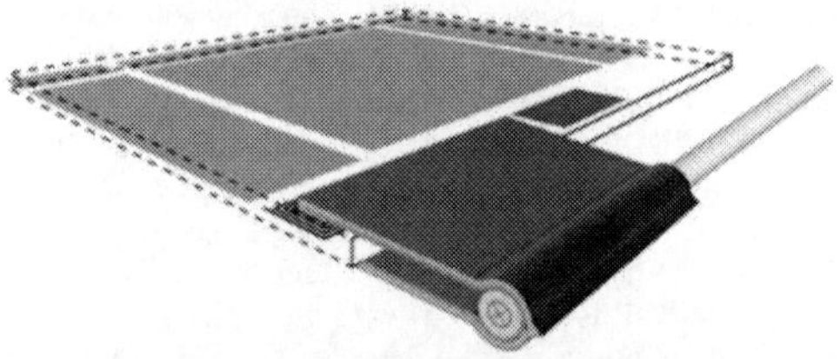

Figure 2: The 3D drawing [3] shows a small section of the new PV module with the solar cells (light grey), the electronics (dark grey) and the planar coil integrated in the laminated module packaging. Outside the PV module, the magnetic clamp needs to be positioned exactly above the planar coil. The common cable is wound around the clamp building the secondary winding. The system is not restricted to only one winding as shown in this drawing.

3 INNOVATION

In recent decades, the focus of public research funding was almost on the optimisation of solar cell efficiency itself and less on another important part, the BOS Components. The reason for this was that the solar cells had the major cost share. In today's PV power pants, the cost share of the crystalline silicon solar cells are roughly one quarter of the total system silicon costs. The rest is divided into module production, installation and balance of system components such as subconstruction, inverter, cabling, installation material and planning [1]. The new PV system concept represents a new solution approach, which affects system components that account for three quarters of today's total system. With the described system in chapter 2, several advantages arise with the goal to reduce initial system costs as well as operating and maintenance costs for large PV power plants (>100 kWp).

Due to the galvanic isolation, there will be no potential induced degradation. Additionally, it allows using even higher system voltages, which is reached by connecting more AC modules in series with lower voltages than state-of-the-art PV modules. The voltage increase will reduce the amount of copper needed for the cabling. The installation costs are going to decrease because there are no electrical connectors. This saves time on wiring and eliminates weather-related connector failures, which are listed as frequently occurring system failures in the IEA PVPS task 13 [2]. Less expertise for electrical connection is needed on-site because the installation of the magnetic clamp is a purely mechanical connection. The IPT technology allows the disconnection under load. This leads to a reduction in the maintenance costs and yield losses because it is not necessary to shut down a whole string. Furthermore, it increases the fire protection and the safety of people (no electric arcs). The new topology has a DC/AC inverter integrated in the AC modules allowing the implementation of an individual MPP tracker. This results in a reduction of module mismatch as well as shading losses. Additionally, the new AC modules have no junction box increasing the reliability of the PV system.

This new solution approach for PV systems is designed for applications in large ground-mounted, rooftop or facade PV power plants. The main reason is that a certain amount of AC modules is needed to reach the higher voltage level and that the potential in labour costs reduction and the benefit of industrial prefabrication are greater for large systems.

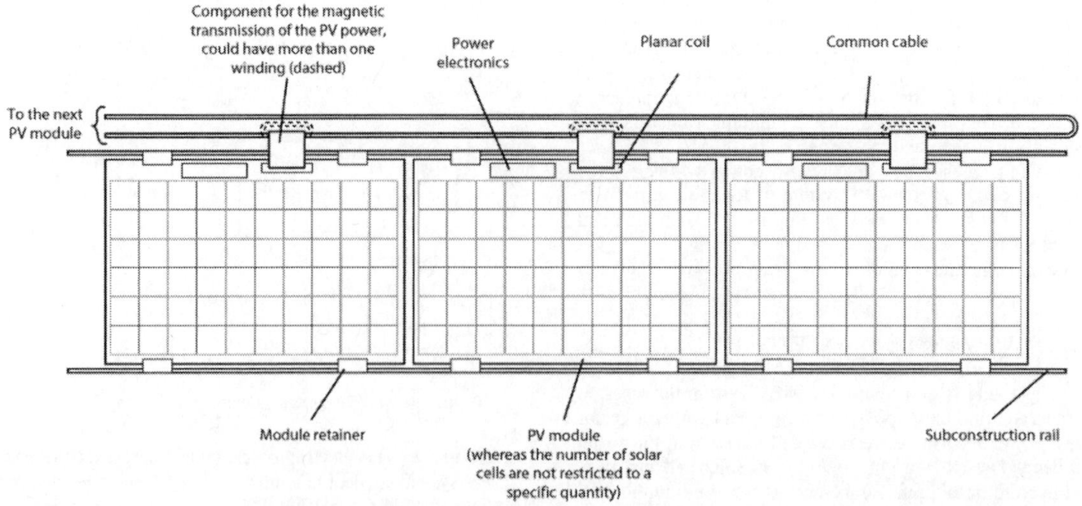

Figure 3: The drawing [3] shows the AC modules installed on the subconstruction rail. In addition to the solar cell matrix, there is the electronic DC/AC circuit and the planar coil integrated in the module package. Both elements are not restricted to be positioned on the long edge of the module. The common cable is wound around each magnetic clamp connecting all modules in series. The voltage on the secondary winding is smaller than the DC voltage of all serial connected solar cells inside the module. This allows adding more AC modules in the PV string compared to a state-of-the-art PV system.

4 CONCLUSION AND OUTLOOK

The new PV system concept is a fundamentally new approach of the electricity transmission in PV power plants. The main idea is not limited to an individual optimisation or exchange of components, but focuses on the further development of the design of the PV modules and the balance of system components including the optimally matched contactless connection technology.

The first task includes the design and development of DC/AC conversion circuit whereas the second task contains the design of the magnetic clamp based on experiments and finite element simulations. Therefore, the optimal system specifications have to be determined including system frequency, module thickness to integrate the electronics, input voltage (serial or parallel cells), input current (cell size), single module output voltage, total system voltage, magnetic material, number of primary and secondary winding, etc.

The goal within the next year is to demonstrate the technical feasibility of the new PV system technology using appropriate switching technology and material combination. In the coming years, this should enable launching a new product that allows a cost-effective installation with minimum maintenance effort.

ACKNOWLEDGMENT

The project (WIPT – Wireless Inductive Power Transfer) receives start-up funding and is monitored by Gebert Rüf Stiftung under the contract number GRS-063/16 [4].

REFERENCES

[1] Ran Fu, Donald Chung, Travis Lowder, David Feldman, Kristen Ardani, and Robert Margolis. U.S. Solar Photovoltaic System Cost Benchmark: Q1 2016. Technical Report NREL/TP-6A20-66532, National Renewable Energy Laboratory (NREL), 2016.

[2] IEA PVPS Task 13, Assessment of photovoltaic module failures in the field, Report IEA-PVPS T13-09:2017

[3] Franz Baumgartner and Fabian Carigiet; Patent Application PCT/IB2017/057228, Solar-Modul zur kontaktlosen Stromübertragung auf eine Sammelleitung sowie Kontaktelement für solche Solar-Module, 17.11.2017

[4] https://www.grstiftung.ch/de/media/portfolio~grs-063-16~.html

LIFE CYCLE ASSESSMENT OF PEROVSKITE SOLAR CELLS IN SINGLE JUNCTION AND TANDEM CONFIGURATION

Juan Camilo Gómez Trillos*, Urte Brand, Sven Peterhammer, Thomas Vogt
DLR Institute of Networked Energy Systems
Carl-von-Ossietzky-Str. 15, 26129 Oldenburg, Germany
*email: Juan.GomezTrillos@dlr.de, phone: +49 44199906-247

ABSTRACT: We present a cradle-to-gate Life Cycle Assessment (LCA) of perovskite solar cells in single junction and in tandem with crystalline silicon to investigate their environmental impacts. We compared both configurations to other conventional technologies such as s-Si, p-Si, CIS, and CdTe regarding cumulative energy demand, energy payback time per m2 of module, and further impact categories included in the CML-IA impact assessment method referred to kWh of produced energy. The results show that the modules containing perovskite absorber in single configuration have the lowest cumulative and energy payback time among the assessed technologies. However, the modules containing tandems with crystalline silicon have the highest cumulative energy demand but lower energy payback time compared to c-Si and mc-Si. Both perovskite technologies exhibit higher global warming potential per kWh of produced energy compared to other technologies due to their short lifetime. Furthermore, additional comparisons concerning CML-IA impact categories show that the improvement via efficiency and lifetime is most critical, e.g. for marine aquatic ecotoxicity. Although the use of perovskite absorbers is favourable from the cumulative energy demand, improvements in efficiency and lifetime are still necessary to get comparable environmental effects in other impact categories in relation to conventional photovoltaic technologies.

Keywords: CO2 footprint, Environmental Effect, LCA, Perovskite, Tandem

1 INTRODUCTION

The adoption of low-carbon energy supplies, including renewable energy, have been acknowledged as a main part of the strategy to mitigate climate change and limit the increase of temperature on the surface of Earth due to the anthropogenic green-house gas emissions [1]. Hence the development of new renewable energy technologies can help to minimize the impact of energy supply. For instance, multiple photovoltaic technologies have been developed to increase the energy conversion, minimize material use, and reduce manufacturing complexity and costs [2]. Among them, perovskite solar cells (PSC) had so far a fast progress. The development of this photovoltaic technology started in 2009 with a power conversion efficiency of 3.8% [3], reaching 22.7% in 2017 in single junction configuration [4]. Additionally, configurations in tandem with other commercial technologies such as crystalline silicon, which have the potential of going beyond the Shockley – Queisser limit for single junction configurations, have already exceeded 25% [5].

Perovskites are materials with the formula ABX3. For photovoltaic applications, A is usually an organic ion, B is lead or tin, although other metals are now in research, and X is a halide. Hence, inexpensive materials are used in the absorber, in comparison to technologies like cadmium telluride (CdTe) or copper-indium-gallium diselenide (CIGS) [6]. Moreover, the possibility of performing the deposition of the different layers via a myriad of processes, such as wet chemistry or vacuum deposition, opens potentially new applications [7]. Furthermore, the light absorption properties can be tuned by changing the composition of the absorbing material, allowing the use together with other photovoltaic technologies [8]. Thus, the future for this technology appears to be promising.

Despite the fact that this new technology has aroused interest due to its power conversion efficiency which is compatible with commercial applications, there are still some challenges left: the reproducibility of the manufacturing processes for commercial scales, the long-term stability, and the environmental effects during the lifecycle of these devices, especially due to the use of lead [7]. Thus, the main purpose of this study was approaching the environmental effects of possible future configurations of PSC from the perspective of a life cycle assessment.

2 METHODOLOGY

In our study, we conducted the methodology of life cycle assessment (LCA) which allows to evaluate the environmental impacts of a system or product throughout the different stages of its life cycle [9]. ISO 14040:2006 and 14044:2006 describe the principles, framework, and requirements to conduct this methodology. These standards divide the methodology in four interconnected phases: goal and scope definition, inventory analysis, impact assessment, and interpretation. The first phase defines the objectives of the study and states its limitations or assumptions. The inventory analysis quantifies the inputs and outputs for a product or service throughout its lifecycle. The impact assessment evaluates the magnitude and significance of the environmental impact of a product or service. Finally, the interpretation evaluates the results in relation to the goal and scope to do conclusions and recommendations [9]. In the following sections, the different phases for this particular study will be described.

2.1 Goal

The main goal of this study was the comparison of the environmental impacts generated by perovskite solar cells in single junction and in tandem configuration with other photovoltaic technologies available in the market. Therefore, we used the life cycle assessment (LCA) complying with the standards ISO 14040 and 14044. Since perovskite solar cells are not currently commercially produced, this study can be catalogued as an ex-ante LCA. This study also aims to identify hot spots of environmental impacts during the life cycle of these systems. This study was commissioned by the DLR Institute of Networked Energy Systems.

2.2 Scope

This study comprised the materials and processes used to produce perovskite solar cells in single junction (hereafter called SJ-PSC) and in tandem with crystalline silicon (tandem PSC/c-Si). Some components, such as encapsulation, connection box, and frame, were included in the analysis. These elements would be required to produce a complete module in case of a practical use.

The approach used here considers PSC in the two mentioned configurations including the raw material acquisition, material processing, and manufacturing of the modules. Since the study is limited to the gate of the manufacturer, it is denominated cradle-to-gate. However, the production of electricity during the use phase is considered to provide results on the base of electricity production.

The geographical system boundary considered here was the European Union and particularly Germany. The electricity used in the manufacturing processes contained in this study was assumed as supplied by the German electric grid. All the products and background systems required for manufacture were selected whenever possible with origin in Germany or the European Union.

Reference information between 2010 and 2018 is used in this study to perform the calculations. Ecoinvent database 3.3 [10] and Ecoinvent 2.2 [11] were used as source of background data for the LCI, along with information from scientific articles and information provided by external research groups. Information and inventories supplied by ESU-services [12] regarding photovoltaics were used to compare different technologies with perovskite solar cell technologies. The software selected for the calculations was Simapro 8.3.0.0.

No mass cut-off rules were applied deliberately, except those already contained in the inventories collected. This was done since some elements with an small mass may have important impact in the overall result, compared to other elements with bigger masses. Allocation was made in a mass basis, especially for the reactants used in the production of the cells. No grouping or weighting were performed to the results for the different impact categories.

2.3. Functional Unit and Impact Assessment Methods

The calculations of the different impact categories were performed using an area unit basis (m^2) and further converted to impacts per energy output unit (kWh). Hence, the functional unit was defined as 1 kWh of electricity produced by the module. For this purpose, the energy output during the lifetime of a module was calculated employing equation 1 [13].

$$E_{out} = \sum_{i=1}^{Lifetime} I \times PCE \times PR \times (1 - DR)^{i-1} \quad (eq. 1)$$

Where E_{out} is the output of energy in kWh during the lifetime, I is the irradiance (1700 kWh/m^2year in southern Europe conditions), PCE is the power conversion efficiency assumed for each technology, PR is the performance rate (assumed 75% in all the cases), and DR is the degradation rate assumed for each technology. The results obtained for an area basis were divided by the energy output to obtain the results in an energy output basis. The assumptions for performing these calculations are summarized in Table I. The assumption for SJ-PSC was done considering the maximum efficiency reported and the difference between the maximum efficiency in research and commercial modules for other technologies, discounting in this case 6%. An assumption of 26% for tandem PSC/c-Si was made, considering the possibilities of increasing the power conversion beyond the Shockley-Queisser limit of single junction. Finally for the commercial technologies, an average of modules from different manufacturers was made and is displayed in Table I.

Table I: Assumed characteristics for the assessed photovoltaic technologies

Technology	Module efficiency (%)	Lifetime (Years)	Degradation rate (%/year)
SJ-PSC	16.7	5	0.7
Tandem PSC/c-Si	26.0	5	0.7
CdTe	17.0	30	1.0
CIS	13.8	30	0.7
p-Si	17.2	30	0.7
s-Si	18.0	30	0.7

The impact assessment was performed using the methods of Cumulative Energy Demand (CED) version 1.09 and CML-IA baseline version 3.04 included in the software SimaPro 8.3.0.0. [14]. The results of Energy Payback Time (EPBT) were obtained applying the equation described by other authors considering the calculated CED [13]. An energy-to-electricity conversion factor for Germany of 31.5% was used to convert the electricity output into primary energy according to information collected from Ecoinvent 3.3 [10].

2.4 Inventory Analysis

The production of perovskite solar cells is an additive process in which each layer is subsequently deposited on top of the previous one. Several different architectures have been assessed by different researchers aiming to increase the efficiency and lifetime of the cells. However, the inventories related to the materials used are limited and, in many cases, not included in background data provided by databases such as Ecoinvent. Hence, different LCAs related to PSC were identified and used as source of information for this LCA, regarding synthesis routes, energy use, and architectures [13][15][16][17][18][19][20][21][22]. In this particular case the configuration used by Bush et al. [23] was used as a basis for the modelling of both SJ-PSC and tandem PSC/c-Si cells.

2.4.1 Inventory of SJ-PSC modules

The substrate for SJ-PSC was considered solar glass with low iron content with a resistivity of 10Ω and hence an indium-tin oxide layer with a thickness of 180nm according to the properties described by the manufacturer [24]. The deposition of indium-tin oxide was considered as a similar process as done for liquid crystal displays, as previously described by other authors [22]. The hole conducting layer was assumed as made of nickel oxide, according with the design considered and previous inventories [22][23].

Regarding the absorber, although the stoichiometry used in the selected design was $Cs_{0.17}FA_{0.83}PbBr_{0.51}I_{2.49}$, a composition $FA_{0.85}MA_{0.15}PbI_3$ was considered for this study as a simplification the stoichiometry. Methylammonium iodide and lead iodide and inventories were obtained from Gong et al. [15], while Formamidinium iodide inventories were obtained from Zhang et al. [19]. The process was modelled as the deposition performed by Yang et al. [25] through intermolecular exchange. Previous mixing, two slot dye coating processes and drying processes were modelled according to the energy intensities described in other studies [16][26]. Slot dye coating was considered with an 80% of material utilization efficiency as a conservative number, given that some manufacturers of this kind of equipment report up to 95% [27]. Finally, an annealing step of 20 min at 150°C was modelled calculating the losses of a horizontal plane in natural convection conditions by means of correlations to calculate the heat transfer coefficient under this condition.

The thickness and materials of the different layers which comprise the electron transporting layer were selected according to the description given in the selected design [23]. Even though, these authors describe the deposition of tin oxide and zinc oxide by atomic layer deposition (a chemical vapour deposition technique). For the purposes of this study, these layers were considered as deposited by thermal evaporation. LiF, SnO2, and ZnO were taken from Ecoinvent 3.3, while the fullerene derivative PCBM was obtained from a LCI previously published [28]. The energy consumption of this process was calculated using as benchmark another study previously published for hard coating deposition processes [29]. The energy consumption in this process was divided by the volume of the layer coated during the experiments obtaining the energy intensity per unit volume of material coated. This parameter was used to calculate the energy used according to the thickness of each layer comprising this element of the cell. The back conducting layer was assumed as made of aluminium evaporated as indicated in the assumed design [23]. A description of the different elements can be observed in Table II.

Encapsulation and module elements were considered using existing systems in the database Ecoinvent 3.3 from crystalline silicon photovoltaic modules. The systems were modified by removing the silicon cells. Front contacts were also removed because are not used in this kind of cell, since a TCO is used instead to convey the current due to the deposition onto a glass.

2.4.2 Inventory of tandem PSC/c-Si modules

As initial material, $1m^2$ of Single-Si wafer was considered. This material was obtained from Ecoinvent 2.2 with datasets offered by ESU-services for photovoltaic data on the LC-Inventories webpage [12]. For this study, a conservative thickness of 190 μm was considered. Additional treatments to produce heterojunction cells were assumed according to previous studies [30]. The deposition of additional 20 nm ITO layers on the front and back of the cell were also modelled in a similar way as afore mentioned. The back layer was topped subsequently with 300nm silicon nanoparticles. This step was modelled as the deposition of a 300nm a-Si layer followed by a thermal phase to anneal and oxidation this layer as depicted by Louwen et al [30]. To complete the cell used as substrate, the deposition of a 200nm silver back electrode was contemplated as done using thermal evaporation, as previously done for other elements in the PSC according to the volume of this layer. This cell was added with perovskite solar cells, encapsulation, and module elements as mentioned in the former section. Table III shows a summary of the different elements contained in the tandem PSC/c-Si cells.

Table II: Architecture, composition, thickness, deposition processes, and deposition efficiency considered for SJ-PSC

Layer		Composition	Thickness	Deposition Process
Substrate		Solar glass	1mm	-
Transparent Conducting Oxide (TCO)		ITO	180nm	SP
Hole transporting layer (HTL)		NiO	25nm	SP
Perovskite absorber (PL)		$FA_{0.85}MA_{0.15}PbI_3$	500nm	SDC
Electron transporting layer		LiF	1nm	TE
		$PC_{60}BM$	10nm	TE
		SnO_2	4nm	TE
		ZnO	2nm	TE
Metal contact (back)		Al	100nm	TE

SP: Sputtering; SDC: Slot Dye Coating; TE: Thermal Evaporation

A deposition efficiency of 15%, 15%, and 80% was assumed for sputtering, thermal evaporation, and slot dye coating respectively.

Table III: Architecture, composition, thickness, deposition processes, and deposition efficiency considered for tandem PSC/c-Si

Layer	Composition	Thickness	Deposition Process
Metal contact	Ag	200nm	TE
Silica nanoparticles	SiO2	300nm	PECVD
Transparent conducting oxide (TCO)	ITO	20nm	SP
p-Type amorphous silicon	a-Si:H (p+)	15nm	PECVD
Intrinsic amorphous	a-Si:H(i)	7nm	PECVD

silicon			
Crystalline Silicon wafer	s-Si (n)	190μm	-
Intrinsic amorphous silicon	a-Si:H(i)	7nm	PECVD
n-type amorphous silicon	a-Si:H(n+)	8nm	PECVD
Transparent conducting oxide (TCO)	ITO	20nm	SP
Hole transporting layer (HTL)	NiO	25nm	SP
Perovskite absorber (PL)	$FA_{0.85}MA_{0.15}PbI_3$	500nm	SDC
Electron transporting layer	LiF	1nm	TE
	$PC_{60}BM$	10nm	TE
	SnO2	4nm	TE
	ZnO	2nm	TE
Transparent conducting oxide (TCO)	ITO	150nm	SP
Front contact	Ag	-	CP
Antireflection (AR) coating	LiF	150nm	TE

SP: Sputtering; SD: Slot Dye Coating; TE: Thermal Evaporation; PECVD: chemical vapour deposition; CP: Contact printing.

2.5. Uncertainty

The quantification of the uncertainty in the different categories was done employing the tool provided with the software SimaPro ® 8.3.0.0. to perform a Monte Carlo analysis for CED and CML-IA impact categories with 1000 runs and 95% of confidence.

2 RESULTS

Figure 1 describes the CED for the different analysed technologies. In this case, modules containing SJ-PSC have the lowest CED per unit area of module, while modules containing tandems PSC/c-Si are the ones with the highest CED. It worth noting that CdTe calculations consider frameless modules, since this element is absent in typical current modules.

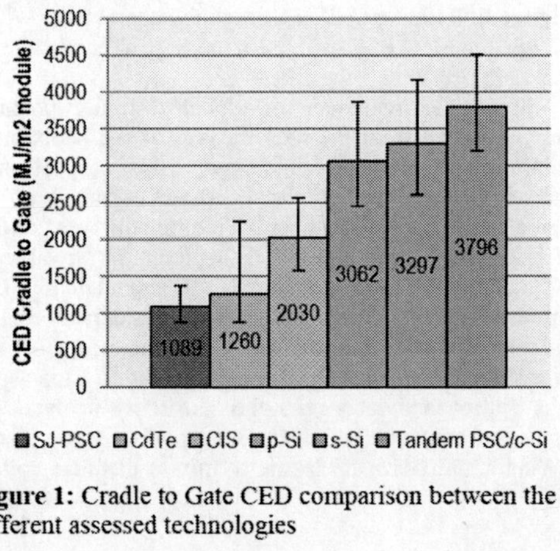

Figure 1: Cradle to Gate CED comparison between the different assessed technologies

Figure 2 shows the contribution of the different module elements to the cumulative energy demand for modules containing SJ-PSC and tandems PSC/c-Si. In the first case, most of the energy is concentrated in the module elements, while the perovskite solar cell only accounts for 18.1% of the CED. Among the elements grouped in encapsulation, frame, and module elements for the SJ-PSC, only the frame gathers 412 MJ of the energy demand representing 37.8% of the total CED of the module. Solar glass concentrates 143 MJ (13.1% of the CED of the module) and ethylene vinyl acetate (EVA) encapsulation material adds 91.6 MJ (8.4% of the CED of the module). In the case of the tandem, the inclusion of heterojunction s-Si cells increases the CED with this element gathering 71.3% of it. Only the s-Si wafers in the previous stage before the production of the heterojunction silicon cell amounts 62.6% of all the CED, while the top PSC only contributes to 5.2% of the total energy demand.

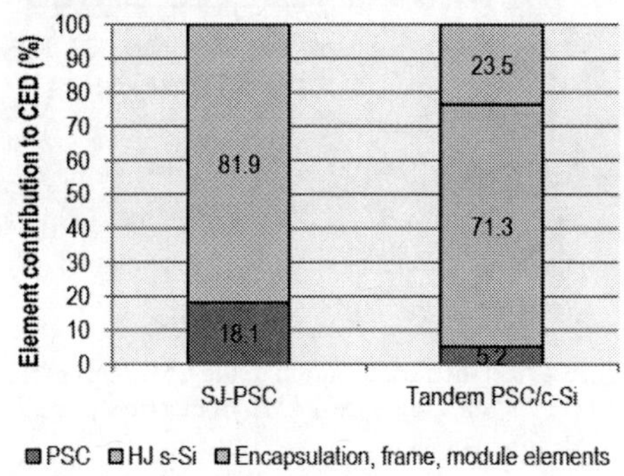

Figure 2: Contribution to CED of different module elements containing SJ-PSC and tandem PSC/c-Si.

Although not depicted in figure 1 or 2, the use of electricity in the manufacturing process contributes considerably to the energy demand. For SJ-PSC, 16.5 kWh of electricity are used during the manufacturing, while this number renders to 233.1 kWh due to the production process of solar grade silicon in PSC/c-Si devices. With a conversion factor of 3.6 MJ/kWh and a grid conversion factor for Germany of 31.5%, this represents 17.3% and 70.2% of the CED for manufacturing of modules containing SJ-PSC and tandem PSC/c-Si respectively.

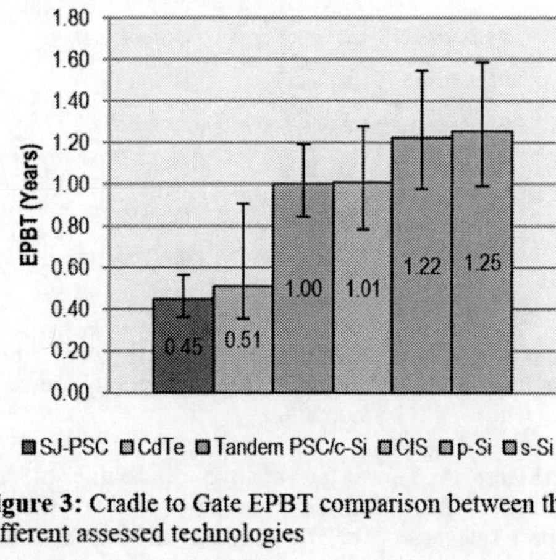

Figure 3: Cradle to Gate EPBT comparison between the different assessed technologies

Figure 3 depicts the comparison between the EPBT of the different technologies according to the previously presented assumptions. In this case, the tandem devices have a lower mean EPBT compared to silicon technologies, but similar to CIS modules. SJ-PSC modules have the lowest EPBT, comparable to the CdTe modules.

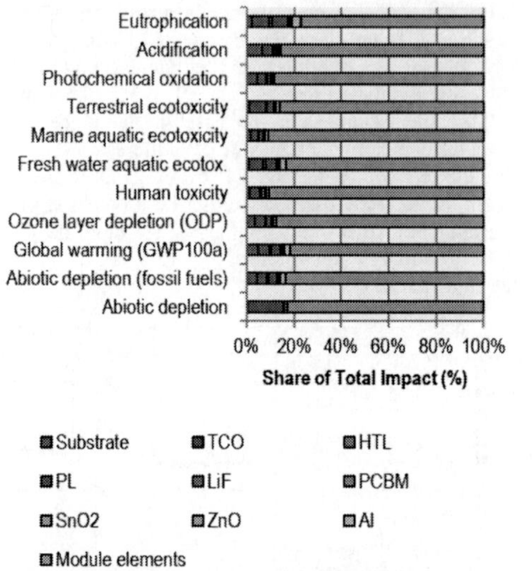

Figure 4: Relative contribution to the CML-IA baseline characterization factors for 1 kWh of electricity produced by a SJ-PSC module.

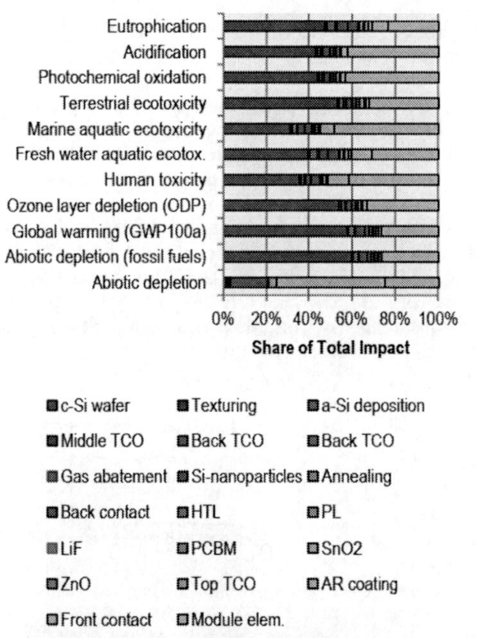

Figure 5: Relative contribution to the CML-IA baseline characterization factors for 1 kWh of electricity produced by a tandem PSC/c-Si module.

Figure 4 shows the relative distribution of the different elements considered for SJ-PSC in the different impact categories. The frame, glass, and encapsulation, grouped here as module elements, gather between 77.3% and 91.0% of the impacts calculated by CML-IA method. The abiotic depletion of elements is particularly affected by the TCO due to the use of indium-tin oxide as layer. This expands the impact of this element to 15.1% of the total for driving most of the abiotic depletion produced by the cell components due to the scarcity of this material. On the other hand, figure 5 shows the same distribution for the modules containing tandem PSC/c-Si. In the latter case the silicon cell produces most of the impacts in all the impact categories, except abiotic depletion. The abiotic depletion of elements is mainly produced by the front and back contacts containing silver and the different transparent conductive oxides containing indium-tin oxide.

Figure 6 depicts the comparison between the different CML-IA impact categories assessed in this work, normalized to the highest value obtained for the technologies per kWh. Perovskite-containing modules have the highest values in all cases except abiotic depletion, mainly due to the shorter lifetime compared to the other technologies. The high scores obtained for CdTe and CIS modules derive from the use of cadmium, tellurium, and indium in the absorbers of these technologies.

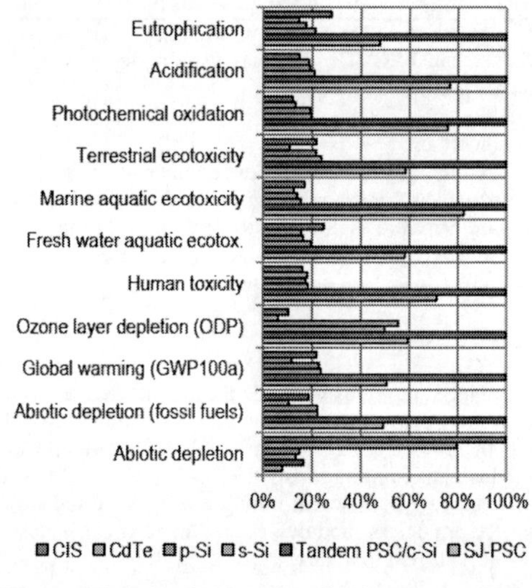

Figure 6: Relative CML-IA baseline impacts per kWh of SJ-PSC, tandem PSC/c-Si, c-Si, mc-Si, CdTe, and CIS.

Since the avoidance of global warming potential emissions is one of the main drivers of the adoption of solar photovoltaic technologies, this aspect was thoroughly explored. Figure 7 shows a detail of the global warming potential per kWh of the different studied technologies in absolute values, depicting mean values of 60 g/kWh and 119 g/kWh for SJ-PSC and tandem PSC/c-Si respectively. These values are considerably higher compared to other photovoltaic technologies, mainly due to a short lifetime assumed for the modules. This aspect was approached later with the sensitivity analysis. As matter of contrast, Ecoinvent 3.3 reports a value of 659 kWh for the German electricity mix at medium voltage [10].

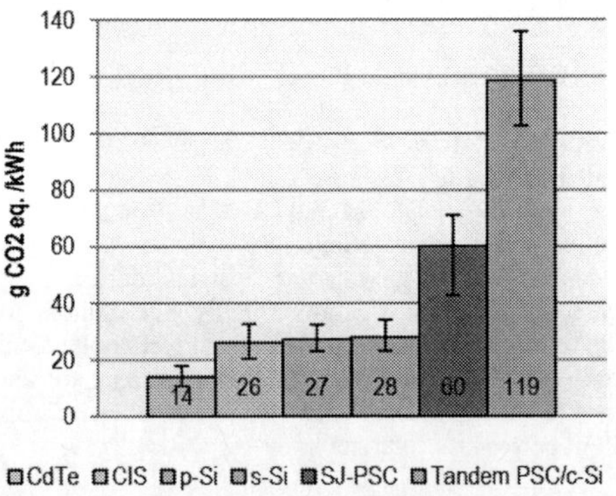

Figure 7: Global Warming Potential per kWh of SJ-PSC, tandem PSC/c-Si, c-Si, mc-Si, CdTe, and CIS.

As a matter of comparison, the results obtained for perovskite-containing modules were contrasted to widest-spread photovoltaic technology: p-Si modules, considering an irradiance of 1700 w/m2year, performance rate of 75%, and degradation rates as summarized in Table I for all the cases. The purpose of this was to see what amount of efficiency and lifetime is necessary to have equal or lower impacts than p-Si devices. The different impact categories were codified as AD (abiotic depletion of elements), ADFF (abiotic depletion of fossil fuels), GW (global warming), OLD (ozone layer depletion), HT (human toxicity), FWAE (freshwater aquatic ecotoxicity), MAE (marine aquatic ecotoxicity), TE (terrestrial ecotoxicity), PO (photochemical oxidation), A (acidification), and E (eutrophication). These results are summarized in figure 8 and 9 showing that the most critical impact category requiring the highest efficiency and lifetime in both cases is marine aquatic ecotoxicity.

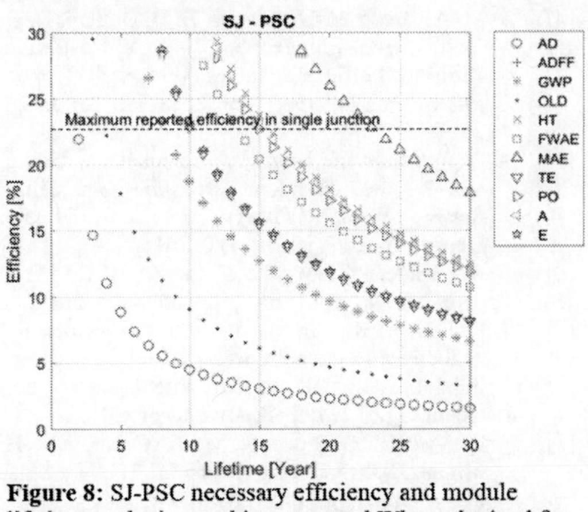

Figure 8: SJ-PSC necessary efficiency and module lifetime to obtain equal impacts per kWh as obtained for mc-Si.

According to the results depicted in figure 8, a lifetime of 30 years and an efficiency of 17.9% are required for SJ-PSC to have a similar value per kWh of marine aquatic ecotoxicity. As shown in figure 9, the efficiency must reach 33.8% with an equal lifetime in the case of tandem PSC/c-Si modules to have similar marine aquatic ecotoxicity exceeding the current maximum reported efficiency for this technology.

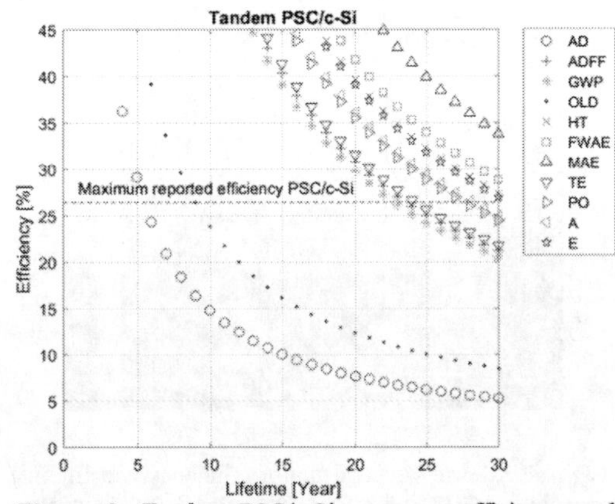

Figure 9: Tandem PSC/c-Si necessary efficiency and module lifetime to obtain equal impacts per kWh as obtained for mc-Si.

The biggest contribution to the marine aquatic ecotoxicity is made by the frame of the module and particularly by the process of prebake for aluminium electrolysis.

3 SENSITIVITY ANALYSIS

A sensitivity analysis regarding the global warming potential was conducted showing the impact of efficiency and lifetime for both perovskite-containing technologies. The results were illustrated in figure 10 and figure 11, and demonstrate that the efficiency and lifetime have an significant role influence in the GWP per kWh. The white-dotted line in both plots is the value obtained for p-Si. Hence, in order to have a similar GWP per kWh compared to p-Si with the assumed efficiencies, this requires a lifetime of 12 and 23 years for SJ-PSC and tandem PSC/c-Si modules respectively. Higher efficiencies or longer lifetimes can further reduce the global warming potential per kWh of produced electricity in comparison to p-Si.

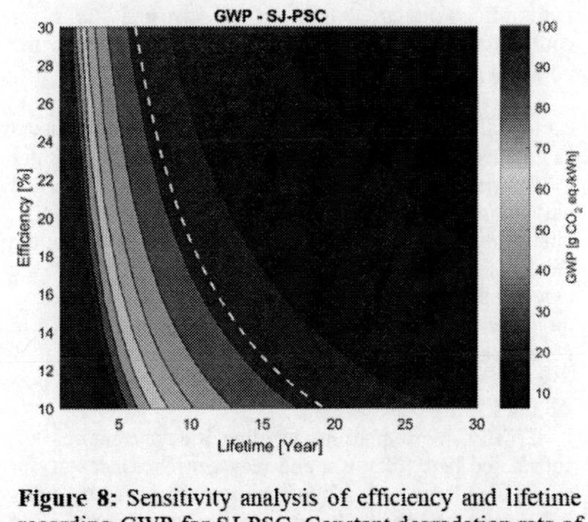

Figure 8: Sensitivity analysis of efficiency and lifetime regarding GWP for SJ-PSC. Constant degradation rate of 0.7% and irradiance of 1700 W/m^2yr. Contours every 10 g CO2 eq. /kWh. White-dotted line representing results for p-Si module (27.2 g CO2 eq. /kWh).

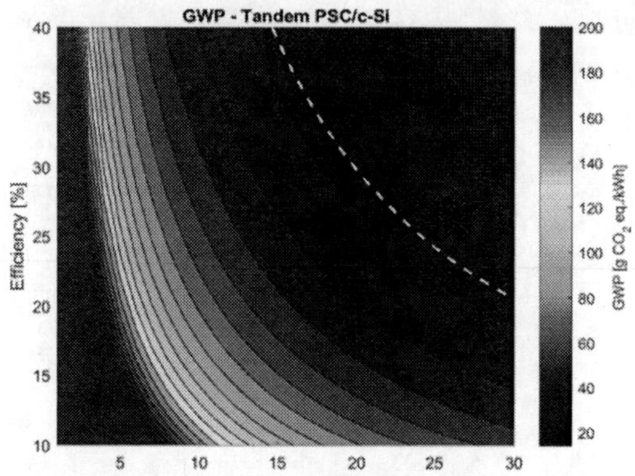

Figure 9: Sensitivity analysis of efficiency and lifetime regarding GWP for tandem PSC/c-Si. Constant degradation rate of 0.7% and irradiance of 1700 W/m^{2}yr. Contours every 10 g CO2 eq. /kWh. White-dotted line representing results for p-Si module (27.2 g CO2 eq. /kWh).

4 CONCLUSIONS AND OUTLOOK

We conducted a Life Cycle Assessment of perovskite solar cells compared to conventional photovoltaic technologies (s-Si, p-Si, CdTe, CIS). The results demonstrate the advantages of PSC regarding the cumulative energy demand and energy payback time especially in the case of single junction perovskite cells. Particularly the mean EPBT for SJ-PSC was 0.45 years, while devices including tandems PSC/c-Si presented a mean value of 1.00 years. The mean result was 0.51, 1.01, 1.22 and 1.25 for modules of CdTe, CIS, p-Si and s-Si respectively. On the other hand, their lifetime and efficiency must improve to reduce their impact per kWh effects in other categories (e.g. global warming potential) compared to conventional technologies. Although most of the environmental impacts are produced by the frame, glass, and encapsulation material in the SJ-PSC and by the silicon cell in the modules containing PSC/c-Si, the indium-tin oxide used in the first case and the silver contacts in the second one concentrate most of the potential abiotic depletion.

The results also revealed that marine aquatic ecotoxicity is the most critical one among the impact categories when compared to p-Si. The main contributor to this impact category is the aluminium frame and particularly the process of aluminium prebake. Finally, our study shows that the encapsulation and module elements have an important environmental impact in the case of single junction cells, mainly due to the energy-intensive production processes of these elements. In general, a lifetime extension beyond the assumption of 5 years is necessary to have results comparable to p-Si modules.

Finally, more analyses in other lifecycle phases not considered here, like use, and recycling are necessary to give a broader picture of the environmental effects of this technology. Additionally, the possibility of leakage of lead to the environment during the use phase must be addressed deeply, since is one of the main concerns regarding these technologies.

5 ACKNOWLEDGEMENTS

The authors would like to acknowledge the members of the Urban and Residential Technologies department of DLR – Institute of Networked Energy Systems for their kind help during this research. The authors would also like to acknowledge Markus Becker from the research group of technical chemistry of the University of Oldenburg, who gave some insights during the first stages of this investigation. Finally, the authors would like to thank Dr Antonio Abate from Helmholtz-Zentrum Berlin, who opened the doors of his labs to see the manufacturing process and shared valuable information about perovskite solar cells.

6 REFERENCES

[1] IPCC, "Climate Change 2014: Synthesis Report. Contribution of Working Groups I, II and III to the Fifth Assessment Report of the Intergovernmental Panel on Climate Change," *Climate Change 2014: Synthesis Report. Contribution of Working Groups I, II and III to the Fifth Assessment Report of the Intergovernmental Panel on Climate Change*, 2014. [Online]. Available: http://ipcc.ch/pdf/assessment-report/ar5/syr/SYR_AR5_FINAL_full_wcover.pdf. [Accessed: 18-Dec-2017].

[2] J. Jean, P. R. Brown, R. L. Jaffe, T. Buonassisi, and V. Bulovic, "Pathways for solar photovoltaics," *Energy Environ. Sci.*, vol. 8, no. 4, pp. 1200–1219, 2015.

[3] A. Kojima, K. Teshima, Y. Shirai, and M. Tsutomu, "Organometal Halide Perovskites as Visible- Light Sensitizers for Photovoltaic Cells," *J Am Chem Soc*, vol. 131, no. October, pp. 6050–6051, 2009.

[4] M. A. Green, Y. Hishikawa, E. D. Dunlop, D. H. Levi, J. Hohl-Ebinger, and A. W. Y. Ho-Baillie, "Solar cell efficiency tables (version 51)," *Prog. Photovolt. Res. Appl.*, vol. 2018, no. 26, p. -20, 2017.

[5] Q. Wali, N. K. Elumalai, Y. Iqbal, A. Uddin, and R. Jose, "Tandem perovskite solar cells," *Renew. Sustain. Energy Rev.*, vol. 84, no. December 2017, pp. 89–110, 2018.

[6] A. Abate, "Perovskite Solar Cells Go Lead Free," *Joule*, vol. 1, no. 4, pp. 659–664, 2017.

[7] L. Qiu, L. K. Ono, and Y. Qi, "Advances and challenges to the commercialization of organic–inorganic halide perovskite solar cell technology," *Mater. Today Energy*, 2017.

[8] M. Anaya, G. Lozano, M. E. Calvo, and H. Míguez, "ABX 3 Perovskites for Tandem Solar Cells," *Joule*, pp. 1–25, 2017.

[9] Deutsches Institut für Normung - NAGUS, *DIN EN ISO 14040 - Environmental Management - Life cycle Assessment - Principles and Framework (ISO 14040:2006) - German and English version EN ISO 14040:2006*. 2009.

[10] G. Wernet, C. Bauer, B. Steubing, J. Reinhard, E. Moreno-Ruiz, and B. Weidema, "The ecoinvent database version 3 (part I): overview and methodology," *International Journal of Life Cycle Assessment*, 2016. [Online]. Available:

http://dx.doi.org/10.1007/s11367-016-1087-8. [Accessed: 15-Mar-2018].

[11] R. Frischknecht, N. Jungbluth, H. J. Althaus, G. Doka, R. Dones, T. Heck, S. Hellweg, R. Hischier, T. Nemecek, G. Rebitzer, and M. Spielmann, "The ecoinvent database: Overview and methodological framework," *Int. J. Life Cycle Assess.*, vol. 10, no. 1, pp. 3–9, 2005.

[12] ESU-Services, "LC-Inventories: Photovoltaics (2012)," 2017. [Online]. Available: http://esu-services.ch/data/public-lci-reports/.

[13] I. Celik, A. B. Phillips, Z. Song, Y. Yan, R. J. Ellingson, M. J. Heben, and D. Apul, "Energy Payback Time (EPBT) and Energy Return on Energy Invested (EROI) of Perovskite Tandem Photovoltaic Solar Cells," *IEEE J. Photovoltaics*, vol. 8, no. 1, pp. 2156–2161, 2017.

[14] Pré Consultants, "SimaPro Database Manual - Methods Library," *PRe'*, 2016. [Online]. Available: https://www.pre-sustainability.com/download/DatabaseManualMethods.pdf. [Accessed: 22-Aug-2018].

[15] J. Gong, S. B. Darling, and F. You, "Perovskite photovoltaics: life-cycle assessment of energy and environmental impacts," *Energy Environ. Sci.*, vol. 8, pp. 1953–1968, 2015.

[16] N. Espinosa, L. Serrano-Luján, A. Urbina, and F. C. Krebs, "Solution and vapour deposited lead perovskite solar cells: Ecotoxicity from a life cycle assessment perspective," *Sol. Energy Mater. Sol. Cells*, vol. 137, pp. 303–310, 2015.

[17] L. Serrano-Lujan, N. Espinosa, T. T. Larsen-Olsen, J. Abad, A. Urbina, and F. C. Krebs, "Tin- and lead-based perovskite solar cells under scrutiny: An environmental perspective," *Adv. Energy Mater.*, vol. 5, no. 20, pp. 1–5, 2015.

[18] I. Celik, Z. Song, A. J. Cimaroli, Y. Yan, M. J. Heben, and D. Apul, "Life Cycle Assessment (LCA) of perovskite PV cells projected from lab to fab," *Sol. Energy Mater. Sol. Cells*, vol. 156, pp. 157–169, 2016.

[19] J. Zhang, X. Gao, Y. Deng, Y. Zha, and C. Yuan, "Comparison of life cycle environmental impacts of different perovskite solar cell systems," *Sol. Energy Mater. Sol. Cells*, vol. 166, no. February, pp. 9–17, 2017.

[20] T. Ibn-Mohammed, S. C. L. Koh, I. M. Reaney, a. Acquaye, G. Schileo, K. B. Mustapha, and R. Greenough, "Perovskite solar cells: An integrated hybrid lifecycle assessment and review in comparison with other photovoltaic technologies," *Renew. Sustain. Energy Rev.*, vol. 80, no. November 2015, pp. 1321–1344, 2017.

[21] M. M. Lunardi, A. Yi Ho-Baillie, J. P. Alvarez-Gaitan, S. Moore, and R. Corkish, "A life cycle assessment of perovskite/silicon tandem solar cells," *Prog. Photovolt Res. Appl.*, vol. 25, no. March 2017, pp. 679–695, 2017.

[22] R. Itten and M. Stucki, "Highly efficient 3rd generation multi-junction solar cells using silicon heterojunction and perovskite tandem: Prospective life cycle environmental impacts," *Energies*, vol. 10, no. 7, 2017.

[23] K. a. Bush, a. F. Palmstrom, Z. J. Yu, M. Boccard, R. Cheacharoen, J. P. Mailoa, D. P. McMeekin, R. L. Z. Hoye, C. D. Bailie, T. Leijtens, I. M. Peters, M. C. Minichetti, N. Rolston, R. Prasanna, S. Sofia, D. Harwood, W. Ma, F. Moghadam, H. J. Snaith, T. Buonassisi, Z. C. Holman, S. F. Bent, and M. D. McGehee, "23.6%-Efficient Monolithic Perovskite/Silicon Tandem Solar Cells With Improved Stability," *Nat. Energy*, vol. 2, no. 4, pp. 1–7, 2017.

[24] Xinjan Technology Ltd, "ITO glass/FTO glass," 2017. [Online]. Available: http://www.xinyan.hk/content/?97.html. [Accessed: 13-Jan-2018].

[25] W. S. Yang, J. H. Noh, N. J. Jeon, Y. C. Kim, S. Ryu, J. Seo, and S. I. Seok, "High-performance photovoltaic perovskite layers fabricated through intramolecular exchange," *Science (80-.).*, vol. 348, no. 6240, pp. 1234–1237, 2015.

[26] N. Espinosa, R. García-Valverde, A. Urbina, F. Lenzmann, M. Manceau, D. Angmo, and F. C. Krebs, "Life cycle assessment of ITO-free flexible polymer solar cells prepared by roll-to-roll coating and printing," *Sol. Energy Mater. Sol. Cells*, vol. 97, pp. 3–13, 2012.

[27] MBraun, "Slot Dye Coating," 2018. [Online]. Available: http://mbraun.de/products/coating-equipment/sloit-die-coater/#specifications. [Accessed: 20-Feb-2018].

[28] R. García-Valverde, J. a. Cherni, and A. Urbina, "Life cycle analysis of organic photovoltaic technologies," *Prog. Photovoltaics Res. Appl.*, vol. 18, no. 7, pp. 535–538, 2010.

[29] M. Gassner, M. Rebelo de Figueiredo, N. Schalk, R. Franz, C. Weiß, H. Rudigier, H. Holzschuh, W. Bürgin, M. Pohler, C. Czettl, and C. Mitterer, "Energy consumption and material fluxes in hard coating deposition processes," *Surf. Coatings Technol.*, vol. 299, pp. 49–55, 2016.

[30] A. Louwen, W. G. J. H. . van Sark, R. E. I. Schropp, W. C. Turkenberg, and A. P. C. Faaij, "Life-cycle greenhouse gas emissions and energy payback time of current and prospective silicon heterojunction solar cell designs," *Prog. Photovolt Res. Appl.*, vol. 23, no. August 2015, pp. 1406–1428, 2015.

A STRATEGY FOR 2ND LIFE C-SI PV MODULES BASED ON FAILURE ASSESSMENT OF SCRAP PV MODULES

E. Kamdje[1], H. Weigand[2], U. Ricklefs[1]
[1]Kompetenzzentrum NanoP, Technische Hochschule Mittelhessen (THM), Wiesenstr. 14, D-35390 Giessen; +49-641-309-1914; ubbo.ricklefs@ei.thm.de
[2]Kompetenzzentrum ZEuUS, Technische Hochschule Mittelhessen (THM), Wiesenstr. 14, D-35390 Giessen

ABSTRACT: End-of-life photovoltaic modules are addressed by the WEEE directive and the pertinent recycling rates need to be fulfilled. The goal of this project is the development of a comprehensive recycling/reuse concept for c-Si PV modules. If possible, defect modules should be repaired for reuse. A reliable failure analysis is basis to the tailored repair/recycling process aimed at. The failure analysis starts with a visible inspection, followed by electrical measurements and ends with analysing images taken with an NIR-enhanced CCD Camera. The method allows for a reliable detection of most types of defects occurring in end-of -life modules. Main part of this paper is a description of the workflow for the image analysis necessary to find these defects.
Keywords: Electroluminescence, Defects, Calibration, PV Modules, Software

1 INTRODUCTION

The end-of-life (EoL) point for large photovoltaic (PV) power plants is not well defined. As long as the total efficiency is high enough, these plants stay at the mains and the loss of efficiency is tolerated. After exceeding a certain number of defect modules, the whole plant may stop its power generation and is dismantled. Techniques of repair for the defect modules will bring these plants back to operation.

The African and South American markets call for small PV-solutions. In these cases low priced reliable second hand modules are preferred. Overall, there is a growing second hand marked for PV components.

Since 2012, PV modules have been included in the EU WEEE Directive [1]. Collection and recycling of discarded end-of-life PV modules is mandatory to an extent of 80%.

The PV industry has started voluntary efforts on PV recycling, particularly regarding the organization of collection, transport and treatment (recycling) of out-of-spec, damaged and end-of-life PV modules by PV CYLCLE.

Today, mainly glass, aluminum and copper are recycled by destructive methods, accepting the loss of components by the destruction of whole modules [2].

In this paper, we discuss methods of detecting and assessing defect cells in modules. These methods are part of an integral concept for repairing and recycling PV modules. The broader framework including the techniques for repair and recycling has been discussed in detail by Glatthaar et al. [3].

2 EXPERIMENTAL

A large number of defects are known in PV c-Si modules. Some of these can be identified by visual inspection [4, 5]:

- Broken glass by volley or mechanical accidents
- Delamination (busbares, EVA foil)
- Broken or deformed frames
- Brownish color caused by hot spots
- Charred electrical connection box
- Defects in connectors
- Environmentally altered surfaces
- etc.

Measurements with photometric testers give additional hints on defect PV modules, especially on defect bypass diodes as well as broken or delaminated connectors and wires. These inspections of modules often can be done while these modules are still installed. However, analysis with photometric testers addresses the modules as a whole and cannot localize the defect within the module. To some degree defects may be localized by thermographic cameras.

In our lab, the modules are measured under defined conditions. We use constant illumination of high power LEDs or halogen bulbs (constant level of illumination but no sun-like spectral distribution) to illuminate the module under test (Fig. 1.a, b). Again, these measurements give hints on certain defects but are not able to localize them.

In a different setup, the modules under test are used as electrical load (Fig. 1.c). When the voltage of the external power source is varied the voltage measured at the module shows if one of the bypass diodes is short or open circuited [6].

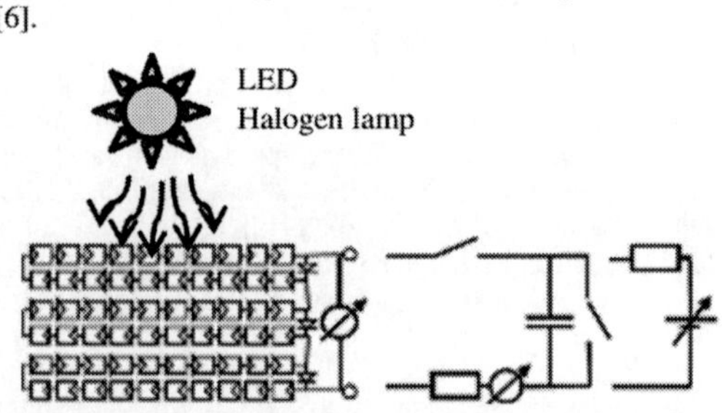

Figure 1: Typical 6x10 PD, 3 bypass diodes, setup for: a. open circuit, b. I/U characteristic, c. test for bypass diodes

Images of the electroluminescence in the NIR-spectral range at wavelengths of 1,1-1,3 μm caused by the recombination of charge carriers are particularly meaningful. Based on this radiation, images of modules can be taken by NIR-enhanced CCD cameras. For the acquisition of these images the module again is operated as an electric load by an external power source (Fig. 2). The current is driven in forward direction (no illumination) through the PV cells (PD). Each PD in a string (1 module consists of typ. 3 strings and typ. 2x10 PDs in a string) is passed by the same current. In PDs with a reduced active area, the density of the current in the remaining area increases. This leads to an increased number of free charges in these remaining areas and thereby to a locally increased electroluminescence. The malfunctioning area of the PD normally show a higher temperature than the intact area and remains darker in the NIR images. The heating up can be visualized directly with an IR camera. The contribution of the thermal IR radiation to the detected NIR images is neglectable small. It does not affect the quality of the NIR-images.

All critical areas that are detectable via the IR camera at long wavelength could as well be detected with the NIR camera. However, NIR-images are more suitable for the defect analysis since they have higher local resolution [7].

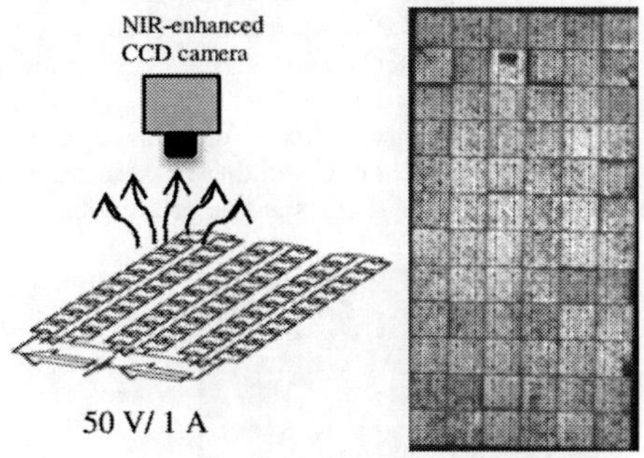

Figure 2: External power source with module as electronic load and NIR enhanced camera for capturing electroluminescence images of modules

Pure silicon shows a spectral sensitivity up to a wavelength λ_c of 1100 nm with a maximal sensitivity at 900 nm. Caused by doping profiles, mostly λ_c is reduced. Therefore, almost no CMOS cameras is able to capture images in the NIR. Some CCD cameras show a small remaining sensitivity at wavelengths greater than 1 µm.

Three cameras were tested. The "Solar Cam for NIR" from MBJ gave best results. However, this camera is expensive (~ 4T€). The "UI-3240LE-NIR-SL" of IDS and the "Raspberry PI 2-V2" (~ 30 €) show a very low sensitivity in the NIR, but may be sufficient for a preliminary result.

The images of the modules were captured in a dark room. Due to the electroluminescence radiation being very weak, the NIR images are very dark and noisy. Caused by the increasing contribution of the dark current to the image it is not possible to enhance the signal to noise ratio of the NIR-images by increasing the exposure time (maximum time of exposure: 20 ms). To compensate for this a larger number of images is taken under same conditions and all these images are summed up. This gives a better signal to noise ratio. To reduce thermal noise, the camera is cooled. Noise in the resulting images than is reduced further by morphological filters.

Typically the NIR images show an affine distortion (Fig. 2). This has to be corrected carefully to an ortho image. Furthermore NIR image are brighter at the center than in the outer regions. This must be numerically compensated. In the next step, the image analysis searches for the edges of the PDs and the busbars in the cells (Fig.3).

Figure 3: Intensity corrected ortho image with identified cell edges (white) and busbars (dark)

A wide variety of specific defects were detected in EoL-PV-modules. The following defects are analysed:

- Dark areas in the cell
- Dark areas at the edges
- Dark areas in the neighborhood of the busbars
- Complete PDs darker than their neighbors
- Cracks and fissures
- Bright spots at the crossing points of the busbar/PD edge
- Bright spots at the busbars
- Strongly structured PDs

Each of these defects has to be addressed by applying specific algorithms in the image analysis. Basis to most of these algorithms are the positions of the edges of the cells and the busbars. Some examples of defects are shown in Fig. 4.

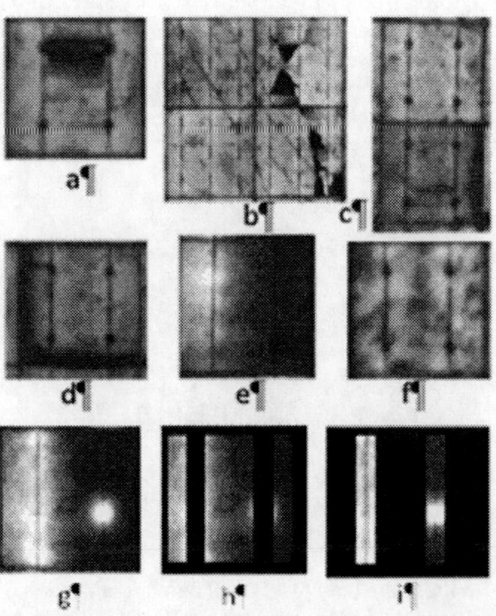

Figure 4: a., b., c., d., e., f. show different types of dark areas, a. caused by defect fingerprints or cell defects, b. shows cracks and clearly defined non contacted areas caused by these cracks, c. completely darkened cell (PID), e. no contact to right busbar, f. strongly structured cell and darkened boarder areas, g. image of a PD segmented in, h. main area of the PD and, i. areas close to the busbars

The mean brightness differs between different types of modules and as well from module to module of the same type. To circumvent the need to define specific thresholds for each individual module, the thresholds are assigned in relation to the measured mean brightness or to the expected standard deviation of the brightness of the module under consideration.

If there are too many defect PDs in a module, module repair would not be efficient. In this case, the goal is to recycle that module by components aiming at the recovery of the maximum number possible non-defect components should be obtained from that module.

Cracks are currently being detected and characterized by their measured length. Future work is related to the assessment of the PD area affected by these cracks.

In some cases the assessment of a PD leads to ambiguous results depending on whether the module as a whole would be assigned to reuse or to repair. For example, if a PD is broken, it cannot be used as a spare part for another module. On the other hand, if it is part of a module with only a low number of defect PDs, it may stay in that module without influencing its total efficiency. Nevertheless, it may be risky to let this broken PD in that module, because the mechanical and thermal conditions for this PD may have changed by the process of repair and reinstallation. That may lead to higher stress and with that to a smaller time of life for the hole module.

Another example of ambiguity is a PD with a hot spot at the busbar. This hot spot may be caused by a singular contact of the busbar to the PD and not by a defect of the cell. So the efficiency of that module is reduced by this PD, but if it is taken out of the module, that PD can be repaired and used as spare part.

Certain defects can be assessed very clearly through image analysis. The efficiency of a string in a module is nearly proportional to the smallest active area of a PD in that string. If there are cracks in the PDs, the inactive areas are easily detected. Other inactive areas in PDs require a more refined approach [8]. Fig. 5 shows an example of a PD with a darkened outer area. The width of the inactive area is not clear defined.

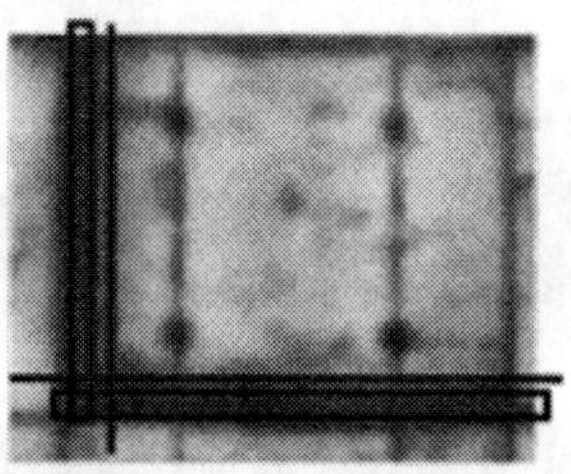

Figure 5: The width and with that the area of the dark region at the borders of the PD are not clear defined.

In literature only for a small number of defects on PDs clear weights for assessment are given. In our analysis eight different types of defects can be detected and assessed. To calibrate the thresholds for a good vs. bad assessment for each of these defects, we searched through our modules and defined borderline PDs by hand. With these PDs thresholds for good - bad assessments are found and the assessments to the other PDs are possible. This set of training data can now be applied for the automated assessment of failure.

3 RESULTS AND DISCUSSION

In many cases, the result of the visual inspection already decides whether a module should get a chance for repair.

Electrical measurements give hints on defects in modules.

Often, analyzing the variance of the brightness of a module's NIR image already shows if it is reasonable to analyze that module for repair.

The NIR images of the modules are analyzed automatically by image analysis. The results are written to an Excel file and assessed in a final image (Fig. 6) of the corresponding module with one of the following marks:

- for material recycling only,
- ready for reuse, requires repair,
- no defect.

In a number of cases different defects are detected for the same PD (Fig. 4. b. c. g.). In these cases, the strongest defect is taken to assess that specific cell in total.

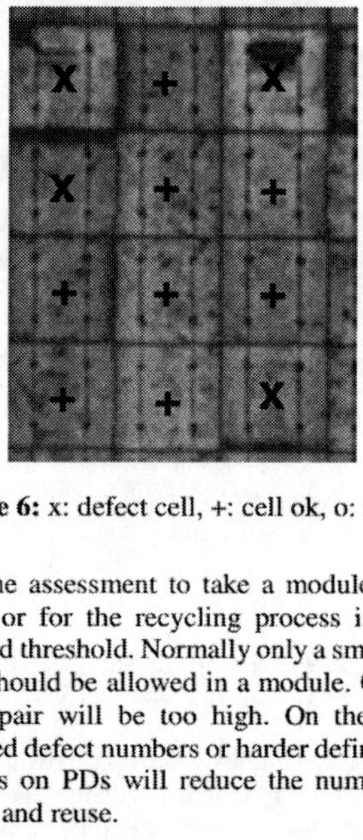

Figure 6: x: defect cell, +: cell ok, o: repair

The assessment to take a module for the repair and reuse or for the recycling process is based on a weak defined threshold. Normally only a small number of defect PDs should be allowed in a module. Otherwise the effort for repair will be too high. On the other hand lower allowed defect numbers or harder defined thresholds to the defects on PDs will reduce the number of modules for repair and reuse.

4 CONCLUSIONS

The acquisition and analysis of NIR-images is a powerful tool for the detection and localization of defects in modules. In combination with a visual inspection and electrical measurements detailed information for the assessment to that module is available. This assessment is basis to the repair or recycling process to be undertaken on that module and its cells.

5 ACKNOWLEDGEMENT

This project (HA project no. 497/16-09) is funded in the framework of Hessen ModellProjekte, financed with funds of LOEWE – Landes-Offensive zur Entwicklung Wissenschaftlich-ökonomischer Exzellenz, Förderlinie 3:

KMU-Verbundvorhaben (State Offensive for the Development of Scientific and Economic Excellence)

6 REFERENCES

[1] European Commission: DIRECTIVE 2012/19/EU OF THE EUROPEAN PARLIAMENT AND OF THE COUNCIL on waste electrical and electronic equipment (WEEE) of 4 July 2012, (2012)
[2] C. Latunussa et al, Solar Energy Materials & Solar Cells, 156 (2016) 101
[3] J. Glatthaar, et al.: Proceedings 33th European Photovoltaic Solar Energy Conference, (2017) 1528
[4] IRENA and IEA-PVPS (2016)
[5] IEA-PVPS: Report IEA-PVPS T13-01, (2014)
[6] A. Wagner: Photovoltaik Engineering, Springer, (2015)
[7] O. Breitenstein et al.: Lock-inThermography-Basics and use for Evaluating Electronic Devices and Materials, Springer, (2010)
[8] U. Jahn: 11. Nationale Photovoltaik-Tagung, (2013), https://docplayer.org/69737026-Qualitaetspruefung-und-zuverlaessigkeit-von-photovoltaik-modulen-und-photovoltaik-anlagen.html (18.09.2018)

PLANNING AND OPTIMIZING THE PV MATERIAL LIFE CYCLE – THE EXAMPLE OF SWITZERLAND

Prof. Urs Muntwyler[1], PD Dr Eva Schüpbach[1], Roman Eppenberger[2]

[1]Bern University of Applied Sciences BFH, Engineering and Information Technology, Photovoltaic Laboratory
Jlcoweg 1, CH-3400 Burgdorf, Switzerland
Tel. +41 34 426 68 37, urs.muntwyler@bfh.ch

[2]Stiftung SENS eRecycling
Obstgartenstrasse 28, CH-8006 Zürich, Switzerland
roman.eppenberger@sens.ch

ABSTRACT: As part of the Energy Strategy 2050, the electricity gap resulting from the phase-out of nuclear energy is primarily compensated with PV modules. The Energy Strategy 2050 plans the expansion of photovoltaics to 12 TWh in 2050. This could rise to 24 TWh due to the elimination of deep geothermal energy (4TWh) and the replacement of cars with combustion engines by electric vehicles (approx. 8TWh) [i]. Based on the current PV market, a huge number of installed PV modules will be seen 2050. These modules can be made up of different technologies such as c-Si, thin-film modules such as Cd/Te, CIS or CIGS. Assuming a service life of 25-40 years[ii], all these modules must be dismantled and disposed of again. Even for PV modules alone the quantities of material that will be disposed of in Switzerland and thus the recycled material flows are considerable. Forward-looking planning of the products employed, and their recycling is important for users and installers. This is where the joint research efforts of SENS eRecycling and the Bern University of Applied Sciences BFH come in.
Keywords: Recycling, power-weight ratio, life cycle

1 INITIAL POSITION

2.1 Brief introduction to the development of photovoltaics in Switzerland

In Switzerland, PV modules have been used for autonomous power supplies since the 1980s[iii]. At that time, the PV systems had small PV capacity, i.e. less than one kWp installed. In 1981, the first two grid-connected PV systems were built in Switzerland and Europe (1.2 kWp Swiss Federal Institute for Reactor Research EIR and 10 kWp SUPSI Lugano). Around 1986, further grid-connected PV systems were installed in Switzerland[iv]. The Tour de Sol in 1987 established a new participant category for "solar vehicles with grid-connected PV systems". This led to the construction of private, grid-connected PV systems in Switzerland. Around 1990, Switzerland was the world leader in the construction of decentralised grid-connected PV systems. However, the installed volumes were less than 1 MWp per year.

2.2 Disposal of PV modules in Switzerland: SENS eRecycling Foundation

With a lifetime of the PV modules of 25-40 years or more, it is to be expected that they will have to be dismantled again in the next few years. In addition, there are cases of fire and guarantee cases in which the modules need to be dismantled prematurely. In concrete terms, the following quantities have been recycled since 2015[v]:

2015: 71 tons

2016: 126 tons

2017: 337 tons

The strong increase in 2017 is astonishing and is primarily due to a warranty recall of a roof-integrated, frameless PV product. A few tons may also be due to disposal after a fire. PV modules that are installed on a building that is set on fire (usually not because of the PV system) usually lose their warranty and are recycled[vi].
The PV modules sent for recycling are accepted by the SENS eRecycling Foundation, the Swiss recycling organisation. SENS is a non-profit organisation founded in 1990 and responsible for the disposal and recycling of electrical waste in Switzerland. The legal framework is the "Ordinance on the Return and Disposal of Electrical and Electronic Equipment (VREG)" of 1 July 1998 (currently under revision). It contains a return obligation for consumers and a take-back obligation for manufacturers, importers and dealers. The costs are covered by the advance disposal fee, which roughly corresponds to the costs of recycling. The recovered material can be recycled to approximately 80% of the market value of the primary material.
The SENS eRecycling network includes manufacturers, importers and wholesalers. Partners and associations ensure coordination. For solar components it is Swissolar, the association of solar entrepreneurs in Switzerland. Recyclers are responsible for the effective processing of the equipment and components disposed of.

2.3 Recyclable material components of PV modules and their further processing

PV modules consist mainly of a cover glass, the embedding material for the interconnected solar cells (standard: 60/72 cells), the embedding material on the back and the foils to protect the solar cells. In most cases, an aluminium frame provides the mechanical stability and a junction box with diodes and two copper cables with plugs ensure the connection to power supply (Fig. 1). Recycling is therefore not very complicated.

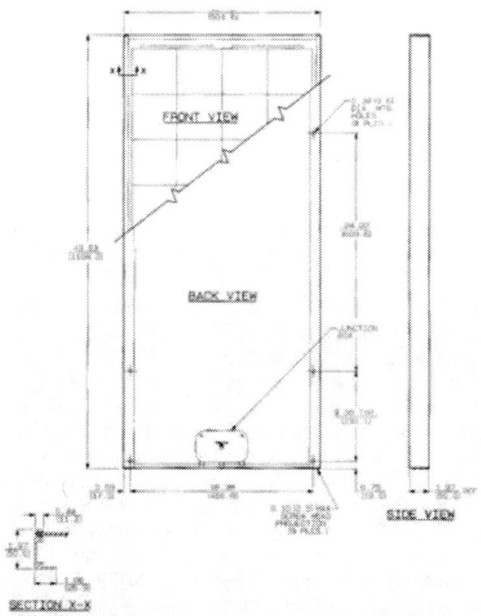

Figure 1: MSX 64 polycrystalline PV module from Solarex (USA) with solid aluminium frame construction, a total weight of 7.2 kg and a large junction box. It is one of the best-selling PV modules from 1988-1998. This module was also very frequently deployed in Switzerland.

Depending on the technology, PV modules need to be recycled separately. All module types used in Switzerland are collected and recycled by the SENS eRecycling Foundation. Crystalline silicon is the dominant type recycled by SENS eRecycling as crystalline SI modules (c-Si) are being the technology primarily used in Switzerland. Modules with cadmium telluride technologies (Cd/Te) are hardly used in Switzerland (<1%). The market leader in this technology (First Solar), which does not sell any modules in Switzerland, has its own disposal line. Copper indium selenide (CIS) / copper indium gallium selenide (CIGS) modules are rarely used in Switzerland and only seen on PV facades.
The shares of materials used in PV modules disposed of by SENS eRecycling in Switzerland include the following materials[vii]. The recycling processes for those present-day small amounts of material are as following (still several hundred tons):

- Glass (including silicon wafers): 88%.
- Metal (especially aluminium): 6%.
- Plastics (foils): 6%.

The aluminium frame is separated and directly transferred to aluminium recycling. The glass is either processed into flat glass or into glass wool and used as insulation material on construction sites. The PV laminate is mechanically shredded in a hammer mill. The particles are mechanically separated in:

- Ferrous metals (screws etc.)
- Non-ferrous metals (e.g. copper)
- Plastics
- Further extraneous materials.

The plastics are either used as substitute fuel in the cement industry or incinerated in a waste incineration plant (energy and heat generation).
All the processes described above are carried out in two factories in Germany, since the current quantities do not justify an own recycling plant in Switzerland[viii]. This means additional transports are required. The analysis of energy expenditure shows energy expenditure for transport is not significant[ix] using large quantities and applying a thought-out choice of means of transport. This even applies for PV modules originating in China or the Far East.
It is conceivable to separate the individual components more precisely and thus increase value creation in the future. It is perceived rather critically (personal communication by Roman Eppenberger, 29.5.2018) whether this could happen in Switzerland as the primary glass produced has little economic value. If silicon could be collected and recycled separately, this added value could improve, which is currently being investigated in various research projects. If chemical separation processes happen to be necessary for this, approval for the process could be a hurdle in Switzerland.

2 FUTURE OF RECYCLING PV MODULES

2.1 Improvements of material

In the future, only half of the material used per kWp of power output will have to be disposed of (Fig. 2; Fig. 3a, 3b). Over the past 40 years PV modules have become larger and more powerful. For example, the power-to-weight ratio of the MSX 64 PV module from Solarex (USA) shown in Fig. 1, with 112.5 kg per 1 kWp, has been halved by today's technology. In some cases, the power output of PV modules has increased 10-fold from approx. 30 Wp to over 300 Wp.

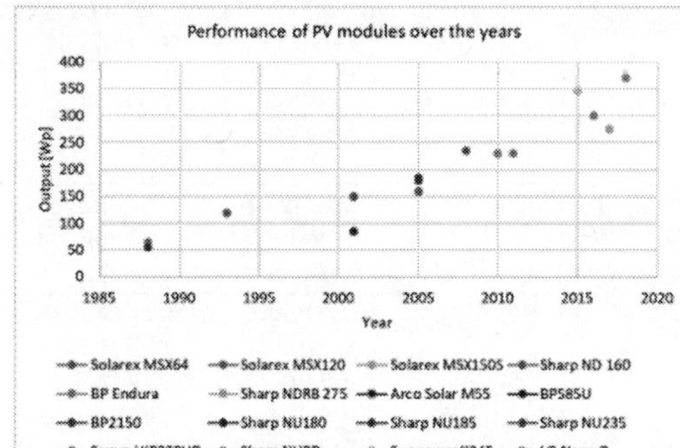

Figure 2: The development of PV modules since 1985 shows the continuous increase in output over the past 30 years.

On the other hand, the efficiency has doubled: Today, a standard module has dimensions of approx. 1x1.6 m. The power-to-weight ratio, i.e. the material quantity for a standard output of 1 Wp, has almost halved. This means that only half as much material needs to be used for a kWh electricity yield as a list of PV modules from the last 30 years shows (Fig. 3a, 3b)

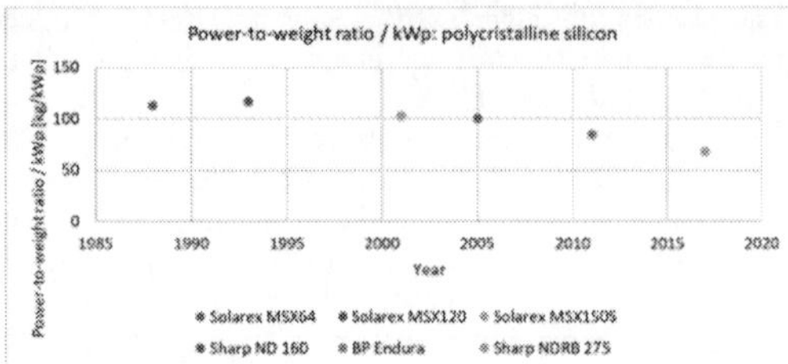

Figure 3a: Development of power-to-weight ratio of polycrystalline PV modules since 1985 (halved in the last 30 years).

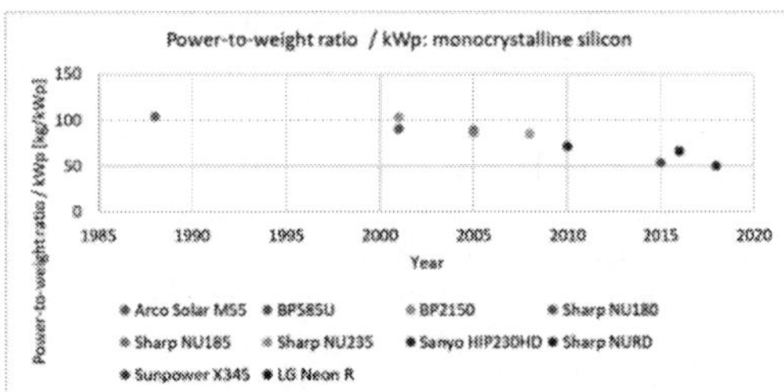

Figure 3b: Development of power-to-weight ratio of monocrystalline PV modules since 1985 (halved in the last 30 years).

Due to the better sensitivity of the solar cells, also the yield per kWp is somewhat higher today. However, this effect is not very pronounced. The longer service life is likely to carry more weight. The fact that old PV modules such as MSX 64 from Solarex (see Fig. 1), KC 50 from Kyocera and first of all M55 from Arco Solar/Siemens last a very long time. This has been demonstrated by long-term PV module measurements by the PV laboratory of BFH[x]. Longevity for new PV modules, has yet to be demonstrated. It is theoretically possible that price pressure could lead to a reduced service life and quality problems.

Table I: Calculated (*) and estimated (**) annually installed PV module volume in Switzerland of 2000-2025.

Year	Market/ Year	Quantity of material (to)/year	Dominant technology
2000*	1 MWp	100 tons	Poly-/ mono-Si (>80% market)
2005*	5 MWp	400 tons	Poly-/ mono-Si (>80% market)
2010*	30 MWp	2'100 tons	Poly-/ mono-Si (>80% market)
2015*	300 MWp	18'000 tons	Poly-/ mono-Si (>80% market)
2020**	500 MWp	25'000 tons	Poly-/ mono-Si (>80% market)
2025**	1'000 MWp	50'000 tons	Poly-/ mono-Si (>80% market)

* Calculation by Muntwyler (based on PV market BFE[xi])/ ** forecast by Muntwyler

The expected cumulative material stock in 2050 (Energy Strategy 2050) is potentially much lower, as a further improvement of the power-to-weight ratio of 20% (40 kg/1 kWp) and an unchanged design of the PV modules can be assumed (Table 2).

Table II: The installed PV material quantities expected in 2050 (**forecast U. Muntwyler).

Year	Installed quantity	Total quantity of material (to)	Dominant technology
2050**	12'000 MWp	480'000 tons	Poly-/ mono-Si (??? market)
2050**	24'000 MWp	960'000 tons	Poly-/ mono-Si (??? market)

Tables 1 and 2 illustrate that the material flows of PV systems in Switzerland today are still relatively small and in the range of several hundred tons per year. Still they increase by a factor of around 1,000 by 2050. The figures suggest that it is worth planning and building up the future recycling for the expected material flows in advance. As explained in section 1.3, current recycling in Switzerland is quite simple and is based on the shredding of c-Si PV modules in hammer mills and further processing of material particles in Germany. The amount of PV modules recycled in Switzerland in 2017 (see section 1.2) corresponds to an output of less than 3 MWp. This quantity will rise to 50,000 tons per year in the coming years (corresponding to 1GWp per year). For example, the oldest and largest PV plant in Europe (Fig. 4) on Mont Soleil (CH) is waiting for disposal (Fig. 4), i.e. an estimated 60 tons of PV module material, plus assembly structures, cables and inverters.

Figure 3b: The free-standing PV installation on Mont Soleil in the Swiss Jura (555 kWp output), built in 1992, is the oldest and largest PV system in Europe.

2.3 Recycling of other components of PV systems

In addition to the PV modules, a PV system requires further components such as PV inverters, copper cables, mounting structures, seals, etc. PV inverters are recycled the same way as a common electrical appliance (e.g. washing machine). An internal study by the PV laboratory at the Bern University of Applied Sciences BFH has shown[xii] the power-to-weight ratio of inverters has also fallen massively. The resulting material flows are becoming smaller and smaller and can be recycled via established recycling organizations such as SENS eRecycling. Today a string inverter like the Delta "Flex" weighs 9.7 kg with 3,800 Wp of power[xiii]. 30 years ago, 39 kg of material were needed for a Solcon HE3400. This improvement by a factor of 4 is enormous as the performance has remained the same. Even more impressive is the comparison between a central inverter that was common 30 years ago and a multi-string inverter that is common today. In 1990, an SMA central inverter required 1,250 kg of material for 50 kWp of power[xiv]. The new multi-string inverter from ABB only 125 kg for 120 kWp[xv]. This is an improvement by a factor of 24 in material efficiency! The efficiency has been improved and prices massively reduced.

2.3 Recycling of PV storage systems

If additional decentralized PV storage systems are used, further considerable quantities of material are installed. In this case not only the battery modules, but also considerable electrotechnical components such as the housing, terminals, cables etc. must be considered. These are easy to recycle. Storage systems in Switzerland are subject to the state obligation (INOBAT) and are processed by the company Batrec in Wimmis (Canton of Berne). Here, Li batteries are recycled according to a hydro-metallurgical process (www.inobat.ch).

3 CLASSIFICATION OF THE QUANTITIES OF MATERIAL TO BE EXPECTED IN THE FUTURE

The total amount of 960,000 tons of PV modules installed in 2050 calculated for the variant with 24 TWh electricity yield in Table 2 seems high. But how high is it in a relative context? For the whole Swiss population this corresponds to 120 kg of material and an output of 3 kWp (approx. 20m^2 area) per capita. This means that electricity can be generated for a 30 years timespan.
Furthermore: The 20m^2 area per capita is equal to an increase of the recycled material by approx. 5 kg (incl. inverters, etc.) per capita or an increase of 50% of the recycled electrical scrap by SENS eRecycling compared to the year 2015. This means that recycling in the Energy Strategy 2050 does not represent an additional order of magnitude.
Compared the amount of recyclable PV material with the total weight of a car, the largest and most expensive product of an average adult living in Switzerland, the total weight of all cars in Switzerland is already 10 times higher (8 million tons) than all PV modules used in Switzerland in 2050 to generate 24 TWh of electricity per year.
The 30 years of electricity production mentioned above are equivalent to 3 generations of cars, i.e. 24 million tons of material. However, every car still needs spare parts, lubricants and almost one ton of petrol per year, i.e. 4.5 million tons of fuel per year or 135 million tons of fuel in 30 years. These 135 million tons of petrol for Swiss cars in 30 years can be replaced by PV electricity, resulting in a recyclable PV module quantity of 960,000 tons. The future belongs to such leverage effects. It goes without saying that this is also more economical. After all, a PV module is much simpler than a car (see Chapter 1) and therefore easier to dismantle and recycle.
Nevertheless, there is still potential in the construction of PV systems to separate PV modules for the recycling process in an even better and faster way and with greater purity. The PV industry can learn a lot from the automotive industry, which has been practicing this for some time. For example, most PV modules today are designed with an aluminum frame. In the future we could see a trend towards glass-glass constructions. This would possibly increase the weight, but the aluminum frame would be omitted. This would improve the energy balance of the PV modules and possibly increase their service life. An increased service life would improve both the economic efficiency of the solar power and the material balance. Completely new designs are also conceivable, possibly based on thin-film technologies, a development that is now more than 20 years old (Solarex Millenia solar modules, flexible Uni-Solar triple modules, etc.). Lightweight modules are another trend.

4 THE MATERIAL OF A PV SYSTEM HAS SEVERAL LIVES (SECOND LIFE)

The material of a PV system has several "lives". It is worth taking this into account when planning PV systems.

4.1 Plan for dismantling and repowering of PV systems

The planning of a PV system also includes considerations on dismantling and recycling. This starts with the selection of the components and solutions (e.g. roof integration or not), continues with the type of assembly, the assembly construction and its aids, through to the planning of the dismantling and disposal of the components. The respective point in time must also be planned. The "repowering" of old plants is a business idea of the future.

4.2 Provisions for the dismantling of plant and equipment

When it comes to financing the deconstruction, the plant operators are in demand and must make provisions for the deconstruction of the plant. This can be done through reserves, e.g. from the "Arbeitsgemeinschaft für dezentrale Energieversorgung ADEV" in Burgdorf. Another strategy is to amortize the plant in 25 years which is the usual period today. The deconstruction could be financed with the extra yield after the regular lifetime of 25 years ("planning by hope..."). It is not very recommendable to pass these costs on to the next plant operator.

4.3 No critical components and materials for CH-PV standard modules

No critical materials are to be expected for the crystalline silicon PV modules predominantly used in Switzerland. This is different for thin-film technologies, which are rarely used in Switzerland for facades and building envelopes. Copper could be a critical component here. It is already an expensive material today and will certainly become even more expensive in the future. Instead of copper, the light metal "aluminium" is available. It is indeed more difficult to use and needs a slightly larger cross-section with a slightly lower weight. At the end the entire electrical industry will face this challenge.

4.4 Energy balances massively positive

Insiders have known for decades that the energy balances of PV modules are very positive, without recycling even in the range of a factor of 30 but have not yet been discussed everywhere. Recycling can further improve energy balances. For example, the main part of calculating energy balances today is whether new or recycled aluminum is used for the frame[xvi].

5 PROSPECTS

Still little agreement and much uncertainty or ignorance is found in the published literature regarding the materials, energy and material flows used in connection with PV modules. The SENS eRecycling Foundation and the PV Laboratory of the Bern University of Applied Sciences (BFH) have therefore established a cooperation to address the existing uncertainties and create the necessary transparency. Joint research efforts are being undertaken on lifetime, material transparency and recycling for both PV systems and PV storage systems. The challenges are easy to solve, because the materials are well known and can already be easily recycled today. The quantities to be expected in the future are large, but in familiar proportions. There is still much potential in the business models for the design, assembly and dismantling of modules, the recycling of materials and objective information on the energy and material flows of recycled PV systems.

6 ACKNOWLEDGMENTS

This research is part of the activities of the Swiss Centre for Competence in Energy Research on the Future Swiss Electrical Infrastructure (SCCER-FURIES), which is financially supported by the Swiss Innovation Agency (Innosuisse - SCCER program). We also gratefully acknowledge funding from Bern University of Applied Sciences BFH, Burgdorf, Switzerland.

References:

[i] Energiestrategie 2050; Urs Muntwyler, www.pvtest.ch, 2017

[ii] Langzeitmessung von PV Anlagen; Urs Muntwyler, PV Labor BFH-TI, www.pvtest.ch, 2018

[iii] Praxis mit Solarzellen, Urs Muntwyler, 1986, 75.ff; Franzis Verlag München

[iv] Muntwylers SolarHandbuch, Urs Muntwyler; 1991, S. 3ff; Muntwyler Energietechnik AG

[v] Persönliche Information, SENS eRecycling, Frühling 2018

[vi] Urs Muntwyler, PV Tagung Staffelstein, 2016

[vii] Persönliche Information, SENS eRecycling, Frühling 2018

[viii] Persönliche Information, SENS eRecycling, Frühling 2018

[ix] Checkliste zum Erstellen einer Energiebilanz, 2014, Unterrichtsmaterialien Wahlmodul „Nachhaltigkeit für Ingenieure; BFH-TI, 2014, PD Dr. Eva Schüpbach

[x] www.pvtest.ch, diverse Beiträge

[xi] Swissolar, PV Markt Schweiz, 2000-2015

[xii] Untersuchung MSE Studenten/ Urs Muntwyler, Sommer 2018 (unveröffentlicht)

[xiii] Tabelle „Wechselrichter", Urs Muntwyler; MSE-Kurs „Photovoltaik" 2018

[xiv] Tabelle „Wechselrichter", Urs Muntwyler; MSE-Kurs „Photovoltaik" 2018

[xv] Tabelle „Wechselrichter", Urs Muntwyler; MSE-Kurs „Photovoltaik" 2018

[xvi] Checkliste zum Erstellen einer Energiebilanz,

5CV.3.75

ECO-SOLAR FACTORY: 40% PLUS ECO-EFFICIENCY GAINS IN THE PHOTOVOLTAIC VALUE CHAIN WITH MINIMISED RESOURCE AND ENERGY CONSUMPTION BY CLOSED LOOP SYSTEMS

K. Wambach[1], R. Peche[1], M. Kroban[1], A. Happach[1], M.P. Bellmann[2], G.S. Park[3], J. Denafas[4], F. Buchholz[5], R. Einhaus[6], G. Noja[7],
B. Ehlen[8], R. Roligheten[9], P. Romero[10], A. Bollar[11]

[1]bifa Umweltinstitut GmbH, Am Mittleren Moos 46, 86167 Augsburg, Germany;
phone: +49 821 7000-198, fax -100, e-mail: kwambach@bifa.de
[2]SINTEF Materials and Chemistry, Trondheim, Norway
[3]Norsun AS, Oslo, Norway
[4]UAB Soli Tek R&D, Vilnius, Lithuania
[5]International Solar Energy Research Center - ISC Konstanz, Konstanz, Germany
[6]Apollon Solar, Lyon, France
[7]Garbo Srl, Cerano, Italy
[8]Boukje.com Consulting BV, Bleiswijk, The Netherlands
[9]Steuler Solar Technology AS, Porsgrunn, Norway
[10]Aimen Laser Application Centre, Porrino, Spain
[11]Ingesea Automation SL, Elgoibar, Spain

ABSTRACT: This conference paper deals gives a summary on the achievement of the targets of the European H2020-FOF-13-2015 Eco-Solar project regarding in material reduction and the environmental impacts like carbon footprint, acidification or resource depletion of the individual processing steps in relation to the environmental impacts of the manufacturing processes of state of the art PV modules (baseline).

By implementing all project developments into the production process chain (crystallization → wafering → cell processing → module design/assembly → enhanced disassembly and module recycling) the project targets reduction demand of aluminium and the reuse of process water can be achieved totally for both, sc-Si and mc-Si based PV modules. The project targets reduction demand of argon gas and reduction demand of ceramics can be achieved totally for sc-Si based PV modules and up to 97 % and 99 % for mc-Si PV modules. Most of the other project targets benefit as well.

Compared to the results of state of the art PV modules (baseline) environmental advantages exist for almost all examined impact categories. The only exception is carcinogenic and non- carcinogenic human toxicity of the improved production process of both, sc-Si and mc-Si based PV modules. In the case of climate change, the greenhouse gas emissions are reduced by more than 17 % for sc-Si based PV modules and more than 22 % for mc-Si based PV modules. The main improvements were achieved in the crystallization process and the module assembly. Improvements can be also found in the wafering in smaller order of magnitude.

Keywords: LCA, c-Si, environmental effect, CO_2 footprint, PV module, material reduction, eco-efficiency

1 AIM AND APPROACH USED

The European H2020-FOF-13-2015 Eco-Solar project that started in October 2015 envisions increasing resource and energy efficiency over the entire photovoltaic value chain, while simultaneously maximising recycling and remanufacturing efficiencies, by introducing design for recovery, repair and reuse, and collaborating for improvements in waste reduction. Reusing materials and reducing the consumption of raw materials will make solar cell panels both cheaper and greener.

One of the main aims of the work package 5 "Environmental impact and repurposing of waste products" is to provide a comprehensive analysis of the progress made with respect to environmental impacts. The implementation of the new eco-friendly production methods requires an ecological validation and has to be economically viable as well. The results of the initial assessment at start of the project (benchmark processes) and the final achievements are presented. bifa carried out an eco-efficiency analysis according to figure 1.

The reference module is a standard module (glass-EVA-Tedlar laminate, nominal power of 270 Wp, 60 cells with an area of 156 x 156 mm each) with front glass and polymer back sheet including aluminium framing.

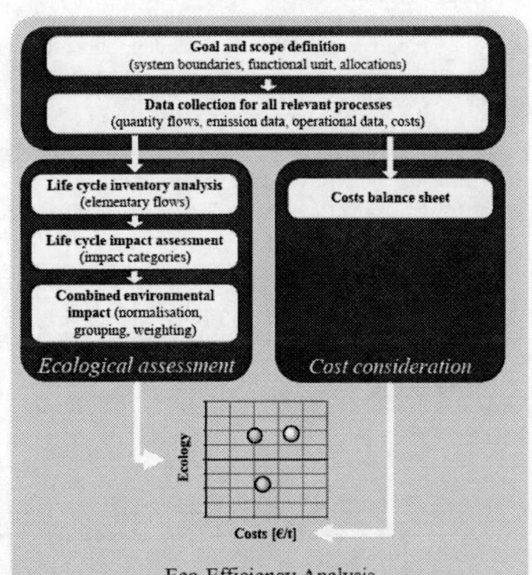

Figure 1: Steps for considering the joint environmental and economic impact within the context of the eco-efficiency analysis

2 SCIENTIFIC INNOVATION AND RELEVANCE

2.1 Life Cycle Inventory Analysis (LCI)

The input data collection for the initial LCI was based on the described reference module. All relevant production steps starting from the ingot casting were modelled using LCI data provided by all partners. Data gaps were filled by using preferably recently published data and expert assumptions. Preliminary baseline data of the PV module production as well as data sets for standard processes like energy supply, transports etc. from professional data bases are used as basis in the material and energy flow modelling [1 - 5].

The LCI baseline of the individual production chain sections crystallization → wafering → cell processing → module design/assembly → enhanced disassembly and module recycling is used as a benchmark for the improvements made throughout the project.

The model structure is designed to facilitate a sectoral evaluation of the results of the LCA. On the one hand it will make it possible to identify the contributions of the individual production chain sections. On the other side the model structure enables to use the results to evaluate direct emissions from sources that are owned or controlled by the reporting entity, indirect emissions from consumption of purchased energy and all other indirect emissions

2.2 Life Cycle Impact Assessment (LCIA): Baseline

Under the framework of LCIA the comprehensive results of the LCI are summarized and prepared for evaluation. The results of the LCI are aligned as far as possible in terms of (potential) environmental impacts and within these impact categories are calculated as accumulated values. Each of the impact categories is related to a more or less complex impact mechanism, at the end of which there are undesired effects on one or more environmental issues. The selection of impact categories for the LCIA has been made by following the ILCD/PEF recommendations [6, 7].

Table 1 shows LCIA results of the reference modules (1.677 m², scope: state of the art production in Europe at project start), used as baseline for evaluation of the project developments.

Table 1: LCIA results of the baseline c-Si PV-modules (60 6-inch solar cells)

Impact category	sc-Si PV-module	mc-Si PV-module
Climate change [CO_2 eq]	360 kg	224 kg
Ecotoxicity, freshwater [CTUe]	13,470	21,500
Particulate matter [PM2.5 eq]	0.38 kg	0.27 kg
Resource depletion [Sb eq]	0.06 kg	0.063 kg
Human toxicity, cancer [CTUh]	2.54E-05	1.58E-05
Ionizing radiation [U235 eq]	78 kBq	33 kBq
Photochemical ozone formation [NMVOC eq]	1.02 kg	0.67 kg
Human toxicity, non-cancer [CTUh]	1.39E-04	1.08E-04
Ozone depletion [CF-11 eq]	0.031 g	0.019 g
Acidification [H+ eq]	2.3 Mole	1.53 Mole
Eutrophication, marine [N eq]	0.41 kg	0.22 kg
Eutrophication, freshwater [P eq]	0.23 kg	0.13 kg
Eutrophication, terrestrial [N eq]	3.3 Mole	2.3 Mole

2.3 Resource consumption: Project targets

The baseline process used as a benchmark includes recent developments (status 2015).

Figure 2 and figure 3 show the reduction of waste and resource consumption per sc-Si PV-module and mc-Si PV-module, respectively, envisaged in the project. The reduction potentials to be achieved are between approx. 40 % for aluminium and approx. 95% for process water.

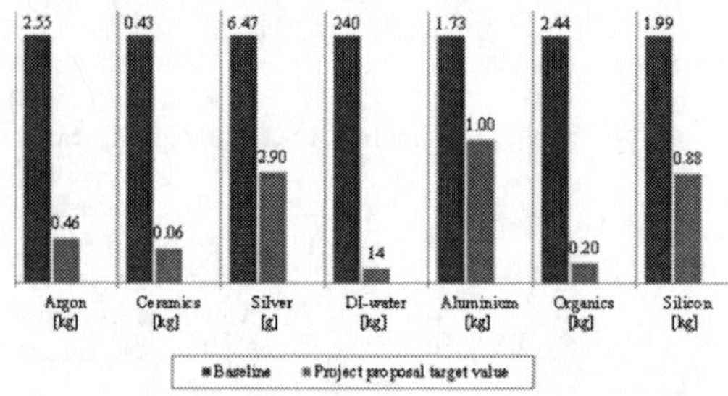

Figure 2: Reduction of waste and resource consumption per one sc-Si PV-module (60 6-inch solar cells) envisaged in Eco-Solar

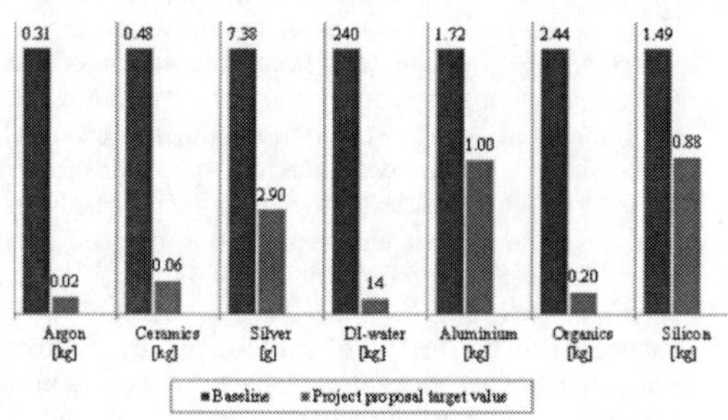

Figure 3: Reduction of waste and resource consumption per one mc-Si PV-module (60 6-inch solar cells) envisaged in Eco-Solar

2.4 Cost consideration: Baseline

Basis for the cost consideration was the current cost consideration of the "International Technology Roadmap for Photovoltaic" and solar PV wafer spot prices [8, 9].

Table 2 shows the resulting proportions of price attributable to individual production chain sections, used as baseline for evaluation of the project developments.

Table 2: Proportions of price at manufacturing of the baseline c-Si PV-modules (60 6-inch solar cells)

Production chain section	sc-Si PV-module	mc-Si PV-module
Production polysilicon:	22 %	23 %
Crystallization and wafering:	22 %	18 %
Cell processing:	21 %	22 %
Module assembly	35 %	37 %

3 RESOURCE CONSUMPTION, LIFE CYCLE IMPACT ASSESSMENT AND COST CONSIDERATON: RESULTS OF SELECTED INDIVIDUAL PROJECT DEVELOPMENTS

3.1 Argon gas recovery by the example of a sc-Si PV module

The approach considers that used argon gas will be treated and afterwards meets the requirement of argon gas for up to 90 % by weight. The data for modelling the argon recovery process are provided by project partner

NorSun.

Figure 4 shows that the project target of reduction of argon gas demand to 0.46 kg per PV module can be achieved totally by implementation the recovery process into the production chain.

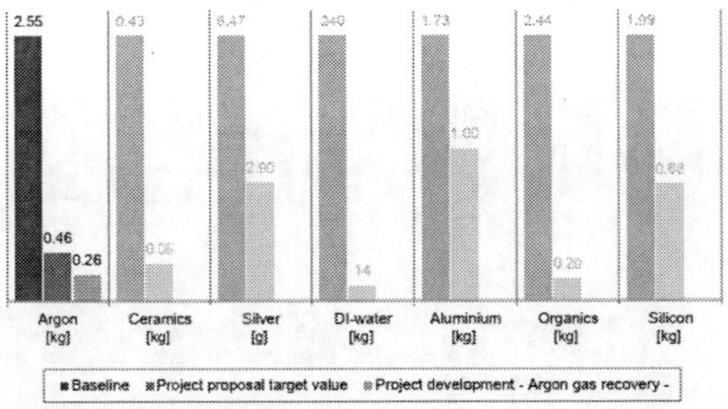

Figure 4: Argon gas recovery: reduction of waste and resource consumption for one sc-Si PV-module (60 6-inch solar cells)

Compared to the LCIA result of the baseline environmental advantages exist for all examined impact categories, even if they are not high. The environmental relief potential ranges from less than 1 % for several impact categories to a maximum of 3 % for particulate matter/respiratory inorganics. All improvements trace back to the crystallization process only.

The estimated cost saving is about 2.1 % for crystallisation process and wafering. That corresponds to approx. 0.5 % of the total manufacturing process costs of a sc-Si PV module.

3.2 New wire sawing process with thinner diamond wire by the example of a mc-Si PV module

The approach considers that a newly wire sawing process developed by the project partner NORSUN makes it possible to use thinner diamond wire (from 80 to 60 µm). This allows more wafers to be sawn out from one silicon ingot. The currently assumed yield plus is 7.92 %. Further the thinner diamond wire leads to a decrease in silicon kerf and wire waste generation.

Figure 5 shows that by implementation the new sawing process into the production chain the project targets of

- reduction of argon gas demand to 0.02 kg per PV module can be fulfilled to 8 %,
- reduction of ceramic demand to 0.06 kg per PV module can be fulfilled to 9 % and
- reduction of silicon demand to 0.88 kg per PV module can be fulfilled to 19 %.

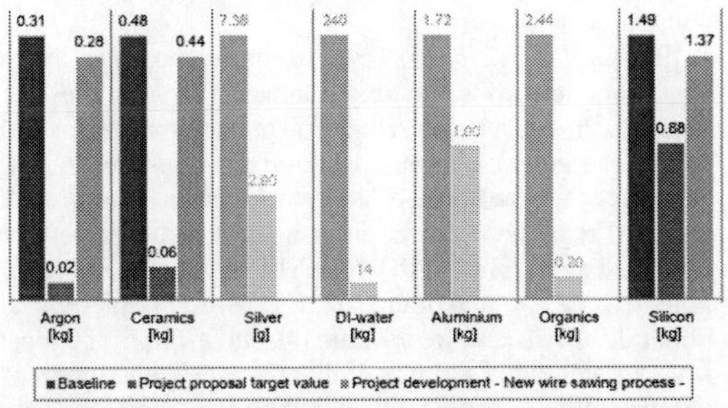

Figure 5: New wire sawing process with thinner diamond wire: reduction of waste and resource consumption for one mc-Si PV-module (60 6-inch solar cells)

Compared to the LCIA result of the baseline environmental advantages exist for all examined impact categories if the new sawing process is implemented into the production chain. The environmental relief potential ranges from less than 0.5 % for depletion of mineral, fossil and renewable resources as well as fresh water ecotoxicity to a maximum of 12 % for carcinogenic human toxicity. The improvements are mainly due to the crystallization process as well wafering on a smaller scale.

The estimated cost saving is about 7.9 % for production polysilicon. That corresponds to approx. 1.7 % of the total manufacturing process costs of a mc-Si PV-module.

3.3 New silicon kerf recovery process from sawing machines coolant by the example of a mc-Si PV module

This approach considers that a newly developed silicon kerf recovery process from sawing machines coolant makes it possible to reuse at least part of the silicon. The current assumption is that for mc-Si wafer 20 % of secondary silicon from kerf can be added to the necessary silicon demand for wafer production without the need for additional preparation of the secondary silicon during melting. In addition, 20% of the silicon content in the crucibles can be substituted as well by the secondary silicon.

The data for modelling the kerf recovery process are provided by project partner GARBO.

Figure 6 shows that by implementation the new sawing process into the production chain the project targets of

- reduction of ceramic demand to 0.06 kg per PV module can be fulfilled by 14 % and
- reduction of silicon demand to 0.88 kg per PV module can be fulfilled by 23 %.

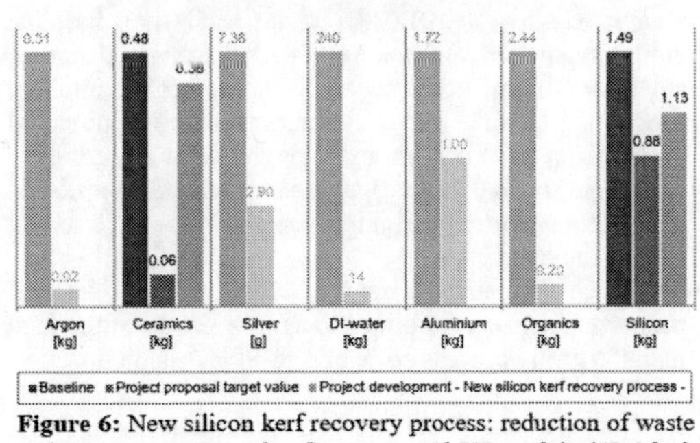

Figure 6: New silicon kerf recovery process: reduction of waste and resource consumption for one mc-Si PV-module (60 6-inch solar cells)

Compared to the LCIA result of the baseline environmental advantages exist for all examined impact categories if the new recovery process is implemented into the production chain. The environmental relief potential ranges from less than 0.5 % for depletion of mineral, fossil and renewable resources as well as fresh water ecotoxicity to a maximum of 12 % for particulate matter / respiratory inorganics. All improvements trace back to the crystallization process only.

The estimated cost saving is about 0.9 % for production polysilicon. That corresponds to approx. 0.2 % of the total manufacturing process costs of a mc-Si PV module.

3.4 Advanced metallization scheme by the example of a sc-Si PV module

According to project partner SoliTek silver metallization paste for the back side (approx. 67 % silver content) can be saved, if the rear side pads are printed with aluminium, so there is a slight increase of aluminium metallization paste consumption. Further experiments regarding the busbars and front side showed that the amount of silver metallization paste for the front side (approx. 84 % silver content) can be reduced by 50 mg per cell. This corresponds to approx. 33 % of the paste quantity. In coordination with project partner a reduction of the total amount of silver metallization paste for the front and back side by about 50 % with a simultaneous increase of aluminium metallization paste by about 2 % was used.

Figure 7 shows that the project target of reduction of silver demand to 2.9 g per PV module can be fulfilled to 86 % by the implementation the advanced metallization scheme into the production chain. The simultaneous increase in aluminium demand is negligible.

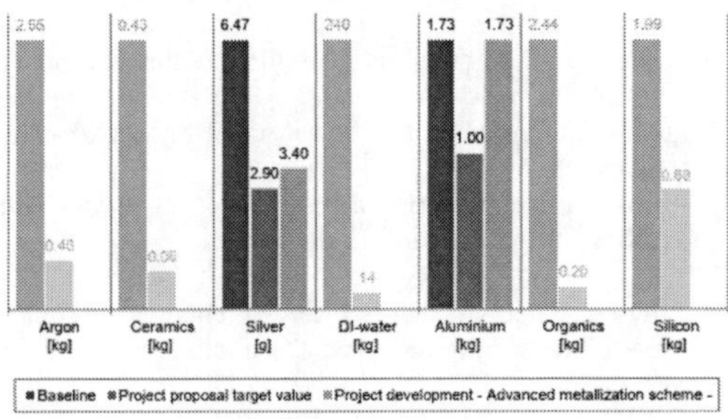

Figure 7: Advanced metallization scheme: reduction of waste and resource consumption for one sc-Si PV-module (60 6-inch solar cells)

Compared to the LCIA result of the baseline environmental advantages exist for all examined impact categories if the new process is implemented into the production chain. The environmental relief potential ranges from less than 1 % for several impact categories to a maximum of 10 % for depletion of mineral, fossil and renewable resources. All improvements trace back to cell processing only.

The estimated cost saving is about 13 % for cell processing. That corresponds to approx. 2.5 % of the total manufacturing process costs of a sc-Si PV module.

3.5 Usage of an EVA-free glass/glass frameless NICE module ("module design for remanufacturing") by the example of a sc-Si PV module

This approach considers that the standard EVA laminated modules with front glass and organic back sheet including an aluminium framing shall be replaced by EVA-free glass/glass frameless NICE modules. The balance model includes both the production and the enhanced disassembly process including treatment and recycling of resulting waste fractions. The regarding material and processing data are provided by project partner APOLLON.

Figure 8 shows that the project target of reduction of aluminium demand to 1 kg per PV module can be achieved totally by using the NICE modules. In addition the project target of reduction of organics demand to 0.2 kg per PV module can be fulfilled by 95 %. All other project targets also benefit to a small extent.

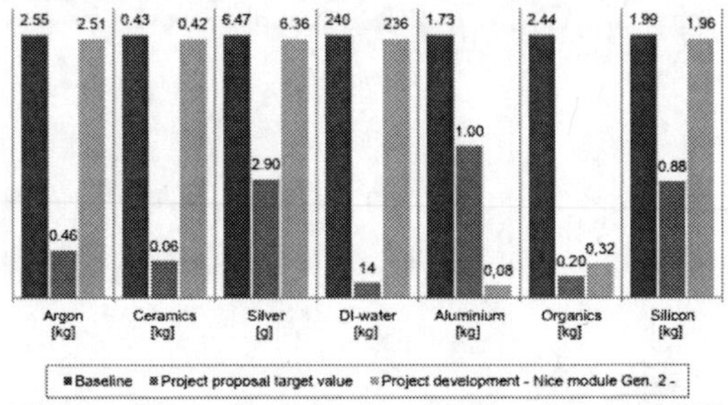

Figure 8: Usage of an EVA-free glass/glass frameless NICE module: reduction of waste and resource consumption for one sc-Si PV-module (60 6-inch solar cells)

Compared to the LCIA result of the baseline environmental advantages exist for ten of the examined impact categories if the standard PV-module be replaced by the NICE PV-module. The environmental relief potential ranges from less than 1 % for acidification to 69 % for freshwater ecotoxicity. At the impact categories carcinogenic and non- carcinogenic human toxicity as well as freshwater eutrophication, however, the use of the NICE module causes environmental burdens between 2 % and 53 %.

The burdens of carcinogenic human toxicity, for example, are a result of chromium(VI) emissions to water, which arise during the production and recycling of the two galvanised steels that replace the standard aluminium frame. The improvements and deteriorations are mainly due to module design/assembly of NICE module and the enhanced module dismantling.

The estimated cost saving is about 5.2 % for module assembly. That corresponds to approx. 1.8 % of the total manufacturing process costs of a sc-Si PV module.

APOLLON demonstrates that module assembly of the Eco-Solar cells result in comparable module output as with current (less eco-friendly) state of the state of the art module technology. APOLLON is working on the disassembly of modules into their major components (glass sheets, copper wires, cells) allowing for their recovery as entire pieces, preserving the value and strong potential for re-use (preferably in PV or other industries) and allowing for repair.

3.5 Other project developments

Each of the other research results show benefits in the processing steps but not as high as the one mentioned above.

Reusable crucibles: Tests at project partners with small silicon nitride crucibles showed that crucibles are reusable. In consultation with project partners Steuler and SINTEF a seven-time reuse of large crucibles was agreed as approach based on the laboratory results. The project target of reduction of ceramic demand to 0.06 kg per PV module can be fulfilled to 97 %. The environmental relief potential has a maximum of 1.5 % for depletion of mineral, fossil and renewable resources and the cost saving potential is estimated roughly with a maximum of approx. 1.1 % of the total manufacturing process costs.

Reuse process water: In consultation with project partner ISC Konstanz it seems to be justified, that waste water from cell processing will be treated and after recycling meets the requirement of process water for up to 97 % by weight. Therefore this project target can be achieved totally. The environmental relief potential has a maximum of 0.2 % for acidification as well as fresh water eutrophication and the cost saving potential is estimated roughly with approx. 0.4 % of the total manufacturing process costs.

New cell process: The approach considers the current cell process at SoliTek. This approach is not recommended for sc-Si wafer. Regarding project targets no changes occurred compared to the baseline-scenario yet though better cell efficiencies due to process modifications were obtained. Compared to the LCIA result of the baseline no environmental advantages exist for the examined impact category The additional environmental pollution ranges from 2 % for fresh water ecotoxicity to a maximum of 33 % for depletion of mineral, fossil and renewable resources. The cost saving potential is estimated roughly with approx. 1.1 % of the total manufacturing process costs. .

"Cell Doctor": The approach considers that a newly cell repair process developed by the project partner AIMEN makes it possible to avoid about 50 % of the cell scrap. The impact on the reduction of silicon demand and the environmental effects are both small. The project target of reduction of silicon demand to 0.88 kg per PV module can be fulfilled to a maximum of 5 %. The environmental relief potential has a maximum of 2.0 % for fresh water eutrophication as well as carcinogenic human toxicity and the cost saving potential is estimated roughly with a maximum of approx. 1.0 % of the total manufacturing process costs.

6.5 Combination of all project developments

The approaches consider that all measured project developments described can be implemented in the production chains, except the new cell process for the mc-Si PV-module.

4.1 sc-Si based PV module

Figure 9 shows that by implementing all project developments into the production chain the project targets of

- reduction of argon gas demand to 0.46 kg per PV module can be achieved totally,
- reduction of ceramic demand to 0.06 kg per PV module can be achieved totally,
- reduction of silver demand to 2.9 g per PV module can be fulfilled to 88 %,
- reduction of process water demand to 14 kg per PV module can be achieved totally,
- reduction of aluminium demand to 1 kg per PV module can be achieved totally,
- reduction of organics demand to 0.2 kg per PV module can be fulfilled to 95 % and
- reduction of silicon demand to 0.88 kg per PV module can be fulfilled to 35 %.

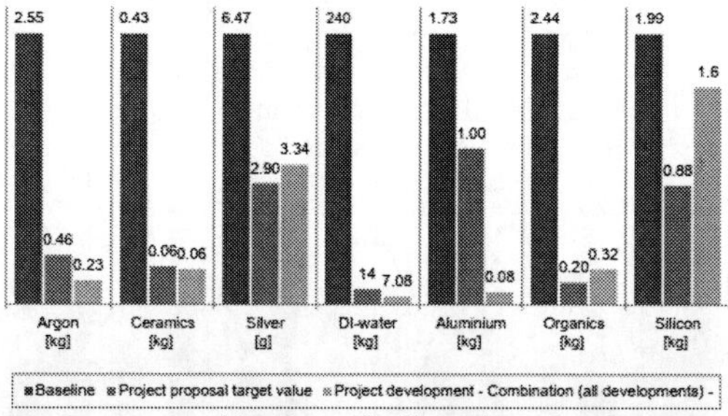

Figure 9: Combination of all project developments: comparison of results for global warming potential of one sc-Si PV-module (60 6-inch solar cells)

Table 3 shows that compared to the LCIA result of the baseline environmental advantages exist for all of the examined impact categories except of carcinogenic and non-carcinogenic human toxicity. The environmental relief potentials range from 10 % for ionizing radiation on human health to a maximum of 78 % for freshwater ecotoxicity. The environmental impact potentials increase by less than 1 % and 37 % respectively. Both deteriorations are a result of the galvanised steels used for NICE module instead of the aluminium frame.

Table 3: Combination of all project developments: comparison of LCIA results of baseline and project developments for one sc-Si PV-module (60 6-inch solar cells)

Impact category	Environmental impact of combination of all project developments compared to baseline	
Climate change [CO_2 eq]	Benefit of 61 kg	reduction 17 %
Ecotoxicity, freshwater [CTUe]	Benefit of 9,802	reduction 73 %
Particulate matter [PM2.5 eq]	Benefit of 0.076 kg	reduction 20 %
Resource depletion [Sb eq]	Benefit of 0.046 kg	reduction 78 %
Human toxicity, cancer [CTUh]	Burden of 9.39E-06	increase 37 %
Ionizing radiation [U235 eq]	Benefit of 8.3 kBq	reduction 11 %
Photochemical ozone formation [NMVOC eq]	Benefit of 0.16 kg	reduction 16 %
Human toxicity, non-cancer [CTUh]	Burden of 1.30E-06	increase < 1 %
Ozone depletion [CF-11 eq]	Benefit of 0.005 g	reduction 16 %
Acidification [H+ eq]	Benefit of 0.29 Mole	reduction 12 %
Eutrophication, marine [N eq]	Benefit of 0.07 kg	reduction 17 %
Eutrophication, freshwater [P eq]	Benefit of 0.024 kg	reduction 10 %
Eutrophication, terrestrial [N eq]	Benefit of 0.46 Mole	reduction 14 %

The estimated cost savings are about

- 8.5 % for production of polysilicon,
- 7.1 % for crystallization process and wafering,
- 18.8 % for cell processing and
- 5.2 % for module assembly.

That corresponds to approx. 9.2 % of the total manufacturing process costs of a sc-Si PV-module.

4.2 mc-Si based PV module

Figure 10 shows that by implementing all project developments (except the new cell process) into the production chain the project targets of

- reduction of argon gas demand to 0.02 kg per PV can be fulfilled to 97 %,
- reduction of ceramic demand to 0.06 kg per PV module can be fulfilled to 99 %,
- reduction of silver demand to 2.9 g per PV module can be fulfilled to 78 %,
- reduction of process water demand to 14 kg per PV module can be achieved totally,
- reduction of aluminium demand to 1 kg per PV module can be achieved totally,

- reduction of organics demand to 0.2 kg per PV module can be fulfilled to 95 % and
- reduction of silicon demand to 0.88 kg per PV module can be fulfilled to 71 %

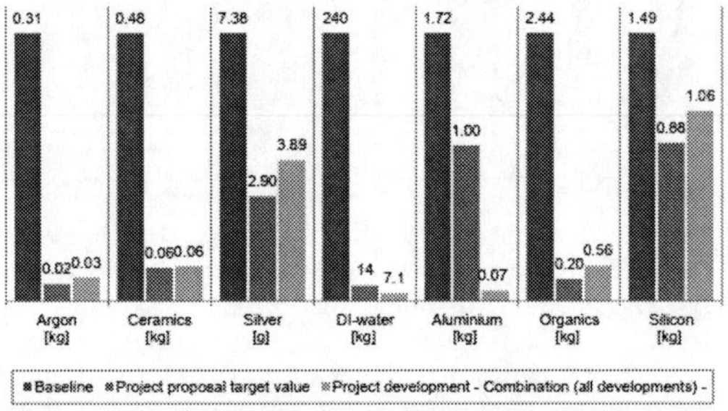

Figure 10: Combination of all project developments: comparison of results for global warming potential of one mc-Si PV-module (60 6-inch solar cells)

Table 4 shows that compared to the LCIA result of the baseline environmental advantages exist for all of the examined impact categories except of carcinogenic and non-carcinogenic human toxicity. The environmental relief potentials range from 11 % for ionizing radiation on human health to a maximum of 78 % for freshwater ecotoxicity. The environmental impact potentials increase by 3 % and 61 % respectively. Both deteriorations are also a result of the galvanised steels used for NICE module instead of the aluminium frame.

Table 4: Combination of all project developments: comparison of LCIA results of baseline and project developments for one mc-Si PV-module (60 6-inch solar cells)

Impact category	Environmental impact of combination of all project developments compared to baseline	
Climate change [CO_2 eq]	Benefit of 49 kg	reduction 22 %
Ecotoxicity, freshwater [CTUe]	Benefit of 9,717	reduction 78 %
Particulate matter [PM2.5 eq]	Benefit of 0.066 kg	reduction 24 %
Resource depletion [Sb eq]	Benefit of 0.047 kg	reduction 75 %
Human toxicity, cancer [CTUh]	Burden of 1.02E-05	increase 61 %
Ionizing radiation [U235 eq]	Benefit of 3.7 kBq	reduction 11 %
Photochemical ozone formation [NMVOC eq]	Benefit of 0.13 kg	reduction 20 %
Human toxicity, non-cancer [CTUh]	Burden of 3.58E-06	increase 3 %
Ozone depletion [CF-11 eq]	Benefit of 0.004 g	reduction 21 %
Acidification [H^+ eq]	Benefit of 0.22 Mole	reduction 14 %
Eutrophication, marine [N eq]	Benefit of 0.05 kg	reduction 24 %
Eutrophication, freshwater [P eq]	Benefit of 0.016 kg	reduction 12 %
Eutrophication, terrestrial [N eq]	Benefit of 0.37 Mole	reduction 16 %

The estimated cost savings are about

- 8.8 % for production of polysilicon,
- 5.5 % for crystallization process and wafering,
- 19.2 % for cell processing and
- 5.2 % for module assembly.

That corresponds to approx. 9.1 % of the total manufacturing process costs of a mc-Si PV-module

5 ECO-EFFICIENCY ANALYSIS (EEA) AND CONCLUSION

The EEA compares the results of the life cycle impact assessment with the specific results of the costs consideration.

5.1 sc-Si based PV module

The portfolio in Figure 11 shows that all project developments achieve a higher eco-efficiency than the baseline. Comparing the project developments with each other the usage of an EVA-free glass/glass frameless NICE module Generation 2 (point 9) obtains the highest increase of eco-efficiency, although this project development does not entail the greatest cost saving.

Then follow the new wire sawing process with thinner diamond wire (point 3) and the advanced metallization scheme (point 5), two project developments which perform both a slightly better ecology index and cost index than the remaining project developments.

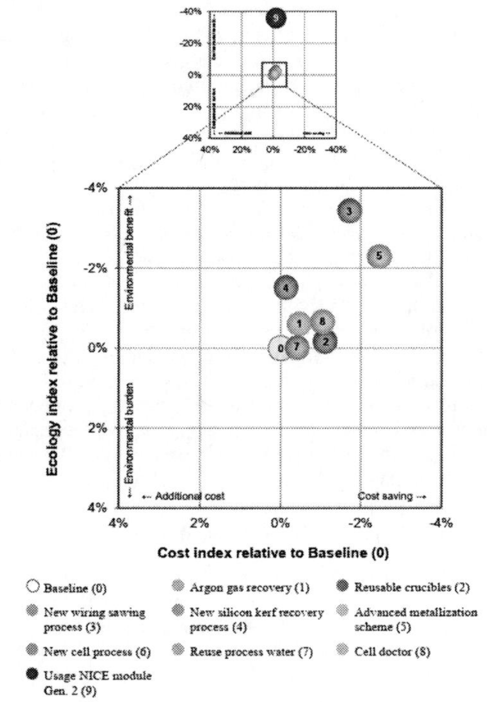

Figure 11: Eco-efficiency portfolio for sc-Si PV modules

5.2 mc-Si based PV module

The portfolio in Figure 12 shows a similar result for mc-Si PV modules. With the exception of the new cell process all project developments achieve a higher eco-efficiency than the baseline.

The usage of an EVA-free glass/glass frameless NICE module Generation 2 obtains the highest increase of eco-efficiency followed by advanced metallization scheme and the new wire sawing process with thinner diamond wire. The new silicon kerf recovery process from sawing machines coolant generates also a good ecology index but the cost index does not improve in the same order of magnitude.

Due to the additional environmental impact for all examined impact categories, the new cell process (point 6) has a significantly worse ecology index than the baseline.

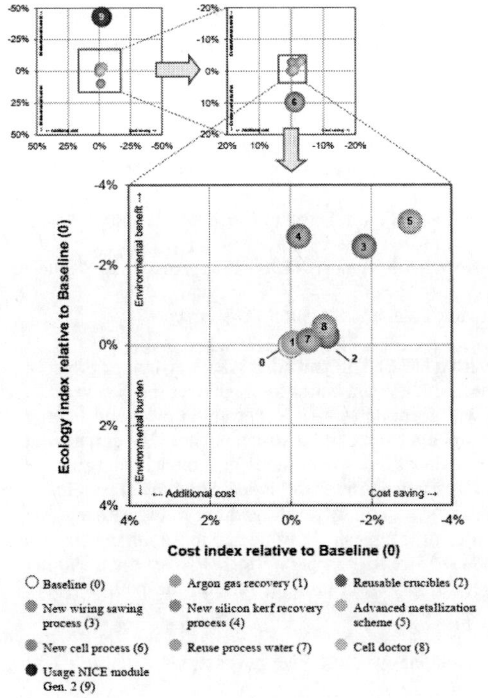

Figure 12: Eco-efficiency portfolio for mc-Si PV modules

5.3 Summary

In general the project developments show a better eco-efficiency with low cost advantages and partly significant environmental relief potentials. The total ecological benefit achieved for sc-Si PV modules is 45 % and for mc-Si PV modules 42 %.

REFERENCES

[1] R. Frischknecht, R. Itten, P. Sinha, M. de Wild-Scholten, J. Zhang, V. Fthenakis, H. C. Kim, M. Raugei, M. Stucki, 2015, Life Cycle Inventories and Life Cycle Assessment of Photovoltaic Systems, International Energy Agency (IEA) PVPS Task 12, Report T12-04:2015.

[2] P. Dold, 2015, Silicon Crystallization Technologies, Semiconductors and Semimetals, Volume 92, Pages 1-61, ISSN 0080-8784.

[3] n.a., 2017, International Technology Roadmap for Photovoltaic (ITRPV) 2016 Results, 8th Edition March 2017

[4] N. Jungbluth, M. Stucki, R. Frischknecht, 2009, Photovoltaics, in R. Dones (Ed.) et al., Sachbilanzen von Energiesystemen: Grundlagen für den ökologischen Vergleich von Energiesystemen und den Einbezug von Energiesystemen in Ökobilanzen für die Schweiz, ecoinvent report No. 6-XII, Swiss Centre for Life Cycle Inventories, Dübendorf 2009

[5] H.J. Möller, 2015, Wafering of Silicon, Semiconductors and Semimetals, Volume 92, Pages 63-109, ISSN 0080-8784.

[6] European Commission-Joint Research Centre-Institute for Environment and Sustainability: International Reference Life Cycle Data System (ILCD) Handbook-Recommendations for Life Cycle Impact Assessment in the European context. First edition November 2011. EUR 24571 EN. Luxemburg. Publications Office of the European Union; 2011

[7] European Commission, Joint Research Centre, Institute for Environment and Sustainability. Characterisation factors of the ILCD Recommended Life Cycle Impact Assessment methods. Database and Supporting Information. First edition. February 2012. EUR 25167. Luxembourg. Publications Office of the European Union; 2012.

[8] International Technology Roadmap for Photovoltaic (ITRPV): 2017 results. Administrated and coordinated by VDMA Photovoltaic equipment, Ninth Edition, March 2018

[9] Solar PV Wafer Weekly Spot Price. http://pvinsights.com, download 24th August 2018

ACKNOWLEDGEMENT

This project has received funding from the European Union's Horizon 2020 research and innovation programme under grant agreement No. 679692.

VISIONS FROM THE FUTURE: THE INTERACTION BETWEEN CURTAILMENT, SPINNING RESERVE SETTINGS AND GENERATOR LIMITS ON AUSTRALIAN PROJECTS WITH MEDIUM TO HIGH RENEWABLE ENERGY FRACTIONS

Bert Herteleer[1], George Dickeson[1], Lachlan McLeod[1], Bradley van Ree[1], Clare Paynter[1],
Dow Airen[2], Phil Maker[4], Sam Latz[2], Anthony Dobb[3], Lyndon Frearson[1]
[1]Ekistica, bert.herteleer@ekistica.com.au, PO Box 8044 (Desert Knowledge Precinct), Alice Springs Northern Territory 0871, Australia
[2]Power and Water Corporation, Ben Hammond Complex, Iliffe Street Stuart Park, Darwin, Northern Territory 0800 Australia
[3]Australian Renewable Energy Agency (ARENA), 2 Phillip Law St NewActon, Canberra City, 2601, Australian Capital Territory, Australia
[4] Enernet Global, Level 3, 169 Fullarton Road, Dulwich, Adelaide SA 5065, Australia

ABSTRACT:. The operational experience and knowledge gained from NT SETuP and other ARENA-funded projects, which typically consist of diesel synchronous generators hybridised with PV and battery storage over the past year are discussed, combined with learnings from other Australian grids with medium to high Renewable Power and Energy Fractions. The operation of PV on these remote grids are used to highlight some of the drivers behind PV curtailment once renewables deliver a large fraction of power and energy, in light of the synchronous generator loading limits, system spinning reserve settings and demand patterns. Based on this experience and worldwide deployment patterns of PV, curtailment will soon become prevalent for grids where renewables routinely reach medium to high Renewable Power and Energy Fractions. A paradigm shift of embracing curtailment as a means of value maximisation, rather than yield maximisation, is identified as a means of ensuring a continued licence to operate for PV over time. Further investment and research needs are discussed for PV to continue to play a meaningful role on the path to 100% Renewable Energy.

Keywords: curtailment, licence to operate, operational limits, Renewable Power and Energy Fractions

1 INTRODUCTION

The cumulative growth of solar photovoltaic (PV) and wind capacity worldwide has surpassed 1 TW in 2018, with wind estimated to reach 592 GW in 2018 [1] and PV exceeding 415 GW by end of 2017 [2]. These numbers indicate the spectacular growth of variable renewable generation on the world's grids, yet in energy and relative power terms, the impact of PV on broader grids is still limited, with PV contributing 7.2% [2] of Germany's annual electricity consumption, and Australia achieving an overall annual Renewable Energy Fraction contribution from PV of 3.9% [3].

By contrast, projects supported by ARENA under its Regional Australia's Renewables (RAR) program [4] aimed to achieve annual Renewable Energy Fractions of 2-30% using PV without battery storage and beyond 50% when PV was combined with battery storage and/or wind and additional enabling technologies. These projects not only can serve as a model for on-grid cases reaching higher Renewable Energy Fractions, they also inform the potential for increased electrification and hybridisation of isolated grids worldwide, with estimates of 1-1.2 billion [5], [6] people living with no or poor supplies of electricity, where renewables such as PV can play an increasing role [7].

For consistency and clarity, the terms Renewable Energy Fraction and Renewable Power Fraction are used, which are defined as [8] in Eqs (1) and (2)

$$\mathrm{REF}_{\mathrm{period}} = \frac{E_{\mathrm{Ren}}}{E_{\mathrm{demand}}}\ [\%] \qquad (1)$$

$$\mathrm{RPF} = \frac{P_{\mathrm{Ren}}}{P_{\mathrm{demand}}}\ [\%] \qquad (2)$$

with the subscript *Ren* denoting renewable, and demand denoting the total demand or load in the system, and the period being the amount of time over which the REF is calculated.

The higher $\mathrm{REF}_{\mathrm{year}}$ aimed for by the projects in the RAR program and other similar isolated grids in Australia result in high RPF values, at which point curtailment of the renewable generation comes into play, particularly in light of the general tendency to depend on diesel-powered[1] synchronous generators which provide a range of services, from the grid reference signal for the grid-connected PV inverters, to reactive power and black start capabilities. In most cases, the amount of PV power permitted on the grid is then restricted to the operational limits of the generators, modified by the spinning reserve safety margins.

2 SPINNING RESERVE

Over short intervals, an excess of generation versus the demand is typically easier to deal with than situations where there is a lack of generation for the given demand. The focus of spinning reserve is to thus primarily deal with demand increases which are harder to accommodate than demand decreases. The use of spinning reserve (SR) permits timely switching to a more appropriately sized generator, given the demand on the grid.

While definitions and interpretations of spinning reserve vary significantly [9], [10], [11] the working definitions employed here are the following:

1. The actual or available spinning reserve is the spare capacity of synchronous generators that can rapidly

[1] Some ARENA RAR projects use gas instead of diesel, yet most of the principles discussed here apply to both.

absorb a large load change while servicing the present demand and maintain system parameters within the grid's technical tolerances. For off-grid situations, the required spinning reserve is the value determined for that grid, based on prior experience or estimate of the probability of load step changes. In practice, the formulation of SR_{req} as per Eq (4) rather than Eq (3) is used.

$$SR_{req} = p\left(\Delta P_{demand,max}(\Delta t)\right) * \Delta P_{demand,max} \quad (3)$$

$$SR_{req} \equiv 1 * \Delta P_{demand,max} \quad (4)$$

$$SR_{act} = P_{gen,nom} - P_{gen,demand} \quad (5)$$

2. Spinning reserve is the spare generation capacity in the system that can rapidly absorb a large load change *with a high likelihood* while servicing the present demand and maintain system parameters within the grid's technical tolerances. The required spinning reserve is determined by the prior experience or estimate of the probability of load step changes, combined with the probability of unplanned outages (typically described as Forced Outage Rate[2] – FOR [12]) of the generator(s) in the system.

$$SR_{req} = \max\begin{pmatrix} p\left(\Delta P_{demand,max}(\Delta t)\right) * \Delta P_{demand,max}, \\ p\left(\Delta P_{PV}(\Delta t)\right) * P_{PV} \end{pmatrix} + \Sigma_i \, p\left(FOR_{gen,i}\right) * P_{gen,nom,i} \quad (6)$$

$$SR_{act} = P_{gen,nom} - P_{gen,demand} + P_{demand,fast-acting} + P_{PV,firm} \quad (7)$$

In the above equations, p denotes the probability, and $\Delta P(\Delta t)$ the step change in power over the next time interval, the subscript *gen* indicates a generator (synchronous or not), and *nom* the nominal or nameplate rating.

3. For the regulated grid in the Northern Territory, the following definition is used: "Spinning reserve is the ability to immediately and automatically increase generation or reduce demand in response *to a fall in frequency*." [13] (Italics added). Given the relationship between the balance of power and frequency in the system, this implicitly also speaks to the available power in the system from generators to deal with net generator loading increases, through demand increases, generator Forced Outage Rates and similar events on the grid.

In all cases, the constraint

$$SR_{act} \geq SR_{req} \quad (8)$$

must be (or attempted to be) met: if constraint (8) will be violated, this results in additional generation to be called up, or may result in demand shedding.

Depending on the interlocutor, the definition of both the actual and required SR is one that focuses only on "firm" synchronous generators (definition 1, above), or SR is viewed (definition 2) as the combination of the spare capacity of all generation sources that is available for the following instant *with a high likelihood*, and may include rapid acting loads that can be ramped down or turned off (e.g. a dynamic resistor or dump load, flywheels or BESSs during charging). Definition 3 is broader, focusing on ensuring that the frequency is maintained, with the priority being given to frequency drops, rather than frequency increases.

Using isolated power systems to describe the following principles, spinning reserve is also influenced by the generator size architecture, whether these are identical in power (e.g. multiple 1 MW diesel generators), or whether generators are sized as small/medium/large, with overlapping operational windows [14]. Further reliability aspects are whether the system is designed for increased resilience in the case of generator failure – e.g. mines typically have at least one additional generator spinning and may operate to N-2 or N-3 contingency events[3] (and thus have a larger spinning reserve requirement), compared to remote communities where the system is typically designed to operate to N-1. This balance of risk versus operational expenses is evaluated on a case-by-case basis. For most of what follows, the described systems operate at N-1.

Contrasting the relative magnitude of spinning reserve used in off-grid contexts versus on-grid situations, on-grid SR_{act} is typically in the range of 3-9% of peak demand, whereas the SR_{act} for isolated systems can be between 10% and 80%. This reflects the benefit of having multiple generators spread over larger areas for (large) on-grid cases and longer lead times available for additional generators to come online when needed. Smaller (isolated) grids (by demand, not by number of customers) have by their nature higher relative spinning reserve needs.

3 STATION CONTROL STRATEGY AND PV CURTAILMENT

A key constraint for the control system of any power station (and, in the on-grid equivalent case) is that the current demand level does not exceed the capacity of the currently online generators:

$$P_{demand} \leq P_{gen,max} \quad (9)$$

Below, two approaches to generator selection in hybrid power station are discussed, demonstrating that distinct strategies can be employed in order to fulfil constraint (9), leading to different levels of PV curtailment. These are the easier-to-describe equivalent of the on-grid counterparts and can thus be used as a model to discuss both off-grid and on-grid cases.

3.1 Net-Demand Generator Selection

To ensure that constraint (9) is satisfied in the event of fast-ramp loads or sudden component failure, a generator control system will seek to satisfy a more stringent constraint by maintaining a level of spinning reserve, as discussed above.

$$P_{demand} \leq P_{gen,max} - SR_{req} \quad (10)$$

If a PV system is installed on an existing system, resulting in reduced generator loading during daylight hours, the most obvious approach to control would have the generator selection be driven by the net load, so that the

[2] The Australian Electricity Market Operator (AEMO) uses FOR statistics to inform SR_{req} for the regions it manages [27].

[3] N-1: Where one large or critical generating component (e.g. a generator) can fail, and yet the grid is expected to operate within its required margins. N-2: 2 simultaneous failures tolerated.

generator control system attempts to satisfy the following constraint.

$$P_{demand} - P_{PV} \leq P_{gen,max} - SR_{req} \quad (11)$$

In this configuration, a sudden reduction in P_{PV} results in sudden increase in generator loading, and SR_{req} may need to be reconfigured to guarantee that constraint (9) is met. Specifically, we must have at all times that

$$P_{PV} \leq SR_{req} \quad (12)$$

This means that we must set SR_{req} to the maximum possible P_{PV}, or equivalently, we curtail the PV power to be less than or equal to the spinning reserve requirement.

This requirement for behind-the-meter PV is currently applied to many isolated grids in Australia (including NT SETuP communities), where the maximum cumulative allowable PV capacity is set by the spinning reserve requirements of that grid – see for example [15], [16].

3.2 Gross-Demand Generator Selection

Rather than using the net demand to dictate the selection of generators, the gross demand may be used instead. This results in the generator selection process remaining unaltered after installing PV, despite the reduced generator loading. Under this approach, the generator control system attempts to satisfy the following constraint

$$P_{demand} \leq P_{gen,max} - SR_{req} \quad (13)$$

This results in a selection of generators that can service the entire demand, so that constraint (7) is met at all times, even in the event of a total loss in PV power.

However, if the P_{PV} is large, the selected generators may be operating outside of the manufacturer's allowable or recommended loading level. To avoid under-loading the generators, a new constraint must be introduced.

$$P_{demand} - P_{PV} \geq P_{gen,min} \quad (14)$$

Where $P_{gen,min}$ represents the minimum loading level for the online generator. This constraint leads to the following PV curtailment strategy

$$P_{PV} \leq P_{demand} - P_{gen,min} \quad (15)$$

Under this approach the level of curtailment depends largely on the value of $P_{gen,min}$. For example, the installation of low-load diesel generators in an existing power station may significantly reduce the level of curtailment required.

4 NT SETUP OVERVIEW AND APPLICATION OF CURTAILMENT IN FUNCTION OF SPINNING RESERVE AND GENERATOR SETTINGS

Among the RAR projects, the Northern Territory Solar Energy Transformation Program (NT SETuP) stands out due to the scale and number of locations involved: NT SETuP aims to deploy 10 MW of PV systems to reduce diesel fuel consumption in 25 remote communities in the Northern Territory by approximately 15% without the need for battery storage, with one additional community – Daly River (Nauiyu) – identified as a high Renewable Power Fraction pilot project, where PV and a Battery Energy Storage System (BESS) are used to achieve an REF_{year} exceeding 50%.

The transformation component of NT SETuP encompasses the organisational change within Power and Water Corporation, from developing the capacity and understanding of PV as part of its electricity generation mix (an ongoing process with prior results being the PV-Diesel Minigrid Handbook [14]), to the establishment of internal supply chains where PV replacement components can be as readily sourced as their diesel generator counterparts. The installation of the PV systems in the SETuP communities without storage was planned to be done in 2 tranches, with the first tranche encompassing 10 communities and the second tranche the remaining 15. The Daly River pilot project was developed separately to the

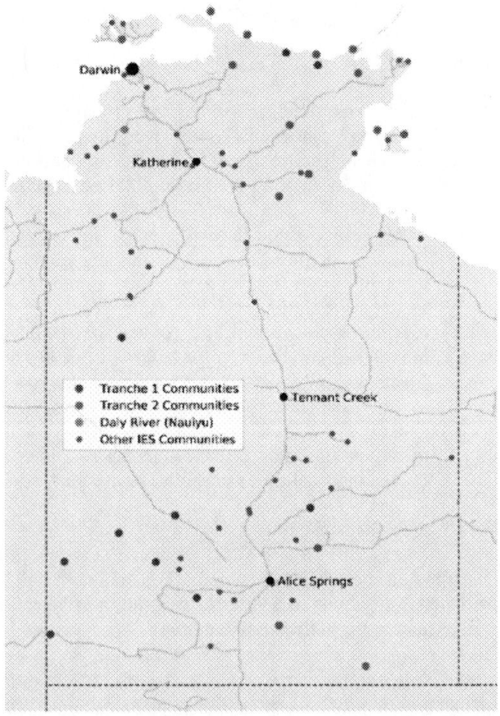

Figure 1: Map of the Northern Territory, with SETuP sites indicated, as well as major roads and the four largest population centres, from [12].

Tranche 1 projects, with the aim of the learnings from this project also informing the delivery of Tranche 2 projects.

The SETuP Tranche 1 projects were completed between February and July 2017, whereas Daly River had the PV generation start in October 2017. Currently (September 2018), the Tranche 2 projects are under construction, with all projects expected to be operational in 2019.

Table 1: Description of power stations for the communities in the Northern Territory's SETuP program Tranche 1 and the Daly River high penetration system, with PV systems commissioned in 2017.

Community	Diesel generators (kW)	PV capacity (kW_p)	Target# fuel saving per year
Maningrida	4: 820, 1000, 2x 1500	800	15%
Ramingining*	3: 400, 640*, 1000	500	15%
Yuendumu	4: 400, 2x 800, 1000	500	15%
Arlparra*	3: 360, 600*, 1000	450	15%
Lajamanu*	3: 420, 650*, 800	400	15%
Kintore	3: 260, 440, 600	225	15%
Nyirripi	3: 240, 270, 320	200	15%
Areyonga*	3: 200*, 360, 400	100	15%
Kaltukatjara	3: 220, 360, 400	100	15%
Amunturangu (Mt Liebig)*	3: 130*, 180, 280	50	15%
Tranche 1 total	**18,325 kW**	**4,325 kW_p**	**15%**
*Daly River * (Nauiyu)*	*3: 450, 600, 800*	*1000 kW_p, BESS: 800 kW/1986 kWh*	*50%*

*: Communities with wider operational window for PV via low-load diesel (LLD) generators or BESS; LLD indicated.
#: The REF_{year} must be larger than this to achieve the target fuel saving

In the case of NT SETuP, the initial aim for the integration of PV on the grids was to minimise the changes to the diesel power station control systems. The engine selection and loading philosophy follow a *gross-demand generator selection* strategy:

1. Determine the SR_{req} based on historical events: this often is the largest observed load step change observed for the community, rounded upwards to the nearest 10 or 100 kW (as appropriate for the demand in the community).
2. Select synchronous generator(s) from available generators based on total demand on the network, while respecting minimum and maximum loading of the generator(s), as per Eq (16). No additional generator operates, given the already high cost and remoteness of operation, and a black-out will occur in the low-probability[4] event of a generator's failure during operation, which then requires an automatic system restart. (The SETuP community power stations have been designed to operate with minimal human intervention, given their remoteness.)
$$P_{gen,min} \leq P_{demand} \leq P_{gen,max} - SR_{req} \quad (16)$$
3. **Given the chosen generators that are online (i.e. based on the demand, and the spinning reserve requirements), set the maximum allowable PV power** as per Eq (17).
$$P_{PV,max} = \min\begin{pmatrix} P_{demand} - P_{gen,online,min} \,; \\ P_{PV}(G,T) \end{pmatrix} \quad (17)$$
This then results in the generator's loading decreasing towards its minimum value, while reducing the risk of reverse power flows in the generator(s). The PV power is thus constrained on the one hand by the size of the operational window determined by the generator and the total demand in the system, and the PV system design and its interaction with the irradiance G and temperature T.
4. Update the SR_{req} to be
$$SR_{req} = \max(SR_{req,no\ PV} \,;\, P_{PV,max} * CR) \quad (18)$$
where CR is the PV cover ratio (currently 100%).
5. The loading of the generator then decreases based on the PV power in the system, while it needs to remain above its mandated minimum loading levels.
$$P_{gen} = P_{demand} - P_{PV,max} \geq P_{gen,min} \quad (19)$$

The rationale employed by Power and Water Corporation to mitigate the risk of the new technology on the operation of the power stations was thus to assume that all of the online or allowable PV generation under its control[5] could be lost at any instant (e.g. by the PV feeder being cut), and that the diesel generator(s) online should therefore have the spare capacity to deal with such an event.

In practice, this means that the expected probability *and* magnitude of a drop in PV power is 100% of the SETuP-controlled PV capacity online (instead of the more likely upper limit of 80% change in PV power during cloud events), and therefore there is *100% back-up of the SETuP-managed PV generation* by the synchronous generation. As some embedded PV generation is possible for the communities [16], it does mean that the SR_{req} covers less than 100% of all the PV power on the grid. While this may still seem conservative, the following factors must be kept in mind:

- The remoteness of the sites and the extant high cost of electricity generation makes running an additional diesel generator to lower the Loss-of-load probability (LOLP) economically infeasible (before, or after PV was installed).
- Operating PV-diesel grids with expected peak Renewable Power Fractions of ~60% reliably *without additional enabling technologies* (such as batteries for smoothing or forecasting of PV generation) at the scale seen for NT SETuP had not yet occurred to convince internal stakeholders. (Recall that the "T" in SETuP stands for Transformation: a need for organisational transformation from understanding diesel-only electricity generation, to generation with PV).

[4] Based on the historical experience of Power and Water
[5] There is some embedded (behind-the-meter) PV allowed in the SETuP communities, see [8]. This results in a mix of both the net-demand and gross-demand control strategies. A detailed discussion is outside the scope of this paper.

- There was little historical weather data available that could be used as inputs for modelling, while the climate (especially in the tropical areas of the Northern Territory) is known for the variability of the solar resource for much of the year, which when combined resulted in more conservative modelling assumptions.
- The (revised) SR_{req} as shown in Eq (18) is a heuristic that covers the drop in SETuP-controlled PV power, the coincident embedded PV power (which is likely to drop a similar amount given the limited geographical area of the grids) *and* the likelihood of a coincident demand increase at that instant.

The SR_{req} drives engine selection and thus has an indirect effect on the potential fuel savings that can be achieved. As such, decreasing the magnitude of SR_{req} to cover less than 100% of the (SETuP or "utility-controlled") PV power increases the risk of blackouts, without materially improving the economics of the system.

5 PV-HYBRIDISED PROJECTS WITH DIFFERENT METHODS OF CURTAILMENT

5.1 Impact of curtailment on the Performance Ratio: fixed limit based on demand, regardless of weather conditions

Figure 2 shows the hourly Performance Ratio versus the hourly Renewable Power Fraction for an isolated grid in Central Australia (this is not an NT SETuP site), where PV has been installed by an additional Independent Power Producer (IPP), upon request from the electricity end user who wished to reduce its electricity costs.

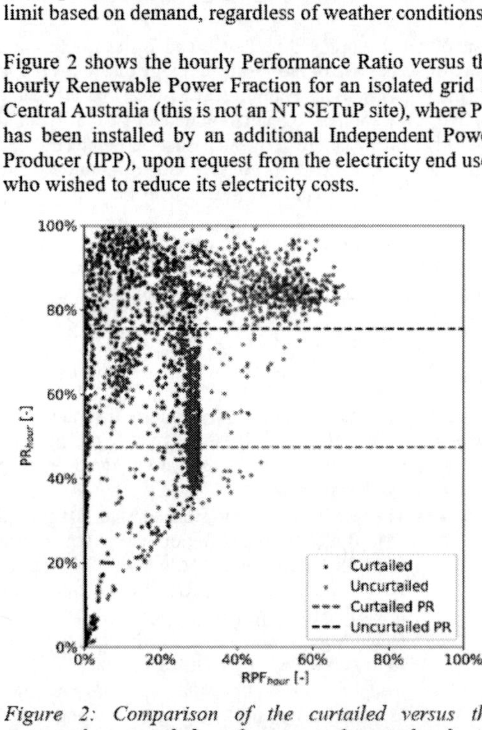

Figure 2: Comparison of the curtailed versus the estimated uncurtailed performance of an isolated grid with synchronous generators and PV. The limitation of the Renewable Power Fraction to 30% is clearly visible.

The Performance Ratio is defined in IEC 61724:1 [17] and is shown in Eq (20), where P_{STC} is the PV system's nameplate power under Standard Testing Conditions (STC), G_{STC} the irradiance at STC, $E_{PV,period}$ the produced PV energy over a certain period and $H_{POA,period}$ the plane-of-the-array irradiation.

$$PR_{period} = \frac{Y_r}{Y_f} = \frac{E_{PV,period}/P_{STC}}{H_{POA,period}/G_{STC}}\ [\%] \qquad (20)$$

The PV system discussed in Figure 2 is configured to operate at a Power Factor between 0.9 and 1.0 and can be curtailed upon instruction from the incumbent IPP who manages the grid.

A fixed upper limit of the RPF to 30% of demand was imposed by the incumbent IPP, resulting in curtailment for much of the year, with most curtailment occurring in the cooler winter months where air conditioning needs are much reduced. Because of the curtailment imposition, the PR drops 28 points, from approximately 75% to 47%, due to curtailment being in place for 38% of the PV operational hours. While curtailment had been anticipated and factored into the financial model of the solar farm, the magnitude of curtailment was not. In part, this was also due to the anticipated load growth for the subsequent years not materialising. The 30% RPF limit thus also speaks to the maximum revenue loss accepted by the incumbent IPP.

Even if the PV system would have been allowed to operate uncurtailed, there would nevertheless have been many instances in the year where the synchronous generators would be under-loaded (increasing both the risk of black-outs and O&M costs to these, due to the less efficient fuel burn). In this instance, the issue of curtailment is not so much related to whether it should occur or not, rather what amount of curtailment is appropriate to result in decreased risks to the power system as a whole, especially on clear sky days.

This case illustrates some of the frustrations experienced by isolated system operators and traditional IPPs (who supply electricity to end-users using gas or diesel synchronous generation) for mines, remote communities and hotels: the consumption of PV-supplied power typically results in a decrease in electricity costs for the end user, whereas the incumbent IPP sees its revenues decline, while its cost per delivered kWh increases. This split incentive between stakeholders is an extant risk for off-grid projects, and is growing in importance for on-grid cases as well, particularly for multiple PV and wind farms that connect to the same transmission network.

The consequence of this is that *not all kWh sold on the grid are equal*: at low RPF values, energy substitution by source is typically of low consequence to the grid's stability and reliability. For increasing RPF values, each additional kWh delivered needs to also contribute more towards the operation of the grid. Anecdotal evidence indicates that some renewable developers have attempted to deliver a (slightly) lower energy price to mines from PV, without considering the wider ramifications and externalities for either the incumbent IPP or the mine operator: PV earning most of the benefits yet not contributing enough to the grid's operation.

5.2 Variable curtailment limit in function of demand and generators

Whereas the previous example illustrated a case where curtailment was driven by economic on top of technical considerations, NT SETuP was designed for the use of PV generation to offset fuel costs in the long term, while ensuring that the quality and reliability of electricity supplied is maintained.

The consequence of this is that curtailment of the PV generation is performed by Power and Water Corporation based on a technical, economic and reliability optimisation while attempting to minimise the risk of black-outs, as the incentives are aligned within one organisation.

The demand profile combined with the diesel generators on-site and the PV generation profile for each community result in the varying normalised efficiency (η_N) versus 1-minute average RPF values as shown in Figure 3. The normalised efficiency is defined as [18],

$$\eta_N = \frac{P_{PV,N}}{G_N} = \frac{P_{PV}/P_{STC}}{G_{POA}/G_{STC}} \; [\%] \qquad (21)$$

where the subscript N indicates normalisation, by the appropriate value (PV nameplate power, or STC irradiance).

While all the NT SETuP power stations are operated according to the same control philosophy and had the PV systems sized to deliver a similar REF_{year} and subsequently fuel savings, the differences in how much the PV generator is allowed to operate unconstrained or not, differ strongly.

Figure 3 clearly illustrates how the operation of PV plants is strongly affected by what happens beyond the limits of the solar farm, ranging from the magnitude and shape of the load over time to how large of an operational window is permitted for PV by other generators.

RPF values, whereas Maningrida (with a demand that is almost twice the size and PV system size to Ramingining – see Table 1) but lacking an LLD sees a hard limit of 60% RPF. Even so, Ramingining has moments where generator 3 operates, which results in the need to curtail the PV strongly.

These communities show the challenge of optimal sizing and selection of synchronous generators, with the addition of PV to the mix making such decisions much more complex. Power and Water Corporation will further improve the operation of the power stations based on these learnings.

In the case of Lajamanu, the minimum allowed loading level of Generator 1 results in the PV being strongly curtailed if both that generator and the PV operate at the same time. Contrasting this with the grid, similar results are likely to occur for (poor) coincidence of PV and wind in the same electrical area.

The interesting aspect for the above variations in the shown NT SETuP communities is that the curtailment that is in place varies strongly, even though all PV systems and generators are managed by the same organisation and operational philosophy.

Overall, the differences between the different community results observed are due to:

- The diesel generators and their operational limits for each community. In many cases, the generators in place today differ from those used for the sizing and modelling exercise, reflecting generator replacements over time.
- The local load shape as a function of climate:e.g. Maningrida and Ramingining are located in the tropical area of the Northern Territory (Köppen-Geiger: Aw), whereas Yuendemu and Nyirripi are in the more desert-like environment of Central Australia (Köppen-Geiger: BWh or BSh), where temperature variations between summer and winter are more pronounced. Even so, the demand shape varies depending on local factors such as the cost of electricity for each community (which reflects remoteness of the site, and the cost of procurement of the diesel fuel), as well as the type of loads in the community.
- The site-specific spinning reserve (SR_{req}) settings.
- The change in the demand pattern, from what was originally modelled, to the actual demand seen in the community
- In some cases, low-load diesel (LLD) generators were installed, outside of the project design for NT SETuP.

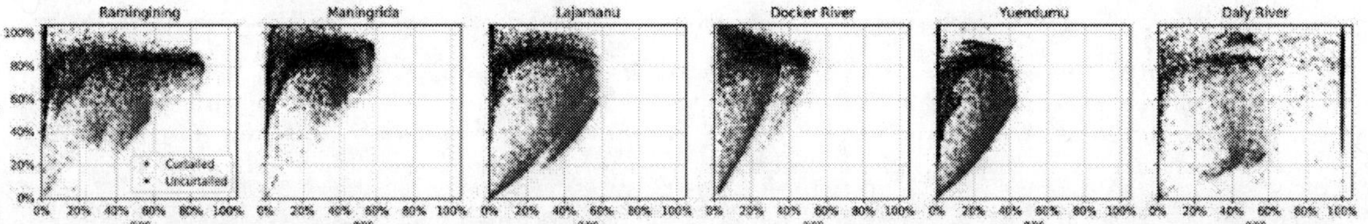

Figure 3: Scatterplot of the normalised efficiency η_N versus the Renewable Power Fraction RPF for 6 of the NT SETuP communities. Data for April-June 2018.

In the case of Docker River, it is possible to observe two curtailment limits appearing as diagonal η_N-RPF lines, each associated with the minimum loading limit of a different synchronous generator. For Yuendemu, the higher minimum loading limit of the diesel generators that are present result in more drastic curtailment (e.g. a PV spill of up to 50% of its power), compared to Maningrida, where curtailment results in less spilled power.

The steepness of the η_N versus RPF in each of the communities reflect the combined interaction of:

- The PV system size;
- The available solar resource;
- The demand profile;
- The generator minimum loading limits; and,
- The spinning reserve limits for that community.

To better understand these points, Figure 4 and Figure 5 are shown below, with Figure 4 illustrating the PV generator's perspective (and providing a view to the magnitude of curtailment), whereas Figure 5 shows how the different generators that are online in function of the demand enable or hamper PV generation.

Ramingining has one low-load diesel genset and thus provides more opportunities for the PV to operate to higher

The last point is worthy of further discussion: the difference between the modelled solar resource and the demand profile, and actual values of these when the PV systems are operational. This is illustrated in Figure 7 for Maningrida.

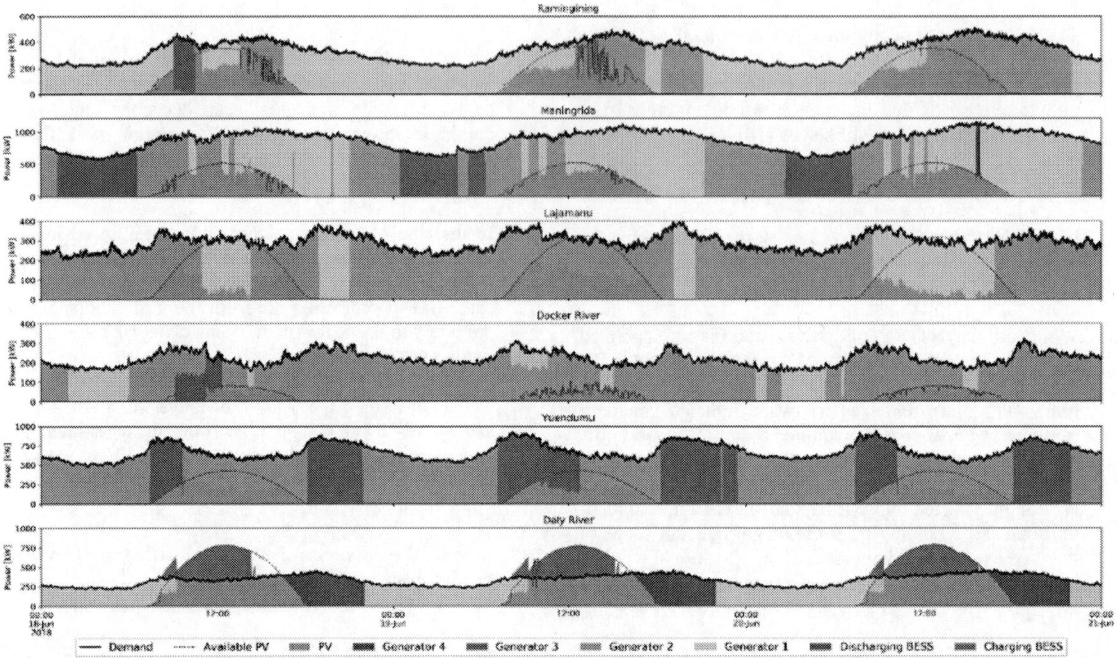

Figure 4: Stacked generation plots for the NT SETuP communities of Figure 3, with PV shown first, for 3 days in winter. The moments of curtailment are generally visible for all communities. PV energy spilled is visible for Daly River, especially on the first afternoon, once the BESS is fully charged. Available PV power estimated from dedicated subsystems or irradiance data.

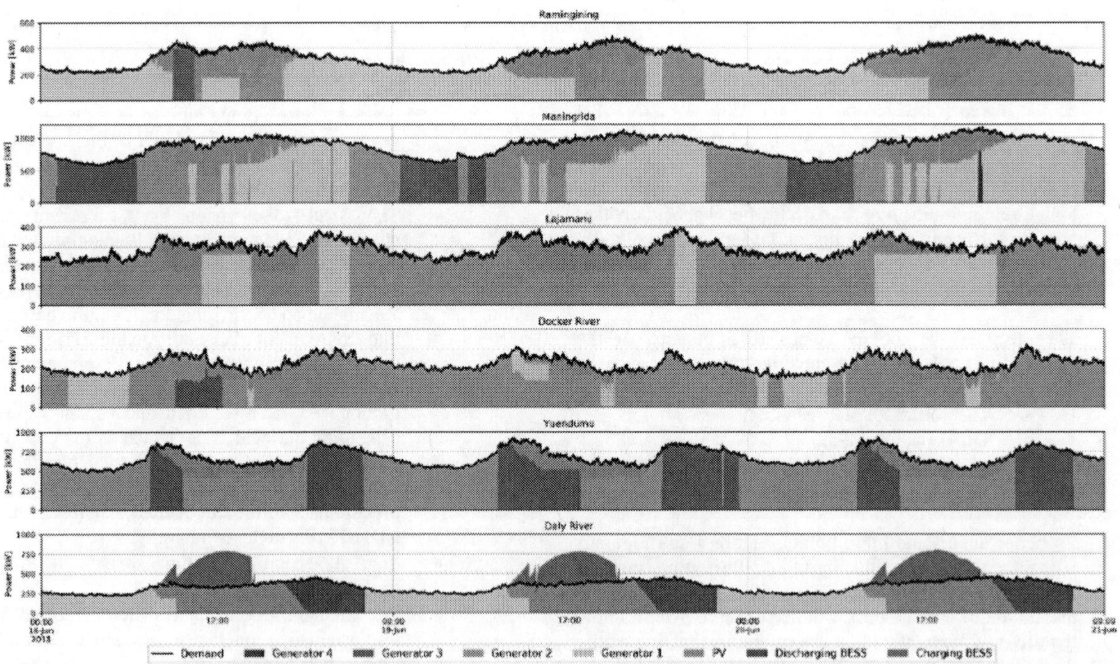

Figure 5: Stacked generation plots for the NT SETuP communities of Figure 3 over 3 days in winter, with the synchronous generators shown first, illustrating the viewpoint of other generators on the grid, or grid managers. The generator minimum loading levels and their effect on the permissible PV are visible in all systems.

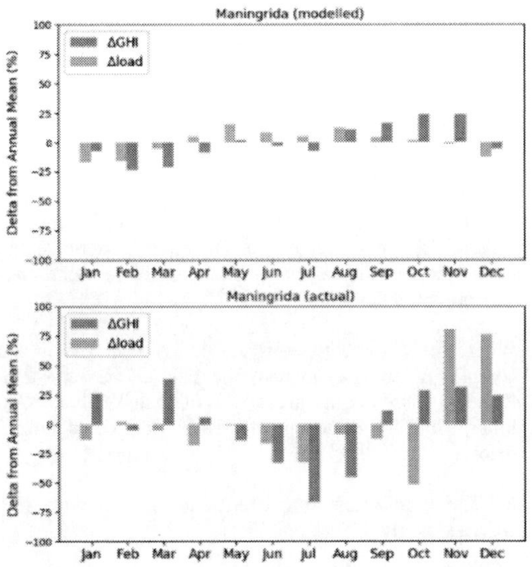

Figure 7: Modelled and actual deviations of monthly global horizontal irradiation (GHI) and load, compared to the annual mean for Maningrida.

Taking the perspective of a PV developer operating in such a location, these differences in load and solar resource are a material risk to the generation that may be achieved, as the change in the demand profile may result in more curtailment occurring, on top of the well-known solar resource risks [19]. Similar impacts will be highly likely in on-grid cases as well, where changes in the net demand (i.e., demand minus behind-the-meter generation) may drive local curtailment. To note though: If PV or a similar technology such as battery storage does become even more affordable to install, some of the resource and demand risks may decrease in importance.

In the case of SETuP, the result of the limited solar seasonal variation throughout the year (when compared to other locations farther from the Equator) is that the impact of other factors can be more easily identified. Thus, the demand variation and the spinning reserve settings, which drive engine selection which *then* drives PV curtailment can be teased apart, as discussed here.

5.3 Consequences of integration of PV and BESS for NT SETuP

Figure 6 shows the daily peak RPF observed for 6 NT SETuP communities and the resulting daily REF achieved. Similar to Figure 3 this shows the limitations of the contribution of PV to the energy mix per day and then over the year, which depend not only on the solar profile, but also on the load in the system and the diesel generator(s) that are online at that instant. The communities Ramingining and Lajamanu achieve high RPF_{peak} values of up to 90% as the LLD generator increases the allowable PV power on the grid, although the number of instances depends on whether the LLD is called upon during the daytime.

Overall, there is an approximate 4:1 relationship of the daily peak RPF_{peak} to the REF_{day} achieved for the NT SETuP communities, for *systems without storage*. This highlights the importance of not only sizing PV

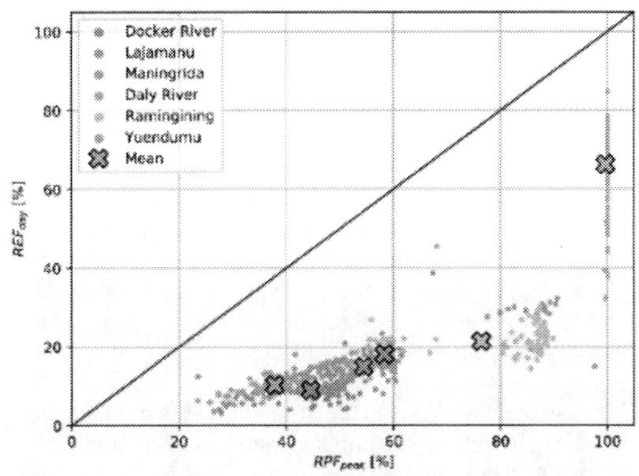

Figure 6: The daily REF observed, versus the RPF_{peak} that day for 6 NT SETuP projects. The red line indicates a project that (theoretically) operates at that peak RPF value for 100% of the time.The cross indicates the mean value for that system.

appropriately for a given load profile, but also the impact that factors outside of the PV generation's control which can affect the resulting energy generation and revenue. It must be noted though, that the RPF_{peak}-REF_{day} relationship is a local value and that the values shown here may not directly apply elsewhere. Nevertheless, systems where the demand and generation profiles are similar to those in the SETuP communities may have comparable high RPF values to achieve target REF_{year} results. Of note, similar RPF_{peak} to REF_{year} (100%:20%) have been observed elsewhere for grids with high amount of wind power such as in Denmark [20].

In the case of Daly River, the BESS widens the resulting operational envelope for the PV energy being delivered to the grid, as it not only permits diesel off operation, smooths PV variability and captures PV energy that would otherwise be spilled. This then results in days where the RPF_{peak} reaches 100% delivering 25% REF_{day}, all the way to exceeding 75% REF_{day}. From the perspective of Power and Water Corporation, the BESS is seen as having the same or better (equivalent) spinning reserve than a diesel generator and alleviates its concerns around the variability of PV.

However, the incorporation of the BESS in the power system did require modifications to the protection settings for the Daly River power system, differentiating among the following modes:

- Diesel only (including night-time operation): no change to previous protection settings.
- Diesel with PV, no BESS (BESS unavailable due to a fault, or lack of stored energy): protection settings same as diesel only, with diesel generator expected to supply the fault current.
- Diesel + PV + BESS, with the BESS in transitional mode: the BESS behaves as a variable load to the diesel generator, allowing the PV to go beyond the diesel engine's spinning reserve. Fault contribution from both the diesel generator and the BESS.
- Diesel off, BESS + PV: reduced fault current detection and action settings, given the limited

fault current capabilities of the BESS and PV inverters.

5.4 Economic drivers, PV system sizing and curtailment

The experience from NT SETuP and similar projects in the RAR portfolio at achieving medium-to-high RPF values proves that PV can deliver a large portion of the power during the day without undue increases in risk to the system. (This may seem obvious with current experience, yet there were doubts in the industry about this, prior to the projects being funded by ARENA.)

This however then leads to the discussion as to how similar results can be achieved without (large) subsidies, and what PV system size is appropriate.

To a certain extent, it can be argued that PV systems that experience much curtailment have been over-sized compared to the local conditions. However, there is an inherent tension between absolute and relative economic returns that can be achieved for energy-displacement systems:

- One design can opt for a small PV system, which results in less spill and achieves a lower REF_{year} and with it, low absolute economic returns.
- On the other hand, larger PV systems will achieve higher REF_{year} values, yet encounter much curtailment and with it a potential decrease in the relative return.

However, apparently sub-optimal curtailment levels can still be chosen in the event of high fuel costs. This optimisation is also often encountered for behind-the-meter PV systems with zero export limitations.

The exploration of whether a BESS can then capture the otherwise spilled energy may then further impact the design decisions, making design decisions more complex as higher REF_{year} values are targeted. The main point here is that unless a holistic approach is undertaken, the likelihood of suboptimal solutions being developed is very high.

6 HOW AUSTRALIAN PROJECTS WITH MEDIUM TO HIGH RENEWABLE ENERGY FRACTIONS INFORM THE PATH TO 100% RENEWABLE ENERGY

Based on the experience of the projects described here as well as that of the authors, challenges and observations are discussed, for the path to 100% renewable energy.

6.1 Impact of other generators and stakeholders on the grid on PV's ability to deliver power and the subsequent financial ramifications

As illustrated in the examples above, the capacity of PV to operate uncurtailed is limited by the permissible operation window dictated by the ability and willingness of the other generators (and occasionally, loads) to accommodate the PV power on the grid.

In some cases seen by the authors, the PV generator has had the ability to deliver the services to the grid (alone, or with a BESS of sufficient capacity), yet this *technical* capability was not deemed to be sufficient by the grid manager to allow a reduced amount of PV curtailment to be instituted.

The inherent difficulty in the above split incentive is that the risk-reward relationship is skewed, as the PV generator or IPP cares about delivering as much power as possible, whereas the grid manager is focused on the continued reliability of the grid. Both careful studies and appropriate engagement (and, where required, education) of stakeholders will become even more necessary and important for increasing amounts of non-synchronous generation on the grid.

Where the above situations are poorly managed or the complexity under-estimated, the risk to the long-term financial viability of projects is not mitigated and may also imperil the licence to operate. This is discussed further below.

6.2 The diminishing value of PR and similar metrics in a curtailed world

For grids that routinely reach 100% RPF, renewables will have to operate curtailed for much of the time to provide the different services that the grid requires. The treatment recommended within IEC 61724:1 [17] of data during moments with curtailment for PR calculations has been to exclude these points throughout the year, based on the premise that the amount of excluded data is sufficiently limited that the resulting corrected PR accurately reflects how well the PV plant operates throughout the year, and its ability to convert irradiance into electricity.

However, the correctness of the PR diminishes as more and more data is excluded, especially as curtailment typically occurs under high irradiance conditions, in which case the largest amount of energy is excluded from the PR calculation. Similarly, determination of the expected normalised efficiency η_N during curtailed moments will also be impacted.

Flow-on consequences from then not being able to calculate the "true" PR and η_N are that their application for fault detection or determination of degradation are no longer possible (or that their use is severely constrained). Alternative approaches such as statistical fault monitoring are described in [21], which would likely still encounter difficulties in delivering the quality of results obtained for low-curtailment PV systems.

Approaches to estimate the yield or determine the magnitude of curtailment for performance and quality determination can be:

- Designate a representative subsystem (e.g. one inverter) to operate uncurtailed in all circumstances and determine the PR for that array and extrapolate to the full system. For NT SETuP, a single module at multiple communities (not connected to any inverter) is measured which provides the maximum PV output, which then also feeds into the power station and PV curtailment control system. A separate example is the Yulara PV system which is divided over multiple arrays, where the largest is curtailed and the remainder is not, providing the opportunity to estimate the spilled energy from the uncurtailed systems.

- Store the inverter setpoint value as either a percentage of possible output or absolute values. This is done by Power and Water for the SETuP arrays and enables detection of curtailment, and when combined with the sub-array data, the magnitude of uncurtailed operation.

Nevertheless, the above descriptions cannot present a comparable level of certainty regarding the operation of PV systems as has historically been expected within the industry. Regardless, more work is required in this field if investors are to determine whether the PV system operates as expected given the climate conditions.

Lastly, the authors have already seen cases where stakeholders for PV systems subject to much curtailment have declined to perform full PR testing (both in commissioning and first years of operation), as the risk to the overall system was deemed to be much higher than the comparative risk reduction in having better knowledge of the PR.

6.3 Changing the paradigm from yield maximisation to value maximisation: embracing curtailment

In on-grid cases, the continued rapid deployment of new variable generation (primarily wind and PV) in each other's vicinity has already resulted in transmission lines reaching their limits, with the incumbent renewable generator being subjected to significantly more curtailment than anticipated in the financial model.

Similarly, the continued uptake of behind-the-meter generation (which is predominantly PV) decreases the net demand on the system. While the current system control settings mandate a minimum amount of synchronous generation in the system, the only remaining generators that can then be curtailed are front-of-meter utility-scale PV and wind farms.

Whereas the (relatively) high initial capital expenditure for PV has historically driven the industry to focus on yield maximisation, this is only possible when variable renewables achieve low RPF values and hence low balancing and grid management costs.

Instead, by embracing curtailment as a necessary requirement for (utility-scale) PV systems to operate, a whole slew of services by PV to the grid become possible.

For example, results from the REservices project [22] showed that PV can deliver various ancillary services, ranging from voltage to reactive power control, as well as the ability for upward power flexibility – taking advantage of the available headroom due to the pre-curtailment of the PV systems. Similarly, [23] discussed Virtual Oscillator Control performed by PV inverters as a method of delivering FCAS (Frequency Ancillary Services). The ability of utility-scale PV plants to deliver FCAS, *including spinning reserves,* was demonstrated in California and is discussed in [24].

Curtailment can also be economically more efficient than energy storage: In the case of the TKLN[6] [25] project in the Northern Territory [26], pre-curtailment of PV through forecasting of solar variability results in a higher system-wide efficiency, as the energy lost due to PV curtailment and higher synchronous generator efficiency is *less* than the round-trip energy storage losses in a BESS, and also results in reduced cycling of the BESS – increasing that asset's longevity.

Furthermore, if PV is to be a meaningful participant in a 100% renewable powered grid, such PV systems must be designed appropriately to not only contribute during normal operation of the grid, but also during contingency events, for example by delivering sufficient short-circuit power. From a PV-only perspective, this may be obtained through over-sizing PV arrays versus inverters, and the inverters over-sized versus the agreed upon grid connection. Alternatives would include the integration of a BESS or synchronous condenser with the PV system. Consequently, it appears unlikely that PV farms as a whole, i.e. including any additional equipment required to be permitted to connect, will continue to decline in cost over time. The "traditional" PV system may become cheaper, yet additional investments as required by the market and/or the regulator will limit long-term cost declines.

6.4 License to operate: social, technical and regulatory

The social license of large-scale PV is impacted by how the development, construction and operation processes evolve, for the given local context. The potential long-term use of high-value land (e.g. farmland, or areas that could become urban over time) by PV has already resulted in utility-scale solar farms in Australia being denied development approval by the local community.

Similarly, Power and Water engaged in deep consultations with the local communities for the installation of the solar arrays. While some of the consultations took a considerable amount of time and resulted in construction being delayed compared to the original plans, the positive view of the PV arrays by the local communities and the perceived long-term benefits have been identified as a crucial component in ensuring that these systems could be installed.

Alternative business or financing models which have community involvement and/or investment in the renewable asset have shown remarkable results [27] in improving the social license of such projects and can address not-in-my-backyard (NIMBY) concerns.

At the technical level, more work is required in educating stakeholders as to the perceived versus actual benefits and costs of PV on the local (area of the) grid. In some cases, this need for continued stakeholder education has stemmed from knowledge imbalances [4], to replacement of decision-makers over the project duration.

Grid operators, regulators and policy advisors such as Power and Water in the Northern Territory [28] and AEMO, the Australian Energy Market Commission, for the Australian National Electricity Market [29] or the North American Electrical Reliability Corporation

[6] TKLN: abbreviation of the three towns in the Northern Territory, Ti Tree, Kalkarindji and Lake Nash, where PV, BESS and forecasting are used to reduce diesel fuel consumption in the power systems.

(NERC) have increasing technical demands [30] for new generators to be connected to their grids. Specifically for utility-scale PV and wind, these requirements include more detailed Electromagnetic Transient (EMT) models done in PSS/E® and/or PSCAD™, to the requirement for PV and wind farms to have sufficient inertia (and where needed, for these to invest in a BESS or synchronous condenser to reach the required levels) [28].

Regulatory acceptance (rather than reform, which is a continuous effort) together with the appropriate market mechanisms will be necessary to accept and understand the capabilities of PV on the grid, as well as the economical valuation of the multiple services provided by PV. Additional learnings and real-world demonstrations are necessary to determine how much spinning reserve or other forms of back-up are required, to obtain cost-effective outcomes.

For this, investments in smart(er) protection equipment and infrastructure upgrades such as circuit breakers that can have their detection and action limits raised or lowered in function of the type of generation online (predominantly renewables with low fault currents or predominantly synchronous generation with high fault currents) may also become necessary. It is then possible that new PV and wind farms will have to contribute in some form to grid upgrades, as some already do today through the construction of new or upgraded transmission capacity.

6.5 Re-evaluating islanding for 100% Renewable Energy and Power Fractions

If a grid is to operate at or close to 100% REF_{year}, this raises the prospect of insufficient or inexistent synchronous generators being available to energise a grid after a blackout, as those synchronous generators would likely be uneconomical to operate for those very few hours in the year. Therefore, asynchronous generators would need to have "black start" capabilities to re-energise a grid. However, current rules for grid-connected PV generators are for these to have anti-islanding detection in place to avoid the possibility of (partial) grid energisation.

While the idea of using PV for grid energisation may appear farfetched and against the developments of the past decades, there are already examples of isolated systems that service thousands of customers where PV combined with storage is used for black start capabilities [31], albeit via an indirect route: the PV is used to charge the BESS, which then forms the grid.

6.6 Ramp rates and impact on other grid actors

The NT SETuP projects discussed here achieve REF_{year} values of 15-20% (and >50% for Daly River), yet have high peak RPF values on a daily basis. For PV, this means that the daily magnitude of variation is between 0% RPF and the peak RPF observed. Both the short-term variability that needs to be covered by spinning reserves or changes in PV operation (e.g. almost permanently curtailed) as well as the daily ramping for morning and evening will require improved coordination with other generators and increases the importance of accurate forecasts.

Moreover, the PV variations for the NT SETuP projects are covered by the synchronous generators or BESS, which are capable of tolerating those net load step changes. While this demonstrates the validity of the spinning reserve philosophy for these projects, achieving 100% REF_{year} is a challenge that requires PV and other renewable generation to deliver power reliably, despite the weather conditions throughout the year.

This will further put a premium on flexibility, not only from PV in delivering more services than has historically been required, but also on other generation and loads on the grid. The transition from current REF_{year} and associated RPF values can be enabled or stymied, depending on how well the various grid actors can cooperate, or not.

For example, inflexible synchronous generators (or hard limits on flexible generators via their minimum generation requirements) as seen by the NT SETuP projects result in PV curtailment, whereas the addition of a BESS as illustrated in the Daly River pilot project introduces a degree of freedom and flexibility to the grid. Similarly, the increased uptake of (more efficient) air conditioning may be used to absorb some of the peak daytime PV generation that otherwise would be spilled, although behaviour and building stock efficiency can still result in missed opportunities.

7 CONCLUSIONS

This article has explored the consequences for PV in Australian grids that achieve medium to high Renewable Power Fractions, where the operational limits of the non-PV actors (primarily the diesel synchronous generators) and system settings drive the amount and frequency of curtailment of PV generation. The described systems are used as a model to discuss likely issues and challenges that may appear in the near future on other grids.

The first conclusion that can be drawn from the experience of the systems described here is that curtailment of PV generation will rapidly become prevalent and have a large impact on individual solar farms in grids worldwide if current deployment trends continue. This has ramifications for existing PV systems seeking to be refinanced, as curtailment is likely to reduce expected revenues. For new PV systems, not only are the recommendations from [19] still applicable, yet developers and financiers must also consider what will happen on the grid, beyond the confines of the PV system proper.

In a curtailed world, the current applications of the Performance Ratio and the Normalised Efficiency and the derivative metrics that these enable may be severely affected, which pose interesting challenges for the PV industry moving forward. On the one hand are issues around performance monitoring and contractual obligations, on the other, there is a need for changes in system designs to provide similar levels of certainty around the quality and performance of PV systems. Further, the impact of curtailing renewable generators with energy production subsidies (such as Large Generation Certificates that are paid for generation sent out, not for theoretical generation) can cause further issues over time, while those subsidies remain in place. (Depending on the jurisdiction, this is likely not going to be a long-term issue.)

The PV industry must also seek to embrace curtailment and maximise the value that PV-generated electricity can provide to ensure the continued deployment of PV on grids worldwide. Failing to do so increases the social, technical and financial risks to PV systems, as the licence to operate becomes harder to obtain, as already seen in the increased technical demands by grid operators and regulators over the past years. However, this paradigm shift of PV operation, where curtailment is the norm rather than the exception, can unlock larger portions of electricity markets, as the additional capabilities made possible through curtailment (such as voltage control, FCAS, improved short-circuit capabilities) address actual and perceived technical concerns by stakeholders on the grid.

The changes to PV system designs to provide the required services then suggests that different DC:AC ratios (modules-inverters) as well as inverter-grid connection ratios may have to be evaluated, and further increases the importance of inverters' capacity to deliver such services, which includes improved EMT models.

As highlighted for the PV systems in the discussed NT SETuP communities, how curtailment eventuates varies strongly, despite these systems having been designed and managed by the same party. From a research perspective, this indicates that teasing out the causal relationships and drivers will require significant effort and cooperation among grid stakeholders for grids where more actors are present.

For the discussed projects, the interaction of the demand on the grid (both its shape and magnitude), the current sizes and operational limits of the synchronous generators present and the spinning reserve requirements together with the "known" PV issues of seasonality result in different curtailment profiles as a function of the Renewable Power Fraction achieved. For NT SETuP, the operational decisions and design of the PV systems were aligned, where the application of curtailment of PV is intended to strike the balance between managing risks of blackouts on the system while using the PV as much as possible to obtain an economic return on investment. Power and Water will further explore changes and enhancements to the operational philosophy of the NT SETuP power stations, to further improve the economic case of displacement of diesel fuel through the use of PV.

A note of caution must be given, as dynamics of split incentives are likely to occur on similar (isolated) grids, where the grid manager or its incumbent IPP can use technical arguments to enforce more curtailment on PV than may technically be required, or allowed for in the original financial model. This lesson is particularly pertinent for isolated power systems where the aim may be to increase the levels of renewables, and builds upon those previously discussed in [4].

The issues encountered by the medium and high Renewable Power Fraction systems further stimulate thinking around potential challenges on the path to reaching 100% Renewable Energy Fractions, and what those may entail for PV. Of these, curtailment is a given, together with the increasing need for timely and accurate forecasts. Further, the question of whether islanding of PV systems will still be a requirement, or whether alternative solutions will need to be developed to permit black start capabilities under specific circumstances can be explored. Similarly, reliable operation of PV systems during severe contingency events and the associated need for capabilities by solar farms to contribute and be seen to contribute positively to the resolution of contingencies is an area where more work is required.
Lastly, more sharing of knowledge and the subsequent application of best practices will be required to further improve understanding among stakeholders on grids. The licence to operate for PV depends on its increased ability to demonstrate that it can carry a larger share of the benefits and burdens of grids with very high Renewable Energy Fractions. This may entail that PV projects may need to co-finance grid upgrades that further enable the uptake of renewables, while business models and market changes are developed that value the additional services provided.

REFERENCES

[1] Global Wind Energy Council, "Global Wind Report - Annual Market Update 2017," 2018. [Online]. Available: http://files.gwec.net/files/GWR2017.pdf. [Accessed 3 September 2018].

[2] S. Philipps and W. Warmuth, "Photovoltaics Report," 2018. [Online]. Available: https://www.ise.fraunhofer.de/content/dam/ise/de/documents/publications/studies/Photovoltaics-Report.pdf. [Accessed 3 September 2018].

[3] APVI, "PV in Australia Report 2017," July 2018. [Online]. Available: http://apvi.org.au/wp-content/uploads/2018/07/PViA-2018-AU.pdf.

[4] B. Herteleer, A. Dobb, O. Boyd, S. Rodgers and L. Frearson, "Identifying risks, costs and lessons from ARENA-funded off-grid renewable energy projects in regional Australia," *Progress in Photovoltaics: Research and Applications,* pp. 1-9, 2018.

[5] OECD/IEA, "Energy access database," IEA Publishing, 2016. [Online]. Available: http://www.worldenergyoutlook.org/resources/energydevelopment/energyaccessdatabase/. [Accessed 23 June 2017].

[6] World Bank Group, "State of electricity access report 2017 (Vol. 2) : full report," 1 May 2017. [Online]. Available: http://documents.worldbank.org/curated/en/364571494517675149/full-report. [Accessed 25 June 2017].

[7] IRENA, "Innovation Outlook: Renewable Mini-grids," Abu Dhabi, 2016.

[8] B. Herteleer, A. Dobb, L. McLeod, G. Dickeson, S. Rodgers and L. Frearson, "Understanding Integration Aspects: Renewable Power and Energy Fractions," in *2017 Asia-Pacific Solar Research Conference*, Melbourne, 2017.

[9] Y. Rebours and D. Kirschen, "What is spinning reserve?," 2005.

[10] North American Electric Reliability Corporation, "Ancillary Service and Balancing Authority Arrea Solutions to Integrate Vriable Generation," March 2011. [Online]. Available: https://www.nerc.com/pa/RAPA/ra/Reliability%20Assessments%20DL/IVGTF2-3.pdf.

[11] Ernst & Young, "Ancillary services parameters - Draft assumptions report," 13 September 2018. [Online]. Available: http://wa.aemo.com.au/-/media/Files/Electricity/WEM/Security_and_Reliability/Ancillary-Services/2018/EY-Draft-Assumptions-Report---2018-09-13---Public-version---Further-Finalised.pdf. [Accessed 13 September 2018].

[12] European Network of Transmission System Operators for Electricity, "ENTSO-E Outage Document Implementation Guide," 19 May 2010. [Online]. Available: https://www.entsoe.eu/fileadmin/user_upload/edi/library/entsoenet-eod-v1r1/outage-document-v1r1.pdf. [Accessed 17 September 2018].

[13] Power and Water Corporation, "System Control Technical Code Version 5.0," May 2015. [Online]. Available: https://www.powerwater.com.au/__data/assets/pdf_file/0006/98493/SYSTEM_CONTROL_TECHNICAL_CODE_-_Approved_by_UC_25May2015.pdf. [Accessed 13 September 2018].

[14] Power and Water Corporation, "Solar/Diesel Mini-Grid Systems," 18 Februrary 2014. [Online]. Available: https://www.powerwater.com.au/news_and_publications/publications/technical/solardiesel_mini-grid_handbook. [Accessed 4 September 2018].

[15] Horizon Power, "Why is there a limit on the solar capacity for each town?," 2018. [Online]. Available: https://horizonpower.com.au/help-support/solar-energy/why-is-there-a-limit-on-the-solar-capacity-for-each-town/. [Accessed 17 September 2018].

[16] Power and Water Corporation, "Photovoltaic (PV) solar systems," 2018. [Online]. Available: https://www.powerwater.com.au/customers/save/photovoltaic_pv_solar_systems. [Accessed 17 September 2018].

[17] International Electrotechnical Commission, "IEC 61724-1 Photovoltaic system performance - Part 1: Monitoring," 2017.

[18] B. Herteleer, B. Huyck, F. Catthoor, J. Driesen and J. Cappelle, "Normalised Efficiency of Photovoltaic Systems: Going beyond the Performance Ratio," *Solar Energy,* vol. 157, pp. 408-418, 2017.

[19] M. Richter, C. Tjengdrawira, J. Vedde, M. Green, L. Frearson, B. Herteleer, U. Jahn, M. Herz, M. Kontges, B. Stridh, K. Berger, E. Janknecht, D. Moser, G. Belluardo and N. Reich, "Technical Assumptions Used in PV Financial Models - Review of Current Practices and Recommendations," IEA PVPS, 2017.

[20] ENTSO-E, "High Penetration of Power Electronic Interfaced Power Sources (HPoPEIPS)," 29 March 2017. [Online]. Available: https://docstore.entsoe.eu/Documents/Network%20codes%20documents/Implementation/CNC/170322_IGD25_HPoPEIPS.pdf. [Accessed 10 September 2018].

[21] M. Green, E. Brill, B. Jones and J. Dore, "Improving Efficiency of PV Systems Using Statistical Performance Monitoring," IEA PVPS, 2017.

[22] P. Kreutzkamp, K. De Brabandere, O. Gammoh, J. De Decker, M. Rekinger, S. Varga and D. Craciun, "Ancillary Services by Solar PV - Capabilities and Costs of Provision," May 2013. [Online]. Available: http://www.reservices-project.eu/wp-content/uploads/REserviceS-D4-1-Final.pdf. [Accessed 3 September 2018].

[23] B. Kroposki, "Can Smarter Solar Inverters Save the Grid?," October 2016. [Online]. Available: https://spectrum.ieee.org/energy/renewables/can-smarter-solar-inverters-save-the-grid. [Accessed 13 September 2018].

[24] C. Loutan, P. Klauer, S. Chowdhury, S. Hall, M. Morjaria, V. Chadliev, N. Milam, C. Milan and V. Gevorgian, "Demonstration of Essential Reliability Services by a 300-MW Solar Photovoltaic Power Plant," NREL, Golden, 2017.

[25] Epuron, "TKLN (Lake Nash)," 2018. [Online]. Available: http://epuron.com.au/solar/tkln/. [Accessed 17 September 2018].

[26] "Cloud detection and prediction for maximising solar PV utilisation in off-grid hybrid power systems - Final report: project results and lessons learnt," 8 November 2016. [Online]. Available: https://arena.gov.au/assets/2016/11/Fulcrum3D-CloudCAM-ARENA-Public-report.pdf. [Accessed 17 September 2018].

[27] S. Caneva, I. Weiss, M. Papapetrou, P. Alonso, O. Gajda, K. Kohl, A. Bergmann, B. Burton, T. Aschenbeck-Florange, A. Dlouhy, T. Drefke, A. de Ferrari, M. Martinoli, J. Wahlmüller, S. Egger, v. M. R. T. Harwood, K. Harder, S. Müller-Windisch and D. V, "CrowdFundRES: A New Opportunity for Financing Renewable Energy Projects," in *33rd EU PVSEC*, Amsterdam, 2017.

[28] Power and Water Corporation, "Northern Territory Generator Performance Standards (GPS) V0.8," 2018. [Online]. Available: http://www.utilicom.nt.gov.au/PMS/Publications/UC-SUB-PWSC-GPS.pdf. [Accessed 18 September 2018].

[29] Australian Energy Market Commission, "National Electricity Amendment (Managing power system fault levels) Rule 2017 No. 10," 19 September 2017. [Online]. Available: https://www.aemc.gov.au/sites/default/files/content/38cbd875-6295-4d8d-acd6-52d5adfc3041/System-Strength-Final-Rule-19-Sept-2017-VERSION-FOR-PUBLICATION.PDF.

[30] North American Electric Reliability Corporation, "Integrating Inverter-Based Resources into Weak Power Systems," June 2017. [Online]. Available: https://www.nerc.com/pa/RAPA/rg/ReliabilityGuidelines/Reliabilty_Guideline_-_Integrating_VER_into_Weak_Power_Systems.pdf. [Accessed 4 September 2018].

[31] F. Neu, *Electrifying African Cities with PV and storage [unpublished EU PVSEC 2017 presentation]*, Amsterdam, 2017.

[32] Ekistica and ARENA, "NT SETuP: A first look at the integration of PV and Diesel Power Stations in remote communities [not yet published]," ARENA, Canberra, 2018.

[33] Australian Energy Market Operator, "Guidebook for Forced Outage Data Recordings: Definitions and Assumptions," 1 July 2010. [Online]. Available: https://www.aemo.com.au/media/Files/Other/electricityops/0240-0014%20pdf.pdf. [Accessed 17 September 2018].

AN OVERVIEW OF FLOATING PV WORLDWIDE

M.M. de Jong, M. Dörenkämper, K. Sinapis, W. Folkerts
Solar Energy Application Centre, High Tech Campus 21, Eindhoven, The Netherlands

ABSTRACT: Floating solar systems can represent a serious alternative to ground mounted solar systems. At present, the international market for floating PV is small compared to the total PV market, but expected to become a multi-billion market in the coming years. A main driver for the growth of the market for floating PV is the mitigation of land cost. Other drivers are the potential added benefits of deploying photovoltaics on water such as a lower module temperature and reflections from the water, leading to higher energy output. Covering hydropower lakes or drinking water reservoirs will decrease evaporation rates. At present, Japan and China are the two driving markets, with Japan hosting a large number of projects and China being home to a number of very large systems, with the largest over 100MWp. In Europe, the UK hosts the largest systems. The yearly installed capacity shows a quickly increasing trend from 3MWp in 2013 to more than 400 MWp in 2017. Research projects are being executed in Singapore, Malta and the Netherlands. Of 20 different available floating PV systems we assessed different properties that affect the energy output, packing density, survivability, ease of maintenance and installation speed. We see large differences in design philosophy between the systems.

Keywords: Floating PV, PV Markets

1 WHY FLOATING SOLAR?

Floating solar systems can represent a serious alternative to ground mounted solar systems. At present, the international market for floating PV is small compared to the total PV market, but growing fast with an estimated multi-billion market in the coming years [1,2]. A main driver for the growth of the market for floating PV is the mitigation of land cost, especially in places where land is scarce, and water is available, such as Singapore, Taiwan or the Netherlands. Moreover, a large fraction of the world's population lives near the coast or other water bodies. Other drivers for the market growth of floating PV are the potential added benefits of deploying photovoltaics on water. Due to cooling effects, the yield per Wp is expected to be higher for floating PV plants, compared to PV plants on land. Also reflections from the water and an improved microclimate above water, resulting in a higher irradiance, may help in a higher electricity output. Covering hydropower lakes or drinking water reservoirs will decrease evaporation rates, which is beneficial in both situations

PRESENT INTERNATIONAL MARKET STATUS

We have compiled a list of known floating PV systems worldwide. We found 157 systems in 24 different countries, ranging in size from 5 kWp to 150 MWp. The first floating systems were already built in 2007, and we have seen an acceleration in deployment since 2015. In the present market, we see two countries that act as forerunners: Japan and China. Analysis of the list shows that the first floating PV system larger than 1 MWp was already built in 2013 in Okegawa, Japan. At present, 84 out of the 157 floating PV installations were built in Japan. In more recent years, a number of much larger projects were realized in China. The top ten largest floating PV plants is almost completely occupied by projects in China, with the largest project larger at 150 MWp. At number ten we find the largest system in Japan, at close to 14 MWp. In Europe, the UK hosts the largest floating PV installations. Two drinking water companies (Thames Water and United Utilities) have commissioned two large floating PV plants on drinking water reservoirs,

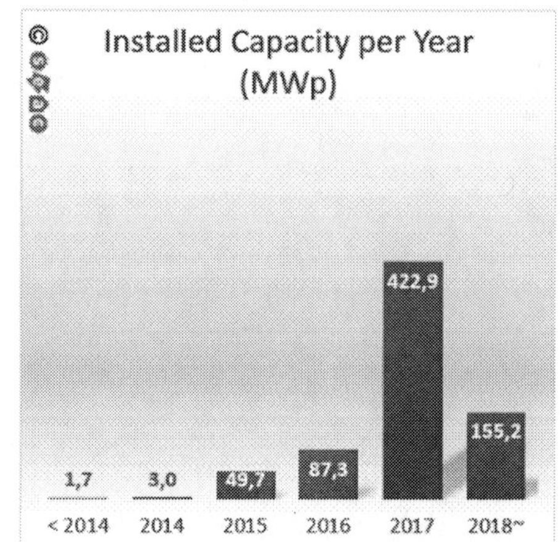

Figure 2: yearly installed capacity of floating PV and cumulative installed capacity. The data for 2018 is incomplete.

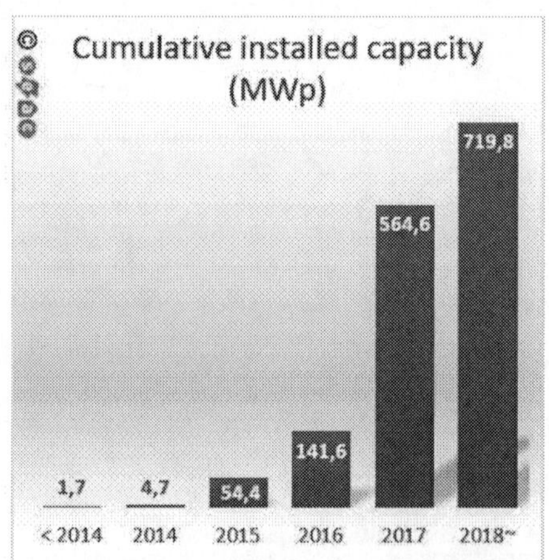

Figure 3: Cumulative installed capacity of floating PV. The data for 2018 is incomplete.

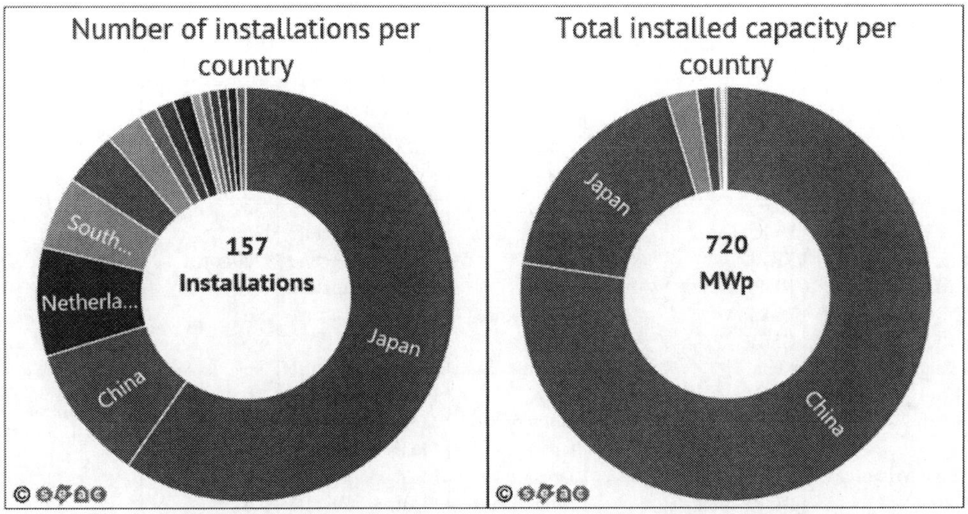

Figure 4: Worldwide number of floating PV installations per country and cumulative installed capacity per country.

which at present are the largest plants in Europe. The yearly installed capacity shows a quickly increasing trend: In 2014, 3MWp was installed, while in 2017 more than 400 MWp was installed. For 2018, we anticipate an installed capacity of more than 1 GWp. Figure 2 shows the yearly installed capacity of floating PV systems, while figure 3 shows the cumulative installed capacity from 2013 to 2017. At the moment, more than 700 MWp of floating PV has been installed worldwide. The data for the running year, 2018, is still incomplete. Figure 4 shows the number of floating PV installations per country and the cumulative installed capacity per country. More than 90% of the installed capacity was built in China and Japan, with China hosting more than 75% of total installed capacity. Third on the list is South Korea.

RESEARCH PROJECTS

Figure 5: A pictures of the SERIS floating PV outdoor research facility, with ten different floating PV designs. (credit: SERIS)

At the Solar Energy Research Institute of Singapore (SERIS) a comparative study is carried out to investigate different aspects of floating PV plants. For this purpose, 10 different systems were deployed at the Tengeh fresh water reservoir in the west of Singapore. These systems represent a large part of the commercial internationally available floating PV systems. A measuring campaign was set up to measure DC and AC output, irradiation, module temperature and general weather information, such as wind speed and direction, ambient temperature and precipitation and their influence on energy performance. As a reference, a roof mounted system was placed on a nearby flat roof. Figure 5 shows an aerial picture of the SERIS floating PV outdoor testing facility.

Figure 5: Pictures of the four different floating PV designs tested in Dutch National Consortium for Floating PV. (credit: zonopwater.nl)

In the Netherlands, a consortium led by Rijkswaterstaat, (the executive body of the Dutch Ministry of Infrastructure and the Environment) and SEAC, brought together more than 40 private and public partners and research organizations together to investigate different aspects of floating PV and initiate large scale construction of floating PV installations in the Netherlands. The shared ambition is to have 2 GWp of floating PV installed by 2023. Research topics include licensability, business modelling and business case analysis. The licensability study includes research on the impact of floating PV systems on their environment. A field test was set up in the Slufter lake near Rotterdam. This 250 hectare artificial lake is only separated from the North Sea by a dike, such that at this location very high wind conditions occur. Due to the large size of the lake, waves grow up to 1 m in height. In these challenging weather and water conditions we perform reliability studies and monitor DC and AC performance. As a

Tabel 1: Overview of floating PV companies and system design properties of their systems

System provider	Float material	Country of Origin	Structure	Tracking
4CSolar	HDPE tubing	USA	Floats + structure	Possible
Adtech	Ferrocement	India	Floats + structure	No
Ciel et Terre	HDPE	France	Floats only	No
Hydro-PV	Unknown	Netherlands	Floats + structure	No
K-Water (Solatus)	FRP	Korea	Floats + structure	Unknown
Koiné Multimedia	HDPE tubing	Italy	Floats + structure	Possible
LG CNS	Unknown	South Korea	Floats + structure	No
Mrac	HDPE	China	Floats only	No
Nemo Eng.	Metal	Korea	Floats + structure	No
NRG Energia	HDPE	Italy	Floats + structure	No
Ocean Sun	Membrane	Norway	Membrane	No
Solaris	HDPE foam	Israel	Floats + structure	Possible
Sumitomo Mitsui	HDPE + styrofoam	Japan	Floats only	No
Sunfloat	HDPE tubing	Netherlands	Floats + structure	No
Sungrow	Unknown	China	Floats only	No
Sunprojects	HDPE tubing	Netherlands	Floats + structure	Yes
Swimsol	Unknown	Maldives	Floats + structure	No
Takiron	HDPE	Japan	Floats + structure	No
Texel4Trading	Concrete + styrofoam	Netherlands	Floats + structure	No
Towa Arc	Unknown	Japan	Floats + structure	No

reference system for studying the differences in performance between ground mounted and floating systems, we built a reference system on land, containing a modules from every system in the field test, at the same orientation and inclination as the floating systems. By monitoring the temperature of the modules both on land and water we can investigate the cooling effect of deploying PV on water.

Several research projects investigate the feasibility of building offshore solar power stations. In Malta, the University of Malta performs research on floating PV systems in the Mediterranean sea [5]. Several floating structure concepts are built and tested. The aim of the so-called SolAqua project is to study and test the technical and financial feasibility of floating solar panels. In the Maldives, Swimsol has built an offshore system [6]. In the Netherlands, two projects aim to build solar field test at sea. The Norwegian company Ocean Sun is developing offshore floating systems on top of floating fish farms [7]. The modules are supported by a floating membrane.

FLOATING SOLAR COMPANIES

We have identified 20 different companies that designed or produce floating PV systems. The maturity of these companies ranges from only having a design but no prototypes, to having commissioned hundreds of MWp of floating systems. Of all system designs we assessed different design properties.

Most designs use high-density polyethylene as a material for the floats, for its robustness and durability, but also some designs incorporate concrete or fibre reinforced plastic (FRP). The structure of the floating systems can be divided into two categories: in the first category, the solar modules are mounted onto the floats directly. In the second category, the solar modules are mounted on a secondary frame that is supported by the floats. These properties affect the flexibility of the system and the installation speed. We also indicated whether the systems can be built as a tracking system. Table 1 shows a summary of the different design properties of the evaluated systems. We see large differences in design philosophy between the systems.

CONCLUSIONS

Evaluation of floating systems worldwide shows a fast emerging market with more than 400 MWp installed in 2017. Most installations are built in China and Japan. Several research projects investigate the performance of floating PV systems, licensability and business modeling. First (research) attempts have been made for deploying floating solar offshore. 24 companies (intend to) produce floating systems for PV deployment. Among these system builders, we see large design differences.

References

[1] Renewable Energy Magazine, http://www.renewableenergymagazine.com/pv_solar/global-floating-solar-panel-market-to-be-20170203, consulted March 2018

[2] http://www.solarquarter.com/index.php/industry-insights/2493-floatavoltaic-floating-solar-panel-market-worth-over-2-5-gw-by-2024, consulted March 2018

[3] http://www.seris.nus.edu.sg/

[4] http://www.zonopwater.nl/en

[5] https://www.um.edu.mt/newspoint/news/features/2016/06/solaquafloatingsolarpanelsprojectnearscompletion

[6] https://swimsol.com/

[7] http://oceansun.no/

INFRARED AND ELECTROLUMINESCENCE IMAGING FOR PV FIELD APPLICATIONS: AN OVERVIEW OF THE LATEST REPORT BY IEA PVPS TASK 13

J. A. Tsanakas[1], U. Jahn[2], M. Herz[2], M. Köntges[3], D. Parlevliet[4], M. Paggi[5], J. S. Stein[6], K. A. Berger[7], B. Kubicek[7], S. Ranta[8], R. H. French[9], M. Richter[10], T. Tanahashi[11]

[1]IMEC, Genk, Belgium [ioannis.tsanakas@imec.be]
[2]TÜV Rheinland GmbH, Cologne, Germany [ulrike.jahn@de.tuv.com | magnus.herz@de.tuv.com]
[3]Institute for Solar Energy Research Hamelin, Emmerthal, Germany [marc.koentges@isfh.de]
[4]Murdoch University, Perth, Australia [d.parlevliet@murdoch.edu.au]
[5]IMT School for Advanced Studies Lucca, Italy [marco.paggi@imtlucca.it]
[6]SANDIA National Laboratory, USA [jsstein@sandia.gov]
[7]Austrian Institute of Technology GmbH, Center for Energy, Austria [karl.berger@ait.ac.at | bernhard.kubicek@ait.ac.at]
[8]Turku University of Applied Sciences, Finland [samuli.ranta@turkuamk.fi]
[9]Case Western Reserve University, Ohio, USA [roger.french@case.edu]
[10]3E sa, Brussels, Belgium [mauricio.richter@3e.eu]
[11]National Institute of Advanced Industrial Science and Technology, Japan [tadanori.tanahashi@aist.go.jp]

ABSTRACT: This paper presents an overview of the latest research and technical reporting activity of TASK13 participants, within the Subtask 3.3 ("Characterization of PV Module Condition in the Field"); and particularly, key findings of the new "Review on Infrared (IR) and Electroluminescence (EL) Imaging for PV Field Applications" TASK13 Report. Goal of the latter is to provide guidelines and recommendations for using IR and EL imaging, in order to identify and assess specific failure modes of PV modules and systems in field applications. As such, the paper provides first a discussion on the relevant state-of-the-art and particularly the new IEC standards, Technical Specifications (TS) and guidelines. It also describes current practices for IR and EL imaging of PV modules and systems, looking at environmental and device requirements and the interpretation of sample patterns with abnormalities. In addition, examples of typical inspection results are given, showing characteristic IR/thermal and EL signatures of different failure modes occurring in fielded PV modules and arrays.

Keywords: operation, performance and reliability of PV systems, infrared imaging, electroluminescence imaging

1 INTRODUCTION

Coordinated by the International Energy Agency (IEA), the Photovoltaic Power Systems (PVPS) Programme is one of the collaborative R&D Agreements established within the IEA and, since 1993, the PVPS participants have been conducting a variety of joint projects in the application of photovoltaic (PV) conversion of solar energy into electricity. Mission of the IEA PVPS Technology Collaboration Programme is to enhance the international collaborative efforts which facilitate the role of PV solar energy as a cornerstone in the transition to sustainable energy systems.

Within the IEA PVPS Programme, TASK13 ("Performance and Reliability of Photovoltaic Systems") engages in focusing the international collaboration on improving the operation and reliability of photovoltaic systems, subsystems and components, by collecting, analyzing and disseminating information on their technical performance and failures; thus, bringing substantial insight into PV systems' technical assessment, and developing practical recommendations for improving their electrical and economic output.

This paper presents an overview of the latest research and technical reporting activity of TASK13 participants, within the Subtask 3.3 ("Characterization of PV Module Condition in the Field"); and particularly, key outcomes of the new "Review on Infrared (IR) and Electroluminescence (EL) Imaging for PV Field Applications" TASK13 Report [1]. The latter describes current practices for IR and EL imaging of PV modules and systems, looking at environmental and device requirements and the interpretation of sample patterns with abnormalities. The goal of this review is to provide guidelines and recommendations for using IR and EL imaging, in order to identify and assess specific failure modes of PV modules and systems in field applications.

2 IR IMAGING AND PV FIELD APPLICATIONS

A significant amount of published research studies has been reported in the literature through the last decade, reflecting the increasing interest in the application of infrared thermography (IRT) measurements and their potential for fault diagnosis of PV installations in the field [2].

In the early steps of such studies, [3] first investigated the applicability and practicality of IRT imaging of PV modules under outdoor conditions. For PV array-scale applications, Moropoulou et al. [4] used IRT to assess the condition of a building integrated, grid-connected PV array. The investigated PV system presented abnormally low performance, which finally exhibited abnormal temperature profiles through the acquired thermal images. These abnormalities are due to either the panel material itself or the overall function. In terms of accuracy and practicality of IRT measurements of PV modules' materials, three different early studies [5-7] discussed the infrared reflection errors induced by the glass front-sheet (or cover), as well as variations between the cell temperature and the measured modules' temperature for different back sheet materials. In addition, [8] presented a procedure for accurate measurement of IR emissivity of the glass surface, in the spectral range of the thermal cameras that were used, proposing a necessary correction (offset) of the temperatures measured by such bolometer detectors.

The importance of further interpretation of the thermal images and the correlation between performance degradation, fault types and thermal signatures of PV modules was introduced by Hoyer et al. and Vodermayer et al. [9,10]. In particular, Hoyer showed that by the combination of several IRT measurement methods, typical defects in crystalline silicon (c Si) PV modules and their influence on the I-V curve (e.g. the series resistance or the short-circuit current) can be explained. In particular, it was found that in the case of thermography of sun illuminated PV modules, several defects can be distinguished by varying the electrical load on the PV module. In the experimental study of Vodermayer, daily profiles of one minute *I-V* curves in the field were acquired from a range of defective PV modules. These were collected together with additional continuous IR imaging and meteorological data sampling. A direct relation between the power output (thus, energy yield) losses and the thermal differences was observed; however, in order to quantify the effective loss in kWh/kW$_p$ per year, further evaluation is needed and is not published yet.

2.1. State-of-the-art on IR imaging

Review of existing standards/guidelines. IEC/TS 62446-3 Technical Specification (TS), for IR Imaging was recently released by the IEC Technical Committee (TC) 82. It describes test procedures, and is not intended to have pass/fail criteria for the PV modules being investigated. IEC/TS 62446-3 describes investigations of PV modules and the entire plant in operation under natural sunlight. Reporting requirements and a Table-"guide" [11] for qualitative interpretation of IR images and failure modes are key conclusions of the latter technical specification.

Overview of existing practice. The classification of different PV module failures by correlating their thermal signature with the I-V characteristics is of utmost importance for assessing the expected power output losses and scheduling the needed maintenance actions for the PV system. Advances in the field of fault analysis and characterization for installed PV modules, based on IR imaging, were recently reported by Moretón et al. and Buerhop et al. [12, 13]. Moretón et al. proposed hot spot evaluation procedures and well-defined acceptance-rejection criteria, based on observations and IR inspections of 200 fielded and defective modules. Buerhop et al. have shown that the PV module power output of areal cell defects behave linear with the solar irradiance intensity, while the temperature development in point-shaped defects exhibits a less predictable behaviour. The demonstrated reliability and effectiveness of IR imaging encourages the implementation of wide-area measurements for large-scale PV plants, by means of unmanned aerial vehicles (UAVs) equipped with IR imaging and PV diagnostics tools [14-16].

2.2 Test requirements

IR camera technical requirements. Today, there is a broad variety of IR cameras with a range of specifications on the market. Key technical characteristics (and thus often selection criteria) include: detector resolution, thermal sensitivity (i.e. from 0.1 up to 0.02 K), accuracy (i.e. ±2°C or ±1°C), emissivity correction and optical resolution (i.e. from 160x120 up to 640x480 pixels). The most suitable type of camera will depend on the application at hand. For example, handheld cameras are suitable for field inspection of PV modules. For unmanned aerial vehicles based surveys of PV plants, other specifications may be of more importance due to weight restriction and the different lenses and distances in use.

Environmental conditions. Stable field conditions are a pre-requisite for high quality outdoor IR inspection of PV modules and arrays. Weather fluctuations must be much slower than the thermal time constant of the PV module. A typical PV module takes 5 to 15 minutes to thermally stabilize for new environmental conditions such as change of global irradiation intensity, temperature, or wind speed. Furthermore, thermal reflections of clouds, buildings, trees, or other objects on the PV modules must be avoided or minimized as they may distort the analysis of the IR image. The impact of different ambient conditions on the characteristics and the temperature profile of defects in PV modules is particularly studied by Buerhop et al. [14].

Apart from the above, in overall, the analysis of the IR images should be done by qualified personnel using relevant IR analysis software. As orientation for personal qualification, the ISO 9712 Level 2 can be named, as also described in IEC 62446-3 TS Ed.1 – Part 3. In any case the personnel should have detailed knowledge of thermal imaging techniques as well as of PV system behaviour and module design.

2.3 Classification and evaluation

Sample patterns with abnormalities. Based on the reported literature and the real PV field experience feedback, main thermal abnormalities in PV power plants can be classified as follows: i) Hot cells (formally hot spots by: breakage of glass front-sheet (front glazing), external shading or internal cell and interconnection problems), ii) heated bypass diodes inside the PV module junction box, iii) heated string fuses in combiner boxes and iv) heated DC and AC cables and connection points.

Figure 1 presents some examples of specific thermal abnormalities. As outlined failures often have not developed yet, the objective interpretation of IR images is not trivial. However, these defects have the potential to cause power degradations in subsequent years. Hot spots with strong local heat gradients can significantly affect the performance of the PV modules and even, in extreme cases, lead to smoldering fires. IR imaging is a valuable tool to detect power relevant, malfunctioning PV modules, module strings, and certain devices in a PV array if the presence of artefacts, like partial shading, dirt, can be excluded.

One example: "Potential induced degradation by shunts" (PID-s) affected PV modules. IRT imaging of PV modules under direct sunlight is an efficient method for getting an estimation of the degree of PID-s during the operation of modules in the PV plant. Typical PID-s IRT patterns (warmer cells close to the bottom frame or patchwork patterns) and module positions close to the negative pole of the module string are strong indications for PID-s effects [1]. A PID affected cell is not hotter than an inactive cell (typically 1 to 5 K but up to 10 K).

To simulate operating conditions, the PID-s affected module was mounted at 30° tilt angle in natural sunlight. A variable resistive load was used to set module current and voltage. IR imaging (see Figure 2b) was carried out with the module operating close to maximum power point (MPP) conditions. The PID-s affected cells show up slightly warmer than the other cells in the PV-module. In contrast, at open circuit conditions (Figure 2c) no pattern appears (uniform temperature distribution); while at short circuit conditions (Figure 2a), certain reverse biased cells with low I_{sc} dominate the pattern. On the other hand, IRT patterns may differ within string-connected modules under MPP operation (Fig. 3), because the modules are never affected homogeneously by PID-s, and therefore have different IV-curves and operating points. Besides, for the PID-affected modules close to the string's negative pole, power loss and number of warm cells in the IRT image are in good agreement. Therefore, the main field method to detect PID-s affected modules is to prepare IRT images under clear sky and MPP conditions [18-20].

Fig. 1: Examples of thermal abnormalities detected by IR imaging of operating PV systems, associated with: front glass breakage (above), shading (middle) and overheated bypass diode (down) [1].

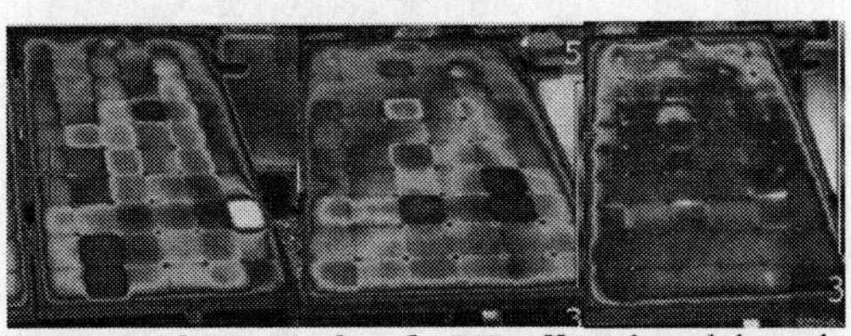

Fig. 2: IR thermography of a PID affected module under sunlight conditions (approx. 900 W/m²); a. left: Short circuit (I=7.3A, V=0V), b. middle: MPP (I=6.6A, V=22V), c. right: Open circuit (I=0A, V=30V).

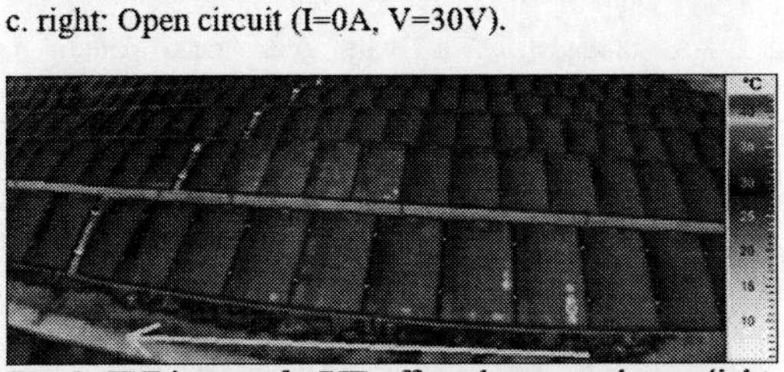

Fig. 3: IRT image of a PID affected array under sunlight, MPP conditions. Modules show PID patterns with warmer cells near their bottom frame. The measured output power of the modules (nominal power 225 W_P) from the negative pole is (see arrow from right to left): 188W_P, 187W_P, 190W_P, 181W_P, 199W_P, 207W_P, 212 W_P, 220 W_P [21].

In the case of dark IRT measurements (Fig. 4), PID-affected modules can also be identified with a "reversed" IR pattern compared to the case of IRT under sunlight. In dark IRT in particular, the cells are heated by an applied current (external forward bias) using a power source. The higher the voltage at each cell, the warmer the cell; and in the case of an occurring PID, i.e. heavily shunted cells (thus, of lower voltage) appear cooler than non-shunted cells. An advantage of dark IRT is also the absence of unwanted reflections and shadows that often disturb IRT measurements under sunlight.

Fig. 4: IRT image of a PID-s affected rooftop system with applied current during the night. In this case, the cooler cells are the PID-s affected ones [22].

Interpretation of abnormal thermal patterns. If the colour scale itself does not provide sufficiently clear information about a thermal pattern, the temperature measuring functions of IR image analysis software should be used to better identify temperature rises and differences. Functions such as the spot measuring tools can be set on local abnormal connection sockets or idling points, for example. Polygon measuring fields can be set on cells, modules or up to whole module strings. The correct setting of the measuring fields on abnormalities and reference areas needs some experience of the IR camera user. In skilled hands these tools can show many complex module or system defects down to just about 1 K of temperature rise. And it can be used for plausibility checks of potentially known defects. In any case, the measuring of temperature differences of just about 1-2 K should be interpreted with good care as there are many misleading effects within a PV array, such as different convective cooling or emissivity effects that may significantly exceed a signal of 1 K.

3 EL IMAGING AND PV FIELD APPLICATIONS

Electroluminescence testing, brought to the attention of the PV community by Fuyuki and co-workers in 2005, has become a very popular technique for inspecting PV modules for minority carrier lifetime [23], cell cracks, shunts, and voltage differences [24]. For these measurements a PV module is typically supplied with a DC current in a dark environment and imaged with a camera sensitive to the emission spectra of the material being analyzed. There are a number of cameras and products available on the market that are suitable for EL measurements of PV modules ranging from modified digital single-lens reflex (DSLR) cameras or the closely related mirrorless interchangeable-lens cameras (MILC) through to dedicated EL systems with a corresponding broad range of specifications.

The key specification for EL camera systems is derived from the need for the camera to be sensitive to the emission spectra of the material being tested. For a crystalline silicon solar module, the emission spectra has a peak at 1150 nm [23], while for other technologies such as organic PV the emission peak resides at a longer wavelength (1200 nm – 1400 nm) [25]. Of the thin-film technologies, CIGS is fairly similar to crystalline silicon, amorphous silicon and micro-morph tandem cells tend to have a broad EL emission spectra, while CdTe has an emission peak at a much shorter wavelength of 830 nm [26].

3.1. State-of-the-art on EL imaging

Review of existing standards/guidelines. The technical specification IEC TS 60904-13 specifies methods to capture electroluminescence images of PV modules, process images to obtain quantitative descriptors, and provide guidance to qualitatively interpret the obtained images. The standard is applicable to PV modules measured indoors in forward bias condition, i.e., by forcing current flow with a power supply where the leads are connected to those of the same polarity on the sample. However, similar technical issues do arise also in the field, where EL imaging of strings of PV modules could nowadays be performed using special tripods or drones.

Overview of existing technology and practices. EL image (camera) detectors are typically charge coupled devices (CCD) or complementary metal oxide semiconductor (CMOS). They may be cooled, usually with thermoelectric cooling, to achieve better signal to noise ratios by means of reducing the device dark current originating from thermally generated charges. Relevant parameters in choosing detectors include image resolution (from <1MP up to >5MP), noise, quantum efficiency (0.3-1.1 μm for Si; 0.7-2.6 μm for InGaAs) at the wavelength of interest, and dynamic range (from 2500:1 up to >10000:1).

There is a range of manufacturers in the market that produce cameras suitable for EL applications. There are two general types of cameras, the line scan camera consisting of a 1D line of pixels and an area camera consisting of a 2D array which is similar to sensors in modern consumer digital cameras.

Quantitative EL image analysis using supervised and unsupervised machine learning. Recently advances have been made to apply computer vision and data science techniques, including image processing and machine learning to EL images [27] These approaches allow for automated pipelines to be applied to thousands of EL images to produce statistically significant measurement of different types of PV module and cell defects, such as busbar corrosion, cracking, between busbar cell darkening and cell edge darkening. In addition, these methods produce EL image based results which quantitatively can predict module power, without requiring direct electrical measurements [28].

3.2 Test requirements

EL testing requirements.

It has been shown in [29] that a relative simple modification (removal of the camera internal IR blocking filters) enables consumer DSLRs and MILCs to perform EL imaging. More recently [30], non-reflex, mirrorless interchangeable lens cameras with large sensors can also be modified for EL applications [31]. The choice of the type of EL camera that is used when imaging PV modules is dependent on a range of factors. As with many types of camera equipment there is trade-off between cost, resolution and sensitivity. For example, silicon CCD cameras tend to have a higher resolution and lower cost than InGaAs CMOS sensors. While their high resolution makes silicon CCD sensors a popular choice, the InGaAs sensor would have a significantly greater sensitivity to crystalline silicon and some thin-film technologies. Despite the differing levels of sensitivity, with appropriate data collection intervals and methodologies, such as longer exposure times, filters and active cooling of the sensor, either camera type can be used for EL imaging of PV modules. This allows a great flexibility in terms of the equipment used, but does make standardization of EL test images and measurements a challenge.

Environmental conditions. High quality images favor dark environment. Precautions should be taken to eliminate stray light entering the imaging studio, e.g. with use of hard walls, curtains and sealing of any gaps with material that are of light absorbing nature (black). Fixed mounting of the camera and a mounting rack for PV module(s) to be imaged are used so that the camera and the module positions are stable. In the field, an option is to use curtains supported by a frame to cover the camera and the photovoltaic module. Alternatively, outdoor EL images have to be taken after sunset or during the night.

3.3 Defect identification and evaluation

Sample patterns with abnormalities. The EL technique can profitably be applied to detect and classify different PV failure modes (Fig. 5). This type of non-destructive inspection includes the possibility to observe defects and attribute them to the different stages of the PV module lifetime. For instance, defects can be caused by production problems not noticed after quality control, wrong handling of the PV module without sufficient care during transport or installation, or can be induced by environmental loading and degradation in the field after installation. However, the amount of time required to take a good quality EL image of a PV module, connecting a PV string to an electrical power source, adjusting the focus and finding the right time of exposure depending on the module type, is often still a limiting factor for large-scale applications of EL imaging in the field.

An example of a wide cell crack pattern in polycrystalline silicon PV modules is shown in Fig. 5 (up), with large portions of solar cells electrically inactive. Other crack patterns are shown in Fig. 6. In the case of potential induced degradation (PID), on the other hand, EL images show the appearance of dark solar cells at the boundaries of the PV module, see Fig. 5 (down).

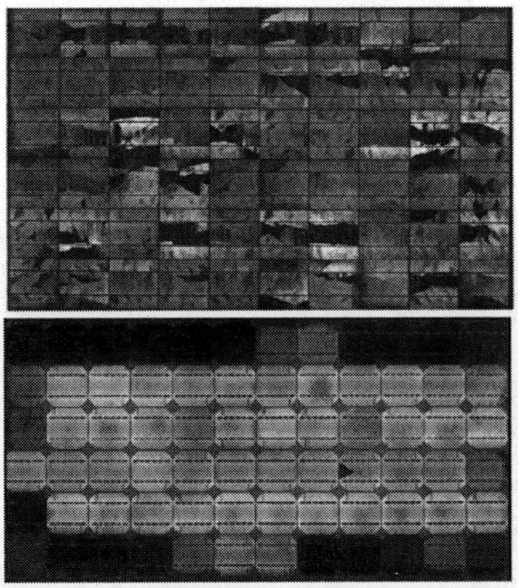

Fig. 5: Examples of EL imaging of PV modules diagnosed with several cracked cells (up) and potential induced degradation (down) [1,32].

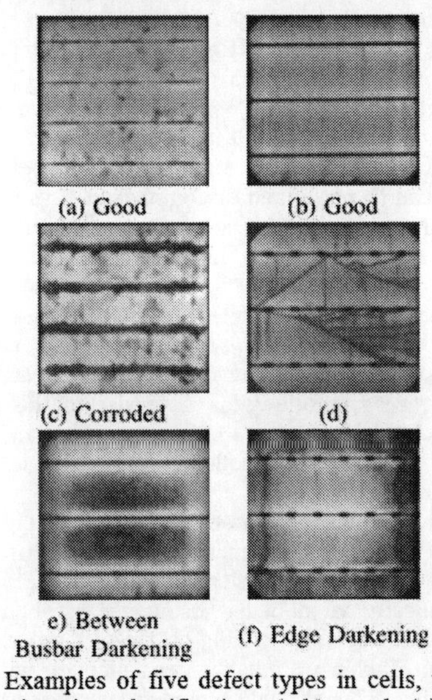

Fig. 6: Examples of five defect types in cells, used for machine learning classification: (a-b) good, (c) busbar corrosion, (d) cracked, (e) between busbar darkening and (f) cell edge darkening.

3.4 Statistical and machine learning methods applied to EL defect classification and quantification

Image characterization techniques capture spatially resolved macroscopic manifestations of microscopic mechanistic behavior. Automated data processing and analytics allow for a large-scale systematic study of PV module health. Image corrections, including planar indexing to align module images, are applied. Subsequently on extracted cell images, defects and degradation states such as busbar corrosion, cracking, wafer edge darkening, and between-busbar darkening can be studied in comparison to new cells using both supervised (where a human labels defects in training images) and un-supervised (where data analysis directly identifies classes in the images) machine learning [27,33] The systematic feature groupings provide a scalable method without bias to quantitatively monitor the degradation of laboratory and commercial systems alike. The evolution of these degradation features through varied exposure conditions provides insight into mechanisms causing degradation in field deployed modules.

The supervised algorithms used in this application are Convolutional Neural Networks (CNN) and Support Vector Machines (SVM). With the increase in data and diversity of features, unsupervised learning can be employed to find relations between inherent image properties. Feature extraction techniques help identify intrinsic geometric patterns formed in the images due to degradation. Principal component analysis is then applied to the extracted set of features to filter the most relevant components from the set, which are then passed to an agglomerative hierarchical clustering algorithm. Google's Tensorflow library was utilized to enhance the computational efficiency of the CNN model by providing GPU-based parallel matrix operations. Using supervised methods on 5 features an accuracy greater than 98% was achieved. For unsupervised clustering, the classification was done into two clusters of degraded and non-degraded cells with the measure of coherence calculated to be 66%.

3.5 Combined EL & IR imaging for comprehensive PV module defect identification.

Night-time EL imaging of PID affected modules installed in the field (Fig. 7), in a black-white-red color scheme and using an exposure time of about four minutes, shows current passing through the decreased parallel resistance instead of the PN-junction for the cells near the edges.

On the other hand, daylight outdoor lock-In thermography of the same modules (Fig. 8) affected by PID (whose night-time EL image is shown in Fig. 7) shows the amplitude of the thermal waves at the glass surface induced by toggling the modules repeatedly between short and open circuit with a period time of 50 seconds. The amplitude of the temperature waves is plotted, white being no thermal change in time, black indicating heat-up of 0.1 K at short circuit, while red indicates heat-up of 0.1 K at open circuit.

Fig. 7: Night-time EL of PID affected modules, using a DSLR type camera, in a black-white-red color scheme. The accumulated exposure time was around four minutes. The darkening of cell edge regions close to the frame can be observed, which is commonly attributed to current passing through the decreased parallel resistance instead of the PN-junction [34].

Fig. 8: Daylight Outdoor Lock-In Thermography of the same modules as in Fig. 7. The modules are repeatedly toggled between short and open circuit with a period time of 50 seconds, while a thermography camera observes transient temperature waves at the glass surface. The amplitude of the waves is plotted, white being no thermal change in time, black indicating heat-up of 0.1 K at short circuit, while red indicates heat-up of 0.1 K at open circuit [34].

4 RECOMMENDATIONS BASED ON IR & EL IMAGING OBSERVATIONS

After detection and evaluation of thermal abnormalities by IR imaging, the following actions are recommended:

Normal temperature behavior

- A temperature difference of a single solar cell, when compared to the average temperature, smaller than 10°C is normally considered as unproblematic and will not be listed separately.

Heated cells

- The temperature differences in a range of 10°C to 20°C are unproblematic in the current stage, although they may increase during the operation of the PV power plant.
- The modules should therefore be observed closely in regular thermographic inspections.

Hot spots

- Temperature differences above 20°C are expected to cause degradations of panel output; in extreme cases (module's absolute temperature >150°C), the material compound may even degrade, resulting in a safety issue during maintenance work. Further increase in temperature gradient is expected during the operation phase of the PV power plant if the modules are not replaced.
- The modules should therefore be replaced.

Activated bypass diodes (modules)

- A permanent activated bypass diode leads to a minimized power output of the affected solar cell string and thus to a reduction of the total power output of the plant. Modules with active bypass diodes should be substituted.

Pattern

- If one module or several modules connected in series have a higher temperature (2K-5K up to 10K), the modules may not be connected to the system and the wiring should be checked.
- Warmer cells close to the bottom frame, and patchwork patterns are strong indicators for PID-effects. The root cause for this pattern should be further investigated as described in chapter 2.3.2.

Due to normal tolerances in cell sorting and module production, thermal abnormalities of less than 10 % of the recorded modules do not indicate a special quality issue regarding the used modules.

If more than 10 % modules show thermal abnormalities, the reason for that behavior should be evaluated. The behavior could be caused by quality problems of the modules as well as planning, installation, maintenance or operation faults of the PV plant.

For warranty claims of the supplied modules the first IRT should be conducted when commissioning the PV plant. If the PV modules of the system are not cleaned and maintained at a suitable frequency, high temperatures of some cells or modules could occur due to bird droppings or power mismatch for a long time which may lead to module damage. At a later stage it might be difficult to evaluate whether the damage was caused by quality problems or by missing cleaning and maintenance procedures.

Electroluminescence (EL) image data provides a rich source of spatially resolvable information pertaining to the health of a PV module. Commonly observed defects and failures contained within these data have been outlined in [11] Sec. 5.4. Using this reference, classification of EL images can be performed by comparison. This qualitative approach has merits for small studies requiring course assessment, however, for large sample quantities or if derived properties are desired, such as electrical properties, techniques for quantitative assessment are needed. Automated analysis pipelines for analysis of thousands of EL images have been developed and include image processing followed by supervised or unsupervised machine learning, and can both classify cell and module defects and predict PV module power losses solely from image analysis.

In the report [1], Section 3.1.2 describes the types of cameras used for EL imaging and provides a discussion of sensors. Sensors with larger numbers of pixels per unit area typically yield better results for resolving image features, however, depending on the material, noise and quantum efficiency limitations can increase with higher resolution. Though with modern signal processing algorithms and image analysis software it is possible to mitigate a portion of the effects of noise on the resultant images.

5 DISCUSSION

Using IR thermography by means of IR measurements of the thermal behavior of cells in a PV module, a number of defects (e.g. short circuits in solar cells, shunts, inactive cell parts etc.) can be clearly identified. Several defects can be distinguished by varying the electrical load corresponding to certain states of the current voltage characteristics. As for IR thermography measurements, it turned out that illuminated thermography is more suited for crystalline modules and dark thermography is a better tool for thin-film modules. Compared to the dark thermography images the illuminated thermography images of crystalline modules supply more details and a better defect resolution. For thin-film PV modules the opposite conclusion applies [1].

EL measurements take advantage of the radiative inter-band recombination of excited charge carriers in solar cells. The solar cells are supplied with a defined external excitation current while the camera takes an image of the emitted photons. Damaged areas of a solar module appear dark or radiate less than areas without defects. The high resolution of the EL images enables resolving some defects more precisely than in IR images. In a study by [35], it is shown very clearly that electroluminescence is a good method to detect cell cracks, interrupted contacts or a number of process failures (e.g. shunts or defects in the antireflection layer), but it is not possible to determine the impact of these defects on the module power output.

6 SUMMARY AND CONCLUSIONS

This paper provides a comparison of the relative merits of EL and IR imaging techniques. These types of optical measurement provide fast, real-time and high-resolution images with a two-dimensional distribution of the characteristic features of PV modules. IR imaging measurements of the thermal behavior of cells in a module allows a number of defects to be easily and quickly identified. These include short circuits in solar cells, shunts, inactive cell parts and PID. However, not all the failures of PV modules lead to an increase in temperature. Taking this into consideration the IR measurements alone cannot identify all the types of failures. On the other hand, PV modules with clear EL-images sometimes provide IR-images with hot areas. A combination of both techniques quickly detects the most common defects in a PV module with high accuracy and provides a good indication of the health and reliability of the PV modules with a plant.

Whether IR or EL imaging is the most appropriate method depends on the specific technological problem to be studied. While EL is superior for lifetime mapping, bulk defect imaging, series resistance imaging and exact junction breakdown site imaging, IR-based inspections are typically faster and superior for all tasks requiring quantitative current measurements [36]. As the origin of a defect is not always located in the identified high temperature area or the remarkable dark EL areas, a combination of EL and IR is often the best solution in order to identify as many defects as possible.

7 ACKNOWLEDGMENTS

Funding was provided by the Federal Ministry for Economic Affairs and Energy (BMWi) under contract no. FKZ 0325786C/A Task13 Projekt, the State of Lower Saxony, and by the Austrian Climate and Energy Fund (KliEn) within the 1st e!MISSION energy research call 2014, project Infinity, under contract no. 850.414, and was partly supported by the Austrian Federal Ministry of transport, Innovation and Technology (bmvit), contract no. 853.028. The work of R.H. French was supported by the U.S. DOE's EERE under Solar Energy Technologies Office (SETO) Agreement Number DE-EE0007140. We like to thank all TASK13 members and experts who contributed to this technical work of ST 3.3.

8 REFERENCES

[1] U. Jahn et al., "Review on Infrared (IR) and Electroluminescence (EL) Imaging for Photovoltaic Field Applications", Report IEA-PVPS T13-10:2018.

[2] J. A. Tsanakas, L. Ha, and C. Buerhop, "Faults and infrared thermographic diagnosis in operating c-Si photovoltaic modules: A review of research and future challenges," Renew. Sustain. Energy Rev., vol. 62, pp. 695–709, Sep. 2016.

[3] C. Buerhop, U. Jahn, U. Hoyer, B. Lerche, and S. Wittmann, "Abschlußbericht der Machbarkeitsstudie zur Überprüfung der Qualität von Photovoltaik-Modulen mittels Infrarot-Aufnahmen," Bayern, Erlangen, 2007.

[4] A. Moropoulou, J. Palyvos, M. Karoglou, and V. Panagopoulos, "Using IR thermography for photovoltaic array performance assessment," in Proc. 4th International Conference on NDT - Hellenic Society for NDT, 2007, pp. 1804–1806.

[5] G. Makrides, B. Zinsser, M. Norton, G. E. Georghiou, M. Schubert, and J. H. Werner, "Error Sources in Outdoor Performance Evaluation of Photovoltaic Systems," in Proc. 24th EU-PVSEC, 2009, pp. 3904–3909.

[6] A. Krenzinger and A. C. de Andrade, "Accurate outdoor glass thermographic thermometry applied to solar energy devices," Sol. Energy, vol. 81, no. 8, pp. 1025–1034, Aug. 2007.

[7] C. Buerhop-Lutz, D. Schlegel, C. Vodermayer, and M. Nieß, "Quality control of PV-modules in the field using infrared-thermography," in Proc. 26th EUPVSEC, 2011, pp. 3894–3897.

[8] C. Buerhop, H. Scheuerpflug, and R. Weißmann, "The role of infrared emissivity of glass on IR-imaging of PV-plants," in Proc. 26th EUPVSEC, 2011, pp. 3413–3416.

[9] U. Hoyer et al., "Analysis of PV modules by electroluminescence and IR thermography," in Proc. 24th EUPVSEC, 2009, pp. 3262–3266.

[10] C. Vodermayer et al., "First results – Correlation between IR images and electrical behavior and energy yield of PV modules," in Proc. 23rd EUPVSEC, 2008, pp. 3134–3137.

[11] M. Köntges et al., "Review of Failures of Photovoltaic Modules", Report IEA-PVPS T13-01:2014.

[12] R. Moretón et al., Proc. 29th EUPVSEC, Amsterdam (Netherlands), 2014, pp. 2722–27.

[13] C. Buerhop et al., Proc. 31st EUPVSEC, Hamburg (Germany), 2015, pp. 1789–92.

[14] C. Buerhop et al., Proc. 31st EUPVSEC, Hamburg (Germany), 2015, pp. 1692–97.

[15] J.A. Tsanakas et al., Proc. 31st EUPVSEC Hamburg (Germany), 2015, pp. 1783–88.

[16] J.A. Tsanakas et al., Renewable Energy, 102 (2017), pp. 224–33.

[17] G. Mathiak, et al., "Potential Induced Degradation (PID) – Comparison of Different Test Methods. Shunt Resistance Measurements of Individual Cells Inside a Module and Infrared Imaging of PV Modules," in Proc. 28th EUPVSEC, 2013, pp. 3332–3335.

[18] P. Hacke and S. Johnston, "All about PID – testing and avoidance in the field PV Tech Power," PV Tech Power, vol. 8, 2016.

[19] C. Buerhop et al., "IR-images of PV-modules with potential induced degradation (PID) correlated to monitored string power output," in Proc. SPIE 9938, 2016, p. 99380J.

[20] T. Kaden, K. Lammers, and H. J. Möller, "Power loss prognosis from thermographic images of PID affected silicon solar modules," Sol. Energy Mater. Sol. Cells, vol. 142, pp. 24–28, Nov. 2015.

[21] B. Weinreich, "Fehlererkennung an Modulen im Feld mittels Thermographie," in Proc. 13th Workshop "Photovoltaik-Modultechnik," 2016.

[22] M. Diehl, "Methoden zur Erkennung von PID," 2015.[Online].Available:http://www.photovoltaikbuero.de/pv-know-how-blog/pid-erkennung-diagnose-methoden/. [Accessed: 22-Dec-2016].

[23] T. Fuyuki, H. Kondo, T. Yamazaki, Y. Takahashi, and Y. Uraoka, "Photographic surveying of minority carrier diffusion length in polycrystalline silicon solar cells by electroluminescence," Appl. Phys. Lett., vol. 86, no. 26, p. 262108, Jun. 2005.

[24] T. Potthoff, K. Bothe, U. Eitner, D. Hinken, and M. Köntges, "Detection of the voltage distribution in photovoltaic modules by electroluminescence imaging," Prog. Photovoltaics Res. Appl., vol. 18, no. 2, pp. 100–106, Mar. 2010.

[25] U. Hoyer et al., "Electroluminescence imaging of organic photovoltaic modules," Appl. Phys. Lett., vol. 97, no. 23, p. 233303, 2010.

[26] T. Weber, A. Albert, M. Roericht, S. Krauter, and P. Grunow, "Electroluminescence investigation on thin film modules," in Proc. 26th EUPVSEC, 2011.

[27] A. M. Karimi, J. S. Fada, J. Liu, J. L. Braid, M. Koyutürk, R. H. French, Feature Extraction, Supervised and Unsupervised Machine Learning Classification of PV Cell Electroluminescence Images, in IEEE WCPEC-7, Waikoloa, HI, 2018.

[28] J. S. Fada, A. Loach, A. J. Curran, J. L. Braid, S. Yang, T. J. Peshek, R. H. French, Correlation of I-V Curve Parameters with Module-Level Electroluminescent Image Data Over 3000 Hours Damp-Heat Exposure, in IEEE PVSC-44, Washington, D.C., 2017.

[29] K. Mertens et al., Proc. OTTI - Symposium Photovoltaische Solarenergie, 2012, pp. 214–19.

[30] J.S. Fada et al., Rev Sci Instruments, 87 (8), 2016 modules after 12 years of operation in Southern Europe," *Prog. Photovoltaics Res. Appl.*, vol. 19, no. 6, pp. 658–666, Sep. 2011.

[31] J.S. Fada, Nicholas R. Wheeler, Davis Zabiyaka, Nikhil Goel, Timothy J. Peshek, and Roger H. French, "Democratizing an electroluminescence imaging apparatus and analytics project for widespread data acquisition in photovoltaic materials," Review of Scientific Instruments, vol. 87, no. 8, p. 085109, Aug. 2016.

[32] F. Spertino et al., "A power and energy procedure in operating photovoltaic systems to quantify the losses according to the causes, " Solar Energy, vol. 118, p. 313-326, 2015.

[33] J. S. Fada, M. A. Hossain, J. L. Braid, S. Yang, T. J. Peshek, R H. French, Electroluminescent Image Processing and Cell Degradation Type Classification via Computer Vision and Statistical Learning Methodologies, in IEEE PVSC-44, Washington, D.C., 2017

[34] B. Kubicek, T. Schlager, and M. Halwachs, "Outdoor Lock-In Thermographie von PID-Modulen," in 31. Symposium Photovoltaishce Solarenergien, 2017.

[35] R. Ebner, S. Zamini, and G. Újvári, "Defect Analysis in Different Photovoltaic Modules Using Electroluminescence (EL) and Infrared (IR)-Thermography," in Proc. 25th EUPVSEC, 2010, pp. 333–336.

[36] O. Breitenstein et al., ""Luminescence imaging versus thermography on solar cells and wafers," in 26th EUPVSEC, 2011, pp. 1031–1038.

PV QUALITY ISSUES APPLYING BUILDING INTEGRATED PHOTOVOLTAICS (BIPV) ON THE FACADE AND ROOF WHEN DEEP RENOVATING A 50 YEAR OLD APARTMENT BUILDING

THE 35TH EU PVSEC 2018

Norwood, Zack[1] and Andersson, Anne[2]
[1]Göteborgs Universitet, Earth Sciences, Box 460, 405 30 Göteborg, E-mail: donkey@berkeley.edu
[2]RISE Research Institute of Sweden, Measurement Technology Company

ABSTRACT:
Kollektivhuset Stacken, a classic, 50-year old Swedish "million program" building with 35 apartments in Göteborg, which has long been run as a cooperative, has demonstrated a cost effective step by step deep retrofit to the international Passive House standard with a building integrated PV (BIPV) façade and roof. The existing concrete façade was externally insulated and covered with thin film amorphous silicon solar modules and more effective crystalline silicon solar modules were installed on the roof. New passive house certified doors and windows, demand-controlled ventilation coupled to a heat recovery unit and vacancy sensor based lighting was installed throughout the building. Finally, sensors to monitor the solar arrays, as well other parameters such as cold and hot water use, ventilation flows, electricity, temperature and humidity in each apartment were also installed.

The research project presented in this article is a collaboration between Göteborgs University, RISE, and Emulsionen that uses knowledge and data gathered from the project towards evaluation of the performance and reliability of the BIPV system.

Keywords: BIPV, renovation, apartment building, solar modelling

1 INTRODUCTION

Sweden has an ambitious political goal to be net zero greenhouse gas emissions (GHG) by 2045 which is in line with the EU's goal that the building sector as a whole should reduce GHG emissions by 80% by 2050. A determining factor in reaching these goals in Sweden is the "million program" buildings, a name given to the (mostly) prefab concrete buildings built in the 60s and 70s to reach a goal of one million additional homes[1]. These buildings are in great need of renovation today, and that poses both a problem and an opportunity[2]. Namely, there is a lack of cost effective renovation solutions that both reduce energy demand and generate renewable energy.

1.1 Background

Building Integrated Photovoltaics (BIPV) applied on roofs and façades of existing buildings during renovations is one such potentially cost-effective solution. If one looks at only the roof area for all buildings in Sweden (approx. 1100 km^2)[3], the contribution to total electricity production would be on the order of 80 TWh per year, which is more than half of Sweden's total electricity use. If this type of solar capacity was installed on the Swedish building stock gradually as roofs and façades were renovated the extra cost to society would be vanishingly small since the price of some solar modules (20-30 €/m^2) are now cost competitive with traditional façade and roof materials[4].

When BIPV is part of a thermal envelope renovation the energy use in the building can also be reduced by 50-80%, and up to 100% of the remaining energy demand can come from the BIPV. So, a BIPV product that can replace traditional building materials such as fiber reinforced concrete, sheet metal, bricks and fiber polymer boards opens up a large potential to improve energy performance of buildings (mostly through insulation and airtightness) and at the same time reduce the environmental impact of the electricity generation system as a whole (with PV). The challenge is to create a BIPV system that is inexpensive to install, fulfills the façade and roofs' functions without compromising the solar modules production, and has a low life cycle embodied energy.

1.2 The pilot project

Kollektivhuset Stacken, a 50-year old building with 35 apartments in Göteborg, Sweden, which has long been run as a cooperative, has demonstrated a cost-effective step by step deep retrofit to Passive House standard with a building integrated PV façade and roof.

The technical goals of the renovation were to reduce space heating demand to below 15 kWh/(m^{2}a) (i.e. to international Passive House Standard) and electricity consumption to approximately 30 kWh/(m^{2}a) while achieving a highly comfortable indoor climate with good air quality and comfortable surface and air temperatures.

To achieve these results the existing concrete façade was externally insulated and covered with thin film solar modules and more effective crystalline solar modules were installed on the roof. New passive house certified doors and windows, and demand-controlled ventilation coupled to a heat recovery unit have been installed in the building. Finally, sensors to monitor the solar arrays, cold and hot water use, ventilation flows, electricity, temperature and humidity in each apartment were installed. Installation of vacancy sensor based lighting is still underway and final ventilation adjustments have not yet been completed as of the time of publication of this article.

The pilot project was led by members of Kollektivhuset Stacken in partnership with Chalmers Tekniska Högskola, Passivhusbyrån, Helhetshus architect studio, Rockwool, i2 Smartwin, Passivhuscentrum, and Ekobanken. Financers for the construction include Naturskyddsföreningen, Energimyndigheten, and Västra Götalandsregionen. Technical support has been provided through inclusion in EuroPHit, a Passive House project of the European Union. Figure 1 shows the completed building in its surroundings next to identical unrenovated buildings on the same street.

Figure 1: Kollektivhuset Stacken with BIPV façade and roof next to neighboring unrenovated "star" buildings. Photo courtesy of Badenfelt and RISE.

This article aims to summarize the testing of solar modules used in the project, technical problems encountered with the BIPV system, and compare actual performance to modelled typical year projected performance. A financial review of this retrofit BIPV concept is also presented.

2 METHODOLOGY

The analysis presented here is a comparison between the modelled and measured production data from the BIPV array, as well as results of IV-curve efficiency testing at STC (so called "flash testing") of the modules. Although the system began producing power in October 2017, and one calendar year is used for simulating performance, measured data from the PV system is only available for a period of about 3 months from the middle of May until the middle of September, 2018, due to a variety of problems with metering the system. Additionally, due to various equipment and design problems including reorganization of the layouts the solar arrays, the arrays did not reach their final state until the beginning of September, 2018. Therefore, the time period with metered production data includes periods when parts of the roof solar array were out of commission.

2.1 The building model

The building energy model is done in the Passive House Planning Package (PHPP)[5] and the solar energy production model in System Advisor Model (SAM)[6] with a 3D modelling scene developed based on a combination of satellite images and on-site measurements of shading objects. Solar irradiance data is from PVGIS5 including calculated horizons for the site[7].

The facade modules are simulated using the IEC 61853 diode model in SAM with module parameters populated from PVSyst, whereas the roof modules are modelled using the CEC model (with user entered parameters from the manufacturer datasheet) in SAM.

21 blocks of modules with expected production of a maximum of 6kW DC are attached to 21 DC-DC optimizers, each with an MPPT, and modelled as separate cases in SAM, each with an appropriate 3d shading scene. The number of strings per optimizer on all but the north facade is 11 on the upper facades, and 13 on the lower facades. The north façade has 24 parallel strings connected to one DC-DC optimizer. To conform to the input requirements for SAM each of the 6kW DC-DC optimizers is modelled as a 6kW inverter, and each string is modelled separately in the 3D scene. In reality, the output from the 21 DC-DC optimizers is transmitted over a 760V bus to a central inverter in the building, with a capacity of 56 kW AC. A typical façade string layout is shown in figure 2.

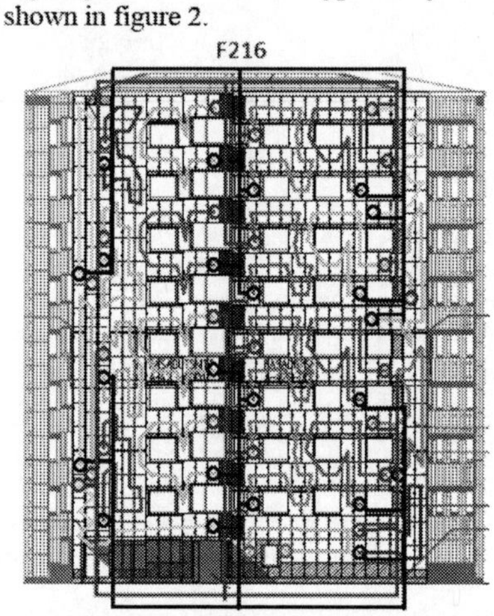

Figure 2: String layout of the southwest facing façade.

2.2 Flash testing of the solar modules

Amorphous silicon GS Solar GS-50 frameless laminated glass modules were chosen for the facades due to their low price per square meter (20 – 30 €/m^2) and the added benefit of having a uniform appearance similar to a standard glass façade.

A combination of monocrystalline silicon Enfinity 185M5-L frameless single glass modules (BIPV) and polycrystalline silicon PPAM Paladium 325 framed modules (BAPV) were chosen for their higher efficiency on the roof where solar irradiance per area is also higher.

The BIPV photovoltaic modules were characterized in a pulsed sun simulator Endeas QuickSun 560Ei class A+,A,A+, according to IEC 60904-1 chapter 4.2. The sun simulator (flasher) was calibrated using the short circuit current on a reference module before the measurements. The reference module was calibrated at Fraunhofer ISE traceable to World PV Scale. The results were corrected to STC standard test condition according to IEC 60891.

3 RESULTS AND DISCUSSION

Table I: GS Solar GS-50 façade module testing.

Facade modules	I_{sc} (A)	I_{mp} (A)	V_{oc} (V)	V_{mp} (V)	P_{mp} (W)
average 10 modules	1,58	1,35	64,2	49,2	66,4
Std. dev. 10 modules	0,07	0,04	0,55	0,55	2,5
max	1,7	1,4	65,6	50,1	72,6
min	1,5	1,3	63,8	48,2	64,8
Deviation from marking	11%	15%	4%	14%	33%

Table II: Enfinity 185M5-L roof module testing.

Roof modules	I_{sc} (A)	I_{mp} (A)	V_{oc} (V)	V_{mp} (V)	P_{mp} (W)
average 19 modules	5,38	5,03	44,2	35,8	180,4
Std. dev. 19 modules	0,07	0,08	0,21	0,23	2,1
max	5,48	5,11	44,6	36,4	184,9
min	5,16	4,75	43,8	35,4	177,5
Deviation from marking	-1%	-1%	-2%	-1%	-2%

3.1 Flash Testing Results

The results of the flash testing (Tables I and II) show that the monocrystalline modules were within the tolerance specified by the manufacturer (3%), but the amorphous silicon modules were significantly above that. One probable explanation for this discrepancy is that the a-Si modules were new and had not been through the rapid "burn in" period that usually occurs during the first year.

Additionally, no spectral correction was applied which can cause an increase of approximately 25% in P_{max} for a-Si but only 1% for mono-Si.

The GS Solar modules was also measured with covers to simulate clamps covering the active PV material at 4 places around the frame. No effect on the IV curve nor on P_{max} was noted.

3.2 Modelled compared to actual results

This project has broken new ground in Sweden by demonstrating that renovations of existing apartment buildings to passive house standard with BIPV facades can be done in an economical way. As such the project has faced much difficulties in both holding timelines and finding builders with the competence needed. Additionally, working with local solar equipment manufacturers has lead to a long debugging process and large amount of down time and decreased production from the solar arrays.

A lack of quality control by the builders and incomplete open circuit voltage tests during installation have made it difficult to debug production problems. Additionally, the fact that string layout has hidden most connections behind the solar panels has led to relatively costly skylifts for repairs.

The result of the first 9 months of solar production after commissioning in November, 2017 have been less than expected production. During the 3 months of June through August 2018, when measurement equipment was working, an approximately 33% reduced production was observed compared to a typical meteorological year (see figure 3).

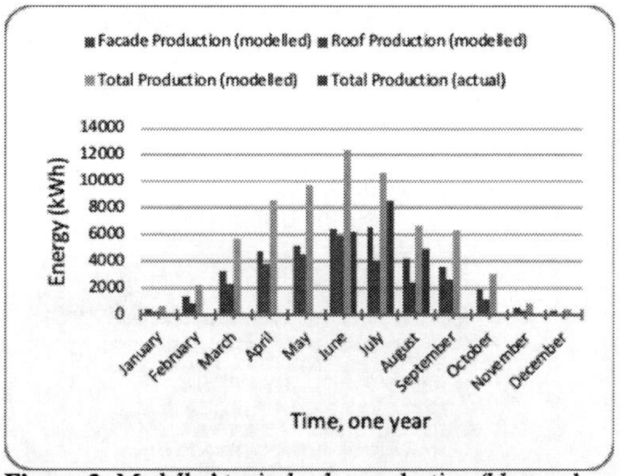

Figure 3: Modelled typical solar production (blue, red, yellow) per month "as installed" excluding parts of the array that were disconnected due to repairs, compared to measured production (green) during the summer.

The main cause to this reduction is believed to be the DC optimizer and inverters which have problems with overheating and with instability at high voltages and ambient temperatures. The equipment manufacturer is actively working on solutions to these problems.

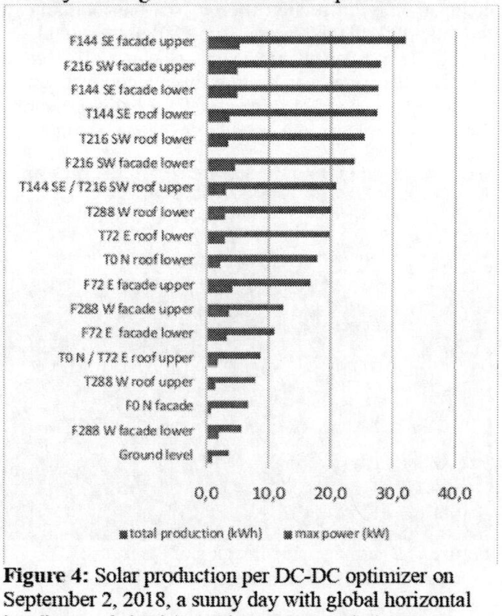

Figure 4: Solar production per DC-DC optimizer on September 2, 2018, a sunny day with global horizontal irradiance of 8kWh/m2. The DC-DC optimizers are named according to the facing direction and whether mounted on the upper or lower parts of the roof and façade. Note that total production was 315 kWh on this day.

Figure 4 shows measurements of the maximum power and energy production for a sunny day in September for each subarray. It can be seen in these results where the most and least productive subarrays are both in terms of peak power and total energy. Figure 5 shows then the comparison of modelled and measured production (both net and brutto) for the system as designed. It was not until the beginning of September, 2018 that the last subarrays were connected and put into operation.

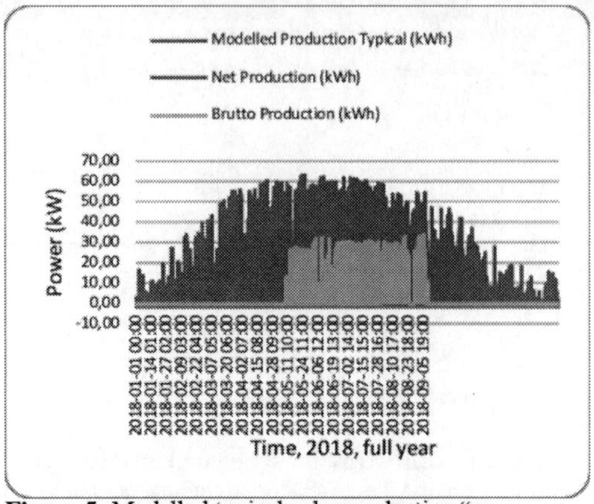

Figure 5: Modelled typical solar production "as designed" during a whole year (blue), compared to measured net production (red) and brutto production (orange) for 2018.

An attempt at using IR thermography with a drone to look for hot spots or incorrectly connected modules has not yet uncovered any results (see figure 6). We hope to continue this inspection nearer to the building to thoroughly debug the façade. The PID effect should also potentially be ruled out as a potential source of problems, although the panel manufacturer installation instructions do clearly state that there is no need for negative grounding for the amorphous silicon solar modules used in this project.

Figure 6: Thermographic imaging of Stackens BIPV facades and roof from a drone mounted IR camera.

4 CONCLUSION

Despite problems with the PV array, the concerted efforts to reduce electricity use by the inhabitants of this cooperative apartment building since the early 2010s has led to significant decreases (see figure 7) in purchased electricity culminating in the last year with the BIPV arrays completion.

With the fixes to the DC-optimizers and inverters the project should be able to attain production inline with the models.

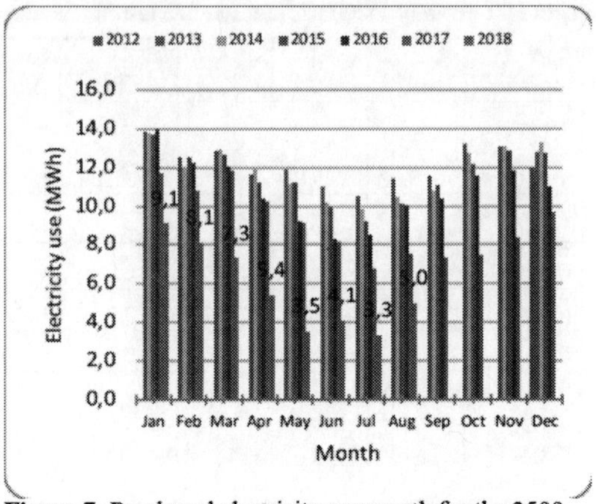

Figure 7: Purchased electricity per month for the 3500 m^2 apartment building from December, 2012 through August, 2018.

5 ACKNOWLEDGEMENTS

The authors would like to acknowledge the funding from Energimyndigheten, Västra GötalandsRegionen and Naturskyddsföreningen for enabling this project. This project was partly financed by the sale of "Bra Miljöval" electricity.

6 REFERENCES

[1] Nationalencyklopedin, miljonprogrammet. http://www.ne.se/uppslagsverk/encyklopedi/lång/miljonprogrammet (accessed 2018-01-07)
[2] Warfvinge, Catarina. "Mycket energi att spara i miljonprogrammet!." VVS-forum-värme och energi. No. April. 2008.
[3] Kamp, Sigrid. "Sveriges potential för elproduktion från takmonterade solceller: Teoretisk, teknisk och ekonomisk analys." Lund University. 2013.
[4] Sukurica, Amer. "Livscykelkostnad för tak och fasad." Högskolan i Halmstad. 2013.
[5] Feist, Wolfgang, et al. "Passive house planning package v. 9.3" Passivhaus Institut, Darmstadt, 2015.
[6] Dobos, Aron, and P. Gilman. System advisor model. Tech. Rep.(National Renewable Energy Laboratory, 2012) NREL Report No. TP-6A20-53437, 2012.
[7] Šúri, M.; Huld, T.A.; Dunlop, E.D. PVGIS: a web-based solar radiation database for the calculation of PV potential in Europe. International Journal of Sustainable Energy 2005, 24, 2, 55-67.

VERSATILE & LIGHTWEIGHT TRANSPARENT COMPOSITE TECHNOLOGY FOR ADVANCED BIPV ARCHITECTURAL SOLUTIONS

Jose M. Vega de Seoane, Oihana Zubillaga, Naiara Yurrita, Jon Aizpurua, Alex Astigarraga, Igor Arrizabalaga, Gorka Imbuluzketa, Maider Machado, Asier Sanz, Eduardo Román, Francisco J. Cano
TECNALIA RESEARCH & INNOVATION
Parque Científico y Tecnológico de Gipuzkoa, Mikeletegi Paselekua 2, 20009 San Sebastián (Spain)

ABSTRACT: Despite their yet low market penetration, building-integrated photovoltaic (BIPV) systems are expected to play a relevant role in the achievement of Nearly-Zero Energy Buildings (NZEBs) related targets. One of the main reasons explaining the low market deployment is the lack of versatility and adaptability of the currently dominating glass-based solutions, which present important restrictions at integration level.
With the ambition of overcoming those limitations and fulfilling architects' demands, TECNALIA R&I has developed Solarface, a transparent fibre-reinforced photovoltaic (PV) encapsulation technology that provides freedom in terms of design, lightness, virtually no shape limitations, mechanical strength and durability, at a competitive cost. The fibre-reinforced PV technology makes possible the fabrication of customized products for various applications: building, urban environment, transport, aerospace. etc.
The technology has been further developed and demonstrated in the FP7 funded BFIRST project (n° 290616), where portfolio of BIPV products were developed and real scale prototypes were manufactured and demonstrated. The development of the technology continues in EnergyMatching (H2020, n° 768766) where the influence of complex geometries on the cell-resin interface, aesthetic finishes and the introduction of additional functionalities to enhance performance are being explored.

Keywords: BIPV, lightweight, composite, architecture, nZEB

1 INTRODUCTION

Over the past decade, the expansion of the building-integrated photovoltaic (BIPV) market predicted by analysts bumped into a context that combined a lack of regulatory and standardisation framework, political support and cost-competitiveness. In addition, the BIPV sector also faced the difficulties derived from merging two different sectors, the photovoltaic sector, thoroughly focused on the manufacturing of standard PV modules and the implementation of economies of scale, and the construction sector, demanding aesthetics, customization, freedom in terms of design, and in general terms reluctant to the introduction of novel active solutions. In that context, the initial approach made by the solar industry was far from being optimal, a lack of customization was perceptible and building aspects were poorly addressed. As a consequence, market forecasts were overestimated, BIPV installations being mainly limited to niche flagship construction projects. Despite that adverse scenario, the BIPV sector has managed to evolve and has steadily consolidated its presence, proving that the bright future depicted by market experts is about to blossom. BCC Research estimated the worldwide installed BIPV capacity during 2016 to nearly 2 GW, confirming the growing trend, increasing by 12.6% compared to 2015, when 1.78 GW were installed worldwide [1].

The ability to integrate PV in the building envelope or other integrated applications (urban environment, transport, etc.) while still complying with requirements related to design, durability, mechanical stability and performance, to name a few of them, ultimately depends on the materials and manufacturing processes and their capacity to adapt themselves to cope with the demands of designers. In a creative context, the manufacturing of complex geometries, including curved shapes and highly customized parts, remains an essential requirement for designers. Ideally, the design of multifunctional PV components should be based on a dialog process, in which the PV design is adapted and customized to meet specifications of designers (shape, dimensions, aesthetics, etc.) rather than the opposite. However, the last situation stands out in most cases. The low customization and integration degree achieved with currently available PV encapsulation solutions forces designers to considerably adapt their approach accordingly, constraining the creative process.

Indeed, traditional PV modules are known to be limited to heavy, simple shapes, typically flat rectangular forms. Other encapsulation materials such as membranes, or textiles lack the required durability and mechanical resistance, and need to be reinforced with additional components (substrates, frames...), leading to poor integration strategies.

Given the constraints shown by conventional PV solutions for integration purposes, TECNALIA Research & Innovation has developed Solarface®, a novel encapsulation solution based on the use of transparent composite materials. The technology shows freedom in terms of design, allowing the manufacturing of fully customized elements in both geometry and size, and leads to lightweight and self-supported solutions, bringing cost reductions at different steps of the value chain (transportation, installation) and presenting advantages in applications demanding fully integrated solutions.

The technology is currently being further developed within the EnergyMatching project (H2020, N° 768766) [2], aimed at the development of new concepts and technologies to optimize the interaction between buildings and energy systems to maximize the RES harvesting in the built environment and district level. Concerning Solarface, the influence of complex geometries on the encapsulant material, such as the possible local stress or cell-resin interface failure are being explored in detail. Compatibility with different cell technologies, the improvement of the aesthetic potential of the technology and the introduction of specific functionalities to enhance BIPV performance complete the objectives of the project related to the composite encapsulation solution.

2 SOLARFACE® TECHNOLOGY

Solarface technology has been developed in-house by TECNALIA and is currently based on two patent applications owned by TECNALIA (WO2016/038000 and EP3006181) [3].

Solarface is a photovoltaic cell encapsulation technology that brings together the advantages of composite materials and those of photovoltaic energy. In hands of architects and designers, Solarface offers a great opportunity to go a step beyond architectural integration of BIPV, approaching the ideal situation in which material and manufacturing solutions are precisely adapted to meet specifications defined by designers, rather than the opposite, where the PV design rules and constrains the creative process.

Key features traditionally linked to composites such as lightness, excellent mechanical resistance, flexibility in terms of design and integrability coexist with the electricity generation, adding a major functionality to final applications. In this approach, the PV cells are completely embedded by the composite, resulting in a monolithic structure offering structural capacity and protection from the environment.

2.1 Materials and manufacturing processes

The basic principle behind Solarface technology relies on embedding the solar cells in a stack of glass-fibres and a transparent resin with coupled refractive indices, thus reducing internal reflections at the front side of the cells. The manufacturing process is based on the adoption of traditional composite manufacturing processes such as vacuum infusion or resin transfer moulding (RTM).

In Figure 1 the vacuum infusion manufacturing process is summarised. A stack of fibres is placed onto the mould cavity, reproducing the exact geometry of the mould. Next the array of solar cells is positioned and electrically connected and finally the back side of the solar cells is covered with glass-fibres until the desired thickness is obtained. Different cell technologies can be used: standard crystalline silicon, back-contact cells, CIGS, etc. A sealant is placed all over the perimeter of the mould and a peel-ply is used to easily remove the auxiliary materials required for the manufacturing (vap, nets, vacuum bag, etc.). Once the vacuum bag has been perfectly glued to the sealant, vacuum is applied and the transparent resin is injected in the cavity, completely impregnating the glass-fibre reinforcement and the solar cells. The manufacturing process ends when the resin has fully filled the cavity. Once the infusion is completed, the piece is left into the mould until the curing phase is finished, in order to provide the module with the resin's optimal mechanical and aging properties.

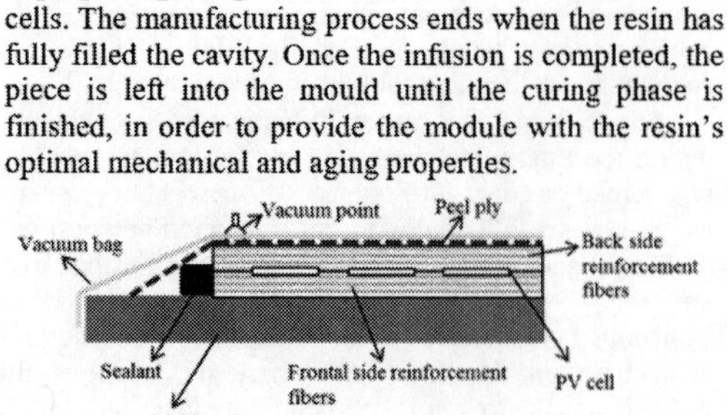

Figure 1: Diagram of the Solarface vacuum infusion manufacturing process

Mechanical and optical performance is optimized through simulation and reinforcement lay-out (number of fibre layers, fabric waving weight) in the back of the cells, together with a thorough control of vacuum, pressure and temperature during the manufacturing process, including curing of the composite material. The resulting elements are monolithic, self-supporting and lightweight (about 30% lighter than glass for an equivalent thickness).

The available range of fibre-reinforcements, resins, coatings and finishes in the market is wide and allows tailoring the properties of final products according to end-users' specifications and final applications. The basic configuration of Solarface technology, designed to comply with the PV standard IEC 61215, fulfils the following requirements:

- Chemical compatibility and stability between composite and photovoltaic cells
- High transparency in the front side
- Excellent adhesion between fibres, resin and PV cells
- Lightweight, adaptable depending on each case (5.2 kg/m^2 for a 3mm thickness module)
- Resistance against moisture
- Excellent integration properties (self-supported, machinable, use of inserts during manufacturing, etc.)

Concerning finishes, solar cells are further protected with transparent coatings that provide protection against UV radiation and humidity. Additional degree of protection can be achieved depending on the requirements of final applications (urban furniture, consumer goods, etc.) by using tailored solutions, such as the use ultra-thin glass, which leads to improved abrasion resistances.

The aesthetical appearance of the PV module can be customized using coloured fibres, veils, paints, and textured finishes, offering a broad range of possibilities to architects. Figure 2 shows some of the results of the work carried out so far in the EnergyMatching project.

Figure 2: Composite minimodules using a black veil and green fibre, as a result of the work carried out in the EnergyMatching project.

2.2 Photovoltaic performance

An optimisation roadmap of the encapsulation technology has been systematically implemented during the past years, encompassing measures focused on the enhancement of the PV performance, the module durability and cost reduction, among others.

Concerning PV performance, the strategy was based on the analysis and optimisation of the optical transmittance of the front side of the encapsulation material, which ultimately depends on the number of glass-fibre layers and the transparency of the resin and protective coating. A minimum number of layers needs to be used in the front face to avoid cell breakage during manufacturing, thus setting constrains in that PV performance optimization.

The photovoltaic performance of Solarface technology has been investigated using different number of glass-fibres and grammage typologies on the front face: 162, 300 and 600 gr/m^2. For performance comparisons, the thickness of the encapsulant on top of the cells was kept constant to the possible extent. Under those conditions, the fibre grammage typology used was not observed determinant, being the obtained differences in power very small in all cases (Fig. 3). The cell-to-module (CTM) values were determined by means of I-V characterisation. A set of individual c-Si cells where characterised before and after encapsulation, leading to CTM values in the range of 92-94% for mono-crystalline cells.

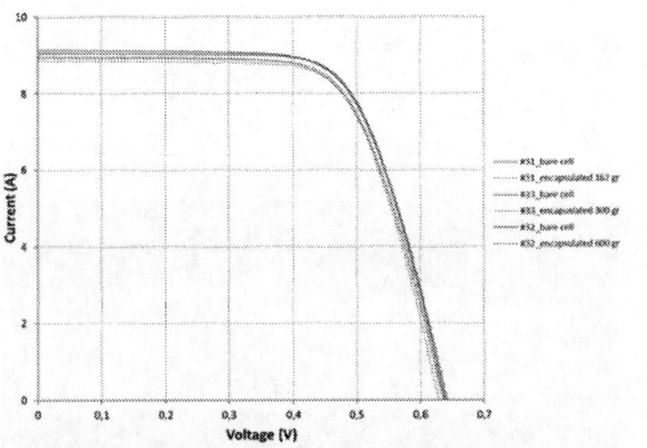

Figure 3: I-V curve of bare cells and after encapsulation with grammages of 162,300 and 600 gr

Concerning durability, the aging test sequences defined in the standard IEC 61215 (UV, thermal cycling, damp-heat and humidity-freeze) were successfully passed.

3 ARCHITECTURAL POTENTIAL OF SOLARFACE TECHNOLOGY

Solarface technology can be a very powerful tool for architects, offering them the possibility to conceive, develop and manufacture fully customized building elements. Unlike architectural glass, which is typically used in rectangular flat format, Solarface allows addressing the manufacturing of complex geometries, including curved ones. Flat and curved transitions can be combined to create elements based on 'saw-tooth' shapes leading, for example, to equivalent aesthetics as those commonly found in standard residential roofs, where flat shingles are overlapped from bottom to top, while concurrently simplifying the installation phase.

3.1 BIPV product examples

Since 2012, a wide range of products have been designed, prototyped and demonstrated using Solarface technology. In BIPV applications, roofing shingles, ventilated facades, shading elements and skylights elements have been developed. A few of them are described in detail herein:

3.1.1 Roofing tiles and shingles

Two roofing elements have been developed so far, using different approaches. The first design was inspired on the curved ceramic tile commonly found in the Mediterranean area. The design was adapted to include two flat surfaces to allow the integration of 20 standard c-Si solar cells, leading to a 1020 x 890 mm panel of 80 Wp and 5 kg of weight.

Figure 4: Lightweight roofing tile based on Solarface technology (SLIGHT project).

A different approach was used in the BFIRST project (FP7, 290616) [4]. In this case, the aim was to replicate the aesthetical aspect achieved with flat shingles. Taking advantage of the Solarface technology, a curved saw-tooth design was implemented to reduce the required number of shingles and overlaps, thus reducing installation costs (Fig. 5). The final design led to a 1395 x 905 x 3 mm roofing shingle and 6.5 kg weight, with 32 solar cells, leading to a power of 132 Wp.

Figure 5: Saw-tooth shaped lightweight roofing shingle

3.1.2 Ventilated façades

Also in the frame of BFIRST project, two ventilated façade systems were developed and demonstrated. The first design (Fig.6) was a flat element with curved transitions of dimensions 1652 x 1209 x 3 mm, 10 kg of weight and 230 Wp of power.

Figure 6: Flat with curved transitions, podium-shaped, ventilated façade element

On the other hand, a standard passive aluminium ventilated façade element from the Spanish company Astrawall was transformed into an active BIPV element by integrating a 1.8 mm thickness and 3 kg of weight composite PV panel based on Solarface technology. The module integrated 20 c-Si solar cells leading to a total approximate power of 80 W.

Figure 7: Integration detail of the composite-aluminium BIPV ventilated façade.

4 CASE STUDIES

4.1 Residential house in Mons (Belgium)

A new residential house in Mons (Belgium) was used for the demonstration of the roofing shingle developed in the BFIRST project. It is a single family detached house, located in Masnuy Saint-Jean, owned by Optimal Computing [5]. The house was designed following Passivhaus principles and a related assessment was done by PHPP (Passivhaus Institute Darmstad), obtaining an A+ level PEB classification.

The BIPV installation was performed with the saw-tooth roofing shingle described in section 3.1.1. A total of 57 units (70 m2) were manufactured and installed on the roof, leading to 7 kWp installed power. The roof is oriented SSW and has a 40° tilt. Two inverters SMA Sunny Boy 4000 TL where installed, connecting 30 and 27 modules each, respectively.

The integration solution was solved using Altuec FL horizontal profiles. A 'U'-shape profile was glued to the roofing shingle's back surface and EPDM joins were used to protect the edges of the panel in contact with the aluminum profiles, as described in Figure 8.

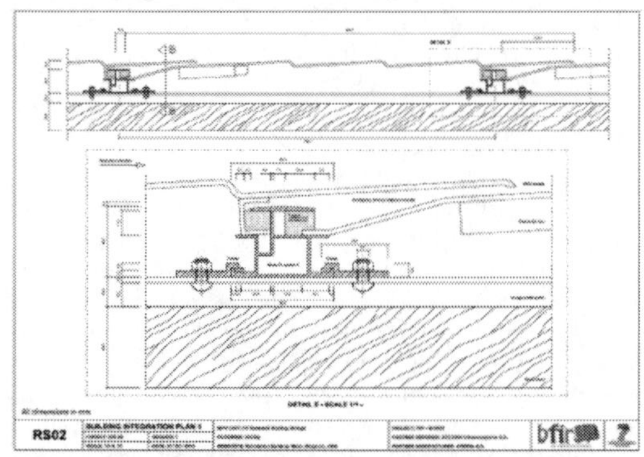

Figure 8: Integration solution implemented

The installation was commissioned during April 2016 (Fig.9) and the monitoring system is gathering data since July 2016.

Figure 9: Demonstration result, where 57 roofing shingles based on the Solarface technology were installed in Mons (Belgium)

The thermal performance and monthly energy generation are described in Figure 10 and 11, respectively. For the thermal performance assessment, three hot and sunny days were chosen (02/08/2018 - 04/08/2018), with maximum ambient temperatures in the range of 30-32 ºC and maximum peak radiation levels close to 1000 W/m^2 on the plane of the array (POA). Wind levels those days where soft, with average wind speeds of 1-1.5 m/s. Under those ambient conditions, the maximum module temperature was 68.9 ºC, in good agreement with the expected temperature of a PV system given by the NOCT model [6]. In BIPV applications, the maximum temperature should typically be higher due to the limited ventilation inherent to integrated solutions, which proves that the implemented integration strategy lead to a good thermal ventilation for the whole system.

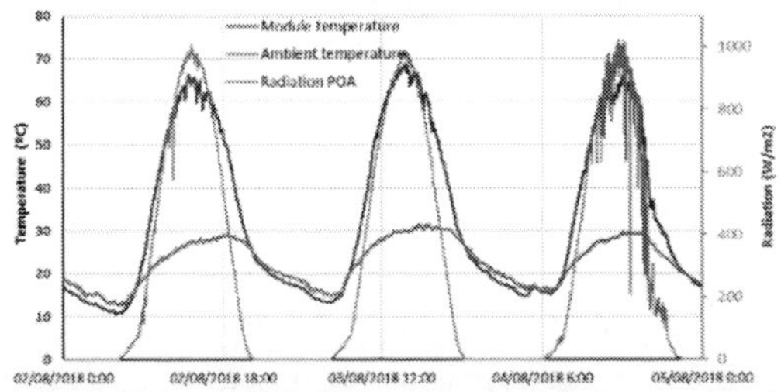

Figure 10: Ambient temperature, radiation and module temperature profile during 02/08/18 to 04/08/18

Regarding energy generation, the performance of the photovoltaic system has been excellent so far, with a total energy generation of 19,245 kWh until September 19th and an estimated Energy Yield of 1025 kWh/kWp for 2017. During the monitoring analysis, a problem with the monitoring data was identified during the December 2017-February 2018 period, which explains the difference between the same period of 2017. The reasons are strictly attributable to lost monitoring data, and no to performance issues.

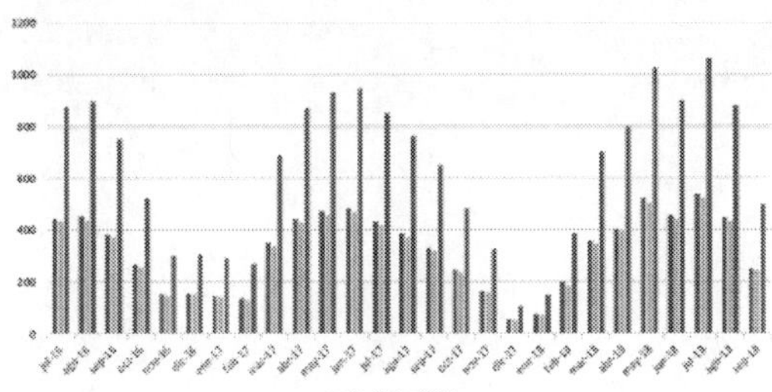

Figure 11: Total energy generation of the BIPV system from July 2016 to September 19th.

4.2 INGRID Smart Grids Building (Spain)

INGRID is the name of TECNALIA's experimental facilities for smart grids, located in Zamudio (Spain). It is composed by two buildings, used as offices and testing laboratories, respectively. A ventilated façade using 34 composite-aluminium BIPV modules was installed (2.42 kWp), placed on the free concrete space under the testing laboratories (SSE orientation).

Figure 12: On the bottom left side, ventilated façade installation carried out with the composite-aluminum module described in section 3.1.2.

5 OTHER PV-INTEGRATED APPLICATIONS

Apart from BIPV, Solarface shows an excellent potential for the integration of multifunctional PV elements into applications such as, urban environment, transport, aerospace, etc. During the past years, TECNALIA has developed several products integrating composite photovoltaic elements: solar streetlights, roofing elements for electric-vehicles and unmanned aerial vehicles, among others.

In Figure 13, FASCOM streetlight is presented, a compact solar streetlight using a dome-shaped composite module integrating flexible c-Si solar cells from Clean Venture 21 [7,8]. The solution allows a flexible power supply mode due to the possibility of feeding the luminaire by means of 3 different sources: (1) domed PV module which is part of the streetlight, (2) associated battery and (3) the electrical grid.

Figure 13: Installation of two FASCOM luminaires in Barcelona

Similarly, a curved CIGS composite module was manufactured for an electric tricycle. The integration took place in two steps. First, the curved composite PV module was manufactured and, in a second step, by means of traditional composite manufacturing processes the final part was manufactured, integrating the PV module during the process.

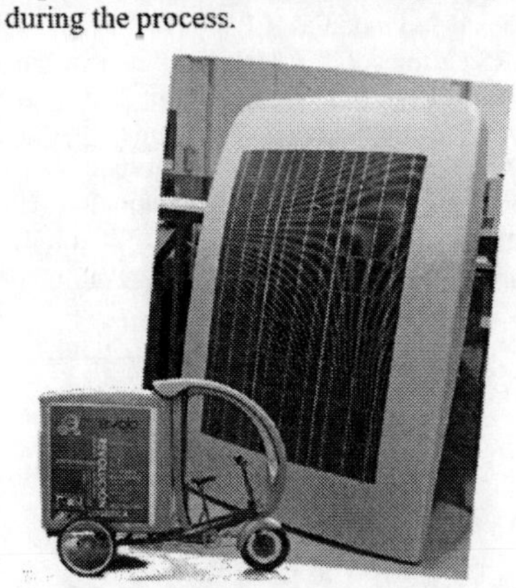

Figure 14: 'Evolo' roofing element for an electric tricycle

6 CONCLUSIONS AND FUTURE WORK

The potential and achievements obtained so far with Solarface technology have been presented. The composite encapsulation solution provides freedom in terms of design, lightness and excellent integration possibilities, going a step further in the deployment of

BIPV solutions. In addition, the technology has the adequate features to successfully address other markets in which electricity consumptions are needed and fully integrated solutions may be pursued, such as urban environment, transport or aerospace sectors.

Concerning future steps, as previously mentioned, the optimisation of the technology continues in the EnergyMatching project, where the study of the influence of complex geometries on the cell-resin interface, the improvement of the aesthetic potential of the composite and the introduction of additional functionalities to enhance performance are pursued. In parallel, TECNALIA's efforts are currently on the manufacturing side, where the impressive cost reductions achieved with automated composite processes (RTM and HP-RTM) are being investigated.

7 ACKNOWLEDGMENTS

The authors want to thank the European Commission for their support on the Solarface technology. This work has received funding from the European Union's Seventh Programme for research, technological development and demonstration under Grant Agreement No 296016, and from European Union's Horizon 2020 Research and Innovation Programme under Grant Agreement No 768766.

8 REFERENCES

[1] BCC Research, EGY072C - Building-Integrated Photovoltaics (BIPV): Technologies and Global Markets, December 2016.
[2] EnergyMatching project (H2020, Grant Agreement Nº 768766). Website: www.energymatching.eu/
[3] Fundación Tecnalia Research & Innovation (2018). Method of encapsulating photovoltaic cells and encapsulated modules. EP3006181 (A2).
[4] BFIRST project (FP7, Grant Agreement No 290616). Website: www.bfirst-fp7.eu; http://bfirst-products.eu/
[5] Optimal Computing: www.optimalcomputing.be
[6] R. G. Ross, "Flat-Plate Photovoltaic Array Design Optimization", 14th IEEE Photovoltaic Specialists Conference. San Diego, CA, pp. 1126-1132, 1980.
[7] http://www.innoenergy.com/innovationproject/our-innovation-projects/fascom/
[8] Clean Venture 21:http://www.cv21.co.jp/en/index.php

MOSAIC MODULE CONCEPT FOR COST-EFFICIENT AND AESTHETIC BIPV MODULES

Max Mittag, Matthieu Ebert, Helen Rose Wilson, Tobias Fellmeth
Fraunhofer Institute for Solar Energy Systems ISE, Heidenhofstr. 2, 79110 Freiburg, Germany
max.mittag@ise.fraunhofer.de

ABSTRACT: We present a module concept based on metal-wrap-through solar cells and conductive copper backsheets that offers a high degree of aesthetic freedom and allows individual designs. Our metal-wrap-through cell design features sub-cells that allow splitting and individual cell sizes enabling a flexible module design. We show results from sample manufacturing and façade mock-up implementation. The manufacturing is performed using automated equipment and low-temperature solder pastes. The process step of "tabbing/stringing" can be omitted compared to conventional PV module manufacturing. We perform IV measurements and a cell-to-module loss analysis. Results show that, considering the aperture area of the cells, competitive cell-to-module power ratios (~95%) are achieved. We estimate the mosaic module costs to be between 57 and 68 €/m² (depending on specific design) compared to 45 €/m² for common industrial, utility scale solar modules.

Keywords: photovoltaic module, Building Integrated PV, BIPV, MWT, electrically conductive backsheets, low temperature solder paste, flexible module design

1 INTRODUCTION

The success of Building-Integrated Photovoltaics (BIPV) strongly depends on the appearance of modules in façades and building envelopes. In addition to aesthetic demands, a high module efficiency and a cost-effective production are necessary [1–3]. These requirements usually represent a target conflict [4, 5].

We develop a module concept which offers design flexibility, high efficiency potentials as well as low production costs [6].

The mosaic module concept is based on metal-wrap-through (AP-MWT) cells [7]. The MWT solar cells are manufactured from common 6" wafers but are designed to be separated into smaller pieces (Figure 1).

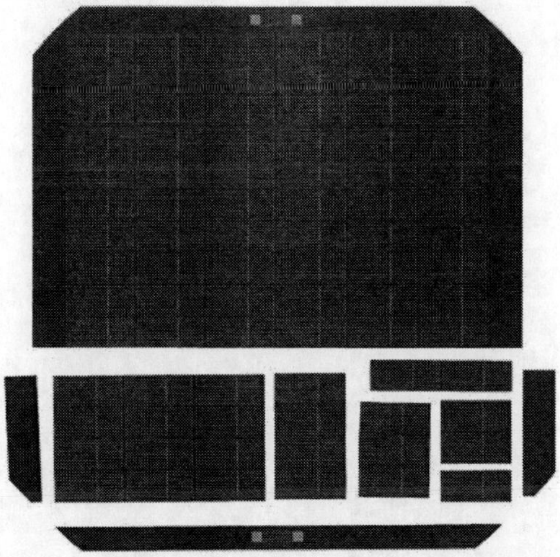

Figure 1: AP-MWT solar cells designed to be separated

The module includes a structured, electrically conductive backsheet that allows flexible positioning of solar cells. The smaller cells together with the variable configuration and different color of the backsheet result in a highly customizable module design (Figure 2).

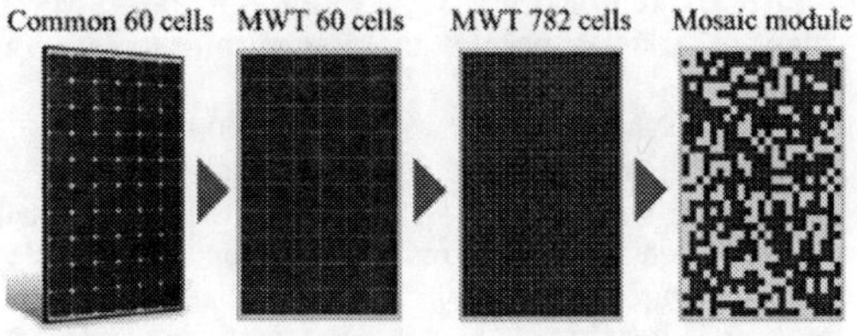

Figure 2: Design flexibility of the mosaic module concept

We apply a cell interconnection based on low-temperature solder paste and a stencil printing process. The suitability for BIPV applications is demonstrated with flexible designs. We present several prototypes of different appearance. A Cell-to-module (CTM) analysis and IV measurements as well as cost estimations are performed.

2 THE ALL-PURPOSE-CELL CONCEPT

The mosaic module concept is based on metal wrap-through (MWT) solar cells [7–9]. These cells are back-contact solar cells, where the electrical front contact is connected to the rear side through laser-drilled holes. Electrical cell interconnection is located on the cell rear surface reducing the shading by interconnector ribbons and thus allowing higher cell power. The visual homogeneity of the solar cell is improved, which is desirable for BIPV modules. The 6" wafers are designed to allow separation into solar cells of different sizes (Figure 1).

The cell concept is named "all-purpose MWT" (AP-MWT) to reflect flexibility. The variability in cell sizes is achieved by adapting the front and back metallization to smaller cell elements providing contact pads for each of those elements (Figure 3). A full-size AP-MWT cell can be separated into sub-cells with a minimal size of 22.5 x 10 mm².

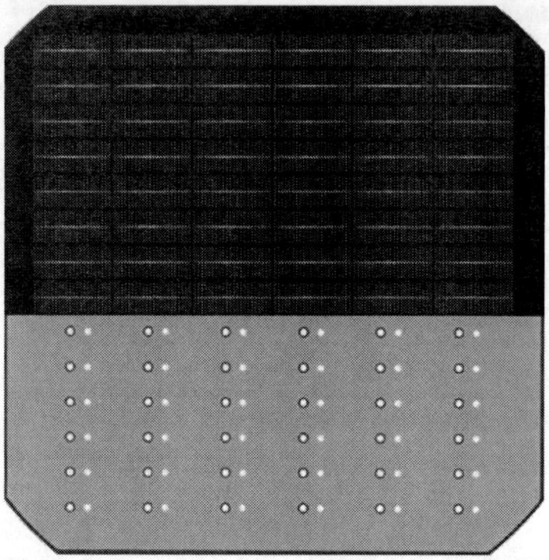

Figure 3: Front and rear view of an all-purpose MWT solar cell for low-concentrating PV applications

The AP-MWT cells can be optimized for variety of different operation conditions such as for facade applications with reduced irradiance due to non-optimal module orientation, for low-concentration PV modules (LCPV, Figure 4) [9] with increased irradiance as well as for device integration [8].

Figure 4: LCPV receiver with AP-MWT solar cells

To determine the optimal cell design for the application in the mosaic module concept, we perform a simulation of the cell efficiency as a function of the irradiance and the number of metal fingers on the front side using Fraunhofer ISE GridMaster [9–11].

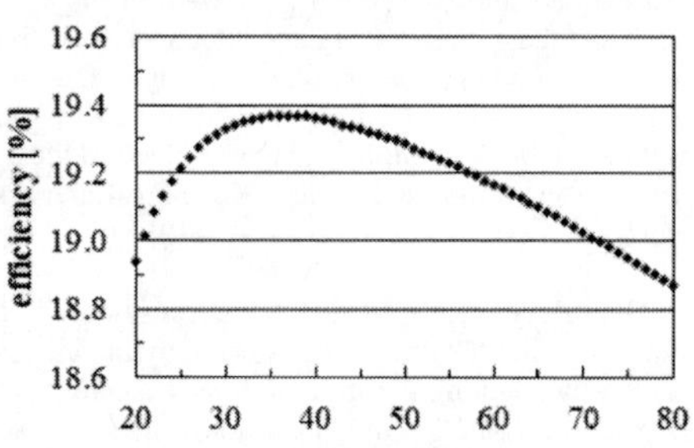

Figure 5: AP-MWT cell efficiency as a function of the number of metal fingers at standard testing conditions calculated using the software GridMaster

A cell design has been specified for mosaic module prototyping with a size of 69.5 x 67 mm^2 featuring 12 pairs of n and p contacts for soldering (Figure 6).

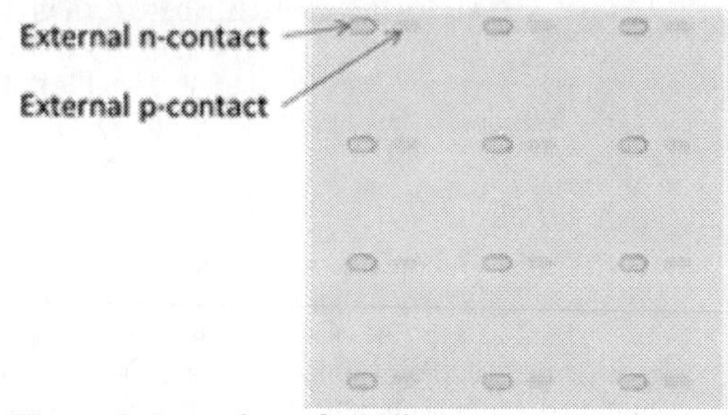

Figure 6: Rear view of an all-purpose MWT solar cell segment with 12 external n and p contacts

The solar cells for this study were manufactured at Fraunhofer ISE PV-TEC and achieve an average efficiency of 19.4% (cell manufacturing in 2015).

3 THE MOSAIC MODULE CONCEPT AND MANUFACTURING

The mosaic module is based on the idea that smaller solar cells provide a higher degree of design freedom for BIPV applications (Figure 2) and that individually structured electrically conductive backsheets (ECB) allow flexibility in solar cell and bypass diode positioning (Figure 7, Figure 8) [12].

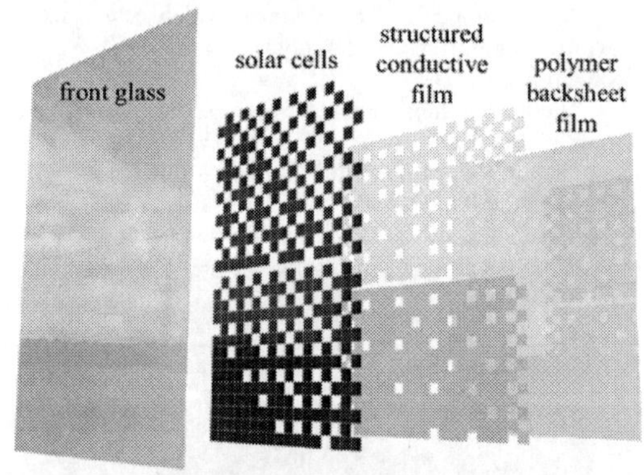

Figure 7: Mosaic module configuration

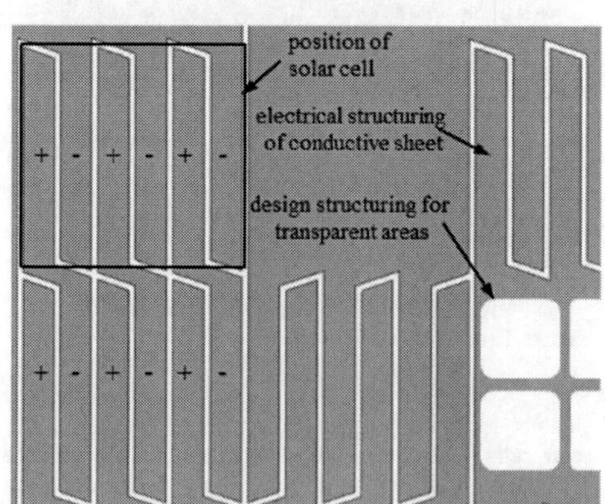

Figure 8: Electrically conductive copper sheet design for flexible cell positioning (example)

The structure of the backsheet provides separate positive and negative electrical contacts. Different structuring of the electrically conductive copper sheet and additional interlayers allow a semi-transparent or colored module appearance (Figure 9, Figure 10).

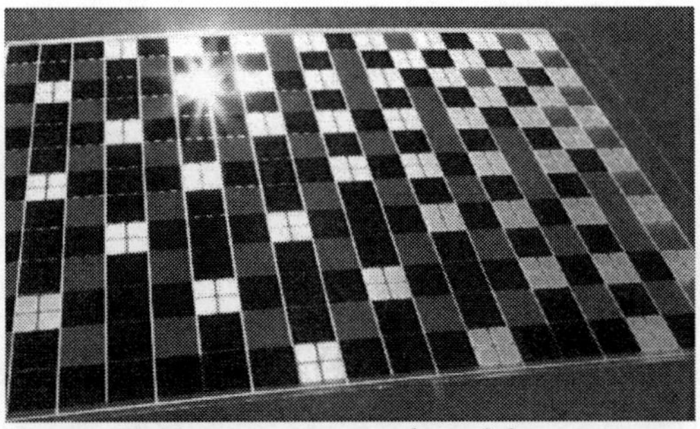

Figure 9: Partly transparent mosaic module prototype

Figure 10: Façade mock-up of mosaic PV modules with different designs and solar cell coverage; top module: 119 AP-MWT cells (38.9% of area photovoltaically active); bottom module: 167 AP-MWT cells (55.0% of area photovoltaically active).

We manufactured modules of different designs with a size of 1460 x 975 mm² (1.42 m²). The first design included 119 AP-MWT cells and the electrically conductive backsheet was combined with a transparent polymer film to allow translucent module areas (Figure 9). The second design includes 167 AP-MWT solar cells and a white polymer film resulting in a totally opaque module (Figure 10). Both modules use 3.9 mm low-iron glass without anti-reflective coating (ARC) and AP-MWT cells with an area of 69 x 67.5 mm² and an average efficiency of 19.4%. Due to the different number of cells, each module configuration has different string lengths. The first design has four strings (34, 31, 27 and 27 solar cells) were the first and last two are connected in parallel and where these parallel blocks are then connected in series. The second design features four strings (47, 42, 37 and 41 cells) all connected in parallel. A series string interconnection and other electrical module topologies are possible but have not been implemented.

While the MWT cells and the ECB would also allow smaller cell spacing, we used a cell distance of 2 mm. The module designs have a relative active area of 38.9% and 55.0%, respectively. Commercially available SnBi58 low-temperature solder paste (LTSP) with 10% flux and EVA encapsulant were used for soldering and lamination, respectively. Cell separation and module manufacturing were performed at Fraunhofer ISE Module-TEC using automated equipment (Figure 11).

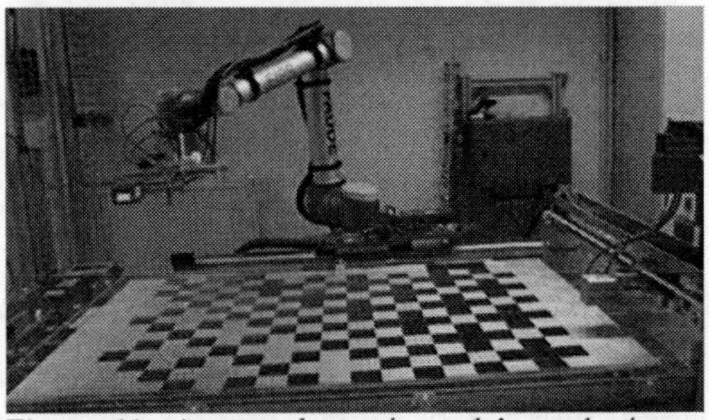

Figure 11: Automated mosaic module production at Fraunhofer ISE Module-TEC

The paste is applied onto the back of the cells using stencil printing before lay-up. The cells with paste are then placed on the electrically conductive backsheet using a six-axis robot system and visual control to increase accuracy in positioning (Figure 12). The electrical cell interconnection forms during lamination using a typical lamination process profile, allowing fast and economic module production. We perform electroluminescence and x-ray inspection of the modules to verify the electrical interconnection between cells and the copper sheet (Figure 12, Figure 13).

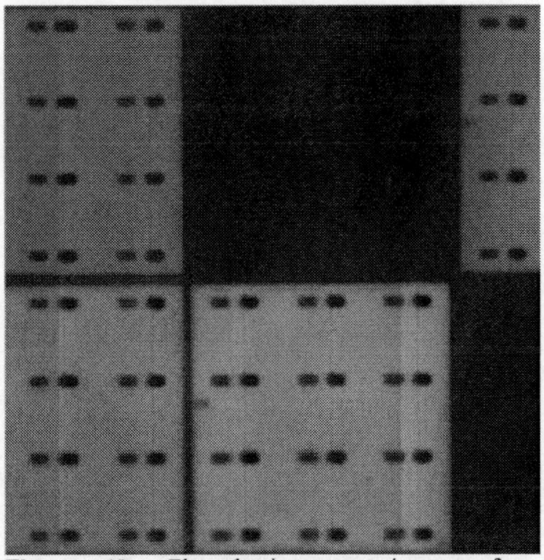

Figure 12: Electroluminescence image of a manufactured mosaic module

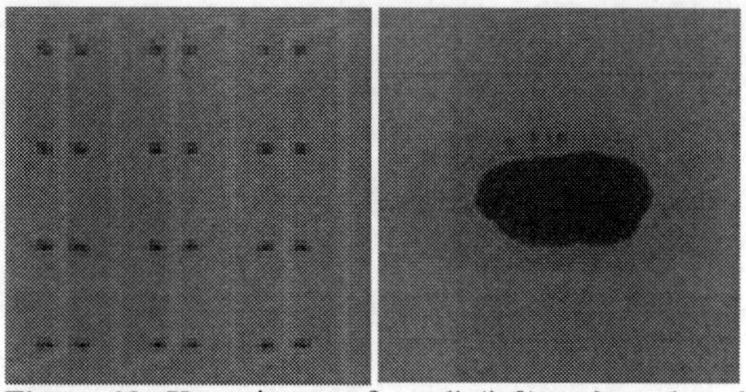

Figure 13: X-ray image of a cell (left) and a selected solder pad (right) showing low-temperature solder paste interconnection

4 POWER MEASUREMENT AND CELL-TO-MODULE (CTM) ANALYSIS

We measure the power of both modules at Fraunhofer ISE CalLab Modules and find the first design to have 102 Wp and the second to have 112 Wp. Due to the uneven string lengths, deviations between the electrical properties of the solar cells and the module topology (mixture of series and parallel string interconnection), significant electrical mismatch losses occur. These losses could not be avoided in the manufactured modules due to the design specifications of the façade mock-up (Figure 10). Specifications requested certain cell patterns, edge margins and mounting tools that made a disadvantageous string interconnection necessary. The focus of this study is on the proof of concept and to demonstrate the possibilities regarding module design and visual appearance. No attention was paid on avoiding mismatch losses but general design rules have been established that allow the reduction of such losses for future designs.

Using only the aperture area of the cells (cell area + 1 mm spacing), we calculate efficiencies of 18.2% for design-optimized versions of the first and 14.4% for the second module. We use the aperture area since the module area would be misleading due to the large inactive area share which is a desired part of the BIPV system design. CTM ratios are displayed in Table I.

Table I: module measurement results and CTM ratios

Module	1	2
Cells	119	167
Total cell power	107.2 Wp	150.5 Wp
Module power	101.7 Wp	112.3 Wp
Module efficiency	7.1%	7.9%
CTM_{power} (aperture area)	94.8%	74.6%
$CTM_{efficiency}$ (module area)	36.6%	40.7%
$CTM_{efficiency}$ (aperture area)	93.8%	74.2%
Active module area share	38.9%	55.0%

CTM analysis shows high losses in nominal efficiency due to the low share of active cell area within the module. CTM_{power} for module 1 is lower than for commercially available, common modules due to lower backsheet gains (copper instead of white backsheet), the lack of an AR coating of the front glass and electrical mismatch losses [13]. Nonetheless, the analysis shows that the mosaic module is capable of effectively generating power and that efficiency losses occur mainly due to aesthetic demands.

5 COST ESTIMATION

We estimate costs of the mosaic modules using the "SCost" model developed at Fraunhofer ISE [14]. The model is based on SEMI E35 and E10 standards for the calculation of costs of ownership. We assume a small industrial production line with 1835 hours manufacturing per year (90% factory uptime for a single working shift, 5 days per week, 51 weeks per year) and 53.5 modules per hour. The total output is approx. 110.000 modules per year.

Table II shows relevant module components and their prices as they were used in the cost estimation. The price assumptions are based on a literature review and market survey and include a surcharge for small order quantities. Costs for engineering and BIPV project management are not included.

Table II: Material prices used for cost estimation

AP-MWT solar cells	0.20 €/Wp
EVA encapsulant	1.00 €/m²
Glass	4.00 €/m²
electrically conductive backsheet	10.00 €/m²
low-temperature solder paste	150 €/kg
junction box	4.00 €/pcs

For cost calculation we assume a module design without mismatch losses (CTM_{power} = 95%). We calculate the module costs for design-optimized versions of design 1 (134 Wp) to be 80.92 € (60 €ct/Wp, 57 €/m²) and of module 2 (187 Wp to be 96.70 € (51 €ct/Wp, 68.1 €/m²). We find the costs higher compared to common industrial, utility scale solar modules (300 Wp, 0.25 €ct/Wp, 1.65 m²) which range at around 45 €/m². The costs of the mosaic module are higher due to the necessary copper sheet, the price of the low-temperature solder paste and the low area coverage of the solar cells (Figure 14). The share of the costs of the solar cells is lower than for conventional modules [13, 15]. Comparing our results to other custom-designed BIPV modules, we find the mosaic module to be competitive compared to other BIPV products [16, 17]. Costs for engineering, design or costs related to specific BIPV projects costs have not been considered.

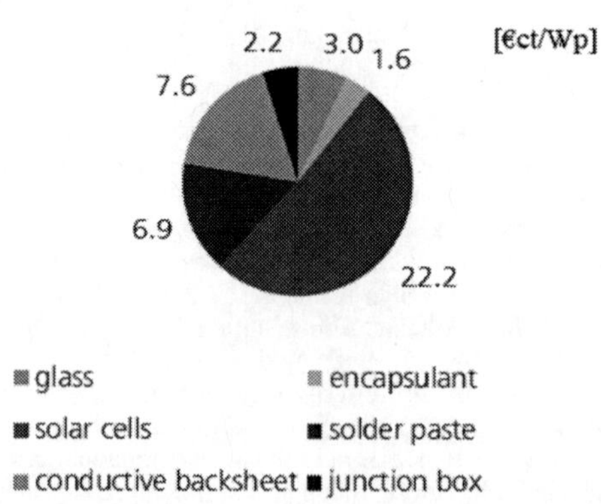

Figure 14: Total Cost of Ownership structure of the mosaic module design 2

6 SUMMARY

We present a module concept for building integration based on MWT solar cells and conductive backsheets that offers a high degree of aesthetic freedom and allows individual module designs. We show prototypes with different visual appearance and façade mock-up implementation. The manufacturing is performed using automated equipment and low-temperature solder pastes, avoiding the process step "tabbing/stringing" compared to conventional PV modules.

We perform a cell-to-module loss analysis. The results show that based on aperture area, competitive cell-to-module power ratios (~95%) are achieved.

We perform a cost analysis and find the mosaic module costs to be between 57 and 68 €/m² (depending on specific design), which is comparable or even lower than other published BIPV products.

7 ACKNOWLEDGEMENTS

This project has received funding from the European Union's Seventh Framework Program for research, technological development and demonstration under grant agreement No. 295981.

We would like to thank all project partners of ConstructPV.

REFERENCES

[1] F. Frontini *et al.*, "From BIPV to Building Component," 28th EUPVSEC, Paris, France, 2013.

[2] T. Minderhoud *et al.*, "Design Innovation from PV-Module to Building Envelope: Architectural Layering and non apparent Repetition," 29th EUPVSEC, Amsterdam, Netherlands, 2014.

[3] L. H. Slooff *et al.*, "An Architectural Approach for Improving Aesthetics of PV," 33rd EUPVSEC, Amsterdam, 2017, pp. 2110–2113.

[4] L. H. Slooff *et al.*, "An Architectural Approach for Improving Aesthetics of PV," 33rd EUPVSEC, Amsterdam, 2017, pp. 2110–2113.

[5] C. Kutter *et al.*, "Decorated building integrated photovoltaic modules: power loss, color appearance and cost analysis," in *35th EUPVSEC 2018*.

[6] *ConstructPV*. [Online] Available: http://www.constructpv.eu/.

[7] B. Thaidigsmann *et al.*, "P-type MWT solar cells: current status and future expectations," 28th EUPVSEC, Paris, France, 2013.

[8] T. Fellmeth *et al.*, "Industrially feasible all-purpose metal-wrap-through concentrator solar cells," 40th IEEE Photovoltaic Specialist Conference, Denver, 2014.

[9] M. Ebert *et al.*, "A low concentrating cell and receiver concept based on low cost silicon solar cells," Aix-les-Bains, France, 2015, 110001-1-110001-7.

[10] T. Fellmeth *et al.*, "Pseudo FF and Voc analysis of Cz-Si Based Low Concentrator Solar Cells," *Energy Procedia*, vol. 77, pp. 572–580, 2015.

[11] M. Ebert *et al.*, "Mosaic Modules for Improved Design Options in Building Integrated Photovoltaic Modules," Duesseldorf, 2016, pp. 415–422.

[12] A. Scognamiglio *et al.*, "Starting Point for a New Cultural Vision of BIPV," 23rd EUPVSEC, Valencia, Spain, 2008, pp. 3222–3233.

[13] ITRPV, "International Technology Roadmap for Photovoltaic (ITRPV): Results 2017, Ninth Edition," 2018.

[14] S. Nold *et al.*, "Cost Modelling of Silicon Solar Cell Production Innovation Along the PV Value Chain," 27th EUPVSEC, Frankfurt, Germany, 2012.

[15] M. Mittag *et al.*, "TPedge: progress on cost-efficient and durable edge-sealed PV modules," 33rd EUPVSEC, Amsterdam, 2017, pp. 48–54.

[16] S. Krawietz *et al.*, "Building Integrated Photovoltaics (BIPV) as a Core Element for Smart Cities: Integrated Research, Innovation and Competitiveness Strategies within the Energy union," 2016. Accessed on: Feb. 09 2018.

[17] University of Applied Sciences and Art of Southern Switzerland, "Building Integrated Photovoltaics: Product overview for solar building skins," 2017. Accessed on: Feb. 09 2018.

AUTHOR INDEX

AUTHOR INDEX

AUTHOR INDEX

AUTHOR INDEX

AUTHOR INDEX

AUTHOR INDEX

AUTHOR INDEX

AUTHOR INDEX

AUTHOR INDEX

AUTHOR INDEX

AUTHOR INDEX

AUTHOR INDEX

AUTHOR INDEX

AUTHOR INDEX

AUTHOR INDEX

AUTHOR INDEX

AUTHOR INDEX

AUTHOR INDEX

AUTHOR INDEX

WIP – Renewable Energies
Sylvensteinstr. 2
81369 Munchen
Germany

ISBN 978-1-5108-7155-7

9 781510 871557

CURRAN ASSOCIATES INC.
proceedings.com

35th European Photovoltaic Solar Energy Conference and Exhibition (EU PVSEC 2018)

Brussels, Belgium
24-28 September 2018

Volume 3 of 3

WIP

35th European Photovoltaic Solar Energy Conference and Exhibition (EU PVSEC 2018)

Brussels, Belgium
24-28 September 2018

Volume 3 of 3

Editors:

P. Verlinden
P. Helm
R. Kenny

ISBN: 978-1-5108-7155-7

Printed from e-media with permission by:

Curran Associates, Inc.
57 Morehouse Lane
Red Hook, NY 12571

Some format issues inherent in the e-media version may also appear in this print version.

Printed by Curran Associates, Inc. (2019)

WIP – Renewable Energies
Sylvensteinstr. 2
81369 Munchen
Germany

Phone: +49 89 72012735
Fax: +49 89 72012791

wip@wip-munich.de

Additional copies of this publication are available from:

Curran Associates, Inc.
57 Morehouse Lane
Red Hook, NY 12571 USA
Phone: 845-758-0400
Fax: 845-758-2633
Email: curran@proceedings.com
Web: www.proceedings.com

TABLE OF CONTENTS
of 35th EU PVSEC 2018 Proceedings Papers

Oral PRESENTATIONS 1CO.1 Advanced Material Development and Analysis for High Performance PV Modules

Oral PRESENTATIONS 1CO.2 Novel Approaches for Special PV Applications

Visual PRESENTATIONS 1CV.4 Fundamental Studies / New Materials and Concepts for Photovoltaic Devices

Silicon Cells

Visual PRESENTATIONS 2AV.2 Homojunction Solar Cells

Visual PRESENTATIONS 2AV.3 Heterojunction Solar Cells

Non Silicon-Based Thin Film Photovoltaics

Visual PRESENTATIONS 3BV.3 Perovskite, Organic and Dye-Sensitised Devices

Concentrator and Space Photovoltaics

Photovoltaic Modules and BoS Components

Visual PRESENTATIONS 5CV.1 PV Module Design, Manufacture, Performance and Reliability

Visual PRESENTATIONS 6CV.2 Design and Installation of PV Systems

PV Economics, Markets and Policies

MEASUREMENT OF SOLAR HEAT GAIN COEFFICIENT FOR SEMI-TRANSPARENT BUILDING-INTEGRATED PHOTOVOLTAICS IN THE TROPICS

Veronika SHABUNKO[1*] and Thomas REINDL[1]
[1]Solar Energy Research Institute of Singapore (SERIS), National University of Singapore (NUS)
7 Engineering Drive 1, Block E3A, #06-01, 117574, SINGAPORE

ABSTRACT: The building envelope in tropical climate conditions continuously receives a high amount of solar radiation throughout the year. Depending on the insulation properties of the building materials used for the roof, wall and fenestration systems, it has a major impact on air-conditioning load and energy utilization in the building. Fenestration systems, in particular, should have a low Solar Heat Gain Coefficient (SHGC) to comply with building codes and regulations such as those of the Singapore's Building & Construction Authority (BCA) in order to achieve the recommended Envelope Thermal Transfer Value (ETTV). This paper investigates the thermal performance of semi-transparent building integrated photovoltaics (BIPV) in a "Japanese bamboo blinds" design using state-of-art indoor laboratory measurements for SHGC in a calibrated Calorimeter Hot Box under steady-state conditions. Results of this study indicate that indoor measurements comply with international standards and provide accurate quantification of the thermal characteristics of semi-transparent BIPV, which has great deployment potential for energy efficient buildings in the tropics. Furthermore, based on ETTV sensitivity analysis, the integration of such glass-glass BIPV into the fenestration system will allow buildings to achieve Gold, Gold Plus, and Platinum Green Mark award with window-to-wall ratios of up to 0.51, 0.40, and 0.36, respectively.
Keywords: Building-Integrated Photovoltaics, BIPV, Solar Heat Gain Coefficient, Thermal Properties.

1 INTRODUCTION

The growing energy demand globally, coupled with increasing environmental concerns call for new applications to utilize sustainable energy generation technologies. This even more so in an urban context where the resources for renewable energies are limited and space is often constraint, like for the case of Singapore, a city-state located close to the equator. In the tropical climate of Singapore, building façades receive a high amount of solar energy throughout the year. In order to harvest the clean energy from the sun and reduce greenhouse gas emissions, one growing trend is to replace conventional materials of the building envelope with building integrated photovoltaics (BIPV). Such advanced building skins (e.g. in the form of solar fenestration systems, curtain walls or shading devices with photovoltaic elements) are also critical for achieving Zero-Energy Buildings (ZEB) or even Positive Energy Buildings (PEB), as they not only enable on-site energy generation but also reduce the solar heat gain into the building. It is pre-requisite though that BIPV elements still provide typical functionalities of the building envelope as divider between the external and internal environment, be it as weather shield, noise protection, heat insulations, as well as aesthetics. More detailed definitions for BIPV as a function of building component or construction product can be found in the two EN standards: EN 50583-1:2016 [1] for BIPV modules and EN 50583-2:2016 [2] for BIPV systems. Both refer to essential building requirements specified in the European Construction Product Regulation CPR 305/2011, such as the mechanical and structural norms, energy economy, sustainable use of natural resources, weather, fire and noise protection, as well as safety aspect on both module and system levels. The application of BIPV technologies is not limited to the roof, and rather expands also to claddings, spandrels, balconies, parapets, shading devices, roof tiles, prefabricated systems, flexible foils, skylight, and fenestration solutions. The latter includes windows, skylights, and curtain walls, as increasingly adopted by architects. Apart from its multi-functionality, sustainable fenestration systems should also consider energy efficiency measures following local building codes, best practices in sustainable designs, and comply with international standards. Nevertheless, up to now, there has been lack of research that quantifies the thermal characteristics of innovative BIPV technologies and compliance with building codes and standards to encourage the building industry to further adopt BIPV solutions.

This paper aims to describe the procedures for the accurate quantification of the thermal properties, in particular the determination of the so-called "Solar Heat Gain Coefficient", SHGC (often also referred to as the "g-value") of semi-transparent building integrated photovoltaic modules under steady-state laboratory conditions using a calibrated Calorimeter Hot Box. The work also addresses the compliance with existing building regulation and Green Mark certification in Singapore. The proposed methodology is meant to assist façade engineers and building professionals to make informed decisions on the adoption of BIPV and its impact on energy performance of the building envelope at an early design stage.

2 EXPERIMENTAL SET-UP

Calorimeter Hot Box Description. The design of the Calorimetric Hot Box at the Solar Energy Research Institute of Singapore (SERIS), National University of Singapore (NUS) was inspired by the requirements of the building industry to test building materials and entire façade elements in real-world dimensions of up to 1.5m x 1.5m. The Calorimeter Hot Box consist of i) a metering box, ii) a guarding box, iii) a surround panel with external air curtain, and iv) a solar simulator. A schematic diagram of the experimental set-up of SERIS' Calorimeter Hot Box is illustrated in Fig. 1.

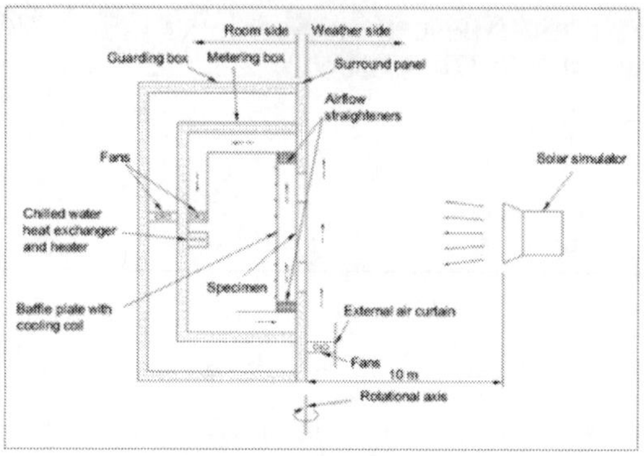

Figure 1: Schematic drawing of the experimental set-up of SERIS' Calorimeter Hot Box for Solar Heat Gain Coefficient (SGHC) measurements.

The diagram also shows details of the "room side" (left) and the "weather side" (right) of the Hot Box. The room side incorporates chilled water heat exchanger, heater, indoor fans, all installed in a metering box and a guarding box. The surround panel is designed for installation of the specimens with various opening dimensions from 30cm by 30cm to 1.5m by 1.5m. The weather side simulates the "weather conditions" with an air curtain (using external fans) and an indoor sun simulator, as shown in Fig 2. The sun simulator (18000ARRIMAX, ARRI 18/12 solution) with a single 18 kW metal halide lamp (HMI 18000W/SE/GX51, Osram) is installed 10m away from the specimen to provide greater uniformity and a small divergence angle. Details of the SERIS in-house procedures for Calorimetric measurements for SHGC are summarized in [3].

Calorimeter Hot Box Calibration. The major components of the Calorimeter Hot Box, for example the high-presicion volumetric flow meters (Emerson) and the pyranometer (CPM11, Kipp&Zonen) are regularly calibrated, following ISO 17025 protocol. Prior to the actual measurement of the fenestration system, the Calorimeter Hot Box is undergoing a number of calibration procedures as outlined in NFRC 201:2010 - *Standard Test Method for Measuring The Solar Heat Gain Coefficient of Fenestration systems Using Calorimetry Methods (NFRC 2001)* [4]. One of the primary calibrations of the calorimeter, which is carried out prior to any facade element tests, is the measurement of a ASHRAE "reference glazing system" (clear single glass) to ensure the accuracy of the apparatus against known values.

Figure 2: Photographs of the metering box with installed specimen (left), and indoor sun simulator (right).

Description of the specimen. The tested specimen is a commercially available, multifunctional glazing element with see-through mono-crystalline silicon solar cells that are cut into stripes and re-arranged in the form of a "Japanese bamboo blinds" design (AGC, "Sudare"), as illustrated in Fig 3.

Figure 3: Original sample of semi-transparent BIPV: view from indoors (left), view from outdoors (right).

This semi-transparent glass-glass PV module consists of the following layers: i) glass; ii) ethylene vinyl acetate (EVA); iii) c-Si solar cells; iv) EVA; v) glass. The module configuration and detailed cell arrangement are shown in Fig. 4. The summary of the technical specifications is given in Table I.

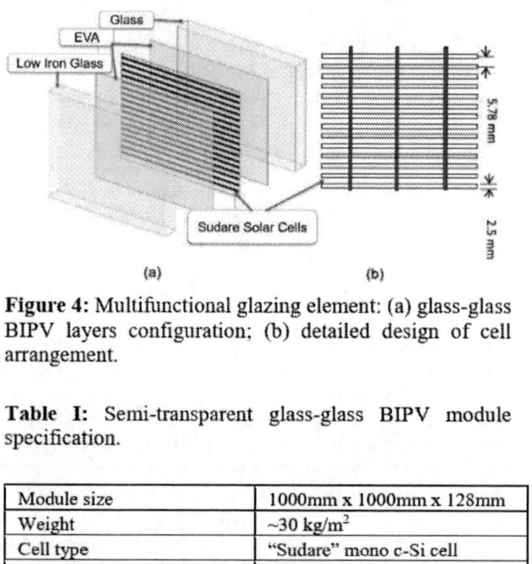

Figure 4: Multifunctional glazing element: (a) glass-glass BIPV layers configuration; (b) detailed design of cell arrangement.

Table I: Semi-transparent glass-glass BIPV module specification.

Module size	1000mm x 1000mm x 128mm
Weight	~30 kg/m^2
Cell type	"Sudare" mono c-Si cell
Cell ratio	30.7%
Number cells per module	56 pcs
Maximum Power, P_{max}	49 W
Fill Factor	84.7

Calculation procedure. Indoor evaluation of the SHGC of the semi-transparent BIPV in a "Japanese bamboo blinds" design as a function of the fenestration system has been carried out at SERIS' Calorimeter Hot Box for hot climates ("summer conditions"). During the test, the indoor side, i.e. the metering box and the guarding box are maintained at steady-state temperature and airflow conditions. On the outdoor side, the specimen is exposed to the external air-curtain and solar radiation from the solar simulator. Steady state heat flow through the specimen due to the transmission of solar radiation heat was measured to derive the SHGC value as follows:

$$SHGC = Q_S/(I \cdot A_s) \quad (1)$$

where Q is the heat flow rate transmitted through the specimen in[W]; A_s is the specimen area in [m²]; and I is the incident solar irradiance in [W/m²].

The heat flow rate is calculated by the following formula:

$$Q_S = -(Q_{CW} + Q_E + Q_{Wl} + Q_{Fl} + Q_{SP} + Q_U) \quad (2)$$

Where Q_S is heat flow through the specimen due to solar heat gain during the steady state in [W]; Q_{CW} is the heat flow discharged by chilled water in [W]; Q_E is heat loss from the electrical devices in [W]; Q_{Wl} is the metering box heat loss, determined during wall loss calibration in [W]; Q_{Fl} is the surround panel flanking heat loss, as detected through flanking loss calibration procedures in [W]; Q_{SP} is the heat loss through the surround panel in [W]; Q_U is the heat flow through the specimen due to U-value (thermal transmittance) in [W]. Table II summarizes the measurement results for determination of SHGC of the "Sudare" semi-transparent BIPV module. The final SHGC for the measured specimen was: 0.45 ± 0.01.

Table II: Measurement results for semi-transparent BIPV.

Specimen	"Sudare" BIPV
Average width (original)	1000 mm
Average length (original)	1000 mm
Average thickness	128 mm
Test duration	8 hours
Laboratory ambient temperature	21.6 – 27.2 °C
Laboratory relative humidity	43 – 52 %
Laboratory barometric pressure	1.00 – 1.01 bar
Metering air temperature	24 °C
External air curtain temperature	27.3 °C
Metering airflow velocity	0.2 m/s
External air curtain velocity	2.2 m/s
Heat exchange with chilled water loops, $Q_{Chilled\ water}$	-279 W
Heat exchange with electrical devices, $Q_{Electrical\ devices}$	100.4 W
Metering box wall heat transfer, Q_{Wall}	8.7 W
Flanking heat transfer, $Q_{Flanking}$	9.6 W
Surround panel heat transfer, $Q_{Surround\ panel}$	8.1 W
Specimen thermal transmission, Q_U	3.7 W
Specimen heat transfer, $Q_{Specimen}$	148 W
Mean solar irradiance, I_{Solar}	589 W/m²
Specimen area exposed to solar radiation, $A_{Specimen}$	0.66 m²
Solar heat gain coefficient (SHGC or G), G	**0.45 ± 0.01**

3 COMPLIANCE WITH BUILDING CODES AND REQUIREMENTS

Prevailing building regulations and codes enable architects, engineers, and building professionals to evaluate the compliance of a proposed building design with local energy codes. In 1975, firstly proposed by ASHRAE, the thermal performance index called "Overall Thermal Transfer Value" (OTTV), was adopted as a regulatory control of air-conditioned buildings by the Building Construction Authority (BCA) of Singapore [5,6]. The original OTTV equation was then modified to the Singapore climate conditions and replaced by the so-called "Envelope Thermal Transfer Value" (ETTV).

Three basic components of heat gain through external building materials, walls and windows, are incorporated into the ETTV equation. These are i) heat conduction through opaque walls; ii) heat conduction through glass windows; and iii) solar radiation through glass windows. These heat components are averaged over the area of the building envelope to accurately characterise the thermal performance of the envelope for air-conditioned non-residential buildings, in form of a single ETTV. The underlying equation is dependent on many factors, including outdoor and indoor environment, SHGC and U-value. The formula is as follows:

$$ETTV = 12(1 - WWR)U_w + 3.4(WWR)U_f + \\ +211(WWR)(CF)(SC) \quad (3)$$

where WWR is the window-to-wall ratio, U_w is the U-value of the opaque wall, U_f is the U-value of the fenestration system with correction factor (CF) for solar heat gain through fenestrations, SC is the fenestration shading coefficient, which is impacted by the SHGC, following this equation:

$$SC = \frac{Solar\ heat\ gain\ coefficient(SHGC)\ of\ any\ glass}{Solar\ heat\ gain\ coefficient\ (SHGC)\ of\ a\ 3mm\ clear\ glass} \quad (4)$$

The physical meaning of the ETTV is the averaged heat flux during air-conditioned active hours, relative to the entire envelope area, expressed in units of [W/m²]. It is important to highlight that the ETTV is now a minimum mandatory standard, regulated under the BCA Green Mark Scheme (GMS) certification. Furthermore, the latest Green Mark requirements for non-residential buildings (NRB), the 5th edition of the GMS, have been tightened with regards to various threshold levels. The acceptable ETTVs now range from 45 W/m² for Gold, 40 W/m² for Gold Plus, and 38 W/m² for Platinum Green Mark [7].

In determining the compliance of the semi-transparent BIPV modules with prevailing building codes in Singapore and the benefits for obtaining Green Mark certifications, further sensitivity analysis of the ETTV has been carried out.

A reference building was defined for the calculation of ETTV for South, North, West, and East orientations with relevant solar correction factors (CF). A representative 20-storey office building with 30m width and 30m depth and a ceiling height of 3m was selected. The window-to wall ratio (WWR) was varied from 10 percent to 90 percent. For simplicity, lightweight wall construction with thermal insulation was assumed for all walls, with a U-value (U_w) of 1.5 W/(m²K). Five typical architectural fenestration systems were compared for the ETTV calculations: 1) 3mm single-pane clear glass; 2) single-pane tinted glass; 3) double-pane (DGU) clear; 4) double-pane (DGU) low-E; 5) semi-transparent BIPV (Sudare). The thermal properties of the 5 fenestration systems are summarised in Table III.

Following equation (3), the ETTV for each fenestration system was calculated, while varying the window-to-wall ratio (WWR), based on their thermal properties i.e. SHGC and U-Value.

The ETTV values of the typical glazing materials was calculated for each of the cardinal directions to determine compliance with Green Mark Certification Scheme with maximum WWR, see Fig 5.

Table III: Thermal properties of 5 glazing systems.

№	Short name	U-Value	SHGC	SC
1	3mm clear glass	5.33	0.87	1.00
2	Single tinted	5.25	0.42	0.48
3	DGU clear	2.83	0.69	0.80
4	DGU low-E	1.59	0.38	0.44
5	BIPV glass	3.74	0.45	0.52

Note: Unit for U-Value is [W/(m^{2}K)] and for SHGC-Solar Heat Gain Coefficient, SC-shading coefficient is [unit-less].

The results show that integrating semi-transparent BIPV as a function of fenestration systems in buildings has comparable thermal properties to low-E glass, see Fig. 5. This, furthermore, indicates that the BIPV modules are able to comply with existing building requirements regarding the ETTV value to achieve Gold, Gold Plus, and even Platinum Green Mark certification at maximum WWR ratios of up to 0.51, 0.40, and 0.36 on North- and South-facing facades, respectively.

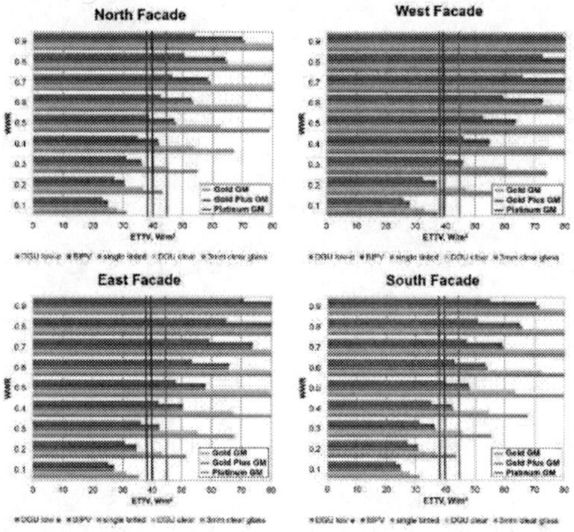

Figure 5: ETTV sensitivity analysis when changing the thermal properties of the glazing (SHGC) and the window-to-wall ratio (from 0.4 as best case to 0.9 fully glazed facade). The Gold and Gold Plus and Platinum Green Mark Award requirement are marked with green, purple and red lines, respectively.

In addition, the GM Scheme also incorporates a numerical scoring methodology, the so-called "Green Mark Score", to assess building design and performance based on degree of compliance with the ideal requirements, as specified in the GM Scheme [7]. According to the total Green Mark Score, including the bonus points under the "Advanced Green Efforts" category, the final Green Mark rating is derived. For example, GM Platinum Rating will be awarded to buildings that score 70 and above points, GM Gold Plus rating for buildings with 60 to 70 points, and GM Gold rating for buildings with 50 to 60 points. Besides, energy efficiency, the deployment of renewable energies is one of the criteria in the building energy performance section. A total of 8 points can be achieved including 0.5 points for a solar energy feasibility study to assess the optimal placement of photovoltaics, 1.5 points for a solar-ready roof (structural, electrical and spatial readiness), and up to 6 points for on-site clean energy generation to offset a building's energy consumption. The Advanced Green Efforts section allows for an additional 5 points for a substantial on-site renewable energy generation. In the bonus section of GM, low heat gain facades, so-called "Tropical Facade Performance" with ETTV equal to 35 W/m^2 an additional point can be scored. This can for example be achieved with semi-transparent BIPV at a WWR of 0.32 on the north facing facade.

4 CONCLUSION

Experimental evaluation of the solar heat gain coefficient (SHGC) of semi-transparent building integrated photovoltaics (BIPV) in a "Japanese bamboo blinds" design as a function of the glazing system has been accurately measured at SERIS' Calorimeter Hot Box for hot climates ("summer conditions"), following international standards and procedures. The results of the study show that the thermal performance of BIPV is comparable with, or better than typical fenestration materials commonly used in the tropics.

This study is meant to primarily assist façade engineers and building professionals to make informed decisions on the adoption of BIPV and its impact on energy performance of the building envelope at an early design stage.

Moreover, the tested specimen complies with BCA Green Mark requirements for ETTV. Based on an ETTV sensitivity analysis, the integration of semi-transparent BIPV into the fenestration system will allow buildings to achieve Gold, Gold Plus and Platinum Green Mark standards with window-to-wall ratios of up to 0.51, 0.40, and 0.36 for South- and North-facing facades, respectively.

Furthermore, the renewable energy generation from the semi-transparent BIPV module becomes even more beneficial for the Green Mark certification, as it allows for additional points, following the GM scoring methodology. BIPV technologies are a critical factor for to achieving the BCA's aspiration of positive-energy low-rise, zero-energy medium-rise and low-energy high-rise buildings in Singapore.

5 ACKNOWLEDGEMENT

This study is supported by the project: "Holistic life-cycle cost analysis and real-world test bedding for Building Integrated Photovoltaics (BIPV) facades", funded by Singapore's Ministry of National Development – Sino-Singapore Tianjin Eco-City Administrative Committee Green Building Related Research and Development Programme and the Building and Construction Authority of Singapore (BCA).

6 REFERENCES

[1] BS EN 50583-1:2016 - Photovoltaics in buildings. BIPV modules," 2016. Retrieved from : https://bsol-bsigroup-com.libproxy1.nus.edu.sg/Bibliographic/BibliographicInfoData/000000000030266507

[2] BS EN 50583-2:2016 - Photovoltaics in buildings. BIPV systems," 2016. Retrieved from

https://bsol-bsigroup-com.libproxy1.nus.edu.sg/Bibliographic/BibliographicInfoData/000000000030328521

[3] Chen, F., Wittkopf, S. K., Ng, P. K. and Du, H., Solar heat gain coefficient measurement of semi-transparent photovoltaic modules with indoor calorimetric hot box and solar simulator. *Energy and Buildings*, **53**, 2012, 74-84.

[4] National Fenestrations Rating Council, NFRC 201-2010, Procedure for interim standard test method for measuring the solar heat gain coefficient of fenestration systems using calorimetry hot box methods, 2010.

[5] ASHRAE (1975) ASHRAE Standard 90-1975, Energy Conservation in New Building Design. American Society of Heating, Refrigerating and Air-conditioning Engineers.

[6] Building and Construction Authority BCA (2004): Guidelines on Envelope Thermal Transfer Value for Buildings, Commissioner of Building Control, Ver 1.01, February 2004.

[7] Building and Construction Authority BCA (2015): Green Mark for Non-Residential Buildings, Ver 5, Dec 2016, retrieved from https://www.bca.gov.sg/GreenMark/others/Green_Mark_NRB_2015_Criteria.pdf.

OUTDOOR CHARACTERIZATION OF COLORED AND TEXTURED PROTOTYPE PV FAÇADE ELEMENTS

Chris Tzikas[1], Roland Valckenborg[1], Maarten Dörenkämper[1], Menno van den Donker[1], David Duque Lozano[2], Ádám Bognár[2], Roel Loonen[2], Jan Hensen[2], Wiep Folkerts[1]

[1]SEAC (Solar Energy Application Centre), Eindhoven, The Netherlands
[2]Eindhoven University of Technology, Eindhoven, The Netherlands

ABSTRACT: The aim of this study is to assess the performance of prototype PV façade elements of various PV technologies, colors and textures. Within this context, a prototype PV façade demonstrator was constructed and monitored at SolarBEAT, Eindhoven. This prototype demonstrator consists of 9 façade PV panels of c-Si and CIGS technologies with flat and textured solar glasses and black, grey and red colors. The field-testing results indicate a limited performance drop of less than 20% for all colors and textures.
Keywords: Photovoltaic Façade, Colored Photovoltaics, Outdoor Characterization, BIPV, Solar Façade

1 INTRODUCTION

Regulations for energy performance of buildings in Europe are becoming more demanding [1]. This requires innovative approaches for integration of solar energy in the building skin. At the same time, especially for façades, the aesthetic appearance is crucial, and the architectural demands are posing a challenge for PV developers [2].

This study has been performed as part of the Interreg PV OpMaat project [2]. The aim of this project is to assess power generating photovoltaic materials, based on newly developed production methods, and tailored in a way that they can be optimally integrated in building elements. In this context, a novel prototype PV façade was constructed at SolarBEAT, the outdoor field-testing facility of SEAC [3]. This research evaluates how various textures and colors affect the performance of c-Si and CIGS PV modules and assesses different coloring techniques in order to obtain aesthetically the best possible result.

2 THE TEST FIELD

The prototype PV façade was commissioned in February 2018 at SolarBEAT, it is facing due south and will be field tested for a complete calendar year. It consists of 9 PV panels of two different PV technologies (conventional crystalline silicon and thin film CIGS), two different textures (normal flat glass and textured glass), and three different colors (black, grey and red). A schematic representation of the demonstrator can be seen in Figure 1 and the constructed PV façade demonstrator in Figure 2.

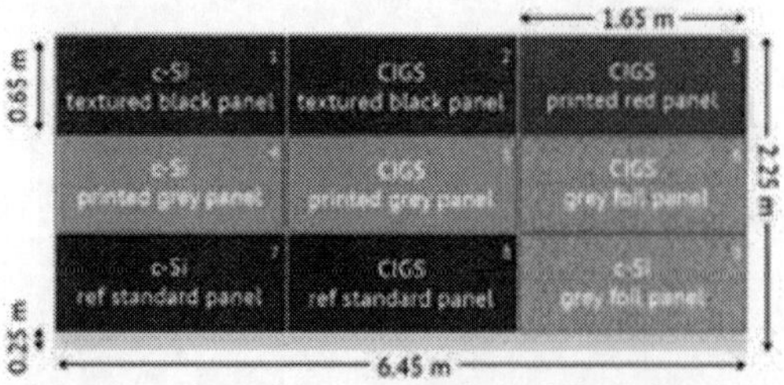

Figure 1: Schematic representation of the colored PV façade demonstrator.

Both the commercial CIGS and the custom-made crystalline silicon PV panels are 1650 x 650 mm in size while the power rating is 185 Wp for the c-Si PV reference panel and 135 Wp for the CIGS PV reference panel.

Figure 2: The outdoor colored and textured PV façade demonstrator installed at SolarBEAT, Eindhoven.

For the grey coloring, two different techniques have been used and are being compared next to each other. At first, there is the standard solar glass dot printing technique which is also being used for the red panel, as reported in other studies [5], [6]. Complementary to it, two standard black panels are equipped with a grey dot printed foil which is glued on the outer surface of the solar glass. This is a quicker and more cost effective solution having the additional advantage that it can be modified in a later stage according to the architectural wishes. Similar foils are already widely used by architects on passive facade elements of commercial buildings. In both coloring techniques used, the color print coverage was 30%.

Figure 3: Close-up photo of the various CIGS panels of the prototype demonstrator. At the top left the black textured glass panel is depicted, at the top right the printed red panel, at the bottom left the printed grey panel and at the bottom right the grey foil panel.

All 9 panels are IV traced with the help of an IV tracer, therefore all DC characteristics of the PV panels can be accurately and individually measured. A secondary standard pyranometer is used for measuring the in-plane irradiance and the panels' temperatures are measured using thermocouples. The individual panels are all connected with a Power Optimizer (PO), to keep the panels in Maximum Power Point (MPP). All the PO's are connected to an inverter. The façade is facing south, and all the panels will be monitored for a complete year.

3 OUTDOOR CHARACTERIZATION RESULTS

The prototype colored PV façade demonstrator is operational since the beginning of March 2018, so already half a year of data have been recorded and analyzed and will be presented below. All data used and analyzed are DC data recorded by the IV tracer to exclude any influence of the power electronics.

The effect that colors and textures have on the performance of the PV panels can be seen in the IV curves shown in Figure 5. Among them, the red color-printed PV panel is performing least optimal followed by the grey colored PV panels of both coloring techniques.

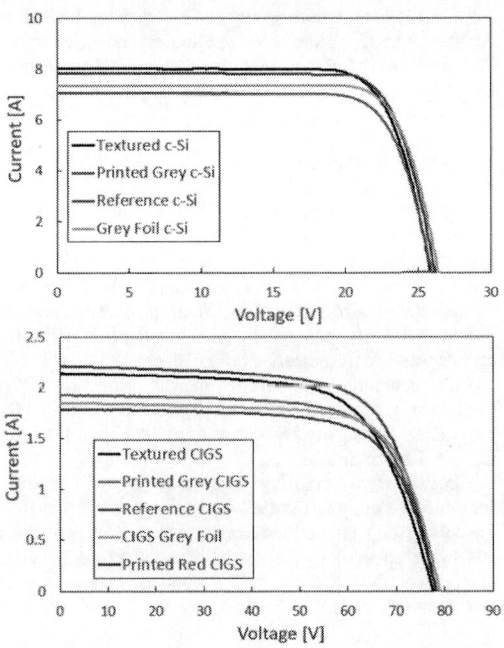

Figure 5: Measured IV curves of the c-Si (top) and CIGS (bottom) panels with various colors and textures, measured on Feb 27, 2018 at 12:00.

In order to compare the performance of the various PV panels, the DC Performance Ratio (PR) was used and compared with the DC PR of the two reference PV panels (one c-Si and one CIGS) installed at the bottom of the demonstrator [5]. The cumulative results of the first 6 months show a relative performance difference of about 9% for the grey colored PV panels of both coloring techniques and 16% for the red colored PV panel when compared to the reference PV panels. The two panels equipped with textured solar glass perform similarly to the reference PV panels if not slightly better. It should be taken into account that there might be a power rating deviation within PV panels of ± 2% according to their manufacturers, which also applies for the performance results shown in Figure 6.

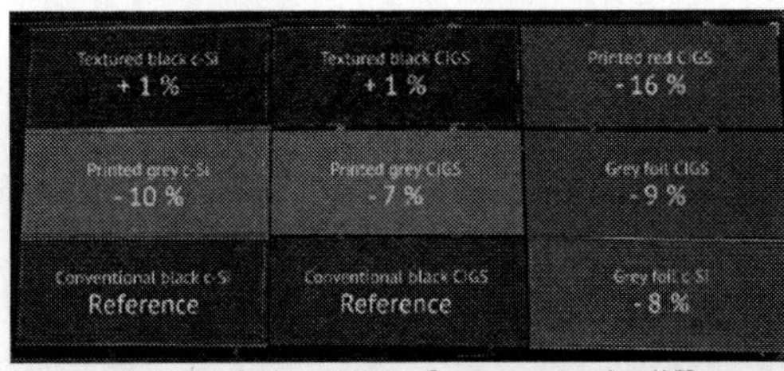

Figure 6: Cumulative DC Performance Ratio differences for all PV panels in regards to the reference ones for the first 6 months of field testing.

The monthly breakdown of the specific yield data for each PV panel is shown in Figures 7 & 8. The performance differences shown above are repeated in the data of every month with the reference and textured black panels performing similarly followed by the grey colored panels while the red colored panel is falling slightly behind.

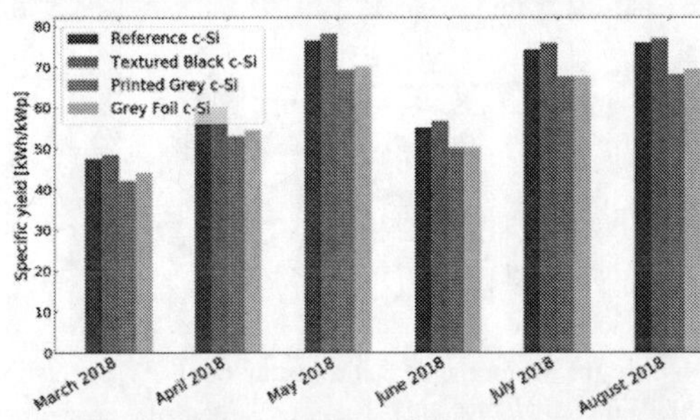

Figure 7: Monthly specific yield of the c-Si PV panels.

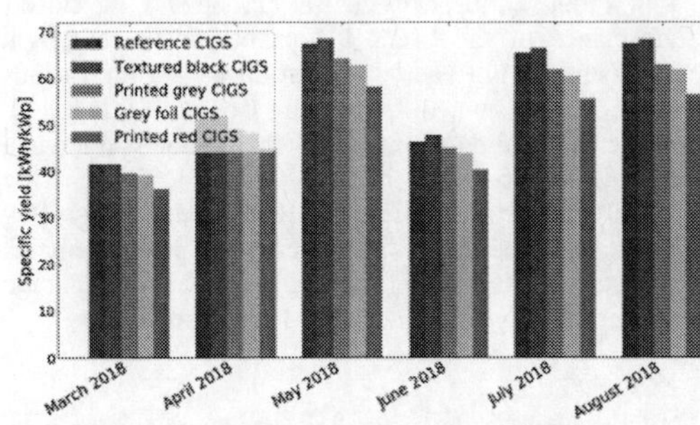

Figure 8: Monthly specific yield of the CIGS PV panels.

Temperature data analysis shows that the CIGS PV panels have higher operating temperatures than the c-Si owing to their lower efficiency, following the trends as reported in other publications [6], [7], [8]. Between them, the red colored and the textured black CIGS PV panels appear to be the hottest, with temperature differences between the 9 PV panels reaching up to 10°C on a warm summer day as shown in Figure 9.

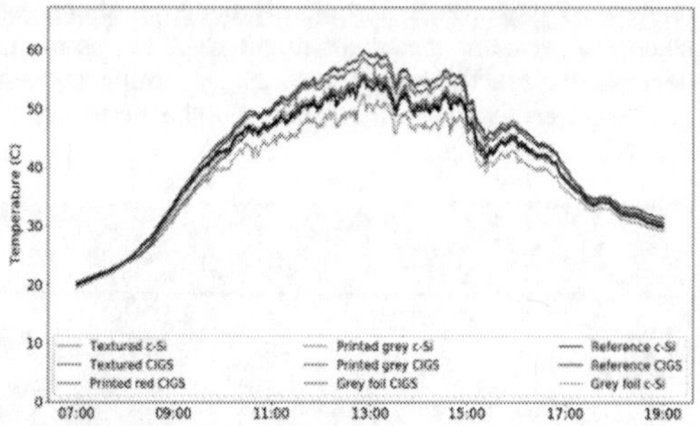

Figure 9: Module temperatures of the 9 PV panels composing the prototype PV facade demonstrator, on July 16, 2018.

More in depth analysis showed a lower performance of all CIGS PV panels under low light conditions. To further investigate this phenomenon, the DC Performance Ratio data were split into irradiance bins of 200 W/m². The results are depicted in Figure 10. It is clear that under low irradiances (<200 W/m²) the DC PR of the CIGS PV panels drops significantly while the c-Si PV panels perform similarly regardless of the irradiance conditions.

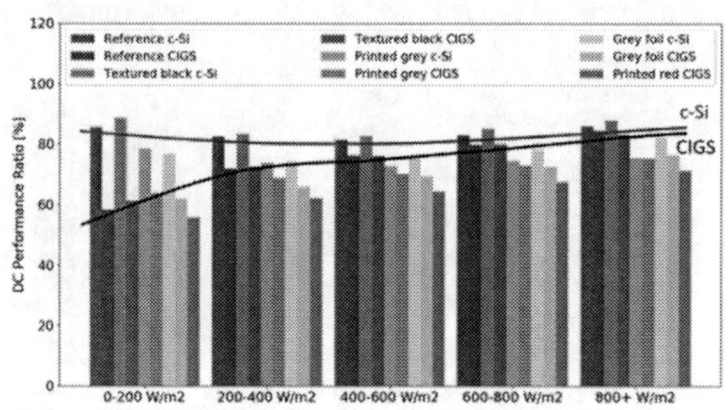

Figure 10: Performance Ratio results of all PV panels in 200 W/m² irradiance bins.

In order to identify the reason behind the lower performance of the CIGS PV panels under low light conditions, the Fill Factor (FF) of all panels for various irradiances was studied. In Figure 11 one can find the FF of the c-Si (blue dots) and CIGS (black dots) reference panels. It can be seen that the CIGS PV panels have generally lower FF which is greatly reduced for low irradiances while the FF of the c-Si PV panels remains unaffected. This FF drop is caused mostly by a drop in the voltage of these panels under low light conditions.

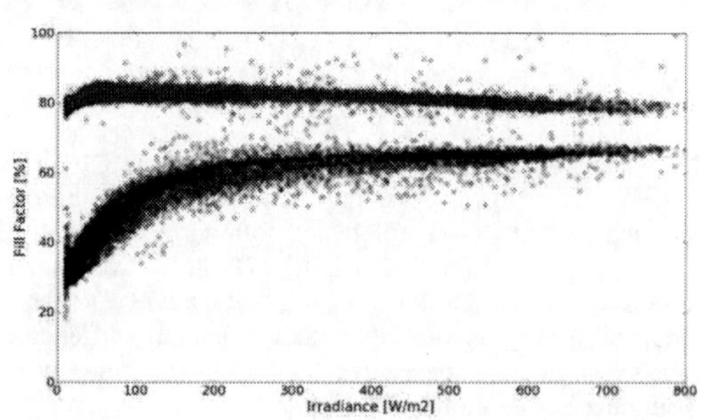

Figure 11: Fill Factor results of CIGS (black) and c-Si (blue) panels for various irradiances.

Finally, a simulation model was created with the PV_LIB tools of the PV Performance Modelling Collaborative (PVPMC) in order to predict the performance of aesthetic PV façades [9]. In Figure 12, indicative results of a simulation are shown for the Amsterdam area and for all PV technologies, textures and colors used in the actual prototype demonstrator. The results include all possible orientations of the PV façade. Such a tool can be a useful tool for architects, assisting them to design aesthetic BIPV projects with sustainability in mind.

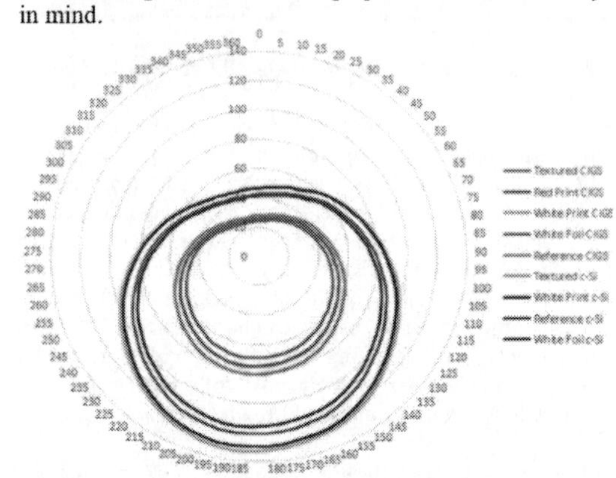

Figure 12: Simulation results for PV façades located in Amsterdam, of various PV techologies and orientations with various colors and textures. The annual yield is in kWh/m²

4 CONCLUSIONS

Within this study, the performance of a prototype colored, and textured PV façade was successfully assessed. Results showed a limited performance drop (below 20% for 30% print coverage) for PV panels equipped with an aesthetic layer (texture, color) while the aesthetic result remained in accordance with the architectural requirements.

In general, the CIGS panels showed lower performance especially under low light conditions than the c-Si panels, owing mostly to lower voltages and fill factors under these conditions.

Lastly, two coloring techniques were assessed (colored foil on top of the solar glass, printed dots behind the solar glass) and yielded similar results. This is offering PV developers an alternative cost-effective way to customize the color of PV panels according to architectural needs.

5 ACKNOWLEDGEMENT

This project is partially funded through a contribution from the European Interreg V Flanders-Netherlands program.

5 REFERENCES

[1] "Directive 2010/31/EU of the European Parliament and of the council of 19 May 2010," Official Journal of the European Union, 18.6.2010

[2] M. Munari and C. Roeckr, "Criteria for architectural integration of active solar systems IEA Task 41, Subtask A," *Energy Procedia,* vol. 30, p. 1195–1204, 2012.

[3] PV OpMaat, "PV OpMaat: Homepage," [Online]. Available: http://pvopmaat.nl/home.

[4] SEAC Solar Energy Application Centre, "SolarBEAT," [Online]. Available: https://www.seac.cc/en/projects/#solarbeat.

[5] M. Mittag, C. Kutter, E. Ebert, H. Wilson and U. Eitner, "Power loss through decorative elements in the front glazing of BIPV modules," *33rd EUPVSEC,* p. 22–26, 2017.

[6] G. Eder, K. Knöbl, L. Maul, M. Aichinger, G. Peharz, W. Nemitz and K. Berger, "Designed BIPV-Elements with Printed Front-Glass: Simulation and Experimental Evaluation of the Effect of Printing on the Electrical Performance," *12th Conference on Advanced Building Skins,* 2017.

[7] A. Woyte, M. Richter, N. Reich, M. Green and S. Mau, "IEA PVPS Task 13, Sub-task 2: Analytical Monitoring of Grid-connected Photovoltaic Systems - Good Practices for Monitoring and Performance Analysis," 2014.

[8] P. H. Choi, H. Baek do, H. S. Park, S. S. Kim and J. S. Yi, "Temperature-dependent electrical characteristics of c-Si and CIGS solar cells," *J Nanosci Nanotechnol,* vol. 14, no. 12, pp. 9206-9209, 2014.

[9] A. Virtuani, D. Pavanello and G. Friesen, "Overview of Temperature Coefficients of Different Thin Film Photovoltaic Technologies," *25th Eur. Photovolt. Sol. Energy Conf. Exhib. / 5th World Conf. Photovolt. Energy Conversion,* p. 4248–4252, 2010.

[10] C. Cañete, J. Carretero and M. Sidrach-de-Cardona, "Energy performance of different photovoltaic module technologies under outdoor conditions," *Energy,* vol. 65, p. 295–302, 2014.

[11] W. F. Holmgren and D. G. Groenendyk, "An open source solar power forecasting tool using PVLIB-Python," *IEEE 43rd Photovoltaic Specialists Conference (PVSC),* p. 0972–0975, 2016.

BIPV MEETS CUSTOMIZABLE GLASS: A DIALOGUE BETWEEN ENERGY EFFICIENCY AND AESTHETICS

E. Saretta[1]*, P. Bonomo[1], F. Frontini[1]

[1] SUPSI-ISAAC, Campus Trevano, CH 6952, Canobbio, Switzerland
* mail to: erika.saretta@supsi.ch; phone +41 052 666 6290, fax +41 052 666 6349

ABSTRACT: The realization of Building-Integrated Photovoltaics (BIPV) facades with black and blue conventional modules has represented a first step of innovation in the process of PV transfer in sustainable buildings. Nowadays, architects are looking for additional BIPV products to design active facades with novel architectural languages. For this reason, BIPV manufacturers started to customize BIPV modules in order to meet architects and clients' requirements. However, changing traditional front solar glass to obtain coloured PV, can influence the final module power output. Even though there are already investigations about the reduction of PV energy output due to the front glass aesthetical treatments, there are not yet clear correlations between alternative design options (glass treatments and colours) and their operative conditions as façade elements.

This paper is firstly aimed at providing a systematic investigation about novel mono-chromatic BIPV products in order to identify the influence of some main design options on the electro-thermal performances of modules. Specifically, this analysis is performed by evaluating, in collaboration with a Swiss industrial partner, a range of customized mono-chromatic BIPV prototypes by means of indoor measurements and outdoor tests. Furthermore, these results have been used to preliminarily assess the electro-thermal behaviour of multi-chromatic BIPV configurations both at STC and outdoor conditions. As a result, it arises that the operative condition should be carefully considered to qualify the behaviour of multi-chromatic BIPV modules with novel aesthetical designs in order to provide reliable data about the real electro-thermal behaviour.

Keywords: coloured BIPV modules, glass aesthetical customization, BIPV efficiency, BIPV operative conditions

1 INTRODUCTION

Building Integrated Photovoltaics (BIPV) provides the opportunity to contribute to the energy sustainability of the built environment since it enables building surfaces to become energy generators, still constituting a building material.

In recent years, the deployment of BIPV facades has been emerging, especially thanks to: (i) their suitability to spread the energy production through the day, when BIPV façades with different orientations are realized, and (ii) their larger surface area in comparison to roofs (in particular for multi-storey buildings), which could potentially improve the building energy target and the self-consumption. However, since building facades represent also the expression of architectural languages, architects are looking for aesthetically appealing materials for realizing active surfaces. In this context, from the collaboration between glass manufacturers and BIPV producers, novel customized BIPV modules are arising thanks to the use of different front glass treatments to obtain various colours, textures or material appearance, different reflectivity, visual/optical effects, etc [1].

If aesthetical improvements are an added value of these BIPV products to support a more and more dynamic market, there can be variations of the energy behaviour in comparison with traditional PV modules, such as the power output reduction and/or the unconventional (e.g. non uniform) operative conditions, which need to be further investigated in order to provide the market with reliable and efficient BIPV systems.

Currently PV modules are optimised, selected and sold on the basis of power produced under standard test conditions (STC). However, this metric does not always reflect real-world conditions. Especially for tailor-made BIPV components, more insight into measurement methodologies, energy rating, safety, degradation rates mechanisms and further knowledge on performances and long-term stability are now requested. On one hand the authors are contributing to develop more accurate measurement methods for emerging solar modules, for extending energy rating to modules applied to or integrated into buildings, with the goal to extend the IEC 61853 standard series with some technical specifications taking BIPV into account [2]. On the other hand, a detailed analysis of power loss and performance under real operating conditions is necessary in order to understand how the special BIPV modules (e.g. coloured, mono or multi-chromatic) behave and, consequently, to laid down a first step in defining characterisation and measurement approaches for special BIPV modules to accurately reflect the real operating environment which is still not fully understood.

Within this framework the paper focuses on analysis of energy performance of coloured modules, by introducing a systemic analysis through an indoor measurement campaign on a range of significant mono-chromatic BIPV prototypes, which take into consideration different colour types and the glass treatments, to define how these design options affect the electrical performances of BIPV modules. Furthermore, outdoor measurements have been carried out in order to also investigate and characterize how the electro-thermal performances of such a products are influenced by real operative conditions.

Besides this, the mismatch among different coloured BIPV prototypes has been analysed by considering both indoor and outdoor conditions with the aim to identify which design parameters can be optimally combined to obtain multi-chromatic BIPV modules, without significantly reducing the electrical performances.

2 APPROACH AND METHODOLOGY

Recently, some aesthetical BIPV novelties, such as coloured modules, has arisen from the encounter between the glass manufacturers and the PV manufacturers know-hows. However, the realization of a coloured coating to hide cells on BIPV modules can involve a different electro-thermal behaviour in comparison to traditional modules, that need to be assessed in order to provide market with reliable BIPV elements.

Some of these mono-chromatic BIPV modules have been already installed in building facades, such as in lighthouse buildings which have become icons of solar architecture. If BIPV modules would become façade elements, it should also offer the option to be customized so that architects and façade engineers can also envision multi-chromatic BIPV modules to develop their own façade designs. Since multi-chromatic modules are not yet in the market as standard products, this till requires glass and PV manufacturers to develop tailor-mades products with specific and complex manufacturing and qualification processes. In order to support the development of custom BIPV products, it would be convenient to investigate "a priori" the main design options and define the most sensitive parameters of customization for ensuring energy efficiency, reliability, in order to avoid time and cost savings.

In this perspective, this paper provides a systematic investigation about some mono-chromatic BIPV prototypes in order to identify the influence of some design aspects (colours and front glass types) on the electro-thermal performances of the module. This analysis has been carried out by means of both indoor and outdoor measurements. Then, on the path to the development of multi-chromatic BIPV modules, the evaluation of the mismatch among different coloured BIPV modules has been performed with the goal to preliminarily assess the electro-thermal behaviour of multi-chromatic BIPV modules both at STC and outdoor conditions.

2.1 Prototypes

All the prototypes are glass/glass modules made of 4 mono c-Si cells connected in series and laminated within two glass panes of 3mm thickness each.

Table 1: Characteristics of the BIPV prototypes and their ID code

Colour	*Satin*	*Float*		*Colour degree*
	Position 2	*Position 1*	*Position 2*	*[%]*
No colour	C1	C2		-
Blue	D1	D3	D2	10
Green	E2	E1	-	10
Light grey	A1	A3	A2	10
Dark grey	B2	B3	B1	10
Terracotta	F3	F2	F1	10

In detail, the front glass of the prototypes has been customized by combining three main design options: the glass type (float/satin finish), the positioning of the colour (outer side – position 1 / inner side – position 2), and the type of colour, as shown in Table 1. For all the prototypes, the coating technique before the glass tempering process has been used for the colour application.

2.2 Methodology

If on one hand front glass treatments can improve the aesthetical perception of BIPV modules by hiding the solar cells, on the other hand they can affect the electrical performances, in comparison to traditional PV modules.

In order to investigate such aspects, this research study has been articulated into two main phases. The first phase regarded the indoor characterization of mono-chromatic BIPV prototypes. In this we evaluated the influences on the electrical parameters of the type of front glass and the type of colour with the aim to compare the coloured modules with reference modules, as well as among each other. Since real operating conditions can affect the electrical behaviour of coloured modules [3], a second phase has been envisioned to characterize the operative thermo-electrical characteristics of the prototypes. Therefore, a group of BIPV prototypes have been exposed in the outdoor test facilities in order to investigate the colour influence on the modules' operative temperature and power output.

The second phase involved the investigation of a set of coloured BIPV prototypes to evaluate how they perform when they are combined in series, in order to preliminarily identify which design options are compatible to form a multi-chromatic BIPV module. Hence, not only theoretical calculations have been carried out, but also indoor and outdoor measurements have been performed with the aim to assess the validity of the method adopted.

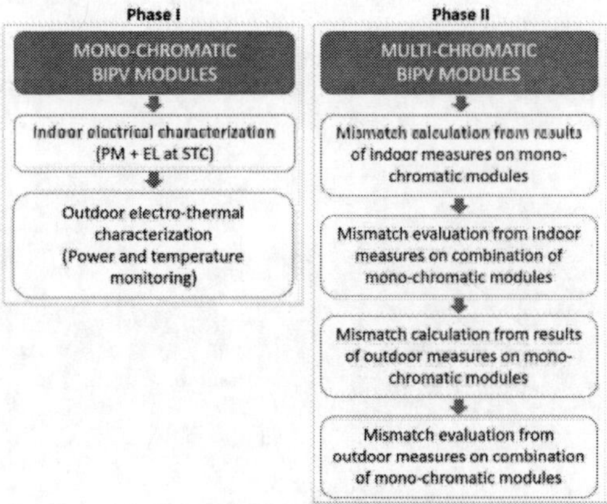

Figure 1: Methodology implemented in the study for the evaluation of the electro-thermal behaviour of mono-chromatic and multi-chromatic BIPV modules.

2.2.1 Indoor characterization

Indoor characterization tests for mono-chromatic modules have been performed in the laboratory of ISAAC-SUPSI with the aim to assess the influence of both the types of glass and the colours on the power parameters at STC. Moreover, electroluminescence (EL) was performed to evaluate the quality of the cells.

Grounding on the resulting IV curves, mismatches among different mono-chromatic prototypes have been calculated. As a consequence, modules with similar electrical characteristics have been combined in series and subjected to power measurements at STC in order to check their electrical compatibility for the development of multi-chromatic modules.

Figure 2: BIPV prototypes. Left picture: PM on reference module “C2”. Right picture: PM on satin finish dark grey module “B2”.

2.2.2 Outdoor characterization

A group of coloured BIPV prototypes have been selected to be exposed in the outdoor test facilities of ISAAC-SUPSI in Lugano (Switzerland) to evaluate their electro-thermal behaviour in real operative conditions. The prototypes have been installed as façade elements and have been equipped with Maximum Power Point Trackers developed by SUPSI, which have been adapted to their voltages and current ranges. Moreover, a pyranometer has been installed to record global irradiance at 90°, and also wind speed and ambient temperature have been monitored. In order to achieve a high level of reliability in the inter-comparison, improved data quality control and an advanced data analysis procedure were used [4]. In addition to electrical parameters, also temperatures have been tracked by means of temperature PT100 sensors placed on the back glass of the prototypes.

Finally, power and temperature measurements have been performed on a couple of mono-chromatic BIPV modules in order to assess their combination in-series.

Figure 3: prototypes installed at ISAAC-SUPSI outdoor test facility in Lugano (Switzerland).

3 MONO-CHROMATIC BIPV MODULES CHARACTERIZATION

3.1 Indoor characterization

As a preliminary step, all the prototypes have been subjected to electroluminescence tests to evaluate the presence of damaged cells that can affect the efficiency. From the emerging results, it arises that there are not relevant damages, hence the power measurement tests at STC have been performed. In detail, the results have been analysed in order to identify the main influencing aspects on the power output of the modules.

When the power output results are analysed for the same colours and different glass types (Figure 4), it arises that differences are in the range between a minimum of 0.3% for the green prototype and 3.3% for the dark grey prototype. Moreover, there is no evidence of efficiency variations when the colour is applied in position 1 or position 2. What is interesting to note is that the satin finish effect of the front glass does not affect significantly the power output of the prototypes. Indeed, in the case of reference, blue and terracotta prototypes, the satin finish modules have a higher power in comparison to float glass modules.

On the other side of the coin, analysing the prototypes with the same front glass treatment (Figure 5), it emerges that the colour type significantly affects the power output. As a result, among the coloured modules, the blue prototypes show less power reduction in comparison to the reference module, while the dark grey prototypes performs the maximum power losses, independently from the glass type.

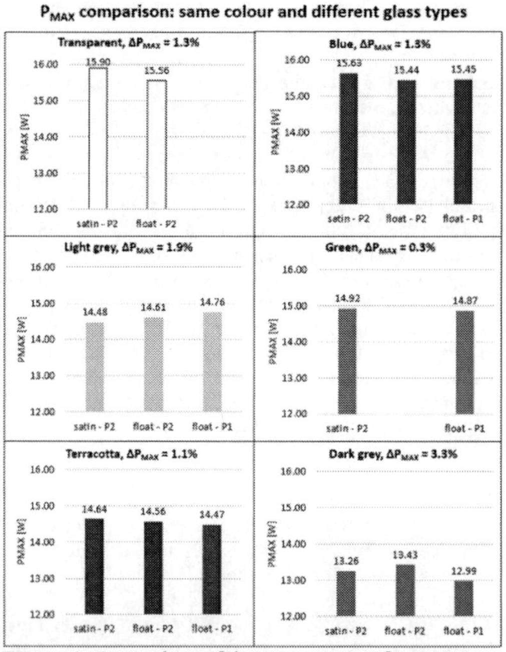

Figure 4: Comparison of the power output for prototypes with the same colour type to analyse the influence of different glass types and colour positioning.

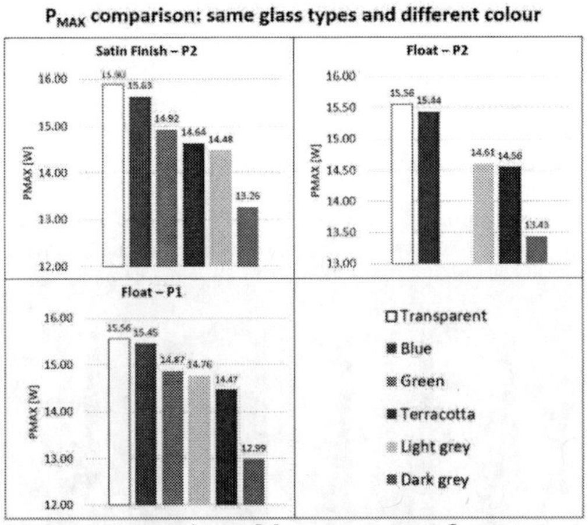

Figure 5: Comparison of the power output for prototypes with the same glass types and colour positioning to analyse the influence of the colour types.

3.2 Outdoor characterization

The outdoor tests have been performed on the mono-chromatic modules with float glass and colour applied in position 2, internally. Specifically, the following prototypes have been mounted in vertical position as façade elements: reference, blue, light grey and terracotta. The power output and temperatures of the prototypes recorded over one day are shown in Figure 6.

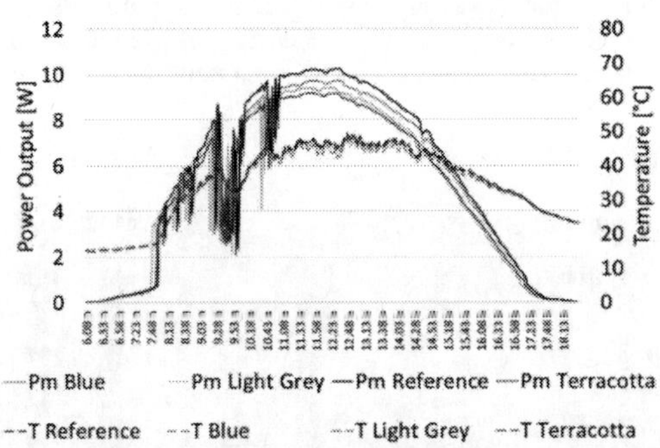

Figure 6: Power output and temperature of the prototype with float glass and colour applied in position 2, internally (data recorded over one day in Lugano)

From the analysis of the power output monitoring, it arises that prototypes have the same electrical behaviour of the indoor measurements. The reference prototype (transparent, non-coloured front glass) performs always better than other prototypes even though it shows higher operative temperature. Among coloured modules, the blue one has a higher power output, while the terracotta and the light grey prototypes are performing worst. In detail, the daily energy losses in comparison to the reference prototype are in the amount of:

- 4.9% for the blue prototype;
- 8.1% for the blue prototype;
- 10.5% for the light grey prototype.

From the evaluation of the recorded data it arises that operating temperature seems to do not significantly affect the modules behaviour since the temperature difference is in the maximum amount of 2 degrees and this is probably due to the low amount of colour that has been applied on the front glass. In detail, this colour degree allows to obtain an aesthetical effect on modules, without affecting in a relevant way their electro-thermal behaviour.

4 MULTI-CHROMATIC BIPV MODULES CHARACTERIZATION

Besides the evaluation of the mono-chromatic BIPV modules, this study is also aimed at investigating how different colored BIPV prototypes can be combined in series for envisioning the design of a multi-chromatic module. For this reason, a calculation method has been defined to evaluate if it is possible to preliminarily identify which design options are compatible to form an efficient multi-chromatic BIPV module. Specifically, this methodology consists of two steps:

- 1st step: evaluation of the mismatch among different mono-chromatic modules, by using data arising from indoor measurements;
- 2nd step: evaluation of the mismatch among different mono-chromatic modules, by the assessment of data retrieved from outdoor monitoring.

4.1 Evaluation of the mismatch using data from indoor measurements

Thanks to the IV curves collected from PM measurements at STC, the mismatch among some mono-chromatic modules has been evaluated starting by plotting the IV curves as shown in Figures 7.

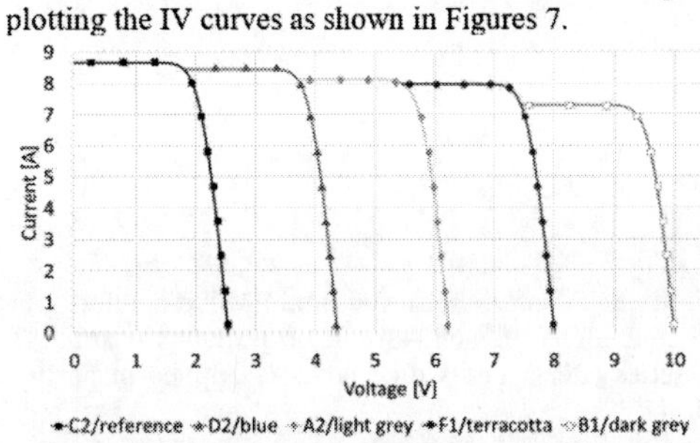

Figure 7: Float glass with colours in applied in position 2, internally. IV curves plotted to visualize better the current mismatch among different colours.

To create multi-chromatic BIPV modules, PV cells are envisioned to be connect in series and the front glass colour should be selected so that there is not a significant mismatch in terms of current intensity. Indeed, thanks to this approach, it can be possible to avoid the activation of bypass diode and avoid significant power reduction. Taking the prototypes with float glass characterized by colours applied in position 2 as examples, the mismatches are calculated by considering the maximum power point of the less performant prototype. The possible combinations for such prototypes are represented in Figure 8, where the comparison with a reference prototype is performed. From this analysis, it arises that the maximum mismatch comes out from the combination of the light grey and the terracotta prototypes, with a power loss in the amount of 6% in comparison to the references.

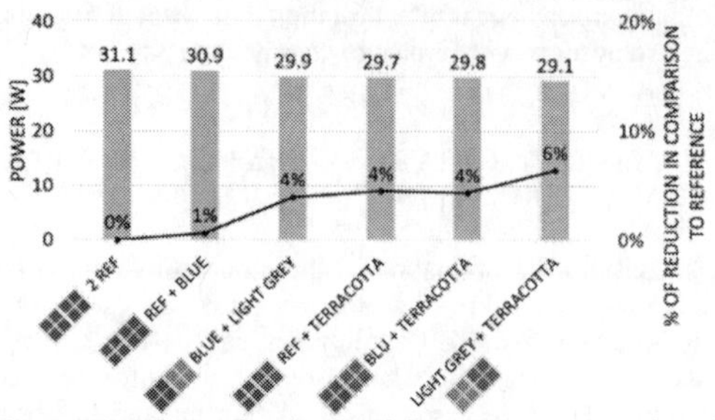

Figure 8: Mismatch calculated from IV curves retrieved from PM at STC and comparison with the two reference modules connected in series.

In order to evaluate the validity of the power output calculations for the combination of coloured BIPV prototypes, a series of prototypes has been subjected to the power measurement at STC. IV curves are shown in Figure 9 and errors are shown in Table 3.

What is interesting to note is that the errors between the calculation method and the real measures are in the range between 0.7% and 1.7%, which are very small errors considering also that the accuracy of the instruments is in the amount of 2.7%.

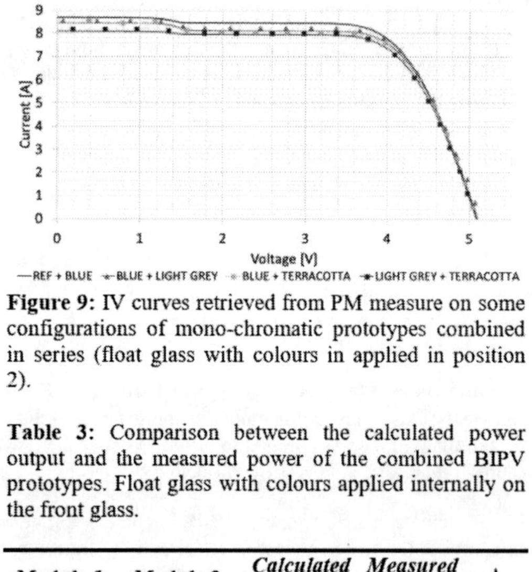

Figure 9: IV curves retrieved from PM measure on some configurations of mono-chromatic prototypes combined in series (float glass with colours in applied in position 2).

Table 3: Comparison between the calculated power output and the measured power of the combined BIPV prototypes. Float glass with colours applied internally on the front glass.

Module 1	*Module 2*	*Calculated P_{MAX} [W]*	*Measured P_{MAX} [W]*	Δ
C2 Ref	D2 Blue	30.97	31.13	+0.6%
D2 Blue	A2 Light Grey	29.93	30.26	+1.1%
D2 Blue	F1 Terracotta	29.79	30.00	+0.7%
A2 Light Grey	F1 Terracotta	28.98	29.46	+1.1%

4.2 Evaluation of the mismatch using data from outdoor measurements

Since a series of mono-chromatic BIPV prototypes has been exposed outdoor, a preliminary evaluation of the mismatch has been performed also when modules are subjected to operative conditions. Thanks to the power output monitoring and IV curves collection, the mismatch among different mono-chromatic modules has been evaluated in three relevant moments during a clear day: with low, medium and high irradiation values, respectively at 200, 400 and 730W/m^2. As shown in Figure 10, the mismatch among the combination of mono-chromatic prototypes in comparison to the reference prototypes is largely higher than the mismatch measured at STC, even though there is the same trend.

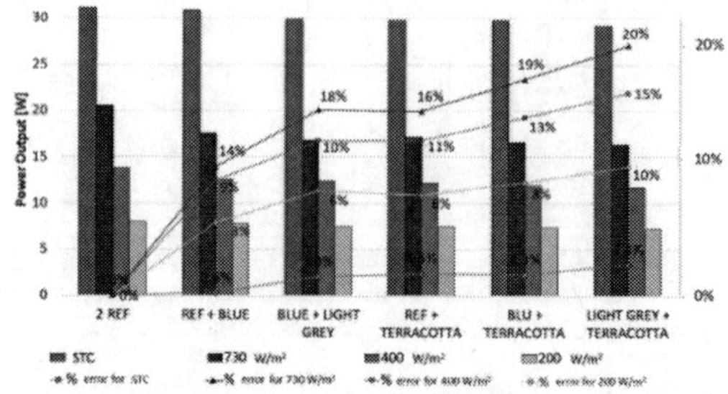

Figure 10: Mismatch calculated from IV curves retrieved from PM at operative conditions and comparison with the two reference modules.

In addition to this calculation, real outdoor measurements on two combinations of prototypes for different irradiation values have been carried out with the aim to evaluate the validity of the calculation method also for outdoor conditions. Tested combinations of prototypes (light grey/blue, and light grey/terracotta) and related errors are shown in Figures 11 and 12.

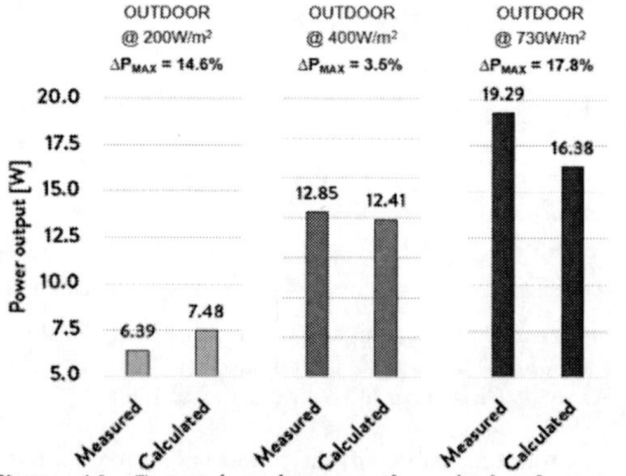

Figure 11: Comparison between the calculated power output and the measured power of the combined BIPV light grey and blue prototypes.

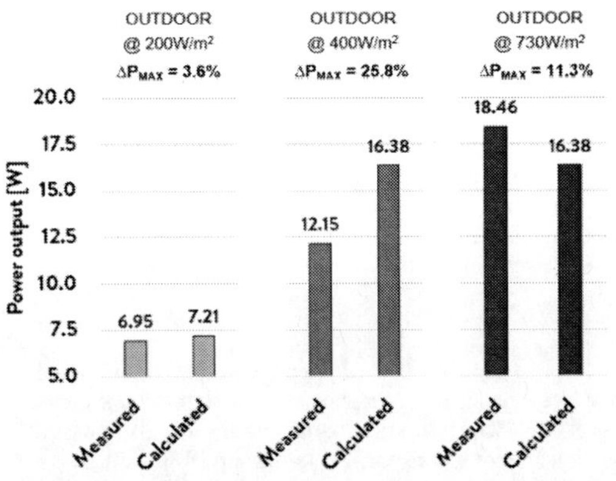

Figure 12: Comparison between the calculated power output and the measured power of the combined BIPV light grey and terracotta prototypes.

From these measures, it arises that errors of the calculation method are up to 25%. This is probably due to a combination of effects, which are not present at STC but only at outdoor conditions, such as the spectral responsivity of the different colours and the angle of incidents.

5 CONCLUSIONS AND FUTURE CHALLENGES

From the indoor and outdoor characterization on mono-chromatic BIPV prototypes, it arises that thanks to the low amount of colour coverage application (10%), the power output is not significantly affected, as well as the operative temperature. Indeed, the higher power loss results for the dark grey prototype in the amount of 17% in comparison to the reference module.

Furthermore, in order to support the development of multi-chromatic BIPV modules, this study has introduced a calculation method to assess the mismatch among two different colours and to determine the resulting power output. Thanks to this calculation method, the power output of two mono-chromatic modules combined in series can be obtained as they form a multi-chromatic element. Hence, the accuracy of this calculation method at STC has been verified by real measurements on combined prototypes. As a result, the errors of this method emerge to be very low, with a maximum error of 1.1%.

However, when the same method is applied to combined prototypes exposed at outdoor conditions, the calculation method provides errors up to 25%, since it is not able to consider real operative conditions (e.g. spectral responsivity of different colours, angle of incidence, low influence of temperature, ...).

Therefore, considering such preliminary results, further investigation should be performed on combined coloured prototypes in order to gather other relevant data and develop a more detailed calculation method that can allow to obtain reliable data about multi-chromatic modules when exposed in real operating conditions. This could be useful for modules manufacturers that can rely on predesign options, thus optimizing costs for producing novel aesthetical BIPV modules. Furthermore, this study has highlighted the importance of considering outdoor conditions, besides STC, when multi-chromatic modules are developed.

ACKNOWLEDGEMENTS

The work has been supported by the Swiss Commission for Technology and Innovation, n°CTI 27973.1 INNO-IW and Advanced PV Energy Rating, "PV-Enerate", funded from the EMPIR programme. Authors would like to thank the industrial partners Sunage (especially Mr. G. Luzi) for the manufacturing of the modules and the support. We also thank a lot the SUPSI team (specifically, Mr. D. Chianese, Mr. G. Bellenda, Mr. R. Roldàn Molinero, Mr. E. Burà and Mr. N. Ostinelli) for their unique and valuable support.

REFERENCES

[1] Saretta, E., Bonomo, P., Frontini, F. (2017) Active BIPV glass facades: current trends of innovation, Proceedings of the Glass Performance Days 2017 - June 2017, Tampere, Finland.

[2] Advanced PV Energy Rating, "PV-Enerate". Project funded from the EMPIR programme co-financed by the Participating States and from the European Union's Horizon 2020 research and innovation programme. https://www.pv-enerate.ptb.de/16eng02-project.html

[3] Saretta, E., Frontini, F., Bonomo, P., Weber, T.& Berghold, J. (2016) Indoor and outdoor characterization of innovative colored BIPV modules for facade application, Proceedings of the 32nd European Photovoltaic Solar Energy Conference and Exhibition, 20-24 June. 2016, Munich, Germany

[4] Ulbikas J., Galdikas A., Stonkus A. (2015) SMART-FLeX – demonstration of integrated approach for design of glass BIPV façade, Proceedings of the 31st European Photovoltaic Solar Energy Conference and Exhibition, September 2015, Hamburg, Germany.

EXPERIMENTAL ANALYSIS OF DIFFERENT CELL AND MODULE TECHNOLOGIES IN A BIPV FAÇADE TEST SET UP

J. Lehmann[1,2], J. Gonçalves[1,2], A. van der Heide[3], G. Yordanov[1,2], H. Goverde[3], K. Baert[1,2], D. Saelens[1,2]
[1] Building Physics, KU Leuven, Kasteelpart Arenberg 10, 3001 Heverlee-Leuven, Belgium
[2] ESAT-ELECTA, KU Leuven, Kasteelpark Arenberg 10, 3001 Heverlee-Leuven, Belgium
[3] EnergyVille, Thor Park 8310, 3600 Genk, Belgium
[4] imec, Kapeldreef 75, 3001, Heverlee-Leuven, Belgium

ABSTRACT: The purpose of this paper is to compare different cell and module technologies comprising monofacial or bifacial, SmartWire or standard busbar interconnection and white or transparent encapsulant in a real conditions BIPV environment. For bifacial modules, the gain obtained in a BIPV configuration using different wall material with contrasting optical properties was investigated. This paper introduces the experimental set-up, followed by an electrical analysis at STC conditions, an electrical and thermal analysis on a sunny winter day and finally a statistical long-term energy yield analysis. Temperatures were measured each second on the back of each module on several relevant location while IV measurements were acquired every 15 seconds, allowing therefore Voc and Isc comparison between the modules. The electrical measurements have evidenced a power production for transparent encapsulant bifacial modules with SmartWire technology up to 16% higher than mono-facial module with standard busbar interconnection. As for the building integration, the results show an important effect of the wall material on the bifacial modules performances, with Isc up to 0.25A higher for white and aluminum wall material, and on the building temperature, with up to 5°C higher temperature observed for black wall material.
Keywords: Building Integrated PV (BIPV), c-Si, Façade, Module Integration, Performance, Thermal Performance, Bifacial

1 SCIENTIFIC INNOVATION AND RELEVANCE

In the European Union, the 2010 Energy Performance of Buildings Directive (EPBD) requires all new buildings to be nearly zero-energy buildings (nZEBs) by 2020. The concept of nZEBs focuses on first improving the energy performance of the building envelope and, then, integrating renewable energy sources (RES) to cover the remaining energy demand [1]. In this context, photovoltaic (PV) systems provide a reliable solution for electricity supply either in existing or new buildings [2]. Building integrated photovoltaic (BIPV) systems have the twofold advantage of increasing the prospects of renewable energy in the built environment, whilst providing savings in materials and construction time by replacing traditional building elements. Recently, bifacial PV cells and multi wire technology (as the SmartWire technology) have attracted interest because of their potential to increase PV cells power production. Moreover, SmartWire technology, due to less visible fingers and more uniform outlook has a greater aesthetic potential for BIPV installations.

In this work, we present the experimental assessment and comparison of new cell and module technologies for BIPV applications. For this study, two experimental BIPV setups based on **mono-facial** or **bifacial** modules and **SmartWire** or **standard busbar** interconnection were installed on a test building in Leuven, Belgium. The installation of BIPV elements in a real building ensures realistic operating conditions for both outdoor and indoor environments. Furthermore, for the bifacial modules, the building integration effect was studied by assessing how the building wall material behind the bifacial module influences the PV performance. Three typical construction materials with contrasting optical properties were employed to cover the surface facing the back of the PV bifacial modules.

This paper is structured as follows. The first section describes the experimental set-up, including the different BIPV configurations and the measurement equipment, the second section focuses on analysing electrical results and the third section displays and interprets thermal results.

2 AIM AND APPROACH USED

The BIPV elements were installed in the vertical plane, on the south-west façade of the KU Leuven test building located in Belgium, with minimum shadowing (Figure 1).

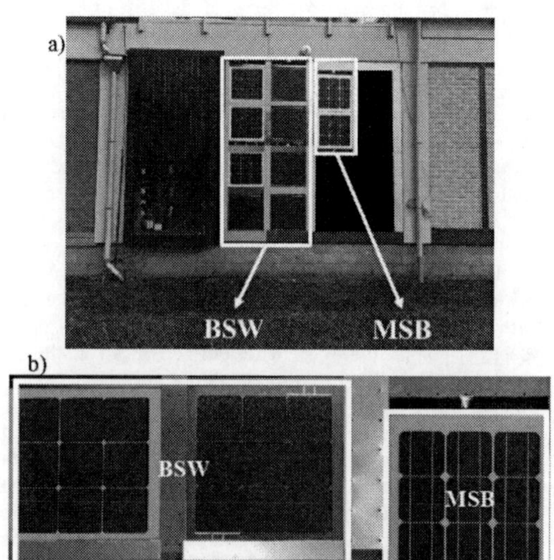

Figure 1: a) Test building with the BIPV setups on the south-west façade. b) PV modules with bifacial cells and SmartWire interconnection (white and transparent encapsulant) and mono-facial cells with white encapsulant and standard busbar interconnection.

The first set-up (Figure 2) is based on eight glass-glass modules; each one composed of nine 6" **bifacial** c-Si solar cell encapsulated on both sides in 0.5mm EVA and covered by a 3.2 mm glass. In order to illustrate the gain obtained using bifacial modules we compared transparent rear encapsulant, allowing bifacial light absorption, to white rear encapsulant, preventing light absorption from the back of the cells (Figure 1b.). The front encapsulant was kept transparent. The cells are interconnected using the **SmartWire technology**. This setup will be referred as BSW: Bifacial SmartWire.

The second set-up (Figure 2) consists of eight glass-glass modules, each one composed of nine 6" **mono-facial** c-Si solar cell encapsulated in a front transparent 0.5mm EVA and rear white 0.5mm EVA and covered on both sides by a 2 mm glass. The cells are interconnected using the **standard busbar** technology. In this work, we will only consider two white ventilated modules. Detailed information and experimental results for this setup can be found in [3]. This setup will be referred as MSB: Mono-facial Standard Busbar.

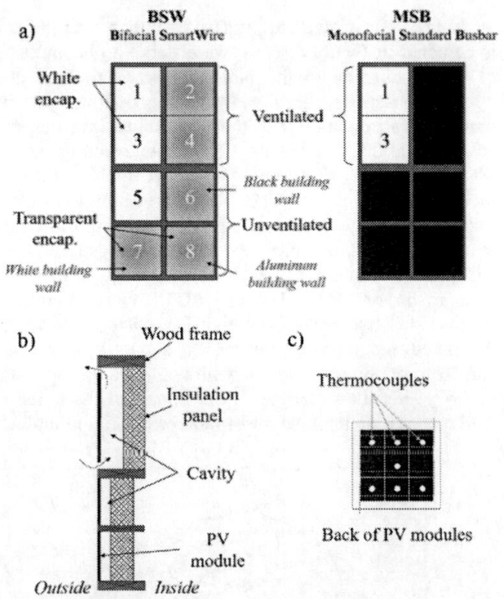

Figure 2: Schematic views: a) front side, b) cross-section view and c) thermocouples location on the back of the PV modules

For both setups, modules 1 and 3 and modules 2 and 4 compose the naturally ventilated configuration (Figure 2b). Modules 5 to 8 are attached to a closed cavity (in this paper, they are referred as unventilated configuration). Additionally, for the BSW setup, in order to assess the influence of the building wall material on the performance of the bifacial cells, the building wall behind modules 6, 7 and 8 was covered respectively with a black, white and aluminum sheet (Figure 2a).

On the back of each module and on the building wall behind the modules we have installed seven thermocouples (Figure 1c.) to measure the temperature at each second. IV characteristics of the PV modules are acquired every 15 seconds. Between each sweep, the PV modules are set at MPP value. Acquiring IV curves instead of solely output power measurements allows to investigate the temperature-related open-circuit voltage [4] and irradiance-related short-circuit current [5] more in detail. A meteorological station records the weather conditions on site. Next to the setups, on the same façade, a pyranometer has been installed to measure the plane-of-array (POA) solar irradiance.

3 RESULTS

The legend in the next sections specifies V for ventilated and U for unventilated; and, W and T for white and transparent rear encapsulant, respectively. The number of the modules corresponds to the indications of Figure 2a.

3.1 Electrical performance

Table 1 shows the results of indoor electrical measurements performed under standard test conditions (STC) for the modules used in this work. For the bifacial modules, we report only the front measurement.

Table I: Electrical performances of the different modules under standard test conditions (STC). For the bifacial cells, we report only the front measurement

Module	Voc (V)	Isc (A)	Power MPP (W)	Module efficiency (%)
1VW BSW	5.91	9.44	41.66	12.38
2VT BSW	5.93	9.25	41.13	12.22
3VW BSW	5.9	9.5	41.68	12.39
4VT BSW	5.93	9.27	41.49	12.33
5UW BSW	5.91	9.51	41.86	12.44
6UT BSW	5.91	9.22	41.3	12.27
7UT BSW	5.94	9.21	41.21	12.25
8UT BSW	5.86	9.22	39.69	11.79
1VW MSB	5.64	8.69	37.19	11.05
3VW MSB	5.64	8.69	37.19	11.05

The results indicate a power increase of 10% at MPP for bifacial compared to monofacial modules. This is caused by an improvement of Voc as well as Isc, allowed by the different technology used. An equivalent result is observed on the module efficiency, with bifacial modules exhibiting efficiencies 1.2 to 1.4% higher. Concerning bifacial modules, white encapsulant modules exhibit higher Isc compared to transparent encapsulant modules. Previous paper has shown that using rear white encapsulant can improve the Isc by increasing light absorption due to photons reflection in the encapsulant [6]. The Voc remains similar for all bifacial modules except 8UT that presents lower Voc causing a decrease of the power at MPP. This is caused by a significant degradation of the fill factor (Figure 3).

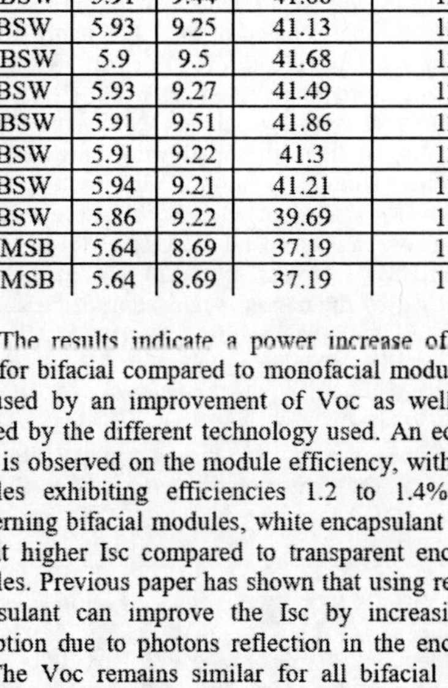

Figure 3: I-V curves of modules 7UT and 8UT BSW

The performances of our setups were studied on a sunny winter day (Figure 4).

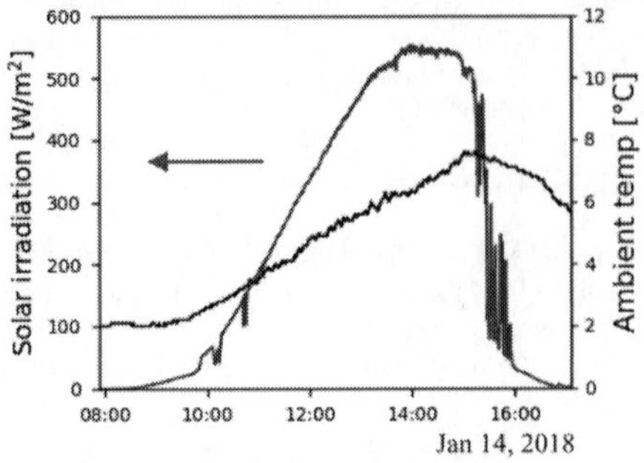

Figure 4: Weather conditions for January 14th

Figure 5 depicts the energy produced by each module during the selected day. Comparing BSW and MSB, we see that BSW modules produce significantly more energy compared to MSB, with differences of up to 21%. Focusing on BSW, we observe that contrary to STC measurements, ventilated bifacial modules with transparent encapsulant (2VT, 4VT) produce more energy compared to modules with white encapsulant (1VW, 3VW). A similar result is observed for unventilated modules (6UT, 7UT and 8UT vs 5VW). Comparing modules with same rear encapsulant color, we oberve an improvement of the total energy produced for ventilated modules (2VT, 4VT or 1VW, 3VW) compared to unventilated modules (6UT, 7UT, 8UT or 5UW). Previous work has shown that this is related to a better cooling of the modules, allowing a higher Voc [3]. It is worth to note that setup MSB has been operational for over two years, while setup BSW has been installed for 1 year. We can consider a decrease of 1% per year for MSB and BSW setups compared to STC measurements to be caused by the normal degradation of the modules. Finally, for both setups, we observe small energy production differences between identical modules in the ventilated configuration (1VW compared to 3VW or 2VT compared to 4VT). Those differences are due to a short shading phenomenon in the morning (Figure 6) associated to different modules temperatures (Figure 7 and Figure 10a).

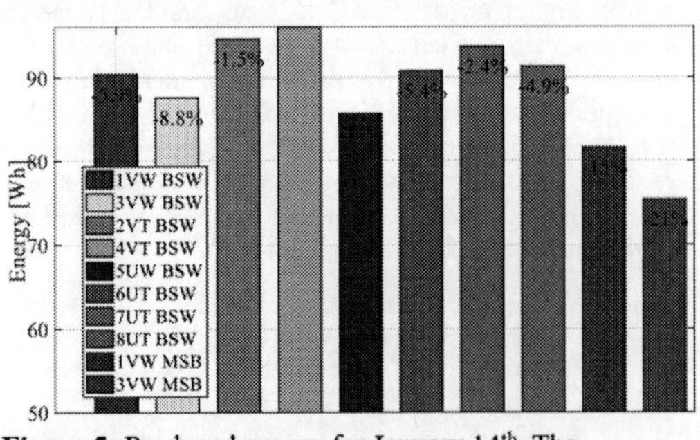

Figure 5: Produced energy for January 14th. The percentages correspond to the energy difference compared to the best performing module (4VT BSW)

Focusing on the building effect on the bifacial modules, Figure 5 shows that 7UT (white wall behind) produces more energy than 8UT (aluminium wall), and both produce more energy than 6UT (black wall). Lastly, the unventilated transparent modules with white and aluminium wall material (7UT and 6UT) produce more energy compared to the ventilated modules with white encapsulant (1VW BSW or 3VW BSW).

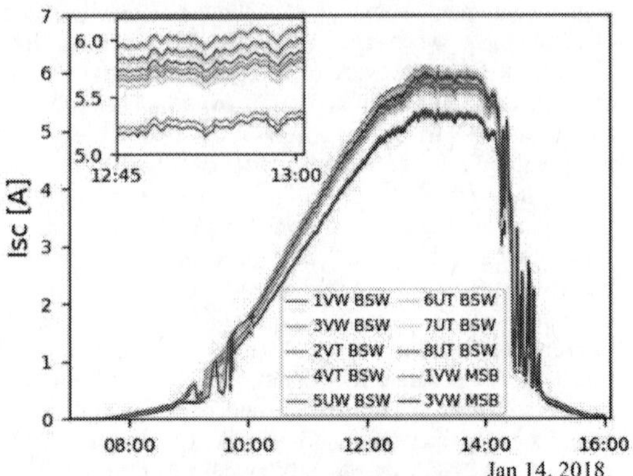

Figure 6: Short-circuit current for January 14th

The short circuit current (I_{sc}) evolution over the day is depicted in Figure 6. First, we observe a phenomenon of partial shading in the morning (between 9h30 and 10h30) caused by the shadow of the building on the panels. In accordance with the STC values, we observe an Isc up to 0.5A higher for BSW modules compared to MSB modules. Concerning the ventilated bifacial modules, as opposed to STC, we observe a higher Isc for transparent rear encapsulant modules compared to white rear encapsulant modules, evidencing the gain in light absorption allowed by bifacial modules. Finally, a comparison of 6UT, 7UT and 8UT shows interesting results: 7UT and 8UT show high Isc, comparable to the ventilated modules, whereas 6UT shows a lower Isc similar to the bifacial module with white rear encapsulant (5UW). This shows the important impact of the building wall material on the light absorption of bifacial modules.

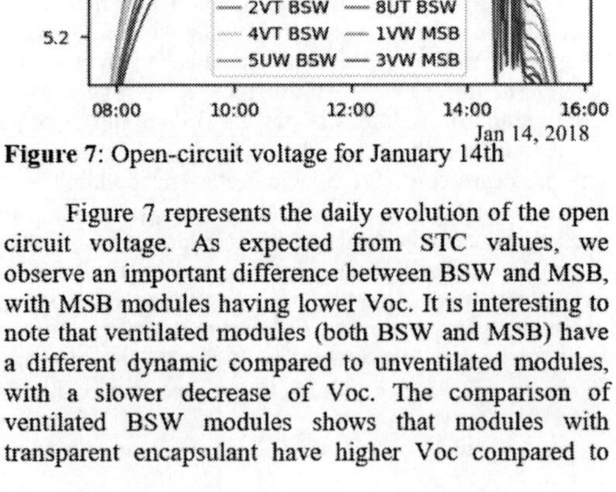

Figure 7: Open-circuit voltage for January 14th

Figure 7 represents the daily evolution of the open circuit voltage. As expected from STC values, we observe an important difference between BSW and MSB, with MSB modules having lower Voc. It is interesting to note that ventilated modules (both BSW and MSB) have a different dynamic compared to unventilated modules, with a slower decrease of Voc. The comparison of ventilated BSW modules shows that modules with transparent encapsulant have higher Voc compared to modules with white encapsulant. Finally, comparing 6UT, 7UT and 8UT, we observe a significantly higher Voc for 7UT. Voc from 6UT and 8UT show similar lower values. It is important to note that 8UT exhibited a lower Voc at STC compared to other bifacial modules (Table 1 and Figure 3).

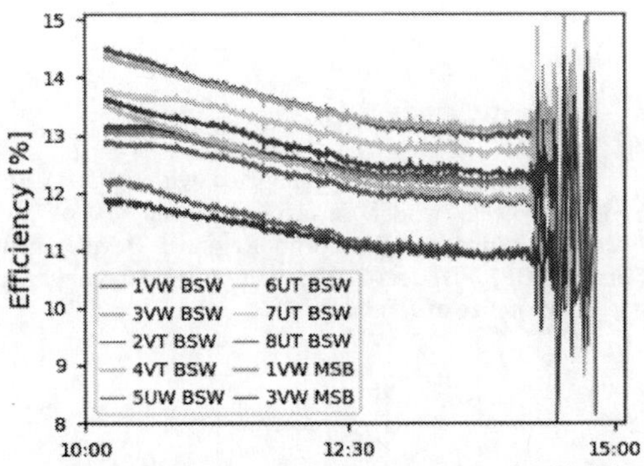

Figure 8: Module efficiency for January 14th

On Figure 8, we have represented the evolution of module efficiency on the studied day. We can observe a decrease of the efficiency during the day: this is a well know effect due to the increase of the modules' temperatures. It is interesting to note that the efficiencies at the beginning of the day are higher than the measured STC efficiencies. This is caused by the low ambient temperature (Figure 4), allowing modules temperature lower than STC (Figure 10a). Comparing ventilated modules, we see an important difference between MSB and BSW modules with BSW modules with white rear encapsulant having efficiencies up to 1.2% higher than MSB modules, which is comparable to STC measurements. However, for ventilated BSW modules with transparent rear encapsulant, we observe efficiencies up to 2.5% higher than MSB modules, evidencing once again the gain in efficiency obtained using bifacial modules. Focusing on BSW modules, we observe that transparent encapsulant ventilated modules (2VT and 4VT) have the highest efficiencies followed by non-ventilated modules associated to white and aluminium wall (7UT and 8UT). Finally, modules with low back of the cell light absorption due to white encapsulant or black building wall (1VW, 3VW, 5UW, 6UT) evidence similar efficiencies.

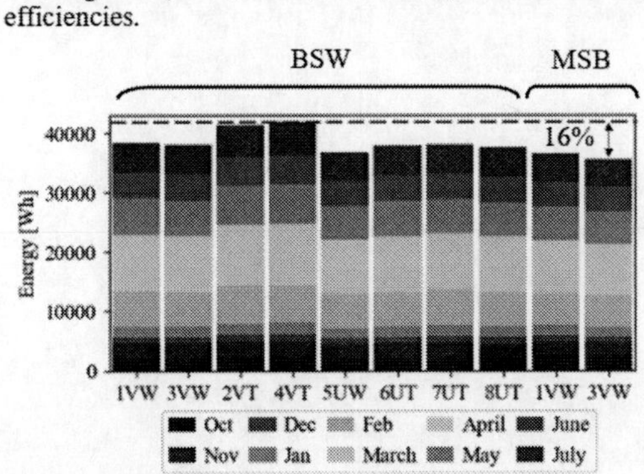

Figure 9: Produced energy on a ten months period

The energy production of each month for a ten months period (Figure 9) confirms the results obtained on the 14th of January with a power production for module 4VT BSW up to 16% higher than for module 3VW MSB. Comparing BSW modules with same encapsulant color (1VW and 3VW compared to 5UW or 2VT and 4VT compared to 6UT, 7UT, 8UT) we see that ventilated modules perform better than unventilated ones. Comparing interconnection technology, we observe a significant gain using SmartWire technology rather than standard busbar interconnection (1VW and 3VW BSW compared to 1VW and 3VW MSB). Comparing both encapsulant color for BSW modules, we observe an important bifacial gain for transparent encapsulant modules (2VT and 4VT compared to 1VW and 3VW). Finally comparing building wall material, we observe a significant improvement with aluminium or white walls compared to black wall (7UT and 8UT compared to 6UT).

3.2 Thermal analysis

Figure 10a shows the PV temperature measured on the back of the module by the middle thermocouple (as presented in Figure 2). Amongst the BSW modules, the unventilated ones experience higher temperatures compared to the ventilated ones, with differences close to 15 K. This confirms the different dynamic between ventilated and unventilated configurations evidenced in Figure 7. Among the ventilated configurations, for all setups, the module located at the bottom of the cavity (modules 3 and 4) presents lower temperatures. This behaviour is due to the buoyancy effect, which drives the airflow from the bottom to the top of the cavity: cold air from the outside enters the cavity through the bottom opening, leading to a better cooling effect on the bottom module. UT8 experiences the highest PV temperature, approximately 5 K higher than its counterparts UT6 and UT7, which may be related to lower efficiency due to lower Voc at STC as shown in Table 1.

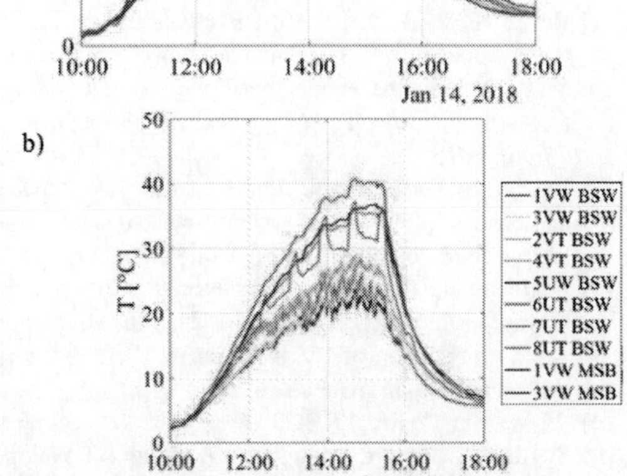

Figure 10 – Temperature evolution over the day for January, 14th: a) PV back temperature, b) building wall (inside the cavity)

On Figure 10b, the wall temperature measured behind each module is presented for the BSW modules. We observe considerably higher temperatures for 6UT (black wall) compared to the other unventilated BSW modules. This result indicates the importance of the building wall material when considering transparent bifacial modules for BIPV applications. Note that the peaks observed for the building wall temperature occur when the sunrays go through the transparent part and reach the thermocouples. When the irradiation is absorbed by the opaque PV cells, lower temperatures are verified. For the ventilated modules, this effect is less pronounced

4 CONCLUSION

Through thermal and electrical analysis, we have compared the performances of mono-facial or bifacial and SmartWire or standard busbar technologies on a BIPV setup. The electrical measurements have evidenced a higher power production for transparent encapsulant bifacial modules compared to white encapsulant bifacial modules or mono-facial modules. It was found that, concerning bifacial modules, the encapsulant color plays an important role with transparent encapsulant modules allowing higher solar absorption (higher Isc) associated to slightly lower temperatures (higher Voc). As for the building integration, the results show an important effect of the wall material on both the PV performance, with Isc up to 0.25A higher for white and aluminum layers, and the building temperature, with up to 5°C higher temperature observed for black layer. Although the white encapsulant helps to conceal the interconnections, the results presented in this paper show the advantage of using transparent encapsulant to fully exploit the bifaciality characteristic of such PV modules.

The selected day had high solar irradiation but low ambient temperatures. To assess the impact of varying weather conditions, future work will focus on extending this evaluation to assess the PV performance under seasonal variability (over a year) and degradation over extended periods.

5 REFERENCES

[1] Recast, E. P. B. D. (2010). Directive 2010/31/EU of the European Parliament and of the Council of 19 May 2010 on the energy performance of buildings (recast). *Official Journal of the European Union, 18(06), 2010.*

[2] Raugei, M., & Frankl, P. (2009). Life cycle impacts and costs of photovoltaic systems: current state of the art and future outlooks. *Energy*, *34*(3), 392-399.

[3] Lehmann, J., Parys, W., Goncalves, J., Govaerts, J., Goverde, J., Baert, K., Saelens, D. (2017). An in-depth comparison of PV modules in a BIPV facade test setup. *In 34th EU PVSEC 2017.*

[4] P. Warren, "IEA ECBCS Annex 32 - Technical Synthesis Report: Integral Building Envelope Performance Assessment", 2003.

[5] G. Van den Broeck, W. Parys, H. Goverde, S. Van der Putten, J. Poortmans, J. Driesen, K. Baert, D. Saelens, Experimental analysis of the performance of façade-integrated BIPV in different configurations *in 33th EU PVSEC, 2016*

[6] Goverde H., Anagnostos D.,Govaerts J., Manganiello P.,Voroshazi E.,Baert K.,Szlufcik J.,Catthoor F., Poortmans J.,Driesen J. (2017). Accurately simulating PV energy production: exploring the impact of module build-up. *In 34th EU PVSEC 2017*

6 ACKNOWLEDGMENT

This work has been conducted within the EFRO SALK project, which receives the support of the European Union, the European Regional Development Fund ERDF, Flanders Innovation & Entrepreneurship and the Province of Limburg.

DUTCH SOLAR DESIGN BIPV: OPTIMIZING POWER OUTPUT AND AESTHETIC PERFORMANCE IN ARCHITECTURAL DESIGN

L.H. Slooff[1], J.A.M. van Roosmalen[1], L.A.G. Okel[1], T. Minderhoud[2], G. Gijzen[2], L.C. Polinder[3], F. Goethals[3], T. Sepers[4]
[1]Energy research Centre of the Netherlands, P.O.Box 1, 1755 ZG Petten, The Netherlands, tel: +31 224 564314, fax: +31 224 56 8214, email: slooff@ecn.nl
[2] UNStudio, PO Box 75381, 1070 AJ Amsterdam, The Netherlands
[3]Design Innovation Group, Nieuwegracht 1, 3512 LB Utrecht, The Netherlands
[4]TS Visuals, Schuit 14, 1724 VD Oudkarspel, The Netherlands

ABSTRACT: The energy transition and more precisely, the contribution of photovoltaics (PV) in that transition, asks for better integration of PV in buildings, infrastructure and landscapes. To realize this, a variation in shape, size and color of PV products is needed. In this paper a technology will be shown that allows variation of all these features. The effect of design variables on the performance is presented as well as stability studies with respect to power output and color. The results show that the performance and visual appearance can be optimized, resulting in a good visual appearance without compromising the power output too much. The outdoor tests show no degradation in color or power output. When subjected to concentrated sunlight, the primary ink colors show limited change, hardly perceivable in a full color print after 25 years. It shows that this technology is able to add the desired variations in shape, size and color to PV modules.
Keywords: module, design, color, façade, BIPV, landscape, solar

1 INTRODUCTION

The need for a more sustainable built environment is urgent and requires cost-efficient, smart and integrated solutions. We envision a (very near) future in which the built environment is efficient and friendly, smart and accessible, high-tech and attractive. The surfaces of the buildings and environments we live and work in represent enormous potential. Incorporating our approach of customized, designed photovoltaic cladding materials in any architectural design adds a novel functionality to façades: generating renewable energy without compromising creative freedom and artistic expression.

Various research groups and companies are developing PV cladding material with a different appearance compared to the traditional photovoltaics [1-14]. These concepts differ mainly in the way the color is applied.

Our aim is to create aesthetic building materials with the added functionality of generating renewable energy including freedom of size and shape and a good performance. We foresee also application in infrastructure and landscape integration.

The concept is based on the back contact foil technology [15-17] in which the back contact cells are interconnected via a conductive foil. Too our knowledge this is the only technology that allows for easy variation of sizes and shapes without the need for large changes in production technology. We developed a technique to integrate a full color, durable print into a façade panel with integrated PV-cells.[18] The desired image or design is processed by an algorithm, which negotiates an optimum between minimal surface coverage and the maximal optical effect. In this way many different attractive designs can be made. The customized prints are engineered out of little dots which 'blend' together to the human eye, creating a subtle optical illusion - while plenty of the surface is actually bare. See Figure 1.
It will be shown that by playing with the print transmission, the power reduction can be minimized further while maintaining the aesthetics. The resulting aesthetic photovoltaic cladding materials are used in first example building and infrastructure integration designs.

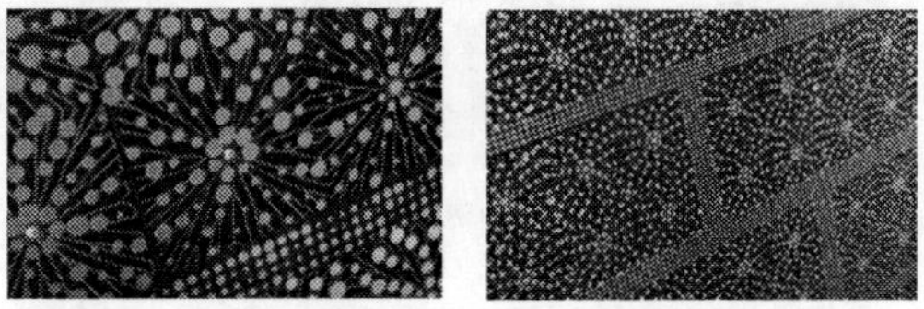

Figure 1: Photographs of the dot pattern details of one of the small solar modules with a brick-like design print.

2 RESULTS

Optimizing power output, aesthetic performance and architectural design while maintaining long product lifetimes are important aspects when developing a new technology and they should be equally addressed. In this work these aspects are all taken into account, in order to develop a product that not only has a nice appearance, but that will also fit into the architectural design of a building and gives a substantial contribution to the renewable energy supply of the building.

2.1 Appearance studies

The applied designs consist of small dots that make up the picture. It was shown before [8] that the size of the dots does not have a large effect on the performance but that it is mainly the coverage of the print and to some extent the colors that influence the performance. So to a large extent, the dot pattern can be tweaked by playing with dot size, coverage and color combination, to optimize the visual appearance of a particular print. To optimize the appearance, different patterns were made based on the same picture to see the effect of the pattern variation.

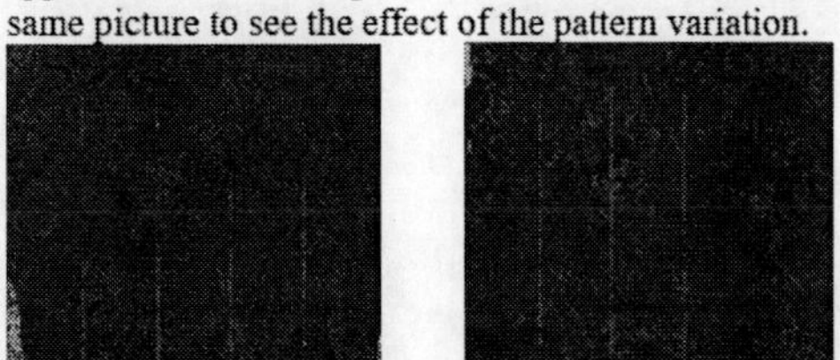

Figure 2: Single cell laminates with printed leaves. On the left small dot size, on the right large dot size (50% larger).

In Figure 2, two laminates with similar leave print are shown where the one on the left has a 50% smaller dot size compared to the one on the right. The print coverage is 35%

As can be seen in Figure 2, the smaller dot size improves the resolution, making the leaves more visible but the appearance darker. The outcome of the trade-off between color and resolution will depend on the viewing distance for the specific situation. The larger the viewing distance, the larger the dot size can be, resulting in pictures with a more vivid color. However, as shown below, the ink transmission also plays an important role in the vividness of the color.

The laminates with small and large dots were measured using a PASAN flash tester. Figure 3 shows the normalized current density for 3 different leave prints, A, B and C, with small and large dots. Currents were normalized to the small dot laminate of the specific print. It shows that the dot size does not influence the short circuit current density. This was seen previously for the primary print colors as well.[19] The cells were measured with a WACOM solar simulator before they were laminated. As such the loss could be determined. For this print the loss in current is 22-24%. Note, that this is the total cell to module loss and includes optical losses in the laminate as well as the optical losses from the applied print.

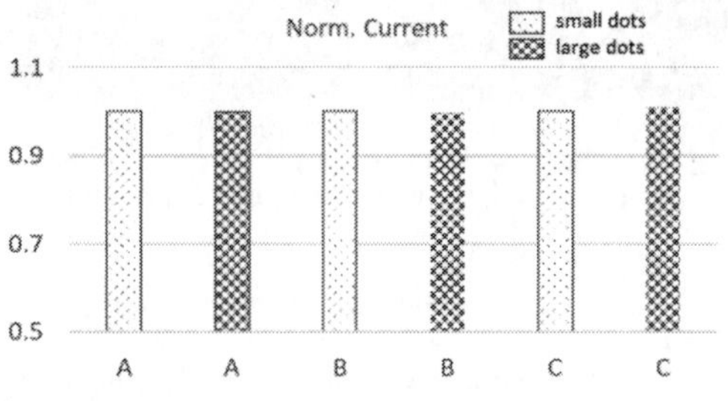

Figure 3: Normalized current density for 3 different leave prints (A, B and C) with small dots or large dots.

2.1 Performance studies

Adding a print to a PV module to improve the appearance of the module will always affect the performance of that PV module. The challenge is to minimize the losses induced by the print. Previously it was shown that a brick laminate as shown in Figure 4, with a print coverage of 44.6%, has a loss in short circuit current density (Isc) of 32% with respect to the Isc of the cell (see Figure 5, blue dots). [19] Although a current loss of 32% for a coverage of 44.6% is reasonably, it is still a large fraction of the current.

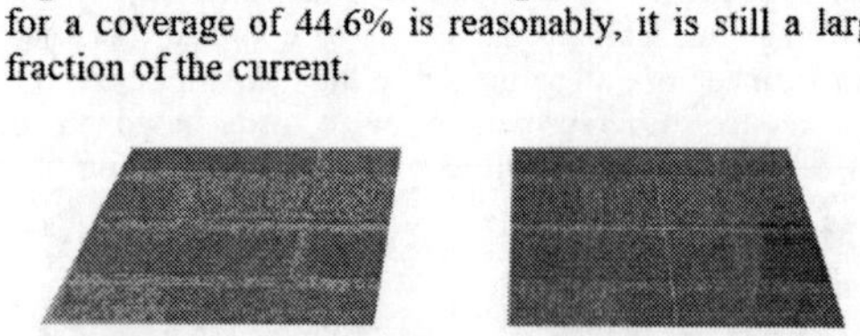

Figure 4: Panels with a different ink transparency: left lower transparency than on the right.

For this reason optimization of the transmission of the ink was performed while aiming at a good visual appearance of the laminates, see Figure 4. There is only a very small difference between the two brick laminates, where the colors of the left panel are slightly lighter than of the panel on the right. The latter is the panel with the increased transparency of the ink. The increased ink transmission reduces the Isc loss to 25%, see the orange crosses in Figure 5.

For aesthetical purposes it often suffices to use lower print coverages, in the order of 25-30%, see e.g. Figure 6. The reduced print coverage is tested for the leave print of Figure 6, which has a print coverage of 30%.

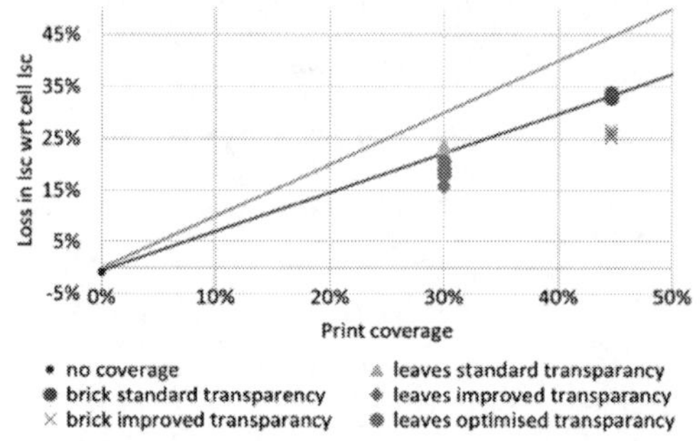

Figure 5: Loss in Isc of the laminate when compared to Isc of the cells in the laminate for different print coverage and different ink transparencies.

On the left the standard ink transmission, in the middle the improved ink transmission, similar to the improved ink transmission of the brick laminate, and on the right an optimized ink transmission. Optimization was done on performance and aesthetics. The results are shown in Figure 5 for the standard ink transmission by the yellow upright triangles, for increased ink transmission by the light blue diamonds and for the optimized ink transmission by the green circles. The solid lines are a guide to the eye. Depending on the ink transparency, the loss of a 30% print coverage is 15-22%. This shows that aesthetic quality can be added to PV panels with acceptable power losses in the order of 15-20% only.

2.3 Lifetime studies

Besides the performance, the lifetime of the modules is also very important. Damp-heat, thermal-cycling, outdoor- and UV irradiance tests were performed to monitor the stability in output and colour on dedicated test modules (Figure 7). The results show no degradation after 1500 hrs of damp-heat and 400 cycles of thermal-cycling. Also, 500 hrs UV testing under a Xenon lamp (Suntest XXL) did not reveal any changes in the reflection and transmission spectra of the prints. Figure 7 shows the latest result on outdoor testing at our test facility in Petten, The Netherlands. Laminates were positioned almost South (170°) oriented at an angle of 30 degrees. Before mounting outdoors, the performance was measured using a PASAN flash tester. These results are indicated by the three individual dot series in Figure 7. The other data was taken from the continuous monitoring in the outdoor test facility. That data was corrected to an irradiance of 1000 W/m2. As can be seen the outdoor and flash test data correspond pretty well and do not show any change in output.

Figure 6: 30% leave print with on the left the standard ink transparency, in the middle the improved ink transparency and on the right the optimized ink transparency.

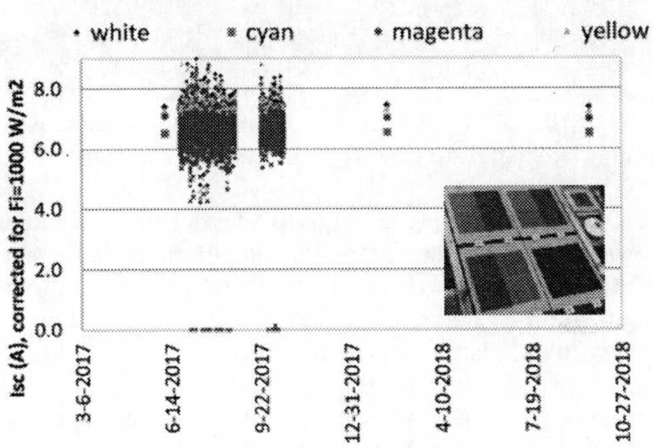

Figure 7 Outdoor measurements on 4-cell laminates where each cell is covered with a different print colour.

However, it must be emphasized that these tests alone do not guarantee an outdoor color stability, even for the UV tests. When looking at the lamp spectra, the UV irradiance is about 10 times higher than for an average yearly irradiance in the Netherlands, so corresponding to 5000 hrs or 208 days. Besides that, the test only contains UV light, whereas some pigments are also sensitive to visible light. Degradation in 4-color inks is mainly due to degradation of yellow and magenta. The reason is that yellow absorbs blue light which has a higher energy and is therefore more destructive, and magenta absorbs green, which is more abundant in the solar spectrum. Besides that, our eye is more sensitive to small changes in magenta compared to yellow. UV testing gives an indication for light fastness, but for a complete test it is thus important that green and blue are also present in the test. To have more realistic outdoor testing we shipped 3 samples of each color to Arizona to be subjected to Q-trac, concentrated sunlight testing, see Figure 9. According to the technical information of the Q-trac [20], the yearly average total irradiance in Arizona is about 8000 MJ/m^2, of which about 333 MJ/ m^2 in the UV. The Q-trac system produces about 1420 MJ/ m^2 in the UV, which is roughly 4 times more. As a rough estimate we can assume that this factor also applies to the other parts of the spectrum. So 1 year of measurements in the Q-trac is, with respect to irradiance, comparable to 4 years under Arizona sunlight. When compared to e.g. the Netherlands, with a yearly average irradiance of 1290 MJ/m^2 [21] this corresponds to 25 years (32300 MJ/m^2). The difference in color perception is measured by measuring the CIElab coordinates [22] using an X-Rite Color i5 photospectrometer.

The International Commission on Illumination (CIE) addressed the topic of color difference by introducing the concept of Delta E (ΔE), the measure of change in visual perception of two given colors:

$$\Delta E = \sqrt{(\Delta L^*)^2 + (\Delta a^*)^2 + (\Delta b^*)^2}$$

Where, L* indicates lightness, a* is the red/green coordinate, and b* is the yellow/blue coordinate, ΔL*, Δa* and Δb* is the difference in respectively L*, a* and b* between the initial color and the color after Q-trac tes-

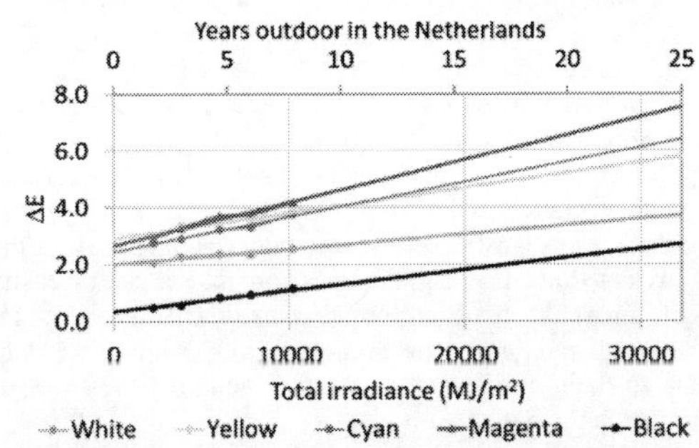

Figure 8: Change in visual perception ΔE as a function of total irradiance. Data points are test results, lines are extrapolated linear fits to the data. Data are an average of 3 samples.

ting. The average ΔE per color of the three samples as a function of the test duration is given in Figure 8. Black shows the lowest initial ΔE, but has the highest change with time. White has a higher initial ΔE but degrades less fast. The other colors show more or less the same initial ΔE. Magenta seems to have a somewhat stronger degradation compared to the other colors. However, the statistics on the data is quite low and as a result conclusions are still preliminary. The extrapolated trend lines indicate a ΔE between 4 and 8 after 1 year of Q-trac testing (as mentioned comparable to 25 years in the Netherlands). But what do these numbers mean for the perception of the color difference? Table I gives a general guideline for the visual perception depending on ΔE. According to this table, the color change is perceptible at a glance. Figure 9 shows a picture of a reference sample and one of the test samples after 10 weeks of testing (corresponding to an irradiance 10206 MJ/m^2). Only when looking carefully one can notice that the colors of the test sample (left) are slightly bleached.

Table I: Visual perception related to ΔE

Delta E	Perception
<= 1.0	Not perceptible by human eyes.
1 - 2	Perceptible through close observation.
2 -10	Perceptible at a glance.
11 - 49	Colors are more similar than opposite
100	Colors are exact opposite

Figure 9: Photograph of one of the test samples after 10 weeks of testing (left) and a reference sample that was kept in the dark (right).

2.4 Inspiration for architects

As mentioned in the introduction, the aim is to introduce these easthestic panels into the building envelope, infrastructure or landscape. For these type of applications variations in the module print, shape and size is important as well. A first series of triangular and quarter circular modules has been made with a leave print, see Figure 10. The panels contained 15 MWT cells each and had an output power of about 50W. The output can be increased by adding half cells to fill the open spaces more. A polymer front sheet with reduced reflection was used in stead of glass for the front side, making the print more visible, especially when looking under an angle. Both the polymer front sheet and the back contact foil can be easily cut into any shape and size. This opens the way for building and landscape architects to design the PV to match their architecture in stead of adapting their design to match the PV.

3. CONCLUSIONS

Appealing photovoltaic cladding material can be made with good performance using an algorithm to convert images into dot patterns. The colored pattern is then integrated in the product. The resulting cladding panels show a loss of about 15-20% in short circuit current depending on the ink formulation and have a good visual appearance.

Figure 10: Two working leave print modules. One with triangular shape and one with quarter circular shape.

The panels do not show signs of degradation of output power and color under outdoor testing for more than a year in the Netherlands. Also 500 hrs UV testing do not reveal degradation of color. Concentrated, accelerated sunlight testing in Arizona, shows a change that is perceptible at a glance in the color readings. When extrapolated to 25 years in the Netherlands these are still small changes and will probably not be observed. Finally it was shown that in combination with the conductive backsheet technology, our approach is suitable to make appealing cladding panels with new sizes and shapes. In fact the conductive backsheet technology is the only interconnection technology that allows freedom of size and shape, thereby offering freedom of design for architects.

ACKNOWLEDGMENTS

This work is financed through the Dutch Solar Design PV (DSD-PV) project with financing from the Topsector Energysubsidies from the Dutch Ministry of Economic Affairs.

[1] L.-E. Perret-Aebi, P. Heinstein, V. Chapuis, S. Pélisset, C. Roecker, A. Schüler, K. Lumsden, Y. Leterrier, J.-L. Scartezzini, J.-A. Manson, C. Balli, Proc. Cleantech for sustainable buildings 1 (2011) 761.

[2] https://www.manz.com/de/maerkte/solar/systemloesungen-fuer-gebaeudeintegrierte-photovoltaik-bipv/manz-cigs-solarmodule/

[3] Swiss Inso: http://www.swissinso.com/

[4] http://www.lofsolar.com/

[5] J.H.Selj, T.T.Mongstad, R.Søndenå, E.S.Marstein, Solar Energy Materials and Solar Cell, 95 (2011) 2576.

[6] V. Neder, S. L. Luxembourg, A. Polman, Appl. Phys. Lett. 111 (2017) 073902.

[7] https://www.sistinesolar.com/

[8] https://www.csem.com/

[9] H.C.A Rooms,. I. Barbu, Z.A.E.P.Vroon, R.Meertens, B.Vermeulen, Proc. SPIE 8110 Thin Film Solar Technology III (2011) 81100V.

[10] C. Tzikas, R.M.E. Valckenborg, M.N. van den Donker, A. Bognar, D. Duque Lozano, R. Loonen, J.L.M. Hensen, W. Folkerts, M. Dörenkämper, 35th EU-PVSEC, Brussels, Belgium, 2018.

[11] E. Saretta, P. Bonomo, F. Frontini, 35th EU-PVSEC, Brussels, Belgium, 2018.

[12] https://www.cnbc.com/2016/10/28/with-teslas-solar-roof-youll-never-have-to-know-these-houses-are-solar-powered.html

[13] *www.kaleo-solar.ch/*

[14] https://kameleonsolar.com/colorblast/

[15] L. Hongfeng, Q. Zhe, C. Kaiyin, Z. Liyan, C. Xianzhi, L. Wei, G. Beaucarne, W. Peng, B. Chislea, Y. Yanghai, A. Zambova, K.M Broek, I.J. Bennett, J. Bakker, N. van Ommen, E. Frederikze, Photovoltaics International (PV-Tech) (2014) 24.

[16] G. Coletti, F. Ishimura, Y. Wu, E.E. Bende, G.J.M. Janssen, K. Hashimoto, Y. Watabe and B.B. Van Aken, Proc. EU PVSEC (2016) 715.

[17] K.M. Broek, I.J. Bennett, M.J.H. Kloos, W. Eerenstein, Energy Procedia 67 (2015) 175.
[18] J.A.M. van Roosmalen, L.H. Slooff, L.A.G. Okel, M.N. van den Donker, T. de Vries, T.Minderhoud, N. Wang, G. Gijzen, T. Sepers, M. Rietbergen, L. Polinder, E.M.B. Heller, A. Versluis, F.Frumau, 12th Conference on Advanced Building Skins, Bern, Switzerland, 2017.
[19] L.H. Slooff, J.A.M. van Roosmalen, L.A.G. Okel, T. de Vries, T. Minderhoud, G. Gijzen, T. Sepers, A. Versluis, F. Frumau, M. Rietbergen, L. Polinder, E.M.B. Heller, F. de Vries, 34th EU-PVSEC conference, Amsterdam, The Netherlands, 2017.
[20]https://www.q-lab.com/documents/public/002697f8-4405-4421-9055-b86f7430e6a2.pdf
[21] De Bilt, The Netherlands: 983 kWh/m2/yr, http://www.allesoverzonnepanelen.nl/voorwaarden/zonnestraling/zoninstraling-in-nederland/
[22] CIE standard ISO 11664-4:2008(E)/CIE S014-4/E:2007 Colorimetry — Part 4: CIE 1976 L*a*b* Colour Space

DECORATED BUILDING-INTEGRATED PHOTOVOLTAIC MODULES: POWER LOSS, COLOR APPEARANCE AND COST ANALYSIS

Christoph Kutter, Benedikt Bläsi, Helen Rose Wilson, Thomas Kroyer, Max Mittag, Oliver Höhn, Martin Heinrich
Fraunhofer Institute for Solar Energy Systems ISE, Heidenhofstraße 2, 79110 Freiburg, Germany
christoph.kutter@ise.fraunhofer.de

ABSTRACT: We evaluate three design solutions for BIPV modules: colored encapsulants, ceramic printed glass covers and spectral-selective photonic Morpho structures, regarding their electrical performance, their optical appearance and their cost. We built single-cell samples on which we perform IV measurements at STC and normal-hemispheric reflectance measurements. We derive color coordinates in CIE L*a*b* space out of the reflectance spectra to quantify color and intensity. Colored encapsulants as well as ceramic prints show low color saturation and high power loss relative to a conventional module (-6 to -31%). The spectrally selective Morpho structure shows strongest color saturation, enhanced brightness and best electrical performance (-3 to -7%). We research module component prices and perform a manufacturing cost analysis for glass-glass modules with the respective decoration and find that decoration of BIPV modules increases the specific module cost, caused by less power output and higher material cost. We find the manufacturing cost of decorated BIPV modules (74-163 €/m²) is within the range of classic cladding materials, such as bricks (60-100 €/m²) or wood (50-180 €/m²) but provide additional benefits regarding power generation.

Keywords: BIPV, facades, power loss, manufacturing cost, color appearance, Morpho structure

1 INTRODUCTION

Since European legislation has tightened the requirements for energy-efficient building [1], the interest in aesthetic, efficient and price-competitive building-integrated photovoltaic (BIPV) solutions has been growing steadily [2–4].

Common silicon PV modules often fail to fulfill the architectural desire for a homogeneous and tailorable appearance. They display a strong contrast between the dark solar cells, the white or transparent background (Figure 1, left) and the shiny interconnectors. Furthermore, their overall appearance is dark. To reduce the contrast between cell matrix and background, a black rear encapsulant can be introduced (Figure 1, right). However, homogeneously colored designs are only possible to a certain extend with this approach.

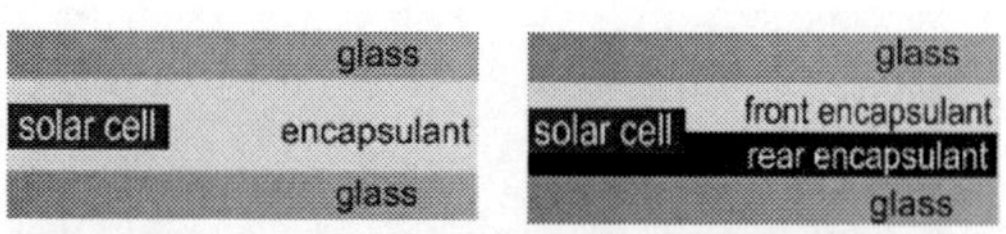

Figure 1: Conventional semitransparent glass-glass module configuration for façade application (left), conventional "black" glass-glass module configuration (right). The top glass cover as well as the encapsulant are suitable design variables for homogenous decoration.

To achieve aesthetic photovoltaic building skins, one approach is to camouflage silicon solar cell appearance by a decorative module component in front of the cell layer, leading to 'Invisible BIPV'.

We focus only on components that create a homogenous color appearance across the whole module area such as colored encapsulants or uniformly coated glass covers.

Each technology leads to a specific power loss (relative to conventional PV modules with fully transparent front glass/encapsulant) and cause additional material costs.

Within this article, we compare three design solutions for colored BIPV modules regarding their optical appearance, electrical performance and economy.

Colored encapsulants are one suitable design option. The most common encapsulant in PV industry is ethylene vinyl acetate (EVA), but no major EVA manufacturer offers colored interlayer films. Within the glazing industry, colored PVB encapsulants are applied in safety glass. Since PVB is a technically feasible encapsulant for PV modules, we use the colored PVB films to manufacture and characterize module samples [5]. The colored PVB replaces the transparent front encapsulant (Figure 2).

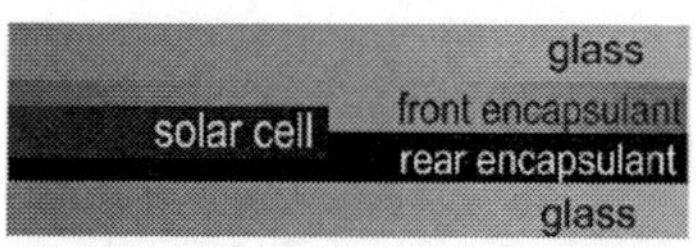

Figure 2 Module configuration with colored encapsulant

Photovoltaic modules with screen-printed or digitally printed glass covers are available on the market. Often partial print patterns are used to obscure the solar cell geometry while keeping transmittance high [6]. The ceramic enamels are stable and UV-resistant. Non-white enamels absorb significant parts of the spectrum, which raises module operation temperature. The print is introduced into the module at the inner side of the glass cover to avoid weathering and soiling exposure. The translucent, light-scattering nature of the applied enamels requires sophisticated measurement equipment to obtain accurate optical data of printed glass panes [7]. Furthermore, the printed glass pane transmittance does not correlate linearly with module performance, as described in [8].

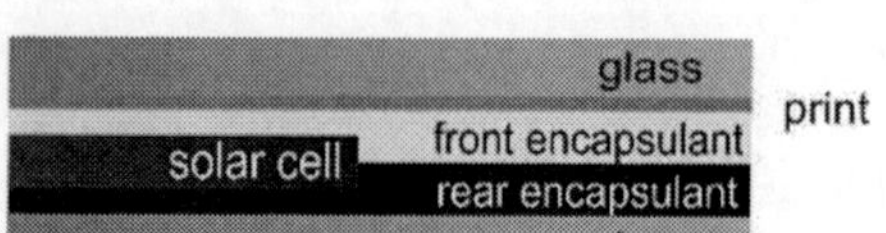

Figure 3 Module configuration with ceramic print

Researchers at Fraunhofer ISE previously presented a spectrally selective photonic structure inspired by the

Morpho butterfly ("Morpho structure") that shows angle-independent saturated colors and low glare [9]. Any color can be obtained and very low losses can be achieved. The photonic structure applied to the front glass reflects narrow bands of the incident spectrum and shows no significant absorbance.

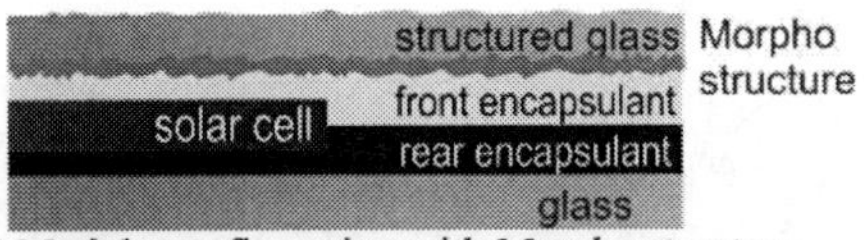

Figure 4 Module configuration with Morpho structure

2 METHOD

2.1 Sample Preparation

To determine the power losses and color saturation associated with each type of decoration, single-cell samples with the respective decorative element were prepared (Figure 5). The lab-scale samples feature no anti-glare layer. Blackened interconnectors increase the homogeneity of the final product appearance but have not been used for the lab-scale samples.

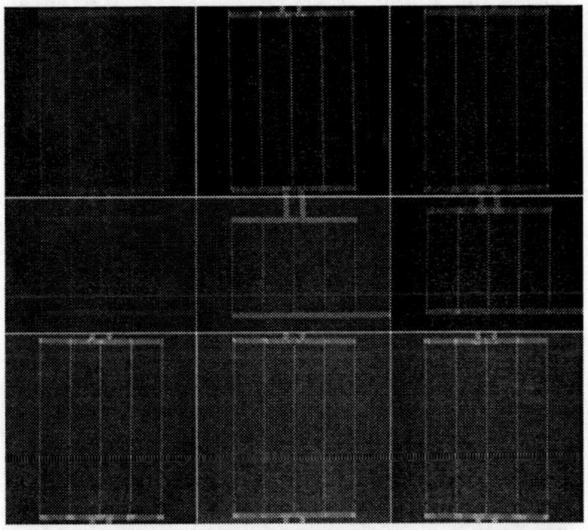

Figure 5: Images of the fabricated single-cell samples under similar illumination conditions of the mentioned design technologies with similar colors (red, green, blue). The metal connectors are deliberately kept visible. From top to bottom: colored encapsulant; full-area digital ceramic print; Morpho structure by Fraunhofer ISE

2.2 Color Analysis

The normal-hemispheric reflectance spectra of the samples in Figure 5 were measured outside the cell area at the TestLab Solar Façades at Fraunhofer ISE with an integrating sphere with a diameter of 220 mm shown in Figure 6 to 8. These spectra represent a lower limit for the reflectance.

The reflectance spectra are plotted in the range from 350 to 800 nm in front of the visible spectrum. The overall reflectance spectra of encapsulants in Figure 6 show low intensities. The red encapsulant shows a reflectance peak with double intensity at 600 nm, providing a better coloration effect compared to the blue and green encapsulant that show no peaks.

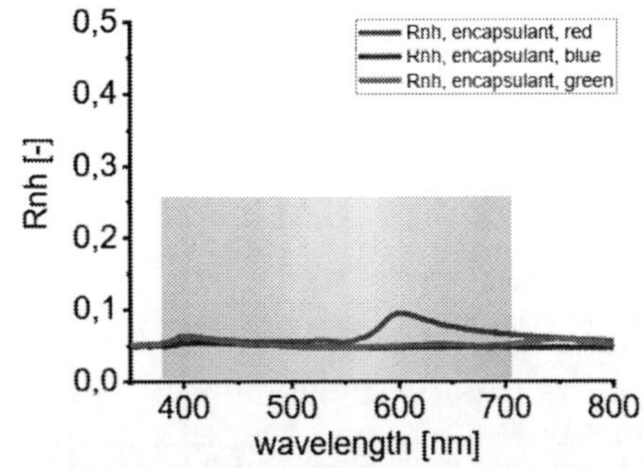

Figure 6 Normal-hemispherical reflectance (R_{nh}) spectra for each module configuration with a colored encapsulant and without solar cell.

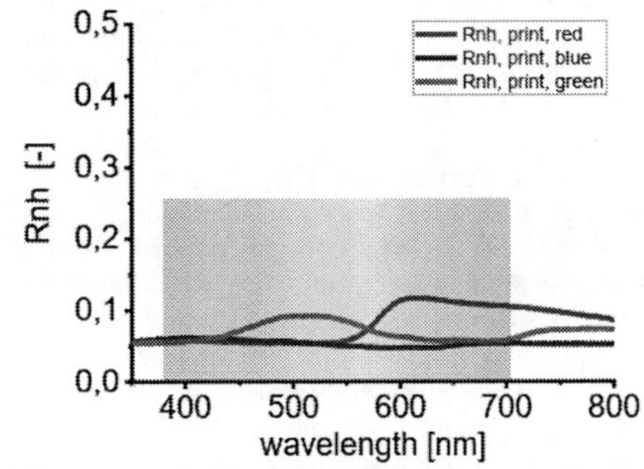

Figure 7 Normal-hemispherical reflectance (R_{nh}) spectra for each module configuration with a digitally printed enamel and without solar cell.

The normal-hemispherical reflectance spectra for red, blue and green digitally printed enamel on glass are shown in Figure 7. The ceramic prints also show low reflectance, with broader reflectance maxima of double intensity for red and green color. The higher reflectance values lead to a better coloration saturation (red>green>blue) of the module (Figure 5).

The overall reflection effect of enamel prints and encapsulants is low compared to the Morpho structure, whose reflectance spectra show narrow and strong maxima of 5-fold increased intensity (Figure 8).

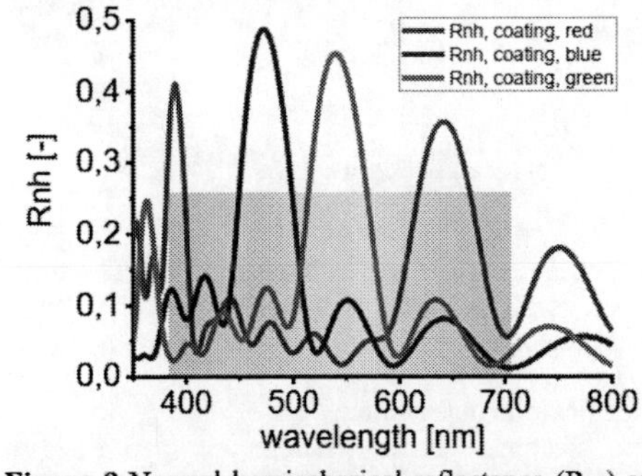

Figure 8 Normal-hemispherical reflectance (R_{nh}) spectra for each module configuration with a spectrally selective Morpho structure from Fraunhofer ISE and without solar cell.

The above plotted normal-hemispherical reflectance spectra are used to derive the CIE L*a*b* color coordinates to quantify brightness and saturation as given in Table 1[10].

Table 1: Color coordinates in CIE L*a*b* space

	L*	a*	b*
Encapsulant, red	30.8	7.4	3.6
Ceramic Print, red	32.5	12.4	7.3
Morpho structure, red	32.9	36.2	0.1
Encapsulant, green	26.5	2.0	-2.6
Ceramic Print, green	33.7	-9.2	2.1
Morpho structure, green	55.3	-42.8	36.3
Encapsulant, blue	26.2	0.4	-1.8
Ceramic Print, blue	27.2	-0.4	-3.5
Morpho structure, blue	39.0	-11.2	-34.7

The L* values show a very similar brightness for the encapsulant and ceramic print system, while the Morpho structure leads to enhanced brightness values for green and blue. The red Morpho Structure sample does not show high brightness due to non-optimized parameters for sample production.

In order to evaluate the color saturation, Figure 9 shows a plot of the a*b* values derived from the reflectance measurements. In this plot, the distance from the origin represents the color saturation. It is evident from this plot that the encapsulant hardly leads to a color effect, the ceramic print shows only a moderate effect, and the Morpho structure features a high color saturation. These results are in good accordance with the subjective impression from Figure 5.

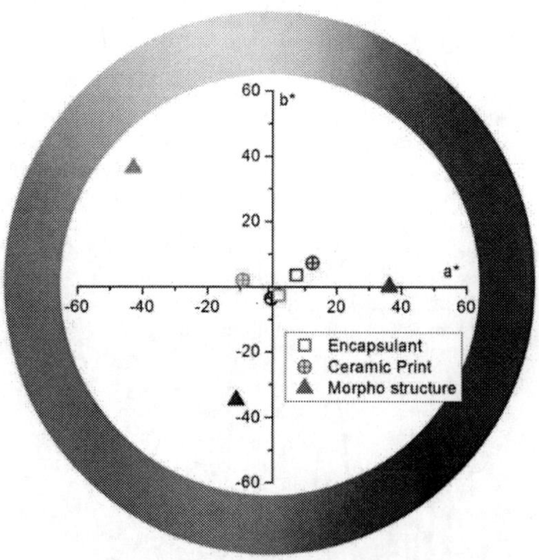

Figure 9 CIE a*b* diagram with the representation of the measured samples. The distance from the origin represents color saturation.

2.3 Electrical Characterization

The electrical characterization was performed at Standard Testing Conditions (STC) at Fraunhofer ISE Module-TEC. The loss in power is determined relative to a reference module, featuring the module configuration in Figure 1 (right) without any decorative components.

The STC Power density $P_{d,MPP}$ [W/m²] is calculated based on the module area and power specified in section 2.4.

Table 2: Measured module power of single-cell samples at STC (Wattpeak Wp) and relative power loss and power density per area

Module	P_{MPP} Wp	loss Δ P_{MPP} %	$P_{d,MPP}$ W/m²
Reference, transparent	4.32	0	194
Encapsulant, red	3.45	-20.1	155
Encapsulant, green	4.07	-5.8	183
Encapsulant, blue	3.59	-16.9	161
Ceramic Print, red	2.97	-31.2	133
Ceramic Print, green	3.46	-19.9.	155
Ceramic Print, blue	3.95	-8.6	177
Morpho structure, red	4.01	-7.2	180
Morpho structure, green	4.15	-4.0	186
Morpho structure, blue	4.19	-3.0	188

We observe that the power loss is strongly dependent on the color for prints and encapsulants. Colored encapsulants achieve 5.8 to 20% less power than a reference module, while whole-area digital enamel prints perform 8 to 30% worse. The spectrally-selective Morpho technology shows by far the lowest power losses on single-cell-modules in the range of 3% to 7%, independent of color.

Calibrated power measurements at Fraunhofer ISE CalLab on full scale modules with the spectrally selective Morpho structures and an additional anti-glare layer confirm these results with a maximum relative loss in efficiency of approx. 7% to a reference module [11].

Significantly more nominal power can be installed on the limited area of the façade using more efficient design technologies. Therefore technologies with higher area specific power feature an advantage when suitable façade areas are limited. In terms of power the Morpho structures clearly outperform the colored encapsulants, while whole-area ceramic prints perform worst.

2.4 Module Cost Analysis

To fulfill structural requirements of BIPV modules for façade applications, glass-glass modules define the calculation baseline for the cost analysis. The 48 cell-modules with an area of 1.32 m x 1 m use 4 mm low-iron front glass cover and 6 mm rear glass. The reference module power is 256 Wp.

Material costs for the respective decorated and standard designs (Figures 1 to 4) have been obtained in a market analysis or estimated based on a sensitivity analysis and are summed up in the tables below.

Table 3: prices of decorative module components

	€/m²
Encapsulant, PVB, colored	12
Ceramic Print	35
Morpho structure	30-100

Table 4: prices of conventional module components

	€/m²	€/Wp
4mm low-iron glass pane	4.5	
4mm structured glass pane	14.9	
6mm rear glass pane	5.5	
Encapsulant, EVA, transparent	1	
Encapsulant, PVB, transparent	5	
Encapsulant, PVB, black	12	
Solar Cells, mono		0.12

The module and manufacturing costs of the respective designs in Figure 1 to Figure 4 are analyzed for a 3-shift production scenario of a single 80 MWp production line in Germany, using the software 'SCost' by Fraunhofer ISE [12]. Labor costs, raw material costs, capital costs and scrap rates are considered along the PV production process.

Using the obtained relative module power decrease (Table 2) and the material cost (Table 3+4), the specific module costs for each decorative design in €/pcs, €/Wp and €/m² are calculated (Table 5).

The conventional glass-glass concept without any decoration achieves manufacturing cost of 0.27 €/Wp. Adding the black encapsulant raises cost to 0.34 €/Wp. Decorated modules increase manufacturing cost from 0.41 to 0.91 €/Wp.

Table 5: Module manufacturing costs for each technological concept

Concept	€/pcs	€/m²	€/Wp
Conventional glass-glass transparent	67.94	51.47	0.27
Conventional glass-glass black encapsulant	88.22	66.83	0.34
Colored encapsulant glass-glass	97.78	74.08	0.41-0.48
Ceramic print glass-glass	134.86	102.17	0.57-0.75
Morpho structure glass-glass 30€-100€/m²	122.2-215.48	92.5-163.24	0.49-0.91

3 RESULTS AND DISCUSSION

We find that decoration of BIPV modules increases the module price due to by power loss and higher material cost.

Colored encapsulants as well as ceramic prints show low color saturation and high power loss relative to a conventional module (7 to 29.5%). The investigated encapsulants were originally designed for solar radiation control of buildings. As a consequence, the main function of these interlayers is to reduce the solar transmittance, which is why absorptance in the encapsulants reaches significant levels and the reflectance of most encapsulants is low. Nonetheless, for PV applications optimized colored encapsulants featuring higher reflectance and lower absorptance could be a feasible design solution for BIPV in the future. The Morpho technology shows strongest color saturation and enhanced brightness. It shows the lowest performance losses (3 to 7%).

With specific module costs from 74 €/m² (0.41-0.48 €/Wp) for colored encapsulants to 102 €/m² (0.57-0.75 €/Wp) for ceramic prints, the decorated modules are within the range of classic cladding materials, such as bricks (60-100 €/m²) or wood (50-180 €/m²) but provide additional benefits regarding power generation [13].

Table 6: Color saturation, module power under STC and manufacturing costs of all concepts.

Concept	Color saturation	loss Δ P_{MPP} %	Cost €/Wp
Conventional glass-glass transparent	none	0	0.27
Colored encapsulant glass-glass	very low	6 to 20	0.41-0.48
Ceramic print glass-glass	low	9 to 31	0.57-0.75
Morpho structure glass-glass	high	3 to 7	0.49-0.91

The sensitivity analysis regarding material costs of the Morpho structure technology (30-100 €/m²) identifies a specific module cost in the range of 93-163 €/m². This is a very attractive price range considering the improved color saturation and higher efficiency. Further due to the low absorptance of the Morpho structure, a lower module temperature and therefore higher yields over the lifetime are expected compared to ceramic prints and colored encapsulants.

4 CONCLUSION

In this paper we analyze three decorative designs for the application in BIPV modules. We investigate colored encapsulants, ceramic printed glass covers and photonic Morpho structures regarding their electrical performance, cost and optical appearance. We perform normal-hemispheric reflectance measurements on single cells laminates and compute CIE L*a*b coordinates. We find the coloration of colored encapsulants and ceramic prints to be low, while the Morpho structure shows high color saturation. Morpho structures additionally feature enhanced brightness compared to the other technologies. Morpho structures show lowest, maximum power loss of approx. 7% compared to a reference module, independent of color. The power loss of colored encapsulants and ceramic prints ranges from 6 to 31% and is highly color dependent. We perform a manufacturing cost analysis and observe that technologies with better color saturation come at increased cost. Higher material cost and power loss are the significant cost drivers. However, we find module manufacturing cost of decorated modules (74-163 €/m²) in the range of classic cladding materials (60-150 €/m²) but point out that PV modules provide economic benefits due to their power generation.

5 OUTLOOK

Within future work, the Net Present Value (NPV) or Life Cycle Costing (LCC) methods can be applied to determine the overall financial benefit or cost of a BIPV installation of the above mentioned technologies over its entire life-cycle, taking the financial effect of generated power into account [14].

6 ACKNOWLEDGEMENT

We would like to thank the German Ministry of Economic Affairs and Energy for their funding (FKZ 0324108 A).

REFERENCES

[1] *Directive of the European Parliament and of the Council of 19 May 2010 on the energy performance of buildings: 2010/31/EU*, 2010.

[2] C. Polo López, F. Frontini, P. Bonomo, and A. Scognamiglio, "PV and Façade Systems for the Building Skin. Analysis of Design Effectiveness and Technological Features," (eng), 2014.

[3] F. Frontini, A. Scognamiglio, G. Graditi, C. Polo López, and M. Pellegrino, "From BIPV to Building Component," (eng), 2013.

[4] Max Mittag, Matthieu Ebert, Helen Rose Wilson, Tobias Fellmeth, "Mosaic Module Concept for Cost-efficient and Aesthetic BIPV Modules," in *35th EUPVSEC 2018*.

[5] I. Hädrich, U. Eitner, M. Wiese, and H. Wirth, "Unified methodology for determining CTM ratios: Systematic prediction of module power," *Solar Energy Materials and Solar Cells*, vol. 131, pp. 14–23, 2014.

[6] M. Mittag, C. Kutter, M. Ebert, H.R. Wilson, U. Eitner, "Power loss through decorative elements in the front glazing of BIPV modules," in *Proceedings of the 33rd European Photovoltaic Solar Energy Conference and Exhibition*, Amsterdam, 2017, pp. 2609–2613.

[7] H. R. Wilson and M. Elstner, "Spot Landing: Determining the Light and Solar Properties of Fritted and Coated Glass," in *Glass Conference Proceedings, Vol 6 (2018): Challenging Glass 6*, pp. 203–212.

[8] Christoph Kutter, Helen Rose Wilson, Andrea Pfreundt, Martin Heinrich, Harry Wirth, "Light Attenuation Model to Predict Nominal Power of Modules with Light-Scattering Ceramic Printed Front Glasses," in *35th European Photovoltaic Solar Energy Conference and Exhibition; Brussels, Belgium (2018)*.

[9] Benedikt Bläsi, Thomas Kroyer, Oliver Höhn, Martin Wiese, Claudio Ferrara, Ulrich Eitner, Tilmann E. Kuhn, "Morpho Butterfly Inspired Coloured BIPV Modules," in *Proceedings of the 33rd European Photovoltaic Solar Energy Conference and Exhibition*, Amsterdam, 2017, pp. 2630–2634.

[10] E. C. Carter, Y. Ohno, M. R. Pointer, A. R. Robertson, R. Seve, J. D. Schanda, and K. Witt, *Colorimetry*, 3rd ed., 2004.

[11] Johannes Eisenlohr, Claudio Ferrara, Benedikt Bläsi, Thomas Kroyer, Oliver Höhn, Martin Heinrich, Ulrich Eitner, Tilmann E. Kuhn, "Highly efficient coloured BIPV modules with anti-glare properties," in *13th Conference on Advanced Building Skins 2018*.

[12] S. Nold *et al.*, "Cost modeling of silicon solar cell production innovation along the PV value chain," in *27th EUPVSEC 2012, Frankfurt, Germany, 2012*.

[13] Isa Zanetti, Pierluigi Bonomo, Francesco Frontini, Erika Saretta and Guus Verberne, Menno van den Donker, Kostas Sinapis, Wiep Folkerts, "Building integrated photovoltaics: Product overview for solar building skins: Status Report 2017," SUPSI – SEAC, 2017.

[14] K. Fath *et al.*, "A method for predicting the economic potential of (building-integrated) photovoltaics in urban areas based on hourly Radiance simulations," *Sol. Energy*, vol. 116, pp. 357–370, 2015.

BUILDING THE WORLD'S LARGEST BIFACIAL SOLAR NOISE BARRIER

M.M. de Jong[1*], J. Kester[2], D. van der Graaf[3], S. Verkuilen[4], W. Folkerts[1]
[1]Solar Energy Application Centre, Eindhoven, The Netherlands
[2]ECN part of TNO, Petten, The Netherlands
[3]Rijkswaterstaat, Ministry of Infrastructure and the Environment, The Netherlands
[4]Heijmans Infra BV, Rosmalen, The Netherlands
*Corresponding author: dejong@seac.cc

ABSTRACT: In the Solar Highways project, knowledge institutes ECN and SEAC, and the executive body of the Dutch Ministry of Infrastructure and the Environment (Rijkswaterstaat) teamed up to build a bifacial fully integrated solar noise barrier. Not only is it the largest solar noise barrier in the Netherlands to date, it will also be the largest bifacial noise barrier in the world. The solar noise barrier will be 400 m long and 5 meters high and will be facing east and west. To achieve the objectives of maximum solar energy output and minimum operation and maintenance costs, we designed and published a procurement based on the principles of 'Most Economically Advantageous Tender' (MEAT), enabling the weighing of monetary alongside non-monetary aspects in the same procurement. Six system designs were submitted by contractors. After assessment of all submitted offers in the MEAT framework, the assignment was granted to Heijmans Infra. By submitting a design that enables a high cell coverage ratio together with minimized construction shading, the output is maximized. Smart design features ensure minimum operation and maintenance effort. After commissioning of the installation in December 2018, we will carefully monitor it with respect to energy output and maintenance costs.

Keywords: Infrastructure Integrated PV, BIPV, Bifacial, PV Noise Barrier, PVNB

Figure 1: An impression of the bifacial solar noise barrier design by Heijmans

1 BIFACIAL SOLAR NOISE BARRIERS

The advantage of *bifacial* solar noise barriers is that they can be used alongside any road, regardless of its direction [1]. More general, solar noise barriers generate renewable energy using existing infrastructure, saving otherwise wasted space. Rijkswaterstaat (RWS) will build a new noise barrier along the A50 highway in Uden, the Netherlands. At this location, the road runs north-south. Therefore the noise barrier sides will face east and west. Using conventional monofacial solar cells would mean a suboptimal electricity output. By using bifacial cells, the output will be increased by capturing light in the morning from the east side, and capturing light in the afternoon using the other side of the barrier.

Combining solar energy systems and noise barriers is an idea that was first executed in the 1980's. In the Netherlands, two systems were built in the 1990's. After these projects no follow-ups were initiated for two decades. This could be attributed to, among other things, the high price of PV, theft, vandalism and the lack of urgency for generating renewable energy. Recently the subject has regained interest in the Netherlands. Several Dutch construction companies have designed and built prototypes of solar noise barriers. In these designs some of the aforementioned problems are addressed. Mostly in Germany and Switzerland, a number of solar noise barriers were built in recent years, including a number of bifacial noise barriers in Zürich along a Highway bridge and in Münsingen along a railway line [2]. The bifacial

Figure 2: A prototype based on the ECN reference design for a bifacial solar noise barrier.

solar noise barrier in Uden will be the largest bifacial solar noise barrier to date.

2 THE PROCUREMENT PROCEDURE

To overcome the problems with theft, the Uden solar noise barrier will have fully integrated solar modules. To tackle problems with vandalism and high maintenance costs, the procurement of the Uden solar noise barrier was set up based on the principles of 'Most Economically Advantageous Tender' (MEAT). In this type of procurement, value is added to aspects other than cost price. In the case of the Uden solar noise barrier procurement, the MEAT tender aimed for maximum electrical output, together with measures to minimize the efforts and cost for maintenance. Six consortia of companies responded to the public tendering procedure, and all submitted a proposal. As guidance for how to maximize energy output, the procurement was accompanied by a reference design made by ECN [3]. The purpose of this design was to minimize the negative effect of self-shading or construction shading of the structural noise barrier components on the solar modules. For this the cell placement over the module was adjusted and a smart stringing and bypass-diode layout was designed. More information on the reference design can be found in the aforementioned reference. A prototype of this design was built and monitored. Figure 2 shows an image of this prototype. The tendering contenders were challenged to come up with a proposal that improves upon the ECN reference design, with regard to the energy output. Furthermore, they were asked to optimize their design to minimize maintenance effort and cost.

3 THE WINNING DESIGN

After assessment of the six designs on total cost, expected energy output and ease of maintenance, the bid by Heijmans Infra came out as the winning design. For the design and construction, Heijmans formed a consortium with Scheuten, Van Campen Bayards and Libra Energy. Heijmans and their partners gained knowledge and experience in the design and building of (bifacial) solar noise barriers in the previously executed SONOB project [4], leading to a very innovative design. A crucial improvement in the design of the PV module is the choice of modules with a height of 2 m instead of the regular height of 1 m, which is the standard height of noise barrier modules in the Netherlands, according to the Modular Noise Barrier (MGS) standards. In this way half the horizontal construction bars are avoided. Therefore the relative effect of construction shading by the horizontal parts of the frames is halved. In addition to that the effects of construction shading [5,6] are mitigated by dividing the active area into two areas. The top part, which suffers most from construction shading in summer, when the sun position is high, has a horizontal string layout, such that when this part is shaded, the output of the main area is not negatively affected. The main part of the module has a vertical string layout, to mitigate the effects of self-shading from the vertical support structure. Using this string layout enables using a very high cell density, without losing much output due to construction shading and thereby maximizing the energy output.

The system is fully equipped with module level power management system by using power optimizers. The main reason for this is safety. In the event of a vehicle crashing into the barrier, the optimizers will return to a safe working voltage, minimizing risk not only for the people in the vehicle, but also for the emergency services. As a bonus less-performing modules, caused by (external) shading, soiling, graffiti or defects, will not influence the rest of the system.

To build a vandalism-proof system, all electronics and cables are integrated into the carrying structure, and are not reachable from outside without specialized tools. For safety reasons it is important that all maintenance can be performed from the safe side of the noise barrier, away from the traffic. This means that cleaning or maintenance can be performed without disturbing traffic. All noise barrier elements are identical. This modular structure means that when a module is irreparably damaged, the modules can be replaced. Four spare modules will be kept in stock by Heijmans Infra.

Preparatory actions for the construction were started in March 2018. In the summer months an old concrete

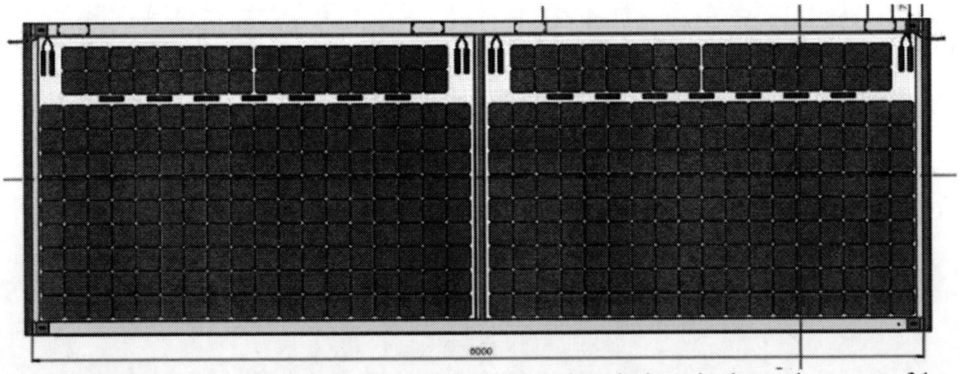

Figure 3: The Heijmans modular solar noise barrier panel design. By using a horizontal string on the top part of the modules and vertical strings in the main part, the effects of self-shading are mitigated.

noise barrier was removed and the foundations for the new solar noise barrier was placed. In September and October the bifacial solar noise blocking elements are placed. Commissioning of the system is planned for December 2018.

4 MONITORING

After finalizing the bifacial solar noise barrier in the last months of 2018, SEAC will carefully monitor the total solar energy output of the installation for 18 months. We will not only monitor the energy output of the bifacial solar noise barrier, also the efforts and costs needed for maintenance will be assessed. To analyze the business case for building and exploiting solar noise barriers, information about the costs of operation and maintenance is crucial. The most important missing information in this respect is the cost of cleaning the installation. A more frequent cleaning regime will enhance the energy output, but will add costs to the maintenance. To find out what is the optimal cleaning frequency with the business case in mind, a part of the installation will function as an O&M research section. This installation is split into 5 identical experimental sections. Each section will have a different cleaning regime, ranging from cleaning of both sides of the barrier every 6 months, cleaning only of the safe side (away from the road) to not cleaning the barrier at all. By assessing differences in energy output and relating them to the cost of cleaning, we can design the optimal maintenance scheme with the business case in mind.

LARGE SCALE ROLLOUT

The Uden bifacial solar noise barrier will be the largest of its kind in the world. Nevertheless we foresee a large and growing market for solar noise barriers in the near future, not only in the Netherlands, but also in the rest of the world. By setting up this project not as a one-time pilot, but through a tendering procedure for a permanent fixture, the procurement of the installation can be copied for future projects. In this way, future projects will be much easier to set up and execute. In this project we show that the technical hurdles for building a vandalism proof bifacial solar noise barrier can be taken. Crucial input for business case calculations is gathered in the monitoring phase, taking away the last barriers regarding the business case towards large scale rollout of building solar noise barriers.

ACKNOWLEDGEMENT

This work has been partially financed by the EU in the framework of the LIFE+ program.

References

[1] G.J.Faturrochman et al., Maximizing annual yield of bifacial photovoltaic noise barriers, Solar Energy, 2018

[2] Th. Nordmann et al., 15 Years Of Practical Experience In Development And Improvement Of Bifacial Photovoltaic Noise Barriers Along Highways And Railway Lines In Switzerland, 27th EUPVSEC proceedings, 2012

[3] J. Kester, M.J. Jansen, M.M. de Jong, D. van der Graaf, Reference Design for a Highway Noise Barrier with Integrated Bifacial PV, 33rd EUPVSEC proceedings, 2017

[4] C. Tzikas, M.M. de Jong, L.H. Slooff, M.G. Debije, S. Verkuilen, W. Folkerts, Graffiti on Solar Noise Barriers - A Case Study, 33rd EUPVSEC proceedings, 2017

[5] https://www.seac.cc/conclusions-on-solar-noise-barriers-sonob-project/

[6] M.M. de Jong, M.N. van den Donker, S. Verkuilen, W. Folkerts, Self-Shading in Bifacial Photovoltaic Noise Barriers, 32rd EUPVSEC proceedings, 2016

APPLICATION OF SEMI-TRANSPARENT PHOTOVOLTAICS IN TRANSPORTATION INFRASTRUCTURE FOR ENERGY SAVINGS AND SOLAR ELECTRICITY PRODUCTION: TOWARDS NOVEL NET-ZERO ENERGY TUNNEL DESIGN

David Y-J Sun[1], Andreas K. Athienitis[1], Katherine D'Avignon[2]

[1]Centre for Zero Energy Building Studies, Department for Building, Civil, and Environmental Engineering, Concordia University, 1455 de Maisonneuve Blvd. W., Montréal, Québec, Canada, H3G 1M8
David.Sun@mail.concordia.ca ; Aathieni@encs.concordia.ca

[2]Department of Construction Engineering, École de technologie supérieure
1100, rue Notre-Dame Ouest, Montréal, Québec, H3C 1K3
Katherine.Davignon@etsmtl.ca

ABSTRACT: Lighting subsystems consume 50% of the energy needs of a typical tunnel. Day-time lighting levels account for over two-thirds of the total system lighting power; their periodic nature creates daily peaks in the tunnel's energy load profile. This paper studies the integration of semi-transparent photovoltaic (STPV) cells into sunscreen structures installed above tunnel entrances to reduce tunnel lighting requirements and offset their day-time lighting loads using energy generated from PVs. The electrical lighting load of a typical 1 km length road tunnel with and without semitransparent photovoltaics (STPV) sunscreen structures was modelled to establish the potential for energy savings. Using a daylighting and energy modeling plug-in called DIVA [1], the transparencies and ratios of photovoltaics (PV) to glass of a STPV sunscreen that are in accordance with the luminance reduction code requirements were determined. Reduced lighting requirements over the whole tunnel length, including the threshold, transition and interior lighting zones of the tunnel were considered, resulting in significant energy savings. The annual power production of the sections covered with STPV was then simulated using the PVsyst software. The integration of photovoltaic cells resulted in annual energy production that reduced annual energy use by up to 7% and the potential to reduce electric lighting loads by up to 60% during the day. Results also demonstrated that STPV sunscreens have the potential to meet luminance requirements if supplemented with an intelligent lighting control system.

Keywords: Semi-transparent photovoltaics, transportation infrastructure, daylighting, net-zero energy systems

1 INTRODUCTION

The research team at Concordia University's Centre for Zero Energy Building Studies (CZEBS) is exploring innovative ways to improve motorist safety and implement net-zero energy usage in critical urban infrastructure located in cold climates. Underground road tunnels fall under the category of high-priority structures since in the event of an emergency or natural disaster, they fulfill the vital role of providing vehicles safe passage through natural barriers. Using a case study of a Canadian tunnel as a baseline for comparison, it was revealed that energy efficiency provisions and local renewable energy generation could significantly improve the tunnel's energy resilience under emergency circumstances in addition to achieving net-zero energy use.

A fundamental requirement in designing safe transportation infrastructure is ensuring that drivers can detect the presence and movement of objects and vehicles on the road. Methods of illuminating underground road tunnels differ from conventional practices of illumination design; tunnel lighting systems must reduce contrast between the outdoor and indoor environment to reduce visual impairment, traffic, and the potential for accidents as drivers transition between environments. Hence, the brightness at the entrance of tunnels must be very high to allow drivers to visually adapt to the shift in lighting conditions. Consequently, lighting subsystems in tunnels can account for up to 50% of the total energy consumed [2][3], presenting a prospective opportunity for energy savings, while proving for increased safety by gradually decreasing lighting levels and avoiding the "black hole" effect at the tunnel entrance.

This paper specifically explores the use of semi-transparent photovoltaic (STPV) sunscreen structures installed at the portals of underground road tunnels which create naturally sun-lit zones that satisfy the recommended visual performance criteria described in international standards [4]. In doing so, sunscreen structures reduce the need for artificial lighting in the most intensely lit segments of underground road tunnels. The additional advantage of having local power generated from the integrated STPVs is that peak power loads —that are caused by intense daytime lighting— can be offset, giving this specific application of photovoltaics exciting potential in the development and implementation of net-zero energy tunnels [3]] that can potentially generate from renewable energy sources as much energy as they consume in a year.

Additionally, implementing sunscreens at the entrances of tunnels can have secondary, but impactful effects on other important subsystems. One important safety hazard that is being studied for tunnels in cold climates is the formation of black ice at the entrance of tunnels. On-going research at Concordia has presented convincing evidence that STPV sunscreens can significantly reduce the occurrence of ice formation by reducing road exposure to precipitation, reducing convective heat losses and raising air temperature through the greenhouse effect, increasing the benefits of STPV sunscreen structures.

2 BACKGROUND

2.1 Lighting in Tunnels

Lighting subsystems are typically the primary energy consumers in tunnels. Based on a study of four tunnels in the Netherlands [3], they can be responsible for up to 50% of the total energy consumed in tunnels. They are activated a higher number of hours and at much higher intensities than conventional road and building lighting systems. This is because the design of lighting systems in underground road tunnels have a different purpose compared to conventional practices of road and building illumination: to ensure the safe visual adaptation of drivers as they approach and circulate through the tunnel. If the contrast between outdoor conditions and lighting immediately at the entrance of the tunnel is too high, drivers will experience what is known as the black hole effect [4], causing disabling glare and visual impairment. This can lead to drivers slowing down as they enter the tunnel, causing higher density traffic, or worse, increasing the potential for accidents [4]. Hence, the brighter it is outdoors, the higher the intensity of the lighting must be at the entrance of the tunnel. As drivers progress further into the tunnel threshold and transition zones, their eyes adapt to the darker environment and lighting intensity can be gradually reduced (Figure 1).

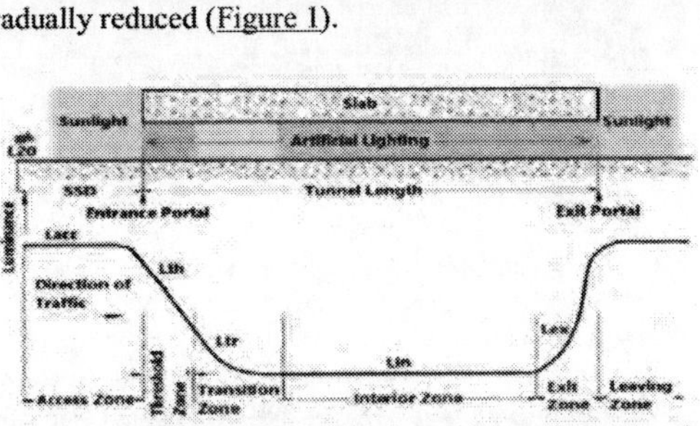

Figure 1: Typical longitudinal section of a tunnel, showing different lighting zones that are active during the day [6].

The International Commission on Illumination's technical report "Guide for the Lighting of Road Tunnels and Underpasses"[4] and the American National Standards Institute's standard practice on tunnel lighting [5] recommend best practices with regards to tunnel lighting. Both use the same nomenclature; they divide tunnels into 5 zones of lighting: the access, threshold, transition, interior and exit zones (Figure 2). The required level of lighting in each zone is dependent on the level of the preceding zone. The threshold and transition zones are calculated using equations (1), (2), and (3), and are responsible for reducing the contrast in brightness experienced by drivers. Their luminance requirements (L_{th} and L_{tr}) are a function of the equivalent veiling luminance (L_{seq}), which is measured at the minimum distance that a driver requires to clearly detect obstructions in their path and immobilize their vehicle. This distance is defined as the safe stopping distance (SSD). The function of the interior zone lighting is to maintain the minimum luminance requirements for drivers in an underground environment after drivers have gradually adapted to the darker environment with the help of the preceding lighting zones [4], [5].

$$L_m = \frac{(\tau_{ws} L_{atm} + L_{ws} + L_{seq})}{(\tau_{ws} \tau_{atm})} \quad (1)$$

$$L_{th} = \frac{L_m}{\frac{1}{C_m}\left(\frac{\rho}{\pi q_c}\right) - 1} \quad (2)$$

Where

- L_{seq} = Total equivalent veiling luminance
- L_{atm} = Atmospheric luminance
- L_{ws} = Luminance of a typical windscreen
- L_m = Real measured luminance
- τ_{atm} = Transmission factor of the atmosphere
- τ_{ws} = Transmission factor of a typical vehicles windscreen
- C_m = Minimum required perceived contrast
- q_c = Contrast revealing coefficient
- ρ = Reflection factor of target object

$$L_{tr} = L_{th}{}^{(1.9+t)^{-1.4}} \quad (3)$$

Where

- t = time spent in transition (in seconds) at the rated speed limit.
- L_{th} is the calculated threshold luminance

2.2 Sunscreen Structure

The application of sunscreen structures to reduce the energy consumed by tunnel lighting systems is a topic that has been studied sporadically. Researchers have tried to reduce energy consumption and daytime lighting loads in two ways:

i) The threshold zone is displaced outside of the tunnel where the transparency of the structure can effectively ensure the luminance requirements are met with the use of filtered, natural sunlight [6]–[10].

ii) The structure acts as a type of sun-shade, reducing the veiling luminance (L_{seq}) and therefore the luminance required in the threshold zone[6], [11], [12].

Point ii) is especially relevant for the tunnel in this study because it is oriented along the East-West axis, bringing sunlight directly into the line of sight of drivers in the mornings and evenings, causing disability glare.

2.2.1 i) Displacing the Threshold Zone

The threshold zone is located immediately after the tunnel portal and must have the highest intensity of lighting to reduce the contrast between outside and inside conditions. A method for reducing the lighting energy use and peak power is to reduce the number of artificial lights activated by placing the threshold zone outside of the tunnel, effectively meeting the required luminance levels passively with sunlight. This can be achieved by installing a semi-transparent structure that acts as an extension of the original tunnel given that it satisfies the following criteria specified in [4]:

i) Achieve an illuminance uniformity (U_o) on the road surface under the structure greater than 0.4 as per minimum requirements. U_o is the ratio of the minimum measured illuminance to the average measured illuminance of a surface for a moment in time.

ii) Achieve an average illuminance (E_{av}) on the road surface under the structure as close as possible to the required average illuminance in the threshold zone (E_{th}).

The required threshold illuminance (E_{th}) incident on the road surface under a sunscreen is calculated using Equation (4) [13] where ρ is the reflectance of the road surface (ρ = 0.3), B is the utilization factor (B = 1), and u is the utilization factor (u = 1).

$$E_{th} = \frac{L_{th} * \pi}{\rho * B * u} \quad (4)$$

Where:

- L_{th} = Threshold luminance
- ρ = Reflectance of the road
- B = Maintenance factor
- u = Utilization factor

Researchers at the University of Granada in Spain have extensively studied tunnel 'pergola' sunscreens for energy savings and tested numerous light diffusive materials. They have recognized the challenges in achieving adequate illuminance uniformity. Their results showed that uniformity could be improved by using light diffusive materials to transmit light but that the average illuminance on the road would be significantly reduced [10].

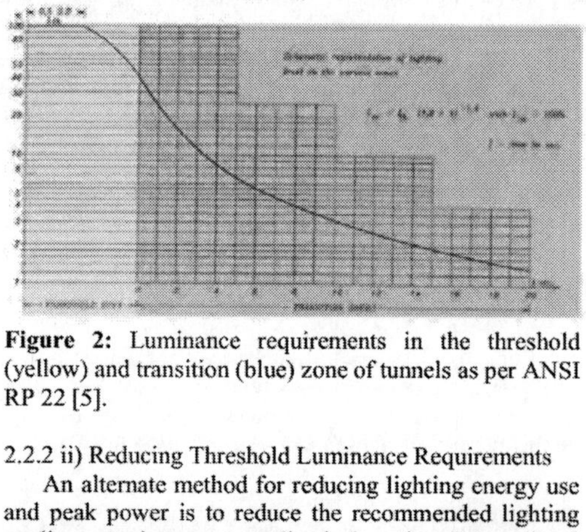

Figure 2: Luminance requirements in the threshold (yellow) and transition (blue) zone of tunnels as per ANSI RP 22 [5].

2.2.2 ii) Reducing Threshold Luminance Requirements

An alternate method for reducing lighting energy use and peak power is to reduce the recommended lighting quality requirements outlined by the International Commission on Illumination [4] and Illuminating Engineering Society [5]. To do this, the equivalent veiling luminance (L_{seq}) experienced by drivers at the beginning of the access zone must be reduced as it determines the lighting requirements of all tunnel zones. The equivalent veiling luminance (L_{seq}) is determined by superimposing a polar diagram over an image that is representative of a driver's perspective as they approach the tunnel as in Figure 3, and summing the average luminances in each section. Equation 1 uses L_{seq} to calculate the real measured luminance and equation 2 calculates the threshold luminance. [4] gives reference data of luminance values for the typical tunnel environments illustrated in Figure 4, which shows that the sky environment is the biggest contributor to veiling luminance. Therefore, by reducing the amount of sky that drivers see at the SSD, the lighting requirements of the tunnel can be reduced significantly.

To this point, a sunscreen structure that can generate electricity while simultaneously providing energy curtailment through either method mentioned earlier has not been explored. Semi-transparent photovoltaics are technologically ready products that are well-suited for this niche application.

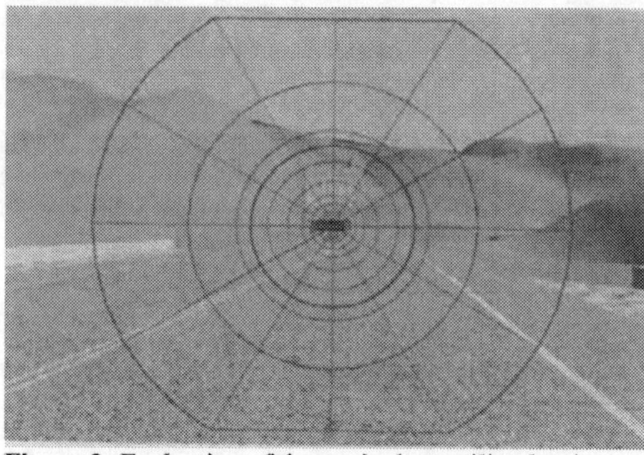

Figure 3: Evaluation of the equivalent veiling luminance, [4].

Driving direction (Northern hemisphere)	L_c (sky) kcd/m²	L_r (road) kcd/m²	L_e (environment) kcd/m²			
			Rocks	Buildings	Snow	Meadows
N	8	3	3	8	15 (V) 15 (H)	2
E-W	12	4	2	6	10 (V) 15 (H)	2
S	16	5	1	4	5 (V) 15 (H)	2

(V) Mountainous country with mainly steep surfaces facing drivers
(H) Flat, more or less horizontal, country
NOTE: In the southern hemisphere N and S should be interchanged.

Figure 4: Examples of typical veiling luminances at tunnel portals, [4]

2.3 Semi-Transparent Photovoltaics (STPV)

Photovoltaics are no longer uniquely installed in empty fields and on top of buildings but are being used to replace standard construction materials and elements. Such Building Integrated Photovoltaics (BIPV) have been used to replace roofs, walls, and windows. Semi-Transparent Photovoltaics (STPV) devices are a sub-category of BIPV devices that can replace building components that have light transmission (and shading) functions, such as overhangs and glazing units, to transmit sunlight for daylighting and electricity generation [14]. There is also on-going research that integrates heat collection functions into STPV glazing units using airflow as a coolant fluid [15], [16]. One crucial characteristic about integrating any type of material into sunscreen structures is its transmissivity – an optical property that represents the amount of visible light that passes through a type of material under standard conditions. Modern day STPV devices are categorized into 3 types (Figure 5) based on how their transparency is achieved and the type of PV technology used [17], [18].

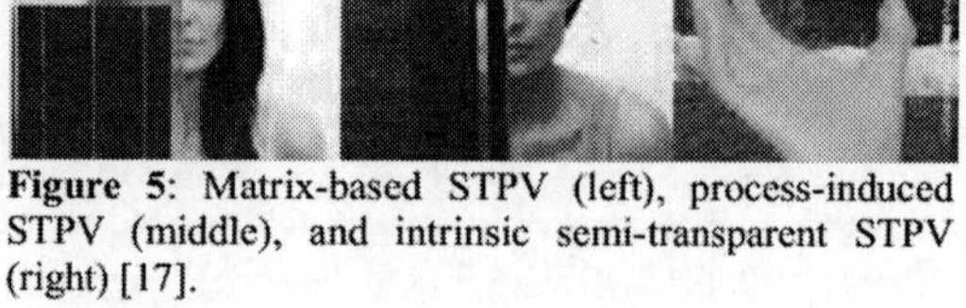

Figure 5: Matrix-based STPV (left), process-induced STPV (middle), and intrinsic semi-transparent STPV (right) [17].

2.3.1 Matrix Based Semi-Transparency

Matrix based semi-transparency (Figure 5a) achieves the intended degree of transparency by alternating opaque crystalline-silicon PV cells with fully transparent gaps on a glazing unit, creating an opaque-transparent pattern [18]. These types of STPV devices are typically designed to resemble shading devices in windows such as fritted glass and shutter blinds.

2.3.2 Process Induced Semi-Transparency

Process induced semi-transparency (Figure 5b) applies the same principle as matrix-based semi-transparent PV devices, achieving its intended degree of transparency with a sequence of opaque PVs and fully transparent spaces. The main difference is that the laser-etching technique can cut micro voids out of the PV material allowing for a finer degree of transparency that gives the illusion of the PV cells themselves being 'see-through' or 'tinted' rather than opaque. This technique has been applied to both crystalline-silicon and amorphous silicon PVs [17], [18].

2.3.3 Intrinsic Semi-Transparency

While Intrinsic semi-transparent PVs are still process induced, their transparency is an innate characteristic (Figure 5c), rather than a physical modification of cells. They are thin-film, organic and perovskite PV technologies that are highly customizable in terms of transparency and shape[17], [18]. Some of the more recent PVs in this category allow for complete transparency and uniform transmission of sunlight, however, they are not as durable or efficient as silicon-based cells [19].

3 RESEARCH OBJECTIVES

The objective of this study is to demonstrate the benefits of applying STPVs in tunnel sunscreen structures. The innovative use of STPVs in this application stems from their dual purposes. The opaque photovoltaic cells act as daylighting shading devices, utilizing daylight transmission to reduce energy use, while simultaneously generating renewable energy. Both aspects play a fundamental role in designing net zero energy systems [20], including tunnels [2][3]. Furthermore, the energy analysis of a representative tunnel in Canada shows that current lighting design is responsible for daily peak loads. Since the tunnel lighting loads and PV generation are both concurrent with daylight hours, STPV sunscreen structures would naturally contribute to balancing the tunnel load profile.

4 SCENARIOS AND MODELS

This section will give an overview of the conditions under which the simulations were done, the models and tools that were used to create said simulations, and the numerical models used to estimate the net-energy use of the system.

4.1 Scenarios

Initially, matrix-based semi-transparent PV technologies were the primary consideration in this study. However, because of the poor uniformity that results from an opaque/transparent sequence of shading [10], intrinsic semi-transparent thin-film PV technologies were also considered. Transparencies ranging from 20% to 50% were considered for both types of STPVs and the impact on load reduction, energy savings, and energy production was quantified. Finally, the threshold luminance during critical hours was considered. We have defined critical hours as moments when the sun is tilted and oriented to be directly in the in the eyes of drivers in the access zone of the east-bound and west-bound tunnel portals (Table). For the east-bound approach, hours when the sun had a tilt angle between 20° to 30° and was oriented 0° to 30° *south of east* were considered critical hours. Similarly, critical hours for the west-bound approach were when the sun had a tilt angle between 20° to 30° and was oriented 0° to 30° *north of west*. Figure 6 is a perfect example of what disability glare and the black hole effect looks like to drivers. Even through a still image, it is difficult to distinguish the distance between the neighboring vehicles, and impossible to tell if there is an obstruction on the road ahead. Consider that the posted speed limits in tunnels can range between 60 to 100 km/h, and it immediately becomes obvious how important it is to address safety concerns during critical glare hours.

Table I: Critical glare hours for the east-bound and west-bound approaches of a reference tunnel.

East-bound critical hours (sun-rise, spring/autumn)	West-bound critical hours (sun-set, summer)
7h00	18h00
8h00	19h00
9h00	20h00
10h00	21h00

Figure 6: An example of the black hole effect that drivers experience when entering underground road tunnels during critical hours.

4.2 Rhinoceros 3D Model

A 3D model of the east-bound and west-bound approaches and their proposed STPV sunscreens (Figure 7) were created separately using a computer-aided design software called Rhinoceros [21]. In order to reduce simulation time, the physical boundaries of the models of each approach started at the SSD from the tunnel portal and ended an appropriate distance into the tunnel where the road was no longer exposed to the sun. The area within these boundaries was modeled using construction plans from the case study. These models were used to determine an appropriate height and shape for the structure to reduce sunlight exposure that drivers receive.

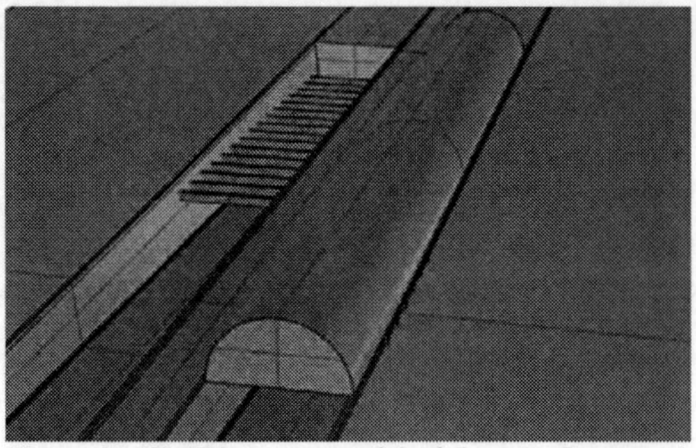

Figure 7: 3D Rhinoceros model of the tunnel approach with a curved surface representation of the STPV cover.

4.3 DIVA4Grasshopper Daylight Simulations

A graphical algorithm editor that is tightly integrated with Rhinoceros' modeling tools called Grasshopper was used in conjunction with a daylighting simulation plug-in called DIVA. These tools were used to simulate the perspective of a motorist in the Rhinoceros model using a camera view placed at the safe stopping distance (SSD) at a height of 1.45m (approximate height of motorist in a vehicle) above the road. False color images showing the luminance reflected off the tunnel portal scene were generated and the equivalent veiling luminances were calculated for every sun-lit hour of the year using a clear sky model. The same model was used to calculate the average illuminance received by the road surface in the access zone of the tunnel. Different scenarios were simulated to evaluate if different transparencies of the STPV sunscreen were in accordance with the luminance reduction code requirements.

4.4 Numerical Lighting Model

A representative 1 km tunnel in Canada was modeled numerically with Python [22]. Using this model, the lighting load profile was simulated and analyzed for the year 2015-2016 with and without the application of an STPV sunscreen to establish the potential energy savings of the application. Energy savings and peak load reduction are primarily achieved by displacing the threshold zone outside of the tunnel, thus eliminating all non-essential daytime lighting in the zone. Secondarily, the geometry of the structure —as viewed at the SSD of the tunnel— reduces excess sunlight during critical hours, thus reducing the veiling luminance and consequently the luminance requirements in the tunnel.

4.5 PV Model

The 2015-2016 annual and hourly energy generation profile of the sections covered in photovoltaics (PV) was simulated using the software package called PVsyst [23]. Since PVsyst is unable to simulate the energy production of PV modules on a curved surface, the portion of the sunscreen covered with PVs was discretized into equivalent flat surfaces using a straight-line approximation (Figure 8). These surfaces were used to estimate the PV generation of the STPV sunscreen structures. The cumulative chordal distance of the straight-line approximation had a percent error of just 2.2% when compared to the actual curved length of the actual arc. Considering that the installed PV capacity for all scenarios is significantly smaller than the base lighting load of the tunnel, it was safe to assume that all energy generated would directly contribute to reducing peak loads and the lighting system energy consumption.

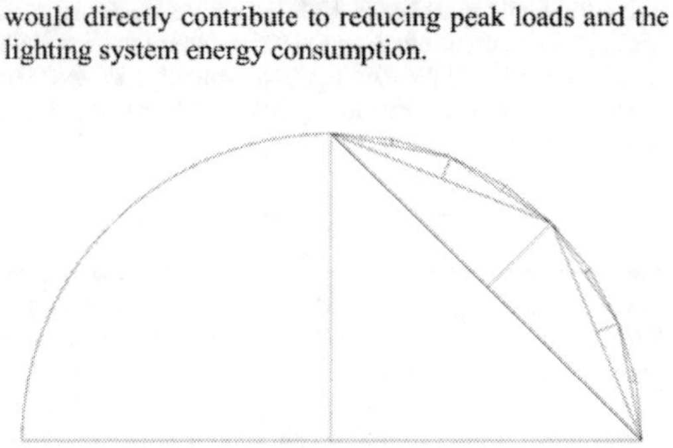

Figure 8: Cross-sectional view of the straight-line approximation method used to create the equivalent flat surfaces. These surfaces were used to estimate the PV generation of the STPV sunscreen structures.

5 RESULTS AND DISCUSSION

In this section, PVsyst was used to estimate the energy generated over a year by the STPV sunscreen structure. The ability of the sunscreen to reduce lighting energy consumption by (i) displacing the threshold zone, and (ii) reducing threshold luminance requirements was also evaluated, both visually and numerically. These three performance criteria, which are directly linked to the sunscreen degree of transparency, were considered in determining appropriate design parameters for the specific Canadian reference tunnel.

5.1 PV Generation

PVsyst estimated that the energy generated over a year by intrinsically semi-transparent PV sunscreens will reduce lighting energy use by up to 2% but that matrix-based semi-transparent sunscreens are much more fruitful, reducing energy use by up to 7.2%. Additionally, under standard testing conditions, the intrinsically based STPVs have the potential to reduce day-time lighting loads by between 53 to 82.3 kW, while the matrix-based STPVs have the potential to supply 100% of the day-time lighting load during the day. It can be tempting to conclude that matrix-based STPVs are the ideal technology to use in sunscreen structures, however, the level of transparency and type of PV technology can hinder both the sunscreen's ability to block the sun, and drivers' ability to carry out crucial visual tasks on the road. This will be discussed in more detail in sections 5.2 and 5.3.

Table II: Summary of the capacity, energy production and energy savings for different types of STPV sunscreens and transparencies.

Name	Transparency (%)	Module Area (m2)	PV Cell Efficiency (%)	Effective PV Cell Efficiency (%)	Installed PV Capacity (kW)	Annual Energy Curtailment (%)
Baseline	100					
Inherent Semi-Transparent PV	40	2420	2.2	2.2	53.2	1.3
	30		2.8	2.8	61.8	1.7
	20		3.4	3.4	82.3	2.1
Matrix Based Semi-Transparent PV	44	1452	15	9.6	217	5.5
	40	1684	15	10.5	254.1	6.3
	20	1936	15	12	290.4	7.2

5.2 Reducing Lth Values

To get a visual idea of what drivers experience during critical glare hours (8h00, 9h00, 10h00 on March 21st) Figure 9 and Figure 10 show visualizations that were generated using daylighting simulations. These figures illustrate the contrast in luminance from the perspective of a driver at the SSD with and without a sunscreen structure shading part of the sun.

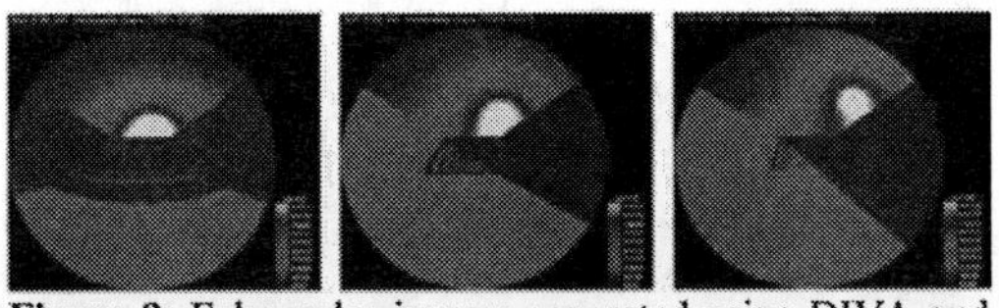

Figure 9: False-color images generated using DIVA and Rhinoceros; illustrates what drivers experience during 'critical hours' of the day when the sun is directly in their line of sight.

Figure 10: False-color images generated using DIVA and Rhinoceros; illustrates the impact that adding an intrinsically semi-transparent PV sunscreen would have on driver experience during 'critical hours'.

While these visualizations are enough to anecdotally understand the impact that sunscreens can have on glare experienced by drivers, it is not enough to make any concrete conclusions. It is important to communicate to designers what this means in terms of lighting requirements; Table 3 summarizes the different threshold luminance requirements as per [4], [5] for the same hours that were visualized in Figure 9 and Figure 10. Additionally, the threshold luminance requirement for the entire year was determined by taking the highest threshold luminance that occurs at least 75 times. This information conveys several important pieces of information:

(i) The application of a sunscreen structure will drastically reduce artificial lighting requirements inside the tunnel;

(ii) Intrinsic STPVs are more effective in reducing glare than matrix-based semi-transparent PVs for this application;

(iii) The range of preliminary threshold luminances estimated in the CIE's technical report on tunnel lighting does not encompass requirements during critical glare hours.

An interesting observation is that when critical hours are omitted from all other day-time hours, the resulting threshold luminance is an order of magnitude lower than the annual values shown in Table 3. These values fall closer to the preliminary threshold luminances estimated using methods outlined by the International Commission on Illumination and Illuminating Engineering Society. This suggests that existing methods that rely on reference data (Figure 4), such as the L20 method [4] may not adequately consider that the sun will be directly in the eyes of drivers during critical glare hours in east-west oriented tunnels.

Table III: This table summarizes the calculated threshold luminances during critical hours on March 21st using the false-color images generated from DIVA for different types of STVP sunscreens and transparencies. A CIE clear sky model was used.

		Time of Day				
Notes	Transparency (%)	7:00	8:00	9:00	10:00	Annual
Baseline	100	[illegible] cd/m²	4241 cd/m²	2704 cd/m²	1673 cd/m²	[illegible] cd/m²
Intrinsic Semi-Transparent PV	40	117 cd/m²	1148 cd/m²	920 cd/m²	287 cd/m²	1100-1200 cd/m²
	30	103 cd/m²	829 cd/m²	[illegible]	[illegible]	1100-1200 cd/m²
	20	103 cd/m²	537 cd/m²	459 cd/m²	151 cd/m²	1100-1200 cd/m²
Matrix-Based Semi-Transparent PV	30	151 cd/m²	2490 cd/m²	1894 cd/m²	812 cd/m²	1500-1600 cd/m²

5.3 Displacing the Threshold Zone

The ability of a sunscreen to save energy and reduce peak lighting loads through displacing the threshold zone is entirely dependent on its ability to replace the function of artificial luminaires; in other words, enough daylight must penetrate the sunscreen to satisfy the minimum required illuminance and illuminance uniformity on the road surface under the structure.

Figure 11 shows a plot of the road surface illuminance (y-axis) at locations starting from the tunnel portal to the end of the sunscreen (x-axis) under a matrix based semi-transparent sunscreen. This examples uses a sunscreen that has an effective transparency of 30%. The erratic spikes in illuminance demonstrates that a matrix-based semi-transparent PV sunscreen will always have a poor illuminance uniformity that can hinder a motorist's ability to detect objects, therefore this option was not investigated further. Figure 12, which shows the same type of plot, shows that intrinsically semi-transparent PV sunscreens will have a much smoother uniformity along the length of the sunscreen.

Figure 13 shows the median illuminance incident on the road over an entire year and compares it to the minimum illuminance requirements that were calculated for each intrinsically semi-transparent PV sunscreen and the minimum illuminance requirement estimated using international standards and reference values from Figure 4. It presents two contradicting conclusions. The first conclusion is that intrinsically semi-transparent PV sunscreens will not always satisfy the minimum illuminance requirements that were determined using the generated false-color images. However, due to the inclusion of critical glare hours in the analysis, these calculated minimum requirements are unrealistically high and would be difficult to satisfy — even with supplemental artificial luminaires. In fact, these requirements are only suitable for the critical glare hours in Table and are excessive to a fault for most of the day. The second conclusion is that the sunscreen structures could satisfy the minimum illuminance requirements *outside* of the critical glare hours most of the time. This is demonstrated in Figure 13 by the preliminary estimate determined by methods outlined in international standards (black line). The major shortcoming is that these standards do not account for the critical glare hours but are adequate for normal operational hours.

These conclusions may leave net-zero tunnel designers with the following dilemma: whether to under-design the lighting system for critical hours to maximize the energy saving potential of the sunscreen structure; or to over-design the lighting system at the cost of peak load and energy expenses. While the logical answer is that the perfect design is a compromise that lies between the two extremes (perhaps with a monitored control system), further research is needed to find an optimal solution that balances safety with energy preservation.

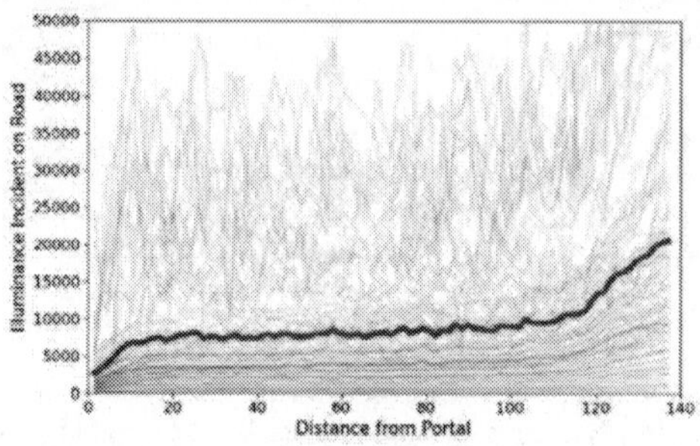

Figure 11: Average road surface illuminance (dark purple) under a matrix-based semi-transparent sunscreen with an effective transparency of 30%. Light purple lines represent hourly data.

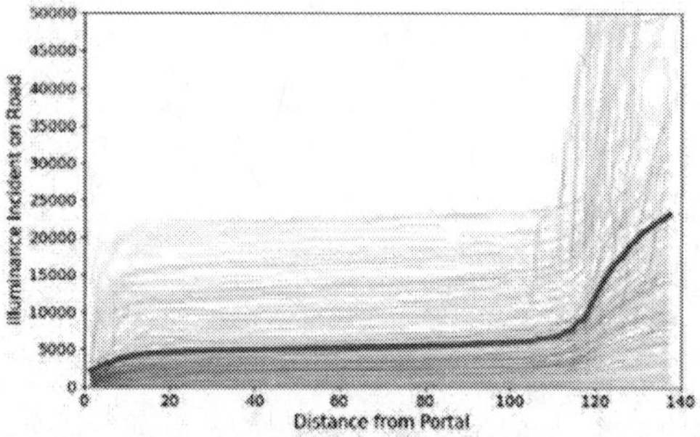

Figure 12: Average road surface illuminance (dark olive) under an intrinsically semi-transparent PV sunscreen with an effective transparency of 30%. The light olive lines represent hourly data.

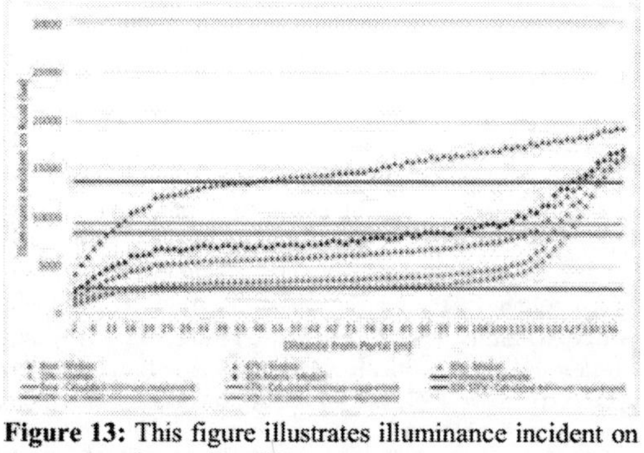

Figure 13: This figure illustrates illuminance incident on the road surface under different sunscreen structures. Each sunscreen's median illuminance over the course of a year is plotted at different distances from the tunnel portal and compared to their calculated minimum illuminance requirements and the preliminary estimate (black line) determined using methods described in tunnel lighting standards.

In the scope of this study, daytime lighting loads in the threshold zone are significantly reduced only if enough daylight is incident on the road to meet minimum illuminance requirements for all hours of the year. The reality is not as binary, and some load reduction can still be attained if the sunscreens are supplemented with artificial lights. While this information is enough to conclude that peak lighting load and energy savings *are possible* through displacing the threshold zone, a more detailed lighting and daylighting model is required to quantify exactly how much can be reduced. The numerical lighting model of a reference tunnel showed that if all artificial day-time lights in the threshold zone were effectively replaced, the potential load reduction and energy savings could be as high as 42% and 16%, respectively, of the total lighting system.

6 CONCLUSIONS

This paper demonstrates that STPV integration into sunscreen structures can provide many benefits to the energy performance and safety in tunnels. The use of a Rhinoceros and DIVA daylighting model enabled us to determine the illuminance conditions on the road surface under matrix based and intrinsic STPV sunscreen structures, and to qualitatively and quantitatively understand the impact of critical glare hours on drivers entering the tunnel. Combined with the numerical lighting model of a reference tunnel, PVsyst was able to determine the potential of STPV sunscreen structures to reduce lighting loads and energy consumption. The most notable findings are:

i) Matrix-based STPV sunscreens are capable of offsetting day-time lighting loads and generating more energy (on an order of magnitude) than intrinsic STPV sunscreens of the same size.

ii) Threshold luminance requirements can be significantly reduced during critical glare hours using STPV sunscreens.

iii) Intrinsic STPV materials are more suited for reducing luminance requirements than matrix-based STPV modules.

iv) There is a gap of knowledge when it comes to designing adequate lighting systems for critical glare hours.

v) Matrix-based STPV sunscreens result in poor illuminance uniformity on the road surface when compared to sunscreens that use intrinsic STPV materials.

vi) Using a STPV sunscreen would most likely require lighting designers to install artificial luminaires in the new threshold zone to supplement transmitted sunlight.

vii) More research needs to be done to quantify how much day-time lighting loads can be reduced, and energy can be saved by displacing the threshold zone.

Concordia University's Centre for Zero Energy Building Studies also has ongoing research involving:

i) The development of a combined lighting, daylighting and PV model that can quantitatively estimate peak load reductions and energy savings induced by displacing the threshold zone.

ii) The development of a singular, generalized model that can estimate peak load reduction and energy savings of STPV sunscreens (which includes inputs for the PV system, lighting system, and daylighting).

iii) Bridging the gap that exists between lighting requirements during critical glare hours and the rest of the day.

iv) The impactful effect of sunscreens on motorist safety and tunnel subsystems other than lighting.

FUNDING

1) Concordia University, School of Graduate Studies
2) Concordia University, Centre for Zero Energy Building Studies

REFERENCES

[1] C. Reinhart, *Daylighting Handbook 1: Fundamentals Designing with the Sun*. Cambridge: Christoph Reinhart, 2013.

[2] K. Janssen, J. Lantinga, and J. Naber, "Working Towards a Zero-Energy Tunnel," Presented at ITA Conference 2017 in Bergen, Norway, 2017.

[3] R. Dzhusupova, J. F. G. Cobben, and W. L. Kling, "Zero energy tunnel: Renewable energy generation and reduction of energy consumption," in *2012 47th International Universities Power Engineering Conference (UPEC)*, 2012, pp. 1–6.

[4] CIE 2004, "CIE 88:2004. Guide for the lighting of road tunnels and underpasses," 2004.

[5] ANSI, "Tunnel Lighting," 2011.

[6] A. O. Abdul Salam and K. A. Mezher, "Energy saving in tunnels lighting using shading structures," *Proc. 2014 Int. Renew. Sustain. Energy Conf. IRSEC 2014*, no. October 2014, pp. 519–524, 2014.

[7] A. Peña-Garćia, L. M. Gil-Martín, R. Escribano, and A. Espín-Estrella, "A scale model of tension structures in road tunnels to optimize the use of solar light for energy saving," *Int. J. Photoenergy*, vol. 2011, no. September, 2011.

[8] A. Peña García, L. M. Gil-Martín, A. Espín Estrella, and F. Aznar Dols, "Energy saving in road tunnels by means of transparent tension structures," in *International Conference on Renewable Energies and Power Quality*, 2010.

[9] D. Drakou, C. Burattini, F. Bisegna, and F. Gugliermetti, "Study of a daylight 'filter' zone in tunnels," in *2015 IEEE 15th International Conference on Environment and Electrical Engineering (EEEIC)*, 2015, no. June, pp. 649–652.

[10] L. M. Gil-Martín, A. Gómez-Guzmán, and A. Peña-García, "Use of diffusers materials to improve the homogeneity of sunlight under pergolas installed in road tunnels portals for energy savings," *Tunn. Undergr. Sp. Technol.*, vol. 48, pp. 123–128, Apr. 2015.

[11] A. Peña-García, J. C. López, and A. L. Grindlay, "Decrease of energy demands of lighting installations in road tunnels based in the forestation of portal surroundings with climbing plants," *Tunn. Undergr. Sp. Technol.*, vol. 46, pp. 111–115, 2015.

[12] Z. Zhang, Y. Zou, and Z. Gou, "Lighting Analysis at Access Zone of Tunnel Entrance of IJEScA," vol. 3, no. May, pp. 97–110, 2016.

[13] G. Parise, L. Martirano, and S. Pierdomenico, "An Adaptive Criterion to Design the Lighting System in the Road Tunnels," in *2007 IEEE Industry Applications Annual Meeting*, 2007, pp. 1244–1248.

[14] S. A. Kalogirou, "Building integration of solar renewable energy systems towards zero or nearly zero energy buildings," *Int. J. Low-Carbon Technol.*, vol. 10, no. 4, pp. 379–385, Dec. 2015.

[15] Z. Ioannidis, "Double Skin Facades Integrating Photovoltaic Panels, Motorized Shades and Controlled Air Flow," Concordia University, 2016.

[16] L. Aresti, R. Agathokleous, P. Christodoulides, and S. A. Kalogirou, "Air flow effect on the temperature of a building integrated PV-panel," 2013, p. 6.

[17] K. Kapsis, "Modelling, Design and Experimental Study of Semi-Transparent Photovoltaic Windows for Commercial Building Applications," Concordia University, 2016.

[18] G. Bizzarri, M. Gillott, and V. Belpoliti, "The potential of semitransparent photovoltaic devices for architectural integration: The development of device performance and improvement of the indoor environmental quality and comfort through case-study application," *Sustain. Cities Soc.*, vol. 1, no. 3, pp. 178–185, Oct. 2011.

[19] M. Makha *et al.*, "Ternary semitransparent organic solar cells with a laminated top electrode.," *Sci. Technol. Adv. Mater.*, vol. 18, no. 1, pp. 68–75, 2017.

[20] A. Athienitis and W. O'Brien, *Modeling, Design, and Optimization of Net-Zero Energy Buildings*. Montreal: Ernst & Sohn, 2015.

[21] R. Mcneel, "Rhinoceros 5 User's Guide Table of Contents Section I: Introduction," 2014.

[22] "The Python Language Reference — Python 2.7.14 documentation." [Online]. Available: https://docs.python.org/2/reference/index.html. [Accessed: 08-Mar-2018].

[23] PVsyst SA, "PVsyst".

PERFORMANCE ANALYSIS OF VERTICALLY MOUNTED BIFACIAL PV MODULES ON GREEN ROOF SYSTEM

Thomas Baumann[1]*, Fabian Carigiet[1], Raphael Knecht[1], Markus Klenk[1], Andreas Dreisiebner[2], Hartmut Nussbaumer[1], Franz Baumgartner[1]
[1] ZHAW, Zurich University of Applied Sciences, SoE, Institute of Energy Systems and Fluid Engineering, Technikumstrasse 9, CH-8401 Winterthur, Switzerland, www.zhaw.ch/=bamt
*phone: +41 (0) 58 934 71 87; e-mail: thomas.baumann@zhaw.ch
[2] Solarspar, Bahnhofstrasse 29, CH-4450 Sissach, Switzerland, www.solarspar.ch

ABSTRACT: A combination of PV and green roof is an ideal fusion in terms of ecology. The green roof improves the water retention in the city, whereas the PV system produces electric power at the place where it is consumed. Flat tilted modules in south or east west direction on green roofs generally require intensive maintenance to prevent them from being shaded by plants and often cover the roof area to a large extent. Because of the space requirement conflict between PV on the roof and green roofs, it is essential to combine these two systems in a smart way. Vertically mounted bifacial modules can be an option to combine PV and green roof and to also allow a cost-effective maintenance.
In this paper we report about the layout and the performance of a corresponding system, subdivided into two groups with differing albedo. Custom made bifacial modules with 20 cells were produced to reduce the wind load and to improve the general appearance. This 9.09 kWp bifacial plant achieved a specific yield of 942 kWh/kWp in one year (11.08.2017 to 10.08.2018). High quality DC power measurement systems are installed to monitor two modules in each bifacial test field and a reference south-facing module. This allows an energy yield comparison between the vertical bifacial test system with east-west orientation and the monofacial south-facing reference over four months of outdoor measurements. The use of plants with good reflective properties, which are also well suited two the ambient conditions on flat roofs, resulted in a yield increase of 17 % compared to a standard green roof planting. The vertically installed bifacial modules obtained an almost identical specific yield (-1.4 %) compared to a stand-alone monofacial south-facing reference module. Due to the increased yield in the mornings and afternoons, the vertical bifacial modules can achieve higher self-consumption depending on the load profile.
Keywords: Bifacial, PV System, System Performance, Green Roof

1 INTRODUCTION

According to the energy strategy 2050 of the Federal Council and Parliament of Switzerland, electricity originated from nuclear power has to be completely covered by electricity from new renewable energies; 16.7 % of the total electricity demand is to be covered by PV electricity by 2050 [1]. Particularly in densely populated countries such as Switzerland, the utilization of flat roofs for PV may be an important means to increase the installed capacity.

Green roofs promise a wide range of environmental, economic and social benefits. Building regulations that support the green roof concept are being implemented around the world at the moment [2]. These regulations aim to reduce rainwater runoff, to improve air quality and to reduce the need for cooling [3]. A comparison of 17 studies shows a cooling at street level between 0.03-3 C° due to green roofs [4]. Even though we could show in a previous study that the green roof cooling effect only has a negligible influence on the PV yield [5], a combined use is desirable with regard to the above considerations.

Today, PV installations on green roofs are rather rare. Because of the space requirement conflict between PV on the roof and green roofs, it is essential to identify options that allow a combination of both approaches.

Vertically mounted bifacial modules may be such an option. According to simulations, vertically mounted bifacial modules can have a higher energy yield than standard bifacial installations depending on the location and the installation conditions [6], [7].

With the presented concept for vertical mounting and corresponding outdoor measurements, we want to show that PV and green roof can be combined and that the maintenance can be done in a cost-effective way.

2 APPROACH

2.1 Project goal

This joint project of green roof experts and photovoltaic specialists [8] was set up to verify the suitability of an adapted vertical bifacial PV system on a flat green roof. One goal was to analyse how vertically east-west-facing mounted bifacial modules perform compared to a monofacial south oriented module and if the vertically mounted bifacial PV panels actually allow a cost-effective maintenance. In addition, the influence of silver-leaved plants on the specific yield should be investigated.

Figure 1 Vertically east-west mounted bifacial modules installed on the green roof in Winterthur, Switzerland with a rated output of 9.09 kWp. The east-west-facing modules have a tilt angle of 90° and an azimuth angle of -65° for the front side and 115° for the rear side. (used azimuth angle definition: east = -90°, south=0° and west = 90°)

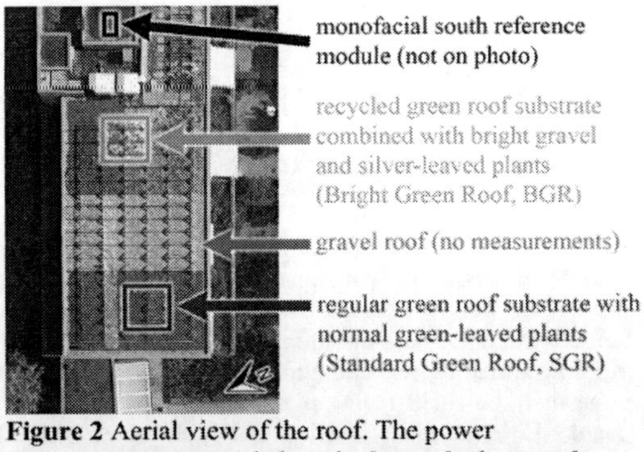

Figure 2 Aerial view of the roof. The power measurements are carried out in the marked rectangles.

Figure 3 Photos of the two green roof areas. On the left side is the standard green-leaved green roof (SGR) and on the right side the "bright" silvered-leaved green roof (BGR). The photos were taken after a long dry and heat period on 21.08.2018. It shows nicely that the silver-leaved plants survive such a period better than a green-leaved plant. The housings on the floor below the modules contain part of the measurement electronics.

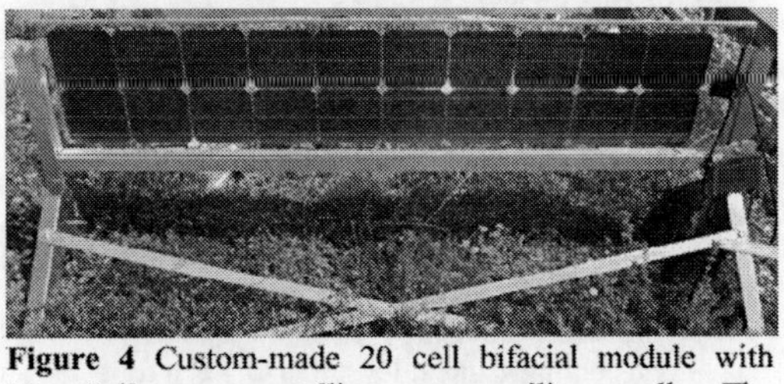

Figure 4 Custom-made 20 cell bifacial module with MegaCell monocrystalline n-type silicon cells. The average STC-power of the four modules measured is 82.2 W on the front side and 71.6 W on the rear side.

Figure 5 Monofacial south facing (25° azimuth, 16° tilt angle) monocrystalline Silicon reference module of the manufacturer Hareonsolar with a measured STC power of 203.3 W.

2.2 Setup

On a flat roof (25° azimuth) located in Winterthur Switzerland, a bifacial PV plant of 9.09 kWp was installed. The modules are mounted vertically in east-west direction (90° tilt angle, -65° azimuth for the front side and 115° for the rear side) as shown in Figure 1. The green roof is divided into two test fields with different soil material. One test field consists of a standard green roof substrate and a standard mixture of green-leaved plants (standard green roof, SGR). The other consists of recycled green roof substrate combined with bright gravel and silver-leaved plants to achieve a higher albedo (bright green roof, BGR). Figure 3 shows the two test areas. The silver-leaved plants are usually found in the Mediterranean region. The bright color of the plants protects them from drying out by partially reflecting the sunlight, accordingly they also have an improved resilience in dry conditions. Plants were selected with a low growth height. Between these two test fields there is a normal gravel roof with installed PV where no detailed measurements are made. A standard monofacial module is installed in south direction (16° tilt angle, 25° azimuth) as reference. The arrangement on the roof is shown in the aerial view in Figure 2.

The bifacial modules are custom made glass/glass modules where 20 monocrystalline n-type silicon cells from the manufacturer MegaCell are embedded. A photo of such a module with the vertically mounting system from the company ZinCo is shown in Figure 4. The modules have a dimension of 36 x 170 cm, whereby the total cell area is 31.5 x 157.5 cm. The distance between the floor and the lower edge of the module is 40 cm. The row distance between the modules is 98 cm, which corresponds to a shading angle of 20.2°. The ground coverage ratio (GCR) is 36.7 %.

The south oriented (25° azimuth, 16° tilt angle) monofacial reference module is a glass/backsheet module from the manufacturer Hareonsolar (type HR-210-24/Aa) with 72 monocrystalline silicon cells as shown in Figure 5. For a fictive installation with south oriented modules on the whole roof, a row distance of 1.4 meters would have been chosen for a shading angle of 20°. The resulting GCR would then be 57.7 %, which is significantly higher compared to the solution using the vertically oriented bifacial modules. The tilt angle of 16° was chosen in order to avoid shading of the bifacial modules in a potential combination of bifacial east-west-facing and monofacial south orientated modules.

The bifacial PV plant was put into operation in March 2017 and the south-facing reference module was installed in spring 2018. The power measurement of the 5 modules (reference module plus 4 bifacial modules) was started on 19.05.2018.

2.3 Surroundings

The test roof is located in Winterthur, Switzerland (coordinates: 47°30'10.7"N 8°43'22.0"E). There is a higher building on the west side of the roof, which causes shading in the evening. Figure 6 shows a panorama photo, including the sun positions over one year. The shading causes more power loss on the bifacial modules than on the reference module. In order to be able to compare the measurements at the different locations on the roof, only the measurement data up to an azimuth angle of 86° are evaluated. In an additional evaluation, the missing data from 86° azimuth onwards were simulated as described in chapter 2.5 and added to the dataset. This allows to make conclusions for roofs that are not affected by shading.

Figure 6 Panorama photo taken between the module rows with drawn sun positions (orange) and horizon line (yellow). The red line marks the 86° azimuth angle. On the left side of the red line measurement data, on the right side of the red line simulated data are used.

2.4 Measurements

High quality DC power measurement systems are installed to monitor two modules in each bifacial test field and the reference module. The measurement uncertainty for the power measurement is below 0.5 % (k=2) and the measurements are logged every 10 seconds. Each module has its own MPP tracker whereas the power measurement is installed between the PV module and the MPP tracker.

The module temperature of the reference module is measured with a Class A PT1000 sensor mounted on the module backside. The measurement uncertainty of the temperature sensor depends on the actual temperature resulting in ± 0.25 °C (k=2) at 50 °C. The used measurement electronics in the data logger add another ± 0.25 °C measurement uncertainty (k=2).

STC power measurements of the monitored modules were performed with the Portable LED Flasher [9] before the installation (each side of the bifacial module was measured separately while the other side was completely shaded).

2.5 Yield simulations

The 10 minute meteo data (global irradiance, diffuse irradiance, air temperature, wind speed and wind direction) of the nearest meteo station of MeteoSwiss (Federal Office of Meteorology and Climatology) are used for the yield simulation. The closest station is in Aadorf (13.9 km linear distance). A low-cost irradiance measurement is installed on the test roof itself, but it is also affected by the shading. Therefore, it cannot be used for the simulation. The yield simulation is structured as follows:

1. sun position (azimuth and elevation) [10]
2. angle of incidence (AOI) [11]
3. extraterrestrial irradiance [11]
4. direct normal irradiance from global and diffuse irradiance [11]
5. global tilted irradiance [12]
6. low-light performance [13]
7. module temperature [11]
8. module power considering low-light performance, module temperature, power temperature coefficient and spectral mismatch [13]

For the bifacial modules, the global tilted irradiance was calculated for both sides and then summed up considering the bifacility. In addition the mutual shading was taken into account for the calculation of the global tilted irradiance.

In order to obtain a uniform time interval with the 10 second measurement data, the 10 minute simulation data is interpolated linearly [5]. The simulated DC power is scaled according to the measurement data during the day where no shading appears. Then, the scaled data are added to the measurements after the shading occurs as shown in Figure 7.

From chapter 4.5 onwards, the results are given for the pure power measurement data up to 86° azimuth and for the combination of measurement data and simulation. In addition, the results from the combination of vertically bifacial (BGR) and south-facing monofacial systems are given. The combination of 40 % vertically bifacial and 60 % monofacial south leads to the blue colored graph shown in Figure 7. Such a combination could be interesting for industrial companies, but also for households. The combination offers a higher yield in the morning and afternoon and a more homogeneous yield at noon than the south facing plant alone. This blue graph could be optimized by rotating 180° half of the bifacial modules, regarding homogeneity over noon.

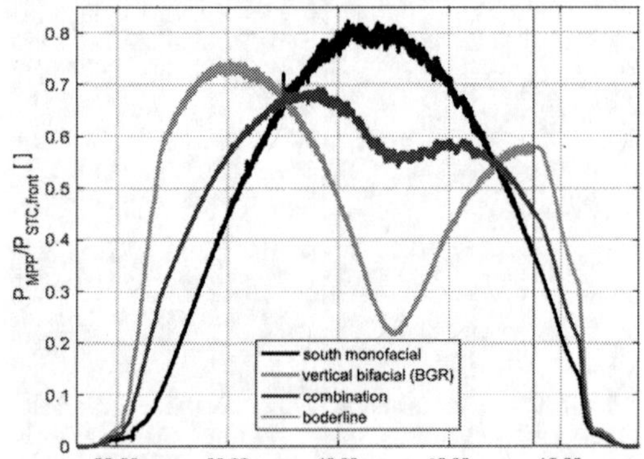

Figure 7 Graph of the measured and simulated powers of the reference south-facing module (black), a bifacial east-west-facing module (orange) and a combination of bifacial and south (blue, 40 % vertically bifacial and 60 % monofacial south). Simulated data was used on the right side of the red border line. The data from 16.08.2018 are shown, where the time is in UTC+1.

3 INNOVATION AND RELEVANCE

Sealed floor surfaces in cities, waste heat of vehicles and houses which cause heat islands enhance the warming of the air during the day and reduce the cooling of the air during night. Thus, summer nights are four to five degrees warmer in cities compared to rural regions [14]. The maximum cooling at street level due to green roofs ranges from 0.03 to 3 °C according to 17 studies that provide data on urban heat island reduction [4].

The economic and ecological advantages of green roofs are undisputed. Today, the decision between PV and green roof has to be often made for flat roofs.

The green roof has several advantages like water retention, reduction of peak water runoff, protection of the roof seal, additional insulation, cooling and air humidification of the ambient atmosphere, habitat for animals (especially insects), filtering air pollutants etc.

Typical PV installations however largely cover the flat roof area, which suppresses the green roof effect to a large extent, if not bitumen or gravel is used from the outset. Plants may also lead to considerable shading of the modules, while the green roof maintenance is hindered by the PV installation.

With vertical bifacial PV on green roofs, a symbiosis of these two approaches can be obtained. In this project, the achieved energy yield is examined for this type of module mounting, which also allows a cost-effective maintenance of the green roof.

4 MEASUREMENT RESULTS AND ANALYSIS

4.1 Data availability

The DC power measurements were started on 19.05.2018 and will be carried out for one year. Outdoor measurements up to 18.09.2018 have been taken into account for this work. Therefore, four months or 123 days of measurement data have been evaluated. Data of the electricity meter of the whole plant are available since July 2017.

4.2 STC power measurement

The STC power of the measured modules was measured in the laboratory with the Portable LED Flasher [9]. Table 1 shows the results of the measurements. Each side of the bifacial module was measured separately while the other side was shaded. The bifaciality (B) in Table 1 was calculated using Formula (1).

$$B = \frac{P_{STC,rear}}{P_{STC,front}} \; [\,] \qquad (1)$$

Table 1 Measured STC powers of all examined modules and calculated bifaciality (SGR = test field with standard green roof with green-leaved plants, BGR = test field with bright green roof with silver-leaved plants and bright substrate)

Module	$P_{STC,front}$ [W]	$P_{STC,rear}$ [W]	B
SGR 1	76.2	56.1	0.74
SGR 2	88.8	81.1	0.91
BGR 1	88.3	81.9	0.93
BGR 2	75.5	67.1	0.89
Sum SGR	165.0	137.3	0.83
Sum BGR	163.8	149.0	0.91
Reference	203.3	-	-

4.3 Albedo measurement

The albedo measurement was performed on a clear sky day (21.08.2018) at a sun position of 25° azimuth and listed in Table 2 (the shadows of the modules were exactly under the modules). The measuring device was placed at the height of the upper edges of the monitored module in the middle of the module rows. Figure 3 and Figure 8 show the photos that were taken at the time of the albedo measurement. As a comparison, the gravel surface was also measured. In addition, the albedo was measured for the areas with more plants where no DC power measurements were performed. These two additional surfaces are shown in Figure 8. The horizontal global irradiance during the albedo measurements was 780 W/m^2.

Figure 8 Gravel (left) and standard green roof (SGR) with more plants (right) at the time of the albedo measurements.

Table 2 Results of albedo measurements carried out on 21.08.2018 of the different surfaces. Measured on the module upper edges in the middle of the module rows.

Surface	Albedo [ ]
SGR (Figure 3 left)	0.09
BGR (Figure 3 right)	0.21
Gravel (Figure 8 left)	0.14
SGR with more plants (Figure 8 right)	0.16

4.4 Specific annual AC energy yield of the entire PV system

An AC electricity meter is installed, which measures the feed-in of the entire PV system. The meter reading was read periodically and evaluated for a period of one year (11.08.2017 to 10.08.2018). The specific energy yield of the 9.09 kWp vertical bifacial PV system in this period is 942 kWh/kWp. A typical value for south-facing plants in our region is 1000 kWh/kWp.

4.5 Specific energy yield of DC power measurement

The resulting specific DC energy yields for the period 19.05.2018 to 18.09.2018 are shown in Table 3.

Table 3 Specific DC energy yields in kWh/kW for the time period of 19.05.2018 to 18.09.2018. The first row contains only the measured data up to 86° azimuth and the second line the combination of measured data and simulation beyond 86° azimuth. The combination of south and vertical bifacial corresponds to the description in chapter 2.5.

Dataset	South monofacial	East-west bifacial SGR	East-west bifacial BGR	Combination South + east-west BGR
Measurement	603.2	476.3	557.9	585.1
Measurement and simulation	658.4	556.9	649.2	654.7

For the two vertical bifacial systems, the bright green roof (BGR) provides a significant increase in yield of 17.1 % for pure measurement data and 16.6 % for measurement data combined with simulation. The lower yield of 7.5 % of the east-west bifacial BGR compared to south-oriented monofacial module for the measurement data and 1.4 % for the combined measured data and simulated data shows that the yield of the vertical bifacial modules is almost identical to that of the monofacial south-oriented module in an environment that does not cause shading by surrounding objects such as buildings.

To estimate the expected specific yield for one year, simulations were performed using PVsyst 6.74. The measured albedo according to Table 2 and the horizon according to Figure 6 (yellow line) were used for this simulation. Table 4 contains the results. The first line contains the simulation results for the measurement period (19.05.2018 to 18.09.2018). For the south oriented monofacial module (-2.2 %) and the SGR (+0.4 %) the simulation results correspond quite well. For the BGR the deviations from PVsyst are comparably higher (-8 %). The second line of Table 4 shows the expected values for one year, whereby the four months measured (Table 3, line 1) were supplemented with the PVsyst simulations for the rest of the year.

Table 4 PVsyst simulation results of specific DC energy yields in kWh/kW for the measurement period (19.05.2018 to 18.09.2018) on the first line and on the second line the combination of the measurement data (Table 3, line 1) with the PVsyst results for the rest of the year to estimate the expected specific yield for one year.

Dataset	**South monofacial**	**East-west bifacial SGR**	**East-west bifacial BGR**	**Combination South + east-west BGR**
PVsyst for measurement period	590	478	513	559
4 months measurement and 8 months PVsyst	1225	978	1090	1171

4.6 Self-consumption

A 10 second load profile [15], [16] was used to make a statement about self-consumption (SC). The load profile was generated for a single-family house (profile name HT20) with a family of two children (profile name CHR57) with an annual consumption of 8535 kWh. The amount consumed in the period under review (19.05.2018 until 18.09.2018) is 2556 kWh. The nominal power of the PV system was set at 8.5 kWp. Table 5 shows the resulting own consumption values and Figure 9 shows the daily course from 20.06.2018.

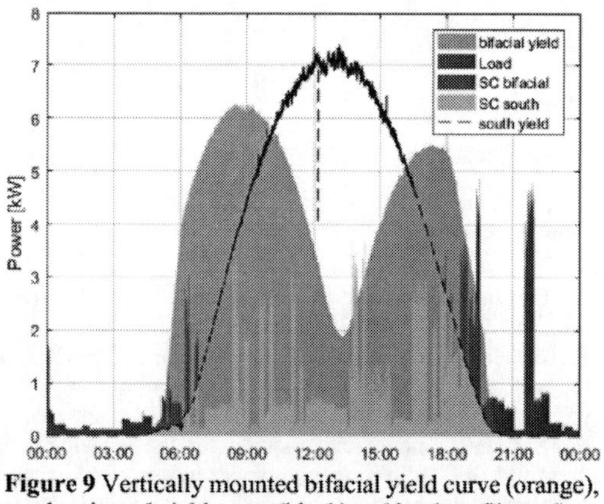

Figure 9 Vertically mounted bifacial yield curve (orange), south-oriented yield curve (black) and load profile (red) on 20.06.2018 (time in UTC+1). Light green shows the self-consumption (SC) of the south oriented system and in dark green the additional self-consumption due to the bifacial east-west orientation. The peak in the afternoon would also be covered by a combination of south and east-west. On this day, self-consumption amounts to 23 % for the south-facing plant and 26 % for the east-west-facing bifacial plant.

Table 5 Self-consumption values in percent for the period under review (19.05.2018 until 18.09.2018)

Dataset	**South monofacial**	**East-west Bifacial SGR**	**East-west bifacial BGR**	**Combination South + east-west BGR**
Measurement	24.2	28.9	26.1	25.4
Measurement and simulation	25.8	29.7	27.0	26.8

5 CONCLUSION

The combination of vertically installed bifacial modules and green roof has been proven very successful. The maintenance of the green roof can be carried out efficiently, because you can walk well between the module rows. Maintenance costs may be further reduced by using a mowing robot [5].

Due to the bright substrate with the silver-leaved plants the yield could be increased by 17 % compared to a standard green roof with green-leaved plants in summer. The additional costs of the bright surface amount to approximately 10 % [17] compared to a standard green roof. Therefore, this investment is worthwhile. The silver-leaved plants also showed very good resistance against heat and drought. The albedo of the standard green roof was further worsened by the drying-out of the plants.

The specific DC energy yield of the bright green roof with vertically mounted bifacial modules is almost identical to that of the south oriented monofacial module. The vertical bifacial installation is expected to produce a good yield in winter compared to the south module, since the vertical modules are not shaded by snow and the snow increases the albedo. The specific yield of the whole 9.09 kWp PV plant over one year was 942 kWh/kWp.

Due to the vertical installation and the associated different power distribution associated with it, there is a potential of increasing the self-consumption rate for specific consumers.

The power measurements will be continued at least until spring 2019. It is planned to install and measure a similar plant in a mountainous region to confirm the expected high yield in winter with snow.

ACKNOWLEDGMENT

Thanks to the founding agency "Klimafonds Stadtwerk Winterthur" and project partners Solarspar, ZinCo, Fenaco, A777 garden design, Stadtwerk Winterthur (public utility Winterthur), Swiss professional association for green buildings SFG, Fritz Wassmann and ZHAW Life Sciences and Facility Management. Thanks to Hanspeter Willi (head housing technology of rentirement home Wiesengrund) for reading the electricity meter.

REFERENCES

[1] BFE, 'Energieperspektiven 2050 Zusammenfassung', Swiss Federal Office of Energy SFOE, Oct. 2013.

[2] J. Snow, 'Green Roofs Take Root Around the World', *National Geographic News*, 27-Oct-2016. [Online]. Available: https://news.nationalgeographic.com/2016/10/san-francisco-green-roof-law/. [Accessed: 05-Mar-2018].

[3] V. Azeñas *et al.*, 'Thermal regulation capacity of a green roof system in the mediterranean region: The effects of vegetation and irrigation level', *Energy and Buildings*, vol. 164, pp. 226–238, Apr. 2018.

[4] L. F. M. Francis and M. B. Jensen, 'Benefits of green roofs: A systematic review of the evidence for three ecosystem services', *Urban Forestry & Urban Greening*, vol. 28, pp. 167–176, Dec. 2017.

[5] T. Baumann, D. Schär, F. Carigiet, A. Dreisiebner, and F. Baumgartner, 'Performance Analysis of PV Green Roof Systems', in *32nd European Photovoltaic Solar` Energy Conference and Exhibition; 1618-1622*, 2016.

[6] H. Nussbaumer *et al.*, 'PV Installations Based on Vertically Mounted Bifacial Modules Evaluation of Energy Yield and Shading Effects', in *31st European Photovoltaic Solar Energy Conference and Exhibition; 2037-2041*, 2015.

[7] S. Guo, T. M. Walsh, and M. Peters, 'Vertically mounted bifacial photovoltaic modules: A global analysis', *Energy*, vol. 61, pp. 447–454, Nov. 2013.

[8] 'Andreas Dreisiebner (solarspar), Ralf Walker (ZinCo), Tobias Probst (Fenaco), Stephan Brenneisen (ZHAW Life Sciences und Facility Management), Fritz Wassmann, project funded by the Klimafonds Stadtwerk Winterthur, Switzerland'.

[9] R. Knecht *et al.*, 'Field Testing of Portable LED Flasher for Nominal Power Measurements of PV-Modules On-Site', in *33rd European Photovoltaic Solar Energy Conference and Exhibition; 2007-2012*, 2017.

[10] 'DIN 5034-2:1985-02 Tageslicht in Innenräumen; Grundlagen'. Beuth Verlag GmbH, Feb-1985.

[11] V. Quaschning, *Regenerative Energiesysteme: Technologie - Berechnung - Simulation*, 8., aktualisierte und erw. Aufl. München: Hanser, 2013.

[12] R. Perez, P. Ineichen, R. Seals, J. Michalsky, and R. Stewart, 'Modeling daylight availability and irradiance components from direct and global irradiance', *Solar Energy*, vol. 44, no. 5, pp. 271–289, 1990.

[13] F. Carigiet, F. P. Baumgartner, J. Sutterlueti, N. Allet, M. Pezzotti, and J. Haller, 'Energy Rating Based on Thermal Modelling of Five Different PV Technologies', in *29th European Photovoltaic Solar Energy Conference and Exhibition; 3311-3315*, 2014.

[14] M. Soukup and S. Häne, 'Mit grünen Dächern gegen die Hitze', *Tages-Anzeiger*, 09-Jul-2015.

[15] N. Pflugradt and U. Muntwyler, 'Synthesizing residential load profiles using behavior simulation', *Energy Procedia*, vol. 122, pp. 655–660, Sep. 2017.

[16] N. Pflugradt, 'LoadProfileGenerator'. [Online]. Available: https://www.loadprofilegenerator.de.

[17] A. Dreisiebner, 'Personal interview with the gardener and project partner'.

SOLAR HYBRID ENERGY POWERING QUADCOPTER

Ching-Fuh Lin[1, *], Hsiang Lan[2], Ta-Jung Lin[3], Wei-Sheng Liao[4], Jiun-Yu Lin[5], Chi-Han Chiu[6], Hong-Jhang Syu[7]
[1] Graduate Institute of Photonics and Optoelectronics, National Taiwan University, 1, Sec. 4, Roosevelt Rd., Taipei 10617, Taiwan., Tel: +886-2-33663540, E-mail: lincf@ntu.edu.tw
[2] Department of Electrical Engineering, National Taiwan University, Taipei, Taiwan, E-mail: footworknuts@gmail.com
[3] Taipei Municipal Jianguo High School, E-mail: metro123star@gmail.com
[4] Department of Electrical Engineering, National Taiwan University, Taipei, Taiwan, E-mail: artliao5559@gmail.com
[5] Department of Electrical Engineering, National Taiwan University, Taipei, Taiwan, E-mail: linjiunyu.tw@gmail.com
[6] Taipei Municipal Jianguo High School, E-mail: ericcm509mh@gmail.com
[7] Graduate Institute of Photonics and Optoelectronics, National Taiwan University, Taipei, Taiwan, E-mail: d02941007@ntu.edu.tw
* The author for all correspondence. He is also a faculty of Graduate Institute of Electronic Engineering and Department of Electrical Engineering

ABSTRACT: Solar power is used to significantly extend the flight time of the multi-copters. So far a flight time of over 3 hours has been achieved. In our field test of flight, the solar module used in the quadcopters is able to provide sufficient power without the battery. However, when the quadcopters flies, its attitudes change all the time, so the flight controller continuously adjusts the current to each motor, leading to the fast fluctuation of the current supplied by the solar module. Thus the voltage and the power also quickly vary. This could possibly damage the electronics for the quadcopters. It could be overcome with a back-up battery to bias the solar cells at a nearly fixed voltage. This back-up battery also has the function of providing emergency power when the sunlight is blocked by the clouds or strong wind suddenly blows. Our experiments show that 1000 mAh charge capacity could make the quadcopter fly for over 2 hours because the major power is provided by the sun. If only this 1000 mAh battery is used, the flight time is 7 minutes and 56 seconds. With the solar power, the flight time is extended to more than 15 times. With hours of flight time, the copters will enable the users to explore the frontiers that people have never gone before.

Keywords: Solar power, quadcopter, hybrid energy

1 INTRODUCTION

Solar cells have many applications, including replacement of conventional power plants and use for portable electronic devices. In recent years, solar energy for vehicles has also caught increasing attention. Solar modules for cars and airplanes have been demonstrated. For example, Solar Impulse 2 had accomplished the longest single flight of 5 days continuously in 2015, from Japan to Hawaii. Using solar panels for fixed-wing airplanes [1] is pretty straightforward because the airplanes have a large horizontal area to install solar panels for harvesting sunlight. In contrast, using solar energies for copters that rely on the propellers to lift their weight is less developed. The reason is mainly two-fold. One is that the propellers typically occupy the most horizontal area of the copter, making few areas available for solar panels. The other is that the power required to lift copter is about five times more than the fixed-wing airplanes of the same weight due to the impossibility of applying Bernoulli's principle. However, the copters are much more convenient to use than the fixed-wing airplanes, particularly for unmanned aerial vehicles (UAV) that are commonly controlled by nonprofessional pilots, so copters have increasing popularity among the UAV markets.

Currently the energy for copters is mostly supplied by batteries or liquid fuels. Because liquid fuels are explosive, people have more concerns in using them. Nonetheless, the battery-powered light-weight copters, typically around 2 kilograms or less, have limited flight time, usually less than 30 minutes. It is hence highly desired to increase the flight time of copters and simultaneously avoid the danger of explosion caused by the liquid fuels. Solar energy then gives the opportunity to increase the flight time with good safety. In this research, we develop solar hybrid energy to power the quadcopters. The longet flight time achieved in our Lab-made quadcopters give flight time of over 3 hours.

2 Quadcopter design and solar modules

2.1 Solar module

We designed and implemented several solar hybrid systems. All used crystalline-Si solar cells of 5 inches with 22 % efficiency. One has 52 cells in series that can give a voltage of above 26 V. Two maximum power point trackers (MPPT) in parallel are used to maximize the output power of the solar module and to convert the high voltage to the output of 25.2 V, which is typical for 6S drones. The second has 72 cells, which are divided to two modules, each with 36 cells. The single module gives the voltage between 18 V and 22 V, which is less than that of 6S batteries, and so is connected to a voltage up-converter to give an output voltage of 26 V, and a diode. The two modules are put in parallel. A 6S battery is then connected in parallel with the output of the MPPTs or the diode. Our test flights show that the flight time could be increased to more than one hour. The ratio of the flight time with the solar power and only the battery is N:1. N could be 3.02, indicating that the power provided by the solar module is more than the battery to effectively increase the flight time.

Because the MPPT and the voltgae up converter cannot give 100 % conversion efficiency, the solar power is wasted. Later on, the third and the forth types of solar modules are developed without those conversin. In the third, we made the solar module for 5S voltage. This module has 76 cells, which are divided to two modules, each with 38 cells. The single module gives the voltage between 19 V and 23 V approximately. The forth type has 86 cells, which are divided to two modules, each with 43 cells. The single module gives the voltage between 21.5 V and 26 V approximately. This type matches the 6S voltage.

Both the third and the forth types give longer flight time than the first and the second types.

2.2 quadcopter design

In order to install the above solar module, the quadcopter needs to have a large frame size, while the propellers are relatively small in order not to block the sunlight incident on the solar module and to avoid the interference of the air flow caused by the solar module. As a result, the frame size (diagonal distance of the motors) is about 140 cm.

In addition, the frame with such a large size needs to be strong and light-weight. Fortunately, the carbon-fiber tubes are common nowadays, so we are able to make the quadcopters with the frame weight of about 1300g, including the motors and all electronics but without the solar module.

18" propellers are used for the quadcopters although much larger propellers are possible due to the large frame size. Limiting the propeller size is based on two major reasons. One is that the propeller area should not be huge to give more areas for the solar cells. The other is that the motors might not be able to efficiently drive large propellers.

To attach the solar module to the quadcopters, many types of small joints are fabricated using the 3D printer. Also, for the convenience of the field test, the solar modules are made foldable, so they can be carried with a usual car.

3 EXPERIMENTS

We have evaluated the power required for the quadcopters and the power possibly generated by the solar panels. Depending on the payload and power efficiency, for a quadcopter with the frame size of 450 mm or less, the current for a motor is usually between 1A and 10A, assuming 6S battery that gives a voltage of around 24V. There are four motors, so the power required is between 96W and 960 W. For the payload of about 2 kilograms, the total current for 4 motors is typically around 10 A, so a brief estimation of power consumption is 240 W.

For 22% efficiency of solar cells, the area required to generate 240 W under the solar intensity of $1kW/m^2$ is $1.09m^2$. Therefore, the usual quadcopter of frame size ≤ 450 mm has too small area to install the solar panels. It is necessary to make a quadcopter with the size that covers an area of about $1m^2$. In addition, because the propellers for quadcopters lie horizontally and cannot be blocked, the area for propellers has to be subtracted, further reducing the available area for solar panels.

3.1 Challenges and consideration to overcome

Such a large area of quadcopter gives rise to several challenges. First, the power required super-linearly increases with the weight of the entire quadcopter. The main reason is that the power efficiency of the motor nonlinearly decreases with the current, as shown in Figure 1.

With the example of the motor performance shown in Fig. 1, the quadcopter with the weight of 1.82 kg will need 4.4 A current. If the weight increases to 4.3 kg, the current required is 14A. The weight and current increase for 2.36 times and 3.18 times, respectively, indicating that the increasing weight consumes power more than the normal proportionality. As a result, the commercially available solar modules are too heavy for the quadcopter to carry. Most commercial solar modules have the area to weight ratio less than $3cm^2/g$. To generate 220 W (assuming 22% efficiency of cells), the area is $1m^2$ under the solar intensity of $1kW/m^2$. Then the weight of the solar modules is more than 3.3 kg. If the motors and the copter body weigh 1kg, then the required power of the motors shown in Figure 1 is 336 W, much more than the power generated by the solar module. This is under optimal condition. In reality, the absorbed solar intensity is less than $1kW/m^2$ because the quadcopter cannot remain at an inclined angle to harvest the maximum sunlight. In addition, the power efficiency of motors is usually worse than the curve shown in Fig. 1. Therefore, the usual solar modules will not be good enough to power the quadcopter and need to be re-engineered. A key factor is the area to weight ratio, which is preferred to be more than 7 cm^2/g under the efficiency condition of current Si solar cells. If the efficiency of solar cells increases, the area to weight ratio can be less restricted.

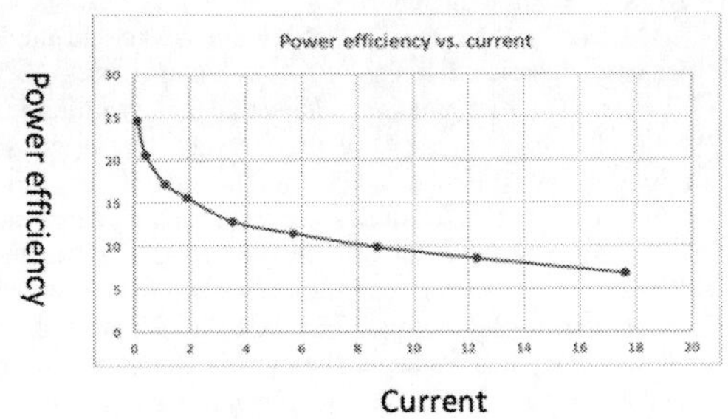

Figure1: Power efficiency of motor vs. current.

Second, because the area of the quadcopter is much more than the usual ones, the air resistance is large, making the quadcopter difficult to take off even though the force to lift the weight is sufficient. We found that this could be overcome by adjusting the flight parameters, i.e., PID values. Also, a battery electrically connected in parallel with the solar module will help because it can provide the extra power required when the quadcopter just takes off. With the battery and proper PID values, the air resistance due to the large area and the weight of the quadcopter together can be overcome.

Third, as the area of solar panels increases to 1 m^2, the torque due to the air resistance is also large. If the motors are placed near the center of the drone, the torque provided by the motors is too small to counter balance the torque from the air even though the motors give enough lifting force to overcome the gravity. A better configuration is to place the motors near the edge of the copter body. Thus, our quadcopter has the configuration shown in Figure 2. In this way, the same force of motors gives more torque to overcome the torque caused by the air resistance.

Fig. 2 only shows the photo of a quadcopter with solar modules. We actually made several types of solar modules that take into account the total power, the voltage, and the current required for the corresponding quadcopter. For a quadcopter, a low current, corresponding to a high voltage, is preferred because the motor usually has better power efficiency when it is running at a low current. In addition, the diameter of the wires that connect the electronic speed controller (ESC) to motors can be small to reduce the weight. This is particularly helpful for our quadcopter with extended frame size and long wires. Hence, we take the 6S or 5S voltage for motors and ESC used in the copter. The solar panels are then electrically configured with 6S or 5S voltage. The 6S voltage is typically 25.2V, while the 5S is 21V. If the voltage from the solar module does not exactly match the 6S or 5S standard, voltage converters are necessary to give a proper output voltage in order to match the battery for connection in parallel.

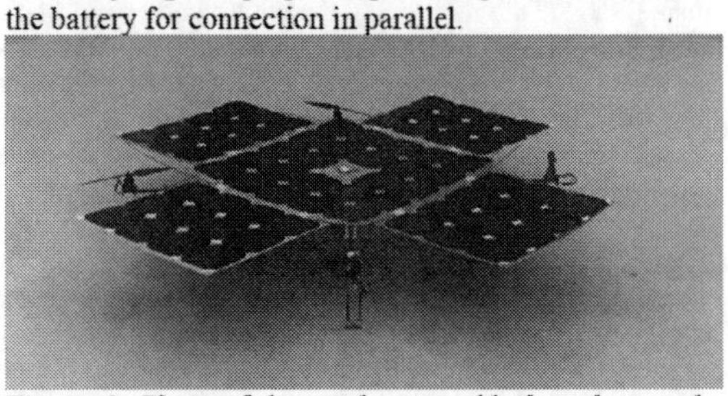

Figure 2: Photo of the quadcopter with the solar panels. The motors are placed near the edge of the body to have sufficient torques.

3.2 Practical implementation

The first type of the solar module has 52 crystalline Si solar cells in series. Each solar cell is 5 inches with 22 % efficiency. The total voltage is above 26 V. Two identical MPPT electronic circuits are used here to optimize the output power of the solar module and to convert the output voltage to 25.2 V. No diodes are used here because the MPPTs already have the function of rectification. The MPPTs have the conversion efficiency of 93.6%. A battery is connected to the output of the MPPTs. We also made monitoring circuits to check the current from the battery and the solar module. The two currents vary due to the variation of the solar power, but the total current is pretty fixed. When the copter flies, its position and orientation vary to remain good balance of forces from the four motors. Then the power of sunlight harvested is not constant. The battery can provide matching current for the quadcopter.

The second type has the same kind of 72 cells to form two modules. Each has 36 cells in series, so the voltage is less than 25.2 voltage and a voltage up-converter is required because a diode is used to prevent the current flowin into the solar cells and the diode will lead to a voltage drop of around 0.8V. The up-converters have the conversion efficiency of 96.2%. The output voltage of the converter is adjusted to 26 V. The two modules are connected in parallel. A diode is also connected to the output of the converters and then connected with the battery in parallel. Because the diode has a voltage drop of about 0.8 V, the voltage to be connected to the battery is 25.2 V. The circuit diagram is shown in Figure 3.

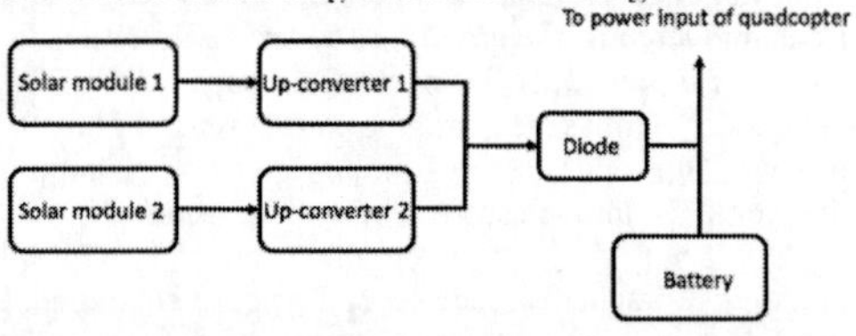

Figure 3. Circuit diagram of the solar modules and the battery connection. The up-converters have the conversion efficiency of 96.2%.

Because the MPPTs and up-converters consume power and they seem to easily burn out, we develop two other types of solar modules. One is with 76 solar cells that are divided to two modules, each with 38 cells in series to give the voltage between 19 V and 23 V that can match the 5S battery. The two modules are connected in parallel with the 5S battery. The other is with 86 cells, also divided to two modules, each with 43 cells. This type of single module gives the voltage between 21.5 V and 26 V, matching the 6S voltage, so these two modules and the 6S battery can be connected in parallel. A diode is also used at the output of each solar module to prevent the current flowing to the solar cells from the battery. The later two types of solar modules give longer flight time than the first and the second types.

4 RESULTS AND DISCUSSIONS

4.1 Battery influences

Several batteries of different capacity (mAh) have been experimented individually with the solar module to form solar hybrid power system for the quadcopters. In general, the high capacity battery can give a longer flight time, but the ratio of solar energy is less. The reason is because the high capacity battery is heavier, so more power is needed. Because the solar energy is limited by the panel area, the increased power required can only be supplied by the battery, reducing the ratio of the energy from the sun. In our experiments, our configuration could extend the flight time to over 20 times. This can be furher improved by changing the configuration and other components. The longest flight time achieved so far is 3 hours and 28 minutes, which cannot be achieved by using only the battery.

Although the battery provides a small portion of the power requirement, it is still necessary. Its main function is to stabilize the voltage to the quadcopter. When the copter takes off, more power is needed to overcome both the gravity and the air resistance from the large area of solar panels. Also, when the copter is in the air, its three-dimensional orientation varies due to the requirement of balance. Then the solar panel is not remained horizontal all the time, so the power from the solar module fluctuates. The battery can match the power variation. In addition, when the quadcopter is adjusting its attitudes, more power is required. In particular, as the wind blows, the attitudes vary significantly. We had recorded the real-time current, voltage, and the power consumption using the data link. The data show that the power consumption can be more than doubled, compared to the well-balanced attitudes. With the solar energy only, the power and the current may not be enough. With the battery in parallel, the current matching is automatically achieved, so the copter can fly well in the air for hours. On the other hand, the battery power is not always consumed. Because the output voltage from the solar module in our design is not below the maximum voltage of the battery, the solar module can charge the battery when the battery voltage is reduced due to the consumption.

4.2 Sunlight influences

In our field test flight, the sunlight may not be stable. When clouds came and partially block the sunlight, the voltage dropped relatively fast. If the sunlight was stable with 95000 lux approximately, the voltage remained almost no change, indicating the quadcopter could continuously fly without consuming the electricity of the battery. If the sunlight was over 95000 lux, the battery could be even charged, implying the solar power was more than what the copter needed. Those tests were done in North Taiwan, where constant sunlight of several hours is rare. If the solar-hybrid quadcopter is used in the areas with excellent sunlight, it should be able to fly nearly from 9 am to 4 pm continuously.

5 CONCLUSION

In conclusion, solar hybrid energy systems are developed for quadcopters. The sunlight can provide more than the power required for the copter. In our experiments, the flight time with the solar power could be extended to over 20 times and can be further improved. The longest flight time achieved now is over three hours, which is much more than the flight time of most commercial quadcopters of the same weight. We expect to have even more hours of continuous flight as the copter configuration and the solar system are further optimized.

6. ACKNOWLEDGEMENT

The authors would like to acknowledge the financial support provided by the project in National Taiwan University with the number of 06HT911001, led by Prof. Ruey-Beei Wu.

•

7. REFERENCE

[1] Gaurav Kumar, Shubham Sepat, Shubham Bansal, "Review paper of Solar Powered UAV," International Journal of Scientific & Engineering Research, Volume 6, Issue 2, February-2015.

INVESTIGATING THE TECHNICAL EFFECTIVENESS OF DIFFERENT PHOTOVOLTAIC CLEANING METHODS IN DUST-INTENSIVE CLIMATES

Feras G. Alzubi, Abdullah Alkandary, and Abdulwahab T. Alasfour
Energy and Building Research Center,
Kuwait Institute for Scientific Research, Safat 13109, Kuwait
Phone: +965 2495 6291, Fax: +965 2498 9099
E-mail: feras.alzubi@gmail.com

ABSTRACT: In climates that are known for frequent dust events, performance of photovoltaic modules can drop significantly due to dust accumulation. Continuous in-field monitoring of PV modules studied in this work shows a soiling loss of 56 % in the power output over six months. This power loss is minimized by frequently cleaning the PV modules; however, this could lead to higher maintenance and operation costs. In this work we focus on investigating the effectiveness and the techno-economics of cleaning PV modules with de-ionized water, detergent, pressurized water, pressurized air, and self-cleaning coating. Amount of time, energy, and water consumed to clean in addition to recovered power after each cleaning event are evaluated. Results indicate that all cleaning methods were effective in recovering lost power within 3% range except pressurized air which could not recover power. Pressurized water cleaning was the fastest method (less than a minute) with the highest energy (average 13 Wh) and water (4.5 L) consumptions per module. Moreover, for residential system with 10KW in size, the cost analyses of all the cleaning methods are dominated by the equipment capital and replacement costs. Pressurized water cleaning method was the least expensive solution after a 25 years operation period based on the electricity and water prices considered in this study. It is possible to optimize the deionized water cleaning method by using a less expensive filter to a point where it is the most financially attractive. Same analyses were depicted for commercial PV system with100 KW size.
Keywords: Soiling loss, PV cleaning methods, cost analysis, power regaining

1 INTRODUCTION

Dust deposition (soiling) can reduce the power output of a 20° tilted PV module up to 24 % over three months in desert environment [1]. The several solutions proposed in literature to soiling problem can be categorized under four groups. The first one is optimizing system installation configuration to minimize dust accumulation [2-4]. The second category is washing the dust with water. In sometimes this includes low surface energy detergent. The third category is direct dust cleaning by mechanical methods such as hand or machine cleaning. The final category of solutions is surface modification by means of coatings [5]. Several in market systems and designs were surfaced as cleaning solutions [6]. These designs use automatic control systems for water washing and mechanical scrapping. The problem with such systems is their case specific cleaning efficiency and cost savings or its debatable effectiveness in cleaning at an extremely dusty environment. In general, the economic feasibility of these proposed solutions is yet to be matured and fully investigated. Comparing gain in energy due to cleaning to the cost associated with parameters involved in the cleaning process such as labor, material, and equipment is a key factor that can favor a cleaning method over another.

In this paper we determine the best combination of five different cleaning methods with respect to cost savings and energy produced. This was done by measuring the percentage of power recovered after a cleaning method was applied. For a cost-effective study to be carried out, water consumption, energy consumption, and time required to apply each cleaning method were measured. The soiling loss (SL) for the modules was calculated and 15 % of SL was set to ignite the cleaning event. The cleaning events were 10 times over a span of one year. Residential and commercial PV system sizes were considered for life cost analysis of each cleaning methods.

2 EXPERIMENTAL SETUP AND PROCEDURE

After the pre-installation and preconditioning processes of the PV modules were executed [1, 7], a PV system of seven modules, each with 250 W, was mounted offshore in natural arid conditions with 20° tilt and connected to an I-V tracer as shown in Fig. 1.

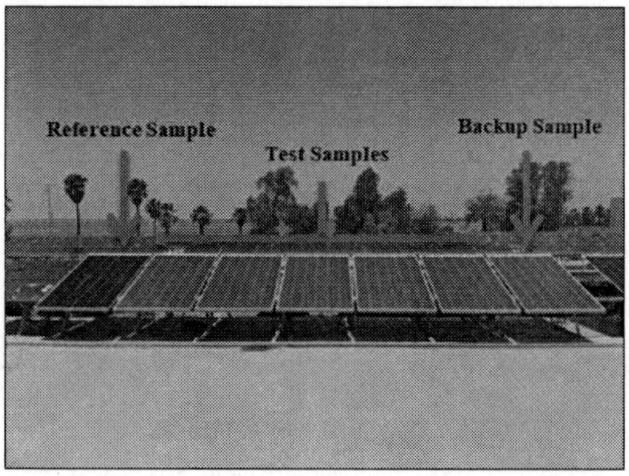

Figure 1: Test setup located at Kuwait Institute for Scientific Research composed of seven PV modules.

Out of the seven test modules, 1 module was a reference sample that is cleaned daily, 5 modules were test samples, and 1 module was a backup test sample. The test samples were left to soil naturally until their power drops by 15%. Each cleaning method was designated to a test sample. As for the backup sample, it was used in case more data needs to be collected for any cleaning method. The purpose of the reference sample was to allow for the monitoring and measurement of dust-induced power losses of the test samples. Once a test sample loses 15% of its power, one of five cleaning methods is applied. Each PV module was cleaned by means of: (1) Dry cleaning (restorative): cleaning directly (manually) with pressurized air up to 90 psi. (2) Cleaning with deionized water (restorative): manual hand cleaning is used with deionized water and squeegee. (3) Cleaning with high pressure water (restorative): 1.5 to 4 bar to wash the surface directly without the need for scrubbing. (4) Cleaning with water and PV detergent (DGT) (restorative): the detergent is mixed with normal, non-pressurized water with a PV brush, microfiber cloth, and squeegee. Finally (5) Self-cleaning coating (preventative by means of surface modification): self-repellent PV coating prevents the adhesion of dust particles on the PV module surface.

3 RESULTS AND DISCUSSION

3.1 Soiling Loss Measurement

During the conducted experiment, we have monitored the dust deposition and calculated the soiling loss in the test site. Figure 2 below shows the loss due to soiling calculated based on the Pmax over duration of nine months (May 16th to February 22nd). SL was 24.1 % after the first three months and reached 56.2 % in six months, which is the maximum SL reached in the year of the conducted experiment. This reflects the harsh weather and the intensive effect that the accumulated dust can cause to the electrical losses of the tested PV modules. The SL 56.2% was reached with no artificial or natural (rain) cleaning which is amongst the highest recorded SL values [8]. After rain events occur, PV module regained 97% of its initial power. After consecutive nine months and five rain events, module was performing at 87% of its initial power state.

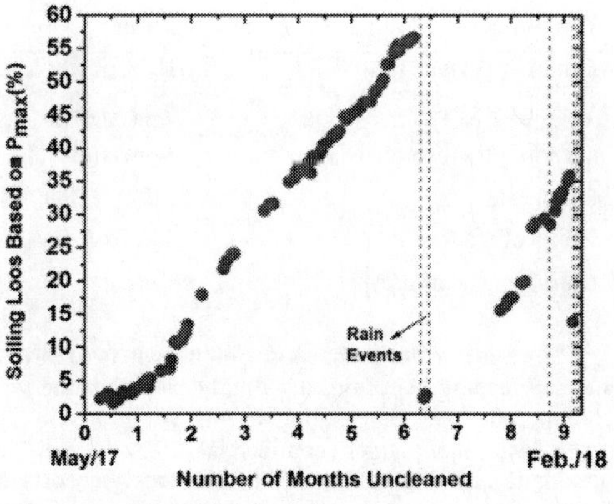

Figure 2: Soiling loss based on the loss in the output power of PV modules for nine consecutive months without cleaning. Blue lines are indicating rain events at certain dates.

3.2 Effectiveness of Cleaning Methods

The main objective of the technical analysis in this work is to assess the cleaning methods' effectiveness in removing the soil collected on the glass surface of the PV samples. This was done by measuring the percentage of power recovered after a cleaning method was applied. For a cost-effective study to be carried out, water consumption, energy consumption, and time required to apply each cleaning method were measured by specialized related meters.

Midway through the experiment, pressurized air was deemed not feasible under the weather conditions of the test site. When this cleaning method was applied for the first time at 15% SL, it was not able to remove any layer of dust from the test sample's surface, and thus did not recover any lost power. Outlet pressures as high as 6 bar were reached, a value that is significantly higher than what has been recommended in literature; however, they were still unable to recover any power. In an effort to determine the effectiveness of this method, it was re-applied at 1% and 5% SL, where dust accumulation was significantly less. In both cases however, power was not recovered. The inability for this method to remove accumulated dust was indicative of the strong particle adhesion that exists on the glass surface. Such strong adhesion forces are likely an outcome of the urban coastline test site, where the soil content might have relatively high organic/minerals and humidity, thus creating heavy cementation layers [9, 10]. Characterizing the soil content in this location will be valuable in understanding the cementation build-up experienced in this study. It was clear that pressurized air was unable to clean the soiled samples after multiple trials and thus was no longer studied. Furthermore, this cleaning solution ought to be re-evaluated under a desert-dry climate to assess its potential in areas of low relative humidity. Furthermore, the hydrophobic (HB) and hydrophilic (HL) self-cleaning coatings also underperformed most of the test year with regard to their ability to mitigate dust accumulation. Consequently, deionized (DI) water and a squeegee were used to clean the coated test samples when they reached an SL of 15%. The ineffectiveness of the pressurized air and coating (to prevent soiling) is believed to be influenced by strong cementation build-up as a consequence of high humidity levels and soil composition. Therefore, it was concluded that these two cleaning methods need to be investigated under dry desert climates for a better understanding of their potential.

Figure 3 presents the technical results of each cleaning method averaged out over ten cleaning events at 15% SL. It shows the (a) consumed time, (b) water, and (c) energy during cleaning in addition to (d) recovered power after applying the cleaning methods. As can be seen, all tested cleaning methods are technically effective in recovering power to roughly 100%. However, each cleaning method requires a different amount of water, energy, and cleaning time. Since pressurized water does not require strenuous labor, cleaning time was three to four times faster than the other cleaning methods. Conversely, it utilized the largest amount of water and energy. Such improvements could lead to lower labor costs despite the large amount of water and energy needed. As for the other three cleaning methods, it was apparent that the HB and HL nano coated samples required less time and water for cleaning. This was a result of the surface modification that altered the water-repellent and water-attraction properties, which enabled the use of less time and water to clean the test samples.

Less amount of water was consumed when applying detergent and coating. However, applying detergent has introduced extra time of cleaning where coating offered quicker cleaning times. With that said, it does not recover as well as other cleaning methods.

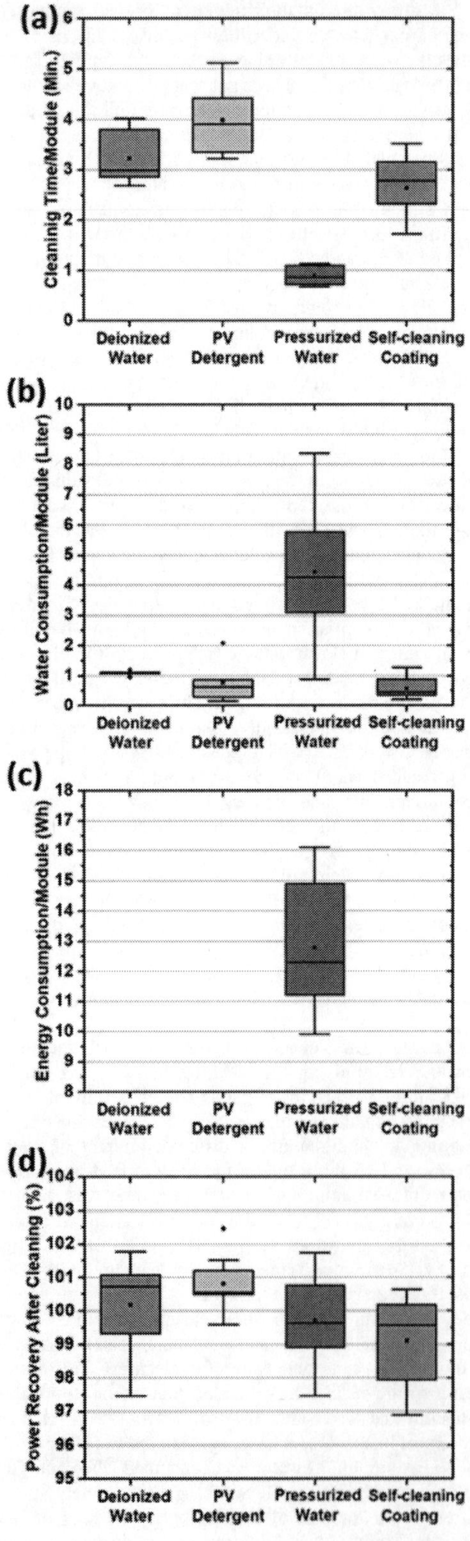

Figure 3: Consumed (a) Time, (b) Water, and (c) Energy per module when applying the four cleaning methods. (d) Power recovery after application of cleaning methods.

3.3 Economic Assessment of Cleaning Methods

After the technical assessment, an economical investigation was carried out for each cleaning solution. Two cost estimation models were developed to assess residential and commercial scale systems. These models utilized technical inputs from the previous results' sections, economic inputs, and other inputs such as PV system size, system life, and cost of water. The reached results showed different economics between the cleaning methods investigated and between residential and commercial scale systems. Thus, different optimization strategies were considered. The lifetime cost estimate of each cleaning method was carried out using the inputs listed Table 1 and 2.

Table 1: Inputs Used to Estimate the Lifetime Costs of Each Cleaning Method

Key Inputs	Value
Pmax of PV module	250 Wp
Area of PV module	1.64 m^2
System lifetime	25 yr
Cost of non-subsidized water	0.006 USD/liter
Cost of non-subsidized electricity	0.12 USD/KWh
Detergent consumption	0.024 Liters/m^2
Cost of detergent	9.6 USD/liter
Cleaning frequency (Residential)	12 events/yr
Emergency cleaning	2 events/yr

Table 2: Additional Inputs Used to Estimate the Lifetime Costs of Cleaning Method for Comercial PV Systems

Key Inputs	Value
Criteria to clean	10 % power loss
Soiling loss rate	0.44 % /day
Cleaning frequency	16 event/yr
Labor rate	3.75 USD/hr
Hours of work	5 hr/day
Cleaning execution speed	1 day

Since it is more practical for homeowners to have a simple cleaning schedule, a monthly cleaning rate was chosen reflecting 12 cleaning events per year. On the other hand, for larger commercial PV systems, an optimized cleaning schedule that balances the costs of energy production and cleaning operation costs is more financially attractive for the owner. Therefore, a performance-based cleaning schedule was chosen where cleaning is only applied when a power loss of 10% is reached. Consequently, 16 cleaning events per year were estimated based on the annual soiling rate observed in this study. As for labor costs, only commercial systems considered these expenses because the homeowners are likely to clean their systems on their own. It is important to note that the model did not consider general expenses and indirect costs.

Fig. 4 summarizes and breaks down the lifetime cost of each cleaning method for a 10 KW residential and 100 KW commercial PV systems. Residential systems usually range in size from 1KW to 10KW. It is concluded from the graph that the expenses for all the solutions are

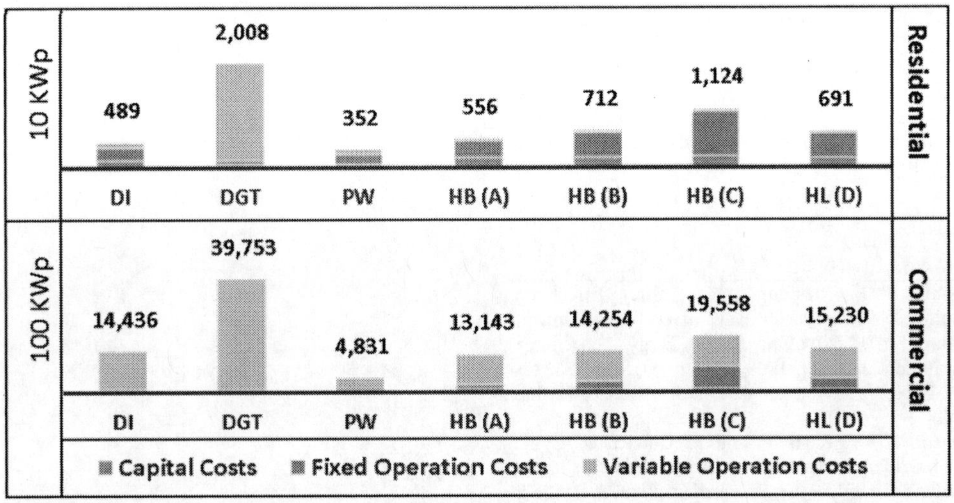

Figure 4: The bar graphs presents the results of the lifetime cost estimation models for each cleaning method for both residential and commercial PV scale systems. These estimations include capital, variable, and fixed costs for a system life of 25 years.

dominated by the equipment capital and replacement costs (i.e. capital and fixed) for the residential systems. Variable expenses on the other hand, which comprise of water, electricity, and consumables are virtually negligible even without government subsidies. As a result, the PW cleaning method was calculated to be the most viable cleaning solution as it requires the least amount of equipment, despite consuming relatively high quantities of water and electricity. After 25 years of operation using PW for cleaning the PV system, water and electricity costs would only be 290 USD (i.e. 0.9 USD per month). Thus, from a homeowner's perspective, as far as finances are concerned, water and electricity consumptions are not a problem.

Unlike the other cleaning methods, the DGT's lifetime cost is strongly influenced by variable costs rather than capital and fixed costs as can be seen in residential figure. High variable costs are a result of the use of expensive PV-friendly (i.e. non-ionic and non-abrasive) detergent which constitutes nearly 90% of the lifetime cost, making it the most expensive cleaning solution. Although this presents an opportunity to financially optimize since small price reductions in the detergent can yield significant price reductions in the lifetime cost, it remains difficult from a practical point of view because the detergent used in this study has a typical market price value for a commercially available product. For this method to enter a relatively competitive lifetime cost range with the other cleaning methods, a two-thirds price reduction in the typical value of the detergents is required (i.e. 6 USD per liter or less). At the present moment however, no such product exists and thus the solution is not economically feasible.

The remaining two cleaning solutions, which are the DI and coating methods, financially lie between the cheapest and most expensive method. As for cleaning with DI, the second most financially attractive solution, optimization opportunities are tangible. The de-ionization filter used in this study accounted for 66% of the lifetime cost. Since numerous less expensive filtration systems exist in the market and considering that the lifetime cost of the DI solution is very close to the lifetime cost of the PW solution, it is possible to optimize the DI cleaning method by using a less expensive filter to a point where it is the most financially attractive cleaning option. Finally, with regards to nanocoatings as a cleaning solution, high lifetime cost fluctuations between the four different coatings have been observed. This is primarily influenced by the large difference in durability amongst the commercially available coatings.

Unlike residential systems, the lifetime cost of all the cleaning methods for commercial systems which usually range from 10KW to 500KW is highly influenced by variable costs and not capital and fixed costs as shown in the Fig.4. This shift in finances is expected when increasing the system size, where the cost of water, electricity, and consumables considerably increased while the fixed costs marginally change.

4 CONCLUSION

Five PV cleaning methods proposed in literature were investigated under Kuwait's environment over a period of one year. All but the DA method have led to full power recovery within a maximum range of 3%. As for the HB and HL coatings, they did not slow down soiling; however, they did reduce water consumption and cleaning time when restored with de-ionized water and a squeegee. PW has shown the fastest time of cleaning.

For the lifetime cost estimation for both residential and commercial system sizes, the PW cleaning method was the least expensive solution after a 25 years operation period. Even though it consumes that highest amount of water and electric energy, this is possible because such expenses are relatively cheap even without government subsidies. As for coatings, lifetime costs fluctuated due to cost variations which resulted from the durability differences amongst the coatings. Finally, cleaning by DGT was deemed not feasible because commercial PV-friendly detergents are expensive.

PW is by far the most user-friendly solution to a homeowner. Ultimately, no other optimization strategy will allow for any of the other cleaning solutions to compete with PW, other than reducing labor expenses, because alone they cost more than the PW's total lifetime cost.

5 ACKNOWLEDGMENT

The presented work is part of project EA048C that is financially supported by the Kuwait Foundation for the Advancement of Sciences (KFAS). Facility and equipment support was provided by the Kuwait Institute for Scientific Research (KISR) under Renewable Energy program.

6 REFERENCES

[1] Alzubi, F.G.; Alasfour, A.T.; Soiling Effect on Photovoltaic Modules' Performance in Arid Environment, 33rd European Photovoltaic Solar Energy Conference and Exhibition. 2017, pp 1809 – 1812. ISBN: 3-936338-47-7, Paper DOI: 10.4229/EUPVSEC20172017-5DV.3.28.

[2] Qasem, H., T. R. Betts and R. Gottschalg. 2011. Effect of shading caused by dust on cadmium telluride photovoltaic modules. Proceedings, 37th IEEE Photovoltaic Specialist Conference, Seattle, USA, pp. 003199-003204.

[3] Qasem, H., T. R. Betts and R. Gottschalg. a. Soiling correction model for long term energy prediction in photovoltaic modules. Proceedings, 38th IEEE Photovoltaic Specialist Conference, Austin, USA, 2012, pp. 003397-003401.

[4] Elminir, H. K.; A. E. Ghitas; R. H. Hamid; F. El-Hussainy; M.M. Beheary; and K.M. Abdel-Moneim. 2006. Effect of dust on the transparent cover of solar collectors. Energy Conversion and Management 47:3192-3203.

[5] Barletta, M.; M. Puopolo; V. Tagliaferri; and S. Vesco. 2014. Retrofitting of solar glasses by protective anti-soiling and –graffiti coatings. Renewable Energy 66: 443-453.

[6] Anderson, M.; Grandy, Hastie, A. J.; Sweezey, A.; Ranky, R.; Mavroidis, C.; and Markopou,Y.P., 2007. Robotic Device for Cleaning Photovoltaic Panel Arrays, Conference on Solar Concentrators for the Generation of Electricity or Hydrogen; San Lorenzo del Escorial, Spain, 12-17 March.

[7] Alasfour, A.T.; Alzubi, F.G., Light-Induced Degradation in Multicrystalline Modules and Systems under Field Conditions, 33rd European Photovoltaic Solar Energy Conference and Exhibition. 2017, pp 1584 - 1587. ISBN: 3-936338-47-7, Paper DOI: 10.4229/EUPVSEC20172017-5BV.4.15.

[8] Maghami, M.R.; Hizam, H.; Gomes, C.; Radzi, M.A.; Rezadad, M.I.; Hajighorbani, S., Power loss due to soiling on solar panel: A review, 2016. Renewable and Sustainable Energy Reviews, Volume 59, Pages 1307-1316.

[9] Biryukov, S. 1998. An experimental study of the dry deposition mechanism for airborne dust. Journal of Aerosol Science 29:129-139.

[10] El-Shobokshy, M. S., and F. M. Hussein. 1993. Degradation of photovoltaic cell performance due to dust deposition on to its surface. Renewable Energy 3:585-590.

BUSINESS CASES FOR ANTI-SOILING COATINGS IN THE NETHERLANDS

Chris Tzikas[1], Marco Cappa[1,2], Menno van den Donker[1], Paul Sommeling[3], Geert Verbong[2], Wiep Folkerts[1]
[1]SEAC (Solar Energy Application Centre), Eindhoven, The Netherlands
[2]TU/e Eindhoven University of Technology, Eindhoven, The Netherlands
[3]ECN, part of TNO, Petten, The Netherlands

ABSTRACT: This paper focuses on the application of coatings with combined anti-soiling and anti-reflective properties in the Netherlands. The main goal is to provide different perspectives and case studies for the application of this new technology in a fast-growing market, identifying the best scenarios from the user's and investor's point of view. A techno-financial analysis of the possible gains achievable through the implementation of this technology will be performed.
Keywords: Anti-Soiling, PV coatings, System Performance, Business Cases, Yield Maximization

1 INTRODUCTION

Energy production gain for PV panels with an anti-soiling (AS) coating has the potential to be significant in desert environments due to high soiling levels [1], but few studies have been performed in north-western European regions such as The Netherlands. The impact of soiling varies considerably in relation to environmental conditions [2]. In particular, rural areas in these regions might be subject to substantial dust formation. The soiling effect could be particularly pronounced when PV modules are installed on farm roofs, noise barriers, intensive industrial areas, or as large-scale systems in open field.

The most common solution to prevent soiling losses is cleaning the solar system. This process can take place through two methods: active and passive procedures. The former is performed through mechanical brushing of the surface, while the latter can be achieved with the use of anti-soiling coatings, which modify the properties of the surface to improve its washability and to reduce adherence of dust particles [2]. The use of anti-soiling coatings can improve the performance of PV systems, increasing the overall system energy production and reducing costs associated with the maintenance of solar installations.

2 THE SOILING EFFECT

Soiling losses are caused by the accumulation of dust over the glass surface of the photovoltaic module. From the technical point of view, soiling reduces the amount of light that can be absorbed by the photovoltaic cells. Phenomena of scattering and absorption of incident light are the main cause of this reduction.

While in desert areas power losses are due to dust deposition, in milder climates they can also be caused by different substances: bird droppings, moss, pollen, and non-combusted carbon particles are among the main soiling sources in these regions [2]. The process of deposition is influenced by external conditions, some of which can aggravate it and increase power losses.

The process of accumulation of dust has been described in detail by Cuddihy and it is named cementation [3].

Soiling rates can be reduced in presence of heavy rains and wind, but the scale of this phenomenon is location-dependent. The material composing the module's surface, the soiling composition, and the module's tilt angle are crucial parameters affecting the impact of natural cleaning [4]. After a certain period of time, the deposition and removal of dust reach a state of equilibrium keeping the soiling level constant.

The types of soiling identified in the Netherlands, relevant for the Dutch solar market along with the main external and system factors affecting the soiling effect are presented in Fig.1.

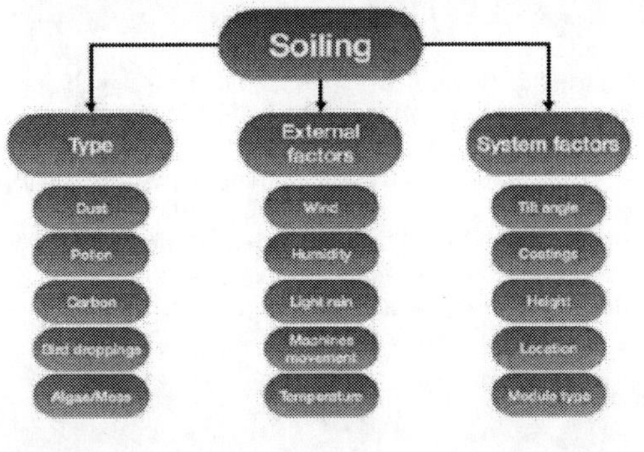

Figure 1: Causes of soiling accumulation on solar installations in the Netherlands.

3 ANTI-SOILING COATINGS FOR PV

Anti-soiling coatings are relatively new products comprising mostly of retrofit solutions but also a few solutions exist that are built in the solar glass, which have a much longer lifetime.

To assess the properties of an anti-soiling material, wetting is the main parameter that needs to be evaluated. Surface wetting represents the ability of a fluid to maintain contact with a surface. The wettability is characterized through the contact angle (θ), which is determined by the balance between adhesive and cohesive forces. Depending upon the value of θ, two different types of anti-soiling materials can be identified: superhydrophilic (where water goes beneath the contaminant) and superhydrophobic coatings (water picks up the contaminant) [5].

4 MARKET ANALYSIS

The solar energy market in the Netherlands is relatively new and small, thus the amount of inputs which can be collected for the techno-financial model is limited. For this reason, the research included reliable inputs from neighboring countries with more mature solar energy markets and similar climate and environmental conditions. The region includes an area of 350 km around the city of Eindhoven.

After the selection of the region in which the study would have been carried out, market segments which could benefit from the introduction of anti-soiling coatings were identified. Two main parameters have been selected to guide the market research: high soiling rates and high cleaning costs. Installations which satisfy one of these two criteria have been identified and they have been represented in Fig. 2.

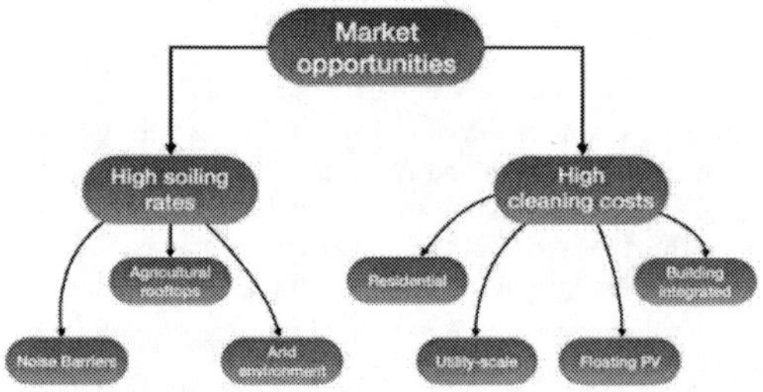

Figure 2: Identified market segments which could benefit from the introduction of anti-soiling coatings.

In order to quantify the soiling rates, three realistic pollution scenarios for a solar installation in the Netherlands have been developed based on data collection from previous studies and interviews with stakeholders. These scenarios are presented in Fig. 3.

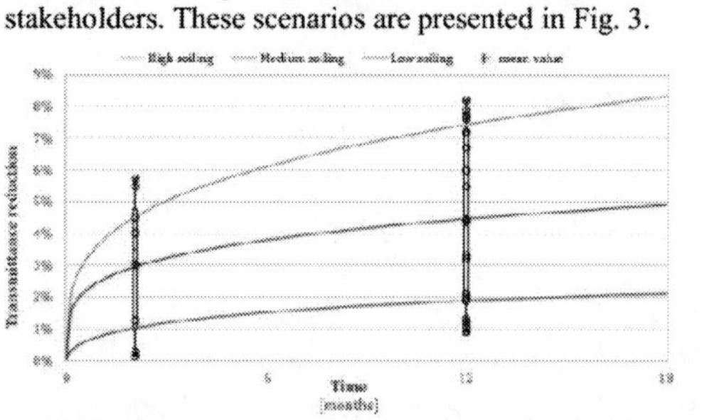

Figure 3: Three pollution scenarios developed including the input values on which they were based.

The second parameter to be defined for the development of the techno-financial model was the cleaning cost for a PV installation. A prediction curve for this parameter has been created implementing data provided from cleaning companies and plant owners. The cost per unit is inversely proportional to the size of the plant and it is dependent to the method used to clean the modules (manual or mechanical). The resulting prediction curve is presented in Figure 4.

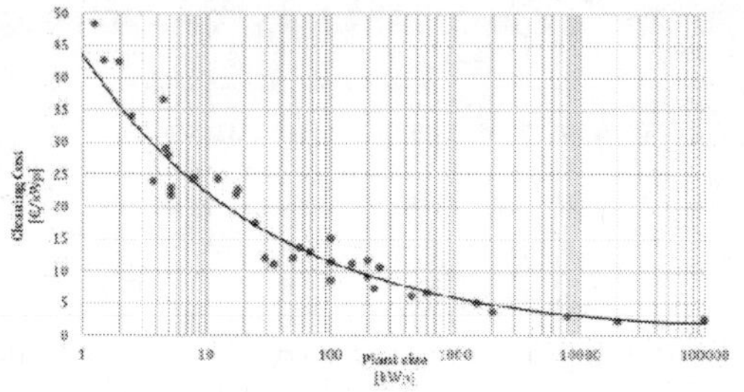

Figure 4: Cleaning cost prediction curve.

5 BUSINESS CASES FOR AS COATINGS

From the data collected through literature studies and interviews conducted with stakeholders of the relevant markets, three market segments were selected for the development of the business cases. The first selected segment was the agricultural rooftop PV systems mostly due to the high soiling rate. The second segment is represented by solar parks as soiling affects their thin profit margins. The final segment selected was the commercial rooftop PV systems, as they represent an increasingly important sector in the densely populated Netherlands.

The business cases developed were based on a constructed techno-financial model which contained inputs of technical and economic nature, which were gathered through literature reviews, stakeholder interviews, and internal inputs. The two main outputs used to assess the benefits of the implementation of this technology were the Net Present Value (NPV) and the Internal Rate of Return (IRR).

For a typical agricultural rooftop PV system of 100 kWp, three soiling rates and three anti-soiling coating performances were simulated and are shown in Fig. 5 & 6. It must be noticed that for a cleaning schedule performed every 6 months, NPV and IRR are negative, thus the vertical axes have been optimized for the remaining cleaning frequencies (less than every 6 months). The maximum point of the three curves has been indicated.

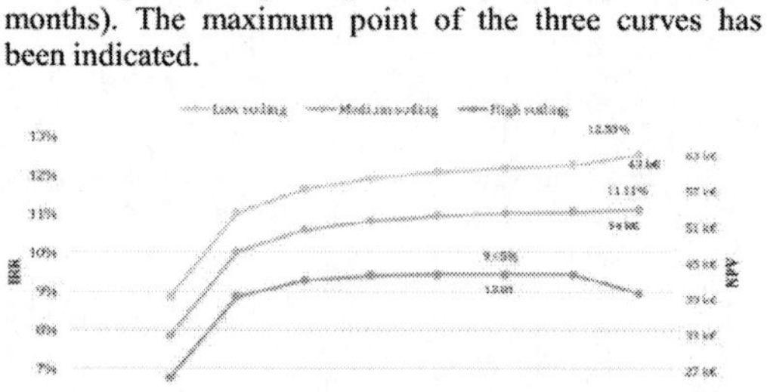

Figure 5: Effect of different soiling levels on the IRR and NPV of a 100 kWp agricultural rooftop PV system.

The economic performance of the system changes radically when changing the soiling scenario. If soiling losses are not significant, the maximum of the curves occurs when no cleaning procedure is performed during the whole lifetime of the solar installation. This effect shows that costs associated with cleaning processes are hardly recoverable with the yield increase, mostly due to the low price of the electricity fed into the grid.

In order to evaluate the anti-soiling performance, a single soiling scenario has been followed. For this case, the high soiling scenario has been selected.

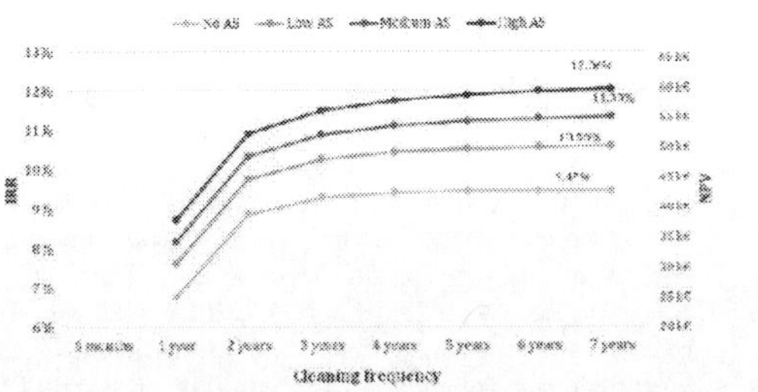

Figure 6: Effect of different AS coating performances on the IRR and NPV of a 100 kWp agricultural rooftop PV system under high soiling conditions.

For the case of solar parks, a typical large utility scale solar park of 5 MWp was selected. Similarly, three soiling rates and anti-soiling coating performances were simulated as shown Fig. 7 & 8.

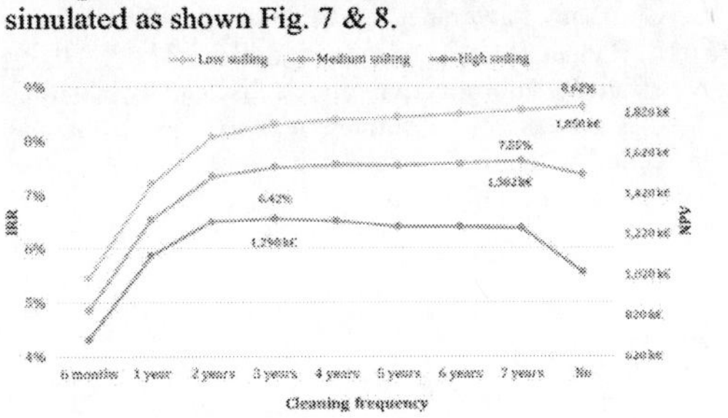

Figure 7: Effect of different soiling levels on the IRR and NPV of a 5 MWp solar park.

Although the three curves present a very similar shape, it is possible to notice that for environments with high soiling losses, the maximum of the curve occurs for more frequent cleaning schedules.

The anti-soiling coating performance has been analysed based on the medium soiling scenario, which has been selected as the most representative for such an application. Although the general performance of the plant is increased with higher anti-soiling performances, the effect is much less remarkable than for agricultural applications. The reason lies in the combination of the two main parameters: lower soiling losses and less expensive cleaning procedures, due to the size of the solar installation.

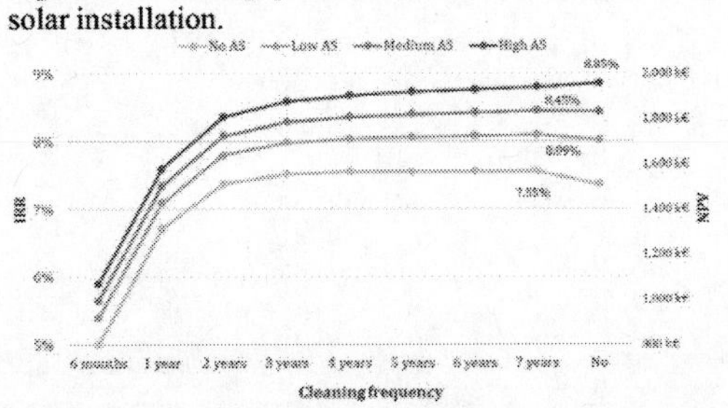

Figure 8: Effect of different AS coating performances on the IRR and NPV of a 5 MWp solar park under medium soiling conditions.

Finally, for the commercial rooftop PV system case, a typical size of 400 kWp was selected. The three soiling rate and anti-soiling coating performance scenarios are shown in Fig. 9 & 10.

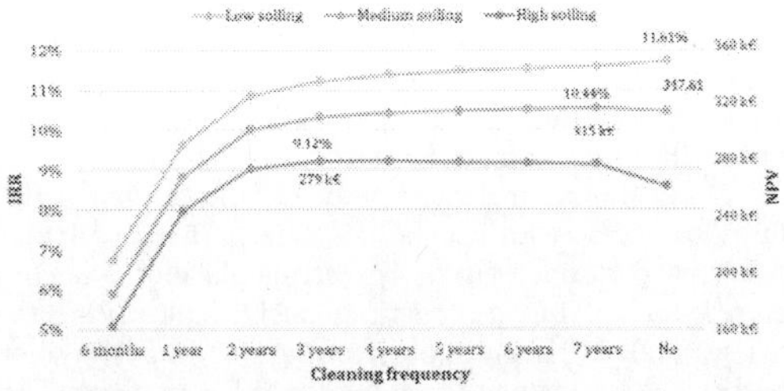

Figure 9: Effect of different soiling levels on the IRR and NPV of a 400 kWp commercial rooftop PV system.

The economic performance of commercial rooftops is largely dependent on self-consumption, which has been set to 40% in our case. Increasing or decreasing the magnitude of self-consumption can change the economic performance of the system in general, but it would not change the effect that soiling and anti-soiling coating would have on the outputs of the model.

The anti-soiling performance has been evaluated implementing the low soiling scenario for commercial applications. The curves reveal again the characteristic shape of the different economic performance over a varying cleaning schedule. In this case as well, as it was observed for solar parks, the optimal performance for high-soiling rates can be obtained with a cleaning frequency of 3 years. The similarity to the previous case makes clear that the slope of the curve representing the cleaning prediction cost reaches a horizontal asymptote very fast. From the analysis of the latter graph, it can be observed that differences between different anti-soiling coating performances are almost imperceptible mostly due to the low soiling scenario chosen for this case.

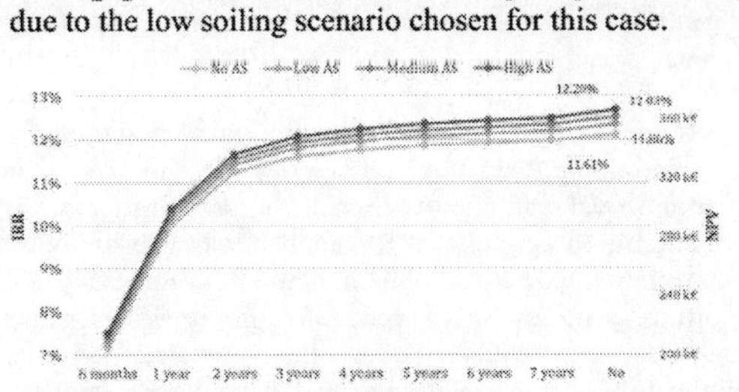

Figure 10: Effect of different AS coating performances on the IRR and NPV of a 400 kWp commercial rooftop PV system under low soiling conditions.

The development of the techno-financial model has been based on the assumption that the anti-soiling coating does not involve any additional cost for the investor. In this way, it has been possible to calculate the real increase in NPV achievable through the use of a anti-soiling coatings in different solar applications in the Dutch market.

The optimal NPV has been extrapolated as the maximum point for every data series; a data series is obtained changing the cleaning schedule over a fixed system size and soiling scenarios, for a total of 27 data series. The added value generated by anti-soiling coatings has been calculated using the equation:

$$Added\ value = \frac{NPV_{max,n} - NPV_{max,i}}{P_x}$$

where: the n subscript indicates the series of values obtained without anti-soiling coatings, while the i subscript indicates the series of values obtained with the anti-soiling coating technology, while Px represents the installed size of the solar system.

Using this normalization process, the results of the three business cases can be compared. The calculation has been performed analysing each application with the corresponding soiling scenario, namely: high soiling for agricultural PV applications, medium soiling for solar parks, and low soiling for commercial PV rooftops. The graphical result for this analysis can be seen in Fig. 11. In the graph, the green bar (maximum) and the red bar (minimum) represent gains observed for reduction of soiling of 70% and 30% respectively (corresponding to minimum and maximum anti-soiling coating performances used before).

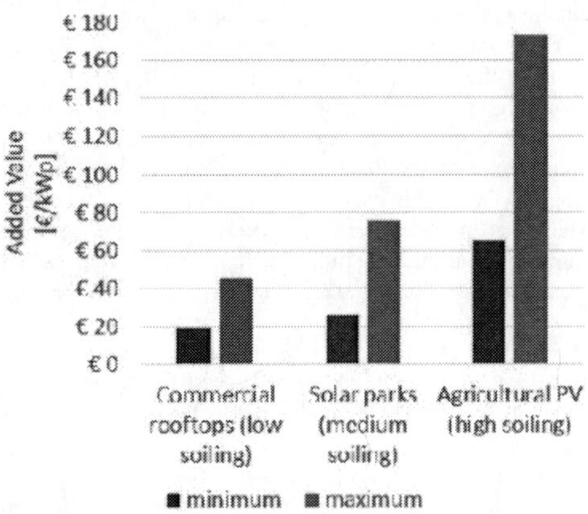

Figure 11: Added economic value in €/kWp for the three business cases.

It can be observed that there is a great potential for the anti-soiling coating technology to improve the economic value for photovoltaic systems; this effect is remarkable in agricultural PV applications. On the other hand, the already favourable situation of commercial PV rooftops presents small room for improvement, thus lowering expectations over the potential of the application of the technology in the whole PV market. The same applies for residential solar systems.

6 CONCLUSIONS

From the results of the techno-economic analysis, it is clear that anti-soiling coatings have the potential to improve the performance of photovoltaic systems even in moderate climates as the Dutch climate.

It is clear that the soiling scenario and the anti-soiling coating performance strongly influence the results of the simulations. Agriculture PV installations present outstanding results, mostly due to the high soiling rates observed in these areas and to the high- cost of cleaning procedures, accentuated by the small PV system sizes.

Finally, a key finding of the analysis, is the fact that it makes financial sense to omit cleaning of the PV system entirely if anti-soiling coatings are applied.

7 REFERENCES

[1] T. Sarver, A. Al-Qaraghuli, L.L. Kazmerski, "A Comprehensive review of the impact of dust on the use of solar energy: history, investigations, results, literature and mitigation approaches", Ren. And Sust. En. Rev. Vol. 22, p.698-733, 2013.

[2] E. Klugmann-Radziemska, "Degradation of electrical performance of a crystalline photovoltaic module due to dust deposition in northern Poland," Renew. Energy, vol. 78, pp. 418-426, 2015.

[3] E. F. Cuddihy, "Theoretical Considerations of Soil Retention," Sol. Energy Mater., vol. 3, pp. 21-33, 1980.

[4] M. P. Rocha et al., "Comparative Analysis of the Dust Losses in Photovoltaic Modules with Different Cover Glasses," 23rd Eur. Photovolt. Sol. Energy Conf., no. September, pp. 2698-2700, 2008.

[5] H. Hanaei, M. K. Assadi, and R. Saidur, "Highly efficient antireflective and self-cleaning coatings that incorporate carbon nanotubes (CNTs) into solar cells: A review," Renew. Sustain. Energy Rev., vol. 59, pp. 620-635, 2016.

MARS™ SOILING SENSOR

Michael Gostein, Stan Faullin, Keith Miller, Jason Schneider, Bill Stueve
Atonometrics, 8900 Shoal Creek Blvd. Suite 116, Austin, TX, 78757, USA, info@atonometrics.com

ABSTRACT: We evaluate a new proprietary optical soiling measurement technology that does not require washing. The compact, water-free, and maintenance-free sensor technology is designed to permit PV system soiling measurement with lower cost and labor requirements compared to traditional soiling measurement systems using a clean/soiled pair of PV reference devices. These attributes make the sensor suitable for a wide range of sites including not only utility but also commercial-scale PV installations. Laboratory test results using soiled glass test coupons demonstrate excellent correlation between the optical sensor and a PV cell, and field testing is underway.

1 INTRODUCTION

Soiling of PV modules is one of the principal energy loss factors for PV systems. Accordingly, in recent years there has been a growing interest in measuring PV soiling at both pre-production prospecting and operational sites [1]. Motivations include performance prediction, monitoring of plant operations, verifying compliance with performance guarantees, and optimizing wash schedules for best return-on-investment of washing expense.

Traditional soiling measurement systems compare the outputs of two PV devices, one of which is allowed to soil naturally and the other of which is routinely cleaned to serve as a reference [1][2]. Daily or weekly cleaning is performed either manually by onsite personnel or by an automated [3] washing system.

Eliminating the requirement for washing a reference device in order to measure soiling would lower the cost of soiling monitoring, and be especially favorable for commercial-scale PV sites that typically have smaller budgets and no onsite personnel.

In this work we evaluate a new proprietary all-optical soiling measurement concept developed by Atonometrics (Mars Soiling Sensor™, [4][5]), conceived as a solution for maintenance-free measurement of soiling loss. The Mars™ sensor is designed to be deployed at a PV installation site in plane-of-array and to operate unattended and without water. Soiling particles accumulate on the sensor window just as on any nearby PV modules and the sensor directly detects the soiling level on the sensor window.

2 OPTICAL MEASUREMENT OF SOILING

Consider a PV device covered by a transparent window upon which soiling particles have accumulated. The soiling particles block incident sunlight through reflection and absorption of light, and therefore result in lost power production by the PV device relative to a clean state of the window. The soiling loss *SL*, the fractional loss in power due to the soiling particles, can be expressed as

$$SL = 1 - T = R + A \quad (1)$$

where *T* is the fractional transmission of light through the layer of soiling particles, *R* is the fractional reflection (or scattering) of light, and *A* is the fractional absorption of light. Here we consider *T*, *R*, and *A* as properties of the layer of soiling particles only, neglecting any reflection or absorption by the glass, i.e. *SL* is only the additional loss due to the soiling particles. For simplicity, in eq. (1) we also neglect the spectral irradiance distribution of incident light, spectral transmissivity of the soiling layer, and spectral response of the PV device, although these can be considered in a more complete treatment.

Our aim is to determine *SL* or, equivalently, *T*.

Recently another optical soiling measurement device has been developed [6] which uses a reflection measurement to measure a signal proportional to the *R* term in eq. (1). A beam of light is directed from inside the sensor to the sensor window and light that is back-reflected/scattered by soiling particles is detected. The *A* term is neglected and *SL* is assumed to be proportional to the reflectance signal or *R* term. The device therefore requires separate calibrations for different colors of dust with different reflectance/absorbance. This is a significant difficulty since the type and color of dust encountered at a particular site are not necessarily known and may even change over time with weather conditions.

In contrast, our sensor concept directly measures both the *T* and *R* terms in eq. (1) using a microscopic imaging approach, similar to the soiling microscope used in [7] to study particle deposition dynamics, although with important differences.

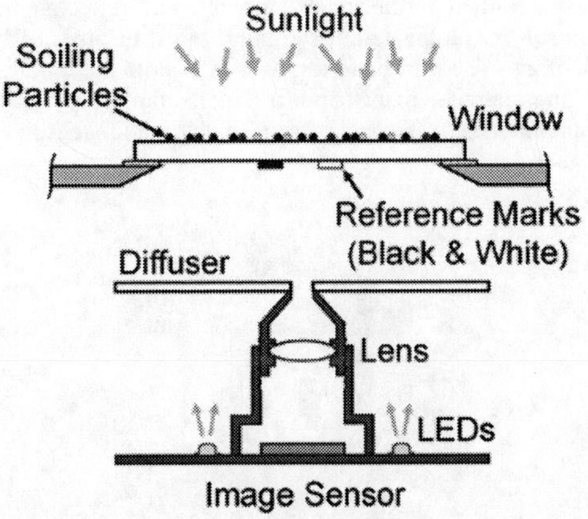

Figure 1: Mars Soiling Sensor™ concept. A microscope system captures an image of a soiling particle layer on a window using either external (sunlight) or internal (LED) illumination. Black and white reference marks provide calibration features for image analysis.

Figure 2: Mars Soiling Sensor™ measurement head, showing reference marks on inside of window.

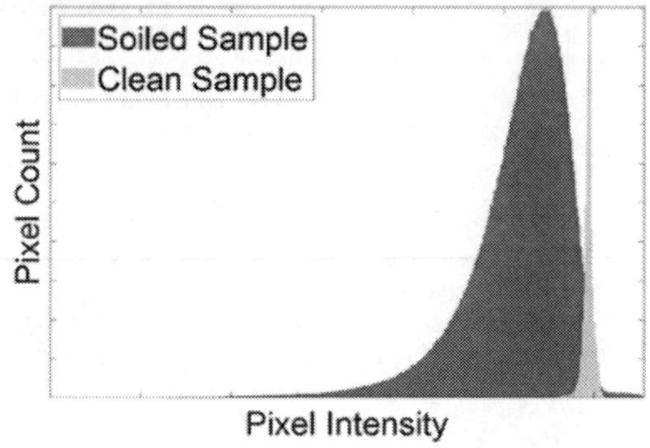

Figure 4: Pixel intensity distribution of Mars™ images of two samples, one clean and one soiled.

3 MARS™ SENSOR CONCEPT

Fig. 1 illustrates the Mars Soiling Sensor™ concept. Soiling particles accumulate on a glass window mounted on the outside of the sensor enclosure. A camera assembly, consisting of a lens and image sensor, is focused on the outer layer of the window and captures a microscopic image of the soiling particle layer, which can be illuminated by sunlight and/or by internal LEDs. A diffuser spreads the LED illumination evenly across the window. Black and white reference marks screen-printed on the inside of the window are used as self-calibration features for image analysis, as well as for detection and rejection of invalid images. Note that the window edge is not covered, to prevent trapping moisture and dust. Fig. 2 is a photograph of the actual sensor head, which better shows the reference marks which are formed as two concentric circles. The sensor is focused to obtain images from a field of view on the window surface with diameter up to approximately 15 mm.

Fig. 3 shows an example acquired image when the sensor head is covered with a layer of ISO 12103-1, A2 fine test dust ("Arizona road dust"), which has a large silica content and is somewhat white and reflective. The sensor is illuminated externally by simulated diffuse sunlight. The dust particles, as well as both the black and white reference marks, appear dark, as they cast shadows blocking light entering from outside the sensor. To analyze the image, we use a proprietary background correction algorithm to correct for non-uniformities in illumination, identify a reference pixel brightness corresponding to local regions of maximum transmission, and then compute the average brightness of all pixels, excluding the reference marks, relative to the reference pixel brightness. The result directly yields the fractional transmission T and therefore SL. Note that this technique does not require that the camera resolve individual dust particles, unlike the method in [7], and that it does not assume any optical properties of the dust itself, which could be either opaque or partially translucent.

The image analysis method is further illustrated in Fig. 4, which shows pixel intensity histograms of Mars™ images, similar to that of Fig. 3, of two samples, one clean and one soiled by dust. Note that pixels corresponding to the circular reference mark pattern are not included in the histograms. For the clean sample the histogram distribution is very narrow, with most pixels having intensity near a value corresponding to clean glass and maximum light transmission. For the soiled sample, the histogram shows a wide range of pixel intensity values, indicating that many of the pixels have reduced light transmission. As described above, soiling loss is calculated by determining from each image a reference pixel brightness corresponding to clean glass and then comparing all pixels to the reference brightness to calculate average transmission and therefore soiling loss.

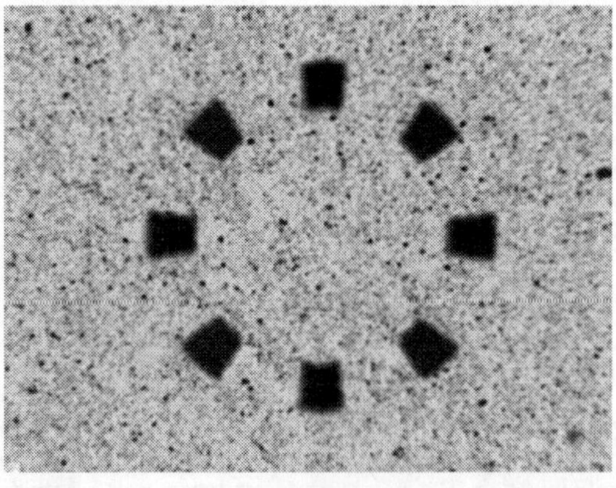

Figure 3: View through the sensor head with external illumination and a layer of dust on the sensor window.

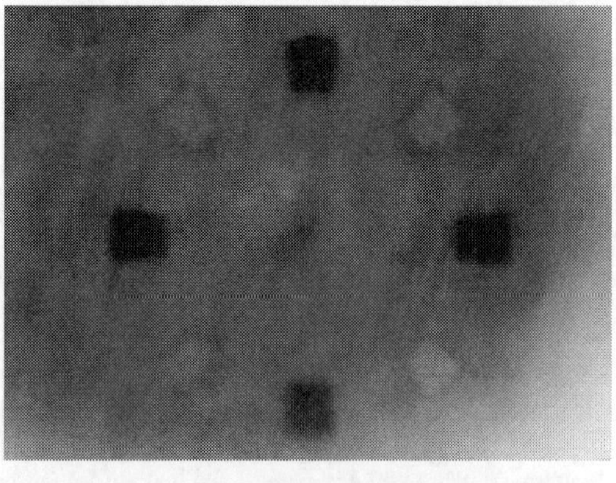

Figure 5: View through the sensor head with internal illumination and simulated snow on the sensor window.

Figure 6: Dust powders (top) and microscope images (bottom) of representative soiled glass test coupons using each of the dust powders, Arizona road dust (left), carbon black (center), and red iron oxide (right). The clear glass soiled coupons were overlaid on a PV cell for the images; although appearance varies due to dust reflectivity, soil levels of each sample are similar (~12%). Image widths are 5 mm.

By using internal illumination from the LEDs, additional information can be gained. When illuminated from the inside (for example, at night), the camera captures an image of the reflection of material on the window. This mode of operation can be used to measure the R term in eq. 1, thus setting a lower limit on SL and providing an independent check. In addition, the measured reflectance can be used to provide a small correction to the external-illumination determination of T to account for daylight which enters the sensor, scatters upwards, and illuminates the particles from below. Thus analysis may combine both externally illuminated (i.e. daytime) and internally-illuminated (e.g. nighttime) images of the soiling layer.

Figure 5 shows another potential application of internal illumination, the detection of snow or other reflective substances (e.g. bird droppings) on the sensor which block external light. The image in Fig. 5 is taken with the sensor covered with simulated snow (a white foam sheet) and illuminated from the inside by LEDs. Note that the white reference marks reflect the internal illumination and that reflection from the simulated snow layer is also apparent in the image and can be referenced to the reflection from the white marks.

4 LABORATORY TESTING

Laboratory evaluation was performed using soiled test coupons. Coupons were prepared using three different dust materials, of different colors, representative

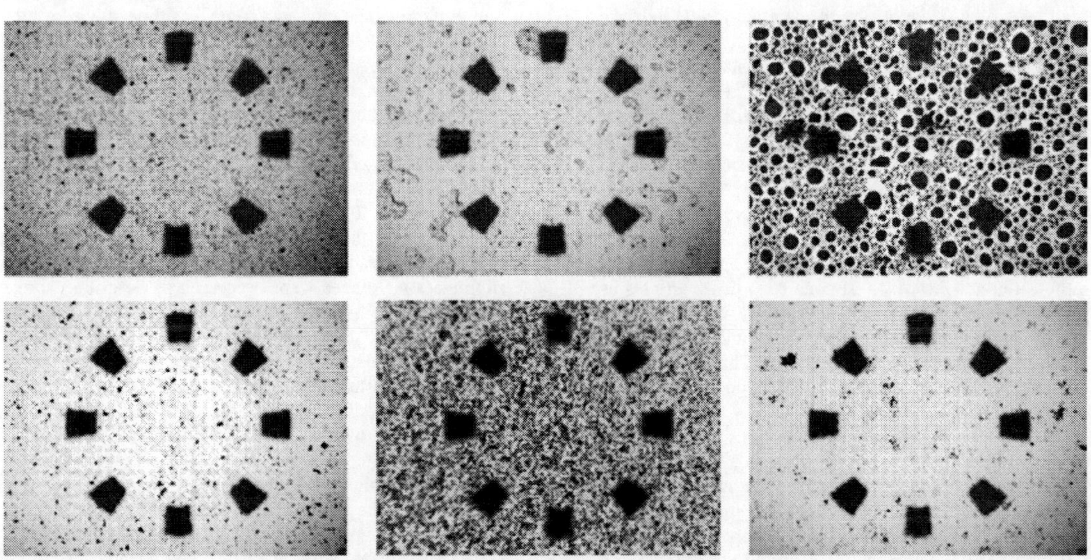

Figure 7: Mars™ images of representative soiled glass test coupons, illustrating the effects of varying deposition conditions. Each image has width ~11 mm.

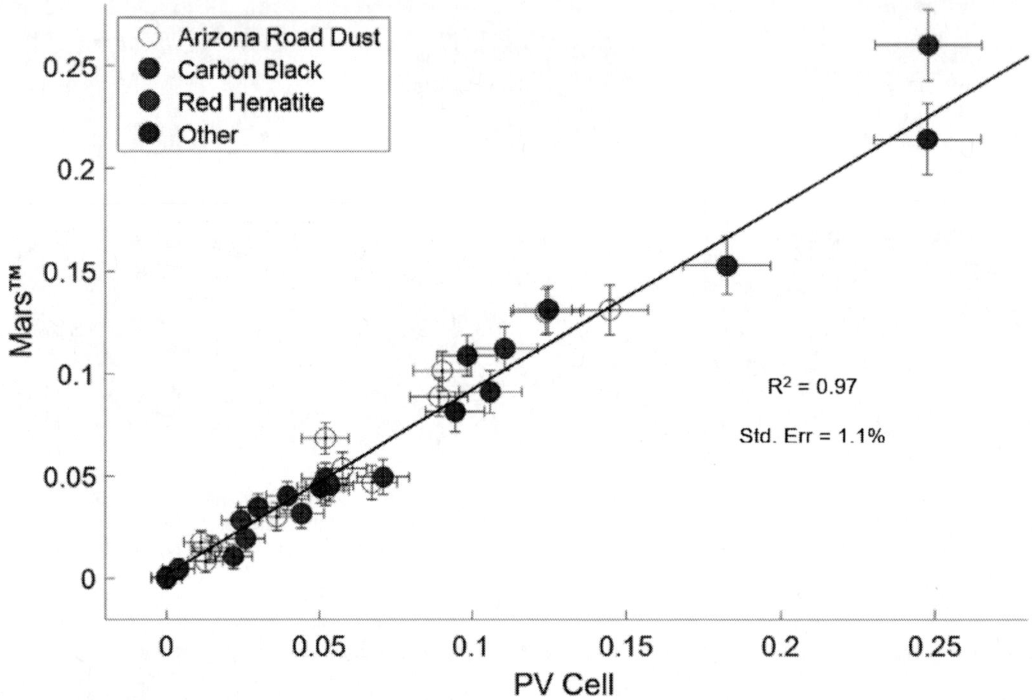

Figure 8: Correlation of soiling loss measured by the Mars Soiling Sensor™ versus results with a PV cell, for various representative dust materials including those with white, black, and red particles as indicated by the colored symbols.

of airborne soiling particles. These included fine Arizona road dust (ISO 12103-1, A2, Powder Technology), 30 um filtered carbon black (Inoxia), and 30 um filtered red iron oxide (Alpha Chemicals). The three materials are respectively white, black, and red in appearance, as shown in Fig. 6. The coupons consisted of 2" x 2" borosilicate glass slides. Dust was uniformly applied to the coupons by placing them, together with a selected dust material, in a custom-built dusting box, which uses a high speed recirculation fan to maintain a dust-laden air stream. Prior to dust exposure, the coupons were chilled to -10 °C to 5 °C, causing moisture condensation during exposure which promotes dust adhesion. Fig. 6 illustrates the varying appearance of the three different materials once applied to the coupons. The figure shows microscope images of coupons with each of the three materials in which the clear glass coupon is overlaid on a PV cell, simulating the appearance of a soiled PV device. Although all three coupons have similar soiling level of ~12% (measured as discussed below) their appearance is quite different: the white Arizona road dust particles are bright and reflective, while the carbon black is barely visible against the dark background and the red iron oxide has an intermediate appearance. This illustrates the difficulty of measuring particle coverage by appearance or by reflectivity.

By varying the deposition conditions, including pre-chilling temperature, exposure time, position within the dusting box, and other parameters, we achieved a wide variety of different dust coverages and morphologies, as illustrated in Fig. 7. The figure shows close-ups of images of various coupons as acquired with the Mars™ sensor, similar to Fig. 3, where each image is an image of the shadows of the dust particles against the external lighting. Note in some images that particles have agglomerated together into larger collections, due to the moisture level, which may be similar to natural soiling processes.

As a control, the soiling loss of each coupon was measured with a PV cell illuminated with a halogen lamp. A 10 mm diameter aperture was placed in front of the cell to limit the spot size of the measurement. The ratio of the PV cell short-circuit current for each coupon to its short-circuit current with clean glass was recorded, and one minus the ratio was logged as the soiling loss for the coupon. A small correction was performed to account for reflection of light from the multiple interfaces of the soiled coupons. The characteristic uniformity of soiling loss within the central region of the samples was estimated by translating selected samples across the PV cell aperture and measuring maximum and minimum soiling loss.

Following PV measurement, each sample was positioned on the window of the Mars™ sensor and images were acquired with both external and internal illumination. Images were processed as described above and the results were correlated with the control measurements using the PV cell.

Results are shown in Fig. 8. Error bars represent the uncertainty in the control measurements due to lamp instability as well as non-uniformity of the soiling in the central region of each sample. White, black, and red symbols indicate the white, black, and red dust particles. Blue symbols show other materials, which include salt, a mixture of the three dusts, and clean glass. The correlation between the optical soiling sensor and the PV cell is very good, with R^2 of 0.97 and standard error of correlation of 1.1%. Furthermore, all three different dust materials fall on the same trend line, suggesting a near-universal response of the sensor independent of dust color.

The results presented in Fig. 8 are for soiling levels up to 25%, which is a practical range for PV system operation. Additional testing has been performed up to 75% soiling levels, showing adequate performance although with larger errors. Work is ongoing in this area.

5 OUTDOOR TESTING

Outdoor testing of the Mars™ sensor is underway, with several units deployed or in process of deployment at locations in Texas and California, together with traditional soiling measurement techniques for data correlation. Due to low soiling and frequent rains at the already installed test sites, data with significant soiling levels for correlation are not yet available, although initial results demonstrate the ability of the sensor to operate properly in the outdoor environment and in diffuse sunlight illumination. Further work is ongoing.

6 DISCUSSION AND CONCLUSIONS

Laboratory data show promising results for the new water-free and maintenance-free Mars Soiling Sensor™ concept. In particular, the results show very little, if any, dependence on dust color, in contrast to [6].

Note that the results presented in Fig. 8 do not include any arbitrary calibration constants. The slope of the best-fit line to all data points is 0.95, indicating that without calibration the technique already measures soiling loss close to the control method. By applying a calibration factor, the results can of course be improved. One potential reason for the non-unity slope of the correlation could be spectral mismatch between the PV cell and the image sensor. Further work is needed to assess this. Also, while to date we have processed acquired images in grayscale only, use of the color sensitivity of the image sensor could in the future allow automatic compensation for a spectral mismatch effect.

Our method for determining T does not depend on dust color, but it does depend on the morphology of the dust layer, since we must be able to observe some gaps between dust particles in order to determine reference intensities corresponding to local regions of clean glass. Therefore the method may be less effective for highly soiled samples with very small and highly uniformly distributed particles. However, in such cases our complementary method of determining R using internal illumination could provide a check and lower limit on SL.

Determining the accuracy of the Mars™ technique in actual practice will require long term outdoor testing with naturally-occurring soiling in a variety of conditions.

Finally, we note that our optical sensor measures soiling loss only at a relatively small region on the sensor itself, while PV module power loss can be significantly affected by the distribution pattern of dust across an entire module [8][9]. We envision potential extensions of the technique that could account for these effects [10].

Overall, the promising results obtained in laboratory testing so far suggest that the Mars Soiling Sensor™ concept could be an effective, compact, low-cost, and unattended soiling sensor practical not only for utility-scale but also commercial-scale PV projects.

7 ACKNOWLEDGEMENTS

We wish to thank Christina Fonda for her help in preparing the dust coupons used for this study.

8 REFERENCES

[1] M. Gostein, J. R. Caron, B. Littmann, "Measuring soiling losses at utility-scale PV power plants," in *40th IEEE Photovoltaic Specialists Conference*, Denver, CO, 2014.

[2] "Photovoltaic system performance – part 1: monitoring," IEC standard 61724-1, 2017.

[3] M. Gostein, et al, "Soiling measurement station to evaluate anti-soiling properties of PV module coatings," in *43rd IEEE Photovoltaic Specialists Conference*, Portland, OR, 2016.

[4] Patents Pending.

[5] M. Gostein, et al, "Mars Soiling Sensor™," in *45th IEEE Photovoltaic Specialists Conference*, Hawaii, 2018.

[6] M. Korevaar, et al, "Novel soiling detection system for solar panels," in *33rd European Photovoltaic Solar Energy Conference*, Amsterdam, the Netherlands, 2017.

[7] B. Figgis, et al, "Outdoor soiling microscope for measuring particle deposition and resuspension," *Solar Energy*, v. 137, pp. 158-164, November 2016.

[8] M. Gostein, B. Littmann, J. R. Caron, L. Dunn, "Comparing PV power plant soiling measurements extracted from PV module irradiance and power measurements," in *39th IEEE Photovoltaic Specialists Conference*, Tampa, FL, 2013.

[9] M. Gostein, B. Stueve, M. Chan, "Accurately measuring PV soiling losses with soiling station employing PV module power measurements," in *44th IEEE Photovoltaic Specialists Conference*, Washington, DC, 2017.

[10] Patents Pending.

PREDICTING FUTURE SOILING LOSSES USING ENVIRONMENTAL DATA

Leonardo Micheli, Michael G. Deceglie
National Renewable Energy Laboratory, Golden, Colorado, United States

ABSTRACT: The correlations between soiling and parameters describing the atmospheric, pollution and geometric characteristic of a site have already been investigated in the literature. In particular, particulate matter and rainfall data have been shown to be the best parameters to estimate soiling losses occurring at photovoltaic sites in the United States. However, previous investigations only considered soiling and environmental data collected over the same time periods. The aim of the present research is, instead, to understand if future soiling losses can be predicted using environmental parameters recorded during previous time periods. To do this, we compared the soiling occurring at 41 U.S. locations with particulate matter and rainfall data recorded 1 to 30 years earlier.
Keywords: Soiling, Reliability, System Performance, Monitoring

1 INTRODUCTION

Soiling consists of accumulated dust and dirt on the surfaces of photovoltaic (PV) modules that reduces energy conversion [1]. This loss leads to a lower and more uncertain energy yield, which means reduced revenues, higher operation and maintenance costs, and more expensive financing.

Investors generally consider an average monthly yearly soiling loss to estimate the energy yield and the economic revenue of a site for a new PV installation. The yearly soiling losses can be estimated by using soiling data available from nearby sites [2] or by considering environmental data available at the site [3–6]. In the first case, the soiling losses can be directly measured using soiling stations [7,8] or by using soiling extraction algorithms to identify soiling trends in PV performance data [9,10]. In the second case, widely available parameters, describing the weather and pollution patterns, can be used to estimate the soiling losses occurring at different sites. In particular, our previous works showed how average particulate matter concentrations and rainfall data correlate with the annualized soiling losses registered at a number of sites in the United States [4,5].

In our previous works, we compared only soiling and environmental data collected over the same time periods. However, when a new PV site is selected, historical environmental data must be used to predict future losses. For this reason, the present work focuses on understanding the correlations between historical pollution and rainfall trends and future soiling losses.

2 METHODOLOGY

2.1 Soiling data

The soiling data used in this work have been extracted from 41 soiling stations installed in the United States between 2013 and 2016 for time periods ranging between 7 and 40 months [11]. We analyzed the same data of Ref. [5] and compared them with environmental parameters that occurred during the soiling data collection period. A soiling station is a system consisting of two PV devices—two cells, two modules or a combination of both—that are mounted outdoor with the same tilt, azimuth, and height. One of the two devices is cleaned regularly (reference device), whereas the second one is left to soil naturally (soiled device). Soiling is quantified by comparing the electrical output of the two devices using a "soiling ratio" metric that is the ratio of the short-circuit current of the soiled device to the short-circuit current of the reference device [12]. The soiling ratio has a value of 1 if both the devices are cleaned, and it decreases as soiling accumulates on the soiled device. The same filters described in [2,4] have been used: we consider only data points occurring between 12 pm and 2 pm and for irradiances > 500 W/m^2.

2.2 Environmental data

Our present work considers only particulate matter and rainfall data as environmental parameters. Indeed, these have been found to be the most relevant parameters for predicting PV losses in the United States [4,5].

The particulate matter consists of the mixture of soiling particles and liquid droplets suspended in the air. It is commonly expressed as PM_{10} and $PM_{2.5}$, which describe the mass concentration of particles less than 10 microns and 2.5 microns in diameter, respectively, suspended in 1 m^3 of air. The values of the annual PM_{10} and $PM_{2.5}$ concentrations are from the U.S. Environmental Protection Agency (EPA) database [13], which records annual average concentrations from monitoring stations installed in the USA. These data have been analyzed using two standard air quality interpolation techniques [5,14]:

- Nearest neighbor: The particulate concentration at a soiling station site is obtained as the mean of the annual values recorded by the closest EPA monitoring station over the data collection period;
- Declustered distance estimation: The particulate concentration at a soiling station site is calculated as the weighted mean of the annual values recorded by the EPA monitoring stations located within a set distance of the soiling site (either 30, 50, 100 or 250 km). As described in [15], we used the inverse of the distance between each EPA monitoring station and the soiling site as well as a parameter describing the distances among the EPA monitoring stations as a weight to give more influence to the closest monitoring stations and to reduce the impact of spatially clustered monitoring stations.

The average and the maximum lengths of the dry periods, which describe the mean and maximum number of days between two consecutive rain events, have been calculated by using the daily rainfall data downloaded from the Oregon State University's PRISM database [16].

2.3 Statistical metrics

The soiling ratios of each site, calculated as the mean of the daily ratios measured over the data collection periods, were compared with the average value of the environmental parameters. These parameters describe the pollution and precipitation trends that occurred at each site for the year and for the 3, 5, 10 and 30 years before the soiling stations were installed. This means that, for example, if a station was installed in 2013, environmental data from 2010 to 2012 were considered for the "prior 3 years" analysis, and from 1983 to 2012 were considered for the "prior 30 years" analysis.

It is important to mention that, while the PM_{10} data have been available since 1982, whereas the first $PM_{2.5}$ concentrations were reported in 1997. Therefore, even if labelled as "30 years before", $PM_{2.5}$ values have been calculated as the average of the concentrations measured between 15 and 17 years before the start of the soiling data collection.

We used a linear single-variable regression analysis to study the correlations between soiling and each parameter. The adjusted coefficient of determination (adjR^2) has been employed to determine the quality of each correlation [4,5]. In addition, we discarded any correlation showing a p-value higher than 0.05.

3 RESULTS

In Ref. [5], we found that PM_{10} and $PM_{2.5}$ concentrations interpolated from the EPA dataset were the parameters with the best correlations to soiling. In particular, we found that the best correlation of PM_{10} occurred for distances between 30 km and 50 km, with R^2 up to 69%, independent of the interpolation techniques used. On the other hand, the correlations between $PM_{2.5}$ and soiling were more robust to the distance, with R^2 ranging between 70% and 63% for distances between 30 km and 250 km, if we used the declustered distance estimation technique.

The same analyses have been repeated in the present work, considering the particulate matter measured over different time periods. The results are shown in Figure 1 and Figure 2, with maximum adjusted coefficients of determination between 68% and 71% for both of the parameters. Overall, we found the same trends of the previous work [5]: PM_{10} performs better at shorter distances, whereas $PM_{2.5}$ seems to be more robust, having limited drops in R^2 at greater distances. In both cases, the nearest neighbor interpolation returns results similar to those obtained with the declustered distance estimation at larger distances (≥ 100 km).

For both indexes, the best results are obtained if data are collected during the same time that soiling was measured. $PM_{2.5}$ returns more consistent correlations to soiling than PM_{10} when different time periods are considered, with adjusted R^2 always above 40%, independent of the method and distance. $PM_{2.5}$ returns higher R^2 values, with better results obtained when we use data collected for long time periods (≥ 10 years).

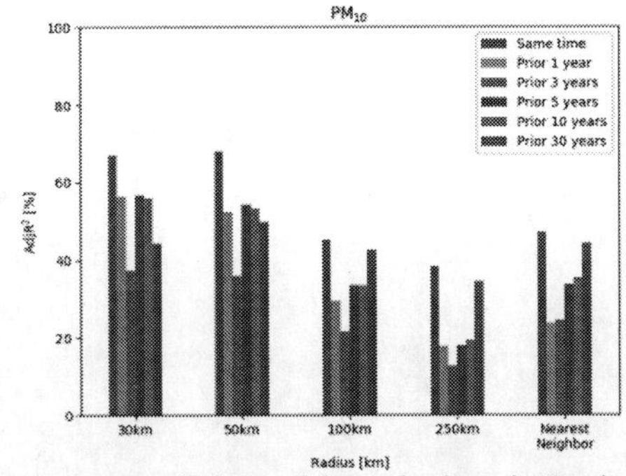

Figure 1: Coefficients of determinations, in %, obtained by interpolating the PM_{10} data using the declustered distance estimation with different radii and the nearest neighbor methods. The bars are colored according to the time the data were collected compared to the soiling data monitoring period.

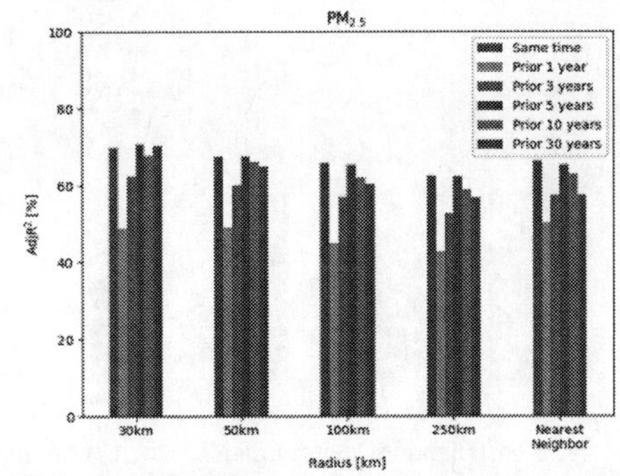

Figure 2: Coefficients of determinations, in %, obtained by interpolating the $PM_{2.5}$ data using the declustered distance estimation with different radii and the nearest neighbor methods. The bars are colored according to the time the data were collected compared to the soiling data monitoring period.

Other than the pollution parameters, the average length of the dry periods and the maximum length of the dry periods were the only significant variables showing correlations to soiling among those investigated in [4,5]. These same parameters were analyzed in this study: their correlations with soiling, depending on the data collection periods, are shown in Figure 3 and Figure 4. Overall, both these parameters return R^2 lower than the pollution concentrations. In particular, the correlation of the average length of the dry period with soiling drops rapidly, with R^2 equal to or lower than 20% if data are collected more than 1 year before soiling occurs. The maximum dry periods are more consistent to the variation in time periods. The best results are generally obtained if 3 years are averaged, and the correlations are slightly worse for longer time period (≥ 5 years). In both cases, the correlations are steady for thresholds ≤ 1mm, whereas they tend to decrease, especially for the maximum length of the dry period, for thresholds of 5 mm.

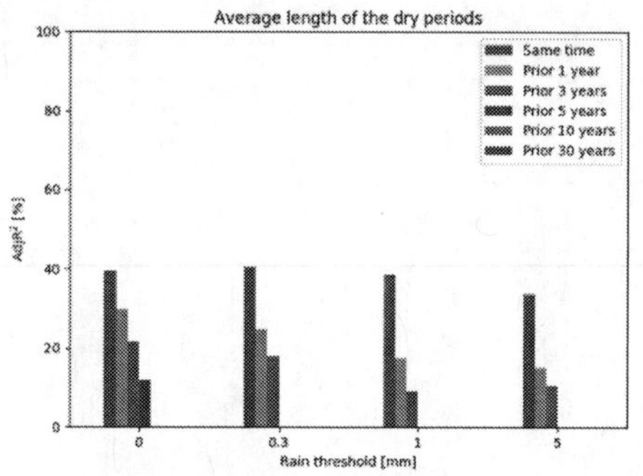

Figure 3: Coefficients of determinations, in %, obtained by comparing the average length of the dry periods with the soiling losses. The bars are colored according to the time the data were collected compared to the soiling data monitoring period.

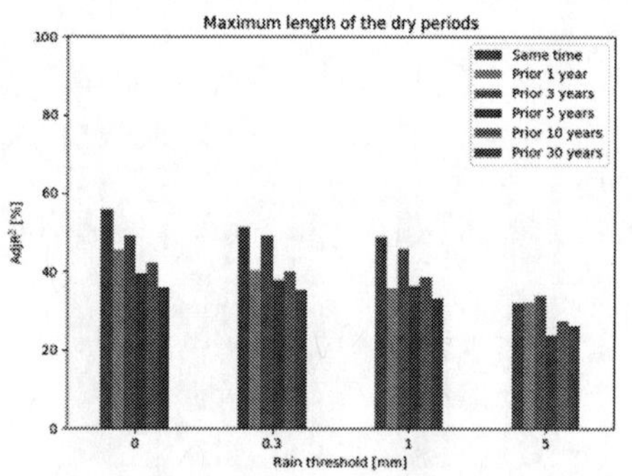

Figure 4: Coefficients of determinations, in %, obtained by comparing the maximum length of the dry periods with the soiling losses. The bars are colored according to the time the data were collected compared to the soiling data monitoring period.

4 DISCUSSION

Among the four parameters analyzed in this study, PM2.5 seems to be the most consistent to the different time periods considered. R^2 values consistently above 50% are obtained if data are averaged over long time periods (≥10 years), independently of the distance considered. This result is because the $PM_{2.5}$ concentrations of the most soiled sites are clearly higher than the rest of the sites (Figure 5). The same distinction cannot be done by instead considering PM_{10} concentrations in Figure 6.

Among rainfall data, the maximum length of the dry periods seems to perform better than the average length of the dry periods. This is probably because, over a 30 years period, the most soiled sites are more likely to experience long summer dry periods, where most of the soiling losses occur (Figure 7). On the other hand, the average dry period of the most soiled locations does not appear to be necessarily longer than those of the cleaner sites.

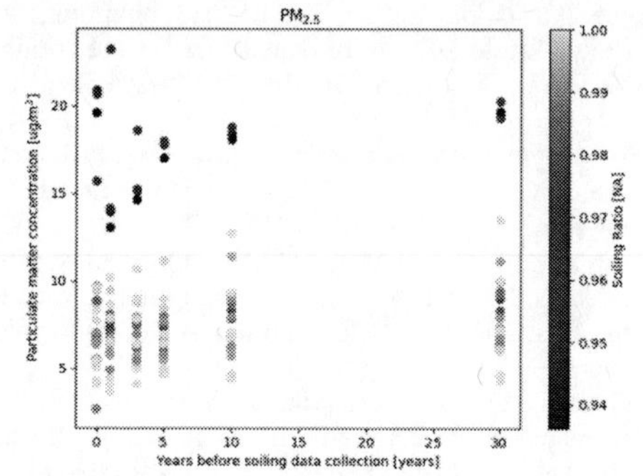

Figure 5: $PM_{2.5}$ concentrations at the 41 sites during the data collections and 1, 3, 5, 10 and 30 years before. The data are colored according to the average soiling ratio of the site.

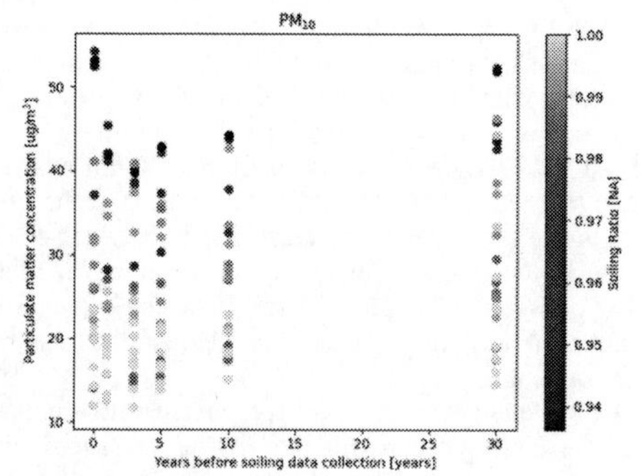

Figure 6: PM_{10} concentrations at the 41 sites during the data collections and 1, 3, 5, 10 and 30 years before. The data are colored according to the average soiling ratio of the site.

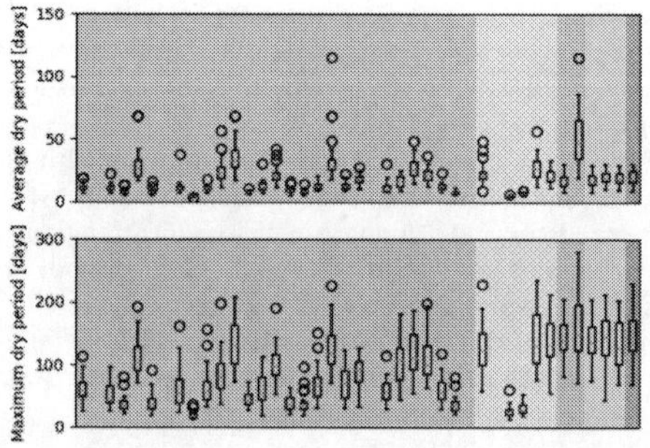

Figure 7: Boxplots representing the average and maximum lengths of the dry periods of each site over a 30 year periods. The background is colored according to the severity of soiling: green for soiling ratio ≥ 99%, orange for soiling ratio ≥ 98%, red for soiling ratio ≥ 96%, light grey for soiling ratio ≥ 94%, dark grey otherwise. Boxplots calculated for rain threshold of 1 mm.

5 CONCLUSIONS

This work analyzes the correlations between average historical environmental data and future soiling losses. We found the parameters describing the particulate matter concentrations to be the best predictors of soiling, even when historical data were used to estimate future soiling losses. In particular, $PM_{2.5}$ was found to be the most consistent predictor, independent of the time period considered, because of the steady distinction in $PM_{2.5}$ concentrations between high and low soiling sites. Among the rainfall parameters, the maximum length of the dry period returned the best results because of the consistent longer summer dry periods experienced by high soiling sites.

ACKNOWLEDGMENTS

This work was authored by Alliance for Sustainable Energy, LLC, the manager and operator of the National Renewable Energy Laboratory for the U.S. Department of Energy (DOE) under Contract No. DE-AC36-08GO28308. Funding provided by the U.S. Department of Energy's Office of Energy Efficiency and Renewable Energy (EERE) under Solar Energy Technologies Office (SETO) Agreement Number 30311. The views expressed in the article do not necessarily represent the views of the DOE or the U.S. Government. The U.S. Government retains and the publisher, by accepting the article for publication, acknowledges that the U.S. Government retains a nonexclusive, paid-up, irrevocable, worldwide license to publish or reproduce the published form of this work, or allow others to do so, for U.S. Government purposes.

REFERENCES

[1] T. Sarver, A. Al-Qaraghuli, L.L. Kazmerski, A comprehensive review of the impact of dust on the use of solar energy: History, investigations, results, literature, and mitigation approaches, Renew. Sustain. Energy Rev. 22 (2013) 698–733. doi:10.1016/j.rser.2012.12.065.

[2] L. Micheli, M.G. Deceglie, M. Muller, Mapping Photovoltaic Soiling Using Spatial Interpolation Techniques, IEEE J. Photovoltaics. (n.d.).

[3] L. Micheli, M. Muller, S. Kurtz, Determining the effects of environment and atmospheric parameters on PV field performance, in: 2016 IEEE 43rd Photovolt. Spec. Conf., IEEE, Portland, OR, 2016: pp. 1724 – 1729. doi:10.1109/PVSC.2016.7749919.

[4] L. Micheli, M. Muller, An investigation of the key parameters for predicting PV soiling losses, Prog. Photovoltaics Res. Appl. 25 (2017) 291–307. doi:10.1002/pip.2860.

[5] L. Micheli, M.G. Deceglie, M. Muller, Predicting photovoltaic soiling losses using environmental parameters: An update, Prog. Photovolt Res. Appl. (n.d.). doi:10.1002/adsc.201.

[6] W. Javed, B. Guo, B. Figgis, Modeling of photovoltaic soiling loss as a function of environmental variables, Sol. Energy. 157 (2017) 397–407. doi:10.1016/j.solener.2017.08.046.

[7] M. Gostein, J.R. Caron, B. Littmann, Measuring soiling losses at utility-scale PV power plants, 2014 IEEE 40th Photovolt. Spec. Conf. (2014) 0885–0890. doi:10.1109/PVSC.2014.6925056.

[8] M. Gostein, T. Duster, C. Thuman, Accurately Measuring PV Soiling Losses With Soiling Station Employing Module Power Measurements, in: IEEE 42nd Photovolt. Spec. Conf., 2015.

[9] A. Kimber, L. Mitchell, S. Nogradi, H. Wenger, The Effect of Soiling on Large Grid-Connected Photovoltaic Systems in California and the Southwest Region of the United States, in: Photovolt. Energy Conversion, Conf. Rec. 2006 IEEE 4th World Conf., 2006: pp. 2391–2395.

[10] M.G. Deceglie, L. Micheli, M. Muller, Quantifying Soiling Loss Directly from PV Yield, IEEE J. Photovoltaics. (2018). doi:10.1109/JPHOTOV.2017.2784682.

[11] National Renewable Energy Laboratory, Photovoltaic modules soiling map, (2018). https://www.nrel.gov/pv/soiling.html (accessed May 18, 2018).

[12] International Electrotechnical Commission, Photovoltaic system performance – Part 1: Monitoring (IEC 61724-1, Edition 1.0, 2017-03), (2017).

[13] US Environmental Protection Agency, Air Quality System Data Mart [internet database], (n.d.). https://www.epa.gov/airdata (accessed May 15, 2018).

[14] D.W. Wong, L. Yuan, S.A. Perlin, Comparison of spatial interpolation methods for the estimation of air quality data, J. Expo. Anal. Environ. Epidemiol. 14 (2004) 404–415. doi:10.1038/sj.jea.7500338.

[15] S. Falke, R. Husar, B. Schichtel, Mapping Air Pollutant Concentrations from Point Monitoring Data I: Declustering and Temporal Variance, (n.d.). http://citeseerx.ist.psu.edu/viewdoc/download?doi=10.1.1.612.5742&rep=rep1&type=pdf (accessed May 18, 2018).

[16] PRISM Climate Group - Oregon State University, PRISM Climate Group - Oregon State University, (n.d.). http://www.prism.oregonstate.edu/explorer/ (accessed June 3, 2017).

PHOTOLUMINESCENCE OUTDOOR MEASUREMENT SYSTEM (PLOMS)

Martin Koch and Benno Bucher
University of Applied Sciences Rapperswil, IET Institute for Energy Technology,
Oberseestrasse 10, 8640 Rapperswil SG, Switzerland,
+41 55 222 43 44, martin.koch@hsr.ch, benno.bucher@hsr.ch

ABSTRACT: Micro cracks are a well-known problem of photovoltaic cells. Different techniques are applied to identify damaged cells. We have developed a portable, lightweight and economical **Photoluminescence Outdoor Measurement System (PLOMS)**. This system allows the detection of micro cracks on **installed PV modules** without switching off, demounting or rewiring the solar plant. It is designed to work on standard glass-foil modules but as well on bifacial **glass-glass** modules on **sunny days**. PLOMS has been compared to electroluminescence imaging and had been tested under real conditions on a solar test installation in the Swiss Alps. The PLOMS revealed high resolution photoluminescence images of PV modules in operation and guaranteed detection of failures such as micro cracks of installed solar plants, which simplifies the maintenance work.
Keywords: Photoluminescence, Micro Crack, Detection, Outdoor, Measurement

1 INTRODUCTION

Photoluminescence (PL) is an established technique for identifying failures in silicon wafers and cells [1]. It is mainly used within the manufacturing of photovoltaic (PV) modules to detect failures such as micro cracks.

However micro cracks do not occur only in the production process but develop during the transport, installation and operation. Studies revealed that 4 - 6% [2, 3] of supplied PV cells suffer from cracks. A long-term study has analyzed 143 cracked cells over 6 years and found that the isolated area increases 3.2% [4] per year. Therefor onsite outdoor measurements are important.

There are three techniques for detecting micro cracks on outdoor installed PV-Modules. While Electroluminescence (EL) [5] needs a modification of the photovoltaic power plant to inject the current, UV Fluorescence and PL doesn't have to alter the installation.

UV Fluorescence [2] excites the EVA, which needs at least one summer of sun exposure to reveal micro cracks. Additionally this technique works only on glass-foil modules, where oxygen diffuses through the foil and reacts with the EVA at the cracked area.

Outdoor PL has been performed with sun light as excitation with a long exposure time [6] and an additional LED light modulated on a control cell [7]. Both Methods need InGaAs cameras which are expensive and have a low resolution.

In this paper we reveal an economical and fast outdoor PL technique, which performs high resolution images with a NIR enhanced CMOS camera.

2 EXPERIMENTAL

PL consists of two successive steps: excitation of the PV cells by an external light source and measuring the subsequent luminescence by a camera. PL has been introduced using laser light [1] as the excitation source. With the advent of high power LEDs a cheaper option [8] became available.

The camera has to be sensitive in the luminance spectra of the silicon solar cell (950 – 1350 nm). InGaAs cameras work within this spectra [6, 7, 8] but are expensive and have a low resolution. CCD cameras are cheaper and have a higher resolution [9]. We have used a NIR enhanced CMOS camera which is also a silicon based camera, but the cheaper technique.

The biggest challenge for outdoor PL measurements is the sunlight interfering with the luminance spectra of the solar cell. A pyramid-shaped design has been applied to prevent any interference, as has been depict in Fig. 1 and Fig. 2. The design blocks the sunlight and assures the luminescence of the solar cell to be gathered by the camera.

To excite the silicon, four passively cooled blue LEDs with an irradiance of approximately 1500 W/m^2 are used. The reflected LED light is blocked from the camera by a set of filters; hence, just the luminance of the solar cell reaches the 4 megapixel sensor of the camera.

The camera itself is connected to a Raspberry Pi used as a user interface and it allows some contrast adjustment of the image by the CLAHE [10] algorithm, to increase the visibility of the micro crack.

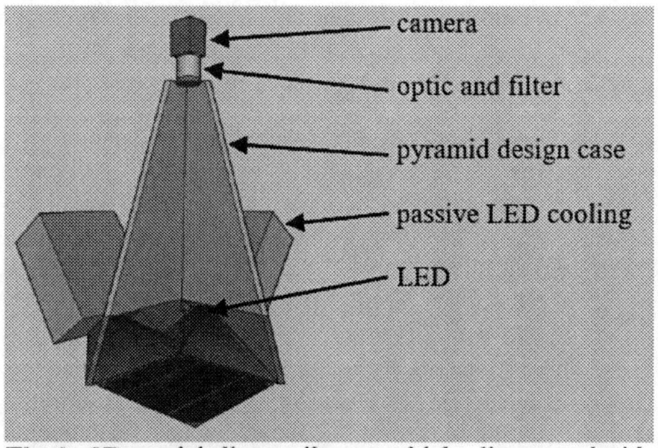

Fig 1: 3D model diagonally cut which allows an inside view of the PLOMS.

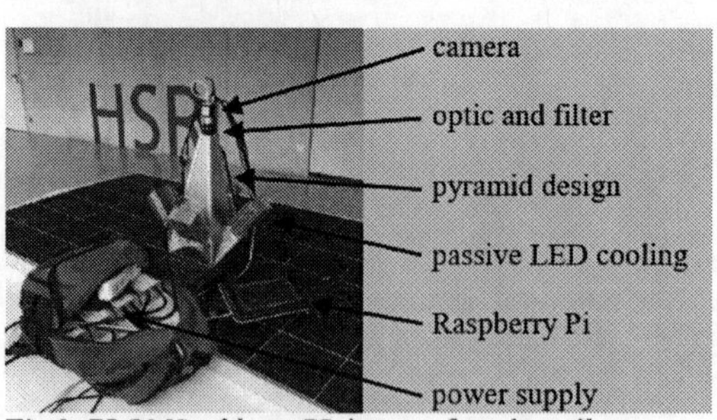

Fig 2: PLOMS taking a PL image of a solar cell

3 RESULTS

We have performed tests to verify that the PLOMS detects all micro cracks. Furthermore this section demonstrates well its outdoor operation in the Swiss Alps.

3.1 Comparison of EL and PL

Electroluminescence measurements have been compared with the PLOMS. Fig. 3 is the result of the PLOMS and shows the same micro crack and polycrystalline cell structure as the EL image on Fig. 4 which was taken with the MBJ Mobile Lab.

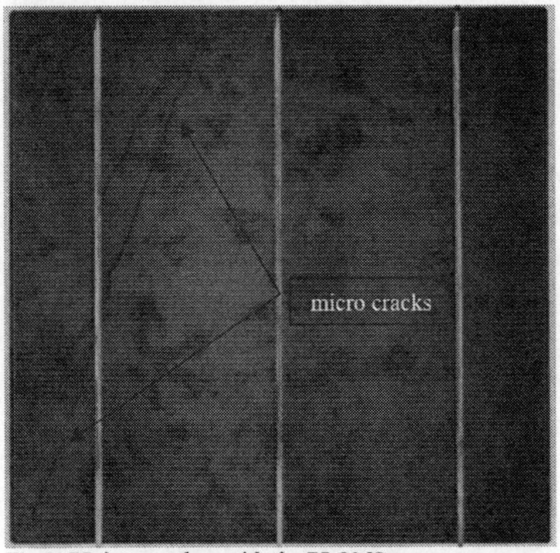

Fig 3: PL image taken with the PLOMS

Fig 4: EL image taken with the MBJ Mobile Lab

3.2 PL Outdoor Measurement

For the verification of its outdoor capability on sunny days, the PLOMS has been applied to make initial PL measurements of a new solar test facility in Davos (operated by the Zürich University of Applied Science (ZHAW) and powered by the "Elektrizitätswerke des Kantons Zürich" (EKZ)) after its installation in autumn 2017 (Fig 5). Measurements were carried out on a bifacial monocrystalline glass-glass module (marked in blue) as well as on a monofacial monocrystalline glass-glass module (marked in brown). Both images were taken with a camera exposure time of 0.5s.

Fig 5: Solar test facility in Davos (operated by the Zürich University of Applied Science (ZHAW) and powered by the "Elektrizitätswerke des Kantons Zürich" (EKZ)). Two modules were measured: a bifacial glass-glass module (blue) and a monofacial glass-glass module (brown).

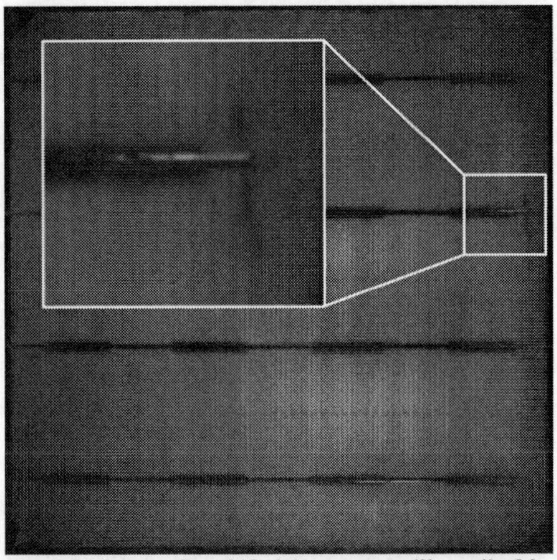

Fig 6: PL image of one bifacial monocrystalline cell of the module marked in blue on Fig 5. Taken with an exposure time of 0.5s.

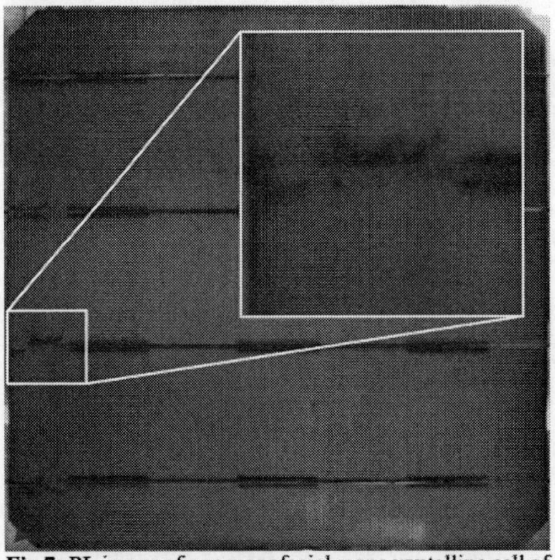

Fig 7: PL image of one monofacial monocrystalline cell of the module marked in brown on Fig 5. Taken with an exposure time of 0.5s.

The back of glass-glass modules must be covered to perform PLOMS. The bifacial module in Fig. 6 was shielded with a black blanket, which is why the background is black (red arrow in Fig. 6). The back of the monofacial module was covered with a cardboard; as the cardboard does not absorb all sunlight, the background in Fig. 7 is bright.

4 DISCUSSION AND CONCLUSION

A Photoluminescence Outdoor Measurement System (PLOMS) has been devised in order to enable PL measurements for failure analysis of PV modules in operation. The measurement results of the PLOMS are comparable to those measured by EL (Fig. 3 and Fig. 4), proofing that PLOMS can be used equivalently for identifying failures of micro cracks on PV modules.

The outdoor measurements can even be performed successfully on glass-glass modules on sunny days (Fig. 6 and Fig. 7) with an exposure time less than a second. It allows the detection of micro cracks on installed PV modules without switching off, demounting or rewiring the solar plant.

Furthermore, it has been demonstrated, that PL measurements can be carried out quite economically with passively cooled LEDs as excitation source and a near infrared optimized CMOS camera.

Possible applications of this device could be:

- Incoming inspection: ensuring quality of received PV modules
- End-control: ensuring quality after installation
- Insurance issues: control after hard weather conditions like strong wind, hail and snow
- Research

5 REFERENCES

[1] T. Trupke. R.A. Bardos, M.D. Abbott, F.W. Chen, J:E. Cotter, A. Lorenz, Fast photoluminescence imaging of Silicon wafers, IEEE, 4th World Convergence on PV Energy Conversion (2006).

[2] M. Köntges, S. Kajari-Schröder, I. Kunze, Cell Cracks measured by UV fluorescence in the field, 27th EU PVSEC, Frankfurt, Germany, 2012, pp. 3033-3040

[3] M. Köntges, S. Kajari-Schröder, I. Kunze, U. Jahn, Crack statistic of crystalline silicon photovoltaic modules, 26th EU PVSEC, Hamburg, Germany, 2011, pp. 3290-3294

[4] B. Jaeckel, T. Franke J. Arp, Long term statistic on micro cracks and their impact on performance, 33rd EU PVSEC, Amsterdam, The Netherlands, 2017, pp. 1396-1401

[5] T. Fuyuky, H. Kondo, T. Yamazaki, Y. Takahaschi, Y. Uraoka, Photographic surveying of minority carrier diffusion length in polycrystalline silicon solar cells by electroluminescence, Applied Physics Letters 86 (2005) 262108

[6] L. Stoicescu, M. Reuter, J.H. Werner, Daysy: Luminescence Imaging Of PV Modules in Daylight, 29th EU PVSEC, Amsterdam, The Netherlands, 2014, pp. 2553-2554

[7] R. Bhoopathy, O. Kunz, M. Juhl, T. Trupke, Z. Hameiri, Outdoor photoluminescence imaging of photovoltaic modules with sunlight excitation, Prog Photovolt Res Appl, 2018, 26:69–73

[8] A. Lawerenz, K. Lauer, M. Blech, A. Laades, M. Zentgraf, Photoluminescence lifetime using LED arrays exitation source, 25th EU PVSEC, Valencia, Spain, 2010, pp. 2486-2489

[9] R. Ebner, G. Újvári, B. Kubicek, M. Halwachs, Concept of a photoluminescence measurement system, 33th EU PVSEC, Amsterdam, The Netherlands, 2017, 1541-1543

[10] Karel, Z.: Contrast Limited Adaptive Histogram Equalization, pp. 474-485. Press Professional, Inc., Randallstown (1994)

DUBIO: A FULLY AUTOMATIC IR INSPECTING SYSTEM FOR LARGE PV PLANTS

Marco Colaprico, Maria Francesca de Ruvo
APIS – Apulia Intelligent Systems
Via P. Fiore 26, 70125 Bari, Italy; tel/fax: +39.080.5963586; [m.colaprico, f.deruvo]@spinoffapis.com
Giuseppe Leotta, Maria Luisa Teresa Lo Trovato, Fabrizio Bizzarri
ENEL Green Power
Zona Industriale - Passo Martino, 95121 Catania, Italy; tel: +39.095.2877026; giuseppe.leotta@ENEL.com
Lungarno Colombo 54, 50136 Firenze, Italy; tel: +39.06.83059859; marialuisa.lotrovato@ENEL.com
Viale Regina Margherita 125, 00198 Roma, Italy; tel: +39.06.83052589, fabrizio.bizzarri@ENEL.com
Silvano Vergura, Francescomaria Marino
Electrical and Information Engineering Dept., Polytechnic University of Bari
Via E. Orabona 4, 70125 Bari, Italy; tel:+39.080.5963710; fax: +39.080.5963410; [vergura, marino]@poliba.it

ABSTRACT: DUBIO is a project jointly developed by APIS and ENEL Green Power.
Its result is an innovative cloud platform (www.dubio.cloud), a pay-for-use service devoted to O&M technicians responsible of the analysis of radiometric infrared videos acquired by drones over large PV plants, which now can only care about uploading these videos onto the platform, for getting an automatic and accurate diagnosis of each PV module.
In fact, thanks to computer vision algorithms, DUBIO automatically detects and crops the PV modules inside each video frame, associates them to the corresponding ones in the topology of the PV plant, and provides an accurate diagnosis of each PV module (whichever the environmental conditions), evidencing the overheating values, and classifying the typology of (eventual) defects. All the numerical and graphical results, are therefore collected in a multimedia database, which thus provides the O&M technicians with significant information, useful to monitor the ageing trend of each module of the plant. Moreover, for those PV modules needing deeper investigations, the platform also embeds tools for performing thermal analysis according to conventional approaches, and for launching an automatic diagnostic protocol, which reveals quantitative information about the thermal behavior of each cell constituting the module.
Keywords: Monitoring, Photovoltaic, Plant Control, Drones, Thermal Analysis

1 INTRODUCTION

Today, thermography constitutes a consolidated and powerful diagnostic method for monitoring the health of PV modules, while unmanned aerial vehicles (UAVs, commonly also called drones) embedding infrared (IR) cameras have been verified as an efficient approach for monitoring large and very large PV plants, where manual inspection would require a very expensive acquisition process. Therefore, computer aided systems for analyzing thermographic videos acquired on PV plants have been already proposed on the O&M market, but, at the best of our knowledge, all of them present certain limits, since:

- they need human to perform the association of the IR images with the modules of the plant (referencing);
- often, no radiometric analysis is achieved, i.e., they provide only a merely subjective analysis, based on the temperature (color in the IR images) of a PV module compared to the other adjacent ones in order to detect hot spots and defective PV modules;
- they do not allow to handle data in order to get evaluations on the ageing trend of the single module during its life (since the data is neither radiometric nor independent from the environmental conditions: i.e. the temperature assumed by a PV module during a given inspection is not comparable with the temperature assumed by the same module under different environmental conditions).

As a result of a long research activity, documented in various scientific publications [1, 5], APIS – APulia Intelligent Systems, under the supervision of the Professor of Electrotechnics at the Polytechnic University of Bari, Silvano Vergura, has developed an automatic diagnostic protocol (named DISS – DIagnostics for Solar Systems) that analytically measures the temperature deviation of a single internal cell of the PV module from the one that it would be (in case of its proper operation), and which depends both on the environmental conditions and on the normal operating cell temperature (NOCT) of the module itself. The diagnosis produced by DISS, being normalized to the environmental conditions, provides much more useful information than a mere absolute temperature evaluation, and it can be effectively used for comparing subsequent measurements on the same PV module, thus allowing to evaluate its ageing trend, suggesting the actions to be undertaken.

The evolution of this research has recently led to project, named DUBIO – Drones and CloUd Based Infrared Monitoring, jointly developed by APIS and the Innovation and Sustainability division of ENEL Green Power.

Goal of this project has been the creation of a cloud platform able to perform the IR monitoring of large-scale PV plants in a complete automatism, processing and analyzing IR videos captured and uploaded by drones patrolling large/very large PV plants. Thanks to its computer vision algorithms, the system:

- automatically detects and crops the PV modules inside each video frame;
- associates them to the corresponding ones in the topology of the PV plant;
- provides an accurate diagnosis of each PV module evidencing the overheating values;
- classifies the typology of (eventual) defects;
- collects all the data in a multimedia data base, for being efficiently queried by the O&M technicians.

Once the testing campaign will be successfully closed (this is expected by the end of October 2018), The system will be world-wide provided as a pay-for-use cloud service for the analysis and management of radiometric IR videos.

This paper is so structured: Section 2 has been devoted to the synergic project that has originated the platform, describing the performed activities and the ongoing testing campaign. Section 3 introduces the idea of evolving the developed platform into a pay-for-use service: it provides not only a punctual characterization of the workflow that a user should follow in order to obtain the diagnoses of his own radiometric IR video, but also a detailed description of the tools embedded in the platform and of their graphical user interface (GUI). Conclusive remarks, in Section 4, end the paper.

2 DUBIO: A SYNERGIC PROJECT

DUBIO – Drones and CloUd Based Infrared Monitoring is the result of a joint project carried out by APIS and ENEL Green Power (EGP).

2.1 The partnership

APIS (www.spinoffapis.com) is a startup, spin off of the Polytechnic University of Bari (Italy), characterized by strong skills in computer vision, and whose main core business concerns design, construction, engineering and installation of intelligent systems aimed to automate industrial processes, provide security and perform diagnostics.

In the mission of the Innovation and Sustainability Division of EGP, what assumes a relevant role is the synergic cooperation with innovative startups and companies, aiming at developing technologies, products and services which might play strategic roles in the core business of ENEL. Such cooperation encompasses technical, economical, and logistical support.

Figure 1: Conceptual representation of the DUBIO project.

In this scenario, a partnership was settled between EGP and APIS, aiming at developing the DUBIO system, basing on the concept depicted in Fig. 1: UAVs (Unmanned Aerial Vehicles, also called drones), embedding thermal cameras and sensors, autonomously patrol large PV plants, aiming at acquiring radiometric IR videos over the PV modules. Once completed the flight, and parked onto suitable docking stations for recharging their batteries, the UAVs upload the acquired IR videos, as well as the measured environmental parameters (solar irradiance, wind speed, etc.) to a cloud server which, thanks to computer vision engines:

- automatically crops all the PV modules pictured in each IR video frame and associate them to the plant topology;
- processes the radiometric values;
- extracts useful diagnostic info, which is collected into a multimedia database, to be accessed by the O&M technicians which thus can be greatly helped in their job.

2.2 Project activities

As a preliminary activity in the development of the computer vision engines on which DUBIO is built, EGP has provided APIS with several IR videos, acquired by drones over PV plants hosting modules of different technologies (conventional or bifacial), framed or not, mounted on solar trackers or not.

These IR videos have been used for developing and testing the algorithms designed to automatically extract and crop single regions of interest (ROIs), each one corresponding to a single PV module, as well as to associate them to a unique identifier related to the topology of the plant (the ordered couple composed by the number of the column and of the row occupied by the PV module in each plant line).

Fig. 2 shows an example of IR videoframe, in which the accuracy of the automatic cropping performed over bifacial not framed PV modules, mounted on solar trackers can be appreciated.

Figure 2: Examples of automatically cropped and labelled PV modules.

Once developed, these algorithms were debugged and tuned, up to reach the following goals in terms of accuracy:

- Portion of PV module area not enclosed in the automatically cropped ROI < 3% (in terms of radiometric pixels), in at least 99.5% of the cases
- Portion of automatically cropped ROI not belonging to the PV module < 1% (in terms of radiometric pixels), in at least 99.5% of the cases
- Number of real PV modules not detected (i.e., false negative) = 0
- Number of automatically cropped ROIs not corresponding to real PV modules (i.e., false positive) = 0
- Number of mismatch between the detected PV module and its location inside the plant topology = 0

Afterwards, several automatic diagnostic procedures were developed. All of them, use as input the environmental meteorological data (as registered during the acquisition) and the punctual temperatures of each PV module (i.e., the radiometric values enclosed in the ROI automatically cropped), once such a ROI has been geometrically transformed to remove the perspective artefacts (see Fig. 3).

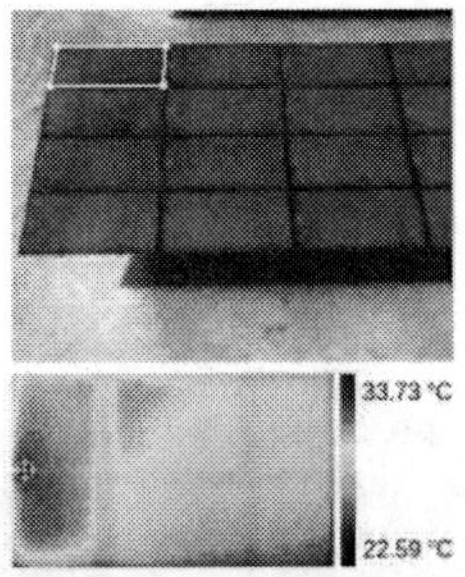

Figure 3: Examples of automatically cropped PV module and its thermographic representation after perspective artifacts removal.

The so called 1st level diagnostic tool employs these data are used in order to:

1. compute and localize the highest temperature inside the PV module;
2. compute the average of the temperatures inside the PV module;
3. determine how much the above values (points 1 and 2) exceed the temperature that the PV module should reach, in its normal behavior, at the registered environmental conditions;
4. use the overheating values (point 3) as macro indexes of the health state of the PV module, independent from the environmental conditions;
5. automatically classify the eventual failure reasons (for those PV modules characterized by significant overheating values), among the most frequent ones, evidencing the expected influence of the detected failure on the electrical output.

In addition, tools for carrying out two other analyses, respectively manual and automatic were also developed and integrated in the platform. These 2nd level diagnostic tools, conceived for being launched on request by the user, only over these PV modules which need deeper investigation, allow:

1. a thermographic analysis according to the usual manual approach focused on histograms, maximum, average and minimum values of temperatures over ROIs specified by the user;
2. an automatic analysis at the level of any single cell constituting the PV module, based on DISS – Diagnostics for Solar Systems [1, 5].

All the implemented diagnostics tools are better described in section *3.2 The platform: features and GUI*, which, showing also some examples of the GUIs, makes clearer the potentialities of the platform.

2.3 Testing campaign

Once developed, the computer vision tools have been integrated onto a cloud platform for being massively tested on the field.

The benchmark sites chosen by EGP were the experimental PV plants at the Innovation Lab in Catania (Italy) and in La Silla (Chile), as well as the industrial PV plant located in Carrera Pinto (Chile):

- the first benchmark site is a 7-years-old plant located in Adrano in the Sicilian island (Italy). It encloses 35,710 PV polycrystalline module (280 Wp), for totally 10 MWp of nominal power.
- the second site is a utility-scale PV plant, activated in September 2016. The plant combines innovative bifacial and smart modules with conventional modules for side-by-side testing, located at 600 km north Santiago, close to a famous astronomical observatory. It produces 1.7 MWp of nominal power thanks to 2,112 bifacial PV modules and 1,884 conventional PV modules.
- the third site was installed in 2014: it employs 312,912 PV modules (72 cells c-Si modules, 310 Wp), for totally 97 MW of nominal power.

All the sites were provided with infrastructures for allowing an autonomous patrolling by means of UAVs: recharging docking stations, with internet connectivity for uploading the acquired data to the cloud server, and UAVs equipped by WIRIS 2 thermal camera by WORKSWELL, able to inspect approximately 3,200 PV module/hour. The campaign is involving approximately 352,600 PV modules, both in the boreal and in the austral hemisphere, and foresees approximately 3 inspections over the same PV module per month, to collect a significant quantity of data to be cross-correlated with the energy production of the plants.

3 NEXT STEP: A PAY-FOR-USE SERVICE

EGP foresees of closing the testing campaign of DUBIO by the end of October 2018. Then, the platform will be worldwide extended to the O&M technicians at the cloud URL www.dubio.cloud, as a service aiming at providing them with a deep and accurate thermal analysis of their radiometric IR videos.

Currently, the use of DUBIO by EGP is completely automatic: the UAVs are suitably configured for acquiring IR videos while they automatically patrol the benchmark plants, following predefined flight routes. Both the features of the PV plants, as well as their topologies, and the flight routes covered by the UAVs, are well known to the DUBIO database, which can thus correctly perform the matching and labelling of each automatically cropped ROI to a real PV module. By this way, human operations are at all avoided, since even the IR data uploading is automatized. In addition, the platform encloses a messaging tool for automatically e-mailing warnings and alarms to selected addresses, in case of events defined by the user: thus allowing, the O&M technicians only to care about those situations automatically signaled by the system to their mail-boxes.

As previously said, the natural evolution of the project is to release a service devoted even to other O&M agencies, different from EGP, for exploiting the cloud platform www.dubio.cloud as a powerful engine for carrying out the thermal analysis of radiometric IR videos: this scenario will be characterized by an automatism partially relaxed, with respect to the above described one, since the task of analyzing IR videos acquired over PV plants unknown to the system, necessarily adds to the DUBIO workflow some additional step, needed for initializing the system, as described in the next paragraph.

3.1 Workflow

For extending the capability of the system to process radiometric videos acquired even over unknown PV plants, the workflow of the service www.dubio.cloud will be structured into the following phases:

- tasks required once for all:
  1. registration
  2. initialization of the PV plants

- tasks required for each monitoring campaign:
  3. data upload and processing;
  4. preview access, check of the diagnoses, acceptation of the service quality, and payment (only if satisfied);
  5. results access and database consultation.

More in detail:

1. Once registered, a user will assume the privilege of administrator of his own data and created entities. In case of structured organizations, the registered USER can also create different accounts, to be assigned to the member of his staff, suitably enabling these USERS to operate over the initialized PV PLANTS. In such a way, different members of the staff will can access to the DUBIO service with different capabilities and purposes.
2. The registered user (or his staff) has to initialize the "PV PLANTS" whose he cares the O&M: this is necessary in order to create all the instances of the multimedia database as well as provide information useful for the processing.

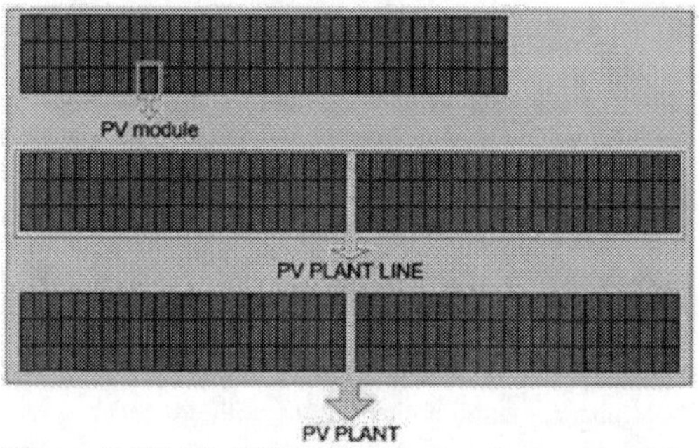

Figure 4: PV PLANT topology example, composed by three PV PLANT LINEs: the first one structured into 3 rows x 36 columns of PV modules, and the other two ones enclosing PV modules arranged into 3 rows x 48 columns.

   In order to accomplish this task, the user has to describe the topological structure of the plant, providing information about the so called "PV PLANT LINES" (i.e., specifying according to which pattern the PV modules have been mounted; in order to clarify, an example is provided in Fig. 4); moreover, several manufacturer's data related to the PV modules of the plant are also needed (e.g., nominal power, temperature coefficient, NOCT).
3. Once an initialized PV PLANT has been inspected, the user creates a "MONITORING CAMPAIGN", which may consist of one or more radiometric "VIDEOS" (acquired during the campaign itself), to be uploaded. At the same time, the user provides data on the environmental conditions registered during the campaign (e.g., solar irradiance, atmospheric temperature, wind speed). Then, as described in *3.2 The platform: features and GUI*, he tags and labels the frames related to the first and last column of each "PV PLANT LINE" (this requires some second for each PV PLANT LINE); afterwards, he can finally submit the data to the system, for being processed and analysed.
4. Though the processing is automatically performed by DUBIO, this step will be supervised by the APIS staff to provide the user with the best accuracy. Accordingly, the results will be available with a short delay since the data submission (no more than five working days, typically one or two). Because of the pay-only-if-satisfied policy, the user will be firstly enabled to preview only a subset of the computed "DIAGNOSES" for checking and evaluating their accuracy; thus, only if their quality will result acceptable, the user will buy the service for having access to the entire set of the results.
5. Once the DIAGNOSES have been bought, the user will access them through a multimedia database, exploiting query and filtering options, having the possibility of downloading printable reports and more deeply analysing the PV modules thanks to the 2nd level analyses tools, if needed. Each data will be stored onto DUBIO database until 18 months after the last bought diagnosis on the same PLANT LINE, meaning that in case of periodical monitoring conducted over the same PLANT LINE, the user will access to all the performed DIAGNOSES, with the possibility of examining the ageing trend of each PV MODULE, thanks to the helpful GUI described in the next section.

3.2 The platform: features and GUI

As already mentioned, once uploaded the radiometric IR videos for being analysed, the user is required to tag and label the frames related to the first and last column of each "PV PLANT LINE". This step is needed in order to allow the system of correctly associate each automatically extracted ROI to a specific PV module of the plant. Anyway, this annotation is facilitated and accelerated by an embedded GUI: the user needs only to browse the uploaded frames, thanks to a suitable player, where he has to mark only those frames where each plant line starts or ends.

At this point, the uploaded videos can be processed by DUBIO, which can now:

- automatically extract and crop the individual PV modules from each IR video frame, correctly associate them with the unique identifier associated with each panel in the plant layout, accordingly to the initialization performed by the user;
- select, among the various frames in which a same PV module occurs, the optimal one for the thermographic elaboration of said PV module;
- eliminate the perspective distortion that will result from the acquisition geometry, presenting to the thermographic analysis a rectangular area in which each contribution in terms of pixels has been equalized; this, in order to avoid perspective aberrations that would distort the average temperature values since the temperature of module areas in the foreground would weigh more than those in background;
- carry out an "automatic radiometric thermographic analysis of 1st level independent of the environmental conditions", being decontextualized from the climatic and environmental conditions present at the time of acquisition. In this analysis, DUBIO computes (for each acquired PV module):
  a) the maximum and average value of over-heating, in absolute differential terms with respect to the temperature that the PV module should have in case of normal operation, based on the climatic and environmental conditions detected during the monitoring campaign;

b) a radiometric map, with the location of the maximum temperature point inside the PV module, useful, for example, to check if it coincides or not with the junction box;
c) a classification of the defects among the most frequent ones, e.g.:
   - module is open circuited / not connected to the system;
   - short circuited (SC) or open sub-string:
     - Bypass diode SC;
     - Internal SC.
   - whole module is short circuited:
     - all bypass diodes SC;
     - wrong connection.
   - massive shunts caused by potential induced degradation and/or polarization;
   - shadowing effects / defect cell / delaminated cell;
   - broken cell / disconnected string interconnect;
   - artifact / partly shadowed (e.g. bird dropping, lightning protection rod);
   - sub-string with missing or open-circuit bypass diode.

Since the computed DIAGNOSES are stored in a multimedia database, the GUI of the platform has been designed for providing the user with different levels of visualization, either for an entire list of diagnoses (Fig. 5) or for a single diagnosis, at a more detailed scale (Fig. 6).

Moreover, for facilitating the consultation, different FILTERS and QUERIES can be defined by the user. In this contest, the COMPARATIVE QUERIES merit particular emphasis: these ones allow of comparing diagnoses performed on the same PV module through different CAMPAIGNS.

In fact, since the service encompasses not only the processing, but also the storage of both numerical and graphical data, in case of ACQUISITION CAMPAIGNS furtherly performed on the same PV plant, thanks to the independence of the diagnostic protocol from the environmental conditions, the overheating values, even occurring in different seasons are effectively comparable.

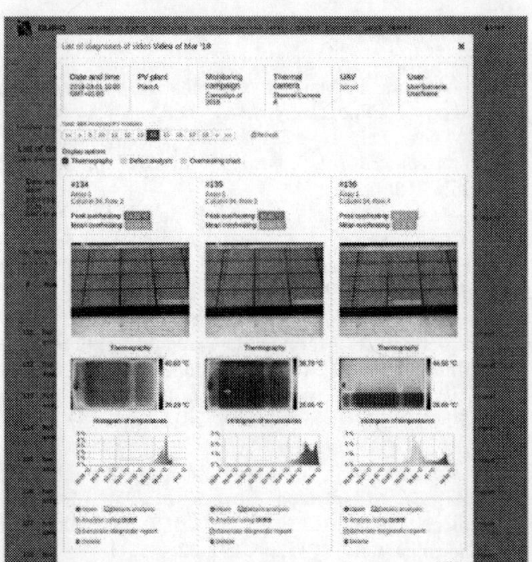

Figure 5: List of diagnoses GUI: panoramic visualization.

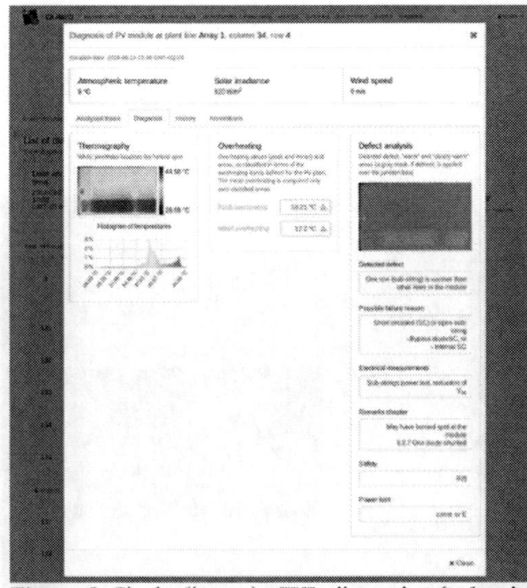

Figure 6: Single diagnosis GUI: diagnosis tab showing the rectified thermography, the overheating values and the defect analysis.

Therefore, the user can exploit a COMPARATIVE QUERY for evidencing those PV modules characterized by a deterioration (i.e., differences between the overheating that were measured in different CAMPAIGNS) higher than a certain threshold.

Similarly, useful deductions about the ageing of each PV module can be extrapolated by useful diagrams provided by the HISTORY panel, which depicts the trends of the recorded overheating (Fig. 7).

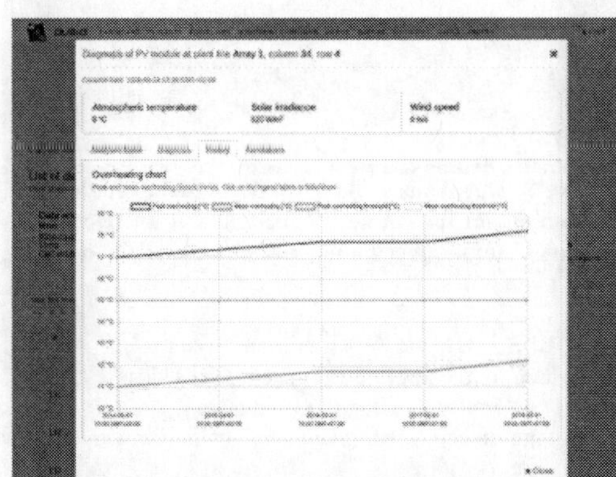

Figure 7: Single diagnosis GUI: history tab, showing the trend recorded by the overheating temperatures in five different campaigns.

Besides the above mentioned 1st level analysis, that the system accomplishes by default over all the acquired PV modules, two other analyses of 2nd level, respectively manual and automatic, can be carried out on request by the user, over these PV modules which need deeper investigation. More in detail, these ones implement:

- a "details analysis" (Fig. 8), i.e., a conventional manual thermographic analysis based on histograms, maximum, average and minimum values of temperatures for ROIs freely identified by the user consisting of:

a) individual points;
b) points of segments;
c) points of polygonal, rectangular, ellipsoidal or circular areas;
d) ROI given by the combination of those previously described.

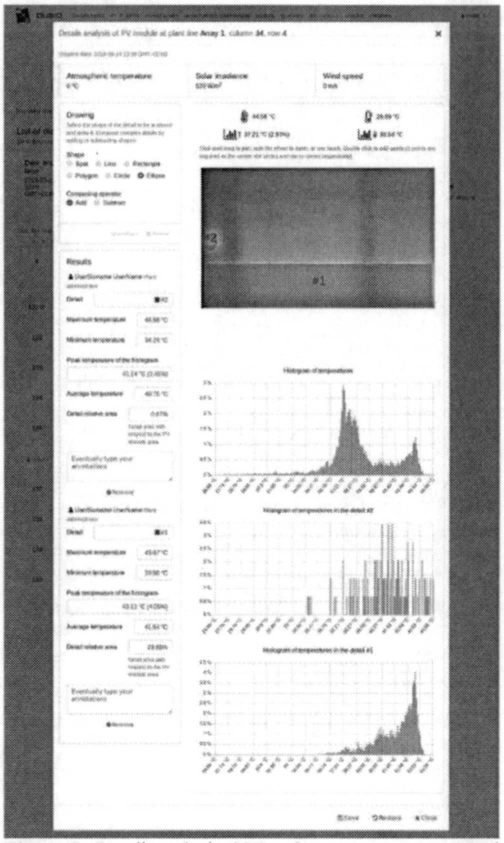

Figure 8: Detail analysis GUI: relevant temperatures and histograms for the whole PV module and for two details (in this example, rectangular and elliptic) defined by the user.

- "DISS – DIagnostics for Solar Systems", a protocol which, in a complete automatism provides:
  a) an analysis of each individual cell;
  b) an analysis of groups of cells having similar temperatures;
  c) a chromatic map with the areas classified according to the punctual overheating, basing on threshold values defined by the user;
  d) tables and diagrams classifying in percentage terms the results of a) and c);
  e) a "health dashboard", which supports the user in writing his prescriptions and scheduling the next inspection.

Finally, the platform allows the possibility of generating printable reports for each one of the monitored PV modules.

4 CONCLUSIONS

This paper has presented DUBIO – Drones and CloUd Based Infrared Monitoring, a project jointly conducted by APIS and ENEL Green Power.

Goal of the project has been to develop a cloud platform for processing and analyzing infrared (IR) videos captured and uploaded by unmanned aerial vehicles (UAVs) patrolling large/very large PV plants. Accurate computer vision algorithms have been developed, able to automatically crop regions of interest (ROIs) corresponding to all the PV modules pictured in each IR video frame, and label them accordingly to the plant topology. The radiometric values associated to the pixels bounded by the so extracted ROIs, are therefore used as input for suitably tuned diagnostic protocols which determine accurate indexes related to the health of the PV modules. These indexes are independent from the environmental conditions, meaning that different diagnoses, obtained over IR data acquired in different moments, can be compared to achieve useful information about the ageing trend of each single PV module.

The platform provides also an automatic defect classifier, and embeds tools for deeper analyses, to be eventually performed on demand over modules showing criticalities. Moreover, all the graphical and numerical data are stored in a data base that the user can manage, thanks to an easy-to-use graphical user interface.

ENEL Green Power is currently testing the platform, and once this phase will be completed (expected data: end of October 2018), a release of the system, designed as a pay-for-use (only if satisfied) service will be installed on the cloud and made available at the URL www.dubio.cloud. This service will allow O&M agencies to upload their radiometric IR videos for being processed and getting (with a very high level of automatism) accurate diagnoses and significant information, useful to monitor the ageing trend of each module of the plant.

Moreover, APIS is going to build a network of UAVs operators, enabled to adopt thermal cameras, for extending the possibility of exploiting DUBIO even to those technicians not equipped with UAV and thermal camera, offering them a key-in-the-hand service.

REFERENCES

[1] S. Vergura, F. Marino, "Quantitative and Computer Aided Thermography-based Diagnostics for PV Devices: Part I – Framework", IEEE Journal of Photovoltaics Vol. 7, Issue 3, May 2017, pp 822-827, ISSN 2156-3381.

[2] S. Vergura, M. Colaprico, M. F. de Ruvo, F. Marino, "A Quantitative and Computer Aided Thermography-based Diagnostics for PV Devices: Part II – Platform and Results", IEEE Journal of Photovoltaics, Vol. 7, Issue 1, January 2017, pp 237-243, ISSN 2156-3381.

[3] S. Vergura, F. Marino, M. Carpentieri, "Processing Infrared Image of PV Modules for Defects Classification", Proceedings of ICRERA 2015, 22-25/11/2015, Palermo, Italy, 2015, pp. 1337-1341, ISBN: 978-1-4799-9981-1.

[4] S. Vergura, M. F. de Ruvo, F. Marino, "A GUI Based Analysis of Infrared Images of PV Modules", Proceedings of IEEE-ICCEP 2015, 16-18/06/2015, Taormina, Italy, 2015, pp. 622-626.

[5] S. Vergura, F. Marino "A Diagnostic workflow and software platform for PV modules", Renewable Energy & Power Quality Journal, No.12, April 2014, pp. 661-664, ISSN 2172-038X.

SOLAR MODULE INSPECTION DRONE

Niklas Treutner*, Sascha Stübing*, Sebastian Hellwig*, Beate Meffert*, Michael Menz#, Jens Killat#
* Institut für Informatik, Humboldt-Universität zu Berlin
Unter den Linden 6, D-10099 Berlin
greateyes GmbH
Justus-von-Liebig-Straße 2, D-12489 Berlin

ABSTRACT: Electroluminescence (EL) is used commonly to detect defects in photovoltaic cells and modules, both during production and in the field. Several different sensors have been proposed to capture the weak EL radiation. We present a novel approach to outdoor inspection of PV modules using an sCMOS sensor that is attached to a drone. Semi-autonomous navigation of the drone improves the speed of the measurement and automatic defect analysis reduces the effort necessary for large-scale evaluation, e.g. entire solar parks. Our field tests show that sufficiently stable and sharp recordings can be made using pre-processing on a FPGA on the drone. This allows for new applications of EL measurements both in academia and business.
Keywords: Electroluminescence, Defects, Monitoring

1 INTRODUCTION

Electroluminescence (EL) is used commonly to detect defects in photovoltaic cells and modules, both during production and in the field. Several different sensors have been proposed to capture the weak EL radiation. We present a novel approach to outdoor inspection of PV modules using an sCMOS sensor that is attached to a semi-autonomous drone.

Common methods to inspect deployed photovoltaic plants using EL include the use of cherry pickers or ladders and tripods, depending on the available money and the number of modules [8]. PV modules can also be demounted in order to inspect them in a mobile lab. Many defects of PV modules can be detected using drone-based thermal imaging or close range InGaAs sensors. However, all of these methods are either slow or cost-intensive or they fail to detect micro cracks and other small defects in the inspected PV modules.

The Solar Module Inspection Drone (SMID) aims to close this gap: It allows a large-scale and affordable drone-based inspection of PV modules while providing a high resolution and photometric sensitivity for defect detection. This is made possible by using and combining several innovations: The sCMOS sensor allows for fast acquisition of the images, an integrated FPGA performs in-flight pre-processing and powerful algorithms control the flight-path of the drone based on the position of the drone and the transmitted images. The images are stored for the automated inspection: Different kinds of defects, including micro cracks, are detected by a set of specialized algorithms.

2 SYSTEM DESIGN

2.1 Electroluminescence

The weak emission of photons of electrical powered photovoltaic cells is called electroluminescence. Sensors with high sensibility can gather this electroluminescence. Defects in the cell structure can be seen in those electroluminescence images.

2.2 sCMOS Sensor

The sCMOS camera used in this project consists of a silicon based CMOS sensor and a Xilinx-Zynq-FPGA. Unlike the normally used CCD sensors in EL measurements the frame rate of a CMOS sensor is much higher and therefore preferable for drone applications. Higher frame rates are required because unavoidable influences make it necessary to deal with blurred or underexposed images. Furthermore, the high frame rate enables the autonomous navigation of the drone. The resolution of the sensor is 2560 x 2160 pixels and its intensity range is 16 bit. A Banana Pi is connected to the FPGA in order to facilitate data transmission to ground control.

Figure 1: AiD-MC8 octocopter during a field test.

2.3 Drone

Drones have been used for PV-inspection before: Infrared (IR)-based sensor systems on-board a drone are commonly used to check the performance of modules and cells. And even EL-inspection has been suggested and tried before, using both CCD [8] and InGaAs Sensors [9].

The drone being used is the AiD-MC8 octocopter by AIDrones. It is a professional drone system with good flight capabilities. The payload for sensor modules is up to 1.5 kg, the flight time with sensor module is approximately 15 minutes, which is enough to inspect at least one string of PV modules. The drone comes with gyroscopes, accelerometers, compass and GPS which supports the autonomous navigation. These sensors are used to stabilize the flight. The drone can be controlled with a simple flight control framework, allowing the use of pre-planned flight plans. The drone can be controlled by a remote controller (via radio) or by software (via Wi-Fi) over MAVLink protocol. For autonomous flight the latter is used.

In figure 1 the drone is shown right before take off during a field test. German law requires positioning lights when flying at night.

3 IMAGE PROCESSING

3.1 Data Transmission

Images recorded by the sCMOS camera are

transmitted to the ground control by Wi-Fi (5 GHz band), using a Banana Pi that is connected to the FPGA. It receives the pre-processed images and transmits them to ground control with a transmission rate of up to 100 Mbit/s. This rate dropped to 30 Mbit/s when using the drone in more realistic environments outside.

Transmission for navigation

A steady stream of images is required in order to perform the autonomous module detection and positioning (3.2). The size of each full-resolution-image is approximately 88 Mbit, which means that an image needs nearly 3 seconds for the transmission in realistic environments. This is insufficient for the navigation task, so in order to transmit more images per second to ground control, the FPGA decimates each image to a resolution of 640 x 539 pixels, which leads to an image size of about 5.5 Mbit. In this case 4 to 8 images per second can be transmitted, depending on the exposure time of the camera.

In figure 2 the different types of data transmission are shown. The low resolution images (thin, light blue arrows) are transmitted to ground control, where the data is analysed in order to compute new control commands. These are verified by the APM Planner software and sent to the drone via MAVLink protocol.

Transmission for defect detection

For detection of small defects like finger defects or micro cracks, full resolution images are required. As stated previously, the low transmission rate prevents a continuous transmission of full resolution images. That's why the transmission rate for full resolution images is reduced to one image every 5-20 seconds depending on exposure time and available bandwidth. With this restriction it has to be ensured that the image is sharp and the module is fully visible so that the transmission isn't wasted. That is achieved by performing a sharpness analysis and primitive module detection on-board the drone for each full resolution image. The full resolution image with the best results within the target time frame is transmitted to the ground station where the defect detection is performed (3.3).

Figure 2 shows how the full resolution image (thick, dark blue arrows) is transmitted to ground control where the module is scanned for defects by image processing algorithms. The results are stored in a database.

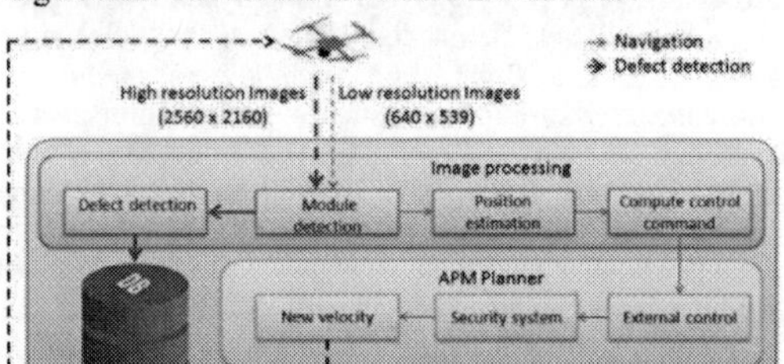

Figure 2: Different data transmissions between drone and ground control.

3.2 Autonomous module detection and positioning

The low resolution image is used in order to perform autonomous module detection and computation of the drone position relatively to the module.

Module detection

Figure 3 shows steps of the module detection. The left image shows the low resolution image as it is recorded by the camera. Due to the weak EL radiation and short exposure time the sensor noise has a high impact on the intensity of the image. In a first step the sensor noise is reduced by using information of the sensor which is stored in first and last rows of the image. These pixels are fully covered and their intensities are directly linked to the gain of each image column, thus a correction by these values reduces errors. After subtracting each image column by its respective intensity a median filter is applied to remove "hot pixel" image noise. The result can be seen in the middle of figure 3. Based on the histogram of the image a threshold is computed which separates the module from the background. A contour finding algorithm detects the outer borders of the module as shown in figure 3 (right). A similar module detection algorithm is used on the full resolution image as a first step to the error detection.

Figure 3: Left: low resolution image transmitted by the drone. Middle: pre-processed image. Right: detected module.

Positioning

After extracting the position of the module the next task is to compute the distance between drone (camera) and the module centre. Therefore some further information is required.

- The Focal length f of the camera (objective) in cm
- The Width b of a pixel in cm
- The Width m of the module in cm (known before inspection)
- The Width $\bar{m}$ of the module in pixel (known after module detection)
- The perspective projection (figure 4) allows to compute the distance of the module from the sensor:

1. From: $\frac{\bar{m} \cdot b}{f} = \frac{m}{z}$ follows: $z = \frac{m \cdot f}{\bar{m} \cdot b}$

With a focal length of 5 cm, a pixel width of 0.00065 cm, a module width of 50 cm and a module width in the image of 507 pixels this leads to:

2. $\frac{50\ cm \cdot 5\ cm}{507 \cdot 0.00065\ cm} = 758\ cm$

In this case the distance between camera and module was 758 cm.

The goal for the autonomous navigation is to position the drone at a point where the module is completely visible within the image and fills it completely. The drone should hold that position so that several images can be recorded to assure a sharp image for defect analysis was transmitted. Once the optimal flight altitude is determined, all further navigation commands can be reduced to move forward, backwards, to the left or to the right. Rotation proved to be unnecessary, once the drone is positioned correctly. The optimal flight altitude depends on camera focus and the angle of the camera.

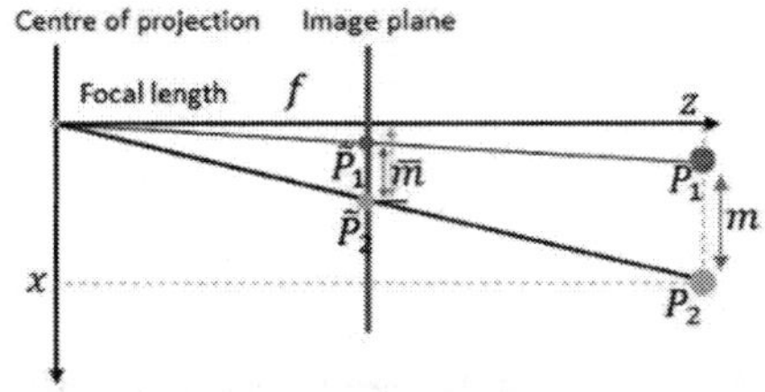

Figure 4: Relation of m and $\bar{m}$ in perspective projection

To centre the module in the field of view of the camera, the centre of the detected module (image coordinates) is compared to the coordinates of the image centre. The difference in horizontal and vertical direction in pixels represents the translation that the drone has to perform in order to centre the module in the field of view. Therefore the translation in image coordinates has to be converted to translation values in world coordinates as follows:

Given:

- Translation in image coordinates $\Delta x, \Delta y$
- Pixel size b
- Focal length f
- Distance z as computed in formula 2

Desired:

- Translation in world coordinates $\Delta X, \Delta Y$

3. $\Delta X = \frac{(\Delta x \cdot b)}{f} \cdot z$ And $\Delta Y = \frac{(\Delta y \cdot b)}{f} \cdot z$

With a focal length f of 5 cm, a pixel width b of 0.00065 cm, z = 758 cm, Δx of 145 pixels and Δy of 227 pixels it leads to:

4. $\Delta X = \frac{(145 \cdot 0.00065\ cm)}{5\ cm} \cdot 758\ cm = 14.2\ cm$

And

5. $\Delta Y = \frac{(227 \cdot 0.00065\ cm)}{5\ cm} \cdot 758\ cm = 22.3\ cm$

As pictured in figure 2, the results of the image processing, in this case $\Delta X = 14.2\ cm$ and $\Delta Y = 22.3\ cm$, are verified in the APM Planner and converted to control commands in form of motor control commands that are sent to the drone via MAVLink.

3.3 Error detection

Different errors can be identified in PV-modules using Electroluminescence. Among those are cracks, micro cracks, finger defects, border decline and contaminations. Since most of those defects are only visible in the active areas of the module, we perform a cell-detection and busbar-localisation beforehand, in order to only search for defects in relevant areas of the module. Then, the defects are detected sequentially and each detection is incorporated in the next detection step in order to avoid double detections.

Cell detection

We assume that the number and alignment of the cells is known. A grid is assumed, but other structures would be possible, too. Therefore we can estimate the coordinate of each cell after we have calculated the position of the module (3.2). These coordinates are then fine-tuned, based on the assumption that the areas between the cells usually appear darker in the image: The centre of the groove between cells is calculated and used as a separation between cells.

In order to avoid detecting multiple instances of the same defect, the location of various structures are stored in a mask, so that these areas will not be analysed in further steps.

Busbar localisation

The location of the busbars can be detected by applying a Hough Transformation to an inversely thresholded image and selecting the best candidates within a certain range. The angles of the candidates are required to be close to each other. The number of busbars n_{Bus} is usually known, so the selection of correct candidates can be simplified by selecting the n_{Bus} candidates. The busbars are then added to the mask.

Crack detection

Reasonably precise crack detection can be achieved by using a straightforward thresholding-approach. If pixels below the threshold can be connected and grouped to a cluster above a certain size, the entire cluster is assigned to be a crack. Smaller patches or single pixels below the threshold are rejected. Good ways to compute the intensity- and size-thresholds are: relative to the mean, median (intensity) and relative to the cell (size).

Border decline

If the intensity increases steadily from the edge of the cell towards the centre, this can be attributed to border decline. In order to prevent misclassification of small structures as border decline, we require for this steady increase to be detectable in a large section of the cell, since border decline usually affects not only a region, but the entire length of the border of the cell.

Micro cracks

Micro cracks are among the most difficult defects to detect in an EL-image. Several approaches have been published over the last years. While some studies focus more on different methods for signal acquisition [2][4], others present techniques to localize defects in EL-images using computer vision [3][5][6].

We employ a two-fold strategy. First we apply different filters on the image in order to separate the micro cracks from the remaining structure. This can be done with simple Laplace-filters and also more sophisticated anisotropic diffusion [6].

The second step uses this filtered image and tries to reduce implausible candidates. This is done by using knowledge about micro cracks: While usually very thin, they are long and can even cross the entire cell. They also tend to start at the edge of the cell. Following these assumptions, we scan for image patches with very high filter-responses, especially among the border of the cell or near the busbars. These points are then considered to be the start- or end-point of a micro crack. A modified dynamic-programming algorithm follows possible ridges in the neighbouring pixels in order to traverse the entire micro crack. All possible paths are compared by their length, accumulated filter-responses and shape. The most likely variants are stored. Further validation allows for combination of different parts of larger micro cracks, once all candidates have been evaluated. Similar results can also be achieved by using graph-cut algorithms instead of dynamic programming.

Results

Results depend on a variety of parameters. Different types of modules require different sets of parameters, both for the defect detection and the data-collection process (exposure time). We tested the algorithms on a variety of data samples; however, we can't provide statistical analysis yet due insufficient annotated images or similar ground-truth.

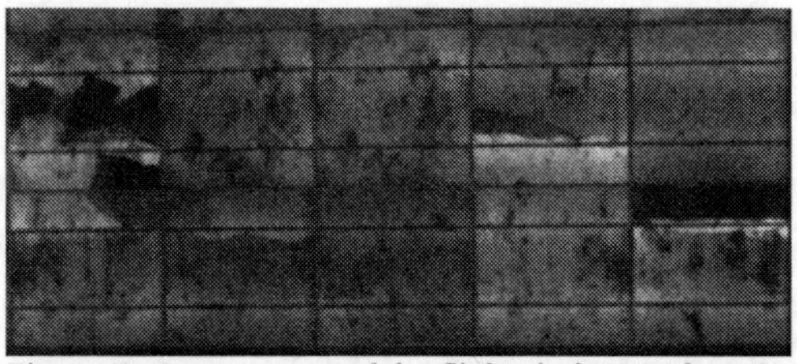

Figure 5: Image captured in flight during testing. Red lines: cell borders

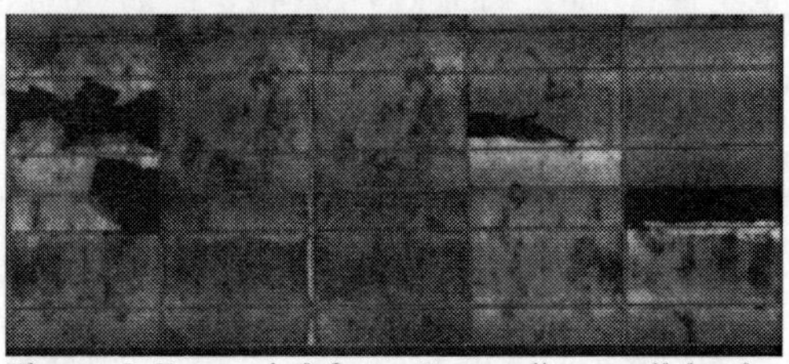

Figure 6: Detected defects. Orange lines: cell borders. Blue: Cracks/Contamination. Green: Border decline. Red (inside cell): micro cracks

4 FIELD TEST

4.1 Location

After several early tests while working on the coordination of drone, camera and software, we performed a larger field test on the test grounds of the ZAE Bayern in Arzberg (Bavaria, Germany). The ZAE Bayern operates several different types of PV-modules in 25 strings for various tests.

4.2 Collected data and key findings

We acquired over 75.000 images during both manual and automatic flights. Among the varied parameters during testing were object distance, exposure time and navigation-modes. However, some restrictions had to be made: The pilot had to navigate the drone into the initial position and due to safety concerns the autopilot was not activated for close-ups (below 3m distance).

The key findings were:

The quality of the detections did not improve with reduced distance. This was probably due to more motion blur. Also, the stability of the drone decreased with increased proximity to the ground. An initial assumption that lower altitude would improve the overall quality could not be confirmed.

Lower exposure time did usually improve the sharpness of the image due to reduced motion blur. The lower light yield seemed to play less of a role for the performance of the algorithms than the motion blur. This was especially true for very efficient modules.

5 CONCLUSION

We have presented a semi-automatic drone-based measurement and defect-detection system, that uses an sCMOS-sensor to capture EL-radiation. While a human pilot due to legal reasons still needs to supervise each flight, the collection of data can be semi-automated. A next step would be to provide a convenient input mechanism for plans of PV-plants.

The automatic detection provides good results, although results on images taken in a Lab still yield better precision and recall. Annotated data samples would be helpful to better compare results.

ACKNOWLEDGEMENT

This project was partially funded by the European Regional Development Fund (EFRE).

EUROPEAN UNION
Investing in our Future
European Regional
Development Fund

6 REFERENCES

[1] E. Sovetkin, A. Steland, Automatic Processing and Solar Cell Detection in Photovoltaic Electroluminescence Images, arXiv:1807.10820 [cs.CV]

[2] M. Abdelhamid, S. Rajdeep, M. Omar, Review of microcrack detection techniques for silicon solar cells, IEEE Journal of Photovoltaics, Vol. 4 (2014), 514-524

[3] S.A. Anwar, M.Z. Abdullah, Micro-crack detection of multicrystalline solar cells featuring an improved anisotropic diffusion filter and image segmentation technique, EURASIP Journal on Image and Video Processing (2014), 1-17

([4] Yih-Chih, C., J.Z. Liu, Y.T. Liang, Micro crack detection of multi-crystalline silicon solar wafer using machine vision techniques, Sensor Review, Vol. 31 (2011), 154-165 --- QUELLE NICHT ONLINE)

[4] M. Israil, S.A. Anwar, M.Z. Abdullah, Automatic detection of micro-crack in solar wafers and cells: a review, Transactions of the Institute of Measurement and Control, Vol. 35 (2013), 606-618

[5] Z. Mahdavipour, M.Z. Abdullah, Micro-crack Detection of Polycrystalline Silicon Solar Wafer, IETE Technical Review (2015)

[6] D.M. Tsai, C.C. Chang, S.M. Chao, Micro-crack inspection in heterogeneously textured solar wafers using anisotropic diffusion, Image and Vision Computing, Vol. 28 (2010), 491-501

[7] AIDrones website: https://www.aidrones.de/english/drone-systems/octocopter-mc8/ viewed on 10.09.2018.

[8] S. Koch, T. Weber, C. Sobottka, A. Fladung, P. Clemens, J. Berghold, Outdoor electroluminescence imaging of crystalline photovoltaic modules: comparative study between manual ground-level inspections and drone-based aerial surveys, 32nd European Photovoltaic Solar Energy Conf. and Exhibition (2016)

BETTER FAULT DETECTION AND DIAGNOSIS WITH ARTIFICIAL INTELLIGENCE: METHODS, EXAMPLES AND BUSINESS CASES

Achim Woyte, Babacar Sarr, Karel De Brabandere, Mauricio Richter, Werner Coppye
3E sa
Quai à la chaux 6, B-1000 Brussels, Belgium
Tel. +32 22 17 58 68, Fax +32 22 19 79 89, E-mail: info@3E.eu

ABSTRACT: Based on a sample of plants monitored by 3E, we see more than 20% of professional photovoltaic plants performing more than 10% below expectation. The main cause for these cases of underperformance are lack of time and experience. Especially not so obvious flaws and failures often persist for weeks to months before they are noticed and fixed eventually. This paper shows how artificial intelligence can be used for PV operations and maintenance to detect faults automatically and formulate recommendations on their most probable root causes. It shows several examples, namely, performance and loss analysis with limit checking and degradation analysis with trend checking. In conclusion, with the methods presented here, PV plant operators and asset managers can identify faults early and draw conclusions on the underlying root causes. By estimating the losses and comparing them to the losses as expected from a simplified model, we can associate them with the different failure events and compute the gains to be expected when the fault had been detected and corrected early. 3E is currently offering the methods presented here under the name PV Health Scan.
Keywords: Big data, Data analytics, Data mining, Monitoring, Operation and maintenance

1 INTRODUCTION

1.1 Problem Description

Professional photovoltaic (PV) plants today are virtually always monitored. Nevertheless, based on a sample of plants monitored by 3E, we see more than 20% of such plants performing more than 10% below expectation. Based on interviews with PV plant operators, the main cause for these cases of underperformance are lack of time and experience. Inverter outages or larger disturbances on plant level are commonly identified and fixed fast. But less obvious flaws and failures often persist for weeks to months before they are noticed and fixed eventually. Many of these flaws can be detected fast, easily and reliably through data analytics.

1.2 State of the Art and Current Trend

Methods for automatic fault detection from monitoring data in PV plants have been explored, published and tested for many years (see e.g.,[1]–[3]). A literature review and presentation of a general

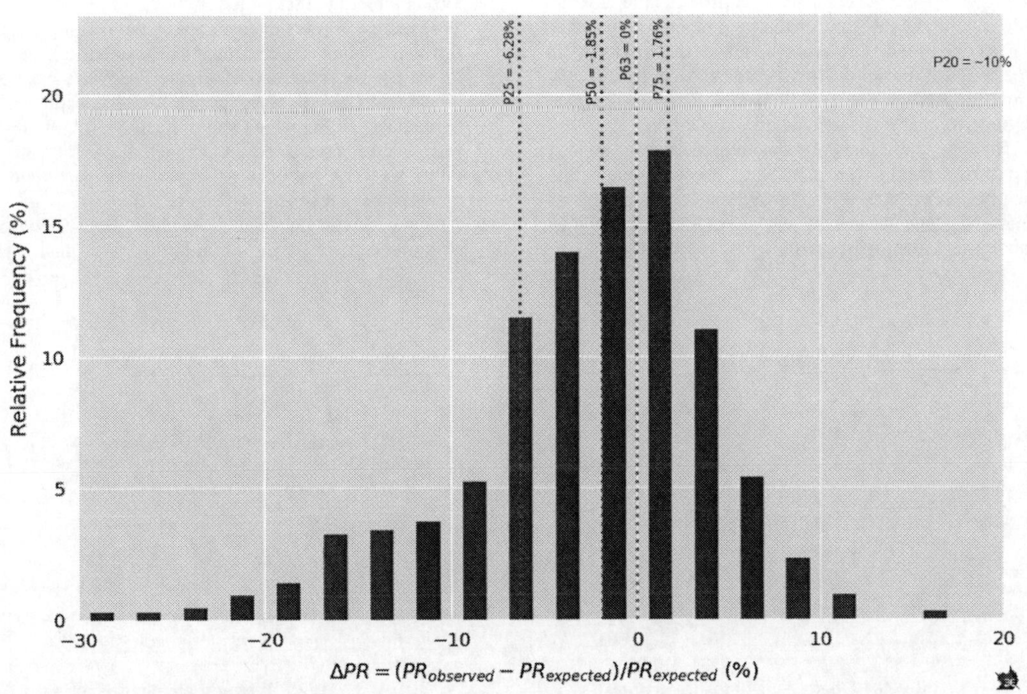

Figure 1: Statistics of performance ratio (PR) as observed compared to the expected PR from a simple model; data set: 644 PV plants as monitored by 3Es monitoring service SynaptiQ in 2016.

methodology based on linear regression was published by the IEA PVPS Task 13 in 2014 [4]. Nevertheless, commercially available monitoring systems in practice did not offer corresponding functions for fault detection and diagnosis in the past.

Today, cloud computing and the renewed attention for artificial intelligence and data analytics have recently led to an increased interest in such functions from the PV asset management and O&M companies. Accordingly, several monitoring providers have started offering or announced fault detection and diagnosis tools of different kinds.

1.3 Outline

Here we will situate the different methods and approaches in the field of data analytics and artificial intelligence. This paper shows how artificial intelligence can be used for PV operations and maintenance (O&M) to detect faults automatically and formulate recommendations on their most probable root causes. It situates automatic fault detection and diagnosis within the field of big data analytics and describes the different analysis steps and use cases (fault detection, fault diagnosis, fault prediction, maintenance optimisation). We present some of the most interesting methods currently applied (feature engineering, limit and trend checking of performance and loss features, time-series based degradation forecasting) and illustrate them with real-world examples. We also discuss the business value of early fault diagnosis for the examples.

2 WHAT ARTIFICIAL INTELLIGENCE CAN DO

Figure 1 shows data from 644 commercial PV plants monitored by 3E. We compare performance ratio (PR) as realized with the PR from a simplified model and weather data as measured. On average, the plants perform as expected. But the bottom tier 25% perform at least -6.3% below expectation and approximately 20% of all plants (P20) perform at least -10% below expectation.

Notably, these plants are monitored and supervised by professional teams. The reason for this underperformance is often not ignorance but a lack of time to dig into the details and a lack of experience in isolating and identifying possible root causes fast.

Automatic fault detection can support the operator to detect the not so obvious faults and locate the fault to a certain inverter, array or string. Automatic diagnosis can suggest a subset of probable root causes. Predictive analytics can estimate whether and when degradation may lead to a warranty violation. In conclusion, artificial intelligence will not replace PV plant operators for a long time, but it will help them to be more effective and bring also many of the low-performing plants up to the top tier.

3 EXAMPLES

3.1 Performance and Loss Analysis with Limit Checking

For checking the operational performance over different energy conversion steps, we define a performance loss ratio per step. The performance loss ratio is computed for a given time span, e.g., a day up to several months. It is the specific loss over the energy conversion step divided by the reference yield. We call these losses normalized to reference yield *performance loss ratio*, in analogy with the definition of performance ratio.

In practice, we compare the performance loss ratios from measurements to the model-based performance loss ratios and thresholds. The model is fed with measured values of plane-of-array irradiance and ambient temperature. The model parameters can be set from data sheet parameters of the devices in the PV plant or identified from measurements from the plant in a healthy state. Accordingly, adequate limits can be derived either from tolerances on the data sheet parameters or from choosing percentiles from the healthy plant. Both the model-based performance loss ratios and their limit values vary depending on the PV plant and the weather during the evaluation period.

Figure 2 illustrates this application of limit checking for a PV plant in Belgium. The current-related array losses (*Array (current)*) in Figure 2a by far exceed the threshold. This triggered a maintenance action end of September, during which approximately 20% of all string fuses had to be replaced. In October, all performance loss ratios are back within their expected ranges, yielding a much higher PR of 82.9% (Figure 2b).

Other typical failure modes are disconnected strings/subarrays with multi-string or small central

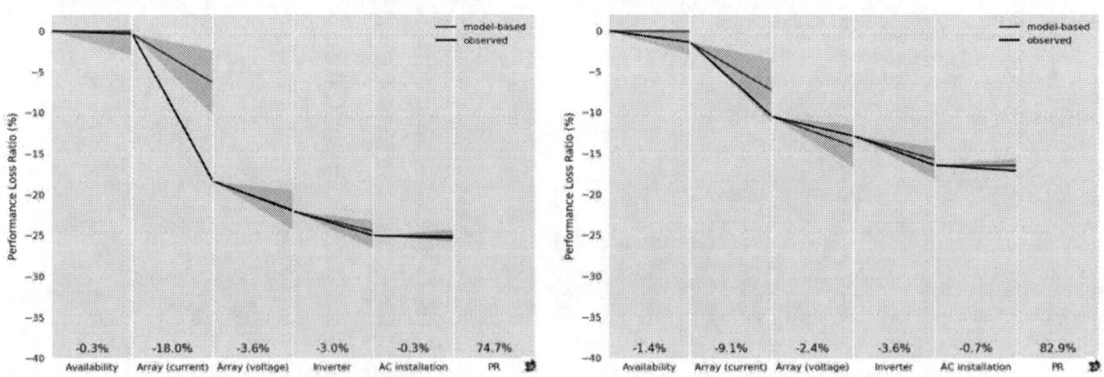

(a) 9/2017, before maintenance action b) 10/2017, after maintenance action

Figure 2: Example of limit checking for the energy conversion process in a 4 MW PV plant; performance loss ratios per conversion step are compared to the model for each conversion step inverters, fast intermittent inverter output disconnections and limitations as they can occur due to grid disturbances or overheated inverters, and different modes of degradation (e.g., caused by soiling, hotspots or potential induced degradation).

3.2 Estimating Business Value from Limit Checking

In order to illustrate the business case of automatic fault detection and diagnosis, we can compute the losses due to the deviation from the expected values, and finally the value of this energy lost. Here we list results from a 300 kW PV rooftop plant in a sub-tropical climate, where several typical failure modes occurred over the year of 2017, and estimate the potential gains:

- Reduced array capture losses due to soiling by 3%, corresponding to an annual yield improvement of 13.8 MWh,
- Reduced array capture losses due to intermittent disconnections of 1 out of 8 inverters by 16.6%, corresponding to an annual yield improvement of 9.5 MWh,
- Reduced array capture losses by 3% by faster detecting when inverters are unavailable, corresponding to an annual yield improvement of 13.8 MWh,
- Reduced array capture losses by 0.73% by faster detecting disconnected strings, corresponding to an annual yield improvement of 3.3 MWh.

3.3 Long-term Degradation with Trend Analysis

Figure 3 shows the deseasonalized trend of current-related array losses for a 2 MW plant with two inverters in Belgium. Both arrays/inverters degrade more or less equally with, on average, approximately 2.4% per year. The degradation appears to be strongest in the beginning and then gradually decreases.

In this case, a diagnosis module can suggest different possible root causes that can explain gradually decreasing PV array currents for the entire plant, as implemented into a knowledge base, e.g., optical degradation, soiling or cell breakage. In this case, the PV plant operator has confirmed to 3E that the plant suffers from heterogeneous soiling by bird droppings, which appear to have caused irreversible damage through hot spots.

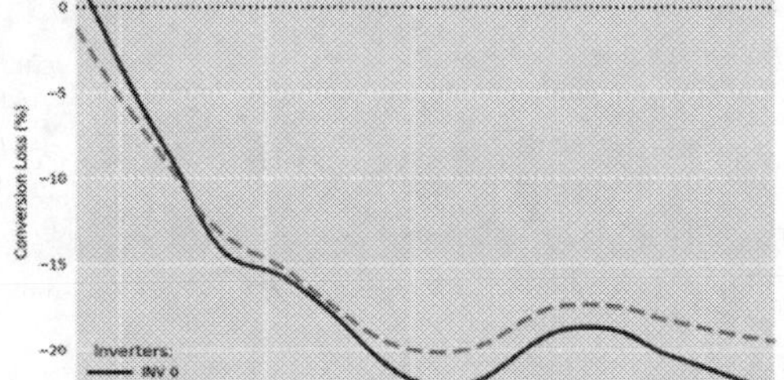

Figure 3: Long-term desesonalized trend of current-based PV array losses for a PV plant with two inverters

In this case, the degradation is clearly identified. However, from the data only, the trend analysis cannot identify its root cause. Nevertheless, such trend analysis is particularly useful when the end of a warranty period is approached.

4 CONCLUSION AND OUTLOOK

In conclusion, with the methods presented here, PV plant operators and asset managers can identify faults early and draw some conclusions on the underlying root causes. By estimating the losses and comparing them to the losses as expected from a simplified model, we can associate them with the different failure events and compute the gains to be expected when the fault had been detected and corrected early.

3E is currently offering the methods presented here under the name PV Health Scan. The PV Health Scan is available as part of a technical due diligence, especially when PV plants are for sale or when retrofitting is considered. Moreover, it has been integrated into 3E's PV monitoring and reporting platform SynaptiQ where it is available through the monitoring tool. 3E is currently further exploring data analytics for benchmarking and diagnosis.

5 ACKNOWLEDGEMENTS

The project FleetAID that led to these results was subsidized by the Brussels Capital Region (het Brussels Hoofdstedelijk Gewest – Innpoviris) under the contract No. BHG/2018-SHAPE-12b.

REFERENCES

[1] N.M. Pearsall and B. Atanasiu, 'Assessment of PV System Monitoring Requirements by Consideration of Failure Mode Probability', in *24th EUPVSEC*, Hamburg, Germany, 2009, pp. 3896–3903.

[2] T. Oozeki, T. Izawa, K. Otani, and K. Kurokawa, 'An evaluation method of PV systems', *Solar Energy Materials and Solar Cells*, vol. 75, no. 3, pp. 687–695, 2003.

[3] S. Stettler *et al.*, 'Failure detection routine for grid-connected PV systems as part of the PVSAT-2 project', in *20th EUPVSEC*, Barcelona, Spain, 2005, pp. 2490–2493.

[4] A. Woyte *et al.*, 'Analytical Monitoring of Grid-connected Photovoltaic Systems - Good Practice for Monitoring and Performance Analysis', IEA PVPS, Report IEA-PVPS T13-03: 2014, Mar. 2014.

ADVANCED DIAGNOSTIC APPROACH OF FAILURES FOR GRID-CONNECTED PHOTOVOLTAIC (PV) SYSTEMS

Andreas Livera[1,*], Marios Theristis[1], George Makrides[1], Juergen Sutterlueti[2] and George E. Georghiou[1]
[1]PV Technology Laboratory, FOSS Research Centre for Sustainable Energy,
Department of Electrical and Computer Engineering, University of Cyprus
Nicosia, 1678, Cyprus
[2]Gantner Instruments GmbH, Montafonerstraße 4, 6780 Schruns, Austria
*corresponding author email: livera.andreas@ucy.ac.cy

ABSTRACT: Real-time identification of failures in photovoltaic (PV) systems is crucial for achieving reactive maintenance schemes that, in turn, will increase the system reliability and guarantee the lifetime output. Following this line, failure detection routines (FDRs) that operate on acquired data-sets of grid-connected PV systems were developed to diagnose the occurrence of failures. The developed FDRs comprise of a failure detection and a classification stage. The detection stage was based on the comparison between the measured and predicted DC power production against set threshold levels (TL). The classification stage was based on data-driven algorithms, which were used to post-process the detected failure patterns through the application of the developed decision trees (DT), k-nearest neighbours (k-NN), support vector machine (SVM) and fuzzy inference systems (FIS). The experimental results showed that the FDRs were capable of detecting all the different types of failures (open- and short-circuited PV module, inverter shutdown, shorted bypass diode and partial shading) that were introduced to the test-bench PV system. Finally, amongst the investigated models, the k-NN model achieved the highest average true negative rate of 91%, when classifying each type of failure used for benchmarking.
Keywords: failure diagnosis, failure classification, reliability, photovoltaics.

1 INTRODUCTION

Over the last decade, online failure diagnosis (detection and classification) for photovoltaic (PV) systems has attracted a lot of attention as it is a fundamental task that can significantly improve and ensure the quality of operation of the systems having a positive impact on the investment cost [1].

Currently, the lifetime performance output of PV installations is guaranteed through monitoring systems, either via the inverter or other proprietary data-acquisition hardware and software dashboards [2]. The vast majority of the existing monitoring systems provide a simplistic view of the energy yield and Performance Ratio (PR) while the most sophisticated of these dashboards can decide upon fault conditions through online comparisons between the measured and predicted PV power production against set threshold levels (TL). For the predicted PV power production, different predictive models (based on parametric modelling and machine learning approaches) are used [3]-[4]. The set TL can be either static (i.e. predefined) or dynamic (i.e. there is feedback from the stream of measured values). The advantages of this signal threshold approach include the ability for real-time failure detection and the simplicity of implementation in a wide range of PV systems. On the other hand, the disadvantages include the requirement of a simulation model and the high false alarm rate.

As most of the reported studies in the field were developed in simulation and laboratory environments, the scope of this work is to develop a diagnosis technique based on field experiments to represent a real-world environment. Consequently, failure detection routines (FDRs) based on the signal threshold technique have been developed for the automated identification of commonly occurred failures (open- and short-circuited PV module, inverter shutdown, shorted bypass diode and partial shading) in grid-connected PV systems.

2 METHODOLOGY

2.1 Experimental setup

The proposed FDRs were developed and experimentally verified using actual measurements from a test-bench PV system installed at the outdoor test facility of the University of Cyprus (UCY) in Nicosia, Cyprus [5].

The test-bench PV system comprises of 5 poly-crystalline Silicon (poly-c Si) PV modules (see Fig. 1), each of nominal power 235 W_p as depicted by the manufacturer's datasheet. The system is connected in series to form a string at the input of a grid-connected inverter. It is installed in an open-field mounting arrangement due South, at the optimum annual energy yield angle for Cyprus of 27.5°.

Figure 1: Test-bench PV system used for the emulation of failures.

The PV system performance and the prevailing irradiance and meteorological conditions were recorded according to the requirements set by the IEC 61724 standard [6] and stored with the use of an advanced measurement platform. This platform comprises of solar irradiance (Kipp & Zonen CM21 pyranometer), temperature (PT 100) and electrical operation (current and voltage transducers, energy meter) sensors connected to a central data acquisition system that stores data every second. The measurements were also averaged and stored every one- and fifteen-minutes. Specifically, the irradiance and meteorological measurements include the incident global irradiance at the POA (G_{POA}), wind direction (W_a), wind speed (W_s) and ambient temperature (T_{amb}). The PV system's operational measurements include the module temperature (T_m), DC current (I_{DC}), voltage (V_{DC}) and power (P_{DC}) of the array at the maximum power point and AC power (P_{AC}).

Additionally, five types of failures (open- and short-circuited PV module, inverter shutdown, shorted bypass diode and partial shading) were emulated at the test-bench PV system and formed the basis of creating failure signatures for the training and testing phase of the developed routines.

2.2 Data quality routines (DQRs)

Initially, DQRs that operate on the acquired data-sets were developed in order to ensure data validity for performance assessment. The recorded data were thoroughly checked for consistency and gaps before any additional analysis was conducted. More specifically, the developed DQRs run algorithms for missing (or erroneous) data, duplicate records, outliers and outages along with data filtering and imputation techniques that can be applied to identify data anomalies and reconstruct the data-set. These periodic data checks are essential to ensure high data quality for the development of FDRs.

2.3 Failure signatures

The acquired data-sets required for the development of the FDRs included the fifteen-minute average irradiance, meteorological and PV operational measurements of G_{POA}, T_m, I_{DC}, V_{DC} and P_{DC}. These measurements were collected in both normal operating and emulated fault conditions. The fault conditions were introduced at different time periods and irradiance levels ranging from 400 to 1100 W/m^2 in order to characterise the effect of the failure on the main electrical parameters at the DC side. Accordingly, labelled data (indicating normal PV operation, open-circuit PV module failure, short-circuit PV module failure, inverter shutdown failure, bypass diode failure and partial shading) were also inserted to the acquired data-sets in order to characterise the operating conditions and failure signatures. This is necessary for the learning phase of the classification stage.

Finally, the acquired data-sets were split into 70:30% train and test set approach (both sets contained failure signatures). In particular, the train set contained 2083 data points; 2062 normal and 21 fault data points (7 failure signatures at irradiance intervals between 400-600 W/m^2, 600-800 W/m^2 and 800-1100 W/m^2). The test set contained 893 data points that included 884 normal and 9 fault conditions (3 failure signatures at irradiance intervals between 400-600 W/m^2, 600-800 W/m^2 and 800-1100 W/m^2).

2.4 Failure Detection Routines (FDRs)

The developed FDRs comprise of a failure detection and a classification stage and make use of the readily available field measurements. More specifically, the FDRs require as input features the acquired data-sets of the PV system, its technical details and specifications (total number of the modules in the string, string nominal power and maximum power temperature coefficient at standard test conditions (STC)). Therefore, the proposed approach does not require any additional hardware /software installation or extra labour cost.

2.4.1 Failure detection

The failure detection stage was based on a comparative assessment between the measured and predicted DC power produced. In particular, the measured DC power production of the PV system was compared to the predicted power production on a daily basis. A failure was detected when the absolute error (AE), defined as the absolute difference between the measured and predicted DC power, exceeded a predefined TL which incorporated the predictive model and power sensor uncertainties. The set TL is calculated by multiplying the power of the array at STC with the combined uncertainty.

For the predicted DC power produced, two modelling techniques were initially compared in order to find the most appropriate prediction model. The performance of the PV system was thus simulated using a parametric model [7] and a machine learning Feed Forward Neural Network (FFNN) model [8]. For this investigation, a one-year data-set containing fifteen-minute average measurements of the test-bench PV system was used. The prediction performance results showed that when the data was split into a 70:30% train and test set approach, the FFNN outperformed the parametric model since the predicted normalised root mean square error (nRMSE) was 1.06% for the FFNN while the parametric model achieved a nRMSE of 1.44% over the test set period.

On the contrary, when a smaller amount of data was used for the training process (10:90% train and test set approach), the parametric model achieved a power prediction nRMSE of 1.47%, while the FFNN achieved a nRMSE of 1.87% over the test set period. Therefore, the parametric model could predict the DC power of the array at the maximum power point (MPP) more accurately compared to the FFNN when a small amount of data (approximately 36 days) was available for the training process reflecting in this aspect real PV system scenarios. Furthermore, since the availability of PV operational data is sometimes scarce, the parametric model was selected for the prediction of the PV performance due to the lowest error exhibited using the smaller data-set.

Finally, by taking into consideration the uncertainties (2% relative error of the power measurements and 4% relative error associated to the power prediction model), the set TL was calculated to be 70.5 W.

2.4.2 Failure classification

For the classification stage, DT, k-NN, SVM and FIS models were constructed for post-processing the detected failure patterns [9]-[10]. The constructed supervised classification models were generated by employing a learning procedure based on data-set partition. In particular, the train set contained 2083 data points (2062 normal and 21 fault data points) and each instance in the set contained one target value (the labelled data) and several attributes (irradiance, meteorological and PV operational measurements). The test set contained 893 data points that included 884 normal and 9 fault conditions. After the training process, the goal of the constructed classification models was to predict target values of instances in the test set, for which only the attributes are known.

In order to evaluate the classification performance of each model, confusion matrices of a binary classifier were constructed (see Table I). The four outcomes of the binary classifier are defined as true positive (TP), true negative (TN), false positive (FP) and false negative (FN). The TN represents the fault data points correctly detected, while the FN represents the normal data points that were incorrectly detected as fault conditions. On the other hand, the TP represents the normal data points correctly detected, while the FP represents the fault data points that were incorrectly classified as normal operating points.

Table I: The confusion matrix of a binary classifier.

	Predicted normal points	**Predicted fault points**
Actual normal points	TP	FN
Actual fault points	FP	TN

To assess the efficiency of each model, the specificity performance metric was used [11]:

$$Specificity = \frac{TN}{TN + FP} \quad (1)$$

The specificity (also called the true negative rate) measures the proportion of fault points that were correctly classified as fault conditions.

3 RESULTS

3.1 Normal operation

Normal operation describes the expected operation of a well performing PV system without any fault conditions and the system is affected only by environmental conditions. During normal operation, the measured DC power exhibited similar behaviour to the predicted DC power, the AE was below the static TL and the FDRs detected the fault-free operation (see Fig.2).

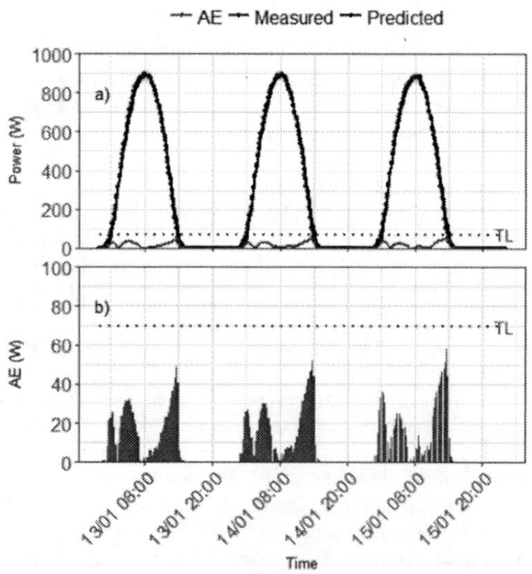

Figure 2: Measured against predicted DC power during normal operating conditions and AE.

For the following failure scenarios, the acquired data-sets contained 2976 data points of which only the 1% (~30 data points) were fault signatures. The classification models were trained using a data-set containing 2083 data points (2062 normal and 21 fault data points), and then presented with 893 data points that included 884 normal and 9 fault conditions.

3.2 Open-circuit PV module failure

The open-circuit PV module failure was emulated by open-circuiting a PV module in the string, while the measurement platform continued to function in order to capture the failure profile. The developed FDRs accurately detected the failure occurrence as the absolute difference between the measured and predicted DC power of the array exceeded the set TL (see Fig. 3).

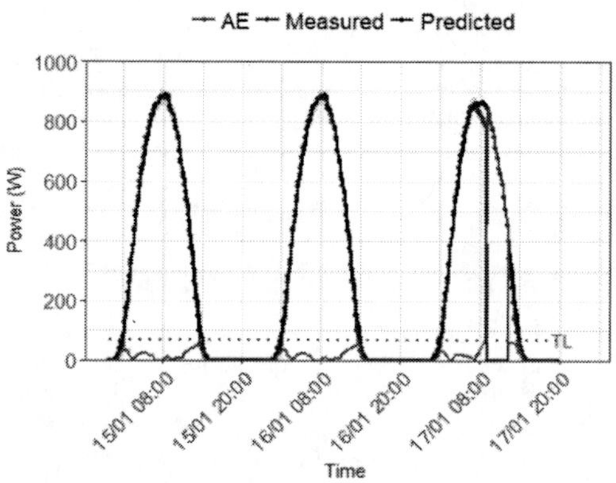

Figure 3: Measured against predicted DC power during an open-circuit PV module failure.

During the failure occurrence, the DC current and power were nearly zero, while the measured voltage was approximately equal to the system's open circuit voltage. The affected electrical parameters were post-processed using the developed classification models and the confusion matrices of the models classifiers are summarised in Table II.

Table II: Confusion matrices of the model classifiers for the open-circuit PV module failure.

	Predicted normal points	Predicted fault points
DT classifier		
Actual normal points	884	0
Actual fault points	0	9
k-NN classifier		
Actual normal points	884	0
Actual fault points	0	9
SVM classifier		
Actual normal points	884	0
Actual fault points	0	9
FIS classifier		
Actual normal points	884	0
Actual fault points	0	9

More specifically, the developed models applied to the test set identified 884 TP normal conditions without producing a single FP. In addition, these models correctly identified 9 TN fault conditions and 0 FN. Thus, the test set was classified entirely correctly (the true negative rate was 100% when classifying the open-circuit PV module fault using the developed DT, k-NN, SVM and FIS models).

3.3 Short-circuit PV module failure

The short-circuit PV module failure was emulated by short-circuiting a PV module in the string, while the measurement platform continued to function in order to capture the failure profile. The developed FDRs accurately detected the failure occurrence as the difference between the measured and predicted DC power of the array exceeded the set TL (see Fig. 4).

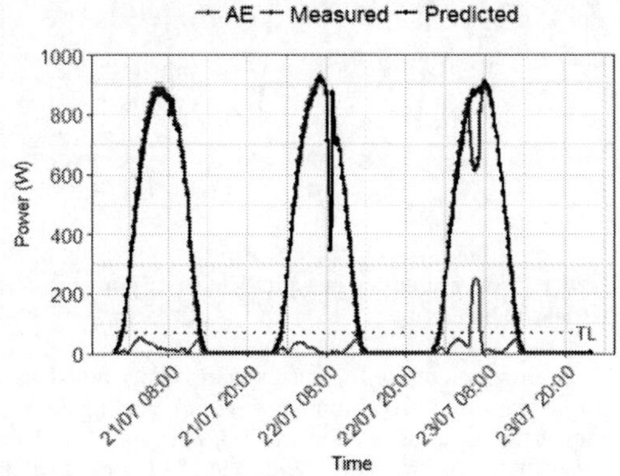

Figure 4: Measured against predicted DC power during a short-circuit PV module failure.

The detected failure pattern (reduced DC voltage and power) was post-processed using the developed classification models for predicting either normal or short-circuit PV module failure condition. The classification performance of the models is summarised in Table III.

Table III: Confusion matrices of the model classifiers for the short-circuit PV module failure.

	Predicted normal points	Predicted fault points
DT classifier		
Actual normal points	884	0
Actual fault points	1	8
k-NN classifier		
Actual normal points	884	0
Actual fault points	0	9
SVM classifier		
Actual normal points	884	0
Actual fault points	3	6
FIS classifier		
Actual normal points	884	0
Actual fault points	7	2

The obtained results showed that the specificity for classifying a short-circuit PV module failure was 88.89%, 100%, 66.67% and 22.22% when using the developed DT, k-NN, SVM and FIS models, respectively. For this failure scenario, the FIS model performed poorly as 7 out of 9 fault data points were misclassified.

3.4 Inverter shutdown failure

Inverter shutdown was emulated by switching off the inverter of the test-bench PV system, while the measurement platform continued to function properly. A failure occurrence was identified by the developed FDRs as the AE exceeded the set TL (see Fig. 5).

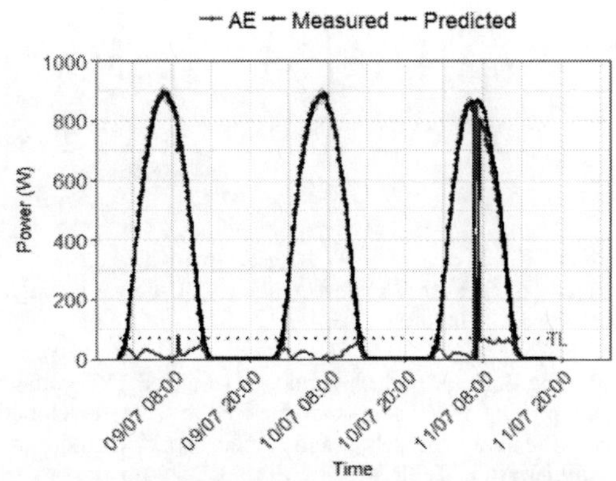

Figure 5: Measured against predicted DC power during the inverter shutdown failure.

To classify the failure occurrence during which the three electrical parameters were affected as in the open-circuit PV module failure scenario, different data-driven models were developed. The performance of the classification models is summarised in Table IV.

Table IV: Confusion matrices of the model classifiers for the inverter shutdown failure.

	Predicted normal points	Predicted fault points
DT classifier		
Actual normal points	884	0
Actual fault points	0	9
k-NN classifier		
Actual normal points	884	0
Actual fault points	0	9
SVM classifier		
Actual normal points	884	0
Actual fault points	0	9
FIS classifier		
Actual normal points	884	0
Actual fault points	0	9

For the given data-set, the specificity was 100%, 100%, 100% and 100% when classifying the inverter shutdown failure using the developed DT, k-NN, SVM and FIS models, respectively.

3.5 Bypass diode failure

The bypass diode failure was emulated by short-circuiting one of the three bypass diodes located at the junction box of a module of the test-bench PV system. The bypass diode failure was accurately detected by the comparative algorithm and the DC measured and predicted power during the fault introduction is depicted in Fig. 6.

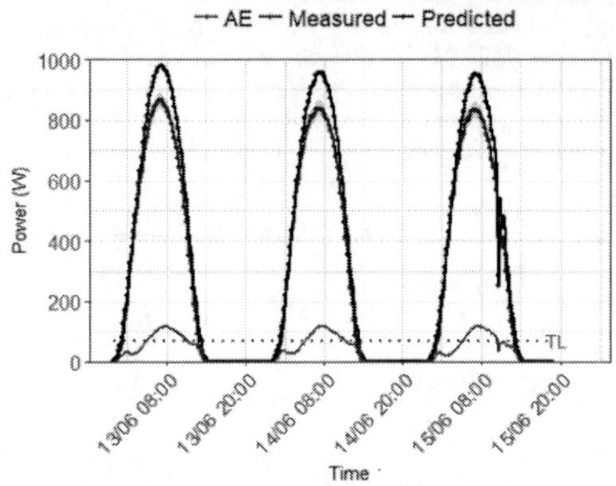

Figure 6: Measured against predicted DC power during a bypass diode failure.

The detected failure signature (reduced DC voltage and power) was again characterised by the developed classification models and their performance is summarised in Table V.

Table V: Confusion matrices of the model classifiers for the bypass diode failure.

	Predicted normal points	Predicted fault points
DT classifier		
Actual normal points	884	0
Actual fault points	3	6
k-NN classifier		
Actual normal points	884	0
Actual fault points	4	5
SVM classifier		
Actual normal points	884	0
Actual fault points	8	1
FIS classifier		
Actual normal points	884	0
Actual fault points	8	1

The obtained results showed that the specificity for classifying a bypass diode failure was 66.67%, 55.56%, 11.11% and 11.11% when using the developed DT, k-NN, SVM and FIS models, respectively. For this failure scenario, the SVM and FIS models performed very poorly as almost all the fault points were misclassified as normal operating points. The failure signature (especially at low irradiance intervals) could not be characterised by these models.

3.6 Partial shading

Partial shading of the array was achieved by placing an object on the surface of a PV module of the array, which covered 30 of the total 300 cells (10% of total amount of cells), for a specific time duration while the solar irradiance was relatively constant. As partial shading is a repetitive type of failure that can be linked to time or solar angle locations, two additional parameters (attributes), the Solar Azimuth Angle (*AzS*) and the Solar Elevation Angle (*AlS*), were calculated as part of the failure classification stage. Consequently, when the DC power difference exceeds the set TL, the algorithm simultaneously checks the persistency of the fault, whether the fault occurs at the same elevation and azimuth angle over a one-week period.

Partial shading was accurately detected by the developed FDRs as the absolute difference between the measured and predicted DC power of the array exceeded the set TL (see Fig. 7).

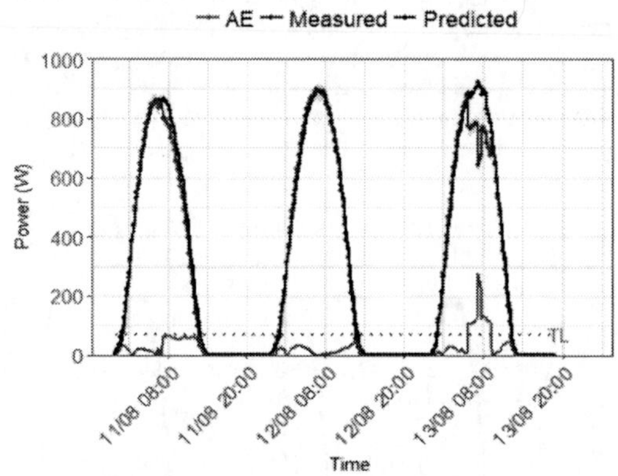

Figure 7: Measured against predicted DC power during partial shading.

A reduction of the DC current and voltage and, hence power was observed during the partial shading period. Four different classification models were developed for characterising this fault and their performance is summarised in Table VI.

Table VI: Confusion matrices of the model classifiers for partial shading.

	Predicted normal points	Predicted fault points
DT classifier		
Actual normal points	884	0
Actual fault points	1	8
k-NN classifier		
Actual normal points	884	0
Actual fault points	0	9
SVM classifier		
Actual normal points	884	0
Actual fault points	2	7
FIS classifier		
Actual normal points	884	0
Actual fault points	9	0

For the given data-set, the true negative rate was 88.89%, 100%, 77.78% and 0% when classifying partial shading using the developed DT, k-NN, SVM and FIS classification models, respectively.

4 CONCLUSIONS

Failures can occur during the PV plant operation due to different factors such as MPP tracking error, shading effects and faulty equipment. Advanced PV monitoring systems encompassing FDRs can provide a clear indication whether the system is performing as expected and therefore safeguard the optimal performance and reliability.

In this work, FDRs based on the analysis of acquired performance data were developed for the online identification of failures in grid-connected PV systems. The results obtained when emulating different failure patterns to the test-bench PV system demonstrated that the proposed methodology was capable of detecting accurately the faults upon their occurrence.

In addition, the developed FDRs could classify the faults by applying different data-driven algorithms. The developed classification models showed good accuracy when classifying each failure occurrence within the test set used for benchmarking. Specifically, for the short-circuit PV module failure and partial shading, the best performing classification model was the k-NN, which accurately classified all instances of the test set (with a true negative rate of 100%). For the bypass diode failure scenario, the best performing classification model was the DT achieving a specificity of 66.67%. Lastly, for the open-circuit PV module and inverter shutdown failures scenarios, all the developed classification models achieved a specificity of 100%. Overall, the k-NN model exhibited the best classification performance with an average true negative rate of 91%. Though, to achieve higher classification accuracies, a combination of classification models is recommended.

An important remark is that the proposed method uses readily available field measurements obtained from a test-bench grid-connected PV system connected to a monitoring system that acquires the main weather and PV operation parameters. However, the supervised learning models require an amount of labelled data (known as fault types) to achieve accurate failure classification. Moreover, data collected both at normal operating and fault conditions were required for the training and testing phase of the developed routines, and such data may be difficult to obtain in a real-world environment. It is also important to note that the developed FDRs worked adequately with the acquired fifteen-minute average measurements; a sensitivity analysis of time resolution will be conducted in future work. Finally, as the developed FDRs are unique to the specific PV installation, future investigation will focus on developing a universal failure diagnosis tool that will be usable for all PV systems.

5 ACKNOWLEDGEMENTS

This work was funded through the IPERMON project which is co-financed by the SOLAR-ERA.NET Transnational Calls of the European Union under the grant agreement number SOLAR-ERA.NET/1214/08. The authors would also like to acknowledge the financial support from the Cyprus Research Promotion Foundation.

6 REFERENCES

[1] Z. Chen, L. Wu, S. Cheng, P. Lin, Y. Wu, and W. Lin, "Intelligent fault diagnosis of photovoltaic arrays based on optimized kernel extreme learning machine and I-V characteristics," *Appl. Energy*, vol. 204, pp. 912–931, 2017.

[2] S. R. Madeti and S. N. Singh, "Monitoring system for photovoltaic plants: A review," *Renew. Sustain. Energy Rev.*, vol. 67, pp. 1180–1207, 2017.

[3] R. Platon, J. Martel, N. Woodruff, and T. Y. Chau, "Online Fault Detection in PV Systems," *IEEE Trans. Sustain. Energy*, vol. 6, no. 4, pp. 1200–1207, 2015.

[4] W. Chine, A. Mellit, A. M. Pavan, and S. A. Kalogirou, "Fault detection method for grid-connected photovoltaic plants," *Renew. Energy*, vol. 66, pp. 99–110, 2014.

[5] A. Phinikarides, G. Makrides, B. Zinsser, M. Schubert, and G. E. Georghiou, "Analysis of photovoltaic system performance time series: Seasonality and performance loss," *Renew. Energy*, vol. 77, pp. 51–63, 2015.

[6] IEC 61724:1998, "Photovoltaic system performance monitoring — Guidelines for measurement , data exchange and analysis," *1st ed. Geneva, Switz. IEC*, 1998.

[7] A. Livera, M. Theristis, G. Makrides, and G. E. Georghiou, "On-line Failure Diagnosis of Grid-Connected Photovoltaic Systems based on Fuzzy Logic," in *12th IEEE International Conference on Compatibility, Power Electronics and Power Engineering*, 2018.

[8] S. Theocharides, G. Makrides, A. Kyprianou, and G. E. Georghiou, "Machine Learning Algorithms for Photovoltaic System Power Output Prediction," in *5th IEEE International Energy Conference, ENERGYCON*, 2018.

[9] A. Livera, G. Makrides, J. Sutterlueti, and G. E. Georghiou, "Advanced Failure Detection Algorithms and Performance Decision Classification for Grid-connected PV Systems," in *33rd European Photovoltaic Solar Energy Conference and Exhibition*, 2017.

[10] A. Livera, M. Theristis, G. Makrides, and G. E. Georghiou, "On-line failure diagnosis of grid-connected photovoltaic systems based on fuzzy logic," in *Proceedings - 2018 IEEE 12th International Conference on Compatibility, Power Electronics and Power Engineering, CPE-POWERENG 2018*, 2018.

[11] B. Lantz, *Machine Learning with R*. Packt Publishing Ltd, 2013.

PERFORMANCE ANALYSIS OF PRE-CRACKED PV-MODULES AT CYCLIC LOADING CONDITIONS

Claudia Buerhop[1], Thilo Winkler[1,2], Felix Enzenberger[1], Tirth Patel[1], Jens Hauch[1], Christian Camus[1,2], Christoph J. Brabec[1,2,3]
[1]ZAE Bayern e.V., D-91058 Erlangen, Immerwahrstraße 2,
[2]i-MEET, FAU Erlangen-Nürnberg, D-91058 Erlangen, Martensstraße 7
[3] Helmholtz-Institut Erlangen-Nürnberg (HI ERN), D-91058 Erlangen, Egerlandstraße 3

ABSTRACT: PV-modules with precracked solar cells are frequently present in PV-plants as current inspection methods reveal. How these modules perform in the future is still unknown. Uniform static loading tests characterize short-term performance well, as presented previously (EUPVSEC 2017). In order to simulate long-term stability, the application of alternating loads, comparable to wind gusts, is important. The load cycle is fed into the improved mechanical loading test-setup by a PLC whilst recording EL-images simultaneously and measuring IV-curves. For the first time, module performance for long-term stability with 100,000 cycles are presented. The results show that with an increasing number of cycles and pressure, the EL-intensity distribution of cracked cells changes but the performance loss is rather small. These findings correlate with analysis of field-exposed modules.

Keywords: cracks, electroluminescence, long-term stability

1 INTRODUCTION

Quality control and inspection of PV-installations is of great interest nowadays. Imaging methods, such as thermography (infrared) IR-imaging and electroluminescence EL-imaging, visualize defective modules and cells. Especially, the degradation of modules with cracked cells is of importance because cracks can electrically isolate cell areas and cause power loss. The timescale on which these losses occur is still unknown.

Minor performance losses of pre-cracked PV-modules after six years of field operation are documented by [1]. Previous analyses and studies showed that pre-cracked modules do not show measurable and detectable changes and performance losses [2-4]. In [2], the field study recorded the performance for one year whilst the mechanical loading tests carried out in the lab only applied static, uniform loads simulating snow and stable wind loads.

For a better understanding of the lifetime of pre-cracked wind-exposed modules in the field, we modified our test facility. The improved test setup supports cycle tests by alternating the load pressure between over- and underpressure and simultaneously recording EL-images, IV-curves and strain data.

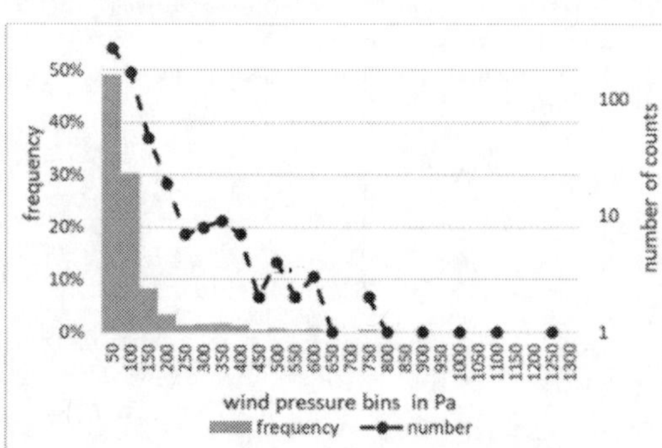

Figure 1: Distribution of maximum daily wind pressure at a PV-site in Southern Germany during one year

Previous analysis revealed that hurricane forces are equivalent to a uniform pressure of roughly 1300 Pa, according to standards transforming wind loads to pressures on structures [5]. The resulting pressure frequency for the maximum daily wind speed of a PV-location in Southern Germany for one year is given in Figure 1. Although single days with strong winds, storms and hurricanes causing high pressures exist, days with fairly low wind pressures are most frequent. Therefore, our test cycles are designed such that after a run of cycles with stronger loading ($p > 400$ Pa), a run of cycles (1,000 cycles) with a small load ($p < 200$ Pa) always follows. Since hurricanes are less frequent than calm breezes and winds, in general, the number of cycles per run decreases with increasing load pressure. That means a run represents a certain number of cycles between constant, defined load pressure and 0 Pa. The loading experiments in the lab are complemented by a highly time-resolved wind speed data and displacement data of PV-modules at operating conditions measured in the field.

This cyclic load test facility, our evolved test procedure, as well as the first results are presented in the paper for the first time.

2 EXPERIMENTAL PROCEDURE

For this investigation, pre-cracked, crystalline PV-modules (230 W) were investigated. The modules were initially loaded up to 5,000 Pa using the static mechanical load test-setup, as described elsewhere [2]. They had differing numbers of cells with cracks and fracture. Many cracks were identified parallel to and near the busbars, mostly in the second and fifth long cell rows.

Several modules were available for artificial load testing. Simultaneous EL-imaging and IV-measurements during loading of the modules are presented. The test facility applies the pressure by vacuum pumps and air blowers. Thus, underpressure as well as overpressure can be used to simulate snow, wind loads and wind suction, respectively. Alternating wind gusts are simulated by alternating the pressure level periodically. A programmable logic controller PLC was used to simulate real wind gusts scenarios. Here, the pressure level goes up to 1,500 Pa. Test cycles with identical pressure were grouped to certain runs. In total, 31 runs were designed.

With increasing pressure level, the number of cycles decreases. The order of test cycles is shown in Figure 2. The number of low loading cycles dominates because after each "stronger pressure event" a run of low loading cycles follows. The period for one cycle is about 5 s. After each run IV-curves were measured at an illumination of approximately 1000 W/m² and the peak power P_{pk} was calculated for standard test conditions (25°C, 1,000 W/m², 1.5AM). Additionally, EL-images were recorded at the unloaded stage after the cycling.

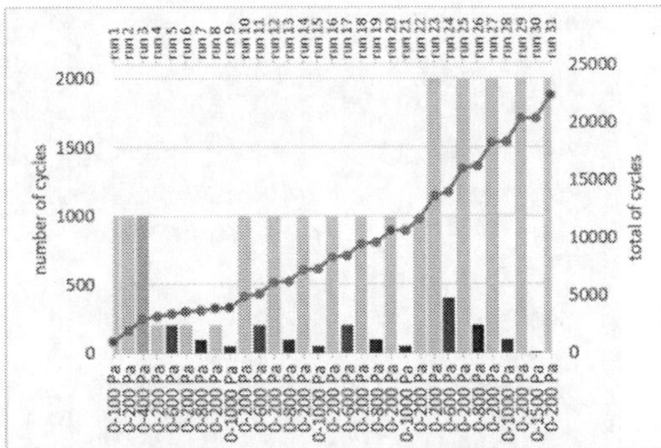

Figure 2: Order of loading tests and number of cycles, colours visualize differing loading pressures

Furthermore, a long-term test of 100,000 cycles at constant, subcritical, low load (pressure) was carried out. The load distribution was as follows: 90% at 200 Pa and 10% at 400 Pa. After every 10,000 cycles IV-curves and EL-images were recorded.

Furthermore, the first measurements of the displacement of installed PV-modules in the field at operating conditions were recorded for three months. The displacement measurement took place at a test setup on the roof of the ZAE building in Erlangen. The deflection was measured in the middle of the module with a Panasonic HG-C1200 microlaser distance sensor, which was placed at its zero point, 200 mm from the module back surface. The given analog signal was recorded with the Driesen+Kern DCXP-16 data logger. Additionally, the air and the module temperature were measured, as well as the wind speed and direction directly above the module. For the temperatures, common 4-wire PT100 sensors from Fühlersysteme were used and the wind was measured with a MESA Systemtechnik ultrasonic anemometer.

3 RESULTS

3.1 Cyclic loading with alternating runs

Up to 23,000 cycles were realized during the cyclic test procedure. 88% of the loading cycles occurred at loads less than 200 Pa. This corresponds to the distribution of maximum daily wind loads found for field installations (see Figure 1). For that specific installation site, about 10% of the recorded maximum wind gusts were in the range between 260 Pa and 1300 Pa (which include singular storm and hurricane events).

In order to study crack growth and propagation in the cells, the EL-images are evaluated. Figure 3 gives an overview of EL images after selected loading cycles. The EL-image at 3,900 cycles is right before the first loading with 1000 Pa and the one at 3,950 cycles directly afterwards. Here, first permanent changes become visible. In general, the cell areas which are electrically separated by a crack from the busbar become darker with increasing number of cycles. Main modifications of the EL-intensity distribution are observed in the second and fifth long cell row. These are the highly stressed cells, as previous analysis and FEM-simulations show [4, 6].

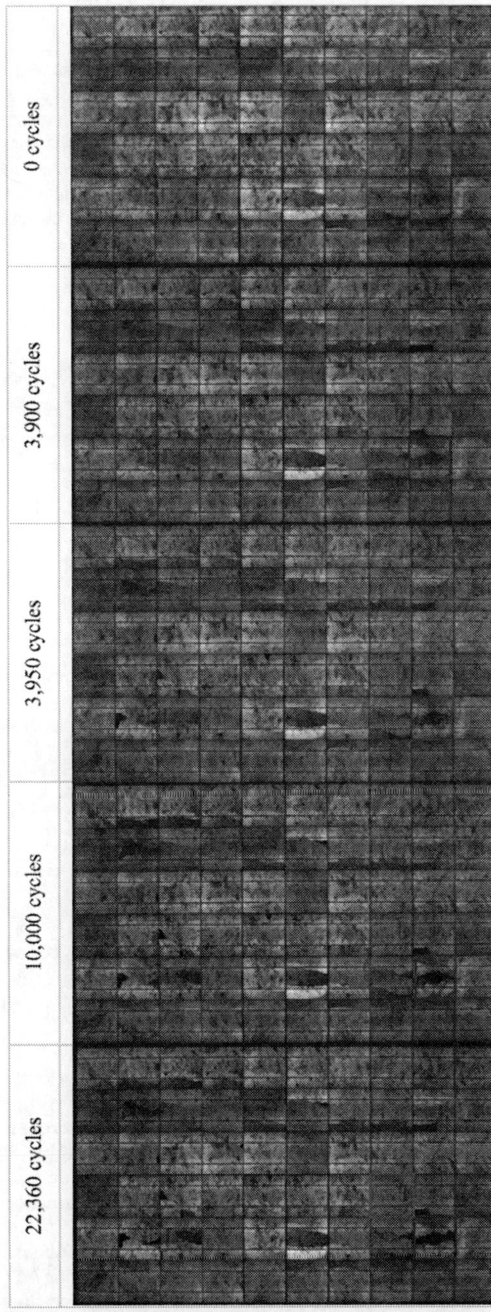

Figure 3: EL-images of a repeatedly mechanically loaded module, ID 148, with increasing number of cycles, red frames mark exemplary cells with modifications in EL-intensity, green frame a reference cell

Automated evaluation of the EL-images indicates how characteristic values like mean, median, quantile (as thresholds characterizing the EL-intensity distribution) and difference between median and quantile as a measure for the cell quality change. The changes and shifts of the EL-intensity distribution (visible in both Figure 3 and Figure 5) are visualized exemplarily for particular cells of module 555 with and without cracks in Figure 4. A constantly high median for all runs describes a high-quality cell, e. g. c8, whose quality does not change. The broken cell b2 has a constantly low median, because the crack structure does not alter. However, the cells, c3, e9, e2, and b7, show the strongest degradation. Their median increases. Simultaneously, their crack structure changes, which is visible in the EL-image after 31 runs and in the increase of the difference between median and quantile. The intensity distributions spread and the quality of these cells worsens.

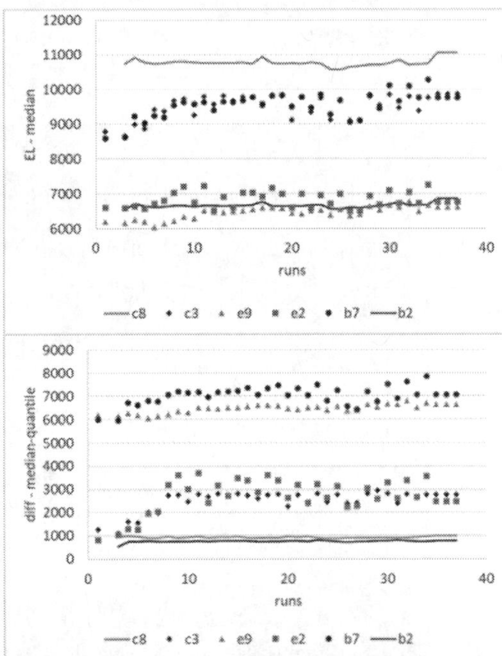

Figure 4: Statistical evaluation of module 555 for particular cells (marked in Figure 5), top: median of the EL-intensity, bottom: difference between median and quantile as a measure of cell quality changes

The EL-images reveal clearly that no new cracks are initiated during the cycling test. Previously (before cycling test) crack-free cells remain unchanged. Furthermore, interruptions of interconnect ribbons, as described by [8-10], were not detected.

In order to determine the impact on the performance, the IV-curves are evaluated. At 0 cycles the module power is measured as P_{pk}(0 cycles) = 223.9 W. Even though changes in the intensity distribution of the differing EL-images exist, the measured IV-curves at identical cycling parameters with respect to the EL-images show only minor differences. The resulting module power at the end of the cycle test is P_{pk}(22,360 cycles) = 221.9 W.

Four modules with more than 20 pre-cracked cells and differing number of cells with open cracks were loaded according to the long-term cycling test procedure (Figure 2) and characterized. Figure 5 shows the EL-images of the tested modules before run 1 and after the final run 31. In general, no new cracked cells are generated (compare Figure 3 and Figure 5). However, after run 31 the previously cracked cells appear much darker due to opening of the cracks. The impact on the electrical performance, the output power at maximum power point, is shown in Figure 6.

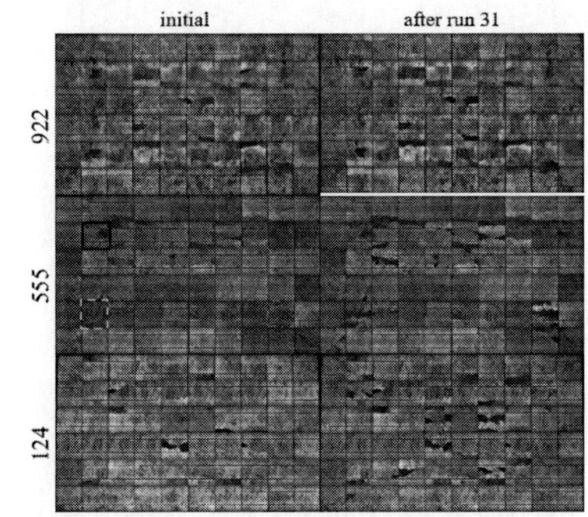

Figure 5: EL-images of module 922, 555 and 124 before and after the cycling procedure with 31 runs at the unloaded state

As Figure 6 exhibits, there is a small, but negative trend for the module power with progression of the loading cycles. The average power loss is 3.6 ± 1.5 W. Three modules (555, 922, 148) lose power between 0.8% and 1.7%. The impact on the power loss of module 124, which was once loaded with 5 kPa, (comparable to an extremely severe storm or snow event) is much stronger. Its power is reduced by 2.5%. The power loss after 22,360 cycles in total is rather small, and within the measurement accuracy of 5%. Similar power loss after accelerated cycle loading test was recently stated by [11].

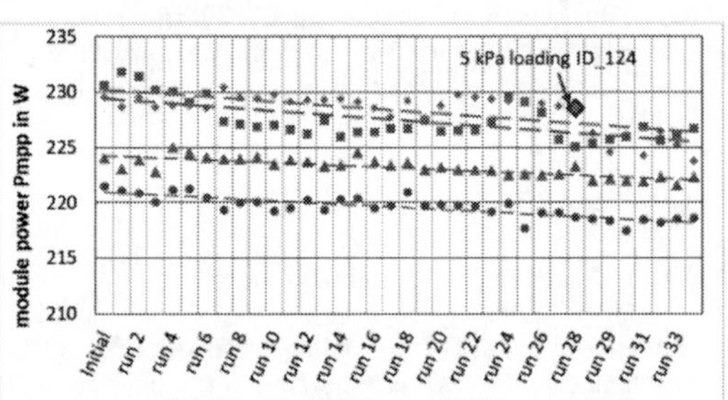

Figure 6: Module power during progression of the cycle loading test

In summary, the power loss is rather small. That corresponds to the observed changes in the EL-images of the investigated modules. In order to detect measurable degradation, significant degradation effects are expected to be seen in the EL-images. The image evaluation revealed that close to 90% of the cells (good and already cracked ones) remain unaffected, 7-9% show quality loss and 3-5% quality increase. Since the cell EL-intensity does not drop below a threshold of 10% of the measured EL-intensity of good cells [7], the power loss is rather small.

3.2 Long-term cycling procedure with identical runs

In order to study the long-term stability of pre-cracked PV-modules, cycling tests with subcritical pressures were carried out. Module ID 124 was loaded 100,000 times, mostly with 200 Pa, and occasionally with 400 Pa. As will be shown in 3.3.2, this load is slightly above the value expected due to daily temperature differences. EL-images before and after this long-term test are shown in Figure 7.

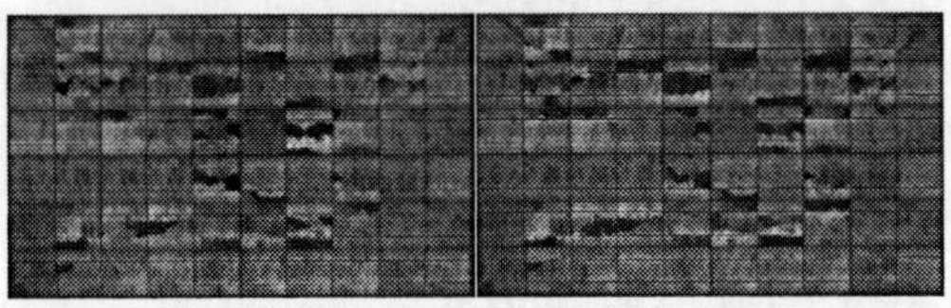

Figure 7: EL-images of module 124 before and after the long-term cycling, left: initial EL-image, right: EL-image after 100,000 cycles

At the beginning, the modules show already a rather high number of cells with open cracks. Due to the long-lasting cycle loading, even with small loads, the cracks open further (increase of darkness) and more cracks appear. In total, seven cracks degraded, while five cells improved in quality (e.g. cell e8 in Figure 7).

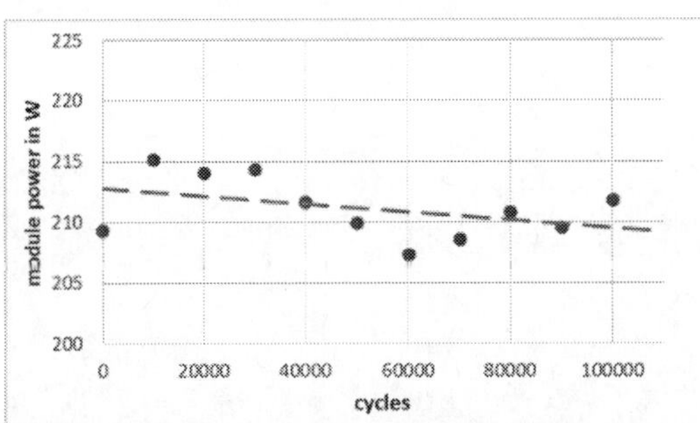

Figure 8: Module power at long-time loading with subcritical loads, 90% of the cycles at 200 Pa and 10% of the cycles at 400 Pa, ID 124

The long-term behavior with respect to the power is plotted in Figure 8. The measured power alternates between 207 W and 215 W, which is a relative power change of 4%. A slight power drop with progression of the long-term test cycle can be noted.

3.3 Outdoor performance

In order to give an idea of the timescale for the cycles carried out (the number of pressure changes) the number of relevant wind gusts per day and year needs to be quantified.

3.3.1 Wind gust analysis

First, the extraction of single wind events is described, and second, the distribution of relevant wind events throughout the entire year is analyzed.

Figure 9 gives a first perception of the wind gust distribution for three different days for Erlangen, located in moderate, continental climate zone in Europe: a calm day, a normal day and finally a very stormy day. On calm days, the wind speed fluctuates between 0 and 1.3 m/s. Corresponding to the rather low wind speeds, no wind pressures are expected. On normal days, the wind speed reaches 5 m/s, several of which are apparent in the data. On the stormy day in Erlangen, wind speeds of up to 12.1 m/s are measured temporally. The wind speed alternates frequently between 6 m/s and 8 m/s.

In order to describe an entire year, every day is at least represented by one wind corresponding pressure. At windy days, wind speeds above 8 m/s are counted and evaluated.

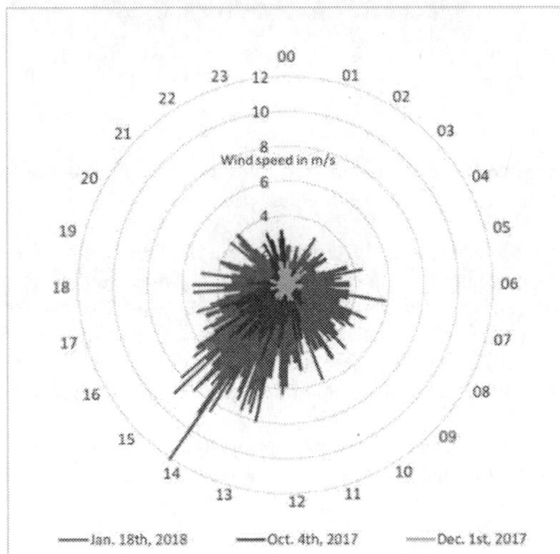

Figure 9: Wind speed distribution during one day in Erlangen, Germany, for different days: a stormy day (hurricane „Friederike"), January 18th, 2018; a normal windy day, October 4th, 2017; and a calm day, December 1st, 2017. Time resolution is 10 seconds.

Therefore, available wind data for two entire years were analyzed. The data were collected in Arzberg, Germany for central European climate zone, where the ZAE test facility is located. The available 2-second wind data are angle-corrected, because the PV-modules are orientated southwards. It results that the number of extracted wind gusts differs strongly for the two analysed years: 2016: 1870 wind gusts, 2017: 5128 wind gusts. In total there were 6998 wind events, and on average 3499 wind loads a year. Also, the calculated maximum wind pressures according to DIN 1991 are different for both years. 2016 the maximum wind gust corresponds to 778 Pa whereas in 2017 it was close to 1300 Pa. These basic data of real ambient weather conditions correspond fairly well to the proposed test procedure presented in Figure 2. According to the recorded weather data, the test procedure simulates approximately 6 years of operating conditions in central Europe.

Further data analysis will provide more clues to the distribution of wind gusts.

3.3.2 Module displacement

On top of the wind speed, the resulting stress on the module is of great interest. In order to measure the influence of wind on the module, the displacement of installed modules is recorded in the module center at operating conditions, see **Figure 10**. During the measurement period the module temperature varied between 50°C at daytime and 5°C at nighttime. The wind speed was mostly less than 4 m/s, which corresponds to a light breeze.

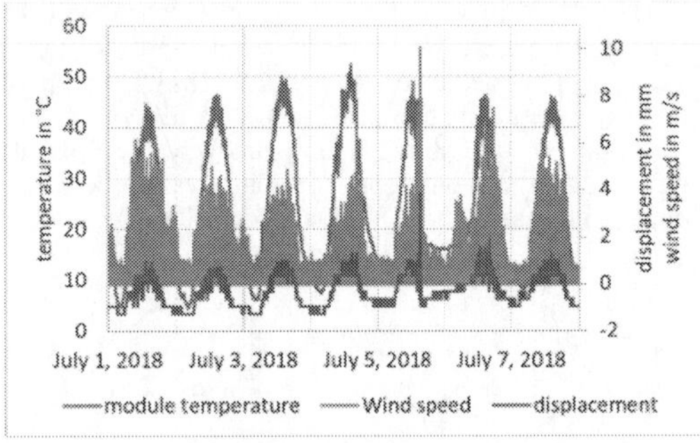

Figure 10: Measured module displacement with respect to module temperature and wind speed

Due to the daily temperature cycles of about 45 K, displacements of 2.5 mm were measured. This corresponds to 0.047 mm/K. These small registered displacements equal to small loads of around 100 Pa according to [2, 12]. One wind gust of 10 m/s was recorded on July 5, 2018 that forced a displacement of roughly 4-5 mm.

Unfortunately, neither stormy nor windy days were present during the current measurement period.

4 DISCUSSION

Summarizing the results shows that first slight changes in the intensity distribution of the EL-images occur above 3,900 cycles, after simulating a first, short hurricane event. Strong reductions in the module performance were not observed. This holds for modules with more cells with open cracks, too. The test cycle reveals a degradation of cracks of 1 – 1.5% for modules with cracked cells for an estimated field exposure of about 6 years in central Europe according to the recorded wind data for two years. The findings agree with the existing 3-year field experience of the degradation of precracked PV-modules.

5 CONCLUSION

The newly introduced cyclic load test gives indication that negative impact on the module performance of precracked PV-modules due to repeatedly applied moderate stresses is rather small. However, the recording of more field data in terms of module degradation and weather scenarios as well as investigating more modules is the lab will manifest the model.

6 ACKNOWLEDGEMENTS

ZAE Bayern gratefully thanks the German Federal Ministry for Economic Affairs and Energy (BMWi) for financial funding of the project AQUAM.

The authors sincerely thank the Allianz Risk Consulting GmbH / Allianz Zentrum für Technik (AZT) in Munich, Germany for supporting the project with a large number of PV-modules.

7 REFERENCES

[1] J. Arp, et al., 33rd EU PVSEC (2017).
[2] C. Buerhop, et al.; Progress in Photovoltaics (2018).
[3] C. Buerhop-Lutz, et al., 33rd EU-PVSEC (2017) 1451
[4] C. Buerhop-Lutz, et al., 44th IEEE (2017).
[5] Standard DIN EN 1991-1-4 2010 Eurocode 1: Actions on structures - Part 1-4: General actions - wind loads, 2010 Allgemeine Einwirkungen - Windlasten.
[6] M. Sander, et al.; Solar Energy Materials and Solar Cells 111 (2013) 82.
[7] R. Amesöder (2017) Master Thesis, FAU Erlangen-Nürnberg.
[8] M. Pander, et al., 32nd EU-PVSEC (2016) 1589.
[9] M. Pander, et al., 33rd EU-PVSEC (2017) 1457 .
[10] J. Käsewieter, et al.; IEEE Journal of Photovoltaics 6 (2016) 28.
[11] H. Seigneur, et al., WCPEC-4 (2018)
[12] S. Wirsching (2016) Master Thesis; FAU Erlangen-Nürnberg

OPTIMUM CONDITION FOR ACCURATE MEASUREMENT OF PHOTOVOLTAIC ARRAY TEMPERATURE

Kazuki Okumoto, Kazuyuki Miyamura, and Kensuke Nishioka
University of Miyazaki
1-1 Gakuen Kibanadai-nishi, Miyazaki 889-2192, Japan

ABSTRACT: For further development of solar power generation, the highly precise power generation forecast and evaluation of solar power generation's performance are extremely important. What important here is a highly accurate temperature measurement technique of PV module, array, and system. However, the PV temperature in outdoor environments is affected by various environmental factors such as wind direction, wind speed, solar irradiance, and outside air temperature. In this study, we have established optimal temperature measurement conditions for PV array. As a result, it was found that in the case of low irradiance, the variation of temperature distribution is small and it is suitable for a temperature measurement condition. Furthermore, it was found that the difference between the temperature in the center of the PV array and the average temperature is the smallest and it is suitable for a temperature measurement point.
Keywords: PV array, Evaluation, Temperature distribution

1 INTRODUCTION

In recent years, energy consumption has continued to increase with the population increase. However, fossil fuels are finite resources and fear of depletion of fossil fuels is a concern. In addition, global warming caused by emission of greenhouse gases when generating electricity using fossil fuels has become more serious. As a method to solve this problem, a power generation method using renewable energy, which is a clean power generation method without using fossil fuel, is required. In particular, solar power generation that does not emit greenhouse gases such as carbon dioxide attracts attention, and demand for solar power generation is expected to grow in the future [1-3].

Also, since PV cells convert solar energy into electricity energy, output fluctuates drastically due to the weather, and a power supply configuration combined with other power generation methods is required. Along with this, the power generation forecast and evaluation of solar power generation's performance are attracting attention. What important here is a highly accurate temperature measurement technique of PV module, array, and system. Since the output of the PV is strongly influenced by the temperature, grasping the temperature with high accuracy leads to an increase in the accuracy of power generation forecast and evaluation of performance [4-9]. However, the PV temperature in outdoor environments is affected by various environmental factors such as wind direction, wind speed, solar irradiance, and outside air temperature, so it is not easy to measure the temperature with high accuracy [10-11]. In particular, temperature distribution occurs in the array at the time of wind blowing, so that a large temperature measurement error occurs due to the difference in the measuring points.

In this study, we have established optimal temperature measurement conditions for PV array.

2 EXPERIMENTAL METHODS

2.1 Measurement method

In order to clarify the variation of the temperature distribution of the PV array, the PV array was exposed outdoors and the temperature was measured.

Fig. 1 shows the PV array used in this study. The PV array used was set at the University of Miyazaki and it was composed of polycrystalline Si PV modules of 3 vertical (2325 mm) × 28 horizontal (43400 mm). The measurement was carried out for a year from July 1, 2017 to June 30, 2018, and the temperature and the solar irradiance were measured every second from 8:00:00 to 16:00:00 on each day.

Next, we describe the temperature measurement method in this study. Fig. 2 shows the temperature measurement points in the PV array. As shown in Fig. 2, thermocouples were attached to the center of the 21 PV modules and temperature measurement was performed. In order to attach the thermocouples to the back sheet of PV modules, Kapton tape with 10 mm x 40 mm was used.

Figure 1: Photograph of the PV array.

Figure 2: Temperature measurement points in the PV array.

2.2 Analysis method

In order to show the typical temperature of the PV array with temperature distribution, we describe analysis method using measurement results.

Table I shows solar irradiance condition. The influence of the solar irradiance is considered as the cause of the large measurement error. This is because solar irradiance is heavily involved in temperature rise of the PV. Therefore, for each range of solar irradiance shown in table 1, a temperature difference distribution was created by using the temperature difference between each measured temperature and the average temperature. First, the difference between the average temperature of the PV array and the temperature of each PV modules was derived every second. Using the temperature difference, the average deviation of the temperature difference was calculated for each PV modules.

Next, based on these results, the range of solar irradiance suitable for temperature measurement was obtained. Based on the created temperature difference distribution, average deviation in the PV array was derived for each range of solar irradiance. Also, the derivation equation is shown in equation (1),

$$M_{dev} = \frac{\sum_{i=1}^{n} |X_i - \bar{X}|}{n} \quad \cdot \cdot \cdot (1)$$

where M_{dev} is average deviation, X_i is temperature difference at each measurement point, $\bar{X}$ is the average temperature difference within the PV array, and n is data points, respectively.

Table I: Range of solar irradiance.

The amount of solar radiation range
200≦Irr＜300
300≦Irr＜400
400≦Irr＜500
500≦Irr＜600
600≦Irr＜700
700≦Irr＜800
800≦Irr＜900
900≦Irr

In order to obtain a more detailed trend, a frequency distribution at each measurement point was created. The measurement results used are the same as that used in the temperature difference distribution. Based on this, a frequency distribution was created for each range of solar irradiance shown in Table 1.

3 RESULT AND DISCUTTION

3.1 Temperature difference distribution

Figure 3 shows the temperature difference distribution in each range of solar irradiance. From Fig. 3, the maximum and the minimum temperature difference from the average temperature of the PV array were 1.08°C, 0.21°C in the case of 200≦Irr＜300 W/m² and 4.48°C, 0.54°C in the case of 900≦Irr, the higher the solar irradiance, the greater the temperature difference. As a cause of this, it is conceivable that the higher the solar irradiance, the higher the temperature of the PV array and the greater the temperature difference from the outside temperature. Generally, factors responsible for cooling the PV array are convective heat transfer and radiative heat transfer, as shown in Eq. (2) and Eq. (3).

$$Q = h(T_o - T_f)A \quad \cdot \cdot \cdot (2)$$
$$E = \varepsilon\sigma A(T_H{}^4 - T_L{}^4) \quad \cdot \cdot \cdot (3)$$

Where Q is the total heat flux from a object, h is the heat-transfer coefficient, $T_o - T_f$ is the temperature difference between object and fluid, A is the surface area of the object, E is radiant energy, ε is the emissivity, σ is the Stefan-Boltzmann constant, T_H is surface temperature, and T_L is the ambient temperature, respectively. From the equations (2) and (3), in the process of cooling of the PV, the difference between the PV temperature and the outside air temperature is greatly related to the amount of heat transfer. Therefore, when the range of solar irradiance is large, the difference between the PV temperature and the outside air temperature becomes large, and it is conceivable that the variation of the temperature difference distribution increases.

Figure 4 shows the average deviation in the array at each solar irradiance. As shown in Fig. 4, the average deviation within the PV array tends to increase as the solar irradiance increases. This is considered to be caused by the high temperature of the PV array with an increase in the solar irradiance.

From the above results, it was found that when the solar irradiance was low, variations in temperature in the PV array were small, and temperatures close to the average temperature of the PV array could be easily measured.

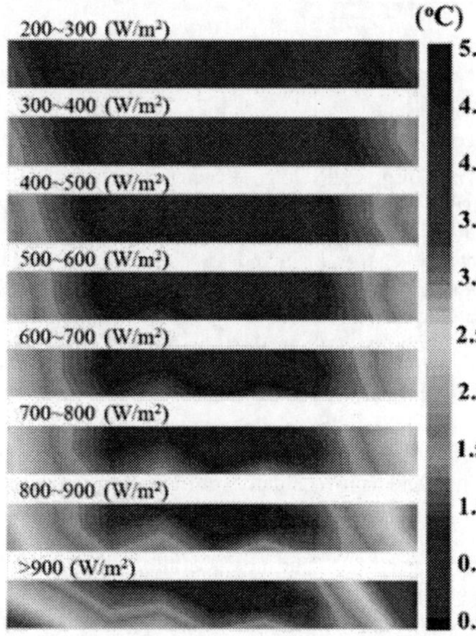

Figure 3: Temperature difference distribution from average temperature in each solar irradiance.

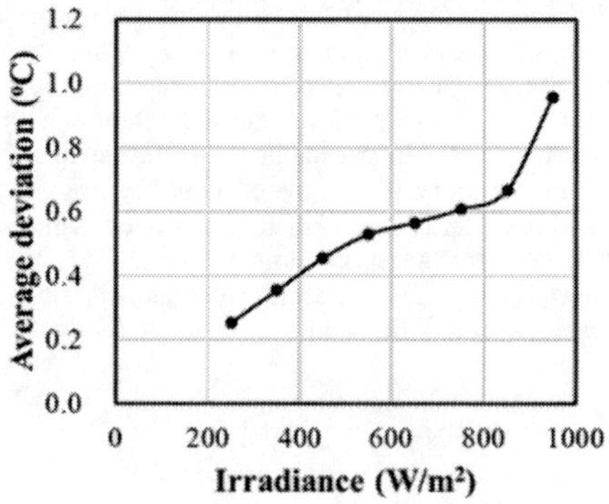

Figure 4: Average deviation in the array at each solar irradiance.

3.2 Frequency distribution

Figures 5 and 6 show the frequency distributions at the center and the left end of the PV array. From Fig. 5 (a) for 200≦Irr<300, the fraction peak was at around 0°C. From Fig. 5 (b) for 900≦Irr, it can be seen that the fraction of the points varies with the increase in solar irradiance. This is considered that the temperature difference between the air and the PV increases due to an increase in solar irradiance, and the influence of convective heat transfer increases, so that the temperature distribution is likely to occur. As a result, a large temperature difference occurs in the PV array, and the number of times that the difference from the average temperature becomes large. As shown in Fig. 6 (a), the peak becomes broad compared with Fig. 5 (a). This is because the influence of wind is larger in the edge of PV array. Furthermore, as shown in Fig. 6 (b), the peak was divided into two. The reason is the difference of wind direction. When the measurement point is on the windward side, since the cooling effect is greatest in the array, the temperature is lower than the average temperature, and the peak moves to the left side. On the other hand, when the measurement point is on the leeward side, the influence of the cooling effect by wind is small and the temperature of the solar cell module is higher than the average temperature, so the peak moves to the right side. Therefore, this tendency was observed at the edge of the PV array.

From the above results, it was found that the influence of the wind could be minimized by measuring the temperature of the PV module located at the center of the PV array

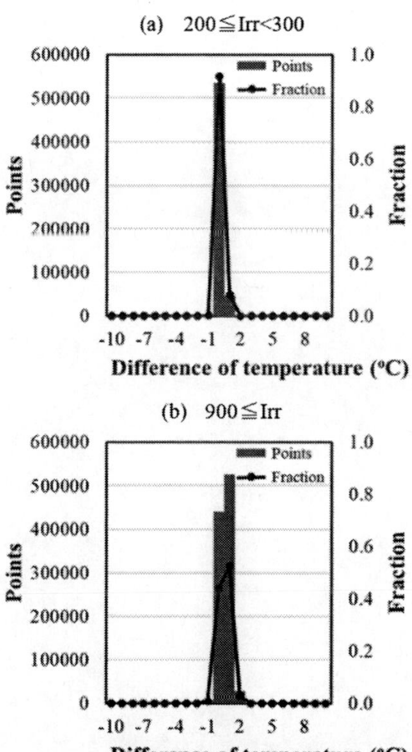

Figure 5: Frequency distribution in the center of the array.

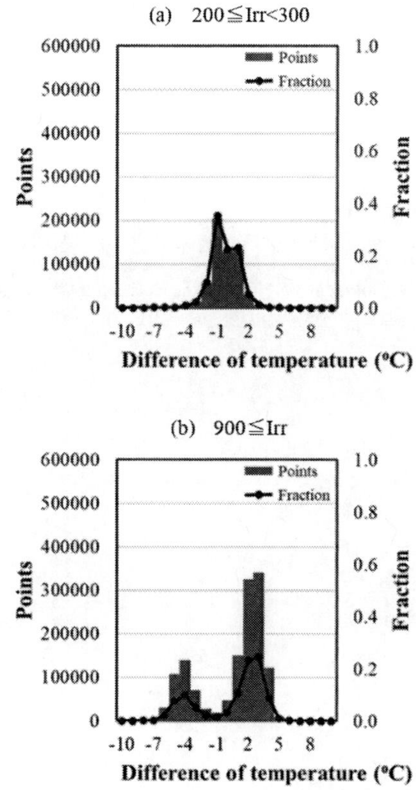

Figure 6: Frequency distribution at the edge of the array.

4 CONCLUSION

In this study, we have established an optimal temperature measurement conditions for PV array base on the 1 year measured data.

The maximum and the minimum temperature difference from the average temperature of the PV array were 1.08°C and 0.21°C in the case of 200≦Irr<300 W/m^2, and 4.48°C and 0.54°C in the case of 900≦Irr. Furthermore, it was found that the dispersion of the temperature difference within the PV array was small at low solar irradiance and it was suitable for temperature measurement. Moreover, the center of the PV array was suitable for temperature measurement because the temperature difference from the average temperature was small and the temperature difference fluctuation was small.

5 REFERENCE

[1] A. Goetzberger, C. Hebling and H. W. Shock, Materials Science and Engineering: Reports 40 (2003) 1.
[2] T. M. Razykov, C. S. Ferekides, D. Morel, E. Stefanakos, H. S. Ullal, and H. M. Upadhyaya, Sol. Energy 85 (2011) 1580.
[3] L. L. Kazmerski, Renewable and Sustainable Energy Reviews, Vol. 1 (1997) 71-170.
[4] J. J. Wysocki and P. Rappaport, Journal of Applied Physics 31 (1960) 571-578.
[5] J. C. C. Fan, Solar Cells 17 (1986) 309-315.
[6] P. Singh, S. N. Singh, M. Lal, and M. Husain, Solar

Energy Materials and Solar Cells 92 (2008) 1611-1616.
[7] D. J. Friedman, Proceedings 25th IEEE Photovoltaic Specialists Conference, (1996) 89-92.
[8] K. Nishioka, T. Hatayama, Y. Uraoka, T. Fuyuki, R. Hagihara, and M. Watanabe, Solar Energy Materials and Solar Cells 75 (2003) 665-671.
[9] K. Nishioka, N. Sakitani, K. –I. Kurobe, Y. Yamamoto, Y. Ishikawa, Y. Uraoka, and T. Fuyuki, Jpn. J. Appl. Phys 42 (2003) 7175-7179.
[10] J. K. Kaldellis, M. Kapsali, and K. A. Kavadias, Renewable Energy 66 (2014) 612-624.
[11] N. Gokmen, W. Hu, P. Hou, Z. Chen, D. Sera, and S. Spataru, Renewable Energy 90 (2016) 283-290.

UNDERSTANDING THE TIME EVOLUTION OF THE PVGIS PERFORMANCE MODEL PARAMETERS AND THE TEMPERATURE COEFFICIENTS

Philip Ingenhoven[1*], Giorgio Belluardo[1], Sascha Lindig[1,2] and David Moser[1]
[1]EURAC Research / [2]University of Ljubljana - Faculty of Engineering
Viale Druso 1, 39100 Bolzano, Italy / Trzaska cesta 25, 1000 Ljubljana, Slovenia
*philip.ingenhoven@eurac.edu

ABSTRACT: The time evolution of the performance of photovoltaic (PV) modules has attracted increasing interest in recent years and many methods exist to estimate the performance loss rates. In this work, we analyze the relative efficiency depending on temperature, irradiance and time. To analyze the temperature and irradiance behavior we use the model adopted by the Photovoltaic Geographical Information System (PVGIS). This might help to improve the PVGIS model to take into account the aging of PV modules. As the analysis of this model yields ambiguous results, we decided to investigate the degradation depending on either temperature or irradiance, separately. We investigated the time evolution of the temperature coefficients of power, current and voltage. With the help of the temperature coefficients, it is possible to find yearly power, current and voltage values at standard test conditions (STC), without taking the modules out of the field for laboratory inspection. With this approach, it is not possible to include spectral, annealing, light soaking and wind effects. The STC values can be used to study the performance loss of the analyzed modules. As a further step, we analyzed the degradation at different irradiance levels, which might help to identify the root causes of the degradation of the module. Whereas, it is not possible to find meaningful trends for the fitting parameters of the model used by PVGIS and the temperature coefficients, we see a definite irradiance dependency of the degradation rates. As a possible cause for this behavior, we identified a decrease of the shunt resistance.

Keywords: Degradation, Durability, Electrical Properties, PV Module, Reliability, Thermal Performance, CdTe, Heterojunction, Silicon

1 INTRODUCTION

The study of photovoltaic performance metrics has been a more or less one-dimensional matter. The model used by Photovoltaic Geographical Information System (PVGIS) [1] describes the efficiency of PV models in dependence of irradiance and temperature. The study of degradation of photovoltaic devices as discussed in [2], [3], [4] or [5] relates different performance metrics of the devices with time. In our approach, we would like to combine these two aspects of the field. This way, looking at 'multi-dimensional degradation', it can become clear under which conditions the degradation modes are most apparent.

PVGIS [6] is an interactive tool that allows evaluating the performance of a PV plant using advanced algorithms based on satellite data. For our purposes, the relevant part of their model is the translation from in-plane irradiance and temperature data into electric power produced by the PV module, which was developed in several steps, see [7], [8], and [9]. The data is fitted yearly to the parameter in the model and the time evolution of the model parameters will be evaluated. Initially a monthly fitting was used. However, in months with low irradiance like January and December, the fitted curves dropped sharply at the highest irradiance level, rendering the parameters useless. Further, the temperature behavior of current, voltage and power will be examined by determining outdoor temperature coefficients each year. Finally, the time evolution at different irradiance levels will be scrutinized.

The data used for this study was collected during an ongoing monitoring campaign at the PV outdoor test facility of Airport of Bolzano (position ca. 46.46N, 11.33E, elevation: 262 m) located in South Tyrol, Italy [4]. The facility is located at the junction of three valleys: Val d'Isarco, Val Sarentino, and Val d'Adige. The climate is classified as *Cfb* (temperate, without dry season, warm summer) according to Köppen classification [10]. Average monthly precipitation in winter in the analyzed period was 37.5 mm, in spring 58.5 mm, in summer 85.1 mm, and in autumn 84.8 mm according to historical data measured by the province [11]. Furthermore, the average daytime temperatures in summer were 25°C with a maximal temperature of 40°C reached; average daytime winter temperatures reached were 5°C with a minimum night temperature of -11°C measured in our weather station. Twenty-six PV arrays from 1 to 7 kWp of different module types (technology, manufacturer, and design) are installed at a fixed tilt of 30° and an orientation of 8.5° West of South. Four arrays of different technologies, namely mono-crystalline (mc-Si) and poly-crystalline (pc-Si) silicon, heterojunction (HIT) and cadmium telluride (CdTe) were chosen for this analysis. All arrays are installed within 100 m of the weather station and are exposed to similar horizon shading. DC-side electrical parameters of each array are measured every 15 min by commercial inverters (SMA) ensuring a good level of accuracy in maximum power point current and voltage.

The remainder of this report is structured as follows. In the following section, the methods used for the analysis are described in detail. In section 3, the results are presented and finally in section 4, some conclusions are reached and future work is discussed.

2 METHODS

2.1 Data preparation

The data were initially filtered for outliers. Irradiance data were used only between 0 and 1500 W/m^2. To exclude incidents in which the irradiance condition at the PV system and the monitoring system were different due to local shading, a performance ratio filtering as described in [4] was used.

2.2 Time evolution of the 6-parameter model

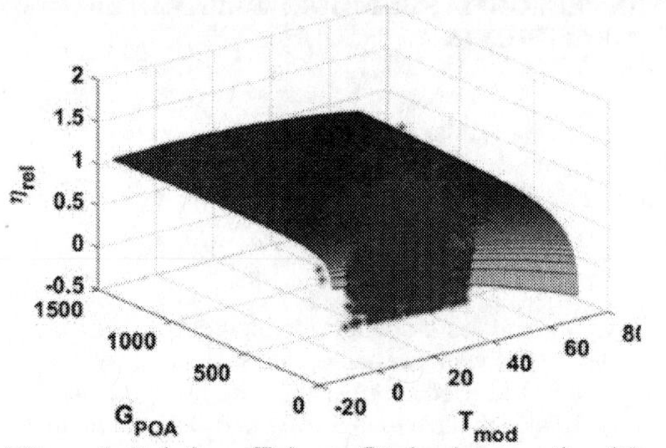

Figure 1: Relative efficiency fitted using equation (1) and (2). For poly-crystalline silicon in the year 2015 plotted against module temperature and in-plane irradiance. The red dots show the measured data and the blue plane shows the three-dimensional fit

To capture the influence of irradiance and temperature on the power output of a given PV system we use the approach described in [9]. The relative efficiency $\eta_{rel}(G_i, T_{m,i})$ at time t_i is defined through

$$P(G_i, T_{m,i}) = \frac{\eta_{rel}(G_i, T_{m,i}) P_{STC} G_i}{G_0}, \tag{1}$$

where $P(G_i, T_{m,i})$ is the electrical power measured at the DC side of the inverter, G_i is the plane-of-array (POA) irradiance at time t_i, $T_{m,i}$ is the module temperature at time t_i, P_{STC} is the PV power output at STC and $G_0 = 1000\,W/m^2$ is the STC irradiance.

The relative efficiency is then fitted using six parameters k_1 to k_6 with

$$\begin{aligned}\eta_{rel}(G_i, T_{m,i}) = 1 + k_1 \ln\left(\frac{G_i}{G_0}\right) + k_2 \left(\ln\left(\frac{G_i}{G_0}\right)\right)^2 \\ + (T_{m,i} - T_{STC})\left[k_3 + k_4 \ln\left(\frac{G_i}{G_0}\right) \right. \\ \left. + k_5 \left(\ln\left(\frac{G_i}{G_0}\right)\right)^2\right] \\ + k_6(T_{m,i} - T_{STC}),\end{aligned} \tag{2}$$

where $T_{STC} = 25°C$. Each year the 15 minute DC power data, the temperature data and the irradiance data are fitted to equation (2). As the power output of the system degrades, it is necessary to adjust the STC power, P_{STC}, in equation (1) each year. Thus, the number of fitting parameters rises to seven. Figure 1 shows an example of this procedure for a poly-crystalline silicon system for the year 2015. The seven parameters were fitted using the nonlinear least squares method [12] implemented in MATLAB.

2.3 Time evolution of the temperature coefficients

In data sheets of PV modules, temperature coefficients of power, open circuit voltage and short circuit current are given. These were measured in postproduction and cannot be assumed constant over time or resistant to aging. In this part of the study, we analyze the time evolution of the temperature behavior. As we are collecting data from grid-connected systems, it is not possible to measure open and short circuit values for voltage and current. Instead, we performed the analysis on maximum-power-point (mpp) data.

To find the temperature coefficients the data were filtered around $G_0 = 1000 \pm 25\,W/m^2$ and normalized with nominal values. The remaining points were then fitted linearly against the temperature. It is worth to mention that earlier studies have shown that the temperature coefficients of the power vary greatly with irradiance [13]. Figure 2 shows the procedure for the normalized power in the year 2012.

2.4 Degradation at different irradiance levels

For the examination of the performance loss rate of PV modules, evaluated at different irradiance levels, the data was sorted by irradiance intervals. The intervals were chosen with a width of $100\,W/m^2$ from 0 to $1200\,W/m^2$. In each of the these intervals we compute the performance ratio for each month using

$$PR_{month} = \frac{Y_a}{Y_r}, \tag{3}$$

where the array and the reference yield are defined as

$$Y_a = \sum_{month} \frac{P(G_i, T_{m,i})}{P_{STC}} \quad \text{and} \quad Y_r = \sum_{month} \frac{G_i}{G_0}. \tag{4}$$

The performance loss rates in each irradiance interval are then computed using the local regression seasonal trend analysis as described in [4]. The yearly performance loss can then be either quantified by taking the gradient of the resulting trend line multiplied by 12 (absolute performance loss) or this values can be divided by the y-intercept of the trend line (relative performance loss). To understand the overall performance loss the same procedure is repeated for the complete data. For more details, see [4]. Figure 3 shows the time evolution of the performance ratio for mc-Si in three selected irradiance intervals.

3 RESULTS

The procedures described in section 2 were performed for four model types representing four different technologies. The results are presented in the following

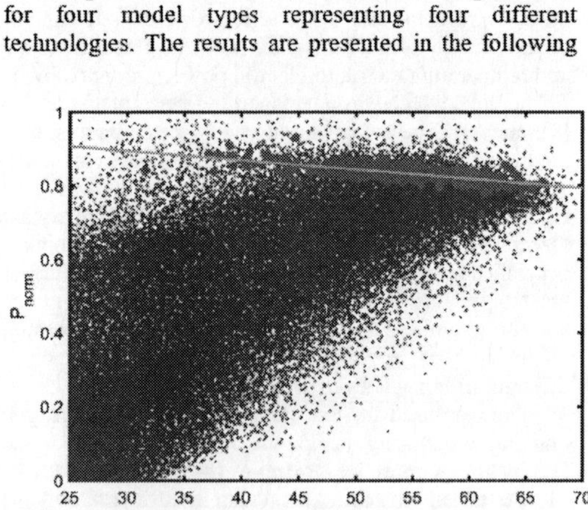

Figure 2: Normalized PV power for the year 2012 plotted versus temperature. Blue: all data, Red: filtered around 1000 W/m², Orange: linear fit.

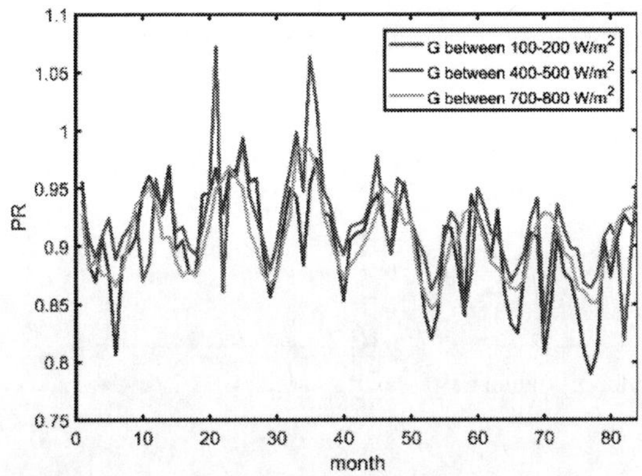

Figure 3: Performance ratio in three different irradiance intervals over 8 years for mono-crystalline silicon.

sections.

3.1 Time evolution of the 6-parameter model

Figure 4 shows the time evolution of the six fitting parameters where none of the four technologies shows a clear trend in any of the parameters k_1 to k_6. The only trend in this analysis was visible in the fitted power at STC, which shows a performance loss compatible with other methods for the determination of the performance loss rates [4]. A small effect can be seen in the fitted STC power, see figure 5. Table 1 shows the linear change rates for all the fitted parameters and technologies.

In all cases the error of the linear fit coefficients are larger than the trend of the parameters. Even for the trends of the STC power values, which show performance loss rates that are within the range of observed performance loss rates [2] the error is larger than the trend itself. As the linear regression only relies on seven points, the error is accordingly large. Further, the fitting space (seven parameters) makes it difficult to find reliable k-parameters and STC power values. Hence, further studies of the time evolution of the temperature and irradiance dependence were performed in separate steps as described in section

Table 1: Change rates of the fitted parameters with the corresponding errors.

pc-Si			mc-Si		
	rate	error		rate	error
k_1	-6.7E-04	1.8E-02	k_1	-1.5E-03	2.1E-02
k_2	-3.6E-04	4.6E-03	k_2	-8.8E-04	5.5E-03
k_3	5.1E-04	1.6E-03	k_3	5.7E-04	1.5E-03
k_4	3.5E-04	1.5E-03	k_4	2.9E-04	1.4E-03
k_5	4.8E-05	2.1E-04	k_5	1.2E-05	1.8E-04
k_6	-1.2E-05	3.9E-05	k_6	-1.6E-05	4.1E-05
	%/year	error		%/year	error
P_{STC}	-1.2	1.8	P_{STC}	-1.0	1.7

HIT			CdTe		
	rate	error		rate	error
k_1	-1.1E-03	2.0E-02	k_1	-7.3E-03	1.8E-02
k_2	-7.8E-04	5.5E-03	k_2	-5.9E-03	7.0E-03
k_3	6.2E-04	1.2E-03	k_3	5.0E-04	1.4E-03
k_4	4.2E-04	1.1E-03	k_4	3.3E-04	1.5E-03
k_5	5.6E-05	1.0E-04	k_5	1.0E-05	2.3E-04
k_6	-1.8E-05	3.7E-05	k_6	-8.2E-06	2.5E-05
	%/year	error		%/year	error
P_{STC}	-1.1	1.3	P_{STC}	-2.2	2.0

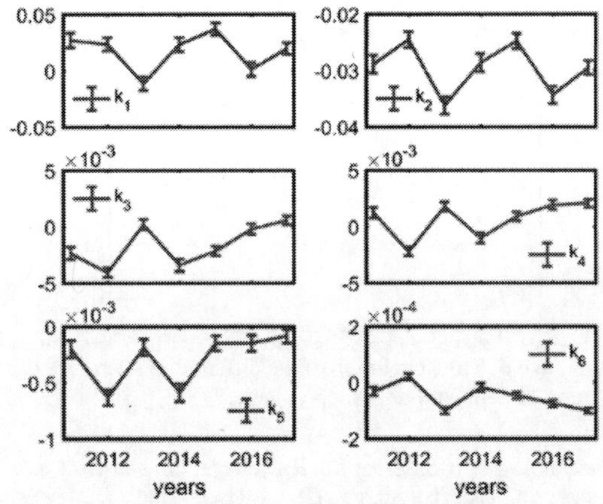

Figure 4: Time evolution of the six parameters k_1 to k_6 from equation (2) for pc-Si. All parameters show minor dependence.

2.3 and 2.4. The results of these analyses follow in the next sections.

3.2 Time evolution of the temperature coefficients

The outdoor temperature coefficients were determined as described in section 2.3. Figure 6 shows the yearly values of the temperature coefficients of the mpp values for CdTe. A slight trend for the voltage coefficient is visible here. However, in general the temperature coefficients do not display a clear trend and stay near the data sheet values with a few outliers. See Table 2 for the average temperature coefficients at maximum power point, their trends and the data sheet values.

3.3 Degradation at different irradiance levels

In this section, we report the results of the analysis of performance loss rates at different irradiance levels. Interestingly we found a strong dependence of the performance loss rates on the irradiance levels for all four analyzed technologies. The effect is visible in absolute as well as relative values and cannot stem from changing overall performance ratios at different irradiance levels. In all technologies, we see that the performance loss rates as defined in section 2.4 are larger at lower irradiance levels. In figure 7, the relative performance loss rates for all

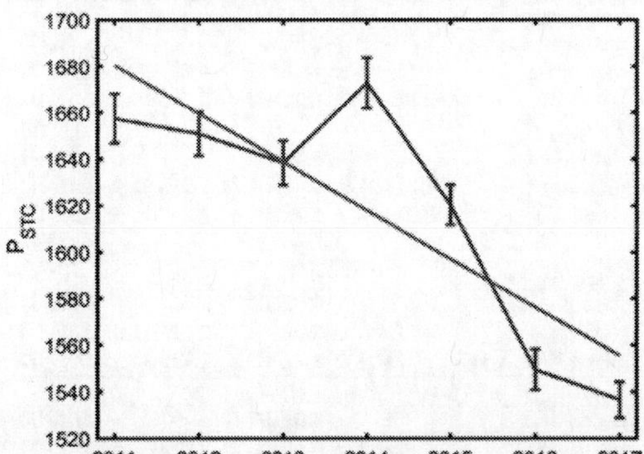

Figure 5: Fitted power output at STC for poly-crystalline silicon. Blue fitted points, red, trend line.

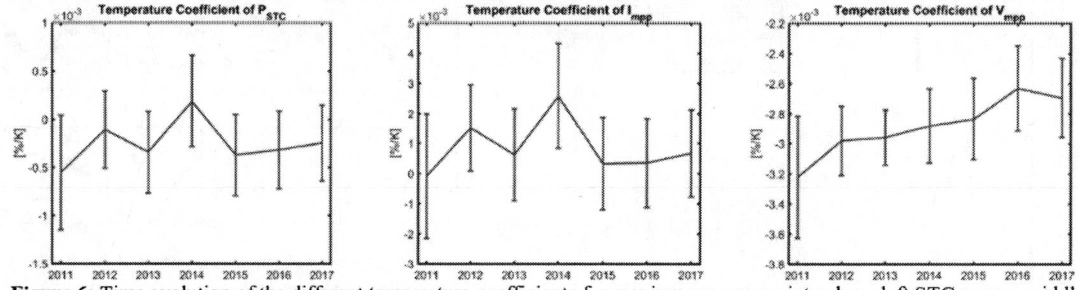

Figure 6: Time evolution of the different temperature coefficients for maximum power point values, left STC power, middle mpp current, and right mpp voltage.

technologies at different irradiance level are shown. CdTe shows an especially large effect at lower irradiance levels. CdTe, however, shows large overall performance loss as well as high performance loss at high irradiance levels. The overall performance loss rates are

- PLR = -0.6 ± 0.2 %/year for pc-Si,
- PLR = -0.8 ± 0.2 %/year for mc-Si,
- PLR = -1.0 ± 0.2 %/year for HIT, and
- PLR = -2.0 ± 0.1 %/year for CdTe.

These values are compatible with previously found values; see [2] or [4]. As the effect is seen in all technologies, the reason for it must be universal. As lower irradiance levels have a different spectrum, it might be possible that spectral effects play a role. However, it is difficult to prove the spectral influence with the present data. Another factor that is easier to understand is the change of the parasitic resistances over time.

We used the generic single-diode model to analyze the effect of changing shunt R_{sh} and series R_s resistances and examine the effect of changing irradiance levels on the efficiency. The two-diode model is defined via the implicit equation

Table 2 Temperature coefficients for power, voltage and current.

	HIT1	mc-Si
$P_{STC}T_{coeff}$	$[1/K]$	$[1/K]$
data sheet	-3.0E-03	-5.0E-03
measured	-2.6E-03 ± 5.0E-04	-3.2E-03 ± 4.5E-04
trend $[\%/year]$	-9.3E-04 ± 5.3E-02	-2.9E-03 ± 4.7E-02
$V_{oc}T_{coeff}$	$[1/K]$	$[1/K]$
data sheet	-2.5E-03	-3.7E-03
measured	-2.9E-03 ± 1.3E-04	-4.6E-03 ± 1.6E-04
trend $[\%/year]$	-2.9E-03 ± 1.3E-02	-5.2E-03 ± 1.3E-02
$I_{sc}T_{coeff}$	$[1/K]$	$[1/K]$
data sheet	3.3E-04	3.0E-04
measured	1.3E-04 ± 4.4E-04	1.2E-03 ± 4.4E-04
trend $[\%/year]$	3.2E-04 ± 4.7E-02	-1.3E-04 ± 4.6E-02
	pc-Si	CdTe
$P_{STC}T_{coeff}$	$[1/K]$	$[1/K]$
data sheet	-4.6E-03	-2.5E-03
measured	-3.1E-03 ± 3.5E-04	-2.5E-04 ± 2.3E-04
trend $[\%/year]$	1.1E-03 ± 3.7E-02	1.7E-03 ± 2.5E-02
$V_{oc}T_{coeff}$	$[1/K]$	$[1/K]$
data sheet	-3.6E-03	-2.5E-03
measured	-4.0E-03 ± 6.1E-05	-2.9E-03 ± 2.0E-04
trend $[\%/year]$	-1.3E-03 ± 5.7E-03	8.6E-03 ± 6.9E-03
$I_{sc}T_{coeff}$	$[1/K]$	$[1/K]$
data sheet	6.0E-04	4.0E-04
measured	7.1E-04 ± 4.0E-04	8.6E-04 ± 9.0E-04
trend $[\%/year]$	-5.8E-04 ± 4.2E-02	-1.2E-03 ± 9.6E-02

$$I = I_l - I_0 \exp\left[\frac{q(V + IR_s)}{nkT}\right] - \frac{V + IR_s}{R_{sh}}, \quad (5)$$

where I_l is the light induced current, which is proportional to the irradiance and is used as the parameter to set the relative irradiance levels, I_0 is the diode current, n is the ideality factor of the diode, k is the Boltzmann constant and T is the cell temperature. The implicit equation was solved numerically for the different tested conditions. As we could not determine any time evolution of the temperature coefficients, we kept the temperature constant for simplicity. This is further useful, as temperature effects can be excluded theoretically this way. In future work a detailed relation between irradiance and temperature can be included. As larger series resistance and lower shunt resistance cause performance loss, we considered these two cases. Large series resistance can be caused by e.g. corroding contacts. Lower shunt resistances are caused by alternative routes for the charge carriers to recombine, which could e.g. be caused by moisture or humidity between the cells and the encapsulation layers. Both of these cases are caused by aging.

In the case of rising series resistance, we found a lower relative efficiency at higher irradiance levels and higher relative efficiency for lower irradiance levels, which does not explain the findings of higher degradation rates for lower irradiance. In turn, the analysis of the shunt resistance yielded that with decreasing (aging) shunt resistance the relative performance is lower at lower irradiance levels, see figure 8. This trend fits well with the found results; with a

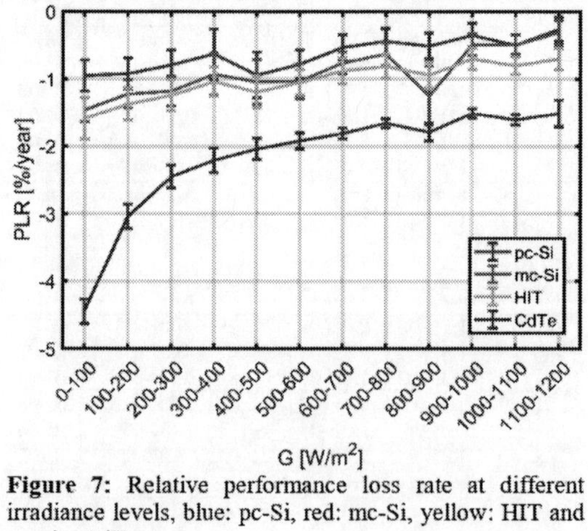

Figure 7: Relative performance loss rate at different irradiance levels, blue: pc-Si, red: mc-Si, yellow: HIT and purple: CdTe.

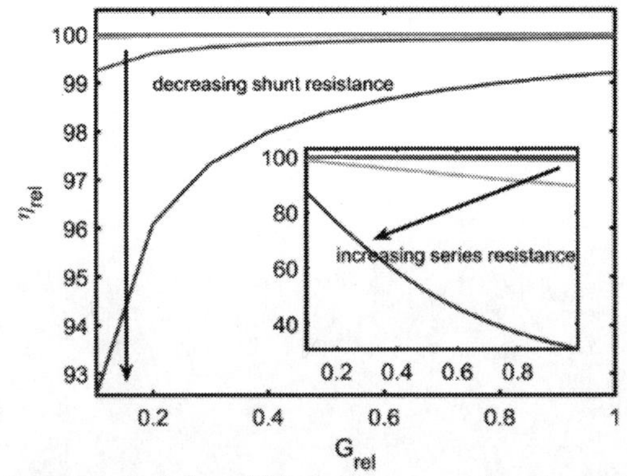

Figure 8: Relative efficiency for a generic solar cell model with changing parasitic resistance and irradiance levels, main figure effect of decreasing shunt on the efficiency at different irradiance levels, inset effect of increasing series resistance.

more detailed less generic analysis it might be possible to quantify the found effects.

4 CONCLUSIONS AND FUTURE WORK

In this study, we investigated the effect of aging on grid connected PV modules. In contrast to previous studies, we did not look at one performance metric in the course of time but evaluated the time behavior of the performance for the influence of irradiance and temperature.

First, we analyzed the PV power output over seven years and fitted a three-dimensional function given in equation (1) and (2) each year. The resulting time evolution of the seven fitting parameters did not yield much insight as the trends in the parameters were not clearly visible and the computed errors surpassed the values in each instance, see table 1. Thus, with this method it was impossible to determine a time evolution for the fitted parameters. The only exception was the fitted power at STC, however even in this case the errors of the degradation rate found were larger than the actual value. To reduce the errors a monthly fitting was performed, however the resulting curves were highly dependent on the available data each month. A possible solution for this would be an analysis after a much longer time such as 20 years. Within the 7 years that we analyzed no trends were visible.

Secondly, an investigation of the temperature coefficients of power, voltage and current was performed. The values of the coefficients reasonably agreed with the datasheet values, however a time evolution was not detectable here either. Longer monitoring periods might help to clarify the picture.

Finally, an analysis of the performance loss rates at different irradiance levels showed clear results. A technology independent trend of higher performance loss at lower irradiances was found. One possible explanation is that the decrease of the shunt resistance over time causes this behavior. These findings could help to understand climatic differences in PV performance loss [5]. An open question here is whether the high degradation seen at low irradiance levels is also visible in regions with low irradiance like northern Europe or if the high performance loss is caused by higher irradiances and only shows at low irradiance levels.

In a future study, we will analyze data from various geographical and climatic regions to understand if the different performance loss rates are visible in all climates, especially low irradiance climates, in which this effect would contribute more to energy production losses. Further, the study of the temperature coefficients and the degradation of the six-parameter model will be repeated later, when more data is available and trend can be found more clearly.

ACKNOWLEDGEMENTS

This project has received funding from the European Union's Horizon 2020 research and innovation programme under the Marie Sklodowska-Curie grant agreement No 721452.

BIBLIOGRAPHY

[1] G. H. Yordanov, "Relative efficiency revealed: Equations for k1–k6 of the PVGIS model," in *IEEE 40th Photovoltaic Specialist Conference*, Denver, 2014.

[2] D. Jordan, "Compendium of photovoltaic degradation rates," *Progress in Photovoltaics: Research and Applications - Wiley Online Library*, 2016.

[3] A. Phinikarides et al., "Review of Photovoltaic Degradation Rate Methodologies," *Renewable and Sustainable Energy Reviews*, 2014.

[4] P. Ingenhoven, G. Belluardo and D. Moser, "Comparison of Statistical and Deterministic Smoothing Methods to Reduce the Uncertainty of Performance Loss Rate Estimates," *IEEE Journal of Photovoltaics*, vol. 8, no. 1, pp. 224 232, 2018.

[5] R. Dubey, S. Chattopadhyay, V. Kuthanazhi, A. Kottantharayil, C. Singh Solanki, B. M. Arora, K. L. Narasimhan, J. Vasi, B. Bora and Y. K. a. S. O. S. Singh, "Comprehensive study of performance degradation of field-mounted photovoltaic modules in India," *Energy Sci Eng*, vol. 5, pp. 51-64, 2017.

[6] "PVGIS," Joint Research Center, [Online]. Available: http://re.jrc.ec.europa.eu/pvgis/. [Accessed 5 September 2018].

[7] M. Šúri, T. Huld, E. Dunlop and H. Ossenbrink, "Potential of solar electricity generation in the European Union member states and candidate countries.," *Solar Energy*, vol. 10, no. 81, pp. 1295-1305, 2007.

[8] T. Huld, M. Šúri and E. Dunlop, "Geographical variation of the conversion efficiency of crystalline silicon photovoltaic modules in Europe," *Progress in Photovoltaics: Research and Applications*, no. 16, pp. 595-607, 2008.

[9] T. Huld, R. Gottschalg, H. Beyer and M. Topić, "Mapping the performance of PV modules, effects of module type and data averaging," *Solar Energy*, vol. 84, pp. 324-338, 2010.

[10] M. C. Peel, B. L. Finlayson and T. A. McMahon, "Updated world map of the Köppen-Geiger climate

classification," *Hydrol. Earth Syst. Sci.*, vol. 11, no. 5, pp. 1633-1644, 2017.

[11] "Historical Data Weather," Autonomous Province of Bozen/Bolzano, [Online]. Available: http://weather.provinz.bz.it/historical-data.asp. [Accessed August 2017].

[12] J. J. Moré, "The Levenberg-Marquardt Algorithm: Implementation and Theory.," in *Numerical Analysis, ed. G. A. Watson, Lecture Notes in Mathematics 630*, Springer Verlag, 1977, p. 105–111.

[13] D. Moser, M. Pichler and M. Nikolaeva-Dimitrova, "Filtering procedures for reliable outdoor temperature coefficients in different photovoltaic technologies," *"J. Sol. Energy Eng*, vol. 136, no. 2, 2013.

AUTOMATED MODULE FAILURE IDENTIFICATION AND PROPOSAL OF REPOWERING IN OPERATING SOLAR PLANTS FOR CONTINUOUS OPTIMUM OPERATION

Rodríguez San Segundo HJ*[1], Sánchez Serrano JS[1], Imbert R[2], Gonzalo-Martín C[2], Robles V[2], Calo López A[1] and de Vicente Suso C[1]
[1]The South Oracle, S.L.
Paseo de las Delicias, 1
41001 Sevilla (Spain)
Tel. +34 954 562 955, Fax +34 954 560 857
[2]Escuela Técnica Superior de Ingenieros Informáticos, Universidad Politécnica de Madrid
Campus Montegancedo, s/n
28040 – Madrid
Tel. +34 913 367 399, Fax +34 913 367 399
*Author for correspondence. E-mail: hugo@thesouthoracle.com

ABSTRACT: Solar PV power plants are composed of thousands of solar modules. It is a known fact that 2% of them will fail after year 10 of operation, causing losses as high as 27% of total income. In a market of plants over 10 years old of 10 Gigawatts peak (GWp) in 2018, and 100 GWp by 2025, this leads to global losses of EUR 750 million and EUR 7.5 billion, respectively. Despite these figures, this problem is currently not being tackled, because, to recover the losses, a continuous repowering of the PV plant must be performed, i.e., a continuous detection and substitution of the failed modules for new ones. This implies at least an annual inspection, and an ulterior analysis of the inspection results in order to make the best decision on the repowering configuration. While the inspection will soon start reducing costs by the increasing penetration of drones, the ulterior analysis is still carried out manually by qualified technicians (supervising, one by one, the thousands of images taken by the drone). This fact escalates costs and makes the a continuous repowering unaffordable, i.e., not worth the recovery of the losses. The only way to make continuous repowering viable is to automate the image processing, analysis and decision-making. The authors, through our experience in the field, have calculated that automation will save up to 90% of current costs, making repowering profitable. Therefore, in the frame of a Spanish R&D project, we are currently developing Optimized Solar Repowering (OSR), a software that carries out an automated image analysis of the images taken by the drone, in which failures are identified, geo-localized and classified, and, also, automatically proposes the optimum repowering configuration, by means of own developed algorithms embedded into a user-friendly software architecture. First results and advances of the project are presented.

Keywords: PV plant O&M, PV Module Failure, Artificial Intelligence, Aerial Inspection.

1 INTRODUCTION

According to international reports[1], at least 2% of the solar photovoltaic (PV) modules in operating solar plants will fail after year 10 of operation. As these modules are series-connected to other "healthy" modules in strings, they can lead to monetary losses as high as 27% of total income, depending on location, type of solar plant and remuneration fee. Table 1 shows several examples of different types of solar plants in Spain – some of which we have analyzed manually at very high analysis costs. Losses are high – in the range of tens of thousands of euros yearly – even for highly subsidized plants.

Table I: Losses due to defect modules in several types of solar power plants in Spain (own calculations).

Plant type*	A subsid.	B subsid.	C subsid.	D subsid.	E non-subsid.
Annual avg. loss (€)	20.890	27.579	3.358	6.071	15.025
% over total income	1,53%	1,62%	3,31%	6,54%	26,62%

*A: Ground-mounted, 1,89 nominal megawatts (MWn), fixed, Spanish Royal Decree (RD) 661/2007, installed in 2007.
B: Ground-mounted, 1,89 MWn, two-axis tracking, RD 661/2007, installed in 2008.
C: Rooftop, 200 nominal kilowatts (kWn), RD 1578/2008, zone 4, installed in 2009.
D: Rooftop, 200 kWn, RD 1578/2008, zone 4, installed in 2013.
E: Rooftop, 200 kWn, non-subsidized, self-consumption, zone 4, installed in 2017.

For non-subsidized plants (case E in Table 1) losses may exceed 25%. This is significant because these plants represent much better the current state-of-the-art as they are much younger plants implying updated costs of retail electricity (higher) and of solar power (much lower, around USD30/MWh in the "sunbelt" regions, and close to USD50-60/MWh in less sunny areas).

Today, there are 10 gigawatts peak (GWp) of solar plants over 10 years old[2]. By 2025, this number will explode until 100 GWp. The potential losses could reach EUR 750 million in 2018 and EUR 7.5 billion by 2025.

The straightforward solution to recover the losses would be continuous repowering. I.e., continually substituting the modules that fail for new ones – for instance, yearly or at lower intervals. This process would imply inspecting the plant in search for failed modules. Failures in modules lead to hot spots that can be detected by visible (VIS)[1], thermography (IR)[3] and electroluminescence (EL)[4] images. This inspection, which is still mainly manual, is rapidly reducing costs by the increasing penetration of drones, which will very probably become state-of-the-art soon[5,6,7].

However, once the images are taken, currently there are no commercially available solutions for the analysis, failure identification and ulterior decision-making, which are completely manual: a specialized technician analyses the thousands of images – one by one – and provides a diagnosis of the power plant and a repowering proposal. The cost of this method usually exceeds the actual losses (stated in Table 1).

Therefore, as of today, there is no economically available solution for continuous repowering. The only way to make it viable is to automate the analysis step. Were this achieved, analysis costs would reduce dramatically and 80 to 90% of the losses due to module failures could be recovered (Table 2).

Most common modes of failure include soiling, glass or cell cracks, encapsulant or backsheet delamination or junction box misconnection[1]. The detection by IR, VIS or EL imaging gives much more precise information – at module level – than any other data source (e.g. production monitoring data, which will be soon deployed also at module level thanks to Internet of Things (IoT)[8,9]). These imaging techniques can be used in combination with any or a variety of those other data sources.

Table II: Average annual losses due to defect modules and recovery after repowering (own calculations).

Plant type (Table 1)	A	B	C	D	E
Annual avg. loss (€)	20.890	27.579	3.358	6.071	15.025
Avg. *net* recovery (€)	16.962	23.187	2.877	5.476	13.793
% recovered	**81%**	**84%**	**86%**	**90%**	**92%**

Our approach combines both automatic image analysis of these images AND automatic plant repowering proposal.

To reach this uniqueness, this work is innovative in several ways:

- It uses the latest advances in technologies of Big Data[10], Artificial Intelligence[11] and Image Processing[12]. Dealing with data volume, variability and veracity is one of the big challenges.
- Also, specific, unique binary models are being developed for module failure identification, geo-localization and classification into several types of failures, depending on their degree of recoverability.
- Furthermore, we are also developing specific algorithms to automatically propose the best repowering configuration, considering both technical and economic data of the plant.

2 DESCRIPTION OF PROPOSED DEVELOPMENT

2.1 Overview

Much has been researched on the modes of failures that PV modules can present while on operation. Task 13, "Performance and Reliability of Photovoltaic Systems", of the International Energy Agency Photovoltaic Power Systems Programme (IEA PVPS) gathers experts worldwide on the matter and provides excellent summaries of the advances in science and technology. In this sense, module failure modes are well known and classified. Their causes are also state-of-the-art, and it is well tabulated how each of them can be identified with different measuring techniques[1].

Furthermore, the effects of each failure mode on performance is also subject to past and on-going research[13], from which performance loss depending on location, climatic conditions and many other parameters can be estimated.

In addition, recent technology developments have allowed massive penetration of aerial inspections with drones. While electroluminescent (EL) cameras are still making their way towards commercial use[5] – certainly available in the near future –, high-quality thermographic (IR) and visible cameras for drones are already being used. Worldwide efforts are currently going on to standardize such imaging techniques, with an important focus on IR and EL cameras[14]. Some of them have already made their way through to national standards, as is the case of the German VATh, the Association for Applied Thermography[15].

However, once the aerial inspections are carried out, there are usually thousands of raw images to analyze. This, which is still carried out manually by experts who go through the many images one by one, is the missing piece that prevents wide deployment of aerial inspections in regular O&M operations.

Our software, Optimized Solar Repowering, consists on an automatic analysis of such a high number of images. Given the current commercial state of development, we have focused on IR and visible (VIS) images, but we will incorporate EL in the future. The reason to add VIS images is to help identify the hot spots seen in the IR images. While some modes of failure are clearly identified only with IR images (PID effects, string disconnect, or substring disconnect, for instance), others require a visual inspection. With the VIS image right at the side of the IR image, this visual inspection can take place right away and save precious time. Furthermore, with these three imaging technologies, more than 95% of most frequent modes of failure can be identified[1].

The software development has two main blocks, which are described below:

Block I: Module failure recognition, identification and classification. OSR must not only locate where a failure is, but also unequivocally identify which mode of failure it is and classify it in diverse categories as a function of their "recoverability" or "substitutability".

Block II: Plant repowering proposal. As a final result, OSR must provide a report of repowering, i.e., which is the optimum solution, both technically and economically, to repair and/or substitute the failed modules.

Block I will be developed in three highly differentiated stages, which will gradually incorporate more and more Artificial Intelligence (AI). These three stages are:

Stage I: Manual module failure detection platform ("Manual annotator"). A web platform will ease the manual work of the PV expert. It will present both the IR and the VIS image, allow the expert to zoom in and out, and annotate the detected failure. It will provide a report of all found failures, their location and their type. All the gathered information will be stored in an in-house designed Data Base with the aim of using it in the next stages.

Stage II. Semi-automatic module failure detection platform ("Semi-automatic annotator"). With machine learning algorithms, the software will increasingly learn from the PV expert's inputs. This second stage will already select where potential failures are and suggest failure modes to the expert. This stage will gradually increase automation from almost always manual to almost automatic.

Stage III. Automatic module failure detection platform ("Automatic annotator"). This final stage will automatically detect, identify and classify the module failures as a result of machine learning from the thousands of images analyzed by the PV expert.

3 ADVANCES IN MANUAL ANNOTATOR

The current stage of development of the OSR project is an almost finished Stage I in Block I. The "Manual Annotator" has simplified the analysis work of PV experts, already reducing their analysis time by 30%.

As seen in Figure 1, it presents the two images of the same part of the PV plant, IR and VIS.

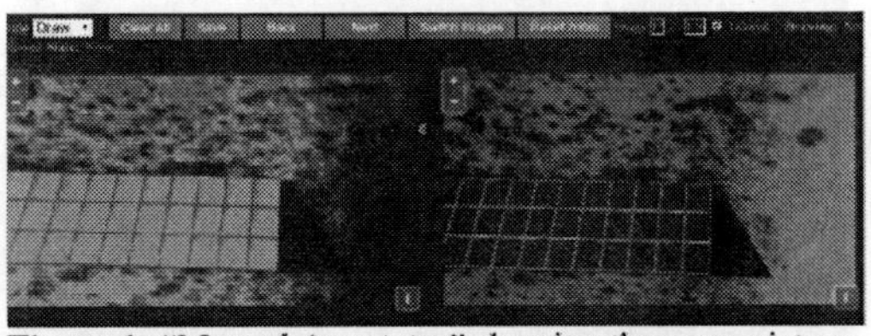

Figure 1: "Manual Annotator" showing the same picture within a PV plant, IR left and VIS right, with a hot spot on the IR picture.

Once a failure is detected, generally a hot spot on the IR image whose temperature is known, the PV expert can zoom in. Both images zoom in at the same time, so that the hot spot is matched with the visual image. They can be marked within a polygon for easier recognition of the area. If visually identifiable, the PV expert will be able to tag the type of defect from a menu. Figures 2 and 3 show this process for a partial shadowing. Behind the scenes, the machine learning algorithm saves these data and uses them for the process automation. All data are saved in a Data Base.

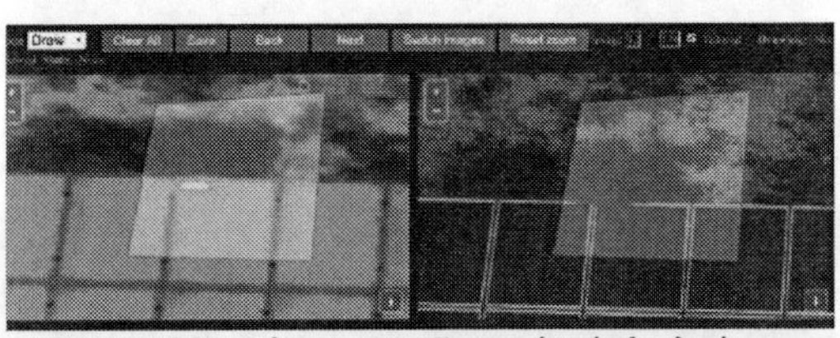

Figure 2: "Manual Annotator" zooming in both pictures of Figure 1, IR left and VIS right. Failure has been framed for easier recognition. Partial shadowing is clearly visible in the VIS image (right).

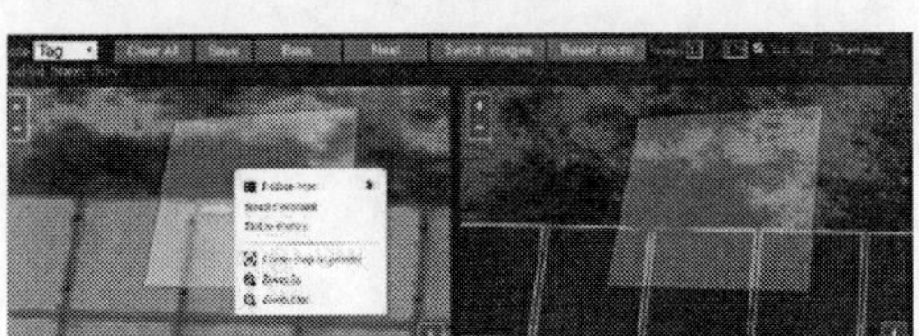

Figure 3: "Manual Annotator" showing failure selection menu, for the picture shown in Figure 1 and zoomed in in Figure 2.

Figures 4 to 6 show further modes of failure that are easily detected by the Manual Annotator. In fact, in Figure 4, the hot spot on the IR image is right on the junction box. This could mislead to the conclusion that there is a failure in the junction box. However, the VIS image clearly shows that the third cell from the top left is burnt and therefore damaged. This module, for instance, must be replaced and not repaired (as would be the case in some junction box failures, if only the junction box is damaged).

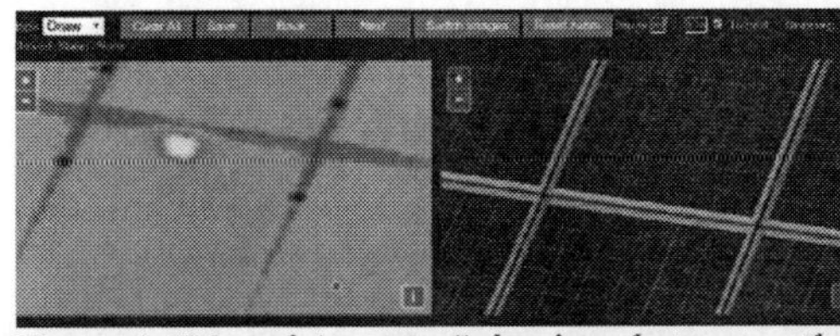

Figure 4: "Manual Annotator" showing a hot spot on the IR image, which turns to be a burnt cell when seen in the VIS image (third cell from the top left of the module).

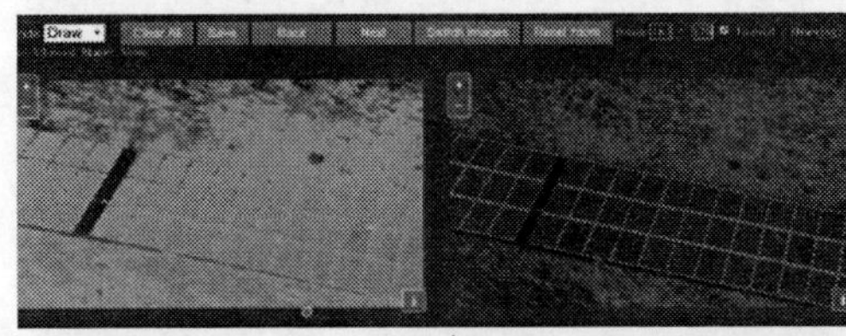

a.)

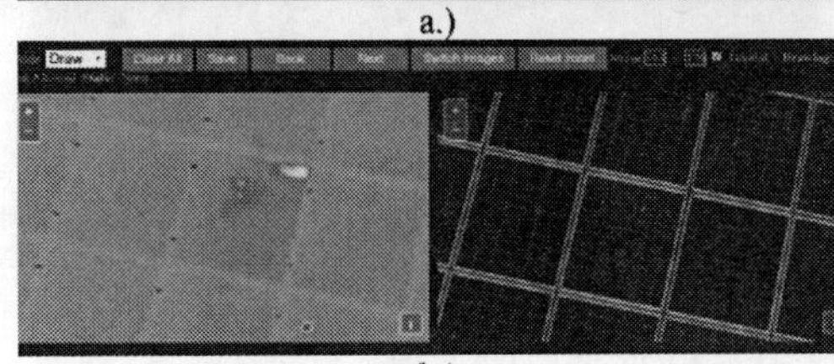

b.)

Figure 5: One module shows several hot spots in the IR picture (a.) left. When zoomed in (b), the VIS image clearly shows that the front glass is broken in several places.

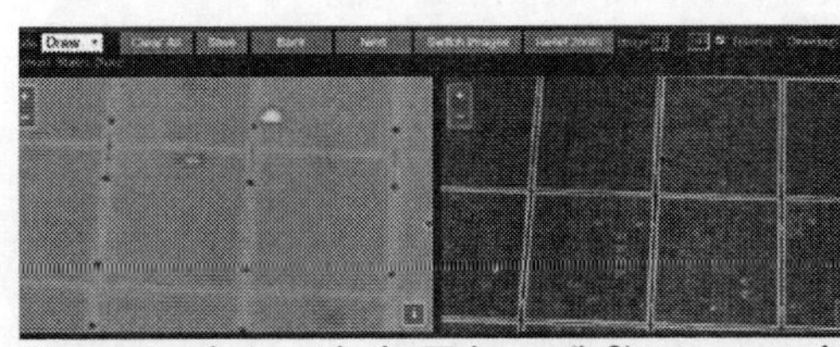

Figure 6: A hot spot in the IR image (left) turns out to be due to snail tracks, clearly recognizable in the VIS image (right).

4 CONCLUSIONS

A novel software of economic continuous repowering of solar PV plants has been presented: Optimized Solar Repowering (OSR). It overcomes the main obstacle for continuous substitution of failed modules in operating plants for new ones: the unaffordable cost of continuous data analysis by experts. I.e., even if data gathering is becoming more and more economic thanks to the ever-growing penetration of drone technologies with embedded IR, EL and VIS cameras, the post-processing of these data is still manual and depends on the criteria of PV experts.

OSR is a patented[16] software currently in development which uses Image Processing, Artificial Intelligence and Big Data algorithms to detect, identify and classify the defects found at drone inspections of PV plants, and makes the decisions on what the most suitable repowering proposal is, both economically and technically. As an automated process, it saves up to 90%

of current manual analysis costs, therefore allowing for a more continuous plant inspection and continuous repowering of the solar plant, keeping its performance at the maximum possible level at all times.

OSR current stage of development is a functional manual annotator platform, in which a PV expert comfortably navigates along the whole photos of a PV plant, both thermal and visible, can zoom in and out, identify the modules that fail and the type of failures. We have calculated that this step alone saves around 30% of analysis time, and therefore of costs. Moreover, all information generated by the PV expert is saved in an in-house designed Data Base, with the aim to use this information for training machine learning algorithms in the next OSR stages.

Next steps towards 90% of cost reduction will include semi-automatic annotator, which will already involve machine learning from the expert's input, and a wholly automatic annotator.

5 ACKNOWLEDGEMENTS

The authors wish to acknowledge the Spanish Centre for Industrial Technological Development for their support through R&D project IDI-20180194.

6 REFERENCES

[1] Köntges, M., et al., "Review of Failures of Photovoltaic Modules", Report IEA-PVPS, T13:01, 2014

[2] Jäger-Waldau, A., "PV Status Report 2016", JRC Science for Policy Report, 2016

[3] Buerhop, C., et al., "Verifying defective PV-modules by IR-imaging and controlling with module optimizers", in Proc. of the 33rd EUPVSEC, pp. 1488-1500, Amsterdam, 2017

[4] Vannier, G., et al., "Classification of photovoltaic defects through type of characterizing methods and faults discriminations", in Proc. of the 33rd EUPVSEC, Amsterdam, 2017

[5] dos Reis Benatto, G.A., et al., "Luminescence imaging strategies for drone-based PV array inspection", in Proc. of the 33rd EUPVSEC, pp. 2016-2020, Amsterdam, 2017

[6] Koch, S., et al., "Outdoor electroluminescence imaging of crystalline photovoltaic modules: Comparative study between manual ground-level inspections and drone-based aerial surveys", in Proc. of the 32nd EUPVSEC, pp. 1736-1740, Munich, 2016

[7] Lanz, M., et al., "Drone-based assessment of cleaning effects on PV installations", in Proc. of the 32nd EUPVSEC, pp. 1960-1963, Munich, 2016

[8] Jana, J., et al., "An IoT Based Smart Solar Photovoltaic Remote Monitoring and Control unit.", 10.1109/CIEC.2016.7513793, 2016.

[9] Basuvaiyan, S. and Rathinasabapathy, V., "IoT based Solar Photo Voltaic Monitoring System", 33rd NCEE & National Conference On "Hybrid AC/DC Power Systems for Effective Utilization of Renewable Energy", Thiruchirappalli, 2017.

[10] Goswami, R., et al., "Materialized view selection using evolutionary algorithm for speeding up big data query processing", Journal of Intelligent Information Systems, Vol. 49, Issue 3, pp. 407-433, 2017.

[11] "Artificial Intelligence", in: Carayannis E.G. (eds) Encyclopedia of Creativity, Invention, Innovation and Entrepreneurship. Springer, New York, NY, 2013.

[12] Gonzalo, C., et al., "Local optimal scale in a hierarchical segmentation method for satellite images", Journal of Intelligent Information Systems, Vol. 46, Issue 3, pp. 517-529, 2016.

[13] Köntges, M., et al., "Assessment of Photovoltaic Module Failures in the Field", Report IEA-PVPS, T13:09, 2017

[14] Köntges, M., et al., "Review on Infrared and Electroluminescence Imaging for PV Field Applications", Report IEA-PVPS, T13:10, 2018

[15] VATh, "VATh-Guideline: Electrical Infrared Inspections • Low Voltage: Planning, execution and documentation of infrared surveys on electrical systems and components ≤ 1kV", Edition February, 2016

[16] Rodríguez San Segundo, H.J., de Vicente Suso, C. and Calo López, A., Spanish Patent Application ESP201730710

HOW TO MAXIMIZE THE KWH/KWP RATIO: SIMULATIONS OF SINGLE-AXIS TRACKING IN BIFACIAL SYSTEMS

G.J.M. Janssen[1] (janssen@ecn.nl), A.R. Burgers[1], A. Binani[1], A.J. Carr[1], B.B. Van Aken[1], I.G. Romijn[1]
M. Klenk[2], H. Nussbaumer[2], T. Baumann[2]
1. ECN part of TNO, Solar Energy, Westerduinweg 3, NL-1755 LE Petten, The Netherlands
2. Zurich University of Applied Sciences, SoE, Institute of Energy Systems and Fluid Engineering
Technikumstrasse 9, 8401 Winterthur, Switzerland

ABSTRACT: The energy yield of PV systems with horizontal single-axis tracking and bifacial panels was calculated using BIGEYE. BIGEYE is a versatile code developed at ECN part of TNO to calculate the yield of bifacial PV. Comparison to recent measured data with the BIFOROT set-up at the Zurich University of Applied Sciences showed very good agreement. Our BIGEYE results show that the bifacial energy gain and the tracking gain are mostly additive, making the combination of bifacial panels and tracking a very attractive option. For two different locations, Doha and Amsterdam, increases of energy yield in the order of 25% compared to monofacial, fixed tilt systems are possible at relatively modest ground cover values.
Keywords: Bifacial, Energy performance, Tracking, Simulation.

1 INTRODUCTION

The yield of a PV system can be improved by making the most of the available sun light. Apart from using cells and modules with a high conversion efficiency, there are two other ways to achieve this. The first is not to waste the light incident on the rear, i.e. use bifacial panels. The second way is to maximize the direct, or beam component of the irradiation by tracking the sun. For large scale applications Horizontal Single Axis Tracking (HSAT) is a very practical, reliable way (compared to 2-axis-tracking) but uses more area than fixed tilt (FT) fields. Monofacial HSAT systems produce in the range of 20% more energy than monofacial FT systems [1].

On the other hand, bifacial PV farms are becoming more and more common. Bifacial cells and modules are now available at limited additional cost. Depending on the location, climate and the albedo values, they can produce 20% or even more energy than monofacial systems [2].

Although no significant technological barriers seem to prevent combining HSAT and bifaciality, there some concerns whether the advantages of bifaciality are maintained in a HSAT configuration, since the maximized beam direct beam will cause a significant shade of the panels on the ground. Recently, a few studies have been published that show promising yield for HSAT with bifacial panels [3-6].

We present a simulation study using BIGEYE, the bifacial energy yield model for PV systems developed at ECN part of TNO. Features of this model are explained in the next section and we present validation of the model by comparison to measured data of the Bifacial Outdoor Rotor Tester at ZHAW. After discussing geometrical parameters that affect the bifacial and HSAT gain, results are presented for two different locations that will show that bifacial gain and HSAT gain are to a good approximation additive. This is especially relevant for higher latitude (above 50°) where monofacial HSAT would probably not be cost effective.

2 BIGEYE SIMULATION CODE

BIGEYE calculates the energy yield of bifacial PV systems at a given location. Minimal inputs are the module specifications and a set of time-dependent meteorological data containing at least the Global Horizontal Irradiance (GHI) and ambient temperature. Diffuse Horizontal Irradiance (DHI) can be supplied or calculated. If wind speed data is provided, it can be included in the thermal model. Parts of this model were published before [7,8].

The latest version of this model, BIGEYE V3, is very flexible in the geometry of the PV field. It can treat fixed-tilt (FT) systems with any surface azimuth angle as well as HSAT configurations. The PV field can consist of single or multiple arrays, each having any number of rows and columns of modules. The model is therefore not limited to infinitely long arrays or stand-alone modules; essentially PV fields of all sizes can be modelled. This means also edge effects, i.e. modules at the edge of an array receiving different irradiance can be treated. Moreover, tilt angles and row-to-row distances in the field can be non-uniform as illustrated by Fig. 1.

The total front and rear irradiance is calculated at (sub)cell level, where the position of the cell in the

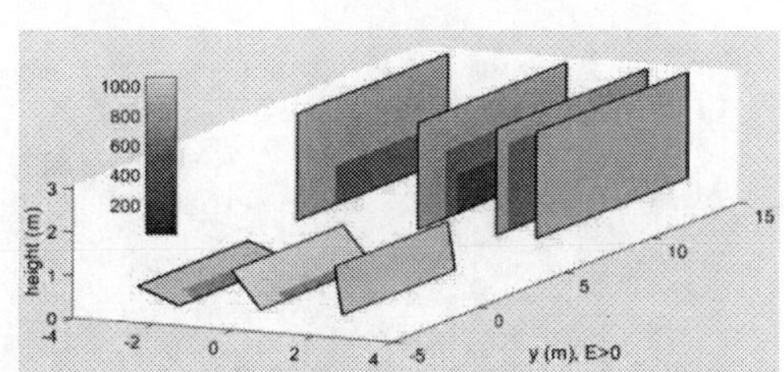

Figure 1: Beam/direct irradiance on module surfaces illustrating the flexibility in field geometry BIGEYE can handle. Note the modules have exaggerated transparency to show shadow on shadow effects.

system, as well as the length, width, ground clearance and pitch of the arrays are determining how much irradiance is received at specific insolation conditions.

The ground-reflected (albedo) irradiance is an essential feature of bifacial yield modelling. In BIGEYE this ground-reflected irradiance considers the geometry of the system: both the direct and diffuse irradiance incident on the ground are limited by adjacent arrays (self-shading), as well as the part of the ground that can be "seen" by the surface of the panel (view factor approach). The self-shading is the main reason why the irradiance on a bifacial panel is non-uniform. In BIGEYE V3 all view factors, from panel the ground as well as to the sky are fully 3D and numerically calculated. Of course, direct shading is also accounted for in the model; but panels can be partly transparent, reducing the direct shading and self-shading. Angle-of-incidence dependent reflectance of the direct component on front and rear is used; effective angles of incidence are used for diffuse and ground-reflected irradiance on front and rear.

The electrical output is calculated with a full electrothermal model based on the single-diode equation describing the module output, with temperature-corrected coefficients. The operating temperature of the cells is obtained from a heat balance, considering the appropriate irradiance absorption coefficient for front and rear irradiance [9]. The generated current I_L in the module is calculated according to the equivalent or compensated current method:

$$I_L \propto G_{front} + \varphi \, G_{rear} \qquad (1)$$

with G_{front} and G_{rear} the irradiance on front and rear, respectively, and φ bifaciality factor of the module (the ratio of the rear-side and front-side efficiency at STC conditions).

Direct shading on cells by adjacent arrays and objects leads to a further reduction of power output depending on the configuration of the diode-protected blocks. This power reduction is calculated with the Rodrigo model [10]. For the much smaller mismatch resulting from the mismatch in current due to non-uniform rear-irradiance a correction is used based on the relative standard deviation in the total current [11].

The major part of the BIGEYE code was written in-house, but use has been made of code available in the library provide by the PV Performance Modelling Collaborative [12].

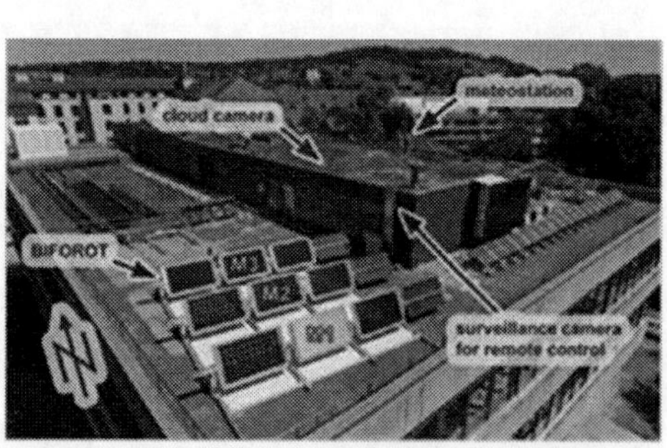

Figure 2: BIFOROT on the roof of ZHAW. The presented results were measured with 0.75 m height of the rotation axis above the ground. The pitch of the rows is 2.86 m and the ground albedo was measured to be 51%. The measured module is indicated as M2.

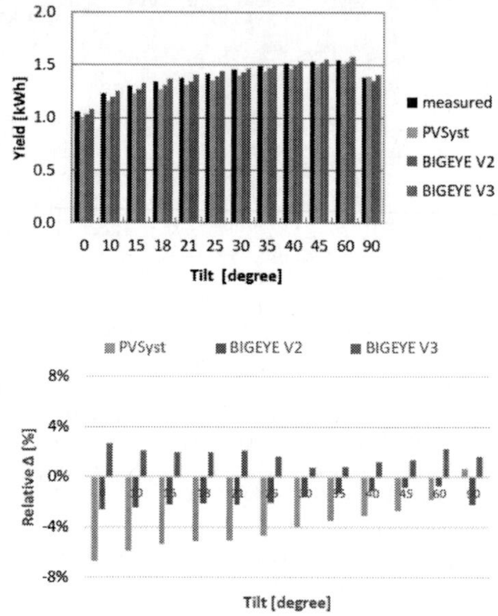

Figure 3: The measured output of the BIFOROT system and simulation results (top). The bottom figure shows the relative deviations of the simulated from the measured data.

3 VALIDATION WITH BIFOROT EXPERIMENTS

Data from the Bifacial Outdoor Rotor Tester (BIFOROT) at the Zurich University of Applied Sciences (ZWAH) were used to validate the BIGEYE approach. The BIFOROT is an experimental tool to understand and optimize bifacial PV systems. It consists of three rows of four modules oriented toward the south and continuously rotating with 12 steps between 0° and 90° tilt angle in a one-minute cycle (Fig. 2). In this way the set-up mimics twelve fixed-tilt configurations at the same conditions. The output of a bifacial module in the center of the set-up (M2) is continuously monitored, as well as the local meteorological data. More details can be found in [13-14].

Fig. 3 shows the measured accumulated energy produced during a single sunny day (15 October 2017) at each angle. We compare this to results from 1) simulations by BIGEYE V3, 2) the previous BIGEYE version (V2) and 3) and simulations with a bifacial version of PVSyst [19]. Also shown are the relative deviations of the simulations from measured data. All simulated yields are close to the measured yield and capture the tilt dependency well, but the BIGEYE V3 data are within 2% accuracy at moderate angles and within 3% at extreme angles like 0° and 90°. It should be mentioned that PVSyst states their code is only suited for tilt angles up to 60°. Further analysis is ongoing, but the main difference in the two versions of BIGEYE was found in the model describing the division of diffuse light into isotropic diffuse and circumsolar irradiation, i.e. the Perez [15] used in V3 and the Hay-Davies [16] model used in V2. More results on the BIFOROT validation will be presented elsewhere.

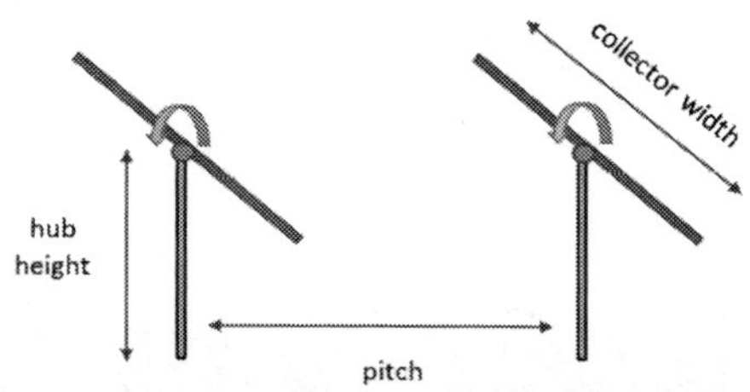

$$GCR = \frac{\textbf{collector width}}{\textbf{pitch}} \qquad H = \frac{\textbf{hub height}}{\textbf{collector width}}$$

Figure 4: Schematic of a HSAT geometry with definition of the normalized parameters *GCR* and *H*.

4 HSAT SYSTEM GEOMETRY

In the remainder of this paper we considered trackers with a North-South oriented axis. To minimize shading effects backtracking was assumed according to the algorithm in [17]. A parameter with a strong effect on the yield is the distance between trackers, expressed by the Ground Cover Ratio (*GCR*) which is the collector width divided by the tracker pitch as illustrated in Fig. 4. In the case of bifacial PV, the distance of the panel to the scattering ground is an important parameter [18]. For HSAT this distance is characterized by the parameter *H*, i.e. the height of the axis divided by the collector width

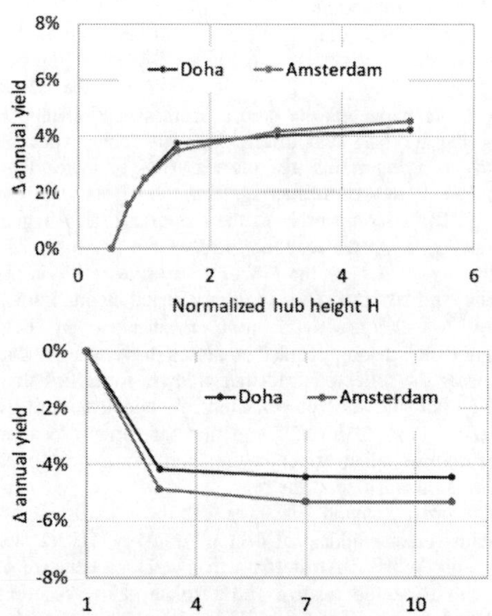

Figure 5: The change in annual energy yield of a bifacial HSAT tracker as function of the normalized hub height (top). Considered is the central tracker in a field of seven parallel trackers. The bottom figure shows the change in energy yield of a tracker in a field with a given number of trackers compared to an isolated tracker.

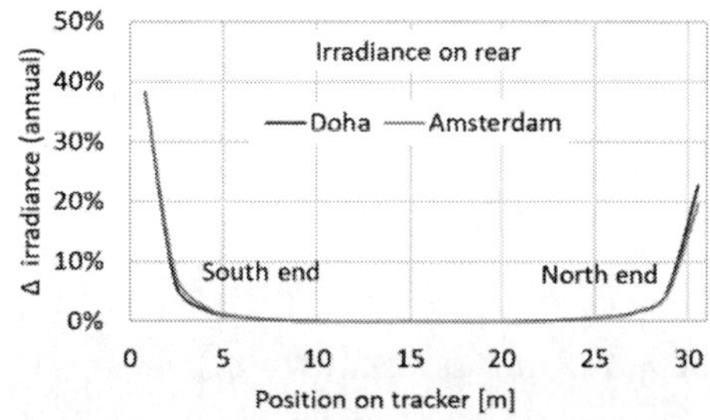

Figure 6: The rear irradiance on a module as a function of the distance to the edge of the tracker, relative to the rear irradiance at the center of the tracker.

(Fig. 4). By using the normalized parameters, the results on irradiance are independent of actual configuration, i.e. whether the panels are in landscape or portrait orientation and how many panels there are in the direction perpendicular to the axis. Of course, the actual layout will influence the electrical system output when the full electrical string is considered. Also, even though back tracking avoids shading on the front, the rear irradiance on the panels will depend on their position on the tracker, making the tracker length another parameter. Finally, the total number of trackers in the field determines the total shading on the ground.

In this study we considered 19 panels with their long side parallel to the tracker axis with the axis (virtually) positioned in the center of the panel. Meteorological data were synthetically generated from monthly average values in PVGIS [20] using PVSyst with one-hour resolution [21]. We considered two locations, one with a high beam component, i.e. Doha at 25° latitude (annual GHI 2273 kWh/m^2, annual DHI 774 kWh/m^2, DHI:GHI=0.34) and Amsterdam at 51° latitude, where the diffuse component is dominant (annual GHI 1067 kWh/m^2, annual DHI 566 kWh/m^2, DHI:GHI=0.53).

Unless stated otherwise, a moderate ground reflection coefficient (albedo) of 0.3 was assumed. The panels had a front-side efficiency of 18 % and a bifaciality factor φ of 0.8. The base case parameters of the simulations considered the center tracker in a field of seven parallel trackers with *GCR*=0.35 and *H*=0.75.

Fig. 5 shows the effect of the normalized axis height at both locations. As expected the yield increases with *H* but the differences between the two locations are rather small. A more pronounced difference is found by varying the number of trackers, compared to the annual yield for an isolated tracker. Next to the reduced view factors to the sky and to the ground for a none-edge tracker, the shaded area on the ground is larger at higher latitudes leading to a larger reduction in the electrical output. Notice that that with seven trackers the yield of the central tracker seems representative for the yield of a tracker in a larger field. The above results are in good agreement with other, recently published simulation results [5,6].

Noteworthy is also that in both locations there is a large increase of the rear irradiance at the ends of the tracker, especially at the south end (Fig. 6), as also observed in [6]. Whether this additional irradiance can be harvested depends on electrical (string) arrangement of the field. When the whole tracker is in one string the non-uniform irradiance will lead to mismatch losses, which could outweigh the irradiance gain. Note that we expect such mismatch losses to be small as the uniform front-side illumination of the tracked panels is dominating [11].

The ground cover ratio *GCR* will have a strong impact on the output of both monofacial and bifacial configu-rations and this will be discussed in the next section.

5 BIFACIAL AND TRACKING GAINS

Fig. 7 compares for two locations the calculated gains with respect to monofacial panels in a fixed-tilt configuration that can be obtained by using, respectively, bifacial panels in a FT configuration, HSAT with monofacial panels, and HSAT with bifacial panels. For the FT systems an optimized tilt was used, i.e. tilt angles varied between 12° and 25° at Doha and 10° to 36° at Amsterdam, depending on the *GCR*. Note, that by considering a central row in a seven row FT field or a central tracker in a seven tracker HSAT field, these results are representative for a large field. The FT bifacial gain increases with *GCR*, reaching values up to 10 % at both locations at an assumed albedo of 0.3. HSAT gains are even more strongly dependent on the *GCR*. For Doha, with a strong direct component in the light, substantial HSAT gains are calculated. At a higher latitude, such as Amsterdam, the HSAT gain is much more modest.

Doha:

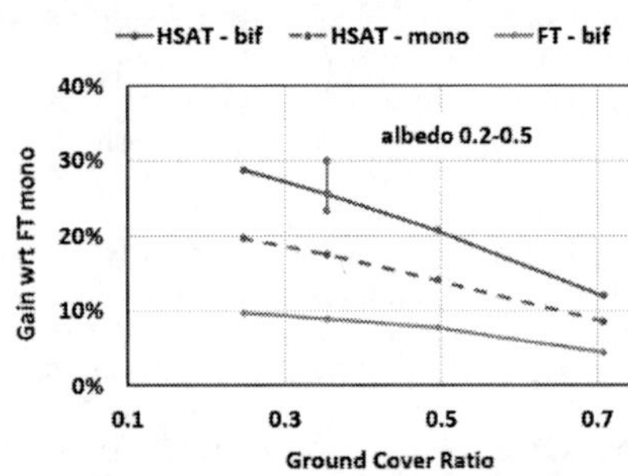

Amsterdam:

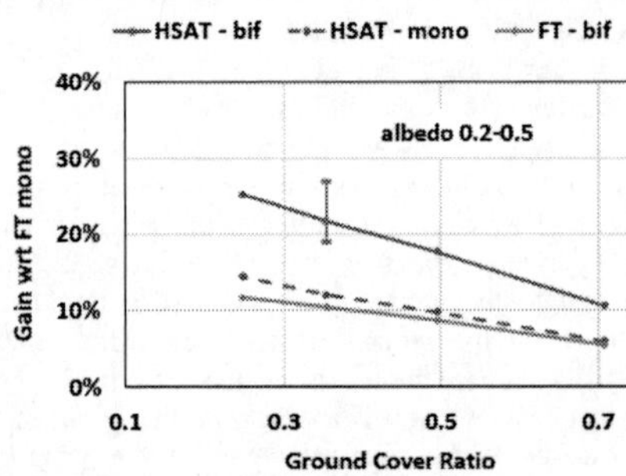

Figure 7: Calculated gains with respect to monofacial panels in a fixed-tilt configuration that can be obtained by using, respectively, bifacial panels in a FT configuration,(orange) HSAT with monofacial panels (blue dashed lines), and HSAT with bifacial panels (blue solid lines). For the FT systems an optimized tilt was used. The assumed albedo was 0.3 but the green lines indicate the variation in gain for albedo between 0.2 and 0.5.

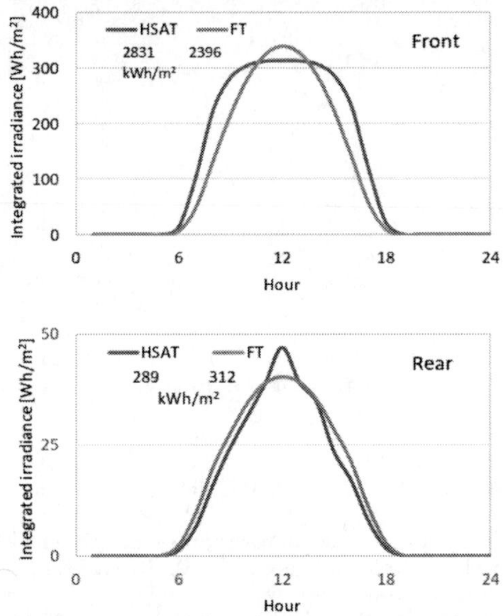

Figure 8: The total annual irradiance on the front and rear of a panel in a HSAT and a FT configuration at Doha.

Moreover, Fig. 5 shows that the extra yield that is obtained when substituting monofacial panels by bifacial panels in a FT system, is also found in the case of HSAT. This implies that the advantages of bifaciality are not lost in a HSAT configuration.

In Fig. 8 the annual irradiance distributions on the front and rear of a central panel in a HSAT configuration are compared to a FT configuration for the Doha location. As expected, the gain in front side irradiance is during the morning and afternoon, with some reduction around noon when the tilt angle is 0°, i.e. lower than optimal. The total irradiance gain on the front is in the order of 18 %, comparable to the monofacial HSAT gain seen in Fig. 7. At the rear the irradiance is much smaller, and there is a loss for the HSAT compared to FT in the morning and afternoon but a gain around noon, leading overall to almost equal rear irradiances in both configurations. More detailed analyses have shown that at the rear the reflected irradiance is lower for HSAT than for FT, but this is compensated by additional diffuse irradiance. For other *GCR* and the Amsterdam location similar compensating effects were found for the reflected and diffuse irradiance at the rear.

Considering again Fig. 7 it can be concluded that depending on the additional cost of trackers, HSAT will be less profitable in Amsterdam than in Doha because of the lower expected relative and absolute gains. Whereas in Doha bifacial panels would add about 50% additional gain to the HSAT gain, deploying bifacial panels in Amsterdam would about double the HSAT gain, making HSAT a viable option in Amsterdam, even at modest *GCR*.

The yield of a bifacial system increases with the albedo, i.e. the ground reflection coefficient. In Fig. 7 it is indicated that with albedo values of 0.5, which are not unrealistic, the gains can go up by 5% absolute which brings bifacial HSAT gains in the order of 30 % within reach for both locations.

6 CONCLUSIONS

The new BIGEYE simulation tool to calculate bifacial energy yield was validated against BIFOROT data. The total yield of bifacial HSAT PV will depend on geometry parameters such as *GCR* and hub height. Based on our simulations we think increases of energy yield in the order of 25% compared to monofacial, fixed tilt systems are possible at relatively modest ground cover ratio.

The BIGEYE simulations show that bifacial energy gain and HSAT gain are mostly additive, making the combination of bifacial panels and HSAT a very attractive option. Especially at latitudes above 50° bifacial panels can double the yield gain of a horizontal single axis tracking system.

ACKNOWLEDGEMENT

BIGEYE was developed in the project BING, Dutch TKI Solar Energy and in the project BIFACE, Netherlands Enterprise Agency, solar-era.net.

REFERENCES

[1] T. Huld, M. Šúri, and T. Cebecauer, Photovoltaics International 6 (2009).
[2] R. Kopecek, Y. Veschetti, E. Gerritsen, A. Schneider, C. Comparotto, V. D. Mihailetchi, J. Lossen, J. Libal, Photovoltaics International 26 (2015).
[3] I. Shoukry, J. Libal, R. Kopecek, E. Wefringhaus, and J. Werner, Energy Procedia 92 (2016) 600.
[4] A. Lindsay , M. Chiodetti, D. Binesti, S. Mousel, E. Lutun, K. Radouane, and J. Christopherson, 32nd EUPVSEC, München Germany, Europe (2016).
[5] D.Berrian, J.Libal, M. Klenk, H. Nussbaumer, R. Kopecek, Proc. 7th WCPEC, Hawaii (2018).
[6] S. Ayala Pelaez, C. Deline, P. Greenberg, J. Stein, R.K. Kostuk, Proc. 7th WCPEC, Hawaii (2018).
[7] G. J. M. Janssen, B. B. Van Aken, A. J. Carr, and A. A. Mewe, Energy Procedia 77 (2015) 364.
[8] G.J.M. Janssen, A.J. Carr, B.B. Van Aken, and I.G. Romijn, 26th Asian PVSEC (2016).
[9] M.W.P.E. Lamers, E. Özkalay, R.S.R. Gali, G.J.M. Janssen, A.W. Weeber, I.G. Romijn, B.B. Van Aken, Solar Energy Materials and Solar Cells, 185 (2018) 192.
[10] P. Rodrigo, E. F. Fernández, F. Almonacid, and P. J. Pérez-Higueras, Sol. Energy (2013) 322-333.
[11] G. J. M. Janssen, R. S. R. Gali, K. M. de Groot, A. J. Carr, B. B. Van Aken, and I. G. Romijn, 33rd EUPVSEC, Amsterdam (2017).
[12] PV Performance Modelling Collaborative, https://pvpmc.sandia.gov.
[13] T. Baumann, M. Klenk, H. Nussbaumer, F. Baumgartner, N. Keller, 3th Bifacial Workshop Miyazaki, Japan, 2016. http://bifipv-workshop.com.
[14] T. Baumann, M. Klenk, H. Nussbaumer, F. Baumgartner, N. Keller, 4th Bifacial Workshop Konstanz, Germany, 2017. http://bifipv-workshop.com/
[15] R. Perez, P. Ineichen, R. Seals, J. Michalsky, and R. Stewart, Solar Energy, 44 (1990) 271.
[16] J.E. Hay, and J.A Davies, 1980. In: J.E. Hay, T.K Won,. (Eds.), Proc. of First Canadian Solar Radiation Data Workshop, 59. Ministry of Supply and Services, Canada.
[17] E. Lorenzo, M. Pérez, A. Ezpeleta, and J. Acedo, Progress in Photovoltaics 10, (2002) 533–543.
[18] U. A. Yusufoglu, T. M. Pletzer, L. J. Koduvelikulathu, C. Comparotto, R. Kopecek, and H. Kurz, IEEE Journal of Photovoltaics 5 (2015) 320.
[19] A. Mermoud an B. Wittmer, 4th Bifacial Workshop, Konstanz, Germany 2017. http://bifipv-workshop.com/
[20] PVGIS, 2013, http://re.jrc.ec.europa.eu/pvgis/apps4/pvest.php#.
[21] PVsyst V6, Photovoltaic System Software, 2013, www.pvsyst.com.

OPERATION, PERFORMANCE, AND MAINTENANCE OF FIRST UTILITY-SCALE SOLAR PHOTOVOLTAIC PLANT IN KUWAIT OIL COMPANY FOR THE OPERATION OF ELECTRIC SUBMERSIBLE PUMPS

Raed A. Sherif, Ahmad Al-Qudaihi, Hani Alsaqabi, Aziz Najaf, Emad Safar, and Rakan Al-Ajmi

Well Surveillance Group, Kuwait Oil Company
P.O. Box 9758, Ahmadi 61008, Kuwait, Phone: +965 98009010, Email: ra3solar@yahoo.com

ABSTRACT: This paper presents real-life performance of Sidrah 500, the first utility-scale solar photovoltaic plant installed inside Umm-Gudair oilfield of Kuwait Oil Company (KOC), and first in the world to operate electric submersible pumps (ESPs) used to lift the oil from under the ground. There has been lots of skepticism about the operation of solar PV plants in the harsh weather environment of Kuwait, where temperatures constantly exceed 50 deg. C and relative humidity can exceed 95% during summer months. This 10 MW solar PV plant uses poly-crystalline silicon panels mounted on single axis trackers, and connected to an 11-kV substation synchronizing the electricity from the solar plant and the national grid. The solar plant was commissioned in October 2016, and has been operational now for almost two years. The plant has met its Performance Ratio and generation targets when adjusted with the weather data. The plant showed that there are no practical issues associated with operation of solar PV plants in Kuwait, and future projects ought to take the experience of operation of Sidrah 500 into account. The project has been registered with the UN Clean Development Mechanism, reflecting its environmental benefits that amount to avoiding emission of 250,000 tons of CO2 over its lifetime, which is equivalent to planting 500 thousand Sidrah trees.
Keywords: Sidrah 500, CO2 footprint, Solar PV Plant in Kuwait, Solar for ESPs, Operation and Maintenance

1 INTRODUCTION

Kuwait has set a target to generate 15% of its energy from renewable energy sources by 2030. Reducing the country's burning of fossil fuels will not only reduce the emissions of greenhouse gases, but will also increase the number of barrels of oil available for export. The oil sector is the largest consumer of electricity in Kuwait, and its consumption is projected to increase significantly over the next few years.

The use of solar energy in Kuwait has been met with a great deal of skepticism because of the perception that solar equipment will not function properly in the harsh desert climate of Kuwait. The weather in Kuwait reaches temperatures above 55 deg. C under the sun in summer months, and yet temperatures can fall below zero deg. C at nights in winters. Relative humidity reaching above 95% with temperature above 40 deg. C happen over a couple of months during the summer. Dust storms that last for a few days in the spring and summer also pose a serious challenge to the operation and maintenance of any solar plant. Lastly, gusting winds can reach speeds of 60 km per hour.

A 15% target in Kuwait corresponds to about 4 GW of installed solar capacity. Before any meaningful installations can be deployed, a clear demonstration of a functioning solar PV plant must be shown; otherwise, it will be too risky for project developers to obtain non-recourse financing for solar projects.

2 PLANT DESCRIPTION

2.1 General

Sidrah 500 is a 10 MW (DC) solar photovoltaic plant located inside Kuwait Oil Company near Umm-Gudair oilfield, and is feeding electric submersible pumps (ESPs) used to extract oil from under the ground.

The plant uses single-axis trackers, mounted on pre-fabricated concrete pedestals designed to anchor the structure in place, since the soil was deemed too sandy to support the structure otherwise. The plant is divided into 6 sections, where each section has one power conversion center that includes two central inverters and one step-up transformer. Sections 1 - 4 are identical, as each section is connected to five trackers. Sections 5 and 6 have one inverter connected to five trackers and the second inverter connected to five-and-half trackers. Each tracker has 28 strings, and each string has 19 solar panels. There is a toal of 32,452 poly-crystalline silicon from Trina, each rated at 310 Watts each (efficiency of 16.2%) under Standard Test Conditions. The conversion centers are provided by SMA, with each inverter rated at 720 kW. An aerial view of the plant is shown in Fig. 1, whereas the layout of the plant is shown in Fig. 2

Figure 1: Aerial view of Sidrah 500

There are 29 ESPs connected to the F-193 substation, which corresponds to a load of approximately 8 MW that the substation is required to meet 24/7; hence the F-193 substation is connected to Minagish-B substation to provide for electricity at nighttime or when the solar plant output is not sufficient to meet the ESPs load demand, see Fig. 3.

Sidrah 500 was called as such to reflect the environematl benefits of the sola rplant. Over its lifetime, it will avoid emission of ~ 250,000 tons of CO2, which is roughly equivalent to planting 500 thousand Sidrah trees (Sidrah being the national tree of Kuwait).

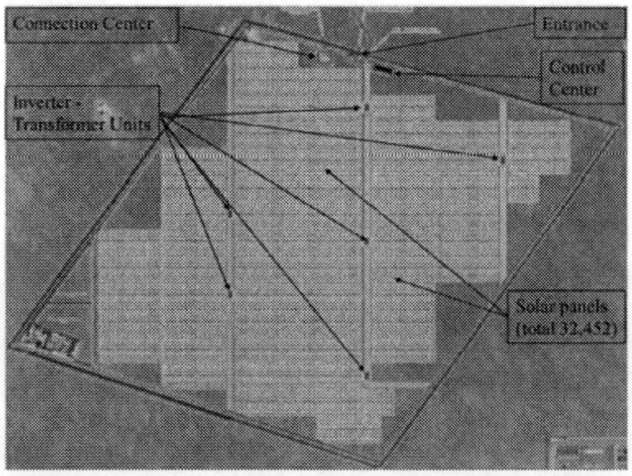

Figure 2: Layout of Sidrah 500

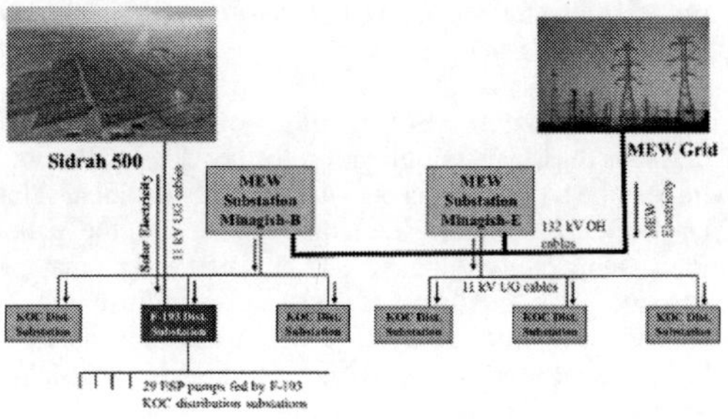

Figure 3: Sidrah 500 connection with National grid

2.2 Contractual Obligation

Unlike solar plants that operate under a Power Purchase Agreement (PPA), Sidrah 500 is built and operated on a lease model. The project has two phases: Phase 1 is for the engineering, procurement, construction, and commissioning, to be followed by Phase 2 for the operation and maintenance. During Phase 1, Kuwait Oil Company (the Client) makes payments to the Contractor against milestones, with payments reaching 80% of the EPC costs by end of Phase 1. The remaining 20% of the EPC costs in addition to the O&M costs are paid by the Client on a quarterly basis over 5 years. At the end of 5 years, the Client will assess how much degradation occurred to the plant as compared to performance when it started. If the degradation is over 10% of initial performance, the Client may ask the Contractor to uninstall the plant at its own costs. If, however, the degradation is within industry standards, the Client has the option of purchasing the plant or extending the O&M contract to the Contractor.

The Contractor has provided Guaranteed Solar Electricity Generation (GSEG) for each month of the O&M phase during the first 5 years of operation. Their GSEG was estimated based on the Typical Meteorological Year (TMY) data for the site, which was obtained by analyzing 17 years' worth of weather data. As the actual weather data will be slightly different from the TMY data, adjustments will have to be made to arrive at the Adjusted Guaranteed Solar Electricity Generation (AGSEG), which will be compared to the actual kWh produced every quarter. If the Actual energy generation per quarter is less than the AGSEG, then deductions are applied for each kWh short of the AGSEG. The Contractor, therefore, has to manage the O&M of the plant in order to avoid deductions being applied by the Client.

3 PERFORMANCE OF SIDRAH 500

Sidrah 500 was commissioned in October 2016 with the supervision of a third party who did 100% inspection of the solar panels to ensure the panels are free from hot spots. The plant entered into Phase 2, the O&M phase, by January 2017.

3.1 Plant Management Philosophy

The Contractor is responsible for managing Sidrah 500 to meet its AGSEG using a Supervisory Control and Data Acquisition (SACAD) system that monitors the performance of each section of the solar plant in addition to the data coming from the weather stations. Weekly performance data and monthly summaries are provided to the Client. The weekly reports show the average weather data every 10 minutes and kWh generated by each section of the plant averaged over 15 minutes. The SCADA system is used to identify if a particular section of the plant is not performing as it should, after which the fault can be attributed to one or more tracker, and ultimately identify which strings need to be physically inspected.

Panel cleaning decisions are made by comparing the output of two calibrated solar panels that are mounted side-by-side on the tracking PV arrays. One panel is left uncleaned while the other panel is cleaned every morning. The loss in power due to soiling is then identified based on which decisions can be made on whether or not to initiate a cleaning cycle.

3.2 Performance Data

Figure 4 shows the Performance Ratio (PR) of the solar plant in the first year of operation, not corrected for temperature. The PR for the first year was 77.5%, which is higher the Contractor's guaranteed PR of year 1 of 73.9%.

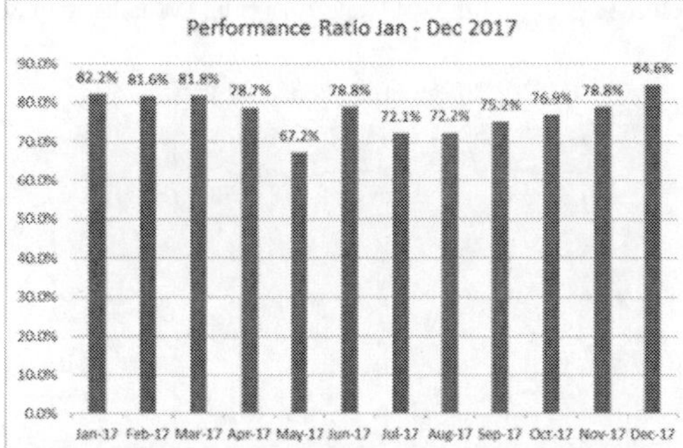

Figure 4: Performance Ratio of first year of operation

Note that the PR drop in the summer months is due to the high temperatures, when the ambient temperature is constantly above 40 deg. C and sometimes reaching over 50-55 deg. C. However, despite the extremely high temperatures of the summer months, the combination of the higher solar intensity and longer daytime during summer translate to higher energy production, as shown in Fig. 5. In fact, the energy production from June – September has amounted to about 40% of the energy production in the first year.

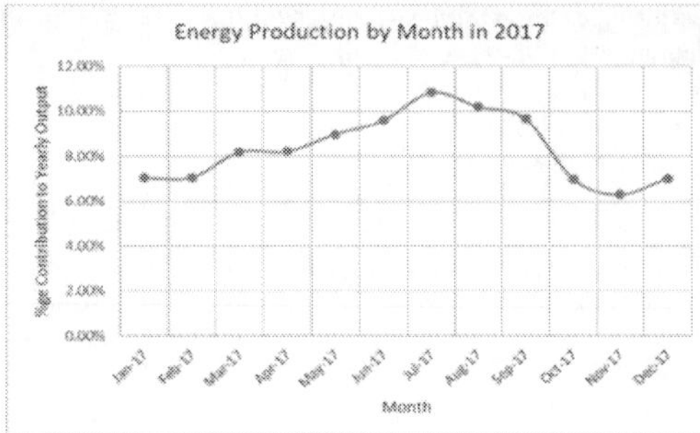

Figure 5: Energy production by month in first year

A key issue in assessing the performance of Sidrah 500 is its seasonal performance given the frequency of dust storms, high relative humidity and high temperatures in the summer months. Figure 6 shows the solar plant after the dust storm that occurred on March 16, 2018, relative to the solar panel that is cleaned every morning, whereas Fig. 7 shows the impact of the dust storm on the PR, dropping from 80% to 50%. The energy production in the week following the dust storm saw the energy fell by 20-25% compared to pre-dust storm energy production.

Figure 6: Sidrah 500 after one dust storm, showing the cleaned panel vs. the uncleaned panel on the same tracker

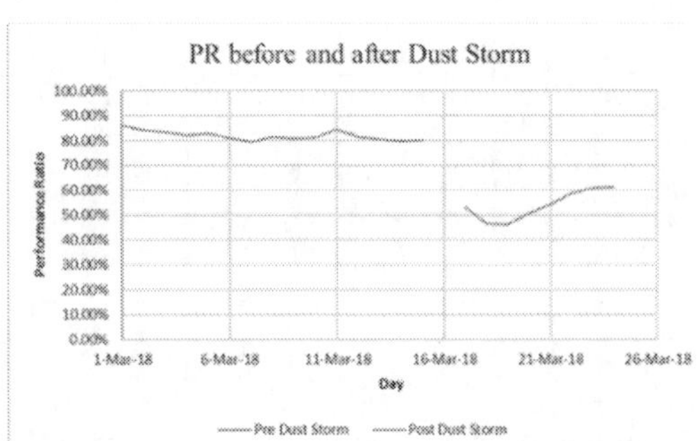

Figure 7: PR before- and after the dust storm of 16-3-2018

A graph of the soiling losses over a period of 9 months is shown in Fig. 8. A first examination of this chart shows that the soiling losses during summer months were surprisingly limited to 2-4%. Apparently, the high relative humidity in the summer results in water condensation on the panel surfaces at nighttime. As the panels move from stow position (horizontal) to vertical position in early morning hours, the condensed water cleans the panel. This self-cleaning mechanism had greatly reduced the soiling losses.

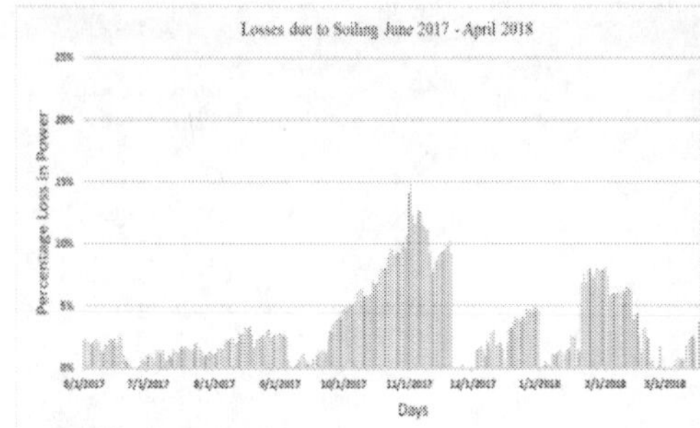

Figure 8: Energy production by month in first year

As the humidity season ended, dust accumulation starting the second half of September on led to power loss at a rate of almost 1% every 3 days, reaching a peak of almost 14% in a span of 1 month. The months of March-April 2018 illustrate the impact of dust storm, where losses of close to 25% can take place.

The actual energy generated in 2017 has been compared to the AGSEG in order to determine if the Contractor has met its obligation to the Client. We note here that the adjustment takes into account the global solar energy incident on the tracking surface and the panel surface temperature against the TMY weather data that the Contractor has used for the site. This adjustment is necessary, as – for example -- solar irradiation data for 2017 was measured at 2,665 kWh/m^2, which is 15% higher than the TMY data of 2,372 kWh/m^2.

As Fig. 9 shows, the actual kWh delivered to the substation has exceeded the AGSEG every month except in May 2017; however, no deductions have been applied to the Contractor since the quarterly energy generation exceeded the contracted amount.

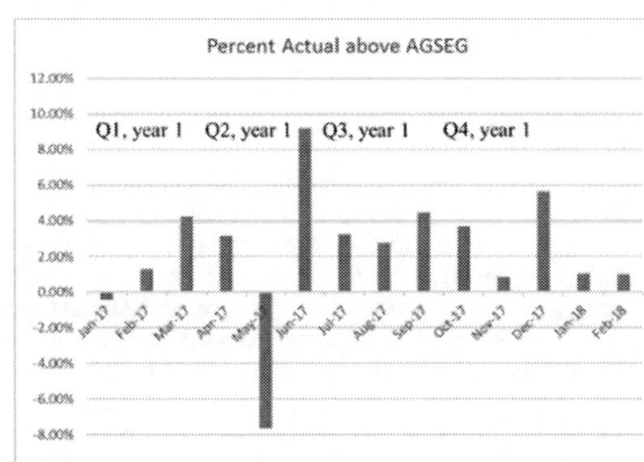

Figure 9: Actual vs. Adjusted Guaranteed Solar Electricity generation in year 1

4 CONCLUSIONS

Sidrah 500 has proved feasibility of operating solar PV plants in the harsh desert weather of Kuwait, with first year PR of 77.5%. Dust storms do take a heavy toll on the performance, reducing the output by up to 25%. Absent of dust storms, soiling losses accumulate at a rate of 1% every 3 days. High relative humidity and tracking panels provided a self-cleaning mechanism during the summer months, limiting soiling losses to under 4%.

INTRODUCING 'PEARL PV': PERFORMANCE AND RELIABILITY OF PHOTOVOLTAIC SYSTEMS - EVALUATIONS OF LARGE-SCALE MONITORING DATA

Angèle Reinders
ARISE, University of Twente / Energy Technology Group, TU Eindhoven
PO Box 217, 7500 AE Enschede, The Netherlands

David Moser
Institute for Renewable Energy, EURAC
39100 Bolzano, Italy

Wilfried Van Sark
Copernicus Institute of Sustainable Development, Utrecht University
3584 CB Utrecht, The Netherlands

Gernot Oreski
Polymer Competence Center Leoben GmbH
A-8700 Leoben, Austria

Nicola Pearsall
NPAG, Northumbria University
Newcastle upon Tyne NE1 8ST, United Kingdom

Alessandra Scognamiglio
ENEA Portici Research Centre
80055 Portici, Italy

Jonathan Leloux
Polytechnic University of Madrid
28040 Madrid, Spain

ABSTRACT: This paper will introduce a recently initiated COST Action titled PEARL PV and show its 4-year research and work plan with the aim to create exposure, receive feedback from various stakeholders and involve new participants to this research network. The aim is i) to improve the energy performance and reliability of photovoltaic (PV) solar energy systems in Europe leading to lower costs of electricity produced by PV systems by a higher energy yield, ii) a longer life time eventually beyond the guaranteed 20 years as specified by manufacturers, and iii) a reduction in the perceived risk in investments in PV projects. This objective will be achieved by analyzing big data of the actual monitored long-term performance, defects and failures in PV systems installed all over Europe to quantitatively determine the absolute influences of components rated performance, key design of systems, installation, operation, maintenance practice, geographic location and weather factors on the performance, performance degradation over time and failure modes of these PV systems. The PEARL PV project aims at the formation of an inclusive network of PV system researchers, data resources that will be analyzed by researchers and experts that can include more-nuanced evidence-based reliability in PV system evaluation methods and simulation and design tools. The COST Action PEARL PV assembles an inclusive partnership network of PV system researchers with experience with PV monitoring and simulation, with solar resources, and with PV operators who collectively bring together their ongoing work to contribute to the collation, analysis and interpretation of an agglomeration of PV system performance data in Europe.
Keywords: PV systems, Monitoring, Energy performance, Durability, Simulation, Building Integrated PV

1 INTRODUCTION

PEARL PV is the abbreviation for 'Performance and Reliability of Photovoltaic Systems: Evaluations of Large-Scale Monitoring Data' and was initiated in October 2017 for the following reasons. PV systems can achieve a long-term pivotal position in the future single European energy market, only by ensuring effective performance, high reliability and prolonged durability [1, 2]. PV solar systems have now become an established part of the electrical energy mix in standalone and grid-connected applications. Their modularity, consisting of PV modules, inverters, other power electronics and support structures, make them applicable at the fullest range of scales from hundred watts to hundreds of megawatts. At present PV systems are the most dominant solar technology in the world with an installed capacity exceeding 400 GW (at the end of 2017) which significantly more than installed solar thermal systems and concentrating PV. This is mainly due to significant costs reductions in the past decades, while the market for PV systems has grown at 44% per annum. Recently Power Purchasing Agreements for PV systems were reported below 2 cents per kWh in Saudi Arabia. In 2016 in Germany and Italy resp. 6 to 8% of the annual electricity was generated by PV systems, while across Europe in 2017 PV systems account for 4% of the electricity demand. Targets for further PV deployment have been set, e.g. the International Energy Agency expects 3 TW of PV systems to be globally installed by 2050 representing about 11% of the global electricity production. With such large numbers, reliability and performance are of paramount importance with a timeline for the next 20 years looking at the horizon of energy scenarios in 2050. As the majority of all PV systems have been installed in the past 10 years, each single installation can provide only a limited insight into the long-term performance and reliability of PV systems. The challenge is therefore to learn from the experience and information as embodied in the monitored performance of many PV systems. Important interlinked non-technical factors driving growth in photovoltaics deployment include:

- Government regulations and mandates and obligations for renewable power,
- Falling prices for all PV system components,
- Take-up of third-party funded PV array ownership models,
- Availability of dedicated feed-in tariffs or other financial (tax) incentives for PV systems,
- Need to decrease the global warming environmental impact of electricity generation,
- Reducing dependence on centralized generation electric utilities,
- Scalable, silent and clean operation enables PV systems to be used readily in smart cities and in building integrated PV systems.

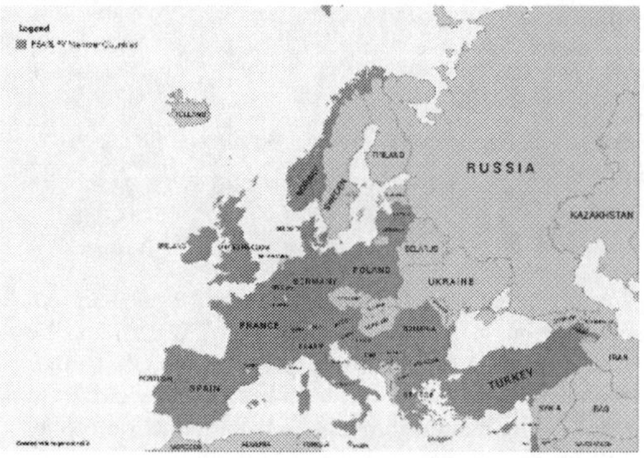

Figure 1: 30 countries that take part in this COST Action PEARL PV by 21 September 2018, indicated in orange.

Well-executed PV system research requires long term collaboration in geographically distributed research networks due to the strong relation between PV performance, geographically distributed climates and related solar resource assessments. Also this will be required due to the global spread of data on energy performance of PV systems and failure statistics of their components. Therefore the COST Action PEARL PV assembles an inclusive partnership network of PV system researchers with experience with PV monitoring and simulation, with solar resources, and with PV operators who collectively bring together their ongoing work to contribute to the collation, analysis and interpretation of an agglomeration of PV system performance data in Europe.

The network includes (i) experts that can make the most-nuanced evidence-based consideration of reliability in PV systems (ii) experts that have the ability to integrate this knowledge in evaluation methods, simulation models and design tools and (iii) the direct participation of end users to ensure there will be knowledge about their insights and perspectives. The project will therefore contribute to the development of a mature research field for PV system analysis in relation to irradiance potentials, long term PV performance, and PV grid interactions in an international context inside and outside Europe, see Figure 1. At present in total 31 countries, namely 30 at the European continent and the USA are a member of the PEARL PV network. The network is aiming at involving all 36 European COST member countries and if possible many more countries in the further world.

2 MAIN OBJECTIVES

PEARL PV aims to increase performance and lower costs of electricity produced by photovoltaic (PV) solar electricity systems in Europe via (i) obtaining higher energy yields, (ii) achieving longer operational life time (beyond the 20 years usually guaranteed by manufacturers) and (iii) lowering the perceived investment risk in PV projects. These objectives will be achieved by a cooperative European COST Action partnership, collating and analyzing a very large aggregated set of monitored long-term PV System operational performance data, with a focus on understanding defect and failure of PV systems installed across Europe, in the context of integration of PVs facilities into grids, and the impact of regional climate characteristics on the generation of PV energy. This data will be used to determine quantitatively:

- the absolute influences of (i) components' rated performance, (ii) system design, (iii) installation type, (iv) operation and maintenance practice, (v) interactions with grids, (vi) geographic location and (vii) weather and climate conditions;
- on (i) performance degradation over time and (ii) failure modes as they affect (i) economic viability, (ii) securing project investment, (iii) environmental sustainability (iv) security and predictability of electricity supply and (v) diversity and distribution of electricity supply;
- to (i) improve the electrical design of PV systems, (ii) achieve optimal sizing via the use of simulation models, (iii) enhance expected system efficiency, (iv) ease maintenance, (v) achieve high reliability and (vi) demonstrate excellent durability.

For this purpose 5 Working Groups have been set up that will conduct research using a shared data bank and shared simulation tools and models to analyze and compare these data that are collected in this data bank, see Figure 2. The 5 Working Groups are focused on (WG1) PV monitoring, (WG2) Reliability and durability of PV, (WG3) PV simulation, (WG4) PV in the built environment and (WG5) PV in grids. In the following sections (2.1 to 2.5) the objectives and activities in each Working Group will be introduced.

2.1 WG1: PV monitoring

The main objectives of WG1 are to:

1) identify relevant data to be collected to properly assess PV performance of installed PV systems in the field and on rooftops,
2) design guidelines for the collection,
3) set-up a database, and
4) guidelines for database access.

These objectives will be achieved by analyzing data of the actual monitored long-term performance, defects and failures in PV systems installed all over Europe to quantitatively determine the absolute influences of components rated performance, key design of systems, installation, operation, maintenance practice, geographic location and weather factors on the performance. This also will allow to study performance degradation over time and failure modes of these PV systems. A first version of the open database is planned for fourth quarter of 2018. At present WG1 consists of 30 members.

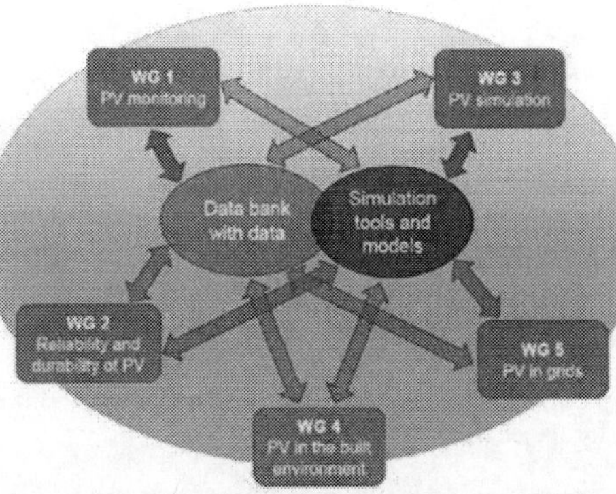

Figure 2: The 5 Working Groups of COST Action PEARL PV in relation to a shared data bank and simulation tools.

2.2 WG2: Reliability and durability of PV

The main objectives of WG2 are to find a proper definition of reliability and durability metrics for PV modules, components and systems as well as to identify relevant data to be collected to measure reliability and durability. WG 2, which currently consists of 30 members, will address the following problems and research questions:

1) Definition of reliability and end-of-life considering different views of PV stakeholders and different technologies.
2) How to test reliability?
3) Can reliability, lifetime or end-of-life be modelled from short term testing?
4) Can available models for single effects or single materials be combined to an comprehensive PV module lifetime assessment method?

The planned outcome is an extensive review paper to be published in 2020. Contributions are still welcome.

2.3 WG3: PV simulation

WG3 considers the use of modelling tools that are used to simulate the performance of PV devices and systems, both for prediction and assessment of performance. The main objectives are to

1) identify and classify the current PV simulation models by scope and usage category, (
2) provide access to information on PV solar electricity simulation models,
3) compare the performance of various models and
4) share the knowledge originating from this study with the PV community.

The initial activity of WG3 has been to issue a software usage questionnaire to establish the tools commonly used by the community and to obtain information on the way in which those modelling tools are utilized. Initial analysis has been based on the response of 33 members of the PEARL PV community, who identified 14 different packages that are regularly used, ranging from solar irradiance models to system simulation packages. The survey is being used to inform the scope of the model comparisons to be performed over the next few months. Further responses to the survey are welcome. The survey can be accessed on the PEARL PV website.

2.4 WG4: PV in the built environment

The use of PV in the built environment is the focus of the WG4, with the main aim of improving the performance of PV systems in such environment, and improving the ability to predict the behavior of PV in operating conditions. The starting point is the systematization of existing knowledge the COST PEARL participants share from previous or on-going projects, publications, and thematic networks they are involved in.

The composition of the participants is strongly multidisciplinary, with focus on: architecture; building technologies and PV integration; technology and PV systems; standardization, norms and regulations; data monitoring, modeling and simulations; reliability, testing, components, materials.

Within the WG4 there are two main tasks:

1) Collection of information in the data server of this Action about PV in the built environment, from realized projects, publications and information retrieved from internet, together with PV experts, architects, installers, construction and building services engineers and city/urban planners.
2) Identification of required data and appropriate simulation models to be used in the framework of PV systems in the built environment given the challenges of (i) shading by neighboring buildings, (ii) building and planning codes and regulations and (iii) energy performance norms that apply to and/or are required for, buildings.

About 20 experts are participating in the activities of the WG4.

Up to now, a review of trends in development of simulation models and tools has been prepared, with particular emphasis on integration in the building information modelling (BIM) environments and on application in early design stages. Considering available reviews [53], recommendations [54] and recent explorations in domain of simulation tools, [55], [56], [57], as well as a preliminary examination performed in first grant period of the PEARL PV Project-WG4 including 18 tools, a set of trends has been identified, clearly indicating a converging development of PV simulation tools and BIM platforms.

2.5 WG5: PV in grids.

WG5, currently composed of 27 participants, aims at contributing towards PV systems that are better integrated into the grid, that perform better, and whose operation under real-world conditions is better understood. This WG will use information from WG1, WG2 WG3 and WG4 on PV monitoring, reliability and durability and PV simulation respectively and information from communications with PV experts, electrical engineers, utilities, smart grid experts and meteorologists. These objectives will be achieved by exploring five different complementary pathways:

1) PV power forecasting
2) PV power fluctuations for grid operators
3) PV power quality in the grid
4) PV energy storage and management strategies
5) PV performance and fault detection

2.6 Training Schools

Adding to the research activities in the Working Groups, COST Action PEARL PV also aims at knowledge dissimination to young researchers by means of Training Schools. The topics that will be addressed by these annual Training Schools are connected and overlapping with the content of the Working Groups. Topics that have been identified for the period of 2018 to 2021 are as following:

1) 2018: Monitoring and simulation of the performance and reliability of PV systems in the built environment.
2) 2019: Evaluation of the performance degradation of PV-systems – influence factors, failure modes and their detectability and affect on economic viability.
3) 2020: Simulation tools and models for the forecast of system efficiencies of PV plants – with focus on environmental and integration aspects.
4) 2021: Potential of monitoring tools and advanced operation and maintenance practice for security and predictability of PV performance.

3 DISCUSSION AND RELEVANCE

Whilst the highest efficiencies for small PV cells in laboratory contexts are near 32%, commercial PV modules have a maximum rated efficiency of close to 22%. In operational PV installations these efficiencies decline further to be in the range of 13 to 17%, because a PV system [3-5] (i) is exposed to solar irradiation intensity, collimation and spectra that vary but are different from those under which standard rating procedures are conducted, (ii) is subject to soiling, (iii) has a lower than expected inverter efficiency, (iv) may have inefficient maximum power point tracking and (v) suffers thermal de-rating of PV modules (e.g. 0.4% loss per °C for silicon PV modules). System efficiency strongly depends on (i) local meteorological conditions, (ii) PV module technology (mono- or multi-crystalline silicon, thin film or organic PV), (iii) other electronic components involved (inverters, batteries etc.), (iv) the system configuration (e.g. local shading) and (v) the operation and maintenance of the system. Key factors determining the optimal performance of a PV system are shown in Figure 3.

There are extensive international IEC and SEMI [6,7] conformity standards to which PV modules and systems must comply to assure users and investors of the long-term value of their systems. However, as relatively few studies have considered the actual realised longterm performance of PV systems in use [8-32], there is a lack of collated comparable data from many installations on actual system and component reliability and durability [33-36]. It is imperative to find relationships between system performance over time and reliability in the long run.

Optimal design of switching converters for the integration and optimal exploitation of photovoltaic solar electricity systems into the electric grid remains a crucial issue often debated in the recent power electronics literature [37-39]. The installed capacity of operational PV systems has increased at a much faster rate than the development of adequate grid codes to effectively and efficiently manage high penetrations of PV into the distribution system.

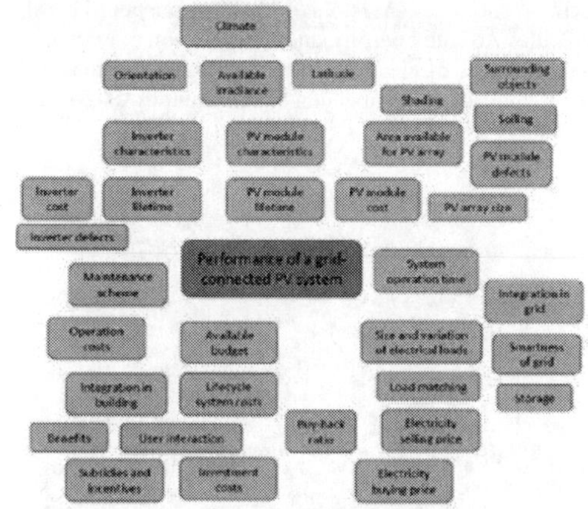

Figure 3: Key factors that determine the optimal performance of PV systems.

Management strategies vary considerably by country, some still have an approach that PV systems should behave as passive as possible, whereas other countries demand an active participation in grid control. For smart grid scenarios with a high penetration of distributed renewable energy generation into the electric grid, PV systems have to supply ancillary services to the distribution grid, in particular voltage and frequency regulation that maintain grid stability [40-43].

Therefore this COST Action compares monitoring data with results from simulation models for multiple photovoltaic installations, to define the range of applicability of each simulation model and make recommendations for improvement of simulation models to include reliability and durability. This COST Action creates a network that will uniquely bring together groups in Europe who have long-term PV system performance measurements with groups that have developed and can further enhance simulation models for PV reliability/durability. The aim is to understand the key factors influencing real-life relations between performance, reliability and durability [44] to ensure that this insight is embodied in (i) design methodologies [3] (ii) conformance standards [6], (iii) monitoring methods, (iv) predictive models for reliability/durability [16], and (v) system performance indicators such as efficiency, final yield and performance ratio [8,23,45,46]. This includes, for examples, replacement of components, and correcting irradiance or temperature mismatches in maximum power point tracking [47] and effects of application of PV systems in buildings [48-52]. It will produce a consolidated input/output performance chart that places all annual systems performance in a common format, to determine trends in system performance with (i) type of installation, (ii) components types (PV, inverters etc.) (iii) location,(iv) date of installation. This will be used to produce evidence-based design recommendations [58] based on (i) a systematic categorization of typical failures, (ii) their statistical distributions and (iii) both estimated and measured their impact on PV performance.

Indicators, which will be used to quantify and compare the performance of PV modules and PV systems, are among others: performance ratio (PR), final yield (Yf) and temperature-corrected efficiencies (η25°C). These indicators will be related to available irradiation (H) and modes of operation and eventual failures. For reliability and durability research, firstly performance data – as indicated above – will be used to determine the rate of degradation of PV module and systems, secondly failure modes will be collected, identified and statistically evaluated. This evaluation will happen in the context of among others, location, hence irradiation, life time of the PV modules and PV systems, typical components and modes of installation, system operation and integration with the grid as well as maintenance schemes.

4 SUMMARY AND CONCLUSION

This paper presents a recently initiated COST Action entitled PEARL PV and shows its 4-year research and work plan with the aim to create exposure, receive feedback from various stakeholders and involve new participants to this research network.

In the forthcoming period PEARL PV will continue organizing events such as Seminars, Workshops and Training Schools to reach out to a large audience of PV experts, young researchers and companies. Before the end of 2018 the data bank containing PV monitoring data should become available as well as a complete workplan for the further implementation of research and dissemination activities in the period of 2018 to 2021.

For more information about this project, registration for participation in this network and for future events of PEARL PV please visit the website of this COST Action at https://www.pearlpv-cost.eu/ and the project page of this Action at the COST Association at http://www.cost.eu/COST_Actions/ca/CA16235.

5 ACKNOWLEDGEMENTS

We would like to thank all 165 participants of PEARL PV for their enthusiasm and efforts.

This abstract is based upon work from COST Action PEARL PV, supported by COST (European Cooperation in Science and Technology).

COST (European Cooperation in Science and Technology) is a funding agency for research and innovation networks. Our Actions help connect research initiatives across Europe and enable scientists to grow their ideas by sharing them with their peers. This boosts their research, career and innovation, see www.cost.eu.

6 REFERENCES

[1] Fraunhofer Institute of Solar Energy Systems, www.ise.fraunhofer.de/de/downloads/pdf-files/aktuelles/photovoltaics-report-in-englischer-sprache.pdf, 2015.

[2] A. Golnas, "PV System Reliability: An Operator's Perspective," IEEE Journal of Photovoltaics, vol. 3, pp. 416 - 421, 2013.

[3] A.H.M.E. Reinders, V.A.P. van Dijk, E. Wiemken and W.C. Turkenburg, "Technical and economic analysis of grid-connected PV systems by means of simulation," Progress in Photovoltaics, vol. 7(1), pp. 71-82, 1999.

[4] N. Reich, J. Zenke, B. Müller, K. Kiefer and B. Farnung, "On-site performance verification services to significantly reduce PV energy yield prediction uncertainties," in 42th IEEE Photovoltaic Specialist Conference, 2015, p.1.

[5] D.L. King, W. E. Boyson, and J. A. Kratochvil, "Analysis of factors influencing the annual energy production of photovoltaic systems," in 29th IEEE Photovoltaic Specialist Conference, 2002, p. 1356.

[6] H. Ossenbrink, H. Müllejans, R. Kenny and E. Dunlop, Standards in Photovoltaic Technology, Chapter 1.39 in PV Technology, Vol. 1 in Comprehensive Renewable Energy, Elsevier, United Kingdom, pp. 787-803, 2012.

[7] M., Pellegrino, G. Flaminio and G. Graditi, "Testing and standards for new BIPV products," in IECON 2013, 2013, p. 1.

[8] J.D. Mondol and B. Norton, "Comparison of measured and predicted long-term performance of a grid connected photovoltaic system," Energy Conversion and Management, vol. 48(4), pp. 1065-1080, 2007.

[9] C. E. Chamberlin, M. A. Rocheleau, M. W. Marshall, A. M. Reis, N. T. Coleman and P. A. Lehman, "Comparison of PV module performance before and after 11 and 20 years of field exposure," in 37th IEEE Photovoltaic Specialist Conference, 2011, p. 101.

[10] E.E. van Dyk, E. Meyer, F. Vorster and A.W.R Leitch, "Long-term monitoring of photovoltaic devices," Renewable Energy, vol. 25(2), pp. 183-197, 2002.

[11] J.A. Eikelboom and A.H.M.E. Reinders, "Determination of the irradiation dependent efficiency of multicrystalline Si PV modules on basis of IV curve fitting and its influence on the annual performance," in 14th EU PVSEC, 1997.

[12] K.B.D. Esmeijer and W.G.J.H.M. van Sark, "Statistical analysis of PV performance using publically available data in the Netherlands,",in 28th EU PVSEC, 2013, p. 4024.

[13] T. Friesen, D. Chianese, A. Realini, G. Friesen, E. Burà, A. Virtuani, D. Strepparava and R. Meoli, "TISO 10kW: 30 years of Experience with a PV Plant," in 27th EU PVSEC, 2012, p. 3125.

[14] E. Hasselbrink, M. Anderson, Z. Defreitas, M. Mikofski, Y.C. Shen. S. Caldwell, A. Terao, D. Kavulak, Z. Campeau, D. DeGraaff, "Validation of the PVLife model using 3 million module-years of live site data," in 39th IEEE Photovoltaic Specialist Conference, 2013, p. 7.

[15] U. Jahn, and W. Nasse, "Operational performance of grid-connected PV systems on buildings in Germany," Progress in Photovoltaics, vol. 12, pp. 441-448, 2004.

[16] D.C. Jordan, and S.R. Kurtz, "Photovoltaic degradation rates - An analytical review," Progress in Photovoltaics, vol. 21, pp. 12-29, 2011.

[17] D.C. Jordan, J.H. Wohlgemuth and S.R. Kurtz, "Technology and climate trends in PV module degradation," in 27th EU PVSEC, 2012, p. 3118.

[18] A.S. Joshi, I. Dinzer and B.V. Reddy, "Performance analysis of photovoltaic systems: A review," Renewable and Sustainable Energy Reviews, vol. 13, pp. 1884–1897, 2009.

[19] J. Leloux, L. Narvarte Fernández and D. Trebosc, "Performance analysis of 10,000 residential PV systems in France and Belgium," in 26th EU PVSEC, 2011, p. 3939.

[20] E. Lorenzo, R. Zilles, R. Moretón, T. Gómez and A. Martínez de Olcoz, "Performance analysis of a 7-kW crystalline silicon generator after 17 years of operation in Madrid," Progress in Photovoltaics, vol. 22(12), pp. 1273–1279, 2013.

[21] B. Müller, W. Heydenreich, K. Kiefer and C. Reise, "More Insights from the Monitoring of Real World PV Power Plants - A Comparison of measured to predicted performance of PV systems," in 24th EU PVSEC, 2009, p. 3888.

[22] D. Polverini, M. Field, E. Dunlop and W. Zaaiman, "Polycrystalline silicon PV modules performance and degradation over 20 years," Progress in Photovoltaics, vol. 21, pp. 1004-1015, 2013.
[23] N.H. Reich, B. Müller, A. Armbruster, W. G. J. H. M. van Sark, K. Kiefer and C. Reise, "Performance ratio revisited: Is PR>90% realistic?," Progress in Photovoltaics, vol. 20, pp. 717–726, 2012.
[24] N.H. Reich, A. Goebel, D. Dirnberger and K. Kiefer "System performance analysis and estimation of degradation rates based on 500 years of monitoring data," in 38th Photovoltaic Specialist Conference, 2012, p. 1551.
[25] P. Sanchez-Friera, M. Piliougine, J. Peláez, J. Carretero M. and Sidrach de Cardona, "Analysis of degradation mechanisms of crystalline silicon PV modules after 12 years of operation in Southern Europe," Progress in Photovoltaics, vol. 19, pp. 658–666, 2011.
[26] J.H. Wohlgemuth, D.W. Cunningham, P. Monus, J. Miller and A.M. Nguyen, "Long term reliability of PV modules," in IEEE 4th World Conference on Photovoltaic Energy Conference, 2006.
[27] F. De Lia, S. Castello, A. Mittiga, R. Sensoli, G. Graditi, M. Pellegrino and G. Flaminio, "Energy rating for different module technologies," in 29th EU PVSEC, 2014, p. 3265.
[28] I.J. Bennett, J. Zhu, R. Gottschalg, M. Köhl, S. Hoffmann, K.A. Berger, S. Zamini, E. Gerritsen, P. Malbranche, P.M. Pugliatti, A. Di Stefano, F. Aleo, D. Bertani, F. Paletta, F. Roca, G. Graditi, M. Pellegrino, O. Zubillaga and P. Cano, "PV module lifetime prediction and quality assurance as addressed by Sophia," in 29th EU PVSEC, 2014, p. 2495.
[29] P. Manshanden, N.J.C.M. van der Borg, M. Bliss, B. Mihaylov, R. Gottschalg, M. Izzi, M. Tucci, F. Roca, M. Pellegrino, A. Romano, G. Graditi, J. Hohl-Ebinger, W. Warta, M. Debucquoy, O. El Daif, I. Gordon, J. Champliaud, A. Jouini, J. Glatz-Reichenbach, K. Bothe and A. Herguth, "Round robins of solar cells to evaluate measurement systems of different European research institutes," in 28th EU PVSEC, 2013, p. 861.
[30] S. R. Williams, T. R. Betts, R. Gottschalg, D. G. Infield, H. De Moor, N. Van Der Borg, A. R. Burgers, G. Friesen, D. Chianese, A. Guerin De Montgareuil, T. Zdanowicz, D. Stellbogen and W. Herrmann, " Accuracy of energy prediction methodologies," in WCPEC-6, Kyoto, 2014, p. 2206.
[31] J.M. Cuadra, A.B. Cueli, J. Díaz, M. Ezquer Mayo, J. Moracho, I. Petrina and A.R. Lagunas., "Comparison of different PV modules evolution after being submitted to special test-to-failure protocol beyond the IEC-61215 standard," in 29th EUPVSEC, 2014, p. 3147.
[32] Abentin I., "Characterization of visual defects detected on c-Si PV modules after certain period of field exposure," in 28th EUPVSEC, 2013.
[33] A. Woyte, M. Richter, D. Moser, S. Mau, N. Reich and U. Jahn, "Monitoring of photovoltaic systems: Good practices and systematic analysis, in 28th EUPVSEC, 2013, p. 3686.
[34] S.V. Spataru, D. Sera, P. Hacke, T. Kerekes and R. Teodorescu, "Fault identification in crystalline silicon PV modules by complementary analysis of the light and dark current–voltage characteristics," Progress in Photovoltaics, vol. 24(4), 2015.
[35] D.C. Jordan, and S.R. Kurtz, "Dark horse of evaluating long-term field performance - Data filtering," IEEE Journal of Photovoltaics, vol. 4(1), pp. 317-323, 2014.
[36] G. Graditi, and G. Adinolfi, "Photovoltaic power optimizers: A comparison in reliability evaluations," in 12th IMEKO, 2013, p. 221.
[37] G. Graditi, G. Adolfini and G.M. Tina., Photovoltaic optimizer boost converters: temperature influence and electro-thermal design, Applied Energy, vol. 115, pp. 140-150, 2014.
[38] G. Aurilio, M. Balato, G. Graditi, C. Landi, M. Luiso, and M. Vitelli, "Fast hybrid MPPT technique for photovoltaic applications: Numerical and experimental validation," Advances in Power Electronics, vol. 2014, pp. 1 – 15, 2014.
[39] F. Spertino and G. Graditi, "Power conditioning units in grid-connected photovoltaic systems: A comparison," Solar Energy, vol. 108, pp 219-229, 2014..
[40] M. Braun, T. Stetz, R. Bründlinger, C. Mayr, K. Ogimoto, H. Hatta, H. Kobayashi, B. Kroposki, B. Mather, M. Coddington, K. Lynn, G. Graditi, A.Woyte and I. MacGill, "Is the Distribution Grid ready to accept large scale photovoltaic deployment?," Progress in Photovoltaics, vol. 20, pp. 681-697, 2011.
[41] R. Bründlinger, C. Mayr, H. Fechner, M. Braun, K. Ogimoto, K. Frederiksen, B. Kroposki, G. Graditi, I.F. MacGill, D. Turcotte and L. Perret, "Bringing together international research on high penetration PV in electricity grids - The new Task 14 of the IEA-Photovoltaic Power Systems Programme," in 25th EU PVSEC, 2010, p. 4971.
[42] S. Favuzza, G. Graditi and F. Spertino, Comparison of power quality impact of different photovoltaic inverters: IEEE International Conference on Industrial Technology, 2004, p. 542.
[43] S.Tselepis, "Literature survey about lifetime expectancy of components, in particular the energy storage systems in existing PV applications," in 19th EU PVSEC, 2004.
[44] S.R. Kurtz, T. Sample, J. Wohlgemuth, W. Zhou, N. Bosco, J. Althaus, N. Phillips, M. Deceglie, C. Flueckiger, P. Hacke, D. Miller, M. Kempe, M. Yamamichi, and M. Kondo, "Moving toward quantifying reliability - the next step for a rapidly maturing PV industry," in 42th IEEE Photovoltaic Specialist Conference, 2015, p. 1.
[45] F. Bonanno, G. Capizzi, G. Graditi, C. Napoli and G.M. Tina, "A radial basis function neural network based approach for the electrical characteristics estimation of a photovoltaic module," Applied Energy, vol. 97, pp. 956-961, 2012.
[46] G.M. Tina, S. Gagliano, G. Graditi and M. Angelo, "Experimental validation of a probabilistic model for estimating the energy output from double axis tracking PV systems," Applied Energy, vol. 97, pp. 990-998, 2012.
[47] W.G.J.H.M. van Sark, N.H. Reich, B. Mueller, A. Armbruster, K. Kiefer and C. Reise, "Review of PV Performance Ratio Development," in World Renewable Energy Forum, 2012.
[48] A. Scognamiglio, G. Adinolfi, G. Graditi, E. Saretta, "Photovoltaics in Net Zero Energy Buildings and clusters: enabling the smart city operation," Energy Procedia, vol. 61, pp. 1171-1174, 2014.
[49] F. Frontini, A. Scognamiglio, G. Graditi, C. Polo López and M. Pellegrino, "From BIPV to multifunctional building component," in 28th EU PVSEC, 2013.
[50] J. Benemann, O. Chehab and E. Schaar-Gabriel, "Building-integrated PV modules," Solar Energy and Solar Cells, vol. 67, pp. 345-354, 2001.
[51] A. Henemann, "BIPV: Built-in solar energy,"

Renewable Energy Focus, vol. 9(6), pp. 16-19, 2008.
[52] B. Norton, P.C. Eames, T.K. Mallick, M.J. Huang, S.J. McCormack, J.D. Mondol and Y.G. Yohanis, "Enhancing the performance of building integrated photovoltaics," Solar Energy, vol. 85(8), pp. 1629-1664, 2011.
[53] C. Eastman, P. Teicholz, R. Sacks and K. Liston, BIM Handbook, A Guide to Building Information Modelling, John Wiley & Sons, Inc., 2011.
[54] The American Institute of Architects (AIA), "An Architect's Guide to Integrating Energy Modeling in the Design Process," 2012. [Online]. Available: http://aiad8.prod.acquia-sites.com/sites/default/files/2016-04/Energy-Modeling-Design-Process-Guide.pdf. [Accessed 4 March 2018].
[55] M. Horvat and M. C. Dubois, "Tools and methods for solar design – an overview of IEA SHC Task 41, Subtask B," Energy Procedia, no. 30, pp. 1120-1130, 2012.
[56] T. Reeves, S. Olbina and R. Issa, "Gudelines for Using Building Information Modelling for Energy Analysis of Buildings," Buildings, vol. 5, pp. 1361-1388, 2015.
[57] N. Jakica, "State of the art review of solar design tools and methods for assessing daylighting and solar potential for building integrated photovoltaics," Renewable and Sustainable Energy Reviews, no. 81, pp. 1296-1328, 2018.
[58] A. Reinders, P. Verlinden, W. van Sark, A. Freundlich (Eds.), "Photovoltaic Solar Energy, From Fundamentals to Applications," Wiley, 2017.

FULLY AUTOMATED PHOTOVOLTAIC SYSTEM MODELLING FOR LOW COST ENERGY MANAGEMENT APPLICATIONS BASED ON POWER MEASUREMENT DATA

Benedikt Hanke [a*], Mariano Bottega [a], Dorothee Peters [a], Nailya Maitanova [a], Jan-Simon Telle [a], Matthias Grottke [b], Karsten von Maydell [a], Carsten Agert [a]

[a] DLR Institute of Networked Energy Systems, Carl-von-Ossietzky-Str. 15, 26129 Oldenburg, Germany
[b] Hammer Real GmbH, Sylvensteinstr. 2, 81369 München / Munich, Germany
[*] Corresponding author: benedikt.hanke@dlr.de

ABSTRACT: Within this work an automated algorithm for the modelling of photovoltaic systems solely based on smart meter power readings of the system is proposed assuming low cost, stationary rooftop photovoltaic systems. As an example system, data from a nearly horizontal (7° tilt) thin film system is used for method validation. The photovoltaic power output modelling results are compared to an individually engineered INSEL® model and a Linear Regression (LR) Model. The resulting auto model output is discussed in the context of day ahead and intraday photovoltaic power forecast using no-cost / low-cost irradiation information. This work presumes that the characteristics of a photovoltaic system can be extracted from its clear sky /best system power curve. The extracted best system curve can be used directly as a model by scaling the clear sky power output with the ratio of the radiation to the clear sky radiation. The clear sky/best system curves are constructed by identifying the power maximum for a given time of the day for a data set of a sample number of previous days.
The auto model is comparable to the fitted INSEL® model and the linear regression (LR). The yearly error metrics for estimation of accuracy of the simulation, concerning 5-minute power values, are between -0.3% for the LR model, 3.7% for the INSEL® model and 4.7% for the auto model. Also the annual RMSE and MAE of the auto model (74.8 W and 27.4 W) are comparable to the LR (66.8 W and 26.2 W) and the INSEL® fit (68.1 W and 25.5 W). The auto model operates with less input data (generated energy only) and no knowledge about the engineering, design or surroundings of the PV system while yielding similar simulation results as conventional methods.
Keywords: PV System, Modelling, Prosumer

1 INTRODUCTION AND MOTIVATION

Energy systems in buildings and districts become more complex and diverse due to the available technologies and the ongoing sector coupling between electricity, mobility and heat. Additionally energy efficiency regulations encourage or even enforce local energy generation using renewable energy technologies [1][2]. To optimize the local electric infrastructure expenditure, load management becomes an integral part of future electricity grid designs, especially in the private sector targeting electrical installations behind the grid connection point. To integrate local photovoltaic production with flexible loads from sector coupling applications in the building, a low cost photovoltaic modelling and prediction is needed to assist the local energy and load management system. To ensure a low cost installation of the energy management system, operation and adaption due to degradation of the solar cells and the system, the modelling of the photovoltaic system needs to be highly automated and based on as few parameters as possible. These need to be configured without special photovoltaic knowledge by electricians without any additional training.

Various commercial (e.g. PV*SOL[1], INSEL®[2], PVsyst[3], HOMER Pro®[4]) and open source tools (e.g. PV_LIB[5] Toolbox) are available to model the behaviour and energy generation of a photovoltaic system mainly before system installation. This allows for building business cases and scaling renewable energy generation systems (e.g. mini grids or battery storage systems). Across these tools, a detailed knowledge about the

Figure 1: Roof top photovoltaic system at DLR Institute of Networked Energy Systems in Oldenburg

photovoltaic system and the geographic lcation is required for the modelling. The pre-installation models are usually not fit to be used for operational optimisation, as installation changes need to be manually changed within the photovoltaic system model.

2 DATA

Data from December 1st 2016 until December 31st 2017 is used for evaluation and training purposes. Due to measurement errors in the photovoltaic system, the period from September 2nd 2017 until September 17th 2017 is replaced with data from 2011 of the same dates to achieve a full year functional dataset. The December 2016 time frame is only used for the automated modelling as a pre-training phase. All available data has been checked for plausibility.

[1] https://www.valentin-software.com/
[2] http://www.insel.eu/
[3] http://www.pvsyst.com/en/
[4] https://www.homerenergy.com/
[5] https://pvpmc.sandia.gov/applications/pv_lib-toolbox/

2.1 Photovoltaic System

The reference photovoltaic system used in this work is located at the DLR Institute of Networked Energy Systems in Oldenburg, Germany[6] (see authors contact for address and Figure 1). The roof top photovoltaic system is facing south-west (237°) with a 7° tilt.

The system uses 12 Schott ASi95 modules for energy generation with a nominal total system power generation of 1140 W_P. The modules are rated with a nominal cell efficiency of 6.6% at standard test conditions (STC). The module size is 1.108 m x 1.308 m = 1.499 m² resulting in an estimated total module area of 17.4 m² - ignoring additional installation gaps between modules. The resulting calculated module efficiency is 6.55%. For the year 2017 782.9 kWh of electricity are generated.

A SMA Sunny Boy 1100[7] is used as inverter for grid connection. A SMA Sunny Sensorbox is mounted in the plane of the photovoltaic modules to collect additional temperature and radiation information. The data is collected using a SMA Sunny Webbox configured to 5 min data interval.

For the manual system modelling, original system component data sheets are available for all system components.

2.2 Weather Measurements

Publically available reference weather measurements from the University of Oldenburg[8] are used as reference weather data for the system modelling. The weather station is located 32 m above sea level atop the main university building[9]. The weather station and the photovoltaic system are 1000 m apart. Weather data is publicly available in 30 min intervals since July 2003.

2.3 Data Resume

The plotted solar radiation of the given data set for the year 2017 versus the measured power of the photovoltaic system is shown in figure 2. The data points fit the simple linear regression of P = 0.0474 G with a coefficient of determination of 85% resulting in a system efficiency of 4.74%.

The correlation between the measured radiation and the photovoltaic system power is given most of the time. The distance between the radiation measurement and the photovoltaic system locations is never the less visible in the wide scattering of the data points.

3 METHODOLOGY

For the Linear Regression Model (LR) the linear regression from section 2.3 is used.

Based on the described data set and system information, a static INSEL® model (manual model) of the photovoltaic system is designed. The manual model is benchmarked against the training data. The received benchmark is therefore reflecting the best case model performance for a full year model scenario.

The algorithm based automatic model (auto-model) is using only power and location information to create a model of the photovoltaic system (details see below). Thirty days of data from December 2016 are used to pre-train the auto-model, allowing for a full evaluation of the 2017 dataset. The same benchmark parameters are calculated as for the manual model. The dataset had some missing values (2nd – 18th of September and 11th – 13th of December) which were replaced by values from the year 2011 from the same PV system.

To describe the performance of the models the mean absolute error (*MAE*, eq. (1)), the root mean square error (*RMSE*, eq. (2)) and the simulation error (*ΔE*, eq. (3), [3]) are calculated using power values from the measurement P_M, the power values from the model simulation P_S and the number of data points n.

$$MAE = \sum_{i=1}^{n} |P_{Mi} - P_{Si}| / n \quad (1)$$

$$RMSE = \sqrt{\sum_{i=1}^{n} (P_{Mi} - P_{Si})^2 / n} \quad (2)$$

$$\Delta E = \frac{E_{model} - E_{measured}}{E_{measured}} \quad (3)$$

The results from the linear regression model and the manual model are used as a reference for the evaluation of the performance of the auto-model. The modelling results of both models are created by independent research teams to reduce the chance of result-biasing.

4 RESULTS

4.1 INSEL® Model

Two INSEL® models where created for comparison: the first solely build on database values and the second by matching the power measurements with the radiation measurements thus analysing the system performance. The block diagram of the created INSEL® models is shown in Figure 3.

- Block FROM contains a reference to the input data with global horizontal radiation (W/m²) and ambient air temperature (°C) for the whole year 2017.
- Block PVAI returns electric current and temperature of the PV modules depending on climate data and PV voltage. Calculation of the electric current is based on a modified one-diode model. The parameters of the block are given in Table 1.

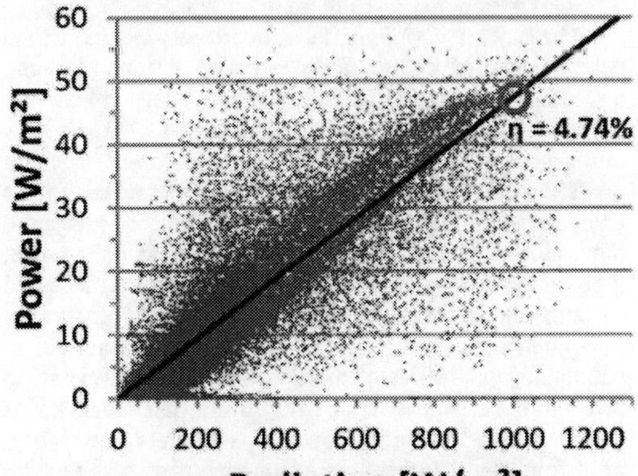

Figure 2: Measured solar radiation versus measured PV system power. The plot shows 105120 data points in 2017 at a resulting time resolution of five minutes.

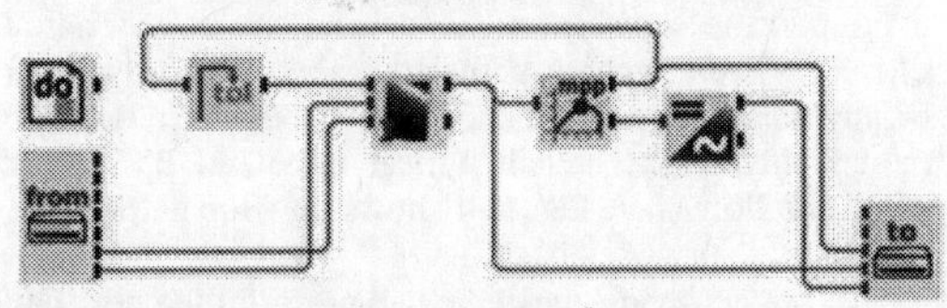

Figure 3: Block diagram of INSEL® model

[6] 53°09'03.8"N 8°09'59.5"E, https://goo.gl/maps/ewzJ1s5zcXy
[7] Serial number 2000689124
[8] http://www.uni-oldenburg.de/wetter/
[9] 53°08'51.4"N 8°10'48.4"E, https://goo.gl/maps/xiDm3Pgxa3m

Table 1: Characterization table of the block PVAI with the fitted parameters

Fitted Parameters	
Number of cells in series per module	24
Number of cells in parallel per module	3
Single cell area	0.0201
Single module area	1.449
Coefficient of short-circuit current density	0.113206+00
Temperature coefficient of short-circuit	0.568527E-04
Merten parameter	0.547033E+01
Built-in voltage	0.120437E+07
Coefficient of saturation current density	0.446160E+01
Temperature exponent of saturation	0.457965E+01
Band gap	0.152050E+01
Diode ideality factor	1.791466
Series resistance	0.44343E-03
Parallel resistance	0.110715E+01
Module tolerance plus	5.0
Module tolerance minus	-5.0
Characteristic module length	1.308
Module weight	18.000
Absorption coefficient	0.70
Emission actor	0.85
Specific heat of a module	900.0
Nominal operating cell temperature	47.0
Single cell voltage error tolerance	0.100000E-02

- Block MPP calculates maximum power points of the PV system under different climate conditions. This block operates in a combination with a block TOL (top of loop)
- Block IVP simulates the operation of the inverter and its losses and returns ac power produced by the PV system.
- Block TO gathers selected results after running the simulation and saves them into a file.

The block PVAI contains standard parameters, which need to be fitted for amorphous silicon solar modules and also for PV systems that have already been taken into operation in order to take a degradation of the modules into account.

The parameters of the block PVAI presented in the characterization table above were fitted on measured data: electric current and voltage of the real PV system, global horizontal radiation and air temperature measured for the whole year 2017. Then this fitted block was integrated in a model (see the block diagram above) to simulate a time series of the PV power over the year. In order to define the effect of the fitting, the block PVAI with the standard values of the parameters and values only from the datasheet, i.e. without fitting, was also be used to simulate PV power over the year.

4.2 Auto Model

The approach of the auto model [8] deals with the assumption to forecast and model the power generation of a PV system without knowledge about the orientation or any parameters of the modules and the inverter. The model only works with historical measured PV power data. The flow chart Figure 4 shows how the auto model works.

To predict/model the generated power of the next day the auto model only uses measured PV power data from the last few days. Therefore, PV data from the 1st December 2016 until the 31st December 2017 were used from the PV system which was described in chapter 2.1.

To find the best system power curve for the next day, the algorithm takes a sample number of PV data from previous days (PD) and calculates the maximum value for every time step (288, 5-minutes resolution) of all chosen previous days. Thereof results the best system power (P_{BS}) as shown, for the example of six previous days, in Figure 5.

To model and scale the PV power (P_{MO}) and make the auto model comparable to the INSEL model and the linear regression, the P_{BS} was scaled by the ratio of the measured irradiance GHI (GHI_{MS}) (Global Horizontal Irradiance) and the clear sky GHI (GHI_{CS}) irradiance. The comparison of the P_{MS} curve, the P_{BS} curve and the P_{MO} curve is presented in Figure 6 for the example of 6 previous days.

To find the optimum of how many days to look back to find the ideal P_{BS}, all calculation steps were repeated thirty times previous days and the calculation of 365 P_{BS}

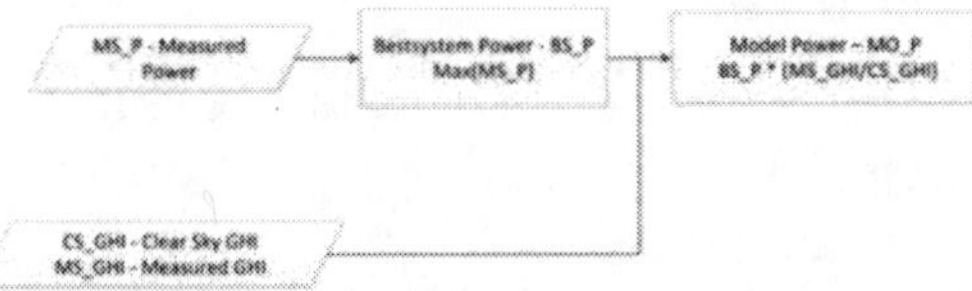

Figure 4: Flow chart of the auto model

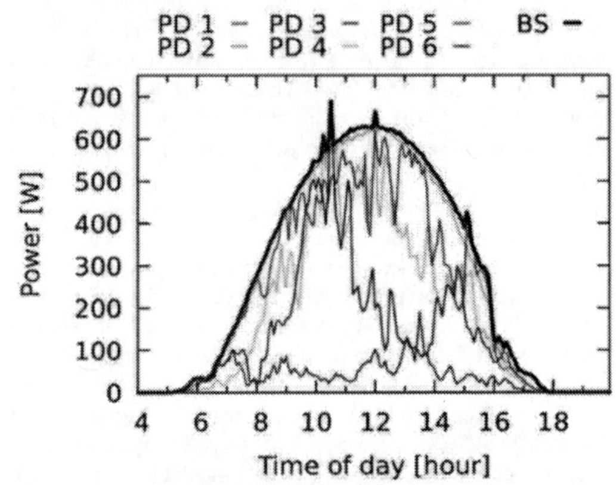

Figure 5: example of building the best system power curve of six previous days

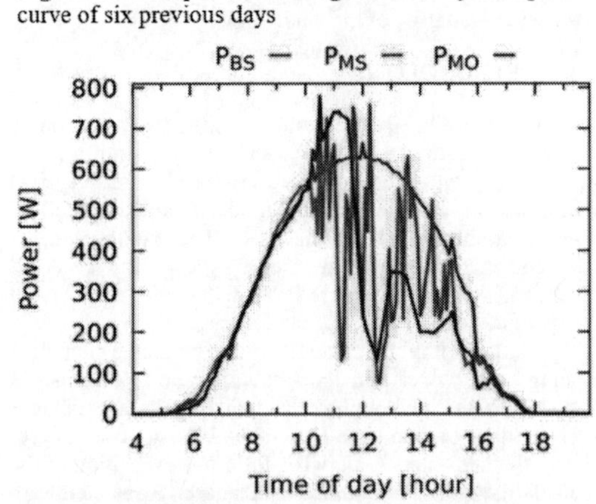

Figure 6: Comparison of the measured power P_{MS} and the auto modelled best system power curves P_{BS} and model power curve P_{MO}.

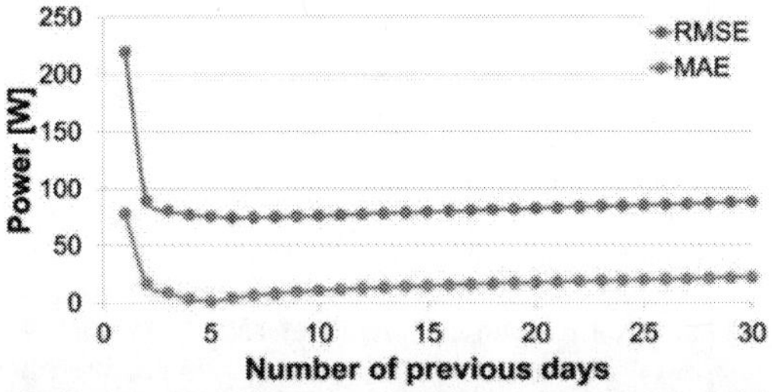

Figure 7: Simulation error from the auto model simulation of one year as selection criteria for the number of PD.

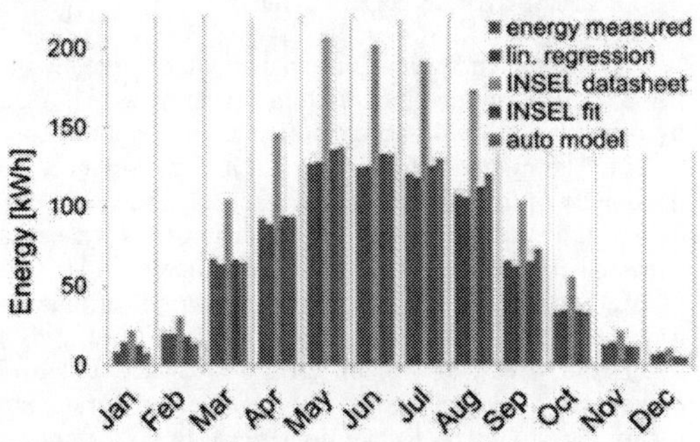

Figure 8: monthly energy output of the PV system simulated with different models in comparison with the measured energy output from PV system for year 2017.

curves. The selection criteria for the optimal number of previous days were the minimum simulation error or rather the *RMSE* (as seen in eq. 1) and *MAE* (eq. 2) between the modelled power and the measured power as shown in Figure 7.

From these results it was suggested that the optimal number of previous days is 6. Another important point is to choose PD > 1 so that persistence should exclude and also not to go too far in the past. This have a negative effect with regards on seasonal impacts of time-dependents increasing or decreasing irradiance/power.

4.3 Linear Regression Model

The linear regression model described in section 2.3 is used as a most simple model for the photovoltaic system. This model requires radiation and output power as input parameters and cannot be created independently of weather data. The model is forced to 0 W power output P_{LR} at 0 W/m² of solar radiation G by omitting a constant factor. The resulting linear regression is shown in eq. (4).

$$P_{LR} = 17.4\ m^2 * 0.0474 * G \quad (4)$$

The *MAE* for this model is 26.3 W and the *RMSE* is 67.2 W based on the training data set of 2017 values.

4.4 Simulation results

The results of the linear regression approach, INSEL simulations with and without fitting of the block PVAI and auto model were evaluated and compared with the measured power of the PV system (see Fig. 8).

When the block PVAI was not fitted, the INSEL software overestimated the power of the PV system dramatically. INSEL didn't take the features of the amorphous Si and the degradation effect into account. A 7 year old PV-system requires manual adaption of the datasheet values to reflect current system behavior.

Table 2: Statistical metrics for estimation of the accuracy of the simulation (concerned power values)

	MAE	RMS	Simulation
Linear	2	66.8	-0.3 %
INSEL	6	130.4	61.5 %
INSEL fit	2	68.1	3.7 %
Auto	2	74.8	4.7 %

Table 3: Statistical metrics for estimation of the accuracy of the simulation (concerned energy values)

	Lin.	INS	IN	Au
Simulation error [%]				
Mean	41,01	115,	27,	14,
Minim	-	5,11	-	-
Maxim	2784,	4610	28	17
MAE [Wh]				
Mean	2,18	5,08	2,1	2,2
Minim	0,14	0,09	0,0	0,0
Maxim	8,53	16,5	9,0	9,7
RMSE [Wh]				
Mean	4,33	8,77	4,3	4,8
Minim	0,22	0,22	0,2	0,1
Maxim	15,59	25,5	16,	18,

An accuracy of the considered simulation approaches was measured with the help of three statistical metrics: mean absolute error (MAE), root mean square error (RMSE) and simulation error ΔE (difference between simulated and measured energy outputs of the PV system in percent).

The values of the simulation error can be positive and negative: positive values mean overestimation and negative values – underestimation. The calculated values of the MAE, RMSE and simulation error ΔE are demonstrated in the Table 2. These results are comparable and better referring to the simulation results of [4] with a mean RMSE of 80.97 W and a MAE of 95.17 W.

The statistical metrics also indicated that simulation values of the INSEL model with non-fitted PVAI block diverged significantly from the measured power of PV system. The fitting of the model improved the simulation results and increased the accuracy of the simulation. Despite the fact that the PVAI block was fitted, the INSEL model continued to overestimate the measured values, but to a far lesser extent. The datasheet based INSEL® model was discarded as not being suitable for benchmarking the auto model.

The annual energy output calculated with the linear regression approach deviated from the measurements o by only 0.3 %. The deviation of the energy outputs delivered by the fitted INSEL® model and auto model amounted to 3.7 % and 4.7 % respectively.

The fitted INSEL® model and auto model had one significant commonality: these two models underestimated the PV energy output in almost all winter months and overestimated it in summer months. The simulation error per day is shown in Figure 9 and per month is shown in Figure 11. During the summer the simulation error is in a much lower range as during the winter month. This can be accounted to the much higher generation. The error distribution is presented in Figure 10. In 73% of the days, the errors are within ±20% and in 90% of the days the error is within ±40% of the generated energy. As a result, 37 days per year have a power prediction error of more than ±40%. Taking the annual distribution (Figure 9) into account, these outliers happen during times of low solar activity and are therefore representing only a small part of the energy generation.

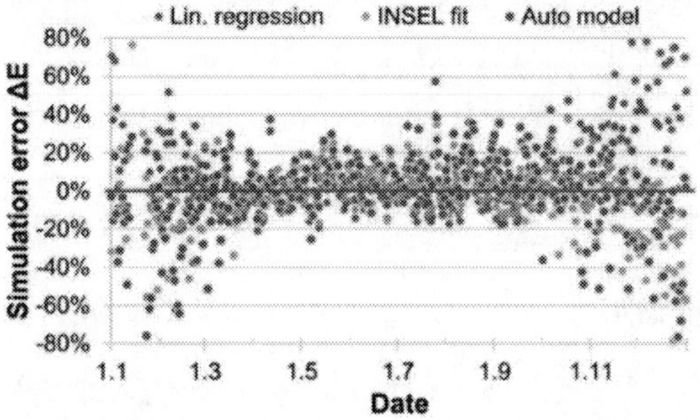

Figure 9: Values of the simulation error ΔE in % per day

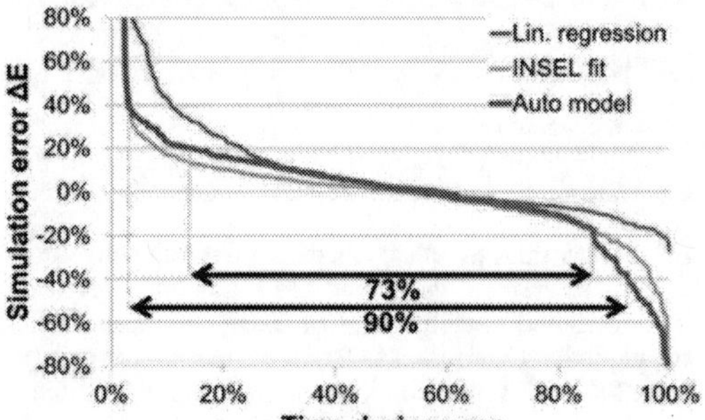

Figure 10: Distribution of daily simulation error ΔE in % over one year (365 days).

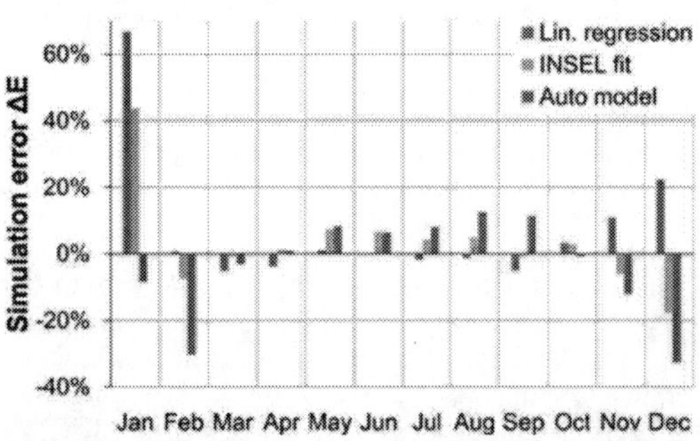

Figure 11: Simulation error in percent ΔE of the fitted INSEL model and auto model

Table 4: Monthly simulation error ΔE

Month	Lin.	INSEL fit	Auto
January	66,9%	43,7%	-8,4%
February	0,7%	-7,5%	-30,3%
March	-5,1%	-0,8%	-3,2%
April	-3,9%	1,4%	0,9%
May	1,1%	7,3%	8,3%
June	0,3%	6,5%	6,3%
July	-1,8%	4,1%	8,1%
August	-1,3%	4,8%	12,5%
September	-5,0%	-0,7%	11,3%
October	3,4%	2,8%	-0,9%
November	10,8%	66,3%	-12,2%
December	22,4%	64,5%	-32,6%

Although the annual simulation error of the auto model was not much higher than the annual error of the fitted INSEL model, the auto model underestimated the PV energy output in winter months up to 30 %.

The simulation error shown in Fig. 11 and Table 3 are comparable to [5] and [6] with simulation errors between 3 % and 5.2 % of simulation and measured PV power data in Germany. The course of the monthly simulation errors are consistent referring to [5] and also the great deviation of the monthly simulation error between summer and winter months are comparable referring to [7].

5 DISCUSSION AND OUTLOOK

The presented approach proved suitable to predict the behavior of a photovoltaic system for an upcoming day, by providing at least two days of measured photovoltaic output. The number of previous has a large influence on the quality of the systems model. Especially in the second half of the year, the better performance of the system is expected for the earliest days in the dataset – therefore ignoring seasonal variations for longer periods of time. In a further optimization of the algorithm, the historic data timeframe used for model creation can be obtained dynamically by analyzing the prediction error and modifying the data frame time span used for model creation. Further the historical data could be scaled by its individual clear sky G_{HI} curve, to reduce seasonal effects on the input data.

Figure 12 plots the PV systems power output, irradiation and the modeled results of 19 days in January 2017. Due to snow on the modules the power generation was far below the expectations from the radiation measurements. Starting Jan. 13th the auto model starts to adapt – as it is not using the radiation readings and only reacting to the changes in power output - and the predictions of the 19th and 20th of January represent very well the electric generation despite high levels of solar radiation. Starting on the 22nd the snow started to melt and the generation got back to normal. The auto model however needed the number of previous days (six) to adapt back to normal operation. This demonstrates how the algorithm is able to react to introduced disturbances (e.g. shadow of growing trees or newly build structures in the area of the generation site, failing strings, dust and dirt) without any interaction of an operator. In case of the snow the very short period of disturbance is not bringing any advantage, as the other techniques do not require re-adaption after the disturbance is gone. The behavior of the fitted INSEL® model is visible in Figure 13 and the one of the linear regression model in Figure 14 respectively. As the output power has no feedback loop to the model, the radiation and temperature measurements are the only indicator for the model calculation. Therefor beginning with the 21st the calculated power is instantly a very good approximation of the generated power. By comparison the linear regression model is working better in low irradiance situations in comparison to the fitted INSEL® model. A dynamic evaluation of the prediction error could reduce the size of the historic data frame after a radical change in the photovoltaic generation, allowing for faster adaption times.

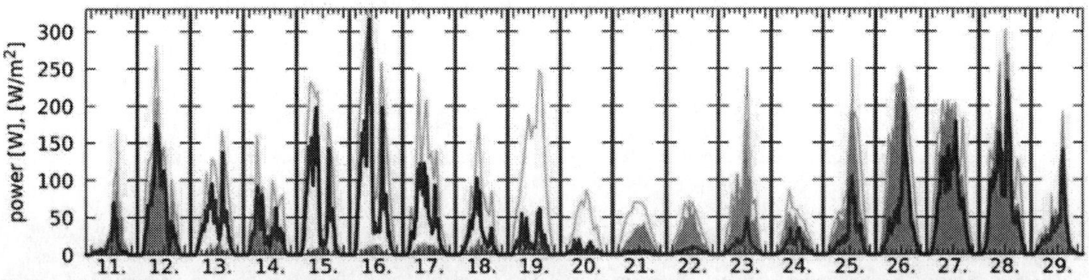

Figure 12: Irradiation (orange), generated power (light green, area) and auto model prediction (dark green) for 19 days in January 2017 (days are given below subgraphs) from 7:00 to 17:00.

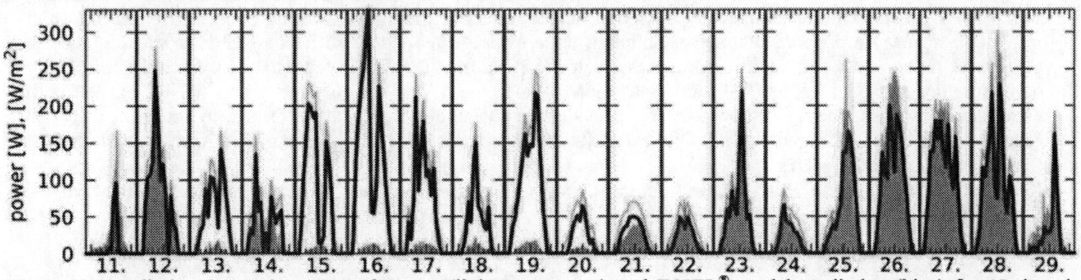

Figure 13: Irradiation (orange), generated power (light green, area) and INSEL® model prediction (blue) for 19 days in January 2017 (days are given below subgraphs) from 7:00 to 17:00.

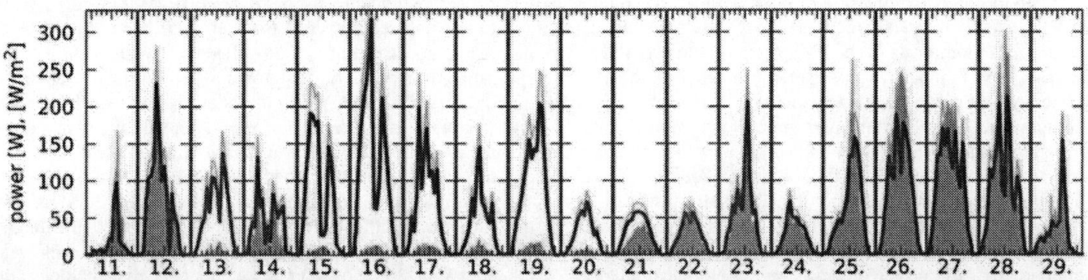

Figure 14: Irradiation (orange), generated power (light green, area) and linear regression model prediction (red) for 19 days in January 2017 (days are given below subgraphs) from 7:00 to 17:00.

6 REFERENCES

[1] die tageszeitung (taz), July 5th 2018, p. 08, "Sonne darf nicht mehr nur so scheinen"

[2] „Verordnung über energiesparenden Wärmeschutz und energiesparende Anlagentechnik bei Gebäuden (Energieeinsparverordnung - EnEV)", status as of October 24th 2015, §5 „Anrechnung von Strom aus erneuerbaren Energien"

[3] Zinßer, B; Dissertation (2010); Shaker Verlag Aachen; „Jahresenergieerträge unterschiedlicher Photovoltaik-Technologien bei verschiedenen klimatischen Bedingungen"

[4] Chouder, A. et al.; Solar Energy (2012); "Monitoring, modelling and simulation of PV systems using LabVIEW"

[5] Becker, G.; et al.; "Energy Yields of PV Systems - Comparison of Simulation and Reality"; Bavarian Association for the Promotion of Solar Energy; Stuttgart University of Applied Sciences, Munich University of Applied Sciences, BEC Engineering GmbH

[6] Matthiss, B., et al. (2015). PV Performance Modelling Workshop; "Comparison and Validation of System and Irradiance Models for Yield estimation"

[7] Schweiger, P. Oberpertinger, R., Petz, A.. Forschungsforum der österreichischen Fachhochschulen (2017); „Performance-Monitoring von PV-Anlagen auf der Basis von Wetterdaten. Ertragssimulation und potenzielle Fehlerquellen"

[8] Bottega, M., Master thesis (2017); "Management of Local Energy Availability and Customer Charging Demand for Battery Electric Vehicles Parking Areas"; University of Oldenburg

7 ACKNOWLEDGEMENTS

The authors acknowledge the financial support of the "Federal Ministry for Economic Affairs and Energy" of the "Federal Republic of Germany" for the project "EG2050: EMGIMO[10]: Neue Energieversorgungskonzepte für Mehr-Mieter-Gewerbeimmobilien" (03EGB0004G). For more details see: emgimo.eu. The presented work is created as part of this project.

Finally the authors acknowledge the paper size limitations for the conference proceedings of the EUPVSEC and regret the figure quality and jpeg artefacts.

[10] https://www.emgimo.eu/

A MORE ACCURATE MACHINE LEARNING PHOTOVOLTAIC SYSTEM PERFORMANCE ANALYSER BY USING FUZZY LOGIC

Sandy Rodrigues[1,3], João Paulo Carvalho[2], Helena Geirinhas Ramos[3], and F. Morgado-Dias[1,4]
[1]Madeira Interactive Technologies Institute, [2]INESC-ID / Instituto Superior Técnico, University of Lisbon, [3]Instituto de Telecomunicações from Instituto Superior Técnico of University of Lisbon, and [4]Universidade da Madeira,

ABSTRACT: Machine Learning Techniques (MLTs) deal well with the non-linear behaviour of the Photovoltaic systems and therefore are ideal to be used in Photovoltaic System Performance and Fault Analysis. The annual historical data of a roof-top Photovoltaic system located in Portugal in years 2015 and 2016 were used to train and test five different MLT prediction models such as the Regression Trees, Artificial Neural Networks, Multi-Gene Genetic Programming, Gaussian Process and Support Vector Machines for Regression. The aim of this work is to determine which MLT provides the most accurate prediction output results of the daily AC solar energy based on the solar irradiation inputs. An experimental setup including three experiments and 12 different scenarios was designed to achieve the aim of this work. Each scenario is associated to a season of the year in order to break down the complexity of the prediction model training. The first and second experiments simply compare various MLTs, but the third experiment describes a novel methodology that provides very accurate MLT estimate results, by means of a Hybrid Machine Learning and Computational Intelligence System that combines a Takagi-Sugeno type Fuzzy Logic Inference System with the various MLTs.
Keywords: Takagi-Sugeno Fuzzy Logic Inference System, Machine Learning, Artificial Intelligence, Photovoltaic System Performance

1 INTRODUCTION

The return of the investment (ROI) of a Photovoltaic (PV) system is maintained as high as possible if continuous monitoring of the PV system output efficiency is made and compared to past measurements of the PV system (during good working conditions). However, this comparison between the solar production output data measurements of the present and the past cannot be done in a simple linear manner as the weather conditions of one year are not the same as another. The solar production outputs of a PV system depend on the combination of many environmental factors such as the solar irradiance, temperature, soiling, and others as well as electrical factors at the solar cell, module, string and array level. Therefore, the solar production outputs of a PV system are nonlinear.

Data-driven approaches, which include Machine Learning Techniques (MLTs), tend to deal well with the nonlinear behaviour of the PV systems and are used to estimate the solar production output. The MLT prediction models should be trained with inputs and outputs of the measured data of a PV system in good working conditions in order to provide the correct estimates of the solar production outputs which can then be used to compare with the new measured solar production output of the PV system. This comparison study between the measured data (PV system) and the estimated data (PV system data-driven approach) is carried out by a PV system performance analyser to detect anomalies in the solar production output so that the maintenance team can act on them as soon as possible and ensure maximum PV system output efficiency.

The key element to correctly detect anomalies in PV systems is related to the accuracy of the estimated values provided by the MLT prediction models. Therefore, the more accurate the estimate values are the more correctly the PV performance can be predicted and the more accurately the PV anomalies and faults can be detected.

PV system monitoring analysis requires annual onsite historical data due to the unpredictable and nonlinear nature of the weather conditions throughout the various seasons of the year that consequently produce nonlinear PV system solar production outputs. For this work both annual historical data and seasonal historical data were considered, since seasonal datasets present closely related trends and relationships that can facilitate the process of machine learning when dealing with PV system production prediction as shown in [1]. The onsite historical data used in this work includes daily solar irradiation measures in kJ/m^2 as the inputs and daily solar energy measures in kWh as the outputs. The historical data used in this work is described in detail in section 2.2. The training dataset is from year 2015 with a total of 361 samples, and the testing dataset is from year 2016 with 209 samples. The total number of samples from each year should be 365, however some data was unavailable.

Based on this information we were motivated to research the behaviour of various MLTs in predicting the daily AC solar energy values from a PV system installation. One of the main challenges of this work is to correctly choose the MLT that provides the most accurate estimates for the daily AC solar energy output of a PV system since there is a vast number of MLTs available. Generally, researchers tend to use the most popular MLTs such as the Artificial Neural Networks (ANN) and Support Vector Machines (SVM), however there are many more MLTs that are unexplored and eventually can provide better results as mentioned in [2]. This work introduces other MLTs that should be considered in the Machine Learning PV system monitoring field of research and analyses its results.

Therefore, the aim of this work is to implement a machine learning PV system performance analyser that is capable of accurately predicting the daily AC solar energy values of a roof-top PV system. In other words, the predicted values should be closely correlated to the values of the measured data. There are a number of ways to predict PV solar AC energy output, however MLT have been mostly used in this type of prediction problem [3]–[6].

Given the vast number of MLTs and methods that are available to achieve the aim of this work, a Research Scheme (RS) divided into two approaches was designed to narrow down the number of MLTs to study in this work and to determine which experiments were needed to study in order to provide accurate prediction values. Therefore, Approach 1 describes the research method used to choose the MLTs and Approach 2 describes the Experimental Setup (ES) used to determine which MLTs provided the most accurate results.

The RS is described in more detail in the methodology section and is used to provide answers to the following main objectives:

- Choose which MLTs to study in this work from the vast number of MLTs that are available for PV System production prediction.
- Determine the type of training dataset needed for accurate results (Annual or Seasonal).
- Determine which MLT presents the most accurate results based on the root mean square error (RMSE) and the normalized root mean square error (NRMSE) of the error analysis.
- Propose a new methodology carried out by a PV system performance analyser to accurately predict the daily AC solar energy values of a roof-top PV system by using all five MLTs and applying them according to their best ability by means of a single input and single constant output Takagi-Sugeno type Fuzzy Logic Inference System (FIS).

The research made in Approach 1 led us to choose five MLTs to study in this work which include the Regression Trees, Artificial Neural Networks, Multi-Gene Genetic Programming, the Gaussian Process and the Support Vector Machine for Regression.

Approach 2 (Experimental Setup) describes the three experiments and twelve associated scenarios that were carried out in an attempt to understand the behaviour of the MLTs and accuracy of the prediction results. All scenarios correspond to a season (Spring, Summer, Autumn or Winter) of the year and each Experiment is associated to four scenarios. All testing datasets from 2016 were the same in all three experiments. Experiment 1 uses the 2015 annual historical data to train the MLT model in each scenario/season. Experiment 2 uses the 2015 seasonal historical data to train the MLT model in each scenario/season. Experiment 3 uses the same concept used in Experiment 2 and adds the Takagi-Sugeno type FIS to choose the MLT that should produce the best prediction results based on the input that is given.

The research made in Approach 1 also led us to the concept of combining various MLTs to create an ultimate Master Algorithm suggested by [7], believed to be capable of providing very accurate prediction results. This concept motivated us to create a new methodology (Experiment 3) for the PV System Performance Analyser which consists in combining various MLTs in order for it to provide even more accurate prediction values.

Here, the challenge lied in finding a logical rule-based system for the Analyser to choose one of five MLTs as the output based on input ranges (normalized ranges of daily solar irradiation values) that presents non-linear behaviour and is never the same from one year to the other. A simple Fuzzy Logic Inference System (FIS) is able to provide various outputs based on the inputs that it is given, and therefore meets the requirement for the Analyser to choose one of five MLTs as an output from a rule-based logic system.

Hence, the reasoning behind the setup of Experiment 3 which combines the use of a Fuzzy Logic Inference System with a Machine Learning System. The Takagi-Sugeno type FIS was chosen for this work, since a single constant output is expected to choose the MLT that should predict the best results. In the context of this work, the Machine Learning System is used to create the prediction models, and the Fuzzy Logic Inference System is used to analyse the inputs and choose one of the MLTs that should produce the best results.

This work contributes to further research in the PV system monitoring field by providing information about how

- Daily AC solar energy production behaves in the different seasons of the year;
- Different MLTs deal with annual datasets for training and seasonal datasets for training;
- A Fuzzy Logic Inference System can be used to provide prediction results that are more accurate.

The results explored in this work contribute to further research in the PV system monitoring field not only in the PV performance analysis but also in the PV fault analysis.

The PV fault study includes methods such as PV fault detection, identification, diagnostics and prognostics that rely on ML estimate model analysis to provide information to the PV system owners, investors and maintenance teams in order to keep ROI payback times on track. The results of this work are of relevance to the PV system diagnostics and prognostics field of research since these two methods include solar production output estimates, where the first analyses the performance of the PV system and the latter analyses the remaining useful lifetime (RUL) of the PV system which is useful to schedule condition-based maintenance trips which lowers the O&M costs and improves the ROI results.

The organization of the paper is as follows: Section 2 provides information about the methodology of the experimental design of this paper as well as information about the historical data used to train and test the MLTs. Section 3 presents and discusses the results obtained from the research scheme and finally section 4 concludes the main findings of the work.

2 METHODOLOGY

This section describes the methodology that was used to determine the MLT with the best results when predicting the daily solar AC energy values.

The following sections describe the research scheme that was used in this work along with the description of the historical datasets and finally the Machine Learning Process.

Section 2.1 describes the two approaches of the research scheme in detail. Section 2.2 describes the historical datasets that were used in the experimental setup. Finally, section 2.3 explains how the machine learning process works in order to train and test the MLTs.

Matlab 2017b software was used to implement the Machine Learning Performance analyser, that trains and tests the different MLTs to then perform the error analysis. All MLT prediction models were trained and tested using default parameter settings from the Statistics and Machine Learning Toolbox as well as the GPTIPS2 Toolbox. The latter toolbox (GPTIPS2) is used to train and test the Multi-Gene Genetic Programming algorithm [8], while the first toolbox trains and tests the rest of the MLTs.

2.1 Research Scheme used for Machine Learning Technique Comparison

A RS was designed in order to achieve the objectives proposed in this work. This Research Scheme is divided into two approaches in which the first approach describes the method that was used to select the MLTs to study in this work. The second approach describes the experimental setup used to compare the various MLTs which include three types of experiments and twelve scenarios. Each scenario is associated to a particular season of the year (Spring, Summer, Autumn, or Winter).

The annual training dataset used in all scenarios are from 2015 and the annual testing dataset used in all scenarios are from 2016. The testing dataset was divided into four seasonal datasets and used this way in all experiments.

The whole annual training dataset was used in Experiment 1, while for Experiments 2 and 3 the annual training dataset was divided into four seasonal training datasets.

The diagram of the Research Scheme can be found in the Supplementary Material section of this work.

The following two sections describe the approaches in more detail.

2.2.1 Approach 1

Approach 1 describes the method suggested in [7] to narrow down the vast number of machine learning techniques studied in this work down to five. This method provides a MLT selection scheme that allows to cover all types of MLT to research the Machine Learning PV system monitoring topic. This method consists in organizing all MLTs into five "tribes" or "schools" in which each one is represented by its own master algorithm, as presented in Table I [7]. The five machine learning techniques used in this work include the Regression Trees, Artificial Neural Networks (ANN), Multi-Gene Genetic Programming, the Gaussian Process and the Support Vector Machine for Regression (SVR).

These five specific MLTs were also chosen due to the fact that they were available in the Matlab 2017b software. An extended list of MLTs according to their tribe can be consulted in [2].

Table I: The Five MLT Tribe Concept [7]

#	Tribe	Master Algorithm	Selected MLT for this work
1	Symbolists	Inverse Deduction / Induction	Regression Tree
2	Connectionists	Backpropagation	Artificial Neural Network (ANN)
3	Evolutionaries	Genetic Programming	Multi-Gene Genetic Programming
4	Bayesians	Probabilistic Inference	Gaussian Process
5	Analogizers	Kernel Machines	Support Vector Machine for Regression (SVR)

2.2.2 Approach 2

Approach 2 describes the experimental setup that was used to achieve the aim of this work, that includes three experiments and twelve scenarios. Each experiment is associated to four scenarios (Spring, Summer, Autumn, Winter) related to a different season of the year. The experiments are used to show that the MLT prediction models can become more accurate and what procedures need to take place in order for the accuracy to be improved. The experimental setup is designed to show how the accuracy of the prediction results keep improving, in which Experiment 1 presents the worst results and Experiment 3 presents the best results.

The First Experiment consists in training only one prediction model per MLT by using the whole 2015 annual training dataset of 361 samples to then be tested by season with the 2016 testing dataset.

The Second Experiment consists in training four prediction models per MLT by using seasonal data from the 2015 annual training dataset by dividing it into 4 seasons to then be tested by season with the 2016 testing dataset.

The Third Experiment is implemented by combining a Fuzzy Logic Inference System (FIS) with various MLTs.

The FIS designed for this work was a single input and a single constant output Takagi-Sugeno (TS) type FIS. The TS method was chosen, since it allows the FIS output to be a constant and therefore the Fuzzy rule output is associated to a single MLT output based on the single input that is given. Each input represents a range of daily solar irradiation values and each output represents one of the five MLTs.

Since the FIS designed in this work presented a single input and output, there was no need to set a particular operator (AND, OR). The rule output methods that were used include the *Weighted Average* ('wtaver') defuzzification method which is the default for TS systems, the product ('prod') implication method as well as the sum aggregation method. The input *Triangular-shaped membership function* ('trimf') was used as well as the range [0 1]. Finally, the output constant membership function was used.

A constant output TS type FIS was created for each season of the year in which the Spring FIS has 29 rules, the Summer FIS has 26 Rules, the Autumn FIS has 30 Rules and the Winter FIS has 49 rules. The number of rules is associated to the number of inputs that a FIS season has. As an example, the Winter test dataset has 72 samples, and these samples were grouped into 49 different ranges, in which each range is associated to a specific MLT. The MLT that is associated to a range presents the best prediction result for that specific input range of solar irradiation values.

The Methodology used in Experiment 3 consists of the following four steps

1. Associating each input/output to a MLT that produces the best result by following the flowchart in Fig. 1. Most of the procedures included in this step were completed in Experiment 2 (Machine Learning System);
2. Grouping ranges of inputs that are associated to the same MLT;
3. Designing the constant Takagi-Sugeno type Fuzzy Logic Inference System (FIS) by following the flowchart in Fig. 2;
4. Using the designed FIS to choose the MLT that will predict the best daily AC solar energy output based on the input by following the flowchart in Fig. 3.

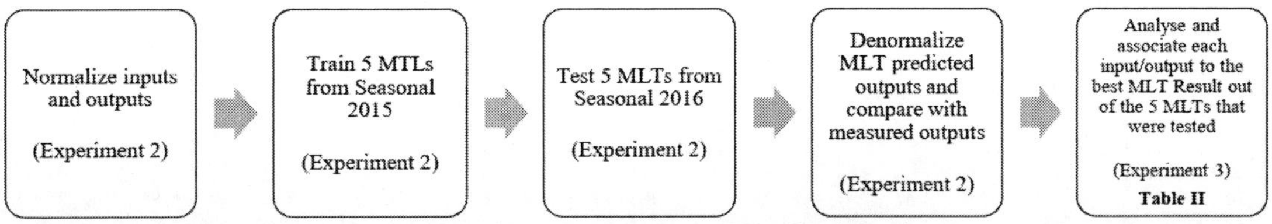

Figure 1: Machine Learning System Flowchart

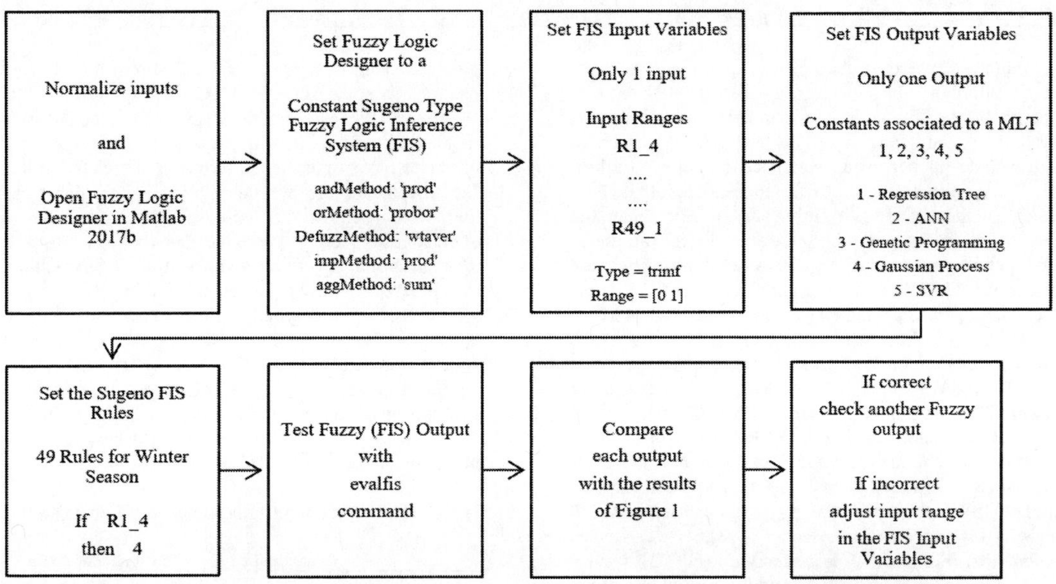

Figure 2: The Constant Takagi-Sugeno type Fuzzy Logic Inference System (FIS) Flowchart

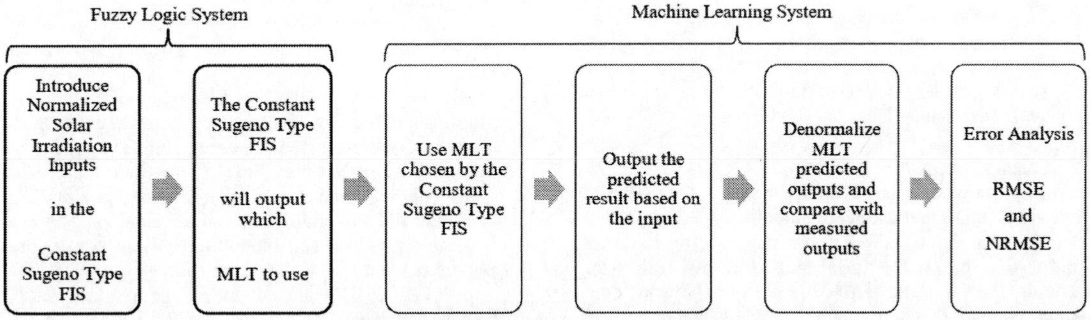

Figure 3: Experiment 3 System Flowchart

Table II: Samples from the Winter season and corresponding best MLT

Date	Measured		Estimated Output					Best MLT
	Input Solar Irradiation (kJ/m^2)	Output Solar Energy (kWh)	Regression Tree	ANN	Genetic Programming	Gaussian Process	SVR	
15-01-2015	3231	2.7816	4.844	5.702	3.873	3.244	5.542	Gaussian Process
08-01-2015	3911	4.2024	4.844	5.768	4.837	3.918	6.464	Gaussian Process
24-01-2015	6265	6.9044	8.861	8.373	7.597	7.483	9.655	Gaussian Process
12-01-2015	6974	9.3728	8.861	8.575	8.651	8.843	10.616	Regression Tree
22-01-2015	7345	8.3668	8.861	8.964	9.275	9.588	11.119	Regression Tree
10-01-2015	7924	10.1324	10.928	10.188	10.348	10.780	11.904	ANN
16-01-2015	8449	11.1324	10.928	11.073	11.416	11.878	12.616	ANN
08-02-2015	8820	9.9072	13.636	13.464	12.215	12.656	13.119	Genetic Programming
26-12-2015	8904	12.1344	13.636	13.503	12.400	12.832	13.232	Genetic Programming
14-01-2015	9423	12.8252	13.636	13.691	13.566	13.909	13.936	Genetic Programming
23-01-2015	10448	13.4492	16.157	16.522	15.899	15.956	15.326	SVR
18-02-2015	10823	13.3812	17.493	17.083	16.723	16.665	15.834	SVR

Table II serves as an example to explain how the input solar irradiation values were grouped and associated to a specific MLT. This table presents 12 samples from the Winter season and the result of the best MLT associated to each sample, in which the estimated output of the best MLT is the closest value to the measured output. It is possible to identify ranges of input values associated to the same MLT. In this case through Table II, an example of the input ranges to apply to the Fuzzy Logic Inference System would be as follows: from 0 kJ/m^2 to 6273 kJ/m^2 associated to the Gaussian Process MLT, from 6974 kJ/m^2 to 7923 kJ/m^2 associated to the Regression Tree MLT, from 7924 kJ/m^2 to 8819 kJ/m^2 associated to the ANN MLT, from 8820 kJ/m^2 to 10447 kJ/m^2 associated to the Genetic Programming MLT, and finally from 10448 kJ/m^2 to 10823 kJ/m^2 associated to the SVR MLT.

The most common methodology to choose the most accurate MLT has been to analyse and compare the RMSE and NRMSE error results of an entire test dataset, in which the MLT that presents the lowest error results is chosen as the one that presents the most accurate results. However, the RMSE and NRMSE results provides information of the "big picture" of an entire test dataset (and changes when using different test datasets) nonetheless does not provide information about the error results of the individual samples of the test dataset. Therefore, the RMSE and NRMSE results associated to the best MLT of an entire test dataset is always different to another test dataset and consequently the associated best MLT is also always different. Thus, the determination of the best MLT depends on the test dataset that is used at the time and thus the determination of the best MLT is a subjective answer. The best solution to determine the best MLT is to analyse the samples of the test dataset individually as explained above through Table II.

2.2 Historical Datasets

The historical dataset of the PV system is composed of one set of inputs and one set of outputs.

The roof-top 4.14kW PV system dataset for training and testing the MLT estimate models in this work were measured in the Island of Madeira (Portugal) in which the input samples (solar irradiation) were gathered by the Portuguese Institute of the Sea and Atmosphere (IPMA) and the output samples (daily solar energy) were gathered by a PV system installation company called *Factor Energia*. The solar irradiation values were only available in daily measurements, and therefore the output solar values were converted into daily solar AC energy values and consequently the prediction results of the MLTs are also in daily solar AC energy values.

The weather station is 440m away from the PV system installation used in this work. It is costly for a roof-top PV system to have a precise pyranometer associated to it and therefore the nearest weather station is considered as an alternative. It should also be noted that pyranometers are in most cases thermal based devices that have long time constants not adequate for identification and modelling of PV systems that have time constants of milliseconds.

Two sets of incomplete annual datasets were used in this work. One of the annual datasets with 361 samples measured in 2015 was used to train the prediction models and the other annual dataset with 209 samples measured in 2016 was used to test the prediction models.

Table III presents information about the Experiments and associated datasets used in this work.

Table III: PV system installation dataset information

	Experiment 1	Experiment 2 and 3
Train Dataset 2015	361 Annual Samples	92 Spring Samples 89 Summer Samples 91 Autumn Samples 89 Winter Samples
Test Dataset 2016	58 Spring Samples	
	32 Summer Samples	
	47 Autumn Samples	
	72 Winter Samples	

2.3 Machine Learning Process

This section describes the pre-processing technique applied to the raw data, the Machine Learning Process, and finally the error analysis.

Before training the prediction models, the raw historical data is pre-processed by normalizing both the inputs and outputs with a pre-processing technique called Min-Max normalization. This normalization procedure speeds up the prediction model training process. The minimum and maximum values of the input samples were respectively 1 kJ/m^2 and 30000 kJ/m^2, while the minimum and maximum values of the output samples were respectively 1 kWh and 30 kWh. These values are the highest and lowest values that the PV system can ever experience. This normalizing technique is used before training the models, and therefore it has to be reversed at the output.

The Machine Learning model is generated by analysing the relationships and trends of the historical data by means of a mathematical algorithm.

The Machine Learning Process of any MLT starts by training a Machine Learning (ML) model, then testing the ML model, to then compare the measured data (targets) with the estimated data (outputs) generated by the ML model and finally performing an error analysis.

In order to compare the performance of the MLTs, an error analysis is conducted by using two indicators namely the Root Mean Squared Error (RMSE) and the Normalized Root Mean Squared Error (NRMSE) after de-normalizing the data. The RMSE is easily interpreted due to having the same units as the targets and outputs (kWh) and computes the average error compared to the target (observed output) value (Equation 1). A large positive RMSE value represents a large deviation scale in the prediction values from the target values.

The accepted RMSE value in this work, is a RMSE value lower that 6 kWh since this value represents a 2.40 € / day loss (6 x 0.40 €/kWh = 2.40 € / day loss) assuming that the grid injection price is 0.40 €/kWh (as once practiced in Portugal). If this loss is an average daily loss over the course of 3 months, the total monetary loss is of 2016€ (2.40€ x 90 days = 216€) which is quite a significant amount and is negatively reflected in increasing the investment payback time period.

The NRMSE provides the average percentage value of the error (Equation 2), in relation to the difference between the unnormalized minimum and maximum output values, which is 29 kWh (30 kWh–1 kWh = 29 kWh) assumed in this work.

$$RMSE = \sqrt{\frac{1}{n}\sum_{i=1}^{n}(Yt - Yp)^2} \qquad (1)$$

$$NRMSE = \frac{RMSE}{Yt_{max} - Yt_{min}} \qquad (2)$$

Where *Yt* is the observed targets, *Yp* is the predicted outputs and n is the number of observations.

The Machine Learning for Regression model parameter settings for this work are mainly the Matlab default Machine Learning model settings. Validation and cross-validation were not performed to any of the models in this work unless automatically done by the fitting command of the Matlab 2017b toolbox. The MLTs were only trained with simple/default parameter settings and then tested.

3 RESULTS AND DISCUSSION

The results of the three experiments and respective scenarios are presented and discussed in this section. Experiments 1, 2 and 3 are followed by a table with the ranking score of each MLT in section 3.4.

3.1 Experiment 1 - Seasonal Comparison with Annual trained models

All the MLT models in Experiment 1 were trained with the same 2015 annual dataset (inputs and targets) of 361 samples from the PV system. This Experiment is divided into four scenarios where each one tests the trained model with a seasonal dataset from the year 2016.

Table IV presents the error analysis results that include the RMSE and NRMSE values. This table shows us the MLTs that best predict the AC daily solar energy values (green highlight) which are the SVR MLT in scenarios 1 and 2 (Spring and Summer), the Gaussian Process in scenario 3 (Autumn) and finally the ANN in scenario 4 (Winter).

The Multi-Gene Genetic Programming and Gaussian Process MLTs present results that are very similar to each other. These two MLTs rank second and third place in all scenarios except for scenario 3 where the Gaussian Process MLT ranks first place. It should be pointed out that the Gaussian Process MLT is much faster and uses less computer resources compared to the Multi-Gene Genetic Programming MLT.

The worst results in scenarios 1, 2 and 3 are presented by the Regression Tree and the ANN MLTs, while the Regression Tree and the SVR MLTs present the worst results in scenario 4.

Overall the results are under the highest RMSE value assumed as acceptable in this work of 6 kWh as explained in the methodology section except for the ANN MLT in scenarios 1 and 2. The corresponding NRMSE value of this RMSE value is approximately 17.2414 %, since the difference between the normalizing minimum and maximum values is 29 kWh.

The paragraphs that follow describes how each MLT reacts in each scenario. Overall, the MLTs react badly in scenario 1 (Spring) except for the ANN that reacts the worst in scenario 2 (Summer). This indicates that the Spring months of year 2015 are not similar to the ones of 2016. This indicates that the MLTs do not deal well when predicting what takes place in Spring since it might have a mixture of results from the whole year. The bad results are due to the small number of testing samples.

Overall the best results of each MLT take place in scenario 4 (Winter) except for the SVR MLT. This indicates that the annual data is very similar to the Winter months or that the Winter months of 2015 are very similar to the ones in 2016. The results are better because there are more test samples.

The Regression Tree MLT presents the worst results in scenario 1 (Spring), while in the rest of the scenarios it presents very similar results regardless of how many training samples are used. This MLT presents the best results in scenario 3 (Autumn).

The ANN MLT presents its worst results in scenario 2 (Summer) which coincides with the lowest number of testing samples. However, the theory of having less samples will produce worse results does not apply when comparing scenario 1 (Spring) with scenario 3 (Autumn) but this comparison shows that even the ANN MLT also has a hard time dealing with the Spring months of scenario 1.

The Multi-Gene Genetic Programming MLT presents its worst results in scenario 1 (Spring) and its best results in scenario 4 (Winter). Scenarios 2, 3 and 4 present results that are very similar to each other regardless of how many test samples are used.

The Gaussian Process MLT results are followed very closely to the results of the Multi-Gene Genetic Programming MLT. The difference between these two MLTs is related to the processing speed and the use of computer resources. The Gaussian Process MLT presents faster speeds as well as has a necessity of fewer computer resources to operate.

Table IV: The Error Results of Experiments 1 of the Scenarios of the Seasonal Comparison

			Experiment 1 - Seasonal Comparison with Annual trained models			
Tribe	**MLT**	**Error Analysis**	**Scenario 1** **Spring** 361 Annual Train Samples 58 Spring Test 2016 Samples	**Scenario 2** **Summer** 361 Annual Train 2015 Samples 32 Summer Test 2016 Samples	**Scenario 3** **Autumn** 361 Annual Train 2015 Samples 47 Autumn Test 2016 Samples	**Scenario 4** **Winter** 361 Annual Train 2015 Samples 72 Winter Test 2016 Samples
1	**Regression Tree (RT)**	**RMSE**	3.9858 kWh	3.2559 kWh	3.2422 kWh	3.3212 kWh
		NRMSE	13.7440 %	11.2273 %	11.1799 %	11.4524 %
2	**ANN**	**RMSE**	6.3825 kWh	12.0830 kWh	4.4680 kWh	1.9359 kWh
		NRMSE	22.0084 %	41.6654 %	15.4070 %	6.6755 %
3	**Multi-Gene Genetic Programming**	**RMSE**	3.2236 kWh	2.6083 kWh	2.6396 kWh	2.4128 kWh
		NRMSE	11.1159 %	8.9941 %	9.1023 %	8.3201 %
4	**Gaussian Process**	**RMSE**	3.2260 kWh	2.6030 kWh	2.6265 kWh	2.4017 kWh
		NRMSE	11.1240 %	8.9758 %	9.0570 %	8.2818 %
5	**SVR**	**RMSE**	2.8665 kWh	2.5552 kWh	2.8410 kWh	3.7598 kWh
		NRMSE	9.8844 %	8.8110 %	9.7965 %	12.9650 %

Table V: The Error Results of Experiments 2 of the Scenarios of the Seasonal Comparison

			Experiment 2 - Seasonal Comparison with Seasonal trained models			
Tribe	**MLT**	**Error Analysis**	**Scenario 5** **Spring** 92 Train Samples 58 Test Samples	**Scenario 6** **Summer** 89 Train Samples 32 Test Samples	**Scenario 7** **Autumn** 91 Train Samples 47 Test Samples	**Scenario 8** **Winter** 89 Train Samples 72 Test Samples
1	**Regression Tree (RT)**	**RMSE**	2.6519 kWh	2.0613 kWh	2.8508 kWh	2.3406 kWh
		NRMSE	9.1446 %	7.1081 %	9.8305 %	8.0711 %
2	**ANN**	**RMSE**	2.5535 kWh	1.8272 kWh	2.5357 kWh	1.9678 kWh
		NRMSE	8.8052 %	6.3007 %	8.7439 %	6.7856 %
3	**Multi-Gene Genetic Programming**	**RMSE**	2.8016 kWh	1.7965 kWh	2.4507 kWh	1.8466 kWh
		NRMSE	9.6606 %	6.1949 %	8.4505 %	6.3675 %
4	**Gaussian Process**	**RMSE**	2.4541 kWh	1.8860 kWh	2.3301 kWh	1.7781 kWh
		NRMSE	8.4624 %	6.5033 %	8.0349 %	6.1315 %
5	**SVR**	**RMSE**	2.5136 kWh	1.9187 kWh	2.3232 kWh	2.1969 kWh
		NRMSE	8.6675 %	6.6161 %	8.0112 %	7.5755 %

The SVR MLT deals differently to the test datasets of the scenarios when compared to the other MLTs even though it presents the best results in scenarios 1 and 2 out of all the MLTs. The worst results presented by the SVR MLT is in scenario 4 (Winter) which is opposite to the all the other MLTs. The best results are presented in scenario 2 (Summer) which is opposite to the results presented by the ANN MLT.

3.2 Experiment 2 - Seasonal Comparison with Seasonal trained models

All the MLT models in Experiment 2 were trained with the corresponding 2015 seasonal dataset (inputs and targets) and therefore there are four different MLT trained models where each one is associated to each scenario. This Experiment is divided into four scenarios where each one tests the seasonal trained model from the year 2015 with a seasonal dataset from the year 2016.

Table V presents the error analysis results that include the RMSE and NRMSE values. This table shows us the MLTs that best predict the AC daily solar energy values (green highlight) which are the Gaussian Process MLT in scenarios 5 and 8 (Spring and Winter), the Multi-Gene Genetic Programming MLT in scenario 6 (Summer) and finally the SVR MLT in scenario 7 (Autumn). The error results are very similar in scenarios 5 and 7 (Spring and Autumn) and very similar in scenarios 6 and 8 (Summer and Winter), in which the latter case presents better results.

It can be observed that the tribe with the second-best results of each scenario is adjacent to the tribe with the best results of each scenario. Therefore, tribes 4 and 5 (Gaussian Process and SVR) present the best results in scenario 5 (Spring); tribes 2 and 3 (ANN and Multi-Gene Genetic Programming) present the best results in scenario 6 (Summer); tribes 4 and 5 (Gaussian Process and SVR) present the best results in scenario 7 (Autumn); and finally tribes 3 and 4 (Multi-Gene Genetic Programming and Gaussian Process) present the best results in scenario 8 (Winter).

Overall, the MLTs that rank in second and third place lie in all tribes except for tribe 1 (Regression Tree) which presents the worst results.

In this experiment all the results are under the highest RMSE value assumed as acceptable in this work of 6 kWh as explained in the methodology section.

The paragraph that follows describes how each MLT reacts in each scenario, and again it can be observed that scenarios 5 and 7 (Spring and Autumn) present very similar results as does scenarios 6 and 8 (Summer and Winter). This indicates that the Summer and Winter seasons from one year to another as well as the weather variations of the season are very similar throughout.

The Regression Tree MLT presents the worst results in scenario 7 (Autumn), while in the rest of the scenarios it presents very similar results regardless of how many training samples are used. This MLT presents the best results in scenario 6 (Summer).

The ANN MLT presents its worst results in scenario 5 (Spring) and best results in scenario 6 (Summer).

The Multi-Gene Genetic Programming MLT presents

its worst results in scenario 5 (Spring) and its best results in scenario 6 (Summer).

The Gaussian Process MLT presents the worst results in scenario 5 (Spring) and its best results in scenario 8 (Winter).

The SVR MLT presents the worst results in scenario 5 (Spring) and its best results in scenario 6 (Summer).

When comparing Experiment 1 to Experiment 2, we can observe that all the MLTs improve the results because the training and testing datasets are very similar to each other in Experiment 2. This is due to the presence of less training data samples in Experiment 2 (the testing samples are the same ones used in Experiment 1).

The season that improves the most from experiment 1 to experiment 2 is the Summer Season.

All the results of Experiment 2 are better than the results of Experiment 1 by approximately 3% of the NRMSE values. This indicates that training the prediction models by season (Experiment 2) improves the estimation outputs when compared to the prediction models that are trained with the annual data (Experiment 1). This way the MLT prediction model training procedure focuses only on the nonlinear data of 3 months (small group variance) instead of 12 months (large group variance) and thus providing more accurate output estimates by minimizing the estimate error.

Therefore, it is best to narrow down the training dataset to a seasonal dataset rather than using an annual dataset for training. This makes the training process focus more on the fluctuations of the weather conditions that is in accordance to the season.

3.3 Experiment 3 - Seasonal Comparison with Fuzzy Logic

The results of Experiment 3 that uses a Takagi-Sugeno type FIS is presented in **Table VI** and they indicate that they are all better than the results of Experiment 2 (by approximately 2% of the NRMSE values) and consequently of Experiment 1. The estimate accuracy improved approximately 5% of the NRMSE values from Experiment 1 to Experiment 3.

These results indicate that the new methodology proposed in this work is a better solution to accurately estimate the daily solar energy output of a PV system. The Winter season in scenario 12 presents the most improvement. This may have various reasons such as the inputs and outputs of the training dataset of 2015 might be very similar to those of the testing dataset of 2016. Another possible reason would be due to the fact that there are more test samples in this scenario 12 than the rest of the seasons (scenarios 9, 10 and 11).

3.4 Ranking Score of the MLTs.

This section Table VII presents a ranking score associated to each of the five MLT tribes within the seven scenarios of only Experiment 1 and 2 carried out in this work.

It should be pointed out that the trained models of Experiment 1 have more samples (annual ~361 samples) and therefore the learning curve is more disperse, while the trained models of Experiment 2 have less samples (Seasonal ~90 samples) and therefore learning curve is more focussed. Given this information and analysing Table VII, we can observe that the ranking of the MLTs also change accordingly. This said, it is possible to determine which MLTs deal better with disperse learning curves (Experiment 1 - SVR, Gaussian Process and ANN) and which MLTs deal best with narrow learning curves (Experiment 2 – Gaussian Process, Multi-Gene Genetic Programming and SVR).

In Experiment 1 the ranking of the MLTs are very similar between Spring and Summer and also between Autumn and winter. On the other hand, in Experiment 2 the ranking of the MLTs are very similar between Spring and Autumn and also between Summer and Winter.

Table VI: The Error Results of Experiment 3 of the Scenarios of the Seasonal Comparison when using Fuzzy Logic

MLT	Error Analysis	Experiment 3 - Seasonal Comparison with Seasonal trained models with Fuzzy Logic			
		Scenario 9 Spring 92 Train Samples 58 Test Samples	**Scenario 10 Summer** 89 Train Samples 32 Test Samples	**Scenario 11 Autumn** 91 Train Samples 47 Test Samples	**Scenario 12 Winter** 89 Train Samples 72 Test Samples
Fuzzy Logic + 5MLTs	**RMSE**	1.8751 kWh	1.5338 kWh	1.7122 kWh	1.1561 kWh
	NRMSE	6.4658 %	5.2888 %	5.9042 %	3.9865 %

Table VII: Ranking Score of the Machine Learning Techniques trained by the Spring season in each scenario

Rank	Experiment 1 - Annual trained models				Experiment 2 - Seasonal trained models			
	Scenario 1 Spring	Scenario 2 Summer	Scenario 3 Autumn	Scenario 4 Winter	Scenario 5 Spring	Scenario 6 Summer	Scenario 7 Autumn	Scenario 8 Winter
1º	SVR	SVR	Gaussian Process	ANN	Gaussian Process	MG Genetic Programming	SVR	Gaussian Process
2º	MG Genetic Programming	Gaussian Process	MG Genetic Programming	Gaussian Process	SVR	ANN	Gaussian Process	MG Genetic Programming
3º	Gaussian Process	MG Genetic Programming	SVR	MG Genetic Programming	ANN	Gaussian Process	MG Genetic Programming	ANN
4º	Regression Tree	Regression Tree	Regression Tree	Regression Tree	Regression Tree	SVR	ANN	SVR
5º	ANN	ANN	ANN	SVR	MG Genetic Programming	Regression Tree	Regression Tree	Regression Tree

Overall, the Regression Tree MLT is very consistent in both Experiments by presenting bad ranking scores (4th and 5th place ranking).

The ANN MLT has a high-ranking score in Experiment 2 when dealing with a trained model with a narrow learning curve when compared to the ranking scores of Experiment 1. This MLT ranks either in a very bad location (5th place) in Spring, Summer and Autumn or ranks in a very good location (1st place) in Winter. This ranking score of the ANN MLT tells us that it does not deal well with a disperse learning curve, however it is capable of successfully learning the Winter patterns from the annual data when compared to all the other MLTs. Curiously in Experiment 2 the ANN MLT ranks in the middle (2nd, 3rd and 4th place), which indicates that this MLT deals better when the learning curve is narrow.

The Multi-Gene Genetic Programming MLT ranks in second and third place in all scenarios except for scenarios 5 and 6 where it ranks at 5th and 1st place respectively.

The Gaussian Process MLT also ranks in second and third place in all scenarios except for scenarios 3, 5 and 8 where it ranks at first place.

The SVR MLT deal better with the disperse learning of the trained model curve (Experiment 1) than the other MLTs and the opposite in Experiment 2 with the narrow learning curve.

Overall the Multi-Gene Genetic Programming, Gaussian Process and SVR MLTs rank between the first and third place, while Regression Tree and ANN rank in fourth and fifth place.

Many researches tend to choose the most popular (ANN and SVR) MLTs to work with in their research and this work is able to show that there are other MLTs that should be considered depending on the training dataset at hand. Interestingly, the most popular MLTs (ANN and SVR) are exactly the ones that present the worst results in this work in some cases/scenarios.

4 CONCLUSIONS

The aim of this work is to accurately estimate the daily solar energy output of a PV system by using a data-driven approach. Five Machine Learning Techniques were tested in three experiments with a total of twelve scenarios, as explained in the methodology section. The five MLTS are the Regression Tree, Artificial Neural Network (ANN), Multi-Gene Genetic Programming, Gaussian Process and the Support Vector Machine for Regression (SVR) MLTs. The results show that it is possible for MLT models to produce very accurate results and is dependent on the type (annual or seasonal) of training datasets and number of samples.

The solution with the best results was associated to Experiment 3 where the test results of Experiment 2 were analysed individually and grouped into input ranges to apply in the Takagi-Sugeno type Fuzzy Logic Inference System. The estimate accuracy improved approximately 5% of the overall NRMSE values from Experiment 1 to Experiment 3.

Each trained prediction model created in Experiment 2 did not have more than 92 training samples. Therefore, the results of this work related to the five MLTS are of great relevance to the PV system monitoring research field, to show that accurate models can be trained using limited data, even though some MLTs present better results than others. Therefore, it is best to narrow down the training dataset to a seasonal dataset rather than an annual dataset for training. This makes the training process focus more on the fluctuations of the weather conditions that is in accordance with the season at hand.

This work shows that the best solution to accurately estimate the prediction outputs of any PV system is to use various MLTs to their best ability instead of using only one MLT. Hence, a new methodology to acquire the most accurate MLT estimate results is presented in this paper by using a Takagi-Sugeno type Fuzzy Logic system combined with a Machine Learning Techniques.

The results would improve if the parameters of each MLT were fine-tuned instead of using the default parameters established by the Matlab 2017b software.

Experiment 2 is better than Experiment 1 should Experiment 3 not be possible for use. Therefore, should the use of a Fuzzy Logic system not be available to combine the MLTs then Experiment 2 should be considered. Overall, the Gaussian Process, Multi-Gene Genetic Programming and Support Vector Machines for Regression presents the best results.

The ranking score of the Multi-Gene Genetic Programming, Gaussian Process and SVR MLTs are between the first and third place, while Regression Tree and ANN rank in fourth and fifth place. Many researches tend to choose the most popular (ANN and SVR) MLTs to work with in their research and this work is able to show that there are other MLTs that should be considered depending on the training dataset at hand. Interestingly, the most popular MLTs (ANN and SVR) are exactly the ones that present the worst results in this work in some scenarios.

The aim of this work involves exploring both the PV system monitoring research field as well as the Machine Learning Technique research field. The PV system monitoring research field provides information about the features that influence the performance of the PV system such as the weather conditions, PV system equipment behaviour and PV system monitoring devices, methods and related issues. The Machine Learning Technique research field provides information about the numerous available MLT and their applications. In recent years, MLTs have been used in the PV system monitoring research field to improve the estimate outputs of the solar production outputs in order to perform PV system performance analysis.

The scientific innovation of this work includes the organizing and narrowing down of the vast number of MLT into 5 MLT tribes suggested by [7], the comparison of 5 MLT that represent all types of MLT, the procedures that need to be taken in order to implement an accurate PV system performance analyser, and finally the presentation of a new methodology to acquire the most accurate MLT estimate results is presented in this paper by using a Takagi-Sugeno type Fuzzy Logic Inference system. Another scientific innovation includes the fact that this work considers samples of almost a whole calendar year, which is uncommon in most research papers that study the PV system performance by using MLTs due to the difficulty in gathering annual data.

Future work would include narrowing the training dataset down with sunny days from each season. Another future work to improve accuracy would include using more than one input feature rather than just solar irradiation.

ACKNOWLEDGMENTS

The authors acknowledge the PV system installation company *Factor Energia* for providing the historical data from the 3 PV systems located in the island of Madeira, Portugal, and the meteorological institution IPMA (IPMA—*Instituto Português do Mar e da Atmosfera*) for providing the daily solar irradiation historical data. The PhD Grant financial support from ARDITI (*Agência Regional para o Desenvolvimento da Investigação, Tecnologia e Inovação*) under the scope of the project M1420-09-5369-FSE-000001, is gratefully appreciated. Acknowledgments to the Portuguese Science and Technology Foundation (FCT) for their support through Projeto Estratégico LA 9 - UID/EEA/50009/2013. This work was also developed under the Instituto de Telecomunicações and it was supported in part by the Portuguese Science and Technology Foundation (FCT) projects: UID/EEA/50008/2013. This support is gratefully acknowledged. This work was also supported by national funds through Fundação para a Ciência e a Tecnologia (FCT) under reference UID/CEC/50021/2013 and grant SFRH/BSAB/136312/2018.

REFERENCES

[1] S. Rodrigues, H. G. Ramos, and F. Morgado-Dias, "Machine learning PV system performance analyser," *Progress in Photovoltaics: Research and Applications*, vol. 26, no. 8, pp. 675–687, 2018.

[2] S. Rodrigues, H. G. Ramos, and F. Morgado-Dias, "Machine Learning in PV Fault Detection, Diagnostics and Prognostics: A Review," in *International IEEE 44th Photovoltaic Specialists Conference*, 2017.

[3] H. T. C. Pedro and C. F. M. Coimbra, "Assessment of forecasting techniques for solar power production with no exogenous inputs," *Solar Energy*, vol. 86, no. 7, pp. 2017–2028, 2012.

[4] R. H. Inman, H. T. C. Pedro, and C. F. M. Coimbra, "Solar forecasting methods for renewable energy integration," *Progress in Energy and Combustion Science*, vol. 39, no. 6, pp. 535–576, 2013.

[5] J. Antonanzas, N. Osorio, R. Escobar, R. Urraca, F. J. Martinez-de-pison, and F. Antonanzas-torres, "Review of photovoltaic power forecasting," *Solar Energy*, vol. 136, pp. 78–111, 2016.

[6] Muhammad Qamar Raza, Mithulananthan Nadarajah, and Chandima Ekanayake, "On recent advances in PV output power forecast," *Solar Energy*, vol. 136, no. 136, pp. 125–144, 2016.

[7] P. Domingos, *The Master Algorithm: How the Quest for the Ultimate Learning Machine Will Remake Our World*. Basic Books, 2015.

[8] D. Searson, D. Leahy, and M. Willis, "GPTIPS: An open source genetic programming toolbox for multigene symbolic regression," *Proceedings of the International of the MultiConference of Engineers and Computer Scientists*, vol. I, pp. 17–20, 2010.

SUPPLEMENTARY MATERIAL

The Supplementary material includes a diagram of the Research Scheme.

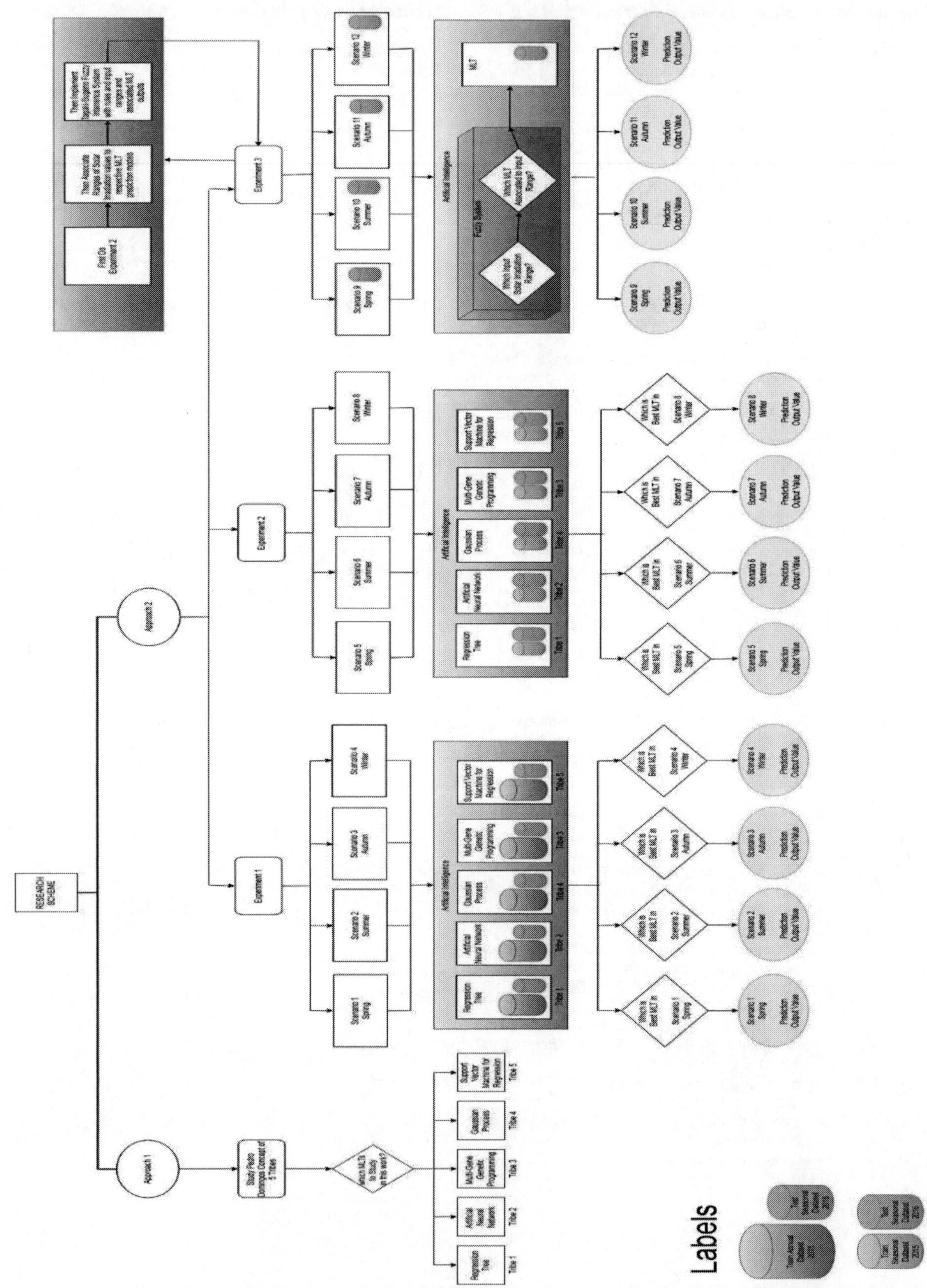
RESEARCH SCHEME
Approach 1
Approach 2
Study Pedro Domingos Concept of 5 Tribes
Which MLTs to Study in this work?
Regression Tree
Title 1
Artificial Neural Network
Title 2
Multi-Gene Genetic Programming
Title 3
Gaussian Process
Title 4
Support Vector Machine for Regression
Title 5
Experiment 1
Experiment 2
Experiment 3
Scenario 1 Spring
Scenario 2 Summer
Scenario 3 Autumn
Scenario 4 Winter
Scenario 5 Spring
Scenario 6 Summer
Scenario 7 Autumn
Scenario 8 Winter
Scenario 9 Spring
Scenario 10 Summer
Scenario 11 Autumn
Scenario 12 Winter
Artificial Intelligence
Which is Best MLT in Scenario 1 Spring
Which is Best MLT in Scenario 2 Summer
Which is Best MLT in Scenario 3 Autumn
Which is Best MLT in Scenario 4 Winter
Which is Best MLT in Scenario 5 Spring
Which is Best MLT in Scenario 6 Summer
Which is Best MLT in Scenario 7 Autumn
Which is Best MLT in Scenario 8 Winter
Prediction Output Value
First Do Experiment 2
Then Associate Ranges of Solar Irradiation values to respective MLT prediction models
Fuzzy System
Which Input Solar Irradiation Range?
Which MLT Associated to Input Range?
MLT
Labels
Train Annual Dataset 2015
Test Seasonal Dataset 2015

SIMULATION OF BIFACIAL PV MODULES IN NORDIC CONDITIONS FOR LOW AND HIGH ALBEDO

Elin Molin [ac*], Bengt Stridh [b], Andreas Molin [c,d], Ewa Wäckelgård [a]
a Dalarna University, School of Technology and Business Studies, Energy Technology, SE-791 88 Falun, Sweden
b Mälardalen University, School of Business Society and Engineering, P.O. Box 883, SE-721 23 Västerås, Sweden,
c PPAM Solkraft AB, Corporate Research, Husarvägen 57, SE-590 72 Ljungsbro, Sweden,
d Linköping University, Division of Energy Systems, Department of Management and Engineering, SE-58183 Linköping, Sweden
* Corresponding author. E-mail: elin.molin@ppam.se, Tel. +46 (0)70 567 66 31

ABSTRACT: In this study simulations of bifacial HIT modules using PVsyst will be compared with measured data from test plants in Linköping, Sweden. The simulations have been validated with measured data in the field study. The deviation between simulated and measured annual energy yields were in the range of 3 % for site 1 and 5% for site 2. Moreover, simulations show that the annual specific energy yield would increase to 1.45 MWh/kW$_p$ compared to 1.10 MWh/kW$_p$, for south orientation at 40° tilt and to 1.47 MWh/kW$_p$, compared to 0.98 MWh/kW$_p$, for vertical east-west orientation by changing the albedo to 0.80 instead of the existing tar paper with albedo 0.05. The annual specific energy yield increased with 32 % for the south facing modules at site 1 and with 50% for east-west facing modules at site 2 when the albedo change from 0.05 to 0.80. Monofacial simulations show that the annual specific yield increased with 7% for south orientated at 40° tilt on an annual basis when the albedo change from 0.05 to 0.80.

Keywords: Bifacial, HIT, monofacial, field study, simulation, Nordic, Sweden.

1 BACKGROUND

This study presents a comparison between energy yield simulations in the software PVsyst 6.68 and the result of an earlier experimental field study of bifacial PV (Bi) in a Swedish location [1]. The Bi systems in this study are based on frameless modules that can be used as solar shields, carports, greenhouses, visual barriers, safety fences or noise barriers.

Bi have earlier been studied both in simulations and in field studies but few were conducted for Nordic conditions. In simulations it was found that depending on the actual albedo vertical east-west (EW) installations may display a higher annual energy yield than south-facing (S) modules [2]. A computation study of plants in Germany and Singapore showed that the Bi power increases linearly with albedo [3]. A linear relation between the annual energy yield and the albedo coefficient was also found in [4.] Furthermore, Bi is more beneficial in regions with higher latitude, because the Bi yields more than M when the sun height is lower [3]. Ref [5] studies a Bi oriented east-west (EW) with a bifaciality factor of 92%, located in Amsterdam, compared to a MS tilted module. This study shows that the Bi annual energy yield is 10 % higher at albedo 0.2 and 30% higher with albedo 0.5. Bifacial modules have also a great potential for locations dominated by more diffuse irradiance conditions like Oslo being able to produce 15% more energy than standard modules even at a common surface albedo coefficient of 0.2 [4]. As these earlier works show, high albedo is of importance in order to reach the full potential of bifacial modules. As pointed out in these studies, also higher latitudes are advantageous but this needs to be studied further to be fully understood and a simulation model that can vary albedo, tilt and asimuth in order to improve the installtions in Nordic conditions is needed.

In a first study the authors of this report compared Bi and M systems in to configurations at two roof top sites in Linköping, Sweden. It was found that the annual energy gain was 5% at site 1 and 1% at site 2 compared to the M modules for the installations on tar paper roof. The difference in power temperature coefficient between Bi (-0.28%/°C) and M (0.42%/°C) influenced the BG$_E$ by +0.4% (site 1) and +0.1% (site 2). The roof tops had tar paper with an albedo of about 0.05 which explains the low energy gain derived in these cases. In the present study the same test site systems have been modelled and validated in order to vary the albedo, elevation from ground and tilt of modules to improve installations in Sweden.

2 AIM

The overall aim is to compare simulated energy yields with experimental results for one full year (Dec 2016 – Nov 2017) under operational conditions for bifacial PV modules in the Nordic climate. The maximum tilt in the PVsyst program for simulations of Bi is 60° [8] but here we also simulated a EW rail. The aim is to specifically address:

- Validate simulated with experimental yields in the two set-ups: 40° tilt, south facing (BiS40) and vertical east-west oriented (BiEW).
- Comparing simulated yields and BG$_E$ for low (0.05) and high (0.80) albedo for BiS40.
- Comparing simulated yields for low (0.05) and high (0.80) albedo for BiEW.

3 METHODOLOGY

This simulation study compares the annual energy yields of the Bi with the measured in [1] for the time-period to validate the used simulation model. The test site systems are described in [1]. The modules were created in PVsyst using recalculated data from RISE and manufactuer, see Table 1. The same type of cell (HIT) is used in the simulations for both Bi and M, in [1] the M was an ordinary mono-crystalline Si-cell. According to SMHI [6] snow covered the ground 37 days in December to March during this one-year period. In the simulations the albedo is 0.05 for each simulation when validating the data and if the modules are covered or higher albedo from the snow is not considered in the simulations.

PVsyst 6.68 can only simulate 2D bifacial PV models, nearby-shadings in 3D can only be simulated for M. All systems were simulated in PVsyst as free mounted modules with air circulation using a thermal loss factor of 29 W/m^2 which is a default value in PVsyst. For the simulations, calculated global horizontal irradiation hourly values (Wh/m^2) was used from STRÅNG [7] with the coordinates lat 58.40 and long 15.58 degrees, and hourly ambient temperature from the open air temperature data at SMHI [8]. The hourly values was used to make an ASCII table including specific yield global horizontal irradiation and ambient temperature, later imported into PVsyst as input data for the simulation. The diffuse irradiation values are estimated in PVsyst from the horizontal global values [7] with the Liu and Jordan´s correlation model.

In order to simulate the system BiS40 with shadows from surrounding area [1] and the bifacial energy gain at site 1, the used method in this study is described in Eq. 1 and 2. The model of the bifacial set-up (called the system in Eq. 1) is a 3D M front side simulation including nearby shadings and roof size and a 2D backside simulation. To get the 2D backside data, see Fig. 1, a simulation of a 2D M front side simulation is subtracted from a 2D Bi simulation (Eq. 2).

$$3D\ M + 2D\ Backside = System \qquad [Eq\ 1]$$

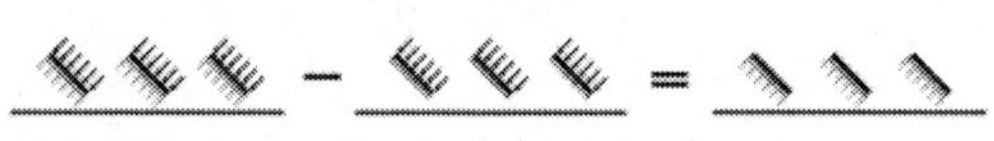

Fig. 1. Illustration of equation 2.

$$2D\ Bi - 2D\ M = 2D\ Backside \qquad [Eq\ 2]$$

The PVsyst supplier [8], states that simulations of a tilt higher than 60° is inaccurate for Bi. To overcome this problem, this study uses one east M array and one west M array (Eq 3) for simulation of vertical EW set-up at site 2. The east M array is simulated as 7 * 278W$_p$ and the west (rear side) as 7*261W$_p$, see Table I.

$$2D\ West\ M + 2D\ East\ M = System \qquad [Eq\ 3]$$

The specific energy yield is calculated using the power at STC given in Table I. In PVsyst the Bi modules were created from the input of RISE data. The specific yield and BG$_E$ values are significantly dependent on flash test method for Bi. There is no common standard for the back-side cover in the flash test method. This study consider black back-side cover and measured the power of the panel at standard test contidions (STC) 1000 W/m^2 (1.0 +0), 25°C, 1,5 AM. A comparison between the manufacturers flash test data to a flash test done on one solar panel at RISE is made. At RISE the solar module was characterized by a pulsed solar simulator Endeas QuickSun 560Ei class A +, A, A +, in accordance with IEC 60904-1, chapter 4.2. The solar simulator was calibrated against the short-circuit current on a reference module before the measurement. The measurement results have been corrected to STC in accordance with IEC 60891. Bifacial measurement was made according to TC82 document 82/1044 / NP which is a working document for a proposed standard IEC 60904-1-2 "Measurement of current-voltage characteristics of bifacial photovoltaic devices" . A black cloth with a 4% measured reflectance was used to cover reflections on the back of the module.

A factor of 0.928 for power was found between the front and backside flash data from RISE black back-covered panels, which was used to recalculate the black back-covered power, see Table I. Similar methodology but other factors were used for I_{sc}, I_{mp}, V_{oc}, V_m to create the module in PVsyst.

TABLE I
SIMULATION SET-UP. BI = BIFACIAL- [1]. AVERAGE POWER AT STC

Site	Name in article	No	Average power @STC (Wp)	Azimuth (°)	Tilt (°)
1	BiS40	6	335	0	40
2	BiEW	7	278 (east)	-90	90
2	BiEW	7	261 (west/rear)	90	90

Albedo in the simulations for tar paper was 0.05, the albedo changed then until 0.8 on annual basis.

4 RESULTS AND DISCUSSION

4.1 Validation of simulation model

4.11 Validation of simulation model for site one, albedo 0.05 (tar paper)

Measured data and simulated data was compared on annual, monthly and hourly basis. It is simulated for the validation model without effect of snow on annual basis. In site 1 the specific annual yield of the bifacial south facing modules was measured to 1.07 MWh/kW$_p$ whereas the simulation with albedo 0.05 showed 1.10 MWh/kW$_p$, see Fig. 2, with an annual deviation of 3 %. This may depend on the lack of considering snow in the model.

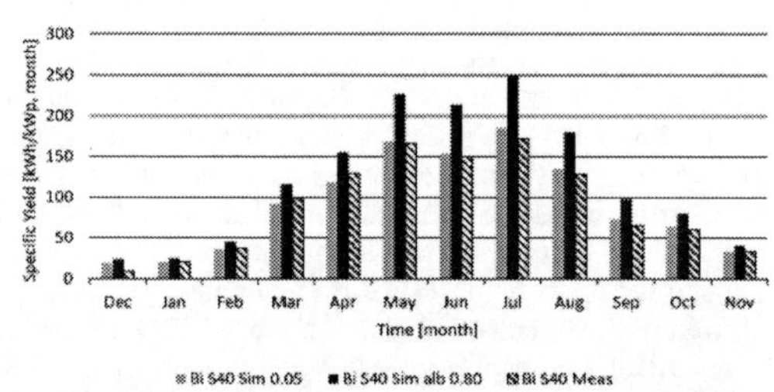

Fig. 2. Site 1. Monthly specific yield for BiS40, simulated with albedo 0.05, 0.80 and measured December 2016- November 2017.

The modules yield depends of the temperature and one cold and one warm day was studied. On February 24th, one cold and sunny day this year with a measured air temperature of up to 0.9°C. The daily specific yield was 5.1 kWh/kW$_p$ when measured and for the simulation with albedo 0.05, 4.8 kWh/kW$_p$ when the global irradiance peak reached 842 W/m^2 [7]. The highest specific power during a 5-minute period was measured to 0.8 kW/kW$_p$ the same as for the hourly value in the simulation with albedo 0.05, see Fig. 3. On May 27th, one warm and sunny day with a measured air temperature of up to 29°C, the measured daily specific yield was 7.0 kWh/kW$_p$ while simulated with albedo 0.05 was 7.4 kWh/kW$_p$ at a global irradiance peak at 1020 W/m^2. The highest specific power during a 5-minute period was measured to 0.87 kW/kW$_p$ compared to 0.89 kW/kW$_p$ hourly value in the simulation with albedo 0.05, see Fig. 3.

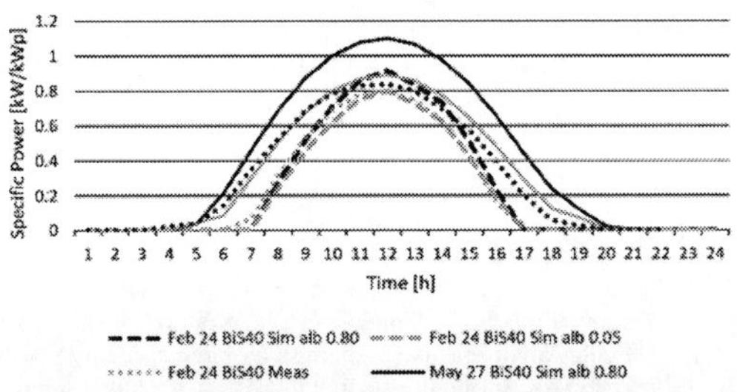

Fig. 3. Site 1. Specific power measured, simulated with albedo 0.05 and 0.80 of BiS40 May 27th, one warm and sunny day in Sweden in this study, and February 24th, one cold and sunny day.

4.12 Validation of simulation model for site two, albedo 0.05 (tar paper)

For site 2 the specific annual yield of the BiEW was measured to 1.03 MWh/kW$_p$ whereas the simulation with albedo 0.05 showed 0.98 MWh/kW$_p$, see Fig. 4, with an annual deviation of 5 %.The annual irradiation on the collector's plane was with an albedo of 0.05, 1.20 MWh/m^2.

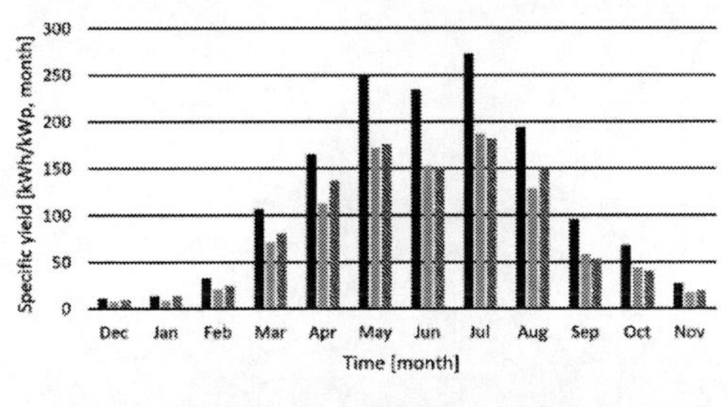

Fig. 4. Site 2. Monthly specific yield for BiEW, simulated with albedo 0.05, 0.80 and measured December 2016- November 2017.

On February 24th, one cold and sunny day this year with an air temperature of up to 0.9 °C. The daily specific yield was 3.4 kWh/kW$_p$ when measured and for the simulation with albedo 0.05, 3.2 kWh/kW$_p$ at the effective global irradiance peak of 348 W/m^2. The highest specific power during a 5-minute period was measured to 0.6 kW/kW$_p$, same as the hourly value in the simulation with albedo 0.05. On May 27th, one of the warm and sunny day this year with an air temperature of up to 26.8 °C. The daily specific yield was 8.0 kWh/kW$_p$ when measured and for the simulation with albedo 0.05, 7.7 kWh/kW$_p$ at the global irradiance peak of 445 W/m^2. The highest specific power during a 5-minute period was measured to 0.75 kW/kW$_p$, the same as for the hourly value in the simulation with albedo 0.05, se Fig. 5.

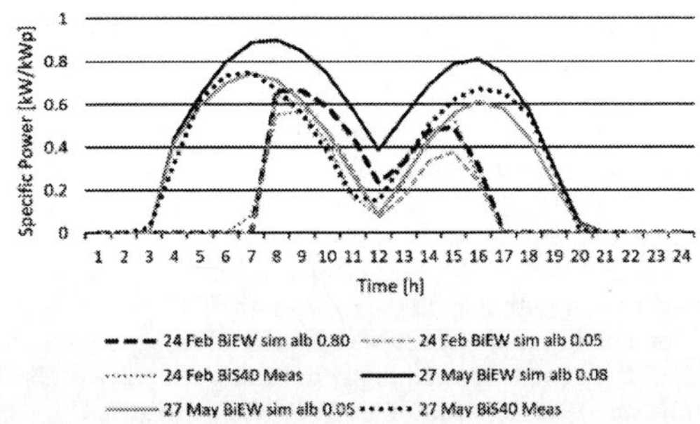

Fig. 5. Site 2. Specific power measured, simulated with albedo 0.05 and 0.80 of BiEW May 27th, one warm and sunny day in Sweden in this study, and February 24th, one cold and sunny day.

4.13 Validation for high albedo 0.80 (snow) for site two,

The day April 15th was a sunny day with fresh snow on the roof. The specific yield for the measured values was 7.8 kWh/kWp, simulated value with albedo 0.80 the specific yield was 7.7 kWh/kWp. It is clear that the power midday does not drop as much as without snow when there was fresh snow on the roof with a high albedo, se Fig. 6. The average daily yield of the 15th April shows high resemblance between simulated and measured data at albedo 0.80. However, the profile is not perfectly following the same pattern, which could be explained by cloud reflections and uneven snow (0.1 m [7]) surface on the roof.

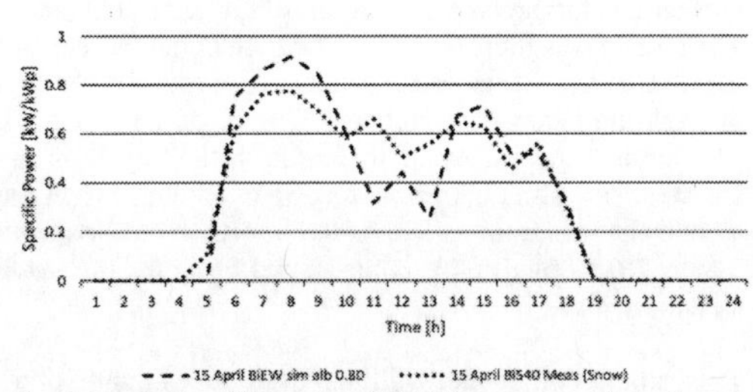

Fig. 6. Site 2. Specific power of BiEW measured with fresh snow (April 15th). Simulated values with albedo 0.80.

4.2 Influence of albedo

4.21 Influence of albedo Site one,

With albedo 0.80 the annual specific yield was simulated to 1.45 MWh/kW$_p$. The simulated annual specific yield increased with 32 % when the albedo change from 0.05 to 0.80. This resulted in a simulated BG$_E$ of 3% for albedo 0.05 and BG$_E$ of 26 % with albedo 0.80 at site 1. With albedo 0.80 the simulation on February 24th shows 5.6 kWh/kW$_p$. 9.8 % more than for albedo 0.05. The highest specific power during a 5-minute period with albedo 0.80 in the simulation 0.9 kW/kW$_p$, 12.5% more than for albedo 0.05. On May 27th, the simulated daily specific yield with albedo 0.80 was 9.8 kWh/kW$_p$. The highest specific power during a 5-minute period was 1.1 kW/kW$_p$ with albedo 0.80, see Fig. 3.

4.22 Influence of albedo Site two,

The simulation with albedo of 0.80 showed an annual specific yield of 1.47 MWh/kW$_p$ for an annual irradiance on collector's plane was 879 kWh/m^2. The annual specific yield increased with 50 % when the albedo change from 0.05 to 0.80.

On February 24th, one cold and sunny day this year with a measured air temperature of up to 0.9°C. When simulated with albedo 0.80 the specific yield was 4.2 kWh/kW$_p$. The highest specific power during a 5-minute period was measured to 0.6 kW/kW$_p$, same as the hourly value in the simulation with albedo 0.05. With an albedo of 0.80 the simulated specific power was 0.7 kW/kW$_p$, see fig. 5. On May 27th, one warm and sunny day this year with a highest measured air temperature of 29°C the specific yield was simulated to 10.7 kWh/kW$_p$ with an albedo of 0.80. The highest specific power during a 5-minute period was measured to 0.75 kW/kW$_p$ the same as for the hourly value in the simulation with albedo 0.05. With albedo 0.80 the value was 0.9 kW/kW$_p$, se Fig. 5.

5 CONCLUSIONS

A simulation method was developed using PVsyst 6.68 that can evaluate potential improvements in energy yield for Bi in Nordic conditions. The simulated annual specific energy yields have a maximum deviation of 5 % in comparison with measured yields in Linköping, Sweden. The model was used to estimate energy yield by changing parameters such as albedo.

The first and most important parameter to study was the albedo since this was very low on the test site (0.05) though it is tar paper, a common roof material in Sweden. The albedo was increased to 0.80 which could be realistic for a reflective coated polyurethane roof (albedo 0.8 [9]) or fresh snow in winter (albedo 0.82 according to PVsyst). The annual specific yield increased with 32 % for the BiS40 at site 1 and with 50% for BiEW at site 2 when the albedo changes from 0.05 to 0.80. The BG_E increased from 3 % to 26 % for BiS40 on an annual basis for the same albedo increase.

This simulation model will be used to estimate other parameters influence of energy yield such as tilt, elevation and different albedo in a coming study.

6 ACKNOWLEDGEMENTS

The work has been carried out under the auspices of the industrial post-graduate school Reesbe, which is financed by the Knowledge Foundation (KK-stiftelsen). The authors thank Rickard Linderström of Byggvesta AB and Peter Alfredsson of Stångåstaden for invaluable support with giving us access to the project sites. I would also like to thank the students Jessica Trossell and Johan Perua from Linköping's university corporate research with PPAM Solkraft for the discussions and experience sharing.

7 REFERENCES

[1] Molin, E., Stridh, B., Molin, A., Wäckelgård, E., "Experimental yield study in Nordic conditions" IEEE Journal of photovoltaic, DOI 10.1109/JPHOTOV.2018.2865168.

[2] Guo, S., Walsh, T.M., Peters, M.," Vertically mounted bifacial photovoltaic modules: A global analysis" (2013) Energy, 61, pp. 447-454.

[3] Wang, S., Wilkie, O., Lam, J., Steeman, R., Zhang, W., Khoo, K.S., Siong, S.C., Rostan, H., "Bifacial Photovoltaic Systems Energy Yield Modelling" (2015) Energy Procedia, 77, pp. 428-433.

[4] Janssen, G.J.M., Van Aken, B.B., Carr, A.J., Mewe, A.A., "Outdoor Performance of Bifacial Modules by Measurements and Modelling" (2015) Energy Procedia, 77, pp. 364-373.

[5] Yusifoglu, U.A., Lee T.H, Plexer M., Hlm a., Koduvelikulathu J., Comparotto C., Kopecek R., Kurz H., "simulation of energy production by bifacial modules with revision of ground reflection" (2014) 4th International Conference on Silicon Photovoltaics

[6] Swedish Meteorological and Hydrological Institute (SMHI), STRÅNG. http://strang.smhi.se/.

[7] Swedish Meteorological and Hydrological Institute (SMHI), STRÅNG.https://opendata-download-metobs.smhi.se/explore/

[8] Mermoud A, Wittmer B, Bifacial shed simulations with PVsyst, Bifacial Workshop 2017, 25-26.10.2016 Konstanz, Germany

[9] Mariseal, test of spectral reflectance done by Center for renewable energy sources, Pikermi Attica. 04.03.2009.

SIMPLIFIED METHOD FOR PARTIAL SHADING LOSSES CALCULATION FOR SERIES CONNECTED PV MODULES WITH EXPERIMENTAL VALIDATION

M. Dallapiccola[1,2,3], P. Ingenhoven[1,3], D. Moser[1], J. S. Stein[4]
[1]Institute for Renewable Energy, EURAC Research, Bolzano, Italy
2 Università degli Studi di Trento, Trento, Italy
3 Free University of Bozen, Bolzano, Italy
4 Sandia National Laboratories, Albuquerque, New Mexico USA
[1] 39100 Bolzano, Italy, Phone: +39 0471 055 717; Fax: +39 0471 055699; E-mail: mattia.dallapiccola@eurac.edu

ABSTRACT: Partial shading strongly affects the power output of photovoltaic systems. It occurs when the system is under non-uniform irradiance conditions due to possible obstacles, e.g. close buildings, trees, dust, clouds and other objects, which affect the incident radiation on the modules. In these conditions, the nominal properties of photovoltaic systems can be strongly reduced resulting in power losses due to current and voltage mismatch. Moreover, shaded cells dissipate power resulting in a dangerous local heating. Modules are usually connected in series or in parallel to form an array. Particularly for the series connection, partial shading can compromise the whole array power output. To avoid these problems bypass diodes were introduced in commercial modules. The aim of this work is to present a fast and accurate method to evaluate the output power and the power loss of a series connected system under partial shading including the effect of by-pass diodes. The method is interesting for optimization processes, annual calculations and for all the applications where the computational time needs to be small.
Keywords: Modelling, PV System, Shading, Energy Performance, Building Integrated PV (BIPV)

1 INTRODUCTION

In this study we present a new model for the calculation of partial shading losses of series connected PV modules. In complex shading scenarios, such as building integrated photovoltaics (BIPV) applications, the effect of partial shading strongly affects the output power of a photovoltaic system and can cause damage due to the power dissipation in the shaded cells. Modern commercial modules are usually equipped with by-pass diodes to reduce the impact of shading on the power output. The effect of by-pass diodes is to modify the I-V and P-V curves of the partially shaded modules introducing multiple peaks on the power curve and causing the need of more advanced MPPT algorithms. The increased complexity is also reflected in simulation phases. Common approach for the simulation of partially shaded modules are mathematical models or simulations using electrical circuit simulators e.g. SPICE [1], Ngspice [2], Qucs [3], LTspice [4] and Xyce [5]. In all cases the computational time is usually too long to be used in optimization processes or long-term calculations due to the complex implicit equations involved.

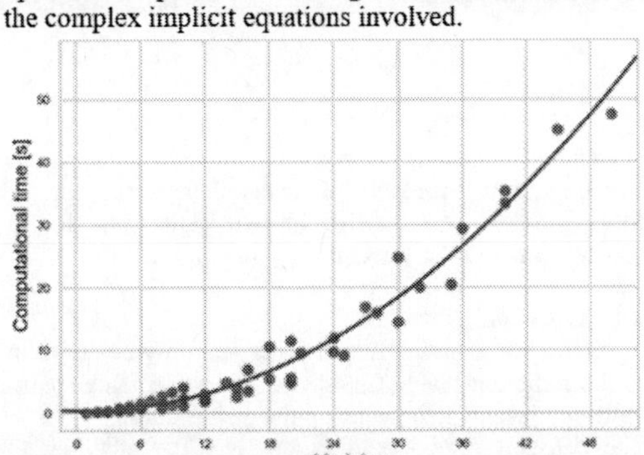

Figure 1: Computational time of an electrical circuit simulator for different plant sizes. The computational time increases nonlinearly with the increasing of the number of modules connected in series.

The model presented in this work aims to reduce the computational time required for the analysis of PV systems under partial shading conditions maintaining an acceptable accuracy.
Finally, the model was experimentally validated and compared with other publications [6], [7], [8], [9], [10].

2 APPROACH

The model was extracted from the results of several detailed simulations of a reference system under different shading and radiation conditions. Simulations were performed with PySpice [11], an electrical circuit simulator with a Python interface based on Ngspice [2]. It was assumed that the radiation of each string of cells (series of cells protected by a by-pass diode) is uniform. In fact, the I-V and P-V curves of a string completely shaded and the ones of a string with only one shaded cell are similar. The consequence of this assumption is that a string of cell can only be considered fully shaded or not shaded. Moreover, it was assumed that the incident radiation can have only two values (shaded and not shaded condition). In real applications, the shaded condition can be approximated with the diffuse component and the not shaded radiation can be set equal to the global radiation in the plane of the array. The module temperature used in the simulations was calculated with the Nominal Operating Cell Temperature (NOCT) formula. It was calculated as [12]:

$$T_{cell} = T_{amb} + \frac{NOCT - 20}{800} G$$

where the ambient temperature was set to T_{amb} = 20°C and G is the radiation in W/m^2.

2.1 Model

Due to the large availability of parameters and data of commercial modules, it was chosen to use the single diode model for the simulation of the solar cells. The equivalent electrical circuit was implemented in PySpice [11], the electrical circuit simulator used.

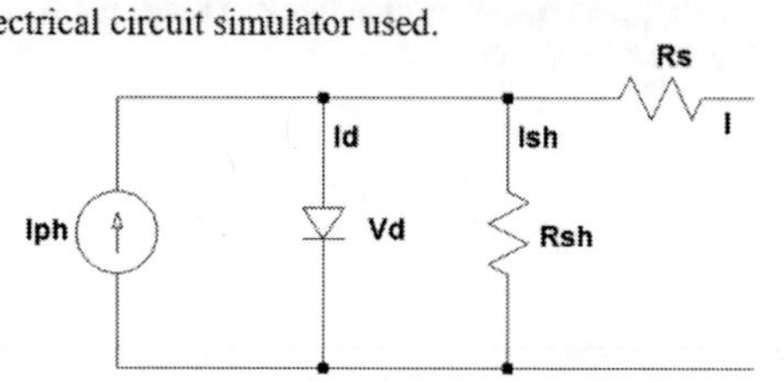

Figure 2: Equivalent electrical circuit of the single diode model used for the simulation of the solar cells.

2.2 Reference system

The reference system used for the simulations is composed by two series connected modules. Modules used are Canadian Solar CS6P 230P, 60-cells, multi-crystalline silicon modules. Each module has three by-pass diodes and is divided in three strings of 20 cells. The properties of the chosen modules and the single diode model parameters at standard test conditions (STC) were reported according to the CEC [13] database implemented in PVLIB [14] see table I.

Table I: Properties of the reference module

Property	Value	Unit
NOCT	43.6	°C
N_{cells}	60	-
$I_{sc,STC}$	8.34	A
$V_{oc,STC}$	36.8	V
α_{sc}	0.003002	A/K
β_{oc}	-0.1277	V/K
$I_{L,STC}$	8.361	A
$I_{0,STC}$	1.233exp(-10)	A
R_s	0.335	Ω
$R_{sh,STC}$	132.91	Ω

2.3 Dimensionless approach

Results of the simulations were expressed in terms of dimensionless variables. The dimensionless parameters introduced are the power ratio *p*, the shading ratio *n* and the radiation ratio *g*. The power ratio can be calculated as the ratio between the power under partial shading conditions P_{sh} and the power of the system under non-shaded conditions P_0:

$$p = {}^{P_{sh}}/_{P_0}$$

The shading ratio is defined as the ratio between the number of shaded strings N_{sh} and the total number of strings in the system N_{tot}:

$$n = {}^{N_{sh}}/_{N_{tot}}$$

The radiation ratio is defined as the ratio between the radiation of the shaded strings and the radiation of the non-shaded strings:

$$g = {}^{G_{sh}}/_{G_0}$$

The model was obtained fitting the results of several simulations (performed sweeping the parameters G_0, G_{sh} and N_{sh}) as function of the shading and the radiation ratio *p=f(g,n)*. The dimensionless approach was chosen because it reduces the need of calibrating the model for different modules and technology, as will be shown later. Further, the computational time required for the calculation of the power ratio does not increase with increasing system size. These two aspects are of great importance for generalizing the model and for its application in optimization processes and in long-term calculations. The set of parameters used in the simulation is summarized in table II.

Table II: Parameters of the simulations

	Start	End	Step
G_0 not shaded radiation	100	1000	100
G_{sh} shaded radiation	10	G_0	10
N_{sh} number of shaded strings	0	N_{tot}	1

In this way, known the shading and the radiation condition, the model allows the calculation of the power ratio.
Once the power ratio is known it is possible to calculate the power under partial shading as:

$$P_{sh} = p \cdot P_0$$

where P_0 can be easily calculated with the common tools used for the calculation of the power under uniform conditions, e.g. PVLIB [14]. Consequently, the power ratio *p* results to be a reduction factor which accounts of the losses caused by partial shading.

3 RESULTS OF THE SIMULATIONS

The power ratio *p* as a function of *g* shows two different trends depending on the radiation conditions. For low values of *g*, it is approximately constant and for higher values of *g* a cubic fit describes the results well. Figure 3 shows the simulation results for all parameter groups.

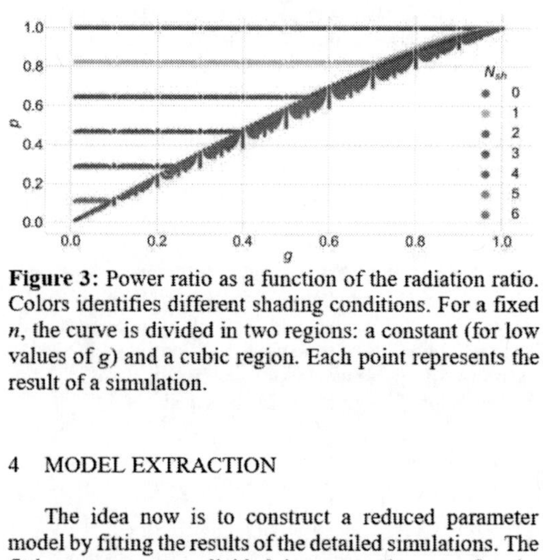

Figure 3: Power ratio as a function of the radiation ratio. Colors identifies different shading conditions. For a fixed *n*, the curve is divided in two regions: a constant (for low values of *g*) and a cubic region. Each point represents the result of a simulation.

4 MODEL EXTRACTION

The idea now is to construct a reduced parameter model by fitting the results of the detailed simulations. The fitting process was divided in two paths: one for the constant and one for the cubic region.

4.1 Constant region

Since we assumed that a string can only be subjected to the radiation levels G_0 or G_{sh}, a module under partial shading presents two peaks on the power curve.
The constant region is originated by the fact that the maximum power point of the system is on the first peak of the P-V curve for low values of *g*, for better understanding an example I-V and V-P curve for *g=0.2* and *n=0.33* is shown in figure 4. Figure 7 shows the case in which the second peak is the absolute maximum.

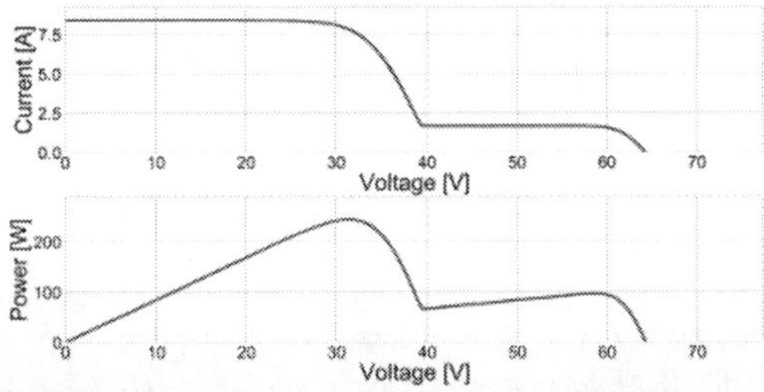

Figure 4: I-V and P-V curve for g=*0.2* and n=*0.33*

By increasing the value of g, the second peak becomes more prominent but is still lower than the first one. The maximum power point of the system stays on the first peak and consequently p is constant. In this region, the power ratio can be calculated as an ideal p minus a correction factor which depends on G_0 and n

$$p_{const} = p_{ideal} - e$$

where p_{ideal} is the idealized reduction caused by the by-pass of the shaded strings assuming that the shaded strings simply do not contribute to the power output and is defined as

$$p_{ideal} = 1 - {N_{sh}}/{N_{tot}} = 1 - n$$

The correction factor e was fitted using a linear function of n whose gradient is a linear function of G_0, see figure 5 and 6 and table III for the fitting coefficients.

$$e = m_e \cdot n$$
$$m_e = m_{me} \cdot G_0 + q_{me}$$

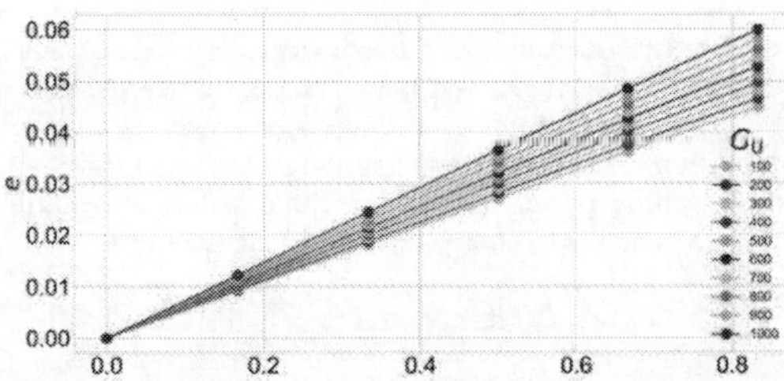

Figure 5: Correction factor e is a linear function of n

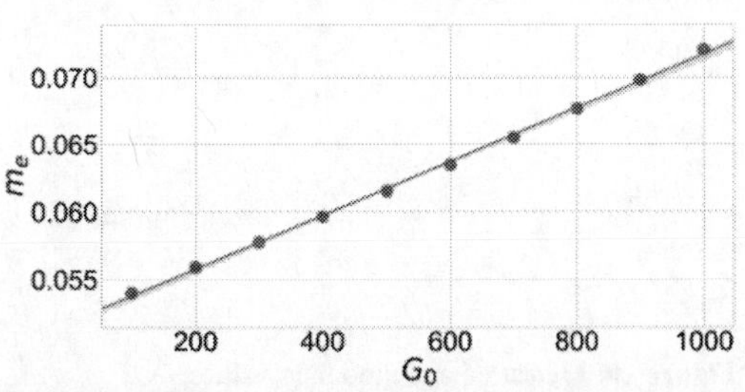

Figure 6: The gradient of the relation is a linear function of G_0

Table III: Parameters of the constant region

Parameter	Value	Unit
m_{m_e} gradient	$2.097 \cdot 10^{-5}$	m²/W
q_{m_e} intercept	0.05169	-

4.2 Cubic region

For a fixed n, increasing g the second peak of the P-V curve becomes more relevant until it becomes the maximum power point of the system, see figure 7. From this point on, increasing g, the power and the power ratio increase.

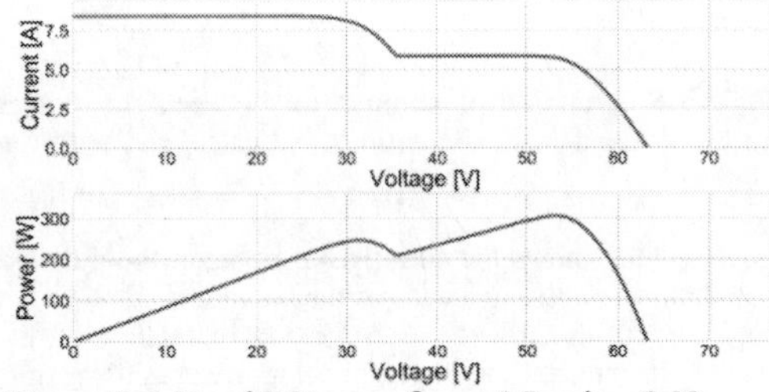

Figure 7: I-V and P-V curve for g=*0.7* and n=*0.33*

Points of this region were fitted using a cubic function of g:

$$p_{cubic} = b \cdot g + c \cdot g^2 + d \cdot g^3$$

where the parameters b, c and d are linear functions of G_0

$$b = m_b \cdot G_0 + q_b$$
$$c = m_c \cdot G_0 + q_c$$
$$d = m_d \cdot G_0 + q_d$$

At the same time, the gradient m_i and the intercept q_i of the linear functions are fitted to quadratic functions of n as

$$m_i = d_{m_i} + b_{m_i} \cdot n + c_{m_i} \cdot n^2 \quad for\ i = b, c, d$$
$$q_i = d_{q_i} + b_{q_i} \cdot n + c_{q_i} \cdot n^2 \quad for\ i = b, c, d$$

where the parameters are reported in table IV

Table IV: Parameters of the cubic region

	b_{m_b}	c_{m_b}	d_{m_b}
m_b gradient	-0.00183	0.00133	0.00091
	b_{q_b}	c_{q_b}	d_{q_b}
q_b intercept	3.92005	-2.65245	-0.44742
	b_{m_c}	c_{m_c}	d_{m_c}
m_c gradient	0.00417	-0.00279	-0.00183
	b_{q_c}	c_{q_c}	d_{q_c}
q_c intercept	-10.06717	6.40570	4.00014
	b_{m_d}	c_{m_d}	d_{m_d}
m_d gradient	-0.00234	0.00146	0.00092
	b_{q_d}	c_{q_d}	d_{q_d}
q_d intercept	6.14712	-3.75325	-2.55272

4.3 Power ratio

Once the power ratio in the constant region p_{const} and in the cubic region p_{cubic} is known, it is possible to calculate the power ratio p simply as the maximum of the two

$$p = max\{p_{const}, p_{cubic}\}$$

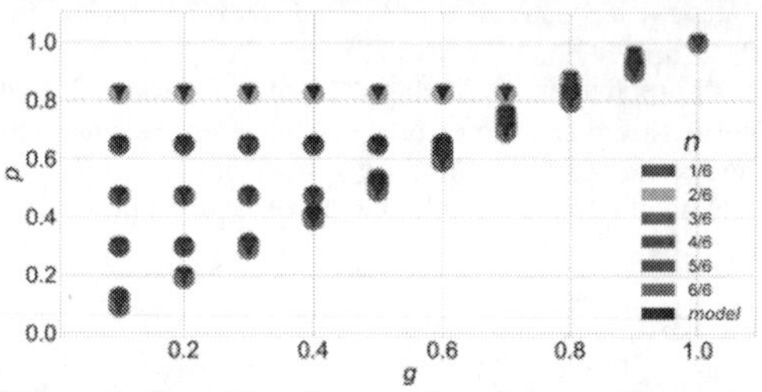

Figure 8: Example of comparison between the power ratios obtained from simulations and calculated with the model.

Figure 8 shows the comparison between the detailed simulation and the reduced parameter model. Due to the complex fitting process used for the extraction of the model, the reduced parameter model reproduces the results of the simulations well. The mean absolute percentage error calculated between the results of the simulations (Figure 3) and the power ratio obtained with the reduced parameter model is equal to

$$MAPE = \frac{100}{N}\sum_{i}^{N}\frac{|p_{i,sim} - p_{i,model}|}{p_{i,sim}} = 0.39\%$$

where N=2700.

5 MODEL VALIDATION

The model presented in this work was experimentally validated and compared with the results presented in other publications (measurements and simulations).

5.1 Comparison with literature

The developed reduced parameter model was compared to nine examples presented by other authors covering different module types, technologies, system sizes and conditions [6], [7], [8], [9], [10]. The analyzed cases include both experimental and simulation results. We reported the main properties, parameters and the absolute percentage error between the results of the experimental/simulation results and the model presented in this work, for each of the analyzed cases, see table V for a summary. Note that the model was calibrated at operating temperature while the temperature of the comparing articles was 25°C or not known.

Table V: Parameters and properties of the experimental (Exp) and simulation (Sim) analysis for mono-crystalline (m-cSi) and poly-crystaline (p-cSi).

Cite	n°	Tech	g	n	Sim/Exp	APE
[8]	2	m-cSi	0.7	0.5	Sim	3.46%
[6]	3	m-cSi	0.12	0.33	Exp	2.89%
[6]	3	m-cSi	0.11	0.17	Exp	4.18%
[7]	1	p-cSi	0.5	0.5	Sim	6.18%
[7]	1	p-cSi	0.21	0.5	Exp	3.95%
[7]	2	p-cSi	0.21	0.25	Sim	2.29%
[7]	2	p-cSi	0.21	0.25	Exp	1.56%
[9]	4	m-cSi	0.5	0.25	Sim	2.62%
[10]	1	p-cSi	0.25	0.5	Sim	0.21%
					Average	**3.04%**

In general, the reduced parameter model predicts the results of the literature used for the comparison well. It is possible to confirm that the dimensionless approach used in this work decreases calibration needs of the model if it is used for the analysis of systems of different technologies, size (number of series connected modules) and module types. Further, due to the dimensionless approach, the computational time does not increase at the increasing of the size of the system. This aspect is of particular interest because it allows the use of the model in optimization processes or long-term calculations also for large systems.

5.2 Experimental validation

To verify the accuracy of the model, a Canadian Solar CS6P 230P module was tested under different shading and radiation conditions. Tests were performed using a Pasan Solar Simulator for G_0 equal to 200, 700 and 1000 W/m^2. Shading conditions were controlled using opaque masks covering different percentages of the strings (1-2-3 strings shaded 20-50-80%). During the tests, the temperature ranged between 25°C and 26.5°C. For each of the shading and radiation conditions, the maximum power was calculated from the measured P-V curves. Each measure was repeated three times to decrease the error during the measurements.

Figure 9: Example of experimental setup.

For each measurement, the power ratio p was obtained dividing the average (over the three measurements) power in shaded conditions by the average power in uniform conditions. The model was applied in the same cases and the resulting power ratios were compared. The resulting mean absolute percentage error was calculated as

$$MAPE = \frac{100}{N}\sum_{i}^{N}\frac{|p_{i,meas} - p_{i,model}|}{p_{i,meas}} = 2.76\%$$

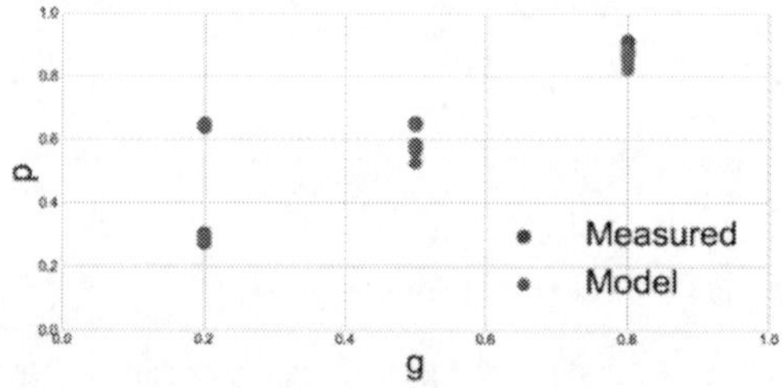

Figure 10: Measured and modelled values.

5.3 Larger systems

Considering that real plants are usually constituted by several modules, the model was compared with the results of simulations of large systems implemented in PySpice [11].

Table VI: Comparison between the model and simulations for a system composed by 24 modules connected in series. Module: Canadian Solar CS6P 230P

n°	n	g	p_{sim}	P_{model}	APE [%]
24	0.25	0.2	0.73195	0.73183	0.016
24	0.75	0.2	0.24351	0.24841	2.01
24	0.2	0.6	0.73267	0.73183	0.12
24	0.75	0.6	0.67849	0.67939	0.13
24	0.2	0.8	0.89244	0.89033	0.24
24	0.75	0.8	0.85828	0.85519	0.36

In this case, the model describes the simulated behavior of partially shaded photovoltaic modules well without increasing the computational time required. For each simulation, the absolute percentage error confirms the good accuracy of the model.

5.4 Different module
The model was also compared with the results of simulations implemented in PySpice [11] for a different module type with respect to the one used for the calibration.

Table VII: Comparison between the model and simulations for a system composed by 8 monocrystalline modules connected in series. Module: SunPower 210-BLK-U

n°	n	g	p_{sim}	P_{model}	APE [%]
8	0.25	0.2	0.73720	0.73183	0.73
8	0.75	0.2	0.23133	0.24841	7.3
8	0.25	0.6	0.73769	0.73183	0.79
8	0.75	0.6	0.65689	0.67939	3.43
8	0.25	0.8	0.86810	0.89033	2.56
8	0.75	0.8	0.84347	0.85519	1.39

In this case the model was used to calculate the power under partial shading for a different module with respect to the one used for the calibration of the model. Even if the module and the size are different, the error was considered acceptable and it is possible to say that the model describes the behavior of the analyzed photovoltaic system well.

5.5 Comments
In this work a new model for the calculation of the power losses related to partial shading of series connected photovoltaic systems was presented. The model was compared with the results of other publications and experimentally validated. In both cases the error was considered acceptable for the aim of this work. In fact, the model describes the behavior of partially shaded systems with series connected modules with good accuracy. The error can be ascribed mainly to temperature effects (the model was calibrated at operating conditions while the results used for the comparison were obtained at ambient temperature) and partly to the non-calibration of the model for the specific case. However, it was observed that the dimensionless approach used decreases the need of calibration of the parameters even if the model is used for analyzing different modules, technology or sizes of the system.

6 CONCLUSIONS

The model presented simplifies the calculations related to partial shading and allows to avoid time-consuming solution of complex implicit equations. It requires simple inputs such as the percentage of shaded strings and the ratio between the radiation in shaded and non-shaded conditions. The model was experimentally validated using a sun simulator and providing artificial shading to the strings and it was also compared with the results of other publications. The mean absolute error calculated during the experimental validation is equal to 2.76% while the one related to the comparison with other publications is equal to 3.04%. It shows good accuracy also in different cases including different module types, technology and size of system. Due to the dimensionless approach used during the model extraction, it can be generalized to different cases of study. Moreover, the model introduces two important improvements with respect to the common approaches used. The model is fast (based on polynomials) and does not required the solution of complex implicit equations (as common simulation tools) and the computational time is independent from the size of the analyzed system. Thus, it can be used in all applications where time is an issue such as optimization algorithms or long-term calculations.

In future, it is planned to extend the simulation for the case of modules or strings connected in parallel or more complex wiring. Due to the large amount of possible connections paths, it will not be possible to find a fit for the resulting reduced parameter model. It is planned to run numerous detailed simulations and then apply neural network tools to find the reduced parameter models.

7 REFERENCES

[1] O. Pederson Donald and W. Nagel Laurence, "SPICE (Simulation Program with Integrated Circuit Emphasis)," EECS Department, University of California, Berkeley, Apr 1973. [Online]. Available: http://www2.eecs.berkeley.edu/Pubs/TechRpts/1973/22871.html.

[2] "Ngspice," [Online]. Available: http://ngspice.sourceforge.net/presentation.html. [Accessed 2018].

[3] "Qucs," 2004-2017. [Online]. Available: http://qucs.sourceforge.net/.

[4] "LTspice," 2018. [Online]. Available: http://www.analog.com/en/design-center/design-tools-and-calculators/ltspice-simulator.html.

[5] Sandia National Laboratories, "Xyce," [Online]. Available: https://xyce.sandia.gov/.

[6] M. Abdulazeez and I. Iskender, Simulation and experimental study of shading effect on series and parallel connected photovoltaic PV modules., In Electrical and Electronics Engineering (ELECO), 2011 7th International Conference on (pp. I-28). IEEE., 2011, December.

[7] B. Alsayid, S. Alsadi, S. Ja'far and M. H. Dradi, Partial shading of PV system simulation with experimental results, Smart Grid and Renewable Energy, 4(06), 429, 2013.

[8] L. Fialho, R. Melicio, V. M. F. Mendes, J. Figueiredo and M. Collares-Pereira, Effect of shading on series solar modules: simulation and experimental results, Procedia Technology, 17, 295-302, 2014.

[9] P. V. Simulator and L. Ardhenta, Photovoltaic Array Modeling under Uniform Irradiation and Partial Shading Condition, International Journal of Applied, 6(3), 144-152, 2017.

[10] D. Sera and Y. Baghzouz, On the impact of partial shading on PV output power, In WSEAS/IASME International Conference on Renewable Energy Sources (pp. 229-234), 2008, October.

[11] F. Salvaire, "PySpice," 2018. [Online]. Available: https://pyspice.fabrice-salvaire.fr.

[12] R. G. Jr Ross, Flat-plate photovoltaic array design optimization, In 14th Photovoltaic Specialists Conference (pp. 1126-1132), 1980.

[13] "CEC database," [Online]. Available: http://www.gosolarcalifornia.org/equipment/pv_modules.php.

[14] J. S. Stein, W. F. Holmgren, J. Forbess and C. W. Hansen, "PVLIB: Open source photovoltaic performance modeling functions for Matlab and python," in *In Photovoltaic Specialist Conference (PVSC), 2017 IEEE 44th (pp. 1-6). IEEE*, 2017.

A LOWER TC:
IN THE FUTURE MAYBE NOT ALWAYS THE BEST IDEA?

Jonathan Govaerts[1], Imre T. Horvath[1], Hans Goverde[1], Bader Aldalali[2], Ian Beausoleil-Morrison[3], Jef poortmans[4]
[1]imec-Energyville, Thorpark 8320, B-3600 Genk
[2]Kuwait University, College of Engineering and Petroleum, Khaldiya, Kuwait
[3]Carleton University, Ottawa, Canada
[4]Also at KULeuven and UHasselt

ABSTRACT: In the future, with costs for PV modules rapidly declining and an expected increase of PV penetration in energy production, the cost for PV installations may be impacted more and more by balancing costs of storage, load management, and curtailment, rather than the currently dominating hardware components (PV modules and BOS components). With a low overall cost in mind, we should consider the interplay of generation (of course also intraday, but in particular seasonal variations are investigated in this work), load (consumption pattern) and storage costs for each location.
In this paper we want to raise awareness about the increasing complexity of technology optimization resulting from the evolving cost structure of PV electricity generation. The latter can lead to counter-intuitive outcomes such as: a lower temperature coefficient may not always be the best in the future.
First we build some first order analytical model to illustrate the concept and indicate trends. Then we apply our physics-based bottom-up modeling approach, taking into account coupled optical, thermal and electrical effects (and published more in detail elsewhere). This can be used to quantify with high accuracy PV production throughout the year as function of climate data and PV module technical data, in particular the temperature coefficient.
Finally, we indicate some locations where the benefits of a higher temperature coefficient may materialize most, and that deserve some further attention.
Keywords: temperature coeeficient, energy yield

1 INTRODUCTION

Up until now, development of c-Si PV technology, the most widely deployed technology, has been mainly focused on increasing rated performance at standardized test conditions (STC) and reducing (investment) costs. To steer technology development to this end, modelling has been used from the start. In first instance this was done through modeling cell performance, using basic models as PC1D [1], evolving to more and more complex simulators taking into account 2D and 3D effects [2]. In a second optimization stage, also module materials came into play, either adding it inside the cell simulation software [3], or working on a higher level looking at cell-to-module losses [4]. Currently, more and more attention is pointed towards maximization of annual energy yield in operational conditions [5], with increasing accuracy (using validated models) to take into account coupled optical, thermal and electrical effects, and allowing performance prediction of different technologies and conditions [6]. Such bottom-up, physics-based modeling allows then to evaluate different technologies for different installations and drive technology optimization for specific conditions. In [7] e.g., the impact of different cell technologies is explored for a specific climate using so-called what-if simulations. In particular the impact of the temperature coefficient (TC) of the cells on the annual energy yield is studied, for 2 different climates.
In the future, with costs for PV modules rapidly declining and an expected increase of PV penetration in energy production, the cost for PV installations may be impacted more and more by balancing costs of storage, load management, and curtailment, rather than the currently dominating hardware components (PV modules and BOS components).
The purpose of this paper is to raise awareness about the increasing complexity of technology optimization resulting from the evolving cost structure of PV electricity generation. The latter can lead to counter-intuitive outcomes such as: a lower temperature coefficient may not always be the best in the future

2 ANNUAL VARIABILITIES

With increasing penetration of renewable energy (wind and PV) in the electricity production mix, it is widely acknowledged that the intermittency, as production is depending on local weather conditions, will be one of the big challenges. Currently, a number of options are under development to tackle this, and boil down to storage and demand-side (load) management. All this depends of course on consumption pattern and is expected to be reflected in electricity pricing that is variable in time. Already market prices can vary significantly over the year, as illustrated in Figure 1.

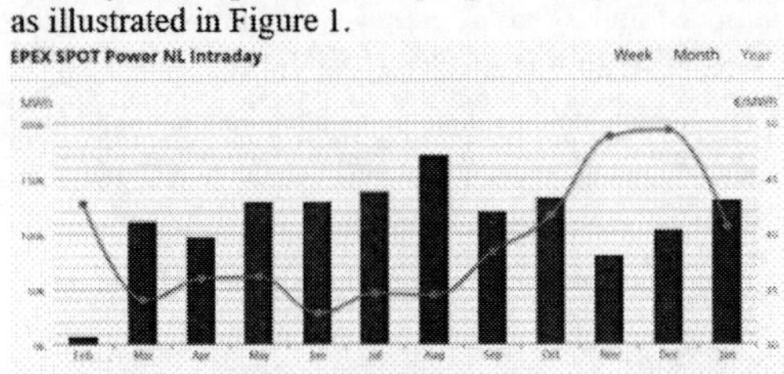

Figure 1: Monthly volume (bars) and price averages (in red) for the Netherlands varying between 33 (June) and 49 (December) EUR/MWh, from [8]

While storage may conceivably cover short-term (day-week) needs of low to moderate capacity, for the longer-term (season) it is probably less suited. Load management is also part of the solution, but is expected to reach limits in societal acceptance/adoption. For this reason, it is interesting to take a closer look at ways to improve performance of PV installations in times with low output (weekly-monthly averaged) to reduce seasonal variations. On a higher level this helps to reduce the needs for longer-term storage and keep costs low despite the interplay of generation, load and storage, whether from a society or a business point-of-view.

3 FIRST ORDER ANALYTICAL MODELING

To get a feeling for trends, first we did some rough simulations using an analytical approach. For this we assumed a simplified evolution of daily mean cell temperature and insolation profile throughout the year: sinusoidal functions closely matching measured data in Kuwait and Oldenburg in 2014 [7]. To check the separate impact of temperature and insolation, we also constructed two hypothetical "hybrid" climates, interchanging those profiles into different combinations, so as to mimick:

(1) Low latitude, high temperature (Kuwait)
(2) High latitude, low temperature (Oldenburg)
(3) Low latitude, low temperature
(4) High latitude, high temperature

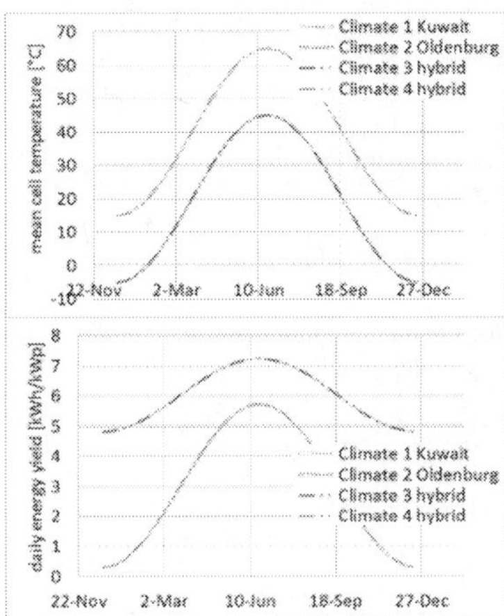

Figure 2: Simplified profiles for mean cell temperatures (top) and "basic" daily energy yield without impact of temperature (bottom) to reflect different climates

Figure 2 indicates these profiles, both for the mean cell temperatures, and for the insolation, that is already translated into a "basic" energy yield, where temperature effects are not yet taken into account.

To get a more or less realistic behaviour, we then applied distinctly different thermal coefficients for the maximum power, namely 0.40 (e.g. [9]), 0.28 (e.g. [10]) and 0.16%/°C, resp. for technologies "tech1", "tech2" and "tech3". These are illustrated in Figure 3.

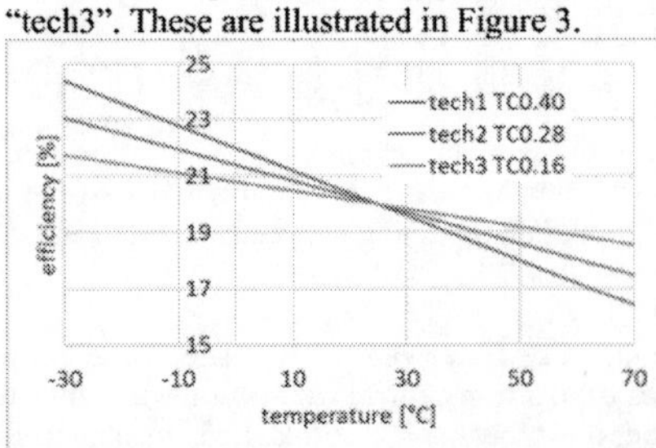

Figure 3: Graphical representation of the assumed temperature coefficients for efficiency or maximum power for the 3 technologies to be simulated

4 SIMULATION RESULTS

With the above assumed conditions, we superimpose the effect of the cell temperatures on the "basic" energy yields, to determine the impact of the different TCs in different climates. The graphical result of this is shown in Figure 4.

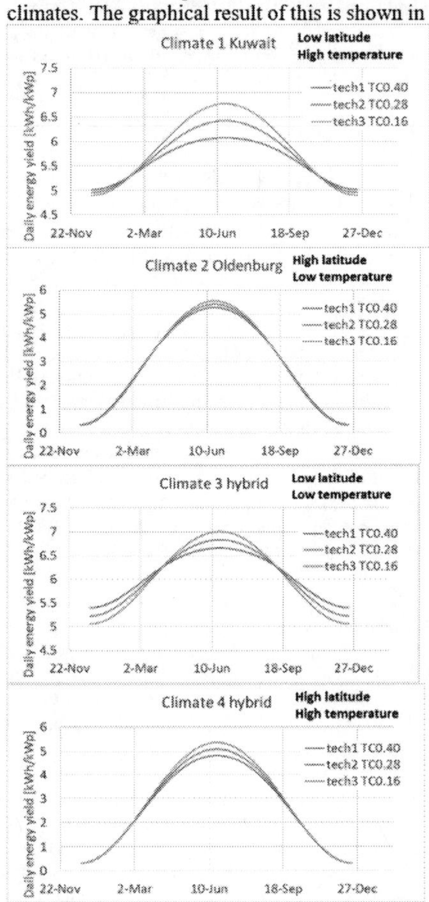

Figure 4: Simulated daily energy yield profiles throughout the year for the different climates

While clearly a lower TC is typically beneficial for overall annual energy yield, an increased performance can be observed in wintertime for cells with a higher TC, in particular more pronounced for lower latitude and lower temperature. Indeed, a higher TC causes an improvement in performance at temperatures below 25°C (used for the STC rating), and this improvement is relative to the insolation, which will be higher in wintertime for climates at low latitude (closer to the equator). Alternatively formulated: a higher TC flattens out energy production throughout the year.

Such graphs can be further extended towards cost gains throughout the year. As an example, Figure 5 illustrates how an assumed simplified electricity pricing profile materializes in cost gains and losses for a technology with a higher TC throughout the year for different climates. Of course the Kuwait results are very much hypothetical as the electricity pricing profile may be very different. In the future, with an increasing penetration of PV, pricing may very well further affect the results here, in terms of (increased) wintertime values, as well as reduced summertime values (and potentially curtailment). The hypothetical "climate 3 hybrid" results indicate how a reduced temperature lifts the up potential gains across the year.

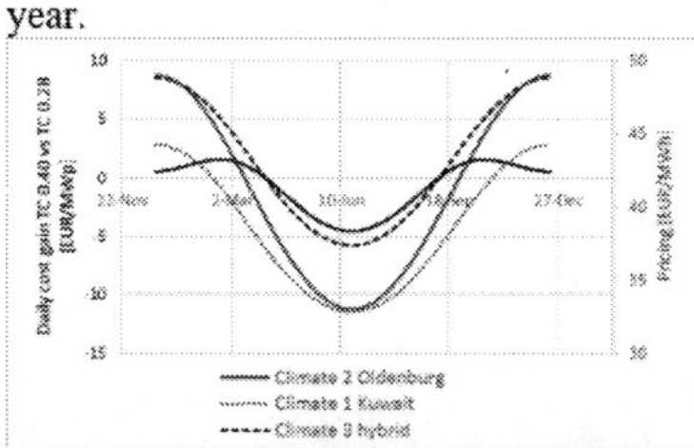

Figure 5: Hypothetical cost gains and losses for a technology with a higher TC throughout the year (left axis) for an assumed electricity pricing (right axis)

5 VALIDATION IN ADVANCED SIMULATIONS

To check if such local improvement also materializes in realistic conditions, we applied our energy yield modeling approach [5] to the measured climate data in Kuwait and Oldenburg in Figure 6, also for technologies with different TCs [7]. The resulting daily energy yield differences throughout the year are plotted in Figure 7.

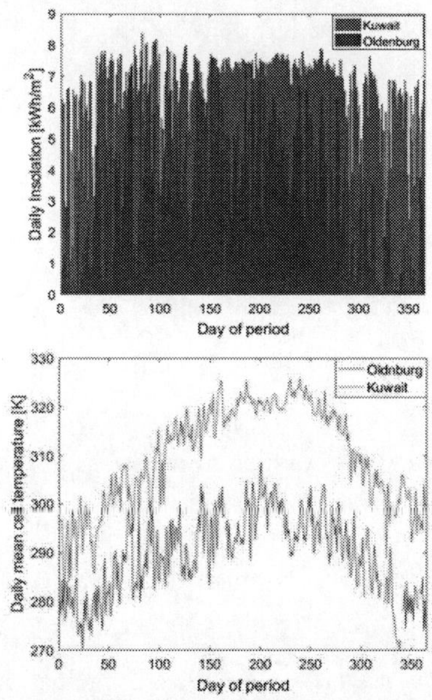

Figure 6: Daily insolation (top) and mean cell temperatures during production hours (bottom) in Kuwait and Oldenburg

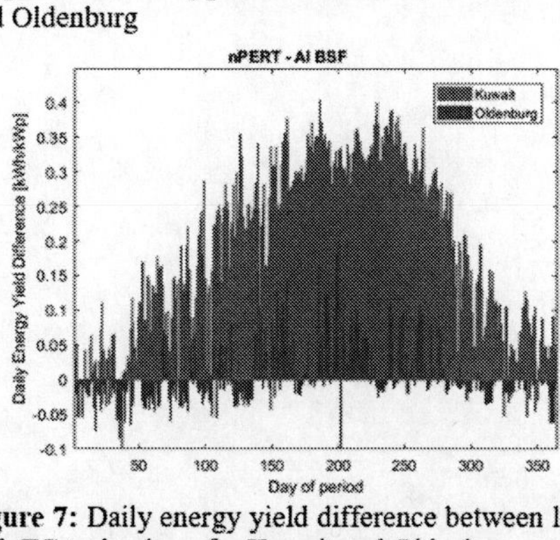

Figure 7: Daily energy yield difference between low and high TC technology for Kuwait and Oldenburg

To be complete, for these simulations the cell models were slightly adapted to rule out the impact of differences in low-light behaviour, as this is not the focus of this paper. For Kuwait it appears that there is never a gain in daily energy yield for the higher TC, which is explainable due to the high temperatures. Even if the mean module temperature gets below 25°C for some days, the majority of the energy is generated during the part of the day that it is hotter than 25°C. Even so, we observe indeed, during wintertime, a relatively better performance for the high TC (Al BSF, 0.32%/°C) vs the low TC (n-PERT, 0.28%/°C) technology. For Oldenburg, we observe indeed even negative values, indicating a better performance for the high TC technology.

6 REGIONS OF INTEREST FOR HIGHER TC

Considering the above understanding, climates with lower temperatures close to the equator would be impacted more favourably from technologies with a higher TC to yield a flatter PV production (increased production in low-production times, decreased production during high-production times). Additionally, also climates that induce cell temperatures below 25°C (used at STC rating) for significant amounts of time, can benefit, even in annual energy yield.

To this end, we have identified in Figure 8 some interesting locations that are promising to check further in detail. The graphs show daily irradiance values (radiation values on equator-pointed tilted solar panels based on NASA database [11]) and daily maximum ambient temperatures for 3 climates in comparison to the Kuwait and Oldenburg "references".

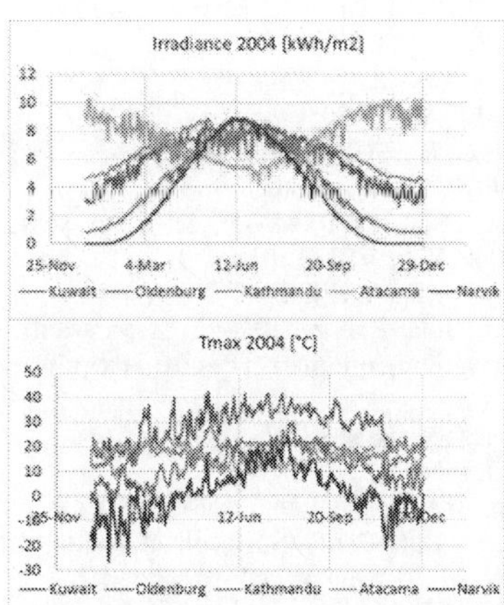

Figure 8: historical irradiance (top) and maximum ambient temperatures (bottom) for climates of interest for further investigation of impact of TC differences

Kathmandu in Nepal is interesting as it lies on roughly the same latitude as Kuwait, and receives similar irradiance levels, but at a significantly lower temperature. The Atacama desert in Chile is of similar interest, in the southern hemisphere, with even somewhat higher irradiance than Kathmandu due to the height (3200 i.o. 1800 m above sea level). Narvik in Norway on the other hand has a high latitude (above the arctic circle) but is interesting for its very low temperature.

7 CONCLUSION AND OUTLOOK

In this paper we want to raise awareness about the increasing complexity of technology optimization resulting from the evolving cost structure of PV electricity generation. The latter can lead to counter-intuitive outcomes such as: a lower temperature coefficient may not always be the best in the future.

First we build some first order analytical model to illustrate the concept and indicate trends. Then we apply our physics-based bottom-up modeling approach, to illustrate how it can be used to quantify with high accuracy PV production throughout the year as function of climate data and PV module technical data, in particular here for the temperature coefficient.

Finally, we indicate some locations where the benefits of a higher temperature coefficient may materialize most, and that deserve some further attention. Apart from temperature coefficient, that is discussed here as a concrete example, other parameters as e.g. tilt and orientation can be similarly used to influence this variability throughout the year, and can also be simulated.

ACKNOWLEDGMENT

The authors gratefully acknowledge imec's SiPV industrial affiliation programme and its partners, the support of the European Union, the European Regional Development Fund ERDF, Flanders Innovation & Entrepreneurship and Province of Limburg through the funded EFRO-SALK project SolsThore, the Kuwait Institute for Scientific Research (KISR) for providing climate data from Kuwait, and the partial funding by the Kuwait Foundation for the advancement of Sciences under project number P115-15EE-01.

REFERENCES

[1] P.A. Basore IEEE TED, 1990.
[2] P.A. Basore et al., IEEE JPV, 2011.
[3] http://www.pvlighthouse.com.au
[4] I. Haedrich et al., SOLMAT 2014.
[5] H. Goverde et al., EU PVSEC 2014.
[6] H. Goverde et al., EU PVSEC 2017.
[7] I. Horvath et al., Solar Energy, 2018 (in preparation).
[8] https://www.apxgroup.com/market-results/apx-power-nl/dashboard/
[9] Trina Solar datasheet TSM-PD05
[10] Sunpower datasheet X22-360
[11] obtained from the NASA Langley Research Center Atmospheric Science Data Center Surface meteorological and Solar Energy (SSE) web portal supported by the NASA LaRC POWER Project

FROM BIPV MODULE TO SYSTEM: A MODELICA-DEVELOPED FRAMEWORK FOR BUILDING ENERGY SIMULATIONS INCLUDING BIPV

Konstantinos Spiliotis[1,3], Juliana Gonçalves[2,3], Kris Baert[1,3], Johan Driesen[1,3], Dirk Saelens[2,3]

[1] ESAT-ELECTA, KU Leuven, Kasteelpark Arenberg 10, 3001 Leuven, Belgium
[2] Department of Civil Engineering, Building Physics Section, KU Leuven, Kasteelpark Arenberg 40, 3001 Leuven, Belgium
[3] EnergyVille, Thor Park 8310, 3600 Genk, Belgium

ABSTRACT: Building integrated photovoltaic (BIPV) systems have been attracting attention as a promising means of complying to an increasingly stricter building energy regulations. Although building energy simulation (BES) tools can be used to assess the BIPV potential contribution at the building level, the existing BIPV modelling in these tools often focuses on either the electrical or the building physics aspect of those systems. Consequently, the existing BES is not fully capable to capture interdependencies between system configurations and the resulting performance. Further to this, the study of complex phenomena, crucial in BIPV applications, such as (partial) shading or temperature gradient over the BIPV façade may not be performed. This paper presents the development and validation of a detailed multi-physics BIPV modelling framework for building energy simulations within the openIDEAS environment. The validation shows that the proposed model capture accurately the absolute values and the dynamics of the real system by comparing the simulation results with measurements from an outdoor flexible façade test facility.

Keywords: building integrated photovoltaic (BIPV), building energy simulation (BES), Modelica, experimental validation, curtain wall.

1 INTRODUCTION

The Energy Performance of Buildings Directive (Directive 2010/31/EU - EPBD) established in 2010 stipulates that all new buildings should be Nearly Zero-Energy Buildings (NZEB) by 2021, while supports in parallel the transformation of refurbished buildings into NZEBs. Next to increasing the building energy performance, the integration of renewable energy generation into the built environment is decisive to cope with this policy. In this context, Building Integrated PhotoVoltaic (BIPV) systems are considered as a promising technology to increase the share of renewable energy in buildings by exploiting additional surface area for energy production.

The integration of PV modules into the building envelope affects the performance on both PV and building sides. On the one hand, BIPV elements drastically alter the building envelope characteristics, including U-values, optical properties and lightning conditions, which will dictate the building overall energy performance. On the other hand, the integration also modifies the boundary conditions imposed to the PV cells and the power converters operation. Therefore, in order to assess the performance of the entire system, both the PV modules and the building have to be considered in a fully coupled and integrated approach. Building Energy Simulation (BES) tools can be used to evaluate the year-round energy performance of such integrated systems. However, current modelling of BIPV systems in BES tools tends to be simplified, and does not cover the wide range of BIPV applications and designs. Additionally, traditional BES programs are based on imperative language, which makes the code hard to maintain and prevents the inclusion of new models [1][2].

The proposed approach takes a step further by considering the BIPV module as part of a greater BIPV system. Next to the module, the electrical components are incorporated within the boundaries of the BIPV system. BIPV modules are connected to power converters so that their output is conditioned at a specific grid operating voltage. A thermal-electrical model of the power converter has been developed to simulate the behavior of the apparatus and provide estimation of the losses that occur. This modeling approach can therefore provide us with a better estimate of the expected energy generation of the BIPV module. Furthermore, it is easy to compare different BIPV module placement and interconnection strategies. The state-of-art BES tools usually do not include the system's electrical components or there are very little possibilities to modify the available models. In practice, that makes it hard for the building designer to simulate a BIPV system in a customized fashion.

The literature indicates the need for dynamic models that take into account the thermal-electrical coupled effects and are able to predict the shading effects raising up from the integration of PV modules into building facades. In this paper, we propose: (1) an integrated approach as depicted on Figure 1 combining the building, the BIPV and electrical components within the same dynamic environment; (2) a circuit-based model to represent the PV electrical behavior, which allows for modelling partial shading effects; (3) the use of the equation-based object-oriented Modelica language, which provides the necessary flexibility for modelling multi physics systems. As a case study, the proposed approach is used to model a curtain wall BIPV element integrated into a BIPV test facility located in Leuven, Belgium. Both the methodology and the results are compared with TRNSYS software, a commonly used BES tool in terms of estimation accuracy but also modularity and easiness to use. The results are further compared to the measured results from the experimental setup.

2 METHODOLOGY

2.1 Modelica language

Modelica is a language for modelling complex and multi domain systems. Contrary to imperative languages such as Fortran or C, where the specific physical knowledge is hard-coded, in Modelica the physics of the models are coded in the language itself while the compiler encapsulates all the mathematical knowledge [3]. This characteristic makes Modelica a general-purpose language. Further, due to the separation between solver and physical modelling, Modelica models are easier to maintain and upgrade. The advantages and suitability of Modelica-based modelling for building energy and control systems simulations is thoroughly discussed in [1][2]. The use of Modelica language for electronic circuit modeling and simulation is considered in [3][4]. Finally, the equation-based object-oriented nature of Modelica provides the necessary flexibility to cover the broad range of BIPV applications. For these reasons, Modelica language has been chosen to model the proposed integrated BIPV framework. The models are further compatible with IEA EBC Annex60 building libraries and, in particular, with openIDEAS environment, a framework developed for integrated district and building energy simulations [5].

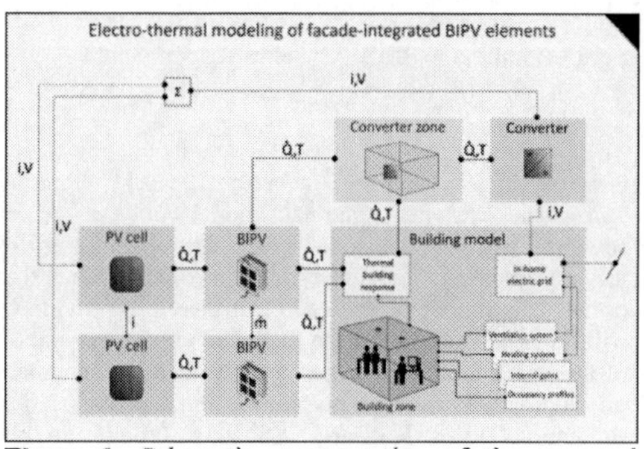

Figure 1: Schematic representation of the proposed integrated electrothermal modelling of BIPV systems

2.2 BIPV system package

A thermal-electrical model has been developed to simulate the BIPV element. Contrary to roof-mounted PV installations, in BIPV systems, the two sides of the modules are exposed to very different boundary conditions. Moreover, BIPV systems are normally cooled by natural or mechanical ventilation. Therefore, a detailed thermal modelling is critical to represent BIPV systems. Making use of the object-oriented approach provided by Modelica, the thermal model is characterized here by a combination of discretized elements (or instances), which represent the heat transfer phenomena governing the BIPV thermal behavior (Figure 2, right). In this way, it is possible to simulate the thermal vertical stratification occurring on the PV, air and building wall layers. Figure 2 shows a schematic representation of the coupling between the PV cells and the BIPV thermal model.

The external heat transfer coefficient (HTC) was calculated from the expressions presented in [6]. The radiative losses to the ground and to the sky are taken into account separately. Radiation exchange is also considered between PV module and the building wall. The airflow circulating inside the cavity behind the PV modules is obtained from the balance between driving forces (buoyancy and wind) and cavity losses. Convection inside the cavity is based on the correlations presented in [7].

Due to a good balance between complexity and accuracy [8], the one-diode model is adopted in this work. The model is used at the cell level, which means that PV cells can be connected to each other either in parallel or in series, according to the user requirements, to compose a PV module. In turn, PV modules can also be connected to each other. The use of circuit elements also allows including resistances to account for losses between cells and strings. A sampling MPP tracker is used to maximize power extraction under varying conditions. To our knowledge, this approach is currently not available in BES tools.

The coupling between the electrical circuit and the thermal model is achieved by using the PV temperature from the thermal model to calculate the power in the electrical model, which is fed to the input of a DC/DC boost converter. The solver iterates the model until convergence is achieved.

The model of a DC/DC boost converter suitable for BIPV applications has been developed. The converter's maximum operating power is 330W and its components are properly dimensioned to handle the current and voltage output of the bespoke BIPV module. A DC/DC boost converter typically consists of passive components (inductor and capacitor) at the input and output respectively and a switching cell (switch and diode) that allows stepping up voltage in proportion to the duty cycle (ratio of on and off time). The average switching model has been adopted in order to avoid excessive simulation steps proportional to the switching frequency. The average switching model is well established in the literature and it is documented that it provides accurate simulation results. However, the passive components still would impose several simulation steps in proportion to the resonance frequency. The power converter is eventually simulated with the use of a voltage and a current source at the input and output side respectively.

Even though the switching cell and the passive components are not physically connected to the converter's circuit, their losses are estimated and deducted from the converter's output. This strategy allows us to have an accurate estimate of the converter losses but not at the cost of computation time. Another concern would be the great difference between the BIPV module thermal time constant and the converter electrical constant. With the proposed approach however, this problem is avoided.

Given the BIPV module output, the hypothetical values for the current of the passive components and the switching cell are derived. With the use of the root mean square (RMS) and average values of the currents the switching and conduction losses are estimated using an established loss estimation methodology [9]. Further, the losses in the inductor and the capacitor are estimated from their RMS currents and their equivalent series resistance (ESR) and DC resistance (DCR) respectively. All of those losses are dependent on the temperature. Therefore, the efficiency and consequently the output of the DC/DC converter is coupled to the whole BIPV system and its operating conditions. Figure 3 depicts this correspondence between the operating temperature, T_{air} and the converter's losses, $Q_{1\text{-}2}$, through the 'Cauer' equivalent electrical circuit where R_{th} represent the thermal resistances and C_{th} the heat capacitance of the concerned air zone, here the converter zone.

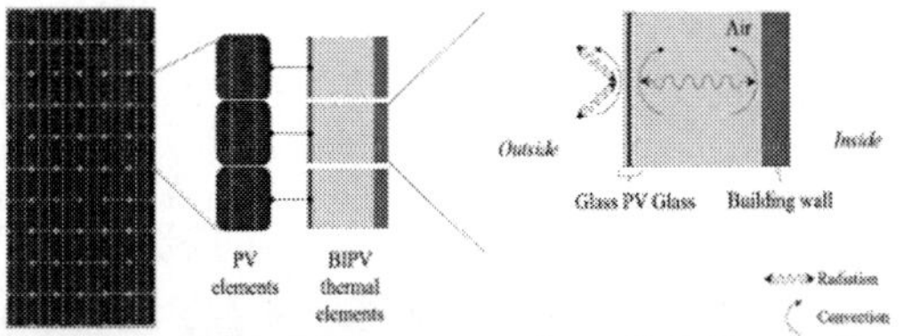

Figure 2: Schematic representation of (left) coupling between PV cells and BIPV thermal model; (right) physical phenomena included in the BIPV model

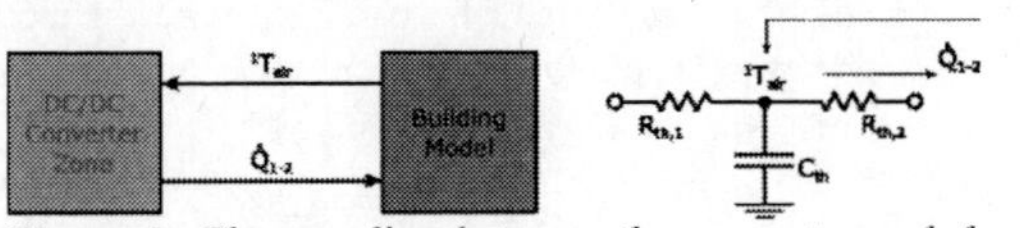

Figure 3: The coupling between the converter and the building/BIPV model

3 CASE STUDY

For the purposes of this study, a BIPV curtain wall is considered. The module corresponds to the experimental BIPV setup installed on the façade of a BIPV test facility in Leuven, Belgium. Detailed information on the BIPV design and performance can be found in [10]. The maximum power output of the PV module is 244.1 W, as estimated at standard conditions. The meteorological measurements of 28th of August 2017 have been used as an input to the developed BIPV model. The weather conditions of that day provide a good opportunity to get a glimpse at the model's functionality. In the full paper a longer period is considered, including diverse weather conditions.

As a benchmark with a state-of-art BES, TRNSYS is used to replicate the case study conditions. TRNSYS is a tool widely used to describe the behavior of transient systems [11]. TRNSYS is one of the most common tools for BES and thermal systems in general. For this case study, a simulation has been set up where the BIPV test facility structure is imitated through a Type56 block. Type567 has been used to model the curtain wall BIPV module. Details on how to implement the coupling between the BIPV and the building in the TRNSYS environment are discussed in the literature [12]. Finally, the TRNSYS BIPV modules require the cavity flow rate as an input. To estimate the flow in a naturally ventilated cavity, both the wind and the buoyance effects are considered [13].

4 RESULTS

Figure 4 shows the estimated output power against the one measured at the test facility. As observed on Figure 4 the peak power is accurately estimated by the proposed modelica-based model. It appears that the modelled dynamics of the BIPV module on the DC/DC converter follow closely the experimental measurements. The deviation maxes out around the peak values and it is equal to approximately 14 Watts of difference. The mean deviation is 3.5 Watts throughout the day. In comparison to the estimation using TRNSYS, the proposed modeling package is more accurate both in terms of deviation and in terms of predicting the dynamic behavior of the real system. The simulated results using TRNSYS have a mean deviation of approximately 14 Watts. As shown in Figure 4, the BIPV model in TRNSYS does not captivate the dynamics of the physical module to the same accuracy as the proposed modelling package. Both the rise and the drop of power generation are quite faster compared to the corresponding measured values.

Further to power estimation, the model has the capability of estimating thermal parameters such as the PV cells temperature, the temperature and heat flux in the ventilation cavity and in general to investigate any thermal coupling between the BIPV module and the building on which it is integrated. By applying a given irradiation input, it is possible to estimate the PV cell temperature as seen on Figure 5 From one perspective, this information is useful in hygrothermal building simulations. In addition to this, the estimated temperature time series can be used to estimate PV cell lifetime and degradation for the given conditions.

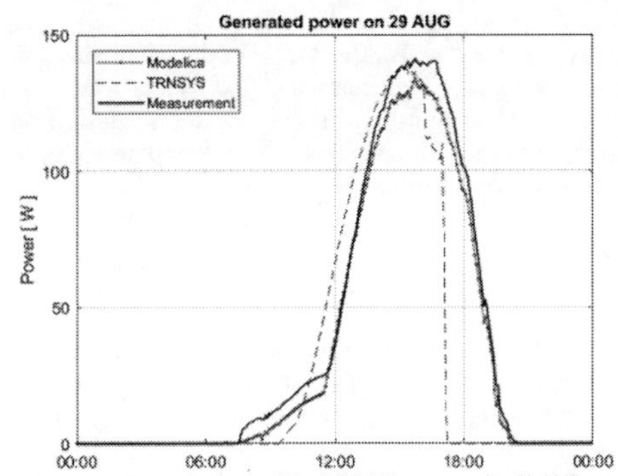

Figure 4: A comparison of the power generation estimation using the proposed modelica-based modelling approach (red) and using TRNSYS (blue) against the measured values from the test setup (black)

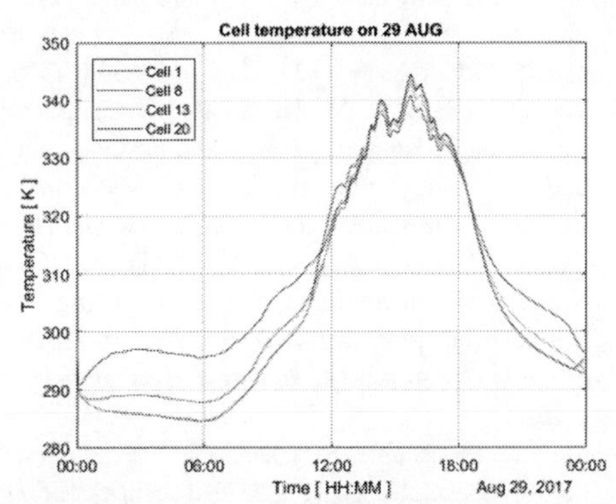

Figure 5: Temperature of cells 1,8,13 and 20 of the simulated 60-cell mono-Si BIPV module.

5 CONCLUSIONS AND OUTLOOK

In this paper, an electric-thermal integrated BIPV model developed in Modelica is presented and benchmarked with outdoor measurements and with TRNSYS, a commonly used BES tool. Besides a high degree of accuracy as indicated by the preliminary results, the proposed approach aims at bringing various benefits. The electric-thermal models are coupled (PV and cavity model; converter and cavity/building model), which allows the investigation of further effects of different design options. The components of the BIPV electrical system are included (power converters, energy storage, loads); in that way power conversion losses can be taken into account when designing a BIPV system or quantifying its benefits. The presented research offers a package to cover a wider range of BIPV module configurations (i.e. non-ventilated, mechanically and naturally ventilated) as well as a wider range of electrical system architectures (i.e. different interconnection regimes). The high modularity of the models developed allows the detailed study of partial shading and their effects; a phenomenon which is rather common in the urban environment. Finally, the models have been developed to be user-friendly through clear and organized documentation. This provides the flexibility needed for the fast prototyping and assessment of difference BIPV systems and energy scenarios. Beyond research, such a framework may assist the work of building designers, architects and engineers by providing an inter-disciplinary insight on BIPV systems design and operation without requiring the constant interlocking of multiple experts.

ACKNOWLEDGEMENT

The results presented in this paper are part of a greater effort towards an integrated BIPV simulation framework supported by the European Union, the European Regional Development Fund ERDF, Flanders Innovation and Entrepreneurship and the Province of Limburg.

REFERENCES

[1] Wetter, M., Bonvini, M., & Nouidui, T. S. (2016). Equation-based languages–A new paradigm for building energy modeling, simulation and optimization. *Energy and Buildings*, 117, 290-300.

[2] Wetter, M. (2009). Modelica-based modelling and simulation to support research and development in building energy and control systems. *Journal of Building Performance Simulation*, *2*(2), 143-161.

[3] Cellier, F. E., Clauß, C., & Urquía, A. (2007, September). Electronic circuit modeling and simulation in Modelica. *In Proceedings of the Sixth Eurosim Congress on Modelling and Simulation* (pp. 1-10).

[4] Majetta, K., Böhme, S., Clauß, C., & Schneider, P. (2009, December). SPICE3 Modelica Library. In *Proceedings of the 7th International Modelica Conference; Como; Italy; 20-22 September 2009* (No. 043, pp. 567-575). Linköping University Electronic Press.

[5] Baetens, R., De Coninck, R., Van Roy, J., Verbruggen, B., Driesen, J.; Helsen, L.; Saelens, D. (2012). Assessing electrical bottlenecks at feeder level for residential net zero-energy buildings by integrated system simulation. *Applied Energy*, 96, 74-83.

[6] Montazeri, H., & Blocken, B. (2017). New generalized expressions for forced convective heat transfer coefficients at building facades and roofs. *Building and Environment, 119*, 153-168.

[7] Churchill, S. W., & Chu, H. H. (1975). Correlating equations for laminar and turbulent free convection from a vertical plate. *International journal of heat and mass transfer*, *18*(11), 1323-1329.

[8] Chin, V. J., Salam, Z., & Ishaque, K. (2015). Cell modelling and model parameters estimation techniques for photovoltaic simulator application: A review. *Applied Energy*, *154*, 500-519.

[9] Drofenik, U., Kolar, J. (2005) A general scheme for calculating switching-and conduction-losses of power semiconductors in numerical circuit simulations of power electronic systems. *International power electronics conference (IPEC)*, Niigata, Japan.

[10] Lehmann, J, Parys, W., Gonçalves, J., Yordanov, G., Baert, K., Saelens, D. (2017) Experimental analysis of the performance of a BIPV curtain wall component. In *Advanced Building Skins Conference*.

[11] TRNSYS (2018) [online] Available at: http://www.trnsys.com/ [Retrieved on 8 March 2018].

[12] Lindig, S., Moser, D., Avesani, S., Lollini, R. (2015) International Building Performance Simulation Association, *International Building Performance Simulation Association (IBPSA)*, Bozen, Italy.

[13] Griffith, B. (2006) A model for naturally ventilated cavities on the exteriors of opaque building thermal envelopes, *Simbuild conference*, Cambridge, USA.

PREDICTING YIELDS OF BIFACIAL PV POWER PLANTS – WHAT ACCURACY IS POSSIBLE?

Matthieu Chiodetti[1], Jinsuk Kang[2], Christian Reise[2], Amy Lindsay[3]
[1]EDF R&D - Avenue des Renardières - 7250 Moret Loing et Orvanne - France
[2]Fraunhofer ISE - Heidenhofstraße 2 - 79110 Freiburg - Germany
[3]EDF Inc., USA
Phone +33 1 60 73 63 42 matthieu.chiodetti@edf.fr

ABSTRACT: Realistic bifacial gains expected for large scale PV systems are in the range of 5% to 15%. This gain is a big game changer when compared with other improvements in PV in the last few years. However, the performance of a bifacial system will be strongly affected by its design and the resulting impact on the irradiance received on the rear side of the modules. To accurately estimate the extra yield generated, new tools able to assess the irradiance reaching the rear side are needed. Different methods are available to estimate the bifacial yield. A very accurate one is to use 3D modeling of all the system components that can interfere with light combined with ray tracing. EDF R&D and Fraunhofer ISE have both developed their own bifacial yield assessment tool based on ray tracing. To validate this approach, both tools have been used for cross comparison simulations. Several validations have been carried out against real bifacial systems – including large scale fixed tilt and horizontal single axis tracking systems. Results of the validation and the tool comparison are presented here.
Keywords: Bifacial, Modelling, Ray tracing, System Performance

INTRODUCTION

Bifacial PV is set to become a mainstream technology in the coming years thanks to the energy gain it can supply to a PV project. Major PV manufacturers are now starting to propose bifacial modules, and with the industry now moving into a period of mono-based PERC (60% forecasted market share by 2027), it is very likely that bifacial PV will be the next step. Indeed there is very little difference between the cell production process of a PERC monofacial or bifacial module. Accordingly, the ITRPV roadmap also forecasts that bifacial modules would represent 30% of the market in 2026 [1]. This shift is of course motivated by the promise of additional energy generated from the rear side of the bifacial modules.

However, the performance of a bifacial PV system not only depends on PV module properties, but is strongly affected by its mechanical design and the resulting impact on the irradiance received on the rear side of the modules. To accurately estimate the extra yield generated, new tools being able to assess the rear side irradiance are needed.

1 MODELING AND PREDICTION TOOLS

The modelling of bifacial yield requires an accurate estimation on the rear side irradiance of the modules. To achieve that calculation, two main approaches exist: the "view-factor" method and the "ray tracing" approach.

A view factor expresses the fraction of radiation emitted from a surface which is received by another. The receiving surface might be a section of or even the whole rear surface of a bifacial PV module, while all other surfaces in the vicinity are assumed as emitting surfaces. In a very simple case, there are at least two emitting surfaces, the unshaded ground below the module and the ground section shaded by the module itself. Examples of such calculations may be found in [2] and [3].

The view factor method is rather easy to implement, but easy to handle for single modules or very small systems only. For large and more complex installations, the individual programming effort grows quickly, and even then, effects like inhomogeneity caused by mounting structures cannot be evaluated.

The approach chosen by EDF and Fraunhofer ISE for rear surface irradiance prediction is based on the ray tracing method. While the system model is easy to define, the ray tracing calculations as such consume big amounts of computer time. Nevertheless, they may solve a number of problems simultaneously: the calculation of rear side irradiance including its inhomogeneity (influenced by the mounting structure) is easily accompanied by the calculation of front side irradiance – including mutual shading effects. Complex geometries, topology, or edge effects can also be taken into account.

Today, several tools – publicly available or not – exist for the simulation of bifacial production. Most of them, such as Pvsyst [4], SAM [5] or the ISC Konstanz tool [6] are based on the view-factors method. Due to its complexity, ray tracing approach is less commonly used, but research institutes such as NREL [7], EDF R&D or Fraunhofer ISE are using it for the more details it can give. Explanations on the two models from EDF and ISE are detailed in the next part of the document.

1.1. EDF's approach

For the irradiance assessment on bifacial modules – and more generally for all PV surfaces – EDF is developing an in-house ray tracing tool. It is using backward ray tracing. i.e. the rays are sent from the PV cells towards the rest of the world. It allows the user to choose different types of sky modeling: a simplified isotropic sky or the more sophisticated Perez model which discretizes the sky in the circumsolar and horizon regions.

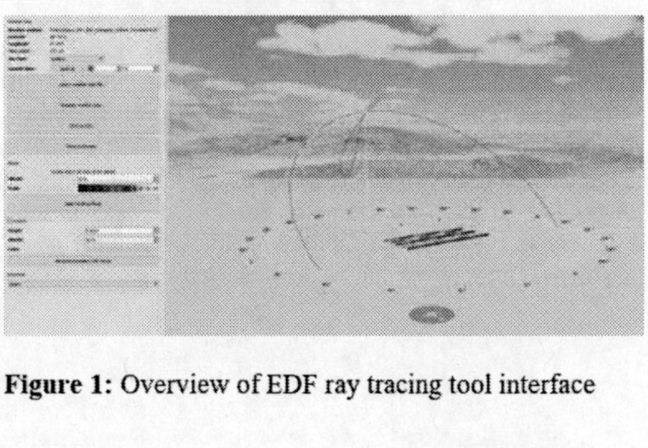

Figure 1: Overview of EDF ray tracing tool interface

Complex 3D models can be imported or generated in the tool. All the objects (including ground) have a reflectivity that can be defined and their impact on the irradiance received on the PV surfaces can be analyzed. Optical properties of the modules such as incident angle modifier are defined in the tool.

1.2. Fraunhofer ISE's approach

The approach of Fraunhofer ISE is based on *Radiance*, a powerful lighting simulation program based on backward ray tracing calculation. A Radiance tool named *Gendaylit* [8] creates a complete sky radiance distribution using Perez model for any pair of global and diffuse irradiance. The defined model, including geometry and surface and ground materials, may be rendered as an image, and numerical values of irradiance may be calculated for arbitrary positions. "Irradiance measurements" in Radiance are performed as "seen" by virtual sensors on front and back side of defined PV cells.

The irradiance values on each sensor are then aggregated as irradiance values on module level and handed over to Fraunhofer ISE's home grown PV system simulation tool Zenit. The remaining calculation steps in Zenit are standard and well known in the community.

The whole bifacial modelling process from optical (Radiance) to system yield simulation (Zenit) is controlled by a Python tool package.

2 BIFACIAL PV SYSTEM DATA SETS

To ensure the validity of the modelling approach, it is necessary to assess the performance of the models against real bifacial installations in outdoor operation. First results of validation were already presented in [9] and [10]. In this paper, more validation cases are considered to cover more types of configuration (array size, fixed-tilt, tracking…), different modules types and different climates. The location and climatic conditions of the installations are presented in Figure 2.

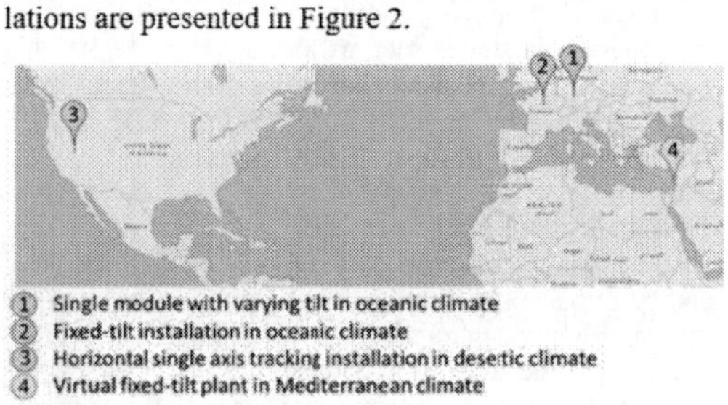

Figure 2: Situation of the 4 bifacial datasets chosen.

2.1. System 1 – Single module

System 1 consists in a setup of a single bifacial PV module on a roof at Fraunhofer ISE. The module is mounted on a test rack with unobstructed rear surface and may be rotated and fixed at any angle between 0° and 90°. The actual tilt angle is recorded by an electronic inclination sensor, two silicon reference cells provide values of front and rear irradiance measured close to one module edge. The module itself may be connected either to an I-V-curve tracer or to a simple short circuit current measurement. The module can be seen on Figure 3.

Figure 3: Single bifacial module on the roof of Fraunhofer ISE

2.2. System 2 – 16 kWp fixed-tilt array

System 2 is located in the Paris region at an EDF R&D test facility. It is a 16 kWp array built with 60 n-type bifacial modules, with a bifaciality of 90%. The array is mounted on an adapted structure to avoid shading on the rear side. The configuration has been chosen to be representative of a real large plant installation with a structure length of 50m and a structure height of 3m, and a tilt of 30°. Another PV stand is behind the bifacial array, resulting in an equivalent Ground Covering Ratio (GCR) of about 50%. The ground is made of clear concrete, with an albedo monitored at about 30%.

The system is monitored with pyranometers on the front and the rear side and thermal sensors on the back of the modules. DC voltage, current, and power are measured for each string of the installation. System 2 can be seen on Figure 4.

Figure 4: Overview of bifacial system 2 - a 16 kWp array at EDF R&D test facility.

2.3. System 3 – Tracking installation

Several bifacial modules were installed on Horizontal Single Axis Trackers (HSAT) in the Californian desert on an EDF test facility. The test area is formed with several rows of trackers. In this document, the studied installation is one row in the middle of the plant with modules. The PERC 72-cells bifacial modules are positioned in portrait on the tracker rows, as shown in Figure 5. Structure of the trackers are standard, they have not been optimized for bifacial modules. Measurements are available for front and rear irradiance, back of modules temperatures, DC and AC voltages, currents and powers.

Figure 5: System 3 is a HSAT installation in the Californian desert

2.4. System 4 – 2.5 MWp fixed-tilt plant

System 4 is not a real installation but rather a test case which has be defined to be as close as possible to a real bifacial PV plant. As there are no experimental measurements, the coherence between the models is studied in this case.

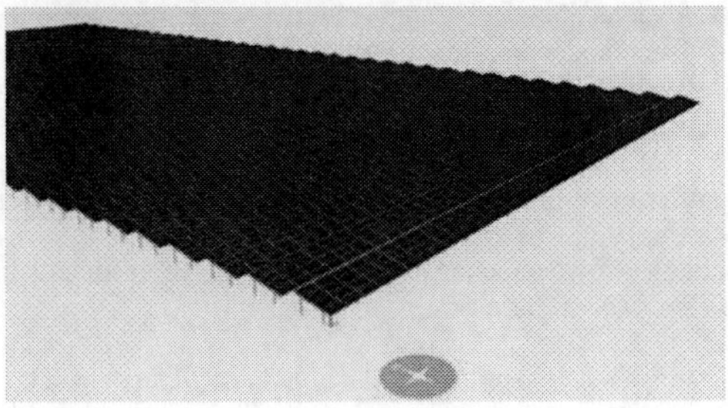

Figure 6: System 4 is a virtual 2.5MWp bifacial plant located in Israel.

The system is a virtual 2.5 MWp plant in Israel, on sand ground. The albedo of sand has been taken as 30%, based on the literature and measurements carried out by EDF. The modules considered are n-type 72-cells bifacial modules with 84% of bifaciality. The structure is tilted at 12° and adapted for bifacial modules. 70% of GCR was considered to be representative of actual PV plant. An illustration of the virtual plant can be seen in Figure 6. The characteristics of the different installations presented are summarized in Table 1.

2.5. Data filtering

System 2 and 3 are operating over several months with below one minute of time interval, which results in large amount of data. To facilitate the modeling without smoothing the power variations, data were aggregated to 10 minutes, according to the following rule:

- Data are aggregated to 10min if more than 90% of the data are available on the time interval

For the modelling, a weather file was constructed from irradiance and temperature measurements on site. Global Horizontal Irradiance (GHI) was measured using CMP10 pyranometers, Diffuse Horizontal (DHI) using a SPN1 pyranometer. Data from the pyranometers were filtered according to the following rules:

- GHI and DHI values below 0 W/m² were removed
- If DHI > GHI, DHI is set to GHI value
- GHI and DHI are set to zero for sun height < 0°

For the array measurements, only positive values of current, voltage and power are kept.

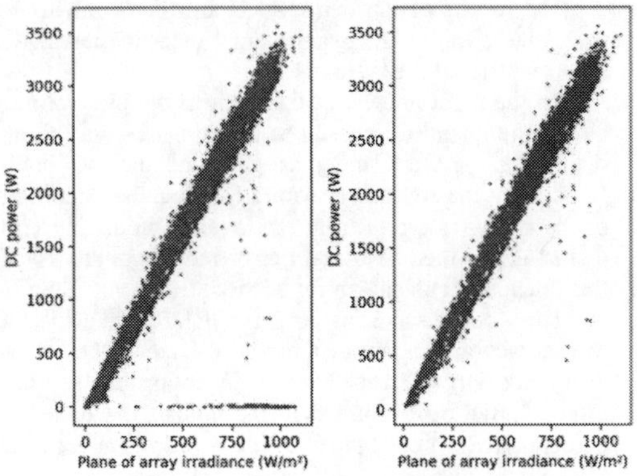

Figure 7: Illustration of filtering on data from system 3. DC power is plotted as a function of the plane of array irradiance before filtering (left) and after filtering (right). 86% of points were kept, representing 99.2% of the energy.

Finally, to remove the effects of snow, power cut and environment shading that are not considered in the simulation, the points for which difference measure / experiments is greater than 50% have not been considered, assuming that such difference are due to external condition, not to the modelling. As an illustration, when applying the filter to the experimental data of system 3, the ratio of accepted points is 86.0%. In term of energy, these points represent 99.2% of the total energy monitored during the time period.

Table 1: Summary of the characteristics of the bifacial installation considered in this study

	System 1	System 2	System 3	System 4
Peak power (kWp)	0.35	16	3.5	2500
Module type	60-cells n-type	60-cells n-PERT	72-cells p-PERC	72-cells n-PERT
Ground albedo	25%	30%	32%	30%
Fixed-tilt / tracking	Varying tilt	30°	East-West HSAT	12°
Ground covering ratio	NA	50%	35%	70%
Location	Freiburg	Paris	California	Israel
Climate	Oceanic	Oceanic	Desert	Mediterranean
Time period	15. July, 2015	27. Jun – 30. Dec, 2016	17. Aug – 18.Feb, 2018	01. Aug, 2017 – 28. Feb, 2018

It can be seen in Figure 7 that removed points are mainly caused by a default of the inverter which was connected but wasn't delivering power.

3 PREDICTION VS. OBSERVATION

For each dataset, EDF and ISE have run independently their own model with the following inputs:

- GHI, DHI, ambient temperature of the site
- Albedo of the ground
- 3D geometry of the mounting structure
- Configuration and orientation of the installation
- Electrical characteristics of the bifacial module from the datasheet or IV curves if available

The results of the two models are then compared to each other, and to the experimental measurements (with the exception of the system 4).

As the main interest of the study is the bifacial modelling, the quantities chosen to be compared are the irradiance on the front and rear side, and the DC power. Looking at the AC power would increase the uncertainty of the study, as errors from the inverter modelling could distort the results. Two main error indicators are used in this document and shown for each system:

The Root Mean Squared Error (RMSE) is a measure of accuracy and is proportional to the size of the squared error thus larger errors have a disproportionately large effect. It can be expressed in the unit of the data or in percentage. It is calculated with the following formulas:

$$RMSE_{absolute} = \sqrt{\frac{\sum_{i=1}^{N}(sim_i - mes_i)^2}{N}}$$

$$RMSE_{relative} = \frac{RMSE_{absolute}}{mean(mes)}$$

where sim and mes are the considered simulated and measured data, and N is the number of samples.

The Mean Bias Error (MBE) is also a measure of accuracy. It gives the average error between the simulation and the true value. It can be expressed in the unit of the data or in percentage. It is calculated with the following formulas:

$$MBE_{absolute} = \frac{\sum_{i=1}^{N}(sim_i - mes_i)}{N}$$

$$MBE_{relative} = \frac{MBE_{absolute}}{mean(mes)}$$

The indicator more relevant to the market is the Mean Bias Error (MBE) of the power, as it represents the amount of energy under or over-predicted by the model. As discussed, the mentioned systems in section 2 are analyzed with above error indicators and the results are following in the next sections.

3.1. System 1

The geometry and the optical properties of the module, rack and ground were implemented both in EDF's and ISE's ray tracing tools, based on the real dimensions of the system (cf. Figure 8). Irradiance is then calculated on the surface of the two sensors, as well as on the cells on the front and the rear of the module.

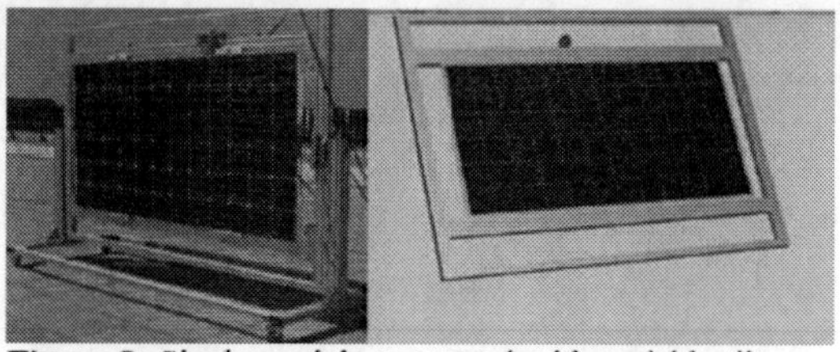

Figure 8: Single module test stand with variable tilt angle and orientation at Fraunhofer ISE, and its model in a ray tracing platform.

The structure allows to change the tilt of the module, so different configurations were tested and compared against the simulations. For a measurement taken in July 2015, Figure 9 compares measured and modelled irradiance values. For all tilt angles below some 70°, the values of front and rear side irradiance are reproduced quite well, above 70°, the front side irradiance is overestimated, which also affects total irradiance.

Both EDF and ISE tools show a similar behavior and similar error on the evaluation of the rear side irradiance varying with the tilt angle. These results show that ray tracing allows an accurate estimation of the irradiance in different configurations.

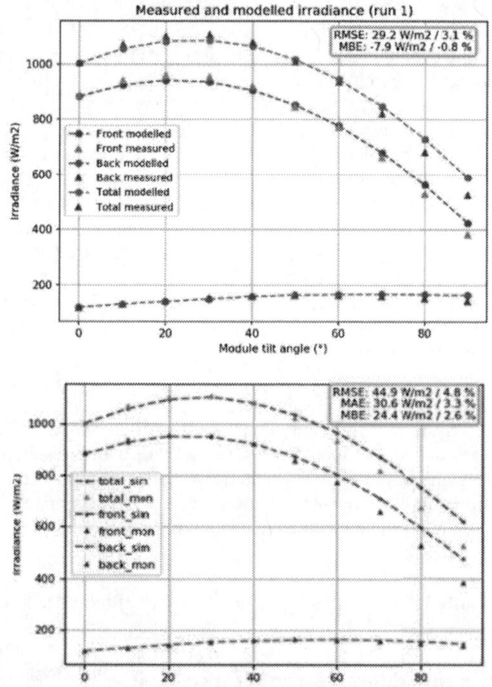

Figure 9: Front and rear irradiance of the single bifacial module compared to the results from EDF model (top) and ISE model (bottom).

3.2. System 2

System 2 is a large-scale ground mounted array installed at EDF R&D, near Paris. Meteorological measurements are made on-site and used for generating the weather inputs of the models. An albedometer was used to monitor the ground reflection over time, and its average value was taken to define the albedo in the ray tracing tools.

The 60 modules of the installation were characterized with front and rear side flash-tests, resulting in an average bifaciality of 90%. The characteristics of the modules were implemented in the models. A 3D model of the bifacial installation and of its surroundings (adjacent PV arrays) was built in the two tools, as shown in Figure 10.

In order to validate the irradiance calculation received on the rear side over time, backside of the array is measured with 6 pyranometers at different heights (only the three not affected by any edge effects are presented in this study).

Figure 11 shows that the irradiance varies strongly between winter and summer, not only in its shape but the hierarchy between the most and the lowest irradiated position is also changed. In winter the lowest pyranometer (Pyrano 3) receives more irradiance, while in summer it is the top pyranometer (Pyrano 6) that gets more lights – except around solar noon. This can be explained by the shadow casted on the ground. Figure 10 shows that the shadow is further away from the lower module in winter compare to in summer.

Sensors in the ray tracing models were placed at the same positions as Pyranometers, in order to evaluate the accuracy of the prediction of these irradiation variations over the months of data available. Results of the irradiance and power comparison are presented in Figure 12 and Figure 13.

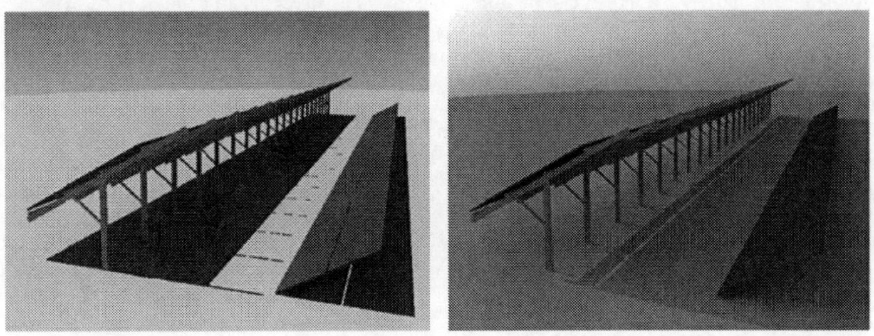

Figure 10: (left) ray tracing visualization with Radiance tool, 'rshow' in one sunny day in summer (19, July) at 12:02; (right) same visualization in winter (21, December) at 12:25 (source: Fraunhofer ISE).

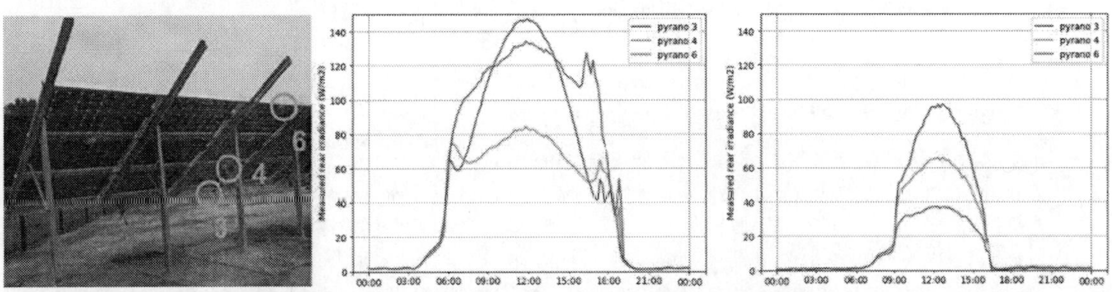

Figure 11: Illustration of the measured irradiance received on the rear of the modules in summer and in winter, at different position in the array (top in green, middle in orange, bottom in blue). It can be seen that the modules that the shape and the ranking of the modules that received the more irradiance are changed with the season.

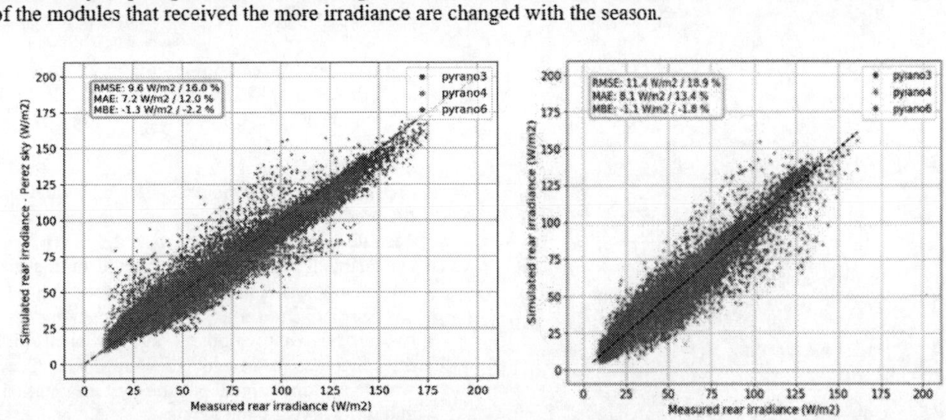

Figure 12: Correlation graph between the simulation (left: EDF (using Perez sky), right: Fraunhofer ISE) and the measurements of rear irradiance for top (purple), middle (green) and bottom (blue) pyranometers.

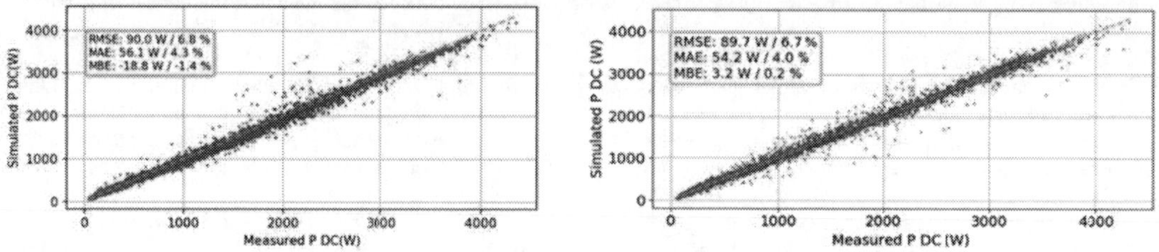

Figure 13: Correlation graph between the simulation and the measurements for the DC power of the system 2 (left: EDF, right: Fraunhofer ISE)

Models are in good agreement with power measurement, with absolute errors under 10 W/m². It can be noted that the absolute errors are smaller for the rear side than for the front side. However as values are significantly lower on the rear side, it results in relative errors slightly higher (RMSE of 18%).

Results in term of power prediction are presented in Figure 13. It represents the output from the electrical models from EDF and ISE, compared to the data from the DC monitoring of the installation. After removing the shading loss from buildings nearby by applying the filters described in 2.5, there is a good correlation with both models (RMSE < 7%, and MBE < 2%).

3.3. System 3

One array of 3.5 kWp of bifacial modules on HSAT was considered for this comparison. As with the other systems, the same inputs were used for the two models. Due to some issue with the backtracking algorithm implemented in the trackers, the models were run directly with the angles measured on the trackers. Therefore the backtracking algorithm was not modelled for this study.

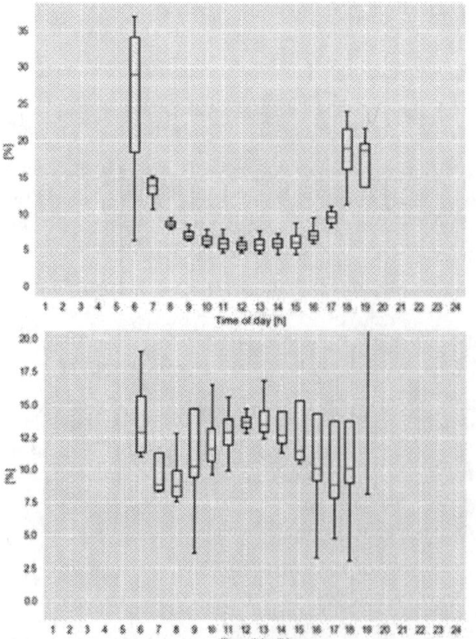

Figure 14: Typical variation of bifacial gains over the course of the day for fixed-tilt (top) and HSAT (bottom)

Again, consistent results were found by both models. The diversity in behaviors of bifacial gains is well represented, in comparison to fixed-tilt simulations. For fixed-tilt installations, bifacial gain has a U-shape, while HSAT shows a W-shape with a maximum at solar noon (Figure 14). Indeed, the shadow cast on the ground is at its minimum size at solar noon under the trackers, the rear side benefits from the sunlit areas between rows.

For the fixed-tilt structures, the rear irradiances are mainly explained by diffuse irradiance (sky diffuse and ground diffuse), so while the front production increases at solar noon thanks to the direct light, the rear production does not in the same extent. Figure 15 shows the results from both models for the prediction of the DC power.

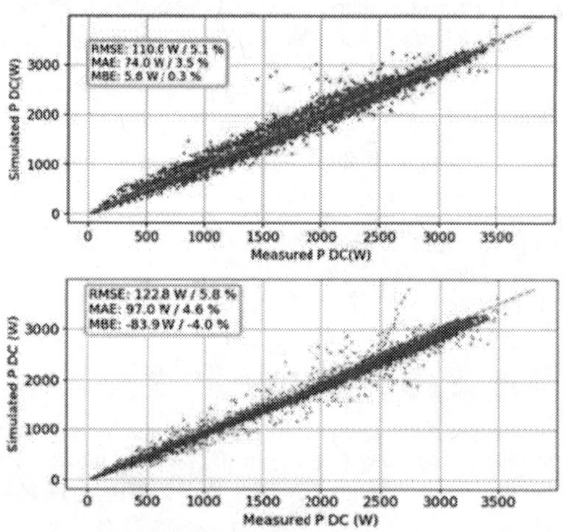

Figure 15: Correlation graph between the simulation with models (top: EDF, bottom: Fraunhofer ISE) and the measurements of DC power

3.4. System 4

After validating the irradiance on a single module and the production of a full bifacial array, the two tools have been tested on a fictive case of a 2.5 MWp fixed-tilt bifacial plant. It represents what would be the yield prediction on a potential industrial project.

As there are no experimental data for this case, the two models were compared against each other to check the coherence of the modelling approach. They were fed with the information about the system detailed in Part 3, and run with the same inputs.

The results presented in Figure 16 that both models are very well correlated and give very close bifacial gain and yield estimation. The RMSE in total difference between models is 30 kW, MBE is 24.2 kW and MBE is -4.9 kW while the total error is between -0.4% and 2.4%.

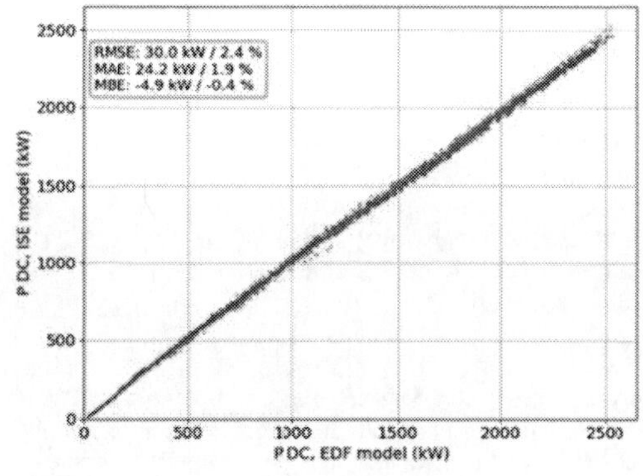

Figure 16: Correlation between the ISE (y axis) and EDF (x axis) models on a 2.5 MWp bifacial plant.

3.5. Summary

The performance of the models is summarized in Table 2. The average total error on yield indicates in both models between 3% and 7% for any designs and climates, from single module system to tracking system. The deviation between two models reveals less than 3%. The errors are much smaller than the typical uncertainty of a PV yield assessment which lies between ±5% to ±10% [11]. These results show that the modelling of bifacial PV yield can be of similar accurate than standard PV modelling with the use of ray tracing approach.

Table 2: Summary of the performance of the two models. Irradiance values are compared for system 1, DC power values for systems 2 to 4. Errors are given with regards to the experimental values with the exception of system 4, where two models are compared to each other.

	Syst 1	Syst 2	Syst 3	Syst 4 (models)
RMSE (%)	3.1 – 4.8	6.7 – 6.8	5.1 – 5.8	2.4
MBE (%)	0.8 – 2.6	-1.4 – 0.2	-4.0 – 0.3	0.4

CONCLUSION

Accurate yield predictions on bifacial PV system became more importance as bifacial PV system depend on many different factors and are not as simple to model as monofacial systems. Each EDF R&D and Fraunhofer ISE applied ray tracing approaches for a respective bifacial system's yield model. Both models were validated with existing single module, fixed array and tracking systems in different climates and on different albedos. Additionally, a 3 MWp fixed system was considered to correlate the two models for large scale PV systems.

The comparisons presented here show a good correlation of both models with the observed results and may be seen as a reliable validation. For a single module, RMSE errors are in a range of 3% to 5%, for the small scale system in a range of 5% to 7%. These total average errors on the DC yield achieved by the two models are smaller than errors typically obtained by modelling monofacial systems with standard PV tools [12].

Comparisons to other bifacial yield prediction tools such as view factor based models could be performed as future research.

REFERENCES

[1] ITRPV, International technology roadmap for photovoltaic (ITRPV): 2016 results, 2017.

[2] P. Dupeyrat, C. Lucas, A. Lindsay, A. Plotton, K. Radouane, Investigations on Albedo Dependency of Bifacial PV Yield, EU PVSEC, 2014.

[3] U. A. Yusufoglu, T. Pletzer, L. J. Koduvelikulathu, C. Comparotto, Analysis of the Annual Performance of Bifacial Modules and Optimization Methods, IEEE Journal of Photovoltaics, vol. 5, pp. 320 - 328 , 2015.

[4] A. Mermoud, B. Wittmer, PVsysts new framework to simulate bifacial systems, PVPMPC Workshop, 2016.

[5] B. Marion, S. MacAlpine, C. Deline, A. Asgharzadeh, F. Toor, D. Riley, J. S. C. Hansen, A Practical Irradiance Model for Bifacial PV Modules, IEEE 44th Photovoltaic Specialists Conference , 2017.

[6] I. Shoukry, J. Libal, R. Kopecek, E. Wefringhaus, J. Werner, Modelling of Bifacial Gain for Stand-alone and in-field Installed Bifacial PV Modules, Energy Procedia, vol. 92, pp. 600-608, 2016.

[7] A. Asgharzadeh, T. Lubenow, J. Sink, B. Marion, C. Deline, C. Hansen, J. Stein, F. Toor, Analysis of the impact of installation parameters and system size on bifacial gain and energy yield of PV Systems, IEEE PVSC, 2017.

[8] Gendaylit, Radiance-online, Online available: http://www.radiance-online.org/learning/documentation/manual-pages/pdfs/gendaylit.pdf.

[9] Ch. Reise, A. Schmidt, Realistic yield expectations for bifacial PV systems – an assessment of announced, predicted and observed benefits, EU PVSEC 2015.

[10] M. Chiodetti, A. Lindsay, Enhancing Bifacial Modelling with Ray-Tracing, PVPMC Workhop, 2016.

[11] M. Richter, Reducing uncertainty in PV yield assessment, PV-TECH, pp. 50-56, 2018.

[12] J. Freeman, J. Whitmore, B. Nate, A. P. Dobos, Validation of Multiple Tools for Flat Plate Photovoltaic Modeling Against Measured Data, 2014.

MITIGATING SNOW ON ROOFTOP PV SYSTEMS FOR HIGHER ENERGY YIELD AND SAFER ROOFS

Bjørn B. Aarseth[1,2,b], Mari B. Øgaard[1,2,b*], Junjie Zhu[2], Tommy Strömberg[3], John A. Tsanakas[2,a], Josefine H. Selj[1,2] and Erik S. Marstein[1,2]
[1]Department of Technology Systems, University of Oslo
Gunnar Randers vei 19, 2007 Kjeller, Norway
[2]Department of Renewable Energy Systems, Institute for Energy Technology
Instituttveien 18, 2007 Kjeller, Norway
[a]Now employed at IMEC
[3]Innos AS, Karoline Kristiansens vei 1H, 0603 Etterstad, Norway
*Corresponding author. E-mail: mari.ogaard@ife.no, Tel.: +47 976 356 08
[b]Both authors contributed equally to this work.

This study investigates the potential and technical limitations of applying a controlled forward bias on photovoltaic (PV) modules for snow mitigation on building-applied photovoltaic (BAPV) systems installed in regions with substantial snowfalls. Mitigating snow on BAPV systems has a twofold purpose: to reduce the mechanical stress on the PV modules and the building structure itself, as well as to limit shading and thus increase the overall PV energy yield in the winter months. The technology is tested in both a climate chamber and under field conditions in an operating, commercial rooftop PV system covered with snow in Oslo, Norway. Payback time calculations for the snow mitigation is done both in terms of energy and economy, with the latter taking the national power tax into account.

Both in the climate chamber and initial full-scale rooftop PV system tests, the system was successfully demonstrated. Energy calculations show that there is large potential for optimization of the snow mitigation process to boost the PV system's energy yield in the spring. Hence, active snow mitigation can be a beneficial solution not only for necessary snow mitigation, but also for boosting the yield of PV systems.

Keywords: PV System, Rooftop, Safety, Snow

1 INTRODUCTION

In Norway, like in the rest of the world, the installed photovoltaic (PV) capacity is growing at an impressive rate. According to the International Energy Agency (IEA), 11.4 MW_p was installed in 2016, corresponding to a 366 % increase compared to the 2015 figures, giving a total accumulated PV capacity of 26.7 MW_p [1]. The largest part (7.4 MW_p) of the installed capacity in 2016 is made up of PV systems installed on roofs and facades of commercial/industrial buildings while the rest refers to residential and off-grid PV systems.

With regard to the Norwegian climate, there are several challenges that PV asset owners and operators are faced with. Snow during the winter is one of these. This is especially an important concern for PV systems installed on flat roofed commercial buildings, since many of these are under-dimensioned for carrying large snow loads during winter. Normally, manual snow removal is required to reduce the load, but when a PV system is installed on the roof, the process becomes more complicated and expensive. Rooftop PV systems in Norway are often installed with low tilt angles (in an east-west configuration), which will reduce the accessibility to large parts of the roof. Additionally, manual shoveling increases the risk of damaging the PV modules. Manual mechanic snow removal will be expensive, and is not recommended for PV systems [2]. In addition to being a risk for the building structure itself, large snow loads can also lead to physical failures and performance losses in PV modules [3].

To solve these challenges the Norwegian supplier Innos [4] has introduced a snow mitigation system, WeightWatcher, which uses the heat generated by forward-biasing the PV arrays for melting and removing the snow from the modules' surface. Today, this system is primarily marketed for reducing the snow loads on the roof and the consequent risks for the building structures. However, it is well-known that snow reduces the power production in winter time significantly [5,6], due to reduced transmission and significant shading effects on the affected PV modules or arrays. The latter is known to increase the risk of follow-up hot spot effects. This raises the question of whether it is possible to optimize the snow mitigation process to reduce PV production losses from shading, and thus increase the net PV energy yield in the winter months.

In this study, the forward-biased PV modules' capability of melting snow is assessed. The temperature increase of a forward biased module is measured for different ambient temperatures and applied currents in a climate chamber. The technology is also tested under field conditions in an operating rooftop PV system covered with snow. To investigate the potential of using the system to increase PV generation in addition to load reduction, the energy consumption for heating up the PV modules and adequately melting the snow is calculated. Based on this, the potential of such snow mitigation practice for increasing the system's energy yield is quantitatively evaluated.

2 METHODS

2.1 Climate chamber tests

The technology was tested in a climate chamber set-up, at different ambient temperatures (+25°C, -1°C, -5°C and -10°C) and at different levels of current bias (10 A, 15 A and 19 A), provided by an external power source, to investigate the increase in temperature as a function of both variables. The tested module was a specially designed mini-module (prototype), consisting of 2×3 c-Si solar cells, with seven individual PT100 temperature sensors, to provide the module's thermal response at different positions (Figure 1).

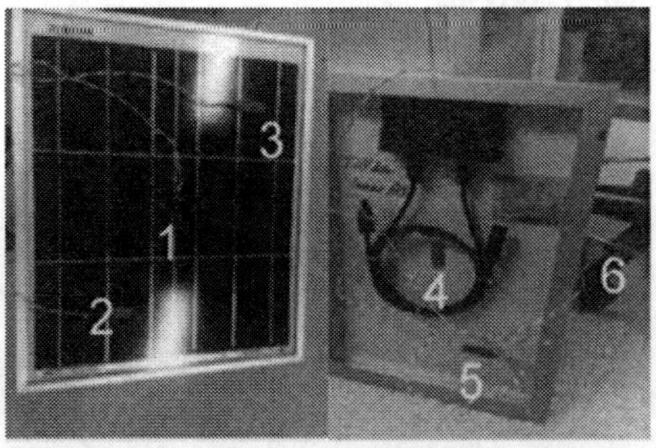

Sensor	Position
1	Between cells, front
2	Middle of cell, front
3	Middle of cell, front
4	Between cells, back
5	Middle of cell, back
6	Frame
7	Ambient

Figure 1: Image of the test module with temperature sensors.

2.2 BAPV snow mitigation system

The WeightWatcher BAPV snow mitigation system was tested at a site in Oslo, Norway, installed at ASKO Norge by INNOS AS and Solenergi FUSen AS. In the WeightWatcher system, the PV modules are heated up by applying forward bias through an external power source. The system consists of 260 W multicrystalline and 300 W monocrystalline silicon modules, and is installed on a flat roof commercial building, where the modules are installed in an east-west configuration with a tilt angle of 10°. The installed capacity of the PV system with WeightWatcher technology is 1 MW_p. The applied current through the modules is 10A, which corresponds to approximately 45V per module depending on the modules' temperature. Heated channels are used to drain off the melting water from the roof. To avoid icing on the roof and on the modules during the melting process, the system is normally used when the ambient temperature is above -5°C.

2.3 Energy calculations

The energy calculations were done using clear sky modelling from the commercially available PVsyst software with the Meteonorm weather database [7] and subsequent analysis of melting energy obtained from field test data. In the PVsyst calculations the modelled PV-system was a miniature of the real system described above, with 20 x 260 W mc-Si modules in total, using an azimuth matching the real system. A clear sky model was used, with additional weather data from the Meteonorm database, using the systems coordinates in the Oslo-area.

To estimate the minimum payback time, the energy and economic calculations use the number of *consecutive clear sky days* (CCSD) needed to break even in terms of energy used and money spent on melting.

2.4 Snow mitigation test under field conditions

A field test of the snow mitigation solution was performed in December 2017, when there was a thin layer of snow on the PV modules (~1cm (1 kg/m²)). The test involved 360 modules, with 20 modules per string. The temperature on the back and front surface of the modules was measured before and after melting. The back-surface temperature was measured with a portable TRI-SEN sensor from TRITEC with an accuracy of ± 3 %, and the front surface temperature was measured using a FLIR E-series IR camera.

An additional run was done in late March 2018, with a larger accumulated snow load, and the snow load sensors in place.

3 RESULTS

3.1 Climate chamber temperature tests

The average increase in temperature in measuring point 2 and 3 for different applied currents at different ambient temperatures is given in Figure 2.

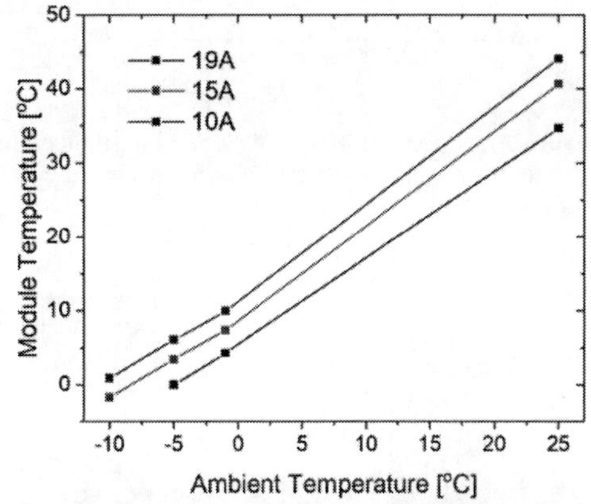

Figure 2: Temperature increase of the test module at different ambient temperatures and at different applied currents.

As expected, forwards bias heats up the module. The module temperature correlates positively with the applied current and increase in ambient temperature. The typical heating process of the module is shown in Figure 3. After a steady increase for approximately 1000 seconds, the module temperature saturates. The cooling of the module happens within the same time frame as the heating.

The temperature increase measured at the positions given in Section 2.1 for the investigated PV module at 10 A and different ambient temperatures, is presented in Figure 4. There are low temperature differences between the front and the back of the module, and in the middle and between cells. The low temperature of the frame might reduce the efficiency of the snow melting, as this might refreeze the melting water at low ambient temperatures.

3.2 Field validation – proof of concept

As shown in Figure 5, during the initial test of forward biasing snow covered modules under field conditions, the temperature of the modules increased, and the snow melted. Before the test, the temperature of the back and front surface was respectively -10.5°C and -10°C. At the end of the test the corresponding measured temperatures were -1°C and 0°C. During the test, the ambient temperature was slightly below what climate chamber tests indicate is necessary for efficient melting, making the process slightly harder than under normal operation. Despite this, most of the snow was removed from the modules within 45 minutes of operation time. Thus, the temperature increase was slightly above what is expected from the climate chamber results for such low ambient temperatures, and is likely due to limited heat transport in the field due to the isolating properties of the snow layer.

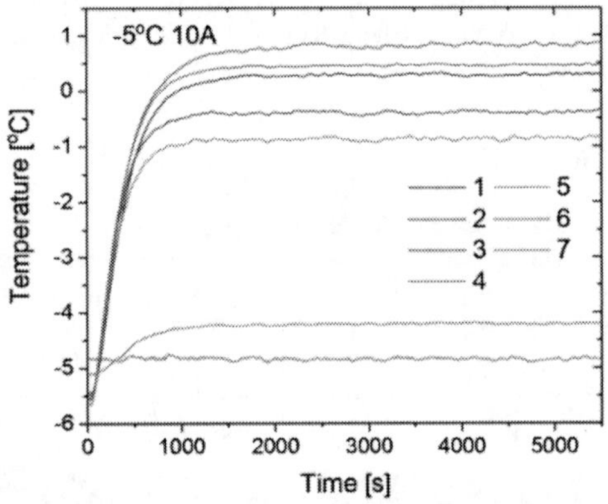

Figure 3: The increase in temperature as a function of time at an ambient temperature of -5°C and an applied current of 10 A.

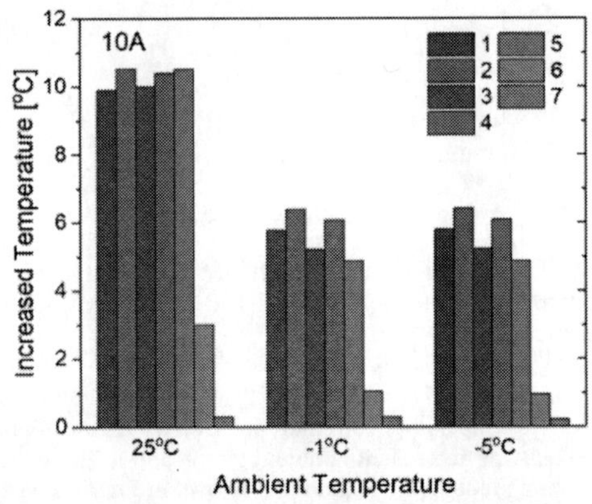

Figure 4: Increased temperature at different ambient temperatures for an applied current of 10 A.

Figure 5: Left: The PV modules in the WeightWatcher system [3] after 30 minutes of forward bias, compared to unbiased modules. Right: IR-photography of forward biased, partly snow covered, modules.

The snow load data recorded by the system before, during and after the melting done in March can be seen in Figure 6. It is compared to snow load estimates from the Norwegian Water Resources and Energy Directorate [8] at the ground level of the site of the system for 2018. The data recording started on February 13th and contains one active melting on March 24th. The measured snow load on the roof is lower than the estimates, which is likely due to an offset in load sensor calibration, and possibly leakage of heat through the roof, causing some of the snow to melt on the hotter days. The active melting makes the roof snow free 25th of March, while the snow remains on the ground until the 15th of April, according to the snow load estimates. This indicates that due to the active melting it was possible for the PV system to produce power three weeks before what would normally be the case. Because the snow load data set only consists of 45 snow days, and system tests have been done on the roof during winter, it is difficult to say how well the snow load estimates from [8] translates to the building roof. It is however, assumed that the estimates are a good approximation of roof snow loads during a normal winter, with no interference from the WeightWatcher system.

The above mentioned melting was a test run of the system, and the roof snow load was not critical. However, should it be critical in months with less sun than March, a melting in e.g. January would still reduce the roof snow load through the spring months, and ultimately result in snow free modules earlier than natural snow melting. Which means that if a melting during mid-winter results in snow free modules earlier in the spring, there would still be a window, like in Figure 6, where it would be possible to return the energy invested in melting – and possibly even produce more energy than invested.

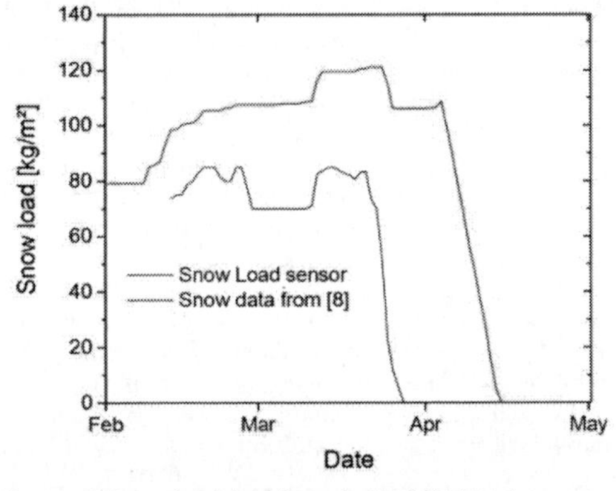

Figure 6: Snow load data from the WeightWatcher system and snow load data from seNorge.no [8] for 2018.

3.3 Energy payback time calculations

The daily clear sky DC energy output per square meter calculated in PVsyst can be seen in Figure 7. Because of the latitude of Oslo, the seasonal variations in production are large, from under 0.025 kWh/(m^2 day) December 22nd, to over 1.1 kWh/(m^2 day) June 1st. From this the *energy payback time* (EPBT) of running the system under different conditions has been calculated below. The metric chosen is *consecutive clear sky days* (CCSD), which, as mentioned above, represents the minimum theoretical payback time. This means that a melting at any given time through the winter and spring will result in a need for a given number of CCSD to return the energy investment. This naturally means that the number of CCSD needed after melting under similar conditions in December and April will result in a larger number for December, due to the low production during a December clear sky day.

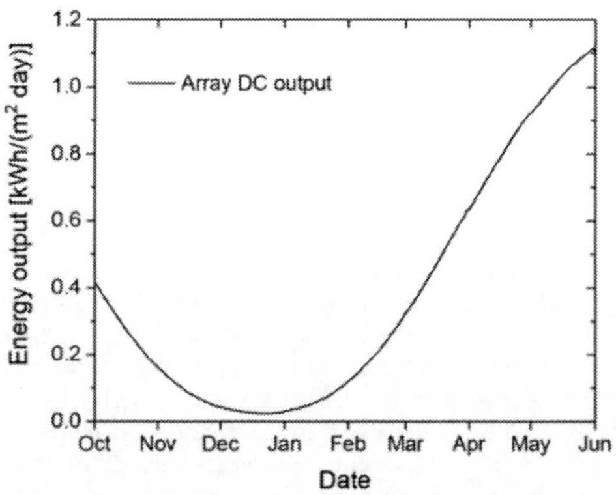

Figure 7: PVsyst clear sky model output through the year.

The number of CCSD needed to return the energy invested in melting 10 kg/m^2 of snow given melting energies of 0.15, 0.1 and 0.05 kWh, a given date from October through May is shown in Figure 8. The melting energy is the amount of energy in kWh needed to melt 1 kg snow. This number is very dependent on conditions when melting. Ambient factors play a large part, where lower temperatures require larger energies like demonstrated by the climate chamber tests above. The same is true for wind, as this will contribute to module cooling. This means that unless the snow load approaches critical values, melting should preferably be done on relatively hot days with no wind, which will also help reduce ice generation on the modules.

Here 0.15, 0.1 and 0.05 kWh/kg is chosen because this is in the range that has been observed in initial melting tests in the field. Since melting energy is directly proportional to the energy needed for melting a given snow load, it largely impacts the CCSD needed to return the energy investment. In the calculations below, 0.1 kWh/kg is used, as this is close to the average observed through the initial testing. More tests should be done under better defined conditions to be able to report more accurate numbers for the melting energy for different conditions.

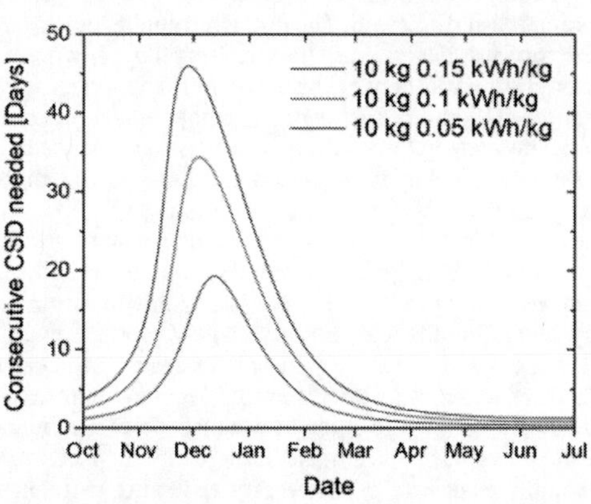

Figure 8: The number of clear sky days with varied melting energy needed per kg.

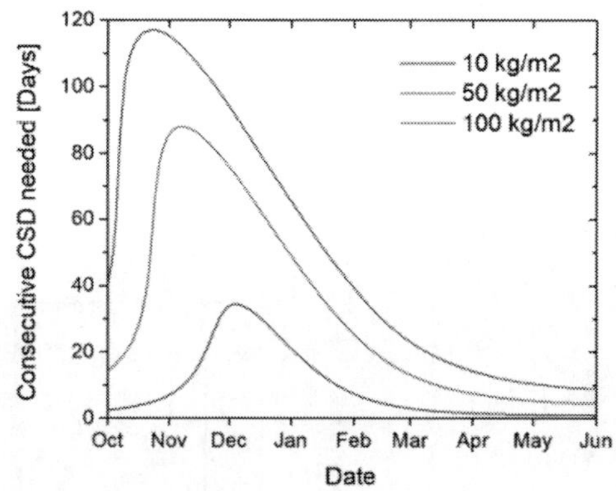

Figure 9: The number of clear sky days needed for melting every day through the winter.

Another important factor in the EPBT calculations is the snow load melted. Figure 9 shows the number of CCSD needed to melt a snow load of 10, 50 and 100 kg/m^2. Naturally, an increase in snow load increases the EPBT and shifts the worst melting date from early December to late October. For 10 and 100 kg/m^2 the number of CCSD needed after melting is just above 30 and 115 days in early December and late October, respectively. However, the 1st of April less than two CCSD are needed to return the energy for 10 kg/m^2. For a larger snow load of 100 kg/m^2 the number is 14 days, which still is a commercially interesting number compared to the test data for the spring of 2018, where the use of the snow mitigation system potentially allowed for increased production for 21 days.

Since the WeightWatcher system first and foremost is a system installed for safety, in order to mitigate snow from under dimensioned roofs before, during or after heavy snowfalls, and keep the roof from collapsing, the need for an active melting event mid-winter is to be expected. This would lead to large CCSD numbers with regards to the EPBT. Potentially, the larger the snow load melted mid-winter, the earlier the roof will be bare compared to the surroundings, and the larger the window for power production in the spring could become. It should also be noted that to have the shortest payback time, weather forecasts for temperature, wind, and snow precipitation should be utilized, to avoid ending up having to melt excessive snow loads in cold or windy weather, and ultimately increase the power investment needed.

3.4 Economic Payback time

The economic payback time of melting snow is a complex calculation and is done here in a somewhat reduced form. The extra installation cost of the power supply for the system is not considered, and neither is the potential savings from not having manual snow removal from the roof. In other words, these calculations only show the potential economic gain due to increased energy generation, given that the system already is installed because of necessary load reduction. In addition, the calculations assume that all the generated power is used in-house, and not exported to the grid. The price for importing power from the grid is much higher than the price received for export, meaning that the payback time could potentially be considerably longer.

The most important factor for economic payback time for melting in Norway is the monthly Power Tariff (PT) cost. This is calculated from the monthly peak power point times the PT rate. The PT rate varies through the year, which is illustrated in Figure 10. Here the y-axis shows the full Power Tariff cost of turning on the melting system, per square meter, while the real PT rate is shown below each step in the graph. The price per square meter is calculated using the system module voltage and current, divided by the module area. As the figure shows the PT rate is at its highest during winter.

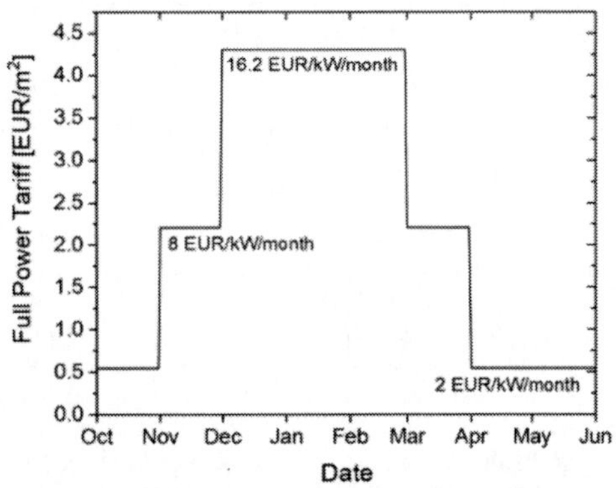

Figure 10: Power Tariff costs through the year. The y-axis represents the full power tariff cost for melting per square meter, while the EUR/kW/month values in the plot represent the base power tariff cost.

How the PT rate affects the cost of melting snow can be seen in Figure 11. It can be seen that the price of just turning on the system is 4.3 and 2.2 EUR/m^2 in Dec-Feb and Mar/Nov, respectively, if the entire power consumption from melting increases the monthly peak load value. Surprisingly the energy cost is close to negligible compared to the initial PT cost, using a power price of 0.063 EUR/kWh (0.6 NOK/kWh). This implies that to cut melting costs it is paramount to have exact control of peak loads and avoid melting when the melting power consumption will increase the monthly peak load.

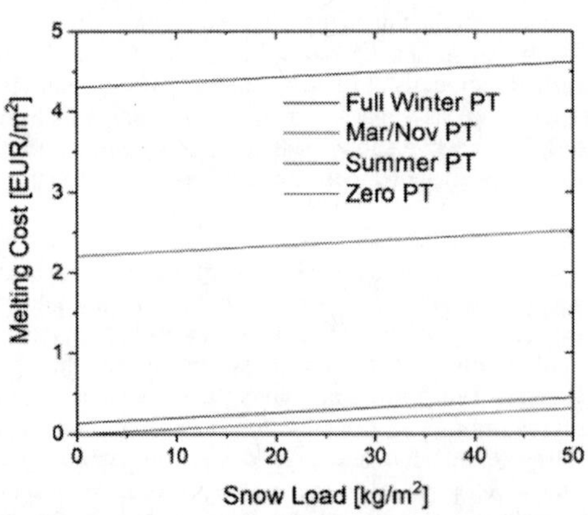

Figure 11: The price of melting snow as a function of snow load, with the different PT rates represented by each line.

Figure 12 shows the number of CCSD needed to pay back the snow mitigation, given the different PT rates though the year. As can be seen, melting done at times where the monthly peak load is increased, increases the CCSD needed. If melting triggers full PT costs during winter (1st Dec-28th Feb), the CCSD needed in early December becomes 170 days. Interestingly, this also gives an CCSD number of over 90 in early March and over 60 in the end of May. This means that the modules would have to become and stay snow free three clear sky months before they naturally would in March. For the Mar/Nov PT rate, the CCSD number is just below 60 in early March, and just above 30 in the end of May, which is better, but still kills a lot of the profitability of both the melting, and the PV system in general. However, if the melting is done without raising the monthly peak load, the CCSD needed becomes the same as in Figure 9, and for 10 kg/m^2 this means 3 days in early March. The short payback time for no PT makes it possible to forecast based on weekly weather forecasts, whether a melting at a given day would be profitable.

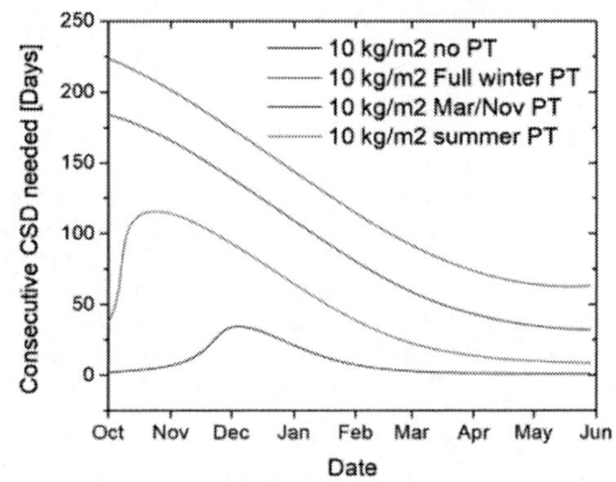

Figure 12: The number of CCSD needed to pay back the money invested in melting snow, given the different PTs through the winter.

4 CONCLUSIONS

In this work a snow mitigation solution and its application potential for rooftop PV systems has been presented and assessed. The discussed application is based on forward-biasing and thus heating up snow-covered modules, until they get snow-free. The concept was validated both in a climatic chamber under controlled ambient temperature and forward-bias (on a mini-module prototype), and under real-field and full-scale conditions, on a rooftop PV system. In addition, calculations on the energy – and economic payback time show that it is possible to return both energy and economic investment of the snow mitigation, and that it is possible to earn money by mitigating the snow to increase PV production, given that the system is already installed because of necessary load reduction. This does, however, require delicate peak load control and utilization of weather forecasts, to hinder that power used to mitigate snow does not increase the monthly peak load of the facility or require more energy than potentially needed.

REFERENCES

[1] Ø. Holm, National Survey Report of PV Power Applications in Norway, IEA PVPS Natl. Surv. (2017) 1–19.

[2] NREL, Sandia, Sunspec Alliance SuNLaMP PV O&M Working Group, Best Practices in Photovoltaic System Operations and Maintenance, 2016. doi:NREL/TP-7A40-67553.
[3] F. Reil, G. Mathiak, K. Strohkendl, S. Raubach, C. Schloth, B. v. Wangenheim, K. Fotak, Experimental Testing of PV Modules under Inhomogeneous Snow Loads, 27th Eur. Photovolt. Sol. Energy Conf. Exhib. (2012) 3414–3417.
[4] Innos, Innos.no, (n.d.). http://www.innos.no/.
[5] B. Marion, R. Schaefer, H. Caine, G. Sanchez, Measured and modeled photovoltaic system energy losses from snow for Colorado and Wisconsin locations, Sol. Energy. 97 (2013) 112–121. doi:10.1016/j.solener.2013.07.029.
[6] B. Stridh, Evaluation of economical benefit of cleaning of soiling and snow in PV plants at three European locations, Conf. Rec. IEEE Photovolt. Spec. Conf. (2012) 1448–1451. doi:10.1109/PVSC.2012.6317869.
[7] A. Mermoud, Pvsyst: Software for the study and simulation of photovoltaic systems, (2012). www.pvsyst.com.
[8] NVE, seNorge, (n.d.). www.senorge.no (accessed September 20, 2018).

LOAD FLOW SIMULATION OF A LOW-VOLTAGE PV-BATTERY BASED DC MICROGRID TO SUPPLY SMALL ISOLATED COMMUNITIES

Pedro Ferreira Torres[1] – pedro.ftorres@itec.ufpa.br
José de Arimatéia Alves Vieira Filho[1]– arimateia.eng@outlook.com
Vilson Lima Chaar Junior[1] – vilsonchaar@gmail.com
Leonam Ferreira de Araújo[1] – leo_into@hotmail.com
Samuel Williamson[2] – sam.williamson@bristol.ac.uk
Marcos André Barros Galhardo[1] – galhardo@ufpa.br
João Tavares Pinho[1] – jtpinho@ufpa.br
Wilson Negrão Macêdo[1] – wnmacedo@ufpa.br
[1]Universidade Federal do Pará, Grupo de Estudos e Desenvolvimento de Alternativas Energéticas
[2]University of Bristol, Electrical Energy Management Group

ABSTRACT: This paper presents the modelling and load flow simulation of a low-voltage distribution DC microgrid which is developed in the test area of the Group of Studies and Development of Energy Alternatives (GEDAE-UFPA). This microgrid is intended to serve as a basis for conducting tests of operation modes, power conditioning devices, control strategies, among other possibilities, as well as applications related to low voltage direct current distribution systems, and to validate mathematical models of related processes. The simulated microgrid structure is composed of three photovoltaic generators of 500 Wp each, a battery bank of 4.8 kWh, three charge controllers forming a 24 V DC bus, and load banks distributed along a grid with a total length of 200 m. The modelling of microgrid components are presented in detail, and the load flow study used is derived from the traditional Newton-Raphson power flow algorithm used in AC systems.
Keywords: Modelling, Rural Electrification, Battery Storage and Control, DC Microgrid

1 INTRODUCTION

The search for electricity generation systems that have less impact on the environment and are independent from fossil fuels is growing in several parts of the world. Renewable energy systems are being used in several applications: grid-connected low-voltage distributed systems, centralized PV plants and wind farms, and off-grid battery-powered applications are examples of the various applications being developed worldwide. In this context, the concept of microgrids for energy distribution aims to create a more flexible distribution system. The high penetration of distributed energy sources, especially those with high variability as is the case of PV, can bring problems to the power quality of the distribution grid [1].

Low-voltage DC power distribution systems (LVDC) can be a more efficient alternative to traditional AC systems [2]. The increase in the share of distributed generation sources (in large part by photovoltaic systems) and the use of more efficient electric loads in direct current applications justify the development of this type of microgrid, since it has a reduced number of stages of conversion, making the system more efficient, reliable, and with lower cost [3]. Several other benefits of LVDC can be listed, such as: reduction of line loading and transmission losses (since there is no reactive power flow), reduction in the complexity of power converters, and reduction in losses due to skin effect.

The use of LVDC to supply energy for isolated communities is a feasible alternative to traditional AC systems, as the higher efficiencies in the conversion processes indirectly represent a reduction in the energy storage capacity, which heavily contributes to the overall costs of such systems. Countries like India and Brazil have high demand for such projects, since a significant number of isolated communities without access to electricity could benefit from such systems [4,5]. In this context, this paper proposes a methodology to evaluate the technical feasibility of isolated PV-battery powered LVDC distribution grids for small communities, where it is possible to simulate the battery bank capacity, PV resource and load curves, and also to evaluate the voltage profile at the grid nodes over a day and for several scenarios.

The simulated DC microgrid has been built in the outdoor test area of the Group of Studies and Development of Energy Alternatives (GEDAE-UFPA) laboratory. It is used to perform several studies and tests on this topic, such as: control and energy management strategies, power converters, DC loads, etc. This microgrid is formed by three distributed generation systems, each comprising a 500 Wp PV generator, a configurable lead-acid battery bank (24 V/132 Ah and 48 V/66 Ah) and an MPPT charge controller [6]. Three independent load banks are distributed along the grid, with a maximum power of 200 W, and with configurable load profile (load steps of 40 W), so that a typical load curve for a small community household can be evaluated. To form the grid, 35 mm² aluminum multiplexed cables are used, together with 3 poles installed in the laboratory test area. The grid layout is presented in Fig. 1. It is important to highlight that this paper deals only with the simulation results of the proposed system, and real operating results will be considered in a future study.

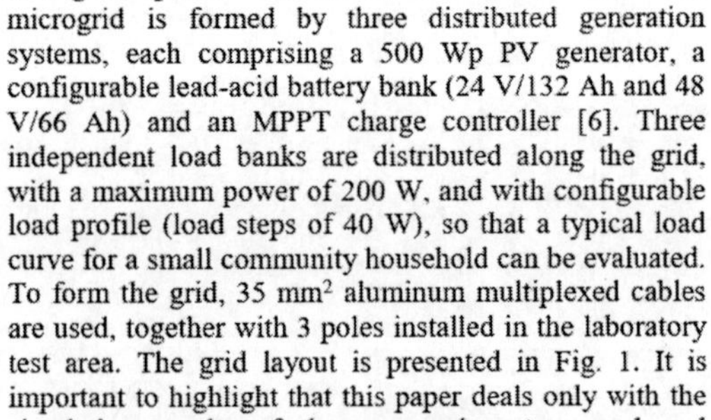

Figure 1: DC Microgrid layout used for simulations.

The following sections describe the modeling of the microgrid components and the algorithm used to evaluate the power flow. Next, simulation results are presented, considering both real operating load curves and real data of solar resource availability. A discussion of the results is carried out as well as perspectives for future works with the developed microgrid and on the simulation tool.

2 MICROGRID COMPONENTS´ MODELS

The microgrid can be divided into three components to be modeled: the distribution cabling, the generation system, and the load banks. The following items present in detail de model used for each component.

2.1 Grid Cabling

Since no reactive power flow is considered in DC distribution systems (only active power is transmitted), inductive and capacitive effects associated to distribution lines are not present. Hence, only the resistive effects are considered, and the distribution lines are modelled as a single fixed value of resistance.

The resistance value, R, of a conducting line between two grid nodes, n_a and n_b, is obtained from the conductor's electrical resistivity, ρ, the total length, l, and the cross-sectional area, A, as shown in Equation 1.

$$R_{a,b} = \rho . l / A \qquad (1)$$

The electrical resistivity is a function of temperature and is usually given considering 20 °C operating temperature, along with a temperature correction factor used to evaluate it for different temperatures. Hence, the distribution line resistance will vary over a day due to temperature changes. However, to simplify calculations, a fixed value for operating temperature of 40 °C was used, which can be considered a worst case scenario. For the aluminum conductors at the considered temperature, $\rho = 3.047 \times 10^{-8}$ Ω.m. The total length, l, considers twice the distance between the nodes, since there is both a positive and a negative conductor.

2.2 Generation System

Fig. 2 presents the connections on each generation system.

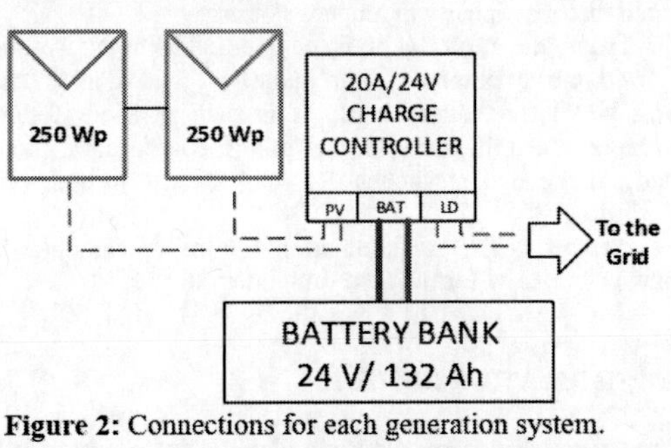

Figure 2: Connections for each generation system.

The charge controller has the function of regulating the charge and discharge of the battery bank, and has maximum power point tracking (MPPT) capability, that ensures the PV generator delivers its maximum available power at a specific operating condition. The charge controller is used to prevent deep discharge of the battery bank, as well as stop any potential overcharging from the PV generation. The output voltage, identified in Fig. 2 as the "LD" terminal, is the same as the battery bank.

The charge controller operates according to two hysteresis curves, shown in Fig 3.

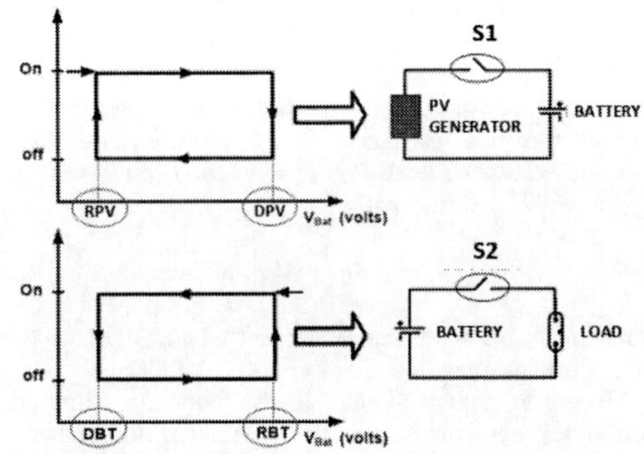

Figure 3: Hysteresis curves for the charge controller (Adapted from [7]).

In the first plot one can identify the regulation to prevent overcharge of the battery bank. The voltage identified as DPV in Fig. 3 is the *Disconnection point of the PV generator*, and when the battery voltage reaches DPV, the charge controller cuts out the PV, as represented by the on/off switch, S1. When the battery starts discharging, the PV generator will be reconnected as battery voltage reaches RPV (*Reconnection point of the PV generator*).

The second plot presents the discharging regulation. The voltage identified as DBT is the *Disconnection point of the battery*, and when the battery voltage reaches DBT, the charge controller cuts out the load (which is connected to the microgrid bus), to prevent deep discharge. This disconnection is represented by the on/off switch, S2, in the figure. When the battery starts recharging, the load is reconnected as battery voltage reaches RBT (*Reconnection point of the battery*).

To model the battery bank, a simple Thevenin equivalent circuit was considered. In this model an ideal voltage source is connected in series with a resistance, R_{BAT}, which represents charge and discharge power losses. The battery open-circuit voltage, V_{OC}, is mainly a function of its state-of-charge, SoC [8,9]. For this simulation, a curve $V_{OC} = f(SoC)$ was obtained from the manufacturer's datasheet and embedded in the model.

For the PV generation, real operating data available from PV systems installed in the laboratory was used. The model only needs the output power from the PV generator, since the MPPT algorithm of the converter works for a broad range of voltages (up to 100 V) so that the voltage of the generator is not considered in the calculations. Fig. 4 shows the model used for the generation system.

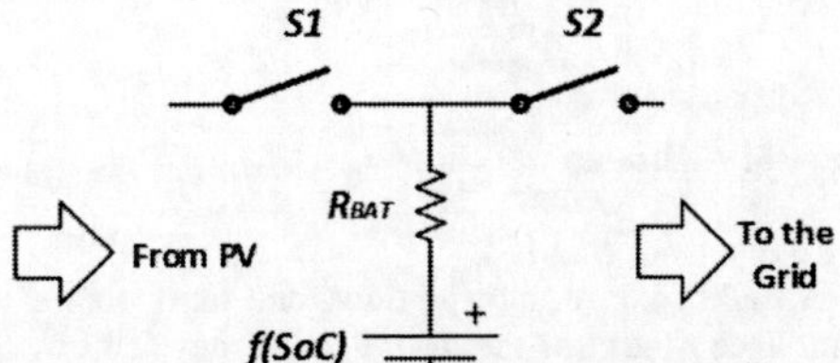

Figure 4: Model used for the generation system.

2.3 Load Banks

Each load bank of the DC microgrid present in GEDAE's laboratory is currently composed of five resistive lamps rated at 24 V/ 40 W. These lamps can be switched independently, so that each load bank can demand powers of 40, 80, 120, 160 and 200 W. These are considered constant resistive loads, since they can be modelled as pure resistances. The power demanded by a constant resistance load, P_{LD}, is a function of its voltage, V_{LD}, as expressed in equation 2.

$$P_{LD} = V_{LD}^2/R \tag{2}$$

where R is the load resistance. Each lamp used in the microgrid has a resistance of $24^2/40 = 14.4\ \Omega$.

However, most of the loads found in common household appliances can be considered as constant power loads. This kind of load demands the rated power independently of the voltage at its terminals (provided that the voltage is within the range specified by the manufacturer).

For the power flow studies presented in this paper, only constant power loads were considered. However, constant resistance loads can be easily implemented in the simulation by using equation 2.

3 SIMULATION METHODOLOGY

The simulation algorithm can be divided into three main sections: setup of initial conditions and grid impedance matrix; Newton-Raphson iterative process to evaluate the voltage at each grid node; calculation of the charge at each battery bank. The simulation performs successive power flows with a specified time step, so that a voltage profile over a day can be evaluated. The algorithm flow diagram is shown in Fig. 5.

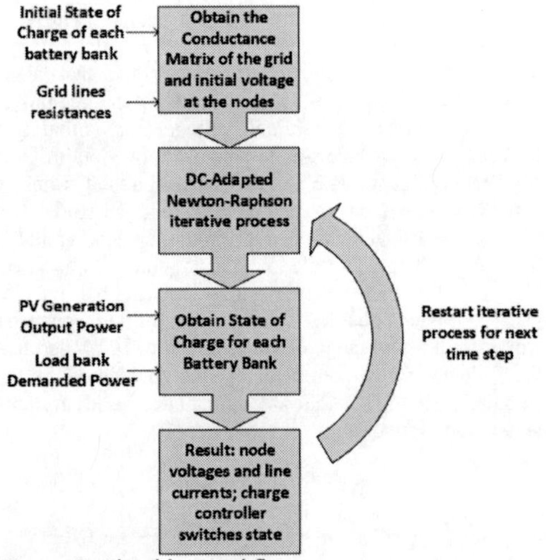

Figure 5: Algorithm workflow.

3.1 Conductance Matrix

In traditional AC power flow, one must obtain the Admittance Matrix of the grid. For DC power flow, the impedance is simplified to resistances (grid cabling is modelled as resistances), and hence, instead of an admittance matrix, Y, a conductance matrix, G, is found [10,11]. Matrix G is formed in a similar manner as the Y matrix. Equation 3 summarizes the calculation process for the elements in matrix G, where k is the number of nodes of the grid, and $R_{i,j}$ is the resistance of the line between grid nodes i and j.

$$G_{i,j} = \begin{cases} -1/R_{i,j}\,, i \neq j \\ \sum_{n=1,n\neq i}^{k} 1/R_{i,n}\,, i = j \end{cases} \tag{3}$$

3.2 Newton-Raphson iterative process for DC grids

For DC systems the power flowing at a given bus, i, is expressed in equation 4.

$$P_i = \sum_{j=1}^{k} |V_i|.|V_j|.|G_{i,j}| \tag{4}$$

In matrix form, the Newton-Raphson algorithm for a DC system can be expressed as in equation 5, where $\Delta P_i^{(k)}$ is the residual power at a given bus, i, after an nth iteration, J is the Jacobian matrix, and $\Delta V_i^{(n)}$ is the change in voltage.

$$[\Delta P_i^{(n)}]=[J].[\Delta V_i^{(n)}] \tag{5}$$

The Jacobian matrix is formed as expressed in equation 6.

$$J_{i,j} = \begin{cases} V_j.G_{i,j}, i \neq j \\ \sum_{j\neq i}^{k} V_i.V_j.G_{i,j} + 2.V_iG_{i,j}, i = j \end{cases} \tag{6}$$

3.3 State of Charge for the battery banks

Once the voltages at each node and the current flow at each line are computed, it its time to calculate the SoC (%) of each battery bank. This is calculated as expressed in equations 7 and 8.

$$SoC_{i,t} = SoC_{i,t-1} + (P_{FV,i,t} - I_{i,t}.V_{i,t}).\Delta t/ V_{i,t} \tag{7}$$

$$SoC_{i,t}(\%) = SoC_{i,t}/C_{BAT} \tag{8}$$

where $SoC_{i,t}$ is the state-of-charge in ampere-hours of a load bank in node i for a given discrete time t, $P_{FV,i,t}$ is the output from the PV generator at the same node and time, $I_{i,t}$ and $V_{i,t}$ are the current and the voltage at node i, Δt is the time step considered in the simulation, and C_{BAT} is the rated battery capacity in ampere-hours.

From the SoC (%) it is now possible to apply the $f(SoC)$ curve obtained from datasheet and derive the voltage of the battery bank. It is then verified if this voltage is within the setpoints of the charge controller, and, if needed, switches S1 or S2 are turned off accordingly.

At this point, the simulation returns to calculate a new power flow for the next time interval.

4 SIMULATION RESULTS

Successive power flow simulations were performed for a day, considering intervals of five minutes. Three buses were considered reference buses for the Newton-Raphson iterative process: N1, N5, and N12 in Fig. 1, which correspond to nodes where the generation systems are connected. The voltage reference adopted for these buses was calculated from the battery model, given an initial known state-of-charge for each battery bank (for the simulation presented in this paper, the initial state of charge for all battery banks is 80 %).

The resistances of the lines for the grid layout presented in Fig. 1 are shown in Table I. Fig. 6 presents the data for PV generation and load profile (both given in watts), for a typical day and with intervals of five minutes for successive simulations of power flow. From Fig. 6 it is possible to see that for this day the maximum output power from PV is close to 350 W at 12 noon, even though the PV generators are rated at 500 Wp. Also, from the generation curve it can be seen that it was a partly cloudy day, with sudden quick drops in the PV generation. From the load point of view, the sudden, repetitive growth that takes about one hour, represents the operation of a freezer, that works at rated power approximately eight hours a day. The load curve presented in Fig. 6 was created considering typical appliances of a household in an isolated community, and a typical use over time of those appliances.

Table I: Resistance of the microgrid lines.

Line (From-To)	2 x Length (m)	Resistance (Ω)
1-2	36	0.031265
2-3	24	0.020843
2-8	24	0.020843
3-4	26	0.022580
3-6	60	0.052109
4-5	78	0.067741
4-10	26	0.022580
8-7	12	0.010422
8-9	20	0.017370
10-11	52	0.045161
9-12	32	0.027791
9-10	20	0.017370

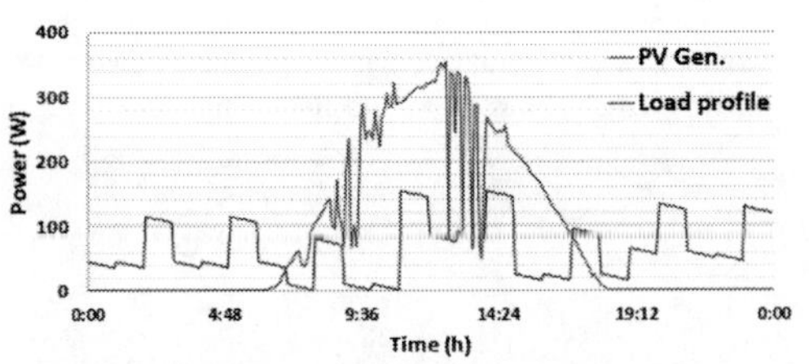

Figure 6: PV generation and load profile for a day with five minutes of data interval.

Fig. 7 shows the voltages at the three load bank nodes, with the results obtained from the power flow algorithm. Two situations were evaluated here: the first, for the implemented 24 V system, and the second for a prospective 48 V bus. Plots in Fig. 7 are given in per unit system (p.u.) with respect to both 24 V, for 24 V bus, and 48 V, for 48 V bus simulations.

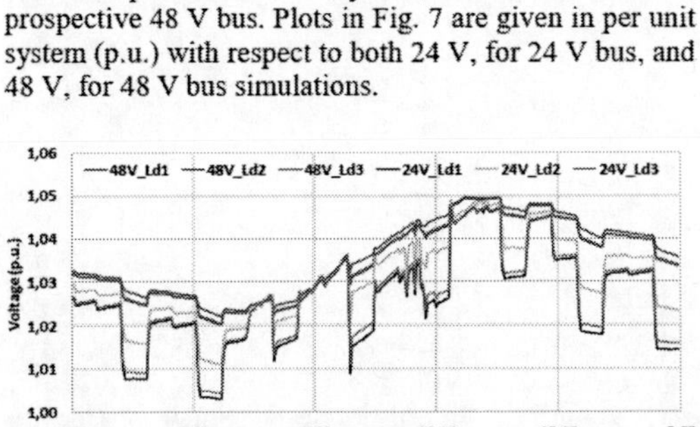

Figure 7: Voltage profile at load buses for 24 V and 48 V.

From Fig. 7 it can be observed that the lower voltages, i.e. higher voltage drops, occur in the 24 V system, which are due to the higher currents circulating at this voltage level. Lower voltages are found in the beginning of the day (around 05h00), when PV generation is not present, the battery bank is partially discharged, and the loads are operating normally (some lamps, freezer, etc.). From an operational point of view, the system will normally supply the loads, since the voltages are always above 1.00 p.u., and thus above rated voltage. The maximum voltage found is around 1.05 p.u., i.e. 5 % above rated voltage, in the afternoon, when PV generation is at its maximum and the battery bank is fully charged.

Fig. 8 shows the SoC (%) for all battery banks. Again, in 48 V system batteries' SoC is slightly higher that in 24 V. This is explained by the fact that power losses at 24 V are higher, and so more charge is demanded from the batteries. The SoC has a profile similar to the bus voltages, since the battery SoC is related to its output voltage. Fig. 8 shows that around 15h30 the battery banks are fully charged by the PV generation, and start discharging around 17h00, when PV power is decreasing. For the simulated situation, batteries end the day at a higher SoC (around 87 – 90 %) than they started (80 %), but these values tend to oscillate according to PV generation and load profile for a given day.

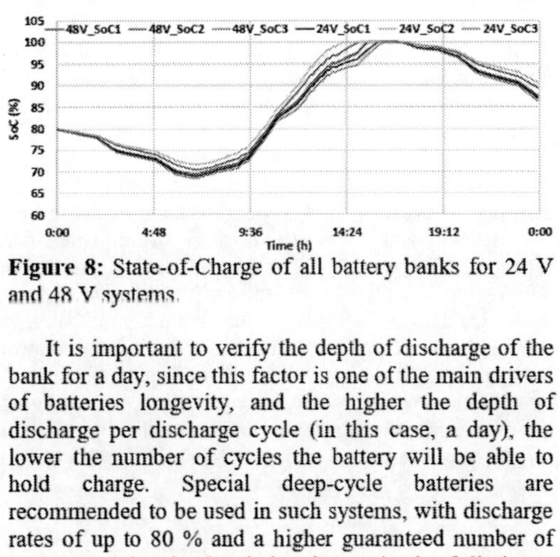

Figure 8: State-of-Charge of all battery banks for 24 V and 48 V systems.

It is important to verify the depth of discharge of the bank for a day, since this factor is one of the main drivers of batteries longevity, and the higher the depth of discharge per discharge cycle (in this case, a day), the lower the number of cycles the battery will be able to hold charge. Special deep-cycle batteries are recommended to be used in such systems, with discharge rates of up to 80 % and a higher guaranteed number of cycles. For the simulated situation, a depth of discharge of 30 % is a common value used for stationary general-purpose batteries.

The power distribution efficiency is a key factor to be evaluated prior to the deployment of a system. System designer must take into consideration the costs associated when selecting the cross section of the distribution cables, bus voltage, etc. Fig. 9 presents the power losses over a day, for both 24 V and 48 V systems. Losses at 24 V are higher and reach up to about 7 % of the total loading of the system. From the data presented in Fig. 9, a detailed economic assessment must be done to evaluate the more feasible configuration, considering both technical and economic perspectives.

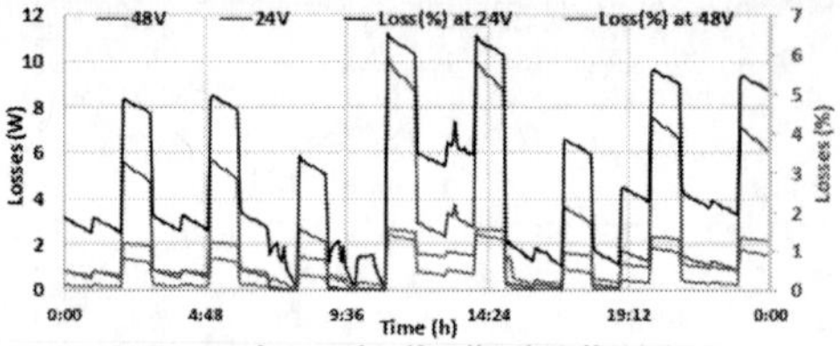

Figure 9: Power losses in distribution lines.

5 CONCLUSIONS

DC distribution microgrids are already being used in specific applications, such as servers and ships, and in most cases present higher efficiency than the traditional AC grids, since they have a reduced number of conversion stages, and their power lines transmit only active power. The increasing number of applications involving distributed generation with renewable energy sources has made DC microgrids gain attention, since efficiency gains are more evident in systems with distributed energy sources and can reduce the complexity of the system and global costs.

This paper has developed a load flow simulation based upon a distributed generation DC microgrid, with PV generators, battery storage and resistive loads. Using models constructed for each component and Newton-Raphson techniques, the system operation for varying PV generation and load has been demonstrated. Two distribution voltages were evaluated, 24 V and 48 V, with the 48 V system showing a 6-times reduction in loss during times of high load, when compared to a 24 V system.

Simulation tools such as presented in this paper are an important way to visualize system performance for several scenarios, helping system designer in the project and development stage. Further work will compare real operating data from the DC microgrid already installed in GEDAE´s laboratory, and improve the simulation tools from this data. Also, other simulations are under way, to evaluate system stability and other operating and control strategies. Also, other works on this simulation tool will include improvements in battery modeling, so that aging processes can be considered during simulations, as well as the implementation of demand-side management and droop control voltage regulation.

REFERENCES

[1] Jenkins, N., Allan, R., Crossley, P., Kirschen, D., Strbac, G. *Embedded Generation*. IET Power and Energy Series, 2008.

[2] Kumar, D., Zare, F., Ghosh, A. "DC Microgrid Technology: System Architectures, AC Grid Interfaces, Grounding Schemes, Power Quality, Communication Networks, Applications and Standardizations Aspects", *IEEE Access*, June 2017.

[3] Elsayed, A., Mohammed, A. A., Mohammed, O. A. "DC Microgrids and Distribution systems: An overview", *Electric Power System Research*, no. 119. 2015.

[4] Nascimento, R. L., "Energia Solar no Brasil: Situação e Perspectivas", *Consultoria Legislativa*. Tech. Report, March 2017.

[5] Jhunjhunwala, A., Lolla, A., Kaur, P. "Solar-DC Microgrid for Indian Homes: A transforming power scenario", *IEEE Eletrification Magazine*, June 2016.

[6] Torres, P. F.; Vieira Filho, J. A. A.; Williamson, S. J.; Chaar Junior, V. L.; Pinho, J. T.; Galhardo, M. A. B.; Macêdo, W. N.; "Concepção de Estrutura Laboratorial para Realização de Estudos em Microrrede em Corrente Contínua de Baixa Tensão", In. Proc. VII Congresso Brasileiro de Energia Solar, Gramado, Brazil. 2018.

[7] Oliveira, L.G. M, "Estratégias de Controle de Carga e Descarga em Sistemas Fotovoltaicos Domiciliares", M.Sc. dissertation, Universidade de São Paulo, Brazil, 2005.

[8] Coleman, M; Lee, C. K.; Zhu, C.; Hurley, W. G.; "State-of-Charge Determination from EMF Voltage Estimation: Using Impedance, Terminal Voltage, and Current for Lead-Acid and Lithium-Ion Batteries", *IEEE Transactions on Industrial Electronics*. Vol. 54, No. 5. October 2007.

[9] Rana, M. J.; Abido, M. A.; "Energy management in DC microgrid with energy storage and model predictive controlled AC–DC converter", *IET Generation, Transmission & Distribution, Special Issue: Smart Grid Voltage Control*. Vol. 11, Iss. 15, pp. 3694-3702. 2017.

[10] Jayaranthna, C.; Binduhewa, P.; Ekanayake, J.; Wu, J. "Load Flow Analysis of Low Voltage dc Networks with Photovoltaic", In. Proc. 9th International Conference on Industrial and Information Systems, ICIIS 2014, art. no. 7036566, 2015.

[11] Farooq, R.; Mateen, L.; Ahmad, M.; Akbar, S. Q.; Khan, H. A.; Zaffar, N. A.; "Smart DC Microgrids: Modeling and Power Flow analysis of a DC Microgrid for off-grid and weak-grid connected communities", In. Proc. (2014) Asia-Pacific Power and Energy Engineering Conference, APPEEC, 2015-March (March), art. no. 7066139.

OPEN SOURCE TOOL FOR A BETTER DESIGN OF BIPV+ BATTERY SYSTEM: AN APPLIED EXAMPLE

Marco Lovati*[1,2], Jennifer Adami[2], David Moser[2]
[1]Dipartimento di Ingegneria Civile, Ambientale e Meccanica (DICAM) –Università di Trento Via Mesiano 77, 38123 Trento, 0461-282621
[2]Institute for Renewable Energy, eurac, 39100 Bolzano, Italy, Phone: +39 0471 055667; Fax: +39 047055699;
*E-mail: marco.lovati@eurac.edu

ABSTRACT: In this paper a case study is examined with an innovative open source tool for designing cost effective BIPV+ battery systems. The aim is to bring awareness of the economic profitability and visual impact of a BIPV system to the designer in the early stage of the architectural design. The main factors influencing the output are the sunlight availability, its variation along the HOY (Hour Of the Year), the electric demand, the cost of different parts of the system and the cost of electricity. The method is applied on a group of three residential high-rise buildings in northern Italy, the system is optimized for maximum NPV (Net Present Value) or for an established IRR (Internal Rate of Return) supposing different prices for the electric storage. The system has been sized using two different demand profiles: one from a well known building standard and a realistic one produced with a stochastic generator and calibrated on a measured aggregated demand profile. Using the realistic curve the coverage of the electric demand of the optimized configuration goes beyond 20%.
Keywords: Building Integrated PV (BIPV), Economic Analysis, System Performance, Storage, building integration, Software

1 INTRODUCTION AND AIM

Photovoltaic energy is seen as one of the main energy sources of the future given the fact that is renewable, abundant and releases low specific CO_2 emissions. Contrary to free field photovoltaics, BIPV (Building Integrated Photovoltaics) do not compete for agricultural land and produces energy in proximity of the demand. Despite its interesting properties and the optimism around it in past forecasts, BIPV is still a niche application within the building industry [3]. Among possible hurdles along its spread is that BIPV is generally more expensive than free field installations, and that it impacts the aesthetic of the buildings. According to numerous relevant studies there are two main axes of improvement: the development of better BIPV technologies [4]–[6] and the improvement in BIPV design methodologies [7], [8]. In this study the use of an advanced design tool for the positioning and dimensioning of BIPV system is presented. The tool, available at [9], has the potential to improve the BIPV design and help the spread of this technology. In this paper for BIPV system is intended the PV modules, the inverters, mounting structures and the electric storage connected to the system.

Relevant concerns when it comes to the installation of photovoltaic system in buildings is the fear of wasting money in a poorly performing technology that does not have a relevant impact on the actual energy consumption of the building. A neglected phenomenon is that the performance of the BIPV system strictly depends on its scale relative to the electric demand of the building and on the positions of the active material on the different facades. The performance of the BIPV system ultimately comes down to the quantity of electricity that it produces during its lifetime and on the fraction of this electricity that is directly used in the building. Energy performance aside, it is also important to ensure that a BIPV system is profitable, or at least that it does not realizes a loss of money. As it is easy to imagine, the economic profitability of the BIPV system is strongly dependent on its scale relative to the electric consumption of the building: in other words, to make a building totally energetically independent is way more expensive than having a fraction of a specific load covered by PV. The economic performance of a BIPV system ultimately depends on the relative capacities of its components amid electricity prices, electric demand, availability and orientation of surfaces, presence of shading bodies and so on. Provided that BIPV is an environmental asset, if the adoption of a BIPV system is economically advantageous, is there a reason not to adopt it? The software POW [9] exists to ensure the economic gain of BIPV systems.

Given the relation between BIPV size and electric demand and the present technological performances and costs: what percentage of the electric load will be covered by the most profitable system? How much the load can be covered without earning nor losing money? Is the electric storage advantageous? Is a façade integrated system advantageous? What is the impact of using standardized load profiles [1] for the dimensioning of the PV system?

The present study does not aspire to answer to these questions in a general way but only to present an example of what can be realistically obtained in a case study.

2 METHOD

2.1 Description of the tool.

The aim of the tool implemented in this study is to design a BIPV system with the primary aim of ensuring profitability. The size of the system is of primary importance because if it is too small it has no impact on the overall electric demand of the building, if it is too big it often exceeds the electric power demand feeding electricity into the grid which is typically poorly valorised. The position of the system is also extremely important as the quantity of the irradiation can vary wildly in terms of magnitude and time of the day on the diverse facades of the building. It is possible to assume that the shape of the production curve of a BIPV system is a function of the relative capacities of the system on the different facades. Size and positions of the BIPV system can therefore become parameters in a problem of optimization as described in [8], the inputs and outputs of this optimization process have been effectively synthetized in ***Figure 1*** taken from [8].

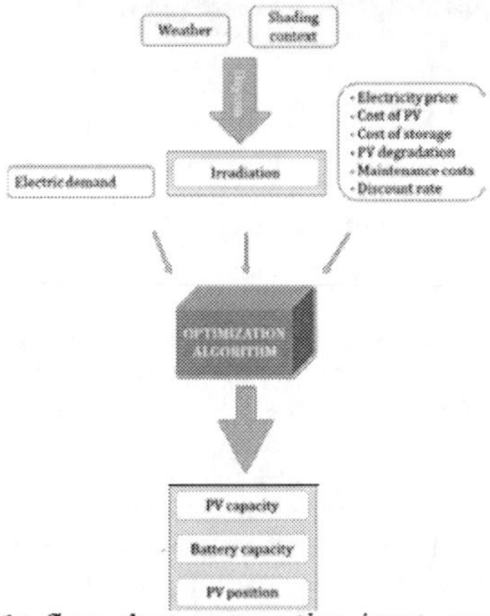

Figure 1: flow chart representing input, process and output of the tool

There are some innovations respect the previous version, in fact in the new version it is possible to optimize the electric storage alongside the PV system, and there is a new fitness function based on ***Equation 1.***

$$\sum_{t=0}^{end\ of\ life} \frac{CF}{(1+i)^t} = CAPEX$$

Equation 1: IRR fitness function. In this fitness function the NPV is forced equal to 0 and the self-consumption is maximized.

Using this fitness function the NPV [10] will not be maximized but assumed equal to 0 while the coverage of the electric demand by the system will be maximized. The discount rate that recover the NPV in the set lifetime is equal to the Internal Rate of Return. .

The optimization algorithm used for this task works similarly to a painter making a painting: assuming that the task is imitating a given image, the solution is not found directly, but each stroke brings the imitation closer to the blueprint. In ***Figure 2*** the algorithm used in POW is applied to the problem of imitating a photography.

Figure 2: application of the optimization algorithm of POW to the task of imitating a picture using discs of color

POW treats the photovoltaic surface similarly to the color on a canvas aimed at reducing the difference between the imitation and the original: the photovoltaic system is applied on roof or façade of a building to maximize the NPV or to maximize the self-production given an IRR (Internal Rate of Return).

2.2 The case study.

The case study presented in this paper is a city district composed by three high rise residential buildings (***Figure 3***) located in Trento city, in the Province of Trento in northern Italy.

Figure 3: 3-Dimensional representation of the buildings in the case study

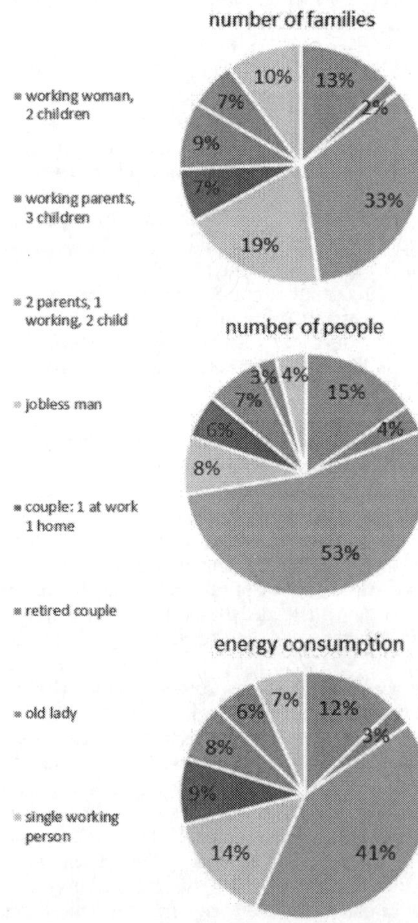

Figure 4: population and energy consumption make-up of the case study by type of family.

Without measured data available, for the sake of a realistic example an electric demand profile was generated with LPG (Load Profile Generator[2]): 8 different type of family were stochastically simulated, and their prevalence was adjusted to fit as closely as possible an aggregated electric demand profile measured in a residential district in northern Italy. The population composition by type of family and their prevalence in cumulative electricity demand is shown in ***Figure 4***.

In the optimization process the following set of inputs was used:

PV module area: **1.44 m²**
PV module efficiency: **0.165**
Performance ratio at STC: **0.85**
Number of years for the NPV: **25**
Price of electricity self consumed: **0.2 €/kWh**
Revenues from the electricity sold: **0.0 €**
Cost of PV modules: **1800.0 €/kWp**
Cost of battery: **908.0 €/kWh**
Linear annual growth of the power demand: **0 - 2 %**
Linear annual growth of the price of electricity for the consumer: **± 2 %**
Linear annual growth of the revenues of electricity for the plant owner: **0 %**
Linear annual growth of discount rate: 0-4 **%**
Annual cost of maintenance: **18 – 36 €/kWp**
Linear annual efficiency losses as a percentage of the initial efficiency: **0.5 - 1 %**

Using the fixed IRR function (***Equation 1***) the stochastic variables must assume a single value to achieve the equality. The mean of the intervals were taken for all the input except the discount rate (which represents the IRR). The IRR was assumed equal to 3.23% a year, this value is equal to the worldwide annual growth of 2016 according to the international monetary fund [11].

Aside from these inputs, as shown in **Figure 1**, the optimization algorithm also needs meteorological data and an electric demand curve. The meteorological file is the one for the city of Trento taken from the Meteonorm database [12]. For the electric demand two different curves were used: one is the realistic demand of the building produced with LPG, the other one is taken from the prEN16798-1 and ISO/FDIS 17772-1 Standards [1].

3 RESULTS

3.1 Comparison of the demand curves.

Using diverse demand curves for the optimization process can lead to different results, in this section the difference in terms of average power between the two curves is shown. The difference between the standard and the realistic curve matters for us as much as it affect the result of the optimization. ***Figure 5*** shows that in this example the curve from the standard suggests a lower electric demand (-26.7%), this would lead the optimization algorithm to choose a smaller capacity compared to the actual optimal one. The standard curve does not possess a seasonal variability, this feature is surely uncommon in real demand curves as the electric consumption for the lights is likely to have more intensity during the winter months. This aspect might praise excessively the annual cumulative irradiation over the irradiation in winter, advantaging the modules with a low tilt (e.g. those integrated in the roof) over the ones with a high tilt (e.g. those integrated in the facade).

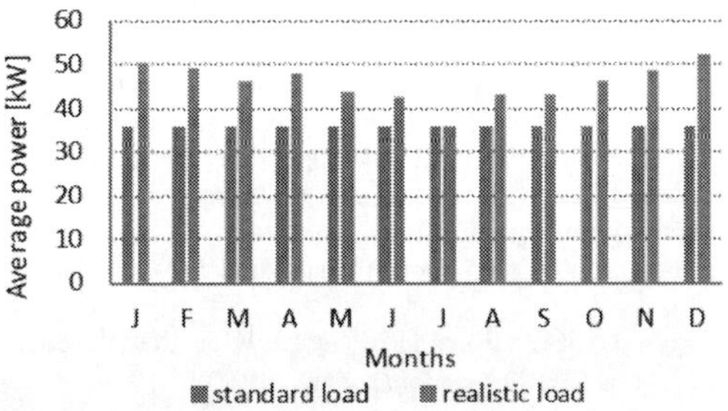

Figure 5: average power demand by month

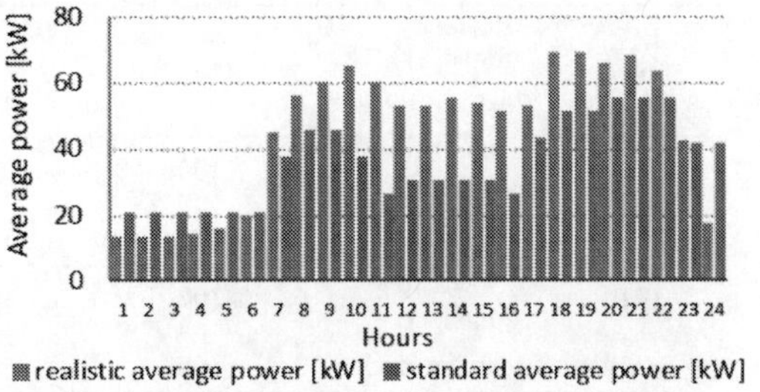

Figure 6: average power demand by hour of the day

Looking at the demand curves along the day is possible to compare the average value of the realistic curve with those of the standard one (in the standard curve every day is the same, therefore the values are not averages but are the actual ones). It is visible how the realistic curve is more constant along the day (the reduction of the demand during the central hours of the day is lower) and both the demand peaks are shifted toward the central hours. In terms of effect on the optimization, the shape of the realistic curve turns out to be way more advantageous for PV: a strongly reduced demand during the central hours of the day (when the solar irradiation is higher) makes it easy for a large system to over-produce electricity sending part of it to the grid and thus reducing the potential savings in terms of energy. A demand curve with a strong reduction of the power during the central hours of the day would push the algorithm into choosing a smaller system, but also to install a larger part of the system in unconventional positions in order to get more energy in the early morning and in the late afternoon. Both these two effect would tend to disappear when the battery price is set very low (more battery capacity is installed).

3.2 Comparison of the optimal systems

The optimization can bring diverse results because it is affected by the demand curve, by the fitness function and by the prices of the components. ***Table 1*** collects the results in terms of capacity and of percentage installed in the southern façade for the two demand curves and the two fitness function. As expected from the analysis of the demand curve, the realistic load curve yields a higher capacity result compared to the standard one. The difference in result due to the shape of the curve is remarkable: being only 26.7% higher in terms of cumulative energy, the realistic demand produces a system that is more than 63% larger both with NPV and IRR. The relative stability of the realistic curve along the central hours of the day (i.e. a lower reduction of the daytime electric demand) allows the optimization algorithm to install a larger capacity without being forced to inject huge shares of the energy produced into the grid for little value. It is important to notice how the gain in capacity due to the load curve shrinks from 63% to 55% with the installation of larger batteries (see IRR row of ***Table 2***). This effect is due to the mitigation effect, provided by the battery, of the strong drop in intensity of the standard curve.

Table 1: optimal installed capacity for the two fitness functions (IRR 3% and maximum NPV)

		PV capacity			
		East (%)	South (%)	West (%)	Total (kWp)
IRR (Batt. cost 908 €/kWh)	Standard load[1]	17	82	1	100
	Realistic load	4	86	10	163
NPV (Batt. cost 908 €/kWh)	Standard load[1]	0	100	0	62
	Realistic load	0	100	0	102
IRR (Batt. cost 250 €/kWh)	Standard load[1]	0	100	0	115
	Realistic load	0	100	0	178

Figure 7 shows the change in optimized PV system from a maximum NPV fitness function to a fixed IRR fitness function, is visible that with both the load profiles the IRR fitness function leads to a higher installed capacity. This is because the maximum NPV obtainable, if transformed to an IRR would lead to a higher rate than the 3.23. In this sense the IRR fitness function is less "greedy" then NPV and more "self-coverage oriented" (i.e. has a tendency to maximize the energy self-consumed accepting lower earnings during the lifetime of the system). The tendency of maximizing self-consumption in the IRR fitness function is also visible in the choice of PV modules in different facades: in ***Table 1*** and ***Figure 7*** is possible to see how with both demand curves ca. 15% of the system is installed outside of the southern façade. It is interesting to notice how these east/west oriented parts of the plant are abandoned in favor of more irradiated southern portions when the electric storage get cheaper. An economic electric storage would allow to cover morning and evening demand peaks using electricity gathered during the central hours of the day.

Figure 7: 3D representation of the optimal PV system for standard load (Left) and realistic load (Right). The modules in red represent those that are installed for a fixed IRR of 3% but not for the maximum NPV fitness function.

Table 2: PV, battery capacity and performance for a 908 €/kWh battery

		PV capacity (kWp)	Battery capacity (kWh)	Self-production (%)
IRR	Standard load[1]	100	0.5	20
	Realistic load	163	0.4	26
NPV	Standard load[1]	62	0	16
	Realistic load	102	0	20

Table 2 and ***Table 3*** show the system components capacities and the self-production (i.e. the coverage of electricity demand) for two different prices for the electric storage. In both the hypothesis, the IRR fitness function causes a larger installation of batteries, although for the 908 €/kWh price the installed capacity is negligible. It is important to notice that the installation of batteries have a positive effect on the PV capacity installed (see also ***Figure 8***), for example the cost of battery equal to 250 €/kWh would cause a 9% and 15% increase in the IRR optimal PV capacity of the standard and the realistic profile respectively. A larger improvement for the standard load profile is not a surprise as the shape of this curve is less favorable for PV and can therefore benefit more by the installation of batteries.

Table 3: PV, battery capacity and performance for a 250 €/kWh battery

		PV capacity (kWp)	Battery capacity (kWh)	Self-production (%)
IRR	Standard load[1]	115	130	27
	Realistic load	178	179	33
NPV	Standard load[1]	65	6	16
	Realistic load	107	16	21

Figure 8: 3D representation of the increase of PV installation due to drop in electric storage price. The optimal system are represented for the standard curve on the left, and for the realistic one on the right.

3.3 Performance of the standard dimensioned system in the realistic demand case

In case of new construction or in case there is not the time to measure the electric demand of the building, a designer would rely on a standard demand curve to proceed in the optimization of the BIPV system. In this study we cannot estimate the average error made by the use of these demand curves as we do not have a large set of examples out of which extrapolate general conclusions. Nevertheless is possible to evaluate what type of economic and energy loss would have happened in this specific case had the designer chosen to apply the standard demand. ***Table 4*** shows the variation in KPIs (Key Performance Indicators) caused by the use of a standard load if the realistic load happens. In other words a perfectly predictive optimization case is compared with the optimization case based on the standard curve available today.

Table 4: variation of KPIs from a standard load optimized system to a realistic load optimized system in a realistic load economic simulation.

	Optimized by standard	Optimized by realistic	Variation [%]
PV capacity [kWp]	62	102	-39
Self-production [%]	15	21	-29
NPV (25 years) [€]	42995	65193	-34

4 CONCLUSIONS

The study presented in this paper gives reason to be optimistic about the future economic performance of BIPV as it shows that a large PV installation on the façade of an high rise building is possible with present prices and technologies. The study shows also that is possible to cover 26% of the cumulative electric demand of the case study without the need for storage and employing economically advantageous solutions.

The demand profile taken from the standard, being slightly lower in magnitude (-27%), led as a result a much smaller system (-63%) compared to the actual optimum. This aspect shows that the shape of the curve along the day, and probably along the seasons, has a strong effect on the size of the optimal BIPV system. This study, being based on a single example, cannot conclude that the standard is too low in magnitude nor give indications on any specific correction that can be done on its shape (e.g. decrease the prevalence of the daytime drop in power demand). Nevertheless it provides a principle to evaluate the demand curves: with growing measured data, the variability of the load profiles will be gradually probed, and the features impacting the optimal solution will be known. It seems therefore feasible to find, at least for what concerns residential buildings, one or more standardized demand curves able to lead to an economically effective system. A set of curves that, though not identical to the real demand of a building, do not differ with this in such a way to cause significant differences in the optimal solution. For what concerns the example presented in this study, the standard curve taken from [1] is significantly affected in terms of performance, it in fact leads to losses about 30% on both lifetime self-production of electricity and NPV of the investment. The optimization has to be run with more examples to assess the real usefulness of the standard curve and ultimately accept it or discard it for our method. Statistical studies on various demand profiles and on their effect over the result of the optimization would lead to propose new ones more fit for the purpose. A higher level of confidence on the input data for this method would lead to more accurate results, therefore to more profitable BIPV system and hopefully a higher penetration of this technology.

5 AKNOWLEDGEMENTS

This work is part of the research activities of the project **EnergyMatching**, funded by the European Union's Horizon 2020 research and innovation programme under **grant agreement No 768766**.

The authors wish to thank Mr. **Mattia Dallapiccola** for the precious work on the tool POW of which he is co-author.

6 REFERENCES

[1] K. Ahmed, A. Akhondzada, J. Kurnitski, and B. Olesen, "Occupancy schedules for energy simulation in new prEN16798-1 and ISO/FDIS 17772-1 standards," *Sustain. Cities Soc.*, vol. 35, pp. 134–144, Nov. 2017.

[2] N. Pflugradt, J. Teuscher, B. Platzer, and W. Schufft, "Analysing low-voltage grids using a behaviour based load profile generator," in *International Conference on Renewable Energies and Power Quality*, 2013, vol. 11, p. 5.

[3] R. J. Yang, "Overcoming technical barriers and risks in the application of building integrated photovoltaics (BIPV): hardware and software strategies," *Autom. Constr.*, vol. 51, pp. 92–102, 2015.

[4] J. Escarré *et al.*, "When PV modules are becoming real building elements: White solar module, a revolution for BIPV," in *2015 IEEE 42nd Photovoltaic Specialist Conference (PVSC)*, 2015, pp. 1–2.

[5] R. Valckenborg, W. van der Wall, W. Folkerts, J. Hensen, and A. de Vries, "ZIGZAG STRUCTURE IN FAÇADE OPTIMIZES PV YIELD WHILE AESTHETICS ARE PRESERVED."

[6] P. Heinstein, C. Ballif, and L.-E. Perret-Aebi, "Building integrated photovoltaics (BIPV): review, potentials, barriers and myths.," *Green*, pp. 125–156, 2013.

[7] H. Kaan and T. Reijenga, "Photovoltaics in an architectural context," *Prog. Photovolt. Res. Appl.*, vol. 12, no. 6, pp. 395–408, 2004.

[8] M. Lovati, R. Albatici, L. Maturi, D. Moser, and others, "EFFECT OF MODULE ORIENTATION AND BATTERIES ON PERFORMANCE OF BUILDING INTEGRATED PHOTOVOLTAIC SYSTEMS.," in *33rd European Photovoltaic Solar Energy Conference and Exhibition*, 2017.

[9] "Photovoltaic Optimization Ware." [Online]. Available: https://github.com/mlovati/POW/releases.
[10] D. Kurt, "Net Present Value - NPV," *Investopedia*, 24-Nov-2003. .
[11] International Monetary Fund, "World economic outlook. Cyclical Upswing, Structural Change.," Washington DC, Apr. 2018.
[12] J. Remund and S. C.Müller, "Solar radiation and uncertainty information of meteonorm 7." .

SIZE OF A BASIC SIMULATION UNIT IN PV SYSTEM PARTIAL SHADING STUDIES

Kari Lappalainen and Seppo Valkealahti
Tampere University of Technology, Laboratory of Electrical Energy Engineering
P.O. Box 692, FI-33101 Tampere, Finland
E-mail: kari.lappalainen@tut.fi, seppo.valkealahti@tut.fi

ABSTRACT: The operation of partially shaded photovoltaic (PV) systems have been studied by many research groups during the past decades; mostly using the well-known one-diode model of a PV cell to model the electrical behaviour of the PV systems. However, the accuracy and correctness of the used simulation tools and approaches have obtained little attention.

This article presents a study of the effect of the size of a basic simulation unit on the accuracy of PV array simulations, implemented using the one-diode model, under partial shading conditions resulting from cloud shadows. The aim was to find out an adequate basic simulation unit size to be used in PV array partial shading studies to minimise the computing time while retaining sufficient simulation accuracy. The study was based on the characteristics of 9097 measured irradiance transitions. The results show that on certain shading situations the used simulation unit size has a notable effect on the obtained results. However, the differences in the produced energy and relative mismatch losses during all the studied irradiance transitions, resulting from the use of a larger basic simulation unit with respect to cell level modelling, were less than 1%. Most of the mismatch losses resulting from cloud shadows take place between PV modules and the mismatch losses between submodules and cells can be regarded as negligible.
Keywords: Modelling, PV system, Shading, Simulation

1 INTRODUCTION

Photovoltaic (PV) systems are prone to irradiance fluctuations, mainly originated from overpassing cloud shadows, which cause fluctuations in PV power production. Partial shading resulting from cloud shadows can also lead to mismatch power losses and to the occurrence of multiple maximum power points (MPPs) leading to failures in MPP tracking thereby causing extra losses. The mismatch losses of a PV system are the difference between the combined maximum power of individual PV cells or modules of the system, as if they were operating independently, and the global MPP power of the whole system. The average length of irradiance transitions on cloud shadow edges has been found to be in the region of 150 m [1] indicating that they may induce problems on the operation of PV systems of all sizes.

During the past years, several studies have been presented regarding the operation of partially shaded PV systems, e.g. [2]–[7]. These studies have mainly been implemented using the one-diode PV cell model. However, the accuracy and correctness of the used simulation tools and approaches have obtained only little attention. In [8], an LTspice simulation model, based on the one-diode model, has been presented and validated. A comparison between the one and two-diode models of a PV cell under partial shading conditions, carried out using MATLAB Simulink, has been presented in [9]. In [10], a study on the effects of the length of a simulation time step, the resolutions of *I–U* curves and irradiance and the size of a basic simulation unit on the accuracy of the simulations of PV system operation under partial shading conditions has been presented. It has been found that the irradiance accuracy of 1.0 W/m^2 and the number of *I–U* curve points of 1000 are sufficient for PV string and PV array level partial shading studies. In that study, a submodule and a PV module were used as the basic simulation unit.

In this paper, the effect of the size of a basic simulation unit on the accuracy of PV array simulations under partial shading conditions caused by cloud shadows was studied more comprehensively than has been done earlier. The main aim of the study was to find out an adequate size for the basic simulation unit to be used in PV array level partial shading studies to minimise the computing time while retaining sufficient accuracy of the simulations. Moreover, mismatch losses between PV modules, submodules and cells were studied. Typically, only mismatch losses existing between PV modules have been studied. The study was based on the characteristics of 9097 irradiance transitions caused by cloud shadows identified in measured irradiance data and was performed using the one-diode PV cell model. The studied PV system was an array of two parallel strings of 20 series-connected PV modules.

2 METHODS AND DATA

2.1 Simulation model for the PV modules

In this study, the electrical behaviour of PV modules was modelled by an experimentally verified MATLAB Simulink model, based on the model presented by Villalva et al. [11]. The model is based upon the widely used one-diode model of a PV cell that provides the following relationship between the current I and the voltage U of the cell:

$$I = I_{\text{ph}} - I_0\left(e^{\frac{U+R_s I}{A U_T}} - 1\right) - \frac{U + R_s I}{R_{\text{sh}}}, \quad (1)$$

where A is the ideality factor, I_0 the dark saturation current, I_{ph} the light-generated current, R_s the series resistance, R_{sh} the shunt resistance and U_T the thermal voltage of the PV cell. In order to obtain the simulation model for a PV module or submodule, the parameter values used in the model of a PV cell were scaled by the number of PV cells in the module or submodule represented by N_s. The thermal voltage of the PV module or submodule can be expressed as $U_T = N_s kT/q$, where the temperature of the cells is represented by T, k is the Boltzmann constant and q is the elementary charge.

Bypass diodes of the PV module were modelled using Eq. (1) by assuming I_{ph} to be zero and R_{sh} infinite. The bypass diodes were assumed to be at the same temperature as the module. The characteristics of the simulation model were fitted to the characteristics of the PV modules (NAPS NP190GKg) used in the solar PV power station research plant of Tampere University of Technology (TUT) [12]. The parameter values of the simulation model are compiled in Table I.

Table I: Parameter values of the simulation model for the PV modules and bypass diodes.

Parameter	Value
$I_{SC, STC}$	8.02 A
$U_{OC, STC}$	33.1 V
A	1.30
R_s	0.329 Ω
R_{sh}	188 Ω
A_{bypass}	1.50
$R_{s, bypass}$	0.020 Ω
$I_{o, bypass}$	3.20 μA

2.1 Irradiance transitions and the shading of PV arrays

Irradiance transitions on cloud shadow edges were modelled mathematically using the equation

$$G(t) = \frac{G_{us} - G_s}{1 + e^{(t-t_0)/b}} + G_s, \quad (2)$$

where G is the irradiance, t is the time and G_{us} and G_s are the irradiances of an unshaded and fully shaded situation, respectively [13]. Parameter b is related to the steepness of the transition and the sign of it defines whether the transition is decreasing (fall) or increasing (rise). Parameter t_0 defines the midpoint of the transition. The mathematical model has been validated with around 40,000 measured irradiance transitions [13]. The shading strength (SS), i.e., the irradiance attenuation due to shading, of an irradiance transition is defined as

$$\text{SS} = \frac{G_{us} - G_s}{G_{us}}. \quad (3)$$

By using the mathematical irradiance transition model of Eq. (2), transitions can be defined by four independent variables: SS, b and apparent speed and direction of movement [1], [14]. Parameter t_0 was calculated from the duration of a transition, which was calculated as the product of the absolute value of b and the experimental regression coefficient of 7.67 [14].

In this study, the data measured in the TUT solar PV power station research plant was used. A total of 9097 irradiance transitions (4438 falls and 4659 rises) were identified in five months (May–September 2013) of data, measured by three sensors (S2, S5 and S6 [12]) by utilising the method offered in [13]. The apparent velocity, i.e., the component of shadow velocity normal to the shadow edge, was calculated from the time lags between the shading of the sensors [1]. The median values of the SS, the apparent speed and the absolute value of b of the identified shadow edges were 57.6%, 7.87 m/s and 1.49 s, respectively. The dominant apparent movement direction of the shadow edges was towards the directions from north-east to east; about 21% of the shadow edges moved towards these directions.

A shadow edge was assumed to be linear across a PV array and the apparent shadow edge velocity was assumed to be constant during each simulation period, which are both justified approximations. A simulation period started when a shadow edge moved over the first PV module of the array and ended when the shadow edge had moved across the PV array. The PV array was chosen to be under standard test conditions (STC) in the beginning of each simulation period and the operating temperature of the array was chosen to stay constant during the simulation periods. The used simulation time step was 0.1 s.

The studied PV array was composed of two parallel strings of 20 series-connected PV modules. The PV strings were located in straight lines without gaps between the modules having a gap of 2.0 m between each other and forming a rectangle. For the studied PV array, a partial shading event resulting from an overpassing irradiance fall is symmetrical to an event resulting from a similar overpassing irradiance rise. The absolute values of parameter b for the identified shadow edges were thus used in the simulations.

The characteristics of MPPs for a PV array at each time step were determined by finding maxima (peaks) of the P–U curve of the array. Peaks that were very close to each other were ignored by restricting the acceptable peak-to-peak separation to 1.0 V. The mismatch losses of the PV array were computed as the difference between the sum of the global MPP powers of the PV modules of the array, as if they were operating independently, and the global MPP power of the array. Relative mismatch losses were computed in relation to the sum of the global MPP powers of the modules.

3 RESULTS

In this study, the effect of the size of a basic simulation unit on the accuracy of PV array simulations under partial shading conditions resulting from cloud shadows was studied. Four cases were studied with the used basic simulation unit sizes of a PV cell, a submodule and a module. The PV submodule is a series-connection of 18 PV cells protected by a bypass diode and a PV module is a series-connection of three submodules. In the three cases, the irradiance at the centre of each simulation unit was used for the whole simulation unit during a time step. In the fourth case, a PV submodule with minimum G was studied, i.e., the lowest irradiance at the centre of a PV cell of the submodule was used for the whole submodule during a time step. The irradiance values were calculated with an accuracy of 0.1 W/m².

The case of a PV submodule with minimum G was included in this study because in a series connection of identical PV cells the cell with the lowest irradiance has the lowest short-circuit current and thus limits the total current of the series connection and, accordingly, the power output of the submodule. For this reason, the use of the central point irradiance of the submodule for the whole submodule may overestimate the power production of the submodule in certain shading situations. If the irradiance transition is monotonic, i.e., irradiance changes monotonically during the transition, the PV cell with the lowest irradiance is on one of the corners of the submodule. Thus, the irradiances of the PV cells in the corners of the submodule were calculated and the lowest irradiance value was used for the submodule during a time step.

To elucidate the studied problem, the average irradiance of a 2 × 2 PV array for all the studied cases during partial shading resulting from a sharp cloud shadow edge moving over the array is presented in Fig. 1. The 95th percentile value of the SS (80.6%) and the 5th percentile values of the apparent speed (2.86 m/s) and the absolute value of b (0.43 s) of all the 9097 identified shadow edges were used in these simulations to represent the sharp shadow edge. The lines of average irradiances for the cases of having PV cells, submodules and modules as the simulation units are almost on top of each other, but the average irradiance is smaller for the submodule case with minimum G. In the case of a submodule with minimum G, the PV array undergoes the irradiance transition a bit earlier than in the other cases.

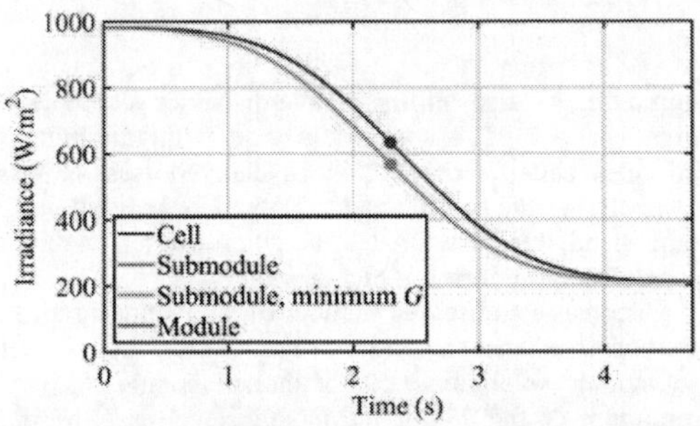

Figure 1: Average irradiance for a 2 × 2 PV array during partial shading resulting from a sharp shadow edge moving over the array with 45° apparent movement direction with respect to the strings of the array.

The irradiance levels falling on the cells of the 2 × 2 PV array at the middle time step of the partial shading event resulting from the movement of the sharp shadow edge over the PV array shown in Fig. 1 are presented in Fig. 2 (a). The corresponding *I–U* and *P–U* curves of the array for all the studied cases are presented in Figs. 2 (b) and (c), respectively. The irradiance differences between the cells of the array during an irradiance transition are the largest when the midpoint of the transition is in the middle of the array. In Fig. 2, this very moment is presented. In this case, the use of large module or submodule basic simulation units overestimated the power production of the array compared to cell level modelling. Due to irradiance differences within the submodules, the global MPP for cell level modelling was also on somewhat higher voltage compared to the other cases.

There are only minor differences in the *I–U* and *P–U* curves between the module and submodule level cases in Fig. 2 since the points of irradiance measurements were close to each other. The largest difference results from the fact that in module level modelling all the three bypass diodes of a PV module begin to conduct at the same point while in submodule level modelling the bypass diodes start to conduct one by one causing small steps in the curves.

Although the average irradiance was almost the same for cell, submodule and module level modelling (Fig. 1), cell level modelling provided clearly lower current and power than submodule and module level modelling. The reason for this is that, in cell level modelling, the cell with the lowest irradiance in a submodule limits the total current of the submodule.

The submodule case with minimum G provided clearly lower power than the others in accord with Fig. 1. The difference with respect to submodule and module level modelling is obvious, because they are based on their midpoint irradiance, which is always higher than the lowest cell level irradiance in a submodule. When a PV cell is the basic simulation unit, the cell receiving a high

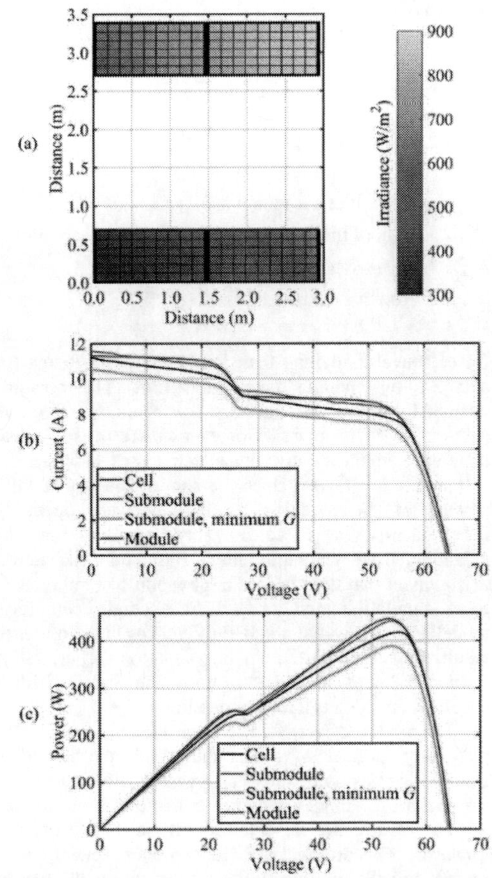

Figure 2: Irradiance levels received by the PV cells (a), *I–U* curves (b) and *P–U* curves (c) of the 2 × 2 PV array at the midpoint of the partial shading event (dots in Fig. 1) resulting from a sharp shadow edge moving over the array.

irradiance in a submodule forces a cell receiving lower irradiance to operate slightly in the negative voltage region and thus pass through higher current than its short circuit current. This is the reason why the overall PV array current is larger in the case of cell level modelling than in the submodule case with minimum G. These findings indicate that the use of submodule as a basic simulation unit with minimum G is not a better approach for PV system modelling than using module or submodule as the basic simulation unit with the midpoint irradiance.

Values of various electrical quantities of the PV array at the same moment as in Fig. 2 are presented in Table II for all the studied cases. The voltage of the global MPP was between 53.1 and 55.0 V in all the studied cases and the differences in the maximum power were mainly caused by the differences in the current of the global MPP. Due to the lower global MPP current, the global MPP power for the submodule case with minimum G was clearly lower than for the other cases.

No major differences took place between the electrical quantities of the module and submodules cases. However, they both underestimated the voltage and overestimated the current and power with respect to the more accurate modelling using PV cells as the basic simulation unit. The absolute and relative mismatch losses were clearly larger

Table II: Values of electrical quantities of the 2 × 2 PV array at the midpoint of the partial shading event (dots in Fig. 1) resulting from a sharp shadow edge moving over the array.

Quantity	Case (simulation unit)			
	Cell	Submodule	Submodule, minimum *G*	Module
Power of the global MPP (W)	418	441	389	446
Voltage of the global MPP (V)	55.0	53.5	53.7	53.1
Current of the global MPP (A)	7.60	8.24	7.24	8.40
Sum of the global MPP powers of the modules (W)	457	473	422	477
Mismatch losses (W)	38.7	32.3	33.1	30.6
Relative mismatch losses (%)	8.48	6.82	7.85	6.42

for cell level modelling than for the other cases and the smallest for module level modelling. The absolute mismatch losses were almost the same for the two submodule level cases. However, there was a clear difference in the relative mismatch losses between the submodule level cases since the sum of the global MPP powers of the modules, as if they were operating independently, was considerably smaller for the submodule case with minimum *G*. Based on these results, it is evident that the use of a large module or submodule level simulation unit instead of accurate cell level modelling may lead to notable differences in the obtained results on certain partial shading situations. The relative mismatch losses were 3.1% during the partial shading event, when using cell level modelling.

The produced energy, relative mismatch losses and the maximum number of MPPs during all the identified irradiance transitions and the largest relative mismatch losses during a single irradiance transition for the studied 2 × 20 PV array are presented for all the studied cases in Table III. The difference in the produced energy with respect to cell level modelling was around 0.3% for the submodule case with minimum *G* and almost negligible in the module and submodule level cases. In the relative mismatch losses, the differences were between 0.5% and 0.9%. In the largest relative mismatch losses during a single irradiance transition, they were from 1.3 to 3.0%. The relatively large differences in the largest relative mismatch losses during a single transition were anticipated since that quantity was computed from a single irradiance transition instead of all the identified transitions. In both of the studied mismatch losses quantities, the difference was the smallest for the submodule case with minimum *G*. The difference in mismatch losses with respect to cell level modelling was the smallest for the submodule case with minimum *G* also in the above-discussed example of Figs. 1 and 2. The submodule case with minimum *G* provides better accuracy for mismatch losses analyses compared to the module and submodule cases, but on the other hand, it underestimates energy production leading to somewhat larger error in that quantity.

The largest observed number of MPPs during all the studied irradiance transitions was 18 for the cell and submodule level cases, 19 for the submodule case with minimum *G* and 23 for the module level case. Module level modelling leads to higher maximum number of MPPs compared to modelling on the level of smaller elements. In module level modelling, the *I–U* and *P–U* curves of adjacent PV modules differ more from each other than when a smaller simulation unit is used. This results from the irradiance differences within individual modules and from the above-mentioned fact that in module level modelling all the bypass diodes of a module begin conducting at the same point while in submodule and cell level modelling the bypass diodes start to conduct one by one. This phenomenon can be seen in Fig. 2 (c), where in the module level case the lower voltage MPP is a clear peak in the *P–U* curve while in the cell level case it is only a slightly prominent peak.

In conclusion, the effect of simulation unit size in long-term PV analysis was found to be moderately small. The use of a larger basic simulation unit than one cell caused only marginal errors. However, on certain partial shading situations, such as the one demonstrated in Figs. 1 and 2, the differences on the obtained results can be perceptible. PV module level modelling is accurate enough for most of PV array level partial shading studies. However, the submodule case with minimum *G* provides better accuracy, compared to the module and submodule cases, for mismatch losses analyses. For analyses where

Table III: Produced energy, relative mismatch losses and the maximum number of MPPs of the 2 × 20 PV array during all the identified irradiance transitions for the studied cases. The differences were calculated with respect to the PV cell simulation unit case.

Quantity	Case (simulation unit)			
	Cell	Submodule	Submodule, minimum *G*	Module
Produced energy (kWh)	249.4	249.6	248.7	249.7
Difference in produced energy (%)		0.076	0.290	0.082
Relative mismatch losses (%)	2.886	2.861	2.871	2.867
Difference in relative mismatch losses (%)		0.890	0.537	0.657
Largest relative mismatch losses during an irradiance transition (%)	19.55	18.97	19.31	18.97
Difference in the largest relative mismatch losses (%)		2.997	1.261	3.002
Maximum number of MPPs	18	18	19	23

high accuracy is needed cell level modelling provides the optimal accuracy while, naturally, it is computationally the most demanding alternative.

The relative mismatch losses between the PV modules during all the identified irradiance transitions, calculated by cell level modelling, were 2.9%. The relative mismatch losses between the submodules of the individual modules and the cells of the individual submodules were only about 0.01% and 0.05%, respectively. In the example of Figs. 1 and 2 they were 0.40% and 1.16% during the whole partial shading. It was an extreme example since the studied PV array was very small (only four modules) and the partial shading was caused by the movement of a sharp shadow edge. It has been found earlier that the mismatch losses resulting from cloud shadows decrease with increasing PV array size [5], [15] and with decreasing shadow sharpness [16]. In conclusion, most of the mismatch losses resulting from the shadows of moving clouds take place between PV modules and the mismatch losses between smaller elements can be regarded as negligible.

4 CONCLUSIONS

In this paper, a study of the effect of the size of a basic simulation unit on the accuracy of PV array simulations, implemented using the one-diode PV cell model, under partial shading conditions resulting from moving cloud shadows was presented. The aim of this study was to find out an adequate size for the basic simulation unit to be used in PV array level partial shading studies to minimise the computing time while retaining sufficient accuracy of the simulations. Moreover, mismatch losses between PV modules, submodules and cells were studied. The study was based on the characteristics of around 9000 irradiance transitions resulting from cloud shadows identified in measured irradiance data. The studied PV system was an array of two parallel strings of 20 series-connected PV modules.

The used simulation unit size was found to have a notable effect on the obtained results on certain partial shading situations. However, the effect of simulation unit size in long-term PV analysis was found to be moderately small. The differences in the produced energy and relative mismatch losses during all the studied irradiance transitions, resulting from the use of a larger basic simulation unit with respect to cell level modelling, were less than 1%. It was found that the submodule case with minimum *G* provides better accuracy for mismatch losses analyses compared to the module and submodule cases. It was also found that most of the mismatch losses resulting from cloud shadows take place between PV modules and the mismatch losses taking place between smaller elements, PV submodules and cells, can be regarded as negligible.

5 REFERENCES

[1] K. Lappalainen, S. Valkealahti, Solar Energy, 138 (2016) 47.

[2] S. Vijayalekshmy, G.R. Bindu, S. Rama Iyer, Solar Energy, 135 (2016) 92.

[3] S.R. Potnuru, D. Pattabiraman, S.I. Ganesan, N. Chilakapati, Renewable Energy, 78 (2015) 264.

[4] N. Rakesh, T.V. Madhavaram, Frontiers in Energy, 10 (2016) 227.

[5] K. Lappalainen, S. Valkealahti, Solar Energy, 158 (2017) 455.

[6] K. Lappalainen, S. Valkealahti, Applied Energy, 190 (2017) 902.

[7] R. Yan, T.K. Saha, P. Meredith, A. Ananth, M.I. Hossain, IET Renewable Power Generation, 10 (2016) 1229.

[8] S. Gallardo-Saavedra, B. Karlsson, Solar Energy, 170 (2018) 828.

[9] O.W. Yin, B.C. Babu, Optik, 169 (2018) 48.

[10] K. Lappalainen, S. Valkealahti, Proceedings of 33rd European Photovoltaic Solar Energy Conference, Vol. I (2017) 1973.

[11] M.G. Villalva, J.R. Gazoli, E.R. Filho, IEEE Transactions on Power electronics, 24 (2009) 1198.

[12] D. Torres Lobera, A. Mäki, J. Huusari, K. Lappalainen, T. Suntio, S. Valkealahti, International Journal of Photoenergy, 2013 (2013) Article ID 837310.

[13] K. Lappalainen, S. Valkealahti, Solar Energy, 112 (2015) 55.

[14] K. Lappalainen, S. Valkealahti, Proceedings of 32nd European Photovoltaic Solar Energy Conference, Vol. I (2016) 1485.

[15] K. Lappalainen, S. Valkealahti, Solar Energy, 144 (2017) 548.

[16] K. Lappalainen, S. Valkealahti, IET Renewable Power Generation, 11 (2017) 248.

ECONOMIC VALIDATION OF LARGE POWER PV IRRIGATION SYSTEMS

R.H. Almeida[1,2], I.B. Carrêlo[1], C. Lorenzo[1], L. Narvarte[1]
[1] Instituto de Energía Solar - Universidad Politécnica de Madrid, 28031 Madrid, Spain
[2] Instituto Dom Luiz, Faculdade de Ciências, Universidade de Lisboa, 1749-016 Lisboa, Portugal
*Phone: +34 913365531; E-mail: isaac.barata@ies.upm.es

ABSTRACT: This paper presents an economic evaluation of large-power PV irrigation systems (from 40 to 360 kWp) installed in Mediterranean countries, evaluating the economic feasibility of substituting or reducing their current energy sources (diesel or electrical grid) with a Photovoltaic (PV) generator and also an economic assessment based on PVIS simulated for ECOWAS region is also presented.
In the first case, five PV irrigation systems, operating in Portugal, Spain, Italy and Morocco were considered and two energetic scenarios were created in order to do the economic analysis (Optimistic and Pessimistic Scenario). The normalized Net Present Value (NPV) values are in the 2.3-3.8 €/W_p range, Internal Rate of Return (IRR) values are in the 10-16% range and Payback Period (PBP) values are in the 7-11 years range. Finally, the Levelized Cost of Energy (LCOE) values for PV irrigation systems the values are in the 0.05-0.20 €/kWh range, which represents a percentage reduction of 34-67% % if compared to the actualized cost of the previous energy source.
Regarding the second case, seven countries from the ECOWAS region were considered and two irrigation operating modes were compared (pumping to a water tank or at constant pressure). NPV values are in the 0.28-35.2x10^5 € range, (IRR) values are in the 8-47% range and (PBP) values are in the 2.1-10 years range. LCOE for PV irrigation systems are in the 0.04-0.15 €/kWh range, which represent percentage savings of 30-84% if compared to diesel-powered and grid-powered systems.
Keywords: Design, Monitoring, Stand-alone PV Systems, Water-Pumping

1 INTRODUCTION

Irrigation for agricultural applications is a very high water and electricity-consuming activity, as most of the water demanded must be pumped from underground reservoirs. Traditionally, water pumps are powered by the local electric grid, if accessible, or by diesel generators in isolated regions or in regions where the grid service is unreliable [1]. Solar photovoltaic (PV) generators to feed irrigation systems represent an attractive alternative for reducing the cost of this electricity consumption. A H2020 project named MASLOWATEN [2] recommended solutions which included the installation of 5 real scale photovoltaic water irrigation systems (PVIS) as a way to answer the irrigators' needs [3], [4], [5], [6] in Spain, Portugal, Italy and Morocco, ranging from 40 to 360 kWp. The economic effectiveness of these large power systems has been evaluated and it is here presented. However these results are site-dependent, so it would be interesting to evaluate the economic feasibility of large power irrigation systems in other regions. One of particular interest is the ECOWAS (Economic Community of West African States) region [7], where institutions like the ECREEE (ECOWAS Centre for Renewable Energy and Energy Efficiency) or the World Bank are promoting renewable alternatives for agricultural applications. Countries from this region are generally characterized by highly decentralized farms, with very limited or unreliable grid-access points. This has favored an abundance of diesel generators for back-up services, with the consequent elevated fuel and transportation costs.

First, an economic validation of the real large scale demonstrators in the Mediterranean Countries is presented. Based on this, it is estimated the economic feasibility for the ECOWAS region.

In Section 2 is presented the methodology used to preform both studies. Section 3 shows the results and finally, in section 4 are summarized some conclusions.

2 METHODOLOGY

2.1 Economic study

The viability of the economic investment required for substituting diesel generators or grid-connection points by a PV generator was evaluated through three indicators (NPV, IRR and PBP). Thirdly, the LCOE of the PV-powered systems was estimated and compared with the LCOE of the previous energy source.

For estimating the values of NPV, IRR and PBP for the investment, the annual cash flows were calculated for the whole lifetime of the system (25 years), considering both the annual profits obtained and the Amortization (AM) of the Initial Investment Cost (IIC). In the particular case assessed in this work, the annual profits are given by the economic savings derived from substituting the national grid or diesel generators by a PV generator.

An investment should be considered as profitable if the NPV is positive, IRR is higher than the local discount rate and PBP is significantly lower than the lifetime of the system.

2.2 Mediterranean Region

During the MASLOWATEN project, 5 PVIS were installed in Portugal, Spain, Italy and Morocco and they have been fully operating for the past two years. It must be emphasized that these systems are intended to be wide-ranging of configurations (stand-alone or hybrid - one with the grid, another with diesel - pumping to a water pool or at constant pressure).

Photographs of the systems can be found in the next figures.

Figure 1: Villena, Spain

Figure 2: Alaejos, Spain

Figure 3: Uri, Italy

Figure 4: Tamallalt, Morocco

Figure 5: Alter do Chão, Portugal

They are all different from each other, regarding configurations, sizes and applications:

- A 360 kWp stand-alone PV pumping system to a water pool (WP) at variable water flow (in Villena, Spain);
- A 160 kWp stand-alone PV irrigation system with one pump elevating water from a borehole to an intermediate tank at variable water flow and other pump irrigating at constant pressure (CP) through pivots (in Alaejos, Spain);
- A 40 kWp stand-alone PV irrigation system with 2 pumps pumping from a borehole at variable water flow to a water tank and one pump irrigating at CP through sprinkles (in Uri, Italy);
- A 120 kWp PV-grid drip irrigation systems, with 2 pumps and hybridization in the electric part of the system (in Tamallalt, Morocco);
- A 140 kWp PV-diesel drip irrigation systems, with 3 pumps and hybridization in the hydraulic part of the system (in Alter do Chão, Portugal).

In order to do the economic validation, 2 different scenarios (optimistic and pessimistic scenarios) were considered, depending on the irrigation needs and water availability which greatly influence the use of the PVIS.

2.3 Simulation for the ECOWAS region

In the case of the ECOWAS region, since there is no real system installed we had to do some assumptions and to design a possible PVIS. Using the simulation tool SISIFO, we had simulated a 380 kWp PV System mounted on a North-South horizontal axis tracker (due to its better match between incident irradiance and irrigation needs). It was also considered thatt it could operate by pumping to a water tank or at constant pressure. For these two kind of systems, the viability of the economic investment was done considering the substitution of the diesel generators or grid-connection points by the PV generator. Simulations were carried out for the 7 locations presented in Table 1. Based on it, SISIFO is able to import the monthly mean values of horizontal daily irradiation, as well as the maximum and minimum ambient temperatures for each place from the PVGIS database [13].

Country	Latitude [°]	Longitude [°]
Benin	12.050	3.032
Burkina Faso	14.881	-0.1
Cape Verde	14.924	-23.533
Guinea	11.222	-10.723
Liberia	8.413	-9.748
Nigeria	12.018	-8.613
Sierra Leone	9.649	-12.225

Table1: Location input used for the simulation of the PV irrigation systems.

3 Results

Figure 6 shows the NPV values for the 5 PVIS (for the two scenarios). It is easy to understand that Villena's system is the one that offers the best profitability because of many reasons: its size, for being stand alone but also because the PVIS operates the whole year.

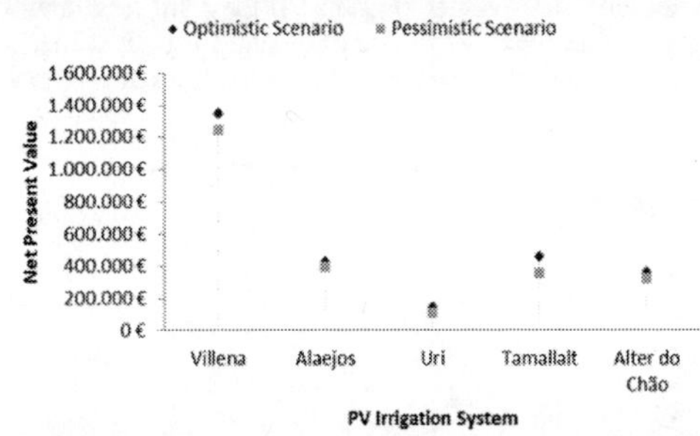

Figure 6: Net Present Value (€) for the 5 PVIS considering two different scenarios (optimistic and pessimistic).

Figure 7 shows the NPV normalized by the PV peak power for the 5 PVIS, regarding the two scenarios. As it can be observed, in this case the profitability is very similar for the 5 PVIS, ranging from 2280 €/kWp (pessimistic scenario of Alter do Chão) to 3770 €/kWp (optimistic scenario of the Tamallalt).

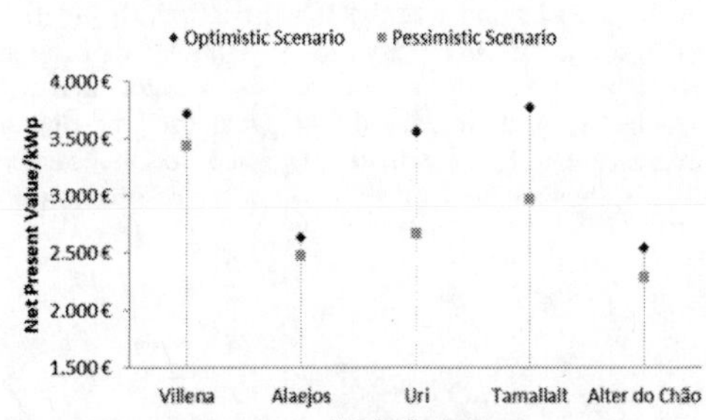

Figure 7: Net Present Value (€) normalized for the size of the PV system for the 5 PVIS considering two different scenarios (optimistic and pessimistic).

In terms of IRR (Figure 8), the values vary from 10% to 16%. Villena's system is the one which offers the best profitability in both scenarios. On the other side, the worst results are obtained for the Uri's system, which presents an IRR of 10% for the pessimistic scenario. It is important to emphasize that an investment is profitable if this tax is higher than the discount rate, in this case the WACC considered for each place.

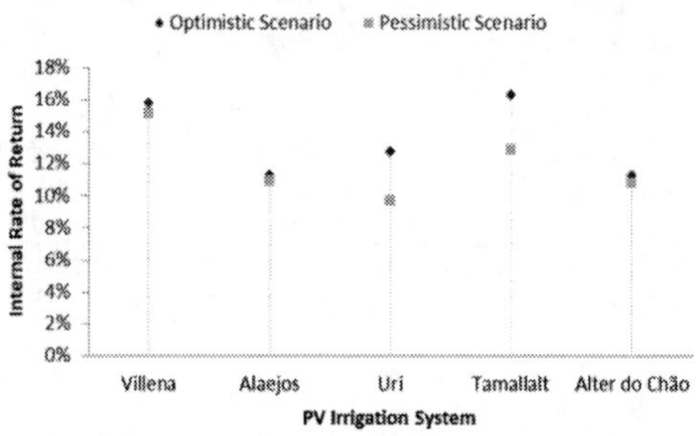

Figure 8: Internal Rate of Return (%) for the 5 PV Irrigation Installations considering two different scenarios (optimistic and pessimistic).

Regarding PBP values (Figure 9), they are less than 11 years for all the cases under study and for both scenarios. For the optimistic scenario, this value goes from 7 to 9 years.

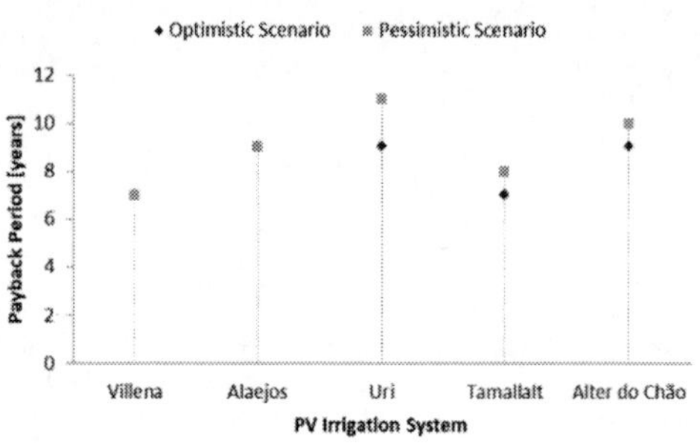

Figure 9: Payback Period (years) for the 5 PVIS considering two different scenarios (optimistic and pessimistic)

Figure 10 shows the values of LCOE obtained for the 5 PVIS, together with actualized electricity or diesel prices for the 5 PVIS under study. It is important to highlight that generating electricity for irrigation applications would be cheaper by far with a PV generator than with the electric grid or with diesel generators: percentage savings are all higher than 30%, and for the case of Villena, they are higher than 60%, for the two scenarios.

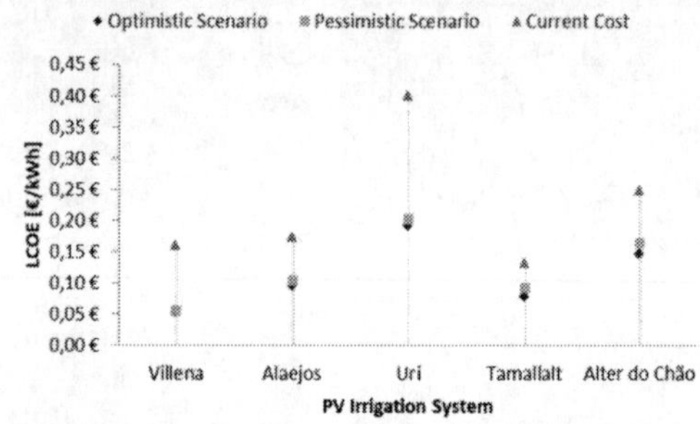

Figure 10: Levelized Cost of Energy (€/kWh) for the 5 PVIS considering two different scenarios (optimistic and pessimistic) compared to the current cost.

Regarding the economic analysis for the ECOWAS countries, if comparing the 7 countries in terms of NPV (Figure 11), Guinea offers the best profitability for the substitution of diesel-powered systems because of its elevated diesel price and its low real discount rate and corporate tax rate (both equal to zero); Burkina Faso presents the highest profitability for the substitution of grid-powered systems due to its low real interest rate, although it has an intermediate electricity price.

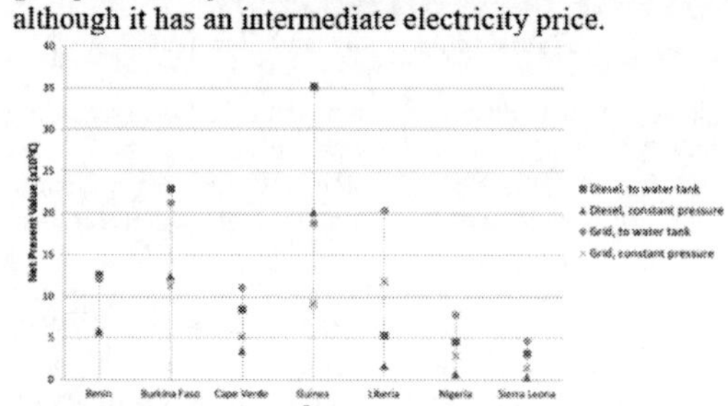

Figure 11: NPC (x10^5€) for the 7 countries and the 4 cases under study: substitution of diesel generators or the grid with PV irrigation systems, for both operating modes (to a water tank or at constant pressure).

In terms of IRR (Figure 12), Guinea also offers the best profitability for the substitution of diesel-powered systems because fuel is very expensive in this country, which implies big savings for a PV powered system, and the income tax rate is zero; Liberia offers the highest IRR for the substitution of a grid-powered system due to the elevated electricity prices, despite it having the highest income tax rate.

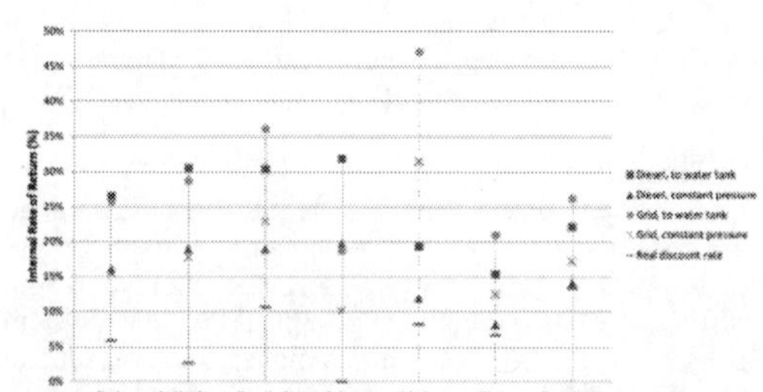

Figure 12: IRR (%) and local discount rate (%) for the 7 countries and the 4 cases under study: substitution of diesel generators or the grid with PV irrigation systems, for both operating modes (to a water tank or at constant pressure)

As regards PBP values (Figure 13), they are less than 10 years for all the cases under study, and lower than 6 years (which represents less than a quarter of the lifetime of the system) except for some systems pumping at constant pressure.

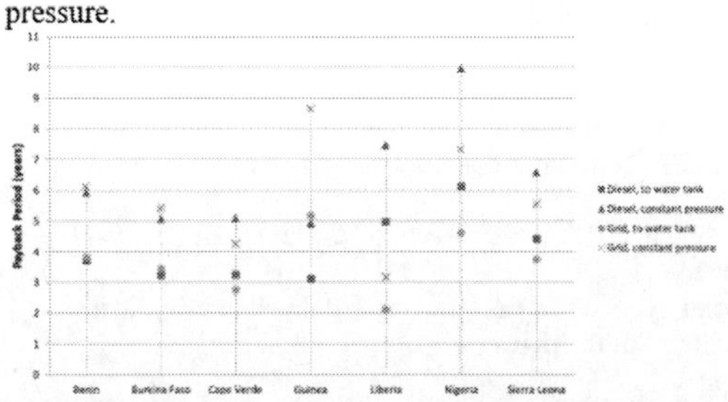

Figure 13: PBP (years) for the 7 countries and the 4 cases under study: substitution of diesel generators or the grid with PV irrigation systems, for both operating modes (to a water tank or at constant pressure)

As regards LCOE values (Figure 14), they are smaller than electricity prices and the LOCE for diesel powered systems.

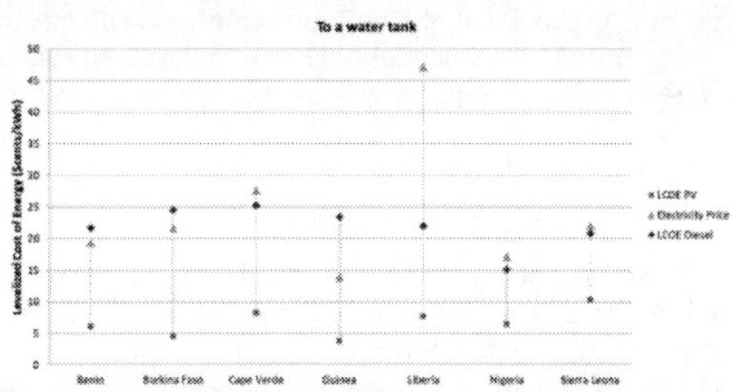

Figure 14: LCOE (€cents/kWh) for PV-powered and for diesel-powered systems and electricity prices for the 7 countries under study, for systems pumping to a water tank.

4 CONCLUSIONS

Regarding the real scale demonstrators, it can be concluded:

- NPV values are all positive and in the range 2279-3771.6 €/kWp.
- IRR values are all higher than the discount rate and in the 10-16% range.
- PBP are far below the lifetime of the system (25 years) and in the 7-11 years range.
- LCOE values are the lowest for PV irrigation systems and in the 0.05-0.20 €/kWh range, representing very high percentage savings in the range 34-67% if compared to the cost of the previous energy source.

In what concerns, the ECOWAS region, the following may be concluded:

- NPV values are all positive and in the range $0.28\text{-}35.17 \times 10^5$ €.
- IRR values are all higher than the local real discount rate (*i*) and in the 8-47% range.
- PBP are far below the lifetime of the system (25 years) and in the 2.1-10 years range.
- LCOE values are the lowest for PV irrigation systems and in the 0.04-0.15 €/kWh range, representing very high percentage savings in the range 30-84% if compared to diesel-powered and grid-powered systems.

5 AKNOWLEDGEMENT

This work has been possible thanks to the funding from the European Union's Horizon 2020 research and innovation programme in the project MArket uptake of an innovative irrigation Solution based on LOW WATer-ENergy consumption (MASLOWATEN), under grant agreement n°640771, as well as the financial support by MIT Portugal Program on Sustainable Energy Systems and the Portuguese Science and Technology Foundation (FCT), grant PD/BD/105851/2014, and IDL project (UID/GEO/50019/2013).

6 REFERENCES

[1] J. Carroquino, R. Dufo-López, J. L. Barnal-Augustin, "Sizing of off-grid renewable energy systems for drip irrigation in Mediterranean crops," Renewable Energy, Vol 76 (2015) 566-574.

[2] Grant Agreement number 640771 – MASLOWATEN, European Commission.

[3] I. B. Carrêlo, R. H. Almeida, L. M. Carrasco, F. Martínez-Moreno and L. Narvarte, "A 360 kWp PV irrigation system to a water pool in Spain," in 33rd European Photovotaic Solar Energy Conference and Exhibition, Amsterdam, 2017.

[4] I. B. Carrêlo, R. H. Almeida, F. Martínez-Moreno, L. M. Carrasco and L. Narvarte, "A 160 kWp constant pressure PV Irrigation system in Spain," in 33rd European Photovotaic Solar Energy Conference and Exhibition, Amsterdam, 2017.

[5] R. H. Almeida, I. B. Carrêlo, F. Martínez-Moreno, L. M. Carrasco and L. Narvarte, "A 140 kW hybrid PV-diesel pumping system for constant-pressure irrigation," in 33rd European Photovotaic Solar Energy Conference and Exhibition, Amsterdam, 2017.

[6] R. H. Almeida, I. B. Carrêlo, L. M. Carrasco, F. Martínez-Moreno and L. Narvarte, "Large-scale hybrid PV-Grid irrigation system," in 33rd European Photovotaic Solar Energy Conference and Exhibition, Amsterdam, 2017.

[7] ECOWAS Energy Sector Website (2017), http://www.ecowas.int/ecowas-sectors/energy/; [accessed 15 November 2017]

DESIGN OF HYBRID MINIGRIDS IN SOUTH AFRICAN RURAL AREAS UNDER CONSIDERATION OF SOCIAL AND CULTURAL ASPECTS

Meike Kühnel[a], Benedikt Hanke[a], Yuliana Baranova[b], Olga Weigel[c], Ingo W. Stuermer[d], Alister McMaster[e], Sander Mabe[c], Karsten von Maydell[a]
[a] DLR Institute of Networked Energy System, Carl-von-Ossietzky-Str. 15, 26129 Oldenburg, Germany
[b] DEULA-Nienburg GmbH, Max-Eyth-Str. 2, D 31582 Nienburg/Weser, Germany
[c] Deutsche Gesellschaft für Internationale Zusammenarbeit (GIZ) GmbH, Dag-Hammarskjöld-Weg 1-5, 65760 Eschborn, Germany
[d] Lower Saxony Ministry for Environmental Affairs, Energy, Building and Climate Protection, Archivstraße 2, 30169 Hannover, Germany
[e] Department of Economic Development, Environmental Affairs and Tourism of Eastern Cape, Palmsquare Business Park, Beacon Bay, East London, South Africa

ABSTRACT: Small scale systems are vulnerable to changes. The development of solution strategies for future changed conditions should be considered from the outset in planning. So far load estimations for first electrification is determined primarily by interviews of the population. However, studies have shown that there are still major differences between the power demand estimated by surveys and the real load after electrification. Reinforcing here is the inadequate assessment of load development through immigration and increased prosperity. On basis of a currently developing mini grid in the Eastern Cape of South Africa we could show that a load increase of approx. 60% within 10 years is to be expected. The following possibilities were outlined as possible counter measures: 1. communicating with the population, i.e. on energy efficiency to raise awareness and understanding, 2. early planning of capacity building and identification of key performance parameters to trigger the expansion based on local socio-cultural conditions and grid supporting qualities. An adequate data base for the initial electrification and power development should also be established and available on open source basis for researchers, developers, communities and installers.
Keywords: hybrid mini grid, LCOE, pv-diesel, first electrification

1 MOTIVATION

Energy use and the access to reliable electric power is an essential cause and facilitator of economic grows and a binding constrain for all sizes of business [1]. In the Development Impact Evaluation – Evidence Review- the commonwealth Development Corporation summarized the impact of access to energy on wealth as follows:

"Energy plays a fundamental part in the economic growth process. More than three quarter of the good quality statistical studies [...] find a positive correlation between energy use and economic growth, and half of the studies find a positive and significant causal link from energy use to economic growth. Energy use is either the cause or the facilitator of economic growth."

With 15.8 % of South Africa's population without access to electricity [2] and poverty head count ratio of around 20 % reducing the backlog in electrification is one aspect in the anti-poverty campaign of the country [3].

In far off rural areas islanded mini-grids based on renewable energy generation offer both, access to electricity on a power grid equivalent quality level and an economic solution for the responsible authorities.

Main challenges for planner, developer and engineers of mini-grids are the appraisal of a residential load profile of former un-electrified areas and the matching with highly fluctuating renewable energy resources. The adjacent increase of the load caused by economic growth, new demands for electric appliances and a growing village size triggered by the availability of electricity adds further challenges and uncertainties.

2 INTRODUCTION

ENGINEER is a sub-project of the "Pilot project for the construction and sustainable operation of local energy grids based on renewable energies in the Eastern Cape" funded by the Federal-State-Government-Project of the Federal Ministry for Economic Cooperation and Development Germany (BMZ) and the federal state of Lower Saxony and implemented by the GIZ. South African institutional funding is given by the Department of Economic Development, Environmental Affairs & Tourism (DEDEAT) and the Eastern Cape Rural Development Agency (ECRDA). Aim of the project is to demonstrate a solution for an economically and ecologically sustainable energy supply to rural communities and test it in Upper Blinkwater, a small non-electrified village in the mountains of Eastern Cape in South Africa. Upper Blinkwater is located 32.85° S and 26.56° E in a height of approximately 900m above sea level and around 17km away from the next main road (cf. Figure 1).

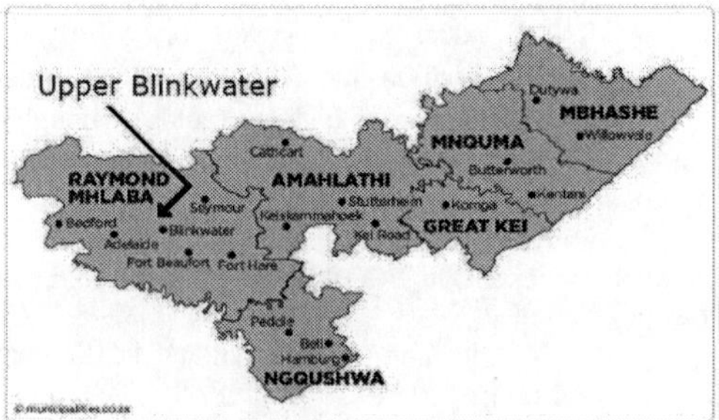

Figure 1: Location of Upper Blinkwater, Raymond Mhlaba Municipality, Eastern Cape, South Africa

Joint monitoring of the institutional integration promotes the development of regulatory frameworks for so-called mini grids and decentralized electricity supply scenarios. The village will be used as a „living lab" by mainly South African universities and research facilities to study the development of power consumption, increasing loads and the desired feedback on local prosperity in the long run. Monitored achievements of load development, necessary cost of operation and maintenance as well as integration in regulatory and institutional processes can be used to develop and define a political and institutional framework for future mini grid installation and integration.

Including retro-fitting structures and a growing renewable power generation plant into upfront considerations is essential for an economical sustainable and persistent mini-grid. An adaptive system will learn from the development of the villages' electric power consumption during the first years of electrification. A recommendation for further installation of renewable energy capacity such as micro wind turbines and additional solar power will be triggered by the recorded demand development and additional onside data using the infrastructure of the generation side. This opens up a new look on business plans and the calculation of levelized cost of electricity and cost reflecting tariffs with a more fluently and dynamic behavior reflecting real life challenges.

3 METHODS

The development of sustainable isolated power systems requires a multi-stage iterative process. In the first step, the ideas, hopes and expectations of the actors are identified and recorded. In the following, these expectations were compared with the underlying local situation. Weather and climate conditions directly influence the chosen technology of hybrid mini grids. Previous projects were analyzed and important findings were gained mainly from their failure.

This analysis showed that it was not the technical decisions that were decisive for the lack of sustainability of the projects, but local acceptance, knowledge and awareness, as well as the implementation in the structure of rural communities. Therefore, the technical decisions follow subordinate to considerations about social acceptance, estimation of the future load profile and climate realities. Cost analyses based on the technical layout and expected load profile close the circle to social acceptance.

3.1 Social acceptance

Early and honest communication with villagers is essential to their acceptance. The biggest threats to mini grids are theft (due to economic constraints) and vandalism of resentment or the perception of discrimination. Lack of understanding of technology often leads to lasting damage to infrastructure in case of small defects. Therefore, already at the beginning of the project a facilitation manager was used to ensure a constant dialog. Facilitation activities help to identify concerns, needs and prospects as well as the demand for workshops and education activities.

3.2 Load profile estimations

For the design of a hybrid mini-grid in a previously un-electrified region with a high percentage of poverty, a good estimate of the expected power consumption in power and energy as well as the consumption behavior over the day is necessary. At the present time there are no reliable household profiles for households after initial electrification. Socio-cultural, financial and political differences complicate the creation and transferability of standardized household load profiles across continental and national borders and climate zones. However, literature shows significant differences between the widely used interviews based method and later consumption measurements [4,5]. Based on this, two approaches were used to estimate the future load profile of Upper Blinkwater.

I. Baseline Survey of Upper Blinkwater

With the help of a villagers questionnaire necessary data on the current use of energy carriers and their consumption patterns, income structure and available financial budget for electric energy were collected. In addition, we surveyed residents about their expectations for a power system, respectively their preferred electrical applications.

II. Multi-Tier Approach

Like the CDC, the WORLD BANK Group has identified access to energy as an important factor and benchmark for the economic development of marginalized communities. The WORLDBANK Group has developed the multi-tier framework to measure and rate amongst others the current state of electrification and reveal possible development scenarios. The five tier-clusters are in detail shown in Table I. Tier 1 to tier 3 can be covered with so called solar home systems whereas tier 4 and 5 are referred as grid-like supply that can be covered with isolated mini grids.

Table I: Multi-Tier framework -WORLD BANK Group

	Tier 1	Tier 2	Tier 3	Tier 4	Tier 5
Capacity [W]	3	50	200	800	2000
Capacity [Wh]	12	200	1000	3400	8200
Durability [h/d]	4	4	8	16	23
Durability Night [h/d]	1	2	3	4	4
Maximum of outages/week				14	3

In order to develop a sustainable mini grid, it is necessary to pay attention to the growing electricity demand after initial electrification.

3.4 Climate conditions

The climate conditions at Upper Blinkwater have been analyzed by the use of synthetic typical meteorological year data provided by METEONORM[1].

[1] https://www.meteonorm.com/de/

3.3 Technical layout

Based on the estimated load profile and climate conditions HOMER Energy pro[2] were used to identify a suitable generation system matching all framework conditions. Multi-year modelling allows the system to be stressed with increasing power requirements.

3.3 Levelized cost of electricity and tariff system

The capital cost (CAPEX) of this South African mini grid is covered by different South African and German funds. For replicability reasons levelized cost of electricity (LCOE) was calculated with CAPEX.

Depreciation is excluded from the LCOE, as well as the interest rate. Cost of diesel is assumed to be constant as most diesel price forecast are been proven wrong. Power tariffs in South Africa a regulated and approved by the National Energy Regulator South Africa (NERSA). In view of the general understanding of access to electricity as a fundamental right, electricity tariffs are highly regulated, in particular for the poor, indigenous and low-income communities, and must be approved by NERSA. For the area of Raymond Mhlaba Municipality approved tariffs for domestic consumption of 0 to 50 kWh per month is 88.21 c/kWh and for 51 kWh to 350 kWh a tariff of 111.99 c/kWh are applied (South African Rand).

4 RESULTS

4.1 Social acceptance

All stakeholders and funding agencies agreed on a targeted supply quality of with a grid like quality compared to the supply quality of ESKOM the national power company. Main points of argument are:

- As stated in the introduction a good quality of electric power and energy supply is considered one main driving factor for economic development.
- Energy access is considered a fundamental right in South Africa
- The mini grid must not be perceived as a second best solution in public perception
- A high rate of acceptance and identification of villagers with the mini grid to protect the installation from vandalism and theft

4.2 Evaluation of the questionnaire

More than 90% of the community members live below the South African poverty line of 2000 Rd/month and their main income is generated by social grant, pension or child benefits. Energy needs are currently met by collecting free wood and using candles and paraffin. Around 16% are using LPG and 37% are using single use batteries. On average, their monthly energy costs amount to 180 Rd/month. Their self-estimated monthly budget for electric energy ranges from 50 Rd/month to 250 Rd/month (cf. Figure 2).

The favored electrical applications are high power consumers like electrical stove, kettle and fridge. These are the essential and applications for a minimal quality live standard. Electric stoves in conjunction with the average daily routine lead to a very high peak to off-peak ration with the maximum power consumption in early evening after sunset. Gas stoves are seen as the second best solution in rural South Africa, in contrast to their perception in many northern European countries. Over 82% of the population would prefer an electrical stove. But 60% would be persuaded to choose a gas stove if their affordable electricity consumption would cover more equipment and for a longer period of time.

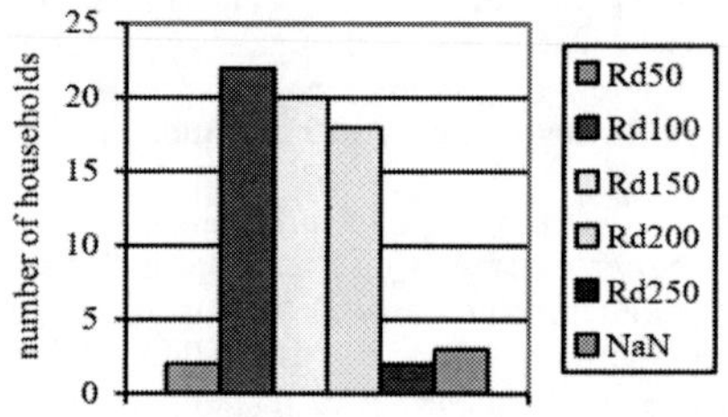

Figure 2: Monthly household budget for electric energy.

Power applications in the field of entertainment and information such as radio and television are second in row after kitchen applications. "Luxury" goods like DVD, DSTV, Laptop, even heater hair dryer and iron are not as important for the residents of Upper Blinkwater (cf. Figure 3).

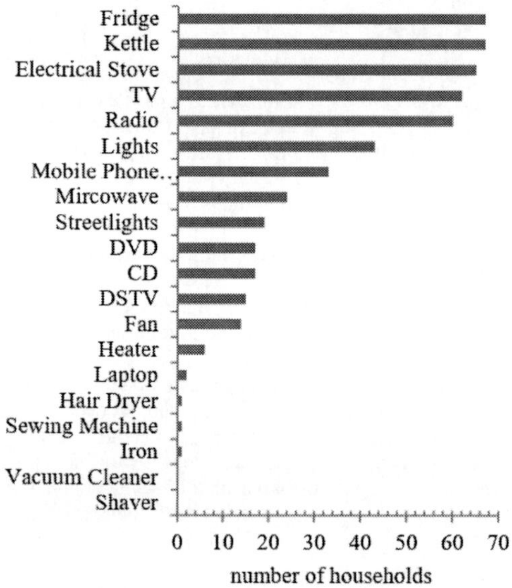

Figure 3: Electrical application with the highest priority against the number of households

In addition to personal, individual application requirements, we also asked about the need for common goods. In particular, the need for an electronic library and for a commonly used cooling container for perishable goods stood out.

4.3 Load profile

In the first year of energy access a mean daily consumption of 3.85kWh per household is aspired, assuming a slight distribution regarding the energy consumption per household caused by different financial scope and including system losses of 5%. The first year daily load

[2] https://www.homerenergy.com/index.html

profiles are exemplary shown in Figure 4. The estimated load profile displays the typical pattern of southern countries with a maximum peak at dinnertimes and two smaller peaks for breakfast and lunch. The annual medium daily load profile is indicated in yellow.

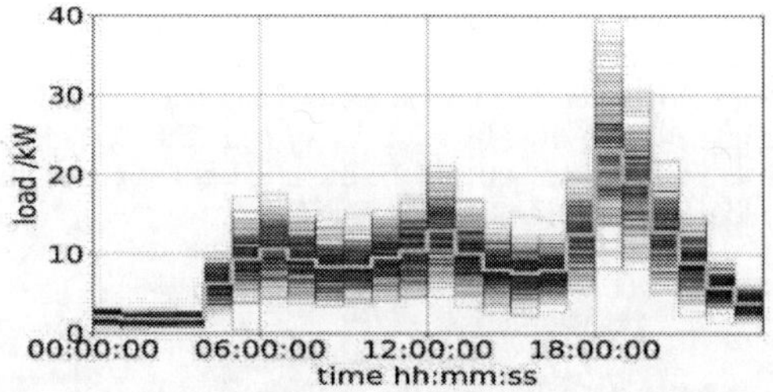

Figure 4: Daily load profile of first years village consumption with the median daily demand indicated in yellow. The typical structure of southern countries domestic load profiles, with a prominent afternoon peak.

Even if the the individual household budget is very limited, most people have access to old and not energy efficient electrical applications from relatives living in bigger cities. Ten years load development was calculated including some system losses (theft of electricity) and an increase of connections and wealth (see Figure 5). Over the years, an increase in total connections due to the pull effect of electrification has to be expected.

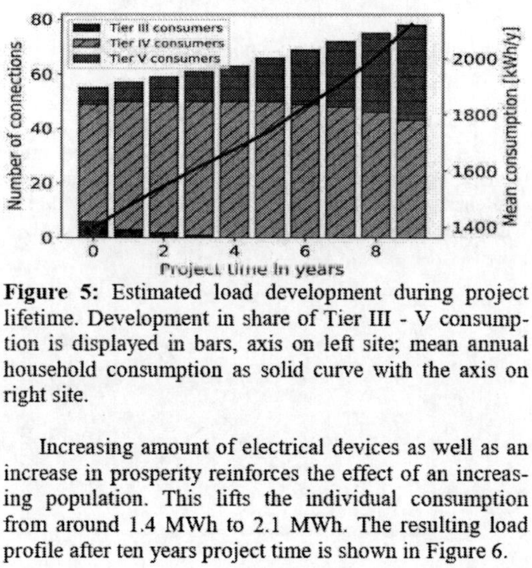

Figure 5: Estimated load development during project lifetime. Development in share of Tier III - V consumption is displayed in bars, axis on left site; mean annual household consumption as solid curve with the axis on right site.

Increasing amount of electrical devices as well as an increase in prosperity reinforces the effect of an increasing population. This lifts the individual consumption from around 1.4 MWh to 2.1 MWh. The resulting load profile after ten years project time is shown in Figure 6.

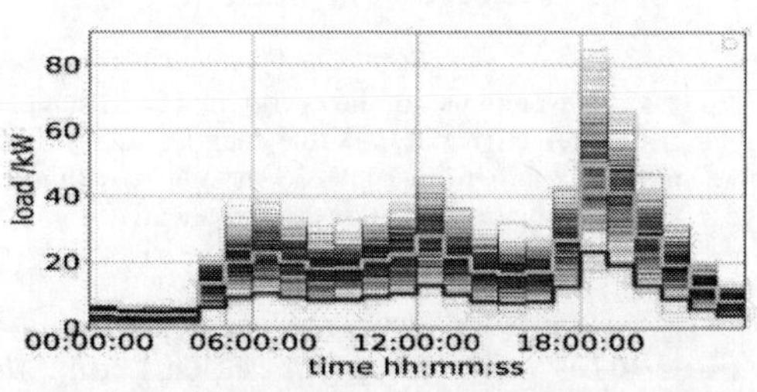

Figure 6: Load profile after ten years project time (median load first year in red, tenths year in yellow).

4.4 Energy system design and response

I. System design

The final design consists of 75 kW_{peak} pv power with around 280 kWh battery storage on the dc site. An additional 40 kW diesel generator for back up and times of high peak demand supports the grid on the ac site. Three converter of 20 kW each will be used to form the grid and enable battery charging with the diesel generator. All converters are supposed to be able to work as inverter and rectifier. The grid will be formed by one of the inverter. A simplified model of the system setup is shown in Figure 7. Illustrated is also a potential retrofit connection of a micro wind turbine (M-WT) either on the DC side of the power generation or on the AC bus.

Pre-paid Smart meter will be installed at all households. They are able to monitor consumption and detect fraud. The second important ability for smart meter in a vulnerable mini grid connection is the ability for load reduction and curtailments following a period frequency scheme. In cases where a household receives more power than the system allows, the consumer is automatically cut off from service for a short time (e.g., 5 minutes). A continuing violation of the performance restriction prolongs the shutdown periods. These kind of smart meters are currently in use for a testing phase by ESKOM, the South African national energy supplier.

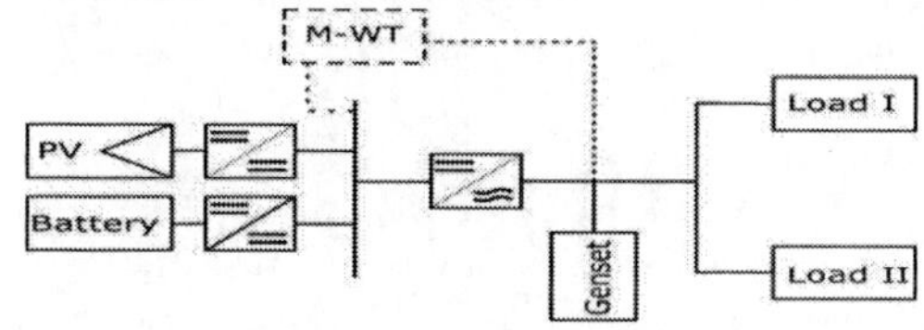

Figure 7: Model of chosen system design

II. First year system response

Annual renewable penetration within the first years is estimated to 98% with a diesel consumption of only 610 liters. The generation system will offer annual excess electrical production of around 57MWh. The highest share of excess power is in the early afternoon, when the battery system is charged and public demand is still low. The system response to a sunny day in the first year is displayed in Figure 8.

The battery systems state of charge (SOC) is still sufficient to supply early morning load until sunrise. During daytime all needs are satisfied with the pv system. The battery system can be charged up to the maximal SOC of around 80%. The high evening peak after sunset will be supplied fully by the battery system. The backup diesel generator is not used during sunny days with adequate battery storage.

Figure 9 shows the maximum hourly load and maximal excess energy. This surplus energy can be used to create and support business models. Thus, the villagers' survey revealed the need for a larger cooling facility to collect and cool the proceeds of gardening such as fruit and vegetables for future sale. In the first year, a mean of around 100 kWh is available for further use.

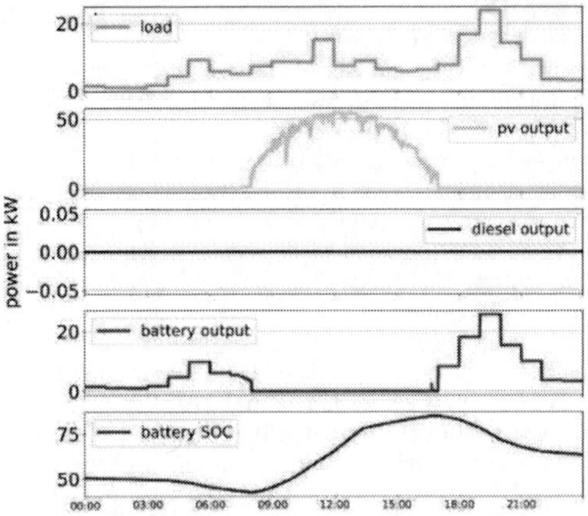

Figure 8: Power system response to a sunny summer day showing from top to bottom the load, pv power, diesel generation, battery output and the battery state of charge

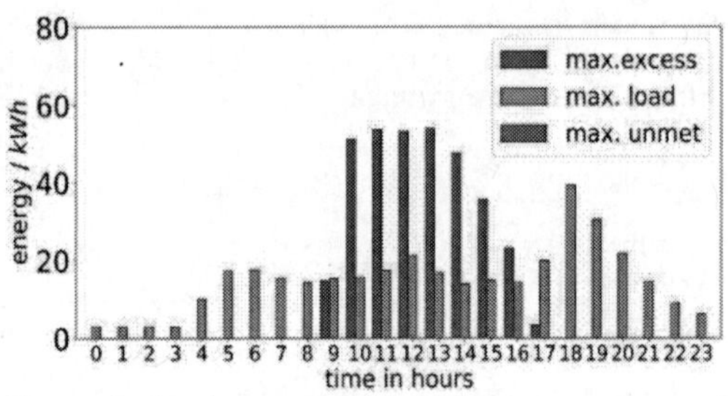

Figure 9: Maximum of daily load, excess energy and unmet load in the first year

III. Tenth year system response

The available surplus energy is strongly reduced and the amount of unmet load is increased (cf. Figure 10). Nonetheless, the share of missing energy is only 1.13 % of the full energy demand. However, the missing power capacity can reach a maximum of 50%. Load shedding is the result.

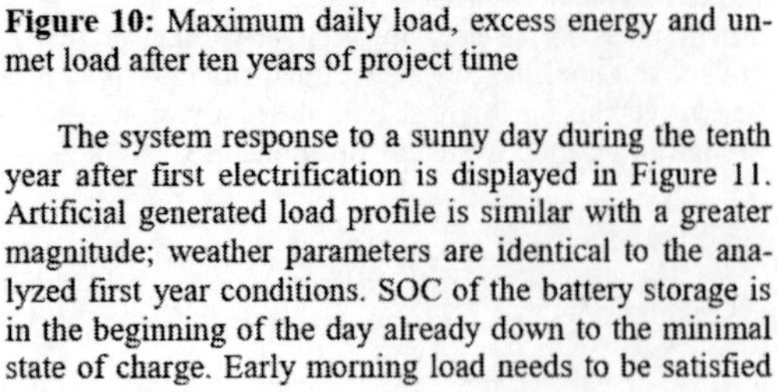

Figure 10: Maximum daily load, excess energy and unmet load after ten years of project time

The system response to a sunny day during the tenth year after first electrification is displayed in Figure 11. Artificial generated load profile is similar with a greater magnitude; weather parameters are identical to the analyzed first year conditions. SOC of the battery storage is in the beginning of the day already down to the minimal state of charge. Early morning load needs to be satisfied by the diesel generator as pv power is not available before sun dawn. The diesel generators minimum power and runtime leads to the spiky pattern in the morning. This can be countered with a smart battery charging system to reduce start and stop cycles. However, nocturnal charging of the batteries can limit the greater use of pv capacities. During day, the pv system is able to meet the power demand and almost fully charge the batteries. The high power consumption in the evening hours cannot be completely covered by the solar system. The SOC drops again to its minimum and the diesel generator covers the additional power and energy required.

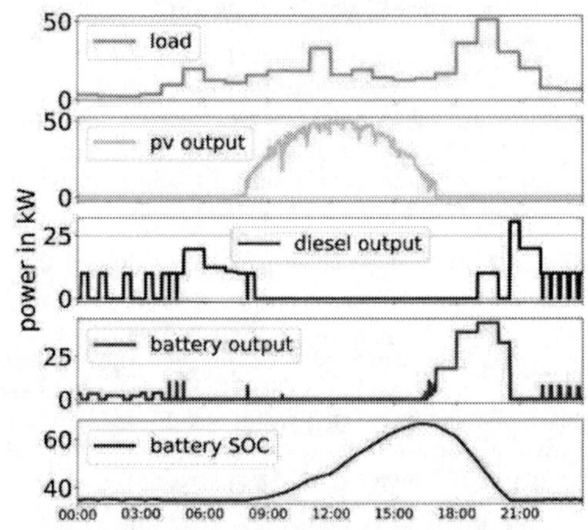

Figure 11: Power system response to a sunny day during the tenths year after electrification.

Increasing load raises the use of fossil fuel to fulfil villagers' needs without further installation of renewable generation capacity. The share of renewable energies in the energy mix is reduced from 90% to 60% within 10 years of the project (cf. Figure 12).

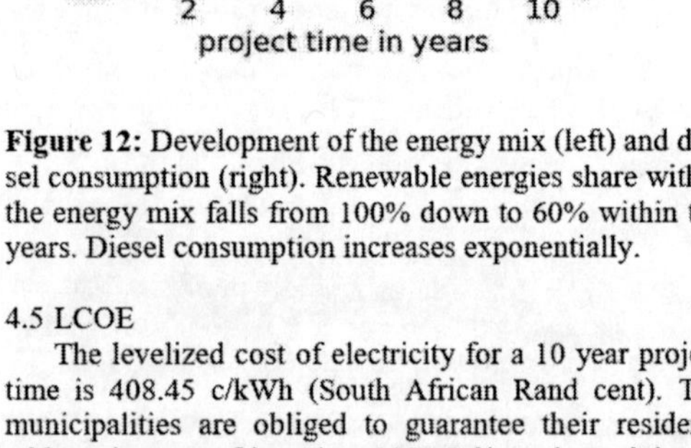

Figure 12: Development of the energy mix (left) and diesel consumption (right). Renewable energies share within the energy mix falls from 100% down to 60% within ten years. Diesel consumption increases exponentially.

4.5 LCOE

The levelized cost of electricity for a 10 year project time is 408.45 c/kWh (South African Rand cent). The municipalities are obliged to guarantee their residents with an income of less than 2000 Rd/month a minimum of 50 kWh/month free allowance. At Raymond Mhlaba Municipality, the cost of electric energy substitutes delivered to rural areas without grid connection exceeded 3 million Rand for 946 households in 2016. Converted to 50 kWh per month, this corresponds to an electricity price of more than 520 c/kWh.

5 CONCLUSION

The designed generation system is able to cover the estimated demand. All parameters given by the "multi-tier framework" of the WORLD BANK Group are successfully covered. But a cost reflecting tariff on basis of LCOE would be too high to be approved by the legislation agency NERSA. Main driving factor on the long term for increasing LCOE is fossil fuel consumption caused by increasing energy demand. Discussions with all stakeholders revealed three possible strategies as counter measures.

5.1 Reduction of energy demand

Access to electrical power introduces a pull effect on the population. This will increase the number of residents in Upper Blinkwater. Ignoring this situation will cause energy fraud and illegal connections to the grid. Usage of smart meters with the ability to curtailment of demand can be used to stabilize the grid and prevent blackouts. An overshot of first consumption will be prevented by the use of prepaid metering. The available power per plot will be reduced in times of very high demand if necessary. A facilitation manager is in contact with the villagers to raise awareness and educate on energy topics like energy efficiency. The use of gas stoves to reduce the peak load on evening is emphasized and supported by a first free batch of funded devices. However, the possibilities to prevent an increase in demand are very low, especially since an increasing demand is also a sign of the desired increase in prosperity.

5.2 Increasing generation capacity

The generation system is designed for the first year demand with some additional capacity for a growing and prospering society. Considering the negligible diesel consumption in the first year and the high surplus energy, a larger design of power generation at the beginning of electrification seems unsuitable. Moreover, the immense investment costs would reduce the chances of replicability to other municipalities or communities without external funds. The definition of key performance parameters to trigger retro-fitting of the generation system is a possible counter measure. Considerations of measured demand profile and mini grid system behavior facilitates the selection of the best power generation technology.

5.3 Adequate framing of a mini grid

LCOE of mini grids are generally higher than those of a main and existing distribution grid with cross-subsidization of generation. A high backlog in electrification and the high cost of extending the distribution network to small and remote locations relocates the considerations also economically to the benefit of mini grids. For public, non-commercial mini grids owned by municipalities we propose to change the view, from an isolated, independent generation and distribution facility towards an small part of the whole municipal distribution system, just with a gab in between the lines. This allows municipalities to fulfil their supply mission to rural areas. At the same time, a non-discriminatory electricity tariff for the poor population can be guaranteed.

All in all, small scale systems are vulnerable to changes. The development of solution strategies for future changed conditions should therefore be considered from the outset in planning.

6 OUTLOOK

Future work in Upper Blinkwater will focus monitoring on the power system, consumption and grid parameters and determination of general load profiles for first electrification of rural villages to close the gap between interviews based demand assumptions and later electrification. Identification of key performance parameters to trigger retrofitting in advance..

Together with SAWEP (South African Wind Energy Program) and the renewable energy group of CSIR (Council for Scientific and Industrial Research, South Africa) a strategy is currently being developed to integrate micro wind turbines into the mini grid. Particular emphasis is placed on supporting the network under development. Especially since wind turbines with the highest energy gain do not offer the best grid support.

7 ACKNOWLEDGEMENT

The authors would like to thanks all local, regional and national authorities in South Africa for the fruitful discussion and support to this pilot project. The huge interest in new solution and willingness to in cooperation is astonishing. We would also like to thank the German project partners and authorities for their practical, financial and ideal support of the Upper Blinkwater pilot project.

7 REFERENCES

[1] CDC Group plc., „Development Impact Evaluation, Evidence Review - "What are the links between power, economic growth and job creation?",“ CDC in partnership with the Overseas Development Institute, 2016.

[2] World Bank Group , „Poverty & Equity Data Portal,“ 2018. Available: http://povertydata.worldbank.org/poverty/country/ZAF. [Zugriff am 21 September 2018].

[3] World Bank Group and IEA, „Access to electricity (% of population),“ August 2018. [Online]. Available: https://data.worldbank.org/indicator/eg.elc.accs.zs. [Zugriff am August 2018].

[4] E. O. A. Elias Hartvigsson, „Comparison of load profiles in a mini-grid: Assessment of performance metrics using measured and interview-based data,“ *Energy for Sustainable Development,* Bd. Volumne 43, pp. 186-195, April 2018.

[5] P. D. H. L. L. K. Courtney Blodgett, “Accuracy of energy-use surveys in predicting rural mini-grid user consumption,” *Energy for Sustainable Development,* vol. Volumne 41, pp. 88-105, December 2017.

IMPROVING THE ACCURACY OF THE NATIONAL SOLAR RADIATION DATABASE (1998–2016)

Manajit Sengupta, Aron Habte, Anthony Lopez and Yu Xie
National Renewable Energy Laboratory, Golden, CO, 80401, USA

ABSTRACT:
The National Solar Radiation Database (NSRDB) provides gridded satellite-based data sets for the United States and other parts of North and South America using cloud properties retrieved from the Geostationary Operational Environmental (GOES) series of satellites. These cloud properties in combination with ancillary atmospheric parameters are used in the Physical Solar Model (PSM) to compute solar radiation. This paper seeks to 1) describe the changes made in the PSM for the recent update to the National Solar Radiation Database (NSRDB 1998–2016) and 2) examine the performance of this update by comparing the data to high-quality ground-based measurements. We summarize the changes to the input data for the radiative transfer models as well as updates for the processing algorithms. We also analyze the performance and accuracy of the NSRDB across the United States. Additional details of the data set and model are described in [1]–[4].

Keywords: NSRDB, PSM, MERRA-2, Aerosol

1 INTRODUCTION

The National Solar Radiation Database (NSRDB) provides a foundational public data set that underpins a large range of research activities that take place within the solar energy industry. The NSRDB data are used as inputs to performance and economic models to improve the accuracy of grid integration studies; facilitate energy modeling using the System Advisor Model (SAM) and PVWatts; enable resource planning using the National Renewably Energy Laboratory's (NREL's) Regional Energy Deployment System (ReEDS), Resource Planning Model (RPM), and Distributed Generation Model (dGen); the U.S. Energy Information Administration's National Energy Modeling System (NEMS) model; as well as production cost modeling using PLEXOS. Therefore, advancing the science used to produce high-quality solar resource data sets is essential to achieve higher penetrations of photovoltaic and concentrating solar power on the grid, reduce integration and deployment costs, and ultimately reduce project risk [5]. As described in [[1]–[4], the current NSRDB, which is V3, is developed using the Physical Solar Model (PSM) V3. The PSM is a unique model that addresses the physics of clouds, aerosol, water vapor, and the primary atmospheric constituents impacting solar radiation. Aerosol and ancillary data sets were obtained from the Modern-Era Retrospective analysis for Research and Applications, Version 2 (MERRA-2) (sample parameters are illustrated in Figure 1).

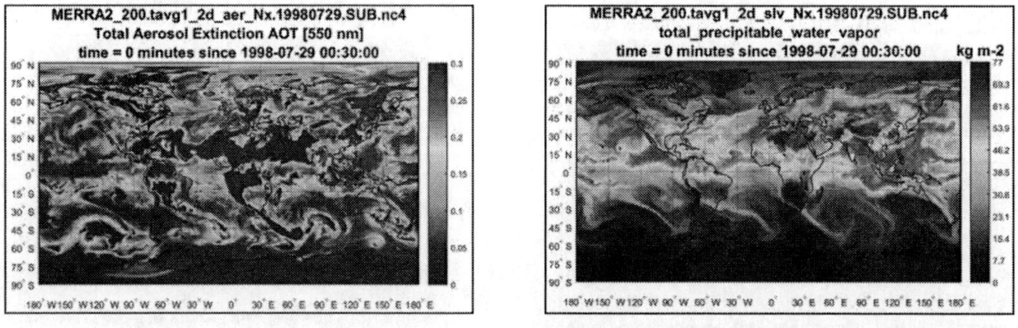

Figure 1: Sample aerosol optical depth and precipitable water vapor data from MERRA-2

2 METHOD

The development of the NSRDB is a computationally intense process because of its high temporal and spatial resolutions and the large volume of diverse input data. Figure 2 demonstrates the complexity of the NSRDB processing and data delivery; however, the use of NREL's high-performance computing capabilities enables the timely processing and delivery of the NSRDB data sets.

The NSRDB update incorporates MERRA-2, which superseded MERRA in 2016. MERRA-2 uses a more advanced data assimilation scheme and better simulates the amount of aerosols and other variables in the atmosphere [6]. In the NSRDB processing, the interpolation and extrapolation methodologies were updated to maintain physical consistency while adapting low-resolution MERRA-2 output to the higher resolution NSRDB. The MERRA-2 variables used in the NSRDB are hourly at a 50-km^2 spatial resolution, which is then interpolated to match the 4-km by 4-km grid and half-hourly temporal resolution. Understanding the impacts of these changes on the solar resource and quantifying the uncertainty are essential for accurately designing utility-scale solar energy conversion systems.

Validation of the NSRDB was carried out by comparing the data to ground-based measurements from the seven National Oceanic and Atmospheric Administration's (NOAA) Surface Radiation Budget Network (SURFRAD) stations. These sites were selected because their ground-based measurements are high quality because the instruments are well calibrated and maintained. The surface data were averaged from 1–3

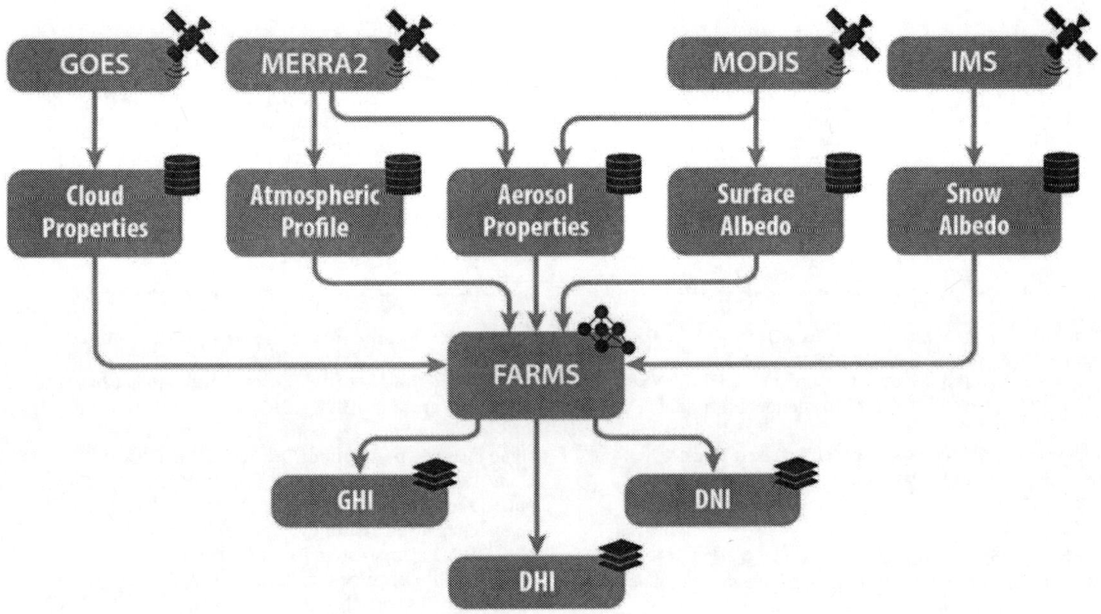

Figure 2: NSRDB processing framework

minutes to 60 minutes (one-half hour before and after the hour) to compare them with the hourly data from the NSRDB.

3 RESULTS AND DISCUSSION

The NSRDB through the PSM model constitutes an important scientific advancement because additional information content provided either from new ancillary data and/or from using newer generation satellites will automatically flow into improvements in the accuracy of the data set. As described in [6], the improvements in MERRA-2 bring observations and numerical models together in a unified standardized framework resulting in high-quality ancillary information. Hence, the PSM model easily assimilated the new MERRA-2 data set. Table 1 shows the MERRA-2 parameters used in the NSRDB and data variables distributed using the NSRDB website (https://nsrdb.nrel.gov).

Validation of the NSRDB 1998–2016 was conducted by comparing it to satellite-modeled and good quality ground-based measurement data. The results showed that satellite-derived hourly data have a systematic error (bias) of less than ±5% for global horizontal irradiance (GHI) and ±10% for direct normal irradiance (DNI) when compared to hourly averaged ground data. Further, as described in [3], we implemented a comprehensive uncertainty determination approach using the Guide to the Expression of Uncertainty in Measurement (GUM) [7]. The approach identifies the source of uncertainties and quantifies standard uncertainties using the measure of statistic for each source. The standard uncertainties are then combined using the root sum squared method (Figure 3). The statistics such as bias and root mean square error were derived by comparing the NSRDB data to the seven measurement stations from NOAA's SURFRAD. The evaluation was conducted for hourly values, daily totals, monthly mean daily totals, and annual mean monthly mean daily totals and demonstrates the quality of the new data sets currently available from the NSRDB (Figure 3).

The hourly data set demonstrated higher uncertainty; however, with increasing average timescales, the uncertainty approaches the measurement uncertainty, e.g., annual. As described in [3], one reason for the high uncertainty in the hourly data set is because the NSRDB pixel represents a 4-km by 4-km area, whereas a ground-based solar measurement represents only a small area above the measuring station. The subpixel variability in clouds appears to contribute to increased differences.

Table 1. Data set in the NSRDB

Data used in the PSM:	Data Delivered to:
• MERRA-2 o Atmospheric pressure o Surface albedo o Aerosols o Aerosol optical thickness o Single-scattering albedo o Aerosol Angstrom parameter o Total ozone o Precipitable water. • GOES (PATMOS-X retrievals) o Cloud effective radius o Cloud optical depth o Cloud type. • Moderate Resolution Imaging Spectroradiometer/National Snow and Ice Data Center o Surface albedo.	• GHI • DNI • Diffuse horizontal irradiance (DHI) • Clear-sky GHI, DNI, and DHI • Cloud type • Dew point** • Air temperature* • Atmospheric pressure • Relative humidity** • Solar zenith angle • Precipitable water • Wind direction** • Wind speed.**

* From MERRA-2
**Recalculated from MERRA-2

The NSRDB V3 demonstrates significant reduction in hourly uncertainty estimations compared to V2, as shown in Fig. 3. These improvements are clearly

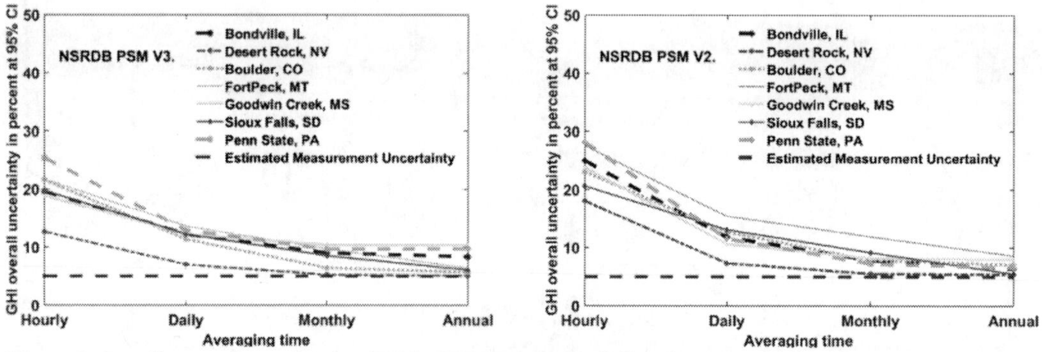

Figure 3: Overall uncertainty estimation GHI for NSRDB PSM V3 (left) and PSM V2 (right) using seven ground-based solar measurement locations. Data used in this figure are from 1998–2015.

noticeable for all locations, especially Desert Rock, where the uncertainty was reduced to ~12% in V3 from ~18% in V2.

The monthly and annual averages for NSRDB V2 appear to be better for the eastern locations; this could be attributed to higher bias in NSRDB V3 where the clear and cloudy condition biases are either both positive or negative. However, NSRDB V2 appears to contain biases of opposing signs, resulting in cancellation of biases (Fig. 4.). This is clearly shown in the mean bias error (MBE) statistics in Fig. 4 for all-sky conditions, which provide some insights in explaining the overall uncertainty estimation described in Fig. 3. The MBE (Fig. 4 left panel) shows a positive bias for the eastern locations and a small negative bias for the western locations in the NSRDB V3. The western locations demonstrate significant improvement under cloud conditions, which are attributed to the changes implemented in NSRDB V3. In contrast, NSRDB V2 shows higher negative bias for the western locations; however, as mentioned, because of a cancellation of errors, the eastern locations showed better results under all-sky conditions.

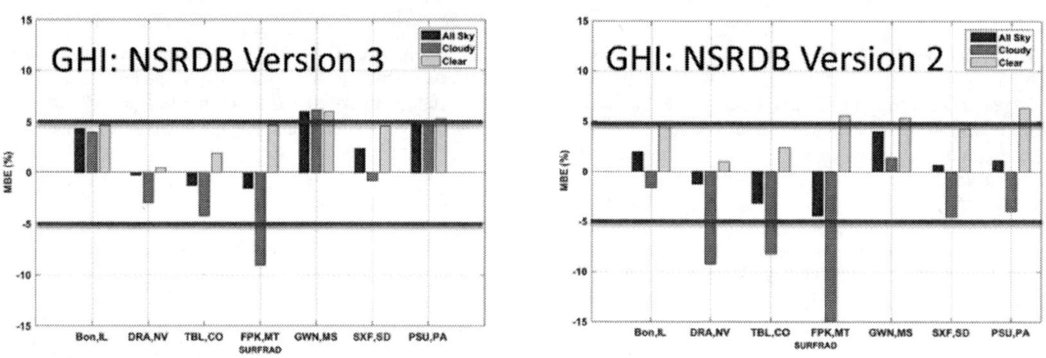

Figure 4: Estimated biases for seven ground-based solar measurement locations for NSRDB V3 and V2. Data used in this figure are from 1998–2015.

4 CONCLUSIONS

Comparisons of the NSRDB (1998–2016) V3 were assessed by comparing the data with selected ground-measured data under both clear- and cloudy-sky conditions and covered the period from 1998–2015. The results show improvement compared to the previous version of the NSRDB. The bias is ~ ±5%, and the overall uncertainty reduced significantly on the hourly biases. Overall, the current NSRDB shows less uncertainty compared to the NSRDB 1998–2015 data (PSM V2). The improvement is assumed to result from a combination of factors, including better downscaling methodologies, particularly in the interpolation and extrapolation used to align the multiple data sets to the same grid, the use of hourly values of aerosol information, and the use of surface albedo instead of interpolated monthly averages. Using hourly precipitable water vapor from MERRA-2 might have marginally contributed to the improvement too.

ACKNOWLEDGMENTS

This work was authored by Alliance for Sustainable Energy, LLC, the Manager and Operator of the National Renewable Energy Laboratory for the U.S. Department of Energy (DOE) under Contract No. DE-AC36-08GO28308. Funding provided by U.S. Department of Energy Office of Energy Efficiency and Renewable Energy Solar Energy Technologies Office. The views expressed in the article do not necessarily represent the views of the DOE or the U.S. Government. The U.S. Government retains and the publisher, by accepting the article for publication, acknowledges that the U.S. Government retains a nonexclusive, paid-up, irrevocable, worldwide license to publish or reproduce the published form of this work, or allow others to do so, for U.S. Government purposes.

We acknowledge the DOE Solar Energy Technologies Office and its Systems Integration subprogram for supporting this research. Specifically, we thank Dr. Tassos

Golnas, Technology Manager of Systems Integration, for his support and encouragement.

REFERENCES

[1] Y. Xie, M. Sengupta, and J. Dudhia. "A Fast All-Sky Radiation Model for Solar Applications (FARMS): Algorithm and Performance Evaluation," Solar Energy 135: 435–445, 2016.

[2] M. Sengupta, A. Weekley, A. Habte, A. Lopez, C. Molling, and A. Heidinger, "Validation of the National Solar Radiation Database (NSRDB) (2005–2012): Preprint," Paper presented at the European PV Solar Energy Conference and Exhibition, Hamburg, Germany, September 14–18, 2015.

[3] A. Habte, M. Sengupta, A. Lopez, "Evaluation of the National Solar Radiation Database (NSRDB): 1998-2015," NREL/TP-5D00-67722, 2017.

[4] M. Sengupta, Y. Xie, A. Lopez, A. Habte, G. Maclaurin, J. Shelby (in press), The National Solar Radiation Data Base, Renew. Sustain. Energy Rev.

[5] M. Sengupta, A. Habte, C. Gueymard, S. Wilbert, D. Renné. (2017). Best Practices Handbook for the Collection and Use of Solar Resource Data for Solar Energy Applications: Second Edition.

[6] Gelaro R, McCarty W, Suárez M, Todling R, Molod A, Takacs L, et al. The modern-era retrospective analysis for research and applications, version 2 (merra-2). J Climate. 2017;30:5419-54

[7] Joint Committee for Guides in Metrology Working Group 1 (JCGM/WG 1). (2008). Guide to the Expression of Uncertainty in Measurement. www.bipm.org/utils/common/documents/jcgm/JCGM_100_2008_E.pdf. Accessed January 08, 2018.

Disaggregation of Local Photovoltaic Generation from Composite Power Flows with Direct Measuring and Satellite Estimations of the Irradiance: a Comparison

Fabrizio Sossan, Enrica Scolari, Mario Paolone
Distributed Electrical Systems Laboratory (DESL)
EPFL, École Polytechnique Fédérale de Lausanne, Switzerland
{fabrizio.sossan, enrica.scolari, mario.paolone}@epfl.ch

***Abstract*—We consider the problem of estimating the unobserved PV generation and demand from aggregated measurements taken at a point of common coupling, e.g., at the grid connection point of a radial distribution system. We outline the formulation of an existing algorithm that achieves the disaggregation task by using measurements of the composite power flow and the local global horizontal irradiance (GHI), and compare its performance when the estimates of the local GHI are from a local pyranometer and when from satellite, considering HelioClim-3v.5 estimations. In the comparison, we use measurements from a real-life setup that includes four houses with rooftop PV installations and battery systems for PV self-consumption. Results show that the training performed with pyranometer measurements is more reliable than using satellite estimations.**

***Index Terms*—Disaggregation, PV generation, Demand, Distribution Systems, Distributed Renewable Generation**

I. Introduction

Although the decreasing price of photovoltaic (PV) panels is fostering the deployment of distributed PV generation, the increasing number of PV installations is a concern for electrical operators because it determines *i)* increased reserve requirements at the system level, and *ii)* overvoltages and line ampacity violations in low voltage and medium voltage distribution systems, which are normally characterized by a limited hosting capacity, e.g., [1], [2]. Solutions to mitigate the effects associated to large PV penetration levels are PV self-consumption with flexible demand and battery systems [3]–[5], coordination of the operation of PV facilities [6]–[8], integrated planning of energy systems and storage [9], [10], and dispatching PV systems according to pre-established and network-safe trajectories [11], [12]. A key prerequisite to enable the efficient monitoring of local distribution systems and the control of the associated distributed energy resources is the availability of real-time estimations of the power output of PV installations. Solutions based on the direct monitoring of all PV installations are pervasive and communication intensive, therefore unlikely. Moreover, factors like conflicts due to the ownership of the metering infrastructures, privacy concerns, lack of standards for monitoring, communication and aggregation of measurements, and the fact of relying on non redundant communication channels subject to failures, are barriers inhibiting the possibility of harvesting real-time information from distributed PV facilities.

To overcome issues related to the limited observability of distributed PV generation, the problem of estimating PV production from aggregated power flow measurements has been considered in the existing literature. To introduce the problem, let us refer to the setup sketched in Fig. 1, i.e. a radial feeder with unobserved distributed PV production and unobserved demand connected to the upper level network through a point of common coupling (PCC), where PV generation is possibly from multiple sites with different tilt and azimuth configurations of the panels. This configuration is commonly encountered in low and medium voltage distribution systems for commercial and residential areas, therefore it is representative of a wide class of networks and worth of being considered as a case study. The problem is estimating the unobserved amount of PV generation.

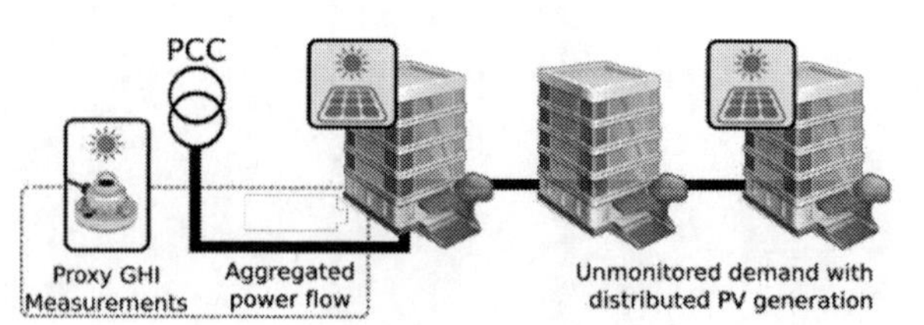

Fig. 1. An electrical grid with unmonitored demand and PV generation from multiple production sites with different azimuth/tilt configurations. Disaggregation algorihms exploit measurements of the active power flow at the point of common coupling and global horizontal irradiance (GHI) to identify PV generation patterns and estimation PV panels' properties.

Algorithms proposed in the literature use measurements of the power flow at the PCC and estimations of the local global horizontal irradiance (GHI) to disaggregate the profile of PV generation from the whole. Estimators described in [13] use measurements of the GHI or the production of monitored PV installations as a proxy signal to explain PV production patterns. The work in [14] proposes tractable estimation algorithms based on convex optimization which, with respect to [13], leverage transposition models to explain PV generation patterns from installation with different tilt and azimuth configurations of the panels. This is a key feature for applications to urban areas, where panels are normally installed in a non uniform way according to roof characteristics. In this paper, we consider one of the disaggregation algorithms from [14] and investigate its performance when the input GHI measurements are from a local pyranometer versus satellite estimations. Satellite estimations are less invasive than local measurements based on pyranometers (or alternative methods for the estimation of the local GHI, like exploiting information from nearby monitored installations [15]–[18] or clear-sky camera [19], [20]) and are commercially available for extended geographical areas although having higher spatial resolution, e.g. 3 km at nadir for Europe. In this paper, we verify the performance of satellite estimations in application to PV disaggregation and how they perform compared to using local estimations.

The paper is organized as follows. Section II summarizes the modelling principles of the PV disaggregation algorithm used in this work, section III and III present and discuss results, respectively, and Section IV concludes the paper.

II. On the Adopted PV Disaggregation Algorithm

A. *General principles and assumptions*

For the sake of clarity and remarking where the information on the GHI is used, we summarize here the principles and formulation of one of the disaggregation algorithms proposed in [14], i.e. so-called Method C. The algorithm uses historical measurements of the active power flow at the point of common coupling and GHI and attempts to estimate the amount of installed PV capacity by identifying PV generation patterns in the aggregated power flow. Reactive power is not considered in the estimation process because it is not representative of the PV production, as PV plants normally operate at unitary power factor and it is difficult to do assumptions on potential reactive power control policies implemented at the PV converter level. The algorithm initially requires a training phase, illustrated in Fig. 2a: historical GHI measurements are transposed onto a number of tilted planes equally spaced on a south facing semi-sphere and used to create exploratory signals of the tilted irradiance (global normal irradiance, GNI), which are representative of potential PV generation patterns from production sites with different tilt and azimuth configurations. This is a key element, especially in urban and suburban distribution feeders, where PV panels are installed according to roof characteristics. Exploratory signals are used in a convex optimization problem, detailed later, together with a model of the electrical demand, that estimates the installed PV capacity at each considered tilt/azimuth configuration. Once the installed PV capacity for each configuration is known, it can be used in real-time to disaggregate the total active power flow in the trajectories of PV generation and demand, as highlighted in Fig. 2. The performance of the disaggregation procedure can be affected by the presence of elements performing self-consumption; in particular, demand patterns (e.g., storage or flexible demand) strongly correlated with the PV self consumption would make PV generation in the aggregated active power flow measurements less prominent, and, therefore, more difficult to be identified. However, this greatly depends on the resolution of the control action of flexible demand/batteries and measurements: e.g., actuations at 15 min and measurements at 10 sec would leave a large proportion of the spectrum untouched. In [14], a slight degradation of the estimation performance was observed when using power flow measurements that included elements implementing PV self-consumption.

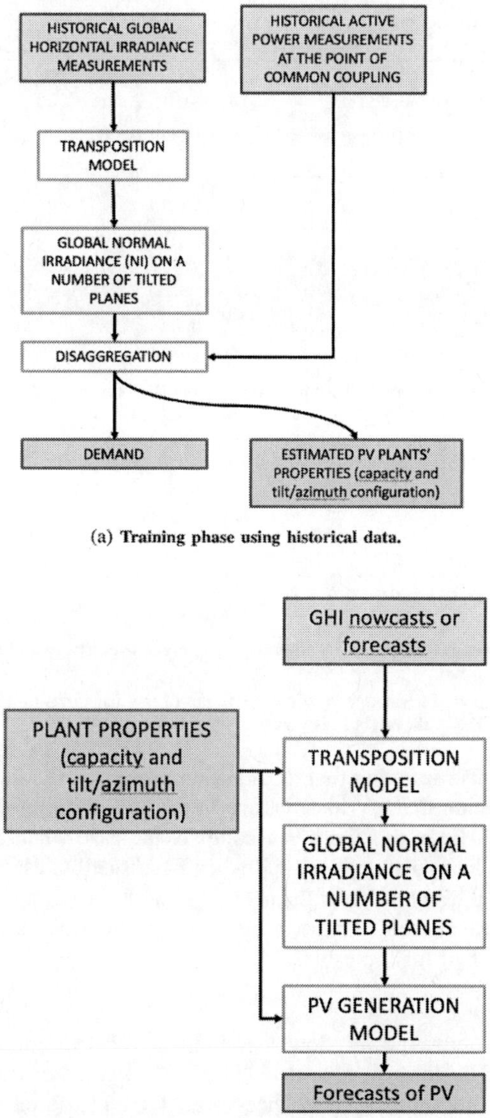

Fig. 2. Operation principles of the PV disaggregation algorithm used in this work.

Algorithm's operations are exemplified in Fig. 3, where night time observations are omitted. Fig. 3a shows the inputs, intermediate results and outputs of the algorithm. The inputs are the aggregated power flow P_k at the PCC (top panel) and GHI (middle panel, solid fill), measured with a pyranometer in this case. Fig. 3b shows the trajectories of the transposed irradiance, used to explore the PV production of plants with different tilt/azimuth configurations. Fig. 3c shows the output of one of the proposed algorithms, with the estimated demand and estimated PV generation (orange line), the latter close to the measured ground truth PV generation (solid gray fill).

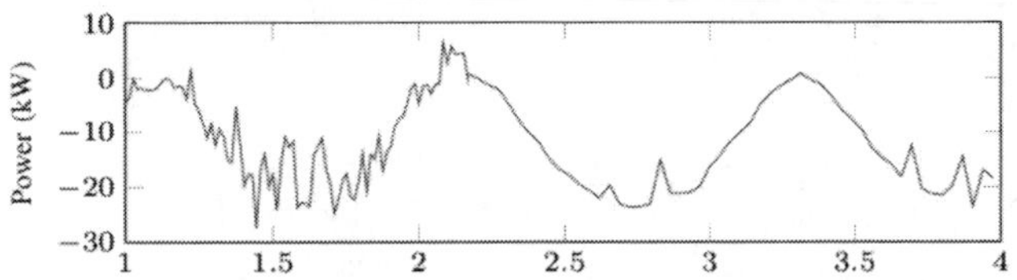

(a) **Observed active power flow at the PCC.**

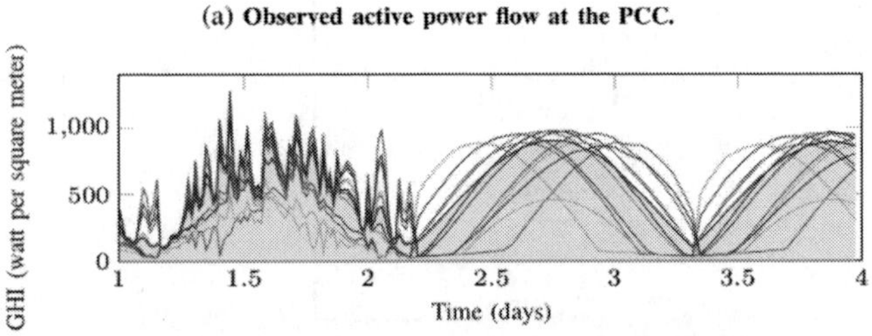

(b) **Exploratory GNI profiles derived from GHI observations (solid).**

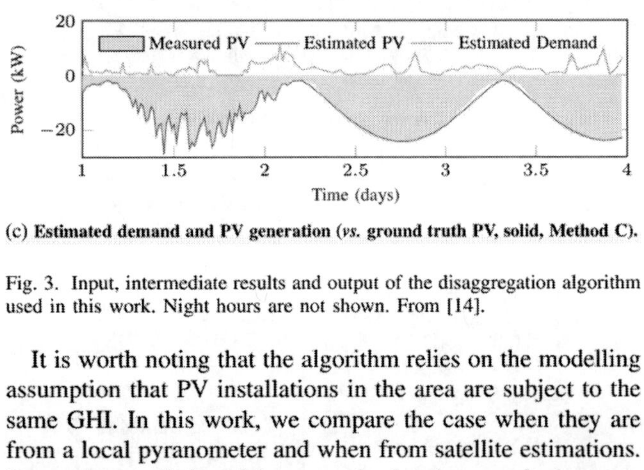

(c) **Estimated demand and PV generation (*vs.* ground truth PV, solid, Method C).**

Fig. 3. Input, intermediate results and output of the disaggregation algorithm used in this work. Night hours are not shown. From [14].

It is worth noting that the algorithm relies on the modelling assumption that PV installations in the area are subject to the same GHI. In this work, we compare the case when they are from a local pyranometer and when from satellite estimations. Due to the small size of the networks that these methods target, grid transmission losses are neglected at this stage due to the short length of the cables.

B. Notation

The active power flow measured at the PCC at the discrete time interval k is denoted as P_k (kW). Positive flows denote consumption and vice-versa. GHI measurements are denoted by $\mathcal{I}_k^{-}$ in $\mathrm{kW\,m^{-2}}$, while $I_{jk}^{\searrow}$ ($\mathrm{kW\,m^{-2}}$) denotes the estimated global normal irradiance (GNI) to a certain tilted plane j, where $j = 1, \ldots, J$ denotes the plane tilt/azimuth configuration. We consider $J = 21$ tilted planes with tilt and azimuth values equally spaced on a south-facing semi-sphere, chosen to have a reasonably representative set of potential configurations of PV installations in the northern hemisphere. GNI estimations are from the Hay-Davies transposition model [21], [22]. The quantity $\widehat{G}_k$ and $\widehat{L}_k$ (kW) respectively denote the estimated PV production and estimated demand, which are to be determined. Series are sampled at resolution T_s.

C. Formulation

a) Model of the PV generation: The estimated global PV generation $\widehat{G}_k$ at the PCC is modelled as the sum over all the tilt/azimuth configurations $j = 1, \cdots, J$ of the transposed irradiance $I_{jk}^{\searrow}$ times J nonnegative scalars $\alpha_j \in \mathbb{R}_+$ (i.e. unknown and to identify) that represent the PV generation capacity available at configuration j. Formally, it is:

$$\widehat{G}_k(\boldsymbol{\alpha}) = \sum_{j=1}^{J} \alpha_j \cdot I_{jk}^{\searrow}, \qquad k = 1 \ldots, K, \tag{1}$$

where $\boldsymbol{\alpha} = \{\alpha_j, j = 1, \cdots, J\} \in \mathbb{R}_+^J$ is the set of variables to identify. Coefficients α_j are kilowatt peak, kWp, and refer to the power produced in standard test conditions (STC, i.e. $\mathrm{kW\,m^{-2}}$ GNI at 20 °C). The temperature effect on the PV conversion efficiency is accounted for by preprocessing the input irradiance time series using the model described in [23], see [14] for further details. By modelling the PV generation as in (1), we assume that PV plants operate in the *maximum power point tracking* (MPPT) mode. Plants which might operate in curtailed mode, are likely to be directly monitored and the amount of their generation can be subtracted from the aggregated demand, still enabling to apply the disaggregation algorithm to estimate the amount of unobserved generation. Partial shading effects, although not explicitly modelled, does not impact significantly the operation of the algorithm, as discussed in [14].

b) Estimating α: Let $\widehat{\boldsymbol{L}} = [\widehat{L}_1, \ldots, \widehat{L}_K]$ be the estimated demand trajectory. The estimated active power flow at the PCC $\widehat{P}_k$ is the sum between the estimated total PV generation and demand

$$\widehat{P}_k(\boldsymbol{\alpha}, \widehat{\boldsymbol{L}}) = \widehat{L}_k - \widehat{G}_k(\boldsymbol{\alpha}), \qquad k = 1 \ldots, K, \tag{2}$$

the former with negative sign because it corresponds to generation. The PV disaggregation algorithm attempts to determine $\widehat{L}_k$ and $\boldsymbol{\alpha}$ by minimizing the norm-2 of the estimation error $P_k - \widehat{P}_k$. This problem is however under-determined since there are more free variables than observations. Therefore, we augment the cost function and consider a piecewise constant model of the demand, i.e. we require the unknown sequence $\widehat{\boldsymbol{L}}$ to be piecewise constant for c consecutive samples, where c is a parameter. For the first c samples, this is achieved by enforcing the following $c - 1$ equality constraints:

$$\widehat{L}_1 = \widehat{L}_2 = \cdots = \widehat{L}_c. \tag{3}$$

Extending to the set of K measurements (K multiple of c) yields:

$$\widehat{L}_{c(i-1)+1} = \cdots = \widehat{L}_{c(i-1)+c} \qquad i = 1, \ldots, K/c. \tag{4}$$

Modelling the demand as piecewise constant is a reasonable assumption when the length of the constant segment does not overlap with typical intra-day dynamics of the demand, i.e. for small c values and densely sampled series. In other words, it is reasonable when considering short periods of time (e.g., seconds), when the persistence model of the demand has unbeaten performance, see, e.g., [24]. When the demand has shorter variations than c (e.g., load inrushes), the estimated demand will have the average value of the waveform and the residuals will be the estimation error. The sampling time of the input time series T_s and c are design parameters: in [14], it was shown that they can be chosen in certain wide ranges, i.e. $T_s \in (30, 400)$ s and $c \in (5, 15)$, to have good and quite stable estimation performance.

Formally, the PV disaggregation problem consists in minimizing the norm-2 of the estimation error $P_k - \widehat{P}_k$ subject to the models of the estimated aggregated power flow and piecewise constant demand:

$$\begin{bmatrix} \boldsymbol{\alpha}^o \\ \widehat{\boldsymbol{L}}^o \end{bmatrix} = \underset{\begin{bmatrix} \boldsymbol{\alpha} \in \mathbf{R}_+^J \\ \widehat{\boldsymbol{L}} \in \mathbf{R}_+^K \end{bmatrix}}{\arg\min} \left\{ \sum_{k=1}^{K} \left(P_k - \widehat{P}_k(\boldsymbol{\alpha}, \widehat{\boldsymbol{L}}) \right)^2 \right\} \quad (5)$$

subject to:

$$\widehat{P}_k(\boldsymbol{\alpha}, \widehat{\boldsymbol{L}}) = \widehat{L}_k - \sum_{j=1}^{J} \alpha_j \cdot I_{jk}^{\searrow}, \qquad k = 1, \ldots, K \quad (6)$$

$$\widehat{L}_{c(i-1)+1} = \cdots = \widehat{L}_{c(i-1)+c} \qquad i = 1, \ldots, K/c. \quad (7)$$

The problem is a the combination of a quadratic cost function and linear constraints, thus convex if the former is convex. The convexity of the quadratic term depends on the input data and can be verified a-priori as discussed in [14].

III. Results and Discussion

A. Case study

Measurements used in this paper are from a real-life setup in the region of Basel, Switzerland, with 4 private households, each equipped with a rooftop PV installation. The experimental setup spans an area of $0.07\,\mathrm{km}^2$. PV converters operate in MPPT mode at unitary power factor. The total PV installed capacity is 35.3 kWp and the peak demand is 12 kWp. Measurements are: the power flow at the PCC, GHI (i.e., from a local pyranometer or from satellite estimations) and PV generation measurements for 1 year. The first 2 are used for the training, while the last as the ground truth value to test the algorithm's estimation performance. Pyranometer measurements, power flow at the PCC and PV production are sampled at 30 s. Satellite GHI estimations are from the HelioClim-3v.5 database and are provided by SoDa Service [25]. They are based on Meteosat images, which are sampled at 15 min intervals and have 3 km spatial resolution at nadir. HelioClim-3v.5 estimations are known to be slightly positively biased, and the commercial provider suggests to apply a non-linear regression model fitted on ground measurements to correct them [26], [27]. In this case, we do not perform any de-biasing as we would invalidate the principle of *unsupervised* training. Satellite estimates are downsampled to 30 s by linear interpolation. The whole datasets is divided into daily sequences to preserve daily dynamics. Night time observations are omitted because not informative about the PV generation. The daily sequences are randomly shuffled (i.e. to avoid to train and test the algorithms on different periods of the year) and divided into 3 datasets to perform three-fold cross-validation, i.e. the algorithm is trained on the first fold and tested on the remaining 2. It was verified that measurements refer to days with a uniform mix of sky conditions: partly cloudy, clear-sky and overcast.

B. Discussion

The disaggregation results from a single run of the three-fold process at 30 sec resolution are shown in figures 4 and 5, for GHI measurements from the local pyranometer and satellite estimates, respectively. Night hours are omitted. As visible, when using satellite estimates, *PV (Estimated)* is substantially different than ground truth measurements *PV (Measured)*, denoting that the algorithm is not able to reconstruct accurately PV generation patterns. Statistics (i.e., minimum, maximum and mean) of the *root mean square error* (RMSE) and of the *mean error* (ME) for the estimation error (i.e., ground truth minus estimated value) are reported in Table I, during the complete three-fold process at 30 sec resolution. All metrics score in favour of using pyranometer estimations. When using pyranometer estimates, the algorithm achieves nearly unbiased estimations, as denoted by metric ME, which scores single percentage digit values. In particular, the mean value of ME scores 3.5%, and the minimum and maximum values are 4.5% and 2.6%, denoting that the interval of variations of ME is small.

A second test was performed by using the original satellite estimations at 15 min and downsampling the rest of the series to 15 min by averaging samples. The disaggregation performance with satellite estimations scores similar average results as the former, whereas it decreases by about 30% (i.e., 13% mean RMSE) with pyranometers. However, local GHI estimations largely achieve better performance than satellite estimations.

The reason for the poor performance of the satellite GHI estimations in the disaggregation process is the lack of temporal resolutions. In particular, by using temporally sparse measurements, the algorithm fails in associating fast fluctuations of the power flow at the PCC with PV generation patterns.

TABLE I
Estimation Performance Summary at 30 s Resolution

Metric	Stat	Satellite Estimations	Pyranometer
RMSE	min	22.2	7.9
	max	58.4	11
	mean	34.8	9.3
ME	min	11.39	2.6
	max	39.8	4.5
	mean	21.1	3.5

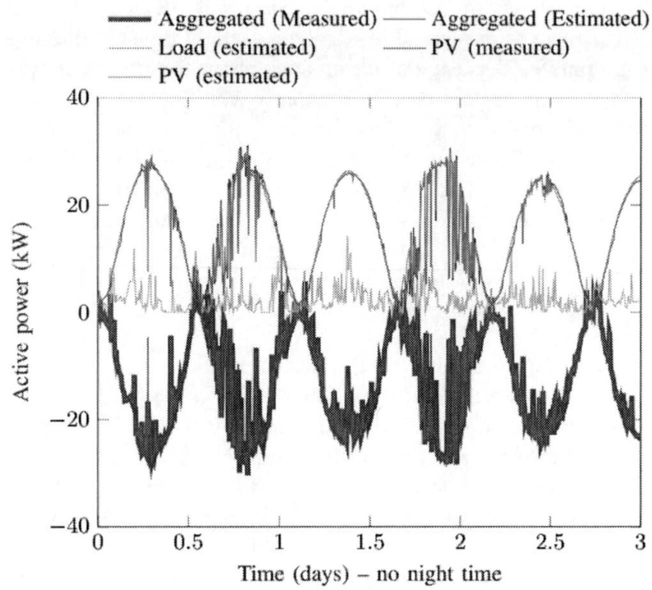

Fig. 4. Performance of the disaggregation algorithm with GHI estimations from pyranometers.

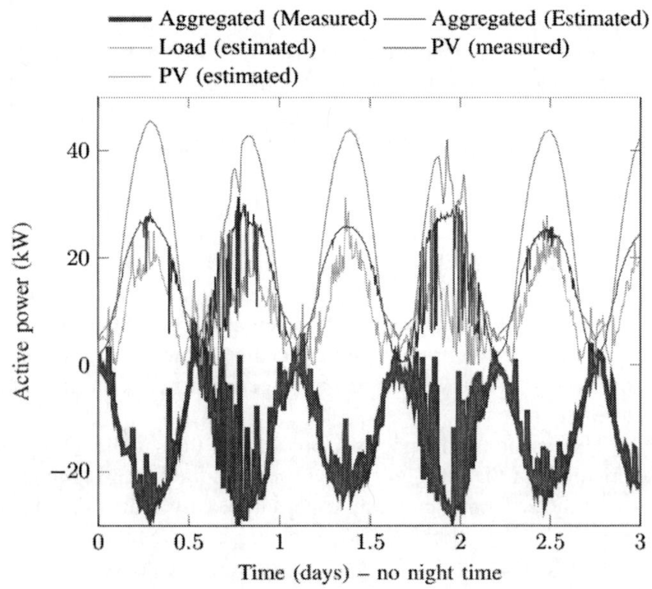

Fig. 5. Performance of the disaggregation algorithm with GHI estimations from satellite.

IV. Conclusions

We considered the problem of estimating the unobserved PV generation and demand from composite power flow profiles measured at a point of common coupling, e.g., at the grid connection point of a radial distribution system. We outlined the formulation of an existing algorithm from the literature that achieves the disaggregation task by using measurements of the composite power flow and of the local GHI. The algorithm uses, first, a transposition model to determine the global normal irradiance (GNI) for a number of tilted planes (i.e., 21) as a function of the GHI, and second, combines power flow measurements and GNI estimations in an optimization problem that determines the installed capacity for each plane.

We have compared the performance of the disaggregation algorithm when the estimates of the local GHI are from a local pyranometer and when from satellite (i.e., HelioClim-3v.5 from Soda service). Results showed that the training performed with pyranometer measurements outperforms the case when satellite estimations are used. In particular, the root mean square and the mean of the estimation error are in average three and eight times larger, respectively, for satellite estimations. When trained with measurements from the pyranometers, the disaggregation algorithm achieves nearly unbiased estimations of the PV generation.

References

[1] A. Ballanti and L. F. Ochoa, "On the integrated pv hosting capacity of mv and lv distribution networks," in *Innovative Smart Grid Technologies Latin America (ISGT LATAM), 2015 IEEE PES*. IEEE, 2015, pp. 366–370.

[2] F. Sossan, J. Darulova, M. Paolone, A. Kahl, S. J. Bartlett, and M. Lehning, "Large scale deployment of pv units in existing distribution networks: Optimization of the installation layout," in *2016 Power Systems Computation Conference (PSCC)*, June 2016.

[3] Y. Zong, L. Mihet-Popa, D. Kullmann, A. Thavlov, O. Gehrke, and H. Bindner, "Model predictive controller for active demand side management with pv self-consumption in an intelligent building," in *IEEE International Conference on Innovative Smart Grid Technologies (ISGT)*, 2012.

[4] F. Sossan, A. M. Kosek, S. Martinenas, M. Marinelli, and H. W. Bindner, "Scheduling of domestic water heater power demand for maximizing PV self-consumption using model predictive control," in *IEEE International Conference on Innovative Smart Grid Technologies (ISGT)*, 2013.

[5] R. Luthander, J. Widén, D. Nilsson, and J. Palm, "Photovoltaic self-consumption in buildings: A review," *Applied Energy*, vol. 142, pp. 80–94, 2015.

[6] O. Gagrica, P. Nguyen, W. Kling, and T. Uhl, "Microinverter curtailment strategy for increasing photovoltaic penetration in low-voltage networks," *Sustainable Energy, IEEE Transactions on*, vol. 6, no. 2, pp. 369–379, April 2015.

[7] S. Ghosh, S. Rahman, and M. Pipattanasomporn, "Distribution voltage regulation through active power curtailment with pv inverters and solar generation forecasts," *IEEE Transactions on Sustainable Energy*, vol. 8, no. 1, pp. 13–22, 2017.

[8] R. Gupta, F. Sossan, E. Scolari, E. Namor, L. Fabietti, C. Jones, and M. Paolone, "An ADMM-based coordination and control strategy for PV and storage to dispatch stochastic prosumers: Theory and experimental validation," in *Power Systems Computation Conference (PSCC) 2018*, 2018.

[9] G. Celli, E. Ghiani, S. Mocci, and F. Pilo, "A multiobjective evolutionary algorithm for the sizing and siting of distributed generation," *IEEE Transactions on Power Systems*, vol. 20, no. 2, pp. 750–757, May 2005.

[10] M. Nick, R. Cherkaoui, and M. Paolone, "Optimal siting and sizing of distributed energy storage systems via alternating direction method of multipliers," *International Journal of Electrical Power & Energy Systems*, vol. 72, pp. 33–39, 2015.

[11] F. Conte, S. Massucco, M. Saviozzi, and F. Silvestro, "A stochastic optimization method for planning and real-time control of integrated pv-storage systems: Design and experimental validation," *IEEE Transactions on Sustainable Energy*, pp. 1–1, 2017.

[12] E. Stai, L. Reyes-Chamorro, F. Sossan, J. Y. L. Boudec, and M. Paolone, "Dispatching stochastic heterogeneous resources accounting for grid and battery losses," *IEEE Transactions on Smart Grid*, pp. 1–1, 2017.

[13] E. C. Kara, C. M. Roberts, M. Tabone, L. Alvarez, D. S. Callaway, and E. M. Stewart, "Disaggregating solar generation from feeder-level measurements," *Sustainable Energy, Grids and Networks*, vol. 13, pp. 112 – 121, 2018. [Online]. Available: http://www.sciencedirect.com/science/article/pii/S2352467717301169

[14] F. Sossan, L. Nespoli, V. Medici, and M. Paolone, "Unsupervised disaggregation of photovoltaic production from composite power flow measurements of heterogeneous prosumers," *IEEE Transactions on Industrial Informatics*, 2018.

[15] A. Laudani, F. R. Fulginei, A. Salvini, M. Carrasco, and F. Mancilla-David, "A fast and effective procedure for sensing solar irradiance in photovoltaic arrays," in *2016 IEEE 16th International Conference on Environment and Electrical Engineering (EEEIC)*, June 2016, pp. 1–4.

[16] M. Carrasco, F. Mancilla-David, and R. Ortega, "An estimator of solar irradiance in photovoltaic arrays with guaranteed stability properties," *IEEE Transactions on Industrial Electronics*, vol. 61, no. 7, pp. 3359–3366, 2014.

[17] E. Scolari, F. Sossan, and M. Paolone, "Photovoltaic model-based solar irradiance estimators: Performance comparison and application to maximum power forecasting," *IEEE Transactions on Sustainable Energy*, 2017.

[18] L. Nespoli and V. Medici, "An unsupervised method for estimating the global horizontal irradiance from photovoltaic power measurements," *Solar Energy*, vol. 158, pp. 701 – 710, 2017. [Online]. Available: http://www.sciencedirect.com/science/article/pii/S0038092X17309076

[19] E. Scolari, F. Sossan, M. Haure-Touze, and M. Paolone, "Local estimation of the global horizontal irradiance using an all-sky camera," *Solar Energy*, vol. 173, pp. 1225 – 1235, 2018.

[20] B. Kurtz and J. Kleissl, "Measuring diffuse, direct, and global irradiance using a sky imager," *Solar Energy*, vol. 141, pp. 311 – 322, 2017. [Online]. Available: http://www.sciencedirect.com/science/article/pii/S0038092X16305722

[21] J. E. Hay and J. A. Davies, "Calculation of the solar radiation incident on an inclined surface," in *Proc. of First Canadian Solar Radiation Data Workshop (Eds: JE Hay and TK Won), Ministry of Supply and Services Canada*, vol. 59, 1980.

[22] E. L. Maxwell, "A quasi-physical model for converting hourly global horizontal to direct normal insolation," Solar Energy Research Inst., Golden, CO (USA), Tech. Rep., 1987.

[23] J. A. Kratochvil, W. E. Boyson, and D. L. King, "Photovoltaic array performance model." Sandia National Laboratories, Tech. Rep., 2004.

[24] N. Hatziargyriou, *Microgrids: Architectures and Control*, ser. Wiley - IEEE. Wiley, 2014.

[25] Soda service, helioclim-3 archive. [Online]. Available: http://www.soda-pro.com/web-services

[26] C. Vernay, P. Blanc, and S. Pitaval, "Characterizing measurements campaigns for an innovative calibration approach of the global horizontal irradiation estimated by helioclim-3," *Renewable Energy*, vol. 57, pp. 339 – 347, 2013. [Online]. Available: http://www.sciencedirect.com/science/article/pii/S0960148113000876

[27] S. S. R. Data. (As seen on July 2018) Validation protocol. http://www.soda-pro.com/help/general/validation-protocol.

MODELING REFLECTED IRRADIANCE IN URBAN ENVIRONMENTS – A CASE STUDY FOR SIMULATION-BASED MEASUREMENT QUALITY CONTROL FOR AN OUTDOOR PV TEST SITE

Ádám Bognár[1], Roel Loonen[1], Roland Valckenborg[2], Jan Hensen[1]
[1]Building Physics and Services, Eindhoven University of Technology, Postbus 513 5600 MB, Eindhoven, the Netherlands
[2]Solar Energy Application Centre (SEAC), High Tech Campus 21, Eindhoven, the Netherlands

ABSTRACT: The increased application of PV in urban environments goes together with a growing awareness of the need to consider the influence of partial shading and reflected irradiance on PV performance. Most of the conventional PV simulation tools do not have the capabilities to account for the possible additional PV yield due to reflected sunlight. On the other end of the spectrum are complex ray-tracing tools that can accurately model reflected irradiance, but require a large amount of input data, are computationally intensive and have limited or no PV modeling algorithms. In this work, as a step towards quantifying uncertainties due to reflected irradiance in urban settings, the potential of a modeling strategy using building energy simulation software, with intermediate complexity is investigated using a case study of an outdoor PV test site.
Keywords: Simulation; Solar radiation; Pyranometer; Building integration, EnergyPlus

1 INTRODUCTION

The growth of the solar market in the 2010s exceeded most expectations. In 2017 (99.1 GW), the installed new PV capacity was almost as large as the total capacity in 2012 (100.9 GW) [1]. The main driving factors for such growth are ground-mounted utility-scale solar applications, which are becoming an increasingly attractive investment due to the decreasing prices of PV modules and system components. State-of-the-art PV simulation software are well-suited to predict the performance of such applications, partly because they are typically installed in an ideal location, chosen to be well-suited for maximizing PV yield. The situation of PV integration in the urban environment, on the other hand, is quite different. The European Energy Performance of Buildings Directive (EPBD) requires all new buildings in the European Union to be "nearly zero energy" buildings by 2020 [2]. Such government-induced requirements are pushing building designers to apply passive (energy demand reduction) and active (on-site energy generation) building energy performance measures. The former often requires the preparation of a building energy model to conduct simulations in order to optimize the energy performance of the building [3, 4]. The latter leads to increased use of Building Integrated Photovoltaic (BIPV) applications in an urban environment, where the irradiance conditions are more complex (due to shadows, diffuse and specular reflections) than in the case of ground-mounted PV farms. Consequently, it is a likely scenario, that a state-of-the-art building design, that satisfies the nearly zero energy requirements, is optimized using building energy models and relies on PV for on-site electricity generation. Therefore it is a logical step, to investigate the capability of building energy simulation software to cope with the complex irradiance conditions of the urban environment.

In this work, EnergyPlus [5], a whole building simulation software is investigated regarding its ability to predict the occurrence of specular reflections. The detection and classification of reflection events is demonstrated on the case study of an outdoor PV test site, utilizing the irradiance models of EnergyPlus and post-processing with Python programming language [6]. Using the reflection event classification and irradiance transposition models, filtering the effect of unwanted specular reflections from measured irradiance data at an outdoor PV test site is demonstrated.

2 IRRADIANCE SIMULATION APPROACH

EnergyPlus is an open-source whole building energy simulation software, primarily developed for the aim of conducting thermal simulations. However, during its development, more detailed solar irradiance models were added to the software, making it capable of conducting the investigation presented in this paper. In this section, the necessary simulation model settings are described in order be able to model the occurrence of specular reflections on irradiance sensors and active surfaces.

In order to decrease the computational time, EnergyPlus by default calculates an average daily solar path and shadow coefficients in every 20 days. This setting can be overridden and set to timestep frequency, that is, the calculation of the solar position and shading factors is executed at every timestep of the dynamic building energy simulation. The maximum timestep frequency is 60 iterations per hour.

The irradiance on a tilted surface (G_{poa}) is calculated according to Equation 1:

$$G_{poa} = G_d * cos\theta * SF + G_{sd} * F_s + G_{gd} * F_g, \quad (1)$$

where G_d is the direct normal solar irradiance, θ is the angle of incidence of direct irradiance on the tilted surface, SF is the shaded fraction calculated by the shading module of EnergyPlus. G_{sd} is the sky-diffuse irradiance, calculated with the Perez-model [7], G_{gd} is the ground reflected diffuse irradiance. F_s and F_g are the surface-sky and surface-ground "angle factors".

EnergyPlus has a simple, single-bounce ray-tracing method to calculate external specular and diffuse reflections on building surfaces, originally developed for daylighting and heat-transfer calculations, and rarely used for solar energy applications. Validation studies of this model for predicting reflected irradiance from exterior building surfaces could not be found in literature. For every exterior surface, EnergyPlus assigns four receiver points (see Figure 1), to calculate the incident solar radiation reflected from neighboring surfaces. The reflecting surfaces can be specularly reflective (e.g. glass surfaces), diffusely reflective (opaque surfaces), and the mixture of the two (shading surfaces). Table I summarizes the settings needed to be set in the EnergyPlus Input Data File (IDF) in order to conduct the simulation case studies described in Section 3.

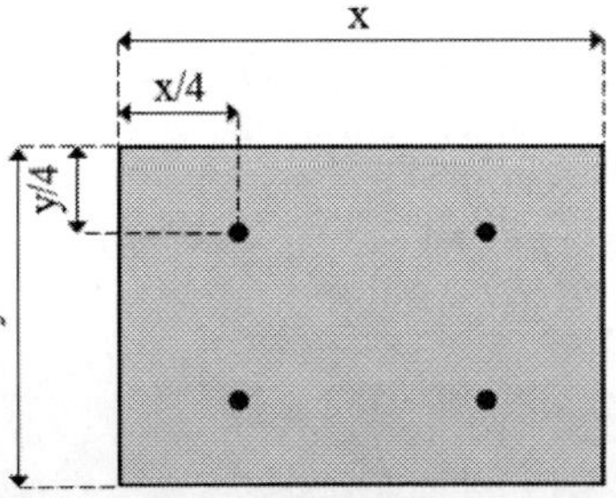

Figure 1: Receiving points on a rectangular surface.

Table I: IDF object settings to model reflected irradiance with EnergyPlus

IDF object	Value
Timestep	60
Building:SolarDistribution	FullExteriorWithReflections
Shad.Calc.:Calc.Method	TimestepFrequency

3 CASE STUDY DESCRIPTION

The SolarBEAT outdoor PV research facility (Figure 2) [8] is located on the top of the Vertigo building at the Eindhoven University of Technology (TU/e) and operated by the Solar Energy Application Centre (SEAC). Similar PV test sites can be found at numerous research institutes across the globe [9]. The site is elevated from the ground, there is no near shading and it has a clear view of the south sky hemisphere. Nevertheless, effects of the built environment can never be eliminated completely. On the North of the test site, there is an elevator shaft whose curtain wall has a reflective coating that is currently covered by a banner, to prevent specular reflections. The curtain wall will be changed in the future to a semi-transparent PV facade that despite the anti-reflection coating, might cause minor specular reflections on the test site. When local reflection events cause differences between irradiance received at PV surfaces and the neighboring irradiance sensors, this can negatively impact the reliability of PV performance metrics calculated from the measurements.

Figure 2: SolarBEAT outdoor PV research facility.

To aid the analysis of experiments, and to maintain the high quality of the measurement data collected at SolarBEAT, a simulation-based method is developed to indicate and classify momentary specular reflection events on the test site. The approach is to simulate the reflected irradiance on the PV surfaces and at the irradiance sensor points on the test site under clear sky conditions using the EnergyPlus software. The geometry of SolarBEAT was generated with Euclid, a SketchUp plugin for EnergyPlus geometry input. The reflective surfaces (elevator shaft and the façade of the Vertigo building) were modeled with shading surfaces (see Figure 3), the active surfaces were modeled as roofs and the pyranometers were modeled as 0.1x0.1m rectangles in plane of the sensors.

Figure 3: SolarBEAT modeled in SketchUp. The reflective surfaces are purple, the active surfaces red and the irradiance sensors are marked with green.

The clear sky weather file for the simulation was generated with the get_clearsky function of the PV_LIB Python library [10], using the Ineichen clear sky model. By simulating the reflected irradiance with a clear sky weather file, all time periods can be identified and classified, when specular reflections can possibly occur.

4 CLASSIFICATION OF REFLECTION EVENTS

4.1 Classification approach

The steps to classify and flag the reflection events on the test site are as follows:

1) Create an accurate geometry of the test site (active surfaces, irradiance measurement points and obstructions) in SketchUp and EnergyPlus with the assumed reflection properties of the nearby surfaces.

2) Create a weather file with only clear sky conditions.

3) Run a yearly simulation with minutely resolution using the clear sky weather file and register the reflected irradiance on the active surfaces and irradiance measurement points.

4) Classify the reflection events with the rule table (see Table II) and flag them in a yearly time series assigned to the individual measurements on the test site.

Measuring plane of array irradiance and generated power of the PV provides the basis to calculate derived performance metrics, such as Performance Ratio (PR). The effect of the reflected irradiance on the calculated metrics can be categorized to three different reflection events (see Figure 4): Reflection on Active Surface (RoAS), Reflection on Irradiance Sensor (RoIS), and Reflection on Active Surface and Irradiance Sensor (RoASIS).

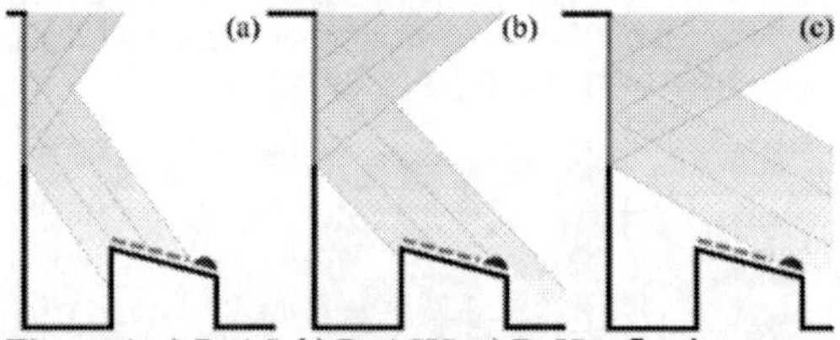

Figure 4: a) RoAS, b) RoASIS, c) RoIS reflection-events

Table II: Reflection-event rule table

Reflection detected		Flagged reflection event		
Active surface	Irradiance sensor	RoAS	RoIS	RoASIS
0	0	0	0	0
1	0	1	0	0
0	1	0	1	0
1	1	0	0	1

In the latter case, the point-like irradiance measurement might represent well the irradiance on the PV surface, but not the weather conditions on the site. It is also difficult to interpret properties like incidence angle, when the active surface sees the sun and the reflected image of it at the same time. In the case of a RoAS reflection event, the power output of the PV might increase, while measured irradiance does not, causing an error in the calculation of performance metrics, such as overestimation of PR. In the case of a RoIS event, PR would be underestimated.

The classification is done in post-processing the simulated and measured irradiance data with Python. Two conditions are needed to be simultaneously met, in order to classify an irradiance sensor or an active surface as being affected by reflections:

$$\frac{DHI_m}{GHI_m} \leq k_{Dt}, \quad (2)$$

and

$$G_{poa\_ref\_sim} \geq G_{poa\_sim} * RTM. \quad (3)$$

Equation 2 is about the irradiance conditions on the site, it expresses the condition, that specular reflections are only possible, if there is direct light. DHI_m and GHI_m are the site measured diffuse and global horizontal irradiance. k_{Dt} is the diffuse fraction threshold. For this investigation k_{Dt} = 0.95 was used. Equation 2 expresses the condition, that a surface can be classified as affected by specular reflections, if the simulated specularly reflected clear sky irradiance ($G_{poa\_ref\_sim}$) surpasses a given value. G_{poa_sim} is the simulated clear sky irradiance on the surface in question, RTM is the reflectance threshold multiplier. An RTM = 0.05 was used in this investigation.

4.2 Case study results

Figure 5 shows measured and simulated clear sky irradiance on an irradiance sensor at SolarBEAT, from August 2015, when the banner covering the reflective elevator shaft was not installed yet. The simulated reflected component is plotted separately as well, with the threshold for indicating a reflection event. Figure 6 shows the classification of reflection events for a whole year.

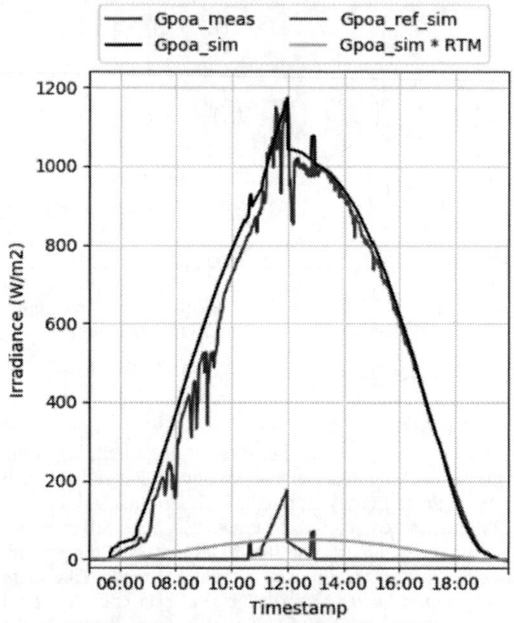

Figure 5: Simulated and measured irradiance of a pyranometer in 22th of August, 2015.

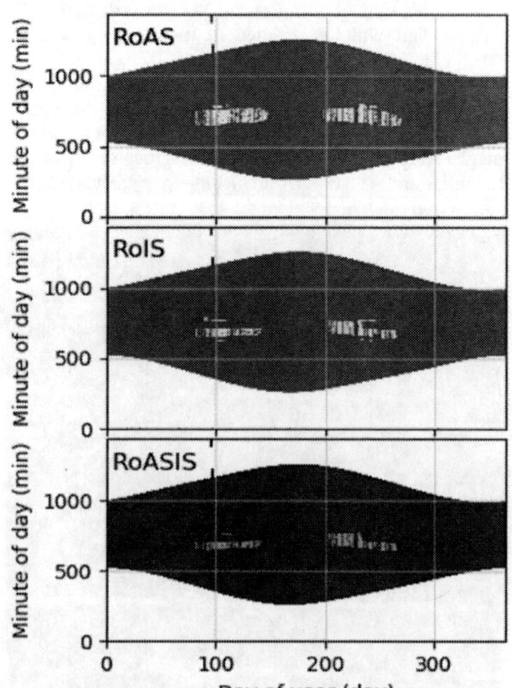

Figure 6: RoAS, RoIS and RoASIS reflection-events on a yearly heatmap. The background color indicates, when the sun is over the horizon, the contrasting colors show the occurrence of a reflection-event.

4.3 Irradiance corrections

A possible application of the flagging of reflection events is to aid the correction of irradiance measurements that are influenced by unwanted specular reflections. On the SolarBEAT test site, each experiment has its own plane of array pyranometer, that measures global irradiance. Moreover, there is a central measuring station, that provides detailed irradiance data, such as global horizontal, direct normal and diffuse horizontal irradiance. To filter unwanted specular reflections from an irradiance measurement, first a reflection-event detection should be conducted on the pyranometer in question and the central measuring station, which are situated approximately 40 meters from each other. It should be verified, that specular reflections do not occur at the same time on the two sensors. Once this is done, the detailed irradiance measured at the central measuring station can be transposed to the tilt and orientation angle of the pyranometer in question. For this the pvlib.irradiance.total_irrad function was used with the Klucher model [11] for transposing the sky-diffuse irradiance, implemented in Python. For the calculation of the ground-reflected component of the irradiance, an albedo value of 0.22 was used. Figure 7 shows, as the tranposed irradiance replaces the measured at times, when it is affected by spcular reflections.

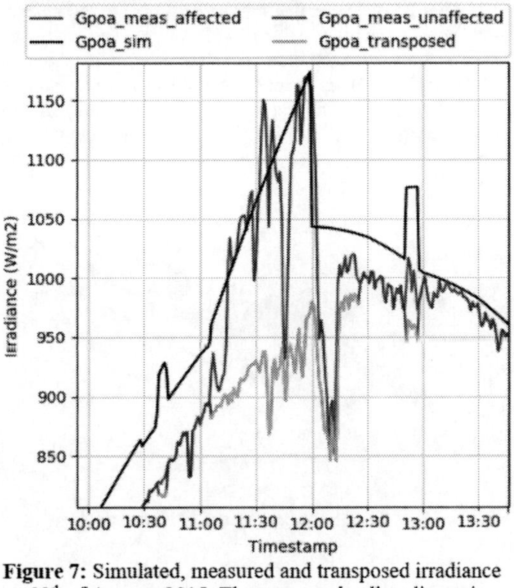

Figure 7: Simulated, measured and transposed irradiance on 22th of August, 2015. The green and yellow lines give the corrected irradiance time series.

5 CONCLUSIONS

In this paper, it was demonstrated, that EnergyPlus is capable of simulating certain aspects of the complex irradiance conditions of dense urban environments. This finding provides supporting arguments for the use of EnergyPlus to simultaneously predict both building energy consumption as well on-site energy generation. With the use of a simulation case-study, it was shown how EnergyPlus can be used for classifying reflection events on an outdoor PV test site. With post-processing the irradiance data, using the flagged reflection events, a method was demonstrated, to filter the effect of unwanted reflections from irradiance measurements. Here we would like to note, that filtering the reflections from the measured irradiance data is not always the best solution. It depends on the utilization of the measured irradiance data, that ignoring, replacing with NaN values or filtering the effect of reflections with the presented method is the best solution. The reflection-event classification presented in this paper can be used, to aid this decision.

6 ACKNOWLEDGEMENTS

This work has been supported by the project "PV OpMaat", financed by the cross-border collaboration program Interreg V Flanders-Netherlands with financial support of the European Regional Development Fund (CCI 2014TC16RFCB046).

7 REFERENCES

[1] SolarPower Europe, "Global Market Outlook For Solar Power / 2018 - 2022," 2018.

[2] EU, "Directive 2010/31/EU of the European Parliament and of the Council of 19 May 2010 on the energy performance of buildings (recast)," *Off. J. Eur. Union*, pp. 13–35, 2010.

[3] T. Østergård, R. L. Jensen, and S. E. Maagaard, "Building simulations supporting decision making in early design - A review," *Renew. Sustain. Energy Rev.*, vol. 61, pp. 187–201, 2016.

[4] J. Hensen and R. Lamberts, *Building Performance Simulation for Design and Operation*. 2011.

[5] D. B. Crawley *et al.*, "EnergyPlus: Energy Simulation Program," *ASHRAE J.*, vol. 42, pp. 49–56, 2000.

[6] Python Software Foundation, "Python Language Reference." [Online]. Available: http://www.python.org.

[7] R. Perez, P. Ineichen, R. Seals, J. Michalsky, and R. Stewart, "Modeling daylight availability and irradiance components from direct and global irradiance," *Sol. Energy*, vol. 44, no. 5, pp. 271–289, 1990.

[8] R. M. E. Valckenborg, J. L. M. Hensen, W. Folkerts, and de A. Vries, "Characterization of BIPV(T) applications in research facility 'SOLARBEAT,'" *Proc. EU PVSEC 2015, 14-18 Sept. 2015, Hamburg, Ger.*, 2015.

[9] G. Eder, L. Maul, P. Illich, and W. Folkerts, *Report IEA-PVPS T15-02. BIPV research teams & BIPV R&D facilities An international mapping*. 2017.

[10] R. W. Andrews, J. S. Stein, C. Hansen, and D. Riley, "Introduction to the open source PV LIB for python Photovoltaic system modelling package," *2014 IEEE 40th Photovolt. Spec. Conf. PVSC 2014*, pp. 170–174, 2014.

[11] T. M. Klucher, "Evaluation of models to predict insolation on tilted surfaces," *Sol. Energy*, vol. 23, no. 2, pp. 111–114, 1979.

DIRECT NORMAL IRRADIANCE MEASUREMENTS USING A TRACKER-LESS SUNSHINE DURATION MEASUREMENT CONCEPT

Mário Pó, Kees Hoogendijk, Isamu Chiba, Akihito Akiyama, Will Beuttell

EKO Instruments Co., Ltd., Hatagaya Shibuya-ku, Tokyo 151-0072, Japan
E-mail: mario.po@eko-eu.com

ABSTRACT: A new concept to measure the direct normal irradiance (DNI) is evaluated. While the state of the art setup to measure the DNI is comprised of a pyrheliometer assembled on a sun tracker, the EKO MS-90 measures DNI based on a sunshine duration concept which doesn't require a tracker to align the sensor aperture with the sun disk, presenting a relatively low-cost solution to measure DNI. In this work, we present a study of the EKO MS-90 accuracy. We evaluate the sensor measurement error by comparing the DNI measurements with the output from a reference pyrheliometer assembled on a sun tracker. The preliminary results for clear sky conditions show deviations from the reference within ±5%, for solar zenith angles smaller than 75° and DNI values higher than 700 W/m^2. Larger deviations practically within 10% are observed for greater angles and lower DNI values.
Keywords: Solar radiation, Characterization, Direct normal irradiance, Ground based

1 INTRODUCTION

The characterization of the solar energy resource is crucial for several solar energy applications [1-2]. With the variable solar resource depending on location and time of the year, it is common practice to perform in-situ measurements, and quantify it in terms of global horizontal irradiance (GHI), diffuse horizontal irradiance (DHI), and direct normal irradiance (DNI). Depending on the acceptable level of uncertainty and cost, different approaches can be pursued to measure the DNI. The state of the art setup to measure the DNI is comprised of one pyrheliometer assembled on a sun tracker.

Lower cost alternatives that do not require a tracker have been introduced [3-6], where an alternating shading band, or a static shading pattern with multiple sensors, is used to measure both GHI and DHI and estimate the DNI by relating the two measurements.

In this work, we introduce and analyze a cost effective new solution for measuring the DNI without the need of a sun-tracker, which is based on the EKO MS-093 sunshine duration measurement concept.

2 SENSOR WORKING PRINCIPLE

Sunshine duration (SD) provides a measure of the amount of time that the brightness of the solar disc exceeds the background of diffuse light from the sky. According to the World Meteorological Organization (WMO), the sunshine duration is defined by the time at which the DNI is above 120W/m^2 [7]. Concepts that measure SD have a way to properly differentiate between the diffuse and direct beam.

The former EKO MS-093 is a reliable and accurate sunshine duration measurement concept, which has a long experience in the field, been deployed for over 10 years within the Japan Meteorological Agency (JMA) Automated Meteorological Data Acquisition System (AMeDaS) network. In this concept, a rotating mirror reflects the sunbeam onto a thermal pyroelectric detector.

An evolution of this sensor was accomplished with the EKO MS-90, in which several optical improvements were implemented. Both pyroelectric sensor and mirror employed have a nearly flat spectral characteristic responding well to changes in the solar spectrum. While a tracker constantly aligns a pyrheliometer aperture with the suns' disk to measure DNI, the MS-90 sensor body is set up with a fixed tilt optimized according to the location latitude. The direct beam is reflected to the sensor head by a 45° tilted rotating mirror. Each revolution of the mirror has a period of 15s, reflecting the direct beam radiation to the sensor head once every 15 sec. A pyroelectric detector then measures the time derivative signal which is proportional to DNI. To make accurate measurements with the MS-90, the sampling criteria must account for this operation principle, in Figure 1 the sensor schematics is shown.

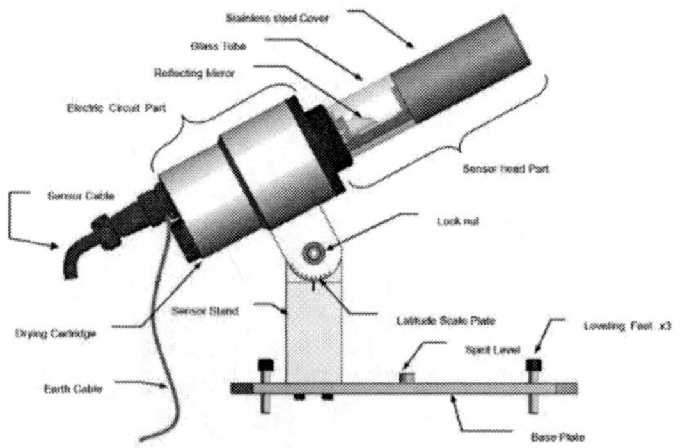

Figure 1. The EKO MS-90 components

A pyroelectric sensor has a slightly different working principle than the traditional thermopile detector. While both are heat sensitive devices, when exposed to constant irradiance the traditional thermopile sensors will reach thermal equilibrium and remain in steady state conditions. Pyroelectric devices have spontaneous temperature dependent polarization. When exposed to constant irradiance the charge rapidly builds up and decays, reacting in transient to the irradiance stimulus. Resulting in a fast response to irradiance changes, which works well in combination with the rotating mirror. In Fig. 2 exemplary voltage signal output from the MS-90 is shown.

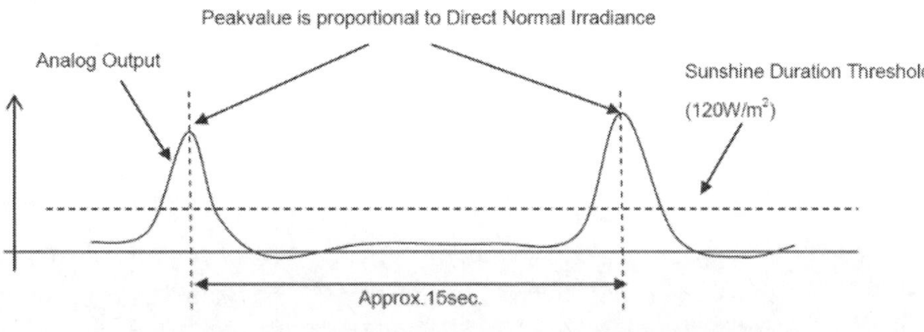

Figure 2. Tipical MS-90 output signal when responding to DNI.

3 SENSOR OUTDOOR EVALUATION

To evaluate the sensor measurement performance we assembled it outdoors. One MS-90 was installed at the EKO roof in Japan with an MS-56 pyrheliometer assembled in an EKO tracker to measure the DNI (see Fig. 2). An MS-80 pyranometer is also used to measure the GHI and estimate the DHI. The sensors are then connected to a CR300 data logger and sampled every 15s. A peak-hold function was employed in the CR300 logger program to sample the MS-90 and detect the maximum signal under the 15s sample rate, related to the 15s mirror rotation period.

Figure 3. Sensors employed for the outdoor comparisson; top) MS-90, bottom) MS-56 pyheliometer and tracker

The tests were performed in an urban environment at the EKO Instruments headquarters, in Shinjuku, Tokyo. Spatial limitations were encountered during the campaign as the sensors were affected by shading from the surrounding buildings at low elevation angles at sunrise and near sunset in the summer time. Shading events were filtered from this analysis based on the solar azimuth angle (80° < Sun Azimuth < 275°). For the resulting data, the deviation between the reference DNI and the MS-90 DNI is quantified according to eq. 1,

$$Deviation = \frac{DNI_{reference} - DNI_{MS-90}}{DNI_{reference}} \quad (1)$$

4 RESULTS & DISCUSSION

Considering all weather conditions, in Fig. 4 the entire filtered data set is employed to plot the MS-90 as a function of the reference DNI. The deviation is also determined for all weather conditions and plotted in Fig 5.

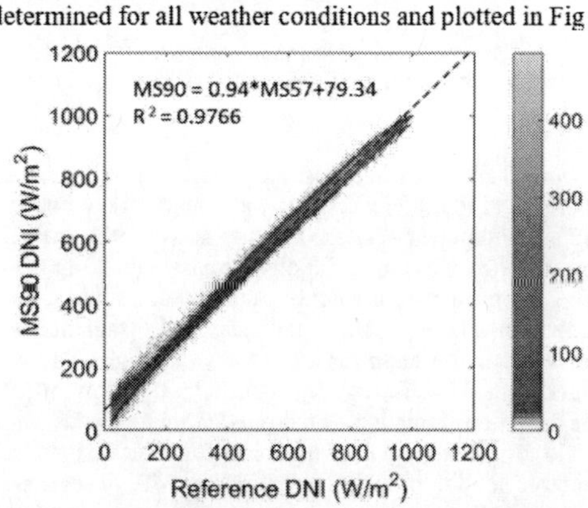

Figure 4. MS90 DNI vs Reference DNI measurements

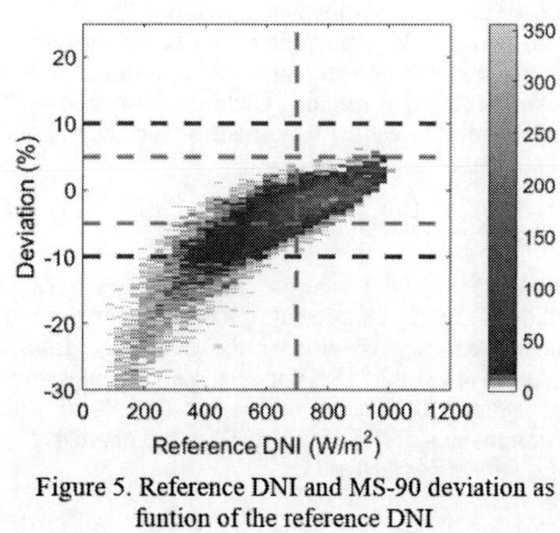

Figure 5. Reference DNI and MS-90 deviation as a funtion of the reference DNI

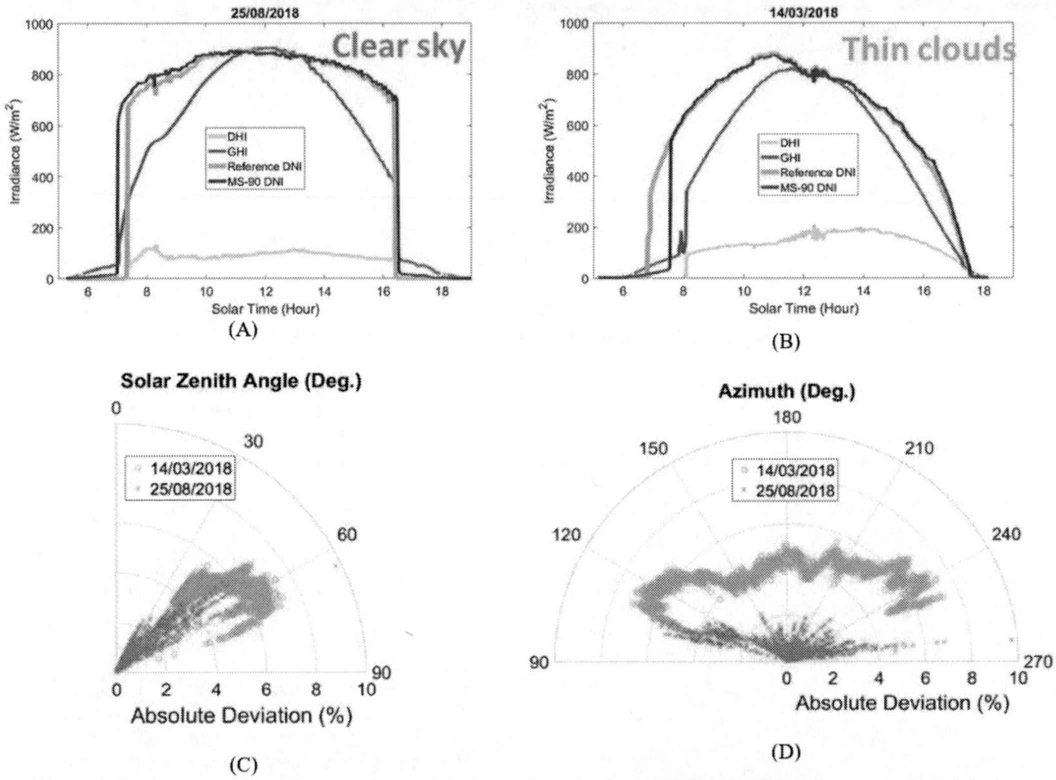

Figure 6. Study cases: A) Irradiances measured during a clear sky day 25/08/2018; B) Irradiances measured during a cloudy day with thin clouds 14/03/2018; C) Absolute deviation determined for both days as a function of the solar zenith angle; D) as a function of the solar azimuth angle.

A good agreement between the DNI measured with the MS-90 and the reference DNI was found. The deviations remain within ±5% for DNI values above 700W/m². For lower DNI values the deviations increase remaining below 10% for most measurements. A particular study case is shown in Fig 6., where the measured irradiance and deviations are plotted for a clear sky day and a day with thin clouds. Under clear sky conditions with low diffuse, the deviations remain well below ±5% up to a solar zenith angle of 75deg. When higher diffuse irradiance and variable conditions are experienced the MS-90 registers a higher DNI values, owed to the sensor measurement principle. Further research is necessary to evaluate these effects in the measurement performance. With the available data, the evaluation of the measurement errors due to seasonal effects was not performed, however possible correction methods could be developed for this sensor and improve the measurement accuracy based on angular dependency.

5 CONCLUSIONS

The EKO MS-90 is able to accurately measure DNI without a sun tracker, presenting a cost-effective solution. Under clear sky conditions, the deviations from the reference are within ±5% for solar zenith angles smaller than 75° and DNI values higher than 700 W/m². Larger deviations practically within 10% are observed for most cases with lower DNI values.

6 REFERENCES

[1] K. Chhatbar.R. Meyer, The influence of meteorological parameters on the energy yield of solar thermal plants. Proc. SolarPACES Symp. Granada, Spain, Sep. 20-24, 8p, (2011)

[2] J. M. Pó, A. Los, and W.G.J.H.M. van Sark, Assessment of STC conversion methods under outdoor test conditions, Proceedings of the 26th European Photovoltaic Solar Energy Conference and Exhibition in Hamburg, Germany, 2011, p.3458–3462.

[3] M. Pó, K. Hoogendijk, W. Beuttell, S. Kazunori, E. Takeuchi, Direct Spectral Irradiance Measurements from Rotating Shadowband EKO Grating Spectroradiometer, 7th WCPEC - 45th IEEE PVSC Conference – Waikoloa (2018)

[4] J. J. Michalsky. R. Perez, and R. Stewart, Design and development of a rotating Shadowband radiometer solar radiation/daylight network, Solar Energy Vol. 41. No 6. pp. 577-581, 38, (1988)

[5] N. Geuder, R. Affolter, O. Goebel, B. Dahleh, M. A. Khawaja, S. Wilbert, B. Pape and B. Pulvermueller, J. Sol. Energy Eng 138(5), 051007 (2016), 10.1115/1.4034070

[6] J. Badosa, J. Wood, P. Blanc, C. N. Long, L. Vuilleumier, D. Demengel, and M. Haeffelin, Atmos. Meas. Tech., 7, 4267–4283, 2014

[7] WMO, Part I, Measurement of Meteorological Variables, Chapter 8, 2008

RADIOMETER RESPONSE TIME AND IRRADIANCE MEASUREMENT ACCURACY

Anton Driesse
PV Performance Labs, Freiburg, Germany

anton.driesse@pvperformancelabs.com

This study evaluates the response times of three types of ISO 9060 First Class pyrheliometers in the laboratory, and more generally, investigates the stabilization of their output signals in response to step changes in irradiance. Observations are similar to, but much more detailed than our previous testing of pyranometers. The link between response time and measurement accuracy in various situations is discussed, and a model of the time-dependent instrument response is developed which can be used to explore the achievable accuracy in specific measurement contexts.

Keywords: Solar radiation, pyranometer, pyrheliometer, response time, calibration, performance, uncertainty

1 INTRODUCTION

Radiometers based on thermal sensing elements cannot respond instantaneously to changes in solar irradiance since the thermal capacity of the component materials delays changes in temperature. Pyranometers and pyrheliometers intended for measuring solar irradiance may require 3-30 seconds to reach 95% of the final value after a rapid change in irradiance. However some newer instruments are now available that offer 95% response times of less than one second. This paper presents measurements and analysis of the transient response of instruments with a range of response times and discusses links to measurement accuracy.

The transient response of radiometers is frequently described as an exponential decay, characterized by a single time constant τ_e:

$$\varepsilon(t) = e^{(-t/\tau_e)} \quad (1)$$

where ε is the error with respect to the final value, which starts at a value of 1 at t=0 and approaches a value of 0 as $t \rightarrow \infty$. The durations needed to reach various response levels are then represented as multiples of τ_e, for example the 95% level (ε=0.05) would be reached at $3\tau_e$, which is referred to in this paper as T95.

Nevertheless it is also mentioned in various sources new and old (for example [1]) that this is a simplified view, and that in reality the transient response is governed by more than one time constant. ISO 9060 states:

> "Owing to the fact that in general the thermal balancing process of pyranometers can only be described by several time constants, the settling behaviour should be characterized by the time during which the instrument reaches 95% of the final value. The value given by the WMO is 99%, which is closer to the final value but has a greater uncertainty owing to the offsets and is more dependent on wind speed. However, it should be emphasized that for accurate measurement of the irradiance of rapidly changing radiation sources, even the time taken for 99.5% of the value should be considered." [2]

This excerpt indicates at the same time the complexities of the topic, the link to measurement accuracy and the limitations of the single T95 specification.

2 RELEVANCE OF RESPONSE TIME

From the point of view of a single irradiance transition, slow response appears to be a significant handicap. But what goes up often also comes down, and as long as the transient response is symmetrical, a slower response effectively produces a weighted moving average of the true irradiance. A longer response time therefore muddles some details, but should not reduce the accuracy of medium- and long-term average values. However there are some situations where this may not be enough.

2.1 Instrument classification

In order for an instrument to be assigned a classification according the ISO 9060 standard [2] or WMO Guide [3], the T95 response time must be measured, reported and fall below specific limit as defined for each class. This formal requirement underscores the importance of the characteristic.

2.2 Instrument calibration

The responsivity (or calibration factor) of an instrument should be determined after the output signal has stabilized sufficiently. Relevant ISO [4] and ASTM [5][6] standards do not agree on a definition of this condition, nor on a procedure for ensuring that it is met. However they do agree that significantly more time than T95 is required for calibration.

While the light source used for calibration is usually quite stable, response time is nevertheless very important because the procedures last only a limited time, and may involve interruptions of the light source, such as the sun-shade method for outdoor pyranometer calibration[3][6]. For a given procedure, the accuracy of the calibration will be better for instruments with faster response; or if the procedure is adapted to the faster instrument, the same level of accuracy can be attained within a shorter time span.

2.3 Measurement system design

Although it is possible to use an analog integrator to perform continuous measurements of solar energy, most modern measurement systems rely on periodic instantaneous measurements. According to the well-known Nyquist criterion, the rate at which such samples are taken should be at least twice the highest frequency contained in the signal. A radiometer with a response time of τ_e actually functions as a single-pole low-pass filter with a cut-off frequency equal to $1/(2\pi\tau_e)$. Thus instruments with slower response times can be sampled more slowly because they have already filtered out some of the high-frequency content of the irradiance signal. This is reflected in design advice given in [3] and [7].

2.4 Specialized studies

Sometimes accurate average irradiance values are not enough. Meteorological studies of cloud enhancement effects, for example, have been shown to require at least 10 samples per second for accurate characterization [8], and of course the radiometer must respond adequately within this short time as well.

Outdoor evaluations of PV modules frequency utilize multiple types of radiometers at the same time, in particular thermal pyranometer in combination with photovoltaic reference cells, which respond much faster. Analyses requiring both measurements must take into account the different response times, and a faster pyranometer provides a useful advantage in this application.

3 PREVIOUS TESTS AND RESULTS

Pyranometer tests were carried out by PV Performance Labs in 2015 in collaboration with JRC Ispra and first reported in [9]. 20 thermopile instruments were exposed to both positive (0→1000 W/m²) and negative (1000→0 W/m²) transitions, after having been preconditioned either in the darkness or full light. Before and after the each transition several minutes of output signal were recorded at 20 samples per second. The transitions themselves required about 0.6 to 0.8 seconds to complete.

Table 1: Pyranometer response time specifications

Manufacturer	Model	ISO Class	95% resp. time (s)
Eko Instruments	MS-802	sec. std.	5
	MS-602	2nd class	17
Eppley	PSP	sec. std.	15
	SPP	sec. std.	5
	GPP	sec. std.	5
Hukseflux	SR20	sec. std.	3
	SR03	2nd class	1
	LP02	2nd class	18
Kipp & Zonen	CMP 10	sec. std.	5
	CMP 3	2nd class	18

The measurements showed that all pyranometers conformed to their specifications and met ISO 9060 classification requirements. We also observed that some of the lines in **Figure 1** cross each other, meaning that some instruments are "speeding up" or "slowing down" relative to others. This would not be possible with a single exponential transient response, and therefore confirms the presence of a more complex transient behaviour governed by two or more time constants.

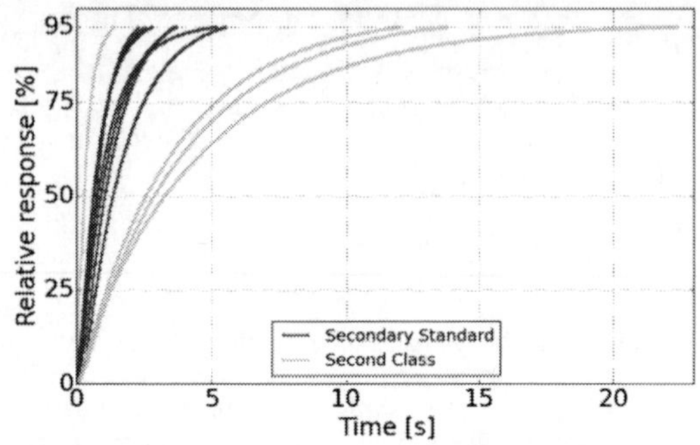

Figure 1 Response of pyranometers to a 0→1000 W/m² step change after pre-stabilization at 1000 W/m²

4 DESCRIPTION OF NEW TESTS

PV Performance Labs carried out a new series of response time tests in Freiburg in 2018. This time we evaluated three ISO First Class pyrheliometers whose specifications are mostly quite similar with the exception of response time. As shown in **Table 2**, the slowest and fastest models differ by a factor of 25.

Table 2: Pyrheliometer response time specifications

Manufacturer	Model	95% response time (s)
Eko Instruments	MS-57	0.2 s
	MS-56	1.0 s
Kipp & Zonen	CHP1	5.0 s

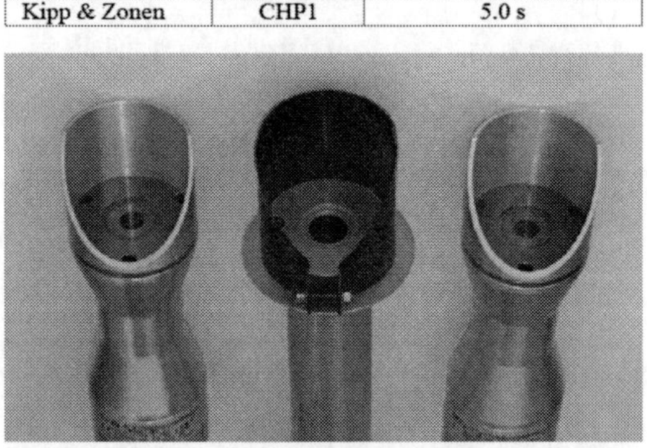

Figure 2: The ISO 9060 First Class pyrheliometers that were tested. From L to R: MS-57, CHP1 and MS-56.

The test bench at PV Performance Labs comprised the following components:

- A halogen light source with optical elements to produce a suitably narrow and uniform beam of light. The maximum intensity received by the instruments is approximately 500 W/m².
- Neutral filters to reduce the intensity of the light.
- A computer-controlled shutter to interrupt the light and thereby create step changes in irradiance.
- A stable power supply for the light source.
- A photodiode detector to monitor both stability of the light source and timing of the shutter.
- A data acquisition system capable of measuring 2 channels at the rate of 250 samples per second.
- A computer and software to execute a planned sequence of light levels and durations autonomously.

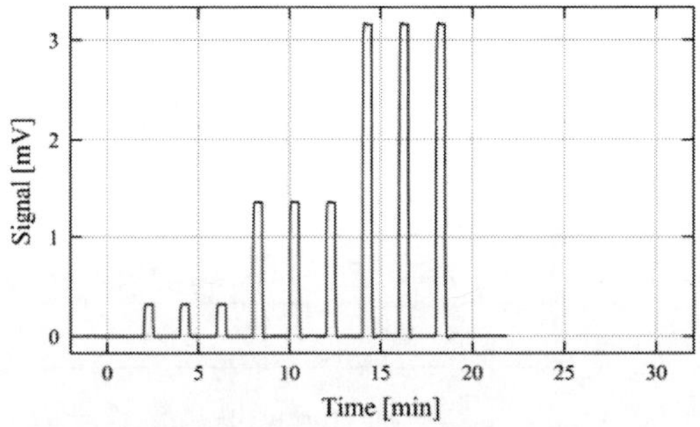

Figure 3: The initial test sequence used short illumination pulses which each produce one dark→light and one light→dark transition.

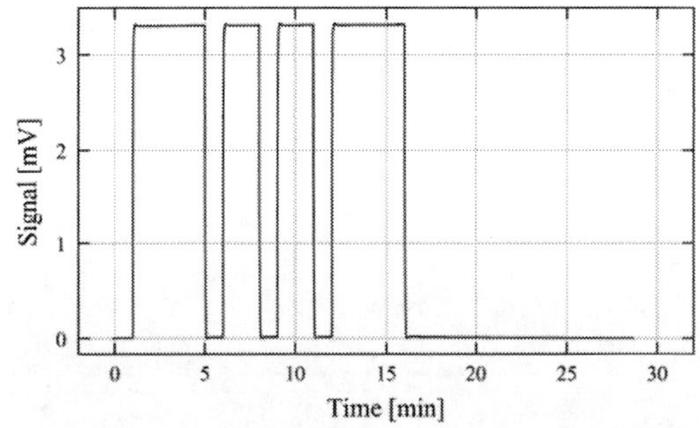

Figure 4: The final test sequence featured longer periods of exposure and stabilization.

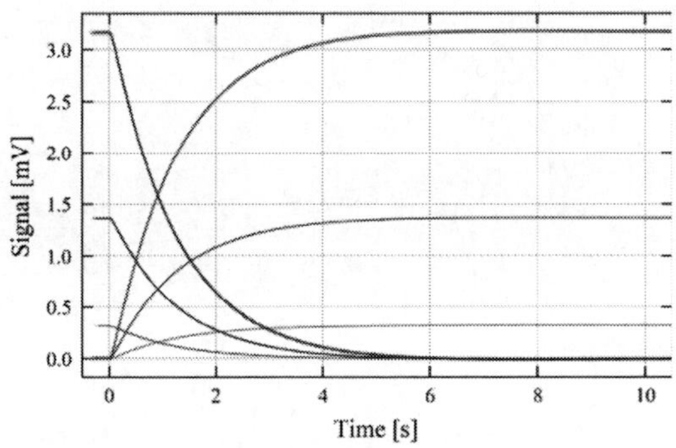

Figure 5: All transitions plotted relative to their respective starting times, dark→light in red and light→dark in blue.

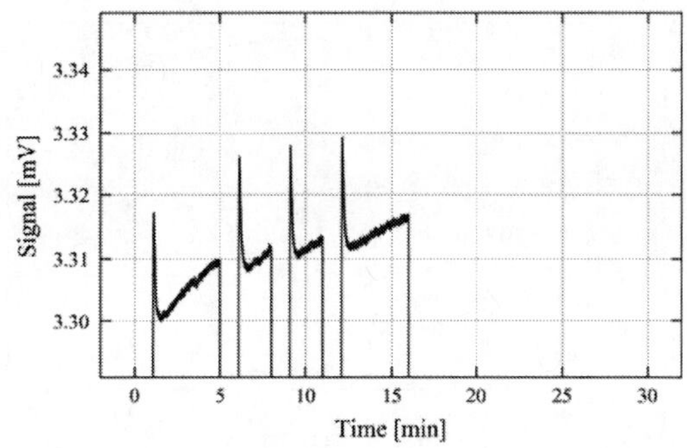

Figure 6: A closer look at the periods of light, showing a very slow approach to the final value lasting more than 15 minutes, with setbacks due to the dark interruptions.

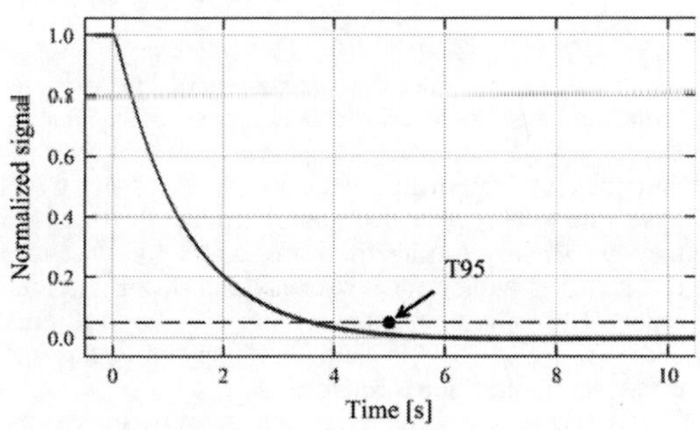

Figure 7: All transitions with magnitudes normalized and inverted where necessary.

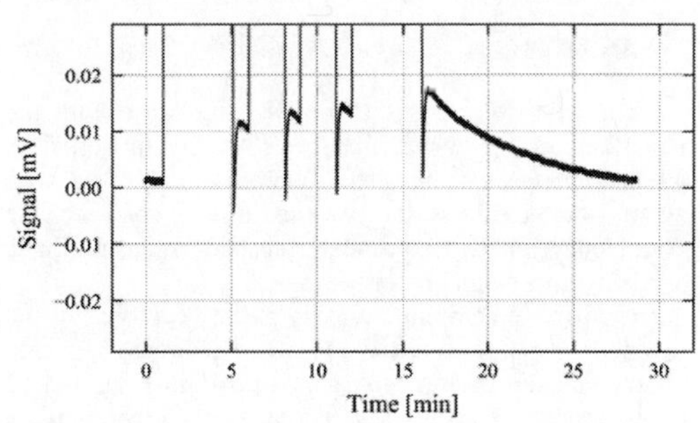

Figure 8: A closer look at the periods of darkness, showing the cumulative influence of the preceding periods of exposure.

5 MEASUREMENTS

To begin, we tested a single instrument of each type using a series of short pulses at multiple light levels (**Figure 3**). Each positive and negative transition was then plotted relative to its own starting time (**Figure 5**), normalized and inverted where necessary to make transitions from 1 to 0 (**Figure 7**). The latter figure demonstrates that all normalized transitions essentially follow the same curve, suggesting that the transient response scales linearly, and is symmetrical.

We carried out further tests on all available instruments using longer periods of high irradiance as shown in **Figure 4**. Looking more closely at the periods of light (**Figure 6**) and darkness (**Figure 8**), we observe that the initial rapid rise and fall of the signal is followed by a slight overshoot for some instruments and, more importantly, a cumulative influence from pulse to pulse that indicates a much longer settling period. Most of the detailed analysis presented in this paper is based on the measurements taken during the final light→dark transitions because it gives the most complete picture of transient response.

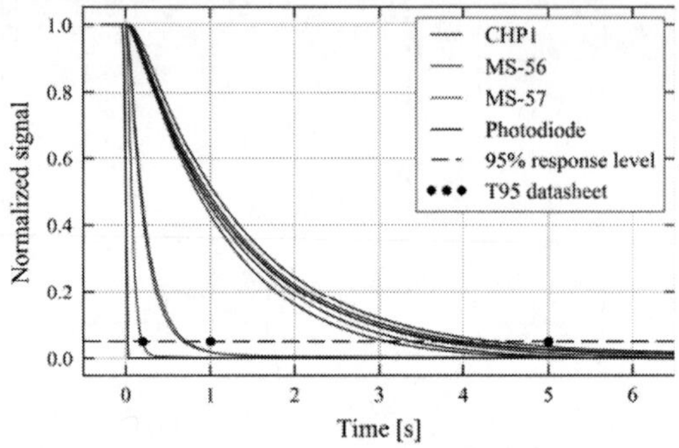

Figure 9: All instruments reach the 95% response level within the T95 interval provided on their datasheets.

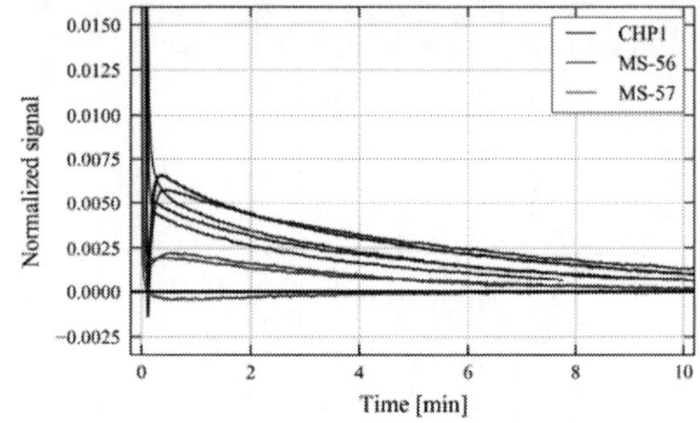

Figure 10: The extended tail ends of the transient responses reveal additional time constants of several minutes for all instruments.

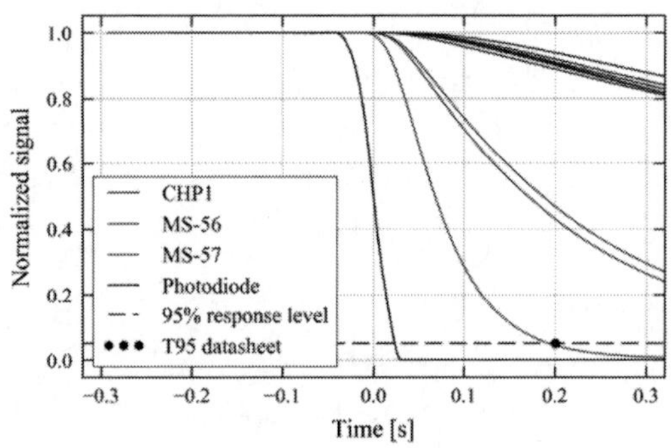

Figure 11: The beginning of the transient response shows a slight delay for all instruments.

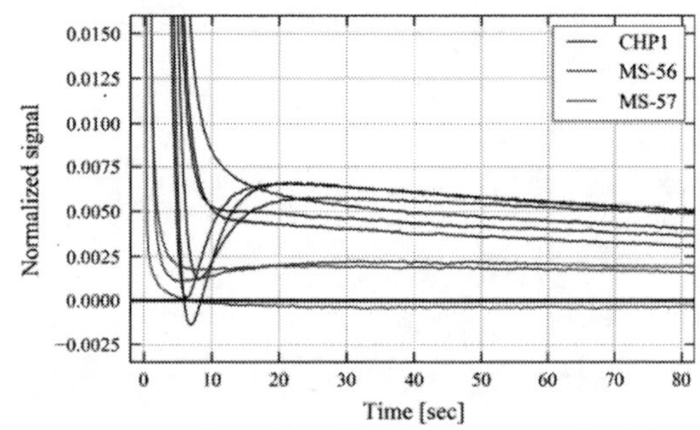

Figure 12: The transition region of the transient response between initial and final decay shows a variety of patterns.

6 ANALYSIS

The first objective is to verify compliance with the manufacturers' own specifications. As seen in **Figure 9**, all instruments pass this test. Some variability is seen among the CHP1's, but we could not compare the consistency for the Eko models because we did not have as many units available for testing.

Next we looked more closely at the beginning of the transient response (**Figure 11**). The photodiode signal shows that the shutter took a non-negligible amount of time to open; however, even after it is fully open all instruments are still "gathering speed", so to speak. This is in contrast to a pure exponential response, which would reach its maximum "speed" as soon as the shutter is completely open. The initial phase of the transient response is therefore better viewed as a *delayed* exponential decay, with delays in the range 27–139 ms.

Moving to the tail of the transient response (**Figure 10**) we observe another set of exponential decays—but with much longer time constants. There is not much difference between these time constants, which range from 2.7 to 5.5 minutes, but what really differentiates the instruments is the signal level at which the final decay begins. For the CHP1 instruments this is between 0.4% and 0.7% of the irradiance change, whereas for the MS-56 and MS-57 this is +0.2% and −0.04% respectively. (The negative value indicates a slight overshoot of the target.)

Finally we consider the transition period between the initial and final decays, which is shown in **Figure 12**. Here we find multiple patterns which cannot be fully described by exponential decays. Three of the CHP1 have a noticeable overshoot, but when they bounce back they are slightly further from the final value than their counterparts without the overshoot. Given that this region is small in magnitude compared to the initial decay, and also short in duration compared to the final decay, its significance is limited.

By taking into account duration we can also compare the significance of the initial decay and the final decay. The integral of the transient irradiance measurement error over an interval represents the error in energy that would be recorded. Interestingly, for all instruments the energy error for the interval 0→T95 was nearly equal to the energy error for the interval T95→10 minutes, with the ratios ranging from 0.5 to 1.5. By this measure, therefore, the final decay is roughly equal in significance to the initial decay.

7 MODELLING

From the foregoing analysis we retain that the main characteristics of the transient step response (excluding the transition region) can be described by the delayed sum of two exponential decays, which is represented by the following equation:

$$\varepsilon(t + T0) = (1 - a)e^{(-t/_{T1})} + ae^{(-t/_{T2})},\ t > 0 \quad (2)$$

where ε is the error with respect to the final value, *T0* is the initial delay, *T1* is the short time constant, *T2* is the long tail time constant, and a is the magnitude of the tail. By various curve fitting procedures we calculated these parameters for each of the instruments, and the median values for each model are provided in **Table 3**.

Table 3: Transient response model parameters

Model	T0 [ms]	T1 [ms]	T2 [min]	a [%]
MS-57	27	58	2.7	−0.07
MS-56	31	211	3.6	+0.26
CHP1	99	1223	5.5	+0.56

8 APPLICATIONS

How can this model be used? In relation to the topics presented at the beginning of this paper, we can suggest the following:

- For instrument classification the model is not useful, although it can be observed that *T95=T0+3T1*.
- For calibration, the model can predict exactly how much time one needs to wait for a specific level of accuracy (disregarding other sources of accuracy).
- For measurement system design the model can be used to simulate the measurement errors under different irradiance conditions, with different sampling rates, and different averaging intervals. For examples of this, see [10].
- And for specialized studies, especially those involving multiple instruments, the model can be used to post-synchronize measurements from instruments with different transient response characteristics.

Or one can use this additional information to make an informed decision when purchasing an instrument.

9 CONCLUSIONS

The output signal of a thermal radiometer responding to a change in irradiance is often represented by a simple exponential function with a single time constant. Analysis of the transient response to step changes in irradiance reveals that the response is more accurately represented by an initial delay, followed by the sum of two exponentials having different time constants. Such a model captures well the final settling from 99% to 100% of the final value, which is especially relevant for calibration procedures.

With regard to the specific instruments that were tested, they were all found to meet or exceed their respective 95% response time specifications. However, the T95 value is not necessarily a good indicator of the response time beyond 99% since all instruments have a long, slow tail response that is governed by a different time constant. The instrument with the fastest initial response, the Eko Instruments MS-57, also came closest to the final value before slowing down. It reached 99.9% of the final value in only 2 seconds, which means that its tail response can generally be considered insignificant.

10 FURTHER WORK

Having developed the apparatus to accurately evaluate radiometer response time as well as the procedures to analyse and model the same, we now plan to evaluate additional instruments, in particular pyranometers. We will also carry forward the results to look at measurement system design in general, but also to make more informed choices about the systems we design and operate.

At our test site in Freiburg we will also compare these instruments in an outdoor setting in order to corroborate the indoor measurements and calculations.

11 REFERENCES

[1] M. R. Riches et al "Proceedings of the IEA Conference on Pyranometer Measurements" Appendix R. SERI/TR-642-1156R. October 1982.

[2] ISO 9060:1990(E) Solar energy—Specification and classification of instruments for measuring hemispherical solar and direct solar radiation.

[3] World Meteorological Organization, Guide to Meteorological instruments and Methods of Observation, No. 8, 2014 edition, WMO, Geneva, 2014.

[4] ISO 9846:1993(E) Solar energy—Calibration of a pyranometer using a pyrheliometer.

[5] ASTM G207-11 Standard Test Method for Indoor Transfer of Calibration from Reference to Field Pyranometers.

[6] ASTM G167-15 Standard Test Method for Calibration of a Pyranometer Using a Pyrheliometer.

[7] IEC 61724-1:2017 Photovoltaic system performance - Part 1: Monitoring.

[8] G. H. Yordanov et al. "100-millisecond Resolution for Accurate Overirradiance Measurements" IEEE Journal of Photovoltaics, vol. 3, no. 4, October 2013.

[9] A. Driesse et al. "Indoor and Outdoor Evaluation of Global Irradiance Sensors", 30th EU-PVSEC, Hamburg, 2015.

[10] A. Driesse, et al, "Sampling and Filtering in Photovoltaic System Performance Monitoring", Sandia Report SAND2014-19137, 2014.

MODELLING AND FORECASTING PV PRODUCTION IN THE ABSENCE OF BEHIND-THE-METER MEASUREMENTS

Tomas Landelius[1], Sandra Andersson[1] and Roger Abrahamsson[2]
Swedish Meteorological and Hydrological Institute[1], Tekniska Verken Linköping Nät AB[2]
601 76 Norrköping, Sweden[1], Brogatan 1, 582 78 Linköping[2]

ABSTRACT: This paper deals with the prediction of the net load from photovoltaic installations. The forecasts were only based on information from a numerical weather prediction model and measured net load. The study thus tackles the problem of estimating the contribution of PV power to the grid without knowing the actual production and consumption "behind-the-meter". The main approach using a physical model was compared with persistence (forecast with yesterday's values) and black box models represented by linear regression and an artificial neural network. The artificial neural network performed best with a normalized (with installed power) RMSE of about 10 %. Linear regression was only slightly inferior while the physical model performed worse (about 15 %). The physical model only predicts the gross production and needs information about the on site consumption in order to predict the net load. Here the consumption was predicted using persistence. This assumption did not hold up for the physical model to match the performance of the black box models. Thus, a better description of the gross consumption is needed in order to make the physical model more competitive.
Keywords: Modelling, Photovoltaic, Solar Radiation

1 INTRODUCTION

Meeting a growing worldwide demand for energy, while addressing climate change, is a challenge. One way of tackling it is to increase the generation of renewable energy, such as solar PV power. Renewable energy depends on weather conditions and high quality forecasts are required by transmission system operators, independent services operator (ISOs), grid owners and energy suppliers for managing the energy mix, maintaining grid balance and trading on the energy market. As an example, reducing the forecast error by just 0.1 percentage points is said to save the California ISO and California ratepayers more than $2 million per year [9].

In order to make a good PV power forecast for an installation, information about its orientation, installed peak power and geographical location is needed along with a reliable prediction of the relevant weather parameters. However, for most behind-the-meter (BTM) systems producing electricity primarily for on-site use, before delivering excess energy to the grid, only data about installed nominal power and the address is recorded.

Moreover, measurements of BTM PV generation are scarce. What is available is the measured net load which does not equal actual electricity consumption/production since some portion of it may come from BTM solar PV generation. This causes a disconnection between the measured load and the electricity demand, when the former is used in models to predict the electricity demand. This problem will be worsened with a growing portion of solar power and adoption of BTM storage, vehicle charging and time dependent electricity rates for instance, introducing more factors that influence the forecast accuracy. Further, the solar power is an unstable and difficult to predict source of energy which increases forecast errors and uncertainties in predictions used for planning and day ahead trading.

Here we describe a method to estimate parameters in a physical PV model (PVLIB Python, [5]). The estimation is based on hourly numerical weather prediction (NWP) data, net load measurements and an average daily cycle of the gross electricity consumption. Information about the latter is assumed to be obtained from net load measurements prior to the PV installation. The model is then used to forecast net electricity production (here defined as the gross production minus the gross consumption) under the assumption that the gross consumption can be predicted using persistence (only weekdays considered here). The result using this model was then compared with persistence and black box models represented by linear regression and an artificial neural network (ANN) using NWP data and measured net load from the previous day as input.

The aim of the study presented in the paper is to develop a model for predicting regional net PV power production and estimating its parameters when BTM PV measurements are unavailable. Ultimately the goal is to predict PV power production at different scales in a regional electricity network, e.g. in terms of total production, production per entry point or per secondary substation.

Previous work (e.g. [4, 12]) has tackled the problem with the general lack of BTM measurements of PV production by up-scaling information from a few well monitored representative sites. However, finding such representative sites could be difficult since there are no requirements on measuring the gross production.

In this study we do use BTM measurements of gross consumption. However, we assume that the daily cycle of the mean hourly gross consumption can be estimated based on data prior to the PV installation when BTM measurements are not available. Such consumption patterns are well-studied and known to be predictable. However, once a PV installation is in place the consumption pattern may very well change [9]. Our approach could be extended to take this into account by also including a model for the daily cycle of the average hourly consumption pattern in the optimization.

2 DATA

In this section we describe the data used for the study. It consists of two parts; measured energy production and consumption, along with NWP forecasts for the corresponding dates. Both data sets consist of hourly data covering the time period March - October 2016.

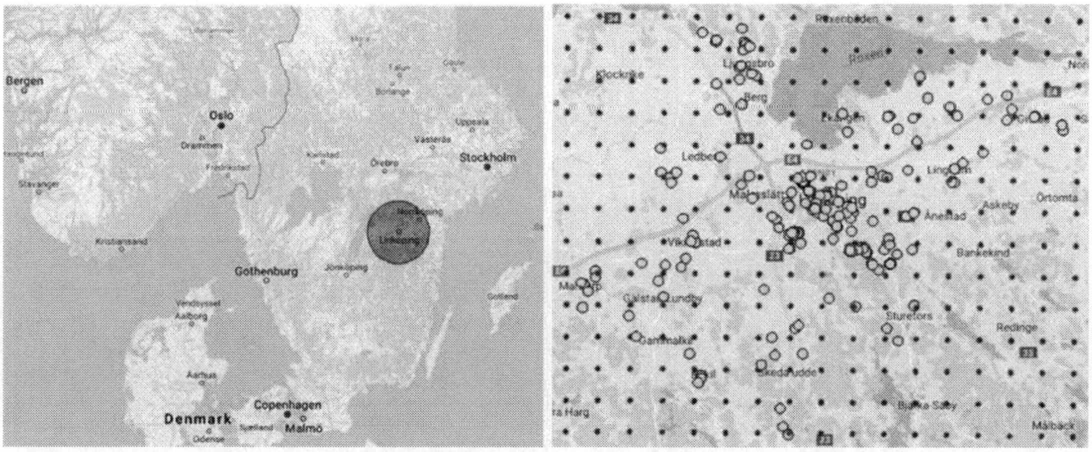

Figure 1: Left: approximate location of the Linköping network. Right: NWP grid points (red) and PV installations (yellow).

The data was divided into one training set, consisting of 80 % of the data (3360 hourly values per site), and one evaluation set with the remaining 20 % (840 hourly values per site). The division was made with a random sampling using a uniform distribution. This was done in order to end up with similar probability distributions for the training and evaluation data sets.

2.1 Measured production and consumption

The electricity measurements were provided by Tekniska verken that is responsible for the electrical grid in the Municipality of Linköping, parts of the Municipality of Mjölby and large parts of the Municipality of Katrineholm in Sweden. Hourly measurements of BTM PV production are available from 11 out of the 220 sites with PV installations in the Linköping net, see Fig. 1. Since we only had data from sites with PV installations and only from the year 2016 we had no information about the gross consumption prior to the installations. Instead we were restricted to use data from the 11 sites where this information was available. Two of these 11 sites were left out due to changes in their installations during the time period. The remaining 9 installations are located on the roofs of five households, four apartment complexes and one office building. For each installation information about the geographical location and the installed effect is also known.

2.2 NWP model data

NWP data was obtained from MetCoOp (Meteorological Co-operation on Operational NWP) where the meteorological services of Sweden, Norway and Finland run a common ensemble prediction system (MetCoOp EPS - MEPS). MEPS is developed in the framework of Aire Limitée Adaptation Dynamique Developpement InterNational (ALADIN) - High-Resolution Limited Area Model (HIRLAM) NWP system. The system can run with different configurations. Here we used the current cycle (40h1.1) of the HIRLAM-ALADIN Regional Meso-scale Operational NWP In Europe-Application of Research to Operations at Mesoscale (HARMONIE-AROME) [2]. The radiation model is based on the radiation scheme by Morcrette [10] and uses the Rapid Radiative Transfer Model of [8]. The model domain contains 900x960 points with 2.5 km grid spacing (see example in Fig. 1) and 65 levels covering a Nordic region. For this study we only used the deterministic MEPS control forecasts started at 00 UTC with a length of 24 hours. The parameters of our interest were accumulated hourly values of GHI and DNI together with instantaneous values of two meter temperature and 10 meter wind speed.

3 PV MODELS

To predict the net load we employed a physical model and included three statistical ones for comparison. By a physical model we here mean a model that simulates the physics of the PV power production based on information about the module characteristics, its orientation and the state of the atmosphere. The statistical models are here represented by persistence, linear regression and an ANN. In order to estimate the parameters of the physical model we need some information about the BTM production. This is not the case for the statistical models which can be fitted directly to the measured net load.

3.1 Physical

As a main approach, PVLIB Python was used to model the BTM PV power production. This is a open source community supported tool that provides a set of functions and classes for simulating the performance of photovoltaic energy systems. It was originally based on a toolbox developed at Sandia National Laboratories. For our purposes we chose a module and inverter that should match common installations in Sweden were selected from the Sandia library in PVLIB; SunPower SPR 220 and ABB MICRO 0 25 respectively.

The installations at each of the 9 sites with BTM measurements were modelled with a scaling factor times the output from one module in a single orientation. This approach could be seen as semi physical since the installation may consist of modules in different orientations and may be affected by shading. Moreover, it needs to be combined with a forecast for the gross error consumption in order to predict the net load. The tilt, azimuth and scaling factor (parameter vector w) was estimated by optimizing the summed squared error for the daily cycle of the mean hourly net production using a Nelder-Mead simplex algorithm (Python function

This is a pre-peer review version of the paper submitted to the journal Progress in Photovoltaics.
It has been selected by the Executive Committee of the 35th EU PVSEC 2018 for submission to Progress in Photovoltaics.

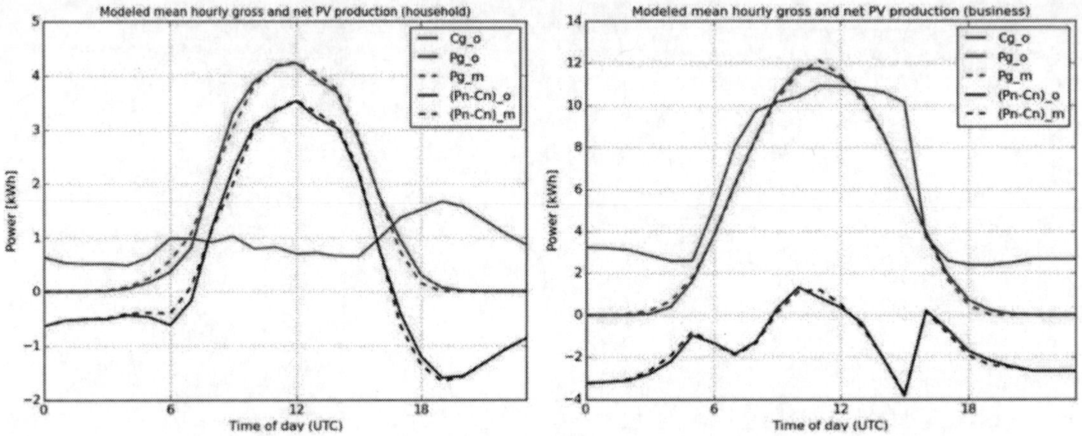

Figure 2: Left: Example of optimization with PVLIB targeting the daily cycle of the net production (solid blue line) and the resulting model values (dashed blue line) at a household installation. Predicted (dashed red line) and observed (solid red line) gross production and consumption (green line). Right: As in left panel but for a site at an office building.

scipy.optimize.fmin):

$$w^* = \operatorname{argmin} \sum (\bar{p}_g^m(w,t) - \bar{c}_g^o(t) - (\bar{p}_n^o(t) - \bar{c}_n^o(t)))^2. \quad (1)$$

Here the observed (superscript o) net (subscript n) and gross (subscript g) production and consumption for a given day and hour is related as:

$$p_g^o(d,t) - c_g^o(d,t) = p_n^o(d,t) - c_n^o(d,t). \quad (2)$$

Now, including the net consumption in the modelled (superscript m) net production (i.e. modelling the negative net load), the daily cycles of the mean hourly values are calculated as:

$$\bar{c}_g^o(t) = \frac{1}{n_d} \sum c_g^o(d,t) \quad (3)$$

$$\bar{p}_n^o(t) = \frac{1}{n_d} \sum p_n^o(d,t) \quad (4)$$

$$\bar{p}_n^m(w,t) = \frac{1}{n_d} \sum p_g^m(w,d,t) - c_g^o(d,t) \quad (5)$$

for the hourly intervals t = 0–1,…,23–24 UTC during the time period March – October 2016.

Here, the mean daily pattern of the gross consumption was calculated with data from the training data set. In practice, this information could be derived from measurements of the load (N.B. not BTM) during a time period before the PV installation was made.

Once the optimal PVLIB model parameters are estimated we have a physical model for the hourly gross (BTM) production. This model can then be used to predict the net PV production by assuming persistence for the gross consumption, i.e. that $c_g(d+1,t) \approx c_g(d,t)$. Hence the gross consumption can be eliminated from the equation and the net production predicted as:

$$p_n(d+1,t) = p_g^m(w,d+1,t) - c_g(d+1,t) \quad (6)$$

$$\approx p_g^m(w,d+1,t) - c_g(d,t) \quad (7)$$

$$= p_n(d,t) + p_g^m(w,d+1,t) - p_g^m(w,d,t). \quad (8)$$

The idea is that the daily pattern in the gross consumption is more stable than either the daily net or the daily gross production. This leads to a model for the net load that is given by persistence modified with the difference in gross production between the day ahead and today.

3.2 Statistical

Another alternative is to turn to statistical or black box models and model the net load directly. Here the idea is to fit a parametric model to the data. The assumption is that the residual is given by some model noise source when the optimal parameters have been estimated.

3.2.1 Persistence

The persistence model does not involve any parameter estimation and the input is the same as the output - the observed value for the corresponding hour from the previous day:

$$p_n^m(d+1,t) = p_n^m(d,t),\ t = 1,\dots,24. \quad (9)$$

3.2.2 Linear least squares

The linear and the ANN model have different parameters but share the same input. As input parameters (x) for these two models we used NWP forecast data (for the next 24 hours as well as for the past 24 hours) together with measured net load from the previous day. The idea behind this is that model should be able to make a connection between yesterday's state of the atmosphere and the net production. We also included the cosine of the solar zenith angle to provide some information about the time of the day and the season. It should also help to account for the effect of PV on tilted surfaces. For the linear model we also added a constant to the input vector in order for the model to be able to add a bias.

The linear model is given by

$$p_n^m(w,d+1,t) = w^T x. \quad (10)$$

We assume that the residual error is described by a normal distribution and employ a linear least squares method to estimate the model parameters (Python function numpy.linalg.lstsq):

Table I: Model performance in terms of mean r^2 values and RMSEn for gross PV production using PVLIB and persistence at the different sites (mean value, H:household, A: apartment block, O: office).

		Mean	H1	H2	H3	H4	H5	A1	A2	A3	O1
PVLIB	r^2	0.80	0.80	0.80	0.78	0.82	0.78	0.81	0.79	0.80	0.82
Persistence	r^2	0.45	0.46	0.45	0.48	0.42	0.44	0.45	0.43	0.43	0.46
PVLIB	RMSEn [%]	11	11	12	10	11	11	11	11	12	9.2
Persistence	RMSEn [%]	21	18	22	17	29	18	18	29	21	17

$$w^* = \text{argmin} \sum \sum (w^T x - (p_n^o(d,t) - c_n^o(d,t)))^2 \quad (11)$$

3.2.3 Artificial neural network

Using machine learning to train non-linear models has a long history within the area of energy forecasting [11, 6]. Here we used an off the shelf ANN from TensorFlow [1] to see if it a non-linear black box model could offer some improvements.

Using the standard feedforward ANN (DNNRegressor) from TensorFlow we set up a network with a three layer feedforward topology with one input, one hidden, and one output layer. In our set up it had 11 inputs, 32, 64 or 96 nodes in the hidden layer and 1 node in the output layer.

Determining the number of hidden neurons in the hidden layer(s) is a trade off between the networks ability to generalize from the training data (not too many neurons) and its representative power (not too few). Here we were guided by the empirical relation for the number of hidden layer neurons proposed by [6]: $n_{hl} = 0.5(n_{in}+n_{out}) + (n_{train})^{1/2}$, where n_{in} , n_{out} and n_{train} denote the number of input, output and the size of the training data set respectively. In our case the number of inputs equals 11 and we have one single output (the net load).

The number of cases in the training data set was 3360. Hence the suggested number of nodes in the hidden layer becomes 64. To check the robustness of this choice we also tried with 32 and 96 neurons in the hidden layer. The network with 32 neurons actually performed slightly better than the others on the evaluation data set so the results presented here are based on the outputs from that network.

In order to harmonize the amplitudes of the input variables we remove the mean (m) and normalize the input vector by multiplying it with the inverse of the square root sample covariance matrix (C), calculated from the training data set. The TensorFlow minimization algorithm then finds the solution to

$$w^* = \text{argmin} \sum \sum (w_o^T f(w_h, C^{-\frac{1}{2}}(x - m)) - (p_n^o(d,t) - c_n^o(d,t)))^2 \quad (12)$$

using an iterative procedure. We ran the minimization for 10,000 iterations (saving the result at each 100:th iteration) at which point the error for the evaluation data sets had started to increase for all sites. The prediction network was then given by the parameters from the iteration for which the evaluation error had a minimum.

4 RESULTS

In order to evaluate the performance of the different models we compared the models by looking at the root mean squared error (RMSE) normalized by the nominal installed power (RMSEn). We also calculated the square of the Pearson's correlation coefficient (r^2) between the modelled and observed net load. Only hours when the sun was over the horizon were included in the calculations.

First we look at the results for the main approach using PVLIB. Fig. 2 shows two examples of the fit of the optimized PVLIB model (blue dashed line) to the measured net production $(P_n-C_n)_o$ (blue solid line) which is the error criteria in equation 1. Using the model to predict gross PV production (dashed red line) results in a good fit to observed values (red solid line). The left and right panels show the results for a household and an office building respectively. Here the daily pattern for gross consumption (green lines) of the household shows a typical structure with peaks during the morning and afternoon while the other installation shows a consumption patterns related to the office hours.

Table I summarizes the error measures for the nine installations when using the PVLIB to model the gross production, which is an essential part of the physical approach. The average RMSEn for the PVLIB and persistence models are 11 % and 19 % respectively. Note that the error does vary much between different type of installations.

The daily cycles of the mean RMSEn when using PVLIB to model the gross consumption are illustrated in Fig. 3. The error corresponding to a persistence model is included for comparison. The error is shown as a function of forecast length. All forecasts are initialized at 00 UTC and the PVLIB based approach outperforms persistence for all forecast lengths.

We then turn to the prediction of the net load, now also including the statistical approaches. Table II summarizes the error measures for the nine installations when using the four different models; PVLIB, linear, ANN and persistence. The errors for both PVLIB and persistence are larger when predicting the net load compared to the errors for the gross PV forecasts. The reason is that the physical model now also needs information about the hourly gross consumption. Assuming persistence for the gross consumption turned out to be problematic. The r^2 values for such predictions varied between 0.10 and 0.83 for the nine sites. This shows that the gross consumption was less stable over time than we had thought.

The statistical models all perform better compared to the physical PVLIB approach. The ANN performs best on average with an r^2 value of 0.76 followed by the linear model (0.71), PVLIB (0.62) and persistence (0.40). The same order goes for the RMSEn where again the ANN is best with a mean value of 11 % followed by the linear model (12 %), PVLIB (16 %) and persistence (20 %). Note that there is a difference in the r^2 values between different sites for the net load predictions. The correlation for the office site is notably smaller for all the models even though only weekdays are included in the data.

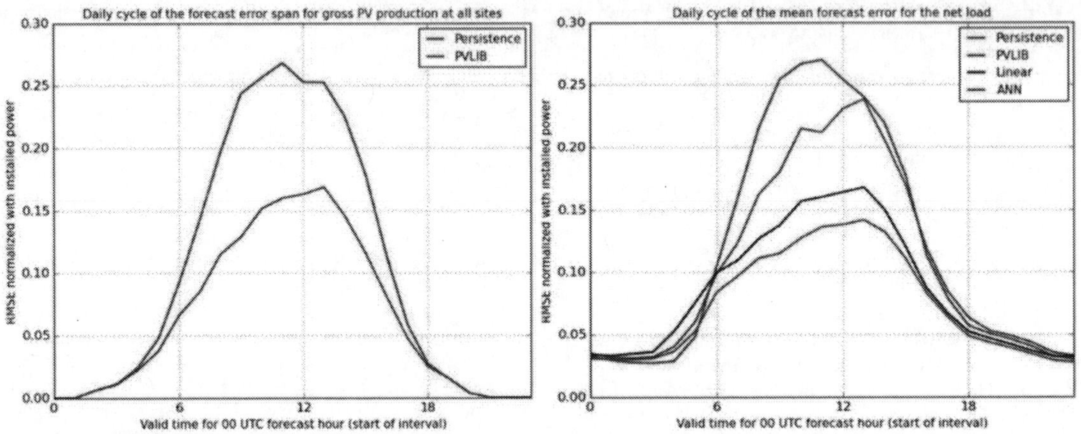

Figure 3: Left: Daily mean RMSEn cycle for the gross PV predictions (persistence: red, PVLIB: green). Right: Daily mean RMSEn cycle for net load predictions (persistence: red, PVLIB: green, linear model: blue, ANN: magenta).

The daily cycles of the mean RMSEn, when using PVLIB and the statistical models to forecast the net load, are illustrated in the right panel of Fig. 3. Again, the error is shown as a function of forecast length and all forecasts are initialized at 00 UTC. The ANN performs best, followed by the linear model, PVLIB and persistence. For the linear and ANN models the error grows somewhat towards the afternoon. This can be explained by the NWP forecast deteriorating with the length of the forecast. In the afternoon the performance of the PVLIB models becomes very similar to that of the persistence model. The reason for this is likely due to its dependence on the persistence forecast for the gross consumption. The error in the latter is high in the late afternoon when there is a peak in the consumption.

5 DISCUSSION AND CONCLUSION

Solar radiation is fluctuating in time and space and is non-trivial to predict. When a day-ahead forecasts is needed NWP models offer the best information [3]. However, in order to turn the solar radiation into a PV production forecast information about the geography and geometry at the installation is needed.

In this paper we studied how one physical and three statistical models performed on the task to predict the net load at a single site for the coming day. The input consisted of information about measured net production and net consumption from the previous day along with a NWP forecast for the next 24 hours. In a real situation the forecast has to be available well before the electricity market closes at about midday and hence the forecast horizon needs to be stretched to 42 hours. Such considerations will be the subject of further studies along with up-scaling of the forecast to an area of a regional electricity network.

The results show that it is possible to estimate the parameters in a physical model (PVLIB) based only on the daily cycle of the mean hourly net production and gross consumption. Information about the latter can be assumed to be available from load measurements during a corresponding time period prior to the PV installation. However, the pattern of consumption may change after the installation has come into place [9], e.g. by the purchase of an electric car or by considering the electricity produced BTM to be available free of charge. On the other hand, our modelling approach using PVLIB opens up for the possibility to include a model for the daily cycle of the average hourly consumption pattern, e.g. by doing a principal component analysis of a data base of consumption patterns and describe the pattern with a few eigenvectors in the optimization along with the PVLIB model parameters. In this way both the BTM production and consumption could be included in the optimization.

The physical model based on PVLIB shows good results when checked against the gross PV production. This means that both the NWP and the PVLIB models perform well. However, here the puropose was to predict the net load and for this purpose the physical model was outperformed by the black box models (linear and ANN). A likely reason for this is that the physical approach calls for an additional model for the hourly gross consumption. The gross consumption was predicted using persistence but this assumption turned out to be too weak.

Using an off the shelf ANN from TensorFlow reduced the overall RMSE, normalized with installed power, with only one percentage unit compared to the linear model. However, looking at the error with respect to forecast length reveals that the ANN actually provides a better prediction. Earlier studies have shown that the choice (e.g. the ANN) between different non-linear models is not critical [7].

Even better performance could perhaps be achieved if a recurrent structure is tried, exploiting the correlation in time between the foretasted values. One could also think of using a hybrid approach where the gross production is modelled with PVLIB and the gross consumption is modelled with an ANN. What to include in the input vector is another question for further investigations. Here we picked information we thought was reasonable. No evaluation was made regarding how useful different input parameters were for the prediction. Future work should also look at using input from probabilistic NWP forecasts. This should be a way to describe and account for uncertainties in the solar radiation forecasts.

To conclude, we have shown that a physical model

Table II: Model performance in terms of mean r 2 values and RMSEn for net load using PVLIB, a linear model, an ANN and persistence at the different sites (mean value, H: household, A: apartment block, O: office).

		Mean	H1	H2	H3	H4	H5	A1	A2	A3	O1
PVLIB	r^2	0.62	0.63	0.66	0.59	0.70	0.67	0.55	0.67	0.68	0.41
Linear	r^2	0.71	0.69	0.78	0.74	0.78	0.78	0.60	0.76	0.77	0.49
ANN	r^2	0.76	0.74	0.81	0.77	0.82	0.81	0.65	0.82	0.81	0.63
Persistence	r^2	0.40	0.44	0.44	0.42	0.47	0.45	0.31	0.45	0.46	0.18
PVLIB	RMSEn [%]	16	16	17	16	15	16	15	16	16	16
Linear	RMSEn [%]	12	13	13	12	11	12	12	12	12	12
ANN	RMSEn [%]	11	12	12	11	10	11	11	10	11	10
Persistence	RMSEn [%]	20	20	23	19	19	20	18	20	20	18

can be estimated and used to predict the gross PV production in the absence of BTM data. However, it was outperformed by statistical black box models such as linear regression and ANN when it comes to forecasting the net load.

ACKNOWLEDGEMENT

The authors greatly acknowledge the Swedish Energy Agency that funded this work via grant 43231-1 as part of the national research program “El från solen”.

REFERENCES

[1] Martin Abadi, Ashish Agarwal, Paul Barham, Eugene Brevdo, Zhifeng Chen, Craig Citro, Greg S. Corrado, Andy Davis, Jeffrey Dean, Matthieu Devin, Sanjay Ghemawat, Ian Goodfellow, Andrew Harp, Geoffrey Irving, Michael Isard, Yangqing Jia, Rafal Jozefowicz, Lukasz Kaiser, Manjunath Kudlur, Josh Levenberg, Dandelion Mane, Rajat Monga, Sherry Moore, Derek Murray, Chris Olah, Mike Schuster, Jonathon Shlens, Benoit Steiner, Ilya Sutskever, Kunal Talwar, Paul Tucker, Vincent Vanhoucke, Vijay Vasudevan, Fernanda Viegas, Oriol Vinyals, Pete Warden, Martin Wattenberg, Martin Wicke, Yuan Yu, and Xiaoqiang Zheng. TensorFlow: Large-scale machine learning on heterogeneous systems, 2015. Software available from tensorflow.org.

[2] Lisa Bengtsson, Ulf Andrae, Trygve Aspelien, Yurii Batrak, Javier Calvo, Wim de Rooy, Emily Gleeson, Bent Hansen-Sass, Mariken Homleid, Mariano Hortal, Karl-Ivar Ivarsson, Geert Lenderink, Sami Niemelä, Kristian Pagh Nielsen, Jeanette Onvlee, Laura Rontu, Patrick Samuelsson, Daniel Santos Mu noz, Alvaro Subias, Sander Tijm, Velle Toll, Xiaohua Yang, and Morten Odegaard Koltzow. The HARMONIE-AROME Model Configuration in the ALADIN-HIRLAM NWP System. Monthly Weather Review, 145(5):1919–1935, 2017.

[3] D. K. Chaturvedi and Isha. Solar power forecasting: A review. International Journal of Computer Applications, 145(6):28–50, Jul 2016.

[4] Lorenz Elke, Scheidsteger Thomas, Hurka Johannes, Heinemann Detlev, and Kurz Christian. Regional PV power prediction for improved grid integration. Progress in Photovoltaics: Research and Applications, 19(7):757–771, 2011.

[5] W. F. Holmgren, R. W. Andrews, A. T. Lorenzo, and J. S. Stein. Pvlib python 2015. In 2015 IEEE 42nd Photovoltaic Specialist Conference (PVSC), pages 1–5, June 2015.

[6] Soteris A. Kalogirou. Artificial neural networks in renewable energy systems applications: a review. Renewable and Sustainable Energy Reviews, 5(4):373 – 401, 2001.

[7] Luis Martn, Luis F. Zarzalejo, Jess Polo, Ana Navarro, Ruth Marchante, and Marco Cony. Prediction of global solar irradiance based on time series analysis: Application to solar thermal power plants energy production planning. Solar Energy, 84(10):1772 – 1781, 2010.

[8] E. J. Mlawer, S. J. Taubman, P. D. Brown, M. J. Iacono, and S. A. Clough. Radiative transfer for inhomogeneous atmospheres: RRTM, a validated correlated-k model for the longwave. Journal of Geophysical Research: Atmospheres, 102(D14):16663–16682, 1997.

[9] F. A. Monforte, C. Fordham, J. Blanco, S. Barsun, A. Kankiewicz, and B. Norris. Improving Short–Term Load Forecasts by Incorporating Solar PV Generation. Technical Report CEC-500-2017-031, California Energy Commission, 2017.

[10] J-J. Morcrette, H. W. Barker, J. N. S. Cole, M. J. Iacono, and R. Pincus. Impact of a New Radiation Package, McRad, in the ECMWF Integrated Forecasting System. Monthly Weather Review, 136(12):4773–4798, 2008.

SUPPORT VECTOR REGRESSION FOR SPATIO-TEMPORAL PV FORECASTING

R. Amaro e Silva[1], L. C. C. Teixeira da Silva[2], M. C. Brito[1]
[1] Instituto Dom Luiz – Faculdade de Ciências da Universidade de Lisboa.
[2] Faculdade de Ciências da Universidade de Lisboa.
Faculdade de Ciências da Universidade de Lisboa, Ed. C8 (8.3.33), Campo Grande, 1749-016 Lisboa, Portugal

ABSTRACT: Spatio-temporal PV forecasting has been showing promising results, detecting weather patterns based on spatially-distributed solar timeseries. Support vector regression (SVR), a robust machine learning approach capable of non-linear modelling and with good generalization ability, was tested for local forecasting with a forecast horizon ranging the 0.5 - 3 hours using data from 57 residential-scale PV systems from the East Midlands region (United Kingdom). The results were compared to a smart persistence and a linear autoregressive model with exogenous inputs (ARX) approaches. ε-SVR led to larger forecast skill improvements for forecast horizons up to 1.5 hours, with absolute median values between 1 – 1.5 % (7.5 – 10.5 % relative). When changing the number of input variables, the ε-SVR handled better larger datasets. A sensitivity analysis showed that CPU demand increased linearly with the number of variables as well as with the number of training examples. Adapting the ε-SVR to an υ-SVR framework led to a substantial 85.7 % reduction with almost no impact in forecast accuracy. From the literature reviewed, this work appears to be the first to test an SVR framework for spatio-temporal PV forecasting.
Keywords: PV integration, PV forecasting, spatio-temporal forecasting, ARX, SVR

1 INTRODUCTION

The inherent variability in PV generation due to the passage of clouds poses a serious challenge for an adequate grid operation [1] in high PV penetration scenarios, with potential negative implications such as curtailment [2]. Accurate PV forecasts have shown to be extreme relevance in coping with this variability, allowing for example for : i) PV operators to maximize their profits [3]; ii) grid operators to reduce costs [4]: iii) an optimal operation of PV-battery systems [5]. Kaur et al. highlight in their work its economical added value for shorter horizons [6].

Support vector regression (SVR) is a statistical framework which is capable of interpreting linear and non-linear patterns and has a good generalizing ability (i.e. is tolerant to overfitting). Several works have been reported which address solar radiation [7]–[13] and PV generation [14]–[20] forecasting. They either address local [3], [7], [21], [8]–[11], [15], [16], [19], [20] or regional [14], [17]–[21] spatial scales; short-term [18]–[21] or day-ahead forecast horizons [7]–[9], [14]–[17]. Most cases base their forecasts on numerical weather prediction (NWP) outputs such as temperature, cloud cover and humidity [7]–[9], [14]–[18]. Other works integrate additional variables such as past irradiance or PV records [19]–[21] or satellite derived irradiance forecasts [20]. Spatio-temporal approaches have been showing promising results by integrating spatially-distributed solar time series in diverse frameworks such as linear linear autoregressive model with exogenous inputs (ARX) [22], artificial neural networks (ANN) [23] or Gradient Boosting (GB) [24]. However, spatio-temporal SVR literature is quite scarce with two cases for irradiance forecasting [8], [11] and a single one for PV forecasting [20].

The present work proposes a short-term spatio-temporal framework for PV forecasting integrating the generation records from a network of residential-scale PV systems in the East Midlands region (United Kingdom). Past generation from 57 PV systems where used to build forecasting models for 12 PV systems, with a 30-minute resolution and for horizons ranging from 30 minutes to 3 hours. The use of small-scale PV records is of particular relevance when the considerable costs of irradiance sensors and the limited spatial and temporal resolutions of satellite and NWP data is taken into account.

2 METHODS

For each framework, an individual forecast model is trained for each PV system and forecast horizon. With only one year of data available (cf. section 3), the train, validation and test sets where distributed across the whole time period. Although not applicable in an operational context, this setup should better indicate the potential performance of the tested forecasting models as the training, validation and testing includes data from different seasons and weather conditions. Thus, for every nine days, the first five days were assigned to the training set (56 % of the data), the following two to the validation set and the remaining two to the testing set (each with 22 %). Additionally, night time (i.e. null) generation values were dismissed.

2.1 Pre-processing

The PV generation records were converted into clear-sky index, which quantifies the atmosphere attenuation on solar radiation or PV generation, when compared to a clear-sky day (Eq. 1). This step isolates short-term weather patterns by removing the data's daily and seasonal trends. Moreover, for PV data the modules' technical characteristics (tilting, orientation, rated power, thermal coefficient, etc.), have also to be taken into account.

$$K_{PV} = \frac{PV}{PV_{CS}} \quad (1)$$

where K_{PV} is the clear-sky index, PV is referring to either PV power or generation and PV_{CS} is the respective clear-sky expectation.

The data-driven clear-sky model suggested in [25] was used in this work, due to its simplicity. The PV clear-sky expectation at any given moment is originally defined as the 80th percentile of the generation records from the previous 10 days and at the same hour. Due to the persistence of cloudy weather in the region of study (cf. section 3), the percentile was empirically increased to 95. According to Lorenzo et al., this model takes into account possible calibration errors and permanent obstacles (e.g. a neighboring building) [26]. As a purely statistical approach, it dismisses any atmospheric or technical information on the location and PV installation, which are required for physical approaches [20].

2.2 Smart Persistence: benchmark model

A smart persistence model assumes that the clear-sky index remains unchanged with time,

$$\widehat{K}_i(t+h) = K_i(t) \quad (2)$$

where $K_i(t)$ is the PV clear-sky index for the ith PV system and instant t and $\widehat{K}_i(t+h)$ is the forecast for the horizon h.

This approach, although very simplistic, performs remarkably well in clear-sky and overcast conditions (i.e. low K variability). The fact that spatio-temporal approaches seek to detect weather patterns and forecast their evolution, makes smart persistence, which greatly underperforms for variable weather conditions, an ideal benchmark.

2.3 Spatio-temporal approaches

2.3.1 Linear ARX

The linear AutoRegressive model with eXogenous inputs (ARX) consists in a multivariate linear regression (Eq. 3) which forecasts PV generation based on autoregressive (past information from the target system) and exogenous (its neighboring systems) components.

$$\widehat{K}_i(t+h) = \sum_p w_{i,p} \times K_i(t-p) + \\ + \sum_q \sum_p w_{q,p} \times K_q(t-p) \quad (3)$$

where i is respective to the ith target system, q is the qth neighbor, p are the considered lagged terms and w the regression coefficients. The coefficients are estimated using the ordinary least squares method, minimizing the sum of the squared errors.

This framework is rather simple to deploy, does not computationally demanding and is easy to interpret.

2.3.2 ε-SVR

This subsection was adapted from [27]. ε-SVR seeks to define a function $f(x)$ which does not deviate from any of its training data points more than a tolerated value ε, while being the flattest possible [28]. In its simplest linear form, it can be written as a standard linear equation (Eq. 4) which assumes an ε-insensitive loss function (Eq. 5), ignoring errors below an ε threshold, and includes an L_2 regularization term (i.e. the squared norm of the weights) in its cost function, which promotes flatness by penalizing larger weights.

$$f(x) = \langle w, x \rangle + b \quad (4)$$

where $\langle \cdot,\cdot \rangle$ is the dot product, w and x weight and input vectors, respectively, and b a bias term.

$$|\xi|_\varepsilon = \begin{cases} 0, & if\ |\xi| < \varepsilon \\ |\xi| - \varepsilon, & otherwise \end{cases} \quad (5)$$

Ideally, the cost function would be described by Eq. 6, where its focus is on minimizing the regularization term.

$$\text{minimize } \frac{1}{2}\|w\|^2 \\ \text{subject to} \begin{cases} y_i - f(x_i) \le \varepsilon \\ -y_i + f(x_i) \le \varepsilon \end{cases} \quad (6)$$

However, it is quite unlikely that such function exists. Slack variables ξ_i and ξ_i^* are, thus, introduced to better cope with the imposed constraints,

$$\text{minimize } \frac{1}{2}\|w\|^2 + C \times \sum_i(\xi_i + \xi_i^*) \\ \text{subject to} \begin{cases} y_i - f(x_i) \le \varepsilon + \xi_i \\ -y_i + f(x_i) \le \varepsilon + \xi_i^* \\ \xi_i, \xi_i^* \ge 0 \\ C > 0 \end{cases} \quad (7)$$

where the cost constant C balances between flatness and accuracy.

Based on the support vector expansion, the weight vector can be re-written as a linear combination of the training data points,

$$w = \sum_j (\alpha_j - \alpha_j^*) \times x_j \\ \text{subject to} \begin{cases} \sum_j(\alpha_j - \alpha_j^*) = 0 \\ \alpha_j, \alpha_j^* \in [0, C] \end{cases} \quad (8)$$

where α_j and α_j^* are Lagrange multipliers, parameters which are optimized during the SVR training stage.

Thus, Eq. 4 can be rearranged to Eq. 9, transferring the SVR complexity dependence from the dimensionality of the input space to the number of support vectors

$$f(x) = \sum_j (\alpha_j - \alpha_j^*) \times \langle x, x_j \rangle + b \quad (9)$$

Non-linear modelling can be achieved by mapping the inputs into a transformed higher-order feature space, where $f(x)$ will be the flattest,

$$f(x) = \sum_j (\alpha_j - \alpha_j^*) \times \langle \phi(x), \phi(x_j) \rangle + b, \\ \phi: \mathbb{R}^n \to \mathbb{R}^{n+k}, n, k \ge 1 \quad (10)$$

where ϕ is the mapping function.

The inner product, however, can become computationally unfeasible for a large number of variables. Thus, it can be replaced by a kernel function K, which directly calculates $\langle \phi(x), \phi(x_j) \rangle$ while skipping the input mapping (Eq. 11). This is only possible because the SVR can be described as a set of dot products and without having to explicitly calculate w.

$$f(x) = \sum_j (\alpha_j - \alpha_j^*) \times K(x, x_j) + b \quad (11)$$

The Gaussian kernel (Eq. 12) was implemented as it was dominant in the SVR works mentioned in section 1.

$$K(x_i, x_j)_{Gaussian} = \exp\left(-\gamma \times \|x - x_j\|\right) \quad (12)$$

where γ an additional hyperparameter for the SVR.

Overall, the ε-SVR framework has three user-defined hyperparameters: the deviation tolerance ε; the cost constant C; and the gaussian kernel constant γ. A grid search procedure is typically used to find which parameter values optimizes model accuracy. Five values, empirically defined, were tested for each parameter (Table 1), resulting in 125 possible hyperparameter sets.

Table I: Set of hyperparameters values for ε-SVR training

Hyper-parameter	Tested values				
ε	0.05	0.10	0.15	0.20	0.25
C	0.5	1	2	3	4
γ ($\times 10^{-3}$)	0.08	0.16	0.31	0.63	1.25

2.3.3 ν-SVR

Later introduced by Smola et al. [29], the ν-SVR framework allows for the user to define with the υ parameter the fraction of data points in the training set to be considered as support vectors, while minimizing the value of ε. An adequate value of ν can lead to a substantially lighter training, as there is less hyperparameter to optimize and the amount of support vectors can be reduced.

2.4 Performance metrics

K_{PV} forecasts are first converted to PV generation by inverting Equation 1,

$$\widehat{PV} = \widehat{K_{PV}} \times PV_{CS} \quad (12)$$

where $\widehat{PV}$ is the PV generation forecast.

To evaluate a model's forecasting accuracy the root mean squared error is calculated (RMSE, Eq. 13), highlighting higher magnitude errors, typically related to changes in cloud cover.

$$RMSE = \frac{1}{n} \times \sqrt{\sum_{i=1}^{n} (\widehat{PV_i} - PV_i)^2} \quad (13)$$

Following common practice, forecast skill (FS, Eq. 14) quantifies a model's relative variation in forecast accuracy over a benchmark smart persistence.

$$FS = 1 - \frac{RMSE_{ARX\ or\ SVR}}{RMSE_{smart\ persistence}} \quad (14)$$

3 Data

3.1 Microgen Database

Developed by Sheffield University, Microgen Database is a web platform where PV owners can upload the generation records from their system in exchange for free monitoring reports [30].

For this work, 30-minute averages of PV generation from 57 2.82 kWp PV systems installed in the East Midlands region (United Kingdom) were considered, covering July 2014 to June 2015. These systems represent a diverse set of module tilts and orientations and their spatial distribution is shown in Fig. 1, characterized by a minimum, median maximum spacing of 0.1, 34 and 86 km, respectively. According to the provider, this dataset was subjected to a quality control procedure [31].

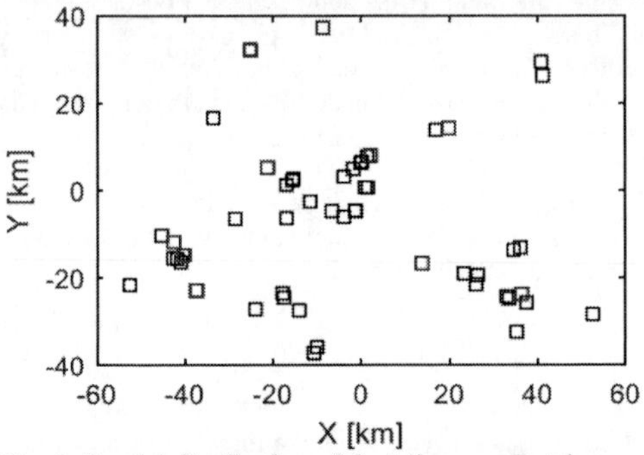

Fig. 1: Spatial distribution of the Microgen Database dataset in the East Midlands region (United Kingdom). Although every system was considered as an input, only those in red were forecasted (cf. section 4).

4.1 Spatio-temporal forecasting accuracy

The range of FS values obtained for the twelve PV systems using either the ARX or ε-SVR are shown in Fig. 2. Both approaches surpass the smart persistence for every forecasting horizon. ε-SVR surpasses the ARX mostly for horizons up to 1.5 hours, with median absolute values between 1 – 1.5 % (7.5 – 10.5 % relative). For farther horizons, the improvements are reduced to 0.6 – 0.8 % (3.5 – 4.6 %).

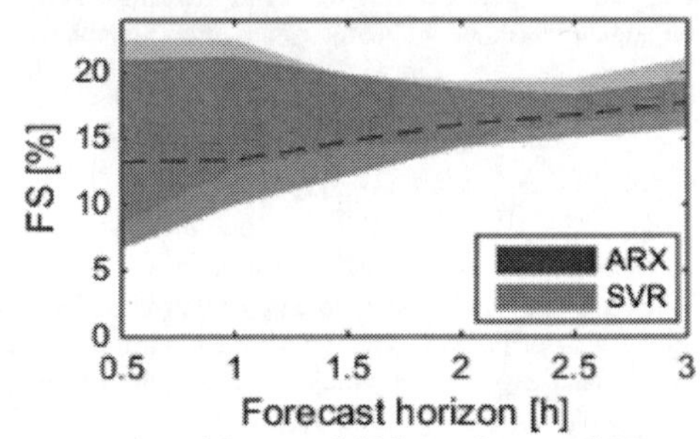

Fig. 2: Range of forecast skill for twelve tested PV systems for forecast horizons between 0.5 and 3 hours either using an ARX or ε-SVR approach (median values indicated by the dashed lines).

4.3 Forecast models' generalization ability

The generalization ability of the ARX and ε-SVR frameworks was tested by comparing the previous results with models which only consider as inputs the first lagged value of each PV system (171 against 57 variables). An increase of FS driven by the reduction of variables would derive from a model's inability to handle datasets with a considerable number of correlated variables (where some redundancy is to be expected). Overall, the results showed that the ε-SVR is less prone to overfit than the ARX. For shorter forecast horizons (up to 1.5 hours), the ε-SVR FS degraded the most, while some improvements where obtained for ARX (Fig. 3). For farther horizons, although the ε-SVR shows slight improvements, it remains more robust than the ARX (a median improvement of +0.34 %, compared to +0.85 % for the ARX).

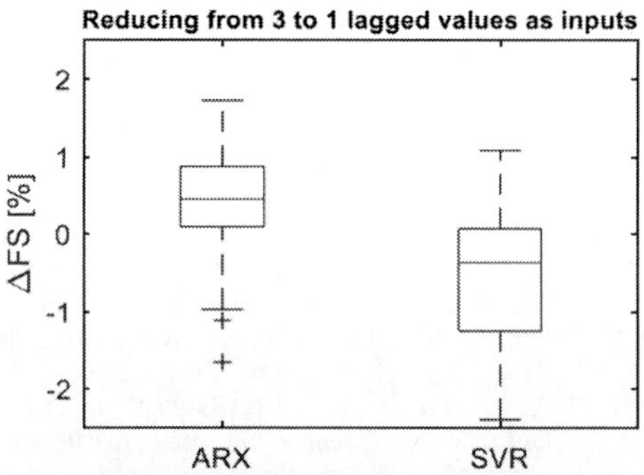

Fig. 3: Variation in forecasting skill for twelve tested systems when the input variables are reduced from three to one lagged term for forecast horizons between 0.5 and 1.5 hours. The fact that forecast skill degrades more for ε-SVR shows that it can better extract useful information from a larger number of variables.

4.4 CPU demand sensitivity analysis and ν-SVR

Despite the advantages of ε-SVR over ARX, it is a significantly more demanding (ARX takes 1% of the time of ε-SVR). Thus, a sensitivity analysis was performed to assess the impact of the number of variables and number of training examples for both ARX and ε-SVR frameworks. For both studies, fixed size and randomly defined subsets of either training data or input variables were used to train the models. Results showed that the ε-SVR computational time increases linearly with the number of variables (Fig. 4) and, with a steeper slope, with the training set size (Fig. 5).

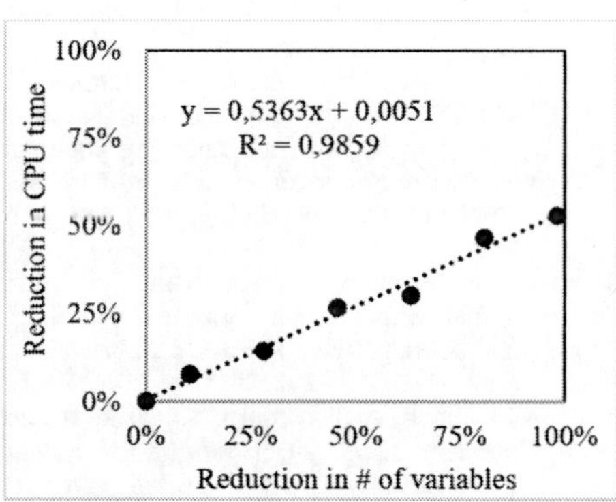

Fig. 4: Impact of the number of variables on the ε-SVR CPU demand.

To lighten the CPU demand, a υ-SVR was tested considering a range of υ values (cf. section 2.3.3): 0.25, 0.50, 0.75 and 1. The results presented in Table II show that a substantial CPU demand reduction can be achieved through this framework. A υ value of 0.5 seems to be the best option in terms of forecast skill compromise.

Table II: Impact on CPU demand and forecast skill of the υ hyperparameter in an υ-SVR framework

υ	0.25	0.50	0.75	1.00
Time reduction over ε-SVR	92.9 %	85.7 %	77.4 %	70.1 %
Median FS variation	-3.84 %	0.05	0.00	-0.37 %
Worst FS variation	-6.35 %	-1.26 %	-1.18 %	-1.62 %

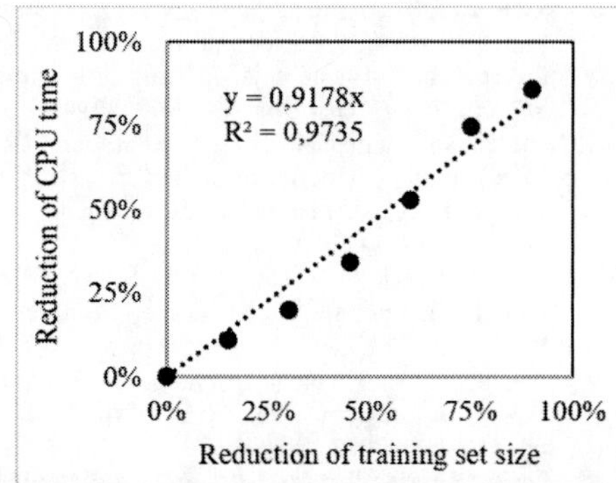

Fig. 5: Impact of the training set size on the ε-SVR CPU demand.

5 CONCLUSIONS

The present work explores SVR frameworks for spatio-temporal PV forecasting. Both ε-SVR and υ-SVR approaches were compared to a benchmark smart persistence and a spatio-temporal linear ARX in terms of forecast accuracy and CPU demand.

larger number of variables. However, this approach is considerably more demanding in terms of CPU demand (100 times more than ARX). From a sensitivity analysis, the CPU time showed to be proportional to the number of input variables and to the training set size, although it is more sensitive to the latter. On the other hand, υ-SVR is more efficient, resulting in a time reduction over 86 %, with little impact on forecast accuracy.

From the literature reviewed, this work appears to be the first to test an SVR framework for spatio-temporal PV forecasting. The results here presented point out the interest in further exploring SVR frameworks for spatio-temporal solar forecasting applications. Computational demand may be further optimized using parallel computing and feature selection or reduction techniques.

6 Acknowledgements

This publication was supported by the Instituto Dom Luiz institute (UID/GEO/50019/2013) and funded by the Fundação para a Ciência e Tecnologia (PhD grant PD/BD/106007/2014).

7 REFERENCES

[1] E. Ela, V. Diakov, E. Ibanez, and M. Heaney, "Impacts of variability and uncertainty in solar photovoltaic generation at multiple timescales," *Natl. Renew. Energy Lab.*, no. May, 2013.

[2] D. Lew, L. Bird, M. Milligan, B. Speer, X. Wang, E. M. Carlini, A. Estanqueiro, D. Flynn, E. Gomez-lazaro, N. Menemenlis, A. Orths, I. Pineda, J. C. Smith, L. Soder, and P. Sorensen, "Wind and Solar Curtailment," *Int. Work. Large-Scale Integr. Wind Power Into Power Syst.*, no. September, 2013.

[3] J. Antonanzas, D. Pozo-Vázquez, L. A. Fernandez-Jimenez, and F. J. Martinez-de-Pison, "The value of day-ahead forecasting for photovoltaics in the Spanish electricity market," *Sol. Energy*, vol. 158, no. September, pp. 140–146, 2017.

[4] C. Brancucci Martinez-Anido, B. Botor, A. R. Florita, C. Draxl, S. Lu, H. F. Hamann, and B. M. Hodge, "The value of day-ahead solar power forecasting improvement," *Sol. Energy*, vol. 129, pp. 192–203, 2016.

[5] J. Weniger, J. Bergner, and V. Quaschning, "Integration of PV power and load forecasts into the operation of residential PV battery systems," *4th Sol. Integr. Work.*, pp. 383–390, 2014.

[6] A. Kaur, L. Nonnenmacher, H. T. C. Pedro, and C. F. M. Coimbra, "Benefits of solar forecasting for energy imbalance markets," *Renew. Energy*, vol. 86, pp. 819–830, 2016.

[7] J. G. da S. Fonseca, O. Takashi, O. Hideaki, S. Ken-ichi, T. Takumi, and O. Kazuhiko, "Analysis of Different Techniques to Set Support Vector Regression to Forecast Insolation in Tsukuba, Japan," *J. Int. Counc. Electr. Eng.*, vol. 3, no. 2, pp. 121–128, 2013.

[8] R. Martin, R. Aler, J. M. Valls, and I. M. Galvan, "Machine learning techniques for daily solar energy prediction and interpolation using numerical weather models," *Concurr. Comput. Pract. Exp.*, vol. 28, no. 4, pp. 1261–1274, 2016.

[9] Y. Gala, Á. Fernández, J. Díaz, and J. R. Dorronsoro, "Support vector forecasting of solar radiation values," *Hybrid Artif. Intell. Syst.*, vol. 8073 LNAI, pp. 51–60, 2013.

[10] P. Lauret, C. Voyant, T. Soubdhan, M. David, and P. Poggi, "A benchmarking of machine learning techniques for solar radiation forecasting in an insular context," *Sol. Energy*, vol. 112, pp. 446–457, 2015.

[11] A. Zagouras, H. T. C. Pedro, and C. F. M. Coimbra, "On the role of lagged exogenous variables and spatio-temporal correlations in improving the accuracy of solar forecasting methods," *Renew. Energy*, vol. 78, pp. 203–218, 2015.

[12] M. Lazzaroni, S. Ferrari, V. Piuri, A. Salman, L. Cristaldi, and M. Faifer, "Models for solar radiation prediction based on different measurement sites," *Meas. J. Int. Meas. Confed.*, vol. 63, pp. 346–363, 2015.

[13] J. G. da Silva Fonseca Junior, H. Ohtake, T. Oozeki, and K. Ogimoto, "Local and Regional Hour-Ahead Forecasts of Solar Irradiance with Training Data Selection and Support Vector Regression," *IEEJ Trans. Power Energy*, vol. 136, no. 12, pp. 898–907, 2016.

[14] J. G. Da Silva Fonseca, T. Oozeki, H. Ohtake, K. I. Shimose, T. Takashima, and K. Ogimoto, "Forecasting regional photovoltaic power generation - A comparison of strategies to obtain one-day-ahead data," *Energy Procedia*, vol. 57, pp. 1337–1345, 2014.

[15] J. G. Da Silva Fonseca, T. Oozeki, T. Takashima, G. Koshimizu, Y. Uchida, and K. Ogimoto, "Photovoltaic power production forecasts with support vector regression: A study on the forecast horizon," *Conf. Rec. IEEE Photovolt. Spec. Conf.*, pp. 002579–002583, 2011.

[16] J. G. Da Silva Fonseca, T. Oozeki, T. Takashima, G. Koshimizu, Y. Uchida, and K. Ogimoto, "Use of support vector regression and numerically predicted cloudiness to forecast power output of a photovoltaic power plant in Kitakyushu, Japan," *Prog. Photovoltaics Res. Appl.*, vol. 20, no. 7, pp. 874–882, 2012.

[17] Á. Fernández, Y. Gala, and J. R. Dorronsoro, "Machine Learning Prediction of Large Area Photovoltaic Energy Production," *Data Anal. Renew. Energy Integr.*, pp. 38–53, 2014.

[18] A. Catalina, A. Torres-Barrán, and J. R. Dorronsoro, *Machine learning prediction of photovoltaic energy from satellite sources*, vol. 10097 LNAI. 2017.

[19] B. Wolff, E. Lorenz, and O. Kramer, "Statistical learning for short-term photovoltaic power predictions," *Stud. Comput. Intell.*, vol. 645, pp. 31–45, 2016.

[20] B. Wolff, J. Kühnert, E. Lorenz, O. Kramer, and D. Heinemann, "Comparing support vector regression for PV power forecasting to a physical modeling approach using measurement, numerical weather prediction, and cloud motion data," *Sol. Energy*, vol. 135, pp. 197–208, 2016.

[21] J. G. da Silva Fonseca Junior, H. Ohtake, T. Oozeki, and K. Ogimoto, "Local and Regional Hour-Ahead Forecasts of Solar Irradiance with Training Data Selection and Support Vector Regression," *IEEJ Trans. Power Energy*, vol. 136, no. 12, pp. 898–907, 2016.

[22] R. Amaro e Silva and M. C. Brito, "Impact of network layout and time resolution on spatio-temporal solar forecasting," *Sol. Energy*, vol. 163, pp. 329–337, Mar. 2018.

[23] A. G. R. Vaz, B. Elsinga, W. G. J. H. M. van Sark, and M. C. Brito, "An artificial neural network to assess the impact of neighbouring photovoltaic systems in power forecasting in Utrecht, the Netherlands," *Renew. Energy*, vol. 85, pp. 631–641, 2016.

[24] R. J. Bessa, A. Trindade, and V. Miranda, "Spatial-temporal solar power forecasting for smart grids," *IEEE Trans. Ind. Informatics*, vol. 11, no. 1, pp. 232–241, 2015.

[25] V. P. Lonij, A. E. Brooks, K. Koch, and A. D. Cronin, "Analysis of 80 rooftop PV systems in the Tucson, AZ area," *Conf. Rec. IEEE Photovolt. Spec. Conf.*, pp. 549–553, 2012.

[26] A. T. Lorenzo, W. F. Holmgren, and A. D. Cronin, "Irradiance forecasts based on an irradiance monitoring network, cloud motion, and spatial averaging," *Sol. Energy*, vol. 122, pp. 1158–1169, 2015.

[27] A. J. Smola and B. Schölkopf, "A tutorial on support vector regression," *Stat. Comput.*, vol. 14, no. 3, pp. 199–222, 2004.

[28] H. Drucker, C. J. C. Burges, L. Kaufman, A. Smola, and V. Vapnik, "Support Vector Regression Machines," *Adv. Neural Inf. Process. Dystems*, pp. 155–161, 1997.

[29] B. Schölkopf, P. Bartlett, A. Smola, and R. Williamson, "Support Vector Regression with Automatic Accuracy Control," *Proc. 8th Int. Conf. Artifcial Neural Networks*, pp. 111–116, 1998.

[30] S. Sheffield, "Microgen Database." [Online].

Available: http://www.microgen-database.org.uk/.

[31] J. Taylor, J. Leloux, A. M. Everard, J. Briggs, and A. Buckley, "Monitoring thousands of distributed PV systems in the UK : Energy production and performance," in *PVSAT-11*, 2015.

ENSEMBLE DETRENDING FOR SOLAR NOWCASTING

Antonio Sanfilippo, Luis Martin Pomares, Daniel Perez-Astudillo, Dunia Bachour
Qatar Environment & Energy Research Institute, Hamad Bin Khalifa University, Qatar Foundation
P.O. Box 34110, Doha, Qatar
{asanfilippo, lpomares, dastudillo, dbachour}@hbku.edu.qa

ABSTRACT: We present a novel treatment of detrending in solar forecasting where several ways of partitioning training data into more cohesive subsets are used in parallel to improve predictions. First, a solar irradiance training set covering a whole year is detrended according to season-based and cluster-based detrending criteria. Then, a classification model is developed which predicts for each choice of time series what is the most appropriate detrending option, including no detrending. A comparative evaluation of the emerging multi-detrending approach with single detrending or no-detrending techniques shows very strong gains in error reduction.
Keywords: solar forecasting, detrending, clustering, classification, ensemble learning

1 INTRODUCTION

Nowcasting, the ability to forecast solar irradiance up to 6 hours ahead [1], is crucial in managing power network operations and regulating electricity markets when solar energy is integrated in power grids. Solar irradiance is subject to abrupt changes due to meteorological change (e.g., clouds, haze, dust storms). With increasing solar energy penetration, this variability can cause critical imbalances in the intra- and inter-hour generation, transmission, distribution, and pricing of electricity that can disrupt power networks and cause uncertainty in energy markets. Solar nowcasting can help redress these imbalances by providing insights about solar variability that can be used to optimize power network operations and normalize electricity markets in near real-time.

Machine learning and stochastics techniques have been shown to be better suited for solar nowcasting than physics-based approaches, such as Numerical Weather Prediction (NWP) models, which lack the temporal resolution needed for intra-hour forecasts [2, 3]. These machine learning and stochastics techniques typically rely on regression methods to learn coefficients that provide the basis for prediction by measuring the relationship between an observation at time t and the observations at previous times. For example, within an autoregressive modeling approach, AR(N), the current term x_t of a time series $x_i, \dots, x_n$, for $N \geq t \geq i \leq n$, can be estimated as the linear weighted sum of previous terms $x_{t-1}, \dots, x_{t-n}$ in the series, as shown in (1) where c is a constant, ε_t is white noise, and $a_i, \dots, a_n$ are the auto-regression coefficients that function as the weights in the sum. Autoregression coefficients can be derived using Yule–Walker equations, or regression techniques such as least squares estimation or maximum entropy [4].

$$x_t = c + \sum_{i=1}^{N} a_i x_{t-i} + \varepsilon_t \qquad (1)$$

In general, the use of larger historical datasets promotes model accuracy through the development of more robust regression coefficients. However, evidence suggests that fitting a solar nowcasting model to a training dataset that presents some form of cohesion can help increase accuracy. For example, the authors in [3] observe that when using statistical analysis for forecasting purposes, it is desirable to detrend and deseason the training datasets. This is because most statistical time series analyses assume that the input time series data are stationary or at least weakly stationary.

Several detrending methods have been developed in the solar forecasting literature. The authors in [5, 6] propose a Gaussian function to fit the experimental ratio of hourly to daily solar radiation data, and an alternative method is presented by [7]. The authors in [8] measure the stationarity of a detrended series using the Augmented Dickey Fuller test to establish when to switch from a linear to a non-linear solar prediction model. The authors in [9] use two partitioning techniques, one based on seasons and the other on the clearness index, to establish whether a linear or an empirical model should be used to predict hourly solar radiation.

So far, no clear winner detrending technique has emerged. Arguing that no single detrending techniques can provide a turnkey solution in solar forecasting, the goal of this paper is to show that an ensemble method where diverse detrending approaches are selectively engaged provide a superior approach. First, a solar irradiance training set covering a year is detrended according to diverse criteria. Then, a classification model is developed which predicts for each choice of time series what is the most appropriate detrending option, including no detrending. A comparative evaluation of the emerging multi-detrending approach with single detrending or no-detrending techniques shows twofold to over threefold order of magnitude reductions in forecasting error.

2 SEASON AND CLUSTER-BASED DETRENDING

Using a dataset of 5-minute averages of normalized clearness index measurements (henceforth KT_p) [10] collected in Education City, Doha, Qatar, in 2014 [11], we generated partitions of the data in five configurations:

1. Single partition (*No Detrending*)
2. Season-based partitions (*Season Detrending*)
3. Cluster-based partitions using K-means clustering Detrending (*K-means Detrending*)
4. Cluster-based partitions using Expectation Maximization clustering (*EM Detrending*)
5. Cluster-based partitions using Learning Vector Quantization clustering (*LVQ Detrending*).

Each data record provided 12 consecutive KT_p values and a date specification, as shown in Figure 1. Data records were provided at 5-minute intervals.

M	D	H	Mn	KT_{p1}	...	KT_{p12}
1	1	6	40	0.527	...	0.674
1	1	6	45	0.718	...	0.666
1	1	6	50	0.725	...	0.666
...	...	...	...	...	...	...

Figure 1: Data record sample.

Normalized clearness index measurements alleviate the dependency of the clearness index on the zenith angle, which changes the traversed air mass during the course of a day [10]. The clearness index (KT) is calculated as the ratio of Global Horizontal Irradiation to the incoming solar radiation, on a horizontal surface at the top of the earth's atmosphere [12, 13]. KT_p is obtained by normalizing KT with respect to a standard clear-sky global irradiance profile for a relative air mass of one, as shown in equation (2), where AM the air mass at the (same) corresponding minute. The air mass is calculated according to the formula in (3), where SZA is the solar zenith angle (measured from the vertical overhead) in degrees at the corresponding minute [14].

$$KT_p = \frac{KT}{0.1 + 1.031 * \exp[-1.4/(0.9 + 9.4/AM)]} \quad (2)$$

$$AM = \frac{1}{\cos(SZA) + 0.50572 * (96.07995 - SZA)^{-1.6364}} \quad (3)$$

Cluster-based data partitions were developed by assigning each data record to a given cluster based on the record's attribute values, according to the three clustering algorithms mentioned above, as shown in Figure 2. We used the WEKA workbench [15] to carry out the clustering task. The three clustering algorithms assume a given number of clusters, which in our case was set to four to match the number of partitions in season detrending. K-means clustering assigns observations to the preset clusters so that the means across clusters (for all variables) are as different from each other as possible. EM clustering computes probabilities of cluster memberships based on one or more probability distributions, by maximizing the overall probability or likelihood of the data, given the (final) clusters. LVQ clustering can be understood as a special case of an artificial neural network; it applies a winner-take-all Hebbian learning-based approach. See [15] and references therein for further details.

M	D	H	Mn	KT_{p1}	KT_{p2}	...	KT_{p12}	Cluster
1	1	6	40	0.527	0.976	...	0.674	0
1	1	6	45	0.718	0.711	...	0.666	2
1	1	6	50	0.725	0.676	...	0.666	1
...	...	...	...	...	...	...	...	...

Figure 2: Sample of clustered dataset.

To assess the relative performance of the three clustering algorithms, we trained a Bayesian net classifiers on 66% of each clustered dataset, and then used the weighted harmonic mean of precision and recall (F-Measure) to evaluate the performance of each classifier on the remaining 34% of the dataset. As shown in Table 1, the EM algorithm outperforms both the LVQ and KMeans algorithms, and LVQ is the worst performing algorithm. Interestingly enough, the performance of a clustering algorithm as detrending method, as defined through classification, has a complex correlation with forecasting accuracy. For example, while LVQ detrending yields the lowest number of superior forecasts (Figure 3), overall it provide better forecasting results than both EM and KMeans detrending (Figure 5). As discussed with reference to Figure 6, this is due to variance across forecasting results, which appears to be a determinant factor in establishing the overall forecasting accuracy of detrending algorithms.

Table 1: Evaluation results for the clustered data sets.

Clustering Algorithms	*Weighted Average*		
	Precision	*Recall*	*F-Measure*
EM	95.9%	95.9%	95.9%
LVQ	79.2%	78.2%	78.0%
KMeans	88.5%	87.9%	88.0%

Using each data partition as a distinct training dataset, we built 13 models using the *Nonlinear Autoregressive Neural Networks with Exogenous Inputs* (NARX) algorithm [16], one for the *No Detrending* configuration, and four each for the *Season, K-means, EM* and *LVQ Detrending* configurations. We then used the relativized Root Mean Square Error (rRMSE) defined in (4) to evaluate model accuracy, and calculated the number of times each detrending method provided the best forecast (lowest rRMSE value).

$$rRMSE = \frac{\sqrt{\frac{\sum_{t=1}^{n}(x_{predicted_t} - x_{observed_t})^2}{n}}}{mean(observed_{1,\dots,n})} \quad (4)$$

As shown in Figure 3, no single detrending technique provides the best solution in absolute. Even the lowest performing detrending method provide the correct forecast about 14% of the times. Therefore, if we could predict the best detrending method for each choice of time series input (e.g. a data record like the one in Figure 1), the resulting forecasts would be significantly more accurate. We used classification as the technique to achieve this objective.

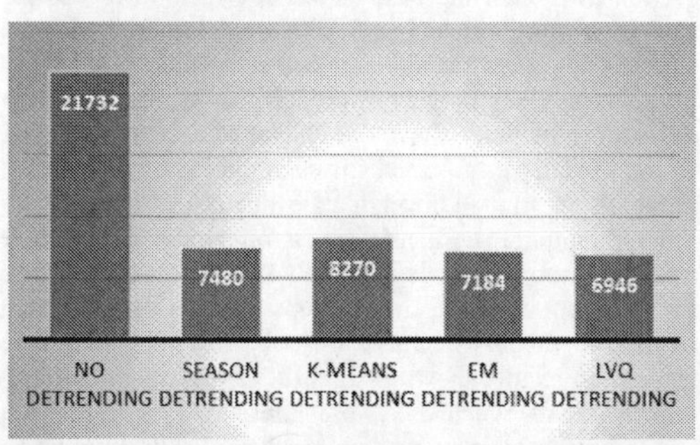

Figure 3: Number of superior forecasts (i.e. lowest rRMSE) by detrending method.

3 USING CLASSIFICATION FOR CONTEXTUAL DETRENDING SELECTION

Using the evaluation results of the forecasting models built with each detrending method (see section 2), we create a dataset where each record consist of a time series data with its best performing detrending method, as shown in Figure 4. Using these data, we trained a classification model that for each input time series predicts the best performing detrending method for the time series.

M	D	H	Mn	KT_{p1}	KT_{p2}	...	KT_{p12}	Detrend method
1	1	6	40	0.527	0.976	...	0.674	Season
1	1	6	45	0.718	0.711	...	0.666	No Detr.
1	1	6	50	0.725	0.676	...	0.666	KMeans
...	...	...	...	...	...	...	...	...

Figure 4: Training data sample used to train the classifier.

As classification method, we selected the *Random Forest Tree* algorithm implemented in the Weka Workbench (see [15] for details). Several trials with diverse classification methods indicated that this algorithm outperformed several other classifiers. Overall, performance of the *Random Forest Tree* classifiers is somewhat underwhelming, as shown in Table 2. Still, even with such mediocre performance, the selection of best detrending method by input time series emerging from the classification model offers impressive improvements when compared to competing detrending approaches, as discussed in the next section.

Table 2: Classification Results

Detrending method	*Precision*	*Recall*	*F-Measure*
No detrending	57.1%	79.8%	66.6%
Season	66.7%	60.2%	63.3%
K-means	45.1%	38.2%	41.4%
EM	48.0%	32.9%	39.0%
LVQ	35.2%	16.9%	22.8%

4 RESULTS

We used the classifier described in the previous section to predict the best detrending method to use for each input time series. The NARX model trained with the selected detrending methods was then applied to the input time series. The emerging approach provides an *ensemble detrending* method where the choice of detrended model is contextually established for each choice of time series input, as determined through classification.

We evaluated *ensemble detrending* alongside the other five detrending methods using rRMSE to assess error rate for one-hour forecasts in 12 steps of five minutes each relative to the 2104 dataset described in section 2. The evaluation results shown in Figure 5 show overwhelming improvements when *ensemble detrending* is used as a detrending techniques as compared to all other methods, with twofold to over threefold reductions in forecasting error rates. Another important advantage with *ensemble detrending* is the relatively low variance in performance from month to month, a performance characteristic only found with the *LVQ Detrending* method.

Additional insights emerging from the results obtained in this study concern the role of clustering as a detrending technique, and the correlation of clustering performance with forecasting accuracy.

4.1 Role of clustering as a detrending technique

Unlike season-based techniques, where the detrended data maintain temporal contiguity, a cluster-detrended dataset may contain discontinuous time series data, which may even belong to diverse seasons. When clustering solar irradiation time sequences, the relevant criterion is time sequence similarity, as determined by solar irradiation values. While there is a significant correlation between solar irradiation levels and time of the year, cloudy and clear sky conditions can occur in any season. This explains why clustering detrending performs no worse (EM, K-Means) or significantly better (LVQ) than season detrending, as shown in Figure 5.

4.2 Clustering performance and forecasting accuracy

Prima facie, it is somewhat surprising that LVQ detrending, the detrending method that has the fewest number of superior forecasts (Figure 3), is the second-best performing detrending method in terms of overall lower forecasting error rate (Figure 5). The question arises then as to what are the relevant clustering performance factors for forecasting accuracy.

The ability to group solar irradiation time sequences according to their similarity regardless of time contiguity (4.1) is one of the advantages of clustering as a detrending technique. However, that alone is not sufficient to outperform season detrending or no detrending. For example, both EM and K-Means detrending offer equivalent performance to season detrending and underperform when compared to no detrending, as shown in Figure 5.

Further inspection indicates that another equally significant factor correlating clustering performance to solar forecasting accuracy is *harmony* across forecasting results. For example, LVQ detrending yields a much lower rRMSE average and standard deviation across forecasting results than all other detrending methods, aside from ensemble detrending, as shown in Figure 6. The ability of LVQ detrending to yield greater harmony across forecasting results when compared to season, clustering and no detrending (Figure 6), makes LVQ detrending a better performing detrending method with reference to solar forecasting (Figure 5), despite its lower performance in predicting number of superior forecasts (Figure 3).

5 CONCLUSIONS

While significant effort has been devoted to detrending in solar forecasting, no clear winner technique has emerged. The study described in this paper suggests that no single detrending technique can provide a turnkey solution. Rather, an approach that enables contextual selection of the best detrending method, including no detrending, appears far better suited to improve forecasting results.

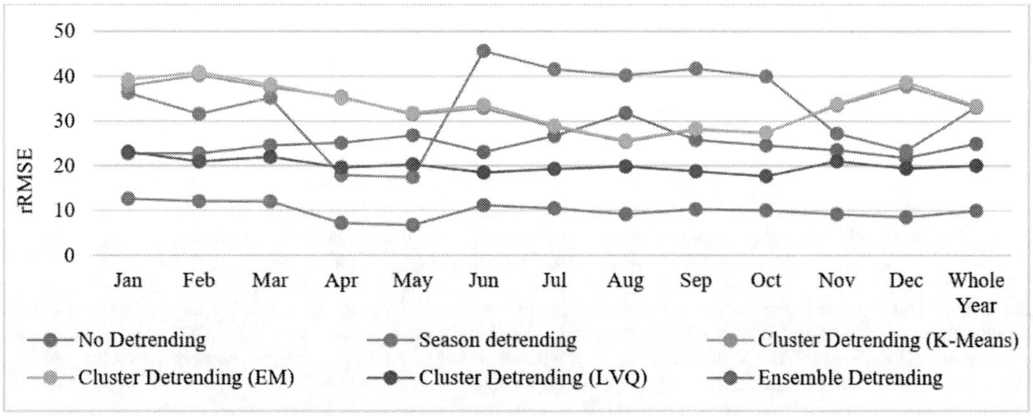

Figure 5: Overall forecasting error rates for diverse detrending methods.

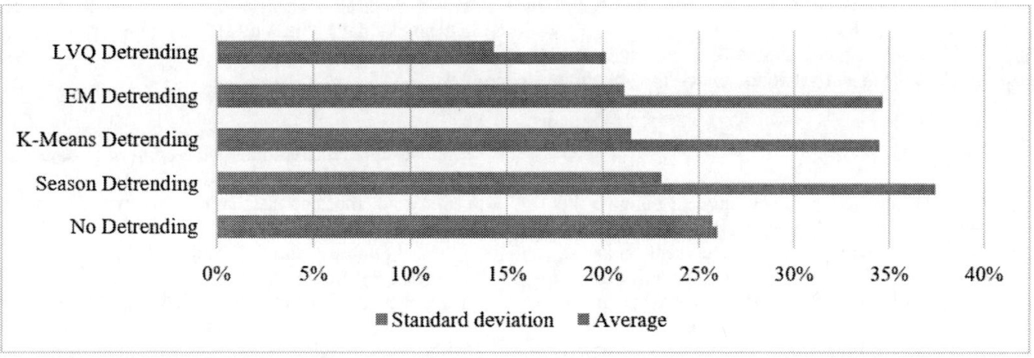

Figure 6: rRMSE average and standard deviation across forecasting results for diverse detrending methods.

6 REFERENCES

[1] http://www.wmo.int/pages/prog/amp/pwsp/Nowcasting.htm.

[2] Diagne, M., David, M., Lauret, P., Boland, J., & Schmutz, N. (2013). Review of solar irradiance forecasting methods and a proposition for small-scale insular grids. Renewable and Sustainable Energy Reviews, 27, 65-76.

[3] Inman, R. H., Pedro, H. T., & Coimbra, C. F., 2013. Solar forecasting methods for renewable energy integration. Progress in energy and combustion science, 39(6), 535-576.

[4] Brockwell, P. J., & Davis, R. A., 2006. Introduction to time series and forecasting. Springer Science & Business Media.

[5] Jain, P. C. (1984), Comparison of techniques for the estimation of daily global irradiation and a new technique for the estimation of global irradiation, Solar Wind Technol. 1: 123-134.

[6] Baig, A., P. Akhter and A, Mufti (1991), A novel approach to estimate the clear day global radiation, Renewable Energy, l:(1) 119-123.

[7] Akarslan, E., & Hocaoglu, F. O. (2016). A novel adaptive approach for hourly solar radiation forecasting. Renewable Energy, 87, 628-633.

[8] Ji, W., & Chee, K. C. (2011). Prediction of hourly solar radiation using a novel hybrid model of ARMA and TDNN. Solar Energy, 85(5), 808-817.

[9] Kaplanis, S. N. (2006). New methodologies to estimate the hourly global solar radiation; comparisons with existing models. Renewable Energy, 31(6), 781-790.

[10] Perez, R., Ineichen, P., Seals, R., & Zelenka, A. 1990. Making Full use of the Clearness Index for Parameterizing Hourly Insolation Conditions. Solar Energy, 45, 111-114.

[11] Perez-Astudillo, D., Bachour, D., 2014. DNI, GHI and DHI Ground Measurements in Doha, Qatar. Energy Procedia, 49(0), 2398-2404.

[12] Black, J.N., Bonython, C.W., Prescott, J.A. (1954) Solar radiation and the duration of sunshine. Q. J. R. Meteorol. Soc. 80 (344), 231–235.

[13] Duffie, J.A., Beckman, W.A. (1991) Solar Engineering of Thermal Processes, second ed. Wiley.

[14] Kasten, F., Young, A., 1989. Revised optical air mass tables and approximation formula. Appl. Opt. 28 (22), 4735–4738.

[15] Eibe Frank, Mark A. Hall, and Ian H. Witten (2016). The WEKA Workbench. Online Appendix for "Data Mining: Practical Machine Learning Tools and Techniques", Morgan Kaufmann, Fourth Edition, 2016.

[16] https://www.mathworks.com/help/nnet/ref/narxnet.html.

SHORT-TERM PHOTOVOLTAIC POWER FORECASTING BASED ON ARTIFICIAL NEURAL NETWORKS: A NUMERICAL WEATHER PREDICTION-FREE APPROACH

Spyros Theocharides[1,*], George Makrides[1], Marios Theristis[1], Florencia Almonacid[2], Eduardo F. Fernández[2] and George E. Georghiou[1]
[1]PV Technology Laboratory, FOSS Research Centre for Sustainable Energy, Department of Electrical and Computer Engineering, University of Cyprus, Nicosia, 1678, Cyprus
[2]Centre for Advanced Studies on Energy and Environment (CEAEMA), IDEA Solar Energy Research Group, Electronics and Automation Engineering Department, University of Jaén, Jaén, 23071, Spain
*corresponding author email: theocharidis.spyros@ucy.ac.cy

ABSTRACT: The increased penetration of photovoltaic (PV) systems introduce new challenges for the stability of electricity grids. In this scope, a machine learning technique utilising artificial neural networks (ANN) was implemented to forecast the hour-ahead (HA) PV power production without the utilisation of numerical weather prediction (NWP) models. Instead, historical PV operational and meteorological data-sets were used for the training and validation stages of the model to calculate HA PV power generation data-sets for time $t + 1h$ while the model input parameters (weather measurements) were applied for time t. The best-performing model comprised of four input parameters (in-plane global irradiance (G_I), ambient temperature (T_{amb}), elevation angle (AlS), azimuth angle (AzS)), a single hidden layer and 22 hidden neurons. Additionally, the results obtained over a test set period of 55 days demonstrated that accurate HA forecasts could be achieved without incorporating NWP since a daily-normalised root mean square error (nRMSE) of 3.63% was achieved. Finally, the model performed very well during clear sky days with a nRMSE close to 1% whereas 57% of all days demonstrated a nRMSE below 5%.
Keywords: artificial neural networks, forecasting, photovoltaic, power.

1 INTRODUCTION

Accurate PV production forecasting can mitigate the power quality effects posed by large shares of distributed systems through active grid management and is, therefore, an important feature that can assist utilities and plant operators in the direction of energy management and dispatchability planning. More specifically, short-term PV production forecasts (intra-hour) are necessary for power ramp and voltage flicker prediction as well as control operations and dispatch management. On the other hand, mid-term PV production forecasting (intra-day and day-ahead) is used for load consumption and production monitoring to control voltage and frequency levels and reduce secondary reserve.

During the last decades, electricity system operation has been upgraded involving PV production forecasts and most commonly adopted PV production forecasting approaches are based on time series analysis techniques [1]. In addition, parametric models for PV production forecasting have already been developed [2], [3] but their ability to forecast the power output of PV systems is not a straightforward process since accurate knowledge of system characteristics and behaviour should be provided. Therefore, a huge share of research is devoted to the development of more sophisticated and flexible prediction techniques using non-parametric models based on machine learning algorithms [4]–[9]. In order to train a PV power forecasting model with identical data, weather classification and machine learning techniques may be performed [10], [11]. Moreover, joint models comprised of combination of features of a physical model combined with artificial neural networks (ANNs) were presented elsewhere [7], [12]. Although a significant number of PV power forecasting tools have been developed, the challenge to provide a global and validated (against large scale data-sets) model for different PV remains unsolved. Additionally, to improve the accuracy of the power prediction of PV systems, adaptive methods that can capture system information and behaviour without the need of datasheet and installation parameters must be employed. This is crucial because a large proportion of PV systems includes de-centralized rooftop installations where knowledge of system information is not always available.

Furthermore, the system's behaviour can be estimated by processing recent PV operational data-sets. using the classical approach of the feedforward neural network (FFNN). This method was inspired by the method of biological neurons and is the simplest type of ANNs. It is a popular and powerful machine learning algorithm widely used in other fields for both prediction and classification purposes. The classical approach of the FFNN with an input, a hidden and an output layer of linear and non-linear activation functions can be viewed as a convenient way to predict the power output of PV systems. FFNN can be trained to develop relational weighted chains between internal nodes to overcome the limitations of traditional methods to solve complex problems, which can be modelled through a supervised learning technique based on historical data. Because of this chain of relationships, theoretically, multi-layered neural networks can be universal approximators and have tremendous potential to perform any nonlinear mapping through a learning process based on historical time-series [13]. In addition, ANNs are efficient for online training due to their capability of reflecting the information of new instances on a model by changing the weight values only.

Another approach is to utilise Numerical Weather Prediction (NWP) models to forecast weather variables [3], [14]–[16]. NWP models deliberates the atmosphere as an fluid thus the concept of weather forecasts is to parametrise the state of the fluid at a certain time (t) by utilising the fluid- and thermo-dynamic equations to forecast the state of the fluid at time (t + n) [17]–[19]. The utilisation of mathematical equations and models to simulate the atmosphere are not accurate, thus the NWP methods are associated with forecasting errors and biases [20], [21].

In this work, a NWP free methodology for hour-ahead (HA) PV power production forecasting utilising data-driven machine learning models and specifically ANNs is proposed in this study. One-year data from a reference poly-crystalline silicon (poly-c-Si) PV system in Cyprus were used for training and testing the model and the forecasting performance results showed that the application of this method improves the HA PV production forecasting accuracy.

2 EXPERIMENTAL SETUP

2.1. Outdoor test facility description

The outdoor test facility of the University of Cyprus includes a fixed plane infrastructure for outdoor performance assessments at both the module and system level. The installed poly-c-Si system was mounted in a portrait arrangement on aluminium mountings, at the optimum annual energy yield plane-of-array (POA) angle for Cyprus of 27.5°.

The PV system was connected to a data-acquisition platform, used for the monitoring and storage of meteorological and PV operational data. The performance of the system and the prevailing meteorological conditions were recorded according to the requirements set by the IEC 61724 [22]. Specifically, the irradiance and meteorological measurements include the in-plane global irradiance (G_I), relative humidity (RH), wind direction (W_a), wind speed (W_s) and ambient temperature (T_{amb}). The PV system operational measurements include the maximum power current (I_{mp}), voltage (V_{mp}) and power (P_{mp}), as measured at the output of the PV array (DC side) [23], [24].

2.2 PV system description

The PV system comprises of five poly-c-Si PV modules. The modules of the system are connected in series to form a PV string at the input of a string inverter. The main technical specifications of the test PV system are summarised in Table I.

Table I: Installed PV system technical characteristics.

Technical characteristic	Parameter
Modules	5 × poly-c-Si
System power (datasheet)	1365 W_p
Installation date	01/06/2015
Efficiency	14.40%

3 METHODOLOGY

The HA PV power production forecasting model was based on a FFNN that employed recent historical data-sets of a reference PV system, as the training parameters of the model. The best-performing neural network was chosen through a series of validation tests performed by identifying optimum combinations of input parameters, sizes and hyper (architectural) parameters of the model (optimal hidden layers, neurons, iterations and learning function). The historical PV operational and meteorological data-sets were used for both the training and validation stages of the model. For each model, the training step yielded the respective HA PV power (output parameter) for time $t + 1h$ while the model input parameters (weather measurements) were applied for time t. The aim of the training step was to develop a model which is capable of accurately forecasts the power generation for time $t + 1h$, based on the relationships established between the weather measurements of time t (without incorporating forecasted weather data) and the actual output PV produced power of the reference system at time $t + 1h$.

In particular, the annual historical data-sets of the reference system and prevailing weather conditions were divided into three subsets, the training, validation and testing set. The training and validation sets were utilised in order to implement the best-performing model, by varying the training period from 30% to 70% of the actual data-set. The model inputs included the meteorological parameters in addition to the solar elevation angle (AlS) and solar azimuth angle (AzS) that were calculated using solar position algorithms [25], [26]. Additionally, the hyper parameters of the neural network were tested by varying the number of hidden neurons, layers, iterations and learning function. The validation set was employed in order to evaluate the performance of the various neural network models and therefore identify the best performing model, while the test set was applied in order to yield forecasting results for the developed model. The complete methodological steps followed for the development of the short-term HA PV power production forecasting model are exhibited in Fig. 1 and Eq. (1).

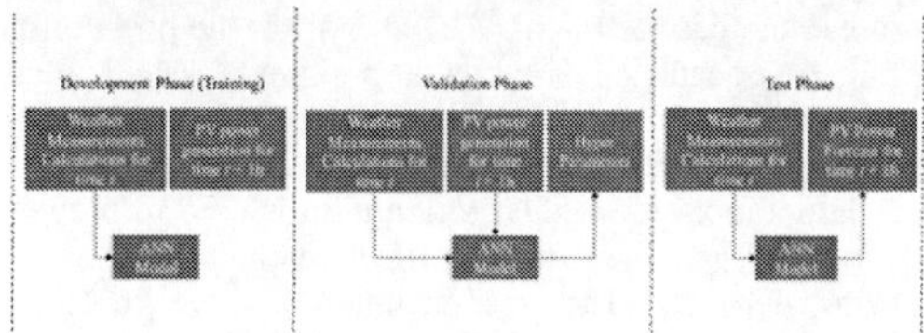

Figure 1: Training, validation and testing phases of the short-term HA PV power production forecasting model.

$$P_{DC}(t+1) = G_I(t) + T_{amb}(t) + ... \quad (1)$$

Where $P_{DC}(t + 1)$ is the power for time t + 1, G_I(t), T_{amb}(t), … , are the weather measurements for time t.

3.1 Neural network

A Bayesian Regularization Neural Network (BRNN) is essentially a simple Multi-Layer Perceptron (MLP), in which a Bayesian regularization has been applied to its training function. The proposed model is given by Eq. (2) [27], [28]:

$$y_i = g(x_i) + e_i = \sum_{k=1}^{9} w_k g_k (b_k + \sum_{j=1}^{p} x_{ij} \beta_j^{[k]} + e_i), i = 1, ..., n \quad (2)$$

where $e_i \sim N(0, \sigma^2 e)$, s is the number of neurons, w_k is the weight of the k-th neuron, b_k is a bias for the k-th neuron, $\beta_j^{[k]}$ is the weight of the j-th input to the net, *and* $gk(\cdot)$ is the activation function:

$$g_k(x) = \frac{exp(2x) - 1}{exp(2x) + 1} \quad (3)$$

The model will minimise according to Eq. (4) [27], [28]

$$F = \beta E_D + \alpha E_W \quad (4)$$

where E_D is the error of sum squares, E_W is the sum of squares of network parameters (weights and biases). E_W

is the sum of squares of network parameters (weights and biases), β and α are the dispersion parameters for weights and biases.

The regularisation term applied was the squared sum of the weights of the neural network [29]:

$$\varepsilon_T = \beta\varepsilon_D + \alpha\varepsilon_R = \frac{\beta}{2}\sum_{k=1}^{N}(y_k - t_k) + \frac{\alpha}{2}\sum_{i=2}^{M} w_i \quad (5)$$

where α and β are coefficients assigned to each term. The second term in Eq. (5) is called weight decay, as it ensures that the weights of the network do not exceed the total error of the network.

After the data were fed into the network, the density function for the weights can be updated according to Bayes' rule [28]:

$$P(w|D,\alpha,\beta,M) = \frac{P(D|w,\beta,M)P(w|\alpha,M)}{P(D|\alpha,\beta,M)} \quad (6)$$

where D represents the data-set, M the model used for the Neural Network and w is the vector of neural network's weights. $P(w|\alpha, M)$ represents the values of weights prior to the data-set input. $P(D|w,\beta, M)$ is the probability of the data occurring based on the weights. $P(D|\alpha,\beta, M)$ is a normalisation factor, which ensures that the total summation of the probability is one.

In addition, the regularisation term is used to prevent overfitting, by controlling the effective complexity of the neural network. The regularisation of the designed networks in this study was performed by adding a penalty equal to the L2-norm of the weights, in order to reduce the value of the weights by the same factor.

3.2 Model architecture

The acquired annual data-set of the PV system was divided into three different sets in order to determine the optimal parameters for the neural network. The annual data-set was separated into the train, validation and test set with the 70:15:15% portion approach [30]. The 70:15:15% data split portion was selected in order to include systematic information for the training but to also preserve sufficient data-points for validation and testing. The data splitting techniques involved the partitioning of data into an explicit training set used to develop the model, the validation set used to tune the parameters of the model and the test set necessary to evaluate the model performance on unseen data. The training, validation and testing set, comprised of the model inputs which included the measurements of G_I, T_{amb}, RH, W_s, W_a and the calculated parameters of AzS and AlS. Finally, the DC P_{mp} was the output parameter of the model.

In order to achieve optimal forecasting performance, different input feature combinations and hidden layer topologies (number of hidden units) were investigated while training the network models. The initially designed FFNN, which was subsequently used in the tuning process comprised of two inputs (G_I, T_{amb}), four hidden nodes and an output layer (P_{mp}) (see Fig. 2). The optimal model parameters were identified by comparing the performance of the developed models based on the training and validation sets. Re-iteration of the process (epochs) was performed until the results from the network became acceptable, indicating low performance error while avoiding over-training the network, that may result in over-fitting.

Lastly, the tuned networks were then evaluated by feeding the test set input variables and assessing the actual measured power against the predicted power according to some performance assessment metrics (that will be discussed later).

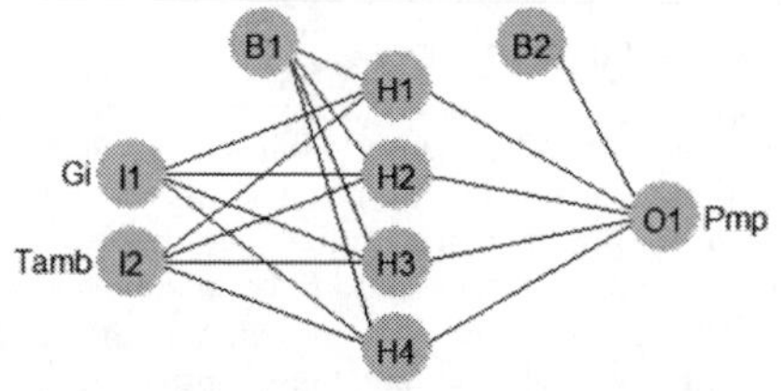

Figure 2: Network interconnection diagram (NID) of a FFNN.

3.3 Model performance assessment

The forecasting performance accuracy was assessed based on several predefined metrics when the test set was applied to the developed algorithms. The metrics commonly used in PV production forecasting applications include the mean absolute error (MAE), mean absolute percentage error (MAPE), root mean square error (RMSE) and normalised RMSE (nRMSE) to the nominal PV system peak power:

$$MAE = \frac{1}{n} \times \sum_{i=1}^{n} \left| y_{actual,i} - y_{predicted,i} \right| \quad (7)$$

$$MAPE = \frac{100}{n} \times \sum_{i=1}^{n} \left| \frac{y_{actual,i} - y_{predicted,i}}{y_{actual,i}} \right| \quad (8)$$

$$MBE = \frac{100}{n} \times \sum_{i}^{n} \frac{y_{actual,i} - y_{predicted,i}}{y_{actual,i}} \quad (9)$$

$$RMSE = \sqrt{\frac{1}{n} \times \sum_{i=1}^{n} (y_{actual,i} - y_{predicted,i})^2} \quad (10)$$

$$nRMSE = \frac{100}{P_{nominal}} \times \sqrt{\frac{1}{n} \times \sum_{i=1}^{n} (y_{actual,i} - y_{predicted,i})^2} \quad (11)$$

where $y_{actual,i}$ and $y_{forecasted,}$ is the actual and predicted power respectively, $P_{nominal}$ is the nominal peak power of the PV system (1365 W).

4 RESULTS

In order to establish an accurate HA PV power production forecasting model, it is necessary to analyse the sensitivity of the output on the input variables. For this purpose, the influence of each input parameter on the output power, was examined by calculating the Pearson Correlation Coefficient (PCC), which measures the direction and strength of the linear relationship between two variables. Table II summarises the results from the PCC between the investigated features.

Table II: Pearson Correlation Coefficient between PV power output and environmental factors under typical weather conditions.

	Pearson Correlation Coefficient
	Maximum Power, P_{mp}
Incident global irradiance, G_I	0.86
Sun elevation angle, AlS	0.77
Relative humidity, RH	-0.74
Ambient temperature, T_{amb}	-0.60
Wind direction, W_a	-0.10
Wind speed, W_s	0.10
Sun azimuth angle, AzS	-0.01

Intending to select only the highly linearly correlated input parameters, a threshold was set at the Pearson Coefficient; i.e. $|PCC| \geq 0.5$. The parameters complying with the set threshold level were the G_I, AlS, T_{amb} and RH. In order to achieve a model optimisation, different input parameter combinations at the input layer of each devised network were investigated comprising not only of the highly correlated parameters but also the remaining input parameters that exhibited non-linear behaviour. However, the highly correlated input parameters were expected to achieve high accuracy and feature importance. In addition, the number of hidden units was set to an empirical value between the total of the number of inputs and output layer nodes. Table III tabulates the validation results based on the nRMSE of the input feature analysis to the FFNN model.

Table III: Input feature parameter models and the corresponding nRMSE.

Input features	Hidden Layer Neurons	nRMSE (%)
G_I, T_{amb}	2	5.50
G_I, T_{amb}, AlS	3	5.27
G_I, T_{amb}, RH	3	5.68
G_I, T_{amb}, RH, AlS	4	5.33
G_I, T_{amb}, W_{alpha}, W_S	4	6.10
G_I, T_{amb}, AlS, AzS	4	4.60
G_I, T_{amb}, W_S, RH	4	5.13
G_I, T_{amb}, AzS, AlS, W_a, W_S, RH	7	5.77

The results clearly demonstrated that the optimum feature combination included all the input variables (G_I, T_{amb}, AzS, AlS), confirming that FFNNs could detect both the linear and non-linear behaviours among the input and output parameters. Varying the number of input features resulted in an nRMSE ranging from 6.10% to 4.60%, which highlights the robustness of neural networks.

To further improve the accuracy of the predictions of the optimal network, the effect of varying the number of hidden units on the prediction accuracy was also investigated. The results presented in Fig. 3, demonstrate that the nRMSE decreases with the increasing number of hidden units, in the case of the training set.

In particular, although the training set was improving by increasing the number of hidden neurons, the nRMSE of the validation set showed a different behaviour when the number of hidden neurons was increased beyond a certain number. The results presented in the plot showed that the best accuracy for the validation set was obtained when the selected number of hidden neurons was in the range of 22 to 25. To minimise the error and the noise of the training phase more hidden units were utilised. However, to decrease the complexity of the model and prone to overfitting [31] the number of nodes was set to 22 in the hidden layer as an additional optimisation step, which yielded a nRMSE of 3.63%.

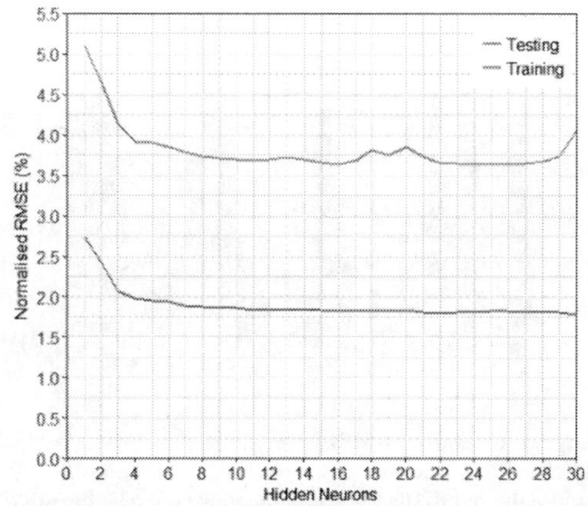

Figure 3: Daily nRMSE (%) against the number of neurons for both the training and validation sets using a single hidden layer.

At this stage, the neural network was designed by specifying the number of input variables and amount of hidden neurons. The optimal model therefore comprised of the input parameters of G_I, T_{amb}, AzS and AlS. The 70% of randomly selected samples and a single hidden layer with 22 neurons (see Fig. 4) developed the training set of the ANN model. Additionally Eq. (12) describes the best-performing model.

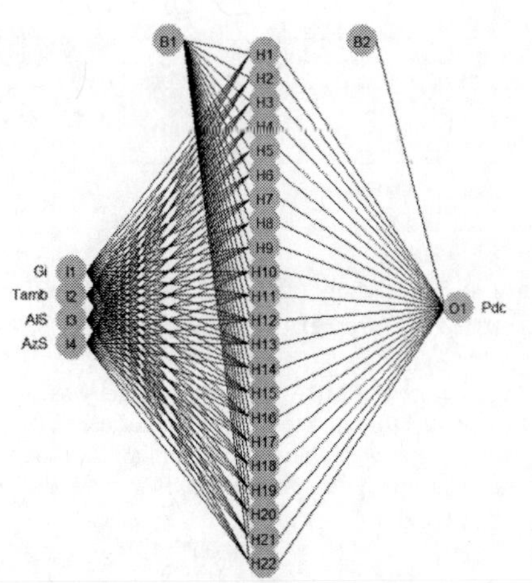

Figure 4: NID of the optimimally designed FFNN.

$$P_{DC}(t+1) = G_I(t) + T_{amb}(t) + AlS(t) + AzS(t) \quad (12)$$

When the testing set data were applied to the model, the daily average nRMSE was 3.60%, while 57% of the days demonstrated a nRMSE below 5% (Fig. 5). Furthermore, when typical weekly data with both clear and overcast days were applied to the model it was found that the systematic behaviour of the PV system was captured (see Fig. 6a). Specifically, the error relative to the capacity of the system for all points during clear sky was in most cases less than 0.1 W/Wp (see Fig. 6b).

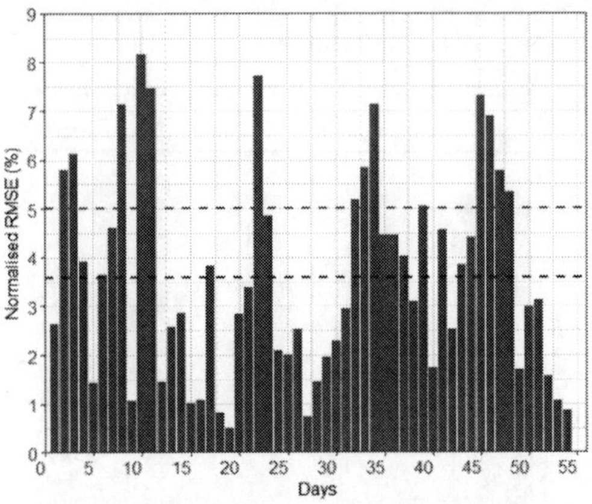

Figure 5: Daily nRMSE over the test set samples. The dashed blue line represents the daily nRMSE of 3.60% while the red dashed line demonstrates the 5% threshold.

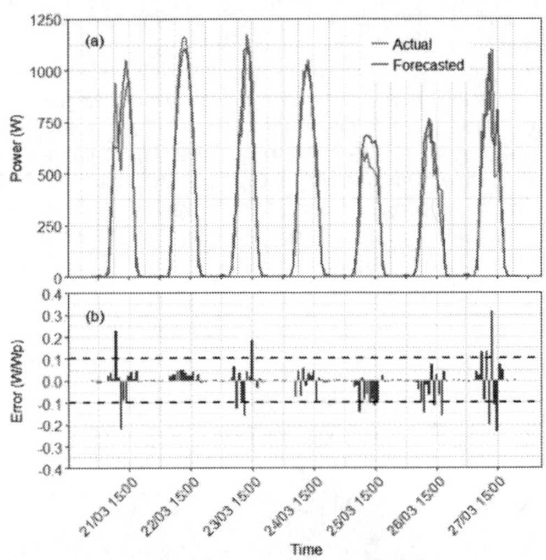

Figure 6: Plot of daily (a) Actual against HA PV power forecast and (b) Error relative to the capacity.

In addition, the MBE of the testing set was found to be 0.1%, indicating that there is no significant bias in the forecasts (neither an overestimation nor an underestimation of the forecasted P_{mp}, over the actual P_{mp}) (see Fig. 7).

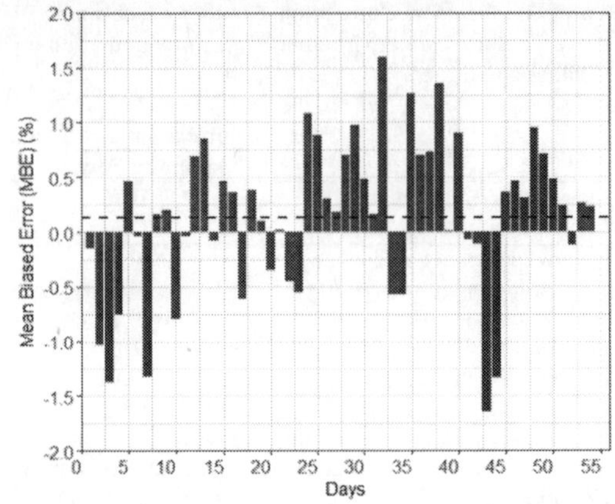

Figure 7: Daily forecasted bias error. The blue dashed line represents the MBE (~0.1%).

5 CONCLUSION

Increased PV penetration requires agile forecasting techniques in order to ensure grid stability and dispatchability. For the development of an accurate ANN model, a procedure with subsequent stages of training and testing steps was followed and the devised ANNs were assessed by tuning their input features and network architecture. More specifically, a conventional FFNN was designed, whose parameters were estimated by a feedforward supervised learning process, trained using the BP technique and regularised by employing L2 regularisation. The final ANN comprised of 4 input parameters (G_I, T_{amb}, *AzS*, *AlS*), 22 hidden neurons and was trained with a random samples of 70:15:15% train, validation and test set approach.

The forecasting results demonstrated that the average daily nRMSE was 3.60% over the tests period and the model performed very well during clear sky days exhibiting a nRMSE close to 1% while the MBE was 0.15% indicating no biases between the forecasted and the actual power.

6 ACKNOWLEDGEMENT

This work was funded through the INFORPV project which is co-financed by the SOLAR-ERA.NET Transnational Calls of the European Union under the grant agreement number SOLAR-ERA.NET/1215/02.

7 REFERENCES

[1] S. Pelland, J. Remund, J. Kleissl, T. Oozeki, and K. De Brabandere, "Photovoltaic and Solar Forecasting: State of the Art," *Int. Energy Agency Photovolt. Power Syst. Program. Rep. IEA PVPS T14*, pp. 1–40, 2013.

[2] E. Lorenz, T. Scheidsteger, J. Hurka, D. Heinemann, and C. Kurz, "Regional PV power prediction for improved grid integration," *Prog. Photovoltaics Res. Appl.*, vol. 19, no. 7, pp. 757–771, 2011.

[3] S. Pelland, G. Galanis, and G. Kallos, "Solar and photovoltaic forecasting through post-processing of the Global Environmental Multiscale numerical weather prediction model," *Prog. Photovoltaics Res. Appl.*, vol. 21, no. 3, pp. 284–296, 2013.

[4] I. Ashraf and a Chandra, "Artificial neural network based models for forecasting electricity generation of grid connected solar PV power plant," *Int. J. Glob. Energy Issues*, vol. 21, no. 1, pp. 119–130, 2004.

[5] D. Caputo, F. Grimaccia, M. Mussetta, and R. E. Zich, "Photovoltaic plants predictive model by means of ANN trained by a hybrid evolutionary algorithm," in *Proceedings of the International Joint Conference on Neural Networks*, 2010.

[6] S. A. Kalogirou, "Applications of artificial neural-networks for energy systems," *Appl. Energy*, vol. 67, no. 1–2, pp. 17–35, 2000.

[7] A. Gandelli, F. Grimaccia, S. Leva, M. Mussetta, and E. Ogliari, "Hybrid model analysis and validation for PV energy production forecasting," in *Proceedings of the International Joint Conference on Neural Networks*, 2014, pp. 1957–1962.

[8] A. Mellit and A. M. Pavan, "A 24-h forecast of solar irradiance using artificial neural network: Application for performance prediction of a grid-connected PV plant at Trieste, Italy," *Sol. Energy*, vol. 84, no. 5, pp. 807–821, 2010.

[9] F. Almonacid, P. J. Pérez-Higueras, E. F. Fernández, and L. Hontoria, "A methodology based on dynamic artificial neural network for short-term forecasting of the power output of a PV generator," *Energy Convers. Manag.*, vol. 85, pp. 389–398, 2014.

[10] C. Chen, S. Duan, T. Cai, and B. Liu, "Online 24-h solar power forecasting based on weather type classification using artificial neural network," *Sol. Energy*, vol. 85, no. 11, pp. 2856–2870, 2011.

[11] J. Shi, W. J. Lee, Y. Liu, Y. Yang, and P. Wang, "Forecasting power output of photovoltaic system based on weather classification and support vector machine," in *Conference Record - IAS Annual Meeting (IEEE Industry Applications Society)*, 2011.

[12] A. Dolara, F. Grimaccia, S. Leva, M. Mussetta, and E. Ogliari, "A physical hybrid artificial neural network for short term forecasting of PV plant power output," *Energies*, vol. 8, no. 2, pp. 1138–1153, 2015.

[13] M. P. Almeida, M. Muñoz, I. de la Parra, O. P. Lamigueiro, and L. N. Fernández, "Comparative study of nonparametric and parametric PV models to forecast AC power output of PV plants," in *EU PVSEC Proceedings*, 2015, pp. 2230–2234.

[14] S. E. Haupt *et al.*, "Building the Sun4Cast System: Improvements in Solar Power Forecasting," *Bull. Am. Meteorol. Soc.*, vol. 99, no. 1, pp. 121–136, 2018.

[15] D. P. Larson, L. Nonnenmacher, and C. F. M. Coimbra, "Day-ahead forecasting of solar power output from photovoltaic plants in the American Southwest," *Renew. Energy*, vol. 91, pp. 11–20, 2016.

[16] X. Sun and T. Zhang, "Solar Power Prediction in Smart Grid Based on NWP Data and an Improved Boosting Method," in *Proceedings - 1st IEEE International Conference on Energy Internet, ICEI 2017*, 2017, pp. 89–94.

[17] P. Lynch, "The origins of computer weather prediction and climate modeling," *J. Comput. Phys.*, vol. 227, no. 7, pp. 3431–3444, 2008.

[18] L. F. Richardson, *Weather prediction by numerical process, second edition*. 2007.

[19] R. Kimura, "Numerical weather prediction," *J. Wind Eng. Ind. Aerodyn.*, vol. 90, no. 12–15, pp. 1403–1414, 2002.

[20] H.-L. Xue, X.-S. Shen, and J.-F. Chou, "A forecast error correction method in numerical weather prediction by using recent multiple-time evolution data," *Adv. Atmos. Sci.*, vol. 30, no. 5, pp. 1249–1259, 2013.

[21] L. Gao, H.-L. Ren, J.-P. Li, and J.-F. Chou, "Analogue correction method of errors and its application to numerical weather prediction," *Chinese Phys.*, vol. 15, no. 4, pp. 882–889, 2006.

[22] British Standard, "IEC 61724:1998. Photovoltaic system performance monitoring — Guidelines for measurement , data exchange and analysis," p. 20, 1998.

[23] G. Makrides, B. Zinsser, M. Norton, G. E. Georghiou, M. Schubert, and J. H. Werner, "Potential of photovoltaic systems in countries with high solar irradiation," *Renewable and Sustainable Energy Reviews*, vol. 14, no. 2. pp. 754–762, 2010.

[24] G. Makrides, B. Zinsser, M. Schubert, and G. E. Georghiou, "Energy yield prediction errors and uncertainties of different photovoltaic models," *Prog. Photovoltaics Res. Appl.*, vol. 21, no. 4, pp. 500–516, 2013.

[25] I. Reda and A. Andreas, "Solar position algorithm for solar radiation applications," *Sol Energy*, vol. 76, no. 5, pp. 577–589, 2004.

[26] NREL, "Solar Position Algorithm (SPA) Source Code," *NREL Renewable Resource Data Center (RReDC)*, 2008. [Online]. Available: http://rredc.nrel.gov/solar/codesandalgorithms/spa/.

[27] D. E. Rumelhart, G. E. Hinton, and R. J. Williams, "Learning representations by back-propagating errors," *Nature*, vol. 323, no. 6088, pp. 533–536, 1986.

[28] D. J. C. MacKay, "Bayesian Interpolation," *Neural Comput.*, vol. 4, no. 3, pp. 415–447, 1992.

[29] F. Dan Foresee and M. T. Hagan, "Gauss-Newton approximation to bayesian learning," in *IEEE International Conference on Neural Networks - Conference Proceedings*, 1997, vol. 3, pp. 1930–1935.

[30] S. Theocharides, G. Makrides, V. Venizelou, P. Kaimakis, and G. E. Georghiou, "Pv Production Forecasting Model Based on Artificial Neural Networks (Ann)," *33rd Eur. Photovolt. Sol. Energy Conf.*, pp. 1830–1894, 2017.

[31] K. P. Burnham and D. R. Anderson, "Model Selection and Multimodel Inference," *Model Sel. Multimodel Inference*, pp. 49–97, 2004.

OPTIMIZATION OF COMPONENT DIMENSIONING FOR A COMBINED HEAT AND POWER SYSTEM WITH SPECIAL FOCUS ON PV GENERATOR SIZE

Angenendt Georg[1,2,3], Zurmühlen Sebastian[1,2,3], Axelsen Hendrik[1,2,3], Wirtz Christian[1], Sauer Dirk Uwe[1,2,3]
[1] Chair of Electrochemical Energy Conversion and Storage Systems, Institute for Power Electronics and Electrical Drives (ISEA), RWTH Aachen University, Germany
[2] Juelich Aachen Research Alliance, JARA-Energy, Germany
[3] Institute for Power Generation and Storage Systems (PGS), E.ON ERC, RWTH Aachen University, Germany
Email: georg.angenendt@isea.rwth-aachen.de
Tel.: +49 241 80 49373, Fax: +49 241 80 92203

ABSTRACT: The German energy sector faces a continuous growth of residential PV battery energy storage systems (PV BESS). These systems are applied to enhance the self-consumption rate of domestic photovoltaic systems. The power generated by the PV system can also be used for space heating and domestic hot water (DHW) to contribute to the decarbonisation of the heating sector. Sector coupling of heat and power can be achieved by using heat pump systems. The combined system is capable of providing renewable energy to the household demand for heat and electricity. The economics of these combined systems are the key to increase their market share. The dimensioning of the different system components heavily influences the economics of PV BESS with power and heat coupling. Capital-intensive components such as the PV generator, the BESS, the heat pump, and the thermal storage units play a major role for profitability. Moreover, the thermal and electrical load profile of the residential household as well as the solar radiation profile must be considered to define efficient or optimal component dimensions. This is why an optimization of component sizing is necessary in order to enhance the economics of the system. This paper presents optimization results based on an evolutionary algorithm.
The annuity is used as the target function of the optimization. Additionally, the levelized cost of energy (LCOEnergy) are calculated, to enable an inter-sectorial comparison and an aggregated economic treatment. The levelized cost of energy are determined by calculating the respective levelized cost of electricity (LCOEle) as well as the levelized cost of heat (LCOH).
First optimization results for a household with combined heat and power system show that a system with relatively small storage components compared to the rated power of the PV generator leads to the lowest LCOEnergy. Diminishing subsidies and feed-in tariffs (FIT) have opposite effects on the optimal value for the rated PV power of the system compared to the additional demand from combined heat and power. If the FIT is lower than the generation cost, the PV generators size is mainly depended on load of the integrated home. Overall, the results show that an optimization of the component sizes can increase the economics, but the optimal sizes have to be calculated for every scenario individually.
Keywords: photovoltaics (PV), battery energy storage systems (BESS), thermal storage, optimization, power to heat (P2H)

Nomenclature	
BESS	battery energy storage system
COP	coefficient of Performance (heat pump)
DC	direct current
DHW	domestic hot water
FIT	feed-in tariff
LCOEle	levelized cost of electricity
LCOEnergy	levelized cost of energy
LCOH	levelized cost of heat
MPP	maximum power point
NMC	nickel-manganese-cobalt-oxide
PV	Photovoltaic
SC	self-consumption
SOC	state of charge

1 INTRODUCTION

One quarter of the energy demand in Germany is consumed in residential households. In 2017, 91% of the energy consumed for residential heating was generated from fossil fuels [1].Therefore, residential households can play a major role in the decarbonisation process [2].

The rising number of PV systems in Germany [3] leads to a rising production of renewable energy in the residential sector. To use this energy in the heating sector, heat pumps can be applied. Heat pumps increase the self-consumption rate of residential PV generators [4]. Further enhancement of the PV self-consumption rate can be achieved by the installation of battery energy storage systems (BESS) [5].

Integrated households combine a PV generator with a BESS and a heat pump for power-to-heat coupling. An optimization of the component sizes of integrated houses can enhance their economics. The optimization of PV BESS is broadly discussed in literature [6–9]. The load transfer potential of heat pumps is optimized in [10]. A cost optimal allocation of heat pumps is analysed in [11]. The storage of thermal energy due to the mass of the building is investigated in [12]. An optimization of an integrated home with the target to enhance the self-consumption is presented in [13].

This paper analyses the economic optimization of integrated households. Furthermore, the influence of feed-in tariffs (FIT) for PV systems on the optimal size of the components of the integrated home is investigated.

2 METHODOLOGY

2.1 Model

To find the optimal system configuration for an integrated home with a PV BESS and a combined power-to-heat system, a detailed model is crucial. The presented model of an integrated home has a PV BESS and a heat pump to couple the electrical and the thermal system.

The electrical model is based on the model presented in [14], with its structure being depicted in Figure 1. The model represents a DC-coupled BESS with a PV power plant. The battery model is parametrized with test results of a cylindrical lithium-nickel-manganese-cobalt-oxide (NMC) 18650 lithium-ion battery cell from LG Chem (LG ICR18650MF1) with a rated capacity of 2.15Ah. The battery model is based on the model presented in [15] and contains a separate calendar and cycle aging model. Input data for the load [16] and PV radiation of the electrical model are based on real data measurement.

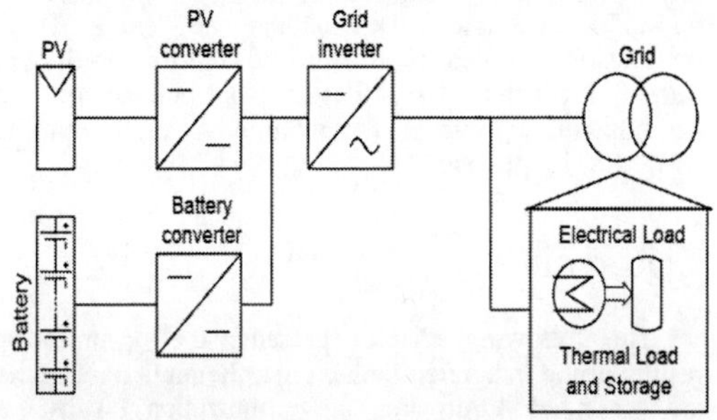

Figure 1. Model structure of an integrated home with a DC-coupled PV BESS and a heat pump for heat and power coupling [4].

The heat pump and thermal storages provide thermal energy to cover the thermal demand of the household. This energy is used for space heating and the domestic hot water supply (DHW). The applied thermal model is presented in [4] and shown in Figure 2. It consists of a heat pump and two thermal storage units: a DHW storage and a buffer storage. The thermal storage units are arranged in parallel [17] and have been modelled according to commercially available systems [18, 19]. The heat pump model is based on a Vitocal 200-S heat pump from Vissmann [20].

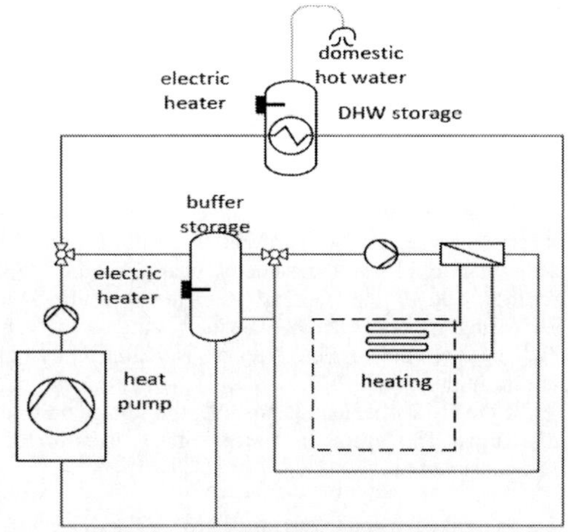

Figure 2: Model structure of the thermal system with a heat pump and two thermal storage units [4].

To optimize the size of the heat pump, a flexible heat pump model is necessary. The COP of the heat pump is reduced with a reduced thermal power [20]. Figure 3 shows the COP over the thermal power with respect to different operation points. Therefore, the COP of the heat pump is scaled to meet this effect. The modified parameters in comparison to the parameters of the heat pump based on the datasheets are shown in Figure 3.

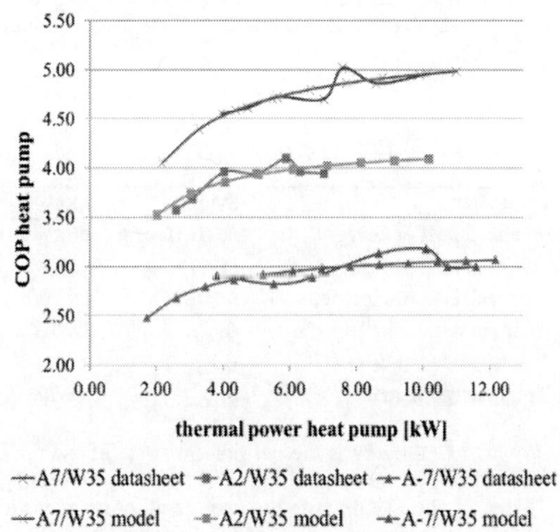

Figure 3: Comparison of simulation results and manufacturer data [20] for the COP in relation to the thermal power of the heat pump. Different operation points of the heat pumps are analysed (air or ambient temperature/water or flow temperature).

2.2 Operation strategies

The introduced integrated home aims to enhance the self-consumption rate of energy from PV. The primary goal is to cover the energy demand of the household to meet the needs of the residents. If PV power is available, it is used to cover the current load demand. Load that cannot be covered by the PV system is called residual load and is covered by the grid or the storage units. PV generated energy that is available after covering the load is called negative residual load and calculated as equation (1).

$$P_{RL} = P_{LH} - P_{PV} \quad (1)$$

Parameter	Symbol	Unit
Residual load	P_{RL}	W
Load household with heat pump	P_{LH}	W
PV feed-in	P_{PV}	W

If the residual load is negative, the BESS and the thermal storages can be charged. The BESS is prioritized according to [4]. If the state of charge of the BESS reaches 100%, the thermal storage is charged by increasing its temperature. The maximum temperature $T_{Buffermax}$ of the buffer storage is defined by the optimization.

If the residential load is positive, the storage units are discharged. This operation strategy aims to maximize the self-consumption.

2.3 Optimization and target function

The target of the optimization is to increase the economics of the presented integrated home. Hence, the total costs are minimized. The annuity is used as the indictor for these costs. It is divided by the annual energy consumption to calculate the levelized cost. The evaluation of the levelized cost enables the potential of a comparison with other supply concepts. The LCOEnergy are the weighted sum of the levelized cost of heat (LCOH) and the levelized cost of electricity (LCOEle). With the separate consideration of the LCOH and the LCOEle, an inter-sectorial evaluation is possible. Equation (2) depicts the calculation of the levelized cost of energy.

$$LCOEnergy = LCOH \cdot \frac{Q_{heat}}{E_{total}} + LCOEle \cdot \frac{E_{elect}}{E_{total}} \quad (2)$$

Parameter	Symbol	Unit
levelized cost of energy	$LCOEnergy$	ct/kWh
levelized cost of heat	$LCOH$	ct/kWh_{th}
levelized cost of electricity	$LCOEle$	ct/kWh_{el}
Thermal demand	Q_{heat}	kWh_{th}/a
Electrical demand	E_{elect}	kWh_{el}/a
Total energy demand	E_{total}	kWh/a

The LCOEnergy is the net present value of the unit-cost of energy over the lifetime of a generating asset [21]. Therefore, the LCOEnergy incorporate all costs over the system's lifetime including initial investment, cost of operation, maintenance cost and cost of capital.

The calculation of these costs are presented in detail in [4]. The LCOEle and the LCOH are calculated with the same method as the LCOEnergy.

2.4 Optimization algorithm and parameters

The optimization parameters have a direct influence on the target function presented in chapter 2.3. These parameters and the corresponding boundary conditions are given in Table I. The algorithm optimizes the sizes of the different components of the electrical and the thermal system and therefore the investment cost. These costs directly influence the economics.

Furthermore, parameters are optimized which influence the behavior of the systems. These parameters are the maximum temperature of the buffer storage and the DHW storage. The storage temperature influences the flow temperature of the heat pump and therefore influences the efficiency (COP) of the heat pump.

The minimum value of the sizes of the thermal storage and the DHW storage are flexible, because they depend on the size of the heat pump. The minimum volume of the buffer storage can be calculated with equation (3). The maximum output temperature of the modelled heat pump is 60 °C. In equation (4), the minimum storage temperature of 32 °C is minded as well as the minimum spread temperature of 5 °C between DHW and thermal storage.

$$V_{min} = \frac{P_{th,N} \cdot 3.600 \frac{s}{h} \cdot t_{min}}{c_p \cdot \rho \cdot \Delta T_{min}} \quad (3)$$

$$\Delta T_{min} = min(T_{Buffermax} - 32\,°C, 5\,°C) \quad (4)$$

Parameter	Symbol	Unit
min. water volume of the storage	V_{min}	m^3
thermal power heat pump	$P_{th,N}$	kW
min. runtime heat pump	t_{min}	h
thermal capacity water	c_p	kJ/(kg·K)
water density	ρ	kg/m^3
min. temperature spread storage	ΔT_{min}	K

The described optimization problem is a non-linear (battery aging), non-convex and a multimodal optimization problem. Due to the complexity of the optimization problem, a probabilistic algorithm is chosen. These algorithms are faster in comparison to deterministic algorithms. An evolutionary algorithm, the CMA-ES, is applied [22].

3 RESULTS

The following chapter presents the optimization results of the integrated home. First, the input parameters are presented. Following, the optimization results are compared to conventional energy supply concepts. Finally, the influence of the feed-in tariffs (FIT) on the system size is examined.

3.1 Input parameters

This chapter provides an overview on the input data of the model. The solar radiation data for the integrated home are based on real-time measurements performed in Lindenberg (Tauche), close to Berlin. The load is based on [16], resulting in an electrical demand of 4,674 kWh_{el}/a and a thermal demand of 12,523 kWh_{th}/a. The economic parameters for the calculation of the LCOEle are depicted in Table II. Table III illustrates the input parameters for the LCOH calculation.

Some of the components costs depend on the component sizes. The corresponding equations are depicted in Table IV. The specific converter costs of one converter are calculated with equation (5) and are based on a market research done in [6]. The electric model contains three converters because a DC-topology is used.

Similar to the calculation of the converter cost, the calculation of the specific cost of the heat pump and the thermal storage units are based on equations (6), (7) and (8). The equations are based on a market research. Therefore, the prices for different sizes of the heat pump [20] and storage (buffer storage [19] and DHW storage [18]) have been analyzed.

Costs are calculated using linear regression. The cost of radiators are not investigated. The costs of material and installation of floor heating depend on the size and geometry of the building. Therefore, a general statement on the costs is difficult to make. Furthermore, neglecting these enables to compare different heating concepts.

Table I: Optimization parameters of the integrated home and the corresponding boundary conditions

Parameter	Symbol	Min.	Max.	Unit
PV generator size	P_{PVPeak}	5	10	kWp
power Maximum Power Point (MPP) tracker	P_{MPP}	5	10	kW
power inverter	$P_{Inverter}$	5	10	kW
power battery converter	$P_{BatConv}$	2	10	kW
capacity battery system	C_{BESS}	0	24	kWh
volume buffer storage	V_{Buffer}	variable	2	m³
volume DHW storage	V_{DHW}	variable	2	m³
power heat pump	$P_{heat\ pump}$	2	15	kW_{th}
max. temperature buffer storage	$T_{Buffer\ max}$	39	60	°C
max. temperature DHW storage	$T_{DHW\ max}$	45	55	°C

Table II: Input data for LCOEle calculation

Parameter	Symbol	Value	Unit
calculation period for investment assessment	t_{calc}	15	a
interest rate	i_{calc}	1.3	%/a
specific battery cost	c_{bat}	250	€/kWh
annual battery cost degression	i_{bat}	7	%/a
specific converter cost of one converter the model contains three converters	c_{conv}	eq. (5)	€/kW
converter lifetime	L_{conv}	20	a
specific PV generator cost	c_{PVgen}	1170	€/kWh
converter lifetime	L_{PVgen}	20	a
feed-in tariff (Sep. 2017)	$c_{feed-in}$	0.122	€/kWh
electricity price (Sep. 2017)	$c_{electricity}$	0.292	€/kWh
annual electricity price increase	$i_{electricity}$	1.85	%/a
annual maintenance cost relative to investment cost	$i_{maintenance}$	1.5	%/a

Table III: Input data for LCOH calculation

Parameter	Symbol	Value	Unit
specific cost heat pump	$c_{heat\ pump}$	eq. (6)	€
specific cost buffer storage	c_{buffer}	eq. (7)	€
specific cost DHW storage	$c_{DHW\ storage}$	eq. (8)	€
cost DHW pump	$c_{pump\ DHW}$	150	€
cost heat system supplying pump	$c_{pump\ heat}$	150	€
cost heater rod buffer storage	$c_{heater\ rod,buffer}$	275	€
cost three-way valve	$c_{3\ way\ valve}$	300	€
cost elec. installation	$c_{elec.installation}$	500	€
annual maintenance cost of the heat system	$i_{maintenance,heat}$	50	€/a
heat pump lifetime	$L_{heat\ pump}$	20	a
heater rod lifetime	$L_{heater\ rod}$	25	a
storage lifetime	$L_{storage}$	25	a
three-way valve lifetime	$L_{3\ way\ valve}$	20	a
elec. installation lifetime	$L_{elec.installation}$	20	a
pump lifetime	L_{pump}	10	a

Table IV: Equations for converter, heat pump, buffer storage and DHW storage costs

$c_{conv} = \max\left(1000, \frac{970{,}3 \cdot (\frac{P_{BatConv}}{kW})^{-1{,}957} + 304{,}5}{2}\right)\frac{€}{kW}$	(5)
$C_{heat\ pump} = 1{,}119.25\ € + 801.11\frac{€}{kW} \cdot P_{heat\ pump}$	(6)
$C_{buffer} = 480.33\ € + 656.53\frac{€}{m^3} \cdot V_{buffer}$	(7)
$C_{DHW} = 62.90\ € + 3{,}337.10\frac{€}{m^3} \cdot V_{DHW}$	(8)

Table V: Optimization results of an integrated home located in Lindenberg (Tauche), close to Berlin, with an electrical demand of 4,674.3 kWh_{el}/a and a thermal demand of 12,523 kWh_{th}/a. Results for a PV BESS system without (elec) and with power-to-heat coupling (heat) is depicted.

Parameter	Symbol	$Value_{elec}$	$Value_{heat}$	Unit
PV generator size	P_{PVPeak}	10.0	10.0	kWp
power Maximum Power Point (MPP) tracker	P_{MPP}	8.05	9.4	kW
power inverter	$P_{Inverter}$	7.41	8.3	kW
power battery converter	$P_{BatConv}$	1.43	1.8	kW
capacity battery system	C_{BESS}	1.15	0.83	kWh
volume buffer storage	V_{Buffer}		0.16	m^3
volume DHW storage	V_{DHW}		0.13	m^3
power heat pump	$P_{heat\ pump}$		2.8	kW_{th}
max. temperature buffer storage	$T_{Buffer\ max}$		47.2	°C
max. temperature DHW storage	$T_{DHW\ max}$		47.8	°C

3.2 Optimal size

Based on the parameters described in Table I and the input parameters for the economic evaluation, the component sizes of the integrated home are optimized. If the annuity of the best individual is not improved significantly (20 €/a) the simulation is stopped. The presented results are based on an optimization, which executed 1417 annual simulations with a one-minute resolution. The lifetime is extrapolated after one year of simulation. The optimum set of parameters found by the optimization algorithm is given in Table V under consideration of a feed-in tariff of 12.2 ct/kWh. Table V illustrates resulting parameters if a PV BESS system without a power-to-heat coupling (elec) as well as a system with power-to-heat coupling (heat) is minded. The sizes of the components of the electric system are slightly higher if a power-to-heat coupling is applied. Overall, the optimization of the combined system shows that a system with relatively small storage components is economical. The component sizes of the *electricity-only* system (elec) are comparable to the sizes presented in [6] and the sizes of the heat system are comparable to the sizes presented in [13].

3.3 Costs

The resulting costs of the presented system, based on the sizes given in Table V, are presented in Table VI. Considering the solar irradiation data for Lindenberg (Tauche) and an optimal PV generator size of 10 kWp, the integrated home generates 11,404 kWh/a. The calculation of the levelized cost of energy is based on equation (2). The modelled integrated home has a thermal demand of 12,523 kWh_{th}/a and an electrical demand of 4,674 kWh_{el}/a.

Table VI: Resulting LCOEnergy, LCOEle and LCOH of the optimization of an integrated home located in Lindenberg (Tauche), near Berlin, with an electrical demand of 4,674 kWh_{el}/ and a thermal demand of 12,523 kWh_{th}/a.

Parameter	Symbol	Value	Unit
Levelized cost of energy	$LCOEnergy$	13.72	ct/kWh
Levelized cost of heat	$LCOH$	12.89	ct/kWh_{el}
Levelized cost of electricity	$LCOEle$	15.93	ct/kWh_{th}

The resulting levelized cost of electricity (LCOEle) of the optimized system is shown in detail in Figure 4 and is compared to a conventional energy supply by the grid (29.2 ct/kWh). The main costs for the electric system are the investment cost for the PV generator. The PV generator due to the FIT generates the revenues. Load, which cannot be covered by PV generator directly or by discharging the storage units, has to be covered by the grid.

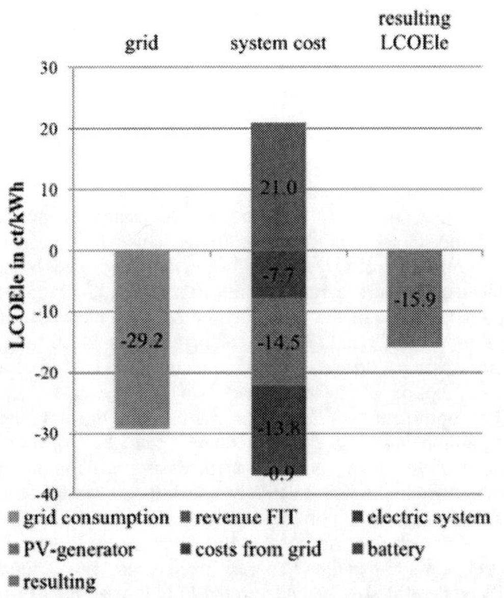

Figure 4: Electricity cost of an optimized integrated home with a PV BESS and power-to-heat coupling in comparison to a household with conventional energy supply.

The LCOH of the optimized integrated home is depicted in detail in Figure 5 and is compared to a fossil heating system. The costs for the fossil heating system are based on [23]. The main shares of the LCOH are the electricity cost. This cost is mainly generated in the winter month. The thermal energy demand is higher in the winter month. In winter, the PV generated electricity is lower.

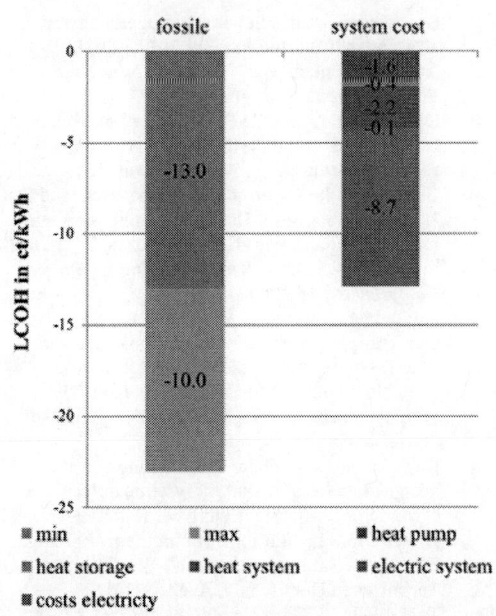

Figure 5: Heat cost of an optimized integrated home with a PV BESS and heat power coupling in comparison to a household with conventional energy supply.

The levelized cost of energy (LCOEnergy) are mainly driven by the cost for the energy consumption. This cost consists of energy cost for grid consumption and diminished feed-in revenues. The resulting LCOEnergy for the integrated home is compared to a conventional supply system with fossil heating and grid supply. The results are illustrated in Figure 6. The energy costs for the integrated home are at the lower end of the cost for a house with a conventional energy supply concept.

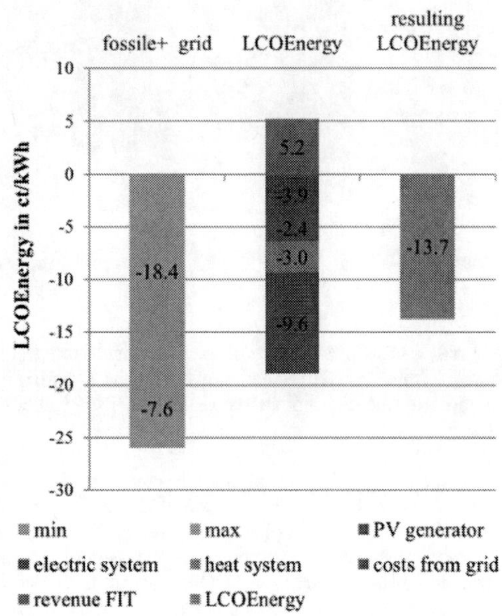

Figure 6: Energy cost of an optimized integrated home with a PV BESS and power-to-heat coupling in comparison to a household with conventional energy supply.

3.4 Analysis of feed-in tariffs for PV generated energy

The economics of the aforementioned systems are highly depended on the economic parameters presented in chapter 3.2. Therefore, the influence of the feed-in tariffs (FIT) for PV-generated energy is examined in the following. Figure 7 depicts the LCOEle of the integrated home under consideration of different FITs. If the FITs are higher than the generation cost of the PV energy, the optimization chooses the upper boundary of the PV generator size (10 kWp). The PV generator size is reduced when the FIT is lower than the generation costs. If no FIT is applied, the revenue for feed-in of PV-generated energy is slightly above the exchange price. Therefore, a price of 6 ct/kWh for PV generated energy is expected. In this case, the PV generator size is mainly driven by the energy consumption of the household, because PV-generated energy reduces grid consumption. This is why the generator size is bigger when a power-to-heat coupling is applied.

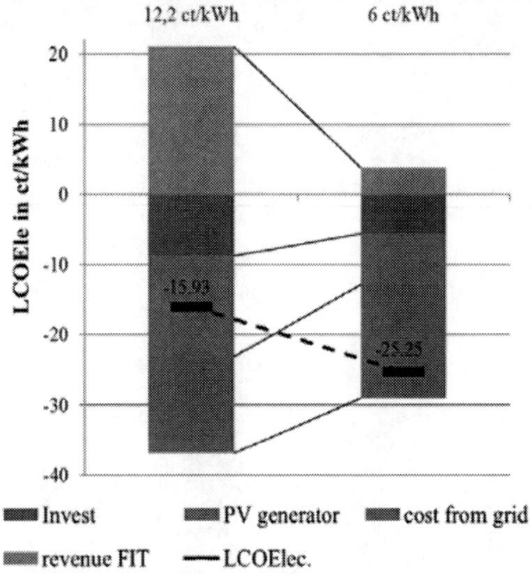

Figure 7: Electricity cost of an optimized integrated home with a PV BESS and power-to-heat coupling in relation to the feed-in tariffs (FIT) for PV generated energy.

If the PV generator is only used to cover the electricity demand, the optimal size is 5.0 kWp. If a power-to-heat coupling system is applied, this value increases to 9.2 kWp. The costs for the increased PV generator size are attributed to the LCOH. The LCOEnergy are increased in this case, because of the falling revenues from the feed-in. The LCOEnergy, as well as the costs for the PV generator are illustrated in Figure 8.

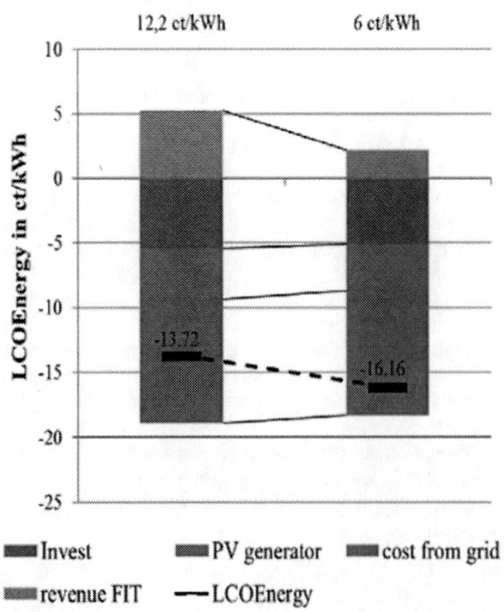

Figure 8: Energy costs of an optimized integrated home with a PV BESS and heat power coupling in dependency of the feed-in tariffs (FIT) for PV generated energy.

4 CONCLUSION AND DISCUSSION

Households that integrate a PV battery energy storage system (BESS) and a heat pump for power-to-heat coupling can contribute to the decarbonisation process. To enhance the economics of these integrated homes, the component sizes have to be optimized. In this paper, a tool for the optimization of an integrated home has been presented, which is able to consider many economic parameters and boundary conditions. This paper presents optimization results for a four-person integrated home located in Lindenberg (Tauche), near Berlin. To evaluate the economics of the integrated home, the levelized cost of energy (LCOEnergy) are calculated. The LCOEnergy are mainly driven by the costs for the energy consumption of the combined heat and power system. The optimization result show that under the assumed conditions an integrated home can be operated beneficially if an optimal system design is chosen. An integrated home with relatively small storage components shows to be an economic solution.

Additionally, the influence of different feed-in tariffs (FIT) for PV-generated energy is analyzed. The optimal PV generator size depends on the FIT. In case the FIT is lower than the generation costs, the PV generator size is mainly depended on the load of the integrated home. Overall, the results show that an optimization of the component sizes can increase the economics, but the optimal sizes have to be calculated for every scenario individually. The results of the optimization are highly dependent on the input parameters as shown exemplary by the FIT sensitivity analysis. This is why the presented results have to be considered as exemplary.

5 REFERENCES

[1] Bundesverband des Schornsteinfegerhandwerks, "Gesamtzahl Wärmeerzeuger 2017," 2018.

[2] AG Energiebilanzen e.V., *Energiebilanz der Bundesrepublik Deutschland 2015*.

[3] D. H. Wirth, "Aktuelle Fakten zur Photovoltaik in Deutschland," Fraunhofer Institut für Solare Energiesysteme (ISE), Freiburg, Aug. 2017.

[4] G. Angenendt, S. Zurmühlen, H. Axelsen, and D. U. Sauer, "Prognosis-Based Operating Strategies for Smart Homes with Heat and Power Coupling," in *12th International Renewable Energy Storage Conference (IRES 2018)*.

[5] J. Weniger, J. Bergner, T. Tjaden, and V. Quaschnig, "Bedeutung von prognosebasierten Betriebsstrategien für die Netzintegration von PV-Speichersystemen," in *Tagungsband des '29. Symposium Photovoltaische Solarenergie'*, Bad Staffelstein.

[6] D. T. Magnor, "Globale Optimierung netzgekoppelter PV-Batteriesysteme unter besonderer Berücksichtigung der Batteriealterung," Dissertation, Institut für Stromrichtertechnik und Elektrische Antriebe, Rheinisch-Westfälischen Technischen Hochschule, Aachen, 2016.

[7] T. H. Pham, F. Wurtz, and S. Bacha, "Optimal operation of a PV based multi-source system and energy management for household application," *IEEE International Conference on Industrial Technology*, pp. 10–13, 2009.

[8] T. Beck, H. Kondziella, G. Huard, and T. Bruckner, "Assessing the influence of the temporal resolution of electrical load and PV generation profiles on self-consumption and sizing of PV-battery systems," *Applied Energy*, no. 173, pp. 331–342, 2016.

[9] B. Claessens, K. De Brabandere, G. Mulder, and F. De Ridder, "Break-even analysis of PV-battery self-consumption systems.,"

[10] C. Verhelst, F. Logist, J. van Impe, and L. Helsen, "Study of the optimal control problem formulation for modulating air-to-water heat pumps connected to a residential floor heating system," *Energy and Buildings*, no. 45, pp. 43–53, 2012.

[11] T. Ikegami, K. Kataoka, Y. Iwafune, and K. Ogimoto, "Optimal demand controls for a heat pump water heater under different objective functions," *IEEE International Conference on Power System Technology (POWERCON)*, pp. 1–6, 2012.

[12] G. Reynders, T. Nuytten, and D. Saelens, "Potential of structural thermal mass for demand-side management in dwellings," *Building and Environment*, no. 64, pp. 187–199, 2013.

[13] T. Beck, H. Kondziella, G. Huard, and T. Bruckner, "Optimal operation, configuration and sizing of generation and storage technologies for residential heat pump systems in the spotlight of self-consumption of photovoltaic electricity," *Applied Energy*, no. 188, pp. 604–619, 2017.

[14] G. Angenendt, S. Zurmühlen, C. Chen, D. Magnor, and D. U. Sauer, "Reducing Cut-Off Energy of PV Home Storage Systems by Using Forecast-Based Operation Strategies," in *Tagungsband der '11th International Renewable Energy Conference (IRES 2017)'*, Düsseldorf, Mar. 2017.

[15] M. Ecker, N. Nieto, S. Käbitz, J. Schmalstieg, H. Blanke, A. Warnecke and D. U. Sauer, "Calendar and cycle life study of Li(NiMnCo)O2-based 18650 lithiumion-Ion batteries," *Journal of Power Sources*, pp. 839–851.

[16] Mark Bost, Dr. Bernd Hirschl, and Dr. Astrid Aretz, *Effekte von Eigenverbrauch und Netzparität bei der Photovoltaik: Beginn der dezentralen Energierevolution oder Nischeneffekt.*

[17] K. Schwamberger, *Modellbildung und Regelung von Gebäudeheizungsanlagen mit Wärmepumpen.* Zugl.: Darmstadt, Techn. Hochsch., Diss. Düsseldorf: VDI-Verl., 1991.

[18] kamdi24, *Pufferspeicher 300 L mit 1x Solarwärmetauscher - Schindler+Hofmann.* [Online] Available: https://www.kamdi24.de/Pufferspeicher-300-L-mit-1x-Solarwaermetauscher.html. Accessed on: Jan. 02 2018.

[19] kamdi24, *Pufferspeicher 300l - Schindler+Hofmann.* [Online] Available: https://www.kamdi24.de/Pufferspeicher-300-L.html. Accessed on: Jan. 02 2018.

[20] Viessmann, *Datenblatt VITOCAL 200-S*, 2017.

[21] K. Branker, M. Pathak, and J. M. Pearce, "A Review of Solar Photovoltaic Levelized Cost of Electricity," *Renewable and Sustainable Energy Reviews*, vol. 15, no. 9, pp. 4470–4482, 2011.

[22] A. Auger and N. Hansen, "Evolution Strategies and Covariance Matrix Adaptation,"

[23] BDEW, *BDEW-Heizkostenvergleich Altbau 2017.* [Online] Available: https://www.bdew.de/media/documents/2017_Januar_Heizkostenvergleich-Altbau.pdf.

PHOTOVOLTAIC ENERGY INTEGRATION WITH HOUSEHOLDS' DEMAND: A CASE STUDY OF A RESIDENTIAL SMART GRIDS PILOT IN THE NETHERLANDS

Cihan Gercek [1], Angèle Reinders [1,2]
[1] Department of Design, Production and Management, University of Twente
Drienerloolaan 5, 7522 NB, Enschede, The Netherlands
[2] Energy Technology Group at Mechanical Engineering, Eindhoven University of Technology (TU/e)
P.O. Box 513, 5600 MB, Eindhoven, The Netherlands

ABSTRACT: This paper presents a data analysis of a residential smart grid pilot in the Netherlands with PV installations. This study has been executed in the ERA-Net Smart Grids Plus project "Co-Evolution of Smart Energy Products and Services" with the aim to evaluate the flexibility and self-consumption of generated PV energy in residential smart grids. The pilot that has been analysed, namely PowerMatching City, has various configurations with PV systems, heat-pumps and micro-cogeneration units which were used as a tool for the efficient integration of renewable power into electricity network. Different energy variables as electricity production, purchase or consumption of the pilot households has been announced monthly and yearly basis and discussed. In conclusion, the combination of μ-CHP and PV systems increases the self-sufficiency (40%) compared to households that have heat-pumps (20%).
Keywords: Prosumer, PV System, Energy Performance, Demand-side, Grid Connected, Grid Management, Energy Options

1 INTRODUCTION

In the framework of the sustainable energy transition, integration of PV installations in residential areas are a perfect solution to produce sustainable energy without occupying additional spaces and avoid transmission losses [1,2]. The main obstacle to the transition is matching the demand and supply, both varying daily and seasonally, especially in Western Europe countries as the Netherlands [3,4]. A smarter grid operation with new energy management systems may match the demand-supply balance, if the network has enough flexibility [5]. Zero energy buildings (ZEB) with 100% renewable energy are achievable with PV systems [6] and batteries [7] or by using community-based storage systems [8]. However, if those *low emission* systems' batteries won't be optimized, they might have an additional detrimental impact on the environment [9] [10]. Those impacts are hopefully expected to decrease with time and technological advancements [11], however, besides storage, other solutions for flexibility are needed for minimizing the impacts.

In our paper, we will only focus on flexibility for electricity. Flexibility may be defined as the ability to provide instantaneously electrical power according to changes in demand, for a time interval and a location of the grid. Diverse storage systems might contribute to the flexibility of the network, (hydroelectricity, fuel cells and hydrogen) [12]. Although, fuel cell technologies using hydrogen may be applied for heating or transport[13], or as an additional electricity supply to networks, such systems are still quite expensive [14]. Therefore, in this study, we would like to analyse the compatibility of micro combined-heat-and-power (μ-CHP) units with residential PV systems, which offers both electricity and heating. In this context, we would like to evaluate the efficient integration of PV systems into a local network. Our case study is a smart grid pilot called PowerMatching City (PMC), which the household configurations are resumed in Figure 1 and Table I. In this pilot, 22 households are equipped with PV and two main DER configuration, either smart hybrid heat pumps(SHHP) or μ-CHP.

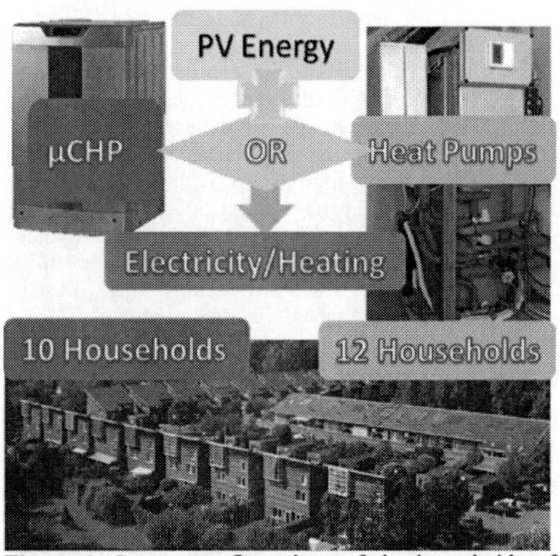

Figure 1. System configurations of the households of PowerMatching City [15]

An energy balance of these households during 2012 will be provided for various variables in order to highlight how well DER increased self-sufficiency and contribute to a sustainable energy usage. This will be done by data that were collected in 2012. The pilot is situated in Groningen-Hoogkerk, in the northern Netherlands.

This paper is structured as follows. In Section 2, the technical and socio-economical properties of the pilot PMC, the methods of data processing and equations used will be detailed. In Section 3, the results will be presented at a yearly and monthly basis. In this section the different system configurations will be compared indicating minimum and maximum values for their consumption and production of electricity. Finally in Section 4 the results will be discussed. The paper will be completed with conclusions in Section 5.

Table I : Households features and PV Details for PowerMatching City in 2012 [16]

Energy Sources		PV systems	Smart appliances
10 Households equipped with μ-CHP (1kW Electric, 6 KW Thermal)	2 Households: PV installed at roof	500 W_p	No smart appliance
		2200 W_p	Equipped with smart app.
	8 Households: PV installed at a distance	1590 W_p	4 Households equipped with smart app.
12 Households equipped with Heat Pump (4.5 kWh Thermal)	2 Households: PV installed at roof	500 W_p	Equipped with smart app.
		3840 W_p	No smart appliance
	10 Households : PV installed at a distance	1590 W_p	5 Households equipped with smart app.

2 MATERIAL AND METHODS

2.1 Household Configurations

μ-CHP systems in PMC have a nominal power of 1 kW for producing electricity. In addition, they have the ability to deliver 6 kW thermal output for heating. The SHHPs have 4.5 kW thermal energy production and adding to that, a condensing boiler of 20 kW thermal output. Both μ-CHP and SHHP systems have a 210 litre hot water buffer. In PMC, ten households use μ-CHP and twelve of them SHHP. In this study, both groups will be compared.

Only four households had rooftop PV installations, while eighteen receive PV infeed from a close by located PV installation at 2.3 km distance. The infeed was shared virtually via a smart grid software called Powermatcher [17]. This software manages the activation of smart applications such as smart washing machines or dishwashers, or SHHP systems [18].The project ran from 2009 until 2015 with two different phases [19]. We have evaluated data from 2012, which makes part of the phase 2 [20, 21].

In Figure 2, an example of daily consumption and production of a household of PMC is shown. The targeted season is winter, as the radiance levels are low. The example shows the electricity generation of PV with a complementary DER, here μ-CHP, aiming to diminish the electricity purchase from the grid and increased the self-sufficiency.

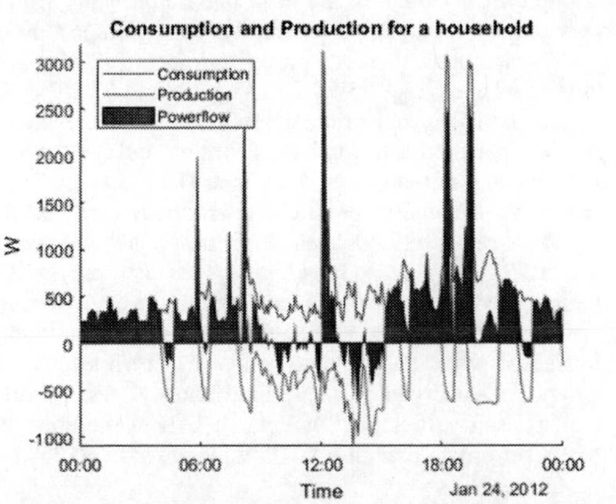

Figure 2. Electricity consumption and delivery to the grid during winter from a PMC household equipped with a PV system and a μ -CHP.

2.2 PMC Households and Residents

PMC households are having more surface than the average of the Netherlands (136 m2). Moreover the average income, education level, and especially the number of residents per household are significantly higher than the Dutch average. The size of the households of PMC according to the number of resident per household, indicates clearly that PMC has more residents per household (3.11) compared to Dutch households in average (2.21) [10],[12]. Table II resumes the household size and compares it to a relatively significant Dutch sample. The higher number of residents and the surface of the households compared to Dutch average expect to result in higher energy consumption in these households. However, the higher level of salary and education make it difficult to estimate the impact on consumption. The participants were conscious of being part of a pilot and that they were under observation. That implicates their awareness and behaviour change, called *Hawthorne* effect [23], so only a slight change of their electricity use is expected [24].

Table II: PMC households' size according to the number of resident per household for the year 2012, compared to a Dutch survey [19].

Household size	PMC	The Netherlands
1 person	9%	37%
2 person	22%	33%
3 person	9%	12%
4 person	31%	13%
5 person	9%	6%
Unknown	18.1%	-
Sample Number	22	7,51 million

2.3 Energy Tariffs

The Dutch electricity prices for households depend on the level of consumption. For the period of 2011 to 2013, the cost of electricity in euro cent per kWh is summarized in Table III [25]. The decrease of the energy price in yearly low consumption cases or negative energy prices under 1000kWh are due to the energy tax that the state is claiming for transmission and other services related to the delivery of the electricity and environment [25]. In other words, the price of electricity (in euro cent per kWh) excluding taxes and VAT decreases with increasing consumption level, contrarily, the same price including tax and VAT increases for higher consumption levels.

Although this article does not aim at an in-depth economic analysis, even at the residential level, the price of electricity is expected to have a major impact on the amount of purchased and delivered electricity. Smart grids will optimise those purchase and delivery times, by

Table III : Dutch Electricity prices in euro cent per kWh including VAT and taxes, between 2011 to 2013.

Energy in kWh	2011	2012	2013
<1000kWh	-16.6	-15.8	-16.2
1 000– 2 500 kWh	10.6	11.5	11.6
2 500– 5 000 kWh	17.8	18.7	19.2
5 000–15 000 kWh	20.8	22.3	22.8

shifting the demand with the smart appliances, activating them in the PV energy production times, or by activing DER when delivery prices are interesting.

Nowadays, projects and pilots are more focused on dynamic pricing effects, however this particular pilot applied the time of use tariff for the simplicity for users. Purchased and delivered electricity are separately metered and the rates are applied for both. The timing of the normal rate was on week days from 7:00h to 23:00h. In other time slots the low rate was applied. In this context, smart appliances would help to self-consume and reducing the peak load. Moreover, because PV systems mostly produce energy during periods with a normal rate (except for the weekends), the delivery of excess PV power to the grid will have an advantage to benefit often from normal rate pricing.

2.4 Data Processing and Equations

In the PMC pilot diverse appliances, as well as electricity and gas consumption have been monitored with smart meters [15]. Those records result in 29 different variables for each 5 minutes for each of the 22 households, which required processing using Matlab algorithms. We cleaned unexpected spikes or illogical values that do not correspond to normal values expected for the physical situation.. The data that were processed were mainly power flows, which were calculated by taking the difference between total production and total consumption. The time sequence of the data was divided according to tariffs: n.r. stands for the normal rate and l.r. stands for the low rate. The equations that were used for our analysis are grouped in Table IV, where E is electricity in kWh, P is power in kW, t is time, n is the number of appliances, m is the number of energy sources:

Table IV: Equations used for the analysis of data of PMC.

	INSTANTENIOUS	**SUM**
Power (P)	$\sum_{i}^{n} P_{app.\,i}(t) - \sum_{k}^{m} P_{gen.\,k}(t)$	$E_P - E_D$
Purchased (E_p)	$P(t) > 0$	$E_{P\_l.r.} + E_{P\_n.r.}$
Delivered (E_d)	$P(t) < 0$	$E_{D\_l.r.} + E_{D\_n.r.}$
Generation (E_g)	$P_{PV}(t) + P_{\mu\text{-}CHP}(t)$	$E_{PV} + E_{\mu CHP}$
Self-Consumption (E_s)	$P_{PV}(t) + P_{\mu\text{-}CHP}(t) - \lvert P_f(t) < 0 \rvert$	$E_G - E_D$
Consumption (E_c)	$\lvert P_f(t) > 0 \rvert + P_{PV}(t) + P_{\mu\text{-}CHP}(t) - \lvert P_f(t) < 0 \rvert$	$E_P + E_S$

3 RESULTS

To respect their privacy, we will anonymise the results of the participants of the pilot, by not showing these results for individual households. Nevertheless, we present the monthly average, maximum and minimum values for consumption and production of electricity, independently from the households. The results are shown in Figure 3, 4 and 5 and explained in detail in further sections.

3.1 Consumption

The yearly electricity consumption for an average household of PMC in 2012 was around 5100 kWh, which is more than 50% higher than the Dutch average which is 3300 kWh [26]. These numbers can be connected to the number of residents per household in PMC, which is about 40% higher than average households in the Netherlands.

The monthly electricity consumption is shown in Figure 3, for the average household of PMC. The monthly average consumption is 420 kWh. The variation of the values in Figure 3 highlight the consumption behaviour differences. The maximum and minimum values mostly are the results of a different household performance for each month.

Considering monthly minimum consumptions, we assume that some residents were away during a certain period of time as the values were as low as 50 kWh per month (Figure 3). This hypothesis could be reinforced by indicating that the lowest value occurred during August, which is an official summer holiday period in the Netherlands. Controversially, the maximum values differ up to 340 kWh between each other, with a peak of 830 kWh, which happened during winter, mostly due to heating.

3.2 Production

Monthly energy provided by PV systems is shown in Figure 4. The peak of the average values occurred in May. The most of average and minimum values are matching with monthly irradiation regimes of Groningen [27]. However, the maximum values differ compared to the irradiation regimes, due to a non-optimised slope and slight location variation for virtual PV installations. An example of it would be the peak maximum value which has occurred in August, incoherently to the average values.

The average PV energy production for April, May and June is respectively 95 kWh, 135 kWh and 95 kWh. In winter, the PV systems production gradually decreased to 25 kWh per month in January, February and November; and felt to a critical level, 9 kWh in December. In 2012 the average PV energy production per month was 70 kWh.

The overall self-sufficiency of households in PMC is also affected by seasonal effects. The decrease in PV energy production during winter results in 2% of self-sufficiency on an average for December for some household with PV systems only. Meanwhile μ-CHP generation leads to 20% self-sufficiency. The yearly average self-sufficiency without μ-CHP generation is 20.3%, meanwhile, with μ-CHP, it increases to 40.7%.

4 DISCUSSIONS

The data sample is limited to one year and 22 households, which might be considered as statistically insufficient for general assumptions. Moreover, the level of education and the household size is above the Dutch averages, and the sample is voluntary based. Therefore the

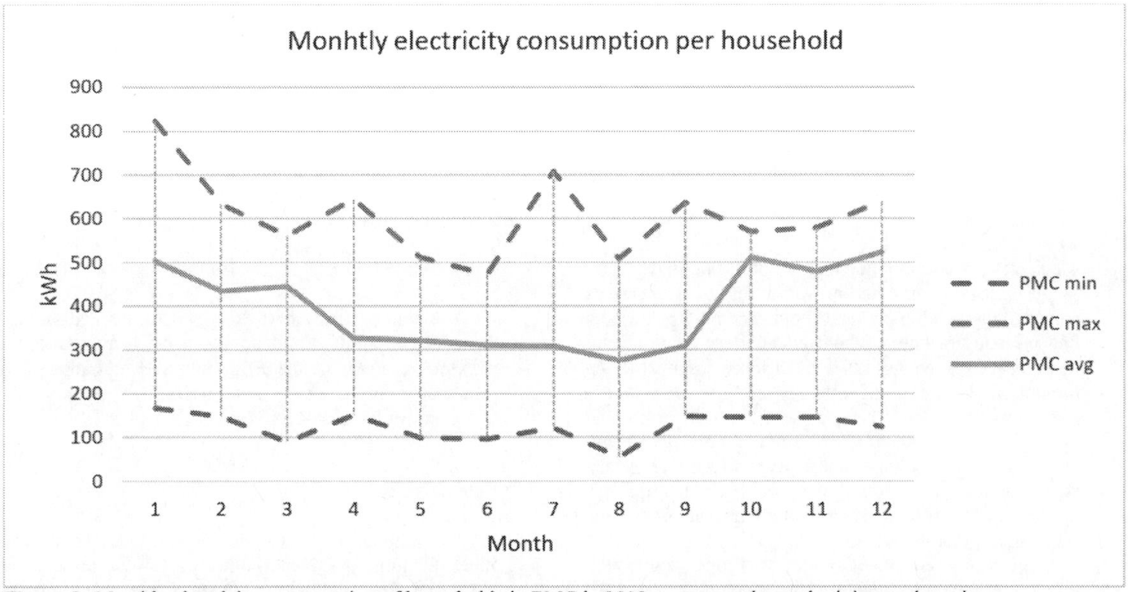

Figure 3. Monthly electricity consumption of households in PMC in 2012: average value and minima and maxima.

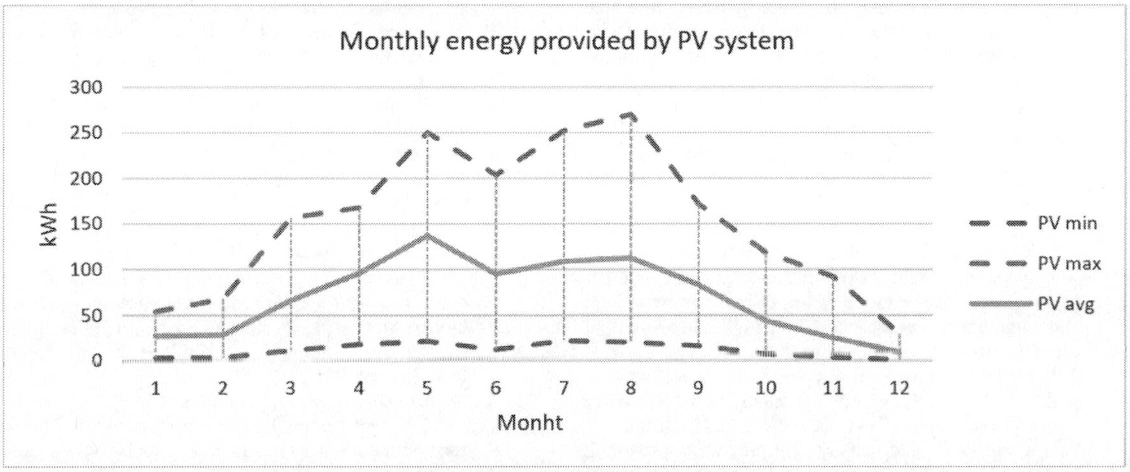

Figure 4. Monthly energy provided by the PV systems of households in PMC in 2012: average value and minima and maxima.

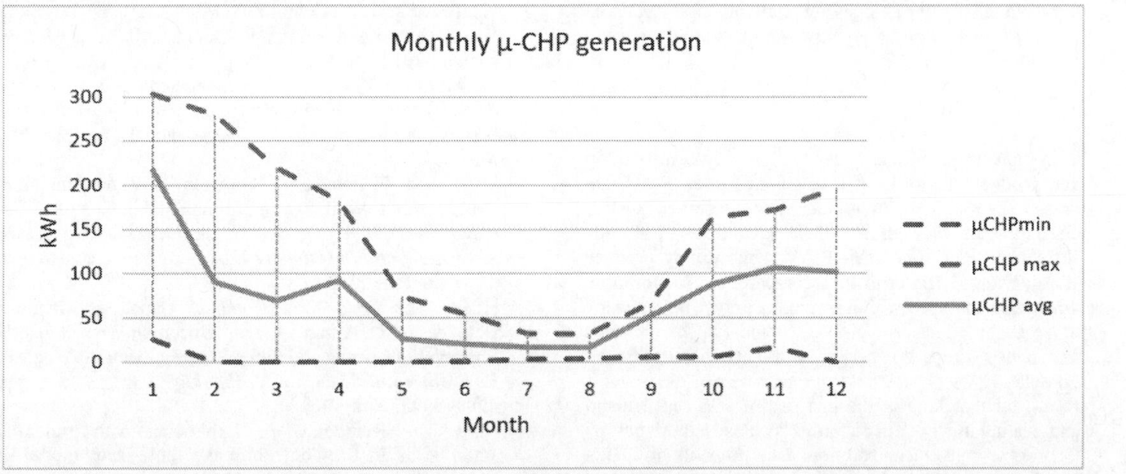

Figure 5. Monthly electricity production by μ-CHP in households in PMC in 2012: average value and minima and maxima.

result should be taken into consideration cautiously. Nevertheless, our results demonstrate a sample of consumption profiles, PV systems, DER applications and degree of self-sufficiency of the pilot. The different types of system configurations in PMC help us to evaluate their complementarity with the smart grid. Monthly based results clarify the seasonal variations of the energy provided by PV systems and the consumption of the households.

4.1 Electricity consumption on a monthly and yearly basis

Electricity consumption over all the year is considerably varying per household. The motivations and the common practices of the users are very important for them to become energy efficient or more sustainable. As the impact assumed to be only 2.7%[22], the *Hawthorne* effect was neglected. The impact of the user profile and their consumption practices were much more significant, which is another reason to not focus on this effect. As the small sample as 22 households, we could not make a cluster of user profiles, which is our perspective with the acquisition of larger data.

The average consumption increases considerably with the winter season, therefore the maximum value of monthly electricity consumption occurred also as expected in winter. In these months people are spending more time in their home because of daylight time and outside-temperature decrease. The months are quite demanding, as they require more heating than in summer. We also see an interesting peak in July for the maximum values, which may be caused either by air conditioners or cooling appliances, or can be simply related to this particular household practices. From the data however it is difficult to indicate the precise reason for this peak.

4.2 PV and µ-CHP : electricity production

Figure 4 shows the maximum and the minimum output of PV panels. The average values differ from maximum and minimum values because the tilt angle and azimuth of the installations vary. As can be seen in Figure 5, the µ-CHP power generation has a complementary performance with PV systems, and jointly, the production of electricity is kept to a certain level, even in winter when the irradiance values are low, additionally, µ-CHP were providing heating. The power generation of µ-CHP could even become more sustainable by the use of biogas or hydrogen, although the prices of biogas are relatively high compared to other actual fuel sources. However this might change in the future.

5 CONCLUSIONS

In conclusion, in this paper the monthly consumption and production profile of a smart grid pilot, PMC, are presented for 2012 We identified different system configurations as well as their advantages. The self-sufficiency increases when PV systems jointly produce energy with µ-CHP, up to 40%, compared to 20% for those household equipped with heat pumps only. The tendency of seasonal effects is also discussed, where the self-sufficiency drops to nearby zero in winter for the households which are not equipped with µ-CHP. As a result, for matching of the increase of seasonal demand and heating in winter in countries such as Netherlands, µ-CHP is a convenient solution. But probably µ-CHP's installation and the maintenance costs, their complexity of installation in existing households and isolation requirements to be energy efficient are barriers to their widespread use.

ACKNOWLEDGEMENTS

This project has received funding in the framework of the joint programming initiative ERA-Net Smart Grids Plus, with support from the European Union's Horizon 2020 research and innovation programme.

DISCLAIMER

The content and views expressed in this material are those of the authors and do not necessarily reflect the views or opinion of the ERA-Net SG+ initiative. Any reference given does not necessarily imply the endorsement by ERA-Net SG+.

References

[1] A. Mahone *et al.*, 'On the Path to Decarbonization: Electrification and Renewables in California and the Northeast United States', *IEEE Power and Energy Magazine*, vol. 16, no. 4, pp. 58–68, Jul. 2018.

[2] M. Miller, 'Electrification: Its Role in Deeply Decarbonized Energy Systems [Guest Editorial]', *IEEE Power and Energy Magazine*, vol. 16, no. 4, pp. 20–23, Jul. 2018.

[4] A. Reinders, P. Verlinden, W. van Sark, and A. Freundlich, *Photovoltaic solar energy: from fundamentals to applications*. 2017.

[5] Y. Xin, B. Zhang, M. Zhai, Q. Li, and H. Zhou, 'A Smarter Grid Operation: New Energy Management Systems in China', *IEEE Power and Energy Magazine*, vol. 16, no. 2, pp. 36–45, Mar. 2018.

[6] B. P. Heard, B. W. Brook, T. M. L. Wigley, and C. J. A. Bradshaw, 'Burden of proof: A comprehensive review of the feasibility of 100% renewable-electricity systems', *Renewable and Sustainable Energy Reviews*, vol. 76, pp. 1122–1133, Sep. 2017.

[7] W. L. Schram, I. Lampropoulos, and W. G. J. H. M. van Sark, 'Photovoltaic systems coupled with batteries that are optimally sized for household self-consumption: Assessment of peak shaving potential', *Applied Energy*, vol. 223, pp. 69–81, Aug. 2018.

[8] T. AlSkaif, W. Schram, G. Litjens, and W. van Sark, 'Smart charging of community storage units using Markov chains', 2017, pp. 1–6.

[9] M. C. McManus, 'Environmental consequences of the use of batteries in low carbon systems: The impact of battery production', *Applied Energy*, vol. 93, pp. 288–295, May 2012.

[10] J. F. Peters, M. Baumann, B. Zimmermann, J. Braun, and M. Weil, 'The environmental impact of Li-Ion batteries and the role of key parameters – A review', *Renewable and Sustainable Energy Reviews*, vol. 67, pp. 491–506, Jan. 2017.

[11] B. Cox, C. L. Mutel, C. Bauer, A. Mendoza Beltran, and D. P. van Vuuren, 'Uncertain Environmental Footprint of Current and Future Battery Electric Vehicles', *Environmental Science & Technology*, vol. 52, no. 8, pp. 4989–4995, Apr. 2018.

[12] A. Reinders *et al.*, 'Experiences with residential smart grids in the context of the three-layer model of stakeholders, markets and technologies', *MDPI Applied Science*, vol. Under Submission, 2019.

[13] C. B. Robledo, V. Oldenbroek, F. Abbruzzese, and A. J. M. van Wijk, 'Integrating a hydrogen fuel cell electric vehicle with vehicle-to-grid technology, photovoltaic power and a residential building', *Applied Energy*, vol. 215, pp. 615–629, Apr. 2018.

[14] C. B. Robledo, V. Oldenbroek, F. Abbruzzese, and A. J. M. van Wijk, 'Integrating a hydrogen fuel cell electric vehicle with vehicle-to-grid technology, photovoltaic power and a residential building', *Applied Energy*, vol. 215, pp. 615–629, Apr. 2018.

[15] F. Bliek *et al.*, 'PowerMatching City, a living lab smart grid demonstration', in *Innovative Smart Grid Technologies Conference Europe (ISGT Europe), 2010 IEEE PES*, 2010, pp. 1–8.

[16] D. V. Geelen, 'Empowering end-users in the energy transition: An exploration of products and services to support changes in household energy management.', TU Delft, Delft, The Netherlands, 2014.

[17] K. Kok *et al.*, 'Dynamic pricing by scalable energy management systems 2014; Field experiences and simulation results using PowerMatcher', in *2012 IEEE Power and Energy Society General Meeting*, San Diego, CA, 2012, pp. 1–8.

[18] A. van den Noort, J. van de Velde, H. de Heer, M. Eijgelaar, B. Roossien, and R. Kamphuis, 'Real-time trade dispatch of a commercial VPP with residential customers in the PowerMatching City SmartGrid living lab', in *22nd International Conference and Exhibition on Electricity Distribution (CIRED 2013)*, Stockholm, Sweden, 2013, pp. 0309–0309.

[19] A. Van Den Noort, 'PowerMatching City: Lighthouse project, Knowing smart grids, and how to deliver consumer-friendly energy services', 2018.

[20] U. Obinna, P. Joore, L. Wauben, and A. Reinders, 'Comparison of two residential Smart Grid pilots in the Netherlands and in the USA, focusing on energy performance and user experiences', *Applied Energy*, vol. 191, pp. 264–275, Apr. 2017.

[21] J. P. Wijbenga, P. MacDougall, R. Kamphuis, T. Sanberg, A. van den Noort, and E. Klaassen, 'Multi-goal optimization in PowerMatching city: A smart living lab', in *IEEE PES Innovative Smart Grid Technologies, Europe*, Istanbul, Turkey, 2014, pp. 1–5.

[22] D. Geelen, A. Reinders, and D. Keyson, 'Empowering the end-user in smart grids: Recommendations for the design of products and services', *Energy Policy*, vol. 61, pp. 151–161, Oct. 2013.

[23] J. G. Adair, 'The Hawthorne effect: A reconsideration of the methodological artifact.', *Journal of Applied Psychology*, vol. 69, no. 2, pp. 334–345, 1984.

[24] D. Schwartz, B. Fischhoff, T. Krishnamurti, and F. Sowell, 'The Hawthorne effect and energy awareness', *Proceedings of the National Academy of Sciences*, vol. 110, no. 38, pp. 15242–15246, Sep. 2013.

[25] Centraal Bureau voor de Statistiek, 'Elektriciteit in Nederland', Feb. 2015.

[26] Energieonderzoek Centrum Nederland, 'Energie Trends 2012', Nov. 2012.

[27] Joint Research Center of European Commission, 'Photovoltaic Geographical Information System (PVGIS), Geographical Assessment of Solar Resource and Performance of Photovoltaic Technology'.

TECHNO-ECONOMIC EVALUATION OF VOLTAGE DEPENDENT ACTIVE AND REACTIVE POWER CONTROL TO REDUCE VOLTAGE VIOLATIONS IN DISTRIBUTION GRIDS

Raphael Knecht*, Fabian Carigiet, Alain Schwab, Petr Korba, F. P. Baumgartner
[1]ZHAW, Zurich University of Applied Sciences, School of Engineering, IEFE
Technikumstr. 9, CH-8401 Winterthur, Switzerland, www.zhaw.ch/=kner
*phone: +41589344576; email: kner@zhaw.ch

ABSTRACT: High penetration of PV plants or numerous electric vehicle (EV) charging station stations connected to the low voltage distribution grids (LVDG) may cause a voltage rise or voltage decrease respectively. There are several measures of maintaining the voltage stability such as grid reinforcement, battery energy storage, line voltage regulator, etc., although they vary in effectiveness and economic viability. This paper focuses on using decentralised voltage-dependent active and reactive power (PQ(V)) control of PV inverters to stabilise the voltage in the grid. Using two grid models in Southern Germany and Switzerland the best PQ(V) control strategy is evaluated using load flow calculations. The weakest node in the first grid exhibits a maximum voltage of 1.072 pu on a sunny day. Due to the implementation of the PQ(V) control the maximum voltage is reduced to 1.024 pu at the same node. Costs considered for PQ(V) control are the PV yield loss and the additional reactive power compensation, which amount to roughly CHF 2'600.- per year. The future installation of EV charging stations may positively interact with PV feed-in. The voltage decrease can further be limited using PQ(V) control. Further grids and means for voltage stabilisation will be analysed in the future.
Keywords: Voltage Stabilisation, Grid Integration, Reactive Power

1 VOLTAGE VIOLATIONS IN LVDG

Voltage stability becomes an increasing concern for distribution system operators (DSO) due to increasing decentralised renewable energy generation and a large number of high power electric vehicle (EV) charging stations connected to the low voltage distribution grid (LVDG). Technical regulation in EN 50160 requires the voltage at a node in the LVDG to be within ±10% of the nominal voltage [1]. The D-A-CH-CZ-Regulations applied in Switzerland and Germany are more restrictively limiting the maximum voltage deviation to ±3% around the nominal voltage [2].

Measurements at a distribuation box in the LVDG of Dettighofen, Southern Germany (see Section 4.1) showed a voltage of over 110% of the nominal voltage for ten minutes on a sunny day in May 2014 [3].

The opposite voltage violation (below 90% of nominal voltage) is possible with an intermittent high power demand due to EV charging.

Distribution system operators need to find appropriate solutions to counteract these voltage violations when the PV penetration is high or a large amount of EV charging stations are going to be installed in their LVDG. Technical measures to improve voltage stability are:

- Grid Reinforcement
- On-load Tap Changer
- (Battery) Energy Storage
- Demand Side Management
- Line Voltage Regulator
- Active Power Curtailment
- Reactive Power Control

In order to provide results for a decision-making tool in the planning phase of a LVDG, ZHAW is evaluating and comparing both the technical as well as the economic feasibility of the above-mentioned measures for different grid classes. These grid classes were identified in collaboration with DSO and should contain the typical LVDGs in Switzerland:[4]

- Medium sized industry
- Small sized industry
- City border with shopping mall
- Urban neighbourhood with apartment buildings
- Urban neighbourhood with business
- Village centre
- Village periphery
- Hamlet

A combination of active and reactive power control using PV inverters is a promising cost-effective tool since no additional hardware costs arise. This method is presented in this work and its technical as well as economic viability is discussed.

2 REACTIVE POWER CONTROL WITH PV INVERTERS

PV inverters usually are programmed to feed-in only active power into the LVDG ($\cos\varphi = 1$). The current induces a voltage drop across the line impedance ($|\underline{V}_{pv}| > |\underline{V}_{grid}|$) when active power is fed in. An additional reactive power injection will effectively reduce the voltage drop and thus reduce the voltage rise at the PV node (see [5]).

2.1 Different Control Strategies

There are several strategies for reactive power control using PV inverters like *fixed cosφ*, *cosφ(P)*, *cosφ(V)* and *Q(V)*. Their respective dis-/advantages are listed in Table I.

The voltage dependent reactive power control (*Q(V)*) is a decentralised approach since it only controls the reactive power according to the voltage at the location where the inverter is installed. Contrary to this, *fixed cosφ* and *cosφ(P)* both may cause excessive reactive power in the LVDG. *Q(V)* control also does not induce costs for real time control from a centralised location, apart from implementation costs.

Similarly to *Q(V)* the active power can be curtailed by the line voltage, henceforth denoted as P(V). The reduced active power feed-in also reduces the voltage at the PV node.

This decentralised method is implemented identically to the Q(V) control without any additional communication infrastructure.

Table I: Dis-/Advantages (-/+) of different methods for reactive power control using PV inverters.

Characteristic	fixed cosφ	cosφ (P)	cosφ (V)	Q(V)
Voltage Reduction	++	++	+	+
Voltage Rise	--	--	++	++
Reactive Power Input	--	-	+	+
Independent of Active Power	--	--	-	++
Effectivity for DSO	-	-	+	++
Effort of Implementation	++	+	-	-

2.2 Voltage Dependent Active and Reactive Power Control

A prospective control method is the combination of both *P(V)* and *Q(V)* control, denoted as *PQ(V)* control. The implementation is done by parametrisation of the PV inverters preferably at commissioning. There are already several inverters available on the market that provide this functionality [6].

The control ramps, as seen in Figure 2, define the relative active/reactive power feed-in as a function of the line voltage at the PV node. While an underexcited reactive power injection (from 0% at 1.01 pu to - 100% at 1.05 pu in example from Figure 2) leads to a reduction of the node voltage, the overexcited mode (from 0% at 0.99 pu to 100% at 0.95 pu in Figure 2) leads to a voltage rise.

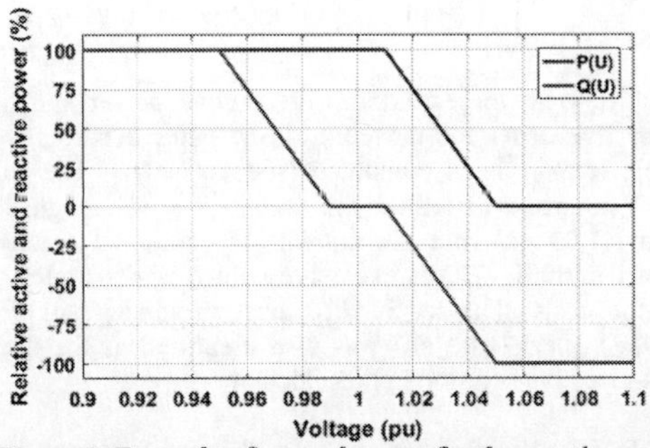

Figure 2: Example of control ramps for the reactive power feed in and active power curtailment as a function of line voltage in per unit (pu). The shown example is for a control strategy where 50% of the respective control bandwidth is reached at 1.03 pu.

3 COSTS OF PQ(V)

The implementation of PQ(V) control requires no communication infrastructure to a centralised station, instead costs for inverter oversizing, active power losses and additional reactive power compensation arise. These costs need to be included in the economic analysis for the comparison to the other methods presented in chapter 1.

3.1 Oversizing of inverter

When deploying 100% of the reactive power capability of the PV inverter (cosφ = 0.9), the rated apparent power output of the inverter has to be 1.11 times the active power output. Thus, the inverter has to be oversized accordingly:

$$S_{Inv} \geq \frac{1}{\cos(\varphi)} * P_n$$

The oversizing does not manifest a direct benefit for the PV plant operator and thus has to be reimbursed.

3.2 Losses

There are two types of active power losses which will arise with PQ(V) control. Active power curtailment will reduce obviously the effective yield of the PV plant. Furthermore, the reactive power feed-in will cause heat dissipation on the lines, resulting in active power losses in the LVDG. Both of these losses need to be included in cost calculations. They were quantified in Switzerland with roughly CHF 60.- per MWh.

3.3 Compensation of reactive power

There might be situations where the activated Q(V) control at multiple PV inverters will cause an imbalance of reactive power at the transformer to the next higher grid level. The compensation of reactive power is usually charged if the cosφ is lower than 0.92 at the transformer. If explicitly attributed to the Q(V) control this compensation needs to be included in the costs for PQ(V). The costs for reactive power were quantified with CHF 41.- per Mvarh.

4 LOAD FLOW CALCULATIONS

The voltage dependent active and reactive power control has been implemented to the load flow calculation software Matpower [5]. Using grid models provided by two DSOs, the two grid classes *village periphery* and *urban neighbourhood with apartment buildings* could be assigned. The technological analysis was performed over one year using the load flow calculation. Based on these results the costs could be identified.

4.1 Grid Class: Village Periphery

The western part of the LVDG in Dettighofen, Southern Germany (Village periphery) is connected to the medium voltage grid with a 400 kVA transformer (see Figure 3) and corresponds to the grid class *village periphery*. The installed PV capacity amounts to 535.5 kWp [3].

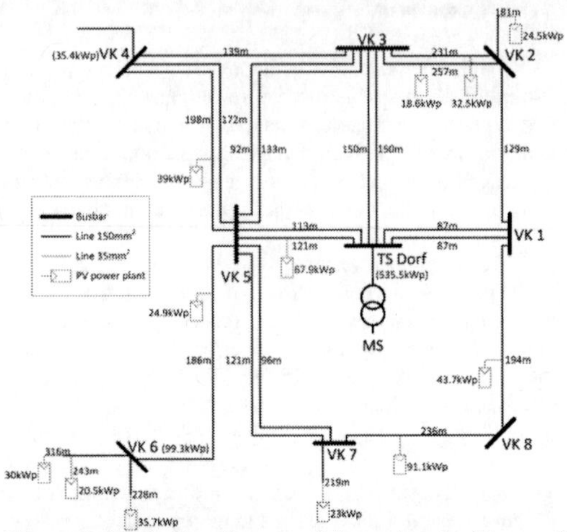

Figure 3: Single line diagram of the analysed grid part of Dettighofen. The transformer pictured in the center has an apparent power of 400 kVA. The grid features 33 PV plants with a total of 535.5 kWp nominal power.[5]

4.1.1 High PV Penetration

Additionally to measurements performed in 2014 [3], the load flow calculations without any control (status quo) identified a critical node at long stub connected to the distribution box *VK6* (see Figure 3). The voltage rise at the critical node reached a maximum of 7.23% above the nominal voltage. The distribution of the voltage in Figure 4 suggests that the load connected to this stub is relatively small compared to the installed PV nominal power. It can be seen that there are multiple instances of voltage violations beyond the 1.03 pu limit.

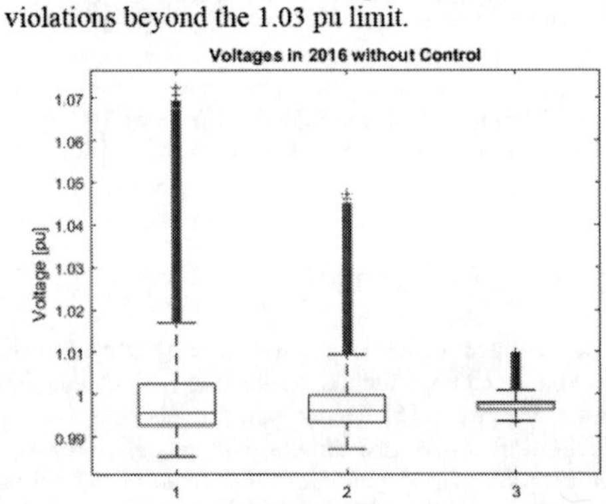

Figure 4: Voltage distribution (based on 10-minute mean values) at 1) the critical node, 2) the connection box of the critical node, 3) the transformer in Dettighofen followed from the load flow calculations without any control.

Several different control methods and ramps have been tested in the load flow calculation (see Table II).

The control strategy with the best performance regarding voltage rise limitation was PQ(V) with a mean of 1.02 pu (definition see Figure 2). With this inverter profile the voltage can be kept within the D-A-CH-CZ technical rules band at any time during the year (see Figure 5).

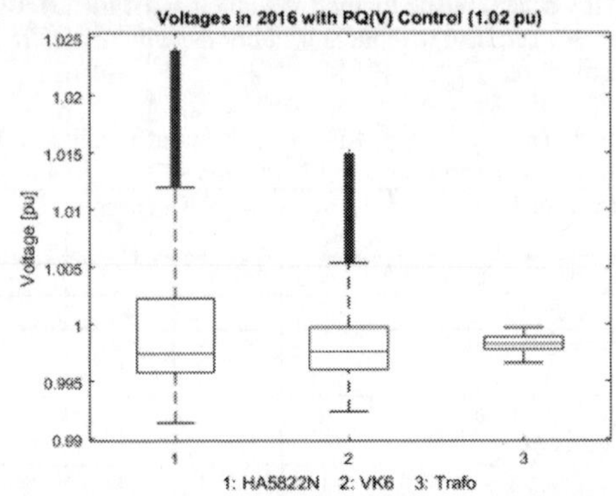

Figure 5: Voltage distribution (based on 10-minute mean values) at 1) the critical node, 2) the connection box of the critical node, 3) the transformer in Dettighofen followed from the load flow calculations with PQ(V) control with a mean at 1.02 pu.

Table II: Maximum Voltages over one year in Dettighofen, results from load flow calculations with different P/Q(V) control strategies.

Control Strategy	Maximum Voltage Critical Node	Maximum Voltage VK6	Maximum Voltage Transformator
Status quo	1.0723 pu	1.0471 pu	1.0098 pu
PQ(V) Mean 1.02 pu	1.0239 pu	1.0149 pu	0.9996 pu
P(V) Mean 1.02 pu	1.0282 pu	1.0203 pu	1.0049 pu
Q(V) Mean 1.01 pu	1.0466 pu	1.0243 pu	0.9962 pu
Q(V) Mean 1.00 pu	1.0430 pu	1.0206 pu	0.9993 pu

The yield losses, additional reactive power and the respective costs for the control strategies according to Table II have been computed (see Table III). The sole Q(V) control will reduce the voltage rise (although not below 1.03 pu), but the amount of additional reactive power in the LVDG is excessive, which is also reflected on the costs. The technically most promising solution (PQ(V), mean 1.02 pu) was also identified as the most cost-effective solution in this calculation.

Table III: Yield losses, additional reactive power and their respective costs over one year for the same control strategies as in Table II.

Control Strategy	Yield Loss	Additional Reactive Power	Cost per Year
PQ(V) Mean 1.02 pu	21.71 MWh	32.54 Mvarh	CHF 2'637.-
P(V) Mean 1.02 pu	46.68 MWh	--	CHF 2'801.-
Q(V) Mean 1.01 pu	--	417.36 Mvarh	CHF 17'112.-
Q(V) Mean 1.00 pu	--	745.75 Mvarh	CHF 30'575.-

4.1.2 EV Charging

Four hypothetical scenarios for EV charging in Dettighofen are assumed:

- No EV
- 1 EV at weakest node
- 50% of PV nodes with EV
- 100% of PV nodes with EV

The scenario 100% EV means that all nodes with a connected PV plant will also connect an EV charging station.

The charging is specified to happen during noon to investigate the interaction between EV charging and high PV feed-in. The charging profile is assumed the same for all EV, according to Figure 6.

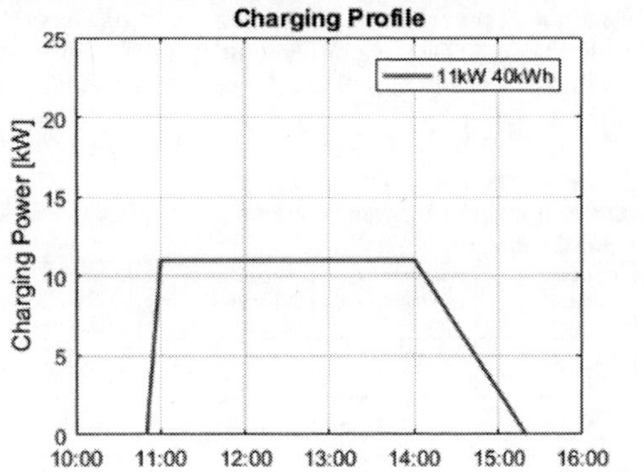

Figure 6: Charge profile used for the load flow calculations. The EVs (40 kWh) are charged during noon at a rate of 11 kW until they are 80% full. After that the charging rate is linearly decreased until SOC of 100% is reached.

Figure 7 shows the impact of EV charging on the voltage at the weakest node. While EV charging can limit the voltage rise due to PV power feed in to some extent, it also will result in a voltage below 0.95 pu for the 100% EV scenario.

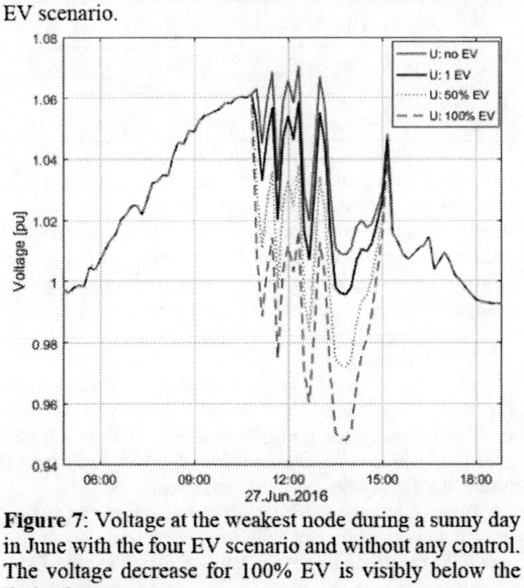

Figure 7: Voltage at the weakest node during a sunny day in June with the four EV scenario and without any control. The voltage decrease for 100% EV is visibly below the limit of 0.97 pu.

The voltage rise due to PV feed-in can be limited to 1.03 pu (see Figure 8) by implementing the PQ(V) control with a mean value of 1.03 pu (see Figure 2).

The voltage decrease due to EV charging is also limited within the limit of 0.97 pu by over-excited reactive power mode of the PV inverters, except for the 100% EV scenario.

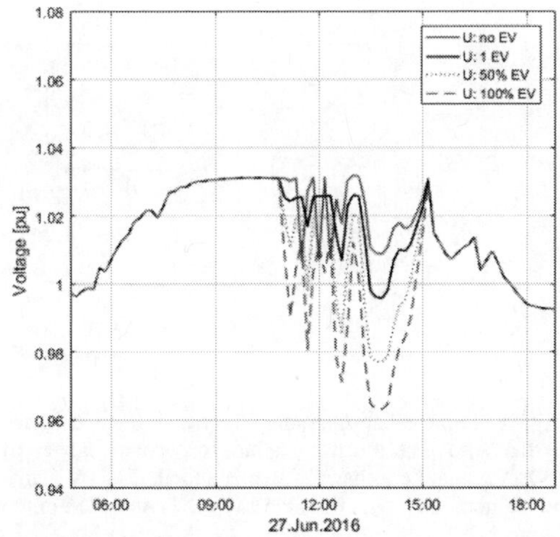

Figure 8: Voltage at the weakest node during a sunny day in June with the four EV scenario and with PQ(V) control with a mean value of 1.03 pu. The voltage decrease is limited compared to no control, although not quite within the limits for the 100% EV scenario.

4.2 Grid Class: Urban Neighbourhood with Apartment Buildings

The second analysed grid in Figure 9 is the current state of the small city of Ilanz, Canton of Grisons (Urban neighbourhood with apartment buildings), corresponding to the grid class *urban neighbourhood with apartment buildings*. The local DSO estimates a maximum of 41 PV power plants with a total nominal power of 788.5 kWp to be connected to their 630 kVA transformer in the future.

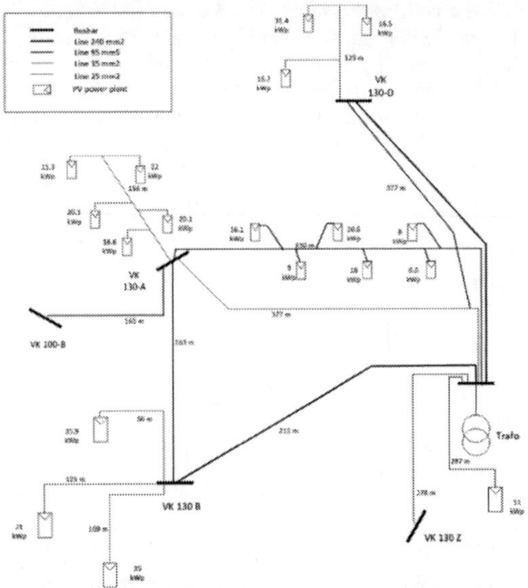

Figure 9: Single-line diagram of the grid in Ilanz, Grisons, Switzerland. The transformer has an apparent power of 630 kVA and the scenario estimates a total of 41 PV plants with a nominal power of 788.5 kWp to be installed in the future.

4.2.1 High PV Penetration

The load flow calculation without any control identified the critical node with a maximum voltage rise of 3.98% over the nominal voltage (see Figure 10).

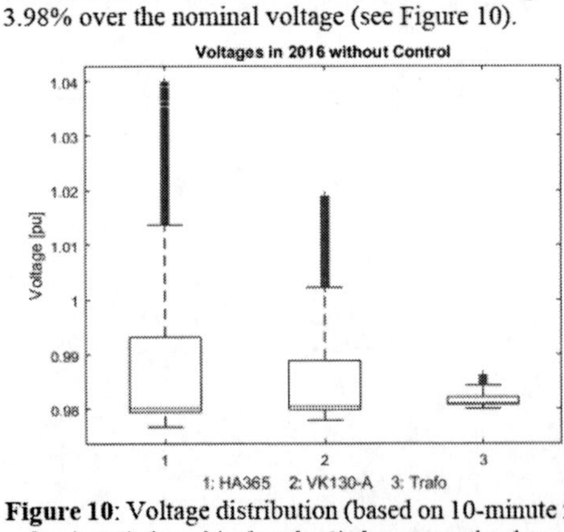

Figure 10: Voltage distribution (based on 10-minute mean values) at 1) the critical node, 2) the connection box of the critical node, 3) the transformer in Ilanz followed from the load flow calculations without any control.

Compared to Dettighofen the grid is more rigid since there are no long stubs with high PV penetration. This results in a lower maximum voltage rise at the weakest node (see Figure 10).

Different analyses with variation of the control methods show that the best performance could be reached with a PQ(V) control and a mean of 1.03 pu (see Table IV and Figure 11).

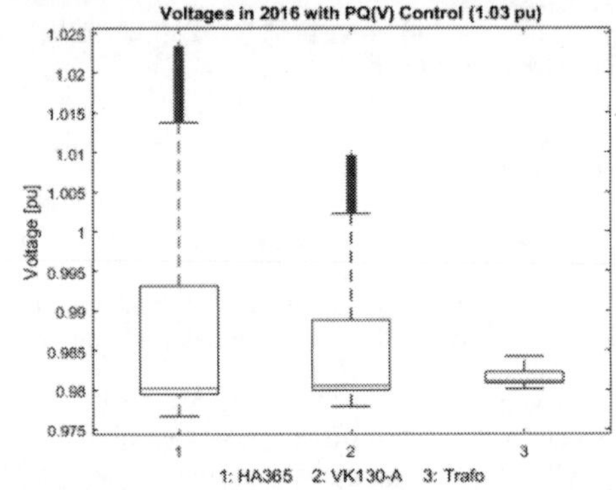

Figure 11: Voltage distribution (based on 10-minute mean values) at 1) the critical node, 2) the connection box of the critical node, 3) the transformer in Ilanz followed from the load flow calculations with PQ(V) control with a mean of 1.03 pu.

Table IV: Maximum voltages over one year in Ilanz, results from load flow calculations with different P/Q(V) control strategies.

Control Strategy	Maximum Voltage Critical Node	Maximum Voltage VK130-A	Maximum Voltage Transformator
Status quo	1.0392 pu	1.0185 pu	0.9861 pu
PQ(V) Mean 1.03 pu	1.0233 pu	1.0095 pu	0.9842 pu
P(V) Mean 1.03 pu	1.0249 pu	1.0116 pu	0.9850 pu
Q(V) Mean 1.02 pu	1.0295 pu	1.0094 pu	0.9832 pu

Table V: Yield losses, additional reactive power and their respective costs for the same control strategies as in Table IV.

Control Strategy	Yield Loss	Additional Reactive Power	Cost per Year
PQ(V) Mean 1.03 pu	4.16 MWh	--	CHF 250.-
P(V) Mean 1.03 pu	6.82 MWh	--	CHF 409.-
Q(V) Mean 1.02 pu	--	--	--

Again, the technically best control method causes the least costs regarding yield loss and additional reactive power compensation (see Table V). Although activated, the reactive power control does not cause the cosφ to go below 0.92, thus not inducing reactive power compensation.

4.2.2 EV Charging

The same scenarios as in Section 4.1.2 were assumed for the EV charging in Ilanz. It can be observed that there is a positive interaction between PV feed-in and EV charging, which complement each other to some extent in terms of voltage keeping (see Figure 12).

By applying the PQ(V) control with a mean of 1.03 pu the voltage deviation can be limited to ±3% of the nominal line voltage, although not for 100% EV (see Figure 13).

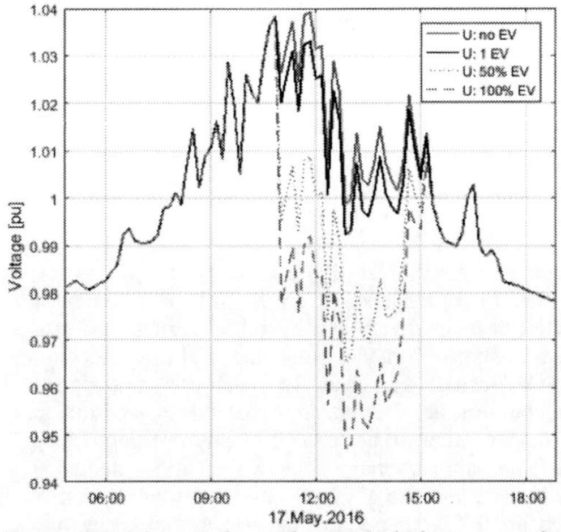

Figure 12: Voltage at the weakest node during a sunny day in May with the four EV scenario and without any control. The voltage decrease for 50% and 100% EV is visibly below the limit of 0.97 pu.

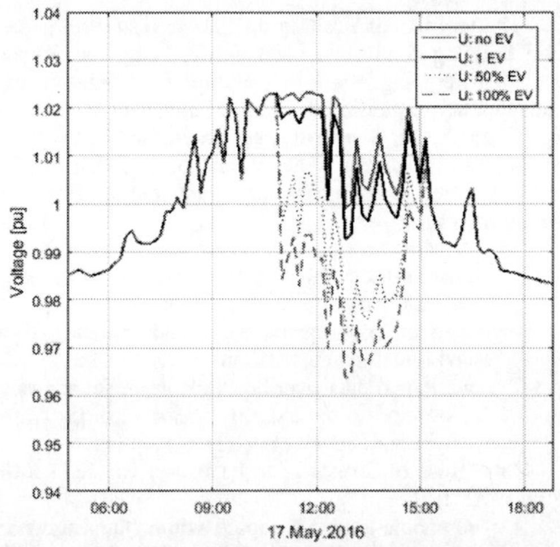

Figure 13: Voltage at the weakest node during a sunny day in May with the four EV scenario and with PQ(V) control with a mean value of 1.03 pu. The voltage deviation limits of ±3% can be met except for the 100% EV scenario.

5 CONCLUSION

The voltage dependent active and reactive power control was applied to two LVDG models using load flow calculation and a techno-economic analysis was carried out. It has been shown that PQ(V) control is a cost-effective and efficient approach to reduce voltage limit violations in LVDG with high PV penetration as well as in case of a large number of installed EV charging stations. The maximum voltage rise of 1.0723 pu could be reduced to 1.0239 pu using PQ(V) control in the case of the LVDG of Dettighofen. The yearly costs amount to CHF 2'637.- for this measure. The interaction between EV charging, PV and PQ(V) control has been proven positive under certain circumstances, reducing both the voltage rise due to PV feed-in as well as the voltage drop due to EV charging. Costs for different measures of voltage control, such as LVR, battery energy storage, etc., need to be acquired in order to compare the PQ(V) method. Furthermore, additional grid models need to be classified and analysed in the load flow calculation. The ultimate goal is to provide a decision-making tool for DSO to evaluate the most techno-economic measure for voltage stabilisation in their grid.

ACKNOWLEDGMENT

This project is funded by the Swiss Federal Office of Energy (SFOE) under the contract number SI/501370-07 and supported by EKS and Repower (DSOs of the two analysed grids).

REFERENCES

[1] "EN 50160: Voltage characteristics of electricity supplied by public distribution networks." DIN, Feb-2011.

[2] "D-A-CH-CZ Technical Rules for the Assessment of Network Disturbances." 2007.

[3] F. Carigiet, M. Niedrist, C. Scheuermann, and F. Baumgartner, "Case study of a low-voltage distribution grid with high PV penetration in Germany and simulation analyses of cost-effective measures," *Prog. Photovolt. Res. Appl.*, vol. 24, no. 12, pp. 1523–1532, Dec. 2016.

[4] Franz Baumgartner, Fabian Carigiet, and Petr Korba, "Project CEVSol," Swiss Federal Bureau of Energy, Public annual report, 2017.

[5] F. Carigiet, F. P. Baumgartner, P. Korba, R. Knecht, M. Koller, and M. Niedrist, "Optimisation of the Load Flow Calculation Method in Order to Perform Techno-Economic Assessments of Low-Voltage Distribution Grids," *33rd Eur. Photovolt. Sol. Energy Conf. Exhib. 2181-2187*, 2017.

[6] "Liste untersuchte Wechselrichter Vorarlberg Netz." VKW AG, 12-Sep-2018.

Spatial representation of low-voltage network hosting capacity for photovoltaic roof-top installations using an open-source tool

M. Joos[1], N. Lebert[1], B. Gaiddon[1], E. Séguin[2], P.E. Gautreau[2]
[1]Hespul, 14 place Jules Ferry, 69006 Lyon France
[2]IGN, 73 Avenue de Paris, 94165 Saint-Mandé, France

ABSTRACT: Just as solar cadastres have become the norm to assess the roof-top PV potential in the last years, assessing the hosting capacity of the low-voltage network for electrical production is on the path to becoming a necessity for an efficient solar production action plan. The model developed here is a hybrid tool composed of a physical model computing results based on a real infrastructure and a statistical model using empirical rules based on frequently encountered cases. The tool, built entirely on PostgreSQL/PostGIS, can be easily launched by non-experts on any set of data and on any computer, and has a low computation time. Results show that the major limit in connecting high shares of PV in existing grids is above all the low summer power consumption at noon. Key messages can be drawn from results: 1) hosting capacity of the existing infrastructure on the two territories studied is important but must be utilized with care by avoiding buildings located far from a substation, 2) on the long run, a deep reflexion on grid settings and planning is compulsory in order to achieve ambitious shares of renewable electricity at low marginal cost.
Keywords: Grid Integration – Grid Management - Strategy - Distributed System Operators (DSO) - Software

1 INTRODUCTION

Today solar production of energy is developing rapidly in France: 875MW were added in 2017, comprising 298MW on the low-voltage (LV) grid [1]. As local planning documents on climate, air and energy have become mandatory for all local authorities with over 20,000 inhabitants, many of them have come to realize through studies evaluating the renewable energy sources (RES) potential that most of the solar potential resides on low-voltage grid, in the form of small to large roof-tops. As an example, on the Valence-Romans conurbation, ground-based photovoltaic parks' potential total 62MW, 8MW of which are already in service, compared to 1362MW on roof-top, amongst which 1007MW are on the low-voltage grid, the rest being on the medium-voltage (MV) grid and marginally on the high-voltage (HV) grid.

Knowing that connection costs in €/kVA are the highest on the low-voltage grid and that infrastructure's hosting capacity for renewable electricity production is largely unknown by producers prior to connection request, improving knowledge on the low-voltage grid's hosting capacity appears to be a key issue in the energy transition, with the risk otherwise to cancel the efforts of local authorities to massively develop photovoltaic production. The present study mirrors similar approaches in European countries that attempt to assess nation-wide distribution network hosting capacity [2]. The exercise could not be generalised nationwide in France since the distribution network topology is not yet fully available in open data [3], and must be asked for territory by territory for a specific purpose.

The purpose of the work is to develop a geographical information system (GIS) tool computing low-voltage network hosting capacity for roof-top photovoltaic systems on large areas (conurbations of over 20,000 inhabitants, departments etc.) based on data readily available and free of charge, that is user-friendly, open-source, and fully transferred to local authorities' technical services. The tool aims above all to highlight levers that would facilitate renewable electricity grid integration, whether they are technical or pertain to the governance structure of network planning and connection.

2 DATA SET

The tool uses the cadastral database (cadastral parcel and building footprints), data delivered by the French National Mapping Agency (IGN) free of charge for any municipality requesting it. The tool also uses a basic set of electrical data, data already communicated by the DSO to its licensing authority every year or data that can be requested free of charge by any municipality. This data set consists in:

- distribution network topology: geographical positions of LV and MV lines, HV/MV and MV/LV substations
- type of lines (overhead or underground), type and function of substations
- additional data may be available in some cases: substations nominal power and loading rate, nature and section of cables.

Two areas of investigation have been chosen in the Rhône-Alpes Region:

- Grenoble-Alps Metropole within the European Commission funded FP7-City-Zen project [4] focusing on innovation in the transition process
- Valence-Romans conurbation, a more rural area, in the context of the preparation of Climate Air Energy strategy ("Plan Climat Air Energie"), which has since being politically approved [5].

The present work has in both cases responded to local authorities' need to better understand network limits with regards to RES integration and to advise decisions on RES development goals and network development.

2.1 Characteristics of the low-voltage network

The two areas of study differ substantially in the terms of network topology (see Table 1). Valence-Romans conurbation is representative of the national average in terms of length of LV lines per substation, whereas Grenoble-Alps Metropolitan area is on average a little less dense, which is probably due to the fact that the city center of Grenoble is out of the scope of the study whereas the cities of Valence and Romans are part of the study. Nonetheless, Grenoble's LV network comprises of a higher percentage of underground LV lines and its average linear meter of LV line per building is half that of Valence-Romans conurbation (20 meter per building versus 40 meters per building). This higher building density in Grenoble must be kept in mind when comparing the two areas.

Table 1: Network characteristics of Valence-Romans conurbation (VRA) and Grenoble-Alps Metropole compared to national averages.

	Nationwide	VRA	Grenoble
% of overhead LV lines over total LV length (overhead + underground)	N/A	57%	35%
Average LV lenght per public MV/LV substation (km/substation)	0,9	0,9	1,1
Average LV lenght per building (km/building)	N/A	0,04	0,02

3 MODEL

The model developed here is a hybrid tool composed of a physical model computing results based on a real infrastructure and a statistical model using empirical rules based on frequently encountered cases. Empirical rules allow gaining computation time over a classical load-flow model computing voltage constraints. The precision thereby lost in the process isn't crucial at this stage since the tool serves primarily to raise awareness and gain a preliminary understanding of grid integration challenges on a large scale. It aims at pointing out how current decision-making processes and tools used for grid investment planning, traditionally focused on consumption growth, must evolve to ensure cost-effective integration of high shares of small-scale photovoltaic systems.

Main steps of the algorithm to perform grid constraints calculation are the following:

1. Evaluate PV potential of each building,
2. Find their grid connection point,
3. Compute the base load power of buildings and substations,
4. Calculate the network distance between buildings connection point and the nearest substation,
5. Estimate hosting capacity of substations.

3.1 PV potential – solar cadastre

Many studies that compute automatically the roof-top solar photovoltaic potential over large areas date back to the years 2010 and were funded through European projects [6]. Today, these solar cadastres have somewhat become the norm in energy planning studies. They may be carried out by consultants or directly by local authorities themselves for several reasons, the main reasons being:

- identify the PV potential on their territories (mandatory input for their energy planning documents),
- raise awareness of elected officials on the feasibility of the energy transition and the opportunity it represents for the territory,
- engage inhabitants in development of private projects by providing them adequate information on the solar potential of their rooftop.

Solar cadastres provide two core indications per rooftop:

- installable PV peak power
- estimation of the annual production of the system.

In the present study, as the low-voltage network hosting capacity for PV plant is constrained by the worst-case scenario, namely maximum PV production and minimum electrical consumption (base load in summer), only PV peak power is required as in input to further steps.

The methodology used is a classical in-house 2-D solar cadastre, based on ratios obtained through Hespul's field experience:

1. Calculate roof surfaces from building ground coverage and estimated roof tilt (the tilt is generally provided by the local Urban Planning Agency which has knowledge of constructions' characteristic in the region),
2. Estimate the usable surface for PV installation, depending on roof size and azimuth (e.g. for gable roof, only the South-facing slope is equipped with PV panels)
3. Calculate PV peak power that can be installed.

Figure 1 shows an example of output from the solar cadastre.

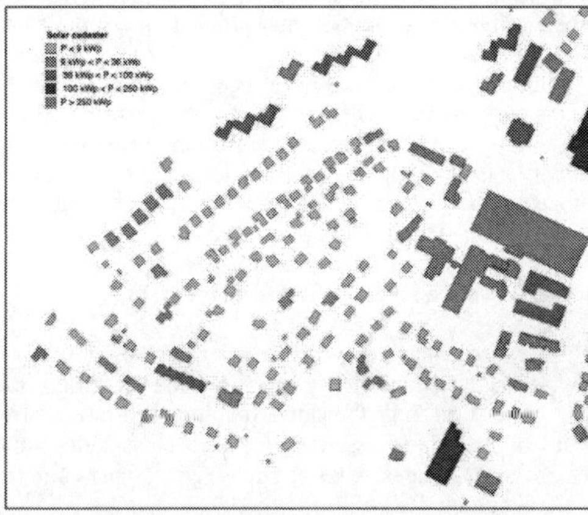

Figure 1: Building ranking based on their PV potential

3.2 Network connection

The tool has access to two different datasets, from two different sources:

- cadastral database by the National Geographical Institute (IGN)
- distribution network topology by the distribution system operator (DSO) via the local authority.

In practice, each building of the cadastral dataset is physically connected to the distribution network. As no information is given on the location of this or these connection points, they must be computed using available data.

There are indeed two connection points to compute: one for consumption and one for production, which location (at which voltage level and to which electrical equipment – line or substation) will depend on the maximum power consumed or supplied.

On the production side, the connection level is defined by the PV peak power installed. System with a peak power less than 250 kVA will be connected to low-voltage network, whereas medium voltage network can receive PV with a maximum power of 17MW (see Table 2).

On consumption side, the connection level depends on the power subscribed. As this latter information is not known (considered confidential in current national rules), it is evaluated based on floor surface computation and a ratio of power consumption per square meter (40VA/m^2) used indistinctively for residential and tertiary buildings since the nature of the building's activity is not known in datasets used.

Table 2: Voltage connection levels of the distribution grid as a function of installed power (generally taken to be the minimum between inverter installed capacity and PV panels' total peak power) [7].

	Value (V)	Maximum installed power
Low-voltage	400	250kVA
Medium voltage	20 000	12-17MW

The following assumption is made for buildings connected to the low-voltage grid: the network connection point of a building is the nearest point of the low-voltage network. On the distribution network, the photovoltaic connection point will most of the time be located right next to the consumption pre-existing network connection point as long as voltage connection levels of both consumption and production are the same (see Table 2) (see 4.5).

Buildings connected to the medium voltage network on the consumption side have a dedicated private MV/LV substation, which is often located on the same cadastral parcel. All buildings on this cadastral parcel will be connected to this dedicated substation on the consumption side, and an option is available to allow this connection point on the production side, corresponding to actual network connection technical rules [8].

3.3 Distance to the electrically nearest substation

Once the connection point has been identified, the next step is to find the corresponding substation. The problem can be expressed as: for each connection point, which is the nearest substation by going upstream the low-voltage sections of the network?

The question looks actually like a simple graph shortest path problem; and it is! Nonetheless, in the datasets considered here are tens to hundreds of thousands of connection points, as well as thousands of substations. As a result, launching a shortest path algorithm for each connection point independently appeared to be time-consuming (more than 20 hours for 70 000 supply points). No shortest path algorithm provided with the classical GIS network library (e.g. pgRouting) could thus be used.

However, multiple connection points (up to hundreds) often lie between the end of an LV line and its substation. When processing the nearest substation of one connection point, it is then also possible to update all connection points crossed over along the path. This fact is at the root of an in-house algorithm, based on a breadth-first search (BFS) principle, whose computation time is more than 20 times less than the GIS solution available off the shelf in the network library.

The result of this algorithm is the association of both a substation and the distance between the connection point and the substation. Figure 2 represents the results of the algorithm on a small urban area : the color of each building is a function of its distance to the electrically nearest substation. It can be noted that for some buildings, no connection point has been found using the algorithm (the percentage of concerned buildings varies from 2% to 10% of the total): in some cases, the building is truly off-grid, and in other cases, the reason that no connection point has been found lies in data quality issues, which are further developped in the next paragraph.

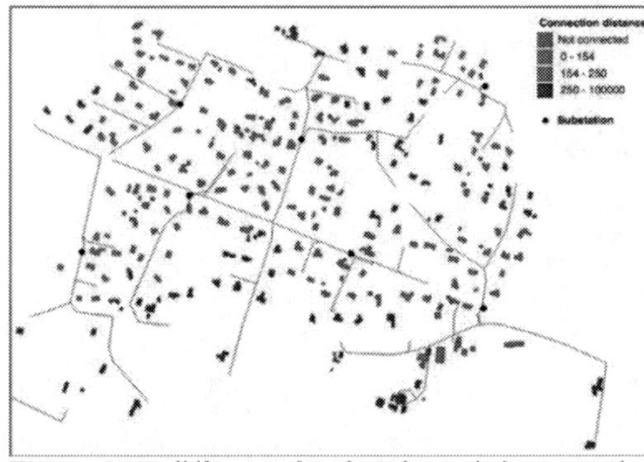

Figure 2: Building sorting based on their connection distance (meters) to the nearest substation showing connection costs inequalities. The distance is computed using an in-house algorithm of shortest path following LV lines.

3.4 Data quality issues

The algorithm can be used with default values when substation power values are missing from the database: default values are attributed using other substation parameters such as the substation's type (e.g. number and size of transformers vary whether they are in a substation up on a pole or in a substation on the ground) or the number of clients served.

The main data quality issue is actually network discontinuity, as displayed in Figure 3, where three discontinuities appear on one of the distribution line. As a result of this data quality issue, all buildings whose connection point are downstream the discontinuity on the electrical line (considering the substation as the starting point), will not be connected to the substation in the constructed graph.

Those discontinuities do not exist of course in the physical network, and are only due to missing information or inaccuracies, but lead to a number of non-connected buildings that is non-negligible.

Data cleaning process can be performed in order to improve the results, but caution is needed in order not to add network parts that in reality do not exist.

Figure 3: Network discontinuities appearing in the GIS network data. The quality of the dataset necessarily affects the precision of the results.

3.5 Hosting capacity

The hosting capacity of a substation is defined as the total PV power that can be connected to it without generating a major constraint which would require important network upgrade (e.g. LV line reinforcement over several tens of meters, transformer size increase, MV/LV substation creation, etc.).

Constraints generated by PV power connection are mainly voltage constraints rather than constraints of current. Voltage rise appears when PV production is higher than local consumption. This imbalance can be supported by the LV network as long as voltage rise remains in the European standard EN50-160 power quality levels of nominal voltage (Un) +10%. French connection requirements specified in technical reference documents [9] also specify that the voltage set on the LV side of the transformer is slightly higher than nominal level: MV/LV transformation ratio without charge is 48.7 (voltage output of the transformer at 410V) rather than 50 (20kV/400V), since historically network planning's major constraint has been to accommodate for high winter electrical consumption.

However, as monitoring campaigns have shown [10], this transformer setting can strongly limit the share of PV that can be connected in existing grids, since the worst case scenario corresponds to a summer day around noon when the consumption is at its lowest (therefore voltage values close to a no-load condition), all PV systems are producing at their peak power (see Figure 4). In the absence of data on minimal power consumption, these values are taken to be 20% of subscribed power in network connection studies performed by DSOs [9].

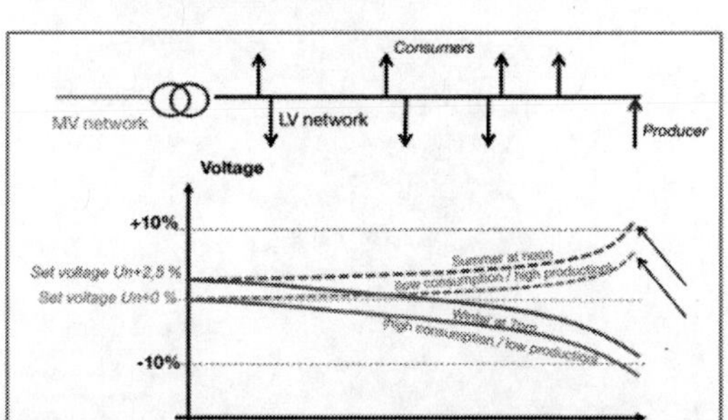

Figure 4: Simplified voltage plan showing the impact of transformer voltage setting (nominal level – Un+0%, or higher level – Un+2.5%, where Un = 400V) on the voltage rise (see arrows on the right hand side) at the end of an LV line where a photovoltaic plant is connected. In the case of a setting higher than nominal level, the power quality standards are not met, with voltage levels higher than Un+10%.

These considerations have led to simplify voltage constraints using the following two criteria:

- Distance: as an empirical fact (e.g. obtained through field experience), PV systems located on buildings that are more than 250 meters away from a substation systematically face prohibitive connection costs.
- Minimum consumption: cumulated PV power higher than minimal consumption power consumed on the same substation will lead to major voltage constraints.

The distance has been previously computed. Minimum consumption constraint requires the calculation of the minimum electrical consumption at the substation level. This value is obtained from the sum of the maximum building consumption level adjusted with a diversity factor which accounts for the fact that not all clients are reaching the maximum power at the same time. Minimum electrical consumption at the substation level is obtained by multiplying the maximum electrical consumption by 20%.

The two criteria are ultimately combined to provide a synthetic view of substation hosting capacity corresponding to the result of the most stringent criteria.

3.6 GIS technology

During the methodology definition phase, the different steps of the procedure were performed manually on test areas. As it was very time consuming, and needed special GIS skills, it was decided to automate the whole process, so that non-GIS-expert users could handle it. The current tool is thus built entirely on PostgreSQL/PostGIS, and can be easily launched on any set of data, on any computer, and by non-experts, as long as the input format is as defined. Efforts made by IGN and Hespul finally allowed to optimize the algorithms developed jointly and thus divide computation time by more than 20.

To be used on a computer, the current tool needs a special environment, with specific libraries installed, and a database management system such as PostgreSQL. This is an important limitation for the tool deployment on non-expert's computers.

Furthermore, the portability of a SQL database from one computer to another one is not straightforward, which makes the teamwork on a project quite complicated. A solution may be to use a database installed on a server, but this solution also has its downsides for non-specialists.

Ongoing research is performed to assess the possibility to transfer the tool from Postgis to Spatialite: this would indeed make the tool easy to deploy and use and improve its database portability. The main advantage of Spatialite is indeed to use ordinary files as database, making the portability as easy as sending a file.

The whole code has been transferred in Spatialite and it appears to be both as stable and as efficient as PostGIS to deal with extended territories. However, further work is required to verify that the improved performance applies to different computer capacities and territories sizes, and that it is possible to integrate the whole process in a QGIS plugin.

3.7 Limits and areas of improvement

The main limits of the present method lie in the following three points:

- connection point location
- base load consumption computation
- simplification rules to compute network constraints.

Concerning connection points' location, the method described does not ensure to get a perfect image of field reality since network development is incremental and cannot systematically be synthesized in pure logical rules such as the ones that are applied here on a wide scale. For example, a building cannot be connected to the network on an electrical line that did not exist when the connection was made. Notwithstanding the errors thereby generated, precision is considered sufficient here considering the purpose of the tool, not to mention the high number of points considered and the radial structure of the grid that make it unlikely to attribute a false connection point to a building.

Base load consumption computation is probably a higher source of error, most probably leading to an overestimation of PV potential that can be connected at low marginal cost. Traditional network constraints study focus on peak consumption power, generally occurring in the winter (especially in the context of a high share of electric heating in France), whereas little attention has been paid to summer middle day consumption, which is crucial to studies of network constraints with high share of PV.

As said previously, the strength of the present method is also its flaw. Indeed, using empirical rules allow processing large areas in minimal time and gaining a first reading of network's limits. However, this first reading must then be completed with an actual voltage plan using the lines' cross-section.

4 RESULTS AND COMPARISON

This section presents first the characteristics of the PV potential computed through the solar cadastre method, and then delves into grid impact with various angles in order to gain the most complete picture of the results' implications for PV and network development policies.

4.1 PV potential characteristics

On Grenoble-Alps Metropolan area, PV potential on roof-top is evaluated at 916 MWp on 69 194 buildings, all voltage connection levels combined. The low-voltage potential (power less than 250kVA) in itself is 684MWp, the rest being on the medium-voltage. 406 PV plants that would be connected to the MV level have been identified. 95 % of potential installations have a power lower than 36 kVA, which represents 50% of the whole PV potential. Figure 5 shows the spatial distribution of the potential on the metropolitan area.

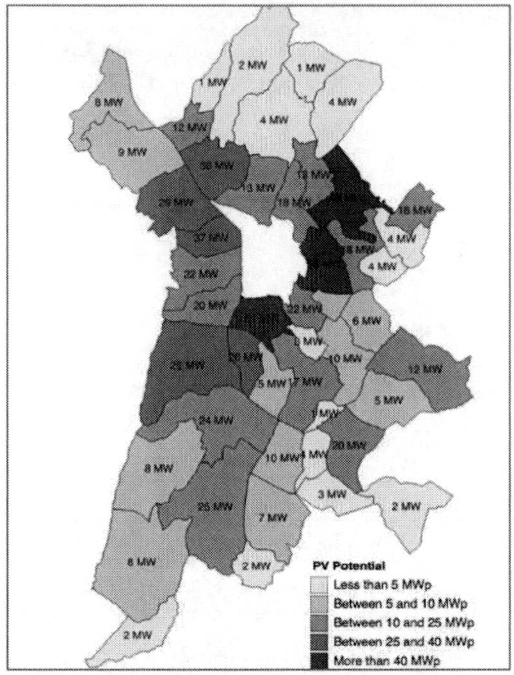

Figure 5: Grenoble-Alps Metropole map showing the low-voltage PV potential in MW by municipality. Potential on the city center has not been analysed, explaining the absence of data in the middle of the map.

On Valence-Romans conurbation, PV potential on roof-tops totals 1362MW on 64 370 buildings, amongst which 1007MW are on the low-voltage grid, the rest being on the medium-voltage (MV) grid. The greatest part of the roof-top PV potential is located in Romans, Valence and peripheral municipalities (see Figure 6).

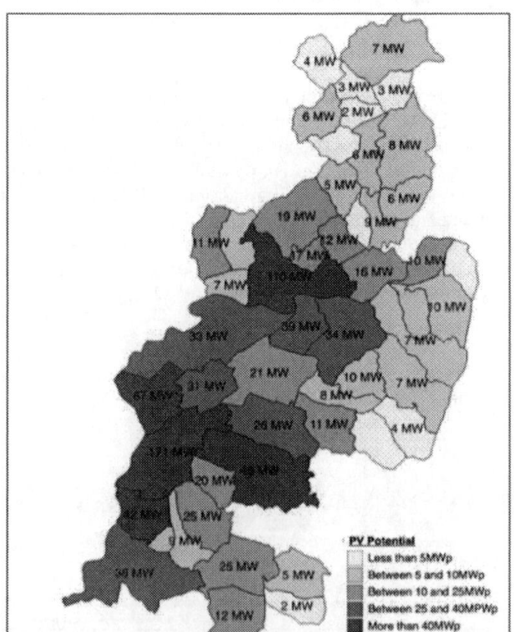

Figure 6: Valence-Romans conurbation map showing the low-voltage PV potential in MW by municipality.

4.2 Network limitations

Common general conclusions can be drawn from the two studies. First of all, it appears that the major limit in connecting high shares of PV is the low summer power consumption on substations at noon rather than the long distance between buildings and the nearest substation (see Table 4). As an illustration, 62% of Valence-Romans's public MV/LV substations (3048 in total) have a PV potential in the vicinity to the substation (electrical distance less than 250 meters) that is higher than what they can absorb considering their minimal power consumption. In the case of Grenoble-Alps Metropole, which is a more urban area, this percentage reaches 95.5% of the 1790 public MV/LV substations (see Figure 7) i.e. less than 5% of substations have a hosting capacity limited by the percentage of PV potential close to the substation. In fact, in the Grenoble-Alps area, 74% of the PV potential power is located less than 250 meters away from the nearest substation.

Learnings can also be drawn from each of the studies due to their geographical and demographical differences and differences in input data. Last but not least, areas of focus, and thus learnings, vary between the two territories based on the local authorities' needs: Grenoble-Alps Metropole has jurisdiction over electricity distribution on 8 out of 47 of its municipalities (the departmental union covers the remaining 39) and thereby the ability to provide guidelines on network development on its area, whereas Valence-Romans conurbation has not yet reached the size to have this legal competence but has competence in energy and the legal obligation to develop a Climate Air Energy strategy.

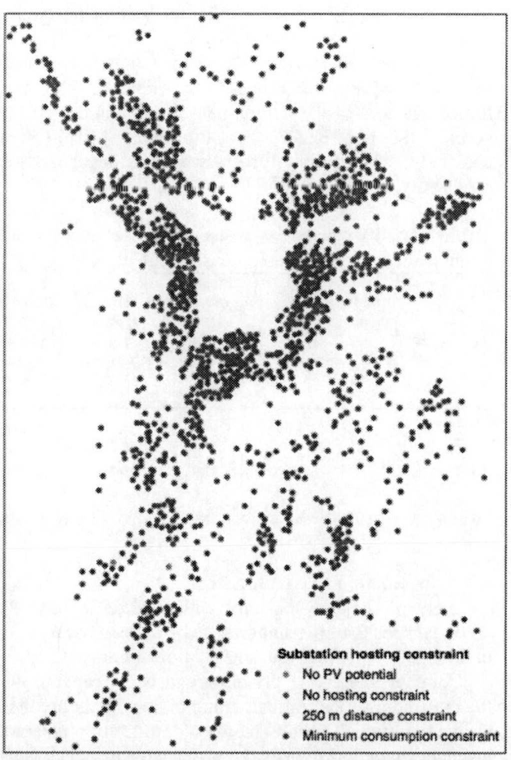

Figure 7: Substation sorting based on the most limiting parameter to the connection of PV potential without prohibitive investment, for each substation on Grenoble-Alps metropolitain area.

4.3 Comparison between urban and rural PV connection potential

Areas on which the tool has been tested always comprise one or two dense urban areas, surrounded by suburban neighbourhoods and then rural areas. Results confirm the well-known correlation between network density and hosting capacity. Figures 8 and 9 display the percentage of PV potential that can be connected to each MV/LV substations, from over 25% in green to less than 10% in red. It can be easily observed that the majority of green dots are substations in or close to city centers, while red dots are rather on the outskirts of the chosen territories, in municipalities of only few hundred inhabitants.

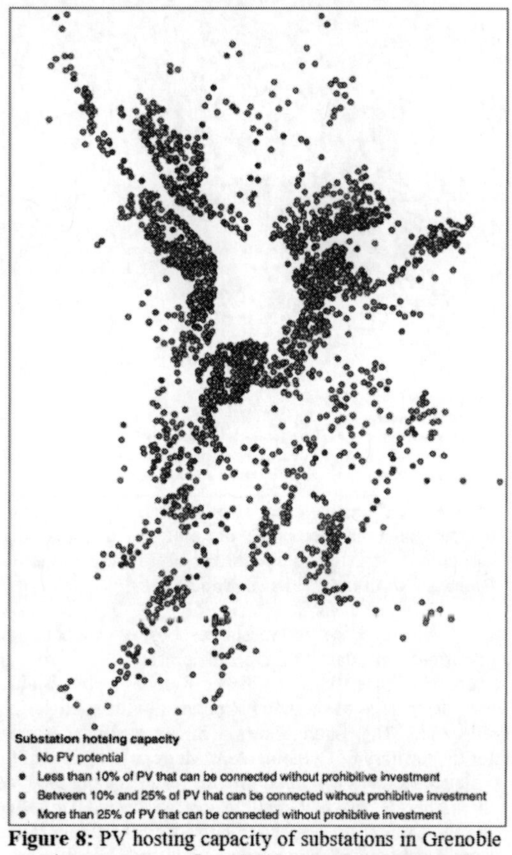

Figure 8: PV hosting capacity of substations in Grenoble area in percentage of total PV potential per substation. The whole in the center corresponds to Grenoble city center, which is not part of the study (different DSO).

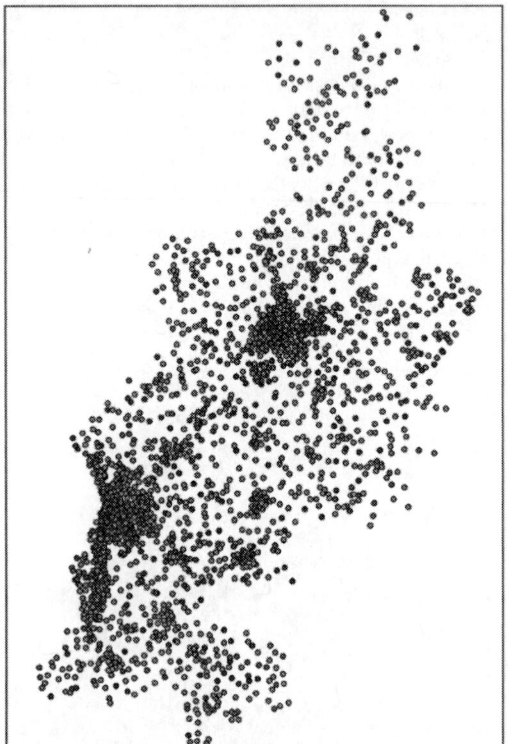

Figure 9: PV hosting capacity of substations in Valence-Romans area in percentage of total PV potential per substation. The two dense urban areas correspond to Romans (North) and Valence (South).

This difference in substation hosting capacity is confirmed by data. Figure 10 presents the hosting capacity of substations of Valence-Romans and Grenoble as a percentage of the LV PV potential attributed to the substation. This figure shows a higher hosting capacity for the territory of Grenoble-Alps Metropole compared to Valence-Romans conurbation. This result is explained by the higher density of Grenoble area which is favourable for PV connection (less buildings are further than 250m from a substation). Furthermore, data for Grenoble area do not include Grenoble city itself for which the density of buildings is higher, and thus the hosting capacity of substations would certainly be much higher if the scope of the study were to include the city center.

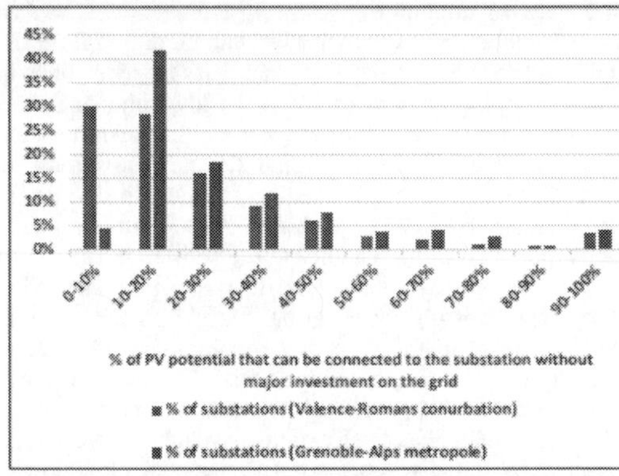

Figure 10: Distribution of substations according to their hosting capacity as a percentage of the LV PV potential attributed to the substation in Valence-Romans and in Grenoble-Alps Metropole.

In summary, three municipalities in Valence-Romans conurbation have been taken as examples of the diversity in table 3: one very rural with less than 500 inhabitants, a suburb a Valence and the city of Valence. As expected, PV potential on rooftops is higher in dense areas than in rural municipalities, and the fraction that can be hosted by the network also. This can be explained both by the higher density of buildings and network, reducing the distance constraint risk, and by the higher consumption area per ground square meter in urban area (e.g. tall buildings versus small houses), reducing the minimum consumption constraint risk.

Nevertheless, the PV potential that can be hosted per inhabitant remains higher in rural areas than in dense municipalities, supporting the assumption that energy transition requires an urban-rural solidarity. The most limiting factors of PV deployment in urban areas will be access to non-shaded roof-tops and architectural constraints whereas it will be network hosting capacity in rural areas.

Table 3: PV hosting capacity in a very rural municipality, a suburb and a dense urban area.

	Total PV potential per inhabitant (kW)	PV potential hosted (kW)	Fraction of total PV potential that can be hosted (%)	PV hosting capacity per inhabitant (kW)	PV hosting capacity per substation (kW)
Rural area	12.6	900	15%	1.9	45
Suburb	4.4	7 700	18%	0.8	87
Dense urban area	2.9	48 300	28%	0.8	120

4.4 Infrastructure adaptation needs

Network hosting capacity with regards to total PV potential speaks of the importance of network connection in the light of an ambitious energy transition.

Another interesting prism though is the capability of the existing physical infrastructure to absorb photovoltaic production, which speaks more of optimization potential through changes in network settings by using classical methods, smart grids or a combinaison of both. Grenoble-Alps Metropole's total PV potential on the low-voltage grid amounts to 915MWp concerning PV potential on both MV and LV levels. This can be compared to an electricity consumption for heating of 383 569 MWh and a maximum power consumption (all usages) of 340 MW. Thus, the reader can have in mind the following order of magnitude: the PV potential is 3 times the peak power consumed.

Concerning more specifically the low-voltage network, PV potential amounts to 681MWp, which incidentally corresponds more or less to the sum of public MV/LV substations' nominal power of 527MVA. That is to say, notwithstanding voltage constraints and cable power bottlenecks, the low-voltage network could host more or less the whole PV potential. Nonetheless, further analysis shows great disparities between rural and periurban municipalities (see Figure 11).

In the current study, which takes into account voltage constraints, the PV potential that can be connected without major network investment, is rather 146MWp (see Table 4), thus 28% of the substation nominal power.

In the realm of distribution network planning, this comparison shows that optimization of the existing network is a promising avenue for cost-effective integration of high shares of PV.

Figure 11: Grenoble-Alps Metropole map showing the sum of MV/LV public substations (MVA) compared to the overall low-voltage PV potential on roof-tops (MWp) by municipality. Colors sort municipality by the total nominal power of MV/LV public substations.

4.5 Levers to increase hosting capacity

Beyond these somewhat predictable results, what is more important is identifying potential levers based on these maps. For example, on Grenoble-Alps Metropolitan area, data on the percentage of electric heating is available and can be crossed with maps of hosting capacity to prioritized households refurbishment. Indeed, electric heating is responsible for much of the voltage drop in winter, which prevents setting the off-load transformer ratio to a lower level which would increase grid hosting capacity. Other examples of levers to increase hosting capacity can be modelled in the tool, notably:

- undersizing inverters following the model of the German regulation, which limits the power injection from PV plants to 70% of installed power of PV modules [11], obligation introduced in the guide VDE-AR-N 4105 [12],
- connecting power plants to private LV grids in the case of buildings that have their own MV/LV substation rather than trying to connect to the public LV grid, according to actual network connection rules.

For the Grenoble-Alps Metropolitan area, Table 4 summarises the impact of those levers on the PV hosting capacity of the LV grid. Overall, the existing LV network can host a little less than a quarter of the PV potential evaluated at that voltage connection level. However, this quite urban area hosts an important number of tertiary and semi-industrial buildings connected to the MV level through private MV/LV substations. Computation of the PV potential on those buildings shows that the percentage of the PV potential is not negligible: 75 MW (11% of the total PV potential), thus liberating hosting capacity on the LV grid and raising the LV PV potential that can be connected overall by that same quantity. Last but not least, undersizing inverters also has an important effect: it increases the PV potential that can be connected without network development by as much as 26%, from 146MWp to 184MWp.

Table 4: PV hosting capacity of LV grid in Grenoble and Valence-Romans area in MW and as a percentage of total PV potential on the territory. Impacts of levers such as inverter undersizing or PV connection to private LV grids, on the PV potential that can be connected at low marginal cost are also displayed

	Power (MWp)	Fraction of total PV potential on the area (%)	% of increase in the hosting capacity compared to the base case (%)
Overall PV potential on the area	681		
PV potential located at an electrical distance lower than 250 meters from a substation	506	74%	-
Base case: PV hosting capacity without network development works	146	21%	-
1. PV hosting capacity without network development works but considering systematic inverter undersizing ($P_{inverter} = 0,7P_{peak}$)	184	27%	+26%
PV potential that may be connected to private LV grid in the case of buildings that have their own MV/LV substation	75	11%	-
2. PV hosting capacity without network development works but connecting LV plants to private MV/LV substations where possible	221	32%	+51%

4 CONCLUSION

First and foremost, the tool reaches its goal in making a preliminary diagnosis of the network's capacity to host PV and thus gaining knowledge on what is one of the main limiting factor of massive PV deployment (other main factors being load-bearing in the case of large surfaces and architectural constraints in the vicinity of historical monuments). Several uses of this diagnosis can be imagined. For example, a quick look allows to discard all roof-top that are at 250 meters or more from a substation as potential projects. This can help gain time in project development, for example in the case of cityzen groups canvassing the territory for adequate roof-tops on which to invest as third party.

Such a method applied nationwide would bring much added value to the current renewable electricity policies. Indeed, such results tend to show that 1) hosting capacity of the existing infrastructure is important (since only a few percent of the total PV potential is often substantial) but must be utilized with care, 2) on the long run, a deep reflexion in the development of further hosting capacity is compulsory in order to achieve ambitious shares of renewable electricity.

Current studies are been led with the national DSO Enedis in order to evaluate the tool's ability to generate a good approximation of network hosting capacity. The objective in this joint work is to develop a methodology to compute more precisely low-voltage network hosting capacity. This work pursues two objectives: 1) to orient better local authorities' renewable energy policies and development efforts, 2) to identify the organizational changes that must be put in place to transform current decision-making processes and tools used for grid investment planning.

8 REFERENCES

[1] Service de la donnée et des études statistiques (SDES), « Tableau de bord photovoltaïque : quatrième trimestre 2017, Évolution du parc national des installations de production électrique solaire photovoltaïque », Sta@t Info énergie, n°79 février 2018, http://www.statistiques.developpement-durable.gouv.fr/publicationweb/79

[2] J. Lassila, V. Tikka, J. Haapaniemi, M. Child, C. Breyer, J. Partanen Lappeenranta (University of Technology), "Nationwide Photovoltaic hosting capacity in the Finnish electricity distribution system", proceedings of the 32nd European Photovoltaic Solar Energy Conference, June 20 – 24, 2016, Munich, Germany

[3] Enedis, "Cartographie des réseaux aériens d'Enedis", https://www.enedis.fr/cartographie-des-reseaux-aeriens-denedis

[4] http://www.cityzen-smartcity.eu/

[5] L. Brard, J. Michelon, J. Vye (Valence-Romans Agglo), "Plan Climat Air Energie Territorial", 2018, Downloadable: http://www.valenceromansagglo.fr/fr/des-services-au-quotidien/developpement-durable/transition-energetique/notre-plan-climat.html

[6] M. Grauthoff, U. Janssen (Climate Alliance)
J. Fernandes (Lisboa E Nova), "Identification and mobilisation of solar potentials via local strategies", Polis project (project funded by Intelligent Energy Europe), 2012, downloadable: http://www.polis-solar.eu/IMG/pdf/polis_guidelines_en_final_web.pdf

[7] French legislation, "Arrêté du 23 avril 2008 relatif aux prescriptions techniques de conception et de fonctionnement pour le raccordement à un réseau public de distribution d'électricité en basse tension ou en moyenne tension d'une installation de production d'énergie électrique"

[8] Enedis, "Enedis-NOI-RES_46E Schémas de comptage V1", 2017

[9] Enedis, "Enedis-PRO-RES_43E-V4", 2017

[10] Sjef Cobben (Continuon), Bruno Gaiddon (Hespul), Hermann Laukamp (Fraunhofer ISE), "Impact of Photovoltaic generation on power quality in urban areas with high PV population", 2008.

[11] EEG 2012 - Erneuerbare-Energien-Gesetz

[12] VDE-AR-N 4105 Power generation systems connected to the low-voltage distribution network

ENERGY CLUSTER MODEL FOR THE HUNSRÜCK-HOCHWALD NATIONAL PARK REGION

Dominik Jung, Henrik te Heesen
Trier University of Applied Sciences, Environmental Campus Birkenfeld, Campusallee, 55768 Neubrücke (Nahe), Germany, h.teheesen@umwelt-campus.de

ABSTRACT: We analyze the energy production and energy demand of ten municipalities in the national park region Hunsrück-Hochwald in Southwest Germany. The energy cluster is characterized by a rural structure with 155,000 inhabitants and just a small number of commercial and industrial companies. The region cannot fulfil its energy demand and thus, we investigate the contribution of renewable energy capacity additions to the energy balance. This will help the region to reduce greenhouse gas emission and to get a zero-emission energy cluster in 2050. From today to 2050, three municipalities are even today an energy export subcluster and five municipalities show a turn-over from an import to an export subcluster. In total, the national park region is changing from a negative annual energy balance to a positive one by using more wind power as well as roof-top and ground-mounted PV systems. The results of the energy model of the national park region can be transferred to other energy cluster and will help these clusters to identify expansion pathways for the integration of renewable energy systems.

Keywords: Energy Rating, Grid Integration, Modelling, PV System

1 INTRODUCTION

An energy cluster represents a spatially delimited region that optimizes itself with the help of a smart grid and a virtual power plant via a control center. The aim is to maximize the supply from fluctuating renewable energies and finally to reduce greenhouse gas emissions. To this end, not only power generation plants but also various flexible consumers will be analyzed, while the continuous energy supply must be granted.

An energy cluster is not an autonomous system within the nationwide electricity supply. Rather, it pursues the goal of optimizing local and regional efficiency potentials in generation and consumption in an environment of increasingly decentralized power generation. Expansion scenarios can be analyzed to support local and municipal authorities to regulate the capacity addition of renewable energy systems. Also, an energy exchange with neighboring energy clusters can be considered.

Figure 1 shows the location of the national park region and the adjected energy cluster. The national park Hunsrück-Hochwald is marked in orange. In total, ten municipalities are combined to the national park energy cluster region, eight from Rhineland-Palatinate and two from Saarland. In total, there are 155,000 inhabitants and the region has an area of 150,00 ha. The region is a mainly rural area with just one city (Idar-Oberstein, 32,000 inhabitants) and nine municipalities which are characterized by small to mid-sized companies and no large or heavy industry.

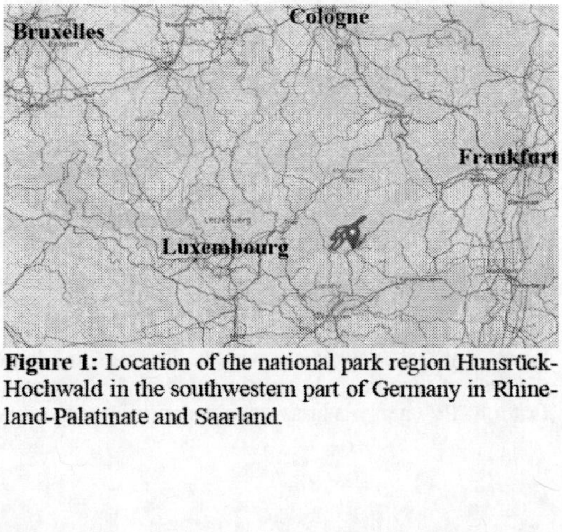

Figure 1: Location of the national park region Hunsrück-Hochwald in the southwestern part of Germany in Rhineland-Palatinate and Saarland.

We focus in the first step to the electric energy production and consumption of the national park region in 2015. We will analyze the development of the region by 2050 regarding the demographic change (reduction of population) and capacity additions of photovoltaic (PV) systems and wind turbines.

2 METHODS

We perform a simulation of energy production by wind turbines and PV systems as well as energy consumption on an hourly basis. The energy production of the wind turbines which are installed within the region is calculated by using the wind speed at 10 m height [1] and the technical parameter (capacity, turbine type, hub height, location) [2]. We use the horizontal insulation [1] and the technical data [2] for the energy production calculation of the PV systems. The insulation in module plane is derived by the isotropic atmospheric model [3]. As there are no information about the orientation and inclination of the PV modules, we take an inclination angle of 30° and an orientation distribution of 25 percent to the east (90°), 50 percent to the south (180°) and 25 percent to the west (270°). The installed capacity of bioenergy systems is also taken from Open Power System Database [2]. Finally, we model the energy consumption by scaling the default load profiles (residential H0, commercial G0, agricultural L0) of BDEW with the population and the annual energy demand in 2015 provided by the regional energy utility innogy SE.

3 RESULTS AND DISCUSSION

The installed capacity of renewable energy systems in the energy cluster Hunsrück-Hochwald is show in Figure 2. 200.5 MW wind energy capacity, 71.8 MWp PV capacity and 14.1 MW bioenergy capacity is installed in the national park region in 2015. Figure 3 shows that the largest total capacity can be found in Birkenfeld (83.9 MWp), the smallest one in Rhaunen (3.0 MW). The year 2015 is selected as the energy consumption data are only available for this year. The installed capacity per capita (PC) is between 0.14 kW/PC in the city Idar-Oberstein and 4.16 kW/PC in Thalfang. The mean installed capacity per capita in the national park region is 1.85 kW/PC.

We use the following parameter to derive the development of the population and the capacity additions listed in Table 1. We consider an increase of installed capacity of wind and PV and a reduction of the population due to the demographic change in the national park region.

Table 1: Parameter for prognosis of the development of the national park region.

Wind energy	2 percent of municipality area is used for wind energy with an area usage factor of 5 MW/ha
Photovoltaic	Annual capacity addition: 3%/a
Bio Energy	No chances
Population	Reduction of 0.2 percent per annum due to demographic change

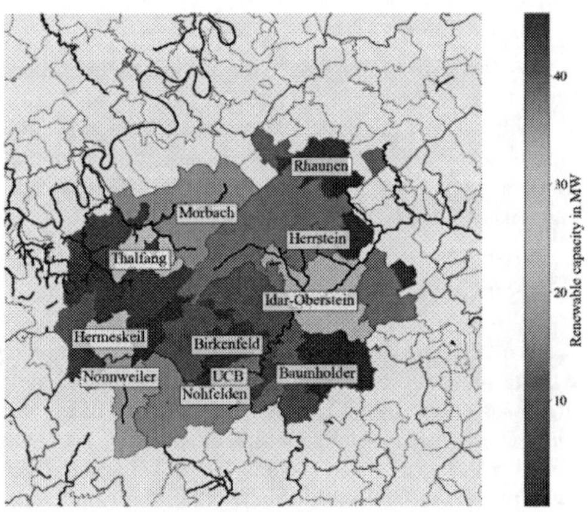

Figure 2: Capacity distribution of renewable energy systems (wind, PV, bioenergy) on the postcode level in the ten municipalities of the national park energy cluster Hunsrück-Hochwald in 2015.

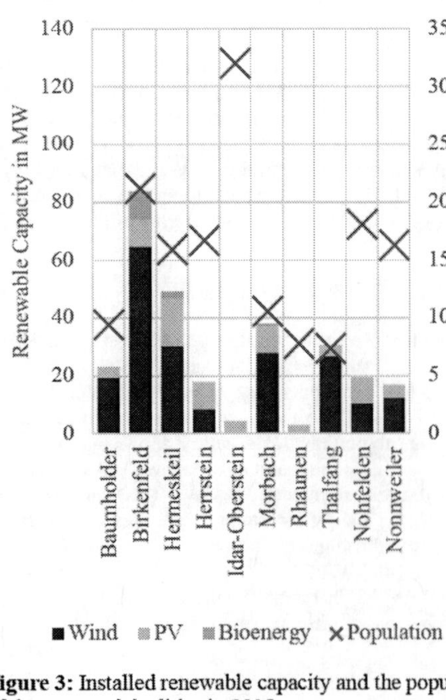

Figure 3: Installed renewable capacity and the population of the ten municipalities in 2015.

The results of the prognosis are shown in Figure 4. The population structure of the ten municipalities keeps unchanged, but the total population drops from 155,087 in 2015 to 144,586 in 2050. The capacity of wind energy rises to 599 MW and of PV to 202 MW. In 2050, 814.8 MW renewable energy capacity will be installed in the national park region under the given parameter.

The simulation of the energy production and consumption in each municipality on an hourly interval for 2015 and 2050 leads to an import demand of electric power in 2015 of -228,6 GWh/a whereas there is an overproduction and thus a possible export of electric energy to neighboring energy cluster of +556.0 GWh/a in 2050. The energy balance for each municipality can be seen in Figure 5. Just three municipalities show a positive energy balance in 2015 – Birkenfeld, Hermeskeil, and Thalfang due to installed bioenergy capacity (see green bars Figure 3) and an electric energy per capita ratio larger than 3 kW/PC. Baumholder, Herrstein, and Rhauen show nearly a balanced energy balance. The two municipalities in Saarland, Nohfelden and Nonnweiler, the municipality Morbach and the city Idar-Oberstein have a large negative energy balance. The installed renewable energy capacity in Nohfelden and Nonnweiler is not sufficient to fulfil the energy demand, the capacity per capita ratio in Idar-Oberstein is just 0.14 kW/PC and Morbach has a large commercial and industrial area. The three municipalities with export capacities cannot handle the energy demand of the seven municipalities, thus the total energy balance in 2015 is negative.

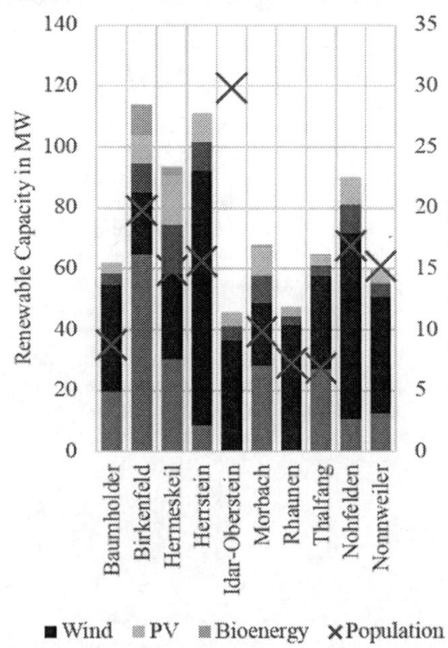

Figure 4: Installed renewable capacity and the population of the ten municipalities in 2050 based on data of Table 1.

The situation changes significantly in 2050. Eight of ten municipalities show a positive energy balance, just Morbach and Idar-Oberstein stay negative. In particular, the capacity additions in the wind sector leads to a large increase of the energy production. Although the growth factor of PV energy is larger than the one of wind energy

(2.5 to 2.1), there is smaller influence of the risen PV capacity to the energy balance compared to the risen wind capacity.

This observation can also be seen in Figure 6. The comparison of the monthly energy production and energy demand in 2015 and 2050 shows that there is a negative energy balance in all months during 2015. In contrast, the energy balance of the national park region in 2050 is positive due to typically larger wind speed in winter. This balance drops to negative values during summer as the wind turbine do not produce sufficient amount of electric energy and the PV systems in the municipalities cannot compensate this drop. The balance recovers to positive values in the beginning of October. The effect of the demographic change can also be identified as the energy demand decreases from 2015 to 2050 by 7.1 percent.

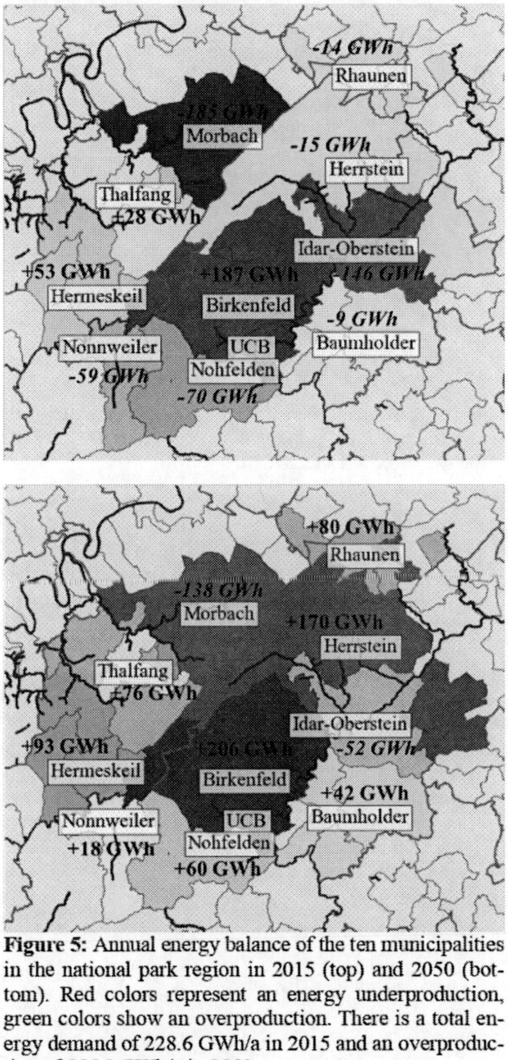

Figure 5: Annual energy balance of the ten municipalities in the national park region in 2015 (top) and 2050 (bottom). Red colors represent an energy underproduction, green colors show an overproduction. There is a total energy demand of 228.6 GWh/a in 2015 and an overproduction of 556.0 GWh/a in 2050.

Taking the hourly energy balance values in 2015 and 2050 into account leads to Figure 7. The assorted annual load line of the electric energy balance shows that the capacity addition of wind turbine and PV systems in the national park region shifts the annual load line to a positive import-export-balance. While the energy balance is negative for more than 6,000 hours in 2015, the balance threshold changes that there are just 2,300 hours with a negative energy balance in 2050. One possible way to reduce the negative load even in 2050 is to use storage systems.

The next step will be to consider electromobility and the influence to the electric energy demand within the national park region. Also, we will analyze the transfer of energy within the region as well as the exchange with neighboring energy cluster. Finally, we will model different expansion scenarios for wind turbines, PV systems and bioenergy to identify rules for the best expansion pathway.

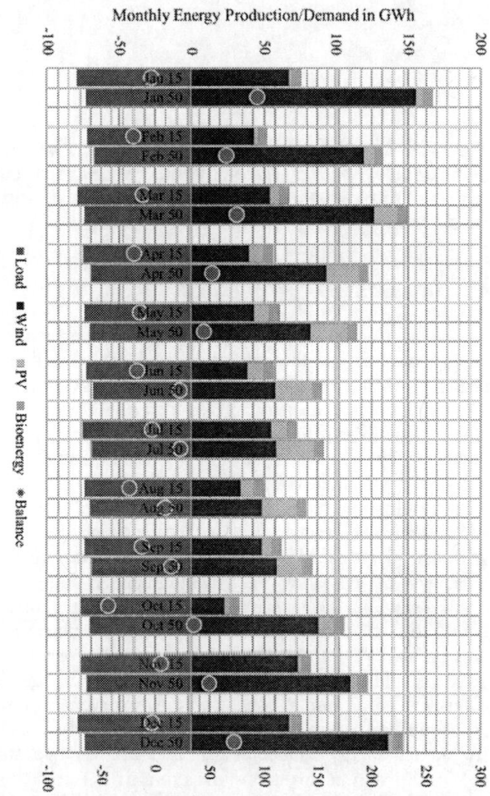

Figure 6: Monthly energy production of the renewable energy systems, the energy demand and the balance each month in the national park region in 2015 (upper bars) and 2050 (lower bars). A positive balance can be seen in winter month. The energy balance is negative in 2015 and 2050 during summer due to smaller wind speed and thus small wind energy production.

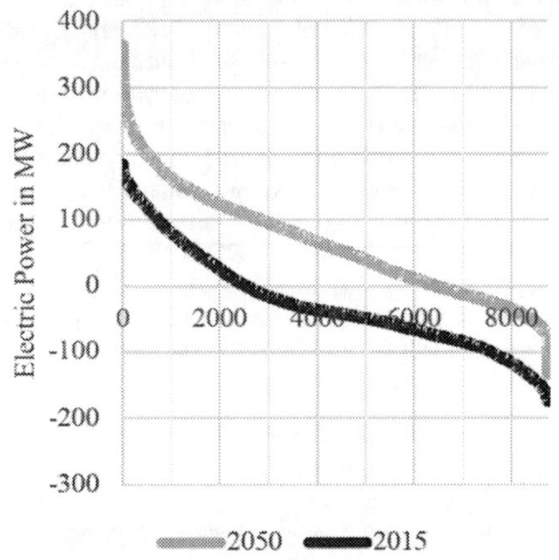

Figure 7: Assorted annual load line of electricity balance in 2015 (blue) and 2050 (green). Positive values represent an energy surplus within the national park region, negative values show an energy undercover.

4 CONCLUSION

We have analyzed the energy production by renewable energy systems and the energy demand for the national park region Hunsrück-Hochwald in Southwest Germany. The energy cluster is characterized by a rural structure of ten municipalities with a population of 155,000 inhabitants and a low-level industrial density. We have taken the actual energy demand based on measurements of the local energy utility and the installed capacity of wind turbines, PV systems and bioenergy plants into account. Also, we have considered an addition of wind and PV capacity and a decrease of population due to demographic change.

We simulate the energy production and demand on an hourly interval, compare the electric energy balance of the national park region in 2015 and make an extrapolation to 2050 to decide, if the region is able to increase their energy autarky ratio. Figure 8 concludes the two scenarios for 2015 and 2050. Three municipalities have already a positive annual energy balance in 2015 and increase their rate in 2050. Five municipalities show a turn-over from an energy import subcluster to an energy export one. Two municipalities still depend on electric energy import although a reduction of import dependency can be observed.

Our model shows that a rural energy cluster can significantly change its energy usage structure by using renewable energy systems. Next, we will investigate the influence of electromobility and an electric energy exchange within the region.

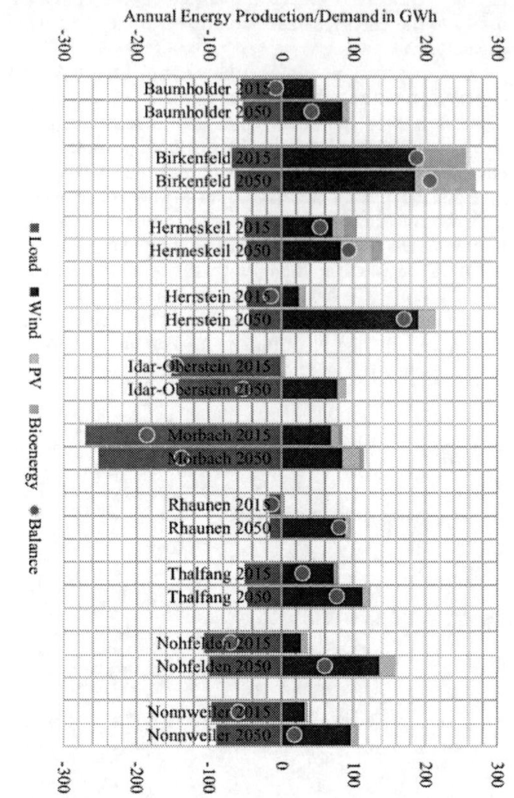

Figure 8: Annual energy production and demand in the ten municipalities in the national park region Hunsrück-Hochwald for 2015 and 2050.

5 REFERENCES

[1] Deutscher Wetterdienst. Hourly wind climate data. ftp://ftpcdc.dwd.de/pub/CDC/observations_germany/climate/hourly/

[2] Open Power System Data. OPSD Data Package: Renewable Power Plants in Germany. Package version 2018-03-08. https://open-power-system-data.org

[3] Loutzenhiser, P.G. and Manz, H. and Felsmann, C. and Strachan, P.A. and Frank, T. and Maxwell, G.M. Empirical validation of models to compute solar irradiance on inclined surfaces for building energy simulation. Solar Energy 81. pp. 254-267. 2007.

PERFORMANCE OF IN-HOUSE LI-ION BATTERY STORAGE SYSTEM BASED ON VARIOUS STRATEGIES

Bhawna Verma, Nina Munzke, and Marc Hiller
Karlsruhe Institute of Technology
Hermann-von-Helmholtz-Platz 1, 76344 Eggenstein-Leopoldshafen, Germany

ABSTRACT: Energy storage systems with Lithium-ion batteries coupled to solar photovoltaic arrays will become a vital pillar in the future energy landscape. An important question is to maximize the self-supply ratio as well as the battery lifetime at. Out of the various critically important aspects affecting the performance and as a consequence the commercial viability, one is the system control software and its ability to predict both the energy needs of the customer as well as the energy supplied by the PV-array, throughout the day. At KIT, an intelligent control software was developed that incorporates online and offline PV forecast to predict the battery charging regime. These different forecasts methods control a 32 kWh AC-coupled Li-ion (NMC) battery energy storage system, using standard household load curves as well as one of the arrays (17.5 kW) of the 1 MW solar-storage facility as PV generation input. The battery is charged in a way that it is fully charged just a short period of time before sunset. The quantification of the level of intelligence of the storage system control software developed at KIT is done based on the following criteria: (a) the software performance based on the predictive module used (online or offline PV-prognosis), (b) the success of delayed charging in reducing the time spent at high state of charge (SOC) levels and (c) whether prediction errors lead to the battery not being fully charged so that the user's self-sufficiency is unnecessarily reduced.
Keywords: Battery Storage and Control, PV System, Strategy, Lifetime, Storage

1 INTRODUCTION

The global shift away from the non-renewable, dispatchable power generation towards the renewable, non-dispatchable sources such as solar and wind has led to an increase in the variability of power supply. This leads to a requisite of cost-effective and efficient energy storage. In the household context, the economics of storage systems depend to a large extent on system lifetime and performance. Li-ion batteries are the current market leader [1] for short-term (minutes to hours) storage applications, and are typically installed in combination with photovoltaic (PV) arrays in household or industry applications. There are several critically important aspects affecting the performance and as a consequence the commercial viability such as system design, efficiency and dimensioning along with control strategy and the battery lifetime. As mentioned one of them is the system control software, which should be regulated in a way as to maximize the self-supply ratio. The underlying algorithms must also ensure the longest possible operational lifespan of the battery. Therefore, a vital component of the software is its ability to predict both the energy needs of the customer as well as the energy supplied by the PV-array, throughout the day. In essence one needs to be able to predict the load as well as the available power, to deduce how the system should best be regulated in order to increase profitability.

One concrete example of this is that excess energy especially on sunny days should not be stored in the battery right at the start of the day, since Li-ion batteries with certain cell chemistries such as NMC, LFP and NCA age faster when their state of charge (SOC) is high [2].

There are several ways to predict the solar irradiation and resulting PV power (see for example the reviews in Refs. [3], [4]); the different methods can be divided into those that use exogenous inputs ("online" forecasts [5]) and those that use only data collected on site ("offline" forecasts [6]).

Calendar ageing of the batteries on system level is so far only discussed in a few publications. Li and Danzer [7] and Angenendt et al. [8] examine calendar ageing of batteries during operation by means of simulations; while Munzke et al. [9] does an evaluation of the influence of the control strategy on calendar ageing of NMC based batteries of exciting commercial systems.

During the concurrent studies of commercially available storage systems, it has been observed that only around 25 % of storage systems on the market are capable of intelligent battery management [9] .

The unique combination of solar PV and storage facilities at KIT allows for in-depth studies of system control, both on the planning (forecast models and predictive analytics) and operational (control algorithms) level. The goal is to examine the effects of intelligent control software more closely.

At Karlsruhe Institute of Technology (KIT) the control strategies using online or offline PV-prognosis are developed and tested on an AC-coupled PV-storage system. In the paper these strategies are briefly presented, compared and evaluated. The next section explains the methodology and system description in greater details.

2 METHODS

2.1 System description

The schematic of the AC-coupled storage system at KIT is represented in Figure 1. The AC-coupled storage system comprises a 32 kWh battery, a 30 kW battery inverter, a 17.5 kWp PV-plant, and a control unit.

The topology of the battery, having NMC based Li-ion cells is known. It consist of that 2 modules of 168 cells each (connected in series) which are connected in parallel. The size of the battery as well as the power electronics has an important impact on the charging strategy throughout the day. For power above 15 kW, the overall round-trip efficiency is 80 %. This includes the losses from battery and inverter. The battery is charged and discharged using the control software developed at KIT. The load data (3 family household) is taken from the summer week of households 2, 8 and 17 from the ADRES data [10]. The next section explains the software description in detail.

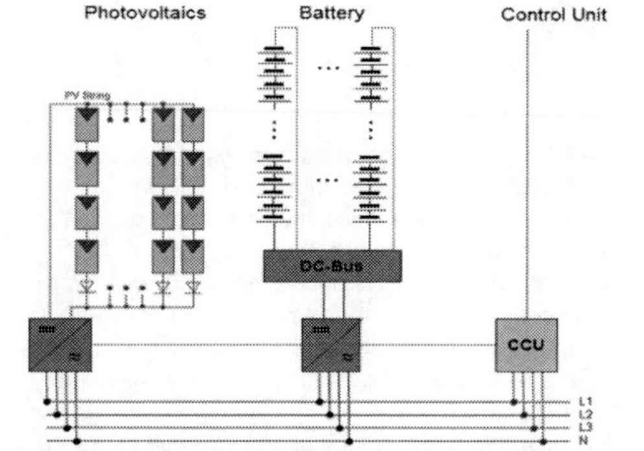

Figure 1: Schematic diagram of the AC-coupled storage system at KIT, Germany [5]

2.2 Control software based on a predictive module

The system control software allows monitoring and control independently of one another. The control algorithm can run at a maximum frequency of 2 Hz, which is the highest sampling rate of the sensors. This means that every 500 ms values such as PV power, grid power, battery power and SOC are measured and used as input values for the control loop. As mentioned the storage system should be controlled in a way as to simultaneously maximize the self-sufficiency of the system operator as well as the battery lifetime. Since NMC-based Li-ion batteries age faster when their SOC is high [2], thus, for a longer lifetime, it should be avoided that the battery is kept at a high SOC for extended periods of time. For peak shaving applications the time spent at high SOC level represents the time when the battery is completely charged before it is discharged in the evening. This means that one should decide in advance when and at which power the battery should be charged in order to simultaneously satisfy the load and guarantee a long battery life. The KIT algorithm, works on the principle that charging starts at sunrise. The strategy used is based on slow charging so that longer times spent at high SOC can be avoided. The charging power of the battery is calculated based on an online or offline PV-prognosis. Therefore, a prediction of PV power and load is needed. For the AC system an online (with web connection) as well as an offline PV-prediction (without web connection) was developed. The charging power in both charging methodologies is recalibrated every hour based on the online or offline forecast as explained below.

In the offline prognosis, the energy generated the day before is used to scale the correction factor as explained by James et al. in [11]. The correction factor represents the value that scales the battery charging power. It is calculated according to the available excess energy for the rest of the day until sunset. A prediction is made based on the simple assumption that the weather of the present day is more or less the same as the day before. The first prediction is made at midnight for the upcoming day with a 5-minute resolution as represented in Equation 1.

$$\hat{P}_{PV}^{offline}(t) = \frac{\int_{yesterday} P_{PV}(t)dt}{\int_{today} \hat{P}_{PV}^{CSC}(t)dt} \hat{P}_{PV}^{CSC}(t) \qquad (1)$$

where, $t \in [00:00, 24:00]$ represents the time from midnight to midnight the next day of the day and $\hat{P}_{PV}^{CSC}$ is the predicted clear sky curve. This prediction is updated every hour throughout the day using the generated PV data obtained on hourly basis. The resolution of the prediction data obtained from the prognosis has an hourly resolution. Therefore, the correction factor is recalculated on an hourly basis. The recalibrated version of the forecast algorithm employed in the storage system is explained in greater detail in [11].

The online prediction is based on the global forecast system (GFS) which in-turn is based on a numerical weather prediction (NWP) model. The online PV prognosis (with an hourly resolution) is obtained every 6 hours, which is then used to scale the correction factor on an hourly basis.

Errors in the prediction can result in unsuccessful battery charging so that the battery is either completely charged too early or not fully charged until sunset.

In the current research, the software performance is studied for the online PV-prognosis, the offline PV-prognosis as well as for the cases where no prediction was done. The results are explained qualitatively and quantitatively in the result section of this paper along with the comparisons of all three control strategies that are under consideration.

2.3 Evaluation of the control strategies

The quantification of the level of intelligence of the storage system control software is done based on following criteria: (a) the software performance based on the predictive module used (online or offline PV-prognosis) that is updated regularly, (b) the success of delayed charging in reducing the time spent at high SOC levels and (c) whether prediction errors lead to the battery not being fully charged so that the user's self-sufficiency is unnecessarily reduced.

During operation, the recording of the measurements is done for several parameters such as voltage, current, SOC, power during charging as well as discharging of the battery. In order to evaluate the time spent at higher SOC, a methodology proposed by Munzke et al. [12] in is used. This requires the ohmic resistance of the battery, which is measured at several SOC levels of 25 %, 50 %, 75 % and 90 % (as represented in Figure 2), respectively, by using a charge pulse for 20 s. Figure 2represents the recorded values of current and voltage (charging from 07:00 am to 01:30 pm and discharging from 01:30 pm to 04:30 pm). The resistance measurement is done using a power of 5 kW, 7.5 kW and 10 kW, since this is the charging and discharging power during a regular day. Figure 3 represents the resistances obtained for the power levels under consideration.

$$U_{BAT,OCV} = U_{BAT,act} + I_{BAT,act} + RO_{BAT} \qquad (2)$$

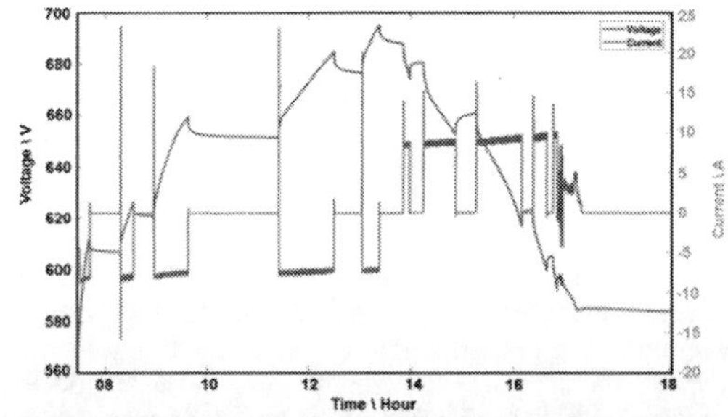

Figure 2: Representation of the variation in current and voltage when the battery is being charged and discharged with 10 kW on a reference day.

The obtained resistance values are then used to estimate the open circuit voltage (OCV) using Equation 2. This OCV can be used to calculate how much time the battery spends at each voltage interval during operation. This method is used to evaluate the success of delayed charging and is discussed in greater detail by Munzke et al. in [12].

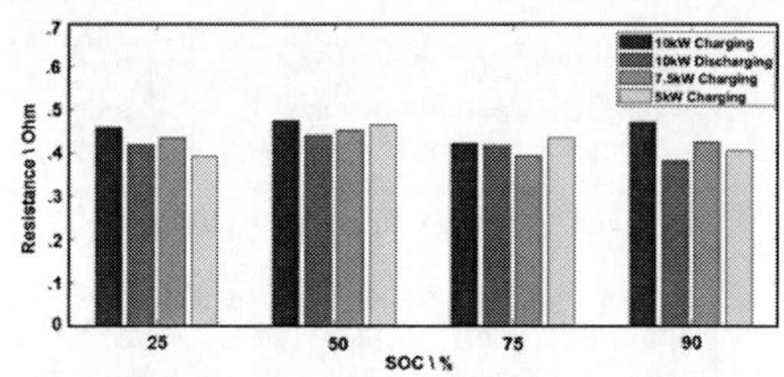

Figure 3: Representation of resistances estimated at various SOC levels and a power of 10 kW (charging and discharging), 7.5 kW (charging) and 5 kW (charging), respectively.

Since the module topology of the battery system under consideration is known as explained in section 2.1, the ohmic resistance per cell can be calculated. The mean ohmic resistances of the different SOC levels are 0.42 Ohm, 0.42 Ohm and 0.43 Ohm (for charging power of 5 kW, 7.5 kW and 10 kW respectively). The estimated values are just a bit higher than the ohmic resistance of pouch cells determined using the single cell measurements [12]. These measurements are considered valid because the BMS is a part of the battery module under consideration which results in an increase in the ohmic resistance.

If prediction errors occur the battery will end up being fully charged too early so that it spends a long period of time completely charged or it will not be full at the end of the day. Due to the iterative nature of the algorithm the battery charging schedule is iterated every hour, which should minimize the effect of an incorrect PV-forecast.

The error estimation as well as the estimation of the success in delayed charging is done for the control strategies namely: online prediction, offline prediction and without prediction. They are further explained in the result section.

3 RESULTS

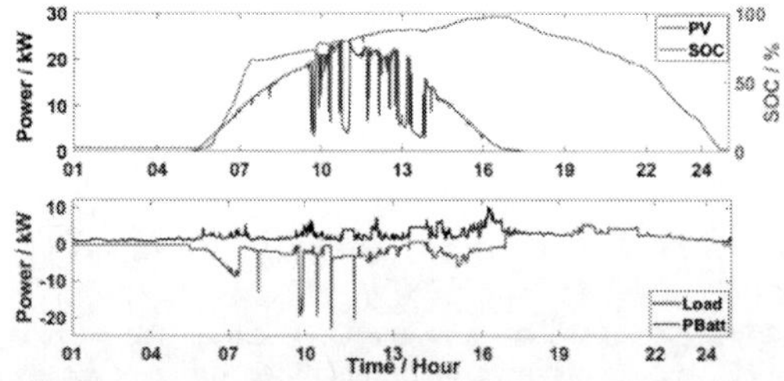

Figure 4: Example of offline persistence forecast with PV data from the15th of September 2018.

Figure 4 shows the results of a measurement with PV data from the 15th of September 2018. In the case shown, the system was controlled using the offline PV persistence forecast. The graph shows the PV power (blue), the load (black), the battery power (green) as well as the SOC of the battery. While negative battery power means charging positive values means discharging. It can be clearly seen that instead of being full early or too late, charging takes place so that the battery is just full around sunset (around 05:00 pm).

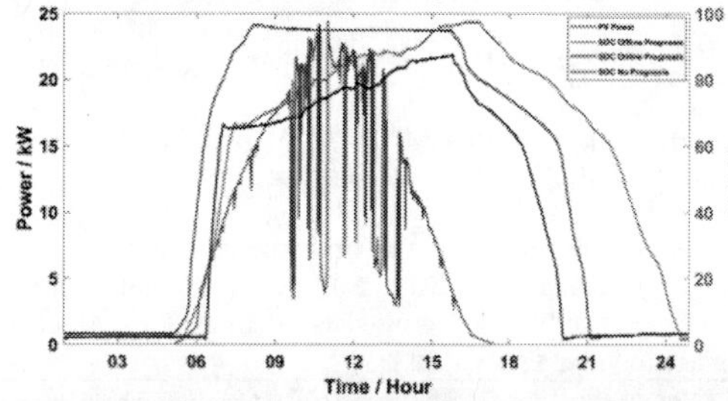

Figure 5: Comparison of the SOC of the three different control strategies (offline persistence forecast, online forecast and without any prediction) with PV data from the 15th of September 2018.

Figure 5 shows a comparison of the SOC for the three different control strategies (offline persistence forecast, online forecast and without any prediction). In all three cases the PV data stems from the15th of September 2018 (blue curve). The different SOC patterns obtained are represented by the orange (offline PV prognosis), the black (online PV prognosis) and the green curve (without any prognosis). It can be seen that during daytime (05:30 am to 05:00 pm) charging of the battery takes place according to the different control strategies while after 05:00 pm the battery is discharged according to the load of the three family household. As explained in the methodology section 2.2, hourly updates of the correction factor are done which consequently influence the charging power of the battery and therefore the battery is charged much slower when one of the predictions are used. It can clearly be seen that the battery is already full after two to three hours after sunrise when no prediction is used. In the example shown in Figure 5 the offline prognosis shows a much better performance than the online one. While the battery is just full at the end of the day, using the offline PV forecast the battery is not completely charged when the online PV forecast is used.

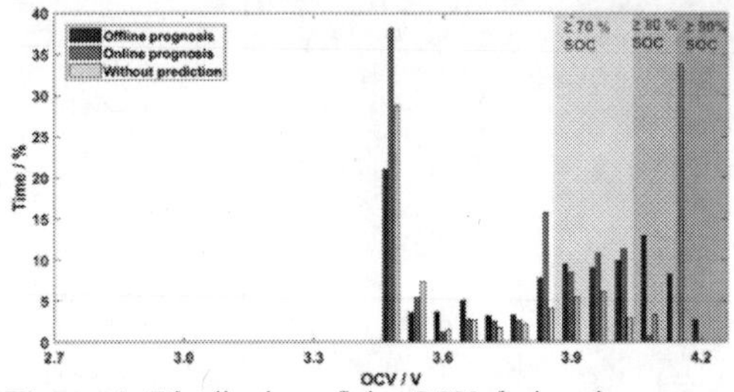

Figure 6: Distribution of the OCV during the curse of one day. A comparison of the three different control strategies is shown for the PV data of the 15th of September 2018. The area marked represents the time spend at SOC greater than 70 %, 80 % and 90 % respectively (this data corresponds to Table 1).

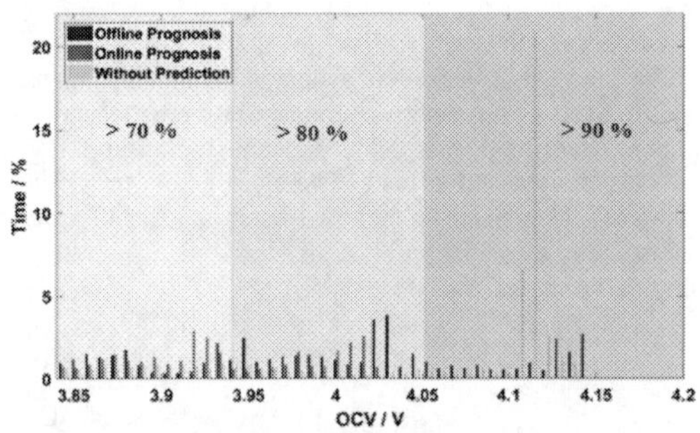

Figure 7: Distribution of the OCV with respect to the time spent at SOC levels greater than 70 %.

Table 1: Percentage of the time spent at various SOC levels. Comparison of the three different control strategies (online PV-prognosis, offline PV-prognosis and without any prognosis).

Strategy	>70 % SOC	>80 % SOC	>90 % SOC
Offline	52.30 %	37.41 %	14.15 %
Online	31.44 %	13.30 %	-
without prediction	51.63 %	41.45 %	35.56 %

To evaluate the time spend at SOC levels above 70 %, 80 % or even 90 % the measured resistances (for different power levels and at specified SOC levels) can be used to calculate the OCV and therefore the SOC during the measurement. Figure 6 shows the distribution of the OCV during the curse of the day and compares the results of the three control strategies under consideration. In addition to that Figure 7 shows a more detailed distribution of the OCV between 3.84 V and 4.2 V. As the module topology of the battery is known, as explained in the section 2.1 the results are shown per cell. The voltage for NMC cells normally lays between 2.7 V and 4.2 V. The obtained voltages during operation vary between 3.45 V to 4.2 V. For NMC an OCV of 3.84 V represents a SOC of 70 %, 3.945 V a SOC of 80 % and 4.05 V a SOC of 90 %.

Table 1 shows the results of the percentage of time which the storage systems spent at a voltage higher than 3.84 V (SOC > 70 %), 3.945 V (SOC > 80 %) and 4.05 V (SOC > 90 %) calculated by using the determined OCV. Figure 6 corresponds to Table 1. By comparing the results of the different control strategies it can be observed that the battery is fully charged during that day using the offline prognosis (97.5 %,) or no prognosis at all (96.5 %) while it is only charged up to 87.5 % SOC using the offline prognosis.

When no prognosis is used the battery is already fully charged at around 08:00 am whereas it only reaches a SOC higher than 90 % in the afternoon when it is charged using the offline prognosis. As explained in section 2.2 the charging power and the charging time depend on the correction factors and the available PV power. Without prognosis the battery ends up spending 41.45 % of time at a SOC higher than 80 % and 35.56 % at a SOC higher than 90 % as represented in Figure 7.

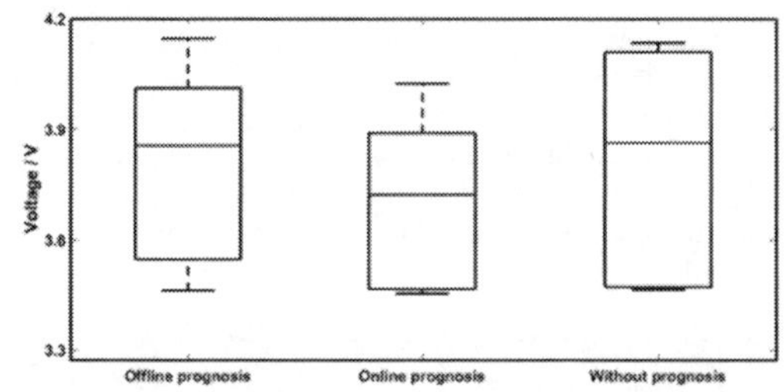

Figure 8: Distribution of the OCV during the measurement for the respective charging strategies in form of a bock plot. The red line represents the median value of the OCV obtained for respective strategy.

Figure 8 represents the distribution of the OCV (in the form of boxplot) of the three different control strategies. The median values of the OCV vary between 3.7 V and 3.87 V depending on the operational strategy. While the OCV during the day varies between 3.55 V and 4.1 V for the offline PV-prognosis, it varies between 3.5 V to 3.88 V for the online PV-prognosis and between 3.5 V to 4.15 V for the charging strategy without any prognosis. The higher the percentage of time spent at high SOC the higher is the OCV median value. Thus it can be concluded that a charging strategy, including a prediction (online or offline), that is updated on regular basis and can therefore alter the charging power of the battery can highly reduce the time spent at high SOC.

4 DISCUSSION AND CONCLUSION

In the current paper a simple method to determine the OCV during operation, which is proposed by Munzke et al. [12] is used and further developed. The method only uses the current and voltage of the battery as well as its ohmic resistance. In comparison to Munzke et al. [12] the ohmic resistance is measured for different power levels and is therefore considered as more accurate. The battery within the storage system used has NMC based cells. The calculated OCV can therefore be used to determine the SOC [9].

Furthermore a comparison of the three control strategies is done. A strategy without any prediction is compared to a strategy with a prediction using either an offline PV-forecast or an online PV-forecast. As shown an intelligent control strategy can reduce the time spent at high SOC. This obviously depends on the configuration of the PV-storage system, including the size of the PV-plant, the power electronics as well as the battery capacity and the time of the year. To show that a control strategy using a PV-forecast can reduce the time spent at high SOC a summer day was chosen for the measurements.

Keil et al. [2] presented that batteries with NMC based cells age faster when their SOC is high. As an intelligent control strategy can reduce the time spent at high SOC it will therefor also reduce calendar aging.

The strategy using the offline PV-forecast reduces the time spent at a SOC higher than 80 % by 4.04 pp and at a SOC higher than 90 % by 21.41 pp in comparison to a strategy without any prediction. These values are valid for a sunny summer day. To extrapolate results for a whole year, how it was done in Munzke et al. [12] further days especially transition and winter day need to be tested as well. By Munzke et al. [12] it was observed that a difference in battery lifetime caused by calendar ageing of up to 1.5 years is possible.

For the day tested, the strategy using the online PV-forecast did not work as it should have, so that the battery was not fully charged at sunset. Even if the time at high SOC was even more reduced, less energy is stored within the battery and can therefore be used during the night if needed. The SOC only reached 87.5 % while it reaches 96.5 % to 97.5 % when one of the other two strategies is used. This can for example happen if one overestimates PV generation and consequently battery charging is reduced too much, which. can lead to a reduction in self-sufficiency of the system operator. If more electricity is needed at night than stored within the battery the system operator will be forced to buy electricity at night. Therefore further improvements of the ganging strategy using an online PV-forecast need to be done.

Furthermore measurement for all different strategies over a longer period of time will be conducted.

ACKNOWLEDGMENT

The authors thank Dr. James Barry (Heidelberg University) for sharing his knowledge and work towards this study.

REFERENCES

[1] D. Sauer, „Lithium-ion batteries become the benchmark for stationary applications – Markets, players, prices, technology“, 15-März-2016.

[2] P. Keil *u. a.*, „Calendar Aging of Lithium-Ion Batteries I. Impact of the Graphite Anode on Capacity Fade“, *J. Electrochem. Soc.*, Bd. 163, Nr. 9, S. A1872–A1880, Jan. 2016.

[3] J. Antonanzas, N. Osorio, R. Escobar, R. Urraca, F. J. Martinez-de-Pison, und F. Antonanzas-Torres, „Review of photovoltaic power forecasting“, *Sol. Energy*, Bd. 136, S. 78–111, Okt. 2016.

[4] R. H. Inman, H. T. C. Pedro, und C. F. M. Coimbra, „Solar forecasting methods for renewable energy integration“, *Prog. Energy Combust. Sci.*, Bd. 39, Nr. 6, S. 535–576, Dez. 2013.

[5] E. Lorenz, J. Kühnert, D. Heinemann, K. P. Nielsen, J. Remund, und S. C. Müller, „Comparison of global horizontal irradiance forecasts based on numerical weather prediction models with different spatio-temporal resolutions“, *Prog. Photovolt. Res. Appl.*, Bd. 24, Nr. 12, S. 1626–1640, Dez. 2016.

[6] H. T. C. Pedro und C. F. M. Coimbra, „Assessment of forecasting techniques for solar power production with no exogenous inputs“, *Sol. Energy*, Bd. 86, Nr. 7, S. 2017–2028, Juli 2012.

[7] J. Li und M. A. Danzer, „Optimal charge control strategies for stationary photovoltaic battery systems“, *J. Power Sources*, Bd. 258, S. 365–373, Juli 2014.

[8] G. Angenendt, S. Zurmühlen, R. Mir-Montazeri, D. Magnor, und D. U. Sauer, „Enhancing Battery Lifetime in PV Battery Home Storage System Using Forecast Based Operating Strategies“, *Energy Procedia*, Bd. 99, S. 80–88, Nov. 2016.

[9] N. Munzke, B. Schwarz, und J. Barry, „The Impact of Control Strategies on the Performance and Profitability of Li-Ion Home Storage Systems“, *Energy Procedia*, Bd. 135, Nr. Supplement C, S. 472–481, Okt. 2017.

[10] A. Einfalt *u. a.*, „EA : ADRES-Concept – Konzeptentwicklung für ADRES - Autonome Dezentrale Regenerative Energie Systeme“, Wien, Aug. 2012.

[11] J. Barry und J. Thomas, „Online and Offline PV Power Forecasts for Optimal Control of Storage Systems“, *33rd Eur. Photovolt. Sol. Energy Conf. Classhighlight SpanExhibition*, S. 2729–2732, Nov. 2017.

[12] N. Munzke und B. Schwarz, „Intelligent control of household Li-ion battery storage systems“, gehalten auf der 12th International Renewable Energy Storage Conference, IRES, Düsseldorf, 2018.

Residential battery storage sizing based on daily PV production and consumption load profile characterization

Stavros Afxentis*, Michalis Florides, Spyros Theocharides, Venizelos Venizelou and George E. Georghiou
Photovoltaic Technology Laboratory, FOSS Research Centre for Sustainable Energy,
Department of Electrical and Computer Engineering, University of Cyprus (UCY)
Panepistimiou 1 Avenue, P.O. Box 20537, 1678, Nicosia, Cyprus

*corresponding author Stavros Afxentis (phone: +357 22 894396, Fax +357 22 895370, safxen01@ucy.ac.cy), Michalis Florides (florides.michalis@ucy.ac.cy), Spyros Theocharides (theocharidis.spyros@ucy.ac.cy), Venizelos Venizelou (venizelou.venizelos@ucy.ac.cy), George E. Georghiou (geg@ucy.ac.cy)

***Abstract*—The integration of battery energy storage systems (BESS) to existing grid-connected residential PV systems can reduce issues stemming from the increased PV penetration and at the same time improve the performance of the existing system. In this paper, a methodology for determining the optimal sizing of BESS is presented based on a clustering method by considering energy profiles of residential prosumers in Cyprus. As such, the daily energy required for residential premises with existing 3 kWp photovoltaic system in Cyprus is determined in three primary clusters. The clustering procedure is carried out based on the daily import electricity profiles recorded for the prosumers over the period of one year which revealed the daily energy needs of each cluster targeting to maximize self-consumption. Additionally, the battery capacity level of each cluster is compared with the daily export electricity of each prosumer within the same cluster to reveal the amount of users with enough surplus electricity to meet the capacity requirement of each cluster. Finally, a power rate dimensioning of the BESS power converter is performed by analyzing the import power measurements over the one year test period.**

Keywords*— *Photovoltaics, Battery Energy Storage Systems, Clustering, K-means.

I. Introduction

Environmental concerns, energy security and the fossil fuel resources have driven towards the integration of renewable energy sources, and more specifically solar photovoltaic technology, into the modern power grid. Despite the intermittent nature of photovoltaics (PV) and the initial high capital cost, in recent years governments around the world have encouraged the integration of PV systems into the electricity network either as residential-scale (rooftop) systems or large-scale (PV parks) systems. The financial incentives and governmental subsidies in conjunction with the recent drop in the cost of PV modules and the increasing electricity prices, have driven to the dramatic uptake of residential PV in some countries [1], [2].

Taking Cyprus as an example, photovoltaic technology has seen a remarkable growth during the last few years, with the installed PV capacity almost doubling in consecutive years, as it can been seen in Fig. 1. Grid parity conditions combined with the implementation of favourable policies such as the net-metering, have contributed to this trend of increasing PV system installations on the island to promote PV penetration towards achieving the 2020 national energy targets.

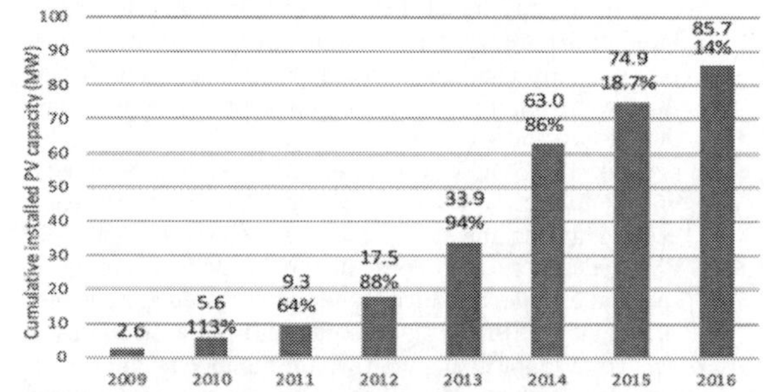

Fig. 1. Cumulative installed PV capacity (power) in Cyprus for the years 2009 till the end of 2016.

Considering the isolated electricity network of Cyprus along with the steadily increasing PV penetration in the energy mix of the island, adverse consequences can impact the performance of the low-voltage (LV) distribution network. For instance, voltage rise can be noticeable in areas with dense rooftop PV systems connected to the grid [3]–[8]. At the same time, voltage deviation occurs mainly from the intermittent generation of PV systems [5], [9], [10] and is primarily affected by the sky clearness conditions, thus compromising the power quality and stability of the grid [3], [11].

Several solutions exist for balancing grid voltage level; one of them is to limit the PV generation that is injected to the grid in order to preserve voltage fluctuations within an acceptable range [12], [13]. However, such an approach is not optimal without incorporating financial incentives since spilling PV generation leads to loss of revenue for the prosumer and also increases the overall integration cost of PV systems. Other solutions include the adoption of energy management strategies to control the flow of PV generation in the LV grid. An example is the use of direct and price-responsive load management by facilitating Advanced Metering Infrastructure (AMI). This approach requires end-to-end communication between the smart utility meter and the utility company [14], [15] and also the utilization of residential Battery Energy Storage Systems (BESS) for storing or delivering electricity to the low voltage (LV) grid. Towards this direction, the utility can respond to a price-responsive load control mechanism in order to push prosumers to schedule battery operation based on time-varying tariffs. For instance, the network peak demand can be reduced by allowing prosumers to interact with the grid based on the electricity cost or by reducing the power flow from the grid when premise demand is higher than the PV production by discharging the local BESS system. Therefore, a battery scheduling mechanism in line with proper battery sizing are important parameters towards achieving a sustainable environment for the residential use of BESS.

Relating the above with the current situation in Cyprus, BESS technology is still at its infancy and technological improvement is needed in order to lift off the barriers before it becomes a reliable and affordable solution. Towards this end, proper battery sizing is required to reduce the aggregated battery cost to the end-user. The objective of this paper is to study the energy profile of residential prosumers with PV systems in Cyprus and reveal the daily energy levels imported from the LV grid. This reflects the additional energy that is required to balance grid demand. Finally, since self-consumed electricity fluctuates between users and week days, the evaluation of surplus PV electricity to achieve the same import energy levels will also be examined.

II. METHODOLOGY

A. Residential Pilots in Cyprus

Sixty-five prosumers in Cyprus have been selected, in the prospect of sizing a BESS unit ideally coupled with an existing 3 kWp PV system. All participating prosumers were geographically spread throughout the island to cover different socio-geographical situations. Two smart metering (SM) devices were installed as depicted in Fig. 2 to acquire energy profiles for each prosumer. In particular, one SM device is placed on the AC side of the PV power converter (inverter) in order to measure the PV generation and the second one is placed on the grid side to measure the energy exchange with the grid (i.e. import and export electricity). The SM device used is the Elster A230, a single phase bidirectional meter which operates as a data logging equipment transmitting 30-minute average power values. The datasets are being post-processed and the premise power and energy exchange is calculated based on the registered power (30-minute interval), before they are stored to a central database platform.

B. Load Characterization

Towards achieving a 100% renewable energy system, several parameters that affect the current electricity network should be taken into consideration. Particular emphasis is given on the energy demand of households, since by achieving dynamic user response could efficiently alleviate issues concerning the gradual increase of PV deployment to the grid. At the same time, electricity consumption depends on the daily habits and societal level [16] of the individuals, and comprises an important step towards achieving resilience to the existing energy system. Towards this direction, the process of collecting energy datasets from the 65 participating prosumers begun in 2015 for the period of one year (reference year). Grid

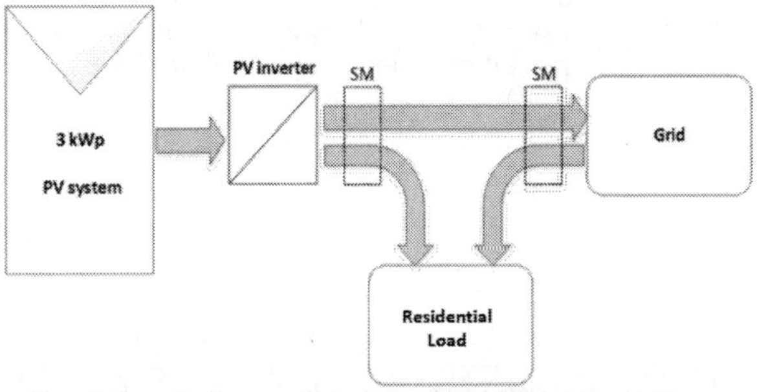

Fig. 2 Schematic diagram of a typical residential PV system in Cyprus.

exchange electricity such as import and export electricity along with PV production data were collected from the installed SMs and were used for the characterization process described next.

To start with, only the imported power samples were taken into consideration as they provide a significant indication of the additional energy that is supplied from the grid. This is the main objective towards achieving 100% self-sufficiency rate. Having this in mind, the cumulative daily import electricity values (in kWh) were extracted for the entire test group throughout the year and used for the grouping process. Then, the *k*-means technique was applied to perform the characterization between different import profiles and obtain an initial knowledge about the interaction of the prosumers with the grid. In particular, the *k*-means method uses fixed input dimensions, in this case the import electricity time series of each user, to separate them based on the input data-series. The algorithm applies an autocorrelation-based procedure to compute a dissimilarity vector between the time series as a function of the weighted Euclidian distance expressed in Eq. 1 below [17], [18].

$$d_{ACF} = \sqrt{\left(\hat{\rho}_{X_T} - \hat{\rho}_{Y_T}\right)^T \Omega \left(\hat{\rho}_{X_T} - \hat{\rho}_{Y_T}\right)} \quad (1)$$

The parameters $\hat{\rho}_{X_T}$ and $\hat{\rho}_{Y_T}$ are defined as the autocorrelation vectors of X_T and Y_T time-series respectively having length *L*, whereas Ω denotes the weight coefficient matrix. For this study, the weight matrix is chosen to be identical such that $\Omega = I$, and let Eq. 1 become:

$$d_{ACF} = \sqrt{\sum_{i=1}^{L}\left(\hat{\rho}_{X_T} - \hat{\rho}_{Y_T}\right)^2} \quad (2)$$

Once the Euclidian distance is calculated, the *k*-means clustering technique is applied. The idea behind clustering is to partition the observations with length L into k sets $S = \{S_1, S_2, \ldots, S_k\}$ such that $L \leq k$. Finally, the k-means clusters are derived as in Eq. 3, where μ_i is the mean of points in S_i.

Recalling the fact that the number of clusters *k* must be a known priori and is user-defined, the elbow method is used in this study to identify the optimum number of clusters that the data will be partitioned to. This is expressed in Eq. 4, where the variable *k* is chosen such that the total intra-cluster variation, also known as the within-cluster sum of squares (WSS) is minimized. The parameters $W(C_k)$ denotes the intra-cluster variant of sequence C_k, where k is the number of clusters in each iteration, ranging from 1 to K clusters.

$$argmin \sum_{i=1}^{k} \sum_{x \in S_i} \|x - \mu_i\| \quad (3)$$

$$wss = \min\left(\sum_{k=1}^{K} W(C_k)\right) \quad (4)$$

C. Battery Storage System Sizing

The dimensioning of the BESS system in terms of power and capacity is an important technical aspect towards achieving an efficient battery system coupled with a grid-connected PV system. The BESS should be able to fully utilize the excess PV generation in terms of storing and supplying energy, fully or partially the residential loads. In light of this, the capacity sizing in this study is initiated by taking the cumulative daily import electricity time series and applying the *k*-means clustering method as described above. This results to the classification of the daily import profiles based on the grid-imported electricity behaviour and also it categorizes the prosumers into groups. However, taking into account the environmental variations throughout the year, which affect the PV generation and hence the power demand of end-users, the battery dimensioning with respect to capacity levels, is performed seasonally. In particular, the reference year is divided into three main periods, the summer, winter and intermediate as shown in Table I. Next, the annual import profile of each cluster is divided into the aforementioned periods and the average seasonal battery capacity of each period is determined. With this, the accumulated daily import electricity of each period is determined, as this reflects to the additional electricity that the BESS needs to compensate during each season.

Additionally, a comparison between the desired battery capacity and the cumulative daily export PV electricity of the prosumers within the same cluster is performed. This provides an indication of the number of end-users that have adequate surplus energy to satisfy the average battery size allocated to each cluster. For this study, deep cycle Lithium-based BESS technology is considered for the battery capacity calculation of each cluster. Assuming that the BESS coupling topology is done on the AC side, the approximated system round-trip efficiency is 90%. The rest 10% of the surplus energy is lost on the AC-to-DC and DC-to-AC conversion, in the battery system itself and the cable losses. Finally, a power analysis of the users is examined as well. The 30-minute import power measurements were used to study the average import behaviour of each cluster seasonally and gain insights about the power dimensioning of the proposed BESS.

TABLE I. SEASONAL PARTITIONING OF DATASET

Season	Days of year	Period Length
Winter	Jan – Feb, Nov-Dec	121
Summer	May – August	123
Intermediate	March – April, Sep-Oct	121

III. Results

The clusters determined from the cumulative import energy time series using the within-cluster sum of squares (WSS) plot is shown in Fig. 3. This is a graphical demonstration of the intra-cluster variation for various cluster values. The optimal *k* value is retained from the elbow point, in this case 3 clusters, and the representative patterns were extracted as depicted in Fig. 4. The majority of prosumers fall in Cluster 3 with 34 out of 65 users (52.3%), Cluster 1 follows with 20 users (30.7%) and finally Cluster 2 with 11 users (17.0%). The clusters revealed a similarity on the daily import energy profiles over the benchmark year, however with different energy levels as this mainly depends on the seasonal and environmental conditions. This was also validated by comparing the import energy profiles with the average ambient temperature profile of each season as depicted from the green dashed line in Fig. 4. It can be observed that the grid imported electricity has significantly higher levels during the summer period, where the ambient temperature peaks (average temperature reaches 29.7 °C) and electricity demand for cooling purposes is increased. Similarly, during the winter period, import electricity remains high but it is lower than the summer period. This stems from the fact that during the winter season, the power demand for heating and domestic hot water is relatively high (average temperature 13.2 °C). On the other hand, grid import electricity levels remain low during the intermediate period. Taking into account that the PV yield peaks during this season, a significant share of power demand is supplied instantaneously from the PV production.

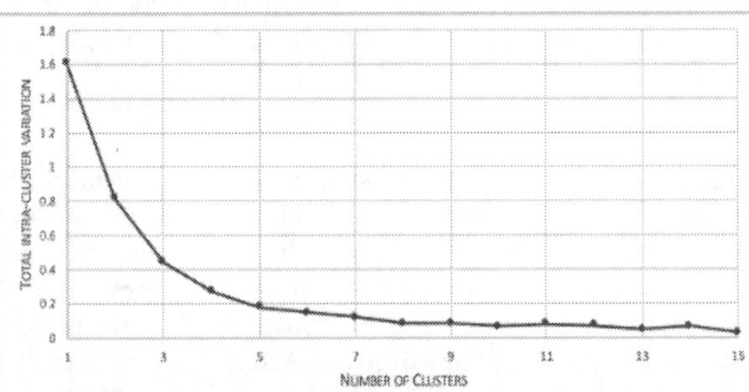

Fig. 3 Scree plot of cumulative daily import energy time series.

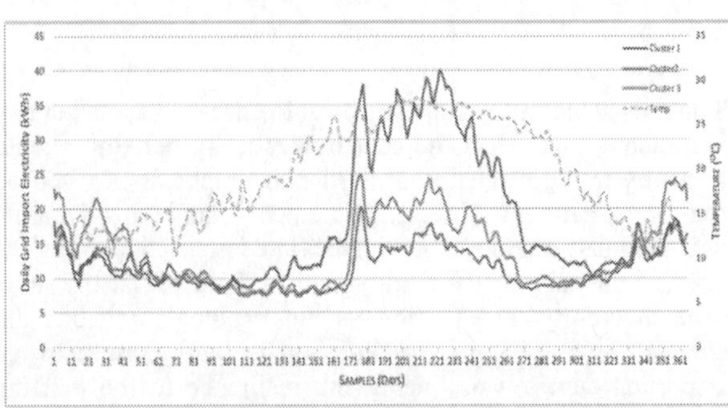

Fig. 4 Clusters of the cumulative daily import energy and average ambient temperature (dashed green line) over the test year.

Following the analysis described, the seasonal import levels of each cluster are shown in Table II. More specifically, it can be noted that Cluster 1 has the lowest import energy needs during the entire test year. On the contrary, prosumers with significantly high import energy levels, especially during the summer period, fall in the second cluster with the average daily grid consumed energy during the summer period to be 24.95 kWh. Also, prosumers that have energy needs in between the other two clusters are categorized in Cluster 3 and represent the majority of prosumers within the test group. Nevertheless, the energy requirements for the entire year reveal the high dependency of residential prosumers in Cyprus to be grid-connected as significantly high energy levels are required to meet the demands. The high energy levels reveal that additional PV capacity is required to cover the on-site demand. Considering that the estimated annual PV yield in Cyprus, is around 1600 kWh/kWp per year (or 4.3 kWh/kWp per day) for a 3 kWp system [19], residential PV system oversize may be required to balance demand and PV production.

TABLE II. DAILY IMPORT ENERGY PER CLUSTER

	Cluster 1	Cluster 2	Cluster 3
Winter	12.76 kWh	12.85 kWh	15.93 kWh
Summer	12.13 kWh	24.95 kWh	15.55 kWh
Intermediate	10.67 kWh	13.62 kWh	12.09 kWh
Average	**11.85 kWh**	**17.14 kWh**	**14.52 kWh**

Further to the daily grid import energy analysis, the BESS power dimensioning comes next as it comprises another important parameter to consider. Ideally, the battery converter power rate should be sufficient enough to supply the load power demand in a way to minimize the power flow coming from the grid side. In this scope, the power rating is determined by analyzing the actual 30-minute grid import power data of the users within the same cluster for each season as shown in Fig. 5. Starting with Cluster 1 and Cluster 3, both clusters follow a similar power distribution profile. During the winter period, the average power of Cluster 1 is at 0.53 kW with the upper quartile (Q3) reaching up to 0.76 kW, where for Cluster 3, the mean power and upper quartile are 0.66 kW and 0.93 kW respectively. It can also be noted that for both clusters, numerous extreme observations (outliers) occur, revealing the fluctuation of import power demand during these seasons. On the other hand, the clustering algorithm classifies users with relatively higher import levels to Cluster 2. The wide interquartile region (IRQ) especially during the summer period, ranging between 0.43 kW to 1.61 kW, confirms the comparably high import energy demand during this period. At the same time, there are very few outliers but with high power demand reaching up to 3.6 kW. Based on the above observations and also considering the system topology, the battery power rate can be chosen to fulfill the maximum seasonal observation (excluding outliers) as shown in Table III. This is to ensure that the battery converter is properly sized to supply the expected power range and also maintain the battery converter operation close to the peak efficiency point. In other words, oversizing

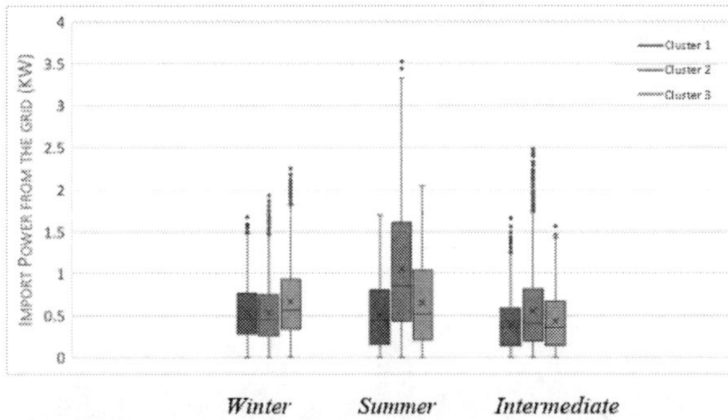

Fig. 5 Box-plot representing the average daily import power of the seasons, divided in Cluster 1 (blue), Cluster 2 (orange) and Cluster 3 (gray).

TABLE III. DAILY IMPORT POWER PER CLUSTER

	Cluster 1	Cluster 2	Cluster 3
Winter	1.48 kW	1.47 kW	1.82 kW
Summer	1.69 kW	3.32 kW	2.05 kW
Intermediate	1.25 kW	1.75 kW	1.42 kW
Average	**1.47 kW**	**2.18 kW**	**1.76 kW**

of the BESS power can cause the battery converter to operate at a lower power point and thus at low efficiency conditions. From the analysis above, it is possible to understand the user behaviour and interaction with the grid over the entire test year. However, for the BESS sizing, the surplus electricity needs to be taken into consideration. As already mentioned, the scope of this study to integrate BESS into existing residential PV systems to increase prosumer self-sufficiency rate. Towards this direction, the export power measurements were used to study the seasonal surplus energy delivered to the grid. In Fig. 6, the daily export electricity of each cluster and season is illustrated using a box-and-whisker plot. The schematic representation clearly shows that the export electricity during the intermediate period is higher than in the summer and winter, whilst winter season has the lowest surplus energy. At the same time, the wide interquartile region during winter reveals the fluctuation of export energy observations due to PV intermittency and variation in environmental conditions. To evaluate the adequacy of users to supply the cluster energy requirements shown in Table II, with the daily PV export energy, various import energy levels were compared with the surplus energy levels of users within the same cluster, reduced by 10% to consider also the BESS system losses. The optimum battery capacity is chosen such that the adequacy of users within the clusters to range between 20% and 80%. As shown in Fig. 7, Cluster 1 battery capacity can be selected between 50%-70% of the average cluster import energy (i.e. 5.9 - 8.3 kWh) since this is the transitional region until none of the users within the Cluster meeting the energy requirements. Even though Cluster 2 achieves high export energy levels, the required import profile limits battery sizing to a maximum usable capacity between 27% - 42% of the average import energy (i.e. 4.6 –

7.2 kWh). On the other hand, Cluster 3 achieves the highest export levels and the battery capacity can be sized in the range of 40% - 55% of the average cluster import energy (i.e. 5.8 – 8.0 kWh). From the battery capacity sizing performed above, it can be noted that Cluster 1 reaches the highest surplus electricity levels allowing the battery to be sized up to 8.3 kWh. At the same time, the quite large energy needs of Cluster 2, this does not a great impact on the battery sizing as the capacity is very close to the other clusters.

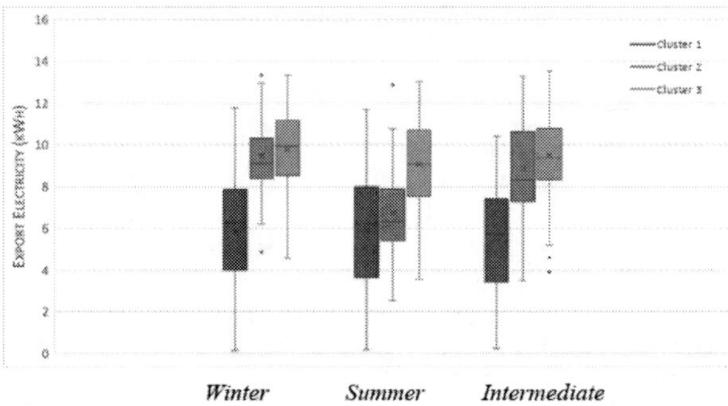

Fig. 6 Box plot representing the average daily export energy of the seasons, divided in Cluster 1 (blue), Cluster 2 (orange) and Cluster 3 (gray).

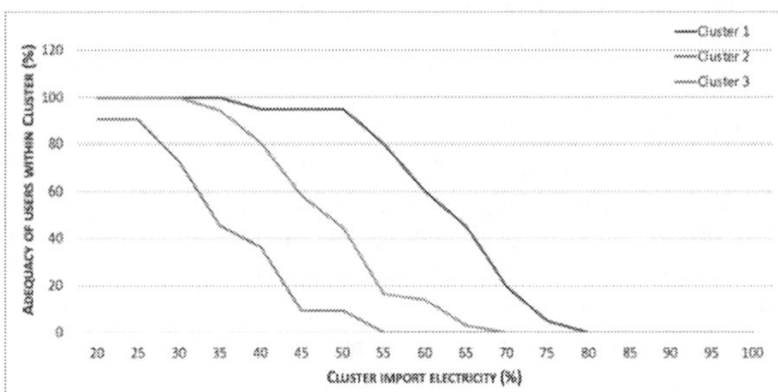

Fig. 7. Adequacy of export PV energy to meet cluster import energy requirements.

IV. Conclusions

A methodology to characterize the import energy profiles of typical prosumers in Cyprus was presented. The annual energy measurements taken from Smart Meters (SMs) were taken for a group of prosumers, and based on the k-means clustering technique, the import energy clusters were determined.

The extracted clusters presented useful information about the prosumer daily import energy levels over the test year as power and energy fluctuations depend on seasonal variations. The clustering technique revealed that 52.3% of the users are predominantly classified in Cluster 3 which represents a typical import profile over the year. On the contrary, Cluster 1, which holds the other 30.7%, has significantly low import needs over the year in comparison to Cluster 2, which has the rest 17% of the users, and has the highest grid import energy needs over the test year.

Next, the BESS sizing in terms of power rating and energy were extracted from the 30-minute power measurements of the import and export profiles. Starting with the energy sizing first, the analysis demonstrated that the PV export electricity can partially supply the premise demands since the relatively high import energy levels cannot be compensated from the daily surplus PV electricity and hence by the energy stored in the BESS unit. In particular, the surplus PV electricity for Cluster 1 is between 5.9 kWh to 8.3 kWh, whereas for Cluster 2 and Cluster 3 is between 4.6 kWh to 7.2 kWh and 5.8 kWh to 8.0 kWh respectively. As for the power dimensioning of the BESS, a statistical analysis of the power measurements was performed for each cluster. The analysis was performed on a seasonal basis over the year and revealed similar import profile for Clusters 1 and 3 with the average power to be 1.47 kW and 1.76 kW respectively. Finally, the summer period is characterized by significantly high power levels with a suitable power rate dimensioning to be 2.18 kW.

V. Future Work

In this study, a BESS unit that is AC-coupled with a PV system is considered and the battery energy and power rate dimensioning were chosen based on the daily import behaviour of three (3) dominant clusters of prosumers in Cyprus. As part of StoRES project scope, the optimum battery energy and power converter utilization will be discovered based on the measurements collected from the project pilot systems. In particular, a 2.5 kW / 9.8 kWh BESS system has been embedded to five (5) residential premises with an existing 3 kWp PV system forming an AC-coupling topology. The primary scope of the project is to group pilot users into the three primary clusters, and also verify the sizing methodology proposed in this study. Finally, the option of allocating a share of residential BESS units to a centralized storage is also considered as a future work of this project. To validate such a scenario, a 30 kW / 50 kWh centralized BESS unit has been installed to the LV distribution feeder that supplies the five residential pilots.

Acknowledgments

This work was funded through the StoRES project which is co-financed by the European Regional Development Fund (ERDF) through the Interreg MED Programme under the grant agreement number 1MED15_2.2_M2_184.

References

[1] J. von Appen, M. Braun, T. Stetz, K. Diwold, and D. Geibel, "Time in the Sun: The Challenge of High PV Penetration in the German Electric Grid," *IEEE Power Energy Mag.*, vol. 11, no. 2, pp. 55–64, Mar. 2013.

[2] K. Ogimoto, I. Kaizuka, Y. Ueda, and T. Oozeki, "A Good Fit: Japan's Solar Power Program and Prospects for the New Power System," *IEEE Power Energy Mag.*, vol. 11, no. 2, pp. 65–74, Mar. 2013.

[3] M. van Werven and M. J. J. Scheepers, "The changing role of distribution system operators in liberalised and decentralising electricity markets," in *2005 International Conference on Future Power Systems*, 2005, p. 6 pp.-6.

[4] T. Stetz, F. Marten, and M. Braun, "Improved low voltage grid-integration of photovoltaic systems in Germany," in *2013 IEEE Power & Energy Society*

General Meeting, 2013, pp. 1–1.

[5] R. Tzartzev, W. Mack Grady, and J. Patel, "Impact of high-penetration PV on distribution feeders," in *2012 3rd IEEE PES Innovative Smart Grid Technologies Europe (ISGT Europe)*, 2012, pp. 1–6.

[6] M. E. Baran, H. Hooshyar, Z. Shen, J. Gajda, and K. M. M. Huq, "Impact of high penetration residential PV systems on distribution systems," in *2011 IEEE Power and Energy Society General Meeting*, 2011, pp. 1–5.

[7] A. S. Masoum, P. S. Moses, M. A. S. Masoum, and A. Abu-Siada, "Impact of rooftop PV generation on distribution transformer and voltage profile of residential and commercial networks," in *2012 IEEE PES Innovative Smart Grid Technologies (ISGT)*, 2012, pp. 1–7.

[8] F. Katiraei and J. Aguero, "Solar PV Integration Challenges," *IEEE Power Energy Mag.*, vol. 9, no. 3, pp. 62–71, May 2011.

[9] R. Tonkoski, D. Turcotte, and T. H. M. El-Fouly, "Impact of High PV Penetration on Voltage Profiles in Residential Neighborhoods," *IEEE Trans. Sustain. Energy*, vol. 3, no. 3, pp. 518–527, Jul. 2012.

[10] C. A. Hill, M. C. Such, D. Chen, J. Gonzalez, and W. M. Grady, "Battery Energy Storage for Enabling Integration of Distributed Solar Power Generation," *IEEE Trans. Smart Grid*, vol. 3, no. 2, pp. 850–857, Jun. 2012.

[11] C. L. Masters, "Voltage rise: the big issue when connecting embedded generation to long 11 kV overhead lines," *Power Eng. J.*, vol. 16, no. 1, pp. 5–12, Feb. 2002.

[12] Y. Ueda, K. Kurokawa, T. Tanabe, K. Kitamura, and H. Sugihara, "Analysis Results of Output Power Loss Due to the Grid Voltage Rise in Grid-Connected Photovoltaic Power Generation Systems," *IEEE Trans. Ind. Electron.*, vol. 55, no. 7, pp. 2744–2751, Jul. 2008.

[13] R. Tonkoski, L. A. C. Lopes, and T. H. M. El-Fouly, "Coordinated Active Power Curtailment of Grid Connected PV Inverters for Overvoltage Prevention," *IEEE Trans. Sustain. Energy*, vol. 2, no. 2, pp. 139–147, Apr. 2011.

[14] M. J. E. Alam, K. M. Muttaqi, and D. Sutanto, "Mitigation of Rooftop Solar PV Impacts and Evening Peak Support by Managing Available Capacity of Distributed Energy Storage Systems," *IEEE Trans. Power Syst.*, vol. 28, no. 4, pp. 3874–3884, Nov. 2013.

[15] E. Vrettos, K. Lai, F. Oldewurtel, and G. Andersson, "Predictive Control of buildings for Demand Response with dynamic day-ahead and real-time prices," in *2013 European Control Conference (ECC)*, 2013, pp. 2527–2534.

[16] G. Walker, "The dynamics of energy demand: Change, rhythm and synchronicity," *Energy Res. Soc. Sci.*, vol. 1, pp. 49–55, Mar. 2014.

[17] P. Galeano and D. Pena, "Multivariate Analysis in Vector Time Series," *Resenhas do Inst. Matemática e Estatística da Univ. São Paulo*, vol. 4, no. 4, pp. 383–403, 2000.

[18] P. (University of A. C. Montero and J. A. (University of C. Vilar, "TSclust: An R Package for Time Series Clustering," *JSS J. Stat. Softw.*, vol. 62, no. 1, pp. 1–43, 2014.

[19] G. Makrides, B. Zinsser, M. Norton, G. E. Georghiou, M. Schubert, and J. H. Werner, "Potential of photovoltaic systems in countries with high solar irradiation," *Renew. Sustain. Energy Rev.*, vol. 14, no. 2, pp. 754–762, Feb. 2010.

A MODULAR STAND-ALONE PHOTOCATALYTIC REACTOR FOR WASTE WATER PURIFICATION: THE HPSOLAR PROJECT

Paolo Bernardoni[1], Micol Boschetti[1], Giulio Mangherini[1], Donato Vincenzi[1], Silvio Fugattini[1], Alfredo Andreoli[1]
Vito Cristino[2], Stefano Caramori[2], Carlo Alberto Bignozzi[2]
[1] University of Ferrara, Department of Physics and Earth Sciences
Via Giuseppe Saragat 1 - Building C, 44122 Ferrara (FE), Italy
[2] University of Ferrara, Department of Chemical and Pharmaceutical Sciences
Via Luigi Borsari 46 - 44121 Ferrara

ABSTRACT: In this work we present the construction and performances of a photocatalytic reactor developed for the HPSolar project, the purpose of this reactor is the treatment of water in order to remove pollutants like drugs or pesticides that are not removed effectively by present water treatments.
The reactor is modular and composed by multiple elementary cells where the photocatalyst is a thin layer of tungsten trioxide biased positively with respect to a metallic cathode; this polarization is switched periodically in order to restore the functionality of the anode.
The cells are completely self-reliant because are powered by four solar cells laying behind the semi-transparent anode while a microcontroller integrated into each cell applies the biasing cycles.
The abatement measurements on atenolol and carbamazepine show that this device can remove 70% of the drugs from a sample of water within 5 and 7 hours respectively.
Keywords: Off-grid, Stand-alone PV Systems, Utilities,

1 PROJECT

The purpose of HPSolar project is the development of reactors based on tungsten trioxide (WO_3) photoanodes whose purpose is the remediation of wastewater or the purification of drinking water.

The aim of the project is to assess the abatement effectiveness of pollutants in a set of samples provided by a local water treatment and management authority (HERA Spa).

WO_3 photoanodes can oxidize most organic compounds (with the exception of fluorinated compounds) but their most interesting application is the abatement of emerging pollutants like drugs or pesticides that are removed with difficulty by the present water treatment plants.

The project covers three possible applications of these photoanodes: unbiased photoanodes for use under concentrated sunlight, biased photoanodes grown on a metallic tungsten substrate used under concentrated sunlight and transparent photoanodes integrated in completely autonomous modular units.

In particular, this work concerns the latter application: a stand-alone photocatalytic reactor powered by solar cells integrated in the reactor itself.

The final deliverable of the project is a modular device that may be integrated in the present water treatment cycle in order to reduce the concentration of emerging pollutants in the treated water.

The HPSolar project is part of the program POR-FESR 2014-2020 funded by Regione Emilia-Romagna and the European Union through the European Regional Development Fund (en:ERDF, it:FESR).

2 PHOTOCATALISYS

Photocatalysis is a process that involves the activation of a catalyst by means of light. In order to be useful for water purification, these catalysts must be able to break down effectively most organic contaminants as well as to reduce the COD (Chemical Oxygen Demand) oxidizing the natural organic matter [1].

It is well known that, in semiconductors, absorption of photons with energy $E=h\nu$ above the bandgap Eg generates electron (e-) hole (h+) couples with an energy difference equal to the bandgap.

At the surface of the semiconductor the charge carriers can interact with the chemicals present in the environment, in particular electrons behave as reducing agents while holes as oxidizers [2].

While effective with no external action, the efficiency of the photocatalytic effect increase dramatically when the semiconductor is externally biased.

What happens in this configuration is that the charge carriers drift according to the external electric field reducing the chance of recombination.

Therefore, the carriers accumulate on the surface depending on the applied field: the net result is an increased reactivity of the catalyst.

The energy of the carriers must be high enough to react with the target chemicals hence typical semiconductors used as photocatalysts for this purpose are metal oxides (TiO_2 [3], WO_3 [4], α-Fe_2O_3 [5]).

Specifically titanium dioxide (TiO_2) is the cheapest and most studied photocatalyst because of its low cost and high bandgap (3.0-3.2eV) but requires an exposure to light with wavelengths lower than 410nm, which represents only 10% of the AM1.5 solar spectrum [5].

In this project, tungsten trioxide (WO_3) was chosen instead of TiO_2 thanks to its lower activation energy (~2.7eV) that makes it able to work under a wider range of natural sunlight (~15% AM1.5 [6]) than TiO_2.

The photocatalytic effect of WO_3, as previously explained, can be boosted by biasing the photoanode with a tandem cell and coupling the system with a metal (Ni, Au, Pt) as cathode.

However, in liquids and in case of high bias, polarization effects occur and the catalytic efficiency is reduced down to a minimum after a few minutes of operation.

This limiting effect can be overcome by applying periodically a small reverse potential (0-0.3 V) to the cell (see Fig.3).

3 THE MODULAR DEVICE

The reactor is composed of an array of PVC frames having a size of 30×30 mm 2 building up a cylindrical structure with a diameter of 140 mm (see Fig. 1).

Figure 1: Picture of the reactor showing the modular structure: the modules are assembled in circular stages that are stacked on top of each other to create a cylindrical shape

Each element is a completely autonomous system composed by a photoanode, four solar cells, the electronics to drive polarization cycles and a metallic cathode.

The shorter wavelength band (from UV to Blue) of the solar spectrum is used for photocatalysis while the longer one passes through the anode and can be converted by the PV cells in order to obtain the energy necessary to bias the photo-electrolytic cell.

The placement of the solar cells, directly embedded in the photoreactor modules, is essential to avoid a complex wiring of the array and to prevent the corrosion issues that may arise with cables in an aqueous environment.

The incapsulation of all these components is made with self-levelling silicone in a watertight plastic frame (see Fig.3).

3.1 Anode

The photocatalytic element consists of a film of tungsten trioxide with an area of 15x15mm2 and a thickness of 1μm deposited on a transparent and conductive layer of FTO (Fluorine-doped Tin Oxide).

The substrate of these two layers is a glass window with an area of 19.6x19.6mm^2 that represents the front element (anode) of the PEC cell.

The layer of tungsten trioxide has been obtained from a colloidal water-based solution, this solution is deposited on a glass substrate with a spin-coating technique then heated to 550°C: the process is repeated until obtaining a final layer of the desired thickness (~1 μm) [7].

Given that the anode deposition is the most time-consuming task of the assembly process, the thickness is balanced in order to obtain a good compromise between the best light absorption and the duration of the deposition process.

The result is an anode that in the region between 350 and 1000nm has an average transparency of 32% (See Fig. 2)

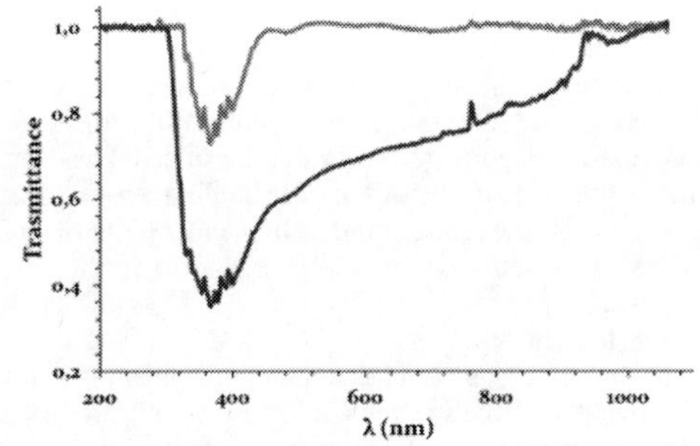

Figure 2: Total transmittance of the anode (blue) that includes the absorption from WO_3, FTO and glass plus reflection from the WO_3 layer. Pure transmittance of the WO_3 layer (orange) without the contribution from reflection and the other layers. Data below 350nm is non-significant.

3.2 Solar cells

Each module is powered by four back-contact c-Si solar cells with an area of 8x8mm^2 each, these cells are connected in series in order to reach the voltage higher than 1.8V that is necessary to bias the tandem cell (See Fig. 3).

Four silicon cells connected in series have exactly the correct voltage to power the microcontroller (that works with a minimum of 1.8V) without reaching a voltage too high (above 3V) that could damage the WO_3 layer.

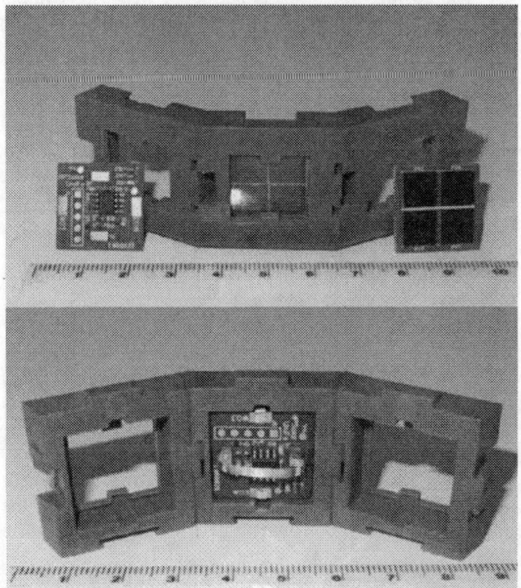

Figure 3: The picture on the top shows a complete cell (without encapsulant) with the transparent photoanode and the two sides of the electronic circuit (cells side and microcontroller side), on the bottom is visible the back of the same cell and the cathode.

Tungsten trioxide is photoactivated only by light with wavelengths below 450nm [8] and is therefore transparent to longer wavelengths.

In order to achieve the best efficiency, the module is assembled stacking the solar cells behind the photoanode in a tandem structure so that the light not absorbed by the anode is used to power the microcontroller that applies the bias and the polarization cycles.

Even with the presence of the anode the current produced by the solar cells is not a limit for the efficiency of the system, the measured current densities under 1 sun illumination are of ~1mA/cm^2 (depending on the final thickness) in the anode and 30mA/cm^2 (without the anode) for the cells.

3.3 Electronics

In the prototype, an electronic circuit integrated into each element of the reactor applies to the electrodes an alternating bias.

This electronics consist of a microcontroller directly powered by the solar cells connected in series, which provides the required voltages (+ 2.0↔ + 2.5 V, -0↔-0.3 V, and timings (+30 min, -3 min) to bias the photoanode and restore periodically its efficiency.

Since the current available from the solar cells is 10 times higher than the current absorbed by the anode a MPPT (Maximum Power Point Tracking) is not necessary simplifying the design of the circuit.

Anode and cathode are directly connected to the pins of the microcontroller thus the applied bias is roughly equal to the Voc of the solar cells minus the voltage drop (typically around 0.1-0.2V) across the microcontroller itself.

In order to limit the bias to -0.3V during the reverse polarization cycles, a Shottky diode is connected in series to the electrodes while a resistor sets the current limit.

The current limitation is necessary because, without MPPT, it prevents an excessive voltage drop from the solar cells.

3.4 Reactor

The PVC frame of the PEC cells have wedges and slots that permit to assemble 14 of these frames into a circle. In the same way, thanks to the modularity of the elements, the height of the reactor column can be tailored according to user needs by simply stacking multiple stages of PVC frames.

The tube composed of the plastic frames has a diameter of 140mm and a thickness of 10mm, this structure is inserted between two PMMA (chosen for UV transparency and cost) tubes of which the outer has a diameter of 160mm and the inner one has a diameter of 120m.

The water to be treated flows in the gap between the tubes from the bottom to the top of the reactor.

Then, upon reaching the top of the reactor, it falls into the inner tube that acts as a sink. At the bottom of the inner tube, a solar-powered high efficiency (2W with a prevalence of 50cm) water pump recirculates the wastewater pumping it again in the gap between the tubes.

4 RESULTS

Measurements carried out on both the bare anodes and on the PEC cells show the effectiveness of this technology in the abatement of micro pollutants.

In particular, the tests focused on atenolol (used to reduce high blood pressure) and carbamazepine (used as a treatment for epilepsy) because present water treatments do not remove effectively (less than 10% [9]) these drugs.

The abatement tests were performed with an illumination of 1-sun AM1.5 and with an area to volume ratio of 1cm^2 of catalyst to 1cm^3 of solution.

Measurements of atenolol at a concentration of 10ppm show a degradation of 75% of the pollutant after 7 hours of exposure to light while, without illumination, the concentration of drug remained stable (See Fig. 5).

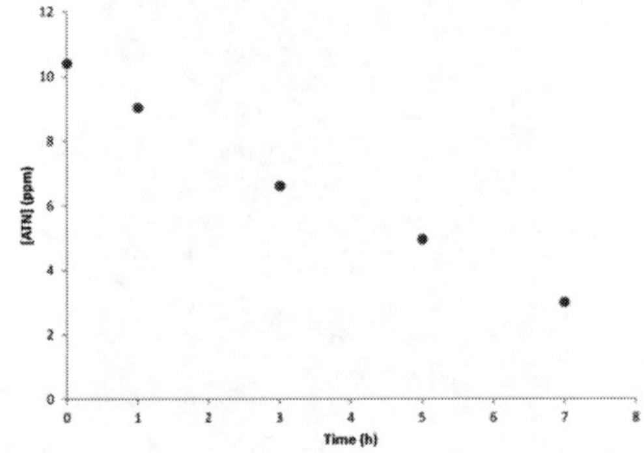

Figure 4: Atenolol abatement in mechanically filtered sewage water illuminated with an intensity of 500mW/cm^2

Carbamazepine is abated more quickly than atenolol since, in the same conditions, the abatement of 75% is reached after 5 hours while the unbiased anode, at the same time, only removed 35% (See Fig.4).

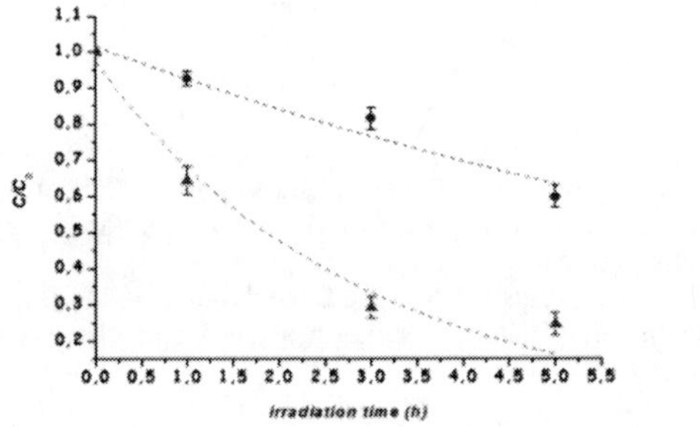

Figure 5: Carbamazepine abatement in mechanically filtered sewage water illuminated with an intensity of 500mW/cm^2

5 BIBLIOGRAPHY

[1] Qin, D. D., Tao, C. L., Friesen, S. A., Wang, T. H., Varghese, O. K., Bao, N. Z. & Grimes, C. A. (2012). Dense layers of vertically oriented WO_3 crystals as anodes for photoelectrochemical water oxidation. Chemical Communications, 48(5), 729-731.

[2] Longobucco, G., Pasti, L., Molinari, A., Marchetti, N., Caramori, S., Cristino, V., ... & Bignozzi, C. A. (2017). Photoelectrochemical mineralization of emerging contaminants at porous WO3 interfaces. Applied

Catalysis B: Environmental, 204, 273-282.
[3] Chen, X., & Mao, S. S. (2007). Titanium dioxide nanomaterials: synthesis, properties, modifications, and applications. Chemical reviews, 107(7), 2891-2959.
[4] Alexander, B. D., & Augustynski, J. (2009). Nanostructured Thin Film WO_3 Photoanodes for Solar Water and Sea-Water Splitting. - On Solar Hydrogen & Nanotechnology, 333-347.
[5] Sivula, K., Le Formal, F., & Grätzel, M. (2011). Solar water splitting: progress using hematite (α-Fe2O3) photoelectrodes. ChemSusChem, 4(4), 432-449.
[6] American Society for Testing and Materials (ASTM) - Terrestrial Reference Spectra for Photovoltaic Performance Evaluation
[7] Meda, L., Tozzola, G., Tacca, A., Marra, G., Caramori, S., Cristino, V., & Bignozzi, C. A. (2010). Photo-electrochemical properties of nanostructured WO3 prepared with different organic dispersing agents. Solar Energy Materials and Solar Cells, 94(5), 788-796.
[8] González-Borrero, P. P., Sato, F., Medina, A. N., Baesso, M. L., Bento, A. C., Baldissera, G., ... & Ferreira da Silva, A. (2010). Optical band-gap determination of nanostructured WO_3 film. Applied Physics Letters, 96(6), 061909.
[9] Castiglioni, S., Bagnati, R., Fanelli, R., Pomati, F., Calamari, D., & Zuccato, E. (2006). Removal of pharmaceuticals in sewage treatment plants in Italy. Environmental Science & Technology, 40(1), 357-363.

ANALYSING THE VOLTAGE STABILITY OF PHOTOVOLTAIC INVERTERS REACTIVE POWER CONTROL IN THE LABORATORY INCLUDING THE DISTRIBUTION GRID TRANSFORMER

F.P. Baumgartner[1]*, C. Messner[2,3], C. Seitl[2], G. Lauss[2], F. Carigiet[1], R. Bründlinger[2], T.I. Strasser[2]

[1]ZHAW, Zurich University of Applied Sciences, School of Engineering, IEFE
www.zhaw.ch/~bauf, Technikumstr. 9, CH-8401 Winterthur, Switzerland
*phone: +41589347232; email: bauf@zhaw.ch

[2]AIT Austrian Institute of Technology, Electric Energy System – Center for Energy
www.ait.ac.at, Giefinggasse 2, A-1210 Vienna, Austria
{christian.messner, christian.seitl, georg.lauss, roland.bruendlinger, thomas.strasser}@ait.ac.at

[3]presenting the talk at the conference

ABSTRACT:
Photovoltaic (PV) inverters increase the line voltage in Distribution Grids (DG) by active power feed in. Today, modern PV inverters are also able to feed in reactive power to mitigate the above voltage rise. The favoured, cost effective implementation is the control of reactive power feed in according to the instantaneous measured line voltage. The stability of this decentralised Q(U) PV inverter closed-loop control is mandatory and analysed in this work. The DG operator must guarantee the voltage limits given in the regulatory framework. This is challenging due to fast changing solar irradiance, load flows and the interaction of an overlying automatic voltage control-loop of a connected sub-station. The performed tests in the AIT SmartEST laboratory resulted in very stable operation even at small Time Constants (TC) below 5 seconds of the Q(U) control parameter. As one test scenario, an abrupt rise of solar irradiance immediately followed by load drop is realized by use of the PV and load emulators. The PV inverter reduces the resulting voltage rise by increasing its reactive power, depending on the Q(U) control time constant.
It was found that even at smaller Q(U) time constants than the typical applied values of 5 or 10 seconds no sign of instability arises. It is recommended to the DG operator to apply TC of 1 or 2 seconds of Q(U) control to minimise the duration of overvoltage condition during the transient voltage adjustment. Applying irradiance conditions of a typical cloudy day in the lab test yields 45% of the time the line voltage was above a given grid voltage limit, while applying Q(U) at TC of 1 second there was no occurrence of overvoltage. In detail at Q(U) TC setting of 20 seconds the over-voltage arises 3.4% of the total period and only 0.3% at TC 5 seconds. Only stable operation conditions were found including the automatic voltage control of the transformer sub-station at a typical setting of 10 seconds delay time of that sub-station control setting. Summarized, in combination with that delay time setting the smaller Q(U) time constant of the PV inverters below 5 seconds are beneficial due to the minimised overvoltage time.

Keywords: PV inverter, grid integration, voltage regulation, qualification and testing, cost reduction

1 MOTIVATION AND MARKET NEEDS

One of the main duties of the DG in the coming years is to deliver power and stay within the permitted grid voltage levels, even during periods of high Photovoltaic (PV) power feed into the grid. Further, costly hardware investments have to be evaluated economically and technically regarding the ability to stabilize the voltage according to the grid code [1].

An effective and low-cost approach is the control of the reactive power Q(V) and the active power P(V) of the PV inverter, by means of the actual grid voltage measured by the individual PV inverter [2,3]. Thus, no additional hardware investment in Information and Communication Technology (ICT) on the grid level is needed. The PV inverter settings of the static characteristics of the Q(V) and P(V) have to be in accordance with the grid operating code. Additionally, the time constants (dead and delay time) of these Q(V) and P(V) control activities have to be specified to guarantee a stable grid operation.

Today, this economically very attractive Q(V) method is not widely used in practice. To convince the DG operator, more effort of sophisticated tests have to be performed, including feedback loops of the tap changer controller of the transformers close to the stability boundary.

A report about such extreme laboratory tests are given in this paper, like results and analyses of stability measurements in the laboratory of PV inverters, powered by DC sources, emulating changing weather conditions, especially abrupt transient trigger by changing solar irradiance. Other transient inputs have been generated by different load, emulating Electric Vehicles (EV) charging stations.

The first published results of this test series here [6], will be extended here and tested including the use of the substation [4]. The transformers medium voltage On-Load-Tap-Changers (OLTC) in the lab will complete the total analysed control loop with the PV inverter in the main focus. It will be elaborated in the paper, if different appropriate settings of the parameters are needed for each category of distribution grids and type of the static Q(V) und P(V) characteristics. A set of stable parameters are given of the times constants of the controllers for each grid category including the demand of highest load transients occurring at EV charging. Finally, a suggestion is made of how these laboratory-based parameter findings should be introduced into a general standardization process.

2 DESIGN OF THE LAB TEST CASES

A commercial PV inverter is the Device-under-Test (DUT) of the experimental set-up in particular his implemented control regime of the line voltage based Q(U)

method is tested as shown in Fig. 1 and 2. The direct current input of the PV inverter was feed by either as a change of active power or according the calculation of the current voltage characteristics of the PV generator according to the measured solar irradiance on a cloud day in the Zurich area. The goal was to measure instability occurring as oscillations of the line voltage.

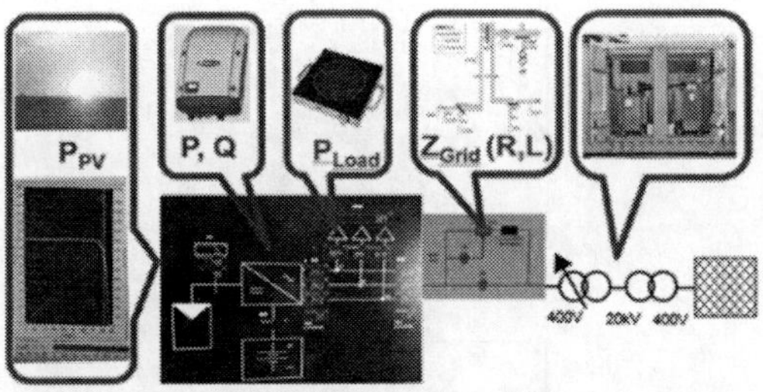

Figure 1: Engaged devices of the SmartEST labs to realise the stability test of the Q(V) feature of the PV inverter as the Hardware-under-Test (Hut).

Figure 2: Commercial smart secondary substation with OLTC as part of the SmartEST lab equipment embedded in the experiments.

3 CONFIGURATION OF TEST SETUP

Different static Q(U) characteristics and values of the TC of the inverters Q(U) setting was tested. Here the results are reported with the static Q(U) settings with a slope of 3% voltage rise resulting in the linear increase of reactive power from zero to the maximum, with a dead band of again 3% in between the positive and negative branches. [7]. The adjusted Q(U) time constants at the DUT is defined as first order filter in which the reactive power set point is reached after 1τ (63 %) of a voltage step. The lab experiment settings are adapted to the voltage drop along the real impedance in the DG of the village Dettighofen, from a farmer's house with a 100kW PV installation, located on the village rand, to the transformer in the centre of that typical village [2].

Thus, the voltage drop on the lab-impedance at the maximum active power of the DUT, which was smaller than 100KW, was comparable with the scaled slope of the static DUT Q(U) characteristics. The tests are performed with different TC settings between 1 to 20 seconds and with and without the connected sub-station transformer equipped with OLTC control.

The DEWE hardware and software solution are used to acquire the electrical measurements of the three-phase power and voltage and first analyses are found by the generated plots of the DEWE Software. Further analyses are performed by the use of MATLAB.

4 ACHIEVED RESULTS

4.1 Results of transient load and irradiance chances

In Fig. 3 results are given of a test applying by a step function of solar irradiance resulting in a nearly immediate rise of active power of the PV inverter followed by an immediate reduction in line voltage. In the linear regime of the static Q(U) characteristics this have to give rise to reactive power, according to the settings given by the TC time constant of the control system.

Smaller values of the Q(U) time constant leads to shorter periods the line voltage is above a given value, before it is compensated. The final value voltage value results from the delivered reactive power according to the DUT's static Q(U) characteristics, at the given grid impedance. All this measured step responses provided stable control conditions of the DUT.

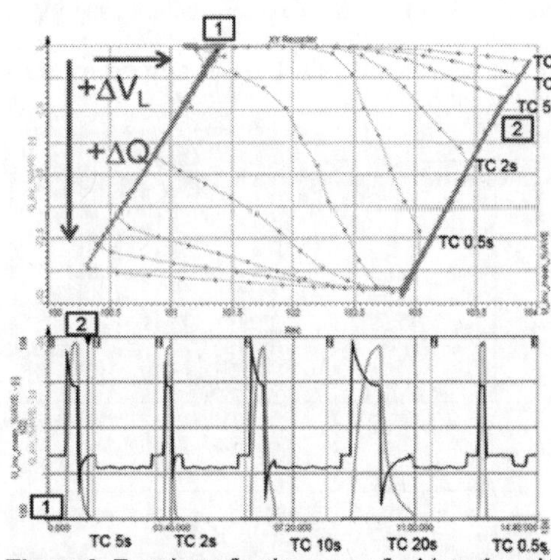

Figure 3: Transient of active power feed into the grid by the DUT forcing reactive power according to demanded Q(U) control at different parameter settings of TC 0.5, 2, 5, 10 and 20 seconds and thus compensate the line voltage rise. While the graphs in the bottom shows the measured line voltage (blue) and the produced reactive power by the DUT, the above graphs represents the same measurement data plotted reactive power Q versus line voltage V_L.

4.2 Results of real weather data performance

The requested active power of the DUT during an excerpt of a typical cloudy day in the Zurich area in April, as shown in Fig. 4 is performed by applying the alteration of the calculated current voltage characteristics of a crystalline silicon PV generator onto a DC amplifier according to the measured solar irradiance characteristics.

In Fig. 5 the measured produced reactive power is given for three TC parameter settings and without Q(U) control. The characteristics of the measured line voltage versus time is shown just below.

How effective Q(U) may reduce the voltage rise is illustrated in Fig. 6 and 7 by sorting the voltage measurement according the total time the line voltage exceeds a certain value shown in % of the nominal voltage, which is given in function of minutes.

In detail the analysis performed at a threshold of 102.8% of nominal voltage yield 45% of the time without Q(U) and 3.4% at TC of 20 seconds, only 0.3% at TC 5 seconds and never above at a TC of one second.

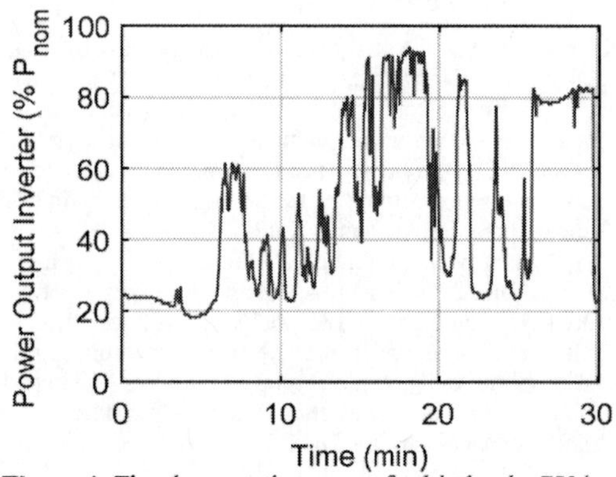

Figure 4: The shown active power feed-in by the PV inverter into the grid is the result of the applied solar irradiance characteristics in the HIL setup as a function of DC amplifier output.

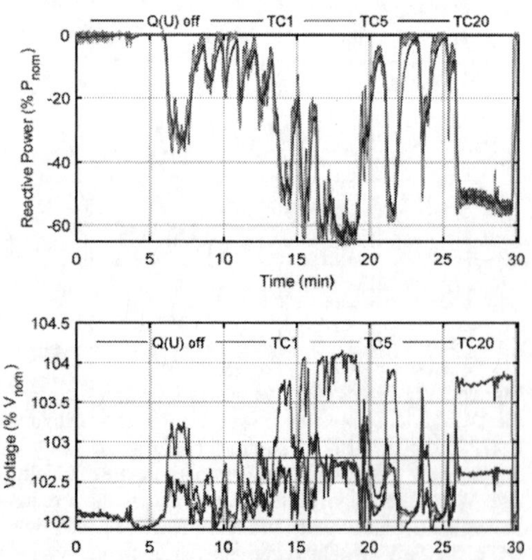

Figure 5: Thirty-minute voltage characteristics as a function of reactive power applied by a PV inverter during the same real solar irradiance conditions with parameter settings of 1, 5 and 20 seconds of the Q(U) controller and without Q(U) control.

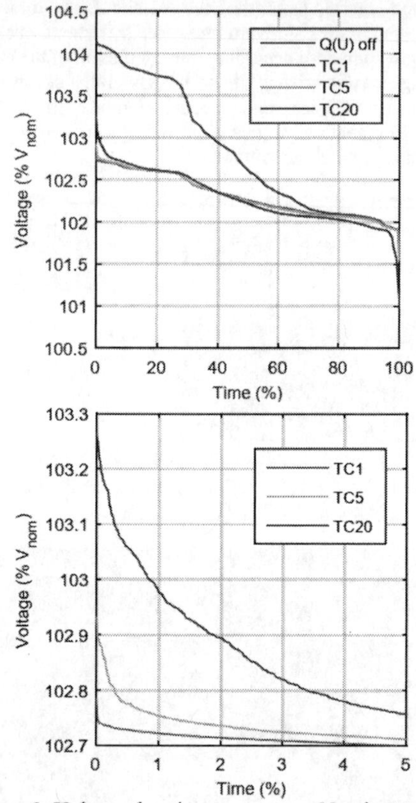

Figure 6: Voltage duration curve over 30 minutes as a function of reactive power applied by a PV inverter during the same real solar irradiance conditions with parameter settings of 1, 5 and 20 seconds of the Q(U) controller and without Q(U) control.

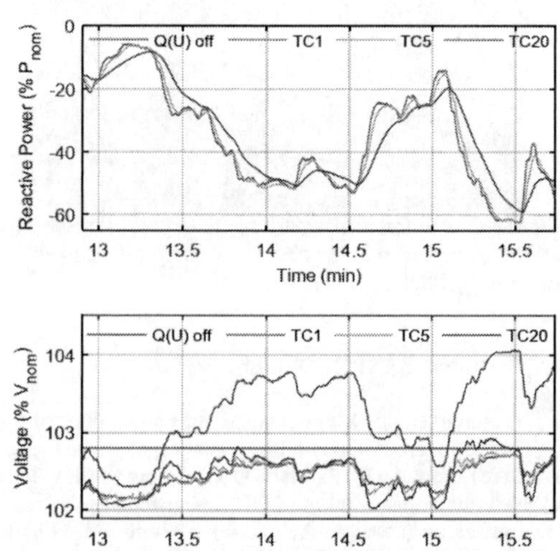

Figure 7: Three-minute voltage characteristics as a function of reactive power applied by a PV inverter during the same real solar irradiance conditions with parameter settings of 1, 5 and 20 seconds of the Q(U) controller and without Q(U) control.

4.3 Results including a substation

One solution to improve the DG voltage quality is the use of transformers to the next grid level with automatic operation of the OLTC to select the appropriate voltage at the output of the substation. If the voltage is higher than a certain threshold the OLTC will step down one interval as shown in Fig. 8. The implementation of a delay time of typ. larger 10 sec and algorithm to change voltage by only one-step, e.g., 1% of nominal voltage was intended as the base of a stable operation, not to activate the OLTC due to regular small fluctuations of irradiance. Larger steps of the tap-changer exceeding a regular smallest voltage interval will only occur at higher voltage changes, which will never be a results of Q(U) control activities. But what happened if the PV installer set the wrong sign of the static characteristics?

These assumptions and the theoretical work dedicated to the stability of Q(U), focusing on the inclusion of a dead time of the Q(U) control [7] leads to the test in the lab.

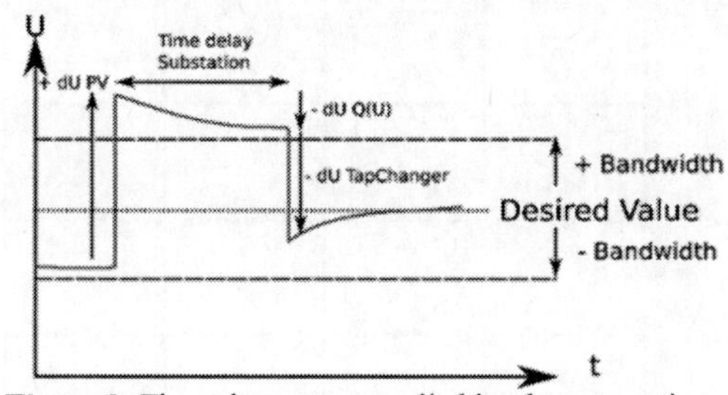

Figure 8: The voltage steps applied by the automatic tap changer after a certain delay time at the substation is superposed by the PV inverters automatic Q(U) control resulting in –dU Q(U).

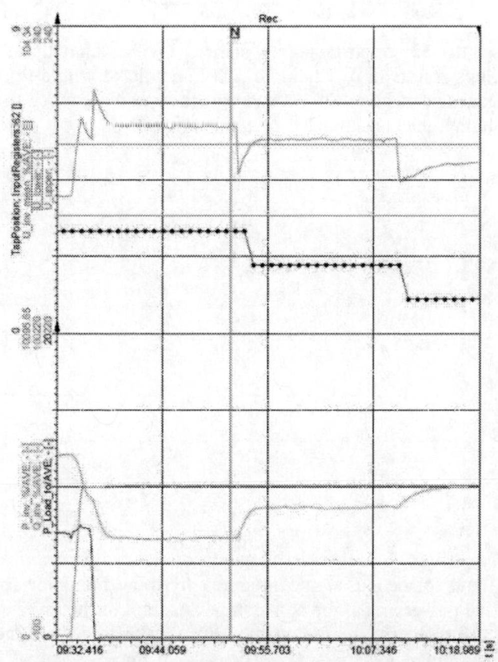

Figure 9: Voltage characteristics (blue) and the position of the tap changer (blue dots) shown in the top and the characteristics of the delivered reactive (green) and active (orange) power by the DUT at a time constant setting of 2 seconds together with a load loss (blue below).

As shown in Fig. 9 if the voltage is above the threshold it needs to step down for the transformer to stay in the voltage band, due to counter action of the Q(U) voltage control if the signs of the static Q(U) settings are chosen well.

In principle in most of the cases both actors the tap-changer and the Q(U) control will both work against voltage changes occurring due to irradiance changes over one day as seen in Fig. 10.

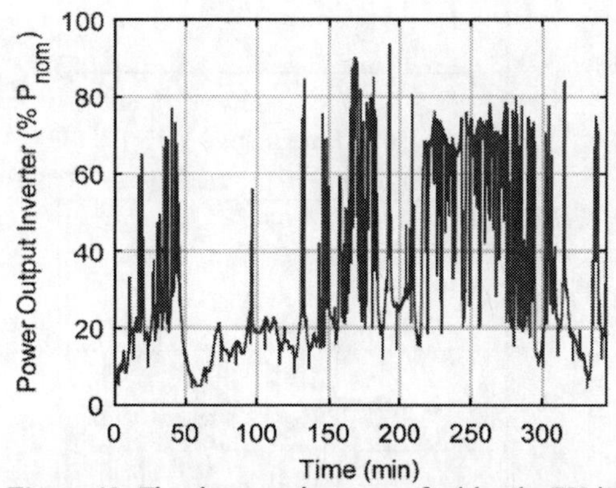

Figure 10: The shown active power feed by the PV inverter into the grid is the result of the applied solar irradiance characteristics over 350 minutes in the HIL setup as a function of DC amplifier output.

The result of this is that about 90% of that period the voltage is within a bandwidth of about 1% as illustrated by Fig. 12. In detail in Fig. 11 also the number of switching actions of the tap changer is seen with the threshold values at 99 and 101%.

Tab. 1 Duration in % of total period to be above/below the given voltage level around 100% including substation OLTC and Q(U) control.

in%	> 100.5%	>101%	<99.5%	<99
Q(U) aus	4.8	4.8	32.3	2.6
TC5	1.5	1.5	4	0
TC25	1.4	1.4	21.3	0.2

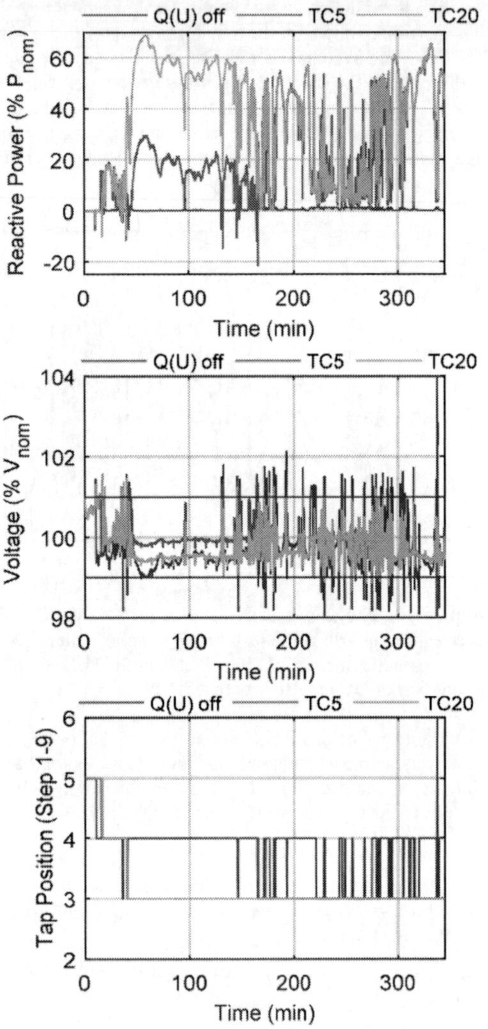

Figure 11: Three-minute voltage characteristics as a function of reactive power applied by a PV inverter during the same real solar irradiance conditions with parameter settings of 1, 5 and 20 seconds time constant TC of the Q(U) controller and without Q(U) control.

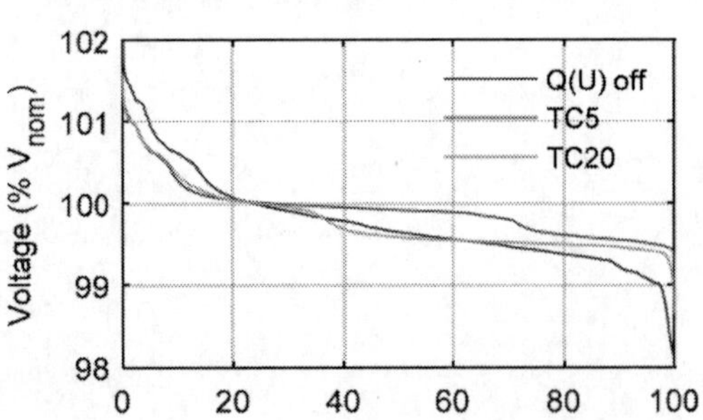

Figure 12: Voltage duration curve over 350 minutes including a substation and reactive power applied by a PV inverter during the same real solar irradiance conditions with parameter settings of 5s and 20s time constant TC of the inverters Q(U) controller and without Q(U) control.

4.4 Measured Oscillation of the line voltage

In principle oscillations expected only if at least to actors in a control loop work in the opposite direction. Like decreasing the voltage by the Q(U) control, due to a step up in irradiance, followed by an action of the OLTC, step down the voltage due to exceeding the threshold limit longer than the delay time.

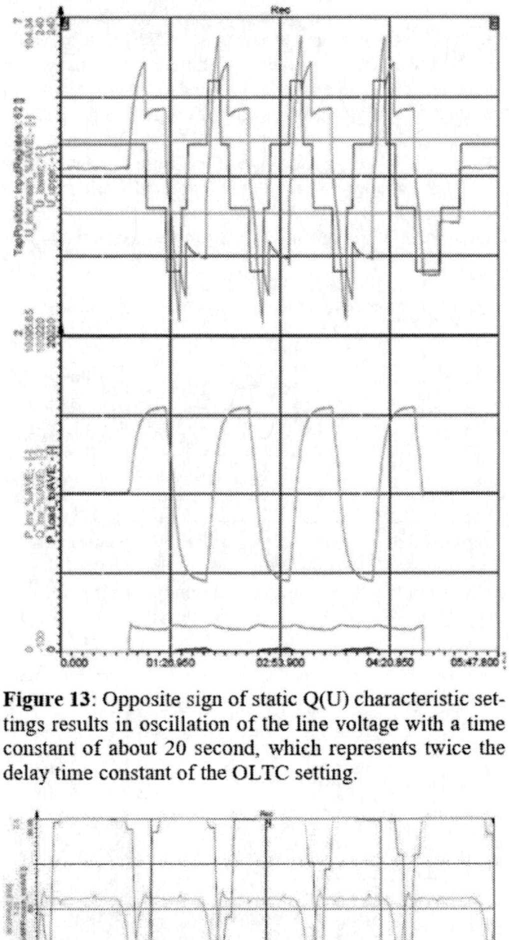

Figure 13: Opposite sign of static Q(U) characteristic settings results in oscillation of the line voltage with a time constant of about 20 second, which represents twice the delay time constant of the OLTC setting.

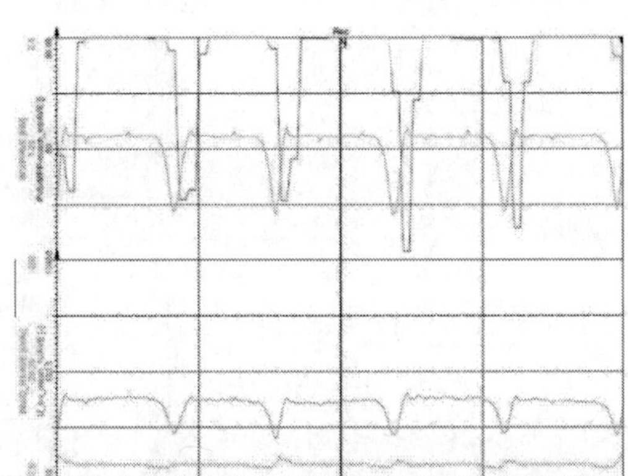

Figure 14: Small line voltage oscillations at a period of about 15 seconds at maximum reactive power and low active power and at the apparent power limit of the inverter performed by increasing the DC voltage to reduce the maximum power tracing efficiency of the PV generator current voltage characteristics.

5 RECOMMENDATIONS AND CONCLUSIONS

The very cost-effective method of reducing the rise of line voltage due to changes of solar irradiance and load was always stable at regular settings of the PV inverter parameters. Oscillation of the line voltage of several percent were found when the sign of the static characteristics of the reactive power demand is mixed up by the PV-installer or the PV inverter settings are hacked. Than the voltage rise is not compensated by the Q(U) but get progressively changed. Later only happen if the PV inverter is connected to a sub-station transformer equipped with an automatic controlled tap-changer interacting with the inverters control regime in a positive feedback mode.

It is recommended to the DG operator grid code to prescribe values of the PV inverter Q(U) time constant to be below 5 seconds resulting in reduced time intervals of overvoltage during the first moments when the Q(U) demands changes in the reactive power production.

It has to be paid attention how the time constant is executed by the inverter, as multiple definitions exist in different national grid codes. The time constant might be defined as the time reaching 1τ (63%) or 3λ (95%) of the required reactive power set point after a voltage step.

ACKNOWLEDGMENT

The authors would like to thank the AIT Austrian Institute of Technology, Vienna for hosting the Transnational Access project "TIPI-GRID" and providing laboratory infrastructure. This research work is supported with the European Union's Horizon 2020 grant 654113 (ERIGrid) under the Trans-national Access programme.

REFERENCES

[1] F. Baumgartner, F. Carigiet, P. Korba; Swiss Federal Bureau of Energy, BFE-Contract No. SI/501370-01, Project CEVSOL; public annual Report 2016, 2017

[2] F. Carigiet, M. Niedrist, C. Scheuermann, and F. Baumgartner, "Case study of a low-voltage distribution grid with high PV penetration in Germany and simulation analyses of cost-effective measures," Prog. Photovolt. Res. Appl., 2015

[3] F. Carigiet, F. P. Baumgartner, P. Korba, M. Niedrist; "Optimisation of the Load Flow Calculation Method of Low-Voltage Distribution Grids", 33rd EU PV Conference; Amsterdam, Sept 2017

[4] F.P. Baumgartner, F. Carigiet, C. Messner, C. Seitl, T. Strasser, R. Bründlinger; "Stability of Photovoltaic Inverters Reactive Power Control by the distribution GRID voltage"; ISENEC conference, Nürnberg, July 2018

[5] F. Andrén, B. Bletterie, S. Kadam, P. Kotsampopoulos and C. Bucher, "On the Stability of Local Voltage Control in Distribution Networks With a High Penetration of Inverter-Based Generation," in IEEE Transactions on Industrial Electronics, vol. 62, no. 4, pp. 2519-2529, April 2015[6]

[6] A. Constantin and R. D. Lazar, "Open loop Q(U) stability investigation in case of PV power plants," in Proc. 27th Eur. Photovoltaic Solar Energy, Conf. Exhib., Frankfurt, Germany, 2012, pp. 3745–3749

[7] O. Marggraf, I. Berber, B. Engel; " Sichere Netzintegration mittels Q(U)-Regelung und rONT im Praxistest "; PV Tagung Staffelstein, März 2018

VALIDATION OF THE METEONORM SATELLITE IRRADIATION DATASET

Stefan C. Mueller and Jan Remund
Meteotest AG
Fabrikstrasse 14, CH-3012 Bern, Switzerland, stefan.mueller@meteotest.ch

ABSTRACT: Global irradiation is the main source for photovoltaic (PV) energy production. The knowledge of the incoming solar short-wave irradiation is of major importance for the planning, monitoring and operation of solar energy power plants and grid operations. In this paper we present a validation of the newly developed Meteonorm Satellite Irradiation (MNSI) dataset for Europe, Middle East and Africa. Uncertainties of 12-25% relative RMSE for hourly values has been found.
Keywords: solar resources, irradiation data, satellite data

1 INTRODUCTION

Global irradiation is the main source for photovoltaic (PV) energy production. A large number of PV power plants produce a significant amount of power penetrating into the electrical grid under sunny conditions. Therefore the knowledge of the incoming solar short-wave irradiation is of major importance for the planning, monitoring and operation of solar energy power plants and grid operations.

Ground station measurements from national weather services still provide the most accurate data for global irradiation. They represent the ground truth, which is the relevant parameter for solar energy applications. However, the network of the measurement stations is not dense enough to cover an entire area in sufficient spatial resolution. In this context, satellites have become a valuable source for solar irradiation data, in particular in areas with sparse distribution of meteorological stations. Satellite images from weather satellites have a horizontal grid resolution of a few km^2. Beside the availability of high resolution gridded time series, the rapid update of satellite images every 15 minutes in real time allows to retrieve information of the actual cloud state. This allows the use for monitoring and nowcasting applications [1], [2].

In this paper we present a validation of the newly developed Meteonorm Satellite Irradiation (MNSI) Dataset for Europe, Middle East and Africa. It is based on satellite images from Meteosat Second Generation (MSG). The dataset was compiled combining the Heliosat method [3, 4] applied to the visible channels 1 (VIS06) and 12 (HRV) with a multichannel retrieval for snow cover detection. Irradiation values from the HRV channel images are calculated on a 1/40 degree grid (Figure 1), those from the VIS06 channel on a 1/16 degree grid (Figure 2). The spatial extension is from 66°N to 66°S and 66°W to 66°E. A regionalized statistical regression approach integrating ground measurements is adopted for bias correction and uncertainty reduction. The dataset covers a time period from 2006 onwards and is updated in real-time.

The dataset is compared to reference ground measurement stations from the Baseline Surface Radiation Network BSRN [5] (http://bsrn.awi.de/). These stations are not included in the statistical bias correction.

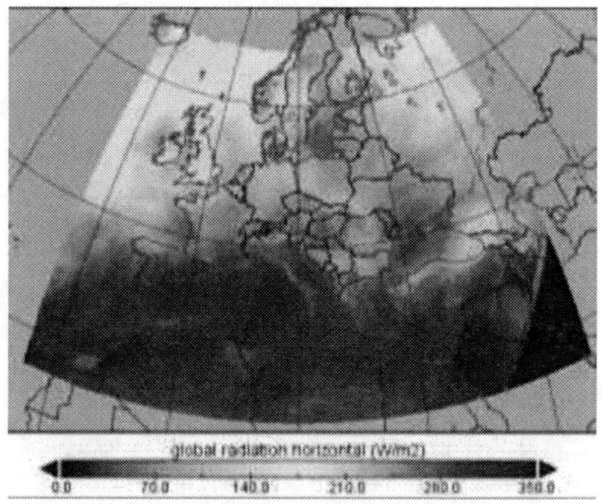

Figure 1: Global horizontal irradiance for May 2017 showing the area covered by the high resolution visual (HRV) channel.

Figure 2: Global horizontal irradiance for May 2017 showing the area covered by the full disk VIS06 channel.

2 METHOD

We validate the satellite dataset by a comparison to reference ground stations from the BSRN. The recommended quality checks [6] have been applied to the BSRN data.

2.1 Validation sites

Table I and Figure 3 shows the list of the validation sites. Eleven sites were used, where nine are located in

Europe from Spain up to Estonia and two are located in the southern part of Africa.

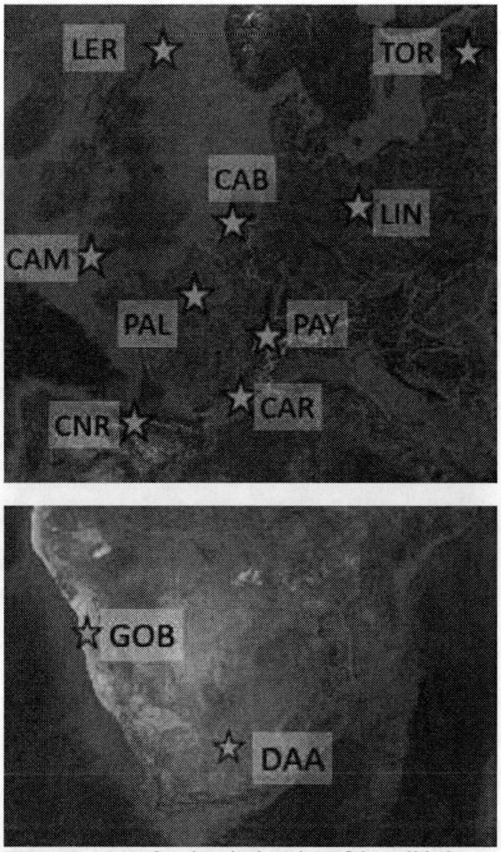

Figure 3: Maps showing the location of the validation sites (top: Europe / bottom: Southern Africa)

Table I: List of the validation sites including the period of data used to validate.

Abbr.	Name	Latitude [°]	Longitude [°]	Period
CAB	Cabauw	51.9711	4.9267	2006-2017
CAR	Carpentras	44.0830	5.0590	2006-2017
CAM	Camborne	50.2167	-5.3167	2006-2017
CNR	Cener	42.8160	-1.6010	2009-2017
LIN	Lindenberg	52.2100	14.1220	2006-2017
LER	Lerwick	60.1389	-1.1847	2006-2017
TOR	Toravere	58.2540	26.4620	2006-2017
PAY	Payerne	46.8150	6.9440	2007-2013
PAL	Palaiseau	48.7130	2.2080	2006-2017
GOB	Gobabeb	-23.5614	15.0420	2013-2017
DAA	De Aar	-30.6667	23.9930	2015-2017

2.2 Validation procedure

The following procedure has been applied to validate the dataset.

- The dataset is validated on a per station basis.
- Hourly and daily values are validated separately.
- The statistical measures bias and rmse are calculated in absolute and relative values. The relative value was gained by division of the absolute value with the average measured value.
- Only daytime values are used for the comparison. We defined daytime when GHI (Global horizontal irradiance) is above 10 W/m^2.

3 RESULTS

Figure 4 and Figure 5 are showing exemplary scatter plots of hourly and daily value comparisons of satellite versus ground data for the location of Carpentras. RMSE is 14.8% for hourly and 9.1% for daily values, which is a relative improvement of 38% in RMSE by averaging to daily values. Figure 6 (hourly values) and Figure 7 (daily values) show absolute and relative values of the statistical measures for all sites. Sites below 55°N show a RMSE of 60-70 W/m^2 or 12-25% relative RMSE. Higher latitude sites show larger uncertainties. RMSE of daily values are in the range of 9-16% for the sites below 55°N. This is a relative reduction of 30-40%.

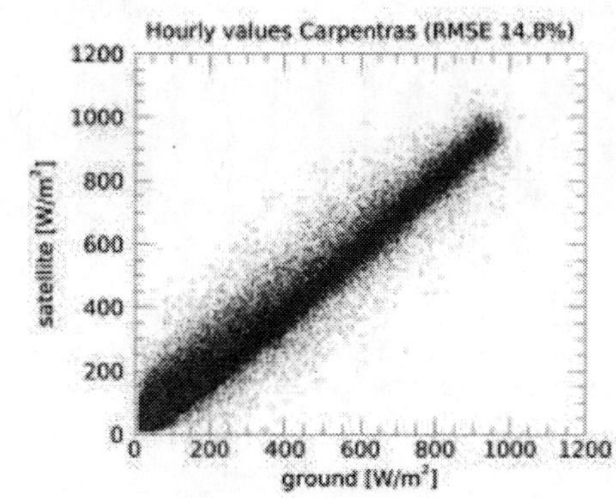

Figure 4: Scatter plot of hourly values of GHI for Carpentras (FR). Relative RMSE is 14.8% for this location.

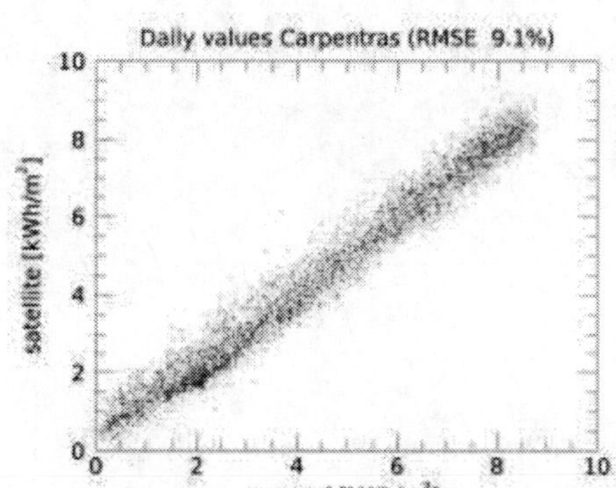

Figure 5: Scatter plot of daily values of GHI for Carpentras (FR). Relative RMSE is 9.1% for this location.

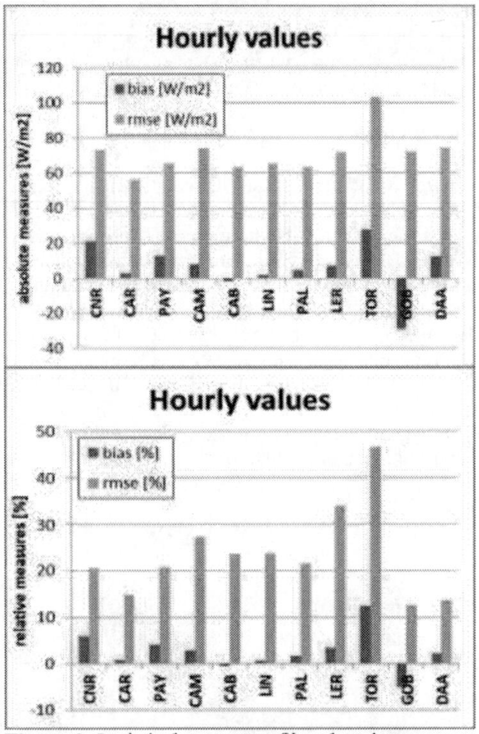

Figure 6: Statistical measures of hourly values per station (green: bias, violet: rmse). Absolute values in top Figure and relative values at the bottom Figure.

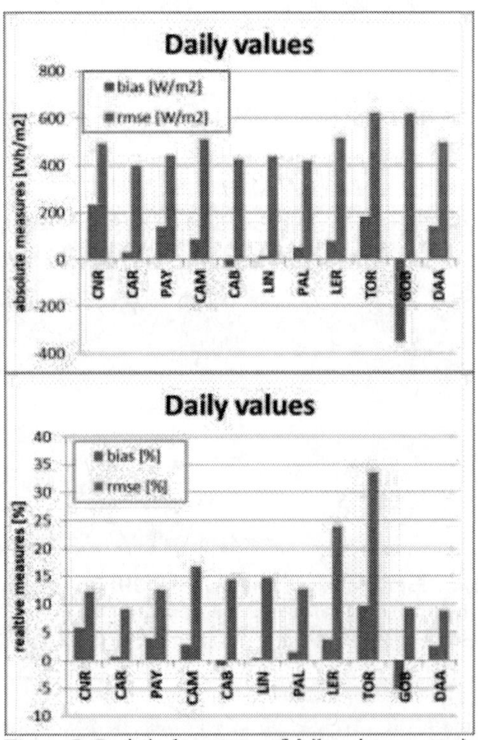

Figure 7: Statistical measures of daily values per station (blue: bias, orange: rmse). Absolute values in top Figure and relative values at the bottom Figure.

4 AVAILABILITY

The dataset is available through the climate database software tool Meteonorm in version 7.3 (www.meteonorm.com).

5 CONCLUSION

We presented the validation of the newly compiled Meteonorm Satellite Irradiation (MNSI) Dataset for Europe and Africa. A representative number of eleven stations, nine across Europe and two in the Southern part of Africa, from the BSRN network served as reference for the validation. A RMSE of 60-70 W/m^2 for hourly values has been found. This is in the order of 20% for the temperate zone in Europe and around 13% for the sunny African sites. Larger uncertainties were found in high latitude sites above 55°N in Northern Europe. Aggregating hourly to daily values reduces the uncertainty. Errors are smoothed out. The gain in relative RMSE is in the order of 30-40%.

6 REFERENCES

[1] Müller & Remund, PVSEC 2014, Satellite based shortest term solar energy forecast system for entire Europe for the next hours

[2] Müller & Remund, PVSEC 2016, Shortest term forecasting of DNI for concentrated solar technologies

[3] Cano, D., Monget, J. M., Albuisson, M., Guillard, H., & Wald, L. 1986. A method for the determination of the global solar radiation from meteorological satellite data. Solar Energy, 37(1), 31–39.

[4] Rigollier C. et al., 2004: The method Heliosat-2 for deriving shortwave solar radiation data from satellite images. Solar Energy, 77(2), 159-169

[5] Driemel A. et al., 2018: Baseline Surface Radiation Network (BSRN): structure and data description (1992–2017), Earth Syst. Sci. Data, 10, 1491-1501

[6] Long C.N. and Dutton E.G., BSRN Global Network recommended QC tests, V2.0 (Document from https://wiki.pangaea.de/wiki/BSRN_Toolbox#Quality_Check)

6 ACKNOWLEDGEMENTS

Thanks to the World Radiation Monitoring Center (https://bsrn.awi.de/) for providing the data from the BSRN station network.

Assessment of Error Sources in the Modeling of POA Irradiance under Clear-sky Conditions

Yu Xie and Manajit Sengupta
Power System Engineering Center, National Renewable Energy Laboratory
15013 Denver West Parkway, Golden, CO 80401, United States

ABSTRACT: A new physics-based model, the Fast All-sky Radiation Model for Solar applications with Narrowband Irradiances on Tilted surfaces (FARMS-NIT), accurately computes plane-of-array (POA) irradiances in both narrow- and broad wavelength bands within the solar region. The evaluation studies suggest the new model provide more accuracy than a model assuming isotropic diffuse solar radiation and an empirical model linking the POA irradiances with long-term surface observations at selected locations. This study anlyzes the surface observations used by FARMS-NIT and is intended to understand the error sources under clear-sky conditions as well as their impact to the performance of FARMS-NIT.
Keywords: solar radiation, resource assessment, POA irradiance

1 INTRODUCTION

Conventional models that estimate the production of solar energy conversion systems compute or measure solar radiation on a horizontal surface and convert it to plane-of-array (POA) irradiance [1, 2]. To improve the accuracy and efficiency of those models, Xie et al. [1] and Xie and Sengupta [3] developed the Fast All-sky Radiation Model for Solar applications with Narrowband Irradiances on Tilted surfaces (FARMS-NIT) to provide physics-based solutions of spectral irradiance in the POA. This new model employs the Simple Model of the Atmospheric Radiative Transfer of Sunshine (SMARTS) [4] to efficiently compute clear-sky atmospheric properties, e.g. the optical thickness of the atmosphere, in 2,002 wavelength bands from 0.28-4.0 μm.

Following the Rayleigh scattering correction technique, FARMS-NIT considers five independent events where photons (1) directly transmit through the aerosol/cloud, (2) are scattered by gas molecules in the upper clear sky and scattered by the aerosol/cloud, (3) are scattered by gas molecules in the upper clear sky and directly transmit through the aerosol/cloud, (4) are scattered by the aerosol/cloud and scattered by gas molecules in the lower clear sky, and (5) directly transmit through the aerosol/cloud and are scattered by gas molecules in the lower clear sky. Then surface radiances from all spacial directions are computed by considering the possible paths for individual photon transmission, and the relevant absorption, scattering and reflection by aerosols, clouds and land surface. For clear-sky conditions, the computation is simplified using the assumption of single scattering within aerosols and the single-scattering phase functions. For cloudy-sky conditions, we developed a lookup table (LUT) of cloud bi-directional transmittance distribution functions (BTDFs) for 97 wavelengths, 39 optical thicknesses, 28 effective particle sizes, 50 solar zenith angles, 25 viewing zenith angles and 18 relative azimuth angles. The POA irradiances are then accurately computed by integrating radiances over inclined surfaces with any tilt angles. The broadband POA irradiance is calculated by integrating the spectral irradiances over the 2,002 wavelength bands.

The FARMS-NIT will be used to provide spectral irradiances in the POA using satellite observations and the estimation of cloud microphysical and optical properties [5-16]. The simulations will be disseminated by the National Renewable Energy Laboratory's (NREL's) National Solar Radiation Data Base (NSRDB) [17]. The operational application of FARMS-NIT requires reliable validations in various scenes to understand the magnitude of the model uncertainties. An analysis is also required to understand the error sources associated with the uncertainty in FARMS-NIT.

2 EVALUATION OF POA IRRADIANCE

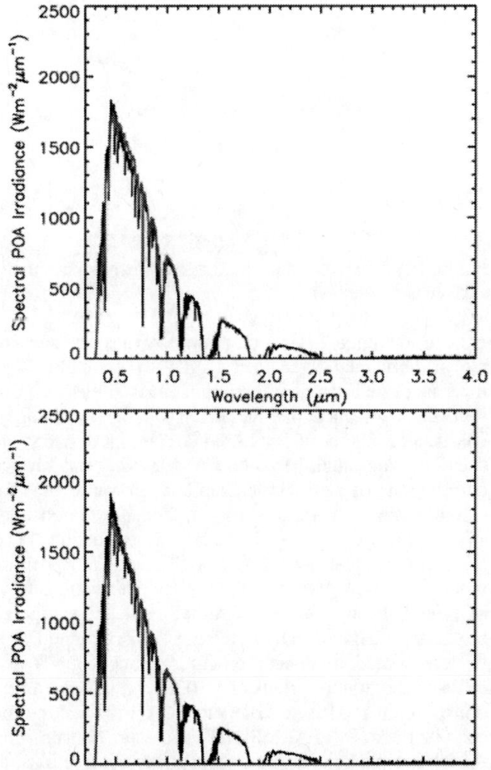

Figure 1 Spectral POA irradiance on a one-axis tracker measured by a LI-1800 (red line) and simulated by FARMS-NIT (black line) at 8:30 am (upper panel) and 12:05 pm (lower panel) on June 18, 2017.

We utilize the surface-based observations at the NREL's Solar Radiation Research Loboratory (SRRL) to validate FARMS-NIT in clear-sky conditions [1]. An EKO WISER spectroradiometer on a one-axis tracker is used to measure spectral POA irradianes from 0.3 to 1.1 μm for every 5 minutes. The aerosol optical depth (AOD), surface albedo, surface air pressure, and solar and PV orientations are measured for every 1 minute. The precipitable water vapor (PWV) is measured for every 30 minutes by a Zephyr Geodetic GPS antenna. The surface observations are collocated to the time steps of the spectral POA irradiance measurements for each 5 minutes, and they are applied to FARMS-NIT to compute spectral POA irradiances from 0.28 to 4.0 μm. The model simulations are compared with the surface observations at NREL's SRRL.

Our results suggest that the percentage error (PE) and absolute percentage error (APE) of FARMS-NIT are 0.615% and 4.93%, respectively, over the 121 5-min intervals on June 18, 2017. These errors are significantly lower than models assuming isotropic diffuse solar radiation [18] and an empirical model linking the POA irradiances with long-term surface observations at selected locations [2].

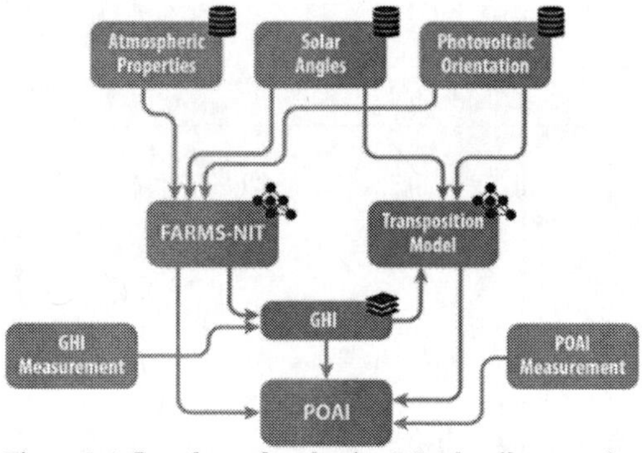

Figure 2 A flowchart of evaluating POA irradiances using surface measurements.

Current model evaluation efforts use measurements of global horizontal irradiance (GHI) to compute POA irradiances and compare them with surface observations [19-22]. To reduce the restriction by the availability, consistency, and accuracy of the GHI measurements, we use a new algorithm based on FARMS-NIT, and surface observations of AOD, precipitable water vapor (PWV), cloud fraction, surface pressure, surface albedo and solar angles. Figure 2 is a flowchart of evaluating POA irradiance computed by FARMS-NIT and transposition models. We first collocated the surface observations to 5-min intervals, input them to FARMS-NIT, and computed broadband GHI and POA irradiance. The computed GHI is decomposed to direct normal irradiance (DNI) and diffuse horizontal irradiance (DHI) and used by transposition models to compute POA irradiances. The computations by FARMS-NIT and the transposition models are finally compared with surface observations.

3 ASSESSMENT OF ERROR SOURCES IN FARMS-NIT AND TRANSPOSITION MODELS

Our analysis indicates the uncertainty of the transposition models comes from the inherent limitations of the models, e.g. limitation of directions and locations of GHI and POA irradiance measurements. The uncertainty of the transposition models may also come from the computed GHI by FARMS-NIT as well as measurements of the atmospheric properties to compute GHI. It is thus important to assess the uncertainty sources of FARMS-NIT for the optimal employment of the models and to further improve their performance.

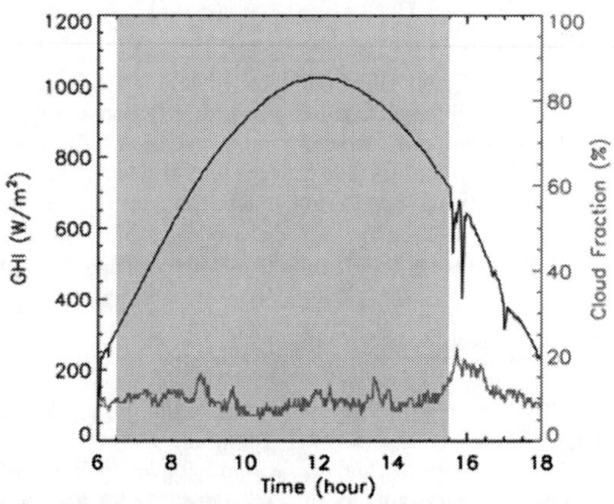

Figure 3 Observation of GHI and cloud fraction at NREL's SRRL on June 18, 2017.

Because FARMS-NIT directly solves POA irradiance from the computed radiances, the error sources come from the uncertainty and limitation of the atmospheric properties used to compute solar radiances. To evaluate FARMS-NIT in clear-sky conditions, we selected relatively clear days at NERL's SRRL based on measured GHI by a Kipp and Zonene CM Pyranometer 22 (CMP22) and cloud fraction by a Yankee Total Sky Imager at NREL's SRRL. Figure 3 shows the GHI and cloud fraction on June 18, 2017 that is selected because the average cloudy fraction is less than 15%. However, the 100% clear-sky condition assumed by FARMS-NIT leads to uncertainties because of the partially covered cloud conditions.

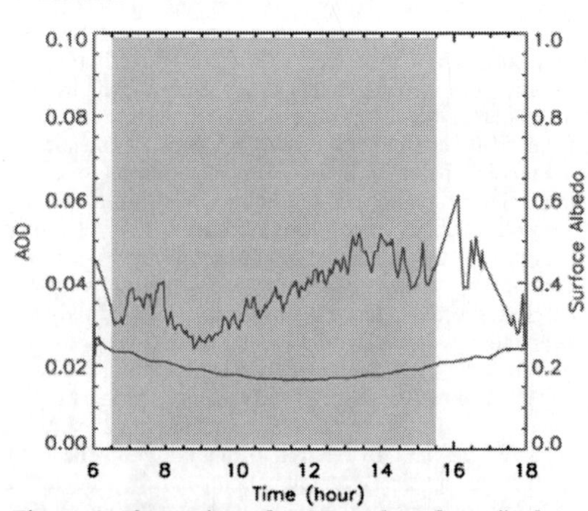

Figure 4 Observation of AOD and surface albedo at NREL's SRRL on June 18, 2017.

Other important error sources include measurements of AOD and land-surface albedo. Figure 4 shows the daily variation of AOD, which indicates the use of daily averaged AOD may lead to significantly bias in a day. FARMS-NIT uses measurements of broadband surface albedo because those in spectral bands are unavailable at

NREL's SRRL. Following previous studies by McKenzie and Kotkamp [23], spectral surface albedo of green grass tends to be smaller and greater than broadband albedo in visible and near-infrared regions, respectively. Therefore, the land-surface albedo may lead to significantly bias when used by FARMS-NIT to compute spectral irradiances. Their impact on the simulation of broadband irradiance requires further sensitivity studies.

4 CONCLUSIONS

FARMS-NIT solves POA irradiance from the optical and physical properties of the atmosphere, and it thus reduces the inherence limitation in the transposition models that relies on long-term observations of GHI and POA irradiance. FARMS-NIT has less than 5% uncertainties as evaluated by the surface observations at NREL's SRRL. Our study suggests that the error sources in FARMS-NIT come from the use of the clear-sky model when a small fraction of cloud exists. The error sources in FARMS-NIT can also be related to infrequent observations of AOD and the representation of spectral surface albedo by observations in the broadband region.

ACKNOWLEDGEMENTS

This work was authored by Alliance for Sustainable Energy, LLC, the Manager and Operator of the National Renewable Energy Laboratory for the U.S. Department of Energy (DOE) under Contract No. DE-AC36-08GO28308. Funding provided by U.S. Department of Energy Office of Energy Efficiency and Renewable Energy Solar Energy Technologies Office. The views expressed in the article do not necessarily represent the views of the DOE or the U.S. Government. The U.S. Government retains and the publisher, by accepting the article for publication, acknowledges that the U.S. Government retains a nonexclusive, paid-up, irrevocable, worldwide license to publish or reproduce the published form of this work, or allow others to do so, for U.S. Government purposes.

We acknowledge the DOE Solar Energy Technologies Program and its PV subprogram for supporting this research. Specifically, we thank Dr. Dave Rench-McCauley, Dr. Lenny Tinker, Dr. Rebecca Jones-Albertus, Dr. Charlie Gay, and the PV subprogram for their support and encouragement.

REFERENCES

[1] Y. Xie, M. Sengupta, and M. Dooraghi, "Assessment of uncertainty in the numerical simulation of solar irradiance over inclined PV panels: New algorithms using measurements and modeling tools," *Sol. Energy*, vol. 165, pp. 55-64, 2018.

[2] R. Perez, R. Seals, P. Ineichen, R. Stewart, and D. Menicucci, "A new simplified version of the Perez diffuse irradiance model for tilted surfaces," (in English), *Sol. Energy*, vol. 39, no. 3, pp. 221-231, 1987.

[3] Y. Xie and M. Sengupta, "A Fast All-sky Radiation Model for Solar applications with Narrowband Irradiances on Tilted surfaces (FARMS-NIT): Part I. The clear-sky model," *Sol. Energy*, submitted.

[4] C. Gueymard, "SMARTS2: a simple model of the atmospheric radiative transfer of sunshine: algorithms and performance assessment," *Florida Solar Energy Center*, vol. Cocoa, FL, 1995.

[5] Y. Xie, "Study of ice cloud properties from synergetic use of satellite observations and modeling capabilities," Ph. D., Department of Atmospheric Sciences, Texas A&M University, College Station, TX, 2010.

[6] Y. Xie and Y. G. Liu, "A new approach for simultaneously retrieving cloud albedo and cloud fraction from surface-based shortwave radiation measurements," *Environ. Res. Lett.*, vol. 8, pp. doi:10.1088/1748-9326/8/4/044023, 2013.

[7] Y. Xie, Y. G. Liu, C. N. Long, and Q. L. Min, "Retrievals of cloud fraction and cloud albedo from surface-based shortwave radiation measurements: A comparison of 16 year measurements," (in English), *J. Geophys. Res. Atmos.*, vol. 119, no. 14, pp. 8925-8940, Jul 27 2014.

[8] Y. Xie, M. Sengupta, and J. Dudhia, "A Fast All-sky Radiation Model for Solar applications (FARMS): Algorithm and performance evaluation," *Sol. Energy*, vol. 135, pp. 435-445, 2016.

[9] Y. Xie, P. Yang, G. W. Kattawar, P. Minnis, and Y. X. Hu, "Effect of the inhomogeneity of ice crystals on retrieving ice cloud optical thickness and effective particle size," (in English), *J. Geophys. Res.*, vol. 114, pp. D11203, doi:10.1029/2008JD011216, Jun 5 2009.

[10] Y. Xie, P. Yang, G. W. Kattawar, P. Minnis, Y. X. Hu, and D. Wu, "Determination of ice cloud models using MODIS and MISR data," *Int. Remote Sens.*, vol. 33, pp. 4219-4253, 2012.

[11] Y. Xie, P. Yang, K. N. Liou, P. Minnis, and D. P. Duda, "Parameterization of contrail radiative properties for climate studies," *Geophys. Res. Lett.*, vol. 39, no. L00F02, p. doi:10.1029/2012GL054043, 2012.

[12] Y. Xie, P. Yang, B. C. Gao, G. W. Kattawar, and M. I. Mishchenko, "Effect of ice crystal shape and effective size on snow bidirectional reflectance," (in English), *J. Quant. Specrtrosc. Radiat. Transfer* vol. 100, no. 1-3, pp. 457-469, Jul-Aug 2006.

[13] Y. Xie, F. Wei, G. Chen, T. Zhang, and L. Hu, "Analysis of the 2008 heavy snowfall over South China using GPS PWV measurements from the Tibetan Plateau," (in English), *Ann. Geophys.*, vol. 28, no. 6, pp. 1369-1376, 2010.

[14] Y. Xie, P. Yang, G. W. Kattawar, B. Baum, and Y. X. Hu, "Simulation of the optical properties of ice particle aggregates for application to remote sensing of cirrus clouds," *Appl. Opt.*, vol. 50, pp. 1065-1081, 2011.

[15] P. Minnis *et al.*, "CERES edition-2 cloud property retrievals using TRMM VIRS and TERRA and AQUA MODIS data, part I: algorithms," *IEEE Transactions on Geoscience and Remote Sensing*, vol. 49, no. 11, pp. 4374-4400, 2011.

[16] R. Lawless, Y. Xie, and P. Yang, "Polarization and effective Mueller matrix for multiple scattering of light by nonspherical ice crystals," (in English), *Opt. Express*, vol. 14, no. 14, pp. 6381-6393, Jul 10 2006.

[17] M. Sengupta, Y. Xie, A. Lopez, A. Habte, G. Maclaurin, and J. Shelby, "The National Solar Radiation Data Base (NSRDB)," *Renew. Sustain. Energy Rev.*, vol. 89, pp. 51-60, 2018.

[18] B. Liu and R. Jordan, "The long-term average performance of flat-plate solar-energy collectors," *Sol. Energy,* vol. 7, no. 2, pp. 53-74, 1963.

[19] C. Gueymard and J. Ruiz-Arias, "Extensive worldwide validation and climate sensitivity analysis of direct irradiance predictions from 1-min global irradiance," *Sol. Energy,* vol. 128, pp. 1-30, 2016.

[20] M. Lave, W. Hayes, A. Pohl, and C. Hansen, "Evaluation of global horizontal irradiance to plane-of-array irradiance models at locations across the United States," *J. PV,* vol. 5, no. 2, pp. 597-606, 2015.

[21] P. Loutzenhiser, H. Manz, C. Felsmann, P. Strachan, T. Frank, and G. Maxwell, "Empirical validation of models to compute solar irradiance on inclined surfaces for buiding energy simulation," *Sol. Energy,* vol. 81, pp. 254-267, 2007.

[22] A. Noorian, I. Moradi, and G. Kamali, "Evaluation of 12 models to estimate hourly diffuse irradiation on inclined surfaces," *Renew. Energy,* vol. 33, pp. 1406-1412, 2008.

[23] R. McKenzie and M. Kotkamp, "Upwelling UV spectral irradiances and surface albedo measurements at Lauder, New Zealand," *Geophys. Res. Lett.,* vol. 23, pp. 1757-1760, 1996.

IEA PVPS TASK 16 AND IEA SOLARPACES V: STATE OF THE PROJECT

Remund Jan[1], Philippe Blanc[2] and Perez Richard[3]
[1]*Meteotest, Fabrikstrasse 14, 3012 Bern, Switzerland*
[2]*Mines Paristech, Sophia Antipolis, France*
[3]*SUNY, Albany NY, USA*
E-mail: jan.remund@meteotest.ch

ABSTRACT: A new joint Task of IEA Technological Collaboration Programmes PVPS and SolarPACES started in July 2017 and will last till 2020. 21 countries and 54 partners are part of the Task. Many existing partners of IEA SHC Task 46 like NREL or Mines Paristech are part but also new partners – like the Swedish met service SMHI or Energinet (TSO of Denmark). In 2018 two workshops where held to get basic information about uncertainty of resources and forecast needs. The results will be shown in this paper.
The Task offers a unique platform of exchange for private and public organizations as well as data providers and science organizations. A strong emphasis on extra gain of collaboration will be promoted. Benchmarks, standardizations, state-of the art descriptions and dissemination will be in the focus. Main added values are: lower resource uncertainties and higher forecast accuracies, which will lead to lower implementation costs of PV and CSP projects.
The first year was used to gather basic information for benchmarks and user needs. Two workshops where held of which the results will be presented. The first workshop (March 2018) was dedicated to the definition of uncertainty in dependence of climate. Up to now uncertainty information was presented differently by each service providers. A common way of definition would help the users to compare the services.
The second workshop was held at Intersolar June 2018 to get requirements concerning solar and PV forecasts for different applications. A focus was set to utilities and grid operation. Enhanced knowledge of user needs will enhance the service.

1 STATE OF THE TASK

The first of three years was used to organise the team and collaboration, to get to know user needs and the state of the groups' knowledge. The Task consists of 56 participants of 21 different countries (http://www.iea-pvps.org/index.php?id=389).
Many groups are working on the diverse topics of the Task. In the following two examples of publications done by group members linked to the work plan are shown.
The team of German Weather Service (DWD) looked at the variability and trends of surface solar radiation in Europe based on CM SAF satellite data records (www.cmsaf.eu). [1].

The following conclusions could be drawn:

- CM SAF SARAH and CLARA reproduce (most) trends derived from surface measurements
- Largest trends in Europe during spring (> 5 W/m2/dec); average trend about 2 W/m2/dec
- There is substantial variability in the linear trend: spatial, decadal, and seasonal!

Fig. 1 shows the trends of global radiation based on surface and satellite measurements.

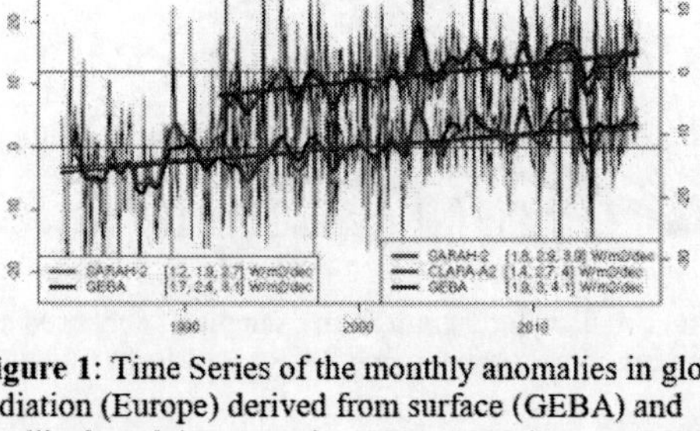

Figure 1: Time Series of the monthly anomalies in global radiation (Europe) derived from surface (GEBA) and satellite based data records, 1983 – 2015 (source: DWD).

JRC published a paper about evaluation of the latest re-analysis datasets [2].

This study examines the progress made by two new reanalyses in the estimation of surface irradiance: ERA5, the new global reanalysis from the ECMWF, and COSMO-REA6, the regional reanalysis from the DWD for Europe. Daily global horizontal irradiance data were evaluated with 41 BSRN stations worldwide, 294 stations in Europe, and two satellite-derived products (NSRDB and SARAH). ERA5 achieves a moderate positive bias worldwide and in Europe of +4.05 W/m2 and +4.54 W/m2 respectively, which entails a reduction in the average bias ranging from 50% to 75% compared to ERA-Interim and MERRA-2 (Fig. 2). This makes ERA5 comparable with satellite-derived products in terms of the mean bias in most inland stations, but ERA5 results degrade in coastal areas and mountains. The bias of ERA5 varies with the cloudiness, overestimating under cloudy conditions and slightly underestimating under clear-skies, which suggests a poor prediction of cloud patterns and leads to larger absolute errors than that of satellite-based products.
In Europe, the regional COSMO-REA6 underestimates in most stations (MBE = −5.29 W/m2) showing the largest deviations under clear-sky conditions, which is most likely caused by the aerosol climatology used. Above 45°N the magnitude of the bias and absolute error of COSMO-REA6 are similar to ERA5 while it outperforms ERA5 in the coastal areas due to its high-resolution grid (6.2 km). We conclude that ERA5 and COSMO-REA6 have reduced the gap between reanalysis and satellite-based data, but further development is required in the prediction of clouds while the spatial grid of ERA5 (31 km) remains inadequate for places with high variability of surface irradiance (coasts and mountains). Satellite-based data should be still used when available, but having in mind their limitations, ERA5 is a valid

alternative for situations in which satellite-based data are missing (polar regions and gaps in times series) while COSMO-REA6 complements ERA5 in Central and Northern Europe mitigating the limitations of ERA5 in coastal areas.

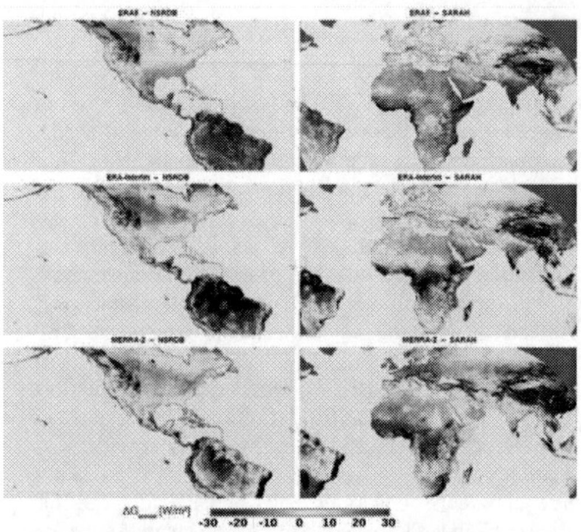

Figure 2: Comparison of global reanalysis with satellite estimates (source: JRC)

2 WORKSHOP ON UNCERTAINTY OF SOLAR DATASETS FROM SATELLITE AND RE-ANALYSIS PRODUCTS

The main objective of the workshop was to initiate, within PVPS Task 16 and SolarPACES Task V, a common effort to revisit the comparison protocols between satellite-derived estimates or re-analyses outputs and in-situ measurements of the downwelling solar radiation at surface.

During this first workshop, a list of the properties related to uncertainty of satellite-derived solar products that could be assessed has been proposed and discussed:

- the estimates should reproduce the time-series of observations on a one-to-one basis;
- the Cumlative Distribution Function (CDF) of the estimates should reproduce the CDF of the observations;
- the estimates should reproduce the temporal variability of the observations for a given range of timescales;
- the estimates should reproduce the spatial variability of the observations for a given range of spatial scales.

The next step will be to complete, if needed, this list and to set up, document and accept a protocol for each property.

For more than 40 years, many scientific articles and reports have been published that compare the satellite-derived estimates to qualified measurements performed at ground stations. A survey of this literature shows that for a given satellite product, the uncertainty is a function of the local climate including effects of the orography, the air mass and its spectral distribution through the solar zenithal angle, cloud presence and their optical properties and the composition of the cloud-free atmosphere. This list is not exhaustive.

The workshop organised by PVPS Task 16 / SolarPACES Task V aimed to set up or contribute to protocols for a better description of the uncertainty of estimates of solar radiation from satellite-based methods in dependence of local climate.

A better description will support a better modelling of the uncertainty and further will contribute to an enhanced uptake of such estimates by the targeted communities: developers, planners, investors, banks, builders, direct marketers and maintenance companies of PV, solar thermal and concentrating solar power installation and operation, utilities, distribution (DSO) and transmission system operators (TSO), academics and researchers.

Furthermore, a better description of the uncertainty supports more accurate methods for solar modelling, satellite-based estimation and fusion of satellite-derived products with ground measurements and other similar products.

During the workshop the following conclusions have been drawn

- The use updated criteria seems adequate
- separate bias and standard deviations for investments / site assessments
- rmse is not a good measure for long term assessments
- complexity vs. inclusiveness is one of the main issues
- user need to understand criteria (not only producers and science)
- Taylor diagram [3] includes many criteria in one plot – helpful or to complex? (Fig. 3)
- Within Task 16 we will write a new user guide for uncertainty measures on basis of Mesor report [4] and cams validation report [5]. It will include a new chapter about how to plot multi-dimensional plots and to interprete them

Figure 3: Taylor diagram as an example of enhanced and complex comparison (source: Mines Paristech)

2 SOLAR FORECAST REQUIREMENTS & VALUE FOR GRID APPLICATIONS

The workshop brought together forecast providers users - - grid operators – from North America and Europe to offer views on forecast products, requirements and value. Over fifty people attended.

The forecast providers, Clean Power Research' Skip Dise, and Fraunhofer ISE's Elke Lorenz, presented applications of their operational solar forecast products, discussed their accuracy, geographic applicability from single plants to regional and national markets, and the influence of current pricing & regulations on value – e.g., the self-consumption market optimization can greatly benefit financially from accurate forecasts with current regulations. Both stressed the importance of model blending in deterministic accuracy and the importance of probabilistic forecasts – e.g. for congestion planning. Examples of grid congestion management and intraday market spikes traceable to forecast uncertainty were presented.

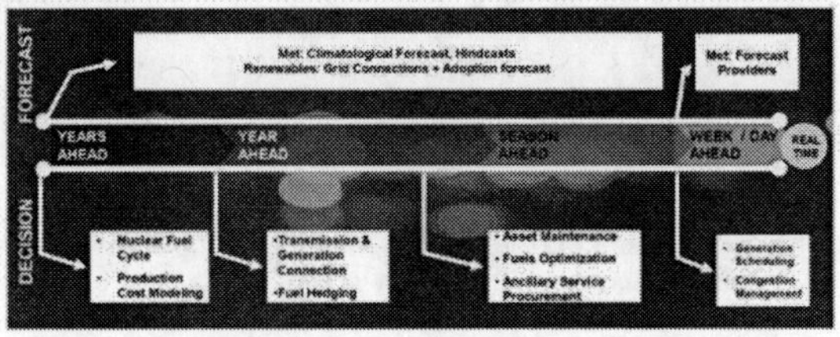

Figure 4: Bulk System Forecasts Uses for Grid Operators (source: EPRI).

On the grid operator front, Eammon Lannoye (EPRI) and Michael Osmann (EnergiNet) placed the solar power forecasts in a TSO time horizon perspective, with time horizons extending from multiple years (production modeling) to one year (fuel hedging) to seasonal (maintenance) to days (market scheduling) up to near real-time (Fig. 4). PV power forecast only pertain to the short operational horizons (< days), although PV deployment forecasts (not weather related) obviously influence multi-year horizons. Examples of comprehensive deterministic forecast evaluation and operational grid balancing with 50% wind and 5% PV in Denmark were presented (Fig. 2).

Michael Osmann of energinet.dk showed, that future has already arrived in Denmark. Situations with very high shares (> 100%) of renewables happen regularly – with 44% average wind power and 3% average PV power (Fig. 5). A mix of pragmatic handling, pro-active measures and good international grid connections helps to tackle the issues linked to the variability.

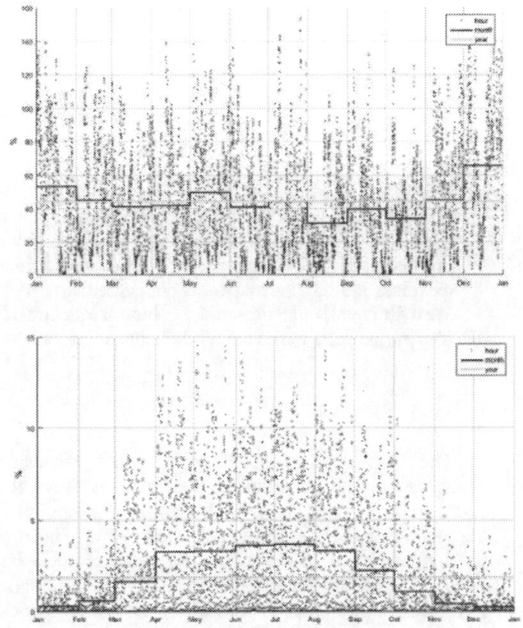

Figure 5: Share of wind power in Denmark (above) and solar power (below) (source: energinet.dk).

References

[1] Pfeifroth, U., A. Sanchez-Lorenzo, V. Manara, J. Trentmann, and R. Hollmann (2018), Trends and Variability of Surface Solar Radiation in Europe based on Surface- and Satellite-based Data Records, Journal of Geophysical Research: Atmospheres, 123, 1735-1754, doi: http://dx.doi.org/10.1002/2017JD027418

[2] Urraca, R., Gracia-Amillo, A. M., Huld, T., Martinez-de-Pison, F. J., Trentmann, J., Lindfors, A. V., … Sanz-Garcia, A. (2017). Quality control of global solar radiation data with satellite-based products. Solar Energy, 158(August), 49–62. http://doi.org/10.1016/j.solener.2017.09.032

[3] Taylor, K. E. (2001), Summarizing multiple aspects of model performance in a single diagram, J. Geophys. Res., 106(D7), 7183–7192, doi: 10.1029/2000JD900719.

[4] Hoyer-Klick, C., Beyer, H.G., Dumortier, D., Schroedter-Homscheidt, M., Wald, L., Martinoli, M., Schilings, C., Gschwind, B., Menard, L., Gaboardi, E., Ramirez-Santigosa, L., Polo, J., Cebecauer, T., Huld, T., Súri, M., de Blas, M., Lorenz, E., Pfatischer, R., Remund, J., Ineichen, P., Tsvetkov, A., Hofierka, J., 2008. MESOR. Management and exploitation of solar resource knowledge. In: Proceeding of the EUROSUN 2008, 1st International Conference on Solar Heating, Cooling and Buildings, Lisbon, Portugal, October 2008, pp. 7–10.

[5] Lefèvre, M., & Wald, L. (2017). CAMS Regular Validation Report Issue # 17 D - J - F 2017 CAMS - 72.

SHORT-TERM SOLAR IRRADIANCE FORECASTING BASED ON SUNSHINE NUMBER

Marius Paulescu[1], Oana Mares[2], Delia Calinoiu[1,2], Eugenia Paulescu[1]
[1]Faculty of Physics, West University of Timisoara, Romania
[2]Department of Physical Foundations of Engineering, Politehnica University Timisoara, Romania

ABSTRACT: In this paper a new procedure for nowcasting solar irradiation is proposed. The procedure is based on an improved version of the two-state model for nowcasting solar irradiance at ground level. Principally, the procedure sums-up the solar irradiance forecasts generated by the two state-model at one-minute sampling of the lead time. The proposed procedure demonstrates a better performance than the main competitor based on ARIMA forecasting of the clearness index. The study was conducted with high-quality radiometric data measured on the Solar Platform of the West University of Timisoara at a high frequency of four samples per minute.
Keywords: solar radiation; modelling; forecasting

1 INTRODUCTION

The equilibrium of an electricity grid is constantly changing with the fluctuation of the load. Nowadays, it is furthermore threatened by the increasing penetration of the renewable energy sources. The inherent variability in the output power of the wind and solar sources may induce significant energy fluctuation into the grid. Forecasting the energy production of these sources represents a crucial task for enabling operators to take control actions to balance the electricity grid. Definitely, the quality of forecasting the PV plants energy production closely follows the accuracy of forecasting the solar resource. In this context, the present paper is focused on forecasting solar irradiance at ground level. Passing clouds may ascribe an erratic character to the solar irradiance variation, and therefore, generating high-quality forecasts is always a challenge. A huge effort has been paid in the last years to increase the accuracy of the solar irradiance forecasts [1-2]. In a recent study [3] based on the results of 1705 accuracy tests reported during twelve years (2006-2017), we have noted a significant progress in forecasting solar resources: on average, the relative mean bias error was reduced by two thirds, while the relative root mean square error was reduced by one third, as compared to the values at the beginning of the considered period.

This paper presents an improved version of the two-state model for nowcasting solar irradiance [4]. On its basis, a novel approach for nowcasting solar irradiation is proposed.

2 DATA

This study was conducted with radiometric data recorded on the Solar Platform of the West University of Timisoara, Romania (http://solar.physics.uvt.ro/srms). The town of Timisoara (45°46°N; 21°25°E and 85 m asl) has a warm temperate climate, (Köppen climate classification Cfb [5]). The built database contains raw and post-processed radiometric data. Raw data consist of global G and diffuse G_d horizontal solar irradiances, measured using Kipp & Zonen SMP10 pyranometers at high temporal resolution (15 seconds). On basis of measured data, the following parameters were calculated.

- Sunshine number defined as a time dependent random binary variable [6]:

$$SSN_t = \begin{cases} 0 & \text{if the sun is covered by clouds at time } t \\ 1 & \text{otherwise} \end{cases} \quad (1)$$

Series of SSN values have been derived using the WMO sunshine criterion [7]: the Sun is shining at time t if the direct-normal solar irradiance exceeds 120 W/m^2:

$$SSN_t = \begin{cases} 1 & \text{IF } \left(G_t - G_{d,t}\right)/\sin h_t > 120 \text{ W/m}^2 \\ 0 & \text{otherwise} \end{cases} \quad (2)$$

where G_t and $G_{d,t}$ denote the measured global and diffuse solar irradiance at the time t and h_t is the Sun elevation angle. The average value of SSN over a given time interval Δt is equal to relative sunshine.

- Sunshine stability number defined as:

$$SSSN_t \equiv \begin{cases} 1 & \text{if } SSN_t > SSN_{t-1} \\ 0 & \text{otherwise} \end{cases} \quad (3)$$

$SSSN$ is an indicator that captures the changes in SSN time series [8]. The average value of $SSSN$ over Δt measures the frequency of changes in the SSN time series, thus being a straightforward indicator of the solar radiative regime stability.

- Clearness index defined as

$$kt_t = G_t / G_{e,t} \quad (4)$$

where $G_{e,t}$ is the global solar irradiance estimated at the top of the atmosphere. Clearness index isolates the stochastic components in solar irradiance time series. Statistical extrapolation of the clearness index time series is a traditional approach for nowcasting solar irradiance.

A total of 65500 solar irradiance measurements were post-processed in this study. Records from the first 12 days of May 2018 were used to fit the models while 19000 measurements from 6 different days of May 2018 were used for testing the models. The test days were selected to cover a large range of categories of solar irradiance variability. Variation of solar irradiance and SSN in these six days are displayed in Fig.1. During the days 05/19 and 05/22 the solar irradiance experienced a high variability. The days 05/20 and 05/28 were sunny and stable with the difference that the solar irradiance curve from 05/22 presents several huge spices due to the sparse clouds.

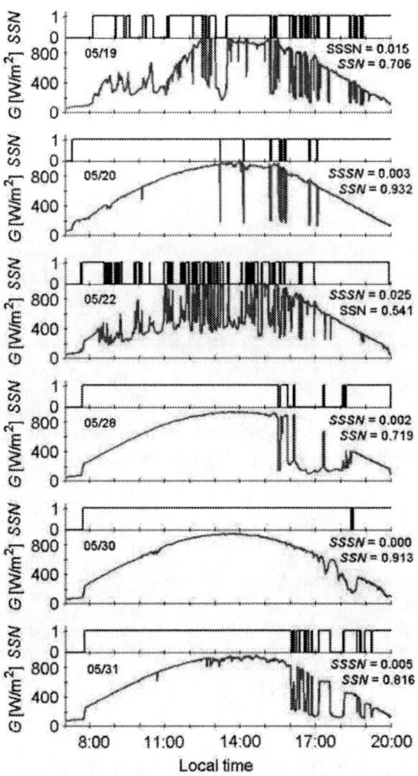

Figure 1: Global solar irradiance G and sunshine number SSN in six days of May 2018. The mean values of $SSSN$ and SSN are indicated on the graphs.

3 FORECASTING MODELS

3.1 Two-state model v.2

The 2-state model for nowcasting solar irradiance was developed by our group in 2014 [4]. Basically, the model connects in an innovative manner an empirical model for estimating the clear-sky solar irradiance with a statistical model for forecasting SSN. A new version, was recently reported [9]. The model equation is:

$$\hat{G}_t = \begin{cases} c_{cs} \cdot G_{0,t} & \text{if} \quad \hat{SSN}_t = 1 \\ \overline{\tau}_c \cdot G_{0,t} & \text{if} \quad \hat{SSN}_t = 0 \end{cases} \tag{5}$$

where $\hat{G}_t$ is the forecasted value of the solar irradiance at time t. $G_{0,t}$ is the estimated solar irradiance under clear sky at time t and c_{cs} is a dynamic correction applied to the mean atmospheric transmittance encapsulated into the $G_{0,t}$ equation. $\overline{\tau}_c$ is the attenuation applied to $G_{0,t}$ according to the cloud transmittance. The sole parameter that needs to be forecasted in Eq. (5) is SSN. The model operates as follows ($\hat{SSN}_t$ denotes the forecasted value of SSN at time t). On the state $\hat{SSN}_t = 1$, the global solar irradiance is being estimated using a clear-sky model. In the first version of the two-state model $c_{cs} = 1$. On the state $\hat{SSN}_t = 0$ the estimate under clear-sky is adjusted with the cloud transmittance $\overline{\tau}_c$.

Table 1: Indicators for the models' performance in nowcasting solar irradiance. Lead time = 5 minutes.

Date	Model	*nMBE*	*nRMSE*	T = 10%
05/19	2-state	-0.006	0.341	57.3
	kt	-0.008	0.357	25.9
05/20	2-state	-0.007	0.303	89.6
	kt	-0.002	0.209	83.2
05/22	2-state	-0.005	0.433	63.0
	kt	-0.006	0.423	40.4
05/28	2-state	0.002	0.209	83.2
	kt	0.003	0.201	86.3
05/30	2-state	0.000	0.030	86.3
	kt	-0.002	0.053	75.9
05/31	2-state	0.000	0.156	79.7
	kt	0.000	0.200	60.2

Both parameters c_{cs} and $\overline{\tau}_c$ are estimated simultaneously by a linear regression applied to data measured in a period Δt prior to the forecasting moment. Regarding Δt , an important finding was reported in [10]: a minimum error is reached if Δt equals the forecasting lead time.

3.2 kt-based model

The autoregressive integrated moving average (ARIMA) model applied to the clearness index time series is considered as one of the main competitors in the race for accurate forecasting solar irradiance. The model equation allows to evaluate kt at the discrete time t as a function of the values recorded at previous time moments. The general form of the model equation ARIMA(p,d,q) is provided by the classical Box-Jenkins theory [11].

3.3 A new model for nowcasting solar irradiation

Solar irradiation $\hat{H}(LT)$ is forecasted by summing-up the solar irradiance values $\hat{G}(LT_i)$ predicted by the 2-state models at one minute sampling of the lead time LT:

$$\hat{H}(LT) = \frac{1}{60}\sum_{i=1}^{N}\hat{G}(LT_i) \quad [\text{Wh/m}^2] \tag{6}$$

For example, if Eq. (6) is applied for nowcasting solar irradiation at 5 minutes ahead, then N = 5 and $\hat{G}(LT_1)$ represents the forecasted value of solar irradiance at one minute ahead, $\hat{G}(LT_2)$ at 2 minutes ahead, and so on.

4 NOWCASTING SOLAR IRRADIANCE

The clear-sky solar irradiance was estimated with the Adnot et al. model [12]. Following on the results from [4], regardless the lag, an ARIMA(0,1,0) model was used for nowcasting SSN. On basis of AIC criterion an ARIMA(1,1,1) model was selected for nowcasting clearness index and it was considered as reference in the models' evaluation process.

Table 1 shows the models' performance in nowcasting solar irradiance during the test days, at a horizon of time of 5 minutes.

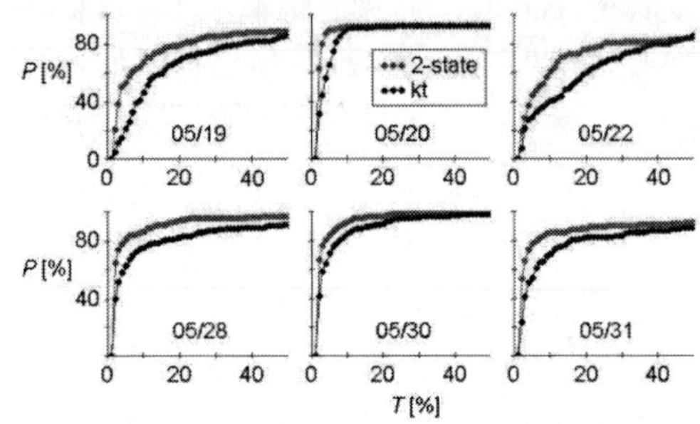

Figure 2: Percentage of forecasts P accurate to within a given tolerance interval T centered on the measurement. Lead time = 5 minutes.

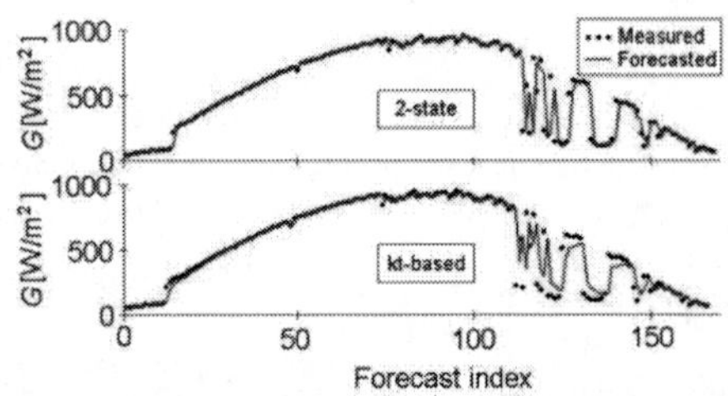

Figure 3: Measured and forecasted solar irradiance in 05/31, Lead time = 5 minutes

Taking into account only the statistical indicators, normalized root mean square error *nRMSE* and normalized mean bias error *nMBE*, the two-state model v.2 and the kt-based model lead to comparable results. A better performance of the two-state model is noted in the days with low-frequency variation of solar irradiance (05/30 and 05/31). The spikes presented in the solar irradiance curve affect the performance of the two-state model (05/20). If the percentage P of the forecasts accurate to within a given tolerance interval T around the measurement is considered, the two-state model performs considerably better than the kt-based model. During the day 05/19 (mostly cloudy in the morning and high variable afternoon), for a tolerance interval T =10% the two-state model precision is double to the precision of the kt-based model ($P_{2\text{-state}} > 2 \cdot P_{\text{kt-based}}$). Figure 2 underline this advantage of the two-state model, showing the percentage P plotted against the tolerance T. In general, for $T < 20\%$, the precision of the two-state model is far greater than the precision of the kt-based model. Roughly, at $T = 40\%$, the success rates of the two models become comparable, but the two-state model still preserves an advance. It is worth to note that difference between the precision of the two models increase with increasing the instability of the solar radiative regime. An explanation of this behavior follows. Figure 3 displays the measured and forecasted solar irradiance values during the day 05/31, Visual inspection shows that the curve generated by the *kt*-based model is smoother than the curve generated by the two-state model. This is a consequence of the fact that clearness index is forecasted by an ARIMA model. A forecasted value is always conditioned by the past measured data, which assign to model a kind of inertia at tracking the changes in the clearness index time series. Differently, the two-state model equation is permanently adapted to actual measurements of solar irradiance by estimating the atmospheric transmittance on the most recent measured data. This is an important feature of the two-state model, which contributes decisive to the increasing of its precision in forecasting solar irradiance.

Table 2: Indicators for the models' performance in nowcasting solar irradiation.

Date	Lead time	Model	*nMBE*	*nRMSE*	*T*= 5%
05/19	5 min	2-state	-0.017	0.266	56.4
		kt	-0.009	0.263	27.5
	10 min	2-state	-0.023	0.277	51.3
		kt	-0.018	0.290	24.3
05/31	5 min	2-state	-0.007	0.133	85.3
		kt	0.000	0.161	64.7
	10 min	2-state	0.000	0.115	79.5
		kt	0.000	0.162	58.9

5 NOWCASTING SOLAR IRRADIATION

Solar irradiation was forecasted with the procedure proposed in this study (Eq. 6). The results were compared with the ones obtained by ARIMA extrapolation of the clearness time series. For each lead time, the clearness index was evaluated as:

$$kt_t = H_t / H_{e,t} \tag{7}$$

The solar irradiation H_t was calculated by integrating the measured solar irradiance G_t over the forecasting lead time interval. $H_{e,t}$ represents the counterpart of H_t, estimated at the top of the atmosphere.

Table 2 presents the models' performance for two different lead times of 5 and 10 minutes, respectively. The values of the statistical indicators emphasize a better performance of the two-state model, both in terms of accuracy and precision. The lead time has no major influence on the models' accuracy. Contrary, the stability of the solar radiative regime influences dramatically the forecasting accuracy. As example, for the two-state model forecasting solar irradiation at a lead time of 10 minutes, in 05/31 ($\overline{SSN} = 0.816$, $\overline{SSSN} = 0.005$) the value of the relative root mean square is $nRMSE$ = 11.5%, while in 05/19 ($\overline{SSN} = 0.706$, $\overline{SSSN} = 0.015$) $nRMSE$ = 27.7%.

Figure 4 displays the forecasted vs. measured values of solar irradiation in two days of May 2018. Visual inspection shows that in case of the proposed procedure (Eq. 6) the most of the dots are placed very close to the first diagonal, emphasizing again the high performance of the model. The dots placed far from the first diagonal are due to the errors committed in forecasting *SSN* by the ARIMA model. For instance, when the actual value of *SSN* is 0, while the forecasted value is 1, the model two-state model will overestimate the value of the solar irradiance, because the correction with the cloud transmittance is no longer applied. The other way around, if the actual value of *SSN* is 1 and the forecasted value is 0, the model will underestimate the solar irradiance. Such large errors in forecasting solar irradiance are further propagated in the evaluation of solar irradiation by Eq. (6).

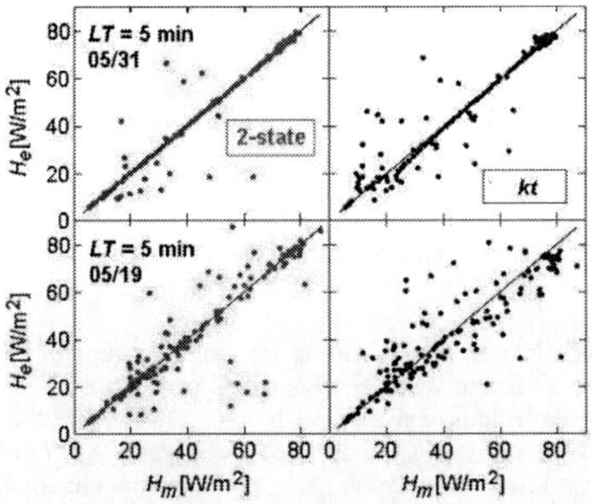

Figure 4: Forecasted vs. measured solar irradiation in two days of May 2018: (left) the proposed procedure (Eq. 6); (right) kt-based model.

When *SSN* is correctly predicted, due to the real-time correction of the atmospheric transmittance ($\bar{\tau}_c$ and c_{cs}), Eq. (6) will predict accurate the solar irradiation. We can deduct from here that the more stable the radiative regime, the better Eq. (6) will work. Without a real-time correction for atmospheric transmittance, the kt-based model even in stable meteorological conditions will predict solar irradiation with a lower accuracy. It is clearly visible in Fig. 4, were the dots corresponding to the kt-based model prediction are dispersed in a larger cloud around the first diagonal than the dots corresponding to the two-state model prediction.

6 CONCLUSIONS

A new version of the two-state model [4, 9] for nowcasting solar irradiance was presented. Based on the two-state model a new procedure for nowcastig solar irradiance was introduced. The two-state model works as follows: (1) if the sun is shining, the solar irradiance is estimated by a clear-sky model whose atmospheric transmittance is adjusted in real-time; (2) otherwise, the clear sky estimation of solar irradiance is adjusted with the cloud transmittance. These two states are differentiated by the sunshine number *SSN* [6], which is the sole quantity directly forecasted. The results show that the two-state model demonstrated better performance than the main competitor, ARIMA forecasting of the clearness index. When *SSN* is correctly predicted, the forecasts trace with accuracy the measurements. Although many efforts were devoted to increasing the accuracy of nowcasting *SSN* (e.g. [13], further researches are required, mainly for removing the persistence in *SSN* prediction. A notable advantage of the two-state model is its modularity, which facilitates to increase the forecasting accuracy by fine tuning its components: the dynamic adjustments and the model for nowcasting *SSN*.

The new procedure of forecasting solar irradiation by integrating the solar irradiance values predicted by the two-state model at one minute sampling, proves to be a reliable alternative to the traditional models. It is very simple and the results reported here shows its remarkable accuracy.

7 REFERENCES

[1] J. Kleissl (Ed.) Solar energy forecasting and resources assessment. Elsevier (2013).
[2] J. Antonanzas, N. Osorio, R. Escobar, R. Urraca, F.J. Martinez-de-Pison, F. Antonanzas-Torres. Solar Energy 136 (2016) 78.
[3] R. Blaga, A. Sabadus, N. Stefu, C. Dughir, M. Paulescu, V. Badescu (2018) Submitted to Progress in Energy and Combustion Science.
[4] M. Paulescu, O. Mares, E. Paulescu, N. Stefu, A. Pacurar, D. Calinoiu, P. Gravila, N. Pop, R. Boata Energy Conversion and Management 79 (2014) 690.
[5] M. Kottek, J. Grieser, C. Beck, B. Rudolf, F. Rubel. Meteorologische Zeitschrift 15 (2006) 259.
[6] V. Badescu, M. Paulescu. Atmospheric Research 101 (2011) 194.
[7] World Meteorological Organization. Guide to Meteorological Instruments and Methods of Observation. WMO-No.8/2008
[8] M. Paulescu, V. Badescu Theoretical and Applied Climatology 103 (2011) 459.
[9] M. Paulescu, O. Mares, C. Dughir, E. Paulescu, International Conference on Renewable Energy, Barcelona, 25-27 April 2018,
[10] D. Costa, O. Mares. AIP Conference Proceedings 1634 (2014) 151.
[11] G.E.P. Box, G.M. Jenkins, G.C. Reisnel, G.M. Ljung. Time series analysis. Forecasting and Control. 5th edition. Wiley (2016).
[12] J. Adnot, B. Bourges, D. Campana, R. Gicquel. In Lestienne R (Ed) Analyse Statistique des Processus Meteorologiques Appliquee a l'Energie Solaire. CNRS Paris (1979) 9-40.
[13] M. Brabec, M. Paulescu, V. Badescu. Solar Energy 101 (2014) 272.

MACHINE LEARNING TECHNIQUES FOR FORECASTING SINGLE-SITE PV PRODUCTION

Max BOEGLI
CSEM PV-Center
max.boegli@csem.ch

Marco PIERRO
EURAC Research
marco.pierro@eurac.edu

David MOSER
EURAC Research
david.moser@eurac.edu

Pierre-Jean ALET
CSEM PV-Center
pierre-jean.alet@csem.ch

Address for correspondence: Rue Jaquet-Droz 1, Case postale, CH-2002 Neuchâtel, Switzerland, +41 32 720 5194

ABSTRACT: The objective of this work is to develop and benchmark data-driven algorithms for the forecasting of PV power production on individual sites. This is necessary for e.g., compliance with regulation and operations on the electricity market. In this paper we present several machine-learning algorithms developed by both CSEM and EURAC Research. Using data from a 662 kW_p CdTe PV plant in Bolzano (Northern Italy), we clarified the respective roles of the statistical models and of the numerical weather forecasts in the resulting accuracy. We also evaluated the impact of the availability of weather parameters. We used a one-year dataset of power output measurements and numerical weather forecasts for training (2012) and another one for validation (2013). For this evaluation, we used numerical weather predictions generated by the Integrated forecasting system global model by the European centre for medium-range weather forecasts (ECMWF). Results with support vector regression (SVR)-based techniques showed an RMS error in PV power estimation of 2% with measured weather data as inputs. With weather forecasts as inputs, the RMS error in PV power prediction was 9% (RMSE) with weather data forecast.
Keywords: Solar Radiation, Photovoltaic, Forecasting, Machine learning.

1 INTRODUCTION

Since power system components have a finite dynamics, forecasting of power production and consumption is necessary to ensure the system is balanced at all times. A critical horizon where forecast is needed is 24 h ahead, which corresponds both to a major trading cycle on the electricity market and to the operating cycle of most building-level energy management systems [1]. Single-site forecasting is the most challenging since it does not benefit from the smoothing effect seen in regional forecasts [2].

The objective of this work is to develop and benchmark data-driven algorithms for the forecasting of PV power production on individual sites. This is necessary for e.g., compliance with regulation (more and more balancing authorities mandate that PV producers provide day-ahead forecasts) and operations on the electricity market.

Here we focus on two-step forecasting, which relies on general-purpose numerical weather forecasts and on statistical models to convert the predicted weather conditions into forecasts of PV power production. In this paper we present several machine-learning algorithms developed by both CSEM and EURAC Research. From a scientific perspective, we wanted to clarify the respective roles of the statistical models and of the numerical weather forecasts in the resulting accuracy. We also evaluated the impact of the training period and of the availability of weather parameters.

The algorithms were evaluated in several steps:

- Accuracy of the statistical model itself by comparing measured power production with its output when fed with measured weather data.
- Impact of variability and bias in the weather forecasts by comparing measured power production with model output when the model was fed with "synthetic forecasts" i.e., measured weather data to which noise and/or bias were added.
- Accuracy of the complete forecasting chain of operational numerical weather forecasts and statistical model.

We validated the algorithms on data from several PV power plants. The results shown here refer to a 662 kW_P CdTe PV plant located in Bolzano (Northern Italy). We used a one-year dataset of power output measurements and numerical weather forecasts for training (2012) and another one for validation (2013). For this evaluation, we used numerical weather predictions generated by the Integrated forecasting system global model by the European centre for medium-range weather forecasts (ECMWF).

2 METHODOLOGY

2.1 Support Vector Regression

We developed a prediction algorithm based on the machine-learning technique called support vector regression (SVR) [3]–[5]. SVR is an extension of support vector machine (SVM), which is commonly used for classification. It uses a non-linear mapping of kernel function to project data into a higher-dimensional space where solving the regression task is easier than in the original space [3], [4]. Standard SVR architecture is given in Figure 1:

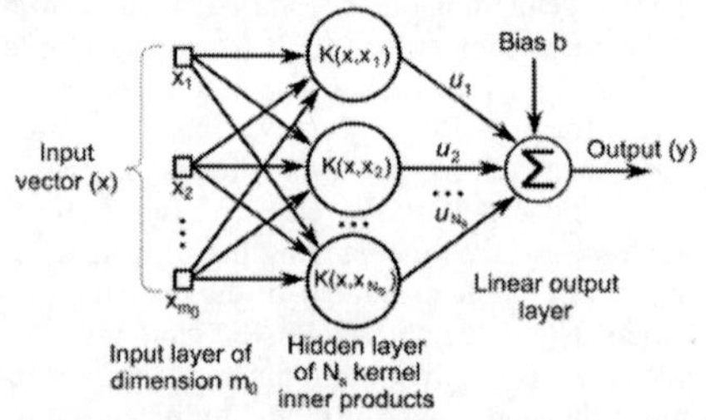

Figure 1: Support vector regression (SVR) architecture with the inner kernel layer in between the input-output layers. (Courtesy of www.mdpi.com, Sensors 2014)

Here, a Gaussian kernel function is used:

$$K(x_i,x_j) = \exp(-|x_i-x_j|^2 / 2h^2),$$

with h the kernel bandwidth. This non-linear mapping makes SVR methods well suited to estimate non-linear models such as the response of PV systems to weather variations. This non-linear regression aims to minimize a training cost function, which is a tradeoff between the estimation error and a regularization term, subject to a linear mapping in the kernel space.

2.2 Training and prediction

The algorithm takes as input various types of weather forecast information, including solar irradiance, air temperature, clear-sky index and solar almanac. In this work, the following five input signals are considered:

- From the numerical weather forecast by ECMWF:
  - Global horizontal irradiance (GHI)
  - Air temperature
- From deterministic calculations based on time and geographical coordinates:
  - Clear-sky GHI
  - Sun elevation
  - Sun azimuth

In the training phase, the output signal is the measured PV power. While in the prediction phase, the output signal is the forecasted PV power.

Figure 2 shows a snapshot example of the five input and one output signal for both training and prediction phases.

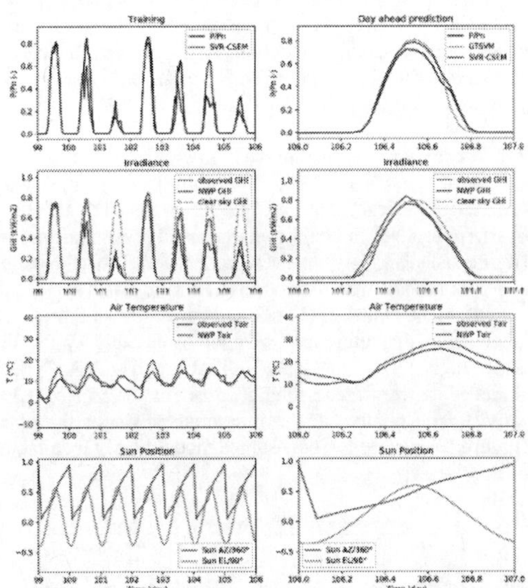

Figure 2: Snapshot of seven training days (left column) with five input signals: NWP GHI, Clear-sky GHI, NWP air temperature, sun azimuth and elevation. The relative measured PV power (*P/Pn*) is the output training signal (upper left graph). A one-day-ahead prediction is shown in the upper right graph for the GTSVM (EURAC)[6] and the SVR-CSEM algorithm. In the prediction phase, the measured PV power is the reference signal used to compute the prediction error.

2.3 Details and performance of the SVR model

Two SVR hyper-parameters are needed: the kernel bandwidth *h*, which defines the regression smoothness or granularity, and the regularization parameter λ, which defines the regression robustness. The impact of their variation is summarized on Table 1.

Table 1: Impact of SVR tuning parameters variations

Increased value	Decreased value
Bandwidth	
Wide kernel bandwidth filtering/smoothing effect	Narrow kernel bandwidth sharp and spiky effects
Regularization	
Better robustness to unpredictable dynamics. Less fitting accuracy. Avoiding overfitting.	Better fit on the training data set. May lack robustness towards unpredictable dynamics. Risk of overfitting.

A third parameter, epsilon, specifies a tube within which no penalty is associated in the training cost function. For the tuning of the SVR hyper-parameters, a manual search is carried, further validated by an extensive grid search.

3 RESULTS AND ANALYSIS

3.1 Impact of accuracy of the implied model

As first possible source of errors in the power forecasts is the finite accuracy of the underlying, implied model of the PV plant. To quantify this accuracy, we ran the algorithms with measured values of global horizontal irradiance and air temperature instead of forecasts as inputs.

In that case, the error metrics of the SVR model (excluding night time) normalized to the measured PV power *P* of the plant were:

- Root-mean-square error (RMSE): 2.16% of *P*
- Mean absolute error (MAE): 1.65% of *P*
- Mean bias error (MBE): 0.70% of *P*

These errors are shown in Figure 3.

3.2 Impact of weather forecast accuracy

Since the output of PV power plants is primarily determined by the irradiance (GHI), the accuracy of weather forecasts has a significant impact on the PV power forecasting accuracy. For a systematic investigation of this impact we synthesized pseudo forecasts at time t-1 by adding noise and/or bias to the GHI values which were measured at time t. The resulting errors are summarized on Figure 3.

Even with significant error levels in the weather forecasts (up to 40% noise on the irradiance), the accuracy of the PV forecasts based on SVR statistical models is consistently better than the reference persistence model. In particular, the SVR models significantly reduce the bias.

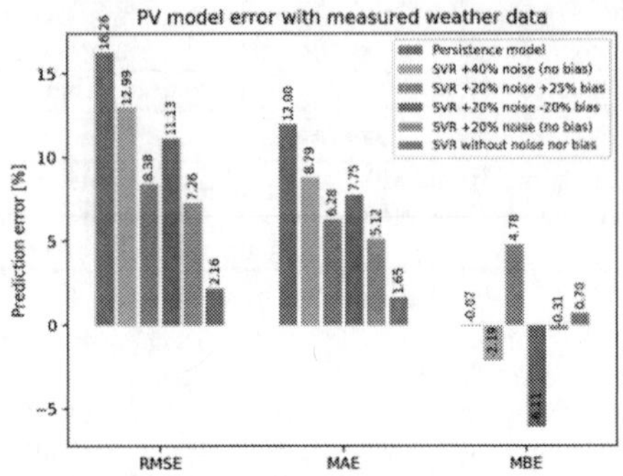

Figure 3: Annual error relative to measured power (excluding night time) obtained for a 662 kWP PV power plant with pseudo weather forecasts (weather measurements with added noise and/or bias) and an SVR statistical model, compared with persistence.

3.3 Impact of different SVR input signals

Input signals are key information for SVR algorithm performance. As mentionned earlier, five input signals are used in the SVR algorithm under study. We investigated the SVR algorihtm with reduced sets of inputs. An overview of input signal conbinations is shown in Table 2.

Table 2: Investigated combinations of SVR input signals

Input signals	NWP GHI only	NWP GHI + GHIcs	NWP GHI + hour	Train: Meas, Pred: NWP	SVR-CSEM w/out Tair	SVR-CSEM
For Training						
NWP GHI	×	×	×		×	×
NWP Tair						×
Measured GHI				×		
Measured Tair				×		
Clear sky GHI		×		×	×	×
Sun Elevation				×	×	×
Sun Azimuth				×	×	×
Daily pattern			×			
For Prediction						
NWP GHI	×	×	×	×	×	×
NWP Tair				×		×
Measured GHI						
Measured Tair						
Clear sky GHI		×		×	×	×
Sun Elevation				×	×	×
Sun Azimuth				×	×	×
Daily pattern			×			

Six different configurations were considered. One with only the NWP GHI, two with the NWP GHI and one extra information, either the clear sky GHI or a daily pattern (1-24h cycle). One with all five input signals and one with all signals except air temperature. In addition, a special case has been investigated in which two signals, i.e. GHI and air temperature, were different between the training and prediction phases. In this case, a common NWP forecast for GHI and air temperature was used for prediction while measured GHI and air temperature were used for training.

Results are shown in Figure 4 for these six different configurations and the persistence model. When all available inputs were included and the global irradiance and the air temperature were taken from the numerical weather forecasts (NWP), the SVR algorithm achieved a root-mean-square error of 9.3% relative to the measured power of the plant (Figure 4), beating the persistence model nearly by a factor of two.

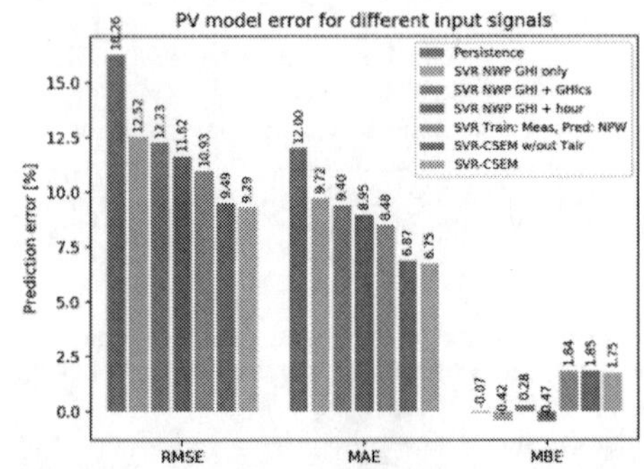

Figure 4: Annual error (excluding night time) relative to measured power obtained for day-ahead forecasting on a 662 kWP PV power plant with different SVR input signal combinations, as per selection in Table 2. It is cleary shown that the prediction error decreases with the amount of relevant input signals.

3.4 Accuracy of the full forecasting chain

Here, the SVR-CSEM algorithm is compared with different statistical models developed in the EURAC study by Pierro et al. [6], including an improved persistence model (K Persistence), a SARIMA model, and two Neural-Network models with different weather forecasts.

When the global irradiance and the air temperature were taken from the numerical weather forecasts (NWP), the SVR algorithms (GTSVM-EURAC and SVR-CSEM) achieved a root-mean-square error of 9.2% to 9.3% respectively, relative to the measured power of the plant (Figure 5), beating the persistence model nearly by a factor of two.

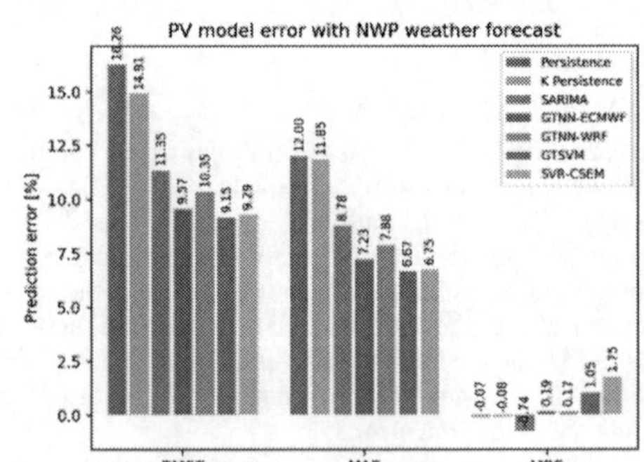

Figure 5: Annual error relative to measured power obtained for day-ahead forecasting (excluding night time) on a 662 kWP PV power plant with persistence models and with several statistical models applied to numerical weather forecasts.

The results show similar performance between the GTSVM-EURAC and the SVR-CSEM algorithms. This comes without surprises as both algorithms uses the same SVR architecture. SVR based prediction techniques are significantly better than other algorithmic schemes in terms of RMS and mean absolute errors. However, they come at the expense of a larger bias than with other neural-network techniques.

4 CONCLUSIONS AND FUTURE WORK

For SVR-based prediction techniques, this study compared the prediction errors for various weather forecast accuracy. First with observed weather data (with and without synthetic noise) and second with forecasted weather data in comparison with different input configurations and statistical models. An RMSE of about 2% is obtained with measured weather data without synthetic noise and an RMSE of 9% is obtained with weather data forecast. Further studies implies multi-models ensemble [6] and analysis on forecast combination [8]. However the results presented here show that the bulk of the error comes from the weather forecasts rather than from the underlying, implied model of the PV plant.

6 REFERENCES

[1] P.-J. Alet et al., 'Forecasting and Observability: Critical Technologies for System Operations with High PV Penetration', in 32nd European Photovoltaic Solar Energy Conference and Exhibition, Munich, 2016.

[2] J. Antonanzas, N. Osorio, R. Escobar, R. Urraca, F. J. Martinez-de-Pison, and F. Antonanzas-Torres, 'Review of photovoltaic power forecasting', Sol. Energy, vol. 136, pp. 78–111, Oct. 2016.

[3] K. D. Brabanter, J. Suykens, and B. D. Moor, 'Nonparametric Regression via StatLSSVM', J. Stat. Softw, vol. 55, no. 2, pp. 1–21, 2013.

[4] J. Smola and B. Schölkopf, 'A tutorial on support vector regression', Stat. Comput., vol. 14, no. 3, pp. 199–222, Aug. 2004.

[5] M. Espinoza, J. A. K. Suykens, R. Belmans, and B. D. Moor, 'Electric Load Forecasting', IEEE Control Syst., vol. 27, no. 5, pp. 43–57, Oct. 2007.

[6] M. Pierro et al., 'Multi-Model Ensemble for day ahead prediction of photovoltaic power generation', Sol. Energy, vol. 134, pp. 132–146, Sep. 2016.

[7] K. E. Taylor, 'Summarizing multiple aspects of model performance in a single diagram', J. Geophys. Res., vol. 106, no. D7, pp. 7183–7192, Apr. 2001.

[8] F. Chan, L. L. Pauwels, 'Some theoretical results on forecast combinations', Int'l J. Forecasting, vol. 34, pp. 64-74, 2018.

COMPARISON OF IRRADIANCE FORECASTING METHODS APPLIED FOR BUILDING SOLAR ENERGY ESTIMATION

Vladislav Martinek, Petr Wolf, Ludivine Dupont
University Centre for Energy Efficient Buildings, Czech Technical University in Prague,
Třinecká 1024, 273 43 Buštěhrad, Czech Republic, vladislav.martinek@cvut.cz

ABSTRACT: The paper is focused on state-of the-art methods for local PV yield estimation in middle and very short time horizon. Both numerical weather prediction as well as statistical approaches are discussed and the accuracy for a chosen place is compared. The developed approach for data acquisition of irradiance sensors is discussed together with the method of image processing. In the end general statements are made based on accuracy and suitability of each prediction method.
Keywords: solar irradiation, forecasting, sky imager, image processing

1 INTRODUCTION

Buildings are worldwide responsible for the highest energy demand – one third of all primary energy and one half of electrical is used there. This makes one third of CO_2 emissions. As the population is expected to increase 2,5 bil. till 2050, the energy demand for buildings will rise as well [1].

On the other hand the buildings offer a potential to implement new technologies and approaches to use the energy more effectively and implement renewable energy sources, especially photovoltaics.

Both residential and commercial loads are equally significant to the energy consumption mix in the global scale. Compared to residential building loads, commercial energy management has features that drive research in the field of forecasting and optimization using weather variable inputs [2]. To use the local generated energy for the building loads in an effective way appropriate design of energy systems and its progressive control is needed. If PV source is implemented, an essential part for decision making by control schemes is the forecasting of the PV power with high accuracy.

2 PV FORECASTING MODELS

The forecasting methods can be divided according to spatial resolution and time horizon, Fig 1, [3]. Long term prediction of several days can be provided using numerical models. Nowcast applies for very short time horizon of several minutes to few hours, but achieves better accuracy than forecast based on the numerical computational model. It also respects the local influencing factors such as near and far shadowing objects or reflections. Therefore it is useful to combine numerical models with outdoor sensors feedback to increase accuracy of prediction in short term horizon.

One of the important factors for prediction of PV output is cloud cover which involves the amount of sun irradiation getting through atmosphere to PV modules. Clouds could be of a great variety. Using sky imagers we can detect and predict cloud movement and their "collision" with sun direct beam. This information can be used for short term prediction together with other, especially statistical models.

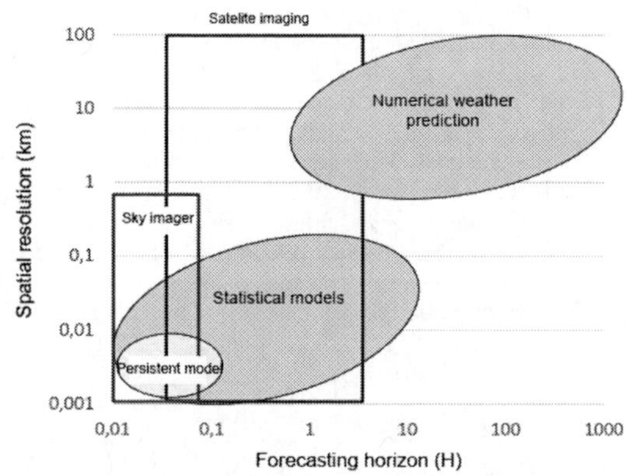

Fig. 1: Types of PV forecasting models. Based on [3].

3 FORECASTING SERVICE FOR BUILDINGS OPERATED AT UCEEB

For better management control of several buildings that are being monitored and controlled by UCEEB CTU a service based on combined model was prepared. As the basis the long term numerical weather prediction is used.

If a building has any irradiance sensor, this can be used to adjust the numerical model as well as to build a short term statistical model. Both models are then combined using weights to get best prediction results for short term as well as long term prediction horizon, Fig. 2.

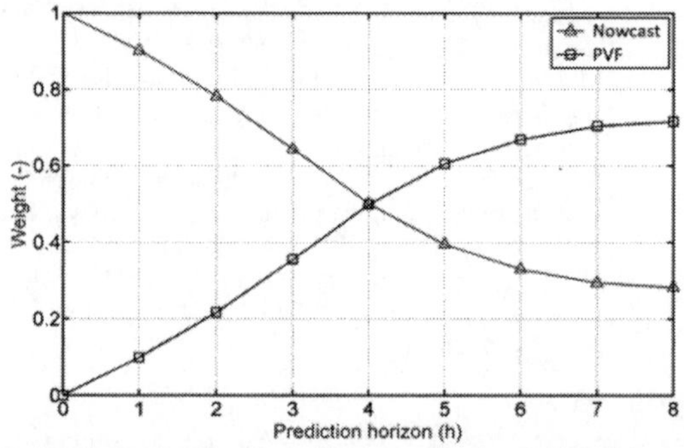

Fig. 2: Example of statistically determined weights of the numerical weather prediction (PVF) and nowcast statistical model.

To achieve more accurate forecasting for very short time horizon, a model based on sky observing was prepared. It is based on taking images of sky using wide angle camera, image recognition and statistical data processing. Unfortunately the sky imager is a costly device and the techniques are computational intensive. This limits its utilization exclusively for large buildings with sophisticated energy control or a pool of nearby buildings. At the moment the sky imager is being operated at 2 locations (UCEEB center in Buštěhrad 30 km away from Prague and Fenix administrative building in Jeseník, northeast part of Czech Republic). The collected data are being used to test and further improve prediction techniques.

4 FORECAST BASED ON NUMERICAL WEATHER PREDICTION

For the long term forecasting using numerical weather prediction free of use models are used as inputs. With the help of local feedback from several places in the Czech Republic a new model is calculated (PVF). Fig. 3 show the accuracy for of the original HAWAI model and PVF model.

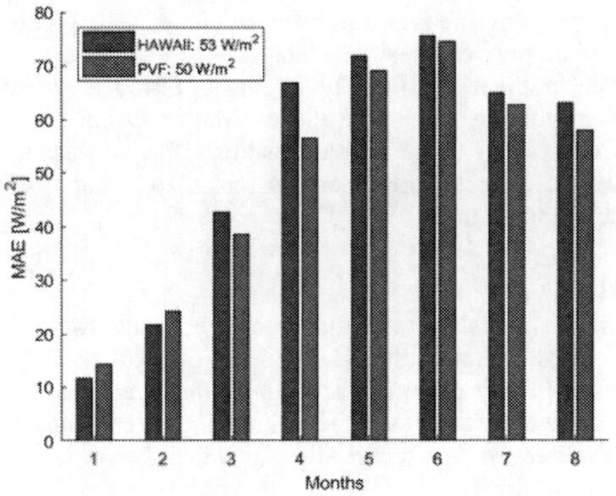

Fig. 3: Accuracy comparison of HAWAI and PVF - monthly RMSE of hourly values, day ahead forecast, UCEEB Buštěhrad

5 NOWCAST USING THE LOCAL IRRADIANCE SENSOR FEEDBACK

For better short term energy control of buildings a simple statistical model is being operated. It is based on cloud cover persistency - assuming that the cloud cover is not changing. The forecast of irradiance G_{T2} can be calculated as

$$G_{T2} = G_{T1} * \frac{G_{TeorT2}}{G_{TeorT1}} \qquad [1]$$

where G_{T1} is the measured current irradiance, G_{TeorT1} and G_{TeorT2} are the theoretical maximal incident irradiances in current time T1 and time of forecast T2. When applying this term for real measured values a limitation of $G_{T2} \leq G_{TeorT2}$ is needed, especially by sunrise and sunset or if sudden reflection from nearby object occurs. Fig. 4 shows the forecasted irradiation of selected days using only local irradiance sensor feedback. The 1 hour forecast unlike the 3 hour forecast still preserves the "daily shape" and achieves higher accuracy. The overall prediction accuracy in 2018 can be found on Fig. 5. This simple nowcast method can be slightly improved using e.g. statistical correction based on historical errors.

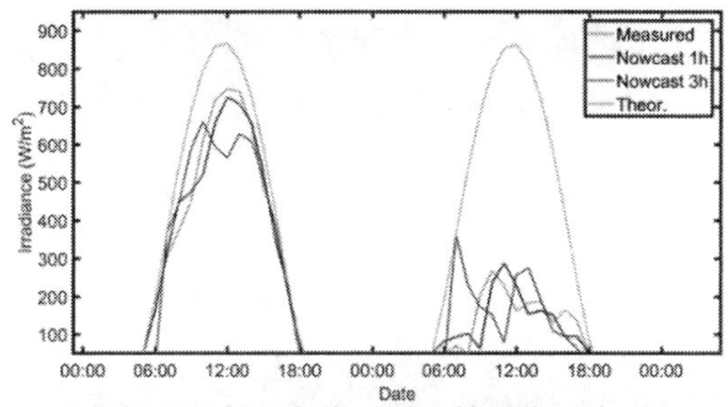

Fig. 4: Nowcast of 1 and 3 hour ahead based on cloud cover persistency, UCEEB Buštěhrad.

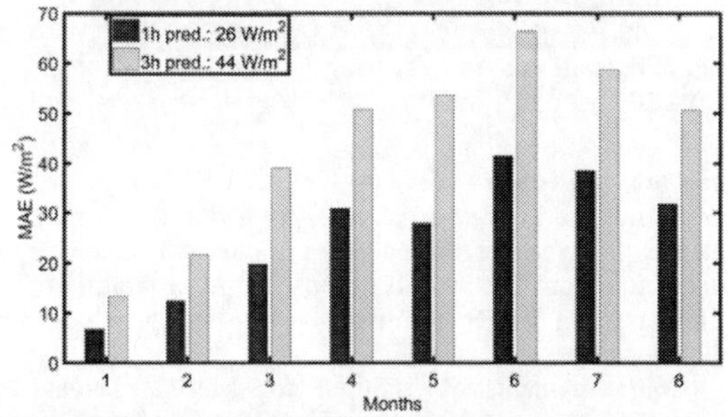

Fig. 5: Accuracy of 1 and 3 hour ahead nowcast based on cloud cover persistency, UCEEB Buštěhrad.

6 SKY IMAGER AND IMAGE PROCESSING

In order to increase the precision of the very short forecasting horizon, an image processing procedure was developed. The image of all sky is taken every 10 seconds during day and processed to create prediction of irradiation for defined prediction horizon.

Sky imaging camera is placed on the roof of UCEEB. Other camera if placed on administrative building of Fenix company in Jeseník where PV modules in connection with battery are used. To assure all functions of the energy system (backup, self-consumption, grid support...) it is crucial to have a precise irradiation prediction.

Captured images are also stored for later analysis of new approaches. Developed prediction methods can be tested on a large set of images and results of prediction can be compared to measured irradiance.

Fig. 6: Sky imager ASI-16/50 on the roof of UCEEB building.

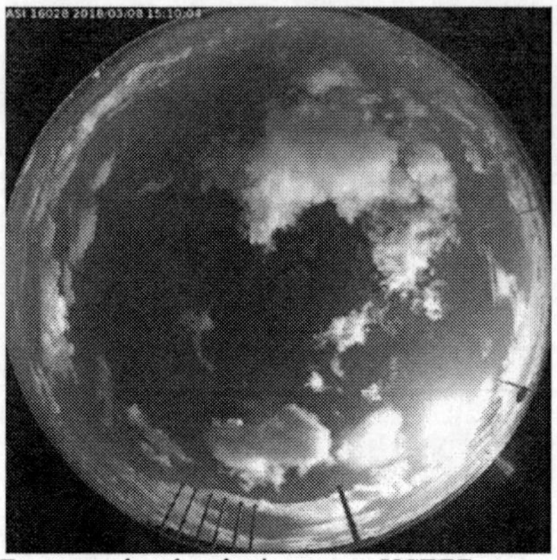

Fig. 7: Footage taken by sky imager at UCEEB, 08/03/2018, 3 p.m.

6.1 Clouds detection

In order to increase irradiation prediction accuracy, it is intended to detect the moments during a day when the energy production from PV modules drops due to coverage of the sun by passing clouds. The first step is to recognize position of clouds in the sky.

Color segmentation is used to determine areas in captured image which represents clouds. Every area representing cloud is then processed separately since clouds move in different heights with different directions and speed. Segmented clouds are marked by different colors. Also image mask is applied to filter out static objects on the landscape horizon, Fig. 8.

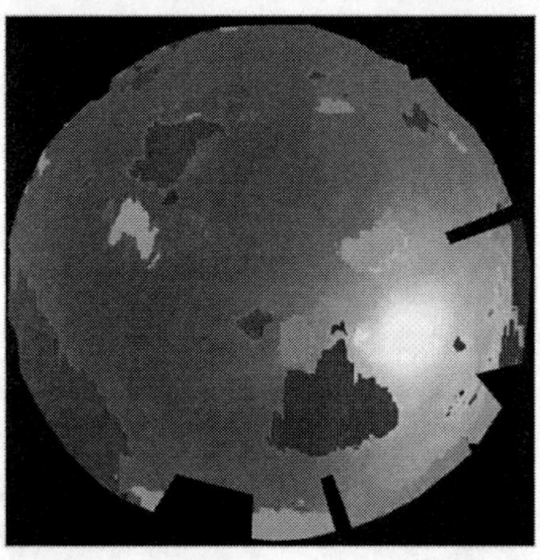

Fig. 8: Color segmentation used on image acquired from Panoramatic camera placed at UCEEB Buštěhrad.

The observations turn out that it is necessary to consider clouds as separate objects since their tendency to move over sky is individual.

6.2 Movement tracking

To be able to recognize the movement direction and relative speed we compared two sequent images captured in the 10 seconds interval. Analyzing features in both images leads to detecting of same points at different position in the sky. This way the relative movement vectors are calculated and assigned to clouds.

Fig. 9 shows feature points detected in a pair of sequent images and also future points predicted by multiplication of the movement vector.

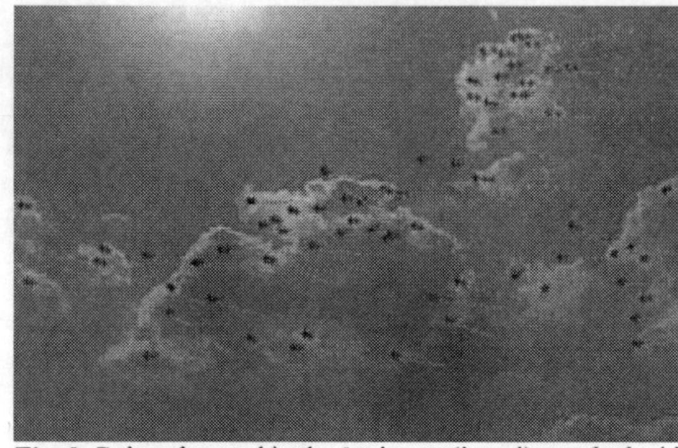

Fig. 9: Points detected in the 1st image (in red) matched with points detected in the 2nd image (in blue) to calculate the future points positions (in green).

6.3 Sun cover

Using cloud detection and their movement tracking we are determining the future position of clouds. The prediction is based on previous position changes.

The moment and duration of sun covering is estimated together with the percentage of direct sun covering.

Using sun covering estimation the influence to theoretical direct irradiation is calculated for defined prediction horizon.

6.4 Sky cover

Since also indirect radiation should be taken into account the overall cloudiness in the sky is calculated and transformed into influence factor for diffused radiation.

The overall sky cover factor comes to great importance to both direct and indirect in situation when the sky is heavily covered by clouds.

6.5 Irradiation prediction

Overall irradiation useable by PV modules consists of direct and diffuse parts. The calculation of irradiation prediction combines both - sun and overall sky coverage and applies this to the theoretical maximal irradiation for given place and time.

An example of applying the described methods is presented on Fig. 10. The image on the left was captured before a cloud covered the sun. On the image on the right the cloud has already shaded the sun. Corresponding irradiance value curves can be found underneath.

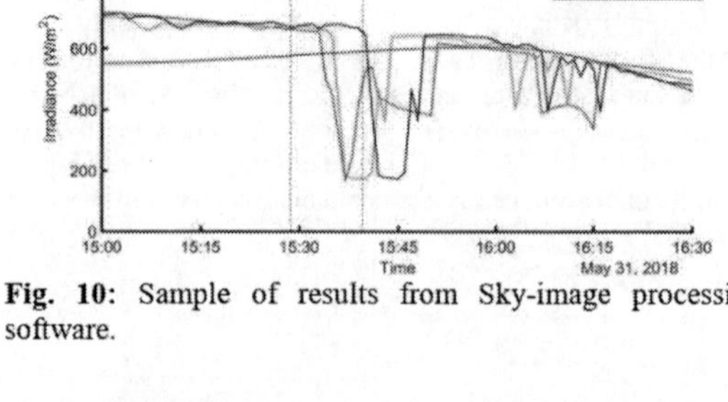

Fig. 10: Sample of results from Sky-image processing software.

Comparison to measured irradiance shows that the cloud movement prediction can handle well situations when isolated clouds are moving approximately constant speed in the sky. So the drops in irradiance can be precisely predicted in advance. Also the duration of drops can be estimated well so the battery management software can be provided with information if the PV production drops for a short time or if the drop duration will be long.

The prediction of cloud movement is not always precise in terms of timeliness since the variation of cloud speed is not taken into account. We expect the clouds move the same speed and direction as they did several seconds ago.

On clear days as well as on cloudy days the prediction using sky imager brings results comparable to other (much simpler and cheaper) methods. So the stability of prediction depends on recent weather conditions.

7 CONCLUSION

Estimation of future sun irradiance is the way to predict PV modules energy production which is vital information for battery storage management systems of energy efficient buildings.

One way is to use statistical models to create prediction based on historical data. It shows that these approaches give sufficient results in long term prediction.

To improve reaction of statistical methods to recent irradiance conditions changes method of local sensor feedback was tested and found useful in overall prediction precision especially in short term horizon.

To go further sky imaging camera in connection with self-developed software for cloud tracking and cloud cover prediction was tested with remarkable results in very short prediction horizon.

Moreover, the resolution of prediction is satisfying for high precision forecasting which is expected to be used by large buildings with sophisticated energy control systems.

Collected data set including sky images of 10 seconds frequency from a large period of time can be precious source of testing data for future development and improvement.

8 ACKNOWLEDGEMENTS

This work has been supported by the Ministry of Education, Youth and Sports within National Sustainability Programme I (NPU I), project No. LO1605 – University Centre for Energy Efficient Buildings – Sustainability Phase and by the Operational Programme Research, Development and Education of the European Structural and Investment Funds, project CZ.02.1.01/0.0/0.0/15_003/0000464 Centre for Advanced Photovoltaics.

9 REFERENCES

[1] "Transition to Sustainable buildings – Strategies and Opportunities to 2050," International Energy Agency, 2013.

[2] D. Lazos, A.B. Sproul, M. Kay, "Optimization of energy management in commercial buildings with weather forecasting inputs: A review," Renewable and Sustainable Energy Reviews, vol. 39, pp. 587-603, 2014.

[3] Diagne, M., David, M. e al. Review of solar irradiance forecasting methods and a proposition for small-scale insular grids. Renewable and Sustainable Energy Reviews. 27, 65-76.

Really building with BIPV

Putting the foundation in place for a successful Dutch BIPV sector (the 'Werkelijk Bouwen aan BIPV' project)

Arthur de Vries[1], Alain Kahn[2], Raul Comuth[3], Anko van Deursen[4], Michiel Arninkhof[5], Guus Verpaalen[6], Cor Maas[7], Sam Kin[8], Paul de Jong[9], Wim van de Wall[10], Zeger Vroon[11], Ando Kuypers[12], Josco Kester[13], Roland Valckenborg[14], Wilfried van Sark[15,1],
Roel Loonen[16], Yael Aartsma[17], Ingrid van Straten[17], Erik Teunissen[17]

[1]Stichting Monitoring Zonnestroom, Arthur van Schendelstraat 550, 3511 MH Utrecht, The Netherlands
[2]4WWWWIE, Ouderkerk aan de Amstel, The Netherlands
[3]Adviesbureau Comuth, Maastricht, The Netherlands
[4]HD Solar BV, Someren, The Netherlands
[5]Holland Solar, Utrecht, The Netherlands
[6]Kameleon Solar Specials BV, Roosendaal, The Netherlands
[7]Chatim BV, Heerlen, The Netherlands
[8]SolarSwing Energy, Delft, The Netherlands
[9]Solinso, Kessel, The Netherlands
[10]Wallvision, Heeze, The Netherlands
[11]Zuyd Hogeschool, Heerlen, The Netherlands
[12]TNO, Delft, The Netherlands
[13] ECN, Petten, The Netherlands
[14]SEAC, Eindhoven, The Netherlands
[15]Utrecht University, Utrecht, The Netherlands
[16]Eindhoven University of Technology, Eindhoven, The Netherlands
[17]Berenschot, Utrecht, The Netherlands

Email: adevries@celstar.nl Phone: +31 6 38751654

SUMMARY

The main result of this project is the organisation of a strong BIPV market and supply chain, to encourage knowledge sharing, and successful business development by collaboration of (SMEs) businesses, R&D knowledge institutions and the solar- and building associations. This should result in a significant increase in the use of BIPV in building projects in Netherlands, and as a result in economic growth. The project runs from 2017 until the end of 2020, and is a follow-up to the national BIPV Roadmap issued in early 2016.

Keywords: BIPV, ecosystem, living lab, market development, innovation process, BIPV Nederland

1. PROJECT OBJECTIVES

The overall goal of a significant increase in the use of BIPV in building projects will be reached by improving the speed and effectiveness of product innovation, increasing the knowledge and skills of BIPV in the building sector, developing tools for easier implementation of BIPV in building projects, removing obstacles in the growth of the BIPV market, and showing the benefits and values of BIPV through demonstration (Living Labs).

2. PROJECT DESCRIPTION

A project team with representatives of 17 parties will be working on the project during a period of 4 years. Each of the parties will contribute with its specific expertise, ranging from building construction technology, design tools, product R&D, all the way to marketing. In effect, the complete knowledge chain from materials and production technology, product development, integration in buildings, up to implementation of projects, is represented in the project team.

We will show results of the project at the PVSEC 2018 conference.

The project is funded by the European Innovation program for the South of the Netherlands (OPZuid), as part of the European Fund for Regional Development (EFRO) of the European Union.

3. EXPLANATORY PAGES

A program of activities (Fig. 1) is structured to deliver the envisioned results, in addition to the increased cohesion in the BIPV community. The impulse to cooperate is driven by benefiting from each other's expertise. The participating partners work on 4 main themes (see in green below). The 4 themes lead to improvement in marketing actions, innovation projects, technological research and demonstration projects (Living Labs).

Fig. 1. Dutch BIPV Ecosystem

The theme 'Connecting (Verbinden)' aims at connecting with the building market to analyse the customer's demand and to be able to convert the demand into valuable BIPV propositions through collective marketing. An easy accessible information sharing point (BIPV information centre) is set up for parties wishing to build projects with BIPV technology.

In theme 'Deepening (verdiepen)', technological challenges will be addressed. In addition to research and development of products and production technology, a human capital agenda will be developed. The latter should lead to better and quicker training of qualified BIPV personnel throughout the solar sector and the building sector, which ranges from architects who will become more enticed to include BIPV in their designs for energy efficient buildings, to scientists who are going to develop the latest technology, and even up to marketeers.

The theme 'Accelerate (versnellen)' helps the many BIPV start-ups in the development of their business. Often SMEs are able to create a potentially successful BIPV product (product development), but struggle with the next step to set up the business (business development). The project will support these SMEs with professional business coaching to guide them in introducing their products in the market and to grow a profitable business.

The last theme 'Strengthening (versterken)' aims at developing the tools and reducing or ideally removing the technical and non-technical obstacles to make the BIPV community as a whole successful. This will be done by collectively working on topics such as certification and standardization and in particular the development of BIPV elements for multidisciplinary CAD/BIM tooling for the building industry.

The activities of the 4 themes will be supported by a streamlined process for innovation and cooperation between solar and building expertise centres in the Netherlands. The outcome of the technical innovations will be shown in BIPV Living Labs (see below).

4. STREAMLINES INNOVATION PROCESS

The technological developments necessary to come to successful BIPV products and design tools are supported by various R&D laboratories in the Netherlands. Solliance is well equipped to apply new technology (in particular thin film) involving innovations in materials, products and production techniques. The Silicon Competence Center offers complementary knowledge and facilities for silicon technology. Both will focus on further improving quality, reducing costs and increasing efficiency. Technologies and BIPV products with a high TRL level will benefit of the SolarBEAT test platform of the Solar Energy Application Centre in Eindhoven. This is where complete BIPV-modules can be tested in an operational environment and statistical research can be performed on the performance, quality and reliability of building integrated modules.

Eindhoven University of Technology contributes with its vast knowledge on building physics, while Utrecht University brings its expertise in performance analyses and modelling to the project. The Prototyping lab at Kameleon Solar offers opportunities for manufacturing new innovative BIPV modules which then can be tested at SolarBeat or at the BIPV test buildings of the 'District of Tomorrow' at the Zuyd Hogeschool, acting as a "validation Lab" where longer time measurements can be made on these innovative BIPV modules. Educational institutions contribute with their expertise on curriculum and training development. The combination of the expertise's and test facilities of these R&D institutions creates a streamlined chain for a fast and tuned innovation process that is unique in Europe, and allows new innovations and the BIPV sector to accelerate!

5. LIVING LABS

The innovation process does not stop in the laboratories and test institutions. An important addition to the innovation process is the introduction of Living Labs, where the BIPV sector will demonstrate the outcome of their innovations in a significant number of homes and buildings, spread out over the Netherlands. The aim is to learn from the users on their experiences on operational, use, quality, and maintenance aspects. In addition, the Living Labs serve the purpose of showing the variety of BIPV solutions and products to architects, design engineers, project developers and other parties from the building sector. For a large-scale rollout of BIPV in the market, these living labs are an important link.

This project is funded by the European Innovation program for the South of the Netherlands (OPZuid), as part of the European Fund for Regional Development (EFRO) of the European Union.

OPZuid

Europese Unie

BIPV Nederland

ENERGY PERFORMANCE EVALUATION OF A PHOTOVOLTAIC WINDOW

F. Serrano-Casares and V. Navas
University of Málaga. Energy Group
Escuela de Ingenieras Industriales. Campus de Teatinos E-29071 Málaga. Spain

ABSTRACT: The energy sector is in a process of radical and unstoppable transformation, motivated by two reasons: the fight against climate change and renewable energies. The two factors are decisive and go hand in hand, in line with the social demands and decisions of the European Union to comply with the emission reduction agreements.
The growing increase in energy consumption in developed and developing countries is an important task that must be addressed by the economic policies of these countries. The percentage of energy consumption of buildings relative to the overall energy use has grown in the last years due to the increasing amount of electrical devices.
The photovoltaic windows as an integrated system in the building façades can contribute to generate clean energy to meet needs of the inhabitants of these buildings.
A photovoltaic window specially built by a manufacturer has been studied. An amorphous silicon photovoltaic module encapsulated between two transparent glass sheets, an air chamber and a second double glass sheet with an air chamber forms the photovoltaic window. Everything is framed in a PVC structure. The effective dimensions of the a-Si photovoltaic module are 0.57 m x 1.17 m, equivalent to a standard measurement of 0.60 m x 1.20 m.
To know the electrical characteristics of PV window in standard test conditions, a test in accordance with IEC standard 61646 it was carried out. A peak power of 50.74 W_p was obtained.
Measurements of energy production in real sunlight were carried out. The window was placed vertically facing south on a test bench. Measurements of the energy produced by the photovoltaic window were made in several sunny days of August and September 2016 from sunrise to sunset. On average, the irradiance received on the plane of the photovoltaic window was 4114 Wh/m^2 and the energy produced 71.2 Wh each day. These results match those obtained using the Malaga radiation databases.
For one square meter of the window studied, 79868 Wh/m^2/year are obtained, when an overall efficiency of 0.8 is considered. Integrating this PV window in a building in Malaga (Spain), an annual electric production of 345030 kWh is obtained when a glazing surface of 4320 m^2 is considered. This energy is enough to meet the annual electricity needs of the 68 household of the building. Furthermore, the saving in electricity bill of the houses of the building would be 59279 €/year and the BIPV system would avoid the emission of 240278 kg of CO_2 into the atmosphere per year.
Keywords: PV window, a-Si, Building Integrated PV (BIPV).

1 INTRODUCTION

Modern buildings are built taller and consume more energy, and efforts are being made in various areas to reduce their energy consumption. Building Integrated Photovoltaic (BIPV) is widely used in modern buildings as part of the façade in recent years. Due to its characteristics, its energy production can be combined with other functional features on buildings. This can not only achieve an aesthetic outlook, but also encourage the development of energy efficient buildings. BIPV system does not require extra installation spacing and can be integrated into the building envelope that replaces the traditional building materials. Semi-transparent photovoltaic (STPV) is one of the BIPV applications which becomes more and more popular. A topic about the energy savings of an office building with the use of semi-transparent BIPV modules is analyzed in [1]. The overall energy performance was studied in terms of heating and cooling loads, daylight availability and electricity generation by using EnergyPlus. The thermal performance of semitransparent BIPV modules is studied in [2] and stated that the majority of total heat gain from PV module was solar heat gain. It also evaluated several important design parameters for the design of solar technology, which included orientation, solar cell area ratio, efficiency of solar cells and module thickness. The energy performance and cost of semi-transparent BIPV modules in office building is analyzed in [3]. Thermal and visual properties were the major components of overall energy performance. The results were conducted by manual calculations in order to estimate the reduction of peak cooling load in addition to the annual electricity consumption.

Thin-film technology based on amorphous silicon (a-Si) offers a range of attractive features that are ideally suited for BIPV installations. Solar modules may be assembled to customer-specific BIPV elements for roofs and façades, and thus may combine various functions, namely electricity generation, thermal insulation, shading, and even satisfy aspects of architectural design. Compared to other PV technologies, a-Si modules show only a minor reduction in power output at elevated temperatures, and in efficiency at lower light levels, and thus offer superior energy yields per peak power.

The aim of this work is to evaluate the energy performance of a semi-transparent photovoltaic window of unknown characteristics, specially built, and to estimate the electricity production potential to meet the energy needs of urban buildings in Malaga (Spain).

2 WINDOW PV CHARACTERISTICS

The semi-transparent PV window is shown in Figure 1. An amorphous silicon photovoltaic module encapsulated between two transparent glass sheets, an air chamber and a second double glass sheet with an air chamber form the photovoltaic window. The effective dimensions of the a-Si photovoltaic module are 0.57x1.17 m^2, equivalent to a standard measurement of 0.60x1.20 m^2. The frame of the window is a PVC structure.

The characteristics of the window were not provided

by the owner, so the experiments necessary to know them had to be carried out.

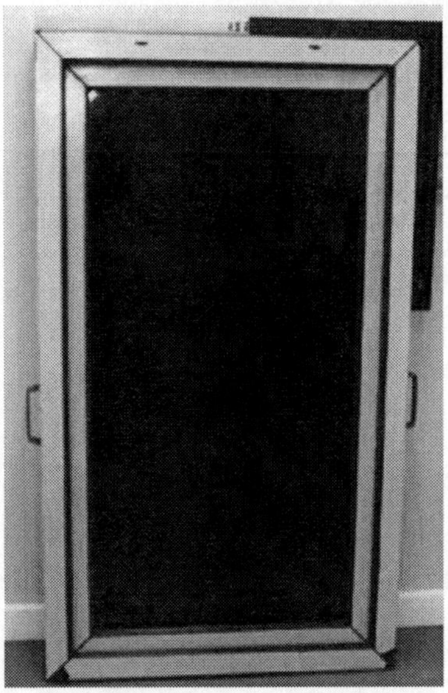

Figure 1. Semi-transparent PV window.

Table 1. Climate test conditions and PV characteristics.

T_{max}(°C)	23,3
T_{min}(°C)	21,3
RH_{max} (%)	54,3
RH_{min} (%)	51,1
RH_{limit} (%)	<70
I_{sc} (A)	0,71
V_{oc} (V)	113,19
V_{max} (V)	87,18
I_{max} (A)	0,58
Peak power (W_p)	50,74

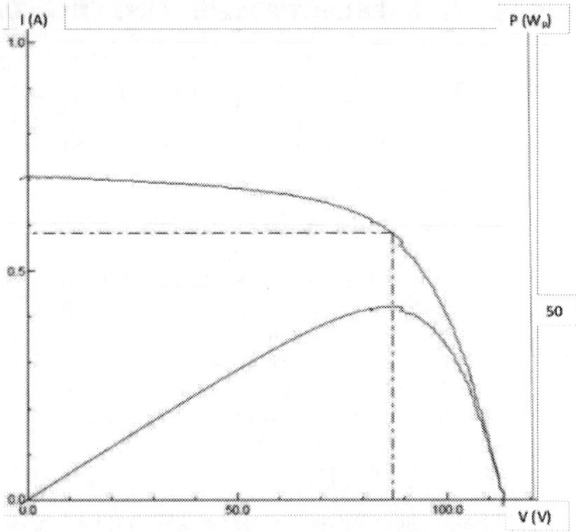

Figure 2. I-V curve of photovoltaic window at standard test conditions obtained in the solar simulator.

The IV characteristic in standard test conditions were determined in a solar simulator according to 61646 IEC standard [4]. The climatic values and the electric results are shown in Table 1. A power uncertainty of ± 6.78% was found. Figure 2 presents the I-V curve of the PV window

Once the window electric characteristics was known, it was placed in a test bench installed on the rooftop of the building of the Andalusian Institute of Renewable Energies, at Málaga (36.73 N, 4.55 W). The PV window was placed on a steel mobile structure. This structure allows an adjustment of both azimuth and inclination. A pyranometer was used to measure the global irradiance on the window plane, and a temperature sensor measures its temperature. The photovoltaic window was placed in vertical position facing South. A protocol for data collection that allows knowing the instantaneous power generated by the photovoltaic window has been developed. The figure 3 shows a scheme of the experimental arrangement.

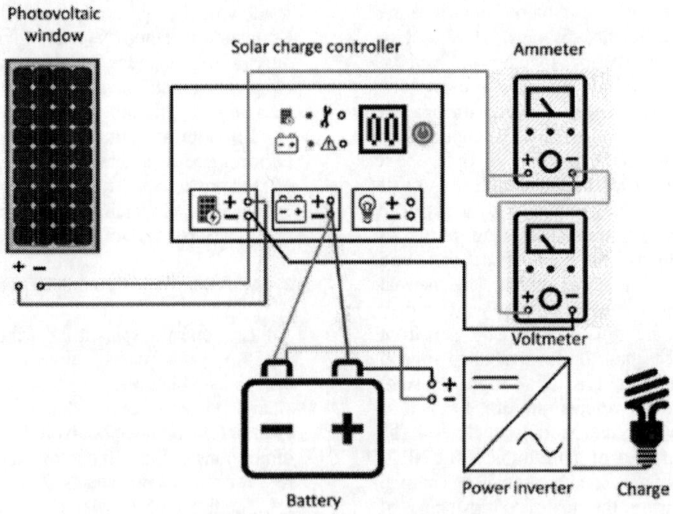

Figure 3. Experimental arrangement.

3 ENERGY PRODUCTION OF PV WINDOW

To know the energy produced by the window in vertical position, the voltage and current generated by the photovoltaic window in sunny days as well as the solar irradiance, module temperature and ambient temperature are collected every ten minutes from sunrise to sunset. The measurements of the energy supplied by the photovoltaic window have been compared with those obtained with an IV photovoltaic meter.

To carry out the measurements, the following devices have been used:

- Fluke model 8842A multimeter: one to measure the voltage and another to measure the current of the photovoltaic window at every moment.
- Fluke Hydra model 2620A (Hydra Data Acquisition Unit) to collect the radiation, the ambient temperature and the temperature of the photovoltaic module by:
  - CM11 pyranometer from Kipp-Zonen located in the same plane of the photovoltaic window.
  - Temperature probe (thermocouple) that is placed on the back of the cell.
  - Temperature probe (thermocouple) for the ambient temperature.
- Fluke-1735 power logger to measure the voltage, current and power consumed by the test load.
- Digital meter for measuring the voltage at the terminals of the photovoltaic battery.
- I-V400 solar I-V meter to verify the instantaneous measurements of the power generated by the photovoltaic window.

In addition, a solar FS-155 battery, a pure sine wave inverter 800W/1600W, a MMPT solar charge controller and a variable charge, form the experimental arrangement to simulate a stand-alone PV system as is shown in Figure 3.

The result obtained for solar radiation and the instantaneous power produced by the photovoltaic window on a sunny day of measurement is shown in Figure 4. The irradiance received on the plane of the photovoltaic window at daytime was 4114 Wh/m^2 and the energy produced was 71.2 Wh. These values are consistent with radiation data obtained from databases [5] and the power value obtained from PV window under standard test conditions.

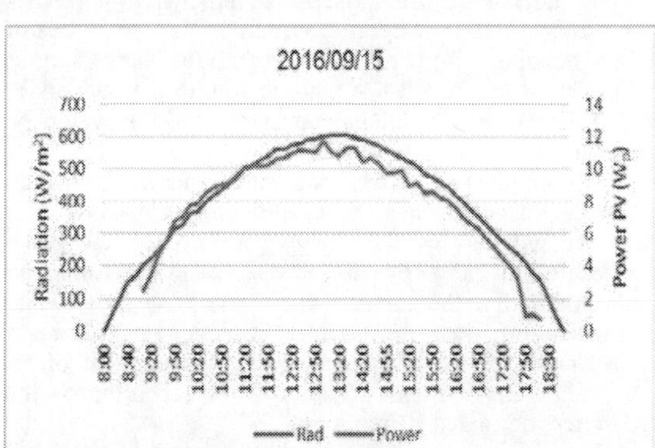

Figure 4. Solar radiation and instantaneous power of PV window in a sunny day.

Considering that the power generated by the photovoltaic window is proportional to the surface, the photovoltaic power of a window of one square meter of useful surface will be 76.1 W/m$^2_{window}$.

The energy yield of a photovoltaic system depends on the power of the system under standard test conditions, on the normalized solar radiation received on the plane of the module and on the efficiency of the system [6]. That efficiency takes into account the total losses that occur in the photovoltaic system like the dispersion of parameters, the actual temperature of the modules, the wiring, the effect of the angle of incidence of the sun on the plane of the module, the performance of the devices of power conditioning (regulator and inverter) and the battery performance. Its value for commercial systems varies between 0.75 and 0.85 usually.

With the PV window placed in vertical position, the monthly yield production for one year has been obtained. Efficiency values of 0.75, 0.80 and 0.85 and radiation data from the Energy Agency of the Junta de Andalucia for Malaga are considered [5]. Figure 5 graphically compares these results.

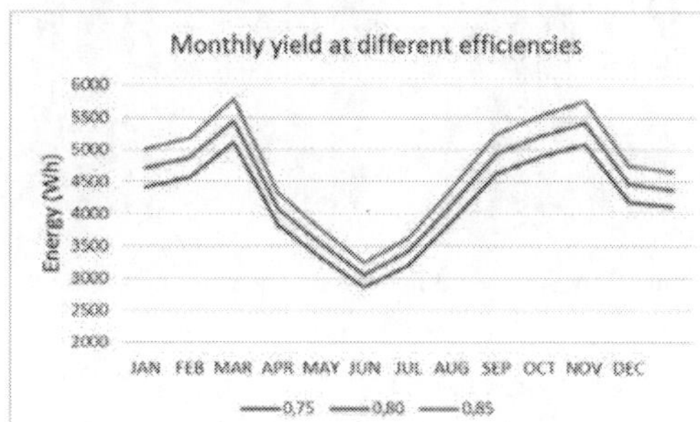

Figure 5. Monthly yield at different efficiencies.

From the analysis of the results, it is observed that when the window is vertically placed (usual position) the energy production in the months from April to September is significantly affected.

The annual production can be improved if instead of using the photovoltaic module as a vertical window, the same semi-transparent photovoltaic module is used as an enclosure element in a building, for example as a skylight for the covering of a courtyard or an atrium. This application would increase the solar energy captured and, therefore, the annual production [7, 8].

To analyze the influence of the photovoltaic window position in a building, the annual yield for 0.85 efficiency has been obtained when the PV window is placed in vertical position, and tilted 60° or 30°. Figure 6 shows the monthly result of one square meter of PV window.

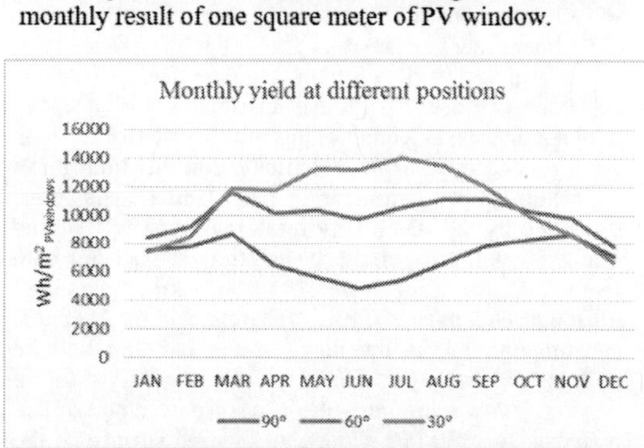

Figure 6. Monthly yield of window at different positions.

As can be seen on Figure 6, the tilt angle affects the yearly energy production of PV window. The bigger production correspond to a tilt of 30°, so the best use of this window in BIPV is as a part of an atrium [9]. Also, the 60° inclination presents a more homogeneous behaviour throughout the year, although the annual production is greater for 30° tilting. This is recommended to obtain maximum production in a photovoltaic system connected to the grid when the installation is placed in a latitude like that of Malaga.

In Figure 7 are presented the annual yield for different efficiencies and different tilting position of the PV window.

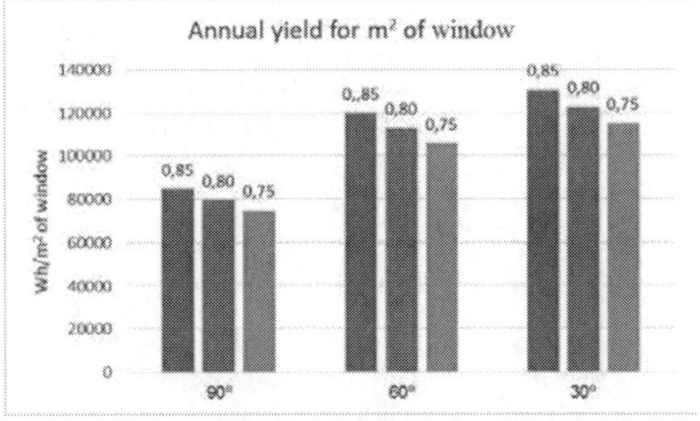

Figure 7. Annual yield for one square meter of window.

4 DISCUSSION

The integration of the photovoltaic window in a building has several advantages over conventional systems: energy saving, economic savings and environmental advantages. In the previous sections, we have analyzed the energy produced per square meter of photovoltaic window. In this section, the production of electricity, the economic savings, and CO_2 emissions avoided will be analyzed if PV semi-transparent windows were installed in an apartment building located in Málaga (Spain).

Energy saving

An apartment building with the following characteristics is considered:

Residential building with 17 floors with 4 apartments per floor (68 apartments):

- 2 apartments of 60 m^2
- 1 apartment of 75 m^2
- 1 apartment of 100 m^2

The glazed surface of the building facing south is supposed to be 4320 m^2. All apartments are equipped with appliances, air conditioning and electric stove.

The average annual electricity consumption for a residential utility consumer in the Mediterranean area, estimated by the IDAE (Institute for Diversification and Energy Saving) according to the study carried out with real data for the years 2010 to 2016, is 3487 kWh/household/year [10]. Therefore, the annual consumption of the dwellings of the building will be 237116 kWh/year.

The PV window provides a peak power per square meter of 76.1 W_p/m^2. If the entire glazed surface of the building is covered by the PV window, a peak power of 328.8 kW_p would be available.

The annual electric production per square meter in Malaga (located in the Mediterranean area) of the window facing South was 79.868 kWh/m^2/year, considering a global coefficient of losses of 0.20 as shown in Figure 7. Therefore, the yield obtained for the BIPV system would be 345030 kWh/year.

As the electric consumption of the apartments is 237116 kWh/year, an electric excess of 107914 kWh/year is obtained that represents an energy surplus of 46%. This surplus of energy could cover the building's common electricity needs, such as elevators, lighting in common areas, pumps for water supply and car park doors, which can be estimated at 6700 kWh/year. The rest of the energy could be sold to the network or to other users, if the legislation allows it.

Economic savings

In March 2014 the Spanish Government introduced a new way of calculating the price of electric energy that consumers must pay: the Voluntary Price for Small Consumers (PVPC in Spanish acronym). The PVPC is the system for fixing the price of electricity for customers on the regulated market (customers who have their contract with a Reference Seller), with a contracted capacity of no more than 10 kW.

Currently, consumers have digital meters so billing is based on the consumption of electricity that occurs every hour, applying the prices established at that time by the energy market. That is, today the price of electricity changes every hour of the day and every day.

However, the small consumer can take advantage of tariffs with a fixed price of electricity, which can be revised annually. They can also choose among different market traders.

On the other hand, the bill of the electric company consists of three concepts: a) the power contracted, b) the energy billed and c) taxes, both the tax on electricity and VAT.

The energy extreme prices of the first week in September 2016, ranged from a minimum of 0.096 €/kWh, corresponding to a Saturday at 5 pm, to a maximum of 0.162 €/kWh which corresponds to a Wednesday at 9:00 p.m. Similar values have been found in other weeks with an uncertainty of 0.005 €/kWh. When taxes are considered a daily average price of 0.25 €/kWh is obtained.

Under these conditions, if an efficiency of 0.80 is considered, the saving in electricity bill of the houses of the building would be 59279 €/year. If the energy surplus produced by the BIPV system is sold to the network at 50 €/MWh, an additional profit of 5396 € would be obtained.

With this type of windows another benefit in relation to the consumption in air conditioning is reached. It is saved not only by the energy generated by the BIPV system but also by the lower energy consumption required by the home when using windows that significantly increase thermal insulation and decrease solar radiation loads, reducing the consumption of air conditioning systems in up to 30% according to the sources consulted [10].

CO_2 emissions avoided

The photovoltaic window not only saves money but also prevents the emission of harmful gases to the environment by generating electricity from renewable sources and, therefore, non-polluting.

If a conversion factor of 0.30 kg of CO_2 emitted per each electric kWh consumed is considered, the BIPV system avoids the emission of 240278 kg of CO_2 into the atmosphere per year.

5 CONCLUSIONS

A PV window has been studied with the aim of being integrated in a building in Malaga (southern Mediterranean area). A peak power of 50.74 W_p was obtained from measurement made in a solar simulator at standard test conditions.

An experimental arrangement was built to obtain the daily production at real sun conditions. A value of 71.2 Wh was obtained when the window is placed in vertically facing south. The production depends on the overall efficiency of the system and on the inclination of the window.

The integration of the PV window in a building allows to obtain energy, economic and environmental advantages.

Considering an apartment building with a surface of photovoltaic windows of 4320 m^2, the BIPV system provides enough energy to cover the annual electric demand of the dwellings of the building, producing an excess of energy that can be sold to the grid and avoiding the emission of 240 tons of CO_2 to the atmosphere annually.

The use of semi-transparent BIPV elements will become more and more popular to combine various functions, namely electricity generation, thermal insulation, shading, and even satisfy aspects of architectural design.

6 REFERENCES

[1] Miyazaki T, Akisawa A, Kashiwagi T. Energy savings of office buildings by the use of semi-transparent solar cells for windows. Renewable Energy 2005, 30: 281-304.

[2] Yang HX, Fung YY. Study on thermal performance of semi-transparent building integrated photovoltaic glazings. Energy and Building 2008, 40: 341-350.

[3] Li HW, Lam NT, Chan WH, Mak HL. Energy and cost analysis of semi-transparent photovoltaic in office buildings. Applied Energy 2009, 86: 722-729.

[4] IEC 61646: 2008 Ed 2- Thin-film terrestrial photovoltaic (PV) modules - Design qualification and type approval.

[5]http://www.agenciaandaluzadelaenergia.es/Radiacion/radiacion1.php. Last access 2018/03/9.

[6] Lu L., Law K. M. Overall energy performance of semi-transparent single-glazed photovoltaic (PV) window for a typical office in Hong Kong. Renewable Energy 2013, 49: 250-254.

[7] Yang HX, Lu L. The optimum tilt angles and orientations of PV claddings for building integrated photovoltaic (BIPV) applications. Journal of Solar Energy Engineering - Transactions of the ASME 2007, 129 (2): 253-255.

[8] Hwanga T., Kangb S., Kim J. K. Optimization of the building integrated photovoltaic system in office buildings-Focus on the orientation, inclined angle and installed area. Energy and Buildings 2012, 46: 92-104.

[9] Maurus H., Schmid M., Blersch B., Lechner P., Schade H. PV for buildings: benefits and experiences with amorphous silicon in BIPV applications. reFOCUS 2004, 5 (6): 22-27.

[10]Informe anual de consumos energéticos. Evolución 2010-2016. http://www.idae.es/estudios-informes-y-estadisticas. Last access 2018/03/09.

[11]Guía Práctica de la Energía. Consumo Eficiente y Responsable. Ed. IDAE (Instituto para la Diversificación y Ahorro de la Energía), 2017. Madrid (Spain).

ACKNOWLEDGEMENT

This work is partially derived from the contract nº 8.06/5.31.4644 Libre Evolución de Energía S.L. The authors want to thank to Universidad de Málaga. Campus de Excelencia Internacional Andalucía Tech the support for presenting this work.

BEYOND WATT PER MODULE AND COSTS PER WATT – NEW WEIGHT RELATED PARAMETERS FOR PHOTOVOLTAIC MODULES

Sebastian Schindler, Dominik Götz, Jens Schneider
Fraunhofer-Center for Silicon Photovoltaics CSP
Otto-Eissfeldt-Str. 12, 06120 Halle, Germany

ABSTRACT: Nowadays, PV modules are almost identical in size and shape and weight worldwide. Standard photovoltaic modules, therefor, show 60 or 72 Si wafer based cell layout, interconnected string-wise and all embedded within an encapsulant material (e.g. Ethylen-Vinylacetat (EVA)) and between a frontglass and a polymeric backsheet. In terms of new applications and up-coming trends in the photovoltaic sector, certain material restrictions as well as size and weight requirements can show up. Examples for such applications are building-integrated photovoltaics (BIPV), vehicle-integrated photovoltaics (VIPV) or comparable mobile implementations, for which low transportation/installation weight is a crucial aspect, e.g. off-grid and solar home system markets.
By this paper, we want to discuss parameters for alternative photovoltaic module classifications, in order to get the ability to compare and separate modules in regards of lightweight and design aspects and underlying technology. The parameters Power-to-Weight Ratio *P2W,* Area density *AD,* Weight-to-Power Ratio *W2P* and Efficiency-to-Weight Ratio *E2W* are proposed to compare the three module groups Silicon, ThinFilm and Lightweight.

Keywords: module technology, lightweight, BIPV, VIPV

1 INTRODUCTION AND MOTIVATION

Common photovoltaic (PV) modules contain 60 or 72 crystalline silicon solar cells, which are interconnected string-wise and embedded between either a frontglass and a polymeric backsheet or two glasses through an encapsulant as e.g. Ethylen-Vinylacetat (EVA). Those stack structures, together with an additional aluminum frame and a junction box usually result in a total module weight in the range of 15 to 27 kg [1-3].

In terms of new applications and trends coming up in the photovoltaic sector, certain material restrictions as well as size and weight requirements can get substantial. Examples for aforesaid applications are building-integrated photovoltaics (BIPV), vehicle-integrated photovoltaics (VIPV) or comparable mobile implementations, for which low transportation / installation weight is a crucial aspect.

To get the ability to compare different photovoltaic modules with regard to their lightweight design quality, suitable parameters have been evaluated. For this purpose the following chapters introduce possible indicators and discuss the aptitude for their use when applied to current photovoltaic modules in the market.

2 APPROACH

2.1 Module classification & weight related parameters

This paper compiles parameters for photovoltaic module classifications in order to get the ability to compare and separate modules in regards of lightweight and design aspects and underlying technology.

For the classification we have chosen major groups of PV modules to distinguish among each other:

- Silicon: modules containing 60 or 72 cells, Si wafer based (Al-BSF, Perc, etc.) with a glass/backsheet or a glass/glass laminate stack, plus framing, etc.
- ThinFilm: containing all thin-film technologies, leading mainly to glass/glass laminates
- Lightweight: these modules take advantage of substituting materials and redesigning manifoldly laminate stacks in varying dimensions and shapes, solar cell technology independent, differenced as foil/foil assembly or rear backing structures with glass or foil front sides

Figure 2: Typical example for silicon wafer-based 60-cell as framed glass-backsheet module configuration, picture taken from q-cells.de

Figure 2: : Example for ThinFilm modules, Solibro CIGS glass-glass laminate, picture taken from solibro-solar.com

Figure 3: Lightweight foil-foil module for VIPV, roof size & performance optimized [prototype by Fraunhofer CSP]

Figure 4: Example of 60-cell lightweight with rigid polymer foam rear side [prototype by Fraunhofer CSP]

So far, no standards exist for parameters allowing comparison of modules for ratios of weight to area, power or cost.

As a first approach to this topic, we suggest the following parameters within this paper:

- Power-to-Weight Ratio *P2W* [W/kg]
- Area density *AD* [kg/m²]
- Weight-to-Power Ratio *W2P* [kg/W]
- Efficiency-to-Weight Ratio *E2W* [%/kg]

2.2 Power-to-Weight Ratio *P2W* [W/kg]

The power-to-weight ratio (P2W) is defined as the STC power rating in watts divided by the weight in kilograms of the module:

$$P2W = \frac{P_M}{m_M} \quad (1)$$

The power output of each module is dependent on the efficiency of the used solar cells as well as sources of gains and losses in the module [4, 5]. The power is measured under standard test conditions (STC: AM 1.5, temperature 25 °C, irradiation 1,000 W/m²).

As a result, the aim will be to achieve the maximum possible *P2W* for given technology and used materials.

Especially for lightweight module concepts, one will significantly reduce material mass, e.g by polymer only laminates without any glasses or excluding frames.

2.3 Area density *AD* [kg/m²]

The area density (AD) is defined as the quotient of module weight to area of the module:

$$AD = \frac{m_M}{A_M} \quad (2)$$

The module weight is defined as the overall weight of the resulting module, including its frame, junction box.

In general, it can be assumed that *AD* shall be lowest as possible, as especially in building integration and mobile installations bearing load limitations exist.

And the reduction of weight will somehow be always beneficial and will address aspects in terms of lower masses for transportation and installation.

2.4 Weight-to-Power Ratio *W2P* [kg/W]

The weight-to-power ratio (W2P) is the reciprocal of the prior defined *P2W* parameter:

$$W2P = \frac{1}{P2W} = \frac{m_M}{P_M} \quad (3)$$

This parameter generates the possibility to compare different types of modules independent of their individual size.

2.5 Efficiency-to-Weight Ratio *E2W* [%/kg]

The efficiency to weight ratio (E2W) is the module efficiency by the overall module weight:

$$E2W = \frac{\eta_M}{m_M} = \frac{P_M/(A_M G_{POA})}{m_M} \quad (3)$$

Where η_M ist the efficiency, P_M the power, A_M the area of the module and GPOA is the irradiance in the plain of area. This parameter can indicate technological improvements, e.g. for increased cell efficiency the so far persistent module manufacturing results in an increasing *E2W* ratio for a given module technology.

The definition of area can be discussed extensively. In our consideration, the outer dimensions of the module were used. More distinct *E2W* can be defined, by differentiation between overall dimensions, cell area only, active module area including cell to cell spaces, etc.

A dimensioning of this parameter does make sense in terms of BIPV, for example. A limited installation area will probably lead to as much as possible watts per square meter and parallel considerations of restrictions of bearing loads of facades and building envelopes.

3 RESULTS

The allocated data for graphs in the result section has been taken from commercially available data sheets [1-3] or if available by in-house measurements of modules. The data represent not a set of completeness of contents of all available PV modules in the market.

Figure 5 illustrates the *P2W* ratio for the 3prior defined groups of PV modules.

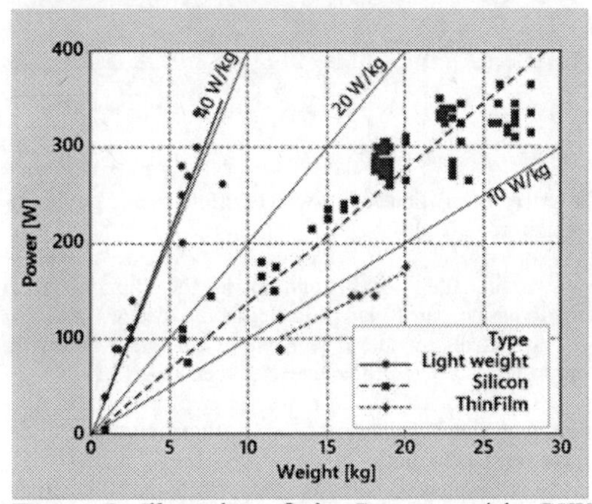

Figure 5: Illustration of the Power-to-weight P2W distribution of the different groups of PV modules

The group Silicon, mainly Si wafer based mainly 60-72 cell glass-glass or glass-backsheet configuration, does have the most data points, as this kind of module technology has to be considered as nr.1 world production standard and highest market share.

This group shows a strong grouping along a contour line of 15 W/kg. This reflects the almost same design and bill of materials for this type of modules. For example, a 260-300 W_P Al-framed 3,2 mm glass - backsheet laminate with 60 cells results in typically 18±x kg module weight.

ThinFilm modules show a less favorable *P2W* ratio with <10 W/kg, as this type of modules are primarily manufactured on glass substrates while having in general lower efficiencies.

The group of Lightweight modules shows a significantly increased *P2W* ratio. The contour line increases more than 2 times up to 40 W/kg for chosen data by us.

Lightweight modules are often laminated as (semi-)flexible module stacks, resulting in a replacement of the front side material glass to polymer sheets or foils. A very representative material is ETFE.

The diagram shows that by advanced development of module technology half the weight for modules within the 280-300 W_P class can be achieved, combining lightweight design and the integration of PV-standard competitive Silicon cell power / efficiency in one module concept.

Figure 6 extends with a face to face illustration of efficiency to weight.

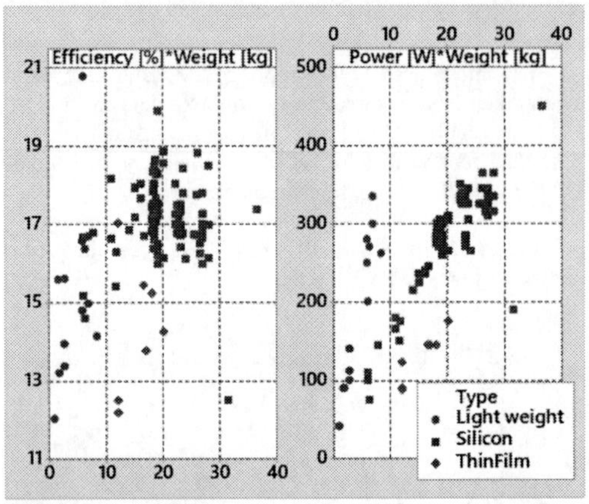

Figure 6: Efficiency-to-weight (left) and power-to-weight ratio (right)

Again, the world standard of the Silicon group agglomerates at 18 kg (60cell module design) and 23-27 kg (72cell module design or 60 cell glass-glass) for nowadays module efficiencies between 16 – 19 %.

ThinFilm modules are weight depended in a 12 – 20 kg span with significant lower efficiencies, due to the given cell technology.

Lightweight modules are very widespread. Due to the change in front side material, very often this type of modules show of higher cell to module losses, resulting in lower module efficiencies for this group of PV.

The next figure addresses the weight and covered area of our 3 chosen PV module groups.

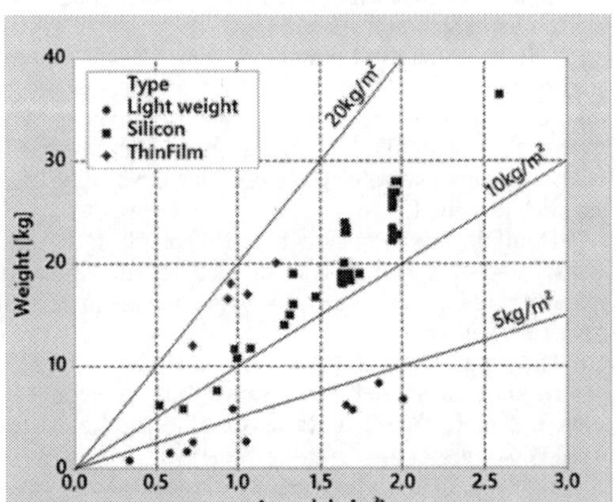

Figure 7: PV module area density: weight over module area

There is a limit of 10 kg/m² for the group of Silicon noticeable. In our opinion mainly due to glass usage and necessary framing solutions, independently of cell technology.

Lightweight modules can be classified below 5 kg/m². These modules differ strong in size and they are very light due to their front side material, often used polymer foil / sheet.

Nowadays ThinFilm solutions build upon glass substrates, leading to certainly higher area densities, see green data points within figure 7.

The following figure 8 illustrates the Power-to-weight over area density. Most modules are on joint hyperbolas with a constant *P2W* and *AD* product.

As a result, this is not too surprisingly since the product is efficiency.

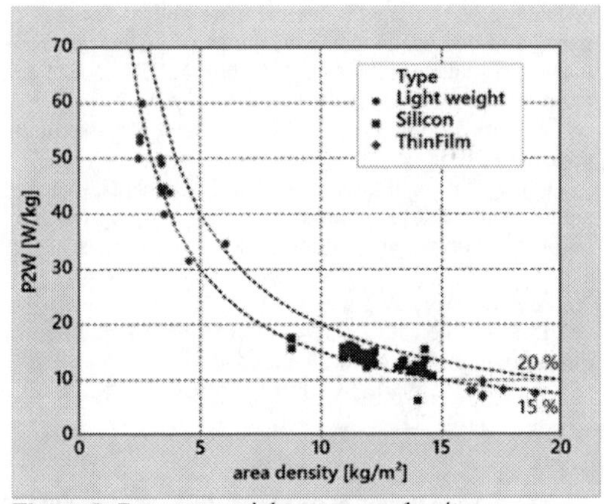

Figure 8: Power-to-weight over area density

Again, Silicon group of PV modules pool due to their weight settings by design and BOM restricted configurations.

In any case, high *P2W* ratios for lightweight modules mean that small area densities are achieved.

The last figure (figure 9) expands the prior considerations with estimated costs for the 3 groups.

We calculated the area-related costs of the modules and set this in ratio to the module power.

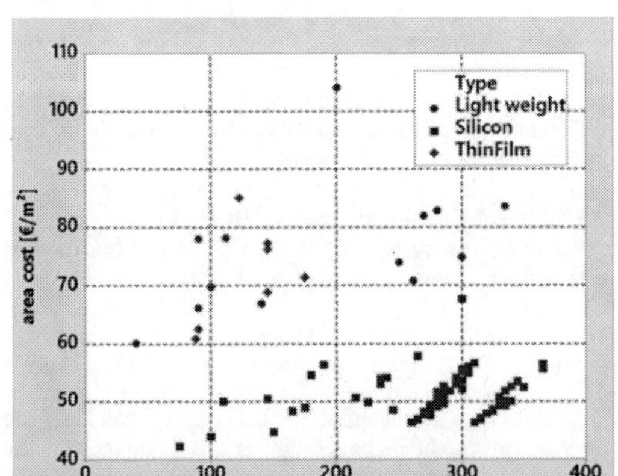

Figure 9: Cost per square meter over module peak power

As Silicon group is the dominant production standard, lowest costs for different module power classes are feasible.

And as today most modules are in the range of 280-310 W_P (60 cell layout), most data points are markable in the diagram.

Lightweight modules are more diversified due to different material stacks. These types of modules are significantly lighter and often thinner. But these advantages are currently often achieved at higher costs.

Costs for solar cells and interconnection can be assumed as high as for Silicon, taking full advantage of achievements in this group.

Based on our expertise in developments of lightweight PV module concepts [6, 7], cost drivers are replacement materials for front and back side of modules, as especially thin and lightweight sheets or foils need to be engineered weather and mechanical protective as well, compared to very cost competitive and stable glasses.

Lower efficiencies for ThinFilm lead to lower power output, while having higher area costs together with lightweight modules.

However, lightweight concepts can already be cost competitive [7] with necessary research in terms reliability of primarily used polymeric materials.

4 CONCLUSIONS AND OUTLOOK

The costs for solar as renewable energy source are decreasing furthermore. This leads to further and new markets for the implementation of PV as an additional feature.

As for BIPV and VIPV applications, today's STC ratings of modules will not be sufficient anymore.

Emerging and fast-growing markets for e-mobility and nearly zero-energy buildings requesting alternative photovoltaic module concepts and PV design interpretation [8-10].

This paper suggests alternative parameters and module classifications at first serve.

Independently of solar cell types, resulting weights of module technology can become dominating, due to transport or bearing load limitations.

The development and availability of lightweight modules will increase market share of such product developments and result in competitive material cost reductions, as seen for standard PV modules (*Silicon group*).

7 REFERENCES

[1] Heckert Solar GmbH: Datenblatt NeMo® 2.0 60 M 295-300 Wp, technical data sheet, pdf version downloaded 08-2018

[2] AE Solar: AE POWERPLUS MODULE AE M6-72 Series 320W-345W, technical data sheet, pdf version downloaded 07-2018

[3] Hanwha Q Cells GmbH: Produktdatenblatt Q.PEAK DUO-G5 Q.ANTUM SOLARMODUL, technical data sheet, pdf version downloaded 07-2018

[4] Hanifi, H. et al.: Investigation of cell-to-module (CTM) ratios of PV modules by analysis of loss and gain mechanisms, 2016, Photovoltaics International, Volume 32

[5] Schneider, J. et al. 2014, "Combined effect of light harvesting strings, anti-reflective coating, thin glass, and high ultraviolet transmission encapsulant to reduce optical losses in solar modules", Prog. Photovoltaics Res. Appl., Vol. 22, pp. 830–837

[6] Götz, D.: Light Weight Solar Modules for Mobile Applications, PV Days 2017, Halle, Germany

[7] Schindler, S. et al.: Development of rigid lightweight PV modules of polymeric based lamination stacks – benefits and challenges for novel module concepts, SNEC 2018, Shanghai, China

[8] Fraunhofer CSP: Scientists present a particularly efficient design solar façade, SOLAR.shell project, Press release, 23.11.2017

[9] Fraunhofer CSP: Concrete can do more – Fraunhofer CSP wants to turn house walls into solar power plants, Press Release, 11.11.2016

[10] Fraunhofer CSP: Fraunhofer CSP and Continental develop flexible photovoltaic modules for trucks, Press Release, 22.9.2016

COMPARATIVE PERFORMANCE MEASUREMENTS OF IDENTICAL BIPV-ELEMENTS IN DIFFERENT CLIMATIC ENVIRONMENTS - A ROUND ROBIN ACTION WITHIN THE IEA PVPS TASK 15 COLLABORATION

Peter Illich[1], Gabriele C. Eder[2], Karl A. Berger[3], Újvári Gusztáv[3], Philipp Rechberger[4], Dieter Moor[5], Martin Aichinger[5], Simon Boddaert[6], Roland Valckenborg[7], Jochem van den Brand[7], Pierluigi Bonomo[8], Francesco Frontini[8], Cristina S. Polo López[8], Matteo del Buono[9], Anne Gerd Imenes[10], Nuria Martín Chivelet[11], Faustino Chenlo[11], Asier Sanz Martinez[12], Maider Machado[12], Jun-Tae Kim[13], Alex Masolin[14], Michiel Ritzen[14]

[1]University of Applied Sciences Technikum Vienna, Giefinggasse 6, 1210 Vienna, Austria - Corresponding author: illich@technikum-wien.at; 0043-1-3334077-577
[2]OFI, Austrian Research Institute for Chemistry and Technology, Austria
[3]AIT, Austrian Institute of Technology GmbH, Austria
[4]University of Applied Sciences Upper Austria, Austria
[5]Ertex Solartechnik GmbH, Austria
[6]CSTB, Centre Scientifique et Technique du Bâtiment, France
[7]SEAC, Solar Energy Application Centre, Netherlands
[8]SUPSI, University of Applied Sciences and Arts of Southern Switzerland, Switzerland
[9]EURAC Research, Italy
[10]UiA, University of Agder, Norway
[11]CIEMAT, Spain
[12]Tecnalia Research and Innovation, Spain
[13]Kongju National University, South Korea
[14]ZUYD, University of Applied Sciences, Netherlands

ABSTRACT: This paper results from an on-going activity within the IEA PVPS Task 15 – Enabling Framework for the Acceleration of Building Integrated Photovoltaics (BIPV). It evaluates the performance of identical and commercially available glass/glass BIPV-elements installed in outdoor test sites at different climatic environments. These c-Si test-modules resemble those modules built in the façade of an existing reference building in Linz, Austria, and are mounted reproducing similar boundary conditions. The reference BIPV façade is currently under operation and has been monitored for more than 10 years. In total, the experiment involves 9 participants from 7 countries in Europe and Asia. Accorded monitoring method and constructive boundaries allow for the analysis of the influence of geographical and climatic conditions on the performance of the BIPV elements and can be related to the pre- and post-installation measurement results under STC conditions in a reference lab. From the experimental design of this work, it is possible to compare the performance results of a commercially available type of glass/glass BIPV elements installed in the façade of a building with those installed and monitored in outdoor-sites and measured under laboratory conditions. Furthermore, the reliability results under natural ageing conditions (high rise building with BIPV façade) are compared to those obtained after accelerated ageing in the laboratory.
At this point, (incomplete) monitoring data of only six modules could be used for the first analysis. Thus, a detailed description of the the experimental design, the technical data of the laboratories, test sites and reference building and the current status of data analysis is presented.

Keywords: BIPV, Monitoring, Performance Measurements, Glass/Glass Modules

1 INTRODUCTION

Within the international technological collaboration program Photovoltaic Power Systems (PVPS) of the International Energy Agency (IEA), a group of over 60 experts from 16 countries are collaborating in Task 15 to create an Enabling Framework for the Acceleration of Building Integrated Photovoltaics (BIPV). As part of this collaboration, a joint action was initiated to compare available outdoor tests facilities in a "Round Robin" test. The test modules selected are identical to the BIPV modules used in an existing BIPV installation on a multi-story building – the Power Tower in Linz, Austria (Headquarter of the Energie AG utility). The BIPV façade of this reference building has already been in operation for more than 10 years. This allows for comparing the performance of the test modules (i) after production (measured in the laboratory under STC conditions), (ii) installed at various outdoor test sites and (iii) integrated in the reference building. In addition, accelerated aging tests are being performed on two of the identically produced test modules in order to compare their reliability with the long term performance of the respective modules in the BIPV façade. After performing an international mapping on BIPV research teams and test facilities [1], 9 participants from 7 countries in Europe and Asia agreed on such a joint round robin action. For this experiment, 13 identical glass/glass BIPV elements with crystalline Si-cells were produced by ertex solartechnik GmbH, pre-characterized in the certified test laboratory of the AIT – Austrian Institute of Technology GmbH and shipped to the 9 selected outdoor test sites. Two of the modules were kept in the laboratory for accelerated testing. A monitoring procedure was developed and harmonized with all participants. The constructive requirements for the test stands are based on the façade setup of the reference building (ventilated curtain wall, cold façade) and were taken into account by all participating institutions. The outdoor monitoring of the round robin action started in November 2017 and is planned to run for at least one year. Based on the acquired data, an annual yield of the test modules in different geographic and climatic environments will be obtained.

2 METHODOLOGY

2.1 Design of experiment

This experiment will allow to directly compare the performance and energy yield of identical and commercially available glass/glass BIPV elements installed and monitored under well-defined conditions in different environments. As the monitoring method and the constructive boundaries for the test stands are agreed by all participants, a direct influence of the environmental and climatic conditions on the performance data and yearly yield of the BIPV elements can be concluded. Obtained and available data are put in relation to:

(i) The measurement results under STC conditions in the reference lab (i.e. pre-characterization) and the irradiance and temperature matrix of the Energy Rating Standard [2] and the module's spectral response [3];
(ii) The monitoring data and reliability results of the BIPV-elements within the reference BIPV-building (already in operation for 10 years);
(iii) The reliability results obtained in the accelerated ageing tests in the laboratory (climate chambers);

After shipping, the test modules were installed at the selected sites using a pre-defined mock-up outdoor test stand with the same constructive and monitoring method requirements. Monthly reports in pre-established data formats of the monitoring data (weather and operational data) are collected at each research institution for comparative data treatment. The BIPV modules used, the reference building as well as constructive requirements and monitoring methodology are described in the following sections.

2.2 BIPV Modules and Pre-Characterization

13 identical 54 (9 × 6) cell glass/glass BIPV modules of the type VSG-L were manufactured at ertex solartechnik GmbH in Austria. The size of the modules is 1.100 mm × 1.600 mm × 11.52 mm. Polycrystalline Si-cells (type Poly TT; 156 × 156mm) were used with a 5 mm by 5 mm distance between the cells, resulting in a module transparency of 25.33%. The edge distance is 78 mm (top and bottom) and 69.5 mm on the sides. As front and backside, tempered 5 mm glass panes are used, with the rear one coated on the outer side with a grey enamel coating. A polyvinyl butyral (PVB) is used as the polymeric encapsulant. The electrical parameters and the corresponding temperature coefficients according to the label at STC are:

- Nominal power (P): 227.26 Wp (+/- 5%); -0.37 %/K
- Short circuit current (I_{SC}): 8.34 A; 0.05 %/K
- Open circuit voltage (V_{OC}): 34.07 V; -0.32 %/K

The pre-characterization of the 13 test-modules in the lab showed a maximum power at STC (after initial light stabilization) in a measured range from 219.0 to 217.9 W and is therefore within the +/- 5% tolerance given by the manufacturer. Electroluminescence (EL) images (at 100% Isc and 10% Isc) showed that all cells were intact, with no cracks or other irregularities and nearly similar brightness.

2.3 Reference building and measurements

The reference BIPV installation is located at the so-called "Power Tower" in Linz, Austria. The building, shown in Figure 1, is located in the center of the town and was built in 2008 as the first high rise office tower according to the passive house standard. Beside a highly efficient building energy system based on solar and geothermal energy, a 66 kWp BIPV system is installed on the south-west façade of the building. The BIPV installation, with an area of 638 m², expands over the upper 17 (of 19) floors. Although the whole back ventilated façade consists of about 60% glass the solar heat gain was minimized by special underlying louvers providing excellent daylighting.

Figure 1: BIPV façade at Power Tower in Linz, Austria (own picture).

All modules span the height between the floors of 3.9 m. However, 18 different sizes of custom-made BIPV-modules are installed. A bypass diode is installed after every 22 cells and the modules are interconnected in two strings per floor. The produced energy is fed into the electricity grid by five inverters. There is no specific monitoring system for the BIPV installation, therefore the only available data is the hourly amount of produced energy monitored by the inverters. Thus, for a more detailed analysis of the performance of the system, an irradiation sensor was installed in May 2018. Furthermore, infrared thermography (IR) and IV-curve measurements were performed in the course of this activity.

2.4 Constructive boundaries and monitoring methodology

The monitoring method and the constructive boundaries were agreed by all participants. Given that differences in the measuring parameters can be then excluded, the direct influence of environmental and climatic conditions on the performance of the BIPV elements can be deduced from the monitoring data.

2.4.1 Constructive boundaries

The constructive requirements for the BIPV outdoor test stand built at each institution use a pre-defined mock-up construction, as shown in Figure 2 and Figure 3. The mock-up construction aims to simulate the ventilated cold façade based on the reference BIPV building skin. This allows for monitoring under equal conditions at all participating institutions.

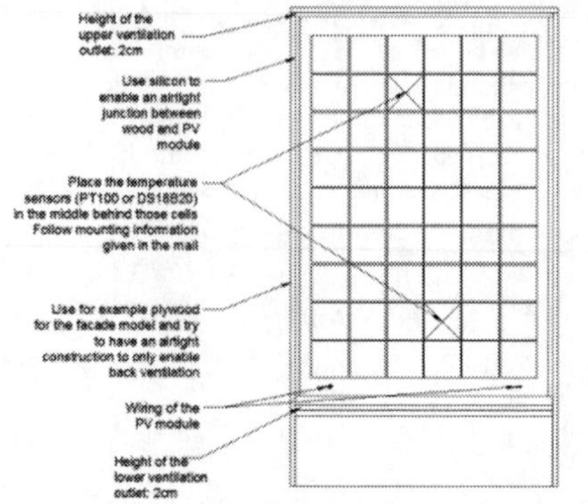

Figure 2: Front view of BIPV mock-up construction elements. (L. Maul, UAS Technikum Vienna).

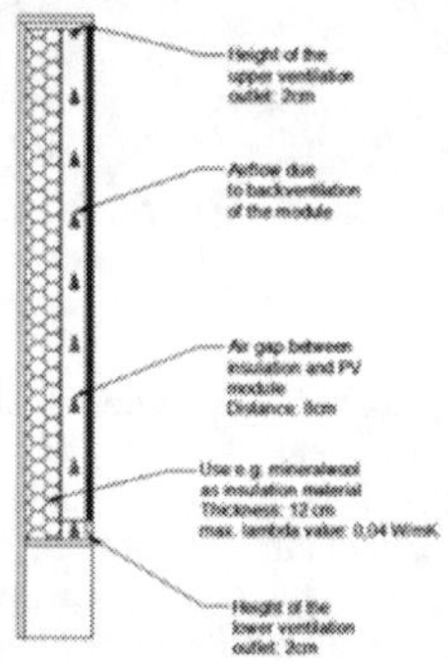

Figure 3: Cross-section of BIPV mock-up construction elements. (L. Maul, UAS Technikum Vienna).

The following constructive details aim to simulate the situation in the BIPV reference building:

- 12 cm mineral wool at the backside backside (or similar insulation with thermal conductivity of 0.04 W/mK)
- air gap between insulation and BIPV element of 8 cm
- upper and bottom ventilation outlet of about 2 cm

The construction avoids the accumulation of rain and/or condensation water through drainage (e.g. via outlets) at the bottom of the mock-up. To avoid direct exposure of the mineral wool to rain water through the ventilation gaps, small rain shields are applied. An example of a test stand is shown in Figure 4.

Figure 4: BIPV test stand – module and mock-up construction at SEAC, Netherlands (R. Valckenborg, SEAC).

2.4.2 Monitoring method

In addition to the definition of constructive requirements of the BIPV outdoor test stands, the monitoring method was also designed by all participants in advance. The minimum requirement of data recording was fixed to be daily morning to evening data when the (in-plane) irradiance is above or equal to 20W/m² (usually 10W/m² is enough for the inverter to operate in normal feed-in mode). Missing data (e.g. due to not operating monitoring and/or down times of individual monitoring equipment) are being reported. Furthermore, date and time of the last cleaning of deployed irradiance sensors are reported as well.

The monitoring raw data files for climatic and PV operational data are provided as csv-files with lists separated by semicolon and time stamps, file headers as well as file naming following a pre-defined format for all participants. The following data are monitored:

- I_{mp} (A),
- V_{mp} (V),
- T_{mod} (°C),
- G_i (W/m²),
- T_{amb} (°C),
- $W_{indspeed}$ (m/s).

Some institutions are capable to provide additional data such as information about shading condition, wind direction, relative humidity, or direct normal irradiance (DNI), diffuse horizontal irradiance (DHI) and global horizontal irradiance (GHI)

The used time stamp format follows the standard ISO 8601 [4], according to the recommendation of the IEC 61724 monitoring standard, [5]. Time stamps are at the end of each minute of 1 minute aggregated, arithmetical averages of instantaneous data points.

A sampling interval of 5 seconds (or below) is chosen to be sufficient for the purposes of this activity – as for e.g. for class A equipment, a sampling interval of max. 3 seconds is recommended, according to the monitoring standard. The instantaneous measurements, taken every 5 seconds, are then aggregated to 1 minute arithmetically averaged values and saved in the corresponding column of the csv-file – according to the pre-defined format.

Separate csv-files are used for the IV curve measurements. Files include also the module temperature at the start time of the IV curve measurement in the header. File headers as well as file naming follow a pre-defined format. Test participants who have the opportunity to measure more frequently than the minimum of one time per week, provide IV curves every 1 hour. The IV curves contain 128 triples of data points as minimum requirement of the following data:

- V (V),
- I (A),
- G_i (W/m²)

With minimum and maximum V_{oc} ranging from -0.5 to 36V. If only I-V pairs are recorded, the header includes the maximum and minimum value of G_i during the IV measurement interval to assure that the irradiance has not changed significantly during the measurement. However, if I-V pairs are recorded together with G_i, the irradiance change during the IV curve sweep should be within a threshold of ±2%. The IV characteristics measurement standard, IEC 60904-1, requires ±0.2% of I_{sc} and V_{oc} at STC [6]. For the current experiment, we suggest to provide a maximum uncertainty of ±2% at 200W/m², i.e. ±0.4% at STC. IV curve measurements performed manually are done at stable weather conditions, while taking care that the modules are not (partly) shaded before and during measurement and that the MPP-tracking is interrupted only for a very short time (less than 1 minute), in order to avoid mismatch between the average cell temperature and the sensors for the module temperature.

Class A sensor equipment is preferably used, as it allows most of the typical applications according to the monitoring system classification of IEC 61724-1. Sensor calibration reports (of irradiance sensors) are provided before and after the round robin monitoring test period.

Thermopile pyranometers are chosen to be best suited for the global horizontal irradiance (GHI) measurements, while matched PV reference devices are best for in-plane (POA) measurement. If pyranometers are used for POA measurement, attention is payed to correction procedures for the 90° tilt angle. For all measurement setup additional temperature correction is applied if required, following the procedures for irradiance and temperature corrections, [6].

2.5 Data processing, organization and analysis

Monitoring started in November 2017 at the first participating institutes. Monthly datasets, containing operational, IV and additional informational data are collected at each research institution and exchanged via online file sharing for further analysis and comparison.
At the University of Applied Sciences Upper Austria, the monitoring data are collected, processed and filtered for further analysis. As software tool, MATLAB is used aiding to build up a database of the acquired measurement data.
Until now, 6 of in total 9 test stands already provided monitoring data. Monitoring of the remaining 3 test stands is currently being processed into the required format. Table 1 shows the amount of collected data so far (until September 2018), giving information of the number of provided days of operational data as well as the number of collected IV curves and temporal resolution as well as other measured parameters. Thus, observations of this paper refer to currently available data.

Table 1: Current status of availability and measurement points (6 out of 9 participants)

Module	Operational Data (days)	Number of Measured IV-Curves	Resolution of Operational Data	Monitored Parameters
M17_04276	87	7	60	I_{mp}, V_{mp}, T_{mod}, G_I, T_{Amb}, Ws, Wd, Rh
M17_02289	83	8	60	I_{mp}, V_{mp}, T_{mod}, G_I, T_{Amb}
M17_02288	192	79213	60	I_{mp}, V_{mp}, T_{mod}, G_I, T_{Amb}, Ws, Wd, Rh, GHI, DHI, DNI, P_{mp}
M17_02281	276	944	60	I_{mp}, V_{mp}, T_{mod}, G_I, T_{Amb}, Ws, Wd, Rh, GHI, DHI, DNI
M17_02286	124	0	60	I_{mp}, V_{mp}, T_{mod}, G_I, T_{Amb}, Ws, Wd, Rh, GHI
M17-04275	65	0	10	I_{mp}, V_{mp}, T_{mod}, G_I, T_{Amb}, Ws, Wd, Rh, P_{mp}

Preliminary analysis of the available measured data has been performed in terms of

(i) performance ratio (PR)
(ii) comparison of parameters calculated with the Loss Factor Model developed by Steve Ransome (www.steveransome.com/) and also mentioned in [8]
(iii) comparison of IV-curves corrected to STC, or to other available irradiance and temperature values, as measured initially according to IEC 61583-1, see chapter 2.1.

The outcome of this first analysis should also provide inputs for the development of potential further evaluation approaches of available monitoring data.

STC corrections are performed based on method 1 in IEC 60891 with irradiance and temperature correction by temperature coefficients and are used for IV-curve translations [7]. Daily module performance ratio values are calculated as the ratio of the specific energy yield measured, and the reference yield derived from the irradiance measurement. With the Loss Factor Model (LFM) physically relevant parameters of the modules are extracted and normalized from IV-curve measurements. They can be easily analyzed in terms of climate and time dependence. The so-called "loss factors" consist of 7 parameters (such as losses at I_{SC}, V_{OC}, FF, etc.) which can be combined to a resulting power factor. In addition, two curvature values can be used to check the shape of the curve [8].

As no outdoor spectral measurements are available, all IV-curve based calculations were performed at irradiations greater than 200W/m². All other data was filtered for physically valid values. Current and voltage were limited to datasheet values plus overhead.

2.6 Accelerated ageing test methodology

In order to measure the influence of enhanced stress impact on the performance of the BIPV-module design used, accelerated ageing in climate chambers was performed:

(i) Damp Heat storage at 85% relative humidity (RH) and 85°C for 1000h (2000h) in accordance with IEC 61215 test 10.13.
(ii) Combined irradiation, temperature and humidity application at 1000W/m² (artificial sunlight), 60°C chamber air temperature, and 50% RH, which corresponds to a module temperature of about 80°C for 1000h (2000h);

3 RESULTS

3.1 Survey of participating R&D organizations

Before the official start of the monitoring and reporting phase of this activity, an initial report including orientation, location and deployed monitoring equipment and sensors, as well as information about the surrounding environment of the BIPV outdoor test stand at each participating organization was provided. In total, the experiment involves 9 institutions from 7 different countries in Europe and Asia – as illustrated in Figure 5.

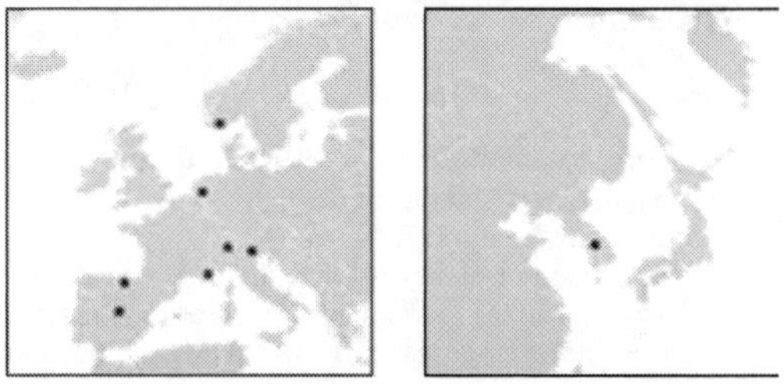

Figure 5: Illustration of the locations of participating institutions from seven different countries in Europe (left) and one in Asia (right hand side).

A summary of the initial survey report of each participating institution is shown in Table 2Table **2**, giving information about the location and general information about the deployed monitoring equipment and test site.

Table 2: Compiled information on participating institutions and deployed monitoring equipment (PY= pyranometer)

Participating Institutions	Orientation / Latitude / Longitude / Climate zone	Weather station / Temperature / Irradiance / Wind sensor	IV curve tracing
SEAC	180° / 51.446° / 5.485° / Cfb	LUFFT WS600/ PT100 class A / MS-802 PY/ wind speed and direction	MPPT 3000
SUPSI	176° / 46.027° / 8.916° / Cfb	VaisalaWXT 536/ PT100 class A / CM21 PY / wind speed and direction	MPPT 3000
ZUYD	180° / 50.830° / 6.016° / Cfb	- DS18B20 digital T / SMP-10 PY / wind speed and direction	Solaredge P305I / BK8502
CSTB	180° / 43.61° / 7.05° / Csb	- / Micro-sensors type T/ CMP21 PY / ENPHASE M250 micro-inverter / wind speed and direction	PVE PVPM 1000C
CIE-MAT	180° / 40.45°/ -3.73° / Csa	Young model / PT100 class A / ATERSA mc-Si cell / wind speed and direction	APS YC205A
EU-RAC	198° / 46.342° / 11.330° / Cfb	- / PT100 class A / CMP11 PY / -	MPPT 3000
UiA	-/-/-/ Dfc	-	-
Tecnalia	180° / 43.293° / 2.862° / Csb/Cfb	Gill GMX Compact / PT100 class A / CM6B PY & Atersa MET Si-cell / wind speed	PVPM 1000CX40
KNU	180° / 36.81° / 127.11° / Dwa/Cfa	- / PT100 class A / MX802 pyranometer / wind speed and direction	TNE Tech MPPT IV Tracer

3.2 Analysis of monitoring data

As stated in section 2.5, data of 6 modules were used for the first analysis. Figure 6 shows an example dataset of power, irradiance and temperature. The unexpected difference between irradiance and power detected in the afternoon is the result of some partial shading of the module which does not affect the irradiance sensor. In this example, the BIPV module suffers from partial shading through a neighbor building on the western side.

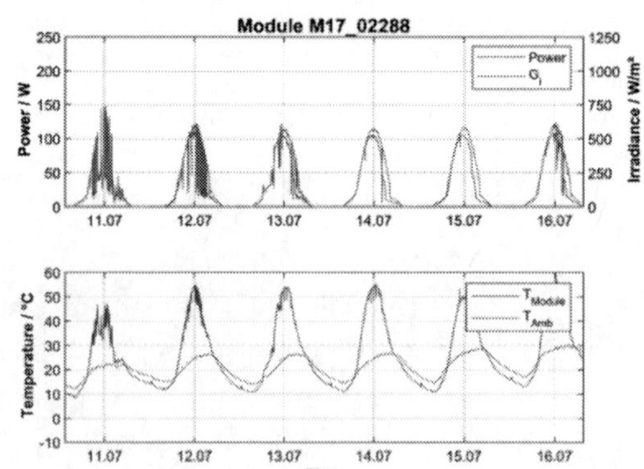

Figure 6: Comparison of power, in plane irradiance (G_i) and temperature values (T) of six example days of module M17_02288.

In Figure 7 daily performance ratios are shown on a timeline, while Figure 8 presents these data, as box-plots for all days with an overall production over 227 W to reduce outliers. The performance ratios (PR) of the modules are mostly between 1 and 0.75 and show differences with time of the year and installation site. Furthermore, the values of M17-04275 seem to be affected by some measuring error, requiring further analysis.

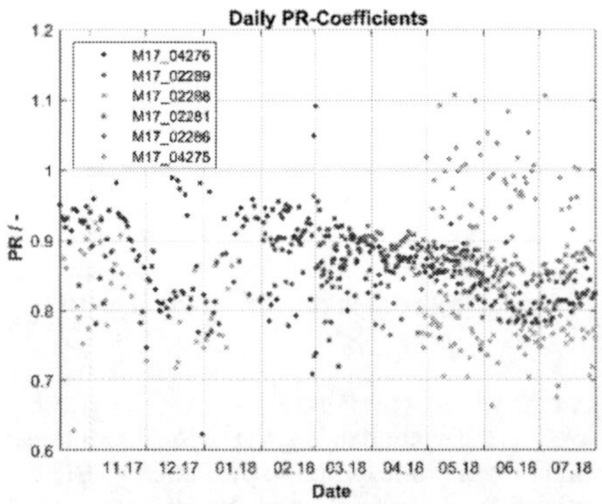

Figure 7: Daily performance ratios of the modules.

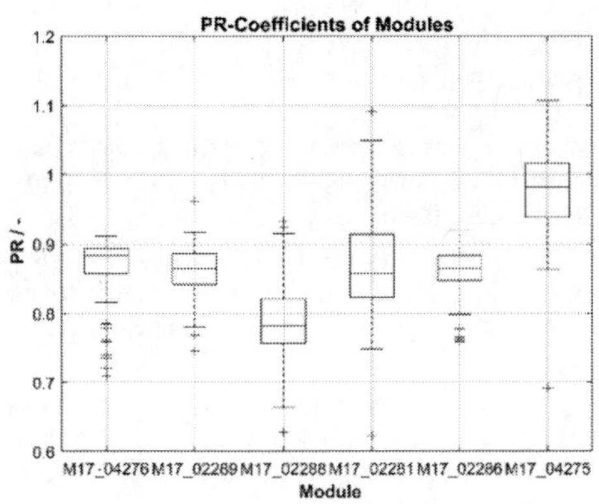

Figure 8: Box-plot of the daily performance ratios.

IV curves were analyzed by comparison of STC corrected curves and the LFM, respectively. Up to now, only for one module (M17_02288) a sufficiently high number of IV-curve measurements were performed to allow for the calculation of significant loss factor parameters (see Figure 9). It can be seen that the normalized loss at I_{SC} shows a wide scattering and needs further analysis.

The comparison of the corrected IV curves shows multiple differences which could be the result of shadings and inaccurate temperature and/or irradiance measurements.

3.3 Comparison to measurements at the reference building

As stated in section 2.3, up to May 2018, the reference installation was only equipped with an inverter-based monitoring of the energy production. Figure 10 shows the monthly energy production of the BIPV façade at the reference building in kWh between 2009 and 2015. As there was no irradiation measurement on-site, a detailed evaluation of the performance regarding overall trends is difficult at this point. It can be seen that, for each month, the energy yield can vary significantly depending on the actual weather/irradiation conditions. A calculation based on available irradiance data for tilted modules, or satellite GHI data to vertical in-plane irradiance values would introduce large errors. Thus, a direct degradation rate of the whole system was hard to derive from currently available data.

However, for a more detailed analysis of the performance of the system, an irradiation sensor was installed in May 2018. The irradiation data of the nearby weather stations, the newly installed irradiation sensor on the building and a higher resolved power measurement will allow to provide a more detailed insight to the performance of the system and to derive potential information about degradation behavior for the current year.

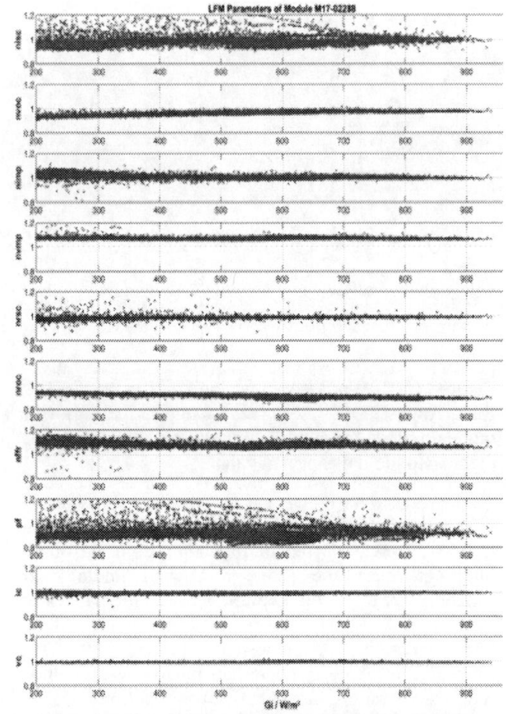

Figure 9: LFM parameters of module M17-42288.

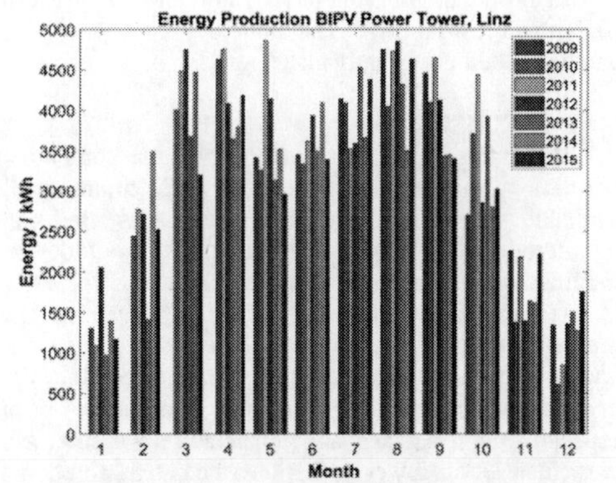

Figure 10: Monthly energy production of the Power Tower BIPV installation.

In August 2018 an on-site visit was made, were detailed infrared imaging (IR) as well as IV curve measurements of all individual strings were taken. A first analysis of the results shows that the majority of the BIPV façade is working very well, but some issues with diode defects and connection problems within two strings were identified. Exemplarily the IV-curves of the problematic strings are shown in Figure 11 in comparison to the expected IV-curve. The effects could also be identified in the IR images, however, as the viewing angle especially to BIPV modules in higher positions are not optimal, this paper does not show any IR images – also, for legal reasons, IR images taken by drones were not possible.

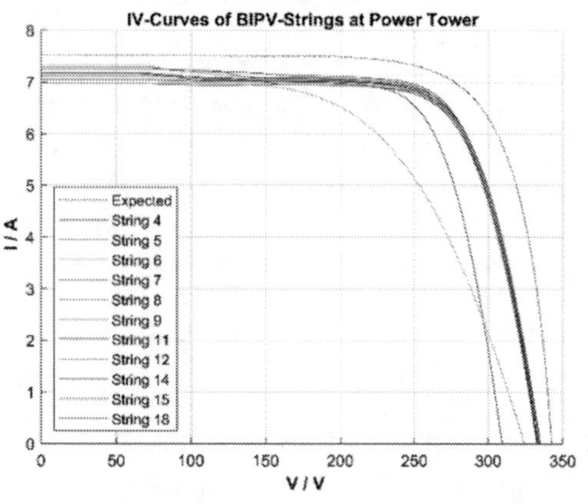

Figure 11: BIPV reference installation at the Power Tower Linz, August 2018: IV curves of selected strings with a diode problem, some of them with decreased fill factor in comparison to the expected values.

As the measured irradiance for correction to STC possibly differs spatially across the building's BIPV façade (e.g. due to different degree of reflection from surrounding buildings, etc.), the above mentioned findings from Figure 11 are to be seen as preliminary qualitative analysis results.

Furthermore, within the visual inspection of the BIPV reference installation, no failure modes or irregularities (e.g. discoloration, delamination, cracks, scratches, etc.) could be identified.

3.4 Accelerated ageing preliminary test results

As expected, DH storage for 1000h did not affect the performance of the BIPV test module. Further storage for another 1000h is currently on-going.

4 CONCLUSION

Within this experiment described, a direct comparison of the performance of identical and commercially available glass/glass BIPV elements installed and monitored under well-defined conditions in outdoor test facilities in different environments will be established.

At the moment, measurement data of seven modules are available of which six datasets were used for a first analysis. Incomplete data sets complicated the data analysis approaches. Furthermore, possible errors especially at irradiance and temperature measurements were identified. However, this intermediate evaluation of the available data resulted valuable information for improved continuation of the international large-scale "round robin" action. Although up to now, no conclusions on the influences of climatic stress factors can be drawn, it is expected that with an increased size and quality of the available datasets better and more significant conclusions can be thrown within the next months.

The comparison of the results (performance) obtained in the laboratory measurements under STC conditions (pre-characterization) and the outdoor measurements yielded good accordance showing satisfying comparability of the test facilities involved.

For the improved analysis of long term data of the reference building, an additional irradiance measurement was set up directly at the façade. The data collected in this way can be directly compared to those from the surrounding sensor stations which will allow for the deduction of correction factors. An on-site visit and measurements for maintenance purposes have revealed that the predominant part of the 66 kWp system of the BIPV façade performs without problems, but same issues with diode and connection problems were identified.

5 ACKNOWLEDGEMENTS

The work was supported by the IEA PVPS Task 15 "Enabling Framework for the Acceleration of BIPV" and the national Austrian funding of the projects nr. 853591 and nr. 853029, funded by the Federal Ministry for Transport, Innovation and Technology.

4 REFERENCES

[1] BIPV research teams & R&D facilities. An international mapping. 93p. see www.iea-pvps.org/index.php?id=458

[2] IEC 61853-1 ED1 (2011): Photo-voltaic (PV) module performance testing and energy rating - Part 1: Irradiance and temperature performance measurements and power rating

[3] IEC 61853-2 ED1 (2016): Photo-voltaic (PV) module performance testing and energy rating - Part 2: Spectral responsivity, incidence angle and module operating temperature measurements

[4] ISO 8601:2004 Data elements and interchange formats

[5] IEC 61724-1:2017 Photovoltaic system performance - Part 1: Monitoring

[6] IEC 60904-1:2006 Photovoltaic devices - Part 1: Measurement of photovoltaic current-voltage characteristics

[7] IEC 60891:2009 Photovoltaic devices - Procedures for temperature and irradiance corrections to measured I-V characteristics

[8] J. Sutterlueti, S. Ransome, R. Kravets and L. Schneider: Characterizing PV Modules Under Outdoor Conditions: What's Most Important for Energy Yield. In: Proceedings of 26th EU-PVSEC, Hamburg 2011.

LIGHT ATTENUATION MODEL TO PREDICT NOMINAL POWER OF MODULES WITH LIGHT-SCATTERING CERAMIC PRINTED FRONT GLASSES

Christoph Kutter, Helen Rose Wilson, Andrea Pfreundt, Martin Heinrich, Harry Wirth
Fraunhofer Institute for Solar Energy Systems ISE, Heidenhofstraße 2, 79110 Freiburg, Germany
christoph.kutter@ise.fraunhofer.de

ABSTRACT: One feasible design for Building-Integrated Photovoltaic modules applies ceramic printed glass covers. Power prediction of such modules is difficult as transmittance of printed glass panes in air was shown to be much lower than the relative power loss in the respective modules for the same print coverage ratio. Applying a matrix-based approach, we developed a light attenuation model, which can describe the relative transmittance decrease of glass panes in air and relative power decrease of modules for glass covers with arbitrary coverage ratios. The model shows good correlation with measured data. The modelled transmittance of glass panes deviate with a root mean square error of 0.55% from the measured transmittance values. The model can compute the module performance loss for investigated coverage ratios with a RMSE of 0.73%.

Keywords: ceramic printed glass cover, enamel, transmittance, PV, BIPV, matrix model, light scattering

1 INTRODUCTION

Ceramic printed glass covers are one design option for Building-Integrated Photovoltaic (BIPV) modules [1, 2]. The print pattern on position 2 of the glass pane in Figure 1 obscures the appearance of silicon solar cells and creates a homogeneous, aesthetic appearance that can satisfy the expectations of architectural design with a prototype shown in Figure 2.

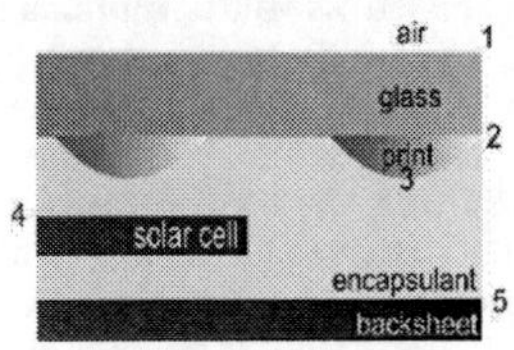

Figure 1: Schematic drawing of module stack with ceramic printed glass cover, print on internal position 2 (not to scale)

Figure 2: PV module prototype with digitally printed glass cover and blackened connectors

To promote this design approach in industrial manufacturing, understanding of the correlation between print coverage ratio and module performance is crucial.

Frontini *et al.* performed indoor and outdoor tests to evaluate the influence of digital prints on module performance [3]. These authors attested the digitally printed patterns a general technical feasibility for PV applications and found daily yield losses from 10% (white prints), through 15% for green prints, to 35% for black prints with 30% coverage ratio compared to an unprinted reference module.

Mittag *et al.* reported that the measured output power of a laminated PV module is higher than predicted from normal-hemispherical transmittance measurements of the glass pane in air [4]. Due to the large refractive index gap between glass and air, much of this diffuse light is trapped inside the glass pane by total internal reflection. After multiple reflections, all light has been either transmitted, reflected, absorbed or laterally lost through the edges of the pane. The measured normal-hemispherical transmittance of a glass pane in air is lower than in the laminated case for two main reasons. One of these is lateral losses, which can be minimized by an appropriate measurement configuration, as described in [5].The second and more fundamental reason for the difference is that once the glass pane is optically coupled by encapsulant within the module stack, reflection at the glass rear surface is strongly reduced, allowing large fractions of the scattered light to reach the solar cell, explaining the lower module power loss. Mittag *et al.* conclude that neither the hemispherical transmittance of glass panes nor the print coverage ratio qualify as sole indicators to predict module performance within a module efficiency analysis.

Within this paper we present an optical model to calculate the module power decrease with ceramic printed glass. We use a simplified version of the matrix model applied in OPTOS [6] to compute the relative decrease in transmittance of glass panes and relative decrease in performance of modules with arbitrary coverage ratios compared to an unprinted reference. The model requires only a single adjustable parameter, the "print transmittance factor", to describe print coatings.

2 OPTICAL PRINT CHARACTERIZATION

To assess the light scattering, we perform normal, diffuse transmittance measurements on whole-area printed glass panes at the TestLab Solar Facades of Fraunhofer ISE. We find that ceramic prints predominantly transmit diffuse light as shown in Figure 3. The light-scattering characteristics is primarily caused by the composition of the enamel and not by surface roughness or the geometry of the printed pattern. Prints are enamels, which consist of a glass matrix filled with scattering pigments giving an opaque appearance. [5].

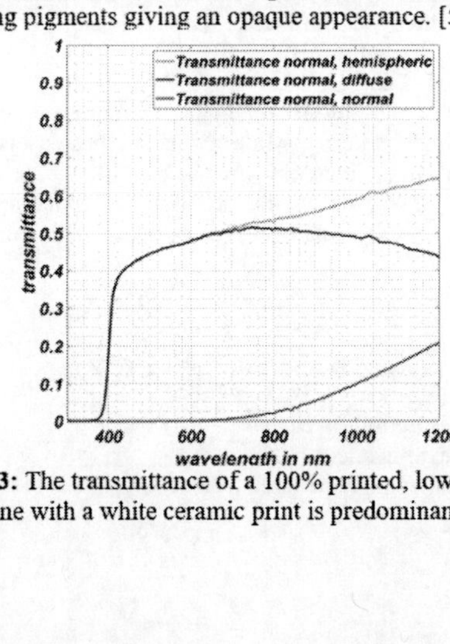

Figure 3: The transmittance of a 100% printed, low-Fe glass pane with a white ceramic print is predominantly diffuse

The light attenuation model needs to evaluate the angle-dependent light propagation at material interfaces and pathlength-dependent absorption within materials to describe light scattering of ceramic prints. The optical model described in the next section is capable of taking both effects into account.

3. MODELLING

3.1 Description of the optical model

The optical model used in this work is a simplified version of the matrix approach also applied in the OPTOS formalism [6, 7]. Unlike OPTOS, it takes multiple reflections only within one of the optically thick sheets into account and it is restricted to unpolarized light of one single wavelength. Within this method, the power distribution of light is described by vectors with finite elements corresponding to a number of discrete angle channels.

$$v_n = \begin{pmatrix} p(\vartheta_1) \\ \vdots \\ p(\vartheta_i) \\ \vdots \\ p(\vartheta_m) \end{pmatrix} \qquad (1)$$

with
v_n= nth power distribution vector
$p(\vartheta_i)$= power fraction for angle i
ϑ_i = polar angle between 0° and 90° in 1° steps

Refraction or scattering of light rays at material interfaces leads to a redistribution of power between angle channels and an angle-specific power loss. A redistribution matrix provides the information on power loss and redistribution. Surface interaction is efficiently modelled by multiplying the incoming power distribution vector with a redistribution matrix.

$$v_{out} = v_{in} \times TM_x \qquad (2)$$

with
v_{out}= outgoing vector
v_{in}= incoming vector
TM_x= propagation or redistribution matrix for layer/interface x

The allocation of the output angle at an interface to the incoming angle within the redistribution matrix is computed with Snell's law while the power fraction transmitted or reflected for each angle channel is computed by Fresnel's law.

The Lambert-Beer law describes the exponential decay of the power of a light ray within homogeneous, absorbing material. Multiplying the power distribution vector of an incoming light ray with the propagation matrix TM_1 allows for accurate computation of the pathlength-dependent absorption within material layers.

$$TM_1 = \begin{pmatrix} e^{-\alpha d/cos\vartheta_1} & \cdots & 0 \\ \vdots & \ddots & \vdots \\ 0 & \cdots & e^{-\alpha d/cos\vartheta_m} \end{pmatrix} \qquad (3)$$

with
α = absorption coefficient
d= glass thickness
ϑ_i = polar angle between 0° and 90°

The scattering print layer is made up of a glass matrix that contains scattering particles and pigments [5]. The print layer is described by a redistribution matrix that redistributes the incoming light intensity between angle channels. The scattering is assumed to be isotropic, however Nitz showed that prints can have an uneven scattering behavior comparing front and rear power distribution [8]. The ratio of reflected and transmitted light is influenced by layer thickness, scattering particle concentration and optical coupling. Therefore, we implemented an adjustable print transmittance factor that allows the forward and backward scattered power fractions to be varied.

As we expect the refractive index of the enamel glass matrix to be close to the refractive index of the glass substrate, we neglect transmittance losses at the glass/enamel interface. Due to the narrow layer thickness of the white enamel (5 µm) and no observed absorption minima in transmittance measurements (see Fig. 3), we assume the white print absorbance to be negligible. Therefore we assumed the absorbance to be zero in our model.

3.2 Application of the model to a (printed) glass pane in air

The light attenuation is modelled by following the light path through the system as shown in Figure 4. The incoming light intensity is defined as 100%, normally incident radiation is assumed (1). The transmittance is calculated at the air/glass interface (2). The internal transmittance for the glass is calculated based on the path length (3). At the rear glass surface, the light intensity is weighted according to the coverage ratio. Light attenuation at the unprinted rear glass/air interface will be processed according to the Fresnel laws (4g). Multiple internal reflections starting from the unprinted rear interface is considered in the effective transmittance until the intensity of the reflected vector sinks below 0.01% of the incoming light intensity.

Light attenuation at the printed rear interface is calculated differently since scattering at the print occurs. The incoming light intensity is divided according to the print transmittance factor into a transmittance and a reflectance vector (T=1-R) labeled with 4p and 5 in Figure 4. The power distribution in both vectors is isotropically distributed into all angle channels.

This reflectance vector is the starting point of an iteration, calculating the transmittance due to multiple reflection within the printed glass pane: The reflectance vector is attenuated by the internal glass transmittance (6), glass/air reflectance (7) and again internal transmittance (6) before being weighted again by the coverage ratio and transmittance is computed for the respective interface labeled as 8g and 8p in Figure 4. Light ray (9) indicates the beginning of a new iteration in which steps (5) to (8) are repeated. Iterations are repeated until the remaining light intensity in the layer stack (9) sinks below 0.01%. After each iteration step, the sum of all light intensities 4g, 4p, 8p and 8g is calculated as the effective transmittance.

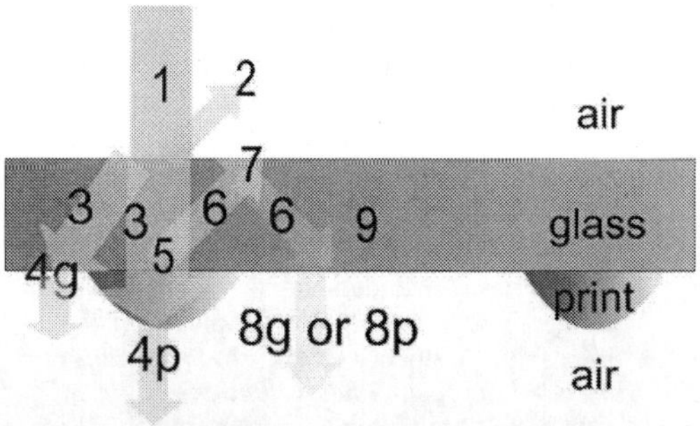

Figure 4: Glass pane in air with light attenuation pathways. "g" refers to glass, "p" to print.

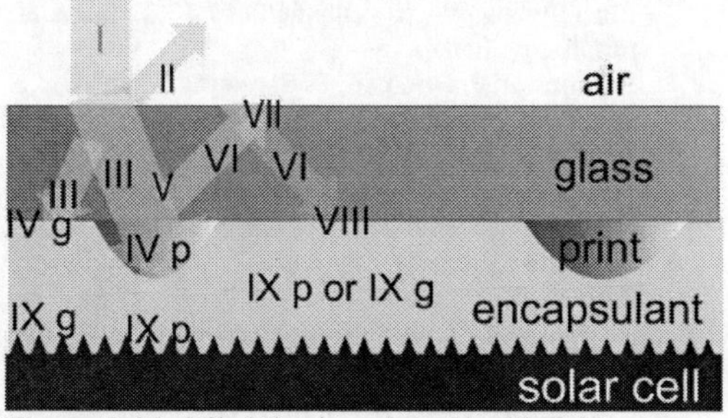

Figure 5: Modelled 'laminated stack' with light attenuation pathways, solar cell is modelled as perfect transmitting and fully absorbing.

3.3 Application of the model to a simplified module stack with printed glass

The module is modelled in a similar way to the glass/encapsulant stack displayed above (Figure 5). The differences to the glass pane setup are:

- Due to optical coupling of the enamel with encapsulant, the transmittance of the rear print surface increases (IVp > 4p)
- Glass and the encapsulant have similar refractive indices, so the reflectance IVg and VIII at the unprinted rear glass/encapsulant interface is negligible.
- The solar cell is assumed to show no reflection at the front surface and perfect absorption in the bulk. This is a sufficient approximation for comparing different print coverages as done within this work, since the reflectance of a random pyramid interface with ARC is low [9].

4. RESULTS

4.1 Results - glass pane in air

To validate the model, we plot the relative, modelled normal-hemispherical transmittance over the relative, measured transmittance of a glass pane in air for coverage ratios 0% to 100% as shown in Figure 6. The print transmittance factor was used as fitting parameter being constant for all coverage ratios.

By definition, the relative normal-hemispherical transmittance of the reference glass pane (print coverage of zero) is set to 100%. For a print transmittance factor of 30%, we find very good correlation of the results for coverage ratios equal and below 98%.

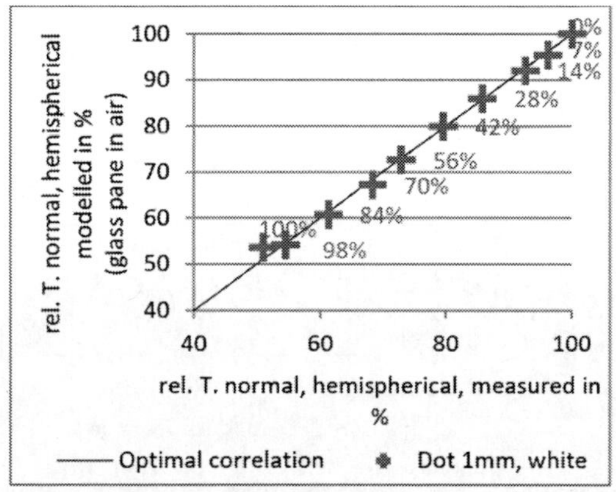

Figure 6: Correlation between computed and measured relative normal-hemispherical transmittance T. for a glass pane in air.

The deviation in the modelled transmittance for 100% can be explained by a different printing process with different droplet positioning of the inkjet leading to a thicker enamel layer and consequently to a lower measured transmittance [5]. In consequence the 100% value is not taken into account when the root mean square error (RMSE) values are calculated below.

4.2 Results - simplified module stack

Figure 7 shows the modelled normal-hemispherical transmittance within the laminated stack over the relative module power for coverage ratios 0 to 100% and a print transmittance factor of 68%. The print transmittance factor is constant for all coverage ratios, but is not equal to that determined from the printed glass panes. The optical coupling of the enamel with encapsulant leads to higher print transmittance.

For lower coverage ratios (14 to 28%) the modelled transmittance in the module stack slightly underestimates the actual transmittance, while for higher coverage ratios (98%) the modelled transmittance is slightly overestimated compared to the measured module power.

We do not consider the angle-dependent solar cell reflectance in our model. The observed deviation between modelled and measured data might be explained by this effect. We expect that with higher print coverage ratios, the share of diffuse light reaching the solar cell will also increase. The more light hits the solar cell at an oblique angle, the higher the solar cell reflectance is. As a result, the modelled transmittance is slightly overestimated for higher print coverage ratios.

At 100% coverage, the change in printing parameters is again assumed to be responsible for the discrepancy between the modelled and measured results.

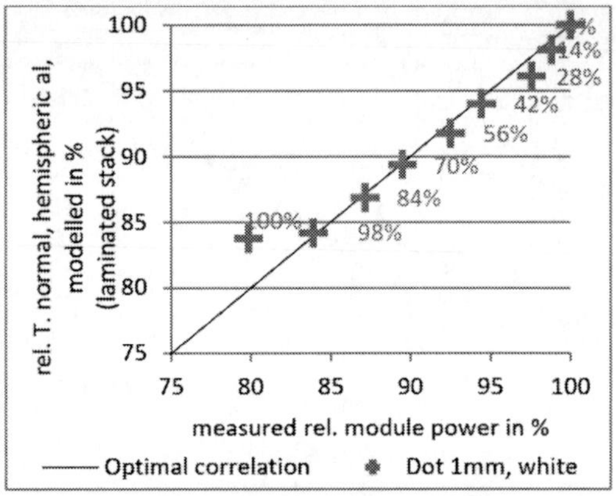

Figure 7: Correlation between modelled relative transmittance of the laminated stack and relative module power.

5 DISCUSSION

The good correlation of the modelled and measured transmittance of the glass pane in air demonstrates that the matrix model is capable of describing the correlation between coverage ratio and transmittance of light-scattering printed glass panes. A RMSE of 0.55% for coverage ratios from 0 to 98% is very low. Hereby, the print transmittance factor plays a key role as a fitting parameter, as it determines the ratio of transmitted and reflected light intensity by the print layer for a given background material.

Likewise, the overall correlation between modelled transmittance in the module stack and relative module power is good. A RMSE of 0.73% for 0 to 98% coverage is good.

By adjustment of the print coverage ratio and the print transmittance factor, modelling of various white prints is now possible. The linear correlation between the modelled and measured data sets indicates that all main optical influence factors are considered within the model.

6 CONCLUSIONS AND OUTLOOK

The introduced matrix model, based on a simplified OPTOS formalism, can accurately describe the transmittance of partially printed glass panes in air as well as the relative power loss of the corresponding module. The print transmittance factors, that determine the ratio of transmitted/reflected light power by the print layer, plays a key role. They are empirically determined by fitting the modelled data points to obtain the lowest RMSE to the measured transmittance data, for a series of partially printed glass panes, either in air or in the module configuration.

It should be possible to determine the print transmittance factor by detailed characterization of the ceramic print enamels, e.g. by determination of the bidirectional scattering distribution functions (BSDF). With this information, the module power could then be predicted by application of the described approach.

By modelling both configurations, the light attenuation processes within printed glass panes and modules are well explained. The optical coupling of the glass pane with encapsulant and thereby the inhibition of total internal reflection of the scattered light at the rear glass surface can be identified as the relevant difference between the two optical systems.

The model can be extended for colored prints by the implementation of an absorbance factor of the print layer.

REFERENCES

[1] L.H. Slooff, J.A.M. van Roosmalen, L.A.G. Okel, T. de Vries, T. Minderhoud, G. Gijzen, T. Sepers, A. Versluis, “An Architectual Approach For Improving Aesthetics Of PV,” in *Proceedings of the 33rd European Photovoltaic Solar Energy Conference and Exhibition*, Amsterdam, 2017.

[2] P. Hannen, “Facade built with photovoltaic modules in six sizes and 7 print patterns,” https://www.pv-magazine.de/2017/10/27/fassade-mitphotovoltaik-.

[3] F. Frontini, P. Bonomo, E. Saretta*, T. Weber, J. Berghold, “Indoor and Outdoor Characterization of Innovative Colored BIPV Modules for Façade Application,” in *Proceedings of the 32nd European Photovoltaic Solar Energy Conference and Exhibition*, Munich, Germany, 2016, pp. 2498–2502.

[4] M. Mittag, C. Kutter, M. Ebert, H.R. Wilson, U. Eitner, “Power loss through decorative elements in the front glazing of BIPV modules,” in *Proceedings of the 33rd European Photovoltaic Solar Energy Conference and Exhibition*, Amsterdam, 2017, pp. 2609–2613.

[5] H. R. Wilson and M. Elstner, “Spot Landing: Determining the Light and Solar Properties of Fritted and Coated Glass,” in *Glass Conference Proceedings, Vol 6 (2018): Challenging Glass 6*, pp. 203–212.

[6] J. Eisenlohr *et al.*, “Matrix formalism for light propagation and absorption in thick textured optical sheets,” (eng), *Optics Express*, vol. 23, no. 11, A502-18, 2015.

[7] N. Tucher *et al.*, “Optical simulation of photovoltaic modules with multiple textured interfaces using the matrix-based formalism OPTOS,” (eng), *Optics Express*, vol. 24, no. 14, A1083-93, 2016.

[8] Dr. Peter Nitz, *Optische Modellierung und Vermessung thermotroper Systeme*. Inaugural-Dissertation. Freiburg im Breisgau, 1999.

[9] I. Geisemeyer *et al.*, “Angle Dependence of Solar Cells and Modules: The Role of Cell Texturization,” *IEEE J. Photovoltaics*, vol. 7, no. 1, pp. 19–24, 2017.

Research Project CIGS-Façade
PV façades - chances and limits

Dieter Geyer[1], Peter Lechner[1], Simon Hummel[1], Dirk Stellbogen[1], Carlos Conejo Gangkofner[2]
[1] Zentrum für Sonnenenergie- und Wasserstoff-Forschung Baden-Württemberg (ZSW), Meitnerstrasse 1, 70563 Stuttgart, Germany, dieter.geyer@ZSW-bw.de
[2] NICE Solar Energy GmbH (NSE), Alfred-Leikam-Strasse 25, 74523 Schwäbisch Hall, Germany

ABSTRACT: ZSW realized a 29 kW_p PV façade based on CIGS technology on parts of its new office and laboratory building in Stuttgart, Germany. For research purposes a detailed monitoring system has been installed. Energy yield measurements based on a one-year period from the PV façade and open rack mounted PV are presented. The article also presents results of a simulation of the specific energy coverage by both PV façades and PV roofs. A detailed cost analysis from the building-owner´s perspective unveils potential for cost improvement in the realized system. Finally, a cost reduction scenario is being discussed.

Keywords: Economic Analysis, CIGS, Façade, Monitoring, Performance, PV System

1 INTRODUCTION

The overall project aim of the research project "CIGS-Façade" is the exploitation of building integrated CIGS thin film PV façades as an important market for the German PV industry to utilize the full PV potential for a successful German Energiewende and the realization of the European Union Energy Performance of Buildings Directive (EPBD) on European level.

Planning and realization of the PV façade at the new ZSW laboratory and office building could be finished with the start of the research project [Figure 1].

Figure 1: PV façade at the new ZSW laboratory and office building in the Stuttgart Engineering Park (STEP).

The research project "CIGS-Façade" addresses the optimization of the PV elements and improvement of the system integration and planning process.

The ZSW building consists of five building blocks touching each other. The façade is realised as a ventilated rain screen. The total opaque façade area of the ZSW building is 1996 m², of which 256 m² are executed as PV façade on the southeast, southwest and northwest side of the highest cuboid of the building. The inactive façade is realized on the same support structure but with anodized aluminium sheets as cover material.

2 APPROACH

In order to evaluate chances and limits of PV facades both a simulation regarding energy potential of BIPV and energy yield, a detailed data monitoring as well as a financial analysis has been carried out at ZSW.

As building owner, ZSW has full access to all costs of both the PV façade and the conventional aluminium façade. This allowed a detailed cost analysis on real data due to the fact that the PV façade was realized as part of an ordinary building process.

At the same time, ZSW was very interested to keep cost low and to reach a high efficiency of the PV modules. Therefore, framed CIGS thin-film PV modules from Nice Solar Energy GmbH (NSE) in almost black colour were chosen. The fixing of the modules happened with the module specific frameline mounting system that had been developed by NSE. Thus, legal aspects of PV modules in the façade could be easily fulfilled.

Furthermore, the legal situation with ZSW as building owner simplifies the monitoring work enormously regarding realization and access for operation and maintenance. Within each of the three façade systems ZSW installed several separated modules, running at own mpp-loads. These modules allow additionally the temperature measurement of both module backside and air gap behind the modules. Leakage current measurement and test of Anti-PID devices are already prepared.

3 RESULTS AND DISCUSSION

3.1 Simulation of energy harvest of PV facades vs. roof

A simulation has been carried out regarding the dependencies from building geometry of the area related energy supply by PV on the roof and a combined PV façade on the east, south and west side, in order to get principal information about the power of both PV roofs and PV facades in bigger buildings (Figure 2).

As the PV roof area and thus the annual energy yield remains the same independent of the building height, the PV energy supply per floor decreases with growing number of storeys. Contrary to that, the PV façades grow proportionally with the number of storeys and hence dominate the energy production at buildings with eight storeys or more.

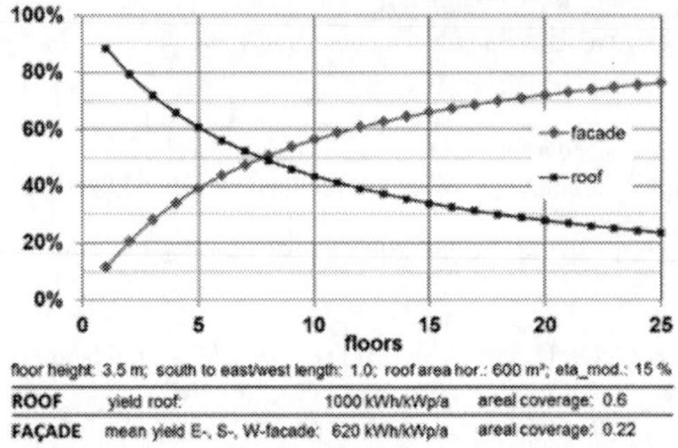

Figure 2: Share of façade and roof on the PV production.

The simplified building with a flat roof consists of stacked cuboids. The cuboid's height is equal to the floor height.

For the simulation, a square area of 600 m² and a storey height of 3.5 m is assumed. The energy yield is set to 1000 kWh/m²/yr. for the roof and 620 kWh/m²/yr. for the combination of east, south and west façade. The areal coverage rate is 60% for the roof and 22% for the façade.

The simulation allows to adjust the length to width ratio of the building as well as its orientation. For this purpose, a south-to-east/west-ratio has been introduced that defines the shape of the ground rectangle and its orientation. A value of the ratio above 1.0 means that the longer side of the building is facing south, a value smaller than 1.0 indicates that the longer sides are facing eat and west. The ground area is equal to the roof area. The number of stacked cuboids is given by the number of storeys.

The ratio of 1.0, as assumed in the example, is the worst case regarding energy supply as a square has the smallest perimeter–to–area-ratio.

A variation of these parameters in a wide but realistic range shows the principally same shape of the curves, the break-even is varying approximately between 5 and 10 storeys.

3.2 Evaluation of monitoring data

The following evaluation of monitoring data is based on a one-year data set from July 2017 to June 2018 in logging intervals of one 1-minute.

Figure 3 and Figure 4 show the specific energy yield of both a CIGS module in open rack, 40° tilted south oriented mounting and a CIGS module in the southwest façade on a daily base.

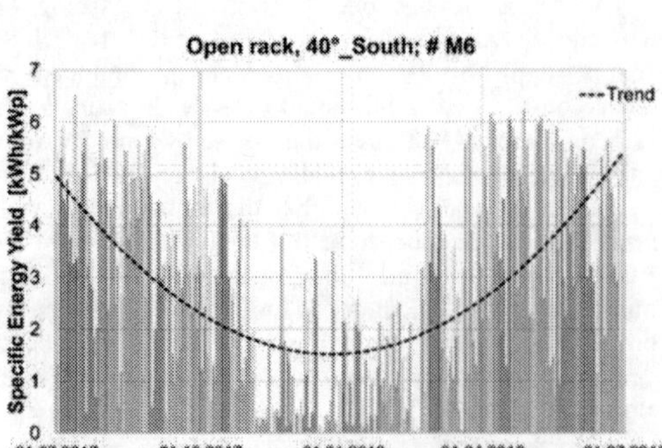

Figure 3: Energy harvest of south oriented, 40° tilted open rack PV, measured 7/2017 – 6/2018.

Comparing the extrema of the trend lines, it becomes evident that the seasonal ratio between mean summer and winter yield is relatively high with 3.5:1 in the case of the 40° tilted mounting, but only 2.4:1 for the SW-façade respectively. This means that the energy supply from a PV façade is distributed more evenly over the course of a year.

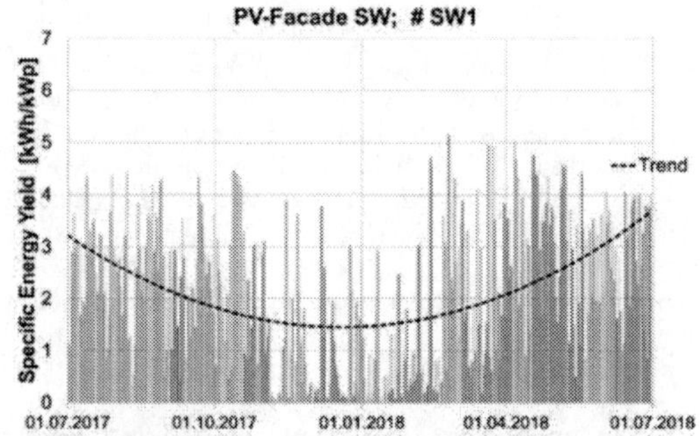

Figure 4: Energy harvest of the SW façade, measured 7/2017 – 6/2018.

The specific yield of PV generators directing to the different façade orientations as well as for a south oriented, 40° tilted open rack generator based on measured data of single modules is given in Table I. The unshaded southwest façade shows a very good yield of still 75%. The high variance between the yields of modules SE3 (659 kWh/kWp due to a one storey higher building at the opposite side of the street in the east) and SE4 (497 kWh/kWp due to an additional building edge in the own façade) shows the problem of reducing a real façade to only one point in a partially shaded façade. This is important in the case of energy forecasts or yield assessments.

Table I: Comparison of individual, MPP-tracked PV modules within the SE-, SW-, NW-façade and a rack mounted module

	S 40° open rack	SE3 façade moderate shading	SE4 façade heavy shading	SW1 façade shade-free	NW6 façade
spec. yield 07/17 to 06/18 [kWh/kWp]	999	659	497	749	380
relative to 40° South	100%	66%	50%	75%	38%

However, the yield of the north-western side is hardly acceptable under economic aspects. Reason for this rather low yield is the fact that the sun reaches this part of the façade only in summer whereas during the rest of the year only indirect solar radiation is usable. The measured mean yield for all facades is determined to 570 kWh/kWp/a.

Furthermore, the course of power output within one day differs fundamentally between tilted south orientation of PV and the vertical orientation in a façade. The need for power limitation of a PV system, as typical for Germany, is rapidly reduced and furthermore energy production is shifted to times where energy is more valuable (morning, afternoon, winter), see Figure 5.

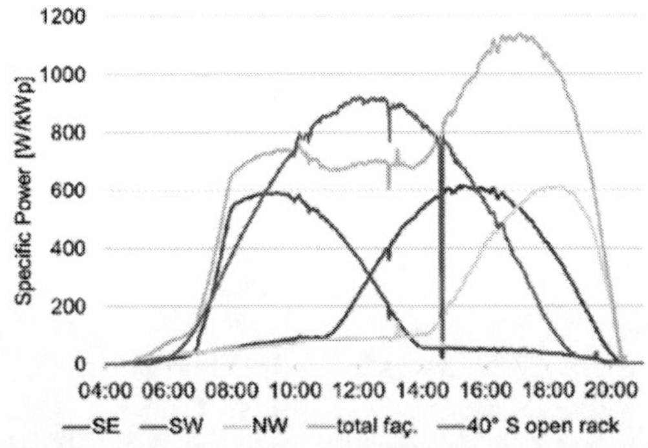

Figure 5: Course of specific power output on 20th June 2018 for the façade modules and for the south oriented, 40° tilted open rack module.

3.3 Costs

For the comparison of the system cost, it has been assumed that all PV elements would have been replaced by conventional elements with an aluminium surface plate. Due to this procedure, it was possible not to use mean cost but precisely the net prices of the specific elements according to the final invoices.

Figure 6 lists all area-related cost on the base of the final invoices to the building-owner for all material and work above concrete including the supporting structure and thermal insulation in the case of façade mounting. The roof-top installation refers to a standard system mounted on the wooden construction under the roof tiles [1] and with a module efficiency of 15%. Both PV scenarios include the complete PV system including mounting and electrical installation, not only the PV modules.

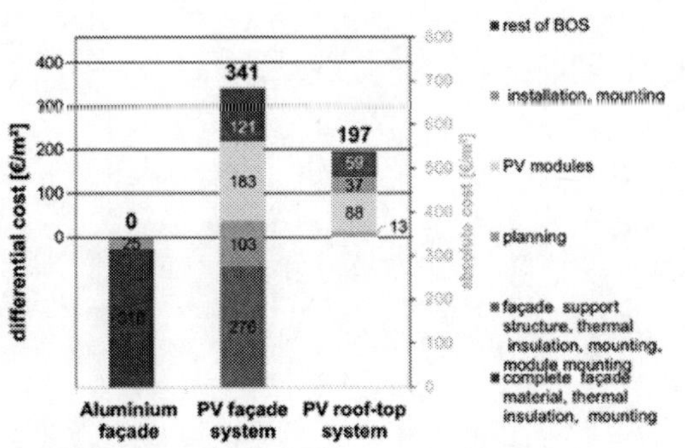

Figure 6: Comparison of the area-related costs of the Aluminium- and PV-façade system (as built). For matters of comparison the cost of a standard roof-top PV system are shown additionally.

In the façade project of ZSW six different module formats are used. This variety became necessary to realize different element heights consisting of three or six modules each in order to serve different building situations such as a neighbour cuboid with a thermal insulation on the flat roof-top or reduced floor heights in the upper floors.

An analysis of module cost unveils that any deviation from the standard module format resulted in module area-related prices that were 2.5 to 3 times higher than those of standard sized modules. The reason for this is that the highly automated production lines are currently not able to handle varying module formats. This leads to many manual production steps that become necessary after the processing of the CIGS module substrate.

In the case of the ZSW building, only 27% of all 357 façade modules could be realized as standard CIGS modules due to architectural demands.

This leads to levelized cost of energy (LCOE) for the PV façade system at the ZSW building as realized of 31 €ct/kWh as shown in Table II. If it would have been possible to use only standard modules, the LCOE would be reduced to 23 €ct/kWh.

The variety of the yield can be explained as following: The measured values from single modules given in Table I result in a mean value of 570 kWh/kWp/a. There the northwest façade with its low yield is included. The values taken on basis of single modules are not fully representative as parts of the façade are self-shading from another building part. The yield of the PV façades systems over a complete year of operation is not yet available as the operation of the grid-connected inverters started in November 2017.

Table II: Levelized Cost of Energy for PV Systems in Southern Germany

PV-System	Invest €/kW	Yield kWh/kWp/a	El. cost €Cent/kWh
PV field xMW	700	1000	5.9
PV roof top xkW	1300	1000	11
PV façade ZSW as built	2270	570 – 750*	26 - 34
PV façade ZSW standard (standard modules only)**	1670	570 – 750*	19 - 25
PV façade potential***	1000 -1300	620 – 750*	10 - 16

assumptions:
- Discount factor 5%, operational cost 1%/a, operational time 25 yrs.
- Façade: PV power 150 W/m²; equivalent to 15% PV efficiency
- Façade as built: differential cost 340 €/m²
-)* moderate shading (SE, SW, NW to none shading (SE, SW)
-)** Façade standard: differential cost 250 €/m€
-)*** Façade potential: differential cost 150-200 €/m²; operational time 40 yrs.

A very simple measure to increase the yield and thus to reduce LCOE could be, to avoid façade parts facing more to north than east or west. In the case of the ZSW building this would mean that by using only the southeast and southwest façades, the yield will raise to values between 620 and 750 kWh/kWp/a, depending on the shading situation.

4 OUTLOOK

Further cost reduction potential towards the level of roof-top PV systems can be reached by the realization of bigger connected module areas, which reduce cabling effort and sheet metal forming drastically. Another important point is the integration of the PV façade design into the planning process from the very beginning. This will lead to a reduction of planning cost. Finally, it is essential to improve the operational time of PV modules in the façade towards 40 years, especially taking into account the hampered access to the PV façade modules for any operation and maintenance measure. On module level, it stays important to further reduce module cost in the future.

5 SUMMARY

ZSW could realize a PV façade at differential cost of 340 €/m².

The resulting LCOE of the built PV system with six different module sizes amounts to 31 €ct/kWh. In an optimised system that only consists of standard modules LCOE would be reduced to 23 €ct/kWh.

The annual energy yield of the SW façade reaches 75% of the open rack PV system facing south at 40° tilt.

The seasonal variation of the energy harvest of this PV façade is 1/3 smaller than that of the south facing system at 40° tilt.

For cost reduction towards the level of roof-top PV systems, bigger connected module areas, the orientation within the east, south and west direction, an early integration of the PV façade design into the planning process and last but not least a raise of the operational time of PV modules towards 40 years will show the highest effect. On module level, it stays important to further reduce module prices in the future.

6 ACKNOWLEDGEMENT

We would like to gratefully acknowledge the support of a part of this work by the Federal Ministry for Economics and Technology (BMWi), contract number FZ 0324156A.

7 REFERENCES

[1] Bundesverband Solarwirtschaft (BSW), web site 2016

OPTIMIZED ORIENTATION AND PROPORTION OF TRANSPARENT COMPONENTS BASED ON THE LEAST ANNUAL HEAT DEMAND

Abbas Rahmani, Prof. Rosemarie Wagner
PhD student in Karlsruhe Institute of Technology (KIT), Department Building Technology (FGB), Karlsruhe, Germany
Phone: +49 721 60844383, Email: abbas.rahmani@kit.edu

ABSTRACT: Transparent components influence every passive house based on their surface proportion compared to opaque surfaces and their main orientation. These components have to provide appropriative potential of sunlight absorption to deliver a considerable proportion of heat demand to building. On the other hand, to reach to an optimized passive house, that be capable in parallel to perform also as a "Zero Energy Building" (ZEB), the very first decision is minimizing total energy demand of building. So, in the very elementary steps of designing of a building it is crucial to include passive strategies to minimize energy demands. These modifications should be applied in the first steps of architectural designing of building and is much wider than only proportion of windows in relation to volumetric and geometrical characteristics of buildings. Based on enormous economic and technical limitations, there is also an unavoidable need to implement photovoltaics or solar thermal panels to compensate the energy demand that passive strategies integrated in buildings are not able to fulfill alone. An accurate implementation of PVs again argues a certain amount of surrounding building's skin and consequently an accurate calculation of the surfaces that should be allocated to PVs is necessary [1]. Also, different geometric proportions result different available surfaces in different orientations for PV integration that influence the final capacity of power plant. In this step, the crucial question is "which one is in priority for geometrical modification of a building? Passive strategies or PV integration?"

In this work, a simple architectural prototype has been evaluated to assess the specific effect of its windows in different orientations. Also, in every orientation, different forms of the same building have been examined to compare the effect of envelop area factor and different "Treated Floor Area" (TFA). So, four architectural variants with the same volumetric amount, but with different TFAs and different geometrical proportions have been developed. The main target of theses comparisons is to assess the annual heat demand of the variants. Also, it would clarify the influence of orientation of windows to reach to different heat demands. In parallel, it would show the effect of different forms of a building with the same orientation and the same surface of transparent surfaces to reach to different heat demands. Design-PH, has been used to determine heat losses and heat gains of these prototypes in different situations. So, in each variant four different parameters have been measured that are; "heat loss of opaque surfaces & thermal bridges", "heat loss of windows", "ventilation heat loss" and "Solar heat gain of windows". The effect of "ventilation heat loss" has been ignored as it always has the same amount in the entire variants with regard to the constant amounts of transparent surfaces. The main principle of the software clarifies the annual heat demand with regard to the difference between the total heat losses and total heat gains. To conclude the final criteria, four different variants have been developed; the entire variants have the same volume of 800 m3 to do the final comparisons fairly. Every variant has been assessed when its windows has been looking toward south, west, north and east. So, sixteen different sub-variants have been calculated and compared to each other. U-values of windows, roof and the ceiling of variants has not been modified as the characteristic of materials has not been topic of this work.

In the second part of theses simulations to assess performance of these variants for integration of PVs, available surfaces of these four variants have been emerged. For emphasis on effect of the main direction of windows, in each variant there is only one facade holding 5 squad windows with a total area of 20 m^2. The effect of main orientation of building has been taken into account, so as subdivision of main variants, they are divided also to south-faced, west-faced, north-faced and east-faced variants. Additionally, "orientation and tilt coefficient" (based on average geographical parameters in Germany) has been included in parts of the comparisons to discuss about the real efficiency of the variants based on their surfaces that resulted a specific parameter of "PV useful area". The proportion of this parametric amount with "TFA", "area of roof" and "total area of the entire components" resulted three different ratios that can clarify the geometrical efficiency of a variant for PV integration. The final comparisons revealed that; In south-faced windows variants there is always least annual heat demand that could be considered as the best situation that is an obvious solar issue. But the point of importance is the amount of the differences. As in elementary steps of architectural design of a passive house the main issues are passive energy performance, necessity of some parallel devices as heat pumps and/or heat exchanger in connection with solar thermal panels or PVs, it should always be a precise estimation of capacity of power plant. In this work, during movement from one-floor variants toward multi-floors variants "PV total area" is almost the same, but after moving from cubic proportional geometry it starts to increase and could result 10% more available surfaces. "PV useful area" softly drops during movement from one-floor variants toward multi-floors variants about 15%. "PV roof area" as the most independence parameter decreases from one-floor variants toward multi-floors variants about more than 65%. The most important comparisons in this work was performed to compare the ratios of above definitions together. Ratio of "PV useful area" to TFA presented a constant amount in all variants except in cubic geometry. Ratio of "PV useful area" to "PV total area" confirmed a constant performance in the entire variants with a little fall in multi-floors. Ratio of "PV roof area" to "TFA presented a dramatic fall in multi-floors variants. These comparisons revealed that maximizing passive characteristics of variant, is not always in favor of increasing potential of variants for mounting PVs.

Keywords: architectural optimization, heating demand, architectural pattern, solar architecture, zero-energy building, passive solar strategies, photovoltaics

1 PURPOSE OF THE WORK

Modification of architectural factors and changing number, dimension or orientation of windows and surface of opaque materials could normally have a significant effect on energy performance of a passive house with regard to its insulation properties, heat gains and heat losses [2]. Through these modifications different heat demand of building also result different annual and monthly energy performances based on comparative statistics [3]. On a very elementary step of designing a passive house, to meet passive house requirements, rather than using well-insulated materials and best orientation of building, considering horizontal and vertical proportion of volumetric design would also have a significant effect in fine-tuning the final performance [4]. As a result, the main focus is determining which factor affects annual heat demand of a building through a smooth movement from one-floor and horizontal buildings toward multi-floors and vertical buildings with the invariable parameter of "total volume" and variable TFAs based on architectural proportions. In the other word, same boundary-volume buildings with different proportional designs have been compared to clarify the effect of area dependent diversity on passive use of solar energy.

2 APPROACHES

To measure and compare effect of proportion of design on annual heat demand of building one key question has been defined; "To which extent energy demand of building would change", if the same volume buildings have;

a- Different dimensional proportions based on height and overall area?

b- Different orientations? [5]

So TFA, as the main characteristic of building has been placed in the focus of these measurements. On architectural aspect, it could be as a specific proportion of building that has to be keep in comfortable conditions to be heated or cooled. In this work, based on geographical location of prototypes, heating is in the focus of comparisons.

For this simulation, a simple quadrangle prototype has been designed in four different dimensional proportions;

2.1 Variants

All of variants are simple cubic forms with exactly the same volumetric size, but with different proportion in dimensions. So, the length, width and height of them change smoothly from variant 01 to variant 04. The point is just keeping the final volumetric size in the entire four variants during conversion from a horizontal concept to a vertical one. For this purpose, multiplying 3 dimensional numbers; length × width × height (Total volume) would always be constant amount of 800 m^3 in the variants. All variants are using flat roofs. They have five windows of 2 × 2 m in one face and one door of 2 × 3 m in the next perpendicular face. So, the total surface of windows is 20 m^2. Area of door is 6 m^2. Total surface of windows and door is supposed to be totally covered with normal glass. All of them have one dwelling unit for a family of 3 persons. Each variant has one kitchen, one bathroom and one WC. The number of bedrooms are the same, but variant-03 and variant-04 have a more vertical design. (Table 01)

	Variant 01	Variant 02	Variant 03	Variant 04
Orientation				
Number of floors	1	1	2	3
Dimension: Length × Width, (m)	20 × 10	16 × 12.5	10 × 10	8.35 × 8
Flat area, (m^2)	200	200	100	66.8
Height, (m)	4	4	8	12
Total volume, (m^3)	800	800	800	800
Treated floor area (TFA), (m^2)	163	164	148	134
Number of windows + doors	5 + 1	5 + 1	5 + 1	5 + 1
Total area of windows + doors, (m^2)	26	26	26	26

Table 01: Dimensional characteristics of variants

2.2 specific characteristics of each variant

Variant-01 is a one-floor house. Beyond the other variants and based on its dimensional proportion, it is the most rectangle and shortest variant with 4m height. Calculated TFA of this variant is 163 m2, as it has no stair box to any other floor.

Variant-02 is a one-floor house. It is almost squad and still shortest variant with 4m height. Calculated TFA of this variant is 164 m2, as it has no stair box to any other floor.

Variant-03 is a two-floor house. Basically, it is still squad with 8m height. Calculated TFA of this variant is 148 m2, as it has one stair box to second floor.

Variant-04 is a three-floor house. It is the most squad variant with 12m height. Calculated TFA of this variant is 134 m2, as it has two stair boxes to second and third floors.

2.3 U-Values

As technical and qualification of materials is not the scope of this work so the total transparent and opaque materials of four variants are supposed to be the same. Normal glasses and other masonry materials for boundary of buildings have been included in software for calculations with usual U-Values. (Table 02)

Building assembly overview	Average U-Value (W/m²K)
Windows	0.915
Door	0.936
External wall - Ambient	0.150

Table 02: U-Values of main components of variants

2.4 Thermal bridges

In principle, effect of thermal bridges has been also automatically included in results. The number of these heat losses have not been represented in final comparisons because in these simple variants, the overall shape is always the same so the difference of heat losses in thermal bridges are not considerable to change sequence of main results. Despite the fact that variant-03 and variant-04 are two-floor variants with more amount of thermal bridges, but software showed that the heat loss differences coming from this part is too little to present it and discuss about.

2.5 Climate, ventilation type

Germany has been selected as geographical location. The climate zone of "Cool-temperate" as the average weather conditions has been used during entre calculations. "Only windows ventilation" has been considered for all variants as ventilation method to reach the simplest output only based on proportional differences. (Table 03)

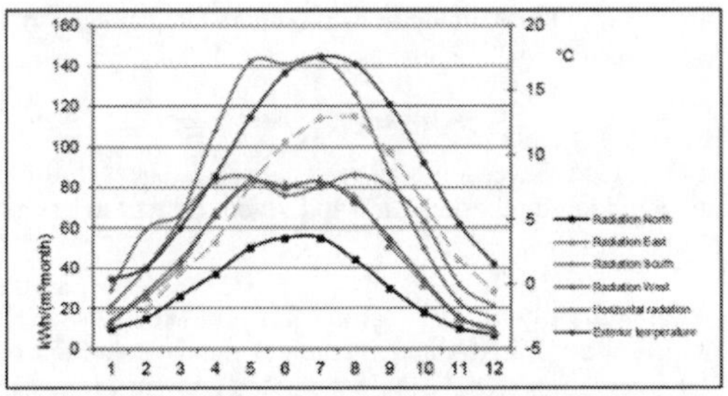

Table 03: Site plan - main accesses and dimensions

2.6 Shading and wind

As no canopy or other external component or planting has been supposed, so the effect of shading has been totally ignored in calculations. Wind effect has been also ignored to simplify the outputs as comparable results with regard to constant geographical location of prototypes.

3 EACH VARIANT, BASED ON DIFFERENT ORIENTATIONS

In principle, the calculations have been done to measure annual heat gains and annual heat losses of prototypes. Heat losses that have been measured are through external walls, roof, floor slab and windows. Effect of windows and door in heat losses has been also included [6]. Heat gains that have been measured are through windows, effect of internal human activities. Difference between annual heat losses and heat gains concluded annual heating demand. (Table 04)

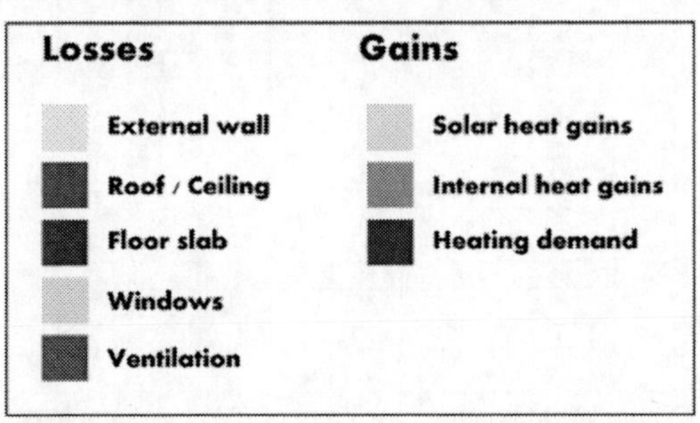

Table 04: Main measured factors of heat losses and heat gains

3.1 Variant-01 orientations

Heat gain of windows is the only different parameter. South orientation has 14.90 kWh/m²a as the most amount of heat gain. North orientation has 6.80 kWh/m²a as the least amount of heat gain. Heat gains of the two other orientations are in between. West has more annual heat demand with regard to place of the door, as in west orientation of windows, door is oriented toward south with maximum heat gain. But in East orientation of windows, door is oriented toward north with minimum heat gain. (Table 05)

	Heat loss			Heat gain	
Orientation	Opaque surfaces	Windows	Ventilation	Windows	Annual heat demand
South	46.20	12.20	23.40	14.90	54.20
West	46.20	12.20	23.40	11.40	57.70
North	46.20	12.20	23.40	6.80	62.30
East	46.20	12.20	23.40	8.50	60.60
Internal heat gain: 12.60					TFA: 163.00 m²

Table 05: Heat losses, heat gains and annual heat demand - variant 01 – Unit: kWh/m²a

3.2 Variant-02 orientations

South orientation has 14.80 kWh/m²a as the most amount of heat gain. North orientation has 6.70 kWh/m²a as the least amount of heat gain. Heat gains of the two other orientations are in between. (Table 06)

	Heat loss			Heat gain	
Orientation	Opaque surfaces	Windows	Ventilation	Windows	Annual heat demand
South	44.80	12.10	23.40	14.80	53.00
West	44.80	12.10	23.40	11.30	56.40
North	44.80	12.10	23.40	6.70	61.00
East	44.80	12.10	23.40	8.50	59.20
Internal heat gain: 12.60					TFA: 164.00 m²

Table 06: Heat losses, heat gains and annual heat demand - variant 02 – Unit: kWh/m²a

3.3 Variant-03 orientations

South orientation has 16.50 kWh/m²a as the most amount of heat gain. North orientation has 7.50 kWh/m²a as the least amount of heat gain. Heat gains of the two other orientations are in between. (Table 07)

	Heat loss			Heat gain	
Orientation	Opaque surfaces	Windows	Ventilation	Windows	Annual heat demand
South	42.00	13.40	23.40	16.50	49.70
West	42.00	13.40	23.40	12.60	53.50
North	42.00	13.40	23.40	7.50	58.60
East	42.00	13.40	23.40	9.40	56.70
Internal heat gain: 12.80					TFA: 148.00 m²

Table 07: Heat losses, heat gains and annual heat demand - variant 03 – Unit: kWh/m²a

3.4 Variant-04 orientations

South orientation has 14.80 kWh/m²a as the most amount of heat gain. North orientation has 6.70 kWh/m²a as the least amount of heat gain. The two other orientations in the case of heat gains are in between. (Table 08)

Orientation	Heat loss			Heat gain	
	Opaque surfaces	Windows	Ventilation	Windows	Annual heat demand
South	46.90	14.80	24.80	18.10	55.60
West	46.90	14.80	24.80	13.90	59.70
North	46.90	14.80	24.80	8.20	65.30
East	46.90	14.80	24.80	10.40	63.20
Internal heat gain: 13.00					TFA: 134.00 m²

Table 08: Heat losses, heat gains and annual heat demand - variant 04 – Unit: kWh/m²a

4 ANNUAL HEAT DEMAND OF FOUR VARIANTS

Alternatively, annual heat demand of each variant in the same orientation has been compared to other variants to address the amount of differences of the results based on different volumetric form and different TFAs. The least annual heat demand belongs to variant-03 in south orientation with annual heat demand of 49.70 kWh/m²a. The most annual heat demand belongs to variant-04 in north orientation with annual heat demand of 65.30 kWh/m²a. (Table 09)

Orientation	Variant 01	Variant 02	Variant 03	Variant 04

South	54.20	53.00	49.70	55.60
West	57.70	56.40	53.50	59.70
North	62.30	61.00	58.60	65.30
East	60.60	59.20	56.70	63.20

Table 09: Annual heat demand of four variants based on main orientation – Unit: kWh/m²a

5 TOTAL AVAILABLE AREA OF COMPONENTS

5.1 Available surface of components

In this step, maximum possibility of integration of PVs in each component has been determined. Main walls with regard to the remaining area after subtracting the area of windows result the amount of possible integration of PVs. As the angle of incidence between the incoming sunlight and a photovoltaic panel results different efficiencies, a coefficient has been multiplied to reach to the useful area of possible integration. This coefficient has been selected with regard to average light intensity and angle of radiation in Germany with latitude of around 51° N. This coefficient for a flat roof with inclination of 0° is 87%, for a south-wall with inclination of 90° is 69%, for a west/east-wall with inclination of 90° is 48%, and for a north-wall with inclination of 90° is 18%. With regard to geographical location of this simulation the ideal efficient component is a south-faced sloped surface with inclination of between 30° to 40° that will reach to a maximum coefficient of 100%. Comparatively a south wall with a total area of 60 m^2 should be multiplied in 69% and will result 41.4 m^2 as;

Total area		Orientation and tilt coefficient for a south wall in Germany	Useful area
60 m²	×	69%	41.4 m²

Table 10: Converting total area of a component to useful area by including orientation and tilt coefficient in a south-faced wall

Similarly, the following charts represent the procedure of converting total area of components to useful area for integration of PVs to reach to comparable number of areas;

Total area		Orientation and tilt coefficient for a south wall in Germany	Useful area
40 m²	×	48%	19.2 m²

Table 11: Converting total area of a component to useful area by including orientation and tilt coefficient in a west-faced wall

Total area		Orientation and tilt coefficient for a south wall in Germany	Useful area
80 m²	×	18%	14.4 m²

Table 12: Converting total area of a component to useful area by including orientation and tilt coefficient in a north-faced wall

Total area		Orientation and tilt coefficient for a south wall in Germany	Useful area
34 m²	×	48%	16.3 m²

Table 13: Converting total area of a component to useful area by including orientation and tilt coefficient in an east-faced wall

After multiplying orientation and tilt coefficient the numbers can fairly added together as they are useful areas for integration of PVs. This separated amount of areas in each variant could be important in elementary architectural steps of design to decide which volumetric concept would result which amount of available surfaces for integration of PVs. Considering different efficiencies of available products in market it will always technically make sense to know the maximum nominal power possibility. On the other hand, as each volumetric concept result different passive properties and as a result different heating and cooling demand, to generate the demand of energy with PVs, this prediction could refine the energy performance of building with having a more accurate estimation of on-site power generation. (Table 14)

Orientation		Variant 01		Variant 02		Variant 03		Variant 04	
		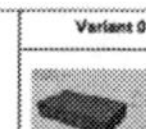		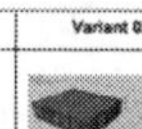		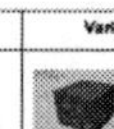		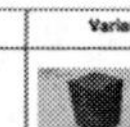	
South-wins		Total area m²	Useful area m²	Total area m²	Useful area m²	Total area m²	Useful area m²	Total area m²	Useful area m²
	South walls	60	41.4	44	30.3	60	41.4	80	55.3
	West walls	40	19.2	50	24.0	80	38.4	96	46.0
	North walls	80	14.4	64	11.5	80	14.4	100	18.0
	East walls	34	16.3	44	21.1	74	35.5	96	46.0
	Roof	200	174.0	200	174.0	100	87.0	66	58.1
	Total	414	265.3	402	261.0	394	216.7	439	223.6
West-wins	South walls	34	23.4	44	30.3	74	51.0	96	66.2
	West walls	60	28.8	44	21.1	60	28.8	80	38.5
	North walls	40	7.2	50	9.0	80	14.4	96	17.2
	East walls	80	38.4	64	30.7	80	38.4	100	48.1
	Roof	200	174.0	200	174.0	100	87.0	66	58.1
	Total	414	271.8	402	265.2	394	219.6	439	228.2
North-wins	South walls	80	55.2	64	44.1	80	55.2	100	69.1
	West walls	34	16.3	44	21.1	74	35.5	96	46.0
	North walls	60	10.8	44	7.9	60	10.8	80	14.4
	East walls	40	19.2	50	24.0	80	38.4	96	46.0
	Roof	200	174.0	200	174.0	100	87.0	66	58.1
	Total	414	275.5	402	271.2	394	226.9	439	233.8
East-wins	South walls	40	27.6	50	34.5	80	55.2	96	66.2
	West walls	80	38.4	64	30.7	80	38.4	100	48.1
	North walls	34	6.1	44	7.9	74	13.3	96	17.2
	East walls	60	28.8	44	21.1	60	28.8	80	38.5
	Roof	200	174.0	200	174.0	100	87.0	66	58.1
	Total	414	274.9	402	268.2	394	222.7	439	228.2

Table 14: Maximum possibility of PV integration in each component of different variants – Useful area of PV integration after multiplying by orientation and tilt coefficient

5.2 Specific ratios between surfaces

Different proportional variants in this simulation result diverse areas for PV integration, even with the same volumetric amount. So, some ratios have been calculated to know the proportion of areas together. These ratios could facilitate the elementary decisions to clarify which amount of surface of building would result the most appropriative integration. Technically, the most efficient integration will produce the most power but not the cheaper, as normally from market point of view, the more efficient PV products as mono-crystalline panels are relatively more expensive. The other limitation is architectural issues as in much cases it is not also possible to allocate best oriented face of building to PVs. To somehow it could be also important to include the surrounding situation of building, scenery values, neighborhoods and other demands. So, in some cases, designers have to switch to a wider integration but with less efficient panels or wise versa. The very critical prediction is always to simulate if the total surface of roof is sufficient for energy demand of building or not. Just relying on available surfaces of building for integration of PVs would not lead to design an accurate energy concept for building as it has to be compared with passive energy performance of building that is still potentially influenced with diverse architectural factors. In this work to reach to more accurate comparisons some specific areas for PV integration have been defined. Treated Floor Area (TFA) also has been included as one impressive volumetric parameter of each variants, because it makes sense to know which amount of floor area gets heat concerning the main place of living in a building.

As following, the main variable parameters of this simulation has been clarified;

- **PV roof area (R), m^2:** Available surface of roof for PV integration

- **PV total area (T), m^2:** Total available surface of roof plus four main walls for PV integration

- **PV useful area (U), m^2:** Total useful surface of roof plus four main walls for PV integration. As a rule, compared to "PV total area (T)", it is always a smaller amount as "the orientation and tilt coefficient" has been multiplied by T. This parametric amount could play a director role during elementary phases of architectural design to know how much is the meaningful surface of building for PV integration regardless of tilt and orientation of building's components.

- **Treated Floor Area (TFA), m^2:** Amount of floor area that gets heat concerning the main place of living in a building. TFA is not much different from the heated living space. The calculation rule has been simplified in some points and adapted to the requirements of the energy balance. Heated auxiliary rooms have been taken into account in this procedure.

- **Ratio of U/TFA:** Ratio of "PV useful area" to "Treated Floor Area". From technical point of view to allocate some components of building for PV integration, the most parametric variable is TFA. As there are usually some parts of building that can be eliminated from building's comfort zone like roofed parking, cellar, store, facility room, etc. For this purpose, it is critical to know which amount of PV useful area could be allocated per capita. For this definition, "PV useful area" has been divided into "Treated Floor Area". Comparing this ratio in different conceptual variants can result a fair comparison to predict if the surface of building's components could be sufficient for electricity demand or not. Expanding this ratio in architectural design phase can result if some components should be added only for extra PV integration, or the building is optimum compared to its energy demand or if there could be some extra electricity production to be fed into grid or be used for electro-mobility. As in each Zero-Energy Building, including electro-mobility possibilities need architectural spaces, so estimation of this ratio could assure if the building could be capable for these additional usages as well.

- **Ratio of U/T:** Ratio of "PV useful area" to "PV total Area". This ratio could statistically explain an "overview efficiency of building's skin" for PV integration, regardless of technical efficiency of PVs. Flat roofs always perform the same performance for PV integration, regardless of orientation of building. But for the roofs with tilt the condition will be totally different mainly for two reasons; firstly, the tilt of roof effects the absorption of PVs, secondly setting the best orientation of PVs is not as simple as in flat roofs and from costs of views not economic. Surrounding walls also perform totally difference not only based on their spread but also based on their direction. In some cases, there is also an inclination in walls that influence the light absorption efficiency. Based on this ratio, the issue becomes more complicated if we want to compare totally different architectural concepts based on their surface efficiency for PV integration. The ratio of U/T accelerate comparing diverse conceptual designs and facilitate modification in architectural layouts based on this ratio.

- **Ratio of R/TFA:** Ratio of "PV roof area" to "Treated Floor Area". This ratio clarifies proportion of the total area of roof into specific portion of building that needs to be in comfort situation. As the very first component that normally mounts PV is roof, so total available surface of roof is crucial, but for comparing potential of different conceptual buildings together, it makes much more sense if its proportion to TFA is taken also into account. Indeed, for diverse technical and architectural reasons, roof is in priority for PV integration. In upcoming sessions, this ratio in different variants will be compared together in parallel to passive characteristics of different variants. (Table 15)

			Variant 01	Variant 02	Variant 03	Variant 04

South-wins	PV roof area (R)	m²	200	200	100	66
	PV total area (T)	m²	414	402	394	439
	PV useful area (U)	m²	265	261	216	223
	TFA	m²	163	164	148	134
	Ratio of U/TFA	coefficient	1.63	1.60	1.46	1.67
	Ratio of U/T	coefficient	0.64	0.65	0.55	0.51
	Ratio of R/TFA	coefficient	1.23	1.22	0.68	0.50
West-wins	PV roof area (R)	m²	200.0	200	100	66
	PV total area (T)	m²	414.0	402	394	439
	PV useful area (U)	m²	271	265	219	228
	TFA	m²	163	164	148	134
	Ratio of U/TFA	coefficient	1.67	1.62	1.48	1.71
	Ratio of U/T	coefficient	0.66	0.66	0.56	0.52
	Ratio of R/TFA	coefficient	1.23	1.22	0.68	0.50
North-wins	PV roof area (R)	m²	200.0	200	100	66
	PV total area (T)	m²	414.0	402	394	439
	PV useful area (U)	m²	275	271	226	233
	TFA	m²	183	164	148	134
	Ratio of U/TFA	coefficient	1.69	1.66	1.53	1.74
	Ratio of U/T	coefficient	0.67	0.68	0.58	0.53
	Ratio of R/TFA	coefficient	1.23	1.22	0.68	0.50
East-wins	PV roof area (R)	m²	200.0	200	100	66
	PV total area (T)	m²	414.0	402	394	439
	PV useful area (U)	m²	274	268	222	228
	TFA	m²	163	164	148	134
	Ratio of U/TFA	coefficient	1.66	1.64	1.50	1.71
	Ratio of U/T	coefficient	0.67	0.67	0.57	0.52
	Ratio of R/TFA	coefficient	1.23	1.22	0.68	0.50

Table 15: Specific amounts of surfaces - defining ratios between surfaces and values

5.3 Comparing and determining relevancy of ratios between surfaces

In this step, the defined ratios have been compared together to determine their logical relevancy and effects on each other. As a strong limitation in integration of PVs is available surfaces, these comparisons (in each project) can narrow the estimation to adopt amounts of PVs and their efficiencies based on provided surfaces.

The blue line describes ratio of "PV useful area" to "Treated Floor Area". From quantity point of view, the larger this ratio is, the more electricity can be generated by PVs in proportion to TFA, as the demand of electricity shows an average number for the total area of building that are included in TFA. Beyond all of variants with different windows-faced orientation, south-faced windows variant has been selected in the graph, as normally in residential building the main preference for windows is toward south. This decision is normally in favor of passive performance and architectural comfort also.

In variant-01 it shows 1.63 that means for each square meter of TFA, there could be equivalent of 1.63 m^2 surface of wall or roof for PV integration. Compared to real buildings the number is much higher that the demand of building. Also, it is much higher than normal standards of demand. The main reason in all of the four variants is that the windows are only utilized in south façade. This decision was taken on purpose to evaluate the maximum possible difference of variable amounts of surfaces in different faced facades. In variant-02 it is almost the same with a small drop and shows 1.60. Variant-03 demonstrates a relative lower ratio of 1.46. The reason is that although TFA drops from 164 m2 in variant-02 into 148 m2 in variant-02, but the "PV roof area" dramatically decreases to 100 m^2. So, despite the fact that variant-03 has less annual heat demand for its more cubic conceptual design (from passive point of views), but it performs a higher active electricity generation per capita of TFA. (49.70 kWh/m²a of annual heat demand in variant-03 compared to 53.00 kWh/m²a in variant-02). In variant-04 it shows 1.67 that is a much higher potential compared to previous variants. Compared to variant-03 its "PV useful area" increases from 216 m^2 to 223 m^2, and on the other hand TFA falls to 134 m^2. Regardless of passive point of view it could be concluded that a three-floors building with this proportion, could perform better for integration of PVs, if the entire facades are available, compared to a two-floors building with the same volumetric design and the total surfaces of windows. Conclusion an absolute result only based on this ratio will only be possible if the passive energy performance of building is not also in focus of investigation. In our example, the three-floors variant performs better for active electricity generation compared to TFA, but the situation is wise versa in comparison of their passive performance. On the other hand, for comparing the variants together from PV potential and passive point of view together, effects of parallel devises that can be integrated in building, clarifies if PVs of a building are sufficient or not. For instance, in variant-04 there is 223 m^2 of useful surface for PV integration. Its sufficiency technically depends on which parallel devises will be integrated in building like; heat pumps, heat exchanger, etc. and comparative efficiency of these devises.

The orange line describes ratio of "PV useful area" to "PV total area". From economic point of view, the larger this ratio is, the more efficient the variant becomes (in the case of quantity of electricity generation that could result a variant with extra electricity generation). Both of variants 01 and 02 show 0.64 and 0.65 respectively. In variant-03, it drops to 0.55, as its "PV useful area" drops to 216 m^2 compared to variant-02 with 261 m^2, but the PV total area is almost the same. In variant-04 it drops again to 0.51. This time the main reason is increasing "PV total area" that shows 439 m^2. A smooth movement from variant-01 toward variant-04 demonstrates a small and constant drop in "ratio of U/T".

The gray line describes ratio of "PV roof area" to "Treated Floor Area". From practical point of view, the larger this ratio is, the more roof PV integration is possible in proportion to TFA that mainly is included in calculation of heat demand of buildings. The influence of this ratio will be more crucial in the case of buildings that have limitations for integration of PV in their surrounding walls for some technical or other reason and the only available component is remaining roof. In this work, ratio of R/TFA in variant-01 and variant-02 is 1.23 and 1.22 respectively with regard to their similar dimensional characteristics. In variant-03 it drops to 0.68, as its R decreases to 100 m^2. In variant-04 it again drops to 0.50, as the R clearly decreases from 100 m^2 in variant-03 to 66 m^2. In this work, from variant-01 toward variant-04, the graph demonstrates decreasing, with a dramatic drop between variant-02 and variant-03. This main reason of this large drop is obviously "PV roof area" that has been half compared to variant-03. This decrease has been happened despite of decreasing TFA from variant-01 toward variant-04. So, the main consequent could be in different variants with the same volumetric amounts, in a movement from one-floor prototypes toward multi-floors prototypes. Although TFA will mainly decrease, but decreasing the final amount of available roof area will mainly result a less possibility for integration of roof-mounted PVs. (Table 16)

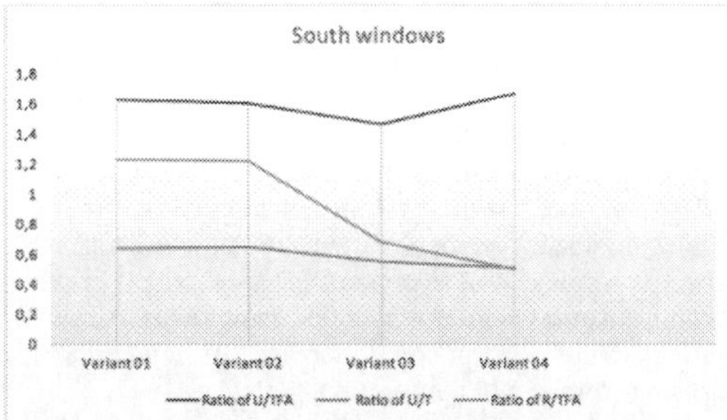

Table 16: Ratios of different variants in south-windows orientation – U: PV useful area, T: PV total area, R: PV roof area, TFA: Treated floor area

5.4 Potential of surfaces in components of different variants

From quantity point of view, it is crucial to estimate the maximum available areas of building's components for integration of PVs and/or solar thermal panels. This issue become more crucial during primary steps of cost estimation of active components, regarding different kinds of photovoltaics. Diverse kinds of crystalline silicon solar cell (c-Si), amorphous Silicon solar cell (a-Si), cadmium telluride solar cell (CdTe), etc. supply different efficiencies and for a certain amount of power generation they mainly differ in two basic parameters; cost and area of utilization. So far, more efficient PVs have been more expensive with regard to the costs of their materials and complexity of their production, but a more efficient panel will have less demand for area of integration. So, with regard to the limitation of construction's costs and available surfaces for PV integration, an immediate decision of type of crystalline panels is not easily feasible. For integration of PVs in north faced facades or other orientations except south, different types of cadmium telluride solar cells (CdTe) are an appropriative selection, as they are not as sensitive as other PV types to orientation. But they need much more surfaces for generation of the same power. For this reason, the buildings with wider component's skins can integrate less efficient PVs to generate a certain amount of electricity.

The graph compares surface quantity of four variants for PV integration. The south-faced alternative has been only selected as the most common orientation of building.

The blue line describes "PV roof area". In variant-01 and variant-02 it is the same amount of 200 m^2, while in variant-03 it drops to 100 m^2 and in variant-04 it reaches to 66 m^2. In this case, regardless of relation to TFA, the nominal potential of one-floor variants for roof-top integration perform a higher statistic availability. In prototypes with the same volumetric amount, from flat concepts toward more vertical concepts, as the dimensional proportion of geometry changes, the available surface of roof certainly decreases. So, in the case of buildings that roof is the only available component for integration, in high-rise buildings, the necessity of cooperation of other components of building in addition to roof is essential. Wise versa, in one-floor buildings with the same TFA, the available surface of roof could be closer to energy demand of building.

The orange line describes "PV total area". In buildings that all surrounding walls will be used for PV integration, prediction of the total available surface will facilitate to reach to a more accurate estimation of power plant. Variant-01 presents 414 m^2, in variant-02 it drops to 402 m^2 and with almost the same decrease in variant-03 it drops to 394 m^2. But in variant-04 it shows a considerable rise and reaches to 439 m^2. The reason is that in a vertical concept like variant-04, the surface of surrounding elements will result a wider surface. Consequently, in taller buildings with possibility of integration of PV in walls, the nominal power plant could perform 6% more in the geometry proportion prototypes that are similar to the selected variants in this test.

The gray line describes "PV useful area". Specific orientation and tilt coefficient has been multiplied by each component to reach to a real nominal power potential. Compared to "PV total area", "PV useful area" could be more reliable in the case of using the same type of PV panels in entire components. Variant-01 performs 265 m2 and in variant-02 it decreases smoothly into 261 m2. But it variant-03 there is a drop and it falls to 216 m2 with regard to geometrical proportion and the orientation and tilt coefficient. Finally, in variant-04 it has a small rise and reaches to 223 m^2. The graph demonstrates that variant-01 and variant-02 that are one-floor prototypes, can result 22% more amount of PV useful area. In the other words, in the case of limited number of PVs, one-floor prototypes could generate more amount of electricity compared to higher prototypes in buildings that their geometrical proportion are similar to this test.

Comparing "orange line" with "blue line" clarifies that in one-floor prototypes there is a higher roof dependence compared to multi-floor prototypes. The amount of this difference is remarkable as the entire variants have a "PV total area" of about 450 m^2 although the "PV roof area" drops in the first two variants from 200 m^2 into 66 m^2. Despite in variant-04 there is a drop in "PV roof area" and it reaches to 66 m^2, but in its "PV total area", there is still a rise compared to previous variant. The quantity of this difference is 214 m^2 and 202 m^2 in the first two variants while I variant-03 it reached to 294 m^2 and in variant-04 it increases to 373 m^2 of differences.

Comparing the three lines together clarifies that "PV useful area" is mainly dependent on "PV roof area" rather than "PV total area". It is a direct outcome of the "orientation and tilt coefficient" that for a flat roof with 87% is relatively higher than roof's coefficient. Comparing "blue line" with "grey line" demonstrates that in multi-floors prototypes like variant-04, despite a considerable drop in "PV roof area", its "PV useful area" increases that could be also an effect of less comparative TFA. (Table 17)

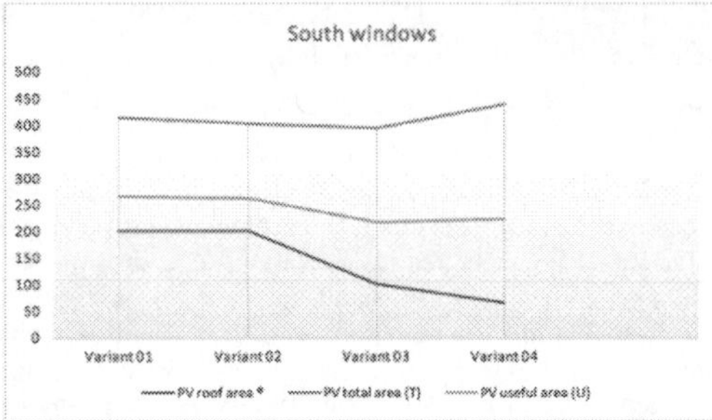

Table 17: Possibility of PV area integration in different variants - R: PV roof area, T: PV total area, U: PV useful area, Unit: m²

5.5 Potential of surfaces in components of different-faced windows in different variants

In this step, the aim was assessing effect of rotating variants in useful amount of the entire components. So, all the four variants have been studied in four main orientations that resulted 16 different alternatives.

This graph presents differences of "PV useful area" in all variants in different faced windows. As a constant effect of "useful surface of roof plus four main walls for PV integration", the procedure of changing is almost the same in all different faced windows. The first two variants perform a higher amount compared to variant-03. Variant-04 also presents a higher value compared to variant-03. The reason of almost the same change in the entire graphs is that interaction of "orientation and tilt coefficient" into the surfaces of each variant always change parallel together in four variants. Comparing the changes in different faced windows shows that the "north-faced windows" have a higher performance than "east-faced windows", "west-faced windows" and "south-faced windows" respectively. The main reason is that when all of the windows of a rectangular building are toward north, then for a southern wall it could be a good opportunity to absorb the light. (In the case of the absence of windows in opposite faces). (Table 18)

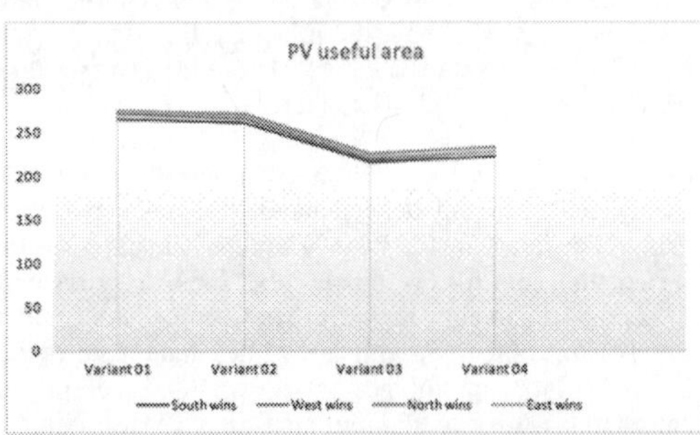

Table 18: Useful area for integration of PV after multiplying orientation and tilt coefficient in different variants with different faced windows

5.6 Possibility of PV integration in variants with two main orientation

This graph presents "PV total area", "PV useful area" and "PV roof area" in all the four variants. As subdivision of four variants, south-faced windows and west-faced windows have been selected to take also the effect of main direction of windows into account. As the graph shows, "PV total area" in the first three variants stay approx. in the same quantity except variant-04 that performs a 10% rise. Accordingly, in every similar geometrical prototype with the same volumetric amount, moving toward vertical proportion of concept could increase "PV total area" after passing from a cubic proportion.

Fluctuations of "PV useful area" is totally different from changes of "PV total area". After moving from one-floor variants toward variant-03 that presents a cubic concept with 2 floors, "PV useful area" drops about 18%. This is partly because of decreasing the surface of roof which normally absorbs more compared to surrounding walls and partly because of decreasing in "PV total area" compared to variant-02. In variant-04 it raises again as (compared to variant-03) TFA falls and "PV total area" rises.

Fluctuations of "PV roof area" is almost the same in the first two variants that have one floor. Obviously, with a movement toward vertical variants with the same volumetric amount, the roof's area decreases based on the new geometry properties. The highlighted quantity difference could clarify that in high rise buildings that the only available component is roof, the sufficiency of this component for the entire building is always critical. So, in the category of vertical conceptual buildings, utilization of at least a proportion of surrounding walls to fulfill electricity demand of building is unavoidable. (Table 19)

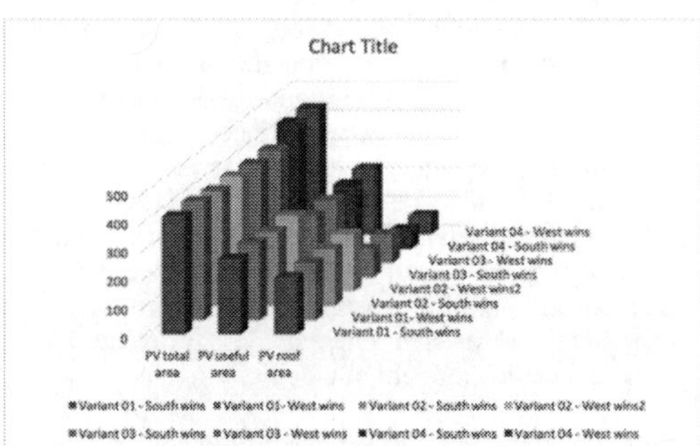

Table 19: Maximum possible integration of PV in south faced windows and west faced windows variants

6 RESULTS AND CONCLUSIONS

• In comparing the variants based on their passive characteristics, it has been revealed that in each geometrical proportion, the south-faced windows alternative could result less annual heat demand. From statistic point of view, the percentage of performance difference in south-faced windows compared to other orientations in not the same. This difference is more moderate in one-floor variants compared to multi-floors variants. Also as a rule, the values of annual heat demands proved that the more cubic variant results a less comparative annual heat demand. But in comparisons of possibilities of integration of PVs in surfaces of components, the graphs proved other kind of superiorities for the tested variants. As The most common component for integration of PVs is roof, one-floor variants proved a remarkable stronger potential for PV integration in the same volumetric geometries.

• In comparing the variants based on the possibilities of their entire components for mounting PVs as well as the available surface of roof, multi-floors variants proved stronger potential, while from passive point of view the multi-floors variants are not absolute the best. In comparing "PV useful area", the numbers proved again that the cubic variant demonstrates the worst potential in the case of available surfaces for PV integration while cubic variants are always the best performance variants from passive point of views.

• In comparing the variants based on the ratios of surfaces and values, with taking the effect of TFA into account again the cubic variant (variant-03 in this test), performed a less efficient performance compared to the other three variants. In comparing the variants based on the ratios of "PV useful area" into "PV total area", the one-floor variants performed a remarkable stronger performance and finally in comparing ratio of "PV roof area" into TFA, again one-floor variants performed absolute stronger performance compared to other variants.

• With regard to specific defined ratios in this work and the results of comparisons, it is revealed that energy performance advantages of a specific proportional variant, compared to other variants with the same volumetric parameters is not always the same. Indeed, it is easily possible that a specific variant (with regard to its geometrical proportions), perform better to use energy of sun to minimize heat demand (passively), but for generating the rest of energy with PVs, it provides less available surfaces for integration. From economic point of view, with regard to enormous types of PVs with different efficiencies and complexity of production, still the question is which proportional variant could perform an absolute higher energy performance? A conclusive answer to this question definitely needs a precise consideration of possible integration of parallel devices in buildings, like heat pumps, heat exchangers, ventilation systems, etc. Also, to include these mechanical devices in preliminary designs of buildings, to estimate energy demand of building, their technical characteristics should also be taken into account, in parallel to effect of geometrical proportion of the architectural design.

• Dealing with this open question, will be strongly influenced by upcoming products in the market of photovoltaics, solar thermal panels, heat pumps, heat exchangers, etc. So, state of the art of technical properties of market, should be precisely considered between different groups of construction professions to assess a more accurate estimation.

7 REFERENCES

[1] R. Krippner, T. Herzog, (2011), Architectural aspects of solar techniques – Studies on the integration of solar energy systems, in Proceedings Eurosun 2000, 3rd ISES-Europe Solar Congress, Cophenagen, Denmark: : International Solar Energy Society.

[2] T.E. Kuhn, S. Herkel, H.M. Henning, (2009), Active Solar Facades (PV and Solar Thermal).

[3] Arbitrary Orientation and Tilt, Retrieved from http://pveducation.org/pvcdrom/properties-of-sunlight/arbitrary-orientation-and-tilt

[4] S. Mende, F. Frontini, J. Wienold, (2011), Comfort and building performance analysis of transparent building integrated silicon photovoltaics, 12th Conference of International Building Performance Simulation Association, Sydney, 14-16 November.

[5] IEA SHC Task 41: Solar Energy and Architecture, Report T.41.A.1, Building integration of Solar, Thermal and Photovoltaics: barriers, needs, strategies

[6] T. OTIS, (2011), Is Solar Design a Straitjacket for Architecture? Conference Proceedings of the 27th International conference on Passive and Low Energy Architecture (PLEA), Louvain-la-Neuve, Belgium.

USING THE HOT AIR UNDER A BUILDING INTEGRATED PV ROOF IN COMBINATION WITH A VENTILATION HEAT PUMP TO REALISE ENERGY EFFICIENT DWELLINGS

Corry de Keizer*[1], Roland Valckenborg[1], René Borro[2], Maarten Laureijssen[3], Wiep Folkerts[1]
*Corresponding author: dekeizer@seac.cc
[1] Solar Energy Application Centre (SEAC), High Tech Campus 21, 5656 AE Eindhoven
[2] Rebor BV, Amsterdam, the Netherlands
[3] Unilin Insulation BV, Oisterwijk, the Netherlands

ABSTRACT: In this paper, we describe and evaluate a ventilated BIPV(T) system that could be used in combination with a ventilation heat pump to realise energy efficient dwellings. European regulations are aiming at (near) zero-energy buildings in the near future. Since a large share of the energy demand in the built environment in continental Europe consists of heat, the combination of both solar heat and solar electricity production in one roof is an interesting option to increase the share of renewable energy. We modeled the ventilated BIPV system and validated the model with a field test. In the field test we also tested the impact of several measures to improve the generation of useful heat, like fins and flow patterns.
The BIPVT/ventilation heat pump concept consists of: a prefab building integrated PV element that includes insulation and a modified ventilation heat pump. Traditionally, a ventilation heat pump uses the source heat from the ventilation air of a dwelling. The heated air under the PV panels could potentially supply a larger share of the source heat that would be required to supply the full heat demand for space heating and domestic hot water.
Keywords: Ventilated BIPVT system

1 INTRODUCTION

European regulations are aiming at (near) zero-energy buildings in the near future. Since a large share of the energy demand in the built environment in continental Europe consists of heat, the combination of both solar heat and solar electricity production in one roof is an interesting option to increase the share of renewable energy. One option is to actively ventilate the air gap between the BIPV modules and the insulated roof. The hot air can be used as an additional source for an air source or ventilation heat pump. The direct use of hot air for space heating is not beneficial in the Netherlands, due to the seasonal mismatch between space heating demand and solar irradiation.
In this paper we will focus on modeling and measuring the thermal energy yield of a ventilated BIPV system. Furthermore, we'll carry out some variations to study the result on the thermal energy yield. The BIPV roof is a prefab roof element that includes insulation as well.

2 FIELD TEST

The field test of the system was running from the end of 2015 to spring 2018 on SolarBEAT, the outdoor research location of SEAC in Eindhoven, the Netherlands. There were two phases of the field test. The first phase consisted of three series / lanes of eight UPT CdTe modules. The difference of these lanes was in the ventilation strategy. The left lane used natural ventilation, the middle column had forced ventilation and in the right column, zig-zag ventilation was created. The second phase of the field test was carried out with three series of three Solibro CIGS modules. Next to a forced ventilation reference lane, the effect of the use of copper fins on the heat production and a reflective foil on the heat production were studied.

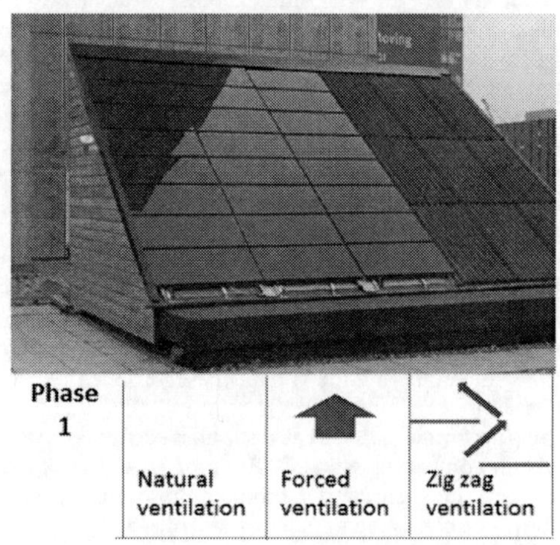

Figure 1: Field test phase 1, three series of 8 modules with natural ventilation, forced ventilation and zig zag forced ventilation

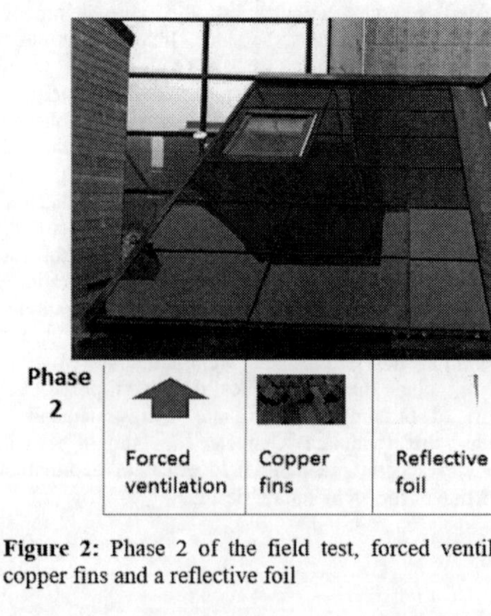

Figure 2: Phase 2 of the field test, forced ventilation, copper fins and a reflective foil

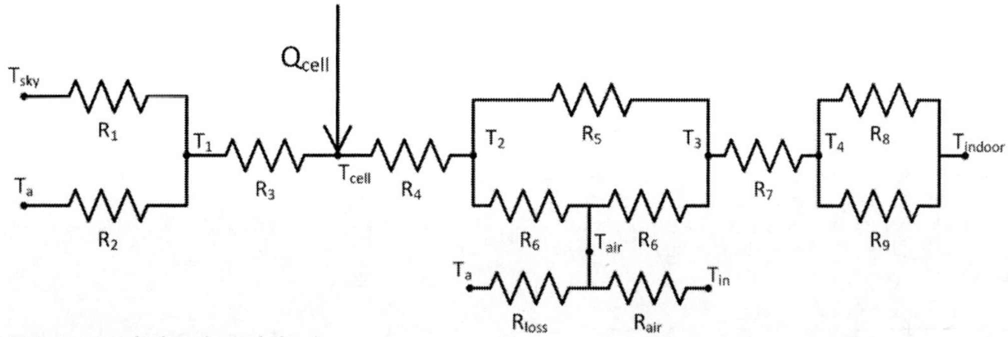

Figure 4: Equivalent thermal circuit

To determine the thermal and electrical performance, the following variables were measured:

- Meteorological measurements: In-plane irradiance, DNI, GHI, winddata from on-site solar measurement station [1]
- Electrical measurements: DC voltage and DC current per string, DC and AC power per string
- Thermal measurements: PV temperature of every module (T-type thermocouples), air inlet and outlet temperatures (Pt100), air flow (pitot method)

3 THERMAL BIPV MODEL

3.1 The BIPV thermal model

A thermal model of a ventilated BIPV roof was set up as an equivalent circuit model. It is an improved version of the model described in [2]. A schematic overview of the modeled layers and temperatures in the BIPV roof are shown in Figure 3.

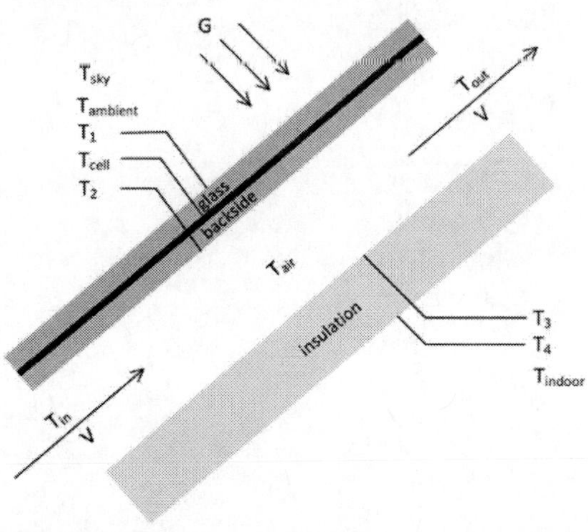

Figure 3: Schematic view of the modeled temperatures in a ventilated BIPV roof

Figure 4 shows the equivalent thermal circuit that was used which includes:

- Radiation (R1, R5, R8)
- Convection (R2, R6, R9)
- Conduction (R3, R4, R7)

Dependent on the PV module characteristics, irradiance and cell temperature, a certain amount of heat (Q_{cell}) and electricity is produced. The heat can then be conducted to the front or the back layer of the module. From there it can be 'lost' to the front by radiation to the sky or convection to the ambient.

The Python model is solved via an iteration process and uses an hourly timestep.

4 RESULTS

4.1 Results of the field test

In the first phase of the field test two ventilation strategies were compared. Figure 5 shows an infrared image of the installed setup. In the left lane there is only natural ventilation, in the middle lane there is forced ventilation going straight up and in the right lane, there is a zigzag forced ventilation flow. The temperature measurements in the airgap show a slightly higher temperature at the end of the zig zag lane in comparison to the middle lane. However, the infrared image shows a large temperature variation in the PV modules and for PV performance these type of temperature differences in the module are not good. Therefore a linear flow is chosen for the final setup.

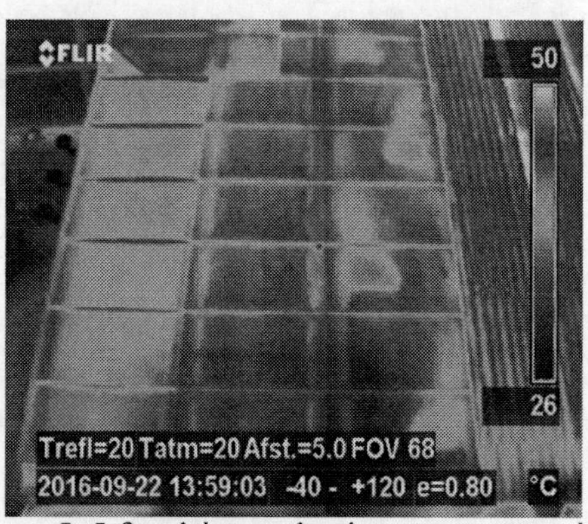

Figure 5: Infrared image showing temperatures in a ventilated BIPV roof

In the second phase of the field test we compared a linear flow with a system with additional fins and a system with a reflecting back layer. We did not find any improvements in the heat transfer. For the additional fins this is most likely due to the manner of installation and possible air pockets.

4.2 Model validation

The thermal BIPVT model has been validated with measurements from the first and second phase fo the field test. The simulation use the measured air temperature as input air temperature. Figure 6 shows the simulated (orange/red) and measured (black) airgap temperatures. The inlet temperature is the black dashed line, while the outlet measured temperature is the continuous black line, which is approximately 2 modules further. The simulated temperature after respectively 1 and 2 modules is shown in red and orange. The visual agreement of measurements and simulations is good.

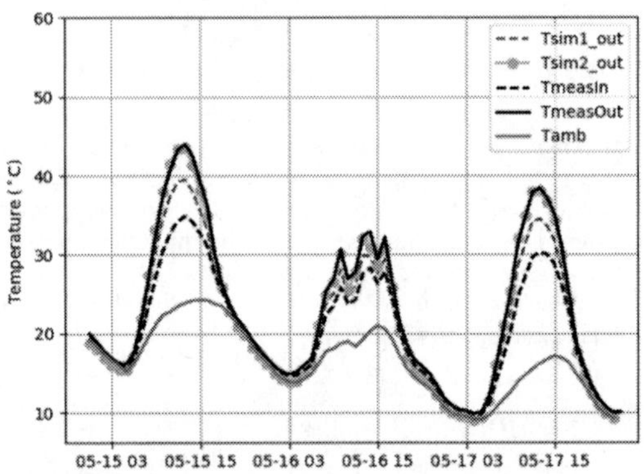

Figure 6: Measured and simulated temperatures in the field test 15 to 17 May 2018. Tmeasin is the measured inlet air temperature under the first module that is also used for simulation inlet temperature, Tmeasout the outlet temperature under the third module. Tsim1_out is the measured outlet temperature after 1 module simulated, Tsim2_out the outlet air temperature after 2 modules.

Table 2 shows a quantitative comparison between simulated and measured air gap temperatures. On average the simulated values are slightly lower than measured values, especially at night.

Tabel 2: Difference and standard deviation between simulations and measurements

	$T_{sim} - T_{meas}$	**Standard deviation** $T_{sim} - T_{meas}$
Phase 1 (CdTe)	-0.75 K	2.31 K
Phase 2 (CIGS)	-0.89 K	1.52 K

5 CONCLUSIONS

We validated a thermal model of a ventilated BIPVT system with a field test. The results show a good agreement between measurements and simulation. The simulation model will be used together with a model for a ventilation heat pump and

6 ACKNOWLEDGEMENTS

This work is supported by the 'Netherlands Enterprise Agency (RVO) and the Dutch Topteam Energy via the project: 'Groene Combi' TKI ZEGO with grant number TEZG214008

7 REFERENCES

[1] R. Valckenborg et al., The BIPV Research Facility 'SolarBEAT' in the Netherlands, 29th EUPVSEC, Amsterdam, The Netherlands, September 2014

[2] K. Sinapis et al, The Glass-Glass Aesthetic Energy Roof: Thermal Behaviour for various Ventilation Levels, 28th EUPVSEC Paris, the Netherlands, 2013

6.4 BUILDING, INFRASTRUCTURE, LANDSCAPE AND OTHER APPLICATIONS OF PV

FINDING THE MOST SUITABLE PV TECHNOLOGY FOR A ZIGZAG-STRUCTURED PV FAÇADE IN NW-EUROPE

Roland Valckenborg[1]*, Shreeganesh Sasidharan[1,2], Chris Tzikas[1], Rudi Santbergen[2], Wim van de Wall[3], Wiep Folkerts[1]
[1] Solar Energy Application Centre (SEAC), High Tech Campus 21, NL-5656 AE Eindhoven, The Netherlands
[2] Delft University of Technology, Delft, The Netherlands
[3]Wallvision, Deken de Bruijnplein 3, NL-5591 CZ Heeze, The Netherlands

*Corresponding author: valckenborg@seac.cc; Tel: +31 652803729

ABSTRACT: Choosing the most suitable PV technology for a ZigZag-structured façade cannot be based on one aspect, but involves four important aspects. First, a maximal energy density (Wp/m^2 panel) of the PV-panel itself is preferred. Secondly, the amount of irradiance from the reflective panels onto the PV-panels should be maximized. However that could lead to a non-uniform irradiance distribution because of self-shading. The development of a custom made irradiance model has turned out to be crucial to predict full year irradiance and energy yield. This model is validated with an experimental setup for three different angles of reflective panels. In this setup, not only irradiance is measured, but also electrical (specific) yield Y [kWh/kWp]. The graphical representation of Y and PR in the IEC 61853-matrix reveals clearly how local climate conditions have an influence on the full year result. Thirdly, the resulting yearly specific yield [kWh/kWp/yr] is depending on the way that the PV-panel can deal with the non-uniformities. Finally, by introducing a new parameter called 'effective façade PV-coverage' [kWp/m^2 façade] the results of specific yield can be modeled, measured and presented in 'effective façade yield' [kWh/m^2 façade]. This parameter is directly of use for project developers, architects, contractors and other stakeholders in the building industry. The full methodology of this paper could in principle be applied not only to ZigZag-structured façades, but also to other complex PV-building skins with any orientation at any location in the world.

Keywords: BIPV, CIGS, Design, Energy Performance, Façade, NZEB, PV System, Solar Radiation

1 INTRODUCTION

The concept of a ZigZag-structured façade with PV-elements has been proposed and built already some time ago [1]. Some variations of the concept have been built and analyzed [2][3]. A novel application of this concept is the Dutch ZigZag-solar architectural façade developed in 2016, see Figure 1 for an example commercial installation in the city Sittard, the Netherlands. The novelty is the emphasis on the aesthetics rather than on maximizing electrical performance.

Figure 1: Commercial application of a ZigZag-structured façade in city Sittard, the Netherlands. Customer has chosen for white decorative/reflective panels to maximize PV yield; the option of a logo or other print was not preferred in this case.

The system uses a combination of a PV system tilted towards the sun and decorative panels facing the observer, see Figure 2-left. In the same figure on the right, one can see the view point from ground level, with the designed effect that the PV panels remain hidden while the decorative panels determine the appearance of the façade. The decorative panels can be designed according to the architectural requirements. Moreover, the decorative panels act as reflecting panels that mirror a part of the incoming irradiance onto the PV-part. In the remainder of this paper we focus on the PV system performance of the system, and therefore we will use the term 'reflective panels' for the decorative panels.

Figure 2: Left: sketch of the ZigZag-cassette geometry. Right: view point from ground level: the PV-panels are hidden, and the decorative/reflective panels show a custom-made logo.

By default the ZigZag-solar system is delivered with custom-sized crystalline silicon (c-Si) PV panels. There are multiple arguments, though, for choosing an alternative PV-technology. Therefore the main research question of this paper is to find the most suitable PV-technology for the ZigZag-system, with the constraint of the climate conditions in NW-Europe. Four sub research topics are addressed in this paper:

1. How to maximize energy density?
2. How to model yield in this complex system? Validated with measurement results of one experimental set-up in 3 different geometries, and 2 different PV-technologies

3. How to understand under which weather condition which amount of specific yield is produced, by adopting the IEC61853-matrix method?
4. How to mitigate for self-shading effects?

This BIPV-system feeds into the grid of the building with AC electricity. Therefore the DC yield coming from the ZigZag-cassettes needs to be converted to AC. Within the project all kinds of power electronics have been investigated and advised according to system size. However, it is beyond the scope of this article to include the results. As a rule of thumb, it is always possible to convert DC yield into AC yield with an efficiency of η_{EU} =92% or higher. Only for very small system sizes (available façade<5 m²), it can become difficult to find power electronics doing the job decently.

2 REALISTIC ENERGY DENSITY [kWp/m² façade]

The design only allows for PV-panels on the plateaus in the ZigZag-cassette. Many geometrical variations are possible, but which version gives the best realistic energy density defined as [kWp/m² façade]? Previous model efforts have proven that the amount of incident irradiance is rather independent of module tilt. Too flat will not benefit from reflected irradiance of the reflective panels. Too steep will cause that the PV-panels become visible from ground level. The compromise was found at a tilt of 32°. The 'depth' of the PV-panels is also adjustable. A 'depth' of about 350 mm has been chosen because it allows for 2 rows of c-Si cells. Finally the angle of the reflective panel, α_r, should be chosen. Too steep will make the façade being covered with mainly decorative panels. Too flat will give a huge amount of self-shading when the solar elevation is above α_r. In Figure x, the variation of α_r in the experimental setup is shown. Consecutively 3 phases have been built with a reflective panel angle α_r of 62°, 55°, 45°. Each geometry has been analyzed for at least half a year from lowest solar elevation to highest solar elevation to include all possible seasonal effects.

Table 1: Effective façade area.

phase	α_r [°]	h [mm]	effective coverage
1	62	731	49%
2	55	610	58%
3	45	488	73%

In phase 1 with α_r = 62°, absolutely no self-shading is occurring because the highest solar elevation in the location of the experimental setup is equal to this value. Drawback from an energy density point of view, is that only roughly half of the façade area is covered with PV, as can be seen from figure 3 and Table 1. In phase 2 the reflective panels have been made more steep to investigate shading effects. This trend has been prolonged in phase 3 to α_r = 45°, giving huge shading effect which will be discussed in chapter 4. Each m² of façade is now effectively occupied by 73% m² of PV on average. One should realize that vertical PV-systems have an effective coverage between 90% and 95% depending on the thickness of the frame; 100% coverage is practically impossible.

For PV-systems that are installed in free fields on cheap land, or other application areas where m² are not too expensive, system evaluation is often expressed in specific yield [kWh/kWp] and rated power [kWp]. For façades in highly populated areas with a huge electricity demand the square meters become rather essential for high-rise buildings to comply with the new European regulations [6]. For these façades we propose to express the BIPV-system label into [kWp/m² façade] in which the area includes the effective coverage. In combination with specific yield [kWh/kWp] the system performance will give a realistic prediction of electrical yield for complete façades.

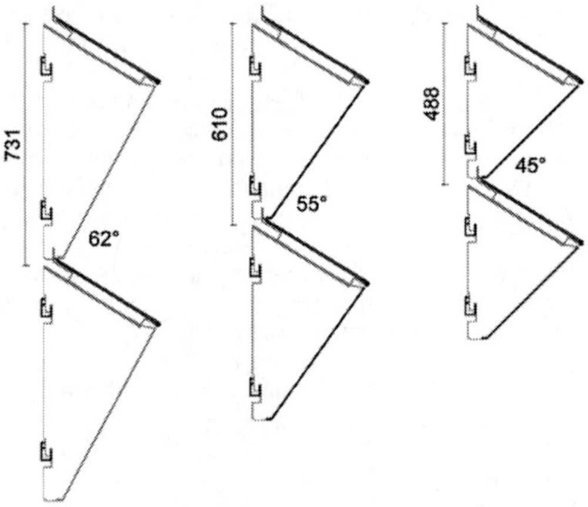

Figure 3: Side view for three different geometries of a ZigZag-cassette.

3 MODELLING AND VALIDATION

3.1 Irradiance model

The irradiance model of the ZigZag-cassette has been developed in-house. It has been introduced in [4][5], and has been sophisticated ever since. At the moment, the time resolution and spatial resolution, as well as the geometry, are highly adjustable. This allows for modelling the irradiance at the precise location of pyranometers and photodiodes in the experimental set-up.

3.2 Experimental Set-up

An experimental setup with 3 rows of 3 ZigZag-cassettes has been built at SEAC-SolarBEAT outdoor research facility [7] in Eindhoven, the Netherlands, facing perfectly South to make analysis easier, see Figure 4. The correlation between photodiodes and a secondary standard pyranometer reading has been investigated in the same facility, in the same period, at just 10 meters away from the ZigZag-façade. Despite the general fact that photodiodes are less accurate than secondary standard pyranometers, for our purpose the accuracy of the photodiodes is good enough. The spatial occupancy of the photodiode (a dot with diameter 22 mm) is much smaller than that of a pyranometer, which enables us to measure the local non-uniformity of irradiance on this small length scale. Especially in phase 3 with a lot of self-shading this is of major importance. The most crucial direction is in the depth of the cassette. Therefore 4 photodiodes have been positioned in this direction as close as possible to the middle (eastern) side of the most western cassette, as this is a position that experiences the most typical shading behavior.

The electrical parameter are measured in the following way:

- Top row: 3 c-Si 22 cells panels are series connected and fed into a micro-inverter. String voltage and current are measured on the DC side of the micro every minute.
- Middle row: 1 c-Si 22 cells panel is connected to an IV-curve tracer that measures every 10 minutes. This panel has one additional bypass diode between the inner 11 cells and the outer 11 cells.
- Bottom row: 2 CIGS panels are parallel connected and fed into a micro-inverter. String voltage and current are measured on the DC side of the micro every minute.

3.3 Irradiance Measurement Results

Phase 1 and phase 2 measurement results (not depicted here) show quite uniform irradiance on the panel, as expected from the fact that solar elevation has never been above the reflector panel; no self-shading occurred in these phases. Phase 3 measurement results show severe non-uniform irradiance and are most important both from an academic point of view and for the understanding of the yield of a realistic ZigZag-system. See Figure 5 for the irradiance for a random clear-sky day with high solar elevation.

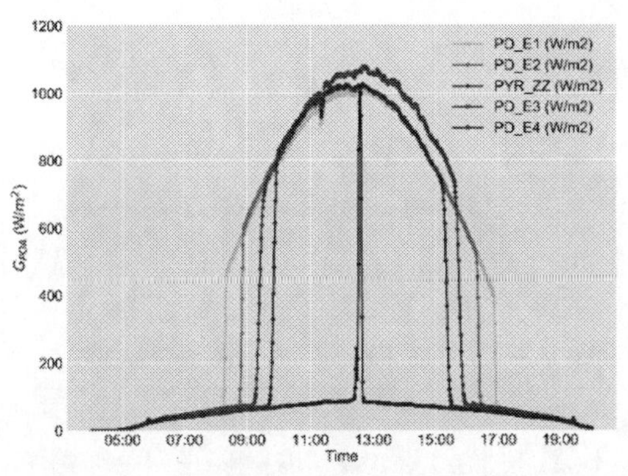

Figure 5: Pyranometer (purple) and photodiode (four shades of green) irradiance measurements on 6th May 2018.

The daily summed irradiance clearly shows how the non-uniformity progresses as the sun climbs higher throughout spring (see Figure 6) and decreases throughout summer (see figure 7).

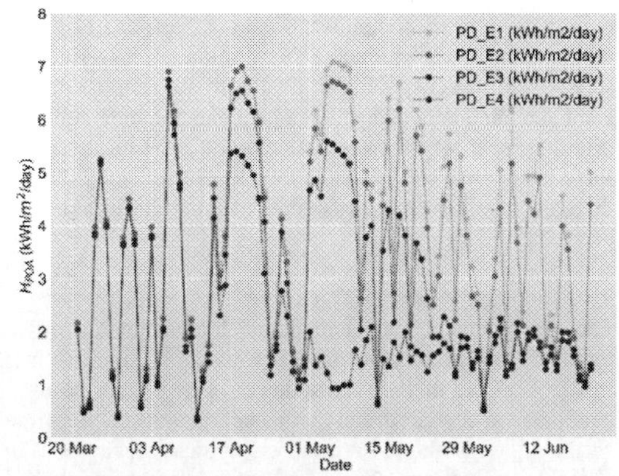

Figure 6: Four photodiode irradiance measurements for period 21st March 2018 till 21st June 2018.

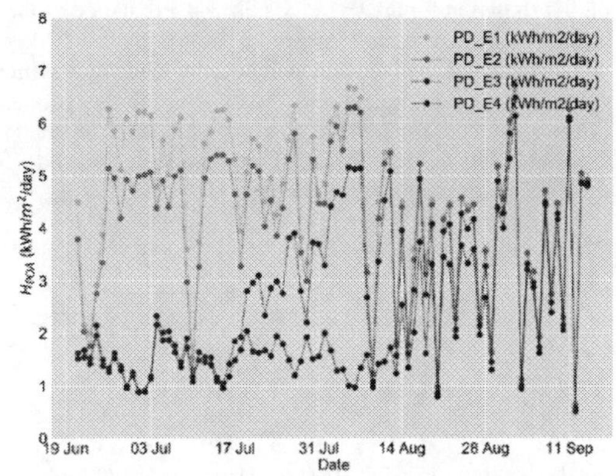

Figure 7: Four photodiode irradiance measurements for period 21st June 2018 till 21st September 2018.

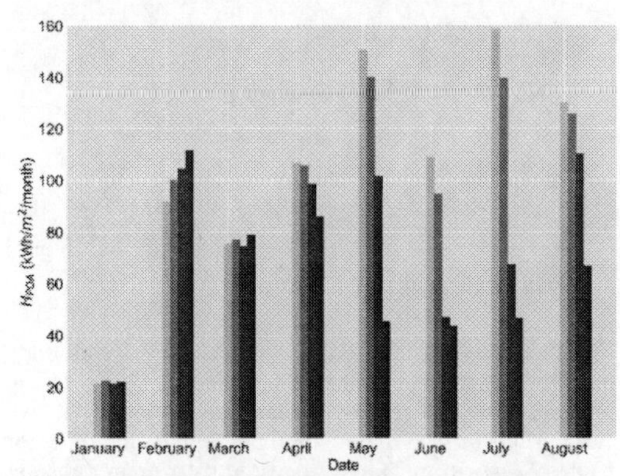

Figure 8: Monthly cumulative irradiance for each photodiode for period 1st January 2018 till 31st August 2018. PD_E1 (lightest green) till PD_E4 (darkest green)

Figure 8 shows the monthly cumulative irradiance for each photodiode. In the spring and summer period we see that the most inner photodiode (E4) gets low amounts of irradiance whereas the outer photodiode (E1) experiences a real good boost. E.g. in July this sum is 158 kWh/m², which shows the significant boost, when comparing to the sum of the vertical irradiance in that same month, 107 kWh/m². Another interesting observation can be made for February 2018, in which the inner photodiode receives the most irradiance due to the lower sun elevation. In combination with complete absence of self-shading this gives the opposite irradiance distribution compared to the spring and summer period. Last but not least, January 2018 was a Dutch record lowest irradiance ever with no direct light at all. Therefore all photodiodes measure the same amount of diffuse light for this month (just a small difference in hemisphere coverage which is too small to observer).

3.4 Irradiance Validation

The measurement results of phase 3 have been used to validate the irradiance model with a 1 minute time resolution. The modeled irradiance follows the measured irradiance quite accurately. Typical deviations in time shifts are less than 10 minutes; typical deviations in absolute value are 25 W/m^2 with some occasional exceptions of 100 W/m^2. Given the tolerances in the built setup, this is acceptable. An overview validation scatter plot is shown in Figure 9. Daily summed irradiance of the model is plotted against measured for all days of May 2018. This is done for each photodiode, using the same color code as Figure 6 and 7. In general the model follows the measurements to an acceptable level. It seems that modeling the low levels of irradiance in the nod of the ZigZag cassette is the most difficult.

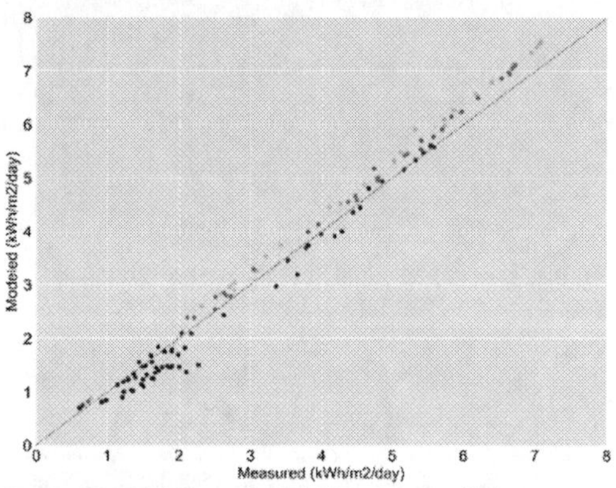

Figure 9: Validation of modelling PD_E1 (lightest green) till PD_E4 (darkest green) for all days of Ma 2018.

4 SPECIFIC YIELD ANALYSIS

For understanding the specific yield of a PV technology, one must exclude self-shading as this complicates the understanding of results. Therefore, in phase 2, we 'waited' for the sun to not climb higher than the reflector angle of our set-up. On 1st August at solar noon the solar elevation in the Netherlands is 56°, and the self-shading just hits the metal structure, but not the PV-panel itself any longer. Therefore the period of analysis has been taken from that day till the day with about lowest solar elevation. As introduced in chapter x, we only have to analyze specific yield in classic terms of kWh/kWp; the translation to realistic yields follows from static conversion constants. Measuring the system performance has be done according to IEC 61724:2017 [8], with a measurement frequency of 1 minute. Analyzing this performance is done by plotting all 15 minute-averaged values of energy yield with the IEC 61853-matrix method, in which the total yearly kWh/kWp is color-plotted with irradiance in the ZigZag-cassette binned on the x-axis and PV-panel temperature binned on the y-axis. It has been experienced that a higher number of bins than prescribed by the norm for both x-axis and y-axis make the interpretation of the measurement results easier; see Figure 10. Still the direct comparison between the two PV technologies is difficult because the color scale is covering the full range between very low irradiance (just above noise level) till very high irradiance. For a better comparison, the PR of each bin is calculated, see Figure 11. Now some significant differences between c-Si and CIGS become visible. First, for the bin of lowest irradiance (G_{POA} < 100 W/m^2) a lot of measurements are done early morning and late afternoon, which experience a triangular shading effect from the sides of the façade. Secondly and most prominent for high irradiance and high temperature the CIGS-panel performs better than the c-Si-panel, which can be explained from the lower temperature coefficient of this CIGS (γ=-0.38%/°C) compared to the c-Si (γ=-0.43%/°C). Finally for low irradiance, somehow surprisingly the c-Si-panel performs better than the CIGS-panel. This can only be understood in combination with full IV-curves; see Figure 12. The voltage of the c-Si is quite independent of the irradiance, but the voltage of the CIGS is dependent of irradiance. For low irradiance, the V_{OC} and V_MPP are significantly lower, and cause a lower P_MPP than expected. A solid explanation for this behavior has not been found yet; it could be related to the internal resistances (R_{series} and R_{shunt}) which become apparently more dominant in the CIGS-panel under low-light conditions.

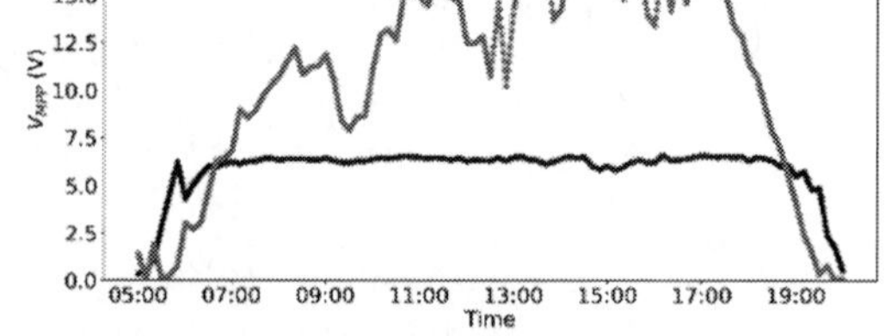

Figure 12: V_MPP for c-Si and CIGS-panel in the middle row of the experimental setup, connected to an IV-tracer; data for 6th September 2018.

5 SELF-SHADING

5.1 Effect of self-shading in IV-curve

Self-shading is observed 24/7 by monitoring the façade with a camera, that has perfect time synchronization with all the other measurement equipment and weather station of the outdoor research facility. However, to quantify the effect of shading on electrical performance, additional IV-curves are measured. The middle row of the experimental setup is equipped with a c-Si panel on the left and a CIGS panel on the right. Moreover, the c-Si panel has a special modification with an additional bypass diode. By observing IV-curves of shaded periods, we see the bypass diode of the c-Si panel kicking in just at the moment one row of cells gets shaded by more than half of the cell. See e.g. 5th May 2018 (Figure 13) when this happens at 8:00. Then at 10:30, self-shading has progressed so far that the full panel is uniform shaded and then the bypass is out. In the afternoon, the reverse process takes place. Of course, I_MPP and V_MPP of that panel flip towards new MPP immediately for every step in this daily process.

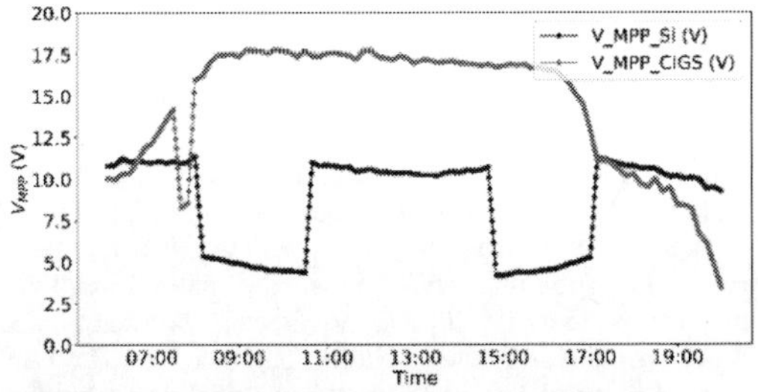

Figure 13: V_MPP for c-Si and CIGS-panel in the middle row of the experimental setup, connected to an IV-tracer; data for 5th May 2018.

5.2 Correlation P_MPP with local irradiance

After having studied I_MPP and V_MPP individually, the P_MPP should be linear dependent on irradiance under the constraint of taking the correct irradiance measurement. Figure 14 shows Y (=P_MPP/Rp) for c-Si as function of average irradiance in the cassette (average of PD_E1 till PD_E4). The correlation fails especially for day numbers around 60 till 120 which represent the two month period around the 21st June, with the highest solar elevation. Figure 15 shows the same P_MPP as function of the average irradiance in the cassette (average of PD E3 and PD E4).

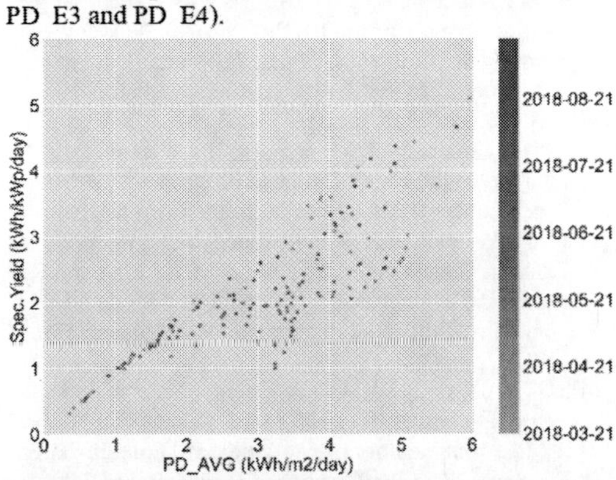

Figure 14: Y_MPP for c-Si as function of average of all photodiodes (PD_E1 till PD_E4). Color represents date.

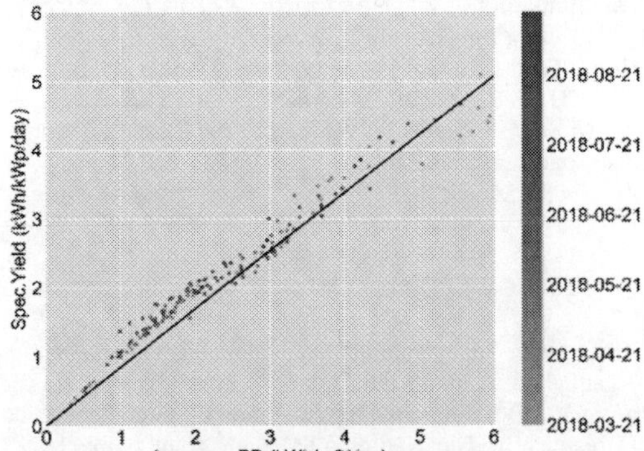

Figure 15: Y_MPP for c-Si as function of average of inner row PD's (PD_E3 and PD_E4). Color represents date. Straight line has slope corresponding to DC_PR* = 84.5%.

Figures 14 and 15 clearly demonstrate the linear dependence of Y_MPP with the lowest irradiance in the non-uniform irradiance distribution in the panel. The same correlation investigation has been done for CIGS, see Figures 16 and 17.

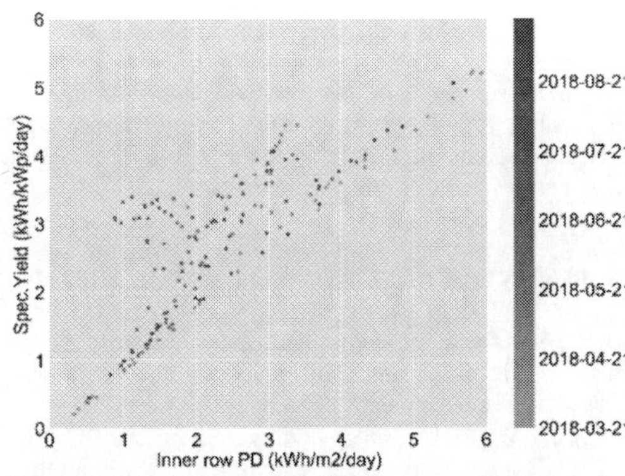

Figure 16: Y_MPP for CIGS as function of average of inner PD's (PD_E3 and PD_E4). Color represents date.

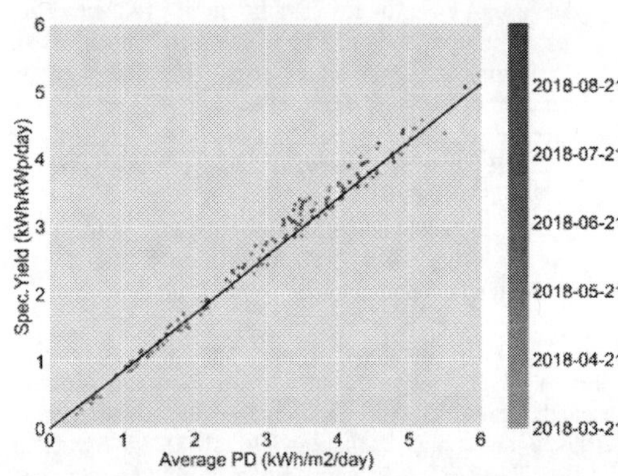

Figure 17: Y_MPP for CIGS as function of average of all photodiodes (PD_E1 till PD_E4). Color represents date. Straight line has slope corresponding to DC_PR* = 85.0%.

For CIGS, the opposite behavior is seen. The best correlation is with average irradiance and not with lowest irradiance of the inner PD's. This behavior has been observed in shading-experiments in the lab before the outdoor measurements (of phase 3) started [9]. Nevertheless the product developers are happy to see the effect now also in realistic outdoor settings, because it proves that this type of CIGS-panels will only loose power proportional to irradiance loss caused by shading.

After these visual proofs of correlation, a full matrix of all possible correlations has been calculated. The best predictor for c-Si turns out to be the average of the two most inward PD's (PD_E3 and PD_E4) with an R^2=0.987. The second best correlation would be with either only PD_E3 (R^2=0.911) or PD_E4 (R^2=0.918). For CIGS, the best predictor power turns out to be the average of all four photodiodes with an R^2=0.992.

Finally, the DC_PR* is actually the slope of Figure X and Figure Y. The additional symbol * is put to remind that not the pyranometer value can be taken. For c-Si performance the lowest amount of irradiance in the non-uniform distribution needs to be taken. Whereas for the CIGS-panels the average amount of irradiance is relevant. The straight line included in Figure 15 has been set on a slope of 84.5% for c-Si. And for CIGS (Figure 17) set to 85.0% as has been retrieved from the PV technology specific yield assessment during phase 2. One can see that for c-Si the data points for the low-light days is actually steeper than this line, which corresponds to the matrix of Figure 11: c-Si has good low-light performance and slightly less PR for high-light and high temperatures.

6 FULL YEAR REALISTIC ELECTRICAL YIELD

All aspects for a ZigZag-system performance prediction have been introduced. In practice, the following procedure can be applied:

- A chosen PV technology will give a certain DC_PR* [%] determined from experiments or modelling efforts.
- Non-uniform irradiance can be measured or modeled. In case of c-Si, the lowest amount of irradiance each time interval should be summed to a H_{POA} [kWh/m²/year]; in case of CIGS, the average amount of irradiance is relevant.
- Specific Yield can be calculated by multiplying DC_PR* with relevant H_{POA} to get Y [kWh/kWp/year]
- Finally, the realistic yield is obtained by multiplying the specific yield Y with the effective façade coverage [kWp/m² façade].

Phase 1 had a duration of one full year. A vertical 'normal' CIGS reference façade on the same outdoor research facility (in Eindhoven, the Netherlands) retrieved an annual realistic yield of 81 kWh/year/m² façade. The ZigZag-system with α_r = 62° retrieved an annual realistic yield of 70 kWh/year/m² façade for that same year. This is somehow lower than the vertical reference system, but one has to remind that not yet the highest energy density was chosen in that geometry. Therefore the authors are very anxious to wait for February 2019 when phase 3 will have one full year outdoor data with the steeper α_r = 45°. Preliminary modeling results show that a close finish is expected for the 'competition' between the c-Si and the CIGS system: the c-Si panels come with a higher power density [kWp/m²], but the CIGS-panels have a much better shade response which is especially beneficial in this BIPV-system with severe self-shading.

7 CONCLUSIONS

In conclusion, choosing the most suitable PV technology for a ZigZag-structured façade cannot be based on one aspect, but involves all four aspects as discussed above. Finding the best compromise is essential. The development of a custom made irradiance model has turned out to be crucial to predict yearly energy yield. Moreover, the graphical representation of kWh/kWp and PR in the IEC 61853-matrix reveals clearly how climate is having an influence on the full year result. Because of the ZigZag-structure, additional self-shading is introduced. Of course, the product developers design the ZigZag-geometry in such a way that the positive effect of the extra reflecting irradiance is much more dominant than the negative effect of self-shading. Moreover, by applying CIGS-panels with an excellent shade-response, independent of shading direction, the effect of shading can be limited. Finally, by introducing a new parameter called 'effective façade coverage' [kWp/m² façade] the results of electrical yield can be modeled, measured and presented in the unit [kWh/m² façade] that is directly of use for project developers, architects, contractors and other persons of the building industry. The full methodology of this paper could in principle be applied not only to ZigZag-structured façades, but also to other complex PV-building skins with any orientation of installation at any location in the world.

8 REFERENCES

[1] I.B. Hagemann, 'Gebäudeintegrierte Photovoltaik', Rudolf Müller (2002), pp. 373

[2] T. Selke, W. Wiedermann, 'A reality check by seven years of monitored energy system performance', 11th EuroSun (2016), doi:10.18086/eurosun.2016.01.03, http://proceedings.ises.org/?mode=list&conference=eurosun2016

[3] A. Schneider, T. Selke, 'BIPV as a Key Technology – On the example of the ENERGYbase', AIT-conf. Shaping the future (14-09-2017) by architect Ursula Schneider of POS Architects, Vienna Austria

[4] R. Valckenborg, R. Loonen, X. Xin, X., D. Verduijn, W. Folkerts and J.L.M. Hensen, 'Performance analysis of the ZigZagSolar BIPV façade system', Proceedings of Advanced Building Skins (2016).

[5] W. Folkerts, R.M.E. Valckenborg, M. van den Donker, C. de Keizer, C. Tzikas, M. de Jong and K. Sinapis, 'Performance assessment of various BIPV concepts', Proceedings of Advanced Building Skins (2017), pp. 1378-1388

[6] EU-Directive 2010/31: The Energy Performance of Buildings Directive (2010/31/EU) recast Oct 2010

[7] R.M.E. Valckenborg, A. de Vries, W. Folkerts, G.P.J. Verbong, 'The BIPV research facility SolarBEAT in the Netherlands', 6AV4.25, EU-PVSEC (2014), pp. 3707-3713

[8] Photovoltaic system performance - Part 1: Monitoring, IEC 61724-1:2017

[9] C. Tzikas, G. Gómez, M. van den Donker, K. Bakker, A.H.M. Smets, W. Folkerts, 'Do Thin Film PV Modules Offer an Advantage Under Partial Shading Conditions?', 5BV4.21, EU-PVSEC (2017), pp. 1593-1596

9 ACKNOWLEDGEMENT

This work is supported by 'Netherlands Enterprise Agency' (RVO) and the Dutch Topteam Energy via the project: 'Building Integrated PhotoVoltaic Panels on Demand - in The Netherlands' with grant number TKI iDEEGO TEID215005.

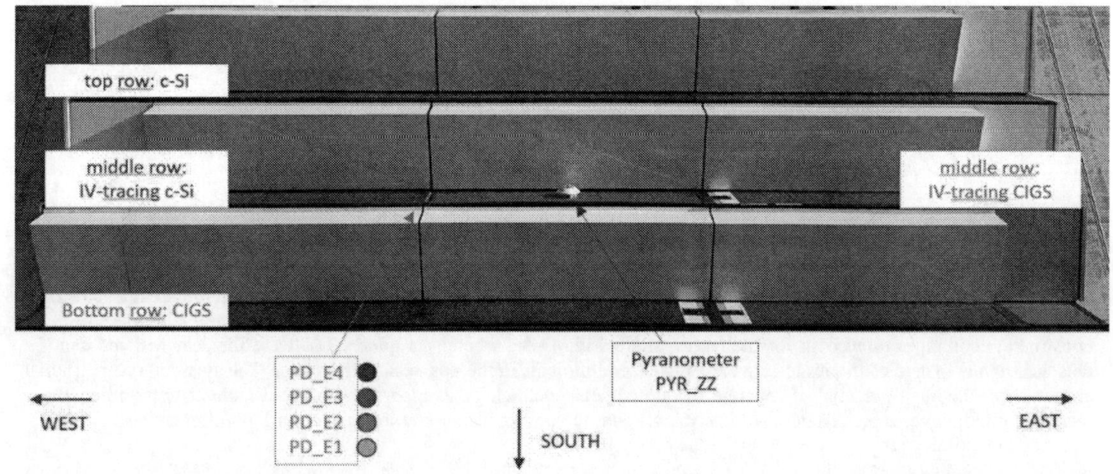

Figure 4: Overview of experimental setup with 3 rows of 3 ZigZag-cassettes. Top row: 3 x c-Si panels with micro-inverter. Middle row: single c-Si (west side) and single CIGS (east side) on IV-tracer. Bottom row: 2 x CIGS panels with micro-inverter. Four PhotoDiode (PD) in a column configuration.

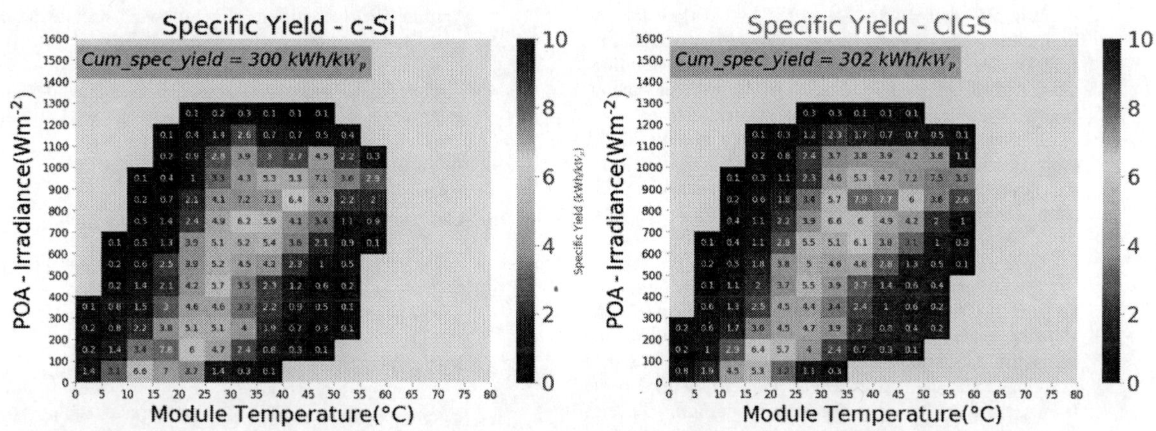

Figure 10: Specific yield as function of Module Temperature and G_POA in a color plot based on the IEC 61853-matrix (with more bins on x-axis and y-axis) for c-Si (left plot) and CIGS (right plot); phase 2 measurement period from 1st August 2017 till 11th December 2017.

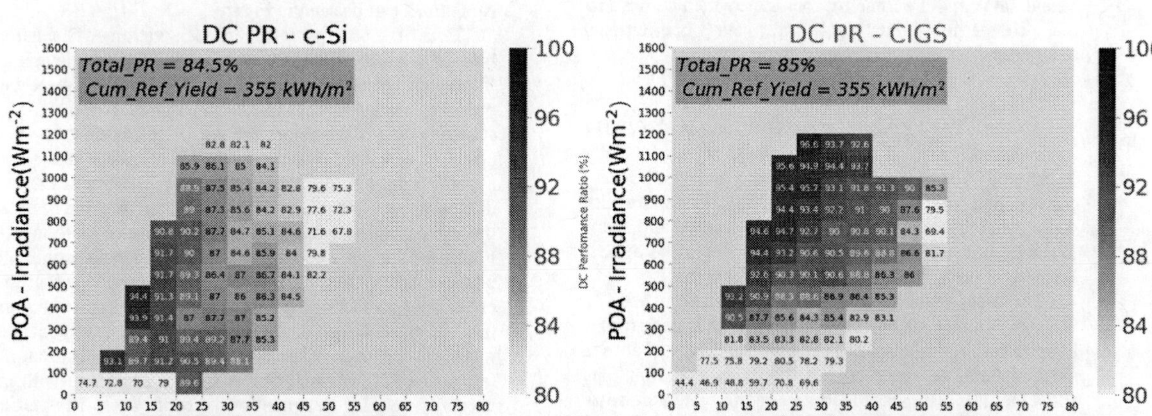

Figure 11: Module Performance Ratio (DC PR) as function of Module Temperature and G_POA in a color plot based on the IEC 61853-matrix (with more bins on x-axis and y-axis) for c-Si (left plot) and CIGS (right plot); phase 2 measurement period from 1st August 2017 till 11th December 2017.

HIGH QUALITY SOLUTIONS OF BUILDING-INTEGRATED PHOTOVOLTAICS (BIPV) – RESULTS OF THE WORLD WIDE COMPETITION IN 2017

Gerd Becker, Fabian Flade, Roland Krippner, Bruno Schiebelsberger, Walter Weber
Bavarian Association for the Promotion of Solar Energy, Friedrich List Strasse 88, 81377 München, Germany
Phone: +49 (0) 89/278134-28, info@sev-bayern.de, www.sev-bayern.de

ABSTRACT: With its award for Building Integrated Solar Technology, the Bavarian Association for the Promotion of Solar Energy honours outstanding contributions to the planning and design of building-integrated solar systems in high-quality architecture. 134 projects from 26 countries were initially registered for the 2017 Architecture Award, 119 projects were presented to the jury. The German-speaking region of the entries makes up roughly 77 %. However, there is a growing number of participants especially from Scandinavia and the Benelux countries. Within the 117 projects evaluated, housing construction continues to account for the largest share with 43 %, whereby a good 60 % were for detached and semi-detached homes. There continues to be a clear increase in projects in the non-residential sector. The awarded competition entries show the huge potential of coherent integrated solar technology as a part of advanced architectural and energy technical building concepts. There were many high quality solutions, an extraordinary contribution to the task field in constructing buildings in rural areas has won the first prize.
Keywords: Building Integrated PV (BIPV), Façade, Solar Architecture

1 INTRODUCTION

With its award, the Bavarian Association for the Promotion of Solar Energy honours outstanding contributions to the planning and design of building-integrated solar systems and raises public awareness about exemplary solutions with high-quality architecture.

The number of participants and the quality of the projects entering the competition, which has been held for the seventh time since 2000, are testimony to the continuing topicality of the theme [1] [2].

1.1 Countries

134 projects from 26 countries were initially registered for the 2017 Architecture Award. 119 projects - that actually entered the competition and fulfilled the competition criteria - were presented to the jury. A closer examination of the entries reveals that only a third come from Germany. Owing to the high proportion of entries from Switzerland and Austria, the German-speaking region nevertheless makes up roughly 77 %. However, a gratifying aspect is the growing number of participants from non-German-speaking countries, especially from Scandinavia and the Benelux countries, which shows that the competition continues to gain international recognition.

1.2 Building Typology

A total of 117 schemes were evaluated in the „2017 Architectural Award for Building-integrated Solar Technology“ competition. Within these projects, housing construction continues to account for the largest share with 43 %, whereby a good 60 % were for detached and semi-detached homes. However, there continues to be a clear increase in projects in the non-residential sector.

The other building types mainly comprise office (21 %), educational (9 %), event (8 %) and industrial / commercial buildings (6 %). The vast majority of entries are new-build schemes, even if the renovation examples show that it is also possible to successfully integrate solar technology into existing roofs and facades. In addition, the entries confirm the continued focus on photovoltaic systems. There are a number of combined solutions in the (small-scale) residential construction field, while pure solar-thermal roof and/or facade systems are the exception.

2 ASSESSMENT CRITERIA AND PRIZES

Within all competitions the following architectural, technical and other criteria had to be fulfilled:

- Architectural aspects (Integral conceptual design, integration of the solar technology in the building envelope, design quality and functionality)
- Innovativeness (energy and construction)
- Energy yield/energy efficiency
- Communicating the system to the general public

Within the competition a first prize and special recognition awards were given. The total prize money amounted to € 25.000.

3 FIRST PRIZE

The first prize - endowed with € 15.000 – went to "Hof 8" in Schäftersheim (Germany),

The construction of buildings in rural areas is characterized by numerous challenges. An extraordinary contribution to this task field in constructing buildings in rural areas can be seen in Figure 1 + 2.

Here the energy efficient renovation of existing buildings using renewable materials is just as relevant as the use of renewable energies for decentralized heat and power generation. With its holistic approach, the „Hof 8“ scheme makes an extraordinary contribution to this complex task field, which also needs to take into account aspects relating to inner-community development and demographic change. In addition to a coherent mix of uses, the use of regional products and the reuse of materials to minimize the share of grey energy in the construction, the photovoltaic system (80 kW) is fully integrated across the entire roof surfaces of the various buildings that make up the former farm ensemble. Although the on-roof installation chosen is somewhat conventional in structural terms, the 550 m^2 solar power system impresses with its carefully detailed treatment of the roof edges. The partial surfaces created by the junctions between different roofscapes are harmoniously resolved with colour-matched panels. The combination with existing rubble stone walls and new timber facades also illustrates the design potential offered by standard solar technology.

Figures 1 + 2: Two views of the award-winning BIPV system of "Hof 8" in Schäftersheim (Germany), a former farm ensemble transformed into an energy-plus building by Architekturbüro Klärle, Bad Mergentheim (D)

„Hof 8“ is a self-assured refurbishment scheme in which a groundwater heat pump, battery storage unit and electromobility form intrinsic components of the energy-plus building ensemble‘s integral energy concept, which also incorporates the neighbours in the medium to long term. The scheme provides valuable impetus for sustainable rural development and shows that photovoltaics can be an entirely self-evident construction material in building renovation.

4 SPECIAL RECOGNITION AWARDS

4.1 "Aktiv-Stadthaus" in Frankfurt

One Special Recognition Award - endowed with € 5.000 - was won by "Aktiv-Stadthaus" in Frankfurt, Germany, see Figure 3 + 4.

Energy-plus buildings in densely populated, inner-city environments have been scarce until now. An outstanding project that could act as a trailblazer in this context is the Aktiv-Stadthaus in Frankfurt. On a confined urban site the architect has succeeded in delivering an 8-storey apartment building with 74 residential units to the Efficient House Plus standard on a 160-metre-long and only 9-metre-wide plot. As part of the ambitious energy concept, photovoltaic systems in the roof (251 kW) and facade (118 kW) are combined with electricity storage (250 kWh), heat pumps and buffer storage tanks along with an innovative energy management system to produce surplus energy in the annual energy balance.

Along the elongated, gently undulating southern elevation, monocrystalline anthracite modules, mounted upright with visible fastenings, alternate with the openings and form an effective contrast to the white bands of Eternit panels and the vertical timber cladding.

Figure 3 + 4: Aktiv-Stadthaus, Frankfurt, above: PV system on roof, below: solar façade (Architect: HHS Planer + Architekten, Kassel; EGS-plan, Stuttgart (D))

4.2 Copenhagen International School

Another Special Recognition Award - endowed with € 5.000 -was won by Copenhagen International School, Nordhavn (Figure 4).

The new school building, prominently located on the waterfront in the Nordhavn docklands area, is characterised by the very differentiated structure of the building form, which is reminiscent of stacked ship containers. Projecting and recessed blocks, whose almost square glass panels create a lively elevational treatment in light blue to turquoise green shades, extend above a partially glazed podium zone.

Figure 4: Copenhagen International School, Nordhavn by C.F. Moeller Architects, Kopenhagen

The technically and creatively innovative approach is illustrated in Figure 5: The specially coated cover glass for the monocrystalline modules enables the photovoltaics to retreat into the background and creates novel, tinted solar facades in which vapour-deposited oxide layers reflect only a specific spectral range of sunlight.

Figure 5: Copenhagen International School - PV Cells, each individually angled to create a sequin-like effect

In order to articulate the 6,000 m^2 of surface area, the 12,000 modules in metal frames are mounted at different angles so that the facades – depending on the solar radiation and angle of view – change their hue although the modules all have the same colour.

4.3 Special Recognition Award – Family Home (2016)

The photovoltaic roof of the house in Tamins (Figure 6) provides an exemplary solution for detached and semi-detached family homes. The elongated and narrow building features a clear design vocabulary with a restrained pallet of materials. The south-facing roof surface (108 m2) is entirely covered with monocrystalline PV modules (17.28 kW) on a timber substructure that is visually expressed along the eaves and projecting verges.

The solar technology also harmonises with the slightly lower roof surface covered in corrugated sheet panelling, as well as with the exposed concrete surfaces and the slotted vertical timber cladding.

Figure 6: Family home in Tamins (CH) by Bearth & Deplazes Architekten, Chur

4.4 Special Recognition Award - Tschagguns (AT), Stall B. (2015)

The refurbishment and conversion of a former farm building into a private house features an entirely southeast-facing solar roof on which solar thermal (20 m2) and PV modules (55 m^2) are arranged in clearly differentiated sections with slightly overlapping, equally sized component dimensions (Figure 7). The solar roof provides about 200 % of the electricity and most of the heat requirement. The project shows the design possibilities offered by solar technology, including on existing buildings, and visibly expands Vorarlberg's timber building tradition with a sustainable energy concept.

Figure 7: Family home (2015) in Tschagguns (AT) by Bernhard Breuer Architektur Schruns

4.5 Special Recognition Award - Energy plus homes

Drab-looking 1930s terraced housing has been successfully transformed into attractive energy-plus homes, see Figure 8. Although the construction costs were dictated by the subsidy requirements for social housing in Bavaria, the new elevations with their timber extensions and projecting elements reflect both ecological and aesthetic values. With the exception of a few flush-mounted skylights, the south-facing roof (205 m^2) is entirely covered with 152 PV modules (31.16 kW). This is a self-evident solution, which also impresses in terms of the construction through the carefully detailed work.

Figure 8: Energy plus homes (1938/2015) in Neu-Ulm (D) by o5 Architekten Frankfurt

4.6 Special Recognition Award - Basel (CH), Grosspeter Tower (2017)

The surrounding facade of the office tower is characterised by a clear grid structure with window openings that become wider with increasing height. The opaque facade surfaces (5,000 m^2) consist of CIGS thin-film solar modules (glass – CIGS cell – glass, 440 kW) from the Manz Company. Two to three solar sub-modules are applied to a rear panel consisting of expanded glass granules in order to achieve the necessary stability.

The different dimensions and architectural requirements have necessitated the need for around 450 different types of facade elements, which are attached to the substructure by means of rear-mounted support channels. This creates a homogeneous surface effect that, combined with the use of string optimizers, provides an elegant and high performing alternative to natural stone and metal facades.

Figure 9: Grosspeter Tower (2017) in Basel (CH) by Burckhardt + Partner – Architekten

4.7 Special Recognition Award - St. Trinitatis Church

The new church building is based on an ambitious sustainability concept that ranges from the choice of durable regional materials to the visible solar power system. On the narrow southern facade of the church tower, a dark band (140 m^2) of CIGS thin-film modules (15.6 kW) is integrated precisely and flush above the first floor. The transversely arranged, frameless PV panels with a 2:1 aspect ratio sit well proportioned within the internal surface and form an effective contrast to the stepped facade cladding made of reddish-violet Rochlitz porphyry. The „solar"church tower on the edge of Leipzig's city centre provides an important urban landmark and symbolises the church community's support for solar energy.

Figure 10: St. Trinitatis Church (2015) in Leipzig (D) by Schulz und Schulz Architekten Leipzig

4. SHORTLIST

Furthermore, seven excellent examples of BIPV systems from the shortlist shall be presented by means of their photos in Figures 11 - 18.

Figure 11: Omicron Campus Klaus(AT) (Dietrich | Untertrifaller Architekten)

Figure 12: District Town Hall in Freiburg/Germany (2017) by Ingenhoven Architects, Düsseldorf

Figure 14: sto office building, Stühlingen by Orange Blu Architects, Stuttgart

Figure 15: FKI Tower in Seoul/Korea by Adrian Smith + Gordon Gill Architecture, Chicago

Figure 16: Church in Rif (AT) by Architect Walter Klasz, Innsbruck

Figure 17: NEW Blauhaus in Mönchengladbach (D) by kadawittfeldarchitektur

Figure 18: Multi Family Home in Riedberg Germany by HHS Planer und Architekten

5 CONCLUSION

The competition "Building Integrated Solar Technology" which has been held in 2017 for the seventh time since 2000, shows by means of the number of participants and the quality of the projects the actuality of BIPV. 134 projects from 26 countries were initially registered for the 2017 Architecture Award, 119 projects were presented to the jury. The German-speaking region of the entries makes up roughly 77 %. However, there is a growing number of participants especially from Scandinavia and the Benelux countries. Within the 117 projects evaluated, housing construction continues to account for the largest share with 43 %.

High quality solutions are possible in all fields. Housing with single-family and two-family houses take the biggest share; however, there is still a significant increase in non-residential construction projects.

There were many high quality solutions, an extraordinary contribution to the task field in constructing buildings in rural areas has won the first prize.

6 REFERENCES

[1] Roland Krippner (Hrsg.): Building Integrated Solar Technology; Architectural design with photovoltaics and solar thermal; Detail green books. Munich: DETAIL Business Information GmbH, June/2017)

[2] Gerd Becker, Fabian Flade, Roland Krippner, Bruno Schiebelsberger, Walter Weber; Building Integrated Photovoltaics (BIPV) over the time – illustrated by means of examples from competitions; Proceedings 33rd European Photovoltaic Solar Energy Conference and Exhibition; PP. 2099 - 2104

EXPERIMENTAL PERFORMANCE OF A CURTAIN WALL BIPV ELEMENT UNDER REALISTIC BOUNDARY CONDITIONS

Juliana E. Gonçalves[1,3], Jonathan Lehmann[1,3], Georgi H. Yordanov[2,3], Wout Parys[1,3]*, Kris Baert[2,3], Dirk Saelens[1,3]
[1] KU Leuven, Department of Civil Engineering, Building Physics Section, Kasteelpark Arenberg 40, 3001 Leuven, Belgium
[2] KU Leuven, ESAT-ELECTA, Kasteelpark Arenberg 10, 3001 Leuven, Belgium
[3] EnergyVille, Thor Park 8310, 3600 Genk, Belgium
*Currently at BAUPHI, Building Engineering Office, Stropkaai 55, 9000 Ghent, Belgium

ABSTRACT: Building integrated photovoltaic (BIPV) technology has experienced a significant development over the most recent years. Yet some barriers to its large deployment remain. One of them is the lack of design consideration regarding the installation of BIPV modules. This paper demonstrates the technical feasibility of a curtain wall BIPV concept that steps towards the plug-and-play concept to facilitate the installation process. A prototype of such BIPV concept was constructed and installed in the façade of a real building, considering the technical requirements for building components. The BIPV module has been operating for over a year. Between August 2017 and July 2018, the BIPV module operated with monthly PR-values between 88 and 99%, and an average of 94.2%. The yearly PR was 92.6%. Furthermore, a Finite Element Method (FEM) model was developed in COMSOL to investigate the thermal behaviour of the module. Comparison against experimental data showed that the model is able to predict the BIPV temperature over time. This model will be useful in future studies to (1) optimize the BIPV module design in terms of thermal performance, and (2) investigate the suitability of integrating power electronic components inside the module's frame in order to achieve a fully plug-and-play system.
Keywords: Building Integrated PV (BIPV), Building Integration, Energy Performance, Thermal Performance, Finite Element Method (FEM).

1 INTRODUCTION

Although a significant progress has been achieved in the field of Building Integrated PV (BIPV) systems, the sector is still considered a niche market [1]. One of the reasons for that is the lack of design consideration on how to facilitate BIPV installation, maintenance and replacement [2][3][4]. Considering these requirements, standardized pre-fabricated plug-and-play BIPV elements may be a promising solution to support the development of the BIPV sector.

Plug-and-play BIPV solutions have a two-fold aspect. The first one concerns facilitating the BIPV installations from a construction point-of-view. The second aspect focuses on the electrical connection. For BIPV systems, the use of DC/DC (micro) converters at the module level is being considered as a more suitable solution to cope with shading effects occurring in the built environment [7][8][9], which are especially critical in façade applications. Compared to their array-level counterparts, module-level converters are reduced in size, which in turn brings on the opportunity to achieve a complete plug-and-play design, by integrating the converter into the BIPV frame. However, a converter integrated into the BIPV frame may experience higher operating temperatures, which leads to reduced performance [8][10]. Not only reduced performance is of concern, but also the accelerated degradation and decreased lifetime, and thus reliability [8][9].

Experimental studies in the BIPV field normally evaluate the performance of the module itself, and do not address the effects of the integrating the BIPV element into the building [5]. In the case of building-size demonstrations, the focus is mostly on the energy performance [6], and detailed thermal information such as temperature distribution and dynamics is not available. Studying the thermal behaviour of BIPV systems is essential as it provides information to assess not only the temperature effect on the PV efficiency, but also the module lifetime and degradation.

For the present work, a BIPV curtain wall prototype was designed and manufactured following industrial standards for building envelope components. The prototype integrates a commercial-size PV module into a conventional curtain wall aluminium frame, as depicted in Figures 1 and 2. As a first step towards the plug-and-play concept, wiring and junction boxes are incorporated inside the frame. Figure 1 shows the BIPV prototype installed in the façade of the Vliet test building in Leuven, Belgium, which ensures realistic boundary conditions on both the indoor and outdoor environments. In a real building, the opaque PV module covers the area between two floors, and the window provides natural lighting to the building interior space. In this way, the BIPV system can be aesthetically integrated into office buildings.

Figure 1: BIPV curtain wall prototype integrated in the façade of the Vliet Building

Lateral view Front view

Figure 2: Lateral and front views of the BIPV prototype

This paper has a twofold goal. The first is to present the one-year performance of a real-size BIPV curtain wall demonstrator operating under realistic conditions. The electrical performance of the BIPV element for a typical summer day is analyzed, and a long-term study of the Performance Ratio (PR) is presented. The second goal is to develop a detailed model to represent the thermal behaviour of the BIPV module. A Finite Element Method (FEM) approach is employed for this purpose, using COMSOL software. Temperature results are compared to experimental data. Since the model can provide the temperature evolution over time at any point of the computation domain, it is useful to investigate the suitability of integrating a converter inside the BIPV frame in future studies.

In the following section, the BIPV curtain wall design and the experimental setup are presented. Next, Section 3 brings the results of the energy performance analysis. Section 4 presents the FEM thermal model and the validation results. The last section outlines the main results and the relevance of the work, and provides the direction for future work.

2 BIPV CURTAIN WALL PROTOTYPE

2.1 Design and characteristics

Figure 3 presents the schematic representations of the BIPV curtain wall design. It consists of a full-size floor-to-floor curtain wall element with a built-in 60 cells monocrystalline-Si PV module at the top. The window installed at the bottom of the frame consists of a 4/14/4/15/4 mm Argon-window. Behind the PV module, a 4-cm cavity was designed to promote air circulation. Eight apertures (1-cm diameter) located at each lateral side of the frame allow the ambient air to circulate inside the cavity. The cavity is insulated from the indoor environment by a 16-cm Rockwool layer and a plywood panel. To reduce the heat losses to the environment, insulation was also provided around the frame.

The prototype showed in Figures 1 and 2 was constructed by Belgo Metal CW taking into account the industrial requirements for building envelope components, such as air and rain tightness, thermal insulation, aesthetics, fire resistance, and durability. To ensure realistic operating conditions, the BIPV prototype was installed in the southwest façade of the Vliet test building in Leuven, Belgium (Figure 1).

The glass-glass PV module, manufactured by Soltech, is composed of sixty 6 inches mono c-Si solar cells encapsulated in a front transparent EVA and rear black EVA, and covered on both sides by a 4 mm glass sheet. Each sub-string has been tested and their STC ratings are listed in Table 1. From these measurements, the maximum power output of the module was estimated at 244.1 Wp by Soltech. Note the inferior performance of sub-string 2 resulting from fill-factor (FF) losses due to an increased series resistance. The PV module is 164 cm × 109 cm, and has an efficiency of 13.6% at STC.

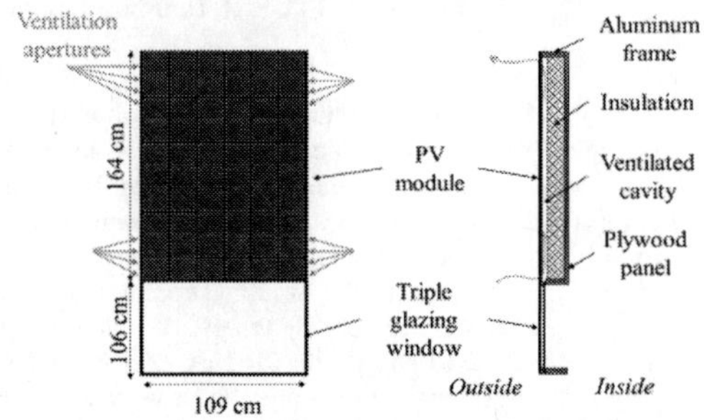

Figure 3: Schematic representation of the BIPV curtain wall design

Table I: PV module specifications

Characteristic	Value/specification
Cell type	Monocrystalline silicon
Cell dimensions	156 mm x 156 mm
Temperature coefficient of power	-0.42 %/K
Number of cells	60
Maximum power	244 W_P
Efficiency at STC	13.6 %
Area	1.79 m^2

2.2 Experimental setup

Figure 4 shows the schematic view of the temperature sensors used to characterize the BIPV thermal behaviour. Twelve type-T thermocouples were attached on the back of module (the insulation and the plywood panel have been omitted for the sake of clarity). The same thermocouples network was attached to the insulation layer inside the cavity and to the plywood layer indicated in Figure 3. The temperatures are recorded every second.

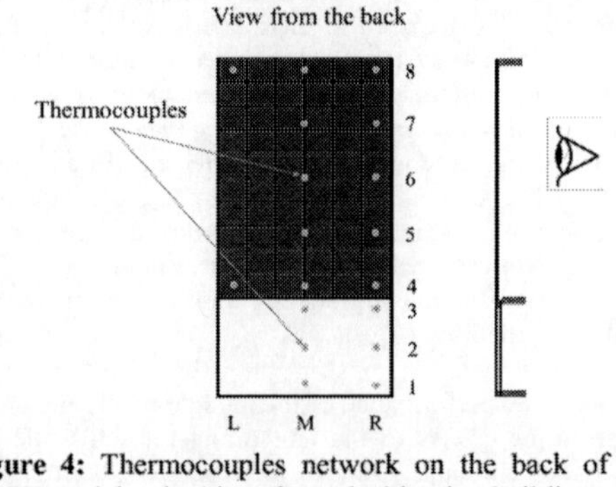

Figure 4: Thermocouples network on the back of the BIPV module, looking from inside the building. For clarity, insulation and plywood panel have been omitted

The electrical measurements were performed using industrial equipment from Femtogrid Energy Solutions. The module operates at maximum power point (MPP) by means of a Femtogrid Solar Power Optimizer PO330, which also increases the voltage to match the 380-V DC bus of the Femtogrid Inverter 2400. Due to the unique BIPV module design, the inverter operates below 10% of the rated input power most of the time, which leads to very low inverter efficiency. Therefore, the AC energy fed into the grid is not considered in this paper, but only the DC energy provided by the BIPV module to the power optimizer. The electrical output of the BIPV module is measured every two seconds.

Finally, two meteorological stations monitor the weather conditions on-site (wind conditions, and ambient temperature), and a pyranometer was installed next to the BIPV element to measure the solar irradiance on the façade.

3 ENERGY PERFORMANCE

Figure 5 presents the solar irradiance and the BIPV module efficiency on a typical summer day. Before 10h, when the module was exposed only to diffuse irradiance, the resulting efficiency is slightly higher, which is a consequence of lower temperatures in the beginning of the day. Between 10h and 18h, when the module was exposed to direct irradiance, the efficiency curve is uniform, varying between 11.1 and 11.7%, which corresponds to a drop of 2-2.5% in relation to the efficiency at STC. For the cells used in this BIPV module, the manufacturer specifies a temperature coefficient of maximum power of -0.42%/K. Using this coefficient, an efficiency decrease of 2.5% leads to a power drop of 46 W, which in turn corresponds to a temperature increase of 45 K.

Furthermore, a drop in the efficiency is observed at the beginning and the end of the daytime period (at around 10h and 18h). This is likely caused by an angular acceptance mismatch between the pyranometer and the BIPV module, with the pyranometer having a better response at flat angles of incidence of direct sunlight.

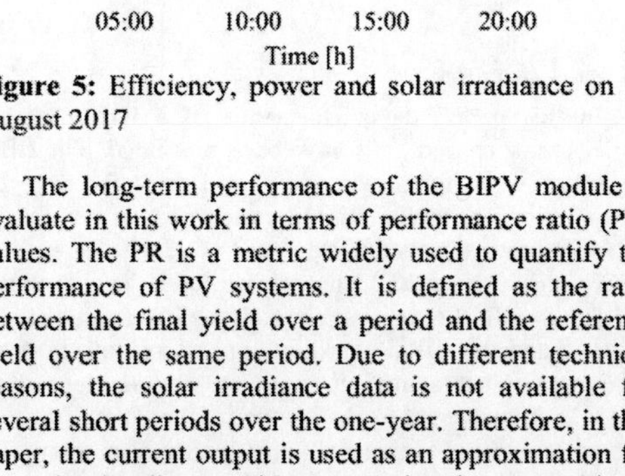

Figure 5: Efficiency, power and solar irradiance on 29 August 2017

The long-term performance of the BIPV module is evaluate in this work in terms of performance ratio (PR) values. The PR is a metric widely used to quantify the performance of PV systems. It is defined as the ratio between the final yield over a period and the reference yield over the same period. Due to different technical reasons, the solar irradiance data is not available for several short periods over the one-year. Therefore, in this paper, the current output is used as an approximation for the solar irradiance. This assumption is reasonable as long as the (BI)PV module does not suffer from shading events. During the first half of August, due to construction work on the same building façade, the BIPV module was operating under slightly shaded conditions. Using the current as an approximation for the irradiance in this case leads to overestimated PR-values. Apart from this period, the BIPV module was operating under non-shading conditions.

Figure 6 presents the monthly PR for the one-year period studied. PR-values between 88 and 99% were obtained, with an average of 94.2%. The BIPV module achieved a yearly PR of 92.6%, slightly below the average. The well-known seasonal effect is also observed, with cold winter months exhibiting higher PR-values compared to summer months. From December to July, the PR decreases by 11%. This situation has been reported in [11] and is a consequence of low temperatures occurring in winter months, when the equivalent temperature of the solar cells can drop below the standard temperature of 25°C, leading to high PR-values, oftentimes superior to 100 %.

For façade BIPV systems, PR values between 0.59 and 0.82 have been reported in the literature [11][12]. Two aspects might explain the slightly higher PR-values reported here. First, in this paper, PR-values were based on the DC energy provided by the BIPV module. This means that the inverter losses are not taken into account. Additional 10% losses would result in a yearly PR of around 83%, still compatible to the results in the literature. Second aspect concerns the BIPV design. The higher PR-values might also be a consequence of lower temperatures resulting from the ventilation effect provided by the cavity and ventilation apertures.

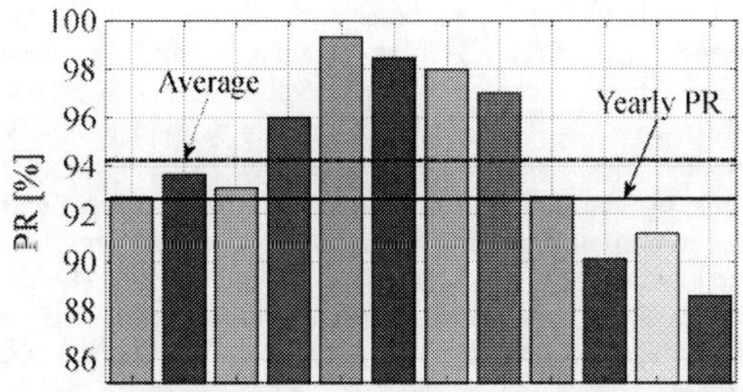

Figure 6: Monthly PR over one year. Results for March have to be considered carefully since data was only available for 13 days

4 FEM MODEL AND THERMAL BEHAVIOUR

4.1 Description of the model

The BIPV module was modelled based on the design presented in Section 2 and on the technical specifications provided by Soltech and Belgo Metal CW. The inputs to the model consist of solar irradiance, ambient temperature, and power produced by the BIPV module, as well as the surface-average temperature measured on the plywood panel, which was imposed as a boundary condition. All input data were measured on site.

Solar irradiance was modelled as a heat source (heat flux) applied to the BIPV module, corresponding to the difference between the amount of energy converted into electricity, i.e. the BIPV produced power, and the amount of solar energy absorbed by the PV module. Because the

PV cells are significantly thinner compared to the glass panels that compose the PV module, the cells were not modelled explicitly. Instead, the heat source was imposed on the surface between the two glass layers composing the BIPV module.

Furthermore, to avoid modelling the complex wind flow around the building, convective boundary conditions were imposed on the BIPV exterior surface. Correlations for natural convection on a vertical wall presented in [14] were employed. The outside environment was considered as far-field temperature boundary condition. Similarly, natural convection occurring inside the cavity was approximated using an enhanced thermal conductivity. The effective thermal conductivity of the air is increased by empirical correlation for vertical rectangular cavities [15][16]. Note that the air domain is still explicitly modelled as a fluid domain, but the airflow fields are not computed and the velocity term is neglected.

Inside the cavity, radiative heat transfer is described by surface-to-surface radiation, which means that the mutual irradiation from the surfaces that can be seen from a particular surface are accounted for. At the outer surfaces of BIPV module, radiation is described by surface-to-ambient radiation, meaning that there is no reflected radiation from the surroundings (i.e. blackbody radiation condition). Long-wave radiation losses to the ground and to the sky were taken into account separately. Finally, adiabatic surfaces were used to represent the insulation around the frame perimeter.

4.2 Validation results

The results presented here focus on a worst-case scenario, where high solar irradiation and high ambient temperature occur simultaneously. Figure 7 shows the weather conditions (solar irradiance and ambient temperature) recorded on two consecutive sunny and warm days in April 2018. In these conditions, the PV module experiences temperatures near 75°C, defining a very critical situation for the electronic components integrated into the BIPV structure. Figure 7 also presents the power produced by the BIPV module, the heat flux on the PV layer, and the average temperature measured on the plywood layer. These are the data used as input to the model.

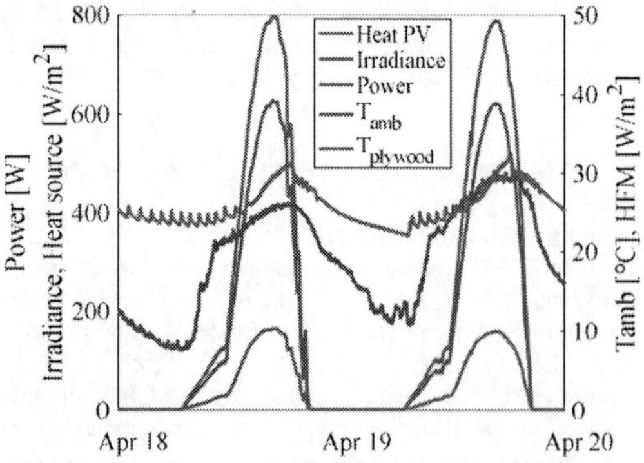

Figure 7: Input data to the model

Figures 8 and 9 compare the average measured temperature of the thermocouples on the right (R) and middle (M) for all heights (4-8) to the predicted temperature at the same positions for the back of the BIPV module and for the insulation layer inside the cavity, respectively. These results indicate that the model is able to predict both the temperature dynamics and the temperature distribution over the BIPV module.

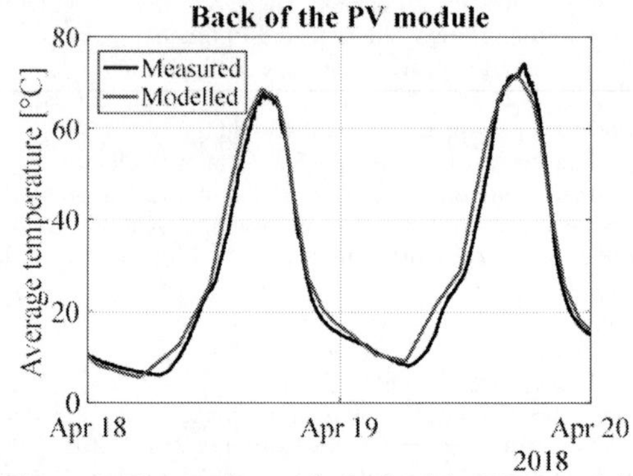

Figure 8: Temperature at the back of the BIPV module: measured vs. modelled

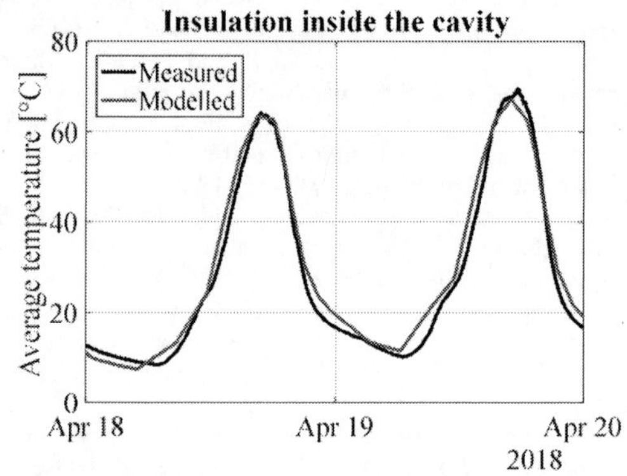

Figure 9: Temperature at the insulation inside the cavity: measured vs. modelled

Figure 10 shows the temperature over the entire BIPV module at solar peak irradiation on the 18th of April. This figure shows that the BIPV temperature is nearly uniform over the surface (at the top, except the glazing part), which is in agreement with the measurements presented in Figure 11. Apart from height 4, where the temperature is slightly lower likely due to ventilation effects, the temperature at heights 5 to 8 is approximately the same.

5 CONCLUSION

In this paper, the performance of a BIPV module under realistic conditions have been presented. The BIPV module is operational over one year. During the first year of operation, the module has operated with an average PR of 94.2%, and achieved a yearly PR of 92.6%.

Furthermore, a FEM model has been developed and validated with the experimental data. The model is able to predicted the BIPV module temperature both in terms of dynamics and spatial distribution. This model will be used in future studies to optimize the BIPV module design regarding its thermal performance, and to support the decision for the most suitable placement for the converter within the BIPV frame in terms of operating temperatures for electronic components.

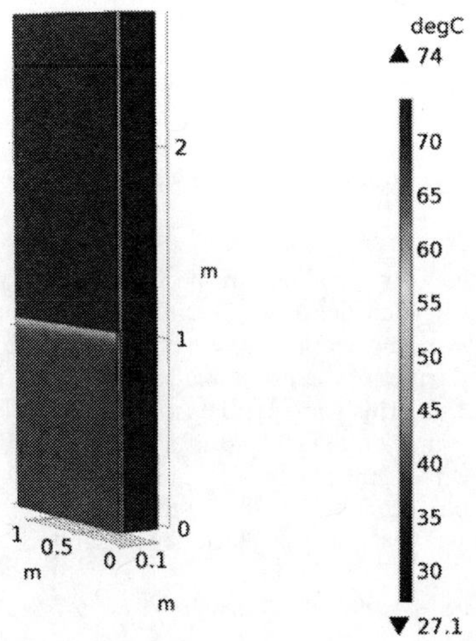

Figure 10: Monthly PR over one year. Results for March have to be considered carefully since data was only available for 13 days

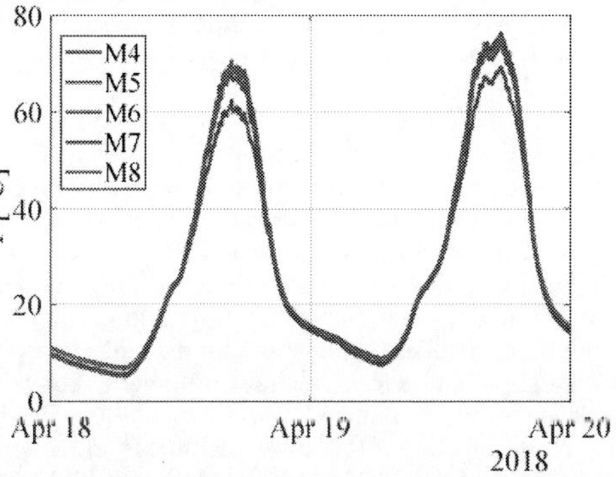

Figure 11: Measured temperature at the back of the BIPV module for all heights at the middle of the module (legend follows Figure 3)

6 REFERENCES

[1] Delponte, E., Marchi, F., Frontini, F., Polo, C., Fath, K., Batey, M. In 31st Eur. Photovolt. Sol. Energy Conf. Exhib (2015) 3046-3050.

[2] Goetzler, W., Guernsey, M., Droesch, M. Research and Development Needs for Building-Integrated Solar Technologies. US Department of Energy Office of Energy Efficiency and Renewable Energy Building Technologies Office Report (2014).

[3] Burgun, F., Snow, M., Sproul, A.B. Research report. CRC for low carbon living (2015).

[4] Yang, R. J. Automation in Construction Vol. 51 (2015) 92-102.

[5] Zogou, O., Stapountzis, H. 2011. Renewable Energy Vol. 36(12) (2011) 3488-3498.

[6] Lee, H. M., Yoon, J. H., Kim, S. C., Shin, U. C. Solar energy, Vol. 145 (2017) 66-77.

[7] Chatterjee, U., Gelagaev, R., Masolin, A., Driesen, J. In Industrial Electronics Society, IECON 2014-40th Annual Conference of the IEEE (2014) 2017-2022.

[8] Ikkurti, H. P., Saha, S. Renewable and Sustainable Energy Reviews, Vol. 47 (2015) 997-1006.

[9] Ravyts, S., Dalla Vecchia, M., Zwysen, J., Van den Broeck, G., Driesen, J. IEEE Texas Power and Energy Conference (2018).

[10] Graditi, G., Adinolfi, G. In Power Electronics, Electrical Drives, Automation and Motion (SPEEDAM) (2012) 1113-1118.

[11] Dierauf T., Growitz A.,Kurtz S., Becerra Cruz J. L., Riley E., Hansen C. In NREL/TP-5200-57991 (2013).

[12] Nordmann, T., Clavadetscher, L. In Photovoltaic Energy Conversion. Proceedings of 3rd World Conference on Vol. 3 (2013) 2243-2246.

[13] Aristizábal A.J.,Páez C.A. Energy and Buildings Vol. 152 (2017) 1-10.

[14] Churchill, S.W., Chu, H.H. International journal of heat and mass transfer. Vol. 18(11) (1975) 1323-9.

[15] Catton, I. In Proceedings of the sixth international heat transfer conference Vol. 6 (1978) 13-31.

[16] MacGregor, R. K., Emery, A.F. Journal of Heat Transfer Vol. 91 (1969) 391-401.

7 ACKNOLEDGMENTS

We are immensely grateful to Jimmy Van Criekingen and Wim Bertels for their technical support with the experimental setup. We would also like to show our gratitude to Patricia Elsen for the assistance with data acquisition.

This work has been conducted within the EFRO SALK project, supported by the EU, ERDF, Flanders Innovation & Entrepreneurship and the Province of Limburg.

METHOD FOR THE ANALYSIS OF TECHNICAL AND ECONOMIC FEASIBILITY OF GRID-CONNECTED PHOTOVOLTAIC SYSTEMS INTEGRATED ON BUILDING FAÇADES

Isadora Custódio & Ricardo Rüther
Centro de Pesquisa e Capacitação em Energia Solar da Universidade Federal de Santa Catarina
(www.fotovoltaica.ufsc.br)
Laboratório Fotovoltaica-UFSC, Sapiens Parque, Av. Luiz Boiteux Piazza, 1302,Florianópolis–SC, 88056-000, Brazil

ABSTRACT: Photovoltaic (PV) energy systems can be better integrated onto buildings, respecting the original architecture or even being considered during the design of a new building. The challenge for designers is to create a compromise between the architecture of the building and the function of generating energy in those buildings. Besides these technical issues, it is important that the projects take into account a study of the economic aspects, since the determination of the value of goods is as relevant as their physical production. This paper defines a method to select appropriated building façades for the integration of grid-connected PV systems, on the campus of a Brazilian public University. In order to be accepted, the façade PV systems must have a positive impact on the buildings' architecture on top of generating energy. In addition, PV systems on buildings should be economic feasible, with a positive Net Present Value (NPV), an Internal Rate of Return (IRR) higher than 3.0% per year and a payback period shorter than the PV systems' lifetime.
Keywords: Building Integrated PV (BIPV), Economic Analysis, Façade, Grid-Connected, Rooftop PV.

1 INTRODUCTION

An architectural or engineering project relies not only on technical, but also on economic, environmental, political and cultural solutions. Therefore, it is not only important to design the technical project of photovoltaic (PV) systems integrated on buildings, but an analysis of their economic feasibility is also essential.

The role of building designers is to use PV to enrich an existing or new architecture. The harmonization of PV technology and buildings can value a rooftop or façade aesthetically, and even improve the building's internal environmental comfort by the creation of PV *brise soleils* or double skin façades. The objective is to show that once the commitment between the architectural space, the building's and the PV installation shape, and the function of generating electricity is established, energy losses must be calculated and, in many cases, considered acceptable [1].

Thus, it is essential to make an economic analysis to ensure that these energy losses are also accepted in financial evaluations. By understanding this, building designers will feel encouraged to start integrating PV technology in the most different ways and make great high-tech architecture.

This paper aims at demonstrating different types of integration of PV modules on building façades, and defining a method to make these PV systems technically and economically feasible. The main focus of this research is to demonstrate how PV modules installed on façades can help improve the architecture of a building, generate energy and be economically feasible at the same time.

2 APPROACHES

Universidade Federal de Santa Catarina is a public university located in Florianópolis-SC, Brazil (27° S, 48° W). Two of the University's buildings were chosen as case studies: a classroom building at the Technological Center (building A) and the University's Main Library (building B).

2.1 Technical feasibility

Both of the selected buildings have façades facing north, where the PV integrations were suggested. Aluminum shutters for air circulation, windows with transparent glass and a big tiled blind wall compose building A's north façade. Building B's north façade is composed mainly by large windows with transparent glass.

The energy generated by the PV systems was estimated using the well-known PVsyst software [2] with input of the climate database from Florianópolis, measured by BSRN's (Baseline Surface Radiation Network) solarimetric station no. 3, which is located at the University itself [3]. Losses due to shading caused by nearby vegetation and neighboring buildings, due to the amount of incident irradiation in the collector plane, temperature, and due to module and inverter conversion efficiencies were estimated.

A degradation of 0.5% was considered in the energy generation for each of the 25 years of the systems' lifetime [4,5].

2.2 Economic analysis

The economic analysis was based on the calculation of Net Present Value (NPV), as shown in Eq. 1, Internal Rate of Return (IRR), as shown in Eq. 2, and payback period, as shown in Eq. 3.

$$NPV = -C_0 + \sum_{n=1}^{T} \frac{C_n}{(1+MARR)^n} \qquad \textbf{Equation 1}$$

$$\sum_{n=1}^{T} \frac{C_n}{(1+IRR)^n} - C_0 = 0 \qquad \textbf{Equation 2}$$

$$NPV = 0 \qquad \textbf{Equation 3}$$

Where:
NPV = Net Present Value, in USD;
C_0 = initial investment, in USD;
T = lifetime, in years;
n = considered duration of the system, in years;
C_n = values referring to the cash flow for the considered duration of the system, in USD;
$MARR$ = Minimum Acceptable Rate of Return, in % per year;
IRR = Internal Rate of Return, in % per year.

In this research, the Minimum Acceptable Rate of Return (MARR) was used to calculate the payback periods, thus changes in the value of money over time were considered.

All values that were adjusted from BRL to USD used the conversion rate of 1 USD = 3.90 BRL from July 2nd, 2018, according to the Central Bank of Brazil [6].

Two economic scenarios were studied: a current one, with a capital expenditure (CAPEX) of 1.28 USD/Wp and an electricity price annual growth rate of 4% (case 1); and an optimistic and futuristic one, with a CAPEX of 1.02 USD/Wp and an annual growth rate of 6% in the price of electricity (case 2).

Since solar energy generation occurs during daylight, the current electric off-peak time tariff of 0.08 USD/kWh, established by the local electric power company, was used [7].

The operating expenditure (OPEX) considered for both scenarios was 1% of the CAPEX every year for maintenance of the systems [8] and 21% of the CAPEX every 10 years for equipment replacement [9]. An interest rate (MARR) of 3.0% per year was applied in both scenarios [10].

In order to be considered economically feasible, the PV systems should have a positive NPV, an IRR higher than 3.0% per year and a payback period lower than 25 years, which is the considered lifetime of each PV system.

2.3 PV modules

The PV modules used in the integrations proposed in this paper are multicrystalline silicon (p-Si) PV modules of 260 Wp. Tab. I shows the technical specifications of the chosen PV modules [5].

Table I: Technical specifications of the PV modules

General characteristics	
Technology	Multicrystalline silicon (p-Si)
Dimensions	1650 x 992 mm
Thickness	35 mm
Weight	18.5 kg
Number of cells	60
Front cover	3.2 mm low-iron tempered glass
Encapsulating	Ethylene Vinyl Acetate (EVA)
Frame	Anodized aluminum alloy
Electrical parameters (STC)	
P_{max}	260 W
Efficiency	15.9 %
V_{mpp}	30.3 V
I_{mpp}	8.59 A
V_{oc}	37.7 V
I_{sc}	9.09 A
Temperature coefficient of V_{oc}	-0.32 %/°C
Temperature coefficient of I_{sc}	0.05 %/°C
Cable lenght	1000 mm

3 RESULTS

3.1 PV integration to the façades

The PV integration suggested for building A's façade is composed by 194 p-Si PV modules, as the layout shown in Fig.1. Its purpose, besides generating energy, was to eliminate the expenditure with the ceramic tiles that would otherwise cover the façade, by substituting this coating material with PV modules. This way, the now double skin façade also becomes more attractive, with a high-tech solution that matches with the University's Technological Center.

A *brise soleil* composed by 54 p-Si PV modules and sheet metals was proposed to building B's façade, as shown in Fig. 2, in order to improve the building's internal thermal and visual comforts. The sheet metals were suggested in areas where there was shading caused by the building's inclined beam and by nearby trees, which would result in considerable energy generation losses.

Tab. II shows the characteristics of the PV integrations for both buildings.

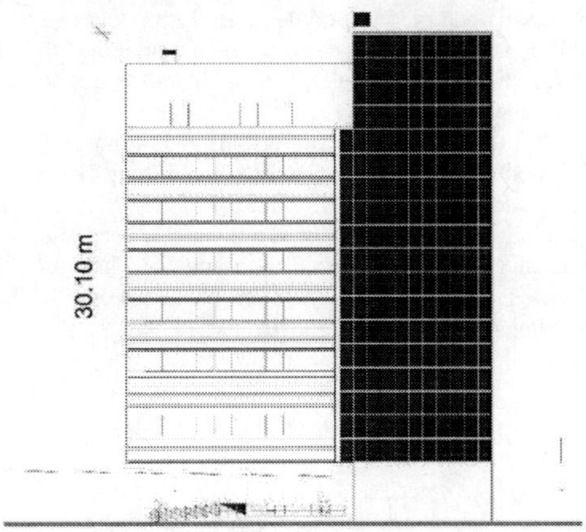

Figure 1: Building A's North façade with PV integration

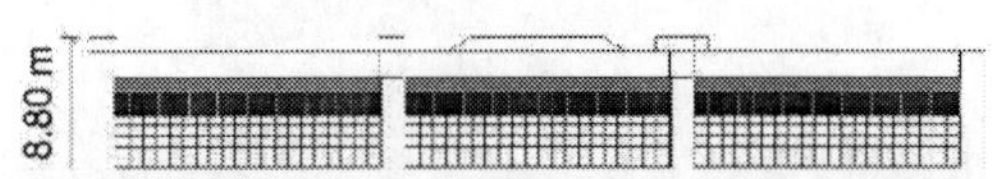

Figure 2: Building B's North façade with PV integration

Table II: Characteristics of the PV integrations to the façades of buildings A and B

Building	Azimuth	Tilt	Installed capacity (kWp)	Energy generation (MWh/year)
A	5°E	90°	50.44	31.90
B	5°E	90°	14.04	8.97

3.2 Economic analysis

Since in building A the coating material was replaced by the PV modules, the cost of the tiles was subtracted from the initial investment for the economic analysis.

The building's façade PV system had, for economic scenario case 1, a NPV of -14,830 USD, an IRR of 0.5% per year and a payback period of more than 25 years. Since the NPV is negative, the IRR is lower than the MARR established and the payback period is higher than the PV system's lifetime, building A's PV façade cannot be considered feasible for this economic scenario.

For economic scenario case 2, the system had a NPV of 20,730 USD, an IRR of 6.2% per year and a payback period of 18 years. Since the NPV value is positive, the IRR is higher than the MARR established, and the payback period is lower than the PV system's lifetime, the PV façade, in this case, can be considered acceptable. However, for many investors the numbers are not attractive.

In building B the cost of the metallic sheets that compose the *brise soleil* with the PV modules was added to the initial investment.

Under both economic cases, the building's façade PV system's NPV was negative (-10,912 USD for case 1 and -963 USD for case 2), the IRR was lower than the MARR required (-1.9% per year and 2.6% per year for cases 1 and 2, respectively), and the payback periods were higher than the PV system's lifetime. Therefore, building B's façade PV system is not economically feasible for any of the cases.

3.3 Optimizing potential

Since none of the façades integrations by themselves were satisfactory, PV modules were also integrated to the buildings' rooftops in order to try making the PV integrations economically viable or more aesthetically attractive.

Rooftop integration for building A is composed of 78 PV modules and is shown in Fig. 3. Building B's rooftop integration has 2,020 PV modules and is shown in Fig. 4. Building B's rooftop area was not entirely used due to the existence of terraces and equipment. The complete systems' characteristics (façades plus rooftops) of both buildings are shown in Tab. III.

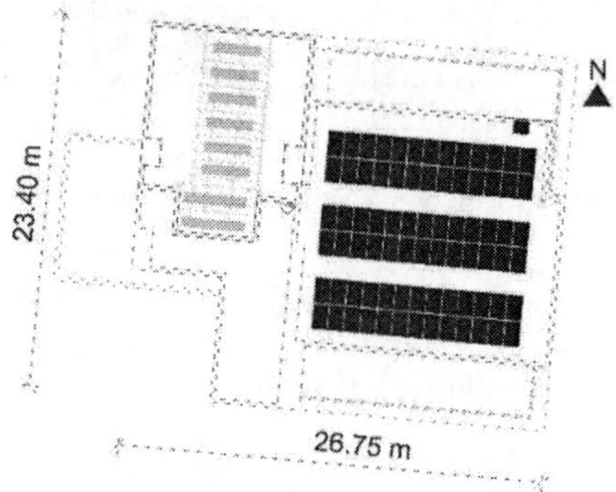

Figure 3: Building A's roofplan with PV integration

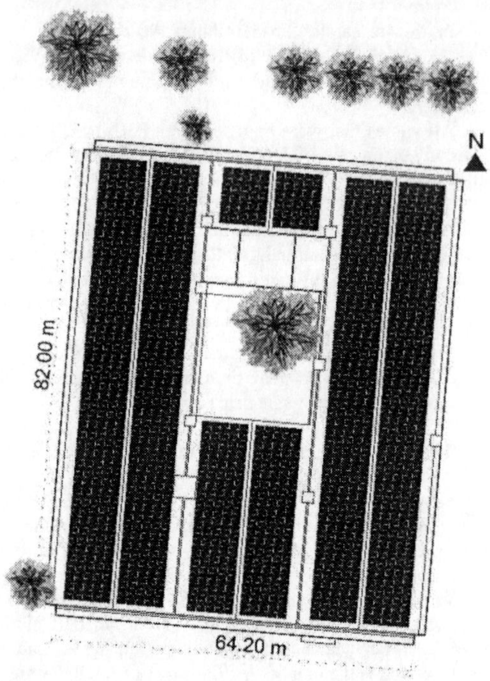

Figure 4: Building B's roofplan with PV integration

Table III: Characteristics of the complete PV integrations to buildings A and B

Building		Azimuth	Tilt	Installed capacity (kWp)	Energy generation (MWh/year)
A	Façade	5°E	90°	50.44	31.90
	Rooftop	5°E	10°	20.28	26.81
	Total	-	-	70.72	58.71
B	Façade	5°E	90°	14.04	8.97
	Rooftop	95°E + 85°W	5°	525.20	756.00
	Total	-	-	539.24	764.97

Building A's economic analysis for case 1 had as result a NPV of 14,012 USD, an IRR of 4.3% per year and a payback period of 23 years. For case 2, the NPV was 69,890 USD, the IRR was 9.4% per year and the payback period was 14 years.

Building B's economic scenario case 1 had a NPV of 862,128 USD, an IRR of 10.5% per year and a payback period of 12 years. For case 2, the NPV was of 1,432,855 USD, the IRR was of 15.2% per year and the payback period was 9 years.

With the incorporation of rooftop PV systems, the PV integrations of both buildings became technical and economic feasible for both economic scenarios. Since these PV systems also play a role in the architecture of the buildings as coating material and as *brise soleil*, the PV systems can be considered suitable.

4 CONCLUSIONS

A method was defined for the analysis of the technical and economic viability of photovoltaic systems integrated to the façades of two buildings. The research showed four viable technical studies, being only one of them economically viable. However, with the addition of rooftop systems, all integrations have become economically feasible and therefore attractive and valuable to build. This demonstrates that it is important for building designers to be aware of the possibilities for functionality and integration of photovoltaic systems and their opportunity to be economically viable.

With the declining cost of PV systems [9] and the increasing cost of electric energy in Brazil [7,11,12,13], PV façades should start to be economic feasible in more flexible ways. Consequently, PV technology could offer attractive solutions of high integration and aesthetic appeal, as well as renewable and pollution-free energy generation.

5 REFERENCES

[1] C. Zomer, A. Nobre, P. Cassatella, T. Reindl, and R. Rüther, "The balance between aesthetics and performance in building-integrated photovoltaics in the tropics," Progress in Photovoltaics: Research and Applications, vol. 22, pp. 744-756, 2014.
[2] A. Mermoud, "PVsyst," PVsyst Photovoltaic Software, 2018. Available in: <www.pvsyst.com>.
[3] S. Colle, "Basic measurements of radiation at station Florianopolis," Brazil, 2016.
[4] A. Limmanee, S. Songtrai, N. Udomdachanut, et al, "Degradation analysis of photovoltaic modules under tropical climatic conditions and its impacts on LCOE," Renewable Energy, vol. 102, pp. 199-204, 2017.
[5] YINGLI SOLAR, "YGE 60 Cell Series 2," Yingli Green Energy Holding Co., Ltd., 2017.
[6] Central Bank of Brazil, "Conversão de moedas," Brazil, 2018. Available in: <www4.bcb.gov.br/pec/conversao/conversao.asp>.
[7] ANEEL, "Resolução Homologatória no 2.286, de 15 de agosto de 2017," Brazil, 2017.
[8] C. Lacchini and R. Rüther, "The influence of government strategies on the financial return of capital invested in PV systems located in different climatic zones in Brazil," Renewable Energy, vol. 83, pp.786-798, 2015.
[9] INSTITUTO IDEAL, "O Mercado Brasileiro de Geração Distribuída Fotovoltaica," Florianópolis, 2018.
[10] BNDES, "Fundo Clima - Subprograma Energias Renováveis" Brazil, 2018. Available in: <www.bndes.gov.br/wps/portal/site/home/financiamento/produto/fundo-clima-energias-renovaveis>.
[11] ANEEL, "Resolução Homologatória no 2.120, de 16 de agosto de 2016," Brazil, 2016.
[12] ANEEL, "Resolução Homologatória no 1.927, de 4 de agosto de 2015," Brazil, 2015.
[13] ANEEL, "Resolução Homologatória no 1.770, de 5 de agosto de 2014," Brazil, 2014.

LAMINATED TEXTILES ENABLING CUSTOM APPERANCE OF BUILDING INTEGRATED PHOTOVOLTAIC MODULES

T. Gewohn[1,2*], S. Blankemeyer[1], M. R. Vogt[1], H. Schulte-Huxel[1], M. Köntges[1], B. Lim[1], C. Schinke[2], R. Brendel[1,2]
[1]Institute for Solar Energy Research Hamelin (ISFH), Am Ohrberg 1, 31860 Emmerthal, Germany
[2]Dep. Solar Energy, Leibniz University Hannover, Appelstr. 2, 30167 Hannover, Germany
*Corresponding author: t.gewohn@isfh.de, Tel: +49-511-762-17253

ABSTRACT: We present a new technique to alter the appearance of photovoltaic (PV) modules for building integration by laminating textiles and other customary materials onto the front of PV modules. This approach is highly customizable, since we can use virtually any thin UV-stable textile, whether it is a plain colored or a patterned textile. The short-circuit current density loss caused by the textile laminate is 6% to 11% for a whitish appearance. Textiles can be laminated to any standard PV module. Using commercially available materials and machines, this method is likely to be cost-efficient and readily available.
Keywords: Building Integrated PV (BIPV), Design, Performance

1 INTRODUCTION

Building-integrated photovoltaics (BIPV) becomes increasingly important as current regulations require enhanced energy standards [1], [2]. The activation of the façade is thus an important topic that requires PV modules, which are attractive for architects, home owners and for the society in general by enabling a customary design of the modules. At the same time, the cost-performance ratio needs to be competitive with other façade materials. Currently, specialized manufacturers produce colored solar modules and offer colored foils to alter the appearance of solar modules [3]. These techniques increase the costs of a module by 20% to 45% [4] and may cause power losses of more than 15% [5]. Moreover, they do very often not allow a customized design.

In this work, we use customary textiles to alter the appearance of standard PV modules. We show that the dark blue appearance can be changed to a whitish, colored or even patterned design such as the one shown in Figure 1.

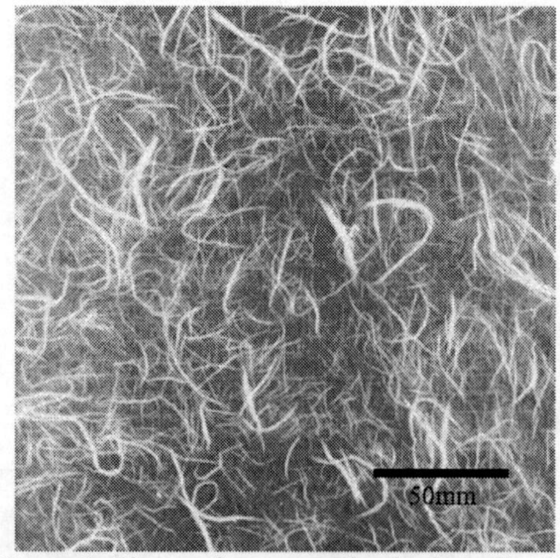

Figure 1: Picture of an encapsulated solar cell, with Japanese paper laminated on the top. The short-circuit current loss is 15%.

2 EXPERIMENTAL DETAILS

2.1 Preparation of the single cell modules

The sample structure is shown in Figure 2. It is a one-cell PV module with an additional top layer. For the module, we use a 156 × 156 mm² aluminum BSF solar cell contacted by the smart wire technology. The module has a three-film transparent backsheet (FPL-FET-T250-T50, Toyal Solar) on the bottom, the cell is encapsulated with standard UV-absorbing EVA (EVASKY S88, Bridgestone) that has a thickness of 460 µm and there is a low-iron glass plate with a thickness of 4 mm on the top.

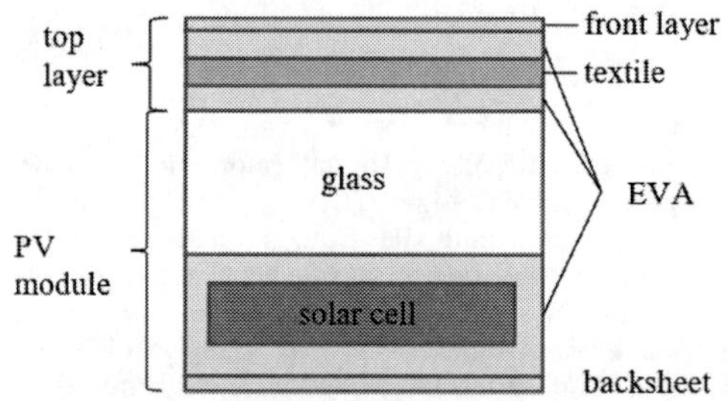

Figure 2: Schematic of the modified PV modules.

The top layer consists of a textile embedded between two EVA layers (EVASKY S88, Bridgestone) and covered with a transparent front sheet (FPL-FET-T250-T50, Toyal Solar).

Beside this setup, we made another attempt, where the textile was placed, embedded between two EVA layers, under the glass. The lamination onto the PV module has various advantages over the embedding under the glass of the module. When it is embedded under the glass, the scattering of the light in the textile is less strong, thus the solar cell is more visible. Furthermore, the textile layer and the PV module must be made in one step, the textile can't be added afterwards.

Figure 1 and Figure 3 show the five samples that we investigate in this paper. The textile materials include Japanese paper (Figure 1), white fleece (Figure 3a), sisal (Figure 3b), white fleece twofold (Figure 3c) and green gauze sixfold (Figure 3d). Japanese paper and white fleece (twofold) hide the cell metallization completely while the other three textiles still show the contours of the cells' metal fingers.

We use a standard PV module laminator with common processes in order to connect the components together. In the first lamination process, we produce the one-cell test module. In a second process, we laminate the components of the top layer onto the test module. We use two steps in

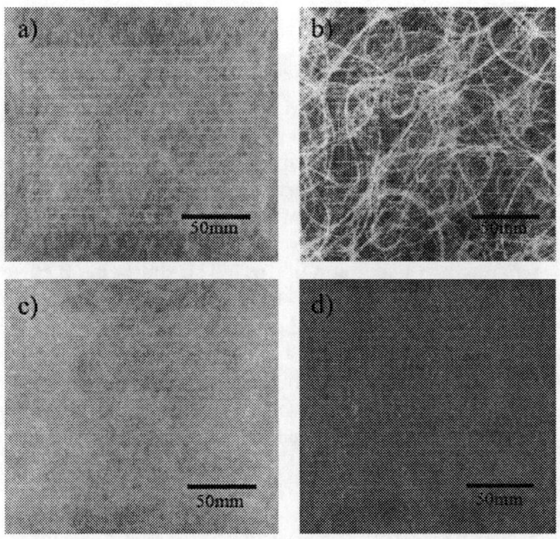

Figure 3: Photographs of four 200 × 200 mm² one-cell PV modules. (a) white fleece, (b) sisal, (c) white fleece (twofold), (d) green gauze (sixfold).

order to be able to characterize before and after applying the textile layer stack. Notably, the textile layer can be laminated onto virtually every standard photovoltaic module.

2.2 Measuring the single cell modules

In order to facilitate the measurement of many textiles with a single and unchanged PV module, we temporarily optically connect the top layer to the PV module using paraffin as a refractive index matching liquid [6].

For the measurement of the short-circuit current density, we use a commercial class AAA flashed solar simulator (flasher) and measure under standard test conditions at an irradiance of 1000 W/m². We relate the current density measured for the five top layers coupled to the PV module to a reference measurement without a top layer. We thus report relative current densities here.

For measuring the external quantum efficiency (EQE), we use the solar cell analysis system LOANA (PVtools). The illuminated area is 20 × 20 mm². The wavelength ranges from 280 nm to 1200 nm.

All samples are examined with an optical microscope before they are laminated into the top layer to reveal the microstructure of the samples.

3 RESULTS

3.1 Measurement of the short-circuit current

Figure 4 shows a photograph of a PV module with a top layer without a textile and compare its short-circuit current to the reference. The short-circuit current loss due to the EVA and front sheet layer is 1.6%. In the following, we report the ratio of the short-circuit current density of measurements with and without a top layer.

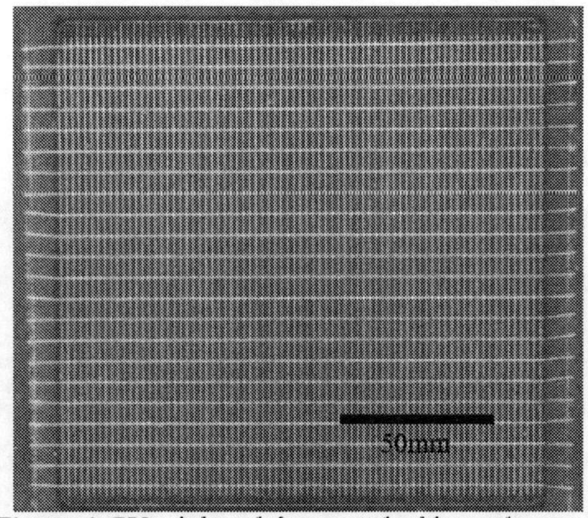

Figure 4: PV mini module covered with two layers of EVA and a front layer.

Figure 5 shows the short-circuit current density of samples with textiles in the top layer varies from 85% to 94% relative to the reference without a textile. The single layer of white fleece (Figure. 3a) shows the smallest loss of 6% but also results in incomplete hiding of the cells grid lines. With the Japanese paper on the front of the PV module (Figure 1), the cell and its grid lines are almost invisible, while the current is still 85% of that of the reference.

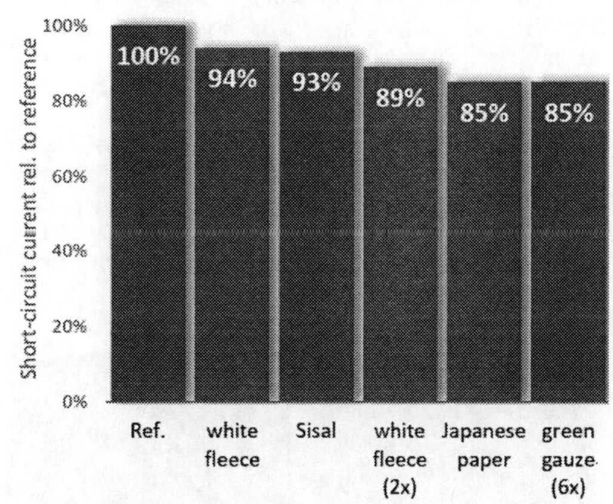

Figure 5: Ratio between short-circuit current density of the textile and the reference samples.

3.2 Spectrally resolved investigation of the EQE

Figure 6 shows the EQE of the reference module (black line), the single-layer white fleece (blue dashed line) and the green gauze (green dotted line). The EQE measurement provides spectrally resolved information. We can thus analyze at which wavelength any power is lost due to reflection or absorption. The green sample loses more light in the wavelength ranges of 400 nm to 500 nm and 600 nm to 650 nm. This is not only due to the green color but also due to the material structure and to the fact that we use six layers here. Measuring a module takes the optical interaction between the top layer and the module into account. This is not the case when simply measuring the optical transmittance of a textile.

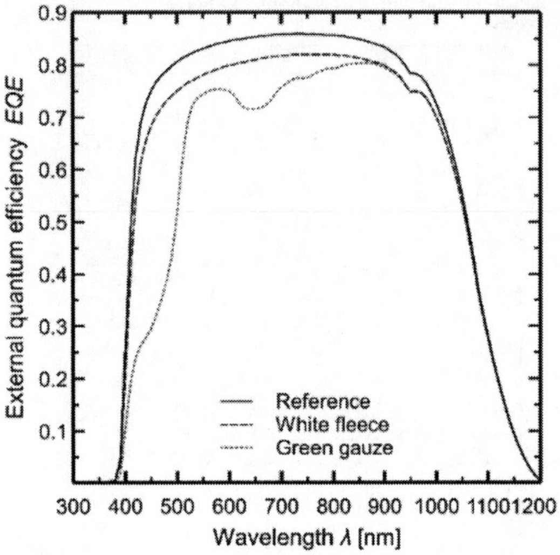

Figure 6: Spectrally resolved EQE of the reference (black line), white fleece onefold (blue dashed line) and green gauze sample (green dotted line).

3.3 Sample structure

Figure 7 shows the optical micrographs. The differences of the microstructures of fleece, Japanese paper, sisal and gauze become evident. The Japanese paper in Figure 7 b) has the densest thread structure. This is in qualitative agreement with losing the larges fraction of 15% of the short-circuit current density with a single layer of Japanese paper and with completely hiding the solar cell. In contrast, the green gauze in Figure 7 d) has a less dense structure and was used sixfold to achieve a strong green appearance. Using a different number of textile layers is a very simple means to alter the color strength without changing any material. The structural differences will affect the optics on the module and will have to be accounted for when modelling modules with textile laminates.

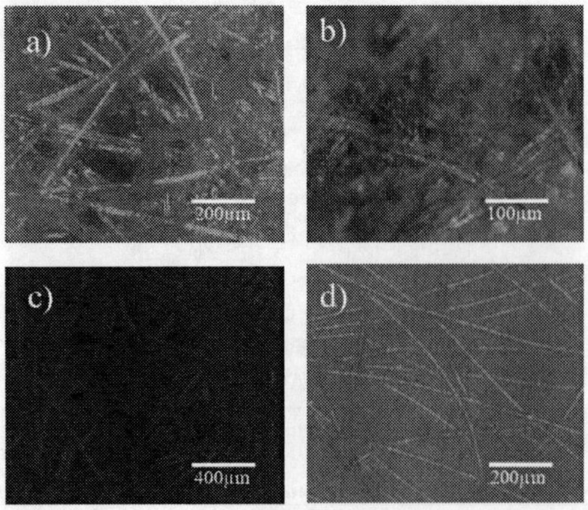

Figure 7: Optical microscope micrographs of a) white fleece (magnification 10x), b) Japanese paper (20x), c) sisal (5x) and d) green gauze (10x).

3.4 Cost estimate

The additional components required are a standard EVA foil, a textile sample and a front sheet. For a cost calculation we assume a PV module price of 0.35 €/Wp, 1.007 €/m² for standard EVA, 2.322 €/m² for the front sheet and 0.5 €/m² to 1.5 €/m² for the textiles. Thus, the additional components increase the cost of a solar module by 7.2% to 10%.

4 CONCLUSION

We investigated an approach for a potentially cost-efficient change of the appearance of PV modules by using customary textiles and standard manufacturing equipment. The additional materials increase the cost of a solar module by a maximum of 10%. The short-circuit current density loss of these PV modules is 6% to 15% which compares favorably to other reported approaches [4], [5], [7].

The top layer containing the textile is weather-resistant and thus we expect the module plus textile to be weather resistant also. The UV stability has to be investigated in future work.

5 ACKNOWLEDGEMENT

The authors would like to thank AMF creation GmbH, who supported our experiments with a selection of nonwoven lines, fleeces and other customary textiles.

8 REFERENCES

[1] B. P. Jelle, C. Breivik, and H. Drolsum Røkenes, "Building integrated photovoltaic products: A state-of-the-art review and future research opportunities," *Sol. Energy Mater. Sol. Cells*, vol. 100, no. 7465, pp. 69–96, 2012.

[2] IEA, *Task 15 Subtask B :* "Inventory on Existing Business Models , Opportunities and Issues for BIPV", no. April. 2018.

[3] I. Cerón, E. Caamaño-Martín, and F. J. Neila, "'State-of-the-art' of building integrated photovoltaic products," *Renew. Energy*, vol. 58, pp. 127–133, 2013.

[4] C. Renken, "Konzepte - und Planungsprozesse für die Gebäudeintegration," 2017.

[5] M. Mittag, C. Kutter, M. Ebert, H. Wilson, and U. Eitner, "Power loss through decorative elements in the front glazing of BIPV modules," *33rd EUPVSEC*, no. September, pp. 22–26, 2017.

[6] M. Köntges, H. Schulte-Huxel, S. Blankemeyer, M. R. Vogt, H. Holst, and R. Reineke-Koch, "Measuring the light recovery factor of backsheets in photovoltaic modules," *Sol. Energy Mater. Sol. Cells*, vol. 186, pp. 175-183, 2018.

[7] F. de V. L.H. Slooff, J.A.M. van Roosmalen, L.A.G. Okel, T. de Vries, T. Minderhoud, G. Gijzen, T. Sepers, A. Versluis, F. Frumau, M. Rietbergen, L. Polinder, E.M.B. Heller, "An architectural approach for improving aesthetics of PV," in *33rd European Photovoltaic Solar Energy Conference and Exhibition*, 2016, no. c, pp. 1–4.

TRANSMISSIVITY AND AESTHETICS OF OPTICALLY UNCOUPLED GLASSES FOR THE APPLICATION IN A PVT ROOF TILE

Nils Reiners[1], Ulf Blieske[1], Peter Hakenberg[2], Julian Münzberg[2]
[1]Cologne University of Applied Sciences, Betzdorfer Str. 2, 50679 Cologne, Germany
[2]Paxos Consulting & Engineering GmbH & Co. KG, Am Kielsgraben 8, 40789 Monheim am Rhein, Germany

ABSTRACT: The purpose of this work was to investigate on different types of solar glass sheets that could be used as front side encapsulant in the PVT (Photovoltaic and Thermal) solar tile that is currently developed by the company Paxos Consulting & Engineering GmbH & Co. KG. Between the glass sheet and the solar cells a gap of several millimeters is foreseen and the space that is created is filled with inert gas to avoid heat loss. Two elements must be considered for the choice of the glass sheet. First the transmission needs to be as high as possible. On the other hand, the aesthetics of the tiles surface is of high importance for the product. For the aesthetics the homogeneity, the color and the potential glazing are of high interest. For the evaluation of the transmissivity of the 9 different glass types a measurement setup was constructed using a Xenon lamp and a large area detector which was constructed for that purpose to avoid the loss of photons that are strongly scattered. For the aesthetics, pictures were taken from the perpendicular view and from 45° with a light source being placed on the opposite side with an angle of incidence of 45°. The coated glasses and the glasses with texture on the backside have a low transmissivity compared to the other samples. The best aesthetics was achieved with coating #1 on Albarino P while the second best was achieved with uncoated Albarino P. The overall best choice for the encapsulation would hence be Albarino P as it gives a good transmissivity and an acceptable aesthetic.

1 INTRODUCTION

To achieve the maximum efficiency from an illuminated surface area, it is meaningful to combine a photovoltaic and a thermal solar system. Such PVT (Photovoltaic and Thermal) system were developed by several companies and the certification procedures were worked out for such modules in the last years [1].

The company Paxos Consulting & Engineering GmbH & Co. KG is currently developing a PVT roof tile that is combining the advantages of a PVT system with those of a roof integrated BIPV (Building Integrated Photovoltaic) system.

A core element of the innovative PVT roof tile is a patented connection system that allows a very fast and reliable connection of the electrical as well as the fluid system.

The liquid flowing through the system is fulfilling two functions. On the one hand, it is cooling the solar cells leading to an increase in efficiency. On the other hand, it is transporting the heat to the thermal storage where it can then be used for different purposes.

To increase the thermal efficiency of the roof tile, the glass pane is optically uncoupled from the solar cell and the space in between is filled with inert gas (e.g. argon) such that the loss due to heat conduction is significantly reduced. In addition, this set-up allows the roof-tiles to be walkable.

But, as the optical coupling is not given, the optical properties of the front side encapsulation are different from the one in a standard solar module and need to be reconsidered.

Two aspects are of importance for the choice of the frontside glass pane: The aesthetics and the optical performance. For the aesthetics the aspects of the color appearance, the homogeneity and the glazing effects need to be considered. For the optical performance it is the transmissivity that is of importance.

2 MEASURING THE TRANSMISSIVITY OF OPTICALLY UNCOUPLED GLASS PANES

For the measurement of the transmissivity a special detector was constructed. It consists of 4 (2 x 2) polycrystalline solar cells with edge lengths of 156 mm. The polycrystalline cell type was chosen because like this no uncovered space is created as it would be the case using monocrystalline solar cells. The cells were connected in parallel and the overall short circuit current was measured using a measurement shunt resistor. The large detector area was chosen to be capable to detect all photons even though they are strongly scattered by the glass sheet. Using a standard transmission measurement setup with an integrating sphere as described for example in [2] would not fully take into account such strong scattering effects as the entrance aperture of the integrating sphere is relatively small. An illustration of the standard method and the method applied in this study is given in Figure 1.

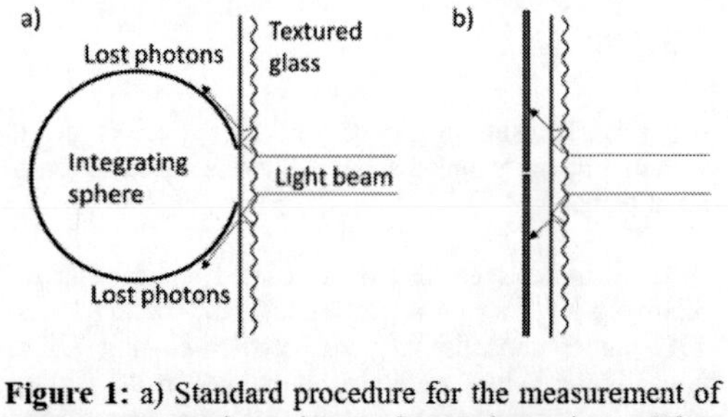

Figure 1: a) Standard procedure for the measurement of transparent samples using an integrating sphere. When thick textured glasses are measured photons can escape from the sphere leading to wrong results. b) Measurement setup applied for this paper using 4 parallel connected solar cells as detector.

The light source that is used for the measurement is a liquid cooled xenon short-arc lamp that provides a constant spectrum comparable to the solar spectrum. Using apertures, the light spot on the sample is reduced to the size of a tennis ball.

The measurement procedure starts by measuring the short circuit current of the detector without a glass sample as reference value ($I_{sc,r}$). In a second step the glass sample under investigation is placed above the detector with a gap of 5 mm and then the short circuit current is measured ($I_{sc,s}$). To calculate the transmission through the glass sample the quotient of the two currents is calculated.

$$T = \frac{I_{sc,s}}{I_{sc,r}} \tag{1}$$

The samples investigated for this study are all using glasses produced by the company Saint Gobain (SG). The two samples 'Albarino P' and 'Albarino G' are textured on one side by a casting process. While in solar modules these textured glasses are only used with the texture on the top side, for this study, also the reverse configuration with the texture on the inner side was investigated. Only in modules with a gap, this configuration is feasible. Two of the samples had an additional coating on their inner side for diffuse scattering. Both companies that provided the coatings claimed that despite the coating a very high transmissivity is preserved. One of the glass samples had an AR-coating on the outer surface to reduce reflection. The measurement results for the nine different glass samples are given in Figure 2.

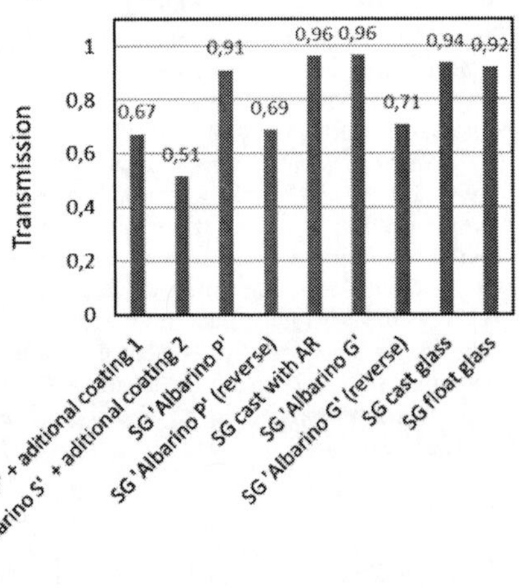

Figure 2: Transmission values of different glass types for an optically not coupled setup. SG is the abbreviation for Saint Gobain

It can be seen that both coated glasses have a relatively low transmissivity. Though the coating 1 has a 15% higher transmissivity compared to coating 2. Low transmission values were also measured for the textured glasses in reverse configuration. All other glasses have transmission values that are higher than 90%. The highest values were achieved by the cast glass with additional AR coating and the 'Albarino G' glass. The values found for the float glass are in agreement with transmission values stated in the literature [3].

Figure 3: Appearance of the nine different glass pane samples photographed from perpendicular view and from a view angle of 45°.

3 AESTHETICS OF OPTICALLY UNCOUPLED GLASS PANES

The aesthetics of the glass encapsulant is of high importance for the development of the PVT solar roof tile. The goal is to achieve homogenous appearance of the surface with a dark blueish color. In future versions also other colors shall be included in the product portfolio. Furthermore, the solar cells beneath the glass shall not be distinguishable from the sheet located behind the solar cells. The third element of the aesthetics is that the glazing on the glass surface should be reduced as much as possible.

In Figure 3 the appearance of the nine different samples is presented from the perpendicular view and from an angle of 45°. Opposite to this view point, a halogen lamp was placed with an angle of 45° for that the general appearance and the glazing effect could be distinguished.

The float glass, the cast glass with AR-coating and the cast glass without AR-coating show a strong glazing effect and the cells can clearly be identified.

For the modules with additional coating, the aesthetics of the sample with coating 1 is really homogeneous, no cell can be identified and the glazing effect is very low. The sample with coating 2 shows positive aesthetic behavior but the glazing is quite strong because of the planar glass that was used as substrate for the coating.

The samples with textures did show a high glazing effect for the reverse configuration. For the standard configuration (texture at the frontside), the glazing effect was low. From the perpendicular viewpoint the cells could still be identified for all textured samples. At the angle of 45 ° the solar cells were barely visible and the appearance was quite homogeneous.

4 CONCLUSIONS

Nine different glass samples were investigated for their suitability in a PVT solar roof tile design without optical coupling between glass and solar cells. Their transmissivities at perpendicular incidence were measured and aesthetical aspects were investigated.

The best aesthetics was reached by coating 1 that was applied on an 'Albarino P' glass sheet. This sample on the other hand showed a transmissivity of only 67% which is not satisfactory for a solar application. Hence, the best choice based on the specified parameters was found to be the 'Albarino P' glass which showed a good aesthetical performance at 45° and a high transmissivity.

5 LITERATURE

[1] Althaus, Jörg; Bott, Matthias; Fritzsche, Ulrich (2014): Qualification of PVT Collectors – Current Status. EuroSun 2014/ ISES Conference Proceedings.

[2] Labsphere (2017): A guide to integrating sphere theory and applications. Hg. v. Labsphere.

[3] Deubener, J.; Helsch, G.; Moiseev, A.; Bornhöft, H. (2009): Glasses for solar energy conversion systems. In: *Journal of the European Ceramic Society* 29

PV BEYOND ELECTRICITY, HEAT PUMPS HYBRIDIZATION FOR A MULTIPLICATIVE EFFECT TOWARDS NZEB

Asier Sanz[1], Raquel Fuente[2], Antonio J. Martín[3], Nekane Hermoso[1] and Jose M. Vega de Seoane[4]

[1]TECNALIA Research & Innovation, Derio, 48160, Spain. Corresponding author.
[2]Faculty of Engineering, University of the Basque Country UPV/EHU, Bilbao, 48013, Spain.
[3]Energy Panel, Lucena, 14900, Spain.
[4]TECNALIA Research & Innovation, Donostia, 20009, Spain.

ABSTRACT: The high energy consuming building sector needs to meet both electricity and heat demands. In a NZEB scenario most of the consumed energy would be generated locally by means of RES that nowadays seem not to provide an attractive cost-competitiveness. Solar based technologies tend to be the most promising ones, but for high densely populated and restricted areas buildings thermal loads are still a major setback concerning zero energy performance. Traditional fossil fuel or single renewable approaches will not be valid from efficiency and/or cost competitiveness. PV seem to have a great future as the build-environment local generation technology, but high capacity scenarios related grid impact is still a serious threat for massive deployment. HPs are an efficient approach for a reduction of building thermal comfort related CO_2 emission and capable of using the electrical infrastructure as back-up source. Therefore, if both technologies are individually suitable, merging them in a unique solution should enable even higher benefits. The current work is focused on the potential analysis of PV and smart combination with HP as entire-lifetime feasible solutions for NZEB. The techno-economic analysis is focused on an 80s residential-use 95-dwelling building, located at 6 different EU sites for a set of 5 scenarios, combining market standard natural gas boilers, conventional HPs and PV systems. The results show that the proposed solution could be a currently efficient, renewable and cost-competitive.

Keywords: photovoltaic, PV, heat pump, HP, hybrid, NZEB, ZEB

1 INTRODUCTION

According to the International Energy Outlook of 2017 [1] the global energy consumption continues rising. The reference for 2015 was of 575 quadrillion British thermal units with a 28% estimated increase by 2040. The European Commission states [2] that buildings are responsible for 40% of the energy consumption and 36% of CO_2 emissions, and pretends to reduce their impact through measures. Most of those buildings need to meet both electricity and heat demands, for domestic hot water (DHW) and space heating/cooling (H&C). In the close nearly zero energy building (NZEB) scenario most of the consumed energy would be generated locally by means of renewable resources. Unfortunately, current renewable solutions seem not to provide an overall, simultaneous, integral and local solution to this need ensuring cost-competitiveness.

However, solar energy is available all over the face of the earth. Thus, buildings should try to take more value form every beam of light reaching their envelopes. Photovoltaics (PV) are becoming today, and solar thermal (ST) applications in the past, widely used for build environment on-site generation. Nevertheless, for high densely populated and shadows restricted areas this kind of single approaches appear to be not enough to satisfy building energy needs. Detailed analysis of solar resource in build environments show that not only roofs but also façades should be considered with higher efficiency solar conversion devices such as photovoltaic-thermal (PVT) [3] or through hybrid systems [4].

Harnessing solar energy should be a must for new and refurbished buildings, but when the sun is not shining and the energy stores are empty solar base systems will always require a back-up that reduces their competitiveness. The electrical grid makes things easier for some loads but thermal needs are still high fossil fuel dependent in great part of Europe. However, heat pumps (HP) seem a promising technology towards a reduction of building thermal comfort related CO_2 emission, and enable the use of the electrical grid for using them as the back-up.

Additionally, PV systems have achieved along this last decade a high drop on cost and present an attractive levelized cost of energy (LCOE) in some countries and will gradually became cost-attractive also in others. However, the current grid use will not be any longer possible in high PV penetration scenarios. The energy generation is hardly full self-consumed and in some grids the overgeneration is polarizing production "duck curves" excessively till becoming a threat for conventional PV massive deployment. Electrical stores seem to be a solution for utility scale or small households power systems, but entire lifetime cost-competitiveness uncertainties still limit them as market standards. In the meantime, build environment thermal loads require non-fossil based RES solutions for enabling the NZEB utopia.

Therefore, if solar and HP are individually suitable for electricity and heat generation, merging them in a unique hybrid system will should enable to obtain even higher benefits. The aim of the current work is to shed some light on this hot topic and try to quantify its potential.

2 TECHNOLOGIES

The hybrid systems coupling solar collection and HPs could lead to multiple solutions that have been widely analysed by different research initiatives.

The current approach towards NZEB is based on the use of solar hybrid systems, coupling simple-collector solar base technologies with conventional vapour compression HPs through advanced overall solution control systems through predictive performance algorithms. These solar and HP combinations (Fig.1) could be done in different ways, where the solar field and the HP are connected:

(i) electrically;
(ii) thermally;
(iii) dually driven solutions.

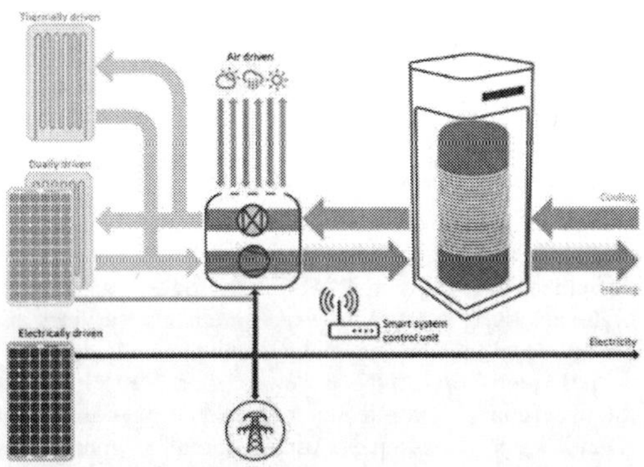

Figure 1: Scheme of different solar coupled HP systems.

The current work analysis is focused on power-to-heat (P2H) alternatives of transferring electricity into heat [5], enabling the potential reduction on fossil energy sources and emissions saving. Thus, the variant to be studied is the electrically PV driven approach, where PV is grid-tie and the air-water HP consumes the energy from it and/or the grid. However, for a broader analysis, different RES and non-RES technological solutions have been considered, due to the cost-competitiveness potential and/or because represent a high market share.

2.1 Non-renewable

Although this kind of solutions will gradually loose relevance and probably will not be part of the future energy ecosystem, nowadays still represent a huge market share in most of the countries [6]. The most common non-renewable solution for satisfying locally building just heat demands are gas boilers. Their simplicity, end-user acceptance and centralized energy grid infrastructure make this solution hard to be beaten despite its high operation cost and service uncertainties. Thus, within the current analysis natural gas based solutions have been considered.

2.2 Renewable conventional HP base

The HP solutions market sector is experiencing a significant growth. The household appliance sales for all heating and cooling (H&C) market technologies increased by 20% in 2015 and show a great potential for near future HPs barometer [7]. The emerging trend over the last years show that:

(i) air-source units are clearly gaining market share to the detriment of the ground-source market;
(ii) the use air as their vector are taking advantage of record temperatures that have given a boost to the cooling market;
(iii) the energy independence and the growth of the self-consumption markets are driving another trend that can turn to their advantage;
(iv) the units must reduce the greenhouse emissions with low global-warming potential (GWP) refrigerant selection, load reductions and efficiency improvements;
(v) a greater performance enhancement should come by combining with solar based solutions, where, the air-water source HPs and their combination with an additional uncoupled parallelly installed PV system has been included in the study in order to analyse also the performance of the solar electrically driven solution.

2.3 Selected solutions

According to the previous considerations, within the study, the following set of 5 technological solutions have been chosen for feeding buildings thermal and electrical loads:

(i) Natural gas boiler & electrical grid;
(ii) Natural gas boiler & PV plus electrical grid;
(iii) HP & electrical grid;
(iv) HP & PV plus electrical grid;
(v) Smart HP & PV maximized plus electrical grid;

3 METHODOLOGY

The assessment of new disruptive technologies or incremental innovations is commonly addressed just from the cost or performance side, when real market success and massive deployment tends to come from a delicate balance of both. In the field of NZEB the energy solutions will come from a technically feasible but economically market-competitive technology and/or systems combinations. Thus, as a previous step to the development of the highest potential system, different solutions comparative techno-economic analysis is needed. The implemented procedure (Fig. 2) is described below.

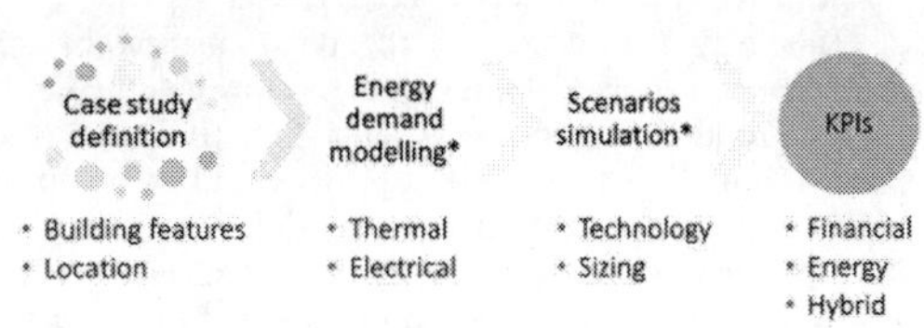

Figure 2: Scheme of the implemented methodology, based on entire-year hourly-step calculations (*).

3.1 Case study definition

The proposed analysis methodology starts with the definition of the case study, where all the building features (location, orientation, composition, use, etc.) are described.

3.2 Building energy demand profiles modelling

Then, the building energy load profiles are obtained, for household appliances, lighting, space heating, cooling and domestic hot water (DHW) demands.

3.3 Scenarios simulation

For each analysed case study five different scenarios are considered, based on the corresponding technological approaches stablished in 2.3:

(i) Baseline: natural gas boiler
(ii) Baseline added PV system
(iii) Heat pump installation
(iv) Combined conventional PV and HP
(v) Hybrid PV & HP solution

Then, for each one of these scenarios the performance of is obtained based on annual operation simulation, for same load profiles and operating conditions. The tool determines the annual performance based on typical meteorological year data for 365 days hourly-profiles. The models used for each technology are listed by category below.

3.3.1 Natural gas boiler

The non-intermittent source for these kind grid dependent technologies makes easier their black-box basic performance modelling. Thus, the models implemented for natural gas based boilers (eq. 2) are just characterized by constant conversion efficiency, where no start-stop energy losses or lifespan degradation is considered. The demand is satisfied by instant operation of units with no thermal store. According to conventional market available electric and natural gas boilers, the following efficiencies have been used $\eta_{NGboiler} = 0.9$.

$$E_{NGgrid} = E_{Load} \,/\, \eta_{NGboiler} \qquad \text{(eq. 1)}$$

3.3.2 Heat pump

Three of the five analysed scenarios are based on market available conventional HPs. The cost and performance of these systems are often high dependent. The scope of the study is focused on the assessment of different technological solution more than in specific products. Thus, in order to minimise the impact of comparing high performance (high cost) solutions versus cheaper (poorer performance) options, an average performance/cost HP has been considered.

This way the model for the determination of the performance is obtained through the following equations. The evaporation temperature is defined for 10 K below the ambient (eq. 2). Then the HP capacity could be obtained as function of the evaporation temperature (eq. 3) and the corresponding polynomic coefficients with an error below 3% (the coefficients depend on the selected machine for each case study and scenario). In the same way the compressor electric consumption is calculated (eq. 4). Then the HP coefficient of performance (COP) could be calculated (eq. 5). Then, the HP heat generation is obtained as function of the thermal load. Finally, the electric grid consumption (eq. 6).

$$T_{eva} = T_a - \Delta T \qquad \text{(eq. 2)}$$
$$Q_{HP} = q_0 + q_1 \cdot T_{eva} + q_2 \cdot T_{eva}^{\ 2} \qquad \text{(eq. 3)}$$
$$P_{HP} = p_0 + p_1 \cdot T_{eva} + p_2 \cdot T_{eva}^{\ 2} + p_3 \cdot T_{eva}^{\ 3} \qquad \text{(eq. 4)}$$
$$COP_{HP} = (Q_{HP} + P_{HP}) \,/\, P_{HP} \qquad \text{(eq. 5)}$$
$$E_{Egrid} = E_{Load} \,/\, COP_{HP} \qquad \text{(eq. 6)}$$

3.3.3 Photovoltaics

The PV electric performance is obtained from satellite based data and PVGIS models, corrected by irradiance profiles. The approach enables easy to handle profiles but with conservative annual yields.

3.4 Scenarios simulation

The holistic techno-economic analysis carried out pretends to address the energy performance and economics as a whole, trying to clarify the traditional gap between the great efficiency results of new research results and real market standard solutions lower cost. In this sense, the implemented methodology is based on a simple quantitative multiple-approach of solutions entire lifetime costs and system performance, in order to conclude with the following different kind key performance indicators (KPI). These KPIs could be classified in the following three different categories:

(i) Financials: investment, operation & maintenance costs, return on investment and internal rate of return.
(ii) Energy: building annual primary energy consumption, seasonal performance factor and self-sufficiency.
(iii) Mixed: levelized cost of energy (LCOE).

The economic assessment of each addressed scenario is based on traditional capital expenditures (CAPEX) and operating expenditures (OPEX). The CAPEX is essentially compound of system cost and turnkey installation. The OPEX is obtained as sum of system grid energy consumption, periodic maintenance labours and probable lifetime component replacement. The market-dependent CAPEX and OPEX items figures are usually high-sensitive, so retail prices, maintenance services and energy costs used in the analysis are a consequence of a deep market study. In the same way, in order to minimise the uncertainty of non-technology related phenomena, no financial cost is assumed but conservative energy cost updates are included.

4 ANALYSIS

The work carried out has been focused on the following parameterization.

4.1 Building

The building considered within the analysed case studies is the same one, although different local enveloped thermal properties have been used for each case. A residential-use and 95-dwelling construction built in 1980, with 4 floors and 8047, 2011 and 1988 sqm gross, roof, free-window façade area respectively. The energy demands are for household appliances, lighting, space heating, cooling and DHW demands.

4.2 Case studies

The same building has been located at 6 different locations, covering a wide-range of solar potential and heat demand across EU. The considered cities have been: Valencia, San Sebastian, Rome, Paris, Berlin and Helsinki.

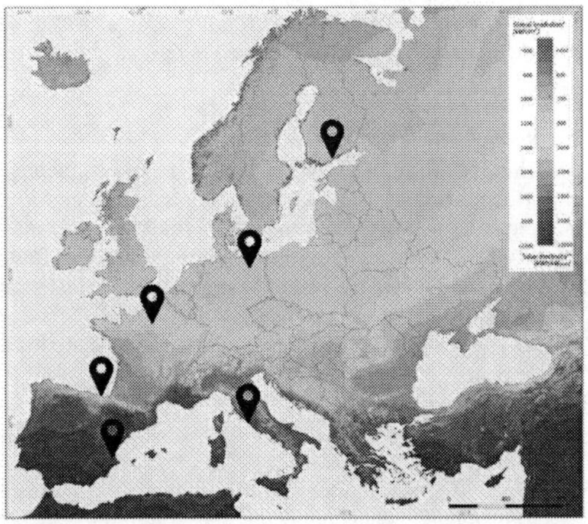

Figure 3: Map representation of the considered 6 different location as function of their solar potential and thermal loads.

4.3 Scenarios

The simulated scenarios have been the previously enumerated 5 different technological approaches, although different system sizes have been considered for each case study. The system sizing has been carried out according to the annual maximum peak demands.

5 RESULTS

The overall energy results show (Fig. 4 and Tab. I-VI) that PV could reduce the annual grid consumption within baseline scenario for different case studies (4%-24%), but the aggregated self-consumption figures (27%-39%) evidence the high decoupled electric generation and consumption. This effect combined with non-favourable regulation driven grid injection costs, decreases its potential profitability although still low paybacks could be achieved (6,4 c€/kWh).

The high annual thermal loads (60%-74%) are the core demands that need to be tackled towards NZEB. In this sense, the power-to-heat approach of HP technology could reduce significantly the whole grid energy consumption (45%-127%), but from an economic perspective these advantages may not be attractive due to required extra investment, cheap gas/boiler, high electricity price or high peak/annual thermal demand ratio.

However, PV & HP combination figures reveals that self-consumption ratios could be significantly enhanced (28%-43%) for conventional and (~80%) for smart couple greater PV fields, reducing the grid consumption and stress. The HP annual grid consumption is significantly reduced. Furthermore, the 20-year period obtained global (electricity & heat) LCOE for such combo systems is the lowest among all case studies. However, the required initial investment (~6 times higher) may reduce its attractiveness for some end-users/investors.

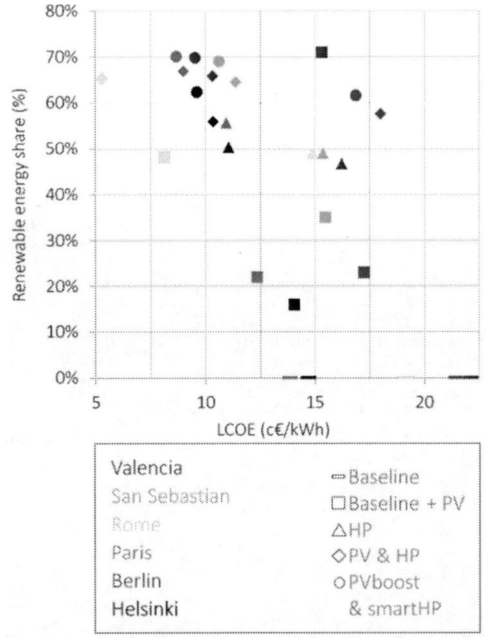

Figure 4: Representation of the obtained results for the analysed 6 case studies and 5 scenarios.

5.1 Valencia

Table I: Summary KPIs for Valencia case study.

Energy demand (MWh/year)		Heating	Cooling	DHW	Electricity	Total
		79	119	71	183	452
		18%	26%	16%	40%	100%
Scenarios		Baseline	Baseline + PV	HP	PV & HP	PV boost & smart HP
Features		NG boiler 2x 160 kW DHW & inertia 4000 l *no cooling	NG boiler 2x 160 kW DHW & inertia 4000 l PV 150 kWp S-40° opt. *no cooling	HP 2x 120 kW DHW & inertia 12000 l	HP 2x 120 kW DHW & inertia 12000 l PV 150 kWp S-40° opt.	HP 2x 120 kW DHW & inertia 12000 l PV 150 kWp S-40° opt. 50 kWp S-90°
CAPEX (k€)		53	203	106	256	331
OPEX (k€/year)		55	32	53	26	21
Grid consumption	**Electricity**	**183**	**116**	**241**	**154**	**136**
(MWh/year)	Natural gas	**167**	**167**	-	-	-
PV generation (MWh/year)		0	238	0	238	289
Self-consumption (MWh/year)		-	67	-	86	104
Grid injection (MWh/year)		-	171	-	152	185
Renewable energy share (%)		0%	71%	47%	66%	70%
Payback (years)		ref.	**6,4**	**22,6**	**7,0**	**8,0**
IRR (%)			14,4%	-	12,2%	9,2%
LCOE* (c€/kWh)		**22,10**	**15,32**	**16,22**	**10,33**	**9,53**

5.2 San Sebastian

Table II: Summary KPIs for San Sebastian case study.

Energy demand (MWh/year)	Heating	Cooling	DHW	Electricity	Total
	191	36	116	183	526
	36%	7%	22%	35%	100%
Scenarios	Baseline	Baseline + PV	HP	PV & HP	PV boost & smart HP
Features	NG boiler 2x 160 kW DHW & inertia 4000 l *no cooling	NG boiler 2x 160 kW DHW & inertia 4000 l PV 150 kWp S-40° opt. *no cooling	HP 2x 120 kW DHW & inertia 12000 l	HP 2x 120 kW DHW & inertia 12000 l PV 150 kWp S-40° opt.	HP 2x 120 kW DHW & inertia 12000 l PV 150 kWp S-40° opt. 50 kWp S-90°
CAPEX (k€)	53	203	106	256	331
OPEX (k€/year)	70	51	59	37	31
Grid consumption **Electricity**	**183**	**120**	**268**	**186**	**163**
(MWh/year) **Natural gas**	**341**	**341**	**-**	**-**	**-**
PV generation (MWh/year)	0	173	0	173	212
Self-consumption (MWh/year)	-	63	-	82	105
Grid injection (MWh/year)	-	110	-	91	107
Renewable energy share (%)	0%	35%	49%	65%	69%
Payback (years)	ref.	**7,8**	**4,6**	**6,0**	**7,0**
IRR (%)		9,8%	22,3%	15,8%	12,3%
LCOE* (c€/kWh)	**18,99**	**15,48**	**15,38**	**11,37**	**10,62**

5.3 Rome

Table III: Summary KPIs for Rome case study.

Energy demand (MWh/year)	Heating	Cooling	DHW	Electricity	Total
	182	65	89	183	519
	35%	12%	17%	35%	100%
Scenarios	Baseline	Baseline + PV	HP	PV & HP	PV boost & smart HP
Features	NG boiler 2x 200 kW DHW & inertia 4000 l *no cooling	NG boiler 2x 200 kW DHW & inertia 4000 l PV 150 kWp S-40° opt. *no cooling	HP 2x 150 kW DHW & inertia 12000 l	HP 2x 150 kW DHW & inertia 12000 l PV 150 kWp S-40° opt.	HP 2x 150 kW DHW & inertia 12000 l PV 150 kWp S-40° opt. 50 kWp S-90°
CAPEX (k€)	56	206	120	270	345
OPEX (k€/year)	65	21	56	11	1
Grid net consumption **Electricity**	**183**	**120**	**264**	**179**	**157**
(MWh/year) **Natural gas**	**302**	**302**	**-**	**-**	**-**
PV generation (MWh/year)	0	217	0	217	264
Self-consumption (MWh/year)	-	63	-	85	108
Grid injection (MWh/year)	-	154	-	132	156
Renewable energy share (%)	0%	48%	49%	65%	70%
Payback (years)	ref.	**3,4**	**6,5**	**3,9**	**4,5**
IRR (%)		31,7%	14,1%	26,9%	23,2%
LCOE* (c€/kWh)	**19,11**	**8,14**	**14,89**	**5,29**	**3,60**

5.4 Paris

Table IV: Summary KPIs for Paris case study.

Energy demand (MWh/year)	Heating	Cooling	DHW	Electricity	Total
	381	53	138	183	754
	50%	7%	18%	24%	100%
Scenarios	Baseline	Baseline + PV	HP	PV & HP	PV boost & smart HP
Features	NG boiler 2x 250 kW DHW & inertia 4000 l *no cooling	NG boiler 2x 250 kW DHW & inertia 4000 l PV 150 kWp S-40° opt. *no cooling	HP 2x 170 kW DHW & inertia 12000 l	HP 2x 170 kW DHW & inertia 12000 l PV 150 kWp S-40° opt.	HP 2x 170 kW DHW & inertia 12000 l PV 150 kWp S-40° opt. 50 kWp S-90°
CAPEX (k€)	61	211	130	280	355
OPEX (k€/year)	73	59	59	42	37
Grid consumption **Electricity**	**183**	**124**	**334**	**250**	**226**
(MWh/year) **Natural gas**	**576**	**576**	**-**	**-**	**-**
PV generation (MWh/year)	0	155	0	155	190
Self-consumption (MWh/year)	-	59	-	84	108
Grid injection (MWh/year)	-	96	-	71	82
Renewable energy share (%)	0%	22%	56%	67%	70%
Payback (years)	ref.	**10,7**	**5,0**	**7,0**	**8,2**
IRR (%)		2,6%	20,6%	12,3%	8,9%
LCOE* (c€/kWh)	**13,83**	**12,33**	**10,95**	**8,98**	**8,67**

5.5 Berlin

Table V: Summary KPIs for Berlin case study.

Energy demand (MWh/year)	Heating	Cooling	DHW	Electricity	Total
	301	50	154	183	587
	44%	7%	22%	27%	100%
Scenarios	Baseline	Baseline + PV	HP	PV & HP	PV boost & smart HP
Features	NG boiler 2x 200 kW DHW & inertia 4000 l *no cooling	NG boiler 2x 200 kW DHW & inertia 4000 l PV 150 kWp S-40° opt. *no cooling	HP 2x 150 kW DHW & inertia 12000 l	HP 2x 150 kW DHW & inertia 12000 l PV 150 kWp S-40° opt.	HP 2x 150 kW DHW & inertia 12000 l PV 150 kWp S-40° opt. 50 kWp S-90°
CAPEX (k€)	56	206	120	270	345
OPEX (k€/year)	87	64	99	72	64
Grid net consumption (MWh/year) Electricity	**183**	**127**	**325**	**249**	**225**
Grid net consumption (MWh/year) Natural gas	**505**	**505**	**-**	**-**	**-**
PV generation (MWh/year)	0	144	0	144	178
Self-consumption (MWh/year)	-	56	-	75	99
Grid injection (MWh/year)	-	88	-	69	79
Renewable energy share (%)	0%	23%	45%	58%	62%
Payback (years)	ref.	**6,4**	**-**	**13,6**	**12,1**
IRR (%)		14,3%	-	-3,5%	-0,5%
LCOE* (c€/kWh)	**21,42**	**17,24**	**22,78**	**17,98**	**16,86**

5.6 Helsinki

Table VI: Summary KPIs for Helsinki case study.

Energy demand (MWh/year)	Heating	Cooling	DHW	Electricity	Total
	375	43	209	216	843
	45%	5%	25%	26%	100%
Scenarios	Baseline	Baseline + PV	HP	PV & HP	PV boost & smart HP
Features	NG boiler 2x 250 kW DHW & inertia 4000 l *no cooling	NG boiler 2x 250 kW DHW & inertia 4000 l PV 150 kWp S-40° opt. *no cooling	HP 2x 170 kW DHW & inertia 12000 l	HP 2x 170 kW DHW & inertia 12000 l PV 150 kWp S-40° opt.	HP 2x 170 kW DHW & inertia 12000 l PV 150 kWp S-40° opt. 50 kWp S-90°
CAPEX (k€)	61	211	130	280	355
OPEX (k€/year)	70	61	67	57	49
Grid net consumption (MWh/year) Electricity	**216**	**181**	**419**	**371**	**317**
Grid net consumption (MWh/year) Natural gas	**649**	**649**	**-**	**-**	**-**
PV generation (MWh/year)	0	131	0	131	163
Self-consumption (MWh/year)	-	35	-	48	101
Grid injection (MWh/year)	-	95	-	83	62
Renewable energy share (%)	0%	16%	50%	56%	62%
Payback (years)	ref.	**16,7**	**26,4**	**16,7**	**14,1**
IRR (%)		-10,3%	-	-10,4%	-4,6%
LCOE* (c€/kWh)	**14,67**	**14,04**	**11,05**	**10,34**	**9,61**

6 CONCLUSIONS

A residential use, 95 dwelling and 1980 constructed building NZEB approach analysis has been carried out for 6 different case studies (Valencia, San Sebastian, Rome, Paris, Berlin and Helsinki) and 5 diverse scenarios (baseline natural gas boiler; baseline added PV system; heat pump installation; combined conventional PV and HP; hybrid PV & HP solution).

The results show that the solar PV potential of the building could change considerably between case studies, although the annual building-external energy need is slightly reduced. However, the high decoupled generation and consumption profiles draw poor self-consumption ratios for the common baseline plus PV conventional systems (non-battery or load shifting) scenario. This phenomenon, added to the already currently usual grid injected/consumed energy cost gap, decrease the real profitability of the PV systems and slows its massive development.

In the same way, the building energy demand is overall significantly increased with latitude. However, the HP approach for satisfying building thermal needs show great figures, regardless different case studies energy costs and climatic conditions. Furthermore, the reversible air-source HP solution offers the added value of cooling while electrifies and decarbonises the high-polluting heat sector.

The combined PV and HP solutions results evidence that PV self-consumption ratios could be easily increased, enhancing the PV field profitability and reducing the HP operational costs at the same time. These double benefits drive the overall entire-lifetime cheapest energy costs, increase significantly the renewable energy share while for reasonable paybacks within all analysed case studies. Additionally, the energy KPIs of the enhanced PV and HP smart operation approach show that even higher energy and financial performance figures may be obtained.

These initial results show that the proposed hybrid PV and HP solution could be a currently efficient, renewable and cost-competitive approach towards the NZEB utopia, expanding the present PV market potential, securing the EU building energy sector energy supply and ensuring a reliable provision of sustainable energy while reducing emissions, pollution, fossil-fuel consumption and dependence.

ACKNOWLEDGEMENTS

The work carried out is comprised under the HANDLE innovation project between Tecnalia Research & Innovation, Energy Panel S.L, Termoidraulica Brotto, Termosol and InnoEnergy, with the collaboration of the University of the Basque Country.

REFERENCES

[1] International Energy Outlook of 2017, https://www.eia.gov/outlooks/ieo/

[2] European Commission EU building needs https://ec.europa.eu/energy/en/topics/energy-efficiency/buildings

[3] Sotehi, V/T Hybrid Solar Water Collector for Net Zero Building Energy (nZEB) in Algeria, 2016.

[4] JC Hadrn, Solar and Heat Pump Systems for Residential Buildings, ISBN: 978-3-433-03040-0, 2015.

[5] Bloess et al, Power-to-heat for renewable energy integration: A review of technologies, modelling approaches, and flexibility potentials, 2017.

[6] Latest trends in the World Traditional & Renewable Heating Markets, BSRIA, 2014. https://www.iea.org/media/workshops/buildingwebinars/webinar4/4BSRIAKrystynaPresentationforIEA

[7] EurObservER HPs barometer, 2016. https://www.eurobserv-er.org/heat-pumps-barometer-2016/

THE POSSIBLE ROLE OF PV IN THE FUTURE POWER SUPPLY OF THE FAROE ISLANDS

Hans Georg Beyer*, Isadora Pauli Custódio#
* Fróðskaparsetur Føroyar, Náttúruvísindadeildini
Tórshavn. Faroe Islands
#Universidade Federal de Santa Catarina, Departamento de Engenharia Civil
Centro de Pesquisa e Capacitação em Energia Solar da UFSC
Florianópolis – SC, Brazil

ABSTRACT: The Faroe Islands form a group of 18 islands located in the North Atlantic at 62° N. They are populated with about 51,000 people. The capital city, Tórshavn has about 21,000 inhabitants. Currently, the energy supply is given on the electricity side by a mix of fossil fired power stations, hydropower plants and wind farms. The heating sectors are mainly based on oil burners. This is the reason to restructure the energy supply to abandon the use of fossil fuels by 2030. There is an ongoing discussion on the best option to reach this goal. First scenarios had been set up. This paper aims to contribute to this discussion by highlighting the options of the inclusion of PV power at this location.
Keywords: Hybrid, Energy Performance, Modelling, System Performance, Utilities

1 INTRODUCTION

On the Faroe Islands - located remotely at 62°N, 7°W in the North Atlantic – an isolated power grid for a population of about 51,000 people is operated. Currently, the generation is based on fossil fired power stations, hydropower plants and wind farms. that the objective is to phase out fossil generation by 2030 [1], [2], whereby the electricity consumption is expected to increase due not only to the replacement of the currently predominant oil based heat supply of the housing stock by electrical heating systems, but also due to the shift to electric mobility. To contribute to the design of the purely renewable based power system, the option of the inclusion of PV capacity is more closely inspected here.

2 THE CURRENT STATE OF THE POWER SYSTEM

Table I shows the rated peak power and the average power generation of the diesel stations, hydropower plants and wind farms in the year of 2017.

Table I: Faroe Islands' power system for the year of 2017.

	Rated peak power (MW)	Average power generation (MWh)
Diesel	36.2	+16.1
Hydro	33.8	+12.3
Wind	18.4	+6.7
Load	53.9	-35.1

In 2017, the generation by wind and hydropower had a share of 54% to the load coverage (previous years had shown higher shares). Instantaneously, the wind power share may reach values that are more elevated. For reasons of system stability it this currently intended to limit the share of this intermittent power to 60%, leading to occasional curtailment (not prominent in 2017). It should be remarked that the "off-shore" wind conditions on the Faroe Islands lead to a good value for the wind capacity factor (average power generated/ rated peak power) of 0.36.

The next session will introduce characteristics of PV generation on the Faroe Islands, followed by an inspection of the temporal patterns of the load consumption, the wind and PV Generation. Monthly scale was chosen to get a first insight in the feasibility of its inclusion.

3 PV ON THE FAROE ISLANDS

First experiences with PV operation under the Faroe Islands' have shown some less favourable conditions, such as high latitude and cloudy skies, and some more beneficial conditions, such as year around modest temperatures. Those experiences were gained by the operation of a small set of kW size grid connected systems, as shown in Fig. 1.

Figure 1: Small PV system installed in 2013 at Tórshavn, Faroe Islands, to gain insight in system performances under the specific meteorological operation conditions at 62°N, 7°W. Blue sky as depicted here is a rare occasion.

Fig. 2 gives an impression of the operation of a system installed in Tórshavn (the Faroe Islands' capital), highlighting the variability of conditions in summer and winter months. The monthly performance ratios (PR) for these sets are 0.95 for July and 0.63 for January. These values were calculated after removing data pairs affected by pyranometer shading.

According to this data set, a simulation scheme as described in [3] is set up. Data for hourly horizontal irradiance and ambient air temperature for Tórshavn are supplied by the Danish Meteorological Institute - DMI. To receive the best annual irradiance, a module tilted at 42° is identified, resulting in an annual irradiance of 834 W/m^2. By this assumption, the pattern of the potential generation of systems installed in this location can be determined. It will be used in the following sessions.

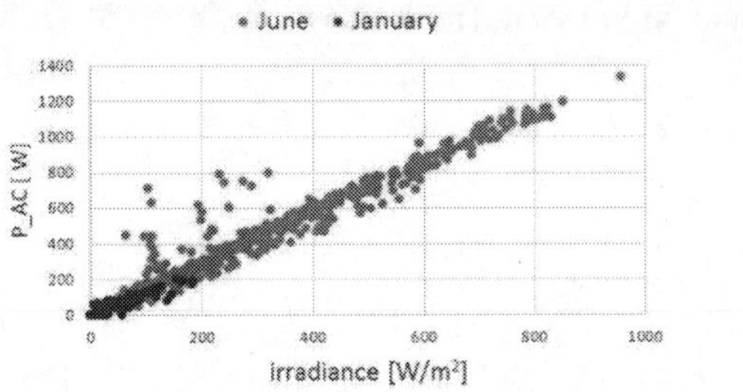

Figure 2: Plot of monthly power versus irradiance characteristics of a section of the generator depicted in Fig. 1 for the months of January and June. Scatter in the lower irradiance region are due to occasional shading of the pyranometer. Large seasonal differences can be seen.

4 MONTHLY PATTERNS OF LOAD, WIND AND PV GENERATION

Data of load and generation by the diesel sets, wind and hydropower (supplied by the system operator SEV) are available with hourly time resolution. The respective monthly means of load and generation by wind, PV and diesel are normalized to the respective annual means and presented in Figs. 3-5.

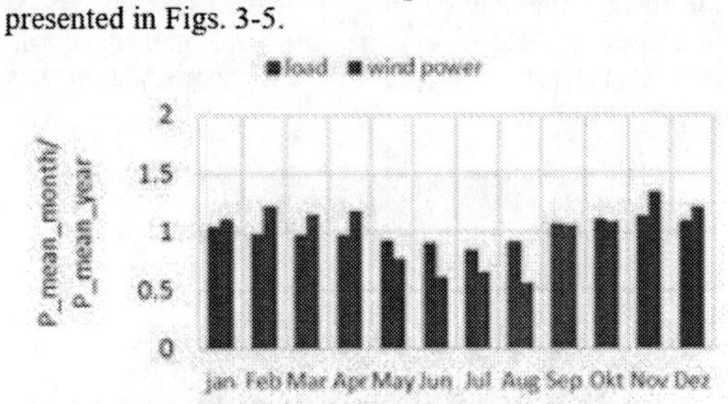

Figure 3: Normalized (monthly mean power relative to annual mean power) monthly load and wind generation in 2017.

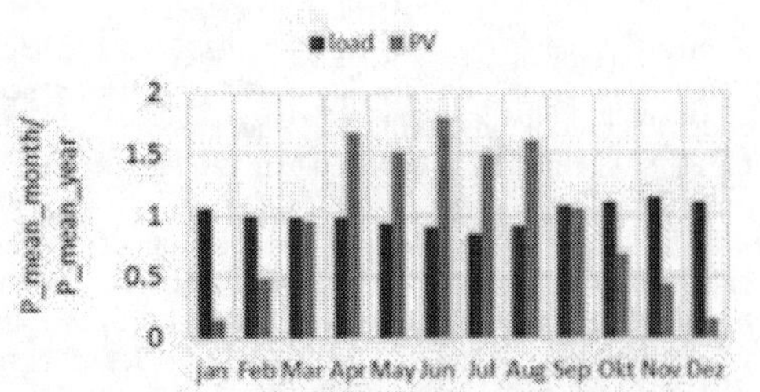

Figure 4: Normalized monthly load and PV generation in 2017.

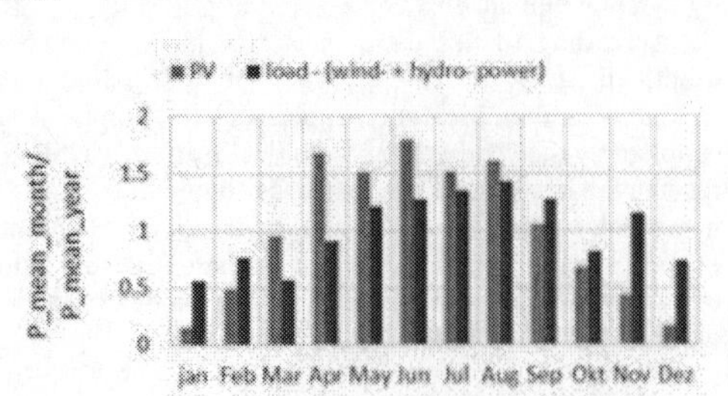

Figure 5: Comparison between PV generation and residual load (load minus wind and hydro generation).

The cross correlations of these sets are given in Table II. A basic good match of wind and load is visible (see also [4]), However, in detail, differences occur. Wind generation scaled to match the mean annual load would result in a surplus in winter and deficit in summer. The PV generation is basically anti-correlated with the load – low PV generation in the winter month with the highest gross load – but shows reasonable correlation to the residual load. Thus, the PV generation may contribute to balance the shortcomings of wind generation in summer.

Session 4 gives an overview of the inclusion of PV into the current system. Session 5 inspects this effect for the future system.

Table II: Cross correlation of monthly generation and load.

Combination	Cross correlation
Wind ↔ Load	0.81
PV ↔ Load	- 0.79
PV ↔ Residual load	0.68

5 INSPECTION OF PV INTEGRATION INTO THE CURRENT SYSTEM

Based on the hourly data of 2017 for load, wind and hydro generation and the simulated time pattern of PV generation, the possible reduction of the generation by the diesel sets due to PV generation is determined. By doing this, requirements for a minimal share of controllable power (see session 2) can be respected.

Fig. 6 shows the fraction of load covered by PV as function of the installed PV power for an acceptable share of wind and solar of 60% and 80%. It can be observed that initial linear increase of load coverage is flatted out with increasing PV capacity. The sensitivity to the required share of controllable generation can be seen.

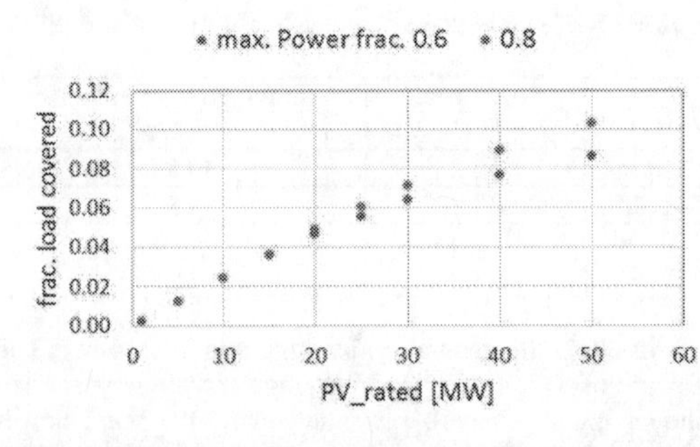

Figure 6: Fraction of load covered by PV as function of PV installed capacity for different acceptable max. instantaneous fractions of wind and PV. The fraction of 0.6 refers to the current situation.

Fig. 7 shows the respective fractions of PV power that has to be shed due to this curtailment. Increased shedding applies for additional wind capacity as well, calling for ways to maintain grid stability other than the operation of active synchronous generators. Respective investigations are under way by the grid operator SEV.

The results for a scenario with both an increased load

and an increased wind and PV generation, presented in the next session are gained with the assumption that grid stability can be secured without active generation by diesel sets.

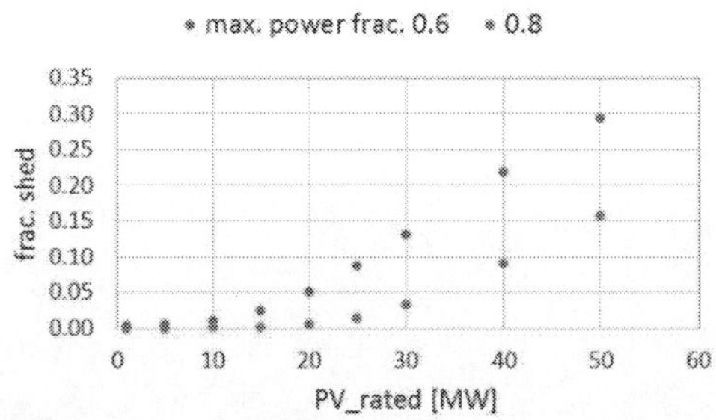

Figure 7: Same presentation as Fig. 6, but for the fraction of possible PV generation that has to be shed due to curtailment.

6 PV AS PART OF A SCENARIO FOR A CLEAN POWER SYSTEM

The scenarios analysed here are based on the skeleton given in [1] where the discussion of different scenarios is given. It is assumed that the load is shared between the current state increased due to coverage of heating loads and electrification in the transport sector. The operation of the diesel sets is discontinued. The actual hydropower scheme operating in pure generation mode is assumed to be unchanged. A pumped hydro scheme is added presenting a storage capacity of about 4.9 days of annual mean load.

Table III: Basics for the scenarios for the future power supply discussed here.

Load	Diesel	Hydro	Wind	PV
Max. power [MW]				
103.7	0	33.8	?	?
Annual mean [MW]				
67.6	0	12.3	?	?
+ pumped hydro cap: 8000 MWh (~4.9 days of load)				

Next, the operation of the system using different wind and PV capacities is analysed, focussing on the state of storage, indicating whether the load can be served or not. The simulation is based on the scheme given above. The time pattern of load and wind generation is modelled by an upscaling of the 2017 data, according to the rating given in Tab. II and the selected installed wind capacity. The same applies for the PV generation. The pumped hydro storage is modelled assuming an efficiency of 90% for both loading and unloading.

For a case a) the wind capacity is scaled for an annual generation equal to the annual load. The time trance of the storage content shown in Fig. 8 indicates several periods of depleted storage i.e. failures in serving the load most notable in late summer.

An increase of the installed wind capacity by both a factor of 2 [case b)] and 3 [case c)] improve the load coverage but cannot give a complete coverage. Reason for this can be traced back to an approximately weeklong periods with low to none wind generation. Comparison with a wind data set from another location on the Faroe Islands indicate a larger scale of low wind situation. Figs. 9 and 10 give the respective evolutions of the storage content and Tab. IV gives the installed wind capacities and generations.

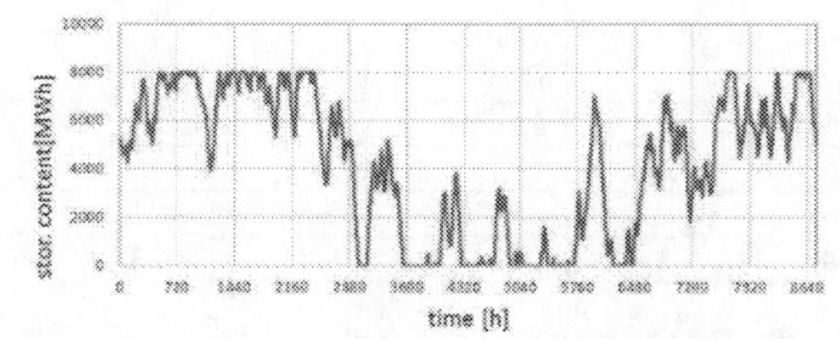

Figure 8: Evolution of the storage content for a system with a mean load of 67.6 MW, mean pure generation hydropower of 12.3 MW and a mean potential wind generation of 55.3 MW (installed wind power: 151 MW).

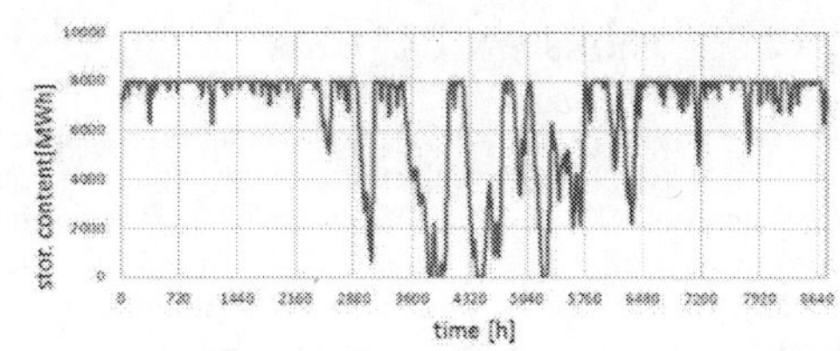

Figure 9: Same presentation as Fig. 8, but for a mean potential wind generation of 110.6 MW (installed wind power 302 MW).

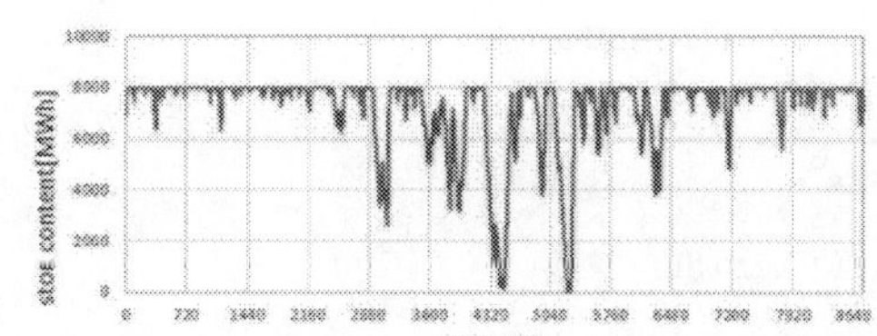

Figure 10: Same presentation as Fig. 9, but for a mean potential wind generation of 165.9 MW (installed wind power 455 MW).

The inclusion of PV is inspected based on case a). The necessary size of a PV generator to ensure the coverage of the load is identified to be 189 MW with a potential generation of 55.3 MW. The evolution of the respective storage content is given in Fig. 11. The lower boundary of the storage is reached, but the storage is refilled in the subsequent time step.

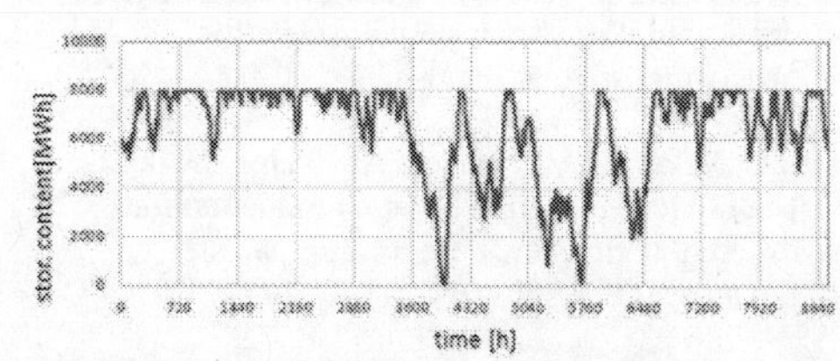

Figure 11: Same presentation as Fig. 10, but for a system using a wind capacity of 55.3 MW and a PV capacity of 189 MW with a potential mean generation of 16,3 MW.

Table IV: Values for installed and annual mean power for the different scenario cases.

Case	Inst. power Wind [MW]	Annual mean power Wind [MW]	Inst. power PV [MW]	Annual mean power PV [MW]
a	55.3	151	0	0
b	110.6	302	0	0
c	165.9	453	0	0
d	55.3	151	189	16.3

As could be expected from the monthly analyses, the addition of the PV can help to overcome the periods of low generation by the wind turbines. Surplus in potential generation is remarkably lower than for the close to sufficient scaled hydro and wind only systems.

7 CONCLUSIONS AND OUTLOOK

The analysis given here has shown that the inclusion of PV capacity can improve the performance of large wind based renewable supply systems even at locations with exceptional good wind conditions. This is one aspect that should be taken into account when designing the most favourable system, together with the assessment of the options of increased generation only hydro capacity and the possible inclusion of tidal stream power as predictable power source.

ACKNOLEDGEMENTS

Thanks are due to the Faroe Islands Energy Agency (Umhvørvisstovan) and the grid operator SEV for data access.

REFERENCES

[1] Anonymous, 100% fornybar kraft, Pumpekraft, vind og sol, Torshavn, Umhvørvisstovan Orka og Elfelagið SEV, Tórshavn, Faroe Islands, July (2018)

[2] Anonymous, Balancing a 100% renewable electricity system, Least cost path for the Faroe Islands, Ea Energy Analyses, Copenhagen, Denmark, July (2018)

[3] A.Drews, A.C.de Keizer, H.G.Beyer, E.Lorenz, J.Betcke, W.G.J.H.M.van Sark, W.Heydenreich, E.Wiemken, S. Stettler, P.Toggweiler, S.Bofinger, M.Schneider, G.Heilscher, D. Heinemann, Monitoring and remote failure detection of grid-connected PV systems based on satellite observations, Solar Energy 81 (2007) 548–564

[4] H G. Beyer, B.A. Niclasen, Analyzing the match of heating load and wind turbine production for different regions in Northern Europe, Bremen, Germany, DEWEK 2017 , 17-18.10 (2017)

MODELLING DIFFERENT PV-BASED COMMUNAL GRIDS ARCHITECTURES FOR RURAL DEVELOPING COMMUNITIES

Dr. Nicholas Nixon Opiyo
Centre for Integrated Energy Research
University of Leeds, Leeds, LS2 9JT, UK
Email: Nicholas.Opiyo1@gmail.com

ABSTRACT: Impacts of grid architectures on temporal diffusion of PV-based communal grids (community owned minigrids) in a rural developing community are modelled and simulated using MATLAB/Simulink and a survey-informed agent-based model (ABM) developed in NetLogo. Results show that decentralised control architectures stimulate minigrid formations and connections by allowing easy expansions of the minigrids, as each decentralised PV system within a minigrid is treated equally and determines its own real and reactive power, eliminating the need for communication links. This also reduces the cost of implementing such a system; fewer connections are realized with centralised controls as such systems require high speed communication links which make them both difficult to expand and expensive to implement. Results also show that multi-master operation modes lead to more communal grid connections compared to single-master operation modes because in the former, all distributed PV systems within a communal grid have the same rank and can act as masters or can be operated as combinations of master generators (VSIs) and PQ inverters, allowing for more design flexibility and easy connections from potential customers.

Keywords: Communal Grids; Control; Architecture; DC/AC Coupling

1 INTRODUCTION

A PV-based communal grid is defined as a locally confined and independently controlled electric power grid in which a distribution architecture integrates distributed loads and distributed energy resources—i.e. local distributed PV systems and energy storage devices—which allows the communal grid to operate in both islanded and utility grid-connected modes. Communal grids are operated in islanded mode when they are situated far from existing utility transmission networks, or if the owners of the grid prefer independence of their own power source. On the other hand, communal grids can be utility-grid connected to improve system economics, to improve operations, and to improve stability, while the utility grid gets to enjoy improved availability, improved stability, and reduced conductor sizes. For a successful interfacing of communal grids to utility grids, converters/inverters should allow net metering (bi-directional flow of power), peak shaving, advanced communications and control, and fast demand response.

Inverters used to interface communal grids with utility grids can be classified according to modes of operation as PQ or V-f (also known as voltage source inverter (VSIs)). A PQ inverter controls the real (P) and reactive (Q) power by adjusting the magnitude of the output real and reactive current. It therefore operates as a voltage controlled current source [1]. A voltage source inverter controls the voltage (V) and frequency (f) at the output terminal, and thus operates as a voltage source [1,2]. The mode of an inverter operation is chosen depending on a communal grid's architecture and control strategy, and may change depending on whether the communal grid is islanded or grid-connected. Unlike synchronous generators, inverters do not have rotors and thus no natural connection between frequency and active power. To achieve stable operation with multiple distributed PV systems, the inverters are controlled so that they mimic the characteristics of synchronous generators with P-f and Q-V droop controls [3]. In P-f and Q-V droop controllers for a PQ inverter, P and Q injections are changed proportionally to frequency and voltage deviations, respectively. On the other hand, P-f and Q-V droop controllers of a VSI changes the frequency and the voltage proportionally to P and Q deviations [3]. These relationships are modelled as [1-3]:

$$P(f) = P_0 - (f_{set} - f)k_f \quad (1)$$

$$Q(V) = Q_0 - (V_{set} - V)k_v \quad (2)$$

where P_0 is the power delivered by the inverter at setpoint frequency f_{set}, k_f is the gradient of the droop, which determines how much the active power P will change in response to a change in frequency f, Q_0 is the reactive power delivered/consumed by the inverter at setpoint voltage V_{set} and k_v is the gradient of the droop, which determines how much the reactive power Q will change in response to a change in voltage V.

The two general control structures used in communal grids are centralised control and decentralised controls [4,5]. In centralised control, a central controller (CC) determines the operating points of the communal grid and sends this information via communication systems to some or all the distributed PV systems and potentially to some of the loads throughout the communal grid network. In decentralised control, a virtual communication system independently determines the operating points of each decentralised PV system and load within the communal grid network, effectively eliminating the need for communication links and thus increasing system reliability as well as reducing cost. This enables easy and cost-effective expansion of the communal grid by enabling decentralised PV systems and loads to have plug-and-play capabilities. The two configurations can be implemented under two main modes of operation, i.e., master slave or multi master [6].

In master slave mode of operation, all the inverters in the network operate in PQ mode when the communal grid is connected to the utility grid, since the latter controls the voltage and frequency. In islanded mode, a single master inverter switches to VSI mode to provide the voltage and frequency references while the remaining distributed PV systems continue to operate in PQ mode [1]. The system can operate in either a centralised configuration or in a decentralised configuration. In multi-master mode of operation, all distributed PV systems within a communal grid can act as masters or can be operated as combinations of master generators (VSIs) and PQ inverters [1]. This mode shares many aspects with single master operation and can also be operated in centralised or decentralised configuration.

In this work, simulation results from different communal grid architecture models developed in MATLAB are fed into an agent-based model developed in NetLogo to determine the most cost-effective option. The novel aspect of this work is that it combines a socio-economic agent-based model with a technical model developed to simulate the most cost-effective communal grid control architecture.

2 METHODOLOGY

2.1 Simulink Models

Different communal grid architectures, revolving about energy storage, are modelled and simulated in MATLAB/Simulink to determine the most cost-effective option for a given transmission and distribution network. Each system modelled comprises of 4 PV arrays, each rated at 100 kW at 1000 W/m^2 radiation and 25 °C.; each array consists of 66 parallel strings, each comprising 5 PV 330 SunPower (SPR-305-WHT-D) modules connected in series (66 x 5 x 305.2 = 100.7 kW). Each array is connected to a 5 kHz boost DC-DC converter with maximum power point tracking (MPPT) and charge-controlling capabilities. Each converter uses the perturb-and-observe method to extract maximum voltage (273V DC) from each array and then boosts it to 500V DC. For utility grid interfacing, a DC-AC inverter inverts the DC bus voltage to 260V AC, while keeping a unity power factor. The inverter uses two control loops: one which regulates DC link voltage to +/-250 V and an internal control loop which regulates active and reactive current. Active current reference is the output of the DC voltage external controller while reactive current reference is set to zero to maintain unity power factor. A 40-kvar capacitor bank is used to filter harmonics produced by the inverter. The inverter is connected to the utility grid through a 400-kVA 260V/25kV step-up transformer.

Each of the communal grids modelled in this work is classified as ether DC-coupled or AC-coupled depending on the common bus voltage. A grid using a DC bus avoids many of the power conversion steps required when using an AC bus leading to higher energy efficiency and cost-saving. Traditionally, AC enabled efficient voltage transformation and high-voltage power transmission over long distances. Recently however, technology advances have led to highly efficient AC/DC and DC/DC converters, making high-voltage DC long-distance bulk power transmission more efficient [7]. Due to lower power and energy rating, stability issues are more prevalent in communal grids than in utility grids. Analyses of stability issues in AC-coupled communal grids follow the same concepts as with utility grids, i.e.:

- Voltage and frequency values need to both be regulated through active and reactive power control
- If a decentralised power source is a traditional synchronous generator with an AC output, and is connected directly to the utility grid without power electronic interfaces, stability is controlled through the machine shaft's torque and speed control

In DC systems, there are no reactive power interactions which suggests that there are few stability issues; system control seems to be oriented towards voltage regulation only. Figure 1 shows a Simulink model of a DC-coupled communal grid with centralized storage.

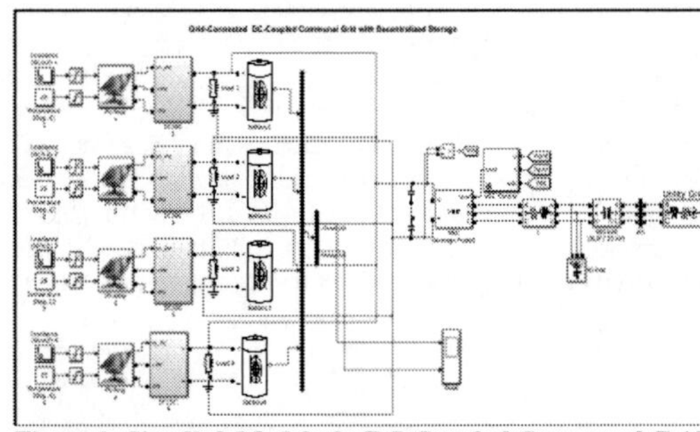

Figure 1: Simulink Model of a DC-Coupled Communal Grid with Centralised Storage

2.2 Agent Based Model (ABM)

Simulated data from the Simulink models are fed into an agent based model (ABM) developed in NetLogo, to simulate effects of control architectures on temporal diffusion of PV-based communal grids in a rural developing community, with the main driving factor being the cost of setting up and connecting to a given control architecture. Kendu Bay survey data are used to inform the ABM [8].

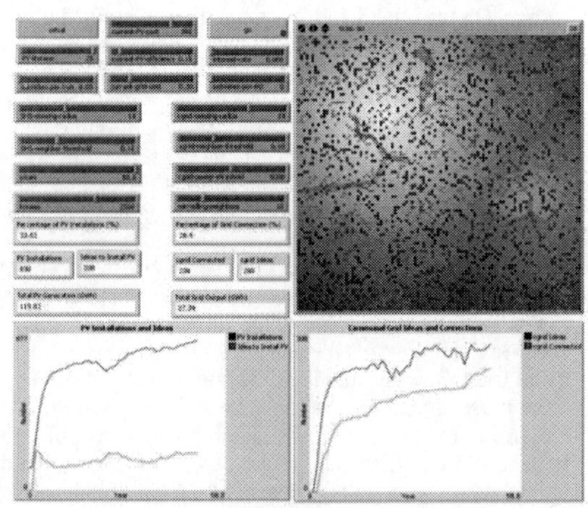

Figure 2: NetLogo Graphical User Interface

Figures 2 shows the graphical user interface of the NetLogo model used to simulate temporal diffusion of the communal grids with the view after 25 years. The landscape is coloured green with the lighter areas being hill tops. Black houses are those that are unelectrified. Houses deciding on installing PV are coloured white while those that have installed PV are coloured yellow. Houses with PV that meet the communal grid-power-threshold and are deciding on joining communal grids are coloured

red. Houses that have joined communal grids are coloured blue and are linked to other houses in the communal grid through grey links (grid lines). From the figure, nearby communal grids join to form even larger regional grids.

The following agents are created in the model: a) a representation of the environment and the solar potential in it, b) the populations in it that require electricity, c) PV seeds that would use the environment to produce electricity, d) links which are used by households with PV to connect to communal grids, and e) a central observer or stakeholder who determines the strategies and preferences for PV diffusions. Through these agents and the rules created for their interactions, the model is used to simulate the way decisions and preferences by human actors affect PV diffusion.

As modelled by Opiyo [8], a household without PV would consider installing PV if

$$LUCE_{PV} < C_{A/kWh} \tag{3}$$

where $C_{A/kWh}$ is avoided cost per kWh, i.e., the prevailing national grid electricity cost per kWh and equals \$0.20/kWh as per current Kendu Bay rates [8], while $LUCE_{PV}$ is the levelised unit cost of delivered electricity and is given by

$$LUCE_{PV} = \frac{ALCC_{PV}}{W_p \times EHFS \times 365 \times CUF} \tag{4}$$

where W_p is the rated peak Watt capacity of the PV module and is based on a household's activity profile and power demand [9], $EHFS$ is the equivalent hours of full sunshine per day and equals 8 [8,9], CUF is the capacity utilization factor which incorporates non-utilization and outages of systems due to various reasons and equals 0.9 based on Kendu Bay survey data [8], and $ALCC_{PV}$ is the annualized life cycle cost which is calculated by summing up the cost of all of its individual components, i.e. the module, battery, charge controller, and appliances multiplied by their respective capital recovery factors plus operations and maintenance costs. It is expressed as

$$\begin{aligned} ALCC_{PV} = (C_{OPV} \times CRF_{PV}) \\ + (C_{Obatt} \times CRF_{batt}) \\ + (C_{Occ} \times CRF_{cc}) \\ + (C_{Oappl} \times CRF_{appl}) \\ + C_{O\&M} \end{aligned} \tag{5}$$

where C_{OPV} is the capital cost of the PV module and is set at \$1.40/Wp based on Kenya's PV costs, C_{Obatt} is the capital cost of the battery and is set at \$0.95/Ah based on the price of a typical 48V 100 Ah battery in Kenya, C_{Occ} is the capital cost of the charge controller and is set at \$28.50 as per Kenyan rates, and C_{Oappl} is the capital cost of appliances and is averaged at \$75 per household based on typical load profiles of 95% of households with PV systems in Kendu Bay [8]. CRF_{PV}, CRF_{batt}, CRF_{cc}, and CRF_{appl} are the capital recovery factors of the PV module, the battery, the charge controller, and appliances, respectively, while $C_{O\&M}$ is the operations and maintenance cost and is set at \$2.50/year based on Kendu Bay survey data [8].

Capital recovery factor (CRF) is calculated using the formula

$$CRF = \frac{i(1+i)^n}{(1+i)^n - 1} \tag{6}$$

where i is the discount rate and is set at 12% based on average Kenya's central bank rates over the past 12 months, while n is the life of the particular component being considered, i.e. 25 years for PV modules/arrays, 5 years for batteries, 5 years for converters/inverters/charge controller, and 10 years for appliances (cumulative load). Based on these figures, $CRF_{PV} = 0.1339$, $CRF_{batt} = 0.2774$, $CRF_{cc} = 0.2774$, and $CRF_{cc} = 0.1770$.

For a communal grid consisting of N households, levelised unit cost of delivered electricity ($LUCE_{CG}$) is given by

$$LUCE_{CG} = \frac{ALCC_{CG}}{PV_{CG} \times EHFS \times 365 \times CUF} \tag{7}$$

where PV_{CG} is the capacity of the communal grid (PV array) in kWp and is based on the activity profiles of all households connected to the communal grid, $EHFS$ is the equivalent hours of full sunshine per day and equals 8, CUF is the capacity utilization factor which incorporates non-utilization and outages of systems due to various reasons and is maintained at 0.9, and $ALCC_{CG}$ is the annualized life cycle cost of the communal grid and is given by

$$\begin{aligned} ALCC_{CG} = [(C_{OPV_{CG}} \times CRF_{PV}) \\ + (C_{Obatt_{CG}} \times CRF_{batt}) \\ + (C_{Opcu} \times CRF_{pcu})] \\ \times [PV_{CG} \times R]^b \\ + [(C_{Odn} \times L) \\ + (C_{Osc} \times N)] \times CRF_{pdn} \\ + C_{O\&M_{CG}} \end{aligned} \tag{8}$$

where $C_{OPV_{CG}}$ is the capital cost of the PV array and associated mounting structures and is set at \$1,400/kWp, C_{OCGrid_batt} is the capital cost of the battery bank and associated structures and equals \$0.95/Ah $\times N$, C_{Opcu} is the capital cost of the power conditioning unit (power electronics) and equals \$28.50 $\times N$, CRF_{pcu} is the capital recovery factor of the power conditioning unit and equals 0.2774, C_{Odn} is the capital cost of the power distribution network per km and equals \$2,500 for Kendu Bay, C_{Osc} is the capital cost of service connections including internal wiring and appliances per household serviced and equals \$125 based on similar networks in Kenya, L is the length of the distribution network in km and is initially set at 1 km for a neighbourhood radius of 500 m, CRF_{pdn} is the capital recovery factor of the distribution network including the service connections and is calculated to be 0.1770, $C_{O\&M_{CG}}$ is the cost of operating and maintaining the communal grid and is estimated to be \$2.50 $\times N$, R is the benchmark unit cost of the communal grid and is set at \$1,400/kWp, and b is a scale factor for incorporating cost reduction in overall cost of the communal grid, without the power distribution network (pdn), due to bulk purchasing of the components used in the grid [10]. Its effect is uniformly distributed over all components of the communal grid, minus the distribution network. It is set at 0.95.

Once cost has been determined, PV is installed by a household if

$$\frac{H_{PV/SR}}{H_{Total/SR}} \times 100 > T_{SR} \quad (9)$$

where $H_{PV/SR}$ is the number of households with PV within a given sensing radius (SR) or neighbourhood, $H_{Total/SR}$ is the total number of households within the same sensing radius, and T_{SR} is the neighbourhood threshold. For Kendu Bay, the initial sensing radius is set at 500m based on population distribution; a sensing radius below 500m limits impacts of neighbourhood influence on a household to one's own clan, based on homestead and farm distributions [9]. The initial neighbourhood threshold is set at 10%; simulation results showed that values below 10% showed no impact (influence) on PV diffusion.

To be allowed to join a communal grid, a household must have electricity needs beyond lighting and must demonstrate this by having installed PV of a given minimum capacity (power-threshold), which is set at 2 KWp. They must also be within a given sensing radius of other houses with PV that meet the power-threshold. A household will join a communal grid if

$$\frac{H_{CG/SR}}{H_{Total/SR}} \times 100 > T_{SR} \quad (10)$$

where $H_{CG/SR}$ is the number of households with PV within the sensing radius that meet the power-threshold and $H_{Total/SR}$ is the total number of households within the same sensing radius.

3 RESULTS AND DISCUSSION

Figure 3 shows the mean power generated by each PV array, bassed on varying irradiances to mimic real environmental conditions. The four power outputs from the PV arrays sum up to the active power injected into the control inverter as shown in figure 4.

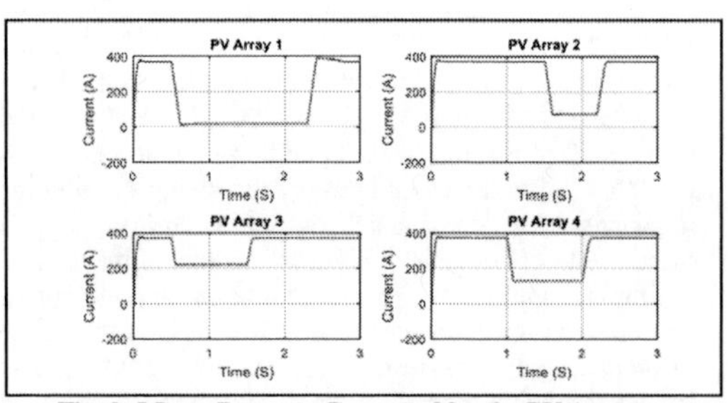

Fig. 3: Mean Currents Generated by the PV Arrays

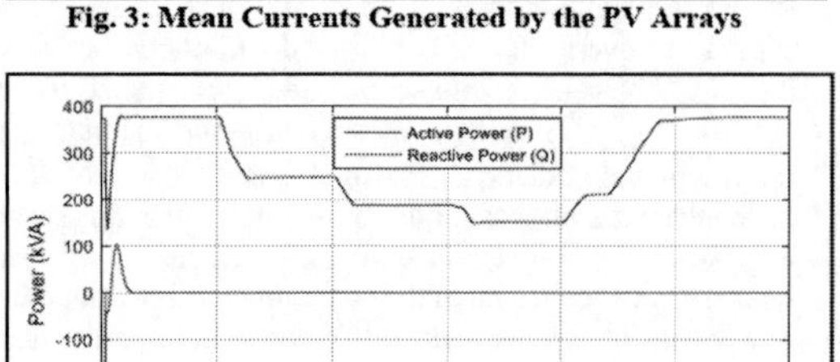

Fig. 4: Active and Reactive Power

Table 1 shows a comparison of additional power electronics required for different control architectures. From the table, DC-coupled networks with decentralised storage are the cheapest options for rural developing communities, with no additional costs beyond costs common to all other control architectures. These are followed by DC-coupled networks with centralised storage which required additional investments into centralised charge controllers. AC-coupled networks with decentralised storage which need 4 DC-AC inverters come in third, followed by AC-coupled networks with centralised storage which need a central charge controller in addition to the 4 DC-AC inverters. Grid connection requires additional investments for all the networks in forms of VSI, 40-kvar capacitor banks to filter harmonics produced by the VSIs, and 400-kVA 260V/25kV step-up transformers. It is assumed that all communal grids are initially islanded, and that investments into utility grid connections are made when it finally arrives at a given location.

Additional Power Electronics	Centralised		Decentralised	
	DC	AC	DC	AC
VSI	1	1	1	1
Central Charge Controller	1	1	0	0
DC-AC Inverter	0	4	0	4
Transformer	1	1	1	1
Filter	1	1	1	1
Total Additional Cost ($)	**250 + GC**	**950 + GC**	**GC**	**700 + GC**

Table 1: Comparison of Additional Power Electronics Required by Networks. (GC is Grid-Connection Cost)

3.1 DC-Coupled Communal Grids:

Figure 5 shows a comparison of households connected to communal grids in centralised and decentralised storage configurations after 25 years while figure 6 shows their corresponding percentages. After 25 years, 2,103 households would have joined communal grids with decentralised storage systems, representing 21.5% of all households. This is higher than the 1,571 households that would have joined networks with centralised storage systems, representing 16% of all households.

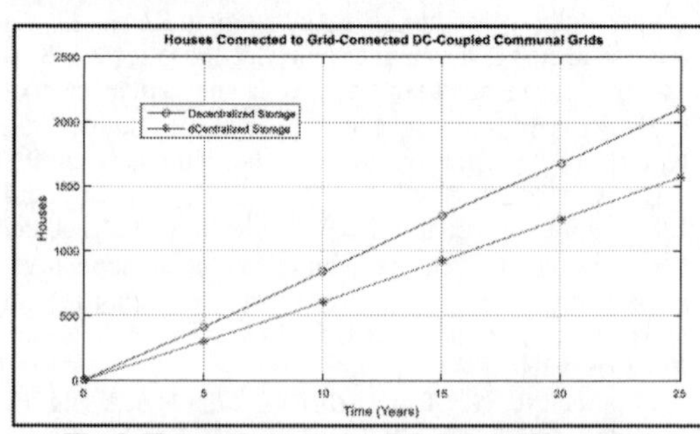

Fig. 5: Houses Connected to DC-Coupled Networks

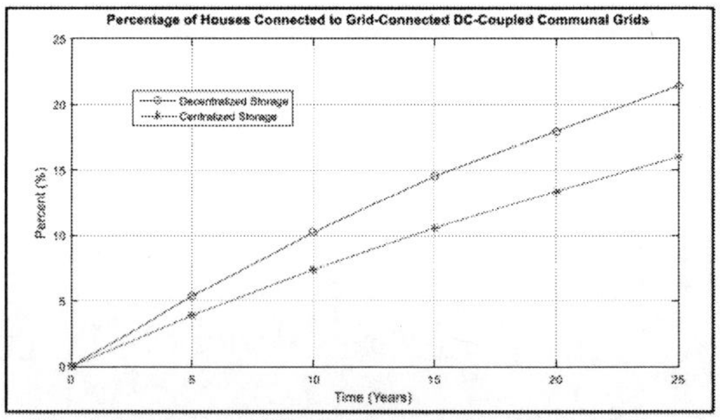

Fig. 6: Percentage of Houses Connected to DC-Coupled Networks

3.2 AC-Coupled Communal Grids:

Figure 7 shows a comparison of households connected to communal grids in centralised and decentralised storage configurations after 25 years while figure 8 shows their corresponding percentages. After 25 years, 1,887 households would have joined communal grids with decentralised storage systems, representing 19.2% of all households. This is higher than the 1,286 households that would have joined networks with centralised storage systems, representing 13.1% of all households

.

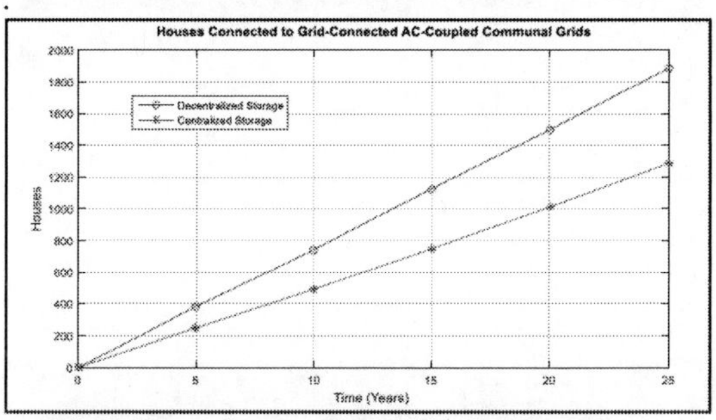

Fig. 7: Houses Connected to AC-Coupled Networks

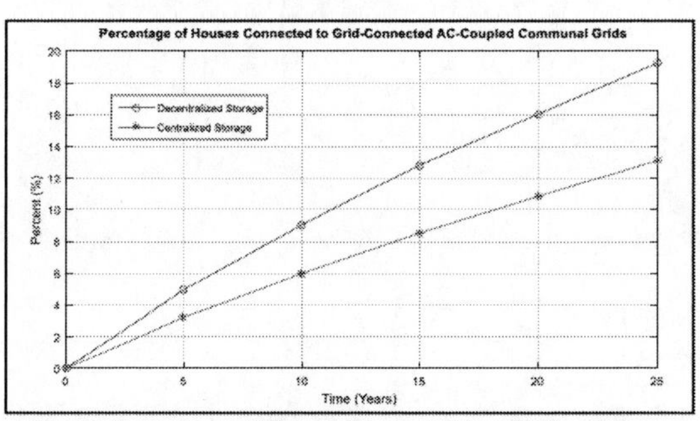

Fig. 8: Percentage of Houses Connected to AC-Coupled Networks

Table 2 shows a comparison of households connected to various grid-connected communal grids after 25 years. Here again it is clear from the table that more households will have joined networks with decentralised storage systems, whether they be DC- or AC-coupled. In the same category, i.e. decentralised or centralised storage, more households would join DC-coupled networks than AC-coupled networks. This is due to additional costs incurred in DC-AC inverter purchases.

Table 2: Comparison of Houses Connected to Communal Grids under Different Control Architectures

Time (Years)	DC-Coupled		AC-Coupled	
	Decentralised Storage	Centralised Storage	Decentralised Storage	Centralised Storage
0	0	0	0	0
5	414	301	383	249
10	842	609	744	493
15	1276	928	1126	748
20	1679	1249	1497	1012
25	2103	1571	1887	1286

4 CONCLUSION

In this work, simulation results from different communal grid architecture models developed in MATLAB are fed into an agent-based model developed in NetLogo to determine the most cost-effective option. The novel aspect of this paper is that it combines a socio-economic agent-based model with a technical model developed to simulate the most cost-effective communal grid control architecture.

Generally, DC-coupled networks seem to fair better than AC-coupled networks in all categories. This is mainly due to cost and ease of set-up of such networks. As explained above, DC systems tend to be more modular and scalable than AC systems because DC converters are easier to control and to parallel. This allows for more flexibility in system design and expansion, and thus more effective capital investment management. In addition, DC system components tend to be more compact than equivalent AC components because of higher efficiency and by not being frequency dependent. This therefore leads to lower capital costs due to fewer electronic components being used. Moreover, most distributed energy sources and storage devices have inherently DC outputs, making DC architectures more natural options for their integrations. Due to lower power and energy rating, stability issues are more prevalent in communal grids than in utility grids. However, in DC systems where there are no reactive power interactions, there are few stability issues; system control seems to be oriented towards voltage regulation only.

Results show DC-coupled networks with decentralised storage stimulate communal grid connections by allowing easy expansions of the communal grids as each decentralised PV system within a communal grid is treated equally and determines its own real and reactive power, eliminating the need for communication links. This also reduces the cost of implementing such a system; fewer connections are realized with centralised controls as such systems require high speed communication links which make them both difficult to expand and expensive to implement.

Results also show that multi-master operation modes lead to more communal grid connections compared to single-master operation modes because in the former, all distributed PV systems within a communal grid have the same rank and can act as masters or can be operated as combinations of master generators (VSIs) and PQ inverters, allowing for more design flexibility and easy connections from potential customers.

In conclusion, for cost-effective communal grids, DC-coupled networks with decentralised storage systems are recommended as the most-cost effective architectures for rural communal grids. Where a communal grid must be AC-coupled communal grids, systems with decentralised storage and multi-master operation modes are recommended for cost-effectiveness.

4.1 Acknowledgement

This research is funded by Leeds International Research Scholarship

4.2 References

1. Pecas-Lopes, J., Moreira, C., Mandureira, A., et al., *Control Strategies for Microgrids Emergency Operation,* International Conference on Future Power Systems, Amsterdam, 2005
2. Pecas-Lopes, J., Moreira, C., Mandureira, A., *Defining Control Strategies for Microgrids Islanded Operation*, IEEE Transactions on Power Systems **21:** 916-924, 2006
3. Ashbani, S., Abdel-Rady, M., *General Interface for Power Management of Micro-Grids Using Nonlinear Cooperative Droop Control*, IEEE Transactions on Power Systems **20:** 2929-2941, 2013
4. Bidram, A., Davoudi, A., *Hierarchical Structure of Microgrids Control System*, IEEE Transactions on Smart Grid **3:** 1963-1976, 2012
5. De Brabandere, K., Vanthournout, K., Driesen, J., et al., *Control of Microgrids*, Power Engineering Society General Meeting, IEEE, Tampa, 2007.
6. Chowdhury, S., Chowdhury, P., Crossley, P., *Microgrids and Active Distribution Networks*, The Institution of Engineering and Technology, London, 2009
7. Vallve, X., Serrasolses, J., *Design and Operation of a 50 kWp PV Rural Electrification Project for Remote Sites in Spain*, Solar Energy **59**: 111–119, 1997
8. Opiyo, N., *A Survey Informed PV-Based Cost-Effective Electrification Options for Rural Sub-Saharan Africa*, Energy Policy **91**: 1-11, 2016
9. Opiyo, N., *Modelling PV-Based Communal Grids Potential for Rural Western Kenya*, Sustainable Energy, Grids and Networks **4:** 54-61, 2015.
10. Bidram, A., Davoudi, A., *Hierarchical Structure of Microgrids Control System*, IEEE Transactions on Smart Grid **3:** 1963-1976, 2012

SIMZUKUNFT: STUDIES ABOUT INTEGRATING A LARGE AMOUNT OF PV INTO THE GRID OF A SMALL SWISS TOWN

Dr.-Ing. Noah Pflugradt, Prof. Urs Muntwyler
Bern University of Applied Sciences, Laboratory for Photovoltaik Systems, Jlcoweg 1, 3400 Burgdorf, Switzerland
Tel: +41 344 266 927, Noah.Pflugradt@bfh.ch

ABSTRACT: The Swiss people decided in a public vote on phasing out nuclear energy and switching to renewable energy until 2050. For this change there is a so-called "energy-strategy 2050". This energy strategy details how exactly the path from the current energy system to the future energy system will look like, but the detailed impact of those changes on the low voltage grids in Switzerland hasn't been researched yet. The "SimZukunft" research project focuses on that. It evaluates the impact of different scenarios from the official planning documents on the grid of the town "Burgdorf". Additionally, two extreme scenarios are evaluated. This paper shows some early results and that it will be for example possible for the town to become a net-zero-energy town.
Keywords: see enclosed list of keywords

1 INTRODUCTION

In 2017 the Swiss population voted to adopt the Energy Strategy 2050 (ES2050), which is a plan for the adoption of renewable energy in Switzerland. This will mean large changes in the energy distribution system in Switzerland. For the ES2050 very detailed scenarios and development paths were created for the entire country. But what is missing so far is an analysis what this would mean for the grid in a typical small town. In the SFOE-funded project SimZukunft, that started in December 2017, we analyze the impact of these plans on the city of Burgdorf, a Swiss town with about 16.000 citizens. The important part is that the scenarios include not only renewable energy adoption, but also electro mobility, climate change with increased air conditioning, increased electric heating with heat pumps and much more. Since the resulting higher energy demand partially cancels out the issues with very high shares of photovoltaic in the low voltage grid, the results are highly relevant for calculating possible PV adoption rates in Switzerland.

2 SCENARIOS

The project uses five different scenarios for Burgdorf. These scenarios are described below. Three of the scenarios are based directly on the document called "Energieperspektive 2050" [1]. This document was originally created by the research institute Prognos and is the foundation of the energy strategy 2050. The other two scenarios evaluate extremes such as a very high penetration of electromobility and photovoltaics, to analyze the sensitivity of the grid to such developments.

2.1. ES-2050 POM-E

This scenario uses the values from the scenario "Politische Massnahmen" (Political Measures) in the variant "renewable energy".

In detail this scenario includes a (from todays perspective) moderate implementation of renewable energy, a fairly low percentage of electromobility and extensive renovation of the existing houses to reduce heating energy demand. The economy is growing slowly and steadily. The nuclear power plants are being turned off and replaced with a mixture of renewable energy and energy purchases from other countries. There will be no smart grid technologies included in this scenario. This scenario will be our reference scenario.

2.2. ES-2050 POM-E Smart

This scenario will use the same basic values as the previous one but add smart grid technology. Technologies such as intelligent curtailment, smart charging, decentralized energy storage and intelligent heat pump control will be included. These technologies have the potential to significantly reduce the need for grid expansion by making sure that the local demand peaks match up with the local generation peaks. Comparing the results of this scenario with the previous one will make it possible to see if it is better to invest in technology or in copper.

2.3. ES-2050 NEP-E

This scenario is rather similar to POM-E but uses somewhat different values for the development path that will enable Switzerland to meet the COP21-targets without getting rid of the nuclear power plants.

2.4. Utopia

This scenario models one of the two extremes: Maximum photovoltaic deployment, very cheap and ubiquitous energy storage, full switch to electric cars, moderate climate change with the corresponding air conditioning deployment, no large population changes and a number of completely grid-independent single-family houses.

This scenario is meant together with the next one for the sensitivity check of the calculations. Therefore, the idea is to set all the parameters to values that are from today's perspective maybe unrealistic, but that will tell if the grid will be able to cope with even those extremes. Additionally, we'll be able to deduce which parameters will have the biggest influence and thus need to be watched closely over the next few years.

2.5. Dystopia

The last scenario models the worst case where everything goes wrong: barely any renewable energy, no decentralized energy storage, large climate change with the associated need for massive air conditioning, barely any electric cars, bad economy and on top of that a lot of climate change refugees.

This scenario models what could happen for example with a shortage of the rare earths, combined with a global trade war with the associated recession and further accelerating climate change.

3 METHOD

To evaluate the impact of the scenarios detailed simulations for the entire year will be performed. For the loads are simulated with an agent-based behavior simulation [2]. The individual agents are controlled by their desires. This enables a very flexible modelling of many different behavior patterns, such as singles, families, retirees or shift workers. Having detailed individual household profiles is especially important to evaluate the impact of decentralized energy storage.
Additionally, the simulations also model the mobility behavior. Because for every simulated person it is clear for every point in time where that person is and what the person is doing, it is possible to create detailed mobility profiles and for example calculate for each car where it is, how far it has driven, if it can be charged and if so, how much energy it will demand.
This enables us for example to compare different smart grid charging control algorithms and the impact of those on the top loads of the grid. The load profiles are then used by Adaptricity to do grid simulations with their proprietary grid simulation software and it is evaluated where problems such as over-voltages, under-voltages or transformer overloads can occur in the grid.
The time plan in the project is that in 2018 the overall sum calculations for the scenarios will be done and the simulation of the presence will be created. In 2019 the scenarios will be implemented into the simulation and the detailed grid planning will be performed.
The project uses the following data sources:

- The local utility company Localnet provides the data about the grid and the consumption of electricity, gas and district heating.
- The building data is from the central Swiss building and apartment registry.
- The potential for photovoltaics is based on studies by the research project Sonnendach.ch. [3] They evaluated for every single building in Switzerland, how high the potential for photovoltaics on the building would be.
- To estimate the energy demand of buildings heated for example with oil or wood, data from the Canton Bern is used. The Canton Bern keeps a database of the estimated energy use of all buildings in the canton.
- To figure out how each building is heated, the data from the local chimney sweeper association is used.
- The population statistics are provided by the local city council.

2.3 Data Merging

One big challenge in the project is creating a comprehensive model from all the different data sources that are as usual not very well integrated. The building registry has a unique identifier, which is used in some of the other data sets and is a great first step. But if, for example, an apartment complex with multiple building is heated with a central heating system, it is not easy to automatically distribute the consumption. Additionally, not all the data sets are referenced to buildings, so some of the data needs to be assigned by address, with all the problems that this brings: one building can have multiple addresses, one address can have multiple buildings and so on.
In this project we used the approach of defining building complexes to deal with these issues. One building complex can have:

- One or more buildings
- One or more addresses
- One or more consumption data entries

The simplest building complex is a single-family house, which is exactly one building with exactly one address and exactly one consumption entry.
After creating the building complexes, the heating and energy demands per building can be summed up per complex and then averaged over the various buildings. This lets us get a reasonable first impression of the building stock energy demand and how much the energy demand might be reduced in future.
Due to privacy reasons the city council couldn't just hand over the data of citizens per address. So here a workaround over statistic values had to be used. For this the citizens were semi-randomly distributed over the building complexes. For this a fuzzy logic engine [4] has proven to be very helpful. We defined a number of heuristics, that try to estimate based on living area and energy consumption how many people are living in each apartment. The fuzzy logic rules were then tweaked iteratively until the sums match up with the statistics from the city council. This data is then the foundation for all future simulations.

4 CURRENT SITUATION

3.4 Energy Demand

The current energy demands in Burgdorf are shown in Figure 1 and Figure 2. It is visible that the largest part of the energy is needed for heating. The main energy sources in Burgdorf are fossil fuels, that is oil and gas. Not included in the chart is the energy use for vehicles.
According to [5] Swiss people drive on average 10'165 km/year and there are about 7000 cars in Burgdorf. That results in an energy demand of gasoline of about 50-60 GWh, depending on which average consumption values one uses and if you consider the energy demand of long-distance drives for such as vacation trips to be part of the energy demand of the city.

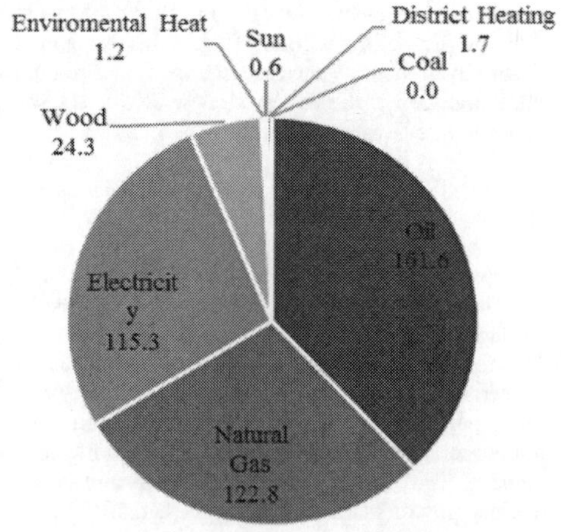

Figure 1: Current energy demand of the buildings in Burgdorf by energy source, based on data from [6]

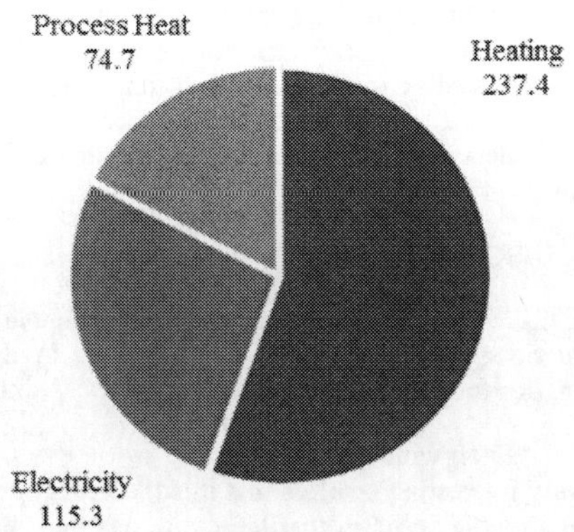

Figure 2: Current energy demand in Burgdorf by sector, based on data from [6]

4.1. Potential for Photovoltaics in Burgdorf

In the project «Sonnendach.ch» [3] an analysis of the potential for photovoltaics for all buildings in Switzerland was performed. The results are occasionally a bit optimistic, because small roof elements have not always been correctly excluded, but the results give a very good first estimate.

The results are shown in Figure 3. It is visible in Figure 4 that it is sufficient to cover the areas with a yearly yield of over 700 kWh/m^2/year to reach 97.6% of the maximum power.

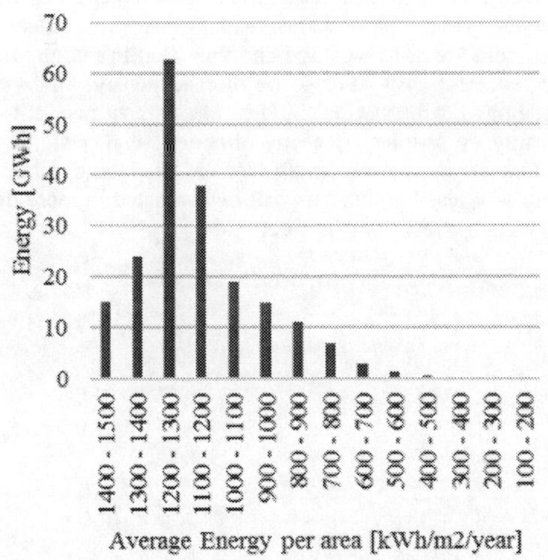

Figure 3: Histogram of the solar energy potential of all roofs in Burgdorf

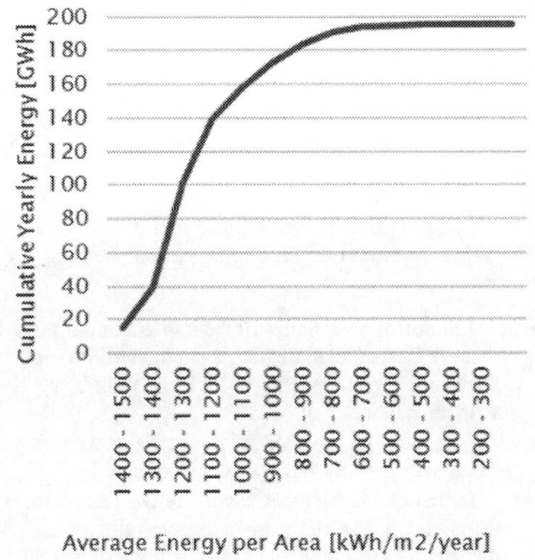

Figure 4: Cumulative energy yield for increasing PV coverage.

4.2. Validation of the building energy data from Canton Bern

To estimate the quality of the heat demand estimates of the Canton, a comparison was performed of the calculated data with the real energy demand of all buildings that are heated with either gas or district heating. The results are shown in Figure 5.

It is visible that the energy demand is partially overestimated and partially underestimated. Mostly these compensate each other. The average over all buildings is 64.3 kWh/m2/year. It seems likely that the extreme differences at the higher and lower end are either data errors or process energy. The average over just the differences between -200 and 200 is -45.2 kWh/m2/year. That means that the cantonal data overestimate the real heat demand slightly, but that they are entire usable for this project. A possible cause of the overestimate might be the increasing adoption of high efficiency condensation boilers.

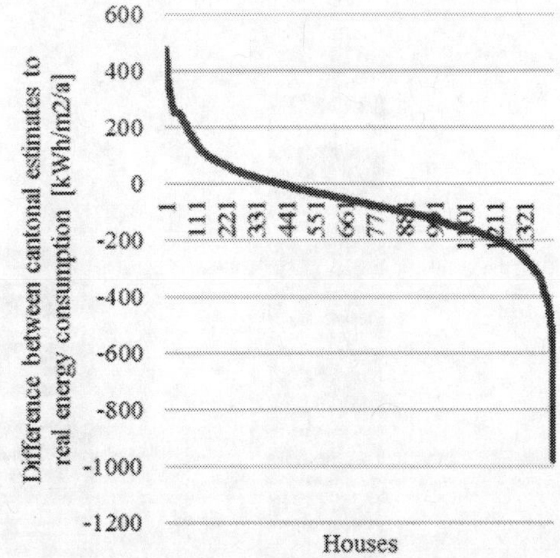

Figure 5: Comparison of the real energy use to the estimates of the Canton

5 RESULTS

The detailed simulations with a 15 min time resolution are only planned for 2019, therefore this paper can only show first, preliminary results. For space reasons this paper will focus on the utopia scenario, since this one yields the most impressive results.

5.1. Assumption in the Utopia Scenario

The most important assumptions in this scenario are as follows:

- A full sector coupling will be implemented, meaning that all fossil fueled heating systems will be replaced. All heating systems will be replaced with heat pumps with an efficiency of 3.
- All cars will be replaced with electric cars. A demand of on average 15 kWh/100 km is assumed.
- All current electricity use will be decreased through efficiency measures such as new lighting and new devices by 10% to 90% of the current value.
- All industrial natural gas heating processes will be optimized for a total efficiency gain of 50%, but they will remain powered by natural gas.

5.2. Air conditioning

The future air conditioning demand is difficult to estimate. The summer 2018 with sold out mobile air conditioning units showed clearly that even in Switzerland there is a large demand for air conditioning. The estimates from [1] show a electricity demand of 13.3 GWh for future air conditioning in Burgdorf. Typical numbers from the US on the other hand seem to be around 3 kW/household and about 1000 hours per year (3-4 months/year with 8-10h runtime per day). Using those numbers, the future electricity demand for Burgdorf would be around 24 GWh, which is almost twice the official estimate. To make the scenarios more comparable though, the utopia scenario will also use the estimates from the "Energieperspektiven"-document, meaning 13.3 GWh/year energy demand.

5.3. Yearly Summary

The results for the utopia scenario are shown in Figure 6. It shows the total energy demand in Burgdorf in 2009 and the final energy demand in 2050 and compares it with the maximum potential for local energy generation.
The first column is the current energy demand of Burgdorf from [6]. The second column shows what would happen with the total energy demand if only the heating is switched over to heat pumps. The third column then adds the efficiency factors. The forth column shows the total generation potential.

6 DISCUSSION

The discussions in this paper mainly focus on the Utopia-scenario again. The results mostly apply to the other scenarios too, but in weakened form.

6.1. Energy independence

A very interesting result is that it is theoretically possible in the utopia scenario that the entire city becomes a net energy neutral city, which would mean that over the year the city generates just as much energy as it consumes.
But to reach actual energy autarky would require a large amount of energy storage, to which there is no good and economical solution yet. The current state of the research seems to suggest that such a solution could be realized for example by the following measures:

- Large scale deployment of facade photovoltaics to maximize energy generation in the winter months.
- A large seasonal heat storage for the entire city, since storing heat is much cheaper than storing electricity.
- A suitable amount of battery storage, to cover up to two weeks of bad weather, especially in winter.

Facade photovoltaics are so far mostly not used, since they generate less energy. But with appropriate financial incentives and a time-dependent feed-in-tariff they could easily become more widespread. Because the differences in yield are not very large and they feed in more at times when electricity tends to be more expensive (mornings, evenings, winter), this could be almost cost-neutral to the utility companies. It seems therefore that a significant increase in the percentage of façade photovoltaics is entirely feasible, but this will be evaluated in more detail during the rest of the project.

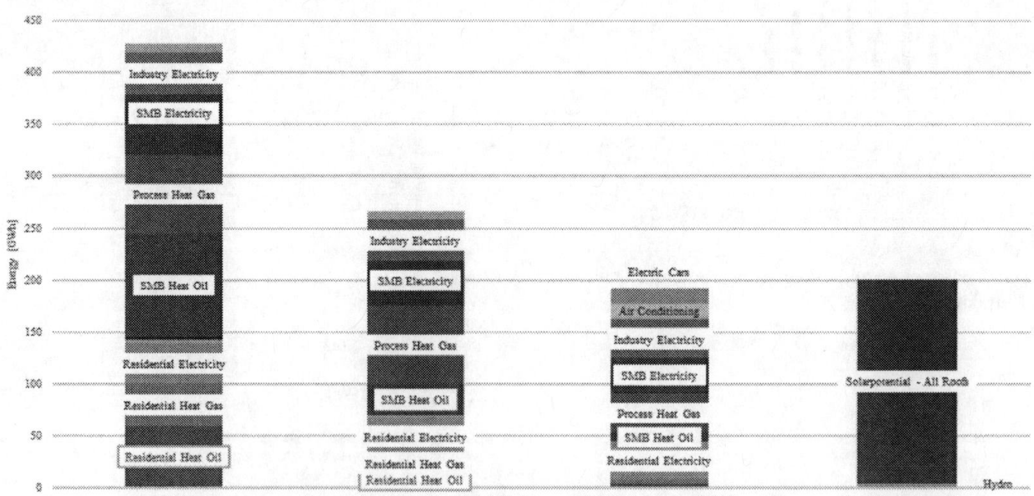

Figure 6: Current and future Energy Balances in Burgdorf in the Utopia scenario

Seasonal heat storages using borehole fields are well-tested and state of the art. Of course, building such a system for an existing, densely populated city would be quite challenging and require large investments.
Especially the battery storage would be quite expensive though. Per day Burgdorf needs about 500 MWh of electric energy. Some people suggest future battery prices of 100 CHF/kWh [7]. This would mean costs of about 50 million CHF per day of battery storage.
Using the electric vehicles as Vehicle-to-Grid battery storage would also not cover this demand. As mentioned there are about 7000 cars in Burgdorf. Even if every car has a 50 kWh Battery that will be drained entirely, that is only 352 MWh and won't cover a single day.
Real energy independence seems therefore with the current technology even in the utopia-scenario not likely.

6.2. Gas-Grid – Tipping Point

One interesting early result is the influence of the adaption rate of heat pumps. As soon as a previously gas-supplied district reaches a certain heat pump adoption rate, then the maintenance of the gas grid suddenly doesn't pay off anymore and the most cost-efficient thing for the utility company is to incentivize all other customers to move away from gas to get rid of the gas grid as soon as possible.

6.3. Smart Grid Technologies

From the early calculations it already becomes obvious that the use of appropriate Smart Grid technologies is able to reduce the grid load significantly, since the load from heat pumps, air conditioning and electromobility is very suited for demand management [8]. In [9] it was shown for example that cars on average tend to be connected to the grid much longer than needed for charging and therefore a huge potential for load shifting exists.

7 OUTLOOK

The project will continue based on the current results and the discussed detailed grid simulations will be performed. The results will be published next year.

8 PARTNER

The project is executed by a cooperation of the Bern University of Applied Sciences (Laboratory for Photovoltaic Systems), the city council of the town of Burgdorf, the local utility company Localnet and the Adaptricity AG, a company that specializes in grid planning.

9 ACKNOWLEDGEMENT

This research project is financed by the Swiss Federal Office for Energy (SFOE). We gratefully acknowledge the co-financing by the Swiss Centre for Competence in Energy Research "SCCER-FURIES" (Future Swiss Electrical Infrastructure), financed by Innosuisse (Schweizerische Agentur für Innovationsförderung - SCCER Programm) as well as the co-financing by the Bern University of Applied Sciences.

10 REFERENCES

[1] Prognos AG. Die Energieperspektiven für die Schweiz bis 2050. Energienachfrage und Elektrizitätsangebot in der Schweiz 2000 2012;2050.
[2] Pflugradt N. LoadProfileGenerator; Available from: http://www.loadprofilegenerator.de.
[3] Bundesamt für Energie Schweiz. Sonnendach.ch; Available from: http://www.sonnendach.ch.
[4] Dmitry Kaluzhny. Fuzzy Logic Library for Microsoft .Net. [August 27, 2018]; Available from: https://sourceforge.net/projects/fuzzynet/.
[5] Bundesamt für Statistik, Bundesamt für Raumentwicklung. Mobilität in der schweiz, Ergebnisse des Mikrozensus Mobilität und Verkehr 2010; 2012.
[6] Stadt Burgdorf. Kommunaler Richtplan Energie. [August 27, 2018]; Available from: https://www.burgdorf.ch/wAssets/docs/Baudirektion/Energie/RPE_Burgdorf_Richtplanbericht_inkl_Gen.pdf.
[7] Mark Chediak. The Battery Will Kill Fossil Fuels—It's Only a Matter of Time; Available from: https://www.bloomberg.com/news/articles/2018-03-08/the-battery-will-kill-fossil-fuels-it-s-only-a-matter-of-time.
[8] Harbrecht A, McKenna R, Fischer D, Fichtner W. Behavior-oriented modeling of electric vehicle load profiles: A stochastic simulation model considering different household characteristics, charging decisions and locations; 2018.
[9] Wolbertus R, Kroesen M, van den Hoed R, Chorus C. Fully charged: An empirical study into the factors that influence connection times at EV-charging stations. Energy Policy 2018;123:1–7.

MODELLING CONTROL METHODS FOR PV-BASED COMMUNAL GRIDS WITH DIFFERENT LINE RESISTANCES AND IMPEDANCES

Dr. Nicholas Nixon Opiyo
Centre for Integrated Energy Research
University of Leeds, Leeds, LS2 9JT, UK
Email: Nicholas.Opiyo1@gmail.com

ABSTRACT: Different droop control methods for PV-based communal grid networks (minigrids and microgrids) with different line resistances (R) and impedances (X) are modelled and simulated in MATLAB to determine the most efficient control method for a given network. Results show that active power-frequency (P-f) droop control method is the most efficient for low voltage transmission networks with low X/R ratios while reactive power-voltage (Q-V) droop control method is the most efficient for systems with high X/R ratios. For systems with complex line resistances and impedances, i.e. near unity X/R ratios, P-f or Q-V droop methods cannot individually efficiently regulate line voltage and frequency. For such systems, P-Q-f droop control method, where both active and reactive power could be used to control PCC voltage via shunt-connected inverters, is determined to be the most efficient control method. Results also show that shunt-connection of inverters leads to improved power flow control of interconnected communal grids by allowing feeder voltage regulation, load reactive power support, reactive power management between feeders, and improved overall system performance against dynamic disturbances.

Keywords: Droop Control; Communal Grid; Inverter

1 INTRODUCTION

A PV-based communal grid can be defined as a collection of distributed PV systems, distributed energy storage devices, and distributed loads, operating as a single and controllable system capable of supplying power to an area of service. They should be operable in both grid-connected and islanded modes. Power electronics interfaces and controllers are used in communal grids to ensure quality, reliable, and independent power supply at all times. A control system must be able to disconnect and reconnect the communal grid from the utility grid, maintain voltage and frequency levels in islanded mode of operation, and facilitate a black start after a system failure [1]. The main drivers of communal grids are demand, cost, technical aspects, and environmental concerns [2]. These factors influence the control strategy applied for a communal grid and will need to give consideration to issues such as load sensitivity, number of distributed generators in the communal grid, power quality requirements, ownership of the communal grid and distributed generators, distances between the distributed generators, the existing communication infrastructure, each distributed generator's energy source, and whether the communal grid is predominantly and exporter or importer of electricity [3-5].

1.1 Droop Control

Droop control method is the most widely used in PV power systems to enable automatic load sharing between different distributed PV systems and to extend operating range of active (P) and reactive (Q) power ratings of a given inverter [6,7]. The method requires P and Q to be measured in order to droop the frequency (f) and voltage (V) accordingly such that the inverter mimics the behaviour of a synchronous generator [6,7]. A reactive power voltage (Q-V) droop control is used to control point of common coupling (PCC) voltage magnitude while an active power frequency (P-f) droop control is used is used to control the frequency of the system in islanded mode [6]. For easy analysis, consider a two bus synchronous generator connected to a high voltage (HV) transmission network as shown in the figure 1 below.

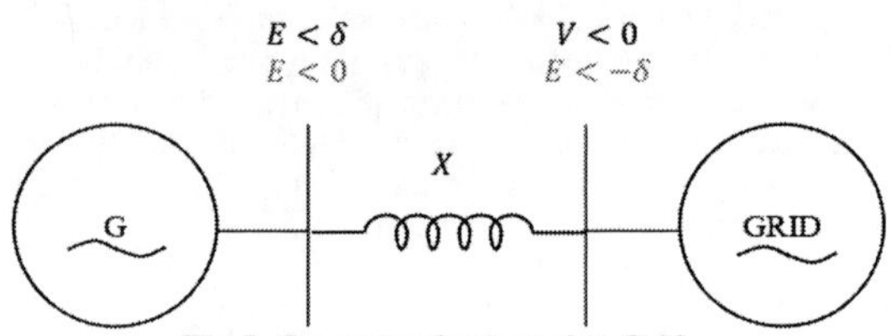

Fig. 1: Generator Connected to Grid

The power produced at the generator terminal is expressed as [8]:

$$P = \frac{E}{R^2 + X^2}[XV \sin\delta + R(E - V\cos\delta)] \quad (1)$$

$$Q = \frac{E}{R^2 + X^2}[-RV \sin\delta + X(E - V\cos\delta)] \quad (2)$$

where δ is the voltage angle.

Since HV networks have high reactance (X) and low line resistance (R), and thus high X/R ratios, the line resistance can be ignored and equations 1 and 2 reduced to [8]:

$$P = \frac{VE\sin\delta}{X} \quad (3)$$

$$Q = \frac{E^2 - VE\cos\delta}{X} \quad (4)$$

Usually, δ is very small (zero) and therefore $\cos\delta \cong 1$ and $\sin\delta \cong \delta$. Equations 3 and 4 can therefore be simplified further to [8]:

$$\delta \approx \frac{PX}{VE} \quad (5)$$

$$E - V \approx \frac{QX}{E} \quad (6)$$

The reactive power Q can therefore be controlled by the difference in voltage between E and V, and the active power P, by the voltage angle δ. The voltage angle δ can be expressed in terms of angular frequency ω (radians/second), and therefore in terms of electrical frequency f (hertz) as [9]:

$$\delta = \int \Delta\omega \, dt = \frac{1}{2\pi}\int \Delta f \, dt \quad (7)$$

The relationship outlined above allows automatic load sharing between many synchronous generators (SGs) when combined with P-f droop control as shown in figure 2a [1,2]; a change in load will cause frequency variation at the terminal of each SG. Active power will flow from regions of higher frequency to regions of lower frequency. Frequency variations within the network will eventually drift to an average steady state value [2]. The new steady state frequency will be proportional to the change in power, as shown in figure 2b [1].

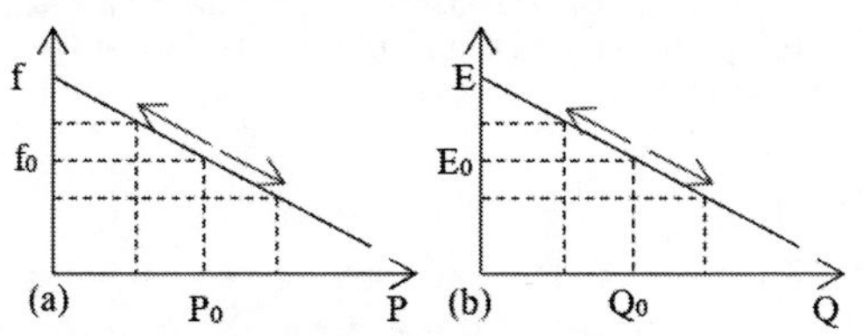

Fig. 2: P-f and Q-V Droops

1.2 Inverter Controls

Inverters are used to interface communal grids with utility grids and can be classified according to modes of operation as PQ or V-f (also known as voltage source inverter (VSIs)). A PQ inverter controls the real (P) and reactive (Q) power by adjusting the magnitude of the output real and reactive current. It therefore operates as a voltage controlled current source [10]. A voltage source inverter controls the voltage (V) and frequency (f) at the output terminal, and thus operates as a voltage source [9,10]. The mode of an inverter operation is chosen depending on a communal grid's architecture and control strategy, and may change depending on whether the communal grid is islanded or grid-connected. Unlike synchronous generators, inverters do not have rotors and thus no natural connection between frequency and active power. To achieve stable operation with multiple distributed PV systems, the inverters are controlled so that they mimic the characteristics of synchronous generators with P-f and Q-V droop controls [11]. For a PQ inverter, these droops are implemented as (f) and (V) functions, whereas a VSI uses (P) and (Q) droops [12].

1.2.1 PQ Inverter with Droop Control

As opposed to a synchronous generator that uses its rotor speed as the frequency input for the P-f droop controller, a PQ inverter does not set the frequency, but rather measures the grid frequency using a phase lock loop (PLL) and then operates at that measured frequency [9-11]. The inverter will adjust its power accordingly, by comparing the measured frequency to a reference (nominal grid frequency) value accordingly. This relationship is modelled as [9-11]:

$$P(f) = P_0 - (f_{set} - f)k_f \quad (8)$$

where P_0 is the power delivered by the inverter at setpoint frequency f_{set} and k_f is the gradient of the droop, which determines how much the active power P will change in response to a change in frequency f. When a Q-V droop is used, the PQ inverter measures the terminal voltage and compares this to the reference value. The reactive power is adjusted by altering the reactive component of the inverter current [10]. This reactive power adjustment is expressed as [9-11]:

$$Q(V) = Q_0 - (V_{set} - V)k_v \quad (9)$$

where Q_0 is the reactive power delivered/consumed by the inverter at setpoint voltage V_{set} and k_v is the gradient of the droop, which determines how much the reactive power Q will change in response to a change in voltage V.

1.2.2 VSI Inverter with Droop Control

A VSI with droop control uses measured active power output to generate the VSI frequency and measured reactive power output to generate the VSI voltage [8]. It thus operates like a synchronous generator. This process can be modelled by equations 10 and 11 below [8]:

$$f(P) = (P_0 - P)k_p - f_{set} \quad (10)$$

where Kp is the gradient of the droop, which determines how much the frequency will change in response to a change in power P.

$$V(Q) = (Q_0 - Q)k_q - V_{set} \quad (11)$$

where Kq is the gradient of the droop, which determines how much the voltage will change in response to a change in reactive power Q.

The conventional droop control relies on the inductive nature of transmission lines and is thus based on the assumption that the line resistance (R) is much less than the reactance (X), i.e., high X/R ratio and therefore active power flow is predominately a function of the voltage angle (δ) [12]. This is the case for high voltage (HV) and medium voltage (MV) lines, but it is not so for low voltage (LV) networks as shown in table 1 below.

Table 1: Typical Line Parameters [13]

Type of line	R (Ω/km)	X (Ω/km)	R/X
Low voltage	0.642	0.083	7.7
Medium voltage	0.161	0.190	0.85
High voltage	0.06	0.191	0.31

For systems with LV transmission networks and thus high line resistance, the low X/R ratio causes a coupling effect between active and reactive power, rendering Q-V droop control method ineffective in achieving required voltage regulation in such systems [14]. A solution to this problem has been proposed by Alatrash et al., in which a control system that incorporates virtual reactance at the terminals of the VSI is used [15]. The idea behind this method is to emulate the higher reactance of MV and HV distribution lines. Abusara et al. proposed a control strategy that employs both virtual inductance to remove harmonic disturbances and a real-time integration filter to eliminate the need for measuring average power [16].

2 METHODOLOGY

Consider a communal grid with two-feeder distribution systems as shown in figure 3 below. The two inverters in the systems are connected in parallel through switch S_4; parallel connection of the inverters enables them to control both active and reactive power flow through shunt-connected transformers, leading to overall improved system performance [17]. The necessary active power for the compensation is drawn from the interconnected feeders via the inverters. The controller then circulates minimum active power between the feeders while the active and reactive power droop coefficients are adjusted online through a look up table based on the PCC voltage level. A real-time integration inductor-capacitor-inductor (LCL)-filter with a virtual inductance is used to actively damp the inverter output without the need for extra sensors to feedback measured signals. Impact of DC link voltage and resonance frequency on stability would then be analysed.

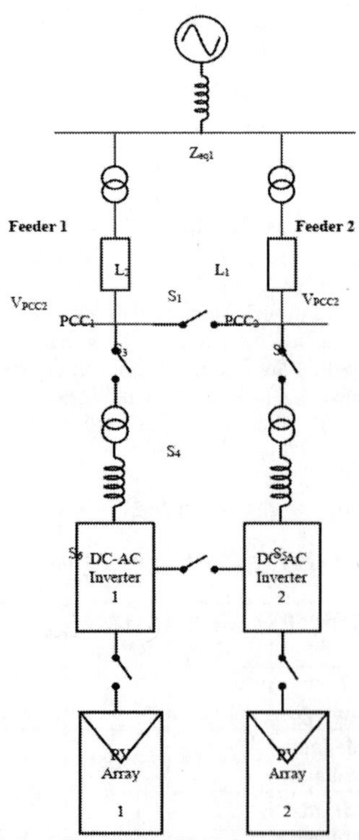

Fig. 3: Interline-PV (I-PV) System Configuration

The PV system does not produce real power during the night and therefore switches S_5 and S_6 can be opened. During the day when the PV system generates power, switches $S_2 - S_6$ are closed, while switch S_1 is kept open in order to operate the system as interline-PV (I-PV), controlled through an interline power flow controller (IPFC) [18]. Let's set the voltage levels of both the feeders at 25 kV and the acceptable range of PCC voltage variation at $\pm 5\%$. Let's consider the loads at the ends of feeder 1 and 2 as P and Q, respectively, and let's assume that they have different values. If inverters 1 and 2 are each rated at 2 MVA then only a maximum of 1 MW active power can circulate between feeders 1 and 2 through the inverters. This is to avoid considerable voltage drop on the other feeder from which the active power is to be taken. The PCC voltages at feeders 1 and 2 are expressed as V_{pcc1} and V_{pcc2}, respectively. The active and reactive powers injected or absorbed by feeders 1 and 2 are expressed as P_{inv1}, Q_{inv1}, P_{inv2} and Q_{inv2} respectively. The reactive powers supported by the inverters are shown as positive quantities since they are leading, while the reactive powers are shown as negative quantities since they are lagging. For simplicity, the simulation results are expressed in per unit (*pu*), with base voltage of 25 kV and base power of 1 MVA. Feeders 1 and 2 line parameters are: 0.08 + j0.04 ohm/km and line lengths: $L_1 = L_2 = 20$ km.

Figure 4 shows a Thevenin equivalent of feeder 2 connected to inverter 2. The figure represents a power source ($E_{PV}\angle\phi$) where Z_{th} represents the feeder as well as inverter coupling impedance, $\emptyset$ is the phase angle difference between PCC and grid voltages, while θ is the impedance angle due to Z_{th}.

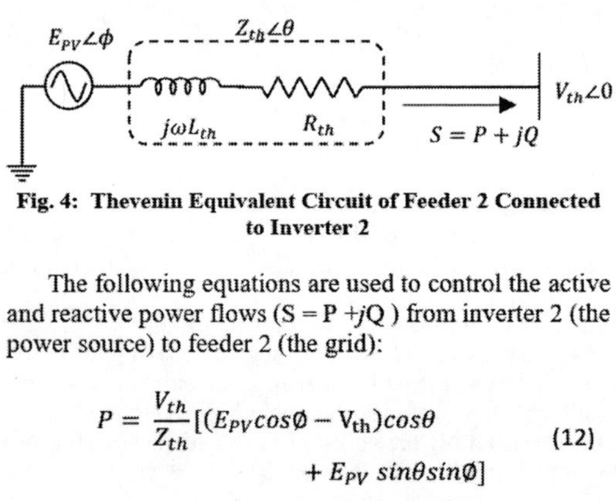

Fig. 4: Thevenin Equivalent Circuit of Feeder 2 Connected to Inverter 2

The following equations are used to control the active and reactive power flows (S = P +*j*Q) from inverter 2 (the power source) to feeder 2 (the grid):

$$P = \frac{V_{th}}{Z_{th}}[(E_{PV}cos\emptyset - \mathrm{V_{th}})cos\theta + E_{PV}\ sin\theta sin\emptyset] \tag{12}$$

$$Q = \frac{V_{th}}{Z_{th}}[(E_{PV}cos\emptyset - \mathrm{V_{th}})sin\theta - E_{PV}\ cos\theta sin\emptyset] \tag{13}$$

Since $\emptyset$ is usually very small, i.e., $cos\ \emptyset \approx 1$, and $sin\ \emptyset \approx \emptyset$, the equations above are reduced to:

$$P = \frac{V_{th}}{Z_{th}}[(E_{PV} - \mathrm{V_{th}})cos\theta + E_{PV}\emptyset\ sin\theta] \tag{14}$$

$$Q = \frac{V_{th}}{Z_{th}}[(E_{PV} - \mathrm{V_{th}})sin\theta - E_{PV}\emptyset\ cos\theta] \tag{15}$$

Equations 14 and 15 show the dependency of delivered active and reactive power on the impedance angle θ and the phase difference angle $\emptyset$.

2.1 P-f Droop Control Method

This method is suitable for networks with high resistive values and low reactance, i.e. low X/R ratios. This makes the impedance angle θ equals to zero hence, from equation 14, the active power delivered by the inverter is proportional to the voltage difference ($E_{PV} - V_{th}$) and thus proportional to the inverter E_{PV}. The reactive power of inverter 2 is proportional to the phase difference angle $\emptyset$, i.e. proportional to the frequency f of the system. It should be noted that $\emptyset$ is varying within a small range. P remains constant irrespective of any change in $\emptyset$ while Q significantly changes with changing $\emptyset$. On the other hand, P significantly increases with increasing E_{PV} while Q is hardly affected by changes in E_{PV}.

2.2 Q-V Droop Control Method

This method is suitable for networks with high X/R ratios, i.e. low line resistance and high reactance and thus the impedance angle θ goes to 90°. The reactive power of the inverter is proportional to the inverter voltage E_{PV} and the active power is proportional to the frequency f. P significantly changes with changing $\emptyset$ while Q remains constant regardless of any changes in $\emptyset$. On the other hand, P hardly changes with changing E_{PV} while Q significantly changes with E_{PV}. The Q-V droop control method is one of the widely used methods for voltage regulation; unlike the P-f droop method where additional provision for real power is required; Q-V droop method does not need such a source of real power for generating the necessary Q for compensation.

2.3 P-Q-V Droop Control Method

This method is suitable for systems with complex line impedances, i.e. where neither the line resistance nor the reactance is more significant over the other and therefore neither can be ignored. In such systems, the X/R ratio is near unity and therefore neither the P-f nor the Q-V droop method is sufficient to regulate the PCC voltage. In this system both the active and reactive power are simultaneously affected by changes in voltage magnitude E_{PV} and the phase difference angle $\emptyset$. Since in these systems both active and reactive power affect the voltage magnitude, the system can be represented by the following equation [18]

$$V = V_{pcc} - (n_L * P) - j\,(m_L * Q) \qquad (16)$$

where n_L and m_L are the electric load droop coefficients while V_{pcc} is the voltage before compensation and is given by

$$V_{pcc} = V_{ref} + (n_d * P) + j\,(m_d * Q) \qquad (17)$$

where n_d and m_d are the active and reactive power coefficients for the proposed P-Q-V droop method while V_{ref} is the desired reference value of PCC voltage, i.e. 1 pu (25kV).

3 RESULTS AND DISCUSSION

For simplicity only feeder 2 in figure 3 is regulated. If PCC voltage rises by more than 5%, another distributed PV generator is used to inject active power into the system while if PCC voltage drops below 5% a heavy load is connected to feeder 2. At t_1 (0.0 S), inverter 2 starts operation to regulate feeder 2 PCC voltage based on P-f, Q-V, or P-Q-V droop control methods with normal loads. At t_2 (0.5 S) feeder 2 PCC is increased above 5% due to active power injected from another PV source. At t_3 (1.0 S) feeder 2 PCC voltage is decreased below 5% limit due to increased load and no additional active power injection.

3.1 System Performance with P-f Droop Control Method

Figure 5 shows the performance of feeder 2 at t_1, t_2, and t_3. The PCC voltage is regulated according to the P-f droop control method. Figure 5 (a) shows the PCC voltage before and after compensation, figure 5 (b) shows the PCC voltage before and after active and reactive powers injected by inverter 2, figure 5 (c) shows the circulated active power between feeder 1 and feeder 2, and figure 5 (d) shows the apparent power S circulated through inverter 2 with respect to its rated capacity of 2 MVA. From t_1 to t_2, the PCC voltage is 1.035 pu, and it is reduced to 1.015 pu using P-f droop method; 0.6 pu of active power is absorbed through inverter 2 and is drawn from feeder 1. From t_2 to t_3, the PCC voltage is increased to 1.089 pu and it is reduced to 1.06 pu. This is done by absorbing 1 pu active power. Even though inverter 2 could absorb up to 2 pu to regulate the PCC voltage, it has been limited to a maximum of 1 pu for active power injection in order not to overload feeder 1 and thus to maintain its voltage within acceptable limits. From t_3 to 1.4 sec, the PCC voltage falls to 0.925 pu due to heavy load on feeder 2. Inverter 2 then injects maximum allowable 1 pu active power to improve the PCC voltage from 0.925 to 0.945 pu. The droop coefficient for this method is 0.02 pu/MW for all the operating conditions.

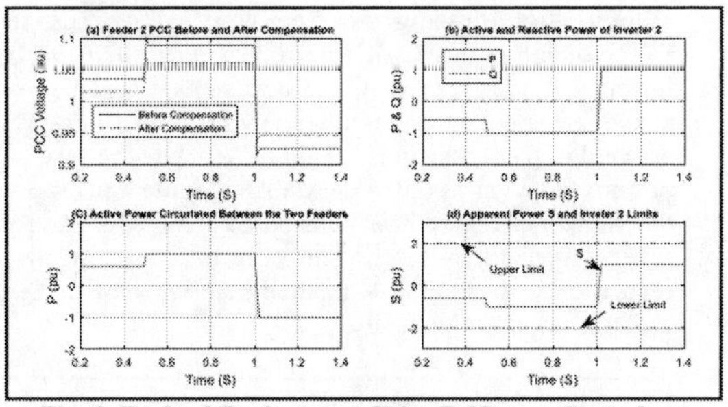

Fig. 5: Feeder 2 Performance Using P-f Droop Control Method

3.2 System Performance with Q-V Droop Control Method

Figure 6 shows the performance of feeder 2 with Q-V droop control method. The conditions for the PCC voltage before compensation are kept identical to that for P-f droop method. From t_1 to t_2 the PCC voltage is reduced from 1.035 pu to 1.02 pu as shown in figure 6 (a) by injecting 1.0 pu inductive reactive power through inverter 2. From t_2 to t_3, the PCC voltage is regulated at around 1.055 pu by absorbing 2 pu inductive reactive power capacity of inverter 2 as shown in figure 86(b). From t_3 to 1.4 seconds, the PCC voltage is raised from 0.925 to 0.948 pu by injecting capacitive reactive power of 2 pu as shown in figure 6 (c). The droop coefficient for this method is 0.01 $pu/MVAR$. Figure 6 (d) shows the total apparent power S handled by feeder 2. The Q-V droop method utilizes the full inverter rating to compensate for voltage rise and voltage drop; a low X/R ratio system would require higher capacity of inverter to regulate the PCC voltage below the set margin of ±5%.

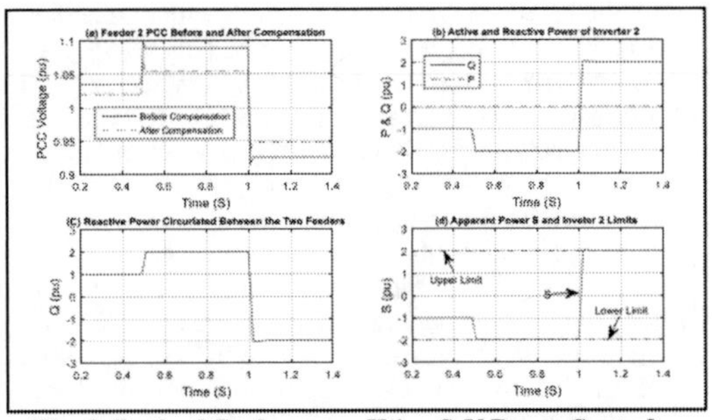

Fig. 6: Feeder 2 Performance Using Q-V Droop Control Method.

3.3 System Performance with P-Q-V Droop Control Method

Figure 7 shows the performance of feeder 2 with P-Q-V droop control method. The conditions for the PCC voltage before compensation are kept identical to those for P-f and Q-V droop method. From t_1 to t_2 the PCC voltage is reduced from 1.035 *pu* to 1.025 *pu* as shown in figure 7 (a) by absorbing 1.0 *pu* inductive reactive power through inverter 2. The active power coefficient is 0 since the controller mainly uses the reactive power for compensation; the reactive power coefficient is 0.04 *pu*/MVAR. From t_2 to t_3, the PCC voltage is regulated at 1.04 *pu* by absorbing 1.4 *pu* inductive reactive power and 0.8 *pu* active power as shown in figure 7 (b). The active droop coefficient is 0.08 *pu*/MW, while the reactive power coefficient is 0.018 *pu*/MVAR. The reactive power coefficient is thus reduced as more reactive power is absorbed from the inverter. From t_3 to 1.4 seconds, the PCC voltage is increased from 0.925 *pu* to 0.96 *pu* by injecting 1.4 *pu* capacitive reactive power and 0.7 *pu* active power simultaneously as shown in figure 7 (c). The active droop coefficient is 0.1 *pu*/MW, while the reactive power coefficient is 0.024 *pu*/MVAR. Figure 7 (d) shows the total apparent power S handled by feeder 2. The inverter capacity is efficiently utilized for voltage regulation with the P-Q-V method than with P-f or Q-V droop methods individually.

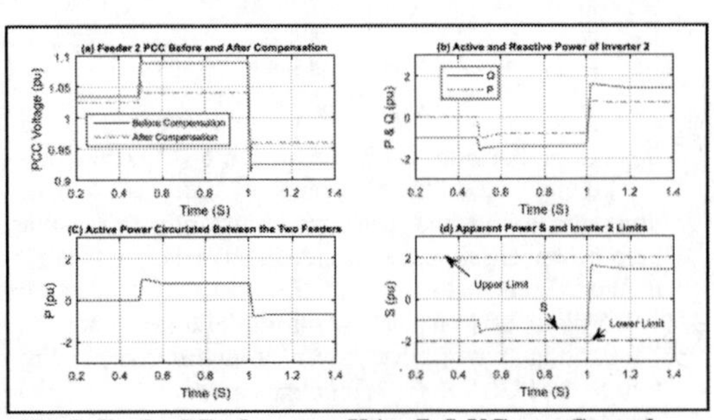

Fig. 7: Feeder 2 Performance Using P-Q-V Droop Control Method.

4 CONCLUSION

In this article, different droop control methods for PV-based communal grid networks with different line resistances (R) and impedances (X) have been modelled and simulated in MATLAB to determine the most efficient control method for a given network. The results show that active power-frequency (P-f) droop control method is the most efficient for low voltage transmission networks with low X/R ratios while reactive power-voltage (Q-V) droop control method is the most efficient for systems with high X/R ratios. Results also show that P-f or Q-V droop control methods cannot individually efficiently regulate line voltage and frequency that for systems with complex line resistances and impedances, i.e. near unity X/R ratios. For such systems, P-Q-V droop control method, where both active and reactive power could be used to control PCC voltage via shunt-connected inverters, is determined to be the most efficient control method. Results also show that shunt-connection of inverters leads to improved power flow control of interconnected communal grids by allowing feeder voltage regulation, load reactive power support, reactive power management between feeders, and improved overall system performance against dynamic disturbances. The necessary active power for the compensation is drawn from the interconnected feeders via the inverters. The controller then circulates minimum active power between the feeders while the active and reactive power droop coefficients are adjusted online through a look up table based on the PCC voltage level. A real-time integration inductor-capacitor-inductor (LCL)-filter with a virtual inductance is used to actively damp the inverter output without the need for extra sensors to feedback measured signals.

4.1 Acknowledgement

This research is funded by Leeds International Research Scholarship

4.2 References

1. Huang, W., Lu, M., Zhang, L., *Survey on Microgrid Control Strategies*, Energy Procedia **12:** 206-212, 2011
2. Hatziargyriou, N., *Microgrids: Architecture and Control*, F. Edition, John Wiley & Sons, New York, 2014
3. Huang, J., Jiang, C., Xu, R., *A Review on Distributed Energy Resources and Microgrids*, Renewable and Sustainable Energy Review **12:** 2472-2483, 2008
4. Basak, P., Saha, A., Chowdhury, S., *Microgrid: Control Techniques and Modelling*, Proceedings of the 44th International Universities Power Engineering Conference (UPEC), Glasgow, 2009
5. Pedrasa, M., Spooner, T., *A Survey of Techniques Used to Control Microgrid Generation and Storage during Island Operation*, 16th Australasian Universities Power Engineering Conference (AUPEC2006), Melbourne, Australia, 2006,
6. De Brabandere, K., Bolsens, B., *A Voltage and Frequency Droop Control Method for Parallel Inverters*, IEEE Trans. on Power Electronics **22:** 1107-1115, 2007
7. Renewable Academy (renac), ReGrid:

Frequency and Voltage Regulation in Electrical Grids, 2003 (http://docplayer.net/6038884-Regrid-frequency-and-voltage-regulation-in-electrical-grids.html)

8. Bevrani, H., Shokoohi, S., *An Intelligent Droop Control for Simultaneous Voltage and Frequency Regulation in Islanded Microgrids*, IEE Transactions on Smart Grid **4**: 1505-1514, 2013
9. Pecas-Lopes, J., Moreira, C., Mandureira, A., *Defining Control Strategies for Microgrids Islanded Operation*, IEEE Transactions on Power Systems **21**: 916-924, 2006
10. Pecas-Lopes, J., Moreira, C., Mandureira, A., et al., *Control Strategies for Microgrids Emergency Operation,* International Conference on Future Power Systems, Amsterdam, 2005
11. Ashbani, S., Abdel-Rady, M., *General Interface for Power Management of Micro-Grids Using Nonlinear Cooperative Droop Control,* IEEE Transactions on Power Systems **20**: 2929-2941, 2013
12. Bergen, A., Vittal, V., *Power Systems Analysis*, Prentice Hall, NJ, USA, 1999
13. Engler, A., Soultanis, N., *Droop Control in LV-grids*, International Conference on Future Power Systems, Amsterdam, 2005
14. Yao, W., Chen, M., *Design and Analysis of the Droop Control Method for Parallel Inverters Considering the Impact of the Complex Impedance on the Power Sharing*, IEEE Trans. on Industrial Electronics **58**: 576-588, 2011
15. Alatrash, H., Mensah, A., Mark, E., et al., *Generator Emulation Controls for Photovoltaic Inverters,* 8th International Conference on Power Electronics - ECCE Asia, Jeju, 2011
16. Abusara, M., Shark, S., Guerrero, J, *Improved Droop Control Strategy for Grid-Connected Inverters, Sustainable Energy*, Grida and Networks **1**: 10-19, 2015
17. Vallve, X., Serrasolses, J., *Design and Operation of a 50 kWp PV Rural Electrification Project for Remote Sites in Spain*, Solar Energy **59**: 111–119, 1997
18. Chaurey, A., Kandpal, T., *A Techno-Economic Comparison of Rural Electrification Based on Solar Home Systems and PV Microgrids*, Energy Policy **38**: 3118-3129, 2010

PERFORMANCE OF A 40 KWP PV IRRIGATION DEMONSTRATOR COMBINING VARIABLE AND CONSTANT PRESSURE PUMPING

I.B. Carrêlo[1], R.H. Almeida[1,2], L. Narvarte[1]
[1] Instituto de Energía Solar - Universidad Politécnica de Madrid, 28031 Madrid, Spain
[2] Instituto Dom Luiz, Faculdade de Ciências, Universidade de Lisboa, 1749-016 Lisboa, Portugal
*Phone: +34 913365531; E-mail: isaac.barata@ies.upm.es

ABSTRACT: This paper describes a 40 kWp PV irrigation system combining variable frequency and constant pressure pumping installed in Sardinia, Italy. The objective of this paper is to present the design, as well as the implementation of the system. Moreover, monitoring data from 2017 irrigation season, from June to the end of October is analyzed and some first results are presented. During this period, the PV irrigation system pumped a total of 20574 m^3.
Keywords: Design, Monitoring, Stand-alone PV Systems, Water-Pumping

1 INTRODUCTION

Nowadays a huge part of the irrigation systems are fed by the national grid or from diesel generators [1]. However, the rise of the conventional energy price [2] increased the need for an alternative by using renewable energies [3], such as the solar energy. A H2020 project named MASLOWATEN [4] recommends solutions, and includes the installation of 5 real scale photovoltaic irrigation systems (PVIS) as a way to answer the irrigators' needs [5], [6], [7], [8]. There are 4 criteria that we consider as crucial in order to be considered as a PVIS. First, the system should be able to surpass passing clouds without destabilize. Second, the PVIS should be designed regarding the preexistent irrigation system. Third, the PV production should be similar to the irrigation demand. And finally, its reliability must be assured in order to work for at least 25 years.
This paper is about one of these PV Irrigation Systems that is installed in Sardinia, Italy.

The aim of this paper is to present the design (Section 2), the installation (Section 3) and first results (Section 4) of the last irrigation season (2017) from June to October.

2 DESIGN

2.1 System configuration

The system is located in the farm Sarciofo, Uri, 15 km Southwest from Sassari, Italy. It is a 5.4 hectares dedicated to the ecological cultivation of artichoke. In this case, a previous system was already installed. This was composed by two wells with 4 and 3 kW submersible pumps (from now on, well n°1 and well n°2) elevating water to the same 2000 m^3 water pool. The irrigation can be done by gravity or by using a third horizontal surface centrifugal pump of 4 kW to irrigate through low-pressure sprinklers.

The new system was designed to feed the three pumps on the farm with a 40 kWp PV generator, organized in 8 strings of 20 modules (160 PV modules). These modules are mounted in a North-South horizontal axis tracker - 2 STI-H160 (single row).

The pump of well n°1 – P1 (originally 4 kW) was replaced by a new 18.5 kW submersible pump (Caprari E6SX50/14A+MACX625B-8V). In order to allow the installation of a larger pump, the owner widened and protected the borehole walls with a new tube, achieving an increase in capacity. In order to get the confidence of the user, the system is allowed to feed the pump both from the PV system and from the previous system (electrical grid), switching manually according to the user decision. In both cases the power is supplied by a 22 kW frequency converter (OMRON 3G3RX A4220-E1F).

In the well n°2, the original pump 2 – P2 (Caprari E4XP35 / 27 + MCH44-8V) has been kept. This pump can be supplied from both the solar system with a 5.5 kW frequency converter (OMRON 3G3RX A4220-E1F) or directly from the grid.

The original irrigation pump – P3 (4 kW) was replaced by a 7.5 kW pump (Caprari MEC-MR65-3 / 2C+ FELM 7.5 kW) able to provide 50 m^3/h at 25 m. This pump can be powered from the solar system by an 11 kW frequency converter (OMRON 3G3RX A4110-E1F) or from the grid with a soft starter.

Figure 1 includes the main components of the system (in green the new system, in orange the previous system).

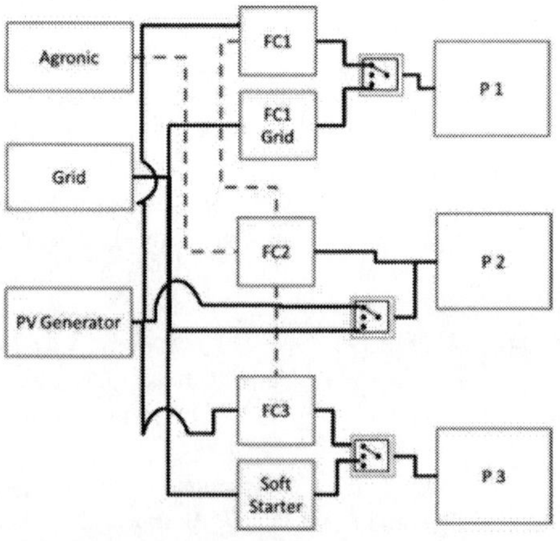

Figure 1: Current irrigation system

2.2 System operation logic

Regarding to the control and operation logic of the system, this system is considered a standalone PV irrigation system. So, it was necessary to create an operating logic control based on the available PV power and the end-user irrigation needs. If there is no irrigation, only P1 and P2 will work. Otherwise, three pumps can work simultaneously according to the available PV power. Table 1 shows the system operation logic according to the user's need to irrigate based on the available PV power.

Table I: Operation of the system when there is an irrigation program

	Increase in PV power	Decrease in PV power
	P1, P2, P3	P1, P2, P3
14.5 kW		
	P2, P3	P1, P2, P3
13.5 kW		
	P2, P3	P2, P3
9.0 kW		
	P2, P3	P2, P3
8.8 kW		
	P2, P3	P2, P3
8.7 kW		
	P3	P2, P3
8.0 kW		
	P3	P3
7.5 kW		
	P2	P3
7.0 kW		
	OFF	P2
2.8 kW		
	OFF	OFF

As it can be seen, hysteresis logic was programed to avoid the continuous stop and start of the pumps due to small variations of the PV power.

For the borehole pumps (P1 and P2) a master-slave configuration has been implemented. FC2 (master) does a Maximum Power Point Tracking routine and FC1 (slave), when is pumping, follows the frequency of the master. FC3 does a pressure control routine, being able to work in three different pressure set-points (2.5, 3 and 3.5 bar) according to the irrigation sector. All frequency converters have passing cloud routines.

3 IMPLEMENTATION

In order to have a better picture of the overall system under analysis, the next figures show the real system. Figure 2 shows the PV generator and its integration on the field.

Figure 3 shows an inside view of the building with the 4 frequency converters (marked in green) and the monitoring (marked in orange).

Figure 4 shows the 2000 m^3 intermediate water tank, between the borehole pumps and the irrigation system, and Figure 5 the low-pressure sprinklers installed in order to save energy.

Figure 2: PV Generator

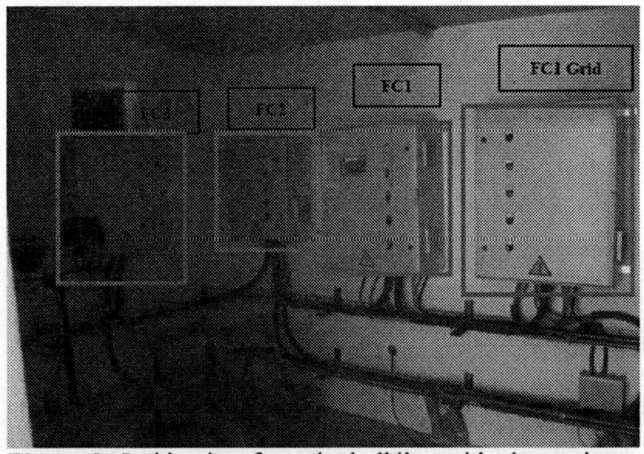

Figure 3: Inside view from the building with electronic, control and monitoring system

Figure 4: Water tank

Figure 5: Low pressure sprinklers

4 FIRST RESULTS

The first results presented are based on monitoring data recorded in the months of June, July, August, September and October 2017 (the last irrigation season). The irrigation period was from June to October, being the highest irrigation needs in July and August (around 6 hours a day). The annual irrigation hours of P3 with PV, were 519, while the boreholes pumps worked for 1902 h. This year was really dry and the nearest lake which feeds most wells in the area was almost without water. This has led local authorities to limit the use of water. Even so, in August, P1 worked, on average, 14.56 hours a day and P2 worked 10.76 hours. These almost 15 hours of use of the P1 cannot be all provided by the PV generator and mainly because of that the PV penetration in this month was 69%.

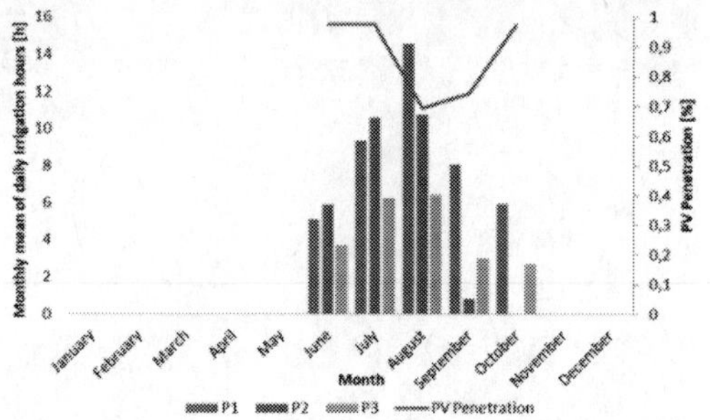

Figure 6: Pumps working hours per day in each month and PV penetration

Figure 6 also shows the PV penetration along the year. In 2017 irrigation period the PV penetration was 87%.

The irrigation pump worked for 683 hours and the borehole pumps for 1323 and 830 hours (P1 and P2), which means a total volume of water of 25695 m^3, 3849 m^3 and 20574 m^3, respectively for P1, P2 and P3.

In table II, it is possible to verify a summary of the measured data.

Based on the measured irradiance presented in Table III, two performance ratios are calculated: the typical one and one that only considers the irrigation hours. The typical PR is 0,16, while the PR during irrigation hours (PR_{I-H}) is 0,37 (see Table IV). These values are low but can be justified by conditioned by the low utilization ratio caused by the lack of water.

Table II: Measured data and first results

Variable	Unit	Value
Total volume of irrigation water	[m^3]	20574
Total volume of pumped water	[m^3]	29545
PV water flow penetration	[%]	87

Table III: Irradiation

Irradiation	Value [kWh/m^2]
Total irradiation	2181
Irradiation during irrigation hours	956

Table IV: Performance ratios

Performance ratio	Value [ ]
PR	0,16
PR_{I-H}	0,37

Until here, only global values were presented. To better know the way the PV system works, the following figure represents a typical daily profile of irradiance and frequency of both pumps.

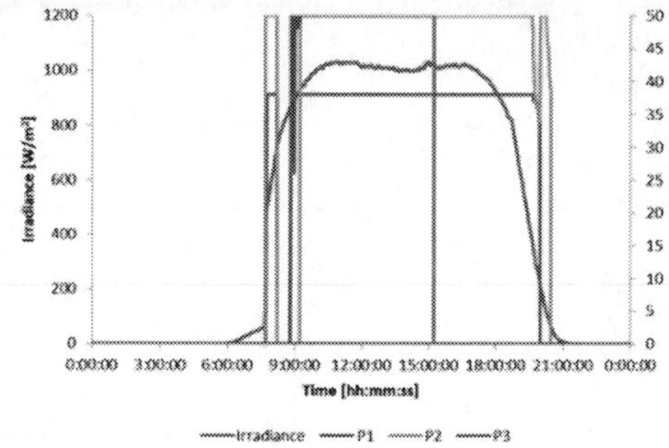

Figure 7: Pumps frequency related with the incident irradiance (3rd June 2017)

As it can be seen both pumps, P1 and P2, started to work at 7h41. They stopped 20 minutes later because they filled the water tank. At 8h49, according to the farmer desire, it starts to irrigate (P3 starts to work) until 15h14. In the meantime the others two pumps worked as well. In fact, P1 works until 19h58 and P2, 30 minutes more. It is interesting to see that at the end of the day P2 starts to regulate the frequency until 34 Hz until P1 stops. After that, having more available PV power, the frequency of P2 returns to 50 Hz.

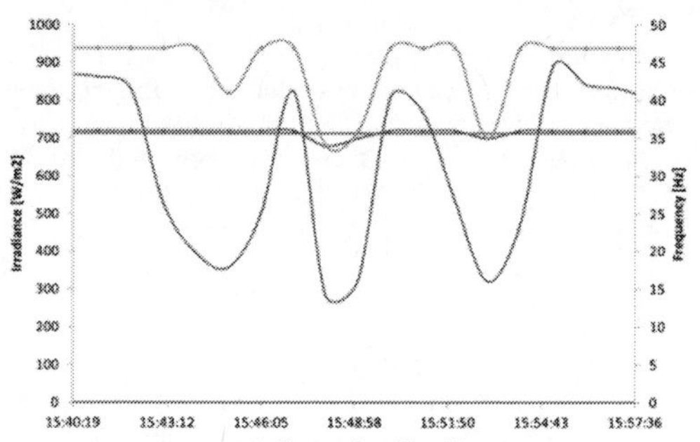

Figure 8: Passing clouds during 17 minutes (22nd September 2017)

In this demonstrator was implemented the algorithm that allows the system to surpass the passing clouds without destabilizing. In figure 8, it can be seen this occurrence. During the 17 minutes the irradiance is varying however the frequency converters do not stop. They are able to surpass these variations.

5 CONCLUSIONS

2017 was a very dry year and, accordingly, the user does not have all the water the user wants to irrigate. For example, July and August were the months in which the irrigation needs were higher.

Regarding the performance of the PV irrigation system, the electric PR during irrigation hours in the highest water-demanding months reached 0.39The PV Irrigation System has shown a good performance - PRs values were conditioned by the low utilization ratio caused by the lack of water.

The control made between the three pumps worked as expected. This allowed a total annual volume of water pumped of 50119 m^3 by the three pumps.

6 AKNOWLEDGEMENT
This work has been possible thanks to the funding from the European Union's Horizon 2020 research and innovation programme in the project MArket uptake of an innovative irrigation Solution based on LOW WATer-ENergy consumption (MASLOWATEN), under grant agreement n°640771, as well as the financial support by MIT Portugal Program on Sustainable Energy Systems and the Portuguese Science and Technology Foundation (FCT), grant PD/BD/105851/2014 to Rita Hogan Almeida.

7 REFERENCES

[1] J. Carroquino, R. Dufo-López, J. L. Barnal-Augustin, "Sizing of off-grid renewable energy systems for drip irrigation in Mediterranean crops," Renewable Energy, Vol 76 (2015) 566-574.
[2] R. Langarita, J. Sánchez Chóliz, C. Sarasa, R. Duarte, S. Jiménez, "Electricity costs in irrigated agriculture: A case study for an irrigation sheme in Spain", Renewable and Sustainable Energy Reviews, Vol 68 (2017), 1008-1019.
[3] M. Abu-Aligah, "Design of Photovoltaic Water Pumping Systems and Compare it with Diesel Powered Pump," Jordan Journal of Mechanical and Industrial Engineering, Vol 5 (2011) 272-280.
[4] Grant Agreement number 640771 – MASLOWATEN, European Commission.
[5] I. B. Carrêlo, R. H. Almeida, L. M. Carrasco, F. Martínez-Moreno and L. Narvarte, "A 360 kWp PV irrigation system to a water pool in Spain," in 33rd European Photovotaic Solar Energy Conference and Exhibition, Amsterdam, 2017.
[6] I. B. Carrêlo, R. H. Almeida, F. Martínez-Moreno, L. M. Carrasco and L. Narvarte, "A 160 kWp constant pressure PV Irrigation system in Spain," in 33rd European Photovotaic Solar Energy Conference and Exhibition, Amsterdam, 2017.
[7] R. H. Almeida, I. B. Carrêlo, F. Martínez-Moreno, L. M. Carrasco and L. Narvarte, "A 140 kW hybrid PV-diesel pumping system for constant-pressure irrigation," in 33rd European Photovotaic Solar Energy Conference and Exhibition, Amsterdam, 2017.
[8] R. H. Almeida, I. B. Carrêlo, L. M. Carrasco, F. Martínez-Moreno and L. Narvarte, "Large-scale hybrid PV-Grid irrigation system," in 33rd European Photovotaic Solar Energy Conference and Exhibition, Amsterdam, 2017.

MAIN FINAL RESULTS OF MASLOWATEN – THE H2020 PROJECT FOR THE MARKET UPTAKE OF LARGE POWER PV IRRIGATION SYSTEMS

R. H. Almeida[1,2], I. B. Carrêlo[1], L. Narvarte[1], J. Fernández-Ramos[3], F. Martinez-Moreno[1], L.M. Carrasco[1]
[1] Instituto de Energía Solar, Universidad Politécnica de Madrid
[2] Instituto Dom Luiz, Faculdade de Ciências, Universidade de Lisboa
[3] Departamento de Electrónica, Universidad de Málaga
[*]Phone: +34 91 336 55 31; E-mail: rita.hogan@ies.upm.es

ABSTRACT: This paper presents the main technical results of MASLOWATEN, a H2020 European Project for the market uptake of large-power photovoltaic irrigation systems (PVIS). Five large-power PVIS, with peak powers ranging from 40 to 360 kWp were installed in real-farms of farmers, cooperatives, agro-industries and irrigator communities (one in Portugal two in Spain, one in Italy and one in Morocco) to show their reliability and economic feasibility in order to introduce them into the market. Furthermore technical specifications, business plan and a simulation tool for PVIS were developed. Finally, the technology developed was transferred to more than 20 SME. The main technical results of the project include: a solution to mitigate the problems associated to PV power intermittences due to passing clouds, the use of North-South horizontal axis trackers in order to match PV production and irrigation needs, the development and application of new performance indices based on the typical performance ratio of a PV system in order to understand the performance of each component of a PVIS.
Keywords: PV irrigation systems, Water pumping, Design, Monitoring

1 INTRODUCTION

The majority of the current agricultural irrigation systems are powered from the grid or diesel generators [1]. However, the increasing price from these energy sources [2], [3] boosted the development of alternative solutions using renewable energies, namely PV solar energy [4], [5]. Large-power photovoltaic irrigation systems (PVIS) are becoming more and more attractive also due to the high energy costs of modernized agriculture (for example, in Spain, 40 to 50% of the production cost of some crops is the energy cost to irrigate).

These large-power PVIS have two main challenges [6]: to match PV production and irrigation needs, and to solve the problems associated with PV power intermittencies due to passing clouds.

MASLOWATEN (a H2020 European Project for the market uptake of large-power PVIS from September 2015 to August 2018 [7]) proposed innovative solutions, less water and energy depend, to face up these problems.

This paper is about the main technical results of MASLOWATEN. Section 2 describes the project, section 3 is about the main innovations of the project, section 4 briefly describes the 5 real-scale large-power PVIS installed, section 5 includes the main results and finally section 6 shows some conclusions.

2 MASLOWATEN PROJECT

MASLOWATEN included the installation of 5 real-scale large-power PVIS, with powers from 40 to 360 kWp, installed in real farms of farmers, cooperatives, agro-industries and irrigator communities to show their reliability in real operating and their economic feasibility to introduce them into the market. The demonstrators cover the different possible configurations of the irrigation systems: water pumping to a pool at a variable water flow and direct pumping to the irrigation network through sprinklers, pivots or drip systems at a constant pressure and water flow; powered by stand-alone PV systems or hybrid systems combining PV with the grid or with diesel generators. The project considered also the development of technical specifications [8] and quality control procedures for PVIS, a simulation tool [9] and business plans, and technology transfer to small and medium enterprises.

3 MAIN INNOVATIONS OF MASLOWATEN

The following technical aspects were studied within MASLOWATEN:

a) The problems associated to PV power intermittences.

Instabilities have been addressed by specific FC tuning control procedures and algorithms. These procedures take advantage of the possibility of power regeneration of the centrifugal pumps and have been patented [10]. This way, instead of a sudden stop of the motor-pump, its frequency is reduced but the motor-pump never stops.

It should be mention that the use of batteries was not considered as an option to solve this problem because their reliability has not been proven yet.

b) The match between PV production and irrigation needs.

The use of a North-South horizontal axis tracker was the solution adopted in the five real-scale large-power PVIS and it seems to be the best solution [11], [12], [13]. This is very interesting both for pumping to a water pool or direct pumping. It presents four main advantages.

First, it maximizes the water pumped during the irrigation period (the match between the yearly profile of irradiance and the yearly water demand of the crops is very good, much better than with the typical static structure facing the Equator [13]).

Second, the daily profile of irradiance is almost flat in this type of tracker [13], which also benefits an irrigation system [14].

Third, this tracker also allows the enlargement of the irrigation time per day when compared to the typical static structure oriented to the Equator - the system will start pumping earlier in the morning and will keep working until later in the afternoon, [11], [15], [13].

Fourth, it requires less nominal power to pump the same water volume than PV static structures [11], [13].

c) The integration of the PV system in the pre-existing irrigation system.

A significant part of the potential PV irrigation market will be the retrofitting of already existing irrigation systems fed by the national grid or diesel generators [16]. Accordingly, it makes sense to keep the already existing irrigation structure (motor-pumps, irrigation controllers, pipe network, drippers, sprinklers, pivots, etc.) and to integrate the PV system on it. In addition to the irrigation system components it is also very important to know the irrigation scheduling and possible restrictions.

d) Tools to guarantee the reliability of the system for at least 25 years.

Technical specifications which include quality control procedures were developed within MASLOWATEN project. These specifications, which should be included in the contracts, will not only guarantee long-term reliability but also the optimization of costs and energy efficiencies.

4. FIVE LARGE-POWER PVIS

The previous solutions have been applied in the design and implementation of the 5 large-power PVIS:

- A 360 kWp stand-alone PVIS pumping water from a deep-borehole to a water pool at a variable water flow (in Villena, Spain) [17] – Figure 1;

Figure 1: PVIS to a water pool in Villena, Spain.

- A 160 kWp stand-alone PVIS with one pump elevating water from a borehole to an intermediate tank at a variable water flow and other pump to irrigate at a constant pressure through pivots (in Alaejos, Spain) [18] – Figure 2;

Figure 2: PVIS through pivots, Alaejos, Spain.

- A 40 kWp stand-alone PVIS with 2 pumps pumping from two different boreholes at a variable water flow to an intermediate tank and one pump to irrigate at a constant pressure through sprinkles (in Sardinia, Italy) [19]– Figure 3;

Figure 3: PVIS through sprinklers, Uri, Italy.

- A 120 kWp hybrid PV-grid drip irrigation systems, with 2 pumps and hybridization in the electric part of the system (in Tamelalt, Morocco) [20] – Figure 4;

Figure 4: Hybrid PV-grid drip irrigation system, Tamelalt, Morocco.

- A 140 kWp hybrid PV-diesel drip irrigation systems, with 3 pumps and hybridization in the hydraulic part of the system (in Alter do Chão, Portugal) [21] – Figure 5.

Figure 5: Hybrid PV-diesel drip irrigation system, Alter do Chão, Portugal.

These 5 PVIS have been working since the summer of 2016 and monitoring data recorded in each system allow:

- The technical validation of the solutions;
- The economical validation of the solutions;
- The environmental validation of the solutions.

This paper pretends to give a general overview of the main technical results of MASLOWATEN and its 5 PVIS.

The economical validation can be found in a concurrent paper of this conference [22], while the environmental validation of the stand-alone PVIS can be found in [23].

5 TECHNICAL RESULTS

5.1 Performance indices

In order to perform the technical validation, there was the need to factorize the typical *PR* used in the PV field in order to have a better understanding of what is happening in a PVIS. This was done since the performance of a PVIS is not only determined by the technical quality of the PV system components and by the efficient use of the available irradiation, but also by the particular characteristics of each irrigation system and by external circumstances such as the lack of water.

So, the following equation (and Figure 6) applies:

$$PR = PR_{PV} \times UR_{IP} \times UR_{PVIS} \times UR_{EF}$$

where:

$PR_{PV} = \frac{E_{PV}}{P^*/G^*} \frac{1}{\int G_{used} dt}$	This is the *PR* considering only losses strictly associated to the PV system itself (i.e., actual versus nominal peak power, dirtiness, thermal and DC/AC conversion losses). It is intrinsic to the technical quality of the PV component and the proper maintenance.
$UR_{IP} = \frac{\int_{IP} Gdt}{\int Gdt}$	This is the ratio of the total irradiation throughout the irrigation period (IP) to the total annual irradiation (Figure 6-a). It is intrinsic to a given crop.
$UR_{PVIS} = \frac{\int G_{useful} dt}{\int_{IP} Gdt}$	This is the ratio of the irradiation strictly required to keep P_{AC} equal to the stable AC power requirement to the total irradiation throughout the IP (Figure 6-b). It is intrinsic to the PVIS design (e.g., it depends on the type of irrigation system, the ratio between the PV peak power and the stable PV power required for irrigation and on the tracking geometry).
$UR_{EF} = \frac{\int G_{used} dt}{\int G_{useful} dt}$	This is the ratio of the irradiation required to keep P_{AC} stable during the irrigation scheduling to the same irradiation during the IP.

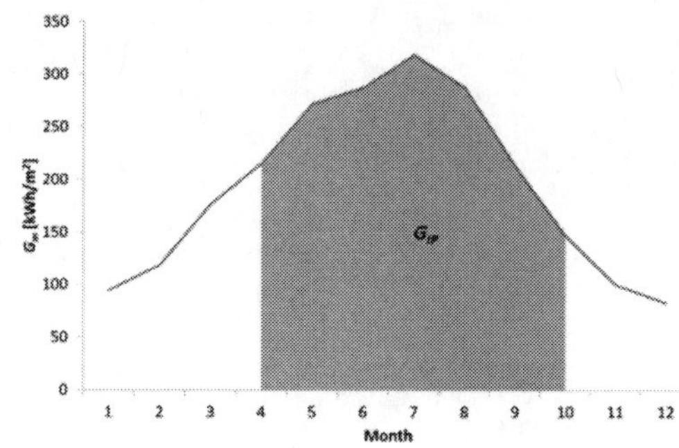

(a)

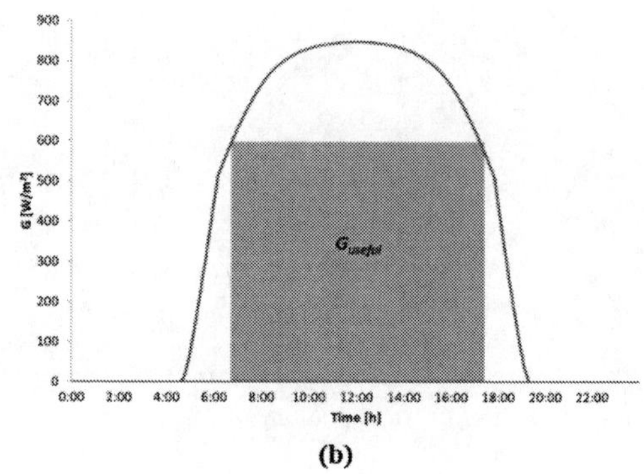

(b)

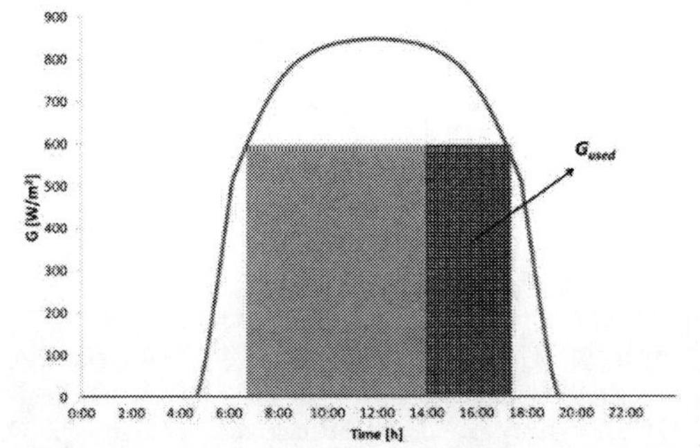

(c)

Figure 6: Graphical representation of the different irradiations considered: (a) $\int G_{IP}$ is the irradiation during the irrigation period, (b) $\int G_{useful}$ is the useful irradiation during the IP determined by the design of the PV irrigation system; and (c) $\int G_{used}$ is the irradiation used effectively by the system.

A representative example of the application of these different components is presented in Table I. It shows the performance indices obtained in August 2018 in the hybrid PV-grid drip irrigation system in Alter do Chão, Portugal. It is interesting to note that the PR_{PV} is 0.83 (similar to the values obtained in a PV grid-connected system), that the UR_{PVIS} is 0.68 (consequence of the type of hybridization and of the power thresholds), and that the global *PR* in this month achieves 0.56.

Table I: Real performance indices in August 2018 – hybrid PV-diesel, Alter do Chão, Portugal

Month	PR	PR_{PV}	UR_{IP}	UR_{PVIS}	UR_{EF}
August	0.56	0.83	1.00	0.68	0.99

5.2 Passing clouds routine

Figure 7 shows the motor-pump frequency and the irradiance received by the PV generator between 10:59:31 and 11:02:24 on the 19th October 2017 in the 360 kWp PVIS in Villena, Spain. The irradiance (in red) decreases intensely due to a cloud, but the stability of the frequency converter is maintained – in 20 seconds, the irradiance falls more than 75%, and the frequency converter responds with a 13% decrease of the frequency without stopping the pump, avoiding all the hydraulic and electric problems that could be caused by an abrupt stop, such as water hammer and overvoltage respectively.

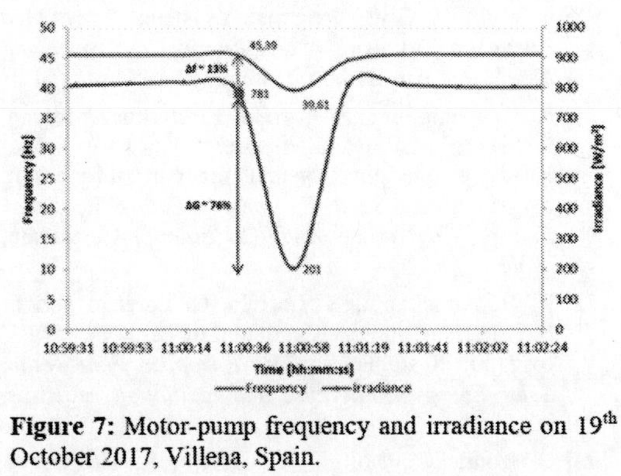

Figure 7: Motor-pump frequency and irradiance on 19th October 2017, Villena, Spain.

5.3 System performance along the day – clear sky day

Figure 8 shows the frequency of the motor-pumps and the irradiance received by the PV generator on the 11th July 2017. It is interesting to observe the variation of the frequency of the borehole pump with the irradiance (in the beginning and end of the day, in blue) as well as the use of two different pressure setpoints (two different levels of frequency in the irrigation pump, in red).

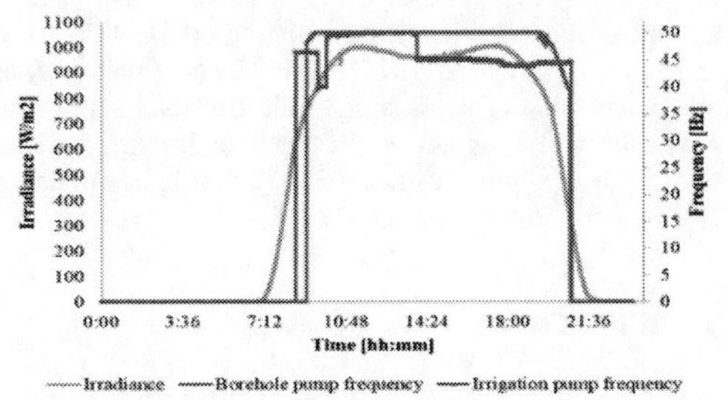

Figure 8: Frequency of the pumps and irradiance on 11th July 2017, Alaejos, Spain.

5.4 System performance along the day – cloudy day

In Figure 9, one can also see the frequency of the motor-pumps and the irradiance received by the PV generator on the 4th October 2017 afternoon. It is interesting to note that the frequency of the borehole pump changes with the irradiance, while the one of the irrigation pump keeps unchanged (two different levels are observed due to 2 different pressure setpoints – a high pressure level is observed in the beginning of the second step due to the filling of the pipes).

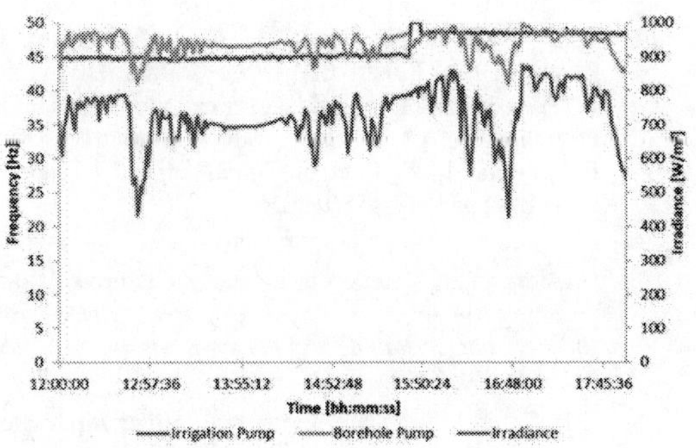

Figure 9: Frequency of the pumps and irradiance on 4th October 2017, Alaejos, Spain.

5 CONCLUSIONS

MASLOWATEN was a H2020 European Project for the market uptake of large-power photovoltaic irrigation systems (PVIS). Five large-power PVIS were installed in real-farms in order to show their technical reliability and economic feasibility. In addition, technical specifications, business plans and a simulation tool for PVIS were developed. Finally, the technology developed under the project was transferred to more than 20 SME.

The project was able to show the technical and economical viability of these large-power PVIS. Particularly, the problems related to PV power intermittences due to passing clouds were addressed and North-South horizontal axis trackers were used in the 5 PVIS in order to match PV production and irrigation needs.

6 AKNOWLEDGMENTS

This work has been possible thanks to the funding from the European Union's Horizon 2020 research and innovation programme in the project MArket uptake of an innovative irrigation Solution based on LOW WATer-ENergy consumption (MASLOWATEN), under grant agreement n°640771, as well as the financial support by MIT Portugal Program on Sustainable Energy Systems and the Portuguese Science and Technology Foundation (FCT), grant PD/BD/105851/2014.

7 REFERENCES

[1] J. Carroquino, R. Dufo-López and J. L. Barnal-Augustin, "Sizing of off-grid renewable energy systems for drip irrigation in Mediterranean crops," *Renewable Energy*, no. 76, pp. 566-574, 2015.

[2] M. Abu-Aligah, "Design of Photovoltaic Water Pumping System and Compare it with Diesel Powered Pump," *Jordan Journal of Mechanical and Industrial Engineering*, vol. 5, no. 3, pp. 273-280, 2011.

[3] R. Langarita, J. Sánchez Chóliz, C. Sarasa, R. Duarte and S. Jiménez, "Electricity costs in irrigated agriculture: A case study for an irrigation scheme in Spain," *Renewable and Sustainable Energy Reviews*, vol. 68, pp. 1008-1019, 2017.

[4] P. E. Campana, S. Leduc, M. Kim, A. Olsson, J. Zhang, J. Liu, F. Kraxner, I. Mc Callum, H. Li and J. Yan, "Suitable and optimal locations for implementing photovoltaic water pumping systems for grassland irrigation in China," *Applied Energy*, vol. 185, pp. 1879-1889, 2017.

[5] C. Gopal, M. Mohanraj, P. Chandramohan and P. Chandrasekar, "Renewable energy source water pumping systems – A literature review," *Renewable Energy and Sustainable Energy Reviews*, vol. 25, pp. 351-370, 2013.

[6] EIP Water, "European Innovation Partnership Water - Strategic Implementation Plan," Brussels, 2012.

[7] MASLOWATEN, "MASLOWATEN project," 2018. [Online]. Available: www.maslowaten.eu.

[8] MASLOWATEN, "Technical Specifications for Photovoltaic Irrigation Systems," 2017.

[9] Instituto de Energía Solar - Universidad Politécnica de Madrid, "SISIFO - Simulación de Sistemas Fotovoltaicos," [Online]. Available: http://sisifo.info.

[10] J. Fernandez Ramos, L. Narvarte Fernandez, R. Hogan Teves de Almeida, I. Barata Carrêlo, L. M. Carrasco Moreno and E. Lorenzo Pigueiras, "Method and control device for photovoltaic pumping systems". Spain Patent ES 2 607 253 B2, 1 3 2018.

[11] L. Narvarte, J. Fernández-Ramos, F. Martínez-Moreno, L. M. Carrasco, R. H. Almeida and I. B. Carrêlo, "Solutions for adapting photovoltaics to large power irrigation systems for agriculture," *Sustainable Energy Technologies and Assessments*, vol. 29, pp. 119-130, 2018.

[12] J. Munoz, J. Carrillo, F. Martínez-Moreno, L. Carrasco and L. Narvarte, "Modeling and simulation of large PV pumping systems," in *Proceedings of the 31th European Photovoltaic Solar Energy Conference and Exhibition*, Hamburg, 2015.

[13] R. H. Almeida, L. Narvarte and E. Lorenzo, "PV arrays with delta structures for constant irradiance daily profiles," *Solar Energy*, vol. 171, pp. 23-30, 2018.

[14] M. Das and R. Mandal, "A comparative performance analysis of direct, with battery, supercapacitor, and battery-supercapacitor enabled photovoltaic water pumping systems using centrifugal pump," *Solar Energy*, vol. 171, pp. 302-309, 2018.

[15] H. Hartung and L. Pluschke, "The benefits and risks of solar-powered irrigation - a global overview," Food and Agriculture Organization of the United Nations and Deutsche Gesellschaft für Internationale Zusammenarbeit, 2018.

[16] I. Kougias, S. S., N. Scarlat, F. Monforti, M. Banja, K. Bódis and M. Moner-Girona, "Water-Energy-Food Nexus Interactions Assessment: Renewable energy sources to support water access and quality in West Africa," European Commission, Luxembourg, 2018.

[17] I. B. Carrêlo, R. H. Almeida, L. M. Carrasco, F. Martínez-Moreno and L. Narvarte, "A 360 kWp PV irrigation system to a water pool in Spain," in *33rd European Photovotaic Solar Energy Conference and Exhibition*, Amsterdam, 2017.

[18] I. B. Carrêlo, R. H. Almeida, F. Martínez-Moreno, L. M. Carrasco and L. Narvarte, "A 160 kWp constant pressure PV Irrigation system in Spain," in *33rd European Photovotaic Solar Energy Conference and Exhibition*, Amsterdam, 2017.

[19] I. B. Carrêlo, R. H. Almeida and L. Narvarte, "Performance of a 40 kWp PV Irrigation Demonstrator Combining Variable and Constant Pressure Pumping," in *35th European Photovotaic Solar Energy Conference and Exhibition*, Brussels, 2018.

[20] R. H. Almeida, I. B. Carrêlo, L. M. Carrasco, F. Martínez-Moreno and L. Narvarte, "Large-scale hybrid PV-Grid irrigation system," in *33rd European Photovotaic Solar Energy Conference and Exhibition*, Amsterdam, 2017.

[21] R. H. Almeida, I. B. Carrêlo, F. Martínez-Moreno, L. M. Carrasco and L. Narvarte, "A 140 kW hybrid PV-diesel pumping system for constant-pressure irrigation," in *33rd European Photovotaic Solar Energy Conference and Exhibition*, Amsterdam, 2017.

[22] R. H. Almeida, I. B. Carrêlo, C. Lorenzo and L. Narvarte, "Economic Validation of Large Power PV Irrigation Systems," in *35th European Photovotaic Solar Energy Conference and Exhibition*, Brussels, 2018.

[23] G. Todde, L. Murgia, I. Carrelo, R. Hogan, A.

Pazzona, L. Ledda and L. Narvarte, "mbodied Energy and Environmental Impact of Large-Power Stand-Alone Photovoltaic Irrigation Systems," *Energies,* vol. 11, no. 8, 2018.

EXPERIMENTAL EVALUATION OF THE INTEGRATION OF MICROGRIDS SUPPLIED BY HYBRID ENERGY SOURCES

João Paulo Alves Veríssimo, Claudomiro Fábio de Oliveira Barbosa, Marcos André Barros Galhardo, Wilson Negrão Macêdo, Victor Parente de Oliveira Alves, Tales Vinícius Pinheiro da Silva, Luana Oliveira de Albuquerque, João Tavares Pinho
Grupo de Estudos e Desenvolvimento de Alternativas Energéticas
Faculdade de Engenharias Elétrica e Biomédica
Instituto de Tecnologia, Universidade Federal do Pará
Av. Augusto Corrêa, 1, CEP 66.050-000 – Belém, PA, Brazil
Tel. / Fax +55 91 3201 7299, verissimo.joaop@gmail.com

ABSTRACT: This work studies the integration between two different microgrids and their interaction with the conventional grid, by performing experimental tests on a real system of the Laboratory of Hybrid Systems and Microgrids, installed at the Group of Studies and Development of Energy Alternatives (GEDAE), Belém, Pará, Brazil. The first test interconnects two different microgrids, the second test connects one microgrid to the conventional grid, and the last test, once more, interconnects two different microgrids, but with a considerable distance between their interconnection points. The experimental results obtained from these tests are presented and explained, showing that the integration of the systems occurred as expected.
Keywords: Hybrid Power Systems, PV Systems, Microgrids

1 INTRODUCTION

Nowadays, microgrids (MG) can be considered key points to: increase the penetration of renewable energy sources; increase energy reliability; increase the efficiency of the whole electrical system; and enable final consumers to gain a certain independence from the conventional grid, since an MG can operate in connected or isolated mode [1]. An MG can be connected to a traditional distribution grid, in order to supply local energy needs and/or provide ancillary services to the grid. Many studies are in the field of interconnection of those MG, forming small clusters, or connected to the main grid. This work aims to contribute to this area by providing experimental results related to the interconnection between two different MG and the connection of one MG to the conventional grid. All experiments were made using the infrastructure of the Laboratory of Hybrid Systems and Microgrids, developed by the Group of Studies and Development of Energy Alternatives (GEDAE) of the Federal University of Pará (UFPA).

2 LABORATORY OF HYBRID SYSTEMS AND MICROGRIDS

In the experimental tests described in this paper, the equipment/infrastructure of the laboratory that were used are briefly described below:

- Cluster A: a group of three bidirectional inverters, each one with rated power of 4.5 kW;
- Cluster B: the same characteristics of Cluster A; however, connected to the other end of the microgrid;
- BB-01: a battery bank of 48 V and 1,000 Ah, connected to Cluster B;
- BB-02: another battery bank of 48 V and 1,000 Ah, but connected to Cluster A;
- Diesel Generator: a backup generation system with nominal capacity of 12 kVA, that can be connected to both clusters;
- PVG-01: a photovoltaic generator of 925 Wp, connected to an inverter with nominal capacity of 700 W, near the output of Cluster A;
- PVG-02 and PVG-03: two identical photovoltaic generators of 1.1 kWp each, connected to inverters with 700 W of nominal capacity, also near the output of Cluster A;
- PVG-05: a photovoltaic generator of 3 kWp connected to the DC bus of Cluster B by means of a charge controller;
- PVG-06 and PVG-07: two identical photovoltaic generators of 2 kWp each, connected to inverters with 2 kW of nominal capacity;
- Distribution microgrid: a three-phase microgrid with multiplex aluminum cables and approximate length of 1 km;
- Loads: resistive and nonlinear loads can be connected to the two clusters. Also, both the lighting and power circuits of GEDAE´s building can be connected near the output of Cluster A by means of a transfer switch. It is important to state that there are also other PV systems connected to the power circuit, thus increasing the generation capacity of the isolated system when GEDAE´s building is inserted in it.

Besides these, there are other components installed in the laboratory that were not used for this specific work. A more detailed description of all components of the laboratory is given in [2].

3 CONFIGURATIONS

According to [3], the bidirectional inverters used in both Cluster A and Cluster B have a specific input for external power sources. Normally, these inverters work in stand-alone mode, setting the parameters of their outputs, which, in this case, are the three phases of the power distribution microgrid. However, if these inverters have an external power source available, they can switch their operation mode, allowing the external source to be responsible for setting their outputs. Additionally, there is a configurable parameter in the settings that allows the inverters to inject power into the external source (either the main grid or other system) in case of generation surplus.

Since these clusters do not communicate with each other and do not synchronize to other power sources connected to their outputs, it became impracticable to put both clusters to supply power to the whole MG at the same time. However, instead of a parallel connection, there is the possibility to connect them in cascade, via the external power source input.

For this work, three different configurations of the laboratory structure were chosen, in order to successfully integrate such systems.

3.1 Configuration 1

This configuration uses Cluster A as an external source for Cluster B, with the distance between their connection points being 10 meters. The output of Cluster A is connected to three PV generators, to some loads and to the input for external sources on Cluster B. At the distribution microgrid, there are only two PV generators connected (PVG-06 and PVG-07). Finally, at the output of Cluster B, some loads and the distribution microgrid are connected. Fig.1 shows the schematic of this configuration.

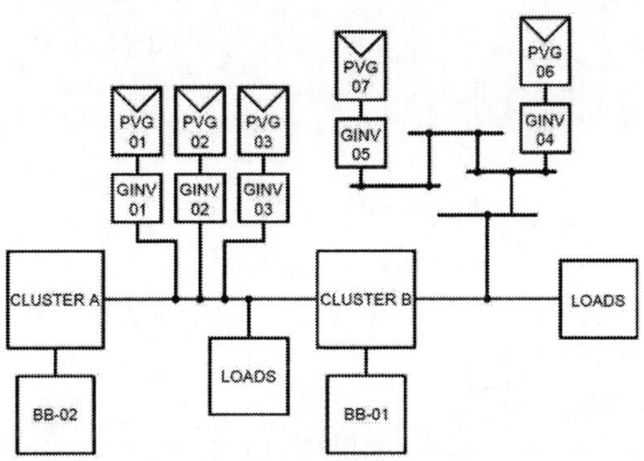

Figure 1: Schematic of configuration 1.

The only devices with significant distances to be taken into consideration are the two PV generators connected to the microgrid. PVG-06 and PVG-07 are connected 95 meters and 570 meters away from Cluster B, respectively.

3.2 Configuration 2

Here, the conventional grid is used as an external source for Cluster B, with the distance between their connections also being 10 meters. At the output of Cluster B the distribution microgrid is connected, which has loads and two PV generators connect to it. PVG-06 acts like a prosumer while PVG-07 acts only as distributed generation (DG). The rest of the loads on the microgrid are connected near each other and far away from Cluster B. Fig. 2 shows the schematic for this second configuration.

The relevant distances between the components are as follows:

- Cluster B to PVG-06: 95 meters;
- Cluster B to PVG-07: 570 meters;
- Cluster B to loads: 570 meters;
- PVG-06 to loads: 475 meters;
- PVG-07 to loads: 120 meters;
- PVG-06 to PVG-07: 475 meters.

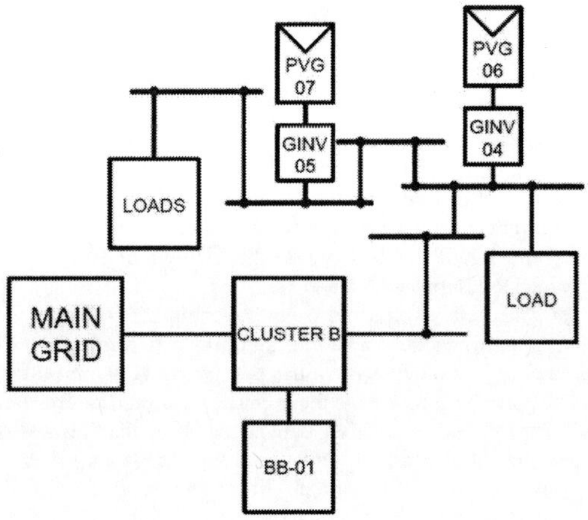

Figure 2: Schematic of configuration 2.

The load connected near PVG-06 is a resistive load with a thermostat, so the demand can vary without the need of operator intervention.

3.3 Configuration 3

In this third configuration, the diesel generator is used as input for Cluster A, which is again used as external source for Cluster B. However, instead of being connected only 10 meters away from each other, the connection between clusters is made through the microgrid, with approximately 570 meters of cable length between them. At the output of Cluster A three PV generators and the distribution microgrid are connected, and it is also possible to connect GEDAE's building by means of a transfer switch. Two PV generators (working as DG) and the input for external sources of Cluster B are connected to the microgrid. Regarding the output of Cluster B, there are only loads connected to it. The only PV generator present in this cluster is connected to the DC bus through a charge controller, which has no direct communication with the bidirectional inverters. Fig. 3 shows the schematic of this configuration.

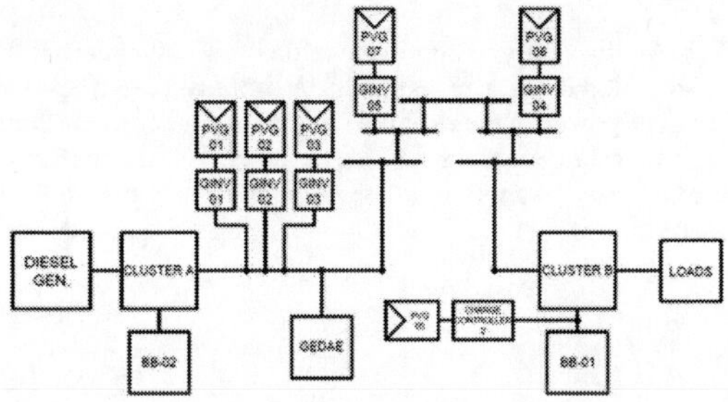

Figure 3: Schematic of configuration 3.

The relevant distances between the components are similar to those of configuration 2:

- Cluster A to Cluster B: 570 meters;
- Cluster A to PVG-06: 475 meters;
- Cluster A to PVG-07: 120 meters;
- Cluster B to PVG-06: 95 meters;
- Cluster B to PVG-07: 570 meters;
- PVG-06 to PVG-07: 475 meters.

4 EXPERIMENTAL TESTS

4.1 First test

The first test started at 7:00 and ended at 18:10 of the same day, and used configuration 1. As an example of the results, Fig. 4 shows the active power flow in phases L2 and L3, at different parts of the system. It is important to note that the inverter of PVG-06 is connected to phases L1 and L2 of the MG, while the inverter of PVG-07 is connected to phases L1 and L3.

The yellow bars represent the output of Cluster B, with positive values meaning that power is coming out of the cluster and negative values that power is going into it. The blue bars represent the input for external sources at Cluster B, with positive values meaning that power is coming into the cluster and negative values that power is going out of it. The black lines represent the power requested by the loads connected to the output of Cluster B. The red dotted lines represent the power on the MG, with positive values meaning that power is being consumed by the MG and negative values that power is being delivered by the MG.

From Fig. 4 it is noticed that at the first hours of the test, Cluster B was requesting a large amount of power from the external source, without having many loads on its output. So, it was mainly being used to charge BB-01. After the cluster fully charged the battery bank, it stopped requesting power from the external source, and, only when a large number of loads were connected to the output of Cluster B it started requesting power again.

It is also possible to see that during the entire test period PVG-07, which is connected to phases L1 and L3 approximately 570 meters away from Cluster B, synchronized only a few times with the microgrid (even when there was a great load demand between 11:40 and 14:20, right next to the output of Cluster B). And since no significant surplus generation happened on phase L3, there was almost no power sent back to the external source.

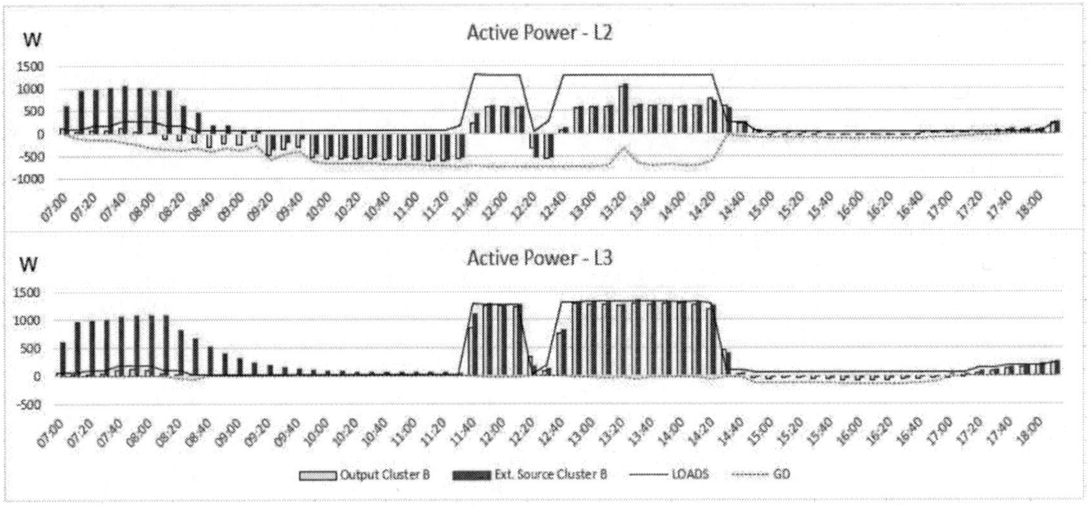

Figure 4: Active Power on phases L2 and L3 with configuration 1.

Another observation made in this first test is that each bidirectional inverter has fully autonomy to request or deliver power to the external source. In Fig. 5, by putting together the graphics of active power at the external source of Cluster B in all three phases, it is possible to see the differences between the power requested and supplied by each inverter.

For example, there are times when only the inverter from phase L3 is requesting energy from the external source, and times when only the inverter from phase L1 is feeding it back.

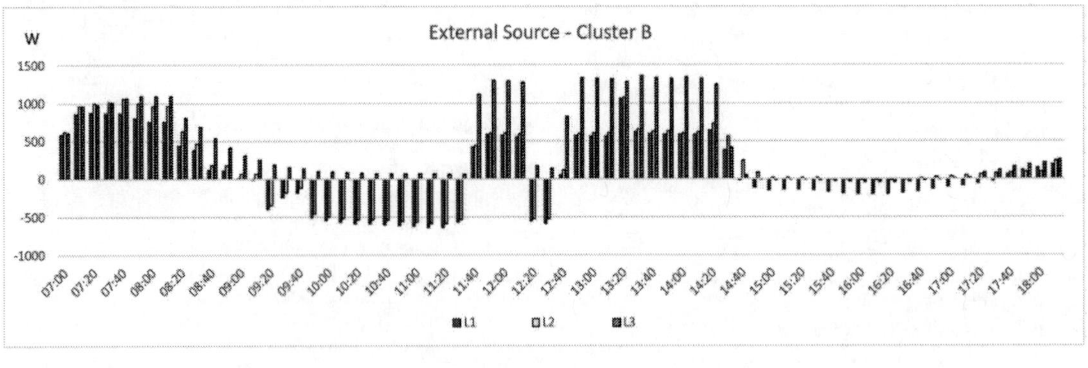

Figure 5: Active power from the external source at Cluster B.

4.2 Second test

In this experiment configuration 2 was used, and the duration of the test was from 7:00 to 18:10. Fig. 6 shows the active power flow on phase L2.

The yellow and the blue bars represent the same variables as in the first configuration, with the same positive and negative values interpretations. The black line represents the power requested by the loads at the end of the MG, and the red dotted line represents the resultant power from the prosumer and the DG, with positive values meaning that the consumption is higher than the generation and negative values meaning the opposite.

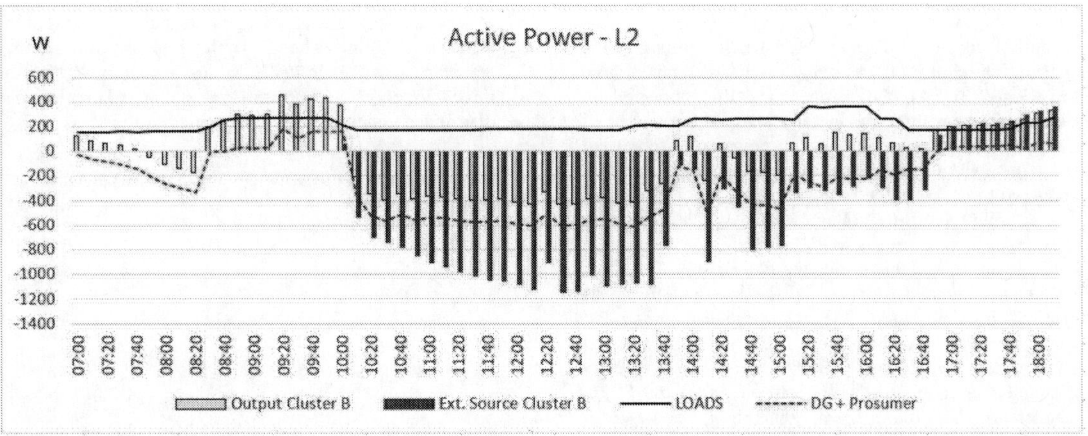

Figure 6: Active power on phase L2 with configuration 2.

In this test the external source was manually connected to Cluster B only at 10:00, and the load used by the prosumer (resistive load with thermostat) was plugged into phase L2 of the MG between 9:20 and 16:30.

Looking at the first two hours of the graphic on Fig. 6 it is possible to see that PVG-06 started working with low power and then tried to synchronize with the MG. Even after the load was connected next to PVG-06 the generator still failed to synchronize. The reason for these events can be explained by the graphics of the battery state-of-charge (BatSoc) and output frequency on Cluster B, shown in Fig. 7.

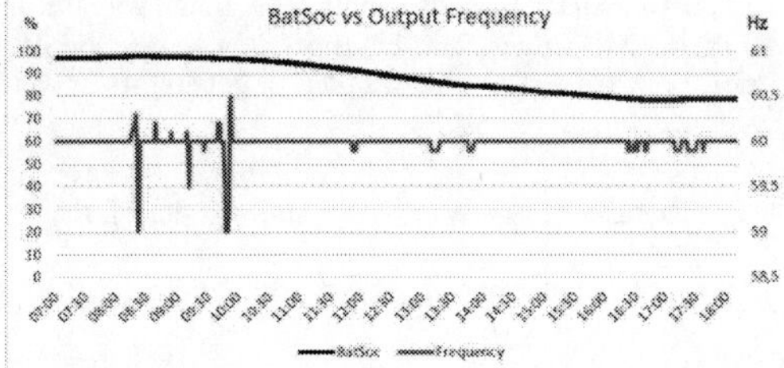

Figure 7: Battery state-of-charge versus Cluster B output frequency.

Since the battery bank was almost full, the inverters changed their frequency, in order to avoid excessive generation. However, when the external source was connected at 10:00, the inverters understood that there was an available route for the surplus generation, thus stabilizing the frequency and allowing the PV generators to synchronize with the MG.

An interesting fact is that at the periods when there was a surplus generation the inverter was feeding its external source with more energy than it was receiving from the PV generator. So, from 10:10 to 16:40 it kept discharging the battery bank (from 97 % to 78 %). Only when the PV generator lost synchronism the system stopped discharging the batteries and started requesting power from the main grid.

4.3 Third test

This final test started at 10h50 and ended at 14h00. Although it had a smaller duration compared to the previous tests, this test used configuration 3, which is more complex than the others presented before, and had a lot of maneuvers that significantly changed the behavior of the system. For this reason, only the active power on phase L1 is shown as an example of the results, and it is divided in seven different parts, in order to facilitate the description of the events that occurred at each time.

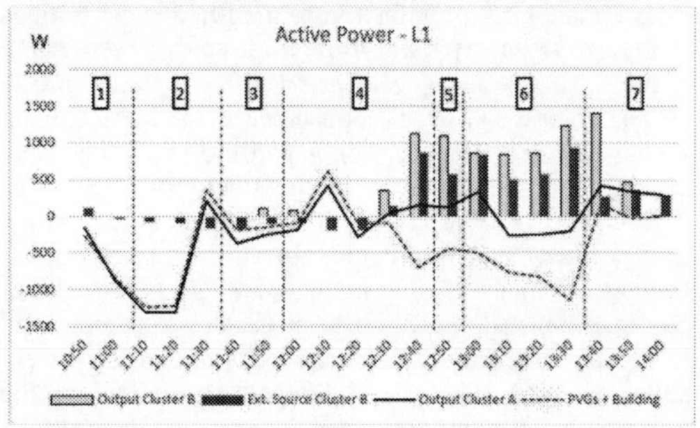

Figure 8: Active power on phase L1 with configuration 3.

Again, the yellow and the blue bars represent the power at the output and at the input of Cluster B, respectively, with the same analysis about positive and negative values made for the previous tests. The black line represents the output of Cluster A, with positive values meaning that power is flowing out of the cluster and negative values that power is coming into it. And the red dotted line is the result of the power requested by GEDAE's building and the power generated by the PV generators, with positive values meaning that consumption is higher than generation and negative values meaning the opposite.

The analyses of each of the seven parts are described below:

- At the first period, the diesel generator was synchronized with Cluster A, which, through the distribution microgrid, was being used as external source for Cluster B. So, they were basically connected in cascade. Even without having any loads at the output of Cluster B, there were times when power was requested from the external source, and times when power was delivered to it, due to the influence of PVG-05 on the DC bus, altering the BB-01 voltage level. Also, it is possible to see that there were a few PV generators synchronized with Cluster A, and they were charging BB-02.
- At the second period, GEDAE's building was inserted in the system, which introduced more PV generators in it, making the diesel generator become unnecessary to supply power, thus, automatically disconnecting it. At some point between 11:20 and 11:30, Cluster A had an excess generation, and so it disconnected the PV generators and started supplying power to the building.
- In the third period, GEDAE's building was disconnected from the system, but since the load demand was low, Cluster A did not request the diesel generator's operation.
- Since the diesel generator was not being requested, GEDAE's building was again connected, but this time without its PV generators, and so the load demand increased and the diesel generator was finally requested, and Cluster A started following the diesel generator parameters. At the end of this fourth period a voltage drop was also noticed between the output of Cluster A and the input of Cluster B, due to the length of the MG.
- Until this fifth period Cluster B was connected to its external source (Cluster A), but when more loads were connected to the output of Cluster B, the voltage on phase L3 dropped to a value below the acceptable limit for external sources by the inverters of Cluster B. Both superior and inferior limits can be modified at the settings of the master inverter, but for these tests they were set to ±10 % of the nominal voltage of Cluster B, which is 127 V. So, when the voltage dropped below the limit, Cluster B disconnected from Cluster A and the loads that were causing the voltage drop in the microgrid were also disconnected from Cluster A. Cluster B then became the grid former, thus stabilizing the voltage level on the microgrid and after one minute (set on timer configuration) the inverters from Cluster B reconnected to Cluster A. However, since the loads of Cluster B were still connected, the voltage dropped again, causing the immediate disconnection from Cluster A. Basically, a cycle of connections and disconnections was created, that occurred twice until Cluster A disconnected the diesel generator. After that, a few loads were manually removed, in order to prevent that cycle, and stabilize the system.
- At the sixth period, the PV generators of GEDAE's building were connected again. At 13:30, the same loads used on the fifth period were connected; but this time the external source was not disconnected from Cluster B.
- At the final period, GEDAE's building was removed from the system and, as shown by the red line, the PV generators lost synchronism and disconnected. After that, the inverters from Cluster B once again entered in that cycle of connections and disconnections, because of the voltage level. Again, after a few cycles some loads were disconnected to stabilize the system.

To better verify the problem with the voltage levels, a few waveform files taken during this test were analyzed to see the instantaneous voltage at Clusters A and B at the same time. Fig. 9 and Table I show the voltage waveforms and their RMS values, respectively, for the three phases of the system, for a case where the two distributed generators were connected to the MG.

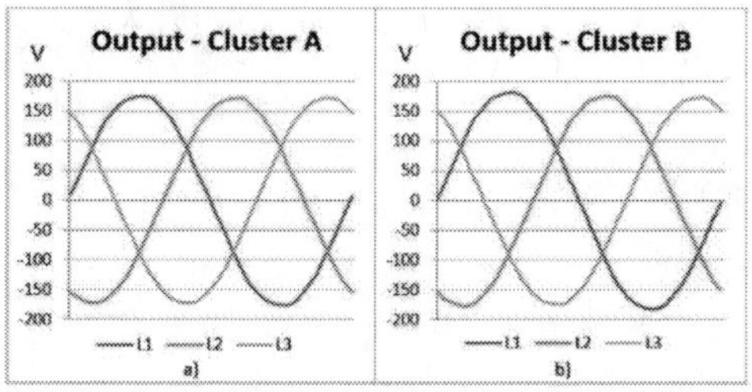

Figure 9: Voltage waveforms on each phase, with distributed generation.

Table I: RMS voltage on each phase, with distributed generation.

	L1 (V)	L2 (V)	L3 (V)
Output of Cluster A	125.50	122.72	123.18
Output of Cluster B	129.32	125.56	124.03

With this comparison it is possible to see that the distributed generators are raising the voltage along the distribution microgrid. It is also noticed that the influence of PVG-06 (which is the one closest to the input of Cluster B and connected to phases L1 and L2) is greater than the PVG-07 (farthest from Cluster B and connected to phases L1 and L3).

Fig. 10 and Table II also show the voltage waveforms and their RMS values, but this time right before the fifth period, when the distributed generators were disconnected.

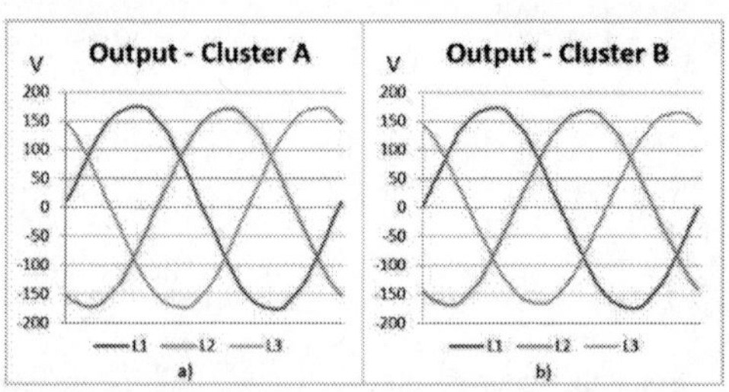

Figure 10: Voltage waveforms on each phase, without distributed generation.

Table II: RMS voltage on each phase, without distributed generation.

	L1 (V)	L2 (V)	L3 (V)
Output of Cluster A	125.13	122.08	123.65
Output of Cluster B	123.70	119.40	118.28

Because there was a high load demand far away from Cluster A and no DG stabilizing the voltage levels on the MG, there is a voltage difference between the two systems. In this case, the load demand on phase L3 was higher than on the others, so that the voltage drop on it was also higher. This explains what happened back on the fifth period, because when more loads were connected, the voltage on phase L3 dropped even more, ending with the disconnection of the external source of Cluster B.

5 FINAL CONSIDERATIONS

In accordance with the initial objectives, the results from the experimental tests confirmed the suitable integration between two MG, and an MG to the conventional grid. Through the analyses of each result it was possible to identify how the logic of the system works for different situations, from simpler to more complex configurations. For example, as seen in the second test, if a cluster has a fully charged battery bank and a low load demand, instead of disconnecting its PV generators (wasting the generation capacity of the system), it could be integrated with other power systems to share with them its energy surplus. Furthermore, as presented in the first and third tests, two systems can also be connected to each other, independent from the main grid. This kind of integration can be very useful for isolated communities near each other, that do not have access to the conventional grid. Still, more tests shall be performed with different load profiles and DG penetration, in order to fully understand the behavior of the systems.

These experiments are only a small step for the future project to implement smart-grids in the laboratory. The considerable lack of communication between the two clusters makes it more difficult to perform system operating strategies (since a cluster makes decisions based on its own necessities, without seeing the needs of the other systems connected as external sources). That is why studies regarding the integration of these laboratory MG must continue, to overcome these barriers and create a unified and intelligent system.

Acknowledgments

The authors would like to thanks the National Council for Scientific and Technological Development (CNPq) for the financial support to development this research.

References

[1] R. Bayindir, E. Hossain, S. Vadi. The Path of the Smart Grid -The New and Improved Power Grid. 3rd International Smart Grid Workshop and Certificate Program (2016).

[2] C. F. O. Barbosa, J. A. L. Correa, J. P. A. Veríssimo, I. H. L. Lemos, M. A. B. Galhardo, E. J. S. Pereira, J. T. Pinho, A. R. A. Manito. Laboratory of Hybrid Systems and Mini-Grids. 33rd European Photovoltaic Solar Energy Conference and Exhibition (2017).

[3] SMA America. Off-grid Inverter Sunny Island 4548-US/6048-US Technical Description, Version 1.2 (2012).

THE DEVELOPMENT AND TEST OF THE PV CONCENTRATOR SYSTEM WITN ELECTRICAL AND THERMAL OUTPUT

Okhorzina A.[1], Yurchenko, A.[1,2], Dinh Wang Tai[1],Mekhtiev A.[3]
[1]National Research Tomsk Polytechnic University
[2] Tomsk State University
Russia, Tomsk, Lenin Avenue 30
[3] Karagandy State Technical University
Karagandy, Kazakhstan

ABSTRACT: The aim of the article is to provide the reader with some information about a design of hybrid system. Hybrid system represents a concentrator photovoltaic solar energy system with electrical and thermal output. This system allows more effective using of photovoltaic modules in region with some solar radiance. Low concentrator adds to a beam solar flow, and cooling system reduces a risk of overheating and premature degradation of photovoltaic modules. Removing heat will be used on water heating in boiler or heat accumulator.
The computer modeling is applied for design of some system elements. Parameters of coil pipe of cooling system are developed. The concentrator optimization is providing more uniformly illumination of photovoltaic module surface.
Keywords: PVT system, PV System, Concentrators, Thermal Performance

1 INTRODUCTION

It is not urgently necessary to use everywhere an alternative energy systems in Russia, because Russia has a large amount and infrastructure for traditional energy resources: oil, coal, gas, hydro and nuclear.
Therefore the authors see a prospect for alternative energy usage for small villages and private houses in distant and remote terrain (example, villages in a taiga, and houses of forestry officers in national parks).
The installation is a hybrid system that converts solar energy into electricity and heat. The sunlight falling silicon photovoltaic modules are converted into electricity. Production of heat occurs at the expense the cooling of photovoltaic modules.
The low concentrator is added to the system to increase energy production. The concentrator is a parabolic reflector.
Coming radiation flux falls on the concentrator and isreflected on the photovoltaic module. The plane of the solar module is located perpendicular to the flow of the reflected radiation. Direct light is not incident on the a photovoltaic module.
For the effective operation work in the cold season it is possible to use the additional cover (transparent to the direct sun rays and impervious to reflected radiation within the system), which would protect the concentrator and photovoltaic modules from precipitation and reduce the outflow of heat into the environment.
To increase the power the tracking system can be used

2 DESCRIBE OF PVT SYSTEM

The Photovoltaic module consists of the two panels arranged in parallel to each other. . The area between the photovoltaic modules is filled with the liquid. The photovoltaic modules are cooled by the liquid flow at the same time the liquid is heating.

The concentrator consists of two parabolic reflecting surfaces. The concentrator can be made of any material that meets the economic and operational tasks, such as metal or plastic with a reflective coating.
The mathematical model takes into account the influence of the concentrator as a multiplier for solar radiation.
The cooling system is a typical thermal solar system. As the coolant can be applied any coolant with a high value of specific heat capacity. The Coolant is glycol, mineral oil or water. Water is pumped from the bottom and moved upward by the pump. A configuration of cooling system is coil.

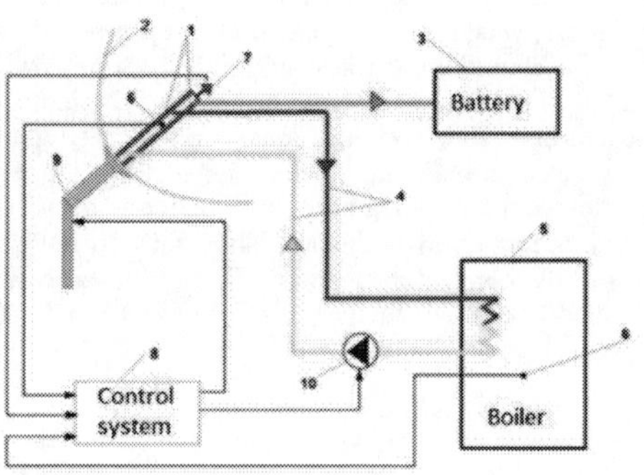

Figure 1: PV concentrator system with a heat sink: 1 is PV module, 2 is Concentrator, 3 is Accumulator Battery, 4 is Copper Tubes, 5 is Boiler, 6 is Temperature Sensors, 7 is Sensor of Tracking System, 8 is Control System, 9 is Rotating device, 10 is Pump

The control system consists of the several subsystems that control individual units. Harmonization of all subsystems is based on a microcomputer BeagleBon. Power for BeagleBon will be provided directly by PV modules as well as other sensors. The microcomputer will form a database of system and save it. That will optimize the system.

Management of PV: Battery control will be based on the MPPT-controller.

The system orientation to the sun is realized on the basis of the active tracking system [1].
This system based on the photoelectric sensor. The Photoelectric sensor can be used in the tracking system. In our system we use three photovoltaic inverters A, B and C [2].
At the heart of the gauge: two obverse of elements A and B define the position of the Sun, the third rear element (C) accepts influence of scattered radiation. The signal of the received element C is subtracted from signals of the elements A and B. The device compares the signals and then generates a control signal to the motor (whether positive or negative), which directs the solar battery.

Control cooling system. The subsystem monitors the temperature of the coolant at the inlet and outlet of the solar panel [3, 4].

If the temperature difference at the inlet and outlet is less than the minimal value (approximately 5°), the pump power is decreased; if greater than the maximum value, it is increased (approximately 20°).

3 THE MATHEMATICAL MODELL

Numerical modeling of a PV concentrator system is designed with the help of MATLAB. MATLAB is a mathematical package of great potential. MATLAB works with matrix data and enables to create a custom calculated program. Simulink can be used together with the MATLAB and other toolboxes, as well as individually. Simulink implements the principle of visual programming: the user uses the library to construct the model, configures the solver and calculation step.

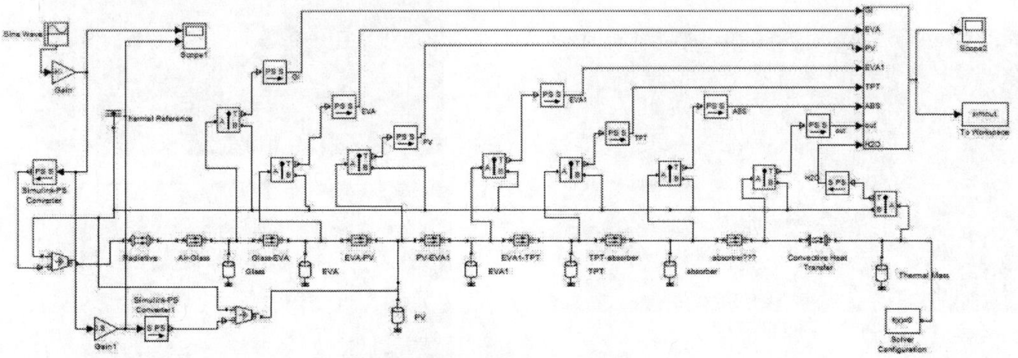

Figure 2: The Simulink model.

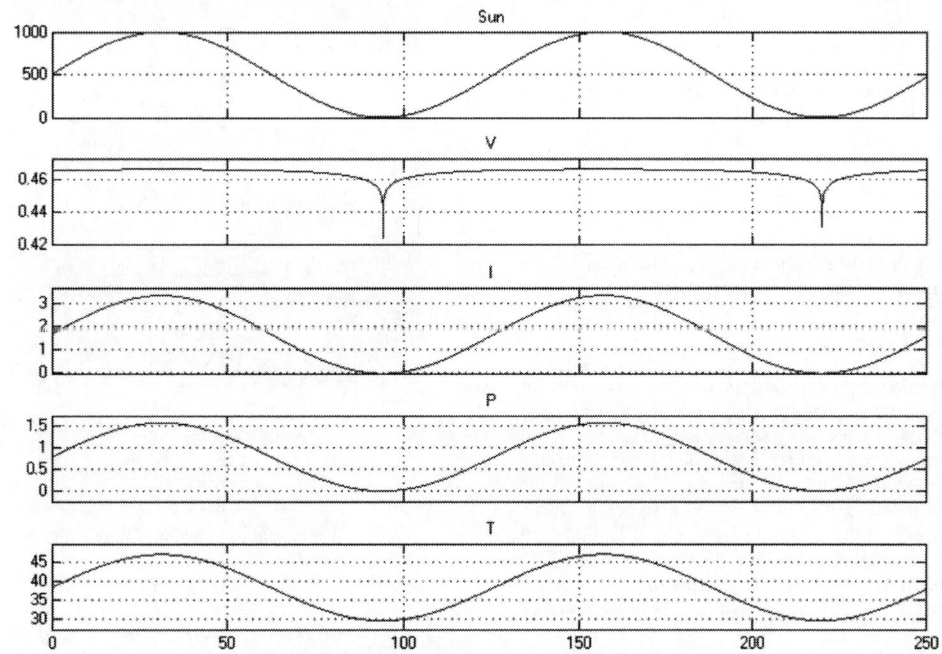

Figure 3: Results of modelling

The electrical part. The electrical model is based on the double exponential model of the photovoltaic cell. The model for the proposed range of the equivalent circuit elements, irradiance and temperature as model inputs, with the corresponding values of voltages, currents, and power as outputs is presented. [3]

The voltage is set by uniformly time-varying signals. The current values are calculated for a definite voltage value at each step. The calculation stops when the current value reaches 0. The temperature effect is taken into account when modeling the photocurrent and diode saturation current. Possible differences in the parameters of the diodes in the equivalent circuit are taken into account, as they have a significant impact on Voc (open-circuit voltage) and power output.

The thermal part. The thermal model bases on the transfer of heat from hot to cold bodies. The model takes into account the heat flow from the heat protective glass flow and heat energy directly to the photovoltaic cells. In the thermal part of the calculated temperature value transmitted to the module and the electric part, wherein on the basis of this temperature is determined by the operating current and voltage. The result of the simulation is shown figure 2 and 3.

3 FIELD TEST

The field experiment was conducted. The experimental results have shown efficiency of the installation. A great influence on the electrical output has a minimal deviation from the direction of the sun.

Figure 4: The prototype

When the temperature setting to 60 ° C, power generation decreased by 8.2% and at a temperature of 90 ° C decline in power output is 18%.

The water temperature is increase 20°C.

As can be seen from the field experiment (fig.5), the heating of the PV surface is a strong (about 85 ° C). When the cooling is turned on there is a significant cooling (about 30 ° C).

To increase the cooling uniformity in the design can be changed or to add absorber, as well to augment the number of turns of the coil.

According to the results of the experiment can be said that the cooling system is functioning well.

4 CONCLUSION

Application of combined system such as the PV concentrator system with heat sink cooling will increase the efficiency of converting solar energy at times. One of the effective ways to increase the solar power collected by PV module is a solar tracker. Position of the maximum power of the solar radiation sometimes does not comply with the position of the Sun. The Designed control system of the optimal PV module orientation allows not only to follow the maximum power of the solar radiation in the sky, but also to maximize received energy.

The second way is the creation of the cooling and concentrating system. The cooling system will take the heat energy that will preserve and extend the life of photovoltaic modules. The concentrator will increase the flow of solar energy on the surface of the photovoltaic module.

The control system monitors the operation of the cooling system and the photovoltaic module orientation. Main control element is necessary for efficient operation, which is implemented on a microcomputer BeagleBon. The platform is easy to use, practically feasible and cheap.

The field experiment was conducted. The experimental results have shown efficiency of the installation.

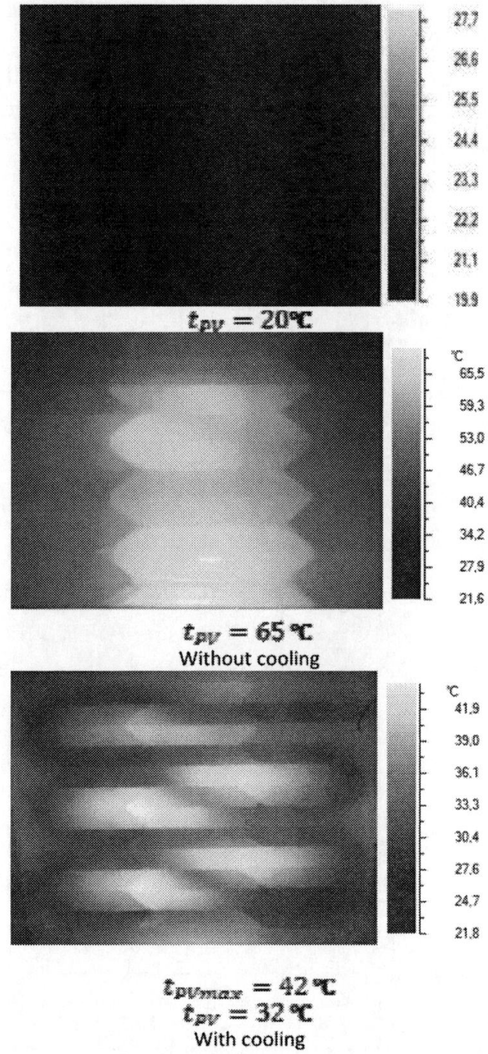

Figure 5: The results of experiments

5 REFERENCES

[1] Chia-Yen Lee, Po-Cheng Chou, Che-Ming Chiang, Chiu-Feng Lin. Sun Tracking Systems: A Review. Sensors 2009, 9, pp. 3875-3890;

[2] Kitaeva M. V., Yurchenko A. V., Okhorzina A. V. Efficiency of PV systems with solar trackers for Russian regions, Proceedings: 7th International Forum on Strategic Technology (IFOST – 2012), Tomsk: TPU Press, Vol. 2 , 2012,p. 103-106,

[3] Dupeyrat, P., Mnzo, C., Rommel, M., Henning, H.M., Efficient single glazed flat plate photovoltaic–thermal hybrid collector for domestic hot water system. Solar Energy, 2011, 85, pp. 1457–1468.

[4] M. Rosa-Clot, P. Rosa-Clot, G.M. Tina TESPI: Thermal Electric Solar Panel Integration. Solar Energy. 2011, 85, pp. 2433–2442

INNOVATIVE SOLAR SPECTRAL BEAM SPLITTING CONCEPTS: ALTERNATIVE FUELS PRODUCTION

Gur Mittelman[1*], Avi Kribus[2]
[1]Agricultural Research Organization, POB 15159, Rishon LeZion 7528809, Israel.
[2]School of Mechanical Engineering, Tel Aviv University, Tel Aviv 69978, Israel
*e-mail: gur.mittelman@gmail.com tel: 972-52-9404011

ABSTRACT: To date solar hydrogen and other solar fuels production is not competitive with fossil fuels. We present an analysis of a solar fuels/chemicals production plant based on spectral beam splitting and high temperature electrolysis. In this system, long wave radiation is reflected to a thermal receiver, while short wave radiation is transmitted to photovoltaic converters located below the splitter. The combined electricity and high temperature heat is then utilized to produce alternative fuels/chemicals in a high temperature electrolysis reactor. Significant advantages of this design include (i) conversion of the infra-red radiation into high grade heat, reducing the reactor power requirement (ii) reduction in reactor energy losses and (iii) simple design combined with reactor operating well below 1000°C. The resulting solar-to-fuel efficiency is higher than reference figures by 25-60%.
Keywords: Hybrid, Hydrogen, Photovoltaic, Storage, Tandem, Spectral Splitting, Solar Fuels, High Temperature Electrolysis, Power-to-Gas

1 INTRODUCTION

It is highly desirable to substitute fossil fuel energy with renewable alternatives. However, if intermittent renewable sources are to become major supplies of energy for electricity generation and motor vehicles, effective storage and transportation of renewable energy must be developed. Therefore, methods to convert raw materials such as H_2O and CO_2 to a fuel using intermittent renewable energy can greatly contribute to the world's climate crisis. A large amount of energy must be supplied in order to chemically reduce or dissociate H_2O into hydrogen and/or CO_2 into carbon fuels. For a renewable fuel production process, this energy must be supplied by a renewable energy source, such as solar, in the form of electricity or heat. Overall solar-to-electricity conversion efficiency of solar power-generation systems (including all the system components) is typically in the range of 10–30%, whereas the production of high temperature heat from solar energy can be performed at 60-90% efficiency. These features make solar-driven dissociation of H_2O and CO_2 potentially more efficient as the temperature increases, because it requires conversion of less solar energy to electricity and more to heat. This will be illustrated in more detail later on.

The two principal parameters indicating the potential of any solar-to-fuel conversion process to become economically feasible are (i) the overall process thermodynamic efficiency – from incident solar radiation to chemical energy and (ii) the rate of the energy conversion reaction. In solar driven processes, increasing the efficiency decreases the size of the collection area (e.g. solar panels, reflectors, etc.) which results in reduction of both land requirements and potentially cost. Because the conversion of solar energy to high temperature heat is 3–4 times more efficient than the conversion to electricity, high temperature solar thermal processes could be, in principle, less expensive. Furthermore, operation at high temperature will increase the rate of the endothermic dissociation reaction. Therefore, it is desirable to operate a solar-driven fuel process at as high temperature as possible while staying within practical limits of sunlight concentration, operating temperature and structural materials. These practical limits can be extended by future developments.

To date, the most promising methods for high temperature conversion of H_2O or CO_2 to fuel include thermolysis and thermo-chemical processes, and high temperature electrolysis. Current solar hydrogen from H_2O dissociation devices is still in the R&D phase and not likely to reach commercialization or cost competitiveness in the short run [1]. Solar thermo-chemical radiation to hydrogen conversion efficiency is rather low, typically about 20% or below for thermal decomposition of water where some of the proposed processes involve very high operating temperatures (e.g. >1600°C in the thermo-chemical zinc-zinc oxide cycle) which greatly limits the plant materials long term stability. Electrical decomposition such as electrolysis at room temperature (no thermal requirements at all), is a robust process but the conversion efficiency from solar to fuel is only about 20% or less, due to the low solar-to-electricity conversion and reactor energy losses.

An alternative approach is High Temperature Electrolysis (HTE) driven by solar cogeneration (simultaneous production of power and heat) where in contrast to solar thermo-chemical processes, the HTE working temperature could be well below 1000°C, allowing more standard reactor materials and a reasonable durability. The HTE has two advantages over room temperature (or 'cold') electrolysis: The conversion efficiency could reach some 80-85% [2, 3] compared to 40-60% [1] and, some of the process work demand (high thermodynamic value) can be replaced by thermal energy (low thermodynamic value). The latter is illustrated in Fig. 1 for the dissociation of H_2O (on the left) and CO_2 (on the right). The reactions (endothermic) are:

$$H_2O \rightarrow H_2 + \frac{1}{2}O_2 \text{ and}$$

$$CO_2 \rightarrow CO + \frac{1}{2}O_2$$

If a mixture of H_2O and CO_2 is to be dissociated (often referred as 'co-electrolysis'), then the outcome is syngas, a mixture of hydrogen (H_2) and carbon monoxide (CO):

$$nH_2O + CO_2 \rightarrow nH_2 + CO + \frac{n+1}{2}O_2$$

The HTE technology was already demonstrated successfully e.g. using Solid Oxide Electrolysis Cell (SOEC) designs [2-6]. Because considerable joule heating is unavoidable when electrolysis is done under practical constraints, with HTE this energy can be utilized in the highly endothermic process, increasing the efficiency (the first advantage).

Combining the advantages of HTE and solar cogeneration, a design based on a dish type, High Concentration Photovoltaics (HCPV) with a hyperbolic 'hot mirror' beam splitter located near the dish focal (secondary optics), was suggested before [7]. Solar irradiation is reflected to the beam splitter, where short waves (SW) are transmitted to PV collectors located behind the splitter, while long waves (LW) are further concentrated into a thermal receiver through an optical fiber. The secondary optics is required to enhance the concentration level at the thermal receiver because only LW radiation is converted into heat. Thus, in this plant the HTE is driven, simultaneously, by DC electricity and high temperature heat. However, the engineering requirements involved in such a design are very demanding and include: very accurate 2-axis tracking and reflector, curved splitter, concentrator PV cells, cooling water system, circulation of raw and product materials through the entire solar field, and other sophisticated elements.

The designs presented in the current paper are arguably simpler than the references above, yet suggesting better conversion efficiency while operating at reasonable reactor temperature.

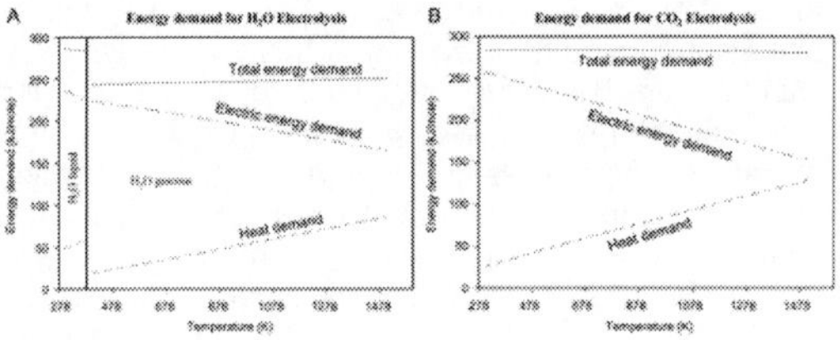

Figure 1: Energy demands for H_2O (A) and CO_2 (B) dissociation [8]. When temperature increases, the total energy demand remains about constant, but the electrical and thermal energy contributions are different, with more energy consumed in the form of thermal.

2 THE CONCEPT

The proposed concept is illustrated in Fig. 2 with a geometry similar to solar Central Receiver Systems (CRS). Using a 'hot mirror' beam splitter, LW radiation is reflected to a thermal receiver, while SW radiation is transmitted to photovoltaic (PV) converters located below the splitter. The combined electricity and high temperature heat is then utilized to produce alternative fuel in a HTE reactor (H_2O electrolysis or co-electrolysis). To allow the implementation of commercial, 1 sun PV modules, only LW radiation is concentrated to obtain high temperature operation at the (central) thermal receiver (e.g. volumetric air absorber), while there is no need to transport heat and materials through the entire solar collection area. For a HTE operating temperature below 1000°C, the electricity (work) demand is 2-3 times higher than the thermal energy demand (see Fig. 1). Thus, two configurations are considered, each based on different PV converter: I – single junction Si (Fig. 2) and II - dual junction (tandem) GaAs/Si, to allow more power from short waves (300-1200 nm), while keeping the same thermal output from waves longer than 1200 nm (Fig. 3). In the first design, excess heat could be stored as will be discussed later.

A critical element in this layout is the large area (in the scale of say, 1 m^2) 'hot mirrors' beam splitters. The technology of such splitters (filters) is already available with several manufacturers worldwide e.g. SHINCRON [9]. Using Physical Vapor Deposition (PVD), it is possible, in principle, to tune and optimize The spectral optical properties of the filters according to the system requirements e.g. the incidence angle where the desired cut-off can be achieved.

The input data for the energetic analysis is given in Tables I and II where the difference in the data is only related to the PV modules performance, along with the power and heat fractions and reactor temperature. The filters reflectance is embedded in the CRS optical efficiency. The resulting conversion efficiencies are given in Table III for the case no Thermal Energy Storage (TES) is considered. For instance, with a GaAs/Si modules with a Standard Test Conditions (STC: 1000 W/m^2, AM1.5G, 25°C) efficiency of 32.7% [10], the ratio between the produced power and heat is 2.5 (70% power, 30% heat). For H_2O electrolysis (generating H_2), the reactor has to be kept at about 900°C while for co-electrolysis of CO_2 and H_2O (generating syngas), the electrolyzer temperature should be about 800°C. To achieve that level of temperature only from the LW radiation (about 18% of the total insolation, AM1.5D), secondary optics may be considered. The peak (nominal, design point) conversion efficiency from solar to fuel (hydrogen or syngas) in the case of GaAs/Si is 33.1%, and the annual efficiency is 28.9%. These figures are considerably higher than reference solar hydrogen production [1]. For example, if the same PV electricity (yet with GaAs/Si modules) was used to drive cold electrolysis at say, 50% conversion efficiency, the overall solar-to hydrogen conversion efficiency would be $0.327 \cdot 0.5 = 16.4\%$, about half the value obtained in the proposed design. In fact, the efficiency limit of Si based dual junction (Si as the bottoming layer) modules is well above 40% which could, in principle, further enhance the proposed layout performance. The performance of the single junction based plant falls below the tandem case due to the lower conversion of solar to electricity. However, single junction modules are more applicable in the short run. Only two PV converters were considered in the analysis but in principle, the performance can be optimized for different module materials and beam splitting cut-off.

Hydrogen, the product of water electrolysis, could be used directly as a fuel or a raw material e.g. for the production of ammonia. However, syngas, which is the product of the co-electrolysis, is usually not considered as a final fuel product but with the process of methanation, it could be converted to methane (CH_4) which is often referred as a Substitute/Synthetic Natural Gas (SNG). Other options are the conversion of syngas into liquid hydrocarbons (e.g. diesel fuel) by Fischer-Tropsch reaction (F-T) or the conversion to methanol. Due to the energy losses in the F-T reactor system, the conversion efficiency from solar to say, methane, is only 23% for the case of GaAs/Si (see Table III).

When the sky is not clear or when beam irradiation is unavailable, a solar plant without TES could only generate electricity which could be inconvenient for the reactor which is designed for high temperature. A possible solution could be decreasing the reactor operating temperature such that some of the (reactor) thermal demand is relaxed and could be therefore stored during clear sky conditions. The other option is using a single junction PV material as in configuration I (Fig. 2), where the power to heat ratio is such, that some of the heat generated during available beam radiation times can be stored and used later along with the PV electricity, when beam is low but diffuse and ground reflected radiation are sufficient for power production. If we consider that 20% of the heat is to be stored annually, then due the high TES efficiency (90%, see Table I), the resulting overall efficiency in configuration I is 23.0%, only 0.2% below a case which storage is not required.

Configuration I (Fig. 2) can be used as an energy storage device, for either solar energy or excess grid electricity during off-peak periods. By storing grid electricity (e.g. power-to-gas-storage), the plant becomes hybrid with the advantage that reactors could operate more continuously (Fig. 4). However, the choice of the PV material as well as the scale of the solar field have to be made carefully, to allow sufficient thermal energy storage.

A cost analysis was performed [11] for the proposed plants. The capital (investment) cost of each configuration is given in Table II in terms of $/Watt peak of the PV field, while the annual costs are given in terms of ¢/kWhe production of the PV field. The resulting hydrogen costs are around 4-5 $/kg (Table III), well below reference solar hydrogen plants (7-10 $/kg), but still above the price of conventional hydrogen produced from coal or local natural gas, which is about 3 $/kg. When local natural gas production is small (e.g. Japan), the price of conventional hydrogen could reach as high as 10 $/kg.

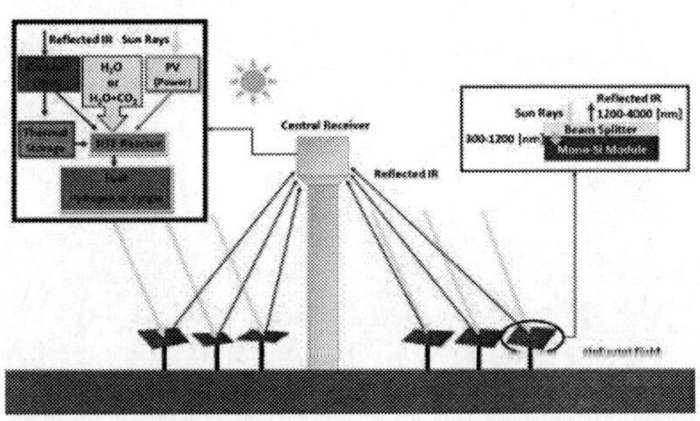

Figure 2: Configuration I - A plant with single-junction, Silicon PV modules.

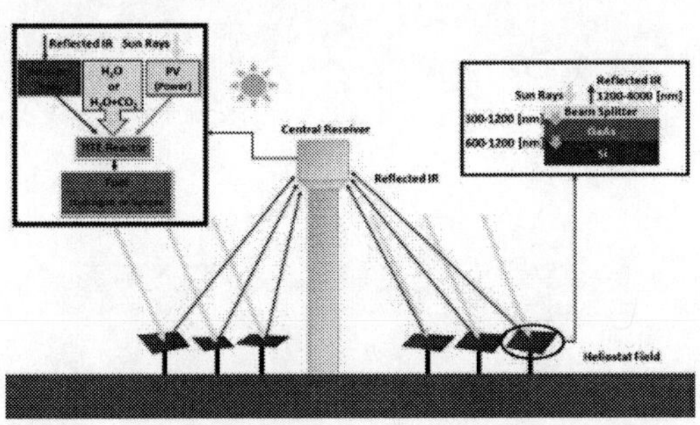

Figure 3: Configuration II - A plant with dual-junction, GaAs/Si PV modules.

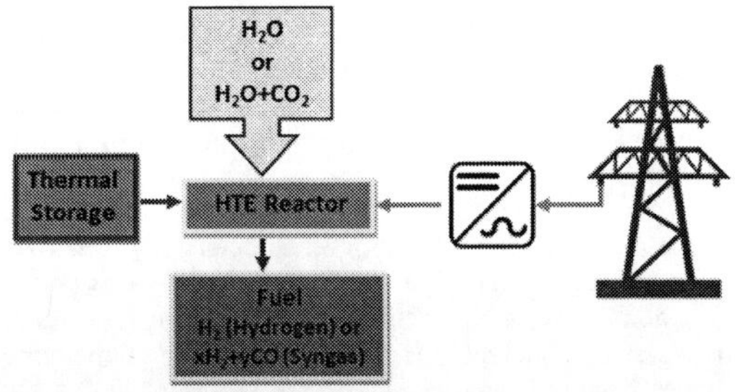

Figure 4: A hybrid plant.

Table I: Input data. POA=Plane of Array.

Parameter	Value
Solar resource global	AM1.5G
Solar resource direct	AM1.5D
Annual G_{POA}, kWh/m^{2}yr	2,300
Annual DNI, kWh/m^{2}yr	2,400
Splitter transmittance (SW) [9]	95%
DC/DC converter	98%
Optical efficiency (CRS)	70%
Thermal receiver efficiency [12]	90%
HTE reactor efficiency [13]	78%
F-T reactor efficiency [13]	70%
PV Performance Ratio (PR)	75%
Thermal storage efficiency [14]	90%

Table II: Configurations key features and input data

Configuration	I	II
PV module material	mono-Si	GaAs/Si
PV module efficiency (II [10])	23.0%	32.7%
Beam splitting cut-off	1200 nm	1200 nm
Fraction of power supply	64%	70%
Fraction of thermal supply	36%	30%
HTE reactor temperature, °C	900-1000	800-900
Investment cost, $/W$_p$ [11]	1.65	3.30
Annual costs, ¢/kWhe [11]	2.4	2.0
Capital Recovery Factor (CRF)	7%	7%

Table III: Results summary. Peak efficiencies are normalized to the incident AM1.5G radiation.

Configuration	Ref. [1, 15-16]	I	II
Solar to H_2 peak efficiency	2-21%	26.0%	33.1%
Solar to H_2 annual efficiency	-	23.2%	28.9%
Hydrogen cost $/kg	7-10	4.4	5.3
Solar to CH_4 peak efficiency	2-15%	18.2%	23.1%

3 CONCLUSIONS

A novel concept for the production of solar fuels was presented. The design is based on solar cogeneration with spectral beam splitting. The approach is motivated by simplicity yet offering a promising solar-to-fuel conversion efficiency and a reasonable cost. In case H_2 is produced, it could be used either as a fuel ("Hydrogen Economy") or, as a raw material for the production of commodity chemicals such as ammonia. If syngas is produced, it could be further converted into more convenient fuels such as SNG, liquid hydrocarbons and methanol. With the help of high temperature thermal storage, the plant could be hybridized using low-cost, off-peak grid power, increasing the annual capacity and most likely the cost effectiveness (not analyzed here). Future research should be also dedicated to performance optimization through the selection of the PV material/s.

The implementation of the proposed concept in large industrial scale depends on several factors: (i) low cost, serial production of the spectral filters (ii) in the case of dual-junction PV, serial production of modules such as GaAs/Si (iii) serial production of HTE reactors capable of operating under dynamic conditions (iv) in case TES is integrated, the applicability of high temperature storage in industrial scale.

4 REFERENCES

[1] P. Nikolaidis, A. Poullikkas, Renewables and Sustainable Energy Reviews 67 (2017) 597-611.
[2] L. Wang, M. Pérez-Fortes H. Madi, S. Diethelm, J. Van herle, F. Maréchal, Applied Energy 211 (2018) 1060-1079.
[3] D. Ferrero, Ph.D. thesis. Politecnico De Torino (2016).
[4] L. Frost, E. Elangovan, J. Hartvigsen, The Canadian Journal of Chemical Engineering 94 (2016) 636-641.
[5] Y. Alioshin, M. Kohn, A. Rothschild, J. Karni, Journal of Electrochemical Society 163(2) (2016) F79-F89.
[6] Y. Seong-Bin et al., Chemical Engineering Journal 355 (2018) 41-51.
[7] J.R. Thompson, R.D. McConnel, M. Mosleh, International Conference on Solar Concentrator for the Generation of Electricity or Hydrogen, Scottsdale, Arizona (2005).
[8] S.H. Jensen, M. Mogensen, International Journal of Hydrogen Energy 32 (2007) 3253-3257.
[9] Wenjuan Huang, Hui Lv, Jun Liao, Yiyuan Shang, Wen Liu, Qinghua Lv, Chunfu Cheng, Yuehong Su, R. Saffa, Journal of Renewable and Sustainable Energy 9 (2017).
[10] D.C. Bobela, L. Gedvilas, M. Woodhouse, K.A.W. Horowitz, P.A. Basore, Progress in Photovoltaics: Research and Applications 25 (2017) 41-48.
[11] D. Cohen, M. Yehoshou, Dept. of Energy Engineering, Graduate Studies, Afeka Tel Aviv College of Engineering (2018).
[12] C.K. Ho, B. D. Iverson, Renewable and Sustainable Energy Reviews 29 (2014) 835-846.
[13] M. Götz et al., Renewable Energy 85 (2016) 1371-1390.
[14] Z. Ma, G. Glatzmair, M. Mehos, Journal of Solar Energy Engineering 136 (2014).
[15] H.L. Tuller, Materials for Renewable and Sustainable Energy 6.1 (2017).
[16] W.A. Jong, M.Sc. thesis. Utrecht University (2018).

Optimal Energy management in microgrid using a particle swarm optimization algorithm

M. AIT BENALI[1], A. OUTZOURHIT[1],
[1]Laboratory of Nanomaterials for Energy and the Environment (LN2E) - Cadi Ayyad University, Faculty of Sciences-semlalia Marrakech Morocco
simo.aitbenali@gmail.com
aoutzour@uca.ma

ABSTRACT: Smart-grid and smart micro-grids technologies can significantly enhance the integration of renewable energies and alleviate the effects of their intermittence. This mostly due to integration of modern communication technologies allowing an intelligent and flexible control of power flow in the grid. The dispatch of electrical power produced by many generators to the loads requires rapid and robust decisions, this process of control is known as **energy management**. The purpose of this work is the research of the optimal energy management solution for a PV-WIND-Micro turbine energy system to minimize the cost of generated energy with the application of the particle swarm optimization (PSO). Matlab software is used for programming and simulation of this work.

Keywords: Smart Micro-grid, energy management, particle swam optimization.

1. INTRODUCTION:

As demand for clean renewable sources of electricity increases, many people are installing their own personal generators, one proposed way to integrate these different distributed generations (DG) sources is to aggregate them into Microgrids. A Microgrid is a mix of DG sources, loads, and storage on a low to medium voltage network [1-3], it can operate as part of the system grid or in islanded mode. So proper control method should be implemented to improve the operation of the MG in an intelligent and coordinated manner. The requirement to assure a proper balance between the powers generated within a Microgrid and the demand load to avoid instability problems in the Microgrid operation is the purpose of using energy management systems in Microgrids. A general structure of an EMS [4-5] for a Microgrid with DERs is shown in Figure 1 :

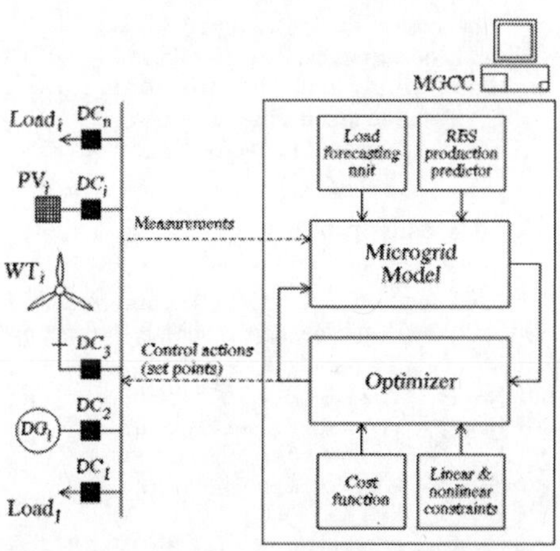

Figure 1: Architecture of an optimal EMS.

The following factors are the functional requirements of EMS:

- Forecasting :

The historical data and other inputs are used by the EMS to forecast the DERs outputs and the loads on different time scales. These data are used as the input for optimization. This is a challenging requirement in a Microgrid setting due to the inherent intermittency and variability of DER and uncertainty in controllable loads.

- Optimizer :

Real time Optimization is the brain of the EMS. It ensures continuity of the load supply an d decreases the overall cost of energy production. Different optimization techniques are followed for different applications which are typically formulated as non-linear optimization problems with different objectives.

- Measurements :

The data of the DERs and the loads are used for analyzing EMS. The data analysis provides better understanding of the characteristics of the DERs and the loads. This can be utilized for adjusting the forecast and the optimization models for better performance and also used for designing control policies for new applications.

2. PARTICLE SWARM OPTIMIZATION :

The search process of the particle swarm optimization method is similar to the social behavior of flying birds when searching for food position [6], these birds are called particles or swarm flies. The particle's position is always adjusted according to its own experience and to the cooperative communication with the other particles. For each particle is assigned a vector position x_i adjusted with velocity v_i.

For a swarm of m particles the velocity and position vectors are:
$P_i = [x_1^i, x_2^i, \ldots\ldots, x_n^i]$; $i = 1,2, \ldots.\ m$, n problem dimension
$V_i = [v^i_1, vi2, \ldots. v^i_n]$

Where i is the index of the particle, V is the swarm velocity and n is the number of the problem's parameters. First, the particles positions and their velocities are generated randomly respecting the limits of every decision variable. Then, the objective function is calculated for every particle.

$OF_i = f(x_1^i, x_2^i, \ldots x_n^i)$; n problem dimension.

The best position of every particle is set at its first position, the best global position is set at the position of the particle with the best objective function. After, the positions and velocities are updated using the process below.

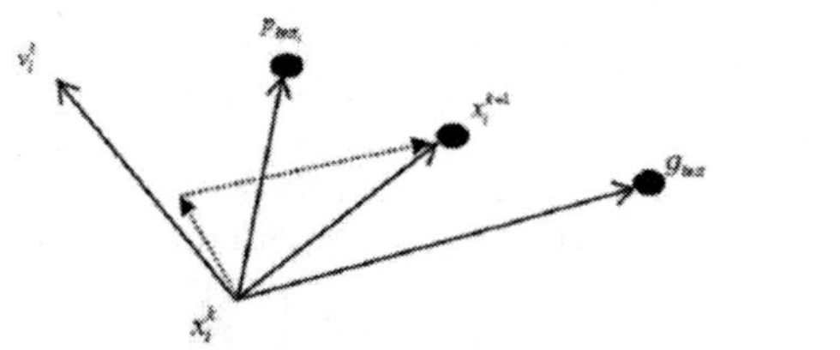

Figure 2: Position and velocity updates.

The updated velocity and the new position vectors for the particle i are :

$V_i^{(k+1)} = wV_i^{(k)} + c_1 r_1 (P_{besti}^{(k)} - P_i^{(k)}) + c_2 r_2 (g_{best}^{(k)} - P_i^{(k)})$

$V_i^{(k)}$ is the previous velocity of particle i

w is the inertia weight.

c_1, c_2 are the individual and social acceleration positive constants.

r_1, r_2 are the random values in the range [0,1].

P_{besti} is the personal best position of particle i.

g_{best} is the global best position of all the particles.

$P_i^{(k+1)} = P_i^{(k)} + V_i^{(k+1)}$

Where :

$P_i^{(k+1)}$ is particle i new position at iteration $k+1$.

$P_i^{(k)}$ is particle i old position at iteration k.

$V_i^{(k+1)}$ is particle i new velocity at iteration $k+1$.

The objective function values are calculated using the new positions and velocities. For every particle, the objective function of this particle is compared with its best position, the best one is taken and its best position is updated. Finally, the best global solution is the one with the best objective function from all the particles. The process of updating positions and velocities, calculating objective function, updating best positions and global best position is repeated for all iterations.

3. PROBLEM FORMULATION:

A microgrid with PV-Wind-Microturbine-Storage Unit connected to the Utility is optimized using particle swarm optimization. The selected decision variables to be optimized are the power output of the microturbine $P_{t,MT}$ and the output power of the battery $P_{t,batt}$.

3.1 Microturbine:

P_{MT} is the output power of the micro turbine limited by the inequality $0 \leq P_{MT} \leq P_{MTmax}$ where P_{MTmax} is the maximum power that will be given by the micro turbine. The emission of the micro turbine is also modeled by the equation:

emission = (emissionSO_2 + emissionCO_2 + emission NOx) * P_{MT}

where the characteristics of the micro turbine are :

emissionSO_2 = 720/10^6 Kg/Wh.

emissionCO_2 = 0.0036/10^6 Kg/Wh.

emissionNOx = 0.1/10^6 Kg/Wh.

3.2 Battery :

In this work, a 25 Ah capacity battery is used with constant voltage 12 volts (300Wh), the total amount of batteries to use to serve a maximum P_{bmax} output power at a time dT is calculated used this equation :

$$K = \frac{Pbmax}{\frac{Vbatt * Ahinit * nb}{dT}} \quad ; \quad Ah_{total} = Ah_{init} * K$$

After power consumption, The state of charge (SOC) of the battery is updated using the formula :

$$SOC = \frac{Ahtotal - Ahtotal\frac{\sum Pbold + Pbnew}{Pbmax}}{Ahtotal}$$

3.3 Objective function :

The objective function in this work is the cost of the generated power from the micro turbine and the utility given by:

$OF = \sum_{t=1}^{24}(F_{t,MT} + F_{t,utility})$ where :

$\sum_{t=1}^{24} F_{t,utility} = C_{utility} * \sum_{t=1}^{24} P_{t,utility}$

$\sum_{t=1}^{24} F_{t,MT} = C_{MT} * F_{MT} \sum_{t=1}^{24} P_{t,MT} * \Delta T + \sum_{t=1}^{24} OM_{t,MT} + \sum_{t=1}^{24} SC_{t,MT}$

$C_{utility}$: Cost of the power utility (\$/Wh).

$P_{t,utility}$: Power output of the utility.

C_{MT} : fuel cost of the MT (\$/m3).

F_{MT} : fuel consumption rate (m3/kWh).

$P_{t,MT}$: Power output from the microturbine.

$OM_{t,MT}$: Operation and maintenance cost.

$SC_{t,MT}$: Start-up cost of the microturbine unit.

3.4 EMS algorithm :

The selected decision variables to be optimized are $P_{t,MT}$ and $P_{t,batt}$, they are generated randomly respecting the constraints of each equipment :

$P_{t,MT} = rand() * (P_{t,MTmax} - P_{t,MTmin}) + P_{t,MTmin}$

$P_{t,batt} = rand() * (P_{t,battmax} - P_{t,battmin}) + P_{t,battmin}$

Then , $P_{t,utility}$ is calculated using the equation :

$P_{t,utility} = P_{t,load} - (P_{t,MT} + P_{t,WT} + P_{pv} + P_{t,batt})$

for $1 \leq t \leq 24$.

The objective function is calculated for each particle, then the lowest one is selected as best solution, the $P_{t,MT}$ and $P_{t,batt}$ are updated and the constraint are verified, the best solution also is updated in every iteration, and the final solution is the global best solution adopted by the PSO algorithm.

The complete process of the EMS algorithm is shown in the chart below :

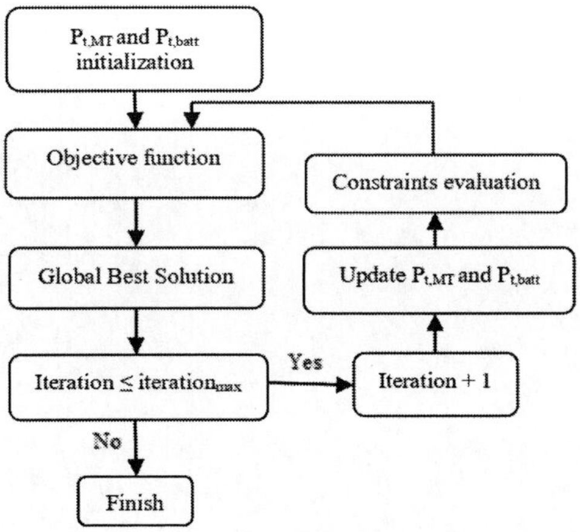

4. SIMULATION & RESULT :

After the prediction of the PV-Wind and load powers (Figure 3) for the following day. The objective is to find the optimale scenario for the Grid and the microturbine that minimizes the cost of the generated power in the microgrid. The simulation was done using the characteristics of Microgrid elements:
Battery : P_{bmax} = 20 kW
PV : P_{pvmax}= 200 kW
Wind turbine : P_{WTmax} = 100 kW.
Microturbine : P_{MTmax} = 20 kW.

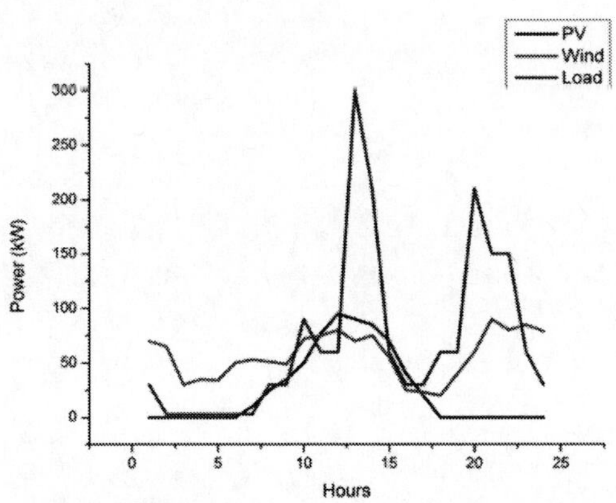

Figure 3: Used scenario

The algorithm was run for 50 iterations with 50 particle swarms, the convergence of the objective function is presented in the figure 4, a negative cost means that the power has been sold to the public grid.
The corresponding power charts of all the Microgrid elements giving the optimal cost is showed on the Figure 5.

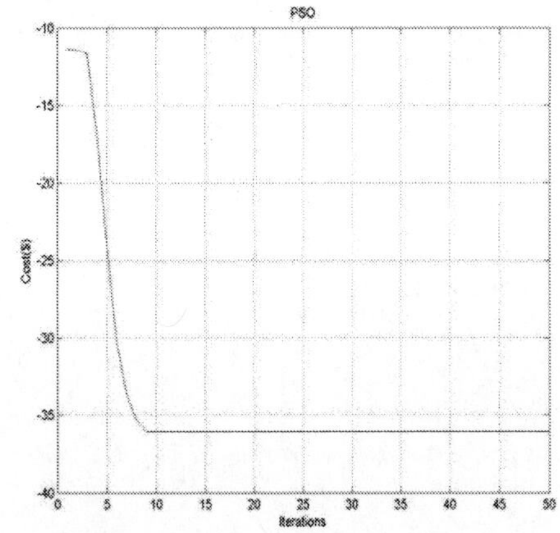

Figure 4: Objective function convergence

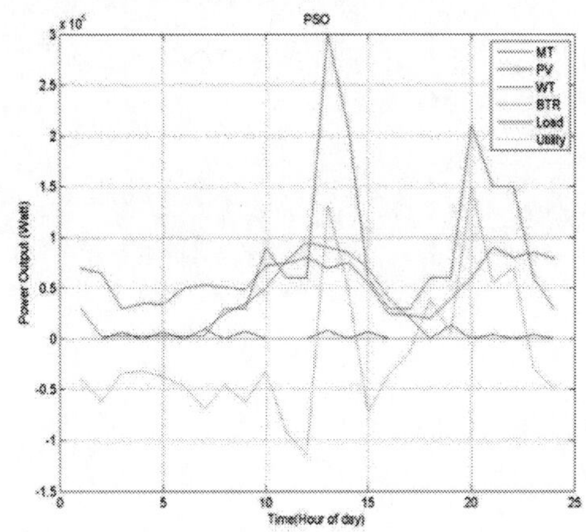

Figure 5: optimal solution.

5. CONCLUSION :

This work focused on the application of the particle swarm optimization for a better and inexpensive management of power inside Microgrids. We tested our programmed algorithm for a given scenario and the results show that the algorithm converges to an optimal solution that minimizes the cost of the generated power of the microgrid. We are currently investigating the performances of a modified PSO compared to the classic explained in this work.

Acknowledgements

This project is funded by the Research Institute on Solar Energy and New Energies (IRESEN, Morocco) in the framework of the InnoPV/Innowind2014 program

References

[1] Committee C, Generation D, Storage E. IEEE Guide for Design, Operation, and Integration of Distributed Resource Island Systems with Electric Power Systems IEEE Standards Coordinating Committee 21 Sponsored by the; 2011.
[2] Hatziargyriou N, Asano H, Iravani R, Marnay C. Microgrids. IEEE Power Energy Mag 2007;5:78–94.
[3] Lasseter RH. MicroGrids. 2002 IEEE Power Eng. Soc. Winter Meet. Conf. Proc. (Cat. No. 02CH37309), vol. 1; 2002

[4] Kanchev H, Lu D, Colas F, Lazarov V, Francois B. Energy management and operational planning of a microgrid with a pv-based active generator for smart grid applications. IEEE Trans Ind Electron 2011;58:4583–92
[5] Chen C, Duan S, Cai T, Liu B, Hu G. Smart energy management system for optimal microgrid economic operation. IET Renew Power Gener 2011;5:258
[6] Yuxin Zhao, Wei Zu, and Haitao Zeng, "A modified particle swarm optimization via particle visual modeling analysis," Computers and Mathematics with Applications, Volume 57, P. 2022-2029, June, 2009.

AUTOMATIC TOPOLOGY IDENTIFICATION FOR ENERGY LOAD MANAGEMENT IN ELECTRO MOBILITY CHARGING APPLICATIONS TO INCREASE SHARE OF LOCAL RENEWABLES

Babak Ravanbach, Peter Klement *, Benedikt Hanke, Karsten von Maydell
DLR Institute for Networked Energy Systems
Carl-von-Ossietzky-Str. 15, 26129 Oldenburg, Germany
* Corresponding author: Tel.: +49 441 999 06-226, Fax: +49 441-999 06-107, E-Mail: peter.klement@dlr.de

ABSTRACT: Existing distribution networks are designed based on certain assumptions. Therefore, standard load profiles and coincidence factors are identified and the power lines are dimensioned with sufficient power reserve. Nowadays, the use cases are changing and will change even more, when sector coupling applications like electro mobility and distributed renewable energy generation systems get into focus to achieve common carbon reduction targets. Then, the power distribution networks can either be strengthened in order to operate the networks with large power reserve or the power applications can be managed in order to achieve higher utilization factors of the distribution networks. In the latter case, detailed information about the distribution network need to be known, for not exceeding any physical limits of the power lines. This detailed information is difficult to maintain and extract from available documentation especially in old or privately operated networks, so that this information needs to be gathered by measurements. An automatic topology identification method has been developed in order to identify the low voltage network topology for distributed loads and sources. The aim of this approach is to get the basic information for a local energy and load management system, which updates topology information under normal operation conditions.
Keywords: Network Topology, Electro mobility, Energy Management

1 INTRODUCTION

Various research efforts have been made to provide a solution to the problem of topology identification for low voltage distribution networks. End-user data, such as active & reactive powers and voltage magnitude measured by smart meter, can be used to detect the network topology [1]. In [2] an optimization based approach to infer phase connectivity and also network topology from time series of energy measurements from meters was used. In [3], a method to identify the phase distribution based on a cross-correlation analysis of time series of voltage measurements in an unbalanced 3-phase network is presented. In [4-6] similar methods were proposed to infer topology from measurements of PMUs in the network. In [7], a topology detection algorithm based on voltage correlation analysis is presented. In [8] a real–time monitoring approach is proposed to detect the underlying network topology based on the status of circuit breakers. In [9] a topology identification technique is proposed using the signals generated by the PLC network laid in a smart grid. In [10] a novel graphical learning based approach is developed to estimate the radial operational grid structure using voltage measurements collected from the grid loads.

Optimal operations in the distribution network depend on the correct estimation of its bus and node states (voltage and power consumption) and its operational radial topology. However, lines in the network still suffer from limited real time metering that hinders the network operator from learning the true topology [10]. Traditionally, utility operators collect voltage, current, and power readings only from a few grid points. To this end and given the currently prohibitive cost of installing synchro-phasors on a wide scale, data from advanced metering infrastructure (AMI) and power inverters can provide useful grid information [11]. One such critical piece of information is the operational structure of a grid. Further, for a reliable state estimation in a distribution network, accurate information of the network topology is essential [12].

A major challenge in topology identification is having access to accurate network data. In most previous work, historical data is collected and used for carrying out the results. In most cases the lack of access to data due to privacy issues or limitations of installing measurement units at every node minimizes the amount of data available for analysis. In this work a voltage correlation method is used, which uses the measured voltage data from every node in the network. Active control over loads will provide the opportunity to isolate single loads (on and off) one at a time and measure the voltages at every node throughout the network. But the active control could also be replaced by a time series of data, so that the topology information will be gained over time by normal operation.

The information of the underlying network topology is essential for efficient and reliable integration of distributed renewable energy systems, specially PV, and the management of controllable loads (i.e. electro mobility) in low voltage distribution networks.

A proposed strategy to increase the share of locally generated renewable energy from distributed PV systems by implementing an intelligent load management system to support power applications, mainly electro mobility, is presented in this work. The underlying concept is to increase the utilization factor of the low voltage distribution networks by absorbing the locally generated renewable energy with controllable electro mobility loads connected to charging points near the source of generation.

The proposed strategy suggests using the existing adjacent public street lighting network to provide the base infrastructure. Retrofitting the existing lighting poles with installed on-pole power sockets would enable power consumption, and implementing an intelligent load management system would further optimize and maximize the power absorption from the locally generated renewable energy while maintaining the reliability and quality of power in the network.

2 THEORY

2.1 Network fundamentals

There are several constraints which should be satisfied during the planning and operation of low voltage distribution networks, including the permissible voltage band and the current capacity of the lines (Fig. 1).

The EN50160 Characteristics of Voltage in Public Electricity Networks is a European Standard that defines and specifies the essential characteristics of the voltage at the network connection point under normal operating conditions. According to this standard the voltage changes should not exceed ±10% of the nominal voltage at 230 V (low voltage) [13].

The current capacity of the cables limits the maximum load that can be connected to the network, and the circuit breakers are placed to avoid overloading of the lines. A circuit breaker is an automatically operated electrical switch designed to protect an electrical circuit from damage caused by overcurrent, typically resulting from an overload or short circuit. The network constraints are illustrated in Fig. 1, followed by equations (1) and (2) showing the relationship between network voltages and currents to minimum allowable voltage (-10% U_{source}) and maximum allowable current (I_{max}) respectively.

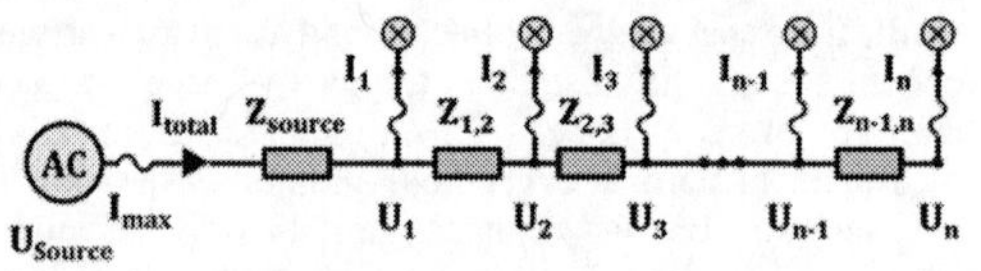

$$U_{source} > U_1 > U_2 > U_3 > \cdots > U_{n-1} > U_n > -10\%\ U_{source} \text{ (Eq. 1)}$$
$$I_1 + I_2 + I_3 + \cdots + I_{n-1} + I_n = I_{total} < I_{max} \text{ (Eq. 2)}$$
$$VD_{i,j} = U_j - U_i = I \times Z_{i,j} \text{ (Eq. 3)}$$

Figure 1: Electrical fundamentals in a network

Voltage Drop is defined as the amount of voltage loss that occurs through all or part of a circuit due to impedance (Fig. 1). The equation (3) below Fig. 1 explains the electrical relationship between the Voltage Drop (VD), Voltage (U), Current (I) and line Impedance (Z) in the network.

Voltage drops occur throughout the distribution network. However, it is necessary to regulate the voltage in order to ensure that these drops stay within the defined permissible range. High voltage drops, beyond the permissible level, can have many negative consequences. Operating electrical equipment below its rated voltage can be dangerous as well as reducing the expected lifetime of the equipment. Resistive loads which are operated at a too low voltage will not be able to produce the desired output [14].

Improving the voltage profile, line loss and system reliability is a main concern in planning of distribution networks. Supporting the load growth and peak load level is another factor which should be considered in the planning procedure.

2.2 Topology

The topology defines how the nodes and branches of a network are connected to each other and the value of line impedance between them. In Fig. 2, notations N_i and N_j represent two arbitrary adjacent nodes, and Z_{ij} represents the impedance of the line connecting them.

Having detailed information about the arrangement of loads on the network is essential to make efficient scheduling of connected loads while maintaining the power quality requirements, including managing permissible voltage drops and avoiding overload of the lines.

The low voltage distribution network is distinguished by its tree-like radial operational topology with the transformer at the root and loads positioned along the non-root nodes of the tree (Fig. 2).

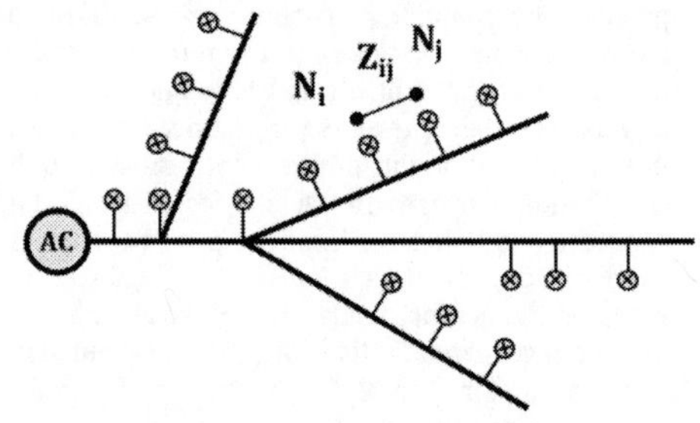

Figure 2: Low voltage radial network topology

The power flow calculation demands certain set of input parameters from the network, including the admittance matrix (Y-matrix), which is derived from the topology information. In this work voltage measurements form the initial input matrix, which will be further manipulated to identify the topology. An example of this matrix is shown below. Each row (U) represents measured voltages (u) at every node in the network when a single node (N) is activated.

$$U_1 = [N_{1,1}: u, N_{1,2}: u, N_{1,3}: u, \ldots, N_{1,n}: u]$$
$$U_2 = [N_{2,1}: u, N_{2,2}: u, N_{2,3}: u, \ldots, N_{2,n}: u]$$
$$U_3 = [N_{3,1}: u, N_{3,2}: u, N_{3,3}: u, \ldots, N_{3,n}: u]$$
$$\vdots$$
$$U_n = [N_{n,1}: u, N_{n,2}: u, N_{n,3}: u, \ldots, N_{n,n}: u]$$

Having access to topology information, a given network can be modelled and various analyses including the power flow and load management can be carried out.

2.3 Voltage behaviour in a radial network

In an arbitrary low voltage radial network (Fig. 3), the nodes are then classified as "In-line" and "Branching".

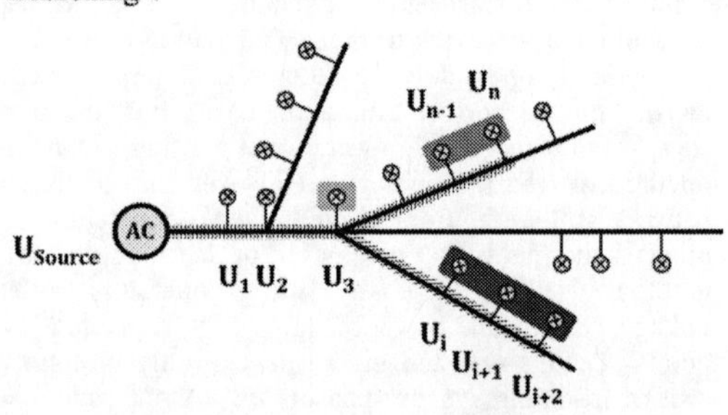

Branching Node ■ In-line Node, ON ■ In-line Node, OFF

Loaded Line ≡ Non-loaded Line

$$U_{source} \gg U_1 \gg U_2 \gg U_3 > U_i > U_{i+1} > U_{i+2} \gg U_{n-1} \gg U_n \text{ (Eq. 4)}$$

Figure 3: Voltage behaviour in a radial network

For a "Loaded" line (Fig. 3, vertical stripes) with a known test load switched on at node n, the measured voltages, denoted as "In-line Node, ON" (U_1, U_2, U_3, $\cdots$, U_{n-1}, U_n), decrease due to VD at every node from the node closer to the source to the loaded node (furthest from the source). However, the measured voltages on the "Non-loaded" (U_i, U_{i+1}, U_{i+2}) branch, denoted by "In-line Node, OFF" (Fig. 3, horizontal stripes), show a different pattern. The measured voltages appear to be very close to each other and to the voltage at the closest branching node. Equation (4) below Fig. 3 shows the relationship between the voltages of the nodes on the "loaded" versus the "non-loaded" lines.

3 TOPOLOGY IDENTIFICATION METHOD

Strong sources or loads need to be managed dynamically due to the power line operation close to the physical limit, and therefore additional infrastructure needs to be implemented. That infrastructure contains voltage measuring devices, which are able to detect voltage increases (supply by photovoltaics) or decreases (load by e.g. electro mobility) over the power lines. The measurements are time-synchronously collected at every managed connection point, at every change of operation within the power grid. Based on the voltage changes over the power lines, the interconnection between these voltage measuring devices is estimated. Therefore, the topology of the power grid is verified or updated during operation. An initial setup procedure with defined test loads is proposed.

The proposed topology identification was evaluated with respect to a single phase PowerFactory model (PowerFactory is a power system analysis software application for use in analyzing generation, transmission, distribution and industrial systems, developed by DIgSILENT [15]). Therefore, a radial (tree-like) low voltage distribution network was designed and loads were located at branching points. The loads were activated as needed by the method and the voltage drops were tracked for each activation of loads over the complete network. These voltage drops were supplied to the topology identification method in order to identify the topology. The results were then compared to the original PowerFactory model for various power loads.

The topology identification method is based on the assumption that the branching is occurring at known loads, which can be activated and deactivated as desired. The loads are activated one at a time for all the network loads. The measurements are collected and using an algorithm based on series of data manipulation (including clustering) and mathematical calculations the network topology is identified. An intelligent device installed at every node carries out the measurements, transmits the data to a database and a central computation application using the validated algorithm dynamically identifies the network topology.

When no renewable power is supplied, the radial network is supplied by a voltage source and the voltage drops over the power lines. The set of measured voltage drops represents a unique pattern which is characteristic for the network as previously describe in section 2.3. The voltage drops along the power line, depending on the strength of the load. Higher loads cause higher voltage drops. Therefore, the voltage drop is increased when an additional load is applied. Branches with no load will show less voltage drops caused by the line impedance, and the measured voltages are close to the next loaded node. If the measured voltages are sorted from high to low, the voltage difference is larger for loaded parts of the network and the difference is less for none-loaded branches.

Jenks natural breaks method is a data clustering method designed to determine the best arrangement of values into different classes (suitable for 1-dimesional set of values). This is done by seeking to minimize each class's average deviation from the class mean, while maximizing each class's deviation from the means of the other groups. In other words, the method seeks to reduce the variance within classes and maximize the variance between classes [13]. This method is found suitable to classify the voltage measurements and is used for the clustering part of the topology identification algorithm (Fig. 4).

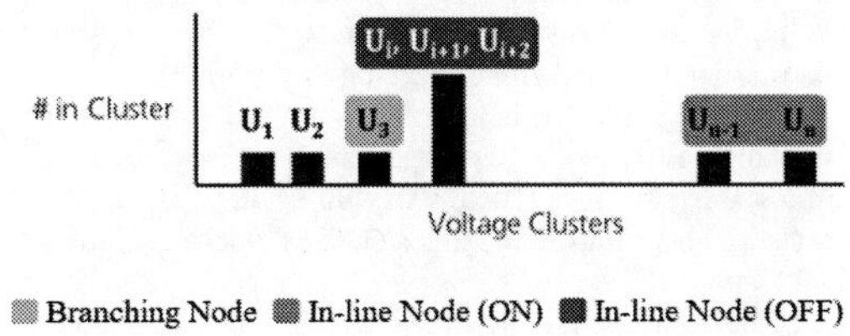

Figure 4: Clustered Voltages by Jenks Natural Breaks

When renewable power is supplied, the origin of the network needs to be shifted for single sources. If several sources are active, the voltage is a superposition of single sources, and can be constructed as such.

4 INCREASING SHARE OF LOCAL RENEWABLES

According to the Agora report, technological innovation and new societal aspirations are transforming the power system toward more decentralization, while the advancements in communication and control technologies is enabling the gird to operate in a bi-directional manner, with real-time flows of electricity and data between consumers and the grid operators [16]. The German Federal Ministry for Economic Affairs and Energy (Bundesministerium für Wirtschaft und Energie, BMWi), states that precise information on the behaviour of millions of distributed generating installations is indispensable in operating an efficient and reliable system and the integration of consumer-side applications such as electric mobility [17].

The new opportunities come with a challenge, which is focused on managing a dynamic combination of generations and loads (often fluctuating, unpredictable or experiencing bidirectional flows) on an existing infrastructure that is designed based on a set of particular standards and specification to support its static original conditions.

New power flow scenarios are pushing the existing networks to their maximum capacity; therefore, the conventional role of distribution system operation will become more critical in maintaining the reliability and safety of the system. While the typical daily energy demand for the additional application can be transported in most situations, the load restrictions require additional management and precautions.

Conventionally, low voltage distribution networks are designed and developed to provide unidirectional flow of energy from the transformer station to the loads. In a typical radial network with a redial topology the transformer is configured to maintain a certain standard voltage level at the root with the ability to handle voltage drops along the line with voltage regulation capabilities.

One of the main issues identified regarding high PV penetration is reverse power flows in the distribution system. In Germany, approximately 70% of the installed PV capacity is connected to the low voltage distribution network, which was not originally designed to host generation (2012) [18]. In distribution networks where the local PV generation exceeds the local load demand, reverse power flows occur toward upstream voltage levels, local transformer and consequently additional power flows to the medium distribution and transmission networks. This phenomenon is often accompanied by voltage rises. This scenario limits the opportunity to increase the share of renewable energy from PV.

One approach to handle this situation is to allow the network to absorb the generated energy from the renewable sources locally (within the same radial network) by controllable loads such as electric vehicles.

4.1 Public street lighting network

Public street lighting networks are fundamentally designed for the specific application of illumination during certain dark intervals of the day. Hence, this unique characteristic provides the opportunity to use the existing infrastructure to provide energy to other applications when the luminaires are turned off during daylight, acting as additional demand on the network, potentially absorbing the energy from locally generated renewable energy. In a special case where the network is designed for conventional luminaires, if the old luminaires are replaced by lower consumption LED lights, the added remaining capacity can also be further utilized for higher levels of power consumption.

4.2 Application: energetic parking lot with PV

The proposed strategy in this work aims to retrofit an existing public street lighting network to create a unique charging platform, where an intelligent hardware and software system capable of measurement, calculation and wireless communication will be developed and installed onto the existing poles, and then a cloud-based load management system will be implemented to manage a host of new electricity applications (Fig. 5).

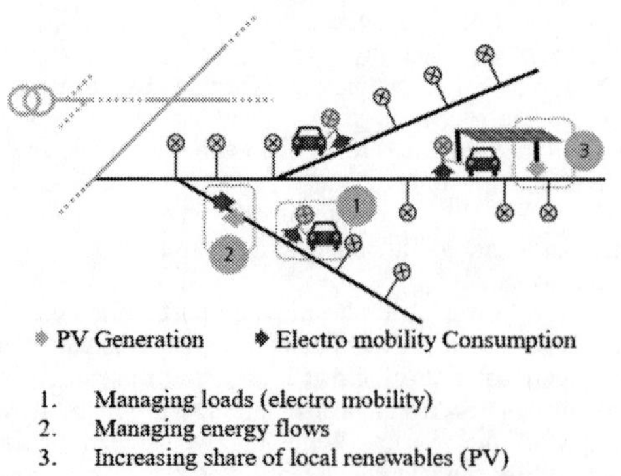

Figure 5: Load management with intelligent light poles

An intelligent load management system capable of carrying out load calculations can provide the solution for scheduling loads in coordination with local renewable energy generation. This solution has the potential to be implemented in an energetic parking lot with PV generation.

This system can measure the PV generation and develop a load profile over time in order to coordinate the power flows between the local renewable energy generation from the PV system installed in the parking lot and the electro mobility loads charging at the adjacent network of intelligent lighting poles in order to maximize the share of renewable consumption and reduce reverse flow while maintaining the network reliability and power quality requirements.

Dynamic control over loads will provide the opportunity to optimize the power flow in the low voltage distribution networks to prevent over voltages in power lines and at the point of connection to the local transformer.

The voltage drop constraint in the network limits the maximum amount of the available charging power at each node; therefore, the topology information is essential to load management. Depending on the topology the maximum amount of power that can be drawn by each load is restricted by the allowable voltage drop band of +/- 10% previously stated in section 2.1.

A study is designed to analyze the influence of topology on voltage drops and maximum available power in a network. A common size of public street lighting network with two different topologies is modelled in PowerFactory to carry out the analysis:

- Topology type 1: Single-line
- Topology type 2: Branched

Table I outlines the network specification, and Fig. 6 illustrates simplified representative models of the two types of topology.

Table I: Public street lighting network specification

Number of poles	50 poles
Topology type	Single-line/Branched
Network Phases	3-Phase
Load Connection	Single-phase
Length	800m
Distance between poles	16m
Nominal Voltage/Frequency	230V/50Hz
Cable type	4-wire NYY, $10mm^2$
Current rating	75A
Allowable voltage drop	-10% nominal voltage

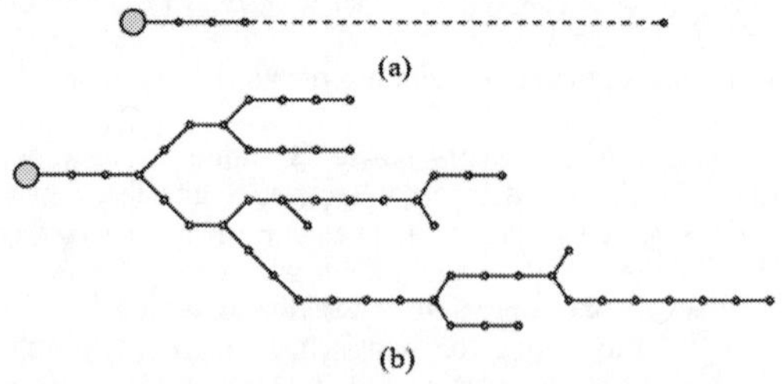

Figure 6: Representative network models with (a) single-line and (b) branched radial topologies

The goal of each analysis is to calculate and compare the maximum amount of load (in Watts) that can be supported by the network infrastructure, or in other words the maximum available power at each node, for both topology types.

The presented method calculates the amount of power that can been absorbed at each pole based on the two network constraints discussed in the theory part:

- Allowable Voltage Drop
- Maximum Current Capacity

The loads are considered to be single-phase connected arbitrarily to phases A, B, or, C of the network. The calculated power values are essentially equivalent to the total amount of local renewable energy that can be absorbed by the adjacent public street lighting network.

5 RESULTS & DISCUSSION

5.1 Topology identification

The algorithm successfully identified the topology of the modelled network, when the resolution of the measurement is at least 2 decimals. Aging effects like corrosion, bad cable connections etc. cannot be taken into account, because they act as impedances, which cannot be distinguished from the cable impedances.

The method is evaluated in comparison to a PowerFactory model and showed good agreement. Low resolution or weak loads/sources lead to lower accuracy in the determination of the topology. The precision needs to be increased, as soon as the power line is operated closer to its physical limits.

The operation in sample scenario applications is discussed and attention is drawn to possible sources of failure, like power line or interconnection corrosion.

The effect of branching between loads is still under discussion.

The measurement resolution is a critical parameter. If the network has to be operated closer to its physical limits, the more critical the resolution is.

5.2 Maximum load calculation

The results from the network modelling and analysis are illustrated in Fig. 6. Light grey and dark grey dots represent the sorted available power at the nodes for the Single-line and Branched topologies respectively. The x-axis shows the pole IDs, and the y-axis represents the power in Watts. Poles are identified by arbitrary numbers, and do not necessarily represent an order or location in the network.

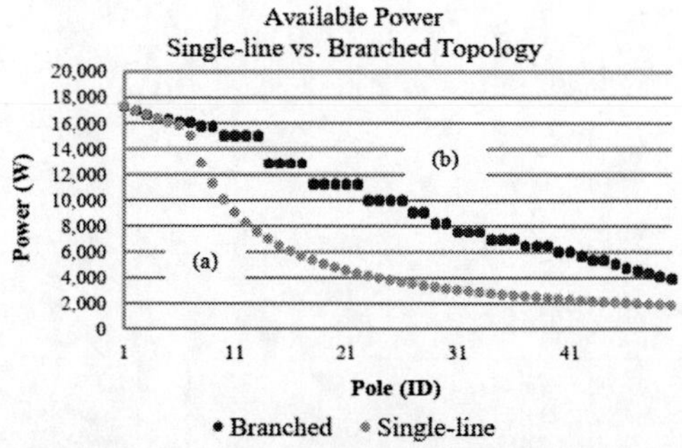

Figure 6: Comparison of available power in (a) single-line vs. (b) branched topology

The average amount of power available at each pole (network node) in the Single-line network is 5,962W compared to 10,166W in the Branched network.

The maximum available power at each node depends on the location of the lighting pole. The poles closer to the source can support higher loads, where the voltage drop has its minimum impact, while the loads connected to the poles toward the end of the line will experience a higher voltage drop; therefor, the available power is less.

In the single-line model the influence of voltage drop increases proportionally along the line from the first pole (1) to the last (50). In this network the voltage drop effect is maximized. In contrast, in the branched model the average amount of voltage drop is lowered, due to shorter lengths of the cable to the loads.

The results reveal that for two networks with equal number of nodes and identical cable length between the two adjacent nodes, the network with branches is capable of supplying or absorbing higher amount of power.

In a special case, for a network with a radial topology so that all the branches share a single common root node, the voltage drop effect is minimized; therefor, the highest amount of power can be absorbed.

6 CONCLUSION & OUTLOOK

In this paper, an algorithm for topology identification is proposed. Using Jenks Natural Break clustering, the topology becomes visible in a matrix illustration of the measured voltage drops. The proposed algorithm was further evaluated and valuated by comparing the output graphical representation of topology with the model. The algorithm is suitable for the identification of single phase low voltage networks and can be used as a sub-module in an intelligent energy management system.

It is concluded that the topology identification can be used as an integral part of the power management in weakly documented radial low voltages power networks.

For more complex applications, the relevance of phase-shift induced cross-phase effects towards successful topology detection needs to be evaluated and quantified. The usability in scenarios with uncontrolled loads that randomly appear at measurement-points using stochastics instead of quasi stationary measurements would open a whole new range of applications.

The analysis of large low voltage city infrastructures based on asynchronous meter readings would be of great interest for future management strategies in smart grids, but would require stochastic analysis on large data sets.

7 ACKNOWLEDGEMENT

This work was carried out as part of the project, "Smarte Pfosten" (smart lighting poles) and is funded by the ZIM program of the Federal Ministry for Economic Affairs (16KN062820).

References

[1] Yizheng Liao, Y. Weng, Meng Wu and R. Rajagopal, "Distribution grid topology reconstruction: An information theoretic approach," 2015 North American Power Symposium (NAPS), Charlotte, NC, 2015, pp. 1-6.

[2] V. Arya, T. Jayram, S. Pal, and S. Kalyanaraman, "Inferring connectivity model from meter measurements in distribution networks," in 4th International Conference on Future Energy Systems, 2013.

[3] H. Pezeshki and H. Wolfs, "Consumer phase identification in a three phase unbalanced lv distribution network," IEEE PES Innovative Smart Grid Technologies, Europe, 2012.

[4] G. Cavraro, "Modeling, control and identification of a smart grid," Ph.D. thesis, University of Padova, 2015.

[5] S. Wiel, R. Bent, E. Casleton, and E. Lawrence, "Identification of topology changes in power grids using phasor measurements," Applied Stochastic Models in Business and Industry, vol. 30, no. 6, pp. 740–752, 2014.

[6] Cavraro, Guido et al. "Data-driven approach for distribution network topology detection." 2015 IEEE Power & Energy Society General Meeting (2015): 1-5.

[7] S. Bolognani, N. Bof, D. Michelotti, R. Muraro, and L. Schenato, "Identification of power distribution network topology via voltage correlation analysis," in 52nd IEEE Conference on Decision and Control, Florence, Italy, 2013.

[8] M. Kezunovic, "Monitoring of power system topology in real-time," in 39th Hawaii International Conference on System Sciences, 2006.

[9] T. Erseghe, S. Tomasin, and A. Vigato, "Topology estimation for smart micro grids via powerline communications," IEEE Transactions on Signal Processing, vol. 61, no. 13, pp. 3368–3377, 2013.

[10] D. Deka, S. Backhaus and M. Chertkov, "Estimating distribution grid topologies: A graphical learning based approach," 2016 Power Systems Computation Conference (PSCC), Genoa, 2016, pp. 1-7.

[11] Guido Cavraro, Vassilis Kekatos, and Sriharsha Veeramachaneni."Voltage analytics for power distribution network topology verification", July 201

[12] Pappu, Satya Jayadev & Bhatt, Nirav & Pasumarthy, Ramkrishna & Rajeswaran, Aravind. (2016). Identifying Topology of Power Distribution Networks Based on Smart Meter Data. IEEE Transactions on Smart Grid. PP. 10.1109/TSG.2017.2680542.

[13] Jenks, George F. 1967. "The Data Model Concept in Statistical Mapping", International Yearbook of Cartography 7: 186–190.

[14] Pandian, Shanmuga S. " Calculating Voltage Drop in Power Distribution Systems", EC&M Electrical Construction & Maintenance;Jun2005, Vol. 104 Issue 6, pC16

[15] DIgSILENT, PowerFactory, Website: https://www.digsilent.de/en/powerfactory.html

[16] Dimitri Pescia, Patrick Graichen, Agora Energiewende (2015): Understanding the Energiewende. FAQ on the ongoing transition of the German power system. October 2015

[17] An electricity grid for the energy transition, (2018, April 24). Retrieved from https://www.bmwi.de/Redaktion/EN/Dossier/grids-grid-expansion.html

[18] Time in the Sun, The Challenge of High PV Penetration in the German Electric Grid, IEEE power & energy magazine, Digital Object Identifier 10.1109/MPE.2012.2234407, Date of publication: 20 February 2013, http:// http://magazine.ieee-pes.org/files/2013/02/11mpe02-vonappen-2234407-x.pdf

IMPACT OF DYNAMIC-MECHANICAL LOAD ON PV MODULES MOUNTED ON SINGLE-AXIS TRACKER

Dirk Stellbogen, Peter Lechner, Olaf Schanz, Simon Hummel, and Jonathan Schmepf
Zentrum fuer Sonnenenergie- und Wasserstoff-Forschung Baden-Wuerttemberg (ZSW)
Meitnerstrasse 1, D-70563 Stuttgart, Germany, dirk.stellbogen@zsw-bw.de

Marius Kaiser
Canadian Solar EMEA GmbH, Landsberger Straße 94, 80339 Munich, Germany

ABSTRACT:
PV modules with crystalline silicon technology in some cases exhibit cells with microcracks which provokes concerns about the impact of such microcracks on the energy delivery performance as well as the reliability of the modules. Further, the increasing application of single-axis solar trackers in PV plants raises the question whether this might cause special performance and reliability challenges due to the increased mechanical stress executed onto the modules.
In order to access the extend of deflections induced onto the PV modules in such installations and the impact these deflections might have on the power and energy yield, a field investigation has been conducted at the ZSW test site Widderstall with complementary laboratory tests. Two PV modules with initial microcracks were monitored mounted on a single-axis tracker and underwent laboratory pre- and post-characterisation as well as power measurements in artificial deflection.
The sample modules showed only a moderate STC power loss after one year of operation, which could not certainly be attributed to the microcracks, and no further power decrease after one and a half years. The module back surface deflections recorded in the field were almost entirely within a few millimeters. In flash simulator measurements with artificial deflection, a significant power reduction was observed only at deflections of 5 mm or more to the back. The combination of these findings with the field observations indicates that energy yield losses induced by deflections from daily cyclic tracker movements and wind load pressure are not expected.

Keywords: PV module, microcracks, field monitoring, tracker

1 INTRODUCTION

PV modules with crystalline silicon technology in some cases exhibit cells with microcracks provoking a strong interest to determine the impact of these microcracks on the energy delivery performance of the modules as well as the future development of the issue. These considerations especially address the impact of module bending under mechanical load from, e.g., wind pressure.

A second current trend in the PV power industry is the application of single-axis solar trackers. This leads to the question whether the combination of PV modules with microcracks and trackers causes special performance and reliability challenges. This is initiated by the suspicion of increased mechanical stress executed onto the modules from the daily tracker moving cycles and varying wind exposure.

In order to access the module deflections occurring in real operation, a field evaluation at the ZSW test site Widderstall was conducted together with laboratory examinations.

2 APPROACH

A set of c-Si PV modules for a single-axis tracker installation at the test site Widderstall [1] has been initially characterized in the ZSW laboratory with flash simulator power measurements and EL imaging. Two modules that already showed micro cracks during the initial EL characterisation were chosen for closer examination.

After mounting the modules under test on the tracker structure (Figure 1) and wiring them into the PV array, they were equipped with laser distance sensors for recording the mechanical movement of the centre of the module rear side (see Figure 2).

Figure 1: Two modules mounted on a single-axis tracker at the ZSW test field Widderstall were monitored for deflections of the module laminate

Figure 2: Laser distance sensor placed at the centre of the PV module backside for measuring the laminate deflection in operation on fielded single-axis tracker

The Laser distance sensors provided the measurement of the module deflection at a distance resolution of

1/1000 mm and with sampling time resolution of about 1/10 s. In order to avoid unnecessarily large data files, samples were only recorded, if the displacement from the previous sample exceeds a certain threshold.

Concurrently, the mean and max wind speed at the test site at 5m height was measured and recorded in 1 min intervals.

A follow-up laboratory characterisation of the test samples with power measurements and EL as well as IR imaging has been conducted after one year and one and a half years of operation, respectively.

Finally, the modules were subject to a laboratory experiment, where the STC power of the samples was determined with imposed deflections of the laminate [2]. For this test series, the modules were mounted vertically into the flash simulator setup. After a reference STC power determination, the power measurements were repeated with the center of the module back plane being mechanically displaced by up to 10 mm to the back or to the front simulating a deflection by pressure or by suction, respectively.

3 RESULTS

3.1 Laboratory evaluations

The revision power measurements with flash solar simulator after one year of operation showed a reduction of the STC power by about 2.5 % with respect to the initial characterisation before outdoor exposure for both modules under test (see Table I). Since the modules were not preconditioned before exposure, this includes the initial LID degradation. Further, there is no direct relation to the observed microcracks, as module A with only one cracked cell behaves similar as module B with 8 cracked cells.

Table I: Results of flash simulator pre- and post-tests

Module / Test	Pmax		Isc		FF	
	W	*rel.*	A	*rel.*	%	*rel.*
Module A: 1 cracked cell						
pre-exposure	259,0		9,06		75,7	
post 1 year	253,0	***97,7%***	8,94	***98,7%***	75,5	***99,7%***
repeat, 1.5 years	254,3	***98,2%***	8,99	***99,2%***	75,4	***99,7%***
Module B: 8 cracked cells						
pre-exposure	257,3		9,03		75,6	
post 1 year	250,1	***97,2%***	8,92	***98,8%***	74,9	***99,1%***
repeat, 1.5 years	251,0	***97,6%***	8,95	***99,1%***	74,9	***99,1%***

Repeat tests of the samples after one and a half years of field operation did not show further degradation although several stormy periods had been experienced by the installation in between.

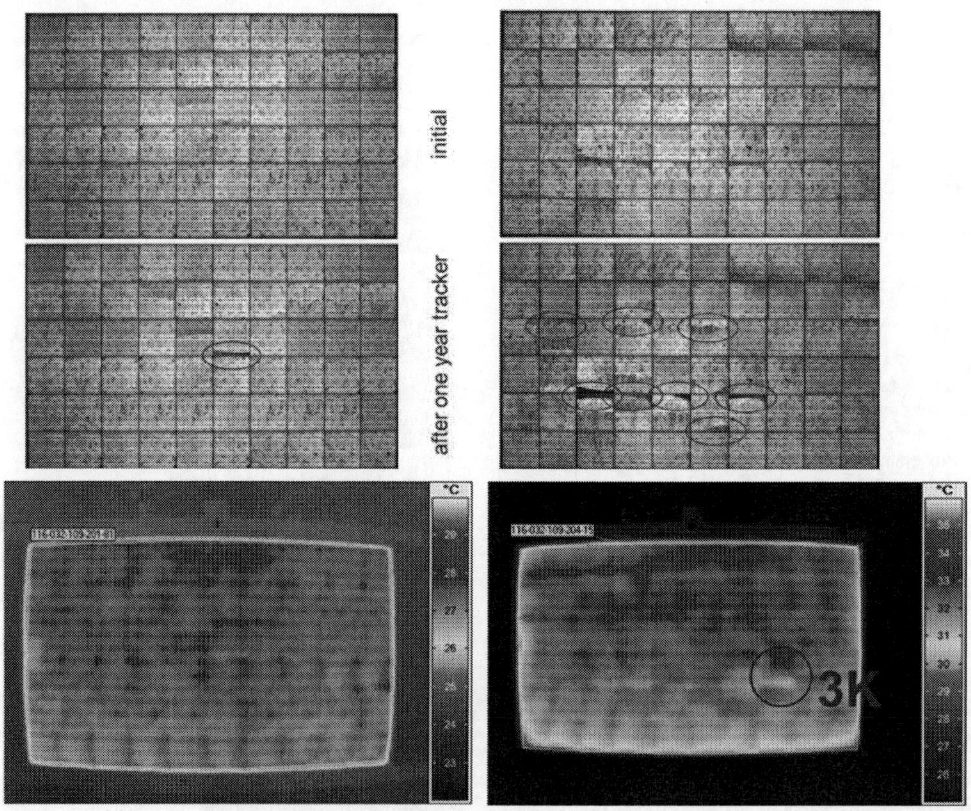

Figure 3: EL images of the two test samples of c-Si PV modules A (left) and B (right) before (top) and after one year of operation on single-axis tracker (middle) showing electrically disconnected/inactive areas; the IR images (bottom) show a temperature raise of only 3K at the most affected cell in module B.

The EL imaging of the test modules after one year (see Figure 3) showed that initial microcracks had developed some electrically disconnected/inactive areas. This was especially the case for the sample module B (right column in Figure 3) that had shown more cracks already initially. It displayed inactive areas at eight cells at the re-assessment but had no higher power loss than the sample module A.

In the IR images (see also Figure 3) there were no peculiarities other than a small temperature increase of +3K at the most impacted cell in sample B.

The results of the power measurement series under deflection are shown in Figure 4, where deflections to the back of the module are counted in the positive direction. Module A exhibited a significant power decrease only at positive deflections ≥ 5mm, with a marked reduction of about 4% at a displacement of the module center by 10 mm to the back.

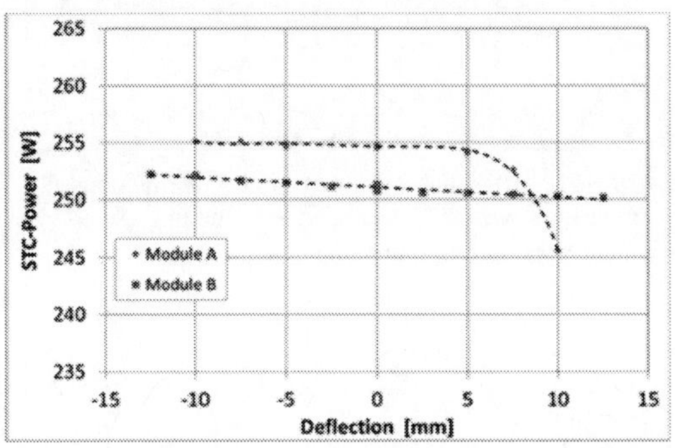

Figure 4: Measured STC power of the test modules at artificial deflection (+: to the back / pressure, -: to the front / suction)

Yet, the more severely cracked module B only showed a very slight variation of the measured power with linear dependence on the module center's displacement. All power changes have been reversible as confirmed by a control measurement at zero deflection.

3.2 Field monitoring results

An example of the measurement data from the laser distance sensor directed to the center of the module back surface is shown in Figure 5, which covers four consecutive days in March 2017. The recordings reveal, that the module deflections are a superposition of wind driven displacements with a gravity driven displacement cycle due to the daily movement of the one-axis solar tracker.

This can be taken clearly from a plot of the observed back surface positions versus the recording time (Figure 6) It shows the daily cycle from the resting position (zero tilt) to East facing in the morning, then moving over to West facing and finally back again to the resting position.

The observed deflections are usually close to the gravity imposed displacement experienced in the tracker resting position and remain almost entirely within a range of a few millimetres (Figure 7).

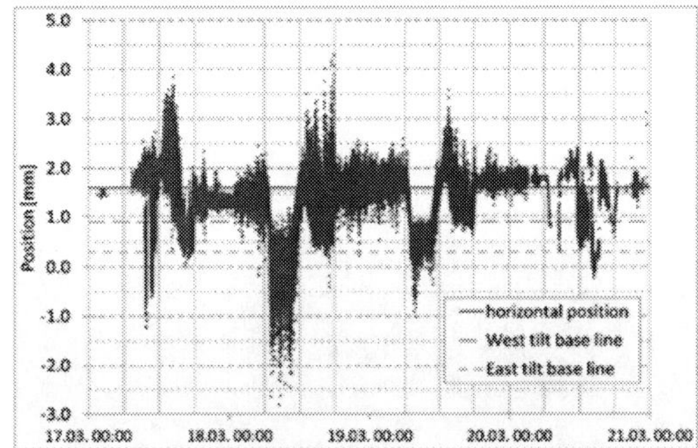

Figure 5: Measured deflection of module back surface on four days in March 2017, showing a superposition of a daily periodic deflection by own weight from the tracker movement and additional deflections from wind impact.

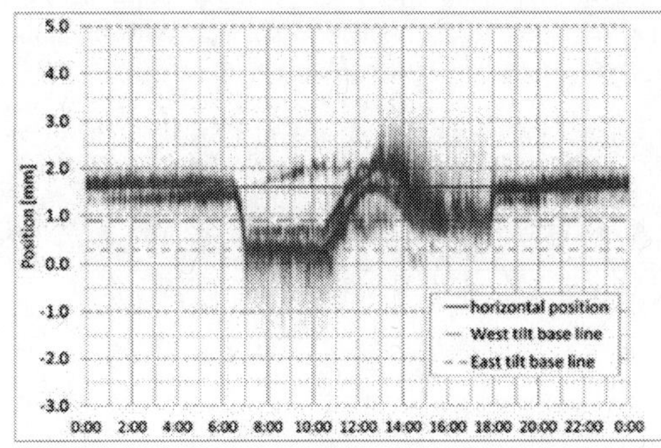

Figure 6: Plot of the measured deflections of the sample module back surface versus the recording time, showing the daily course induced by the tracker movement

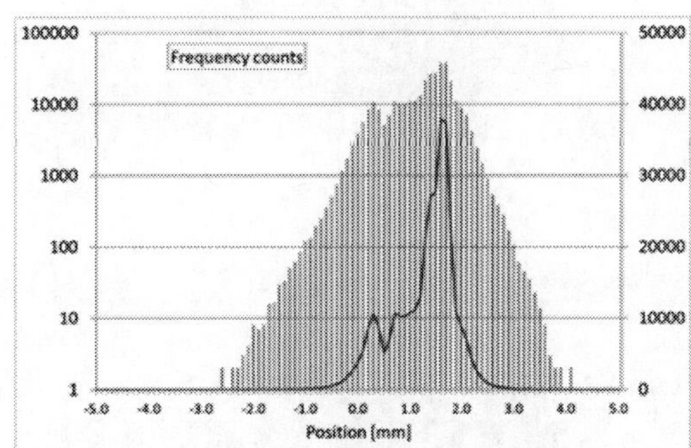

Figure 7: Frequency distribution of the module deflections recorded with the distance sensor during the test period in March 2017 (in logarithmic and in linear scaling)

The ZSW test site Widderstall has a moderate wind regime. Figure 8 displays the frequency distributions and cumulative fraction curves of the observed mean and max. wind speed per minute for a full year period from September 2016 to August 2017. A few records of the wind speed reach values of more than 60 km/h, but the mean and max. wind speed values remain below 25 km/h and 45 km/h, respectively, in 99.9% of the instances.

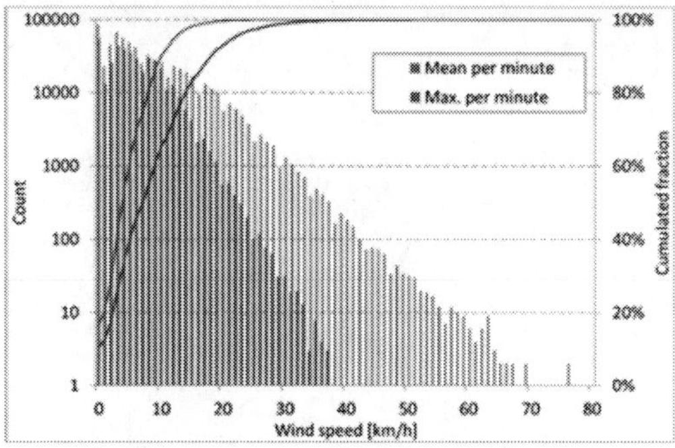

Figure 8: Distribution of one minute measurement values of mean and max. wind speed in 5 m height at the test site Widderstall from 2016/09 to 2017/08.

The comparison of the module deflection data with the concurrent on-site wind measurements shows, that the movements of the module laminate are clearly related to the wind velocity. This is illustrated in Figure 9 which joins the course of the mean and max. wind speed values with the movements of the module laminate between observed positions for the test period already discussed above. Indeed, wide displacements induced by wind remain extremely rare; the vast majority of recorded module movements is in the sub-millimetre range.

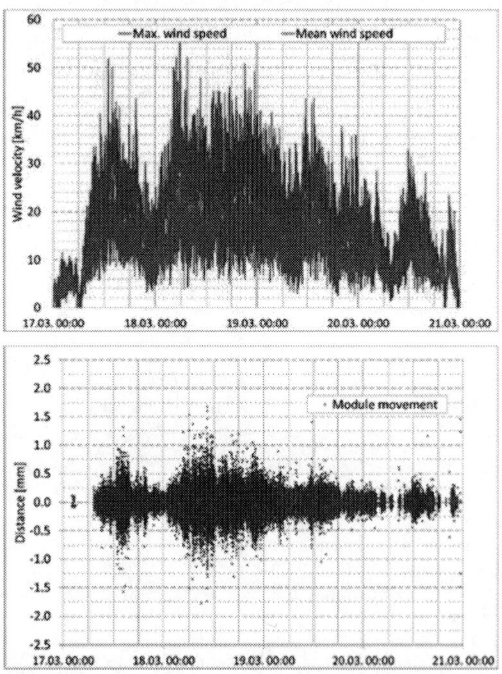

Figure 9: Correlation of wind speeds recorded on site (top) with the extension of mechanical displacements of the module back surface (bottom) for four consecutive days in March 2017.

The correlation of module deflections and wind speed is given in a more quantitative way in Figure 10, showing the data of module position and absolute module movement, respectively, from the same period in dependence of the recorded max. wind velocity in the minute time interval. The range of positions and the movement distances increase almost linearly with the wind speed up to about 30 km/h, however remaining within a few millimeters.

For higher wind speeds, the number of observations gets much smaller. However, even when extrapolating the trend form the lower wind speeds to the full range of the one-year wind data described above, almost all deflections would not exceed about ± 5 mm. This finding should be considered, when devising laboratory cyclic mechanical load tests for accelerated PV module testing.

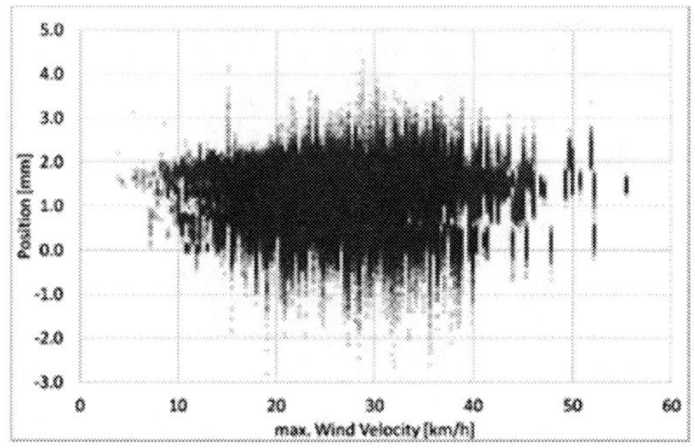

Figure 10: Distribution of test module deflections in the observation period in dependence of the max. wind speed in the respective time interval.

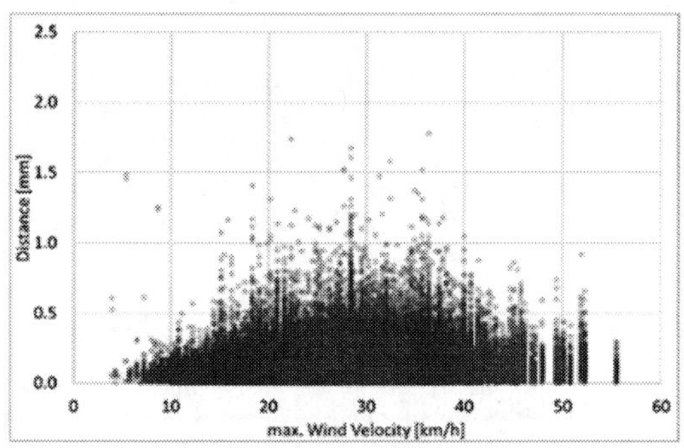

Figure 11: Distribution of test module deflection movements between position observations in dependence of the max. wind speed in the respective time interval.

3.2 Combination of Laboratory and Field monitoring results

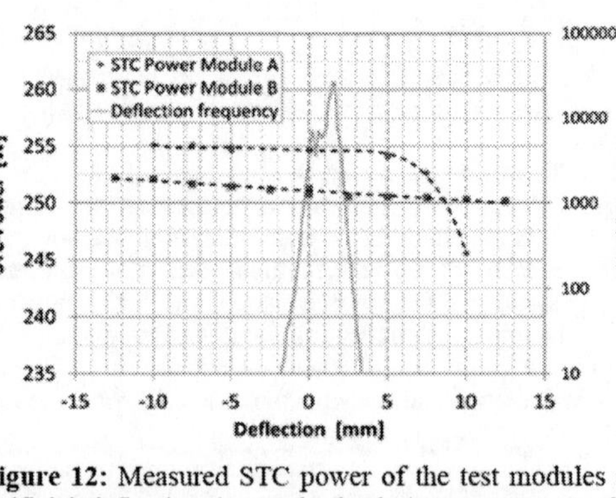

Figure 12: Measured STC power of the test modules at artificial deflection (+: to the back / pressure, -: to the front / suction) together with the frequency distribution of observed module positions (test period in March 2017)

The results from the power measurements with artificial deflection can be evaluated combining them with the frequency distribution of module deflections from the field monitoring, as it is given in Figure 12. This shows, that the observed deflections are in a range where almost now power reduction from the reflection is to be expected. Hence, no significant energy loss should be experienced.

4 SUMMARY AND CONCLUSIONS

Two PV modules with initial microcracks were monitored in the field mounted on a single-axis tracker and underwent laboratory pre- and post-characterisation and experimental power measurements in artificial deflection. This revealed interesting insights into the extension of module deflections in real operation as well as the impact these deflections might have on the power and energy yield of PV modules with microcracks.

The two sample modules showed a moderate STC power loss of 2.5% after one year, which could not certainly be attributed to the microcracks; no further decrease was detected after one and a half years.

The observed module back surface deflections from tracker tilt movements and wind loads were within a few millimeters. This should be taken in consideration when designing laboratory cyclic mechanical load tests for accelerated PV module testing.

In STC power measurements with artificial deflection, one test module exhibited a significant power reduction only at deflections of ≥ 5 mm to the back, whereas the other module sample showed only a very slight variation of the measured power.

The combination of field observations with the laboratory power measurements of deflected modules indicates that energy yield losses induced by deflections from daily cyclic tracker movements and wind load pressure are not expected.

REFERENCES

[1] D. Stellbogen, P. Lechner, H.-D. Mohring, Long term performance characterisation of PV generators, 28th European Photovoltaic Solar Energy Conference and Exhibition Proceedings, 2013

[2] J. Käsewieter, F. Haase, M.H. Larrodé, M. Köntges, Cracks in solar cell metallization leading to module power loss under mechanical loads, Energy Procedia 55 (2014) 469 – 477

TECHNICAL EVALUATION OF A STAND-ALONE PV HEAT PUMP SYSTEM FOR SPACE HEATING/COOLING APPLICATIONS WITHOUT BATTERIES

C. Lorenzo*, L. Narvarte, F. Martínez-Moreno
Instituto de Energía Solar - Universidad Politécnica de Madrid, 28031 Madrid, Spain
*Phone: +34 91 0673485, mail: celena.lorenzo@ies.upm.es

ABSTRACT: In the last years, the concept of solar photovoltaic (PV) powered heat pumps (HP) has become very attractive, in order to match the increasing heating/cooling demand with a renewable and environmentally-friendly energy source. The objective of this work is to technically validate a stand-alone PV-HP system for space heating/cooling applications. This paper describes the prototype system installed in Madrid (Spain) for this purpose, together with the definition of the validation tests to be performed. Finally, some modifications are proposed for the performance indicators that are traditionally used for comparing HP and PV systems, in order to achieve a characterization that is more specific to PV-powered systems.

Keywords: Evaluation, Heat Pump, Stand-alone PV system, Performance

1 INTRODUCTION

Space heating and cooling, both for residential and industrial applications, is a highly energy consuming activity that needs to meet increasing electricity demands along the whole planet. Heat pumps help to reduce significantly the electricity consumption required to cover thermal necessities, so they are becoming an increasingly important technology. In Europe, the heat pump market increased by 12% in terms of units sold in 2015 (which means an installed electrical power of 2.11 GW) [1] and by 12% in 2016 (2.37 GW) [2]. This market growth is expected to be maintained in the near future, with estimations of the compound annual growth rate in the range 6.02% [3] to 10.25% [4] for the period 2016-2021. However, in most cases the electricity consumed by a heat-pump system is obtained directly from the national grid or from stand-alone generation sets (mostly diesel generators). Solar thermal heat pumps, that use solar collectors as the evaporator of the system, have been widely explored and demonstrated: they reduce the electricity consumption of the heat pump, but they do not guarantee that such electricity is generated by a renewable source. Consequently, latest research on solar heat pumps is focused on the use of Photovoltaic (PV) generators for powering the variable-speed compressor that circulates the refrigerant loop [5], [6], [7].

The objective of this work is to technically validate a stand-alone PV Heat Pump (PV-HP) system for space heating and cooling applications, where the variable-speed compressor of the heat pump is directly powered by a PV array through a frequency inverter. Such validation implies to manage the power intermittences due to cloud passing and to minimize the day/night generation cycle problem. The validation test scheduling is presented. Finally, some modifications of the traditional performance indicators are proposed, in order to adapt them to PV-powered systems.

2 SYSTEM DESCRIPTION

For performing the technical validation of a stand-alone heat pump system without batteries, a PV-HP prototype has been installed at the IES-UPM building. Figure 1 shows the schematic of this prototype. It consists of a PV generator with two possible configurations (one of 1.6 kW and another of 3.6 kW) that powers a 1.5 kW air-to-air heat pump through a 4 kW frequency inverter. The heat pump external unit is placed at a small thermally isolated room, which simulates the ambient. The internal unit is placed at a similar room that simulates the space to be heated or cooled. Two auxiliary grid-powered heat pumps (also air-to-air) have been installed in these two rooms for temperature control: the effect of the PV-HP can be accelerated or countered for simulating different testing conditions (i.e winter/summer ambient conditions or different thermal loads). Figure 2 shows the main components of the system described: PV generator, frequency inverter, interior room and exterior room.

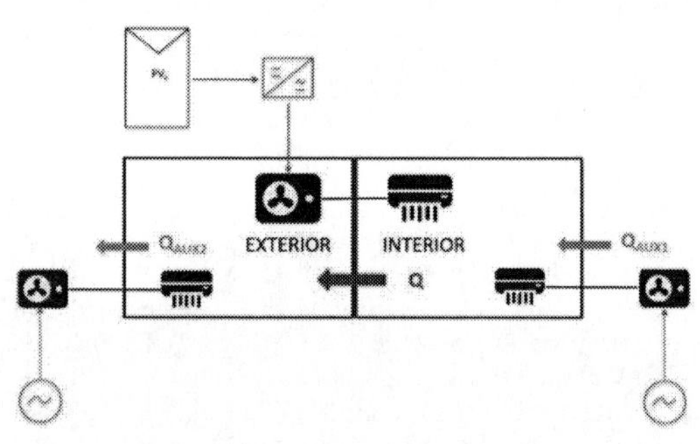

Figure 1: Schematic of the PV-HP prototype installed at the IES-UPM building in Madrid, Spain.

(a) (b)

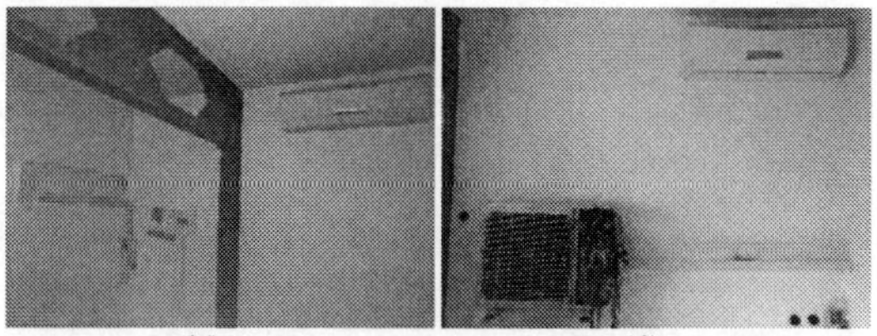

(c) (d)

Figure 2: Images of the PV-HP prototype installed in Madrid (Spain), composed of the PV generator (a), frequency inverter (b), interior room (c) and exterior room (d).

3 VALIDATION TESTS

3.1 Characterization with the electric grid

Before powering the compressor of the HP with a PV generator, it was characterized using the electric grid and the internal control plate of the HP. The objective of this test was to learn the frequency/temperature characteristics of the compressor, in order to replicate the logic of the control plate with the frequency inverter.

Figure 3 shows the results of a test performed at constant T_{ext}, both in cooling (a) and heating modes (b). The input frequency of the compressor depends on ΔT, which is the difference between T_{int} and the temperature set point. It can be observed that the frequency increases almost instantly to its maximum value at the beginning of the test. As ΔT is reduced, so is the frequency of the compressor, until the temperature set point is reached.

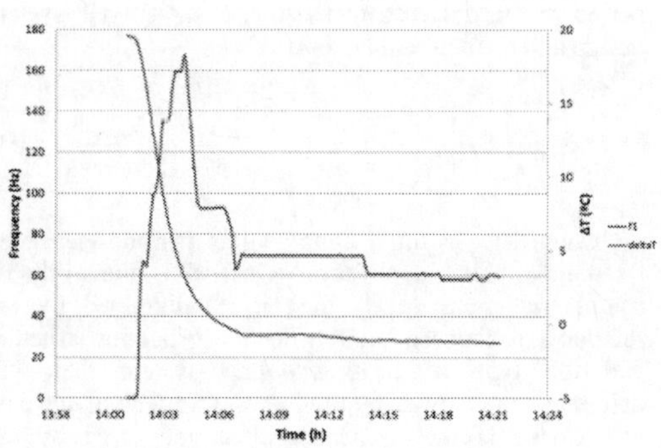

(a)

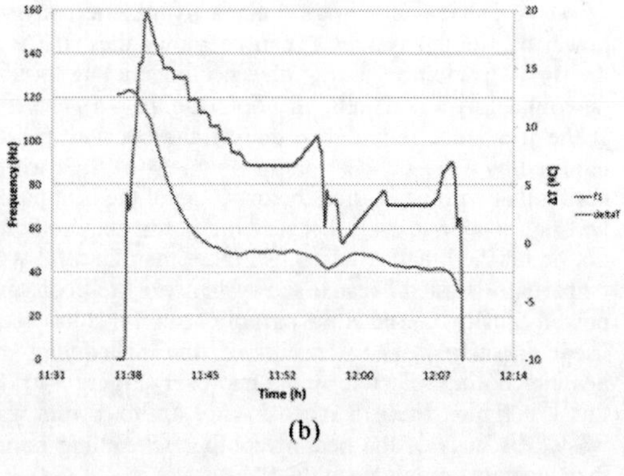

(b)

Figure 3: Results of the frequency/temperature characterization of the compressor using the electric grid and the internal control plate of the HP, for the cooling (a) and heating (b) modes.

3.2 Technical validation with the PV generator

The technical validation of the PV-HP system described in section 2 will be performed through a series of experiments, designed for simulating different operating conditions:

- The PV generator can be programmed to operate always at its Maximum Power Point (MPP), using all the available PV electricity for powering the compressor of the HP. Another possibility is to operate at a point of the I-V curve where the PV output is exactly the electricity demanded by the compressor for reaching a certain temperature in the interior room. This operation point will not be necessarily the MPP.
- There are two possible configurations for the PV array, resulting into two different nominal powers of 1.1 kW (undersized generator) and 3 kW (oversized generator).
- The HP performance is highly affected by the temperature conditions in both the interior and exterior rooms (T_{int} and T_{ext} respectively). The auxiliary grid-powered heat pumps can be used for controlling these temperatures. Usually, Tint is kept constant at a given set value (different for heating and cooling modes). On the other hand, Text can be kept constant or follow a certain daily profile (different for winter or summer conditions). When operating in heating mode, winter conditions will be simulated and vice versa.

The combination of different operating conditions results into the following 8 validation tests:

Table I: Operating conditions for all the validation tests here proposed.

	PV Operation Point	PV generator configuration	Temperature conditions
T1	MPP	Undersized	T_{int} and T_{ext} constant
T2	MPP	Undersized	T_{int} constant, T_{ext} variable
T3	MPP	Oversized	T_{int} and T_{ext} constant
T4	MPP	Oversized	T_{int} constant, T_{ext} variable
T5	Not MPP	Undersized	T_{int} and T_{ext} constant
T6	Not MPP	Undersized	T_{int} constant, T_{ext} variable
T7	Not MPP	Oversized	T_{int} and T_{ext} constant
T8	Not MPP	Oversized	T_{int} constant, T_{ext} variable

4 PERFORMANCE INDICATORS

4.1 Traditional performance indicators

The performance of HP systems, in terms of how efficiently they transform electrical energy into thermal energy, is typically evaluated through the Coefficient of Performance (COP) when operating in heating mode or through the Energy Efficiency Ratio (EER) when operating in cooling mode. These indicators are defined as follows:

$$COP = \frac{Q_c}{W_{com}} \quad (1)$$

$$EER = \frac{Q_e}{W_{com}} \quad (2)$$

where Q_c is the heating power of the condenser in heating mode, Q_e is the cooling power of the evaporator in cooling mode and W_{com} is the electric power consumed by the compressor.

The performance of a PV generator is usually evaluated through the Performance Ratio (PR), which represents the ratio between the PV energy that is actually generated and the PV energy that could have been ideally generated for a certain period of time:

$$PR = \frac{E_{PV}}{\frac{P^*}{G^*}\int G(t)dt} \quad (3)$$

where E_{PV} is the useful (meaning that it is either used for powering the compressor, for charging a battery or exported to the grid) DC energy produced by the PV generator, P^* is the nominal power of the PV generator, $G^*$is the solar irradiance under Standard Test Conditions (STC) and G is the solar irradiance received in the plane of the PV generator.

Finally, when a PV generator is connected to any electric load different from the grid (i.e. a heat pump), the Self-Consumption Ratio (SCR) and the PV Penetration Rate (PVPR) evaluate the quality of the coupling between both systems. For a PV-HP, these indicators are given by the following equations:

$$SCR = \frac{E_{PV.com}}{E_{PV}} \quad (4)$$

$$PVPR = \frac{E_{PV.com}}{E_{com}} \quad (5)$$

where $E_{PV.com}$ is the PV energy consumed at the compressor of the HP, both directly or after being stored in a battery, and E_{com} is the total electric energy consumed by the compressor of the HP, regardless of how it is generated.

4.2 Modified performance indicators for PV-HP systems

When approaching the performance evaluation of a PV-HP system, a combination of the aforementioned indicators would help to integrate three different characteristics of the system: the quality of the heat pump (characterized by the COP or the EER), the quality of the PV generator (characterized by the PR), and the quality of the integration of the two systems (characterized by the SCR and the PVPR).

In this work, an adapted COP for PV-HP systems is proposed:

$$COP_{PV-HP} = COP(1 + PR \times SCR \times PVPR) \quad (6)$$

$COP_{PV\text{-}HP}$ will be equal to the traditional COP if there is no PV energy generation and/or use (PR=SCR=PVPR=0); $COP_{PV\text{-}HP}$ will double the traditional COP if the PV generator performs ideally and all its electricity production is used for powering the compressor, with no excess to be exported to the grid (PRPV-=SCR=PVPR=1).

Similarly, the adapted $EER_{PV\text{-}HP}$ is defined as follows:

$$EER_{PV-HP} = EER(1 + PR \times SCR \times PVPR) \quad (7)$$

This indicator will also present values in the range of 1 to 2 times the value of the traditional EER.

Finally, some considerations must be accounted when calculating the PR As traditionally defined, this indicator is highly affected by factors that do not depend on the quality of the PV system, such as the heating/cooling period, the daily heating/cooling demand profile or possible failures of the compressor or any of the components of the heat pump. For separating these external factors from those that are strictly related to the PV generator performance, the PR can be factorized by distinguishing between irradiation losses for three essentially different reasons: the non-heating/cooling period, the intrinsic characteristics of the PV-HP system design and the external factors:

$$PR = \frac{E_{PV}}{P^*/G^*} \text{x} \frac{1}{\int Gdt} \text{x} \frac{\int_{HCp} Gdt}{\int_{HCp} Gdt} \text{x} \frac{\int G_{useful}dt}{\int G_{useful}dt} \text{x} \frac{\int G_{used}dt}{\int G_{used}dt} \quad (8)$$

where HCp is the heating/cooling period determined by the particular application, Guseful is the available useful irradiance during the HCp determined by the relationship between the P*, the PV generator structure and the type of heating/cooling system that will determine the power requirements throughout the day, and Gused is the irradiance effectively used by the system. To clarify these concepts, GHCp, Guseful, and Gused are shown in Figure 4 for a hypothetic constant power heat-pump system. It can be shown that GHCp is the total irradiance during the heating/cooling period determined by a certain application (Figure 4-a). Guseful is the irradiance required to deliver the constant power required by a certain heat pump (Figure 4-b). It is worth noting that with irradiances below Guseful the heat pump will not be able to work because the power required will not be reached, and irradiances higher than Guseful will be partially wasted because the system works at constant power. Finally, Gused is the part of Guseful that has been used effectively as a result of the scheduling of heating/cooling selected by the end-user (Figure 4-c). In this last figure, the irradiance from 7 am to 14 pm was wasted because of the heating/cooling scheduling rather than technical problems in the PV system.

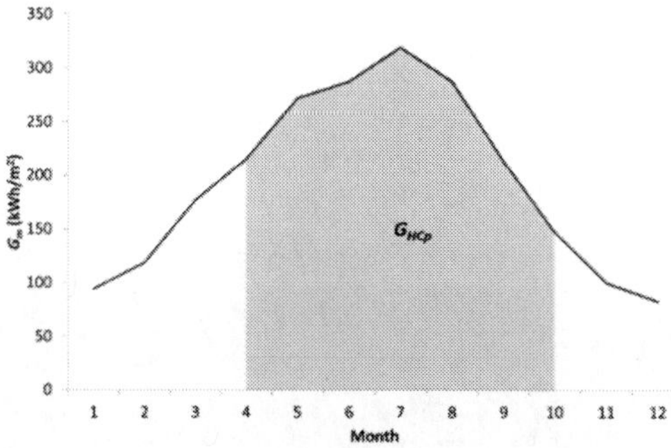

(a)

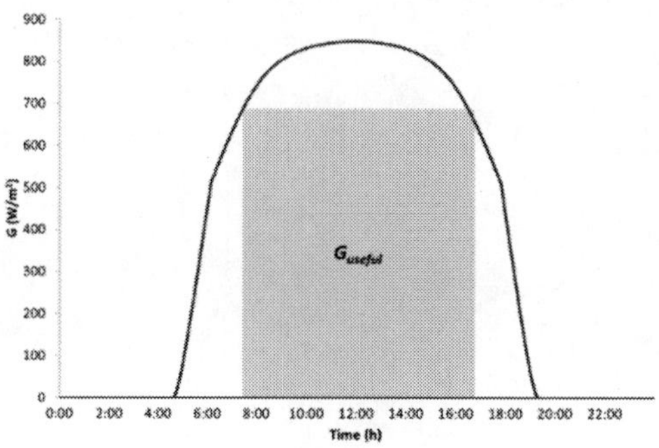

(b)

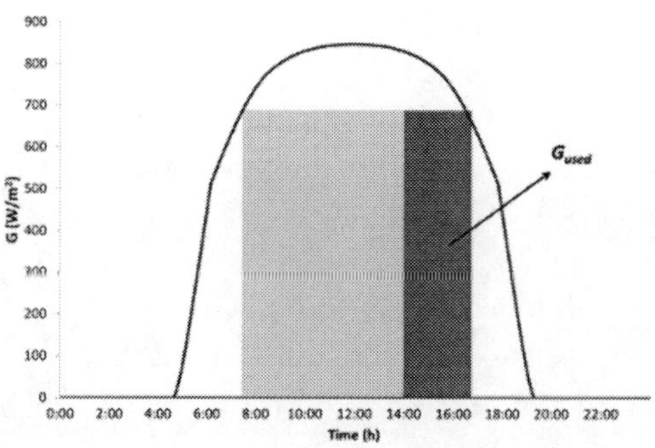

(c)

Figure 4: Graphical representation of the different irradiations considered: (a) $\int G_{HCp}$ is the irradiation during the heating/cooling period (HCp), (b) $\int G_{useful}$ is the useful irradiation during the HCp determined by the design of the PV HP system; and (c) $\int G_{used}$ is the irradiation used effectively by the system.

So, it is possible to rewrite equation (8) as:

$$PR = PR_{PV} \times UR_{HCp} \times UR_{PVHP} \times UR_{EF} \qquad (9)$$

where:

$PR_{PV} = \frac{E_{PV}}{P^*/G^*} x \frac{1}{\int G_{used} dt}$	This is the PR considering only losses strictly associated to the PV system itself, i.e., actual versus nominal peak power, dirtiness, thermal and DC/AC conversion losses. It is intrinsic to the technical quality of the PV component and its maintenance.
$UR_{HCp} = \frac{\int_{HCp} G dt}{\int G dt}$	This is the ratio of the total irradiation throughout the HCp period to the total annual irradiation (Fig. 4-a). It is intrinsic to a given application of heating/cooling. Note that it can only be applied to the annual period.
$UR_{PVHP} = \frac{\int G_{useful} dt}{\int_{HCp} G dt}$	This is the ratio of the irradiation necessary to deliver the power required by the heat pump to the total irradiation throughout the HCp (Fig. 4-b). It is intrinsic to the heating/cooling system design; specifically it depends on the type of heating/cooling system (at constant or at variable power), the ratio between the PV peak power and the PV power required for heating/cooling, and on the tracking geometry.
$UR_{EF} = \frac{\int G_{used} dt}{\int G_{useful} dt}$	This is the ratio of the irradiation required by the heat pump during the heating/cooling scheduling to the same irradiation during the HCp.

The modified performance indicators here proposed ($COP_{PV\text{-}HP}$, $EER_{PV\text{-}HP}$ and PR factorized) will be evaluated for the PV-HP prototype described in section 2, for all the validation tests specified in section 3. It must be noted that PVPR=1 for this study, as corresponding to a stand-alone system. However, this indicator must be taken into account for hybrid PV/grid configurations.

6 REFERENCES

[1] Heat Pump Market and Statistics Report 2016. Executive Summary. European Heat Pump Agency; 2016 Sep.

[2] Heat Pump Market and Statistics Report 2017. Executive Summary. European Heat Pump Agency; 2017 Nov.

[3] Heat Pumps Market - Forecast from 2016 to 2021. Executive summary. Knowledge Sourcing Intelligence LLP; 2016 Aug.

[4] Global Heat Pump Market 2016-2020. Executive summary. TechNavio; 2016 Sep.

[5] Aguilar F. J., Aledo S., Quiles P. V. Experimental analysis of an air conditioner powered by photovoltaic energy and supported by the grid. *Applied Thermal Engineering 123 (2017), 486-497.*

[6] Romaní J., Belusko M., Alemu A., Cabeza L. F., Gracia A., Bruno F. Control concepts of a radiant wall working as thermal energy storage for peak load shifting of a heat pump coupled to a PV array. *Renewable Energy 118 (2018), 489-501.*

[7] Toradmal A., Kemmler T., Thomas B. Boosting the share of onsite PV-electricity utilization by optimized scheduling of a heat pump using buildings thermal inertia. *Applied Thermal Engineering 137 (2018), 248-258.*

Acknowledgements

This work has received funds from the European Union's Horizon2020 research and innovation programme under Grant Agreement 787289.

AN EMPIRICAL TECHNO-COMMERCIAL ASSESSMENT OF INVERTER DC LOADING OF PHOTOVOLTAIC ASSETS IN ASIAN LOCATIONS

André M. Nobre[#], Jason Tan, Shravan Karthik, Rupesh S. Baker, Raghav Malhotra and Alan Khor
Cleantech Energy Corporation Limited
25 Church Street, #03-04, Capital Square Three, Singapore 049482
andre.nobre@cleantechsolar.com

ABSTRACT: PV system performance of Asian assets is a novel frontier. Established markets in Europe, USA and Japan are no longer dominating most new installations coming online. With new systems deployment comes learning curve mistakes on the design front, which could include overloading the DC inputs of inverters to minimize capital expenditure on site. This work addresses real-world examples of "inverter clipping" due to DC overloading situations at PV systems deployed in Asia. It is demonstrated that the use of design techniques from temperate climate PV systems cannot be directly implemented in other parts of the world without considering local conditions, which for Asian locations includes high irradiance levels year-round. DC-to-AC loading ratios of 1.2-1.4 can cause unwanted losses of revenue beyond typical business expectations. Singapore and Cambodia tested PV systems demonstrated inverter "clipping" losses of a few percentage points. For the tested Singapore case, deploying more inverters would have made financial sense only after a payback period of ~7.5 years, which nearly approximates the lifetime of inverters in business plan simulations. Business model strategies may dictate PV system design outcomes irrespective of high-performing PV system performance optimization at sites.

Keywords: PV System, Inverter, Performance, Solar Radiation, Economic Analysis

1 INTRODUCTION

Photovoltaic (PV) systems deployment in Asia has increased considerably in recent years, especially after China became the number one market in the world in 2015. Asian new facilities have dominated the global PV penetration rates, increasing cumulative solar capacities in the region progressively (37% in 2014 – out of a 177 GWp worldwide market, 43% in 2015 – out of 228 GWp, and nearly 50% in 2016 – out of 303 GWp). After the deployment figures of 2017, the global market reached 403 GWp, out of which ~55% of the systems are in Asia [1]. India has just recently crossed 25 GWp of deployed PV systems [2]. Thailand has deployed 2.7 GWp to date, with Malaysia picking up the speed [1]. And as a summary, although a small market, Singapore, in Southeast Asia, serves as an illustrative example of such rapid solar PV deployments in the region, with a compound annual growth rate of ~80% in the decade spanning from 2008-2018 [3].

As multiple GWp are deployed in a short period of time, lack of expertise in novel markets could yield a variety of installation mistakes. One typical flaw is poor system sizing (whether string sizing or inverter over/under-loading). Additionally, lack of proper weather instrumentation and lack of long-term database of system performance in these new markets further hinder lack of knowledge distribution & scientific publications for this booming region of the world (both PV- as well as economically speaking). Trial and error become the norm.

With the tremendous growth of PV systems in the region and outstanding investment conditions (due to excellent irradiation rates combined with high conventional electricity costs), developers & financiers alike are heavily present, with capital injection widely available. Inverter loading decisions then start to venture beyond technical high-performing system decisions, rather towards techno-commercial choices of equipment physical limitations against their ability to return its investment in the most optimum way. Fig. 1 shows the AC power output for a PV system in Singapore on a typical irradiance day, with high cloud variability. After a certain level of irradiance, the inverter is not able to cope any further beyond its nameplate capacity in converting DC to AC power. The red dots indicate the potential AC power output had the PV section performed freely.

Therefore, this investigation intends to assess, considering real-world conditions, PV system power AC output behavior based on DC loading strategies and its consequences to returns on investment. Costing boundary conditions are to be considered when adjusting levels of loading in tropical and subtropical climate settings. Various tropical locations are aimed at being tested such as Singapore, Cambodia, and Thailand. Farther away from Southeast Asia's tropical climate, temperate system behavior is also tested with the participation of systems in South and North of India (Chennai & Delhi respectively).

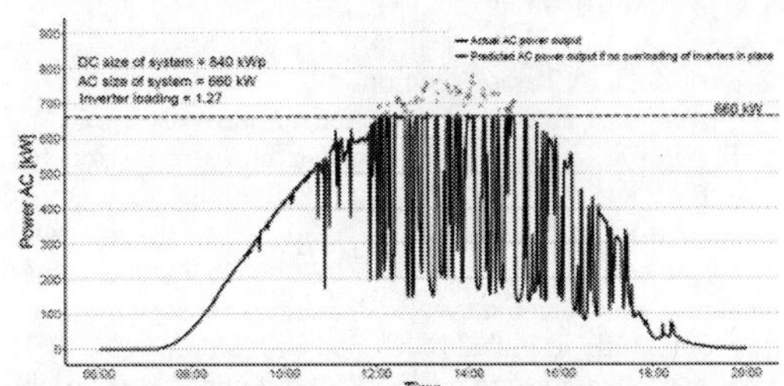

Figure 1: Minutely power AC output of a PV system with a 1.27 inverter DC loading on a typical day in Singapore with high cloud cover variability (blue continuous line). The "clipping" of the inverters can be visualized at peak sun-hours (12:00 to 15:00 pm). Red dots are a representation of the theoretical output of the PV system had inverter DC overloading not occurred.

2 MOTIVATION

This work aims at investigating the technical and commercial implications of inverter DC loading (the amount of DC power deployed at the PV array versus the

AC inverter nominal capacity [4]). Such design decisions are studied for photovoltaic (PV) systems in Asian locations, where tropical conditions are predominant. Indian PV systems are also part of the evaluation, allowing then a contrast of mostly monsoon-dominated locations against subtropical weather ones. Findings arising from this work could be used as a guide for developers and other stakeholders alike on the DC loading topic.

A massive number of Asian-based PV systems have been deployed in the past half-decade [1-3], catapulting this region of the world as the top location for PV systems globally. Further, most of these systems are present in China with a big portion simply translated into sizeable centralized facilities (also the case for up-and-coming solar farms in the Middle East). Yet, the growth of Commercial & Industrial (C&I) rooftop PV systems continues to take place. At those systems, where multiple roofs in a facility are present, string inverters are usually deployed. Inverter loading strategies in novel markets usually happen through trial-and-error or use of experience from mature markets, whose locations do not possess correspondingly equivalent climatic conditions.

With the proposed pool of systems in this investigation, which encompass rooftop installations tailored for the C&I space, this work aims to be an initial empirical analysis for Asian PV systems, with a decent geographic spread of the sites across the region. Although only a selected pool of systems is included, it is a good starting point in revealing intricacies of this topic for PV systems across in these new parts of the globe.

3 METHOD

The study sees the evaluation of the behavior of field PV systems deployed with certain DC loading ratios at various Asian geographies. Weather characteristics (here irradiation patterns) of the cities are investigated to assess the energy amount which takes place at different irradiance binning levels, a critical information when determining the amount of potential electricity being curtailed when inverters "clip" the excess energy due to DC input overloading.

3.1 Irradiance Characterization

Irradiance sensors (silicon sensors) are used for data collection at sites, both for weather characterization but also for PV system performance calculations. Accuracy of such sensors is ±2.0%. Data logger availability throughout the year at sites is greater than 99.0%. Irradiance values are used only when above 5 W/m^2 and sensors mounted predominantly at global horizontal position due to mostly shallow tilt angles found in PV systems closer to the equator [5]. For higher latitudes such as Delhi, sensors are mounted on the plane-of-array of the system under study. Additionally, sensors are cleaned regularly to avoid false readings as reported also in [5].

Fig 2. shows the annual frequency distribution (in %) of global horizontal irradiance values recorded at a PV site in Singapore, where 3.2% of irradiance values are greater than 1,000 W/m^2. This value (1,000 W/m^2) was chosen as it matches the Standard Test Conditions (STC). Additionally, most inverter "clipping" episodes studied took place near those levels. Such frequency distribution graphs are performed for all locations of the study.

The next step considers the amount of solar energy occurring in those levels of concern where financial loss would likely present itself. Fig. 3 shows such an annual irradiation distribution plot (in kWh/m^2), highlighting that once the sun's energy is considered, 8.8% of the generation potential quantum arises at inverter "clipping" instances for the example of Singapore here.

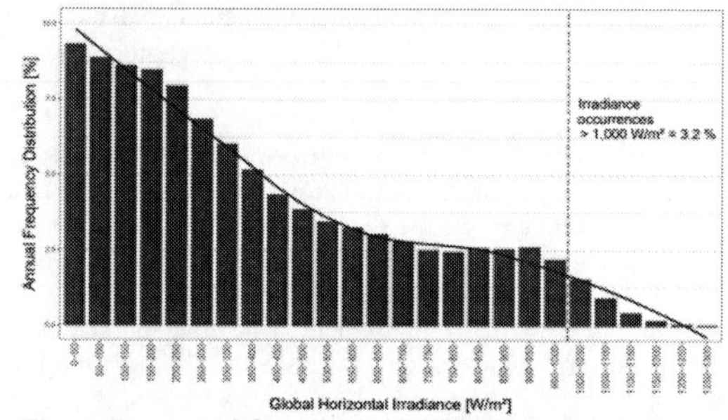

Figure 2: Annual frequency distribution (in %) of global horizontal irradiance values recorded at a PV site in Singapore, where 3.2% of occurrences fall in values greater than 1,000 W/m^2.

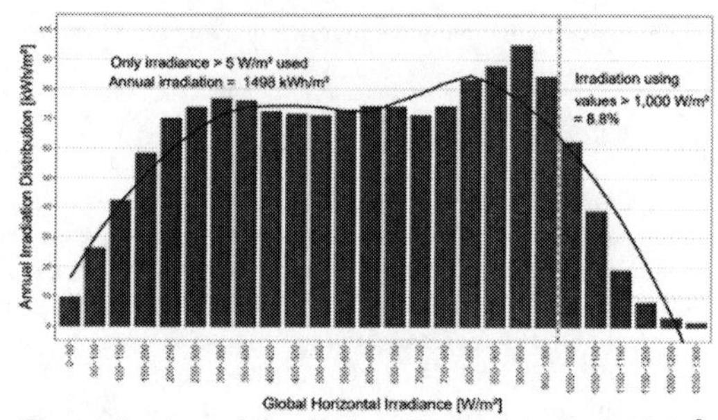

Figure 3: Annual irradiation distribution (in kWh/m^2) highlighting that once the sun's energy is considered, 8.8% of the energy occurs at values greater than 1,000 W/m^2.

3.2 Power AC Assessment

Revenue meters (±0.2% accuracy class) are used for AC power and energy parameters which are necessary for the analysis. The plotting of power AC output against irradiance allows the understanding of curtailment points and system behavior at various conditions. The estimation of loss of energy due to "clipping" is obtained via a method where the difference from the operating point of the inverter without any overloading is predicted, based on known system performance at sites. Secondary methods, such as for example comparing the performance of systems against known benchmark levels, are tested.

3.3 Commercial Analysis

As a final step of this investigation, a commercial analysis is executed, considering general market rates for inverters and solar electricity rates observed by developers alike in Asia. Such step is deemed necessary to close the gap on a thorough investigation, considering not only the technical aspects of levels of inverter "clipping" but also if such strategies make sense across Asian markets. An example of such extra step in the evaluation is the fact that PV modules degrade through time, thus the associated amount of "clipping" on a system is bound to reduce, as less available DC power is available for conversion as time passes. Such model levers are captured and do highlight the importance and necessity of a complete techno-commercial analysis when dealing with this topic.

4 RESULTS & DISCUSSIONS

4.1 Irradiance Characterization

Fig. 4 portrays the annual frequency distribution of global horizontal irradiance values at five Asian locations (Chennai and Delhi in India, Phnom Penh in Cambodia, Rayong in Thailand and Singapore). For Delhi, the values represent the plane-of-array irradiance at the PV system due to a much higher latitude. Phnom Penh and Singapore had higher occurrences of irradiance values above 1,000 W/m^2 (4.0% and 3.2% respectively) which preliminarily could indicate that PV systems in those locations having high DC-to-AC loading ratios would suffer "clipping" losses.

When analyzing the annual irradiation distribution, Fig. 5 highlights that high irradiation concentration (within levels of irradiance above 1,000 W/m^2) occurs for Phnom Penh (9.1% of annual energy levels) and Singapore (8.8%) as the top two locations with such inverter "clipping" potentials. Both Rayong, Thailand (3.1%) and Chennai, India (2.0%) had considerably lower loss potential activity. Coming last, Delhi in India contained only 0.7% of solar irradiation totals with the values of irradiance greater than 1,000 W/m^2. Such results are a strong indication that PV systems in Delhi can be set to high DC-to-AC rations (i.e. 1.3-1.35) without much fear of losses.

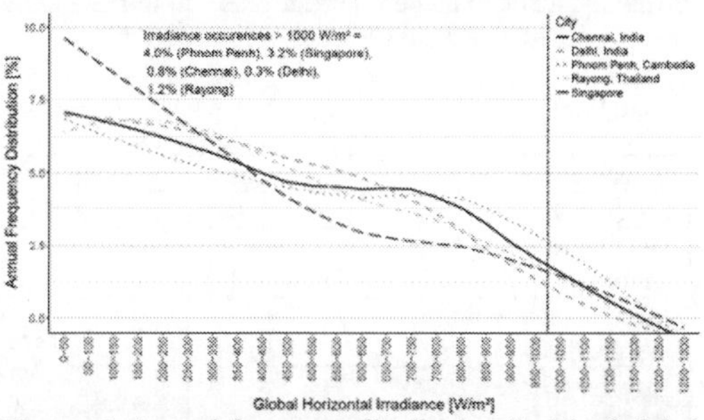

Figure 4: Annual frequency distribution (in %) of global horizontal irradiance values at five Asian locations (Chennai and Delhi in India, Phnom Penh in Cambodia, Rayong in Thailand and Singapore).

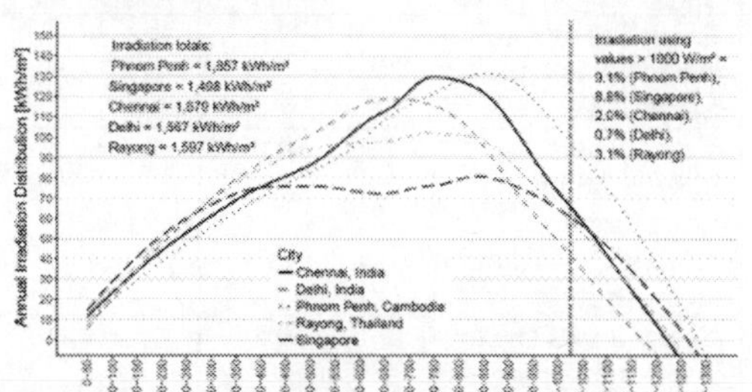

Figure 5: Annual irradiation distribution (in kWh/m^2) for the same four locations. It can be observed a high concentration of occurrences above 1,000 W/m^2 for Phnom Penh and Singapore, highlighting potential "clipping" challenges at tropical locations.

4.2 Power AC Assessment

Fig. 6 shows minutely normalized power AC scatter distribution for one month (Jul/17) for the PV system in Singapore versus global horizontal irradiance on site. The PV system tilt angle is nearly flat at this location, thus the sensor was chosen to be mounted at a global horizontal irradiation position. Scattering of some of the data points seen is attributed to the system rooftop size (here 190 x 65 m) against the position of the irradiance sensor (in middle of roof) and thus, varying cloud cover across the PV field. A linear fit is conducted as a method to estimate the amount of extra power AC which could have been produced had inverter "clipping" not occurred (here ~3.5% of losses for this particular month of the year). Taking all 12 months of this analysis into account, losses for the Singapore PV system amounted to ~4.0%.

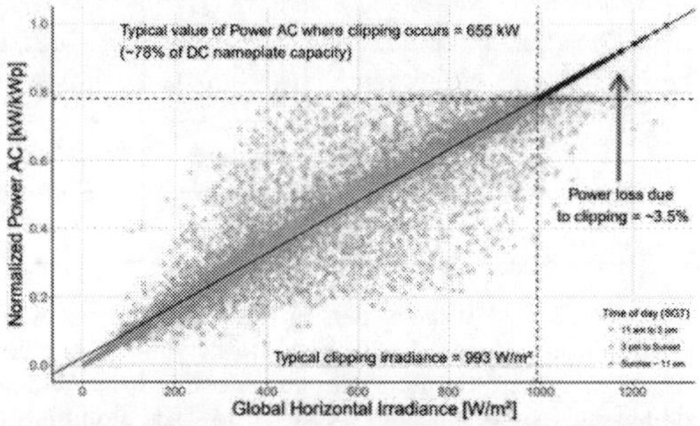

Figure 6: Minutely normalized power AC distribution for one month (Jul/17) for a PV system in Singapore versus global horizontal irradiance (the PV system tilted angle is nearly flat). A linear fit is conducted as a guide to the eyes and additionally as an alternative method to estimate the amount of extra power AC which could have been produced had inverter "clipping" not happened (here ~3.5% of losses for this month).

Another method of gauging potential "clipping" losses can be seen when plotting daily performance ratios against irradiation. In an average day (~4.10 kWh/m^2 of irradiation for Singapore, based on the annual irradiation on site), this PV system performs at ~83% which is circa 3-5% lower than state-of-the-art PV systems in Singapore [6]. This alternative method brings confidence to the primary methodology of losses proposed in this work.

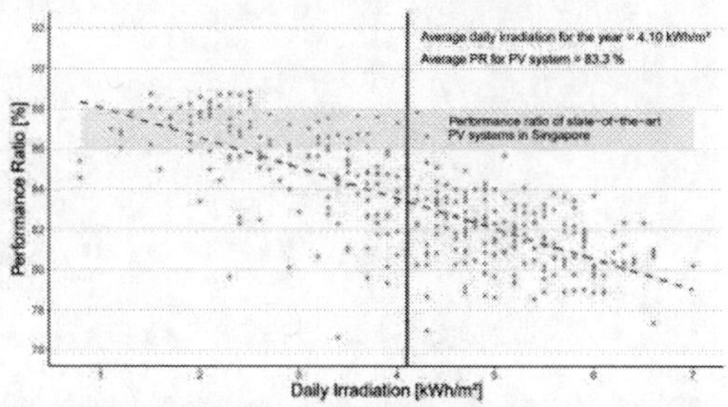

Figure 7: Another method for gauging potential "clipping" losses can be seen when plotting daily performance ratio against irradiation. In an average day (~4.10 kWh/m^2 of irradiation, based on a measured annual total), this PV system performs at ~83% which is usually 3-5% lower than state-of-the-art PV systems in Singapore can obtain [6].

Based on the proposed method and using annual data for the five locations under study, Table I showcases the annual "clipping" losses occurred. It is important to emphasize that the five PV systems of this study had good performance records for the entire year of the investigation, thus no major episodes, such as data loss, technical faults, grid faults and the like had to be filtered out from the analysis. All in all, relevant losses were detected for PV systems in Singapore and Phnom Penh (Cambodia) only.

Table I: DC-AC loading ratio, estimated annual electricity losses and tilt/azimuth for five PV systems in Asia.

Location/Latitude	DC-AC Ratio	Losses [%]	Tilt/ Azimuth
Tuas, Singapore/1°N	1.27	~4.0%	3/N-S
Phnom Penh, Cambodia/12°N	1.29	~1.5%	10/E-W
Rayong, Thailand/13°N	1.30	Negligible	18/N-S
Chennai, India/13°N	1.26	Negligible	12/N-S
Delhi, India/29°N	1.26	None	13/S

From Table I, it can be derived that the less occurrence of irradiance values > 1,000 W/m^2 is likely the reason the Thai PV system did not suffer considerable losses when compared to the Singapore system. In fact, cloud-edge effects, a typical condition found in Singapore, which can induce higher occurrences of 1,000 W/m^2-plus irradiance levels, is a good candidate to explain the higher losses for the Singapore PV system against other systems. A typical sky cloud cover condition is seen in Fig. 8, which helps accentuate irradiance spikes.

Figure 8: Sky camera snapshot for a typical cloud cover condition found in Singapore, with broken clouds all around and which happen virtually daily. Photo courtesy of the National Solar Repository of Singapore [7].

For the system in Phnom Penh, a likely reason for less losses when compared to the Singapore PV system could be linked to the fact that a nearly flat PV system in Singapore faces more situations of high irradiance than the tilted East-West system in Cambodia. This could be an area for further evaluation in future studies, which would require the use of more irradiance sensors at different tilt/azimuth configurations to collect real-world conditions for other combinations. The fact that the losses in Cambodia are smaller than in Singapore is a good indication that tilting the systems will avoid greater losses for shallow tilt angles. However, more data collection from other PV systems in the region is necessary for a definite conclusion.

4.3 Commercial Analysis

As three of the systems did not show any major losses of electricity (i.e. revenue) as per Table I, further analysis on the commercial aspect are not necessary for them. For those systems, the DC-AC loading strategy was successful and proved to be the correct design decision, giving developers/installers the confidence to operate with such inverter loading levels at those areas.

For demonstration purposes in this work, the case study for Singapore is further evaluated on a commercial basis. The following parameters and conditions as shown in Table II were run to determine the payback period in which it would have made sense to deploy more inverters to reduce the DC loading ratio of a PV system. In the simulation, the loading ratio was brought down from 1.27 (actual) to 1.13 (simulated) which then reduced the losses in the first year of system operation from ~4.0% to ~0.5%. A solar electricity rate is used assuming the simulation to be for a developer, thus under the "leasing" or "power purchase agreement" model. For the given set of values, it would take ~7.5 years for the increase on number of inverters on site to make financial sense, in relation to the reality of the PV system showcased in this study.

Table II: Commercial levers utilized for the cost-benefit calculation for a typical Singapore PV system.

Parameter	Value	Unit
Specific yield	1,250	kWh/kWp
Degradation rate (per annum)	0.8	%
Module (share of total system cost)	36	%
Inverter (share of total system cost)	8	%
Balance of systems (share of total system cost)	56	%
System size	1	MWp
System specific cost	0.98	USD/Wp
Solar electricity rate	0.066	USD/kWh
Escalation of solar electricity rate (per annum)	1	%
Year 1 losses due to inverter "clipping" (actual from study)	~4.0	%
Inverter DC-AC ratio (actual from study)	1.27	-
Year 1 losses due to inverter "clipping" (simulated for lower DC-AC ratio)	~0.5	%
Inverter DC-AC ratio (simulated)	1.13	-
Payback period	~7.5	years

Changing certain values of the levers in Table II do highlight areas of interest on this topic. Three examples follow:

a) A reduction in system cost by 10%, which has been the case for PV systems in Asia for the past years, would bring the payback period down by

~0.5 year, highlighting that as time passes and systems become cheaper, inverter loading strategies should be further revised;

b) Increasing yield by 5%, which could be the case on a certain part of Singapore with higher irradiance levels, brings payback period down by ~0.3 year;

c) Increasing the solar electricity rate by 10%, which is a strong aspect for consideration, given countries with rising conventional electricity prices (countries with high content of fossil-fuel-based generation), which face customary price swings due to oil price volatility, would bring the payback period down by ~0.5 year.

All in all, for the sake of a multi-parameter simulation, the deployment of a similar PV system with a cheaper cost (-10%), combined with the location being of higher irradiation resource (+5%) and with a higher solar electricity rate (+10%) would represent a payback period for the analysis of ~6.3 years.

For the PV system in Phnom Penh, Cambodia, the payback period ended up being greater than 10 years, which is higher than the theoretical lifetime of the inverters (commonly assumed in the business to be 10 years with business models of 20 years set such that a full inverter swap would be needed during the lifetime of the deal). Therefore, for the location Cambodia, with the given parameters chosen, the inverter loading strategy is successful.

Finally, if such payback periods do approach levels such as ~5 years, it becomes a strategic decision of developers/installation companies as having a higher system cost (i.e. deploying more inverters to avoid "clipping") would likely mean tougher chances of winning deals against competitors, who could be overloading inverters to reduce system cost and reduce the overall proposed system price to a final client (under direct sale model, "CAPEX"), or reducing the solar electricity rate (under the "leasing" or "OPEX" or "power purchase agreement" model). Such aspects highlight that the overall strategy of a company is linked to engineering as well as business decisions alike.

5 CONCLUSIONS

Inverter loading strategies implemented in mature markets (mostly temperate climates of the world) such as USA, Japan and Europe not necessarily yield the most cost-competitive strategies for new markets in Asia. Pitfalls do exist if lack of proper design is implemented, especially with extremely high DC-to-AC inverter loading ratios implemented. Characterizing annual irradiation data for various Asian locations allows better understanding of real-world conditions which a PV system will experience, coupled with the permissible thresholds businesses can withstand.

Empirical information from the field for five PV systems plus commercial inputs for a case study in Singapore were studied in this work. All in all, inverter loading strategies taken for the PV systems under investigations were proven to be successful, with areas of risk highlighted with regards to PV system business model parameter selection/conditions expected on a deal.

Falling levelized cost of electricity combined with high electricity rates in Asia represent a "perfect storm" for rapid deployment of PV systems. Such high volume of installations not always will possess the best information at hand at the design stage jeopardizing maximum returns to stakeholders, e.g. overloading of inverters to levels near 1.4, a common figure found in competitive markets.

With more deployment of PV systems in Asia happening, a region of the world with notably high irradiation rates, the availability of data will allow further investigations and conclusions to be made. Inclusion of novel markets such as the ones showcased in this study, plus new regions of Asia such as Malaysia, Vietnam, Myanmar, Indonesia, to name a few, will yield new data insights which will dictate the path ahead for mass deployment of renewables in the Asian continent.

ACKNOWLEDGEMENTS

The authors would like to acknowledge the support of other Cleantech Solar team members across multiple geographies: Monika Bieri (Singapore), Rohit Jaswal & Abhishek Puri (Delhi), Jeeva Govindarasu (Chennai), Rahul Patil (Pune), Lucas Ferrand & Visal Yang (Cambodia), Jeffrey Olan (Philippines) and Pichet Prasomsri (Thailand).

REFERENCES

[1] International Energy Agency (IEA). *Snapshot of Global Photovoltaic Markets*. Report IEA PVPS T1-33:2018, 16 pages.

[2] Mercom India. *India crosses yet another milestone: 25 GW of Solar Capacity*. https://mercomindia.com/india-25-gw-solar-capacity-installed/. 18th, September 2018.

[3] Energy Market Authority (EMA). *Installed Capacity of Grid-Connected Solar Photovoltaic (PV) Systems by User Type, 2008-2018*. https://www.ema.gov.sg/cmsmedia/Publications_and_Statistics/Statistics/47RSU.pdf. Accessed on 27th September 2018.

[4] V. Sheldon. *Optimizing array-to-inverter power ratio*. Issue 7,6, Oct/Nov 2014. SolarPro Magazine.

[5] A.M. Nobre, S. Karthik, C. Liu, R. Jaswal, R. Baker, R. Malhotra and A. Khor. *Irradiance measurement considerations for system performance assessment when managing fleets of photovoltaic assets across Asia*. 44th Photovoltaic Specialists Conference (PVSC), Washington DC, United States (2017).

[6] Y.S. Khoo, A. Nobre, R. Malhotra, D. Yang, R. Rüther, T. Reindl and A.G. Aberle. *Optimal Orientation and Tilt Angle for Maximizing in-Plane Solar Irradiation for PV Applications in Singapore*. IEEE Journal of Photovoltaics, Vol. 4, No. 2, March 2014.

[7] National Solar Repository of Singapore (NSR). *Sky Cameras*. http://www.solar-repository.sg/sky-cameras. Access on 24th June 2018.

PHOTOVOLTAIC SYSTEM EQUIPPED WITH FLAT REFLECTORS: NEW MPPT MODEL IN CASE OF NON-UNIFORM ILLUMINATION ON PV MODULES

Christine Abdel Nour[a,b,c], Anne Migan-Dubois[a], Jordi Badosa[c], Vincent Bourdin[d], Claude Marchand[a], Tilda Akiki[b]

[a] GeePs ; CNRS UMR 8507 ; CentraleSupélec ; Univ Paris-Sud ; Sorbonne Université ; 11 rue Joliot-Curie, Plateau de Moulon, F-91192 Gif-sur-Yvette Cedex, FRANCE
[b] Université Saint Esprit de Kaslik, USEK, Département de génie électrique et électronique, BP446 Jounieh, LEBANON
[c] LMD, Institut Pierre-Simon Laplace, CNRS, École Polytechnique, 91128 Palaiseau Cedex, FRANCE
[d] LIMSI, UPR 3251 CNRS, Bâtiment 508, Rue John von Neumann, 91405 Orsay Cedex, FRANCE

ABSTRACT: This study presents a new Maximum Power Point Tracking (MPPT) model developed with MATLAB/SIMULINK based on P&O algorithm. This MPPT model takes into account the activation of bypass diodes since it will be used for the simulation of photovoltaic (PV) power plants equipped with flat reflectors.
In order to validate the proposed simulation tool, we have compared our model to experimental measurements realized without and with flat reflectors. The resulting absolute error can be lower than 3.6% for single days either with or without reflectors under stable sky conditions, whereas during one full month (November 2017), for the installation without reflectors, the absolute error was about 4.76%.
In addition to these low errors found, the proposed MPPT model allows simulation for long periods of time, months to years, with a simulation time considerably fast.
Keywords: Photovoltaic modelling, MPPT modelling, Flat reflectors, Low concentration

1 INTRODUCTION

PV system's production can be increased by increasing the incident irradiance on the modules. This gain can be achieved by adding flat reflectors in front of the PV arrays. Recently, several research studies are conducted in the flat reflectors technologies showing the advantages and drawbacks of fixed and automated mirrors topologies [1], [7]. However, adding flat reflectors in front of the PV panel will add a non-uniform illumination over the PV array cells. Therefore, in this case it's crucial to study the irradiance mismatch in order to determine the efficiency of our system.

2 AIM AND APPROACH USED

We have shown in previous work that the energy yield of a photovoltaic (PV) power plant can be enhanced by almost 20% by installing flat reflectors in front of PV modules [1]. This paper focuses on the modelling of the Maximum Power Point Tracking (MPPT), using the toolbox SIMULINK of MATLAB. The MPPT model is based on a Perturb & Observe (P&O) algorithm [2] [3].
We compared models taken from the literature with the model that we developed. The simulation results are compared also with experimental data and error indicators are calculated during one month.
This paper presents first the experimental characterization platform for PV modules equipped with flat reflectors, then, the proposed model is compared to existing models and is validated with measurements realized without reflectors. At the end, the installation with flat reflectors is simulated and compared to measurements, this is the case of non-uniform illumination on PV modules. The main contribution of the model that we propose, is a large reduction of the computation time, for an error close to that of existing models. Moreover, this model is optimized for non-uniform illumination conditions and validated in both cases (uniform and non-uniform).

2 MODEL AND METHODOLOGY

2.1 Experimental characterization platform

The experimental platforms are installed at the SIRTA observatory [4] located in Palaiseau (France, 48.7N, 2.2E), on the campus of École Polytechnique. The SIRTA observatory is a reference meteorological and climate observatory with more than 150 remote sensing and in-situ instruments. In terms of radiometric measurements, the site is part of the Baseline Surface Radiation Network [5] since 2003.

An experimental PV platform was installed in 2014 and is composed of 5 commercial PV modules issued from different technologies as we can see on Fig. 1 [6]. In this paper, we only considered the PANASONIC N240 (2nd in Fig. 1 and see Table I). The purpose of this experimental installation was to form non-uniform illumination conditions in order to validate our modelling, with activated bypass diodes. It is not an optimized installation used to maximize the gain in electricity production.

Fig. 1: Characterization test bench equipped with flat reflectors.

Table I: PANASONIC N240 technical parameters

Electrical data (at STC)

Max. power (Pmax) [W]	240
Max. power voltage (Vmp) [V]	43.7
Max. power current (Imp) [A]	5.51
Open circuit voltage (Voc) [V]	52.4
Short circuit current (Isc) [A]	5.85
Max. over current rating [A]	15
Production tolerance power [%]	+10/-5*
Max. system voltage [V]	1000

Note: Standard Test Conditions: Air mass 1.5; Irradiance = 1000W/m²; cell temp. 25°C

Temperature characteristics

Temperature (NOCT) [°C]	44.0
Temp. coefficient of Pmax [%/°C]	-0.29
Temp. coefficient of Voc [V/°C]	-0.131
Temp. coefficient of Isc [mA/°C]	1.76

At NOCT

Max. power (Pmax) [W]	182
Max. power voltage (Vmp) [V]	41.1
Max. power current (Imp) [A]	4.44
Open circuit voltage (Voc) [V]	49.4
Short circuit current (Isc) [A]	4.71

Note: Nominal Operating Cell Temp.: Air mass 1.5 spectrum; Irradiance = 800W/m²; Air temperature 20°C; wind speed 1 m/s

At low irradiance

Max. power (Pmax) [W]	45.9
Max. power voltage (Vmp) [V]	41.7
Max. power current (Imp) [A]	1.10
Open circuit voltage (Voc) [V]	49.0
Short circuit current (Isc) [A]	1.17

Note: Low irradiance: Air mass 1.5 spectrum; Irradiance = 200W/m²; cell temp. = 25°C

The current-voltage characteristics are measured with CHROMA DC electronic loads, each minute from sunrise to sunset. The maximum power point is derived from these characteristics. PV module temperature is measured with 4-wired class A platinum sensors (Pt100) and global irradiance in the plane of array is measured with a solar radiometer (Hukseflux SR01) and reference PV cells (SOLEMS RG100) (Fig.2). An average during each 10 minutes is calculated.

Fig. 2: Experimental platform equipped with sensors

2.2 Evaluation indicators

In order to compare measurements to simulations, we calculated error indicators. These indicators are absolute and relative mean bias error (MBE and RMBE), and absolute and relative mean absolute error (MAE and RMAE), are as defined by the following equations:

$$MBE = \frac{\sum_{i=1}^{N}[P_{max,meas}(i) - P_{max,model}(i)]}{N}$$

$$MAE = \frac{\sum_{i=1}^{N}|P_{max,meas}(i) - P_{max,model}(i)|}{N}$$

$$RMBE = \frac{\sum_{i=1}^{N}[P_{max,meas}(i) - P_{max,model}(i)]}{\sum_{i=1}^{N}[P_{max,meas}(i)]}$$

$$RMAE = \frac{\sum_{i=1}^{N}|P_{max,meas}(i) - P_{max,model}(i)|}{\sum_{i=1}^{N}|P_{max,meas}(i)|}$$

3 MPPT Model

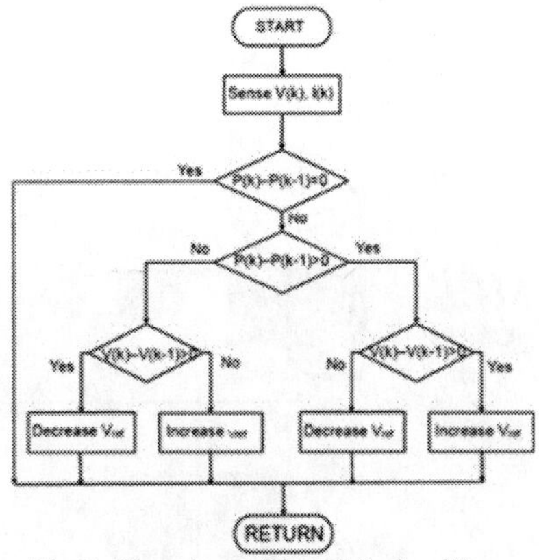

Fig. 3: Flowchart of the P&O algorithm.

The MPPT model proposed in this paper relies on P&O algorithm (Fig. 3 and [2] [3]).

At the first time step, the operating voltage of the PV module is fixed to 80% of its open-circuit voltage given by the manufacturer (Table I). Then, the voltage increases slightly by a constant step. If the PV output power increases, then the algorithm continues to increase the voltage, otherwise, the voltage decreases (always of the same voltage step). The output condition of the algorithm is reached when the variation of power from one voltage step to another is less than a given threshold.

3.1 MPPT Model Description

The proposed MPPT algorithm has been optimized for non-uniform illumination conditions since its aim is to simulate a PV power plant equipped with flat reflectors. Thus, it can exit from a local maximum when bypass diodes are activated in order to converge to the real maximum operating power point.

The presented model is able to simulate, at least, one year of 10-minutes points with a significantly reduced simulation time. The existing MPPT methods can be listed as follow: voltage feedback, P&O, linear approximation, incremental conductance, fuzzy logic control and so on. Our system is based on P&O concept introducing small perturbations in order to vary the PV module's power. Thus, the direction of the perturbation (increased or decreased) will be specified by the power variation (increasing or decreasing). This method is simple to implement compared to the other cited methods and operates well under normally varying atmospheric conditions.

3.2 Simulation

The PV module considered in this study is composed of 3 strings in series and each string is composed of 24 cells in series. Each string is connected to an anti-parallel bypass diode. In order to simulate the non-uniform illumination arriving on the PV module, we model it by 3 strings with their bypass diode, as presented in Fig. 4. In our Simulink model, the photovoltaic array block is based on an electric modelling from the NREL System Advisory Model Database [8].

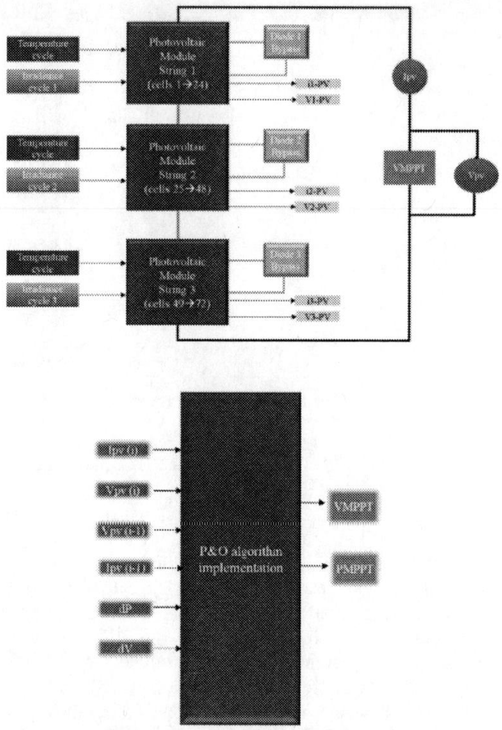

Fig. 4: Block-diagram representation of the model that takes into account a non-uniform illumination.

3.3 Flat reflectors modelling

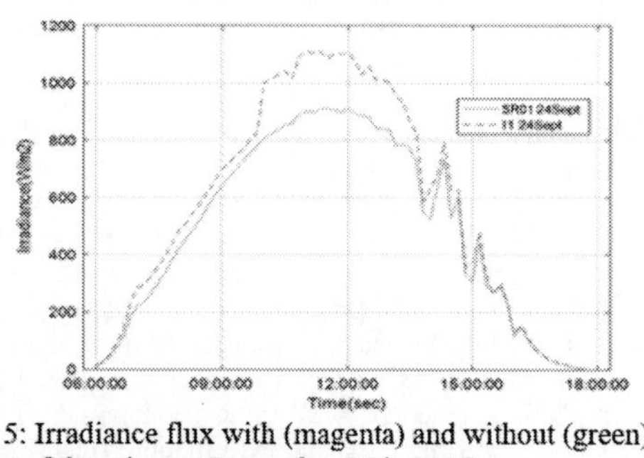

Fig. 5: Irradiance flux with (magenta) and without (green) effect of the mirrors; September 24th, 2017

Fig. 5 illustrates the irradiance fluxes measured by 2 different sensors: where I1 sensor is receiving the mirror's irradiance and SR01 is not.

Non-uniform illumination is modelled in our system by different plane of array irradiances measured on the strings levels (Fig. 4).

The flat reflectors model is just introduced in our system by the increased irradiance coming from the mirrors.

4 PRELIMINARY RESULTS

4.1 Validation under uniform illumination

In order to validate the proposed MPPT modelling, we have plot in Fig. 6a and Fig. 6b, the maximum power point measured and simulated during typical days of the considered period with different weather conditions. There were no reflectors in front of PV modules in the installation shown in Fig. 1.

In the early morning, the simulation deviates from the measurement because there are buildings that cast shadows on PV modules, but not on the irradiance sensor. Error indicators are quite satisfying with a RMAE of 1.5% and 3% for these 2 particular days.

A simulation performed during the whole month of November 2017 with a time step of 10 minutes takes 5 minutes and the error indicators between measurements and simulation are MBE = -0.38W; RMBE = -0.8%; MAE = 2.24W and RMAE = 4.7% (Fig.7) . The simulation time is 5 minutes.

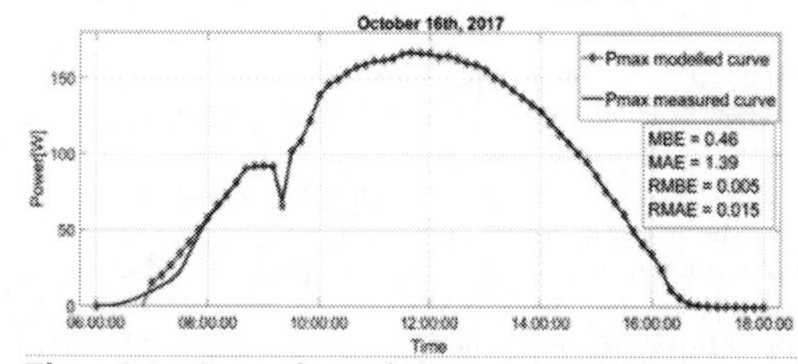

Fig. 6a: Comparison between measurements and simulations, October 16th 2017, installation without reflector.

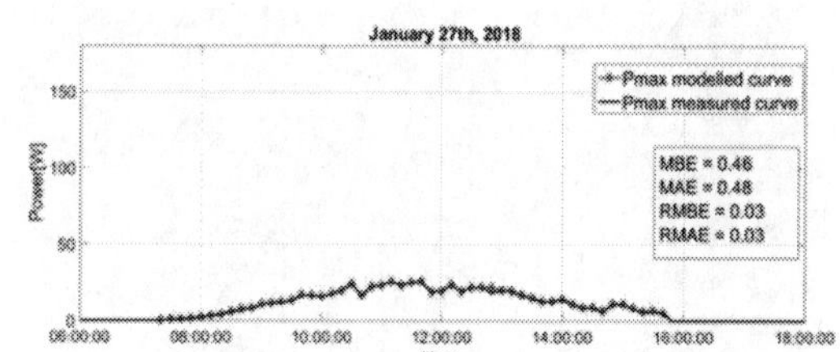

Fig. 6b: Comparison between measurements and simulations, January 27th 2018, installation without reflector.

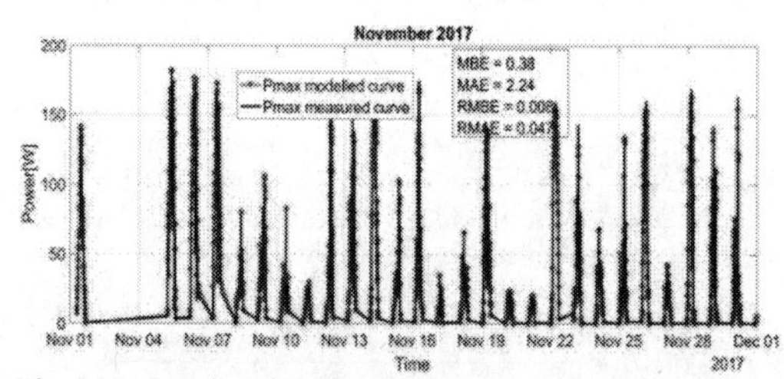

Fig. 7: Comparison between measurements and simulations, November 2017, installation without reflector.

Table II: Error indicators over 10 months, from November 2017 until August 2018

Month	RMBE(%)	RMAE(%)
Nov-17	0.81	4.76
Dec-17	2.33	7.15
Jan-18	3.06	5.95
Feb-18	1.57	5.87
Mar-18	0.75	4.79
Apr-18	1.42	5.13
May-18	4.55	5.15
Jun-18	7.48	7.91
Jul-18	10.83	11.22
Aug-18	8.36	8.86

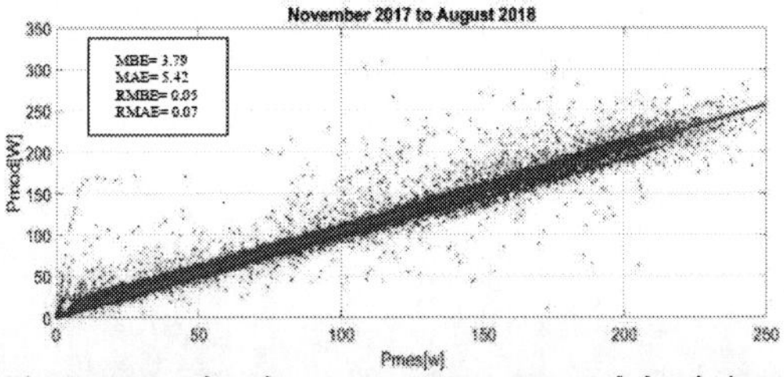

Fig. 8: Comparison between measurements and simulations, for the period: November 2017 to August 2018, installation without reflector.

Fig. 8 and Table II show the experimental validation results over a period of 10 months. As we can notice, the relative mean bias and relative mean absolute errors are of an average of 2.07% and 5.54% respectively for the period of 7 months (Nov-17 to May-18). On the other hand, we obtained an increased error during the summer period (Jun-18 to Aug-18).
This is probably due to the response of our model to high temperature values that needs to be optimized.

3.4 Validation under non-uniform illumination

We tested our MPPT method under non uniform illumination, as illustrated in Fig. 9 and 10.
Fig. 9 and 10 show that the proposed MPPT model works quite well, whether there are 2 (10:40), 1 (11:20) or no (14:27) bypass diodes activated.

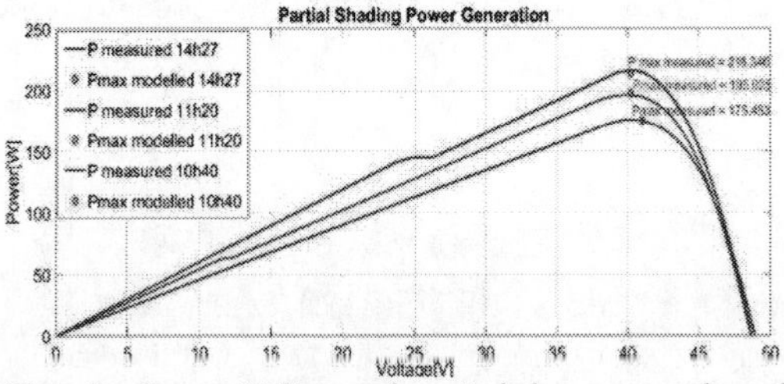

Fig. 9: Power-Voltage characteristic measured and maximum power point simulated, installation with flat reflectors.

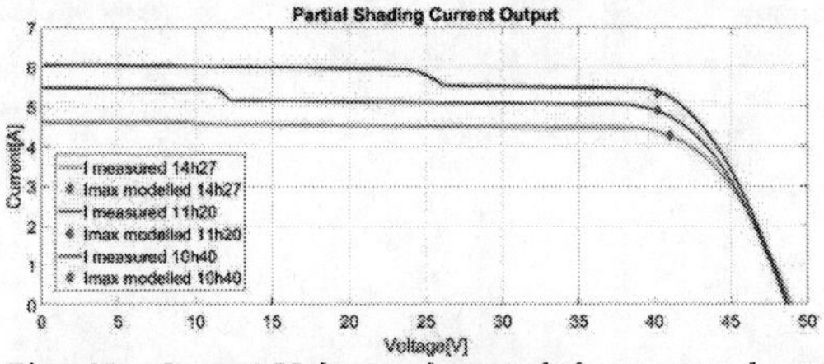

Fig. 10: Current-Voltage characteristic measured and simulated, installation with flat reflectors.

Fig. 11: Photo illustrating the non-uniform illumination on 2 PV modules. The same happens for the PANASONIC.

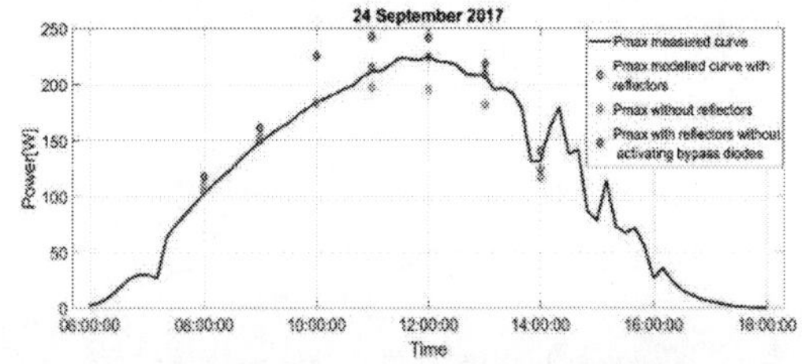

Fig. 12: Comparison between measurements and simulations for September 24th, 2017, installation with reflectors.

In this case, since the PV module will be exposed to non-uniform illumination (Fig.11), we will experience irradiation mismatch levels across the panel. Thus, that mismatch is illustrated by the activation of bypass diodes in Fig. 9.

Fig. 12 shows the comparison between maximum power point measured (blue curve) and simulated with bypass diodes activation (magenta curve) for the whole day of September 24th, 2017. In this figure, we have also indicated, for information, the maximum power point simulated without flat reflectors (green curve) as well as the power simulated with reflectors and without bypass diodes activation (red curve) indicating the over-estimation of power in this case. The error indicators between measurements and simulation are: MBE = 0.07W; RMBE = 0.04%; MAE = 2.32W and RMAE = 1.35% (Fig.12). As we can observe in the figure, ignoring bypass diodes activation will add an over estimation of the power output. This over estimation is around 11.2% for the 24th of September 2017. The power output gain added by the reflectors for the same day is about 6.4%, but as we have said before, the reflectors are installed to form non-uniform illumination and not optimized for the maximum gain yet.

CONCLUSION AND OUTLOOKS

This study presents a new MPPT algorithm developed with MATLAB/SIMULINK that takes into account the activation of bypass diodes. This model is used for the simulation of PV power plants equipped with flat reflectors. In order to validate the simulation tool, we have compared our algorithm to existing models, on one hand, and to measurements realized without and with flat reflectors, on the other hand.
The simulation results are quite similar to those of existing models with a faster simulation time. The error indicators are low considering the previous results. Finally, this model will be coupled to a storage and distribution system. The simulation and experimental results for different geometrical configurations of PV-Reflectors systems will be published in a future work.

REFERENCES

[1] Pavlov M., doi: 10.1109/JPHOTOV.2015.2478057
[2] O.Ezinwannea, doi.org/10.1016/j.egypro.2016.12.156
[3] I.W Christofer, http://www.ajer.org/papers/v2 (12) T.C
[4] Haeffelin M., doi: 10.5194/angeo-23-253-2005
[5] BSRN, http://bsrn.awi.de

[6] Badosa J., doi: 10.4229/EUPVSEC20152015-5BV.2.23
[7] Belhadj, Mohammed. (2018). Analyse et Conception d'un Système autonome de captage solaire hybride (photovoltaïque/thermique) à Concentration.
[8] P. Gilman. SAM Photovoltaic Model Technical Reference NREL/TP-6A20-64102 May 2015. National Renewable Energy Laboratory

YIELD SIMULATIONS FOR HORIZONTAL AXIS TRACKERS WITH BIFACIAL PV MODULES IN PVSYST

André Mermoud, Bruno Wittmer
PVsyst SA
Route du Bois-de-Bay 107, 1242 Satigny, Switzerland

ABSTRACT: The simulation of the energy yield of Bifacial PV modules is not straightforward, since most of the light reaching the back side of the PV modules is scattered back from the ground. For tracking systems this is particularly challenging since the geometry is changing as the sun moves through the sky. In PVsyst, a simplified 2D model was introduced to describe bifacial horizontal single axis trackers with regular spacing. The approach uses view factors to model the fraction of light that is scattered back to the back side of the PV modules. The bifacial calculation includes direct and sky diffuse contributions on the back side, as well as ground scattering to the front and back side of the modules.
Keywords: Bifacial, Tracking, PV System, Modelling, Simulation

1 INTRODUCTION

Bifacial PV modules are being used more and more commonly in PV installations. They increase the energy yield by collecting additional light that reaches the back side of the PV modules. Over the past years, many publications have calculated and measured the bifacial gain of PV modules for particular conditions, giving a wide range of results [1, 2, 3]. The expected increase in energy yield depends strongly on many parameters, like ground reflection properties, module orientation, mounting height, number and distance of surrounding modules, geographical location, climate, etc. Therefore, it cannot be predicted in a general way with simple estimations. A full simulation is required, to get a realistic value for the bifacial gain. The simulation of the energy yield of bifacial PV modules is not straightforward, since most of the light reaching the back side of the PV modules is scattered back from the ground. This is not part of a typical PV simulation, and additional steps have to be introduced to model this behavior. For tracking systems this is particularly challenging, since the orientation of the PV modules and the ground illumination are permanently changing as the sun moves through the sky.

2 BIFACIAL SIMULATION

In PVsyst, a model describing bifacial modules for fixed tilt installations with regular rows was introduced in early 2017, in the Version 6.6.0 [4]. The approach first calculates the irradiance reaching the ground, and then uses view factors to model the fraction of light that is scattered back to the rear side of the PV modules. Similar models have also been described elsewhere [3, 5, 6, 7]. The bifacial calculation of PVsyst includes ground scattering to the front and back side of the modules, as well as direct and sky diffuse contributions on the back side. The current simplified model assumes that the rows of PV modules have all the same orientation and are spaced equally. In this way, the model can be fully described in a two-dimensional cross section of the rows, as shown in figure 1. This approximation is well suited for long regular rows as they occur in large scale ground mounted PV installations, or on flat rooftop commercial installations. It cannot be applied directly to small experimental bifacial PV systems.

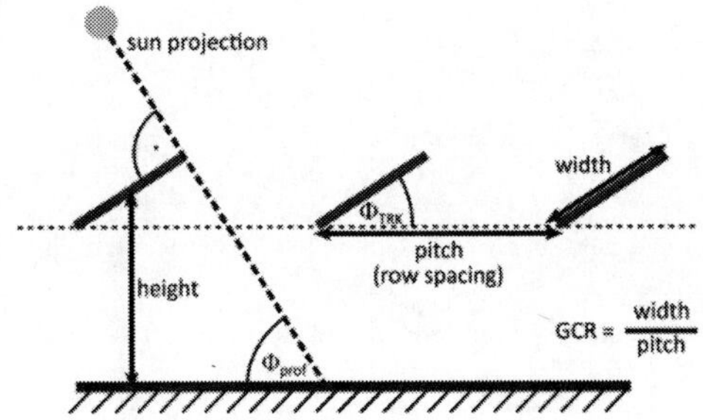

Figure 1: Two-dimensional model used for calculating the irradiance that is scattered back from the ground.

2.1 View Factor Model

The modeling of the bifacial PV modules proceeds in several steps, that are shown in figure 2. The front side irradiance is calculated like in a normal monofacial simulation. The rear side irradiance is composed of the part arriving directly from the sky (direct and diffuse), and the part that is scattered back from the ground. The direct sky contribution, which is only present in fixed tilt rows and does not occur with trackers, follows from straightforward geometrical considerations. The diffuse sky contribution on the back side is calculated as average over three points of the rear side, considering the different opening angles towards the sky. For the part that is scattered back from the ground, it is necessary to first determine the irradiance distribution on the ground. The part coming from the direct sunlight can be easily computed from geometrical considerations. The result are homogeneous stripes of irradiance that run parallel to the rows. The diffuse contribution at each point on the ground is an integral over the visible part of the sky, weighted with the cosine of the incidence angle. The sum of these two contributions gives the total irradiance distribution as function of the ground position. The scattering off the ground surface is assumed to follow the Lambertian law, meaning that the surface has the same apparent brightness from all viewing angles. The fraction of the irradiance that is scattered back is described by the ground albedo factor, which ranges typically between 0.2 and 0.6. For each point at the ground there is a view factor, describing how much of the scattered irradiance reaches the rear side of a row. This view factor is calculated as the integral over all scattering angles for which the rear side is visible, weighted with incidence angle modifier function (IAM). To get the contribution of the entire ground, the integral over all ground points along one row spacing is performed. When the result is normalized to the row width, we get the irradiance in W/m^2 for the additional irradiance reaching the rear side of the PV modules. This is an average value, since the integration was performed over the entire row width, which has a non-uniform irradiance distribution. The rear side irradiance contains inter-row shadings for direct and diffuse contributions, but it does not include shadings coming from obstruction very close to the rear side, like mounting structures or junction boxes. This shading is accounted for in the 'rear shading' factor, which has to be estimated by the user.

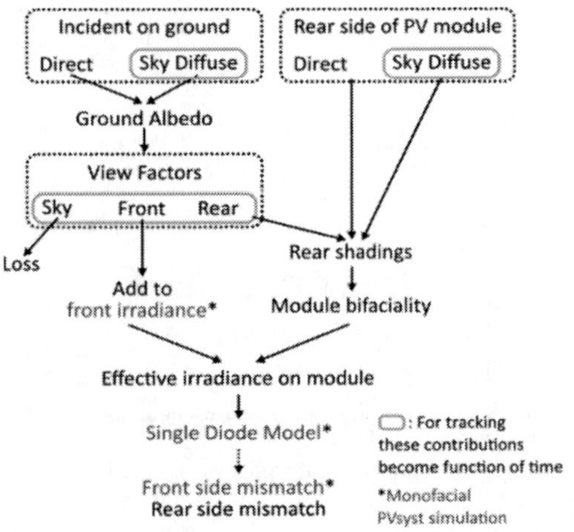

Figure 2: Simulation steps for the bifacial model. The blue steps are the interface to a normal non-bifacial simulation. The green steps are the contributions that are constant with fixed orientations and become functions of time with tracking installations.

Once the rear side irradiance has been calculated, the bifaciality factor of the PV module is applied to it. This factor ranges typically between 60% and 96%, and it describes the reduced efficiency of the module rear side. The reduced rear side irradiance is then added to the front side irradiance and used as input for the single diode model that describes the PV conversion in PVsyst. This approach results in an average value for the PV power, which neglects the mismatch arising from the non-uniform irradiance distribution. Since this non-uniformity is much more pronounced on the rear side of the modules, an additional parameter called 'rear side mismatch' is multiplied to the PV module power contribution from the rear side.

2.2 Generalization for Trackers

The model used in PVsyst to simulate bifacial PV modules has been generalized, so that it can also be applied to single axis trackers [8]. Like for the rows before, it is also assumed that the trackers have a regular orientation, width and spacing. Therefore, the two-dimensional model described in the previous paragraph, has to be changed to allow for a changing tilt. This means, that certain variables that were constant for fixed sheds, become functions of the sun position and thus of date and time. This is shown as green colored contributions in figure 2. It starts with the irradiance reaching the ground, where now both direct and diffuse contributions will depend on the tracker angle Φ_{Trk}. The view factors also need to become a function of Φ_{Trk}, as well as the diffuse contribution on the back side. The tracking algorithm in the simulations is always minimizing the angle of incidence for direct sunlight on the front side of the PV module. It does not aim to maximize irradiance for cloudy sky conditions, where smaller tracking angles might be more favorable. There is also a backtracking mode, that avoids mutual row shadings by tilting back to smaller angles when the sun is low in the sky.

2.3 Simulation Results

When performing a bifacial simulation, there are additional result variables. These include the irradiance on ground, the scattering losses, the view factor losses, direct and diffuse sky irradiance on the rear side, rear side shadings and bifacial mismatch.

After performing the bifacial simulation, PVsyst stores all the losses and intermediate results as hourly values. A summary of these values can be read off the final report that is generated after a simulation run. An example of a loss diagram for a bifacial tracking simulation is shown in figure 3. There is an additional branch on the right side of the diagram, describing all the contributions that lead to the additional irradiance on the rear side. It is important to note, that the irradiance on ground and scattering losses are normalized to the ground surface, while the view factor and subsequent losses are normalized to the PV module surface.

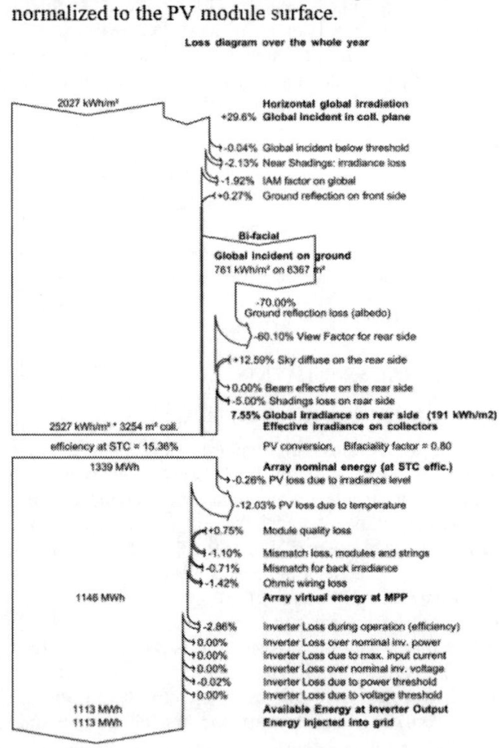

Figure 3: Loss diagram for a simulation with bifacial single axis trackers.

3 PARAMETRIC STUDIES

The PVsyst software is not only designed to simulate single cases of PV installations, but allows also to perform more general studies to understand the behavior of PV installations. It is possible to store hourly values of the simulations in text files, allowing a detailed custom analysis of the results. Furthermore, there are tools to perform parametric studies by running many simulations in an automated way, while changing sets of simulation input parameters. In the following we will present some results obtained with this kind of parametric scans. The parameters that were varied are the row spacing (pitch), the mounting height, the latitude and the climate. The parametric studies will give some insight on the general behavior of tracking and bifacial PV systems. This can be useful to understand under which conditions significant benefit can arise from these two technologies.

3.1 Gain factors

To study the potential gain of bifacial PV modules for tracking systems, it is useful to compare it to the gain obtained by passing from a fixed orientation to a tracking system. In this paper, we focus on the gain in irradiance in the collector plane. This includes row-to row shadings and reflections on the PV module surface. It does not consider any losses in the PV conversion step and all subsequent stages of power generation. This means that also the bifaciality factor of the PV modules and potential mismatch due to non-uniform illumination, is not treated here. This choice was made to get the results independent of any specific bifacial PV technology and electrical design details.

PV systems with bifacial trackers will boost the energy production by both increasing the front side irradiance by tracking the sun, and by collecting an additional irradiance contribution on the rear side of the modules. To make it clearer, how the total irradiance gain is related to tracking or bifacial modules, we consider four different scenarios as shown in figure 4. The basic reference is a fixed tilt non-tracking PV installation, which can be enhanced either by adding trackers, or by moving to bifacial PV modules. This is shown by the light green and orange arrows. The resulting two scenarios can both be converted into bifacial tracking installations, by switching to bifacial PV modules or adding trackers respectively. This gives raise to four different gain factors as depicted in the figure, named BG for bifacial gain and TG for tracker gain. The four quantities to be considered are then:

$BG_{irr}FT$: Bifacial gain for fixed tilt system
$BG_{irr}TR$: Bifacial gain for a tracking system
$TG_{irr}MF$: Tracker gain for a monofacial PV installation
$TG_{irr}BF$: Tracker gain for a bifacial PV installation

From these definitions it follows, that

$BG_{irr}FT \times TG_{irr}BF = TG_{irr}MF \times BG_{irr}TR$,

assuming that all other properties, like ground covering ratio or height, stay the same.

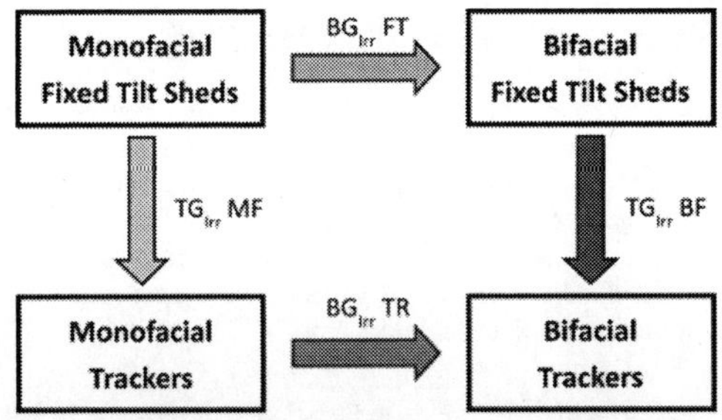

Figure 4: Gain factors for irradiance. Green arrows denote transitions from fixed orientation to tracking. Orange arrows denote transitions from monofacial to bifacial PV modules.

3.2 Geometric parameters

In a first analysis, we look at the different gain factors, when changing the row spacing (pitch) and mounting height. The system for this study is located in Albuquerque, USA, the weather data was generated by the Meteonorm 7.1 software, that is built into PVsyst. It yields hourly values that describe a typical year, and is based on satellite and ground data. The row width and pitch were chosen as 3m and 6.6m respectively, giving a ground covering ratio of 45%. The mounting height of 3m was chosen rather large, to maximize the bifacial contribution. For the ground scattering, an albedo value of 30% was chosen, which corresponds to a reasonably well reflecting surface. Higher albedo values can be reached when specially preparing the ground surface by painting or spreading bright white gravel.

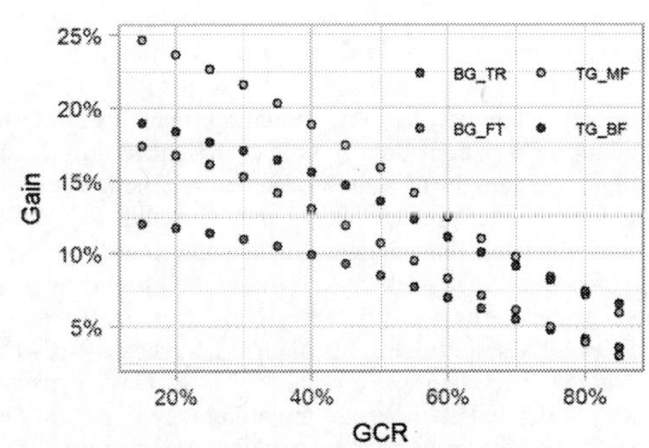

Figure 5: bifacial gain as function of ground covering ratio GCR for the different gain factors.

Ground covering ratio

To express the pitch in a general unit-free way, we use the ground covering ratio GCR, which is defined as the ratio between PV module surface and the total land surface of the installation. For regular row spacing and constant row width, this can be approximated by the ratio of row width over pitch: GCR = row width / pitch. Typical values for GCR are in the order of 40-60%. In the following study the GCR was varied from 10-90% by changing the pitch and keeping the width constant. For the fixed tilt scenarios, the best tilt was determined for each pitch value. The results of the GCR variation are shown in figure 5. As expected, a denser packing in the PV system leads to a decrease of bifacial and tracker gain. It can also be seen, that for this location, configuration and albedo value, the tracking gain always exceeds the bifacial gain. The largest increase in yield is achieved, when switching from a fixed orientation with monofacial PV modules to a tracking system. The bifacial gain in a tracking system is lower than in a fixed tilt system, because the tracker strongly boosts the front side irradiance, thus reducing the impact of the rear side contribution. A more detailed picture of the bifacial gain for different conditions of diffuse and direct irradiance will be presented in section 3.5.

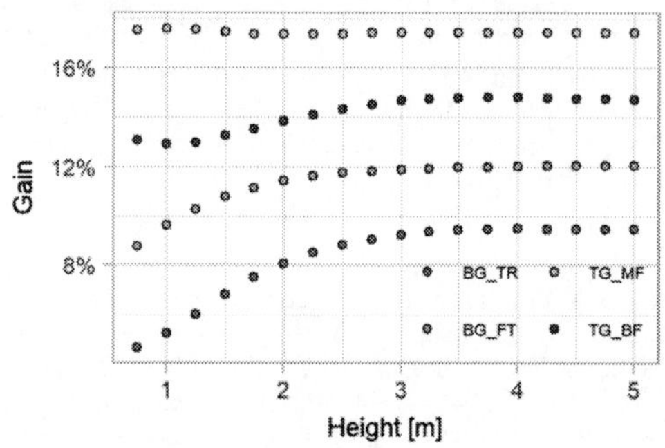

Figure 6: Bifacial gain as function of mounting height for the different gain factors.

Height over ground

The mounting height is a crucial parameter for bifacial PV systems. If the PV modules are mounted very low, the ground close to the back side of the PV modules will be shaded. At the same time, this is the region that has the most favorable view angle towards the rear side. Similarly, the ground region close to the front side of the modules will get most of the irradiance, but it has a very narrow view angle towards the module's rear side. This imbalance leads to a significant scattering loss towards the sky, and it will get attenuated as the mounting height increases. Although the pattern from the direct irradiance on ground will not change, the diffuse sky irradiance passing between the rows will be spread out more evenly, leading to a more uniform ground illumination. At the same time, the view angles towards the back side of the modules, will tend to become the same for different ground regions. Together this leads to an increase in bifacial gain, as shown in figure 6. As expected, the tracking gain for monofacial systems does not depend at all on the mounting height. Again, the benefits coming from having a tracking system, exceed the bifacial gain.

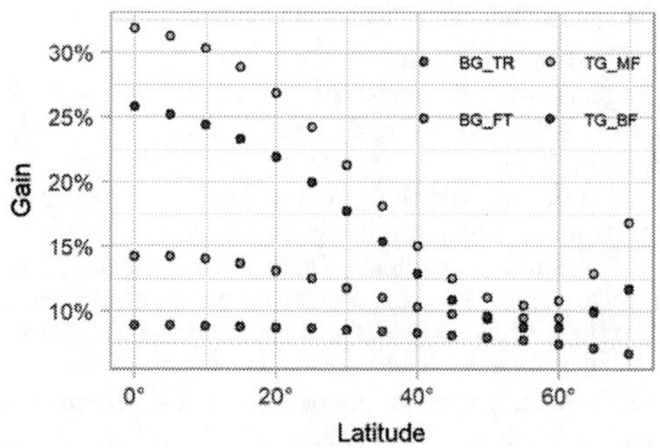

Figure 7: Bifacial gain as function of latitude, for clear sky conditions. For each latitude, the orientation of the fixed tilt scenarios was optimized.

3.3 Latitude

Next, we study how the tracker and bifacial gain is impacted by the latitude. To remove any impact coming from altitude, artificial locations were generated, all placed at sea level and latitudes ranging from 0° at the equator up to 70° northern latitude. The weather data used for this study is based on a clear sky model. This does not describe a realistic climate, but it makes the different latitude results comparable. The results are shown in figure 7. The highest gain factors can be expected at the equator and low latitudes. Since for these latitudes the horizontal tracking system is particularly well suited, the tracker gains highly exceed the expected bifacial gains. One can also see, that equipping a tracking installation with bifacial modules leads to a rather small gain. This is largely due to the fact that only clear sky conditions have been used, which have a smaller bifacial gain than conditions with a strong diffuse contribution. This will become clearer in the following section.

Site	**Sharorah**	**Atacama**	**Stockholm**	**Kuala Lumpur**
Latitude	17.5°N	23.42°S	59.35°N	3.12°N
Diff/Glob Irr.	26.1%	28.6%	49.5%	58.8%
PoA Irr	2999	2889	1225	1753
Ground Irr.	1059	1008	435	804
Rear Irr.	286	276	137	236
$BG_{Irr}TR$	**9.5%**	**9.5%**	**11.2%**	**13.5%**

Table I: Results for different geographical locations with distinct climates. Irradiances are giwen in W/m^2.

3.5 Climate

The benefit from a bifacial PV module is different for direct and diffuse light. Therefore, the ratio between cloudy and clear sky conditions for a given climate will impact the total bifacial gain that can be expected over one year. The ground illumination conditions are different for direct and diffuse light. The direct light creates sharp-edged shadows on the ground, giving rise to regular stripes alternating between full irradiance and complete shadow. On the other hand, the diffuse irradiance from the sky will create a blurry pattern of illuminated ground. For fixed tilt systems, this pattern is constant, while for tracking systems, the tilt changes with the sun position, and the pattern becomes a function of time. The different ground illumination patterns, cause also the amount of light that is scattered back to the rear side of the module, to be different for direct and diffuse light. The impact of this difference for different climate conditions is summarized in table I. Four different locations were compared:

Stockholm: Cold climate at high latitude with average diffuse content.

Sharorah: Dry and hot desert climate with small diffuse content

Atacama Desert: Dry and cool climate with small diffuse content

Kuala Lumpur: Tropical climate with a high diffuse content

The table shows the bifacial gain for tracking installations $BG_{Irr}TR$. The diffuse contents for the entire year ranges from 26% for Sharorah in Saudi Arabia up to almost 60% for Kuala Lumpur. The locations with higher diffuse content have a larger bifacial tracker gain. The reason for this behavior can be seen in figure 9.

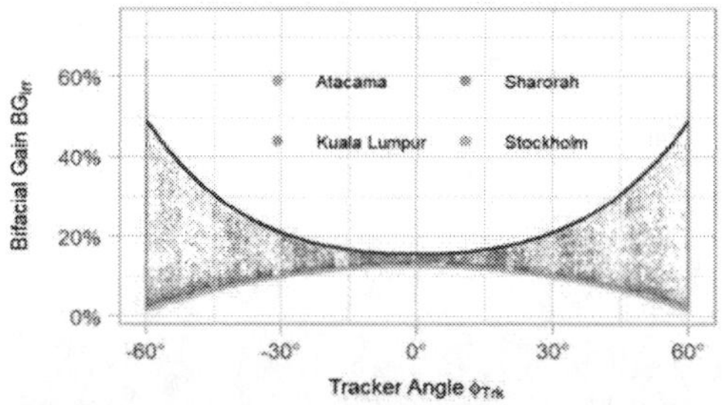

Figure 9: Distribution of bifacial gain factor for tracking systems. The blue and yellow curves are approximate, and represent the bifacial gain factor for the diffuse and direct component only.

In this figure, the hourly bifacial tracker gain is plotted against the tracker angle Φ_{Trk}. A tracker angle of 0° means that the sun is in the zenith, and the PV modules are horizontal. The different colors of the points code the four different locations. As can be seen, the points are all distributed between the blue and the yellow curve, which represent the bifacial gain for purely diffuse and purely direct light respectively. Any given tracker angle means, that the sun position in the projection perpendicular to the tracker axis is fixed. Therefore, the ground illumination pattern for direct light is always the same for a given tracker angle. The same is true for the diffuse illumination pattern. Both patterns are modulated by the absolute value of the direct normal and the diffuse horizontal irradiance respectively. We see, that the bifacial gain for direct light is highest for a tracker angle of 0°, which can also be deduced from figure 10. This plot shows the fraction of direct light reaching the ground in between the rows, and the view factors towards the rear and front side of the PV modules. When the trackers are horizontal, the stripes of light on the ground are widest, and therefore a larger fraction of direct light is reaching the ground. At the same time, the average view angle is also highest for this tracker position, since the back side of the PV modules are facing straight down. As the trackers move to higher angles, the stripes of illuminated ground will become narrower. On top of that, the front side of the PV modules will still point to the sun (in the 2D projection), meaning that the angle of incidence on the front side will be smaller than the one on the ground. This will increase the ratio between the irradiance reaching the front side of the PV modules and the irradiance reaching the ground. Furthermore, the view angle of the module rear side towards the ground will decrease as the modules tilt away from the horizontal position. These combined effects lead to the decrease of the bifacial gain for direct light as the tracker angle increases. The value will reach zero at the limit angle, where the inter-row shadings set on, and no more direct light reaches the ground. The drop of the bifacial gain for larger tracker angles is accentuated by the small contribution of light scattered back from the ground to the front surface of the PV modules, as shown in figure 10.

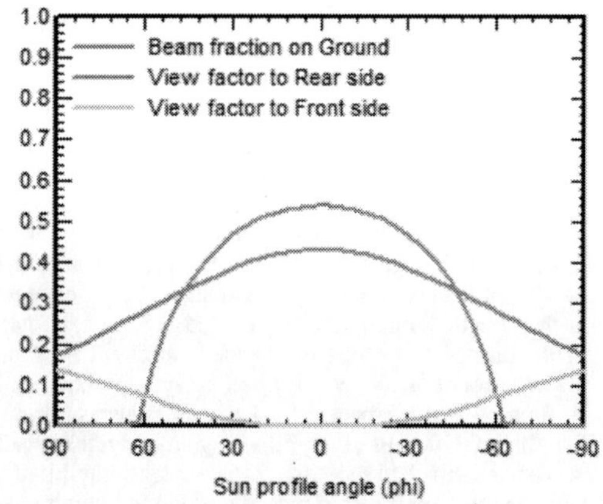

Figure 10: Bifacial model for the beam component.

The diffuse component, on the other hand, displays exactly the opposite behavior. The situation here is more complex than for the direct light. First of all, the fraction of diffuse sky light that reaches the front side of the PV modules, decreases as the tracker tilt increases, while the back side gets more and more diffuse irradiance from the sky. The fraction of diffuse light reaching the ground will also increase with the tracker angle, but the view factor of the module back side towards the ground will decrease, leading to less irradiance scattered back to the module's rear side. Overall, the decrease of irradiance on the front side and the increase of sky diffuse irradiance on the rear side, dominate the trend, and lead to an increased bifacial gain as the tracker angle increases.

For the case Φ_{Trk}=0° the bifacial gain for the direct component is maximal, and the one for the diffuse component is minimal. At this tracker position, the gain for the diffuse is always the larger of the two values, and thus the yellow and blue line in figure 9 never intersect. This can be explained by the fact that the ratio of irradiations reaching the front and rear side of the modules, is the same. Also, the light pattern on the ground is roughly similar for the two components, there are bright stripes between the tracker rows and more shaded stripes underneath the PV modules. For the diffuse component however, some of the light leaks right underneath the PV modules, where the view factor is much more favorable for the rear side. Therefore, the bifacial gain at this tracker position is slightly higher for the diffuse component.

From the above we see, that the bifacial gain is always higher for the diffuse light contribution than for the direct light contribution. This statement is only valid for horizontal axis tracking systems. For rows with fixed orientation, the picture becomes much more complex for direct light, and no general rule can be given.

4 SUMMARY

In this paper we described the bifacial model used in the PVsyst software, that is applied to horizontal axis tracker systems. The model is based on a simplified view factor approach, that can be reduced to a two-dimensional calculation. This is a suitable approximation for large fields of trackers with regular row spacing and width. The model captures the main bifacial contributions, namely the direct and diffuse light scattered back from

the ground, and the direct and diffuse light reaching directly the rear side of the PV modules.
The model was used to study the benefit arising by passing from fixed orientation to tracking systems (tracker gain) and by switching from monofacial to bifacial modules (bifacial gain). The gain values are a function of the geometrical layout, namely the row spacing expressed here in GCR, and the mounting height. Furthermore, the gain depends on the latitude at which the system is placed, and on the climate. The dependency on the climate comes from the fact that the bifacial gain for the diffuse part of the irradiance is different from the bifacial gain of the direct contribution. As a general result we found that the expected tracker gain is always higher than the bifacial gain. Both, bifacial gain and tracker gain are very sensitive to the GCR, whereas only the bifacial gain changes significantly with the mounting height. We found also that the tracking gain is very sensitive to the latitude, locations close to the equator benefitting much more from trackers than those at higher latitude. The bifacial gain is also slightly higher at low latitude. Finally, the trackers in climates with a large diffuse irradiance component tend to have a higher bifacial gain.

5 REFERENCES

[1] U. A. Yusufoglu, T. M. Pletzer, L. J. Koduvelikulathu, C. Comparotto, R. Kopecek and H. Kurz, "Analysis of the Annual Performance of Bifacial Modules and Optimization Methods," in IEEE Journal of Photovoltaics, vol. 5, no. 1, pp. 320-328, Jan. 2015. doi: 10.1109/JPHOTOV.2014.2364406

[2] Gaby J.M.Jansse, nBas B.Van Aken, Anna J.CarrAgnes, A.Mewe, "Outdoor Performance of Bifacial Modules by Measurements and Modelling", Energy Procedia, Volume 77, August 2015, Pages 364-373.

[3] C. W. Hansen *et al.*, "Analysis of irradiance models for bifacial PV modules," *2016 IEEE 43rd Photovoltaic Specialists Conference (PVSC)*, Portland, OR, 2016, pp. 0138-0143. doi: 10.1109/PVSC.2016.7749564.

[4] B.Wittmer, A.Mermoud, "Bifacial shed simulations with PVsyst", 4th bifiPV workshop, Konstanz, Germany, October 25/26 2017 http://bifipv-workshop.com/index.php?id=konstanz-2017-program

[5] Marion, Bill, MacAlpine, Sara, Deline, Chris, Asgharzadeh, Amir, Toor, Fatima, Riley, Daniel, Stein, Joshua, and Hansen, Clifford. "A Practical Irradiance Model for Bifacial PV Modules: Preprint". United States. doi:. https://www.osti.gov/servlets/purl/1364146.

[6] System Advisor Model (SAM) Website. Bifacial PV Modules in SAM. National Renewable Energy Laboratory. Golden, CO. Accessed September 20, 2018 https://sam.nrel.gov /node/80086 .

[7] Ismail Shoukry, Joris Libal, Radovan Kopecek, Eckard Wefringhaus, Jürgen Werner, "Modelling of bifacial gain for stand-alone and in-field installed bifacial PV modules", 6th International Conference on Silicon Photovoltaics, SiliconPV 2016, Energy Procedia 92 (2016) 600 – 608.

[8] B. Wittmer, A.Mermoud, "Horizontal Axis Trackers with Bifacial Modules in PVsyst", 10th PV Performance Modeling Workshop, Albuquerque, NM, USA, 2018 https://pvpmc.sandia.gov/resources-and-events/events/2018-10th-pv-performance-modeling-workshop/

COMPARISON BETWEEN CENTRAL AND STRING INVERTERS PERFORMANCE FOR THE UTILITY-SCALE PV PLANT IN NOVA OLINDA BRAZIL

Giovanni Nobile[1], Mario Cacciato[1], Giuseppe Scarcella[1], Giacomo Scelba[1],
Agnese Giuseppa Federica Di Stefano[2], Giuseppe Leotta[2], Paola Maria Pugliatti[2], Fabrizio Bizzarri[2]
[1]University of Catania Department of Electrical, Electronics Engineering and Computer Science
[2]Enel Green Power SpA
Contact: gnobile@dieei.unict.it

ABSTRACT: One of the most considered solutions to improve the performance of PV plants is the adoption of distributed power converters. The evaluation of the performance increasing by using string inverters or optimizers instead of central inverters in utility-scale PV fields is usually a difficult task due to several reasons such as numerous partial unavailabilities, monitoring system issues, operating constraints imposed by the Power Plant Controllers, etc. To support the data analysis and to forecast the behavior of the PV field under several operating scenarios, a novel modeling approach addressed to large PV plants with multiple conversion stages has been introduced. This work is focused on the comparison among central and string converters installed for testing purpose in two different subfields of the Enel Green Power 300 MW PV plant in Nova Olinda (Brazil). The elaboration of data acquired by the PV plant monitoring system in some months using the developed model highlights that string inverters ensure a remarkable gain in terms of produced energy. Anyway, the accurate estimation of such gain requires data for a larger time period in order to detect the variations due to seasonality. Moreover, the identification of the causes leading to mismatch phenomenon compensated by string inverters requires further investigations in future works.
Keywords: Utility-scale PV fields, State-space average modeling, Simulation, String Inverters, Multi-stage Converters

1 BACKGROUND

The use of distributed power converters in PV fields represents a key topic in solar value chain both for new constructions and retrofitting purpose. Many tests have been carried out in laboratory as well as in small size and medium size PV plant in order to compare string and central inverters performance [1],[2]. On the contrary, in literature a comprehensive analysis about test results at utility-scale level is mostly missing. A first reason for this lack of information is related to the fact that the installation of central inverters represents the most common choice in large plants. Moreover, in those PV fields having distributed power converters, some issues can occur in data analysis due the large number of data to be elaborated leading to a difficult assessment of performance. The monitoring system can be affected by multiple faults in dataloggers, sensors, etc. getting missing or wrong historical and real-time data. Also the number of partial unavailabilities is quite high due to the high number of components in which a fault can occur. Moreover, in some cases there are also additional operating constraints set by the Power Plant Controller (PPC) or by the inverters themselves in terms of maximum power that a specific PV subfield can deliver to the grid.

These issues can be overcome by exploiting suitable models capable to track the expected behavior of the PV plant even in case of operating thresholds and to fill the gaps due to faults, missing data, etc. Unfortunately, the detailed models shown in many papers addresses to represent the main components of a PV plant are not the best choice due to the computational effort occurring for long-term analysis in PV systems having thousands of PV modules, hundreds of distributed converters and so on. The modeling strategy introduced in this paper has been developed focusing on these typical issues.

The case study in this work is the performance evaluation of central and string converters installed in two different subfields being part of the Enel Green Power 300 MW PV plant in Nova Olinda (Brazil). The comparison process is based on energy production and it is supported by a suitable state-space average model implemented in Matlab Simulink environment described in section 3.

This kind of approach based on a behavioral model is very different than the common analysis tools used in PV asset management. Indeed, the latter are mostly based on daily or monthly statistical data reported in spreadsheet files. On the contrary, the implementation of a state-space average model gives the opportunity to analyze the behavior of the entire system with a satisfactory accuracy while avoiding a too large computational effort. Moreover, in case of issues caused by wrong data from monitoring system, partial unavailabilities, external constraints, etc. occurring for one or more conversion systems, data provided by the model can be used to replace the real PV plant data.

2 CASE STUDY

The PV field is divided into subfields, two of those are used as benchmark.

In the 2 MW subfield having central converters there are two inverters whose rated power is 1025 kVA, fan cooled, embedded in an electric cabin.

In the 2 MW subfield with string inverters each converter has a rated power of 60 kVA but can reach 66 kW in case of unity power factor and ambient temperature below 30° C. String converters are mounted on field without using outer cabinets and without fan cooling. They are connected using AC parallel switchboards named QPCA to a dedicated cabin close to the transformers.

The sum of PV modules power values connected to each conversion system is exactly the same (2570.4 kWp) so that the comparisons can be performed in the same conditions.

The configuration of the subfields under test is represented in Fig. 1 and Fig. 2. Table I summarizes main technical data.

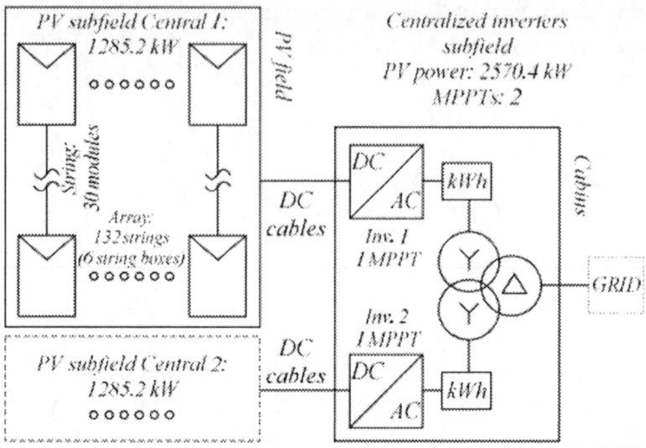

Figure 1: Power system configuration of the 2 MW subfield with central inverters

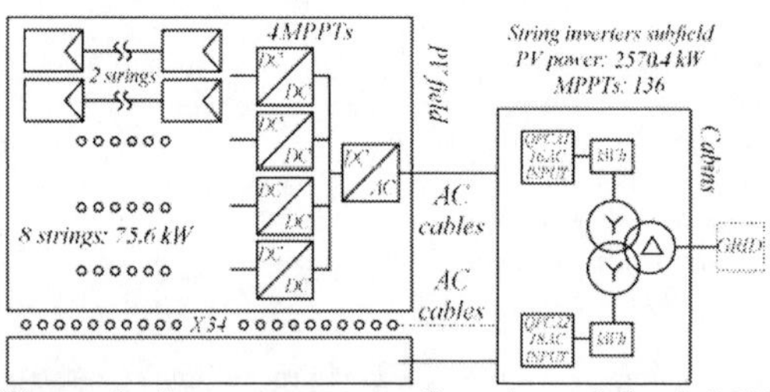

Figure 2: Power system configuration of the 2 MW subfield with string inverters

Table I: Main technical data

PV modules	
P (W)	315
V_{oc} (V)	46.2
I_{sc} (A)	9.01
V_{mpp} (V)	37.2
I_{mpp} (V)	8.48
NOCT (°C)	45±2
P/Temp (%/°C)	-0.40
V_{oc}/Temp (%/°C)	-0.30
I_{sc}/Temp (%/°C)	+0.06
Number of cells	72
Central Inverters	
Rated AC power (kVA)	1025
AC output (V, Hz)	400±10%, 50/60
MPPT input DC voltage range (V)	675-1320
Maximum Efficiency (%)	98.9
MPPTs per power converter	1
String Inverters	
Rated AC power (kVA)	60 (up to 66)
AC output (V, Hz)	800, 50/60
MPPT input DC voltage range (V)	600-1450
Maximum Efficiency (%)	99.0
MPPTs per power converter	4

3 UTILITY-SCALE PV FIELDS MODELING

3.1 Modelling approach

A scheme of a utility-scale PV plant is shown in Fig. 3. Such a configuration is of general-purpose therefore it can be used to simulate the central inverters subfield as well as the string inverters subfield or part of them (strings, string boxes, etc.) with minimal modifications.

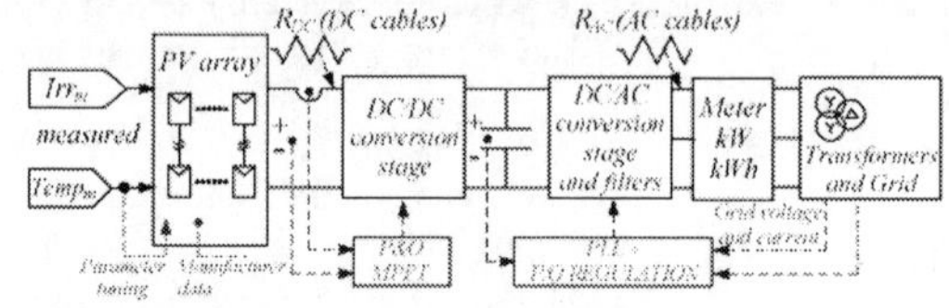

Figure 3: A generic PV plant configuration

This generic PV field has been implemented in Matlab Simulink environment using a novel behavioral model based on a modified state-space averaging method. Since the power converters in large PV fields with distributed inverters normally include both DC-DC (e.g. first stage of string inverters, string optimizers, etc.) and DC-AC (grid connected inverters, sometimes multilevel inverters), the implementation of a single comprehensive model is often a difficult task.

A common approach used to model DC-DC converters while keeping a low computational effort is the state-space averaging method, some examples are in [3]-[5]. On the contrary, the grid connected inverters are usually modeled using analytic relationships coming from energy balances [4],[6],[7] or from equivalent circuits [8], in the latter case the electrical quantities are sometimes expressed as phasors [9],[10].

In literature there is a lack of information about modeling approaches for large PV fields having multiple conversion stages. Accordingly, this works is mainly aimed to fill this gap. The main novelty is related to the creation of a comprehensive model in which each component of the PV field has been considered by using its state-space representation. Such approach gives the opportunity to get a simple identification process by exploiting data coming from the monitoring system of the PV plant. At the same time, all the conversion components are mixed into a single state-space average model obtaining a complete representation for the whole PV plant. In other words, the PV plant becomes a single state-space system whose inputs are irradiance and cell temperature and whose output is the energy production or other electric quantities to be known.

3.2 Conversion system modeling

To better explain the novelty of this paper with regard to multi-stage converters, this section is focused on a specific multi-stage topology represented in Fig. 4.

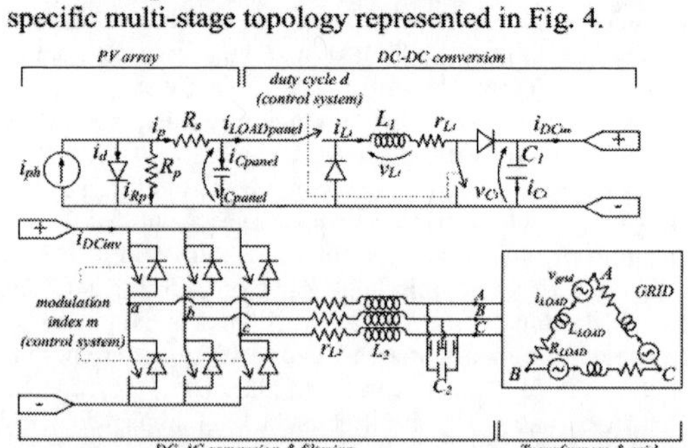

Figure 4: Block diagram of a generic grid-connected PV system with multi-stage conversion system

Considering the DC-DC converter as a single block, Fig. 5, its state-space averaged form can be obtained as in [3]. The variation of duty cycle d due to the implementation of the MPPT strategy can be introduced as a perturbation $\tilde{d}$ [5], leading to:

$$\begin{cases} \dot{X} = \overline{A}_{DC} X + \overline{B}_{DC} u \\ Y = \overline{C}_{DC} X + \overline{D}_{DC} u \end{cases}$$

$$\begin{cases} \dot{\tilde{x}} = \overline{A}_{DC}\tilde{x} + \left[\left(A_{DC_{ON}} - A_{DC_{OFF}}\right) \ \left(B_{DC_{ON}} - B_{DC_{OFF}}\right)\right]\begin{bmatrix} X\tilde{d} \\ v_{in}\tilde{d} \end{bmatrix} \\ \tilde{y} = \overline{C}_{DC}\tilde{x} + \left[\left(C_{DC_{ON}} - C_{DC_{OFF}}\right)X\right]\tilde{d} = \overline{C}_{DC}\tilde{x} \end{cases} \quad (1)$$

with:

$$u = U = v_{in} \quad x = X + \tilde{x} \quad y = Y + \tilde{y} \quad d = D + \tilde{d}$$

$$x = \begin{bmatrix} i_{L1} \\ v_{C1} \end{bmatrix} \qquad \overline{A}_{DC} = A_{DC_{ON}} d + A_{DC_{OFF}}(1-d)$$

$$A_{DC_{ON}} = \begin{bmatrix} 0 & 0 \\ 0 & -\frac{1}{RC_1} \end{bmatrix} \qquad A_{DC_{OFF}} = \begin{bmatrix} 0 & -\frac{1}{L_1} \\ \frac{1}{C_1} & -\frac{1}{RC_1} \end{bmatrix} \quad (2)$$

$$\overline{B}_{DC} = B_{DC_{ON}} d + B_{DC_{OFF}}(1-d)$$

$$B_{DC_{ON}} = \begin{bmatrix} \frac{1}{L_1} \\ 0 \end{bmatrix} \qquad B_{DC_{OFF}} = \begin{bmatrix} 0 \\ 0 \end{bmatrix}$$

$$\overline{C}_{DC} = C_{DC_{ON}} = C_{DC_{OFF}} = \begin{bmatrix} 1 & 0 \\ 0 & 1 \end{bmatrix} \qquad \overline{D}_{DC} \text{ null}$$

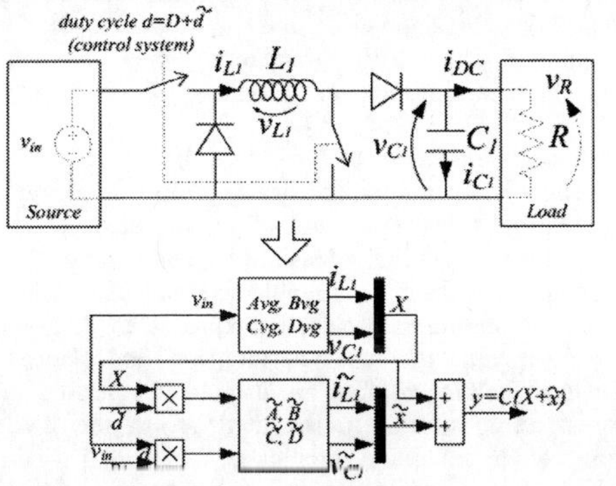

Figure 5: DC-DC converter basic topology and its state-space form

About the inverter and filtering stage in Fig. 6, in this first part of analysis the load is assumed to be three-phase inductive-resistive in delta connection without grid. A valid modelling approach is in [10] exploiting a generalized state-space averaging method based on Fortescue symmetrical components and on Fourier transform. The state-space form can be obtained from:

$$\begin{cases} \frac{d}{dt}\begin{bmatrix} i_{ab} \\ i_{bc} \\ i_{ca} \end{bmatrix} = -\frac{r_{L2}}{L_2}\begin{bmatrix} i_{ab} \\ i_{bc} \\ i_{ca} \end{bmatrix} - \frac{1}{3L_2}\begin{bmatrix} v_{AB} \\ v_{BC} \\ v_{CA} \end{bmatrix} + \frac{\sqrt{3}m}{6L_2}\begin{bmatrix} v_{DC}\cos\left(\omega t - \varphi_a + \frac{\pi}{6}\right) \\ v_{DC}\cos\left(\omega t - \varphi_b + \frac{\pi}{6}\right) \\ v_{DC}\cos\left(\omega t - \varphi_c + \frac{\pi}{6}\right) \end{bmatrix} \\ \frac{d}{dt}\begin{bmatrix} v_{AB} \\ v_{BC} \\ v_{CA} \end{bmatrix} = \frac{1}{C_2}\begin{bmatrix} i_{ab} \\ i_{bc} \\ i_{ca} \end{bmatrix} - \frac{1}{C_2}\begin{bmatrix} i_{LOAD_{AB}} \\ i_{LOAD_{BC}} \\ i_{LOAD_{CA}} \end{bmatrix} \\ \frac{d}{dt}\begin{bmatrix} i_{LOAD_{AB}} \\ i_{LOAD_{BC}} \\ i_{LOAD_{CA}} \end{bmatrix} = -\frac{R_{LOAD}}{L_{LOAD}}\begin{bmatrix} i_{LOAD_{AB}} \\ i_{LOAD_{BC}} \\ i_{LOAD_{CA}} \end{bmatrix} + \frac{1}{L_{LOAD}}\begin{bmatrix} v_{AB} \\ v_{BC} \\ v_{CA} \end{bmatrix} \end{cases} \quad (3)$$

where i_{ab}, i_{bc} and i_{ca} are virtual line currents.

The AC currents and voltages can be represented as the sum of their symmetrical components:

$$\begin{cases} i_{ab} = \langle i_{ab}\rangle_{-1} e^{-j\omega t} + \langle i_{ab}\rangle_0 + \langle i_{ab}\rangle_1 e^{j\omega t} \\ i_{bc} = \langle i_{bc}\rangle_{-1} e^{-j\omega t} + \langle i_{bc}\rangle_0 + \langle i_{bc}\rangle_1 e^{j\omega t} \\ i_{ca} = \langle i_{ca}\rangle_{-1} e^{-j\omega t} + \langle i_{ca}\rangle_0 + \langle i_{ca}\rangle_1 e^{j\omega t} \\ v_{AB} = \langle v_{AB}\rangle_{-1} e^{-j\omega t} + \langle v_{AB}\rangle_0 + \langle v_{AB}\rangle_1 e^{j\omega t} \\ v_{BC} = \langle v_{BC}\rangle_{-1} e^{-j\omega t} + \langle v_{BC}\rangle_0 + \langle v_{BC}\rangle_1 e^{j\omega t} \\ v_{CA} = \langle v_{CA}\rangle_{-1} e^{-j\omega t} + \langle v_{CA}\rangle_0 + \langle v_{CA}\rangle_1 e^{j\omega t} \\ i_{LOAD_{AB}} = \langle i_{LOAD_{AB}}\rangle_{-1} e^{-j\omega t} + \langle i_{LOAD_{AB}}\rangle_0 + \langle i_{LOAD_{AB}}\rangle_1 e^{j\omega t} \\ i_{LOAD_{BC}} = \langle i_{LOAD_{BC}}\rangle_{-1} e^{-j\omega t} + \langle i_{LOAD_{BC}}\rangle_0 + \langle i_{LOAD_{BC}}\rangle_1 e^{j\omega t} \\ i_{LOAD_{CA}} = \langle i_{LOAD_{CA}}\rangle_{-1} e^{-j\omega t} + \langle i_{LOAD_{CA}}\rangle_0 + \langle i_{LOAD_{CA}}\rangle_1 e^{j\omega t} \end{cases} \quad (4)$$

Then, assuming:

$$\begin{cases} \langle i_{ab}\rangle_1 = x_1 + jx_2 \\ \langle i_{bc}\rangle_1 = x_3 + jx_4 \\ \langle i_{ca}\rangle_1 = x_5 + jx_6 \\ \langle v_{AB}\rangle_1 = x_7 + jx_8 \\ \langle v_{BC}\rangle_1 = x_9 + jx_{10} \\ \langle v_{CA}\rangle_1 = x_{11} + jx_{12} \\ \langle i_{LOAD_{AB}}\rangle_1 = x_{13} + jx_{14} \\ \langle i_{LOAD_{BC}}\rangle_1 = x_{15} + jx_{16} \\ \langle i_{LOAD_{CA}}\rangle_1 = x_{17} + jx_{18} \end{cases} \qquad \begin{cases} \langle i_{ab}\rangle_0 = x_{19} \\ \langle i_{bc}\rangle_0 = x_{20} \\ \langle i_{ca}\rangle_0 = x_{21} \\ \langle v_{AB}\rangle_0 = x_{22} \\ \langle v_{BC}\rangle_0 = x_{23} \\ \langle v_{CA}\rangle_0 = x_{24} \\ \langle i_{LOAD_{AB}}\rangle_0 = x_{25} \\ \langle i_{LOAD_{BC}}\rangle_0 = x_{26} \\ \langle i_{LOAD_{CA}}\rangle_0 = x_{27} \end{cases} \quad (5)$$

the AC currents and voltages become:

$$\begin{aligned} i_{ab} &= 2x_1\cos\omega t - 2x_2\sin\omega t + x_{19} \\ i_{bc} &= 2x_3\cos\omega t - 2x_4\sin\omega t + x_{20} \\ i_{ca} &= 2x_5\cos\omega t - 2x_6\sin\omega t + x_{21} \\ v_{AB} &= 2x_7\cos\omega t - 2x_8\sin\omega t + x_{22} \\ v_{BC} &= 2x_9\cos\omega t - 2x_{10}\sin\omega t + x_{23} \\ v_{CA} &= 2x_{11}\cos\omega t - 2x_{12}\sin\omega t + x_{24} \\ i_{LOAD_{AB}} &= 2x_{13}\cos\omega t - 2x_{14}\sin\omega t + x_{25} \\ i_{LOAD_{BC}} &= 2x_{15}\cos\omega t - 2x_{16}\sin\omega t + x_{26} \\ i_{LOAD_{CA}} &= 2x_{17}\cos\omega t - 2x_{18}\sin\omega t + x_{27} \end{aligned} \quad (6)$$

Exploiting simple relationships, the state-space equations of the DC-AC conversion system, for which $(x_1,....x_{27})$ are the state variables, can be integrated in the PV plant model described in section 3.5.

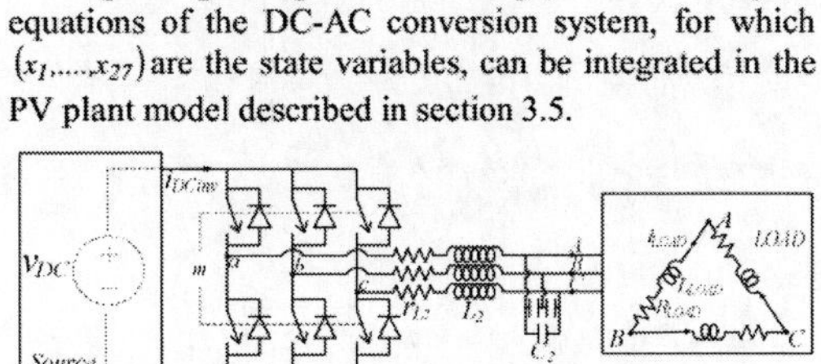

Figure 6: Grid connected inverter and filtering stage, basic topology

The presence of zero-sequence components in the AC currents and voltages is neglected in the following. Anyway, it can be easily introduced as in [10].

From above, considering the complexity for defining the conversion stages, with the aim to obtain a comprehensive representation for the whole system it could be useful to match the different modeling approaches. Thus, the DC-link current i_{DCinv} quantity assumes a key role.

In previous works the average value or RMS value of this current has been calculated from power balances [11] or from integral calculation [11],[12], in some cases the latter has been applied by means of the reduced-order Fourier transform. One of the main drawbacks of these methods is the large computational time due to data storage in integral terms.

This issue has been overcome in this work finding an analytical technique to calculate the average value of the DC-link current as a function of the AC current components using simple algebraic formulas. To achieve this result, the first step is to consider that:

$$\begin{aligned} i_{DCinv} &= s_a i_a + s_b i_b + s_c i_c = \\ &= s_a i_{ab} - s_a i_{ca} + s_b i_{bc} - s_b i_{ab} + s_c i_{ca} - s_c i_{bc} = \\ &(s_a - s_b) i_{ab} + (s_b - s_c) i_{bc} + (s_c - s_a) i_{ca} \end{aligned} \tag{7}$$

Using prosthaphaeresis formulas, such relationship becomes:

$$\begin{aligned} i_{DCinv} = \langle i_{DCinv} \rangle_0 = \frac{\sqrt{3}m}{2} \Bigg[\left\langle i_{ab} \cos\left(\omega t - \varphi_a + \frac{\pi}{6}\right) \right\rangle_0 + \\ + \left\langle i_{bc} \cos\left(\omega t - \varphi_b + \frac{\pi}{6}\right) \right\rangle_0 + \left\langle i_{ca} \cos\left(\omega t - \varphi_c + \frac{\pi}{6}\right) \right\rangle_0 \Bigg] \end{aligned} \tag{8}$$

in which the search for the zero-sequence component of DC-link current is pointed out. Thanks to some Fourier transform properties, the first term into square parenthesis can be rewritten as:

$$\begin{aligned} &\left\langle i_{ab} \cos\left(\omega t - \varphi_a + \frac{\pi}{6}\right) \right\rangle_0 = \\ &= \langle i_{ab} \rangle_1 \left\langle \cos\left(\omega t - \varphi_a + \frac{\pi}{6}\right) \right\rangle_{-1} + \\ &+ \langle i_{ab} \rangle_0 \left\langle \cos\left(\omega t - \varphi_a + \frac{\pi}{6}\right) \right\rangle_0 + \\ &+ \langle i_{ab} \rangle_{-1} \left\langle \cos\left(\omega t - \varphi_a + \frac{\pi}{6}\right) \right\rangle_1 \end{aligned} \tag{9}$$

In (9), the second term on the right is null. On the contrary, the other terms require some further calculations using Euler formulas leading to:

$$\begin{aligned} &\left\langle i_{ab} \cos\left(\omega t - \varphi_a + \frac{\pi}{6}\right) \right\rangle_0 = \langle i_{ab} \rangle_1 \left\langle \cos\left(\omega t - \varphi_a + \frac{\pi}{6}\right) \right\rangle_{-1} + \\ &+ \langle i_{ab} \rangle_{-1} \left\langle \cos\left(\omega t - \varphi_a + \frac{\pi}{6}\right) \right\rangle_1 = \\ &= (x_1 + jx_2) \frac{1}{2} e^{j\left(\varphi_a - \frac{\pi}{6}\right)} + (x_1 - jx_2) \frac{1}{2} e^{-j\left(\varphi_a - \frac{\pi}{6}\right)} \end{aligned} \tag{10}$$

and:

$$\begin{aligned} &\left\langle i_{ab} \cos\left(\omega t - \varphi_a + \frac{\pi}{6}\right) \right\rangle_0 = (x_1 + jx_2) \frac{1}{2} e^{j\left(\varphi_a - \frac{\pi}{6}\right)} + \\ &+ (x_1 - jx_2) \frac{1}{2} e^{-j\left(\varphi_a - \frac{\pi}{6}\right)} = \\ &= \frac{1}{2} \left\{ (x_1 + jx_2) \left[\cos\left(\omega t - \varphi_a + \frac{\pi}{6}\right) + j \sin\left(\omega t - \varphi_a + \frac{\pi}{6}\right) \right] \right\} + \\ &+ \frac{1}{2} \left\{ (x_1 - jx_2) \left[\cos\left(\omega t - \varphi_a + \frac{\pi}{6}\right) - j \sin\left(\omega t - \varphi_a + \frac{\pi}{6}\right) \right] \right\} \end{aligned} \tag{11}$$

so that:

$$\begin{aligned} &\left\langle i_{ab} \cos\left(\omega t - \varphi_a + \frac{\pi}{6}\right) \right\rangle_0 = \frac{1}{2} 2x_1 \cos\left(\varphi_a - \frac{\pi}{6}\right) + \\ &- \frac{1}{2} 2x_2 \sin\left(\varphi_a - \frac{\pi}{6}\right) = x_1 \cos\left(\varphi_a - \frac{\pi}{6}\right) - x_2 \sin\left(\varphi_a - \frac{\pi}{6}\right) \end{aligned} \tag{12}$$

Repeating the same elaboration for the other zero-sequence components in (8), the final formula linking the DC-link current to the AC current components is:

$$\begin{aligned} &i_{DCinv} = \langle i_{DCinv} \rangle_0 = \\ &= \frac{\sqrt{3}m}{2} \Bigg[x_1 \cos\left(\varphi_a - \frac{\pi}{6}\right) - x_2 \sin\left(\varphi_a - \frac{\pi}{6}\right) + \\ &+ x_3 \cos\left(\varphi_b - \frac{\pi}{6}\right) - x_4 \sin\left(\varphi_b - \frac{\pi}{6}\right) + \\ &+ x_5 \cos\left(\varphi_c - \frac{\pi}{6}\right) - x_6 \sin\left(\varphi_c - \frac{\pi}{6}\right) \Bigg] \end{aligned} \tag{13}$$

In such a way, it has been obtained a straightforward approach to calculate the DC-link current from the direct-sequence components of AC currents, the latter being state variables.

3.3 PV array model

The accuracy of the PV array model has a large impact for the accuracy of the whole system representation. An in-depth literature review has been performed and several comparisons between the most popular existing models have been carried out focusing on both accuracy and complexity. The basic formulation of the selected model is in [13]. Since in PV plant of Nova Olinda large temperature variations are expected to take place, a sensitivity analysis has been conducted in order to detect the parameters which have the bigger influence on the I-V and P-V characteristics and to identity the main temperature-dependent parameters, which are on-line tuned by means of look-up tables.

3.4 Power transformer and grid models

The implementation of a detailed equivalent circuit for the LV-MV transformer and of a distributed parameters model for the grid is not feasible for computational effort reasons. Nevertheless, a suitable accuracy can be reached using a minimum order model represented in Fig. 4 in which the values of load resistors, load inductances and also grid voltage sources are calculated referring to the primary side of the transformer i.e. to the LV side connected to the output filter stage.

3.5 PV plant model

Combining the models built for each component of Fig. 4, it is possible to achieve a state-space representation for the whole PV plant or subfield as reported in (14) useful to easily calculate the power and energy quantities in AC side from irradiance and cell temperature data. The rms value of phase-to-phase grid voltage V_{grid} is considered as an input. Alternatively, it could be assigned as a constant term into matrices. The actual value of control variables d and m are set by the control system described in the following section.

The state-space representation of the whole PV system in Fig. 4 is:

$$\begin{aligned} \dot{x} &= \overline{A}x + \overline{B}u \\ y &= \overline{C}x + \overline{D}u \end{aligned} \tag{14}$$

where:

$$x = \begin{bmatrix} v_{C_{panel}} \\ i_{L_1} \\ v_{C_1} \\ x_1 \\ \cdots \\ \cdots \\ x_{18} \end{bmatrix} \quad u = \begin{bmatrix} i_{ph} \\ i_d \\ V_{grid} \end{bmatrix} \quad y = x \quad \text{or} \quad y = \begin{bmatrix} x_1 \\ \cdots \\ \cdots \\ x_{12} \end{bmatrix} \tag{15}$$

Matrix $\overline{A}$ is:

$$\overline{A} = A_{ON} d + A_{OFF} (1-d) \tag{16}$$

with:

$$A_{ON} = \begin{bmatrix} A_{ON\,1.1} & \cdots & A_{ON\,1.21} \\ \cdots & \cdots & \cdots \\ A_{ON\,21.1} & \cdots & A_{ON\,21..21} \end{bmatrix}$$
$$A_{OFF} = \begin{bmatrix} A_{OFF\,1.1} & \cdots & A_{OFF\,1.21} \\ \cdots & \cdots & \cdots \\ A_{OFF\,21.1} & \cdots & A_{OFF\,21..21} \end{bmatrix} \tag{17}$$

Non zero elements in A_{ON} are listed here:

$$A_{ON\,1.1} = -\frac{1}{C_{panel}(R_s + R_p)} \quad A_{ON\,1.2} = -\frac{1}{C_{panel}}$$
$$A_{ON\,2.1} = \frac{1}{L_1} \quad A_{ON\,2.2} = -\frac{r_{L1}}{L_1}$$
$$A_{ON3.4} = -\frac{\sqrt{3}m}{2C_1} cos\left(\varphi_a - \frac{\pi}{6}\right)$$
$$A_{ON3.5} = \frac{\sqrt{3}m}{2C_1} sin\left(\varphi_a - \frac{\pi}{6}\right)$$
$$A_{ON3.6} = -\frac{\sqrt{3}m}{2C_1} cos\left(\varphi_b - \frac{\pi}{6}\right)$$
$$A_{ON3.7} = \frac{\sqrt{3}m}{2C_1} sin\left(\varphi_b - \frac{\pi}{6}\right)$$
$$A_{ON3.8} = -\frac{\sqrt{3}m}{2C_1} cos\left(\varphi_c - \frac{\pi}{6}\right)$$
$$A_{ON3.9} = \frac{\sqrt{3}m}{2C_1} sin\left(\varphi_c - \frac{\pi}{6}\right) \tag{18}$$
$$A_{ON4.3} = \frac{\sqrt{3}m}{12L_2} cos\left(\varphi_a - \frac{\pi}{6}\right)$$
$$A_{ON5.3} = -\frac{\sqrt{3}m}{12L_2} sin\left(\varphi_a - \frac{\pi}{6}\right)$$
$$A_{ON6.3} = \frac{\sqrt{3}m}{12L_2} cos\left(\varphi_b - \frac{\pi}{6}\right)$$
$$A_{ON7.3} = -\frac{\sqrt{3}m}{12L_2} sin\left(\varphi_b - \frac{\pi}{6}\right)$$
$$A_{ON8.3} = \frac{\sqrt{3}m}{12L_2} cos\left(\varphi_c - \frac{\pi}{6}\right)$$
$$A_{ON9.3} = -\frac{\sqrt{3}m}{12L_2} sin\left(\varphi_c - \frac{\pi}{6}\right)$$

and:

$$A_{ON4.4} = A_{ON5.5} = A_{ON6.6} = A_{ON7.7} = A_{ON8.8} = = A_{ON9.9} = -\frac{r_{L2}}{L_2}$$
$$A_{ON4.10} = A_{ON5.11} = A_{ON6.12} = A_{ON7.13} = A_{ON8.14} = = A_{ON9.15} = -\frac{1}{3L_2} \tag{19}$$
$$A_{ON5.4} = A_{ON7.6} = A_{ON9.8} = A_{ON11.10} = A_{ON13.12} = = A_{ON15.14} = A_{ON17.16} = A_{ON19.18} = A_{ON21.20} = -\omega$$
$$A_{ON4.5} = A_{ON6.7} = A_{ON8.9} = A_{ON10.11} = A_{ON12.13} = = A_{ON14.15} = A_{ON16.17} = A_{ON18.19} = A_{ON20.21} = \omega$$
$$A_{ON10.4} = A_{ON11.5} = A_{ON12.6} = A_{ON13.7} = A_{ON14.8} = = A_{ON15.9} = \frac{1}{C_2}$$
$$A_{ON10.16} = A_{ON11.17} = A_{ON12.18} = A_{ON13.19} = A_{ON14.20} = = A_{ON15.21} = -\frac{1}{C_2}$$
$$A_{ON16.10} = A_{ON17.11} = A_{ON18.12} = A_{ON19.13} = A_{ON20.14} = = A_{ON21.15} = \frac{1}{L_{LOAD}}$$
$$A_{ON16.16} = A_{ON17.17} = A_{ON18.18} = A_{ON19.19} = A_{ON20.20} = = A_{ON21.21} = -\frac{R_{LOAD}}{L_{LOAD}}$$

The elements in A_{OFF} being different than A_{ON} are:

$$A_{OFF\,1.2} = A_{OFF\,2.1} = 0 \quad A_{OFF\,2.3} = -\frac{1}{L_1}$$
$$A_{OFF\,3.2} = \frac{1}{C_1} \tag{20}$$

Matrix $\overline{B}$ is:

$$\overline{B} = B_{ON} d + B_{OFF} (1-d) \tag{21}$$

with:

$$B_{ON} = \begin{bmatrix} B_{ON1.1} & B_{ON1.2} & B_{ON1.3} \\ \cdots & \cdots & \cdots \\ B_{ON21.1} & B_{ON21.2} & B_{ON21.3} \end{bmatrix} \tag{22}$$
$$B_{OFF} = B_{ON}$$

whose non zero elements are:

$$B_{ON1.1} = -\frac{1}{C_{panel}\left(1 + \frac{R_s}{R_p}\right)}$$
$$B_{ON1.2} = -\frac{1}{C_{panel}\left(1 + \frac{R_s}{R_p}\right)}$$
$$B_{ON16.3} = -\frac{\sqrt{2}}{L_{LOAD}} cos(\varphi_{grida})$$
$$B_{ON17.3} = \frac{\sqrt{2}}{L_{LOAD}} sin(\varphi_{grida}) \tag{23}$$
$$B_{ON18.3} = -\frac{\sqrt{2}}{L_{LOAD}} cos(\varphi_{gridb})$$
$$B_{ON19.3} = \frac{\sqrt{2}}{L_{LOAD}} sin(\varphi_{gridb})$$
$$B_{ON20.3} = -\frac{\sqrt{2}}{L_{LOAD}} cos(\varphi_{gridc})$$
$$B_{ON19.3} = \frac{\sqrt{2}}{L_{LOAD}} sin(\varphi_{gridc})$$

Each state variable can be selected in order to be extracted from the system as output thanks to a proper identification of matrix $\overline{C}$. The latter can be set as an identity matrix. Anyway, to calculate the energy produced by the PV plant, it is enough to apply simple formulas using variables from x_1 to x_{12} i.e. using AC voltage and current components. Matrix $\overline{D}$ is null.

3.6 Control system

The adopted MPPT strategy for the DC-DC converter is the Perturb & Observe method. The control system of the grid connected inverter is implemented in *qd* reference frame using a suitable PLL subsystem. The DC-link voltage is fixed by *d*-axis current acting on the delivered active power.

To maintain acceptable the computational time, time sampling of quantities in control system can be set larger than time sampling used for the plant e.g. 10x or more.

Referring to the specific case study discussed in this paper, the design of the control system takes into account additional constraints: converters maximum power internal threshold, internal and ambient maximum operating temperature and Power Plant Controller (PPC) limitation. The latter is a threshold fixed by the local utility company as the maximum value of active power that the PV plant can deliver to the grid, 300 MW for the considered case.

3.7 Parameters identification

The values of PV array parameters can be easily assigned using data listed in the PV modules datasheet and considering the number N_s of modules per string and the number N_p of strings per string box. Some parameters e.g. R_s and R_p need to be calculated. For this purpose, a suitable identification method is in [13]. The look-up tables used to tune the main temperature-dependent parameters implement simple linear or quasi-linear relationships coming from the temperature coefficients of P, V_{oc} and I_{sc} reported in PV module datasheet and in Table I.

About the converters, due to know-how protection policies detailed information about their topology, hardware components and control system are usually not available. From this point of view, the modeling approach described in this paper allows to overcome this lack of data giving the opportunity to identify an equivalent behavioral model in a simple but effective way. Accordingly, one of the many methods presented in the literature can be used, useful to identify systems in the form of state-space or transfer function. Similar procedures can be followed to identify the parameters of transformers and grid.

4 PERFORMANCE COMPARISON BETWEEN CENTRAL AND STRING INVERTERS

4.1 Relevance of the introduced model

The usefulness of the proposed modelling approach can be highlighted with an example. In Fig. 7 the power measured at the meter of the string inverters subsection QPCA 1 is compared to the power curve obtained in simulation for the same subfield in a sunny day during which a power limitation occurs due to the inverter internal maximum power threshold. Thanks to the simulation, it can be estimated the theoretical extra-energy with a good accuracy.

This example shows that the introduced model gives the opportunity to track the behaviour of the plant also in case of missing data, PPC limitation, etc. This circumstance represents a significant advantage for comparison processes as in this work.

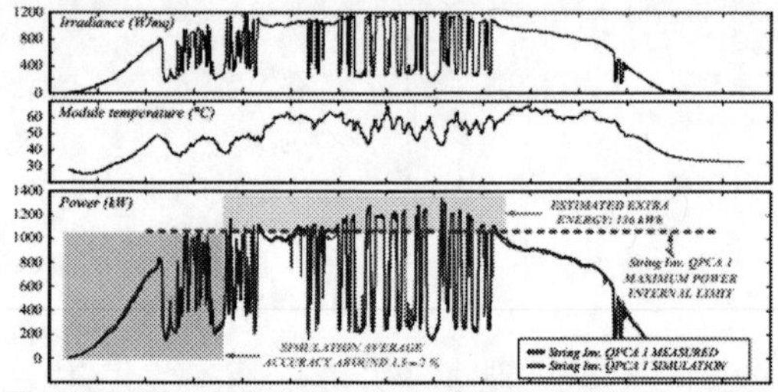

Figure 7: Simulation vs measured data of string inverters subsection QPCA 1 during a sunny day

4.2 Comparison results

The performances of central and string inverters in the case study have been compared looking at the post-commissioning period (5 months long) from December 2017 to April 2018 i.e. during summer and autumn in Brazil. Data for following time periods will be provided in future works. The basic formula used to evaluate the gain obtained by string converters G_{string_inv} is:

$$G_{string\_inv} = \frac{E_{string\_inv} / P_{string\_inv} - E_{central\_inv} / P_{central\_inv}}{E_{central\_inv} / P_{central\_inv}} \quad (24)$$

where E_{string_inv} is the daily energy produced by the string inverters (or by a QPCA), $E_{central\_inv}$ is the daily energy produced by the central inverters (or by a single inverter), P is the whole power of the PV modules connected to each conversion system. Such comparison criterion is equivalent to the comparison between the performance ratio values.

To better clarify the comparison criterion, Fig. 8 shows the energy produced in subfields with centralized and in subfields with string inverters measured during a cloudy day, in absence of power derating and of PPC limitation or other constraints. In this specific case, the gain of string inverters subfield in comparison to centralized inverters is around +2.8%, calculated by formula (24).

The aggregated results for the considered period are reported in Fig. 9 in the form of data distribution charts including a normal density function fitting curve.

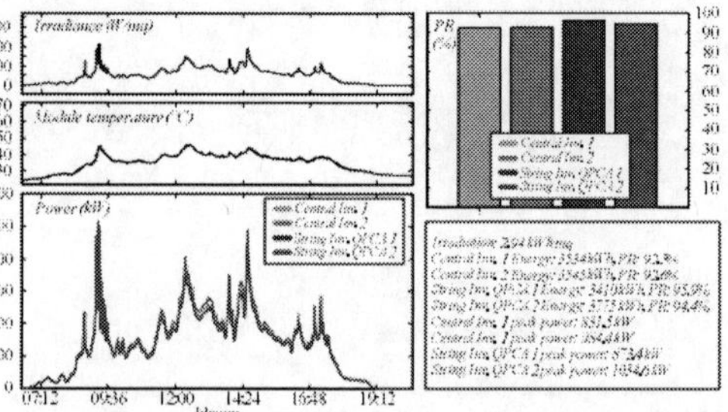

Figure 8: Performance comparison among centralized and string inverters during a cloudy day

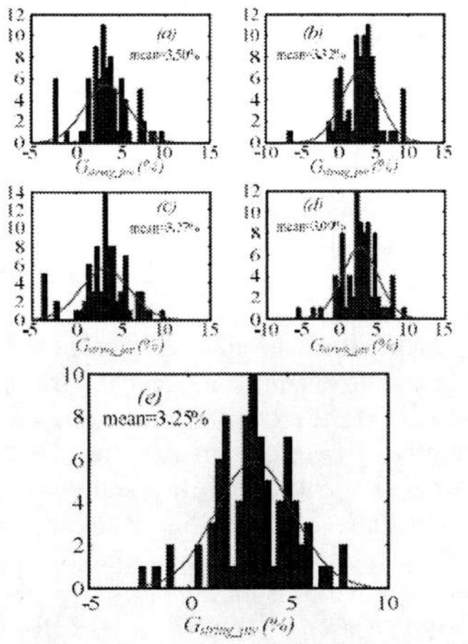

Figure 9: String inverters subfields gain distribution charts and normal density functions for the considered time period: (a) QPCA 1 vs central inverter 1, (b) QPCA 1 vs central inverter 2, (c) QPCA 2 vs central inverter 1, (d) QPCA 2 vs central inverter 2, (e) string inverters subfield (QPCA 1+QPCA 2) vs central inverters subfield (central 1+central 2)

Finally, referring to the established limited period and with no further investigation regarding the causes leading to mismatch phenomenon compensated by string inverters, the average gain obtained by the string converters is around 3% in terms of daily production.

It is worth highlighting that in case of issues caused by wrong data, PPC limitation, partial unavailabilities, etc. occurring for one or more converters, data provided by the model introduced in this paper have been used to replace data coming from the monitoring system of the real PV plant as shown in the previous section.

5 CONCLUSIONS AND FUTURE WORKS

In this paper the comparison among the performance obtainable using central or string inverters for utility-scale PV plants have been addressed. A real case study has been presented considering the results achieved by the central and string converters installed for testing purpose in two different subfields of the Enel Green Power 300 MW PV plant in Nova Olinda, Brazil.

To better investigate the performance of the two considered conversion systems while largely reducing the impact of incorrect data, unavailabilities or external constraints, a novel state-space average model addresses to utility-scale PV plants has been introduced getting a satisfactory tradeoff between computational effort and simulation accuracy.

The process of the data acquired by the monitoring system on the real PV plant, supported by the developed model, highlights that string inverters ensure an improvement in terms of produced energy of about 3%, although the accurate assessment of such value requires data for a larger time period in order to detect the variations due to seasonality. Moreover, an appropriate study of the causes leading to mismatch phenomenon compensated by string inverters requires further investigations in future works.

6 REFERENCES

[1] S. Strache, R. Wunderlich and S. Heinen, "A Comprehensive, Quantitative Comparison of Inverter Architectures for Various PV Systems, PV Cells, and Irradiance Profiles," IEEE Trans. Sustainable Energy vol. 5, no. 3, pp. 813-822, July 2014.

[2] S. Harb, M. Kedia, H. Zhang and R. S. Balog, "Microinverter and String Inverter Grid-Connected Photovoltaic System – A Comprehensive Study," in Proc. 2013 IEEE Photovoltaic Specialists Conference.

[3] R. H. G. Tan and L. Y. H. Hoo, "DC-DC Converter Modeling and Simulation using State Space Approach", in Proc. 2015 IEEE Conference on Energy Conversion.

[4] S. Chiniforoosh, J. Jatskevich, A. Yazdani, V. Sood, V. Dinavahi, J. A. Martinez and A. Ramirez, "Definitions and Applications of Dynamic Average Models for Analysis of Power Systems," IEEE Trans. Power Delivery, vol. 25, no. 4, pp. 2655-2669, Mar 2010.

[5] G. Kanimozhi, J. Meenakshi and V. Sreedevi, "Small Signal Modeling of a DC-DC Type Double Boost Converter Integrated with SEPIC Converter Using State Space Averaging Approach," Energy Procedia, vol. 117, pp. 835-846, June 2017.

[6] X. Liu and A. M. Cramer, "Three-Phase Inverter Modeling using Multifrequency Averaging with Third Harmonic Injection," in Proc.2016 Energy Conversion Congress and Exposition.

[7] J. Guerrero-Perez, E. De Jodar, E. Gomez-Lazaro and A. Molina-Garcia, "Behavioral modeling of grid-connected photovoltaic inverters: Development and assessment," Renewable Energy, vol. 68, pp. 686-696, Aug 2014.

[8] H. A. Pereira, F. D. Freijedo, M. M. Silva, V. F. Mendes and R. Teodorescu, "Harmonic current prediction by impedance modeling of grid-tied inverters: A 1.4 MW PV plant case study," Electrical Power and Energy Systems, vol. 93, Dec. 2017.

[9] A. Coronado-Mendoza, M. A. Perez-Cisneros, J. A. Dominguez-Navarro, V. Osuna-Enciso, V. Zuniga-Grajeda and K. J. Gurubel-Tun, "Dynamic phasors modeling for a single phase two stage inverter," Electric Power Systems Research, vol. 140, pp. 854-865, Nov. 2016.

[10] Z. Lin and H. Ma, "Modeling and Analysis of Three-phase Inverter based on Generalized State Space Averaging Method", in Proc. 2013 Annual Conference of the IEEE Industrial Electronics Society.

[11] J. W. Kolar and S. D. Round, "Analytical calculation of the RMS current stress on the DC-link capacitor of voltage-PWM converter systems", IEE Proc. – Electric Power Applications, vol. 153, issue 4, pp. 535-543, July 2006.

[12] X. Pei, W. Zhou and Y. Kang, "Analysis and Calculation of DC-Link Current and Voltage Ripples for Three-Phase Inverter With Unbalanced Load," IEEE Trans. Power Electronics, vol. 30, no. 10, pp. 5401-5412, Oct 2015.

[13] M. G. Villalva, J. R. Gazoli and E. R. Filho, "Comprehensive Approach to Modeling and Simulation of Photovoltaic Arrays," IEEE Trans. Power Electronics, vol. 24, no. 5, pp. 1198-1208, May 2009.

SIZING OF PHOTOVOLTAIC SYSTEMS FOR SELF-CONSUMPTION IN COMMERCIAL AND INDUSTRIAL APPLICATIONS

Thomas Roessler
Yingli Namene Service Company, Landshuter Allee 8-10, Munich 80637, Germany
Phone: +49 162 1362156; E-mail: thomas.roessler@yinglinamene.com

ABSTRACT: The integration of photovoltaic (PV) systems with the power supply from the electricity grid (and usually Diesel generators as backup power source) is becoming increasingly popular in commercial and industrial (C&I) applications in developing countries. In the absence of a feed-in tariff scheme and net metering, as is common in such markets, self-consumption of the power generated by the PV system is required. Therefore, its size must be carefully chosen to enable a high PV Utilization, thus ensuring a low effective cost of the PV energy, while reaching the highest possible PV Coverage of the energy demand. The work presented here outlines a methodology to determine an appropriate PV system size in this context. Basic inputs are (1) a so-called "load profile", i.e., a time series of the power consumption during a representative period of operation, (2) time-resolved information about the power generated by the PV system, and (3) insights about the seasonality of the power consumption. Based on this, the percentage of both, the PV power consumed instantaneously by the loads and the energy demand covered by PV, can be predicted for a given PV system size, allowing an optimization of the latter. Using example data from actual C&I consumers in Ghana, the methodology is demonstrated in practice.
Keywords: PV System, Grid-Connected, Sizing, Design, Self-Consumption

1 INTRODUCTION

Due to strong solar irradiance and correspondingly high specific yields, power generated by PV systems is economically competitive with the electricity grid in many developing countries in the sunbelt region [1]. This has created increasing interest by local C&I consumers to supplement their power supply from the grid with PV power to lower their utility bill.

In these markets, feed-in tariff schemes and net metering are not common, therefore the application of PV is restricted to self-consumption. More clearly, PV energy can be either self-consumed directly or stored for later, asynchronous consumption. Since storage adds technical complexity and often significantly reduces the economic advantage, it will not be considered here.

In the context of direct self-consumption, an important quantity is the "PV Utilization", i.e., the percentage of the available PV energy which can be used instantaneously. The "Unutilized PV Power", i.e., the portion which exceeds the momentary power demand, simply goes to waste. Capital and operational expenditures for a PV system are tied to the nominal PV system size, hence a high PV Utilization corresponds to a low effective cost of every kilowatt hour from PV. From this point of view, a conservative PV system sizing leading to a high PV Utilization is preferable.

On the other hand, a higher "PV Coverage", i.e., the percentage of the energy demand contributed by PV, results in more of the low-cost kilowatt hours being used, suggesting that an aggressive PV system sizing is commercially advantageous. This paper addresses the question how the right balance can be found.

The basic methodology outlined in Section 2 allows the prediction of PV Utilization and PV Coverage for a given PV system size. This is demonstrated with a variety of examples. Section 3 describes how the seasonality can be included. A simple model to consider the economics is then presented in Section 4. All aspects are combined to analyze the data for one concrete case in Section 5, followed by a brief summary and discussion.

2 METHODOLOGY

2.1 Basic Approach

The underlying simple idea is a direct comparison of demand and supply. If the power demand of the loads at the C&I consumer and the power generated by the PV system are both known as a function of time, they can be easily used to calculate the PV Utilization and the PV Coverage.

This is exemplified by Figure 1, showing a load profile (red) and the PV power (orange) for a single day. Clearly, only the portion of the PV power below the load curve (dashed green) can be used directly. Hence, the PV Utilization is the ratio of the areas under the green and orange curves, corresponding to the used and total PV energy, respectively. Similarly, the PV Coverage is the ratio of the area under the green dashed and red curves, corresponding to the used PV energy and total energy demand, respectively.

Load profiles can be determined by connecting a data logger to the main internal power line at the facility of an interested C&I consumer and measuring the relevant electrical parameters (voltage, current, power) for an

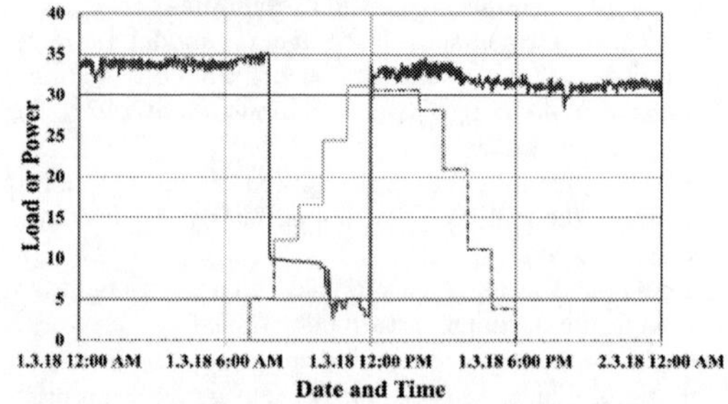

Figure 1: Visualization of the basic approach. Only the portion (dashed green) of the PV power (orange) below the load profile (red) can be used directly. The text explains how the fundamental quantities PV Utilization and PV Coverage can be derived from this.

extended time period. To obtain data representative of normal operation, the measurement should cover at least one full week, including the weekend. The corresponding standard developed by VDE (the association of German Electrical Engineers) even requires a full month [2]. The same standard also recommends a time step of maximally 15 minutes, but clearly the time resolution of the measured data should be fine enough to capture all major load variations.

Since the determination of the PV system size takes place during the planning phase, the PV power cannot yet be measured. However, a prediction of the approximate time dependence can be obtained by performing a simulation with, for instance, the software PVsyst [3]. To the largest extent possible, the simulation should consider the actual circumstances on site, e.g., approximate size, relevant meteorological data as well as orientation and inclination of roofs available for PV. From the simulation output, the time dependence of the PV power for the same period over which the load profile was measured can then be extracted.

With this, for a fixed initial PV system size all the ingredients required for a determination of the PV Utilization and PV Coverage as outlined above, are available. To obtain the results for PV systems of different sizes, in principle the PVsyst simulation would have to be repeated. However, if the changes in size remain reasonably small, it seems a good first approximation to simply scale the PV power by the ratio of considered and simulated PV system sizes.

In principle, the basic approach is comparable to the system sizing which takes place for off-grid PV applications, where loads and solar energy resources also need to be assessed carefully [4]. Apart from the absence of storage, which was deliberately chosen in this work, there are two main differences: On the one hand, the existing grid connection provides an unlimited source of power to immediately make up for an undersupply from the PV system, which therefore does not need to be sized to cover the entire demand. And on the other hand, the grid electricity constitutes a hard cost reference against which the economics of the PV system need to be evaluated.

2.2 Example Results

The usefulness of the approach presented above can be most easily illustrated with a few real-life examples.

At first sight, a department store in Accra (Ghana) seems like the type of C&I consumer that would be able to benefit significantly from PV power. The extensive use of air conditioning causes most of the energy demand to occur during the day and therefore the time dependence roughly follows the intensity of sunshine. Indeed, the analysis (illustrated by Fig. 2) shows that a good PV Coverage of more than 40% can be achieved. However, due to the shift of the store opening hours (9:30 am to 7:30 pm on weekdays) relative to the daytime hours, PV Utilization does not exceed 80%, regardless of how much the PV system size is reduced.

In this example, even if the bare cost of a kilowatt hour from PV power based on plain considerations of capital and operational expenditures were a considerable 20% below the grid utility cost, no economic savings could be achieved due to the low PV Utilization.

Similarly, multi-shift operations in a petrochemical production facility in Tema (Accra), corresponding to a load profile which is almost constant for the entire 24h weekday, would suggest good boundary conditions for significant savings from PV Power. However, a closer look at the one-week load profile reveals that production activity is reduced on Saturdays and basically halted (apart from a base load) on Sundays (see Fig. 3). Aggressive sizing of the PV System considering only the weekday load level leads to a PV Coverage of 26%, but only 94% PV Utilization can be achieved. With a more conservative PV system size considering weekend loads, PV Coverage is reduced to 20%, but PV Utilization reaches 97%.

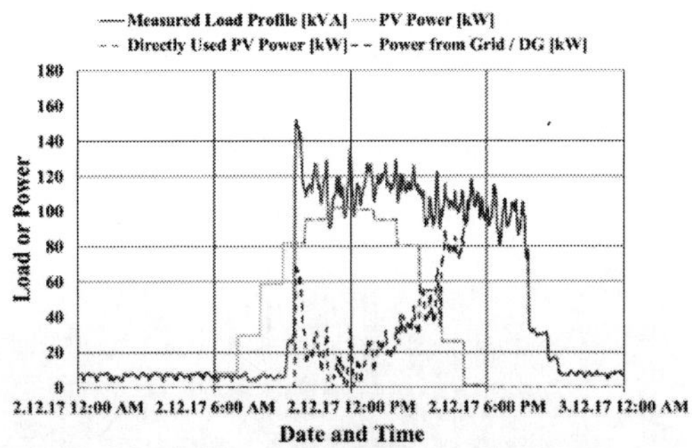

Figure 2: Example single day graph for a department store in Accra (Ghana). The load (red) is coming mostly from air conditioning and occurs during the day, allowing for a high PV Coverage. However, the shift of weekday store opening hours relative to daylight leads to a significant amount of unutilized PV power in the morning (orange line above red line). Consequently, PV Utilization does not exceed 80%.

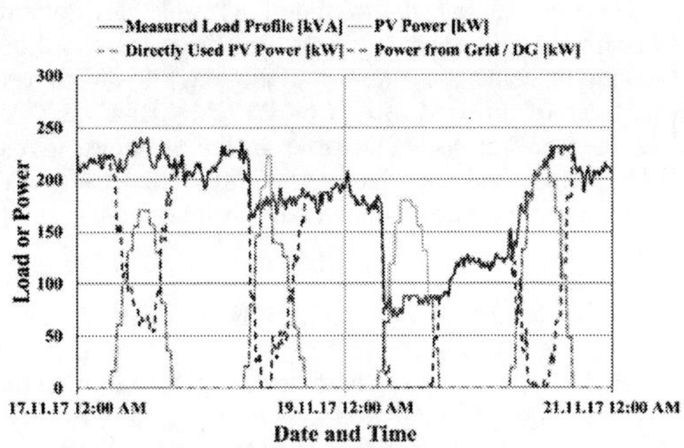

Figure 3: Example graph for a petrochemical factory in Tema (Ghana) with a stable high load during weekdays but exhibiting significantly reduced activity on Saturday (Day 2) and halted operation (apart from a base load) on Sunday (Day 3). With the PV system aggressively sized to the weekday load, PV power is wasted during the weekend (orange curve above red curve) and only a PV Utilization of 94% can be achieved, while PV Coverage reaches 26%. A more conservatively sized PV system increases the PV utilization to 97% but lowers PV Coverage to 20%.

Since there are only a few percent difference between the different choices of PV system size, in this case a full, detailed analysis of the economics is required. An obvious metric for this is the expected percent savings of the C&I consumer on their annual utility bill.

Finally, for a fruit juice producer in Accra (Ghana) the two-week profile shows that the load varies only slightly around a constant value, except for one full and one partial day with the load level reduced to about 25% (see Fig. 4, displaying the first week only), which are related to fruit supply and do not follow a regular pattern. On these days, a considerable amount of PV power is not utilized.

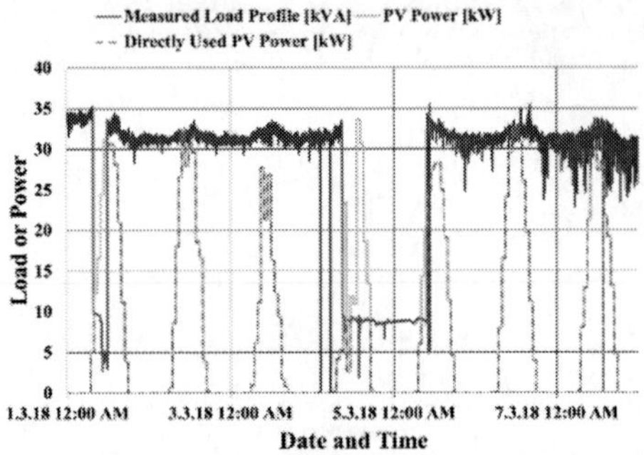

Figure 4: Example graph for a fruit juice producer in Accra (Ghana) with mostly constant load. The full two-week load profile (of which only the first week is shown here) exhibits two dips to about 25% of the normal load that do not follow a regular pattern (Days 1 and 4/5 in the graph). On these days, a considerable amount of PV power is not utilized (orange curve above red curve).

Considering only the average load values for normal weekdays, e.g., from utility bills or estimations based on installed machinery, the impact of intraday and day to day variations of the load profile on PV Utilization and PV Coverage do not become visible. In the examples given above, this would most likely result in a choice of a PV system size that is economically disadvantageous.

3 CONSIDERATION OF SEASONALITY

However, a measured load profile covering an entire work week or even a full month, as recommended by VDE [2], will correctly capture only the short-term variations in the energy demand of the consumer. If demand and supply are just compared on these time scales, the important aspect of seasonality is not considered when determining PV Utilization and PV Coverage.

The two most relevant long-term variations in the present context are the supply seasonality of the PV power and the seasonality of the energy demand.

3.1 Supply Seasonality

The electrical power generated by a PV system is highly dependent on the level of irradiance which reaches the plane of the PV modules. This is the main reason for the typical "clear sky" bell curve of PV power as a function of time during a bright sunny day (see green line in Fig. 1).

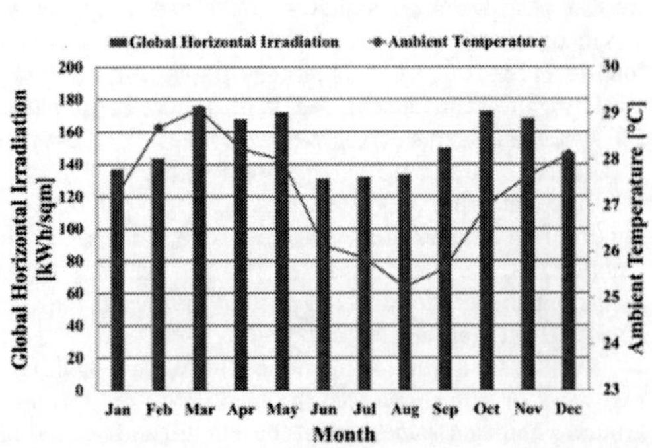

Figure 5: Variation of the monthly global horizontal irradiation (red bars, left axis) and average ambient temperature (blue curve, right axis) over the course of a typical meteorological year for Accra (Ghana).

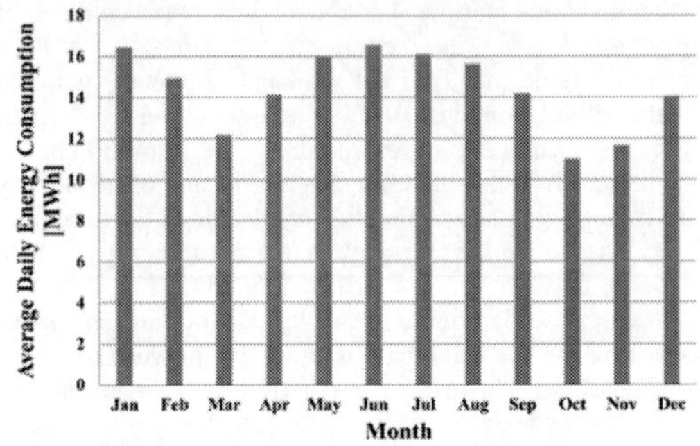

Figure 6: Variation of the monthly average daily energy consumption of a fruit juice producer over the course of one year for Accra (Ghana).

In addition, the ambient temperature has a considerable effect on the PV power production. Obviously, the level of both quantities varies significantly over the course of a year. Using data from a typical meteorological year, this is shown for the example of Accra (Ghana) in Figure 5.

Energy yield simulations, as for instance with PVsyst, must consider the variation of irradiation and temperature over an entire year. Hence, the time dependence of PV power obtained from such simulations already includes the corresponding effects. For a standard PV system design in Accra (Ghana), the monthly peak PV power deviates up to 7.5% from the annual average value.

It is quite straightforward to take this seasonality into account for the determination of PV Utilization and PV Coverage by replicating the measured load profile with its limited time extent of a few weeks on the time axis, i.e., by shifting the start date by one month, two months, and so on until every month is at least partially covered. With the more extensive load profile constructed in this manner and the PV power simulation for the full year, the basic approach as outlined above can then be used again.

3.2 Demand Seasonality

Not only the level of sunshine has seasons, operations in companies also often vary over the course of a year. This is particularly true for businesses in tourism, but also common in the agricultural and food industry or, in a different way, for schools and universities.

Information about the corresponding seasonality of the energy demand can, for instance, be directly obtained from utility bills collected for 12 consecutive months. To illustrate this, Figure 6 shows the average daily energy consumption of a fruit juice producer in Accra (Ghana) for each month of the year. In the given case, the monthly values deviate up to 25% from the annual average.

One way to include the demand seasonality is to use the value for the month during which the load profile was measured as reference and calculate a scaling factor for the other months [2]. Combining this with the temporal shift to account for supply seasonality as described above, these factors are then simply applied to the shifted full-month or partial-month load profiles.

4 BASIC ECONOMIC CONSIDERATIONS

As already indicated in Section 1, PV Utilization and PV Coverage have opposite effects on the economics of the PV system and need to be balanced for an optimal result. This can be seen in a very simple financial model.

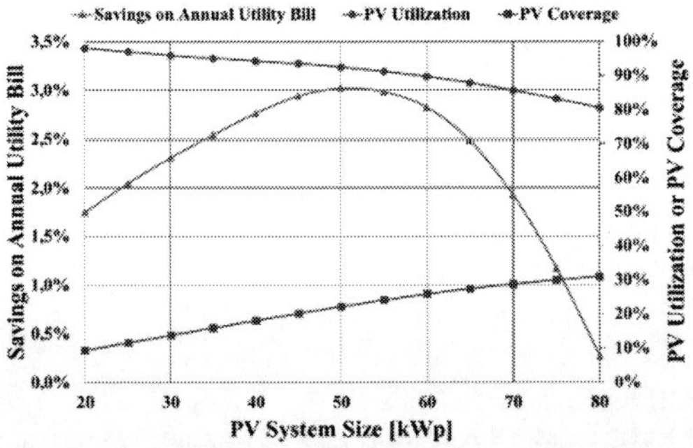

Figure 7: Economics based on a simple financial model for the case of a fruit juice producer in Accra (Ghana). PV Utilization (blue) decreases with PV system size, while PV Coverage (red) increases. The resulting percent savings on the annual utility bill (green) show a clear maximum.

The savings on the annual utility bill of a given C&I consumer result from the difference in cost for the part of the energy demand delivered by PV power, i.e., the PV Coverage. To calculate the difference in cost for every kilowatt hour, the ratio of the bare cost of a kilowatt hour from the PV system and the cost of a kilowatt hour from the grid can be used as basis. "Bare cost" here refers to the case of 100% PV Utilization, i.e., directly derived from capital and operational expenditures over the assumed lifetime of the PV system. For the real cost differential, this ratio, denoted below as BCR, needs to be divided by the PV Utilization. With this, the percent annual savings on the utility bill (PAS) are determined from

$$PAS = PVC*(1-BCR/PVU), \quad (1)$$

where PVC and PVU are the PV Coverage and PV Utilization, respectively. From this equation, the opposite effect of these quantities is obvious, and it is also clear that savings are only possible at all when the BCR divided by PV Utilization is less than unity.

5 EXAMPLE APPLICATION

Data from a fruit juice producer in Accra (Ghana) are now used to demonstrate an application of the full methodology described in the last three sections. The corresponding load profile, supply and demand seasonality are given already in Figures 4, 5 and 6, respectively.

Table 1 shows the results of the analysis for the three different cases of considering (1) no seasonality, (2) only supply (i.e., PV power) seasonality and (3) both, supply and demand seasonality, for a conservatively chosen PV system size. Since the reference month March has both, a higher PV power supply (see Fig. 5) and lower energy demand (see Fig. 6) than most of the other months, PV Coverage decreases significantly when considering seasonality. However, PV Utilization does not change much with the extent to which seasonality is included. Since only 1.5 days of the full two-week period show the reduced load affecting the PV Utilization strongly, this depends mostly on the relatively statistical distribution of high and low PV power days over the year.

Seasonality Considered	None	Supply	Supply & Demand
PV Utilization	95.4%	93.6%	94.3%
PV Coverage	22.7%	21.3%	18.5%

Table 1: Comparison of the PV Utilization and PV Coverage for a fixed PV system size and different types of seasonality considered.

Assuming a bare cost ratio of 80%, the economics discussed in Section 4 are presented for this C&I consumer in Figure 7. As expected, PV Utilization and PV Coverage show the opposing trends as a function of PV system size. The resulting dependence of the percent annual savings on the utility bill has a maximum and approaches zero as PV Utilization reaches a value of nearly 80% from above.

Hence, based on this analysis a clear recommendation for the PV system size can be given.

6 SUMMARY AND DISCUSSION

The work presented here demonstrates that it is indeed possible with reasonable effort to use the quantities "PV Utilization" and "PV Coverage" to determine the economically optimal size of a PV system for self-consumption by a C&I consumer with connection to the utility grid. It should be noted, that the basic ingredients of the proposed methodology, such as the instantaneous comparison of supply and demand, the consideration of seasonality, and the optimization according to an economic metric are in principle already available in existing commercial software, such as HOMER Pro [5].

Limitations of the validity of the results are mainly related to the question if the input data are representative of the long-term behavior. The analysis is likely to be misleading when the typical meteorological year used for the PV power simulation was not up to date, the measured load profile has not captured all the relevant characteristics of operation, or the seasonal variation of energy demand was not well reflected in the available utility bills. The last two points, for both, present and future, are best clarified in communication with the responsible operations manager.

Final note: Experience has shown that C&I consumers who have been made aware of the impact of particularities of the measured load profiles on the economics of a PV system are to some extent open to adjusting the way their facilities are operated.

7 ACKNOWLEDGEMENTS

Many thanks to Dennis Boateng, Frank Atta-Poku, and Firmin Nkamleu Ngassam for making the load profile data available.

[1] IRENA (2016), Solar PV in Africa: Costs and Markets.
[2] Verband Deutscher Elektroingenieure (Association of German Electrical Engineers) (2015), Methods for load measurements and extrapolation and minimum technical requirements for the definition of simulation input parameters for hybrid power generation systems. VDE-PB-0014: 2015.
[3] A. Mermoud, PVsyst (version 6.67) [software]. Geneva, Switzerland: ISE, University of Geneva, 1994-2018, retrieved from www.pvsyst.com.
[4] See, for instance, P. Mohanty et al. (2015) in Solar Photovoltaic System Applications: A Guidebook for Off-Grid Electrification. 10.1007/978-3-319-14663-8.
[5] For more information refer to HOMER Energy at https://www.homerenergy.com/products/pro/index.html.

PV BATTERY SYSTEMS SIZING TO ACCOUNT FOR THE PROVISION OF ANCILLARY SERVICES UNDER DIFFERENT SCENARIOS FOR ENERGY POLICIES AND TARIFFS

Alex Renan Arrifano Manito[1]
Kauê J. F. Novaes Cândido de Souza[1]
Marcelo Pinho Almeida[1]
Gilberto Figueiredo[1]
Roberto Zilles[1]
João Tavares Pinho[2]

1 – Laboratório de Sistemas Fotovoltaicos, Universidade de São Paulo
2 – Grupo de Estudos e Desenvolvimento de Alternativas Energéticas, Universidade Federal do Pará

ABSTRACT: This paper discusses the implications that several aspects, such as cost, types of services to be provided, and policies, may have on the sizing of PV systems with batteries, and presents a framework for decision making on the adoption of such systems, evaluating the profitability and the risks involved. The methodology is based on a sensitivity analysis carried out by simulating several scenarios of electricity prices, system costs, and the possible existence of incentives. The results show that nowadays the viability of PV systems with batteries is still very dependent on a price drop of the battery systems. This makes the adoption of such systems a risky investment, since long payback periods are needed. Moreover, ancillary services such as power factor compensation are only viable to increase the profitability of other services such as energy trade with the grid. If power factor compensation is the sole purpose of the system, the battery would be unnecessary since the storage capacity would be very small, and solutions like capacitor banks would be better suited.

key-words: Storage systems, Economic evaluation, Distributed resources.

1 INTRODUCTION

As smart grid technologies will be more and more adopted, the distribution grid is expected to become a platform for the supply and acquisition of products and services that go beyond the mere electrical energy purchase and/or sale from/to the utilities. This fact has implications for all the stakeholders involved with electricity distribution, and it could drastically change the business model of utilities, giving even the consumers connected to the low voltage grid a more active role in selling energy and/or ancillary services. The possibility not only to generate electricity but also to time-shift its consumption, for example, plays a major role to address issues such as feeders overload, distribution losses, power quality, and the increase of the admissible level of non-dispatchable renewable energy resources such as photovoltaics (PV) and wind.

The increased operational possibilities allowed by an enhanced manageability of energy resources are usually associated with the existence of energy storage. In this context, PV systems with batteries are gaining recognition as a valid approach for distributed generation, due to the maturing of battery technology for grid-connected applications and the cost drop tendency that is expected for batteries, especially with the emergence of niches such as electric vehicles. PV systems with batteries could profit from different policy arrangements for the distribution grid and thus increase the value of the produced energy. Such systems can be operated as controlled units that are able to meet the local needs of the distribution grid, capable of flexible operation (energy dispatch, and isolated operation, among other aspects).

Many studies address the use of batteries to achieve increased value for the energy generated, since this constitutes a promising subject for studies. As the prices for batteries drop and distributed resources become more of a reality in the distribution grid, increasing attention is being given to the research on batteries. He *et al.*[1] discuss the aggregation of services offered by the storage system to increase its value. Hoppmann *et al.*[2] investigate the economic viability of PV-storage systems in Germany under eight electricity price scenarios. Battke *et al.*[3] investigate the uncertainties of input parameters on the life cycle cost of four different battery technologies. Tang *et al.*[4] discuss the use of retired electric vehicles' batteries in PV systems.

However, the sizing of the system may vary greatly depending on the policies, and the type of service to be provided. The benefits achieved with increased flexibility do not present a linear relation with the size of the system, but rather one of saturation. Thus, the benefits with the increase of the nominal power of the PV generator become saturated for a given capacity of the battery bank. The same behavior occurs when increasing the capacity of the battery bank keeping the nominal power of the PV generator constant. Consequently, the amounts of generation and storage capacity are not decoupled and need to be appropriately adjusted to each other, otherwise additional investment in the capacity of a component may result in low profitability or none at all.

This paper discusses the implications that several aspects such as cost, types of services to be provided, and policies may have on the sizing of PV systems with batteries, and presents insights for decision making on the adoption of such systems, evaluating the profitability and the risks involved. The methodology is based on a sensitivity analysis that evaluates the effects on the cost of the system that an increase in the system's capacity would produce. The results show that nowadays the viability of PV systems with batteries are still very dependent on a price drop of the battery systems. This makes the adoption of such systems a risky investment, since long payback periods are needed. Moreover, ancillary services such as power factor compensation are only viable to increase the profitability of other services such as energy trade with the grid. If power factor compensation is the sole purpose of the system, the battery would be unnecessary since the storage capacity would be very small and solutions like capacitor banks would be better suited.

2 METHODOLOGY

The methodology estimates the costs associated with the provision of different services. The ancillary services considered in this paper are: arbitrage, reactive power compensation, load-shift, loading limits on feeders, self-consumption, and phase balancing. Such services were chosen due to their likelihood to be offered by residential and commercial prosumers. Figure 1 presents an overview of the methodology.

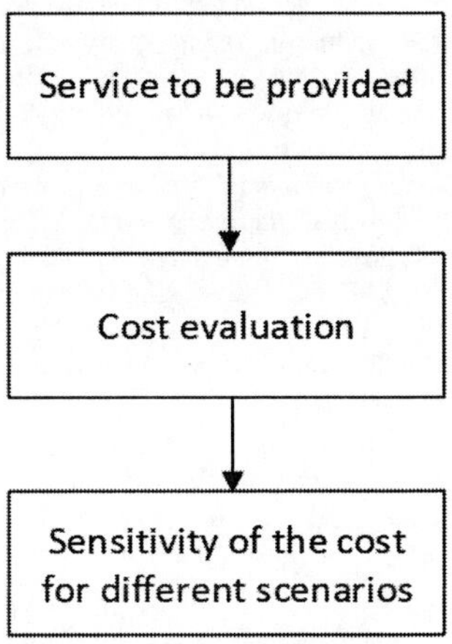

Figure 1: Overview of the methodology.

2.1 Cost of storage per unit of storage size

The following analysis evaluates the cost of the storage system for a given scenario of battery prices and numbers of cycles per period.

The output of the algorithm consists of a period value that represents the cost of having 1 kWh of storage available under a given cycle profile defined for the period. The algorithm for calculating the period cost is shown in the following set of equations and the inputs needed are presented in Table I.

Table I: Input variables for the algorithm.

Variable	Description
$C_{EndLife}$	Capacity at the end of life
$Dep_{PerCycle}$	Depreciation of the battery per cycle
$N_{DailyCycles}$	Number of daily cycles
$DoD_{DailyAverage}$	Average depth of discharge
$Proj_{LifeSpan}$	Project period
$BatCost_{PerkWh}$	Cost of the battery per kWh of storage
i	Interest rate

The values for depreciation per cycle and capacity at the end of life can be acquired using the number of cycles per Depth of Discharge curves from the manufacture's datasheet. Table II presents the variables of the algorithm.

Table II: Algorithm variables.

Variable or parameter	Description
$N_{LifeCycle}$	Number of Cycles during life spam
$Bat_{LifeSpan}$	Life spam of the battery
DEF	Effective Depreciation Factor
E_{Efec}	Effective Energy
$FracRepEvent$	Fractional Storage Replacement Event
$RepEvent$	Storage Replacement Event
$ResValue$	Residual Value
i_{Efec}	Effective interest rate
CRF_{Efec}	Effective Capital Recovery Factor
NPV_{PerkWh}	Net Present Value per kWh
CRF	Capital Recovery Factor
AC_{PerkWh}	Annualized Cost per kWh

$$N_{LifeCycle} = \frac{1 - C_{EndLife}}{Dep_{PerCycle}} \quad (1)$$

$$Bat_{LifeSpan} = \frac{N_{LifeCycle}}{365 \times N_{DailyCycles}} \quad (2)$$

$$DEF = \frac{Dep_{PerCycle} \times \left(1 + Dep_{PerCycle}\right)^{N_{LifeCycle}}}{\left(1 + Dep_{PerCycle}\right)^{N_{LifeCycle}} - 1} \quad (3)$$

$$E_{Efec} = \frac{DEF \times N_{LifeCycle}}{DoD_{DailyAverage}} \quad (4)$$

$$FracRepEvent = \frac{Proj_{LifeSpan}}{Bat_{LifeSpan}} \quad (5)$$

$$RepEvent = floor(FracRepEvent) \quad (6)$$

$$ResValue = [1 - (FracRepEvent - RepEvent)] \times BatCost_{PerkWh} \quad (7)$$

$$i_{Efec} = (1 + i)^{Bat_{LifeSpan}} - 1 \quad (8)$$

$$CRF_{Efec} = \frac{i_{Efec} \times \left(1 + i_{Efec}\right)^{RepEvent}}{\left(1 + i_{Efec}\right)^{RepEvent} - 1} \quad (9)$$

$$NPV_{PerkWh} = \left(BatCost_{PerkWh} + \frac{BatCost_{PerkWh}}{FRC_{Efec}} - \frac{ResValue}{(1+i_{Efec})^{RepEvent}} \right) \times E_{Efec} \quad (10)$$

$$CRF = \frac{i \times (1+i)^{Proj_{LifeSpan}}}{(1+i)^{Proj_{LifeSpan}} - 1} \quad (11)$$

$$AC_{PerkWh} = \frac{NPV_{PerkWh}}{FRC} \quad (12)$$

2.2 Arbitrage and self-consumption

For the arbitrage and self-consumption operation to be profitable, the gains with the purchase and resell or self-consumption should be higher than the period cost, given by equation 13.

$$C_j = \frac{AC_{PerkWh}}{365} \quad (13)$$

The output of the algorithm is a function of the number of cycles per day. A higher number of cycles would increase the cost per day of the storage system, due to the decrease in battery's lifetime, but could also increase the daily income of the arbitrage operation. In this situation, the advantage of doing a subsequent cycle with the battery has to be evaluated. Depending on the Time of Use (ToU) tariff values, it may not be the best option to perform as many cycles as possible. For a subsequent cycle to be profitable, the following condition has to be met:

$$\sum_{k=1}^{j} G_{k,j} > C_j - C_{j-1} + \sum_{k=1}^{j-1} G_{k,j-1} \quad (14)$$

The above condition states that the sum of all gains (G) due to cycle k for an operation with j cycles has to be higher than the variation of the costs between an operation with j and $j-1$ cycles plus the gains with the operation with $j-1$ cycles.

The sizing of the system depends on the tariff posts for maximum profitability to be found. The bi-directional converter should be sized in such a way as to have a net energy balance equal to zero at the end of the considered period, as described in equation 15.

$$S_n = \frac{E_{storage} \times DoD}{N_{hour\,min}} \quad (15)$$

where S_n is the capacity of the bi-directional converter, $E_{storage}$ is the size of the storage system, DoD is the depth of discharge considered for the system and $N_{hour\,min}$ is the minimum number of hours needed to charge or discharge the storage system for arbitrage or self-consumption operation.

2.3 Power factor compensation

Theoretically, power factor compensation would not consume energy either from the battery or from the utility grid, since all the power supplied in half of the cycle returns to the source in the next half cycle. Therefore, the amount of storage per kVAr needed for power factor compensation would be given by the following equation:

$$E_{size}/kvar = \frac{\int_0^{T/2} Q(t)dt}{kvar} \quad (16)$$

In the case of a system operating at 60 Hz, the amount of storage per kVAr would be given by:

$$E_{size}/kVAr = \frac{\frac{1/freq}{2 \times 3600} h}{1kvar} = 2.31 \times 10^{-6}\ kWh/kVAr$$

Since the amount of storage is quite small, other technologies would be better suited. Many converters perform such operation using energy from the grid without the need for the storage system. Thus, using the storage for the sole purpose of supplying reactive power would not be cost effective.

Though the supply of reactive power does not theoretically consume energy, in a real system there are losses in the converter, associated with the provision of reactive power. These losses could be modeled for a given apparent power (S) and power factor (FP) by equations 17 to 22. Table III presents a description for the variables.

$$P = S \times FP \quad (17)$$

$$Q = \sqrt{S^2 - P^2} \quad (18)$$

$$I = \frac{V - V_{cte}}{2 \times R_p} - \sqrt{\left[\left(\frac{V - V_{cte}}{2 \times R_p}\right)^2 - \left(\frac{-Q}{V - V_{cte}}\right)^2 - \frac{P}{R_p}\right]} + j \times \left(\left(\frac{-Q}{V - V_{cte}}\right)\right) \quad (19)$$

$$I_a = Real(I) \quad (20)$$

$$I_r = Imag(I) \quad (21)$$

$$P_{loss} = P_{cte} + |V_{cte} \times I_a| + R_p \times (I_a^2 + I_r^2) \quad (22)$$

Table III: Power factor compensation variables.

Variable or parameter	Description
P	Active Power
Q	Reactive Power
V	Nominal Voltage
V_{cte}	Constant Voltage
R_p	Converter resistance
I	Current
I_a	Active current
I_r	Reactive current
P_{cte}	Losses due to constant voltage
P_{loss}	Converter losses

Typical values for the parameters R_p, V_{cte} and P_{cte} are respectively 0.0355 p.u., 0.5 % of V and 1 % of the nominal power of the converter. The energy for power factor compensation would be supplied either by the battery or by the utility grid, which would incur into costs from the purchase of energy.

For power factor compensation, the size of the bi-directional converter is not dependent on the storage size, being a function only of the amount of reactive power needed to compensate the power factor.

2.4 Postponement in capacity investment and load shift

Storage systems can be used as an alternative for investments in the capacity of the distribution system. The amount of storage, E_n, needed to decrease the peak demand from a value D_a to a value D_b during a period is given by the following equation.

$$E_n = \int_0^T a(t) \times D(t)dt \qquad (23)$$

where T is the period (which can be a day, or a week, for example), $D(t)$ is the demand, and $a(t)$ is a dummy variable given by:

$$a(t) = \begin{cases} 0, & D(t) < D_b \\ 1, & D_a > D(t) \geq D_b \end{cases} \qquad (24)$$

The minimum value for the new peak demand, D_b, is the average demand for the period T. Below the average demand, it would be impossible to decrease the peak demand with storage alone. Once the amount of storage is found, equations 1 to 12 can be applied to determine the cost of the system considering a cycle over period T.

It is important to point out that the size of the storage system needed to avoid reinforcements in the capacity of the distribution system does not increase linearly with the avoided capacity. In this case, the storage system should be sized only to account for periods of overload; otherwise the cost of the system would be too high.

2.5 Phase balancing

The bi-directional characteristic of battery inverters for grid connection could allow phase balancing, which could constitute an ancillary service provided to the distribution grid. The methodology for evaluating the cost of phase balancing uses equations 1 to 12 presented in subsection 2.1.

3 RESULTS

The following subsections present the results and discussion for each of the services presented in section 2. The scenarios consider the case for the city of São Paulo as the base case.

3.1 Arbitrage

Figure 2 presents the result of a sensitivity analysis for the equivalent energy tariff for the battery bank energy considering different storage costs and interest rates, when considering one daily cycle of charge-discharge. Therefore, energy dependent services such as arbitrage, self-consumption, and load shift would only be profitable if the daily monetary gain for a given interest rate were above the curve for the respective interest rate. The profitability of systems with batteries could be improved by providing power dependent services such as power factor compensation. Such services are less dependent on the storage capacity of the battery bank, and increases in profitability could be achieved by only increasing the nominal power of the grid interface inverter.

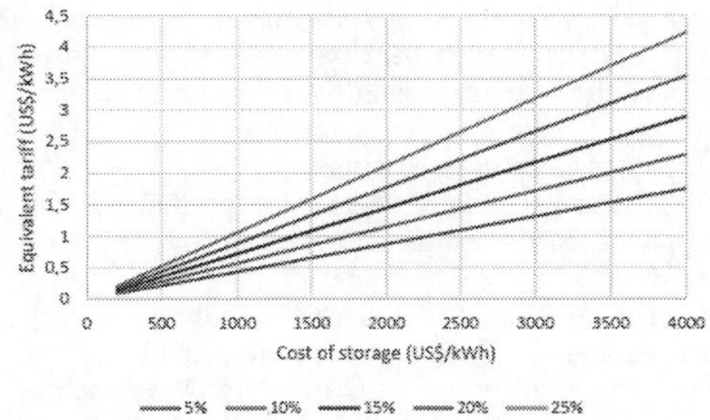

Figure 2: Daily operation cost as a function of the battery cost for different interest rates.

In Brazil, PV-battery systems are still very expensive due to the current prices found in the Brazilian market for lithium-ion batteries, which is around US$ 900/kWh and therefore would need daily gains ranging from US$ 0.5 to 1.0/kWh to break even.

Nowadays, consumers in Brazil may participate in the time of use tariff called Tarifa Branca (TB). Figure 3 presents the profile for TB. The values presented in the figure refer to the values used by Eletropaulo, the utility that supplies the city of São Paulo. For the presented case, there would only be the possibility for one cycle in the arbitrage case between the off-peak period and the peak period. The gain for each kWh would be US$ 0.0927. In such situation, the cost per day of storage availability would have to be at least equal to this value for the storage system to break even, which means that the cost per kWh of the storage system should drop to around US$ 200/kWh.

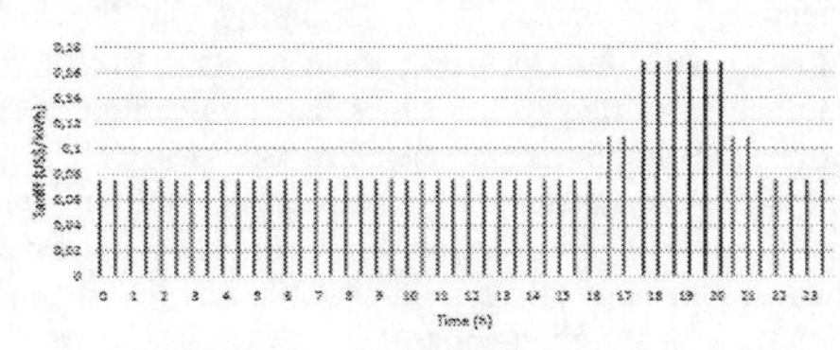

Figure 3: Time of Use tariff in Brazil.

In the case where a distributed generator is used to charge the battery, it is assumed that the battery is discharged at the time in which the energy is more expensive. Presently in Brazil, there would not be profitability even if the batteries were charged by a PV generator. The levelized cost of energy (LCOE) with PV in Brazil ranges from US$ 0.125 to US$ 0.15.

A diesel generator running under an adequate loading would operate with a specific consumption in a range between 0.25 and 0.3 L/kWh. In Brazil the cost of one liter diesel falls within the range of US$ 0.87 to US$ 1. In this situation, the price of producing energy with diesel generators would range from US$ 0.18/kWh to US$ 0.3 /kWh. Some companies run diesel generators during part of the day to prevent the purchase of energy from the utility grid. This could be an entrance market for battery storage in Brazil. Though the use of diesel is still cheaper today, it has an unstable price in Brazil and the lack of price stability in the fuel market could be an attractive characteristic of battery systems.

Nowadays, investment in storage without polices that assure at least a minimum profitability to protect the investor would be risky. Daily gains would have to be a lot higher for profitability to be found, and, even in this case, it needs long payback periods, which increases the risks on the investment and drives consumers away.

3.2 Power factor compensation

In Brazil, currently, there is no profitability for the supply of reactive power. However, it could be a way to avoid fines due to low power factor at the point of common coupling of the consumer with the utility. The size of the bi-directional converter should allow the correction of the power factor to values above 0.92.

Power factor compensation could also be a way to increase the value of an existing storage system due to the possibility of supplying ancillary services. This increases the available market, since it could avoid power quality issues at the point of common coupling, which would otherwise hamper the maximum allowable penetration level of storage and renewable non-dispatchable energy in the distribution grid.

In the case of future profitability associated with power factor compensation, this service should at least pay for the energy needed to provide the service. Otherwise, the owner of the battery system would have an expense for providing the service.

3.3 Postponement in capacity investment

The capacity of the distribution system has to account for moments of peak demand which seldom occur, leaving the distribution system full capacity underused most of the time, and yet having to face periods of overload. Figure 3 presents the typical demand profile for residential, commercial and industrial consumers in the city of São Paulo used by its utility [5], along with their respective Load Factor (LF). Assuming that the feeder supplying such load is using most of its capacity (Use Factor close to unity), the same line could theoretically supply consumptions 49 %, 69 %, 118 %, higher, respectively for residential, commercial, and industrial demand profiles. In practice, however, such increase in the consumption would not be possible due to the time at which the consumption occur.

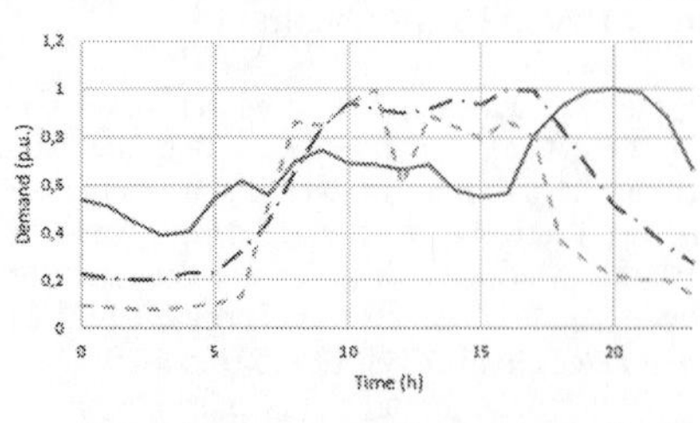

Figure 4: Evaluation on capacity investment.

Moreover, increases in the demand in a certain part of the grid would affect all upstream lines, while storage systems may be able to allow the supply of an increased demand by shifting energy under low loading and supplying it to the loads during peak time. In this manner, all upstream lines would not face increased loading, since the storage system could be charged during moments of low loading.

The cost for capacity reduction due to the use of storage systems would not increase linearly, since it would be influenced by the demand profile. For instance, in Figure 5 areas A1 and A2 represent the amount of energy needed to reduce the same amount of capacity in two different situations. From this figure it is noted that area A2 is larger than area A1, even though the decrease in demand is the same.

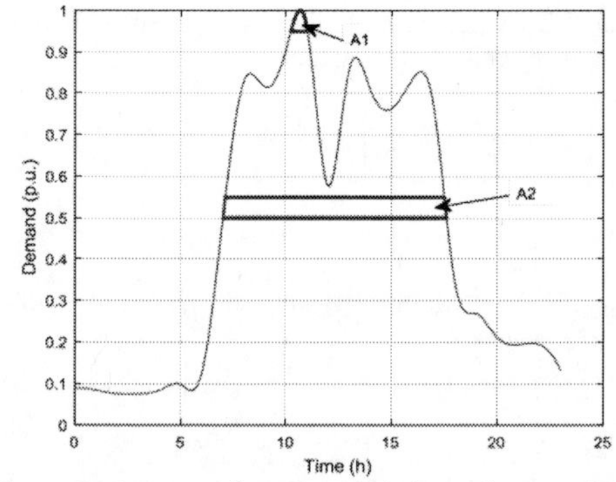

Figure 5: Amount of storage needed to reduce needed capacity for the industrial profile.

This leads to saturation in the ability to avoid increasing the installed capacity with the usage of storage systems. Figure 6 presents the amount of storage needed as the avoided demand increases. Especially for the commercial and industrial profiles, it is noticed that above a certain value (0.07 and 0.15, respectively for the commercial and industrial profile) there is a fast increase in the amount of energy needed to avoid line capacity increase.

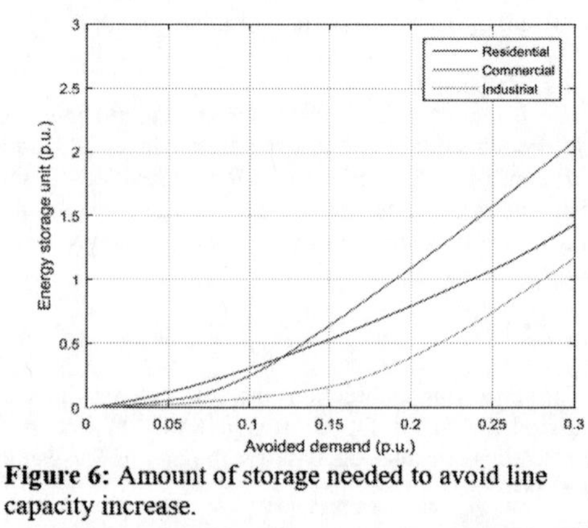

Figure 6: Amount of storage needed to avoid line capacity increase.

Figure 7 presents the behavior of avoided demand unit per storage capacity unity for the three-abovementioned demand profiles as the avoided demand increases. The figure indicates, as mentioned before, saturation in the ability to avoid capacity increase. As the avoided demand increases, the aforementioned relation becomes smaller, which indicates little reduction in the line capacity per added storage unit. This suggests that storage may be suitable to avoid overloads during peak hours, since the amount of energy would not be large, but would not be very effective to significantly avoid capacity increase of the system, since the size of the storage system would increase exponentially.

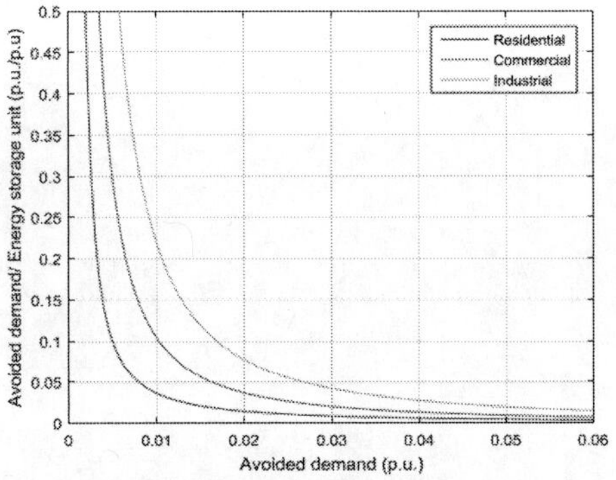

Figure 7: Avoided demand unit per storage capacity unity as a function of avoided demand.

In the case of the commercial and industrial profiles, a PV-battery system could benefit from the coincidence between generation and demand. This would affect the amount of energy needed to avoid capacity increases. In the case of the residential profile, the use of a PV generator affects little the needed storage size, due to the non-coincident peak values of generation and demand.

3.4 Phase balancing

Unbalanced phases may cause several problems, especially in weak grids, due to the uneven efforts imposed on the generator. This is an important issue, as microgrids are more and more becoming a reality. In islanded operation, a microgrid can be a weak grid and unbalanced phases could have stronger outcomes, that otherwise would go unnoticed when connected to the bulk system.

4 CONCLUSIONS

This paper presented the use of storage systems for different applications. The applications selected involve both energy and power applications.

The operation of power distribution systems would become more complex in a scenario with active agents negotiating with the grid, at least initially, because despite the greater flexibility of operation, if clear and well-defined rules are not established on how the agents must interact, unwanted situations may occur. It is noteworthy that the objectives of the various agents that interact with the distribution system are not necessarily the same, they will not necessarily be interested in cooperating, and may at times compete for the purchase or sale of energy, or the provision of services. The distribution grid cannot accommodate any penetration amount of active elements and the disordered operation of such elements may produce unwanted effects such as: reductions in the life of system's components by overload, overvoltage in distribution feeders, creation of new demand peaks and/or their displacement.

In Brazil, although some forms of distributed resources can be connected to the distribution grids according to the conditions established by ANEEL's Normative Resolution No. 482/2012 [6], there are no established conditions for coupling storage systems with bi-directional inverters. This task, however, is not trivial since storage systems for energy exchange with the grid would allow new modes of operation that would affect how the various agents would interact with the grid, and even generate conflicts of interest between these agents.

For example, private utilities might not be interested in a situation where consumers with storage would buy energy at a lower price to sell the same energy to the utility at a higher price. However, in another scenario the same situation could be seen by utilities as the provision of an ancillary service that would reduce thermal stress on lines and transformers that would otherwise be overloaded. This example illustrates how different agents may have different interests in the same situation, and thus it is necessary to create clear conditions for all agents to interact.

Currently, for arbitrage application, the use of storage is still expensive. The gains with the purchase and eventual energy sale are still below the minimum value to break even. However, storage could be an option for consumers that have maximum demand contracts, to avoid fines due to maximum allowed demand surpass. In such cases, since the battery will only be used for peak shaving, the size of the storage system would not necessarily be large.

In the case of power factor compensation, the storage system would only be effective if the provision of reactive power were a complementary activity for the storage system, since other technologies would be better suited. In Brazil, currently, there is no profitability for the supply of reactive power. However, it could be a way to avoid fines due to low power factor at the point of common coupling of the consumer with the utility.

In the case of postponement in capacity increase of the system, storage units are better suited to avoid overloads of short duration. As the avoided capacity increases, the amount of storage needed increases exponentially, which, in turn, would also increase exponentially the cost of the system. One of the main advantages of storage systems is that they do not increase the loading throughout the distribution system, while increasing the capacity of a feeder to supply an increased demand would affect all the upstream lines. If properly placed, storage systems can avoid overload, not only in the feeder in which they are installed, but also avoid overload in other feeders, due to the increased energy management capacity.

Acknowledgment

This paper was supported by Companhia de Transmissão de Energia Elétrica Paulista (CTEEP) through Project PD-00068-0038/2016 under the R&D Program of the Agência Nacional de Energia Elétrica (ANEEL).

5 REFERENCES

[1] Xian He, Erik Delarue, William D'haeseleer, Jean-Michel Glachant. A novel business model for aggregating the values of electricity storage. Energy Police 29 (2010), pp 1575 – 1585. DOI: 10.1016/j.enpol.2010.12.033

[2] Joern Hoppmann, Jonas Volland, Tobias S. Schimidt, Volker H. Hoffmann. The economic viability of battery storage for residential solar photovoltaic

system – A review and a simulation model. Renewable and Sustainable Energy Reviews, 39 (2014), pp 1101 – 1118. DOI: 10.1016/j.rser.2014.07.068

[3] Benedikt Battke, Tobias S. Schimidt, David Grosspietsch, Volker Hoffmann. A review and probabilistic model of lifecycle costs of stationary batteries in multiple applications. Renewable and Sustainable Energy Reviews, 25 (2013), pp 240-250. DOI: 10.1016/j.rser.2013.04.023

[4] Yanyan Tang, Qi Zhang, Benjamin Mclellan, Hailong Li. Study on the impacts of sharing business models on economic performance of distributed PV-Battery systems. Energy, 161 (2018), pp 544 – 558. DOI: 10.1016/j.energy.2018.07.096

[5] AES Eletropaulo. NT-2018: Cálculo da demanda Rede para acréscimo de novas cargas na rede de distribuição de baixa tensão. Technical Report. 2007. Available at: https://www.eletropaulo.com.br/padroes-e-normas-tecnicas/manuais-normas-tecnicas-e-de-seguranca/Documents/Padr%C3%B5es%20e%20Normas%20T%C3%A9cnicas/NT-2018.pdf

[6] ANEEL. Resolução Normativa 492. Agência Nacional de Energia Elétrica. 2012. Available at: http://www2.aneel.gov.br/cedoc/ren2012482.pdf

NEW PHOTOVOLTAIC SYSTEM BASED ON SOLAR RADIATION SPLITTING TECHNOLOGY

M. Tonezzer[1], P. Bernardoni[1], D. Vincenzi[1], P. Decarli[2]

[1] University of Ferrara, Department of Physics and Earth Sciences
Via Giuseppe Saragat 1 - Building C, 44122 Ferrara (FE), Italy
[2] Trentino Rainbow Energy, Altopiano della Vigolana, Italy

ABSTRACT: The University of Ferrara (UNIFE) and the company Trentino Rainbow Energy S.r.l. (TRE) have developed and patented [1] a concentrating photovoltaic prototype (CPV) based on solar spectrum splitting (SSS). The focused beam produced by the concentrating optics is split in two by a short pass dichroic mirror; the resulting infrared beam is reflected on a monocrystalline silicon solar cell while the visible part of the spectrum is focused on a gallium arsenide (GaAs) cell. The experimental results show that this approach is less sensitive than triple junction solar cells (3J) to the changes in solar spectrum typically associated to the last hours of the day or to the presence of haze.
Keywords: see enclosed list of keywords

1 INTRODUCTION

Concentrating PhotoVoltaics (CPV) is a technology that consists in focusing sunlight on solar cells by means of an optical system typically represented by lenses or mirrors. Because of the smaller area of solar cells needed for each square meter of collecting surface, high efficiency III-V devices can be used. Moreover, cells operating under concentration show an increased efficiency, due to an increase in current and voltage at Maximum Power Point (MPP).

CPV systems typically employ MultiJunction (MJ) solar cells because of their high efficiency and their capacity to manage the current density needed to operate under highly concentrated sunlight; as a drawback, these cells are extremely expensive and the number of manufacturers is still limited. MJ solar cells rely on complex stacks of epitaxially grown semiconductors composing the active junctions and the tunnel diodes needed to connect them in series. The choice of materials of the stack is restricted by the lattice matching condition required in epitaxial growth.

As addressed, in a MJ device, the junctions are series connected and any change in the incoming spectrum causes a current mismatch through the junctions, reducing the efficiency of the cell.

Solar Spectrum Splitting (SSS) technology represents a promising technological alternative to MJ-based systems potentially solving the current matching issue. This approach relies on the use of a dichroic mirror that separates the solar spectrum in bands, which are delivered to single junction cells made of semiconductors engineered to operate on specific wavelength bands [2-4].

This approach, unlike triple junction cells (3J), is by far less sensitive to changes in the composition of the incident spectrum because the single-junction cells are independently operated. Daily and seasonal changes of the incoming spectrum do not lead to any current mismatch across the junction.

2 THE WORKING PRINCIPLE

In this work a CPV system with solar spectrum splitting is presented, the prototype has been developed by the Photovoltaic Laboratory at the Department of Physics and Earth Sciences of the University of Ferrara

A schematic of the concentrator is shown in Fig. 1; the concentrator has a Cassegrain configuration with a 45° dichroic mirror to achieve spectrum splitting.

This configuration allows to reduce the thickness of the concentrator, to increase its angular acceptance and to improve light uniformity on the solar cells.

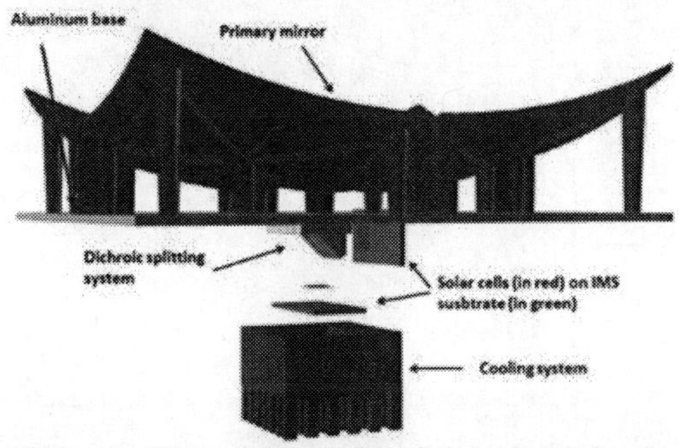

Figure 1: 3D exploded model of the system showing both the concentration optics and the spectral separation device

The primary optics is a freeform concave mirror made of four subcomponents with an area of 100x100mm2 each and a central aperture of 50x50mm^2, while the secondary optics consists of a convex spherical mirror with a semi-diameter and a radius of curvature respectively of 25 mm and 51.6 mm.

The system is characterized by a focal length of 200mm, a geometric concentration ratio C_{geom} = 375 and an optical concentration factor C_{opt} = 300.

The focus of the Cassegrain system doesn't lay on the seating plane of the primary mirrors but 18.8mm behind that in order to accommodate the spectrum splitting optics.

A short-pass dichroic mirror is placed behind the seating plane of the primary concentrator and kept at 45° by a mechanical frame. The cut-off wavelength is 830nm: most of the infrared portion of the incoming light is reflected to the cell with a lower bandgap while the shorter portion of the spectrum is transmitted the cell with a higher bandgap.

An aluminum heatsink has been designed with the dual purpose of cooling the cells and protecting the dichroic filter (See. Fig. 2), the same structure can be used with no dichroic mirror in case the concentration system has to be used with 3J cells.

Figure 2: Detail of the heatsink that acts also as a support for the cells and the dichroic mirror

The numerical simulations performed to optimize a system with two sets of cells show that the optimal choice of semiconductors is Gallium Arsenide (GaAs) for visible light up to 800-900nm plus monocrystalline Silicon (c-Si) for the infrared part of the spectrum.

The simulations are made taking into account the AM1.5 spectrum [5], the external quantum efficiencies of the cells and the transmittance and reflectance values of the optical filter (see Fig. 6): the best theoretical efficiency is obtained with a mirror having a cut-off wavelength of 870nm.

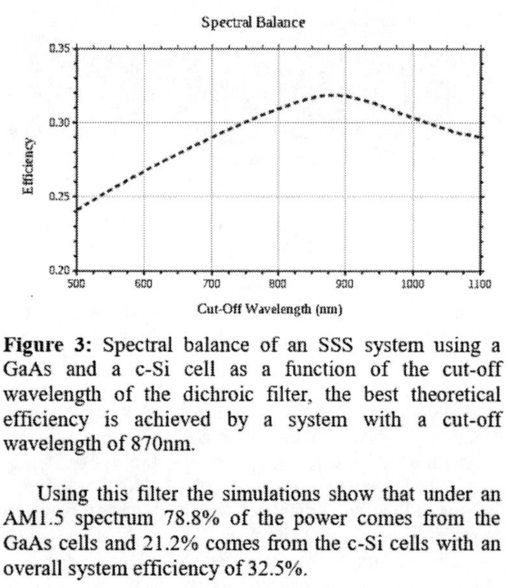

Figure 3: Spectral balance of an SSS system using a GaAs and a c-Si cell as a function of the cut-off wavelength of the dichroic filter, the best theoretical efficiency is achieved by a system with a cut-off wavelength of 870nm.

Using this filter the simulations show that under an AM1.5 spectrum 78.8% of the power comes from the GaAs cells and 21.2% comes from the c-Si cells with an overall system efficiency of 32.5%.

3 EXPERIMENTAL SETUP

In order to assess the performances of the system, a small module with four optics have been assembled in order to compare SSS and MJ technologies:

- With SSS and concentrating optic (400 suns)
- With SSS with no optic (1 sun concentration)
- Without SSS with concentrating optic (400suns)
- Without SSS and concentrating optic (1 sun)

The module was placed on a solar tracking system and the IV curves were acquired as well as the DNI measurement, the power of the SSS system is calculated as the sum of the power outputs of the cells.

The IV curves were acquired with a Keithley 2400 SourceMeter and a software, internally developed, written in LabView.

4 RESULTS

The experimental results of the four configurations are shown in Tab.1, both at 1 sun and under concentration the SSS system show a lower efficiency than the system with the triple junction cell, the results regarding the power balance are far from the values obtained from the simulations (see Fig.3).

In particular the contribution of the GaAs cell to the system power output (65-67%) is significantly lower than expected (79%); this is probably caused by a sub-optimal splitting of the spectrum.

The comparison of the current balance gives a similar information since under direct irradiance at AM1.5 the expected values (obtained from the solar spectrum and the quantum efficiencies) are 73%/27%.

Given that the current is proportional to the number of photons absorbed by the cell this result is coherent with the hypothesis of a sub-optimal optical balance of the spectrum splitting filter.

Table I: Comparison of the four different setups, DNI 822W/m^2, 11-12 UTC.

Configuration	**Efficiency**	**GaAs/Si Power Balance**	**GaAs/Si Current Balance**
3J (1 sun)	32.2%	-	-
SSS (1 sun)	23.4%	65%/35%	56%/44%
3J (400 suns)	34.0%	-	-
SSS (400 suns)	24.9%	67%/33%	58%/42%

The same systems were tested periodically during the day in order to evaluate the possible variations in efficiency caused by the atmospheric conditions, the maximum efficiency of each configuration is taken as a reference and the other values are plotted as a percentage of this value, these measurements were taken without the concentrating optic in order to avoid any possible influence caused by misalignment or mechanical play.

Regarding the variations in relative efficiency, as expected, the spectral separation approach shows a low sensitivity to changes in the composition of the solar spectrum, in this work this change is associated to the longer path of sunlight through the atmosphere at dawn and dusk.

In particular the efficiency exhibits variations within 5% and no time dependent trend (see Fig.4).

Under these same conditions, the increased absorption of the shorter wavelengths caused by the longer path through the atmosphere cause a drop in the current produced by the top junction (typically InGaP) of the 3J cells, the resulting current mismatch is associated with a decrease in efficiency up to 14% (see Fig.4).

With a variation in efficiency limited within 5% the spectrum splitting technology has a significant advantage over triple junction cells in situations of hazed sky and in late afternoon with the maximum advantage at dusk (see

Fig.4).

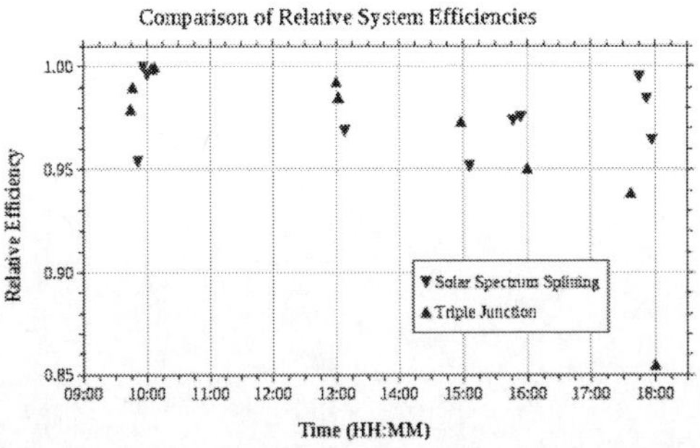

Figure 4: A comparison of the relative system efficiencies (July 2013, DNI>700W/m^2, 1 sun, Ferrara, Italy) show that the SSS system efficiency in late afternoon is almost constant despite the change of the spectral composition of sunlight.

However, the measurement of the absolute efficiency of the system show that the triple junction cells are more efficient and thus produce more power than the cells in the spectral separation system, while the triple junction cells reach an efficiency of 28.9% the SSS system reach a maximum of 21.5% (see Fig.5).

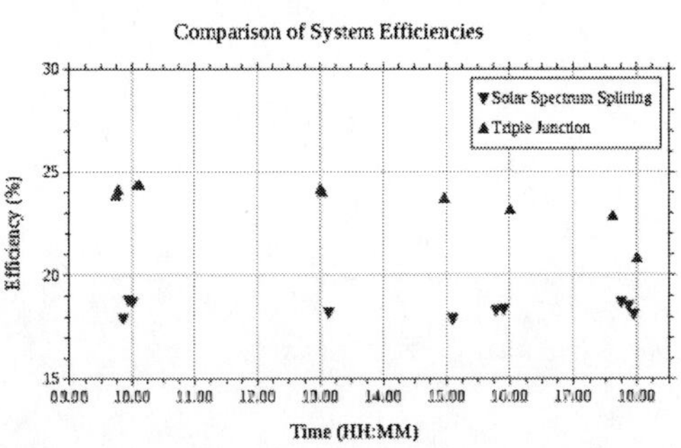

Figure 5: Absolute system efficiencies (July 2013, DNI>700W/m2, 1 sun, Ferrara, Italy) show that the SSS system is less efficient than the one with 3J cells.

Since the efficiency of the SSS system is significantly lower than expected the various components of the system have been re-tested in order to find the cause of this behavior.

The main problem seems related to the particular filter used to separate the spectrum since the actual cut-off wavelength of 830nm is significantly different than the nominal one of 875nm (see Fig. 6) moreover the transmittance of the filter in the visible spectrum ranges from 78% to 94% further limiting the efficiency.

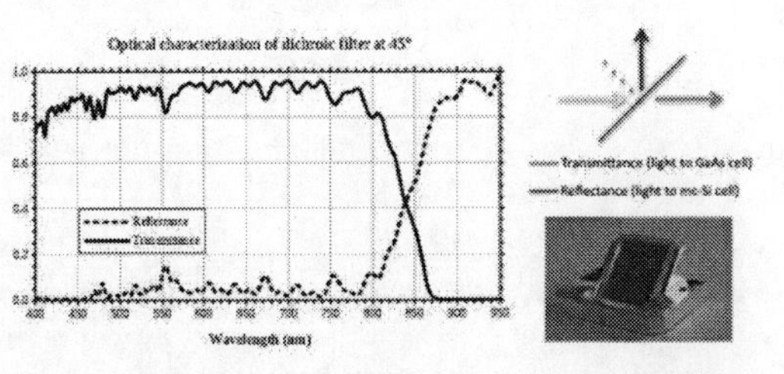

Figure 6: The short-pass filter used for spectral splitting shows a cut-off wavelength (50% transmittance) of 830nm instead of the required value of 875nm, this difference causes a significant drop of the efficiency.

5 CONCLUSIONS

A CPV system with Solar Spectrum Splitting technology has been developed

The main features of this system are:

- The lower cost of the single junction cells compared to the more expensive 3J
- A lower sensitivity to changes in the composition of the solar spectrum such as the presence of haze, at dawn or at dusk.

The short-pass filter used in the spectral separation system is the main cause of the poor efficiency of the SSS system that is highly dependent on the optical quality of this component.

6 WORK IN PROGRESS

Two larger size prototypes of the system were made each one consisting of 20 concentrators with an improved optical design (see Fig.7) based on hexagonal primary mirrors, flat secondary mirrors and a waveguide in order to reduce the divergence of the beam entering the spectral splitting system.

One of the systems has the spectral splitting optics using GaAs and c-Si cells while the second has been fitted with standard triple junction cells without the dichroic mirror and is taken as a reference.

The two systems are currently field-tested at Vigolo Vattaro, Autonomous Province of Trento (Italy).

Figure 7: A picture of a system installed at Vigolo Vattaro, this design uses hexagonal primary mirrors, a flat secondary reflector and a waveguide

7 REFERENCES

[1] D. Vincenzi, P. Decarli, M. Tonezzer, "Solar Concentrator", Italian Patent, BO2012A000348.

[2] A. Mojiri, R. Taylor, E. Thomsen, G. Rosengartend, "Spectral beam splitting for efficient conversion of solar energy - A review", Renewable and Sustainable Energy Reviews, 28, 2013, 654-663.

[3] N. Mohammad, P. Wang, D. J. Friedman, R. Menon, "Enhancing photovoltaic output power by 3-band

spectrum-splitting and concentration using a diffractive micro-optic", Optics Express, 22, 2014, 1519-1525.
[4] D.Vincenzi, A. Busato, M. Stefancich, G. Martinelli, "Concentrating PV system based on spectral separation of solar radiation", Physica Status Solidi A, 2, 2009, 1-4.
[5] American Society for Testing and Materials (ASTM) - Terrestrial Reference Spectra for Photovoltaic Performance Evaluation

PHOTOPUR – PV POWERED WATER DECONTAMINATION FOR WINE GROWERS

Jesus da Costa Fernandes, Michael Schmidt, Elmar Bollin, Florian Ziebold, Simone Lapp
Hochschule Offenburg – University of Applied Sciences / INES – Institute for Energy System Technology
24, Badstraße, 77652, Offenburg, Germany
Phone: +49 781 205 354, Fax: +49 781 205 45354
E-mail: costa@hs-offenburg.de
Internet: fgnet.hs-offenburg.de, ines.hs-offenburg.de, www.hs-offenburg.de

An energy oriented design concept was developed within the research project PHOTOPUR which has the development of a PV powered water cleaning system as main focus. During a wine season Plant Protection Products (PPP) are several times sprayed on plants to protect them of undesired insects and herbs or avoid hazardous fungus types. A work package of the project partner INES in Offenburg led to a design concept introducing an energy profiling already in the early beginning of a product design. The concept is based on three pillars respecting first the requirements of the core process making up filtering and cleaning and secondary aspects which run, support, maintain and monitor the system to secure availability and product reliability. Finally the composed energy consumption becomes a product specific load profile. After analyzing product process and the energy supply conditions assumptions mainly using data sheets were taken to define media and energy flows. Single components and functional groups compose the load profile and are therefore monitored. In a parallel step local conditions of a photovoltaic energy supply were gathered to be included in the system model of a design tool. The tool should help to evaluate technical design parameters and viability. It also enables a view on economic efficiency of competing power supply configurations i.e. solar-battery-combinations. The presented paper shows that the results of the design tools guided the developers to assemble a functional model of the water decontamination unit which was manually tested with its concatenated steps of the water cleaning process.

Keywords: PV-Battery-Systems, PV-powered applications, Renewable energy supply, Smart PV applications

1 INTRODUCTION

1.1 THE PHOTOPUR PROJECT

When wine growers clean vehicles and equipment used for spraying Plant Protection Product (PPP) the water for cleaning gets contaminated by pesticides and is then mostly released to surface waters.

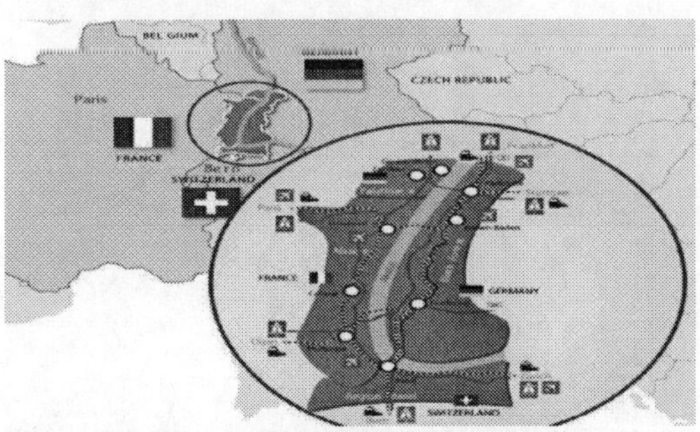

Figure 1a) Wine region TMO (Trimetropolitan Upper Rhine region, www.upperrhinevalley.com) located in the triangle between Germany, France and Switzerland.

Figure 1b) Wineyard in the region Baden/Germany. Photo:J.daCosta

Research works conducted by the Institute of Environmental Sciences at the University of Koblenz-Landau show that these active PPP substances can appear very concentrated and have therefore a high toxic potential to damage ecosystems. At present there is no widespread application or method known to effectively and permanently reduce the impact of pesticides to surface waters.

Figure 2: High pressure cleaning of tools and vehicles.Source:www.kärcher.com.

Currently used methods to avoid entries of pesticides into waters such as activated carbon treatment have so far not proven ecologically and economically effective. Recent publications on the continuous degradation of the water quality forces actors of the viticultural and agricultural market to take measures.

Based on works of Robert, Keller, Malato, Helali and Romero [1-5] pesticide molecules can be cracked down to their ionic components by photocatalytic processes. A concept to decontaminate the content of water tanks was developed within the PHOTOPUR research group as shown in **Figure 3a** using photocatalytic processes shown in **Figure 3b** as reaction principle. As final target the concept will be brought to a marketable product.

The corresponding purification process is developed and validated by project partners in Strasbourg and Landau, who bring in the corresponding environmental, ecotoxicological and process engineering competence.

The Institute of Energy Systems Technology (INES) at Offenburg University contributes the self-sufficient energy supply and the automation concept for the process and can rely on many previous projects in this field. This paper focuses on the design method for the energy supply part of the water decontamination system that strongly interacts with the process automation concept.

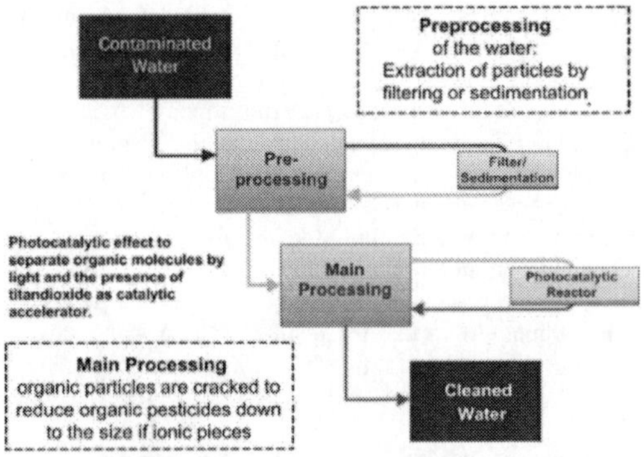

Figure 3a: Schematic view of the process chain

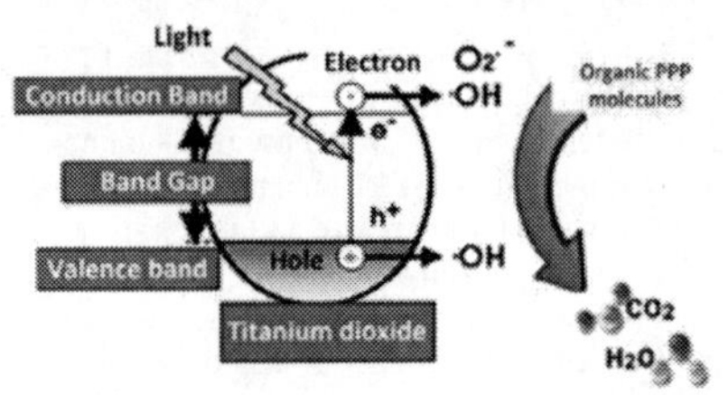

Figure 3b: Reaction scheme of the photocatalytic process

1.2 DESIGN METHOD FOR PV APPLICATIONS

Integrating renewable energy sources already in an early phase of the design of technical systems and their control and automation concept at INES has proven to be a very successful way to finally get sustainable and energy self-sufficient products.

In the following we will show such an integrated design process of energy supply and automation using the example of a photocatalytic water decontamination system that is powered by photovoltaic electricity in combination with a battery.

The design mainly relies on three pillars as presented in Figure 4. First the he design and control of the main process steps lead to a corresponding main energy load profile. Second the consideration of necessary secondary processes lead to a corresponding secondary energy profile. Third the energy supply profile of the renewable energy source is mainly depending on the site-dependent weather conditions and the size and design of the photovoltaic generator with the battery capacity.

The total demand profile, i.e. the sum of loads of main and secondary processes and the supply profile then have to be matched by making use of flexibility in the load design and by shifting energy in time by means of a battery.

Since the process steps and its automation and control are not yet completely determined in the first design phases, the modeling of the process and the estimation of corresponding load profiles should be flexible and open to later changes in the concept and in process parameters. For instance the estimation of the load profile is done by means of an own-developed flexible EXCEL/VBA tool, combining the power demand of significant loads like pumps and valves according to the automation process.

The automation process is defined by functional steps that are processed sequentially or in parallel within certain time slots. Length and power demand of single steps depend on the process and the automation concept. Some details of the control and automation concept are given in Section 2. Some details of its modeling with respect to energy demand within the EXCEL/VBA tool as well as the calculation of expected power supply are given in Section 3.

For a later stage of the project, an energy-related optimization of the process steps is planned. This optimization will, however, take special care of process-related constraints. In particular, dealing with fluids, many processing steps are very time sensitive, i.e. they need a minimum processing time and breaks between processing steps might influence flow properties and mixing of fluids etc. Again, the EXCEL/VBA-tool will be helpful to study the effect of optimizations on the energy supply solution. However, the optimization loop is not included in this paper.

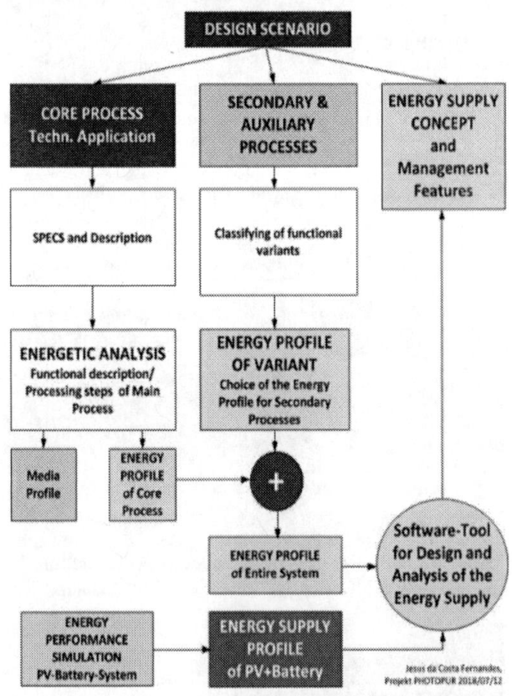

Figure 4: PHOTOPUR - Design process as flowchart.

With respect to practical implementation, a first functional assembly was set up and manually operated. One goal was to see the transfers of water and proper operation of the feed in and the filtering and cleaning process. Short monitoring campaigns measure energy demands of significant electrical devices and were included as real values of the load profile. Previous load profile analysis was done in projects already presented at EU PVSEC 2017 [8].

2 AUTOMATION AND CONTROL CONCEPT

The treatment process of pesticide-contaminated water should be carried out directly at the place of origin or collection, so an automated process of the system is required. Based on the general sequence of the treatment process in Figure 3 this automation was realized according to Figure 5 showing the basic flow diagram of the process.

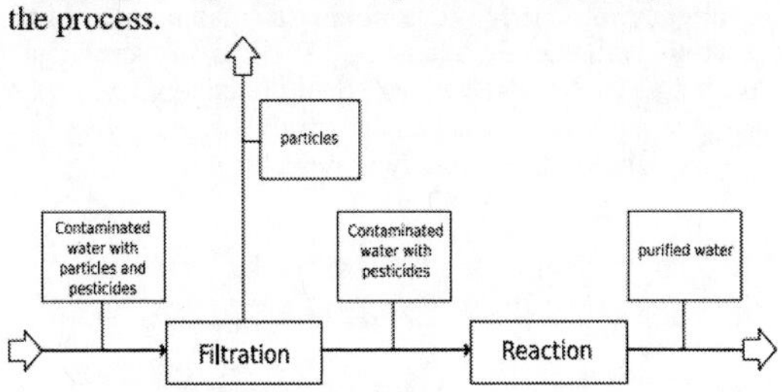

Figure 5: Basic flow diagram

The treatment of the water is divided into two sub-processes. A preprocess which cleans the process medium of suspended matter by filtration/sedimentation as well as a main process in which the water is cleaned by a photocatalytic reaction.

The two sub-processes work with different states of the process medium. Therefore each one of these sub-processes requires its own container for its respective liquid. In the pre-process a closed loop is needed which contains the filtration/sedimentation unit. The main process also needs a closed loop which however contains the photocatalytic reactor unit. The structure of the sensors and actuators for the control can be configured identically for both circuits.

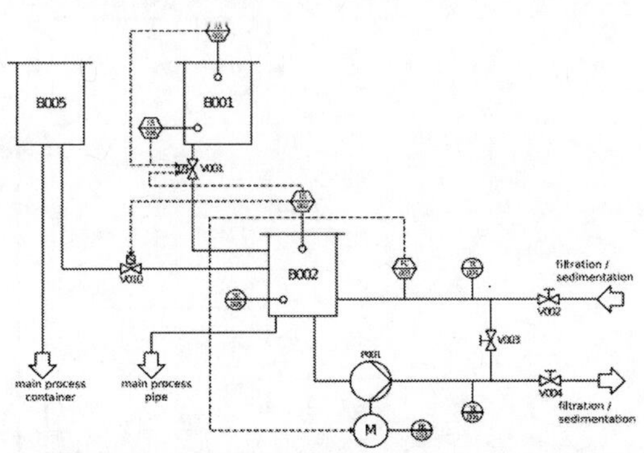

Figure 6: Piping and instrumentation chart of the pre-process

Figure 6 shows the piping and instrumentation diagram of the pre-process, with tank B001 for the input volume, the pre-process medium tank B002 and a clean/fresh water tank B005 as a surge tank. The level switch (LS) and the solenoid valves V001 and V010 control the process flow of the pre-process. In addition, the flow rate of the process medium in the circuit can be regulated with a flow control between the flow meter FC and the pump P001. The temperature measurements (TE) and the current measurements (EE) are initially intended only for data acquisition and are not considered for control purposes.

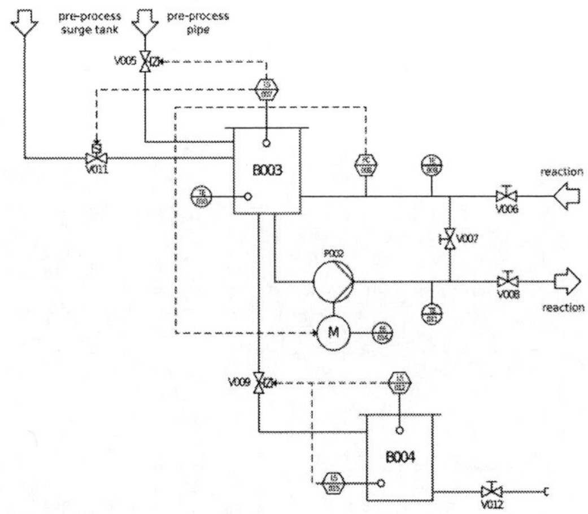

Figure 7: Piping and instrumentation chart of the main process

Figure 7 describes the piping and instrumentation of the main process with the tank B004 containing the final product.

With this system configuration an automation of the system can be realized on basis of a sequence control. The process is divided into three process steps.

- Startup process: Only the pre-process is active. This process step is executed after the system starts.
- Iteration process: The medium to be processed is available for the pre-process and the main process. Both sub-processes are carried out at the same time until the pre-process has completely emptied the starting container B001.
- Discontinuation process: Only the main process is active. The pre-process can no longer be carried out.

Between these three process steps transition phases take place in which the process medium is transferred from one tank to the next. By considering these additional states, the flow control of the system can be derived.

Figure 8 shows the sequence function chart of the controller with possible states and required positions of the actuators. The translations show the integrated sensors of the controller. Actually there are no sensor signals serving as process quality indicators for turbidity of the pre-process or contamination indicator for remaining pesticides during the main process. Therefore the pre-process and the main process are controlled by a defined process time according to experienced decontamination values of the project partners in Landau.

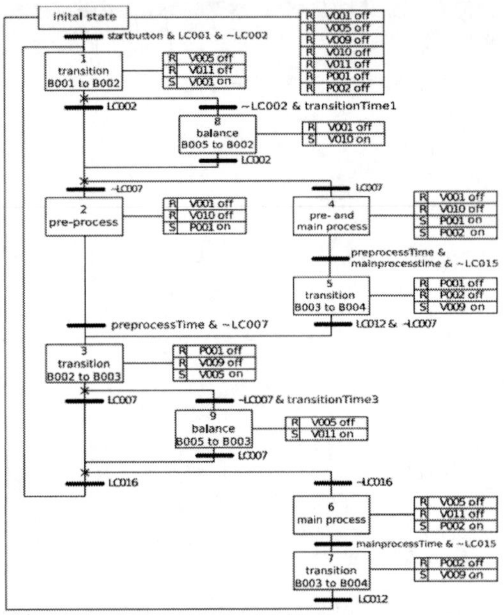

Figure 8: Sequence function chart of the flow control

3 ENERGY SUPPLY CONCEPT

As another work package of the project PHOTOPUR a tool for the design of an energy supply for stand-alone water purification systems was developed. It simulates the power production of a given photovoltaic system, the state-of-the-Charge (SOC) of the battery storage and the energy requirements of the water purification system. Thus the required PV power and battery capacity for a water purification system at a desired size can be determined. The tool was implemented in Excel using VBA-Code. This section gives an overview on the models that were used in the tool.

The load profile of the water purification system is characterized by the sequence of the cleaning as main process. Several process steps are combined for a single run. A run includes all the operations necessary to purify the water of a deposit filled with contaminated water.
For the creation of the load profile consumers of the water purification system are assigned to the individual process steps. It is assumed that during a process step always the active consumers are running at nearly constant power.

The duration of a process step and the performance of the individual components can be changed individually. The load profile then follows the chart of Figure 9:

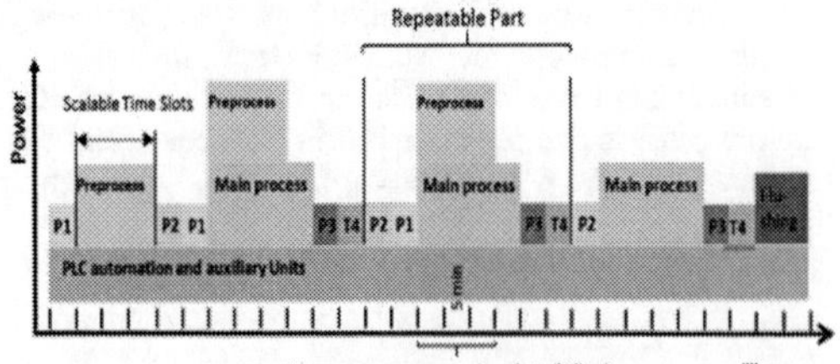

Figure 9: Time chart of the logic controlled process steps to define the expected load profile.

The simulation of the battery is based on the battery model of Quaschning [14]. The model has been slightly extended to consider additional charge and discharge states of the battery which are required for proper simulation. Apart from the cases "load" and "unload" the cases "residual discharge", "trickle charge" and "overcharge/deep discharge protection" were included. The "residual discharge" case occurs when the battery cannot deliver all of the energy demand within the time interval but the battery is still being depleted of residual energy. The trickle charge event occurs when the battery is already full enough to hold only a small amount of the energy available for charging. The case "overcharge / discharge protection" occurs when the battery is either to be charged but can no longer absorb any energy or is to be discharged but no energy is available.

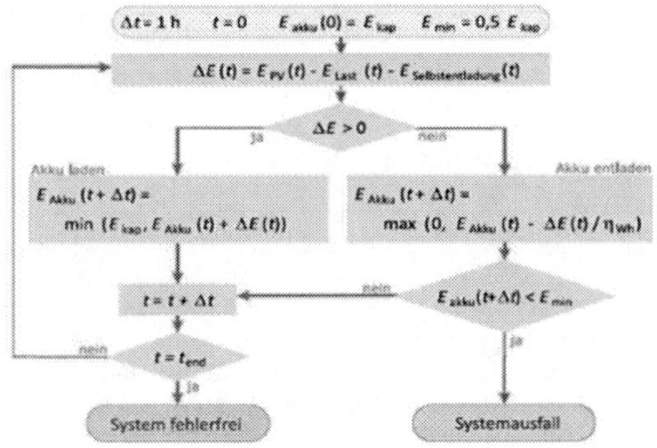

Figure 10: Model to charge and discharge a battery [12]

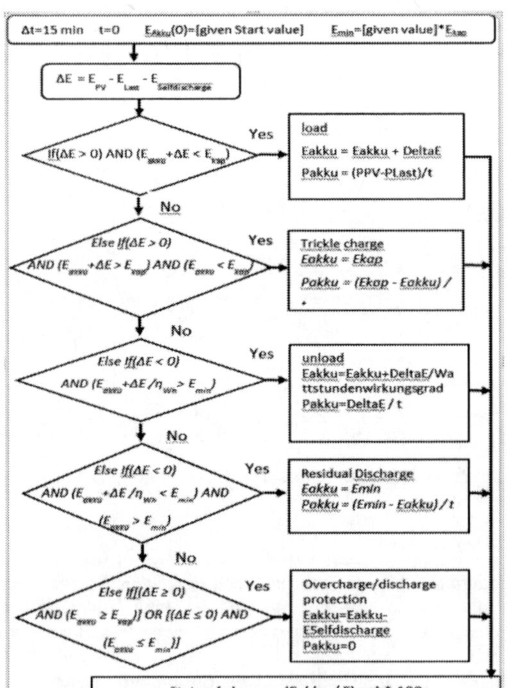

Figure 11: Flowchart to determinate the SOC of a battery [14]

To calculate the power of photovoltaic modules, first the irradiance on the inclined plane of the modules is calculated. The simulation is realized in such a way that the orientation of the modules (azimuth and inclination angle) can be specified and the irradiation on the module surface is calculated accordingly. Then the produced power is calculated with the help of the module efficiency. The efficiency is calculated using the STC efficiency of the modules and a correction for a different spectrum of the irradiation ($\Delta\eta_{\mathrm{irr}} \approx -0{,}04 \cdot \eta_{\mathrm{StC}} \cdot \ln(s)$ with $= \frac{E_{G,\mathrm{gen}}}{1000\,\frac{\mathrm{W}}{\mathrm{m}^2}}$) and a correction for the temperature deviation of the module ($\Delta\eta_{\mathrm{T}} = T_{\mathrm{K}} \cdot (25\,°C - \theta_{\mathrm{M}})) \cdot \eta_{\mathrm{STC}}$). Both models were extracted of the guidelines "Photovoltaic Systems" [16].

The visualization of the simulated data gives an overview about the system configuration and performance. Figure 12 shows the demand shares of the electrical devices or functional groups.

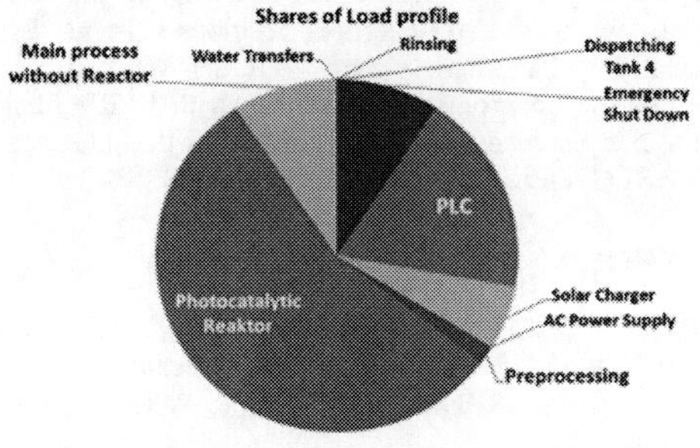

Figure 12: Energy demand shares making up the load profile

As the design tool should also produce some information about economic aspects and reliability of the PV System the charts of **Figure 13** give an overview about the solar coverage (left) and the possible supply gap (right) indicating if there were times where the control of power supply leads to a pausing of the main process.

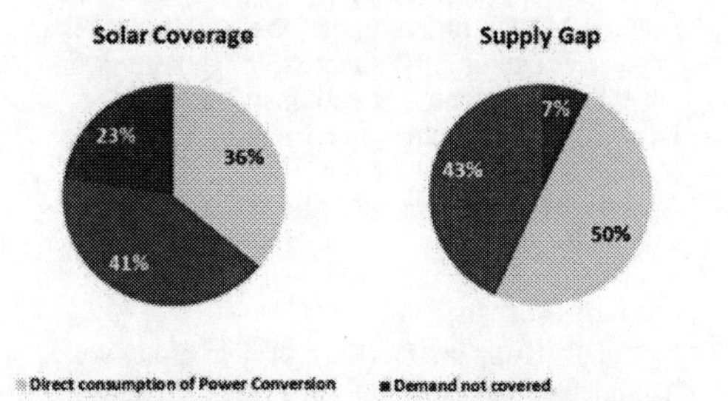

Figure 13: Solar energy coverage and supply gap

Finally the historic graph offers the well-known timeline graph of the simulation run. The left axis of **Figure 14** shows the power supply shares (load curve highlighted in red) and the load profile while the right axis indicates which SOC value (dashed line) is given according to the used balancing model of **Figure 11**.

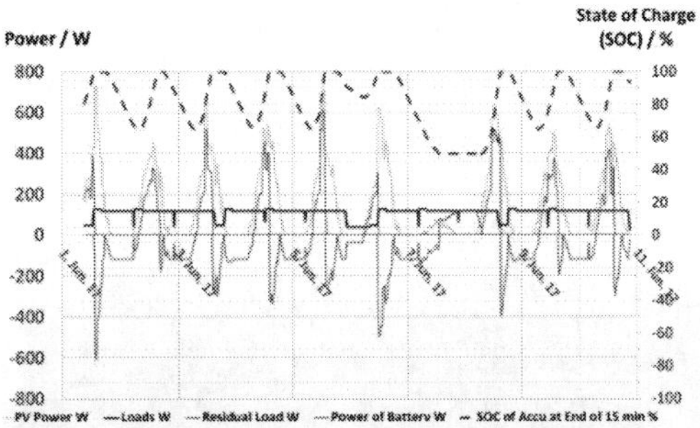

Figure 14: Historical timeline graph of simulated data, not including the preprocessing.

4 RESULTS AND DISCUSSION

The developed EXCEL-VBA tool in combination with the current control and automation concept was used to generate a preliminary design of the supply system. However, the unprecise definition of the automation processes lead to large uncertainties in the load profile of the entire water processing. Hence, the design process must remain flexible. Parameter identification and optimization runs and corresponding data collection will be necessary throughout the whole development process to reduce these uncertainties and to finalize the supply design.

For this reason, already the current functional model assembly has been equipped with additional shunts to reveal specific current profiles of sub-circuits. Detailed acquisition of load data helps to correct previous profiles generated by the design tool and makes simulation runs more precise.

In the current functional assembly, a ratio of 3:7 between auxiliary and useful power for main processes seems to be realistic. This could be considered as moderate but could certainly further optimized in the future.

Also the battery capacity is typically prone to large uncertainties, with respect to uncertainties in the load profiles but also in the practical implementation due to deviations from data sheet values. Therefore, first fully performed charge-discharge-cycles are used as reference to define the battery state levels "deep discharged" or "fully charged". The levels are helpful limits to calculate the SOC value of the battery.

Using the simulation run of the design tool the charts of Figure 13 reveal a supply gap and the solar coverage of the PV powered energy supply. A minimum function for the combination of accumulator cells and PV panels show that the supply gap can be reduced to zero if the solar coverage and a probable surplus of not-used power are respected.

The analysis of the SOC curve gives an idea how the battery cycling is done for the chosen configuration as deep cycling reduces significantly the battery lifetime.

A comparison between the simulation runs with the commercial software PV-SOL Expert 5.5 (R6) and the own design tool of Lapp [12] comes to a difference of about 2 % for the daily energy data. This difference is already quite acceptable and could probably further reduced by further tuning of the tool's solar power model's handling of direct and diffuse sunlight. The weather data were generated for the location Strasbourg (F) using the software module METEONORM© [17] in TRNSYS© [18].

The setup of the water decontamination system undergoing the presented design concept led to the design of Figure 15. The final assembly given in Figure 16 is planned to be completed and tested at the end of the year.

Figure 15: Design sketch of a functional model.

Figure 16: Assembly of the functional model at INES.Photo: J.daCosta

With the ambitious aim to bring a project idea close to market a design concept was followed that allows acceleration of product design for technical applications powered by solar panels. Not focusing only on the power supply the used concept leads to a smart energy management for the technical device. The integrated design of a technical application offers a detailed and analytic monitoring of the product performance and allows energy optimization for remote systems running as stand-alone units but also being able to run as powerline-connected elements.

5 SUMMARY AND OUTLOOK

The reduction of PPP products of surface waters in viticulture is the key motivation of the Project PHOTOPUR. For this aim a water cleaning and decontamination unit using the principle of a photocatalytic process will be developed. This unit requires an appropriate process automation and shall come with a self-sufficient energy supply.

Another project objective is a short time to market. This requires a development of the process automation and energy supply solution in parallel to the development and optimization of the principle decontamination processes. The challenge of such an approach is that the design of the automation and energy supply has to be very flexible at dealing with a large number of uncertainties in the key design parameters.

Therefore a quite generic automation and control concept has been developed and even been implemented in form of an early functional assembly that allows to program already PLC-controls and to make first measurement campaigns to further improve the concept and corresponding system models.

Also the energy supply design has to be very flexible. Therefore an EXCEL/VBA tool has been developed to model the process energy demand profiles and to match it with the supply of a PV-battery system.

These works have been supported by two thesis works on the automation and control concept [11] and on the energy supply concept [12]. The chosen design process helped to significantly move the project forward. The result is the assembly of a fully working functional model.

The base functions of the water processing were soon setup and manually tested while the electric design and setup remains to be completed.

The next steps will be the completion of the electrical and control function by implementing the PLC and visualization. After commissioning the complete system fine adjustment and hardware and sensor optimization are expected mainly related to main processing.

To underline the analytical character of the functional model a monitoring system will be overlaid on the system to get historical data of the performance of the processing and the energy supply. A remote access to the system via GSM or wideband communication may further complete the communication and remote management. These features may open further analytical functions by allowing an online quality control of the water processing.

Acknowledgements

Water conservation is an important issue in the European Union. The project PHOTOPUR is therefore funded by the TMO-neighbor regions Grand Est, Rhineland Palatinate and Baden-Württemberg. Additionally the project is supported with EFRE funds within the Sciences Offensive 2016 of the European INTERREG V Programme under the Ref. 2385/3.5.

References

[1] Pichat P. (Ed.), Photocatalysis and Water Purification: From Fundamentals to Recent Applications, ISBN 9783527331871, Wiley Online Library, 2013

[2] Malato S. et Al.: Photocatalytic decontamination and disinfection of water with solar collectors, Elsevier ScienceDirect Catalysis Today 122 p. 137–149, 2007

[3] Helali S.: Photocatalysis for decontamination and disinfection of water using different types of suspended catalysts irradiated by artificial (UV-A/UV-B) or natural sunlight, Dissertation, Univ. Lyon, 2012

[4] Romero M. et Al.: García E., Solar photocatalytic degradation of water and air pollutants: Challenges and perspectives. Solar Energy, 6(2), 169-182

[5] Lindner M.: Optimierung der photokatalytischen Wasserreinigung mit Titandioxid: Festkörper- und Oberflächenstruktur des Photokatalysators, Dissertation im Fachbereich Chemie, Univ. Hannover, 1997

[6] Bechtloff J.: Regelungstechnik. 1. Aufl., Vogel Buchverlag,.http://www.contentselect.com/index.php?id=bib_view&ean = 9783834361561. – ISBN 978-3-8343-6156-1, 2012

[7] Bindel T., Hofmann D. (Ed.): Projektierung von Automatisierungsanlagen: Eine effektive und anschauliche Einführung. 3., akt. und verb. Aufl., Springer Vieweg, 2017. – 273 S. http://dx.doi.org/10.1007/978-3-658-16416-4. – ISBN 978-3-658-16416-4, Wiesbaden 2017

[8] da Costa Fernandes J., Schmidt M., Feldmann T., Bollin E.: Use of load profiles to optimize microgrids in non-residential environments, Session 6BV.3.90 at 33rd European PV Conference and Exhibition in Amsterdam/ NL, Paper DOI 10.4229 /EUPVSEC20172017-6BV.3.90, p.2736-2746, Amsterdam/NL, 25.-29.09.2017.

[9] Da Costa Fernandes J.: Forecast based energy optimization for solar photovoltaic and hybrid energy systems, Session 6AV.4.6 at 29th European Photovoltaic Solar Energy Conference and Exhibition, ISBN: 3-936338-34-5, p.3648-3653, Amsterdam/NL, 2014.

[10]Schmelas M., Feldmann T., da Costa Fernandes J., Bollin E.: Photovoltaics Energy Prediction Under Complex Conditions for a Predictive Energy Management System, Journal of Solar Energy Engineering, June 2015, Vol. 137 / 031015-3.

[11] Ziebold F.: Automatisierung und Regelung eines mehrstufigen Wasserreinigungssystems, Bachelor-Thesis at Faculty of Electrical Engineering and Information Technology, Univ. of Appl. Sciences Offenburg, Hochschule Offenburg 2018.

[12]Lapp S.: Entwicklung eines Auslegungstools f für eine PV-basierte Energieversorgung mit Batteriespeicher für ein photokatalytisches Wasserreinigungssystem, Bachelor-Thesis at Faculty of Electrical Engineering and Information Technology, Univ. of Appl. Sciences Offenburg, Hochschule Offenburg 2018.

[13]Hinrichsen, G., Programmierung und Visualisierung mit Beckhoff und LabVIEW eines auf erneuerbaren Energien basierenden Smart Grids und einer Raumautomation als Verbrauchsstruktur, Bachelor-Thesis vorgelegt im Studiengang Mechatronik Plus, Hochschule Offenburg, 2011.

[14]Quaschning V., Regenerative Energiesysteme: Technologie ; Berechnung ; Simulation, mit 119 Tabellen (9th ed.). München: Hanser-Verlag, 2015

[15]Wagner, A., Photovoltaik Engineering: Handbuch für Planung, Entwicklung und Anwendung (4th ed., VDI-Buch), Heidelberg, Dordrecht, London, New York, Springer Vieweg, 2015.

[16]Haselhuhn, R. (Ed.), Leitfaden Photovoltaische Anlagen: Leitfaden für das Elektro- und Dachdeckerhandwerk, für Fachplaner, Architekten, Bauherren und Weiterbildungsinstitutionen, 4th Ed., Berlin: DGS Deutsche Gesellschaft für Sonnenenergie, 2010

[17]Meteonorm Weather Data, www.meteonorm.com, **Meteotest AG,** Fabrikstrasse 14, 3012 Bern, Switzerland.

[18 TRNSYS, www.trnsys.com, Transsolar Energietechnik GmbH, Curiestraße 2, D-70563 Stuttgart

CLOUD ENHANCEMENT PHENOMENON AND ITS EFFECT ON PV GENERATORS

Markku Järvelä*, markku.jarvela@tut.fi, phone: +358 50 366 6194
Kari Lappalainen*, kari.lappalainen@tut.fi
Seppo Valkealahti*, seppo.valkealahti@tut.fi
*Tampere University of Technology, Laboratory of Electrical Energy Engineering, P.O. Box 692, FI-33101 Tampere, Finland

ABSTRACT: Due to the cloud enhancement (CE) phenomenon, global horizontal irradiance can exceed the clear sky irradiance value. Enhanced irradiance areas are formed when photons are scattering from the water droplets in clouds near the direct irradiance path. The CE is well-known as a phenomenon, but its effect on photovoltaic (PV) generators has not been thoroughly studied. Output power of a PV generator is almost directly proportional to the irradiance. Therefore, due to the CE phenomenon, output power of a PV generator can exceed the nominal power, which affects the requirements set to the inverter. We used irradiance measurement data of 440 days from an array of pyranometers to calculate the speeds and movement directions of cloud shadows to deduce the diameters of enhancement areas. During the CE events, the irradiance values were occasionally more than 150% of the expected clear sky irradiance value. We discovered that the diameters of these extreme irradiance areas are tens of meters and, therefore, they can have a full effect on residential or commercial scale PV systems. Utility scale PV generators are in practice affected more by less extreme CE events with larger diameters, since spatial smoothing by the large generator area decreases the effects of the extreme events.
Keywords: Solar Radiation, PV System, Design

1. INTRODUCTION

Global horizontal irradiance (GHI) is defined as the irradiance on a horizontal plane on the surface of the earth. On a clear sky summer day, the GHI is approximately 1000 W/m^2 slightly depending on the latitude. On partly cloudy days, the GHI can exceed the clear sky irradiance value. This phenomenon is generally known as cloud enhancement (CE), but some sources refer to it as irradiance enhancement or over-irradiance [1].

The GHI can be divided into direct normal irradiance (DNI) and diffuse horizontal irradiance (DHI). The DNI is the direct irradiance from the sun and it equals to the amount of extraterrestrial irradiance minus the losses from absorption and scattering in the atmosphere. The DHI is the indirect irradiance scattered in the atmosphere and from the clouds reaching the Earth surface. Under clear sky conditions, the DHI can be measured to be around ten percent of the GHI. [2]

Numerous measurements of extreme irradiance values have been reported all around the world [3]. For example, an irradiance value of 1832 W/m^2 was measured in the southern Ecuadorian Andes Mountains at an altitude of 3400 m [4]. The CE phenomenon can also be observed close to the sea level and at high latitudes. A CE event with irradiance reaching 1600 W/m^2 was recorded in Grimstad, southern Norway at an altitude of 60 m [5].

The CE is well known as a phenomenon. It is due to photons scattering from the water droplets in the clouds in a way that creates enhanced irradiance areas on the ground. The effect of CE has also been studied using computer simulations [6], [7]. The CE is the strongest when an edge of a cloud is near the direct irradiance path. Therefore, extreme irradiance values can be measured when the GHI is enhanced from all sides of the sun disk [5]. This can occur when an opening in a cloud aligns with the sun or two clouds are close to each other.

PV power plants are typically constructed of one or more blocks consisting of a PV generator, an inverter and a transformer. Residential and commercial plants are typically build using string inverters and utility-scale power plants using central inverters. Nominal powers of string inverters are typically from kilowatts up to a hundred kilowatts and central inverters from some hundreds of kilowatts up to some megawatts. The land areas of PV generators are almost directly proportional to their nominal powers. Therefore, the effects of the CE phenomenon on PV power production also depends on the generator size.

The effect of the CE phenomenon on PV power plants has not been thoroughly studied. The short circuit current of a PV cell is almost directly proportional to the irradiance. The maximum power of a PV generator depends on the irradiance on all PV cells and the configuration of PV modules. Therefore, due to the CE phenomenon, the maximum power of a PV generator can be higher than the nameplate value under 1000 W/m^2 STC irradiance. In addition, to minimize the levelized cost of electricity, an industry practice is to design the PV power plants so that the nominal power of a PV generator is higher than the nominal power of an inverter [8]. Due to the oversizing of PV generators, the DC to AC power ratio of an inverter can be much higher than anticipated during the CE phenomenon and the inverter is operating against its power limit.

We have investigated the effect of the CE phenomenon on PV generators by analyzing the incidence of CE events and the diameters of the enhancement areas from the irradiance measurement data of 440 days. Accordingly, the relation between the CE land area and the maximum realistic power of the PV generator has been deduced.

2. METHODS

2.1 Measurement data

The analysis is based on irradiance measurement data from the solar PV power research plant on the rooftop of Tampere University of Technology (TUT) in Finland in Northern Europe [9]. The measurements were done by an array of photodiode-based pyranometers with 10 Hz sampling rate.

The layout of PV modules and pyranometers and dimensions of the TUT research plant are presented in Fig. 1. The pyranometers are installed in fixed 45° angle from the horizontal plane facing nearly due south. The TUT solar PV power research plant consists of 21 pyranometers that are marked as yellow triangle and all of them are located inside the area marked with dashed red line in Fig. 1. The width and height of the research power plant are 45 m and 48 m, respectively. The maximum distance between two pyranometers is 56 m. The land area inside the dashed red line is approximately 1300 m^2, which is a typical land area of a 65–100 kW PV generator [10].

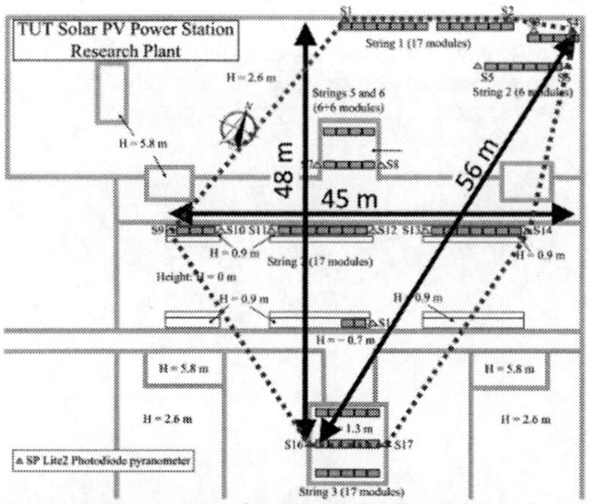

Figure 1: Pyranometer layout and the dimensions of TUT solar PV power research plant.

The measurements were performed for 15 summer months in 2011–2014. Examples of the measured irradiances on a partly cloudy day and a clear sky day are presented in Fig. 2. Due to the CE phenomenon, the irradiance exceeded the clear sky irradiance several times throughout the partly cloudy day.

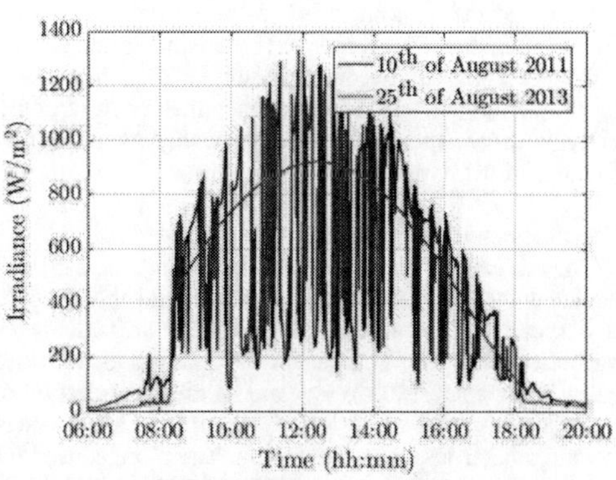

Figure 2: Unfiltered irradiance measurement data on a partly cloudy day on 10th of August 2011 and on a clear sky day on 25th of August 2013.

Maximum irradiance value measured during the observation period was 1438 W/m^2 on 20th of June 2012. It corresponds to almost 60% enhancement, because the maximum clear sky irradiance is 903 W/m^2 for the used setup of fixed 45° tilt angle of the pyranometers oriented nearly towards due south.

2.2 Analysis of the cloud enhancement events

Measuring the durations of enhancement events was straightforward. The duration of an enhancement event is the time that the irradiance exceeds a certain value. Measurement sampling frequency of 10 Hz was used in the analysis. To remove the small ripple from the measurement data, a running average filtering of 1 s was used.

The smoothed irradiance measurement data from a 3-minute period on 10th of August 2011 is presented in Fig. 3. During this period, three cloud shadows passed over the pyranometer resulting in two enhancement events lasting approximately 68 and 85 s. The first enhancement event has a clear cup-like shape, which is typical for the CE phenomenon. It is due to the CE phenomenon being strongest near the edge of a cloud shadow. The second enhancement event consists of three separate peaks. The irradiance drops to approximately 900 W/m^2 in between the peaks. Because of the used method, these individual peaks are counted as separate CE events. Consequently, the number of events in Fig. 3 where irradiance was over 900 W/m^2 was counted to be only two while the number of events with irradiance over 1000 W/m^2 was four, three for irradiance over 1100 W/m^2 and again two for irradiance over 1200 W/m^2. Further, the small ripple in the middle of the of the second CE period of events will cause a series of short enhancement events. The number and durations of enhancement events for 440 summer days are systematically analyzed as a function of irradiance enhancement limit in a similar manner than demonstrated in Fig. 3.

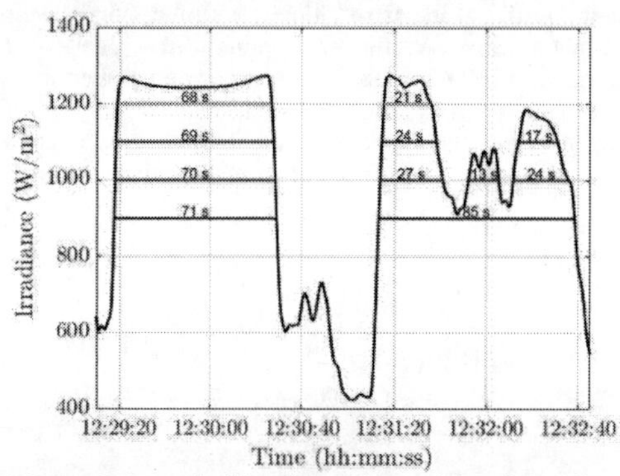

Figure 3: Smoothed irradiance data and durations of enhancement events with four different irradiance limit values during a short period on 10th of August 2011.

2.3 Estimating the diameters of the enhancement areas

Clouds can be classified according to their altitude and optical thickness. The base height of low clouds is under 2 km, middle clouds 2–7 km and high clouds 5–13 km [11]. Because of wind shear, clouds on different altitudes can have different speeds and directions of movement. Therefore, the speeds and movement directions of moving cloud shadows may not be consistent, and correspondingly, enhancement areas on the ground do not always have consistent speeds, movement directions and shapes. In this case, estimating their diameters based on irradiance measurements on a relatively small area is not feasible as such. However, if the cloud shadows are consistently moving in the same direction with similar speed, it can be assumed that the enhancement areas are also moving in the same direction with similar speed. For events that last only tens of seconds, an assumption can made that the clouds are static enough so that the shapes of the enhancement areas do not change during the event.

The speeds and movement directions of cloud shadows were calculated by parametrizing the cloud shadow edges and calculating the time differences of shadowing events between three pyranometers by applying the linear cloud edge method and the modified sigmoid function based mathematical model for irradiance transitions [12].

The analyzing method produced occasionally distinctly anomaly cloud velocity results with respect to the prevailing trend. To ignore these and to get consistent reliable results, a criterion for the analysis was set that at least 15 recognized cloud shadows must take place in an analyzed day. In addition, the standard deviation of 85% of the recognized cloud shadow speeds must be within 20% of the mean shadow speed and circular standard deviation of 85% of the shadow movement directions must less than 25°. The days that did not fulfill these requirements were rejected from the analysis. Finally, the average daily speed of cloud shadows was used as the speed of CE events to estimate their diameter.

Fig. 4 presents irradiance measurement data on 10th of August 2011 aligned with the calculated speeds and movement directions of the recognized cloud shadows. During this six-hour period, the cloud shadow speeds and movement directions are consistent except for a few instances. The average speed of cloud shadows during this day was 8.0 m/s with a standard deviation of 1.3 m/s. The circular mean direction of movement of the cloud shadows was 110° with a standard deviation of 23°. Therefore, it can be assumed that the moving shadows are caused by clouds moving nearly at the same altitude, with the same speeds and in the same direction. Accordingly, the speeds and movement directions of the CE areas can be assumed to be of the same order of magnitude and the average speed of moving cloud shadows can be assumed to give a good estimate for the average speed of the enhancement areas.

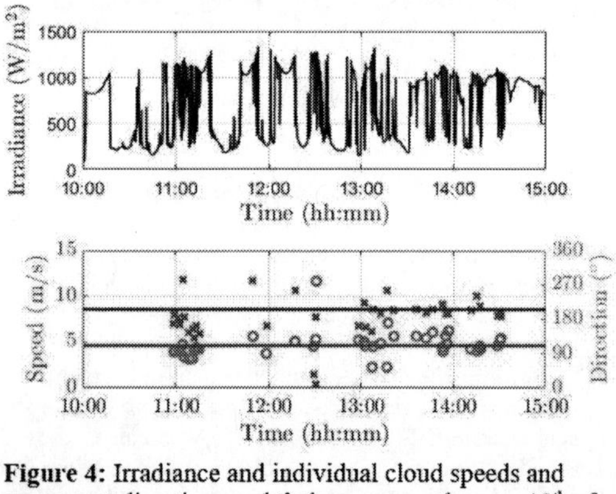

Figure 4: Irradiance and individual cloud speeds and movement directions and their average values on 10th of August 2011.

The total number of analyzed days was 440 and cloud shading events were recognized in 326 days. The rest of the day were clear sky or overcast days. The number of days with more than 15 recognized cloud shadows was 154. The number of days that had enough recognized cloud shadows and fulfilled the consistency requirements for cloud shadow speeds was 67 and for movement directions 87. The days that fulfilled all requirements and were eventually included in the analysis was 56.

3. RESULTS

3.1 Number of cloud enhancement events

The number of the CE events as a function of irradiance is presented in Fig. 5. Red line presents the total number of CE events during the observation period and blue line the number of CE events that met the set consistency criteria and, thereby, were included in the final analysis. The STC irradiance of 1000 W/m² was used as a practical limit for genuine CE events. During the 440-day observation period, irradiance exceeded the STC value altogether 12996 times on 281 separate days. The number of CE events decreases fast with increasing irradiance. During the observation period, irradiance exceeded 1200 W/m² 1414 times on 109 separate days and 1300 W/m² just 188 times on 51 separate days. Extreme irradiance values of over 1400 W/m² were observed only 10 times on six separate days.

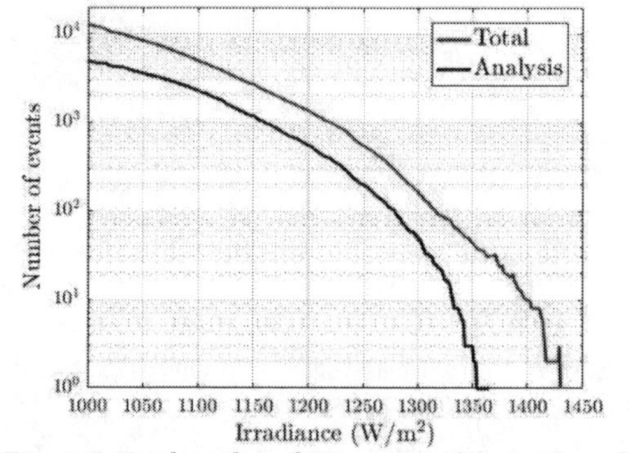

Figure 5: Total number of CE events and the number of CE events that were included in the final analysis fulfilling the set criteria plotted as a function of irradiance.

Approximately 35% of the observed CE events were included in the final analysis up to 1300 W/m² irradiance. The share of the days included in the analysis decreased rapidly when the CE events exceeded the irradiance of 1300 W/m². All CE events exceeding 1365 W/m² were discarded from the final analysis since they occurred only on days with too much variation in cloud shadow speeds and directions of movement.

3.2 Diameters of cloud enhancement areas

The diameter of each individual CE event area has been calculated by using the known duration and the average daily speed of the CE areas. The CE area diameter distributions and their number for the irradiances of 1000, 1100, 1200 and 1300 W/m² are presented in Fig. 6. Approximately 10% of the events on all irradiance values have a diameter less than ten meters. These are caused by the irradiance peaking at a certain value or by the small ripple in the irradiance measurement data, which is also visible in Fig. 2 at 12:32:00. The 1 s running average filtering does not smooth out the ripple in this time scale. A typical diameter of a CE area is some tens of meters and increases with decreasing irradiance up to few hundreds of meters for irradiances close to the STC value having only 10 to 20% irradiance enhancement. The longest observed CE area diameters are of the order of several kilometers. In practice, this means that the CE event has lasted for several minutes at the measurement point. The highest CE irradiances close to 1300 W/m² have typically areas with a diameter of only some tens of meters up to 100 meters.

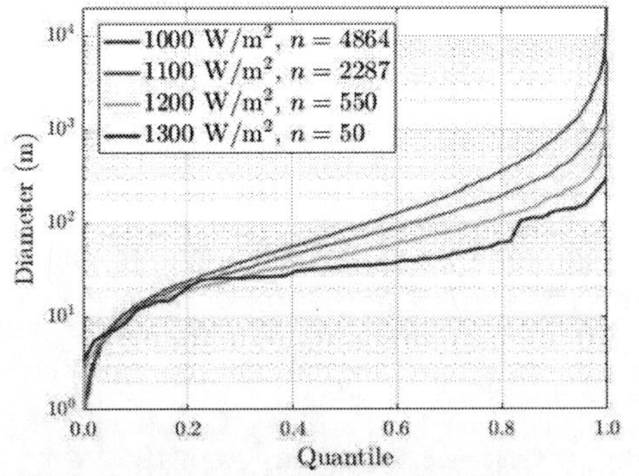

Figure 6: Distributions of cloud enhancement area diameters exceeding different irradiance limits. The number of analyzed cloud enhancement events n is also provided, accordingly.

To further illustrate the results, the diameters of the CE areas of different quantiles are presented in Fig. 7. The top most 100% quantile line presents the maximum diameters of the CE areas as a function of irradiance. The step-like shape is caused by individual events. When considering 98% quantile, the effect of the rare single events is not visible anymore and the diameter of the CE areas decreases more smoothly when irradiance increases. The big gap between the 100% and 98% quantile indicates that the largest CE areas are quite rare. Most of the CE events in the 90% quantile have a diameter of some hundreds of meters. In addition, the decrease of diameters with increasing irradiance is less dominant for lower quantiles, such as for the 80% quantile. This indicates that the events with considerable diameters with respect to the PV generator sizes occur quite often.

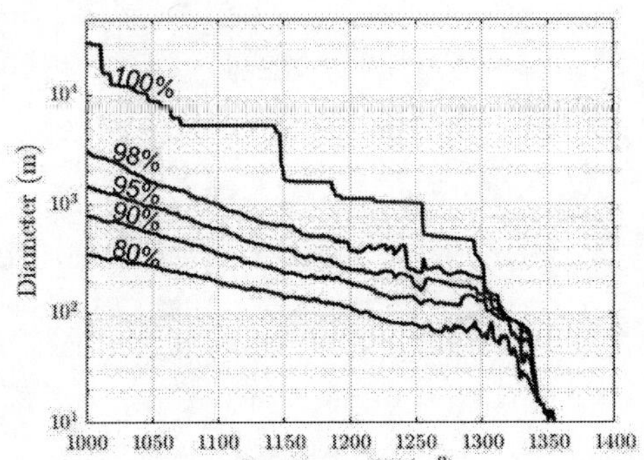

Figure 7: Diameters of cloud enhancement areas as a function of irradiance for 100%, 98%, 95%, 90% and 80% quantiles.

3.3 Validation of the results

The maximum distance between two pyranometers on TUT solar power research plant is 56 m. The CE events with irradiance exceeding 1300 W/m^2 and diameter exceeding 56 m occurred 11 times on six separate days. To validate the results, we compared these 11 events against the actual average irradiance values over the TUT research power plant by analyzing the data from all pyranometers. The average irradiance of all the pyranometers exceeded 1300 W/m^2 in nine cases and was approximately 1280 W/m^2 in two cases. The minimum irradiance value of all the pyranometers exceeded 1250 W/m^2 in all these events. The average irradiance value remained close to or over 1300 W/m^2 for several seconds for the nine events depending on the CE area diameter. Fig. 8 presents the irradiance profile of one of these 11 events on 16th of July 2013. During this event, the maximum irradiance peaked at 1330 W/m^2 and average irradiance exceeded 1300 W/m^2 for 6.2 s.

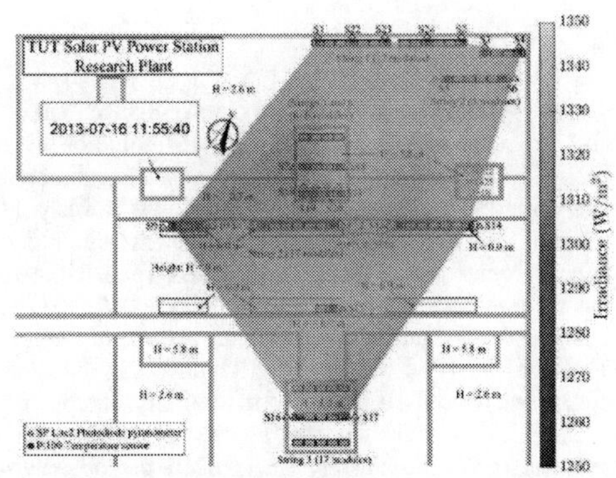

Figure 8: CE event that covers the TUT research plant completely on 16th of July 2013.

By analyzing the data from all pyranometers, it was verified that the used methodology gives a plausible prediction for the diameters and average irradiances of the CE events. Thus, the usage of daily average speed of cloud shadows to estimates the speed of CE areas appears to be reliable and the criterions for selecting the analyzed days appropriate.

3.4 Effect on PV generators

The effect of the CE phenomenon on PV generators can be evaluated by comparing the diameters of the CE areas to the diameters of the PV generators. If the diameter of a CE area is shorter than the diameter of a PV generator, some parts of the PV generator are not covered by the CE area and mismatch losses will limit the maximum power below the value corresponding the enhanced irradiance. When the diameter of the PV generator is equal to or smaller than the diameter of the CE area, the available PV generator maximum power can be directly proportional to the enhanced irradiance.

For the sake of simplicity, let us assume that the PV generator shape is a square. Then the approximate diameters of a 20 kW PV generator is 30 m, a 100 kW PV generator 60 m, a 1 MW PV generator 200 m, and a 4 MW PV generator 400 m [10]. According to Fig. 6 almost 80% of the CE events can cover the 20 kW generator fully and more than half of the events with irradiance below 1300 kW/m^2 can cover the 100 kW generator area fully, but only about 10% of the events with higher irradiance can do that. About 30% of all the analyzed CE events can cover the MW scale PV generators fully and about 20%, 10% and one or two of the events with irradiances over 1100, 1200 and 1300 W/m^2, accordingly.

Based on the analysis, PV power plants constructed using string inverters are affected even by the extreme CE events. During the observation period, a CE event with irradiance exceeding 1335 W/m^2 and diameter of 70 m occurred once. This is 1.5 times the expected maximum clear sky irradiance on the PV plant. Utility-scale PV power plants are also affected by the extreme CE events occasionally. We recorded one CE event of over 1280 W/m^2 with a diameter of 400 m. It is possible that such event can cover a 4 MW PV generator completely. When considering the frequency of the CE events, small residential and commercial PV power plants are affected by the CE phenomenon considerably more often than larger utility-scale power plants.

4. CONCLUSION

We systematically identified cloud enhancement events and measured their durations during 15-month observation period. Diameters of the CE areas were estimated by analyzing the movement of cloud shadows. It was possible to estimate the speeds and movement directions of enhancement areas and consequently calculate the enhancement area diameters for days with consistent enough cloud shadow speeds and directions of movement. The final analysis consisted of 38 days that fulfilled the consistency requirements set for the daily number of CE events and cloud shadow speeds and directions of movement. Approximately 25% of all the observed CE events were included in the analysis.

The results were validated by analyzing the data from all the pyranometers in the TUT research power plant for the recognized extreme CE events with irradiances and diameters of over 1300 W/m^2 and 56 m, respectively. The validation confirmed that the irradiance of the whole PV generator was enhanced accordingly during the CE events and the used methodology gave an accurate prediction for the average irradiance and diameter of the CE events.

Residential and commercial PV power plants, which are typically constructed using string inverters, are affected by most of the CE events. The diameters of even the extreme CE events were of the order of several tens of meters. These can easily cover PV generators in the power range of some tens of kilowatts. During these events, irradiance can be up to 1.5 times higher than the expected clear sky irradiance. Large utility-scale power plants can also be affected by the extreme CE events, but only occasionally. We recorded a CE event with a diameter of 400 m, where irradiance was constantly over 1280 W/m^2. This is still 1.4 times the expected clear sky irradiance value of the setup. However, the frequency of the CE events that could affect utility-scale PV generators is considerably smaller than for residential and commercial PV generators.

5. REFERENCES

[1] C. A. Gueymard, Solar Energy 153 (2017) 776.
[2] T. Stoffel, Solar Energy Forecasting and Resource Assessment (2013) 1.
[3] M. P. Almeida, R. Zilles, E. Lorenzo, Solar Energy 110 (2014) 168.
[4] P. Emck, M. Richter, Journal of Applied Meteorology and Climatology 47 (2008) 2828.
[5] G. H. Yordanov, T. O. Saetre, O. M. Midtgård, Solar Energy 115 (2015) 68.
[6] G. H. Yordanov, Solar Energy 122 (2015) 954.
[7] Z. K. Pecenak, F. A. Mejia, B. Kurtz, A. Evan, J. Kleissl, Solar Energy 136 (2016) 675.
[8] K. Zipp, Solar Power World (2018).
[9] D. Torres Lobera, A. Mäki, J. Huusari, K. Lappalainen, T. Suntio, S. Valkealahti, International Journal of Photoenergy (2013).
[10] S. Ong, C. Campbell, P. Denholm, R. Margolis, G. Heath, NREL/TP-6A20-56290 (2013).
[11] R. A. Houze, International Geophysics 104 (2014) 3.
[12] K. Lappalainen, S. Valkealahti, Solar Energy 138 (2016) 47.

NEW APPROACH FOR THE DESIGN OF LARGE SCALE PHOTOVOLTAIC PLANTS WHICH TAKES INTO ACCOUNT THE INTEGRATION OF PROJECT DATA, TOOLS AND COST ANALYSIS THROUGH THE DIGITALIZATION OF THE PROCESSES

D. Guida, M. Carbone, S. Latella
ENEL GREEN POWER S.p.A.
Via Della Bufalotta 255 - 00139 Rome - ITALY
Phone: +39 3209675167 - E-mail: daniela.guida2@enel.com

ABSTRACT: Digitalization goes beyond business and is entering in various technical and social activities through tools and technologies. The aim of this paper is to show how the digital transformation can be applied also in the photovoltaic business describing what we call Conceptual Design. The first part of the paper describes the main steps of the conceptual design process: the evaluation of the main project data such as site constraints and main equipment, definition of the main parameters of PV (Photovoltaic) configuration like DC/AC and Ground Cover Ratio (GCR) and the research of the final configuration with minimum Levelized Cost of Electricity (LCOE). The second part explains how the optimization tools will be digitalized and automatized describing criticalities, risks and opportunities. The last part shows the benefits of digital transformation in term of design timing reduction and better control of project assumptions and results checking their coherence and feasibility.
Keywords: PV, Design, Tool, Digitalization, LCOE

1 INTRODUCTION

Enel Green Power has started a path to face the important challenges of the new business models developing the Open Power vision based on four pillars: proactivity, innovation, trust and responsibility.

This new strategy needed to address some of world's biggest challenges putting innovation and sustainability at the heart of its corporate strategy.

The role of the new technologies, in particular digitalization, is fundamental in the Open Power Strategy because the development and the application of these new technologies supports the generation and distribution of a more sustainable energy.

In an industrial word where the amount and variety of the information data is huge and complex to manage, the digitalization could create new ways to elaborate these data and to use it in real time in the new innovative applications built for improving the quality of the results of the core business models.

In fact, only organizations that are able to collect and control the data will be in a position to use this power for improving the results and becoming leader in the market.

Within the development and construction of photovoltaics systems, one of the main important process inside the engineering activities is the Conceptual Design phase. Generally this phase is characterized by an initial entropy depending on the necessity to elaborate data of different quality and level, and choosing the best combination of equipment and design parameters for the specific site constraints. Finally the creation of digital platform, studied ad hoc, could be the disruptive solution to edge this inevitable entropy and improve the quality of the results saving time and costs.

2 CONCEPTUAL DESIGN DESCRIPTION

The Conceptual Design phase is the very early stage of design in which step by step all the elements that have an impact in the final configuration of the PV plant are analyzed. It is also the most sensitive and crucial phase because a new, "unknown", project is taken and studied and the risks involved in underestimating some elements, as well as the risks to take wrong decisions for an incorrect choice of the design parameters, could affect the final profitability and in some cases the "survival" of the project itself.

There are two important stages of the Conceptual Design process, the first includes the identification and analysis of constraints and the definition of the preliminary configuration, the second consists of the selection of the equipment and the finalization of the configuration.

2.1 Constraints analysis and preliminary configuration

There are different aspects, technical-economical, commercial and financial to take into account when we start the Conceptual Design of a PV Plant. In fact, the complexity of the Conceptual Design is to verify all these aspects separately and then to check jointly which are the mutual impacts between them. Impact that means defining the limitations to apply in order to determine the development of the PV plant.

Below a not exhaustive list of the main constraints is reported:

- Climate data, mainly including the solar and ambient temperature data. The importance to have accurate data is vital for the profitability of the investment. The main source is from satellite and generally they are adjusted with limited series of measured data through a long-term analysis which gives as results a TMY (Typical Meteorological Year) used for the energy production assessment. A part of the solar radiation data, also wind, rain and humidity data are equally important, in fact they influence the design and consequently the operation and lifetime of the main equipment, such as module, tracker and conversion unit;
- Environmental prescriptions that define the actual land available for installing the PV modules. In order to define that, it is important to identify the areas critical according to geotechnical, topographical, and hydrological criteria, as well as all the areas that present restriction based on environmental impact assessment ruled by the local Authorities which includes all the prescriptions for saving and preserving the local flora and fauna. Furthermore, there will be areas where the presence of

transmission lines, roads, gas pipeline and other infrastructures reduces even more the space available; in fact in this case it is necessary to respect a buffer for the installation according to the local norms. Only the areas identified as "free and safe" will be used for the PV plant;

- Interconnection capacity and infrastructures. Often there are limitations in the capacity of the interconnection and the infrastructures are old and need a technical upgrade for the correct operation. Usually an accurate analysis will be carried out for obtaining the correct value which will be used for the assessment;
- Commercial and financial constraints reported in the tender or on PPA documents which bind the tariff. These indications are related on:
  - Volume of energy to delivery,
  - COD (Commercial Operation Date),
  - AC capacity to be accomplished,
  - Performances of the plant to be guaranteed
  - Liquidated damages to be paid in case of missing the target.
  - Taxes
  - Local content
  - Subsidiary
- The external availability that is the information on the curtailment and the outage of the grid interconnection.

The analysis of the constraints allows to define the perimeter within we can move for choosing the best technical and economical combination that respect our RoI (Return of Investment).

2.2 Equipment selection and final configuration

After checking all the constraints and verifying the interferences that they could generate each other, it is necessary to study the configuration choosing the main equipment (module, structure and conversion unit) and the main parameters of the layout which impact on the BoS (Balance of System). The methodology applied for the selection is LCOE criteria that means finding the combination of Capex (Capital Expenditure), Opex (Operating Expenditure), life time of the plant, actualization rates and EOH (Equivalent Operating hours) that minimize the LCOE function.

In the market a variety of photovoltaic technologies are present for each of the main equipment that have different performances depending on (modules and inverter) that could have a different behavior respect defined environmental conditions, such as radiation, temperature, wind, humidity, etc. and thus change project by project, site by site. The complexity is increased due to the huge numbers of solutions available that give variables results depending on the site and project constraints.

The as-is of the current processes that are in place implies that the resource specialist in charge to perform the main steps of the Conceptual Design is in contact with all the other technical specialist and developer that have the data for running the simulations and use these data for obtaining the results in a iterative and serial process with the use of the tools internal and external (commercial) step by step, one to one without possibility to do a integration of them. The main issues related to this approach is the time consuming to do each passage between the tools (manually) and the fact that the control and quality of the results is delegated to the specialist.

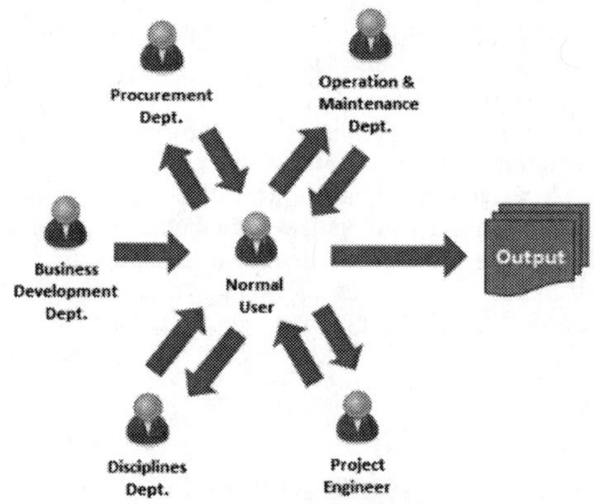

Figure 1 – Project design process (As is)

With the new approach proposed with the digitalization and automation of the processes the expectation is to have a rationalization of the processes implemented in the platform, making them more efficient and quality controlled.

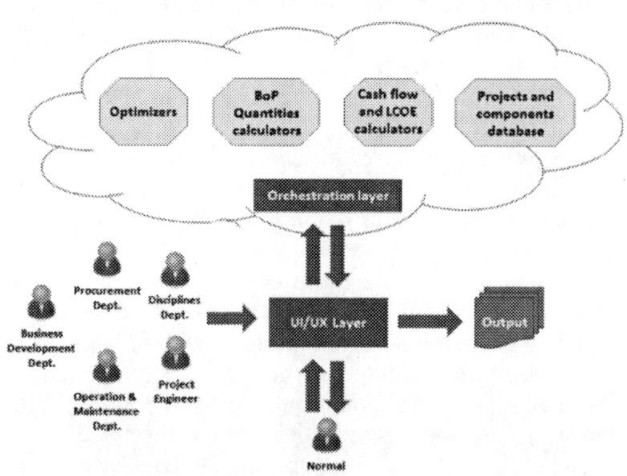

Figure 2 – Project design process (To be)

3 TOOL DIGITALIZATION

The integrated platform collects several specialized in house tools as well as some external software which are dedicated to the evaluation of every part of the plant layout optimization process as shown in Figure 3.

To achieve this goal, a collection of all the available tools or modules currently used in the as is process was done. After this recognition, an in depth analysis was carried out in order to optimize, interconnect and streamline the single parts of the process. In order to integrate the single tools, a low level analysis of all the inputs and outputs of each single module which compose the process (i.e. a single stand-alone operation that provides a consistent output) was performed, so that they could be recalled by the process workflow that is implemented and stored inside a cloud environment.

In this way, most of the information required by the majority of the modules is automatically fetched by the integrated platform during the process workflow, while all the other info that are project dependent, are inserted by the user through an editable graphic user interface before the beginning of the workflow.

One of the complexity of the entire process is that the tools realized in house and the external commercial software have been originally created to "live" and work stand alone. It has been necessary to re-engineering the IT (Information Technology) structures of each of them in order to allow the integration with the other tools/modules and software. In fact one of the main goal of the platform is to have a real time interactions between each component of the platform that avoid the dead times and the arbitrarily of some choices due lack of time.

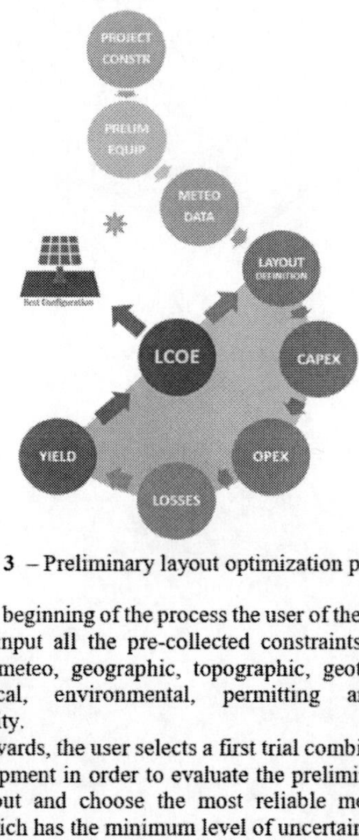

Figure 3 – Preliminary layout optimization process

At the beginning of the process the user of the platform gives as input all the pre-collected constraints such as climatic, meteo, geographic, topographic, geotechnical, hydrological, environmental, permitting and grid connectivity.

Afterwards, the user selects a first trial combination of main equipment in order to evaluate the preliminary best plant layout and choose the most reliable meteo data source which has the minimum level of uncertainty.

At this point the Integrated Platform comes alive: it starts to investigate by seeking the combination respectively of Capex, Opex and yield production which optimize the aforementioned equipment and respect all the constraints through the research of minimum LCOE.

Once the process is finished, the Integrated Platform evaluates all layout details and provides them to the User to enable him to go in a further analysis.

4 INNOVATIVE METHODOLOGY

The Conceptual Design process aims to optimize PV plant design by implementing a series of forefront logics.

In the first phase of the process, on the base of the project constraints and of the first-trial main components choice, the integrated platform uses an optimization algorithm called Layout Research Algorithm (LRA). The LRA, relying on an analytic approach, defines the optimum values of the PV plant layout main parameters (as for example ground covering ratio, DC/AC ratio and so forth), looking for the absolute minimum of an overall function (that takes into account all the project's fixed input parameters) in terms of minimum LCOE. After a certain number of iterations (that depends on the algorithm speed of convergence), the LRA presents to the user the optimum plant layout configuration and all the correlated parameters, which define the preliminary layout. These information are therefore translated into the equivalent bill of quantity and main equipment features and they are transmitted to Procurement team in order to receive the technical and commercial offers from the main equipment and BoP suppliers.

In the second phase of the process, the integrated platform receives as input the technical and economical details of the main equipment and BoP offers and aims to find the best feasible combination of equipment in terms of LCOE. This means that the choice of the best main equipment is not done separately for each component, but instead it takes into account the impact of all the possible complete combinations of equipment and the overall behavior of the PV system, in terms of energy production, capital expenditure and operative costs. The optimal equipment combination is chosen by a module called Optimum Combination Research Algorithm (OCRA): this module evaluates the LCOE of each possible equipment combination and present to the user the best feasible choice of components.

When the equipment selection phase is completed, as a final step, the integrated platform runs once again the LRA in order to verify if the recalculated optimal PV plant layout main parameters are different from those evaluated in the first phase of the process, since they were associated with different equipment. The best components and the final layout parameters settle the final PV plant configuration, which is now mature for the investment committee approval.

5 BENEFITS

Nowadays digitalization is entering in various technical and social activities through a huge amount of tools and technologies. The aim of the Conceptual Design tool is to set up an integrated platform that is able to find the best feasible layout configuration of a PV plant according to geographical, meteorological, economical and technical constraints, relying on many internal specialized tool and through the contribution of some external software.

In this way, on one side the integrated platform will act as a "black box" for the user, taking as inputs all kinds of relevant constraints and equipment features and providing as output detailed information about the PV plant layout main parameters. On the other hand, the user is able to perform specific kinds of sensitivities and to go in depth in the analysis by recalling each single part of the process independently.

This digital integration will lead to several improvements respect to the common approach of PV project design:

- An increase in terms of productivity of each activities, by reducing the time to conclude a project systematically relocating the effort of all the trivial and repeatable tasks on the digital platform and minimizing the interaction among

different units.
For each combination, the timing useful for the activities done for calculating Capex, Opex and EOH is expected to be reduced of 30-40%.

- Saving costs for the Electromechanical and Civil BoP, through the definition of the quantities according the automatic layout and costs estimation based on real offers, it has been estimated a saving between 5-10% of the total BoP Capex which means have a reduction of the LCOE between 0,5 and 1,5%.

- A higher quality of the process results by reducing the chances of occurrence of errors in the manual data processing steps necessary in the standard procedures and of general accidentally data corruption of 90%.

- Conceptual Design will allow to come up with more homogeneous results, by standardizing the whole process and ensuring the same output reliability, reducing its dependence from the specific user skills. The users will also be able to track the process from a high level point of view, fully focusing on the core aspects of the projects.

- Time saving of the resources needed for the development phase is expected to be between 25-35%. Ideally the same resource could work on more projects in parallel without losing the quality and time.

6 CONCLUSIONS

The idea behind the work on the digitalization of the Conceptual Design is based in taking the processes as they are and transform them in a software platform capable to concentrate in a common IT environment managed by the users all the tools/modules/software. In this way the users will be able to evaluate an extremely higher number of possible scenarios and to focus their efforts on the core choices which require personal experience and sensibility that cannot be faced by a digital software.
Although the Conceptual Design process approach was initially created in order to come up to need of speed up the solar photovoltaic utility scale plant development, it can be applied to any other kind of technology or process thanks to its global methodology. The integrated platform will become the keystone of the working procedure in forthcoming future. This cutting-edge approach is borne out of the company perspective in terms of innovation and sustainability principles.

DETAILED CALCULATION OF ELECTRICAL MISMATCH LOSSES FOR CENTRAL- AND STRING-INVERTER CONFIGURATIONS ON UTILITY-SCALE PV ARRAYS

Martín Herrerías, Hugo Capdevila
Capdevila-ITE e.K. 70191 Stuttgart, Germany. +49 711 859 9202
mha@capdevila-ite.com, hdc@capdevila-ite.com

ABSTRACT: The object of this study is to provide an objective answer to the question of whether significant performance improvements can be expected when using distributed (i.e. string-, and 2-string-) inverters in the design of utility-scale PV power plants, as opposed to a central inverter configuration. The study focuses solely on the effects of irradiance mismatch (series and parallel), and is based on 24 test case simulations that consider the effects of system geometry (direct shading at sub-cell resolution, anisotropic diffuse shading and IAM at module level), and detailed calculation of electrical mismatch losses (using full IV curve solution to include the potential effects of MPPT voltage limitations). Three inverter configurations are evaluated for six ~1 MWp layout variations at two distinct locations. The diverse matrix of design conditions is helpful in providing qualitative design guidelines, and in identifying conditions in which a detailed comparative study ought to be performed.

Keywords: inverter, system performance, simulation, large grid-connected PV systems, shading, mismatch.

1 INTRODUCTION

The competitive advantages of PV plant configurations based on decentralized inverters have been a topic of heated discussion in recent years. Aside from differences in cost, logistics, O&M, and reliability; and despite critical studies in the literature (e.g. Woyte et al. [1], García et al. [2]); an incessant claim in favour of string-inverters is the expected improvement in system performance due to a reduction of mismatch losses (see e.g. the recommendations of the IFC [3]). As García and his colleagues point out, the claim is possibly rooted in the intuitive assumption that a system with more independent MPPTs can better adjust to localized conditions that would otherwise lead to electrical mismatch. The assumption seems to be confirmed by analysis of small arrays (~5 kW) subject to irradiance mismatch, such as those of Mäki & Valkealahti [4] and James et al. [5]. These studies, however, cannot — and should not — be readily extrapolated to the long-string, large arrays with uniform, limited shading which are representative of utility plants to date.

At the MW system level (utility scale arrays) there seems to be agreement about the difficulty associated with the relative performance evaluations based on experimental methods (Webber [6]; Evans et al. [7]), although a dual string/central plant is being evaluated by Nobile et al. [8]. On the other side, commercially-available simulation tools typically lack the sufficient accuracy and detail to model subtle, plant-wide effects like diffuse irradiance- and secondary- (i.e. parallel) mismatch.

In the literature some authors have recently analyzed the mechanisms and effects of mismatch losses for larger arrays, and under more detailed conditions: Wurster & Schubert [9] perform a parametric simulation of parallel-mismatch losses for strings with unequal module numbers at STC; Goss et al. [10], Capdevila et al. [11], and Appelbaum, [12] focus on irradiance-mismatch including diffuse components; Vargas et al. [13] analyze manufacturing mismatch for simulated arrays of 10 kW to 1 MW; Lappalainen & Valkealahti, [14] evaluate mismatch losses due to smooth cloud shades on a simulated 32 kW array; Abdin et al. [15] focus on mismatch due to temperature gradients on a 298 kW flat-roof array; and Evans et al. [7] analyze cell and module mismatch for a 280 kW tracking array due to manufacturing, irradiance and temperature variability under no-shading conditions.

The present work is devoted to study the differences in performance of utility-scale photovoltaic power plants using central-, "block"- (2 parallel strings) and string-inverter technologies under changing environmental- and design conditions. The work expands on the results of García et al. [2] and Abdin et al. [15], as it evaluates system configurations much more representative of current utility-scale, fix-tilt PV plants. It also complements the findings of Evans et al. [7] and Vargas et al. [13], as it focuses *strictly* on irradiance and temperature mismatch due to partial shading (direct and diffuse) while excluding manufacturing- and wiring-resistance variability. The level of detail of the simulation (further described in Capdevila et al. [11] [16]), and the flexibility to evaluate a handful of design variations is unprecedented for full-year simulations of systems of this scale, representing a unique opportunity to gain insight on a phenomenon that offers no straightforward experimental validation, and which has profound implications on the optimization of PV systems.

2 METHODOLOGY

Performance models for a test 1.0 MW power plant are developed using proprietary software, introducing a total of 24 design cases. These include two distinct site locations, six layout variations (10° and 20° fix-tilt mounting structures, 2 and 4 module rows, and portrait / landscape module orientation), and three inverter configurations (central, 2-string, 1-string).

Two sites were chosen randomly from regions in Europe with very different weather. Hourly TMY data sets were downloaded from PVGIS [17]. General statistics for both sites are included in Table I and Figure 1. Site D is meant to be representative of high-latitude, high-diffuse-fraction, low-irradiance conditions, whereas site E has a warmer climate with higher direct irradiance; its latitude is lower, but not so far away from D that a completely different layout (i.e. tilt-angle, ground-cover ratio) would be required.

Site Location (key)	D	E
Latitude (°N)	53.14	42.36
Longitude (°E)	8.22	-5.77
Altitude (m)	6	806
GHI (kWh/m²·a)	1018.4	1680.8
BNI (kWh/m²·a)	895.6	1981.1
DHI (kWh/m²·a)	554.1	565.2
Mean Temperature (°C)	9.5	10.8
Mean Wind Speed (m/s)	4.2	3.1

Table I: Site location and meteorological data

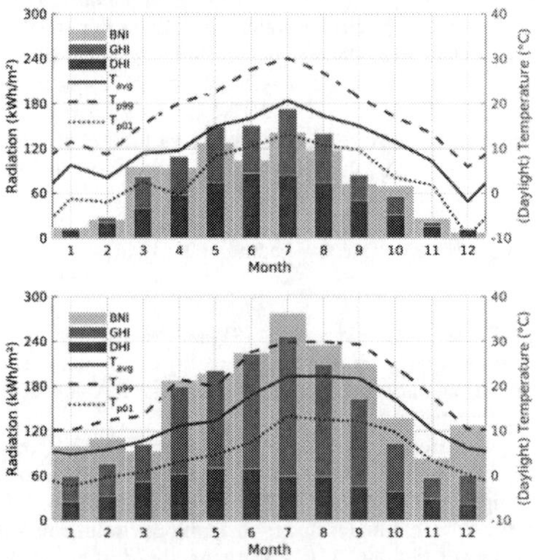

Figure 1: Monthly mean radiation and temperature values for sites D (top) and E (bottom). Note the higher diffuse fraction for D.

Layouts were defined parametrically for a constant number of modules (3072) of 1960 x 996 mm, using a constant ground-cover-ratio (2/3), gaps of 2 cm between modules and 20 cm between *mounts* (sheds or tables). Mount heights are adjusted to achieve a clearance of 50 cm to the ground. All layouts have zero azimuth (i.e. modules are tilted South). Terrain is considered flat, with no external obstacles. The specific details of each layout are listed in Table II. Figure 2 provides an example of the module arrangement on one of the 4H (i.e. 4-row, landscape module) cases.

Layout Variation (key)	4H20	2V20	4H10	2V10	4V10	2H20
Module orientation	H	V	H	V	V	H
Table tilt (degrees)	20	20	10	10	10	20
Module rows per mount	4	2	4	2	4	2
Mod. columns per mount	6	12	6	12	6	12
Mounts per row	8	8	8	8	16	4
Number of mount rows	16	16	16	16	8	32
Mount Length (m)	11.9	12.2	11.8	12.0	6.0	23.5
Mount Height (m)	4.0	3.9	4.0	3.9	7.8	2.0
Row pitch (m)	5.7	5.5	5.9	5.8	11.6	2.8
Diffuse-shading points*	3x3	3x3	4x2	4x2	2x3	5x2

Table II: Detailed layout configurations.

Note that layout configurations pairs 4Hxx and 2Vxx are (at the mount level) geometrically interchangeable, with the aim of capturing the effect of module-orientation for the same irradiance conditions. 10° and 20° tilts modify the overall intensity of beam- and diffuse-shading and thus the effect of irradiance mismatch.

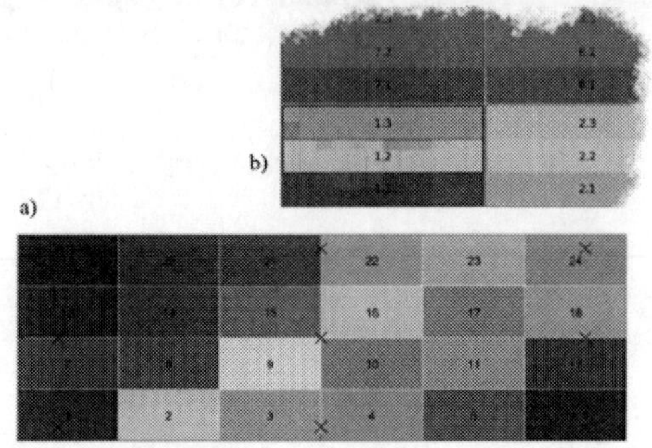

Figure 2: a) Module layout in 4Hxx configurations, and b) Cell/bypass-diode layout inside modules. Black crosses mark the 3x3 grid of points for diffuse-shading analysis.

The methodology for the irradiance transposition and shading calculation is further detailed in [11]. The geometrical framework of the Perez et al. 1990 model [18] is used to calculate independent shaded 3D view-factors for four diffuse irradiance components (isotropic, albedo[1], circumsolar, and a horizon-brightening-band) for a matrix of Chebyshev points distributed over the rectangular mounts (Table II details the number of nodes along each tracker dimension for each layout configuration). Shading factors are estimated as the area of each (orthogonally projected) sky region divided by the same area on an obstacle-free case (Figure 3b). A physical refraction model, as described by De Soto et al. [19]) and implemented by the PVLIB library of the PVPMC [20] is applied to each diffuse component individually (k = 4.0/m, L = 2.0 mm, n = 1.25). The resulting diffuse irradiance at the nodes of analysis is then linearly interpolated to calculate diffuse irradiance at module resolution. Beam shading is calculated at sub-cell resolution for every simulation step (Figure 3c).

Figure 3: Shading analysis step for 4H20/2V20 layouts.
a) Geometrical modeling (masking of obstacles).
b) Diffuse shading analysis for a given mount point.
c) Beam shading analysis for a given mount.

A modified One-Diode-Model as described by Mermoud & Lejeune [21] was used to model photovoltaic modules. A single module is considered for all simulations: a state of the art 340 W pSi module, with 72 cells arranged into three bypass-diode-protected cell blocks (U_{oc} = 47.1 V, U_{mp} = 37.6 V, $\mu_{pmp} = -4\times10^{-3}$/K

[1] Unshaded albedo component was assumed to be uniform and equal to 0.2 of GHI.

and μ_{uoc} = -2.9×10^{-3}/K at STC). The SAPM cell temperature model described by King et al. [22], was used, as implemented in the PVLIB libraries of the PVPMC [20], with the recommended parameters for glass/cell/polymer-sheet modules on an open-rack mount ($a_w = -3.56$, $b_w = -0.08$, $\Delta T = 3.0$K).

Three interconnection schemes (C = central inverter, 128 strings, B = 64 inverters with 2-string, and S = 128 single-string inverters) were evaluated for each layout. Within each of these schemes, three different *physical* string configurations (1-row, 2-row, and 4-row) are included, in order to account for the frequent homogeneity found in real plant layouts due to space constraints. In all layouts, 3/4 of the plant were connected into single-row strings, 1/8 into double-row strings, and 1/8 into 4-row strings (see Figure 4). All strings connect 24 polycrystalline 340 W modules. In the 2-string configuration, strings were connected in pairs to their closest neighbor in the vertical or North-South direction, i.e. 4-row layouts include two-pairs of inverters on each mount-row of 24 modules, whereas 2-row layouts include a single pair.

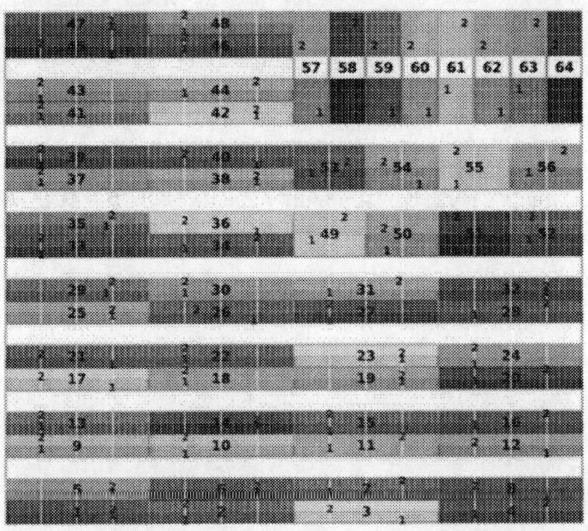

Figure 4: Connection of 4Vxx layout into 64 MPPT units with 2 strings of 24 modules each. Note 2-row strings and 4-row strings on the upper-right quarter.

Commercial inverters with efficiency curves and MPPT voltage windows that were considered representative of each technology were selected, and modeled as fully-*independent MPPT units*.

Inverter MPPT unit (key)	C	B	S
Max. Output Power (kVA)	990	16.5	7.7
Max. Input Power (kW)	1350	16.8	7.9
Max. Input Current (A)	1392	22.0	16.0
Threshold DC Power (W)	5000	20.0	13.3
Lowest MPPT voltage (V)	596	600	200
Highest MPPT voltage (V)	950	1450	950
Max. Efficiency	98.7%	99.0%	98.6%
Euro. Efficiency (800V)	98.2%	98.2%	97.9%

Table III: Inverter specifications

Operational voltage, current, and temperature-dependent power limits were included, and split evenly among independent MPPTs. Note that this is not exactly accurate, as neither power limits nor shared-losses on the AC side are strictly proportional to each independent MPPT's input power. Further on, the imbalance between different MPPT input currents of the same inverter can be limited, which was also not included in the model. For the purposes of this study, however, where the focus lies on the DC side, we consider all of these effects to be of secondary importance.

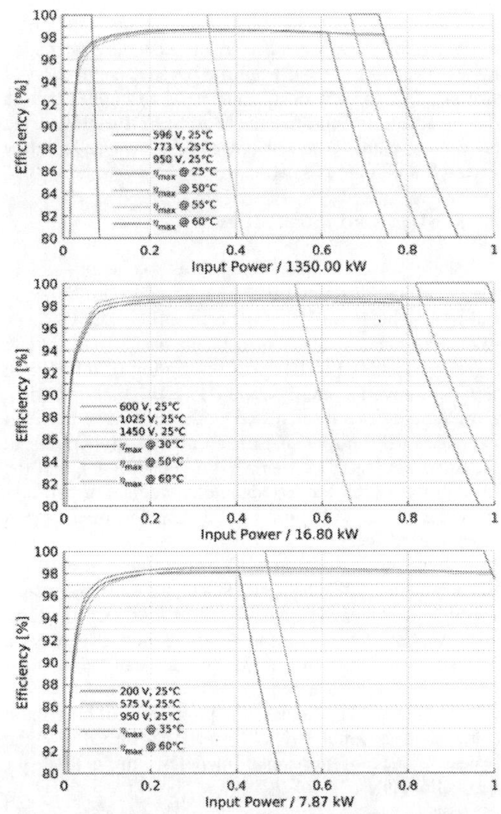

Figure 5: From top to bottom: Central-, Block- (2-string), and String-Inverter models. Efficiency drop at low voltages is due to DC current limitation (i.e. minimum voltage to reach peak power output).

The methodology for the electrical calculation is further detailed in [15]. Module and bypass-diode IV curves are represented with piece-wise-linear approximations (set to ensure a relative tolerance of 1×10^{-4} on both current and voltage). In the general case, IV curves for each cell's operating condition are added in series (and in parallel to a bypass diode) to form cell-blocks; these are added in series to form strings; and depending on the inverter topology, these resulting string IV curves are added in parallel to another string (2-string case) or the remaining 127 strings (central-inverter). In order to reduce the computational expense of the calculation, array *elements* (cell-blocks, modules, and strings) are considered *uniform* when the irradiance over every cell of the element lies within 1% of any other. Detailed calculations proved that electrical mismatch within uniform elements (according to this criterion) lie within the tolerance value of 1×10^{-4}.[2]

[2] There seems to be a general tendency for small-irradiance-mismatch losses to have an electrical effect

An important advantage of the full-IV-curve calculation is that the voltage at MPP (that affects inverter efficiency) is calculated for each operational point. When this voltage lies outside the operational window of the MPPT, a local maximum (and a new operating voltage within MPPT boundaries) can be easily extracted from the curve.

The main focus of this study is on irradiance-mismatch losses, and specifically on secondary-mismatch losses (i.e. those arising from the parallel connection of strings). For this reason a number of additional factors (e.g. spectral correction, soiling, manufacturing tolerances, ohmic wire resistance, etc.) have deliberately been left out of the simulation.

3 RESULTS & DISCUSSION

The results for each simulation case are listed in Table IV, and illustrated in figures Figure **6** to Figure **8**. Cases are labeled following the convention NXTTS for N rows per modules with X orientation, and TT° tilt at site S (refer to tables I to III for a description of the layout, site, and inverter configuration keys). Total electrical mismatch losses are calculated as the difference between the added MPP power of isolated modules (operating at the same irradiance and temperature conditions), and the MPP power of the connected parallel arrays. For partially shaded modules, MPP power is calculated using the mean module irradiance.

Calculated losses (figure 6) seem largely determined by geometrical shading (and much more so by beam-shading, as evidenced by much larger losses on the low-diffuse-fraction site E). Module orientation also plays a dramatic role (e.g. compare the geometrically equivalent cases 2V20E / 4H20E and 2V10D / 4H10D). As will be discussed later, this is due to the prevalence of horizontal beam shades, which are effectively 'absorbed' by the bypass diodes of horizontal modules, but not só for vertical modules.

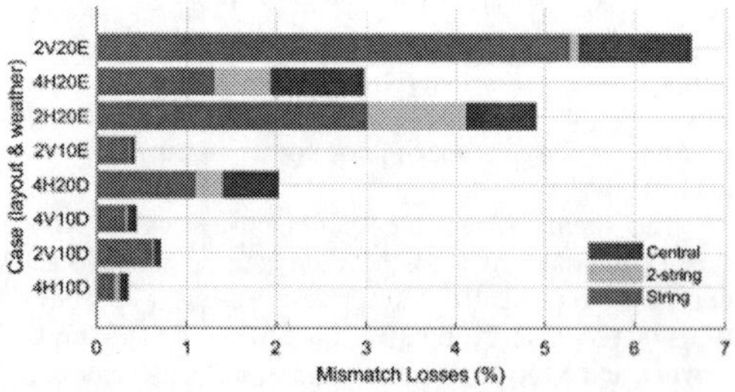

Figure 6: Mismatch losses for each simulation case.

Figure 7 is an attempt to isolate the effect of the different inverter configurations from both module orientation and the intensity of linear (i.e. irradiance) shading. The shading "strength" factor is the ratio of electrical mismatch (electrical loss minus irradiance loss) to a "worst-case" shading loss — calculated by reducing each string's current to that of its cell with the lowest irradiance. In other words, it is a metric of the sensitivity of the array to irradiance mismatch: *zero* is an ideal case where each cell would have its own micro-inverter; *one* is a string of modules with flat reverse-bias response and no bypass-diodes.

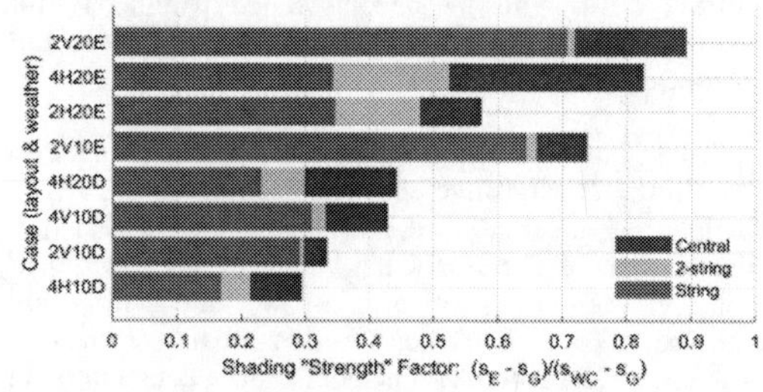

Figure 7: Shading "strength" for each simulation case.

The results suggest that the relative potential advantages of string- and 2-string configurations are also higher for increasing beam-shading fractions, and notably higher for landscape module layouts. Figure 8 provides further insight into the mechanisms underlying this behavior. For two representative cases, the figure indicates where in the array (i.e. in which strings) the highest losses take place. For the landscape module layout, all strings that include the lowest row (L1, U1, and W) — and are thus subject to the highest beam shading — show significant performance improvements when using single-string and 2-string inverters. However the lowest-row single-line strings L1 of portrait-mounted layout 4V10D show no improvement whatsoever (despite shading) when using string or 2-string inverters.

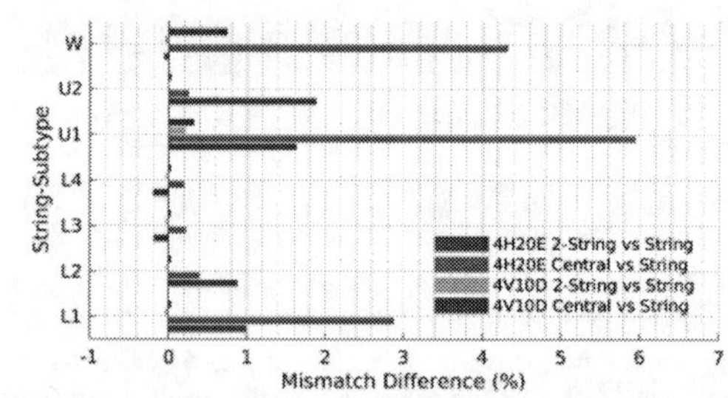

Figure 8: Mismatch losses classified by string *subtype* on 4H20E and 4V10D cases (from mount-bottom to -top, single-row strings can take positions L1 to L4, and two-row strings can be U1 or U2; W are 4-row strings).

The reason for this is that horizontal shades from neighboring mounts ensure that most (if not all) cell-blocks on the string will have at least one shaded cell. This effectively limits the current of the complete block, thus yielding a "flat" IV curve for the shaded string (figure 9) and an MPP voltage B which is not too far away from the central inverter MPP voltage (typically close to A). Bypass-diodes in these circumstances do not help reduce primary (i.e. current) mismatch losses, as all cell-blocks in the string provide roughly the same current. On the other hand, the secondary (i.e. parallel) mismatch losses that are addressed by string-inverters are limited to the small difference B-B', or in the 2-string-inverter case to the negligible C-C'.

For strings that occupy several rows — and for single-line strings of landscape modules — bypass-diodes effectively... well, bypass shaded cell-blocks, and the resulting IV curves present the characteristic "steps" of shaded PV arrays, with multiple local maxima, and

$\Delta p/p \sim (\Delta g/g)^{1/2}$, where p and g are the MPP power and plane-of-array irradiance, respectively; but investigations with other design conditions are required.

MPP voltages potentially far from the non-shaded string MPP (figure 10). In this case the differences C-C' and B-B' become much more important, and distributed inverter configurations become advantageous.

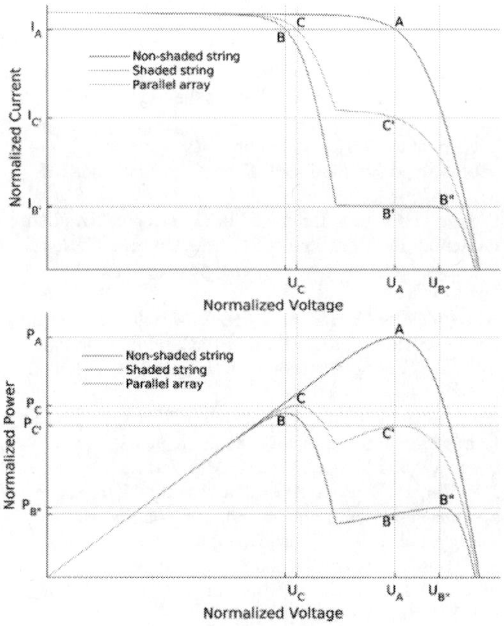

Figure 9: Shading case I (IV and IP curves for a partially shaded array consisting of two strings). Effect of shading is representative of landscape module layouts and/or multi-row strings, i.e. a mixture of effectively bypassed (shaded) cell-blocks, and fully unshaded blocks.

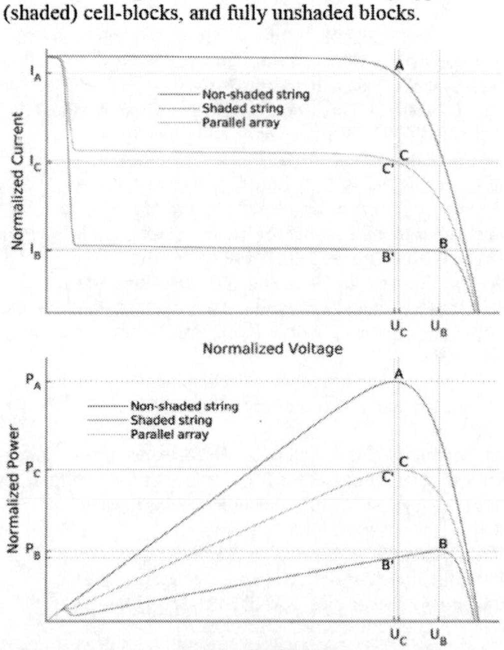

Figure 10: Shading case II (IV and IP curves for a partially shaded array consisting of two strings). Effect of shading is representative of single-string, portrait module layouts: shading affects most cell-blocks simultaneously, leading to internal (sub-module) mismatch losses.

It's worth noting, however, that in these cases the width of the MPPT voltage window (and the efficiency of the MPPT algorithm) become critical, as the MPP of heavily mismatched strings can either lie below the MPPT lower voltage limit, or just not be reachable by e.g. gradient-based MPPT algorithms. In both cases, the string might just be forced to operate on a local maximum (e.g. B") closer to the non-shaded MPP, reducing the potential benefits of string-inverters.

Another caveat is that, for all cases in this study and as a result of the simplified design choices, strings are always coupled with another string of the same *type* (i.e. with the same number of physical rows), although not always in the same *position* (e.g. lower/upper row). This is important when assessing the performance of 2-string inverters, since connecting two strings which do not share the same shading conditions will necessarily result in increased secondary mismatch losses.

Layout	MeteoData	Tilt-gain	Irradiance Shading	Effective Irradiance	Worst-case Shading (E)	Inverter	Mismatch Losses	Inverter Losses	Performance Ratio
2H20	E	14.6%	4.9%	1792	10.6%	C	**4.7%**	1.9%	90.0%
						B	**4.0%**	1.7%	90.9%
						S	**2.9%**	2.1%	91.5%
2V10	D	7.6%	1.2%	1047	2.4%	C	**0.7%**	2.2%	98.1%
						B	**0.7%**	1.8%	98.6%
						S	**0.6%**	2.1%	98.2%
	E	8.6%	0.9%	1760	0.7%	C	**0.4%**	1.8%	98.0%
						B	**0.4%**	1.7%	98.1%
						S	**0.4%**	2.1%	97.7%
2V20	E	14.6%	4.5%	1800	9.4%	C	**6.5%**	1.9%	88.7%
						B	**5.2%**	1.8%	89.0%
						S	**5.1%**	2.2%	89.6%
4H10	D	7.6%	1.2%	1047	1.4%	C	**0.4%**	2.0%	98.6%
						B	**0.3%**	1.8%	99.0%
						S	**0.2%**	2.2%	98.7%
4H20	D	12.9%	5.1%	1063	5.9%	C	**1.9%**	2.2%	93.4%
						B	**1.3%**	1.8%	94.4%
						S	**1.0%**	2.2%	94.3%
	E	14.6%	4.6%	1799	5.6%	C	**2.8%**	1.9%	92.2%
						B	**1.8%**	1.7%	93.3%
						S	**1.2%**	2.3%	93.2%
4V10	D	7.6%	1.1%	1049	1.2%	C	**0.5%**	2.0%	98.7%
						B	**0.4%**	1.8%	99.0%
						S	**0.3%**	2.2%	98.7%

Table IV: Summary of results

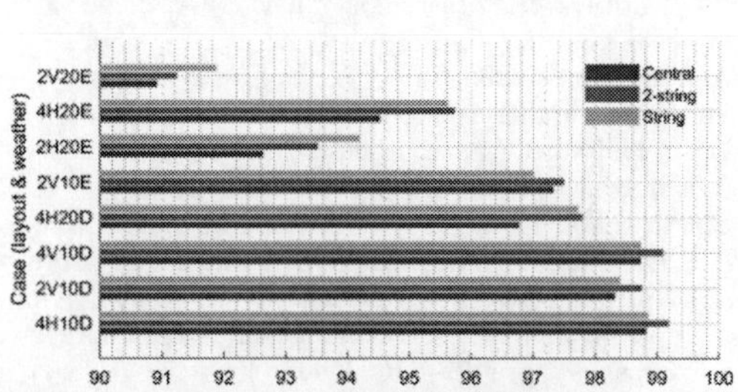

Figure 11: Performance ratio of simulation cases.

Figure 11 considers the effect of inverter efficiency which is usually — but not necessarily — higher for larger inverters, and thus tends to partially compensate the advantages of distributed technologies. Note that a number of loss factors (spectral effect, soiling, wiring, manufacturing mismatch, degradation, etc.) are not included in the calculation. As such, Performance Ratios are not meant to be interpreted as realistic values, but compared in relative terms.

3 CONCLUSIONS

For the particular inverter choices included in this study, and under the simulation assumptions discussed along the paper, string- and 2-string inverter configurations tend to outperform central inverter topologies by a small margin (−0.3 to 1.5%). Differences strongly depend on the layout and precise connection arrangement and of course on the particular choice of inverter.

The performance improvement (secondary mismatch reduction) when using string- and 2-string configurations is higher for increasing beam-shading factors, and notably higher for landscape modules or multi-row-string layouts. Portrait-mounted modules connected on single-row strings show not only much higher primary (sub-module) mismatch losses, but also negligible (or even negative) differences when using distributed inverters. The mechanism responsible for this sensibility on module orientation and string layout is explained by the prevalence of horizontal shades and the effect of module bypass diodes.

Mismatch factors not considered in this study (manufacturing, ageing, cell-defects, temperature, wiring resistance variability and others) might yield additional performance advantages for string inverters. Further research is required in this direction.

It should be emphasized that performance is only one of the many criteria that should be considered when selecting an inverter configuration (along with equipment and BoS costs, logistics, reliability, grid compliance, etc.).

4 REFERENCES

[1] A. Woyte, J. Nijs, and R. Belmans, "Partial shadowing of photovoltaic arrays with different system configurations: literature review and field test results," Sol. Energy, vol. 74, no. 3, pp. 217–233, Mar. 2003.

[2] M. García, J. M. Maruri, L. Marroyo, E. Lorenzo, and M. Pérez, "Partial shadowing, MPPT performance and inverter configurations: observations at tracking PV plants," Prog. Photovolt. Res. Appl., vol. 16, no. 6, pp. 529–536, Sep. 2008.

[3] IFC, "Utility-Scale Solar Photovoltaic Power Plants: a Project Developer's Guide," 2015.

[4] A. Mäki and S. Valkealahti, "Power Losses in Long String and Parallel-Connected Short Strings of Series-Connected Silicon-Based Photovoltaic Modules Due to Partial Shading Conditions," IEEE Trans. Energy Convers., vol. 27, no. 1, pp. 173–183, Mar. 2012.

[5] P. a. B. James, A. S. Bahaj, and R. M. Braid, "PV array <5kWp+single inverter=grid connected PV system: Are multiple inverter alternatives economic?," Sol. Energy, vol. 80, no. 9, pp. 1179–1188, 2005.

[6] J. Webber and E. Riley, "Mismatch Loss Reduction in Photovoltaic Arrays as a Result of Sorting Photovoltaic Modules by Max-Power Parameters," ISRN Renew. Energy, vol. 2013, pp. 1–9, 2013.

[7] R. Evans, M. Boreland, and M. A. Green, "A holistic review of mismatch loss: From manufacturing decision making to losses in fielded arrays," Sol. Energy Mater. Sol. Cells, vol. 174, pp. 214–224, Jan. 2018.

[8] G. Nobile et al., "Comparison between Central and String Inverter Performance for the Utility-Scale PV Plant in Nova Olinda Brazil," presented at the EU PVSEC 2018, Brussels, 2018.

[9] T. S. Wurster and M. B. Schubert, "Mismatch loss in photovoltaic systems," Sol. Energy, vol. 105, pp. 505–511, Jul. 2014.

[10] B. Goss, I. Cole, T. Betts, and R. Gottschalg, "Irradiance modelling for individual cells of shaded solar photovoltaic arrays," Sol. Energy, vol. 110, pp. 410–419, Dec. 2014.

[11] H. Capdevila, M. Herrerías, and A. Marola, "Anisotropic Diffuse Shading Model for Sun-tracking Photovoltaic Systems," Energy Procedia, vol. 57, pp. 144–151, Jan. 2014.

[12] J. Appelbaum, "Current mismatch in PV panels resulting from different locations of cells in the panel," Sol. Energy, vol. 126, pp. 264–275, Mar. 2016.

[13] J. P. Vargas, B. Goss, and R. Gottschalg, "Large scale PV systems under non-uniform and fault conditions," Sol. Energy, vol. 116, pp. 303–313, Jun. 2015.

[14] K. Lappalainen and S. Valkealahti, "Photovoltaic mismatch losses caused by moving clouds," Sol. Energy, vol. 158, pp. 455–461, Dec. 2017.

[15] Z. Abdin, C. J. Webb, and E. M. Gray, "Simulation of large photovoltaic arrays," Sol. Energy, vol. 161, pp. 163–179, Feb. 2018.

[16] H. Capdevila, A. Marola, and M. Herrerías, "High resolution shading modeling and performance simulation of sun-tracking photovoltaic systems". 9th International Conference on Concentrator Photovoltaic Systems: CPV-9, Miyazaki, Japan, 2013, pp. 201–204.

[17] "PVGIS, JRC's Directorate C, Energy, Transport and Climate." URL: http://re.jrc.ec.europa.eu/pvgis/. [Accessed: 25-Sep-2018].

[18] R. Perez, R. Seals, P. Ineichen, R. Stewart, and D. Menicucci, "A new simplified version of the Perez diffuse irradiance model for tilted surfaces," Sol. Energy, vol. 39, no. 3, pp. 221–231, 1987.

[19] W. De Soto, S. A. Klein, and W. A. Beckman, "Improvement and validation of a model for photovoltaic array performance," Sol. Energy, vol. 80, no. 1, pp. 78–88, 2006.

[20] "PV Performance Modeling Collaborative." [Online]. Available: https://pvpmc.sandia.gov/. [Accessed: 25-Sep-2018].

[21] A. Mermoud and T. Lejeune, "Performance assessment of a simulation model for PV modules of any available technology," in 25th European PV Solar Energy Conference, Valencia, 2010.

[22] D. L. King, W. E. Boyson, and J. A. Kratochvil, "Photovoltaic Array Performance Model," Sandia National Laboratories, SAND2004-3535, Aug. 2004.

LOCAL VARIABILITY IN PV SOILING RATE

Michael Gostein[1], Kendra Passow[2], Michael G. Deceglie[3], Leonardo Micheli[3], Bill Stueve[1]
[1]Atonometrics, 8900 Shoal Creek Blvd. Suite 116, Austin, TX, 78757, USA, info@atonometrics.com
[2]First Solar, 135 Main St., San Francisco, CA, 94105, USA
[3]National Renewable Energy Laboratory, 15013 Denver W Pkwy., Golden, CO, 80401, USA

ABSTRACT: We explore local and hyper-local variability in PV soiling rate by examining the correlation between PV system soiling rates over a multi-year period within two groups of First Solar soiling stations in separate regions of California. The results demonstrate significant seasonal variations in monthly-averaged soiling rates, with soiling varying from near 0 to ~0.1%/day in different months and with seasonal patterns that differ between sites. Monthly patterns at nearby sites were more highly correlated, suggesting that soiling rates can be strongly affected by local geography and weather. For the two regions studied, soiling rate pattern correlation diminished beyond a distance on the order of 50 km. In addition, for selected sites we demonstrate within-site variation in soiling rate of up to ~1.5-2x, which may be correlated with proximity to nearby agricultural activity.

1 INTRODUCTION

Soiling of PV modules is one of the principal energy loss factors for PV systems and has a significant impact on PV power plant economics. Accordingly, in recent years there has been a growing interest in measuring, predicting [1]-[2], and monitoring PV soiling rates, for pre-production prospecting as well as operations and maintenance (O&M) purposes, especially at utility-scale PV power plants [3].

An important question concerns the local variability of soiling rates. Is it sufficient to use isolated soiling measurements as representative of a larger region? If so, what are the appropriate distance scales for which a given station's results are considered representative? Conversely, on a very small scale, to what degree does soiling rate vary within a single large PV power plant?

First Solar has deployed an extensive network of soiling measurement stations, at both prospecting and power plant locations, providing multi-year data. In this study, we examine historical soiling data at sites within two regions of California, in order to elucidate trends in local and regional soiling variations.

2 DATA SOURCES

Table I lists the 8 sites used as data sources in this study, each of which is an operating PV power plant. Region 1 and Region 2 are located, respectively, in southern and central California, and are separated by approximately 600 km. Listed distances are measured from the first site listed for each region, and are rounded to the nearest 10 km. Each site has stations collecting both meteorological (met) and soiling data, with varying number of stations per site as indicated. The study includes data collected from July 2013 to December 2017. Each soiling station comprises a pair of PV modules, one of which is cleaned every 1-2 weeks, and the other of which is allowed to soil naturally. A soiling ratio is calculated which quantifies the electrical output of the soiled module versus the cleaned module [3][8], and each soiling station is paired with a met station which collects irradiance and rain data. Table I lists the annual insolation-weighted soiling ratio measured at each site, details of which are provided below. We also list in Table I a "land type" that indicates the nature of the land surrounding the site, using a combination of climate and land use classifications. Although sites 1A-1C and 2B-2D are in areas with desert/arid climate classifications, there is nonetheless extensive agricultural activity in their surrounding areas so we note this in our land type as it may affect soiling rates and seasonal patterns.

Table I: Soiling station sites in the study

Region	Site	Land Type	Stations	Distance (km)	Annual Soiling Ratio
1	1A	Farming	6	0	0.976
	1B	Farming	12	10	0.983
	1C	Farming	2	50	0.956
	1D	Desert	3	120	0.995
2	2A	Arid	9	0	0.995
	2B	Farming	1	30	0.996
	2C	Farming	1	30	0.997
	2D	Farming	1	150	0.987

3 AUTOMATED DATA ANALYSIS

Completely automated analysis techniques based on [4]-[6] were used to process the data from the 35 stations. To illustrate the data analysis method, we show in Fig. 1 a 12-month segment of data from one of the stations located at site 1A.

The upper panel in Fig. 1 shows the soiling ratio, defined in this case as the ratio of temperature-corrected short-circuit current between the soiling station dirty module and clean module [3][7][8]. Light gray points in this plot are minute-by-minute measured soiling ratio values for the period of noon +/- 2 hours on each day. The minute-by-minute points for each day are filtered to remove non-representative points (using a 1σ filter) and values at irradiance less than 500 W/m^2; a daily irradiance-weighted average, shown in red, is then calculated as described in [3]. Note that the daily average soiling ratio values in this plot typically exceed 1.00 during clean periods. This is because the stations in this study were not on-site calibrated; rather, the measured values are normalized during analysis as will be described below.

During dry periods the daily average soiling ratio steadily decreases as soiling accumulates and losses increase. The soiling ratio appears to decrease in a step-

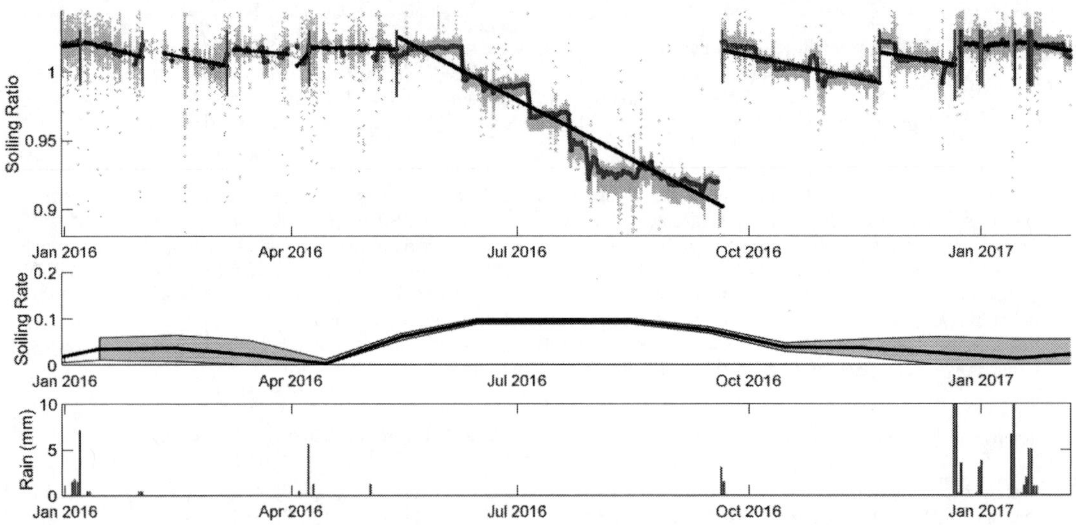

Figure 1: Illustration of automated data analysis method, based on segment of data from a station at site 1A. Upper panel: minute-by-minute measured soiling ratio (gray), daily average soiling ratio (red), detected soiling ratio reset events (blue), and linear fits between resets (black). Middle panel: soiling rate in %/day (black) and its confidence interval band (green). Lower panel: rain events.

wise fashion, due to the fact that cleaning of the clean module in the soiling station module pair was not automated and occurred only every 1-2 weeks. During rain events, plotted in the bottom panel, the soiling ratio value recovers or "resets", due to cleaning of the entire system. Note that in some instances in Fig. 1 the daily average soiling ratio recovers to a clean-state value at times when there is no measurable rainfall. This can be caused by dew, non-measured rainfall, or even in a few cases washing of the entire power plant.

In order to quantify soiling rate in between reset events, reset events must first be detected from the soiling ratio data. We accomplish this by detecting outlier positive shifts in the 14-day moving median of the daily soiling ratio, as per [5], as well as designating any measured rain event as a reset event. Detected reset events are shown in the upper panel of Fig. 1 by vertical blue lines.

We assume soiling rate to be linear in between the reset events. Soiling rate is calculated by determining the slope of the daily soiling ratio values for each data segment in between resets, using the robust Theil-Sen method for slope estimation [4]-[6]. Ninety-five percent confidence intervals of the Theil-Sen slope are determined using bootstrapping [4]; for simplicity, we assumed confidence intervals are symmetric about the mean value. The middle panel of Fig. 1 shows determined soiling rates in %/day (solid line) and their associated confidence intervals (green shading). Soiling rates were determined for each segment of data in between reset events, and were then binned into monthly values by calculating the mean and combined confidence interval of all segments overlapping a given month, weighted by the fraction of the month corresponding to each segment. Note that the width of the confidence interval is primarily determined by the length of time between cleaning events; during the long dry period in the summer of 2016 the confidence intervals are very narrow, indicating a precise determination of the soiling rate, while during the winter months with their more frequent rains the confidence intervals on the soiling rate determination are wider.

Detection of reset events was also used to automatically normalize the daily average soiling ratio values by determining the soiling ratio measured when the station was clean. For each soiling station, the daily average soiling ratios measured 2 days after each reset event were collected into a distribution. This distribution was filtered with a 1σ filter to remove non-representative points, corresponding, for example, to resets that resulted in only a partial cleaning, and the mean value of the resulting distribution was used as the measured clean-state soiling ratio for the station, typically differing from 1.00 by up to ~2%. This value was then used to normalize soiling ratio values for the station.

4 RESULTS

4.1 Site-to-site variability

Fig. 2 shows the soiling rates and their confidence intervals for each of the 8 sites over the multi-year period. Stations within a given site generally showed similar trends, although with differences that will be explored below. For Fig 2, all stations at a given site were averaged together for each month of data using a Monte Carlo resampling method to determine both the mean soiling rate at the site and its confidence interval, taking into account the soiling rates and confidence intervals of the individual stations.

For the nearby sites 1A and 1B, it is apparent that soiling rates display a strong seasonal trend, peaking around 0.1 %/day in the summer months and falling to near 0 in the winter. Possibly this seasonal pattern relates to agricultural activity in the surrounding region, which could generate airborne particulates, as will be explored further below. Sites 1C and 1D, which are farther away, show different patterns. Site 1D, which is located in a

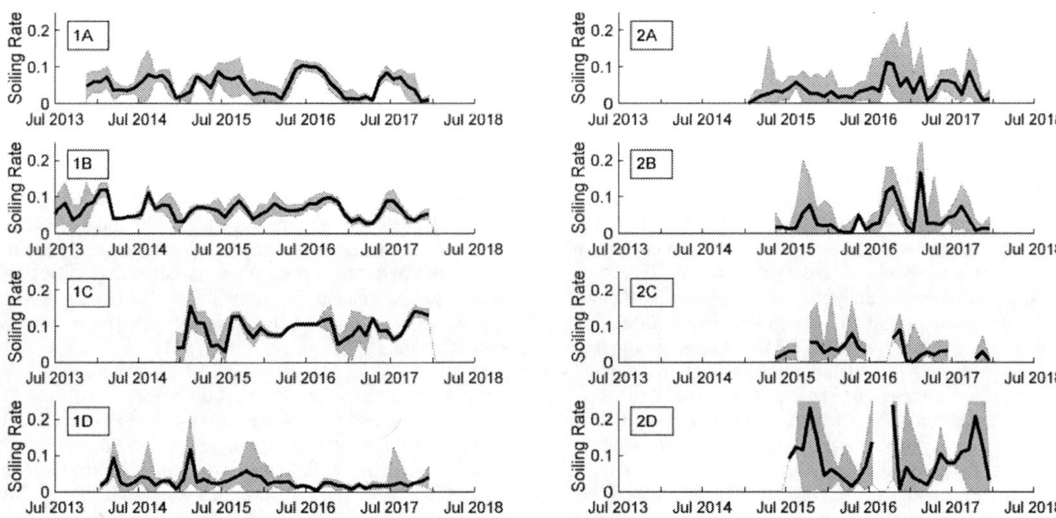

Figure 2: Monthly soiling rates for the 8 sites in %/day (black) and their confidence interval bands (green).

desert area across a mountain range from the other three sites, shows a completely different pattern.

In region 2, sites 2A and 2B appear to show soiling rates peaking to ~0.15 %/day in late summer or early autumn, although their pattern is less regular and contains inter-annual variations, and this pattern changes somewhat in moving to the farthest site, 2D.

We also determined the average soiling ratio of each site. To do this, we first normalized the measured daily average soiling ratios of each soiling station by the clean-state soiling ratio value for that station determined from post-reset measurements, as described above. We then used daily insolation values to calculate an insolation-weighted average of all the daily soiling ratio values for each station during the years 2016 and 2017. Values for all stations within a site were averaged, in order to obtain the site-average soiling ratios shown in Table I. The overall soiling related losses of the sites in this study were in the range of <5%.

4.2 Correlation distance

To quantify the correlation of the monthly soiling rate patterns, we show in Fig. 3 the R-squared correlation coefficient for the linear regression of the monthly soiling rates for site A in each region (i.e. site 1A or 2A) versus the other sites in the same region, plotted as a function of distance. The point at distance 0 corresponds to the A site versus itself and is identically 1. Site 2C was excluded from the plot because unfortunately its one soiling station was offline during the summers of 2016 and 2017 and consequently displayed only very limited range in its soiling rate data (as seen in Fig. 2). Fig. 3 illustrates that the monthly soiling rate patterns of the closest sites are correlated, and that the correlation between sites diminishes with distance. The nearest sites have an R-squared coefficient of approximately 0.35, indicating at least partial correlation. In fact this value is similar to the correlation coefficients between stations within the large multi-station sites, which in good cases were in the range 0.3-0.5. However, note in Fig. 3 that the R-squared values decrease beyond a correlation distance on the order of ~50 km. These results suggests that, for the regions considered, a soiling station must be located within at least ~50 km of a PV plant for its data to be meaningful in soiling prediction at the plant. Of course this distance cannot be regarded as universal for other regions, but would be specific to the local geographic features of any region. We note however that the finding of a correlation length on the order of tens of kilometers is in agreement with a previous study that correlated soiling rates to airborne particulate matter density ($PM_{2.5}$ and PM_{10}) measured at varying distances from soiling stations [1].

Note that the R-squared coefficient provides us only a relative measure of correlation and does not directly quantify prediction accuracy. Ideally, we wish to know what would be the uncertainty in predicting PV power plant losses using soiling station data measured at varying distances from a PV plant. Analysis of our results to determine this prediction uncertainty will require further work.

4.3 Correlation to particulate matter density

We also attempted to correlate the monthly soiling rate of each site in the present study with the nearest ground-based measurements of airborne particulate matter density

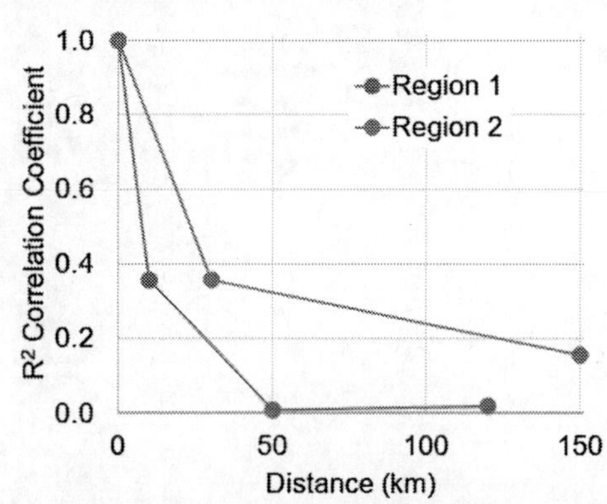

Figure 3: Correlation between monthly soiling rate patterns versus distance for the sites in Table I.

($PM_{2.5}$ and PM_{10}). The particulate matter data were downloaded from the U.S. Environmental Protection Agency (EPA) database [9], which contains data from an EPA network of particulate matter monitoring stations, measuring PM_{10} and $PM_{2.5}$ daily or every 6th day. We calculated the monthly mean PM concentrations for each PV soiling site from the PM concentrations recorded at the closest monitoring station each month. In order to get the most complete data, only stations that took at least 75% of the yearly scheduled measurements have been considered.

At first look, we observed only a very weak correlation between monthly soiling rates and monthly particulate matter for each site, in contrast to the strong correlation observed in a previous study [1]. This is not surprising, as the previous study correlated yearly-average data rather than monthly-average and other researchers [2][10] have already discussed the challenges of predicting short-term soiling using single variable regression compared to longer-term predictions, due to the more complex interactions among the environmental parameters. In addition, in the present study, only data from the closest stations were considered; considering that the closest station could be as far as 93 km from a site, different spatial interpolation techniques could be employed in the future for better correlations. Moreover, the large confidence interval bands in some cases might have affected the results. Indeed, we noted that the correlations between soiling rates and PM_{10} could be enhanced for all the sites in Region 2, with a maximum R^2 higher than 0.7, if high-uncertainty soiling rate data were removed. Despite that, the same improvement could not be found in Region 1. Further work is needed to elucidate the correlation between particulate matter concentration and soiling rate, and to understand the impact that different weather patterns and soil characteristics can have on this correlation.

4.4 Within-site variability in soiling rate

To examine within-site variability in PV soiling rate, we plotted the average soiling rate versus position within the site and identified trends related to nearby potential dust sources. Figure 4 illustrates results for the two sites with the largest number of soiling stations, sites 1B and 2A. The figure shows the average daily soiling rate

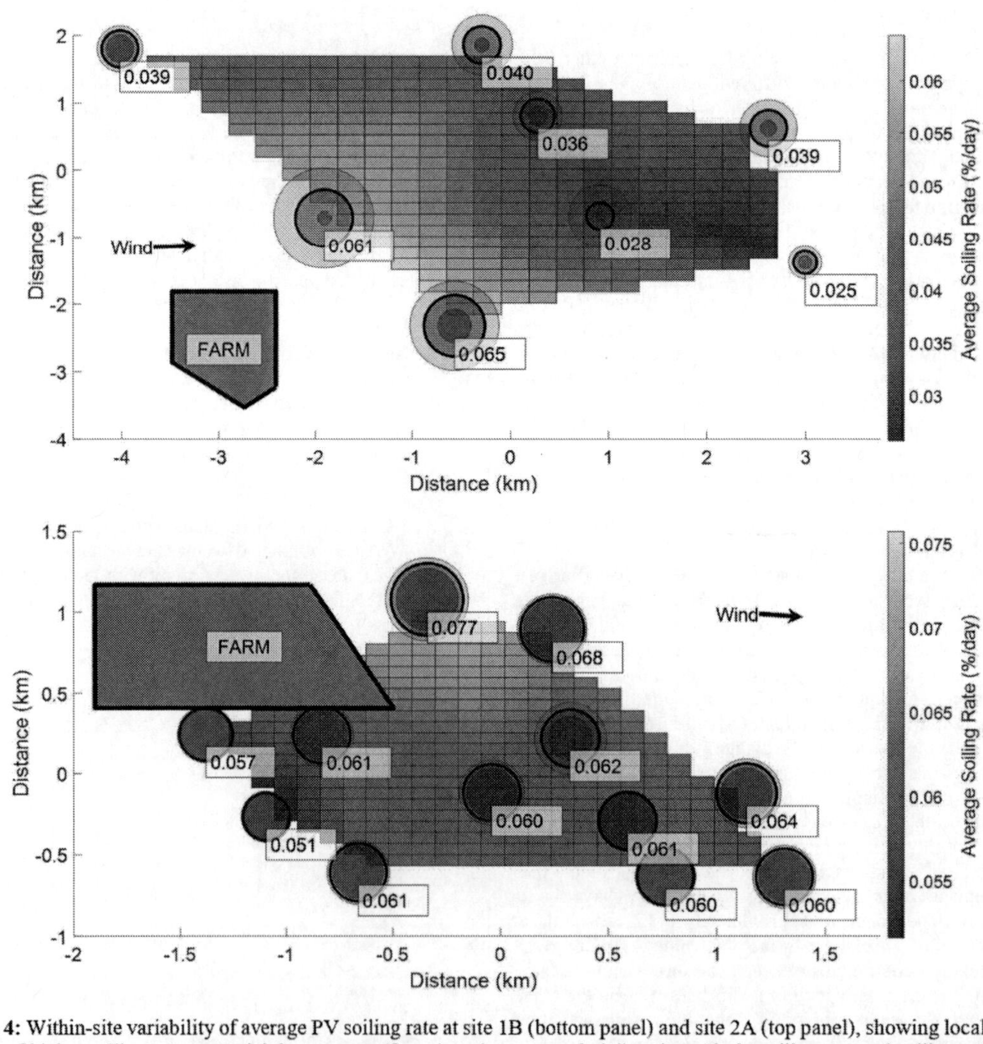

Figure 4: Within-site variability of average PV soiling rate at site 1B (bottom panel) and site 2A (top panel), showing localized regions of higher soiling rate, potential dust sources (farms), and average wind direction. Circle radii: measured soiling rates at each station; inner and outer circles: lower and upper uncertainty limits; color map: linear interpolation between stations.

measured at each station over the entire reporting period. To eliminate potential bias between the stations due to unequal counting of seasonal variations, the average used only 30 months in which all included stations were reporting data; one station was therefore excluded altogether from the analysis of site 2A because it was offline much of the time. The radii of the purple concentric circles indicate the average daily soiling rate and uncertainty at each station; the outermost and innermost circles show, respectively, the upper and lower uncertainty limits, while the average value is shown by the intermediate circle and in text. The colormap is a linear interpolation between the station measurements on arbitrary grids, not related to PV system block size or PV system boundary shape. The positions of nearby farms are indicated. Note that although site 2A is listed in Table I as being situated in an "arid" region, there is nevertheless an isolated farm very close to a boundary of the site.

The results in Figure 4 demonstrate significant variation in soiling rates within the two sites. At site 1B soiling rate varies by up to ~1.5x while at site 2A the rates vary by up to ~2x. The regions of highest soiling rates appear to be correlated with proximity to the nearby farm operations. These results are consistent with the seasonal trends shown in Figure 2 which indicate generally higher soiling rates in the summer or fall in these regions. We have also computed the average speed-weighted wind direction at each site using the on-site met station data. For both sites monthly average wind direction was typically from the west or south, while the overall averages were primarily from the west, as indicated on the figure. Note that for site 1B, only the northernmost soiling stations show elevated soiling rates, although the westernmost stations are equally close to the nearby farm; however, the observation of average westerly winds appears to explain the specific pattern observed. Site 1A (not shown in Figure 4) had similar variations which may also be correlated with wind direction and local farm operations.

5 CONCLUSIONS

Our results demonstrate examples of variability in soiling rates in two nearby regions, with seasonal patterns that differ markedly between sites. Monthly patterns at two pairs of sites within approximately 50 km of each other were more strongly correlated, while farther sites showed significantly lower correlation. Even within individual sites, we observed variations in soiling rates of ~1.5-2x potentially caused by nearby farms, indicating that a larger number of stations is beneficial for determining soiling at a given site more accurately to account for the effects of local dust sources. The results suggest that soiling rates can be strongly affected by local geography, land use, and weather, resulting in specific regional and hyper-local patterns.

6 ACKNOWLEDGMENTS

This work was authored in part by Alliance for Sustainable Energy, LLC, the manager and operator of the National Renewable Energy Laboratory for the U.S. Department of Energy (DOE) under Contract No. DE-AC36-08GO28308. Funding provided by the U.S. Department of Energy's Office of Energy Efficiency and Renewable Energy (EERE) under Solar Energy Technologies Office (SETO) Agreement Number 30311. The views expressed in the article do not necessarily represent the views of the DOE or the U.S. Government. The U.S. Government retains and the publisher, by accepting the article for publication, acknowledges that the U.S. Government retains a nonexclusive, paid-up, irrevocable, worldwide license to publish or reproduce the published form of this work, or allow others to do so, for U.S. Government purposes.

7 REFERENCES

[1] L. Micheli and M. Muller, "An investigation of the key parameters for predicting PV soiling losses," *Prog. In Photovoltaics*, v. 25, Apr. 2017, pp. 291-307.

[2] W. Javed, B. Guoa, and B. Figgis, "Modeling of photovoltaic soiling loss as a function of environmental variables," *Solar Energy*, v. 157, Nov. 2017, pp. 397-407.

[3] M. Gostein, J. R. Caron, B. Littmann, "Measuring soiling losses at utility-scale PV power plants," in *40th IEEE Photovoltaic Specialists Conference*, Denver, CO, 2014.

[4] M. G. Deceglie, M. Muller, S. Kurtz, Z. Defreitas, "A scalable method for extracting soiling rates from PV production data," in *43rd IEEE Photovoltaic Specialists Conference*, Portland, OR, 2016.

[5] M. G. Deceglie, L. Micheli, M. Muller, "Quantifying soiling loss directly from PV yield," *IEEE Journal of Photovoltaics*, v. 8, March 2018, pp. 547-551.

[6] L. Micheli, D. Ruth, M. G. Deceglie, M. Muller, "Time series analysis of photovoltaic soiling station data," NREL/TP-5J00-69131, September 2017.

[7] M. Gostein, B. Littmann, J. R. Caron, L. Dunn, "Comparing PV power plant soiling measurements extracted from PV module irradiance and power measurements," in *39th IEEE Photovoltaic Specialists Conference*, Tampa, FL, 2013.

[8] IEC 61724-1:2017, "Photovoltaic system performance – Part 1: Monitoring".

[9] US Environmental Protection Agency, "Air quality system data mart [internet database]." [Online]. Available: https://www.epa.gov/airdata. [Accessed: 15-May-2018]

[10] L. Boyle, H. Flinchpaugh, and M. Hannigan, "Assessment of PM dry deposition on solar energy harvesting systems: measurement–model comparison," *Aerosol Sci. Technol.*, vol. 50, no. 4, pp. 380–391, Apr. 2016.

ANALYSING THE EFFECT OF SNOW ON THE PV REGULATOR RESPONSE IN A SIMPLE PV SYSTEM

Jorge Solis*,**,***,****, Sahaphol Hamanee*, Magnus Nilsson*****
*Department of Engineering and Physics, Karlstad University, Sweden
**Advanced Research Institute for and Engineering, Waseda University, Japan
***Mechanical Science and Engineering, Tokyo Institute of Technology, Japan
****Faculty of Engineering, University of Southern Denmark, Denmark
*****Galva Energy Center, Sweden
solis@ieee.org

ABSTRACT: During the winter seasons, there is not only the challenge of removing snow downfall on PV solar cell roofs in order to maximize the solar energy efficiency, but it is also to indicate when the effect of snow significantly affects the operation in order to decide to take an action. Up to know, the study of the optical properties of snow have been examined, including shading patterns on photovoltaic solar panels. However, while the snow cover might reduce electricity, it might not have been significant effect due to the lack of solar radiation in the Nordic countries. In this paper, the authors have focused to propose an experimental methodology to analyze the effect of the thickness of the snow on a PV regulator response based on the frequency, statistical and production analyses in a simple PV solar cell module. In order to study the effect of the electrical production in relation to the thickness of the snow and light conditions, a set of experiments were proposed with the LEGO® PV system.
Keywords: PV system, experimental methods, monitoring, environmental effect

1 INTRODUCTION

In the last recent decades, the development of photovoltaic (PV) systems has become feasible to cover an increasingly larger energy demand with solar energy production over the world, even in the Nordic countries. While the solar radtion in the Nordic countries is of comparatively low intensity and energy production is greatly reduced during the winter. Fjedlheim et al. has demonstrated that buildings outfitted with photovoltaic panels can achieve a net surplus of energy over its lifetime, even as far as 60th parallel north. At this latitude, the sun is visible for 18 hours, 52 minutes during the June solstice and 5 hours, 52 minutes during the December solstice.

Several researchers have been introducing notable studies on the effects of snowfall on PV systems. The first was performed in 1979 at the Natural Bridges National Monument, by utilizing a simple linear empirical correlation to determine expected PV output, and used this metric to determine snowfall losses [1]. Townsend and Powers proposed a model to estimate snow losses based on meteorological data from two test sites were obtained, one which had a set of modules that were consistently cleared of snow, and the other was operating fixed tilt system with no module clearing regimen [2]. A model was developed to predict monthly energy loss but it has not been validated yet.

On the other hand, the main influencing factor of snow on PV systems is the lockage of solar radiation on the photovoltaic cells [3]. In order to quantify and assess the importance of this, some understanding of the optical properties of snow is required. Andrews et al. have proposed a methodology to analyze snow-shedding patterns from image data and showed the variability of snow distribution patterns depending on module type and orientation [4]. However, the proposed method requires collecting weather data to predict the effects of snowfall.

As for the optimization of the operation of PV parks, due to the fact that the output power of photovoltaic solar cell has non-linear characteristics, up to know different kinds of control algorithms have been proposed [5], e.g. maximum power point tracking (MPPT), PID, etc.

However, the optimization of the operation of the PV depends strongly on the control parameters which can have a good performance on one solar park might not be good for the other. One main reason for this is that the existent control methods do not take into account the historical response of the system behavior in contrast to the issues that will affect the annual production of a PV parks, such as irradiation, ambient temperature, degradation of modules, etc.

The authors propose to develop an adaptive control system for photovoltaic systems that adapts after changing environmental conditions. One of these conditions is the presence of snow or other form of pollution. Due to its complexity, as a first approach, our research aims to detect those conditions through analyzing power electronic control parameters without extra sensors, in order to reduce their effect on the performance of the system.

In this paper, the authors focused to understand the effect of different kinds of environmental factors such as light source and snow tickness in relation to the electrical efficiency of a photovoltaic (PV) device. In particular, we aims to understand how the snow affects the voltage and power output of a LEGO solar panel by using the renewable energy add-on set with the LEGO® EV3. The proposed experimental setup is shown in Fig. 1. In order to analyze the response of the PV regulator, we have proposed to apply three different methods: frequency analysis, statistical analysis and production analysis.

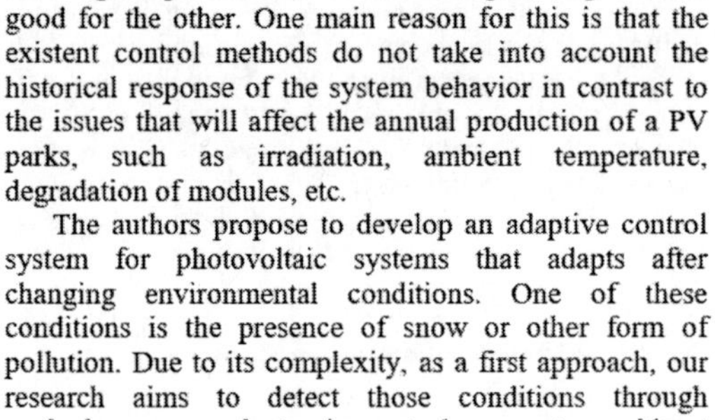

Figure 1: Experimental Setup

2 PROPOSED METHOD

The frequency analysis was considered due to its capability to decompose a complex signal into simpler parts, to facilitate analysis differential equations, and convolution operations in the time domain become algebraic operations in the frequency domain. For this purpose, the fast Fourier transform (FFT) was used in order to analyze the response of the PV regulator in the presence of different kinds of thickness of snow covering the solar panel. The FFT is a very efficient algorithm for computing the discrete Fourier transform (DFT). The DFT is calculated as shown in Eq. 1, where $y[n]$ is the discrete number sequence, N is the sample amount, n is the discrete variable and k is the variable that is converted to different frequencies. The FFT only needs $N\log_2(N)$, where N is the sequence of points of the signal. In our case, we have used 2,400 points for each experiment (output voltage) and the FFT was computed using Matlab 2015a, which it computes the FFT based on the algorithm proposed in [6].

$$Y[k] = \sum_{n=0}^{N-1} y[n] * e^{-ik\frac{2\pi}{N}n} \quad (1)$$

The statistical analysis was considered due to its capability to describe the nature of the data to be analyzed in order to identify trends. For this purpose, the two-way ANOVA was used in order to determines how a response is affected by two factors (e.g. light source and snow thickness). The two-way ANOVA is a parametric test that assumes that all the samples were drawn from normally distributed populations with the same variances [7]. A two-factor analysis of variance tests three hypotheses:

1. There is no difference among the levels of the first factor;
2. There is no difference among the levels of the second factor;
3. There is no interaction between the factors, i.e., if there is any difference among groups within one factor, the differences are the same regardless of the second factor level.

A full two-way ANOVA report displays an ANOVA table describing the variation associated with each factor and their interactions. ANOVA table displays the degrees of freedom, sum of squares, and mean squares for each of the elements in the data table, as well as the F statistics and the corresponding P values. The P value is the probability of being wrong in concluding that there is a true difference between the groups. The smaller the P value, the greater the probability that the samples are drawn from different populations. Traditionally, you can conclude there are significant differences if $P < 0.05$. The two-way ANOVA was computed using Excel.

Finally, the production analysis was considered due to its capability to visualize how the average of the power output drops in relation to the thickness of the snow.

3 EXPERIMENT & RESULTS

In order to carry out the different proposed analyses, the following conditions were considered.

- A strong light source (150 W) was placed at 50 cm away from the LEGO solar panel.
- A low light source (35 W) was placed at 20 cm away from the LEGO solar panel.
- Direct lighting to the LEGO solar panel in order to maximize the voltage output.
- The solar panel temperature should be approximately equal for each experiment.
- The thickness of the snow should be approximately equal for each experiment.
- The measurement time is 4 minutes (because the snow begins to melt) for each experiment.
- The sampling frequency is 10 Hz, where 10 measure samples were taken per second.

Based on the above controlled conditions, the following experiments were carried out.

- Strong and weak light without snow at -5 °C
- Strong and weak light with 14 mm at -5 °C
- Strong and weak light with 5 mm at -5 °C

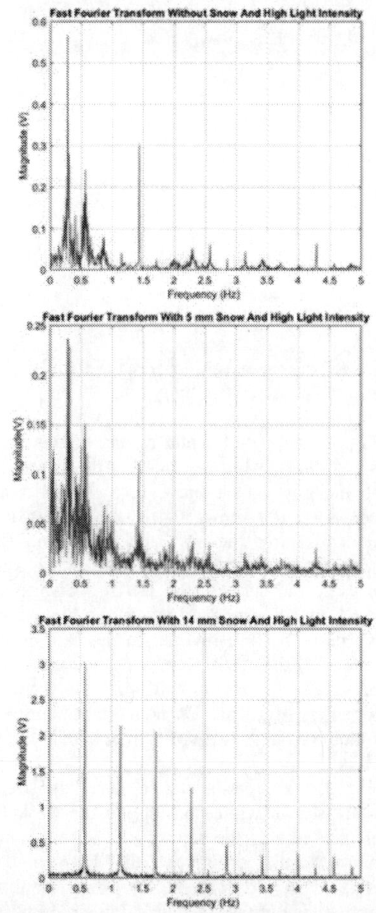

Figure 2: Experimental results from the frequency analysis in the case of strong light

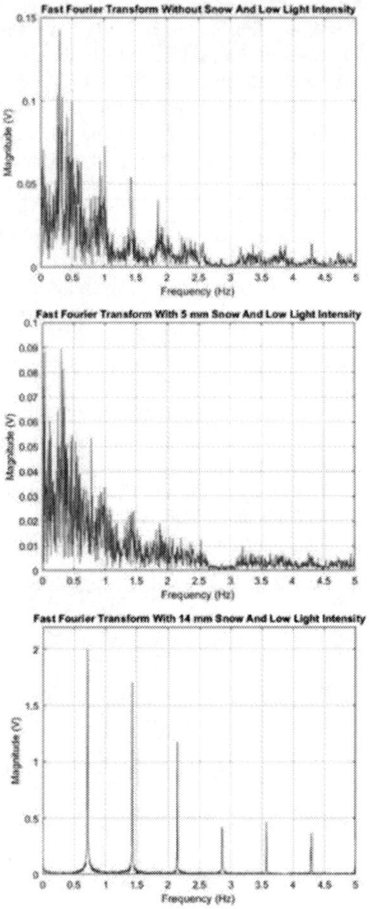

Figure 3: Experimental results from the frequency analysis in the case of weak light

In Fig. 2, the experimental results of the frequency analysis are shown while considering the strong light at different thickness of the snow (0, 5 and 14 mm). The experimental results shows that that between 0 to 14 millimeters of snow thickness, the power amplitude of the output voltage in the frequency domain drops and the frequencies spread between 0-5 Hertz. While comparing the 0 and 5 mm thickness of snow, the highest peak is found within 0.25 Hz. However, in the case of 14 mm, the location of the peak is moving towards a higher frequency (above 0.5 Hz) in comparison to the case of 5 mm. Moreover, it can be also notice that harmonics are more pronounced in the case of 14 mm than in the case of 0 and 5 mm.

In Fig. 3, the experimental results of the frequency analysis are shown while considering the weak light at different thickness of the snow (0, 5 and 14 mm). The experimental results shows that that between 0 to 14 millimeters of snow thickness, the power amplitude of the output voltage in the frequency domain drops and the frequencies spread between 0-5 Hertz. While comparing the 0 and 5 mm thickness of snow, the highest peak is found within the 0.25 Hz. However, in the case of 14 mm, the location of the peak is moving towards a higher frequency (above 0.5 Hz) in respect to the case 5 mm. Moreover, it can be also notice that harmonics are more pronounced in the case of 14 mm than in the case of 0 and 5 mm.

In Table I, the experimental results of the two-way ANOVA are shown. As we can notice, no significant effect of the light source was detected ($P = 0.117$) in the frequency analysis of the PV regulator due to the fact in both cases a direct light to the LEGO solar panel was placed in order to maximize the power output. On the other hand, a significant effect of the snow thickness was detected ($P < 0.05$) in the frequency response of the PV regulator.

Table I: Experimental results from the statistical analysis

Source of Variation	*SS*	*df*	*MS*	*F*	*p*
Light	0.294	1	0.294	2.459	0.117
Snow	0.783	2	0.391	3.272	0.038
Interaction	0.043	2	0.022	0.180	0.835

Finally, the experimental results for the production analysis are shown in Fig. 4. As it can be observed, the average of the power output efficiency drops as the thickness of the snow increases (6.7% and 2.0% reduction of the relative power output per mm for the strong and weak light respectively).

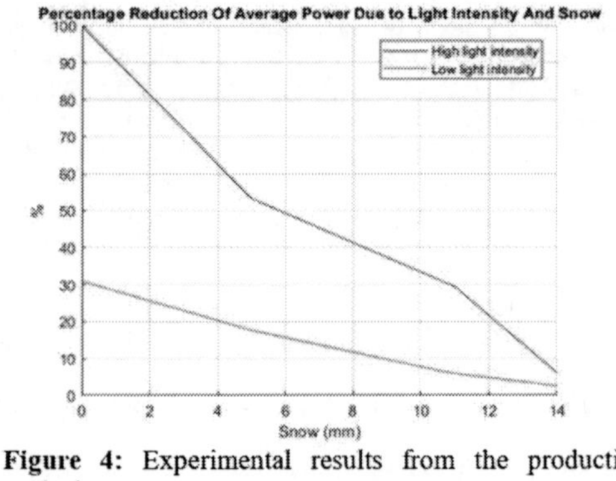

Figure 4: Experimental results from the production analysis

4 CONCLUSIONS & FUTURE WORK

In this paper, the authors have focused to understand how the snow affects the voltage and power output of a LEGO solar panel by using the renewable energy add-on set with the LEGO® EV3. For this purpose, the frequency analysis, statistical analysis and production analysis were considered.

From the experimental results, we could observe the effect of the thickness of the snow in the frequency analysis of the response of the PV regulator, determine a statistical significant effect of the snow thickness as well as visualize how the average of the power output drops in respect the increase of the snow thickness.

As a future work, following the proposed method, experiments will be carried out at the Solar Park 2 in Glava Energy Center, where other parameters will be also considered.

5 REFERENCES

[1] B.L. Brench. Snow-covering effects on the power output of solar photovoltaic arrays. DOE DE-AC02-76ET20279 (1979).

[2] T. Townsend, L. Powers. Photovoltaics and snow: An update from two winters of measurements in the sierra, 37th IEEE PVSC (2011)

[3] Mehrtash, M., Rousse, D.R., Quesada, G. Effects of surroundings snow coverage and solar tracking on photovoltaic systems operating in Canada, Journal of Renewable and Sustainable Energy, Vol. 5, (2013)

[4] R.W. Andrews, A. Pollard, J.M. Pearce. The effects of snowfall on solar photovoltaic performance. Solar Energy 92, 8497 (2013).

[5] S. Yang. Solar energy control system design, Master Thesis, Royal Institute of Technology, pp. 13-25 (2013).

[6] M. Frigo, S.G. Johnson. FFTW: An Adaptive Software Architecture for the FFT, in Proceedings of the International Conference on Acoustics, Speech, and Signal Processing, Vol. 3, pp. 1381-1384 (1998)

[7] R.V. Hogg, J. Ledolter. Engineering Statistics. New York: MacMillan (1987).

Unique soiling detection system for PV modules

Marc Korevaar[1], Joop Mes[1], Ahmed Alami Merrouni[3], Thijs Bergmans[1] Pramod Nepal[2], and Xander van Mechelen[1]

[1]Kipp & Zonen B.V., Delftechpark 36, 2628 XH Delft, the Netherlands
[2]Delft University of Technology, PVMD/EKL, Mekelweg 4, 2628 CD Delft, The Netherlands
[3]Institute of Research in Solar and New Energies (IRESEN), Green Energy Park, Benguerir, Morocco
Email: marc.korevaar@kippzonen.com, Telephone: 0031-152755242

ABSTRACT: In recent years, photovoltaic panels have emerged as one of the leading renewable energy technologies. PV plant performance is, aside from hardware, determined by three main parameters: irradiance, temperature and soiling. For soiling measurement, existing systems use cleaned and uncleaned PV panels or cells. These systems require manual cleaning or moving parts or liquids. We present the DustIQ for accurate and effective soiling measurement without these drawbacks. The DustIQ uses the Optical Soiling Measurement (OSM) technology to determine the soiling. It uses a LED and photodiode to measure the scattering signal of soiling on the top of the PV glass. We investigated the effect of grain size on the system and no influence of the grain size on the linear relation of the DustIQ diode signal and transmission loss was found.

The DustIQ also contains a PV cell, which allows the user to characterize for the local dust using the local dust calibration. Test results from as test locations in Morocco show a stable operation and an increase in soiling in late spring and summer. A preliminary comparison with a cleaned and soiled PV panel is favourable. The DustIQ is a good solution for measuring the soiling on PV panels.

Keywords: Environmental effect, Monitoring, Optical losses, Performance, Plant Control

1 INTRODUCTION

In recent years, photovoltaic (PV) technology has emerged as one of the leading renewable energy technologies currently available. The PV plant performance is, aside from hardware, determined by three main parameters: irradiance, temperature and soiling. Irradiance and temperature are well understood and high quality measurements are available. However, for soiling there is still limited understanding and only basic measurements. The extent of the effect of soiling on the PV plant performance is shown by the accumulation of soiling per day for different locations, shown in fig. 1, reproduced from previously published work [1]. Given as well is the average annual solar insolation for these locations showing their high potential for PV.

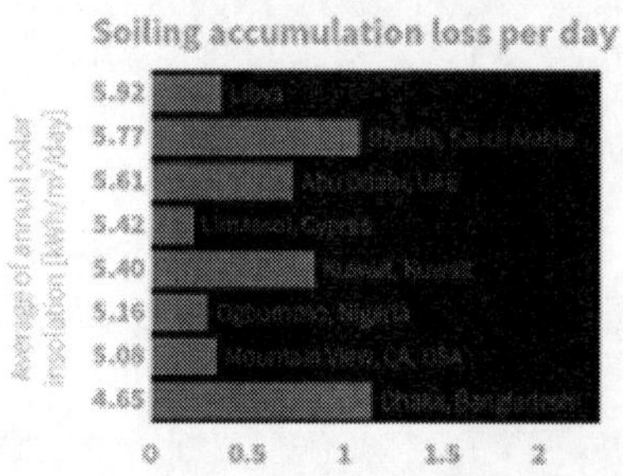

Figure 1: Soiling loss accumulation per day for different locations with average annual solar insolation given (recreated from [1]).

The accumulation of soiling is dependent on the location. A map showing the world distribution, from previously published work [2], is shown in fig. 2. The darker colours indicate a higher level of dust than the lighter colours and the number represents the μg/m³ of PM10. Furthermore the soiling is also dependent on the time of year [3].

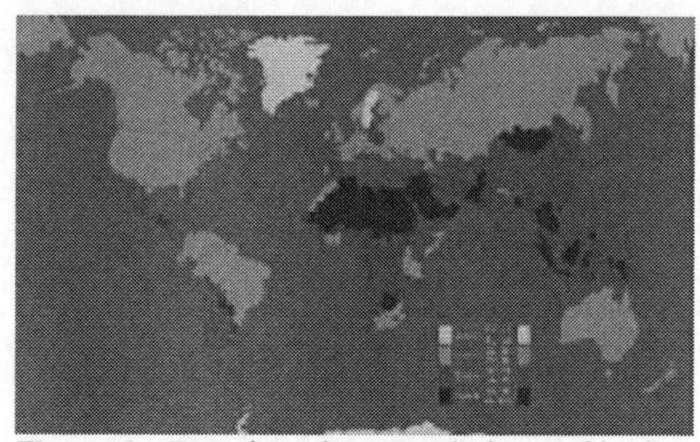

Figure 2: Dust intensity around the world, the number represents the μg/m³ PM10. (recreated from [2])

From the abovementioned work by other groups, it can be seen that the PV plant performance can be significantly influenced by module soiling. But when and where to clean the modules? To answer this question there is a need for accurate measurements of the amount of soiling on PV modules.

Existing commercially available systems all use the principle of a clean and not cleaned solar panel. It requires daily manual cleaning [4] or cleaning with moving parts/liquids.

This need for accurate measurements and the drawbacks of existing systems has been the inspiration for Kipp & Zonen to develop the DustIQ, a novel soiling sensor that avoids daily cleaning, moving parts/liquids, and provides multiple sensors across a plant for a similar budget. This information can be used by an O&M company to agree with stakeholders an optimal cleaning schedule and to trigger action after an event such as a dust or sand storm.

Here we will shortly describe the measurement system, a more thorough analysis of the grain size than previously and show an addition to the system that enables the local dust calibration. Furthermore, results of the measurement campaigned in Benguerir Morocco are also described.

2 SYSTEM & METHODS

2.1 System

The DustIQ system consists of glass plate, the same as in a PV panel, mounted next to or in between the PV modules to ensure that the instrument follows the same soiling and cleaning patterns as the PV panels themselves. Below this glass plate, there are two sensor heads that use the Optical Soiling Measurement (OSM) technology to measure the scattered light they receive from the soiling on top of the glass panel, shown in fig. 3.

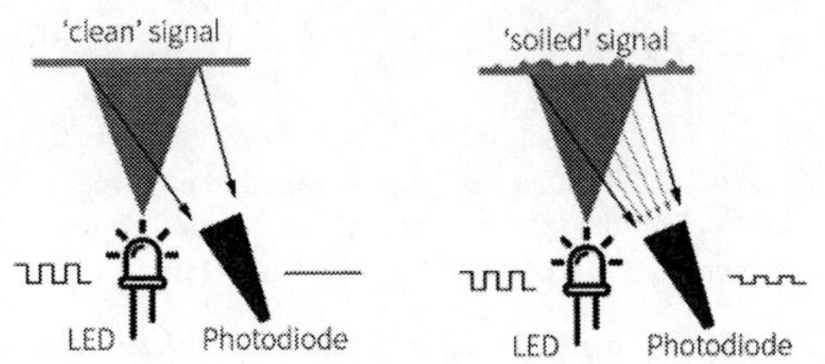

(a) (b)

Figure 3: diagram of DustIQ function with pulsed LED and photodiode with **(a)** clean glass and **(b)** soiled glass, giving rise to detection of scattered light.

With an on-board calculation, the transmission loss due to soiling is determined from the two OSM sensors. It can be shown that this transmission loss gives an equivalent loss in short-circuit current and power. This loss in power is directly related to the Soiling Ratio (SR), as defined as the ratio in short circuit current, I_{sc}, or maximum power, P_{max}, between a soiled and a cleaned PV panel in the IEC 61724-1 [5].

This SR is not constant over the day but is largest around solar noon, and gets smaller closer to sunset and sunrise, as is also reported in literature [3,6,7].

The SR as it is at solar noon is delivered by the DustIQ every minute. A unique feature is that the soiling measurement does not use the irradiance of the sun, therefore it is independent of the sun position and sky conditions. The system communicates using the Modbus® RTU protocol like all Kipp & Zonen Smart instruments with suitable inverters and data acquisition systems. Connectivity is through two-wire RS-485 connection.

2.2 Methods

The dust colour can vary per location and can influence the amount of scattered light measured with OSM and the corresponding transmission loss. Out of the factory, the DustIQ is calibrated for Arizona Test Dust and to accommodate a different dust colour in the field the customer can perform the local dust calibration. To do this, the DustIQ is fitted with an on board polycrystalline silicon cell as shown in figure 4. The local dust calibration can be performed when the DustIQ is soiled, as is shown here at the test facility in Morocco. The short-circuit current in the silicon cells and the scattered light are measured by the on board electronics and the photodiode by pushing the local dust calibration button for 10 seconds. Then the left sensor and the silicon cells are cleaned and the button is pushed a second time. As a last step, also the right sensor is cleaned and the button is pushed a third time. A possible resulting plot that is calculated internally is shown in figure 5. The internal electronics determine the dust slope for the local dust that is on the DustIQ. After pushing the button a fourth time the values are stored internally and the DustIQ measures the transmission loss for the local dust.

Figure 4: A naturally soiled DustIQ. The small silicon-cell module (center) sits between the two round OSM sensors (left and right). Below the DustIQ there is a cleaned PV module as well as a non cleaned PV module.

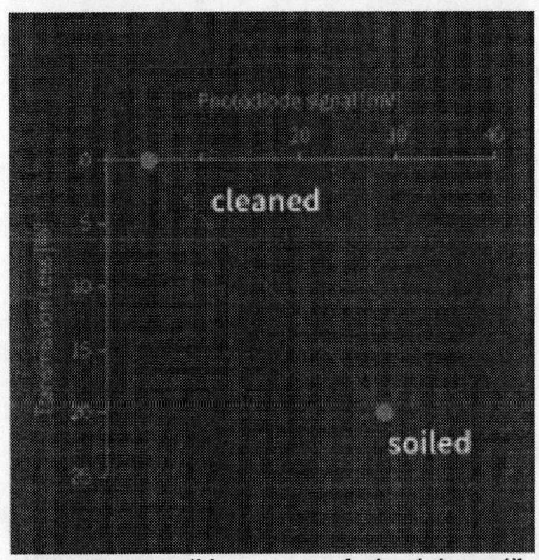

Figure 5: A possible outcome of a local dust calibration result is shown in the plot with the measurement point in the soiled state and in the cleand state.

3 RESULTS

3.1 DustIQ Characterization

The DustIQ was characterized more thoroughly than previously [4] for the effect of grain size on the scattering measurements. Fig. 5 (a) shows the relation between transmission loss and mass loading of different dust grain size distributions of Arizona Quartz in size distributions A1, A2, A3 and A4 according to ISO 12103-1 from KSL Staubtechnik gmbh. For a fixed mass loading different dust particle sizes give different transmission loss. However, if we look at fig. 5 (b) which shows the transmission loss as a function of DustIQ signal for different grain sizes, we see a linear relationship with minimal influence of the grain size.

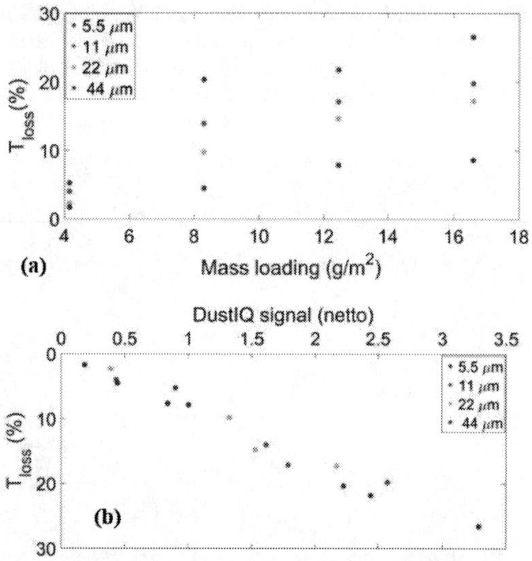

Figure 6: **(a)** Experimental data of transmission loss as function of mass loading on DustIQ. **(b)** Experimental data of transmission loss as a function of DustIQ signal, showing a linear relationship independent of grain size.

Weather data were also collected at each site, including rainfall. The blue curve represents the raw SR value that is influenced by soiling and by short term influences such as by dew or rain. An algorithm generated trendline (red), which is also implemented internally in the DustIQ, shows the trend of the soiling data, minimizing the daily dew/rain signal noise. The DustIQ sensor shows an increase of soiling over time, with the site in Morocco soiling at a rate of up to 0.4% SR per day.

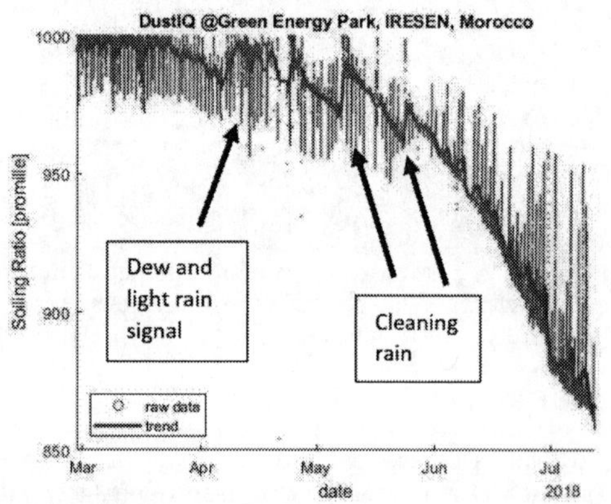

Figure 7: First results of Soiling Ratio measurement with the DustIQ at PSA

Figure 8 shows a comparison of the DustIQ trendline with the ratio of short circuit current of a cleaned and soiled PV panel (figure 5). The comparison shows a good agreement between the 2 measurement methods.

4 CONCLUSIONS

We demonstrated the relevance of soiling for PV plant performance and locations particularly prone to soiling.

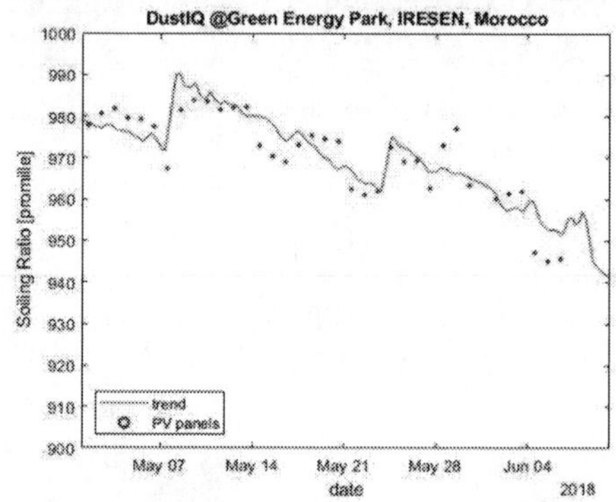

Figure 8 Preliminary DustIQ field comparison with the ratio of 2 PV modules (one cleaned daily, the other soiled). For the limited time there is good agreement between the PV panel ratio and the DustIQ trend.

The DustIQ, using the OSM technology, allows for local dust calibration with the on board polycrystalline silicon cells. Although the transmission loss for a given mass loading of soiling depends on grain size, we have shown that the DustIQ signal versus the transmission loss is not dependent on the grain size.

The stable field test result from the measurement campaign at Green Energy Park in Morocco show a loss in power of up to 14%. The soiling ratio information that is available real-time, day and night can identify when to clean or not.

The DustIQ data identifies morning dew and rain, and this dew signal can be filtered away with an internal algorithm. A preliminary comparison of the DustIQ with the ratio of a soiled and a clean PV panel shows a good agreement.

5 ACKNOWLEDGEMENTS

We would like to thank Green Energy Park Iresen and CIEMAT PSA Almeria for their help with testing.

6 REFERENCES

[1] A. Sayyah, M. Horenstein, and M. Mazumder, "Yield loss of photovoltaic panels caused by depositions", Solar Energy 107, 2014.

[2] M. Maghami et al., "Power loss due to soiling on solar panel", Renewable and Sustainable Energy Reviews 59, 2016.

[3] J. Zorrilla-Casanova, M. Philiougine, J. Carretero, P. Bernaola, P. Carpena, L. Mora-Lopez, and M. Sidrach-de-Cardona, "Analysis of dust losses in photovoltaic modules", World Renewable Energy Congress, 2011.

[4] M. Gostein, T. Duster, and C. Thuman, "Soiling Station Employing Module Power Measurements", Proc. IEEE PVSC, 2015.

[5] International Standard IEC 61724-1.

[6] M. Gostein, J. Caron, and B. Littmann, "Measuring Soiling Losses at Utility-scale PV Power Plants", IEEE Photovoltaic Specialists Conference, 2014.

[7] M. Korevaar, J. Mes, P. Nepal, G. Snijders, X. van Mechelen, "Novel Soiling Detection System for Solar Panels", Proc. EU PVSEC, 2017.

DESIGN AND INDOOR VALIDATION OF "DUSST": A NOVEL LOW-MAINTENANCE SOILING STATION

Matthew Muller[1,2], Joshua Morse[1], Florencia Almonacid[2], Eduardo F. Fernandez[2], Leonardo Micheli[1]
[1] National Renewable Energy Laboratory, Golden, Colorado, United States
[2] University of Jaen, Jaen, Spain

ABSTRACT: Soiling can cause significant losses to photovoltaic systems, and therefore it has to be carefully monitored. Currently, the most common soiling monitoring technologies are soiling stations that use the electrical outputs of a regularly cleaned PV device and of a naturally soiled PV device to quantify soiling. A new class of low-maintenance soiling stations has now been launched in the market to lower the cost and increase the reliability of soiling monitoring. In this work, the design, and the indoor validation of "DUSST" (Detector Unit for Soiling Spectral Transmittance), a new low-maintenance soiling station, are presented. DUSST projects a collimated monochromatic light through a glass surface and on to a light detector to measure the intensity of the transmitted light. As the glass surface naturally soils, the losses are quantified by comparing this soiled reading with a calibrated reading in the baseline clean condition.
Keywords: Soiling, Detector, Reliability, Monitoring, Optical Losses

1 INTRODUCTION

The accumulation of contaminants on the surface of photovoltaic (PV) modules (i.e. soiling) is an issue affecting PV systems worldwide, where it can decrease the annual energy yield by up to 7% [1–3]. In some regions, soiling can cause power drops as high as 70% [4]. Soiling can cause significant economic losses because it lowers the energy yield of PV systems, raises the operations and maintenance costs, and increases the uncertainty on PV performance, which can lead to higher finance rates from the investors. Therefore, it is important to monitor soiling, in order to plan the most accurate cleaning schedule for a system, i.e. only executing cleanings when the associated expenses are justified by the economics of the projected energy gains [5].

So far, the community has been monitoring soiling at PV sites mainly by using soiling stations [6,7]. These systems are generally made of two PV devices (cells or modules), one of which is left to soil naturally, while the other is regularly cleaned. Soiling is quantified by comparing the electrical outputs of the two PV devices. Soiling stations are cleaned either by an operator or through an automated rotating brush or high-pressurized water spray. As discussed in [8], soiling stations require frequent cleanings and careful maintenance in order to produce low uncertainty and reliable data. For example, in the right location one missed cleaning can result in uncertainties ballooning to values that are greater than the reported soiling loss. In this light, we propose an innovative cleaning-less soiling station, named DUSST (*Detector Unit for Soiling Spectral Transmittance*), designed to monitor soiling without moving parts and the need of water [9].

In the present work, the design, and the measurement procedure of DUSST are detailed followed by the results of an indoor validation conducted using artificially soiled coupons.

2 DUSST

2.1 Design

A 2D drawing of the DUSST design, described in the provisional patent submitted to the U.S. Patent and Trademark Office [9], is shown in Figure 1. A collimated light source is used to emit a monochromatic light perpendicularly onto a glass coupon where soiling is collected. The glass coupon is exposed to the same conditions and mounted at the same tilt, azimuth, and height as the PV modules to be monitored. The intensity of the light transmitted through the soiled glass is measured using a light detector and compared to the intensity of the light transmitted through the glass in clean reference conditions. The ratio between the two intensities (*Light Intensity Ratio*) can be used to estimate the *Soiling Ratio*, the most common metric to quantify the impact of soiling on the PV performance, defined as the ratio between the electrical output of a soiled PV module and the same electrical output of the module in clean conditions [10].

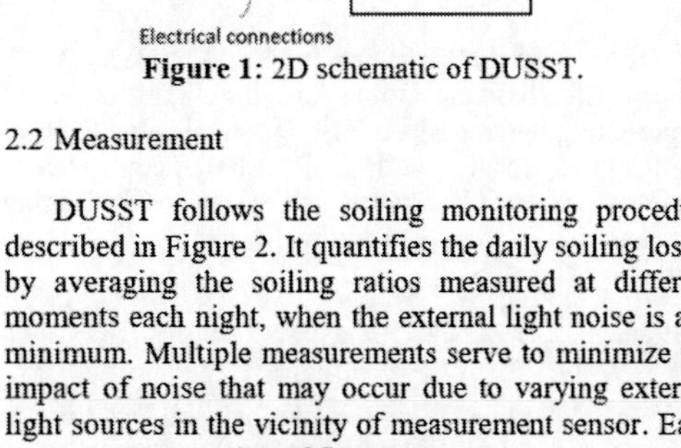

Figure 1: 2D schematic of DUSST.

2.2 Measurement

DUSST follows the soiling monitoring procedure described in Figure 2. It quantifies the daily soiling losses by averaging the soiling ratios measured at different moments each night, when the external light noise is at a minimum. Multiple measurements serve to minimize the impact of noise that may occur due to varying external light sources in the vicinity of measurement sensor. Each measurement consists of five steps:

1. **Zero Measurement**: the reading of the light detector is recorded while the collimated light source is turned off. This way, potential external light sources, which might affect the soiling measurement, are quantified.
2. **Soiling Measurement**: the collimated light source is switched on. Once the signal is stable, the intensity of the light measured by the light detector is recorded. A second zero measurement can be conducted to re-check the absence of

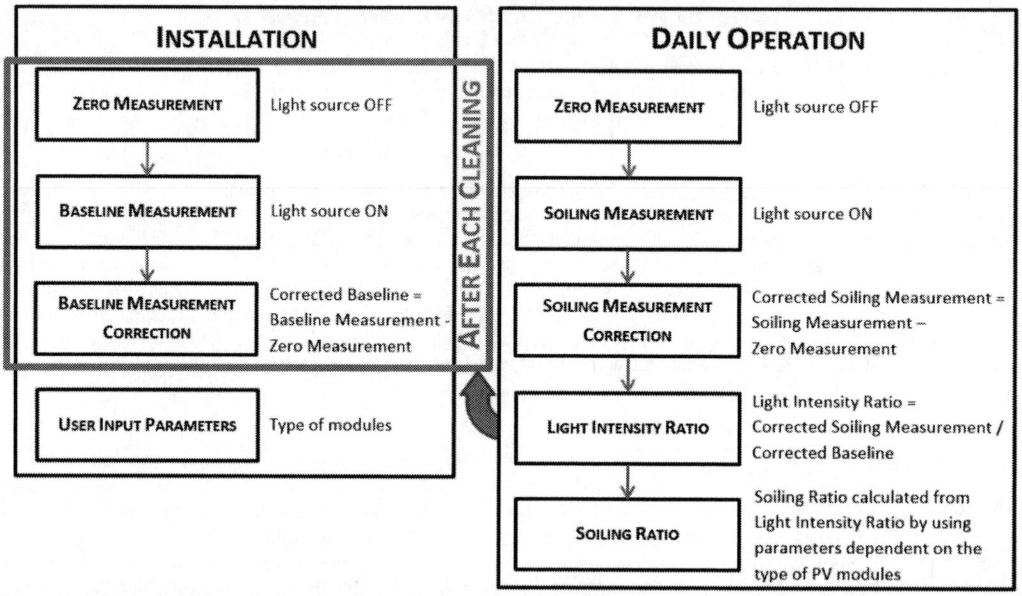

Figure 2: Schematic of the measurement process steps and of the calibration process required after the first installation and after each cleaning.

external light sources and/or to verify the same signal obtained when performing the initial zero measurement.

3. **Soiling Measurement Correction**: the soiling measurement is corrected by subtracting the zero-measurement to remove part of the light signal due to external sources.
4. **Light Intensity Ratio**: the ratio between the intensity of the light transmitted through the soiled glass and its *baseline* (intensity of the light transmitted through the clean glass) is determined.
5. **Soiling Ratio**: the soiling ratio is calculated by correcting the light intensity ratio according to *pre-determined coefficients* that take into account the type of modules under investigation [11].

DUSST should be cleaned at the same time as the PV array. The baseline value of the light intensity, representing the intensity of the light transmitted through the glass in clean condition, is determined (i) at the moment the device is installed, and (ii) after each cleaning. The baseline is measured following steps 1 to 3 of the procedure above.

2.3 Advantages

When light hits soiling particles or a soil layer, part of it is transmitted and reaches the PV material, while part of it is either absorbed or reflected. DUSST replicates the behavior of the solar cell, where the sunlight is replaced by a beam of known intensity, converting the photons into a signal. It measures the transmitted component of the light: transmittance is often larger than the sum of absorbance and reflectance, and its measurement is generally less noisy than a reflectance signal, meaning that DUSST is expected to achieve an excellent signal-to-noise ratio.

DUSST uses a stable monochromatic light source to quantify soiling, instead of the constantly varying sunlight used for typical soiling stations. The need of determining the intensity of the sunlight makes it necessary to regularly clean the reference cells in a typical soiling station. In DUSST, the intensity of the incoming light is known due to the stable monochromatic light source used, thus eliminating the need for cleaning. In addition, the lack of movable parts and the employment of standard components make DUSST reliable, durable, and a low-cost option. The re-calibration, automatically conducted after each cleaning, reduces the potential effects of component aging and degradation on the soiling measurement.

3 INDOOR VALIDATION

3.1 Prototype

The choice of the monochromatic light emitter's wavelength is based on the analysis presented in [11]. A monochromatic diode emitting light at 530 nm (green light) has been chosen for the prototype, because within the waveband that returned the best soiling estimation for any of the PV technologies investigated in [11]. An optical structure has been used to collimate the light onto the light detector.

An encapsulated solar cell has been employed as light detector and a Fluke 289 multimeter has been used to measure the current generated by the cell. In the fielded prototype, a data logger will replace the Fluke multimeter. The on/off switching of the diode is controlled by a timer. Power for the entire station is supplied by a Li-ion USB power bank.

3.2 Artificially Soiled Coupons

Five 4cm × 4cm sized and 3mm thick Diamant low-iron glass coupons have been artificially soiled by using different amounts of Kaolinite. The deposition was performed by using a deposition chamber developed at NREL. The hemispherical transmittances of the coupons were measured between 250 and 1300 nm, at 1-nm steps, using a Cary 5000 dual-beam ultraviolet-visible-near infrared (UV-VIS-NIR) spectrophotometer (Agilent Technologies Inc.) equipped with a DRA-2500 integrating sphere (Agilent Technologies Inc.).

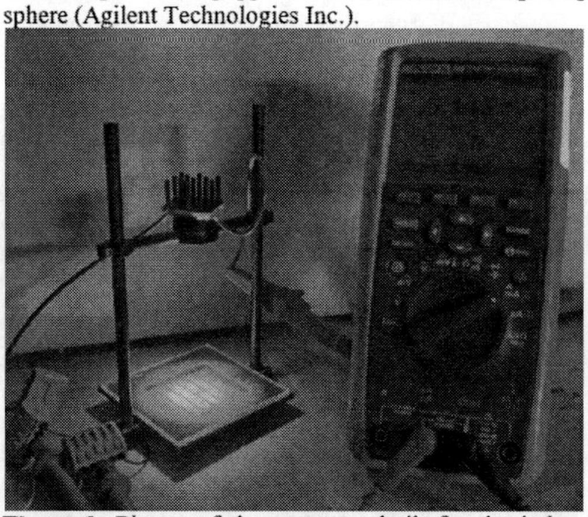

Figure 3: Picture of the prototype built for the indoor validation process.

3.3 Indoor Validation

The validation of the prototype has been conducted by powering the green diode at 510 mA and 3.025 V. The measurements started 30 minutes after the light source had been switched on to let the light beam stabilize. In order to minimize the impact of external light sources, the test was conducted in the dark and a zero calibration was performed (zero reading of the detector: 0.011 mA).

The intensity of the light emitted by the diode and recorded by the light detector under a clean glass resulted in a *baseline* of 5.127 mA. The intensities of the light transmitted through each soiled glass coupon were measured and, along with the baseline, corrected according to the zero measurement. The readings of the multimeter (*soiling measurements*) were compared with the hemispherical transmittance of the same glass at 530 nm, previously measured using the spectrophotometer. The results of the test are shown in Figure 4: the two measurements have a coefficient of determination (R^2) of 97.6%, proving a strong linear correlation between the measurement of DUSST and the actual hemispherical transmittance of the glass coupons.

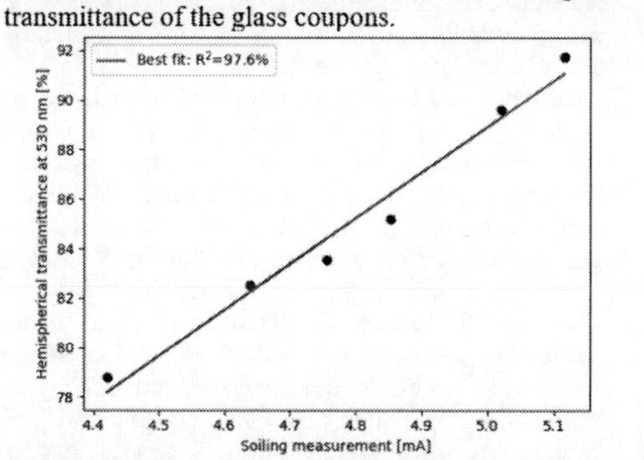

Figure 4: The readings of DUSST (x-axis) compared to the hemispherical transmittance at 530 nm, measured with the spectrophotometer, of the soiled and clean PV glass coupons.

When measuring the intensity of the light transmitted through by each soiled glass coupon and comparing it with the baseline measurement, DUSST functions in a similar manner as a spectrophotometer. Therefore, the quality of the measurement can be determined as shown in Figure 5 by comparing the *Light Intensity Ratio* (ratio of the intensity of the light transmitted by a soiled glass coupon to the intensity of the baseline measurement) with the *Soiling Relative Transmittance* (ratio of the hemispherical transmittance at 530 nm of the soiled glass coupon to the clean glass coupon). Similar to the soiling ratio, the *Soiling Relative Transmittance* and the *Light Intensity Ratio* have values of 100% in clean conditions and decrease with soiling. In ideal conditions, the *Light Intensity Ratio* for a monochromatic light and the *Soiling Relative Transmittance* at that same wavelength should be the same. In our test, the best fit between *Light Intensity Ratio* and the *Soiling Relative Transmittance* (blue dashed line in Figure 5) demonstrates the excellent linear correlation between the two measurements and proves therefore the reliability of the DUSST prototype.

Soiling has a non-linear impact on the irradiance spectra, with the blue region of the light being more affected than the red region [12,13]. This means that the *Soiling Ratio* is the result of a complex interaction among the irradiance spectrum, the soiling transmittance spectrum, and the spectral response of the photovoltaic material. This interaction can modeled for reference conditions as follows [11]:

$$r_s = \frac{\int_{\lambda 1}^{\lambda 2} E_G(\lambda, t)\tau_{soiling}(\lambda, t)SR(\lambda)d\lambda}{\int_{\lambda_1}^{\lambda_2} E_G(\lambda, t)SR(\lambda)d\lambda} \quad (1)$$

where E_g is the spectral distribution of the incident global irradiance, $\tau_{soiling}$ is the hemispherical transmittance of soiling, SR is the spectral response, and λ_1 and λ_1 are the lower and the upper limits of the absorption band of the cell. In this work, the spectral response and the absorption band of a mono-crystalline silicon cell are considered. The global reference spectrum reported in [14] has been considered in this study.

The green continuous line in Figure 5 shows the relation between the *Light Intensity Ratio* and the *Soiling Ratio* (modeled per eq. 1) for the glass coupons under investigation. The slope is substantially less than 1 due to the fact that DUSST measures only a single wavelength, while the *Soiling Ratio* is calculated over the whole absorption band of the cell. This discrepancy can be corrected by applying a calibration to the light intensity ratio, in order to consider the spectral response of each PV material.

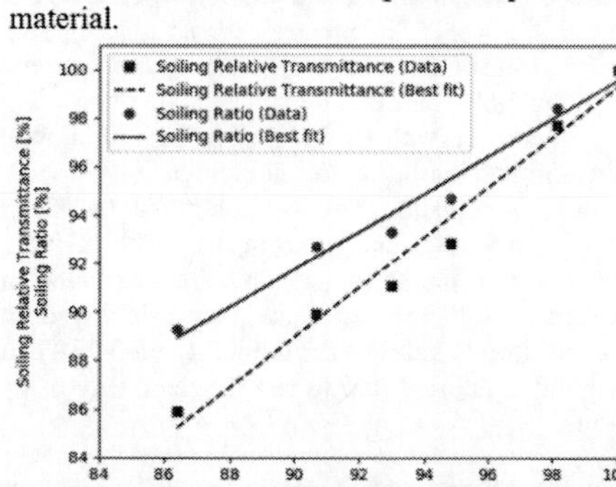

Figure 5: Blue data compare the *light intensity ratio* per DUSST against the *Soiling Relative Transmittance* at 530 nm. Green data compare the *Light Intensity Ratio* per

DUSST against the *Soiling Ratio* of a mono-crystalline cell placed behind each glass, modelled according to eq. 1, under a reference global irradiance spectrum [13]

Applying a calibration factor for mono crystalline cells determined per the line of best fit in Figure 5 (green continuous line) results in accurate DUSST Soiling Ratios. Overall, a mean absolute percentage error (MAPE) of 0.6% is found between the actual soiling ratios and the values of the best fit line calculated at the DUSST Soiling Ratios (see Figure 6). The MAPE expresses the average absolute error between actual *Soiling Ratios* and best fit line. This error is within the uncertainty ranges of the spectrophotometer and is lower than the expected errors of irregularly cleaned traditional soiling stations [8].

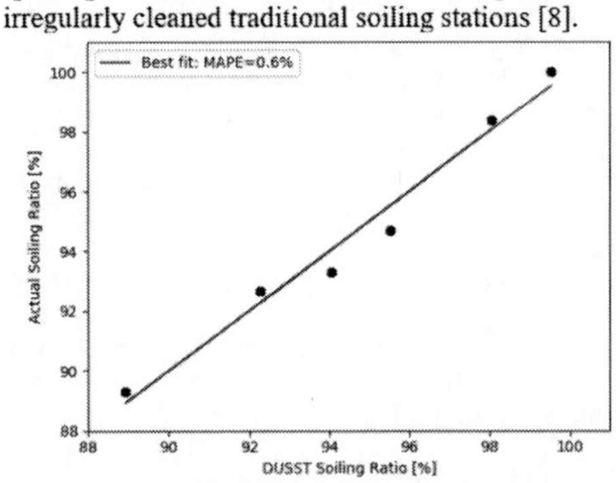

Figure 6: Comparing the *Soiling Ratios* calculated by DUSST (DUSST Soiling Ratio) and the *Soiling Ratios* of a mono-crystalline cell placed behind each glass, modelled according to eq. 1, under a reference global irradiance spectrum [14].

4 CONCLUSIONS AND FUTURE WORK

This work presents the design and the indoor validation of DUSST ("Detector Unit for Soiling Spectral Transmittance"), a novel soiling station that does not need to be cleaned. DUSST, which does not require water to operate and has no movable components, measures monochromatic light emitted by a known collimated source and transmitted through a naturally soiled glass coupon to quantify the soiling ratio of a PV array. The measurements are taken each night, under conditions with minimal external light. DUSST is re-calibrated after each cleaning to take into account any potential degradation of the components.

The indoor validation of a first prototype of DUSST is presented in this work. The analysis shows that DUSST measures the transmittance of soiled glass with extreme accuracy (R^2=97.6%) and that it can calculate the *Soiling Ratio* of a silicon cell with a MAPE of 0.6%, in line with the current soiling monitoring technologies.

The indoor validation will be continued by using different types of dusts and by comparing the DUSST measurements with the soiling losses of actual cells. In addition, more prototypes are now being deployed in outdoor conditions to validate the reliability of DUSST in natural soiling conditions and to test the durability of its components.

ACKNOWLEDGEMENTS

This work was authored in part by Alliance for Sustainable Energy, LLC, the manager and operator of the National Renewable Energy Laboratory for the U.S. Department of Energy (DOE) under Contract No. DE-AC36-08GO28308. Funding provided by the U.S. Department of Energy's Office of Energy Efficiency and Renewable Energy (EERE) under Solar Energy Technologies Office (SETO) Agreement Number 30311. The views expressed in the article do not necessarily represent the views of the DOE or the U.S. Government. The U.S. Government retains and the publisher, by accepting the article for publication, acknowledges that the U.S. Government retains a nonexclusive, paid-up, irrevocable, worldwide license to publish or reproduce the published form of this work, or allow others to do so, for U.S. Government purposes.

Part of this work has been funded through the 2018 NREL Accelerating Inventions to Market Award, duly acknowledged.

This study is partially based upon work from COST Action PEARL PV (CA16235), supported by COST (European Cooperation in Science and Technology). COST (European Cooperation in Science and Technology) is a funding agency for research and innovation networks. Our Actions help connect research initiatives across Europe and enable scientists to grow their ideas by sharing them with their peers. This boosts their research, career and innovation, see www.cost.eu.

REFERENCES

1. National Renewable Energy Laboratory. Photovoltaic modules soiling map 2018. https://www.nrel.gov/pv/soiling.html [accessed May 18, 2018].
2. Appels R, Muthirayan B, Beerten A, Paesen R, Driesen J, Poortmans J. The Effect of Dust Deposition on Photovoltaic Modules. *Photovoltaic Specialists Conference (PVSC), 2012 38th IEEE* 2012: 1886–1889. DOI: 10.1109/PVSC.2012.6317961.
3. Appels R, Lefevre B, Herteleer B, Goverde H, Beerten A, Paesen R, *et al.* Effect of soiling on photovoltaic modules. *Solar Energy* 2013; **96**: 283–291. DOI: 10.1016/j.solener.2013.07.017.
4. Sarver T, Al-Qaraghuli A, Kazmerski LL. A comprehensive review of the impact of dust on the use of solar energy: History, investigations, results, literature, and mitigation approaches. *Renewable and Sustainable Energy Reviews* 2013; **22**: 698–733. DOI: 10.1016/j.rser.2012.12.065.
5. Shrestha S, Taylor M. Soiling Assessment in Large-Scale PV Arrays. *SolarPro Magazine* 2016. http://solarprofessional.com/articles/operations-maintenance/soiling-assessment-in-large-scale-pv-arrays?v=disable_pagination.
6. Gostein M, Faullin S, Dunn LR, Stueve W. Soiling measurement system for photovoltaic arrays 2014.
7. Gostein M, Duster T, Thuman C. Accurately Measuring PV Soiling Losses With Soiling Station Employing Module Power Measurements. *IEEE 42nd Photovoltaic Specialist Conference (PVSC)*, 2015.
8. Muller M, Micheli L, Martinez-Morales AA. A Method to Extract Soiling Loss Data From Soiling Stations with Imperfect Cleaning Schedules. *2017 IEEE 44th Photovoltaic Specialist Conference (PVSC)*, Washington, D.C.: IEEE; 2017.
9. Fernández EF, Muller MT, Micheli L, Almonacid FM. Soiling Spectral Deposition Detector. USPTO 62/652,955, 2018.

10. International Electrotechnical Commission. Photovoltaic system performance – Part 1: Monitoring (IEC 61724-1, Edition 1.0, 2017-03) 2017.
11. Micheli L, Fernandez EF, Caballero JA, Smestad GP, Nofuentes G, Mallick TK, *et al.* Waveband investigation on the impact of soiling on various photovoltaic technologies. *Submitted.*
12. Micheli L, Fernández EF, Smestad GP, Alrashidi H, Sarmah N, Hassan IAI, *et al.* A unified global investigation on the spectral effects of soiling losses of PV glass substrates : preliminary results. *IEEE 44th Photovoltaic Specialist Conference (PVSC)*, Washington, D.C.: IEEE; 2017.
13. Qasem H, Betts TR, Müllejans H, AlBusairi H, Gottschalg R. Dust-induced shading on photovoltaic modules. *Progress in Photovoltaics: Research and Applications* 2014; **22**(2): 218–226. DOI: 10.1002/pip.2230.
14. International Electrotechnical Commission. Photovoltaic devices - Part 3: Measurement principles for terrestrial photovoltaic (PV) solar devices with reference spectral irradiance data (IEC 60904-3:2016) 2016.

I-V CHARACTERISTICS OF BROKEN BYPASS DIODE ON PV MODULE

Shinichiro Oke[1], Homare Sakai[2], Hisanori Tottori[2], Yuhei Shimizu[2], Nicole Choo[2], Ikuo Nanno[3], Toshiyuki Hamada[3], Norio Ishikura[4], and Masayuki Fujii[5]
[1] National Institute of Technology, Tsuyama College, 624-1 Numa, Tsuyama, 708-8580 Japan
E-Mail: oke@tsuyama-ct.ac.jp, Tel. +81-868-24-8272
[2] National Institute of Technology, Tsuyama College, 624-1 Numa, Tsuyama, 708-8580 Japan
[3] National Institute of Technology, Ube College, 2-14-1 Tokiwadai, Ube, 755-8555 Japan
[4] National Institute of Technology, Yonago College, 4448 Hikona-cho, Yonago, 683-8502 Japan
[5] National Institute of Technology, Oshima College, 1091-1 Komatsu, Suo-oshima, 742-2193 Japan

ABSTRACT: Surge voltage or current induced by lightning strike around a PV system often causes short circuit failure of a bypass diode (BPD) on the PV module. We investigated the characteristics of BPDs broken by the induced lightning and evaluated the overheating and degradation of the broken BPD. The BPDs on the broken PV modules with slight damage were often shorted. The BPDs on the broken PV modules with serious damage had previous phase that was a short circuit failure. Current was injected into a BPD with separated cell string by constant current power supply to reproduce degradation of BPD with bypass current. The BPD shorted and had characteristic like a resistance. The shorted BPD was heated with current from a cell string of the shorted BPD in case the broken PV module did not operate. If a PV module had a partial shade in a cell string with the shorted BPD, the shorted BPD was heated by the current from other cell strings without the shade. These results show that if the shorted BPD is left unattended, it can cause serious accidents of PV system such as an overheating or a fire.
Keywords: Degradation, PV module, bypass diode, lightning, short-circuit failure, overheating

1 INTRODUCTION

A bypass diode (BPD) is widely used in photovoltaic (PV) modules to keep electricity generating during shading and current mismatching. The silicon Schottky barrier diode which has a lower maximum peak repetitive reverse voltage relative to surge reliability, a lower thermal resistance, and high leakage current to a P/N junction diode is widely used as the BPD for the PV module [1]. Surge voltage or current induced with lightning strike around a PV system often causes a short circuit failure of the BPD on the PV module [2]. When a current flowed through the damaged bypass diode, the temperature inside the junction box would increase to a level that might be high enough to cause a fire [3]. In this paper, we investigate the characteristics of PV modules and BPDs broken by the induced lightning and a process of its failure. A simple method for evaluating the shorted BPDs is presented based on the data of many damaged BPDs. Moreover, the fire risk of the short circuit failures of BPD is evaluated experimentally.

2 COLLECTION OF BROKEN PV MODULES & BPDs

Broken PV modules and many broken BPDs were retrieved to investigate failures of BPDs from the 3 sites (site A, B, and C) in our research group.

Site A, which was damaged by the induced lightning, had broken PV modules with slight damage. Some broken PV modules in site A emitted smoke after the induced lightning. Fig.1(a) shows inside of a junction box on the backside of the broken PV module in site A. The potting material in the junction box was discolored to a burnt brown. Rated open circuit voltage and rated short circuit current were 37 V and 8.8 A respectively.

PV modules in site B were burned due to the reverse current from other strings, since some BPDs and blocking diodes were shorted by the induced lightning. In this case, current generated by some strings flew into a string consist of some broken modules with shorted BPDs and a shorted blocking diode. The reverse current which was more than 60A, overheated the shorted BPDs which had a low resistance. Fig.1(b) shows the inside of a junction box of the broken PV module at site B. The junction box did not have a potting material. A cell string had 2 parallel BPDs. Rated open circuit voltage and rated short circuit current were 37 V and 8.4 A, respectively. Site B, where the BPDs of all PV modules were replaced with new BPDs before detailed tests, provided many replaced BPDs to our research group for measuring characteristics.

Site C did not receive the induced lightning, but many of its modules had failed. The PV modules at site C had many open circuit failures in their cell strings. The cell strings broke off in their bus bars. Fig.1(c) shows the inside of a junction box of the broken PV module at site C. The potting material around a broken BPD burned black and had cracks. Rated open circuit voltage and rated short circuit current were 25 V and 7.7 A, respectively.

The PV modules of the site A and B had 3 BPDs of Schottky barrier type. The PV modules of site C had 3 BPDs of P/N junction type. All PV modules of site A, B, and C had 3 series clusters. A cluster contained a cell string consists of some series PV cells with paralleled BPD.

3 EXPERIMENTS

3.1 I-V characteristics measurement

I-V characteristics of the broken PV modules, their clusters, and individual BPD were measured by bipolar power supply (PBZ40-10, KIKUSUI) and data logger in dark state. The current and the voltage were measured in the 4-terminal method to exclude a resistance of wire and a contact resistance of probes of measuring devices.

3.2 Overheating of shorted BPD

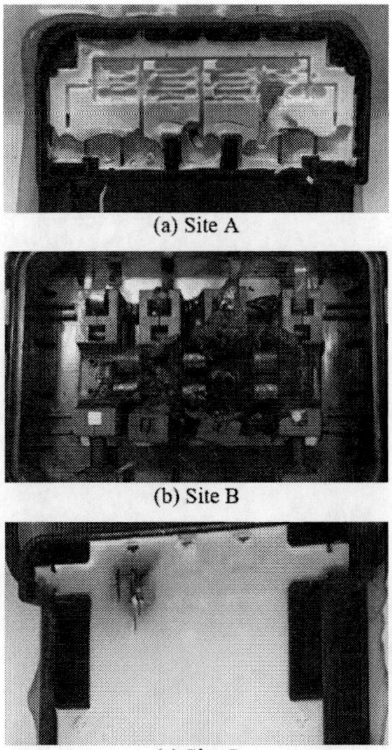

(a) Site A

(b) Site B

(c) Site C

Fig.1 Inside pictures of junction box on the backside of the broken PV modules.

One of the 3 BPDs of a PV module (open circuit voltage: 30 V, short circuit current: 8.1 A) was exchanged with a 1.0 Ω resistor to simulate a short circuit failure of the BPD. The PV module with a resistor instead of the BPD was exposed under the Sun, and temperature of its junction box was measured by thermography. We had 3 kinds of experiment conditions, which were an operating case, a non-operating case, and an operating with partial shade case. For the operating case, connectors of the PV module were connected to each other (short circuit) to flow the short circuit current similar to the maximum power point current. For the non-operating case, connectors of the PV module were opened. For the operating with partial shade case, connectors of the PV module were connected to each other, and the cell string with the 1 Ω resistor instead of a broken BPD was shaded. The shaded area was 1/3 of the PV module surface.

3.3 Degradation of BPD by bypass current

A BPD with a separated cell string is operating at all times, current from other cell string or other PV modules flows through the BPD. Bypass current may cause degradation of the BPD. The PV modules from site C had many separated cell strings, so some BPDs with separated cell strings had operated long time. Rated short circuit current of the PV module was injected into the BPD with separated cell string by constant current power supply (PS40-30A, TEXIO). The rated short circuit current was 7.7 A. The voltage and current were measured in the 4-terminal method. The temperature of the BPD was observed by IR camera (885-2, Testo).

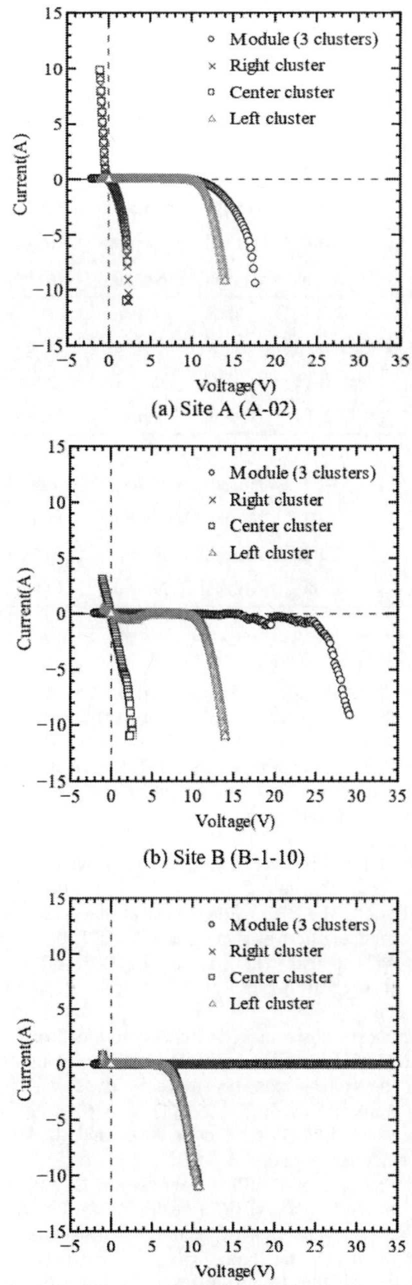

(a) Site A (A-02)

(b) Site B (B-1-10)

(c) Site C (C-01)

Fig.2 I-V curves of broken PV module and its clusters from site A, B, and C.

4 RESULTS & DISCUSSION

4.1 I-V characteristics of broken PV modules and BPDs

Fig.2 (a) shows the I-V curves of the broken PV module A-02 from site A. Open circuit voltage of the module decreased until 1/3 of rated value, since 2 BPDs were shorted. I-V curves of the module and its left cluster did not have a rise caused by forward characteristic of the

Table 1. Conditions of BPD of PV modules from site A, B, and C.

Site	Module	Conditions of BPD		
		Left	Center	Right
A	A-01	0.53 Ω	Normal	Normal
	A-02	Open	2.14 Ω	1.25 Ω
	A-03	0.54 Ω	0.05 Ω	Open
	A-04	1.14 Ω	Normal	Normal
	A-05	Normal	Normal	Normal
B	B-1-6	Open	Open	0.02 Ω
	B-1-10	Open	0.22 Ω	Open
	B-1-13	0.26 Ω	0.13 Ω	Open
	B-2-14	Open	Open	Open
C	C-01	Normal	Normal	Open
	C-02	Normal	Open	Normal
	C-03	Normal	Normal	Open
	C-04	Normal	Normal	Normal
	C-05	Normal	Normal	Open

BPD in the second quadrant, since the BPD of left cluster was opened. I-V curves of center and right clusters in the PV module A-02 did not have a threshold voltage on the voltage axis, since the BPDs of these clusters were shorted. However, they had nonlinear characteristics in the fourth quadrant.

Fig.2 (b) shows the I-V curves of the broken PV module B-1-10 from site B. Open circuit voltage of the module decreased until 2/3 of rated value, since 1 BPD were shorted. The BPD of the center cluster which had a linear characteristic was shorted and the BPD of the right cluster was opened. The state of the BPD of the left cluster changed from short circuit to open circuit in our experiment.

Fig.2 (c) shows the I-V curves of the broken PV module C-01 from site C. I-V curve of the module was flat on the voltage axis, because any BPDs or any cell strings were opened. Both the BPD and cell string were opened in the right cluster. On the other hand, the left and center clusters were normal.

The conditions of BPDs from site A, B, and C are shown in table 1. Many BPDs of the broken PV module from site A were shorted, and their resistance were 0.05~2.14 Ω. On the other hand, many BPDs of the broken PV modules from site B were opened. They were, thus, destroyed by fire. Some BPDs from site B which did not catch fire, were shorted. The shorted BPD B-1-10 changed from short circuit to open circuit during a measurement of I-V curve. Hence, it was found that the induced lightning sometimes short the BPD on PV module. Some shorted BPDs were opened, due to a current from other modules or strings flowed and heated the broken BPDs. At site C, all opened BPDs connected to opened cell strings, and some opened cell strings had normal BPDs. We guessed that the normal BPDs with the opened cell strings accumulated damage from bypass current. If the bypass current is flowed in the BPDs continuously, the BPDs would become short circuit and eventually opened. As a result, the opened BPD

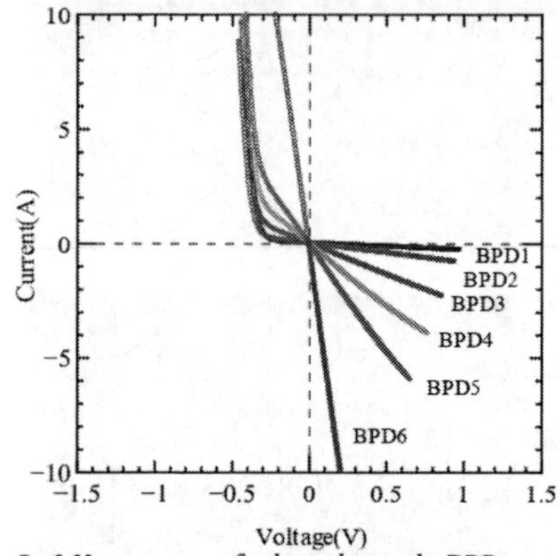

Fig.3 I-V curves of the shorted BPDs on the modules of site B.

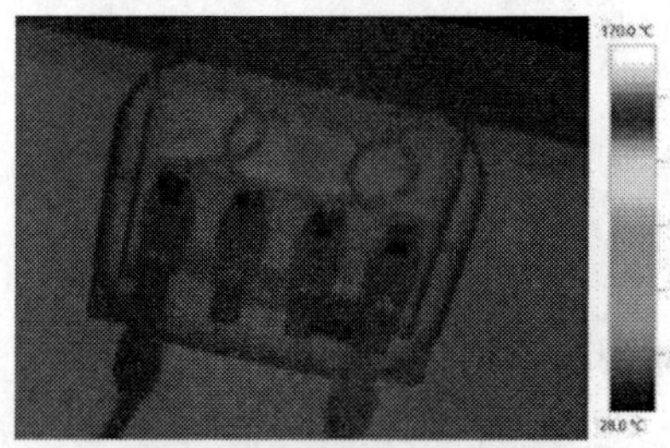

(a) Operating case

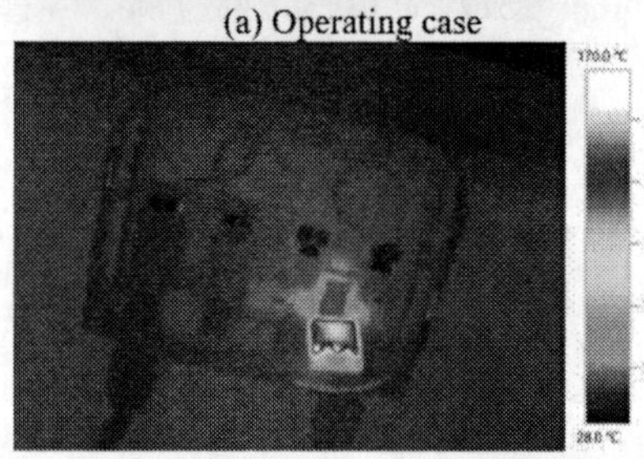

(b) Operating with partial shade case

Fig.4 Thermal images of a resistance instead of a broken BPD in junction box on PV module after 50 seconds from the start of the experiment.

Experienced a short circuit failure prior to the final stage of being opened unless there was an initial failure such as bad soldering or contacts.

Fig.3 is the I-V curves of the shorted BPDs on the PV modules of site B. The shorted BPDs had different characteristics, individually. BPD1, 2, and 3 still had the current rising like a diode in the second quadrant. On the other hand, BPD6 had the characteristics close to a resistance. These graphs show that the extent of damage of a BPD can be determined using the gradient of the I-V curve (i.e. a resistance) around the origin especially in the fourth quadrant. The resistances of the BPD with serious damage and that of the BPD with slight damage were 0.021 Ω, and 3.36 Ω, respectively. Therefore, the extent of damage of BPD can be evaluated by the broken resistance measured by a handy multimeter at on-site.

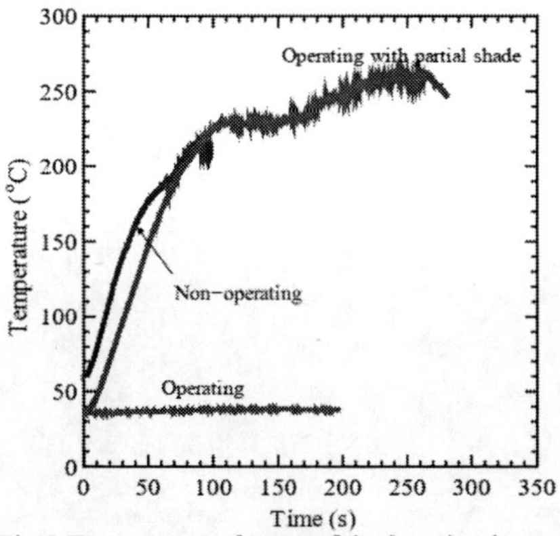

Fig.5 Temperature change of the junction box on a PV module with 1.0 Ω resistance instead of a shorted BPD.

4.2 Overheating of shorted BPD s

Fig.4 shows thermal images of a resistor instead of a broken BPD in junction box on PV module after 50 seconds from the start of the experiment. In the operating case, temperature of the resistance was as low as around components. The temperature of the resistance in the non-operating case and the operating with partial shade case were higher than that of the operating case. The temperature of the resistance was monitored to determine the risk of a fire of the PV module with the broken BPD in the 3 cases.

Fig.5 reveals that the temperature was almost constant at 35°C in the operating case, since current did not flow through the resistance instead of the broken BPD. In contrast, in the non-operating case and the operating with partial shade case, the temperatures rose rapidly to over 200°C. If connectors of the module were not connected the PV module, a current from the cell string with the resistor instead of the broken BPD flowed in that resistor. In case of the operating with partial shade, the current from other cell strings without the shade flowed into the resistor instead of the broken BPD to bypass the shaded cell string.

4.3 Degradation of BPD by bypass current

Fig.6 (a) shows transition of voltage of a BPD with separated cell string from site C. That is minutely data. Some spikes are not important, for it was generated by the turning on of the power supply. Voltage was nearly constant immediately after starting current injection. 33 hours later, voltage had increased slowly. Voltage increase means increasing forward voltage drop i.e. forward resistance of BPD. 73 hours later, voltage rising started suddenly from 0.9 V to 1.3 V. After that, voltage rose and dropped rapidly, since the BPD shorted. Voltage change around a moment of short circuit of the BPD is shown in Fig.6 (b). Voltage rose from around 2 V to 3.8 V in 2 seconds. The drop after the rising was caused by a safety function of the power supply. I-V characteristics of the shorted BPD is shown in Fig. 7. Resistance of the shorted BPD around the origin of the I-V curve was 91 mΩ. This result shows that bypass current flowing in a BPD can decline performance of the BPD and cause short circuit failure of the BPD. However, PV modules from site C had no shorted BPD. We guess that if the bypass

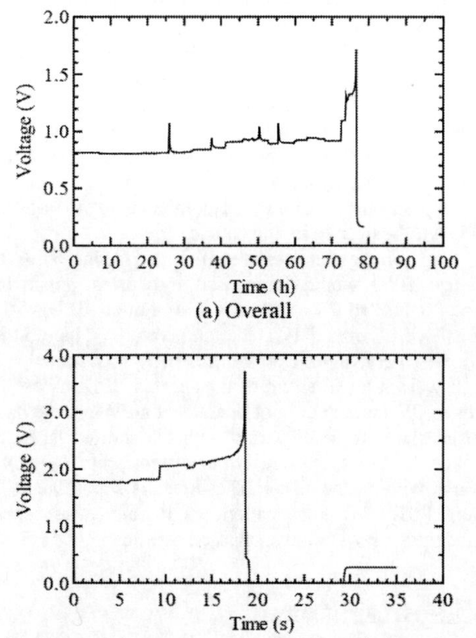

(b) Voltage change around a moment of short circuit of BPD

Fig.6 transition of voltage of a BPD with separated cell string from site C.

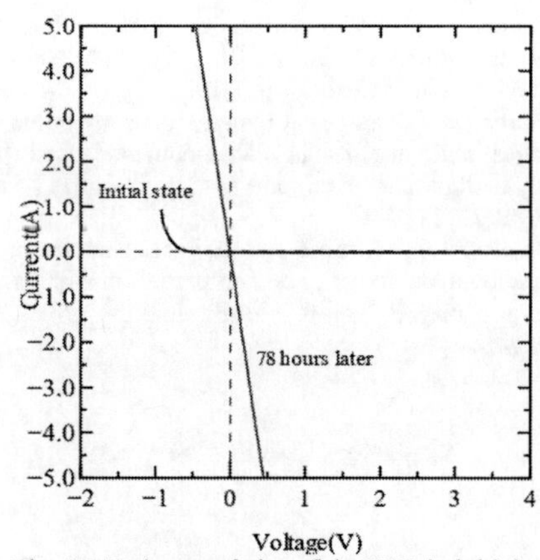

Fig. 7 I-V characteristics of the BPD in initial state and after 78 hours of current injection.

current is flowed continuously, the shorted BPD will be opened in due course.

5 CONCLUSIONS

Characteristics of the broken PV modules and BPDs from 3 sites were measured and evaluated: they were damaged due to the induced lightning or the open circuit failures in their cell strings. The BPDs on the broken PV modules with slight damage were often shorted. The

BPDs on the broken PV modules with serious damage were often opened. Thus, the opened BPD had previous stage that was a short circuit failure except an initial failure such as bad soldering or contacts.

Current was injected into a BPD with separated cell string by constant current power supply to reproduce degradation of BPD by bypass current. The BPD shorted and had characteristic like a resistance. We guess that if the bypass current is flowed continuously, the shorted BPD will be opened in due course.

The raising temperature of a cell string with the opened BPD would be caused by current mismatch. When a broken PV module had a shorted BPD with a resistance of around 1 Ω, the temperature of its junction box was very high. The shorted BPD was heated with current from a cell string of the shorted BPD in case the broken PV module did not operate. If a PV module had a partial shade in a cell string with the shorted BPD, the broken BPD was heated by a current from other cell strings without the shade. These results show that if the shorted BPD is left unattended, it can cause serious accidents of PV system such as an overheating or a fire.

ACKNOWLEDGEMENT

This study was partly supported by the New Energy and Industrial Technology Development Organization (NEDO).

REFERENCES

[1] G. Yumu et al., "Comparison of different bypass diode technologies", Solar 2011, the 49th AuSES annual conference (2011), pp. 1-10.

[2] H. Harberlin, "Damages at bypass diodes by induced voltages and currents in PV modules caused by nearly lightning currents", Proceedings of EUPVSEC22 (2007).

[3] S. W. Ko et al., "Electric and thermal characteristics of photovoltaic modules under partial shading and with a damaged bypass diode", Energy, Vol.128, pp.232-243 (2017)

INFLUENCE OF THE IRRADIANCE ON THE DETECTION AND PERFORMANCE OF PID-AFFECTED PV-MODULES

Claudia Buerhop[1], Tobias Pickel[1], Franziska Wenz[1], Cornelia Zetzmann[2], Jens Hauch[1], Christian Camus[1,3], Chr. J. Brabec[1,3,4]
[1]ZAE Bayern e.V., D-91058 Erlangen, Immerwahrstraße 2,
[2]Rauschert GmbH, D-96332 Pressig, Bahnhofstraße 1,
[3]i-MEET, FAU Erlangen-Nürnberg, D-91058 Erlangen, Martensstraße 7
[4] Helmholtz-Institut Erlangen-Nürnberg (HI ERN), D-91058 Erlangen, Egerlandstraße 3

ABSTRACT: Potential induced degradation (PID) is still a severe problem in many PV-installations and a huge financial risk for investors and plant-owners if not detected early. In previous investigations it was shown that thermography, conducted with handheld or drone mounted IR-cameras, is a fast and reliable tool to detect PID affected PV-modules on site and estimate their remaining electrical power during normal plant operation. In this work we show how to improve the sensitivity of IR- and various other imaging techniques by using the characteristic behavior of PID affected cells regarding the different excitation forces, i.e. solar irradiance for IR- and electric current for DLIT- and EL-measurements. Furthermore, we address the implications of this specific behavior by simulating the energy yield of several PID affected PV-strings based on real monitoring data for two inspected sites and suggest carrying out the measurements on site- and under failure- relevant conditions rather than standard test conditions.

Keywords: EL-imaging, thermography, potential induced degradation, performance, energy yield.

1 INTRODUCTION

Many PV-installations suffer from potential induced degradation (PID) which causes severe performance losses. Early detection is important to initiate suitable actions early enough.

We present a new approach for failure-specific, respectively PID-specific measurement conditions. It takes into account PID-specific properties and performance characteristics as well as site-specific data of PV-installations.

Here, the essential parameter is the solar irradiance. We show how the solar irradiance influences the PID-ratio k, which describes the percentage of affected cells [1],

$$k = \frac{n}{N}$$

with n the number of PID-affected cells in a module or string and N the total number of cells in the module or string.

We present for the first time how irradiance and analogously the current influence the PID-pattern in IR- and EL-images, as well as the performance.

2 EXPERIMENTAL PROCEDURE

A 314 kWp PV-plant installed in 2013 and located in the Atlantic coastal climate (in Portugal) was examined, as well as a 3.8 MWp PV-system in Southern Germany.

The field inspections were carried out with a measurement system consisting of a drone equipped with an IR- and VIS-camera and various sensors. Details about the field inspection and the results can be found elsewhere [1, 2].

In order to study the irradiance-dependent IR-appearance and the electrical performance of PID affected PV-modules, a sample of 27 PV-modules from the Portuguese plant, with different degrees of PID or PID-ratios, were carefully analyzed in the lab. Here, highly resolved EL-images as well as dark lock-in thermography (DLIT)-images were taken at high and low currents, and IV-curves were measured at standard test conditions (STC) and down to 200 W/m². Furthermore, IR-images were taken at different irradiances at an outdoor test facility. A full string of 21 modules was available for lab measurements, so that modules with different degrees of PID or PID-ratios could be analyzed. Their data were compared with the field findings and results.

Finally, monitoring data of the two sites and simple yield simulation completed the study.

3 RESULTS AND DISCUSSION

Detailed results and images for two modules are presented in **Figure 1**and **Figure 2**. These figures show IR-images for fairly high and rather low irradiance, EL- and DLIT-images for low and high excitation currents, as well as IV-curves for 5 irradiances. While **Figure 1** shows a presumably PID-free module at high irradiance and high current, **Figure 2** depicts clearly a PID-affected module.

Both figures reveal that by decreasing the excitation force (solar irradiance or electrical current), the number of dark cells in EL and bright cells in IR, i.e. PID-affected cells, increases. Significantly more cells show the typical PID-pattern of homogeneously increased intensity across the cell area at low irradiances. Studying the DLIT-images at the same measurement conditions, reveals locally increased temperatures at the cell edges for the PID-affected cells, also described in [3]. This is particularly pronounced at low current. Transferring this information about edge degradation to the EL-images, dark areas at the edges can be identified (red arrows in **Figure 1** and **Figure 2** mark one cell exemplarily).

These images clearly show that at low excitation forces more PID-affected cells are visible and relevant for the performance. As the images demonstrate, this is valid for electrical as well as for solar activation. The results of IR-images at low irradiance are comparable to those of DLIT and EL at low electrical activation, and respectively for high irradiance. The outstandingly strong presence of PID at low currents compared to high currents may be explained by shunt shielding or screening effects of "*weak*" PID-cells [4-6]. "*Weak*" PID-cells are defined as cells which are dark at low injection currents and unsuspicious at high injection currents. At low currents, it is beneficial for the electrons to flow through the shunted area of a PID-affected cell [5]. At high currents, this is not advantageous therefore the cell appears bright. Whether a cell behaves as a strong or *weak* PID-cell depends on the ratio of the resistances of the electrical front and back contacts with respect to the shunt.

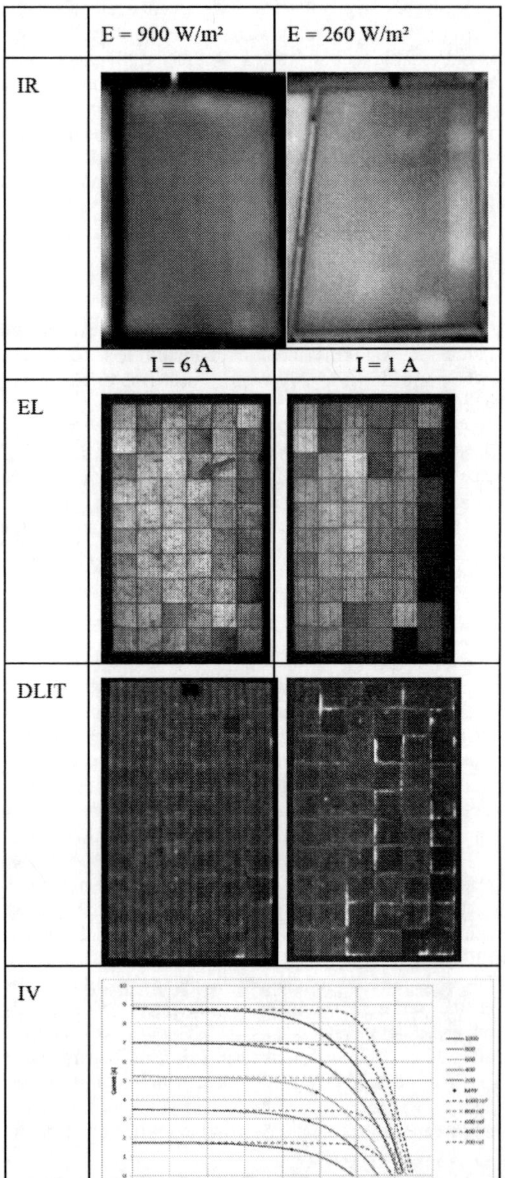

Figure 1: EL- and IR-images PID-affected PV-module, H0318 (M18), P_{STC} = 229.7 W [7]

A correlation between the number of PID-affected cells at high current and low current could not be found, yet. This process seems to be driven by other factors than only the solar or electrical excitation, e. g. material properties and material inhomogeneities.

It has to be highlighted that the apparently PID-free module at high irradiance, and high current, respectively, is not PID-free in all cases, as measurements at low irradiances and currents clearly visualize.

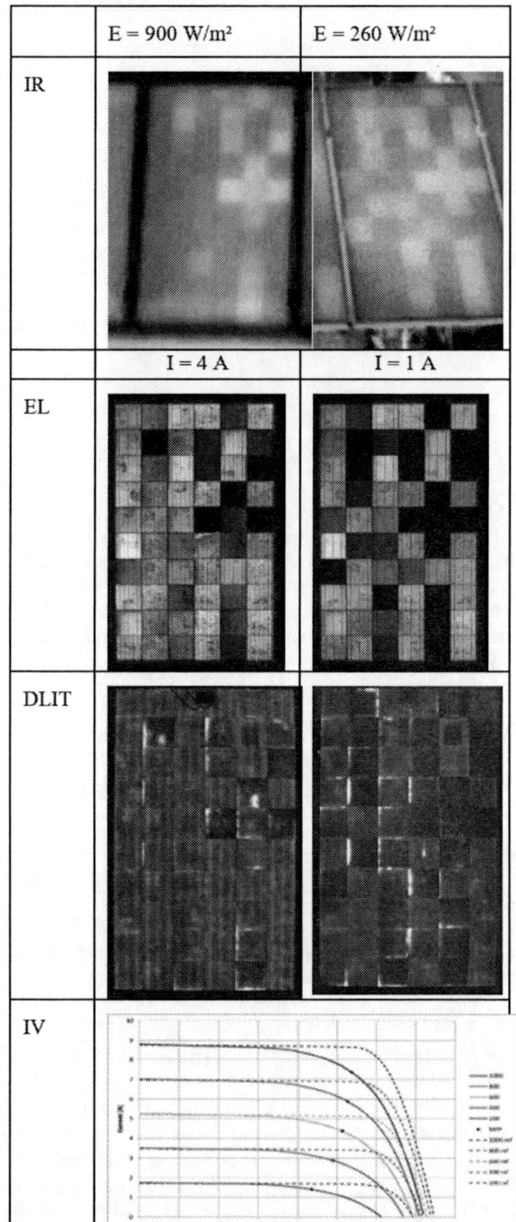

Figure 2: EL- and IR-images PID-affected PV-module, H0025 (M14), P_{STC} = 197.4 W [7]

The IV-curves show not only the decrease of V_{oc} and I_{sc} with increasing irradiance, but also the decrease of the power at maximum power point (mpp). It is apparent that the performance decreases disproportionally with decreasing irradiance. **Figure 3** illustrates the disproportional performance.

For analyzing the influence of the PID-ratio, a threshold of the EL-intensity needed to be determined. Therefore, a threshold of 25% EL-intensity of a good, PID-unaffected cell was found as a reasonable value. Thus, cells with an EL-intensity less than 25% EL-intensity of a good cell at the same excitation level were considered to be PID-active. **Figure 4** illustrates the impact of the PID-ratio on the performance. Non-PID-affected modules have a constant rel. power of 98.6 to 100% (see module H0205 in **Figure 4**). With increasing PID-ratio, the performance decreases linearly with respect to irradiance or current, respectively. The data points highlight clearly that the same module at high irradiance performs better than at low irradiance. Also, the data point out, that PID-affected modules show much more PID-affected cells at lower irradiance since the PID-ratio decreases.

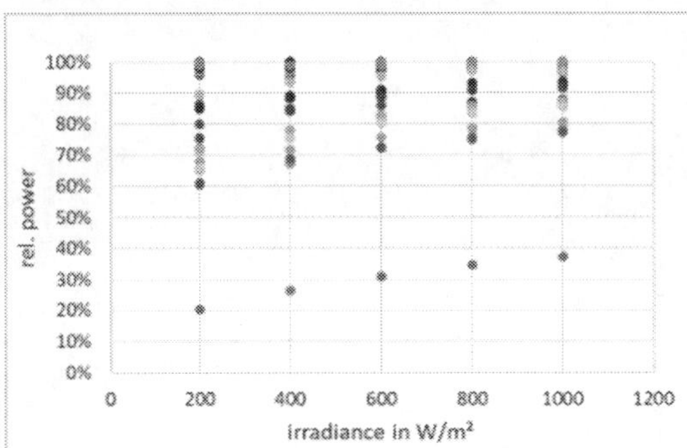

Figure 3: Change of relative P_{mpp} of PID-affected modules with respect to P_{mpp} of a PID-free module at the same irradiance, different colour indicates different module

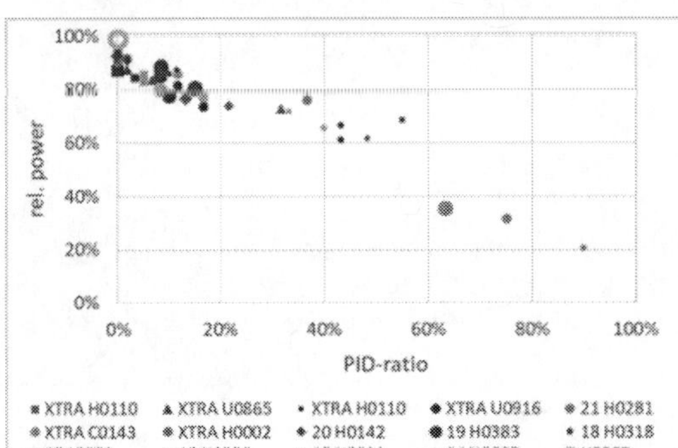

Figure 4: Change of relative power as a function of PID-ratio, large symbol sizes correspond to 800 W/m², medium symbols to 600 W/m², and small symbols to 200 W/m², respectively.

To illustrate this specific issue, a simple yield simulation was conducted based on the irradiance measured by the monitoring systems of the two sites that were inspected and several modeled PV-strings with different PID-ratios. The PV-strings were synthesized based on the measured IV-curves at 25°C. For the yield simulation, the influence of ambient conditions, i.e. shading, wind and temperature, as well as inverter losses were neglected. **Fehler! Verweisquelle konnte nicht gefunden werden.** shows the typical irradiance distribution of the two sites in plane of array. In Portugal, a maximum at a high irradiance of 1000 – 1,200 W/m² exists. For southern Germany, the distribution is flatter and a pronounced maximum is not existent.

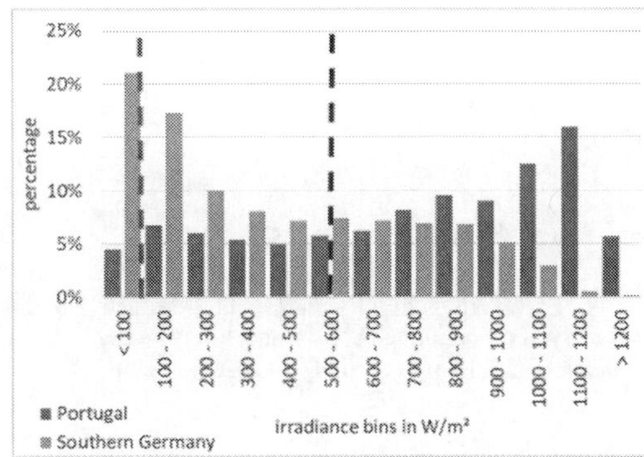

Figure 5: Solar irradiance distribution for the year 2016 for two locations, one in Portugal near the Atlantic coast the other in Southern Germany. Irradiances less than 200 W/m² are neglected.

Table 1 shows the specific yield losses for one year of operation for strings with PID-ratios of 5%, 10%, and 20%, referring to a defect-free string. Based on that simple simulation approach, the specific yield losses in Germany were on average 2% higher than in Portugal. The results in general are confirmed by a study published by Arp et. al. [4], in which a similar approach was used to conduct a more sophisticated yield simulation of PID-affected PV-strings at locations in Germany and Greece.

Table 1: Simulated yield losses modeled PV-string with differing PID-ratios in Portugal and Southern Germany

PID-ratio of one string	**Portugal**	**Southern Germany**
5%	12.7%	14.5%
10%	17.7%	19.8%
20%	25.9%	28.1%

According to the presented findings, the negative effects of PID on the electrical performance of affected modules are just as bad under low irradiation as under higher irradiation conditions. Furthermore, the findings imply that the detection of PID-affected modules can be enhanced by carrying out the measurements in an optimum irradiance range with respect to the local irradiance conditions at the PV-site. The lower the irradiance, until a certain lower threshold irradiance is reached (we suggest at least 500 W/m² under non-cloudy conditions), the more likely is the detection of PID. This is of special importance, since the scaling is not possible because the phenomena is not sufficiently understood yet. Based on the irradiance distribution shown in **Fehler! Verweisquelle konnte nicht gefunden werden.**, we recommend a PID-specific inspection at moderate irradiance of 600 – 800 W/m², whereas in Portugal measurement conditions above 900 W/m² may be advantageous to detect the relevant number of PID-affected modules and the correct PID-ratio for estimating the resulting performance.

4 CONCLUSION

PID-appearance and its impact on performance strongly depend on the irradiance level as field, as well as lab tests demonstrate. Consequently, it is necessary to carry out failure-specific, i. e. PID-specific, locally adapted inspections in order to recognize the impact on the module performance and the yield of the PV-system as early as possible.

5 ACKNOWLEDGEMENTS

ZAE Bayern gratefully thanks the German Federal Ministry for Economic Affairs and Energy (BMWi) for financial funding of the project AQUAM.

6 REFERENCES

[1] C. Buerhop, et al., SPIE Optics and Photonics, Optics + Photonics for Sustainable Energy (2016) 99380J.
[2] C. Buerhop, et al.; IET Renewable Power Generation 11 (2017) 1253.
[3] S. P. Harvey, et al.; Progress in Photovoltaics: Research and Applications 10.1002/pip.2996.
[4] J. Arp, et al., EU-PVSEC (2016).
[5] F. W. Fecher, et al.; Solar Energy 105 (2014) 494.
[6] S. Besold, et al.; Solar Energy Materials and Solar Cells 124 (2014) 133.
[7] F. Wenz (2018) Master Thesis, Karlsruher Institut für Technologie (KIT).

IMAGE PROCESSING FOR DAYLIGHT ELECTROLUMINESCENCE PV IMAGING ACQUIRED IN MOVEMENT

Gisele A. dos Reis Benatto[1], Claire Mantel[1], Adrian A. Santamaria Lancia[1], Michael Graversen[1]; Nicholas Riedel[1], Sune Thorsteinsson[1], Peter B. Poulsen[1], Søren Forchhammer[1], Harsh Parikh[2], Sergiu Spataru[2] and Dezso Sera[2]
[1] Department of Photonics Engineering, Technical University of Denmark, Frederiksborgvej 399, 4000, Roskilde, Denmark,
[2] Department of Energy Technology, Aalborg University, Pontoppidanstraede 101, 9220, Aalborg, Denmark.

ABSTRACT – Regular photovoltaic (PV) system inspections have become a challenge with the significant growth in the number of modules and peak power capacity of PV installations. Image acquisition using drones, based on visual, thermographic, and more recently luminescence, can be a viable solution for large-scale PV inspections. As luminescence can provide a highly detailed and accurate PV module failure diagnosis, the development of a daylight electroluminescence (EL) imaging capability is of high importance. EL imaging performed in the field during the day requires the enhancement of the relatively weak luminescence signal over the noise from the sun. This is accomplished by image averaging and background subtraction, which requires the highly accurate registering of the of individual module images. A sequential EL image acquisition at high frame rates in continuous motion at different angles will be the real-world situation for a drone-based PV inspection in daylight, and to account for this movement while maintaining high quality images, several image processing steps must be developed. With this motivation, here we describe and perform EL image processing on a module with different faults to assure quality of the EL images in different motion speeds.
Keywords: Daylight electroluminescence, Field inspections, Image processing.

1 INTRODUCTION

Imaging techniques such as infrared thermography (IRT) and electroluminescence (EL) are excellent tools for fault detection of both small and large-scale PV installations. At the same time, such inspections are highly important to ensure the expected return on investment (ROI). Drone-based IRT surveys of solar plants are a reality today, and are expected to become an automated inspection method in the near future [1]. The accuracy of thermographic fault detection though, presents limitations, which can be surpassed when performed in combination with EL imaging of the panels. EL can be used to rapidly and accurately detect a large range of major and minor faults in PV modules such as cell cracks, broken interconnections, potential induced degradation (PID) among others [2], [3]. State-of-the-art EL image acquisition however; faces crucial technical challenges, including long exposure times and low signal-to-noise ratio under sun irradiation.

The most commonly used cameras for EL have silicon-based sensors (such as CCD), which present a weak spectral sensitivity in the EL emission range of crystalline silicon (c-Si) solar cells. Although a high resolution (>10MP) can easily be achieved with these detectors, their required long exposure times (of the order of seconds) limit the inspection speed and their application in unmanned aerial vehicle (UAV). In contrast, cameras with InGaAs-based detectors have a good spectral response to the silicon luminescence emission and therefore short integration times, which can be as short as 1 μs [4].

While nighttime EL imaging presents few acquisition challenges, applying high power to PV strings can pose a safety risk to the operators and drone flights can be restricted by national and local laws for UAV night flights. To perform daylight EL in movement, the capability to acquire many images per scene, with high frame rates is essential to perform denoise by averaging and increase signal-to-noise ratio (SNR). No only for EL images, but high frame rates are needed for the obtainment of background images to remove ambient light, required even with the use of filters.

Together with the current efforts to build up a full automated solar plant inspection method including visual; IRT and EL in the future with UAVs [2], [5]–[7], here we describe and perform EL image processing on a module with different faults to insure quality of the EL images acquired in daylight and in movement.

2 EXPERIMENTAL DETAILS

2.1 Sample and electrical signal

For this experiment, a module of 36 multi-crystalline 15.6 x 15.6 cm cells arranged in a 6 x 6 cells matrix was prepared. This module was mechanically stressed and contained cells with several mode A, B and C cracks. The experiment was performed outdoors, under bright daylight with irradiance indicated in the results section. The module was installed facing south with correspondent global horizontal irradiance (GHI) and plane of array (POA) irradiance measured just before an image sequence acquisition using a silicon reference cell from ReRa Solutions connected to a multimeter.

For this outdoor EL setup, a new concept using an AC power supply unit (PSU) was tested. To perform daylight EL imaging, a modulated or lock-in electrical bias is required. In previous works, this modulation was performed by a modulation box synchronized with a camera trigger. This approach required additional hardware to control the power supply unit (PSU) signal and synchronize to camera. In contrast, the AC PSU used in this work (Keysight AC6804A) did not require the additional hardware for synchronization of the PSU and camera, and thus simplified the ground-based system and avoided additional hardware attached to the camera. The forward bias applied to the PV consists of an AC+DC coupling signal (AC signal with a DC offset). The AC+DC electrical signal and applied to the PV module was monitored using a multiphase power analyzer (Tektronix PA4000) in a 4-wire configuration. The AC+DC sinusoidal signal frequency was 60 Hz, while image acquisition was performed with a frame rate of 120 Hz. These frequencies were chosen to accommodate the acquisition of the EL and background (BG) images (Fig. 1). The camera was not actively synchronized with the bias signal, therefore the mean of the pixel numbers of the frames was verified before image processing. We looked at the mean pixel values of the 16 bits image frames acquired in sequence during electrical modulation and compared the differences to determine when the image was EL (high mean) or BG (low mean). The frequencies of the electrical bias and framerate results in collecting an EL and background (BG) image one after the other. This approach avoids the interference of rapid fluctuations in solar irradiance due to different levels of cloud cover during the measurement, as it was verified in our tests that non-modulated EL and BG images acquisition do not offer the same level of image quality after denoising by averaging and EL/BG subtraction.

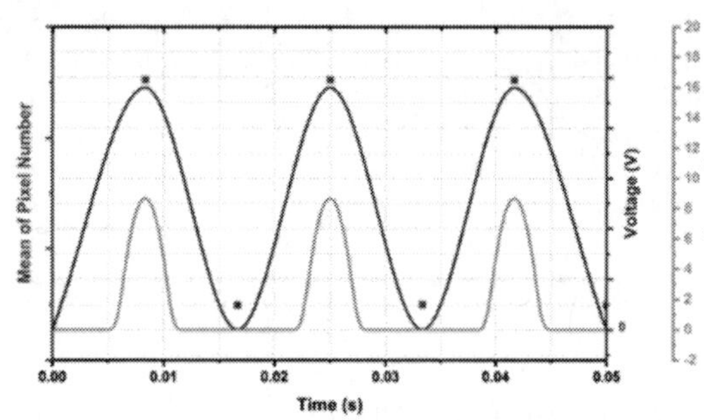

Figure 1: Electrical signal applied to the PV panel and mean of pixel numbers of the images acquired.

2.2 Testbed and moving camera

An InGaAs camera with 640x512 resolution and 14 bits dynamic range from Raptor Photonics is used for the experiment. A PC connected to the camera was required for image data acquisition and settings control. The EL videos were acquired at 120 frames per second (fps). To improve the SNR in the EL images, 16 to 50 EL/BG image pairs are required and the high frame rate allows obtaining this range of images in movement [7]. The exposure time was auto selected by the camera and kept fixed at 0.26 ms for the two image sequences presented here. This exposure time was adequate to avoid over or under exposure for the slightly different irradiances observed during the image sequences (775 W/m^2 and 761 W/m^2 GHI).

The luminescence emission peak for silicon-based solar cells at ambient temperature is centered around 1150 nm [8], therefore, to avoid detecting light from the sun and surroundings, an OD>4.0 1150 nm band-pass filter with 50nm FWHM was used. A 25 mm fixed focus sapphire lens without aperture control was used, and in order to photograph the entire area of the panel, the camera was placed approximately four meters from the test string.

The camera tripod was fixed on a dolly, equipped with proper wheels to be placed over a rail (Fig. 2). This mount allowed the acquisition of EL images at a fixed distance from the PV module, while the camera moved in parallel along horizontally. The tripod was moved manually, permitting speeds of up to 2 m/s and regular speeds up to 1.5 m/s. The speed of the motion during image acquisition was calculated from the video/image sequence acquisition. Two different speeds were studied in similar irradiance conditions: 0.8 m/s (Speed 1) and 1.5 m/s (Speed 2).

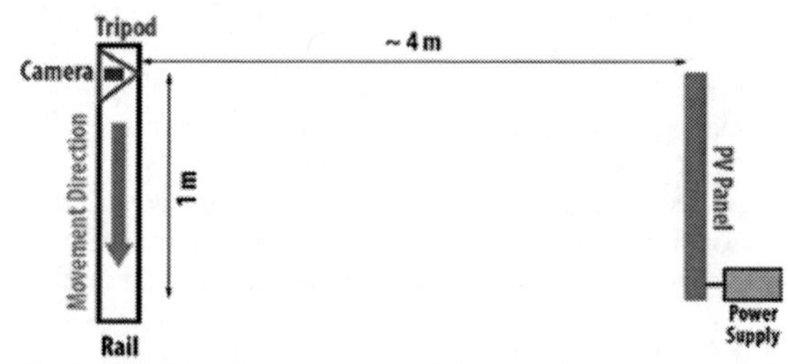

Figure 2: Test bed top view. The camera was mounted on a tripod over a rail to replicate the moviment performed by an UAV.

3 RESULTS

3.1 Image processing

While image processing for stationary EL corresponds to EL/BG image separation, averaging, BG subtraction and finally quality control, images acquired in movement add a much higher level of complexity to this process. At first, the acquired images had to be checked for motion blur. At 2 m/s no visible blur was detected by visual inspection.

The diagram of Fig. 3 shows the image processing steps performed in this work. The colors indicate high (red), medium (orange) and low (blue) complexity of the image processing performed by the algorithm. In each frame, the panel is firstly detected and roughly segmented. That allows compensating the motion between different frames and register across all images of the sequence. After the denoising by averaging and BG subtraction steps, the

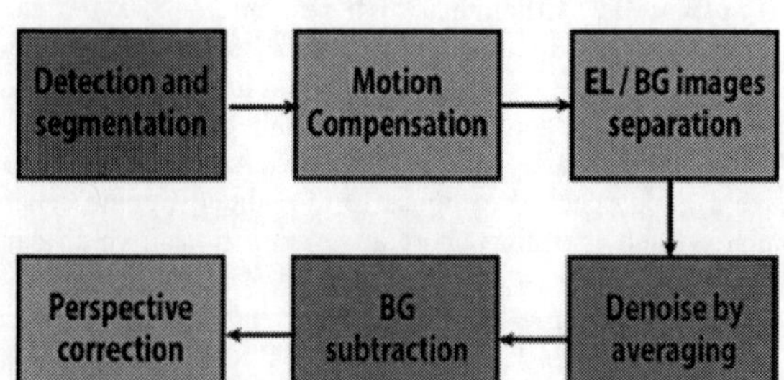

Figure 3: Diagram of image processing steps for daylight EL acquired in motion.

perspective deformation is corrected to obtain a perfectly square panel.

For accurate perspective distortion correction, the homography to determine, i.e. the transformation projecting the imaged PV panel back into the panel plan, has 8 degrees of freedom. As each corner of the PV panel determines two of those degrees of freedom, knowing the localization of the four corners of the considered panel allows to fully determining the homography parameters, which is done using the Direct Linear Transform (DLT) algorithm. The four corners are determined as the intersection of the four edges of the panel, themselves identified by rotating and projecting the image at various angles [9].

The computed motion compensation performed to align the module images is shown in Fig. 4, with "motion in the x" corresponding to the horizontal displacement and "motion in the y" corresponding to vibrations of the test bed construction. Vibrations are highly expected in a UAV

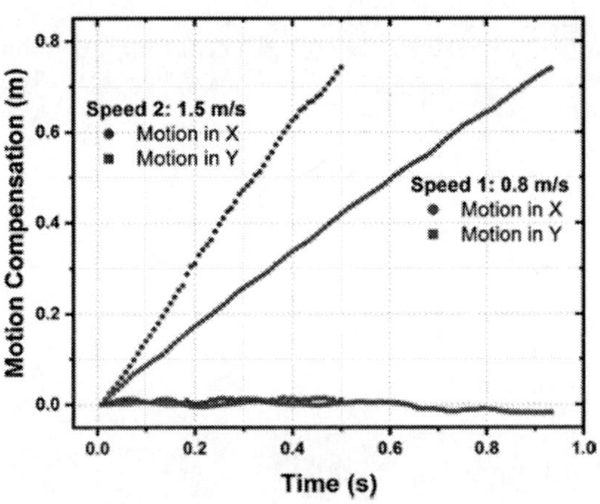

Figure 4: Motion compensation of EL images acquired in motion in two different speeds.

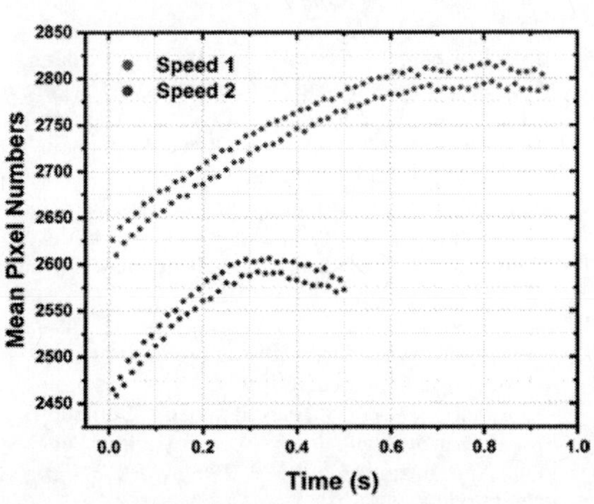

Figure 5: Mean of pixel numbers of EL image sequences used for EL / BG separation.

system specially if the flight is performed in windy conditions, hence the algorithm was prepared to cover this scenario.

Fig. 5 shows the mean of pixel numbers of the image sequences used for EL/BG separation at the two different speeds. The algorithm separates the images with a higher mean as EL images and lower mean as BG images, regardless of the curve shape observed. This curve shape indicates sun irradiance variations during the image acquisition (most likely due to clouds passing).

3.2 EL images

For comparison, Fig. 6 shows the stationary EL images acquired indoors in the dark (top) and outdoors under 849 W/m^2 at the POA (bottom). The outdoor EL image was

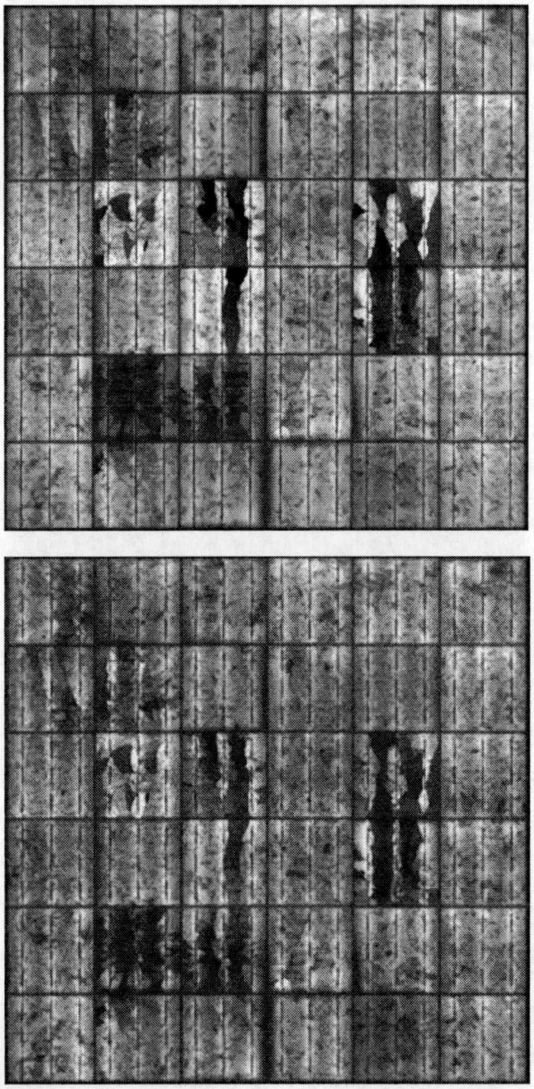

Figure 6: Stationarry EL. Top: Indoors; bottom: Outdoors at 849 W/m^2 POA.

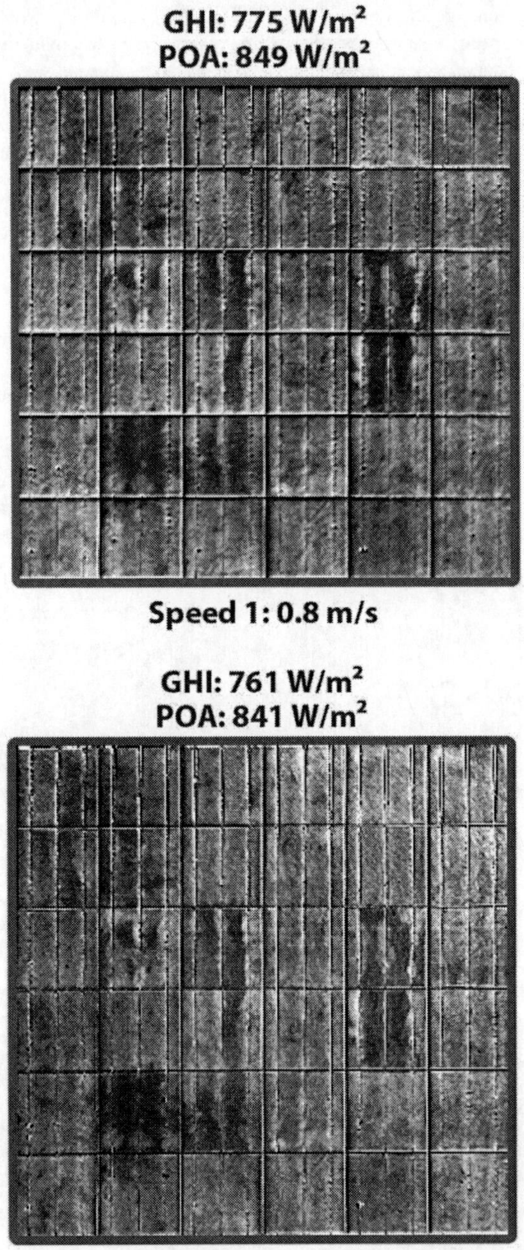

Figure 7: EL in motion. Top: Speed 1; Bottom: Speed 2.

obtained using EL/BG separation, denoise by averaging, BG subtraction and perspective correction image processing steps. The stationary outdoor image had $SNR_{50} > 5$ as recommended by the EL standard [10], where mode A, B and C cracks can be identified.

Fig. 7 shows the processed EL images acquired at speed 1 (top) and speed 2 (bottom). The images presented average quality, with B and C mode cracks well identified. No special treatment was required for the different speeds and the differences observed in the image alignment could be observed in different sequences with similar speeds. The remaining misalignment, especially visible at the bus bar level, is due to perspective distortion as a result of changing viewpoint of the camera. This indicates the need to correct perspective distortion to all images separately first, before averaging and BG subtraction.

4 CONCLUSION AND OUTLOOK

We obtained in this work for the first time outdoor EL images of a PV panel with mode A, B and C cracks mounted in bright sunlight (> 800 W/m² in the POA) while the camera was in motion. The images obtained by motion compensation have an average quality, however it is still possible to recognize major cracks such as B and C cracks.

To improve from average to high image quality, the next step is to apply perspective correction to all acquired images separately and to register them better before averaging and subtraction. This new approach is challenging, as it needs the perspective correction to be perfectly robust across images. Additionally, the blur in EL images must be confirmed mathematically, especially for higher speeds in the future.

The application of such image corrections will support the development of UAV-based PV plant EL inspections in daylight, where each image has a different angular position from the camera viewpoint, and several images from each module have to be registered, averaged, and BG subtracted.

These results confirm the application of daylight measurement of lock-in EL in movement suggested by Adams et al. [5] and is a step further towards fast EL acquisition in the field.

5 REFERENCES

[1] S. Dotenco *et al.*, "Automatic detection and analysis of photovoltaic modules in aerial infrared imagery," *2016 IEEE Winter Conf. Appl. Comput. Vis.*, pp. 1–9, 2016.

[2] J. B. S. Koch, T. Weber, C. Sobottka, A. Fladung, P. Clemens, "Outdoor Electroluminescence Imaging of Crystalline Photovoltaic Modules: Comparative Study between Manual Ground-Level Inspections and Drone-Based Aerial Surveys," *32nd Eur. Photovolt. Sol. Energy Conf. Exhib.*, pp. 1736–1740, 2016.

[3] M. Köntges *et al.*, "Review of Failures of Photovoltaic Modules," *IEA-Photovoltaic Power Syst. Program.*, pp. 1–140, 2014.

[4] A. Vision, "Goldeye CL-033," *https://www.alliedvision.com/en/products/cameras/detail/Goldeye/CL-033 TEC1.html.* .

[5] J. Adams *et al.*, "Non-Stationary Outdoor EL-Measurements with a Fast and Highly Sensitive InGaAs Camera," *32nd Eur. Photovolt. Sol. Energy Conf. Exhib.*, 2016.

[6] C. Buerhop and H. Scheuerpflug, "Field inspection of PV-modules using aerial, drone-mounted thermography," *29th Eur. Photovolt. Sol. Energy Conf. Exhib.*, pp. 2975–2979, 2013.

[7] G. A. dos Reis Benatto *et al.*, "Development of outdoor luminescence imaging for drone-based PV array inspection," *44th IEEE Photovolt. Spec. Conf.*, 2017.

[8] T. Fuyuki, H. Kondo, T. Yamazaki, Y. Takahashi, and Y. Uraoka, "Photographic surveying of minority carrier diffusion length in polycrystalline silicon solar cells by electroluminescence," *Appl. Phys. Lett.*, vol. 86, no. 26, pp. 1–3, 2005.

[9] C. Mantel *et al.*, "Correcting for Perspective Distortion in Electroluminescence Images of Photovoltaic Panels," *IEEE 7th World Conf. Photovolt. Energy Convers.*, 2018.

[10] IEC, "PNW/TS 82-901 Ed. 1.0 Photovoltaic Devices - Part 13: Electroluminescence of Photovoltaic Modules (proposed future IEC TS 60904-13)." http://www.iec.ch/.

LOW-COST ELECTROLUMINESCENCE SYSTEM FOR INFIELD PV MODULES

M. Abdullah Eissa *1, J.A. Silva 2, J. M. Serra 2, A. M. Bassiuny 1 & K. Lobato 2
1 Faculty of Engineering, Helwan University. Egypt.
2 Instituto Dom Luiz, Faculdade Ciências, Universidade Lisboa, Portugal.
*Corresponding email: magdy.abdallah@h-eng.helwan.edu.eg

ABSTRACT: As the installation of PV systems continues to decrease, the weight of the operation and maintenance in the cost structure of systems will become an ever increasing factor. The significant challenges are to maximise electrical energy production by reducing system downtime for planned and unplanned maintenance. Electroluminescence (EL) is becoming an important on-site/infield characterization technique for PV modules. However, proposed high-end EL systems may be too prohibitively expensive, eg, for small systems or low-income areas. This paper demonstrates the principles a low-cost EL system for crystalline PV modules. A consumer grade camera system (camera and lens) was modified by removal of the IR filter to allow for the band-to-band luminescent spectra to arrive at the CCD sensor. To remove background visible light, a bandpass filter tuned to approximately the luminescent spectra of silicon (1100+/-10nm) was placed before the camera lens. Open-source digital imaging processing techniques were then employed to correct geometrical image distortions and to enhance image quality. The effectiveness of the proposed system is demonstrated under three controlled conditions; indoor completely dark, indoor with controlled background lighting and, outdoor urban night-time. The system allowed for acquisition of high resolution EL images of PV modules in a matter of minutes. The quality of the images under the differing conditions are comparable. Easily visible were cracks in individual cells and also non-functioning cells of a purposefully damaged PV module.
Keywords: Silicon PV, EL, infield testing, low cost.

1 INTRODUCATION

Imaging techniques allow for short acquisition times even for large areas (up to arrays of solar modules). However, the high cost of the sensors represents the main obstacle for this approach [1], either by use of high sensitivity & low noise silicon based sensors or high cost InGaAs sensors. For infield testing of PV modules, infrared thermography and EL imaging are the most popular approaches [2, 3].

IR thermography is limited to detecting hotspots [4] at also fairly low sub-megapixel resolutions. In contrast, luminescence imaging allows for the detection of cracks, series resistance effects, bypass diode failures, light-induced degradation, potential induced degradation, and carrier induced degradation mechanisms [2].

This paper reports our new development for infield EL characterization of PV modules. In this work we achieved the following:

1) 1) Development of EL test setup employing a modified consumer-grade digital camera coupled with bandpass filter to remove background stray light;
2) 2) Validation of the effectiveness of the proposed setup in different environments: indoor dark, indoor controlled ambient background lighting and, infield night-time measurements in an urban setting;
3) 3) Application of simple open-source imaging processing techniques to geometrically correct and enhance images by denoising [5].

Our developed EL system is relevant when compared to other existing systems, such as the friendly daylight EL system [6], the non-stationary outdoor EL system [7], the DAYSY system [8], and the system presented by M. Frazao et al. [9]. Table 1 summarizes and compares the referenced systems with the one presented introduced here. The friendly daylight EL system [6] and the Non-stationary outdoor EL systems [7], employ costly high sensitivity industrial grade camera systems to perform outdoor daytime EL inspections. M. Frazao et al. [9] employ a consumer grade camera, however EL measurements are restricted to zero background light conditions. Our system presented here is shown to work in ambient indoor lighting conditions and also in an urban outdoor night-time environment.

The remainder of this paper is organized as follows: Section 2 introduces the proposed EL test setup and methodology, section 3 discusses the result, and conclusion are presented in Section 4.

2 METHODOLOGY

2.1 The proposed electroluminescence setup

Figure 1 shows the experimental setup. The PV modules are connected to a DC power supply by using standard solar cabling and connectors. The digital camera used was a modified Nikon D40 whereby the internal IR filter was removed. A bandpass filter (Melles Griot 03 F11 024, 1100 nm +/- 5 nm) was mounted in front of the lens to remove background ambient light. Automatic image acquisition was operated via a PC with propriety camera software. This limited automatic acquisition times to 30 s. For longer acquisition times, the camera shutter had to be depressed manually operating the via the bulb mode. The EL imaging is undertaken in three conditions: 1) a dark room; 2) a room with normal working indoor lighting conditions and; 3) outdoors under urban night conditions.

Figure 2 is a representation of the intersection of the EL, camera sensitivity and, bandpass spectra. The represented camera spectrum is that of a scientific grade camera. Data is confidential and as such camera model cannot be divulged. To this date, it has not been possible to obtain quantum efficiency data of the consumer grade camera used in here. Although the Si CCD sensor is not particularly sensitive where Si EL is strongest, if background stray light can be removed, then it is simply a

Table 1: Comparison between other existing and our EL system

System[ref.]	A friendly daylight EL system[6]	Non-stationary outdoor EL system [7]	DAYSY system[8]	System A[9]	Our system
Medium	Outdoor-Sun	Outdoor -Sun	Outdoor -Sun	Dark	Outdoor - Night
Camera type	Cost-effective InGaAs CCD- 320x256 pixels	InGaAs CCD-320x256 pixels	-InGaAs CCD*- 320x256 pixels	a consumer digital camera CCD- 3000x2000 pixels	a consumer digital camera CCD- 3000x2000 pixels
Imaging	EL & PL	EL	EL & PL	EL	EL
Exposition time	Less than 20 s.	1 ms - 20 ms	60 s- 120 s	6 min	30 s -120 s
Filter	Appropriate filter setup	-	digital filtering techniques	NO	Hardware bandpass filter

*Information not available, but is inferred by the general characteristics presented by the manufacturer of the DaySy system.

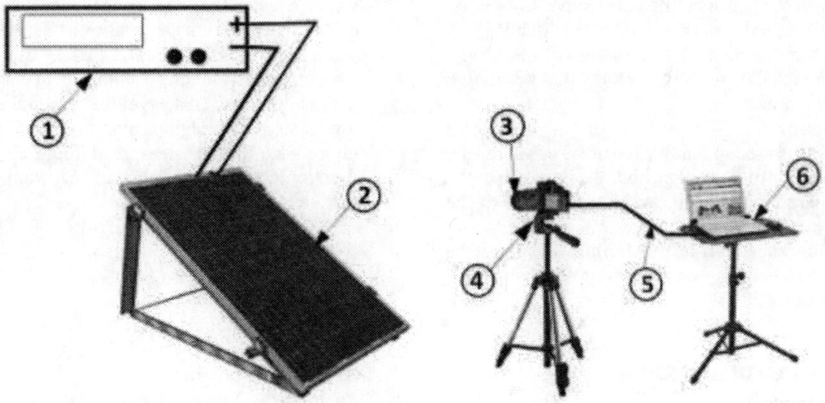

Figure 1: The proposed Electroluminescence setup, the setup consists of the (1) power supply, (2) PV module, (3) bandpass filter, (4) digital camera, (Nikon D40 with a 18-55mm f/3.5-5.6G ED II lens) (5) USB cable (5), and (6) PC.

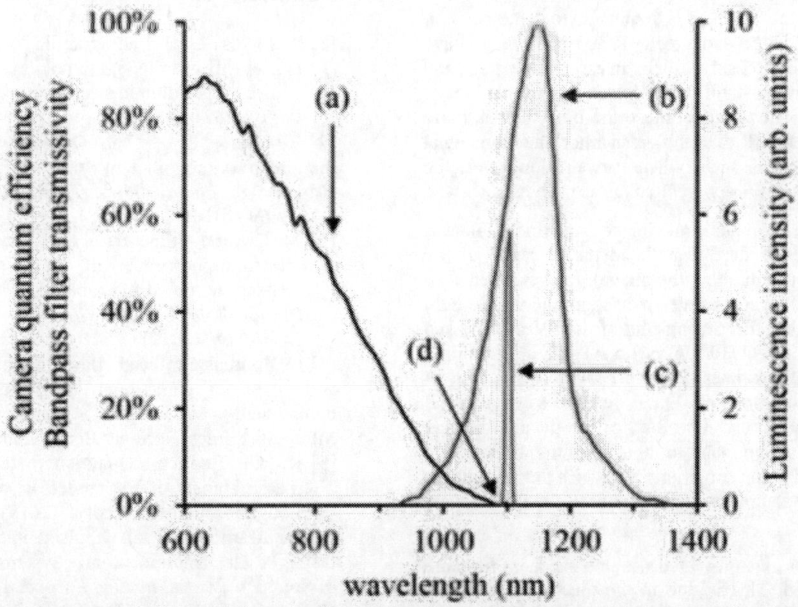

Figure 2: (a) Spectral quantum efficiency of a scientific grade silicon CCD camera. (b) Example of typical luminescence spectrum of a silicon solar cell [1]. (c) Transmission spectrum of the 1100 nm bandpass filter, Melles Griot 03-F11-024. (d) Spectral overlap of camera quantum efficiency, expected luminescence and bandpass filter.

question of extending acquisition times to obtain usable images. The bandpass filter employed has a peak transmissivity of ca. 50%, and a fairly narrow bandwidth (ca +/- 5 nm FWHM), and as such will also block a significant fraction of the EL emission. Tests were undertaken with a 1050 nm bandpass filter with the expectation that in this spectral region the image acquisition would be enhanced. However, the opposite was observed, indicating that at this wavelength the spectral intensity of the tested modules is significantly lower than expected from the example spectrum shown.

2.2 The EL image processing algorithm

Upon acquisition of the EL images, a two-step algorithm was employed for image geometric correction and image quality enhancement.

Step 1 geometric correction: Perspective Correction algorithm which introduced described by Bedrich [5] was applied to correct the position and rotation of the image. This algorithm includes two stages, image grid generation and image cell rectification.

Step 2 denoising: A non-local means denoising algorithm [10] was used to remove the noise of the image. The main concept of this algorithm is to replace the intensity of a target pixel with a weighted average of the neighbouring pixels.

Error! Reference source not found. illustrates the implementation of the imaging processing steps on an EL image of a PV module.

3 RESULTS AND DISCUSSION

Shown in table 2 are EL images (as taken and then digitally processed) of a PV module taken under different background lighting conditions. The images are angled so as to demonstrate the capabilities of the geometric correction of the image processing algorithms employed.

Although the IR filter is removed from the camera, individual red, blue and green (RGB) pixels still have broadband colour selective filters. In essence the green and blue pixels have a significantly lower signal than that of the red pixels. However, because the blue filter is more transparent to the IR than the green filter, the blue signal is stronger and, as such, when viewed along with the strong red signal, the apparent colour is purple.

Bias of the PV module and image acquisition time was adjusted so as to maximise the dynamic range of the images (without saturating the image). Images taken with the bandpass filter to remove the background visible light required a higher forward biasing of the PV module and hence higher currents (0.5 A vs 1.5 A) and, also required longer acquisitions times (30 s vs 120 s). This is evident from the histograms presented, where a large pixel intensity count occurs at ca 80% of the maximum pixel intensity. It is important to note that the background ambient illumination conditions were such that the images would quickly saturate without the use of the IR bandpass filter.

The images here presented with the bandpass filter are slightly blurred. This is due to the mechanical shaking suffered by the camera during the long expositions (>30 s) where manual depressing of the shutter button was required.

Overall the images obtained are similar irrespective of the conditions. However, highlighted is a cell which appears darker in the dark room EL image when compared with the other two conditions which employed a bandpass filter and hence required higher current densities.

The automatically digitally processed images are also shown in table 2 and demonstrate that the algorithm is successful at correcting the image geometry and enhancing the image so as to allow for more facile defect identification on the module.

4 CONCLUSION

The acquisition of high quality infield night time EL images has been shown to be possible with recourse to a lightly modified consumer grade Si CCD camera coupled with a standard bandpass filter along with the enhancement of image quality using open-source image processing algorithms. The images are shown to be of similar quality to those obtained in controlled absolute dark conditions. This paves the way to the development of low-cost EL imaging systems that maybe of vital importance for the widespread implementation of reliable functioning of PV systems for, eg, rural electrification in low income areas.

ACKNOWLEDGEMENRS

Mr. M. Abdullah Eissa acknowledges the student bursary provided by the Erasmus + Medsol student and staff training program.

Prof. K. Lobato, Dr. J. Silva and Prof. J. M. Serra acknowledge the Portuguese Science & Technology Foundation (FCT) for funding through the R&D strategic project UID/GEO/50019/2013.

REFERENCES

[1] T. Fuyuki et al. Photographic surveying of minority carrier diffusion length in polycrystalline silicon solar cells by electroluminescence. *Applied Physics Letters* v.86, p.262108 (2005).

[2] R. Bhoopathy, et al. Outdoor photoluminescence imaging of photovoltaic modules with sunlight excitation. *Progress in Photovoltaics: Research and Applications*, v.26, p.69 (2018).

[3] M. Camino-Villacorta et al. Comparative analysis of electroluminescence, IR thermography and visual inspection on c-SI modules. *In 27th European Photovoltaic Solar Energy Conference and Exhibition*, (2012).

[4] O Breitenstein, et al, Lock-In Thermography-Basics and use for functional diagnostics of electronic components. Springer Series in Advanced Microelectronics, Springer Berlin Heidelberg. (2003).

[5] K. G. Bedrich, Quantitative electroluminescence measurements of PV devices (Doctoral dissertation, Loughborough University). (2017).

[6] O. Martinez. et al. Implementation of a Friendly Daylight Electroluminescence System for the Inspection of Solar PV Panels, *In 33rd European Photovoltaic Solar Energy Conference and Exhibition*.(2016).

[7] J. Adams et al. Non-Stationary Outdoor EL-Measurements with a Fast and Highly Sensitive

InGaAs Camera, *In 32nd European Photovoltaic Solar Energy Conference and Exhibition*, (2015).
[8] L. Stoicescu et al. DaySy: luminescence imaging of PV modules in daylight. *In 29th European photovoltaic solar energy conference and exhibition proceedings (EU PVSEC)* p.2553 (2014).
[9] M. Frazao et al. Electroluminescence of silicon solar cells using a consumer grade digital camera. *Measurement*, v.99, p.7 (2017).
[10] A. Buades et al. Non-local means denoising. *Image Processing On Line*, v.1, p.208 (2011).

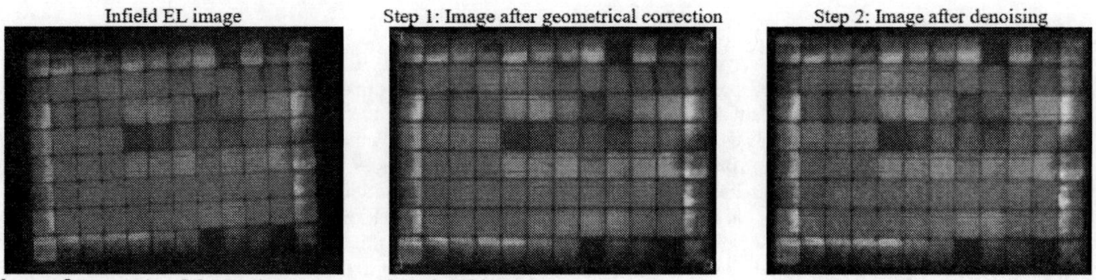

Figure 3: Example of the EL image processing algorithm.

Table 2: EL images before and after digital processing for a PV module under different background lighting conditions. The red square box highlights a dark cell in the darkroom EL image, which is not dark under the other two lighting conditions. The forward bias

	Dark room Without bandpass filter	**Room with ambient light With bandpass filter**	**Infield night-time With bandpass filter**
Original Image		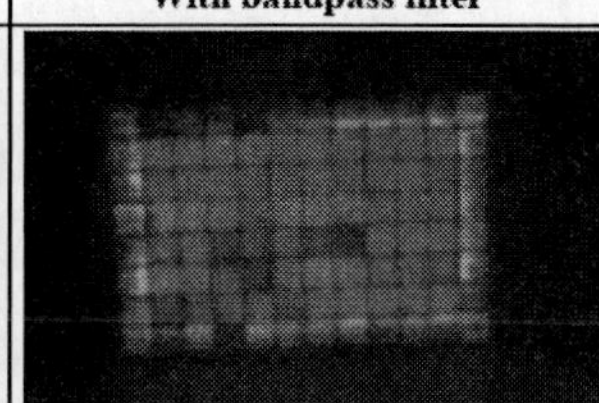	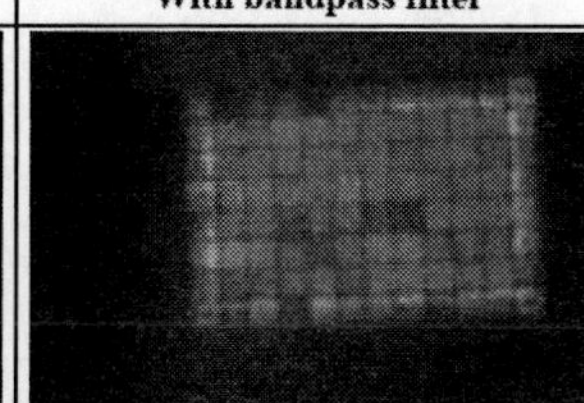
RGB histogram			
Processed image			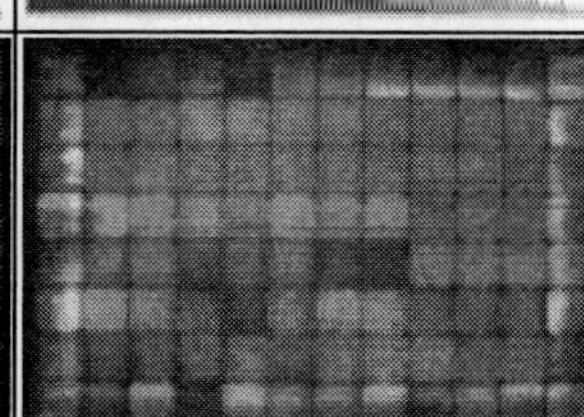
Bias current	0.5 A	1.5 A	1.5 A
Exposure time	30 s	120 s	120 s
Aperture	f/4	f/4	f/4
ISO	200	200	200

A HEALTH CHECK OF THE ITALIAN SOLAR PHOTOVOLTAIC PARK USING SATELLITE-BASED SOLAR RESOURCE DATA

Alessandro Virtuani[1], Juraj Betak[2], and Artur Skoczek[2]

[1] Officina del Sole (O'Sole), Milan, Italy
[2]Solargis, Bratislava, Slovakia
* Corresponding author: Tel. +39 333 7476 939, e-mail: alessandro.virtuani@o-sole.eu

ABSTRACT: At country level, the average Performance Ratio PR for Italy is presently low (68.6% in 2016). At regional level we observe a strong dispersions in the results and the presence of some outliers. Regions with the lowest PR are Lombardia, Liguria, and Campania. Puglia, the region with largest average size of solar plants exhibits the highest PR. We then refine our analysis to province level calculating PR_{2016} values for the 108 Italian provinces using 2016 data, a year for which irradiation data are well aligned to long-term averages. With some outliers, the vast majority of PR_{2016} for the different provinces lies between 60-70%. We then look for some correlations that can describe the strong scatter in the data presented: (1) it is difficult to observe a clear correlation between average system size in the province and PR for average sizes below 45 kW. On the other hand, above this threshold, system size is generally a good indicator of good performance; (2) the residential (and building) density in the province (which can be correlated to some extent to increased shading losses, particularly for small/medium-size rooftop installations) has a strong impact on PR and can be described by a hyperbolic relation; (3) the effect of province elevation on PR is not clear, nevertheless, not necessarily provinces located at higher altitudes (and potentially increasingly affected by far shading) have lower PRs. These correlations are sometimes weak, but we believe that this analysis could be improved considerably by increasing the spatial granularity of the data and focusing on smaller geographical units, such as communes or even at the scale of the single PV plant.
Keywords: PV systems, operation, performance, reliability

1 INTRODUCTION

With a share of 8.4% of electricity consumption covered by solar electricity, in 2016 Italy has become – and potentially still is - the country with the highest penetration of photovoltaics (PV) into its energy mix [1, 2].

In order to investigate the health status of the Italian PV park as a whole and potential signs of early decline, we plot first the evolution over time (2005-2017) of performance ratio PR data calculated for Italian regions, and notice a significant discrepancy between PR values calculated for different regions. PR values are calculated by combining solar electricity data (for regions/provinces) made available by the GSE (*Gestore Servizi Energetici*, the Italian public company in charge of managing subsidies to renewables) with highly accurate satellite-based solar resource data provided by Solargis.

In order to describe the strong discrepancy in PR between difference regions, we increase the spatial resolution to the province level, and plot 2016 PR data (a year for which solar resources have been in line with long-term averages) for the different provinces as a function of a set of parameters. Including: (1) morphological parameters (e.g. province elevation), (2) average size of PV systems, (3) building and residential density in the province.

2 SOURCES AND APPROACH

2.1 Input data

In this paper, we make use of official figures released by *GSE* for statistics relative to Italian solar PV (solar PV production and installed capacity [3, 4]), and insolation data from 2005 up to 2017 and long-time averages (1994-2017) by *Solargis* [5, 6, 7, 8].

The inputs to the Solargis model are multispectral satellite images from Meteosat satellite and atmospheric parameters like aerosols and water vapour. Spatial resolution of Meteosat data considered in the calculation scheme is approximately 3 km at sub-satellite point. Model results are served in 2 arc-minutes (app. 4x4 km) regular grid in geographical coordinate system. For the purpose of this study, the spatial resolution of solar data products is enhanced to 15 arc-seconds (nominally 500 m) using digital terrain model comprising a combination of data from the Shuttle Radar Topography Mission. The primary temporal resolution of satellite data is 15 min. The atmospheric parameters are updated daily.

2.2 Key performance indicators

We make use of two KPIs, which are generally used at the solar plant/array level, and that, utilizing aggregate average values, we use to compute country, regional and province performance indicators. These are:

(1) the *final yield* Y_f (kWh/kWp): i.e. the annual net AC energy output E_a [kWh] of a system divided by the peak DC power P_{STC} [kW] of the installed PV array at standard test conditions (STC: 1000 W/m², 25°C, AM1.5):

$$Y_f = E_a / P_{STC} \quad (1)$$

Y_f is a direct function of the cumulative irradiation (i.e. higher Y_f should correspond to sunnier regions) and, therefore, it generally depends on the latitude as well.

(2) the *performance ratio PR* (%) is an index illustrating the overall effect of losses on the system/array's rated output due to device temperature, incomplete utilization of the irradiance (e.g. due to shading or non-optimal orientation), and system component inefficiencies or failures [9]. It provides an indication of how a system effectively performs under real operating conditions, compared to how it would be operating if it were constantly exposed to standard laboratory (STC) ones.

$$PR = (E_a \times G_{STC}) / (H_a \times P_{STC}) \quad (2)$$

with G_{STC} [kW/m^2] the irradiance at STC (1000 W/m^2), and H_a [kWh/m^2/y] the yearly cumulative irradiation. In this work, for H_a we use the Global Tilted Insolation at optimal tilt (GTI_{opta}) for the whole country and for each separate region and province. As an example, the optimal tilt would be 36°, 34° and 30°, respectively, for a south-facing installation in Milan, Rome and Palermo.

3 RESULTS

3.1 Performance at regional and country level

In order to investigate the health status of the Italian PV park as a whole and potential signs of early decline, we plot first the evolution over time (2005-2017) of PR data calculated for the Italian regions. Fig. 1 shows only data for the period 2011-2016 for which the data look more consistent and stable. For earlier years data seam very noisy, a fact that could be related to some uncertainties in the data published (e.g. newly added PV capacity effectively connected to the grid). At country level, the average PR for Italy is presently below 70% (68.6% in 2016), a value that is well below the 80-85% value exhibited by well-functioning and properly maintained PV plants. However, this is a value that averages all plants irrespective of their size, and whether they are well or poorly performing, or properly maintained.

Further, at regional level we observe a strong dispersions in the results and the presence of some outliers. Regions with the lowest PR are Lombardia, Liguria, and Campania. Puglia, the region with largest average size of solar plants (~58.8 kW in 2016), exhibits the highest PR [1].

Furthermore, starting from 2013 (the last year of the *Conto Energia* feed-in FiT incentive scheme) moderate signs of decline in PR can be observed for most regions and for the country as a whole.

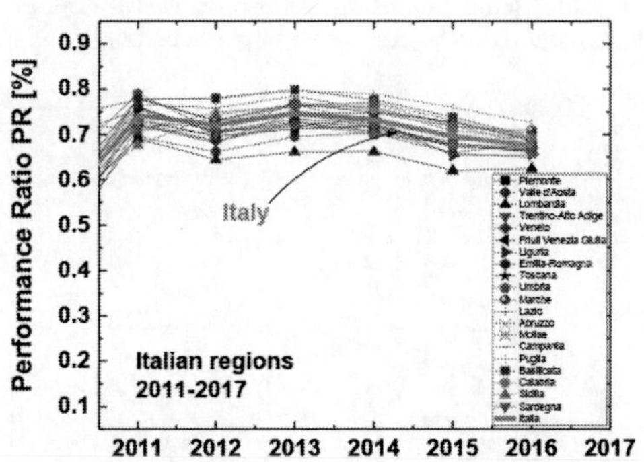

Fig. 1: Evolution of the PR (2011-2016) for the different Italian regions and the whole country (solid orange line).

To get a better understanding of the picture, we increase the spatial resolution of our analysis by focusing on data from provinces (~108) rather than regions (20) and select data from 2016, a year for which, with the exception of Valle d'Aosta, solar resources are well aligned to long-term averages (LTA: 1994-2017), as can be seen in Fig. 2.

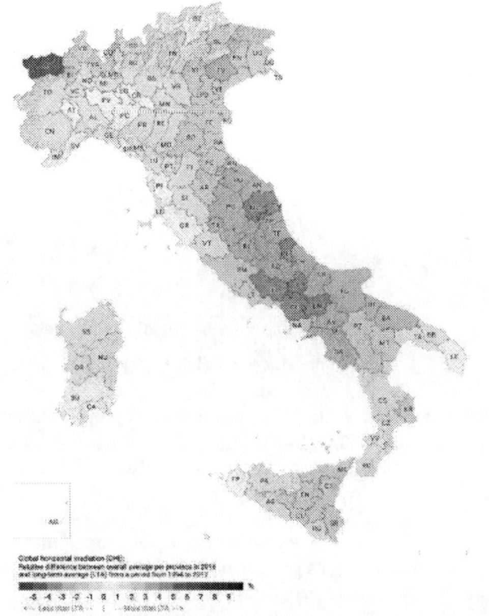

Fig. 2: Global Horizontal Irradiance (GHI): relative difference between the overall average per province in 2016 and long-term average from the period 1994 to 2017.

3.2 Performance at province level

PR_{2016} and Final Yield Y_f data (as a function of GTI_{opta}) for 2016 and each province are shown in Fig. 3 and 4, respectively.

Surprisingly, a couple of provinces have average PR_{2016} values below 60% (MI, MB, PV, LO, NU). The highest values is reported for the provinces of BR in Puglia (PR> 75%) and TP in Sicily. PR values for all other provinces are scattered between these two extremes.

Fig. 3: PR values for Italian provinces in 2016 calculated using GTI_{opta} data.

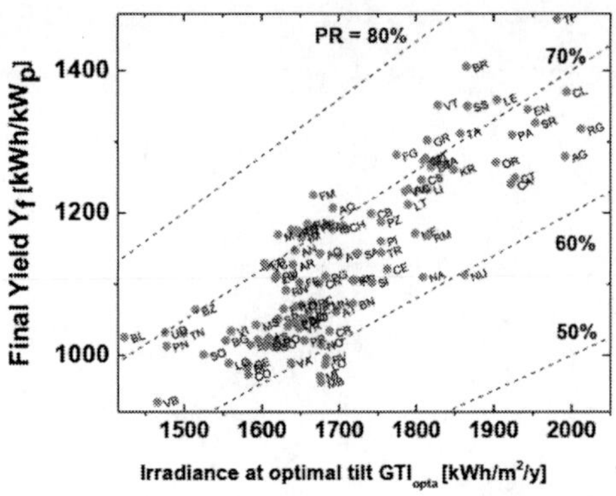

Fig. 4: Final Yield Y_f as a function of irradiance at optimal tilt GTI_{opta} for the different provinces in Italy in 2016. Dashed lines correspond to different levels of PR.

3.3 Performance at province level and correlations

For the different provinces, we plot PR_{2016} data as a function of a set of parameters. These include:

1. the average size of PV systems;
2. building and residential density in the province;
3. morphological parameters such as province elevation (Fig. 6), for which both median and mean province elevation values are used.

As Fig. 5 shows, no clear correlation can be observed between PR and the province average plant sizes below 45 MW. On the other hand, above this threshold, system size is generally a good predictor of good performance, particularly for provinces in Central and Southern Italy that host the majority of large-scale ground-mounted PV installations in Italy. On average, larger plants are better maintained and suffer less from shading losses, which may have a strong impact on plants in highly urbanised areas.

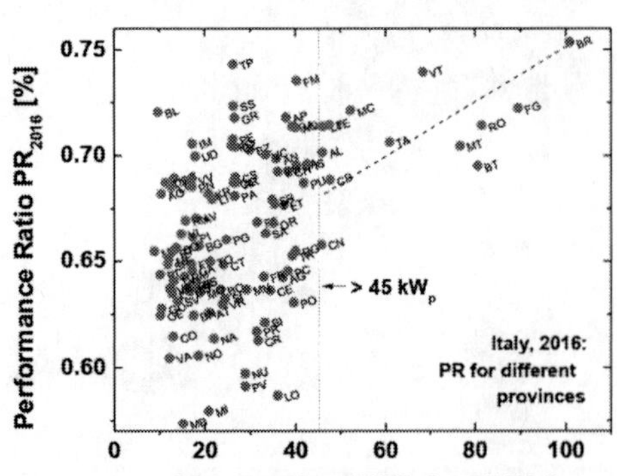

Fig. 5: PR values for Italian provinces in 2016 as a function of average PV plant size.

This is shown in Fig. 6 that displays a correlation between PR_{2016} values for Italian provinces as a function of province residential density, color coded by plant average size. An analogous trend – not shown here - can be observed if we use as parameter the province building density.

Apparently, the residential (and building) density (see as wellcRef. [10]) in the province has a strong impact on PR that can be described by a hyperbolic relation.

To some extent, this can be linked to increased shading losses that will increase particularly for small/medium-sized rooftop installations and in urban environments.

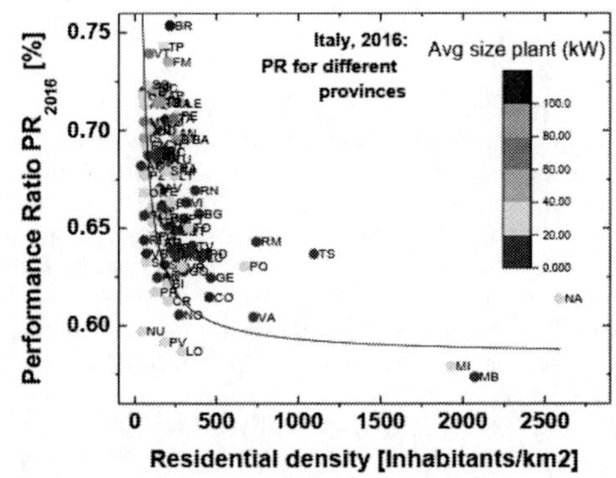

Fig. 7: PR values for Italian provinces in 2016 as a function of province residential density, color coded by plant average size.

The effect of elevation (mean and median values) on province PR_{2016} values is not clear. Nevertheless, not necessarily provinces located at higher altitudes (AO, BZ, BL, TN, AQ, VB) exhibit lower PR values.

In general, solar plants in mountain areas can to some extent be affected more severely by far shading during the whole year and particularly in wintertime. In these areas, however, this effect can be partly compensated by higher insolations, and on average by lower ambient and module operating temperatures. Additionally, GTI_{opta} values in mountain regions come with a larger uncertainty.

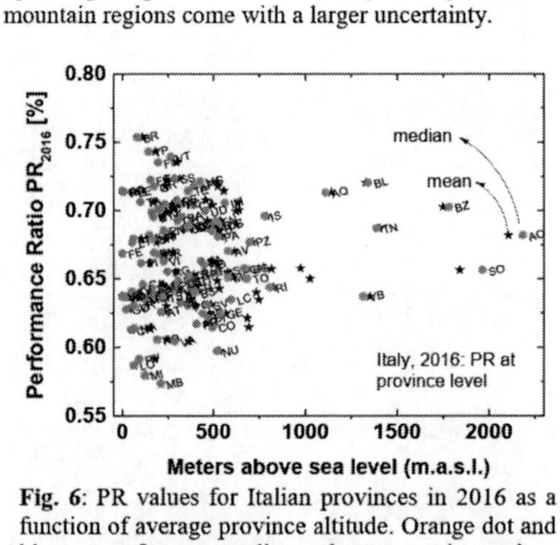

Fig. 6: PR values for Italian provinces in 2016 as a function of average province altitude. Orange dot and blue stars referrer to median and mean province values for the altitude, respectively.

These correlations are sometimes weak, but we believe that this analysis could be improved considerably by refining it. This could be achieved by increasing the spatial granularity of the data and focusing on smaller geographical units, such as communes or even at the scale of the single plant. Of course, the amount of data to deal with would become much larger, and these data (PV generation and installed capacity) are not directly in the public domain. Still, they could possibly be accessed upon request. The present work lays down the foundation for future work.

A detailed analysis of the uncertainties of the PR and Y_f data presented in this work will be part of a subsequent study.

5 CONCLUSIONS

At country level, the average PR for Italy is presently below 70% (68.6% in 2016), a fairly low value. Further, at regional level we observe a strong dispersions in the results and the presence of some outliers. Regions with the lowest PR are Lombardia, Liguria, and Campania. Puglia, the region with largest average size of solar plants, exhibits the highest PR.

We therfore refine our analysis to **province level** calculating PR_{2016} values for the 108 Italian provinces using 2016 data, a year for which irradiation data have been well aligned to long-term averages (LTA) for the whole country and most regions/provinces. With some outliers, the vast majority of PR_{2016} for the different provinces lies between 60-70%.

We then look for some correlations that can describe the strong scatter in the data presented:

(1) It is difficult to observe a clear correlation between **average system size** in the province and PR for average sizes below 45 kW. On the other hand, above this threshold, system size is generally a good indicator of good performance;

(2) Apparently, the **residential (and building) density** in the province (which can be correlated to some extent to increased shading losses, particularly for small/medium-size rooftop installations) has a strong impact on PR and can be described by a hyperbolic relation.

(3) The **effect of province elevation** (mean and median values) on PR is not clear; nevertheless, not necessarily provinces located at higher altitudes (and potentially increasingly affected by far shading) have lower PRs.

These correlations are sometimes weak, but we believe that this analysis could be improved considerably by increasing the spatial granularity of the data and focusing on smaller geographical units, such as communes or even at the scale of the single PV plant. The present work lays down the foundation for future work.

6 ACKNOWLEDGMENTS

We gratefully acknowledge Luca Morganti and Luca Parini (*Officina del Sole*,), as well as Marcel Suri and Tomas Cebecauer (Solargis) for fruitful discussions and comments.

REFERENCES

1 IEA Report: Trends 2016 in Photovoltaic Applications - Survey Report of Selected IEA Countries between 1992 and 2015, Report IEA PVPS T1-30:2016, ISBN 978-3-906042-45-9

2 A. Virtuani, M. Pravettoni, L. Parini, L. Morganti, A. Skoczek, J. Betak, Where Has All The Power Gone? A Health Check of Italian Solar Electricity in 2016, EUPVSEC 2017, Amsterdam;

3 GSE Report "Solare Fotovoltaico - Rapporto Statistico 2015": *idem*;

4 GSE Report « Solare Fotovoltaico - Rapporto Statistico 2016 »: *idem*;

5 Cebecauer T., Šúri M., Perez R., 2013. Semi-Empirical Satellite Models. In Kleissl J. (ed.) Solar Energy Forecasting and Resource Assessment, Academic press, pp. 21-48

6 Cebecauer T., Šúri M., Skoczek A., 2011. SolarGIS: Solar Data And Online Applications For PV Planning And Performance Assessment. 26th European Photovoltaics Solar Energy Conference, September 2011, Hamburg, Germany

7 Cebecauer T., Šúri M., Skoczek A., Fantík J., 2011. Analysis of PV Power Generation in the Czech Electricity Transmission System CEPS. 26th European Photovoltaics Solar Energy Conference, September 2011, Hamburg, Germany;

8 Cebecauer T., Šúri M., Perez R., 2013. Semi-Empirical Satellite Models. In Kleissl J. (ed.) Solar Energy Forecasting and Resource Assessment, Academic press, pp. 21-48

9 International Electro-technical Commission, IEC-61724:1998(E), I.S, Photovoltaic system performance monitoring – guidelines for measurement, data exchange and analysis. International Electrotechnical Commission;

10 P. Moraitis, B. Kausika, W. Van Sark, Effects of urban Environment on solar Performance, Proc. EUPVSEC 2017, Amsterdam.

SYSTEM FAILURE ANALYSIS TOOL FOR PV PLANTS
A SOFTWARE CONCEPT FOR NON-SPECIALISTS

A. Horn[1], W. Mühleisen[1], C. Hirschl[1], R. Ebner[2], M. Spielberger[3], H. Sonnleitner[4]

[1] CTR Carinthian Tech Research AG, Europastr.12, 9524 Villach/St. Magdalen, Austria
(Author and Contact: arnold.horn@ctr.at; T: +43 4242 56300, wolfgang.muehleisen@ctr.at, christina.hirschl@ctr.at)
[2] AIT Austrian Institute of Technology GmbH, Giefinggasse 2, 1210 Vienna /Austria (Rita.Ebner@ait.ac.at)
[3] PVSV e.U., Unterer Markt 7, 9334 Guttaring, Austria, (markus.spielberger@pvsv.at)
[4] ENcome Energy Performance GmbH, Lakeside B08, A-9020 Klagenfurt, Austria, (Horst.Sonnleitner@en-come.com)

ABSTRACT:
Large knowledge about failure possibilities of PV plants has been accumulated during the research Project INFINITY. This paper intends to develop tools for optimising the maintenance strategies of PV systems. The developed software, presented in this article, provides a tool for responsible people to overcome the lack of expert knowledge in that area in a way to enable them to find these potential failures. It represents a support for laypersons as well as professionals.

Keywords: PV maintenance, software analysis tool

1 INTRODUCTION

The majority of photovoltaic systems are still built up of crystalline PV panels. What they all have in common is the need of operation and maintenance strategies, which is often underestimated. For private PV plants (up to 10 kWp) mostly utilized for self-consumption purposes and for huge commercial power plants in the range of several MWp the maintenance strategies are very different speaking of costs.

Especially for smaller power plants, at best, owners pay attention to the annual yield and compare it to the simulated prediction. Due to financial reasons, often monitoring efforts (energy yield, temperature, irradiation, etc.) are not made at all. Besides that one major motivation is that laypersons have a lack of knowledge interpreting such data or suspicious effects of the PV plant.

Within the joint research project INFINITY a focus is put on operation and maintenance strategies. For the evaluation of PV failures, several characterization techniques like thermal/electroluminescence imaging, UV fluorescence measurements or U-I characteristics are available. For all these tools and their results the interpretation of the data is crucial. In many cases a lot of expert knowledge is necessary to interpret those failures. However, plant operators and local service staff (service technicians or electricians) normally do not have the expert knowledge to correctly rate the state of a panel or pending failures.

The developed software, presented in this article, provides a tool for responsible people to overcome the lack of expert knowledge in that area in a way to enable them to find potential failures. It represents a support for laypersons as well as professionals.

2 APPROACH

2.1 Data Collection

Over two years, data of PV plant failures have been collected from literature, internal project partner sources as well as from internal experiments (see Figures 1and 2).

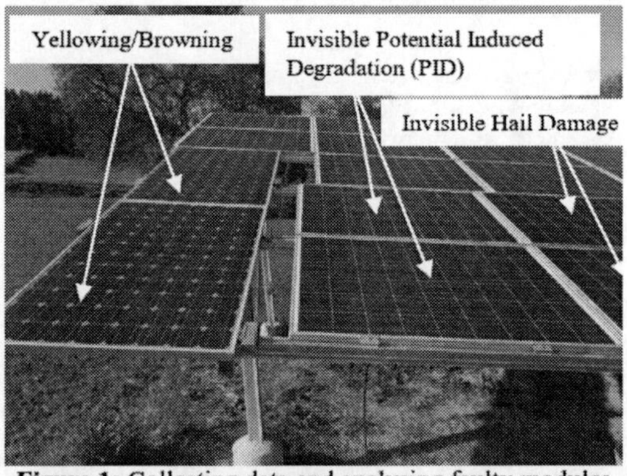

Figure 1: Collecting data and analysing faulty modules as information/data input for the software

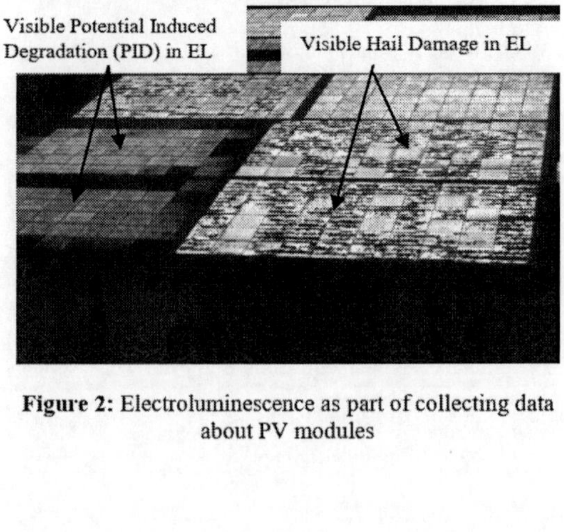

Figure 2: Electroluminescence as part of collecting data about PV modules

2.2 Software Concept / Web Interface

The software presents a summarizing description of common failure modes, provides corresponding exemplary measurement data (see Figures 3 -5) as well as possible solutions for each case. Additionally maintenance suggestions and possible actions in terms of counteracting strategies can be found. The goal was to create a software tool which is easy to use and to understand. Existing failures can be compared to depicted cases and understanding of the background can be enforced. Help is provided to solve the occurring problems or, in the case of more sophisticated cases, suggestions for additional laboratory tests are proposed. These direct guidelines can lower costs and decrease invested time until the plant is working properly again and necessary actions can be taken.

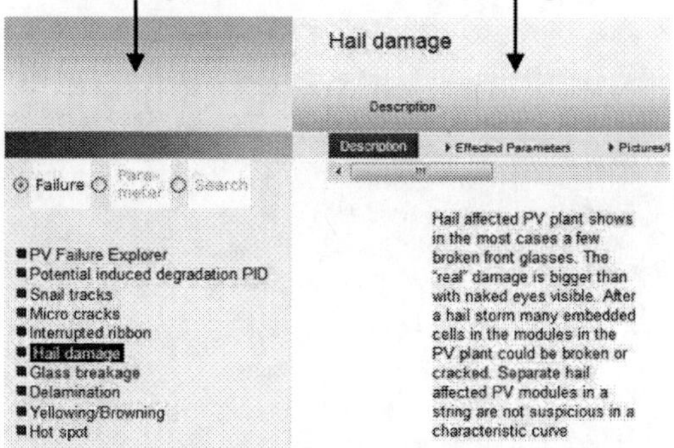

Figure 3: Web Interface Detail: Hail Damage, Failure Scope and Info Scope

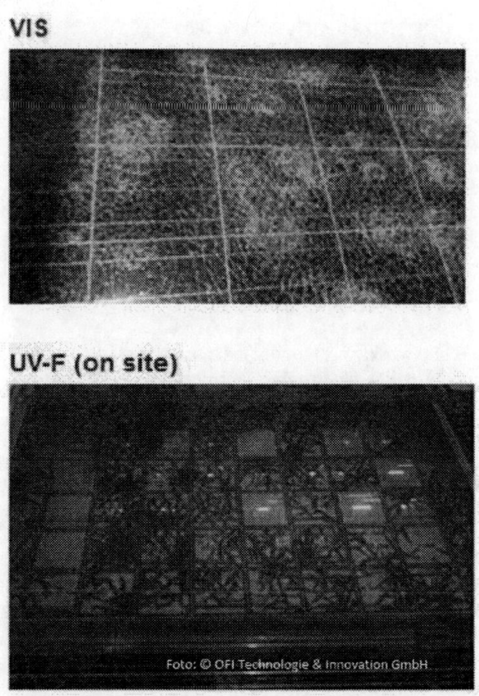

Figure 4: Web Interface Detail: Hail Damage: Visible and UV fluorescence picture

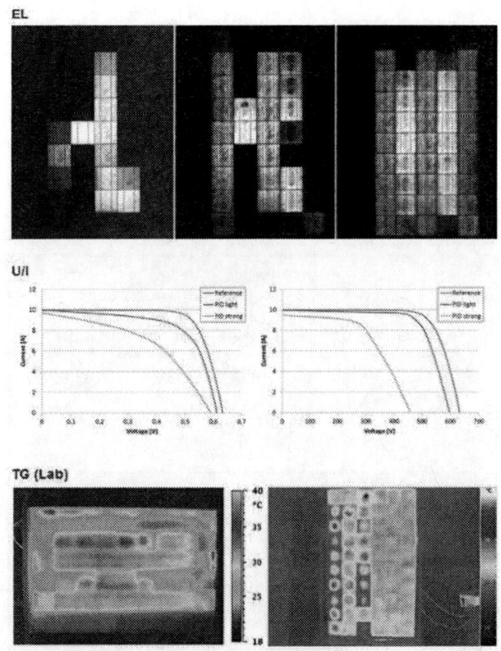

Figure 5: Web Interface Detail: Example PID Electroluminescence pictures, U/I diagram showing power loss and thermographic pictures

3 RESULTS AND DISCUSSION

Obtained research results have been included into a platform-independent, browser-based application which can be accessed via the web. Additionally local installation of the software is possible in order to ensure isolated operation.

A better understanding of PV plants, modules and components can be established and failures can be identified already at an early stage which leads to an effective avoidance of power loss of the system. Furthermore it should be mentioned, that the software tool is designed as a living, dynamic instrument, which will be regularly enhanced and updated.

4 SUMMARY & OUTLOOK

The technical expertise won by collecting different failure types all over the project time indicate that there are some features in common with failures PV panels.

The influence of other conditions in Austria as well as in other climate zones is still to be observed and may be extended to the existing knowledge.

Furthermore it should be mentioned, that the software tool is designed as a "living & dynamic instrument", which will be regularly enhanced.

The following short glimpse will show a roughly idea about what is prepared for mobile devices.

A draft package for android is currently under development, a preview to some views used by the app is shown in Figures 6 and 7.

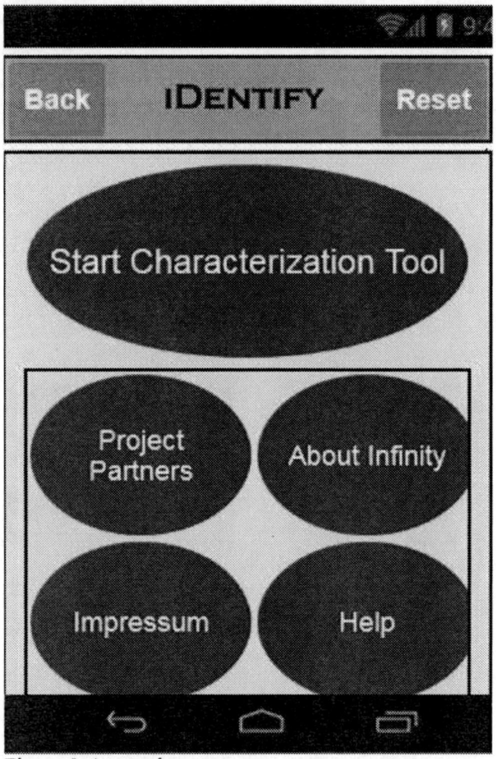

Figure 6: App tool start screen

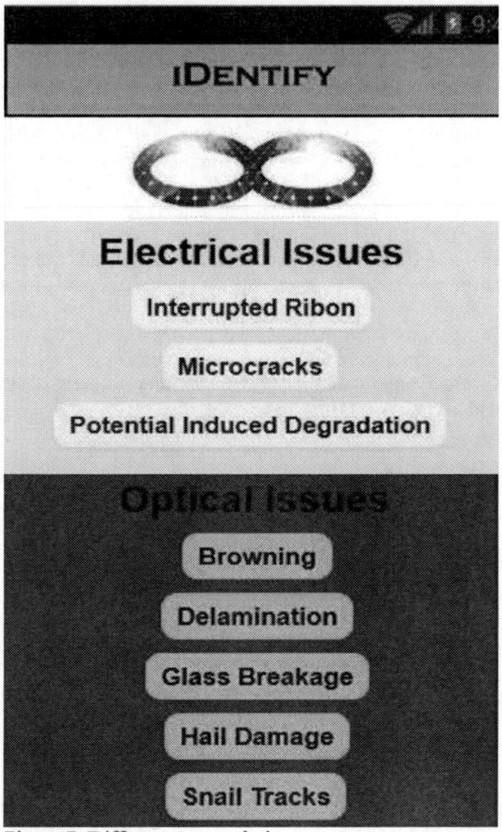

Figure 7: Different App tool views

5 DEVELOPMENT ENVIRONMENT

Web Development:
OfficeHelp (Extension software for Microsoft Word)
App Development:
MIT App Inventor (http://appinventor.mit.edu)

6 ACKNOWLEDGEMENTS

This work was conducted as part of the Austrian "Energy Research Program" project INFINITY. Funded by the Austrian Climate and Energy Fond and the Austrian Research Promotion Agency (FFG), both are gratefully acknowledged.

EVALUATION OF IRRADIANCE SENSOR TECHNOLOGIES FOR PLANT MONITORING OF PV SYSTEMS WITH CIGS THIN FILM MODULES

Stefan Grünsteidl, Peter Borowski, Thomas Dalibor
AVANCIS GmbH, Otto-Hahn-Ring 6, 81739 München, Germany
phone: +49 89 219620-458, email: stefan.gruensteidl@avancis.de

ABSTRACT: Selected data of a network of 20 small-scale reference PV plants in different climatic locations was analyzed with respect to the irradiance sensor technologies used. The modules of the PV systems solely consist of Copper-Indium-Gallium-Selenide (CIGS) thin film modules. The respective sensors, being commercially most commonly used sensors, were further individually analyzed with different data acquisition setups and specific test stands under outdoor (location Munich/Germany) and indoor conditions. Angular, spectral and temperature behavior was analyzed in order to quantify relative difference between the sensors under different irradiance conditions. Two CIGS modules with differing spectral response were investigated separately. The results show good agreements between CIGS and c-Si sensors (temperature compensated and uncompensated), while a pyranometer and a-Si sensor lead to larger discrepancies under certain light and temperature conditions. As normative standards do not imply strict technological requirements, this can lead to severe discrepancies in plant monitoring and high deviations in predictability. Suitable sensors and high calibration standards are strongly recommended.
Keywords: Reference Cell, Performance, CIGS

1 INTRODUCTION

Irradiance sensors are the key elements for PV performance monitoring. Small-scale PV systems are often monitored with only a single irradiance sensor as provided by inverter manufacturers. Normative standards [1] allow for a variety of sensor technologies, as well as a relatively wide calibration uncertainty. Prior studies show the influence of different sensor technologies [2][3][4][5] and module technologies [6][7] on performance evaluations and plant monitoring. In this work, we evaluate selected monitoring data obtained from 20 small-scale reference PV plants in different climatic locations (**Figure 1**) with respect to the irradiance sensor technologies used. Modules solely consist of Copper-Indium-Gallium-Selenide (CIGS) thin film modules by AVANCIS. The applied sensor technologies include standard commercially available monocrystalline silicon sensors with and without temperature correction (c-Si corr/uncorr), amorphous silicon sensors (a-Si), pyranometers and CIGS reference modules of AVANCIS' own production.

The present article is organized as follows: In this section, we motivate our work by showing yield data from a few PV plants and briefly describe the test equipment used for the study of irradiance sensors. We then present the results of these studies separately for effects of angle of incidence, spectral effects, and effects due to sensor stabilization. With this, we are able to interpret the original yield data. We conclude with a summary and a recommendation for improved irradiance sensing in PV plants.

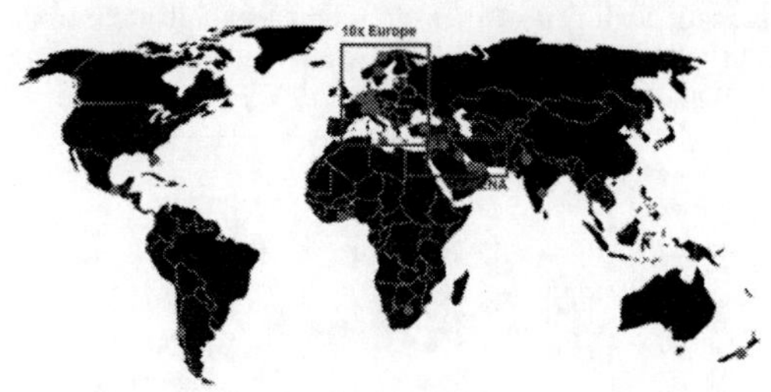

Figure 1: World map showing the locations of AVANCIS reference PV plants (green circles).

1.1 Motivation

Comparative data between retrospectively simulated yield (using PVsyst [8]) based on sensor data [9] and real yield show seasonal variations for some plants with a-Si irradiance sensor technology. These seasonal variations are less prominent or absent in the analysis of systems monitored with c-Si sensors (**Figure 2**). The seasonal variation can also be observed in the daily performance ratio (PR) values (**Figure 3**).

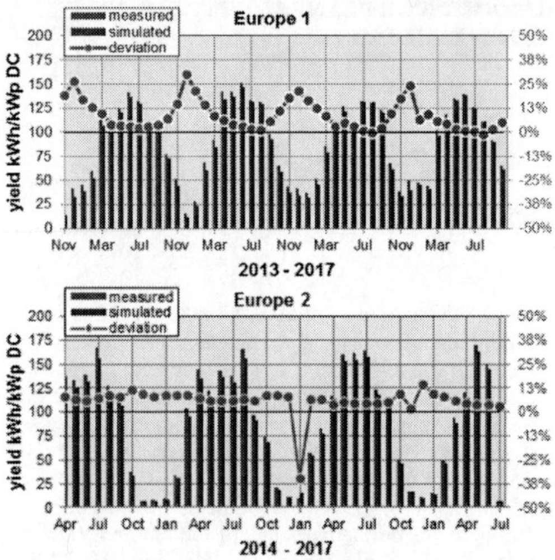

Figure 2: Retrospectively simulated vs. achieved ("measured") monthly yield (left axis) and the deviation between both (right axis). Data is shown for two PV plants. Irradiance of plant "Europe 1" (top panel) is measured with an a-Si sensor, irradiance of plant "Europe 2" (bottom panel) with a c-Si sensor without temperature compensation.

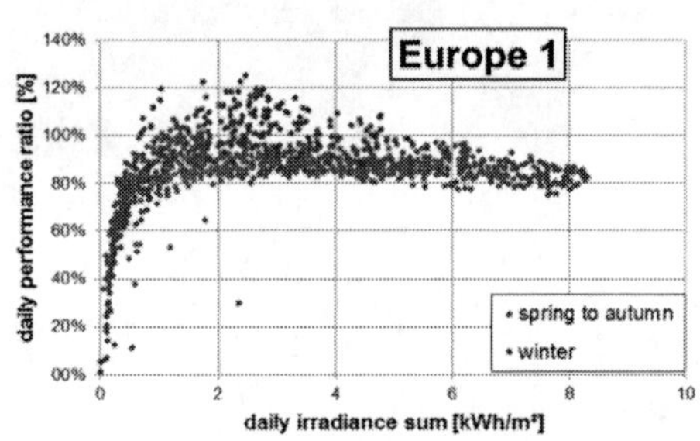

Figure 3: Daily performance ratio values of the system shown in the top panel of **Figure 2** (a-Si irradiance sensor; data shown over the course of 20 months).

Since the PV plants are all built with the same module technology, it is likely that the nature of the irradiance sensing is accountable for the seasonal variations. Module-specific explanations as well as flaws in the yield simulation (e.g., an incorrect modelling of the temperature or low light behavior) could be excluded by extensive simulation studies. Different kinds of irradiance sensors differ in their angle of incidence (AOI), spectral and temperature behavior with respect to both the PV modules and different sensors employed. For the assessed reference PV plants, these factors cannot be easily split individually, as spectral and sensor temperature data is hardly available. Yet, AOI data can be derived from time of measurement via simulation software [8], which resembles in practice a mixture of the different sources of influence (spectral mismatch, temperature correction) and shows mainly the amount of deviations encountered for the different sites (**Figure 4**). Here, the DC current during operation (I_{mpp}) is used to demonstrate different light induction into the module in comparison to the sensor. To extract the deviation from a linear relation, the measured I_{mpp} is compared to a current derived from linear regression of the respective irradiance sensor data.

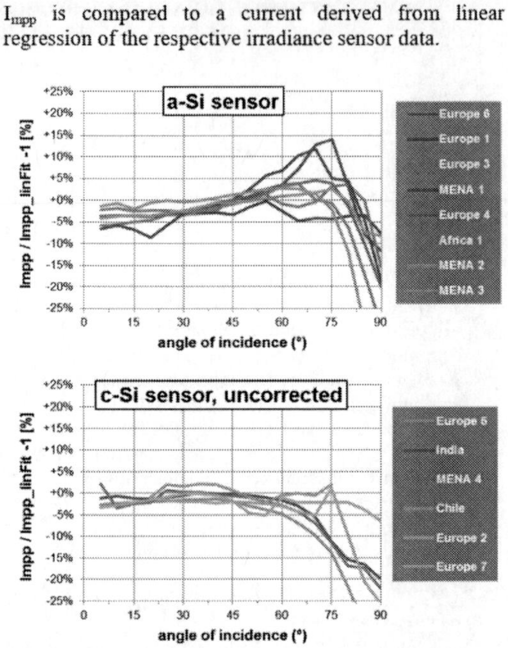

Figure 4: Deviation of PV system DC current (I_{mpp}) from linearly fitted current of irradiance sensor data over angle of incidence for different plants. Irradiance is measured either with a-Si sensors (top panel) or c-Si sensors without temperature compensation (bottom panel).

Figure 4 shows a clear dependence on sensor technology causing stronger deviations at lower sun angles. Effects of soiling were visible for the plants in the MENA region which is beyond the scope of this research. However, the effects shown are not necessarily rooted solely in varying AOI behavior. In order to discriminate between further effects (temperature dependence and effects of differing spectral response), further tests are necessary.

1.2 Test equipment

A test stand was built for deriving comparative data of the individual sensors in usage at the different reference PV plants. The test stand has been in operation in Munich, Germany since summer 2017. The sensors (listed in Table I) include commercial a-Si and c-Si sensors, and CIGS reference modules connected to 0.1 Ohm precision shunt resistors. Analog data (voltages) is taken in 5 seconds intervals with a commercial logging system. Digital data is read out via a Raspberry Pi microcomputer interface. Pyranometer data is measured simultaneously (**Figure 5**). The CIGS modules can be used as irradiance sensors by measuring their short circuit current (approximated by a voltage measurement across the 0.1 Ohm shunt resistor).

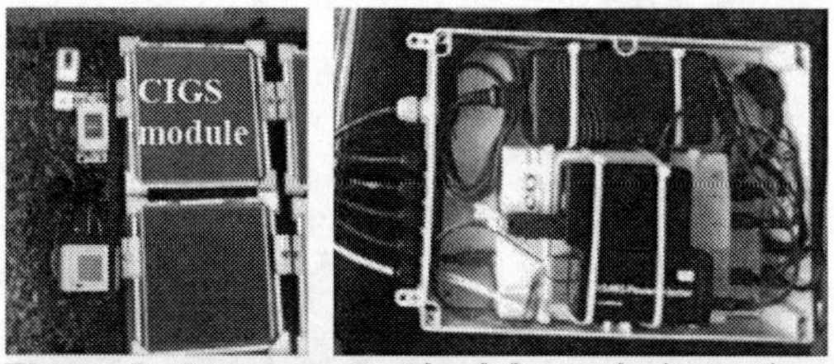

Figure 5: Sensor setup. The left panel shows three commercially available irradiance sensors (red marking: two based on c-Si and one based on a-Si technology) and two CIGS reference modules mounted in the same plane. The pyranometer is not shown. The electronic equipment used for logging the sensors' data is shown in the right panel.

Angular measurements were performed with a fixed and a tiltable test stand. In order to assess the influence of varying spectra, data from a spectrometer was recorded simultaneously to the sensor measurements. However, the data recorded by the spectrometer proved to be of questionable quality, mostly due to misaligned calibration. Indoor light soaking measurements were performed for comparing the sensors under constant conditions.

Table I: Sensors analyzed within this study. The a-Si and the c-Si sensor without temperature compensation are sensors offered by inverter manufacturers.

	size [cm x cm]	uncertainty P95 [%]
a-Si	12 x 9	8%
pyranometer	5 x 5	2% (daily)
c-Si corr.	15.5 x 8.5	4.1%
c-Si uncorr.	11 x 4	5% (annual)
CIGS	30 x 30	(own production)

2 MEASUREMENTS

The recordings of the sensor test stand (as described in 1.2) with a fixed tilt angle on a nearly clear sky day are shown in **Figure 6**. The daily course shows an overall offset between the curves which is typically encountered between different irradiance sensors [3]. The differences usually lie within the calibration uncertainty of the respective sensor, i.e., in a range between 2% and 8% (Table I), highest for a-Si sensors. The specific a-Si sensor used shows an unusually high constant deviation from the other sensors. As the constant differences between the sensors are attributed to calibration offsets, they will be ignored in the following evaluations.

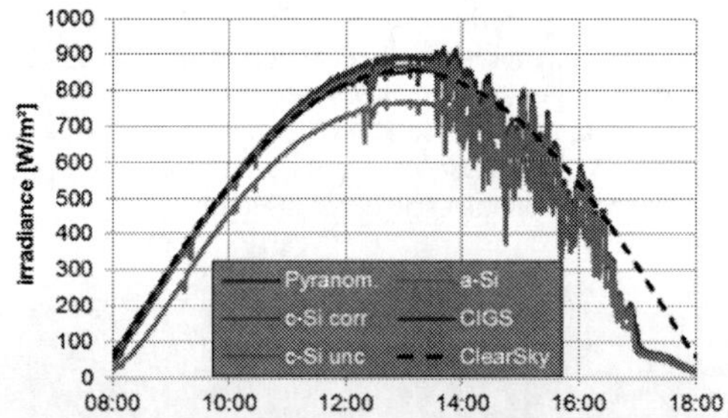

Figure 6: Daily course of irradiance as measured by the different sensor technologies on a nearly clear sky summer day in Munich, Germany for a fixed tilt angle (theoretical clear sky data derived from PVsyst).

2.1 Angular measurements

Angular measurements were performed on a clear sky summer day using a tiltable frame to which all sensors were attached to. The sensors were aligned to the sun (AOI = 0°, maximum irradiance) and measurement data was taken within a timeframe of 30 minutes in steps of 5° tilt angle (inclination), while azimuth was kept constant. The solar elevation angle at time of measurement was 45°. At AOI = 90°, the sensors therefore pointed at an angle of 135° towards North. The data was normalized to the irradiance at AOI = 0° and averaged over two consecutive cycles of changing inclination. The results are shown in **Figure 7** and show a decrease of irradiance along the cosine angle [10], with a larger deviation towards larger angles before settling at a diffuse irradiance of about 20%. The larger deviation is attributed to the reflection of an adjacent wall and the different positions of the sensors on the setup plane.

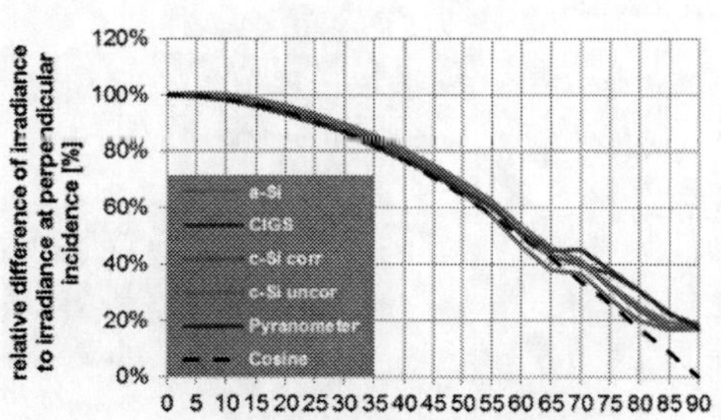

Figure 7: Irradiance as measured by the different sensors with varying AOI on a clear sky day. Data is normalized to irradiance measured at direct alignment to the sun.

To analyze technology-specific effects on the AOI behavior of the various sensors, **Figure 8** shows the data of **Figure 7** with the cosine of the AOI subtracted. The course of all graphs in **Figure 8** show an increasing influence of diffuse and reflected (wall) light for larger angles of incidence, as the direct irradiance decreases. For angles up to 65°, before reaching the region of wall reflection, differences between all sensor technologies are always less than 5%. Excluding the sensor based on a-Si, the differences between the CIGS module and the other sensor technologies were smaller than 3% for this range. As the vast majority of yearly yield is generated at angles smaller than 65° for all reference plants investigated in this study, the influence of deviating AOI behavior is rated as comparably small. E.g., the plant whose data is shown in **Figure 3** generates more than 70 percent of its yearly yield at AOI less than 65°, even during the winter months. AOI dependence is anticipated to depend rather on encapsulation than sensor technology.

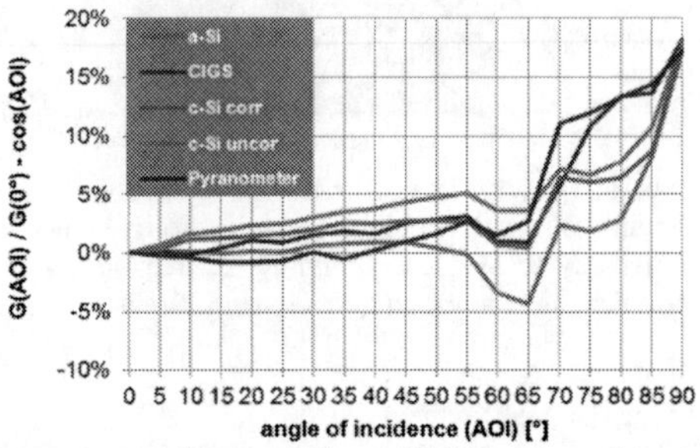

Figure 8: As of **Figure 7** but with the cosine of the AOI subtracted from the data.

2.2 Spectral mismatch

In order to judge on the effect of spectral mismatch between the sensors, outdoor data under varying irradiance situations was investigated. The evaluated data is influenced by a combination of effects of the angle of incidence (AOI) and spectral mismatch. Over the course of three and two averaged consecutive days with clear sky irradiance in October 2017 (location Munich/Germany, azimuth 0° South, tilt 24.5°), the sensors show minor differences in the group of CIGS and c-Si sensors, and stronger differences to the pyranometer and a-Si sensor when it comes to lower sun angles at lower irradiances (**Figure 9**).

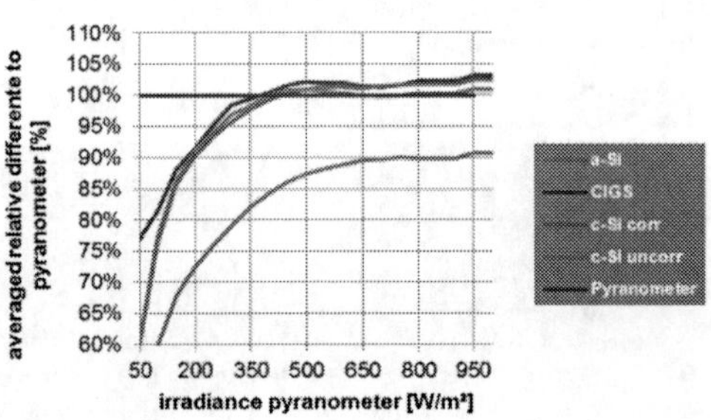

Figure 9: Irradiance as measured by the various sensors, relative to the measurements of the pyranometer.

Within classes of 50 W/m² and above 150 W/m², the difference in irradiance measured within the CIGS/c-Si group was always lower than 2.6%, well within the measurement uncertainty. This difference towards the CIGS module when compensating for calibration differences at 1000 W/m² is depicted in **Figure 10**, which shows the relevant contribution for monitoring a CIGS reference plant. Yet, the differences of the a-Si sensor towards CIGS are not high enough to fully explain the discrepancies shown in subsection 1.1.

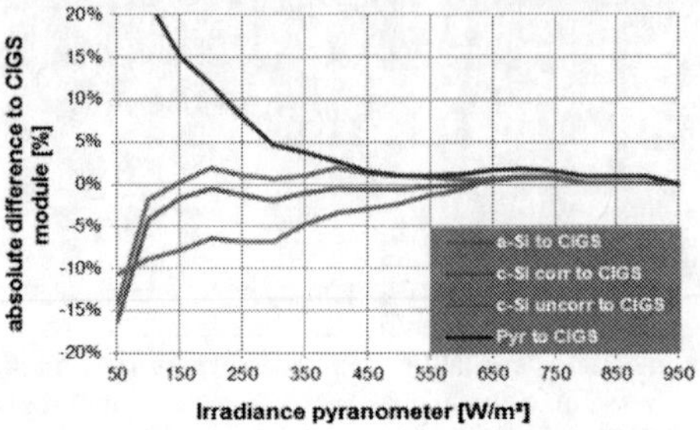

Figure 10: As of **Figure 9** but with CIGS data subtracted from sensor data.

The test setup was enhanced in July 2018 with one CIGS module with a high Gallium (highGa) content, known to significantly vary the spectral response of the device. With the sensor test setup, the influence of this altered spectral response was analyzed. Within the irradiance classes analyzed in the present work, no significant difference could be observed between the two different CIGS modules for all data sets taken in August and September 2018. The influences of air mass and differing spectral behavior relative to the pyranometer can be shown when manually choosing clear sky days for assessment. **Figure 11** shows the difference between choosing clear sky days to choosing all available days. No significant difference can be observed between the two CIGS modules with different Gallium content in the absorber. However, the difference relative to the pyranometer is bigger for lower irradiances on clear sky days.

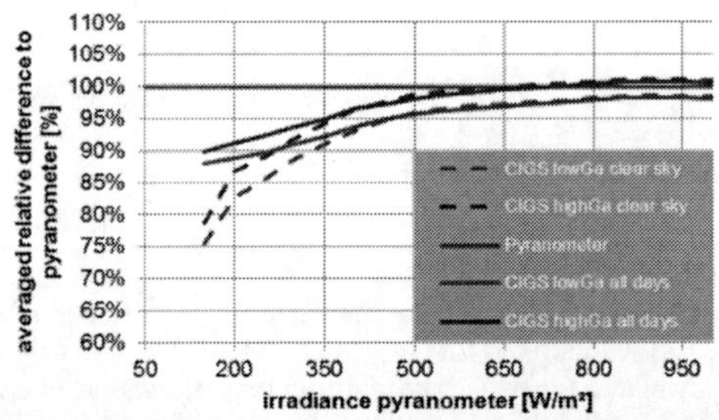

Figure 11: Irradiance as measured by two CIGS modules with differing spectral responses due to variations in the absorber layers (high vs. low Ga content) on clear sky and all days in August and September 2018, relative to the measurements of the pyranometer.

This behavior can be explained with the infrared content in the incident sunlight. While a higher air mass (AM) at lower irradiances results in a higher content of infrared light, this content gets lowered by water vapor on cloudy days. This difference affects the pyranometer stronger, as it absorbs infrared light very efficiently. For the CIGS module, the infrared region beyond near-infrared lies outside the bandgap. Therefore the difference to the pyranometer rises at lower irradiances and rising AM. But this effect is significantly weaker on cloudy days with reduced infrared content.

To better understand the spectral influence, we computed spectral mismatch factors between different sensors under two spectra. **Figure 12** shows the relative spectral responses (SR) for the technologies investigated. While measuring the SR of the a-Si sensor, we discovered that the a-Si sensor under investigation was in fact a tandem a-Si/a-Si PV module with the known difficulties of obtaining a realistic SR in a lab setup [11][12]. The data shown in **Figure 12** was measured internally on individually contacted cells (a-Si, CIGS), provided by manufacturer (pyranometer) or is representative for the specific technology (c-Si).

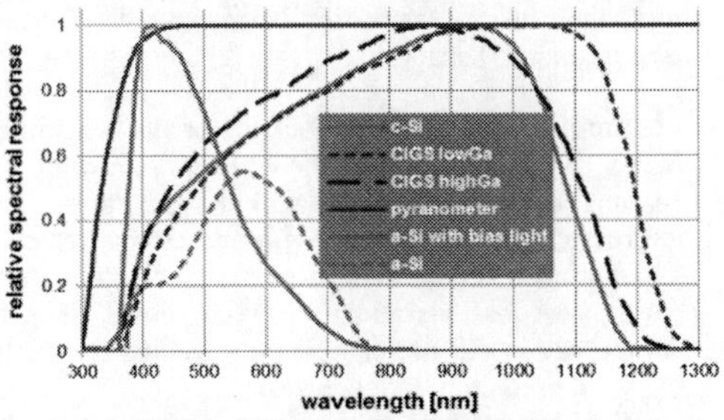

Figure 12: Relative spectral response (SR) of the technologies investigated in this work. The SR data for a-Si shows internal measurements of the sensor with and without bias light (no voltage bias applied).

The spectral mismatch factor (SMM) quantifies the difference in short circuit currents of two solar cells under varying spectra relative to the short circuit currents under a reference spectrum (AM1.5). Lacking data of sufficient quality on the spectral composition during the measurements with the sensor test stand, we exemplarily compute SMM for spectra with varying air masses, generated by SMARTS 2.9.2 [13][14] (**Figure 13**). The AMs chosen should represent the situation on clear sky days in September in Munich at irradiances above 200 W/m².

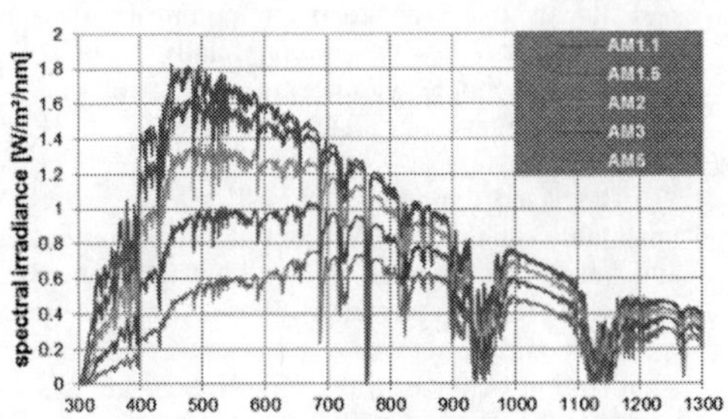

Figure 13: Spectra (E_{meas}) used exemplarily for the analysis of the influence of differing spectral response of the sensors.

Spectral mismatch factors were computed relative to AM1.5 (E_{ref}) according to IEC 60904-7 [15] for two scenarios: the reference sensor (S_{ref}) being a c-Si sensor or alternatively a pyranometer. SMM are given for the remaining sensors, respectively (S_{meas}). The results of the computation are shown in **Figure 14**. With the a-Si sensor being an a-Si/a-Si tandem PV module, a corresponding computation of the SMM for this technology under varying spectra is not meaningful without detailed knowledge of the optical properties of the device and is therefore not being shown.

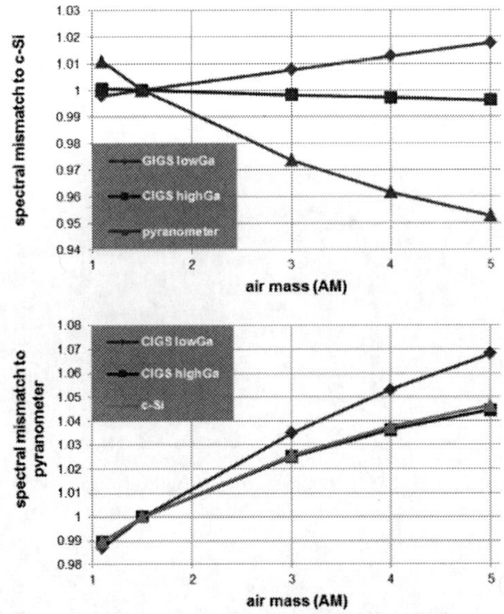

Figure 14: Spectral mismatch factors (SMM) for the sensors investigated. Top panel: The c-Si sensor being the reference device. Bottom panel: Reference device is a pyranometer.

Even though the spectral response curves of c-Si and low-Gallium CIGS show a significant difference in the near-infrared, the difference in sensor reading at the varying spectra considered is still in acceptable ranges. The small differences between these technologies were also observed in outdoor data (**Figure 10**). From this, we conclude that the usage of standard c-Si sensors for monitoring irradiance in CIGS PV plants is acceptable considering typical calibration uncertainties in standard irradiance sensors. This, however, is not valid for high precision measurements of CIGS PV modules in sun simulators, where a calibrated CIGS irradiance sensor is recommended or a correction for spectral mismatch in combination with spectral measurements beyond the limits given in the IEC standard [16].

From **Figure 14** we also observe the well-known discrepancy between c-Si sensors and pyranometers that can lead to systematic differences in performance assessment of PV plants independent of the PV module technology employed [2][3][4]. With the data measured and presented, we are not able to assess the influence of varying spectra on the reliability of irradiance sensing when using the a-Si/a-Si tandem sensor.

2.3 Stabilization

In addition to differing AOI behavior (little effect) and effects of spectral mismatch (magnitude unknown), stabilization effects can contribute to the seasonal variations observed for the PV systems equipped with an a-Si irradiance sensor. Therefore, further measurements were performed on the a-Si sensor to investigate the stabilization behavior and the temperature compensation of the sensor under bright light exposition.

For this, the sensor setup was installed in a

continuous sun simulator (light soaking setup) with an irradiance of roughly 1100 W/m² (laterally varying). The light soaking setup ran for several days, followed by several days of storing the sensors in the dark (dark storage; DS). **Figure 15** shows the measurement results of one sample c-Si sensor and the a-Si sensor after different time periods of DS.

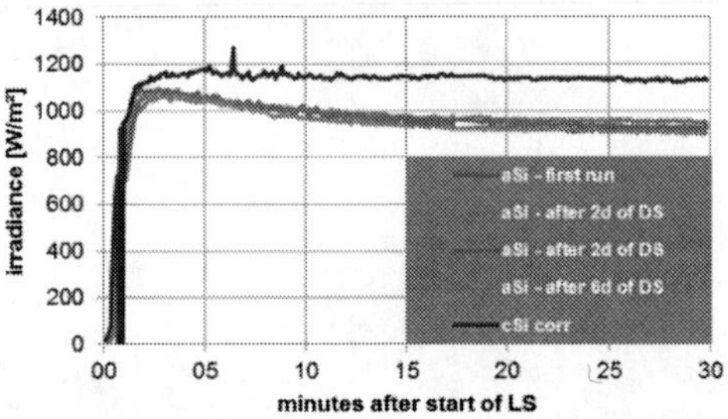

Figure 15: Start phase of the light soaking setup within the first minutes, after varying duration of prior dark storage (DS).

The recording of the c-Si sensor shows the stabilization of the sun simulator's lamp within the first roughly 3 minutes. After this period, the sun simulator provides a constant irradiance within a few percent (as can be concluded from the reading of the temperature-corrected c-Si sensor). While none of the other sensors (including the CIGS modules) showed a significant stabilization behavior, the a-Si sensor shows a constant reading only after about 25 minutes with only about 85% of the initial irradiance reading. The time constant and steepness of the stabilization curve appears to be independent of the prior length of dark storage. As the sensor was equipped with an onboard microprocessor and proprietary firmware, neither cell temperature nor the algorithm could be recorded or analyzed. Therefore it could not be concluded whether the stabilization shown in **Figure 15** is an effect of insufficient or absent temperature compensation within the sensor or is due to the semiconductor technology employed (e.g., light-induced annealing).

2.4 Effect on yield prediction

An attempt was made to retrospectively correct the irradiance data recorded by the a-Si sensor. As a substitute for the absent temperature of the sensor, separately recorded module temperature was used to compute a 'true' sensor reading, using an additional temperature compensation with a constant coefficient (0.8 %/°C). Retrospective yield simulation was repeated with this corrected data set and it was found that the deviations between real and simulated yield can be greatly decreased with this technique. The result of this corrected yield simulation for the PV plant of **Figure 2** (top panel) is shown in **Figure 16**.

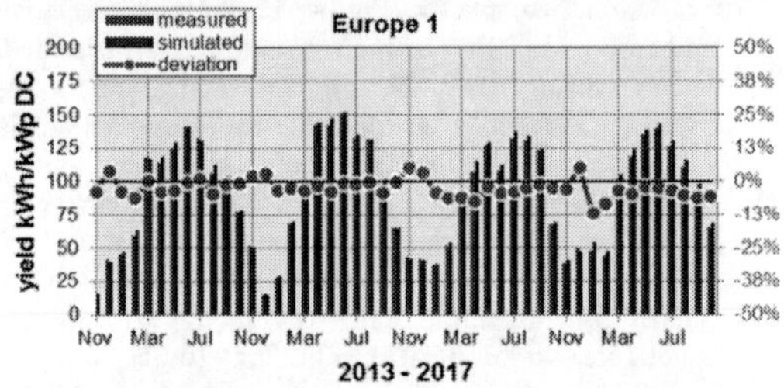

Figure 16: Reference plant monitoring (as of **Figure 2** - top panel) using temperature-corrected recordings of the a-Si irradiance sensor.

Similar results were obtained for the remaining PV plants with a-Si irradiance sensors showing strong seasonal deviations between real and simulated yield. As a cross-check, we applied the described temperature adaption also on the data of plants that do not show the strong seasonal variations. The additional temperature compensation did not significantly change the quality of the yield prediction in these cases.

Similarly to the temperature compensation of the irradiance reading of the a-Si sensor, we attempted to compensate the irradiance reading by potential other mismatches. Neither a realistic consideration of AOI effects nor a rough linear correction of irradiance at instances with high air masses (AM) was sufficient to significantly reduce the seasonal effects as shown in **Figure 2**.

3 CONCLUSION

By carefully analyzing the monitoring data of 20 AVANCIS reference PV systems operating since several years in various climatic regions, the significance of the nature of the irradiance sensor for performance assessment could be shown. In order to produce field-realistic and customer-relevant monitoring data, all PV plants operate with commercially available and comparably simple monitoring equipment and sensors rather than high-quality R&D equipment.

The used tandem a-Si/a-Si sensor proved being unsuitable for precise PV plant monitoring due to strong and unrealistic seasonal effects both in the performance ratio and yield interpretation. By testing the sensors both outdoors and in the lab, we identified a pronounced stabilization behavior after dark storage for this specific sensor technology as a likely cause for the discrepancies that adds to the uncertainties in irradiance measurement due to spectral mismatch. The observed discrepancies in yield monitoring could be decreased by retrospectively performing a temperature compensation of the measured irradiance data. In conclusion, the usage of this sensor technology for plant monitoring is not advisable.

Care should be taken when comparing to pyranometers when performing plant analyses, as reported in prior studies [4]. Our results show good agreement between CIGS modules and c-Si sensors (temperature compensated and uncompensated), even when considering CIGS modules with significantly differing spectral responses. As normative standards do not imply strict technological requirements, applying unsuitable sensors could lead to severe discrepancies in plant monitoring and lead to uncertainties in predictability and in failure analysis. Suitable sensors and high calibration standards are strongly recommended. For the performance assessment of AVANCIS CIGS modules, c-Si irradiance sensors should provide sufficiently good monitoring data for typical PV systems.

7 REFERENCES

[1] IEC 61724-1:2017 Photovoltaic system performance - Part 1: Monitoring (2017)

[2] S. Reinhardt et al., "Influence of Technology Differences in Performance Ratio Calculations", Proc. of the 27th EUPVSEC (2012) 4025

[3] A. Driesse et al., "Indoor and Outdoor Evaluation of Global Irradiance Sensors", Proc. of the 31st EUPVSEC (2015) 1704

[4] B. Kirn et al., "Performance of Different Types of ETSC Solar Irradiance Sensors", Proc. of the 28th EUPVSEC (2013) 3196

[5] B. Schulz et al., "Evaluation of Calibrated Solar Cells and Pyranometers Regarding the Effective Irradiance Detected by the PV Modules", Proc. of the 25th EUPVSEC (2010) 4797

[6] M. Gilg et al., "Stabilized Performance Ratio Calculation on Photovoltaic Parks", Proc. of the 28th EUPVSEC (2013) 4000

[7] M. Schweiger et al., "Spectral Analysis of Various Thin-Film Modules Using High Precision Spectral Response Data and Solar Spectral Irradiance Data", Proc. of the 27th EUPVSEC (2012) 3284

[8] PVsyst V5.74 (16.6.2014), PVsyst SA.

[9] P. Borowski et al., "Energy Rating of CIGS Thin Film Modules and Systems", Proc. of the 27th EUPVSEC (2012) 2294

[10] H. Häberlin, "Photovoltaik", AZ Verlag (2007)

[11] J. Trigo et al., „Progress in Optical Characterization of Amorphous Silicon Tandem Cells", Proc. of the 26th EUPVSEC (2011) 2544

[12] C. Ulbrich et al., "Matching of Silicon Thin-Film Tandem Solar Cells for Maximum Power Output", International J. of Photoenergy (2013) 314091

[13] C. Gueymard, "Parameterized Transmittance Model for Direct Beam and Circumsolar Spectral Irradiance", Solar Energy 71:5 (2001) 325

[14] C. Gueymard, "SMARTS, A Simple Model of the Atmospheric Radiative Transfer of Sunshine: Algorithms and Performance Assessment", Professional Paper FSEC-PF-270-95. Florida Solar Energy Center, 1679 Clearlake Rd., Cocoa, FL 32922 (1995)

[15] IEC 60904-7:2008 Photovoltaic devices - Part 7: Computation of the spectral mismatch correction for measurements of photovoltaic devices

[16] A. Avellan et al., "Challenges in Measuring the True Efficiency of CIGS Modules", Proc. of the 28th EUPVSEC (2013) 2118

DELTA STRUCTURE FOR CONSTANT DAILY POWER PROFILE IN PV IRRIGATION SYSTEMS

R. H. Almeida[1,2], I. B. Carrêlo[1], L. Narvarte[1], E. Lorenzo[1]
[1] Instituto de Energía Solar, Universidad Politécnica de Madrid
[2] Instituto Dom Luiz, Faculdade de Ciências, Universidade de Lisboa
*Phone: +34 91 336 55 31; E-mail: rita.hogan@ies.upm.es

ABSTRACT: This paper presents a different type of photovoltaic (PV) structure called delta, ΔS(60). This was developed in order to achieve a constant irradiance daily profile similar to the one obtained with a North-South horizontal axis tracker (1xh) since this constancy is essential for a set of PV applications, namely for PV irrigation systems because the dynamic of the well usually requires constant water flow. The performance of ΔS(60) in a PV irrigation system (PVIS) pumping to a water pool in Spain is compared with the one of 1xh and with the one of a typical static structure oriented to the Equator, S(25). Results show that ΔS(60) is of keen interest in PVIS since it requires a peak power lower than the one needed in the case of S(25) in order to achieve the same water volume along the irrigation period.

Keywords: PV irrigation systems, PV pumping, constant daily profiles, PV structure

1 INTRODUCTION

The constancy of in-plane irradiance profile is an advantage for some PV applications [1], [2], namely PV irrigation systems (PVIS) [3].

The North-South horizontal axis tracker seems to be the best solution to achieve this constancy [4]. Nevertheless, sometimes trackers are subject to reliability and cost issues when small and isolated systems are concerned [3], [5].

Since the traditional static structure facing the Equator does not fulfill this constancy requirement [1], this paper briefly presents a new type of structure called delta.

The performance of delta, ΔS(60), is compared with two representatives of the current state-of-art: a static structure oriented to the South and tilted 25^0, S(25), and a single North-South horizontal axis tracker, 1xh, on the frame of PVIS to a water pool. This performance is assessed in terms of water pumped along two different periods: the whole year and from May to September, which is a typical irrigation period (IP) for many crops in the Mediterranean area.

A 1xh PV system is simulated and optimized. Then the needed peak power for ΔS(60) and S(25) to achieve the same water volume of 1xh is obtained and the constancy of the profile is analyzed.

This paper is structured as follows: section 2 introduces the structure and section 3 the case study of the PVIS under analysis, section 4 includes the main results and section 5 the conclusions.

2 DELTA STRUCTURE

The delta, ΔS, is a static ground-mounted structure where half of the PV array is oriented to the East and the remaining to the West, both parts with the same tilt angle, β=60^0. With this angle, the PV array surface seen by the Sun is equal at three moments of any day: at morning when the Sun is perpendicular to the East oriented surfaces, at Midday and at afternoon when the Sun is perpendicular to the West oriented surfaces. Roughly, these moments occur about 4 hours before midday, at midday and 4 hours after midday. In-plane direct irradiance tends to be equal on these moments, which leads to in-plane global irradiance reasonably approaching constancy during 8 hours per day. A more detail description of this structure can be found in [6].

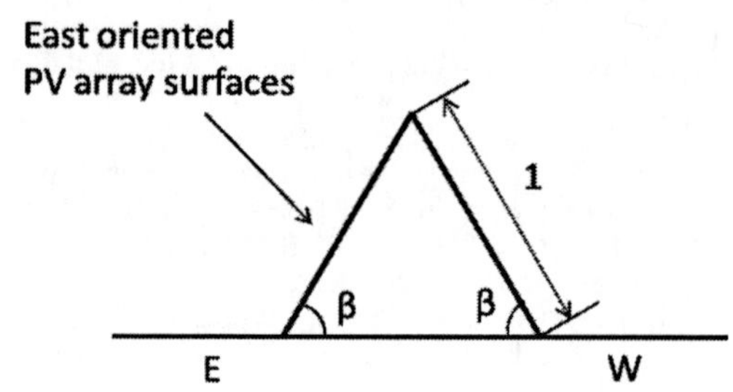

Figure 1: The delta structure, ΔS: The PV array is distributed into two halves – one half is oriented to the West while the other is oriented to the East.

The ΔS has been incorporated in SISIFO (a PV system simulation tool developed at IES-UPM) [7].

3 STUDY CASE: PVIS TO A WATER POOL IN VALLADOLID, SPAIN

A simulation exercise was performed in a PVIS pumping from a borehole to a water pool in Valladolid, Spain (41.307 N, 5.215 W). The monthly mean daily horizontal irradiation values, $G_{dm}(0)$, have been obtained from PVGIS [8] and the simulation of the PVIS has been done using the SISIFO tool [7]. Details of the parameters are out of the scope in this work but can be found in [6].

The system includes a Caprari motor-pump – E8P95/7ZC MAC 870-8V – and a OMRON frequency converter – RX75. The main hydraulic characteristics of this particular PVIS are shown in Table I and Fig. 2.

Table I: Characteristics of the PVIS

Parameter	Value
Type of PV irrigation system	Stand-alone
Type of pumping	Water pool
Pumping head [m]	105
Working flow [m³/h]	Variable

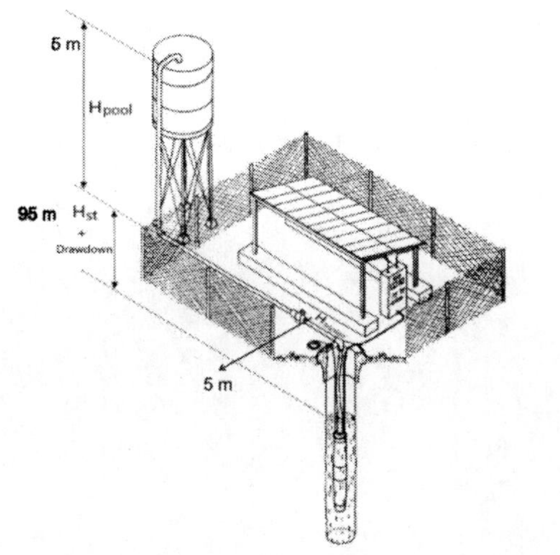

Figure 2: The pumping head of 105 m (see Table I) is the sum of the static head plus drawdown (95 m), the height of the water pool (5 m) and the friction losses (5 m).

4 RESULTS

The peak power of the 1xh system is 80 kWp. If this power is used in the other two structures under analysis, the profiles of the water pumped along a typical year can be seen in Figure 3. The water pumped with 1xh is 336681 m^3 per year and 195235 m^3 during the IP. If the same 80 kWp are used in the other 2 structures, much less water is pumped – ΔS(60) can reach only 42% of the annual water of the 1xh and 54% if ones consider the IP; in the case of S(25) these values are 63 and 70% respectively.

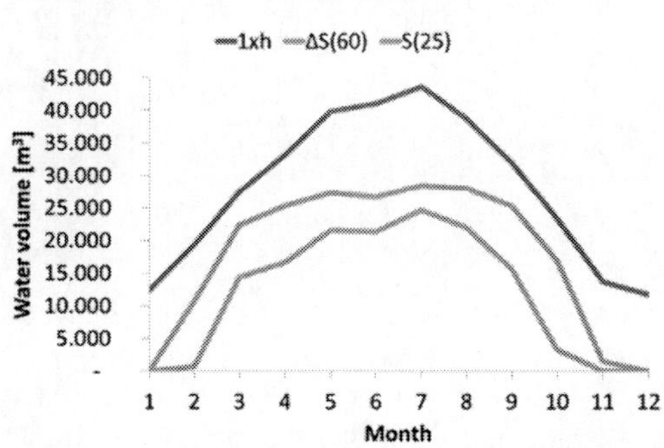

Figure 3: Water pumped along a typical year with the three structures under analysis. The peak power of the 3 structures is 80 kWp.

Since the water pumped with ΔS(60) and S(25) with 80 kWp are lower than the one pumped with 1xh, the next step should be the increase of the peak power of ΔS(60) and S(25) in order to achieve the performance of the 1xh. This is shown in Figure 4, where both the water volume and the peak power are normalized by the values of the 1xh (195235 m^3 and 80 kWp respectively). It is possible to conclude that to guarantee the same water volume during IP, it is necessary to install 1.75 the peak power of the 1xh in the case of the ΔS(60), and 2.75 in the case of the S(25).

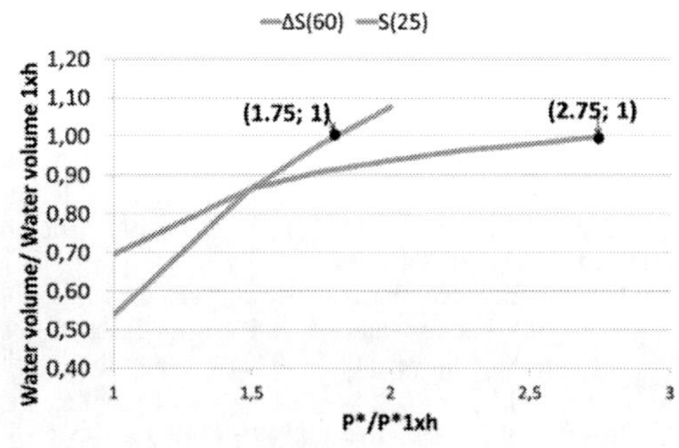

Figure 4: Water pumped during the IP with ΔS(60) and S(25) normalized by the water pumped by a 80 kWp 1xh as a function of its PV peak power normalized by the 80 kWp of the 1xh.

The same exercise was performed considering the whole year. The e obtained results, in addition to the ones obtained during IP, are in Table II. It is interesting to verify that the needed power in S(25) considering only the IP is higher than the one needed if the whole year is analysed.

Table II: PV peak power needed to guarantee the same water volume than 1xh

Period	P^*/P^*_{1xh}	
	ΔS(60)	S(25)
Yearly	1.75	2
IP	1.75	2.75

In order to have a better visualization of the constancy of the daily profile, Figure 5 shows the water pumped during the typical day of June by the three structures under analysis. The similarity between the profiles of the 1xh and ΔS(60) is seen. In the case of the S(25), the system worked less hours a day – it starts pumping latter and stops earlier than the other 2 structures. Furthermore, since the system is oversized in this case, the 100% constant profile in the central hours of the day is a consequence of the motor-pump maximum frequency (50 Hz) – PV energy is being wasted due to the characteristics of the irrigation system.

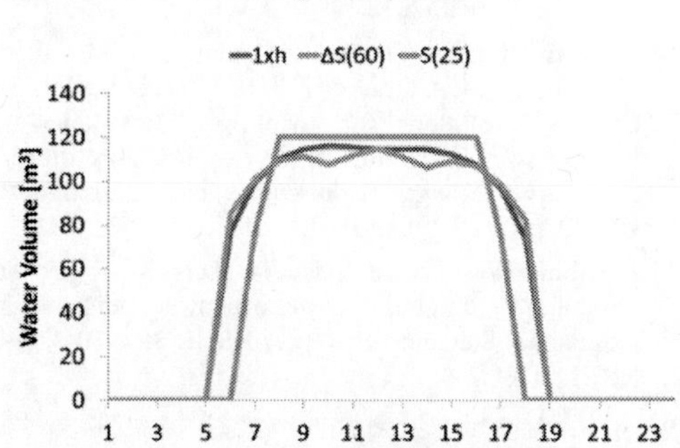

Figure 5: Water pumped during the typical day of June by the 3 structures under analysis.

5 CONCLUSIONS

A new type of structure called delta, ΔS(60), is briefly presented and applied to a study case – a PV irrigation system pumping from a borehole to a water pool in Valladolid, Spain.

The results obtained were always compared to the ones obtained with two typical PV structures: a North-South horizontal axis tracker, 1xh, and a static structure oriented to the Equator, S(25).

It is remarkable that the hypothesis of using the ΔS(60) is of particular interest for PV irrigation systems, being the needed power in ΔS(60) lower than the one in S(25).

In this study case, if the peak power of 1xh is 80 kWp, the one of ΔS(60) and S(25) needs to be 140 and 220 kWp respectively.

6 AKNOWLEDGMENTS

This work has been possible thanks to the funding from the European Union's Horizon 2020 research and innovation programme in the project MArket uptake of an innovative irrigation Solution based on LOW WATer-ENergy consumption (MASLOWATEN), under grant agreement n°640771, as well as the financial support by MIT Portugal Program on Sustainable Energy Systems and the Portuguese Science and Technology Foundation (FCT), grant PD/BD/105851/2014.

7 REFERENCES

[1] G. Litjens, E. Worrell and W. van Sark, "Influence of demand patterns on the optimal orientation of photovoltaic systems," *Solar Energy,* vol. 155, pp. 1002-1014, 2017.

[2] J. Remund, C. Calhau, L. Perret and D. Marcel, "Characterization of the spatio-temporal variations and ramp rates of solar radiation and PV," International Energy Agency - Photovoltaic Power Systems Programme, 2015.

[3] H. Hartung and L. Pluschke, "The benefits and risks of solar-powered irrigation - a global overview," Food and Agriculture Organization of the United Nations and Deutsche Gesellschaft für Internationale Zusammenarbeit, 2018.

[4] L. Narvarte, J. Fernández-Ramos, F. Martínez-Moreno, L. M. Carrasco, R. H. Almeida and I. B. Carrêlo, "Solutions for adapting photovoltaics to large power irrigation systems for agriculture," *Sustainable Energy Technologies and Assessments,* vol. 29, pp. 119-130, 2018.

[5] J. Shingleton, "One-Axis Trackers - Improved reliability, durability, performance and cost reduction," Subcontract Report NREL/SR-520-42769 NREL, 2008.

[6] R. H. Almeida, L. Narvarte and E. Lorenzo, "PV arrays with delta structures for constant irradiance daily profiles," *Solar Energy,* vol. 171, pp. 23-30, 2018.

[7] Instituto de Energía Solar - Universidad Politécnica de Madrid, "SISIFO - Simulación de Sistemas Fotovoltaicos," [Online]. Available: http://sisifo.info.

[8] European Commission, Joint Research Centre, "Photovoltaic Geographical Information System (PVGIS)," [Online]. Available: http://re.jrc.ec.europa.eu/pvgis/.

INVESTIGATION ON RESPONSE CHARACTERISTICS OF PV-PCS AT STEP-CHANGE IN SOLAR IRRADIANCE

Hiroo Konishi*, Jun Hashimoto, Kenji Otani
National Institute of Advanced Industrial Science and Technology (AIST)
2-2-9 Machiikedai, Koriyama-shi Fukushima 963-0298 Japan
*Tel:(+81) 24-963-0827, Fax:(+81) 24-963-0828, E-mail: h.konishi@aist.go.jp

ABSTRACT: The installation of solar power generation system was limited to places with good sunshine conditions, but it tends to be built in various places and areas. Generated electric power is influenced by various factors such as weather condition and installation environment, and it depends greatly on the performance of power conditioning system (PV-PCS) cooperating with the grid. For evaluating the PV-PCS conversion efficiency, measurement of static conversion efficiency and MPPT efficiency under certain conditions is specified by IEC and UL. However, against the factors such as the change of weather conditions and the installation environment, the method of performance evaluation of the PV-PCS for changing in solar radiation is not clear. Comprehensive evaluations of PV-PCS against both static and dynamic change in solar radiation are needed. For this reason, we investigated the performance evaluation method of the PV-PCS by analysis which can comprehensively evaluate solar power generation system easily.

Keywords: power conditioning, performance, evaluation, solar radiation, HILS test

1 INTRODUCTION

The use of renewable energy is being promoted in each country as a measure to prevent global warming and energy savings. In Japan, the FIT system was enforced from July 2012 and the introduction of renewable energy has been promoted. As an energy source, solar power generation systems are rapidly introduced with short lead time, relatively easy construction, and low cost. And the installed capacity until 2016 is over 41.6GW. In 2030, 53 GW of government introduction target is planned which is 40 times the introduction amount of 2005, further introduction is necessary for achieving the target. The installations of solar power generation system were limited to places with good solar radiation conditions until now, but they tend to be built in various places and areas. Generated electric power is influenced by various factors such as weather conditions and installation environment and so on, and it depends greatly on the performance of power conditioning system (PV-PCS) in cooperation with the grid. The conversion efficiency of PV-PCS is important as an performance index. It is presented as multiplying of efficiency of maximum power point tracking (MPPT) control generally performed by a DC/DC converter and conversion efficiency of the inverter if the inverter does not include the DC/DC converter function. In order to increase the generated electric power, it is necessary to make the conversion efficiency as higher as possible. Regarding the conversion efficiency, measurement of static conversion efficiency and MPPT efficiency under fixed operating conditions has been specified by IEC and UL [1]-[6]. However, the method of evaluating the performance of PV-PCS with respect to the change in solar radiation taking into account the operating point which varies depending on the weather conditions and enviroment is not clear. Comprehensive evaluations of the PV-PCS against both static and dynamic irradiation changes is needed. In this paper, we investigated static and dynamic performance evaluation method of the PV-PCS by analysis which can comprehensively evaluate solar power generation system easily.

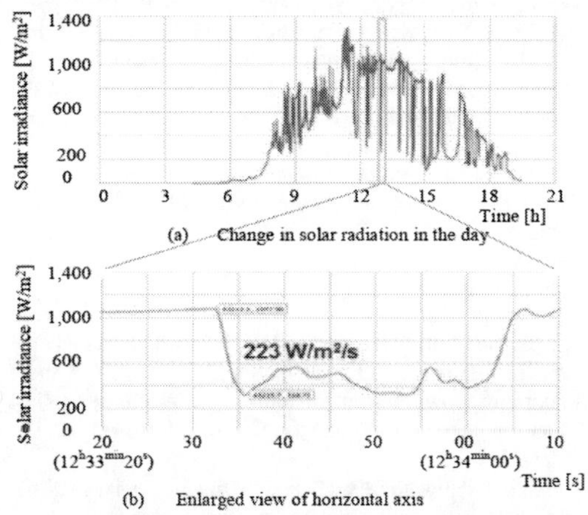

Fig. 1: Measured solar radiation at FREA (2016.8.11)

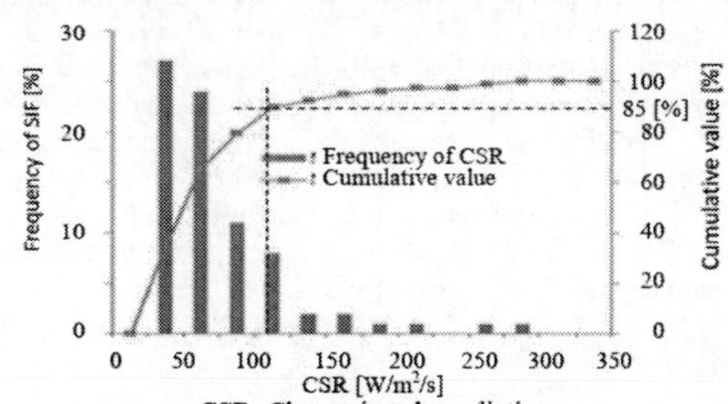

CSR: Change in solar radiation

Fig. 2: Histogram of frequency of change in solar radiation

2 VARIATION AND FLUCTUATION SPEED OF SOLAR RADIATION

The Tsukuba meteorological observation site, where AIST is located is cited as one of the good points of sunshine conditions in Japan. Looking at the past statistics of solar radiation measured data, even in such a

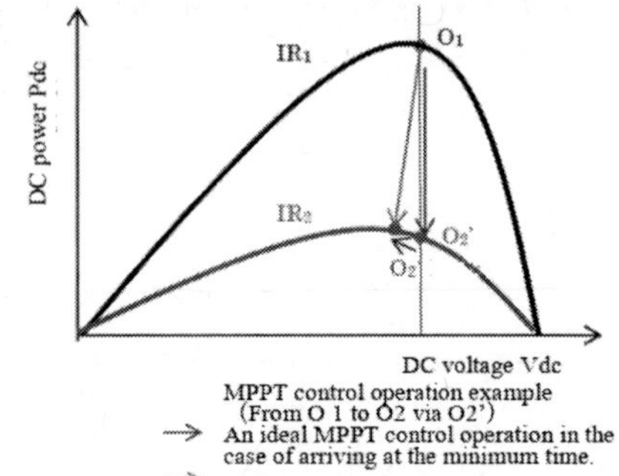

Fig. 3: MPPT behavior at change in solar radiation amount

place, about 60% of amount of solar radiation in a year is fluctuating greatly as with following mentioning data [7]. Fig. 1 (a) shows the solar radiation data [W/m^2] measured every 1 second sampling of a cloudy day at the Fukushima Renewable Energy Laboratory of AIST (FREA) in August 11th 2016 and their time expanded waveform is shown in Fig. 1 (b). A fluctuation speed is reaching at 223 [W/m^2/s]. The falling fluctuations of the solar radiation are generally faster than the rising. Fig. 2 shows their histogram of fluctuation speed [W/m^2/s]. In the figure, their rising and falling in a range from 0 to ± 350 [W/m^2/s] are displayed in percentage. On the measured day, About 85% of the fluctuation speed of the solar radiation amount is less than ±125 [W/m^2/s], and faster fluctuations are 15% remaining. For PV-PCS used in PV systems where change in solar radiation occurs frequently, it is important to apply the PV-PCS with fast response equipped with appropriate dynamic performance characteristics in order to maximize the amount of power generation.

In the MPPT control of the PV-PCS, when the amount of solar radiation suddenly changes, a new maximum power point is found, for example by hill climbing method, by changing in DC voltages or DC currents.

Fig. 3 shows the PV characteristics of a solar cell, the horizontal axis shows the DC voltages, and the vertical axis shows the DC power. In the figure, A MPPT control operates as follows. When the amount of solar radiation suddenly drops from IR_1 to IR_2 ($IR_1 > IR_2$), the maximum power point moves from O_1 to O_2. Depending on the PV-PCS, for example, it reaches the maximum power point O_2 via the point O_2'. Transient losses occur until reaching the new maxim power point O_2. Therefore, how fast the PV-PCS reaches to the new maximum power point after changing in solar radiation is most important. This depends on dynamic performance characteristics of MPPT control.

3. PERFORMANCE EVALUATION TEST METHOD AND TEST CIRCUIT

3.1 Performance evaluation test method

In order to evaluate the dynamic performance characteristics under solar radiation fluctuations, we proposed an evaluation method based on the indicial response method, which is used for evaluation of conventional linear control systems, to obtain the dynamic performance characteristics of the PV-PCS [8].

Fig. 4 shows the step-change of solar radiation S_R as

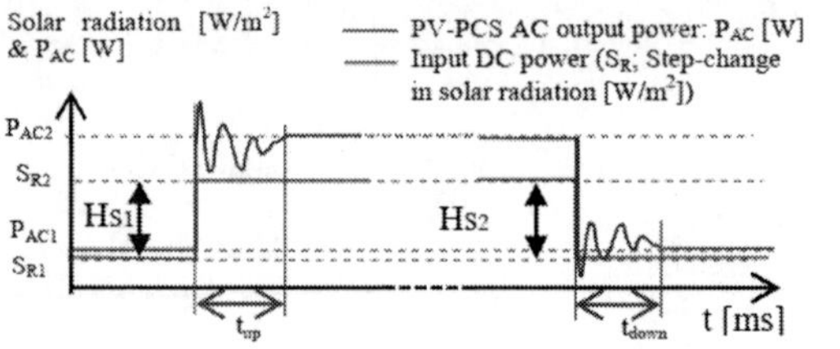

Fig. 4: Test method for evaluation of PV-PCS performance

input and example waveforms of AC output power P_{AC} of the PV-PCS at that time. As shown in the figure, the step-change in the amount of solar radiation is converted into step-change in generated power by a PV simulator that simulates an I-V characteristic of a solar cell. Against step-change in solar radiation, the new maximum power point is tracked by the MPPT control of PV-PCS. It is considered that the P_{AC} reaches in a steady-state with some delay time. The response of the P_{AC} was evaluated by measuring rising time (t_{up}) and falling time (t_{down}). Here, Hs1 and Hs2 are the step-up height and the step-down height of solar radiation and these are parameters in the response time evaluation.

3.2 HILS test circuit

We considered applying the proposed performance evaluation test method to many kinds of PCS. The tests using physical models require testing equipment having larger capacity than the models. They need much cost and spaces, and moreover much time for testing. On the other hand, the tests by digital simulation need to built detailed hardware models and control and protection systems and so on by software. It is not always possible to obtain such detail information from the manufacturer. Even if they can be got, detailed modeling involves difficulty, it takes a lot of time to get satisfactory analysis results. For this reason, we examined the tests combine with physical models and digital simulation. The proposed performance evaluation test method is a HILS (hardware in loop simulation) test method that simulates a test environment using a real-time digital simulator and performs a test by combining a physical model including PV-PCS and peripheral connecting devices. According to the test method, it is possible to carry out the performance evaluation tests of physical model in detail, even if their hardware is unknown and modeling is difficult.

Fig. 5 shows the HILS test system for performance evaluation tests of PV-PCS. We use the RTDS (real-time digital simulator) of RTDS Technologies Inc. [9] which is used generally as a real-time simulator. As for a solar cell of DC circuit of the PV-PCS, the PV array model prepared in RSCAD, which is an exclusive analysis software of RTDS, can be used. By designating solar radiation and temperature, I-V characteristics of a PV cell is created, and the output DC power P_{DC} is amplified by the DC amplifier with a predetermined setting gain and is connected to the physical model of PV-PCS (EUT to be tested). On the RSCAD, the output DC voltage V_D of the PV array model is short-circuited by a variable resistor R_{PCS} which is simulated the behaviors at performance evaluation of physical model of PV-PCS. The variable resistance R_{PCS}[Ω] is obtained from a following equation.

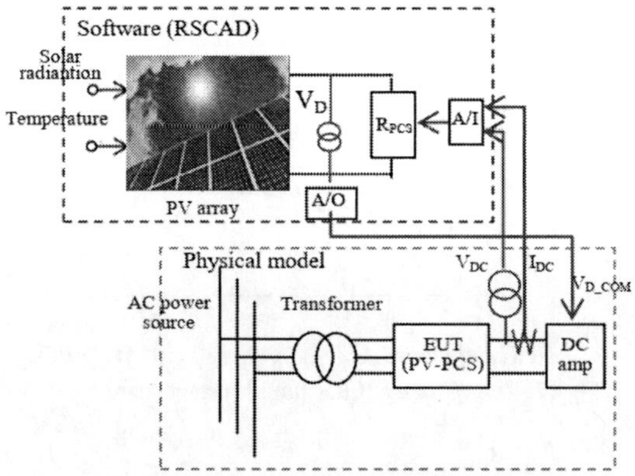

Fig. 5: Configuration of HILS test system for PV-PCS performance evaluation

$$R_{PCS}=V_{DC}/I_{DC} \quad \cdots\ (1)$$

V_{DC} [V] and I_{DC} [A] are DC voltages output and DC currents output of physical model of the DC amplifier and feedbacked to the RSCAD.

AC output voltage V_{AC} of the PV-PCS (EUT) is connected to an AC power source via an insulated transformer. The PV simulator is constructed of the PV array modeled by RSCAD and the DC amplifier for creating the DC power of the PV array. The amplified DC voltages are connected to DC input of the PV-PCS. Between the physical model and RSCAD, output signals V_D, V_{DC} and I_{DC} feedback signals are exchanged every computation time step of 50 μs, and real time analysis is performed in synchronization with both. The HILS performance evaluation test system is constructed by these components and input signals for evaluating performance characteristics use the step-changing command of solar radiation amount into the RSCAD mentioned above.

4. PV-PCS PERFORMANCE EVALUATION BY DIGITAL SIMULATION

In the paper, instead of constructing the HILS test system, the test method for getting the PV-PCS behavior characteristics and response against a step-change of solar radiation was carried out by digital simulation and the effectiveness of the test method was verified.

Fig. 6 shows the simulation model by the RSCAD. As the PV-PCS including a standard control circuit, a PV system model in the RSCAD was used. Any step-changes of solar radiation amount shown in Fig. 4 is input to the PV array model and the temperature set to 25℃ constant. For the I-V characteristics of a solar cell, a default value is used prepared in the PV system libraly. The maximum power point voltage (MPPT voltage) is selected by PV-PCS and the DC maximum power output is converterd to an AC power by PV-PCS and connects to an AC power source.

As the analysis conditions of the performance evaluation tests, the solar radiation amount of the PV array was varied between 50 and 1,200 [W/m^2].

5. SIMULATION RESULTS

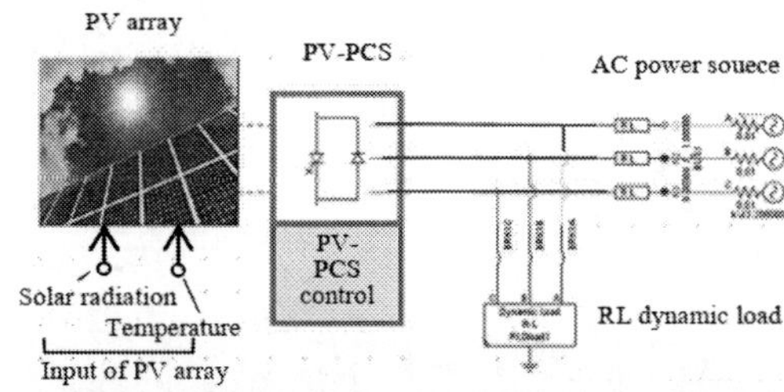

Fig. 6: Configuration of PV-PCS evaluation test circuit

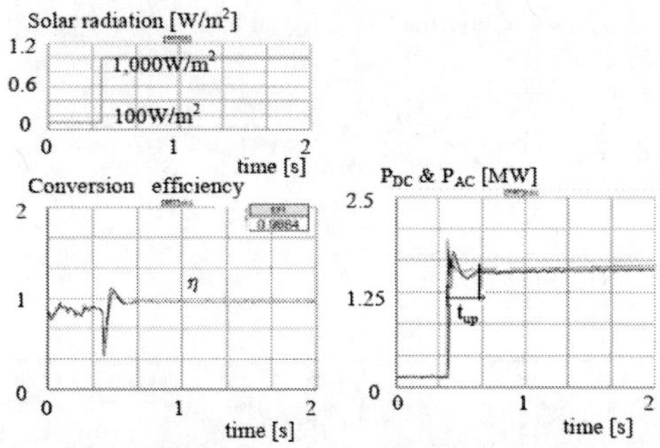

Fig. 7: Test results at step-up change in solar radiation

One of the analysis results when the solar radiation amount is step-up-changed from 100 [W/m^2] to 1,000 [W/m^2] is shown in Fig. 7. The figure shows the input waveform of solar radiation step-up-change, calculated conversion efficiency η, input DC power P_{DC} [MW] and output AC power P_{AC} [MW] of the PV-PCS. The conversion efficiency η is calculated from a following equation for each time step.

η= AC power of PV-PCS output P_{AC}[MW]/ theoretical DC input power P_{DC} of PV-PCS input determined by solar radiation [MW] $\cdots$ (2)

From the figure, in the steady-state of 100[W/m^2], P_{AC} = 0.11[MW], P_{DC} = 0.14[MW], V_{DC} = 1.70[kV], η = 0.88, and in the 1,000[W/m^2], P_{AC} = 1.68 [MW], P_{DC} = 1.74[MW], V_{DC} = 1.98[kV], η = 0.97. In this case, the measured result of rising time t_{up} of the Pac is about 300 [ms].

5.1 Characteristic analysis in constant solar radiation amount

The steady-state of AC output power P_{AC} [MW], input DC power P_{DC} [MW], DC voltage V_{DC} [kV], and DC current I_{DC} [kV] of the PCS are analyzed when the solar radiation amount is varied from 50[W/m^2] to 1,200 [W/m^2].

The results are shown in Fig. 8. The V_{DC} is almost constant but gradually increasing as the amount of solar radiation increases, and the I_{DC} and the P_{AC} are increasing nearly in proportion to the amount of solar radiation.

Fig.9 shows the analysis result of conversion efficiency η. It becomes higher as the amount of solar radiation increases. When it exceeds 500[W/m^2], the η is almost constant and then becomes nearly $\eta \fallingdotseq 0.97$. As the solar radiation increases, the V_{DC} increases slightly. Consequently, the η also increases. When the V_{DC} exceeds a certain DC voltage, the V_{DC} saturates, and the η becomes almost constant. Since the V_{DC} and the I_{DC}

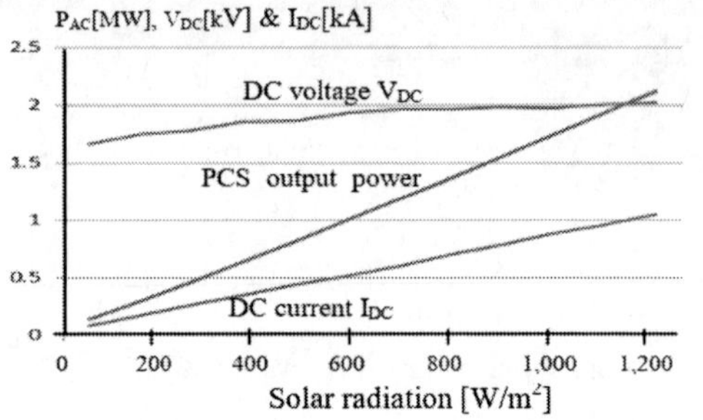

Fig.8: Solar radiation vs. V_{DC}, I_{DC} and P_{AC} in steady-state

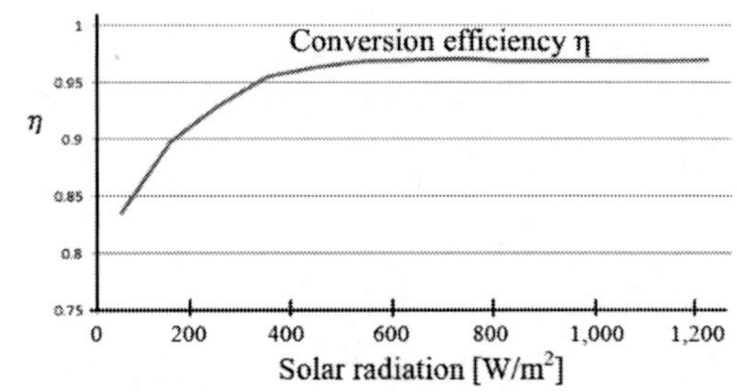

Fig.9: Solar radiation versus PCS conversion efficiency at steady-state

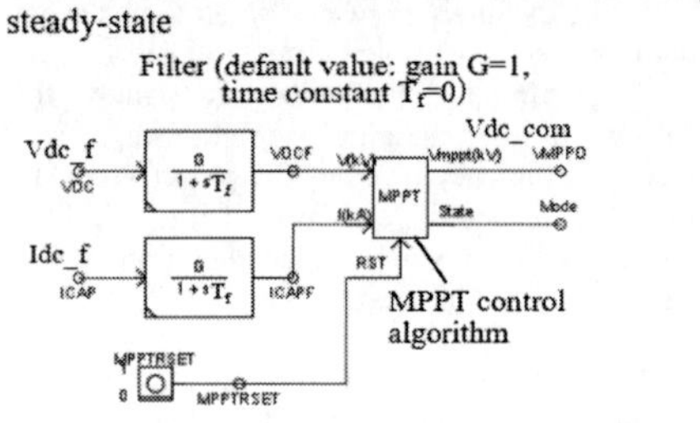

(a) MPPT control brock

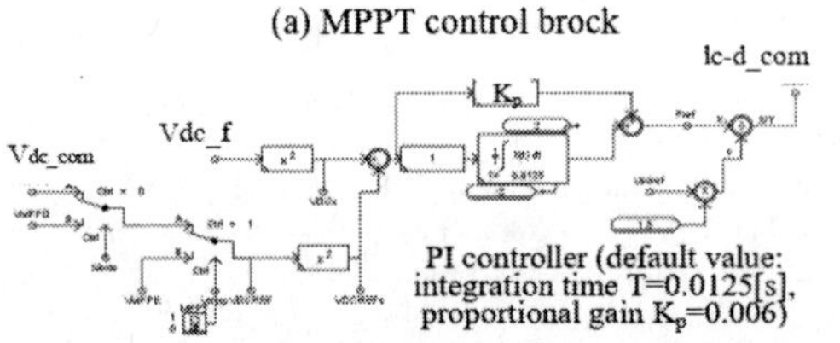

(b) DC-AVR control block

Fig.10: Control block of MPPT and DC-AVR for PV-PCS

fluctuate by searching control of the MPPT, the η also fluctuates (see analysis example in Fig. 7).

Similarly, the performance evaluation analysis was carried out by changing in the step height Hs of solar radiation amount and response times of P_{AC} were measured and evaluated.

5.2 Performance analysis in step change of solar radiation amount

In the analysis on response time, the solar radiation was step-changed when a filter time constant for DC voltage and DC current detection in the MPPT control and constants of a PI controller of DC-AVR were changed.

Fig. 10 shows (a) MPPT control block and (b) PI controller of DC-AVR. Both are prepared in the RSCAD.

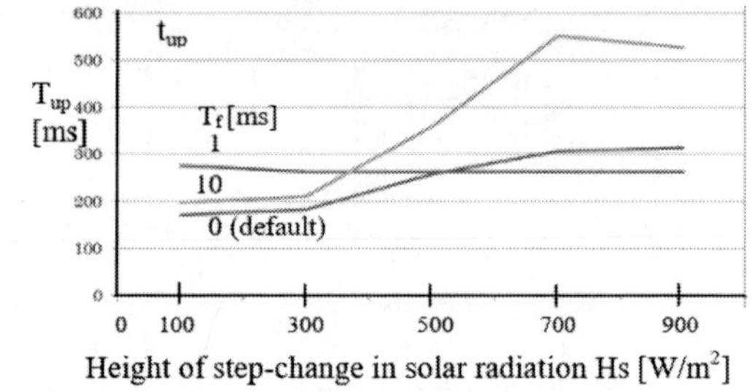

Fig.11: Rising time at filter time constant change

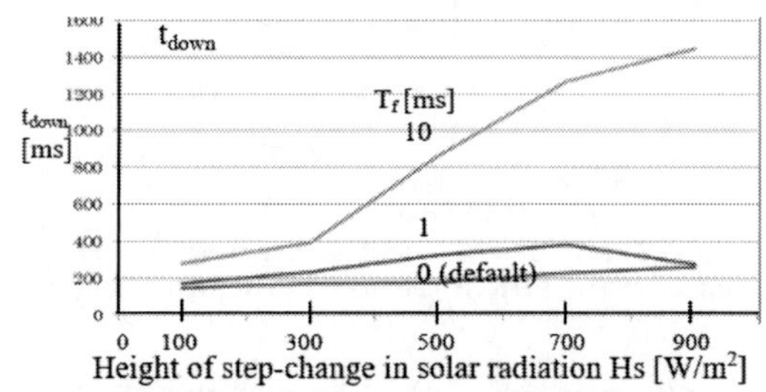

Fig.12: Falling time at filter time constant change

In the MPPT control block, Vdc_f and Idc_f are feedback signals of DC voltage and DC current and are guided to the MPPT control block via filter circuits. T_f shows a filter time constant and is a parameter. Output DC voltage of the MPPT becomes a DC voltage command value Vdc_com and is connected to the DC-AVR control block. In the DC-AVR control block, a DC current command value of d-axis component Idc-d_com is created from the DC voltage command value and DC voltage feedback value Vdc_f by the PI controller. Constants of proportional gain K_p and integration time T of the controller are as parameters. These parameters influence on the PSC dynamic performance.

Here, the solar irradiance amount change for evaluating the response time is based on 100W/m^2, and is stepped up from 100W/m^2 when step-up and is stepped down to 100W/m^2 when step-down. At this time, the step height of solar irradiance amount is being varied.

(1) Step Response in change of VI detecting filter time constant

Fig. 11 and 12 show when the time constant T_f of V_{DC} and I_{DC} detecting filter, which are the input signals of the MPPT control, is changed to 0[ms], 1[ms], and 10[ms]. In addition, 0[ms] indicates the default value. Fig. 11 is the case of step-up and Fig. 12 is step-down. As the T_f increases, both the t_{up} and the t_{down} become long as same as the step-response of conventional linear control systems. Also, they becomes long when height of step-changing Hs in solar radiation amount increases. As the filter time constant becomes the operation delay of the MPPT control, they are preferable to be as small as possible. In the analysis, the default value of the T_f is zero, but since the influence on the MPPT control operation by noise in the detecting signals was not seen, it seems that there is no problem even if the filter time constant is zero. Due to the nonlinearity of the MPPT control algorithm, the response time does not increase linearly according to the magnitude of the filter time delay. The response time becomes larger as Hs becomes higher and the response time of t_{up} and t_{down} in default

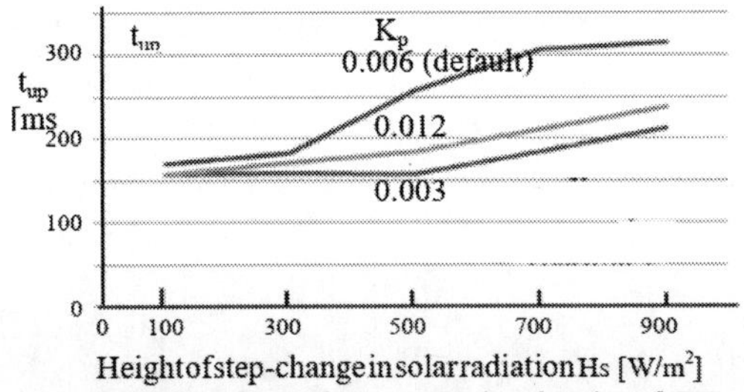

Fig. 13: Rising time when proportional gain of DC-AVR is changed

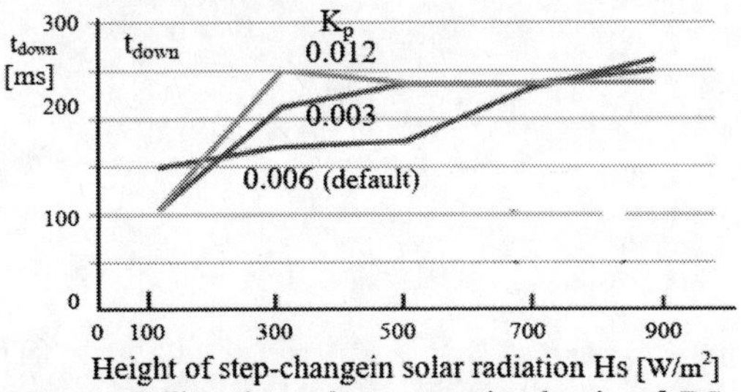

Fig. 14: Falling time when proportional gain of DC-AVR is changed

value is within about 300ms for Hs from 100[W/m²] to 900[W/m²].

(2) Step response in change of proportional gain of DC-AVR

Fig. 13 and Fig. 14 show the response time analysis results of step-up change and step-down change in solar radiation, when the proportional gain K_p of the PI controller of DC-AVR is changed to 0.003, 0.06 and 0.012. As for the t_{up}, they are being smaller even if the proportional gain is larger or smaller than the default value (K_p = 0.006), but as for the t_{down}, they are the smallest at the default value. These characteristics are due to nonlinearity of MPPT control. The response time becomes larger as Hs becomes higher. Considering both the t_{up} and the t_{down}, as the differences in response time against the different proportional gains are small, the default value is seemed to be a reasonable for the PV system.

(3) Step response in change of integration time of DC-AVR

Fig. 15 and 16 show the results of the response time analysis for the solar radiation amount step-change when integration time T of the DC-AVR is increased to 0.0125s, 0.025s and 0.05s. In addition, 0.0125s indicates a default value. Fig. 14 shows the t_{up} and the t_{down} in Fig. 15. The response becomes longer as the integration time becomes longer as it is said theoretically. The default value shows better results among parameter changes. Similarly to the results so far, the response time becomes larger as Hs becomes higher. And although a difference is small, the rising time t_{up} is being longer than the falling time t_{down}. it seems that the MPPT algorithms and nonlinearity of the MPPT control are causing.

(4) Consideration of performance evaluation result

As a result of performance evaluation analysis using PV-PCS prepared with the RSCAD, the conversion efficiency was about η = 0.97 at the standard solar radiation dose of 1000[W/m²]. On the other hand, from

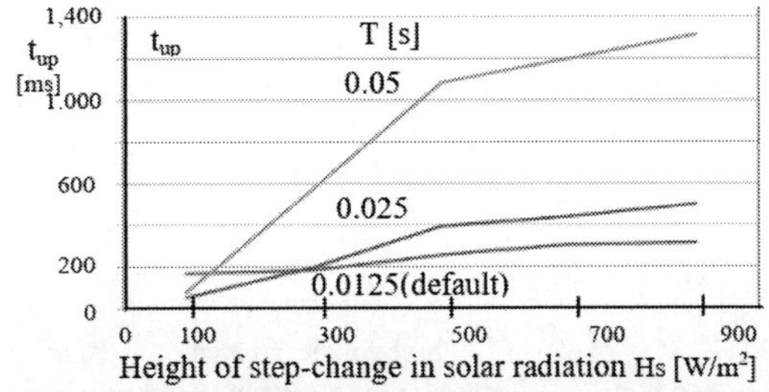

Fig. 15: Rising time when integration time of PI controller is changed

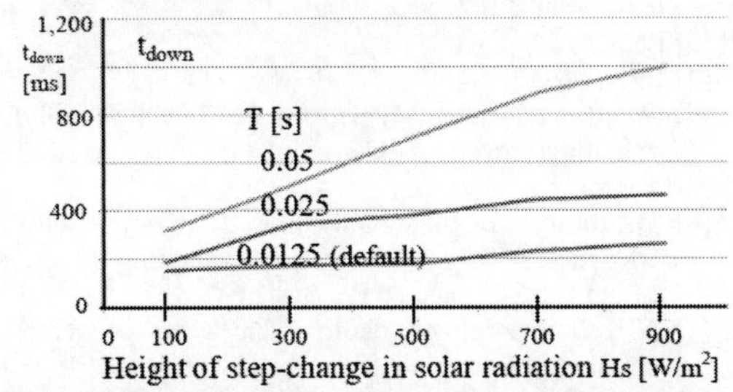

Fig. 16: Falling time when integrationl time of PI controller is changed

the analysis results shown in Fig. 11 and Fig. 12 for example, the rising time t_{up} and the falling time t_{down} at the default value of the PV-PCS in RSCAD were within about 300ms when the height of step-change Hs in solar radiation amount is within ±900 [W/m²]. As 85% of the fluctuation speeds is within ±125 [W/m²/s] according to the example of the solar radiation fluctuation measurement shown in Fig. 2. The PV-PCS is thought enough to follow to such changing speeds because the speed of step-change in solar radiation is much faster than 125[W/m²/s].

In the paper, the conversion efficiency was evaluated at the standard solar radiation amount of from 50[W/m2] to 1,200[W/m2] and the response time of rising and falling was evaluated at height of step-changing in solar radiation within ±900 [W/m²]. In conclusion, it is important to evaluate considering both the operating conditions based on the place and the environment where the PV system is installed.

6. SUMMARY

It is necessary to promote the introduction of further PV generation systems. Unlike in the past, it may be considered to install them in a place where solar radiation conditions are not necessarily good. In order to maximize the amount of PV generation installed in such a place, it is important to use PV-PCS having the characteristics suitable for the installation site operating conditions. In the paper, an evaluation method for dynamic performance of PV-PCS was examined.

We proposed a dynamic performance evaluation test method of PV-PCS by digital simulation. The method is similarly a step-response evaluation which has been used for conventional linear control systems. We analysed dynamic behaviors of PV-PCS in step-changed solar radiation inputs. For evaluation and verification of the method, it was applied to the PV system prepared in RSCAD. From the results, it is considered that the PV system operates efficiently for solar radiation fluctuations as shown in Fig. 1. And the tests were satisfactory.

From performance evaluation tests, it is important to evaluate in consideration of the weather conditions of the installation site and the system environment.

In future, we will apply the proposed test method to the HILS test system and evaluate various PCSs.

REFERENCES

[1] IEC 61683 「Photovoltaic system – Power conditioners – Procedure for measuring efficiency」
[2] IEC 62891 「Overall efficiency of grid connected photo- voltaic inverters」
[3] EN50530
[4] IEC TS63156
[5] W. Bower, et. al., "Performance Test Protocol for Evaluating Inverters Used in Grid-Connected PV Systems"
[6] B. Brooks, et. al., "Guideline for the use of the Performance Test Protocol for Evaluating Inverters Used in Grid-Connected PV Systems"
[7] NEDO meeting on photovoltaic power generation technology development and related projects," R & D of autonomy enhancement type photovoltaic power generation system, (2006)
[8] H. Konishi, et. al. "A study of efficiency evaluation method for PV-PCS with fluctuated irradiation", 7-025, IEEJ annual meeting 2017
[9] RSCAD– Power system simulator software, https://www.rtds.com/the-simulator/our-software/about-rscad/

BIOGRAPHY

Hiroo Konishi (Dr.)

He received the M.S. degree in electrical engineering from Osaka University in 1972. He joined Hitachi Research Laboratory, Hitachi Ltd. in the same year and he worked in the research and development of control and protection system of HVDC then. In 2006, he worked in the development of large-scale solar systems for Solar Project Headquarters of NTT Facilities Inc. Presently; he has been working for advanced industrial science and technology/Fukushima renewable energy research center since 2014. He interests in research and development of power electronics applications to power systems, especially renewable energy systems, and interests in analysis and simulation technology in transmission and distribution systems, too.
He received the Ph. D degree from Osaka University in 1989.
He is a Fellow of Institute of Electrical Engineers of Japan.

Jun Hashimoto

Since joining AIST in 2017, Jun has contributed to PV research, especially PV system performance testing and characterization. More recently he has expanded his research to study smart grid systems, especially advanced inverter interoperability for PV and energy storage. Jun is currently involved in IEC TC82 WG2, 3, 6, and 7 to develop international standards as an expert on PV modules, systems, and CPV.

Kenji Otani

He currently leads Energy Network Team of Renewable Energy Research Center at Fukushima Renewable Energy Institute, AIST. He has researched the system integration technologies of PV and hybrid power systems.

REVIEW OF GUIDELINES FOR PV SYSTEMS PERFORMANCE AND DEGRADATIONS MONITORING

Basant Raj Paudyal[1], Anne Gerd Imenes[1], Tor Oskar Saetre[1]
[1]University of Agder, Faculty of Engineering and Sciences, Grimstad, Norway
[1]Tel: +4737233103; e-mail: *basant.paudyal@uia.no*

ABSTRACT: The recent growth in the PV system deployment and the higher scale of their production make it necessary to pay more attention towards the performance of those systems. System monitoring will provide the status of the systems showing possible degradations and any impending system failures. It also helps with maintaining system quality and ensuring optimal performance of the system for its lifetime. COST Action PEARL PV aims to formulate such monitoring guidelines for the PV systems applicable for all types of PV technologies, depending upon the end use and duration of the monitoring need. The objective of the present paper is to quantify and provide a review of the guidelines and standards for monitoring of PV systems performance and degradation in practice, and also some results of preliminary work done regarding important information needed to be included in the monitoring guideline.
Keywords: PV Systems, Performance, Degradation, Monitoring

1 INTRODUCTION

Solar PV is one of the renewable energy technologies which converts sunlight into electricity. The PV industry and the research domain have had a remarkable global progress regarding the performance and installation volume of PV systems in the last few decades. With the laboratory efficiency crossing 40% [1] together with considerable rise in the outfield module efficiencies, PV has become an integral producer of electrical power globally. Performance monitoring of the PV systems provides a unique and adept solution to understanding the root causes of degradation while also indicating the possible measures to improve the performance of the system. In the present context, with over 400GW of global PV installation, the PV monitoring mechanisms provide a sense of security to investors, researchers and system owners alike.

Different types of PV modules are being designed based on the different materials and characteristics. The performance monitoring of such systems then becomes a necessity since they have different spectral response, sensitivity to temperature, shading, and other parameters in the outdoor environment. In this regard, monitoring can provide the essential information about the operation of such systems and could also provide insight on the necessary measures to be taken to improve their performance. Monitoring of PV systems may draw substantial costs depending upon the degree of accuracy of measured data, so the design of it must be considered carefully. For the data, appropriate performance parameters need to be chosen and their measurements should be continuously saved and updated. Depending upon the requirements, the values of different parameters could be logged every second, every minute or every hour resulting in accumulation of a huge amount of data. Standard sampling interval of measurement has been defined by the IEC61724 standard, but the monitoring frequency could be adjusted depending upon the need and budget of the user. It is better to evaluate the data that has incorporated seasonal variations as this gives a better idea of system performance along with weekly or monthly averages.

The present paper comes out as a preliminary work done within COST Action PEARL- PV, Work Group 1 PV monitoring with the objective of developing generally acceptable approaches and guidelines for the collection of data about the performance of PV modules and systems using advanced monitoring through various techniques [2]. The paper includes the literature review of the different performance parameters of PV systems and the monitoring guidelines formulated globally.

In the present paper, the authors review the monitoring guidelines formulated and implemented by various institutions. They review the current state of the art on monitoring guidelines and based on the initial feedback from the experts on the field, summarize the results on the degree of requirement of the parameters for monitoring.

2 PV PERFORMANCE AND DEGRADATION

2.1 Performance and loss factors

The performance ratio (PR) is the most important quantity to be measured for evaluating the overall behavior of a PV plant. The performance of the power plant depends on several parameters including the site location, the climate and several loss mechanisms[3]. The overall losses of a system are categorized broadly into two types of loss mechanisms, Capture Losses (LC) and System Losses (LS). Captures losses are the losses which affect the performance of PV systems with conversion of available energy into useful/output power of PV systems e.g. module and ambient temperature, obstruction or attenuation of incident radiation, module mismatches, nonoptimal MPP tracking and the parasitic resistances present in the cell and module level. System losses on the other hand are those types of losses that happen within the system and are caused by system components like DC to AC conversion losses, wiring losses and transformer conversion losses[4].

2.1.1 Capture losses

Capture losses are those types which occur at the DC side of a PV conversion chain and are attributed to operating temperature, tilt angle of modules, shading, temperature dependency of PV modules efficiency and most importantly dependence on solar irradiance level. While the PV array itself encounters some losses like MPP tracking error, module parameters dispersion (mismatch), wiring losses and aging[5]. These can be broadly classified into thermal capture losses and other miscellaneous losses[6].

2.1.1.1 Thermal capture losses

Module efficiency of crystalline modules at standard test condition (STC) is defined at 25°C temperature. Depending on the various parameters like wind speed and the type of mounting of the PV modules, there could be a temperature rise of the modules with respect to the ambient of 20–40°C at 1000 W/m^2. The corrected temperature DC output power, at real working irradiance and a standard temperature of 25°C, can be obtained by the simulation model where the input are: the monitored irradiance and a fixed temperature of 25°C[6]. The normalised capture losses can be then determined by the following expression:

$$Lct = Ya\ (G, 25°C) - Ya\ (G,Tc) \qquad (1)$$

where Lct are the thermal losses, Ya (G, 25°C) is the normalised energy yield at real working irradiance and 25°C of temperature, and Ya (G,Tc) is the array yield at real working irradiance and real module temperature Tc. Normalized thermal losses can give us the amount of power losses due the rise of temperature above 25°C.

2.1.1.2 Miscellaneous capture losses

All the inherent losses such as non-uniformity of the irradiance, low irradiance, module/cell mismatch, MPPT errors, wiring, string diodes, dirt accumulation and losses caused by faulty operation at the DC side such as faulty strings, faulty modules, partial shadowing and short circuit of modules are grouped in this category[7].Since the inputs to the simulated model of the PV plant are the effective irradiance and module temperature, the simulated capture losses can be calculated by the following expression, after the calculation of the reference yield:

$$LC = Yr(G, Tc) - Ya(G, Tc) \qquad (2)$$

where LC are the capture losses, Yr(G, Tc) is the measured reference yield and Ya(G, Tc) is the array yield at irradiance and temperature at the site. Finally, the reference miscellaneous capture losses are given by:

$$LCm = LC - Lct \qquad (3)$$

where LC is the capture loss and Lct is the thermal capture loss of modules.

2.1.2 System losses

These losses are mainly referred to the power conditioning units, which are the DC-DC converters responsible to extract maximum available power from the PV plant, the wiring losses, transformer conversion losses as well as the DC to AC conversion losses. It can be obtained by the difference between the array yield and final yield.

$$Ls = Ya - Yf \qquad (4)$$

The losses in a PV system are categorized in Figure 1.

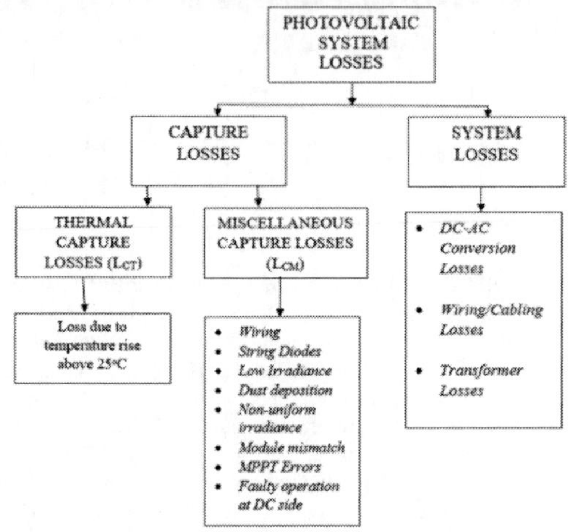

Figure 1: Categorization of losses in PV systems.

2.2 PV system performance

The reference yield is an expected yield from a system considering the energy of incoming light and modules' nominal efficiency whereas the final yield is the total yield after encountering different losses i.e. electricity delivered to the AC. The losses in plant are categorized broadly in two sections namely capture losses and system losses. Capture losses incorporate the various conversion losses in the array due to environmental parameters like irradiance, temperature, shading, soiling and design faults in DC side whereas the system losses include losses in power conditioning units, wirings and transformer conversion.

The power output of a PV system depends primarily on the latitude, outdoor operating conditions as well as the system design. Irradiance and module temperature are known to be the major operating conditions to determine the PV output, where module temperature is a function of solar irradiance, ambient temperature, other thermally related conditions such as wind speed and direction, array mounting and thermal behavior of the module [8]. Other influencing factors are the incident spectrum, angle of incidence of the irradiance and system level losses including soiling, shading and power conditioning losses. These factors in inclusion with the site-specific climate, location and the loss mechanisms are used to calculate a parameter which is the most widely used quantity for evaluation of healthiness of a PV system called Performance Ratio (PR) [4], which is defined as the ratio of final yield to reference yield.

$$PR = \frac{Y_f}{Y_r} = \frac{P_{DC}/P_o}{H/G} \qquad (5)$$

Production and performance of grid connected systems are often measured in Performance Ratio (PR) and sometimes annual energy yield with both of them incorporating the degradation rates within the guarantee periods [8]. Energy yield may have some benefit of consideration in weather conditions, as the measured output can be compared with the predicted output under the similar weather. This requires constant monitoring of the parameters like solar irradiation, ambient and module temperatures while reliable simulation software should also be used.

Performance ratio in comparison is a simpler alternative as it takes the different losses in account and while the irradiation needs to be measured adjacent to the array, it gives a better output response to irradiation variation. But the seasonal change of ambient temperature will have a simultaneous impact on the module temperature. As the ambient temperature during winter gets colder this drops the temperature of the module, for the same amount of irradiance the reduced temperature will help in increased yield and hence the performance ratio. The performance characteristics used to calculate the PR is based on STC, which translates a module temperature of 25°C. Since the modules operate at higher temperatures in normal conditions, it is possible to do a temperature correction of the power to STC. This calculation of the temperature corrected performance ratio (PR) is based on the measured back panel temperature.

$$\tau_{corr} = (T_{module} - 25°C) \cdot \gamma P_{MPP} \quad (6)$$

Here, τ_{corr} is the temperature correction coefficient, T_{module} is the temperature on the backside of the module, and γP_{MPP} is the temperature coefficient of the efficiency. Temperature coefficients for each type of module technologies can be found in corresponding data sheets provided by the manufacturer. Then using the temperature correction coefficient, it is possible to find the temperature corrected performance ratio as:

$$PR_{TC} = PR/(1+\tau_{corr}) \quad (7)$$

2.2 Current trends in PV system performance

We already discussed how PR has been the main parameter to determine the healthiness of a solar PV system. The trend of using PR as a reference of a system performance started as far back as in the 1980's. It has continuously rose from 0.5 to 0.65 in the late 1980s to 0.7 in the 90s to more than 0.8 nowadays [4]. Early PV systems did not generate the expected energy yield due to various constraints related to poor systems design and installation coupled with inefficient devices [9].

Monitoring for one year with measurement of performance ratio and different losses of a 171.36 kWp system in Crete, Greece found that the PV system supplied 229 MWh to the grid during 2007, ranging from 335.48 to 869.68 kWh. The final yield (YF) ranged between 1.96 to 5.07 h/d, and the performance ratio (PR) varied between 58 to 73%, with an annual PR of 67.36% [10]. Meanwhile, one-year measurement and performance analysis of two real-life grid connected PV power plants in Sardinia Italy [11], which receives more than 1700kWh/m2 irradiation per year provided very encouraging results. A ground mounted system of 395.61kWp had a performance ratio of 87.3% and another building integrated system of 1042.29kWp had a performance ratio of 83.2%. Interestingly, the measured performance values were better than the expected PR simulations in the design phase. The performance analysis of a 190 kWp grid interactive solar PV system in Punjab India, the parameters like final yield, reference yield and performance ratio, were found to vary from 1.45 to 2.84 kWh/kWp-day, 2.29 to 3.53 kWh/kWp-day and 55–83% respectively. The annual average performance ratio was found to be 74% and the average measured annual energy yield was found to be 812.76 kWh/kWp, which was very close to the average annual predicted energy yield found to be 823 kWh/kWp using PVSYST. Regular monitoring of the system found the total system losses due to irradiance, temperature, module quality, array mismatch, ohmic wiring and inverter to be 31.7% [12]. In a study of crystalline silicon modules in four different climates [13], the result showed that the failure mechanisms of the modules are also location dependent along with weather. Interestingly, the same study found that the highest performance degradation rate was observed in the polar/alpine climate, which could be caused by high wind and snow loads. Regarding the performance of PV systems in higher latitudes, a study [14] of a 236kWp system in Piteå, Sweden reports a performance ratio of 0.850, specific production of 897 kWh/kWp, and an annual yield of 208 kWh. But the measured results showed inverter efficiency of 89.1% against the 96.3% as suggested by the manufacturer which was unexplained. Similarly, one of the first grid connected PV systems in Norway, installed in 2011 delivered specific yield reaching 950 kWh/kWp and PR approaching 0.8 in 2014[15]. Improvement in PR was evident with the increased operational experience and reduced system downtime.

2.3 System degradation

Degradation of PV modules in a system is known by the gradual deterioration of the initial characteristics which may hamper its operation within the acceptable limits and which is caused by the operating and environmental conditions. Production of a degraded module may not completely stop as it may still produce electricity from sunlight, although with the significant reduction from the initially labelled value [16]. But when the degradation rate exceeds some predefined threshold level, it could be problematic [17]. From a study of around 2000 degradation rates, derived from the measurements of individual modules or entire systems, a median of 0.5% degradation per year was found [18]. When a module's output power reduces below 80% of the initial power during the service period, the module is said to be degraded but the degraded module may not last the entire lifetime and to wait for the module to be tested for degradation after life time may be unacceptable for the PV power plant. Most of the modules go through the wear out scenario, which is the basis of best case yield analysis and determines the cost efficiency of well operating modules [3]. So, the degradation of modules can be classified in short term, medium term and long-term categories. The PV module performance can be degraded due to several factors such as: temperature, humidity, irradiation, and mechanical shock [19, 20]. Each one of these various factors may induce one or more types of module degradation such as corrosion, discoloration, delamination and cell cracks.

The IEA-PVPS task 13 report [3] gives an outline presented in Figure 2 of the major identified degradations of the crystalline silicon modules. There are different failure modes in the lifetime of a PV module. The early life failures appearing at the beginning of the working period are called the infant failures, the one occurring at the mid-years are called the mid-life and the failures occurring at the end of working period are called wear out failures[3].

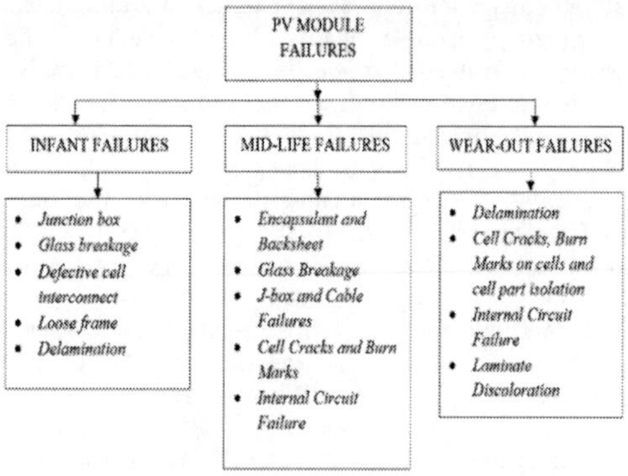

Figure 2: Major identified degradation modes of Crystalline PV modules [3]

2.4 Failures in PV systems

The PV array is the main component of a PV power system which will be affected by the degradation and failures. The PV modules undergo different modes of failures which is discussed in brief below.

2.4.1 Backsheet adhesion loss

The backsheet of a PV module provides protection of the electronic components from external factors as well as shielding from high DC voltages. This failure can be caused by delamination leading to the exposure of active electrical components which would result in insulation faults presenting safety concerns[21]. The backsheet may be constructed with various materials like glass, polymer and metal foils, and each material is susceptible to their own failures. Glass breakage can cause a failure and this could produce an electric arc. When this happens in conjunction with bypass diode failure, there's a possibility of a fire. Metal foils provide more robust electrical insulation layers in order to prevent modules from being charged at the system voltage. Multiple layer polymer laminates have multiple interfaces behaving differently in response to heat, thermal cycling, mechanical stress, humidity, UV light and others. There's a chance of bubble formation which is a more serious issue when it is formed around junction box and module edges compared to other parts[3].

2.4.2 Bubbles

Degradation caused by bubbles formation is much similar to delamination but in this case, the loss of EVA adhesion only affects a small area and the surface gets swelled after loosing the adhesion. The bubbles at the backside of modules are formed as a result of backsheet adhesion loss, which is generally due to chemical reactions that emit gases trapped in the PV module. When this happens on the back side of a module it causes difficulty for heat dissipation in cells, increasing their overheating and reducing then their lifetime[3]. Bubbles appearing in the center are not as harmful as those which are formed in the area around the junction box which may cause the box to become loose, thus adding stress on components and potentially breaking them. Bubbles on the module edges on the other hand may provide a pathway to liquid penetration during rain or dew formation, thus providing a direct electrical pathway to ground, which may be a serious safety concern. Their formation can be due to poor adhesion of the cell caused by the high temperature [20]. Bubbles located on the module front side can produce a reduction of the radiation reaching the module since they can cause a decoupling of the light and increase reflection[20].

2.4.3 Burn marks

These are one of the most common failures observed in Silicon modules. This failure is associated with some module parts becoming very hot because of localized heating caused by reverse current flow. This happens due to some cells operating in reverse bias by partial shading or cell cracks[3], solder bond or ribbon breakage failures caused by thermal fatigue. These marks can be seen through visual inspection but infrared imaging under illuminated or partially shaded conditions can be used to identify if the module needs to be changed or not[3].

2.4.4. Cell cracks and breakage

Cell cracks are failure modes which can be encountered at every lifetime level of the PV module. Silicon cells are prone to breakage during the manufacturing process due to its brittle nature and can avoid inspection from human eye. The stringing process of solar cells is known to have high risk of creating cell cracks[3]. These faults can be generated during packaging or transportation due to mishandling and vibrations. Small cracks or microcracks in cells due to thermo-mechanical stresses, always pose a risk of developing into longer and wider ones during the operation[22, 23].

2.4.5 Corrosion

Moisture penetration through lamination edges in PV modules is known to cause corrosion resulting in increment of electrical conductivity of the material [24]. Corrosion is known to degrade the adhesion between cells and metallic frame and attack the metallic connections of PV cells causing a loss of performance by increasing leakage currents[16]. In [25], the impact of humidity and temperature on PV module degradation is studied thoroughly. A accelerated testing named 85/85 (T = 85 ^{0}C/RH = 85%) was carried out according to the IEC 61215 standard. It was found out that corrosion appeared after 1000 hours of exposure of PV module under 85 ^{0}C and 85% relative humidity.

Authors in study [26] argue that corrosion and discoloration are the generally prevalent modes of photovoltaic modules degradation. Sodium contained in the glass which is reactive with moisture is a major factor of the corrosion of PV modules' edges. Sodium contained in the glass which is reactive with moisture is a major factor of the corrosion of PV module edges[27] while faster silicon PV module degradations are caused by oxygen which is the main factor of the corrosion of silicon junctions[28].

2.4.6 Defective bypass diode

Bypass diodes are integrated into the PV module in parallel to certain number of cells. The main role is to reduce the power loss and avoid reverse bias caused by partial shading on PV modules. Bypass diodes avoid reverse biasing of single solar cells higher than the allowed cell reverse bias voltage of the solar cell[3]. If a cell is reversed with a higher voltage than it is designed for, the cell may evolve hotspots[29] that may cause browning, burn marks or even fire. Bypass diodes are Schottky diodes which are very susceptible to static high voltage discharges and mechanical stress and this demands a careful handling[3].

2.4.7 Delamination

Delamination is also a type of adhesion loss which occurs between the encapsulating polymer and the cells or between cells and the front glass. Delamination is easily visible but if it cannot be identified visually, pulse and lock-in thermography methods may be used while X-ray tomography and ultrasonic scanner can also be used[3]. Different factors like UV, moisture and temperature affect the durability of lamination interfaces in a PV module. Study [3] points out that new pathways and subsequent corrosions following delamination reduce module performance but do not pose a safety threat, study [20] argues delamination can be major problem in the long term because it increases the light reflection and water penetration inside the module structure. Delamination is known to be most severe when it occurs on the edges of the module because, in addition to the degradation of power, it causes electrical risks for the module and the entire installation. Delamination is more frequent in hot and humid climates causing moisture penetration in the module and therefore resulting in various chemical and physical degradations such as metal corrosion of the module structure[16]. Delamination may also be caused by salt accumulation and moisture penetration into the PV module.

2.4.8 Disconnected cells and string ribbons

Poor soldering between string interconnect and cell interconnect ribbon is the main reason for the disconnections. Hot spots by partial shading, stresses during transportation, thermal cycle or repeated mechanical stress during long term operations, force weak ribbon interconnects to break[21]. Short distance between cells develops this kind of failure resulting in short circuited or open circuited cells and resistance increment[30]. These failures can be identified by electroluminescence, infra-red imaging, ultraviolet imaging or the signal transmission methods[3].

2.4.9 EVA discoloration (yellowing and browning)

Discoloration is a widely reported defect in solar modules which usually results in a degradation of the encapsulant module, EVA (Ethylene Vinyl Acetate) or adhesive material between the glass and the PV cells. Module discoloration is a change in color of material which turns yellow and sometimes brown. This is noticeable during visual inspection[16, 31]. It modifies transmittance of light reaching PV cells and therefore the power generated by the module is reduced. Main causes of discoloration are encapsulant quality and UV rays combined with water under temperatures higher than 50°C[32]. Sometimes insufficient adhesion between the cells and glass materials is also known to contribute[33]. Discoloration may appear at different and not necessarily neighboring zones of a PV module, meaning it is more dependent encapsulating polymers. Since the discoloration doesn't lead to any fatal failures, it can be termed as degradation rather than failure. When very high incident UV radiation of 4000 W/m^2 was applied, there was an increase in photosensitivity after 400 h exposure and an increase in transmissivity between 280 nm and 380 nm[34]. This is mainly due to the depression of the UV absorber which must protect PV cells of photodegradation. Besides, there appears a weak yellowing at the EVA films that causes a loss of the photovoltaic module power[34]. However, for a 1000 Wm^2 radiation, no change occurred in the wavelengths range between 280 nm and 380 nm after 500 h exposure. In UV tests on PV modules carried out under a 60 °C temperature, it was found that the discoloration of the encapsulant only appears when global UV irradiation reaches 15 kWhm^{-2} in a wavelength range included between 280 nm and 385 nm without exceeding a 250 Wm^{-2} exposure[25]. Research shows that the slow module degradations in the long term are linearly correlated to the exposure of PV modules under UV radiation[28]. From various experiments, it was found that the discoloration degrades the short-circuit current (Isc) of the PV module; this degradation of Isc may vary from 6% to 8% below the nominal value for a partial discoloration of the PV module surface and from 10% to 13% for complete discoloration[16]. The module discoloration degrades maximum power (Pmax) of the PV module as shown in Figure 3.

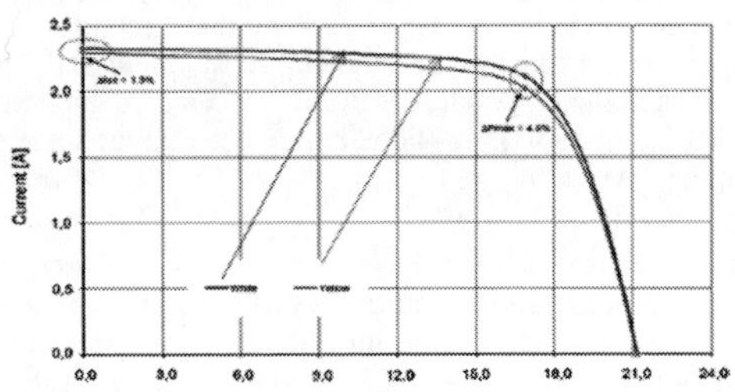

Figure 3: I-V Characteristics of normal and discolored PV modules [35].

2.4.10 Hot-spots:

A hot spot is an area of a PV module that has a very high temperature that could damage a cell or any other element of the module[16]. Hot spots result from the generation of less current in a cell than string current in modules, possibly due to reasons like cell failures, including partial shadowing, cells mismatch or failures in the interconnection between cells [36, 37]. In short circuit conditions, when a PV cell is defective, its voltage is reversed. This defective cell becomes both a load for other cells and a place of a relatively high thermal dissipation constituting thus a hot spot[16]. The risk of hotspots can be reduced by using shorter solar cell strings and by using advanced in-line quality control tools for cell testing[38]. Bypass diodes normally protect against this failure mode.

2.4.11 Junction box failure

Junction box failure is a failure mode generally seen in every type of solar module. Junction box is basically a box attached at the back side of PV modules which also houses a bypass diode to protect cells in the strings during partial shadowing and which protects the connection of cell strings of modules to outer terminals. Generally, poorly manufactured, junction box fixed with poor adhesion to backsheet, and moisture penetration in the modules leading to corrosion of connections are the failures observed in junction boxes[3]. Improper soldering and bad wirings can cause arcing in the box which may lead to fire.

2.4.12 Light induced degradation

Light induced degradation is a natural degradation phenomenon when crystalline silicon modules are initially exposed to light. It is caused by physical reactions causing permanent reduction in output power. It is reflected as a loss in the silicon solar cell efficiency, and seen as a reduction of the short circuit current and open circuit voltage of the solar cell[39]. Boron doped silicon solar cells are found to degrade in efficiency as much as 10% when exposed to light or when minority charge carriers are injected in the dark until a stable level of performance is reached, and this effect is related to the presence of boron and oxygen in different concentrations in the material. More recently it has been related to the presence of copper as well[39, 40]. On average, LID for crystalline silicon modules ranges from 0.5% to 3%, with some modules exhibiting a loss of up to 5%. LID is an unavoidable failure type and manufacturers take this into account by factoring in a 3% power loss (typically) during the first year of the module warranty as the rated module power is printed on PV modules after adjusting the expected standardized saturated power loss[3, 41]. To ensure that LID does not jeopardize the conclusions of the chamber testing, all PV modules in the PV Module Reliability Scorecard are light soaked for at least 40 kWh/m2 before entering the testing chambers[41].

2.4.13 Module/frame breakage

Glass breakage is a typical degradation mode of PV modules generally occurring during transportation, installation and maintenance at sites, and are easily detected by visual inspection [25]. Modules broken or with cracks may keep functioning correctly, however, the risk of an electrical shock and of a moisture infiltration increases[16]. Sometimes the cracks may be present from the manufacturing process due to thermochemical and mechanical stresses. These cracks, if not inspected and treated in time, are usually followed by other severe degradation types such as corrosion, discoloration and delamination. For saving silicon and reducing production costs, the thickness of silicon PV cells has been reduced from 300µm to less than 200µm and sometimes to less than 100µm, whereas the cell surface has been increased to 210mm x 210mm[42]. This makes PV cells more fragile and more susceptible to breakage during handling (rolling and storage). Since it is often impossible to detect cracks on the already operational PV module to the naked eye, various optical methods, in particular electroluminescence can be employed to detect the cracks [43].

2.4.14 Potential induced degradation

Individual modules in PV systems are often connected in series to increase the voltage of the system. The potential difference of such a connection may sometimes reach several hundred volts[44]. In order to protect people against electrical shocks, all metallic structures of modules are often grounded. Because of this electrical voltage between PV modules and their structure, it is possible that electrons in materials used for PV modules escape via the grounded framework when the insulation between structure and active layers is not perfect, thus creating leakage currents[44, 45]. From this phenomenon subsists a reversible polarization that may degrade the electrical characteristics of the photovoltaic cells. This phenomenon known as PID (Potential Induced Degradation) is characterized by the progressive performance deterioration of crystalline silicon photovoltaic modules, due to the presence of an electrical current induced in the module[46]. If some cells in a module remain at the original short-circuit value, the module's short circuit value is almost unchanged. A study showed that PID is more likely to happen in humid climates than in hot and dry ones[45]. In early stages, power degradation due to PID at high irradiation conditions is small and it is only pronounced in low irradiation conditions, making it difficult to detect within the power plant's monitoring data. Even a high degree of PID may not be seen through visual inspection[46, 47]. The electrochemical degradation was descried as the migration of ions from front glass through the encapsulant to the anti-reflective coating (SiN_x) at the cell surface driven by the leakage current in the cell to ground circuit[48]. This leakage current is typically in order of µA and its value strongly relies on material properties, surface conditions, applied voltage and environmental conditions dependent parameters like module temperature and applied voltage[45].

2.4.15 Shadowing

Two types of shading exist, (i)hard shading; which occurs if PV panels are shaded by a solid material, e.g. buildings or dust and (ii) soft shading; which can be caused by smog in the air[49]. The first one results in a voltage decrease whereas the second one affects current only, hence both are known to affect the performance of PV modules[21]. Shaded cells behave as a resistance to generated current within a module and as a result the heat builds up creating a hot spot which may lead to permanent module damage.

2.4.16 Snail tracks

PV modules exposed to outdoor environment for a few months can develop a degradation effect known as snail tracks, which effectively is a grey/black discoloration of the silver paste of the front metallization of screen printed solar cells. The discoloration usually starts at edges of cells and along usually invisible cell cracks and take up irregular dark stripe shapes in cells[37]. Choice of EVA and back sheet material is important for snail track to show up as the modules affected by it show a tendency of high leakage currents[3]. General IEC61215 testing is not competent enough to generate snail tracks as the module under investigation should have preexisting cell cracks existing in the module itself. Combined mechanical load, UV and humidity freeze tests could be applied in conjunction to each other for testing of snail tracks[50].

2.4.17 Soiling

Soiling is an important degradation mode which affects the performance of PV modules or systems by obscuring the irradiation through the layer of dust present at the face of the module. If proper cleaning mechanisms are not applied, soiling can starkly reduce the power production of PV systems. Various articles from around the world have suggested significant efficiency reduction from the dust accumulation as a function of exposure time[51]. Power output and efficiency are drastically reduced due to soiling in various locations as some regions like Kuwait and Nepal experienced power reductions up to 17% and 29.76% respectively [52, 53]. The corresponding reduction of output efficiency varied from 0% to 26% when dust deposition density increased from 0 to 22 g/m^2

[49]. PV modules require frequent cleaning and sometimes the orientation and tilt angle of modules also play a crucial role in soiling[54].

3 MONITORING OF PV SYSTEMS

Solar PV fortunately is a simple technology with no moving parts and requires very low maintenance. However, there is a high chance of hidden problems being overlooked during operation [8]. The power consumer may not know the difference between what is being achieved and what it was supposed to be. Hence, there should be available valid and regular data of electric outputs with other desired or required parameters at suitable intervals. To be applicable, a monitoring report should include information on relevant aspects of operations in such a way that a third party could understand[55].

The main purpose of monitoring a PV system is to determine and validate the required level of agreement between system performance and expectation. For this it is mandatory to have measurement systems with sufficient accuracy, a standard method to convert or compare the output to the output at standard test conditions and finally an expected value which should be maintained [56]. So, it is desired to have some monitoring guidelines to provide instruction on how to make measurements, how to analyze those measurements and how to determine whether the system is performing as expected. If the user wishes to ensure that the optimum output is obtained, then this process must be carried out regularly over the operating period with any shortfalls in output investigated and rectified. Monitoring guidelines are thus expected to provide guidance on the frequency of both measurement and analysis and some diagnostic capability in terms of loss mechanisms. A robust and dynamic Operation and Maintenance Program enables PV system production to reach and maintain the desired efficiency levels [21].

It is difficult to regularly judge the PV systems' performance with the expected production. A more scientific way is to measure and inspect the electrical as well as other parameters at regular intervals [8]. The main reason for monitoring a PV system is to constantly measure the energy yield, verify the performance of the system to required standards and sometimes identify if there is a problem in the system. Monitoring of the PV system generally includes certain number of parameters to be measured and stored in a data acquisition system.

Presently, the most accredited reference document for the monitoring of PV systems is the IEC61724-1 standard, but the older IEC 61724 standard, guidelines of DERlab, and other guidelines being implemented in the different countries are also reference literatures for the PV system monitoring which will be discussed later. These standards and guidelines provide the instructions for measurement and analysis of the measured data and determine the performance of the system.

3.1 Monitoring procedures and parameters

The required level of monitoring in terms of parameters to be measured, duration of the measurement and precision vary according to the size and complexity of the PV system along with the timely need to identify faults [8]. The faster the system faults are identified; the lower the system downtime, which boosts the overall output of the system and eventually the revenue. Monitoring the system performance is often a balance between the costs of instrumentation with the loss of revenue due to output loss.

Monitoring of the PV system can be done in two different ways in general, known as global and analytical monitoring[8]. The global monitoring includes measurement of limited set of parameters, sometimes just the output energy of the system through inverters to ensure the operational status of the system is at a reasonable level. This is usually implemented in systems where the monitoring costs need to be low. However, this approach usually requires additional information to determine the cause of any losses observed. Analytical monitoring on the other hand provides a more comprehensive dataset of operating parameters and allows investigation of performance trends and problems. Table 1 shows the list of parameters that need to be monitored in a PV system.

Table 1: Basic parameters required for monitoring [57, 58]

PV System Type	Parameters		
	Metrological	Electrical	
Grid Connected	i) Total irradiance, in plane of Array ii) Ambient Temperature iii) Module Temperature iv) Wind Speed v) Wind direction vi) Humidity vii) Barometric pressure	*PV Array:* i) Output Voltage ii) Output current iii) Output Power iv) Output energy	*Utility Grid* i) Grid Voltage ii) Current to utility grid iii) Current from utility grid iv) Power to utility grid v) Power from utility grid vi) Utility grid impedance
Stand Alone			*Load:* i) Output Voltage ii) Output current iii) Output power

3.2 Basic parameters for measurement

For the grid connected utility scale PV systems, monitoring provides the comparison between performance and expected energy yield. So, for the performance assessment purpose, the monitoring should include electrical energy generated as well as the incident irradiation. Electrical parameters include current, voltage and power. These parameters are sometimes available in the inverters, but these inverter integrated measurements lack required precision, so dedicated energy meters are prescribed.

3.2.1 Irradiance

Irradiance is the major parameter to be monitored. Irradiance simply is defined as the solar power received in a unit area denoted by W/m^2. There are two possible measurement instruments for solar radiation, pyranometers and reference cells, with their own differences. Pyranometers being thermopile devices, have a slow response time and are affected by the angle of incidence, tilt angle as well as ambient and dome temperature. Reference cells on the other hand, suffer from stability as only crystalline silicon sensors provide the required stability [4]. It is also a general norm to use the same type of reference cell as the modules in PV array. Pyranometers have a slow response time compared to the reference cells and hence cannot detect the sudden change in irradiation values due to a passing cloud or a partial shade.

For the standard measurement of irradiance, the reference device or pyranometer should be calibrated and aligned on the same plane as the photovoltaic array. The accuracy of the irradiance sensors, including signal condition should be better than 5%. Choice of irradiation sensors is based on how the measured data would be used. Despite the thermopile pyranometers having a slower response time, [59] propose them for the long-term monitoring, while the reference devices would be preferred to measure irradiance when investigating the variability effect of PV. For the highest level of accuracy on irradiance, the data should be measured on site, ensuring appropriate calibration, installation and maintenance of sensors. The sensors should be co-planar with the modules with maximum deviation not more than 20 degrees. Regardless of shadowing effects on PV-arrays, the sensors should be placed in such a way that they remain unshaded as well as remain accessible for frequent cleaning. For the small-scale systems with relatively low budget on performance monitoring, low standard sensors or various data sources provide another option.

3.2.2 Ambient temperature

Ambient temperature values represented by ^{0}C are measured at location exactly supplying the photovoltaic array conditions. As the ambient temperature influences module temperature, it is monitored to estimate the losses compared to the STC conditions. The sensors need to be housed in radiation shields, ensuring temperature sensor measures the air temperature only and is not heated by direct irradiance from the sun or even cooled by the wind effects. These sensors should not be placed near any kind of heat sources, being 1m above the ground as prescribed by IEC 61724 while DERlab TG 100-01 guidelines recommend ground clearance of 2m [60].

3.2.3 Wind speed

Wind speed is known to affect the performance of PV systems by triggering convective heat losses thus reducing the module temperatures. As recommended by IEC 61724, for the high standard measurement, wind speed should be measured at the same conditions as the arrays. Anemometers position as defined by IEC 61215 should be 1.2m on the east or west side of the system and 0.7m high above the upper edge of the system.

3.2.4 Module temperature

Similar to the ambient temperature, module temperature is also represented by °C and is measured at locations exactly depicting the photovoltaic array conditions by the help of temperature sensors attached on the back of one or more modules in such way that the temperature of cell in front of sensor is not altered. It is recommended for the large systems to measure module temperatures from various parts within the array as temperature fluctuations can occur within the array [60]. Such measurements can be averaged to provide better understanding of temperature effects on the entire PV system.

3.2.5 Current, voltage and power

Measurement of AC and DC voltage, current and power are the most important parameters for the PV system performance monitoring. These parameters can be used, in combination with the irradiance measurements, to directly determine when a PV system is underperforming [60]. The DC parameters help in diagnosing system faults at system level as reduction in DC quantities means problem in array and its yield, which is a fundamental quantity for degradation analysis.

Voltage and current values may be measured either on DC or AC side. Sensors should be selected to ensure they have a measurement range that is compatible with the output of the PV array (i.e. upper voltage limit > 1.3*V_{OC} and upper current limit > 1.5*I_{SC}) and selected to have minimal impact on the electrical operation of the array. Accuracy of these sensors, including signal conditioning, should be higher than 1% of the reading [57]. For large PV systems containing huge number of arrays and sub-arrays, it is recommended to individually measure both the AC and DC parameters of each sub-array which thus simplifies the location and isolation of faults.

Power data can also be available on DC or AC side or both sides. It can be measured directly by means of power sensors or calculated in real time as the product of sampled voltage and current values. The general recommendation is to measure power directly, but if the direct measurement of power is not available it can be derived from the product of current and voltage. The current and voltage quantities must be the simultaneously sampled ones rather than the averaged quantities. Power sensors accuracy, including signal conditioning, should be greater than 2% of the reading.

3.3 Monitoring standards and guidelines

With the growing deployment of the PV systems, there arose a need to properly measure and verify the degree of production from the PV systems. Some research institutes and some joint projects have worked together in different parts of the world to develop a general norm to guide through the monitoring of those systems. Some of those guidelines are discussed here.

3.3.1 IEC61724 standard

The IEC 61724 standard is a set of international standards for the parameters for PV systems measurements and monitoring presenting guidelines for analysis and monitoring of PV systems performances. Although this standard has been replaced by the IEC61724-1, it still remains one of the influential standards as the many monitoring systems and instruments in practice today are based on this standard. Figure 4 presents an example for basic monitoring parameters required and illustrates the electrical and environmental data to be measured. These data are listed in Table 1. Other parameters can be calculated from these measurements in real time, using data acquisition systems. Monitored parameters are classified under two groups: environmental or meteorological parameters, and electrical parameters.

This standard has requirements to measure total irradiance, module and ambient temperature, wind speed and electrical parameters (current, voltage and power).

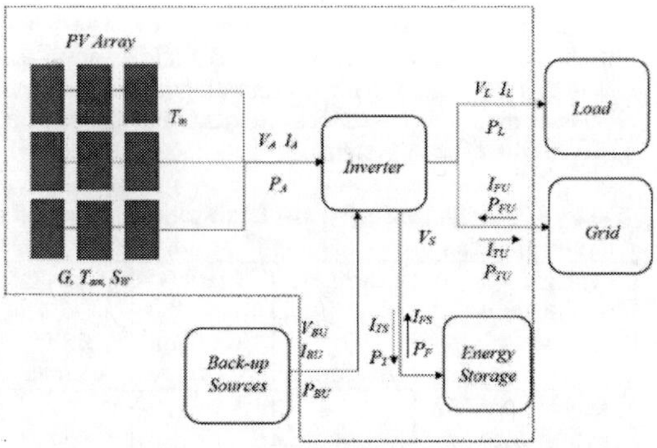

Figure 4: Diagram of basic real time parameters to be measured[57]

The irradiance should be measured in the plane of the array with an uncertainty including the instrumentation to be less than 5%. This should be done either by use of a pyranometer or a reference cell. The irradiance sensors installed according to this standard have to be calibrated every second year. The ambient temperature should be representative of the array location, but the module temperature should be measured on the center of the back of the module or more modules, or the center of arrays while uncertainty including the instrumentation should be less than 1%. Wind speed should be measured at the height and location representative of the array conditions with an uncertainty including the instrumentation less than $0.5ms^{-1}$ for speeds less than $5ms^{-1}$ and less than 10% of the reading for speeds greater than 5 ms^{-1}. For current and voltage both AC and DC parameters have to be measured with the level of uncertainty including the instrumentation to be less than 1% of the reading. For power, the DC power is to be calculated based upon instantaneous measurements but not the averaged reading or directly measured with a wattmeter. AC power meanwhile accounts for the power factor and harmonic distortions and the uncertainty including the instrumentation has to be less than 2%.

3.3.2 Australian guidelines

The Australian guideline is a national technical guideline developed for the Australian PV Institute (APVI) by various research institutions with the support of the Australian government. This guideline offers a guidance on need to measure parameters, measurement mechanisms of those parameters and the frequency of those measurements based on the seven different uses of PV performance and reliability data [60]. This guideline is primarily based on the IEC61724 standard but with a more detailed approach to the use of measured data. The potential use of the data has been categorized in seven different uses which are briefly presented in Table 2.

This guideline has defined conditions for the need or demand of the monitoring on the different uses along with the instrumentation and financial requirement of sensors. The general flow of the guideline is based on the pre-determination of the end use of the measured data. When the end use has been finalized, it then requires selecting the parameters as not all parameters defined in Table 2 are required in each of the seven end uses, but if the end use is not known, all the parameters are suggested to be measured.

Table 2: Categorization of recording interval according to the uses[60]

Use	Use of measured data	Recording interval (τ_r)	Monitoring period
1	Performance assessment under outdoor conditions	Max 15min	Min 1 year
2	Performance diagnostics	Hourly	System Lifetime
3	Degradation & uncertainty analysis with time	Hourly	Min 3-5 years
4	Understanding system losses via comparison to modelled data	Hourly	Min 1 year
5	PV Performance forecasting	5min, 30min or hourly	Min 1 year
6	Grid interaction of PV systems	1 to 15min	Min 1 week
7	Grid integration of distributed generation and load control	1 to 15 min	Min 1 week

Next is the confirmation of required level of accuracy, as the instrumentation requirements are already stated in the IEC61724. To follow those requirements is a best option but for limited budget, monitoring instruments with less accuracy and less costs can also be used. Finally, the recording and monitoring intervals as mentioned in Table 2 can be applied based on again the end use. The whole process is shown in Figure 5.

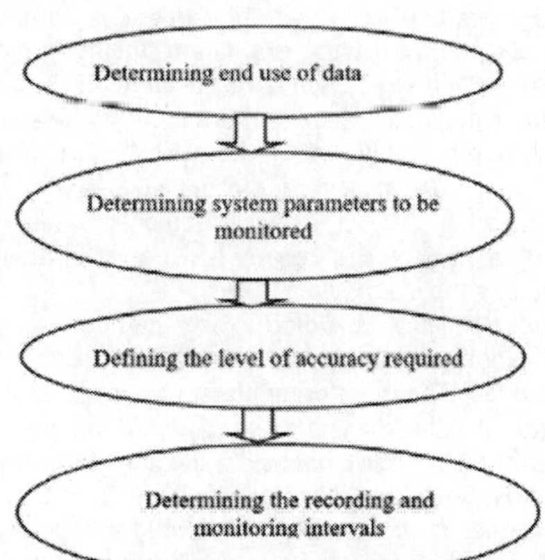

Figure 5: Flow diagram of monitoring decision making from Australian Guideline[60].

3.3.3 DERlab guidelines

This guideline was developed to harmonize the outdoor testing procedures that would be useful for energy yield measurement of different PV technologies. It is meant to provide basic guidelines and procedures regarding testing locations, placement of sensors and accuracy of the whole measurement equipment. Developed as a complimentary to the international standards, this guideline discusses only the measurements on the DC side and it does not include data analysis and evaluations. The DERlab guideline mentions that a maintenance plan consisting of visual inspection and cleaning of the measuring instruments should be well documented along with the proper calibration of the used sensors on a regular basis.

The parameters suggested by this guideline to be measured are the basic ones also prescribed by the IEC61724. Current and voltage are to be taken at the maximum power point (MPP), but the recording time interval is set 15 seconds. The secondary class pyranometer is prescribed for the irradiance measurement while comparing different technologies, owing to its sensitivity over the wide range of spectrum. But for the single technology modules, a reference cell of same technology is preferred. The installation plan of this standard suggests the modules to be placed 1m above the ground and the meteorological sensors to be placed 2m from the ground while being at least 1.2m from the PV array and with such location that it would not shade the modules. For the module temperature measurement, the sensor is suggested to be installed at the middle of the modules rear side. Also, for the long-term PV yield measurement this guideline proposes that an IV curve is retrieved once each month, on a clear day with light wind and corrected for STC, to analyze PV module degradation over time. The matter of operativity of the sensors and subsequent data logging is also quite important as this guideline proposes the availability of 98-99% of data from its potential operation per year necessary to avoid uncertainties in forecasting of the annual energy yield.

3.3.4 IEC 61724-1 standard

The IEC 61724 standard[61] is now replaced with the IEC 61724 series consisting of three separate parts, each part with specific purpose as opposed to the former one. The first part IEC61724-1 is named as monitoring, IEC61724-2 as capacity evaluation method and IEC 61724-3 as energy evaluation method. IEC 61724-1 standard categorizes the monitoring of PV systems on three different scales with the specific emphasis on accuracy and frequency of sensors and measurements, recommending number of measurements based on the scale of a project and requirements for servicing and maintaining the instruments [62].

With the new standard, each monitoring system classification provides a guideline of what measurements can be done. The monitoring classes are divided in three parts with A being the one with highest accuracy and class C being the most basic one and a medium accuracy level of class B. Solar Irradiance and ambient temperature are the parameters commonly required for all level of measurements. Table 3 presents the outline of monitoring system classifications as per the required applications.

The standard has requirements not only for the accuracy and parameters, but these are also related to the number of sensors. This is based on the different classification from the system less than 5 MW to the system greater than 750MW. The cleaning and recalibration of sensors is also an important issue. The irradiance sensor needs to be calibrated annually for high class accuracy and in every two year for other accuracies. It also lays requirements to at least have an annual inspection every year for sensors in class A and B monitoring systems. The power consumed by the tracking and monitoring systems installed in the site are now categorized as the power loss of PV plants as opposed to the loads of the plant. The sampling and recording intervals are different for each of the classes as class A require 3 seconds sampling and 1 min recording interval of the basic parameters whereas class B and C have 1 min sampling for basic parameters and while the class B has 15 min of recording interval, an hourly recording interval is enough for the system with basic accuracy. The newly introduced parameters like soiling, rain, snow and humidity are required to have a sampling rate of 1 min for all classes in ground based measurements and also satellite based for the class A systems.

Table 3: Monitoring system classifications as per the required applications

Typical Applications	Class A High Accuracy	Class B Medium Accuracy	Class C Basic Accuracy
Basic System performance assessment	×	×	×
Documentation of a performance guarantee	×	×	
System loss analysis	×	×	
Electricity network interaction assessment	×		
Fault localization	×		
PV technology assessment	×		
Precise PV system degradation measurement	×		

This standard has the recommendations for all the sensors to be used according to their accuracy including uncertainty levels. Thermopile pyranometers, reference cells and photodiode sensors are prescribed along with the locations and angular alignments based on the accuracy required for irradiance measurements. For both the ambient and module temperatures, the measurements should represent the array conditions and recalibration of sensors should be done every two years for high accuracy systems. The module temperature sensors require the uncertainty of ≤ 2°C while for the ambient temperature this should be ± 1°C. Wind sensors require to measure the wind speed and direction from position representing the array conditions but the recalibrations is dependent on the manufacturers recommendations. Soiling ratio(SR) is a new parameter introduced in this standard which is the ratio of actual power output of PV array under soiling conditions to the expected power when the array had no soiling effects. The prescribed method to find SR is to calculate maximum power loss reduction due to soiling among other alternative methods. Rainfall, shading effect of snow and effects on incident spectrum through humidity are also the parameters that need to be measured in the high accuracy monitoring system. The electrical measurements are also defined according to the accuracy requirements where the uncertainty levels are stated as

±2% for the class A systems, ±3% for the class B systems and the basic systems have no such requirements on inverter level measurements.

The data processing should not include the night hours and should strictly include the hours between sunrise and sunset with irradiance being ≥ 20 W/m^2. The measured data requires to be filtered and checked to ensure there are no missing or invalid data points. The missing or invalid data can be processed in various ways which are discussed in detail in the part 2 and 3 of the now IEC 61724 series.

4 DISCUSSION AND CONCLUSIONS

Regular measurements and analysis of the performance ratio (PR) over the life span of a PV system allows the detection of the serious malfunctions or the degradation and through prompt operation and maintenance action, facilitates the reliable power production. Normally the bigger scale PV systems are equipped with the sophisticated monitoring devices whereas the small scale residential and commercial systems are not given the same level of attention. This could be due to financial constraints. The performance of PV systems mainly depends on the solar irradiation, ambient temperatures and local weather conditions as well as the power conditioning units employed. These are the fundamental components of any PV system. The various monitoring guidelines discussed are based on the IEC61724 standards and with the revision of this standard it can be concluded that with growing investments and grid injections, the monitoring needed to be more precise. IEC61724-1 standard fills the void in many such conditions with defined level of precision for sensors and monitoring mechanisms.

The IEC61724 standard created an understanding of the basic relation of different variables such as irradiance, temperature with power output as well as their combined effects on one another, but the understanding was limited on the variation in the performance and of the instruments/sensors as well as the data those sensors reported. Performance monitoring is significantly dependent on the accuracy of the measuring instruments and the installation procedures. Regular calibration is the best way to ensure the greater significance of the monitored data and subsequent modelling. The guidelines formulated by DERlab contains the basic parameters to be measured and the procedures to be measured. Guideline adopted in Australia has defined the frequency and complexity of measurements as per the length of monitoring need within the IEC61724 standard. The IEC 61724-1 will be a benchmark for a long time to come as it addresses the range from simple residential systems to the utility scale systems. The guidelines formulated under the PERFORMANCE project have been mentioned in different literatures, but it is not accessible and available at the present.

The performance of a PV system depends primarily on the parameters discussed above but is not confined to just them. Parameters like shading, soiling, angle of incidence, tilt and spectral response among others are also crucial for the performance monitoring as different module technologies have a varying response to these parameters. For instance, amorphous silicon modules have a better temperature response but suffer more on spectral response than the crystalline modules, indicating that the same monitoring regime may not be useful to compare the performance of these modules. Statistical methods of monitoring is also applied in the places unable to install the monitoring equipment, but the interpretation of such monitoring needs to be done clearly as the correlation between power produced by the system and actual weather conditions could lead to unrealistic PR values.

The initial work done under the COST Action PEARL PV is summarized in Table 3. The monitoring regime has been categorized under different headings and all the necessary topics and measurements have been enlisted under them. Experts were provided with the questionnaire which contained the classification of variables under the sections mentioned on the Table 3 and their response is being tabulated. The experts have provided their opinions regarding each parameter that are necessary. Some of the parameters are desired to have, but not necessary to include in the monitoring. These types of parameters like cabling, lightening protection and the details of battery storage and surveillance system should be included in the detailed monitoring which may be required when investigating the PV system in detail. But in normal conditions where the monitoring is to be done for the life time of a PV systems, these minute informations may not be a preferred one due to the storage issues of huge amount of data and need to continuously monitor these factors too. The guidelines will be prepared based on the parameters under categories summarized in Table 3, below.

Table 4: Summary of the sections formulated for the development of monitoring guidelines

Section	Contents	Elaboration of Contents
Quality Control	General requirements to ensure the data and equipment quality	• Instrumentation/sensor Accuracy • Data accuracy and Calibration procedures • Data quality control and handling process
File Format	File format required for the storage and transfer of the data.	• XML/ASCII/CSV Formats • Date and Time stamp format. • Reporting of the missing data and required format
Metadata: Site	General information regarding the PV site	• Site name and address • GPS Coordinates with site elevation and Time Zone (UTC)
Metadata: PV system and Components	General information regarding the PV system	• Fixed or Tracking installations and details • Installation type (BIPV/BAPV/Free Standing) • PV Module technology and the electrical/mechanical parameters • PV String design, PCU and Inverters • Shading and Soiling losses • Balance of System components
Metadata: Sensors	Instrumentation, sensors and data logging	• Irradiance, Temperature and Wind Sensors • Other Meterological sensors (Rain, Humidity and others) • Current and Voltage transducers • Power meters • IR and Electroluminescence Imaging • System status
Monitored data: Meteorological	Meterological informations	• Solar Radiation • Climate and Weather • sampling and recording rate of data
Monitored data: Yield and Durability	Information regarding the yield of PV systems and degradation analysis through various methods	• Time series of the available data • PV module / String yields • sampling and recording rate of data • PV module/String degradation 1. IV curves 2. Soiling 3. Other degradation or failure indicators
Monitored Data: PV in the Built Environment	Information about the integration of PV system as a part of buildings or other energy systems	• Building integration / Architectural integation • Hybrid Energy Systems • Environmental footprint • Economical consideration
Monitored Data: PV in Grids	Grid integration of PV systems, parameters and measurements	• Inverter power, phase, voltage, power factor, harmonics and frequencies • Load Characteristics • Utility characteristics • Energy Storage • Protection and Management systems • sampling and recording rate of data
Modelled Data: PV Simulation	Simulation of the PV output to be compared with the measured data	• Weather and Irradiance data files • Simulation models • Comparison of modelled vs measured data • Modelling system performance external influences and thermal properties • Modelling grid interaction and forecasting models for PV production

5 REFERENCES

1. King, R.R., et al., *Solar cell generations over 40% efficiency*. 2012. **20**(6): p. 801-815.
2. Sark, W.v., *Detailed Workplan of Working Group 1, PV Monitoring COST ACTION PEARL PV-CA16235*. 2017: Brussels.
3. Kontges, M., et al., *IEA-PVPS {Task 13}: Review of failures of PV modules*. Technical report, 2014.
4. Woyte, A., et al. *Monitoring of photovoltaic systems: good practices and systematic analysis*. in *Proc. 28th European Photovoltaic Solar Energy Conference*. 2013.
5. Nordmann, T., U. Jahn, and W. Nasse. *Performance of PV systems under real conditions*. in *European Workshop on Life Cycle Analysis and Recycling of Solar Modules, The" Waste" Challenge, Brussels, Belgium*. 2004.
6. Chouder, A. and S. Silvestre, *Automatic supervision and fault detection of PV systems based on power losses analysis*. Energy Conversion and Management, 2010. **51**(10): p. 1929-1937.
7. Guasch, D., S. Silvestre, and R. Calatayud. *Automatic failure detection in photovoltaic systems*. in *Photovoltaic Energy Conversion, 2003. Proceedings of 3rd World Conference on*. 2003. IEEE.
8. Pearsall, N.M., *6 - Prediction and measurement of photovoltaic system energy yield*, in *The Performance of Photovoltaic (PV) Systems*, N. Pearsall, Editor. 2017, Woodhead Publishing. p. 183-208.
9. Decker, B. and U. Jahn, *Performance of 170 grid connected PV plants in Northern Germany—Analysis of yields and optimization potentials*. Solar Energy, 1997. **59**(4): p. 127-133.
10. Kymakis, E., et al., *Performance analysis of a grid connected photovoltaic park on the island of Crete*. 2009. **50**(3): p. 433-438.
11. Ghiani, E., et al., *Evaluation of photovoltaic installations performances in Sardinia*. 2013. **76**: p. 1134-1142.
12. Sharma, V. and S.J.E. Chandel, *Performance analysis of a 190 kWp grid interactive solar photovoltaic power plant in India*. 2013. **55**: p. 476-485.
13. Bogdanski, N., et al. *PV reliability (cluster II): results of a German four-year joint project—part II, results of three years module weathering in four different climates*. in *Proceedings of the 25th European Photovoltaic Solar Energy Conference*. 2010.
14. Lockertsen, E.O., *The Performance and Potential for PV systems at Northern Latitudes*, in *Department of Electric Power Engineering*. 2015, Norwegian University of Science and Technology: Trondheim. p. 215.
15. Imenes, A.G., et al. *Performance of grid-connected PV system in Southern Norway*. in *2015 IEEE 42nd Photovoltaic Specialist Conference (PVSC)*. 2015.
16. Ndiaye, A., et al., *Degradations of silicon photovoltaic modules: A literature review*. Solar Energy, 2013. **96**: p. 140-151.
17. Charki, A., R. Laronde, and D. Bigaud, *The Time-Variant Degradation of a Photovoltaic System*. Journal of Solar Energy Engineering, 2012. **135**(2): p. 024503-024503-4.
18. Jordan, D.C. and S.R. Kurtz, *Photovoltaic Degradation Rates—an Analytical Review*. Progress in Photovoltaics: Research and Applications, 2013. **21**(1): p. 12-29.
19. Osterwald, C.R. and T.J. McMahon, *History of accelerated and qualification testing of terrestrial photovoltaic modules: A literature review*. Progress in Photovoltaics: Research and Applications, 2009. **17**(1): p. 11-33.
20. Munoz, M.A., et al., *Early degradation of silicon PV modules and guaranty conditions*. Solar Energy, 2011. **85**(9): p. 2264-2274.
21. Triki-Lahiani, A., A. Bennani-Ben Abdelghani, and I. Slama-Belkhodja, *Fault detection and monitoring systems for photovoltaic installations: A review*. Renewable and Sustainable Energy Reviews, 2018. **82**: p. 2680-2692.
22. Kajari-Schröder, S., et al., *Spatial and orientational distribution of cracks in crystalline photovoltaic modules generated by mechanical load tests*. Solar Energy Materials and Solar Cells, 2011. **95**(11): p. 3054-3059.
23. Sander, M., et al. *Investigations on crack development and crack growth in embedded solar cells*. in *Reliability of Photovoltaic Cells, Modules, Components, and Systems IV*. 2011. International Society for Optics and Photonics.
24. Kempe, M.D., *Modeling of rates of moisture ingress into photovoltaic modules*. Solar Energy Materials and Solar Cells, 2006. **90**(16): p. 2720-2738.
25. Wohlgemuth, J.H., Kurtz, S., *Reliability Testing Beyond Qualification as a Key Component in Photovoltaic's Progress Toward Grid Parity.* , in *IEEE International Reliability Physics Symposium* 2011.
26. Vázquez, M. and I. Rey-Stolle, *Photovoltaic module reliability model based on field degradation studies*. Progress in Photovoltaics: Research and Applications, 2008. **16**(5): p. 419-433.
27. Carlson, D.E., et al., *Corrosion effects in thin-film photovoltaic modules*. Progress in Photovoltaics: Research and Applications, 2003. **11**(6): p. 377-386.
28. Osterwald, C.R., et al. *Degradation analysis of weathered crystalline-silicon PV modules*. in *Conference Record of the Twenty-Ninth IEEE Photovoltaic Specialists Conference, 2002*. 2002.
29. Herrmann, W., W. Wiesner, and W. Vaassen. *Hot spot investigations on PV modules-new concepts for a test standard and consequences for module design with respect to bypass diodes*. in *Conference Record of the Twenty Sixth IEEE Photovoltaic Specialists Conference - 1997*. 1997.
30. Sharma, V. and S. Chandel, *Performance and degradation analysis for long term reliability of solar photovoltaic systems: a review*. Renewable and Sustainable Energy Reviews, 2013. **27**: p. 753-767.
31. Jordan, D.C., et al., *PV degradation curves: non-linearities and failure modes*. Progress in Photovoltaics: Research and Applications, 2017. **25**(7): p. 583-591.
32. Oreski, G. and G.M. Wallner, *Evaluation of the aging behavior of ethylene copolymer films for solar applications under accelerated weathering conditions*. Solar Energy, 2009. **83**(7): p. 1040-1047.
33. Quater, P.B., et al., *Light Unmanned Aerial Vehicles (UAVs) for cooperative inspection of PV plants*. IEEE Journal of Photovoltaics, 2014. **4**(4): p. 1107-1113.
34. Berman, D. and D. Faiman, *EVA browning and the time-dependence of I−V curve parameters on PV modules with and without mirror-enhancement in a desert environment*. Solar Energy Materials and Solar Cells, 1997. **45**(4): p. 401-412.
35. Realini, A.J.F.R., Federal Office for Education and B. Science Tech. Rep., *Mean time before failure of photovoltaic modules*. 2003. **99**.
36. Molenbroek, E., D.W. Waddington, and K.A. Emery. *Hot spot susceptibility and testing of PV modules*. in *The Conference Record of the Twenty-Second IEEE Photovoltaic Specialists Conference - 1991*. 1991.

37. Kumar, M. and A. Kumar, *Performance assessment and degradation analysis of solar photovoltaic technologies: A review.* Renewable and Sustainable Energy Reviews, 2017. **78**: p. 554-587.
38. Wendlandt, S., et al. *Operating principle of shadowed c-Si solar cell in PV-modules.* in *ISES*. 2011.
39. Schmidt, J. *Light-induced degradation in crystalline silicon solar cells.* in *Solid State Phenomena*. 2004. Trans Tech Publ.
40. Lindroos, J. and H. Savin, *Review of light-induced degradation in crystalline silicon solar cells.* Solar Energy Materials and Solar Cells, 2016. **147**: p. 115-126.
41. Meydbray, J. and F. Dross, *PV module reliability scorecard report 2017*. DNV-GL Google Scholar, 2017.
42. Dallas, W., O. Polupan, and S. Ostapenko, *Resonance ultrasonic vibrations for crack detection in photovoltaic silicon wafers.* Measurement Science and Technology, 2007. **18**(3): p. 852.
43. Rueland, E., et al., *Optical u-crack detection in combination with stability testing for in-line inspection of wafers and cells.* Proceedings of 20th EU PVSEC Barcelona, 2005: p. 3242-3245.
44. Schütze, M., et al. *Laboratory study of potential induced degradation of silicon photovoltaic modules.* in *2011 37th IEEE Photovoltaic Specialists Conference*. 2011.
45. Hacke, P., et al. *System voltage potential-induced degradation mechanisms in PV modules and methods for test.* in *2011 37th IEEE Photovoltaic Specialists Conference*. 2011.
46. Pingel, S., et al. *Potential Induced Degradation of solar cells and panels.* in *2010 35th IEEE Photovoltaic Specialists Conference*. 2010.
47. Mathiak, G., M. Schweiger, and W. Herrmann. *Potential-induced degradation—Comparison of different test methods and low irradiance performance measurements.* in *27th European Photovoltaic Solar Energy Conference and Exhibition*. 2012.
48. Hoffman, A. and R. Ross. *Environmental qualification testing of terrestrial solar cell modules.* in *13th IEEE PVSC*. 1978.
49. Jiang, H., L. Lu, and K. Sun, *Experimental investigation of the impact of airborne dust deposition on the performance of solar photovoltaic (PV) modules.* Atmospheric Environment, 2011. **45**(25): p. 4299-4304.
50. Berghold, J., et al. *Electrochemical corrosion within solar panels.* in *27th European Photovoltaic Solar Energy Conference and Exhibition, Frankfurt*. 2012.
51. Siddiqui, R. and U. Bajpai, *Deviation in the performance of solar module under climatic parameter as ambient temperature and wind velocity in composite climate.* International Journal of Renewable Energy Research, 2012. **2**(3): p. 486-490.
52. Paudyal, B.R. and S.R. Shakya, *Dust accumulation effects on efficiency of solar PV modules for off grid purpose: A case study of Kathmandu.* Solar Energy, 2016. **135**: p. 103-110.
53. Wakim, F., *Introduction of PV power generation to Kuwait.* Kuwait Institute for Scientific Researchers, Kuwait City, 1981.
54. Mani, M. and R. Pillai, *Impact of dust on solar photovoltaic (PV) performance: Research status, challenges and recommendations.* Renewable and Sustainable Energy Reviews, 2010. **14**(9): p. 3124-3131.
55. Tahri, A., T. Oozeki, and A. Draou, *Monitoring and Evaluation of Photovoltaic System.* Energy Procedia, 2013. **42**: p. 456-464.
56. Pearsall, N. and B. Atanasiu. *The European PV system monitoring guidelines-Modernisation under the PERFORMANCE project.* in *Photovoltaic Specialists Conference (PVSC), 2009 34th IEEE*. 2009. IEEE.
57. IEC61724, *Photovoltaic System Performance Monitoring Guidelines for Measurement, Data Exchange and Analysis*. 1998: Geneva, Switzerland.
58. S. R. Madeti, S.N.S., *Comparative analysis of solar photovoltaic monitoring systems.* AIP Conference Proceedings, 2017. **1859**(1): p. 020116.
59. Reinhardt, S., Eggers, C., Grünsteidl, S., Vodermayer, C., Becker, G. *Influence of Technology Differences in Performance Ratio Calculations.* in *27th European Photovoltaic Solar Energy Conference and Exhibition*. 2012. Frankfurt, Germany.
60. Copper, J., et al., *Australian Technical Guidelines for Monitoring and Analysing Photovoltaic Systems*. 2013, Australian PV Institute.
61. 61724-1, I., *Photovoltaic System performance - Part 1: Monitoring*. 2017. p. 57.
62. Beuttell, W., *IEC 61724-1: What's it all about?*, in *PV Tech*. 2017.

OBSERVATIONS IN PV MODULE OPERATION VOLTAGE DISTRIBUTION ALONG A PV ARRAY. AN IN-DEEP LOOK ON MISTMACH LOSSES.

C. H. Rossa, E. Lorenzo, F. Martínez-Moreno*

Instituto de Energía Solar – Universidad Politécnica de Madrid. Grupo de Sistemas Fotovoltaicos (IES-UPM).
Campus Sur UPM. Ctra. Valencia km. 7. ETSI Sistemas Telecomunicación. 28031 Madrid, Spain.
*Corresponding author: francisco.martinez@ies.upm.es

Considering that the current through all the PV modules of a same string is constant, any differences in the electrical behaviour of the modules give raise to differences in the corresponding operation voltages. That opens the door for careful observations of the PV array functioning carried out under normal operation conditions, that is, with the inverter delivering energy and, thus, without the need for disconnection as required for more standard PV array analysis methods: measuring I-V curves, open-circuit voltages and so on. This in-routine operation observation procedure is relatively easy implemented in practice, by means of "T" shaped quick connectors to connect PV modules to each other, which are widely commercially available. In fact, this has been successfully applied to in-the-field PID detection on commercial PV plants, by monitoring the voltage differences between the modules at the extreme poles of the string. This time, we have applied such observation procedure to all the 21 PV modules comprising a 5 kW PV array connected to the grid at the IES-UPM terrace. Moreover, we have also glued a PT1000 in the back of each module. Together with a PV reference module for measuring in-plane irradiance, this set-up allows for continuous and simultaneous monitoring of, both, the operation conditions (irradiance and temperature) and the electrical response (current and voltage) of all the PV modules.

Keywords: voltage dispersion, real time monitoring PV generators, mismatch losses, power loss estimation

1 INTRODUCTION

In a real situation, where the PV modules are interconnected in a same string, the current through all of them is the same. So, any differences in the electrical behaviour of the modules give raise to differences in the corresponding operation voltages. In fact, these differences already occur on the level of the cells, due to manufacturing variance (each individual cell has similar but not identical electrical and thermal characteristics than the other cells of the module) and variation on the operation conditions (especially those related to temperature [1]). So, at string level, the sum of the differences between modules leads to a deviation of the energy delivered into the grid [2] in comparison to the expected one. This is the so called mismatch losses. Even if the mismatch losses in an array may be typically less than 1 % in normal operation conditions [3] [4] [5], this value may increase along the years to values up to 5 % after 15 years exposed to the same conditions [6]. In large PV plants, the instantaneous relative mismatch loss can reach values up to 60 %, due to clouds passing, which causes shading effects on some panels. Even so, these type of losses are below 1 % in total electricity produced by the arrays [7]. In fact, most of the time the relative mismatch losses were less than this value in normal operation conditions, e.g. without shades caused by any obstacles. To avoid these problems, many authors have suggested methods like different types of interconnection between the modules [8] [9] or sorting cells according to its current at the maximum power point [10].

The differences in the corresponding operation voltages between the modules opens the door for careful observations of the PV array functioning carried out under normal operation conditions, i.e., with the inverter delivering energy and, thus, without the need for disconnection as required for more standard PV array analysis methods: measuring I-V curves, open-circuit voltages and so on. This work presents the measurements performed to monitor in real time, the operation voltage and temperature from every single module of the PV generator and the operation current from each string along many days with different situations like soiling and different weather conditions while the generator goes on delivering energy into the grid. This in-routine operation observation procedure is relatively easy to implement in practice, by means of "T" shaped quick connectors to connect PV modules to each other, which are widely commercially available. In fact, this has been successfully applied to in-the-field PID detection on commercial PV plants, by monitoring the voltage differences between the modules at the extreme poles of the string [11].

Thanks to these measurements, the mismatch losses may be quantified by the ratio between the standard deviation to the mean of the observed individual voltage values ($\Delta V = \sigma_V/\bar{V}$). Then, a quadratic relation between voltage dispersion and power loss ($\Delta P = a \cdot \Delta V + b \cdot \Delta V^2$) is empirically adjusted to the I-V curve of a constituent module.

2 METHODOLOGY.

The measurements performed to analyze the operation voltage and temperature distribution along a PV system have been done in a PV generator with a maximum power peak of 5.1 kW_p, which is installed at the roof-terrace in the south campus at Universidad Politécnica de Madrid (Fig. 1). This PV generator has 21 modules, divided in 3 strings connected in parallel, each one with 7 modules interconnected in series.

Figure 1: PV generator composed by 21 c-Si modules with 245 Wp each one.

2.1 Voltage and temperature dispersion measurements.

To assess the operation voltage in each module, "T" shaped connectors were used between each interconnection of them. Besides that, a PT1000 sensor class B (± 0.3 °C of tolerance) was glued on the back of each module, positioned right in its centre (Fig. 2) to measure its temperature. The values of operation voltage and temperature, as well as the in-plane irradiance from a PV reference module and the operation current of each string measured with shunt resistors, were simultaneously measured and continuously stored with a data logger each 5 minutes during a period of 17 months from February 2017. These data have been used to analyze the evolution of mismatch losses in the broadest sense, i.e., caused not only by differences in the PV modules STC characteristics but also by differences in their operation conditions, especially cell temperature.

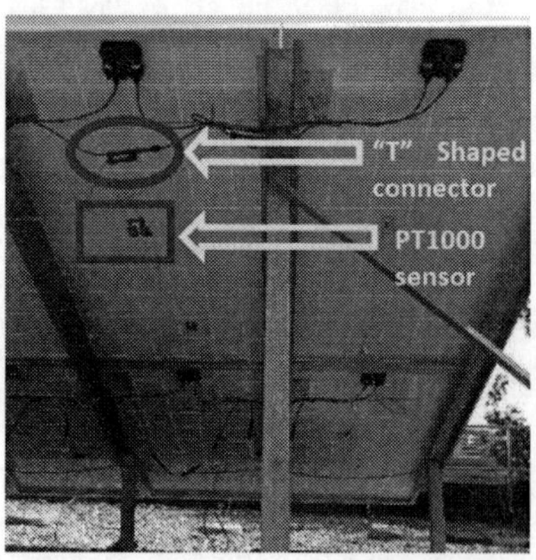

Figure 2: "T" shaped connectors with the PT1000 sensor glued on the back of each module.

2.2 I-V curves measurements.

The I-V curves of each module, string and PV generator were measured with capacitive loads [12] [13].

The measurements of the 21 modules were done outdoors in a "Solar Box", a device which allows to control the values of irradiance and temperature [14]. Each module has been measured at two temperature conditions: 25°C and temperatures beyond 50°C, both with an irradiance of 1000 W/m^2. The temperature is obtained as the average of 9 PT1000 sensors glued on the back of the module.

The curves of the strings and PV generator have also been measured outdoors, but now at ambient temperatures (greater than 50°C) and at 1000 W/m². Now the temperature is obtained as the average of PT1000 sensors glued on the back of the modules of the system under test (string or generator).

From the I-V curves, a relation between power and voltage can be represented; that is, the P-V curve. This curve allows to estimate the power loss, ΔP, related to any voltage deviation, ΔV, from the maximum power point voltage, (V_m, P_m). So, different values of power loss are obtained when the voltage is shifted to values greater or smaller than this point (Fig. 3). As can be noticed, the growth below V_m is almost linear, while the growth above this point is quadratic. So, the analysis presented in the next section is divided in these two areas separately, for more accurate results.

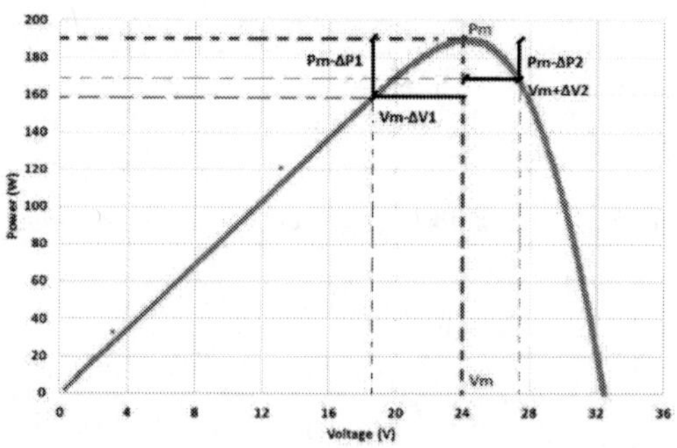

Figure 3: P-V curve measured from a module installed at the generator under study, showing the difference in power – or losses – when the voltage is shifted for greater or lower values than the voltage on the maximum power point.

3 RESULTS AND DISCUSSION.

3.1 Temperature and voltage dispersion along the PV generator.

Fig. 4 shows the temperature dispersion along the PV generator ($\Delta T = T_{max} - T_{min}$), measured along 17 months. As can be seen, the temperature dispersion increase with the irradiance, reaching values up to 14 °C for 1000 W/m^2.

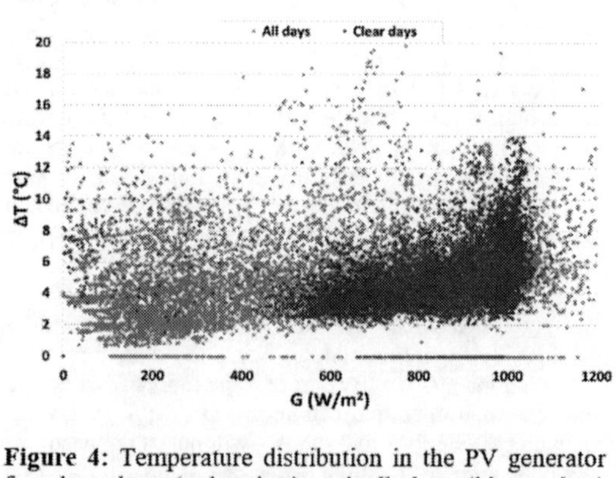

Figure 4: Temperature distribution in the PV generator for clear days (red points) and all days (blue points) observed along 17 months.

Fig. 5 shows the operation voltage dispersion along

the PV generator, measured in the same period ($\Delta V = V_{op,max} - V_{op,min}$). This dispersion decreases from values greater than 2 V for low irradiance values (below 500 W/m^2) to values lower than 2 V at irradiances greater than 700 W/m^2. Therefore, this time the voltage dispersion increases when irradiance decreases, which can be explained by the dominance of shunt resistance effects at low irradiances. Interestingly, such dominance overtakes the temperature dispersion effects which act in the opposite way.

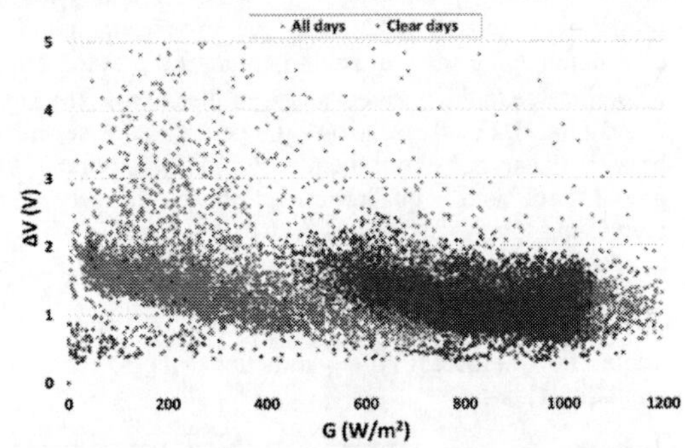

Figure 5: Operation voltage distribution in the PV generator for clear and all days observed along 17 months.

3.2 Power loss estimation from P-V curves.

Fig. 8 shows the power loss estimation for all modules of the PV generator. The power loss estimation is done as follows: considering the point (V_m, P_m) the origin of the losses (that is, the power loss is $\Delta P = 0\%$ when there is a shifting $\Delta V = 0\%$ from V_m), any variation on the operation voltage ΔV leads to a decrease ΔP in the power. The curves (a) and (b) are for voltage values below V_m, and the curves (c) and (d) are for voltage values above V_m. It should be noticed that ΔV and ΔP are always positives (see Fig. 3).

According to Fig. 8a and Fig. 8b, the power loss grows up with a behaviour almost linear when values less than V_m are evaluated. Besides, the temperature seems not to considerably affect the power losses on the modules: there is light differences between the values from curves obtained at 25°C or the values from curves obtained at temperatures greater than 50°C. In both graphics, the values of $\Delta V < 7\%$, which are related to the maximum voltage variation between 800 and 1000 W/m^2 (see Fig. 5, where $V_m \approx 27\ V$), lead to $\Delta P < 5\%$.

When the power loss is evaluated for values greater than V_m (Fig. 8c and Fig. 8d), the losses tend to grow up faster with the variation from V_m (quadratic). Besides, in this case the temperature seems to have some influence: the losses evaluated for the curves measured at 25 °C grow up faster than the curves evaluated for temperatures greater than 50 °C. Here, the values of $\Delta V < 7\%$, at the same interval between 800 and 1000 W/m^2 evaluated above, lead to $\Delta P < 6.5\%$. at values of temperature of 25 °C and to $\Delta P < 5.5\%$ for values of temperature above 50 °C.

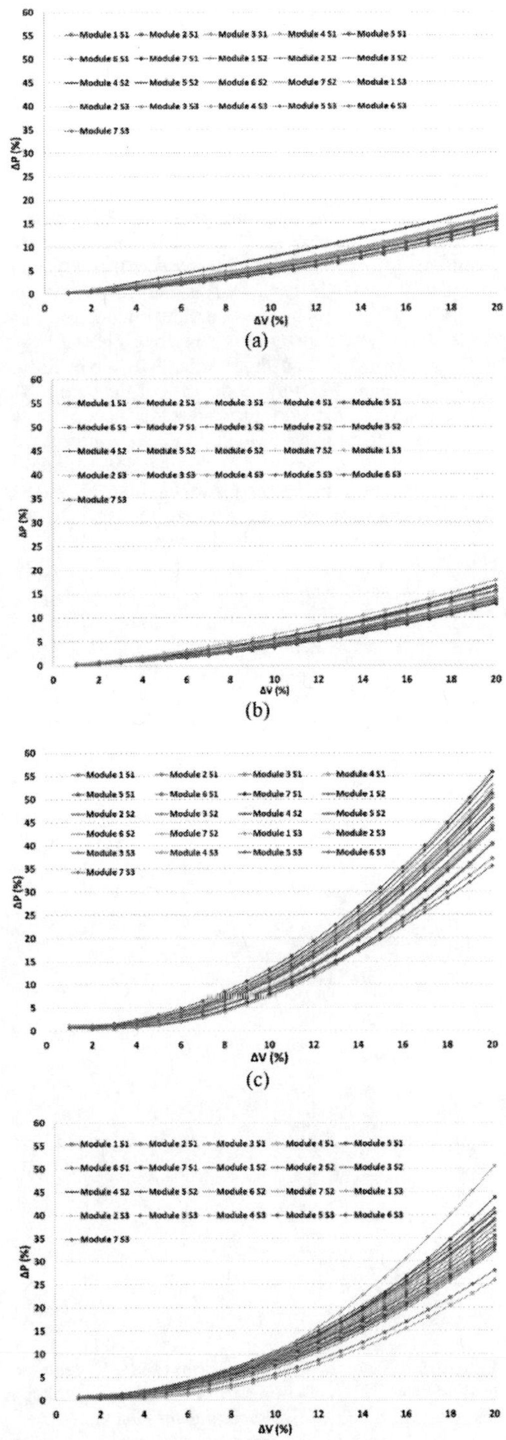

Figure 8: Power losses (%) of each module respect to the voltage operation variation (%) from V_m. Graphics (a) and (b) are for voltages below V_m and graphics (c) and (d) are for voltages above V_m. Graphics (a) and (c) are related to measurements at 25°C while graphics (b) and (d) are related to temperatures greater than 50°C. All the curves were measured at 1000 W/m^2.

Finally, the graphics show clearly that not only the values of the power losses increase quicker for voltages higher than V_m as operation voltage moves away from their optimal point: the dispersion of these power losses also increases considerably. That is, mismatch losses will be considerably higher when the operation voltage is above V_m.

The same analysis of power losses but now related to the P-V curves of the three strings and the PV generator is shown in Fig 9. Now, the curves have been measured at different operation temperatures greater than 50ºC and at 1000 W/m². The behaviour is similar to the one observed previously in the individual modules: it is almost linear when the power losses are evaluated below V_m (Fig. 9a) and it is quadratic when they are evaluated above this value (Fig. 9b). In the case of the strings, the dispersion observed and the figures achieved as the operation voltages move away from V_m are coherent with the obtained in the analysis of the individual modules. Interesting, the array power losses far from V_m (higher voltages) are lower than the obtained for the individual strings. This could be explained because the strings are connected in parallel.

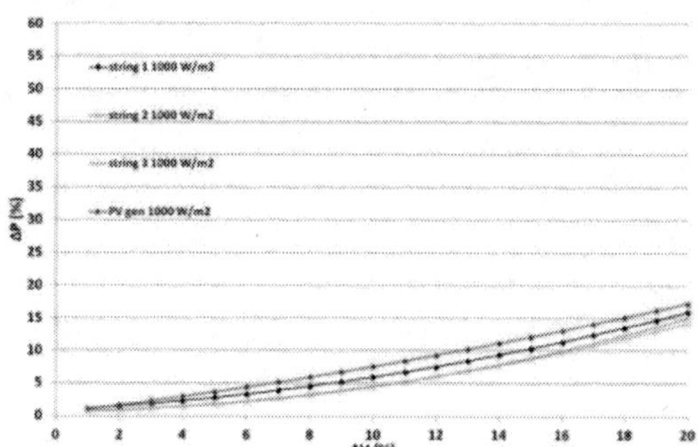

(a)

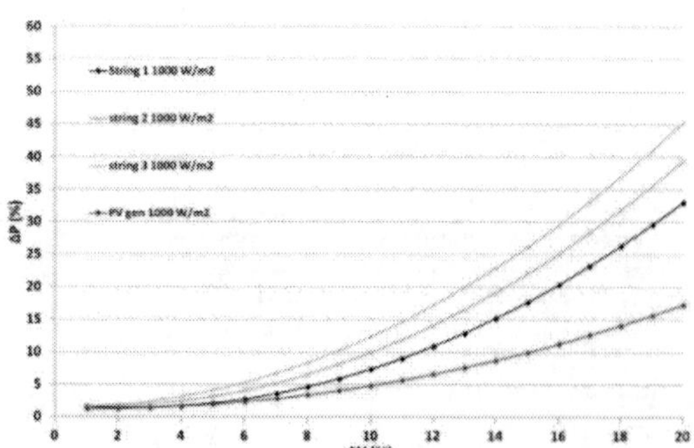

(b)

Figure 9: Power loss of each string and PV generator evaluated at different irradiance values (a) for voltages less than V_m and (b) for voltages greater than this value.

3.3 Modelling of power losses.

The power losses observed from the PV curves of modules, strings and generator can be adjusted with a quadratic equation which allows us to estimate them with a good accuracy (Equation 1)

$$\Delta P = a \cdot \Delta V + b \cdot \Delta V^2 \quad (1)$$

In the previous equation ΔV is the difference –always positive– between the voltage at the maximum power point and the evaluated voltage.

Table I shows the coefficients obtained for voltages below V_m (a_- and b_-) and for voltages above V_m (a_+ and b_+) when mean values of the power losses of all the measurements from each system (module, string or array) are considered. As can be seen, the coefficients at both sides of V_m are quite different. So, the proposed strategy of evaluate separately these regions leads to better power losses estimations. Besides, as $b_+ > b_-$ the contribution of the term related to ΔV^2 to the final estimation is considerable higher for operation voltages $V_{op} > V_m$. This is logical due to the different shape of the curves (see Fig. 8 and Fig. 9). Finally, although the power losses seem to have a linear behaviour below V_m, it is important to model them with a quadratic equation as the proposed one: if the term related to ΔV^2 is removed ($b_- = 0$) the goodness of the estimation (correlation coefficient R^2) is deteriorated at least a 5%.

Table I: Coefficient for estimation of power loss (equation 1).

		$V_{op} < V_m$		$V_{op} > V_m$	
		a_-	b_-	a_+	b_+
T=25ºC	**Module**	0.356	0.022	-0.264	0.130
T>50ºC	**Module**	0.220	0.025	-0.087	0.095
	String	0.275	0.023	0.094	0.092
	Array	0.671	0.009	0.159	0.035

4 CONCLUSIONS.

This work has presented a deep look into the mismatch losses by means of the observation in real time of the individual operation voltage and the temperature of each module in a 5kWp PV generator during 17 months. This monitoring has reported that there is a mismatch loss between the modules inversely proportional to the irradiance. Interestingly, the operation voltage dispersion overtakes the temperature dispersion observed between modules, which is proportionally to the irradiance, i.e. it increases as the irradiance increases.

Taking into account the I-V curve of each module, as well as the curves of the strings and the PV generator, the mismatch power loss can be estimated by means of the voltage variation around the voltage at the maximum power point, V_m. The power loss estimation has been divided in two parts for better accuracy of the results: below V_m the power loss tend to be almost linear, independently the temperature and the size of the system considered (module, string or generator). Above this value the power loss has a quadratic behaviour. This time, it seems to have a dependence of the loss with the temperature: it is greater at lower temperatures. Besides the dispersion of power loss is higher at this side of V_m. The results from the I-V measurements of strings and generator are similar to the obtained for single modules. There is any difference only when the power losses are evaluated in the array for values above V_m: its behaviour is also quadratic, but now the dispersions reached are lower than those reached by each module/string individually. This could be explained because the strings are connected in parallel.

Anyway, the power losses on both sides can be estimated by adjusting a quadratic equation, which brings results with good accuracy. A linear adjust could be used for voltages below V_m, but goodness of the adjust is then deteriorated.

As far as we know, this is the first time the relation between these losses and irradiance is experimentally demonstrated.

5 ACKNOWLEDGEMENTS

Authors would like to thank to CNPq (Conselho Nacional de Desenvolvimento Científico e Tecnológico), for the financing which made possible this research.

6 REFERENCES

[1] Rossa C.H., Martínez-Moreno F., Lorenzo E. *Reducing uncertainty in outdoors PV module characterization*. 35th EUPVSEC 2018.

[2] Evans, R., Boreland M., Green M. A. *A holistic review of mismatch loss: From manufacturing decision making to losses in fielded arrays*. Solar Energy Materials and Solar Cells. 174, 214-224 (2018). doi:10.1016/j.solmat.2017.08.041

[3] Iannone F., Noviello G., Sarno A. *Monte Carlo techniques to analyse the electrical mismatch losses in large-scale photovoltaic generators*. Solar Energy. 62 (2) 85-92. (1998). doi:10.1016/S0038-092X(97)00085-6

[4] Herrmann W., Kämmer S., Yusufoglu U. *Circuit losses in PV arrays caused by electrical mismatch of PV modules e impacts of temperature gradient and a variation of irradiance*. 28th EUPVSEC 2013, pp. 4127-4131. doi:10.4229/28thEUPVSEC2013-5BV.7.20

[5] Pavan A.M., Mellit A., De Pieri D., Lughi V. *A study on the mismatch effect due to the use of different photovoltaic modules classes in large-scale solar parks*. Prog. Photovoltaics Res. Appl. 22 (3) (2014) 332-345. doi:10.1002/pip.2266

[6] Han C., Lee H. *Investigation and modelling of long-term mismatch loss of photovoltaic array*. Renewable Energy. 121, 521-527 (2018). doi:10.1016/j.renene.2018.01.065

[7] Lappalainen K., Valkealahti S. *Photovoltaic mismatch losses caused by moving clouds*. Solar Energy. 158, 455-461 (2017). doi:10.1016/j.solener.2017.10.001

[8] Picault D., Raison B., Bacha S., de la casa J., Aguilera J. *Forecasting photovoltaic array power production subject to mismatch losses*. Solar Energy. 84, 1301-1309 (2010). doi:10.1016/j.solener.2010.04.009

[9] Karatepe E., Boztepe M., Çolak M. *Developing a suitable model for characterizing photovoltaic arrays with shaded solar cells*. Solar Energy. 81, 977-992 (2007). doi:10.1016/j.solener.2006.12.001

[10] Forniés E., Naranjo F., Mazo M., Ruiz F. *The influence of mismatch of solar cells on relative power loss of photovoltaic modules*. Solar Energy. 97, 39-47 (2013). doi:10.1016/j.solener.2013.08.004

[11] Martínez-Moreno F., Figueiredo G., Lorenzo E. *In-the-field PID related experiences*. Solar Energy Materials and Solar Cells. 174, 485 – 493 (2018). doi:10.1016/j.solmat.2017.09.037

[12] Muñoz J., Lorenzo E. *Capacitive load based on IGBTs for on-site characterization of PV arrays*. Solar Energy 80 (11), 1489-1497 (2006). doi:10.1016/j.solener.2005.09.013

[13] Muñoz J., Lorenzo E., Carrillo J.M., Moretón R. *Design of a twin capacitive load and its application to the outdoor rating of photovoltaic modules*. Prog. Photovoltaics Re. Appl. 23, 247-252 (2015). doi:10.1002/pip.2425

[14] Carrillo J.M., Martinez-Moreno F., Lorenzo C., Lorenzo E. *Uncertainties on the outdoor characterisation of PV modules and the calibration of reference modules*. Solar Energy 155, 880-892 (2017). doi:10.1016/j.solener.2017.07.028

ESTIMATION OF THE PV OUTPUT POWER IN THE VARIOUS SETTING DIRECTIONS AND ANGLES FROM THE IV CHARACTERISTICS OF THE INCLINED PV MODULE

Kimihiko Saito[1], Michio Sato[1], and Michio Kondo[1,2]
[1] Faculty of Symbiotic Systems Science, Fukushima University, Kanayagawa 1, Fukushima, Fukushima, 960-1296, Japan
[2] Fukushima Renewable Energy Institute, National Institute of Advanced Industrial Science and Technology (FREA), Machiikedai 2-2-9, Koriyama, Fukushima, 963-0215, Japan

ABSTRACT: We tried to estimate the global horizontal irradiance (GHI) and the panel temperature (T_{panel}) from the inclined modules. The GHI was converted from the global solar irradiance on the inclined mini-module ($GSI_{mimi-module}$) which was calculated by its short circuit current, however, this $GSI_{mimi-module}$ should be underestimated unlike the case using a pyranometer due to the reflection of direct beam radiation (DBR) on its surface. Therefore, the correction of DBR component by taking into account the incident angle dependence of its reflection on the mini-module's surface was applied in the conversion of $GSI_{mimi-module}$ to the GHI. The power generation output of another photovoltaic power system, which locates about 20km away from the location of mini-module and has the different setting direction from the mini-module one, was evaluated by this monitored GHI and monitored T_{panel}, and it showed good agreement with the actual one on a clear day.
Keywords: c-Si, Monitoring, Pyranometer, Solar Radiation

1 INTRODUCTION

Installation of photovoltaic power systems (PVPSs) for commercial use, residential use and so on has been rapidly increasing all over the world as a major source of renewable energy. In this circumstance, the number of small scale PVPSs whose power generation capacity is under 50kW including the residential use occupied about 98% of the total (about 46% of the total capacity) as of the end of Sep 2017 in Japan [1], and the simple and low-cost failure diagnosis for these small scale PVPSs has been desired. As one of possible method for it, the system providing PV output power estimated from the global horizontal irradiance (GHI) measured by a pyranometer through the internet has been proposed for the failure monitoring of PVPSs in the area around the pyranometer by comparison with those actual power generation, where the estimated output power of PVPS is calculated by the global solar irradiance on the tilted surface converted from the measured GHI depending on its facing azimuth and tilt angle [2]. Considering increase of PVPSs installation in these days and in the future, enabling the estimation of GHI from the inclined PV panel in a general setup should further simplify the installation of this failure monitoring method in any area by unnecessity of the additional installation of pyranometer.

In this study, we tried to calculate the GHI and the panel temperature (T_{panel}) from the current-voltage (IV) characteristics of the mini-module set up on the inclined plane. Furthermore, this appropriateness and usefulness were demonstrated by comparison between the actual output power of another PV system which is about 20km away from the location of mini-module and its evaluated one from the GHI and T_{panel} obtained by the mini-module measurement.

2 EXPERIMENTS

We prepared the PV mini-module as a monitor of GHI and of T_{panel}, composed of the series connection of 10 c-Si solar cells of 125 mm x 15 mm strip, the tempered glass plate with the size of 240 mm square, the PET-based back sheet and the Al-based frame, which were encapsulated not by the commonly used EVA but by the silicone resin [3]. Its maximum output power (P_{max}) and 'deemed' short circuit current (I_{sc}) were obtained every 1 minute from the IV characteristics measured by the simple circuit for outdoor use with its tracing time of 7 seconds, whose minimum load resistance was not zero but 1Ω, therefore, the saturated current in the low voltage region was regarded as I_{sc}. These mini-module and measurement system were installed on the roof of the building in Motomiya City in Fukushima Prefecture as shown in Fig.1, in which the mini-module was equipped parallelly to the roof surface whose facing azimuth and tilt angle are 51 degrees from south to west and 14 degrees, respectively. The two 2nd-class pyranometers were also installed to measure the global solar irradiance on the tilted surface with the same facing azimuth and tilt angle as the mini-module ones ($GSI_{pyranometer}$) and the GHI ($GHI_{pyranometer}$) in order to check the validity of the global solar irradiance obtained by the inclined mini-module ($GSI_{mini-module}$) and of the GHI ($GHI_{mini-module}$) converted from the $GSI_{mini-module}$. It should be noted that the response speed against the solar irradiance of the mini-module is much faster than that of pyranometer and the intensity of solar radiation by the pyranometers was monitored every 1 minute as the average of its values measured at every second.

Figure 1: Photograph of the setup of the mini-module and pyranometers on the roof.

Verification of the correspondence of expected power generation output (EPGO) evaluated from the $GHI_{mini-module}$ and $GSI_{mini-module}$ to the actual power generation output (APGO) was carried out at the PVPS in Kawamata Machi, which locates about 20km away from the location of the mini-module and is composed of the c-Si PV panels whose facing azimuth and tilt angle are south-

southeast and 15 degrees, respectively. This PVPS consists of the four systems with the rated panel output of 12kW each and of the system with it of 9kW and the two systems of 12kW in these were used for this verification. The APGO of these systems were monitored in DC. The GHI was also measured by the 2nd-class pyranometer at the side of this PVPS.

3 RESULTS AND DISCUSSION

3.1 Global Horizontal Irradiance (GHI)

The GSI by the mini-module ($GSI_{mimi\text{-}module}$) was calculated by the following equation by using its I_{sc}.

$$GSI_{mini\text{-}module} = I_{sc}/I_{sc@std} \times 1000[W/m^2] \quad \cdots(1)$$

$I_{sc@std}$: I_{sc} measured under the standard condition (with AM1.5 irradiance of 1kW/m^2 at 25°C)

Figure 2 shows the correlation between the daily amount of $GSI_{mimi\text{-}module}$ and it of $GSI_{pyranometer}$ from July/2017 to July/2018, in which the $GSI_{mimi\text{-}module}$ has a good linier relationship with the $GSI_{mimi\text{-}module}$ with its proportionality factor of 0.95. The spectral response of the mini-module is different from it of the pyranometer, which should influence on this few % deviation of the factor from 1 as an intrinsic error. However, this deviation should also come from the tilt error of sensitivity of the pyranometer (<±2% in its specification) and from the difference between the response speed against the solar irradiance of the mini-module and it of the pyranometer, therefore, we have concluded that the evaluation of solar irradiance by the I_{sc} of the mini-module with the equation (1) has no serious problem in this work.

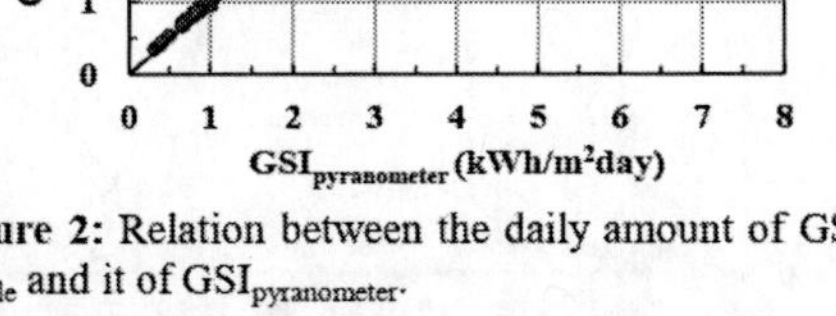

Figure 2: Relation between the daily amount of $GSI_{mini\text{-}module}$ and it of $GSI_{pyranometer}$.

Consequently, the $GHI_{mini\text{-}modul}$ was calculated by solving the simultaneous equations expressed below, where the I_{sc} of the mini-module was divided into the three components I_{scdir}, I_{scdiff} and I_{scref} generated by the direct normal irradiance (DNI), the diffuse horizontal irradiance (DHI) and the ground-reflected diffuse irradiance, respectively.

$$I_{sc} = I_{scdir} + I_{scdiff} + I_{scref} \quad \cdots(2)$$
$$I_{scdir}/I_{sc@std} \times 1000[W/m^2] = E_{hdir} \times \cos\rho/\cos\theta_z \quad \cdots(3)$$
$$I_{scdiff}/I_{sc@std} \times 1000[W/m^2] = E_{hdiff} \times (1 + \cos\theta)/2 \quad \cdots(4)$$
$$I_{scref}/I_{sc@std} \times 1000[W/m^2] = alb \times E_h \times (1 - \cos\theta)/2 \quad \cdots(5)$$

$$GHI_{mini\text{-}module} = E_h = E_{hdir} + E_{hdiff} \quad \cdots(6)$$

E_{hdir}: direct horisontal irradiance [W/m^2]
E_{hdiff}: diffuse horisontal irradiance [W/m^2]
E_h : global horisontal irradiance [W/m^2]
alb : albedo coefficient
θ : tilt angle of the mini-module
ρ : incident angle of sunlight to the mini-module
θ_z : solar zenith angle

In this calculations, the data reported by Erbs et.al. [4] were employed for the determination of the ratio of E_{hdiff} to E_h. Furthermore, the DHI on the mini-module was calculated by the isotropic model as expressed in the equation (4) and the ground albedo coefficient in the equation (5) was fixed at 0.2. One of its examples on a clear day is shown by the red solid line in Fig.3, which has the good agreement with the actual measurement value of $GHI_{pyranometer}$ denoted by the black solid line during the afternoon, however, its deviation from the $GHI_{pyranometer}$ is observed during the morning. The exposure of the mini-module to the DNI started from 6:14AM 54 minutes after the sunrise on this example's day according to the simulation, therefore, this deviation should be caused not by the geometric configuration of the mini-module against the DNI but by the incident angle dependence of direct beam on the I_{sc} of the mini-module. The upper right figure in Fig.3 shows the incident angle dependence of the normalized I_{scdir} generated by the straight light derived from the optical simulation of surface reflectance of the mini-module and recalculation of the $GHI_{mini\text{-}modul}$ by taking into account this dependence as a function of $f(\rho)$ was carried out with the modification of eq.(3) to the one as follows,

$$I_{scdir}/I_{sc@std} \times 1000[W/m^2] = f(\rho) \times E_{hdir} \times \cos\rho/\cos\theta_z \cdots(7)$$

The blue dashed line in the Fig.3 shows this recalculated one, which becomes consistent with the $GHI_{pyranometer}$ in a whole day with disappearance of the deviation during the morning and supports appropriateness of the modification represented by eq.(7).

Figure 4 shows the daily amount of $GHI_{mimi\text{-}module}$ from July/2017 to July/2018 plotted against it of $GHI_{pyranometer}$, which has a good linier relationship with its proportionality factor of 0.98. Furthermore, the small standard deviation value (σ) of 3.05% was obtained in its difference ratio defined as ($GHI_{mimi\text{-}module}$- $GHI_{pyranometer}$)/ $GHI_{pyranometer}$.

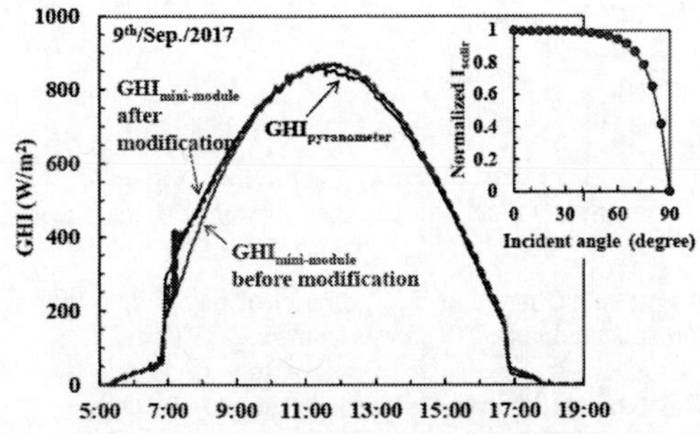

Figure 3: Daily spectrum of the GHI on a clear day. Upper right figure shows the incident angle dependence of the normalized I_{scdir}.

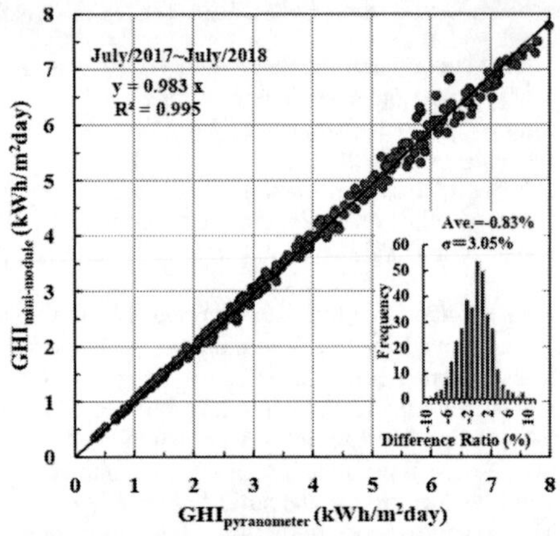

Figure 4: Relation between the daily amount of modified $GHI_{mini\text{-}module}$ and it of $GHI_{pyranometer}$. Bottom right figure in the graph shows the histgram of difference ratio $(GHI_{mimi\text{-}module}$- $GHI_{pyranometer})/ GHI_{pyranometer}$.

3.2 Panel Temperature (T_{panel})

The T_{panel} of the mini-module ($T_{mini\text{-}module}$) was calculated by the following equation with the typical temperature coefficient (α) of 0.45%/°C.

$$T_{mini\text{-}module} = (100\%/\alpha)\times[1-(P_{max}/P_{max@std})\times(I_{sc@std}/I_{sc})]+25°\mathrm{C} \quad \cdots(8)$$

$P_{max@std}$: P_{max} measured under the standard condition (with AM1.5 irradiance of 1kW/m^2 at 25°C)

The large noise of $T_{mini\text{-}module}$ was observed especially at the time with the low solar irradiance such as dawn and dusk and with the rapid fluctuation of solar irradiance by flowing cloud, therefore, smoothing treatment was employed to determination of $T_{mini\text{-}module}$ based on the large difference of time constant between the change of solar irradiance and of the panel temperature. Figure 5 shows the one of calculated examples of $T_{mini\text{-}module}$ on a clear day. This calculation is applicable only under the solar radiation, therefore, $T_{mini\text{-}module}$ at no solar radiation was treated as 25°C.

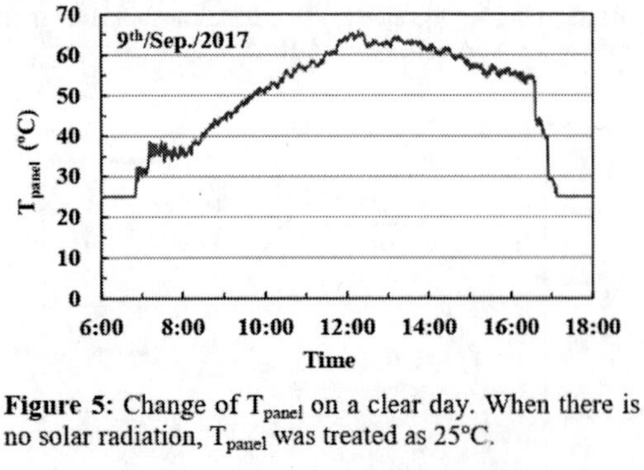

Figure 5: Change of T_{panel} on a clear day. When there is no solar radiation, T_{panel} was treated as 25°C.

3.3 Expected Power Generation Output (EPGO)

The EPGO of the PVPS in Kawamata Machi was estimated by the following equation (9) using the $GHI_{mini\text{-}module}$ and T_{panel} obtained in Motomiya City as described above. The temperature of the panels in the PVPS was regarded as the same as it of the mini-module ($T_{mini\text{-}module}$) and the typical temperature coefficient of 0.45%/°C was used in this estimation.

$$EPGO = 12[\mathrm{kW}] \times GSI_{panel}[\mathrm{W/m^2}]/1000[\mathrm{W/m^2}] \times\{1- 0.45/100\times(T_{mini\text{-}module} - 25[°\mathrm{C}])\} \quad \cdots(9)$$

Here, the global solar irradiance on its inclined panels (GSI_{panel}) was calculated by eq. (6) and (10) ~ (13) as expressed below, where the ratio of E_{hdiff} to E_h reported by Erbs et.al. and the albedo coefficient of 0.2 were used in the same way as the conversion of $GSI_{mini\text{-}module}$ to $GHI_{mini\text{-}module}$.

$$SR_{tdir} = E_{hdir} \times \cos\rho_{panel}/\cos\theta_z \quad \cdots(10)$$
$$SR_{tdiff} = E_{hdiff} \times (1 + \cos\theta_{panel})/2 \quad \cdots(11)$$
$$SR_{tref} = alb \times E_h \times (1 - \cos\theta_{panel})/2 \quad \cdots(12)$$
$$GSI_{panel} = SR_{tdir} + SR_{tdiff} + SR_{tref} \quad \cdots(13)$$

SR_{tdir}: direct solar radiation on the tilted panels [W/m^2]
SR_{tdiff}: diffuse solar radiation on the tilted panels [W/m^2]
SR_{tref} : ground-reflected diffuse radiation on the tilted panels [W/m^2]
θ_{panel} : tilt angle of the panels
ρ_{panel} : incident angle of sunlight to the panels

Figure 6 (a) and (b) show the comparison between the $GHI_{mini\text{-}module}$ estimated in Motomiya City and the $GHI_{pyranometer}$ measured at this PVPS in Kawamata Machi and between its PVPS's EPGO and APGO on a clear day, respectively, which represent the good agreement of those except the parts influenced by the shade of trees and demonstrate the appropriateness of the GHI and T_{panel} obtained from the mini-module.

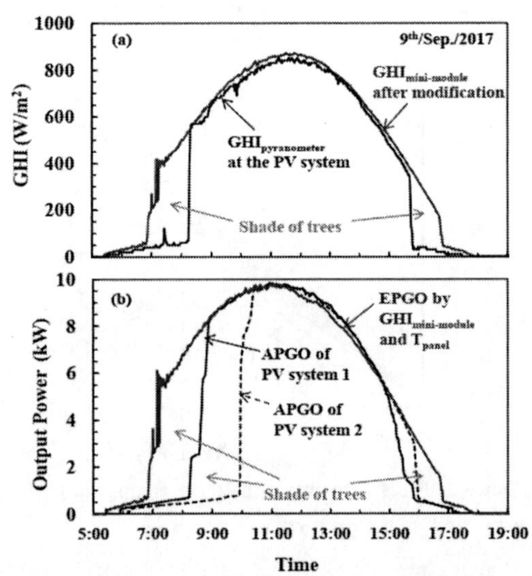

Figure 6: Results of the daily spectrum at the PVPS in Kawamata Machi which locates about 20km away from the location of mini-module on a clear day. (a) comparison between the $GHI_{mini\text{-}module}$ and the $GHI_{pyranometer}$, (b) comparison between the EPGO and APGO (DC output) of the two 12kW systems (PV system 1 and 2) in the PVPS.

Figure 7 shows the correlation between the amount of EPGO and the amount of APGO during the time without any influence of the shade of trees on the PV output

(9:30~13:00 for PV system 1 and 11:00~14:00 for PV system 2) on clear days from Sep/2017 to July/2018. This 'clear days' means the days which showed the sine-curve-like daily septra of GHI at both sites in Motomiya City and in Kawamata Machi and the other days were excluded from this comparison, because the time of change in GHI caused by the cloud shading was different between these both sites due to its long distance of 20km. Even though the exsitence of this long distance, the good correspondence between the EPGO and the APGO on the clear days is observed, which suggests the effectiveness of the estimation of EPGO for the failure monitoring of PVPSs by using the IV characteristics of the inclined PV module nearby.

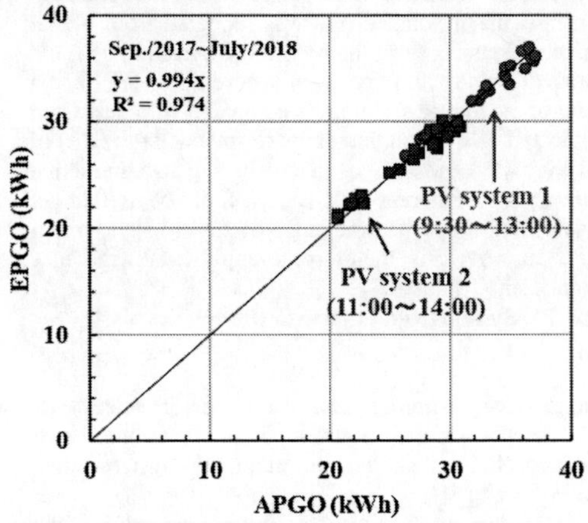

Figure 7: Relation between the amount of EPGO and the amount of APGO (DC output) on clear days. Red circle and blue square indicade those amount of the PV system 1 from 9:30 to 13:00 and of the PV system 2 from 11:00 to 14:00, respectively.

4 SUMMARY

The estimation method of the GHI and T_{panel} from the current-voltage (IV) characteristics of the PV module installed on the inclined plane was proposed and the appropriateness of its GHI was demonstrated by comparing the GHI estimated from the I_{sc} of the mini-module with the GHI measured by the pyranometer nearby the mini-module. Furthermore, the effectiveness of this method was demonstrated by the accurate estimation of EPGO of another PVPS locating about 20km away from the location where the GHI and T_{panel} were obtained, which promises well for providing the simple, widely-usable and cost-effective failure monitoring of PVPSs.

5 ACKNOWLEDGEMENTS

We thank to the Nihon Rokki Ind. Co., Ltd. for their support of the outdoor measurement.

6 REFERENCES

[1] http://www.enecho.meti.go.jp/category/saving_and_new/saiene/statistics/

[2] Y. Ueda, S. Kawamoto, M. Saijyo, and N. Abe, Annual Conference of Power and Energy Society IEE of Japan, proceedings, No.127 (2010) pp.06-11-06-12

[3] Saito K, Yamazaki J, Yoshino D, Higuchi N, Kondo M, 32nd European PV Solar Energy Conference and Exhibition, proceedings, (2016) 5BV.2.8.

[4] D. G. Erbs, S. A. Klein, and J. A. Duffie, Solar Energy Vol.28, No.4 (1982) 293

PERFORMANCE OF A REMOTE HYBRID PV SYSTEM BASED ON REAL AND MODELLED DATA IN INDONESIA

Kun Kunaifi[1,3], A.H.M.E. Reinders[1,2]
[1] University of Twente, Faculty of Engineering Technology, Department of Design, Production and Management, P.O. Box 217, 7500 AE Enschede, The Netherlands; k.kunaifi@utwente.nl; a.h.m.e.reinders@utwente.nl.
[2] Eindhoven University of Technology, Department of Mechanical Engineering, Energy Technology Group, P.O. Box 513, 5600 MB Eindhoven, The Netherlands; a.h.m.e.reinders@tue.nl.
[3] UIN Suska Riau University, Faculty of Science and Technology, Department of Electrical Engineering, Jl. H.R. Soebrantas No. 115, Pekanbaru 28292, Indonesia: kunaifi@uin-suska.ac.id.

ABSTRACT: This paper is aimed at assessing the performance of a remote hybrid PV/diesel system located on the Island of Kri in Papua, Indonesia. For this purpose, monitoring data of electrical variables at 5 minutes recording intervals were used. Using the INSEL simulation environment, in-plane irradiance (G_i) was modeled from global horizontal irradiance (G_h) data from five weather stations at various distances from the hybrid PV system. Also, PV module temperature (T_m) was predicted from ambient temperature (T_a) and wind speed (v) according to the Sandia model for module temperature. The performance of the PV element of the hybrid system was compared with data from another PV system in Jayapura (also in Papua) which is grid-connected. We found that the performance ratio (*PR*) of the PV systems in Kri and Jayapura during the observation period were 41% and 90%, respectively. We also found that data from remote weather stations with a distance of up to 100 km from the reference PV system could be used to get the same *PR* values. However, with greater distance, the *PR* deviates non-linearly in the range of 20% to 50%, which is in line with previous findings indicating no correlation occurs. Our advice is, therefore, to apply local irradiance monitoring at PV sites instead of extrapolating irradiance over significant distances.
Keywords: Grid-Connected, Modelling, Monitoring, Performance, PV System, Rural Electrification.

1 INTRODUCTION

For many years, diesel generators (gensets) have been used for producing electricity in remote areas of Indonesia. However, they have high operation and maintenance costs, create noise, cause pollution, and emit carbon dioxide (CO_2) [1]. Given the increasing costs of diesel fuel and decreasing costs of PV modules, hybrid PV/diesel have become favorable alternatives for diesel gensets. They are economically viable for 15-20 years of project lifetimes [2], especially in rural areas and remote islands [3].

Highly performing hybrid PV systems are required and therefore, understanding their performance is important for further improvements. However, the essential variables that are needed for analyzing their performance are not always measured on-site. This especially applies to small hybrid PV systems [4] for which monitoring systems are considered to be too expensive. Alternatively, their performance can be assessed using data from other locations, such as meteorological stations or other PV systems that are monitored [5].

Based on that background, this study evaluates the use of remote data to estimate the performance of a selected hybrid PV system in Indonesia. This study proposes a modeling and statistical approach for performance analysis of a PV system with incomplete on-site datasets. The procedure involves two PV systems and seven automatic weather stations (AWS) of the Indonesian Agency for Meteorological, Climatological and Geophysics (BMKG) which are available online (Figure 1).

This study is relevant for Indonesia because Indonesia lacks best practices in implementing PV systems while these could be a proper alternative power source for thousands of communities in rural areas or small islands.

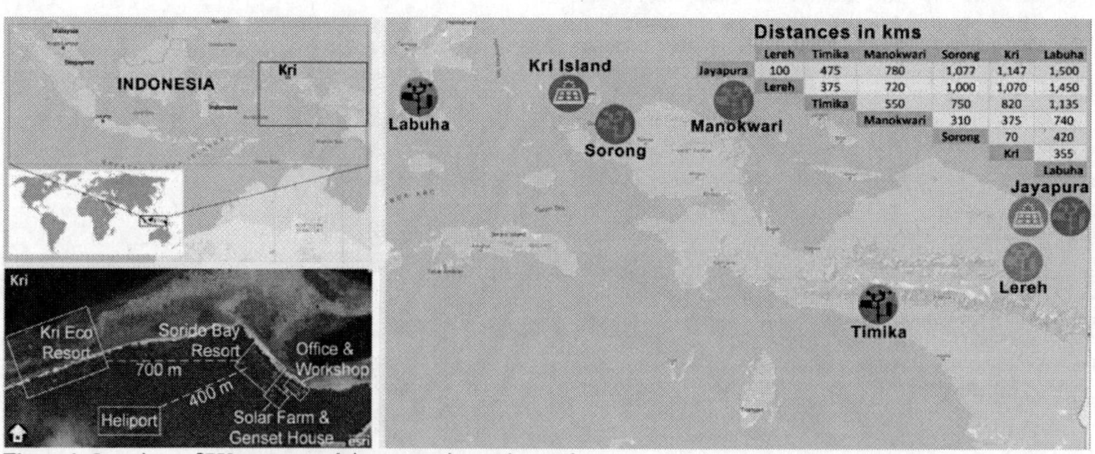

	Lereh	Timika	Manokwari	Sorong	Kri	Labuha
Jayapura	100	475	780	1,077	1,147	1,500
	Lereh	375	720	1,000	1,070	1,450
		Timika	550	750	820	1,135
			Manokwari	310	375	740
				Sorong	70	420
					Kri	355
						Labuha

Figure 1: Locations of PV systems and the automatic weather stations.

2 MONITORING AND METHODS

2.1 Hybrid PV/diesel system on Kri Island

The hybrid system on the island of Kri (0° 557 'S, 130°685 'E) in the Raja Ampat Regency, the Province of West Papua, Indonesia, consists of three PV arrays with a total capacity of 28.8 kWp and two 42 kVA and a 60 kVA Yanmar gensets connected in an AC-coupled configuration (see Figure 2). The PV arrays are ground-mounted close to the equator with 15° tilted angle to the North (two arrays) and to the South (one array). Each array consists of 6 panels which contain 8 monocrystalline PV modules of 200 Wp from Sky Energy. Every two panels are connected to one inverter from Murata.

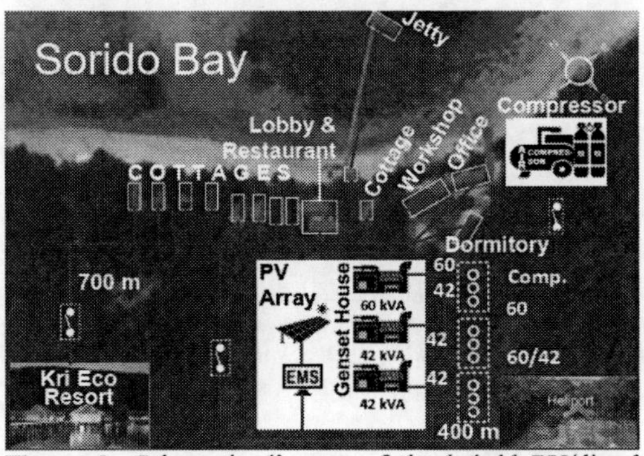

Figure 2: Schematic diagram of the hybrid PV/diesel systems on Kri Island, reproduced with permission of Murata. Photo credits: Kri Eco Resort (Herbert Innah).

Being a diving resort, two main electrical loads on Kri include the resort load and the compressor load. The resort load comes from guest accommodations, restaurant, offices, and workshop located at two locations namely Kri Eco Resort and Sorido Bay Resort. The two resorts are about 700 m apart. The compressor load comes from three compressors for filling the compressed air tanks for divers. The energy management system (EMS) of Murata measures power and energy from the PV panels, gensets, and loads with 5 minutes recording intervals. However, global in-plane irradiance (G_i) and PV module temperature (T_m) were not measured on-site.

2.2 Grid-connected PV system in Jayapura

As a reference, a 34 kWp grid-connected PV system has been operating since 2012 in the City of Jayapura (2°562 'S, 140°692 'E) for research purpose and has been extensively monitored and analyzed [6-8]. The PV modules are tilted 10° to the North. It consists of four arrays of different sizes and types of PV modules. Due to the availability of data, in this study, we use only the 7 kWp array of micro-amorphous silicon modules (Ample Sun) which are connected to an SMA Sunny Mini Central (SMC) inverter.

All variables for performance analysis were measured on-site at a recording interval of 5 minutes for electrical variables and 1 minute for meteorological variables. They include the following variables: total energy production, currents, power, voltage, global horizontal irradiance (G_h), global in-plane irradiance (G_i), ambient temperature (T_a), and PV module temperature (T_m).

2.3. Automatic weather stations (AWS)

Data from seven AWSs were used as modeling inputs. Three variables monitored by the AWSs used in this study include G_h, T_a, and wind speed (v). The data sampling varies from one sensor to others, but the recording intervals were the same, namely 10 minutes.

Because the timeframe of the available data from the PV systems at Kri and Jayapura do not match one and each other, we selected only AWSs that provide data in similar time frames. Therefore, the AWS stations used for the performance analysis of PV system in Jayapura are different to the AWS stations used for the analysis of PV system in Kri. As shown in Figure 1, stations with green background refer to Kri and orange refer to Jayapura. To model the G_i and T_m for the performance analysis of power system at Jayapura, data from AWSs at Lereh, Timika, and Labuha with the distance of 100 km, 475 km, and 1,500 km, respectively, from PV system at Jayapura, will be used. Also, the AWSs at Sorong, Labuha, Manokwari, Timika and Jayapura with the distances of 70 km, 355 km, 375 km, 820 km, and 1,147 km respectively, from Kri island, will be used to model the G_i and T_m for the performance analysis of power system at Kri, see Figure 1.

The big distances among the PV systems and the AWSs are proportionate to the size of Indonesia which lies over more than 5,000 km length from its easternmost to its westernmost.

2.4. Modeling G_i from G_h

Some models offer validated approaches of estimating G_i from the G_h. Each model differs from others depending on how the diffuse fraction of the solar irradiance is handled. The Liu and Jordan model [9], for example, assumes an isotropic distribution over the complete skydome, while Hay [10] assumes a brightening of the horizon band and the circumsolar region [11].

In general, modeling the G_i from G_h involves two steps. First, the decomposition of G_h to its direct and diffuse components. Second, the transposition of these components to G_i. The conversion has a typical uncertainty of 2% to 5% [12].

In this study, we evaluated 60 combinations of decomposition and transposition models available in INSEL® software. INSEL is a simulation tool that uses block diagram for engineering programming, including applications of the renewable energy sector, which is mainly intended for use in research and education [13].

2.5. Modelling T_m from T_a and v

Because the PV module temperature (T_m) is often not measured on-site in many PV systems, including in Kri, we use the Sandia Module Temperature Model (SMTM). SMTM uses T_a and v as inputs to predict the Tm.

The T_m according to the SMTM is given by:

$$T_m = G.(e^{a+b.v}) + T_a \quad (1)$$

where a and b are parameters that depend on the module construction and materials as well as on the mounting configuration of the module. A more detailed explanation about the SMTM including the values of a and b can be found at the Sandia Laboratories website [14].

2.6. Performance assessment

Performance assessment of the hybrid PV system and the reference system was be conducted according to monitoring standard IEC 61724 [5]. Figure 3 shows the complete procedure applied in this study which involves the modeling of the G_i and Tm, and the calculations of the performance of the PV systems. The performance assessment will involve two main operations. First, using the G_h as input, the G_i will be calculated using INSEL. Then, using the T_a and v, T_m will be modeled according to the Sandia Temperature Model [14]. This step will be first applied to the reference PV system in Jayapura.

Second, based on the results from the first step, the performance of the PV system in Jayapura and Kri be calculated. This process will use the modeled G_i and T_m and measurement data from each PV system.

The procedure used to calculate the performance analysis according to IEC 61724 was developed in Python environment.

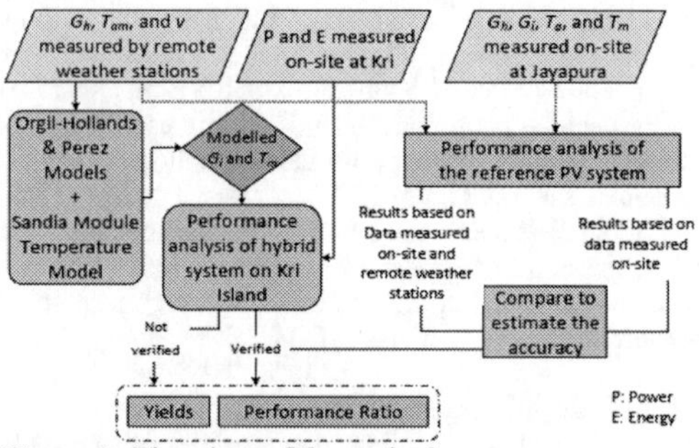

Figure 3: Flowchart of the method.

2.7. Treatment to the datasets

All irradiance values, either measured or modeled, which lie beyond 10 W/m^2 and 1500 W/m^2 have been removed from the datasets. The reason for this is because the former is prone to error due to extraneous night-time data values and the latter is because it is too high. Further, all the daytime data, between 06:00 and 18:00 local time, were kept for the analysis. All missing data were excluded from the analysis.

For the ambient temperature, only values between -40 and 60 °C were used. For the module temperature, values between ambient and ambient plus 40 °C were used because the studied PV systems are open rack-mounted systems. As such, any T_m values lower than T_a that might occur in the early morning due to irradiation to the sky were not included in the analysis.

For the electrical data of the PV system, the values used for the array voltage only those between 0 and 1.3 x V_{oc} of the array under STC. Also, the array current values between 0 and 1.5 x I_{sc} under STC were used in the analysis. Various Python codes were written for managing the data.

3. RESULTS AND DISCUSSION

As shown in Table I, weather data from seven remote weather stations and electrical data from two PV systems were used in this study. Weather data four stations were used for performance analysis of PV system at Jayapura (τ:15 January to 15 March 2015), while data from five stations were used for PV system at Kri (τ:20 September 2017 to 20 February 2018). Data from Labuha and Timika were commonly used for performance analysis of both PV systems.

All data have been checked for consistency and gaps to identify obvious anomalies. All nighttime data (between 18:05 and 15:55) have been excluded from the analysis, including outliers and discovered missing data.

Table I: Data locations and measured variables

Location	Measured Variables	Record Interval (min)	Time Period
1. Jyp *(AWS)*	G_h, T_a, v	10	20/9/17-20/2/18
2. Jyp *(PV)*	E_{Tot}, I_{ac}, I_{PV}, P_{ac}, V_{ac}, V_{PV}	5	15/1/15-15/3/15
3. Jyp *(AWS_{PV})*	G_h, G_i, T_a, T_m	1	15/1/15-15/3/15
4. Kri *(PV)*	L_R, L_C, P_{DG}, P_{PV}, E_R, E_C, E_{DG}, E_{PV}	5	20/9/17-20/2/18
5. Labuha	G_h, T_a, v	10	15/1/15-15/3/15 20/9/17-20/2/18
6. Lereh	G_h, T_a, v	10	15/1/15-15/3/15
7. Manokwari	G_h, T_a, v	10	20/9/17-20/2/18
8. Sorong	G_h, T_a, v	10	20/9/17-20/2/18
9. Timika	G_h, T_a, v	10	15/1/15-15/3/15 20/9/17-20/2/18

Jyp: Jayapura; L: Load (kW); P: Power output (kW); R: Resort; C: Compressor; DG: Diesel genset; PV: PV array; E: Energy (kWh); AWS: automatic weather station of BMKG, AWS_{PV}: automatic weather station included in a PV system.

3.2. Dealing with missing parameters

3.2.1. Tilted irradiance

An INSEL model was developed to convert G_h to G_t using 60 combinations of decomposition and transposition models. First, the horizontal extraterrestrial irradiance (I_o) was extracted using G_h measured by the AWS during the τ as input. Second, using the G_h and I_o, the diffuse component (I_d) of the irradiance was generated.

Third, the G_t was produced with inputs of G_h, I_d, tilted angle, azimuth degree, and ground reflectance. The last step can also generate other elements of irradiation, namely tilted beam radiation, tilted I_d, and tilted ground reflected radiation, but they were not used in this study.

Only 46 of 60 model combinations have produced acceptable results, while all models containing the Gueymard transposition model [15] were excluded because they gave very small or very large values in this particular study.

In the analysis, we found that the combination of Reindl Beckman and Dufie (RB) [16] and Willmott (WM) [17] models performs best in this study. However, their capability cannot be generally justified over the other model combinations because we tested them only at one location. Therefore, for further analysis in this study, we use the commonly accepted Orgill & Hollands (OH) [18] decomposition model and Perez (PZ) [19] transposition model instead [20]. Some other model combinations give small variation in their results.

The comparison of all modeled G_i with ground measurements was assessed using scatter plots, and on the basis of the coefficient of determination of linear regression (R^2) and root mean square error (RMSE), where the measured G_i was used as the reference data. Figure 4 shows small differences of R^2 ranging between 0.70 and 0.99 with the mean value equal to 0.97. A wide range of RMSE from 16.98 to 201.24 takes place with the

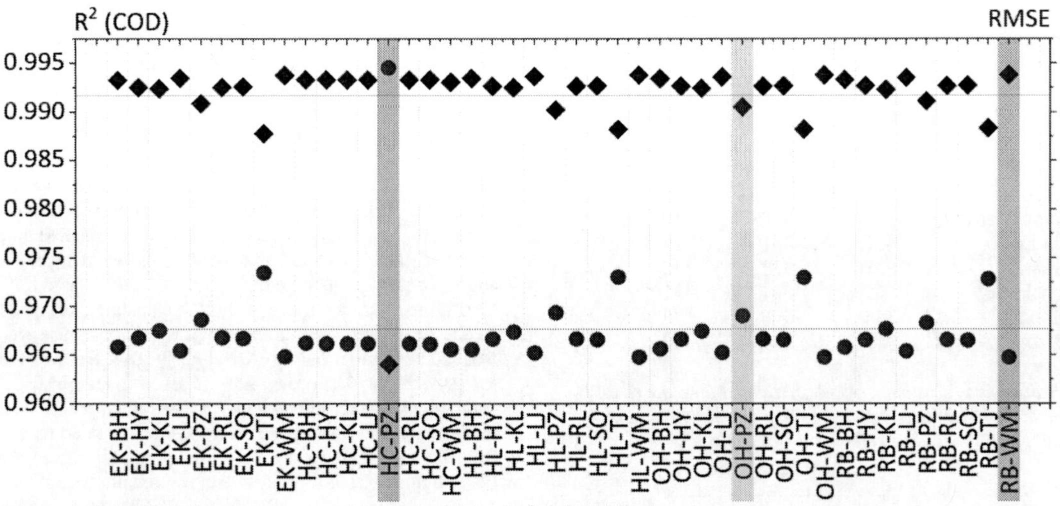

Figure 4: Modelled G_i versus measured G_i on the basis of R^2 and RMSE from Jayapura.

contribution from the Gueymard transposition model. However, the mean RMSE was 31.36.

Figure 5 shows a simple linear fitting between the measured G_h and measured G_i from Jayapura. It shows a small difference between the two data sets with the R^2 of 0.98284. This occurs because the study location is near the equator. This also indicates that modeling G_i from G_h in the equatorial region would be interesting only if a detailed evaluation is of importance. However, for further analysis in this study, we use the modeled G_i to give a better result.

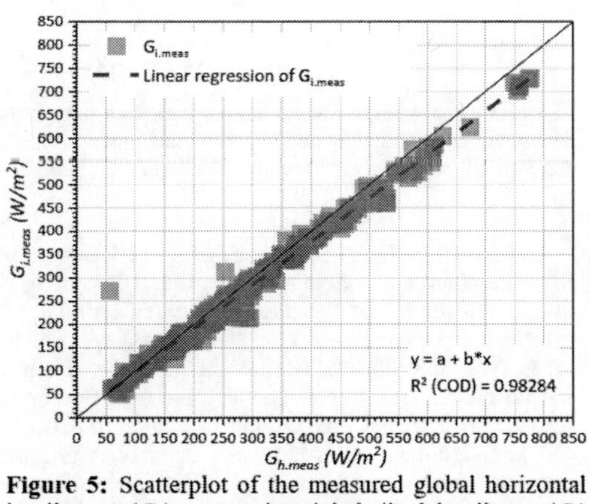

Figure 5: Scatterplot of the measured global horizontal irradiance (G_h) versus the global tilted irradiance (G_i) based on data from the PV system in Jayapura (15 January – 15 March 2015).

Next, the comparison statistics of the measured G_i in Jayapura to the modeled G_i from remote AWSs were separately calculated to assess whether these statistics were dependent on location. The statistical significance of the differences in the comparison statistics derived for the weather stations was tested using the two-sided paired linear fitting tests. Figure 6 shows significant reductions in the R^2 values. Referring to G_i in Jayapura, R^2 value for Lereh was 0.48411, 0.32156 for Timika, and 0.30392 for Labuha (a). The distances, however, do not correlate with the fitness of the G_i datasets (b and c).

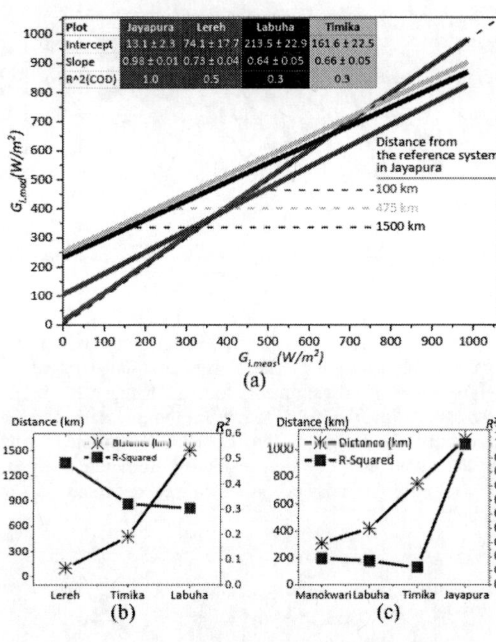

Figure 6: The measured versus the modeled tilted irradiances respectively based on data from PV system in Jayapura and remote AWSs; (a) R^2 of measured G_i and modeled G_i, (b and c) relation of R^2 and the distance.

3.2.2. PV module temperature

The SMTM has been applied to predict the T_m using the T_a and v as inputs. Further, the similar statistical approach used for analyzing the G_i (see Sec. 3.2.1) was also applied to analyze the T_m.

Although the previous datasets showed for the G_i were not linearly fit, the hourly means of G_i (daytime) were close to one and each other. This also applies to T_m as shown in Tabel II and Figure 7.

3.3. Performance of the PV systems

The performance of the PV system in Jayapura has been extensively monitored and analyzed as appear in the publications [6-8]. In general, the present analysis produces the same performance ratios (PR) as those publications.

Table II: The hourly averaged values of G_h, G_i, and Tm

Location	G_h	G_i	T_m
Jayapura *(PV)*	368	353	39
Jayapura *(AWS)*	385	348	39
Labuha	452	441	39
Lereh	338	332	36
Manokwari	405	384	41
Sorong	394	348	38
Timika	419	394	40

R^2 values from Jayapura (AWS), Labuha, Manokwari, and Timika are referred to Sorong. R^2 values from Labuha, Lereh, and Timika are referred to Jayapura.

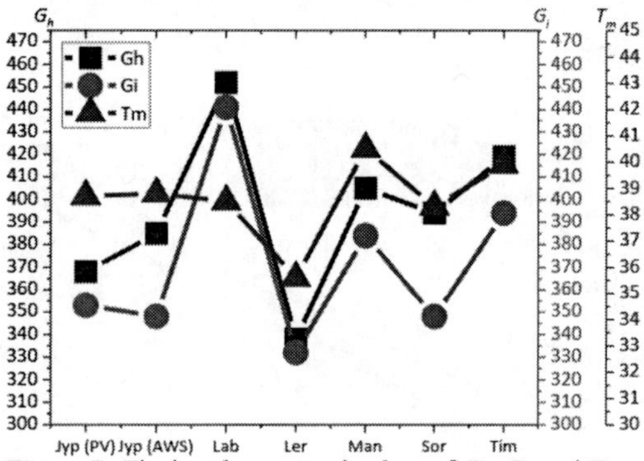

Figure 7: The hourly averaged values of G_h, G_i, and T_m

Figure 8 shows the weekly *PR* calculated using five meteorological stations. They were calculated based on eight weeks of measurements from 22 January 2015. For the calculation, the PV system in Jayapura was used as the reference. According to the IEC 61724 [5], data during system unavailability were excluded from the analysis. Therefore some missing points can be observed on the graph.

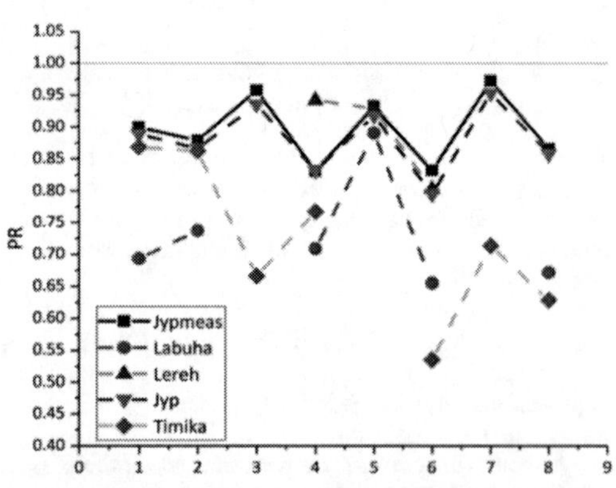

Figure 8: The weekly PR calculated based on data from the reference system in Jayapura and five remote meteorological stations (22 January - 15 March 2015).

As shown in Figure 8, the measured data show PR of 90% which was very close to PR 91% as reported in previous studies [6-8]. We found that, as expected, the PRs calculated using data from AWSs in Jayapura and Lereh were very close to 90% because they are relatively closely located, namely 3 km and 100 km from the reference PV system, respectively. However, with the distance above 100 km, the same PR of 70% was observed, although the great difference of distance from Timika and Labuha to the reference system exists (475 km and 1500 km for Timika and Labuha, respectively). Therefore, it can be concluded that with the distance greater distance 100 km, the correlation between the G_h and G_i cannot be observed.

Figure 9 shows the weekly *PR* calculated using 22 weeks data from five meteorological stations measured from 20 September 2017 to 20 February 2018. Because the PV system at Kri does not measure meteorological parameters on-site, the G_i and Tm were taken from AWS measurement in Sorong which is 70 km separated by sea from Kri. The data regarding power production from the PV system at Kri, however, was measured on-site.

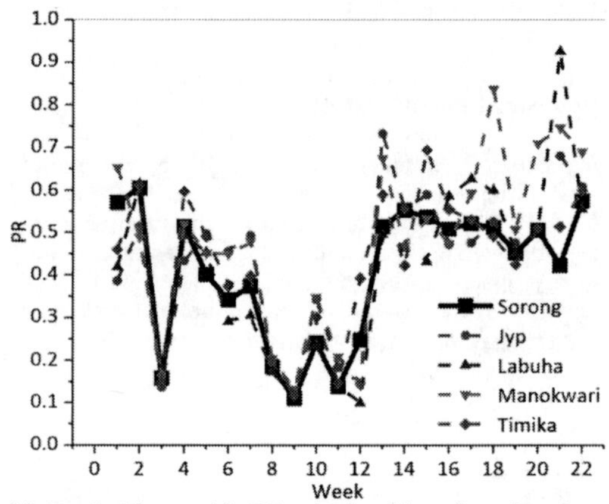

Figure 9: The weekly PR calculated based on data from the reference system in Sorong and four remote meteorological stations (20 September 2017 - 20 February 2018).

As shown in Figure 9, the mean *PR* of PV system in Kri using the reference meteorological data in Sorong is 41% (solid line). The *PR* calculated based on modeled data from other remote AWSs (dash lines) were 48%, 42%, 43%, and 50% for Manokwari, Labuha, Timika, and Jayapura, respectively. The distance from Sorong to Manokwari, Labuha, Timika, and Jayapura are 310 km, 420 km, 750 km, and 1077 km, respectively. Again, distances from the source of data and the reference system do not show any correlation in the calculation of PR.

There are three possible reasons for the low PR of the PV system at Kri. First, the value of the PR depends on the local energy consumption on the island. It means that the loads do not always utilize the produced energy from the PV array. Second, the efficiency of the system component, particularly the inverter. The inverter used at Kri is a new model with no information about its efficiency. Third, the measurement accuracy issues both from the PV system and from the AWSs.

However, if compared to the PR of other stand-alone PV systems, the PR of the PV system at Kri lies in the correct range. The majority of stand-alone PV systems have PR ranging from 0.3 and 0.4 [21].

4. CONCLUSION

In the absence of in-plane irradiance and PV module temperature, the performance of the PV system can still be estimated using the global horizontal irradiance and ambient air temperature and wind speed, respectively. This approach has been applied in this study under the tropical climate of Indonesia. A 34 kWp grid-connected PV system in Jayapura, Papua, was used as a reference system. The analysis results from the reference PV system was applied for predicting the performance of a 28.8 kWp PV array, which is part of a hybrid power system on a remote island of Kri in Raja Ampat, West Papua.

It has been found that the PR of PV system in Jayapura was 90%. We also found that using data from remote AWS with the distance up to 100 km from the reference PV system produce the same result. However, with distance more than that, the PR deviate in the range from 20% to 50%, which is in line with previous findings indicating no correlation occurs. Our advice is, therefore, if possible, to always apply local irradiance monitoring at PV sites instead of extrapolating irradiance over significant distances.

The PR of the hybrid PV system at Kri was 41% which is characteristics for stand-alone PV systems including in Indonesia [21].

5. ACKNOWLEDGMENT

We highly appreciate the financial support provided for this project by the Indonesia Endowment Fund for Education (LPDP). Expert support came from the Advanced Research on Innovations in Sustainability and Energy (ARISE) of the University of Twente which also supported the field study in Kri financially. Makoto Fujita of Murata provided invaluable data support, also Max Ammer and Andreia Nogueira of the Raja Ampat Research and Conservation Centre (RARCC). Special thanks also go to Frengki Numberi of the Jayapura City Government, Suroto of BMKG Papua, and EnReach of UIN Suska Riau University. We also acknowledge Juergen Schumacher for granting a lifetime INSEL license.

7. REFERENCES

1. Kusakana, K. and H.J. Vermaak, *Hybrid renewable power systems for mobile telephony base stations in developing countries*. Renewable Energy, 2013. **51**: p. 419-425.
2. Vetter, M. and G. Bopp, *Stand-Alone and Hybrid PV Systems*, in *Photovoltaic Solar Energy: From Fundamentals to Applications*, Angèle Reinders, et al., Editors. 2017, John Wiley & Sons, Ltd: West Sussex, United Kingdom. p. 539-552.
3. Sen, R. and S.C. Bhattacharyya, *Off-grid electricity generation with renewable energy technologies in India: An application of HOMER*. Renewable Energy, 2014. **62**: p. 388-398.
4. The Australian Photovoltaic Institute, *Australian Technical Guidelines for Monitoring and Analysing Photovoltaic Systems* in *Version 1*. 2013, The Australian Photovoltaic Institute.
5. IEC, *IEC 61724-1 IEC standard for photovoltaic system performance*, in *Part 1: Monitoring*. 2017, The International Electrotechnical Commission: Geneva, Switzerland.
6. Veldhuis, A.J. and A.H.M.E. Reinders, *Performance of a 34 kWp Grid-Connected PV System in Indonesia- A Comparison of Tropical and European PV Systems*, in *2014 IEEE 40th Photovoltaic Specialist Conference (PVSC)*. 2014, IEEE: Denver, CO, USA.
7. Veldhuis, A.J., et al., *An Empirical Model for Rack-Mounted PV Module Temperatures for Southeast Asian Locations Evaluated for Minute Time Scales*. IEEE Journal of Photovoltaics, 2015. **5**(3): p. 774-782.
8. A.H.M.E. Reinders, et al., *PV Electricity in Indonesia – A Sustainable Transition with Great Potential*, in *29th European Photovoltaic Solar Energy Conference and Exhibition*. 2014, WIP Wirtschaft und Infrastruktur GmbH & Co Planungs KG: Amsterdam, The Netherlands. p. 4109 - 4112.
9. Benjamin Y. H. Liu and R.C. Jordan, *The interrelationship and characteristic distribution of direct, diffuse and total solar radiation*. Solar Energy, 1960. **4**(3): p. 1-19.
10. Hay, J.E., *Calculating solar radiation for inclined surfaces- Practical approaches*. Renewable Energy, 1993. **3**(4-5): p. 373-380.
11. Schumacher, J., *INSEL 8 Tutorial: Simulation of Renewable Energy Systems*. 2014, Doppelintegral GmbH: Stuttgart.
12. Richter, M., et al., *Technical Assumptions Used in PV Financial Models: Review of Current Practices and Recommendations*. 2017, International Energy Agency: Paris, France.
13. Doppelintegral. *INSEL 8: Software for simulation, monitoring, and visualization of energy systems*. n.d. [cited 2018 16 June]; Available from: http://www.insel.eu/en/what-is-insel.html.
14. Sandia National Laboratories. *Sandia Module Temperature Model*. n.d. [cited 2018 23 February]; Available from: https://pvpmc.sandia.gov/modeling-steps/2-dc-module-iv/module-temperature/sandia-module-temperature-model/.
15. Gueymard, C.A., *Direct and indirect uncertainties in the prediction of tilted irradiance for solar engineering applications*. Solar Energy, 2009. **83**(3): p. 432-444.
16. D. T. Reindl, W. A. Beckman, and J.A. Duffie, *Diffuse fraction correlations*. Solar Energy, 1990. **45**(1): p. 1-7.
17. Willmott, C.J., *On the climatic optimization of the tilt and azimuth of flat-plate solar collectors*. Solar Energy, 1982. **28**(3): p. 205-216.
18. J. F. Orgill and K.G.T. Hollands, *Correlation equation for hourly diffuse radiation on a horizontal surface*. Solar Energy, 1977. **19**(4): p. 357-359.
19. Perez, R., et al., *Modeling daylight availability and irradiance components from direct and global irradiance*. Solar Energy, 1990. **44**(5): p. 271-289.
20. Lave, M., et al., *Evaluation of Global Horizontal Irradiance to Plane-of-Array Irradiance Models at Locations Across the United States*. IEEE Journal of Photovoltaics, 2015. **5**(2): p. 597-606.
21. Mayer, D. and M. Heidenreich. *Performance analysis of stand alone PV systems from a rational use of energy point of view*. in *The 3rd World Conference on Photovoltaic Energy Conversion*. 2003. Osaka, Japan: IEEE.

YIELD OF SMALL ROOF-TOP PV SYSTEMS IN GERMANY 2017

Henrik te Heesen*, Volker Herbort#, Martin Rumpler*
*Trier University of Applied Sciences, Environmental Campus Birkenfeld, Campusallee, 55768 Neubrücke (Nahe), Germany, h.teheesen@umwelt-campus.de
#Ulm University of Applied Sciences, Albert-Einstein-Allee 55, 89081 Ulm, Germany

ABSTRACT: We provide an analysis of public yield data of more than 23,000 photovoltaic systems with a capacity up to 30 kWp in Germany for 2017. We have used daily yield data including the configuration of the PV systems. We have aggregated the yield data to monthly values. The data have been cleansed by a two-step algorithm to consider only PV systems without any malfunction in this time period. The evaluation of the yield represents the energy production of small PV systems in Germany. 2017 was a below average year for photovoltaic systems in Germany. The yield in 2017 was about three percent smaller compared to the reference period from 2012 to 2016. The specific yield in June was larger than in previous years, the other months show a specific yield which is just in the final third of long-term monthly yield values. The highest specific yield of more than 1,100 kWh/kWp was observed in the southwestern part of Germany. In contrast, the yield in Lower Saxony is the smallest one with just 900 to 950 kWh/kWp. The regional spread of the energy yield is large compared to the previous years. The analysis in this study is based on statistical methods and algorithms. Therefore, the results are objectively verifiable and have no subjective influence. The significance analysis shows that the statistical results will reflect the real income of the solar systems very well if a sufficiently large number of PV systems is evaluated.

Keywords: Monitoring, Rooftop, System Performance

1 INTRODUCTION

There are several publications which have shown that the yield production and the performance of photovoltaic (PV) systems is typically smaller than the technical possibilities. This observation can be found for all European countries [1, 2, 3, 4, 5, 6, 7, 8, 9].

In particular, a lack of quality can be identified for small roof-top PV systems in Germany which are the majority of PV systems (see Figure 1) [10, 11]. There are more than one million roof-top PV systems with a capacity of up to 30 kWp installed in Germany. Very often, the performance of these systems which are not maintained or repaired by the system operator/owner keeps below the expectation due to technical issues. We are continuously addressing the owners of the small roof by publishing yield surveys to show the expected yield within a regional cluster.

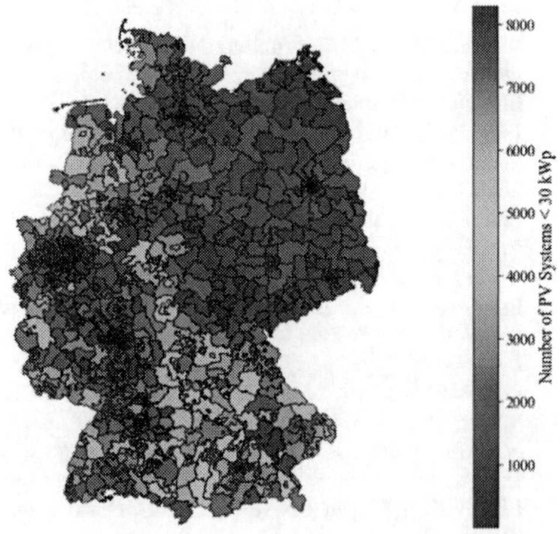

Figure 1: Distribution of installed PV systems with a capacity up to 30 kWp in Germany. Data source: Bundesnetzagentur.

2 METHODS

We use a modification of our algorithm to analyze the yield of PV systems [12]. The original algorithm considers the yield and the insolation as well. If the insolation data is neglected the results do not differ significantly from the previous ones.

Thus, raw yield data and system information (location, capacity, inclination and orientation) are downloaded by screen scraping scripts from monitoring websites providing public data on a daily time interval. Implausible data are eliminated in a primarily step. Statistical values like median and quartiles are calculated for all systems within each one-digit postal code in Germany. PV systems with outlining values are neglected in the following steps. This data cleansing step is repeated once. Next, the yield of systems in each two-digit postal code and the neighboring two-digit postal code is cleaned. Finally, the remaining PV systems are used for further analysis.

The number of PV systems within the capacity classes between 0 and 30 kWp is shown in Figure 2. Most PV system have a capacity between 4 and 10 kWp and represent the group of small roof-top residential systems. The peak at 30 kWp relates to feed-in-tariff restrictions in former EEG (German Renewable Energy Law) versions.

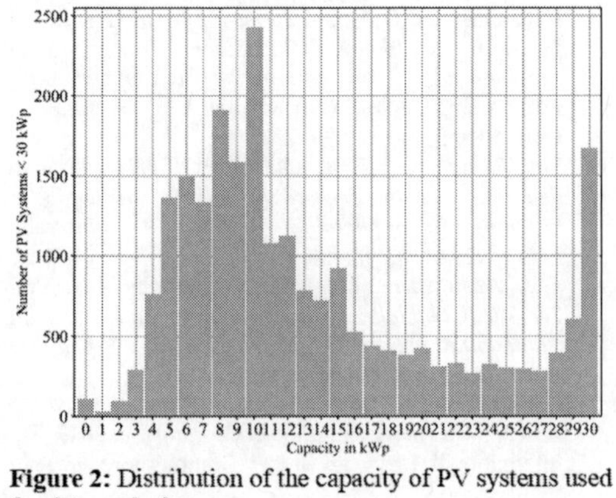

Figure 2: Distribution of the capacity of PV systems used for data analysis.

3 RESULTS AND DISCUSSION

The spatial specific yield distribution for 2017 in Germany is shown in Figure 3. The values represent the yield of well-operated and maintained roof-top PV systems with a southern orientation and tilted modules. There is a strong south-north-gradient of the specific yield. PV systems in North and Northwest Germany have reached only specific yields of 900 to 950 kWh/kWp. A band with a specific yield of around 1,000 kWh/kWp runs along the 50th and 51st latitudes in Germany. South of the line from Karlsruhe to Nuremberg, yields of over 1,075 kWh/kWp can be observed. The highest yields in 2017 are in Breisgau with values larger than 1,100 kWh/kWp.

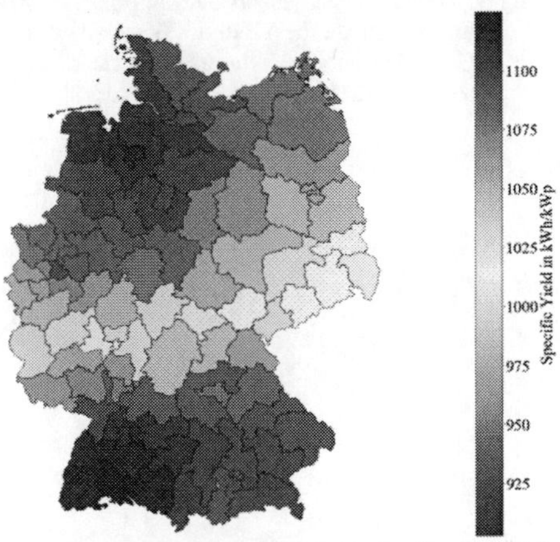

Figure 3: Specific yield in kWh/kWp in tilted module plane (Southern orientation, 30° inclination angle) for small PV systems in Germany 2017.

Figure 4 compares the average specific monthly yield of roof-top PV systems in Germany for the years 2012 to 2017. While the first half of 2017 still shows slightly above average yield, the specific yield in Germany in the second half of the year is below the long-term average. Thus, the mean annual yield of PV systems in Germany for 2017 of 1,009 kWh/kWp is about three percent smaller than the long-term average of 1,040 kWh/kWp.

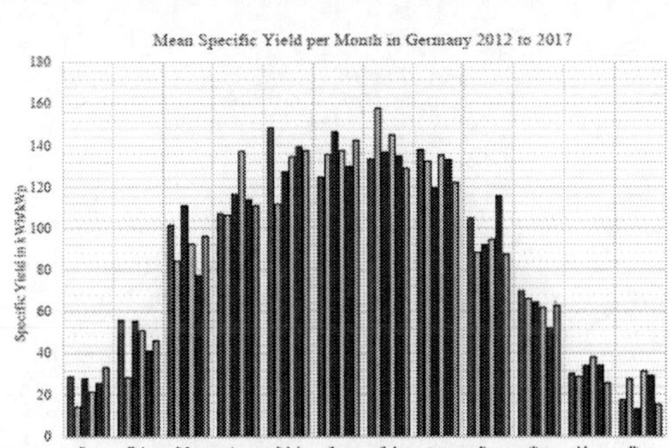

Figure 4: Comparison of the mean specific yield per month of PV systems in Germany from 2012 to 2017 [1, 10].

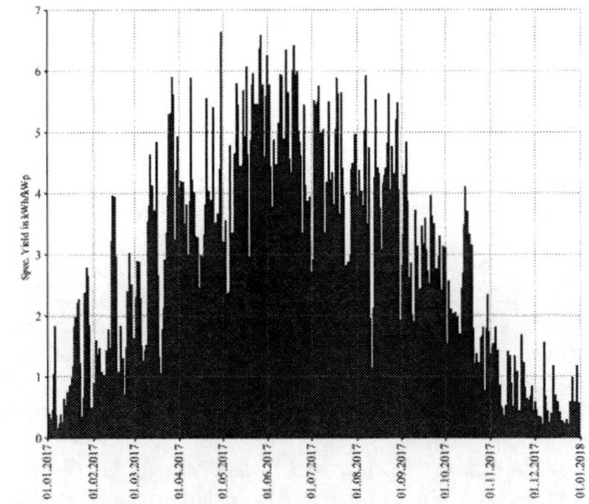

Figure 5: Distribution of the mean specific yield of PV systems 2017 per day in Germany.

The temporal distribution of monthly yields can also be seen in the analysis of specific daily yield (see Figure 5). While in January and February 2017, specific yield values of more than 2 kWh/kWp per day are already being achieved on average throughout Germany, specific daily yields of more than 3 kWh/kWp and even 4 kWh/kWp respectively can be seen regularly from the end of March. In May and June 2017, the yield is almost consistently larger than 4 kWh/kWp and reaches values of up to 6 kWh/kWp. However, the specific yield starts to drop from July. In September 2017, specific yield values larger than 4 kWh/kWp are achieved on a few days. In October, there is another sunny week (week 42) with specific yields above 3 kWh/kWp, otherwise the autumn of 2017 in Germany shows small yield values with a daily yield of less than 2 kWh/kWp.

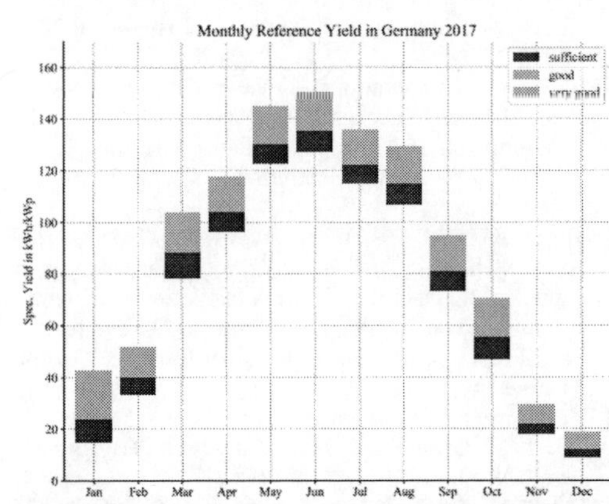

Figure 6: Monthly specific reference yield 2017 in Germany.

The monthly reference yield for PV systems in Germany is shown in Figure 6. The colored bars represent the specific yield for roof-top systems which are running sufficient (red bar), well (yellow bar) and very well (green bar). Thus, if a PV system has reached a yield of 150 kWh/kWp in June 2017, this system will to be rated "very good". If the specific yield is 140 kWh/kWp, the system can be rated a "good" PV system. A specific yield of 130 kWh/kWp represents a "sufficient" PV system. The monthly values highly depend on the location of the system. A more detailed analysis of the data can be found in [11].

4 CONCLUSION

We have analyzed daily yield data of small roof-top PV systems in Germany 2017, provided by public monitoring websites. In total, data of more than 23,000 PV systems with a capacity up to 30 kWp have been collected by screen scraping scripts. We have calculated a reference specific yield by taken only well-operated systems into consideration. PV systems with technical problems due to faults or maintenance issue have been eliminated by a data cleansing algorithm.

The mean specific yield in Germany 2017 was 1,009 kWh/kWp and is thus about three percent smaller than the long-term average. The first half of 2017 was an average one, the second half stays below the long-term average yields. Highest annual yield values can be observed in Baden-Wurttemberg of more than 1,100 kWh/kWp, whereas the annual specific yield in Lower Saxony just reaches 925 kWh/kWp in 2017. A more detailed view regarding the spatial and temporal resolution can be found in [11].

5 REFERENCES

[1] H. te Heesen, V. Herbort, R. Pfatischer, Survey on yield of PV systems in Germany 2013, in: 29th European Photovoltaic Solar Energy Conference and Exhibition, 2014, pp. 2198-2201.

[2] B. Kausika, W. Folkerts, G.J.H.M. van Sark, B. Siebenga, P. Hermans, A big data approach to the solar PV market: design and results of a pilot in the Netherlands, in: 29th European Photovoltaic Solar Energy Conference and Exhibition, 2014, pp. 4030-4033.

[3] J. Leloux, L. Narvarte, D. Trebosc, Performance analysis of 10,000 residential PV systems in France and Belgium, in: 26th European Photovoltaic Solar Energy Conference and Exhibition, 2001, pp. 3939-3946.

[4] E. Lorenz, J.W.H. Betcke, A. Drews, A.C. de Keizer, S. Stettler, M. Schneider, S. Bofinger, D. Heinemann, Intelligent Performance Check of PV System Operation Based on Satellite Data (PVSAT-2), final technical report. Tech. rep., Copernicus Institute, Utrecht University, 2004.

[5] T. Nordmann, L. Clavadetscher, W.G.J.H.M. van Sark, M. Green, Analysis of Long-term Performance of PV Systems, IEA PVPS Task 13, 2014. Subtask 1.

[6] M. Gromaire, H. te Heesen. Survey on Specific Yield of Photovoltaic Systems in France 2014. 30th European Photovoltaic Solar Energy Conference and Exhibition. pp. 2307 - 2309. Sept. 2015.

[7] O. Tsafarakis, W.G.J.H.M. van Sark. Analysis and Comparison of Performance of Photovoltaic Systems from the "Counting the Sun" Public Awareness Campaign on PV Performance. 31st European Photovoltaic Solar Energy Conference and Exhibition. pp. 2245-2248. June 2016.

[8] P. Moraitis, B. B. Kausika, N. Nortier, W. J. G. H. N. van Sark, Urban Environment and Solar PV Performance: The Case of the Netherlands. Energies, Volume 11, Number 1333, May 2018

[9] O. Tsafarakis, , W. J. G. H. N. van Sark, Analysis and Comparison of Performance of Photovoltaic Systems from the 'Counting the Sun Public' awareness Campaign on PV Performance. 31st European Photovoltaic Solar Energy Conference and Exhibition, pp. 2245-2248, June 2015.

[10] H. te Heesen, V. Herbort, M. Rumpler, Survey on Yield of PV Systems in Germany 2014 to 2016. 33rd European Photovoltaic Solar Energy Conference and Exhibition. Amsterdam, The Netherlands. P. 1901-1903. Sept. 2017.

[11] H. te Heesen, V. Herbort, M. Rumpler, Studie zum Ertrag von Photovoltaikkleinanlagen in Deutschland 2017. Hochschule Trier. März 2018

[12] H. te Heesen, V. Herbort. Development of an Algorithm to Analyze the Yield of Photovoltaic Systems. Renewable Energy. Volume 87, Part 2, March 2016, 2016, Pages 1016–1022.

DATA-FILTERING-DEPENDENT VARIABILITY OF LONG-TERM DEGRADATION RATES OF MW-SCALE PHOTOVOLTAIC POWER PLANTS FROM "NON-IDEAL" MONITORING AND WEATHER DATA

Christian Camus[1, 2]*, Maximilian Hüttner[1], Daniel Lassahn[3], Christian Kurz[3], Jens Hauch[1, 2], Christoph J. Brabec[1, 2, 4]
[1] Bavarian Center for Applied Energy Research (ZAE Bayern), Immerwahrstr. 2, 91058 Erlangen, Germany, [2] Materials for Electronics and Energy Technology (i-MEET), Friedrich-Alexander-Universität Erlangen-Nürnberg, Martensstr. 7, 91058 Erlangen, Germany, [3] meteocontrol GmbH, Spicherer Strasse 48, 86157 Augsburg, Germany, [4] Helmholtz Institute Erlangen-Nürnberg for Renewable Energy (HI ERN), Egerlandstraße 3, 91058 Erlangen, Germany
*Corresponding Author: Christian Camus; christian.camus@zae-bayern.de

ABSTRACT: In contrast to research photovoltaic (PV) systems, commercial PV systems are usually not equipped with sophisticated weather metrology and quite frequently even lack an irradiance sensor. Still, the determination of their degradation rate is of paramount importance for the plant owners. This paper compares various approaches to determine the degradation rates of commercial MW-scale PV systems from such "non-ideal" data sets as they are frequently obtained from monitoring data. In particular, the use of model-based irradiance and temperature data is evaluated and several novel methods for data filtering are introduced. It is investigated, for which weather conditions the use of externally weather station data is superior to on-site data. By reducing the data set to mostly sunny conditions, reliable DRs can be determined. Furthermore, the data scattering can be reduced significantly so that sudden performance drops or changes in the DRs become visible, which were previously masked by scattering and data drifts. These results demonstrate that significantly different DRs and uncertainties are obtained depending on the method of data filtering. Hence, care must be taken, when deriving DRs from "non-ideal" monitoring and weather data and when comparing DRs determined by different methods.
Keywords: Outdoor performance, yield analysis, PV technologies, temperature behavior

1 INTRODUCTION

Nowadays, photovoltaic (PV) power plants are an important part of the energy mix and their contribution is expected to grow rapidly in the future. Hence, the short-, mid- and long-term performance of these plants is no longer only relevant for investors but also for utilities and consumers, who depend on the energy provided from these plants. Therefore, the determination of degradation rates (DR) of system-relevant PV systems is of paramount importance for predicting electrical yield and to identify the need for additional capacity or repowering measures. Usually, DRs are determined by assessing a plant's performance ratio (PR) in order to exclude the influence of the volatility of the solar irradiance. The DR is then expressed as a linear rate in the unit of %/year as determined from the PR even though PR data is influenced by various non-linear effects such as snow cover in winter, irregular soiling removal by rain or intentional cleaning etc. Whilst in research, PV systems are usually equipped with extensive weather metrology, electrical yield data is obtained at high frequency and the systems themselves as well as the metrology systems are well-maintained, the monitoring and operation and maintenance of MW-scale commercial PV power plants is usually cost-limited. Hence, these systems are usually not equipped with sophisticated weather metrology and quite frequently even lack an irradiance sensor. Furthermore, commercial plants are usually monitored remotely, so that visual inspections and on-site visits are rather rare. Hence, many of the yield-determining factors are unknown even though reliability is of paramount importance for the operation of these plants.

2 APPROACH

This paper investigates the variation of DRs which are determined from "non-ideal" data of MW-scale PV power plants. Due to price pressure such PV systems are usually – if at all - equipped with rather simple weather metrology systems and electrical yield data is gathered at rather low frequencies of the order of 10 min or more on the AC string level. This situation is very different from research PV systems [1, 2], for which usually complete and highly reliable data sets are available. We investigate several approaches for a reliable determination of DRs of two large-scale mono-crystalline silicon PV systems – one (3.8 MWp) in Southwest Germany (PV#1) and one (1.0 MWp) in Southern Italy (PV#2). Irradiance data in the module plane and AC inverter data were recorded at a frequency of 10 min constitute the only data obtained from the plant itself. Based on this data, the PR is calculated and a first DR is derived, for which only measurement failure data was removed. Due to the volatility of weather this data is potentially influenced by various factors such as variations in module temperature, snow covering, soiling, partial shadowing, hardware drifts etc. All these (among others) factors contribute to the PR data. Hence, the DR and its uncertainty can be deceiving. In order to obtain the most reliable DR, several approaches are made to improve the DR determination. Firstly, the irradiance in the module plane and the module temperature were calculated from the irradiance, temperature and wind speed data of the closest official weather station based on Sandia's PV Performance Model [3, 4]. For PV#1 data from a German Weather Service (DWD) weather station was used [5]. These stations are well-maintained and data reliability is assured. Then a temperature-corrected daily PR was calculated and the corresponding DR was derived. Since the DWD weather station was several kilometers away from the PV site the advantage of high quality weather data goes along with the disadvantage of potentially differing weather conditions at the station and the PV site for volatile weather conditions. Hence, several approaches for excluding such conditions were made. Additionally, data was filtered based on a modification of the K-POP method introduced by Kang et al. [6], which allows determining classes with respect to total irradiance and its persistence for weather data of arbitrary time intervals. For all methods, DRs and their uncertainties were compared. Based on this comparison the most reliable method for determining a DR of a large-scale PV plant from "non-ideal" data is deduced.

3 SYSTEM DETAILS & METHODOLGY

3.1 PV Systems

In this study 2 different PV systems were investigated. The first system (PV#1) is a roof-top system located in southwest Germany It consists of 8 separated sub-systems of a total nameplate capacity of 3.8 MWp. The sub-systems are installed with a tilt angle of 15°C. Seven sub-systems are oriented with an azimuth angle of ~250°, whilst one exhibits an azimuth angle of 68°.All sub-plants are equipped with monocrystalline PV modules with a nameplate power of 180 Wp. DC electrical yield data was available for teach sub-plant, separately.

The second system (PV#2) is a field-installed system which is located in southern Italy and has a nameplate capacity of 1 MWp. Here, the modules are installed with a tilt angle of 30° and an azimuth angle of 180°C. The plant was equipped with monocrystalline PV modules with a nameplate power of 185 Wp. DC electrical yield data was available for three input channels of 35 invertes totaling to 105 channels.

For both PV systems, electrical yield and irradiance data was analyzed with a time resolution of 15 minutes.

3.2 Temperature-corrected performance ratio

Since none of the systems investigated here was equipped with module temperature sensors, the module temperature was calculated according SANDIA's PV performance model [3, 4] from the respective irradiance data set used for the given evaluation method. Based on this temperature, the PR was temperature-corrected based on the formalism in [3, 4].

3.3 Time-correction

Firstly, monitoring and irradiance data was checked for consistency. Therefore, the switch-on and -off times of the inverters were analyzed (see **Figure 1**). Analogously, the times at which irradiance data exceeded the nighttime dark level for the first and last time for each day were analyzed. It can be seen from Figure 1 that two kinds of time shifts were present in the data. The first one being time shifts in spring and fall of each year which could be attributed to the respective time change. The second time shift only occurred once for PV#1. Here, a time shift of two hours occurred after a gap of data. Possibly, this shift may have been caused, when the system went operational after a downtime due to failure or maintenance. This shift was corrected by aligning the dataset such that the switch-on and -off were balanced to noon. Similar kinds of time shifts were observed in the irradiance data as well as in the electrical yield and irradiance data of PV#2.

It needs to be mentioned that even small time shift of the order of 15 min can have a detrimental impact on the quality of the performance ratio data deduced from the irradiance and electrical yield data. This is due to the non-constant irradiance profile during the course of the day. Hence, even small shifts lead to significant mismatches between the amount of energy impinging on the PV modules and the amount of generated power for a given time interval as it is illustrated in **Figure 2**.

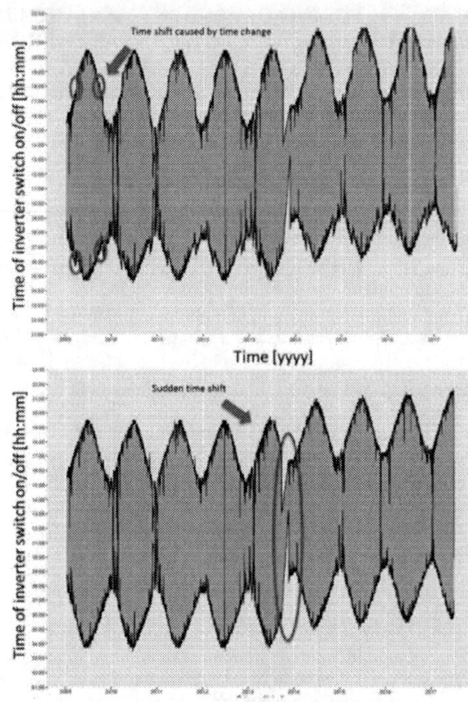

Figure 1: Times of inverter switch on/off for the PV plant PV#1.Upper image: Effect of time change in spring and fall. Lower image: Effect of a sudden time shift.

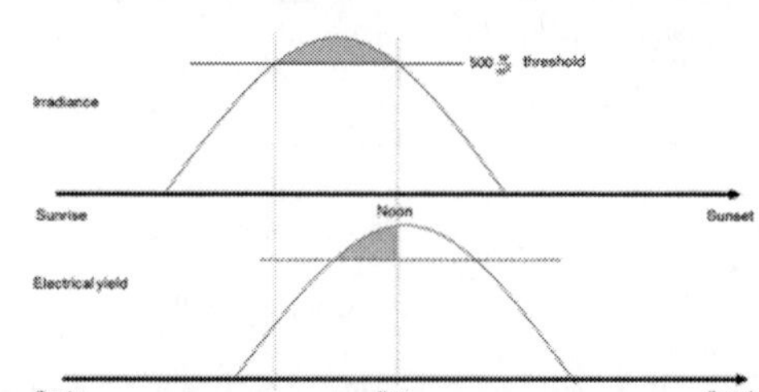

Figure 2: Schematic sketch of the impact of a time-mismatch between electrical yield and irradiance data

3.4 Data-filtering

Various data filters were applied in order to reduce the data set to constantly sunny conditions as these can be considered to be most representative for the actual technological capability of a PV system. Otherwise, e.g. periods with a volatile cloud situation can lead to lower PR values due to the fact that PV modules were heated up in a sunny period and remain at elevated temperature for a certain period of time after clouds have suddenly reduced the irradiance level. Since in this study the module temperature is solely determined from the irradiance data, this effect cannot be accounted for. Similarly, snow covering the modules in winter even under sunny conditions can lead to unrealistically low PR values. At least for PV#1 the latter effect needs to be taken into account. Furthermore, it is known, that PV module efficiency is also slightly reduced for weak light conditions. Therefore, also periods of lower irradiance (i.e. cloudy conditions) can lead to an apparent reduction of the PR, even though the modules did not really degrade. .

Table I: Scheme of stepwise refinement of data-filtering in order to determine the most realistic degradation rates. The colored squares refer to the respective columns in **Figure 7** and **8.**

<table>
<tr><td>Case 1:</td><td colspan="4">• El. yield data
• irradiance measured in module plane
• mod. temp. modelled acc. SANDIA</td><td>Case 2:</td><td colspan="4">• El. yield data
• irradiance modelled acc. SANDIA
• mod. temp. modelled acc. SANDIA</td></tr>
<tr><td colspan="3">PR as-is</td><td colspan="2">Temp.-corr. PR</td><td colspan="3">PR as-is</td><td colspan="2">Temp.-corr. PR</td></tr>
<tr><td colspan="5">All days considered</td><td>All days considered</td><td colspan="4">Assessment of completeness of days: neglect days with <5h of el. yield / irradiance data</td></tr>
<tr><td>Degradation rate as-is</td><td colspan="2">Use all data</td><td>Selection of K_{POP} classes 1, 4</td><td>Selection of K_{POP} classes 1, 2, 4, 5</td><td>Degradation rate as-is</td><td colspan="2">Use all data</td><td>Selection of K_{POP} classes 1, 4</td><td>Selection of K_{POP} classes 1, 2, 4, 5</td></tr>
<tr><td></td><td>$T_{ambient}$>3°C</td><td>$T_{ambient}$>3°C; E_{POA}>500 W/m²</td><td>$T_{ambient}$>3°C; E_{POA}>500 W/m²</td><td>$T_{ambient}$>3°C; E_{POA}>500 W/m²</td><td></td><td>$T_{ambient}$>3°C</td><td>$T_{ambient}$>3°C; E_{POA}>500 W/m²</td><td>$T_{ambient}$>3°C; E_{POA}>500 W/m²</td><td>$T_{ambient}$>3°C; E_{POA}>500 W/m²</td></tr>
</table>

In this study, several data filters were tested in order to reduce the weather conditions to sunny and snow-free conditions. These filters are summarized in **Table I.** Firstly, all data for ambient temperatures, $T_{ambient}$, below 3°C were excluded in order to exclude periods with (partially) snow-covered modules. Since it was not clear, if this condition sufficiently exclude snowy conditions an irradiance level in the module plane (E_{POA}) of E_{POA} > 500 W/m² was added.

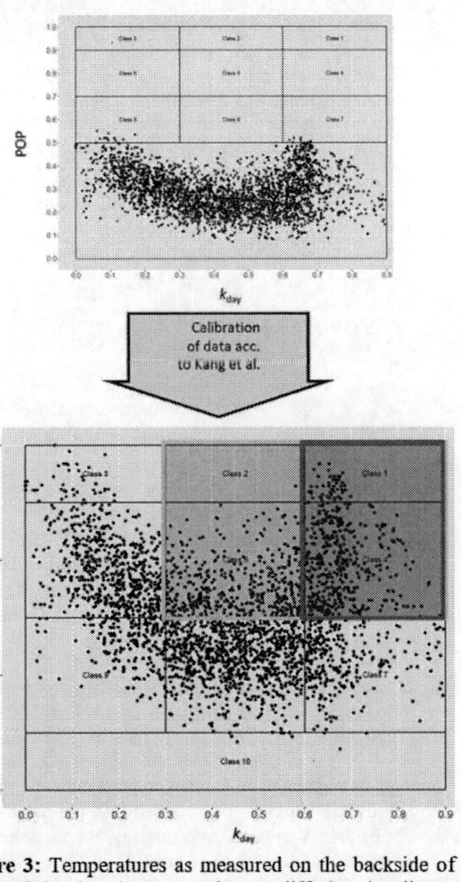

Figure 3: Temperatures as measured on the backside of the modules by thermocouples at differing irradiance levels. The ambient temperature is also shown (dark blue trace).

In a second approach, the irradiance condition was further specified based on the K-POP formalism introduced by Kang et al. [6]. This formalism classifies the irradiance situation based on the persistence of irradiance as well as on the ratio of the irradiance measured at the site of interest and the extraterrestrial irradiance [6]. This leads to ten different K-POP classes, of which 1 is persistently sunny and 10 corresponds to complete overcast. The K-POP formalism originally requires 1-min-resolved irradiance data. Consequently, if it is directly applied to the irradiance data of this study, almost all data points are assigned to class ten, whilst their internal distribution is qualitatively comparable to the distribution observed by Kang et al. Therefore, we calibrated our data distribution such that it quantitatively agrees with the distribution observed by Kang et al. (see **Figure 3**). Based on this calibrated distribution two further data filters were deduced: Firstly, only data of classes 1 and 4 with $T_{ambient}$>3°C and E_{POA} > 500 W/m² was considered. As a second case only data of classes 1, 2, 4 and 5 with $T_{ambient}$>3°C and E_{POA} > 500 W/m² was considered.

Generally, the filtering classes described above were applied to two different data sets (Class 1 and 2 in **Table I**): The first data set consists of measured electrical yield and irradiance data, whereas the second class uses measured electrical yield data but irradiance data was modelled (with respect to the module plane) from meteorological irradiance of a close-by weather station.

4 RESULTS

4.1 Influence of time shifts and temperature correction

Due to the effects described above, the correction of the time shift had a significant effect on the quality of the PR data obtained for both PV systems. In **Figure 4** it can be seen that, for PV#1, the correction practically removed all PR values which had remained after applying the $T_{ambient}$>3°C and E_{POA} > 500 W/m² filtering. After correction of the time shifts the degradation rate from -0.21±0.16%/a (without temperature correction, tc) and 0.14±0.20%/a (with tc) to 0.13±0.09%/a (without tc) and 0.11±0.10%/a. Thus, the correction reveals that practically nod degradation was measurable for this system.

At the same, time **Figure 4** also demonstrates the effect of the tc itself. Upon applying a tc based on simulated module temperatures the apparent rise in PR during winter is practically removed. However, it needs to be noted that for some of the sub-plants of PV#1, the tc also led to a certain "over compensation" (not shown here) so that PR actually appeared to reduce during winter. This is likely due to the fact that modelling yielded the same module temperatures for all sub-plants with the same irradiance (i.e. orientation). However, the actual module temperature may differ for the different sub-plants due to local effects (reflections, building temperature etc.).

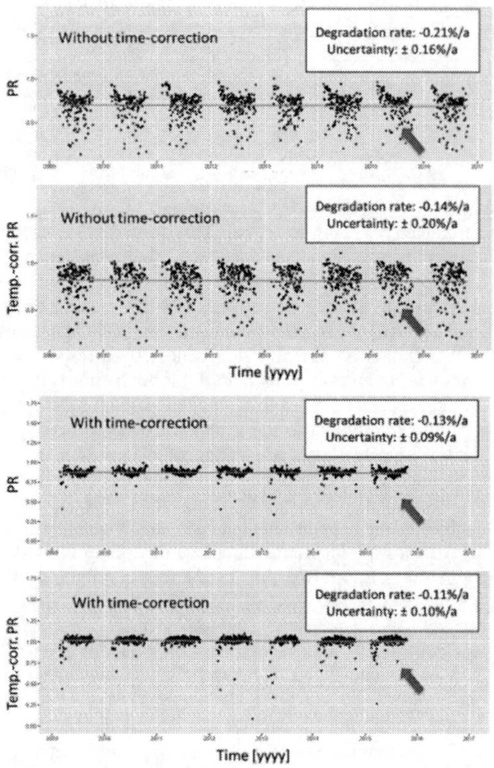

Figure 4: Effect of a time mismatch of ~15min between measured electrical yield and measured irradiance data, resulting in numerous data points with PR << 70%

4.2 Influence of data-filtering

By applying the various data-filtering approaches stated in Table I, the various DR could be defined for the two PV systems. **In Figure 5**, the stepwise effect of applying the various filters becomes clear. Here, the filtering was applied to data of a sub-plant of PV#1 with measured electrical yield and irradiance data. Whilst the fraction of data points was reduced from 100% to 24% the DR deduced from the data changed from ~-0.5%/a for unfiltered data to 0.0%/a when only data for $T_{ambient}$>3°C and E_{POA}>500W/m² and only K-POP classes 1, 2, 4 and 5 are considered. Here it becomes clear that in spite of losing ~75% of the data, the DR deduced from the filtered data can still be considered to be more representative.

In **Figure 6**, the PR plot corresponding to the lower image of **Figure 5** is shown for modelled module plane irradiance data, which was modelled based on irradiance data of a close-by weather station. It becomes clear, that the scattering significantly increased for the modelled data, which is likely due to the fact that weather station was about 10 km away from the PV system. Hence, the irradiance rather constitutes a good approximation of the irradiance situation at the PV system, but cannot fully represent the actual weather conditions.

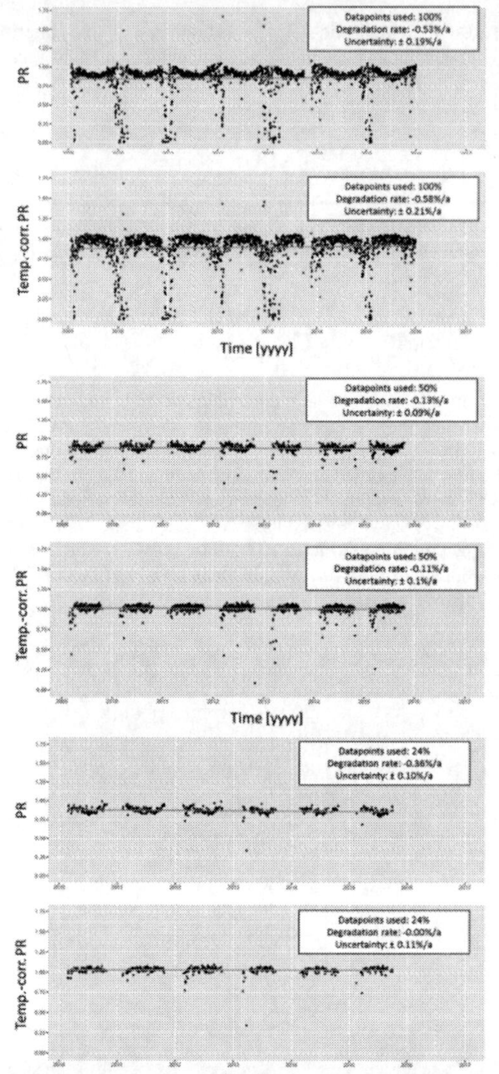

Figure 5: Sequence of stepwise refinement of data-filtering for PR-data of a sub-plant of PV#1. Upper image: PR calculated form irradiance data in the module plane, no additional filtering; center: only data for $T_{ambient}$>3°C and E_{POA}>500W/m² is considered; lower image: additionally, only K-POP classes 1, 2, 4 and 5 are considered.

It can also be seen that the DR is slightly higher for the modelled data, which is likely due to the fact that for the measured data, irradiance data was not available for 2016 due to a technical issue at the site. Hence this year was included in **Figure 5**, whereas it could be included for the modelled data which did not depend on the hardware at the site of the PV system. Additionally, the modelled data set contains some more low PR outliers than the measured data. However, both DR are well below the benchmark value of -0.5%/a published by Jordan et al. [7].

In **Figure 7**, the various degradation rates determined for the different filtering approaches are visualized. The colors of the columns refer to the squares in Table I.

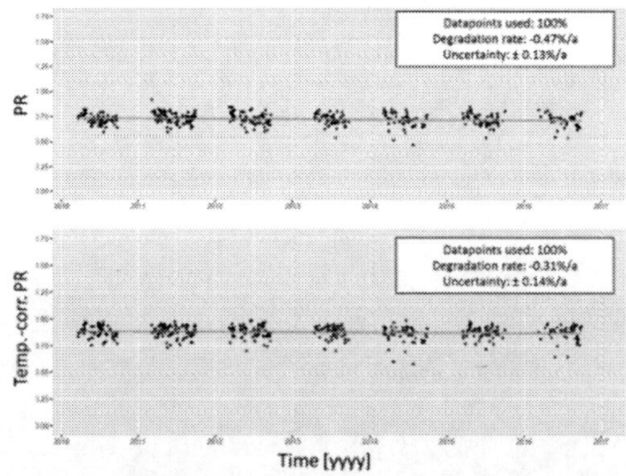

Figure 6: Visualization of the same electrical yield data as shown on the lower image of **Figure 5**, but here, modeled data for the irradiance in the module plane was used.

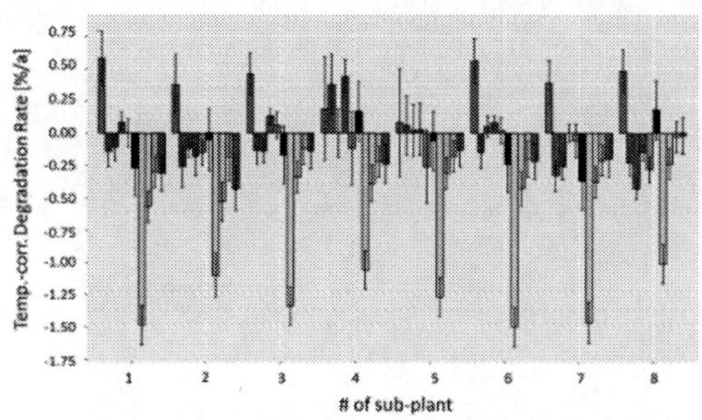

Figure 7: Variety of temperature-corrected DR deduced for the different sub-plants of PV#1 for different levels of data-filtering refinement.

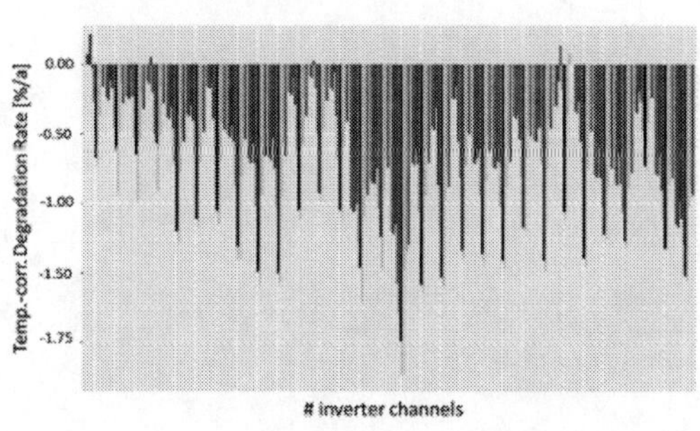

Figure 8: Variety of temperature-corrected DR deduced for 30 different inverter channels (out of 105; 1 sub-plant) of PV#2 in southern Italy for different levels of data-filtering refinement.

Dependent on the method of data-filtering, DR between -1.5%/a and +0,75%/a were deduced indicating the importance of appropriate filtering. Generally, a rather strict filtering resulted in more realistic DR values in spite of a reduction of the number of data points by ~75%. It can be seen that for all sub-plants of PV#1 by far the most negative degradation rates were obtained for modelled irradiance data and a filtering for $T_{ambient}$>3°C only. The root-cause for this extreme value is not yet completely clear. However, the PR data obtained from modelled irradiance was found to be very noisy. At the same time the filter $T_{ambient}$>3°C was not able to exclude all snowy days. Hence, the data contained a significant number of low-PR outliers. Additionally, the last years happened to exhibit more snow days than the previous years. Hence, the number of outliers increased for the most recent years, which may have caused the tendency towards higher DR.

In general it can be seen that almost all more rigorous filtering approaches yield comparable DR rates well below 0.5%/a for all sub-plant but sub-plant 4. Within the uncertainty of the degradation rates most of these include 0%/a so that it can be concluded that the DR for the sub-plants of PV#1 is negligible. For sub-plant 4, the rates for measured and modelled irradiance differ strongly, which can be explained by the fact that no measured data was available after 2013. Therefore, bot data sets were not comparable.

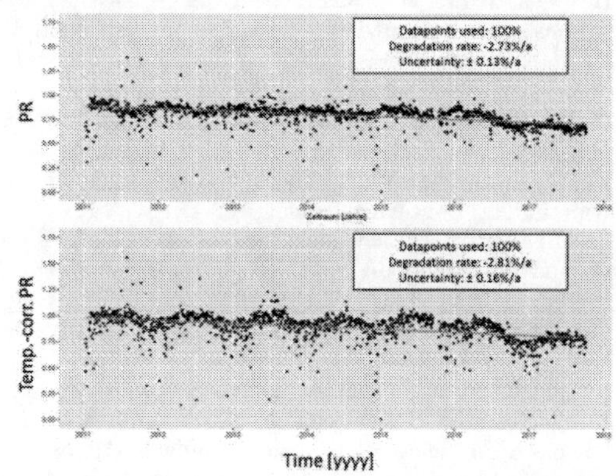

Figure 9: PR data for an inverter channel of PV#2 in southern Italy showing an extraordinary high degradation rate >2%/a

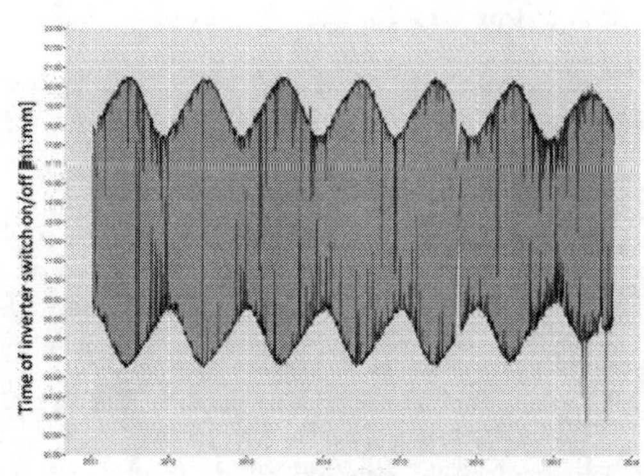

Figure 10: Inverter switch-on and -off times of the inverter channel for which PR data is shown in **Figure 9**.

For the PV system in southern Italy the situation was slightly different. The variety of DR depending on the filtering approach is depicted in Figure 8 for 30 out of 105 inverter channels. Again, it can be seen that the different approaches yield very different DR. Similarly to PV#1, the highest rates are obtained for modelled irradiance without any filtering or with the $T_{ambient}$>3°C filter only. For more rigorous filtering modelled and measured data yield mostly comparable DR. However, in contrast to PV#1, the results of the various channels vary significantly. **Figure 9** exemplarily shows data of one channel with a high degradation rate above 2%/a. Here it can be seen that the systems seems to exhibit a steady decrease in performance during the last year. In order to verify that this was not due to a so far undetected time shift the switch-on and -off where analyzed (see **Figure 10**). Indeed, it can be seen from **Figure 10** that both times shifted steadily during the last year so that the time of power production was reduced. Thus, this result indicates that a true degradation of the PV system took place.

5 DISCUSSION

The results of this work clearly reveal the importance of an appropriate choice of data-filtering for monitoring data from utility-scale PV systems. In particular it has been shown, that a DG cannot be assumed to be reasonable without temperature-correction of PR data. However, it has also been shown that – not only for regions prone to snowfall – an additional restriction of the data to persistently sunny conditions is needed in order to obtain reliable DR. Furthermore, it was demonstrated that, if rigorous filtering is applied, DG obtained from measured and modelled irradiance data yield comparable results. These finding largely agree with findings from Jordan et al. [10].

It has also been shown that the analysis of the switch-on and –off times of the inverter data as well as of irradiance data can be a very powerful tool for assessing the quality of monitoring data as it can reveal hidden time shifts and creeping reductions in power generation.

In conclusion it can be stated that with rigorous filtering even for PV systems for which hardly any data but electrical yield data is available reliable DR can be determined with acceptable uncertainties by using irradiance data calculated from close by weather stations.

6 ACKNOWLEDGMENTS

This research was co-financed by the Free State of Bavaria.

7 REFERENCES

[1] M. Schweiger, J. Bonilla, W. Herrmann, A. Gerber, U. Rau, Prog. Photovolt: Res. Appl., 25 (2017) 968
[2] J. Bogenrieder, M. Hüttner, P. Luchscheider, J. Hauch, C. J. Brabec., Prog. Photovolt: Res. Appl., 26, (2018) 74
[3] PVPMC website, accessed: October 25th, 2017: https://pvpmc.sandia.gov/modeling-steps/,
[4] D. L. King, W. E. Boyson, J .A. Kratochvill., SANDIA REPORT (2004) SAND2004-3535
[5] German Weather Service (DWD, Deutscher Wetterdienst) website; accessed: February 6th 2017: https://www.dwd.de/DE/leistungen/cdcftprasterdaten/cdcftpraster.html
[6] B. O. Kang, K.-S. Tam., Solar Energy 94 (2013) 102
[7] D. C. Jordan, S. R. Kurtz, K. VanSant, J. Newmiller., Prog. Photovolt: Res. Appl., 24 (7) 978–989, 2016
[8] D. C. Jordan, S. R. Kurtz, IEEE J. Photovolt., 4 (2014) 317
[9] D. C. Jordan, C. Deline, S. R. Kurtz, G. M. Kimball, M. Anderson, IEEE J. Photovolt., 8 (2018) 525

PERFORMANCE OF SI-HETEROJUNCTION MODULES WITH DIFFERNENT CELL INTERCONNECTION AND MODULE TECHNOLOGIES

A. Titov[1,2], K. Emtsev[1], D. Andronikov[1], A. Abramov[1], B. Bulygin[3], A. Dubrovskiy[3], E. Terukov[1,2], D. Orekhov[1], I. Shakhray[4]
[1] R&D Center for Thin Film Technologies in Energetics (TFTE), St.-Petersburg, Russia
[2] Ioffe Institute, St.-Petersburg, Russia
[3] Hevel LLC, Novocheboksarsk, Russia
[4]Avelar Solar Technology, Russia

ABSTRACT: The results of outdoor performance monitoring of different Si-heterojunction (HJT) modules is presented. We estimate energy gain of production HJT modules over commercial mc-Si module. Also, bifacial gain of HJT modules over monofacial HJT modules for extended period of monitoring for almost 2 years is presented with clear influence of seasonal and irradiance effects. In the research performance of HJT modules with different cell interconnections (smart wire, ECA busbar) and modules designs (mono- and bifacial) are compared. In addition, HJT modules were tested under extremely low temperatures in the lab and outdoor.

Keywords: heterojunction, bifacial, performance, monitoring, reliability

1 INTRODUCTION

According to the ITRPV roadmap, silicon heterojunction (Si-HJT) technology [1] is expected to grow over the next years as a promising new technology for high efficiency modules [2]. New production facility operated by Hevel LLC for Si-Heterojunction (HJT) modules with annual capacity of 160 MW is in operation in Russia since 2017. The HJT technology was developed at the TFTE R@D Center, Saint-Petersburg based on the a-Si micromorph equipment [3]. There are ongoing activities to further develop HJT cells and modules at the TFTE R&D center. HJT 6" cell efficiency above 23% on production equipment as well as HJT modules with >320W output power have been achieved. In addition to standard production modules by Hevel, modules with different cell types, cell interconnections (smart wire, ECA busbar (BB), half cells) and module designs (glass-glass (GG), glass-backsheet (G/BS)) are investigated. This work extends previous analysis [4] of the outdoor performance of production and the RD modules. The aim of the study is to monitor long-term performance of the HJT modules at various weather conditions in the central Russia climatic conditions. The system has been in operation for 2 years. In addition, HJT modules are being tested under extremely low temperatures outdoor and in the lab.

2 EXPERIMENTAL SETUP AND WEATHER CONDITIONS

The presented data were obtained from monitoring station located in Novocheboksarsk, Russia (N56°07′, E47°30′) and includes overall 9 modules (see Fig. 1). The collected data for each module are P_{mpp}, I_{mpp}, V_{mpp} DC values at the maximum power point, in plane solar irradiance via Kipp&Zonen pyranometer, module and ambient temperatures. The data are collected every 30 sec. The monitoring begun in August 2016 and continues thereafter.

Meteodata (in-plane solar irradiation and ambient temperature) collected for the monitoring period are summarized in Fig. 2. The coldest months of monitoring also exhibits the highest temperature variation. Daytime temperatures down to -30°C occurred in winter 2017 and down to -20°C in winter 2018. At the same time, the winter warmest days were close to 0°C for both years. Generally, however, winter season is of very low irradiance and, in addition, the performance was affected by snow. In this respect, the period from March till October appears most useful for data evaluation. In particular, weather in March with snow coverage and many sunny days is of interest for bifacial modules operation in winter.

The overall description of modules under study is summarized in Table 1. There are 8 HJT modules of different types as well as 1 reference mc-Si module. Three standard Hevel production modules have been monitored for 18 months by now. These are G/BS modules with output power of up to 300W as at the time of the ramp up stage of Hevel in April 2017. The power rating of the modules was independently confirmed by Fraunhofer ISE. Hevel production modules acquired the TUV certification for IEC 61215 and IEC 61730 standards. Prior to the production stage several modules were assembled at the Meyer Burger Research from the cells made at the TFTE (Gen0-2016 in Table 1, see also [4]). These are G/BS and GG modules with smart wire interconnection (SWCT) [5]. In addition to production modules, non-production research modules at the monitoring are: HJT module with Electrically Conductive Adhesive (ECA) busbar interconnection, HJT module with half cells and bifacial HJT module with glass-transparent backsheet design.

Figure 1: Outdoor monitoring stand located in Novocheboksarsk, Russia (N56°07′, E47°30′).

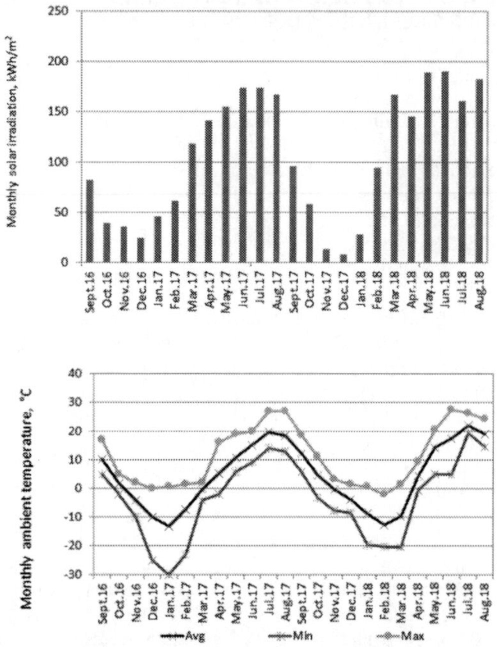

Figure 2: Monthly solar irradiation (top) and ambient temperatures (bottom) for the monitoring period.

Table 1: Monitoring modules with different cell and module designs.

Module	Module type	Cell type	Cell connection	Pmpp, W	Qty	Monitoring, month
Multi mc-Si reference	G/BS	Multi	3 BB	240	1	24
HJT Production	G/BS	Bifacial	SWCT	300	3	18
HJT Half	G/BS	Half cells	SWCT, halfcells	300	1	6
HJT Bifacial	G/BS transparent	Bifacial	SWCT	298	1	10
HJT BB	G/BS	BB cells	4 BB	294	1	10
HJT Gen0	G/BS	Bifacial	SWCT	262	1	24
HJT Bifacial Gen0	G/G	Bifacial	SWCT	268	1	24

3. RESULTS AND DISCUSSION

Comparison of the HJT and mc-Si modules

Fig. 3 compares monthly performance ratio (PR) wich defined as energy yield in kWh/kW divided by total solar irradiation for Hevel production HJT G/BS modules and standard mc-Si module. The HJT data for production modules represents average of the three modules as very small difference in energy yield between modules occurred. In addition, data for the early stage research HJT module (Gen0) is shown for comparison. As can be seen in Fig.3, Hevel HJT modules of production classes outperform mc-Si module for the monitoring period. For the hottest month (July) the relative gain of HJT over mc-Si module of about +5% is detected (Table 2, left). To a great extent, the HJT gain is due to lower temperature coefficients of HJT modules over mc-Si. The Hevel HJT modules have a temperature coefficient of -0.305%/K as measured by the Fraunhofer ISE.

At the same time, production HJT modules show higher PR values than HJT-Gen0 modules as result of the technology and production process improvements. The data for the HJT gain are summarized in Table 2 (left).

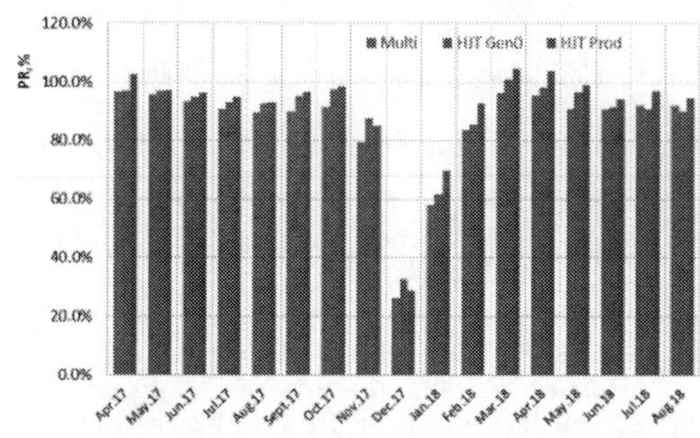

Figure 3: Monthly performance ratio (PR) for the production HJT modules in comparison with commercial mc-Si module and early stage research HJT-Gen0 module. Winter data are affected by snow.

Table 2: The energy gain of the HJT over mc-Si on the left and energy gain of the bifacial HJT modules over monofacial G/BS HJT on the right for the monitoring period. Winter data are affected by snow.

HJT gain

Month	HJT gain, %
Apr.17	6.4%
May.17	1.8%
Jun.17	3.4%
Jul.17	4.6%
Aug.17	3.6%
Sept.17	7.7%
Oct.17	7.9%
Nov.17	7.0%
Dec.17	9.6%
Jan.18	20.1%
Feb.18	10.9%
Mar.18	8.5%
Apr.18	8.6%
May.18	9.3%
Jun.18	3.6%
Jul.18	5.3%
Aug.18	2.9%

HJT bifacial gain

Month	Bifacial gain, %
Apr.17	6.8%
May.17	12.3%
Jun.17	11.9%
Jul.17	9.5%
Aug.17	10.1%
Sept.17	7.0%
Oct.17	6.5%
Nov.17	12.2%
Dec.17	103.0%
Jan.18	34.2%
Feb.18	12.8%
Mar.18	14.5%
Apr.18	11.0%
May.18	8.5%
Jun.18	12.3%
Jul.18	8.6%
Aug.18	7.9%

Comparison of bifacial and monofacial HJT modules

In Fig. 4 we plot monthly performance ratios of monofacial glass/backsheet vs. bifacial glass/glass HJT modules for the full monitoring period. The bifacial modules outperform G/BS HJT modules at any time of the year at all weather conditions. In total, annual energy yield of the bifacial module exceeded that of the monofacial HJT by approximately 10% (see Table 2, right). Data for March is the most representative of the winter period with presence of snow coverage and many sunny days. Bifacial gain of around 15% is obtained for this period due to higher reflectance from snow. Extremely high bifacial gain in some winter months is caused by snow shadowing of the front side of the modules, while there is still some energy yield for bifacial modules due to backside collection.

As in our previous study [4], in Fig.5 we plot bifacial gain as function of daily solar irradiation for different seasons. At any weather conditions there is substantial bifacial gain due to backside light collection. The gain increases strongly for low light overcast cloudy or mixed conditions due to stronger diffused light scattering on the module's backside. The gain values at low light of the order of 15% to 25% can be expected depending on the season. For clear sunny days the bifacial gain of the order of 5% to 12% is observed. At the same time, there is clear seasonal dependence of the bifacial gain that is mostly governed by the condition of the ground. The highest gains are obtained for winter snowed ground, while the lowest gain values are observed in autumn on the bare ground. The data compares well with the literature [5]. The most striking fact from the Fig. 5 is that in winter time extremely high bifacial gain > 25 % has been observed not only for days with low irradiation value, i.e. in cloudy weather, but also for the days with irradiation >4.5 kWh/m^2 that, taking into account short daylight hours corresponded to sunny days. This finding correlates with different reports on the bifacial gain study for the modules installed in area with high albedo level and allow one to expect bifacial gain of HJT modules similar to that obtained in winter, if special measures are applied to increase albedo of a solar plant site.

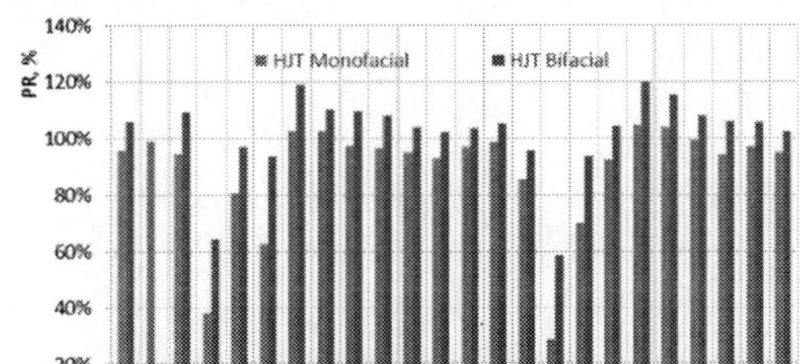

Figure 4. Monthly performance ratio of the monofacial G/BS and bifacial HJT modules. In total, ~10% annual gain of bifacial HJT over monofacial HJT is observed. Winter data are affected by snow.

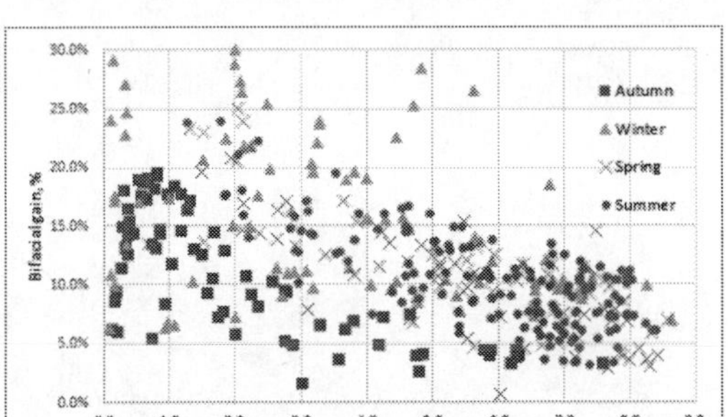

Figure 5: HJT bifacial gain over monofacial HJT module as a function of daily irradiation for different seasons and ground conditions.

Comparison HJT modules with different module design

Performance of the research grade HJT modules with different interconnection and cell types are compared to the production HJT modules in Fig.6. The HJT module with Electrically Conductive Adhesive (ECA) busbar interconnection is marked as "HJT BB". The modules are of the same output power class. Data for several summer months is shown as the most representative for comparison. As can be seen, busbar HJT module perform equally with the SWCT contacted module. Half cells module used for monitoring purposes has been chosen from very first modules made with this technology at Hevel, so its performance might be affected by initial stage of this technology development at Hevel.

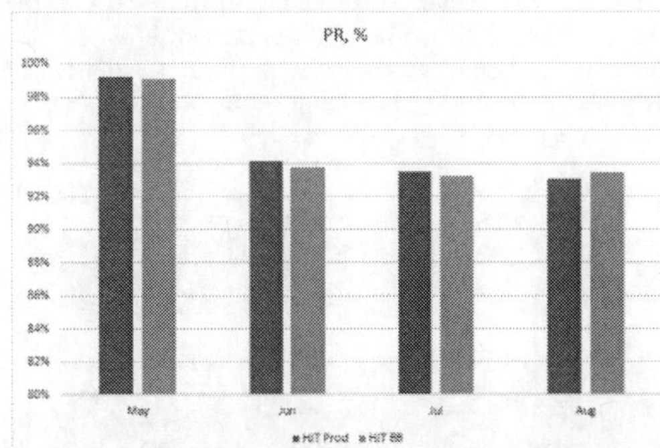

Figure 6: Monthly performance ratios of the modules with different cell interconnections. HJT production with SWCT contacting is compared with HJT busbar interconnection (BB) and half cut HJT cells with SWCT.

STC measurements after outdoor exposure

Modules from the monitoring stand are periodically measured on the solar simulator indoor at STC conditions to check degradation rate during outdoor testing. Current degradation rate of the output power with respect to its initial values for several types of modules is presented in Table 3. Production HJT modules after 18 months of outdoor exposure show degradation of around 0.7% with respect to initial values without additional drop of efficiency in the first months of exposure characteristic of other technologies. Research type HJT module with busbar interconnection after about one year of operation exhibits similar degradation of 0.5%.

Table 3: HJT modules output power degradation after outdoor exposure. Research modules are not assembled from the same materials as production ones.

Module	P_{nom}, W	Monitoring period, month	Degradation for the monitoring period, %
HJT Production module	300	18	-0.7%
HJT - Busbar Research module	294	10	-0.5%
HJT - Bifacial (glass-backsheet) Research module	298	10	-1.4%

For research type bifacial module assembled from alternative materials with transparent backsheet, detected degradation rate was higher than 1%, that is higher than the values demanded by the PV market at the moment. The reason for this could be higher water ingress for transparent backsheet material as compared to the standard one. So, further optimization of this class of bifacial HJT modules with transparent backsheets are needed to improve their reliability.

HJT modules under extreme weather conditions

For solar applications in many regions in Russia, there is need to test module's reliability and performance under extremely low temperatures. Up to now, HJT modules were subject to the thermocycling test (TC200) with the lowest temperature of -60°C. HJT modules successfully passed TC200 test with the very moderate degradation of about 1-1.5%.

At the same time, outdoor performance of Hevel HJT modules is monitored at the site with extreme weather conditions with winter temperatures down to -50°C at the location: (61°12, 151°56). Modules are part of the remote autonomous road emergency station (Fig. 7). After the 1st half year of monitoring including winter period there is no detectable degradation observed.

Figure 7: Remote autonomous road emergency station with Hevel HJT modules, Russia (61°12, 151°56).

4. SUMMARY

In this work the outdoor performance of different HJT modules is evaluated in direct comparison with each other and with respect to the reference mc-Si module. Hevel HJT modules outperform standard commercial mc-Si due to lower temperature coefficients of the HJT cells. Furthermore, bifacial HJT modules show energy gain over monofacial HJT modules at all weather and seasonal conditions. The annual energy gain of about 10% is obtained. Energy gain for clear sunny days of 5- 10% in summertime and up to 20-25% in wintertime with ground covered by snow is observed. No difference in the performance ratios of SWCT and BB interconnected modules with similar nominal power is observed. Average annual power degradation for the HJT production modules is estimated around -0.6%/year. In addition, HJT modules are being tested under extreme low temperatures outdoor and in the lab.

REFERENCES

[1] M. Taguchi, A. Yano, S. Tohoda, K. Matsuyama, Y. Nakamura, T. Nishiwaki, K. Fujita, E. Maruyama, IEEE Journal of Photovoltaics, 4, 96(2014).

[2] The International Technology Roadmap for Photovoltaic (ITRPV): http://www.itrpv.net/

[3] D. Andronikov et al. "A successful conversion of silicon thin-film solar module production to high efficiency heterojunction technology", 33rd European Photovoltaic Solar Energy Conference and Exhibition, p.732(2017).

[4] K. Emtsev et al. "Outdoor performance analysis of the si-heterojunction modules with different cell and module designs". 33rd European Photovoltaic Solar Energy Conference and Exhibition, p.1628(2017).

[5] A. Faes et al., "Smart wire solar cell interconnection technology", Proc. of the 29st EUPVSEC, pp. 2555–2561, 2014.

[6] T. Soderström et al., "Low cost high energy yield solar module lines and its applications", Photovoltaic Specialist Conference (PVSC), IEEE 42nd, 2015. DOI: 10.1109/PVSC.2015.7356431.

PRIORITIZATION OF TEST EXECUTION WITH OPERATIONAL PROFILE MECHANISM FOR SOFTWARE RELIABILITY OF SOLAR ENERGY MONITORING SYSTEM

Woo Sung Jang, Bo Kyung Park, and R. Young Chul Kim
Software Engineering Laboratory, Sejong Campus, Hongik University, 30016, South Korea
{jang, park, bob}@selab.hongik.ac.kr

ABSTRACT: Generally, our solar monitoring system consists of web monitoring server, data collection server, data collection client, heterogeneous inverters, and junction boxes. In this system, each device uses different protocol and interface with different data processing algorithms, which have some problems for communication between different equipment. To prevent this problem, we should require to test it with many test cases. But we have a limited time to release it, that is, difficult to execute all tests. To solve these problems, we propose prioritization of test execution with Musa's operational profile mechanism for software reliability of the solar energy monitoring system. This method identifies elements of the system through incremental steps, which generates operational profile using identified elements. Then we calculate execution probability of operation through quantitative measurement. As a result, we can prioritize quantitative tests, which can perform intensive testing against critical and fatal things.
Keywords: Operational Profile, Solar Energy Monitoring System, Test Case Priority

1 INTRODUCTION

Solar energy research is a representative renewable energy with a technology-intensive mechanism that requires development of various technologies in many scientific and engineering fields. Recently, many researches are focused on power generation forecasting, real-time monitoring, and environmental prediction [1,2]. Also interested on fault detection and diagnosis of solar energy generation system [3,4]. But there is a lack of research on monitoring system failure analysis. The monitoring system gets data with different protocols from heterogeneous inverters. But test period of the monitoring system is insufficient on system development period.

To solve this problem, we propose prioritization of test for this monitoring system using operational profile. This is to quantify the system resources and function execution probabilities, and generate the operational profile using quantified values. Then calculate the execution probability of the generated Operational Profile, and generate test execution probability based on Operation Profile. As a result, it can perform intensive testing of functions with high criticality.

Section 2 describes operational profile and our monitoring system based on metamodel. Section 3 describes generation process of operation profile of software reliability for our monitoring system. Section 4 describes prioritization of test using operational profile. Finally, describes the conclusions and future work.

2 RELATED WORKS

2.1 What is Operational Profile?

The Operational Profile describes how customers use the system and perform quantitative measurements of system usage [5]. This is an essential method in software reliability engineering, which shows how to increase productivity, stability, and speed of development. Operational Profile-based tests can take a full advantage of test schedules limited.

The process of developing the Operational Profile is as follows: 1) find Customer Profile, 2) set User Profile, 3) define System-mode Profile, 4) determine Functional Profile, and 5) determine the Operational Profile. Detailed options for each step are shown in Figure 1.

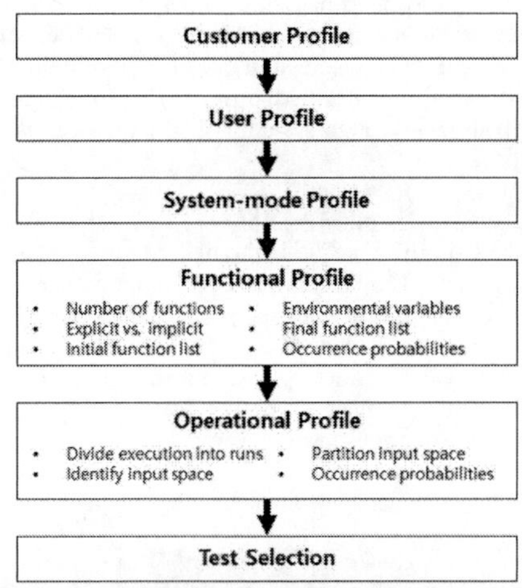

Figure 1: Operational Profile on Development Process

2.2 Our Solar Energy Monitoring System using Metamodel Mechanism on Heterogeneous Protocol

The metamodel-based solar energy monitoring system is a solar monitoring system that supports easy communication between different types of inverter [6,7]. This approach supports an automatic conversion between different protocol models. That is, if new inverter is added, we just design metamodel to perform system maintenance with minimal cost [8].

The DC electricity produced by the solar cell is converted into AC electricity by the inverter. The inverter measures the generated power and transfers the measured power to the Local Server. Inverters manufactured by various companies use various kinds of protocols. Local Server converts data transmitted by various protocols into integrated XML data using metamodel-based conversion method. Converted data is transferred to the Integrated Server using TCP/IP or LoRa network. The integrated Server stores the received data in the DB. The web server reads the data in the DB and displays the web browser. The overall structure of the system is shown in Figure 2.

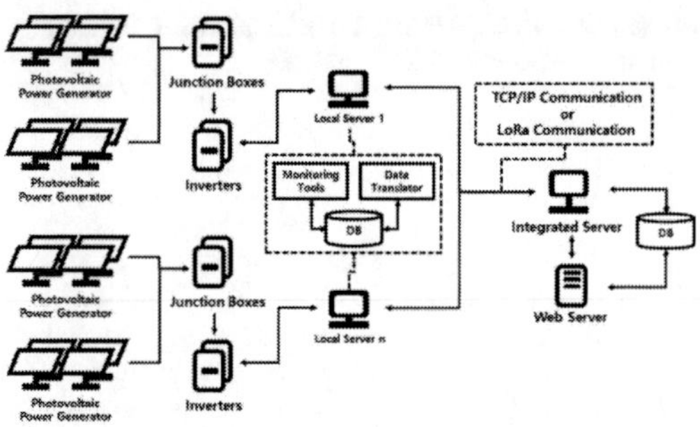

Figure 2: Our Solar Energy Monitoring System

3 OPERATIONAL PROFILE OF SOLAR ENERGY MONITORING SYSTEM BASED ON METAMODEL

First, collect system information for generating operational profile of the photovoltaic energy monitoring system based on metamodel. Then check and collect user type, system menu connection rate, and power plant type. Classify collected data. Measure probability of use by classified data through system log. Quantify all data and start profiling.

3.1 Generate Customer Profile

The Customer profile is the type of power plant used by customers. The metamodel-based photovoltaic energy monitoring system has power plants based on LoRa and TCP/IP communication. Therefore, Customer is classified into two types of power plants. Probability of customer profile is the number of power plants. The generated Customer Profile is shown in Table I.

Table I: Generated Customer Profile

Customer Name	Probability
LoRa Plant	0.77
TCP/IP Plant	0023

3.2 Generate User Profile

The User profile is type of system users. System user groups include System Administrator, System engineer, Plant Behavior, and Maintenance Personnel. The probability of a user profile is probability that each user belongs to a customer group.

For example, LoRa Plant consists of System Administrator 6%, System Engineer 15%, Plant Behavior, 71% and Maintenance Personnel 8%. Each probability is the number of users using the monitoring system. For calculate probability of users using LoRa's plant system, multiply probability of users of the LoRa's Plant by 0.77. Probability of the generated User Profile is shown in Table II.

Table II: Generated User Profile

User Group	LoRa plant probability = 0.77		TCP/IP plant probability = 0.23		Total User-group probability
	User-group probability within customer group	Overall user-group Probability for customer group	User-group probability within customer group	Overall user-group probability for customer group	
System administrator	0.06	0.0462	0.12	0.0276	0.0738
System engineer	0.15	0.1155	0.29	0.0667	0.1822
Plant behavior	0.71	0.5467	0.41	0.0943	0.641
Maintenance personnel	0.08	0.0616	0.18	0.0414	0.103

3.3 System-mode Profile

System-mode profile is usage of system. System-mode profile includes Business use, Personal use, System administration, and Maintenance. Usage of system is classified into the types of users of the system. Sales employee user is classified as Business use, General customer user is referred to as Personal use, Total administrator as System administration, and Maintenance user as Maintenance. The probability of the generated System-mode Profile is shown in Table III.

Table III: Generated System-mode Profile

System mode	Occurrence probability
Business use	0.16
Personal use	0.68
System administration	0.1
Maintenance	0.06

3.4 Generate Functional Profile

Functional Profile performs 6 procedures. Each procedure is as follows.

First, create Initial Function List. Initial Function List defines functions for each system-mode. Generated initial function is shown in Table IV.

Table IV: Generated Initial Function List

System mode	Initial Function
Business use	Inverter Addition, Inverter Change, Inverter Removal, View Sensor Data, View Power Data
Personal use	View Current Power Data, View Hour Power Data, View Day Power Data, View Month Power Data, View Year Power Data, Print Report
System administration	User Addition, User Change, User Removal, Plant Addition, Plant Change, Plant Removal
Maintenance	Inverter Addition, Inverter Change, Inverter Removal, View Sensor Data, View Power Data, View Current Power Data, View Hour Power Data, View Day Power Data, View Month Power Data, View Year Power Data, Print Report, User Addition, User Change, User Removal, Plant Addition, Plant Change, Plant Removal, LoRa Server Addition, LoRa Server Removal, LoRa Server Change

Second, generate an environmental variable. Environmental variable is all environment variable that can be input to the function list. User is type of system user. Plant Communication is type of power plant communication. Generated Environmental Variable is shown in Table V.

Table V: Generated Environmental Variables

Type	Environmental Variable
User	A : Admin
	M : Monitor
	B : Behavior
Plant Communication	W : Wi-fi
	L : LoRa

Select environmental variable for input in the function list. Selection results for the function list in the Administration Mode are shown in Table VI.

Table VI: Generated Final Function List of System Administration Mode

Function	Environmental Variable
User addition	A/M/B
User removal	A/M/B
User change	A/M/B
Plant addition	W/L
Plant removal	W/L
Plant change	W/L

Third, calculate the probability of initial function list. Each probability is calculated based on the number of function execution. The number of function execution refers system log. Function list probability of Administration Mode is shown in Table VII.

Table VII: Generated Occurrence Probability of System Administration Mode

Function	System-administration-mode occurrence probability	Overall occurrence probability
User addition	0.23	0.023
User removal	0.006	0.0006
User change	0.02	0.002
Plant addition	0.69	0.069
Plant removal	0.013	0.0013
Plant change	0.032	0.0032

Fourth, calculate occurrence probability of environmental variable. Occurrence probability of a user type is the type of system user. Occurrence probability of plant communication type is number of communication methods of plant. Probability of the generated environmental variable is shown in Table VIII.

Table VIII: Generated Probability of Environmental Variable

User type	Occurrence probability
A	0.03
M	0.06
B	0.91
Plant Communication type	**Occurrence probability**
W	0.23
L	0.77

Fifth, calculate probability when the environmental variable is entered in function list. Generated probability is shown in Table IX.

Table IX: Generated Occurrence Probability of System Administration Mode per Environment Variable

Function	Environment type	Overall occurrence probability
User addition	A	0.00069
	M	0.00138
	B	0.02093
User removal	A	0.000018
	M	0.000036
	B	0.000546
User change	A	0.00006
	M	0.00012
	B	0.00182
Plant addition	W	0.01587
	L	0.05313
Plant removal	W	0.000299
	L	0.001001
Plant change	W	0.000736
	L	0.002464

Sixth, generate implicit functional profile. Implicit functional profile is the menu of software in the system used by Users. Calculate probability using number of menu execution used by users. Number of menu execution refers system log. Figure 3 shows implicit functional profile of System Administration Mode.

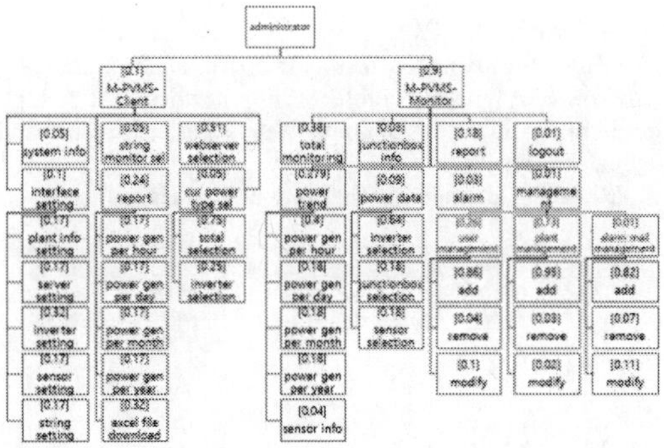

Figure 3: Implicit Functional Profile of System Administration Mode

3.5 Generate Operational Profile

Operational profile defines operation and probability of function. Operation is a combination of functions in implicit functional profile. Probability of operation is multiplication of probabilities defined in the implicit functional profile. Generated operational profile of system administration mode is shown in Table X.

Table X: Generated Operational Profiles of System Administration Mode

Function	operation	Occurrence probability
User addition	M-PVMS-Monitor, management user_management, add	0.0020124
User removal	M-PVMS-Monitor, management user_management, removal	0.0000936
User change	M-PVMS-Monitor, management user_management, change	0.000234
Plant addition	M-PVMS-Monitor, management plant_management, add	0.0062415
Plant removal	M-PVMS-Monitor, management plant_management, removal	0.0001971
Plant change	M-PVMS-Monitor, management plant_management, change	0.01314

4 PRIORITIZE TEST EXECUTION BASED ON OPERATIOAN PROFILE

4.1 Generate Operational Profile Probability per Environment Type

Metamodel-based photovoltaic monitoring system is granting to be accessed by each user. If the user does not have access, it cannot be access to system menu.

Environment types A, M, and B are types of users. Operation of operational profile is a combination of system menu. System menu is used by the user only. Therefore, calculate operation access probability for each user using Environment type. Probability of the operation per user of environment type is shown in Table XI.

Table XI: Generated User Addition Function with Operational Profile Probability per Environment Type

Function	operation	Environment type	occurrence probability
User addition	M-PVMS-Monitor, login, solar_plant_selection, management user_management, add	A	0.000001388556
		M	0.000002777112
		B	0.000042119532

4.2 Prioritize Test Execution

Enter the expected result of operation execution by environment type, and preferentially perform test of high probability operation. Compare expected test result and actual test results.

Table XII: Generated Probability of Test Execution

Function	operation	Environment type	occurrence probability	Expected Result
User addition	M-PVMS-Monitor, login, solar_plant_selection, management user_management, add	A	0.000001388556	SUCCESS
		M	0.000002777112	FAILED
		B	0.000042119532	FAILED

4 CONCLUSION

Our monitoring system manages to communicate heterogeneous inverters, and controls inputs of various users. Therefore, this system includes the possibility of various errors, and requires sufficient testing. But test period is insufficient because of short development period.

We propose prioritization of test for solar energy monitoring system using operational profile, which first identifies commonly used functions in the system, prioritizes tests using identified quantitative data, and performs intensive testing of functions with high importance.

In future work, we will design and create optimal test cases for solar monitoring systems.

Acknowledgments

This research was supported by Basic Science Research Program through the National Research Foundation of Korea(NRF) funded by the Ministry of Education(NRF-2017R1D1A3B03035421), and supported by Development of Information Communication and Broadcast R&D(2018 Open OS Environment Development and Diffusion) through NIPA(National IT Industry Promotion Agency)(R-20180402-002888).

References

[1] Korea Energy Agency New and Renewable Energy Center, "2016 Understanding of New and Renewable Energy", 2016.

[2] R. Leyva, C. Alonso, I. Queinnec, A. Cid-Pastor, D. Lagrange, L. Martinez-Salamero, "MPPT of photovoltaic systems using extremum-seeking control", IEEE Transactions on Aerospace and Electronic Systems, Vol.42, pp.249-258, 2006.

[3] D. Sera, R. Teodorescu, "Photovoltaic module diagnostics by series resistance monitoring and temperature and rated power estimation," Annual Conference of IEEE, pp.2195-2199, 2008.

[4] K.H. Chao, S.H. Ho, M.H. Wang, "Modeling and fault diagnosis of a photovoltaic system," Electric Power Systems Research, Vol.78, pp. 97-105, 2008.

[5] John D. Musa, "Operational Profiles in Software-Reliability Engineering", Journal IEEE Software, Vol.10, pp.14-32, 1993.

[6] Woo Sung Jang, Hyun Seung Son, Bo Kyung Park, Byung Kook Jeon, R. Young Chul Kim, "Implementation of Effective Web Integrated Monitoring System for Small Renewable Energy business Industries", Korean Institute of Smart Media, Vol.5, No.1, pp.303-305, 2016.

[7] Woo Sung JANG, Hyun Seung SON, Chae Yun SEO, R. Young Chul KIM, "Metamodel based Photovoltaic Monitoring System for Heterogeneous Renewable Energies", Asia Pacific Society for Computing and Information Technology, pp.44-47, 2017.

[8] Son, H.S., Kim, W.Y., Kim, R. Y. C., "A Study on Modeling Tool for Convergence of Smart Appliances", The Journal of IIBC (The Journal of The Institute of Internet, Broadcasting and Communication), Vol.8, No.4, pp.119-125, 2008.

METHODS FOR QUALITY CONTROL OF MONITORING DATA FROM COMMERCIAL PV SYSTEMS

Mari B. Øgaard[1,2*], Halvard Haug[1,2] and Josefine Selj[1,2]
[1]Department of Technology Systems, University of Oslo
Gunnar Randers vei 19, 2007 Kjeller, Norway
[2]Department of Renewable Energy Systems, Institute for Energy Technology
Instituttveien 18, 2007 Kjeller, Norway
*Corresponding author. E-mail: mari.ogaard@ife.no, Tel.: +47 976 356 08

The aim of this work is to develop and test new methods for quality control of data from commercial monitoring systems for small and medium sized PV installations. Such installations often have limited or non-existent maintenance of their monitoring systems. Quality issues in e.g. irradiance and temperature measurements will cause errors in the analysis of the PV system performance and might lead to non-optimal maintenance of the system. To determine the condition of the sensors and the monitoring system based on the measured data itself is therefore essential to improve performance analysis algorithms and to understand historical data from these types of PV systems, and consequently this is of significant economical and practical value. In this work, we use data from both commercial and research systems in Norway to assess the robustness of the methods in a real-world scenario. We demonstrate that drift and deviations in the sensitivity of irradiance sensors, in addition to misalignment of the sensors, can be accurately quantified and detected based on comparison with clear sky irradiance modeling. Furthermore, we show that analysis of temperature data potentially can be used to detect snow cover of modules, in addition to identification of detached temperature sensors.

Keywords: monitoring, PV system, data quality

1 INTRODUCTION

Regulations and irradiation conditions greatly influence the size and type of PV systems installed in a given market. In Norway, as in other northern climates, moderate solar irradiance and incentives for self-consumption has resulted in a market dominated by relatively small PV systems with installed capacity less than 1 MWp. Moderately sized commercial systems constitute a significant fraction. Typically, these systems have a simple monitoring system, measuring both electrical and environmental data. Most commonly, the electrical data is collected from the inverter, the plane of array (POA) irradiance is measured by a reference cell, and there are sensors measuring ambient temperature and the module temperature. In addition to the typically low accuracy of commercial monitoring solutions [1,2], often very little maintenance is performed on these systems, as the cost of maintenance of distributed monitoring systems may exceed the expected benefit if local personnel are not available. In most cases, necessary maintenance like cleaning of the irradiance sensors and visual inspection of the system and sensors is not performed, and the sensors are not regularly recalibrated. Drift in sensors, dirty irradiance sensors and detached module temperature sensors can lead to significant misinterpretation of the PV system performance, and consequently also suboptimal or unnecessary maintenance of the PV system. The distribution of the systems, combined with little local competence and small economic incentives, makes increased manual supervision an unrealistic solution. A more realistic path ahead is quality control of the sensor data based on data analysis. In addition to saving money by reducing the need for maintenance, this approach also enables validation of the quality of historical data for any system, independent of previous monitoring system maintenance routines.

Data-based evaluation of the condition of the sensors and the monitoring system is not widely discussed in the literature. The quality control work done today is mostly limited to detecting abnormal points (i.e. outliers and data exceeding physical possible limits) [3–5], as opposed to detecting permanent changes in the measurement over longer periods. In this work we suggest a new data-based method for detecting changes in the quality or accuracy of the irradiance measurements. We also discuss a method for quality control of module temperature measurements and how this method can be used to identify snow cover on the modules. Analysis is performed using data from commercial PV systems in addition to scientific test sites to assess the robustness of the methods in a real-world scenario.

The challenge of missing maintenance and supervision of irradiance sensors is also discussed by Jordan et al. [6], who proposed to calculate PV system degradation based on clear sky irradiance simulations instead of measured irradiance. However, ground measurements provide valuable additional information on the performance of the PV system under cloudy conditions and enable a more comprehensive analysis of PV systems. This is particularly important for locations with few clear sky days.

In this work we use clear sky irradiance simulations to assess the reliability and quality of the measured irradiance. This has previously been suggested by Reno et al. [7], but is to our knowledge not tested before. The method is assessed for both high and low-quality irradiance sensors.

A relatively common problem with measurements of module temperature, is detachment of temperature sensors, which are normally attached to the back sheet of the module. This could be detected by monitoring the difference between the module and the ambient temperature. To our knowledge, this method has not been tested beyond the observations presented by Woyte et al. [8]. In this work we test this method for two different types of PV installation, and assess it for a new application; snow detection. The possibility of detecting snow coverage using measurement equipment that already exists on-site, could improve both the monitoring algorithms for PV system fault detection and the energy generation forecasts by separating snow events from other failures. Additionally, it will also simplify the analysis of historical data, with respect to estimation of performance and validation of snow loss models. In previous studies, the probability of snow cover on PV modules is estimated by detecting when the PV production is low relative to irradiance or estimated irradiance in combination with evaluation of other parameters, including ambient temperature [9], predictions of snow depth and temperature [10], and satellite observations [11]. Using ambient temperature alone will not separate snow events from e.g. total black outs or other serious failures [12]. The aim of this work is to increase the accuracy of snow detection at a specific location by also using the module temperature sensor data.

2 METHODS

2.1 Test of irradiance measurements based on clear sky irradiance simulations

Global horizontal clear sky irradiance is simulated using the Ineichen and Perez clear sky model based on zenith angle, air mass, elevation and Linke Turbidity [13]. The model error has been shown to have low dependency of time of the day and day of the year compared to other clear sky models [7]. The simulated clear sky irradiance is then transposed to a tilted plane by calculating beam, reflected and diffuse irradiance in the plane. The sky diffuse irradiance is calculated using the isotropic sky model [14], where the sky diffuse irradiance in the plane of the PV array is found using the diffuse horizontal irradiance, the tilt angle of the plane of array, and the assumption that the sky is a uniform source of irradiance. Periods with measured irradiance equivalent to clear sky conditions, i.e. a smooth irradiance curve, was detected using the algorithm proposed by Reno and Hansen, which compares GHI time series statistics to the Ineichen clear sky model [15]. The length of the clear sky time periods selected are at least two hours, to optimize between amount of data and correct selection of clear sky periods. All models are implemented in compliance with the methods in the Matlab version of the PV_LIB Toolbox [16], using the default Link turbidity values provided by SoDa in the clear sky modeling.

The modeled clear sky irradiance is in this case used as a reference, and the measured data is compared to the modeled results in the periods with irradiance equivalent to clear sky conditions. The relative difference (ΔI) between measured (I_{meas}) and modeled irradiance (I_{CS}) is given by:

$$\Delta I = (I_{meas} - I_{CS})/I_{meas} \qquad (1)$$

To evaluate how the measurements change relative to the model from year to year, a scaling factor (α) is estimated for every year by minimizing RMSE between measured and clear sky irradiance, given by [15]:

$$RMSE(\alpha) = \sqrt{\frac{\sum_{i=1}^{n}(\alpha \times I_{meas} - I_{CS})^2}{n}} \qquad (2)$$

2.2 Method for detection of sensor detachment and snow

The difference in module and ambient temperature is strongly correlated with the irradiance (Eq. 3) [17]. Hence, by monitoring this correlation, signatures of temperature sensor detachment can be identified [8]. We have tested this method for two different installation configurations, by investigating the changes in linear regression fits of hourly values of temperature difference and irradiance for different weeks. Additionally, we have tested if the same approach can be used for snow cover detection.

$$T_{mod} - T_{amb} = I \times e^{a + b \times Wind\ speed} \qquad (3)$$

Where a and b are coefficients compensating for site specific configurations.

2.3 Measurement sites

The proposed method for quality control of irradiance sensors was tested for the uncorrected raw data, given as 10-minute averages, measured by an old pyranometer installed at the supervised weather station at the Norwegian University of Life Sciences. The method was also tested on a commercial flat roof PV system, where the POA irradiance was measured by tilted reference cell and given in 5-minute averages. The tilt of the system and the reference cell is 10°, and the orientation of the modules is south-east and north-west.

The method for detection of sensor detachment is tested for two different types of PV installations: South-orientated modules installed in an open-rack configuration with a tilt of 28°, and modules installed on a roof with a tilt of 35°. The module temperatures are measured by a resistant thermometer attached to the back sheet of the modules, and the POA irradiance is measured by reference cells. For the rack installation, the ambient temperature sensor is installed under the modules, and for the tilted roof installation the ambient temperature is measured by a weather station at the same location.

For both systems, the same approach is tested for snow cover identification. Additionally, snow cover identification is tested for the flat roof system used in the irradiance quality control tests. For this system, the cell temperature of a reference cell is used as an estimation of module temperature, the ambient temperature is measured by a PT-1000 element, and irradiance is measured by a ventilated pyranometer. Normally, the reference cell is covered by snow in the same period as the modules because of the low tilt angles, and the pyranometer is less affected, most likely because of the ventilation and a more elevated installation position. Both roof installations are commercial systems, while the open rack system is a scientific test site. Hourly averages are used in the analysis.

3 RESULTS AND DISCUSSION

3.1 Quality control of irradiance measurements

The irradiance measurement control method was tested on raw data from the pyranometer at the Norwegian University of Life Sciences. As presented in Figure 1, the relative difference between measured and modeled irradiance is scattered and high, especially in periods with low irradiance. In periods with low irradiance and high angle of incidence, the relative errors of both the measurements and the modeled results is probably higher [7,18].

To improve the analysis, the periods with low irradiance were filtered out. The difference between measured irradiance and modeled clear sky irradiance when the measured irradiance was more than 500 W/m² is given as a function of time in Figure 2. The sensor was recalibrated and adjusted in 2008, after this there were no significant changes in sensitivity for two years, followed by a decrease in sensitivity by 1% per year in a four-year period (2010-2014). The pyranometer was replaced in 2014, which lead to an increase in the sensitivity of the irradiance measurements at the site of 6%. All these alterations were detectable through comparison with clear sky modeling. The shift in measured irradiance relative to the modeled irradiance at the time of the sensor adjustment and the pyranometer replacement is clearly shown in Figure 2. After the sensor replacement, the scaling factor α also increased by 6 percentage points. The analysis indicates a degradation of the sensor of 1% per year in the period 2010-2014, the same as the independent instrument calibration, based on a reduction of α of 1 percentage point per year.

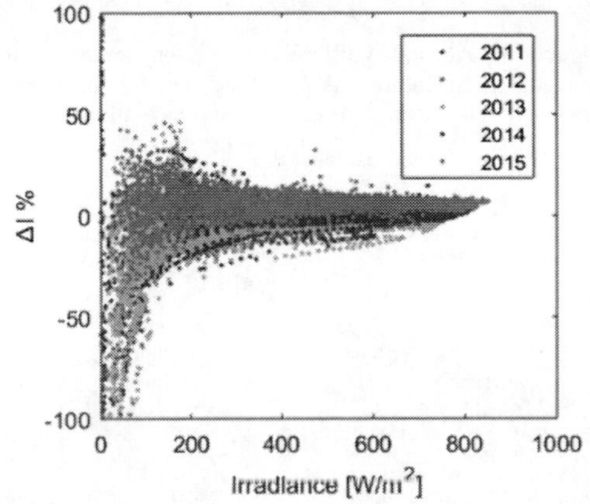

Figure 1: Relative difference between GHI measurements and clear sky irradiance modeling at different irradiance levels.

We observe that the modeled irradiance is slightly biased and most of the time is lower than the measured irradiance, as expected for higher latitudes when using the Ineichen and Perez model [7]. Additionally, the variations between single points is high, indicating that this method cannot be applied for quality control of individual measurements. However, the ability of the method to indicate drift or other permanent changes in the irradiance measurements over time is remarkable.

It is well known that pyranometers are subject to a thermal offset during clear sky periods [19]. This introduce a bias in the comparison between the clear sky model and the measurements. However, as it is the change and not the absolute value that is of importance, this effect is assumed negligible in this context. A more relevant challenge is changes in atmospheric conditions with respect to transmission and scattering. Changes in air pollution are one of the major contributions to what is referred to as global brightening and dimming, which may have an effect of a magnitude that might influence long time PV performance analysis [18]. More practical challenges related to this method for commercial systems, is the low time resolution data these types of systems often have, and potentially also the number of clear sky hours through the year at the specific location.

3.2 Quality control of tilted reference cell measurements

The quality control method for irradiance measurements was also tested for tilted reference cells measuring the POA irradiance of a PV system in a north-west, south-east configuration with a tilt of 10°. Reference cells are, as mentioned earlier, common for medium sized PV systems with monitoring. The uncertainties are however higher than when pyranometers are used. The relative difference between the measurements and the clear sky modeling for irradiance above 500 W/m² is presented in Figure 3. The scaling factor between the measured and modeled irradiance is approximately constant from year to year for the irradiance measured in the north-west direction, with a reduction of less than 0.1 percentage points. This is lower than expected degradation for a c-Si reference cell [8].

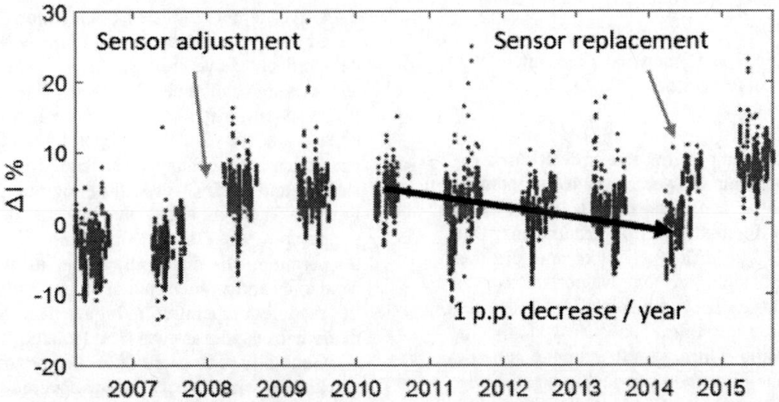

Figure 2: Relative difference between GHI measurements and clear sky irradiance modeling at measured irradiance above 500 W/m². Time of adjustment of the sensor, pyranometer replacement, and the period where 1 % decrease per year was measured is marked.

It should be noted that the north-west orientation results in lower irradiance and fewer data points with irradiance values above 500 W/m². This could potentially affect the robustness of this result. For the irradiance measured in the south-east direction, the difference between measured and modeled irradiance appeared to be season dependent. It was found that this sensor was misaligned, which could be an explanation for this behavior.

This misalignment also gave a shift between the position of the measured and modeled irradiance curves (as well as the PV power curve), as shown in Figure 4. A consequence of this was unlikely high performance ratios for the PV system in the morning, as the irradiance sensor measured less irradiance than what the PV modules received, and opposite in the evening.

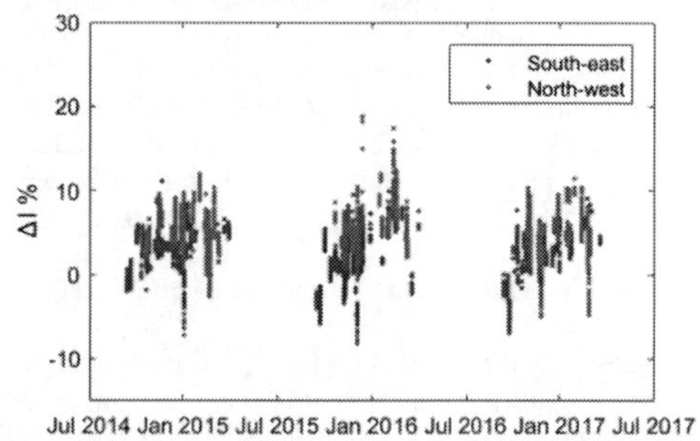

Figure 3: Relative difference between POA measurements and clear sky irradiance modeling at measured irradiance above 500 W/m².

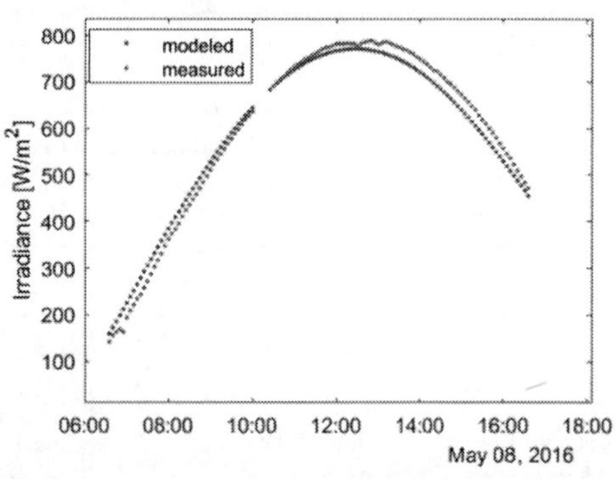

Figure 4: Measured and modeled irradiance with temporal shift in curves because of misalignment of sensor.

3.3 Control of module temperature sensor detachment

The linear relationship between module and ambient temperature difference and irradiance is presented in Figure 5-6 for two different types of installations. Data was selected from four different weeks and different months and years to highlight some important aspects: The correlation between the temperature difference and the irradiance is almost linear and it typically has relatively small variations from year to year and month to month for the ground mounted open rack system, and to some degree also for the roof mounted system. This indicates that it is possible to detect detachment of the module temperature sensor by monitoring changes in the slope. If detached, or partially detached, the module temperature sensor measure a value closer to ambient temperature, and the slope would be less steep or zero. The scattering of the data points is greater for the close roof mounted system and this system also experience the highest temperature difference. One natural explanation for this is the cooling effect of the wind, which is not taken into account in the analysis. Lack of rear side ventilation or heat leakage from the building itself may also be influential. The slope of the regression lines of the close roof mounted system is twice as steep as for the rack mounted system, in agreement with the experimental results of King et al. [17].

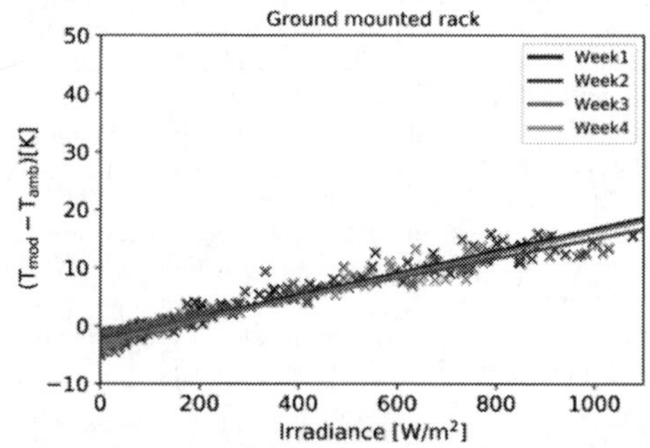

Figure 5: Absolute difference between module and ambient temperature as a function of measured irradiance. Data from four different weeks is shown.

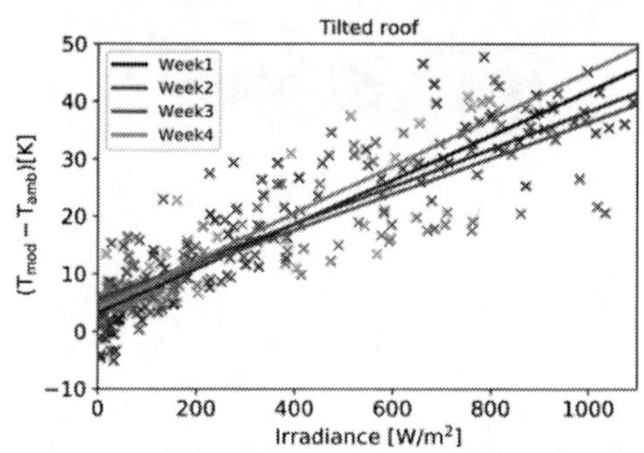

Figure 6: Absolute difference between module and ambient temperature as a function of measured irradiance. Data from four different weeks is shown.

3.4 Use of temperature sensors for snow detection

Snow cover on the PV modules will, like the sensor detachment, have an effect on the slope of the temperature/irradiance regression line. As presented in Figure 8-10, when snow is covering the modules, the correlation between the temperature difference and the irradiance will change substantially, as the module surface and back sheet will be significantly less heated. For the rack mounted system the measured module temperature is lower than the measured ambient temperature. The difference is increasing with increasing irradiance and ambient temperature. For the roof systems, the module temperature is higher than the ambient, most likely because the snow has an isolating effect.

Generally, it might not be possible to separate situations with partly snow-covered modules from situations with low irradiance, or situations when snow is covering the irradiance sensor. The figure showing the snow-covered modules at the tilted roof system, illustrates how it can be challenging to separate periods with low irradiance from periods with snow cover.

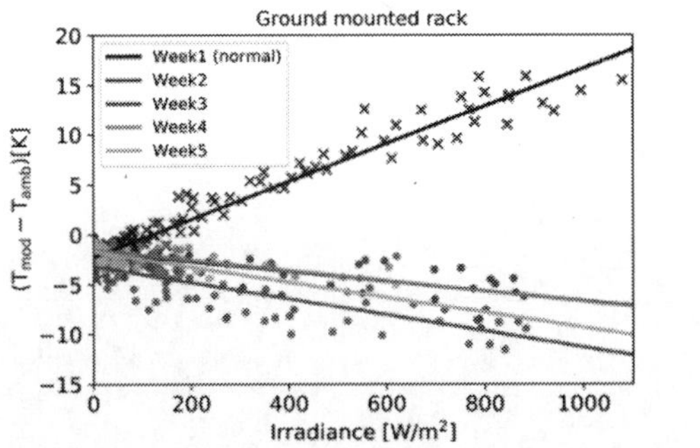

Figure 7: Absolute difference in measured module temperature and ambient temperature as a function of irradiance for weeks where snow is covering the modules compared to a week with normal production.

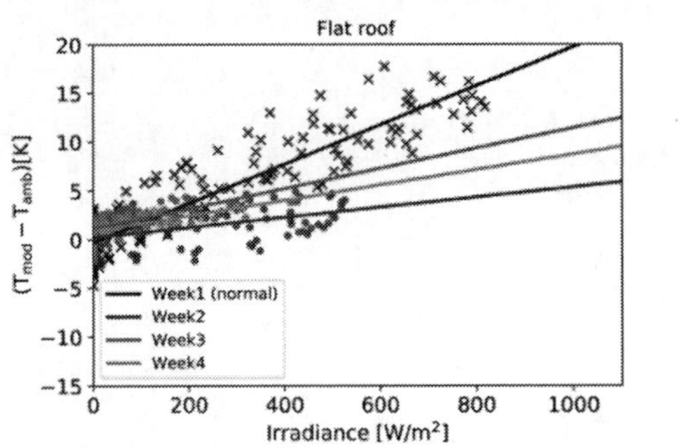

Figure 8: Absolute difference in measured module temperature and ambient temperature as a function of irradiance for weeks where snow is covering the modules compared to a week with normal production.

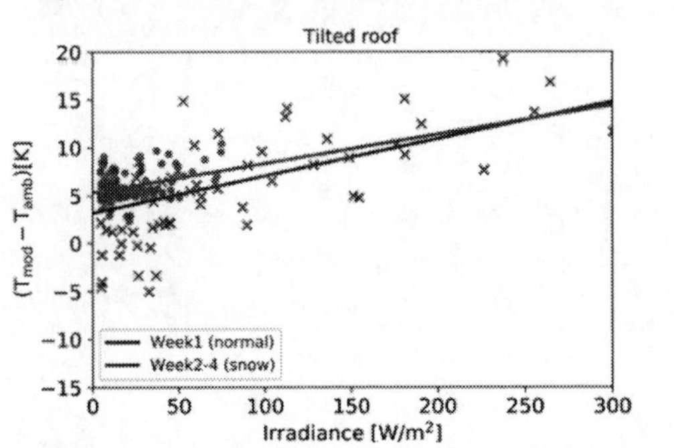

Figure 9: Absolute difference in measured module temperature and ambient temperature as a function of irradiance for weeks where snow is covering the modules compared to a week with normal production.

The main limitation with this method is its dependency of undisturbed irradiance measurements. Low irradiance situations might also be challenging, as the modules do not heat up as much as under normal operation. To give more robust snow detection algorithms, the difference in module and ambient temperature should be combined with PV efficiency data, and potentially also estimated irradiance to both indicate if the sensor is covered by snow and replace the irradiance measurement data.

The placement of the temperature sensors is of importance, as this will define the ability to detect partly covered modules, and give different temperature difference characteristics (i.e. positive, negative or zero). The temperature difference characteristics can also be studied through night time values.

4 CONCLUSION

In this work, a method for detection of permanent changes in irradiance measurements has been tested, as well as a method for detection of module temperature sensor detachment and how the latter can be used to improve detection of snow cover on the PV modules.

Quality control of irradiance measurements based on data anlaysis has been performed for different PV systems with high and low-accuracy irradiance sensors. We show that by using this approach it is possible to accurately detect both abrupt changes, as well as slow gradual changes such as the yearly drift of 1%.

A method for detection of module temperature sensors detachment and snow cover was tested for typical commercial installation configurations in Nordic climates, including open rack mounted, close roof mounted and tilted flat roof mounted systems. The linear relationship between temperature differences and irradiance proved to be stable enough to indicate sensor detachment and snow cover. The robustness of this method might depend on reliable irradiance measurements, sensor placement and installation configuration. Periods with low irradiance and partly snow-covered modules appeared to be challenging. However, in combination with other methods, the presented results show that the use of module temperature sensor increase the confidence and accuracy of forecasting and fault diagnostics with respect to snow cover events.

REFERENCES

[1] M. Fanni, L., Giussani, M., Marzoli, M., & Nikolaeva-Dimitrova, How accurate is a commercial monitoring system for photovoltaic plant?, Prog. Photovoltaics Res. Appl. 22 (2014) 910–922.

[2] M.B. Strobel, T.R. Betts, G. Friesen, H.G. Beyer, R. Gottschalg, Uncertainty in Photovoltaic performance parameters - dependence on location and material, Sol. Energy Mater. Sol. Cells. 93 (2009) 1124–1128.

[3] C. Ventura, G.M. Tina, Utility scale photovoltaic plant indices and models for on-line monitoring and fault detection purposes, Electr. Power Syst. Res. 136 (2016) 43–56.

[4] I. Moradi, Quality control of global solar radiation using sunshine duration hours, Energy. 34 (2009) 1–6.

[5] IEC 61724:1 Photovoltaic system Performance - Monitoring, 2017.

[6] D.C. Jordan, C. Deline, S.R. Kurtz, G.M. Kimball, M. Anderson, Robust PV Degradation Methodology and Application, IEEE J. Photovoltaics. (2017) 1–7.

[7] M.J. Reno, C.W. Hansen, J.S. Stein, Global horizontal irradiance clear sky models : implementation and analysis., (2012).

[8] A. Woyte, M. Richter, D. Moser, M. Green, S. Mau, H.G. Beyer, Analytical Monitoring of Grid-connected Photovoltaic Systems, 2014.

[9] S. Stettler, P. Toggweiler, E. Wiemken, W. Heydenreich, a de Keizer, W. van Sark, S. Feige, M. Scheider, G. Heilscher, E. Lorenz, Failure detection routine for grid-connected PV systems as part of the PVSAT-2 project, 20th Eur. Photovolt. Sol. Energy Conf. (2005) 2490–2493.
[10] E. Lorenz, D. Heinemann, C. Kurz, Local and regional photovoltaic power prediction for large scale grid integration: Assessment of a new algorithm for snow detection, Prog. Photovoltaics Res. Appl. 20 (2011) 760–769.
[11] G. Wirth, M. Schroedter-Homscheidt, M. Zehner, G. Becker, Satellite-based snow identification and its impact on monitoring photovoltaic systems, Sol. Energy. 84 (2010) 215–226.
[12] E. Lorenz, J. Betcke, A. Drews, A.C. de Keizer, S. Stettler, M. Scheider, S. Bofinger, H.G. Beyer, W. Heydenreich, E. Wiemken, W. van Sark, P. Toggweiler, G. Heilscher, D. Heinemann, Intelligent performance check of pv system operation based on satellite data, 2007.
[13] P. Ineichen, R. Perez, A new airmass independent formulation for the linke turbidity coefficient, Sol. Energy. 73 (2002) 151–157.
[14] P.G. Loutzenhiser, H. Manz, C. Felsmann, P.A. Strachan, T. Frank, G.M. Maxwell, Empirical validation of models to compute solar irradiance on inclined surfaces for building energy simulation, Sol. Energy. 81 (2007) 254–267.
[15] M.J. Reno, C.W. Hansen, Identification of periods of clear sky irradiance in time series of GHI measurements, Renew. Energy. 90 (2016) 520–531.
[16] J.S. Stein, W.F. Holmgren, J. Forbess, C.W. Hansen, PVLIB: Open source photovoltaic performance modeling functions for Matlab and Python, Proc. 43rd IEEE Photovolt. Spec. Conf. (2016) 3425–3430.
[17] D.L. King, J.A. Kratochvil, W.E. Boyson, Photovoltaic array performance model, Online. 8 (2004) 1–19.
[18] C. Reise, B. Müller, Uncertainties in PV System Yield Predictions and Assessments, 2018.
[19] G.S. Hernandez, A. Serrano, M.L. Cancillo, J.A. Garcia, Pyranometer thermal offset: Measurement and analysis, J. Atmos. Ocean. Technol. 32 (2015) 234–246.

PHOTOVOLTAIC CLIMATE CLASSIFICATION FOR THE DEGRADATION STUDY

Pratik Mundle, Narendra Shiradkar, Juzer Vasi, S. Patwardhan,
National Center for Photovoltaic Research and Education (NCPRE),
Indian Institute of Technology Bombay, Powai, Mumbai, 400076, India

ABSTRACT: The degradation of photovoltaic modules is dependent on climate conditions in the end-use environments. Conventionally, Köppen-Geiger (KG) climatic classification is used for such studies. But KG classification does not consider the solar irradiance, the most important climate factor regarding the performance of PV technology. This paper proposes a new classification scheme devised specifically by considering the idiosyncrasies of photovoltaic to get a more refined correlation. For classification three parameters - module temperature, strain accumulation caused due to thermal cycling and relative humidity are considered. The analysis is carried out using meteorological data of locations all over the world and 18 classification zones are defined after the analysis of the data.

1 INTRODUCTION

The degradation of various device parameters of PV module – short-circuit current (I_{sc}), open-circuit voltage (V_{oc}) and fill factor (FF), are dependent on PV technology and climatic conditions. Studies to understand and predict the degradation of these parameters can help in the improvement of the design and selection of PV technology for a chosen location. Previous studies [1-2] have tried to correlate the climate condition with degradation of PV module of various technologies mostly using Köppen-Geiger (KG) climatic classification [3].

KG classification splits the globe into 31 different climate zones based on ambient temperature and precipitation. The scheme has five main climate zones based on the vegetation observed in the region as equatorial zone (A), the arid zone (B), the warm temperate zone (C), the snow zone (D), and the polar zone (E). These main climate zones are further subdivided into subzones to identify the microclimatic condition based on temperature and precipitation. However, the KG classification is focused on factors related to vegetation and does not take into account important climate factors related to PV technology. For instance, photovoltaic modules operate at a temperature above ambient temperature owing to incident solar radiation. Therefore, a location may be marked as 'Cold' in KG classification but PV module operates at a significantly higher temperature due to high irradiance.

A previous study [4] identifies this lacunae correctly and proposes a new classification considering solar irradiance as a new parameter but the parameter which directly affects degradation is the module operating temperature, which is a function of irradiance and wind speed. Therefore it is more accurate to consider module temperature as a parameter over irradiance for studies related to degradation. Additionally, daily variation in module operating temperature (thermal cycling) may result in degraded/broken interconnects leading to power loss. Moreover, it is known that relative humidity can accelerate certain failure mechanisms such as corrosion. Therefore these two parameters were also included in addition to module temperature as the relevant criteria for generating different climate zones for PV degradation studies.

A new classification is proposed in this paper incorporating module temperature, cyclic strain accumulation due to thermal cycling and relative humidity as a criterion of classification. The methodology and criteria are explained in the following sections in detail.

2 METHODOLOGY

2.1 Dataset:

The Typical Meteorological Year (TMY) data are the sets of hourly values of solar radiation and meteorological parameters for a period of 1 year which are assembled from measured data after examining multiple years of data and selecting ''typical'' data for each month. TMY data of 2590 locations around the world are considered for this analysis. The datasets are downloaded from the open source website [5].

2.2 Calculation of Plane of array irradiance from Global horizontal irradiance:

TMY data gathered had GHI (Global Horizontal Irradiance), DNI (Direct Normal Irradiance) and DHI (Diffuse Horizontal Irradiance), components of irradiance and therefore transposition model is used to convert this GHI to plane of array irradiance (POI). For conversion, model suggested by Perez et al. [6] (equation 1) is used, because as suggested in [7], it is one of the best performing models.

$$POI = DHI * [(1 - F1)\left(\frac{1 + \cos Q_T}{2}\right) + F1\left(\frac{a}{b}\right) + F2\ \sin(Q_T)\,] \ldots (1)$$

Where F1 and F2 are complex empirically fitted functions that describe circumsolar and horizon brightness, respectively. a = max (0, cos(AOI)) and b = max (cos(85), cos(θ_z)).

Angle between sun and plane of array (AOI)

$$= \cos^{-1}[\cos\theta_z \cos\theta_T + \sin\theta_z \sin\theta_z \cos(\theta_A - \theta_{A,array})] \ldots (2),$$

Solar zenith angle $(\theta_z) = \frac{\pi}{2} - \sin^{-1}(\sin\delta\ \sin\varphi + \cos\delta\ \cos\omega\ \cos\varphi) \ldots (3)$

where δ is sun's declination angle, α is the angle of solar elevation, ω is the current hour angle, and φ is observer's latitude.

Azimuth angle θ_A

$$= \cos^{-1}[\frac{\sin\delta \cos\varphi - \cos\delta \cos\omega \sin\varphi}{\cos\alpha}] \ldots (3)$$

θ_T is the angle tilt from horizontal which is assumed as latitude angle of that particular location.

2.3 : Calculation of module operating temperature:

The module temperature is calculated using empirically derived model published by et al king [8] (equation 5).

$$T_m = T_a + POI * [e^{a+b*ws}] \dots (5)$$

Where T_m is the module operating temperature, T_a is the ambient temperature, ws is wind speed. Quantity a and b are empirically derived quantities. For this study, values of a = -3.47 and b = -0.0594 are considered for open rank mount and glass/cell /polymer sheet module type as given in [8], considering commonly observed condition in power plant in most places in India.

2.4 : Calculation of equivalent temperature for various location:

The operating temperature of the module varies depending on meteorological parameters. To find out the effect of temperature on degradation, rate of degradation of material can be approximated by Arrhenius equation, considering single mechanism dominates degradation within the temperature range of interest [9]. If material follows Arrhenius relationship, then it is possible to define an equivalent temperature T_{eq}, to represent the degradation that would have taken place if the module had been aged for the same time, but at a constant (equivalent) temperature. T_{eq} can be calculated from the thermal history of the module using equation 7.

$$e^{(\frac{-E_a}{kT_{eq}})} = \frac{1}{t_1 - t_2}\int_{t_2}^{t_1} e^{(\frac{-E_a}{kT_m(t)})} dt \dots (7)$$

Where t is time, $T_m(t)$ is the time-dependent module temperature, and limit of integration are the beginning and ending times t_1 and t_2. E_a is activation energy. For simplicity of comparison $E_a = 1.1\ ev$ is considered, as a review of polymeric materials [10] found it the most frequently observed when the degradation of mechanical and electrical properties was considered. Also, in [9] where different locations are compared using T_{eq}, calculated for different E_a values, it can be observervedthat increase or decrease in E_a proportionally increases or decreases T_{eq} and therefore the conclusions and classifications in the study are relatively robust with respect to small changes in the E_a.

2.5 : Calculation of strain accumulation due to thermal cycling:

The results of the study in [11] suggests that the main factor that determines degradation due to temperature variations is the stain at the solder layer caused by thermal cycling. The basic assumption for this calculation is that the damage in solder joint during temperature cycling is primarily due to steady state creep strain accumulation. Although the primary creep and time independent plasticity may play a role, it is assumed that these effects are minor. In [12] there are various models suggested to calculate the rate of creep strain. For this study model suggested by Zhang et al [13] is used.

$$Creep\ rate\ (\dot{\varepsilon}_{cr}) = A_1[\sinh(\alpha\sigma)]^n \exp(\frac{-H_1}{kT})$$

Where H_1 is activation energy, k is Boltzmann constant, T is temperature, σ is sheer stain, α prescribes the stress level at which the power law dependence breaks down. To calculate accumulation (W_{acc}), the integration of creep rate ($\dot{\varepsilon}_{cr}$) with respect to temperature with limits from the minimum temperature observed on that day to maximum temperature observed on the same day is added up for all 365 days.

$$W_{acc} = \sum_{n=1}^{365} \int_{t2}^{t1} ([A_1 \sinh(\alpha\sigma)]^n e^{(\frac{-H_1}{kT})}) dT$$

For sake of convinence in comparison, an equivalent (W_{eq}) is defined

$$W_{eq} = \frac{W_{acc}}{A_1[\sinh(\alpha\sigma)]^n * 10^7}$$

2.5 Calculation of weighted-average relative humidity (RHwa):

KG classification uses precipitation as a classification criterion to classify weather zones. However, relative humidity is known to be more relevant stressor in influencing certain degradation modes. Humidity ingression to the solar module packaging affects degradation mechanisms such as backsheet hydrolysis, corrosion of metallization, snail trail etc. Therefore to accommodate this effect, as suggested in [14], a weighted-average relative humidity (RHwa) based on second-order kinetics (n = 2) is calculated to find a humidity value from the climate history of a particular location using equation 9 which is obtained after simplification of equation 8.

$$RHwa^n e^{\frac{-E_a}{kT_{eq}}} = \frac{\sum_{T=0}^{N} RH^n e^{\frac{-E_a}{kT}}}{N} \dots (8)$$

$$RHwa = \left(\frac{\sum_{T=0}^{N} RH^n e^{\frac{-E_a}{kT}}}{N e^{\frac{-E_a}{kT_{eq}}}}\right)^{\frac{1}{n}} \dots (9)$$

The T_{eq} value obtained from equation 7 is used for calculation in equation 9.

3 CLASSIFICATION

3.1 Data Visualization:

After calculation of all three parameters, all 2590 data points are plotted with T_{eq} on x axis, W_{eq} on y axis and data points are shaded with color specified on color bar according to their RHwa value, as shown in fig 1.

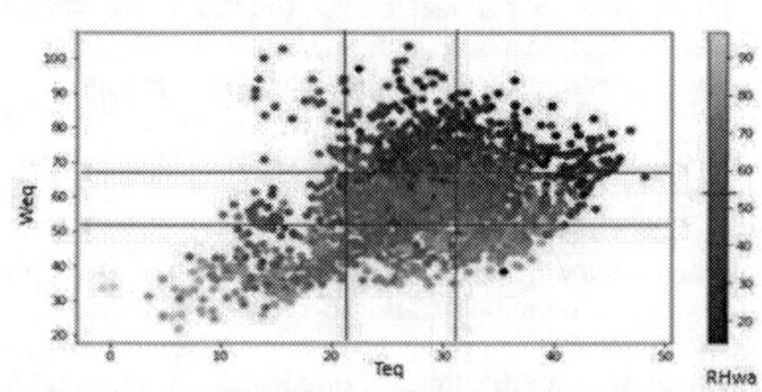

Figure 1: Scatter plot of data points representing all considered location.

It can be observed from the plot that low RHwa data points are crowded in the top right quarter of the plot which implies an inverse relationship of $T_{eq}$and W_{eq} with RHwa which also conforms with the general obeservation as relative humidity is found to reduce at elevated temperature.

3.2 Clustering of Data points into zones:

The density distribution curves of all three parameters are plotted to check for the localized congregation of data points so as to divide them at the point where congregation density is low. But as seen in fig. 2, 3 and 4, except for RHwa, there is no multiple localized congregation in data to divide it into more than one groups.

Therefore to group data points in multiple zones, k-mean [15], hierarchal [16] and fuzzy c means [17] clustering algorithms are used to find clusters in the data. Hierarchal algorithm is unsupervised and therefore clusters data into optimal number of clusters but K-mean and fuzzy c mean are supervised and need number of cluster as an input. Therefore to find the optimal number of cluster as an input for these algorithms, silhouettes method of validation [18] is used. Fig 2, 3 and 4 shows the boundaries found via these clusters. Boundaries defined for W_{eq} and RHwa by all three algorithms are close to each other.

For further analysis boundaries defined by k-mean algorithm is used. T_{eq} is divided into 3 zones, W_{eq} is divided into 3 zones and RHwa is divided into 2 zones. The ranges of each zones found via k-mean are shown in Table 1.

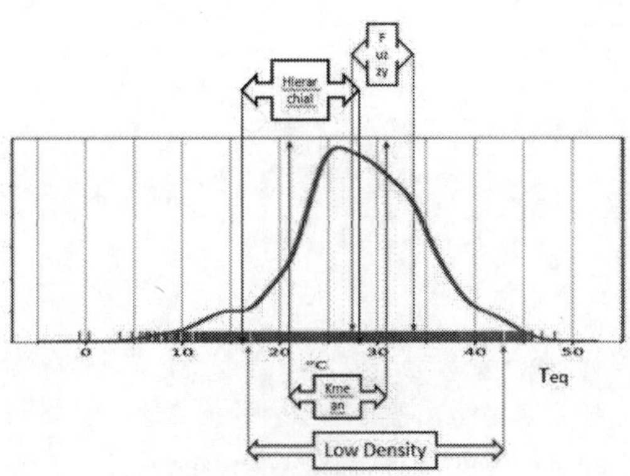

Figure 2: Density distribution plot for T_{eq}.

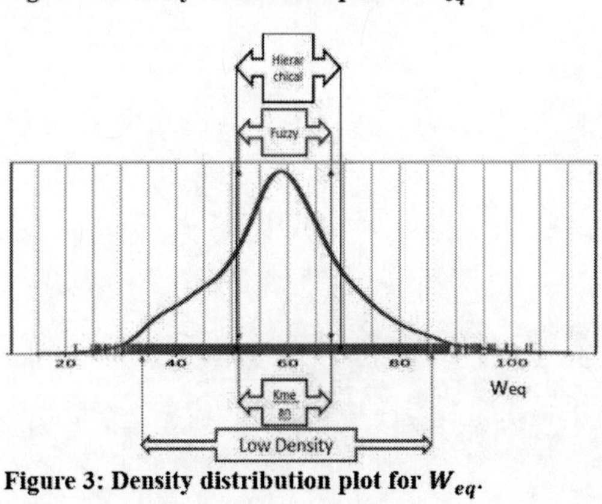

Figure 3: Density distribution plot for W_{eq}.

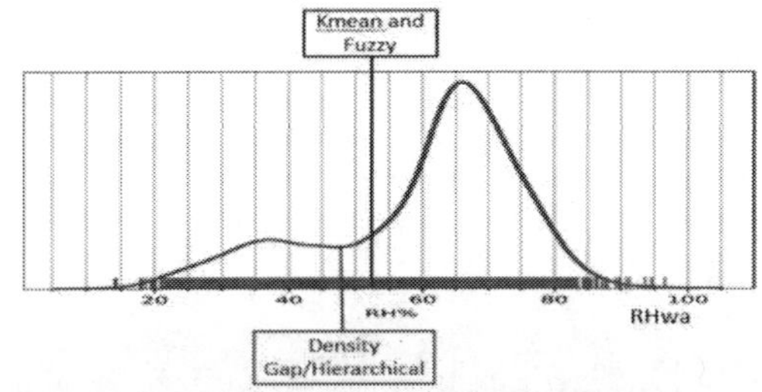

Figure 4: Density distribution plot for RHwa.

Table 1: Range for the parameter for different zones found using k-mean algorithm.

	1st Zone	2nd Zone	3rd Zone
T_{eq}	0 to < 21° C	≥ 21° C to < 31° C	≥ 31° C
W_{eq}	0 to < 51units	≥ 51units to < 68 units	≥ 68 units
RHwa	0 to < 53%	≥ 53%	

3.3 Nomenclature:

In the proposed classification scheme, there are total 18 possible zones and therefore to address each zone a three letter nomenclature scheme is used in which first letter indicates the temperature zone, second letter indicates the cyclic stain accumulation and third letter indicates relative humidity. Table 2 shows the designated letter for each zone.

Table 2: Classification scheme symbol description.

	Name	Range	Designated Letter
T_{eq}	Cool	0 to < 21° C	L
	Hot	≥ 21° C to < 31° C	M
	Very Hot	≥ 31° C	H
W_{eq}	Low Cyclic strain	0 to < 51	l
	Medium Cyclic strain	≥ 51 to < 68	m
	High Cyclic strain	≥ 68	h
RHwa	Dry	0 to < 53%	D
	Humid	≥ 53%	W

For instance, the 'HhD' zone means a very hot (H) temperature zone, high cyclic strain accumulation (h) and dry climate (D).

4 RESULTS AND DISCUSSION:

Locations are classified and are plotted on the map in fig 5. Table 3 shows the number of locations in each zone of classification. From the table, it can be observed that '_lD' locations (with low cyclic strain and dry climate) and 'L_D' locations (with cold and dry climate) are very less in number. Also, LhW and HhW locations are less in number with respect to the size of the dataset.

When compared with KG classification, 521 locations

from D zone (snow, Tmin < -3C) & E zone (polar, Tmax < 10C) are switching to hot zone (M) of new classification scheme. 8 Locations are shifting from D zone and E zone of KG to very hot zone of the new classification scheme. These locations are generally found to have low ambient temperature but high solar irradiance and therefore higher module temperature. Fig 6 shows these shifted loations on map. 503 locations are classified as High cyclic strain (l) in the new classification.

5 CONCLUSION:

The methodology proposed in this paper to classify the climate zones considers the parameters that directly influence PV degradation. This would enable the future studies to derive a more accurate correlation between climate conditions and degradation. The proposed methodology considers module temperature rather than ambient temperature and weighted average relative humidity rather than precipitation and also includes the additional factor of cyclic strain accumulation to accommodate the effect of thermal cycling and therefore is better suited than KG classification for the studies in the area of PV module degradation.

6 ACKNOWLEDGMENT:

This work was supported by the National Centre for Photovoltaic Research and Education (NCPRE) funded by the Ministry of New and Renewable Energy of the Government of India through the Project No. 31/09/2015-16/PVSE-R&D dated 15th June 2016.

Table 3: Number of locations in a the particular zone

	High Cyclic Strain and Humid	Medium Cyclic Strain - Humid	Low Cyclic Strain - Humid	High Cyclic Strain - Dry	Medium Cyclic Strain - Dry	Low Cyclic Strain - Dry
Cool	20 (LhW)	71 (LmW)	246 (LlW)	7 (LhD)	5 (LmD)	2 (LlD)
Hot	76 (MhW)	779 (MmW)	221 (MlW)	182 (MhD)	132 (MmD)	7 (MlD)
Very Hot	30 (HhW)	373 (HmW)	140 (HlW)	188 (HhD)	109 (HmD)	1 (HlD)

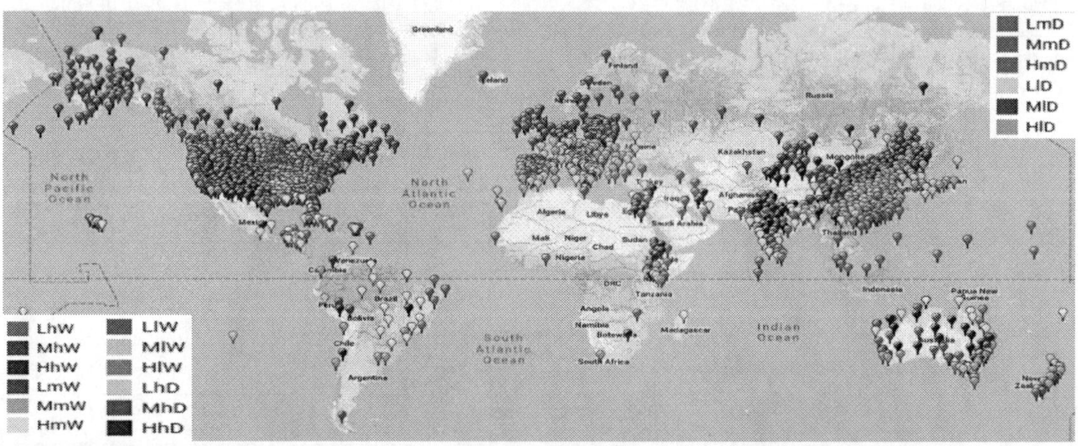

Figure 6: Map of locations which switched from cold KG classification to Hot classification of new scheme.

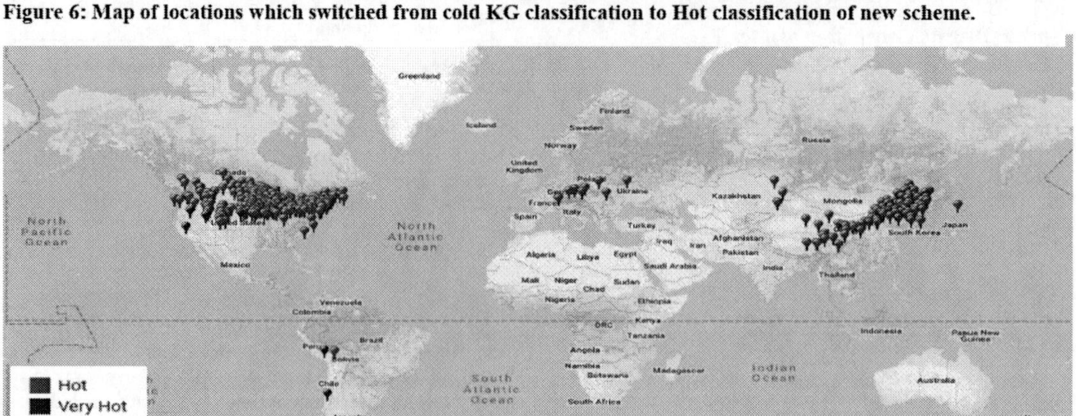

Figure 5: Classification Map of the locations considered

6 REFERENCES:

1. Dirk C. Jordan, John H. Wohlgemuth, Sarah R. Kurtz, "Technology and Climate Trends in PV Module Degradation", at the 27th European Photovoltaic Solar Energy Conference and Exhibition Frankfurt, Germany.
2. Jesús Polo, Miguel Alonso-Abella, Jose A. Ruiz-Arias, José L. Balenzategui, "Worldwide analysis of spectral factors for seven photovoltaic technologies", Elsevier, Solar Energy 142 (2017) 194–203.
3. W. Koppen, "The thermal zones of the Earth according to the duration of hot, moderate and cold periods and to the impact of heat on the organic world", *Meteorologische Zeitschrif*, vol. 30, no. 3, pp. 351-360, 2011.
4. Julián Ascencio-Vásquez, Kristijan Brecl, Marko Topič, "Koppen-Geiger-Photovoltaic Climate Classification", at the 7th World Conference on Photovoltaic Energy Conversion, Hawaii, USA.
5. https://energyplus.net/weather/sources.
6. R. Perez, P. Ineichen, R. Seals, J. Michalsky, and R. Stewart, "Modeling daylight availability and irradiance components from direct and global irradiance," Solar Energy, vol. 44, no. 5, pp. 271–289, 1990
7. Matthew Lave, William Hayes, Andrew Pohl, and Clifford W. Hansen, "Evaluation of Global Horizontal Irradiance to Plane-of-Array Irradiance Models at Locations Across the United States"
8. D.L. King, W.E. Boyson, J.A. Kratochvill, "Photovoltaic Array Performance Model", SANDIA REPORT, Sandia National laboratories.
9. Sarah Kurtz, Kent Whitfield, G. TamizhMani, Michael Koehl, David Miller, James Joyce, John Wohlgemuth, Nick Bosco, Michael Kempe and Timothy Zgonena, "Evaluation of high-temperature exposure of photovoltaic modules", Prog. Photovolt: Res. Appl. 2011; 19:954–965, Wiley Online Library.
10. Dixon RR, "Thermal aging predictions from an Arrhenius plot with only one data point", IEEE Transactions on Electrical Insulation 1980; EI-15: 331.
11. Nochang Park, Jaeseong Jeong, Changwoon Han, "Estimation of the degradation rate of multi-crystalline silicon photovoltaic module under thermal cycling stress", Microelectronics Reliability, Elsevier, 2014
12. Syed A., "Accumulated creep strain and energy density based thermal fatigue life prediction models for SnAgCu solder joints",In: Electronic Components and Technology Conference, 54th, IEEE, 2004. p. 737–46.
13. Qian Zhang, Abhijit Dasgupta, Peter Haswell, "Viscoplastic Constitutive Properties and Energy-Partitioning Model of Lead-Free Sn3.9Ag0.6Cu Solder Alloy", 53rd ECTC 2003, pp. 1862-1868.
14. Michael D. Kempe, John H. Wohlgemuth, "Evaluation of Temperature and Humidity on PV Module Component Degradation", at 39th IEEE Photovoltaic Specialist Conference, Tampa, FL, USA
15. https://en.wikipedia.org/wiki/K-means_clustering
16. https://en.wikipedia.org/wiki/Hierarchical_clustering
17. James C. Bezdek, Robert Ehrlich, William Full, "FCM: The Fuzzy c-Means Clustering Alogorithn", Computers & Geosciences Vol. 10, No. 2-3, pp. 191-203, 1984.
18. Peter J. ROUSSEEUW, "Silhouettes: a graphical aid to the interpretation and validation of cluster analysis", Journal of Computational and Applied Mathematics 20 (1987) 53-65.

IMPACT OF ENVRONMENTAL FACTORS ON THE EFFICIENCY OF A PV SYSTEM: CASE STUDY IN OUARZAZATE, MOROCCO

K. Ettalbi [a, c, *], Z. Naimi [a, c], H. Bouzekri [a], I. Munoz Morales [b], M. Ezquer Mayo [b], A.R. Lagunas [b], M. Maaroufi [c]
[a] Moroccan Agency for Sustainable Energy (Masen), Department of R&D and Industrial Integration, Rabat. Morocco
[b] Centro Nacional de Energias Renovables (CENER), Department of Solar Photovoltaic Energy, Navarra. Spain
[c] Ecole Mohammadia d'Ingénieurs (EMI), Department of Electrical Engineering, Rabat. Morocco
* Corresponding author: k.ettalbi@masen.ma

ABSTRACT: A 10kWp polycrystalline photovoltaic (PV) system was installed in Masen's R&D platform in Ouarzazate, Morocco, in order to validate the performance and reliability of crystalline PV modules under real outdoor conditions, due to the fact that the performance under standard test conditions (STC) can be significantly different from the one measured under outdoor conditions. This difference is mainly caused by various environmental factors (temperature, irradiation, dust, wind) that affect the overall performance of the PV system. The present study focuses on evaluating the effect of environmental factors, particularly solar irradiance, module's temperature and wind speed on the efficiency of the PV system. Recorded data for the past two years, during different seasons, were used in this outdoor performance analysis.

Keywords: Solar energy; Photovoltaics; PV efficiency; Outdoor analysis

1 INTRODUCTION

Manufacturers usually specify PV modules at standard test conditions, which may not be attainable anywhere in the world, due to the fact that performances of PV modules are strongly affected by the environmental conditions which affects their lifetime [1].

In order to predict long term performance of large scale PV plants in arid and desert regions, and to evaluate their average drop of efficiency mostly caused by environmental factors (solar irradiance, module temperature, rain, wind and relative humidity), a 10kW solar PV demonstrator based on a conventional PV technology was installed in Masen research and development platform located in Noor Solar Complex in Ouarzazate, in south of Morocco.

2 PV SYSTEM DESCRIPTION

The PV system was installed in Masen R&D platform in Noor Solar Complex, located in the region of Ouarzazate city, in the south of Morocco, with the GPS coordinates 31°2' North and 6°50' West, at 1260m above sea level. The capacity installed is 10kWp, composed of 40 polycrystalline PV modules with a total effective area of 65,5 m², connected in two strings (20 panels connected in series in each string). The PV modules have a capacity of 250Wp with 60 solar cells in each PV panel, and a module efficiency of 15,3% at Standard Test Conditions (STC). The PV arrays is ground mounted, with a 20° inclination towards the south.

2.1 Data monitoring system

As climatic data is the basis for the performance analysis of PV modules and systems, a meteorological station has been installed on site with the main ojectives of studying the yield performance in correlation with the climatic specific site conditions. The weather station is also used for forcasting the solar resource in Ouarzazate and in similar locations worldwide. The measured parameters by different sensors include:

- Plane-of-array (POA) irradiance, with a pyranometer mounted on the same inclination as the PV panels,
- Global horizontal irradiance,
- Diffuse horizontal irradiance measured with a pyranometer with a shadow band,
- Wind speed and direction mounted on a height of 10m,
- Ambiant air temperature,
- Barometric pressure,
- Relative humidity,
- Rain amount.

These parameters are connected to a high accuracy data monitoring system that has been recording these measurements since september 2015 at 10 seconds

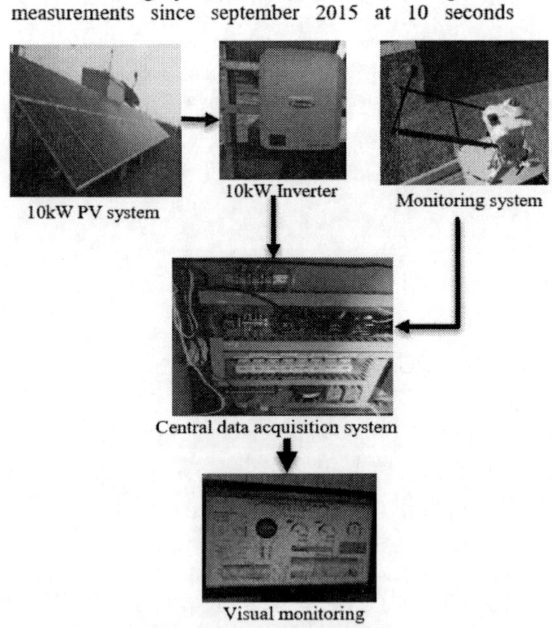

Figure 1: Architecture of the 10kW PV system

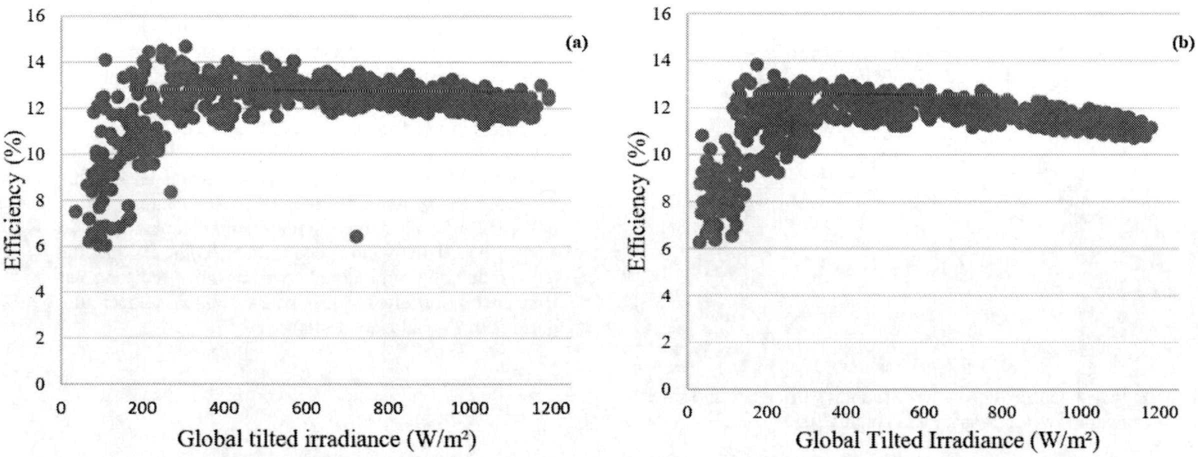

Figure 2: Seasonal correlation between efficiency and solar irradiance during winter2016(a) and summer 2017(b)

sampling rate . The monitoring system is based on a central data acquisition sytem (DAS) for data storage, system control and modeling. The DAS is composed of a data logger and a software interface compatible with the data logger installed. Other than the climatic measurements, the data set monitored also records electrical data which include: DC current, DC voltage, DC power, AC power and AC energy. Module temperature is also recorded and measured by a thermocouple. All the sensors are cleaned on a weekly basis to make sure that the data used in the analysis is accurate. Figure1 shows the architecture of the PV system installed on the site, where the PV modules are cleaned once per month based on a pressurized water cleaning method, due to high soiling effect in the desert conditions.

3 METHODOLOGY

According to the IEC standard 61724, there are several parameters applied to evaluate the overall performance of a solar PV system. Due to the availability of measured AC power data and global tilted irradiance data, this study was based on a system efficiency evaluation, by including various environmental parameters.

The system efficiency represents the overall performance of the PV system; it is calculated as the ratio of the output energy (E_{AC}) on the global-on plane irradiation (H_{POA}) at the module area (A_{MOD}).

$$\eta_{syst} = E_{AC} / (H_{POA} \cdot A_{MOD} \cdot N_{series} \cdot N_{paral}) \ [2]$$

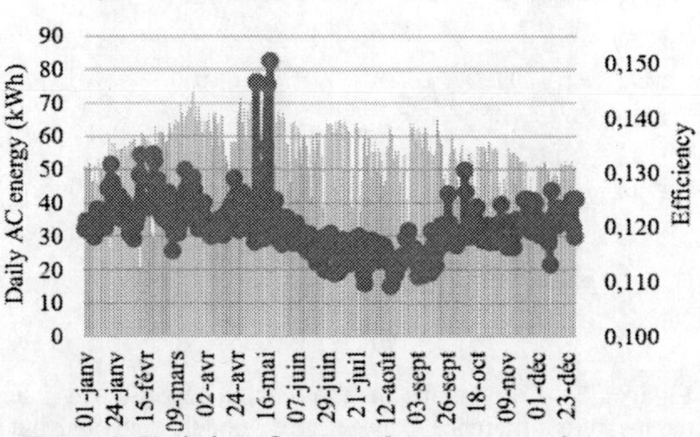

Figure 3: Variation of generated AC energy and efficiency during the year 2017

The PV system performance is evaluated by analyzing the effect of solar irradiance, ambient and module temperature and wind speed on the system efficiency. First, the effect of solar irradiance is studied by evaluating the seasonal correlation between global-in-plane solar irradiance and system efficiency, during summer and winter season. Next, the influence of PV module temperature on efficiency is analyzed. And last, wind speed contribution on the cooling of the PV panels is evaluated.

4 RESULTS AND ANALYSIS

This paper shows results obtained from data recorded during the year 2017. Figure 3 shows daily variation of generated AC energy and efficiency, the daily efficiency varies between 10 and 15%, with a lower efficiency during summer compared to winter, regardless of the higher solar irradiance on summer.

4.1 Global on-plane irradiance effect

Solar irradiance is one of the major parameters influencing the output energy of a PV system and thus affecting its efficiency. The seasonal correlation between efficiency and global-on-plane solar irradiance during winter and summer is shown in figure 2. Results demonstrate an increase in efficiency by the increase of irradiance, followed by a decrease at very high irradiance values with a more significant decrease during summer compared to winter. An average efficiency of 12,10% on winter vs 11,48% on summer has been calculated. A decrease mainly caused by the increase in ambient temperature and thus the increase of modules temperature under operation in desert conditions.

Figure 4 shows the results of the month of January 2017 known for its cold ambient temperature vs the month of August 2017 when the ambient temperature reaches 40°C. The system efficiency remains steady at irradiance level higher than 800W/m² during January, on the other hand, a decrease in efficiency is observed for solar irradiance exceeding 800W/m² during August.

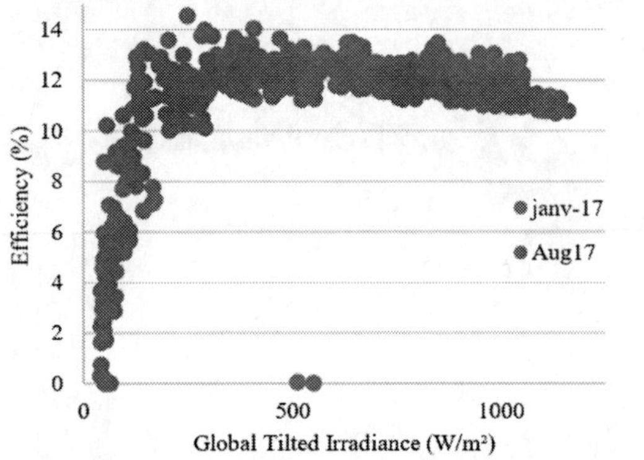

Figure 4: Correlation between monthly efficiency and solar irradiance during January and August 2017

4.2 Module temperature effect

To demonstrate the impact of PV module temperature on the decrease of efficiency, even at high irradiance levels, the seasonal correlation between efficiency and solar irradiance at varying PV module temperatures is reported in figure 8. A high PV module temperature over 60°C is reached during summer, and consequently, significantly decreasing PV system efficiency.

Figure 5 shows the correlation between efficiency and PV module temperature at varying values of solar irradiance during the whole year of 2017, the efficiency starts decreasing from 600W/m² of global-in-plane irradiance corresponding to PV module temperatures exceeding 45°C.

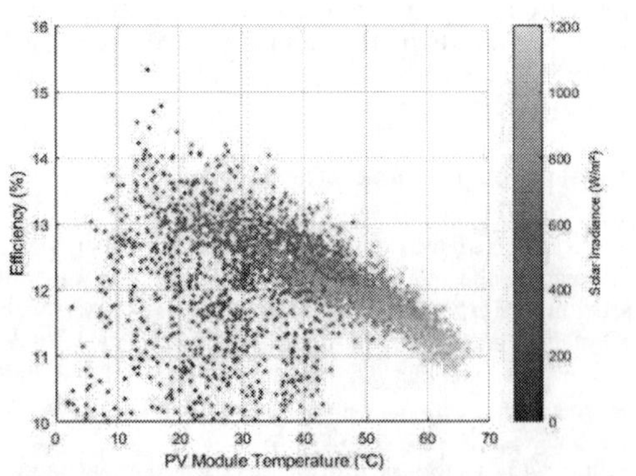

Figure 5: Correlation between efficiency and PV module temperature at varying levels of solar irradiance during the year 2017

To quantify the percentage of decrease in system efficiency due to high PV module temperature, the annual correlation between efficiency and module temperature has been calculated at fixed interval levels of solar irradiance. At an irradiance comprised between 600W/m² and 800W/m², there is a 0,034%/°C decrease, according to the outputs from the trend line's equation. At an irradiance between 800W/m² and 1000W/m², the increase in PV module temperature reduces by 0,045%/°C the system efficiency. And at an irradiance higher than 1000W/m², there is a 0,06%/°C decrease in efficiency, which means that if the PV module temperature increases, for example, by 10°C, there is a drop of efficiency of 0,6% which is relatively significant for a maximum system efficiency of 15%, as it represents 4% from the maximum efficiency.

4.3 Effect of wind speed

Wind speed is one of the environmental factors that can affect the system efficiency through a natural cooling of the PV modules operating under high temperatures. To study whether the wind speed influences the PV module temperature, the correlation between wind speed and solar irradiance at different values of PV module temperature during 2017 is indicated in figure 6.

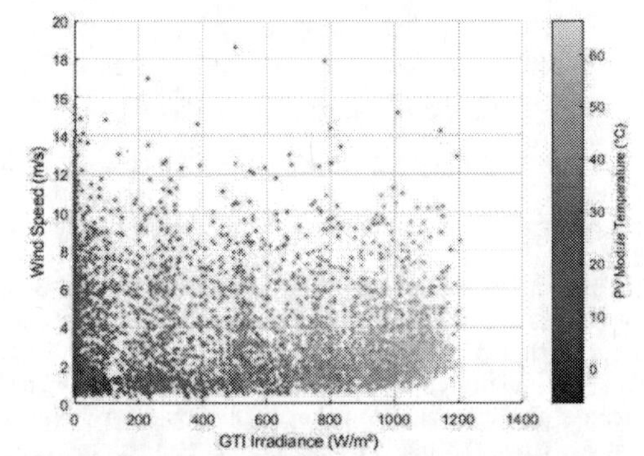

Figure 6: Annual correlation between wind speed and solar irradiance at varying PV module temperatures

Results show that for wind speed values higher than 8m/s, PV modules temperature is significantly lower at different solar irradiances. To quantify the decrease in temperature, figure 7 presents the correlation between the difference in temperature between PV module and ambient temperature vs wind speed. The results clearly show that T_{mod}- T_{amb} decreases with the increase of wind speed.

Through comparing the wind speed influence in Ouarzazate site (annual average wind speed of 2,9m/s) with the ones found in the case study of Ligourio, Greece (annual average wind speed of 1,6m/s) [3], results of both cases demonstrated a decrease of (T_{mod} – T_{amb}) with increasing wind speed values. Based on the trend line equation for (T_{mod} – T_{amb}) vs wind speed in each site, the decrease in temperature is 0,48 times higher in the case of

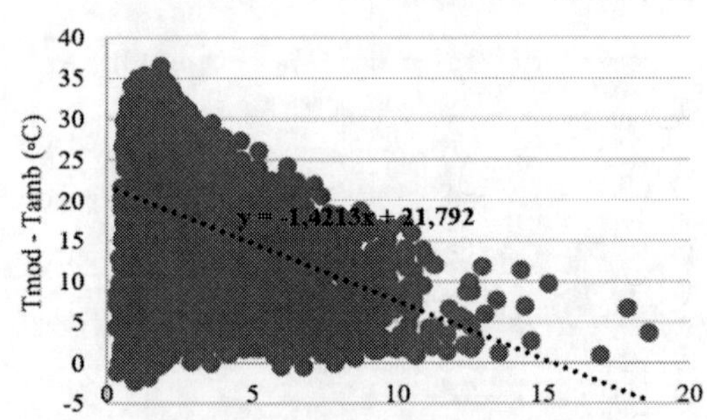

Figure 7: Correlation between wind speed and the temperature difference between PV module and ambiant temperature

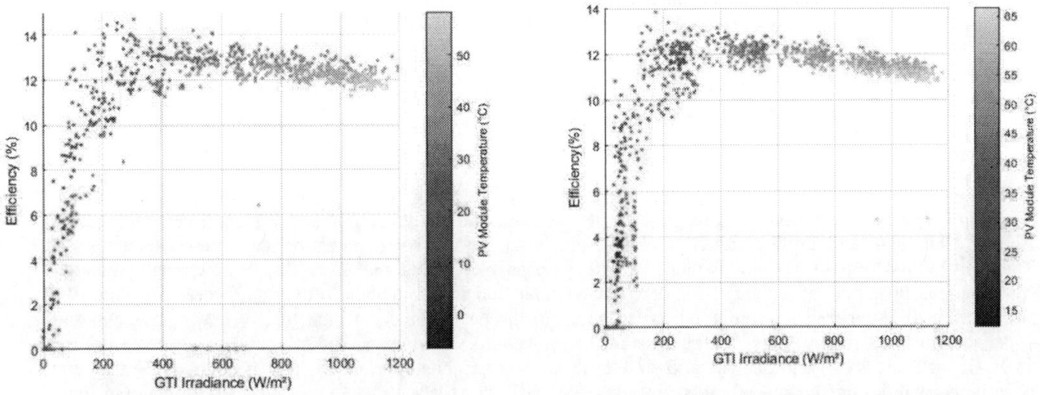

Figure 8: Seasonal correlation between efficiency and solar irradiance during winter 2016 (a) and summer 2017 (b)

Ouarzazate where the wind speed values can reach up to 17m/s in comparison with Ligourio where the wind speed can reach up to 4,2m/s.

5 CONCLUSION

The performance of the 10kW PV system was studied through analyzing the daily and seasonal variation of efficiency in the field under varying meteorological conditions. An annual energy produced of 19,93MWh was calculated.

The daily efficiency throughout the year varies between 10 and 15% with a lower efficiency during summer compared to winter regardless of the higher solar irradiance on summer. An average efficiency of 12,10% was calculated during winter vs 11,48% during summer.

The efficiency decrease caused by high PV module temperatures has been calculated, results showed an efficiency loss in the order of 0,034%/°C at an irradiance level comprised between 600W/m² and 800W/m², going up to 0,06%/°C at an irradiance higher than 1000W/m². Based on these results, a future research will be carried out to compare the decrease in efficiency of polycrystalline modules with thin film modules known for their lower temperature coefficient.

Wind speed is also an important factor contributing to the cooling of PV modules temperature at high solar irradiance levels, and consequently increasing the PV system efficiency. Results demonstrated that an increase in wind speed leads to a 1,42°C decrease in the difference between PV module and ambient temperatures. Further researches will be carried out to study the impact of wind direction and wind speed on the cooling and particles deposition on the PV panels.

REFERENCES

[1] E.D. Dunlop, D. Halton. The Performance of crystalline silicon photovoltaic solar modules after 22 years of continuous outdoor exposure, Progress in Photovoltaics: Research and Applications, Vol. 14, No. 1, Jan. 2006

[2] Vikrant Sharma, S.S. Chandel. Performance and degradation analysis for long term reliability of solar photovoltaic systems: A review. Renewable and Sustainable Energy Reviews. 2013

[3] John K. Kaldellis, Marina Kapsali, Kosmas A. Kavadias. Temperature and wind speed impact on the efficiency of PVinstallations. Experience obtained from outdoor measurements inGreece. Renewable Energy 66. 2014

PERFORMANCE ANALYSIS AND EVALUATION OF DIFFERENT GRID-CONNECTED PHOTOVOLTAIC SYSTEM TECHNOLOGIES IN BOLIVIA, CHILE AND GERMANY

Lionel Clasing, Marvin Raabe, Ulf Blieske, Rudolf Gecke
Cologne University for Applied Sciences, Cologne Institute for Renewable Energy
Betzdorferstr. 2, 50679 Köln

ABSTRACT: The main objective of this study is to compare and analyze different PV system performances in Bolivia, Chile and Germany with different climate preconditions like solar irradiation and ambient temperature. The analyzed PV systems in Chile allow a comparison of poly-crystallin silicon- (p-Si), mono-crystallin silicon- (m-Si) and copper indium gallium selenide- (CIGS) solar cell technologies. Furthermore, there are four PV-strings with different inverter configurations to compare the performance of high to low power inverters and undersized to oversized inverters. The analysis of the evaluated PV systems includes monthly average daily final yield, capacity factor, performance ratio and ratio of self-sufficiency. The highest performance ratio was determined with a combination of oversized inverter and CIGS-cells with 89 % for the overall operation time. Average overall PV system performance ratios in Bolivia, Chile and Germany for the analyzed time-periods in 2015 and 2017 in Bolivia, Chile and Germany were determined with 83 %, 87 % and 69 %, respectively. Overall, the optimal operation conditions result a in energy balance solar coverage ratio of 1.16 in Bolivia and 0.53 in Chile. The climate preconditions of the different countries are one of the key factors for the performance of a PV system, as shown in the results of this paper.
Keywords: Performance, Monitoring, Energy Rating, PV System, Solar Radiation

1 INTRODUCTION

One key factor for the performance of PV systems are the climate conditions of the location. In the following, the performances and yields of three different PV systems in Chile, Bolivia and Germany were compared. In addition, the impact on the yield and performance of various cell technologies, tilt angles and inverter designs are examined.

2 APPROACH

The performance analysis and evaluation are based on three rooftop grid-connected photovoltaic (PV) systems with an installed power of 7.4 kWp in Bella Vista, Bolivia, 40.8 kWp in Santiago, Chile and 20.9 kWp in Cologne, Germany. The energy supply of the systems in Santiago and Bella Vista are part of an overall professional education and training concept with the main objective to increase independence from the national grid and reduce the energy costs of these facilities. Therefore, a self-sufficiency analysis of the facilities in Santiago and Bella Vista is included.

2.1 Climate Conditions

Many studies show the feasibility and performance analysis of PV in European countries [1, 2, 3, 4, 5, 6]. The number of scientific analysis for countries like Chile and especially Bolivia is much lower [7, 8, 9, 10, 11, 12]. In addition, there are few studies comparing the performance of PV systems with different conditions. The climate conditions of the three analyzed locations or rather countries differ greatly, as shown in Table I. The annual irradiation in Bella Vista is much higher than in Santiago and reaches more than double of the one in Cologne. The climate classification for Bella Vista is subtropical highland climate (Cwb), warm-summer mediterranean climate (Csb) for Santiago and oceanic climate (Cfb) for Cologne.

Table I: Climate conditions of Bella Vista, Cologne, Santiago [13, 14]

	Bolivia, Bella Vista	Chile, Santiago	Germany, Cologne
Latitude	- 17.3 °	- 33.4 °	50.9 °
Longitude	- 66.2 °	- 70.6 °	6.9 °
Irradiation [$kWh/m^2/a$]	2,250	1,773	1,062
Elevation [m]	2,800	534	37
Climate classification	Cwb	Csb	Cfb
Average temperature max./min.	26 / 4 °C	30 / 15 °C	24 / 0 °C

Figure 1 shows the distribution of irradiation over a typical year for Bella Vista, Cologne and Santiago. Although, Santiago offers highest values for irradiation, the constant course of irradiation in Bella Vista leads to a higher annual total income. Therefore, Bella Vista offers good conditions for operating PV systems, but the number of installed systems is comparatively low.

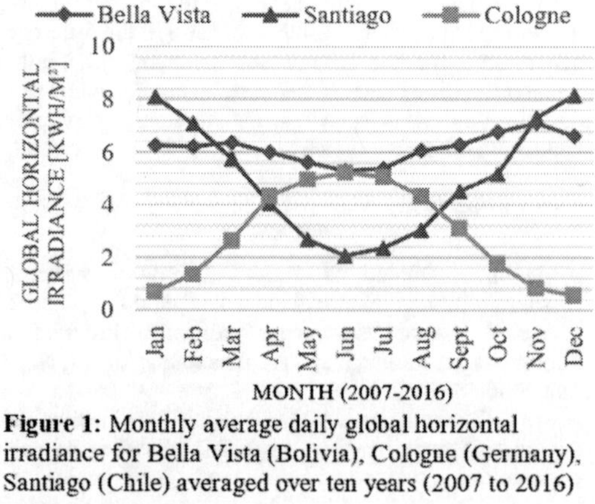

Figure 1: Monthly average daily global horizontal irradiance for Bella Vista (Bolivia), Cologne (Germany), Santiago (Chile) averaged over ten years (2007 to 2016)

[15, 16]

2.2 Technical specification of the PV systems

The three PV systems considered differ in the number and type of modules, in the power capacity and design of modules and inverters and in the tilt angle and azimuth orientation. Table II shows the technical information for the systems in Santiago and Bella Vista. The system in Santiago allows to compare two different cell technologies like p-Si and CIGS with different nominal power capacity and performance. Furthermore, there are four inverter configurations (Figure 3), for a comparison of undersized to oversized inverters (2 x Tripower 5000) based on the nominal module power and inverters with big (Tripower 200000) to small power capacity (Tripower 10000). The PV system in Bella Vista consists of three inverter strings, each with 8 of the mentioned modules, shown in Figure 2.

Table II: Technical conditions of the PV System in **Santiago (Chile)** and in **Bella Vista (Bolivia)** [13], [17]

	Bella Vista	Santiago	
Tech-nology	m-Si	p-Si	CIGS
Model	Sun Power SPR X-19 310 BLK	Astronergy Violin ASM6610P	Avancis PowerMax SMART 120
Power	310 Wp	270 Wp	120 Wp
Number	24	112	84
Installed Power	7.44 kWp	30.24 kWp	10.08 kWp
Incli-nation angle	17 °	23 °	23 °
Azimuth	-90 °	180 °	180 °
Inverter	3 x SunnyBoy 3000 HF	Tripower 10000 TL Tripower 20000 TL	2 x Tripower 5000 TL

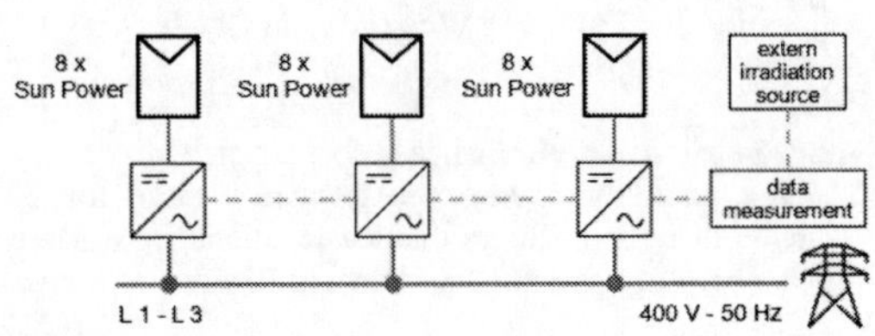

Figure 2: Schematic PV System structure in **Bella Vista (Bolivia)**

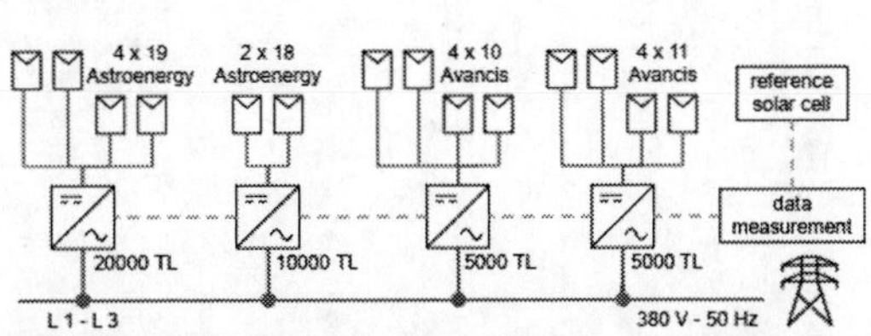

Figure 3: Schematic PV System structure in **Santiago (Chile)**

Table III shows the technical details of the PV system in Cologne with the schematic structure in Figure 4. The cell technologies used here are p-Si and m-Si. Due to the different tilt angles of the modules, depending on the PV string, a comparison of the influence of the inclination on the yield is possible.

Table III: Technical conditions of the PV System in **Cologne (Germany)** [18]

	Cologne		
Tech-nology	p-Si	m-Si	m-Si
Model	SOLON P220/6+	Solar Fun 160-24-M	Solar Fun 160-24- M
Power	205 Wp	175 Wp	180 Wp
Number	12	14	14
Installed Power	2.46 kWp	2.45 kWp	2.52 kWp
Tilt angle	30 °	15 °	15 °
Azimuth	-10 °	-10 °	-10°
Inverter	SunnyBoy 3000 TLHC	SunnyBoy 5000 TLHC	SunnyBoy 5000 TLHC

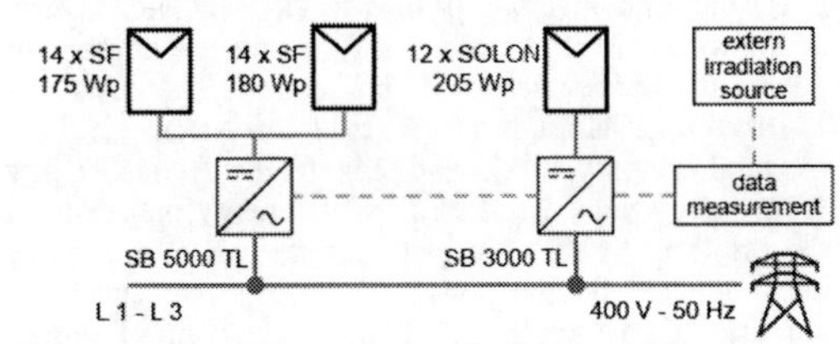

Figure 4: PV System structure in **Cologne (Germany)**

3 METHODOLOGY

3.1 Climate Data Monitoring

The data sets to determine reference final yields for Santiago, Bella Vista and Cologne are based on different sources. In Santiago an in-plane irradiation measurement with a reference solar cell is used.

For Bolivia in-plane irradiation is simulated in *Photovoltaic Geographical Information System* (PVGIS) and for Cologne in *PV_LIB Toolbox for Python* (PVLIB). **PVGIS** is a web application to simulate PV systems with different tilt angles and azimuth orientation. It is a free and user-friendly software which provides also the in-plane irradiation data used for the simulation [16]. **PVLIB** is an open-source Library for *PYTHON* and *MATLAB* to simulate PV system performances with external weather data [19]. The used climate data are of satellite- and weather station-based database as shown in Table IV. 10-minute-values for Bella Vista in 2015, the first year of operation and for Cologne in 2017 were available at different sources [20, 21].

Table IV: In-plane irradiation data sources of the presented performance analysis of different PV systems [19, 20, 21].

Location	Source	Format	Time
Santiago	measurement	hourly	2016-2018
Bella Vista	NSRDB/PVGIS	10 min. interval	2015
Cologne	DWD/PVLIB	10 min. interval	2017

3.2 Performance Analysis

The data of the respective systems to evaluate the described performance indicators are obtained by the *Sunny Portal* from *SMA*. Most important indices, to describe the performance of a grid connected PV system, are final energy output, final energy yield and performance ratio. These indices offer the possibility to evaluate the overall performance of a system and to compare the performance of systems with each other. [7, 8]

The final energy output (1) describes the generated energy (alternating current – AC) of a system over a period of time.

$$E_{AC} = \sum_{t=1}^{N} E_{AC,t} \quad (1)$$

Where t depends on the considered period of time, like minutes, hours, days or month. The generated energy at timestep t is represented as $E_{AC,t}$ and the number of determined data values as N. [7, 8]

Two important types to describe the system yield are the final energy yield (2) and the reference yield (3). The final energy yield is defined as the energy output from the inverter (AC) for a period of time (years, months or days) divided by the power of the PV array at STC-conditions. This factor enables the comparison of similar PV systems. [7, 8]

$$Y_F = \frac{E_{AC}}{P_{PV,rated}} \quad (2)$$

The reference yield (3) is the total in-plane solar irradiation divided by the array reference irradiation (generally with $G_R = 1$ kW/m^2). [7, 8]

$$Y_R = \frac{H_t}{G_R} \quad (3)$$

The performance ratio (*PR*) (4) allows the comparison of PV systems regardless of location, azimuth orientation, tilt angle and their power capacity. The PR of a PV system describes the overall effect of losses on a PV array's power output and is defined by the ratio of a factor for the real operation (final yield) to the ideal performance operation (reference yield). [7, 8]

$$PR = \frac{Y_F}{Y_R} \quad (4)$$

Another index used in this study is the capacity factor (*CF*) (5). The CF is the generated energy from a PV system over a time-period divided by energy output if the system would operate at the full power capacity for the whole given period of time. [7]

$$CF = \frac{100 \cdot E_{AC}}{P_{PV,rated} \cdot t[h]} \quad (5)$$

3.3 Energy balance solar coverage ratio analysis

A main objective for installing the PV systems in the technical school of Bella Vista and vocational school of Santiago is to enhance the independence from the national grid and lower the energy costs of these facilities. Chile has defined laws which include a net-metering-concept for grid-connected PV systems ≤ 100 kWp. Bolivia has no clear legislation for grid-connected PV systems, therefore, a net-metering is only possible in agreement with the local utility [17]. Since the PV systems analyzed in this study are grid-connected and feeding their total final energy output in the local grid, the energy balance solar coverage ratio analysis is based on net-metering. Electrical energy load data for the facilities were only available as monthly based consumption data for 2015 in Bella Vista and 2018 in Santiago. Energy consumption and yield data were analyzed in order to determine the energy balance solar coverage ratio of the facilities. Energy balance solar coverage ratio(a) (6) is defined in this calculation as the PV systems energy fed into the grid divided by the total electricity consumption E_C over a given period [22, 23].

$$a = \frac{E_{AC}}{E_C} \quad (6)$$

4 RESULTS

In this section the results of the performance analysis are presented. Initially, the measured and simulated in-plane irradiation data were used in this study. Furthermore, the performance ratio for entire PV systems in each location and for each inverter and solar cell technology in Chile are calculated. Monthly average daily final yield and capacity factor for each solar cell technology in each location for summer and winter are shown. In Cologne, the monthly average final yield for one year were determined for different tilt angle and solar cell technologies. Finally, the energy consumption and energy output of the technical schools were analyzed to determine the energy balance of these facilities.

4.1 In-plane irradiation data

In Cologne the monthly average daily in-plane irradiation for different tilt angle was calculated by a simulation in Python with PVLIB for 2017 ranges between 0.39 kWh/m²/day and 0.43 kWh/m²/day for 15 ° and 30 ° in December. Maximum irradiation in Cologne was determined with 6.76 kWh/m²/day for 15 ° and 6.64 kWh/m²/day for 30 ° in June. In Santiago, the reference cell measured a radiation of 6.47 kWh/m²/day in January 2018 and 2.42 kWh/m²/day in June 2017. The comparison clarifies again the difference between the two locations. For Bella Vista, a PVGIS simulation estimates values for 2015 of 6.96 kWh/m²/day in October and 4.85 kWh/m²/day in April.

4.2 Comparative performance ratio analysis

The monthly average performance ratio for PV systems in three different climate locations from March to October 2015 and 2017 is shown in Figure 5.

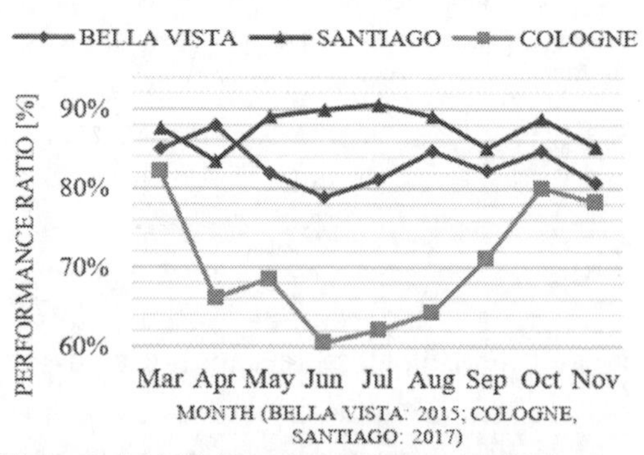

Figure 5: Monthly average performance ratios in three different climate locations during the analyzed time-period.

Due to different climate data sources the irradiation data were only available for 2015 in Bella Vista. Monthly average PR for these periods of time were determined with 69.3 % in Cologne, 83.0 % in Bella Vista and 87.7 % in Santiago. In addition, seasonal PR are represented by the monthly average in February 2018 for summer in Bella Vista and Santiago and July 2018 for summer in Cologne. The monthly average PR in summer ranges from 60.5 % in Cologne and 80.0 % in Bella Vista up to 85.5 % in Santiago. Winter is represented with February 2018 in Cologne and July 2018 in Bella Vista and Santiago. Monthly average PR in winter is between 90.3 % in Santiago, 81.1 % in Bella Vista and 82.2 % in Cologne.

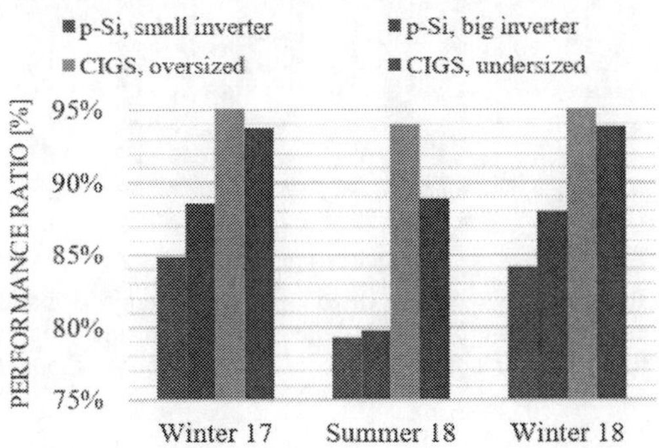

Figure 6: Performance ratio analysis for different solar cell technologies and inverter size in Santiago, Chile.

In Santiago a solar cell and inverter comparison for the same tilt angle as well as irradiation and climate conditions are included. It is shown in Figure 6 that the choice of cell technology has a big impact on PR. Values are presented over the seasons winter and summer, with Bella Vista and Santiago considering July as winter and February as summer, while for Cologne the values from July are displayed for summer and from February for winter. CIGS modules have the best overall PR in winter and summer compared with p-Si. Thus, the systems with p-Si cells provide up to 10 % lower average PR. Related to the differences in inverter design a configuration with an oversized inverter power capacity perform better. Finally, the CIGS in combination with an oversized inverter provide the best results, especially in summer, but less in winter month. Due to the seasonal temperature differences of the location, the plants perform better in winter than in summer. This behavior can be explained by a lower self-adjusting module temperature of the cells in winter. The influence factor of the module temperature on the performance of the systems is better with the CIGS cells than with the p-Si cells.

4.4 Comparative final yield

The monthly average daily final yield of the PV Systems with different configurations is shown in Figure 7. The highest final yield of all systems is calculated as expected for the summer months. Bella Vista shows an incursion of the final yield by comparatively low irradiation data of the month. A comparison of the final yield for the two systems in Santiago as well as in Cologne for summer and the yield of the system for Bella Vista in winter 17 shows that the energy yield output per installed module capacity is comparably high. However, the highest final yield is shown for the CIGS cells in Santiago followed by the p-Si cells with 30 ° tilt angle in Cologne. For Santiago and Cologne, the winter months show a decrease in the final yield, due to the lower irradiation, while Bella Vista delivers almost consistently high values. The systems for Cologne show in winter the lowest final yield due to local low levels of irradiation of this season.

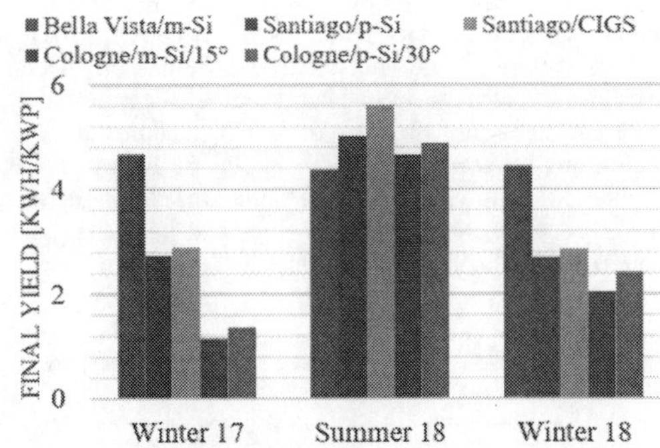

Figure 7: Monthly average daily PV system final yield for summer (July=Cologne; February=Bella Vista, Santiago) and winter (February =Cologne; July=Bella Vista, Santiago) in different locations.

Furthermore, overall PV system monthly average daily final yields for winter 2018 in Bella Vista, Santiago and Cologne are 4.44 kWh/kWp/day, 2.71 kWh/kWp/day and 2.23 kWh/kWp/day, respectively. For comparison, monthly average daily final yields in summer 2018 for Bella Vista, Santiago and Cologne are 4.39 kWh/kWp/day, 5.1 kWh/kWp/day and 4.77 kWh/kWp/day, respectively. In Cologne a final yield analysis for different tilt angle and solar cell technologies is presented over a period of one year. Figure 8 shows the final yield for p-Si modules with an tilt angle of 30 ° and m-Si modules with 15 ° tilt over the time-period of one year. We can see that the p-Si modules with 30 ° tilt angle perform slightly better over the whole year. In the analyzed time-period the p-Si modules with 30 ° tilt angle in Cologne gain over 10 % more final yield compared to the m-Si modules with 15 ° tilt angle.

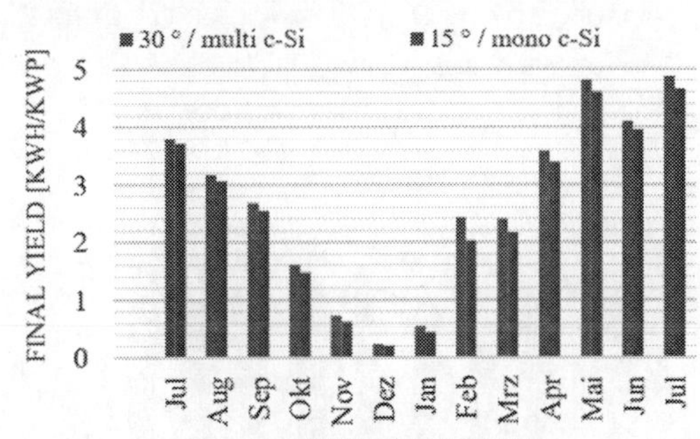

Figure 8: Annual distribution of monthly average daily final yield for the PV system strings in Cologne with different tilt angle and solar cell technologies in comparison.

Generally, p-Si has the lower efficiency, but in this set-up, it takes advantages from chosen tilt angle. Optimal tilt angle over the whole year for the location in

Cologne with azimuth orientation of -10 ° were determined with 38 ° in a PVGIS simulation. Another optimal tilt angle can be found in literature for Berlin with 35 ° [24].

4.3 Comparative capacity factor analysis

It is shown in Figure 9 that the capacity factor varies much in Santiago and Cologne which corresponds to the described differences in the annual irradiation in 2.1. In Bella Vista the capacity factor is almost constant over a year-time-period. The final yield and the capacity factor consider the same ratio of energy output for a period of time and the nominal power of the PV modules at STC-conditions as seen in the comparison of Figure 7 and Figure 8. The values of the individual systems are therefore proportional to each other and keep the same course for the CF and the final yield.

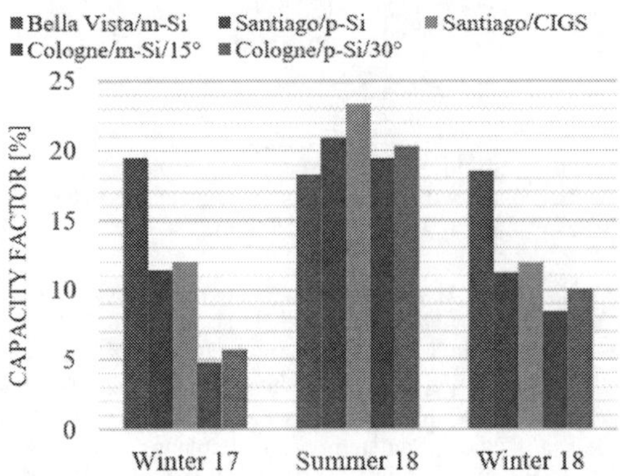

Figure 9: Monthly average capacity factor of the analyzed PV systems for summer (July=Cologne; February=Bella Vista, Santiago) and winter (February =Cologne; July=Bella Vista, Santiago) in 2017 and 2018.

4.5 Energy balance solar coverage ratio analysis

Energy balance solar coverage ratio is calculated in consideration that both systems, Bella Vista and Santiago take part on net-metering according to section 3.4.

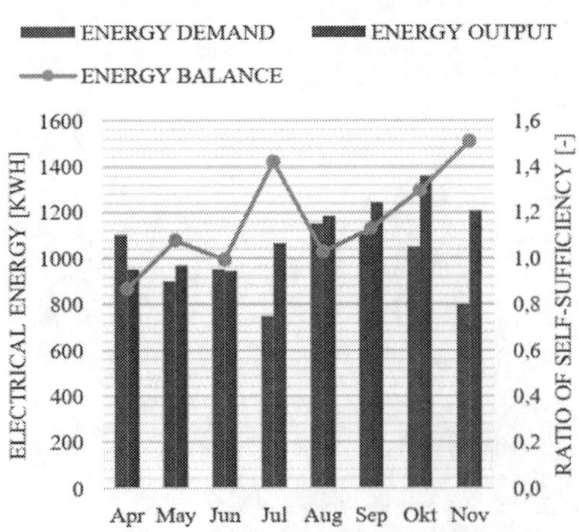

Figure 10: Monthly distribution of energy demand of the technical school, output by the installed PV system and ratio self-sufficiency in Bella Vista, Bolivia.

In Figure 10 is shown that in Bella Vista the energy consumption is constant, while then the energy output of the PV system increases from autumn to summer month gradually. Overall this results in a energy balance solar coverage ratio with values ≥ 1.

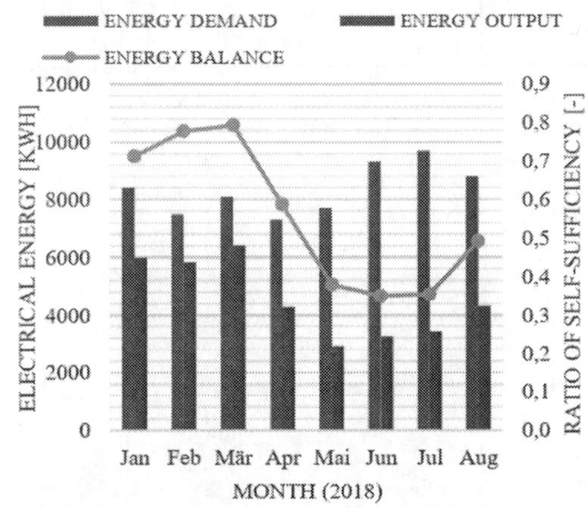

Figure 11: Monthly distribution of energy demand, energy output by the installed PV system and energy balance solar coverage ratio of the vocational school in Santiago, Chile.

Figure 11 shows that in Santiago the energy demand exceeds the energy output of the PV system especially in winter month the energy balance solar coverage ratio reaches minimum values of ≤ 0.35. Overall the optimal operation conditions and scaling of the PV system size lead to an average energy balance solar coverage ratio in Bella Vista of 1.16 from April to November 2015. In Santiago an average of 0.53 from January to August 2018 were determined. The high monthly solar coverage in Bella Vista corresponds to the constant irradiation availability in Bella Vista.

5 SUMMARY & CONCLUSION

PV-systems with different climate conditions, tilt angle, azimuth orientation, solar cell type, inverter configurations and installed power capacity were analyzed to determine and evaluate the performance and ratio of self-sufficiency. Annual, seasonal and monthly average daily results for final yield, capacity factor, performance ratio and energy balance solar coverage ratio were determined and presented.

- Maximum overall specific power output in this study was reached in Bella Vista with 1.1 kW/kWp on 20.12.2017.
- A combination of oversized inverter and CIGS solar cells in Santiago with 89 % average performance ratio for overall operation time provided the best results. It can be observed that the lower temperature sensitivity of CIGS solar cells and a higher operation time in the nominal power capacity of the inverter will result in higher amounts of converted solar energy.
- Overall monthly average PV system performance ratio for the analyzed time-periods were determined from March to November. For Bella Vista a value of 83 % in 2015, for

Santiago and Cologne values of 87 % and 69 % in 2017, respectively.

- Moreover, the calculation of average capacity factors from January to August 2018 results for Cologne, Santiago and Bella Vista are 10.8 %, 15.5 % and 19.8 %, respectively.
- Overall PV system monthly average daily final yields for winter 2018 in Bella Vista, Santiago and Cologne are 4.44 kWh/kWp/day, 2.71 kWh/kWp/day and 2.23 kWh/kWp/day, respectively. For comparison, monthly average daily final yields in summer 2018 for Bella Vista, Santiago and Cologne are 4.39 kWh/kWp/day, 5.1 kWh/kWp/day and 4.77 kWh/kWp/day, respectively.
- In Cologne the annual average final yield in 2017 shows 10 % better results for p-Si modules with 30 ° tilt angle compared with m-Si modules with 15 ° tilt angle.
- High degrees of monthly balance solar coverage ratios were achieved in Bolivia due to high irradiation in summer and winter, as well as optimal operation conditions and sizing of the PV system.

It can be concluded that the climate conditions provide best performance for operating PV systems in the selected area in Chile and Bolivia. Especially in Bolivia the high amount of solar irradiation distributed over the whole year offers a high potential for self-sufficiency.

ACKNOWLEDGEMENTS

The project has been financed by the *German Federal Ministry for Development and Cooperation (BMZ)* through *Misereor* and *Begeca*, the companies *Renusol*, *Sunpower* and *F&S Solar* and the foundation *"Cristo Vive Europe"*. Furthermore, the authors would like to acknowledge the company *Avancis* for donating the CIGS modules for the project in Santiago.

REFERENCES

[1] J. Leloux, L. Narvarte, D. Trebosc, Review of the performance of residential PV systems in France, Renewable and Sustainable Energy Reviews 16 (2012) 1369-1376

[2] N. Allet, F. Baumgartner, J. Sutterlueti, L. Schreier, M. Pezzotti, J. Haller, Evaluation of PV system performance of five different PV module technologies, 26th European Photovoltaic Solar Energy Conference (2011) Hamburg

[3] F. degli Uberti, R. Faranda, S. Leva, E. Ogliari, Performance ratio of a PV power plant: different panel technologies comparison, Solar Energy Tech (2010)

[4] M. S. Adaramola, E. E. T. Vagnes, Preliminary assessment of a small-scale rooftop PV-grid tied in Norwegian climatic conditions, Energy Conversion and Management 90 (2015) 458-465

[5] W. G. J. H. M. van Sark, N. H. Reich, B. Müller, A. Armbruster, K. Kiefer, Ch. Reise, Review of PV performance ratio development, Utrecht (Netherlands), Freiburg (Germany), n.a.

[6] M. Drif, P. J. Pérez, J. Aguilera, G. Almonacid, P. Gomez, J. de la Casa, J. D. Aguilar, Univer Project. A grid connected photovoltaic system of 200 kW_p at Jaén University. Overview and performance analysis, Solar Energy Materials & Solar Cells 91 (2007) 670-683

[7] D. A. Quansah, M. S. Adaramola, G. K. Appiah, I. A. Edwin, Performance analysis of different grid-connected solar photovoltaic (PV) system technologies with combined capacity of 20 kW located in humid tropical climate, International Journal of Hydrogen Energy 42 (2017) 4626-4635

[8] L. M. Ayompe, A. Duffy, S. J. McCormack, M. Conlon, Measured performance of a 1.72 kW rooftop grid connected photovoltaic system in Ireland, Energy Conversion and Management (2011)

[9] M. E. Ya'acob, H. Hizam, T. Khatib, M. A. M. Radzi, A comparative study of three types of grid connected photovoltaic systems based on actual performance, Energy Conversion and Management 78 (2014) 8-13

[10] K. Attari, A. Elyaakoubi, A. Asselman, Performance analysis and investigation of a grid-connected photovoltaic installation in Morocco, Energy Reports 2 (2016) 261-266

[11] R. Sharma, S. Goel, Performance analysis of a 11.2 kWp roof top grid-connected PV system in Eastern India, Energy Reports 3 (2017) 76-84

[12] T. Khatib, K. Sopian, H. A. Kazem, Actual performance and characteristic of a grid connected photovoltaic power system in the tropics: A short term evaluation, Energy Conversion and Management 71 (2013) 115-119

[13] P. Krämer, Bachelor Thesis – Design, implementation and monitoring of a 40 kWp photovoltaic system in Santiago de Chile, Cologne (Germany), 2017

[14] WeatherSpark.com, Cedar Lake Ventures Inc. URL: https://de.weatherspark.com, Excelsior (USA), 2018

[15] J. Remund, S.C. Müller, Meteonorm 7

[16] PVGIS, European Comission, Joint Research Centre, URL: http://re.jrc.ec.europa.eu/pvgis.html, Ispra (Italy), 2017

[17] U. Blieske, N. Reiners, R. Gecke, A. Maixner, P. Schorpp, N. Schwarze, N. Wolf, D. Korber, J. Muenzberg, PV- Systems for demonstration and training purposes in South America, Cologne Institute for Renewable Energy, 5th International Energy and Sustainability Conference 2016, Cologne

[18] Technical Documentation, PV system, Cologne University of Applied Sciences

[19] PV_LIB Toolbox for Python, Photovoltaic Performance Modeling Collaborative, National Technology and Engineering Solutions of Sandia, 2018

[20] NSRDB, National Solar Radiation Data Base, 2018

[21] DWD, Deutscher Wetterdienst, 2018

[22] J.Weniger, V. Quaschning, Begrenzung der Einspeiseleistung von netzgekoppelten Photovoltaiksystemen mit Batteriespeichern, 28.Symposium Photovoltaische Solarenergie, Bad Staffelstein, 2013

[23] T. Schäfer, M. Egler, M. Mack, Impact of different system layouts and -components on the direct consumption and autarky of utility-scale PV power plants, 28th European Photovoltaic Solar Energy Conference and Exhibition 2013, Paris

[24] K. Mertens, Photovoltaik: Lehrbuch zu Grundlagen, Technologien und Praxis, Carl Hanser Verlag, Steinfurt, 2013, S. 53

SECTOR COUPLING IN EUROPE TO REACH THE CLIMATE CHANGE MITIGATION GOALS BY 2050

Marie-Christin Leonhard, Milot Kamberaj, Henrik te Heesen
Trier University of Applied Sciences, Environmental Campus Birkenfeld, Campusallee, 55768 Neubrücke (Nahe), Germany, h.teheesen@umwelt-campus.de

ABSTRACT: The 2015 United Nations Climate Change Conference negotiated a global agreement on the reduction of greenhouse gas emissions. Europe must transform its energy system from mainly fossil power plants to renewable ones in order to reach the climate change mitigation goals by 2050. In particular, wind turbines, photovoltaic systems and the use of biomass can contribute to fulfill the demand of residential, commercial and industrial consumers regarding electric power, heat and transportation. We have developed a simulation model to derive the energy production, exchange and demand in 39 European countries taking different scenarios of bioenergy potential usage into account. Annually, 31.7 GW/a of onshore wind turbines, 35.9 GW/a of offshore wind turbines and 100.2 GW/a photovoltaic systems (roof-top and ground-mounted) have to be installed to transform the European energy systems by 2050. The total installed power plant capacity must increase from 0.9 TW in 2017 to 5.7 TW in case of 40 percent usage of the biomass potential. Out model shows that the transformation process of the energy system in Europe must be accelerated significantly. Then, the climate change mitigation goals can be reached.

Keywords: CO_2 Footprint, Environmental Effect, Modelling, Strategy

1 INTRODUCTION

Using fossil fuels to produce energy leads to an emission of greenhouse gases (GHG) into the environment. These emissions cause continuously increasing temperatures in the atmosphere. Europe's emissions contribution is 12 percent to the global CO_2 emissions. In order to limit global warming to 2 degrees Celsius by the year 2100, the emissions of GHG must be virtually zero until 2050 [1]. In this case, the energy transition has a key role and must be implemented within the next decades. The purpose of this study is to show how renewable energies can completely cover Europe's energy demand with regard to sector coupling to decrease the GHG emissions by 2050.

Quaschning determines in his study the installed capacity of renewable energies in Germany [2]. Different scenarios are carried out in order to consider the development of the annual electrical energy demand. The determination of the electrical energy demand of the Quaschning study is based on the substitution of fossil energy sources by climate-neutral, predominantly electricity-based energy sources which can cover the energy demand in the various sectors. To compare the installed capacity of renewable energies across Europe, the study of Ram [3] is taken into account. The study carries out the assessment of the required installed renewable energy capacity in all European countries. The calculation of the installed capacity in Europe in 2050 is based on the development of electrical energy demand from the electricity sector. The energy demand from the heat and transport sectors is not considered in the study. In contrast to Ram's study, Jacobson's study considers the climate-neutral coverage of energy needs in all sectors [4]. The study from Teske [5] can be used for a Europe-wide comparison of the results. In the publication, various scenarios for covering the energy demand in 22 European countries have been carried out.

2 METHODS

The first step is to model the energy demands in 2050 for the energy sectors electricity, space heating and hot water, process heating and transportation for each European country. The results are summarized to different regions in Europe, which are color-coded in Figure 4. Next, the technical supply potential of renewable energy systems in each country in Europe must be determined. This is followed by the selection of climate-neutral technologies that are used to support sector coupling. Finally, different scenarios are modeled in order to determine the expansion of renewable energies.

2.1 Energy demand

The European energy demand in 2050 for the sectors electricity, space heating and hot water, process heating and transport is expressed in the final energy level. We assume that the population in Europe does not change appreciably. We determine the energy demand in the sector space heating and hot water by the average effective energy per capita which is a specific value for a certain region in Europe. We assume that the energy demand can be reduced by 50 % in 2050 due to more energy efficient construction methods and other improvement in efficiency [3].

The process heating for households, trade and commerce is mainly covered by electric energy. Therefore, it is considered in the electricity sector throughout the study.

Because of the diversity of the technical utilization, we separate the process heating in industry in three different temperature levels

- low temperature level (< 100 °C),
- medium temperature level (100-500 °C),
- high temperature level (> 500 °C).

We set the share of low temperature heat at 30%, medium and high temperature heat at 35%.

The energy demand for the transportation sector is separated in street transportation, air transportation, rail and domestic shipping. We subdivide the street transportation in passenger transportation which constitutes 70% of the energy and freight transportation. We do not need to model the electric energy demand, as there are detailed data available for all the regions in Europe [6].

2.2 Technical supply potential and expansion scenarios

The available area in each country needs to be determined in order to calculate the potential of a certain energy sources. We derive the installed capacity and the possible yield by using region-specific values for the full-load hours for photovoltaic (PV) systems and wind parks.

We consider wind energy and PV systems for the expansion of renewable energy systems. The potential of other energy sources (hydropower, concentrates solar power (CSP), geothermal energy) keeps unchanged to the current capacity. We also take the biomass potential derived by Thrän [7] into account in three different expansion scenarios. The scenarios base on the calculated final energy demand in 2050 and differ in the use of the biomass (0 %, 40 %, 80 %). The biomass can be converted into solid, liquid or gaseous fuels by different transformation processes. We assume no temporal and spatial restrictions in the exchange of energy within Europe in order to analyze the different potentials in each country effectively.

The following technologies are considered to provide thermal energy for the space heating and hot water sector: heat pumps, solid fuel boiler, gas boiler, district heating, combined heat and power plants (CHP) and solar collectors. The share of heat pumps in this sector amounts to 50% whereas the use of the other technologies (except for the solar collectors) depends on the provision of fuels. Those fuels are either made out of biomass or provided by power-to-gas (PtG) processes.

The thermal energy demand for the low temperature level in the process heating sector is covered by the same technologies used for the space heating and hot water sector. Different technologies are used in the medium temperature level because there is a high demand in process steam. The steam can be provided by CHP and electrode boilers, which are run by renewable produced electricity. Additionally, solid fuel boiler, gas boiler, and biomass boiler are used. For the high temperature level different industrial furnaces which run by solid fuels or gas as well as electric arc furnaces are used.

In the transportation sector, we assume that the share of electric cars in the passenger transportation amounts to 50 %. The missing amount is covered by biofuels, partly provided by power-to-liquid (PtL) processes. We do not consider an electrification of the freight transport. The fuel for air transport and shipping is substituted by biofuels. The rail transport, in turn, is considered as fully electric.

The expansion of renewable energies, such as wind energy and PV, depends on the use of the biomass potential.

3 RESULTS AND DISCUSSION

First, we present the prognosis of the energy demand in all three sectors and the total potential of renewable energy production in 2050. Second, we compare three different scenarios regarding the usage of biomass potential. Finally, we select the middle biomass scenario (usage of 40 % of biomass potential) and derive the expansion pathway for Europe to fulfill the energy demand by just using renewable energy systems.

3.1 Energy demand and supply potential

The modeled final energy demand for all sectors (electric power, heat and transportation) amounts to 15 PWh in the year 2050 which is shown in Figure 1.

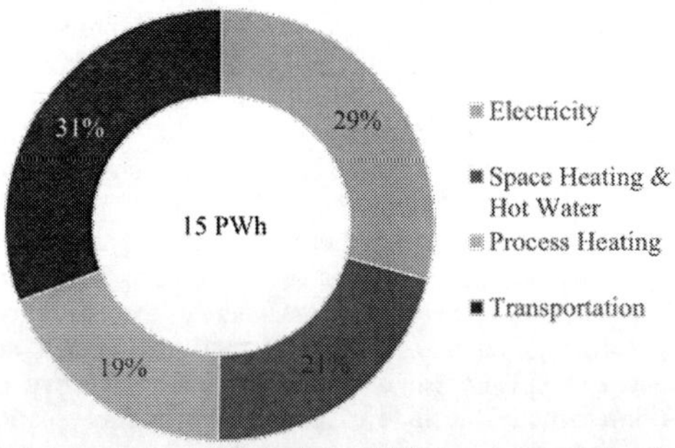

Figure 1: Energy demand in 2050 by sectors electric power, heat (space heating and process heating) and transportation. In total, the energy demand in 2050 is 15 PWh.

Figure 2 shows the total energy production potential of renewable energy systems in Europe amounts which reaches an annual energy production of 35 PWh. Offshore wind turbines have the largest potential followed by photovoltaics.

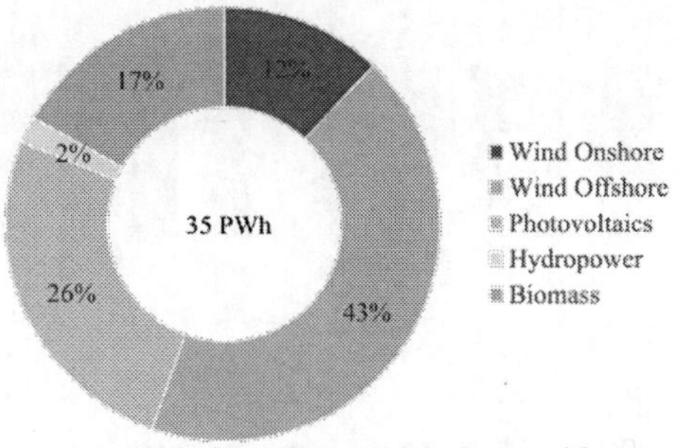

Figure 2: Technical supply potential of renewable energy systems in Europe by 2050. The annual energy production capacity of all systems is 35 PWh.

3.2 Climate-neutral energy system

Figure 3 shows the model of the energy coverage by renewable energy systems and sector coupling. The installed capacity as well as the provided energy for every energy source refer to the scenario "Biomass 40" which means that 40 % of the biomass potential [7] is used. All biomass scenarios mean that there is no displacement of agricultural use for food production. The light green boxes are the renewable energy systems producing electric energy throughout Europe. Additionally, three different types of bioenergy production take part in the energy production – solid biomass (yellow), liquid biomass (green) and gaseous biomass (cyan). The energy flow is represented by the colored arrows. The black arrows are the electric energy flow from the production systems to the transformation systems and finally to the end energy consumption. The red arrows show the heat flow from heat production systems like heat pumps, solar thermal collectors as well as combined head and power systems (CHP). Also, power-to-x (PtX) systems are considered in the energy sector coupling model by the transformation of electric power to natural gas (PtG) or to liquids (PtL). The renewable fuels feed energy transformation systems like CHP or fuel-bases combustion engines (shown by the brown arrows). Finally, the demand of the four different end energy types electricity, space heating, process heating and transportation are covered by the renewable energy production.

The energy flow chart demonstrates the situation of all European countries in 2050. The resulting energy production is free of CO_2 emissions. There is an instantaneous energy exchange within Europe without spatial or temporal restrictions. First, each country tries to fulfil its own energy demand. Second, an exchange of the energy types (electricity, biomass, synthetic natural gas) is considered between all European countries. We use an iterative algorithm to model the trade and exchange of energy resources.

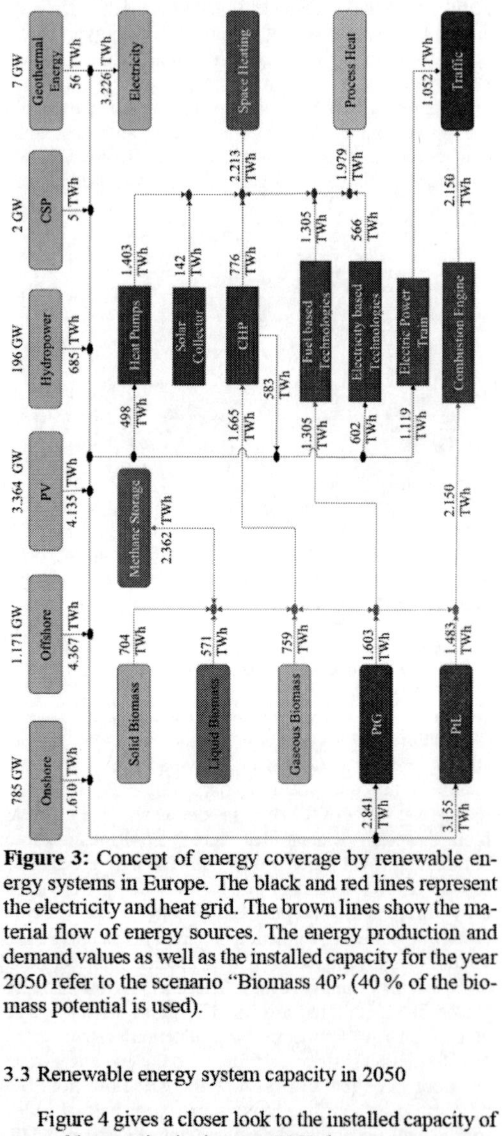

Figure 3: Concept of energy coverage by renewable energy systems in Europe. The black and red lines represent the electricity and heat grid. The brown lines show the material flow of energy sources. The energy production and demand values as well as the installed capacity for the year 2050 refer to the scenario "Biomass 40" (40 % of the biomass potential is used).

3.3 Renewable energy system capacity in 2050

Figure 4 gives a closer look to the installed capacity of renewable energies in the year 2050 that are necessary to reach the goals of climate change mitigation. The 39 European countries are grouped in eight regions with similar industrial levels and climate conditions regarding wind speed, solar irradiation and agricultural land use.

It is noticeable that the capacity of PV has a higher share in the southern regions. In northern regions, the benefits of the coast lines lead in a higher use of offshore wind energy. The values again refer to the scenario with an average biomass use of 40 %. An increasing use of biomass leads to a decreasing energy exchange within Europe which reduces the need of renewable energy production capacity. If less biomass potential is used, more renewable capacity must be installed in all regions in Europe.

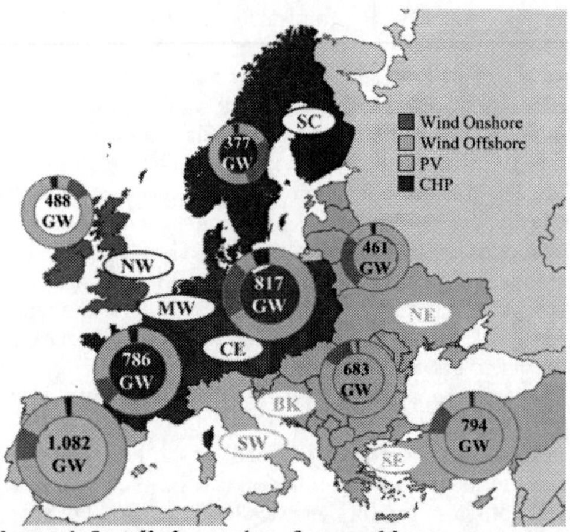

Figure 4: Installed capacity of renewable energy systems in the eight European regions according to scenario "Biomass 40" in the year 2050.

The use and expansion of renewable energy systems is focused on wind energy and PV due to the high potential in Europe. The installed capacity will be achieved by an annual expansion of renewable energy systems from 2014 to 2050 as shown in Figure 5. There is a clear differentiation of the different regions in Europe as the northern regions like Scandinavia (SC) and Northwest Europe (NW) have a large wind energy potential due to high wind speed. The southern regions like Southwest (SW) or Southeast Europe (SE) have a large PV potential due to high solar irradiance values. There is a renewable energy mix of offshore and onshore wind systems as well as PV in Central Europe (CE).

The annual capacity addition varies between 9.8 GW/a in Scandinavia (SC) and 27.7 GW/a in Southwestern Europe (SW) due to the wind and PV potential in the countries. In total, 18.4 GW/a of onshore wind turbines, 32.3 GW/a of offshore wind turbines and 91.0 GW/a of PV systems must be annually added to the energy system from now to 2050 to fulfill the European energy demand and do reduce the greenhouse gas emissions to zero in the "Biomass 40" scenario.

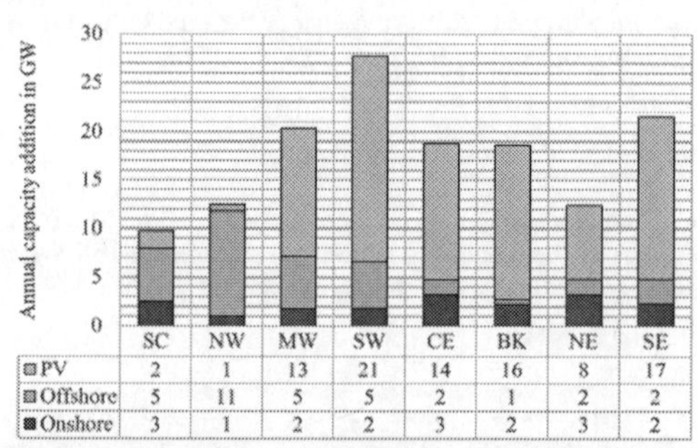

	SC	NW	MW	SW	CE	BK	NE	SE
PV	2	1	13	21	14	16	8	17
Offshore	5	11	5	5	2	1	2	2
Onshore	3	1	2	2	3	2	3	2

Figure 5: Annual capacity additions of photovoltaic systems (PV) and wind energy (onshore and offshore) in the eight different regions in Europe from 2014 to 2050 for the

"Biomass 40" scenario. See Figure 4 for the definition of the regions.

Figure 6 compares the total installed capacity of wind power (onshore and offshore) as well as photovoltaics in 2050 for three different biomass scenarios: Biomass 0 means that no biomass is used for energy production, biomass 40 represents the scenario that 40 % of the biomass potential is used and Biomass 80 uses 80 % of the biomass potential. Additionally, the capacity of combined head and power production (CHP) by biogas is shown as CHP is a highly efficient system for electric power and heat production by the use of biogas or synthetic natural gas. In total, between 4.5 TW ("Biomass 80" scenario) and 7.4 TW ("Biomass 0" scenario) of renewable capacity must be installed in Europe. In comparison, in 2017 932 GW of fossil and renewable (electric) power capacity is installed the European Union (EU-28) [8].

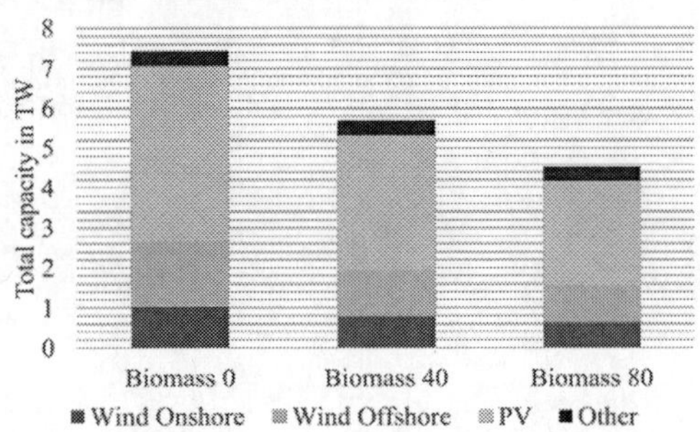

Figure 6: Installed capacity of electrical power of renewable energy systems in Europe for the different scenarios. "Biomass 40" means that 40 % of the biomass potential [7] is used for energy production.

3.4 Evaluation of the scenarios

The comparison of our results with data of different publications which have also analyzed the transformation of energy systems show a common understanding.

Table 1: Comparison of the installed capacity in GW based on the publications of Ram [3] and Teske [5] with our data of the "Biomass 40" scenario. Ram and Teske just considered the electric power sector and neglected the heat and the transportation sectors. Thus, our capacities show twice the values compared to Ram and Teske.

Installed Capacity in GW	Ram [3]	Teske [5]	Our Data Biomass 40
Wind onshore	560	594	785
Wind offshore	0	237	1,171
PV	1,956	926	3,364
Other renewables	358	522	371
Fossil power plants	28	181	---
Total	**2,902**	**2,460**	**5,691**

Table 1 compares our data with the results of Ram [3] and Teske [5]. Both authors have focused on the electric power sector neglecting the heat and transportation sectors. Our model leads to an installed capacity which is twice larger than the capacity of Ram and Teske. Ram does not consider offshore wind parks and concentrates mainly on PV. Teske uses a mix of wind power and PV systems. Both authors still use a small capacity of fossil power plants. Thus, a greenhouse-gas-free energy system is not achieved by both authors as the supply of heat energy and the transportation sector has not been taken into account.

Quaschning [2] analyses just the German energy system considering all three energy sectors electric power, heat and transportation. Table 2 shows the comparison of the installed capacity of Quaschning with our results for Germany. Quaschning's energy system focuses on PV and onshore wind energy. There is a small energy import of electric energy from other European countries to Germany (137 TWh). This is the main difference to our energy model as we are considering a large energy exchange between all European countries which leads to an import of electric energy of 1,412 TWh and a smaller capacity of renewable energy systems in Germany with 411 GW compared to 717 GW by Quaschning.

Table 2: Comparison of the installed capacity in Germany in 2050 based on the study of Quaschning [2] with our results also just for Germany.

Installed Capacity in Germany	Quaschning [2]	Our Data "Biomass 40"
Wind onshore	199 GW	57 GW
Wind offshore	76 GW	35 GW
PV	415 GW	277 GW
Other renewables	27 GW	42 GW
Total capacity	**717 GW**	**411 GW**
Import of electric energy	137 TWh	1,412 TWh

4 CONCLUSION

Our energy model for Europe shows that the transformation of the energy systems is possible but demanding. An annual addition of renewable energy systems of 120.6 GW/a is necessary in Europe to reduce the greenhouse gas emission to zero and to reach the climate change mitigation goals compared to the capacity addition of 26.0 GW in the European Union in 2017. The annual expansion of the renewable energy capacity must be increased by a factor 4.5 compared to today's annual renewable capacity addition.

5 REFERENCES

[1] T. Stocker, G. Plattner, A. Nauels, M. Tignos, Y. Xia, D. Qin et al. (2013): Climate Change 2013. The Physical Science Basis.

[2] V. Quaschning (2016): Sektorkopplung durch die Energiewende. Anforderungen an den Ausbau erneuerbare Energien zum Erreichen der Pariser Klimaschutzziele unter Berücksichtigung der Sektorkopplung. HTW Berlin.

[3] M. Ram, D. Bogdanov, A. Aghahosseini, S. Oyewo, A. Gulagi, M. Child et al. (2017): Global Energy System based on 100% Renewable Energy - Power Sector.

[4] M. Z. Jacobson, M. A. Delucchi, Z. A. F. Bauer, S. C. Goodman, W. E. Chapman, M. A. Cameron et al. (2017): 100% Clean and Renewable Wind, Water, and Sunlight All-Sector Energy Roadmaps for 139 Countries of the World. In: *Joule* 1 (1), S. 108–121.

[5] S. Teske, S. Sawyer, O. Schäfer (2015): Energy Revolution. A Sustainable World Energy Outlook 2015.

[6] Eurostat (2016): Energy balance sheets – 2014 data – 2016 edition. European Union.

[7] D. Thrän, M. Weber, A. Scheuermann, N. Fröhlich, J. Zeddies, A. Henze et al. (2005): Nachhaltige Biomassenutzungsstrategien im europäischen Kontext. Institut für Energetik und Umwelt gGmbH
[8] WindEurope. Installed power capacity in the European Union (EU-28) in 2005 and 2017, by generation type (in gigawatts). Statista. Accessed September 19, 2018. Available from https://www.statista.com/statistics/807675/installed-power-capacity-european-union-eu-28/

ESTABLISHED AND EMERGING SOLAR MARKETS IN 2017
OVERVIEW ON GLOBAL SOLAR MARKET DEVELOPMENT

Chris Werner[1], Alexander Gerlach[2], Christian Breyer[3], Gaëtan Masson[4]

[1] Chris Werner Energy Consulting, 06846 Dessau, Germany
Phone: +49 (0) 176 64093993, e-mail chris.werner@cw-energyconsulting.de
[2] Alexander Gerlach New Energy Consulting, 99755 Ellrich, Germany,
[3] Lappeenranta University of Technology, Skinnarilankatu 34, 53850 Lappeenranta, Finland
[4] Becquerel Institute, Rue Royale 146, 1000 Brussels, Belgium

ABSTRACT
The installed capacity of photovoltaics (PV) shows a constant growth. The major part of PV is installed in highly developed countries. Further, there is an increasing number of emerging markets that show high growth rates of cumulative and annual PV installations. However, reliable installation rates for PV are available only for a limited number of countries. Not only in emerging markets, even for some developed countries exact numbers of cumulative PV installation are rarely available. For the end of 2017 IRENA reports 384.6 GWp of globally installed PV capacity, giving clear national specific data for 156 countries. IEA PVPS provides a number of 402.50 GWp installed by providing detailed data for 24 countries.
This paper gives an overview of installed PV capacities for all countries in the world based on the examination of publicly accessible data. Furthermore, an analysis of the development of cumulative PV installations in recent years is given. Resulting from this evaluation, PV installations are localised in 196 countries, representing 402.15 GWp.

Keywords
PV Markets, Emerging Markets, Installed Capacity, PV Installation, Market Growth

1 MOTIVATION

The global photovoltaic (PV) installations constantly show high growth rates [1]. A variety of countries has reached large cumulative PV installations contributing relevant shares to their annual power generation. Further, numerous markets that today are not among the leading countries per cumulative or annual PV installations, show a high potential to become the new growth markets of photovoltaics. Since reliable and officially traced data on PV installations are only available for a small number of countries, public reports do not usually show any data for installed capacities in those new and emerging PV markets. These markets are usually named as Rest of World (RoW). However, they often show a constant growth of installed PV capacity. With rising photovoltaic capacities and growing economies in these markets, they are about to become important photovoltaic markets in the years to come. Thus, RoW markets have to be better understood.

As a result, an analysis of cumulative installed PV capacity for all countries in the world is strongly required. This paper provides an analysis of photovoltaic capacities and growth rates ranging from 2009 to today for every market globally. Further data starting in 2001 are available, but exceed the scope of this paper. On this basis, an estimation of future growth rates and PV capacities expected to be installed in all markets can be calculated.

2 EXAMINATION METHOD

For many PV markets no reliable estimation of installed PV is available due to a lack in public authority monitoring. This is especially true for emerging markets, but a number of developed countries as well. To improve on a reliable valuation of the photovoltaic markets of all countries, the examination of several sources is necessary to ensure a credible estimation.

A basic source is the international customs database, monitored by the 'Market Analysis and Research' section of the International Trade Centre, an agency of UN's World Trade Organization. This database contains the customs data of all countries worldwide since the year 2000 and receive the data from the particular national customs authorities. The database provides the opportunity to evaluate the total value of imported and exported products for each country per year. Furthermore, a possibility of evaluating bilateral trades for a specific product is given. Products are classified in several specific product numbers, so-called HS Codes, obligatory for more than 200 countries, customs and economic unions, representing more than 98% of world's trade [2]. HS Code group 854140 represents 'photosensitive semiconductors devices, photovoltaic cells and light emitting diodes'. More specific descriptions with detailed determination of codes between light emitting diodes and PV are available for approximately 95% of the data. Otherwise, an experienced ratio of 80% PV and 20% light emitting diodes is used. Surprisingly, a large number of developing countries' customs agencies have reported detailed data. It is valid that: *'imports A from B' = 'exports B to A'*. Data have been assigned for 190 countries in a period of 2000 until 2017. The basis of the calculation of PV capacities per country is the following assumption: *'import A' – 'exports A' + 'production A' = 'market A'*. The production volumes refer to the publications of the magazine 'Photon' on the topic of production output [3]. From this, attention is paid to the fact that approximately 25% of annual production will be installed in the following year as significant quantities of

PV products are either being in transit to their destination or warehoused before their installation.

Since the data include the monetary value of the products, a conversion to PV capacities is necessary. For this purpose a reliable estimation of PV prices per Wp for every country is needed. These price data refer to a worldwide annual average PV module price per Wp [4-7]. Since prices depend on countries' locations and market sizes, there is a significant ratio of divergence. Therefore, all countries are classified in five main groups with one of them having two subgroups, each representing a group of countries that are nearly comparable in market size and economic conditions. The world's average price is multiplied by a factor quoting the ratio of countries' prices for each of these groups. The ratio is based on experience and statements of several countries' experts that have been contacted in the past [8-10]. Those ratios have been validated in previous publications [8, 11-16]. Since the world's given average annual PV prices are selling prices of modules' producers, a factor of 10% is added considering distributor margins and shipping, etc. Therewith, prices for the year 2017 vary between 0.42 USD/Wp for advanced PV market countries and 1.02 USD/Wp for small PV applications in African and Asian low-income countries as can be seen in Table 1.

To lower the probability of errors, further sources with installation rates estimated by IRENA, International Energy Agency Photovoltaic Power Systems Programme (IEA PVPS), journal du photovoltaique, Solar Power Europe and other public resources are used [1, 8, 16-28]. Some countries report PV installation data by giving DC capacities while others refer to their installed AC capacities. Thus a small probability of mismatch between reported data and those number calculated by using the DC capacity according to the customs database is possible. However numbers referring to DC capacities are favoured.

Table 1: PV module price for conversion of customs values into PV per category of countries [4-7]. The category of countries with annual installs below 1 MWp per year is subcategorized per region. Abbreviations: North America (NA) and Latin America (LA).

[USD/Wp]	2000	2001	2002	2003	2004	2005	2006	2007	2008	2009	2010	2011	2012	2013	2014	2015	2016	2017
average world price:	5.17	4.96	4.36	4.13	4.24	4.55	5.07	4.57	4.38	2.53	2.05	1.65	1.21	0.89	0.85	0.74	0.62	0.46
countries >100 MW/year	4.66	4.47	3.92	3.72	3.82	4.09	4.56	4.11	3.94	2.28	1.85	1.49	1.09	0.80	0.76	0.66	0.55	0.42
countries >10 MW/year	5.69	5.46	4.79	4.55	4.67	5.00	5.57	5.02	4.82	2.78	2.26	1.82	1.33	0.98	0.93	0.81	0.68	0.51
countries >5 MW/year	7.76	7.44	6.54	6.20	6.36	6.82	7.60	6.85	6.57	3.79	3.08	2.48	1.82	1.34	1.27	1.11	0.92	0.69
countries <5 MW/year	9.31	8.93	7.84	7.44	7.63	8.19	9.12	8.22	7.89	4.55	3.70	2.97	2.18	1.60	1.52	1.33	1.11	0.83
countries <1 MW/year	10.3	9.93	8.71	8.27	8.48	9.10	10.1	9.14	8.76	5.06	4.11	3.30	2.42	1.78	1.69	1.47	1.23	0.92
Africa & Asia	11.4	10.9	9.59	9.10	9.33	10.0	11.1	10.1	9.64	5.57	4.52	3.63	2.66	1.96	1.86	1.62	1.36	1.02
Europe & NA & LA	9.83	9.43	8.28	7.86	8.06	8.64	9.62	8.68	8.33	4.81	3.90	3.14	2.30	1.69	1.61	1.40	1.17	0.88

Furthermore, all sources are weighted in a certain order: For countries that have their own cell or module production, the arithmetic mean of all sources except customs data is taken for estimation of installation rates, due to production margins and local PV value chain structures influencing the reliability of the customs database. In case of countries having no production, the average of all sources is used.

Due to statistical impacts of additional databases, it is possible in individual cases that estimated installed PV capacities at the end of 2017 can be less than those estimated at the end of previous years.

3 OVERVIEW ON PV INSTALLATIONS 2017

Within the use of this examination, the world's total installed PV capacity is estimated to be up to 402.15 GWp at the end of 2017. It is expected that there is a further 487 MWp that had been installed before the year 2000. PV installations are found in almost all countries worldwide. Appendix Table 1 includes the results for PV installation rates for all countries in the world at end of the year 2017.

As in previous years, China is the country with leading PV installations and the first country worldwide which has surpassed the threshold of 100 GWp. Further six markets show more than 10 GWp of PV capacity installed. Another 22 markets claim more than one GWp of PV capacity by the end of 2017. Thus, there are some 29 countries are among those with more than one GWp deployed. These countries represented by China (130.78 GWp), the United States (51.45 GWp), Japan (49.21 GWp), Germany (43.01 GWp), Italy (19.61 GWp), India (18.35 GWp), the United Kingdom (12.73 GWp), France (8.05 GWp), Australia (7.24 GWp), the Republic of Korea (5.41 GWp), Spain (5.33 GWp), Belgium (3.72 GWp), Turkey (3.43 GWp), Canada (3.01 GWp), the Netherlands (2.76 GWp), Thailand (2.73 GWp), Greece (2.64 GWp), Czech Republic (2.18 GWp), South Africa (1.93 GWp), Switzerland (1.91 GWp), Chile (1.83 GWp), Taiwan (1.73 GWp), Romania (1.41 GWp), Austria (1.29 GWp), Pakistan (1.22 GWp), Israel (1.17 GWp), Ukraine (1.15 GWp), Brazil (1.11 GWp) and Bulgaria (1.04 GWp) in total amount to 96% of global PV installations. Turkey, Brazil and Ukraine are new representatives of this group of markets. Furthermore, there are 37 additional countries claiming more than 100 MWp per country (Denmark, the Philippines, Slovakia, Portugal, Mexico, Honduras, United Arab Emirates, Jordan, Malaysia, Algeria, Morocco, Russia, Hungary, Peru, Slovenia, Luxembourg, Bangladesh, Iran, Poland, Sweden, Yemen, Reunion, Uruguay, Indonesia, Saudi Arabia, Puerto Rico, Singapore, Ghana, Panama, Egypt,

El Salvador, Senegal, Nigeria, Dominican Republic, Belarus, Cyprus and Malta). Remarkable is the increasing share of emerging markets among those countries from Latin America and MENA region. Further, other emerging and developing countries show a significant installed PV base, in particular Kenya (99 MWp), Guatemala (95 MWp), Myanmar (92 MWp) or Cuba (91 MWp). In total 23 countries claim less than 100 MWp but more than 50 MWp, of which the majority is represented by developing countries. Especially in least developed low electrified countries in Sub-Saharan Africa and South Asia like Bangladesh, Tanzania or Myanmar high ratios of installed PV capacity are realised as off-grid systems, i.e. 5 MWp solar home systems with an average size of 50 Wp represent a solar power solution for 100,000 families [29]. On a regional perspective Asia-Pacific and Europe, by far, show highest cumulative PV installations, with 219.25 GWp and 115.1 GWp, respectively. North America represents 55.0 GWp, Latin America 5.4 GWp, Africa 4.4 GWp and the MENA region 3.1 GWp.

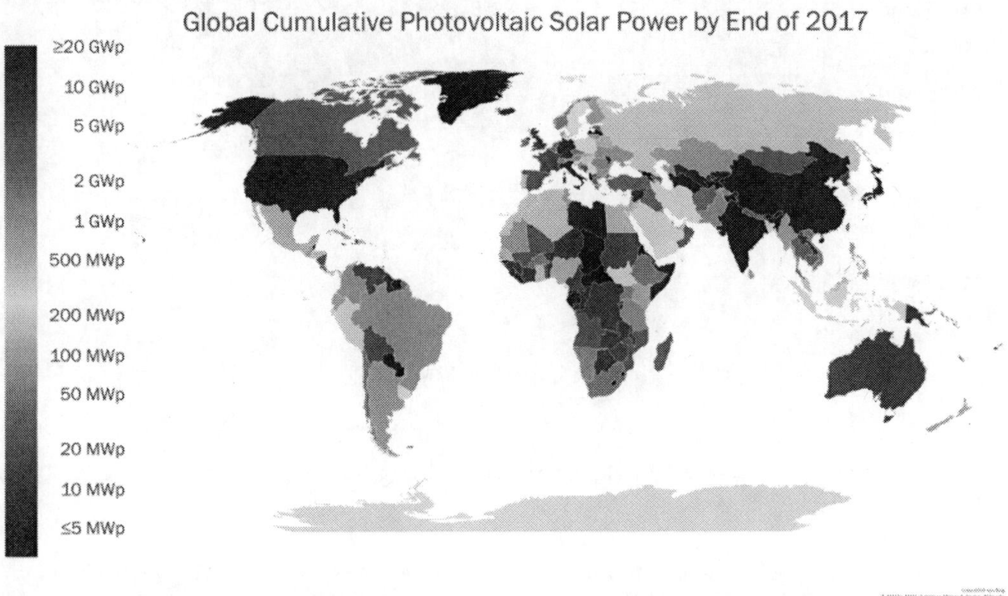

Figure 1: Total installed PV capacity per country by end of 2017.

Approximately 53% of global installed PV capacity (212.8 GWp) is installed in countries that have a gross domestic product (GDP) per capita higher than 30,000 USD per year. Low income countries (GDP/capita <10,000 USD) claim 166.4 GWp, with a high share represented by China. Countries with a GDP per capita between 10,000 USD and 30,000 USD represent only 20.4 GWp. Especially the two groups low and medium income countries with a GDP/capita below 10,000 USD and between 10,000 and 20,000 USD show high growth rates having increased their cumulative installed capacity from 2016 to 2017 by 65% and 86% respectively [16, 31].

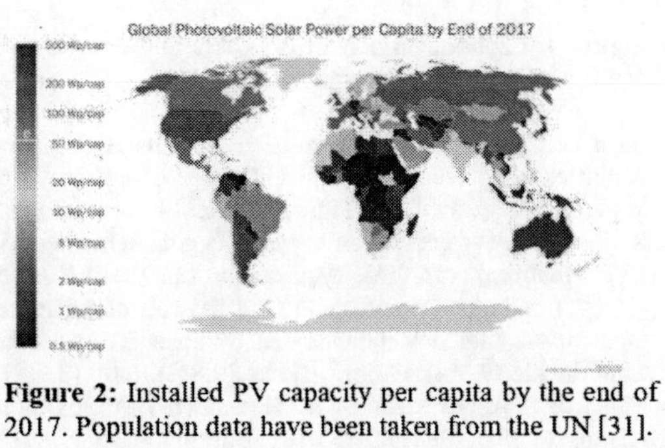

Figure 2: Installed PV capacity per capita by the end of 2017. Population data have been taken from the UN [31].

For three years in a row, the majority of 207 GWp is located in countries with medium irradiation (1,500-2,000 kWh/(m²·a)) [32]. Countries with less than 1,500 kWh/(m²·a), which are mainly highly developed countries, represent 80.2 GWp at low growth rates. Those countries predestined to use PV due to their very high insolation (>2,000 kWh/(m²·a)) claim 29.7 GWp being installed, but show highest growth rates of country groups evaluated.

Figure 2 shows the PV installations per capita and country. Germany (526 Wp/cap.), Luxembourg (511 Wp/cap.), Japan (384 Wp/cap.), Italy (329 Wp/cap.), Belgium (329 Wp/cap.), Australia (304 Wp/cap.), the French overseas department of Reunion (283 Wp/cap.), Malta (238 Wp/cap.), Greece (235 Wp/cap.), Switzerland (229 Wp/cap.), Guam (216 Wp/cap.), the Czech Republic (205 Wp/cap.), the United Kingdom (194 Wp/cap.), the French overseas department of Martinique (173 Wp/cap.), the Netherlands (162 Wp/cap.), Denmark (161 Wp/cap.), the United States (160 Wp/cap.) and the French overseas departments of French Guiana (156 Wp/cap.) and Guadeloupe (155 Wp/cap.) are the ones with more than 150 Wp/capita. Another nine countries show more than 100 Wp/capita (Austria, Israel, Bulgaria, Slovenia, France, Slovakia, Spain, the Republic of Korea, Chile). Especially, highly developed countries and island states show significantly high PV installations per capita.

In total, 15 of the top 50 countries ranking in PV per capita are small islands or states with large parts of their territory located on small islands. This might be due to PV being a clear cost competitive energy source that enables those remote island countries to be decoupled from high prices for conventional electricity generation by diesel power plants [29, 33-37]. As a world average, every human claims 54.6 Wp of PV by end of 2017, an increase of 31%.

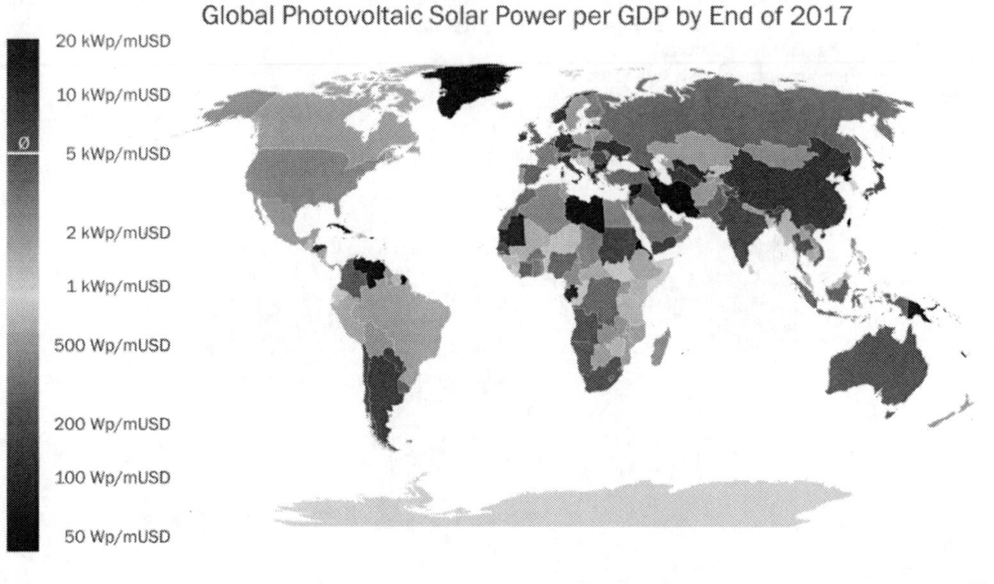

Figure 3: Global overview on cumulative PV installations per million USD of GDP by the end of 2017. Data for GDP have been taken from the World Bank [30].

An indicator of how much of a country's financial capabilities are invested relatively in PV is recognisable in installed PV capacities per countries' GDP (Figure 3). Using this comparative parameter, markets can be categorized depending on their economic possibilities. Some countries have reached significant ratios of PV installations per one million USD of GDP. Honduras (22.7 kWp/mUSD), Bulgaria (19.8 kWp/mUSD), Mauritania (16.3 kWp/mUSD), Kiribati (13.8 kWp/mUSD), Greece (13.6 kWp/mUSD), Germany (12.4 kWp/mUSD), Ukraine (12.3 kWp/mUSD), China (11.7 kWp/mUSD), Jordan (11.7 kWp7mUSD), the Czech Republic (11.3 kWp/mUSD) and Italy (10.6 kWp/mUSD) are those countries with more than 10 kWp installed per million USD of GDP. A large number of smaller countries that often are rarely recognised due to having installed only a few MWp can be distinguished. Thus, countries like Malta (9.3 kWp/mUSD), Yemen (9.2 kWp/mUSD), Cape Verde (9.0 kWp/mUSD), Senegal (8.2 kWp/mUSD), Micronesia (7.1 kWp/mUSD), Namibia (5.3 kWp/mUSD) or Gambia (4.7 kWp/mUSD) invest more of their often limited economic power in PV than the bulk of highly developed countries. In comparison to this, highly developed countries such as Norway (0.117 kWp/mUSD), Ireland (0.117 kWp/mUSD) or Latvia (0.112 kWp/mUSD) still have a vast potential of investing more capital in PV although some of them may not have the need for large PV investments due to already claiming high shares of renewable electricity. Especially, emerging markets, like the African countries or island states, have identified PV as being able to improve significantly people's standards of living and save enormous financial resources in the mid- to long-term. Due to a fast amortisation of PV systems in rural areas of developing countries [29, 33-37], these markets have no need for subsidies and make off-grid PV countries foremost in proportion of PV expenditures to GDP.

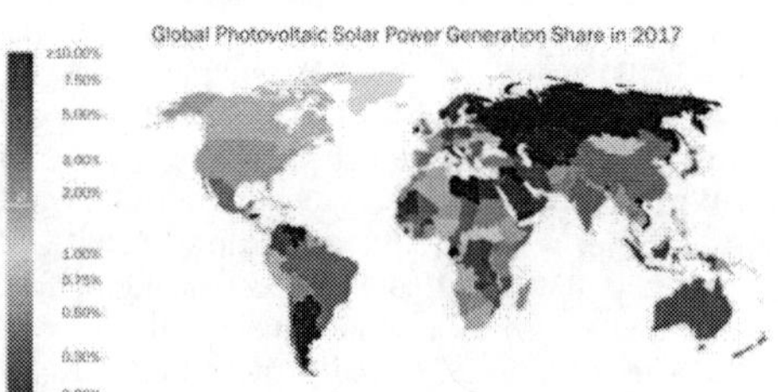

Figure 4: Global overview of PV share in the respective countries total power generation [38].

As a consequence of high ratios of PV to GDP, many countries have reached a relevant share of PV in their total power generation (Figure 4). In 2017 16.6% of Kiribati's power generation capacity is complemented by PV. Honduras (15.0%), Mauritania (13.8%), Yemen (12.0%) and the Comoros (11.0%) as well claim more than 10% of its power generated by solar PV. Further Cape Verde (9.3%), Guinea-Bissau (9.8%), Italy (8.9%), Malta (8.7%), the island of St. Helena (8.6%), Solomon Islands (8.3%), Senegal (8.0%), Burkina Faso (7.8%), Germany (7.8%), Equatorial Guinea (7.0%), Japan

(6.7%), Greece (6.4%), Sierra Leone (6.4%) Djibouti (5.7%) and Jordan (5.2%) saw more than 5% of their power generation provided by PV. In total 99 countries show more than 1% of PV share in their power production with a global average of 2.59% of all global power generated by solar PV. Especially, European, Western African, Asian countries and island states show high ratios of PV in their power generation. While regarding the African and insular countries that represent high ratios of PV in their power generation, it is obvious that they are the ones using a bulk of diesel power plants for generation of electricity. Countries leading in usage of diesel power, due to their remote location usually lead in shares of PV in their total power generation. Thus, in addition to small off-grid appliances, PV is also used as a sustainable alternative for larger solutions as applied in hospitals, schools, supply of commericial appliances and in more and more cases as an upgrade for existing diesel/ oil fired power plants. This is especially prevailing in the environment of high diesel prices and remote areas to be supplied with affordable electricity [29, 33-37].

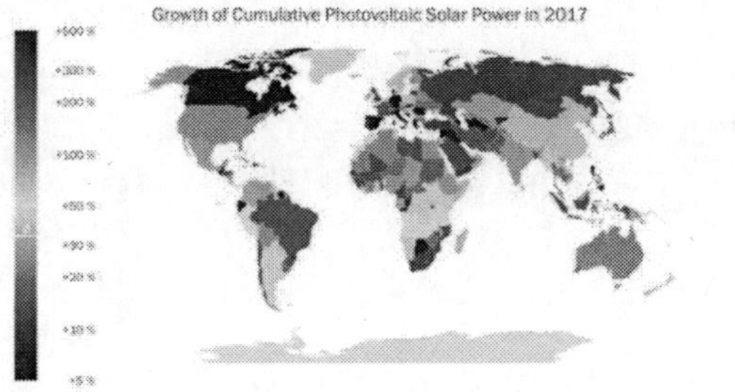

Figure 5: Ratio of added PV capacity in 2017 to cumulative PV installations in 2016, representing growth in PV markets' cumulative capacity.

Figure 5 gives an overview of PV market growth in 2017. Regularly it can be seen that some countries show huge growth rates in a short period of time. In Yemen, for years affected by civil war, the cumulative installed PV capacity has grown by more than 293% from 63.9 MWp to 251.0 MWp, enhanced by low availability of local power grids triggering investments in small scale off-grid PV. Further, other emerging countries have shown high growth rates in 2017. In Turkey, PV capacity by the end of 2017 was 290% higher than the end of 2016, reaching 3,430.0 MWp installed. Mongolia added some 18 MWp in 2017 reaching 25.0 MWp of cumulative capacity, equivalent to a growth rate of 257%. Comparable growth rates can be identified for Burkina Faso (+242%), Brazil (+240%), Afghanistan (+190%), Iran (+186%), El Salvador (+174%), Cambodia (+163%) or Uruguay (+156%). This indicates largely growing PV markets especially in emerging countries all over the world. In Europe, Finland tops PV growth rates with adding 115% last year reaching 80.4 MWp. In addition Belarus (+106%) or Sweden (+73%) have also almost doubled their installed PV. With the Czech Republic (+0.0%), Romania (+0.0%), Greece (+1.0%) or Spain (+1.0%) some of the countries with least growing cumulative PV installations are located in Europe. In contrast, further African markets like Egypt, Mauritania or Burundi showed high growth of their installed capacity per country. Chinese PV installations growth remains at very high growth by adding 53.1 GWp in just one year, equivalent to a growth of 68% and thus has been the first time ever market to surpassed the 50-GWp-threshold of annual new installations. With 9.0 GWp during 2017, new installations in India gained speed, nearly doubling cumulative capacity with a growth of 93%. Apart from Eastern Asia, some South and South-East Asian markets show significant annual installations. The MENA region, apart from high growth rates in Yemen, Egypt, Iran and Oman remains well below its huge solar resource potential. In the Americas, Latin American countries like Cuba, Bolivia, Nicaragua, Colombia and Costa Rica are the ones with high relative growth rates. The Brazilian market finally has taken off during 2017 with annual installations of 784 MWp deployed, whereas Canada has been underperforming with below 100 MWp. The dominant US market added some 10,310 MWp in 2017, a third less than 2016.

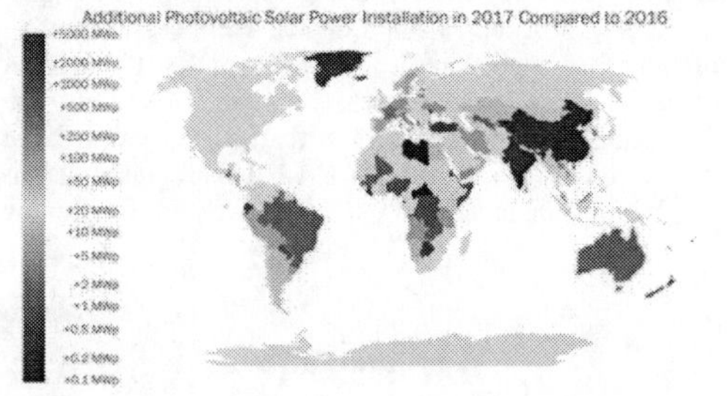

Figure 6: Additional new annual installations in 2017 compared to the annual installations of 2016. Countries marked grey show a decline of annual installs from 2016 to 2017.

Figure 6 shows results of research on the development of annual installations on a year-over-year comparison, trying to give an answer on the question where the additional 20.4 GWp in 2017 compared to 2016 annual installs are based on. With its 53.1 GWp installed in 2017, China installed 19.8 GWp more than the 33.3 GWp installed during 2016. Comparable trends can be identified for 108 other countries that increased their annual installations from 2016 to 2017. India increased the annual installs by 4.9 GWp and thus reached 9.0 GWp of newly installed PV during 2017. Turkey that installed 1,396 MWp during 2016 significantly increased its installations by 2,034 MWp to 3,430 MWp in 2017. Brazil (+632 MWp 2016/17 YoY), Australia (+460 MWp 2016/17 YoY) and France (+350 MWp 2016/17 YoY) show comparable developments. With these countries adding 31.2 GWp (11.4 GWp without China) more during 2017 that they added in 2016, they compensated the loss of 10.8 GWp of annual installations in 86 other countries (marked grey in Figure 6), among other the US (-4,840 MWp 2016/17 YoY), Japan (-2,020 MWp 2016/17 YoY) or the United Kingdom (-1,050 MWp 2016/17 YoY).

The development of PV in emerging markets can be more different compared to most developed PV markets. PV often starts to progress in several emerging countries through tenders for large-scale PV systems, rather than rooftop PV. This development will be even stronger in places where neither the population, nor companies will have the ability to finance PV systems for self-consumption. This trend is already visible since 2014 with the share of large-scale PV increasing in all regions except in Europe. In parallel, small off-grid applications are starting to develop for the electrification of households with very low electricity consumption, as for instance Bangladesh. This development implies the possible cohabitation in emerging countries between Pico PV and Solar Home Systems and utility-scale plants, basically leaving the small and medium rooftop market aside.

Figure 7 shows a segmentation of PV markets. Furthermore, an attractiveness of PV markets is represented. A portfolio analysis chart is applied to PV markets with the use of a logarithmic order. 'Cash cows' are mature markets that claim significant installed PV capacities (more than 0.1% of global installed PV at the end of 2017) with relatively low growth rates, but still remaining the most important driver of the global PV industry. Germany, Italy, Belgium, France, Portugal, Denmark, Slovakia, Japan, Canada, Honduras, Israel or Australia are typical representatives of this PV market segment. The United States, Mexico, Chile and the Philippines are new members of that group, having been categorized as 'rising stars' in 2016. This segment claims 31% of total installed PV capacity. Even with very low growing markets, like Germany or Italy, standing for annual installation of a few 100 MWp due to having an installed PV capacity base of several GWp, this is a major PV market segment. In 2017 'cash cow' countries added 15.9 GWp.

'Rising stars' are probably the most interesting market segment, as respective installed PV capacities (>0.1% of global PV in 2017) and growth rates are above the world's long term average (>46%) and led to significant installations in the relevant year. Brazil, China, India, Jordan, Morocco, Turkey and the United Arab Emirates are those that represent this market segment in 2017. Therefore, this segment stands for 65% of total global PV installations. An interesting question is whether these markets will stay in this segment. In the short- to mid-term, 'rising stars' will become 'cash cows', as such high growth rates will not be able to keep up with increasing cumulative PV capacities in the years to come. It is remarkable that China managed to increase the annual PV installations in 2016 and 2017 at a such high scale, which resulted in moving from 'cash cows' to 'rising stars' again.

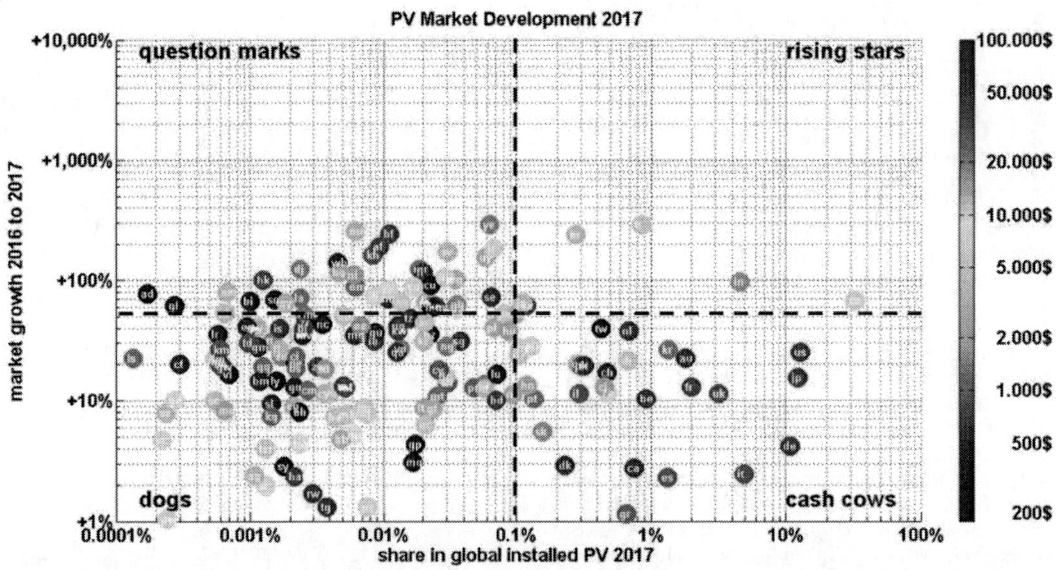

Figure 7: Cumulative installed PV by end of 2017 (x-axis) and PV cumulative capacities growth rate in 2017 (y-axis). Every bubble represents one country. The color code of the bubbles refers to the GDP/capita (left axis). The code inside of the bubble represents the internet top-level domain (ISO 3166-1 alpha 2) of the relevant country.

Markets that show high growth rates, but relatively low installed PV can be named as 'question marks'. Among 46 others, this segment includes markets such as Yemen, Iran, Uruguay, El Salvador, Sweden, Peru, Egypt, Belarus, Burkina Faso, Mauritania, Finland, Ghana, Panama and Cuba that show relatively high growth rates, but because of their low installed cumulative PV capacity, they are so far marginal in the global PV perspective. If these markets keep their high growth rates for several years, becoming a 'rising star' can be a valid opportunity. Most interesting in this segment are markets close to the threshold of becoming a 'rising star'. Yemen, Iran, Uruguay, El Salvador, Sweden, Peru and Egypt are among those that are expected to become new 'rising stars' in the years to come.

At first view 'dogs' seem to be markets of less interest since only 2.0% of total PV is installed in those countries. But often claim a high share of off-grid PV applications, such as Bangladesh. Further markets in this category are among other Russia, Hungary, Slovenia, Luxembourg, Poland, Indonesia or Saudi Arabia.

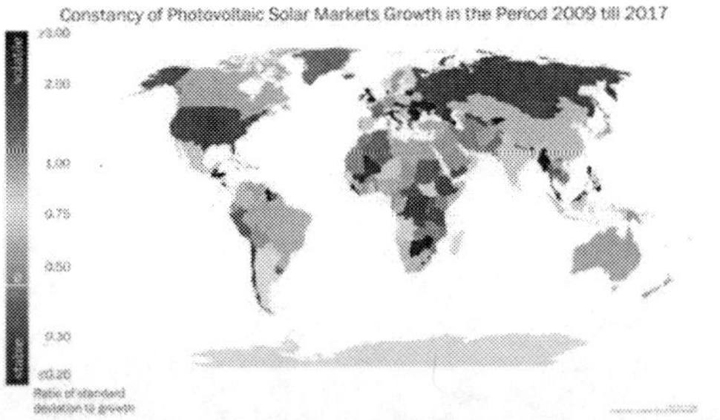

Figure 8: Ratio of standard deviation to average growth of PV capacity in the period from 2009 to 2017. The blue color refers to very unstable PV markets, while red refers to relatively constant growing photovoltaic capacities.

High growth rates of 'rising star' or 'question mark' countries are not always an indicator for a sustainable and constant growth. Deviations in PV markets' annual growth rates are often significant, due to markets often being driven by politically motivated programs, like feed-in tariffs or other policies. Thus, average annual growth rates have to be regarded in the context of annual growth rate standard deviation. Thus, sustainable growth can be defined as significant but constant growth. Very constant growing PV markets can be identified by the ratio of standard deviation of the annual growth rates to average growth rates.

As can be seen in Figure 8, such significant but constant growth in recent years can be identified especially in the Democratic Republic of Congo, the United States, Iceland, Tanzania, Sudan, Ireland, Zambia, the Republic of Korea, Kenya, Ivory Coast or New Zealand and a variety of further emerging markets. All of them show a ratio of standard deviation to average growth of less than 0.50 (refer to Figure 8). Sweden represents an average annual growth of its cumulative PV of 51% in the period from 2009 to 2017. In 2010 Tanzania's PV capacity grew by 81%, in 2011 by 86%, in 2012 by 78%, in 2013 by 50%, in 2014 by 33%, in 2015 by 51%, in 2016 by 51% and in 2017 by 48% to 64.3 MWp at the end of the year. This is an example of a very constant growth in PV and can definitely be regarded as sustainable. If this constant and high growth continues, Tanzania could reach 90 MWp by the end of 2018. Another example is the United States, the PV capacity grew by 53% in 2010, 65% in 2011, 78% in 2012, 65% in 2013, 51% in 2014, 41% in 2015, 58% in 2016 and 25% in 2017, which leads to an average annual growth rate of 54% with a relatively low standard deviation of only 15%. If this constant growth goes on steadily, more than 70 GWp might be cumulatively installed by the end of 2018 in the US. Markets such as Denmark, Slovakia, Ukraine, Czech Republic or Romania have shown high annual growth rates in specific years, but high standard deviations, due to variable growth with boom years being followed by years of very little installations [12-16]. Thus, they show ratios of standard deviation to average growth of more than two. By following this approach, the most volatile markets are among others Cape Verde (ratio of 20.2), Kyrgyzstan (ratio of 15.9), Belize, Botswana, Slovakia, Denmark and Bhutan showing ratios of more than five.

With the use of this evaluation of average annual growth, the constancy of growth rates, the help of publications and the knowledge of seven years of analysis of cumulative PV capacities [11-16], a rough forecast of cumulative PV capacities for the years 2018 and 2019 is possible. In 2018, the global cumulative installed PV capacity is likely to increase to 505 GWp by the end of 2018 (new installations of about 103 GWp in 2018) and by the end of 2019, 630 GWp (new installations of about 125 GWp in 2019) of cumulative installed PV capacity can be expected.

4 CONCLUSIONS

In total, 402.15 GWp can be allocated to 196 countries by end of 2017. Out of those, 96.38 GWp have been installed in 2017. Even, newly installed capacity grew by 26.7% during 2017, the global trade volume of PV modules slightly decreased with 30.0 bnUSD in 2017, after 31.9 bnUSD in 2016.

The most significant amounts of PV are still installed in mature markets in Europe, East Asia and North America. However, a relevant paradigm shift to developing and emerging markets as the new growth regions can be seen. Some low and middle income countries already show remarkable numbers of newly installed PV.

In the years to come, new growth markets in Central and South America (Uruguay, Colombia, El Salvador, Peru, Panama), Asia-Pacific (Mongolia, Myanmar, Sri Lanka), MENA (Iran, Jordan, Oman, Morocco, Egypt) and Sub-Saharan Africa (Mauritania, Nigeria, Senegal, Ghana) will increasingly play a major role within the next years of global PV markets development. In Europe countries will expand their growth perspectives on the short to mid-term again. Enhanced by fast development of competitiveness of photovoltaic solar power, even orthern markets like Sweden or Finland will see increasing numbers of PV installations.

One is able to assess that every world region might represent a great PV market potential. Enormous growth rates in specific regions of the world show that PV is on the way to becoming an important part of global electricity supply. Some developing countries are on the road to solar electrification, others use PV as an opportunity to decrease suffering from lacking access to electricity in remote locations. Lower energy costs enabled by reduced dependence on energy imports will contribute to leading these countries to sustainable development and stable growth.

REFERENCES

[1] [IRENA] – International Renewable Energy Agency, 2018. Renewable Capacity Statistics 2018, Dubai

[2] International Trade Centre Market Analysis Tool, 2018. Product Map, Market Analysis and Research Section, International Trade Centre, Geneva, www.trademap.org

[3] Photon International, 2012. Zellproduktion 2011, Photon Magazin, 2012(4), [plus respective data for the years 2000-2011 either of Photon Magazin or Photon International]

[4] Breyer Ch., Birkner C., Meiss J., Goldschmidt J.C., Riede M., 2013. A top-down analysis: determining photovoltaics R&D investments from patent analysis and R&D headcount, Energy Policy, 62, 1570-1580

[5] Photon, 2016. Factory Gate c-Si Module Prices, Photon Consulting, Aachen

[6] Hanwha Q CELLS Market Intelligence, 2011-2016. Intern Price Snapshot, Hanwha Q CELLS, Bitterfeld-Wolfen

[7] Gerlach A., 2018. private communication, Insights on PV module and cell prices

[8] Werner Ch., Gerlach A., Adelmann P., Breyer Ch., 2011. Global Overview on Cumulative Installed Photovoltaic Power, Proceedings 26th European Photovoltaic Solar Energy Conference, Hamburg

[9] Szewzcyk W., 2018. private communication, PV insights for Poland

[10] Horn S., 2018. private communication, PV insights for Malaysia and South East Asia

[11] Werner Ch., Gerlach A., Adelmann P., Breyer Ch., 2011. Global Overview on Cumulative Installed Photovoltaic Power, 2nd Symposium Small PV-Applications, Ulm

[12] Werner Ch., Gerlach A., Adelmann P., Breyer Ch., 2013. Analysis and Forecast of Global Installed Photovoltaic Capacity and Identification of Hidden Growth Markets, Proceedings 28th European Photovoltaic Solar Energy Conference, Paris

[13] Werner Ch., Gerlach A., Breyer Ch., Masson G., 2014. Global Installed Photovoltaic Capacity and Identification of Hidden Growth Markets, Proceedings 29th European Photovoltaic Solar Energy Conference, Amsterdam

[14] Werner Ch., Gerlach A., Breyer Ch., Orlandi S., Masson G., 2015. Latest Developments in Global Installed Photovoltaic Capacity and Identification of Hidden Growth Markets, Proceedings 31st European Photovoltaic Solar Energy Conference, Hamburg

[15] Werner Ch., Gerlach A., Breyer Ch., Orlandi S., Masson G., 2016. Global Photovoltaics in 2016 – Analysis of Current and Future Solar Energy Markets, Proceedings 32nd European Photovoltaic Solar Energy Conference, Munich

[16] Werner Ch., Gerlach A., Breyer Ch., Orlandi S., Masson G., 2017. Growth Regions in Photovoltaics 2017 – Update on Latest Solar Market Development, Proceedings 33rd European Photovoltaic Solar Energy Conference, Amsterdam

[17] [PVMA] – PV Market Alliance, 2017. The PVMA Global PV Market Report, Brussels

[18] [IEA PVPS] - International Energy Agency Photovoltaic Power Systems Programme, 2018. Snapshot of Global PV Markets 2017, Paris

[19] Journal des Énergies Renouvelables, 2018, Photovoltaic Barometer, EurObserv'er, 2018 (April), Paris

[20] [REN21], 2018. Renewables 2017 – Global Status Report, REN21, Paris

[21] Solar Power Europe, 2018. Global Market Outlook – For Solar Power 2018-2022, Brussels

[22] Bellini E., 2018. Chile's PV capacity surpasses 2 GW, PV Magazine, Berlin, www.pv-magazine.com/2018/02/05/chiles-pv-capacity-surpasses-2-gw//

[229] Maisch M., 2018. Balkan update: Croatia, Bosnia and Herzegovina ready for their biggest PV projects, Macedonia inaugurates first module fab, PV Magazine, Berlin, www.pv-magazine.com/2018/07/05/balkan-update-croatia-bosnia-and-herzegovina-ready-for-their-biggest-pv-projects-macedonia-inaugurates-first-module-fab/

[24] Bellini E., 2018. Cuba's PV capacity reaches 65 MW, PV Magazine, Berlin, www.pv-magazine.com/2018/01/05/cubas-pv-capacity-reaches-65-mw/

[25] [BNetzA], 2018. Installierte Solarleistung EEG-Registerdaten und -FördersätzeG, Bundesnetzagentur, Bonn

[26] Lazaridis V., 2018. private communication, PV insights for Greece

[27] Breyer Ch., 2018. private communication, PV insights for Finland

[28] Bellini E., 2017. Uruguay's installed PV capacity tops 88.4 MW, PV Magazine, Berlin, https://www.pv-magazine.com/2017/06/23/uruguays-installed-pv-capacity-tops-88-4-mw/

[29] Breyer Ch., Werner Ch., Rolland S., Adelmann P., 2011. Off-Grid Photovoltaic Applications in Regions of Low Electrification: High Demand, Fast Financial Amortization and Large Market Potential, Proceedings 26th European Photovoltaic Solar Energy Conference, Hamburg

[30] World Bank, 2017. World Development Indicators, Gross Domestic Product 1960-2016, World Bank, New York

[31] [UN] – United Nations, 2017. World Population Prospects: The 2017 Revision, Department of Economic and Social Affairs, United Nations, New York

[32] Breyer Ch. and Schmid J., 2010. Population Density and Area weighted Solar Irradiation: Global Overview on Solar Resource Conditions for Fixed Tilted, 1-Axis and 2-Axis PV Systems, 25th EU PVSEC

[33] Breyer Ch., Gerlach A., Schäfer D., Schmid J., Fuel-Parity: New Very Large and Sustainable Market Segments for PV Systems, IEEE EnergyCon, Manama, 2010, December 18–22

[34] Blechinger P., Cader C., Bertheau P., Huyskens H., Seguin R., Breyer Ch., 2016. Global Analysis of the Techno-Economic Potential of Renewable Energy Hybrid Systems on Small Islands, Energy Policy, 98, 674-687

[35] Cader C., Bertheau P., Blechinger P., Huyskens H., Breyer Ch., 2015. Global cost advantages of autonomous solar-battery-diesel systems compared to diesel-only systems, Energy for Sustainable Development, 31, 14-23

[36] Werner Ch. and Breyer Ch., 2012. Analysis of Mini-grid Installations: An Overview on System Configurations, Proceedings 27th European Photovoltaic Solar Energy Conference, Frankfurt, DOI: 10.4229/27thEUPVSEC2012-5DO.9.6

[37] Bertheau P., Cader C., Huyskens H., Blechinger P., 2015. The Influence of Diesel Fuel Subsidies and Taxes on the Potential for Solar-Powered Hybrid Systems in Africa, Resources, 4, 673-691

[38] [CIA], 2016. The World Factbook. Central Intelligence Agency, Washington

Appendix Table 1: Alphabetical list of all countries in the world and their cumulative installed PV capacity at the end of 2017. Data are compiled as documented in the paper and based on IRENA, IEA PVPS and further publications plus ITC data [8, 11-31].

	PV installation by end of 2017	PV installation by end of 2016 [16]	PV installation by end of 2015 [15]	PV installation by end of 2014 [14]	PV installation by end of 2013 [13]	PV installation by end of 2012 [12]	PV installations per capita by end of 2017	PV installations per GDP by end of 2017	PV share in power generation 2017
	[MWp]	[MWp]	[MWp]	[MWp]	[MWp]	[MWp]	[Wp/cap]	[Wp/mUSD]	[%]
Afghanistan	37.40	12.90	7.30	5.30	4.60	3.10	1.11	1921.0	1.66%
Albania	2.60	1.70	1.10	0.99	0.30	0.14	0.89	218.0	0.05%
Algeria	419.00	339.00	285.00	78.00	17.60	17.60	10.51	2684.5	1.56%
Andorra	0.69	0.39	0.14	0.12	0.07	0.05	8.84		0.00%
Angola	20.60	13.70	13.00	9.80	9.50	6.70	0.74	229.8	0.71%
Antigua and Barbuda	4.30	4.20	3.60	0.51	0.29	0.17	43.03	2967.2	0.00%
Argentina	82.50	57.00	45.00	35.20	32.70	29.90	1.90	151.1	0.11%
Armenia	2.70	1.50	0.76	0.37	0.26	0.20	0.93	256.0	0.08%
Australia	7240.00	5900.00	5020.00	4080.00	3200.00	2380.00	304.21	6010.2	4.98%
Austria	1290.00	1080.00	940.00	794.00	639.00	420.00	148.64	3338.3	2.08%
Azerbaijan	30.80	30.40	10.10	1.10	0.52	0.24	3.20	813.8	0.23%
Bahamas	8.90	7.50	2.26	1.59	1.33	1.16	23.01	983.8	0.91%
Bahrain	9.50	8.80	8.70	6.80	5.70	4.90	6.92	298.2	0.00%
Bangladesh	284.00	258.00	198.00	172.00	140.00	83.60	1.76	1282.7	1.04%
Barbados	19.70	18.80	9.20	6.00	3.90	2.20	69.31	4294.2	0.00%
Belarus	117.00	56.90	20.80	13.10	7.60	6.60	12.33	2466.6	0.31%
Belgium	3720.00	3370.00	3220.00	3120.00	2990.00	2620.00	329.56	7976.6	4.37%
Belize	5.20	5.20	3.10	1.30	1.10	0.97	14.47	2945.7	1.24%
Benin	10.20	7.30	4.90	4.40	1.60	1.10	0.96	1188.4	1.70%
Bermuda	4.70	4.10	3.30	2.40	1.40	0.97	75.80		0.00%
Bhutan	0.25	0.25	0.24	0.22	0.20	0.19	0.32	111.8	0.02%
Bolivia	18.90	8.80	6.60	5.60	4.50	4.10	1.76	559.1	0.48%
Bosnia-Herzegovina	17.60	16.40	10.10	5.70	4.00	1.90	4.98	1062.8	0.17%
Botswana	9.50	9.10	9.10	5.70	4.90	4.40	4.30	621.9	0.54%
Brazil	1110.00	326.00	174.00	154.00	109.00	78.90	5.39	618.0	0.35%
Brunei	2.40	2.00	1.80	1.30	0.95	0.14	5.75	210.5	0.11%
Bulgaria	1040.00	1040.00	1030.00	1020.00	1000.00	920.00	144.90	19849.2	4.69%
Burkina Faso	44.80	13.10	8.40	6.90	5.80	4.90	2.47	3697.8	7.87%
Burundi	4.00	2.40	1.70	1.40	1.20	0.97	0.39	1330.2	2.04%
Cambodia	33.70	12.80	10.30	9.30	7.70	5.60	2.17	1683.6	1.47%
Cameroon	11.00	9.80	6.90	5.40	3.30	2.50	0.48	454.5	0.30%
Canada	3010.00	2930.00	2530.00	1910.00	1320.00	877.00	83.73	1967.6	0.71%
Cape Verde	14.60	13.10	10.50	10.20	9.90	9.20	27.40	9026.5	9.34%
Centr. Afr. Republic	1.20	1.00	0.92	0.34	0.28	0.25	0.26	683.3	1.16%
Chad	3.90	3.00	2.10	1.40	1.00	1.00	0.28	406.2	3.64%
Chile	1830.00	1620.00	847.00	404.00	134.00	13.00	103.02	7408.1	4.91%
China	130780.00	77690.00	44420.00	28470.00	18310.00	6890.00	93.61	11677.7	3.09%
Colombia	68.80	36.70	26.20	20.00	17.60	12.10	1.43	243.6	0.19%
Comoros	2.40	1.90	1.10	0.50	0.25	0.19	3.09	3892.0	10.98%

Congo Dem. Republic	10.70	7.40	5.10	3.80	3.10	1.90	0.14	305.7	0.22%
Congo Republic	8.50	7.80	6.10	4.10	3.60	2.30	1.70	1085.1	1.51%
Costa Rica	44.50	24.20	15.60	11.50	8.50	6.10	9.26	774.8	0.69%
Croatia	64.80	64.80	47.00	31.80	23.60	12.30	15.30	1285.1	0.48%
Cuba	90.90	47.60	22.80	9.40	8.00	3.20	7.93		0.89%
Cyprus	105.00	89.00	79.10	69.40	34.00	23.70	90.44	5302.6	4.39%
Czech Republic	2180.00	2180.00	2170.00	2170.00	2140.00	2060.00	205.59	11299.8	3.60%
Denmark	920.00	894.00	801.00	654.00	573.00	376.00	161.72	3005.1	2.96%
Djibouti	9.60	4.30	3.40	1.80	1.80	1.80	10.35		5.71%
Dominica	2.20	1.80	1.10	0.81	0.13	0.08	30.07	4187.1	4.68%
Dominican Republic	118.00	102.00	75.30	31.00	21.50	11.50	11.21	1648.4	1.58%
Ecuador	84.20	79.20	74.60	34.00	14.50	10.20	5.22	860.9	0.59%
Egypt	140.00	69.60	40.00	26.60	23.20	11.00	1.49	416.3	0.19%
El Salvador	123.00	44.90	6.60	4.40	3.10	1.10	19.49	4590.0	3.83%
Equatorial Guinea	5.00	4.20	3.90	2.70	2.40	2.10	4.25	491.2	7.02%
Eritrea	3.80	2.70	2.40	1.60	1.40	0.63	0.78		2.37%
Estonia	27.20	19.20	14.90	12.80	8.50	4.80	20.68	1175.6	0.34%
Ethiopia	46.60	46.60	17.70	15.00	13.40	12.30	0.47	643.9	1.57%
Fed. St. o. Micronesia	2.30	1.90	1.90	1.40	0.51	0.15	22.02	7142.9	2.08%
Fiji	6.70	5.30	4.60	3.20	2.20	1.30	7.51	1446.6	1.36%
Finland	80.40	37.40	19.00	17.00	12.60	9.34	14.67	339.5	0.09%
France	8050.00	7120.00	6540.00	5620.00	4690.00	3990.00	124.89	3265.1	2.06%
French Guiana	42.00	42.00	34.00	34.00	34.00	0.00	156.31		
Gabon	1.20	1.20	0.95	0.67	0.47	0.40	0.62	84.4	0.10%
Gambia	4.60	3.60	2.80	2.00	1.40	0.94	2.33	4768.8	3.58%
Georgia	0.71	0.71	0.36	0.36	0.36	0.36	0.18	49.5	0.01%
Germany	43010.00	41270.00	39700.00	38250.00	35680.00	32360.00	526.39	12406.4	7.78%
Ghana	144.00	88.80	73.20	23.80	15.70	5.00	5.22	3373.2	2.02%
Greece	2640.00	2610.00	2610.00	2590.00	2570.00	1530.00	235.34	13569.2	6.41%
Greenland	1.10	0.68	0.56	0.48	0.21	0.06	19.51		0.00%
Grenada	1.10	1.00	0.87	0.35	0.24	0.10	10.30	1082.5	1.15%
Guadeloupe	70.00	67.10	67.10	0.00	0.00	0.00	155.41		
Guam	35.00	25.60	25.60	0.13	0.13	0.13	216.32		3.96%
Guatemala	95.90	88.30	82.70	31.40	3.60	2.00	5.90	1394.6	1.88%
Guinea	8.70	7.70	7.70	5.10	3.20	2.20	0.72	1381.2	1.58%
Guinea-Bissau	2.50	2.10	1.70	1.20	0.99	0.82	1.41	2220.0	8.99%
Guyana	7.40	4.50	4.50	4.30	4.00	2.90	9.63	2147.3	1.89%
Haiti	8.70	8.50	8.30	5.80	4.10	2.50	0.81	1084.4	3.30%
Honduras	489.00	431.00	409.00	73.60	5.00	2.80	54.57	22726.3	15.01%
Hong Kong	5.00	2.50	2.50	2.50	2.50	0.90	0.69	15.6	0.01%
Hungary	354.00	257.00	88.90	20.40	14.40	3.90	36.18	2847.0	1.11%
Iceland	6.70	4.80	3.90	2.50	1.50	0.90	20.29	334.2	0.00%
India	18350.00	9310.00	5180.00	3270.00	2610.00	1460.00	14.02	8106.8	3.45%
Indonesia	233.00	207.00	145.00	83.00	46.70	34.70	0.90	249.9	0.20%
Iran	272.00	95.00	34.96	9.14	1.43	1.43	3.43		0.23%
Iraq	30.50	28.30	10.40	7.20	4.50	4.00	0.84	177.9	0.09%

Ireland	34.50	26.30	9.70	4.00	2.00	1.00	7.34	117.3	0.12%
Israel	1170.00	1050.00	895.00	690.00	455.00	243.00	145.08	3670.7	3.51%
Italy	19610.00	19140.00	18910.00	18320.00	17780.00	16300.00	329.56	10600.2	8.90%
Ivory Coast	9.00	7.30	4.20	2.80	2.30	1.60	0.39	248.9	0.28%
Jamaica	28.80	26.50	11.00	6.60	2.70	0.43	10.03	2053.2	1.63%
Japan	49210.00	42550.00	33870.00	23420.00	13500.00	7230.00	384.53	9962.8	6.74%
Jordan	451.00	274.00	117.00	31.70	4.80	1.10	49.24	11667.4	5.21%
Kazakhstan	81.50	62.10	54.80	25.20	12.80	4.70	4.59	609.8	0.14%
Kenya	99.00	61.70	54.90	31.60	25.40	17.80	2.10	1403.7	2.54%
Kiribati	2.30	2.30	1.70	0.94	0.51	0.34	20.46	13875.1	16.58%
Korea DPR	14.70	14.70	12.40	7.70	5.30	3.30	0.58		0.14%
Korea Rep.	5410.00	4270.00	3360.00	1900.00	1350.00	974.00	106.93	3833.5	1.59%
Kuwait	51.70	37.50	35.50	15.70	9.30	6.00	13.14		0.18%
Kyrgyzstan	5.80	5.40	5.10	5.00	4.10	1.20	0.99	885.3	0.09%
Laos	9.60	5.60	5.60	3.00	2.70	1.80	1.44	603.6	0.49%
Latvia	3.10	3.10	2.40	2.30	1.50	1.20	1.56	112.0	0.05%
Lebanon	34.60	20.00	6.80	4.10	2.10	0.74	5.91	727.9	0.46%
Lesotho	0.54	0.44	0.30	0.29	0.26	0.18	0.25	245.5	0.00%
Liberia	2.30	1.70	1.30	0.92	0.87	0.67	0.51	1094.7	1.18%
Libya	6.30	5.50	4.80	4.20	3.90	2.80	1.01		0.04%
Lithuania	82.10	75.50	72.10	60.70	37.00	8.80	28.00	1921.0	0.87%
Luxembourg	290.00	249.00	243.00	233.00	104.00	50.90	511.70	4837.5	0.00%
Macedonia	21.30	19.80	18.10	15.70	11.20	6.50	10.24	1954.2	0.42%
Madagascar	24.60	18.20	11.20	5.90	5.70	3.50	1.02	2462.3	2.19%
Malawi	10.10	7.50	1.80	1.40	1.30	1.00	0.57	1856.0	0.83%
Malaysia	428.00	336.00	250.00	229.00	73.70	61.90	13.93	1444.2	0.51%
Maldives	9.20	6.80	4.40	4.20	2.30	1.80	21.99	2561.9	0.00%
Mali	21.20	18.80	17.60	16.20	15.50	13.80	1.21	1509.4	4.20%
Malta	102.00	92.00	55.60	32.80	21.60	14.60	238.53	9315.8	8.66%
Martinique	67.00	65.00	62.50	0.00	0.00	0.00	173.65		
Mauritania	75.60	33.70	18.70	15.60	12.60	7.80	18.08	16312.1	13.84%
Mauritius	24.40	23.20	15.50	12.10	11.00	4.10	19.37	2005.9	1.65%
Mexico	526.00	410.00	254.00	187.00	125.00	51.70	4.18	502.9	0.38%
Moldova	2.60	2.40	2.30	1.20	0.91	0.63	0.64	385.2	0.06%
Mongolia	25.00	7.00	5.00	4.50	3.50	3.50	8.40	2240.1	0.91%
Morocco	403.00	265.00	67.10	57.40	36.90	23.70	11.58	3972.6	2.60%
Mozambique	20.60	18.20	11.90	11.80	9.60	3.50	0.74	1870.2	0.30%
Myanmar	92.40	57.90	53.90	4.00	2.00	1.30	1.76	1370.3	1.85%
Namibia	55.20	33.40	23.20	15.90	8.50	3.20	22.76	5376.4	2.45%
Nepal	37.70	37.70	20.40	6.70	6.20	3.80	1.32	1783.0	2.03%
Netherlands	2760.00	2000.00	1440.00	1050.00	695.00	262.00	162.94	3580.5	2.35%
New Caledonia	14.20	9.90	6.80	4.30	3.80	3.40	52.77		0.89%
New Zealand	90.10	66.60	43.70	30.10	23.50	14.50	19.53	487.0	0.29%
Nicaragua	24.30	11.60	10.90	7.50	4.70	3.60	4.00	1836.6	1.10%
Niger	8.80	7.30	4.60	3.30	2.00	1.80	0.44	1171.9	1.80%
Nigeria	120.00	93.60	68.10	45.60	37.80	28.70	0.66	296.2	0.77%

Norfolk Island	1.40	1.40	0.24	0.24	0.24	0.24			
Norway	43.50	25.60	16.10	13.00	11.10	8.40	8.37	117.4	0.03%
Oman	25.00	13.30	12.80	5.90	4.70	2.20	5.95	377.1	0.22%
Pakistan	1220.00	1020.00	780.00	121.00	45.50	18.20	6.44	4300.9	2.59%
Panama	143.00	90.60	48.30	19.90	8.90	6.00	36.03	2591.2	2.77%
Papua New Guinea	11.40	11.40	9.10	6.40	4.20	3.00	1.44		0.53%
Paraguay	4.50	3.20	2.70	1.30	0.74	0.37	0.68	164.0	0.08%
Peru	327.00	215.00	183.00	163.00	100.00	49.20	10.42	1702.3	1.47%
Philippines	903.00	898.00	215.00	109.00	21.00	16.20	8.88	2961.6	2.17%
Poland	270.00	192.00	81.60	27.60	7.90	5.80	7.06	575.1	0.19%
Portugal	543.00	492.00	461.00	405.00	284.00	231.00	52.12	2654.4	1.78%
Puerto Rico	193.00	171.00	113.00	96.00	96.00	1.90	52.54		1.78%
Qatar	51.30	41.00	41.00	30.80	13.10	5.00	20.67	336.5	0.28%
Republic of Palau	0.97	0.96	0.90	0.32	0.19	0.14	45.57	3310.6	
Reunion	245.00	245.00	180.00	0.00	0.00	0.00	283.77		
Romania	1410.00	1400.00	1350.00	1230.00	1050.00	69.60	70.94	7552.6	3.40%
Russia	378.00	340.00	268.00	83.00	13.00	10.00	2.63	294.6	0.04%
Rwanda	12.00	11.80	11.80	11.20	2.60	2.10	1.03	1432.7	4.81%
Sao Tome a. Principe	0.37	0.35	0.18	0.14	0.08	0.02	1.89	1054.0	0.00%
Saudi Arabia	233.00	205.00	137.00	81.60	55.20	33.40	7.38	360.4	0.18%
Senegal	122.00	107.00	38.70	27.60	19.10	15.60	8.15	8262.5	8.02%
Serbia	16.30	14.60	11.00	10.40	5.70	3.50	1.84	431.8	0.08%
Seychelles	2.20	2.00	1.20	0.85	0.53	0.16	23.47	1541.3	1.30%
Sierra Leone	5.80	5.30	4.80	4.40	4.00	2.80	0.80	1580.9	6.40%
Singapore	151.00	115.00	68.40	37.00	14.30	6.40	27.28	508.5	0.00%
Slovakia	628.00	595.00	589.00	576.00	559.00	531.00	115.46	7012.7	2.25%
Slovenia	300.00	300.00	285.00	283.00	270.00	208.00	144.59	6819.6	2.73%
Solomon Island	4.20	3.00	3.00	2.40	2.20	1.30	7.15	3493.8	8.30%
Somalia	6.10	3.60	2.10	1.10	0.32	0.16	0.44	981.2	3.50%
South Africa	1930.00	1730.00	1260.00	915.00	252.00	118.00	34.91	6545.9	1.58%
Spain	5330.00	5210.00	5180.00	5180.00	5020.00	4790.00	114.88	4326.0	3.31%
Sri Lanka	86.50	51.40	23.00	18.00	15.30	10.10	4.18	1063.7	1.23%
St. Helena	0.37	0.35	0.35	0.02	0.01	0.01	91.72		8.55%
St. Lucia	1.00	1.00	0.80	0.33	0.07	0.07	5.64	725.4	0.00%
St. Vintcent a. Grenad.	0.89	0.85	0.65	0.13	0.06	0.06	8.13	1154.6	1.27%
Sudan	14.80	12.50	9.50	6.80	5.80	4.20	0.38	154.8	0.47%
Suriname	5.20	5.10	4.60	3.30	1.10	0.79	9.40	1436.2	0.50%
Swaziland	0.96	0.89	0.66	0.60	0.54	0.50	0.73	257.6	0.12%
Sweden	259.00	150.00	127.00	71.50	40.40	21.90	26.53	506.8	0.19%
Switzerland	1910.00	1630.00	1360.00	1050.00	748.00	414.00	229.57	2894.7	3.86%
Syria	7.20	7.00	4.60	2.00	1.70	1.50	0.38		0.05%
Taiwan	1730.00	1240.00	975.00	590.00	342.00	207.00	73.66		1.01%
Tajikistan	3.60	3.60	3.40	2.80	2.50	1.70	0.42	517.9	0.04%
Tanzania	64.30	43.40	28.80	19.10	14.40	9.60	1.19	1355.7	2.31%
Thailand	2730.00	2250.00	1690.00	1400.00	926.00	439.00	39.76	6710.3	2.67%
Togo	15.40	15.20	15.20	7.50	3.30	2.00	2.08	3500.0	2.44%

Trinidad and Tobago	5.90	4.50	4.50	3.90	3.30	1.80	4.34	281.1	0.12%
Tunisia	84.00	51.10	41.30	28.30	25.00	15.90	7.45	1997.0	0.97%
Turkey	3430.00	880.00	364.00	70.70	49.40	33.40	43.82	3998.8	2.56%
Turkmenistan	5.20	5.00	4.60	2.60	1.70	0.85	0.93	143.7	0.07%
Uganda	52.20	36.90	20.90	13.20	9.40	6.90	1.30	2044.8	2.93%
Ukraine	1150.00	957.00	886.00	855.00	692.00	416.00	25.75	12329.7	0.80%
United Arab Emirates	481.00	298.00	233.00	182.00	118.00	118.00	52.54	1379.2	0.93%
United Kingdom	12730.00	11430.00	9080.00	5260.00	3370.00	1810.00	194.66	4860.8	3.60%
United States	51450.00	41140.00	25990.00	18370.00	12200.00	7400.00	160.82	2770.7	1.93%
Uruguay	235.00	91.80	66.40	22.40	6.00	4.50	68.48	4483.0	3.64%
Uzbekistan	6.90	5.60	5.60	3.50	2.70	1.40	0.22	102.6	0.02%
Venezuela	14.40	14.40	14.40	12.90	10.10	6.60	0.46		0.02%
Vietnam	54.00	42.50	30.00	15.00	11.80	11.80	0.58	266.5	0.07%
Virgin Islands	2.80	2.40	2.40	0.00	0.00	0.00			0.00%
West Bank	18.60	7.70	1.20	0.00	0.00	0.00	3.99	1388.4	0.58%
Western Sahara	0.25	0.25	0.25	0.25	0.25	0.25	0.48		0.53%
Yemen	251.00	63.90	7.30	6.60	1.30	0.73	9.33	9188.2	12.01%
Zambia	10.30	6.80	5.00	4.80	4.00	2.70	0.64	526.8	0.22%
Zimbabwe	13.10	11.00	6.70	4.30	3.90	3.30	0.83	804.2	0.34%

DEVELOPMENT OF INNOVATIVE SELF-CONSUMPTION AND AGGREGATION CONCEPTS FOR PV PROSUMERS TO IMPROVE GRID LOAD AND INCREASE MARKET VALUE OF PV: THE PV-PROSUMERS4GRID PROJECT

Luz Alicia Aguilar[1], Maria Roos[1], Manuel Battaglia[1], David Krehan[1], Christian Grundner[2], Moira Jimeno[2], Dirk Hendricks[3], Dörte Fouquet[3], Georg Lettner[4], Riccardo Battisti[5], Karl Moosdorf[6], Daniela Velte[7], Eduardo Roman[7], António Joyce[8], Gaëtan Masson[9], Carlotta Cambiè[9], Aída González[10], Jose Donoso[10], Christoph Winter[11], Nicole Diewald[11], Ulrich Winter[11], Wouter Schram[12], Wilfried van Sark[12]
[1]Bundesverband Solarwirtschaft (BSW), Berlin, Germany, [2]eclareon GmbH,
[3]European Renewable Energies Federation (EREF), [4]Technische Universität Wien, Energy Economics Group,
[5]Ambiente Italia, [6]Associação Portuguesa de Empresas do Sector Fotovoltaico (APESF),
[7]Fundacion Tecnalia Research & Innovation, [8]Laboratorio Nacional de Energia e Geologia,
[9]Becquerel Institute, Brussels, Belgium, [10]Unión Española Fotovoltaica (UNEF), [11]Fronius International GmbH,
[12]Utrecht University, Copernicus Institute, The Netherlands
Email: w.g.j.h.m.vansark@uu.nl, Phone: +31 30 253 7611

ABSTRACT: In the changing electricity markets in Europe emphasis is put by the European Commission on enlarging the role of prosumers, i.e. consumers that also generate electricity using photovoltaics (PV). The aim of the recently started PV-Prosumers4Grid project is to develop and implement innovative self-consumption and aggregation concepts and business models for PV prosumers that will help integrating sustainable and competitive electricity from PV in the electricity system. This paper presents the concepts of prosumerism, and identifies three different PV prosumers options based on system boundaries: group 1, single direct use; group 2, local collective use; group 3, district power model. A country analysis has been made showing that prosumption in Europe is legally feasible for all countries for single direct use, while this is not the case for collective use, either local or in a district. Finally, a cash-flow model has been developed that allows stakeholders in all countries to perform a cost-benefit analysis taking into account energy demand and sources (heatpump, EV) and PV electricity generation.
Keywords: Photovoltaic, prosumer, markets, policy analysis, economic analysis

1 INTRODUCTION

The European electricity sector is still transforming from public monopolies into competitive private companies in liberalized markets, and, most recently, into models at the customer domain level. This transformation is expected to deliver more efficient and cheaper energy therefore and more competitive. Due to its cost and growth perspective, photovoltaics (PV) will be a key driver of this development throughout Europe because PV has reached a level of competitiveness that allows moving to self-consumption schemes in many European countries and eventually to peer-to-peer transactions of the energy produced by PV systems. In addition, the updated Energy Performance of Buildings Directive [1,2] will drive local generation of PV and will increase self-consumption and self-sufficiency. We define a prosumer to be a natural or legal person that not only consumes energy but also produces energy (e.g. with a PV system). Such a "prosumption" role will empower consumers to participate actively in the electricity market (in compliance with the EC 2016 legislation "Clean Energy for all Europeans" [3]) with the aid of technical developments such as storage systems (electrical/thermal), smart meters, and the use of Demand Response and Demand Side Management (DSM) tools. Moreover, prosumption even creates a new electricity market segment "behind the meter", dealing with energy amounts which are not accounted for in the standard (i.e. grid-based) market models. Advanced business models are necessary to promote self-consumption and aggregation concepts, which will improve the technical design of electricity systems. The success of these developments depends, however, on the regulatory and administrative framework in terms of energy policy and regulation, support schemes, grid financing, taxation and legal relationships amongst the involved parties. This requires innovative solutions coupled with suitable business and energy management models. Thus, innovative business and energy management concepts for residential and commercial prosumers are a key element to enable PV market and system integration.

The aim of the PV-Prosumers4Grid Coordinated and Support Action [4] is to develop and implement innovative self-consumption and aggregation concepts and business models for PV prosumers that will help integrating sustainable and competitive electricity from PV in the electricity system. The PV-Prosumers4Grid project therefore will 1) identify best practice regulatory schemes and business opportunities for prosumers and grid operators, 2) support deployment of PV systems and their combination with other components such as PV storage systems, smart meters and e-mobility with focus on physical and financial grid interactions, and 3) provide PV Prosumers (residential and commercial sector) with competitive and sustainable electricity in a clear 5-step approach based on today's and potential PV prosumer markets.

2 PROJECT APPROACH

The project is following a 5-step approach:
Step 1: Quantitative analysis of the PV prosumer market potential with focus on home owners, landlords, housing cooperatives, commerce, retail, services, as well as industry and real estate companies. Qualitative assessment will be made of technical benefits, grid environments, market options and barriers for PV prosumer business models and distortions for market-compatible PV market integration.
Step 2: Identification of existing management and business models for PVP4Grid concepts and improvement

of these concepts by taking the interdependence with different market frameworks into account. For this, an in-depth analysis of the PV competitiveness of the implementation of the developed improved PVP4Grid concepts into European electricity markets will be performed by means of implementation scenarios for different market design.

Step 3: Investigation of the market readiness of PVP4Grid concepts with real-life testing for market implementation with an increased level of renewable energy shares. Life Cycle Analysis (LCA), socio- economic impact evaluation and transferability of the tested PVP4Grid concepts.

Step 4: Synthesis of the different results and derivation of hands-on guidelines and recommendations for the 3 core target groups and actors of the PVP4Grid project: 1) potential PV prosumers, 2) grid operators, in particular DSOs and 3) policy makers on European as well as on national level.

Step 5: Comprehensive communication and dissemination activities of all the deliverables carried out in the course of previous implementation steps will be directed towards target groups and key actors in order to increase the impact of PV-Prosumer4Grid activities all over Europe. For this, the results will be disseminated at the EU level and also in non-participating European countries next to the 8 partner countries (i.e. Austria, Belgium, France, Germany, Italy, Portugal, Spain and the Netherlands).

3 PROJECT RESULTS

3.1 Prosumer concepts

As the project has started on 1 October 2017, only first results will be presented. At this stage, the current status of PV prosumer concepts in the 8 partner countries has been identified, using three distinct groups related to the use of PV generated electricity based on system boundaries, i.e., single direct use, local collective use, district power model. These are depicted schematically in Figure 1. The groups were identified as follows:

1. Single direct use (behind the meter). One consumer directly uses the generated PV electricity on-site. Two cases are conceivable in this context: (i) generator legally identical to consumer, (ii) generator legally not identical to consumer). The public grid is only used for the residual electricity consumption and possible feed-in of excess electricity. PV self-consumption can be increased due to the implementation of energy storage systems, heat pumps, boilers, demand side management, e-mobility, etc.
2. Local collective use of PV in one place (e.g. in one building). A further distinction can be made based on the grid being private or public:
   a. Public grid: several consumers share the generated PV electricity using the public grid. The public grid is used for both sharing locally generated PV amongst several consumers, and for residual electricity in terms of residual load or feed-in of excessive energy. Each consumer can increase the share of self-consumption by certain measures (storage, demand side management, e-mobility, etc.). In this case only some consumers must agree in participating in this business model. This means that each consumer has its individual grid connection point.
   b. Private grid: several consumers share the generated PV electricity using a private grid which is connected to the public grid through one connection point. The public grid is only used for the residual electricity consumption and possible feed-in of excess electricity. Each consumer can increase the share of self-consumption by certain measures (storage, demand side management, e-mobility, etc.). In this case all consumers must agree in participating in this business model. This means that the connection point to the public grid is common for all consumers of this place (building).
3. District power models. Several consumers directly consume locally generated PV. Also within this group a further distinction can be made based on the grid being private or public:
   a. The PV energy is shared using the public local grid on low voltage level (limitation could be the same substation). To increase the share of self-consumption e.g. district storage devices can be used. In this case only some consumers must agree in participating in this business model. This means that each consumer has its individual grid connection point.
   b. The PV energy is shared using a private local grid on low voltage level. To increase the share of self-consumption e.g. district storage devices can be used. In this case all consumers must agree to participate in this business model. This means that the connection point to the public grid is common for all consumers of this area. This includes the possibility to have an own transformer substation.

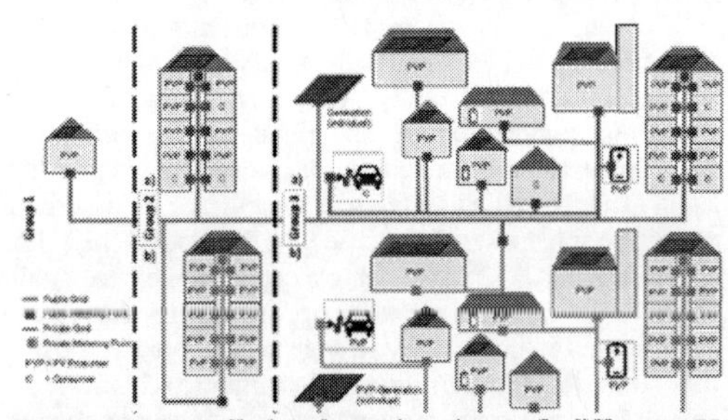

Figure 1: Detailed schematic view of different PV prosumers options based on system boundaries. Group 1: single direct use, group 2: local collective use, group 3: district power model.

In addition, identified benefits of prosumption are: efficient match of demand and supply of renewable energy at lowest (i.e. local) levels due to a high number of consumers and producers, lowest transportation costs of energy, creating energy efficiency awareness in a large part of society and thus contributing to match the EU's climate target on energy efficiency.

3.2 Country status

The status of PV prosumption in the 8 partner countries is shown in Figure 2, in which also information collected from other EU countries is shown. The main conclusions are that individual self-consumption (group 1) is allowed and/or feasible in all countries analysed, while collective self-consumption is currently not feasible in most of the countries, independently of the scale (individual building – residential, commercial- and at larger scale in districts / group of buildings). We remark here that in Austria the model of "gemeinschaftliche

Country	Group 1	Group 2	Grou
Austria	YES	YES	YES
Belgium	YES	NO	NO
France	YES	YES	YES
Germany	YES	YES	YES
Italy	YES	NO	NO
Netherlands	YES	YES	YES
Portugal	YES	YES	YES
Spain	YES	NO	NO
Bulgaria	YES	NO	NO
Croatia	YES	N/A	N/A
Cyprus	YES	N/A	N/A
Czech Republic	YES	N/A	N/A
Denmark	YES	N/A	N/A
Finland	YES	N/A	N/A
Greece	YES	N/A	NO
Hungary	YES	N/A	N/A
Poland	YES	N/A	N/A

Figure 2: Overview of grouped PV prosumer concepts legally feasible in selected European countries (status Spring 2018). Note, this does not guarantee economic feasibility.

Erzeugungsanlagen" actually is situated between group 2a and 2b. Every participant (consumer and generator) has an individual official smart meter. However, the in-house electicity lines are private property and are therefore not part of the public grid, and can be considered as a kind of a private grid. So, for all energy amounts (accounted in 15 minute averages) simultaneously generated and consumed in-house (allocation subject to agreement between the participants), no grid charges apply. The residual load is imported from the grid, where we note that every participating consumer can actually have its individual retailer/supplier for its grid-imported residual load). Any possible surplus (i.e. residual generation) is accounted as fed into the public grid and it is sold to a trader the generator has an agreement with.

Further, PV Prosumer (PVP) concepts that use the public grid to sell excess PV electricity to third parties are legally allowed, but rarely seen, as a result of various economic, administrative, and regulative barriers, with some exceptions ("Mieterstrommodelle" (neighbour PV supply model, which belongs to group 2) in Germany, or collective schemes in France, see for details Ref. [5]). It is also clear that PV prosumer concepts applied to small-scale PV plants are an excellent opportunity for distributed generation. Finally, battery storage is promoted in most of the countries through financial support schemes, tax exemptions and other measures in order to increase self-consumption ratios and reduce contracted power capacity. At present, virtual models (metering, demand aggregation) are more developed outside Europe (e.g. Australia, USA), but are likely to become a future opportunity in Europe.

A preliminary qualitative study on the potential of PV presumption in Austria and Spain has clearly revealed that such a potential is country-dependent. These were selected based on the share of PV generation in these countries. We also distinguish PV prosumer potential with regard to urban, rural, residential, commercial and industrial projects. All in all, the potential of group 3 is the highest in both countries. Regarding group 1, the potential in Austria is much larger than in Spain for single/double/multifamily housing. In the commercial sector group 3, the potential in Spain is mainly found in small businesses, while this potential is high for large businesses and offices in Austria. Further work is needed here.

3.3 Cash-flow model

A cash-flow model has been developed in order to support stakeholders in all countries in assessing costs and benefits for a PV prosumer concept. The economic assessment allows to combine integration of storage, e-mobility, heat pumps, shared use of PV generation, trade of excess storage capacity at wholesale markets and grid services.

The model involves two parts (see screenshots in Fig. 3): 1) PV system planning, and 2) financial analysis. In the

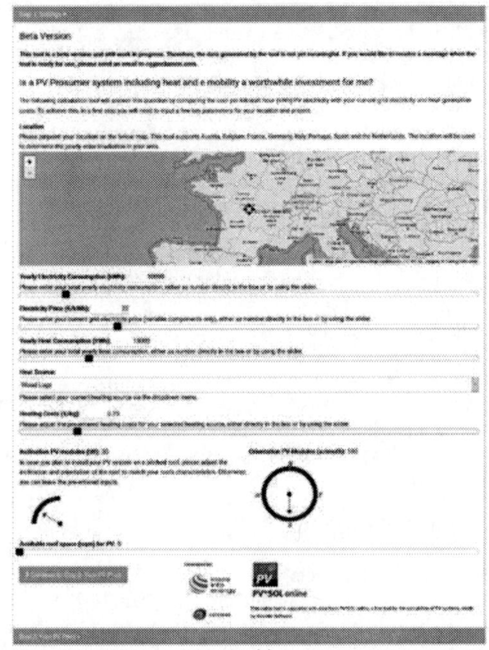

(a)

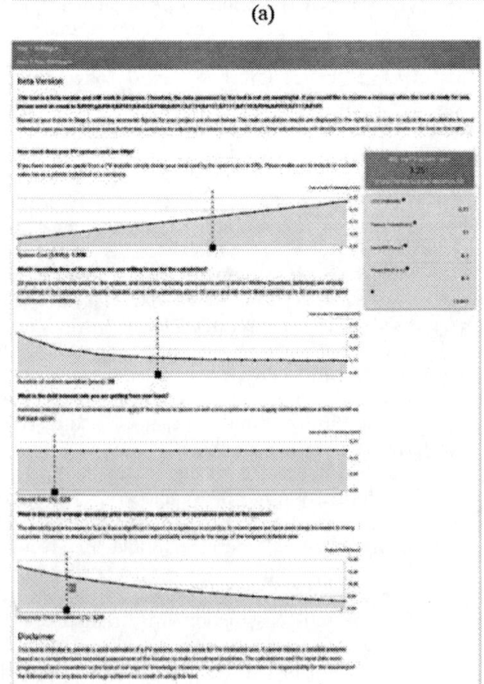

(b)

Figure 3: Screenshots of the two parts of the cash flow model. (a) part 1, PV system planning, (b) part 2, financial analysis.

first part a simple PV system dimensioning (including location) and performance calculation is done, followed by selecting a consumption profile (electricity, heat, e-mobility). The second part makes a profitability assessment including levelized cost of electricity, payback period, internal rate of return and net present value. The user can dynamically adjust the results.

The cash flow model is available free-of-charge from www.pvp4grid.eu/cmt. It should be noted that it is presently in BETA version.

4 CONCLUSION

This paper has presented the first results of the PV-Prosumers4Grid project. We have identified three different groups in PV prosumption, i.e., individual, collective, virtual. We found that in all countries studied individual self-consumption is allowed and/or feasible, while collective self-consumption is not feasible in most countries. Also, a preliminary qualitative study showed that the potential of prosumption is country-dependent. Battery storage is promoted in most countries through financial support, tax exemptions to support increase of self-consumption which is believed to assist in stabilizing the grid.

The next phase of the project is to identify novel prosumption models and to test these in actual or virtual environments.

5 ACKNOWLEDGMENT

This project has received funding from the European Union's Horizon 2020 research and innovation programme under grant agreement No 764786.

6 REFERENCES

[1] European Commission. Directive 2018/844 of the European Parliament and of the Council of 30 May 2018 amending Directive 2010/31/EU on the energy performance of buildings and Directive 2012/27/EU on energy efficiency, Official Journal of the European Union L 156, 19.6.2018, p. 75–91.

[2] The European Parliament and the Council of the European Union, European Commission (2010): Directive 2010/31/EU of the European Parliament and of the Council of 19 May 2010 on the energy performance of buildings (2010) pp. 13–35.

[3] European Commission, Clean Energy for All Europeans, Winter Package 2016, https://ec.europa.eu/energy/en/topics/energy-strategy-and-energy-union/clean-energy-all-europeans.

[4] https://cordis.europa.eu/project/rcn/211874_en.html

[5] G. Lettner, H. Auer, A. Fleischhacker, D. Schwabeneder, B. Dallinger, F. Moisl, E. Roman, D. Velte, A. Huidobro, "Existing and Future PV Prosumer Concepts", Public Report D2.1., 2018, downloadable from https://www.pvp4grid.eu/wp-content/uploads/2018/08/D2.1_Existing-future-prosumer-concepts_PVP4Grid_FV.pdf.

IMPACT OF BATTERIES AND ELECTRIC VEHICLES ON THE COMPETITIVENESS OF SOLAR PV

Eero Vartiainen
Fortum Growth Oy
POB 100, FI-00048 Fortum, Finland, eero.vartiainen@fortum.com

Gaëtan Masson
Becquerel Institute
Rue d'Arlon 69-71, Brussels 1040, Belgium, g.masson@becquerelinstitute.org

Christian Breyer
Lappeenranta University of Technology
Skinnarilankatu 34, FI-53850 Lappeenranta, Finland, christian.breyer@lut.fi

David Moser
Eurac Research
Via G.Di Vittorio 16, I-39100 Bolzano, Italy, david.moser@eurac.edu

ABSTRACT: As the share of PV in the energy system increases, electricity storage becomes more important. In addition to stationary storages, electric vehicles (EVs) offer a means for storing solar electricity. With storage, the economically viable sizing of a PV system is considerably larger. Currently, the LCOE for a residential PV + storage system investment is already lower than the value of the generated PV electricity in European countries with high retail electricity prices like Germany and Italy with reasonable interest rates. In countries with low retail electricity prices like Finland, a PV investment without storage starts now to be profitable. The economically optimal sizing of the storage system in relation to the PV system (kWh of storage per kWp of PV) will increase in the future because storage system prices are decreasing faster than PV system prices.
Keywords: Economic Analysis, LCOE, PV System, Storage, Electric Vehicles, Self-consumption

1 INTRODUCTION

Solar photovoltaics (PV) is becoming the cheapest form of electricity generation in the sunny parts of the world, manifested in the record low PV power purchase agreements (PPAs) in several countries around the world [1]. In many European countries, PV electricity is already cheaper than electricity bought from the grid in all consumer segments [2], or even the least cost electricity source [3]. However, in some European countries with moderate annual solar irradiation, true competitiveness of solar PV has not yet been reached. The main problem with PV generation is that it does not always match the electricity consumption. Especially in residential PV installations, a large share of the self-produced solar electricity cannot be immediately consumed locally, and needs to be fed into the grid. Often the remuneration for the surplus feed-in electricity is very low compared with the full value of retail electricity including energy cost, distribution fees and taxes. This leads to a non-optimal business case for the prosumer. But with electricity storage, the self-consumption ratio of PV could be increased. Whether this increase will enhance the prosumer business case is the topic of this paper.

To determine the true competitiveness of PV without storage is relatively straightforward. It can be done by comparing the levelised cost of electricity (LCOE) of PV with the average PV electricity value [4]. PV electricity value is the weighted average of the variable retail price of electricity which directly replaces consumption and the value of the surplus PV electricity fed into the grid. Only the variable part of the retail electricity price is considered, since for example the fixed annual grid fee cannot be avoided by self-consumption, as long as the prosumer is grid-connected. It is assumed here that prosumers will stay grid-connected, although technical solutions already exist which allow prosumers to go off-grid.

The situation becomes more complex when the PV system is connected to an electricity storage. The main driver for the storage is the price difference between the retail electricity and feed-in electricity. This price difference has to cover the extra cost of the storage system in order to the storage investment be profitable. In essence, it becomes a sizing issue where the PV and the storage system must be optimised in relation to the electricity consumption.

For example, an unprofitable PV system with a low self-consumption ratio could become profitable with a storage system if the self-consumption increases a lot. On the other hand, a profitable PV system with relatively high self-consumption could become unprofitable if the increase of self-consumption with storage is too little. The form of the consumption profile also has an effect: if most of the electricity is consumed around midday, the storage has less impact compared with a situation where morning or evening consumption dominates [5].

To determine which is the most suitable system depends on which economic metric is used. One can, for example, try to maximise autarchy, minimise the combined LCOE of PV + storage, maximise the revenue relative to one's electricity bill, or maximise net present value (NPV) or internal rate of return (IRR). The IRR is a discount rate which makes the NPV zero. IRR is very commonly used and is especially useful in sensitivity analysis. There is also a difference between making a new PV + storage system investment or just adding a storage to an already existing PV system.

In previous works by the authors [2,4], the year of true grid parity, i.e. when PV LCOE equals the average value of PV electricity has been determined for various countries and market segments. This was done with the help of historical learning curves and future projections for the PV system component prices. In the most recent publication [6], the optimal PV + storage system size was analysed from the maximum IRR point of view. In this paper, LCOE for the PV + storage investment is used as the main metric. PV + storage LCOE is compared with the average value of electricity for various system sizes in different markets.

2 INPUT DATA

2.1 Electricity consumption and price

The main consumer segment studied here is the residential. A standard consumption profile for a German household (Standardlastprofil H0 - Haushalt) [7] by German Association of Electricity Industry is used. This is a daily profile with 15 minutes time steps with slightly different values for winter, summer and spring/autumn days. Because the simulation model used here for the analysis has hourly time steps, the 15 minutes values were averaged hourly. The standard household profile gives a 1 MWh annual electricity consumption which in turn is scaled up according to the value used in the simulation. Most simulation cases were made for Berlin in Germany, but the same profile was used also for Rome in Italy, Brussels in Belgium and Helsinki in Finland.

To assess the value of the PV and stored electricity, average retail electricity prices gathered by Eurostat [8] were used. Retail electricity includes both energy, grid, and taxes and fees components. However, the fixed annual and monthly fees were deducted from the total retail price because they cannot usually be saved by own consumption [2]. Peak load shaving and subsequent savings in the fixed power fees were not considered here.

For the feed-in surplus electricity, 2017 average hourly spot market electricity prices were used with a typical 10% deduction for the administrative fee which is usually charged for the feed-in. No special feed-in tariffs or net-metering schemes were considered here, apart from hourly net-metering. The electricity prices used in this study can be seen in Table I.

Table I: Residential retail electricity prices used in the study (derived from Eurostat 2015 data) and feed-in prices for surplus electricity (average spot market price in 2017 - 10%)

	Retail €/kWh	Feed-in €/kWh
Germany	0.273	0.0308
Italy	0.266	0.0486
Belgium	0.178	0.0401
Finland	0.119	0.0299

The reasoning behind country selection was that for moderate annual irradiation countries, Germany had high, Belgium average and Finland low electricity retail prices, whereas Italy represents a country with high irradiation and electricity price.

2.2 PV system

All input data related to the PV system is according to the May 2018 ETIP PV factsheet [9]. For example, the module prices have been projected according to the historical learning rate [10,11], which is close to 25%. Fig. 1 shows the annual volume growth development for three different scenarios. Annual market was 99 GWp in 2017 [12]. The base case scenario assumes a 10% CAGR, slow growth 5% CAGR and fast growth 15% from 2017 to 2020. In the base case, CAGR decreases to 2.5% by 2035 and stay at that afterwards. In slow growth CAGR would decrease to 0% by 2030 and in fast growth to 5% by 2040.

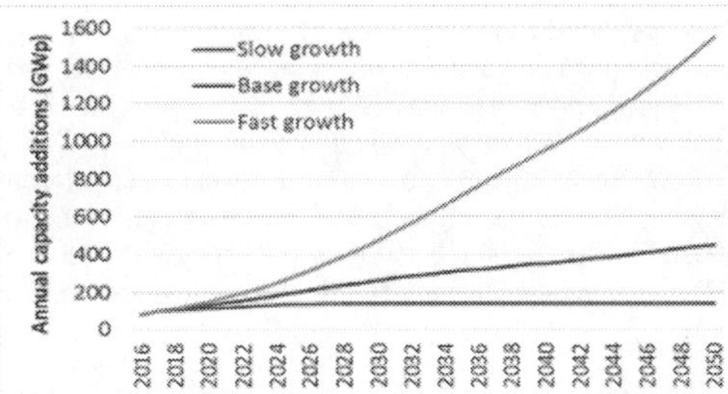

Figure 1: Global annual PV capacity additions 2018-2050 for three different scenarios

Fig. 2 shows the global cumulative capacity growth for the same scenarios. Note that to get cumulative shipments, replacement investments after 30 years for each system have to be added to these figures.

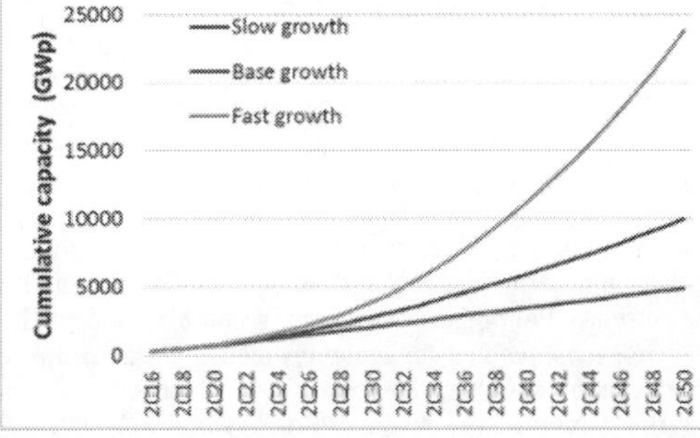

Figure 2: Global cumulative PV capacity development 2018-2050 for three different scenarios. Replacement investments after 30 years for each system have to be added to get the cumulative shipments.

Table II summarises the CAPEX and OPEX prices used for all countries. It is true that these prices may differ from country to country according to the market situation, but in the long term they should converge in Europe.

Table II: CAPEX and OPEX price for a residential PV system for the base volume growth scenario

	CAPEX €/Wp	OPEX €/kWp/a
2018	1.21	19
2020	1.10	18
2025	0.93	16
2030	0.81	15
2035	0.72	14
2040	0.65	13
2045	0.60	12
2050	0.55	11

PV system lifetime is 30 years and power degradation 0.5% per year. For PV generation profiles, European Commission Joint Research Centre's EMHIRES dataset [13] was used. This data includes an hourly time series of capacity factors for years 1986-2015 in country, bidding zone, NUTS-1 and NUTS-2 level for most European countries. The country level dataset gave extremely and unrealistically low values for the Northern European countries like Finland. For this reason, the NUTS-2 level data was used here. For each city, the closest NUTS-2 area was looked up and then a year with an annual yield close to the average of years 1986-2015 was selected for each area. Table III shows the annual yields and comparison with Solargis [14] irradiation data for a 30 degrees tilted and south-oriented PV system for each city. It can be seen that the Performance Ratio (PR) is between 79% and 84% for each city which seems quite realistic [15].

Table III: Annual Solargis irradiation, EMHIRES NUTS-2 level yield and Performance Ratio for a south-oriented 30 degrees tilted PV system

	Irradiation kWh/m^2	Yield kWh/kWp	PR
Berlin	1240	1037	83.6%
Rome	1860	1496	80.4%
Brussels	1240	1022	82.4%
Helsinki	1160	924	79.7%

2.3 Storage system

One of the biggest uncertainties in this analysis concerns the future development of the storage system prices. In this report, Li-ion batteries are used for the storage system. The battery storage market is still immature in many countries and currently there is not one price for a certain battery system. As the German market is currently the biggest in Europe, it can be assumed that it is the most mature one and representing the price levels that can be achieved also in other countries when the market becomes competitive. For example, the average price for small (8-10 kWh capacity) residential turn-key Li-ion battery systems in Germany in early 2018 was just over 600 €/kWh without VAT [16].

Bloomberg New Energy Finance (BNEF) reported at the end of 2017 [17] that the benchmark CAPEX for a fully installed residential storage system (7 kWh capacity) in 2017 was 936 $/kWh, i.e. 790 €/kWh with 1.18 $/€ exchange rate, corresponding to September 2018 rate and also 5-year average 2014-18 and is used throughout this report. According to the same report [17], the residential system price would decrease to 726 $/kWh in 2018, i.e. to 610 €/kWh. That is used as the starting point for the battery system price development here.

The price learning curve for battery systems, especially Li-ion batteries, has been a topic of a lot of discussion in the recent years. One of the first estimates made by Hoffmann [18] was a learning rate (LR) of 15-20% for Li-ion batteries. Schmidt et al. [19] reported that LR for residential Li-ion battery systems in the past would have been 12% ± 4% whereas LR for consumer electronics batteries has been as high as 30% ± 3%. Kittner et al. [20] reported a learning rate of 15.5% for Li-ion batteries. BNEF [21] estimated the LR for Li-ion battery pack as 18%.

It is advisable to break down the battery system price into different components since they might have different learning rates. In this report, the battery system is divided into battery pack, inverter and other Balance of System (BoS) which is analogous to the method used for PV systems here. A LR of 20% is used for the battery pack and the inverter, other BoS has 15% LR here.

As the price will decrease by 20% or 15% every time the global cumulative volume of the respective component doubles, a projection for the battery volume growth is also needed. The cumulative volume of stationary battery storage systems was about 20 GWh at the end of 2017 according to a BNEF market survey [22]. At the same time, the cumulative volume for electric vehicle (EV) batteries, including buses, 2/3-wheelers and trucks, was already more than 200 GWh. Since both stationary and EV batteries use similar modules or packs, it is possible to consider their volumes together in the same learning curve. However, since consumer electronics use often individual battery cells instead of modules or packs, they have been excluded from the learning curve volume here. Annual global battery market was close to 90 GWh in 2017, excluding consumer electronics [22].

The next thing to consider is the future volume growth of battery systems. The task of replacing the fossil fuels in transport sector will be huge, and accordingly, the battery volumes needed will be enormous. According to BNEF projection [22], global annual stationary battery storage market will be less than 5% of the EV battery market during the next decade. Only about 1% of the new passenger cars sold last year [23] were EVs. This is expected to grow to about 28% by 2030, to 55% by 2040 [23], and to 75% by 2050 in the base scenario. For stationary battery systems, the current annual global market is only about 0.04 kWh per 1 kWp installed new PV capacity [22, 12]. This could conceivably grow to 1-2 kWh/kWp by 2050 if all PV systems had an optimally sized storage system [24, 25]. Moreover, assuming conservatively that a storage system will have to be replaced at least once during the lifetime of a PV system, the annual stationary battery market could eventually be about 10-20% of the EV battery market.

Assuming that the average EV battery system size is 50 kWh, the annual EV battery volume growth can be estimated according to the number of EVs sold which was about one million in 2017. To that is added the smaller stationary storage system volume. It is projected here that the annual battery market will grow by 50% in 2018, by 40% in 2019 and by 30% in 2020 in the base case. By 2030 the compound average growth rate (CAGR) would decrease to 15%, by 2040 to 5% and by 2050 to 2.5%. These growth rates would lead to about 1800 GWh annual market in 2030, a figure projected also by BNEF [22]. By 2050, the annual global battery market would grow to about 6300 GWh with these CAGR rates.

Because the uncertainty of the volume growth is very high, two other scenarios in addition to the base case have been devised. In the slow growth scenario, CAGR is 40% in 2018, 30% in 2019 and 20% in 2020. By 2030 CAGR would decrease to 10%, by 2040 to 5% and by 2050 to 2.5% in the slow growth scenario. In the fast growth scenario, CAGR is 60% in 2018, 50% in 2019, 40% in 2020 and decreasing to 20% by 2030, to 5% by 2040 and to 2.5% by 2050. The slow growth scenario would lead to about 800 GWh annual market in 2030 and 2200 GWh in 2050, whereas the fast growth would have about 4000 GWh in 2030 and 17 000 in 2050. Slow growth would mean that only about 50% of the new sold passenger cars would be EVs in 2050 with also a smaller total number of cars and slower PV market growth. Fast growth would require that almost all new cars, including heavy traffic vehicles, would be EVs in 2050. Fig. 3 shows the projected annual volume growth in the different scenarios.

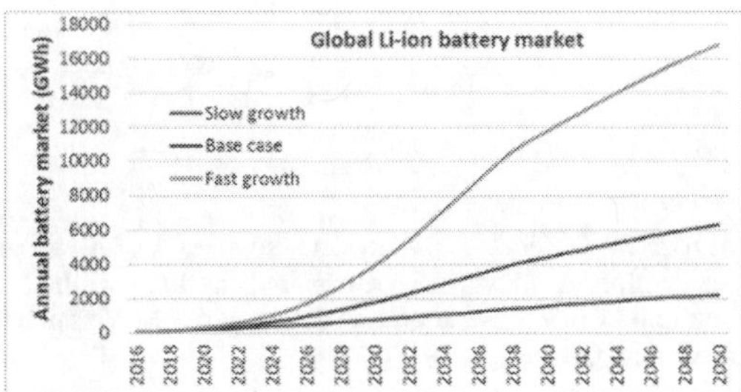

Figure 3: Global annual Li-ion battery system market volume development 2018-2050 for three different scenarios. Consumer electronics is excluded from the figures.

Cumulative volume growth is shown in Fig. 4 for the different scenarios. It must be noted that the figures include the replacement of every stationary battery system after 15 years. In the base scenario, the cumulative market grows from 230 GWh in 2017 to about 10 000 GWh in 2030 and to about 100 000 GWh in 2050. Slow growth will lead to about 60% smaller and fast growth to about 150% larger cumulative market than the base scenario. It must be noted that such high market volumes will probably require very efficient recycling of batteries since there could be scarcity of materials like Lithium.

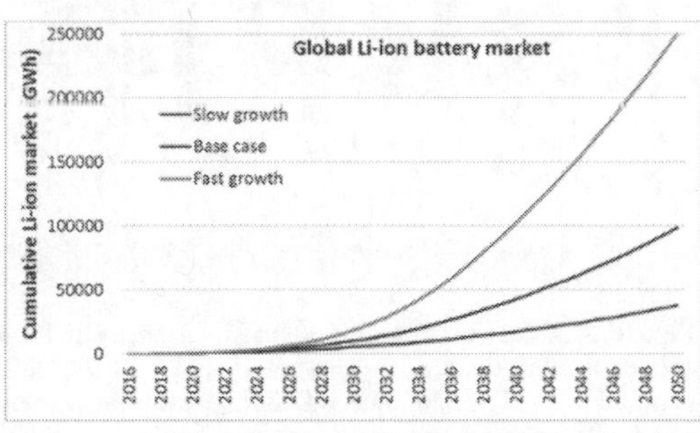

Figure 4: Global cumulative Li-ion battery system market volume development 2018-2050 for three different scenarios. Consumer electronics is excluded from the figures.

After establishing the volume growth scenarios, it is possible to estimate the future price development for battery systems of different sizes. In addition to price, the battery system lifetime is a very important parameter influencing the profitability of the storage investment. It is assumed here that the battery calendar life is 15 years. This means that the battery needs to be renewed once during the lifetime of the PV system. In all simulated cases to be presented, the storage was used for less than 4000 full cycles or 5000 80% deep cycles, since the average maximum usable capacity during the lifetime of the battery was set to 80% of the nominal capacity. This means that the end of calendar life is more probable than the end of cycle life for the battery systems simulated here. In addition, the round-trip efficiency of the storage was set to 90%, leading to a less than 5% loss of the PV generation in all simulated cases. The round-trip efficiency depends very much on the number of DC/AC conversions in the system.

OPEX price for the 1 MWh industrial battery system was set at 1% of CAPEX price per year in 2017, and decreasing with 10% learning rate thereafter. Residential system OPEX is 30% higher, commercial 15% higher and utility-scale 15% lower than the industrial system OPEX. Table IV shows the CAPEX and OPEX price development for the residential battery system for all countries. As storage market is still very small in many countries, there are currently some differences. The current price used here represents the German market [16], the prices are expected to converge in the long run.

Table IV: CAPEX and OPEX price for a residential storage system for the base volume growth scenario

	CAPEX €/kWh	OPEX €/kWh/a
2018	0.61	6.1
2020	0.49	5.6
2025	0.33	4.4
2030	0.24	3.7
2035	0.19	3.3
2040	0.16	3.0
2045	0.14	2.8
2050	0.13	2.6

2.4 Electric vehicles

In addition to stationary storage system, an electric vehicle (EV) battery storage is considered here. In effect, an EV increases the electricity consumption of the household and increases the PV self-consumption ratio. The impact of the EV on self-consumption depends on the charging scenario of the car, i.e., when the car is at home and available for additional storage of surplus PV generation.

It is assumed here that the average daily electricity consumption of the EV is 10 kWh. This would allow an average 50 km daily driving range, or at least 18 000 km per year. Table V lists the different charging scenarios used in this analysis. The first scenario means that the EV is never at home during daytime. The second implies that the EV is used to drive to work on weekdays. The third would mean that public transport is used for commuting and the car would be in use during the weekends. The fourth would probably mean that there are two EVs of which either one is parked at home during daytime. The fifth is same as the fourth with double average EV consumption, i.e. 20 kWh per day. In practice, the charging is also limited by the electrical connections in the house. Slow 1-phase 10 A charging is assumed here.

Table V: Different charging scenarios used in the analysis

1) Every night from 17 to 7
2) Saturday and Sundays from 7 to 17
3) Every weekday from 7 to 17
4) Every day from 7 to 17
5) Every day from 7 to 17 with double consumption

3 RESULTS

3.1 PV without storage

Fig. 5 shows the effect of PV system size on LCOE for a residential installation in Germany in 2018 with 8% nominal WACC and 5 MWh annual electricity consumption. The blue column is the PV LCOE, red column is the average value of the generated PV electricity and grey column is the profit of the investment in €/MWh. It can be seen that a residential PV system without a battery is a profitable investement with 8% nominal WACC up to about 5 kWp currently in Germany. For larger sizes, the PV systems generates too much surplus electricity which has to be fed in to the grid for a very low price. No special feed-in tariffs nor any other forms of subsidies are considered in this paper.

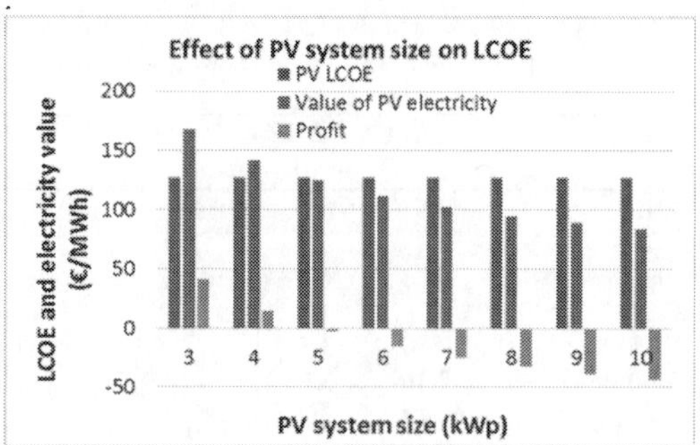

Figure 5: Effect of PV system size on LCOE for a residential installation in Germany in 2018 with 8% nominal WACC, 2% annual inflation and 5 MWh annual electricity consumption.

However, with 5% nominal WACC, the investment would be profitable up to about 8 kWp, as seen in Fig. 6. With 2% nominal WACC, the investment would be profitable for all PV system sizes up to at least 10 kWp, as seen in Fig. 7. In practice, the available roof area could be the limiting factor on the system size. A 60 m^2 roof area could host about 10 kWp of PV.

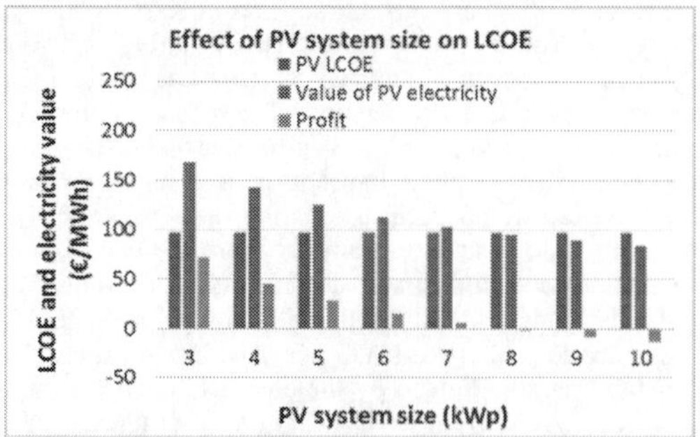

Figure 6: Effect of PV system size on LCOE for a residential installation in Germany in 2018 with 5% nominal WACC, 2% annual inflation and 5 MWh annual electricity consumption.

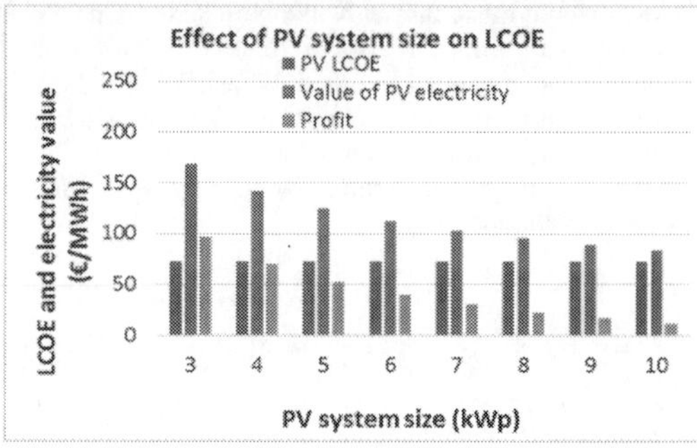

Figure 7: Effect of PV system size on LCOE for a residential installation in Germany in 2018 with 2% nominal WACC, 2% annual inflation and 5 MWh annual electricity consumption.

3.2 PV + storage

With battery storage, the PV self-consumption can be increased. Fig. 8 shows the effect of storage system size on LCOE for a residential installation in Germany in 2018 with 8% nominal WACC, 5 kWp PV system and 5 MWh annual electricity consumption. As seen, PV+battery is not yet a profitable investment in Germany with 8% nominal WACC. The increased PV self-consumption cannot quite compensate the added investment cost of the battery storage.

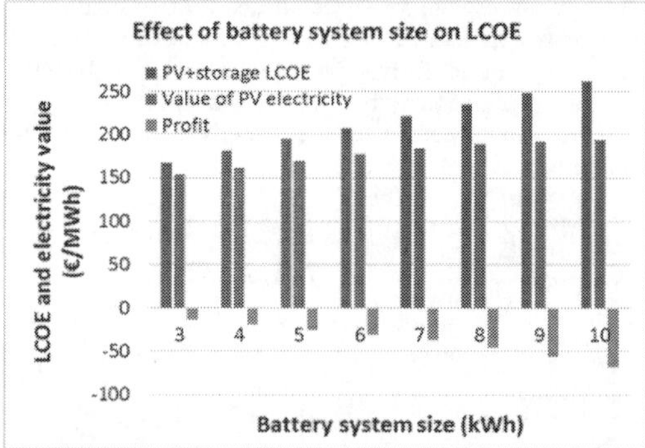

Figure 8: Effect of storage system size on LCOE for a residential installation in Germany in 2018 with 8% nominal WACC, 2% annual inflation, 5 kWp PV system and 5 MWh annual electricity consumption.

Fig. 9 shows the same case with 5% nominal WACC where the investment is profitable up to about 9 kWh storage size. In Fig. 10 the nominal WACC is 2% and storage sizes over 10 kWh would be profitable. However, a storage of about 4 kWh would give the maximum profit in this case.

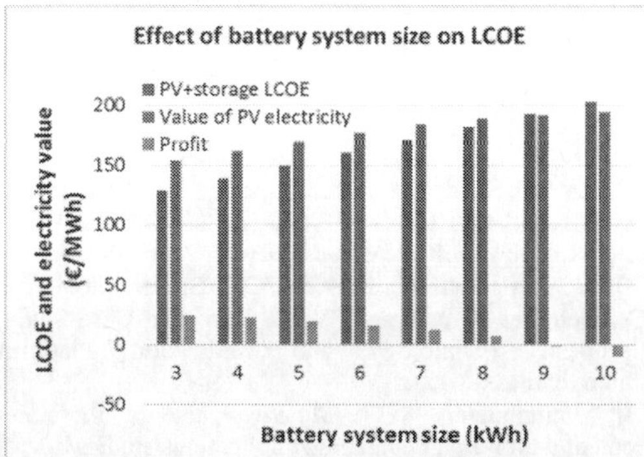

Figure 9: Effect of storage system size on LCOE for a residential installation in Germany in 2018 with 5% nominal WACC, 2% annual inflation, 5 kWp PV system and 5 MWh annual electricity consumption.

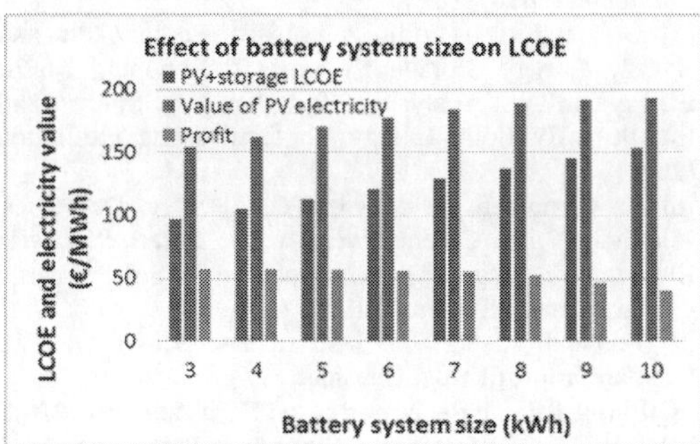

Figure 10: Effect of storage system size on LCOE for a residential installation in Germany in 2018 with 2% nominal WACC, 2% annual inflation, 5 kWp PV system and 5 MWh annual electricity consumption.

Fig. 11 shows the effect of storage system size on LCOE for a residential installation in Italy in 2018 with 8% nominal WACC, 5 kWp PV system and 5 MWh annual electricity consumption. Storage systems up to at least 10 kWh are already profitable because the annual yield in Italy is almost 50% higher than in Germany and there is more surplus electricity with the same PV system size available to be stored with the battery. This would imply that the optimal PV system size is smaller and storage size relatively larger in Italy than in Germany.

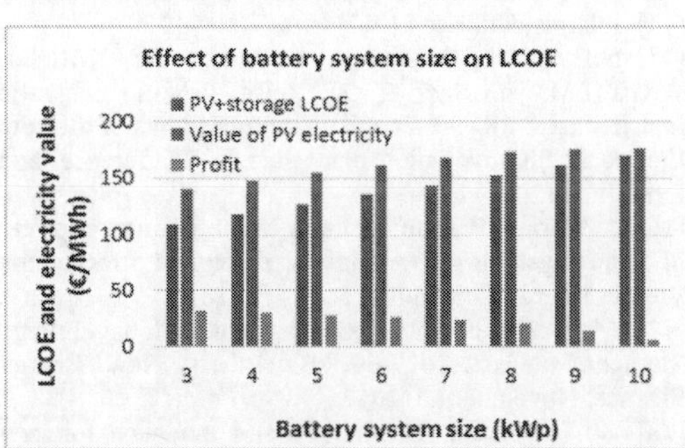

Figure 11: Effect of storage system size on LCOE for a residential installation in Italy in 2018 with 8% nominal WACC, 2% annual inflation, 5 kWp PV system and 5 MWh annual electricity consumption.

In Belgium, the battery investment is not yet profitable with 5% or 8% nominal WACC. But as seen in Fig. 12., with 2% nominal WACC a storage system up to about 7 kWh gives a small profit.

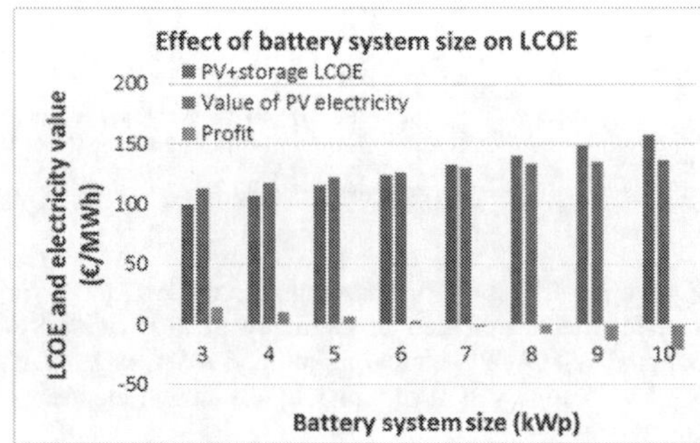

Figure 12: Effect of storage system size on LCOE for a residential installation in Belgium in 2018 with 2% nominal WACC, 2% annual inflation, 5 kWp PV system and 5 MWh annual electricity consumption.

In Finland, the storage investment is not yet profitable even with 2% nominal WACC. The retail electricity price in Finland is more than 50% lower than in Germany. As seen in Fig. 13, the storage investment in Finland starts to create real profit after year 2030.

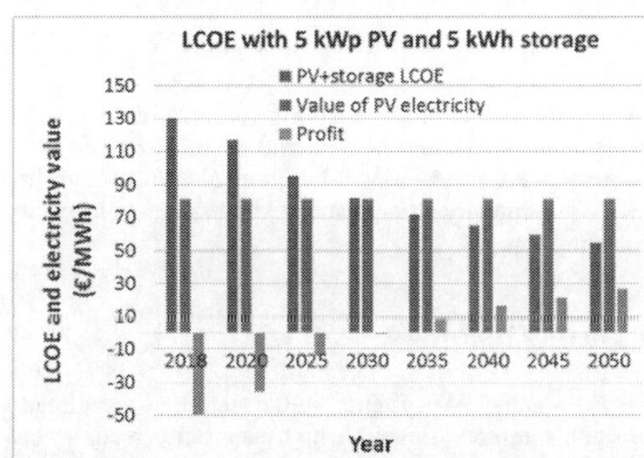

Figure 13: LCOE for a residential installation in Finland in 2018-50 with 2% nominal WACC, 2% annual inflation, 5 kWp PV system, 5 kWh storage system and 5 MWh annual electricity consumption.

3.3 PV + storage + EV

Fig. 14 shows the effect of different EV charging scenarios on the LCOE for a residential installation with 5 kWp PV and 5 kWh stationary storage system in Germany in 2018 with 8% nominal WACC. The EV increases the self-consumption so that the stationary storage investment becomes profitable if the car can be charged at least on weekends during daytime. No extra cost is assumed for the EV. With the EV charged every day during daytime, even a system with 10 kWp and 5 kWh stationary system could be profitable today, as seen in Fig. 15. However, this would probably require two EVs in practice which is maybe not yet realistic.

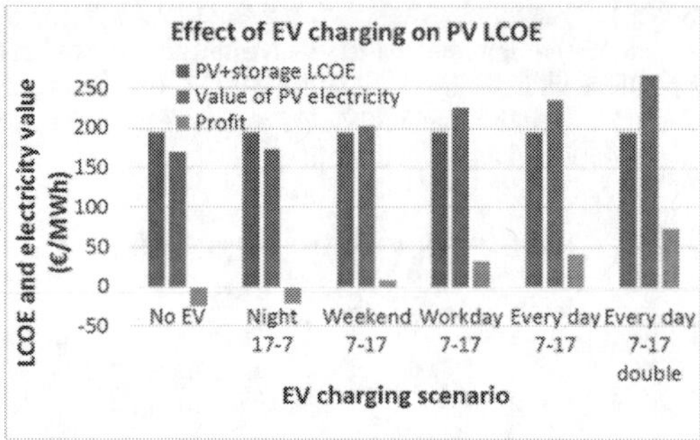

Figure 14: Effect of EV charging scenario on LCOE for a residential installation in Germany in 2018 with 8% nominal WACC, 2% annual inflation, 5 kWp PV system, 5 kWh stationary storage and 5 MWh annual electricity consumption.

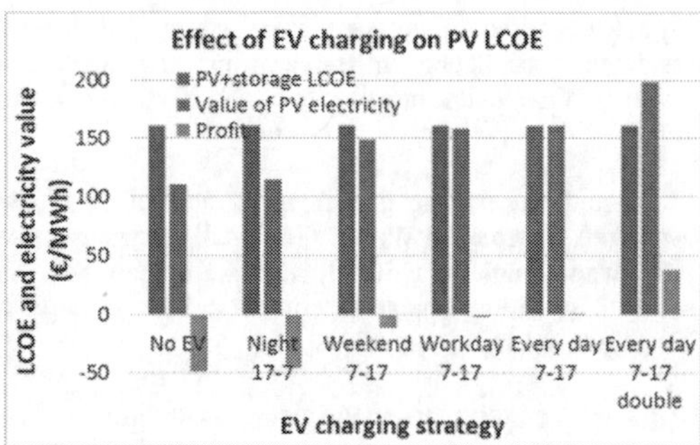

Figure 15: Effect of EV charging scenario on LCOE for a residential installation in Germany in 2018 with 8% nominal WACC, 2% annual inflation, 5 kWp PV system, 10 kWh stationary storage and 5 MWh annual electricity consumption.

4 CONCLUSIONS

Residential PV systems without storage are already profitable in countries with high electricity retail prices like Germany and Italy with all reasonable interest rates. A PV + stationary storage system is profitable in Italy with 8% nominal WACC, in Germany with 5% WACC and in Belgium with 2% today. In Finland, the retail electricity is so low that the storage investment will not become profitable before 2030.

With electric vehicles, the self-consumption of PV and the profitability of PV + storage investment can be increased. However, the EV or EVs must be at home for a significant time during daytime.

5 ACKNOWLEDGEMENTS

The work for this paper has been carried out under the framework of the EU Technology and Innovation Platform Photovoltaics (ETIP PV). We also gratefully acknowledge Solargis for the irradiation data, European Commission Knowledge Management Unit, Directorate C Energy, Transport and Climate, Joint Research Centre for the dissemination of EMHIRES hourly capacity factors, Chris Werner for providing the German consumption profiles and battery storage CAPEX data, and Bloomberg New Energy Finance for storage market information.

6 REFERENCES

[1] IEA PVPS, http://www.iea-pvps.org/fileadmin/dam/public/report/statistics/IEA-PVPS_Trends_2017_in_Photovoltaic_Applications.pdf (2018).

[2] E. Vartiainen, G. Masson, Ch. Breyer, The True Competitiveness of Solar PV – a European Case Study, European Technology and Innovation Platform Photovoltaics (2017).

[3] Fraunhofer ISE, https://www.ise.fraunhofer.de/content/dam/ise/en/documents/publications/studies/EN2018_Fraunhofer-ISE_LCOE_Renewable_Energy_Technologies.pdf (2018).

[4] E. Vartiainen, G. Masson, Ch. Breyer, The True Competitiveness of Solar PV – a European Case Study, 32nd European PV Solar Energy Conference and Exhibition (2016).

[5] G. Barchi, R. Lollini, D.Moser, Photovoltaic and Battery Energy Storage Systems in Shopping Malls: Energy and Cost Analysis of an Italian Case Study, 32nd European PV Solar Energy Conference and Exhibition (2016).

[6] E. Vartiainen, G. Masson, Ch. Breyer, D. Moser, Improving the Competitiveness of Solar PV with Electricity Storage, 33rd European PV Solar Energy Conference and Exhibition (2017).

[7] Verband der Elektrizitätswirtschaft e.V., Standardlastprofil H0 – Haushalt

[8] Eurostat database, ec.europe.eu/Eurostat/data (2016).

[9] PV LCOE factsheet, European Technology and Innovation Platform Photovoltaics (2018).

[10] ITRPV, International Technology Roadmap for Photovoltaic, Nineth Edition, 2017 Results, March 2018.

[11] I. Theologitis, G. Masson, Potential for Cost Reduction of PV Technology – Impact of CHEETAH Research Innovations, 31st European PV Solar Energy Conference and Exhibition (2015).

[12] SolarPower Europe, Global Market Outlook For Solar Power 2018-2022 (2018)

[13] I. Gonzalez-Aparicio, T. Huld, F. Careri, F. Monforti, A. Zucker, EMHIRES dataset - Part II: Solar power generation. European Meteorological derived HIgh resolution RES generation time series for present and future scenarios, Part II: PV generation using the PVGIS model; EUR 28629 EN; doi: 10.2760/044693, JRC Science for Policy report (2017).

[14] Solargis database (2017)

[15] N.H. Reich, B. Müller, A. Armbruster, K.Kiefer, W.G.J.H.M. van Sark & Ch. Reise, Performance ratio revisited: Are PR>90% realistic?, Proceedings of the 26th European Photovoltaic Solar Energy Conference and Exhibition (2011).

[16] C. Werner, Private communication on market prices of battery systems in Germany based on real offers, Chris Werner Energy Consulting, May 4, 2018.

[17] Y. Sekine and L. Goldie-Scot, Global Energy Storage Forecast 2017-30, Bloomberg New Energy Finance, November 15, 2017.

[18] W. Hoffmann, Importance and evidence for cost effective electricity storage, Proceedings of the 29th European Photovoltaic Solar Energy Conference and Exhibition (2014).

[19] O. Schmidt, A. Hawkes, A. Gambhir, I. Staffell, The future cost of electrical energy storage based on

experience rates, Nature Energy, Vol. 2, Article Nr 17110 (7/2017).
[20] N. Kittner, F. Lill, D. Kammen, Energy storage deployment and innovation for the clean energy transition, Nature Energy, Vol. 2, Article Nr 17125 (7/2017).
[21] L. Goldie-Scot, 1H 2018 Energy Storage Market Outlook, Bloomberg New Energy Finance, February 13, 2018.
[22] J. Frith, 2017 Lithium-Ion Battery Price Survey, Bloomberg New Energy Finance, December 5, 2017.
[23] BNEF, Long-Term Electric Vehicle Outlook, May 2018.
[24] J. Weniger, T. Tjaden, V. Quaschning, Sizing of residential PV battery systems, Energy Procedia, 46, 78-87 (2014).
[25] D. Keiner, Ch. Breyer, Modelling of PV Prosumers using a stationary battery, heat pump, thermal energy storage and electric vehicle for optimizing self-consumption ratio and total cost of energy, 33rd European PV Solar Energy Conference and Exhibition (2017).

PRICE-BIFACIALITY RELATIONSHIP OF BIFACIAL MODULES IN VERTICAL EAST-WEST ORIENTED PV SYSTEMS

L. Bodlak, H. Hernandez, J. Bierbaum, J. Kang, S. Nold, R. Preu
Fraunhofer Institute for Solar Energy Systems ISE
Heidenhofstr 2, 79110 Freiburg, Germany
lara.bodlak@ise.fraunhofer.de

ABSTRACT: Bifacial vertical east west orientated PV systems (bifiVEW) are a new approach in PV system application and seem to have a promising future. This is a technology that can provide an increased energy yield compared to conventional PV systems, as well as a shift of peak production from one at noon to two peaks in the morning and in the afternoon. The latter can lead to a more balanced electricity yield throughout the day and therefore allows a better integration of solar energy into the grid. As with this vertical east west orientated mounting configuration, the solar irradiation hits both module sides with almost similar quantity. Thus, a high module bifaciality rate is crucial. Currently, above 80% bifaciality rates are only commercially available with high efficient n-Type Silicon Heterojunction (SHJ) and Passivated Emitter and Rear (n-PERT) technologies, having generally a significantly higher module price than e.g. bifacial PERC (Passivated Emitter and Rear Cell) technologies. The presented work analyzes modules' price-bifaciality relationship, by investigating the bifaciality rate to the module watt peak price, with respect to the achieved LCOE (Levelized cost of electricity) in BifiVEW applications. The results show that for a BifiVEW PV system located in southern Germany, a difference in module bifaciality of 20%abs allows for a higher module price of up to 6.5 or 9.5 €ct/Wp for LCOE of 4 or 6 €ct(kWh, respectively. The relation between a higher percentage of bifaciality and an increased energy gain allows more efficient modules to be more expensive, the rate of this tradeoff is calculated and discussed in this report.
Keywords: bifacial, Economic Analysis, PV System, vertical east-west installation, LCOE

1 INTRODUCTION

When it comes to renewable energies, ambitious goals have been set in Europe and other countries around the world to achieve an increased share in the energy production during the next years. Solar energy technologies play a crucial role in the reach of these goals and it is fundamental that more efficient and lower cost options are not only developed but also produced in an industrial scale and finally, installed. The first practical silicon solar cells were invented at the Bell Labs and presented in 1954 [1]. Just a few years later, also at Bell Labs, a concept of a bifacial solar cell was developed [2]. Bifacial cells are capable of harvesting energy from both the front and rear side. When the modules are installed facing south (or north, when in the southern hemisphere), they are capable of absorbing not only the direct sunlight, but also some of the reflected light by the ground. The effectiveness of the rear side compared to the front side is thereby defined as the bifaciality (BF) which is $BF = \frac{\eta_{Rear\ side}}{\eta_{Front\ side}}$ [%] [4].

Although the concept of bifacial cells has existed for more than 50 years, their real deployment has just starting recently and it is believed to be the future for high-efficiency solar cells [2]. From a market share of under 5% in 2017, a growth to more than 30% is expected by year 2027 [3]. In the meantime, a suitable measurement and standardization framework has to be established to allow a proper comparison between different modules [4]. In the last years, a new concept for the system integration of bifacial cells has come up: The Vertical Bifacial East-West facing modules, from now on just mentioned as BifiVEW. This concept allows, through a vertical mounting, to harvest the most energy when the sun is still low in the East or West, with the draw-back only utilizing diffuse radiation at noontime [5]. An energy production simulation done with Zenit [21] can be seen in Figure 2. The characteristics of the BifiVEW system bring interesting advantages compared to conventional technologie such as a higher electricity value, production at off-peak times and alternative or shared land use [6].

First experiences obtained from a test site are already available since the installation of one by the company Next2Sun in 2015 (see Figure 1). The experiences on the test site were very positive: Despite mutual row shading, the BifiVEW plant delivered 10% more energy as a standard plant with the same rating, and additionally the grassland can still be utilized as a grazing field for livestock [5].
However, BifiVEW systems are still a new application and the previously mentioned lack of standards present an impediment for their commercial deployment. The prices of preferable modules with a high bifaciality rate (above 90%) are still significantly higher as bifacial PERC modules with lower bifaciality rates (of approx. 75%). But for BifiVEW systems the higher modules price can be payed back through the increased module energy yield [5]. In the following we present investigations on the relationship between module price and its bifaciality rate within a BifiVEW system configuration.

Figure 1: Vertical east-west oriented PV System [4]

2 METHODOLOGY

For the electrical yield calculations, a grassland location in Gundelfingen (near Freiburg im Breisgau), Germany was adopted. To achieve the most accurate results, simulations were performed using Fraunhofer ISE's in-house PV system simulation tool (Zenit) adapting ray-tracing tool (Radiance) for the accurate irradiation simulations reaching bifacial modules. For the economic assumptions, a project lifetime of 25 years was taken and an discount rate of 5%, as well as Balance of System costs (BOS) of 45 €ct/Wp for all calculations.

2.1 Predicting bifacial PV system yield

2.1.1. Model description

Fraunhofer ISE developed a PV system simulation tool, Zenit [21] and adapted the suite of ray-tracing tools, Radiance [22] for the accurate irradiation simulations for bifacial modules (Figure 2). A powerful lighting simulation program Radiance uses absolute properties of radiance and irradiance in adequate physical units of W/m2sr-1 or W/m2. The model of sky and sun as a natural light source is provided via one of the Radiance tools gendaylit [23]. This tool creates a complete sky radiance distribution using Perez model for any pair of global and diffuse irradiance. The defined model is rendered as 3D images and numerical values of local irradiance are provided via Radiance as "seen" by virtual sensors located on the solar cells. Such single irradiance values for each PV cell are then aggregated to front and rear side module irradiance values and handed over to Zenit. The remaining calculation steps in Zenit towards system yield are straight forward and well known in the community. The whole bifacial modelling process from optical (Radiance) to system yield simulation (Zenit) is controlled by a python tool package.

2.1.2. Simulation description

The bifacial PV system yield is done for the location Gundelfingen (latitude: 48.04, longitude: 7.86), with is located next to Freiburg, Germany. The module peak power is 100 MWp and the bifaciality varies with 5% increment from 70% to 100%. A ground albedo of 20% is considered as the location is assumed to be covered with typical green grass.

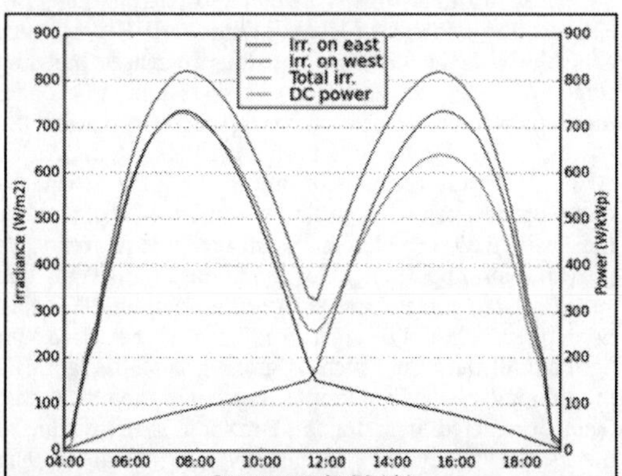

Figure 2: Irradiance and DC plot for BifiVEW system on sunny summer day in Gundelfingen [21, 22].

2.2 LCOE Calculations

The income that can be received from an installed PV plant or module depends on the amount of kilowatt hour (kWh) that it produces and the price at which this kWh can be sold (€ct/kWh). Considering only the Watt-peak price of the installation, without including other costs, is not sufficient for valid economic viability calculations. Hence, finding the best-suited levelized cost of electricity (LCOE) concept is of great importance [17]. The LCOE considers the total expenditures and total energy production of a project for a given lifetime, along with a discount rate.

$$LCOE = \frac{\sum_{t=1}^{n} \frac{I_t + OM_t}{(1+r)^t}}{\sum_{t=1}^{n} \frac{E_t}{(1+r)^t}}$$

With $I_t = P_{mod} + P_{BOS}$

Where:
n = Project Lifetime
t = Steps of project lifetime (year)
I_t = Investments at time t [€ct/Wp]
OM_t = Operation and maintenance costs at time t [€ct/Wp]
E_t = Energy production in year t [kWh/kWp]
r = discount rate per year [%/year]
P_{mod} = Price of module [€ct/Wp]
P_{BOS} = Price of Balance of System [€ct/Wp]
E_{t_front} = Energy produced on the front side per year t [kWh/kWp]
E_{t_back} = Energy produced on the back side in year t [kWh/kWp]

Formula 1: Standard LCOE formula [17]

Simplified, the LCOE represents the minimum remuneration per kWh at which the project would break even at the end of the defined lifetime. For the investigation of the modules' price-bifaciality relationship Formula 1 is transformed to Formula 2, to calculate P_{mod} regarding the Energy yield on the back side E_{t_back} for a given LCOE level:

$$P_{mod} = LCOE * \frac{\sum_{t=1}^{n}(E_{t\_front} + E_{t\_back})}{(1+r)^t} - \frac{\sum_{t=1}^{n} OM_t}{(1+r)^t} - P_{BOS}$$

Formula 2: Adaptation from formula 1 used for calculations

Under this consideration, three different LCOE values were assumed, in order to find the maximum price per Wp to make BifiVEW modules an economically viable option.

Indifference curves for three different LCOEs (4, 5 and 6 euro cents) with fixed BOS costs were calculated, for bifaciality values from 70 to 100%.

As a result, the module price sensitivity for a constant LCOE can be calculated. Based on these LCOE indifference curves, the maximum allowed prices for the specific cell technologies with relating bifaciality rates can be obtained.

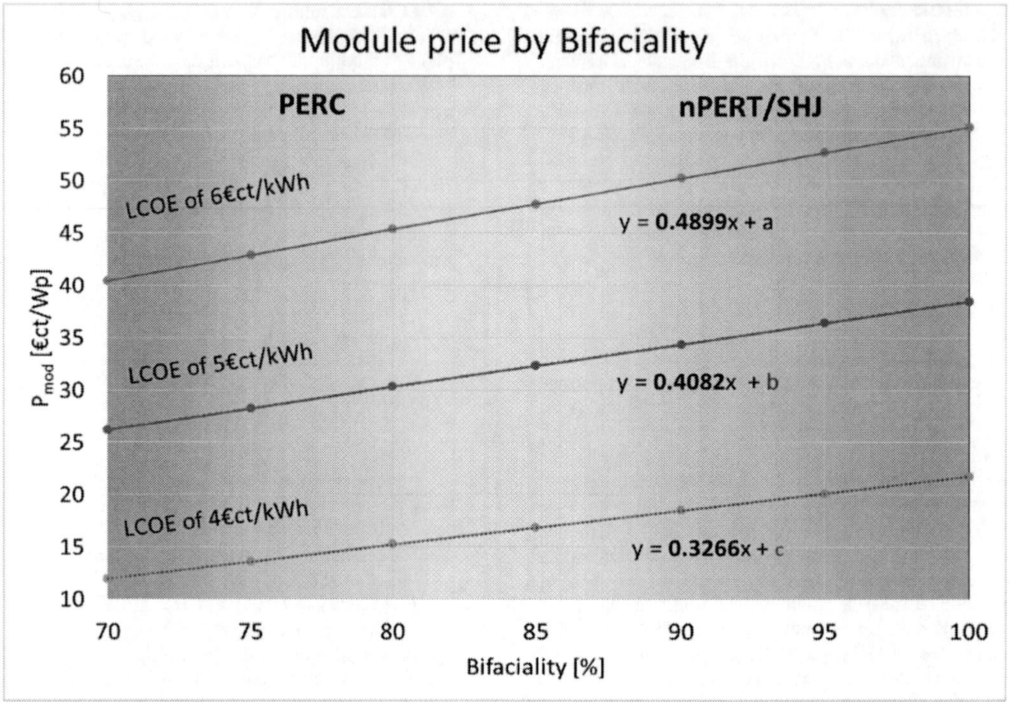

Figure 3: Allowed module price in [€ct/Wp] for the respective bifaciality [%] to reach certain LCOEs

3 RESULTS

The calculated LCOE indifference curves, shown in Figure 3, with the presented approach reveal the trade-off between the PV modules' watt peak price and its bifaciality rate – the module price-bifaciality relationship – for a bifacial east-west oriented PV system configuration. Figure 3 also portrays the maximum prices allowed for the assumed BifiVEW system configuration in order to be competitive against conventional technologies.

For the herein analyzed system location and configuration, the results show, that a decrease of the module bifaciality by $10\%_{abs}$ requires a decrease in the modules' price of 3.3, 4.1 and 4.9 €ct/Wp for equivalent LCOE of 4, 5, and 6 €ct/kWh, respectively. A higher price per percent of bifaciality elevation leads to a non-economic business case, while a lower price would be beneficial.

Thus, assuming a difference in bifaciality of 20% between PERC (with ~75%) and PERT or SHJ (with ~95%) and LCOE of 4 or 6 €ct/kWh, a difference in the modules' price of 6.5 or 9.8 €ct/Wp respectively, will lead to the same LCOE. With regard to current PERC module prices reaching values below 30 €ct/Wp [14], significant price margins for PV technologies with a high bifaciality (as PERT and SHJ) of up to 20% for LCOE of 4 €ct/kWh and margins above 30% for LCOE of 6 €ct/kWh are economically viable within BifiVEW applications.

4 CHALLENGES & OPPORTUNITIES

By now bifacial modules tend to have a considerably higher price than monofacial modules. There are several reasons for that. For once, the market for bifacial technologies is still minor, compared to traditional modules. Due to the expected additional energy gain by the rear-side collection, as well as a longer module lifetime (due to glass covers on both sides), a decrease of LCOE is anticipated. These factors lead the International Technology Roadmap for Photovoltaic (ITRPV) to estimate an increase of the global market share of bifacial technology from 5% in 2016 to 30% in 2026 [3]. This growing market will considerably help to reduce prices of bifacial modules.

Another challenge is the module mounting system for BifiVEW systems, which is not standardized yet. There are several aspects, which need to be considered. First, to reduce the mutual shading of the modules, they need to be placed within 10 meter distance to one another [13]. Second, the vertical structure of the BifiVEW systems acts like sails in the wind and thus increases the drag force on the construction. This results in higher mechanical requirements and material consumption (80 kg steal/kWp) compared to traditional PV-systems [20]. Also, as already mentioned, due to the greater required distances between the module rows, more overall area for the BifiVEW installation is needed. However, this presents an opportunity for additional usage of that specific land, since about 90% of the total area can still be utilized [5]. For instance, it can be used for agricultural purposes, such as grazing lands for animals or growing of smaller crops. Another option is to use (semi-)protected areas for the BifiVEW systems, due to the fact that they are ecologically more valuable than conventional PV farms [20]. Furthermore, it has been demonstrated experimentally that compared to conventional tilting, vertical installations are virtually immune to soiling degradation, which leads to a significant reduction of cleaning costs and water usage [3].

5 CONCLUSIONS

In order to achieve a high share of PV into the electrical grid, BifiVEW installations could play an important role given their shifted peak yields during morning and afternoon compared to the conventional PV production peak at noon time. This feature helps to smoothen out fluctuations in energy production and leads to a more balanced electricity yield throughout the day.

BifiVEW systems are still a niche configuration, but their importance will rise, with increasing PV market share and the need of increased production peak shifting and the dual use of agriculturally used land.

The results of the price-bifaciality assessment with in this paper show that high module bifaciality rates are especially important within BifiVEW applications and significant PV module price margins of up to 30% for technologies with higher bifaciality rates can still be economically beneficial. Thus, PERT and SHJ technologies are seem to be the beneficial choice over PERC for BifiVEW system applications.

6 REFERENCES

[1] A. Chodos, "APS News," 04 2009. [Online]. Available: http://www.aps.org/publications/apsnews/200904/physicshistory.cfm. [Accessed 16 01 2018].

[2] R. Kopecek, Y. Veschetti, E. Gerritsen, A. Schneider and C. Comparotto, "Bifaciality: One small step for technology, one giant leap for kWh cost reduction," Photovoltaics International, pp. 32-45, 2014.

[3] X. Sun, M. Khan, C. Deline and M. Alan, "Optimization and Performance of Bifacial Solar Modules: A Global Perspective," National Renewable Energy Labaratory, Colorado, 2017.

[4] S. Chunduri and M. Schmela, "Bifacial solar module technology," Taiyang News, Munich, 2017.

[5] H. Hildebrandt, "3 MWp vertical E-W oriented system in Germany," in 5th PV bifacial Workshop, Konstanz, 2017.

[6] L. Kreinin, N. Bordin and N. Eisenberg, "Bifacial solar cells: design principle and last achievements in industrial production," bSolar GmbH, Heilbronn, 2016.

[7] L. Guang, L. Cheng-Ching and C. Mattson, "Day-Ahead electricity price forecasting in a grid environment," IEEE Transactions on Power Systems 22, vol. 1, pp. 266-274, 2007.

[8] F. Sensfuß, M. Ragwitz and M. Genoese, "The merit-order effect: A detailed analysis of the price effect of renewable electricity generation on spot market prices in Germany," Energy policy 36(8), pp. 3086-3094, 2008.

[9] J. Stein, D. Riley, M. Lave, C. Deline and F. Toor, "Outdoor field performance from bifacial photovoltaic modules and systems," in 44th IEEE Photovoltaic Specialist Conference, Washington, USA, 2017.

[10] C. Heuberger, I. Staffell and N. Shsh, "Levelised Value of Electricity-A Systemic Approach to Technology Valuation," Elsevier, vol. Computer Aided Chemical Engineering, no. Vol. 38, pp. 721-726, 2016.

[11] D. Brearly, "Solarprofessional," 03 2017. [Online]. Available: http://solarprofessional.com/articles/design-installation/bifacial-pv-systems#.WmBYAqjia00. [Accessed 18 01 2018].

[12] F. Fertig, S. Nold, N. Wöhrle, J. Greulich, I. Hädrich, K. Krauß, M. Mittag, D. Biro and S. a. P. R. Rein, "Economic feasibility of bifacial silicon solar cells," Progress in Photovoltaics, 2016.

[13] H. Hildebrant, Personal Interview. [Interview]. 15 01 2018.

[14], "PVinsight," 01 2018. [Online]. Available: http://pvinsights.com/. [Accessed 17 01 2018].

[15] E. Union, "PVGIS," [Online]. Available: http://re.jrc.ec.europa.eu/pvgis/apps4/pvest.php#. [Accessed 18 01 2017].

[16] P. D. Guide, "http://www.atlassolarproducts.com," atlassolarproducts, [Online]. Available: http://atlassolarproducts.com/files/5714/3776/5993/Prism.Design_guide.pdf. [Accessed 16 01 2017].

[17] J. Libal, D. Berrian and R. Kopecek, "Overview: energy yield simulations and calculation of LCOE for bifacial PV systems," in International Solar Energy Research Center (ISC), Konstanz, 2017.

[18] D. o. I. Energy, "US Department of Energy," [Online]. Available: https://energy.gov/sites/prod/files/2015/08/f25/LCOE.pdf. [Accessed 18 01 2018].

[19] A. Energiewende, "Current and Future Costs of PV," Fraunhofer ISE, Freiburg, 2015.

[20] H. Hildebrant, "Eine Insel mit zwei Bergen," Photon, no. Juni, 2017.

[21] W. Heydenreich, B. Müller, Ch. Reise: Describing the World With Three Parameters: A New Approach to PV Module Power Modelling. Proc. 23. EUPVSEC 2008

[22] G.W. Larson, R.A. Shakespeare: Rendering with Radiance. Morgan Kaufmann Publishers, 1998.

[23] Radiance Manual, 2018. [Online]. Availble: http://www.radiance - online.org/learning/documentation/manual-pages/pdfs/gendaylit.pdf [Accessed 26 06 2018]

[24] Bundesnetzagentur, 2017. [Online]. Available: https://www.bundesnetzagentur.de/DE/Sachgebiete/ElektrizitaetundGas/Unternehmen_Institutionen/Ausschreibungen/Solaranlagen/BeendeteAusschreibungen/BeendeteAusschreibungen_node.html. [Accessed 19 01 2018].

DIVERSIFYING LAND-USE OPTIONS FOR THE FUTURE LARGE SCALE EUROPEAN PV DEPLOYMENT

Thomas Huld, Nigel Taylor, Arnulf Jäger-Waldau, Sandor Szabo, Katalin Bódis
European Commission Joint Research Centre, Directorate C - Energy, Transport and Climate,
via Fermi, 21027 Ispra, Italy

ABSTRACT: Decarbonising the EU energy system will require that renewables provide most, if not all, of electric power, with photovoltaics (PV) contributing a substantial share. At terawatt-scale deployment levels, the surface area used for PV installations will be large in absolute terms, but not necessarily a large percentage of available area. The study uses a unique combination of energy system scenarios, PV energy yield modelling and high resolution land-use data to analyse the impact for the European Union member states. The results show that future PV land requirements are a small percentage of overall area in most countries, even under deep-decarbonisation scenarios. Rooftops can contribute to some extent, but a small percentage of land currently classified as agricultural is likely to be needed. Use of set-aside land and degraded lands as well as dual-use approaches can mitigate the impact.
Keywords: photovoltaics, performance, integration.

1 INTRODUCTION

In a future where all or nearly all electricity is produced by renewable energy sources, photovoltaics are expected to play a major role and produce a very significant percentage of electrical energy. Such large-scale deployment will require areas for PV installations that are large in absolute terms though not necessarily a large percentage of available area. In order to prevent misconceptions about the impact of PV deployment, it is useful to quantify these area needs and to study how the surface area requirements of large-scale PV deployment can best be met.

This study uses a unique combination of energy systems analysis, PV energy yield modelling and land-use data with high geographic granularity and covering all countries in the European Union. Transparent data on the consequences of large-scale PV deployment is critical to building public support and acceptance for Europe's energy transition.

2 ANALYSIS

2.1 Scenarios for Energy Systems in 2050

Techno-economic studies of the European energy system provide estimates of possible levels of PV deployment in 2050, by which time when substantial decarbonisation is expected to have taken place. For the study we consider scenarios from two different models (see Table I).

A report [1] from the Lappeenranta University of Technology and the Energy Watch Group describes scenarios for 100% renewable power by 2050 for all regions of the world (the implications of including the transport sector in this model are presented in [2]). For the "Europe" region, the electric power consumption (6221 TWh) is approximately double the present value, with PV providing 42% of the total. By disaggregating the results, estimates of total and PV power could be obtained for almost all the countries presently in the European Union (the exceptions being Cyprus and Malta). As shown in Figure 1a, the PV share varies from about 22% in Sweden to 74% in Italy.

To fully decarbonize the energy system, the heating and cooling and transport sectors also need to be addressed. In this regard the JRC-EU-TIMES model [3] represents the entire energy system of the present EU-28 plus Switzerland, Iceland and Norway, with each country constituting one region of the model. It has been developed as a technology-rich tool for assessing the possible impact of innovation and cost developments. For the present study the "RES7-near zeroCO2" scenario [4] was chosen. It envisages a 95% decarbonisation of the energy system by 2050, with no nuclear power and no CO2 storage. Electricity demand reaches almost four times today's levels, reflecting the need to produce alternative fuels for transport and for heating and cooling. PV has a 29% overall share of production. Figure 1b shows the distribution for the EU-28 countries.

Table I: Energy system models and scenarios for the PV share of electricity production 2050.

Model/Scenario	Geog. Cover-age	Electricity Demand TWh	PV Output TWh
Lappeenranta University Technology/ EarthWatch 100% renewable power	Europe (incl. NO, IS, CH, UKR & TR)	6 221	2 634
JRC-EU-TIMES RES7: near-zero CO2 for full energy system, no nuclear, no carbon capture and storage	EU-28 + CH, IS &, NO	12 188	3 520

2.2 PV Area Requirements

To estimate the PV energy productivity we have used the methods described in Huld and Gracia Amillo [5], in which the instantaneous PV power at a specific location is calculated taking into account the in-plane irradiance, spectral content of the sunlight, and the module temperature which again depends on air temperature, wind speed and irradiance. The PV system mounting configuration is assumed to be free-standing racks facing south at an inclination angle of 20 degrees (40 degrees for locations north of 60 deg. N). The area required is calculated assuming 5.5 m^2 per kWp of PV modules, i.e. 18.2 % efficiency. The distance between the module racks is calculated so the shadows of one rack will just avoid hitting the modules on the rack behind at noon at winter solstice

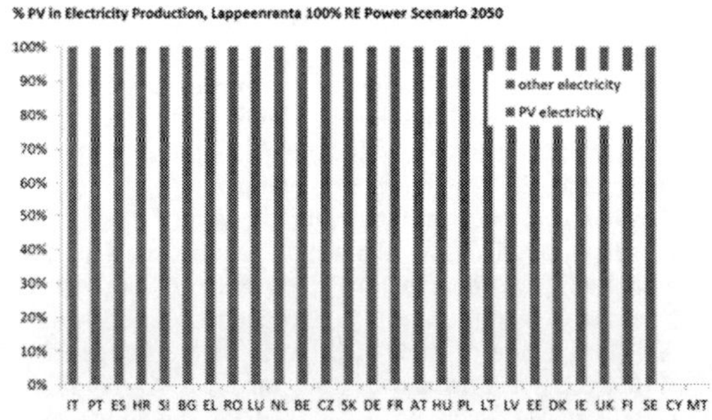

(a) Lappeenranta 100%RE Power

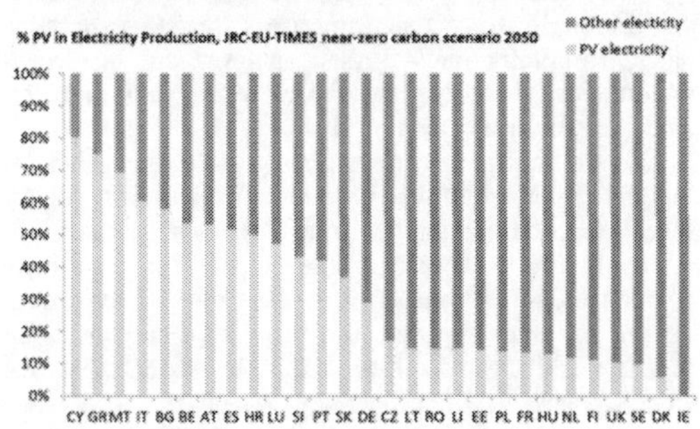

(b) JRC-EU-TIMES near-zero CO2 energy system

Figure 1: Scenarios for the PV share of electricity production 2050 in current EU countries according to two energy system models.

With this configuration the ratio of land area to PV module area is about 1.7 in Southern Europe rising to 3.3 at 55deg. N. Above this latitude we assumed a constant ratio of land to PV module area equal to 3.3. Overall, this variation corresponds to a direct PV area requirement of about 9 m^2 per kW_p in Southern Europe rising to ~18 m^2 in Northern Europe.

The PV energy yield calculation has been performed using the PVGIS methodology [6], using hourly solar radiation data for the period 2005-2016. The spatial resolution is 3 arc-minutes. North of 60°N the solar radiation data have been obtained from the ECMWF ERA-5 reanalysis [7], while for the rest of the study region the solar radiation data are from the CM SAF SARAH data set, calculated from satellite data [8]. The calculation assumes crystalline silicon modules, with balance-of-system losses of 10%. The resulting map of energy yield per unit area of land is shown in Figure 2. Values vary from 47 kWh/m^2 of land in Finland to184 kWh/m^2 in Cyprus.

2.3 Land Availability

The CORINE Land Cover data set [9] provides data on the type of land cover for the EU and candidate countries at a resolution of 100 m. The land cover is divided into four overall classes:

1. Artificial surfaces (urban areas, buildings, road and rail networks, ports, airports, mineral extraction sites, sports facilities etc.),
2. Agricultural areas (arable lands, rice fields, vineyards, pastures, agro-forestry areas etc.)
3. Forest and semi-natural areas (scrub and/or herbaceous vegetation, bare rocks, dunes etc.)
4. Glaciers, wetlands and water bodies

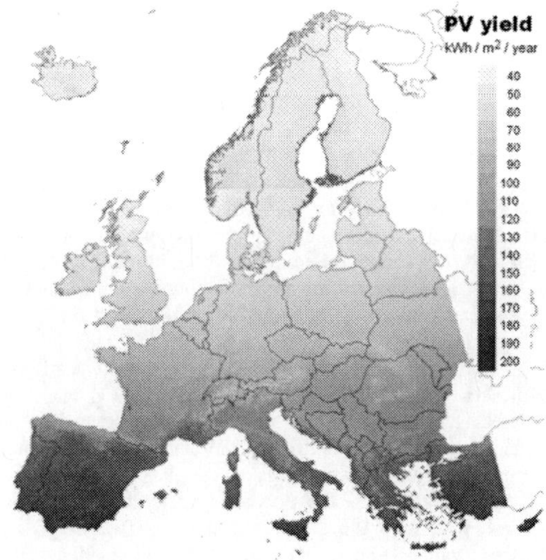

Figure 2: Variation of PV system energy yield for Europe, using the PV-GIS methodology and the system configuration explained in the text.

Two further *ad hoc* categories were considered:

a) Buildings offer considerable potential for deployment of PV and can allow better geographic correlation of supply and demand. A harmonized database on the EU building stock with the required level of detail is lacking, To overcome this, a muti-layer approach [10] was applied to determine the total detectable building footprint area, using the mainland cover dataset and the European Urban Atlas to validate information on EU built-up areas (to a resolution of 10 m x 10 m) derived by the European Settlement Map.

b) Set-aside agricultural land: this refers land set-aside under the Common Agricultural Policy to help reduce surpluses and deliver environmental benefits to agricultural ecosystems and wildlife. In the present context it is used as a proxy measure of agricultural areas area potentially available for non-agricultural purposes.

Table II shows the total areas of each of these categories for the EU as a whole. The high resolution land use data can be combined with the location-specific PV energy yield model to arrive at aggregated values per land cover type for a given geographical entity (in this EU member states, but the same approach could be applied at regional or county levels).

Table II: Total EU surface areas for different categories.

Land Cover	**Surface area, km^2**
CLC1 Artificial surfaces	209 196
CLC2 Agricultural areas	1 998 142
CLC3 Forest and semi natural areas	1 942 049
CLC4 Glaciers, wetlands & water bodies	273 386
Set-aside agricultural lands	62 415
Total detectable building footprint (modelled [10])	30 474

3 RESULTS AND DISCUSSION

The final stage of the analysis compares the surface area needs for the two modelled scenarios of PV deployment per country with the available surface area in several categories. N.B. At EU level, the surface area requirement for PV arising from the Lappeenranta 100% RE power scenario is calculated to be approximately 20,000 km^2, while for the JRC-EU-TIMES near zero CO2 scenario the value is approximately 36,000 km^2. The latter value reflects the far higher level of electrification needed for decarbonizing the entire energy system, including the heating and cooling and transport sectors.

Taking first the total surface area of both artificial surfaces and agricultural areas (CORINE Land Cover classes 1 and 2), Figure 3 shows the percentage of that area that would be required for PV installations in each country under the two energy system scenarios. For the Lappeenranta 100% RE power scenario the average is 1.1%, with only the Benelux countries (BE, LU, NL) giving values above 2%. For the higher electricity production needed in the JRC-EU-TIMES near-zero CO2 scenario, the average usage of CLC1&2 areas is 2.32%. In this case, the land cover share is above 4% for six countries: the Benelux countries plus Austria, Cyprus and Malta.

The next step was to consider the extent to which buildings can be used for the PV installations needed. As described in section 2, the location and building footprint area across the EU has been extracted using a multi.-layer approach. These building roof area values for were reduced by a factor of 0.26 (see ref. [10]) to take account of the technical limitations in using such surfaces for PV. The location-specific PV energy yield values were left unchanged. Although a higher area power density may be possible on roofs and facades could also contribute to production, other factors such as non-optimal orientation and inclination, higher module temperatures and shading can reduce power output. Figure 4 shows the percentage of the modelled PV electricity production that could potentially be covered by PV on buildings in each EU country. Overall there is a significant shortfall, with an EU-28 average of 39% in the case of the Lappeenranta 100% RE scenario and of 33% for the JRC-EU-TIMES near-zero CO2 scenario.

Other artificial surfaces or inland water bodies could also be exploited e.g. parking areas, roads, waste sites, lakes, reservoirs, and further analyses are needed to systematically address these. One example is a recent JRC study on use of waste fill sites [11].

Economic factors also play a role: in general - integrated PV solutions require a substantially higher investment than ground-mounted systems. With this in mind, Figure 5 considers the impact on land areas deemed as agricultural. In this case the EU average is 1.37% of land required for the Lappeenranta model scenario and 2.84% for the JRC-RU-TIMES near-zero CO2 scenario. As a comparison of impact, these values are below the share of land under the CAP set-aside scheme (3.12% overall for the EU).

A dual-use approach combining PV electricity and agricultural production [12] may offer interesting opportunities, and not least provide an additional means of keeping rural communities viable.

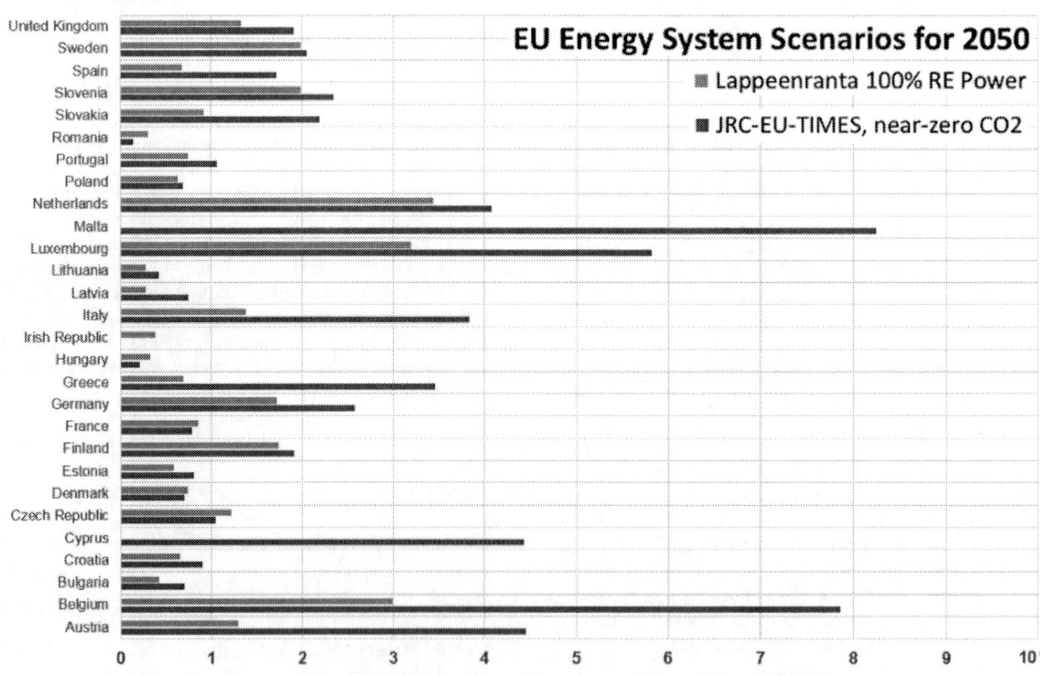

Figure 3: The share of land cover required for PV installations as a percentage of the combined area of CORINE land cover classes 1 (artificial surfaces) and 2 (agricultural areas) for the EU-28 countries under two scenarios: the Lappeenranta 100% RE Power scenario and the JRC-EU-TIMES near-zero CO2 scenario for the full energy system.

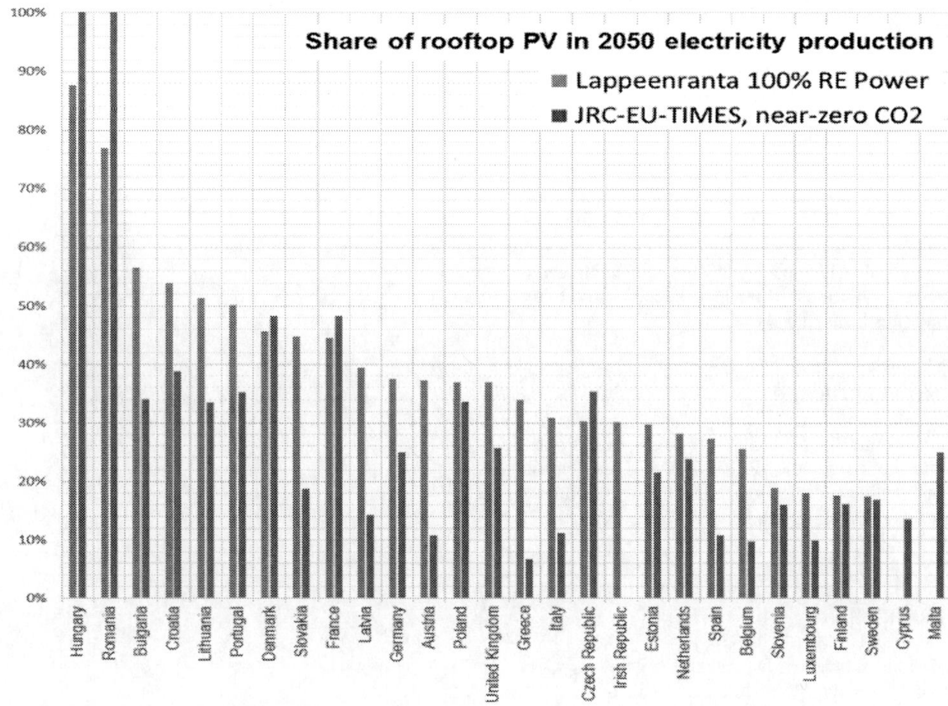

Figure 4: The potential contribution of rooftop installations for PV production to the requirements of the Lappeenranta 100% RE Power and JRC-EU-TIMES near-zero CO2 scenarios.

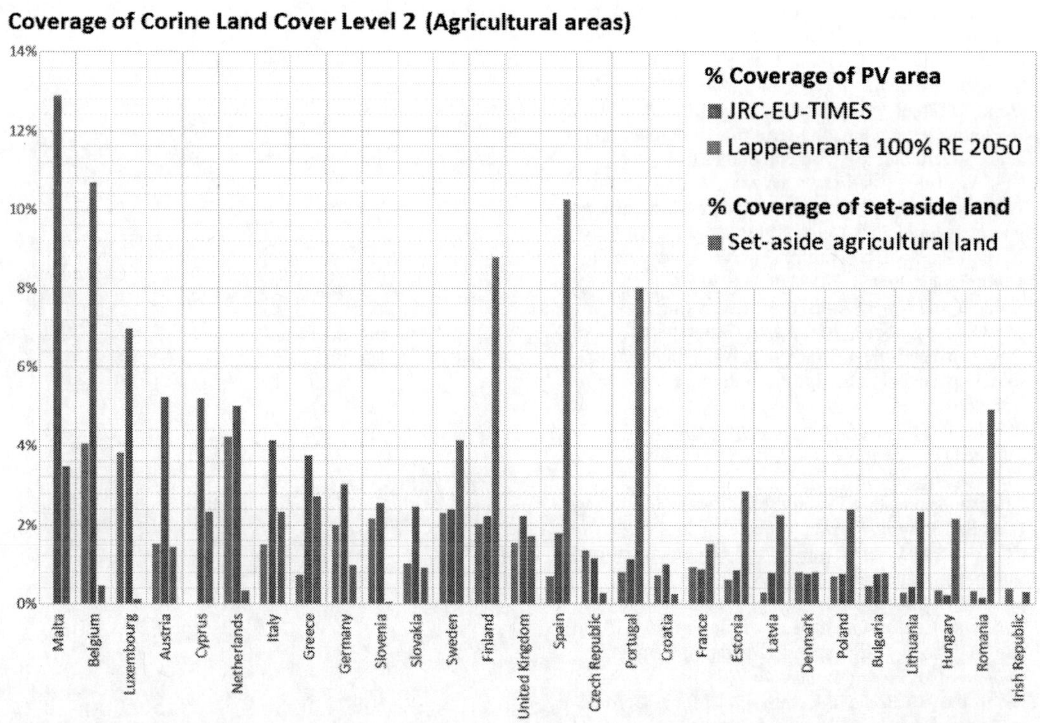

Figure 5: The share of land cover required for PV installations as a percentage of the combined area of CORINE land cover class 2 (agricultural areas) for the EU-28 countries under two scenarios: the Lappeenranta 100% RE Power scenario and the JRC-EU-TIMES near-zero CO2 scenario for the full energy system. The share of set-aside land is also shown.

4 CONCLUSIONS

- Future PV land requirements are a small percentage of overall area in most EU countries, even under deep-decarbonisation scenarios.
- Rooftops can contribute, but a small percentage of land currently classified as agricultural is likely to be needed. Use of set-aside land, degraded lands and dual-use (agro-PV) approaches can mitigate the impact.
- TW-scale PV will impact our lives and landscapes: balanced solutions are needed, involving communities from an early stage.

5 ACKNOWLEDGEMENTS

The authors would like to thank their colleagues Wouter Nijs for the input from the JRC-EU-TIMES model and Ana Gracia Amillo for disaggregating the Lappeenranta University of Technology model data for individual EU countries.

6 REFERENCES

[1] Ram, M. and Bogdanov, D., Aghahosseini, A., Oyewo, A.S., Culagi, A., Child, M., Fell, H-J. and Breyer, C., Global Energy System Based on 100% Renewable Energy – Power Sector. Technical report, LUT, Lappeenranta, Finland, 2017.

[2] C. Breyer, S. Khalili, E. Rantanen, M. Fasihi & D. Bogdanov Solar Photovoltaic Capacity Demand for a Sustainable Transportation Sector to Fulfil the Paris Agreement by 2050, Proc. EU PVSEC 2018

[3] S. Simoes, W. Nijs, P. Ruiz, A. Sgobbi, D. Radu, P. Bolat, C. Thiel and E. Peteves, "The JRC-EU-TIMES model. Assessing the long-term role of the SET Plan Energy technologies," Luxembourg, 2013.

[4] Nijs W., Ruiz Castello P., Tarvydas D., Tsiropoulos I., Zucker A., Deployment scenarios for low carbon energy technologies Deliverable - D4.7 for the Low Carbon Energy Observatory (LCEO), European Commission, Petten, 2018, JRC112915

[5] Huld, T. and Gracia Amillo, A., Estimating PV Module Performance over Large Geographical Regions: The Role of Irradiance, Air Temperature, Wind Speed and Solar Spectrum, Energies, 8, 5159-5181 (2015)

[6] PVGIS: http://re.jrc.ec.europa.eu/pvgis.html

[7] European Centre for Medium-Range Weather Forecast: http://ecmwf.int

[8] Satellite Application Facility for Climate Monitoring (CM SAF): http://www.cmsaf.eu

[9] CORINE Land Cover: https://land.copernicus.eu/pan-european/corine-land-cover/clc-2012

[10] K. Bodis, I. Kougias, A. Jaeger-Waldau, N. Taylor, S. Szabo, Quantifying the role of rooftop solar PV in the 2030 energy targets: an EU-wide assessment, *submitted to Energy, September 2018*

[11] S. Szabo , K. Bodis, I. Kougias, M. Moner Gerona A. Jaeger-Waldau, G. Barton, L. Szabo, A methodology for maximizing the benefits of solar landfills on closed sites, Renewable and Sustainable Energy Reviews 76:1291-1300, 2017

[12] Research Project Agrophotovoltaics - Resource-Efficient Land Use, http://agrophotovoltaik.de/

SOLAR ELECTRICITY IN AFRICA: OVERCOMING BARRIERS AND LESSONS (WHICH MAY BE) LEARNT FROM PREVIOUS EXPERIENCES IN EUROPE

Alessandro Virtuani[1*], and Guido Agostinelli[2]
[1] Officine del Sole (O'Sole), Via Durando 38 A, I-20158 Milan, Italy
[2]International Finance Corporation (IFC), 2121 Pennsylvania Ave NW, DC20433, Washington, DC, USA

* Corresponding author: alessandro.virtuani@o-sole.eu, phone: +39 333 74769 39

ABSTRACT: In this work, by addressing the different phases in the realization of a solar project, we use images to make concrete examples and drive a parallel between the market dynamics happened in Europe in the recent past (15-5 years) and what is presently happening in several African countries today. A strong focus on *quality* should enable the involvement of financial institutions (banks, microcredit and insurances) in the solar sector. Therefore, promoting a virtuous cycle leading to project risks' mitigation, facilitation of access to credit and increase perception and awareness in solar electricity. In short, making solar projects *bankable*. In our vision, a multi-step approach to the *bankability* of solar projects in the rooftop sector should be articulated as follows: (1) focus on **international standards** (and best practices) whenever is possible; (2) adoption/development **of high quality training** to achieve a minimum level of knowledge/skills for professionals and trigger the **learning processes** that are of great relevance to the industry; (3) implement at **local level** institutions able to perform **minimal quality checks** of components (e.g. *testing laboratories*) and systems acting as a trustable and respected third party; (4) design **incentive schemes** or clear regulations aimed at avoiding the unhealthy market dynamics that have characterized the early adoption of this technology in Europe.
Keywords: reliability, testing, international cooperation, emerging markets.

1 INTRODUCTION

Affordable and Clean Energy (i.e. ensuring access to affordable, reliable, sustainable and modern energy for all) is one of the 17 Sustainable Development Goals (SDGs), officially known as *Transforming our world: the 2030 Agenda for Sustainable Development*, set by the United Nations in 2015 [1]. At today's prices, and with a steadily declining price trend, solar photovoltaic (PV) electricity is increasingly perceived as one of the key-enabling technologies potentially leading to the achievement of this single target in those parts of the word where the access to electricity is still extremely limited. As for example in sub-Saharan Africa where 70% of the population (i.e. 600 million person) has no access to this form of energy, and where the availability of solar resources is remarkable (see Fig. 1).

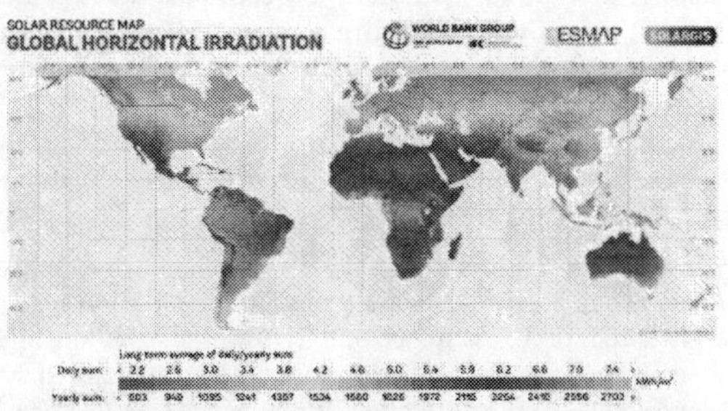

Fig.1: availability of solar resources worldwide (Global Horizontal Irradiation, GHI [2]).

Nevertheless, the diffusion of solar PV installations in developing countries is often hindered by:

(a) a poor perception about the reliability of this technology;

(b) the relatively high up-front capital and financing costs, combined with a difficult access to credit;

(c) a considerably poor focus on quality.

To tackle this, a promotion and diffusion of solar electricity in developing countries cannot take place if adequate control mechanism and quality checks (at component and system level) are not set in place.

Additionally, in this paper we drive a parallel between the current market situation in several African counties and what has happened in some European countries during the early adoption of this technology, when sometimes overgenerous incentives schemes have created unhealthy market dynamics in the first decade of the century (and in more recent years). Dynamics, which have led in some cases to a very poor focus on quality.

Presently, several international players (EPCs, IPPs, etc.) are directly – or through joint ventures - involved in the realization of utility –scale solar plants in some African countries. These players have generally matured the required experience in Europe or elsewhere. On the other hand, these international players face some difficulties in entering some specific markets in the continent. In this paper, we focus primarily our attention on (1) the rooftop market (i.e. residential and commercial/industrial systems, with plant sizes in the range 1 kW to ~5 MW), a market segment that is generally out of attention of the large international players, and on (2) Africa.

Nevertheless, similar observations/findings can be applied to other market segments (including utility-scale plants and micro/pico-solar installations) and other markets, particularly in sun-belt countries presently experiencing a boom in the adoption of this technology.

A strong focus on **quality** should enable promoting a virtuous cycle leading to risk mitigation, increasing access to credit and perception, and, in short, making solar projects ***bankable.***

2 ADOPTION OF SOLAR PV IN EUROPE 2000-2015

To address the issue of quality we initially focus our attention on quality. In some European countries in fact, overgenerous feed-in tariffs, combined with non-optimal timing and sharp year-end rate reduction mechanism have generated distortions in the market, and often caused non-healthy temporal dynamics. In the booming years these dynamics have created a rush to get solar plants connected to the grid by year's end in order to avoid reductions in the awarded incentives. This often at the expense of overall system quality.

As an example, Fig. 1 shows the evolution of the Italian solar market in the years between 2005 and 2017, showing numbers for cumulative and newly added PV capacity (GWp). The Italian market has been characterized by a large increase of installations until 2011 (with a newly added capacity of 9GW in the year) followed by a strong contraction in the following years and a market stagnating at 0.4-0.5 GW of new installations after the end of the FiT incentives in 2013. With some differences, analogous dynamics have been experienced by other European countries such as Spain, Germany, etc.

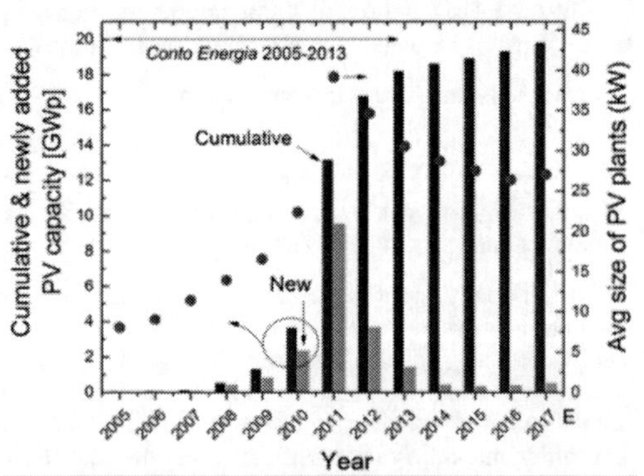

Fig.2: Cumulative and newly added capacity in Italy in the years 2005-2017 (from Ref. [3]).

In parallel, a high demand has often led to considerably higher local market prices for solar components; and, not ultimately, caused the flow of often below-standard components to the domestic markets.

In addition, in booming markets, the lack of trained and skilled professional all along the value chain (systems developers, installers, engineers, electro-technicians, as well as lower qualified personnel) has had a remarkable impact on the overall quality of the solar systems realized. The argument that quality of solar installations clearly follows learning processes is depicted in Fig. 3 (taken from [4]), which shows how, on average, the performance (PR) of the plants realized in later recent years is generally higher. The results were part of a monitoring program realized in the framework of the German 1000-Roofs PV Program.

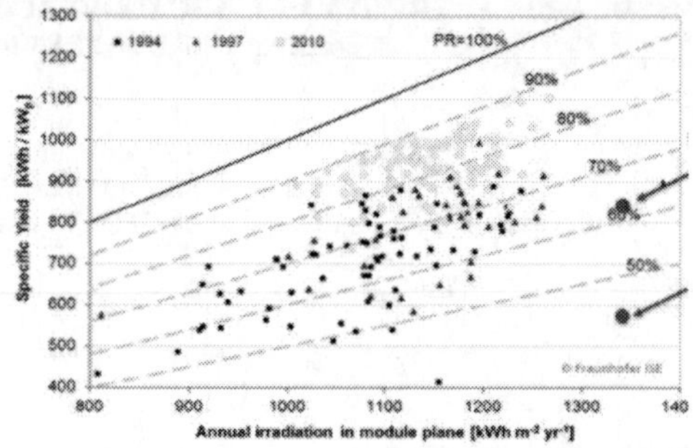

Fig. 3: Specific Yied Y_f (and Performance Ratio PR lines) as a function of plane-of-array irradiance for a selection of PV systems installed in Germany in 1994, 1997, and 2010, and monitored as part of the 1000-Roofs Programs (from Ref. [4]). PV systems installed in more recent years exhibit on average a higher PR, indicating a learning process.

3 THE QUEST FOR QUALITY

In general, quality is a vague concept. When applied to solar projects we can articulate better this concept using a set of complementary words, such as reliability, durability, availability, long-term performance and safety in operation. As such the attention for quality should cover the whole phases of a solar project, including:

a) The selection of state-of-the art components (modules, inverters, batteries, BOS (Balance of System), etc.);

b) State-of-the-art system design (including siting) and planning;

c) Installation and commissioning;

d) Operation and maintenance (O&M), including monitoring;

e) Safety in operation.

In the following, by addressing these different phases in the realization of a solar project, we use some pictures to reinforce our parallel between past and present situation in Europe and Africa, respectively. This gallery of images should help the reader in finding a proper balance between empirical and anecdotal evidence.

3.1 Selection of components

Figure 4 shows different pictures relative to examples of failure modes that are experienced by PV modules installed in solar fields and rooftop installations. All modules were certified products manufactured by different manufacturers in Europe and Asia. The use of certified products is not per-se a guarantee of properly functioning products, but should be considered as a minimum qulity requirement.

Additionally, in the booming years in Europe, a high demand for components has led to evident cases of fraud, with examples of modules (and other components) sold as certified products, but which in reality were carrying falsified labels/certificates. In some cases, European manufacturers of solar panels were selling as own products (and with own labels) modules manufactured in Asia by third party suppliers.

Fig. 4: examples of failure modes on modules that are experienced by PV modules installed in solar fields and rooftop installations. All modules were certified products manufactured by different manufacturers in Europe and Asia. Pictures have been taken in Italy (2017 -2018) on systems installed in the years 2007-2014 (credits: LP).

Presently, several cases of theft of solar modules are reported in different European countries. Allegedly, preferential markets for these stolen products are Africa (in particular North-African countries) and Eastern Europe. A non-careful handling and transportation of these modules (as evident in Fig. 7 and 8) may significantly mine the long-term performance, and sometimes safety, of the devices.

Fig. 5 shows a picture of solar PV modules sold on a local market in Dakar (Senegal). Very often products sold on African markets do not offer any guarantee in terms of **traceability.** Similarly, modules discarded as sub-standard products by manufacturing lines in China, are reported to have easy access to African markets, leading in some cases to serious performance (and in some cases safety) threats.

Fig. 5: PV modules available for sale at a local market (Marché Sandaga) in Dakar (see Ref. [5]). Currently in Africa, most of the PV modules available for individual users or small installers are distributed through retailers with often a very limited knowledge about PV, who cannot offer any traceability about the origin of the products (credits: EA).

Besides the solar modules, Fig. 6 highlights two examples of a poor selection of other components (i.e. the mounting structures) in Europe and Africa.

Fig. 6: poor seletion of mounting structures in Italy (Sicily, absence of ballasts and displacement after strong winds) and in South Africa (Oudtshoorn, non-rigid mountng structure) (credits: MM and RM).

3.2 System design

The proper siting (i.e. site selection) or design of a PV plant are of great relevance to ensure its long-term performance. Particularly in heavily urbanized areas, but not only, one common failure that may lead to significant underperformance is the presence of shading, as shown in Fig.7.

Fig. 7: small shadows, high losses! **Top**: rooftop PV installations in Milan (Italy, 60 kW). Buildings on all sides surround the plant. This reduces considerably the hours of direct sunshine in both summer and winter. Modules rows are so close, casting shadows on adjacent rows. **Bottom**: similarly in Johannesburg, (SA: ~20 kW) strong shading losses will affect plant performance in winter when the sun is much lower over the horizon than in summer (credits: AV and RM).

3.3 Installation and commissioning

Similarly, the installation phase is critical to ensure the quality of a solar project. As an example, often installers, as well as lower qualified personnel (as the workers in charge of transporting and handling the modules), have not received an adequate training on how to handle the modules. Poor handling or transport of the modules may serious endanger their long-term performance. Analogously, stepping on the modules without extreme care during the installation, transport or unloading phase can have very serious consequences on module performance.

Fig. 8: old habits die hard! **Top**: Electro-luminescence (EL) image taken on a module from a residential (5 kW) installation in Switzerland. The dark spots in the image is related to presence of cracks, which have very likely originated during the installation phase and can – to some extent – be correlated to footprints. **Bottom**: footprints on top of a solar rooftop installations in *The Philippines*.

Fig. 8: old habits die hard! South Asia. Extreme care should be taken when stepping on solar panels. The front glass cover may still provide a minimum level of protection to the solar cells. Not the rear (white) polymer cover.

3.4 Operation and Maintenance

Operation and maintenance (O&M) is similarly critical to ensure the long-term performance, durability and – sometimes- safety in operation of the all systems and its components. Similar to shading, dirt and soiling will create strong losses in performance and in some cases, by generating hot-spots, may lead to severe safety threats.

Fig. 9: Dirt and soiling. **Top**: PV module part of a residential rooftop installation in Switzerland covered by bird droppings. **Bottom**: ground-mounted installation covered by a thick layer of sand in Jordan.

3.5 Safety in Operation

Analogously, the safety in operation of a PV system is critical to ensure that anyone (operators, O&M technicians, etc.) is injured during the operation of the plant. Due to the proximity to residents and workers, in the case of rooftop installations, and particularly of Building Integrated PV (BIPV), this requirement becomes even more critical.

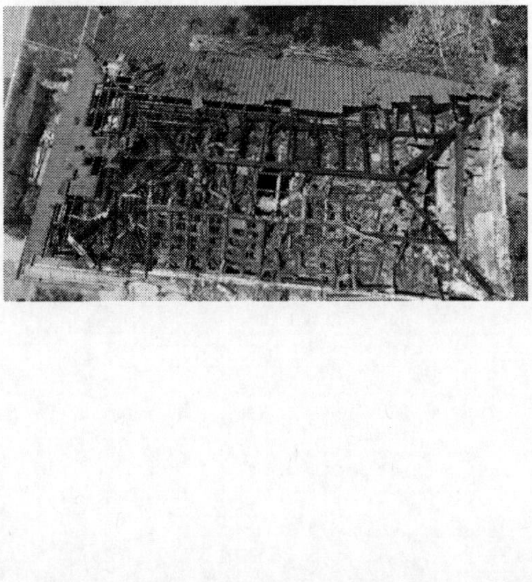

Fig. 10: Safety in operation. **Top**: fire originated by the use of a non-certified Building-Integrated PV (BIPV) product, poor system design, and poor selection of roofing materials (Northern-Italy, 7 kW). **Bottom**: absence of a firebreak, seriously endangering the existence of a groun-mounted PV plant in South Africa (Montague, SA) (credits: RM).

5 RECIPES FOR BANKABLE SOLAR PROJECTS

A strong focus on **quality** should enable the involvement of financial institutions (banks, microcredit and insurances) in the solar sector. Therefore, promoting a virtuous cycle leading to project risks' mitigation, facilitate access to credit and increase perception and awareness in solar electricity. In short, making solar projects *bankable*.

In our vision, a multi-step approach to the **bankability** of solar projects in the rooftop sector should be articulates as follows:

1. Use of **international standards** (and or best practices) whenever is possible. IEC (international Electro-Technical Committee) or *ieee* (Institute of Electrical and Electronic Engineers) standards cover most aspects transversal to the implementation of the whole project: from the selection of certified components, to the design, installation and commissioning phases, as well as to the service life of the plant, including monitoring, and calibration of sensors. Presently, an international standard on O&M is still missing, but international practices – such as the NREL or Solar Power Europe guidelines on O&M [6, 7] - can be used until consensus is reached to draft an international standard. In the case of grid-connected PV plants, country-specific standard covering the "*last mile*" (i.e. the connection of the plant to the grid) will most often be relevant as well.
2. As mentioned previously, **high quality training** is a pre-requisite to achieve a minimum level of knowledge/skills and trigger the **learning processes** that are of great relevance to the industry (see Fig. 3). High-quality training should involve and target professionals all along the value chain, including: systems developers, installers, engineers, electro-technicians, as well as lower qualified personnel, as the workers in charge of transporting and handling the modules. The development of adequate training/qualifications schemes for professionals is of primary relevance.

3. When possible, implement at **local level** institutions able to perform **minimal quality checks** of components (e.g. *testing laboratories*) and systems (e.g. commissioning and site inspections) acting as a trustable and respected third party. Some international aid and cooperation projects are starting to address this with the aim to fill this void (see e.g. the Swiss *REPIC project* (2017-2020) [5] and [8]).
4. When applicable, design **incentive schemes** (FiT, net-metering, tax credits, or other forms of incentives) or clear regulations aimed at avoiding the unhealthy market dynamics that have characterized the early adoption of this technology in Europe (see Sec. 2). Incentives, should rather focus on quality and not on quantity and promote a sustainable take-off of markets, avoiding stop-and-go dynamics.

5 CONCLUSIONS

In this contribution, by addressing the different phases in the realization of a solar project, we use images to make concrete examples and drive a parallel between past and present situations in Europe and Africa, respectively. This gallery of pictures should help the reader in finding a proper balance between empirical and anecdotal evidence.

A strong focus on **quality** should enable the involvement of financial institutions (banks, microcredit and insurances) in the solar sector. Therefore, promoting a virtuous cycle leading to project risks' mitigation, facilitation of access to credit and increased perception and awareness in solar electricity. In short, making solar projects *bankable*.

In our vision, a multi-step approach to the **bankability** of solar projects in the rooftop sector should be articulates as follows:

(1) Focus on **international standards** (and or best practices) whenever is possible. International standards in fact cover most aspects transversal to the implementation of whole project.

(2) Adopt or develop **high quality training** to achieve a minimum level of knowledge/skills and trigger the **learning processes** that are of great relevance to the industry.

(3) When possible, implement at **local level** institutions able to perform **minimal quality checks** of components (e.g. *testing laboratories*) and systems acting as a trustable and respected third party.

(4) When applicable, design **incentive schemes** or clear regulations aimed at avoiding the unhealthy market dynamics that have characterized the early adoption of this technology in Europe.

6 ACKNOWLEDGEMENTS

The authors are grateful to Matteo Marzoli (RSE), Luca Parini (Officina del Sole) and Luca Morganti (Officina del Sole) for fruitful discussions and feedbacks.

We gratefully acknowledge Riaan Meyer (GeoSun Africa), Eleonora Annigoni (EPFL), LP, and MM, for sharing some of the images displayed in this work.

REFERENCES

1 Sustainable Development Goals: United Nations Resolution A/RES/70/1 of 25 September 2015, par. 54;

2 Global solar Atlas: http://globalsolaratlas.info/

3 Alessandro Virtuani1, Mauro Pravettoni1, Luca Parini1, Luca Morganti1, Artur Skoczek2, Juraj Betak2, Where Has All The Power Gone? A Health Check of Italian Solar Electricity in 2016, EUPVSEC 2017, Amsterdam;

4 N. Reich, B. Mueller, A. Armbruster, W.G.J.H.M. van Sark, K. Kiefer, C. Reise, Performance Ratio Revisited: Is PR>90% Realistic?, Prog. Photovolt: Res. Appl.2012;20:717–726

5 E. Annigoni, A. Virtuani, A. Ndiaye, N. Wyrsch, M.L. Ndiaye, C. Ballif, C. M. Fadel Kebe, *Promoting A Sustainable Diffusion Of Solar Pv Electricity In Africa: Results Of The Codev Project,* Proced. EUPVSEC 2017, Amsterdam;

6 Keating, A. Walker, and K. Ardani, SAPC Best Practices in PV Operations and Maintenance, NREL report 2015, www.nrel.gov/docs/fy15osti/63235.pdf

7 O&M Best Practice Guidelines Version 2.0, www.solarpowereurope.org/priorities/operations-maintenance

8 www.repic.ch/repic-de/projekte/laufend/photovoltaik/epfl-imt-senegal/

IMPLEMENTATION OF BUSINESS MODELS FOR RENEWABLE ENERGY AGGREGATORS: EXPERIENCE FROM THE EUROPEAN PROJECT BESTRES

TOPIC 7: PV ECONOMICS, MARKETS AND POLICIES
SUB-TOPIC 7.1: PV ECONOMICS AND MARKETS

Author for correspondence
Silvia Caneva, WIP – Renewable Energies, Sylvensteinstrasse 2, 81369 Munich, Germany
Phone: +49 89 720 12 733, E-mail: silvia.caneva@wip-munich.de

Co-Authors
P. Alonso, S. Challet & I. Weiss, WIP – Renewable Energies, Germany

ABSTRACT

The electricity market landscape is changing because the growth in renewable energy is increasing the share of intermittent electricity and price volatility in the power system. In this framework, the market integration of intermittent electricity from renewable energy sources cannot be achieved by single individual, commercial or domestic consumers since they would only have a limited impact. It is only through a coordinated steering of vast amounts and types of consumers and producers that the use of distributed generation, demand side management and energy storage can be effective. This coordination action can be performed by facilitators such as renewable energy aggregators, which can accelerate the integration of intermittent electricity sources, enhance demand flexibility and decrease the reliance on renewable energy support schemes.

"Renewable energy aggregators are legal entities that aggregate the load or generation of various demand and or generation / production units and aim at optimizing energy supply and consumption technically and/or economically"

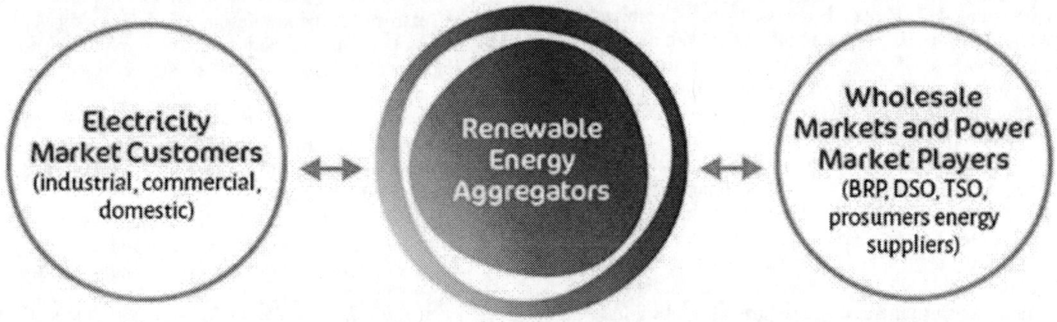

In that context, the project BestRES "Best practices and implementation of innovative business models for renewable energy aggregators" has developed innovative business models for integration of renewable energy sources by aggregating distributed generation such as wind, PV, biogas, biomass, hydro, Combined Heat and Power (CHP) and combining this with demand side management and energy storage. The development of the business models has been based on an analysis of the current barriers for aggregators and the suggestion of ways for improving the role of aggregators in future electricity market designs. Existing European aggregation business models found within and outside the project consortium have been analysed, considering technical, market, environmental and social benefits. This analysis has highlighted the importance of aggregators, who act on behalf of consumers and use technological solutions and ICT (Information and Communication Technology) for optimization, to bring aggregation to the market. Based on the analysis undertaken, the existing European aggregation business models have been improved and, at the time being, some of them are under implementation in real life condition.

The first results related to the implementation of the BestRES business models under real life condition will be presented during the EUPVSEC in Brussels, Belgium, in September 2018.

The BestRES project is a European project, co-financed by the European Commission under H2020 Programme. The project started on 1st March 2016 and will end on 28th February 2019. It involves 11 European partners across 9 different European Countries: United Kingdom, Belgium, Germany, France, Austria, Italy, Cyprus, Spain and Portugal. The main target group, the RES aggregators, are involved directly in the project as partners and they are: Good Energy from UK, Next Kraftwerke Germany (targeting Germany, Italy and France), Next Kraftwerke Belgium targeting Belgium, oekostrom from Austria and EDP (targeting Spain and Portugal). The Consortium includes renewable energy experts (3E and TUW-EEG), 1 legal expert (Foundation for Environmental Energy Law / Stiftung Umweltenergierecht – SUER), 1 dissemination-communication expert (youris.com) and 1 research centre (FOSS). The project is coordinated by WIP – Renewable Energies.

Detailed information on the BestRES project is available at the project website: http://bestres.eu

Keywords: aggregators, business models, policies, renewable energy

1. PURPOSE OF THE WORK AND APPROACH

In a changing electricity market landscape, where the share of intermittent renewable energy in the energy mix is increasing, system flexibility becomes crucial.

In this framework, the market integration of intermittent electricity from renewable energy sources cannot be achieved by single individual, commercial or domestic consumers since they would only have a limited impact. It is only through a coordinated steering of vast amounts and types of consumers and producers that the use of distributed generation, demand side management and energy storage can be effective. This coordination action can be performed by facilitators such as renewable energy aggregators. As part of the solution, the aggregation of renewable energy can significantly accelerate the integration of intermittent electricity sources, complement demand flexibility and decrease the reliance on renewable energy support schemes. Aggregators of demand and/or generation are therefore expected to have an increasingly important role to play in the future.

The BestRES project aims at implementing innovative business models for integration of renewable energy sources by aggregating distributed generation such as wind, PV, biogas, biomass, hydro, Combined Heat and Power (CHP) and combining this with demand side management and energy storage. In order to achieve the overall objective, the work has been structured for achieving the following objectives achieved or to be achieved during the project:

- **Analysis of the existing business models**. The consortium has focused on technical, market, environmental and social benefits that aggregators bring to power markets and systems and the technical, regulatory, and legal barriers preventing optimal deployment of current business models. The consortium has subsequently looked into possibilities for improving business models.
- **Testing of improved business models**. The consortium has improved the business models of selected aggregators in the 9 target countries and will analyze the competitiveness of the improved business models for selected aggregators. A Life Cycle Analysis (LCA) will be also carried out for determining non-economic benefits.
- **Selection of business models for real-life implementation.** The consortium has assessed the feasibility of implementing the business models for each aggregator and defined a pilot implementation plan to identify and attract project participants and implement the business model using real customer data. Special attention will be given to the monitoring and evaluation of the implementation.
- **Recommendations for the uptake of the existing market barriers and for the replication**. This includes an assessment of the impact on RES electricity generation and a development of legal and regulatory recommendations for the target countries and for the replication in countries outside the consortium.

This abstract provides a detailed description of the results achieved until now in the project and on the results to be achieved in the upcoming months to be presented at the EU PVSEC in Brussels, Belgium, in September 2018.

2. SCIENTIFIC INNOVATION AND RELEVANCE

The BestRES projects aims to better integrate renewable energies into the power system with a special focus on market integration. New renewable energy business models that allow aggregators to successfully participate in the market by combining different RES technologies, energy storage, flexible demand and other technologies into a commercially viable product have already been assessed and improved in the first part of the project. At the time being, they are under implementation to proceed then with their evaluation in real life condition. BestRES will foster the full market integration of renewable energy sources, which will increase profitability and reduce reliance on today support schemes. The project will also support the establishment of a robust regulatory framework that encourages investments in the renewable energy sector.

The expected impacts of the project are:

- Successful implementation of business models and full market integration of renewable energy sources beyond support schemes;
- Increase of the hosting capacity of the transmission grid and ancillary services provided to the TSOs through direct implementation of the improved renewable energy business models;
- Assessment of market design for ancillary services across the 9 partner countries, and in particular for implementation of renewable energy aggregators business models;
- Recommendations on market design and on the legal and regulatory framework for creating power systems with flexible demand and improved demand side management;
- Support to policy makers for creating new policy strategies that support the renewable energy aggregator business models and the integration of renewable energy overall.

3. RESULTS AND CONCLUSIONS

During the first phase of the BestRES project, the consortium has assessed existing aggregator business models available at the time being in the European Union highlighting technical, market, environmental and social benefits. Barriers and opportunities for renewable energy aggregation have also been investigated. The existing business models were improved (Table 1) considering the future market designs in the nine target countries: United Kingdom, Belgium, Austria, Italy, Cyprus, Germany, France, Spain and Portugal (Figure 1).

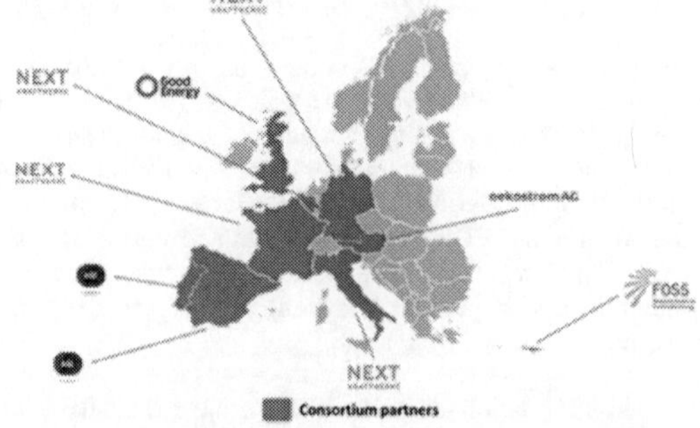

Figure 1 - Countries covered by the BestRES consortium

Table 1 - Description of improved business models

Aggregator	Improved business model
Good Energy (UK)	Automation and control
	"Peer-to-peer" (local) energy matching
Next Kraftwerke Germany (Germany)	Dispatch flexible generation under changing market design on multiple markets
	Supplying „mid-scale" customers with time variable tariffs including grid charges optimization
Next Kraftwerke Germany (France)	Providing decentralized units access to balancing markets
Next Kraftwerke Germany (Italy)	Market renewables on multiple market places
Next Kraftwerke (Belgium)	Trading PV and Wind power
	Using flexibility of customers as third party
Oekostrom AG (Austria)	Demand Side flexibilization of small customers
	Invest and market distributed generation of customers in apartment houses
EDP (Portugal)	Activation and marketing of end user's flexibility.
EDP (Spain)	Activation and marketing of end user's flexibility.
FOSS (Cyprus)	Pooling flexibility for local balancing market and energy service provision

Explanation of the improved business models

- **Automation and control**: Provision of hard- and software solutions allowing customers to automate their devices.
- **"Peer-to-peer" (local) energy matching**: Unite customers and generators on a local level and create value for both.
- **Dispatch flexible generation under changing market design on multiple markets**: Increase value for generators by trading on multiple market places under changing market design, e.g. spot, balancing and reserves markets.
- **Supplying „mid-scale"customers with time variable tariffs including grid charges optimization**: Time variable tariffs (especially grid charges) including monitoring will help consumers to benefit from market signals.
- **Providing decentralized units access to balancing markets**: Distributed generators benefit from portfolio effects.
- **Market renewables on multiple market places:** Forecasting quality can be increased by using live data and portfolio effects. Valorising pooled generation at dispatch and balancing markets.
- **Trading PV and Wind power:** Market the generation of renewable generators. As an effect of portfolio effects, benefits could be achieved.
- **Using flexibility of customers as third party:** This improved business model aims for customers whose supplier is not marketing flexibility. Flexibility can be valorised by an aggregator without changing the supplier.
- **Demand Side flexibilization of small customers:** Activates demand side potential of customers. Allows a shift to a more integrated energy service provider.
- **Invest and market distributed generation of customers in apartment houses:** this improved business model aims in the market integration of solar generation of customers in apartment houses.
- **Activation and marketing of end user's flexibility**: Flexibility of customers will be activated and valorised on energy markets.
- **Pooling flexibility for local balancing market and energy service provision:** because the Cypriot markets are not opened yet, this improved business model uses aggregation for providing DSO services.

The core of all aggregator business models is creating value and revenues by aggregation. Most aggregator business models (9 out 13 improved business models) aim to valorize aggregation at wholesale, balancing and reserve markets. As Figure 2 illustrates, major changes in the electricity market design can be expected from the "Clean Energy for All Europeans Package": active customers, local energy communities, demand response, balancing markets, dispatching of generation and demand response, redispatching and curtailment. To achieve this transformation, new market participants and business opportunities are desirable.

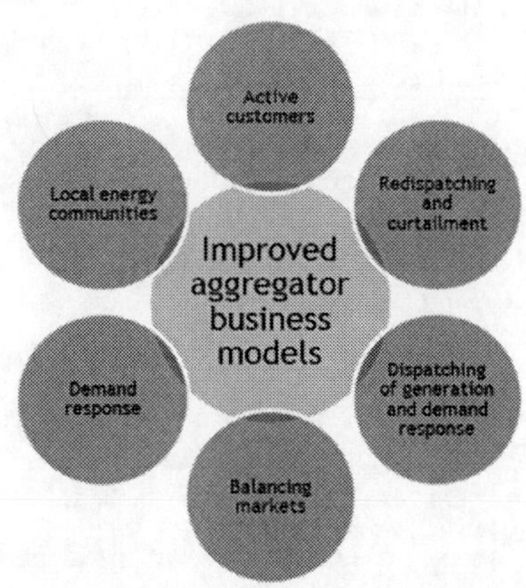

Figure 2 - Future opportunities related to the "Clean Energy for All Europeans Package"

The BestRES consortium has then investigated for each of the improved business models if they are ready for implementation on the basis of an economic feasibility study which led to allocate the business models into three groups:

- **Group 1**: Aggregators with economically viable business models and no substantial legal, social and technical barriers should proceed with real-life implementation with consortium support.
- **Group 2**: Aggregators with economically viable business models but with feasibility issues should proceed with virtual implementation, involving scenario modelling with actual customer data.
- **Group 3**: Aggregators with business models that are not economically viable will not proceed with implementation; however, some commentary on these cases will be provided. Barriers hindering implementation of these models will be identified and discussed.

At the time being, the improved business models ready for implementation are under implementation with real data and monitored. Recommendations will be developed for the replication of the business models in Group 1 and for overcoming the barriers hindering the implementation of the business models in Group 2 and 3. The allocation of the improved business models (BMs) in Group 1, 2 and 3 is shown in Figure 3, 4 and 5.

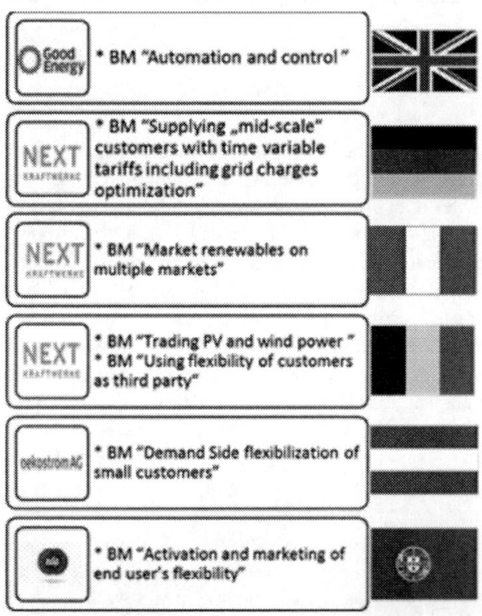

Figure 3 – Allocation of the improved business models (BMs) in Group 1

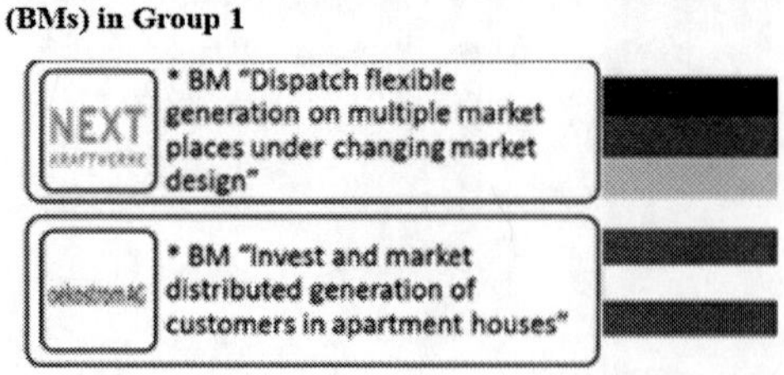

Figure 4 – Allocation of the improved business models (BMs) in Group 2

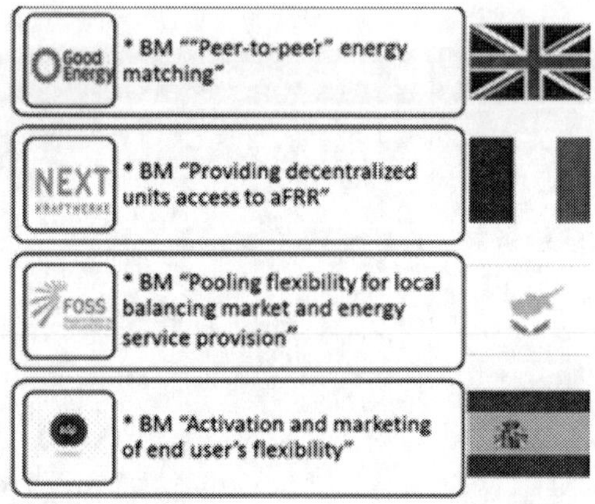

Figure 5 – Allocation of the improved business models (BMs) in Group 3

Business models (BMs) ready for implementation (7 out of the 13 BMs) are identified in the United Kingdom, Austria, Germany, Italy, Belgium and Portugal. For group 1 BMs in the United Kingdom and Austria, the aggregator manages to decrease sourcing costs whereas costs (and turnover) to end customers decrease. In Germany, Next Kraftwerke optimizes wholesale and network tariffs. In both Italy and Belgium, Next Kraftwerke generates revenues from fees on reserve power markets. Next Kraftwerke, with another improved BM, also trades weather dependent renewables such as PV and wind on spot markets in Belgium. Finally, in Portugal, EDP uses flexibility from loads to decrease imbalance penalties of the own portfolio.

Two BMs face barriers that prevent direct implementation in the short to medium term (group 2) and four BMs are not feasible (group 3). The research carried out by the BestRES consortium also underlined that almost all BMs that have no significant barriers are ready for implementation. EDP with its BM "Activation and marketing of end user's flexibility" in Spain is the only exception. In this case, although there are no significant technical, legal, regulatory or other barriers for implementation, the BM is not feasible due to low imbalance tariffs. For all other BMs that are not yet ready for implementation, the main hurdles are related to regulation. Therefore, aggregators will only be able to implement these BMs in the medium (group 2) to long (group 3) run.

The proposed improved business models are designed to address future market and regulatory needs.

ACKNOWLEDGEMENT

The BestRES project has received funding from the European Union's Horizon 2020 research and innovation programme under grant agreement No. 691689.

DISCLAIMER

The sole responsibility for the content of this publication lies with the authors. It does not necessarily reflect the opinion of the European Commission. The European Commission is not responsible for any use that may be made of the information contained therein.

COOPERATIVELY-OWNED BATTERIES AS A CONCEPT TO PREVENT LOCAL GRID CONGESTION

Giulia Leghissa[1,2], Menno van den Donker[1], Chris Tzikas[1], Geert Verbong[2], Wiep Folkerts[1]
[1]SEAC (Solar Energy Application Centre), Eindhoven, The Netherlands
[2]TU/e Eindhoven University of Technology, Eindhoven, The Netherlands

ABSTRACT: There is a growing interest in community-owned large-scale solar projects. However, if large amounts of electricity generated by these PV installations are not directly consumed by the local loads, grid congestion can arise, due to constrains in the local electricity network. This paper explores when does grid congestion start to occur and whether battery storage systems can be an economically feasible solution to mitigate this problem, as an alternative to grid reinforcement measures. In addition, a proper sizing of the battery storage with respect to solar park nominal capacities is assessed, based on simulations and benchmark study of current practices.
Keywords: Batteries, Economic Analysis, Grid Congestion, Large Grid-connected PV systems

1 INTRODUCTION

In the last decades, worldwide solar photovoltaics market has seen an exponential growth, with a cumulative installed capacity of approximately 400 GW in 2017 [1]. A similar increasing interest in this technology can be observed also in community-driven solar projects, typically established by energy cooperatives interested in boosting the local sustainable energy generation. One main advantage of these types of projects is that they provide the opportunity to expand the adoption of solar power to consumers that have financial or structural constraints (such as no available space on their own rooftop or not enough capital to invest in a private home system) [2]. Moreover, they are usually seen as bottom-up initiatives, gaining a higher social support and consent.

However, when large amounts of electricity from the solar projects are not directly consumed by the local community, grid congestion due to a limited network capacity may occur. The current typical solution to prevent this local grid congestion is to take expensive grid extension and reinforcement measures [3]. Nevertheless, recent years have seen a rising interest in alternative ways to address these problems. Among others, Lithium-ion Battery Energy Storage Systems (BESSs, also simply called batteries) have gained momentum as the most promising storage solution to be implemented in combination with solar PV installations [4] that can be used for a variety of different applications, among which also mitigation of grid congestion problems.

This paper describes the financial feasibility of adding a cooperative district battery storage system to a large-scale grid-tied PV installation as an alternative to overcome local grid congestion. Firstly, the evaluation of when grid congestion starts to occur are presented in Section 2. Following that, the possible solutions to mitigate this problem are shown in Section 3, together with the assessment of a proper sizing of the battery storage system. Section 4 describes the financial analysis of the proposed solutions. Lastly, conclusions are given in Section 5.

2 GRID CONGESTION EVALUATION

For this research, a case study linked to a specific Solar-Plus-Storage project in the Netherlands, was selected as basis for the analysis. In particular, the project, called Cooperative Sustainable Business Models for Storage (COOP-Store project), takes as a "living lab" the implementation of a cooperative district storage facility in combination with a community solar park, located in the neighborhood of Altweerterheide (Limburg, The Netherlands). The project is a collaboration between Solar Energy Application Centre (SEAC), Scholt Energy Services, Soltronergy, WeertEnergie and Chematronics, and is financially supported by the Dutch Topsector Energy of the Ministry of Economic Affairs.

To study when grid congestion starts to occur, electricity consumption and production patterns were analysed for different solar PV installation capacities. The former was obtained via the online dataset available from the NEDU (Nederlandse Energie Data Uitwisseling) platform, which provides the average consumption profiles for Dutch private houses and businesses. Instead, the latter has been obtained by simulations of solar PV production profiles with the software PVSyst for a typical meteorological year, using different solar plant sizes.

Lastly, the maximum grid load (i.e. the maximum power that the local grid can handle, above which grid congestion starts) was estimated using the maximum peak for the entire community (defined here as the case when the aggregated household peak coincides with the aggregated business properties' peak load).

The information regarding consumption and production patterns was then combined with the grid characteristics. The presence of grid congestion was defined as the case when the excess electricity (corresponding to the difference between the production from the solar park and the demand of electricity from the community, in other words what gets fed back into the grid) is greater than the maximum grid load.

The results of this analysis show that local grid congestion starts to occur for a Production-to-Consumption (P/C) ratio of 0,5 (i.e. when the solar panels produce on a yearly basis half the electricity consumed by the entire community). This problem is further increased for bigger solar park capacities (corresponding to higher P/C ratios, given the fact that the consumption is kept constant).

Taking a sunny summer's day as an example, Fig. 1 shows how the higher P/C ratio is, the more excess energy is produced. Since this electricity is not consumed directly, it is injected into the grid, where it can cause grid congestion. Assuming that no measures are taken to mitigate these effects, the excess electricity above the maximum grid load is lost since the grid cannot handle it;

thus, it counts as curtailed electricity.

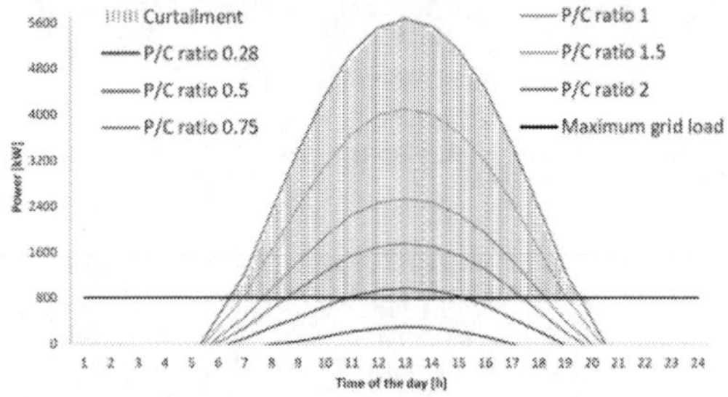

Figure 1: Simulation results of excess electricity and electricity loss (curtailment) for a sunny summer's day, for different production-to-consumption ratios.

On a yearly basis, the limited network capacity can lead to significant energy losses, as shown in Table I. For small P/C ratios, the amount of yearly lost electricity is small, with grid congestion occurring approximately 135 hours throughout a year (representing only 1,5% of the total hours in one year). On the other hand, for higher P/C ratios, this problem is enhanced. For example, when the solar panels produce on a yearly basis the same amount of electricity that is consumed by the village, 20% of their combined production is lost. Lastly, if twice the energy needed in the rural area is produced from PV panels, a remarkable waste of electricity is taking place, since curtailment almost reaches 50%.

Table I: Percentage of time (over one year) for which grid congestion occurs and the corresponding electricity loss (curtailment) as percentage of the total yearly electricity output, for different production-to-consumption ratios.

P/C ratios	% time grid congestion	% electricity curtailed
< 0,5	0	0
0,5	1,5	0,8
0,75	8,1	9,9
1	12,9	20,3
1,5	20,3	36,9
2	25,0	48,4

3 GRID CONGESTION SOLUTIONS

In order to overcome the problem of grid congestion and avoid the curtailment of electricity, the current Business As Usual procedure is to increase the network capacity by strengthening the grid, i.e. additional underground cables or transmission lines are installed to allow a smooth flow of electricity from the utility to the load.

In general, the costs associated with grid reinforcement and extension are site specific, depending on several factors: the most influencing are the power capacity of the power plant, the current structure of the electricity network and the impact of the newly connected utility on grid stability and power quality [5]. While the connection costs of the solar park to the grid (which are always present) are entirely borne by the generator, grid reinforcement costs allocation varies. In this regard, there are currently two charging schemes employed: the deep connection costs scheme, which imposes all the infrastructure-related costs (connection to the network, grid extension and reinforcement) on the developer of the solar park, and the shallow connection costs scheme, where the power station developer is directly responsible only for the costs of the connection to the electricity grid, whereas the other costs within the network, related to the grid enhancement, are borne by the system operator (who recovers the costs via use-of-system charges applied to the network customers) [3].

However, another way to solve this problem consists of installing a battery storage system in proximity of the solar park. Several studies [2, 6-9] have already pointed to the fact that battery systems can be used for different applications; among others, this technology can be used for supply peak shaving purposes. In this regard, the peak power produced by the solar park during times of high renewable resource availability (such as at midday during sunny summer days), which is not directly met by the loads, can be stored into the battery storage and injected into the grid at a later time.

3.1 Sizing of the battery storage system

In order to assess the applicability and feasibility of battery storage as a way to solve grid congestion, it is important to determine an appropriate size of the battery system in relation to the size of the solar system (nominal capacity). To do so, a two-fold analysis was used in this research: simulations and benchmark study of current Solar-Plus-Storage projects.

In case of simulations, beginning with the situation without a battery system, the storage capacity is gradually increased while at the same time curtailment percentages are monitored. The increase of the battery storage's size is interrupted only when 0% of curtailment is reached. The results of this analysis are depicted in Fig. 2, where the different storage systems' capacities are shown in relation to the solar systems' nominal capacities (Battery to PV, or simply B/PV, ratios).

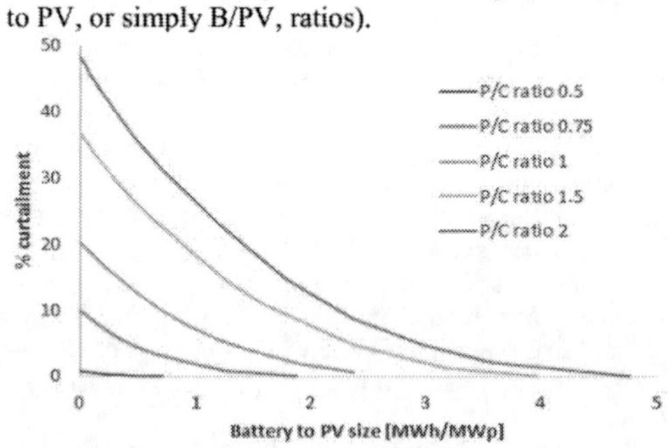

Figure 2: Curtailment percentages as a function of battery storage sizes (in relation to solar systems' nominal capacities), for different production-to-consumption ratios.

As Fig. 2 indicates, for a P/C ratio above 1, a B/PV ratio above 2 MWh/MWp is required to completely solve grid congestion (meaning that a battery storage with a capacity twice as big as the solar park is needed); for higher P/C ratios, even bigger battery systems are required (up to B/PV ratio of 4,7 MWh/MWp).

Focusing instead on current practices of worldwide Solar-Plus-Storage projects obtained by a benchmark study, from the 36 large-scale projects identified (of which 23 smaller than 10MWp shown in Fig. 3), it has been observed that the services that batteries are assigned to provide influence the sizing of the system. In particular, three trends have been determined for battery-to-PV ratios: if the battery system is used in microgrids or off-grid projects, a large B/PV ratio is typically used (approximately B/PV ratio of 3 MWh/MWp); for time shift of electricity, a B/PV of 0,6 MWh/MWp is used; lastly, if the battery system provides ancillary services to the grid, a smaller battery to PV size is usually considered.

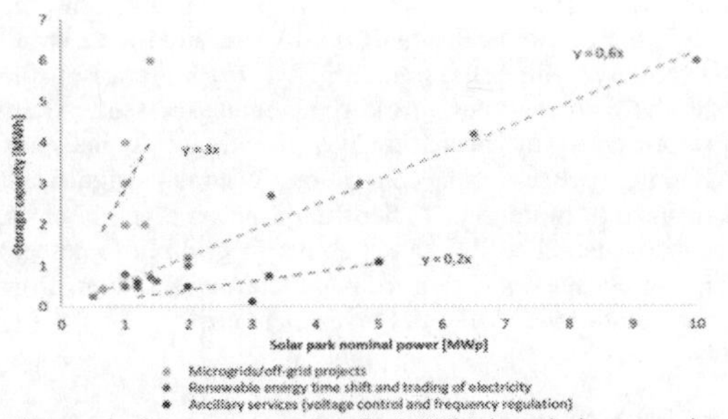

Figure 3: Storage capacity (in MWh) in relation to solar park nominal power (MWp) from current worldwide Solar-Plus-Storage projects.

A first conclusion that can be drawn from this analysis is that there is no "one-size-fits-all" solution: the optimal sizing of the battery storage highly depends on the application that the battery storage system is designed to provide. In general, this ranges from a battery-to-PV size (B/PV) ratio of 0,2 MWh/MWp up to 3 MWh/MWp for current practices, whereas comparing these numbers with the B/PV ratios needed to solve grid congestion obtained via simulations even bigger battery systems might be necessary, depending on the production-to-consumption ratios and the grid characteristics.

4 FINANCIAL ANALYSIS

In order to compare the costs and benefits for the battery storage as an alternative solution to overcome grid congestion, a financial analysis is performed, both if the total investments are calculated as well as if only the costs per time of use are considered.

4.1 Total investment costs

Firstly, the total investment costs required for a solution are considering. In this regard, three different scenarios are compared:

• Scenario 1: large-sized battery scenario (grid congestion is solved by means of a battery storage with a B/PV ratio of 3 MWh/MWp, following the results from the benchmark study; if there is still excess electricity that can cause grid congestion, curtailment is performed).

• Scenario 2: medium-sized battery scenario (also here a combination of battery storage and curtailment is used to overcome grid congestion; however, here the sizing of the battery storage is derived with the B/PV ratio of 0,6 MWh/MWp, again as obtained from the benchmark study).

• Scenario 3: one-solution scenario (here grid congestion is entirely solved by means of one solution, i.e. grid reinforcement, implementation of a battery storage or curtailment of excess electricity).

In this way, not only the investment costs for one-solution strategies are considered, but they also compared with common practices from current projects around the world.

Every scenario considered in this analysis is constructed in such a way that the revenues from selling the electricity produced by the solar panels are the same. For this reason, in order to include curtailment as one of the possible solutions and compare its financial feasibility on the same basis as grid reinforcement measures and battery storage systems (i.e. by looking only at the upfront expenditures needed when a solar park is implemented), the investment costs for additional PV panels that would be needed to recover all the lost electricity are calculated in this case.

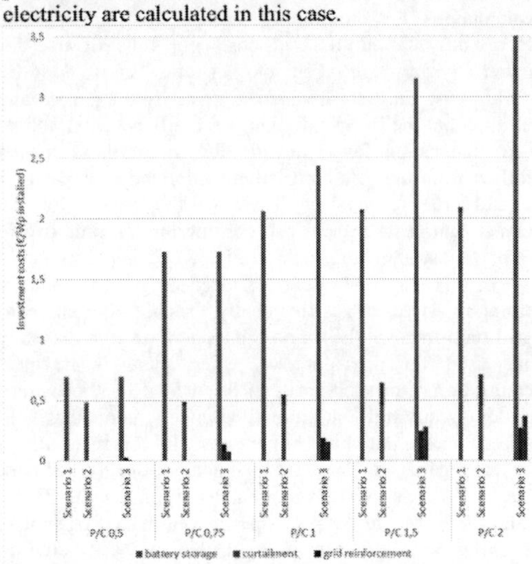

Figure 4: Investment costs for each scenario, for different P/C ratios. The costs are expressed in €/Wp of installed solar power capacity.

The financial analysis indicates that curtailment is the cheapest solution for smaller P/C ratios, or better when the amount of excess energy is not significant. Instead, at higher percentages of extra electricity being produced by the solar panels, grid reinforcement is the most cost-effective strategy. Therefore, it is not surprising that this strategy is the dominant current practice (or Business As Usual measure).

Overall, the battery storage presents the highest investment costs: the bigger the battery storage system, the more expensive a solution is. In addition, the bigger the P/C ratio, the larger is the difference between the costs in case of Large-sized battery scenario, Medium-sized battery scenario and One-solution scenario with only battery storage. For example, for a P/C ratio of 2, the investment costs for the Medium-sized battery scenario are almost one third of the costs for the Large-sized battery scenario and approximately one fifth of the storage One-solution scenario. This results clearly emphasize the impact of the size of the battery on the total investment costs.

However, battery storage systems implemented merely as a solution to solve grid congestion are needed only during peak times during sunny days. As shown in Table I, this corresponds to less than 10% of the time throughout the year for P/C ratios below 0,75. Therefore, when the battery storage systems is not used to mitigate grid congestion problems, it can be used for the provision of other services such as energy trading, frequency regulation or increased self-consumption of the produced solar output; this is typically known as benefit stacking. This means that additional revenues are possible if a battery storage system is implemented to solve grid congestion, whereas this is not the case for grid reinforcement measures or curtailment. To take into account this characteristic, the time of use of the battery storage are included in the financial analysis.

4.2 Investment costs per time of use

In this case, the same financial analysis as is Sub-section 4.1 is considered, the only difference being in the calculations for the battery storage investment costs. Here, only the investment costs per time of use for battery systems are taken into account. This is done by multiplying the total investment costs by the percentage of time that the battery storage is actually needed to solve grid congestion for a specific P/C ratio (these results follow from the grid congestion evaluation from Section 2 and are presented in Table I); the results for the investment costs comparison considering the time of use of battery system are shown in Fig.5.

In this way, it is possible to check whether battery storage systems are a financially feasible alternative to grid reinforcement based on the amount of time that they are used for this purpose; when there is no grid congestion, other services can be tackled by the battery system, resulting in additional revenue streams that help to recover the initial investment costs for this technology. In this regard, a techno-financial analysis for a solar park with a P/C ratio of 0,5 and a battery storage with a B/PV ratio of 0,6 MWh/MWp was performed in case of benefit stacking, with inputs from the Dutch market (including revenues from selling the electricity and from the various applications tacked by the battery system). The battery storage was partially used to solve grid congestion (which is needed approximately 1,5% of the time throughout the year), while the rest of the time it was used for energy arbitrage (in particular on the Dutch imbalance electricity market) and frequency regulation (tendered on the primary reserve capacity market). The results indicate that the total initial investments (for both the solar park and the battery system) can be recovered within 10 years, providing that frequency regulation is tackled for most of the time when there are no grid congestion problems.

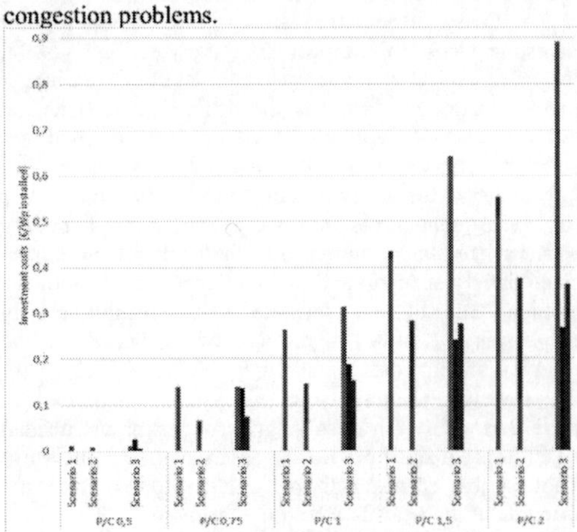

Figure 5: Investment costs (expressed in €/Wp of installed solar power capacity) for each scenario for different P/C ratios, in case of time of use investment costs for the battery storage system.

When the time of use of the battery storage is considered in the investment costs, the outcomes of the financial analysis suggest that this solution is not the most expensive for every P/C ratio. For a P/C ratio of 0,5, the highest investment costs for solving grid congestion are associated with grid reinforcement, while the least expensive solution is to curtail the excess electricity, due to the limited amount of excess electricity above the maximum grid load present in this case. For P/C ratios up to 1, the economically most feasible solution is to install a battery storage system with a B/PV ratio of 0,6 MWh/MWp and curtail the remaining excess electricity (represented by Scenario 2 in the financial analysis). Finally, when grid congestion becomes significant (occurring more than 13% of the time on a yearly basis, corresponding to P/C ratios above 1) grid reinforcement is the cheapest solution to overcome this problem, from the point of view of initial investment costs.

5 CONCLUSIONS

The widespread of PV development, also in the case of community projects, can pose challenges to the electricity grid. In this regard, our results suggest that grid congestion starts when the PV installation produces on a yearly basis half than the aggregated consumption of a community (P/C ratio of 0,5). If the solar PV capacity is further increased, the problems on the electricity network are enhanced. If no measures are taken to mitigate these effects, energy losses due to curtailment can become significant.

To solve this, grid extension and reinforcement measures are typically taken. However, cooperatively-owned batteries also present an interesting and innovative option to prevent local grid congestion. In this regard, a properly sized battery is essential. Although the battery-to-PV ratios can vary from a ratio of 0,2 MWh/MWp up to 3 MWh/MWp or even higher (as obtained via the benchmark study and simulations), the outcomes of the financial analysis indicate that a battery-to-PV ratio of 0,6 can be the most feasible solution in terms of investment costs, provided that the system can be used successfully also for other services.

6 REFERENCES

[1] IRENA, "Renewable capacity statistics 2018", International Renewable Energy Agency (IRENA), Abu Dhabi, 2018.

[2] E. Takata, "Analysis of Solar Community Energy Storage for Supporting Hawaii's 100% Renewable Energy Goals", The University of San Francisco, 2017.

[3] T. Cottier, I. Espa, "International Trade in Sustainable Electricity: Regulatory Challenges in International Economic Law", Cambridge University Press, 2017.

[4] IRENA, "Battery Storage for Renewables: Market Status and Technology Outlook", International Renewable Energy Agency (IRENA), 2015.

[5] R. Barth, C. Weber, D. J. Swider, "Distribution of costs induced by the integration of RES-E power", Energy Policy, Vol. 36 (2008) 3107-3115.

[6] C. A. Hill, M. C. Such, D. Chen, J, Gonzalez, W. M. K. Grady, "Battery energy storage for enabling

integration of distributed solar power generation", IEEE Transactions on Smart Grid 3(2) (2012) 850–857.

[7] R. Luthander, J. Widén, J. Munkhammar, D. Lingfors, "Self-consumption enhancement and peak shaving of residential photovoltaics using storage and curtailment", Energy 112 (2016) 221–231.

[8] Roland Berger, "Energy storage can bring utilities back into the game", Paris, 2017.

[9] J. Tant, F. Geth, D. Six, P. Tant, J. Driesen, "Multiobjective battery storage to improve PV integration in residential distribution grids", IEEE Transactions on Sustainable Energy 4(1) (2013) 182–191.

IMPACT OF PV POWER LOSS/GAIN ON PV POWER COST AND PV ADOPTION

Nassma Mohandes, Ali Elrayyah, Antonio Sanfilippo, Ameni Boumaiza
Qatar Environment & Energy Research Institute, Hamad Bin Khalifa University, Qatar Foundation
P.O. Box 34110, Doha, Qatar
{nsalim, aelrayyah, asanfilippo, aboumaiza}@hbku.edu.qa

ABSTRACT: We evaluate power losses and gains resulting from the integration of Rooftop Photovoltaic (RPV) systems into the power grid, within scenarios where the RPV power is either fed back to the grid or consumed "behind the meter". Power losses resulting from feeding RPV power to the grid are calculated relative to the distance of the site of power generation to the distribution feeder, using GIS data. When the power generated by RPV systems is consumed "behind the meter", this loss is turned into an equivalent power gain. Early results for 65,536 buildings in Qatar with the assumed integration of a 5kW PV system per building show yearly power losses/gains of 7020MWh. Simulations of residential PV adoption show that these power gains have a mild but distinctive effect, leading to a 1.3% increase in adoption rate. The study suggests that overall power losses/gains are modest, providing mild support for a "behind the meter" PV integration solution.
Keywords: Solar energy, power loss, power transmission, residential PV adoption, GIS, agent-based modeling.

1 INTRODUCTION

Many factors contributing to a household's decision to adopt PV have been discussed, including payback period, incentives, PV system and electricity prices, innovation diffusion, rooftop PV potential, and carbon tax [1]-[6]. In this study, we examine the potential impacts of power losses or gains from distributed generation through Rooftop Photovoltaic (RPV) systems on PV power cost, and the ensuing effect on a household's decision to adopt PV. This is a factor that existing studies examining economic drivers of residential PV adoption have not yet considered, to the best of our knowledge. Understanding the impact of PV power losses/gains with RPV systems on PV power cost and adoption can help policymakers and users alike to assess the risks associated with adoption more comprehensively, learn how to address challenges, and take advantage of available opportunities.

We first calculate potential power losses for RPV systems based on the distance from the site of power generation to the distribution feeder, using GIS information about the location of residential buildings with reference to the power network layout. This loss is turned into an equivalent power gain when the power generated by RPV systems is consumed "behind the meter". The measured power losses (gains) are then quantified into their monetary equivalent and added to (subtracted from) the cost of RPV within a multi-agent simulation model of residential RPV adoption to evaluate its impact on RPV adoption.

2 MEASURING POWER LOSS IN GRID-INTEGRATED RESIDENTIAL PV SYSTEMS

The first step in measuring power loss for residential PV systems is to create a dataset that combines information about building location in relation to the low voltage power line that links each building to its distribution feeder. To create such as dataset, we used a 2014 GIS dataset of the power network in Qatar, kindly provided by the Qatar Ministry of Development Planning & Statistics and Kahramaa (Qatar General Electricity & Water Company). We combined three GIS shapefiles contained in this dataset representing the cut-out point layer, where each cut-out point corresponds to a building, the medium voltage layer providing the location of the substations, and the low voltage layer connecting substations to cut-out points through low voltage cables. A sample of the resulting dataset is shown in Figure 1, where black lines represent low voltage lines, red dots are the cutout points connecting buildings to substations, and blue rectangles represent substations.

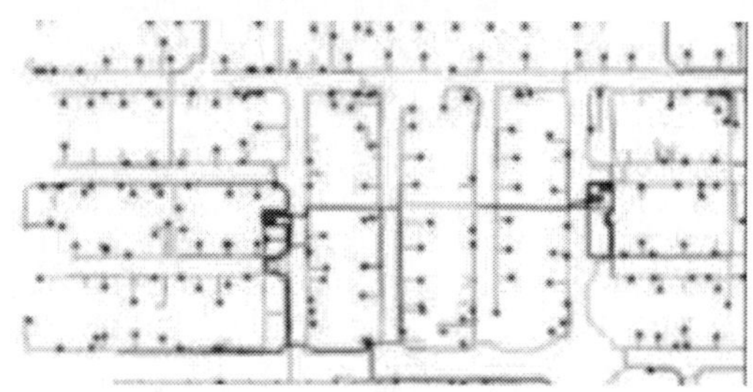

Figure 1: Example of GIS data.

In estimating power loss using GIS data, we assume that the generated PV power is entirely fed to the utility grid. The estimated power loss is considered as an equivalent power gain in a scenario where the power generated by the RPV system is consumed "behind the meter", i.e. within the building, instead of being fed to the grid.

As shown in Figure 2, the segments on a distribution line within a feeder can be combined to characterize power flow using the GIS data. These GIS data contain information about the start and end points of each line segment represented by their geographical coordinates, and the cable type and size. GIS data with such characteristics enable an accurate calculation of the segment resistance factor needed to determine power loss.

The power that flows over each line segment (punctuated by red dots in Figure 2) is determined by summing up the power of all PV sources in the segment. Let the power that pass over specific segment seg_x be P_x, and let the resistance of that segment be R_x. The power loss $P_{loss,x}$ over seg_x can be determined by the equation in (1), where V_L is the standard low voltage line to neutral (240 volts in Qatar), and R_x is calculated by multiplying the segment length by the line resistance per meter (Ω/m) as determined by the cable type and size.

$$P_{loss,x} = \left(\frac{P_x}{V_L}\right)^2 R_x \quad (1)$$

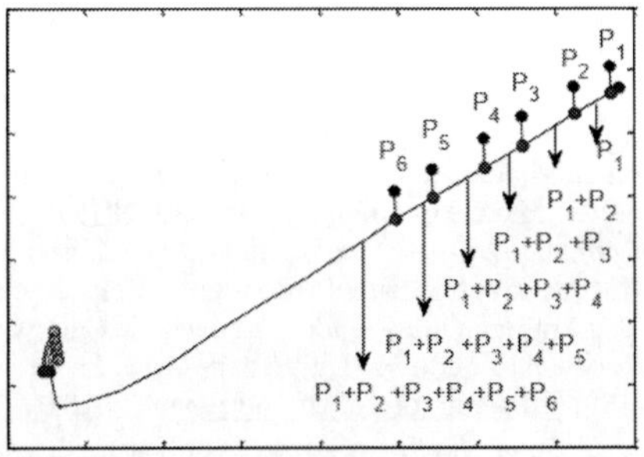

Figure 2: Power flow over various segments in the same distribution line.

3 POWER LOSS/GAIN RESULTS

We calculated power losses and gains for reference 5kW RPV systems installed at each of 65,536 cut-out points, each cut-out point corresponding to a building, assuming that each RPV system would be linked to a substation through a low voltage cross-linked polyethylene insulated power cable with a resistance of $0.669\,\Omega/km$ [7]. According to (1), the power loss resulting from feeding back to the grid power generated from each RPV system is $\frac{\left(\left(\frac{5000}{240}\right)^2 * 0.000669\right)}{1000}$ kW per meter. For example, power generated from a 5kW RPV system that is 430m away from its closest substation, where the PV power would be fed back, would incur a loss of 0.125kW. Assuming an average 5.8 sun peak hours for Qatar [8], the yearly power loss would be 264.32kWh, which at the average electricity tariff of 4.50375¢/kWh for residential, commercial and government buildings [9] is equivalent to \$11.9. Overall, the estimated yearly power loss/gain is 7020MW and its USD equivalent is \$316,172, with an average of \$4.72 per RPV system. Table 1 provides a breakdown of PV power loss percentage for the 65,536 buildings[1] analyzed in five clusters. Figure 3 shows the distribution of estimated yearly power losses/gains in USD.

Table 1: PV power loss percentage for the 65,536 buildings in five clusters.

PV Power Loss Range	*Number of cut-out points*
0%-0.99%	36335 (55.4%)
1%-1.99%	25146 (38.4%)
2%-2.99%	3611 (5.5%)
3%-3.99%	368 (0.59%
4%-8%	75 (0.11%)

[1] In the 2014 dataset we used, 42% of cut-outs points were not connected to substations and could not be included in our RPV power loss/gain calculations.

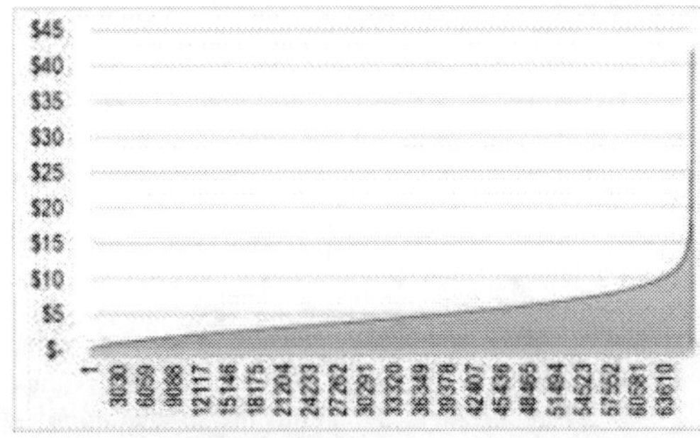

Figure 3: Estimated yearly power loss/gain in USD for 65,536 5kW RPV systems (x-axis) in Qatar.

4 IMPACT OF PV POWER LOSS/GAIN ON RESIDENTIAL PV ADOPTION

We demonstrate the impact of power loss/gain on PV adoption in Qatar using a simplified version of the agent-based model described in [5], where agents represent residential households whose decision to adopt PV is driven by cost. Households adopt solar PV with a probability established by the logistic function in (2), where L is a scaling constant, e is the natural logarithm, x is cost of PV minus the ongoing cost of electricity, and k is a parameter that determines the slope of the adoption curve. We set $L=1$ to normalize the output of the logistic function as a probability. For the k parameter, we select a value that yields a PV market share of 2.5% at the end of the simulation in a scenario where the cost of PV stays constant through time. This market share is equivalent to the number of *innovators* in Rogers' innovation adoption curve [11], who by disposition are willing to adopt novel technology at a premium price.

$$f(x) = \frac{L}{1+e^{-k*x}} \quad (2)$$

At each simulation tick, each household agent that has not adopted yet is presented with the opportunity of doing so. Adoption is determined randomly in terms of the output of the logistic function in (2): a random probability p is generated, and if the probability of adoption as calculated by (1) is greater or equal to p, adoption occurs.

Using the Netlogo [12] environment, we simulated two scenarios: one with power loss and one without. Each scenario was simulated for 14 years starting from 2016, with each simulation tick corresponding to a year. Simulations were iterated 200 times for each year, and the adoption rate for the year was calculated as the average adoption for the year.

To quantify the falling cost of PV, we used the US Department of Energy (DOE) SUNSHOT Initiative targets [13], according to which residential PV costs in 2020 and 2030 are respectively expected to be 50% and 70% less than PV costs in 2016. We adopt the SUNSHOT targets as reference points and use interpolation to calculate PV costs in the intervening years.

Following [14], we set residential PV cost by doubling a utility PV solar plant in the 300MW scale. Given our focus on the Arabian Gulf Region, we used the we selected the 2015 award by the Dubai Electricity and

Water Authority to Acwa Power for a 200MW solar plant, at a fixed rate of 5.84c/kWh over 25 years [15]. The price of residential PV in 2016 was therefore set at 11.68c/kWh. The price of electricity was calculated as the average ongoing KAHRAMAA tariff for residential villas, of 3.555c/kWh [9].

Table 2 provides percentages of total (cumulative) and new adoptions for each simulation year. This simulation results indicate that the integration of PV power losses (gains) as a factor increasing (decreasing) the cost of PV has a mild effect on PV adoption. With PV power gains, the total adoption rate at the end of the simulation (year 14) increases of about 1.3% as compared to a scenario with PV power losses (57.32% vs. 56.05%), with average year by year differences of 0.68% in total adoption rates and 0.09% in new adoption rates.

Table 2: Simulation results.

	Power Gain		Power Loss	
Year	*Total Adoptions*	*New Adoptions*	*Total Adoptions*	*New Adoptions*
1	0.19%	0.19%	0.16%	0.16%
2	0.38%	0.19%	0.34%	0.18%
3	0.82%	0.44%	0.72%	0.38%
4	1.63%	0.81%	1.54%	0.82%
5	3.56%	1.93%	3.37%	1.83%
6	7.71%	4.14%	7.35%	3.99%
7	12.44%	4.73%	11.85%	4.50%
8	17.72%	5.28%	16.94%	5.09%
9	23.63%	5.91%	22.45%	5.52%
10	29.72%	6.09%	28.59%	6.13%
11	36.36%	6.65%	35.11%	6.52%
12	43.25%	6.88%	41.99%	6.88%
13	50.33%	7.08%	49.06%	7.07%
14	57.32%	6.99%	56.05%	6.99%

5 CONCLUSIONS

The results of this study show modest power losses/gains resulting from the integration of RPV system into the Qatari power grid, which provide mild support for a "behind the meter" PV integration solution. The emerging approach can be integrated with a model of residential PV adoption such as the one presented in [5] and [16] to factor in the financial impact of power transmission losses/gains on PV adoption. Preliminary results for a simple model of residential PV adoption in Qatar indicate that the integration of PV power loss (gain) as a factor decreasing (increasing) the cost of PV has a mild but distinctive effect that may lead to a 1.3% increase in adoption rate.

6 ACKNOWLEDGMENTS

This work was supported by the Qatar Environment & Energy Research Institute, and the Qatar National Research Fund [grant number NPRP9-063-5-007].

7 REFERENCES

[1] Zhao, J., E. Mazhari, N. Celik, Y.-J. Son. "Hybrid agent based simulation for policy evaluation of solar power generation systems". Simulation Modeling Practice and Theory, 19: 2189-2205.

[2] Paidipati, J., L. Frantzis, H. Sawyer, A. Kurrasch. "Rooftop photovoltaics market penetration scenarios". National Renewable Energy Laboratory, Golden, CO, 2008.

[3] Drury, E., P. Denholm, R. Margolis. "Modeling the US rooftop photovoltaics market". Solar Conference 2010, American Solar Energy Society (ASES), 2010.

[4] Graziano, M. and K. Gillingham. "Spatial patterns of solar photovoltaic system adoption: the influence of neighbors and the built environment". Journal of Economic Geography, 15: 815-839, 2015.

[5] Mohandes, N., A. Sanfilippo, M. Al Fakhri, "Modeling Residential Adoption of Solar Energy in the Arabian Gulf Region", Renewable Energy, Elsevier, Volume 131, February 2019, Pages 381-389, https://doi.org/10.1016/j.renene.2018.07.048.

[6] Kwan, C.L. 2012. "Influence of local environmental, social, economic and political variables on the spatial distribution of residential solar PV arrays across the United States", Energy Policy, 47: 332–344.

[7] Low Voltage XLPE Cables, http://www.tcisb.com.my/wp-content/uploads/2017/03/low-voltage.pdf.

[8] Sunrise, sunset, dawn and dusk times around the World, https://www.gaisma.com.

[9] Tariff - Qatar General Electricity & water Corporation, https://www.km.com.qa/CustomerService/Pages/Tariff.aspx.

[10] https://www.iea.org/weo/energysubsidies/

[11] Rogers, E.M. Diffusion of innovations. New York, Free Press of Glencoe, 1962.

[12] Wilensky, U. NetLogo. http://ccl.northwestern.edu/netlogo/. Center for Connected Learning and Computer-Based Modeling, Northwestern University, Evanston, IL, 1999.

[13] The SunShot Initiative's 2030 Goal: 3¢ per Kilowatt Hour for Solar Electricity, DOE/EE-1501, 2016, https://www.energy.gov/sites/prod/files/2016/11/f34/SunShot%202030%20Fact%20Sheet-1.pdf.

[14] Tsuchida, B., S. Sergici, B. Mudge, W. Gorman, P. Fox-Penner, J. Schoene. "Comparative generation costs of utility-scale and residential-scale PV in Xcel energy Colorado's service area". The Brattle Group, 2015.

[15] Graves, L., UAE beats renewables cost hurdle with world's cheapest price for solar energy, http://www.thenational.ae/business/energy/uae-beats-renewables-cost-hurdle-with-worlds-cheapest-price-for-solar-energy, 2015.

[16] Bean, P., R. Fuentes, S. Kimbrough, and M. Muaafa. Residential Solar PV Adoption model, version 1.0, http://modelingcommons.org, 2015.

STATUS OF BUILDING INTEGRATED PHOTOVOLTAICS (BIPV) IN LATIN AMERICA AND THE CASE OF SURINAME

Amrita Raghoebarsing[1, 2*] and Angele Reinders[1, 3]
[1] Department of Design, Production and Management, Faculty of Engineering Technology, University of Twente, P.O. Box 217, 7500 AE Enschede, the Netherlands
[2] Department of Electrical Engineering, Faculty of Technology, Anton de Kom University of Suriname, Leysweg 86, Paramaribo, Suriname
[3] Energy Technology Group at Mechanical Engineering, Eindhoven University of Technology (TU/e)
*Correspondence:
Email: a.raghoebarsing@utwente.nl

ABSTRACT: In the recent years Building Integrated Photovoltaics (BIPV) has gained a lot of attention, but little information is available about BIPV in Latin America and Suriname, therefore this paper aims at presenting a status overview of Building Integrated Photovoltaics (BIPV) in the Latin America and Suriname. The application of BIPV happens currently only on small scale in some countries of Latin America, namely in Brazil, Columbia and Chile. BIPV has still not been applied in the built environment in Suriname while the application of rooftop Building Added PV (BAPV) is growing currently. However given the fact that the high initial installation cost of rooftop BAPV is still a financial barrier at this moment for the citizens in Suriname BIPV will not emerge at present but hopefully in the future.
Keywords: Photovoltaics, Building Integrated Photovoltaics, Building Added PV

1 INTRODUCTION

The awareness to use renewable energy is increasing nowadays due to the continuously increasing energy demand, the environmental concerns due to climate change and demand for less dependency on fossil fuels. According to the International Energy Agency (IEA), the built environment is responsible for 24% of greenhouse gas emission and accounts for 40% of the world's total primary energy [1]. This numbers are increasing each year due to increasing world population and improving living standards. Solar PV energy applied in the built environment is a possible solution for this problem. Building Integrated PV (BIPV) systems are PV systems which are integrated into the building envelope as a part of the building structure for example PV modules integrated on the rooftop or on the wall of a building. These systems can be stand-alone or grid-connected PV systems. With the application of BIPV, conventional building materials are replaced with PV materials. However, the PV product can also be added on the building envelope for example PV modules added above the rooftop of a building, this application is called Building Added PV (BAPV).

Two main applications can be distinguished for BIPV namely Roof systems and Façade systems. BIPV makes it possible for the consumer to use the generated electricity directly because it is closely located to the consumer avoiding transportation and storage losses. The application of BIPV can be placed in the context of energy efficiency and zero-energy buildings [2].

In the recent years, Building Integrated Photovoltaics (BIPV) has gained attention due to its many advantages as an application of solar energy. Although the application of BIPV has huge potential, it has still a small market share. It is necessary to cope with some barriers in order to increase the market share of BIPV. These are regulatory-, economic-, environmental-, technical- and communicational barriers [1].

The BIPV market has been growing in the last years and according to it will continue to grow till 2020 [3]. The global market was estimated at 2.3 GW in 2015 which was 2 times more compared to 1.5 GW in 2014 (Table 1). The global BIPV market in terms of installed capacity is dominated by developed countries especially Europe and the United States. In the recent times, Europe has a huge growth and is leading the BIPV market of installation capacity [3] [4]. It is expected that within five years, the BIPV product will get more attention in other parts of the world including Europe, Asia Pacific, Latin America, and South Africa [2]. Table 1 shows results of a BIPV market analysis for the period 2014-2020 by region or country and it gives also the compound annual growth rate (CAGR). From table 1 it can be seen that the Compounded Annual Growth Rate (CAGR) is 27.5% and above, over the analysis period 2014-2020 and the annual installation capacity of BIPV market is projected to further surpass 11 GW by the year 2020.

2 BIPV APPLICATION IN LATIN AMERICA

Table I: Global market BIPV development and forecast from 2014- - 2020 in MW and compounded annual growth rate (CAGR) [4].

Region/Country	2014	2015	2016	2017	2018	2019	2020	CAGR (%)
Asia/Pacific	300	492	772	1,159	1,672	2,329	3,134	47.8
Europe	650	967	1,441	2,103	2,929	3,807	4,838	39.7
Rest of World	81	125	184	263	355	451	561	37.9
USA	319	476	675	917	1,200	1,491	1,766	33.0
Canada	42	61	86	119	157	190	228	32.6
Japan	143	201	268	349	434	520	612	27.5
Total (GW)	1.5	2.3	3.4	4.9	6.7	8.8	11.1	

(Note: Asia-Pacific excluding Japan)

There are 26 countries within the Latin America which are namely Argentina, Bolivia, Brazil, Chile, Colombia, Costa Rica, Cuba, Dominican Republic, Ecuador, El Salvador, French Guiana, Guadeloupe, Guatemala, Haiti, Honduras, Martinique, Mexico, Nicaragua, Panama, Paraguay, Peru, Puerto Pico, Saint Barthélemy, Saint Martin, Uruguay and Venezuela. Suriname is not part of the Latin America.

Since 2012, solar photovoltaic (PV) has emerged as a significant focus of investment in Latin America, mainly in Chile, Brazil and Mexico. In 2015, 20% of $16.4 billion was invested in the application of solar energy in Latin America [5]. Several ground-mounted PV systems (on small- and large scale) have been installed and commissioned in the Caribbean region and in Latin America. Apart from this, rooftop PV installations have also been implemented in these countries but the statistics has not been published internationally. Some studies have been conducted in Brazil regarding the application of BIPV and BAPV, and have been published internationally [6] [7] [8] [9] [10] [11]. Two studies have been conducted in Columbia regarding BIPV [12] [13].Some of the applications of BIPV as rooftop PV systems in the Latin America, are as following:

- 3 MW decentralized rooftop PV system At PepsiCo's bottling plant in San Pedro Sula, Honduras, beverage company Embotelladora de Sula, S.A. (EMSULA). This also the largest rooftop PV system in Latin America.
- 900 kW PV rooftop installation on Caribbean island St. Maarten, on Westin Hotel.
- 102.5 kW rooftop solar power plant for the Champ-de-Mars in Haiti. The solar panels as well as the electrical accessories are secured on the roof of the Ciné Triomphe.

Up till now, there has been few application of facades BIPV in Latin America. For example, in Brazil, a study has been conducted of a building with glass façade [14]. For this study, a hypothetical office building was simulated but it has not been realized yet in practical.

3 BIPV AND BAPV SURINAME

There exist several ground-mounted and rooftop Building Added PV (BAPV) systems in the urban areas and in the interior of Suriname (Fig. 1).

Figure 1: Rooftop BAPV application at a school in Suriname

In the past decade, there has been a demand of small home PV systems (also rooftop BAPV) especially in the interior as there is no electricity available for 24 hours in the interior of Suriname. The existing grid does not expand till the interior of Suriname due to the remote locations which are difficult accessible because of the absence of proper infrastructure.

At this moment, rooftop or facades BIPV systems are not being applied in the built environment in Suriname.

3.1 Survey

A survey has been conducted in Suriname to assess the perception of the citizens regarding the application of rooftop BAPV. The purpose of the survey was to assess:

- The awareness regarding rooftop Building Added PV systems and its benefits, within the local people in Suriname
- The willingness to adopt rooftop PV among the citizens in Suriname.
- Whether the high installation cost of rooftop PV system has a negative impact on the development of the application of PV in Suriname.

For this survey a list of questions were prepared and local citizens in Suriname were interviewed. The survey was also online available, so some of the respondents answered the questions online.

3.2 Results of the survey

According to the results approximately 70% knew the benefits of PV systems (Fig. 2). The respondents were aware that PV systems generates clean energy and it is environment friendly. They also stated that it will be saving money on the long run. Sixty two percent of the respondents indicated that the initial investment cost of PV systems is considered to be high.

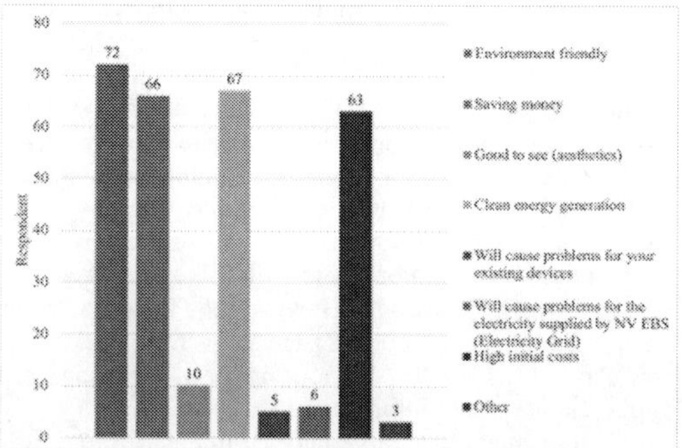

Figure 2: Awareness regarding rooftop BIPV or BAPV and its benefits

According to the obtained results, 64% of the respondents does not want a rooftop PV system installation at their home while 29% does want a PV system (Fig. 3). Five percent of the respondent did not answer this question.

From Fig. 2 and Fig. 4, it can be seen that 62% of the respondents find the initial investment cost for the installation of a rooftop PV system high. With this, it can be concluded that the high initial investment cost at this moment is a financial barrier which makes it difficult for the citizens to have a rooftop PV system at their houses.

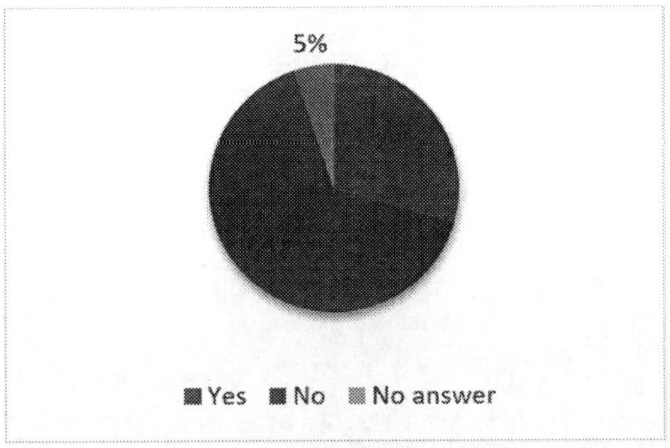

Figure 3: Willingness to adopt rooftop PV by the citizens

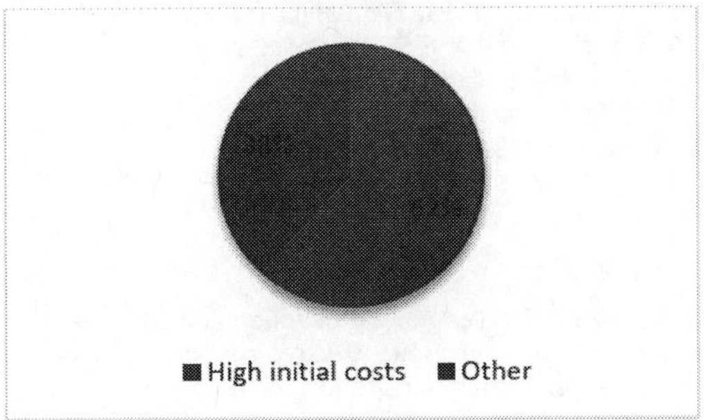

Figure 4: High initial costs for the installation of rooftop PV system

CONCLUSIONS

The application of BIPV happens only at a very small scale in Latin America. At this moment however ground-mounted PV and rooftop BAPV are being applied in several countries and on huge scales (in MW). In few countries, like Brazil and Columbia studies have been conducted regarding BIPV and BAPV. Statistics regarding BIPV and BAPV in Latin America is scarce. But, it is expected that in the near future the market share of BIPV will increase in Latin America because of the growing awareness regarding renewable energy.
In Suriname, there exist no application of BIPV in facades and rooftops, while the application of rooftop BAPV is growing at this moment. The statistics of exact total installed capacity and number of rooftop BAPV systems are not available from Suriname.
According to the survey results held in Suriname, most of the citizens (70%) of Suriname are aware of the benefits of rooftop BAPV. But the high installation cost is a financial barrier at this moment in Suriname which makes it difficult for citizens to adapt to rooftop BAPV.

REFERENCES

[1] IEA International Energy Agency, "PVPS Annual report 2016," IEA International Energy Agency, 2017.
[2] A. Reinders, P. Verlinden, W. v. Sark and A. Freundlich, Photovoltaic Solar Energy. Form Fundamentals to Applications, West Sussex: Joh Wiley & Sons, Ltd, 2017.
[3] A. K. Shukla, K. Sudhakar, P. Baredar and R. Mamat, "Solar PV and BIPV system: Barrier, challenges and policy recommendation in India," Renewable and Sustainable Energy Reviews, vol. 82, no. 3, pp. 3314-3322, 2018.
[4] M. Tabakovic, H. Fechner, W. v. Sark, A. Louwen, G. Georghiou, G. Makrides, E. Loucaidou, M. Ioannidou, I. Weiss, S. Arancon and S. Betz, "Status and outlook for building integrated photovoltaics (BIPV) in relation to educational needs in the BIPV sector," Energy Procedia, vol. 111, pp. 993 - 999, 2017.
[5] IRENA International Renewable Energy Agency, "RENEWABLE ENERGY MARKET ANALYSIS LATIN AMERICA", International Renewable Energy Agency, 2016.
[6] E. Biyik, M. Araz, A. Hepbasli, M. Shahrestani, R. Yao, L. Shao, E. Essah, A. C. Oliveira, T. d. Caño, E. Rico, J. L. Lechón, L. Andrade, A. Mendes and Y. B. Atlı, "A key review of building integrated photovoltaic (BIPV) systems," Engineering Science and Technology, an International Journal, vol. 20, pp. 833-858, 2017.
[7] J. Urbanets, C. D. Zomer and R. Ruther, "Compromises between form and function in grid-connected, building-integrated phtovoltaics (BIPV) at low-atitude sites," Building and Environment, vol. 46, pp. 2107-2113, 2011.
[8] I. P. dos Santos and R. Ruther, "The potential of building-integrated (BIPV) and building-applied photovoltaics (BAPV) in single-family, urban residences at low latitudes in Brazil," Energy and Buildings, vol. 50, pp. 290-297, 2012.
[9] R. Ruther and P. Braun, "Energetic contribution potential of building-integrated photovoltaics on airports in warm climates," Solar Energy, vol. 83, pp. 1923-1931, 2009.
[10] J. Cronemberger, E. Caamano-Martin and S. V. Sanchez, "Assessing the solar irradiation potential for solar phtotvoltiac applications in buildings at low latitudes-making the case of Brazil," Energy and Buildings, vol. 55, pp. 264-272, 2012.
[11] C. da Silva Jardim, R. Ruther, I. T. Salamoni, T. de Souza Viana, S. H. Rebechi and P. J. Knob, "The strategic siting and the roofing area requirements of building-integrated photovoltaic solar energy generators in urban areas in Brazil," Energy and Buildings, vol. 40, pp. 365-370, 2008.
[12] L. F. Mulcue´-Nieto and L. Mora-Lopez, "A new model to predict the energy generated by a photovoltaic system connected to the grid in low latitude countries," Solar Energy, vol. 107, pp. 423 - 442, 2014.
[13] A. Aristizabal and G. Gordillo, "Performance monitoring results of the first grid-connected BIPV system in Colombia," Renewable Energy, vol. 33, pp. 2475 - 2484, 2008.
[14] I. G. d. Santos and M. Takata, "Application of the concept of BIPV on glass facades – Case study in Brazil," in *Glass performance days*, Sao Paulo, 2014.

INTRODUCING THE SUPER PV PROJECT – COST REDUCTION AND ENHANCED PERFORMANCE OF PV SYSTEMS

Juras Ulbikas[1] , Vaidvile Ulbikaite[1], Julius Denafas[2], Robert Witteck[3], Marc Köntges[3], Marko Topič[4], F. Frontini[5], P. Bonomo[5], Philippe Macé[6] , Pieter Jan Bolt[7] , Alexander Ulyashin[8] , Torstein Haarberg[9], Wolfram Palitzsch[10] , Barbara Terheiden[11], Ingrid Weiss[12]
[1]Applied Research Institute for Prospective Technologies (Protech), Vismaliukų g. 34, Vilnius LT 10243, Lithuania
[2]SOLITEK R&D, Mokslininku g. 6A, Vilnius 08412, Lithuania
[3]Institute for Solar Energy Research Hamelin (ISFH)
Am Ohrberg 1, 31860 Emmerthal, Germany
[4]University of Ljubljana, Faculty of Elec Eng, Trzaska 25, 1000 Ljubljana, Slovenia
[5]SUPSI-ISAAC, Campus Trevano, CH 6952, Canobbio, Switzerland
[6]Becquerel Institute, Rue Royale 146, 1000 Brussels, Belgium
[7]Netherlands Organisation for Applied Scientific Research (TNO),
Thin Film Technology, High Tech Campus 21, 5656 AE Eindhoven, The Netherlands
[8]Foundation for Scientific and Industrial Research (SINTEF), Strindevegen 4, 7465 Trondheim, Norway
[9]BNW-Energy, Sildråpevegen 28 B, 7048 Trondheim, Norway
[10]Loser Chemie, Bahnhofstrasse 10, 08134 Langenweissbach, Germany
[11]University of Konstanz, Universitatsstrasse 10, 78457 Konstanz, Germany
[12] WIP Renewable Energies, Sylvensteinstrasse 2, 81369 Munich, Germany

ABSTRACT:
Today photovoltaics (PV) has become one of the most cost-effective forms of electricity production globally and in some regions is already the most competitive unsubsidised form of electricity. Despite positive cost and growing developments, European PV manufacturers are facing a decline in production due to competition from third countries. This decrease in the competitiveness of European PV actors is due to several factors, including a greater fragmentation of the value chain compared to competing actors. Chinese manufacturers have in particular been able to achieve synergies, economies of scale and information sharing throughout the value chain.
SUPER PV is a collaborative European-funded project initiated in 2018 by 26 partners in reaction to this trend. Together, they target a significant LCOE reduction (26%-37%) for European-made PV by adopting a hybrid approach combining technological innovations and data management methods.

Keywords:

1 INTRODUCTION

PV deployment growth is an unmatched success story in the energy sector over recent years. In just a decade the world's cumulative solar capacity increased by over 4,500% - from merely 6.6 GW in 2006 to 307 GW in 2016[1]. Today PV has become one of the cheapest forms of electricity production globally and in some regions is now the most competitive unsubsidised form of electricity with emerging business models based purely on levelized cost of electricity (LCOE) estimations. Despite the positive cost and growth developments, due to competition from third countries European PV manufacturers are facing decline in production (utilization of the module factories in the EU continued to decline to 40% in 2016, from 46% the year before) and without fast measures can completely disappear from the market.

Competitiveness decrease of the European PV actors is due to several factors: Europe PV landscape is today characterized by **fragmentation of the value chain,** while major solar actors are vertically integrated. This fragmentation is believed to be the major threat for the competitiveness of the European PV sector. In fact, substantial share of cost for the PV electricity (kWh/Eur) is related to a fragmented process inefficiency that affects the final cost.

In the SUPER PV project we are proposing a combined solution addressing both, product technological quality and business operation, targeting a significant reduction of LCOE. This reduction will be achieved through demonstration of a careful selection of innovations, which will improve the main components of the LCOE (CAPEX, OPEX and PR[2]). Introducing **SUPER**ior quality **PV** systems will create conditions for accelerating large scale deployment of PV in Europe for both utility (non-urban) and residential (urban) scenarios and help EU PV businesses to regain leadership and competitiveness on world market.

--**SUPER PV is based on a hybrid combination of technological innovations integrated into PV systems and utilization of Big Data Analytics and Data management along the PV value chain. Implementation** into PV systems will be ensured by (i) innovations requiring lowest possible investment into upgrade of manufacturing lines, (ii) compatible with state of the art PV module and Power Electronics (PE) products (iii) while for business operation improvements Digitalization and Data Management solutions based on Industry 4.0 approach will be used through the whole value chain. We will explore several of these applications by adopting already available solutions from the

[1] **International Technology Roadmap for Photovoltaic Results 2016**

[2] CAPEX – capital expenditure, OPEX – operation expenditure, PR – system performance.

construction sector (BIM) for implementing and operating PV plants identifying it as **PV4.0 approach.**

2 PROJECT CONCEPT AND METHODOLOGY

2.1 Scope

To achieve ground breaking impact on cost reduction, **project concept is tackling in integral way following three cornerstone steps impacting PV system performance and, thus, LCOE: module, power electronics (PE) and system integration/O&M (*Figure 1*).**

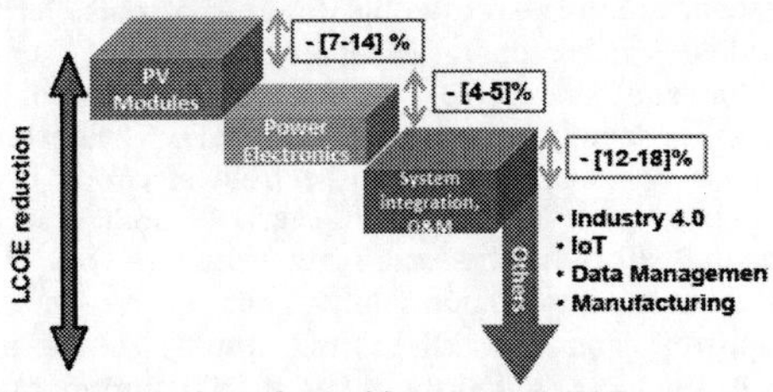

Figure 1 Project planned impact on LCOE

While module costs still represent a major part, the tendency is that this share is decreasing. In 2008, 67 % of an average project's total cost was in the PV module, but today 68 % of total cost resides in the Balance of System (BOS), which includes a variety of structural and electrical components, labour and soft costs. It is becoming clear that there is a significant room for cost reduction at the BOS meanwhile, achieving this target is challenging due to the fragmented value chain that characterizes the EU Solar landscape. Therefore, in addition to hardware level innovations, modern information management tools for PV plant implementation integrating all relevant production and plant implementation steps are needed. Consequently, based on this cost structure assessment, our project concept is addressing main value chain steps with focus on improvements of PV systems manufacturing, implementation and operation performance to achieve expected reduction of LCOE.

-- **PV module innovations**: Cost reduction on the PV module level can be achieved by three main factors: i) increase the PV module efficiency/yield with moderate/no additional production costs, ii) reduce the material cost by using less or cheaper materials, iii) increase the service lifetime by producing more reliable PV modules. The first factor (efficiency/yield increase) also reduces costs on system level because less BOS costs are expected per installed kWp with higher efficiency/yield. The second factor (material saving) reduces the production costs of the PV module. The third factor (increase reliability) is often difficult to address in cost calculations since typical assumed PV service lifetimes are equal or higher than typical financial life models for PV systems. However, we address the lifetime by choosing bifacial PV modules, which have due to its rigid structure an expected service lifetime of at least 30 years as main development basis for our innovations. Furthermore, the bifacial PV modules allow to gain 5%-25% more energy yield for the same installed capacity by additionally collecting diffuse light from the rear [1] [2].

Within the PV modules innovations, we address all three reduction factors. We increase the efficiency/yield of the PV module by increasing the optical and thermal properties of the modules. The applications of nanoparticle-based materials will improve the service lifetime of the PV modules. Furthermore, we address material cost savings for the module production by substituting and recycling expensive materials. The manufacturer SOLITEK, APOLLON and FLISOM will transfer the successful innovations on module level to a test production. The successful transferred innovations will be tested on demonstration level and the cost efficiency will be assessed to reach the final goal of LCOE cost reduction

-- **Power Electronics development**: innovations for PE are focusing on module level power electronics (MLPE) and are forming integral solution ensuring significantly higher power output on the system level by ability to ensure performance monitoring and data collection on string level and long term stability of operation independently of geographical and operation conditions. Proposed hardware level innovations are also ensuring efficient utilization of innovations targeted at O&M cost reduction.

-- **PV system integration and process innovation:** Nowadays, PV installer dealing with PV manufacturing, invest significant amount of resources (with an impact on the final electricity cost from PV) to convert potential clients requests for a PV plant into a project proposal with a technical layout, economic offer (efficiency-performance / cost threshold) and feasibility analysis for production/installation. This process could be significantly optimized. To increase the process quality reducing costs, an optimized **digital process** based on the **information management** would reduce efforts, time, repetitive work, risk of mistakes, information losses, etc., transforming an almost "manual" and **fragmented** work into a digitized and interoperable workflow along all the value chain. The challenges to future large-scale deployment of PV could be overcome by a new methodology that we call **PIM (Photovoltaic information modelling/management)**: a digital and holistic process, supported by innovative digital tools, methodologies and platforms, ensuring clear structures, efficient processes, less time, lower costs and higher quality across the entire PV plant lifecycle.

Starting from this approach, the digitalization that we introduce in this project is expected to be replicable for the whole value chain of solar PV, to be tailored from raw material to electricity, including also building applied and integrated-PV.

2.2 Methodology

Project overall methodology is based on three phases approach:

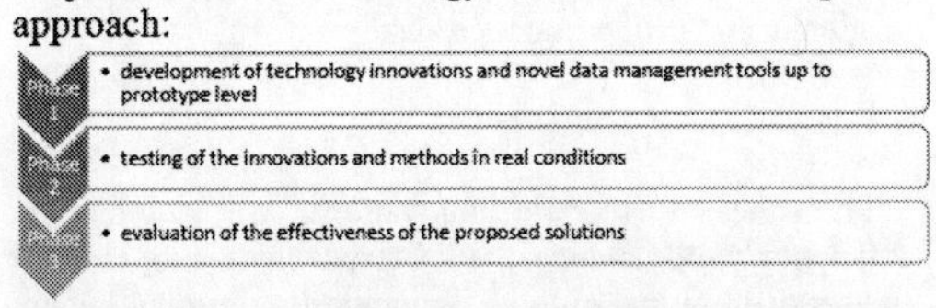

The innovative technologies, suggested in this project, concerns both module and power electronics. All will contribute to significant cost reduction of the PV system cost. To avoid a combined effect of innovations, the testing will be made where only one Innovation Action is

implemented. To ensure fastest possible uptake of the technologies in the production and reduce impact on system cost through increase of CoO, innovations were carefully selected to be compatible with production routing cards. For the SUPER PV systems state of the art module production technologies available for industrial partners were selected to provide possibility for comparable analysis of different technologies and their prospects for cost reduction.

Project results will be tested in demonstration sites in the **markets of primary importance** for European PV producers:

I. European market where roof top market is recently becoming more and more important.
II. Regions where solar irradiation and average temperature are high (South Europe, South USA, Africa and large part of Asia). This market has shown rapid growth with no limits due to already achieved cost competitiveness to other energy sources.

Testing will be performed in three different climatic regions of extreme conditions: tropical equatorial climate: i) hot, humid and salty, ii) hot desert climate (in Morocco and/or Tunisia) and iii) cold climate (winter in Lithuania).

For demonstration of the proposed innovations small systems up 1.5 kWp will be implemented for demonstration of the each PV module technology. Systems will be instrumented with sensors and data loggers. In addition, power electronics and data management innovations will be tested in field.

3 PROPOSED INNOVATIONS

Innovations developed by participating research centres will be applied to state-of-the-art modules (c-Si and flexible CIGS) and power electronics to ensure fast uptake of the project results by the industry. For cost reductions in PV system integration and operation, digitalization and data management solutions based on an Industry 4.0 approach will be adopted following the successful example of the Building Information Modelling approach in the construction sector. The developed innovations will range from nano-coatings to digital tools and will be compatible with existing manufacturing processes, thus reducing the impact on the cost of ownership and ensuring the attractiveness of the proposed technologies. Prototype SUPER PV systems will be tested in different climate conditions, from northern Europe to North Africa, to validate their cost efficiency and competitiveness. Business cases will be prepared on the basis of these demonstration results. These project activities will be complemented by a training and dissemination campaign ensuring the highest visibility and replicability of the project outcomes in both academic and professional circles.

3.1 Module level innovations

Within the PV module innovations, we increase the efficiency/yield, reduce the material costs and increase the service lifetime of the PV modules by introducing and combining five PV module innovations. We apply the innovations to crystalline silicon (c-Si) based bifacial modules and lightweight flexible Copper-Indium-Gallium-(di)Selenide (CIGS) modules. The first innovation is a combination of anti-Soiling (AS), anti-reflection (AR) and infrared (IR) reflection coating based on nanoparticles, which aims to increase the annual yield of the PV modules. The second innovation applies a white reflector to the module's rear glass within the cell gaps, which improves the light harvesting for the bifacial PV modules. This increases the optical gain for bifacial modules, which enhances the modules power output and the annual yield. Furthermore, we employ a deeply structured rear glass to increase light collection from the module rear side and to increase the thermal convection cooling to the surrounding air, which will also improve the efficiency/yield [3]. For the third innovation, we implement in-laminate bypass diodes made with solar cell processing machines, which allow omitting the junction box and requiring fewer materials for the module production. The fourth innovation applies Aluminum oxide (Al_2O_3) gas barrier coatings deposited by spatial atomic layer deposition (SALD). The coating is deposited either directly on the front electrode of the CIGS or on a low cost front-sheet. In both cases the coating will fulfill the same requirements as state of the art flexible encapsulation solutions and achieve similar or improved damp heat (DH) testing results. For the fifths innovation, we will demonstrate a laboratory-recycling tool for all considered module types (c-Si and CIGS) to evaluate the possibility to recycle and re-use the module materials. The manufacturers Solitek, Apollon and Flisom will implement the innovations in a test production. We test reference and innovative PV modules at demonstration sites with different climates, varying from cold and humid, hot and humid to hot and dry. Finally, we evaluate the cost efficiency of the innovations to reach the final goal of LCOE cost reduction.

Recyclability can be planned - that is why we use the many years of experience of Loser Chemie GmbH in the field of photovoltaic recycling in order to be able to offer a recycling strategy with the newly developed products. The focus should be on a recycling all the used materials. Especially the glass of these new solar modules should be used for the production of the next generations of solar cells and relevant modules.

3.2 Power Electronics innovations

Innovations will be focused on module level power electronics (MLPE) to form integral solution ensuring significantly higher power output and to ensure performance monitoring and data collection on string level and long term stability of operation. Proposed hardware level innovations are (i) micro-inverters with advanced switching elements based on GaN, (ii) micro-inverters with smart functionality such as active/reactive power generation and (iii) fault-tolerant converter topologies and converter algorithms, (iv) MPP Optimizers or Smart Boxes with Rapid Shut Down (RSD) functionality that represent dramatic increase of Power Plants fire safety, long-term energy yield, module reliability and PV power plant design flexibility, modularity and longer operation times without maintenance. MLPE will be coupled with wireless communication solutions - gateways (GW) as the heart of the intelligent system of two way connectivity, exchange of information, analytical function, and security assurance towards a cloud and provide the optimum functioning of the PV power plant as a whole.

3.3 Integrated Process and Information Management

A hybrid combination of technological innovations integrated with Data management along the PV value chain with the introduction of PV Information Management (PIM) is the main goal of WP5 for reducing time and therefore costs for PV system design, manufacturing, installation and maintenance. As a general approach, the methodology BIM (Building Information Modeling) widely used in construction industry will be "tailored" and transferred to PV process within this project and we will call it "PIM (Photovoltaic Information Modeling/Management)". The main objectives to implement innovation will be:

- To adopt digital software and hardware tools ensuring integrated information flow through the PV value chain this way reducing costs related to the PV projects implementation and operation:
- Development of a digital platform for design, simulation and operation supporting the cost reduction of PV systems. The platform will allow reducing the cost of PV electricity along the value chain by enhancing the process and by integrating PIM models with LCOE.
- Development of a tool for extending the PIM-based platform functionalities for O&M operations through a Digital Twin Model, continually updated to include the events sustained while in use, thanks to a sensor-enabled digital model that simulates the object in a live setting.

As a first step we will define a map of the PV process and its informative categories and contents, in order to find current bottlenecks, detect strategies to optimize the workflows and reduce costs thanks to the digital implementation. A generic framework of data mining in combination with the PIM will support this stage. An innovative digital platform for the design, co-simulation and operation of PV system and for the implementation of collaborative approaches through the PV value chain, will allow analyzing PV design scenarios against criteria of the minimal cost-to-power ratio. To address also a cost reduction in O&M, the platform developments will also embrace the Internet of Things (IoT) and the adoption of Digital Twin Models for PV plants, by real-time monitoring with the possibility of adding living data and smart knowledge to the digital model to improve reliability and cost reduction.

3.4 Demonstration

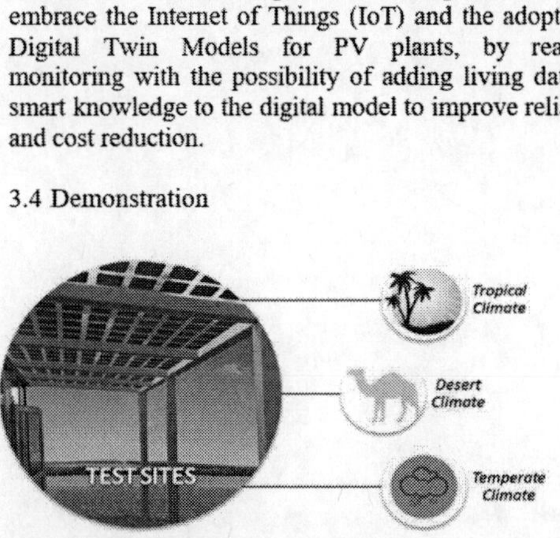

Figure 2: Demonstration sites selection

Conditions of the demonstration sites cover the **markets of primary importance** for European PV producers:

I. European market where roof top market is recently becoming more and more important.

II. Regions where solar irradiation and average temperature are high (South Europe, South USA, Africa and large part of Asia). This market has shown rapid growth with no limits due to already achieved cost competitiveness to other energy sources.

Testing will performed in three different climatic regions of extreme conditions: tropical equatorial climate: i) hot, humid and salty, ii) hot desert climate (in Morocco and/or Tunisia) and iii) cold climate (winter in Lithuania).

4 RESULTS ASSESMENT

To assess the innovation potential from innovations to be developed in SUPER PV, a set of *Key Performance Indicators* (KPI's) are defined:

- ✓ KPI 1: Solar module cost reduction
- ✓ KPI 2: BoS components costs (EUR/Wp)
- ✓ KPI 3: Total installation cost (EUR/Wp)
- ✓ KPI 4: Performance Ratio (PR) over the lifetime
- ✓ KPI 5: Reduction of lifecycle costs per kWh, expressed by the Levelized Cost of Energy (LCOE)
- ✓ KPI 6: Recyclability and environmental footprint of solar modules.

Reaching the defined KPI's will be instrumental for the success of SUPER PV. In Figure 3 calculated LCOE improvement from proposed innovations is shown.

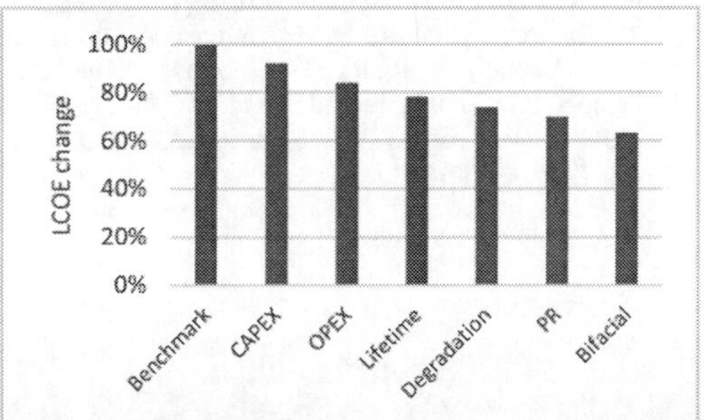

Figure 3: Estimated LCOE change from successive SUPER PV innovations.

The stated ambition in LCOE reduction of 26-37 % refers to the identified state-of-the-art situation as described in the SUPER PV proposal, submitted in September 2017. Hence, to maintain consistency in the SUPER PV Project, benchmark values from the SUPER PV proposal will be used as state-of-the-art at project start. Nonetheless, the technology and market evolution of the PV sector will be closely monitored, to ensure relevance of the results. This crucial task will not only be conducted from a market competition perspective, but also from a SUPER PV technology development point-of-view, as SUPER PV will not cover the entire supply chain needed to produce PV electricity at the lowest possible cost. Hence, SUPER technologies, in combination with relevant technologies outside the scope of the SUPER PV project, may form winning combinations on the market.

5 SUMMARY and CONCLUSIONS

Overall the project will have a twofold impact on the European PV sector: It will on one hand contribute to creating conditions for an accelerated large scale deployment of PV in Europe for both utility (non-urban) and residential (urban) contexts, and on the other hand

will help European PV businesses regain leadership on the world market.
Besides the general project overview, results available for presentation at the time of the conference will include a first estimation and sensitivity analysis for the action's impact on PV LCOE. Technical specifications for the innovations to be developed are also expected.

6 ACKNOWLEDGEMENT

The SUPER PV project has received funding from the European Union's Horizon 2020 Research and Innovation Programme under Grant Agreement
No. 792245.

7 REFERENCES

[1] B. V. Aken, L. Okel, J. Liu, S. Luxembourg and J. v. Roosmalen, "White Bifacial Modules – Improved STC Performance Combined with Bifacial Energy Yield," *32nd EUPVSEC,* pp. 42 - 47, 2016.

[2] U. A. Yusufoglu, T. M. Pletzer, L. J. Koduvelikulathu, C. Comparotto, R. Kopecek and H. Kurz, "Analysis of the Annual Performance of Bifacial," *IEEE JPV,* vol. 5, pp. 320-328, 2015.

[3] M. R. Vogt, H. Schulte-Huxel, M. Offer, S. Blankemeyer, R. Witteck, M. Köntges, K. Bothe and R. Brendel, "Reduced Module Operating Temperature and Increased Yield of Modules With PERC Instead of Al-BSF Solar Cells," *IEEE JPV,* vol. 7, pp. 44-50, 2017.

GRID CONNECTED PV SYSTEMS IN SPAIN: AN ECONOMIC ASSESSMENT WITH SENSITIVITY ANALYSIS

R. Peña[1], A. M. Diez-Pascual[2], P. García Díaz[1], J.A. Luceño Sánchez[2], P. Caño Fernández[3]
[1]Escuela Politécnica Superior, Universidad de Alcala, 28805, Alcalá de Henares, Spain
Tel: +34-91 885 69 68. rafael.pena@uah.es, pilar.garcia@uah.es
[2] Faculty of Sciences, Universidad de Alcalá, 28805, Spain; am.diez@uah.es, jose.luceno@uah.es
[3] Instituto de Energía Solar, Universidad Politécnica de Madrid, 28040, Madrid, Spain; pablo.cano@ies.upm.es

ABSTRACT: This work analyzes the current situation in the Spanish photovoltaic market in terms of profitability of investments for different system configurations: small plants (< 10 kW, typical in residential installations), large plants (some MW) and other systems (commercial, industrial, etc). In all the cases, the calculations assume the PV plants are grid-connected under free market conditions, with no incentives.
Parameters such as return on investment and payback period have been calculated with the data collected from several sources.
A software tool has been used in order to calculate the aforementioned figures. It can carry out sensitivity analysis, making it possible to study the influence of input parameters on the final results. Particularly, the influence of electricity price and system cost is discussed.
Keywords: Economic Analysis, PV Market, Funding and Incentives.

1 INTRODUCTION

2018 could be the first year in history in which a large photovoltaic (PV) plant was installed in Spain with no economic incentives. In this sense, the german group Luxcara recently anouncced the acquisition of a 45 MWp PV project in southern Spain [1]. The plant could be connected to the grid in the second half of 2018, and would sell the generated power under free market conditions.

This project could confirm that investments in PV plants can be profitable under free market conditions in some regions in Spain as well as in other sunny places. The main reason for this being the huge reduction in PV panel price over the last years (figure 1), which has been estimated as 85 % since 2007 [2].

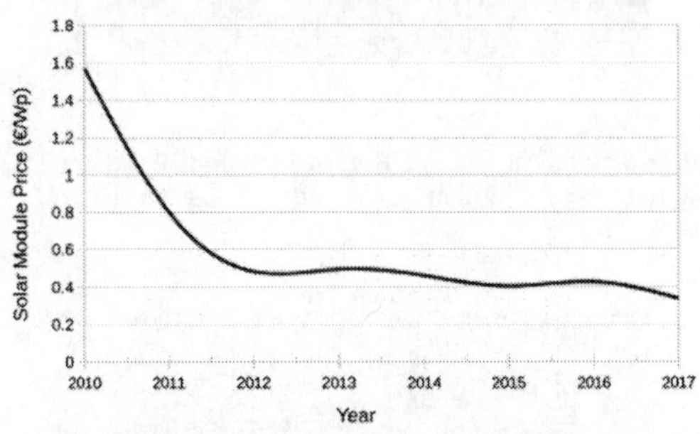

Figure 1: Evolution in the average solar module price over the last years. The prices correspond to crystalline Silicon modules in Spain.

In this regard, PV modules are now available at a price of about 0.40 €/Wp, while in 2008 their price was over 3 €/Wp. This has led to a drastic reduction in the price of PV installations (figure 2), which is currently in the range of 0.7 – 1.5 €/kWp, depending on the system configuration [3].

This work analyzes the current situation in the Spanish PV market in terms of profitability of investments for different system configurations: small plants (< 10 kW, typical in residential installations), large plants (some MW) and other medium size systems (commercial, industrial, etc). Parameters such as return on investment (IRR) and payback period have been calculated with the data collected from several sources within Spanish industry and public institutions.

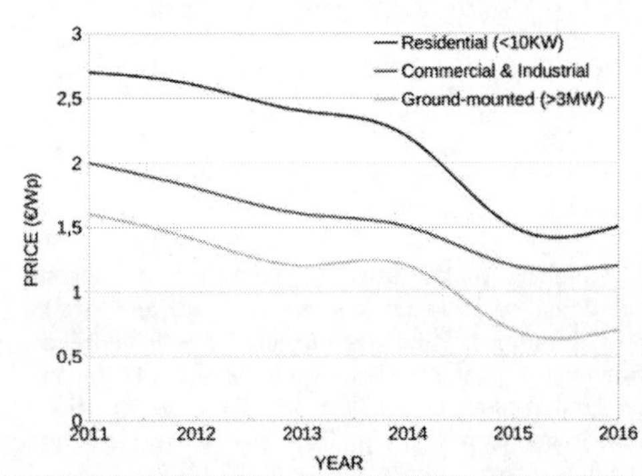

Figure 2: Evolution in the average PV installation price over the last years for different configurations.

2 APROACH

A software tool has been used [4], which calculates the aforementioned figures. It can also carry out sensitivity analysis, making possible to study the influence of input parameters on the final results. Particularly, the influence of electricity and system price has been analyzed. This software, called PVTIR, was designed and registered at the Universidad de Alcalá in a previous project.

In the calculations, the average electricity price in the wholesale market last year in Spain has been used, 50 euros per MW·h. And an inflation of 7% has been assumed for the price of electricity, according to the actual value in last decade in the country.

In the same way, PV plant price has been swept around the average price for the year 2016 for every system configuration. This way, the output parameters are calculated as a function of plant price.

In all cases, it has been assumed that the capital needed for the initial investment is obtained through a 10-year loan, with a 10% down payment. The interest rate of the loan has been fixed to 6%.

3 RESULTS

In the following, the results obtained for Internal Rate of Return and payback period are described.

3.1 Internal Rate of Return

Figure 3 shows calculated Internal Rate of Return for residential PV plants as a function of system price. System price has been varied between 1.2 and 1.9 €/Wp. These are realistic values, since small rooftop PV installations can be implemented in Spain at a price below 1.3 €/Wp. IRRs between 6 and 12 % are obtained for these prices, assuming an average electricity price of 0,05 €/kWh.

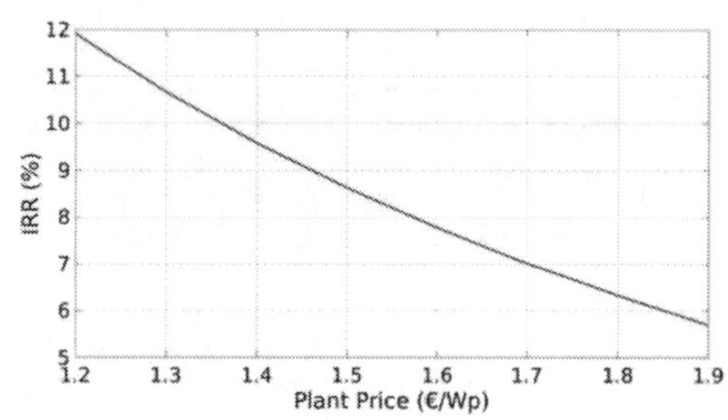

Figure 3: Internal Rate of Return for residential PV plants as a function of system price.

A higher profitability is possible in larger installations, such as in commercial and industrial plants. In these, IRR between 8 and 17 % have been calculated for plant price between 0.9 and 1.6 €/Wp. In larger ground-mounted systems, in which system price can be lower than 1 €/Wp, IRR is well over 15 %, exceeding 24 % for a system price of 0,7 €/Wp.

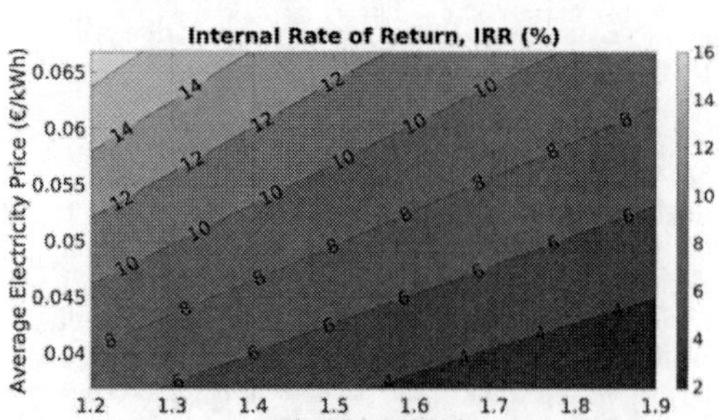

Figure 4: Internal Rate of Return for residential PV plants as a function of system price and average electricity price.

As shown above, the aforementioned values have been calculated for the average price of the electricity last year in Spain (50 €/MW·h). In addition, the influence of electricity price variations has also been considered in this work. This is shown in figures 4, 5 and 6. In these figures, electricity price has been varied between 35 and 70 €/MWh. For residential PV plants (figure 4), for a system price of 1.5 €/Wp, calculated IRR is over 5 % for an electricity price of 40 €/MWh. And values over 12 % are obtained for an electricity price of 65 €/MWh.

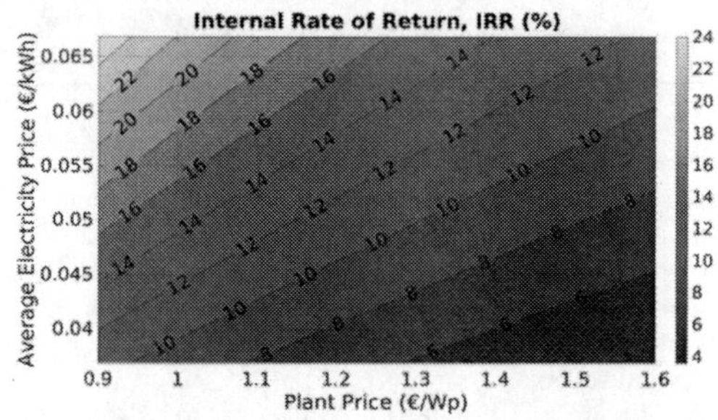

Figure 5: Internal Rate of Return for commercial and industrial PV plants as a function of system price and average electricity price.

For medium size PV plants (figure 5), for a system price of 1,2 €/Wp, calculated IRR is about 8 % for an electricity price of 40 €/MWh. And values close to 15 % are obtained for an electricity price of 65 €/MWh.

For large size PV plants (figure 6), for a system price of 0,7 €/Wp, calculated IRR is about 17 % for an electricity price of 40 €/MWh. And values as high as 35 % are obtained for an electricity price of 65 €/MWh.

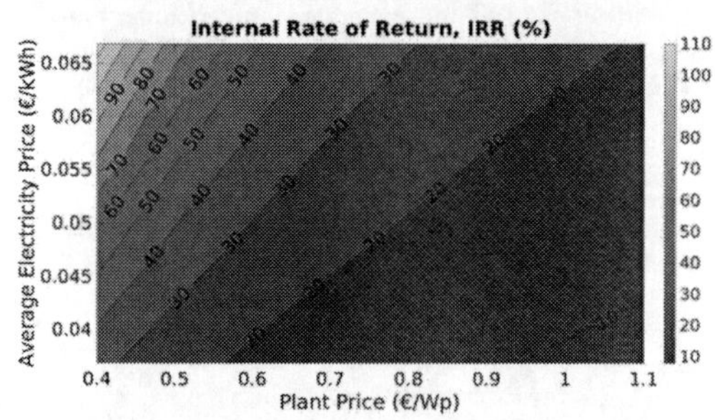

Figure 6: Internal Rate of Return for large PV plants as a function of system price and average electricity price.

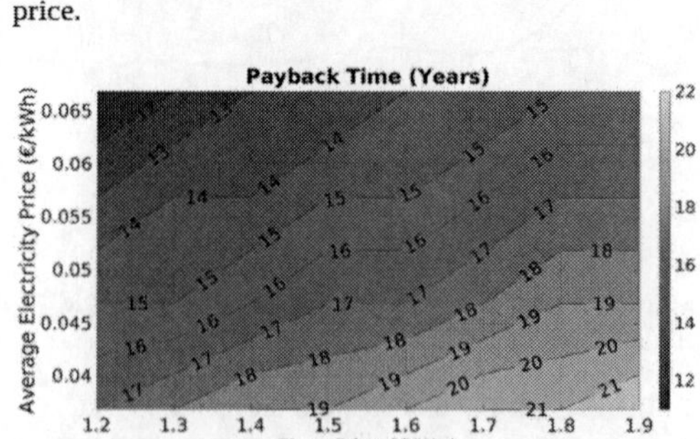

Figure 7: payback time for residential PV plants as a function of system price and average electricity price.

3.2 Payback period

Payback period has been calculated in all the considered cases in this work (small, medium and large plants). It has been calculated as a function of system price and average electricity price.

For residential PV plants (figure 7), for a system price of 1.5 €/Wp, calculated payback period is 18 years for an electricity price of 40 €/MWh. And values close to 14 years are obtained for an electricity price of 65 €/MWh. For the average price of the electricity last year in Spain (50 €/MW·h), a period of 16 years is obtained.

Figure 8: payback time for commercial and industrial PV plants as a function of system price and average electricity price.

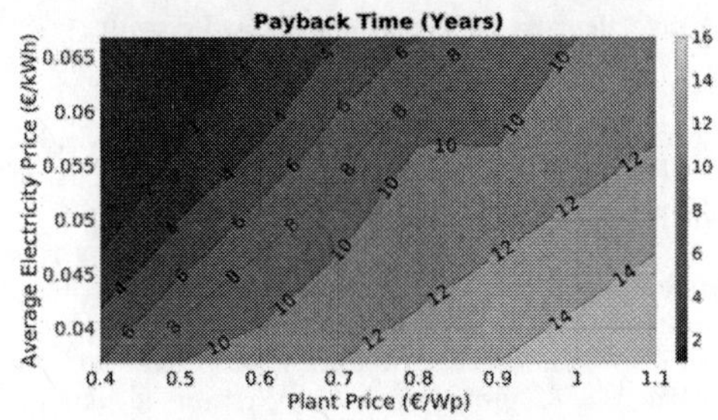

Figure 9: payback time for large PV plants as a function of system price and average electricity price.

For medium size PV plants (figure 8), for a system price of 1.2 €/Wp, calculated payback period is about 16 years for an electricity price of 40 €/MWh. And values about 12 years are obtained for an electricity price of 65 €/MWh. For the average price of the electricity last year in Spain, a period of 14 years is obtained.

For large size PV plants (figure 9), for a system price of 0.7 €/Wp, calculated payback period is 12 years for an electricity price of 40 €/MWh. And values about 4 years are obtained for an electricity price of 65 €/MWh. For the average price of the electricity last year in Spain, a period of 9 years is obtained.

4 SUMMARY AND CONCLUSION

Current prices for electricity in Spain make it possible to connect PV plants to the grid under market conditions. Particularly, in medium and large size PV plants, due to their lower price associated to economies of scale. So, IRRs values well over 10 % are possible with current electricity prices. For smaller plants (such as in residential installations), IRRs over 6 % have been calculated.

With regard to payback period, for the average price of the electricity last year in Spain (50 €/MW·h), a period of 16 years has been calculated for small plants. For medium and large plants, a period of 14 and 9 years has been obtained, respectively.

The influence of electricity price and system cost has also been considered in this work, since the calculations strongly depend on these parameters. So, electricity price has been varied between 35 and 70 €/MWh. And PV plant price has been swept around the average price for the year 2016 for every system configuration.

Even for the lowest price considered of 40 €/MWh, the calculated IRR for large size PV plants is as high as 17 %, with a payback period of 12 years.

REFERENCES

[1] http://luxcara.de/en/news/detail/luxcara-acquires-45mwp-solar-project-in-southern-spain-for-institutional-investors.html

[2] Own elaboration, data from www.iea.org.

[3] National Survey Report of PV Power Applications in SPAIN – 2016, Unión Española Fotovoltaica (UNEF), https://unef.es/.

[4] R. Peña, P. Díaz, and B. Amante, 'Economic Analysis of Grid-Connected Photovoltaic Systems in Spain', WIP-Munich, 2010.

CALCULATING THE COST OF DISTRIBUTION GRID UPGRADES REQUIRED TO ACCOMMODATE CURRENT AND FUTURE LEVELS OF PV DEPLOYMENT ACROSS RURALITY CONTEXTS IN THE UK

Sheridan Few†, Predrag Djapic ‡, Goran Strbac ‡, Jenny Nelson†,§, Chiara Candelise ⊥
† Grantham Institute – Climate Change and the Environment, ‡ Department of Electrical Engineering,
§ Department of Physics, ⊥ Centre for environmental Policy, Imperial College London, Exhibition Road, London, SW7 2AZ

ABSTRACT: In this study, we use a geographical information systems (GIS)-framework—the United Kingdom Photovoltaics Database (UKPVD) to analyze spatial PV deployment, domestic and nondomestic electricity demand, rurality (as a proxy of building and population density), at a resolution of around 600 households. We use this framework to identify four representative rurality contexts which account for ~95% of UK households. Using a Statistical Distribution Network Design and Investment model and making use of network asset data for the Southwest of England as a representative case, we calculate additional network upgrade costs due to load growth and increased PV penetration in representative networks in each of these contexts. We calculate additional costs for a range of PV deployment scenarios and sets of flexibility measures, including storage and demand side management. Trends in network upgrade costs and savings associated with storage and demand side management are broadly similar across rurality contexts, although there are quantitative differences. Up to 2030, network upgrade costs are chiefly associated with avoided investment in low voltage network infrastructure to mitigate voltage impacts of PV. Post-2030, as peak demand rises, upgrades to high voltage networks to accommodate higher power flow begin to dominate. The majority of unavoided network upgrade costs after flexibility measures are installed from the period from 2030 are associated with voltage regulation on the low voltage network to accommodate PV deployment, which modelled levels of storage and demand side management are unable to mitigate.
Keywords: PV integration, electricity distribution system, energy storage, demand side management, load growth

1 INTORDUCTION

Small scale (<50 kW) domestic and non-domestic solar photovoltaic (PV) deployment in Great Britain has increased rapidly over the past decade, rising from 930 MW at the end of 2011 to 4.4 GW at the end of 2016. Deployment in subsequent years has slowed, chiefly due to sudden changes in the feed-in-tariff paid to owners of PV panels per unit of energy generated. However, deployment is expected to accelerate once again over coming years [1], predominantly driven by the falling cost of PV installations alongside increases in electricity price [2]. This is in line with expected global trends [3].
Unlike traditional sources of electricity based upon the combustion of fossil fuels, variable renewable energy sources such as PV do not generate electricity on demand, but rather when the energy source they rely on, in this case solar irradiation, is available. Additionally, traditional sources generate electricity at centralised locations, which is transferred through transmission and distribution networks to consumers. The deployment of solar PV at a distribution level raises the possibility of power flow in the reverse direction, which distribution networks may not be designed to accommodate. Modelling studies also indicate relatively few regions in which challenges due to reverse power flow are expected at current levels of UK PV deployment, both a distribution level [4–6] and a transmission level [7].

Possible technical challenges associated with integration of PV are reviewed in [8] and include voltage rise, reduced power factor, reverse power flow, frequency drift, and phase imbalance. [9] model these impacts in an urban network with high PV penetration, and a series of reports funded by the Low Carbon Network Fund from UK distribution network operators (DNOs) reports on these impacts in case study regions of the distribution network [10–13], finding impacts to be relatively minor at this stage, and in most cases less significant than expected. Reverse power flow represents a significant concern, however, and studies agreed that longer feeders, more PV and higher busbar voltages at substations will make networks more susceptible to over-voltage.

[14,15] carry out a comprehensive review of literature on cost and impacts of intermittency, incorporating that on costs associated with grid upgrades at a distribution and a transmission level, and that associated with providing backup power in order to ensure sufficient energy is available to meet peak demand, and respond to rapid fluctuations in supply associated with variable renewables. Cost of distribution network upgrades to accommodate up to 18% increased penetration of solar PV have been calculated for Europe as a whole and case study countries based upon a combined set of generic urban and rural networks [16]. [17] develop a modelling approach capable of simulating transmission and distribution networks and calculating upgrade costs required to accommodate changes in demand and/or generation. [18] apply this model to calculate low voltage network upgrade costs in one semi-rural and two semi-urban networks in a range of PV and heat pump deployment scenarios (up to 100%). Additionally, a range of other mechanisms may be used to reduce grid impacts of photovoltaic deployment, including the deployment of distributed energy storage to increase self-consumption and reduce reverse power flow [6,19] and demand side management to shift demand from times of peak demand towards those of peak supply [20,21].

However, the challenge of understanding when, where, and how impacts on the grid will manifest themselves as PV deployment increases, how best to ensure a consistent electricity supply in a cost-effective manner in light of these impacts, and how each of these factors depends upon local context, represent important questions which have not yet been fully resolved. Here, we use the UK Photovoltaic Deployment (UKPVD) database [4] to identify key representative rurality contexts and to generate representative data for each of these contexts, and use a statistical distribution network design and investment model [17] to calculate the cost of upgrades in each of these contexts in a range of PV, storage, and demand growth scenarios. We constrain our analysis here to SW England due to data availability on grid assets, but expect conclusions to be of relevance across the UK and in an international context.

2 METHODS

Figure 1 schematises the approach taken in this study.

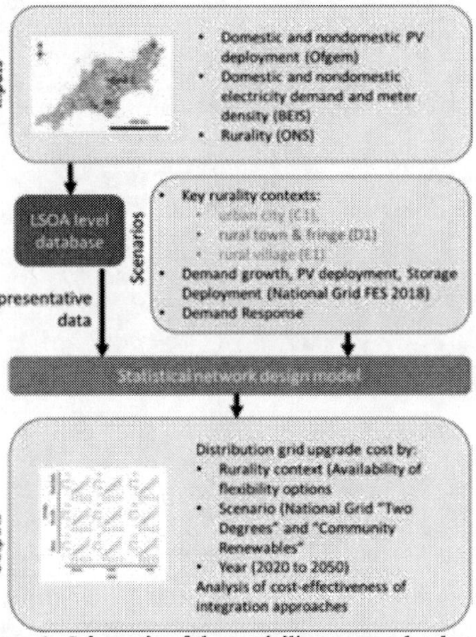

Figure 1: Schematic of the modelling approach taken in this study.

2.1 UK Photovoltaics Deployment Database

We use an updated version of the UK Photovoltaics Database (UKPVD) developed by [4] containing lower level super output area (LSOA, a geographical area containing approximately 600 households) level data from across the UK on:

- PV deployment from FiT statistics produced by OfGem [22]
- Domestic and nondomestic electricity demand and meter density [23,24]
- Population density, spatial area, and rurality [25,26]

This is combined with data on number and type of substations per LSOA from the distribution network operator (DNO), Western Power Distribution (WPD), for the Southwest of England, for which the location of each transformer is known and mapped to an LSOA area. Analysis of distributions of meter density, electricity demand, solar PV deployment, and substation density amongst LSOAs is used to determine three key rurality contexts. Median values of key parameters amongst all LSOAs in the Southwest of England in each of these contexts is fed into the statistical network design model, described below, to model representative networks, before applying load and PV growth scenarios, with and without flexibility from storage and demand side management.

2.2 Statistical Network Design and Investment Model

The statistical network design model [17,18] is a tool capable of simulating a network to accommodate a specified meter density, electricity demand per meter (both magnitude and profile shape), PV and storage capacity, and percentage of households using smart appliances for which demand side management is available. By comparing current infrastructure and infrastructure projected future values for these parameters, we are able to calculate the infrastructure requirements to meet supply and demand profiles at a given date, and, by using infrastructure costs from [27], we are able to calculate the cost associated with these upgrades. The model is also capable of indicating the maximum voltage on the network, and number of feeders where voltage limit is exceeded in future scenarios if networks were not upgraded. Median values from the UKPVD model for each rurality context are fed into the statistical network design and investment model in order to design fifteen realistic networks of each rurality context. Future scenarios are applied to each of these, and average values taken to explore network requirements in future scenarios from a realistic basis.

2.3 Scenario Design

We develop scenarios for peak load growth, PV growth, and storage deployment up to 2050 based upon National Grid's "Two Degrees" (TD) and "Community Renewables" (CR) scenarios [1] (Figure 2a-c). These scenarios contain both highest PV deployment and highest microgeneration capacity (defined as <1MW by National Grid), and as such we consider these most relevant for considering network implications of future PV deployment. However, the "Community Renewables" scenario is significantly more ambitious in terms of microgeneration deployment, reaching a growth factor of 4.6 by 2030, compared to 1.8 in "Two Degrees".

Peak load has been in decline over recent years, predominantly due to more efficient appliances. However, it is projected to rise again in the future as demand from electric vehicles and electric heating increases. As such, growth in peak load is taken with respect to the 2005 peak (which the distribution network must already be capable of accommodating) rather than 2017 values. National Grid do not explicitly break microgeneration capacity down by technology type, but indicate that a large proportion of this is solar PV in 2030 [1]. We assume here that PV growth at an LSOA level occurs with the same growth factor as total UK microgenerating capacity.

We assume half of National Grid's projected non-transmission level battery storage is deployed to balance PV microgeneration. We calculate total micro-PV capacity in a given year by multiplying total 2017 deployment of microgenerating PV (<1MW) based on FiT data [22] by the growth factor in Figure 2b. We then calculate storage per kW of PV deployed by dividing the first quantity by the second.

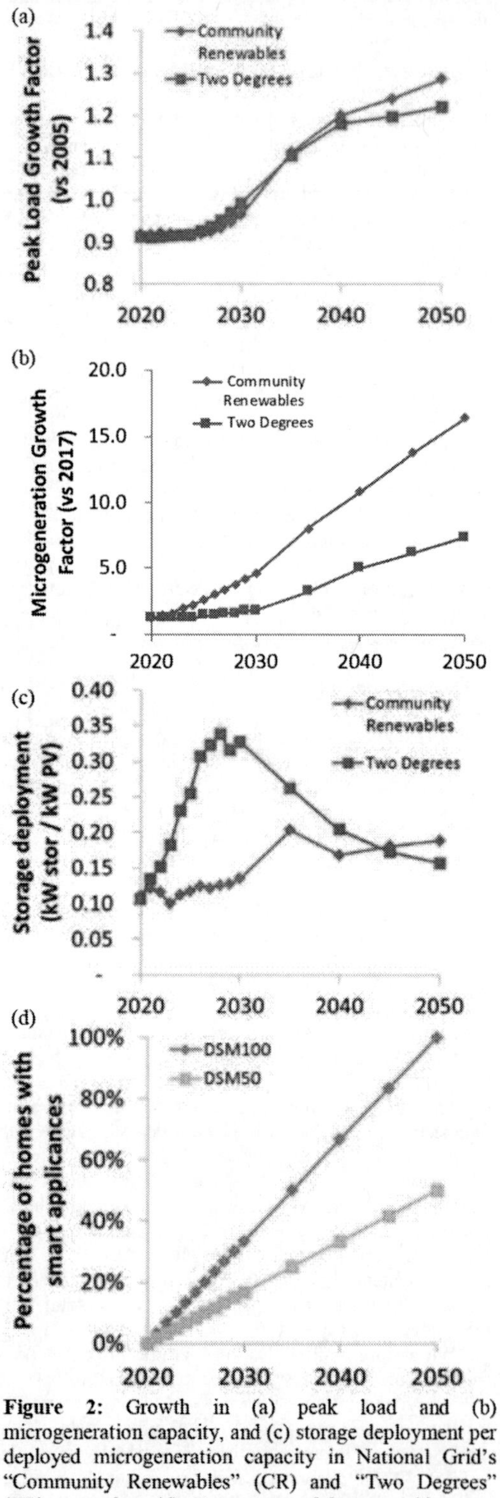

Figure 2: Growth in (a) peak load and (b) microgeneration capacity, and (c) storage deployment per deployed microgeneration capacity in National Grid's "Community Renewables" (CR) and "Two Degrees" (TD) scenarios. (d) Percentage of homes with smart appliances in DSM50 and DSM100 scenarios.

There is relatively little literature on projected uptake of smart technologies and demand side management measures, as such we conduct a sensitivity analysis, developing two growth scenarios with the expectation that reality will fall between these two. These scenarios reach 50% and 100% of households with smart appliances installed by 2050 with linear growth from 0% in 2020 up to this point (Figure 2d). These scenarios are somewhat arbitrary, but illustrate the magnitude of impact which demand side measures could have if such deployment levels can be achieved.

3 RESULTS

3.1 Defining Rurality Contexts

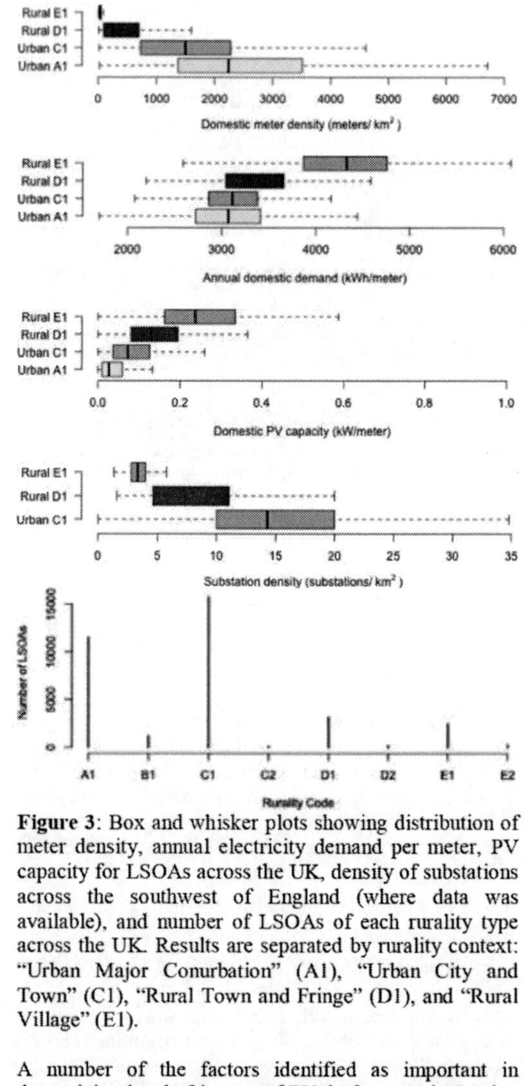

Figure 3: Box and whisker plots showing distribution of meter density, annual electricity demand per meter, PV capacity for LSOAs across the UK, density of substations across the southwest of England (where data was available), and number of LSOAs of each rurality type across the UK. Results are separated by rurality context: "Urban Major Conurbation" (A1), "Urban City and Town" (C1), "Rural Town and Fringe" (D1), and "Rural Village" (E1).

A number of the factors identified as important in determining level of impact of PV deployment in Section 1 (feeder length, busbar voltages) may be expected to typically depend upon meter density and substation density, both of which may be expected to be lower in rural than urban areas. Level of PV deployment and demand by sector (domestic and nondomestic) are also important in determining grid requirements. The UK Office for National Statistics classifies LSOAs into ten rurality contexts based upon density and distribution of dwellings ranging from "urban major conurbation" to "rural hamlets and isolated dwellings". 95% of LSOAs fall into one of four categories: "urban major conurbation" (A1), "urban city and town" (C1), "rural town and fringe" (D1), and "rural village" (E1), and distribution of key parameters for each of these rurality contexts is presented in Figure 3.

Due to the significant difference in distributions for C1, D1, and E1, we consider these contexts most relevant for study. Distributions overlap more significantly for A1 and C1, and PV deployment is relatively low, making these two contexts more similar. Additionally, at the time of the study, we have access to grid asset data only for a region of the southwest of England containing no LSOAs falling into the A1 category. As such, we restrict our attention to C1, D1, and E1 for the remainder of the study.

An example of a modelled network in each rurality category is shown in Figure 3.

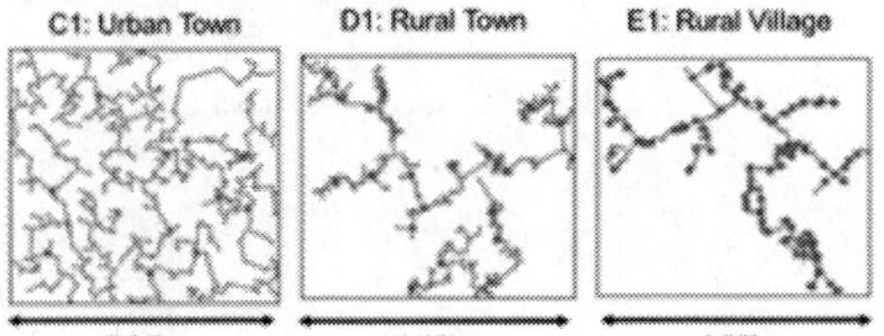

Figure 4: Representative generated networks in each rurality context: C1, D1, E1.

3.2 Distribution Network Upgrade Costs to 2050

Figure 5 compares upgrade costs per meter across rurality contexts and scenarios for load growth with and without PV deployment upto 2050 and no additional flexibility measures installed.

Up to 2030, the only costs are associated with upgrades to the low voltage networks to keep voltage within statutory limits as more PV is deployed. Beyond 2030, significant upgrades to network and transformers are required to avoid thermal constraints as increased peak load growth reaches levels higher than the historical peak in 2005, resulting in to higher power flows at both low and high voltage network levels. Differences in cost between scenarios with load growth alone and load growth and PV deployment are almost all attributable to low voltage network upgrades to mitigate overvoltage caused by solar PV deployment.

For both TD and CR scenarios, when considering load growth alone, costs are highest in the most urban context, C1, and lowest in the most rural context, E1. The increased cost associated with adding PV growth is surprisingly similar in C1 and E1, but significantly higher in the intermediate rurality context, D1. This may be due to large numbers of substations and overengineered infrastructure in particularly rural E1, which is required to reach remote areas, but not operating at full capacity, and relatively dense population in urban C1, leading to small lengths of cable upgrades required. The balance is such that D1 does not benefit to as great an extent from either of these factors, resulting in significantly higher costs associated with PV deployment. Additionally, distribution lines are typically overhead rather than underground in more rural contexts, and transformers polemounted rather than groundmounted, each of which results in lower upgrade costs per unit replaced.

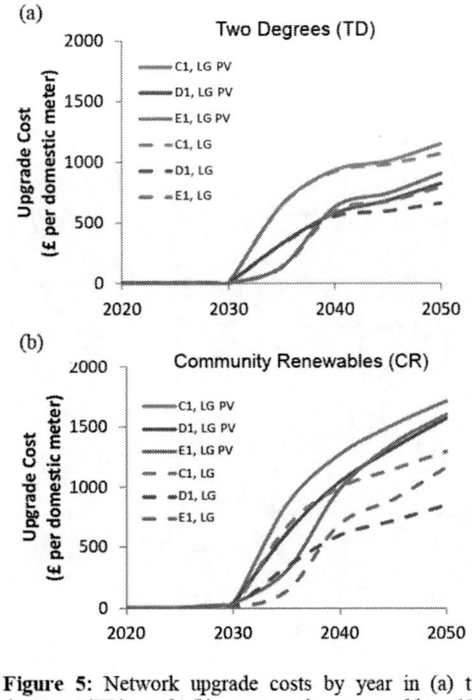

Figure 5: Network upgrade costs by year in (a) two degrees (TD) and (b) community renewables (CR) scenarios. Solid lines are for demand growth plus PV deployment, while dashed lines are for load growth only. (Note that costs are in part associated with nondomestic demand. PV, and storage, although normalisation is to only domestic meters)

3.3 Cost Savings due to Storage and Demand Side Management

Network infrastructure cost savings associated with adding storage and demand side management across scenarios and rurality contexts are shown in Figure 6. For both storage and DSM, savings are higher in C1 and D1 in early years, but savings become larger in E1 between 2035 and 2040. Storage deployment brings much higher cost savings in E1 than C1 or D1, whilst savings associated with DSM are relatively similar. This is likely due to the deployment of storage being in proportion to that of PV in our scenarios, and the higher deployment of PV per meter in rural contexts (Figure 3).

In each scenario and most years, the cost savings associated with going from 50% to 100% DSM are much smaller than those associated with going from 0% to 50%. Cost savings with both flexibility measures are still higher, but the marginal benefit again decreases. When DSM is implemented alongside storage (Figure 6c), there is a significant degree of complementarity and increase in savings in C1 and D1, but very little impact on savings in E1. This implies that, at modelled levels of storage deployment, DSM may not be as important in a highly rural context, although may again be a result of a high level of storage deployment in E1.

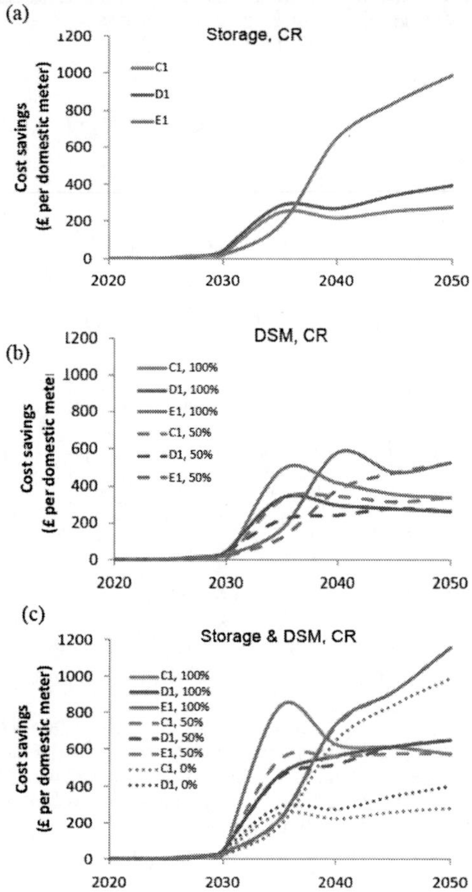

Figure 6: Network upgrade cost savings per domestic meter relative to demand and PV growth alone (LG PV) in rurality contexts C1, D1, and E1 with (a) storage deployment, (b) DSM, and (c) both storage and DSM in CR scenario. (Note that normalisation is to only domestic meters, but costs are for a network also containing nondomestic demand. PV, and storage).

In the CR scenario, savings from both storage and demand side management are exclusively in the LV-V category in early years, when impacts are due to PV deployment, peaking in 2029-2030 for all ruralities before falling steadily to close to zero by 2050 (Figure 7). From 2030 onwards, peak demand reaches levels higher than historically reported, and savings associated with upgrades due to thermal constraints becoming dominant. In each rurality, HV-I becomes the largest source of cost savings between 2035 and 2040, with DT and EHV+ also making significant contributions.

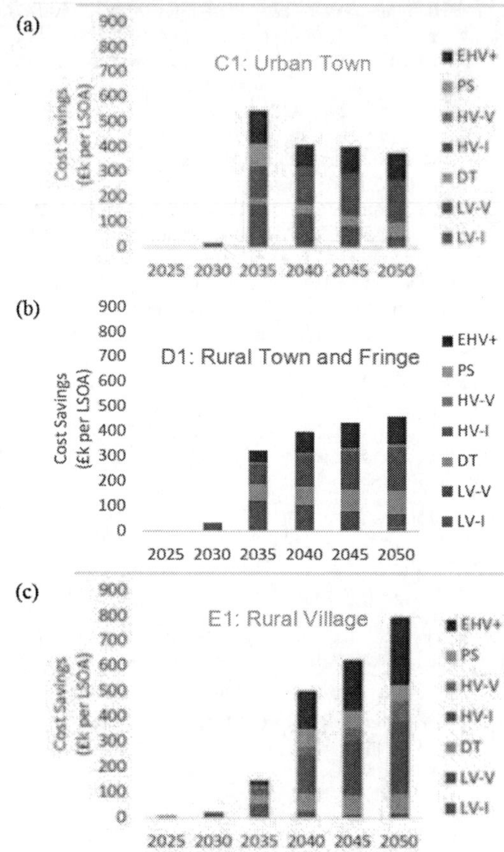

Figure 7 Cost savings associated with adding storage and 100% DSM to a scenario with load and PV growth broken down by upgrade cost category in rurality contexts (a) C1, (b) D1, and (c) E1 in CR scenario. Trends in sources of savings are broadly similar for storage or DSM alone. Costs are broken down into network upgrades at the low voltage (LV), high voltage (HV) and extra high voltage (EHV+) to avoid violation of thermal constraints (I) or voltage constraints (V), and into upgrade costs for primary substations (PS) and distribution transformers (DT).

4 CONCLUSIONS AND FUTTURE WORK

This work has calculated the difference in cost and impacts of increased load growth and PV deployment across rurality scenarios based on real data from across the southwest of England. Results have been compared for two credible PV deployment scenarios up to 2050, and calculations of cost savings in future storage and demand side management scenarios have also been performed.

Further work could include economic analysis of the most cost-effective approach to accommodating future load growth and PV deployment and how this varies by LSOA in a range of future technology cost scenarios. The extension of this analysis to the rest of the UK is also a goal, which could be facilitated by further network infrastructure data. Further work is required in fully interpreting results and how differences between rurality contexts contribute to differences in total cost, breakdown by upgrade type, and reduction associated with different flexibility mechanisms. The consideration of additional flexibility options, such as voltage regulators, could also be a fruitful avenue of research.

ACKNOWLEDGEMENTS

SF, JN, and CC acknowledge the support of the UK Engineering and Physical Sciences Research Council through the Joint UK-India Clean Energy Centre (EP/P003605).

REFERENCES

[1] National Grid, Future Energy Scenarios, 2018.

[2] P. Pearce, R. Slade, Feed-in tariffs for solar microgeneration: Policy evaluation and capacity projections using a realistic agent-based model, Energy Policy. 116 (2018) 95–111. doi:10.1016/j.enpol.2018.01.060.

[3] International Renewable Energy Agency, Renewable power generation costs in 2014, 2015.

[4] P. Westacott, C. Candelise, A novel geographical information systems framework to characterize photovoltaic deployment in the UK: Initial evidence, Energies. 9 (2016). doi:10.3390/en9010026.

[5] P. Westacott, C. Candelise, Assessing the impacts of photovoltaic penetration across an entire low-voltage distribution network containing 1.5 million customers, IET Renew. Power Gener. 10 (2016) 460–466. doi:10.1049/iet-rpg.2015.0535.

[6] C. Candelise, P. Westacott, Can integration of PV within UK electricity network be improved ? A GIS based assessment of storage, Energy Policy. 109 (2017) 694–703. doi:10.1016/j.enpol.2017.07.054.

[7] D. Palmer, E. Koubli, T. Betts, R. Gottschalg, The UK Solar Farm Fleet: A Challenge for the National Grid?, Energies. 10 (2017) 1220. doi:10.3390/en10081220.

[8] M.A. Eltawil, Z. Zhao, Grid-connected photovoltaic power systems : Technical and potential problems — A review, Renew. Sustain. Energy Rev. 14 (2010) 112–129. doi:10.1016/j.rser.2009.07.015.

[9] M. Thomson, D.G. Infield, Impact of widespread photovoltaics generation on distribution systems, IET Renew. Power Gener. 1 (2007) 33–40. doi:10.1049/iet-rpg.

[10] Western Power Distribution, Suburban PV impact, LCNF Closedown Rep. (2016).

[11] UK Power Networks, Validation of Photovoltaic (PV) Connection Assessment Tool: Closedown Report, 2015.

[12] Electricity North West, Low Voltage Network Solutions, Summ. Rep. (2014).

[13] Scottish and Southern Energy, Demonstrating the Benefits of Monitoring LV Networks with embedded PV Panels and EV Charging Point, LCNF Tier 1 Close-Down Rep. (2013) 1–74.

[14] P. Heptonstall, R. Gross, F. Steiner, The costs and impacts of intermittency – 2016 update: a systematic review of the evidence on the costs and impacts of intermittent electricity generation technologies, (2017). doi:10.13140/RG.2.2.17246.02888.

[15] P.H. and M.L. J. Skea, D. Anderson, T. Green, R. Gross, Intermittent renewable generation and the cost of maintaining power system reliability, Gener. Transm. Distrib. IET. 1 (2007) 324. doi:10.1049/iet-gtd.

[16] D. Pudjianto, P. Djapic, J. Dragovic, G. Strbac, Grid Integration Cost of PhotoVoltaic Power Generation, PV Parit. (2013).

[17] C.K. Gan, D. Pudjianto, P. Djapic, G. Strbac, Strategic assessment of alternative design options for multivoltage-level distribution networks, IEEE Trans. Power Syst. 29 (2014) 1261–1269. doi:10.1109/TPWRS.2013.2290103.

[18] R. Mckenna, P. Djapic, J. Weinand, W. Fichtner, G. Strbac, Assessing the implications of socioeconomic diversity for low carbon technology uptake in electrical distribution networks, Appl. Energy. 210 (2018) 856–869. doi:10.1016/j.apenergy.2017.07.089.

[19] A. Lever, D. Sanders, N. Lehmann, M. Ravishankar, M. Ashcroft, G. Strbac, M. Aunedi, F. Teng, D. Pudijanto, Can storage help reduce the cost of a future UK electricity system?, Carbon Trust. Imp. Coll. London. (2016) 121.

[20] G. Strbac, Demand side management: Benefits and challenges, Energy Policy. 36 (2008) 4419–4426. doi:10.1016/j.enpol.2008.09.030.

[21] B. Drysdale, J. Wu, N. Jenkins, Flexible demand in the GB domestic electricity sector in 2030, Appl. Energy. 139 (2015) 281–290. doi:10.1016/j.apenergy.2014.11.013.

[22] Ofgem, Feed-in Tariff Installation Report, September 2017, 2017. https://www.ofgem.gov.uk/environmental-programmes/fit/contacts-guidance-and-resources/public-reports-and-data-fit/installation-reports.

[23] Department for Business Energy & Industrial Strategy (BEIS), LSOA domestic electricity 2016, 2016.

[24] Department for Business Energy & Industrial Strategy (BEIS), MSOA non-domestic electricity 2016, 2016.

[25] Office for National Statistics (ONS), Lower Super Output Area Population Density, 2016.

[26] Office for National Statistics (ONS), Rural Urban Classification (2011) of Lower Layer Super Output Areas in England and Wales, 2011. https://ons.maps.arcgis.com/home/item.html?id=9855221596994bde8363a685cb3dd58a.

[27] Ofgem, Electricity Distribution Price Control Review Final Proposals - Allowed revenue - Cost assessment, 2009.

BUILDING ON CROWDFUNDRES: CROWDFUNDING THE ENERGY TRANSITION

Pablo Alonso, Silvia Caneva, Ingrid Weiss, Sofía Arancón, Antonio Fuentes
WIP – Renewable Energies
Sylvensteinstr. 2, 81369 Munich, Germany

ABSTRACT: European countries have embraced a transition to renewable sources of energy at unprecedented scale. In the 1990s, few achieved a renewable share exceeding 5% of their gross energy consumption. Today, all 28 EU member states exceed that figure, in many instances by a considerable margin, as in the case of Austria, Denmark, Finland, Latvia and Sweden, ranging over 30 % of gross final energy consumption. However, despite the EU remains on track to reach its 20% renewable energy share target for 2020, achieving more ambitious longer-term target of at least 27% across the Union by 2030 will continue to require persistent and joined up efforts directed at seeking out innovative approaches, even more when the rate of renewable energy growth is slowing down in 2016 (to 5.9%) compared with the average annual rate of growth between 2005 and 2014 (6.7 %)[1].

This deceleration is partly attributed to the challenges of renewable energy project developers to finance their projects. Reduced access to conventional financing options over the past few years has triggered innovative financing schemes to emerge, with crowdfunding attracting a lot of attention.

The CrowdFundRES project, identified the vast potential of crowdfunding for filling part of the funding gap that RES project developers have been experiencing by bringing them closer to citizens who are willing to participate as contributors through the use of online crowdfunding platforms.

Crowdfunding can play an important role in the energy transition. With sufficient awareness of its potential among all stakeholders, crowdfunding is well matched to help address resulting funding gaps, provided we attract and sustain high-quality crowdfunding platforms targeted at renewable energy investment, raise awareness among project developers of the suitability of crowdfunding as an additional source of finance, better understand the behavior of individual retail investors and their interest in crowdfunding renewable energy projects, and ensure that the regulatory governance of the range of marketplaces that have emerged is fit for purpose to allow further scaling up across Europe, and without stifling market forces or ongoing innovation in the sector.

The project ran successfully for 3 years, co-funded by the EU under the Horizon 2020 programme. During the 3 years of project implementation, from February 2015 to January 2018, CrowdFundRES successfully achieved the following objectives:

- Deepening understanding of the adoption of crowdfunding for financing RES projects through a qualitative and quantitative analysis and online surveys.
- Performing regulatory analysis and case studies to identify the challenges faced by the application of crowdfunding for RES projects in Europe.
- Developing and applying guidelines aiming at supporting more effective practices for crowdfunding platforms, project developers and investors.
- Formulating policy recommendations to help improve the market and regulatory framework for crowdfunding and renewables in Europe.
- Promoting the crowdfunding concept and its advantages among those who could contribute or raise funds.

CrowdFundRES involves a consortium of renewables project developers, crowdfunding platforms, industry organisations, legal specialists and experts in the sector in order to help unleash the potential of crowdfunding for renewables development. The partnership has been led by WIP Renewable Energies (DE) and involved fourteen partners across six EU member states, representing stakeholders and market actors that are either directly or indirectly involved to crowdfunding. The project partners include the European Crowdfunding Network (BE) and four of the most active crowdfunding platforms for RES projects in Europe, Abundance (GB), Green Crowding/Bettervest(DE), Lumo(FR) and OnePlanetCrowd(NL). Three project developers with projects of different technological and geographical focus around Europe, BNRG Renewables (IE), Valorem (FR) and REG (JE). The consortium was completed by Solar Power Europe (BE), the University of Dundee (GB), Osborne Clarke (DE), Global2000 (AT) and youris.com (IT), leader in the dissemination and communication activities.

The EU PVSEC will represent a unique opportunity to get informed about the results achieved during the CrowdFundRES project in terms of crowdfunding as innovative financing models for projects financing and, at the same time, providing instruments to increase the awareness and public acceptance of renewable energy projects.
Detailed information on the CrowdFundRES project is available at the project website: www.crowdfundres.eu

Keywords: crowdfunding, citizens, investors, awareness, financing, innovative business models, policies, public acceptance

1 AIM AND APPROACH

The overall aim of the CrowdFundRES project is to contribute to the acceleration of the renewable energy growth in Europe by unleashing the potential of crowdfunding for financing renewable energy projects.

> *Crowdfunding is an emerging alternative form of financing that connects those who can give, lend or invest money directly with those who need financing for a specific project. It usually refers to public online calls to contribute finance to specific projects.*

In order to achieve the overall objective, the work has been structured for achieving the following objectives:

- Gain a deep understanding of the public's perception of crowdfunding
- Analyse the challenges faced by the application of crowdfunding for renewable energy projects in Europe
- Develop guidelines that support accepted practices in crowdfunding renewable energy projects
- Apply the guidelines and review them based on practical experience
- Develop policy recommendations to improve the market and regulatory framework
- Promote the crowdfunding concept and its advantages among those who could contribute or raise funds

This paper provides a detailed description of the results achieved during the project to be presented at the EU PVSEC in Brussels, Belgium, in September 2018.

2 SCIENTIFIC INNOVATION AND RELEVANCE

The growing difficulties of renewable energy project developers to finance their project using the conventional avenues have led to the deceleration of renewable energy growth. This is putting significant pressure on independent RES developers, from which only the well-capitalised ones are able to maintain control of their projects at the pre-construction stage, most often by increasing their own equity participation in their projects[2]. There is a need for making use of innovative financing options, such as crowdfunding, that will cover some part of the increased funding requirements. This alternative financing method shows a vast potential for financing renewable energy projects and can help bridging the existing funding gap.

According to the 2nd European Alternative Finance Industry Report, the European alternative finance market as a whole, covering a range from equity-based crowdfunding to peer-to-peer business lending and from reward-based crowdfunding to debt-based securities, grew from €2,833m in 2014 to €5,431m in 2015 [3] .

The fast development of crowdfunding in Europe has attracted the interest of the European Commission and the European Parliament in this relatively new form of financing. The Parliament and commission provided an overview of the current status of crowdfunding in Europe in the briefing 'Crowdfunding in Europe: Introduction and state of play' published in January 2017. Moreover, the Commission published a working document 'Crowdfunding in the EU Capital Markets Union', which discloses the potential of crowdfunding as a key source of financing for SMEs over the long term. Overall, In March 2018, as part of its Fintech action plan, the European Commission presented a proposal for a regulation on crowdfunding service providers[4].

Crowdfunding can therefore play an important role in supporting the effective renewable energy growth in Europe. The project has been designed to target wind, solar, hydro and energy efficiency projects in the range between 100 kW and 5 MW, as these are the technologies and project sizes more suitable for crowdfunding at its current stage of development. However, other case studies analysed have proven that crowdfunding can be effective in ranges from 5MW up to 64MW.

CrowdFundRES project has gained practical experience by analysing 18 crowdfunding renewable electricity projects that has been used as input to the development of targeted guidelines and policy recommendations. As a direct outcome of the 18 projects, there is around 110MW of additional installed capacity. As an indicator of energy savings and CO_2 reduction, one energy efficiency (CHP) crowdfunded project made possible to reach a total energy savings of 37%, leading to an approximate of 270 CO_2 tons savings each year. Overall, through these projects around EUR 29,5 Mio has been successfully crowdfunded.

Key to further support this trend is harnessing the fast-growing crowdfunding market where projects and businesses are financed through contributions from a large number of sources rather than through just a few or one.

The idea of CrowdFundRES is to exploit the potential of crowdfunding for filling part of the funding gap that RES project developers have been experiencing by bringing them closer to citizens who are willing to participate as contributors through the use of online crowdfunding platforms. While the idea of reaching out to the crowd for solutions or for collecting funds isn't new, the phenomenon of using the internet to directly connect with funders has emerged only recently. However, this approach is becoming more relevant as information technologies have been developing and market structures have been evolving.

CrowdFundRES identified and addressed limiting factors and promoted more effective practices among them. This supported further market integration of renewable energy by securing, at acceptable terms, the financing for renewable energy projects that might have difficulties in using conventional financing options; for example when the profile, expected returns or scale of the project do not match the portfolio requirements for bank financing, or when the applied technology is relatively new/complicated and the perceived risk is too high.

Moreover, when members of the public become investors in renewable energy projects the sense of ownership grows, and it contributes to changing mentalities, reducing public opposition that is sometimes associated with renewable energy developments.

For that reason, despite project developers often having access to conventional financing at good terms, they still choose to cover part of the investment through crowdfunding, usually targeted at the local population.

There have been several other projects with a focus ranging from community-owned energy projects with the involvement of cooperatives and communities to the creation of an online aggregating platform for renewable energy projects. CrowdFundRES brought together for the first time all kinds of project developers (incl. traditional RES developers, cooperatives, ESCOs, venture capitalists, etc.) with all citizens in a broader sense (e.g. without excluding those from urban areas who have the largest potential as contributors or even citizens from other countries). Also, the project worked rather as an enabler for the creation of new crowdfunding platforms and the use of existing ones for RES project development than just an aggregator of projects hosted in existing platforms.

3 RESULTS AND CONCLUSIONS

Six main objectives have been identified in order to achieve overall aim of the CrowdFundRES project in contributing to the acceleration of the renewable energy growth in Europe. For each of these objectives an overview on the results is provided.

Gain a deep understanding of the public's perception of crowdfunding

CrowdFundRES performed a qualitative and quantitative analysis through surveys offering a unique insight on the elements of crowdfunding that are positively or negatively received by the public. It also highlighted how the results of this analysis change depending on geographical or other socioeconomic classifications of the public. More specifically, the surveys analyzed perceptions about crowdfunding for renewable energy with the first one focusing on citizens' perceptions as potential contributors, the second one examining the views of crowdfunding platforms and the third one those of RES project developers. The findings provided a detailed picture of the current state of opinion across the EU regarding the present state and future potential of the sector. It was done by conducting a more detailed examination of contributors' preferences on types of investments. Geographical or socioeconomic aspects in the surveys were further analyzed where the sample size allowed.

The results of these activities fed into the formulation of guideline for platforms and developers as well as contributed to the formulation of policy recommendations developed in the second half of the project.

Analyse the challenges faced by the application of crowdfunding for renewable energy projects in Europe

All practical experience of crowdfunding for RES projects has been evaluated in close cooperation with the involved parties in order to identify challenges and limiting factors. The project generated 9 case studies, examining RES crowdfunding projects from France (LUMO / VALOREM), Netherlands (OPC), Germany (GC / Bettervest), United Kingdom (ABUNDANCE / REG / BNRG) and Austria (GLOBAL2000) however this last one it is not strictly a crowdfund and was therefore dealt separately.

The case studies considered a wide range of issues such as the regulatory drivers and barriers, platform technology, as well as the characteristics of each individual fund raise, including investor data and marketing strategy. The case studies aimed to contribute to a better understanding of how platforms operate, identify the challenges facing platforms but also the drivers of growth. The work done helped to analyze the challenges faced by the application of crowdfunding for renewable energy projects in Europe from a regulatory, market, socioeconomic and technical perspective.

Develop guidelines that support accepted practices in crowdfunding renewable energy projects.

The improved understating of the public's perspectives and the identification of challenges for RES developers and platforms have be used to develop three sets of guidelines that will support the wider and more effective development of crowdfunding.

The set of guidelines includes: guidelines for RES project developers interested in crowdfunding, guidelines the crowdfunding platforms interested in hosting RES projects and guidelines for investors in Clean Energy Projects via Crowdfunding.

Apply the guidelines and review them based on practical experience.

The partners LUMO, OnePlanetCrowd, ABUNDANCE and GreenCrowding/Bettervest identified 8 renewable energy projects where the guidelines developed were taken into account.

During the last 18 months of the project, the selected projects were listed and presented in the platform with a specific target for raising funds to finance their realization.

The experiences of preparing and delivering the associated crowdfund raise in each of the cases studies were compared with the Developer and Platform guidelines and, where relevant, small refinements were suggested and incorporated into the guideline documents.

Overall the 8 case studies reported in highlighted the key strength of crowdfunding, which is that for an "investor" the platform is not governed by an investment or lending mandate in the way a fund or bank might be. Instead it can look at each project on a case by case basis and create the right finance solution for the project on the premise that it remains an appropriate investment for its "crowd". This makes crowdfunding a very agile new addition to the funding landscape having as added value the engagement of the public and bringing transparency to project development. The guidelines draw this message out and also provided useful sign posting for new platforms wanting to enter the RES crowdfunding sector.

Develop policy recommendations to improve the market and regulatory framework.

The project involved European and national decision makers and regulators in the development of the policy recommendations from an early stage in order to make sure that they already have a good understanding of the issues and the approach of the project.

The recommendation formulated were intended to improve access to this innovative form of finance for small investors and to develop better policy, regulatory and market support frameworks.

The policy recommendations were validated and promoted at European and national level. The validation process shown that a stable and predictable regulatory framework for project developers is essential to ensure the viability of renewable energy projects, securing return on investments planned by crowd investors. The issue with cross-border crowdfunding appears also to be a major obstacle for crowdfunding in general and for RES project financing. With regards to the aspect of incentivizing RES project developers to boost the use of crowdfunding consensus was found that municipalities are key for promoting crowdfunding and providing reliability to a

RES project on a local level. This in turn could contribute to an increase of citizen's participation, as long as the community is brought together with RES project developers through a local authority.

Overall, the project formulated concrete recommendations for improvements in crowdfunding and renewables regulation in terms of harmonization, support schemes and cross-border investments with increased chances that they were taken into account in the proposal for a regulation on crowdfunding service providers presented by the European Commission in March, 2018.

Promote the crowdfunding concept and its advantages among those who could contribute or raise funds

More than 660.000 stakeholders including crowdfunding experts, different types of renewable energy developers and professionals, NGOs, environmental organisations, policy makers and the wider public have been reached through the projects dissemination activities (Press releases, website, videos, newsletters, leaflets, workshops, articles). Moreover 1.023.411 accounts has been potentially reached through the #CrowdFundRES twitter discussion group resulting in more than 1.650.000 individuals reached through the promotion campaign.

In addition, an eBook has been released which gathers important insights and recommendations for platforms, RES developers, investors and policy makers, drawing on a wealth of analysis of relevant data, industry and investor surveys, and stakeholder engagement across the sector. The ebook is complemented with a range of case studies that amply illustrate how far renewables crowdfunding has come in recent years. Individual platforms have raised funding for individual projects exceeding €10M, spread among 1,000 investors and more. The ebook is available on www.crowdfundres.eu .

4 REFERENCES

[1] European Environment Agency, Renewable energy in Europe – 2017 Update.

[2] Taylor Wessing, "Bridging the Funding Gap: The financing challenge for European cleantech and renewable energy", 2010

[3] B. Zhang, R. Wardrop, T. Ziegler, A. Lui, J. Burton, A. James, K. Garvey, "Sustaining Momentum: The 2nd European Alternative Finance Industry Report", September 2016

[4] European Commission ,"Proposal for a REGULATION OF THE EUROPEAN PARLIAMENT AND OF THE COUNCIL on European Crowdfunding Service Providers (ECSP) for Business" March, 2018

5 ACKNOWLEDGEMENTS

The CrowdFundRES project is a European project, co-financed by the European Commission under the H2020 Programme. The project started on 1st February 2015 and ended on 28th February 2018. The logo of each partner is shown below.

DISCLAIMER

IMPACTS OF SOCIO-ECONOMIC POLICIES ON TEMPORAL DIFFUSION OF PV-BASED COMMUNAL GRIDS IN A RURAL DEVELOPING COMMUNITIY

Dr. Nicholas Nixon Opiyo
Centre for Integrated Energy Research
University of Leeds, Leeds, LS2 9JT, UK
Email: Nicholas.Opiyo1@gmail.com

ABSTRACT: An Agent-Based Model (ABM) is developed in Netlogo as a tool for evaluating impacts of subsidies on temporal diffusion of PV microgeneration systems in a rural developing community. ABMs model individual micro entities within a complex system and the rules that govern the interactions between them within the system, to capture the macro-effects of such interactions. The model simulates how introduction of subsidies would influence households' electrification choices within a given rural developing community over a given period, given various electrification options. Survey-gathered data from Kendu Bay area of rural Western Kenya is used to inform the model. Results show that introduction of capital-based and energy-based subsidies would lead to significant increases in PV installations, which in turn translate into more formations of communal grids (minigrids and microgrids). Observations made from this study can be applied to many similar locations in sub-Saharan Africa. But most importantly, the model could be used by policy-makers in formulating rural electrification policies.

Keywords: Policies; Energy-based subsidies; Capital-based subsidies; Rural electrification

1 INTRODUCTION

Access to grid-level electricity stimulates rural industrial development by enabling productive use of electricity by small manufacturing enterprises such as welding and carpentry shops and by micro-home enterprises such as mobile phone chargers, or by allowing consumers to switch to cheaper, cleaner, and more convenient sources of electricity for their businesses [1]. Even on a smaller scale, the benefits of electricity abound; basic electric lighting alone, usually a replacement for harmful kerosene lanterns, brings with it extended benefits to health and education as improved lighting enables extended hours of studying for school children and extended hours of operation for small businesses, while also providing greater sense of security [2]. Also, basic electricity improves access to information and communication technologies; TVs and radios are major sources of news, business and health information, and distance education [2].

There are two possible routes to rural electrification i.e., through extensions of existing national utility grids from large central power generation systems or through decentralized systems namely communal grids (minigrids and microgrids) and stand-alone microgeneration systems. The International Energy Agency (IEA) estimates that only 30% of the people currently without access to electricity in sub-Saharan Africa can be cost-effectively served through national grid extensions due to the sparseness of the rural populations, rough terrains, low economic activities, and low load densities. The remaining 70% would be most cost-effectively served through decentralized systems, i.e. communal grids (52.5%) or stand-alone systems (17.5%) [1].

The main factors that influence paths to rural electrification by developing communities are costs, demand, availability of resources, and socio-economic factors and policies such as subsidies and neighbourhood influence. Various studies show that positive socio-economic policies implemented by governments of developed nations are largely responsible for the rapid growths in PV installations seen in those nations over the past decade or so; A study by Beise on diffusions of new technologies across national boundaries concluded that positive government policies and incentives significantly stimulated diffusion of PV across many countries, underscoring the importance of government intervention in adaptations of new technologies [3]. A comparable study by Guidolin et al also found that positive government policies in form of incentives and subsidies positively influenced PV diffusions across 11 countries [4]. Wustenhagen found that positive government policies, especially feed-in-tariffs (FiT), were largely responsible for the boom in PV installation in Germany [5]. Zhang et al found that in addition to positive regional governments' policies, costs also played significant roles in PV diffusions across Japan [6]. On the other hand, limited financial resources in developing nations have restricted governments' abilities to subsidize new technologies as priorities are given to crucial projects such as hospitals and schools. In these nations, therefore, PV diffusion largely occurs without government help or influence.

Different modelling tools and techniques have been applied in planning rural electrification paths in many developing countries. However, these often view this problem as a question of expansion of existing national utility grids, and seldom address the unique and regionally-specific challenges presented by each developing nation. In most of sub-Saharan Africa for example, grid electricity is often unreliable, plagued with frequent blackouts, poor maintenance, and low quality of service. In these regions, expansions of the national grids often result in further strain on the systems and thus in further reduction in the quality of services provided to those already grid-connected [7]. The model developed in this work simulates how, given various electrification option, households in a typical rural developing community would make their electrification choices, and how introduction of subsidies would impact on those decisions. Observations made from this study can be applied to many similar locations in sub-Saharan Africa. But most importantly, the model can be used by policy-makers in formulating rural electrification policies.

2 METHODOLOGY

A household without a PV system must first develop the idea to install one given the costs and socio-economic factors such as subsidies. Based on Kendu Bay survey data, 92.2% of households do not have PV systems due to cost. Cost is therefore the dominant factor in a PV installation decision [8]; PV is only installed if a household can afford it and if:

$$LUCE_{PV} < C_{A/kWh} \quad (1)$$

where $C_{A/kWh}$ is avoided cost per kWh, i.e., the prevailing national grid electricity cost per kWh and equals \$0.20/kWh as per current Kenyan rates, while $LUCE_{PV}$ is the levelized unit cost of delivered electricity, and is given by

$$LUCE_{PV} = \frac{ALCC_{PV}}{W_p \times EHFS \times 365 \times CUF} \quad (2)$$

where W_p is the rated peak Watt capacity of the PV module and is based on a household's activity profile and power demand [9], $EHFS$ is the equivalent hours of full sunshine per day and equals 8 [9], CUF is the capacity utilization factor which incorporates non-utilization and outages of systems due to various reasons and equals 0.9 based on Kendu Bay survey data, and $ALCC_{PV}$ is the annualized life cycle cost which is calculated by summing up the cost of all of its individual components, i.e. the module, battery, charge controller, and appliances multiplied by their respective capital recovery factors plus operations and maintenance costs. It is expressed as

$$\begin{aligned} ALCC_{PV} = {} & (C_{0PV} \times CRF_{PV}) \\ & + (C_{0batt} \times CRF_{batt}) \\ & + (C_{0cc} \times CRF_{cc}) \\ & + (C_{0appl} \times CRF_{appl}) \\ & + C_{O\&M} \end{aligned} \quad (3)$$

where C_{0PV} is the capital cost of the PV module and is set at \$1.40/Wp based on Kenya's PV costs, C_{0batt} is the capital cost of the battery and is set at \$0.95/Ah based on the price of a typical 48V 100 Ah battery in Kenya, C_{0cc} is the capital cost of the charge controller and is set at \$28.50 as per Kenyan rates, and C_{0appl} is the capital cost of appliances and is averaged at \$75 per household based on typical load profiles of 95% of households with PV systems in Kendu Bay [8]. CRF_{PV}, CRF_{batt}, CRF_{cc}, and CRF_{appl} are the capital recovery factors of the PV module, the battery, the charge controller, and appliances, respectively, while $C_{O\&M}$ is the operations and maintenance cost and is set at \$2.50/year based on Kendu Bay survey data [8].

Capital recovery factor (CRF) is calculated using the formula

$$CRF = \frac{i(1+i)^n}{(1+i)^n - 1} \quad (4)$$

where i is the discount rate and is set at 12% based on average Kenya's central bank rates over the past 12 months, while n is the life of the particular component being considered, i.e. 25 years for PV modules/arrays, 5 years for batteries, 5 years for converters/inverters/charge controller, and 10 years for appliances (cumulative load). Based on these figures, $CRF_{PV} = 0.1339$, $CRF_{batt} = 0.2774$, $CRF_{cc} = 0.2774$, and $CRF_{cc} = 0.1770$.

For a communal grid consisting of N households' systems connected together, levelized unit cost of delivered electricity ($LUCE_{CG}$) is given by

$$LUCE_{CG} = \frac{ALCC_{CG}}{PV_{CG} \times EHFS \times 365 \times CUF} \quad (5)$$

where PV_{CG} is the capacity of the communal grid (PV array) in kWp and is based on the activity profiles of all households connected to the communal grid, $EHFS$ is the equivalent hours of full sunshine per day and equals 8, CUF is the capacity utilization factor which incorporates non-utilization and outages of systems due to various reasons and is maintained at 0.9, and $ALCC_{CG}$ is the annualized life cycle cost of the communal grid and is given by

$$\begin{aligned} ALCC_{CG} = {} & [(C_{0PV_{CG}} \times CRF_{PV}) \\ & + (C_{0batt_{CG}} \\ & \times CRF_{batt}) \\ & + (C_{0pcu} \times CRF_{pcu})] \\ & \times [PV_{CG} \times R]^b \\ & + [(C_{0dn} \times L) \\ & + (C_{0sc} \times N)] \\ & \times CRF_{pdn} + C_{O\&M_{CG}} \end{aligned} \quad (6)$$

where $C_{0PV_{CG}}$ is the capital cost of the PV array and associated mounting structures and is set at \$1,400/kWp, $C_{0batt_{CG}}$ is the capital cost of the battery bank and associated structures and equals \$0.95/Ah $\times N$, C_{0pcu} is the capital cost of the power conditioning unit (power electronics) and equals \$28.50 $\times N$, CRF_{pcu} is the capital recovery factor of the power conditioning unit and equals 0.2774, C_{Odn} is the capital cost of the power distribution network per km and equals \$2,500 for Kendu Bay, C_{Osc} is the capital cost of service connections including internal wiring and appliances per household serviced and equals \$125 based on similar networks in Kenya, L is the length of the distribution network in km and is initially set at 1 km for a neighbourhood radius of 500 m, CRF_{pdn} is the capital recovery factor of the distribution network including the service connections and is calculated to be 0.1770, $C_{O\&M_{CG}}$ is the cost of operating and maintaining the communal grid and is estimated to be \$2.50 $\times N$, R is the benchmark unit cost of the communal grid and is set at \$1,400/kWp, and b is a scale factor for incorporating cost reduction in overall cost of the communal grid, without the power distribution network (pdn), due to bulk purchasing of the components used in the grid. Its effect is uniformly distributed over all components of the communal grid, minus the distribution network. It is set at 0.95.

Since currently there are no functional PV-related subsidies in Kenya, the values are initially set to zero. To investigate impacts of subsidies on temporal diffusion of PV systems in Kendu Bay area, we simulate impacts of increasing subsidies/kWh of PV electricity generated and increasing subsidies/m² of PV installed. Subsidies per kWh (S_{kWh}) is given by [10]

$$S_{kWh} = C_{PV/kWh} - C_{A/kWh} \quad (7)$$

where $C_{PV/kWh}$ is the PV generation cost per kWh and is given by

$$C_{PV/kWh} = \frac{C_{PV-Total}}{(E_{PV-Total} \times L_{PV})} \quad (8)$$

where $C_{PV-Total}$ is the total PV cost, $E_{PV-Total}$ is the total PV output, and L_{PV} is PV lifetime and is set at 25 years based on Kendu Bay data [8].

Total PV cost ($C_{PV-Total}$) is given by

$$C_{PV-Total} = (C_{OPV} - S_{m^2}) \times \varepsilon_{PV} \times A \times AF \quad (9)$$

where C_{OPV} is the capital cost of the PV module per Wp and is set at \$1.40 based on Kendu Bay data, S_{m^2} is subsidies per m² and is initially set to zero, ε_{PV} is PV efficiency and is set at 15% based on Kendu Bay data, A is roof area available under PV Installation and is randomly set between 10m² and 50m², and AF is an accounting factor given by

$$AF = (1 + r)^{L_{PV}} \quad (10)$$

where r is the lending rate; if a household purchases PV on credit, lending rate (r) on the loan is used for accounting factor AF calculations. For cash buyers, $r = 0$.

Total PV output $E_{PV-Total}$ (kWh) is given by

$$E_{PV-Total} = I_{PV} \times PR \times \varepsilon_{PV} \times A \quad (11)$$

where PR is the performance ratio which considers losses from components, environmental factors, etc., and is initially set at 0.75, while I_{PV} in kWh/m² is the irradiation on tilted panels and is given by [10]

$$I_{PV} = I_{incident} \sin(\alpha + \beta) \quad (12)$$

where $I_{incident}$ is the solar radiation measured orthogonal to the sun rays, α is the elevation angle, and β is the tilt angle of the module measured from horizontal [9].

3 RESULTS AND DISCUSSION

Figure 1 shows a graph of households with PV installations, households connected to communal grids, and households connected to the national grid after 25 years. Currently (at year zero) there are 347 households with PV installations in Kendu Bay area, representing 4.7% of all households. Similarly, there are 229 households connected to the national grid, representing 3.1% of all households; there are currently no communal grids in Kendu Bay and thus zero connections. After 25 years, 4,325 households would have installed PV, representing 44.1% of all households. This massive growth is attributable to many factors with the main ones being falling PV costs coupled with increasing households' incomes, increasing neighbourhood influence due to increasing PV installations, increasing awareness of the socio-economic and environmental benefits of PV systems, and generally increasing PV efficiencies and thus increasing outputs for similarly sized systems.

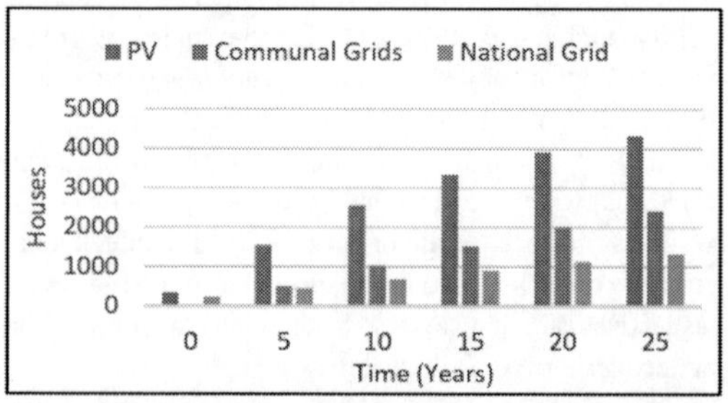

Fig. 1: Kendu Bay Electrification Topologies After 25 Years

As more households install PV over the years, and as households' electricity demands increase, more and more households will start looking for ways of increasing their access to electricity; therefore, households will start coming together to form communal grids or to join already existing communal grids. Specifically, households connected to communal grids are estimated to grow from zero to about 2,410 after 25 years, representing 24.6% of all households. This massive growth is attributable to the same factors as those that influence PV installations. Moreover, communal grids are sized to match demands and can therefore be set up with small initial investments. Also, they are modifiable with increasing demands or with changing technologies, and could thus be expanded with increasing incomes. Even though households connected to the national grid will also grow from 229 to 1,323 after 25 years, representing 13.5% of all households and 23.4% of all electrified households, electrification through PV systems and PV-based communal grids will grow at a faster pace, representing 76.6% of all electrified households.

3.1 Impacts of Subsidies/kWh of PV Power Generated on PV Diffusion

Figure 2 shows a graph of households with PV, households connected to communal grids, and households connected to the national grid as functions of increasing subsidies/kWh of power generated, after 25 years. Without subsidies 4,325 households would have installed PV after 25 years, representing 44.1% of all households. If a subsidy of $0.10/kWh is introduced, the number of households with PV after 25 years jumps to 5,067, representing 51.7% of all households, and an increase of 17.2% in PV installations. A subsidy of $0.20/kWh would lead to 5,591 PV installations, representing 57% of all households and an increase of 29.3% in PV installations. Similarly, introduction of a subsidy of $0.30/kWh would lead to 5,827 PV installations, or 59.4% of all households, and an increase of 34.7% in PV installations. As PV installations increase, so do communal grid formations and/or connections. Without subsidies, 2,410 households would have connected to communal grids after 25 years, representing 24.6% of all households. This figure increases to 2,791 with introduction of a subsidy of $0.10/kWh, representing 28.5% of all households, and an increase of 15.8% in communal grids connections. Similarly, subsidies of $0.20/kWh and $0.30/kWh would lead to 3,217 (32.5%) and 3,685 (37.6%) communal grids connections, respectively and representing increases of 33.5% and 52.9%, respectively.

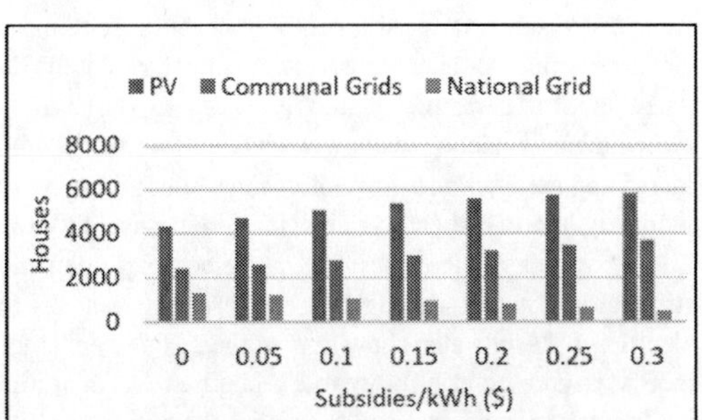

Fig. 2: Impacts of Subsidies/kWh on Electrification Topologies after 25 Years

As PV installations increase, and as more households opt for PV-based communal grids, connections to the national grid fall. Specifically, without subsidies, 1,323 households would have connected to the national grid after 25 years. This falls to 1,069 with introduction of a subsidy of $0.10/kWh, and even further to 802 and 499 with introductions of subsidies of $0.20/kWh and $0.30/kWh, respectively.

3.2 Impacts of Subsidies/m^2 of PV Installed on PV Diffusion

Figure 3 shows a graph of households with PV, households connected to communal grids, and households connected to the national grid as functions of increasing subsidies/m^2 of PV installed, after 25 years. Without subsidies 4,325 households would have installed PV after 25 years, representing 44.1% of all household. If a subsidy of $50/m^2 is introduced, the number of households with PV after 25 years jumps to 5,023, representing 51.2% of all households, and an increase of 16.1% in PV installations. A subsidy of $100/m^2 would lead to 5,601 PV installations, representing 57.1% of all households and an increase of 29.5% in PV installations. Similarly, introduction of a subsidy of $150/m^2 would lead to 6,012 PV installations, or 61.4% of all households, and an increase of 39% in PV installations. As PV installations increase so do communal grids formations and connections. Without subsidies, 2,410 households would have connected to communal grids after 25 years, representing 24.6% of all households. This figure increases to 2,829 with introduction of a subsidy of $50/m^2, representing 28.9% of all households, and an increase of 17.4% in communal grids connections. Similarly, subsidies of $100/m^2 and $150/m^2 would lead to 3,254 (33.2%) and 3,689 (37.6%) communal grids connections, respectively, which represents increases of 35% and 53.1% in connections, respectively.

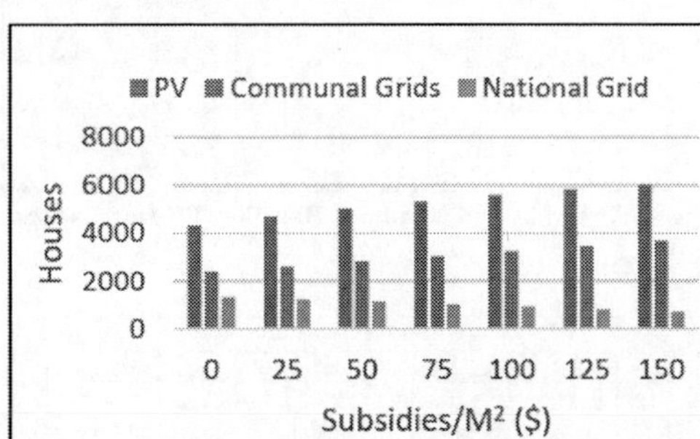

Fig. 3: Impacts of Subsidies/m^2 on Electrification Topologies after 25 Years

As PV installations increase, and as more households opt for PV-based communal grids, connections to the national grid fall with increasing subsidies/m^2. Specifically, without subsidies, 1,323 households would have connected to the national grid after 25 years. This falls to 1,129 with introduction of a subsidy of $50/m^2, and even further to 923 and 701 with introductions of subsidies of \$100/m^2 and \$150/m^2, respectively. In percentage terms, national grid connections fall from 13.5% with no subsidies to 7.2% with subsidies of \$150/m^2.

4 MODEL VALIDATION

Empirical PV installations data from 1985 to 2010 obtained from the Kenya's Ministry of Energy and Petroleum for two separate locations with similar geolocation data to Kendu Bay, i.e. Homabay and Ahero, are used to validate the model [11]. The two towns were also chosen because in 1985, their populations of 31,063 (Homabay) and of 26,322 (Ahero) mirrored that of Kendu Bay (29,518) in 2015 (Central Bureau of Statistics Kenya, 2011). Moreover, both areas are mainly occupied by people of the same ethnic and cultural backgrounds as Kendu Bay and would thus respond similarly to given socio-economic factors. The PV installations data are obtained by the government from retailers who also perform the tasks of installing and repairing the purchased systems.

PV installation data for Homabay at year zero (1985) was input into the model and then simulated over 25 years. Figure 4 shows a comparison of simulated and empirical data for Homabay after 25 years. For empirical data, year 1985 is taken as year zero while year 2010 is taken as year 25. It is assumed that PV costs in 1985 were lower than current costs but that household income levels then were also lower than income levels today by a similar ratio. The errors range between 0.01 and 0.05, making the model reasonably accurate. In all simulations, increasing population and falling PV costs are considered using available census data from the Kenyan Government [12].

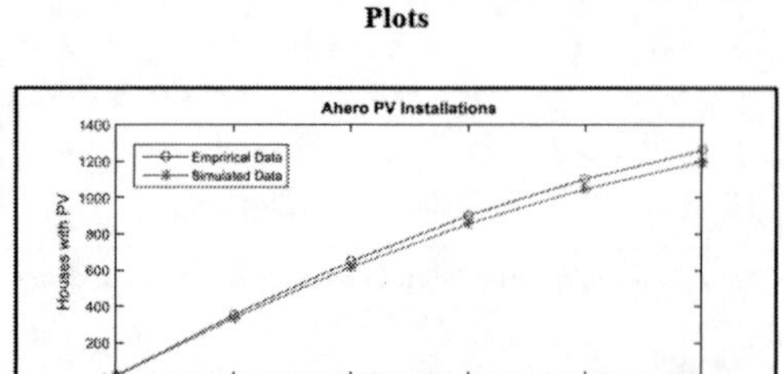

Fig. 4: Empirical and Simulated Homabay PV Installations Plots

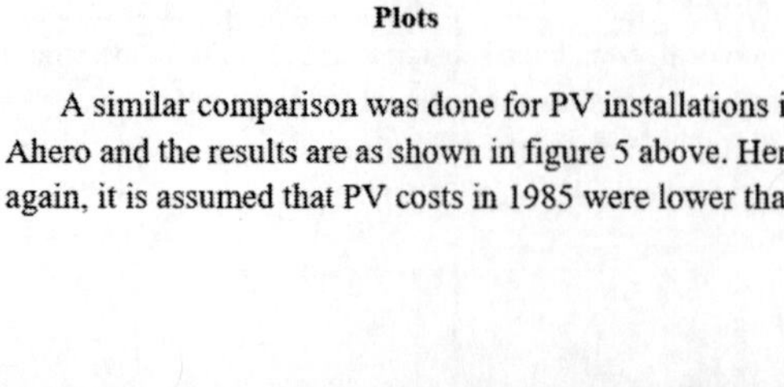

Fig. 10: Empirical and Simulated Ahero PV Installations Plots

A similar comparison was done for PV installations in Ahero and the results are as shown in figure 5 above. Here again, it is assumed that PV costs in 1985 were lower than current costs but that household income levels then were also lower than income levels today by a similar ratio. The error is below 0.05, further validating the model.

Figure 6 shows comparisons of simulated PV installations for Homabay, Ahero, and Kendu Bay after 25 years while figure 7 shows a plot of empirical (Homabay and Ahero) and simulated (Kendu Bay) PV installations data after 25 years. The plots show similar trends in PV installations, with Kendu Bay installations being greater than the other two because of rapidly falling PV costs and increasing awareness of their benefits. The plots further validate the model.

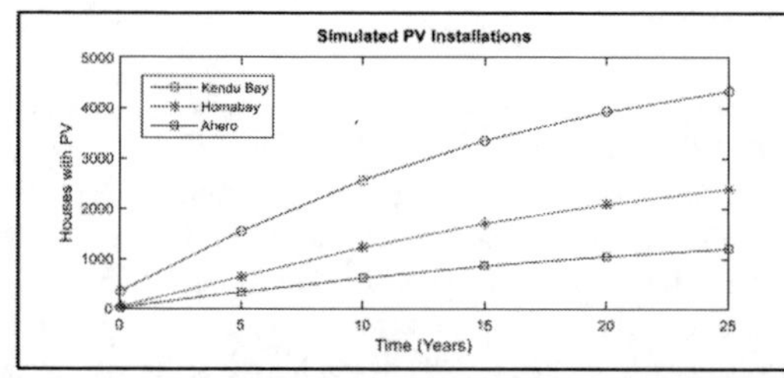

Fig. 6: Comparison of Simulated PV Installations Plots

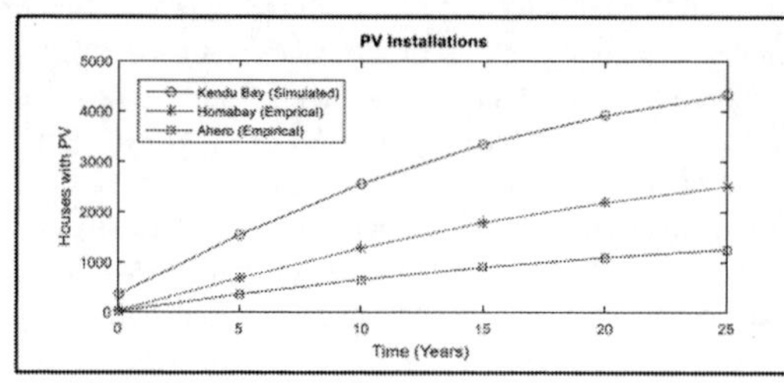

Fig. 7: Comparison of Empirical and Simulated PV Installations Plots

4 CONCLUSION

In this work, temporal diffusion of PV microgeneration systems in Kendu Bay area is modelled and simulated using an agent-based model developed in Netlogo. Results show that after 25 years, more households will be electrified through PV systems than through the national grid. These trends are attributable to rapidly falling PV costs coupled with the availability of small PV systems tailored for rural households with basic electricity needs. These systems enable households to make modest investments into PV systems, tailored to meet their basic power demands. Moreover, with increasing incomes and power demands, households could modestly modify their systems in a step-wise fashion. These systems therefore act as electricity demand stimulators, by introducing households to the benefits of electricity. Results also show that, as their power demands increase, more households with PV choose to form or join communal grids as opposed to joining the national utility grid, mainly due to greater affordability and reliability. Furthermore, results show that the introduction of either capital-based or energy-based subsidies significantly lead to increases in PV installations, which would in turn lead to increased connections to communal grids. Table 1 below summarizes the above information.

Table 1: Impacts of Subsidies on Electrification Topologies after 25 Years

Houses after 25 years	Subsidies/kWh ($)		
	0	**0.15**	**0.30**
Houses with PV	4325	5374	5827
Houses Connected to Communal Grids	2410	2994	3685
Houses Connected to National Grid	1323	937	499
Houses after 25 years	**Subsidies/m^2 ($)**		
	0	**75**	**150**
Houses with PV	4325	5321	6012
Houses Connected to Communal Grids	2410	3046	3689
Houses Connected to National Grid	1323	1027	701

The above findings demonstrate that the Government of Kenya, and by enlarge similar developing communities, could increase its rural electrification rate by introducing favourable subsidies to entice more households to install PV. The merits and demerits of both types of subsidies could be debated, but energy based subsidies enable the government to avoid direct capital investments in a household's own PV system. It could therefore be more affordable for developing countries with less financial capital. It is important to note that subsidies commit future governments to financial burdens and there may be concern over lack of commitment to promises made by previous governments.

4.1 Acknowledgement

This research is funded by Leeds International Research Scholarship

4.2 References

1. International Energy Agency (IEA), 2016. *World Energy Outlook*
2. Ministry of Foreign Affairs of Netherlands, 2013. Renewable Energy: Access and Impact; a Systematic Literature Review of the Impact on Livelihoods of Interventions Providing Access to Renewable Energy in Developing Countries, *IOB Study*
3. Beise, M., 2004. The International Adoption of Photovoltaic Energy Conversion: Is Japan a Lead Market? *Discussion Paper Series* **153**, Research Institute for Economics & Business Administration, Kobe University
4. Guidolin, M., Mortarino, C., 2010. Cross-country diffusion of photovoltaic systems: modelling choices and forecasts for national adoption patterns, *Technological Forecasting and Social Change* **77**: 279-296
5. Wustenhagen, R., Bilharz, M., 2006. Green energy market development in Germany: effective public policy and emerging customer demand, *Energy Policy* **34**: 1681 – 1696
6. Zhang, Y., Song, J., Hamori, S., 2011. Impact of subsidy policies on diffusion of photovoltaic power generation, *Energy Policy* **39**: 1958 – 1964
7. Foster, V., Steinbuks, J., 2009. Paying the Price for Unreliable Power Supplies In-House Generation of Electricity by Firms in Africa, *WPS*
8. Opiyo, N. N., 2016. A Survey Informed PV-Based Cost-Effective Electrification Options for Rural Sub-Saharan Africa, *Energy Policy* **91**: 1-11
9. Opiyo, N. N., 2015. Modelling PV-Based Communal Grids Potential for Rural Western Kenya, *Sustainable Energy, Grids and Networks* **4**: 54-61
10. Green, M. A., 1992. Solar Cells: Operating Principles, Technology and System Applications, *University of New South Wales*
11. Ministry of Energy and Petroleum, Republic of Kenya, 2012. Solar Energy, Nairobi, Kenya
12. Central Bureau of Statistics, Office of the Vice President, and Ministry of Planning and National Development, 2011. *Economic Survey*, Nairobi

SOLAR-ERA.NET COFUND
EUROPEAN NETWORK OF NATIONAL AND REGIONAL RESEARCH AND INNOVATION PROGRAMMES - RECENT DEVELOPMENTS OF JOINT TRANSNATIONAL CALLS

Stefan Nowak, NET Nowak Energy & Technology, Waldweg 8, 1717 St. Ursen, Switzerland,
T: +41 26 494 00 30, F: +41 26 494 00 34, stefan.nowak@netenergy.ch
Marcel Gutschner, NET Nowak Energy & Technology, marcel.gutschner@netenergy.ch
Thomas Biel, NET Nowak Energy & Technology, thomas.biel@netenergy.ch
Stefan Oberholzer, Swiss Federal Department for Environment, Transports, Energy and Communication,
stefan.oberholzer@bfe.admin.ch
Christoph Hünnekes, Forschungszentrum Jülich – PtJ, ch.huennekes@fz-juelich.de
Kambulakwao Chakanga, Forschungszentrum Jülich – PtJ, k.chakanga@fz-juelich.dede
Renate Horbelt, Forschungszentrum Jülich – PtJ, r.horbelt@fz-juelich.de
Melanie Schulte, Project Management Organisation Energy, Technology, Sustainability (ETN), me.schulte@fz-juelich.de
Severino Falcón Morales, Agencia Estatal de Investigación / Ministerio de Economía, Industria y Competitividad
(MINECO-AEI), severino.falcon@mineco.es
Daniel Ruiz, Fundación Española para la Ciencia y la Tecnología (FECYT), daniel.ruiz@fecyt.es
Gema del Rio, Centro para el Desarrollo Tecnológico Industrial (CDTI), gema.delrio@cdti.es
Pierre-Jean Rigole, Swedish Energy Agency, pierre-jean.rigole@energimyndigheten.se
Tobias Walla, Swedish Energy Agency, tobias.walla@energimyndigheten.se
Otto Bernsen, Rijksdienst voor Ondernemend Nederland (RVO), otto.bernsen@rvo.nl
Laurence Polain, Service Public de Wallonie (SPW), laurence.polain@spw.wallonie.be
Elsie Declercq, Hermesfonds / Agentschap Innoveren & Ondernemen (VLAIO), elsie.declercq@vlaio.be
Evi Afentaki, Geniki Grammatia Erevnas Kai Technologias (GSRT), pafe@gsrt.gr
Pavlos Leptos, Research Promotion Foundation (RPF), pleptos@research.org.cy
Tristan Carrere, Agence de l'Environnement et de la Maîtrise de l'Energie (ADEME), tristan.carrere@ademe.fr
Pascal Bain, Agence nationale de la recherche (ANR), Pascal.Bain@agencerecherche.fr
Kaan Karaösz, Turkiye Bilimsel ve Teknolojik Arastirma Kurumu (TUBITAK), kaan.karaoz@tubitak.gov.tr
Aldo Covello, Ministero dell'Istruzione, dell'Università et della Ricerca (MIUR), aldo.covello@miur.it
Alice Goodbrook, InnovateUK, Alice.Goodbrook@innovateuk.gov.uk
Jan Osinski, Narodowe Centrum Badan i Rozwoju (NCBR), jan.osinski@ncbr.gov.pl
Elvira Lutter, Klima- und Energiefonds, elvira.lutter@klimafonds.gv.at
Anita Hipfinger, Österreichische Forschungsförderungsgesellschaft (FFG), Anita.Hipfinger@ffg.at
Ulrike Rohrmeister, BMVIT Ministry of Transport, Innovation and Technology, Ulrike.Rohrmeister@bmvit.gv.at
Gideon Friedmann, Ministry of National Infrastructure, Energy and Water Resources, gideonf@energy.gov.il

ABSTRACT: SOLAR-ERA.NET is a network that brings together more than 20 RTD and innovation programmes in the field of solar electricity technologies in the European Research Area. The network of national and regional funding organisations has been established in order to increase transnational cooperation between RTD and innovation programmes and to contribute to achieving the objectives of the Strategic Energy Technology Plan (SET-Plan) through dedicated transnational activities (especially transnational calls). Through the support of the funding organisations, some 60 MEUR have been mobilised for transnational RTD and innovation projects through the first four joint calls carried out. SOLAR-ERA.NET was an EU funded FP7 project running from 2012 to 2016. Within the H2020, the network continues to work in the ERA-NET Cofund scheme. In this contribution, results are presented for the transnational SOLAR-ERA.NET calls.

Keywords: R&D and Demonstration Programmes, Photovoltaic, Funding and Incentives, Strategy

1 INTRODUCTION AND CONTEXT

SOLAR-ERA.NET is a European network of national and regional funding organisations and RTD and innovation programmes in the field of solar electricity generation, i.e. photovoltaics (PV) and solar thermal electricity (STE) / concentrating solar power (CSP). SOLAR-ERA.NET contributing to reaching the objectives of the European Strategic Energy Technology Plan (SET-Plan), by carrying out the coordination and support actions between national and regional research and innovation programmes.

The SET-Plan aims to increase, coordinate and focus EU support on key low-carbon energy technologies in order to achieve Europe's 2020 and 2030 energy and climate objectives in the future.

2 NETWORK AND MISSION

2.1 Network

As the largest ERA-NET (network in the European Research Area) ever in the solar power field, SOLAR-ERA.NET involves more than 20 national and regional RTD and innovation programmes dealing with PV and CSP. This high level of involvement of most relevant stakeholders provides excellent outreach and allows for a solid coordination needed for an efficient and coherent approach in the highly diverse RTD landscape.

The network has been composed of some 20 organisations being programme owners and managers. Most countries and regions belonging to the SOLAR-ERA.NET consortium also participate in most calls.

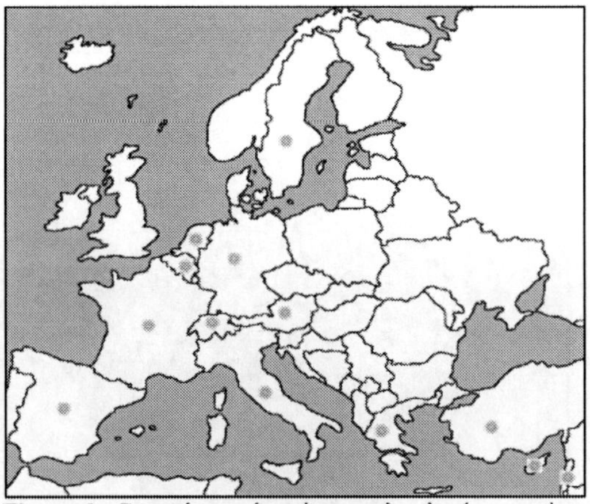

Figure 1: Countries and regions active in the ongoing SOLAR-ERA.NET Cofund 2 call.

2.2 Missions and goals

SOLAR-ERA.NET has two fundamental missions. As a network supported by the EC within the ERA-NET (2012-2016) respectively ERA-NET Cofund (since 2016) scheme, the mission is to improve the coordination and cooperation between national and regional programmes. In the context of the Strategic Energy Technology Plan, the mission of the network is to implement central parts of the SET plan on a transnational level and thus contribute to achieving the goals defined in the SET plan.

The more specific and essential goals and activities of SOLAR-ERA.NET are i) to launch joint calls for RTD proposals by national and regional RTD and innovation programmes and ii) to define and support the best joint activities, strategic information exchange and use of implementation tools.

By identifying and choosing priority topics based on the Implementation Plans for PV respectively CSP for transnational calls, the SOLAR-ERA.NET network selects and funds industrially relevant transnational RTD and innovation projects in the field of solar electricity technologies.

In quantitative, financial terms, SOLAR-ERA.NET shall result in a total funding volume by the participating national and regional programmes of 10 to 15 MEUR per call for innovative projects.

3 ACTIVITIES, FACTS AND FIGURES

3.1 Set up of transnational calls

SOLAR-ERA.NET started in November 2012. Four sets of transnational joint calls were launched until 2016, with the participation of up to 17 countries covering up to 8 PV and 4 CSP topics per call.

Call topics in photovoltaics (PV) and concentrating solar power (CSP) are commonly defined by the SOLAR-ERA.NET consortium and adopted by the participating programmes and countries according to their national / regional priorities. The topics can be found in Table 1.

The transnational call is based on a 2-step / stage-procedure with i) a preproposal and ii) a full proposal stage. Preproposals are checked according to the national / regional regulations and then discussed in the SOLAR-ERA.NET consortium setting up lists of preproposals recommended respectively not recommended for stepping on to the full proposal phase.

Once the full proposals are evaluated by independent international experts as well as by national experts, a list of projects suggested for funding is established.

The first four transnational SOLAR-ERA.NET joint calls are each allocated with a total public funding budget of approximately 10 to 12 MEUR provided by the participating national and regional programmes and agencies.

Within the ERA-NET Cofund scheme, SOLAR-ERA.NET calls come along with an even higher funding budget. The ongoing Cofund Call gathers some 15 MEUR of public funding from national and regional agencies. Furthermore, the EC provides the so-called Top-Up presenting 33% of the total funding effectively used, i.e. the total funding budget for the ongoing Cofund Call amounts to some 22 MEUR. The selection procedure includes a ranking list which has to be strictly followed from top downwards.

Applicants may also come from other countries but they do have to provide their own funding for their participation in the projects.

3.2 Participation in transnational calls

Corroborated data is available for the first four transnational SOLAR-ERA.NET calls. They found good interest in the solar power industry sector and research community. Some 180 preproposals were submitted with a total budget of 227 MEUR (of which 155 MEUR funding requested) involving 804 partners (multiple counting – some partners have been involved in several preproposals).

A majority (54%) of the organisations involved in the preproposals submitted belong to the categories of small and medium sized enterprises and large enterprises; research organizations have a share of 43%. Looking at the 45 projects finally started, the share of small and medium sized enterprises and large enterprises climbs up to 57% (see Figure 2).

Among the 45 projects funded, 29 projects involve partners from Germany and 19 projects have partners from Spain. Figure 3 shows the number of projects by country / region as well as by call.

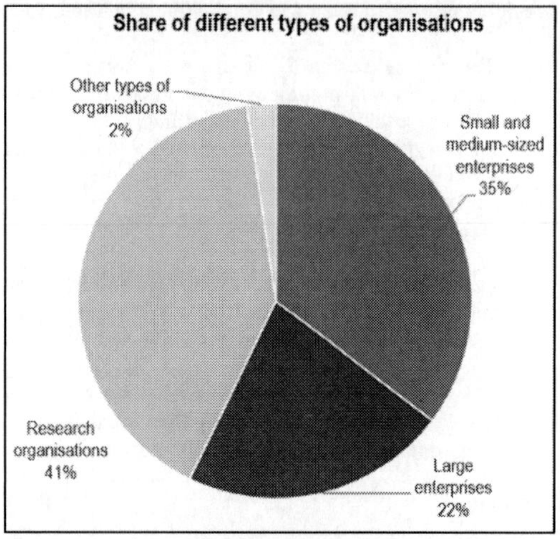

Figure 2: Share of different types of organisations involved in projects funded.

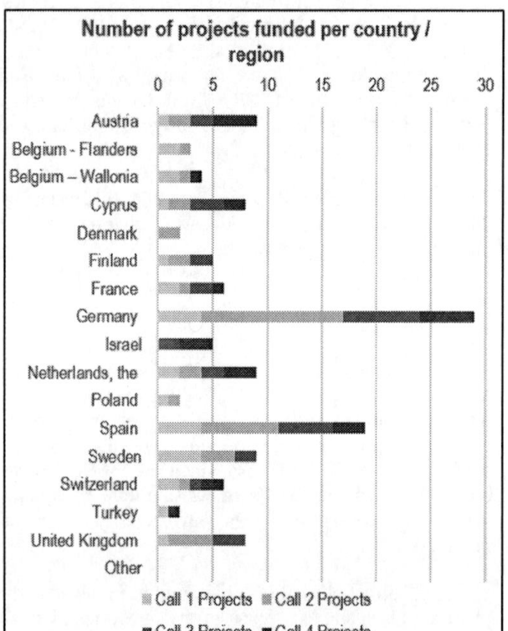

Figure 3: Number of projects by country / region as well as by call resulting from the first four SOLAR-ERA.NET calls.

The 45 projects funded have a total volume of 60 MEUR including 39 MEUR of public funding. The funding share per country / region is displayed in Figure 4.

The success rates of preproposals vary strongly between countries and regions, reaching from around 10% for projects with participants from Poland to 60% for projects with participants from Finland. Reasons are manifold. To quote a few: consulting for parties interested, availability of funds and quality and industrial relevance of proposals.

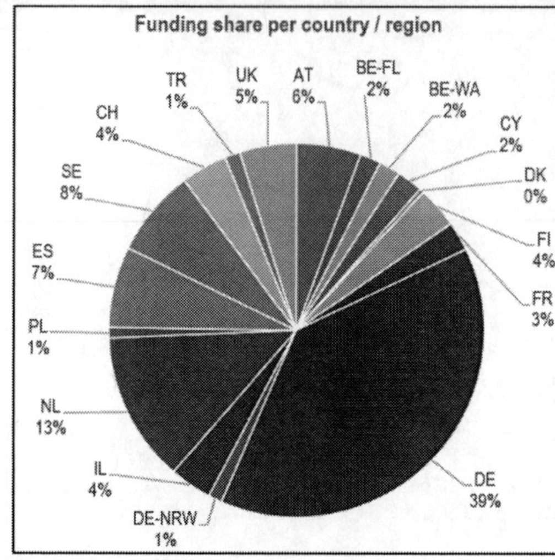

Figure 4: Funding share per country / region for the first four joint calls carried out by SOLAR-ERA.NET.

Figure 5 shows the different success rates (preproposals being finally funded as projects) by countries and regions.

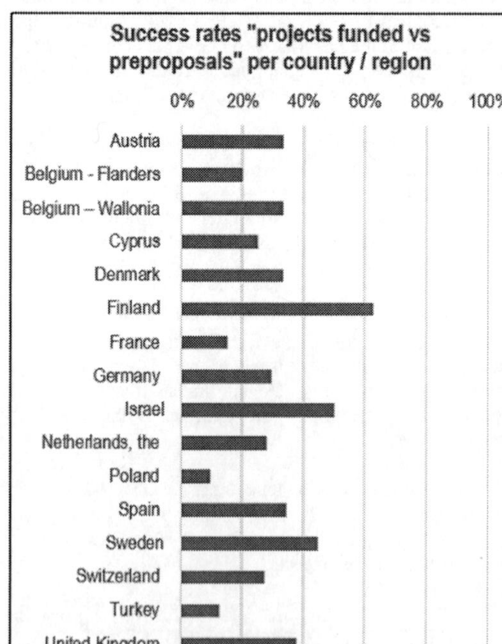

Figure 5: Success rates (i.e. preproposals being finally funded as projects) by countries and regions.

4 RESULTS AND OUTLOOK

4.1 Selected results

The transnational SOLAR-ERA.NET calls found good interest and participation of the sector. Some key figures:

- 180 preproposals submitted with total project costs of 227 MEUR and funding requested of 155 MEUR
- 804 partners / contributions involved in preproposals (multiple counting)
- 17 countries and regions participating in the joint calls
- 83 full proposals submitted
- 45 projects funded
- Funding committed: approx. 39 MEUR public funding for a total project volume of around 60 MEUR

Table 1 shows the number of projects funded per call topic. Some topics have been added in the second and third call.

Table 1: Four transnational SOLAR-ERA.NET calls - topics and numbers of projects funded (number of projects funded per call and topic as well as total sum)

Transnational SOLAR-ERA.NET topics in calls I, II, III, IV and all four together	I	II	III	IV	Σ
PV1 Innovative processes for inorganic thin-film cells & modules	2	2	1	2	7
PV2 Dedicated modules for BIPV design and manufacturing	3	2	1	0	6
PV3 Grid integration and large-scale deployment of PV	2	1	1	3	7
PV4 High-efficiency PV modules based on next generation c-Si solar cells	1	5	3	2	11
PV5 Solar glass and encapsulation materials	1	2	0	1	4
PV6 Concentrator PV technology	0	0	2	0	2
PV7 Si feedstock, crystallization and wafering	0	1	0	0	1
PV8 Organic solar cells and other emerging concepts	0	0	2	0	2
CSP1-4 All four CSP topics together	1	2	1	1	5
Total	*10*	*15*	*11*	*9*	*45*

. The success rates of preproposals differ between call topics, reaching from around 13% for projects on PV concentrator technology to 50% for projects on Organic solar cells and other emerging concepts. Figure 6 shows the success rates (preproposals being finally funded as projects) by call topics.

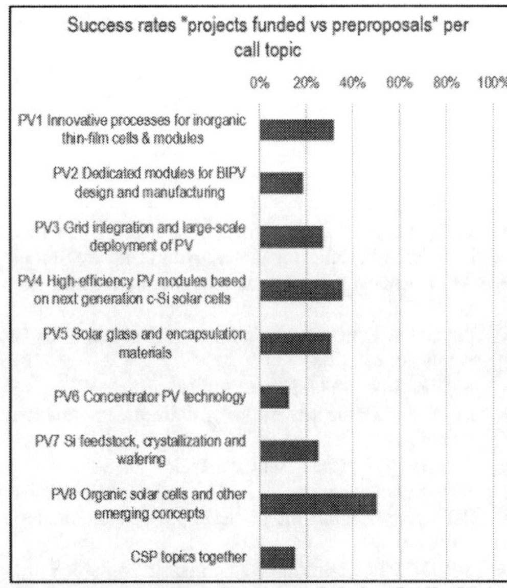

Figure 6: Success rates (preproposals being finally funded as projects) by call topics.

The 45 projects (40 PV projects, 5 CSP projects) funded are:

- SLAGSTOCK: Low-Cost Sustainable Thermal Energy Storage Systems Made of Recycled Steel Industry Waste
- LIMES: Light Innovative Materials for Enhanced Solar Efficiency
- BLACK: Black Silicon and Defect Engineering for Highly Efficient Solar Cells and Module
- INTESEM: Intelligent Solar Energy Management Pipeline from Cell to Grid
- NOVACOST: Non Vacuum Based Strategies for Cost Efficient Low Weight Chalcogenide Photovoltaics
- NovaZOlar: All-non-Vacuum Processed ZnO-based Buffer and Window Layers for CIGS Solar Cell Technology
- HyLighT: Design, Development and Application of a Technologically Advanced System of Natural Day-light and Artificial PV Lighting - Hybrid Light Tube
- InnoModu: Leadfree Modules with Low Silver Content and Innovative Busless Cell Grid
- AER II: Industrialization and System Integration of the Aesthetic Energy Roof Concept
- SNOOPI: Smart Network Control with Coordinated PV Infeed
- PV4FACADES: Photovoltaics for High-Performance Building-Integrated Electricity Production Using High-Efficiency Back-Contact Silicon Modules
- ACCESS-CIGS: Atmospheric Cost Competitive Elemental Sulpho-Selenisation for CIGS
- PV me: Organic PhotoVoltaic Systems Integrated in Manufactured Building Elements
- PV2GRID: A Next Generation Grid Side Converter with Advanced Control and Power Quality Capabilities
- THESEUS: Tandem High Efficiency Solar Cells Utilizing III-V Semiconductors on Silicon
- U-light: Ultra Lightweight PV Modules and their Applications in Innovative PV Systems Achieving Lowest Levelized Cost of Electricity (LCOE)

- HESiTSC: High Efficiency Silicon Based Tandem Solar Cell PV Module
- InGrid: High Efficiency PV Modules Based on Back-Contact Cells and Novel Interconnecting Grid
- Monoscribe: Roll-to-Roll Monolithic Interconnection of Customizable Thin-film Solar Modules
- HighCast: High Performance Silicon Casting and Wafering
- EDITOR: Evaluation of the Dispatchability of a Parabolic Trough Collector System with Concrete Storage
- APPI: Atmospheric Pressure Processing for Industrial Solar Cells
- FunGlass: FunGlass – Multi-Functional Glass for PV Application
- SITEF: Silicon Fluid Test Facility
- PROOF: Competitive Industrialized Photovoltaic Roofing
- SolFieOpt: Optimal Heliostat Fields for Solar Tower Power Plants
- BIPVpod: Building Integrated Photovoltaics Panels on Demand
- SPRINTCELL: Sulfide-Based Ink for Printable Earth-Abundant Solar Cell
- DINAMIC: Dilute Nitride Based Concentrator Multijunction Solar Cells, with Efficiencies over 47%
- ALCHEMI: A Low Cost, High Efficiency, Opto-electronic HCPV Module for 1000 Sun Operation
- IPERMON: Innovative Performance Monitoring System for Improved Reliability and Optimized Levelized Cost of Electricity
- HVolt-PV: High Voltage IBC Photovoltaic i-Cells and Modules
- Bifalo: Bifacial PV Modules for Lowest Levelized Cost of Energy
- CNT-PV: Carbon Nanotube Hole-Transporting and Collecting Layers for Semi-Transparent, Flexible and Low-Cost Solid-State Photovoltaic Cells
- HESTPV: High-Efficiency and Stable Tin-Based Perovskite Solar Cells
- HIPPO: High-Efficiency Poly-Si Passivated Contact Solar Cells and Modules
- INFORPV: Innovative Forecasting PV Energy Yield Solution for Sustainable Large Scale Deployment
- Liquid Si 2.0: Liquid Phase Deposition of Functional Silicon Layers for Cost-Effective High Efficiency Solar Cells
- FrontCIGS: Re-Designing Front Window in Flexible CIGS Modules for Cost-Effective Moisture Protection
- SIMON: Silicone Fluid Maintenance and Operation
- DURACIS: Advanced Global Encapsulation Solutions for Long Term Stability in Industrial Flexible Cu(In,Ga)Se2 Photovoltaic Technology
- Refined PV: Reduction of Losses by Ultra Fine Metallization and Interconnection of Photovoltaic Solar Cells
- ENHANCE: Enhanced Rooftop PV Integration through Kinetic Storage and Wide Area Monitoring
- HIPER: High-Efficiency Si Perovskite Tandem Solar Cells
- PEARL TF-PV: Performance and Electroluminescence Analysis on Reliability and Lifetime of Thin-Film Photovoltaics

SOLAR-ERA.NET not only allows for supporting innovative research projects on a transnational level but also provides added value in terms of networking and programming, namely:

- New industry-led innovation projects and partnerships in a strongly competitive environment
- Common initiatives and implementation activities
- Networking and collaboration between countries across Europe
- Overview on European research and insight into trends
- Best practice for supporting projects on the transnational level

4.2 Outlook

The successful projects demonstrate i) potential commercial impact / relevance to industrial and market needs / contribution to the Strategic Energy Technology Plan (SET-Plan) and added transnational value, ii) scientific and technological excellence and iii) high quality and efficiency of the implementation and the management. Further information on running projects from the transnational joint calls can be found on the solar-era.net website.

Within the Horizon 2020 ERA-NET Cofund scheme, the network continues its activities and launched the 5th joint call in December 2016. The SOLAR-ERA.NET Cofund call addressed 4 topics, i.e. innovative and low-cost PV manufacturing, advanced PV products and applications, PV system integration and CSP cost reduction and system integration. Projects shall aim at Technology Readiness Levels of 5 to 7.

Within this 5th call, SOLAR-ERA.NET is likely to fund another 15 projects with a total project volume of around 15 MEUR of which around 10,5 MEUR is from public funding. Most projects are about to be started by the time of writing this contribution and public information will be soon available.

The 6th call launched by SOLAR-ERA.NET (Cofund 2) comes along with a total funding budget of 22 MEUR (including the EU-Top-Up). The first stage ends on 2nd of October 2018 with the deadline for submission of preproposals. The topics are similar to the previous call, plus a topic dedicated to Emerging PV technologies. The range of Technology Readiness Levels (TRL) is extended, reaching from TRL 3 to 7. An additional call (7th Call) will be launched in spring 2019.

News on ongoing and future SOLAR-ERA.NET activities are channelled through the solar-era.net and funding agencies involved.

REFERENCES

Further information is available on the project website www.solar-era.net.

ACKNOWLEDGEMENTS

The support of the SOLAR-ERA.NET joint activities by the EC under grant agreement N° 321571 within the 7th Framework Programme, SOLAR-ERA.NET Cofund under grant agreement N° 691664 and SOLAR-ERA.NET Cofund 2 under grant agreement N° 786483 within Horizon 2020 is gratefully acknowledged.

LESSONS LEARNED FOR STAND-ALONE PHOTOVOLTAIC SYSTEM PROJECTS FINANCED BY GOVERNMENTS IN DEVELOPING COUNTRIES: THE CASE OF THE RÍO IBÁÑEZ COMMUNE IN THE AYSÉN REGION, CHILE

Juan Carlos Osorio-Aravena (✉)[1,3] and Emilio Muñoz-Cerón[2,3]
(✉)[1] *Corresponding author.* Energy and Environment Group, Campus Patagonia, Universidad Austral de Chile
km 4 camino a Coyhaique Alto s/n, Pabellón Científico. 5950000 Coyhaique, Región de Aysén, Chile
phone.: +56 67 2 526951, e-mail: juan.osorio@uach.cl
[2] Projects Engineering Area. IDEA Research Group (Research and Development in Solar Energy), University of Jaén
Campus Las Lagunillas s/n. 23071, Jaén, Spain, e-mail: emunoz@ujaen.es
[3] CEAEMA. Centre for Advanced Studies in Energy and Environment, University of Jaén

ABSTRACT: This study presents a descriptive analysis, from a technical perspective, of a rural electrification project managed by sub-national government entities in developing countries. It was conducted through a case study applied to a project involving 26 energy access solutions implemented through photovoltaic technology. The analysis is based on information gathered from technical inspections performed three years after the commissioning of the project, in consideration of the guidelines for guaranteeing the quality of stand-alone photovoltaic systems proposed by the International Energy Agency Photovoltaic Power System Programme. The article reveals that despite the fact that most of the solutions have functioned within acceptable margins, a need to improve this type of project in all of the sectors involved; public, business and citizen, is detected. The lessons learned and proposals for improvement are also presented, with the overall aim being for projects of this type, current and future, to move towards success and sustainability with public management, the competence of local professionals and the socio-technical aspects of the inhabitants playing an essential role.
Keywords: Energy Access, Stand-alone PV Systems, Rural Electrification, Off-grid

1 INTRODUCTION

The lessons learned following the conclusion of the Millennium Development Goals period recognize the importance of energy in improving the living conditions of people, reducing poverty and combating climate change [1]. In 2016, *"Goal 7: Ensure access to affordable, reliable, sustainable and modern energy for all"* was established as one of the 17 Sustainable Development Goals with the aim of providing universal access to energy [2]. Attainment of Goal 7 is not simple, however, especially when there are significant differences between the use of modern energy services in urban and rural areas [3]. Multiple challenges exist and thus, without well-designed programmes, it will be difficult for developing countries to successfully eliminate the barriers associated with electricity access and energy poverty [4].

Of the 1.06 billion people who live without electrical energy in their homes –14.0% of the global population–, 84.0% are located in rural areas [5]. In Latin America, there are 17 million people who live in areas of difficult access and without electricity [5]. In Chile, the current government expressed as one of its *"Mega Commitments: To create a map of the country's energy vulnerability, identifying families without electricity and other energy services, with a view to bridging the existing gaps"* [6].

For inhabitants of rural areas, decentralised projects through micro-grids and stand-alone systems continue to be one of the most feasible options for providing access to electrical energy [7]. A growing trend in 2017 was the establishment of associations among multinationals, local companies and/or governments in order to implement renewable systems distributed for energy access [8]. Some of the business models that have arisen as projects for this purpose are: Public-Private Alliance for Development; Micro-franchising; Pay-As-You-Go (PAYG); and, Third Generation Photovoltaic Systems and Quality Control [1]. Isolated solar systems, particularly those marketed through the PAYG business model, were the most significant in the sector, providing electricity access to more than 360 million people around the world [8]. Also observed in 2017 was a growing number of national governments that improved the situation for advancing towards renewable systems that provide electricity access [8]. However, apart from the business models, governments in developing countries have created Rural Electrification Projects (REP) as a strategy for electrifying those more remote populations; large amounts of money have been invested for this purpose. REPs are managed by sub-national governments and are a type of initiative for giving access to electricity without evidence of being studied.

On the other hand, stand-alone photovoltaic (SAPV) systems have traditionally and historically been a technically and economically feasible solution for providing electrical energy to single-family homes isolated from distribution grids. In Latin America and the Caribbean, photovoltaic (PV) solar energy has played an important role in energy access [9]. However, and despite the fact that SAPV systems enjoy a high level of technological maturity –according to a 2004 report by the *International Energy Agency Photovoltaic Power System Programme* (IEA PVPS) significant problems in this type of system do not exist [10]–, there are numerous cases in which the success rate and sustainability of this type of system are still unresolved. In fact, there have been experiences with a variety of results which, in many cases, have become problems that undermine PV technology and its multiple benefits, tending to be more severe in isolated areas.

In the case of the Aysén Region –an area separated from the rest of Chile with conditions of isolation, a population density of 1.00 inhabitant per km^2 and 19.5% rural population– electrical energy in rural sectors is, to a certain extent, an ongoing historical demand. Until 2016, there were more than 440 small-scale stand-alone systems: among them, solar-wind hybrids, mini-hydraulics and PV, 417 of these in the installed capacity range of between 0.50

and 1.50 kWp; implemented under REPs [11]. They were first installed in 2012 with financing coming from public funds, both from the Regional Government (GORE) and the Undersecretary for Regional Development (SUBDERE) of Aysén. In addition, between the two entities, there are a total of 504 SAPV systems designed for rural electrification coverage in the region to be implemented in the short and medium-term [12].

Based on the information gathered in visits to the REP of the "El Claro, Las Ardillas y Lago Alto" committee in the Río Ibáñez Commune, Aysén Region, Patagonia, Chile –a project of 26 SAPV systems of 1.38 kWp each–, after three years in operation, this study presents a technical descriptive analysis considering all of the initiative's sectors; public, business and civic (users/beneficiaries). The aim was to reveal technical and managerial flaws that have arisen in this REP, draw lessons and propose necessary improvements in this type of project for the remainder of their operational lifetime, as well as to ensure that those next to be installed do not end in failure.

Following a description of the background to the REP in question, it presents the observations of the technical inspection, in which flaws in all of the sectors were revealed. In technical terms, problems were detected in the design, the Technical Specifications (TS) and the quality of installation. Improper practises on the part of users were also identified, highlighting a socio-technical problem. In terms of management, bureaucratic complexity and the non-existence of plans for operation and maintenance (O&M) and battery replacement, among other issues, were detected. The lessons learned and the improvements proposed in this study are therefore based on the aspects mentioned in this paragraph.

In short, along with revealing the flaws in the sectors involved in the REPs, this study provides an informational foundation for proposing improvements in both current and upcoming projects. The lessons learned and the improvements presented in this study may also be considered for other types of projects conducive to offering universal access to energy in a successful and sustainable way.

2 METHOD

The procedure used was a case study carried out in two phases. The first phase involved gathering information about the REP in question, followed by an analysis phase from a technical perspective. In the information gathering phase, the public entity in charge of the general management of the REP was asked for background information about the project; the objective of this was to understand the context and its management structure, as well as to identify the location of the beneficiaries. The technical documentation prior to the execution of the project, that is, the TS and the winning technical bid, were also requested. With the prior technical information, an analysis was conducted of the design of the SAPV systems installed through the REP under study. This was followed by visits to each of the 26 SAPV systems installed. These took place between July and August 2016, the winter season in the southern hemisphere. During the visits, a visual and empirical technical inspection was conducted between 10:00 and 16:00 in order to take measurements with instruments such as: a solar power meter, compass, clinometer, digital multimeter, clamp multimeter and thermal imaging camera. In summary, a logical monitoring of the energy, from the radiation available in the Photovoltaic Generator (PVG) to the electricity-consuming components in the homes of the users/beneficiaries, was conducted. The aim of the technical inspection was to assess the performance and condition of the systems, as well as to detect flaws, in order to obtain a general diagnosis of the SAPV systems of the REP under study after three years of operation. In addition, as part of the visit, an open interview was conducted with the beneficiaries in order to determine: what electrical appliances they use and the nature of their usage habits; what problems they have encountered, and; their impression of the project and of PV technology. The second stage consisted of an analysis of the information gathered. This was carried out from a technical perspective, considering the theoretical framework proposed by the IEA PVPS for quality assurance in SAPV system projects [10], associating it with the different sectors. These were classified into three. The first sector was assigned to the public management, responsible for the design of the SAPV systems and their corresponding TS. It also functions as the technical counterpart (quality control of the installation) and financially governs the project. The second sector was related to the company that performed the installation of the systems. The third sector was associated with the beneficiaries of the REP, that is, the final users of the systems and their involvement with the technology. Finally, based on the analysis, the lessons learned and proposals for improvements for this kind of REP were identified as the findings of the study.

3 RESULTS

3.1 Context

The Aysén Region is disconnected from the national electric system. It has three Medium-sized Systems (MS) and nine Isolated Systems (IS) [12], all with different percentages of installed capacity among technologies based on diesel, hydraulics and, to a lesser extent, wind. In Chile, MS are classified as electrical systems with an installed capacity of between 1.50 and 200 MW, and IS as those with less than 1.50 MW. MS cover the parts of the territory with a greater population agglomeration. IS supply electrical energy to villages and small human settlements; in some of them, just hours per day [12]. Among the 19.5% rural population –and considering that the area has 1.00 inhabitant per km^2–, the region has inhabitants in a situation of extreme rurality. To reach these homes, in addition to having to cross rivers at a specific time of year, there are no roads for motorised vehicles. Due to the logistical complexity and its resulting budgetary repercussions, there had not been any possible way to bring electricity to them.

In principle, the REPs were created for the expansion of the distribution grids. REPs based on residential SAPV systems have arisen as a strategy of sub-national governments to provide electrical energy access to the most dispersed and vulnerable inhabitants. They have been implemented in the Aysén Region since 2012.

3.2 Technical information prior to implementation

The development of the initial project was entrusted to an engineer outside of the Municipality hired via Technical Assistance (TA) in electrical energy. It was revealed here that there was no technical report on the design of the SAPV system. This would account, for example, for the sizing of the components of the SAPV system according to the expected consumption. All that was found –in the initial project– was a section titled "General Description of the Solution" and a section titled "Technical Specifications"; the second of which was one page. The solution described consisted of two lists; one corresponded to what the interior electrical installation for the homes of the beneficiaries should contain, associated with the expected consumption, and the other with the components of the SAPV system. Fig. 1 presents a schematic diagram of the solution drawn from the initial project. The diagram summarises what was described in the lists mentioned above, as well as a large part of what was contained in the section "Technical Specifications".

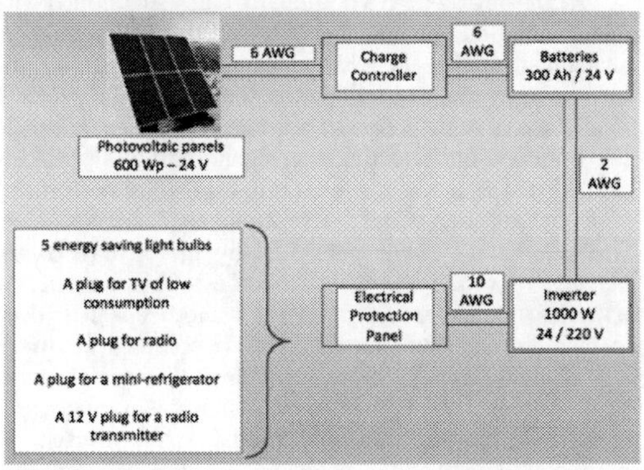

Figure 1: Schematic diagram of the initially proposed solution

The public bidding of the REP in question contained some TS which, in addition to the general description of the solution considered in the initial project, provided information referring to each component of the solution. Particularly significant in this is that, on the one hand, it made references to national and international guidelines for certifying the quality of the components to install and, on the other, it requested a calculation report for the structure supporting the PV modules to resist winds of up to 120 km/h. The TS also included that the offer must propose a training programme, both for the officials of the Municipality's technical unit and for the users/beneficiaries. An additional score was also assigned to the bidder who, as part of the proposed solution, expanded the SAPV system to be able to operate a washing machine.

The company that was awarded the implementation of the REP, in addition to meeting all of the requirements, expanded some of the components of the solution. Fig. 2 shows the SAPV system offered, and ultimately installed, which increased the capacity of the PVG to 1.38 kWp and the inverter to 2.50 kW, also ensuring that the beneficiaries could use a washing machine. This shows that the initially planned system was overvalued.

3.3 Implementation of the REP

Regarding the installation of the SAPV systems, aspects related to the following were observed: the number of hours that the PVG receives unobstructed solar radiation (and if there are obstructions, whether they can be eliminated), the orientation and inclination of the PVG and the installation of the Balance of System (BoS) components.

Figure 2: Photographs showing the components of an SAPV system of the REP under study

In relation to the number of hours that the PVG is directly exposed to the sun, it was observed that nineteen of them have a minimum of six hours –between 10:00 and 16:00–, three have between three and six hours –until 16:00–, two have less than four hours –between 11:00 and 14:30–, and one, in winter, does not receive direct radiation for more than three months. It is worth mentioning that the locations were chosen with technical criteria but the beneficiaries also had to agree on the placement location. With respect to the PVG that does not receive direct radiation for three months, there was no other option due to the geography of the area.

Depending on the orientation measurements of the PVGs, with the exception of seven cases, they were oriented to the north indicated by the compass (magnetic north). The deviations in orientation measured with respect to magnetic north were: one case at 22.0°, two cases at 20.0°, two cases at 10.0° and two cases at 5.00°. While it is true that these deviations in orientation do affect energy production, their significance depends on the user's consumption patterns. In addition, correcting them involves challenging work due to the concrete bases on which the supporting structure of the modules is installed.

Regarding the inclinations of the PVG measured with a clinometer, it was observed that: one was at 65.0°, three at 66.0°, eleven at 67.0°, nine at 68.0° and one at 69.0°. The average inclination of the PVG had a value of 67.0°. The current inclination is configured for winter mode but its average value exceeds the optimum angle for winter by 10.0°.

Furthermore, it was observed that the connections between the module conductors, and these with the conductors that go to the charge controller, did not have connectors. Although this is not supposed to have caused difficulties, it is necessary to have them since overheating can occur and affect the longevity of the system. They also facilitate the handling of the modules when inspecting and making revisions in cases of failure.

Regarding the existing length from the PVG to the charge controller –considering the cross section of the installed conductors and a loss of 3.00% of the maximum voltage in direct current–, it was determined that the maximum must be 3.00 m. None of the SAPV systems meet that distance. The shortest length measured was 4.00 m, followed by 5.00 m, another 6.00 m and another 7.00 m. The others exceeded 10.0 m, with a maximum length of 35.0 m. In addition, solar cables were not used for the interconnection of modules or for the connection of the PVG to the charge controller. Only the 10 AWG outdoor type conductor was used; a different conductor than the one stipulated in the TS. Additionally, it was detected that the cross section of the conductors connecting the charge controller with batteries in the different SAPV systems was undersized. The conductors installed have a maximum electric current capacity of 20.0 and 35.0A and the maximum that can pass through the charge controller is 40.0A. This may explain the fact that there were charge controllers showing evidence of having overheated at the connection points and, in fact, there was one that was burnt. Despite the fact that the type of conductors to use was mentioned in the initial project, it was not stipulated in the TS. Furthermore, it was indicated in the initial project that protections in the DC circuit must be installed –but was not indicated in the TS of the bidding–, however, no SAPV system had them installed.

Moreover, after revising the configuration of the charge controllers, it was discovered that they were configured to operate in a way suitable for liquid electrolyte batteries, the ones installed being the AGM type. It was also found that, for this climate zone, it is advisable to use a charge controller with MPPT technology so as to be able to improve energy production.

With respect to the inverters, several conditions were detected that not only impact the operational lifetime of the batteries but also affect the safety of the systems. The condition affecting the operational lifetime of the batteries stems from the fact that the inverter is configured in voltage ranges different from those of the charge controller. This means that even if the charge controller detects that the charge of the batteries has reached a level at which they can be disconnected from the system to prevent total discharge, the inverter continues drawing energy from them until they run out. This leads to deeper discharge cycles than the charge controller would allow, shortening the operational lifetime of the batteries.

The riskiest condition to the safety of the systems is the fact that the inverter is oversized. The fact that the inverters installed are 2.50 kW means that appliances of more than 1.00 kW can be connected to them in the interior of the installation, the latter figure corresponding to the maximum capacity of direct current that can travel through the charge controller (40.0 A). In fact, this is likely the reason why one charge controller burnt and two show signs of overheating at their connection points.

Finally, it was discovered that neither a technical inspection plan nor an O&M plan considering battery replacement and management existed.

3.4 Overall condition of the SAPV systems

Three years after the installation of the SAPV systems, 24 were functioning, one was not functioning because the charge regulator was burnt and one had been removed because the beneficiary had sold their land. Of the 24 in operation, one was abandoned (the beneficiary no longer lived at the location), one was operating with problems (the batteries ran out at the end of the day) and one only functioned when there was direct radiation.

Regarding the SAPV system that was operating with problems, this was found to be due to a mixture of few hours of direct exposure to the sun, losses in direct current due to the distance between the PVG and the charge controller and high consumption by the users. This reflects a flaw in the technical criteria for the installation and a lack of knowledge of the technology on the part of the users. Nevertheless, all of the sectors bear responsibility. On the one hand, it is the responsibility of both the design and the TS, as well as of the installation company, to know that the distance between the PVG and the charge controller must be reduced as much as possible. This, however, was not stipulated anywhere. On the other hand, it was also the responsibility of the beneficiary for having used other appliances that had not been considered among those suggested in addition to the recommended electrical appliances.

Regarding the SAPV system that only functioned with direct radiation, this was due to the presence of permanent shade caused by the closure of the perimeter by the beneficiary (likely installed after the start of the project). While the dissatisfaction of the beneficiary was clear, to the point of his rejecting the technology and speaking critically of it, it was made clear to him –after performing some measurements and demonstrations– that the problem was due to the shade generated by the closure of the perimeter around the PVG modules. This caused the batteries to never fully charge, with repeated deep discharges, and thus shortened their operational lifetime.

Shade over the PVG was something that was observed in more than half of the SAPV systems. Fourteen cases were found with avoidable partial shade caused by perimeter closure and/or vegetation or other elements that can be removed from the location. In other words, energy production in these systems can be improved without the investment of money.

4 DISCUSSION

Based on the issues included in the procedure suggested to guarantee the quality of the SAPV systems [10], several things can be said regarding the REP under study. With respect to the "*Accreditation of testing laboratories*" of the system components, it was at least indicated in the TS that they must have a certification and that it be presented in the offer. Regarding "*Training and certification*" and "*System installation*", to ensure that quality systems are provided, it is essential that the technicians who design and install the systems have received certified training on the technology. This is because many technicians assume that a general background in electricity is sufficient to be able to correctly design and install PV systems. However, experience shows that this is not the case [10] and, in fact, the REP under consideration is an example of this. In light of the results obtained in this study, in can be inferred that there was a lack of knowledge on the part of the electricity industry professionals. This was the case both in the design and development stage of the TS and in the installation of the SAPV systems, signifying an absence of certified training in PV technology. In addition, it was found that accreditation of specific training in PV systems was not required. This is a significant point that was not considered in the REP in question but which can be improved in future projects. It is also an important element for preventing the mistrust elicited towards technology as the consequence of a poor design and the result of an improper installation of SAPV systems.

With respect to the "*Commissioning tests*", which

should be done at the time of installation, it is likely that they were performed since all of the SAPV systems were operational throughout the first year. Even so, doubt remains as to whether a standardised procedure was followed, given that several details to improve in the installation were observed. In addition, in relation to the "*PV system monitoring*" and the "*Warranties and performance guarantees*", it can only be mentioned that they are not being carried out. It is therefore suggested that these issues be incorporated in some way into the current project and considered for those implemented in the future.

Other issues that are relevant and necessary to improve are: "*Operation and maintenance*" and "*User involvement*". These will be key to the ultimate success of the REP under study –and hence that of the others in operation and those implemented in the future–, due to the fact that there was no O&M plan. Furthermore, particularly in this REP, technical inspections carried out with appropriate frequency and some type procedure to be followed by the users must be ensured. Battery replacement must also be considered comprehensively, in other words, in a way that addresses both a strategy for guaranteeing financing and their management as hazardous waste. In both cases, the direct involvement of the beneficiaries is key.

Furthermore, the only way to provide electrical energy access to the residents of the territory under consideration appears to be through REPs based on SAPV systems. It is likely that the geographical difficulties involved in reaching these homes, coupled with the remoteness of the Aysén Region, make it an unattractive market for companies dedicated to energy access under some of the business models being implemented in other developing countries like those described by Eras-Almeida et al. [1]. REPs based on SAPV systems are a type of initiative that, in addition to being administered and managed by the sub-national public sector, allow for the involvement of the business sector of the territory. They can therefore also contribute to the local economy whilst improving the quality of life of the population through access to energy.

While the bureaucratic complexity of this type of project has not been an impediment to their operation, the greatest difficulties are still to come. Specifically, this relates to future technical inspections and battery replacement. The technical inspections depend on the funds that the Municipality is successful in leveraging to hire a TA to perform them. Battery replacement, as there is no comprehensive plan, not only runs the risk of not being accomplished in all of the SAPV systems but also, in addition to requiring funds for the change to be performed by a company or TA, must include consideration of its handling as a hazardous waste, entailing an additional cost. It is likely that what is expressed in this paragraph is the most crucial point for the success and sustainability of REPs. It also makes a difference in the business models described by Eras-Almeida et al [1], who consider these aspects critical due to their being part the service they provide. However, the solution of the REP in question provides superior benefits with regard to the amount of consumption. In any case, it is suggested that a comparative analysis be conducted between the REPs and the existing business models in which it is additionally recommended to calculate the real cost of the kWh in the different forms of providing energy access.

Regardless of the fact that 22 of the 26 SAPV systems were operating, and with a relative degree of approval on the part of the beneficiaries, the parameters for sustainability are being affected by most of the problems that were identified in this study. It is clear that knowledge gaps exist in all of the involved sectors. It is therefore suggested that specialised training be conducted in all of these sectors. Furthermore, this last point is key to accelerating the sustainable energy transition.

5 CONCLUSIONS

The benefits that energy access can provide to people's quality of life are indisputable. However, even if such access is provided through technology with a high level of maturity, success is not guaranteed. Identifying lessons and possible improvements to energy access projects, especially in those driven by sub-national governments and implemented by local companies, is therefore key for these types of initiatives to be able to provide their multiple benefits.

Although the REP in question has functioned, flaws in all levels of performance were discovered. While it is true that there were technical inconsistencies that have not had a significant impact on the functioning of the SAPV systems, all of them could have been prevented through specialised training for the professionals involved, mainly to obtain a proper design and TS, a rigorous technical counterpart and appropriate training for the users. Nevertheless, many of the problems identified are improvable in the short term.

Finally, the need to improve REPs in all sectors is clear. In turn, it is expected that with the lessons learned and the improvements proposed, it will be possible for these types of initiatives to be both successful and sustainable, contributing to the improvement of the living conditions of the society, the local economy and the environment.

6 ACKNOWLEDGEMENTS

This study is sponsored by the Vicerrectoría de Investigación, Desarrollo y Creación Artística de la Universidad Austral de Chile.

7 REFERENCES

[1] Eras-Almeida A., Fernández M., Eisman J., Martín J., Egido-Aguilera M., 2018. Modelos de negocio para acceso a la energía: Estudio de caso en Perú, México y Bolivia. Libro de actas del XVI Congreso Ibérico y XII Congreso Iberoamericano de Energía Solar (CIES 2018), Madrid, España.

[2] United Nations, n.d. Sustainable Development Goals: 17 Goals to Transform Our World. https://www.un.org/sustainabledevelopment/

[3] Herington M.J., van de Fliert E., Smart S., Greig C., Lant P.A., 2017. Rural energy planning remains out-of-step with contemporary paradigms of energy access and development. Renew. Sustain. Energy Rev. 67, 1412–1419.

[4] Sovacool B.K., 2012. The political economy of energy poverty: A review of key challenges. Energy Sustain. Dev. 16, 272–282.

[5] [IEA] – International Energy Agency, 2017. Energy Access Outlook 2017: From poverty to prosperity.

[6] [ME] – Ministerio de Energía del Gobierno de Chile, 2018. Ruta Energética 2018-2022. http://www.energia.gob.cl/rutaenergetica2018-2022.pdf

[7] [REN21], 2017. Renewables 2017 – Global Status Report. Paris: REN21 Secretariat

[8] [REN21], 2018. Renewables 2018 – Global Status Report. Paris: REN21 Secretariat.

[9] [REN21], 2017. Subastas de Energía Renovable y Proyectos Ciudadanos Participativos: América Latina y el Caribe. Paris: REN21 Secretariat.

[10] [IEA PVPS] – International Energy Agency Photovoltaic Power Systems Programme, 2004. Report IEA PVPS T3-07:2004: Survey of national and international standards, guidelines and quality assurance procedures for stand-alone PV systems, 2nd Edition.

[11] Osorio-Aravena J.C., 2018. Capítulo VI: Sistemas Aislados. Una mirada descriptiva de las experiencias en la región de Aysén. Libro: ¿Conectar o desconectar? Energía y comunidad para las transiciones energéticas. Colección IDEA, Santiago de Chile.

[12] [ME] – Ministerio de Energía del Gobierno de Chile, 2018. Política Energética 2050 Región de Aysén del General Carlos Ibáñez del Campos. http://www.energia2050.cl/wp-content/uploads/2018/02/POLITICA-ENERGETICA-AYSEN-2050.pdf

ENHANCING SOLAR RESEARCH BY USING ICT AND EXPLORATIVE WEB-BASED METHODS FOR COMMUNICATION, EDUCATION AND TRAINING

C. S. Polo López, F. Frontini, P. Bonomo
SUPSI-ISAAC, Campus Trevano, CH 6952, Canobbio, Lugano, Switzerland
*e-mail to: cristina.polo@supsi.ch; phone +41 052 666 6314, fax +41 052 666 6349

ABSTRACT: Within a collaborative research project relating to solar technology and specifically to BIPV, building integrated PV systems (project Construct PV, FP7-ENERGY-2011-2), an important part awaits the effective communication of research results to overcome non technological barriers and the transfer of training and information to a wide range of users and target audience. In this context, new communication strategies, tools and above all new methods of education and training can be implemented in a new way, recognizing an added value of ICT, information and communication technologies, and Web-based methods to stimulate a discussion on the important role of solar energy in buildings and to promote two-way exchange of knowledge, from research to public and from public to research process. Computer-mediated activities along with real-time participatory ITC tools were used in an interactive Webinar to stimulate and encourage participation and, in turn, to gather useful feedback for the project. Furthermore, Web-based utilities (a BIPV Web-library and Web-design tool), were developed as useful tools dedicated to architects, engineers, manufacturers, installers and other interested stakeholders. This paper presents an applied example of successful collaboration between research and education, aimed at creating greater links strengthening the knowledge and competence of relevant stakeholders in solar energy.
Keywords: Education and Training; Dissemination; Information and Communication Technologies, ICT; Building Integrated PV (BIPV), Building Integration; Solar education

1 INTRODUCTION

An effective dissemination of research contents on solar energy and specifically on solar photovoltaic integrated solutions for buildings (BIPV) is fundamental to overcome non-technical barriers that still today slow down the wide diffusion of these solar systems in the built environment. To reverse this trend, as well as, for an effective dissemination of the project results to relevant target groups, influencing policymakers and other stakeholders, a good strategy for communication would be advisable. For the same reason, it will be always necessary to imagine specific methods for training devoted to architects and designers to facilitate market uptake.

Past efforts of different international entities, such as the activities on solar energy in architecture carried out within the framework of the International Energy Agency, Solar Heating and Cooling Programme, IEA SHC Task 41, Subtask A. Criteria for Architectural Integration [1], pointed out that lack of knowledge of architects, end-users and clients were an important barrier to widespread the technology. Clients, contractors, facility managers and owners are even less familiar with these solar systems and consequently make resistance to the implementation of these technologies. Often in architecture, buildings and solar technologies are still seen as two separate aspects even today. Other aspects, just like available useful information on BIPV systems in the market, the lack of common language between various building professionals linked to the need for specific tools in early architectural design process, economic issues and low cost or more aesthetical appeal products to increase acceptance have been also highlighted in some research studies [2-5]. It is clear today that even more available information regarding products and best cases are accessible to everyone through web platforms to reach architects and other target audiences. The BIPV Swiss Competence Center at SUPSI University (www.bipv.ch) has reported for many years about product and example databases by providing a reference platform for BIPV market and innovation trends at European level [6]. The role of dissemination, training and educational activities for new generations in solar BIPV research are essential to identify the obstacles architects are facing when incorporating solar design in their projects, as key figure contributing to its use in buildings, in order to provide resources for overcoming existing barriers and to gather alike useful information for better research developments in return.

2 APPROACH, INNOVATION AND RELEVANCE

Today in Switzerland and abroad, changes in the structural economy of countries and in the labor market are generating the expectation of new and better qualified careers. To guarantee the ongoing development of vocational education and training in such a dynamic context require new pathways for the transfer of knowledge [7]. Universities and also universities of applied sciences have now the important role of contributing to a more in-depth understanding needed for dialogue between research and education, to effectively communicate research findings and to ensure long-term consolidation of knowledge. Quality education and training using innovative approaches with different set of technological tools and resources is associated today with the use of ICT that stands for Information and Communication Technology. For most European countries, the use of ICT in education and training has become a priority changing the face of education over the last few decades. EU policies state that close, effective links between education, research, and innovation – the three sides of the 'knowledge triangle' - are necessary to enhance the contribution of higher education to jobs and growth [8]. ICT and Web-based learning environments allows performing adaptive skills for qualified jobs but also contribute to change the teaching and learning processes, and the interaction mechanisms or the possibility of taking advantage of a two-way exchange of knowledge for research purposes. New types of cooperation that increase flows of knowledge between education institutions, research organizations and business are requested.

3 USE OF ICT AND EXPLORATIVE WEB-BASED METHODS WITHIN THE CONSTRUCT-PV RESEARCH ACTIVITIES

Within the European project Construct PV (FP7-ENERGY-2011-2, www.constructpv.eu) new solutions have been developed for a cost-effective integration of photovoltaic technologies in the building envelope: roof and façade. Innovative, multifunctional and customizable building integrated photovoltaic (BIPV) solutions using new promising technologies with an advanced design were demonstrated in different small mock-up and large scale buildings. The demonstrative target of the project makes tangible the results and the potential transfer of PV technology in the built environment. For these reasons, single small mock-ups to show main research developments (Figure 1) and two real scale buildings has been used as large-scale demonstrators in Athens and Stuttgart [9-10]. Moreover, specific efforts during the project have been made to raise awareness among targeted groups, with a special focus on architects and public authorities, but also including BIPV manufactures and public in general, considering the BIPV market and by evaluating the attractiveness of the innovative BIPV solutions developed within the project [11].

Figure 1: Façade mock-up developed within Construct PV FP7 Project with variants of architectural and technological design/manufacturing of BIPV systems for the building skin (Source: ISAAC-SUPSI).

The importance to reach a huge target audience and to further stimulate potential stakeholders on BIPV specially concerned in novel research developments motivate project partners to perform specific activities within Construct PV project aimed to pursuit innovative communication and dissemination schemes combined with educational activities by using ITC and Web-based tools. The target audience varies from the final user, to the world of professionals and architects who have to implement new architectural systems, to the industry subject to innovation up to the new generations, students and professors which should be aware of innovative and new market developments and the vanguard of innovation. For this purpose, as described in the following sub-sections, an interactive Webinar using an ITC participatory method in education, was organized by SUPSI, in collaboration with other partners to give high visibility to the project outcomes but, at the same time, to gather useful information from end-users, public and technicians (architects, manufacturers, etc.) regarding the new products developed (sub-section 3.1). Besides, more specific practical tools for professionals in the sector were developed using Web-based utilities, such as a BIPV Web-library (sub-section 3.2) with BIPV products and a specific Web-design tool (sub-section 3.3), a user-friendly tool useful for designing of BIPV customizable components.

3.1 Interactive and participatory Webinar using ITC for solar education

With the aim to present the main achievements on solar BIPV technologies reached within the project an innovative and interactive Webinar using ICT in education, was organized by SUPSI at university Campus in Lugano, Switzerland, during an event for students (bachelor course of Architecture of SUPSI), architects and professionals. By combining the traditional method of a frontal live conference with computer-mediated activities in "blended learning", it has been possible promoting the interaction with other remote participants and learning environments for a new e-learning concept.
The in-site live conference was specifically designed to introduce the process of blended learning, intended as combination of virtual and physical environments (i.e. online and face-to-face environments) [12-14] but also to put the basis for an interactive and participatory Webinar with the aim of involving, with the IT e media support, other university bodies, researches or professionals at local level and abroad in a questionnaire on research findings. From the origin, the implementation of ICT in education transformed the teaching and the learning process from the traditional instructional teacher-centered to a learner-centered or student-centered learning approach [15]. This Webinar was aimed at foster the integration of ICT into learning activities in an efficient and fruitful way. A focal point of this Webinar has been to include and to use participatory systems, driven by mobile systems (such as smartphones and tablets, PCs), in live via Wi-Fi among the people attending conference and also remotely on the web platform to enable the participation of a large external public. Adobe Connect Meeting platform was used, using the SWITCHinteract Web (https://www.switch.ch/interact/) with the support of Swiss E-Learning Community and with the collaboration of e-Lab SUPSI members. This software has been used to create a customizable virtual classroom and to deploy learning/research contents (Figure 2).

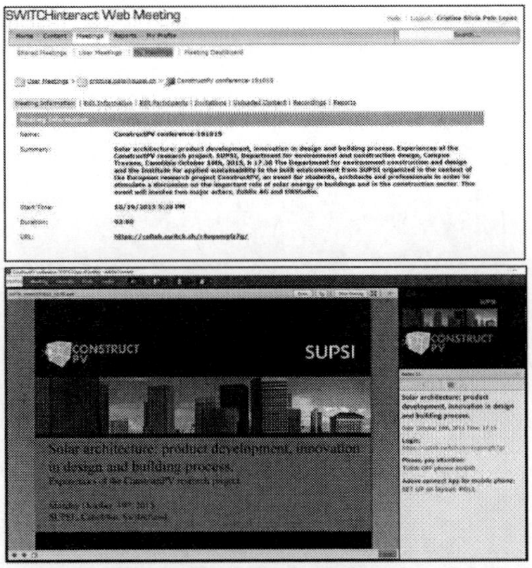

Figure 2: SWITCHinteract web platform for e-learning activities, trainings and meetings on-line. Screenshot of conference meeting room for webinar participants (Source: ISAAC-SUPSI).

The main goal on using these real-time tools, during the Webinar, was to stimulate and encourage the participation of users and to gather useful information for the research project, by making a direct survey of all the topics presented at the conference. A shift of written or visual elements (scrolling titles, flash elements, pods or notes) together with a live survey polls appeared at time of the speakers' presentations. A real-time direct visualization of the flow of participants and its responses to the on-line questionnaire about the project has been possible and the webinar users were able to interact with the speakers by using the chats and post comments available, that was a very interesting aspect. In this way, through this participatory ITC methods by using mobile phones, tablets or a PC workstation has been possible to follow the webinar contents and to see the direct response of the public, encouraging participation in such a way that it was stimulated the fair competition among the participants in order to obtain a greater response in the survey (Figure 3). Besides, it was possible to track the answers for all specific questions and to achieve results for the survey simultaneously to presentations (Figure 4 and Figure 5).

An open discussion of research findings considering the survey poll results was made at the end of the lecture session between the in-site attendees and researchers. Pod and chat allowed the same interaction with the Webinar participants. It has been stated the effectiveness of participatory methods which combine scientific insights with subjective knowledge resources to improve communication between the parties involved [16]. The interaction of stakeholders and researchers facilitates communication and dissemination of main research findings but it is also useful for a bidirectional exchange of knowledge that should allow the integration of scientific data and judgements of experts and stakeholders to improve these research results.

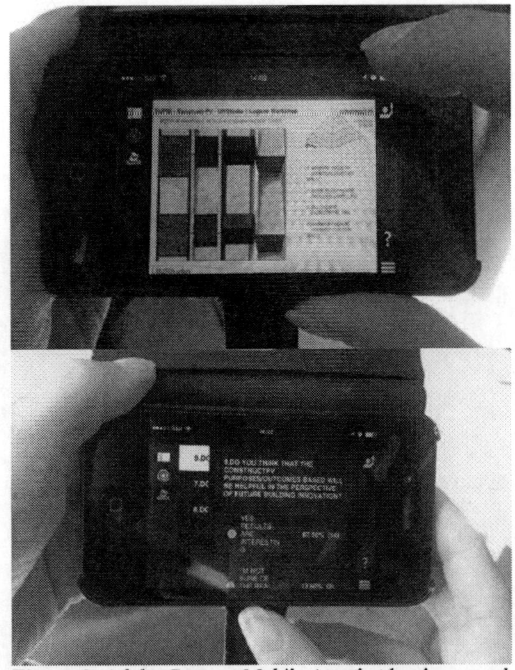

Figure 3: Adobe Connect Mobile App that has been used to make the interactive live survey during the Webinar. (Source: ISAAC-SUPSI).

Figure 4: Lecture session and open discussion to analyze the survey results. (Source: ISAAC-SUPSI).

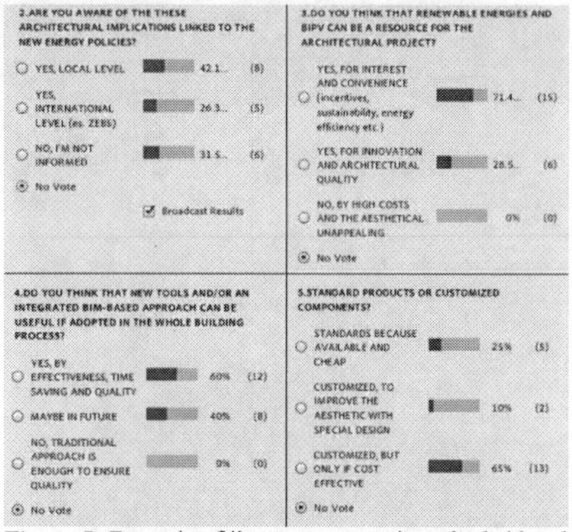

Figure 5: Example of live-survey questionnaire held and results for some interactive polls. (Source: ISAAC-SUPSI).

3.2 Web-based library

Furthermore, other web-based tools, as the Virtual Demonstrator tool, and web-based library of BIPV components have been developed during the project as a prototype to raise attention on Construct-PV research developments. ICT in these cases serves for professional development, increasing opportunities for formative learning or individualized instruction, accessing to online resources and digital tools through the Web.

The web-library of BIPV components developed (http://www.constructpv.eu/weblibrary/) make available to the users the chance to explore, in an interactive and friendly way, different Building Integrated Photovoltaic (BIPV) products currently existing in the market as well as the solutions developed within the Construct PV project, dedicated to architects, engineers, manufacturers, installers and other interested stakeholders. This web-based library of BIPV components, implemented in the website of the project intended to be a useful part dedicated for architects, engineers, manufacturers, installers and in general whoever is interested in the topic. The main goal of the web-library will be giving to the user the chance to explore the different Building Integrated Photovoltaic (BIPV) products that currently exist in the market. The main purpose is to make available to the user an useful innovative and interactive on-line catalogue through which the available products will be accessible, not only according to a market typological classification, but also by starting from different perspectives regarding solar design (visual aspects) and envelope engineering (constructive features of building envelope), at building and component scale.

The Web-library divides contents using smooth scrolling animations and drop down menus by areas of interest and subtopics (Figure 6): 1) *Typological features,* for façade and roof systems (i.e. cold façade, double skin, polyvalent systems all-in-one, warm façade, window, flat-solar floor, membranes, metal sheets, shading systems, etc.); 2) *Aesthetic and architectural features* (cell colour, cell dimension, cell edges, cell shape, cell transparency/perforation, electrical contacts, PV material/technology, frame, module colour, module dimension, module flexibility, module shape, recognisability design potential or special features/layers); 3) *Constructive features* (maintenance system, management/operating, mounting method and special features); 4) *Technological features* (e.g. acoustical protection, mechanical resistance and stability, safety in case of fire, thermal protection or weather protection).

Figure 6: Page screen of the Web-library tool (Source: http://www.constructpv.eu/weblibrary/)

Each main category (*typological, aesthetical, constructive or technological features*) will be divided into more detailed subcategories and choices. Through the web library, the user will have the possibility to select each parameter and arrive to a list of already existing products that correspond to these requirements. Each category should be able to be individually used. In this sense, the user should be able to visualize all available products by selecting only the "typological product classification" (Category 1), or the "aesthetical and architectural features" (Category 2), the "constructive classification" (Category 3) or "technological features" (Category 4) or combinations of them or all of them. For example, in the early design phase, when searching a BIPV component with particular aesthetical features or for a specific envelope element, an architect will be able to select the features of interest and through the web-library; the relative compatible products will appear on the user's screen.

3.3 BIPV Web-design tool, the Virtual Demonstrator

Within the project, virtual test environments (namely Virtual Demonstrator) where new BIPV solutions can be tested have been developed. The Virtual Demonstrator tool (http://www.constructpv.eu/webtool/) is a BIPV Web-design tool for early design and evaluation of customized BIPV components in connection with a BIM-based process, acronym for 'Building Information Modelling' process, in the perspective of an integrated approach ensuring quality, transparency and time/cost benefits (Figure 7) [17].

The BIPV Web-design tool is a software unit aimed to define a customizable BIPV product, through a digital 3D modelling (useful for visualization, rendering, architectural design) and, thanks to BIM technology holds in parameters concerning information, materials, energy performance, costs, etc.. The benefit is to make available a digital representation of physical and functional characteristics of the BIPV product/system that, thanks to the BIM technology, become part of the whole PV and AEC (Architecture, Engineering and Construction) process/workflow.

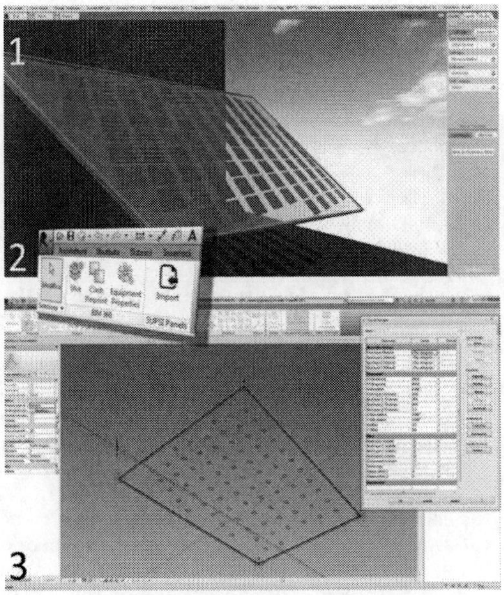

Figure 7: BIPV design tool web-tool for EarlyDesignPhase. BIM interoperability workflow: (1) An *xml* file containing the information is exported as output from the design web-tool; The Revit plug-in (2) allows importing the *xml* and updating the BIM family with the design configuration; An adaptive family (3) is predisposed as the "host" object with geometry and parameters in BIM environment. (Source: http://www.constructpv.eu/webtool/)

Thanks to BIM interoperability the model could be integrated since the design phase, till the production/construction/operating phase considering the whole building life-cycle (e.g. thanks to the interoperability with other tool for time planning, production, cost management, energy analysis, etc.). Furthermore, this design tool could be a simpler yet more extensive method of testing new façades and roof which would complement physical testing very useful too during the manufacturing process. The Virtual Demonstrator digitalize the whole value chain and therefore optimize the flow of information. This will lead to a faster manufacturing and assembly on site which will reduce the construction costs significantly.

Benchmarks for standard building types (new and existing) and climate regions would be established in order that the new option could be comprehensively testing to verify which building categories the façade option would be optimal. Users of any skill levels (i.e. novice to expert) will be able to optimize their building with the intended virtual demonstrator. All the phases that are at the basis of a retrofitting or construction intervention that aim to introduce PV elements in a building can be recreated with this virtual tool: from the planning of the building Re-Construction, to the design and evaluation of the BIPV components, and evaluation of costs and installation on site will be analyzed in such a simulation tool.

4 CONCLUSIONS

Today Computer technologies and Information and Communication Technologies (ICT) that provide access to information through the Internet, wireless networks, cell phones, and other communication paths, as part of the current digital culture have changed how people live, work, play, and learn, changing the way of share knowledge. This paper presents three different approaches based on ICT and Web-based explorative methods (Interactive Webinar, Web-library and Web-design tool) to enhance personal learning skills and to ensure qualified well-trained professionals and at the same time, to improve the result of research in the field of solar BIPV energy.

These ITC and Web-based explorative methods intended to support solar PV innovation and the market growth to generate sustainable processes and practices within Construct PV research project. An interactive bidirectional learning process (from research to public and from public to research) that encourages participatory attitudes helped to discover target markets, foster interest and commitment of stakeholders and to contribute to share and disseminate information.

The participatory approach by using an interactive process in the front-live conference combined with the Webinar event has been demonstrated to be appropriate for consulting several sources (such as architects and manufacturers associations, policy makers, scientists and researchers, etc.) in order to gather feedback about the effectiveness of research findings and to synthesize useful information required to develop new steps by implementing feedback. Furthermore, the development of specific Web-based tools (Web-library and BIPV Web-design tool) has been explored as well within the research process, aiming to support architects or technicians in the early design process and could be potentially exploitable to manufacturers in post-production of customized BIPV solutions. In this case the use of ITC and Web open source tools to enhance solar research, has demonstrated the advantage of digital media to achieve the following aspects:

- To promote the dissemination of knowledge, as well as the exchange of views on scientific research, using the two-way process - with the interaction and listening - to generate/improve dialogue among researchers and the public;
- To encourage the participation of the public (students mainly, but also professionals) by achieving applied direct feedbacks, with possible immediate impact on research activities;
- To track and monitor research interested target audiences (e.g. Webinar event tracking) by allowing knowing the most interesting information (custom dimensions, class of participants and users, answers to specific questions, etc.) and, at the same time, to facilitate the analysis of the information gathered (real-time data & reports dashboards) by using indicators and filters, according to the main topics and goals scheduled;
- To create new forms of interaction between research outcomes and professionals (architects, engineers, installers, students, etc.) through open access digital libraries and more intuitive design tools (e.g. Web-library and Virtual Demonstrator design tool);
- To increase learning opportunities that update

professional skills with innovative research content by accessing to on-line resources and digital tools;
- To implement new methods of teaching, promotion and dissemination of research supported by the integration of ICT (information and communications technology) in education approach.

As result, ICT and explorative web-based methods has been used to enhance solar research, in this case on BIPV, to present the research results, to support technicians and companies of the sector and to create new communication and dissemination pathways opening a new learning paradigm.

AKNOWLEDGMENTS

The Construct-PV project has received funding from the European Union's Seventh Framework Programme, managed by the European Commission under Grant Agreement No 295981.

The authors would like also to thank Mr. Stefano Tardini and Mr. Christian Milani, eLearning Lab members from USI-SUPSI University and to Ms. Erika Saretta from ISAAC, for the technical support in the Webinar preparation and for their contribution in the organizational phases. Besides, special acknowledgement are given to professors and staff in charge for the degree courses and basic education of the Department for environment construction and design DACD CAD services and LCV (visual culture laboratory) of SUPSI.

REFERENCES

[1] Farkas K., Horvat M., (2012). Report T.41.A.1: Building Integration of Solar Thermal and Photovoltaics – Barriers, Needs and Strategies IEA SHC Task 41: Solar Energy and Architecture. Subtask A: Criteria for Architectural Integration, p. 76 <http://task41.iea-shc.org/publications>. Date accessed: 08 mar. 2018.

[2] Prieto, A. et al. Solar façades (2017). Main barriers for widespread façade integration of solar technologies. Journal of Facade Design and Engineering, [S.l.], v. 5, n. 1, p. 51-62, Jan. 2017. ISSN 2213-3038. Available at: <https://journals.library.tudelft.nl/index.php/jfde/article/view/1398>. Date accessed: 08 mar. 2018. DOI: https://doi.org/10.7480/jfde.2017.1.1398.

[3] Bonomo, P., Chatzipanagi A., Frontini F., (2015). Overview and analysis of current BIPV products: new criteria for supporting the technological transfer in the building sector. VITRUVIO- International Journal of Architectural Technology and Sustainability (1). pp. 67-85. ISSN 2444-9091

[4] Saretta, E., Frontini F., Bonomo P., (2016). Laminated BIPV glass: approaches for the integration in the building skin. In: Engineered Transparency 2016: Glass in Architecture and Structural Engineering Engineered Transparency. Edited by Jens Schneider, Bernhard Weller.

[5] Polo López C.S. et al., (2014). PV and Façade Systems for the Building Skin. Analysis of Design Effectiveness and Technological Features. Conference Proceedings of 29th EU PVSEC 2014, pp.3613-3618.

[6] Zanetti I., van den Donker M. et al., (2017). Building Integrated Photovoltaics: Product overview for solar building skins. Status Report 2017 SUPSI-SEAC, SUPSI, <http://www.bipv.ch/images/Report%202017_SUPSI_SEAC_BIPV.pdf> Date accessed: 01 mar. 2018

[7] Vocational Education and Training Research in Switzerland. An OPET programme to promote VET research. OPET, Federal Office for Professional Education and Technology. DFEA, OECD (2007), National Review of Educational R&D, Switzerland, p. 29.

[8] EU Policies for higher education. Education and training: Supporting education and training in Europe and beyond <http://ec.europa.eu/education/policy/higher-education/knowledge-innovation-triangle_en> Date accessed: 07 July 2018

[9] Van Berkel B., Minderhoud T., Piber A., Gijzen G., (2014). Design Innovation from PV-Module to Building Envelope: Architectural Layering and Non Apparent Repetition. 29th EUPVSEC 2014, Amsterdam, The Netherlands. 3606-3612. ISBN: 3-936338-34-5;
Paper DOI: 10.4229/EUPVSEC20142014-6DO.7.2

[10] Polo López C.S., Frontini F., Bonomo P., Cais F., Salvador D., (2016) Outdoor Characterization of Innovative BIPV Modules for Roof Application. 32nd EUPVSEC 2016. Munich, Germany. 2770-2778. ISBN: 3-936338-41-8.
Paper DOI: 10.4229/EUPVSEC20162016-6AV.5.25.

[11] Delponte E., Marchi F., Frontini F., Polo López C. S., Fath K., Batey M., (2015). BIPV in EU28, from Niche to Mass Market: An Assessment of Current Projects and the Potential for Growth through Product Innovation. 31st EUPVSEC 2015, Hamburg, Germany. 3046 - 3050. ISBN: 3-936338-39-6
Paper DOI: 10.4229/EUPVSEC20152015-7DO.15.4

[12] Graham, C., (2006). Blended learning systems. Definitions, current trends and future directions. In C. Bonk & C. Graham (Eds.), The handbook of blended learning: Global perspectives, local designs. San Francisco: John Wiley and Sons.

[13] Vaughan, N., (2007). Perspectives on blended learning in higher education. International Journal on ELearning, 6(1), 81-94.

[14] Mason, R., Rennie, F., (2006). Elearning: The key concepts. London: Routledge Key Guides, 1st Edition. Pages: 200. ISBN-13: 978-0415373074. ISBN-10: 0415373077

[15] Voogt, J., Fisser, P., Pareja Roblin, N., Tondeur, J. and van Braak, J., (2013), Technological pedagogical content knowledge – a review of the literature. Journal of Computer Assisted Learning, 29: 109–121.

[16] Geurts Jac.L.A., Joldersma C., (2001). Methodology for participatory policy analysis, European Journal of Operational Research, Volume 128, Issue 2, 2001, Pages 300-310. ISSN 0377-2217. <https://doi.org/10.1016/S0377-2217(00)00073-4>. Date accessed: 08 August 2018.

[17] Bonomo P., Frontini F., Hecker R. (2016) Supporting integration of photovoltaics in the building skin. A software tool for customized BIPV in a BIM-based process. In: Engineered Transparency 2016: Glass in Architecture and Structural Engineering Engineered Transparency. Edited by Jens Schneider, Bernhard Weller. Pages: 383-388.

AUTHOR INDEX

AUTHOR INDEX

AUTHOR INDEX

AUTHOR INDEX

AUTHOR INDEX

AUTHOR INDEX

AUTHOR INDEX

AUTHOR INDEX

AUTHOR INDEX

AUTHOR INDEX

AUTHOR INDEX

AUTHOR INDEX

AUTHOR INDEX

AUTHOR INDEX

AUTHOR INDEX

AUTHOR INDEX

AUTHOR INDEX

AUTHOR INDEX

AUTHOR INDEX

WIP – Renewable Energies
Sylvensteinstr. 2
81369 Munchen
Germany

ISBN 978-1-5108-7155-7